# Syndromes of the Head and Neck

## OXFORD MONOGRAPHS ON MEDICAL GENETICS



1. R. B. McConnell: *The genetics of gastro-intestinal disorders*
2. A. C. Kopeć: *The distribution of the blood groups in the United Kingdom*
3. E. Slater and V. A. Cowie: *The genetics of mental disorders*
4. C. O. Carter and T. J. Fairbank: *The genetics of locomotor disorders*
5. A. E. Mourant, A. C. Kopeć, and K. Domaniewska-Sobczak: *The distribution of the human blood groups and other polymorphisms*
6. A. E. Mourant, A. C. Kopeć, and K. Domaniewska-Sobczak: *Blood groups and diseases*
7. A. G. Steinberg and C. E. Cook: *The distribution of the human immunoglobulin allotypes*
8. D. Tills, A. C. Kopeć, and R. E. Tills: *The distribution of the human blood groups and other polymorphisms: Supplement I*
9. M. Baraitser: *The genetics of neurological disorders*
10. D. Z. Loesch: *Quantitative dermatoglyphics: classification, genetics, and pathology*
11. D. J. Bond and A. C. Chandley: *Aneuploidy*
12. P. F. Benson and A. H. Fensom: *Genetic biochemical disorders*
13. G. R. Sutherland and F. Hecht: *Fragile sites on human chromosomes*
14. M. d'A. Crawfurd: *The genetics of renal tract disorders*
15. A. E. H. Emery: *Duchenne muscular dystrophy*
16. C. R. Scriver and B. Childs: *Garrod's inborn factors in disease*
17. R. J. M. Gardner and G. R. Sutherland: *Chromosome abnormalities and genetic counseling*
18. M. Baraitser: *The genetics of neurological disorders, second edition*
19. R. J. Gorlin, M. M. Cohen, Jr., and L. S. Levin: *Syndromes of the head and neck, third edition*

OXFORD MONOGRAPHS ON MEDICAL GENETICS NO. 19

# Syndromes of the Head and Neck

Third Edition

ROBERT J. GORLIN, D.D.S., M.S., D.Sc.(Athens)
Regents' Professor and Chairman
Department of Oral Pathology and Genetics
School of Dentistry
Professor of Pathology, Pediatrics, Obstetrics and
Gynecology, Otolaryngology, and Dermatology
School of Medicine
University of Minnesota
Minneapolis, Minnesota

M. MICHAEL COHEN, JR., D.M.D., Ph.D.
Professor of Oral Pathology, Faculty of Dentistry
Professor of Pediatrics, Faculty of Medicine
Dalhousie University
Halifax, Nova Scotia, Canada

L. STEFAN LEVIN, D.D.S., M.S.D.
Late Associate Professor
Department of Otolaryngology—Head and Neck Surgery
Johns Hopkins University
School of Medicine
Baltimore, Maryland

New York Oxford
OXFORD UNIVERSITY PRESS
1990

Oxford University Press

Oxford New York Toronto
Delhi Bombay Calcutta Madras Karachi
Petaling Jaya Singapore Hong Kong Tokyo
Nairobi Dar es Salaam Cape Town
Melbourne, Auckland

and associated companies in
Berlin Ibadan



Published by Oxford University Press, Inc.,
200 Madison Avenue, New York, New York 10016



Library of Congress Cataloging-in-Publication Data
Gorlin, Robert J., 1923–
Syndromes of the head and neck / Robert J. Gorlin,
M. Michael Cohen, Jr., L. Stefan Levin.—3rd ed.
p. cm.—(Oxford monographs on medical genetics)
Includes bibliographies and index.
ISBN 0-19-504518-1
1. Head—Abnormalities, 2. Neck—Abnormalities. 3. Genetics, disorders.
I. Cohen, M. Michael (Meyer Michael), 1937–
II. Levin, L. Stefan. III. Title. IV. Series.
[DNLM: 1. Head—abnormalities. 2. Neck—abnormalities.
3. Nomenclature. WB 15 G669s]
RC936.G67 1989
617.5′1—dc20 DNLM/DLC for Library of Congress
89-8553 CIP

2 4 6 8 9 7 5 3 1

Printed in the United States of America
on acid-free paper

# Contributors

Michael Baraitser, B.Sc., M.B., Ch.B.
The Hospital for Sick Children, London, England

Peter H. Byers, M.D.
University of Washington, Seattle, Washington

David E. C. Cole, M.D., Ph.D.
Dalhousie University, Halifax, Nova Scotia

Cynthia J. R. Curry, M.D.
Valley Children's Hospital, Fresno, California

Robert J. Desnick, M.D., Ph.D.
Mount Sinai School of Medicine, New York, New York

Rosalie B. Goldberg, M.S.
Montefiore Medical Center, Albert Einstein College of Medicine, Bronx, New York

John M. Graham, Jr., M.D., Sc.D.
Cedar-Sinai Hospital, UCLA School of Medicine, Los Angeles, California

Eric Grutzner, D.D.S.
University of Minnesota School of Dentistry, Minneapolis, Minnesota

Bryan D. Hall, M.D.
University of Kentucky, Lexington, Kentucky

James W. Hanson, M.D.
University of Iowa Hospital, Iowa City, Iowa

Martin Ritzau, D.D.S.
Royal Dental College, Aarhus, Denmark

Beverly Rollnick, Ph.D.*
Center for Craniofacial Anomalies, University of Illinois Medical Center, Chicago, Illinois

Edward H. Schuchman, Ph.D.
Mount Sinai Hospital, New York City School of Medicine, New York, New York

Heddie O. Sedano, D.D.S., Dr.O.
University of Minnesota School of Dentistry, Minneapolis, Minnesota

David O. Sillence, M.D.
Children's Hospital, Camperdown, Australia

Mark J. Stephan, M.D.
Madigan Army Medical Center, Tacoma, Washington

Helga V. Toriello, Ph.D.
Butterworth Hospital, Grand Rapids, Michigan

Chester B. Whitley, Ph.D., M.D.
University of Minnesota Medical School, Minneapolis, Minnesota

*Deceased

To the pioneers of dysmorphology

Victor A. McKusick
F. Clarke Fraser
John M. Opitz
David Poswillo
David W. Smith*
Josef Warkany
Hans Rudolph Wiedemann
Hans Zellweger*

*Deceased

# Foreword

Syndromologists, medical geneticists, and (even more so) other health workers have found it increasingly difficult to keep up with the vigorous growth of knowledge about syndromes. The distinguished teratologist Josef Warkany wrote "with the increasing interest in congenital malformations a syndrome fever is spreading through many specialties, and it is difficult for editors of medical journals and readers to separate spurious from durable and meaningful syndromes." That was in 1971 (Congenital Malformations: Notes and Comments). Thus, there is a need for judicious sifting, organizing, and synthesis of the plethora of syndromic literature into meaningful patterns. The first edition of this volume was welcomed by those who were even then beginning to feel this need. Its breadth and depth reflected the encyclopedic knowledge and judgment of Robert Gorlin, an extraordinary phenomenon. And it was by no means limited to the head and neck! To keep up with the burgeoning growth of knowledge, the second edition added another extraordinary repository of syndromic lore, in the form of Michael Cohen. In the present edition, the breadth and depth of the syndrome data base have been further extended by the otolaryngologic knowledge of L. Stefan Levin and by specific chapters from no fewer than 18 collaborators. The result is a truly encyclopedic work, containing descriptions of the phenotypic spectrum, epidemiology, mode of inheritance, and pathogenesis of nearly 700 syndromes. McKusick's Catalogue has more entries, since it covers all Mendelian disorders, not just syndromes, but this volume includes non-Mendelian syndromes as well, and in far more depth. The several computer-aided syndrome-diagnosis systems now available are useful as diagnostic aids, but do not provide as exhaustive descriptions of phenotype or as critical analyses of the literature. This book will be welcomed as an essential component of the knowledge base for the clinical geneticist and others confronted with syndromes. It is an honour to be associated with it, even in this peripheral way. I am certainly looking forward to having it on my bookshelf, and hope that I will be able to persuade my students to let me read it now and then!

F. Clarke Fraser, O.C., M.D., Ph.D., FRS(C)
*Emeritus Professor of Medical Genetics*
*McGill Centre for Human Genetics, Montreal, Canada*

## Foreword to the Second Edition (1976)

The first edition of this book was the pioneering text for this segment of medicine. I vividly recall the excitement upon first reading the book in 1964. At last!—here was a cohesive authoritative text which portrayed the majority of syndromes which had been recognized at that time. Though entitled *Syndromes of the Head and Neck,* it covered all known features of each disorder in a nonspecialized and balanced manner, including the natural history, etiology, differential diagnosis, and pertinent references for each disorder. The field of syndromology has expanded since that time. The number of recognized disorders set forth in the book has more than doubled, and the knowledge has been updated on the original syndromes. A third person, M. Michael Cohen, Jr., has been added to the authorship of this expanded work on syndromes. Thus, this second edition is a most welcome addition for all those who work with, or are interested in, syndromes of malformation. Many children and their parents will be the indirect beneficiaries of this text. For our own patients, we sincerely thank the authors for this monumental work.

David W. Smith, M.D.*
*Professor of Pediatrics, Dysmorphology Unit, Department of Pediatrics*
*University of Washington, School of Medicine, Seattle, Washington*

*Deceased

## Foreword to the First Edition (1964)

The authors of this monograph are keen observers and ardent students of disease in the best tradition of Jonathan Hutchinson, Parkes Weber, and other clinicians of an earlier generation. Many of the diseases on which they have concentrated their attention are rare, but for several reasons no less important. Although these diseases occur infrequently, they constitute in the aggregate a significant portion of medicine. Most of them are congenital malformations or disorders loosely called constitutional; many of them are genetic either in the classic Mendelian sense or as chromosomal aberrations. It is a truism that this body of diseases has come to represent a main challenge to medicine now that infectious and nutritional diseases are better understood and controlled. That the disorder from which he suffers is rare is no consolation to its victim. The first step in the understanding of these conditions must be an accurate and full clinical description.

The careful study of exceptional cases can contribute importantly to medicine and to biology in general. Bateson, a famous early geneticist, said "Treasure your exceptions!" In 1657 William Harvey, of blood circulation fame, eloquently expressed the usefulness of the study of rare diseases:

> Nature is nowhere accustomed more openly to display her secret mysteries than in cases where she shows traces of her workings apart from the beaten path; nor is there any better way to advance the proper practice of medicine than to give our minds to the discovery of the usual law of Nature by careful investigation of cases of rarer forms of disease. For it has been found, in almost all things, that what they contain of useful or applicable nature is hardly perceived unless we are deprived of them, or they become deranged in some way.

Professors Gorlin and Pindborg have done a valuable service to oral pathology and medicine in general by collating their extensive personal experiences and the widely scattered reports of the literature. With skill they have synthesized and interpreted. The fundamental relationship between disorders separately reported, usually under diverse labels, has been carefully explored and convincingly demonstrated in a number of instances. For this, all medicine is in debt of the authors. Their monograph is a particularly valuable addition to the English-language medical literature because they have mined much ore previously not available. With their linguistic prowess they have been able to study in the original a considerable body of literature previously unknown to most of us.

Specialists such as ophthalmologists, dermatologists, and dentists have always been in the enviable position of being able to study disease with simple clinical methods. Fortunately many of these specialists, appreciating the relationship of systemic and constitutional disorders to the manifestations which fall within their purview, have made worthwhile contributions to pathology. With this monograph, Gorlin and Pindborg have joined this company. Their work will be valued, not only by dentists, but by all the numerous medical specialists who are called on to care for patients with these disorders.

Victor A. McKusick, M.D.
*William Osler Professor of Medicine*
*Johns Hopkins University, School of Medicine*
*Physician-in-Chief, Johns Hopkins Hospital*
*Baltimore, Maryland*

# Preface

This third edition of *Syndromes of the Head and Neck* brings many changes and a plethora of new syndromes. We invited L. Stefan Levin of Baltimore to join us as coauthor on the third edition. He wrote many excellent chapters that appear in this book. During the later stages of preparation of this text, he met an untimely death. It is particularly tragic because he succumbed at such an early age. His many contributions to oral pathology, his research into osteogenesis imperfecta, his special interest in hearing loss, and his knowledge of malformation syndromes will long be remembered in the fields of oral pathology, otolaryngology, and medical genetics.

We also invited a number of contributors to participate in this edition: Michael Baraitser (London), Peter H. Byers (Seattle), David E. C. Cole (Halifax), Cynthia J. R. Curry (Fresno), Robert J. Desnick (New York), Rosalie B. Goldberg (New York), John M. Graham, Jr. (Los Angeles), Eric Grutzner (Minneapolis), Bryan D. Hall (Lexington), James W. Hanson (Iowa City), Martin Ritzau (Aarhus), Beverly Rollnick* (Chicago), Edward H. Schuchman (New York), Heddie O. Sedano (Minneapolis), David O. Sillence (Sydney), Mark J. Stephan (Tacoma), Helga V. Toriello (Grand Rapids), and Chester B. Whitley (Minneapolis). We feel honored that Clarke Fraser agreed to write the foreword for this edition. For continuity with earlier editions, we are reprinting the foreword from the first edition, by Victor A. McKusick, and the foreword from the second edition, by David W. Smith.

We puzzled over whether to retain the title *Syndromes of the Head and Neck* for this edition because so many syndromes that appear in the book go beyond the head and neck. Clinicians tend to consult the book for syndromes with prominent craniofacial features, yet often do not think of it for hamartoneoplastic disorders, such as neurofibromatosis and tuberous sclerosis, or for disorders with strong dysmetabolic overtones, such as the mucopolysaccharidoses and the Zellweger syndrome. In fact, we have tried to document these kinds of disorders as carefully and as extensively as the ''pure'' malformation syndromes. For better or for worse, we decided to continue using the same title simply because the book has been known by this title for so long.

There have been several changes in organization with this edition. Most noticeably, conditions are now grouped under headings such as syndromes with craniosynostosis, syndromes with postnatal onset obesity, chondrodysplasias and chondrodystrophies, and hamartoneoplastic syndromes, rather than being listed alphabetically. This reorganization does not solve the problem of syndrome overlap entirely. For example, is craniofrontonasal dysplasia a craniosynostosis syndrome or a hypertelorism syndrome? In this edition, we arbitrarily assigned it to the hypertelorism section of the chapter on syndromes with unusual facies. We hope this new format will help our readers identify specific syndromes more easily. In many instances, other syndromes that make up the differential diagnosis appear on pages adjacent to the syndrome under consideration.

The process of syndrome delineation has progressed at an extremely rapid pace during the past decade and a half. In 1971, for example, 72 syndromes with orofacial clefting were known. Now over 250 such syndromes are recognized. Even with complex craniofacial anomalies such as holoprosencephaly (and arhinencephaly), 66 syndromes are known today compared with 13 in 1971. The rapid advances in syndromology are reflected in the total number of pages of the three editions of *Syndromes of the Head and Neck.* The 1964 edition contained 580 pages; the 1976 edition had 812 pages; and the present edition (with larger format, smaller print and double columns) has about 1,000 pages.

All this added complexity means, of course, that it has become increasingly difficult—in

*Deceased

fact, impossible—for us to have expertise in all the syndromes, disorders, and conditions discussed in this edition. Thus, we have relied on contributing authors and we owe them a great deal of thanks for their expertise, their hard work, and their timely delivery of manuscripts. We especially wish to thank David Poswillo (London) and Kathy Sulik (Chapel Hill) for their careful critical review of several chapters.

We have enjoyed working with Oxford University Press. Jeffrey House, Ellen Fuchs, and Edith Barry have been most helpful and we appreciate their efforts in making this book possible.

Finally, Carol Bauer Church did a truly outstanding job in helping us organize the book during each phase of its production. We are particularly grateful to her for bringing her skills to bear on this project with enthusiasm, love, and dedication. We are also grateful for the superb typing and perseverance of our secretaries Angela M. Faulkner, Ruth E. MacLean, Joy E. Love, and Elfrieda Schneider.

*Minneapolis*
*Halifax*

R. J. G.
M. M. C.

# Contents

# Syndromes of the Head and Neck

# Chapter 1
# Deformations and Disruptions

## Craniofacial deformations

Congenital deformations of the head and neck are common, and most resolve spontaneously within the first few days of postnatal life. When they do not, further evaluation may be necessary to plan therapeutic interventions (1–4,8) that may prevent long-term consequences. The distinction between deformations and malformations and its implications are discussed elsewhere (5,7,8).

Approximately 2% of infants are born with extrinsically caused deformations that usually arise during late fetal life from intrauterine constraint. Approximately 30% of deformed infants have two or more deformations. Deformed infants tend to show catch-up growth toward their genetic potential during the first few postnatal months after release from the intrauterine constraining environment (7). Table 1–1 indicates some of the known causes, both extrinsic and intrinsic, of deformations. Deformations considered here include nasal, auricular, and mandibular deformities, torticollis, nonsynostotic plagiocephaly, craniosynostosis caused by intrauterine constraint, and abnormal fetal presentations, which may result in craniofacial deformation.

Table 1–1. Causes of deformations[a]

| Causes |
|---|
| Extrinsic |
| Mechanical |
| Unstretched uterine and abdominal muscles |
| Small maternal size |
| Amnionic tear |
| Unusual implantation site |
| Uterine leiomyomas |
| Unicornuate uterus |
| Bicornuate uterus |
| Twin fetuses |
| Intrinsic |
| Malformational |
| Spina bifida |
| Other central nervous system malformations |
| Bilateral renal agenesis |
| Severe hypoplastic kidneys |
| Severe polycystic kidneys |
| Urethral atresia |
| Functional |
| Neurologic disturbances |
| Muscular disturbances |
| Connective tissue defects |

[a]From MM Cohen Jr, *The Child with Multiple Birth Defects.* Raven Press, New York, 1982, p 10.

**Nasal deformation.** Deformation of the nose may be associated with face presentation, transverse lie, oligohydramnios, or severe fetal crowding. In most instances, the nose will be compressed and/or deviated (Fig. 1–1), which may make it appear short. Occasionally the nasal cartilage is dislocated from the vomerine ridge, resulting in nasal asymmetry with slanting of the columella. The naris on the side toward the dislocated cartilaginous septum will appear small (8).

**Auricular deformation.** Overfolding of the superior rim of the helix and other alterations in the cartilaginous auricle are frequently caused by fetal constraint. The ear may be flattened against the head by oligohydramnios, and prolonged pressure against the auricle may result in overgrowth. Enlarged, flattened ears have frequently been associated with renal agenesis as part of Potter sequence. When one ear is exposed to more pressure than the other, they may be asymmetric. Occasionally, with prolonged breech presentation, particularly when associated with tilted head position *in utero,* the lower auricle may be lifted by pressure from the shoulder (2).

**Mandibular deformation.** Micrognathia may result from limitation of mandibular growth caused by late gestational constraint, which compresses the chin against the chest. With prolonged compression, there may be a pressure indentation on the superior aspect of the anterior thorax. Marked compression may also lead to pressure necrosis

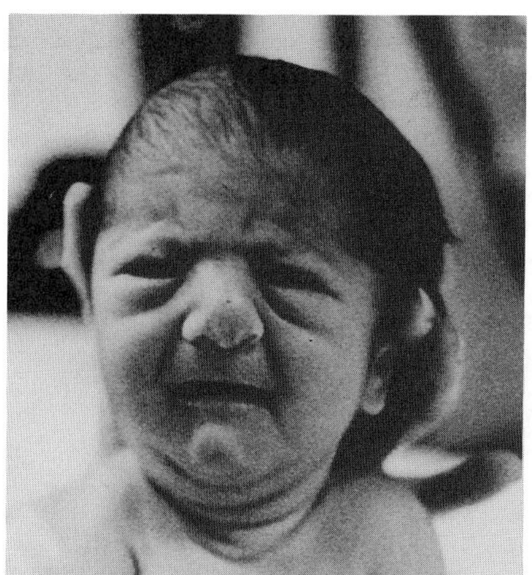

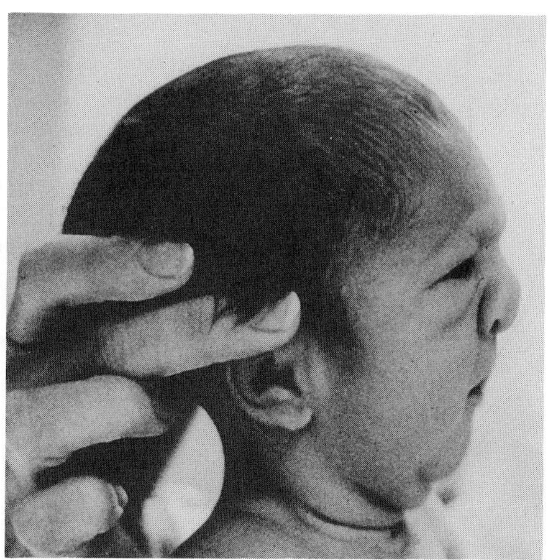

Fig. 1–1. *Craniofacial deformations.* Compression of the face, particularly the nose, from prolonged transverse presentation with the head retroflexed and the face compressed against the lateral wall of the uterus. (From *JM Graham Jr,* Smith's Recognizable Patterns of Human Deformation, W. B. Saunders, Philadelphia, 2nd Ed, 1988.)

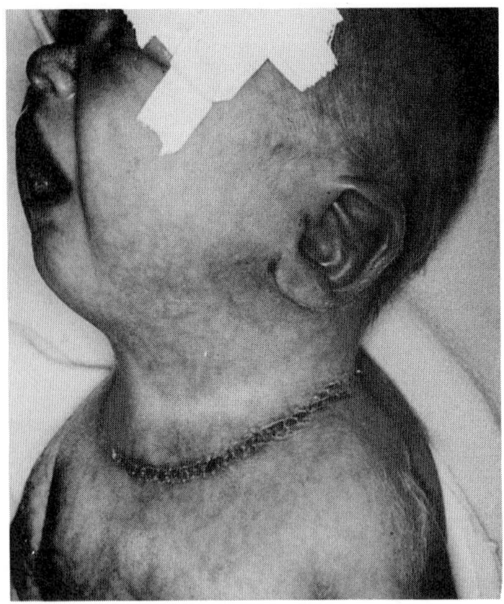
A

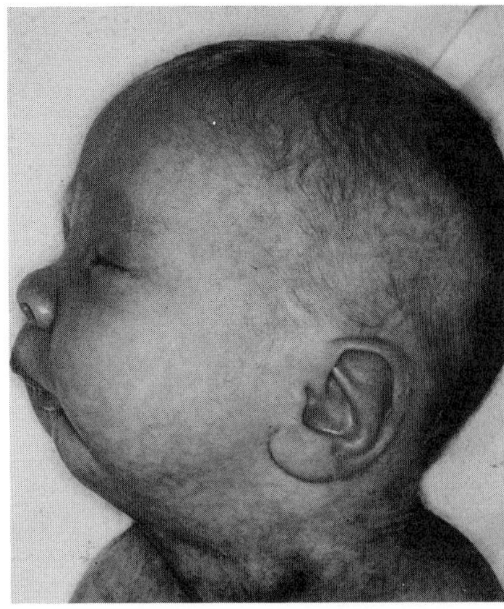
B

C

Fig. 1–2. *Craniofacial deformations.* Extensive compression from persistent leakage of amnionic fluid documented by ultrasound from 17 weeks gestation to birth. (A) Marked mandibular deformity at 3 days of age. Note necrosis along neck creases (which healed with scarring). (B) Extensive mandibular deformation still evident at 3 weeks of age. (C) Partial resolution of mandibular deformity by 2.5 years of age. (From *JM Graham Jr,* Smith's Recognizable Patterns of Human Deformation, W. B. Saunders, Philadelphia, 2nd Ed, 1988.)

along the edges of the anterior neck creases (Fig. 1–2). Such necrosis may extend deeply enough into the dermis to result in scarring. When micrognathia occurs on a deformational basis (rather than on a hypoplastic, malformational basis), there is usually catch-up growth postnatally once the fetus is no longer in the constraining intrauterine environment (5,8).

When mandibular compression is asymmetric, it can produce mandibular asymmetry. Most commonly, this results from the shoulder being thrust up under the mandible with prolonged breech or oblique presentations. A prominent sulcus impression may occur along the neck from shoulder compression (5,8).

**Torticollis.** Congenital torticollis usually occurs together with an obliquely shaped plagiocephalic head and mandibular asymmetry. Often this results from the head being caught askew prenatally. Muscular torticollis may result from constraint on one side of the neck causing ischemia to the central portion of the sternocleidomastoid muscle followed by secondary fibrosis. A fusiform fibrous mass may sometimes be palpable within the muscle, a sternocleidomastoid "knot" or "tumor." Asymmetric shortening of one sternocleidomastoid muscle may lead to aberrant head posture. With persistent torticollis and resultant head posture, craniofacial deformation may be progressive (3,4,8).

Fig. 1–3. *Craniofacial deformations.* Four and one-half-month-old infant with persistent head turn to the left caused by cervical and thoracic vertebral anomalies. Note resultant marked plagiocephaly. (From *JM Graham Jr,* Smith's Recognizable Patterns of Human Deformation, W. B. Saunders, Philadelphia, 2nd Ed, 1988.)

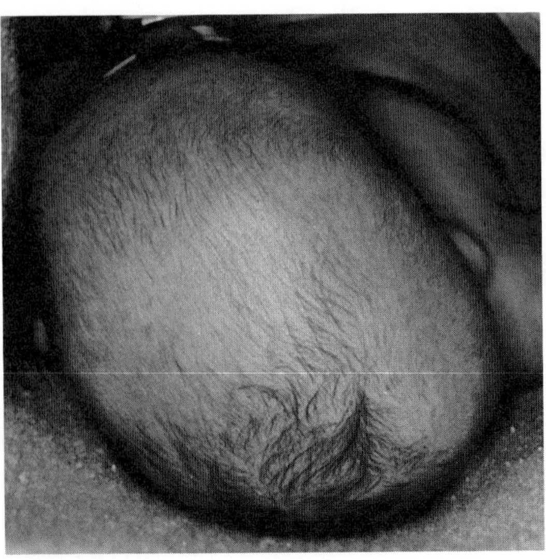

**Torticollis-caused plagiocephaly.** When persistent sternocleidomastoid torticollis or asymmetric cervical vertebral anomalies result in the head resting in an asymmetric position, oblique molding results in the head being rhomboid-shaped, with frontal prominence on the preferred side together with contralateral prominence of the occiput (Fig. 1–3). With marked cranial distortion, the eyes and ears may be asymmetrically placed and the mandible may be asymmetrically deformed. If torticollis remains uncorrected, plagiocephaly may be progressive (3,4,8).

**Deformation-induced craniosynostosis.** Fetal head constraint has been implicated as an important cause for some instances of sagittal, coronal, and metopic craniostenosis. In such instances, craniosynostosis is not associated with other malformations and the condition is not familial. With deformational synostosis, fetal head constraint results in lack of growth stretch across the suture, enhancing liability toward synostosis (6,8,10–12,14). Experimental evidence for this interpretation was provided by Koskinen-Moffett (15), who produced prenatal synostosis of the coronal and squamosal sutures in mouse pups by closing the uterine cervix with a surgical clip to delay birth for several days, which crowded the fetuses.

Fetal head constraint as a cause of isolated craniostenosis helps to explain sex predilection and the prevalence of sidedness. For example, many more males than females are affected by sagittal synostosis and this may be related to the more rapid growth and larger head size of males during the last trimester of pregnancy (11). Constraint-induced plagiocephalic craniofacial deformation is also much more common in male than in female infants. Unilateral coronal synostosis is more commonly right-sided and this nonrandom predilection may be related to the fact that vertex presentations occur most commonly in the left occiput transverse position with the left coronal suture against the sacral prominence. If the fetal head descends early and is maintained in the most common position, the right coronal suture may be more con-

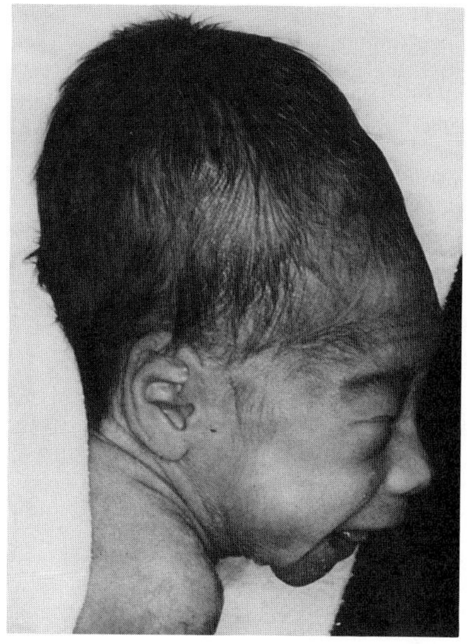

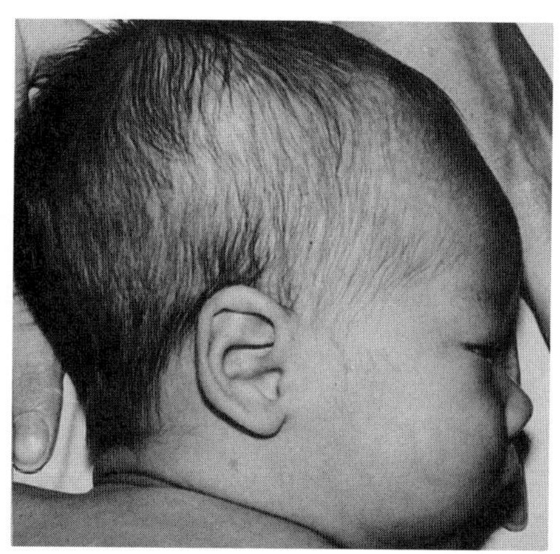

A B

Fig. 1–4. *Craniofacial deformations.* (A) Extensive vertex molding at birth. (B) Self-resolution by 2–3 months of age. (From *JM Graham Jr,* Smith's Recognizable Patterns of Human Deformation, W. B. Saunders, Philadelphia, 2nd Ed, 1988.)

strained than the left, thus explaining the more common right-sided involvement. Most large series of coronal synostosis show approximately equal unilateral and bilateral involvement, but this ratio shifts strikingly toward bilateral involvement when only familial cases are considered. Such a shift would be compatible with the symmetric impact of a mutant gene and the asymmetric impact of intrauterine mechanical forces. Associated malformations suggestive of a mutant gene are also more common with bilateral than with unilateral coronal involvement (8,12).

Fig. 1–5. *Craniofacial deformations.* Prolonged vertex presentation at least 4 weeks prior to delivery. Large area of craniotabes represented by cross hatching. Self-resolution by 2 months of age. (From *JM Graham Jr,* Smith's Recognizable Patterns of Human Deformation, W. B. Saunders, Philadelphia, 2nd Ed, 1988.)

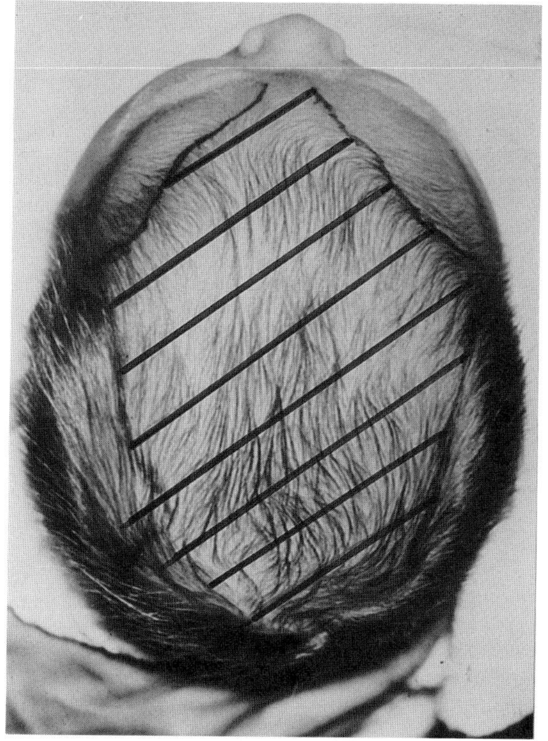

**Abnormal fetal presentation.** A vertex presentation is most common and transient molding of the head occurs in the birth canal. With prolonged engagement, normal vertex birth molding may be accentuated (Fig. 1–4). The prognosis for a complete return to normal form in such instances is excellent (8).

With prolonged pressure on the vertex region, particularly with a first-born infant, a large fetus, or a small mother, diminished mineralization of the cranium may result. Mild vertex craniotabes occurs in 2% of neonates with more extensive cranial softening occurring less frequently (Fig. 1–5). The presence of normal cranial bone along the sides of the calvaria helps distinguish craniotabes from defective calvarial mineralization resulting from inherited metabolic or connective tissue disorders such as hypophosphatasia or osteogenesis imperfecta. Vertex craniotabes, resulting from late fetal head constraint, usually resolves spontaneously within the first few postnatal months (9).

Breech presentation is the most common abnormal fetal presentation, occurring in 6% of all pregnancies; one-third of all deformations occur in breech babies. In Dunn's series (7), 32% of all deformations in newborns were associated with breech presentation. Genu recurvatum (100%), hip dislocation (50%), postural scoliosis (42%), mandibular asymmetry (20–25%), torticollis (20–25%), and talipes equinovarus (20–25%) are all related to breech presentation.

Prolonged breech presentation may give rise to unusual molding of the fetal head, resulting in dolichocephaly with a prominent occipital shelf (13) (Fig. 1–6). Vaginal delivery of a breech fetus with hyperextended head is associated with significant morbidity and mortality (1). Serious consequences include trauma to the spinal cord or brachial plexus, compression of the vertebral artery with cerebral ischemia, or severance of the pituitary stalk with consequent hypopituitarism. Thus, the current trend is toward cesarean delivery of fetuses with breech presentation.

Transverse lie is less frequent than breech presentation, occurring once in every 300–600 deliveries. It is more common in multiparous women, and it results in problems similar to those associated with breech presentation. Thus, unresolved transverse presentation is also an indication for cesarean delivery. Lateral constraint of the fetus may flatten the face, limit mandibular growth, and cause cephalic retroflexion and truncal scoliosis with postional foot deformities. Figure 1–1 demonstrates remarkable facial compression associated with prolonged transverse lie. Such compression is usually only temporary.

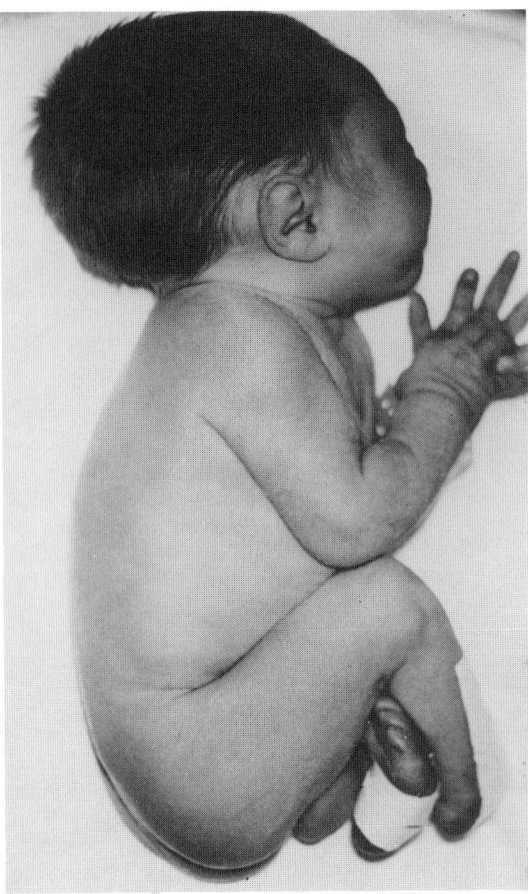

Fig. 1–6. *Craniofacial deformations.* Prolonged breech presentation resulting in prominent occipital shelf. Resultant equinovarus foot deformities treated by taping. (From *JM Graham Jr,* Smith's Recognizable Patterns of Human Deformation, W. B. Saunders, Philadelphia, 2nd Ed, 1988.)

Face and brow presentations (Fig. 1–7) may also temporarily compress the presenting part. Face presentations occur about once in 500 births and brow presentations are even less common. Only anterior positions can be delivered vaginally because of the inability of the neck to further extend in the posterior position. Compression of the mandible may temporarily restrict its growth, and compression of the neck against the pubic ramus during vaginal delivery may cause fracture of the trachea or larynx. Thus, cesarean section should be given consideration. It should also be noted that 90% of babies born in face or brow presentation are infants with major malformations, with anencephaly being the most common. For otherwise normal infants with facial features such as those shown in Figure 1–7, the prognosis for complete return to normal form is excellent.

#### References (Craniofacial deformations)

1. Abrams IF et al: Cervical cord injuries secondary to hyperextension of the head in breech presentations. *Obstet Gynecol* **41**:369–378, 1973.

2. Brown FE et al: Correction of congenital auricular deformities by splinting in the neonatal period. *Pediatrics* **78**:406–411, 1986.

3. Clarren SK et al: Helmet treatment for plagiocephaly and congenital muscular torticollis. *J Pediatr* **94**:43–46, 1979.

4. Clarren SK: Plagiocephaly and torticollis: Etiology, natural history, and helmet treatment. *J Pediatr* **98**:92–95, 1981.

5. Cohen MM Jr: *The Child with Multiple Birth Defects.* Raven Press, New York, 1982.

6. Cohen MM Jr: *Craniosynostosis: Diagnosis, Evaluation, and Management.* Raven Press, New York, 1986.

7. Dunn PM: Congenital postural deformities. *Br Med Bull* **32**:71–76, 1976.

8. Graham JM Jr: *Smith's Recognizable Patterns of Human Deformation.* Philadelphia, WB Saunders, 2nd Ed., 1988.

9. Graham JM, Smith DW: Parietal craniotabes in the neonate: Its origin and relevance. *J Pediatr* **95**:114–116, 1979.

10. Graham JM, Smith DW: Metopic craniostenosis as a consequence of fetal head constraint: Two interesting experiments of nature. *Pediatrics* **65**:1000–1002, 1980.

11. Graham JM et al: Sagittal craniostenosis: Fetal head constraint as one possible cause. *J Pediatr* **95**:747–750, 1979.

12. Graham JM et al: Coronal craniostenosis: Fetal head constraint as one possible cause. *J Pediatr* **65**:995–999, 1980.

13. Haberkern CM et al: The 'breech head' and its relevance. *Am J Dis Child* **133**:154–156, 1979.

14. Higginbottom MC et al: Intrauterine constraint and craniosynostosis. *Neurosurgery* **6**:39–44, 1980.

15. Koskinen-Moffett L: In vivo experimental model for prenatal craniosynostosis. *J Dent Res* **65**, Special Issue, abstr. 980, 1986.

Fig. 1–7. *Craniofacial deformations.* Prolonged face presentation during the last 2 months of gestation. (A) Nasal and mandibular compression. (B) Partial self-resolution by 6 weeks of age. (From *JM Graham Jr,* Smith's Recognizable Patterns of Human Deformation, W. B. Saunders, Philadelphia, 2nd Ed, 1988.)

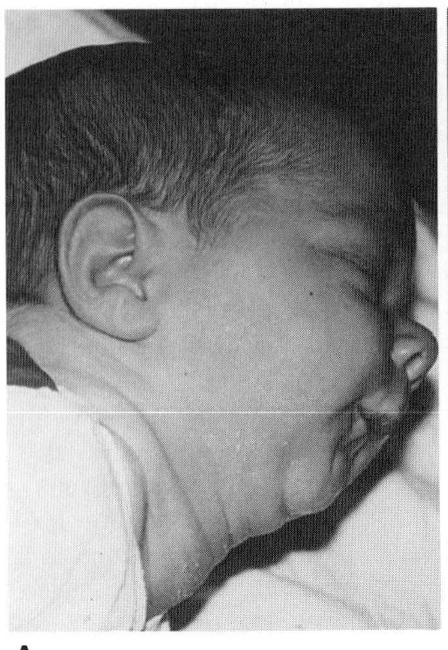

A

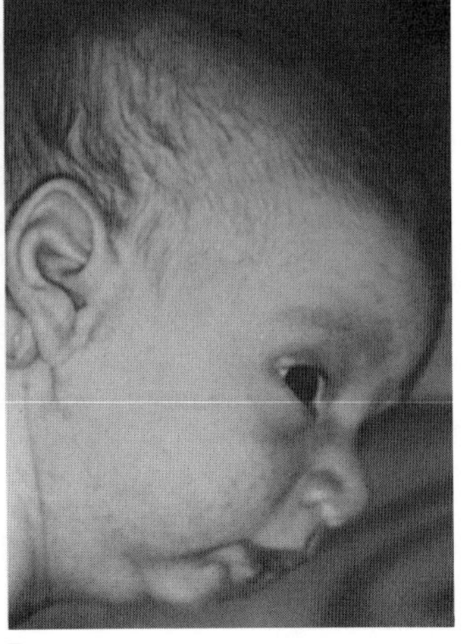

B

## Potter sequence (oligohydramnios sequence)

Potter sequence is characterized by compression deformities of the face and limbs (Fig. 1–8A), pulmonary hypoplasia, wrinkled skin, and growth restriction resulting from any pathologic condition that leads to oligohydramnios (42,95,152), such as bilateral renal agenesis (110,113), cystic dysplasia (11), severe polycystic kidneys (43,60, 103,114), urinary tract obstruction (4,79), or amnionic leakage (20,79, 151). In addition to diverse renal pathology, more extensive malformations may occur in some instances (35), the severe caudal axis defect sirenomelia (Figs. 1–8B–D) being the most extreme example (15,52,77,80,143,146,158). Several unusual experiments of nature have been recorded in which the extrarenal features of Potter sequence are not manifest (4,79,94,151) (Fig. 1–9).

Originally described by Potter (113–116) in association with bilateral renal agenesis, Potter sequence is now known to be extremely heterogeneous, both etiologically and pathogenetically. The term Potter sequence or oligohydramnios sequence best describes the condition, with other terms, such as Potter syndrome (44,45,84,89,110,129, 165,167) (the condition is not a true syndrome *sui generis,* but may occur as a component part of many different syndromes or may occur nonsyndromically), bilateral renal agenesis (48,113–115,122,124,149) (too restrictive), and renofacial dysplasia (18,52,58,60,64,105,106,147) (not specific enough), now being obsolete.

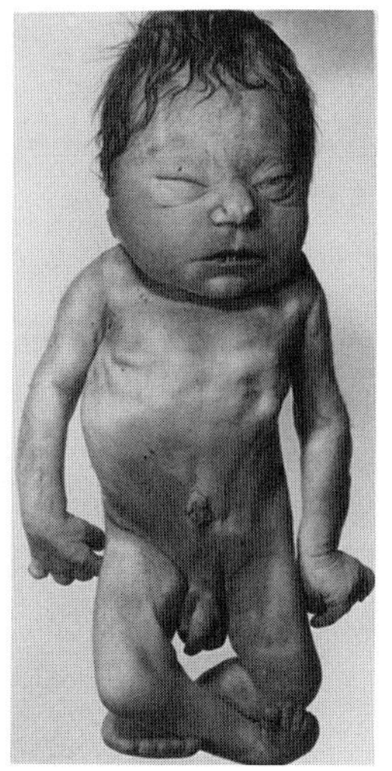
A

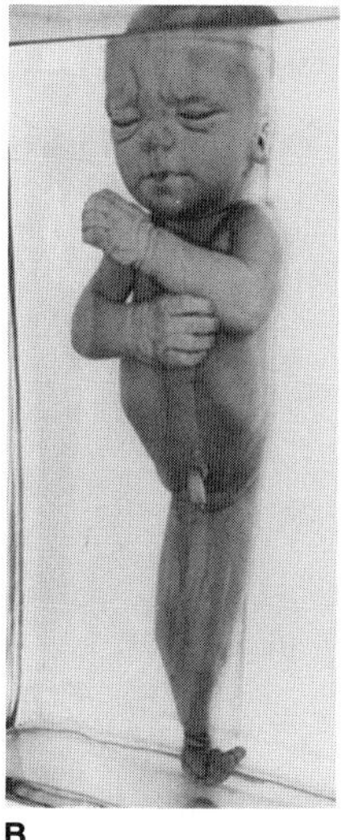
B

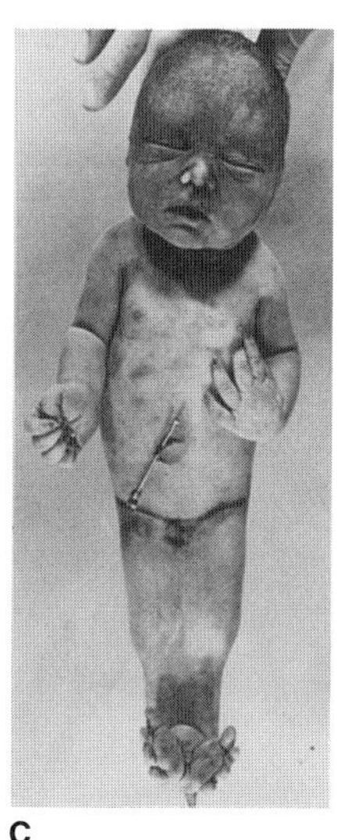
C

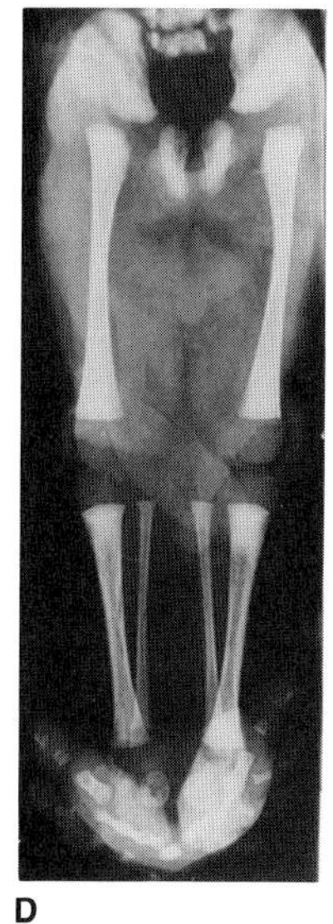
D

Fig. 1–8. *Potter sequence.* (A) Potter sequence with compressed facial appearance and limb positioning deformities. (B) Sirenomelia, a severe caudal axis malformation sequence in which kidneys and genitals are missing. Note Potter facies and upper limb deformities. (C,D) Another example of sirenomelia. (A from *DW Smith,* Recognizable Patterns of Human Malformation, 3rd Ed, W.B. Saunders, Philadelphia, 1982. B courtesy of the Warren Anatomical Museum, Harvard University. Montage from *MM Cohen Jr,* The Child with Multiple Birth Defects, Raven Press, New York, 1982.)

With Potter sequence, malformations of the genitourinary tract, lumbosacral spine, and lower intestinal tract are common (35) and constitute, in Opitz's terminology (107), a developmental field defect. Duncan et al (38) noted the frequent association of Müllerian duct anomalies, renal agenesis, and cervical and thoracic vertebral anomalies, and coined the term MURCS association.

Extensive reviews have been carried out by Curry et al (35) who analyzed 80 cases, Schmidt et al (140) who studied 23 families, and Roodhooft et al (126) and Al Saadi et al (3) who studied 41 and 21 families, respectively. Approximately 80% of Potter sequence cases are nonsyndromic (35), with the remaining 20% occurring in at least 56 different syndromes and associations to date, including 17 chromosomal, 6 autosomal dominant, 21 autosomal recessive, 1 X-linked, 2 teratogenic, 1 disruptive, and 8 of unknown genesis (Table 1–2).

The complex and variable renal pathology found with Potter sequence has been reviewed by Bernstein and his associates (9–11). In 80 retrospectively ascertained cases, Curry et al (35) found bilateral renal agenesis in approximately 21%, cystic dysplasia in approximately 48%, obstructive uropathy in approximately 25%, and other forms of renal pathology in the remaining 6%.

The prevalence of bilateral renal agenesis has been reported to vary from 1 per 3000 to 1 per 9000 births, predominating in males by a ratio of 2:1 or 3:1. The prevalence of unilateral renal agenesis at necropsy varies from 1 per 600 to 1 per 1000 (4,27,47,77,110,116,157).

Genetic evidence is accumulating to indicate that bilateral renal agenesis and cystic dysplasia should be nosologically grouped together. In several documented instances, bilateral renal agenesis has been observed in one sib and cystic dysplasia with or without unilateral renal agenesis has been found in the other sib (22,27,35,61). Affected sibs with either bilateral renal agenesis or cystic dysplasia have been observed on numerous occasions (7,20,22,23,32,61, 89,94,96,97,102,109,124,126,127,132,135–137,161,162). Autosomal dominant inheritance with reduced penetrance and variable expressivity is consistent with affected families in which a parent with unilateral renal agenesis has children with either unilateral or bilateral renal agenesis (12,16,20,23,59,82,103,134,168); Buchta et al (20) called

Table 1–2. Syndromes in which Potter sequence has been reported

| Syndromes | Renal pathology | References |
|---|---|---|
| Chromosomal | | |
| 46,XX, −3, +der(3)t(3;11)(p25;ql3.2)mat | Hypoplastic kidney | Curry et al (35) |
| Deletion short arm 4 | Bilateral renal agenesis | Mikelsaar et al (100) |
| Deletion short arm 5 | Bilateral renal agenesis | Egli and Stalder (40) |
| Duplication long arm 6 | Unilateral renal agenesis, unilateral cystic dysplasia | Curry et al (35) |
| Trisomy 7 | Cystic dysplasia, enlarged kidneys | Yunis et al (167); Pflueger et al (111) |
| Trisomy 8 | Cystic dysplasia, enlarged kidneys | Juberg et al (73) |
| Trisomy 9 | Other | Curry et al (35) |
| Trisomy 13 | Cystic dysplasia, small kidneys | Gilbert and Opitz (53) |
| Deletion long arm 15 | Cystic dysplasia, small kidneys | Clark (30) |
| Monosomy 16 mosaic | Obstructive uropathy | Harley et al (65) |
| Trisomy 18 | Obstructive uropathy | Frydman et al (49) |
| Duplication short arm 20 | Cystic dysplasia, small kidneys | Char et al (29) |
| Trisomy 21 | Bilateral renal agenesis | Egli and Stalder (40) |
| Duplication long arm 22 | Bilateral renal agenesis, obstructive uropathy | Egli and Stalder (40) |
| Turner | Cystic dysplasia, small kidneys, obstructive uropathy | Lubinsky et al (92); Wolf et al (164) |
| XYY | Cystic dysplasia, small kidneys | Côté et al (33); Machin (93); Sutherland et al (148) |
| Familial marker chromosome | Bilateral renal agenesis | Ferrandez and Schmid (44) |

Table 1–2. Syndromes in which Potter sequence has been reported (continued)

| Syndromes | Renal pathology | References |
|---|---|---|
| Autosomal dominant | | |
| Adult type polycystic kidney disease | Other | Proesmans et al (119); Shokeir (142) |
| Anal, ear, renal, and radial anomalies | Obstructive uropathy, other | Kurnit et al (85) |
| Beckwith–Wiedemann syndrome | Obstructive uropathy | Knight et al (78) |
| Branchio-oto-renal syndrome | Bilateral renal agenesis, cystic dysplasia, small kidneys | Fitch and Srolovitz (46); Melnick et al (99); Carmi et al (25) |
| EEC syndrome | Obstructive uropathy | Ivarsson et al (70) |
| Müllerian duct and renal anomalies | Renal agenesis | Biedel et al (12); Schimke and King (134) |
| Autosomal recessive | | |
| Abnormal renal tubular differentiation, microcephaly, and joint hypermobility | Obstructive uropathy | Allanson et al (2) |
| Acrorenal mandibular syndrome | Bilateral renal agenesis | Halal et al (63) |
| Bilateral renal agenesis, lens prolapse, and cataracts | Bilateral renal agenesis | Biedner (13) |
| Cerebro-oculofacio-skeletal syndrome | Bilateral renal agenesis | Preus et al (118) |
| Cryptophthalmos syndrome | Bilateral renal agenesis | Burn and Marwood (21); Codère et al (31); Kahler et al (75) |
| Cystic dysplasia, CNS malformations, and liver abnormalities | Cystic dysplasia, small kidneys | Miranda et al (101) |
| Elejalde syndrome | Cystic dysplasia, enlarged kidneys | Elejalde et al (41) |
| Familial renal tubular dysgenesis | Other | Saunders et al (133) |
| Glutaric aciduria, type II | Cystic dysplasia, enlarged or small kidneys | Kahler et al (76) |
| Infantile polycystic kidney disease | Other | Bernstein (9) |
| Meckel syndrome | Cystic dysplasia, enlarged kidneys | Mecke and Passarge (98) |
| Medullary dysplasia and cerebral dysgenesis | Other | Bernstein and Kissane (10) |
| Mesomelic dysplasia | Other | Rutledge et al (128) |
| Neonatal polycystic dysplasia and brain defect | Other | Goldston et al (54) |
| Prune belly, pulmonic stenosis, retardation, and hearing deficit | Obstructive uropathy | Lockhart et al (88) |
| Renal cystic dysplasia and cerebellar dysgenesis | Cystic dysplasia, small kidneys | Kornguth et al (83) |
| Renal dysplasia and asplenia | Cystic dysplasia, enlarged or small kidneys | Crawford (34) |
| Renal, genital, and middle ear anomalies | Bilateral renal agenesis | Schmidt et al (138); Winter et al (163) |
| Saldino–Noonan syndrome | Cystic dysplasia, enlarged kidneys | Spranger et al (145); Saldino and Noonan (130) |
| Smith–Lemli–Opitz syndrome, type II | Renal agenesis, cystic dysplasia | Curry et al (36) |
| Thymic aplasia, growth retardation, fetal death | Other | Shepard et al (141) |
| X-linked | | |
| Lenz microphthalmia syndrome | Bilateral renal agenesis | Hoefnagel et al (68) |
| Teratogenic | | |
| Diabetes mellitus | Bilateral renal agenesis or cystic dysplasia with small kidneys or obstructive uropathy | Grix et al (57) |
| Thalidomide | Bilateral renal agenesis or obstructive uropathy or other pathology | Pliess (112) |
| Disruptive | | |
| Amnion rupture sequence | Normal kidneys | See amnion rupture sequence section |
| Unknown genesis | | |
| Agnathia, tracheoesophageal fistula, duodenal atresia, and renal agenesis | Bilateral renal agenesis | Saito et al (129) |
| Congenital cystic adenomatoid malformation and bilateral renal agenesis | Bilateral renal agenesis | Krous et al (84) |
| Ear anomalies, cataracts, and cystic dysplasia | Cystic dysplasia, small kidneys | Wright et al (165) |
| Renal agenesis, cardiac anomalies, and skeletal defects | Bilateral renal agenesis | Holzgreve et al (69) |
| Renal dysplasia, mesomelia, and radiohumeral synostosis | Cystic dysplasia, small kidneys | Ulbright et al (155) |
| Renal dysplasia, pancreatic fibrosis, meconium ileus, and situs inversus | Cystic dysplasia, small kidneys | Yoshikawa et al (166) |
| Renal hypoplasia and ectrodactyly | Renal hypoplasia | Fitch and Lachance (45) |
| VATER association | Bilateral renal agenesis or cystic dysplasia with small kidneys or obstructive uropathy, or other pathology | Uehling et al (154) |

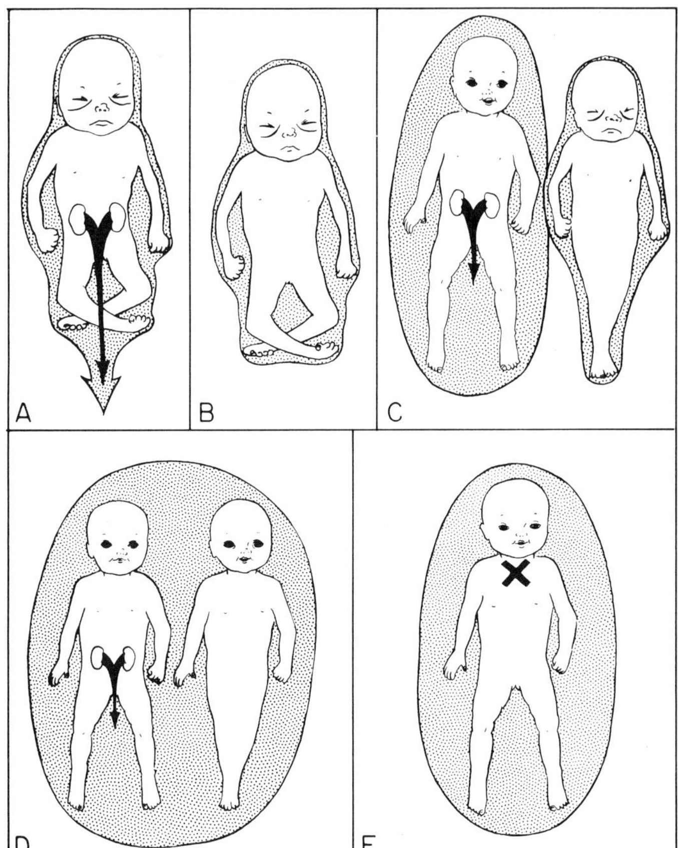


Fig. 1–9. *Potter sequence.* Oligohydramnios has different causes and, except under unusual circumstances, leads to facial and limb deformities of Potter sequence. Normally, small amounts of amnionic fluid cross the amnion as a transudate, but most amnionic fluid results from fetal urination. (A) Amnionic tear with chronic leakage of fluid leading to oligohydramnios, Potter facies, and limb positioning defects. Both kidneys are present and urination is normal. (B) Bilateral renal agenesis. (C) Monozygotic twins with separate amnions. Fetus on the left has kidneys, and enough fetal urine is contributed to amnionic fluid to protect fetus from deformities of Potter sequence. (D) Monozygotic twins sharing common amnionic sac. Note that although the fetus on the right has sirenomelia, Potter deformities are not present because the fetus on the left provides enough urine in amnionic fluid to protect co-twin from deformities of Potter sequence. (E) Fetus has bilateral renal agenesis and therefore does not contribute fetal urine to amnionic fluid. Potter sequence based on neurologic swallowing deficit; amniotic fluid crossing the amnion remains external to the fetus, protecting it from extrinsic, deforming forces. (From *MM Cohen Jr,* The Child with Multiple Birth Defects, Raven Press, New York, 1982.)

the condition hereditary renal adysplasia. In some of the affected families, second and third degree relatives as well as first degree relatives of probands with bilateral renal agenesis had unilateral renal agenesis (12,23,67,82,97). Roodhooft et al (126) ultrasonically investigated the parents and siblings of 41 probands with bilateral renal agenesis and/or cystic dysplasia and found renal abnormalities in 9%. McPherson et al (97a) suggested that hereditary renal adysplasia was more common than previously supposed and may account for most recurrences of bilateral renal agenesis, even when the parents are normal. They calculated penetrance to be between 50 and 90%. Offspring and affected or obligate heterozygotes have an empiric risk for bilateral severe involvement of 15–20%. Other manifestations of the gene may include cryptorchidism in males and Müllerian duct abnormalities in females (12,20,23,136).

Al Saadi et al (3) obtained family histories and performed renal ultrasonography on parents and sibs of 21 probands with renal dysplasia. Among the probands were 16 bilateral and 5 unilateral cases. Specifically excluded from this study were cases of bilateral renal agenesis and known syndromes, both chromosomal and nonchromosomal, with renal dysplasia. Empiric recurrence risk calculated from this family study was only 2.1%. Al Saadi et al (3) concluded that multicystic and aplastic types of renal dysplasia are usually sporadic and only rarely familial in contrast to other types of renal dysplasia identified in the literature as familial.

Obstructive uropathy as a cause of Potter sequence is also complex because of clinical and pathologic heterogeneity. Several possible sites for obstruction exist, particularly in the male. The marked male preponderance for obstructive uropathy is related to the more complex development of the male urethra (108). Clinical variation extends from early fetal lethality to survival with minimal morbidity. Recurrence of prune belly secondary to obstructive uropathy has most frequently involved affected brothers (19,51,56,82,123,159). Three affected sibs—two males and one female—have been reported in one instance (50). Affected male cousins (1,51) and three pairs of discordant monozygotic twins (71) have also been observed.

Family studies of probands with vesicoureteral reflux or duplication of the collecting system have shown that affected sibs have the same abnormality as the proband in 26–34% of cases with reflux (39,72) and in 19% of cases with ureteral duplication (160). Twenty-three percent of parents with children who had ureteral duplication had similar findings (5,160). Finally, six pairs of twins and three pairs of sibs with posterior urethral valves were reviewed by Livne et al (87). Their data suggested that obstructive urinary tract abnormalities may result from an autosomal dominant gene with reduced penetrance.

**Prenatal and perinatal factors.** Amnion nodosum, breech presentation (40%), intrauterine death (25%), and antepartum hemorrhage (15%) have been found in a large series. Cesarean delivery is required in about 10% of cases. About half the infants are small for date of birth. Respiratory insufficiency is common (6,116,121).

**Renal pathology.** Renal pathology is quite variable. Bilateral renal agenesis is a valid diagnosis only when no renal tissue is found on gross histopathologic examination. In such instances, the renal arteries and ureters are also absent and the bladder is hypoplastic, rudimentary, or absent (10).

In renal cystic dysplasia, abnormal renal organization may result from arrested development. Pathologic changes include primitive ducts, nests of cartilage, and primitive glomerular and tubular structures with aberrant relationships to one another. Kidneys may be enlarged, of normal size, or reduced in size. When involvement is unilateral, the other kidney is absent. Various abnormalities of the renal arteries, ureters, and bladder are common (10,11).

In infantile polycystic kidney disease, both kidneys are greatly enlarged with minute cysts on the surface. Medullary ductal dilatation is characteristic, and hepatic lesions with an increased number of portal bile ducts are observed (9).

Hypoplastic kidneys are distinguished from cystic dysplasia. The kidneys are extremely small but structurally normal with normal differentiation (35).

With intrauterine obstructive uropathy, the kidneys may be large and cystic, but may also be small, cystic, and dysplastic when destruction is a significant factor. Obstruction may occur in the distal urethra or the prostatic urethra. Atresia of the ureters is noted less commonly at the level of the bladder or the posterior urethral valves. Massive urinary ascites with distended abdomen may occur on occasion (35).

Medullary dysplasia is uncommon, but, when present, is reported to be accompanied by abnormalities of cortical tubular differentiation (35).

When Potter sequence results from amnionic leakage, both kidneys are normal (152).

**Craniofacial features.** A prominent semicircular skin fold extends from the inner canthus onto the cheek. When this fold is absent, some functioning kidney tissue may be present. Ocular hypertelorism has been observed. The nose may be blunted, with a turned-down nasal tip. A prominent crease is often present on the chin. The ears are low set, posteriorly angulated, large, and floppy with cartilaginous deficiency. Micrognathia has been observed in most cases. Cleft lip and/or palate have been noted in a few instances (110,113–115).

**Skin.** The skin is very dry, loose, and wrinkled, giving a prematurely senile appearance (43,48,114). Markedly hypoplastic nails have been observed in some instances (35). Less commonly, neck webbing may be noted (35). Prune belly may be observed in some instances and does not necessarily correlate with the pathologic diagnosis of obstructive uropathy, nor does lack of abdominal wall distension exclude the presence of urinary tract obstruction. The wrinkled, prune-like appearance is found only in those without abdominal distension. Variable degrees of abdominal distension with abdominal muscle deficiency may be observed (35).

**Skeletal and limb anomalies.** Large fontanels and wide sutures have been observed in some instances (35). Flexion contractures at the knees and hips, spadelike hands, genu varum, and talipes equinovarus are common (106,114). Other findings are quite variable and may include hyperextensibility of the knees and other joints, camptodactyly, hypoplastic arm, hypoplastic leg especially with obstructive uropathy, thoracic hemivertebrae, ischial and sacral aplasia or hypoplasia, other vertebral anomalies, and sirenomelia (1,4,24,35,46, 55,66,86,104,106,108,114–116,120,153).

Pathogenetic hypotheses about sirenomelia are particularly well reviewed by Stevenson et al (146). Their study indicated that sirenomelia and its commonly associated defects are produced by a vascular steal that diverts blood flow and nutrients from caudal structures of the embryo to the placenta. Arteries below the level of the steal vessel are underdeveloped and tissues dependent on them for nutrient supply fail to develop, become arrested in some incomplete stage of development, or are malformed. Thus, the single lower extremity in sirenomelia arises from failure of the lower limb bud field to be cleaved into two lateral masses by an intervening allantois.

**Genital anomalies.** Abnormalities may include cryptorchidism (33% of affected males and 100% of affected males with prune belly), gonadal hypoplasia, absent ductus deferens, absent seminal vesicles, rectovaginal fistula, absent uterus, bicornuate uterus, unicornuate uterus, blind-ended vagina, absent vagina, and masculinization of the external genitalia with 46,XX karyotype (4,12,20,26,35,91,115,116,124, 134,137).

**Lungs.** The lungs are hypoplastic with primitive or absent alveoli (67,114,116).

**Gastrointestinal system.** Imperforate anus, esophageal atresia, duodenal atresia, malrotation, Meckel diverticulum, and omphalocele have been observed (3,35,162).

**Cardiovascular anomalies.** Reported congenital heart defects include atrial septal defect, ventricular septal defect, patent ductus arteriosus, hypoplastic left ventricle, pulmonic stenosis, tetralogy of Fallot, pulmonary atresia, abnormal tricuspid valve, coarctation of the aorta, and single umbilical artery (3,35,162).

**Other findings.** Other findings may include brain malformations, diaphragmatic hernia, asplenia, polysplenia, adrenal hypoplasia, and accessory adrenal glands (3,35).

**Diagnosis, differential diagnosis, and laboratory aids.** Because of the heterogeneous causes of Potter sequence, diagnosis and recurrence risk counseling should be based on (a) careful clinical examination, (b) complete autopsy with special attention to renal histopathology, (c) chromosome analysis, (d) family history, and (e) renal ultrasound examination for parents of infants with renal agenesis and/or cystic dysplasia and more distant relatives if indicated by history or by demonstrated parental involvement based on ultrasound. Prenatal ultrasonography has identified oligohydramnios (35), renal agenesis and/or cystic dysplasia (8,28,37,62,74,125,139,140), and obstructive uropathy (17,131,144).

Syndromes with Potter sequence are listed in Table 1–2. The association of ear anomalies with genitourinary defects independent of Potter sequence has been discussed by several authors (14,90,150,156).

### References [Potter sequence (oligohydramnios sequence)]


1. Adeyokunnu AA, Fimilusi JB: Prune belly syndrome in 2 siblings and a first cousin. *Am J Dis Child* **136**:23–35, 1982.

2. Allanson JE et al: Possible new autosomal recessive syndrome with unusual renal histopathological changes. *Am J Med Genet* **16**:57–60, 1983.

3. Al Saadi AA et al: A family study of renal dysplasia. *Am J Med Genet* **19**:669–677, 1984.

4. Ashley DJ, Mostofi FK: Renal agenesis and dysgenesis. *J Urol* **83**:211–230, 1960.

5. Atwell JD et al: Familial incidence of bifid and double ureters. *Arch Dis Childh* **49**:390–393, 1974.

6. Bain AD, Scott JS: Renal agenesis and severe urinary tract dysplasia. A review of 50 cases, with particular reference to the associated anomalies. *Br Med J* **1**:841–846, 1960.

7. Baron C: Bilateral agenesis of the kidneys in two consecutive infants. *Am J Obstet Gynecol* **67**:667–670, 1954.

8. Bartley JA et al: Prenatal diagnosis of dysplastic kidney disease. *Clin Genet* **11**:375–378, 1977.

9. Bernstein J: Infantile polycystic disease, in *Nephrology,* Hamburger J et al (eds), Wiley, New York, 1979, pp. 1023–1031.

10. Bernstein J, Kissane JM: Hereditary disorders of the kidney, in *Perspectives in Pediatric Pathology,* Rosenberg HS, and Bolande RP (eds), Year Book Medical Publishers, Chicago, 1973, pp. 117–187.

11. Bernstein J, Gardner KD: Cystic disease of the kidney and renal dysplasia, in *Campbell's Urology,* Harrison JH et al (eds), Vol 2, W.B. Saunders, Philadelphia, 1979, pp. 1399–1442.

12. Biedel LW et al: Müllerian anomalies and renal agenesis: Autosomal dominant urogenital adysplasia. *J Pediatr* **104**:861–864, 1984.

13. Biedner B: Potter's syndrome with ocular anomalies. *J Pediatr Ophthalmol Strabismus* **17**:172–174, 1980.

14. Blanc WA, Baens G: Ear malformations, abnormal facies, and genitourinary tract anomalies. *Am J Dis Child* **100**:781–782, 1960.

15. Bloch B: Sirenomelia (sympodia or mermaid deformity). *S Afr Med J* **52**:196–200, 1977.

16. Bound JP: Two cases of congenital absence of one kidney in the same family. *Br Med J* **2**:747, 1943.

17. Bovichelli L et al: Prenatal diagnosis of the prune belly syndrome. *Clin Genet* **18**:79–82, 1980.

18. Braun O: Weitere Beiträge zum Erscheinungsbild der Dysplasia renofacialis. *Zentralbl Allg Pathol* **107**:175–182, 1965.

19. Bronzini E, Moscatelli P: L'aplasia della muscolature della parete addominale associata a malformazioni genitourinarie in due fratellini. *Pathologica* **65**:127–136, 1973.

20. Buchta RM et al: Familial bilateral renal agenesis and hereditary renal adysplasia. *Z Kinderheilkd* **115**:111–129, 1973.

21. Burn J, Marwood RP: Fraser syndrome presenting as bilateral renal agenesis in three sibs. *J Med Genet* **19**:360–361, 1982.

22. Cain DR et al: Familial renal agenesis and total dysplasia. *Am J Dis Child* **128**:377–380, 1974.

23. Carey JC et al: Bilateral and unilateral renal agenesis in the same family. Abstract. National Foundation March of Dimes Birth Defects Meeting, San Diego, California, 1981.

24. Carey JC et al: Lower limb deficiency and the urethral obstruction sequence. *Birth Defects* **18**(3B):19–28, 1982.

25. Carmi R et al: The branchio-oto-renal (BOR) syndrome: Report of bilateral renal agenesis in three sibs. *Am J Med Genet* **14**:625–627, 1983.

26. Carpentier PJ, Potter EL: Nuclear sex and genital malformation in 48 cases of renal agenesis with especial reference to nonspecific female pseudohermaphroditism. *Am J Obstet Gynecol* **78**:235–248, 1959.

27. Carter CO et al: A family study of renal agenesis. *J Med Genet* **16**:176–188, 1979.

28. Cass A et al: Prenatal diagnosis of fetal urinary tract abnormalities by ultrasound. *Urology* **18**:197–202, 1981.

29. Char F et al: Trisomy 20p with features of oligohydramnios sequence.

Abstract. National Foundation March of Dimes Birth Defects Meeting, San Diego, California, 1981.

30. Clark RD: Letter to the editor: del(15)(q22q24) syndrome with Potter sequence. *Am J Med Genet* **19**:703–705, 1984.

31. Codère F et al: Cryptophthalmos syndrome with bilateral renal agenesis. *Am J Ophthalmol* **91**:737–742, 1981.

32. Cole BR et al: Bilateral renal dysplasia in three siblings: Report of a survivor. *Clin Nephrol* **5**:83–87, 1975.

33. Côté GG et al: Oligohydramnios syndrome and XYY karyotype. *Ann Génét* **21**:226–228, 1978.

34. Crawford MP: Renal dysplasia and asplenia in two sibs. *Clin Genet* **14**:338–344, 1978.

35. Curry CJR et al: The Potter sequence: A clinical analysis of 80 cases. *Am J Med Genet* **19**:679–702, 1984.

36. Curry CJR et al: Smith–Lemli–Opitz syndrome—type II: Multiple congenital anomalies with male pseudohermaphroditism and frequent early lethality. *Am J Med Genet* **26**:45–57, 1987.

37. Dubbins PA et al: Renal agenesis: Spectrum of in utero findings. *J Clin Ultrasound* **9**:189–193, 1981.

38. Duncan PP et al: The MURCS association: Müllerian duct aplasia, renal aplasia, and cervicothoracic dysplasia. *J Pediatr* **95**:399–402, 1979.

39. Dwoskin JY: Sibling uropathology. *J Urol* **115**:726–727, 1976.

40. Egli F, Stalder G: Malformation of kidney and urinary tract in common chromosomal aberrations. *Humangenetik* **18**:1–32, 1973.

41. Elejalde BR et al: Acrocephalopolydactylous dysplasia. *Birth Defects* **13**(3B):53–67, 1977.

42. Fantel A, Shepard T: Potter's syndrome—nonrenal features induced by oligoamnios. *Am J Dis Child* **129**:1346–1347, 1975.

43. Feinzaig W: Multicystic dysplastic kidney. A clinical and pathological study of 29 cases. Thesis, University of Minnesota, 1964.

44. Ferrandez A, Schmid W: Potter-syndrom (Nierenagenesie) mit chromosomaler Aberration beim Patient und Mosaik beim Vater. *Helv Pediatr Acta* **26**:210–214, 1971.

45. Fitch N, Lachance RC: The pathogenesis of Potter's syndrome of renal agenesis. *Can Med Assoc J* **107**:653–656, 1972.

46. Fitch N, Srolovitz H: Severe renal dysgenesis produced by a dominant gene. *Am J Dis Child* **130**:1356–1357, 1976.

47. Fraga JR et al: Association of pulmonary hypoplasia, renal anomalies and Potter's facies. *Clin Pediatr* **12**:150–153, 1973.

48. François J, Marchildon A: Facies de Potter et agénesie renale. *Ann Ocul (Paris)* **197**:347–354, 1964.

49. Frydman M et al: Chromosome abnormalities in infants with prune belly anomaly: Association with trisomy 18. *Am J Med Genet* **15**:145–148, 1983.

50. Gaboardi F et al: Prune belly syndrome: Report of three siblings. *Helv Pediatr Acta* **37**:283–288, 1982.

51. Garlinger P, Ott J: Prune belly syndrome—possible genetic implications. *Birth Defects* **10**(8):173–180, 1974.

52. Gärtner H et al: Beitrag zum Problem der renofacialen Dysplasie (Potter-Syndrom). Beziehungen zwischen dem typischen Bild und der Sirenomelie. *Pädiatr Pädiol* **9**:209–216, 1974.

53. Gilbert E, Opitz J: Renal involvement in genetic hereditary malformation syndromes, in *Nephrology,* Hamburger J et al (eds), Wiley, New York, pp 909–943.

54. Goldston AS et al: Neonatal polycystic kidney with brain defect. *Am J Dis Child* **106**:484–488, 1963.

55. Greene LF et al: Urologic abnormalities associated with congenital absence or deficiency of abdominal musculature. *J Urol* **52**:217–299, 1952.

56. Grenet P et al: Aplasie congénitale de la paroi abdominale—À propos d'un cas familial. *Ann Pédiatr* **19**:523–528, 1972.

57. Grix A et al: Patterns of multiple malformations in infants of diabetic mothers. *Birth Defects* **18**(3A):55–77, 1982.

58. Gross H et al: Familiäre Balkenhypoplasie bei Dysplasia renofacialis. *Zentralbl Allg Pathol* **99**:587–592, 1959.

59. Grovoy JD et al: Unilateral renal agenesis in two siblings. *Pediatrics* **29**:270–273, 1962.

60. Habedank M: 18 Beobachtungen von Dysplasia renofacialis. *Z Kinderheilkd* **88**:531–547, 1963.

61. Hack M et al: Familial aggregation in bilateral renal agenesis. *Clin Genet* **5**:173–177, 1974.

62. Hadlock FP et al: Sonography of fetal urinary tract anomalies. *Am J Radiol* **137**:261–267, 1981.

63. Halal F et al: Acrorenal mandibular syndrome. *Am J Med Genet* **5**:277–284, 1980.

64. Hammar I, Roggenkamp K: Augenveränderungen bei Dysplasia renofacialis (Potter–Syndrom). *Klin Monatsbl Augenheilkd* **15**:534–538, 1967.

65. Harley LM et al: Prune belly syndrome. *J Urol* **108**:174–176, 1972.

66. Henley WL, Hyman A: Absent abdominal musculature, genitourinary anomalies and deficiency of pelvic autonomic nervous system. *Am J Dis Child* **86**:795–798, 1958.

67. Hilson D: Malformation of ears as a sign of malformation of genitourinary tract. *Br Med J* **2**:785–789, 1957.

68. Hoefnagel D et al: Heredofamilial bilateral anophthalmia. *Arch Ophthalmol* **69**:760–764, 1963.

69. Holzgreve W et al: Bilateral renal agenesis with Potter phenotype, cleft palate, anomalies of the cardiovascular system, skeletal anomalies including hexadactyly and bifid metacarpal. A new syndrome? *Am J Med Genet* **18**:177–182, 1984.

70. Ivarsson S et al: Coexisting ectrodactyly–ectodermal dysplasia–clefting (ECC) and prune belly syndromes. *Acta Radiol Diag* **23**:287–292, 1982.

71. Ives EJ: The abdominal muscle deficiency triad syndrome: Experience with 10 cases. *Birth Defects* **10**(4):127–135, 1974.

72. Jerkins GR, Noe HN: Familial vesicoureteral reflux: A prospective study. *J Urol* **128**:774–778, 1982.

73. Juberg RC et al: Trisomy C in an infant with polycystic kidneys and other malformations. *J Pediatr* **76**:598–603, 1970.

74. Kaffe S et al: Prenatal diagnosis of bilateral renal agenesis. *Obstet Gynecol* **49**:478–480, 1976.

75. Kahler SG et al: Bilateral renal agenesis in the Fraser cryptophthalmos syndrome. Abstract. David W. Smith Conference on Malformations and Morphogenesis, San Diego, California, 1981.

76. Kahler SG et al: Glutaric acidura type II (multiple acyl-CoA dehydrogenase deficiency): A teratogenic, carcinogenic inborn error of metabolism. Abstract. David W. Smith Conference on Malformations and Morphogenesis, Vancouver, British Columbia, 1983.

77. Källén B, Winberg J: Caudal mesoderm pattern of anomalies: From renal agenesis to sirenomelia. *Teratology* **9**:99–111, 1974.

78. Knight JA et al: Association of the Beckwith–Wiedemann and prune belly syndromes. *Clin Pediatr* **19**:485–488, 1980.

79. Kohler H: Fetal abnormality. *Lancet* **1**:946, 1961.

80. Kohler HG: An unusual case of sirenomelia. *Teratology* **6**:295–302, 1972.

81. Kohn G: Agénesie des muscles de la paroi abdominal chez un nouveau-né. *Prog Med* **32**:1328–1332, 1985.

82. Kohn G, Borns PK: The association of bilateral and unilateral renal aplasia in the same family. *J Pediatr* **83**:95–97, 1973.

83. Kornguth S et al: Defect of cerebellar Purkinje cell histogenesis associated with type I and type II renal cystic disease. *Acta Neuropathol* **40**:1–9, 1977.

84. Krous HF et al: Congenital cystic adenomatoid malformation in bilateral renal agenesis: Its mitigation of Potter's syndrome. *Arch Pathol Lab Med* **104**:368–370, 1980.

85. Kurnit DM et al: Autosomal dominant transmission of a syndrome of anal, ear, renal and radial congenital malformations. *J Pediatr* **93**:270–273, 1978.

86. Lattimer JK: Congenital deficiency of the abdominal musculature and associated genito-urinary anomalies: A report of 22 cases. *J Urol* **79**:343–352, 1958.

87. Livne PM et al: Genetic etiology of posterior urethral valves. *J Urol* **130**:781–784, 1983.

88. Lockhart JI et al: Siblings with prune belly syndrome and associated pulmonic stenosis, mental retardation and deafness. *Urology* **14**:140–142, 1979.

89. Loendersloot EW et al: Bilateral renal agenesis (Potter's syndrome) in 2 consecutive infants. *Eur J Obstet Gynecol Reprod Biol* **8**:137–142, 1978.

90. Longenecker CG et al: Malformation of the ear as a clue to urogenital anomalies. *Plast Reconstr Surg* **35**:303–309, 1965.

91. Lubinsky MS: Female pseudohermaphroditism and associated anomalies. *Am J Med Genet* **6**:123–136, 1980.

92. Lubinsky M et al: The association of "prune belly" with Turner's syndrome. *Am J Dis Child* **134**:1171–1172, 1980.

93. Machin GA: Urinary tract malformation in the XYY male. *Clin Genet* **14**:370–372, 1978.

94. Madisson H: Über das Fehlen beider Nieren. *Zentralbl Allg Path* **60**:1–8, 1934.

95. Marras A et al: Oligohydramnios and extrarenal abnormalities in Potter syndrome. *J Pediatr* **102**:597–598, 1983.

96. Mauer SM et al: Unilateral and bilateral renal agenesis in monoamniotic twins. *J Pediatr* **84**:236–238, 1974.

97. McPherson E: Unilateral and bilateral renal agenesis: Implications for genetic counseling. Abstract No. 277. Am Soc Human Genet, Detroit, Michigan, 1982.

97a. McPherson E et al: Dominantly inherited renal adysplasia. *Am J Med Genet* **26**:863, 1987.

98. Mecke S, Passarge E: Encephalocele, polycystic kidneys and polydactyly as an autosomal recessive trait simulating certain other disorders: The Meckel syndrome. *Ann Génét* **14**:97–103, 1971.

99. Melnick M et al: Familial branchio-oto-renal dysplasia: A new addition to the branchial arch syndromes. *Clin Genet* **9**:25–34, 1976.

100. Mikelsaar AVN et al: A 4p− syndrome: A case report. *Humangenetik* **193**:345–347, 1973.

101. Miranda D et al: Familial renal dysplasia. *Arch Pathol* **93**:483, 1972.

102. Morillo-Cucci G et al: Two sibs with bilateral renal agenesis. *Birth Defects* **10**(4):169–170, 1974.

103. Müntefering H, Schluter I: Beitrag zur Aetiologie der doppelseitigen Nierenagenesie. *Z Morphol Anthropol* **58**:253–285, 1967.

104. Obrinski W: Agenesis of abdominal muscles with associated malformation of the genitourinary tract. *Am J Dis Child* **77**:362–373, 1949.

105. Oppermann J: Beitrag zur Dysplasia reno-facialis. *Monatsschr Kinderheilkd* **114**:397–400, 1966.

106. Oppermann J: Vier Beobachtungen von Dysplasia reno-facialis. *Zentralb Gynäkol* **89**:705–710, 1967.

107. Opitz JM: The developmental field concept in clinical genetics. *J Pediatr* **101**:805–809, 1982.

108. Pagon RA et al: Urethral obstruction malformation complex: A cause of abdominal muscle deficiency in "prune belly." *J Pediatr* **94**:900–906, 1979.

109. Pashayan HM et al: Bilateral absence of the kidneys and ureters. *J Med Genet* **14**:205–209, 1977.

110. Passarge E, Sutherland JM: Potter's syndrome. *Am J Dis Child* **109**:80–84, 1965.

111. Pflueger MV et al: Trisomy 7 and Potter syndrome. *Clin Genet* **25**:543–548, 1984.

112. Pliess G: Thalidomide and congenital abnormalities. *Lancet* **1**:1128, 1962.

113. Potter EL: Bilateral renal agenesis. *J Pediatr* **29**:68–76, 1946.

114. Potter EL: Facial characteristics of infants with bilateral renal agenesis. *Am J Obstet Gynecol* **51**:885–888, 1946.

115. Potter EL: Bilateral absence of ureters and kidneys. *Obstet Gynecol* **25**:3–12, 1965.

116. Potter EL: *Normal and Abnormal Development of the Kidney*. Year Book Medical Publishers, Chicago, 1972.

117. Pramanik AK et al: Prune-belly syndrome associated with Potter (renal nonfunction) syndrome. *Am J Dis Child* **131**:672–674, 1977.

118. Preus M et al: Renal anomalies and oligohydramnios in the cerebro-oculofacio-skeletal syndrome. *Am J Dis Child* **131**:62–64, 1977.

119. Proesmans W et al: Autosomal dominant polycystic kidney disease in the neonatal period: Association with a cerebral arteriovenous malformation. *Pediatrics* **70**:971–975, 1982.

120. Rahs Z, Forbes M: Intrauterine atrophy and gangrene in the lower extremity of the fetus caused by megacystis due to urethral atresia. *J Pathol* **104**:31–35, 1971.

121. Ratten GJ et al: Obstetric complications when the fetus has Potter's syndrome. I. Clinical considerations. *Am J Obstet Gynecol* **115**:890–896, 1973.

122. Regnier C et al: Agénesie renale bilaterale. *Toulouse Méd* **64**:539–547, 1963.

123. Riccardi VM, Grum CM: The prune belly anomaly, heterogeneity and superficial X-linkage mimicry. *J Med Genet* **14**:266–270, 1977.

124. Rizza JM, Downing SE: Bilateral renal agenesis in two female siblings. *Am J Dis Child* **121**:60–63, 1971.

125. Romero R et al: Antenatal diagnosis of renal anomalies with ultrasound. *Am J Obstet Gynecol* **1**:38–43, 1985.

126. Roodhooft AM et al: Familial nature of congenital absence and severe dysgenesis of both kidneys. *N Engl J Med* **310**:1341–1345, 1984.

127. Rosenfeld L: Renal agenesis. *JAMA* **170**:1247–1248, 1959.

128. Rutledge JC et al: Ultrasonographic in utero detection of mesomelic dwarfism with renal hypoplasia. Abstract. National Foundation March of Dimes Birth Defects Meeting, San Diego, California, 1981.

129. Saito R et al: Anomalies of the auditory organ in Potter's syndrome. *Arch Otolaryngol* **108**:484–488.

130. Saldino RM, Noonan CD: Severe thoracic dystrophy with striking micromelia, abnormal osseous development, including the spine, and multiple visceral anomalies. *Am J Roentgenol* **114**:257–263, 1972.

131. Sanders R, Graham D: Twelve cases of hydronephrosis in utero diagnosed by ultrasonography. *J Ultrasound Med* **1**:341–348, 1982.

132. Sangal PR et al: Recurrent bilateral renal agenesis. *Am J Obstet Gynecol* **155**:1078–1079, 1986.

133. Saunders B et al: Potter phenotype secondary to an inherited defect in renal development. Abstract. March of Dimes 14th Annual Birth Defects Conference, San Diego, California, 1981.

134. Schimke RN, King CR: Hereditary urogenital adysplasia. *Clin Genet* **18**:417–420, 1980.

135. Schinzel A et al: Bilateral renal agenesis in 2 male sibs born to consanguineous parents. *J Med Genet* **15**:314–316, 1978.

136. Schinzel A et al: Monozygotic twinning and structural defects. *J Pediatr* **95**:921–930, 1979.

137. Schlegel RJ et al: An XX sex chromosome complement in an infant having male-type external genitals, renal agenesis, and other anomalies. *J Pediatr* **69**:812–814, 1966.

138. Schmidt ECH et al: Renal aplasia in sisters. *Acta Pathol* **54**:403–406, 1952.

139. Schmidt W, Kubli F: Early diagnosis of severe congenital malformations by ultrasound. *J Perinat Med* **10**:233–241, 1982.

140. Schmidt W et al: Genetics, pathoanatomy and prenatal diagnosis of Potter I syndrome and other urogenital tract diseases. *Clin Genet* **22**:105–127, 1982.

141. Shepard MK et al: Familial thymic aplasia with intrauterine growth retardation and fetal death: A new syndrome or a variant of DiGeorge syndrome. *Birth Defects* **12**(6):123–125, 1976.

142. Shokeir MHK: Expression of "adult" polycystic renal disease in the fetus and newborn. *Clin Genet* **14**:61–72, 1978.

143. Smith DW et al: The Duhamel anomalad, from imperforate anus to sirenomelia including VATER association, caudal regression syndrome, and Rokitansky syndrome. Unpublished manuscript.

144. Smythe AR: Ultrasonic detection of fetal ascites and bladder dilation with resulting prune belly. *J Pediatr* **98**:978–980, 1981.

145. Spranger J et al: Short rib-polydactyly (SRP) syndromes, types Majewski and Saldino–Noonan. *Z Kinderheilkd* **116**:73–94, 1974.

146. Stevenson RE et al: Vascular steal: The pathogenetic mechanism producing sirenomelia and associated defects of the viscera and soft tissues. *Pediatrics* **78**:451–457, 1986.

147. Stockenhausen HB von: Beitrag zur Problematik der Dysplasia renofacialis. *Z Kinderheilkd* **105**:303–323, 1969.

148. Sutherland GR et al: XYY males in Victoria. *Med J Aust* **1**:1249–1252, 1972.

149. Sylvester PE, Hughes DR: Congenital absence of both kidneys. *Br Med J* **1**:77–79, 1954.

150. Taylor WC: Deformity of ears and kidneys. *Can Med Assoc J* **93**:107–110, 1965.

151. Ten Berg BS, Wildervanck LS: Familiarie congenitale afwijkingen van uropoetisch en genitaal systeem. *Ned Tijdschr Geneeskd* **95**:2389–2395, 1951.

152. Thomas IT, Smith DW: Oligohydraminios, the cause of the non-renal features of Potter's syndrome, including pulmonary hypoplasia. *J Pediatr* **84**:811–814, 1974.

153. Tuch BA, Smith TK: Prune belly syndrome: A report of twelve cases and review of literature. *J Bone Joint Surg* **60A**:109–111, 1978.

154. Uehling DT et al: Urologic implications of the VATER association. *J Urol* **129**:352–354, 1983.

155. Ulbright CE et al: New syndrome: Renal dysplasia, mesomelia and radiohumeral fusion. *Am J Med Genet* **17**:667–668, 1984.

156. Vincent RW et al: Malformation of ear associated with urogenital anomalies. *Plast Reconstr Surg* **28**:214–220, 1961.

157. Waardenburg PJ: Einseitige Aplasie der Neire und ihrer Abfuhrwege bei beiden eineiigen Zwillingspaarlingen. *Acta Genet Med (Roma)* **1**:317–320, 1952.

158. Warkany J: *Congenital Malformations*. Year Book Medical Publishers, Chicago, 1971, pp 1037–1039.

159. Welling P et al: Beobachtungen zum Bauchmuskelaplasie-Syndrom. *Z Kinderheilkd* **118**:315–335, 1975.

160. Whitaker J, Danks DM: A study of the inheritance of duplication of the kidneys and ureters. *J Urol* **95**:176–178, 1966.

161. Whitehouse W, Mountrose U: Renal agenesis in non-twin siblings. *Am J Obstet Gynecol* **116**:880–881, 1973.

162. Wilson RD, Hayden MR: Brief clinical report: Bilateral renal agenesis in twins. *Am J Med Genet* **21**:147–152, 1985.

163. Winter JS et al: A familial syndrome of renal, genital, and middle ear anomalies. *J Pediatr* **72**:88–93, 1968.

164. Wolf EL et al: Diagnosis of oligohydramnios-related pulmonary hypoplasia (Potter's syndrome): Value of portable voiding cystourethrography in newborns with respiratory distress. *Pediatr Radiol* **125**:769–773, 1977.

165. Wright CG et al: Ear anomalies in an infant with Potter's syndrome. *Am J Otol* **3**:134–138, 1981.

166. Yoshikawa Y et al: Bilateral renal dysplasia accompanied by pancreatic fibrosis, meconium ileus, and situs inversus totalis. *Acta Pathol Jpn* **31**:845–852, 1981.

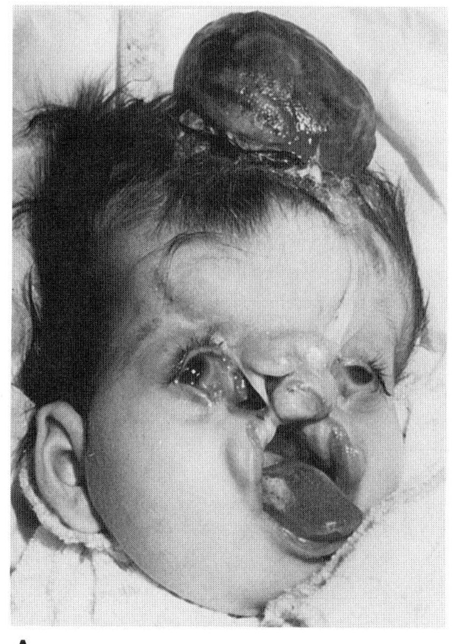
A

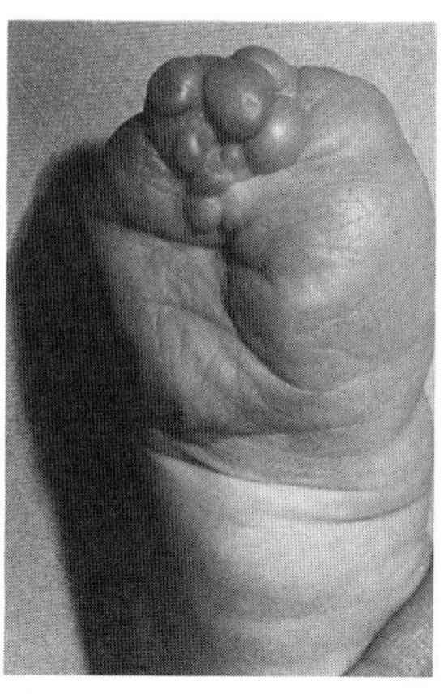
B

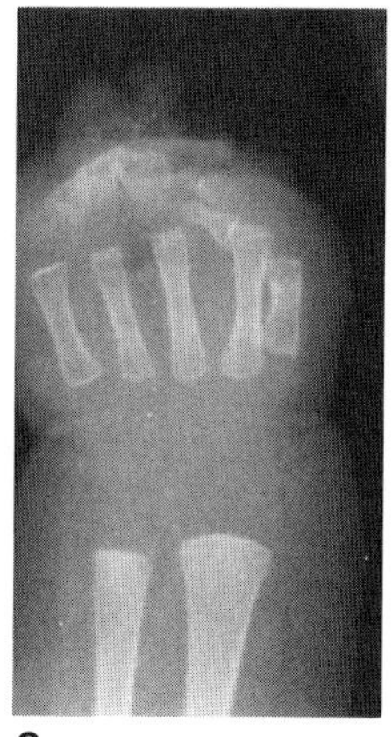
C

Fig. 1–10. *Amnion rupture sequence.* (A) Agenesis of membranous skull bones with protrusion of cranial contents. Bilateral facial clefts extending through premaxilla, lips, lateral nose, medial orbits, and skull. Note tissue band traversing forehead. (B,C) Hand from patient shown in A. Note severe digital amputations. (From *MS Granick et al,* Plast Reconstr Surg **80:**829, 1987.)

167. Yunis E et al: Full trisomy 7 and Potter's syndrome. *Hum Genet* **54**:13–18, 1980.

168. Zonana J et al: Renal agenesis—a genetic disorder. *Clin Res* **10**:420 (abstract), 1976.

## Amnion rupture sequence

Amnion rupture sequence may result in limb reduction defects, amputations, ring constrictions, distal syndactyly, talipes equinovarus, short umbilical cord, craniofacial disruptions and clefts, neural tube defects, limb/body wall deficiency, gastroschisis, extrathoracic heart, scoliosis, growth restriction, and various other anomalies (Figs. 1-10 to 1-14). Severity of amnionic rupture sequence varies from a single ring constriction or amputation to such severe involvement that viability is not possible. The condition has also been known as amnionic band syndrome, aberrant tissue band syndrome, amnionic band lesions, and ADAM complex (3,5,5a,16,18,29,30,34,42). Large series have been reported by numerous authors (1,3,5,5a,16,18,30,36,39–41). Craniofacial involvement has been emphasized by several authors (3,16,18,20,24).

Amnionic rupture may lead to all three basic types of anomalies—disruptions, deformations, and malformations. Disruptions are caused by adhesions, by tearing, and by constriction by amnionic bands. Shepard et al (37a) indicated that facial clefting, anteriorly located encephaloceles, and pseudoanencephaly might be caused by membranes adhering to the perforating buccopharyngeal membrane and lat-

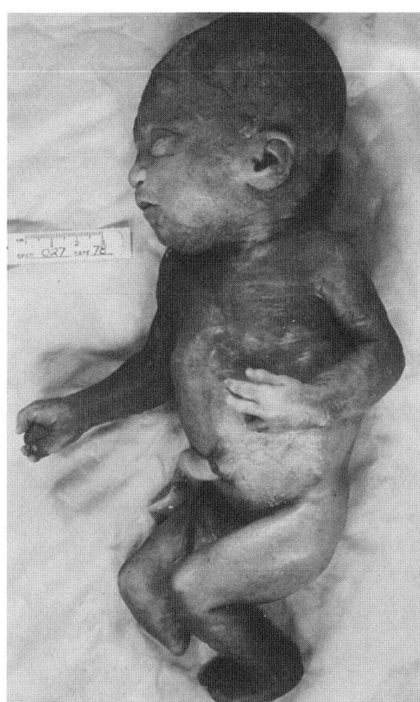
A

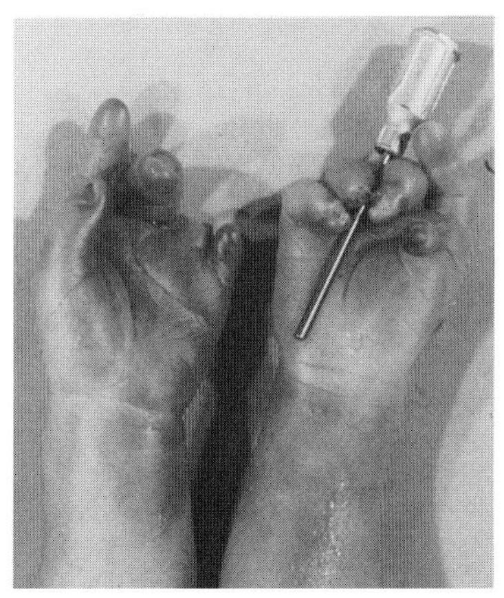
B

Fig. 1–11. *Amnion rupture sequence.* (A) Amputations involving fingers of right hand and right leg. Note Potter facies resulting from oligohydramnios. (B) Hands from patient shown in A. Note amputations and distal syndactyly. (Courtesy of *M Barr,* Ann Arbor, Michigan.)

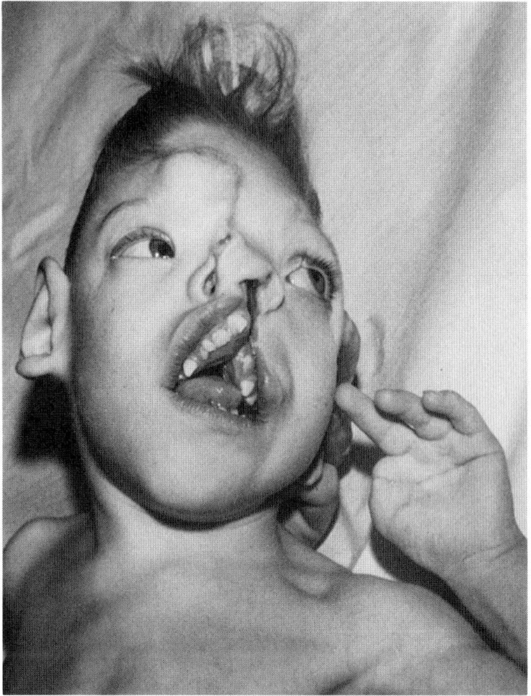

Fig. 1–12. *Amnion rupture sequence.* Bizarre facial clefting and asymmetric encephaloceles. Note amputation of thumb and index finger. (From *MM Cohen Jr,* The Child with Multiple Birth Defects, Raven Press, New York, 1982, p 22.)

eral edges of the closing neural folds between 23 and 35 days of development. If bands are present during the embryonic period, they may interfere with normal embryogenesis, resulting in malformations. Deformations result from oligohydramnios, which leads to intrauterine crowding and tethering of fetal parts. Severe constraint leads to vascular engorgement, hemorrhage, edema, and tissue necrosis resulting in severe disruptions such as limb/body wall defects. Finally, some

Fig. 1–13. *Amnion rupture sequence.* Cloverleaf skull in infant with bilateral oblique facial clefts. Patient also had amputation of digits. (From *A Schuch* and *HJ Pesch,* Z Kinderheilkd **109:**187, 1971.)

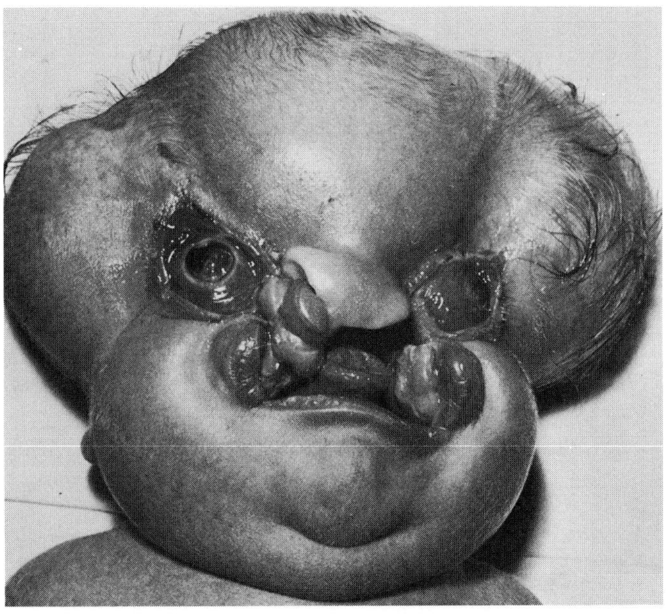

malformations that cannot be explained by bands, constraint, or compression may occur with amnionic rupture sequence anomalies. In such instances, disruptive and nondisruptive abnormalities may have a common primary etiology in some cases and may occur by chance in others (1,14–16,19,22,25,31,39–41).

Baker and Rudolph (1) estimated that congenital ring constrictions and amputations occurred with a frequency of 1/10,000 live births. Herva et al (13) noted a cluster of severe amnionic band cases (9 in 17 months) in Finland, Ossipoff and Hall (30) reported that 1 in 1300 pregnancies had amnionic rupture sequence, and Bryne et al (5) gave a frequency of one per 5,000–15,000 births, but indicated that the condition was relatively common among abortuses. On the basis of ultrasound at 18 weeks *in utero,* Papp et al (32) reported amnionic bands without signs of amnionic rupture or other fetal malformations in 5 of 12,131 consecutive pregnancies. They suggested that "innocent bands" may exist. Kalousek and Bamforth (16a) found amnion rupture sequence was common in previable fetuses, the prevalence being 1 in 56.

The best epidemiologic study is that of Garza et al (9). Using amputation or ring constriction as a minimal criterion, of 388,325 live births the birth prevalence for amnionic rupture sequence was 1.17 per 10,000. The prevalence for males was 0.91 and for females was 1.44. Defects occurred 1.76 times more often in blacks than in whites. Infants of young, black multigravidas showed the highest rate (6.2) and infants of older black multigravidas showed the lowest rate (0.5). Maternal age effect was not observed in black primagravidas or in white mothers. Birth weight was below 2,500 g in 49% of amnionic rupture sequence cases compared with 6.8% among all live births in the Atlanta area. The known case fatality rate was 30%.

The overwhelming majority of cases are sporadic. However, a few examples have been familial (7,21,22). Although most reported identical twins have been discordant (6,17,30,39), concordant twins have also been noted (8,44). Gellis (10) reviewed two families in which father and son had ring constrictions on the terminal phalanges of the same fingers. Such cases are probably best understood on the basis of Streeter dysplasia of genetic origin. Amnionic disruption sequence has been associated with amniocentesis (27).

**Limb defects.** Limb defects are the most commonly observed abnormalities with amnionic rupture sequence (Figs.1–10B,C and 1–11A,B). Findings may include ring constrictions, lymphedema below the ring constriction, congenital amputations of one or more limbs or digits, distal syndactyly, and talipes equinovarus (1,5a,28,33,39). Less common and more unusual abnormalities have included absent limb, oligodactyly, arthrogryposis, single forearm bone, single lower leg bone, radial and ulnar hypoplasia, ectrodactyly, and preaxial polydactyly (40). Proximal syndactyly has also been noted (11). Hall (12a) noted a transposed arm secondary to amnionic band disruption.

**Craniofacial anomalies.** With increasing severity of disruption, craniofacial anomalies may include severe microcephaly with deficiency of the anterior calvaria; asymmetric, usually anteriorly located, and sometimes multiple encephaloceles, microphthalmia; distortion and disruption of the palpebral fissures; various nasal disruptions; cleft lip, cleft palate, and bizarre facial clefts; and aberrant tissue bands about the face (Fig. 1–10A). Potter deformities (Fig. 1–11) and anencephaly have been observed. Various malformations have been reported including holoprosencephaly, septo-optic dysplasia, cloverleaf skull, hydrocephaly, ocular hypertelorism, uveal coloboma, choanal atresia, unilateral proboscis, Robin sequence, and other malformations (2,3,5a,14–16,18,20,22a,24,30a,37,40,41). Neurologic manifestations in survivors have been discussed by Chen and Gonzalez (5a).

**Thoracic, abdominal, and other defects.** Body wall deficiency may include thoracoschisis, abdominoschisis, thoracoabdominoschisis, ectopia cordis, gastroschisis, omphalocele, and even extrophy of the bladder. Many other malformations have been reported including various cardiovascular anomalies, abnormal lung lobulation, ab-

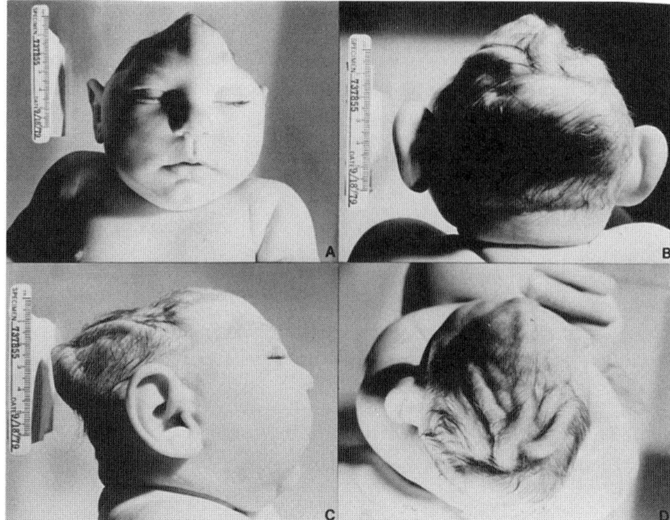


Fig. 1–14. *Fetal brain disruption sequence.* (A–D) Severe microcephaly, ridged and overriding sutures, collapse of skull, prominent occipital bone, and scalp rugae. (From *LJ Russell et al,* Am J Med Genet **17:**509, 1984.)

sent or abnormal diaphragm, abnormalities of intestinal rotation, anal atresia, absent gonads, and anomalous external genitalia (14–16, 25,31,38,40,41). Scoliosis may be observed (25). Low frequency malformations have been tabulated by several authors (15,40,41). Placenta and membranes have been discussed by Van Allen et al (40). Sachdev et al (35) reported a case of amnionic rupture sequence associated with an incompetent cervix.

**Differential diagnosis.** Amnionic band defects may mimic frontonasal dysplasia, ocular hypertelorism, branchial arch dysplasias, Meckel syndrome, cryptophthalmos, arthrogryposis, pterygium-related entities, scalp defect/limb reduction, and ectrodactyly (12). Genetic type ring constrictions of the terminal phalanges (10) may mimic amnionic band constrictions and amputations. Differences between amniogenic and genetic or teratogenic limb anomalies have been discussed by Baker and Rudolph (1). On occasion, postnatal hair strangulation of toes during infancy may mimic amnionic band ring constrictions or amputations of digits (26). Amnionic bands have been observed with severe *osteogenesis imperfecta* (42,43), *type IV Ehlers-Danlos syndrome* (43), and *epidermolysis bullosa* (23).

**Laboratory aids.** Elevated amnionic $\alpha$-fetoprotein levels are often found (4). Prenatal diagnosis by ultrasonography is possible in patients with severe craniofacial defects (5a).

### References (Amnion rupture sequence)


1. Baker CJ, Rudolph AD: Congenital ring constrictions and intrauterine amputations. *Am J Dis Child* **121**:393–400, 1971.

2. BenEzra D, Frucht Y: Uveal coloboma associated with amniotic band syndrome. *Can J Ophthalmol* **18**:136–138, 1983.

3. Broome DL et al: Aberrant tissue bands and craniofacial defects. *Birth Defects* **12**(5):65–79, 1976.

4. Burck U et al: Congenital malformation syndromes and elevation of amniotic fluid alphafetoprotein. *Teratology* **24**:125–130, 1981.

5. Byrne J et al: Amniotic band syndrome in early fetal life. *Birth Defects* **18**(3B):43–58, 1982.

5a. Chen H, Gonzalez E: Amniotic band sequence and its neurocutaneous manifestations. *Am J Med Genet* **28**:661–673, 1987.

6. Donnenfeld AE et al: Discordant amniotic band sequence in monozygotic twins. *Am J Med Genet* **20**:685–694, 1985.

7. Etches PC et al: Familial congenital amputations. *J Pediatr* **101**:448–449, 1982.

8. Fiedler JM, Phelan JP: The amniotic band syndrome in monozygotic twins. *Am J Obstet Gynecol* **146**:864–865, 1983.

9. Garza A et al: Epidemiology of the early amnion rupture spectrum of defects, Atlanta, 1968–1982. *Proc Greenwood Genet Ctr* **6**:127–128, 1987.

10. Gellis SS: Constrictive bands in the human. *Birth Defects* **13**(1):259–268, 1977.

11. Granick MS et al: Severe amniotic band syndrome occurring with unrelated syndactyly. *Plast Reconstr Surg* **80**:829–832, 1987.

12. Hall BD: Syndromes and situations simulated by amniotic bands. March of Dimes Birth Defects Conference, Chicago, June 24–27, 1979.

12a. Hall BD: Transposed arm secondary to amniotic band disruption sequence. David W. Smith Conference on Malformations and Morphogenesis, Greenville, South Carolina, August 15–19, 1987.

13. Herva R et al: Cluster of severe amniotic adhesion malformations in Finland. *Lancet* **1**:818–819, 1980.

14. Higginbottom MC et al: The amniotic band disruption complex: Timing of amnion rupture and variable spectra of consequent defects. *J Pediatr* **95**:544–549, 1979.

15. Hunter AGW, Carpenter BF: Implications of malformations not due to amniotic bands in the amniotic band sequence. *Am J Med Genet* **24**:691–700, 1986.

16. Jones KL et al: A pattern of craniofacial and limb defects secondary to aberrant tissue bands. *J Pediatr* **84**:90–95, 1974.

16a. Kalousek DK, Bamforth S: Amnion rupture sequence in previable fetuses. *Am J Med Genet* **31**:63–73, 1988.

17. Kancherla PL et al: Intrauterine amputations in one monozygotic twin associated with amniotic band. *Am J Obstet Gynecol* **140**:347–348, 1981.

18. Keller H et al: "ADAM complex" (amniotic deformity, adhesions, mutilations)—a pattern of craniofacial and limb defects. *Am J Med Genet* **2**:81–98, 1978.

19. Kennedy LA, Persaud TVN: Pathogenesis of developmental defects induced in the rat by amniotic sac puncture. *Acta Anat* **97**:23–25, 1977.

20. Kubacek V, Penkava J: Oblique clefts of the face. *Acta Chir Plast* **16**:152–163, 1974.

21. Lubinsky M et al: Familial amniotic bands. *Am J Med Genet* **14**:81–87, 1983.

22. Lubinsky M: Familial amniotic bands. *J Pediatr* **102**:323, 1983.

22a. Malinger G et al: Pierre Robin sequence associated with amniotic band syndrome. Ultrasonographic diagnosis and pathogenesis. *Prenatal Diag* **7**:455–459, 1987.

23. Marras A et al: Letter to the editor: Epidermolysis bullosa and amniotic bands. *Am J Med Genet* **19**:815–817, 1984.

24. Mayou BJ, Fenton OM: Oblique facial clefts caused by amniotic bands. *Plast Reconstr Surg* **68**:675–681, 1981.

25. Miller ME et al: Compression—related defects from early amnion rupture: Evidence for mechanical teratogenesis. *J Pediatr* **98**:292–297, 1981.

26. Miller PR, Levi JH: Hair strangulation. *J Bone Joint Surg* **59A**:132, 1977.

27. Moessinger AC et al: Amniotic band syndrome associated with amniocentesis. *Am J Obstet Gynecol* **141**:588–591, 1981.

28. Moses JM et al: Annular constricting bands. *J Bone Joint Surg* **61A**:562–565, 1979.

29. Neuhäuser G, Sitzmann F: ADAM-Komplex—Anomalien des Gesichtsschädels und der Extremitäten infolge amniotischer Abschnürungen. *Dtsch Zahnärztl Z* **34**:546–550, 1979.

30. Ossipoff V, Hall BD: Etiologic factors in the amniotic band syndrome: A study of 24 patients. *Birth Defects* **13**(3D):117–132, 1977.

30a. Pagon RA: Congenital anomalies and aberrant tissue bands. *Am J Med Genet* **27**:491, 1987.

31. Pagon RA et al: Body wall defects with limb reduction anomalies: A report of 15 cases. *Birth Defects* **15**(5A):171–185, 1979.

32. Papp Z et al: Letter to the editor: Are there "innocent" amniotic bands? *Am J Med Genet* **24**:207–209, 1986.

33. Patterson TJS: Congenital ring-constrictions. *Br J Plast Surg* **14**:1–31, 1961.

34. Rohr M et al: Kraniofaziale, amniogene Fehlbildungen im Rahmen des ADAM-Komplexes. *Klin Pädiatr* **192**:474–480, 1980.

35. Sachdev RK et al: Amniotic band syndrome associated with an incompetent cervix. *Am J Obstet Gynecol* **148**:110–111, 1984.

36. Seeds JW et al: Amniotic band syndrome. *Am J Obstet Gynecol* **144**:243–248, 1982.

37. Schuch A, Pesch HJ: Beitrag zum Kleeblattschädel-Syndrom. *Z Kinderheilkd* **109**:187–198, 1971.

37a. Shepard TH et al: "Amniotic band" disruption syndrome: Why do their faces look alike? David W. Smith Conference on Malformations and Morphogenesis, Mills College, Oakland, California, August 3–7, 1988.

38. Tanaka O et al: Amniogenic band anomalies in a fifth-month fetus and in a newborn from maternal oophorectomy during early pregnancy. *Teratology* **33**:187–193, 1986.

39. Torpin R: *Fetal Malformations Caused by Amnion Rupture during Gestation*. Charles C Thomas, Springfield, 1968.

40. Van Allen MI, Myhre S: Ectopia cordis thoracalis with craniofacial defects resulting from early amnion rupture. *Teratology* **32**:19–24, 1985.

41. Van Allen MI et al: Limb body wall complex: I. Pathogenesis. *Am J Med Genet* **24**:840–859, 1986.

42. Van der Rest M et al: Lethal osteogenesis imperfecta with amniotic band lesions: Collagen studies. *Am J Med Genet* **24**:433–446, 1986.

43. Young ID et al: Amniotic bands in connective tissue disorders. *Arch Dis Childh* **60**:1061–1063, 1985.

44. Zionts LE et al: Congenital annular bands in identical twins. *J Bone Joint Surg* **66A**:450–453, 1984.

## Fetal brain disruption sequence

In 1984, Russell et al (3) reported three infants with severe microcephaly, overlapping sutures, prominence of the occipital bone, and scalp rugae. Four additional cases were reported by Moore et al (2).

The fetal brain disruption sequence is postulated to arise from partial brain disruption during the second or third trimester with subsequent skull collapse secondary to decreased intracranial pressure. Possible mechanisms include interruption of the blood supply to selected areas of the brain such as disruption secondary to co-twin demise, prenatal viral infection, and hyperthermia.

Severe microcephaly with severe to profound mental deficiency is characteristic. Diminished intracranial pressure leads to calvarial collapse with overlapping sutures, scalp rugae, and prominent occipital bone (Fig. 1–14).

**Differential diagnosis.** Cutis verticis gyrata occurs as a separate condition, typically having its onset after puberty; collapse of the calvaria does not occur, although microcephaly may be present. Associated endocrine problems are common. Scalp rugae have been observed with severe microcephaly and with hydranencephaly. The fetal brain disruption sequence differs from atelencephaly (1).

### References (Fetal brain disruption sequence)

1. Garcia CA, Duncan C: Atelencephalic microcephaly. *Dev Med Child Neurol* **19**:227–231, 1977.

2. Moore CA et al: The fetal brain disruption sequence: Report of four additional cases. *Proc Greenwood Genet Ctr* **6**:162–163, 1987.

3. Russell LJ et al: In utero brain destruction resulting in collapse of the fetal skull, microcephaly, scalp rugae, and neurologic impairment: The fetal brain disruption sequence. *Am J Med Genet* **17**:509–521, 1984.

# Chapter 2
# Teratogenic Agents

## General considerations

Teratogens are agents that may cause birth defects when present in the fetal environment. Included under such a definition are a wide array of drugs, chemicals, and infectious, physical, and metabolic agents that may adversely affect the intrauterine environment of the developing fetus. Such factors may operate by exceedingly heterogeneous pathogenetic mechanisms to produce alterations of form and function (including growth, learning, and behavior disorders) as well as embryonic and/or fetal death.

The mechanisms of teratogenesis are selective in terms of the target and effect. Thus, characteristic patterns of abnormalities can be expected to be associated with particular teratogenic agents. However, the extent to which an individual may be adversely affected by exposure to a given teratogen varies widely. This variability in the range of effects among members of an exposed population is primarily the result of differences in individual clinical situations among the following four factors:

1. *Differences in dose.* In general, the greater the exposure to a given agent the more likely an effect is to occur, and the more likely the effect will be severe. In addition, some agents are associated with thresholds below which effects are not readily ascertainable.

2. *Developmental timing of exposure.* Most target tissues have relatively specific periods of susceptibility during which a particular tissue is vulnerable to damage by a particular mechanism. Thus, the stage of embryonic or fetal development at which exposure to a teratogen occurs may be of critical importance in determining outcome.

3. *Differences in susceptibility.* Humans vary widely in their genetic control of drug metabolism, resistance to infection, and other biochemical and molecular processes. Thus, some individuals may be at strikingly greater risk from given exposures than the population as a whole. Furthermore, the fetal genome may be substantially different from the maternal genome and even those factors shared between mother and fetus may be differently expressed because of developmental regulatory processes. Hence, during pregnancy the susceptibility of two different individuals, mother and fetus, must be considered.

4. *Interactions among environmental exposures.* It is common during pregnancy for exposures to more than one agent to occur. There is significant potential for interactions among these agents that may enhance or suppress effects on the fetus.

As a result of these four factors most well-studied teratogens are known to produce a continuum of adverse outcomes. Therefore, it seems inappropriate to restrict consideration only to the more severe end of the spectrum, in which a readily recognizable pattern—a syndrome—often emerges. Thus, clinical teratologists have begun to use terminology that encompasses a range of outcomes. In this chapter, the term "fetal effects" more appropriately describes this range and includes the mild end of the spectrum not usually regarded as "syndromic."

This section discusses primarily drugs that have been shown to have teratogenic effects on craniofacial growth and development. Following a discussion of several important agents with well-documented major adverse effects, a tabular summary of other less frequently encountered drugs, chemicals, physical agents, and maternal metabolic factors or less well-documented associations of uncertain significance is presented. References are provided (1–26) which deal with the entire spectrum of teratogenic agents, several being putative.

### References (General considerations)


1. Brent RL, Holmes LB: Chemical and basic science lessons from the thalidomide tragedy. What have we learned about the causes of limb defects? *Teratology* **38**:241–251, 1988.

1a. Briggs GG et al: *Drugs in Pregnancy and Lactation, a Reference Guide to Fetal and Neonatal Risk.* Williams & Wilkins, Baltimore, 1983.

2. Chung CS, Myrianthopoulos NC: *Factors Affecting Risks of Congenital Malformations.* Stratton Intercontinental Medical Book Corp, New York, 1975.

3. Cohen MM: *Craniosynostosis, Diagnosis, Evaluation, and Management.* Raven Press, New York, 1986, pp. 67–69 and 577.

4. Gal P, Sharpless MK: Fetal drug exposure—behavioral teratogenesis. *Drug Intell Clin Pharm* **18**:186–201, 1984.

4a. Greenberg F: Choanal atresia and athelia: Methimazole teratogenicity or a new syndrome? *Am J Med Genet.* **28**:931–934, 1987.

5. Hanshaw JB, Dudgeon JA: *Viral Diseases of the Fetus and Newborn.* Saunders, Philadelphia, 1978.

6. Hanson JW: Teratogenic agents, in *Principles and Practice of Medical Genetics,* AH Emory and DL Rimoin (eds), Churchill Livingstone, Edinburgh, 1983, pp. 127–151.

7. Harper P: *Myotonic Dystrophy.* Saunders, Philadelphia, 1979.

8. Heinonen OP et al: *Birth Defects and Drugs in Pregnancy.* Publishing Sciences Group, Littleton, MA, 1977.

9. Hill LM: Effects of drugs and chemicals on the fetus and newborn (second of two parts). *Mayo Clin Proc* **59**:755–765, 1984.

9a. Jones KL: *Smith's Recognizable Patterns of Human Malformations.* WB Saunders, Philadelphia, 1988.

10. Kalter H, Warkany J: Congenital malformations. *N Engl J Med* **308**:491–497, 1983.

10a. Laegreid L et al: Teratogenic effects of benzodiazepine use during pregnancy. *J Pediatr* **114**:126–131, 1989.

11. Marion RW et al: Human T-cell lymphotropic virus type III (HTLV-III) embryopathy. *Am J Dis Child* **140**:638–640, 1986.

12. Miller P et al: Hyperthermia as one possible etiology of anencephaly. *Lancet* **1**:519, 1978.

13. Newman CGH: Teratogen update: Clinical aspects of thalidomide embryopathy—a continuing preoccupation. *Teratology* **32**:133–144, 1985.

14. Nyhan WL: *Heritable Disorders of Amino Acid Metabolism.* Wiley, New York, 1974.

15. Pexider T: Teratogens, in *Genetics of Cardiovascular Disease,* Pierpont MEM, Moller JH (eds), Martinus Nijhoff, Boston, 1987, pp. 25–68.

16. Potts B et al: Studies of patients and controls for possible role of pestiviruses in the development of microcephaly. *Teratology* **35**:51A, 1987.

17. Remington JS, Klein JO: *Infectious Diseases of the Fetus and Newborn Infant.* Saunders, Philadelphia, 1976.

18. Schardein JL: *Chemically Induced Birth Defects.* Dekker, New York, 1985.

19. Schinzel AA: Cardiovascular defects associated with chromosomal aberrations and malformation syndromes. *Prog Med Genet* **5**:303–379, 1983.

20. Sever JL, Brent RL: *Teratogen Update, Environmentally Induced Birth Defect Risks.* Alan R. Liss, New York, 1986.

21. Sheikh T et al: Lack of evidence for cranio-facial dysmorphism in children with AIDS. *Pediatr Res* **21**:230A, 1987.

22. Shepard TH: *Catalog of Teratogenic Agents,* 5th ed. Johns Hopkins University Press, Baltimore, 1986.

23. Smith DW: *Recognizable Patterns of Human Malformation.* Saunders, Philadelphia, 1976.

24. Stephens TD: Proposed mechanisms of action in thalidomide embryopathy. *Teratology* **38**:229–239, 1988.
25. Stevenson RE: Prenatal AIDS: A cause of postnatal progressive dysmorphism. *Proc Greenwood Genet Ctr* **5**:22–25, 1986.
26. Wilson J, Fraser FC (eds): *Handbook of Teratology*, Vols. 1–4. Plenum Press, New York, 1977.

## Ethyl alcohol

Fetal alcohol syndrome is a term applied to a pattern of abnormalities that lies near the extreme end of a spectrum of defects attributable to prenatal maternal ethanol abuse (77–79,94). Other more mildly affected children may be said to display fetal alcohol effects (65), which are thought to be considerably more frequent than recognizable fetal alcohol syndrome.

Even so, it has been currently estimated that approximately 2 per 1000 children in the United States have fetal alcohol syndrome, which would put the annual U.S. occurrence at over 7000 children (3,61,76a,77,82,129,153,167). Various epidemiologic studies have suggested that approximately 1 of 30 pregnant women abuses alcohol (3) and that approximately 6% of children born to women within this group have clinically recognizable fetal alcohol syndrome. These figures make fetal alcohol syndrome the leading known cause of mental retardation in the western world, exceeding Down syndrome, fragile X-associated mental retardation, neural tube defects, and cerebral palsy (3,64,153).

The amount of alcohol necessary to produce significant damage in the fetus has not been well established (4,163). Mills and Graubard (116), in a prospective study, indicated that total malformation rates were not significantly higher among offspring of women who averaged less than one drink per day or one or two drinks per day. However, they noted that for some malformations, possibly no safe drinking level exists. An average of two or more drinks per day during early pregnancy is associated with increased risk for adverse fetal outcome (65,96,197), and even small amounts of alcohol consumption are associated with increased risk for fetal wastage (165). However, most studies suggest that the risk for a recognizable pattern of abnormal growth and development of serious consequence usually requires somewhat greater levels of alcohol consumption in pregnancy, typically in the range of five or more drinks per day (65). The effects of other concomitant potentially hazardous environmental exposures and of genetic factors affecting alcohol metabolism are not yet clear, though some studies have suggested an interaction with tobacco smoking (53,58,83,102).

The fetus appears to be susceptible to various adverse effects of alcohol consumption throughout the major portion of pregnancy. Recent studies by Clarren suggest that "binging," even early in pregnancy, may have significant implications for learning and behavior (30a). The pathogenesis is uncertain at present (103). Alcohol has many effects on cellular metabolism. There is conflicting evidence about whether ethanol or one of its metabolites, such as acetaldehyde, or other induced effects, which may occur in chronic alcoholics such as malnutrition or zinc or folate deficiency, contributes more to the adverse fetal outcome (26,103,126,145).

Table 2–1A. Principal features of the fetal alcohol syndrome observed in 245 persons affected[c]

| Feature | Manifestation |
|---|---|
| Central nervous system dysfunction | |
| Intellectual | Mild to moderate mental retardation[a] |
| Neurologic | Microcephaly[a] |
| | Poor coordination, hypotonia[b] |
| Behavioral | Irritability in infancy[a] |
| | Hyperactivity in childhood[b] |
| Growth deficiency | |
| Prenatal | <2 SD for length and weight[a] |
| Postnatal | <2 SD for length and weight[a] |
| | Disproportionately diminished adipose tissue[b] |
| Facial characteristics | |
| Eyes | Short palpebral fissures[a] |
| Nose | Short, upturned[b] |
| | Hypoplastic philtrum[a] |
| Maxilla | Hypoplastic[b] |
| Mouth | Thinned upper vermilion[a] |
| | Retrognathia in infancy[a] |
| | Micrognathia or relative prognathia in adolescence[b] |

[a]Feature seen in >80% of patients.
[b]Feature seen in >50% of patients.
[c]From SK Clarren and DW Smith, *N Engl J Med* **298**:1063, 1978.

**Clinical manifestations.** Clinical features consist of prenatal onset growth deficiency, characteristic facial appearance (Fig. 2–1), abnormalities of central nervous system performance, and increased frequency of certain anomalies (Table 2–1A,B). Children having such a pattern of defects may be recognized as having fetal alcohol syndrome (32,33,64,77–79,95,101a,107,120,167,169). Prenatal management has been discussed by Benkendorf et al (18a).

Microcephaly is extremely common. Particular features include, in particular, abnormal neuronal migration resulting in neuroglial heterotopias, polymicrogyria, pachygyria, abnormalities of the corpus callosum, and defects in the limbic system (34,81,112). There is often cerebellar hypoplasia. Occasionally, defects of neuronal migration obstruct cerebrospinal fluid flow resulting in secondary hydrocephalus (75). It has been suggested that alcohol may also contribute to failure of neural tube closure resulting in anencephaly and spina bifida, though the magnitude of this risk is presently unclear (10,26,30,51,55).

Holoprosencephaly has been reported in several infants born to women who drank heavily during pregnancy (103,132,139). Though the significance of this observation is uncertain, it is noteworthy that an animal model of holoprosencephaly has been produced with prenatal maternal alcohol exposure (175,176). The observation of holoprosen-

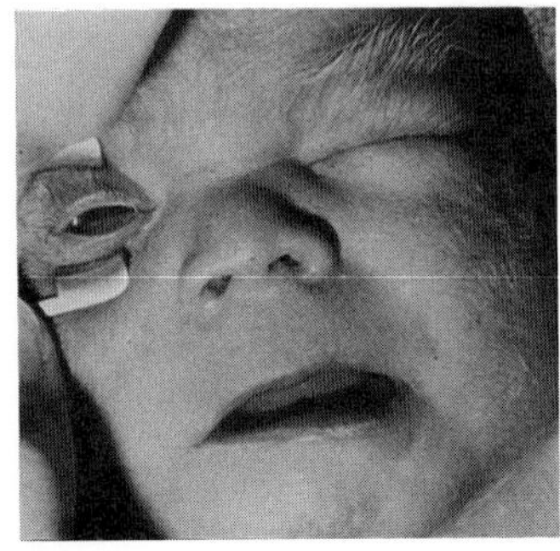
A

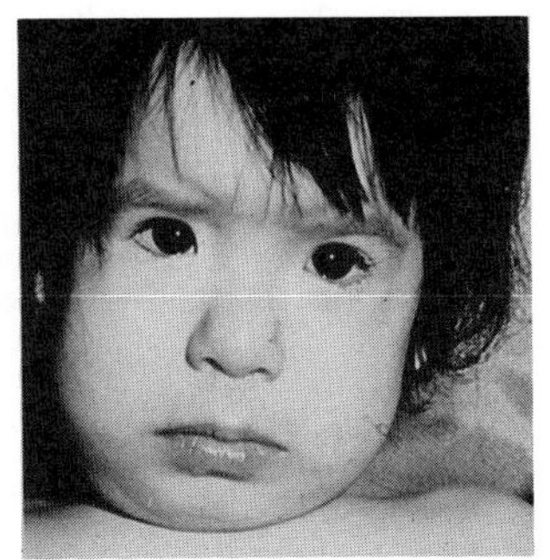
B

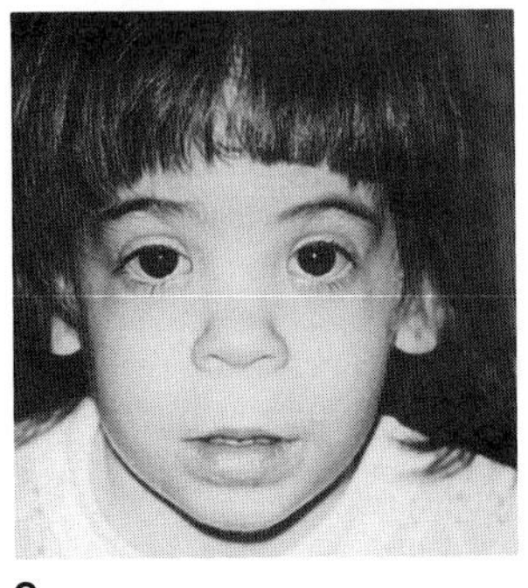
C

Fig. 2–1. *Fetal alcohol syndrome.* (A) Note narrow palpebral fissures, indistinct philtrum, and hirsutism. (B) Compare with A. (C) Short nose, indistinct philtrum, and thin upper lip. (A courtesy of *KL Jones*, San Diego. B from *KL Jones* and *DW Smith*, Lancet **2**:299, 1973. C from *SK Clarren* and *DW Smith*, N Engl J Med **298**:1063, 1978).

cephaly in the infant of a woman who herself had fetal alcohol syndrome is probably coincidental (63).

Neuropsychological studies have demonstrated that alcohol is a behaviorial teratogen (168,199). In addition to a wide range of learning handicaps extending from moderate to severe mental retardation, fine motor dysfunction, including tremulousness, poor eye–hand coordination, and various self-stimulating behaviors have been observed (58,168,171,173). Hyperactivity is common (156,172). In the newborn period, poor sucking, opisthotonus, abnormal muscle tone, irritability, and hyperacusis are often noted (38,87,142,146,151). Abnormal electroencephalographic (EEG) findings have been described (27,66,67,72) and seizure disorders have been identified in 10–20% of some of the children studied. Linguistic deficits appear to be common at later ages (154,158), as are problems with socialization and memory, sleep disturbances, impulsiveness, and personality disorders (166,168,170,171,174).

Facial features are typified by short palpebral fissures sometimes associated with clinical microphthalmia, strabismus, ptosis, epicanthic folds, and a short upturned nose with a broad low bridge (Table 2–1A,B). Midfacial hypoplasia is common (49). The philtrum appears long and smooth. The upper lip commonly displays a thin vermilion border (32,76a). Mild micrognathia may be present and cleft palate, occasionally as part of Robin sequence, may be observed. Hypoplasia of the optic nerve head and increased tortuosity of the retinal vessels are particularly common (174a,b). Other eye anomalies such as myopia, severe microphthalmia, and colobomas are seen less frequently (7,11,52). Relatively little information is available on the dental status of such children, although disturbances in facial growth and malocclusion have been documented (127,138,141,192,195,196). Hearing loss, both conductive and sensorineural, may be more frequent in affected children (28a,47,76).

A wide range of cardiac, renal, and skeletal defects have been reported. ASD and VSD appear to be the most frequent cardiac defects (Table 2–2). However, more serious cyanotic and acyanotic lesions have often been documented (16,43,99,124,145,147,152,179). Skeletal defects include vertebral anomalies (100,111,112), camptodactyly, clinodactyly, mild distal phalangeal hypoplasia, decreased range of joint motion (12,32,33,64,75,161), and possibly major limb reduction defects (9,131). Of renal defects, hypoplastic kidneys occur most commonly (40,42,55,136). Malignant neoplasms, mostly embryonal in origin, have been reported in 10 children, 2 of whom also had prenatal phenytoin exposures. Tumors have included medulloblastoma and neuroblastoma, among other (21,26,35,71,200).

Table 2–1B. Associated features of the fetal alcohol syndrome observed in 245 persons affected[a]

| Area | Frequent[b] | Occasional[c] |
|---|---|---|
| Eyes | Ptosis, strabismus, epicanthal folds | Myopia, clinical microphthalmia, blepharophimosis |
| Ears | Posterior rotation | Poorly formed concha |
| Mouth | Prominent lateral palatine ridges | Cleft lip or cleft palate, small teeth with faulty enamel |
| Cardiac | Murmurs, especially in early childhood, usually atrial septal defect | Ventricular septal defect, great-vessel anomalies, tetralogy of Fallot |
| Renogenital | Labial hypoplasia | Hypospadias, small rotated kidneys, hydronephrosis |
| Cutaneous | Hemangiomas | Hirsutism in infancy |
| Skeletal | Aberrant palmar creases, pectus excavatum | Limited joint movements, especially fingers and elbows, nail hypoplasia, especially 5th, polydactyly, radioulnar synostosis, pectus carinatum, bifid xiphoid, Klippel–Feil anomaly, scoliosis |
| Muscular | | Hernias of diaphragm, umbilicus or groin, diastasis recti |

[a]From SK Clarren and DW Smith, *N Engl J Med* **298**:1063, 1978.
[b]Reported in between 26 and 50% of patients.
[c]Reported in between 1 and 25% of patients.

Table 2–2. Cardiovascular anomalies associated with selected teratogens

| Agent | Anomalies reported (15,19) |
|---|---|
| Alcohol | Frequent: VSD, ASD, tetralogy of Fallot, PDA<br>Occasional: Pulmonic stenosis, subaortic stenosis, endocardial cushion defects, dextrocardia, double outlet right ventricle, coarctation of aorta, peripheral pulmonary stenosis, cardiomyopathy, patent foramen ovale, persistent left superior vena cava |
| Phenytoin | VSD, ASD, pulmonic stenosis, tetralogy of Fallot, valvular pulmonary stenosis, coarctation of aorta, endocardial cushion defects, aortic stenosis, superior vena cava duplex |
| Trimethadione | VSD, ASD, PDA, aortic stenosis, pulmonic stenosis, hypoplastic left heart, endocardial cushion defect, transposition of great vessels, tetralogy of Fallot |
| Valproic Acid | VSD, PDA, coarctation of aorta, hypoplastic left heart, peripheral pulmonic stenosis, levocardia, right bundle branch block |
| Coumarin | PDA, peripheral pulmonic stenosis, transposition of great vessels, total anomalous pulmonary venous return |

**Differential diagnosis.** A variety of dysmorphic syndromes may occasionally be confused with fetal alcohol effects. These include *de Lange syndrome, Noonan syndrome* (65), *Smith–Lemli–Opitz syndrome, trisomy 18 syndrome,* familial blepharophimosis syndrome, *fetal hydantoin effects,* fetal phenylketonuria (PKU) effects (158), and possibly fetal benzodiazepine effects (86a).

### References (Ethyl alcohol)


1. Abel EL: Prenatal effects of alcohol. *Drug Alcohol Depend* **14**:1–10, 1984.
2. Abel EL: Sex ratio in fetal alcohol syndrome. *Lancet* **2**:105, 1979.
3. Abel EL, Sokol RJ: Incidence of fetal alcohol syndrome and economic impact of FAS-related anomalies. *Drug Alcohol Depend* **19**:51–70, 1987.
4. Abel EL, Sokol RM: Fetal alcohol syndrome: How good is the criticism? *Neurobehav Toxicol* **5**:491–492, 1983.
5. Adickers ED, Shuman RM: Fetal alcohol myopathy. *Pediatr Pathol* **1**:369–384, 1983.
6. Adler R, Raphael B: Children of alcoholics. *Aust NZ J Psychiat* **14**:3–8, 1983.
7. Altman B: Fetal alcohol syndrome. *J Pediatr Ophthalmol* **13**:255–258, 1976.
8. Amankwah KS, Kaufmann RC: Ultrastructure of human placenta: Effects of maternal drinking. *Gynecol Obstet Invest* **18**:311–316, 1984.
9. Aro T et al: A multivariate analysis of the risk indicators of reduction limb defects. *Int J Epidemiol* **13**:459–464, 1984.
10. Arulanantham K, Goldstein G: Neural tube defects and fetal alcohol syndrome. *J Pediatr* **95**:329, 1979.
11. Bader DA: Fetal alcohol syndrome. *Am J Optom Physiol Opt* **60**:542–545, 1983.
12. Badois C et al: Maladie des épiphyses ponctuées (MEP) associée à une foetopathie éthylique (FE). *Ann Radiol* **26**:244–250, 1983.
13. Bark N: Fertility and offspring of alcoholic women: An unsuccessful search for the fetal alcohol syndrome. *Br J Addict* **74**:43–49, 1979.
14. Barr HM et al: Infant size at 8 months of age: Relationship to maternal use of alcohol, nicotine, and caffeine during pregnancy. *Pediatrics* **74**:336–341, 1984.
15. Barry RGG, O'Nuallain S: Case report: Foetal alcoholism. *Ir Med Sci* **144**:286–288, 1975.

16. Barth H et al: Über die Kombination von Ventrikelseptumdefekt mit einseitigen Pulmonalarterienstenosen und kontralateraler pulmonaler Hypertonie. *Z Kardiol* **73**:710–716, 1984.

17. Barth PG: Prenatal clastic encephalopathies. *Clin Neurol Neurosurg* **86**:65–75, 1984.

18. Beattie JO et al: Alcohol and the fetus in the west of Scotland. *Br Med J* **287**:17–20, 1983.

18a. Benkendorf JL et al: Maternal fetal alcohol syndrome: Diagnosis and management prenatally. 39th Ann Meeting, *Am Soc Hum Genet*, New Orleans, October 12–15, 1988, Abstract No 152.

19. Beyers N, Moosa A: The fetal alcohol syndrome. *S Afr Med J* **54**:575–578, 1978.

20. Bierich JR et al: Über das embryo-fetale Alkoholsyndrom. *Eur J Pediatr* **121**:155–177, 1976.

21. Bostrom B, Nesbit ME Jr: Hodgkin disease in a child with fetal alcohol-hydantoin syndrome. *J Pediatr* **103**:760–762, 1983.

22. Campbell R, Sullivan O: Alcohol consumption in pregnancy. *Arch Dis Childh* **58**:474–478, 1983.

23. Casey PH: Environment, genes, and alcohol. *Pediatrics* **71**:989–990, 1983.

24. Cate JC et al: Fetal alcohol syndrome and lactic acidosis. *Clin Chem* **29**:1320, 1983.

25. Castro-Gago M et al: Maternal alcohol ingestion and neural tube defects. *J Pediatr* **104**:796–797, 1984.

26. Cavdar AO et al: Fetal alcohol syncrome, malignancies, and zinc deficiency. *J Pediatr* **105**:335, 1984.

27. Chernick V et al: Effects of maternal alcohol intake and smoking on neonatal electroencephalogram and anthropometric measurements. *Am J Obstet Gynecol* **146**:41–47, 1983.

28. Christoffel KK, Salafsky I: Fetal alcohol syndrome in dizygotic twins. *J Pediatr* **87**:963–967, 1975.

28a. Church MW, Gerkin KP: Hearing disorders with fetal alcohol syndrome. *Pediatrics* **82**:147–154, 1988.

29. Clarren SK et al: Using facial protographs to diagnose children with fetal alcohol effects. *Proc Greenwood Genet Ctr* **5**:104, 1986.

30. Clarren SK: Neural tube defects and fetal alcohol syndrome. *J Pediatr* **95**:328, 1979.

30a. Clarren SK et al: Once per week gestational consumption of alcohol and its teratogenic impact on a non-human primate. David W. Smith Conference on Malformations and Morphogenesis, Greenville, South Carolina, August, 1987.

31. Clarren SK: The diagnosis and treatment of fetal alcohol syndrome. *Comp Ther* **8**:41–46, 1982.

32. Clarren SK: Recognition of fetal alcohol syndrome. *JAMA* **254**:2436–2439, 1981.

33. Clarren SK, Smith DW: The fetal alcohol syndrome. *N Engl J Med* **298**:1063–1067, 1978.

34. Clarren SK et al: Brain malformations related to prenatal exposure to ethanol. *J Pediatr* **92**:64–67, 1978.

35. Cohen MM Jr: Neoplasia and the fetal alcohol and hydantoin syndromes. *Neurobehav Toxicol Teratol* **3**:161–162, 1981.

36. Coles CD et al: Neonatal ethanol withdrawal: Characteristics in clinically normal, nondysmorphic neonates. *J Pediatr* **105**:445–451, 1984.

37. Collins E, Turner G: Six children affected by maternal alcoholism. *Med J Aust* **2**:606–608, 1978.

38. Crain LS et al: Nail dysplasia and fetal alcohol syndrome. *Am J Dis Child* **137**:1069–1072, 1983.

39. Davis A, Lipson A: A challenge in managing a family with the fetal alcohol syndrome. *Clin Pediatr* **23**:304, 1984.

40. Debeukelaer MM et al: Renal anomalies in the fetal alcohol syndrome. *J Pediatr* **91**:759–760, 1977.

41. Dehaene PH et al: Le syndrome d'alcoolisme foetal dans le nord de la France. *Rev Alcool* **23**:145–158, 1977.

42. Dunigan TH et al: Extrahepatic biliary atresia and renal anomalies in fetal alcohol syndrome. *Am J Dis Child* **135**:1067–1068, 1981.

43. Dupuis C et al: Les cardiopathies des enfants nés de mère alcoolique. *Arch Mal Coeur Vaiss* **71**:565–572, 1978.

44. Edwards G: Alcohol and advice to the pregnant woman. *Br Med J* **286**:247–248, 1983.

45. El-Guebaly N, Offord DR: On being the offspring of an alcoholic: An update. *Alcoholism Clin Exp* **3**:148–157, 1979.

46. Ferrier PE et al: Fetal alcohol syndrome. *Lancet* **2**:1496, 1973.

47. Flint EF: Severe childhood deafness in Glasgow. *J Laryngol Otol* **97**:421–425, 1983.

48. Forbes R: Alcohol-related birth defects. *Public Health* **98**:238–241, 1984.

49. Frias JL et al: A cephalometric study of fetal alcohol syndrome. *J Pediatr* **101**:870–873, 1982.

50. Fryns JP et al: The foetal alcohol syndrome. *Acta Paediatr Belg* **30**:117–121, 1977.

51. Fuster JS et al: Neural tube defects and fetal alcohol syndrome. *J Pediatr* **95**:328–329, 1979.

52. Garber JM: Steep corneal curvature: A fetal alcohol syndrome landmark. *J Am Optom Assoc* **55**:595–598, 1984.

53. Gibson GT et al: Maternal alcohol, tobacco and cannabis consumption and the outcome of pregnancy. *Aust NZ J Obstet Gynaecol* **23**:15–19, 1983.

54. Golding J, Peters T: Alcohol consumption in pregnancy. *Arch Dis Childh* **58**:474–478, 1983.

55. Goldstein G, Arulanantham K: Neural tube defect and renal anomalies in a child with fetal alcohol syndrome. *J Pediatr* **93**:636–637, 1978.

56. Govoni S et al: Immunoreactive met-enkephalin plasma concentrations in chronic alcoholics and in children born from alcoholic mothers. *Life Sci* **33**:1581–1596, 1983.

57. Grisso JA et al: Alcohol consumption and outcome of pregnancy. *J Epidemiol Commun Health* **38**:232–235, 1984.

58. Gusella J, Fried PA: Effects of maternal social drinking and smoking on offspring at 13 months. *Neurobehav Toxicol Teratol* **6**:13–17, 1984.

59. Gutzke DW: The cry of the children: The Edwardian medical campaign against maternal drinking. *Br J Addict* **79**:71–84, 1984.

60. Habbick BF et al: Liver abnormalities in three patients with fetal alcohol syndrome. *Lancet* **1**:580–581, 1979.

61. Hagberg B, Kyllerman M: Epidemiology of mental retardation—a Swedish survey. *Brain Dev* **5**:441–449, 1983.

62. Hall BD, Orenstein WA: Noonan's phenotype in an offspring of an alcoholic mother. *Lancet* **1**:680–681, 1974.

63. Hanson JW: Personal communication, 1987.

64. Hanson JW et al: Fetal alcohol syndrome—experiences with 41 patients. *JAMA* **235**:1458–1460, 1976.

65. Hanson JW et al: The effects of moderate alcohol consumption during pregnancy on fetal growth and morphogenesis. *J Pediatr* **92**:457–460, 1978.

66. Havlicek V, Childiaeva R: EEG component of fetal alcohol syndrome. *Lancet* **2**:477, 1976.

67. Havlicek V et al: EEG frequency spectrum characteristics of sleep states in infants of alcoholic mothers. *Neuropaediatrie* **8**:360–373, 1977.

68. Hayden MR, Nelson MM: The fetal alcohol syndrome. *S Afr Med J* **54**:571–574, 1978.

68a. Hersh JR, Buchino JJ: Umbilical cord torsion/constriction sequence. David W. Smith Conference on Malformations and Morphogenesis, Greenville, South Carolina, August, 1987.

69. Hogh B, Stenhammar L: Coeliac disease coexistent with fetal alcohol syndrome. *Eur J Pediatr* **143**:74–75, 1984.

70. Hollstedt C et al: Outcome of pregnancy in women treated at an alcohol clinic. *Acta Psychiatr Scand* **67**:236–248, 1983.

71. Hornstein L et al: Adrenal carcinoma in child with history of fetal alcohol syndrome. *Lancet* **3**:1292–1293, 1977.

72. Ioffe S et al: Prolonged effects of maternal alcohol ingestion on the neonatal electroencephalogram. *Pediatrics* **74**:330–335, 1984.

73. Iosub S et al: Fetal alcohol syndrome revisited. *Pediatrics* **68**:475–479, 1981.

74. Iosub S et al: Maternal drinking and fetal clubfoot. *Pediatr Res* **20**:228A, 1986.

75. Johnson CAC et al: Fetal alcohol syndrome with hydrocephalus. *S Afr Med J* **65**:738–739, 1984.

76. Johnson KG: Fetal alcohol syndrome: Rhinorrhea, persistent otitis media, choanal stenosis, hypoplastic sphenoids and ethmoid. *Rocky Mount Med J* **76**:64–65, 1979.

76a. Jones KL: The fetal alcohol syndrome. *Growth Genet Hormones* **4**:1–3, 1988.

77. Jones KL et al: Outcome in offspring of chronic alcoholic women. *Lancet* **1**:1076–1078, 1974.

78. Jones KL et al: Pattern of malformation in offspring of chronic alcoholic mothers. *Lancet* **1**:1267–1271, 1973.

79. Jones KL, Smith DW: Recognition of the fetal alcohol syndrome in early infancy. *Lancet* **2**:999–1001, 1973.

80. Kaminski M et al: Alcohol consumption in pregnant women and the outcome of pregnancy. *Alcoholism Clin Exp Res* **2**:155–163, 1978.

81. Kennedy LA: The pathogenesis of brain abnormalities in the fetal alcohol syndrome: An integrating hypothesis. *Teratology* **29**:363–368, 1984.

82. Kessel N: The fetal alcohol syndrome from the public health standpoint. *Health Trends* **8**:86–89, 1977.

83. King JC, Fabro S: Alcohol consumption and cigarette smoking: Effect on pregnancy. *Clin Obstet Gynecol* **26**:437–438, 1983.

84. Koranyi G: Embryopathia alcoholica. *Orv Hetil* **118**:504–507, 1977.

85. Krous HF: Fetal alcohol syndrome: A dilemma of maternal alcoholism. *Pathol Annu* Part 1, **16**:295–311, 1981.

86. Kruse J: Alcohol use during pregnancy. *AFP* **29**:199–203, 1984.

86a. Laegreid L et al: Teratogenic effects of benzodiazepine use during pregnancy. *J Pediatr* **114**:126–131, 1989.

87. Landesman-Dwyer S et al: Naturalistic observations of newborns: Effects of maternal alcohol intake. *Alcoholism Clin Exp Res* **2**:171–177, 1978.

88. Larsson G: Prevention of fetal alcohol effects. *Acta Obstet Gynecol Scand* **62**:171–178, 1983.

89. Larsson G et al: Evaluation of serum $\alpha$-glutamyl transferase as a screening method for excessive alcohol consumption during pregnancy. *Am J Obstet Gynecol* **147**:654–657, 1983.

90. Lausecker CH et al: A propos de syndrome dit "d'alcoolisme foetal." *Pédiatrie* **31**:741–747, 1976.

91. Lefkowitch JH et al: Hepatic fibrosis in fetal alcohol syndrome. *Gastroenterology* **85**:951–957, 1983.

92. Leiber B: Alkohol-Embryopathie. *Dtsch Med Wochenschr* **103**:880–881, 1978.

93. Leiber B: Schwere multiple Missbildungen beim Kindern von Alkoholikerinnen nehmen Erschreckung zu! *Schleswig-Holsteinisches Ärzteblatt* **2**:89–91, 1977.

94. Lemoine P et al: Les énfants de parents alcooliques: Anomalies observees a propos de 127 cas. *Ouest Med* **25**:477–482, 1968.

95. Little RE: Fetal alcohol effects: An overview. *JAMWA* **38**:46–48, 1983.

96. Little RE: Moderate alcohol use during pregnancy and decreased infant birth weight. *Am J Public Health* **16**:1154–1156, 1977.

97. Little RE et al: Change in obstetrician advice following a two-year community educational program on alcohol use and pregnancy. *Am J Obstet Gynecol* **146**:23–28, 1983.

98. Lipson AH et al: Fetal alcohol syndrome. *Med J Aust* **1**:266–269, 1983.

99. Loser H, Majewski F: Type and frequency of cardiac defects in embryofetal alcohol syndrome. Report of 16 cases. *Br Heart J* **39**:1374–1379, 1977.

100. Lowry RB: The Klippel–Feil anomalad as part of the fetal alcohol syndrome. *Teratology* **16**:53–56, 1977.

101. Madden JJ et al: Increased rate of E-rosette formation by T lymphocytes of pregnant women who drink ethanol. *Immunol Immunopathol* **33**:67–79, 1984.

101a. Majewski F: Die Alkoholembryopathie: Fakten und Hypothesen. *Ergeb Inn Med Kinderheilkd* **43**:1–55, 1979.

102. Martin JC et al: Maternal alcohol ingestion and cigarette smoking and their effects on newborn conditioning. *Alcoholism Clin Exp Res* **1**:243–247, 1977.

103. Majewski F: Alcohol embryopathy: Some facts and speculations about pathogenesis. Neurobehav Toxicol Teratol **3**:129–144, 1981.

104. Majewski F et al: Diagnose: Alkoholembryopathie. *Dtsch Ärzteblatt* **17**:1133–1136, 1977.

105. Majewski F: Untersuchungen zur Alkohol-Embryopathie. *Fortschr Med* **96**:2207–2212, 1978.

106. Majewski F: Über schädigende Einflüsse des Alkohols auf die Nachkommen. *Nervenarzt* **49**:410–416, 1978.

107. Majewski F et al: Zur Klinik und Pathogenese der Alkohol-embryopathie. Bericht über 68 Fälle. *Münch Med Wochenschr* **118**:1635–1642, 1976.

108. Malcolm MT: Foetal alcohol syndrome: Historical aspects. *Alcohol Alcoholism* **19**:261–262, 1984.

109. Manzke H, Grosse FR: Inkomplettes und komplettes fetales Alkohol-Syndrom: bei drei Kindern einer Trinkerin. *Med Welt* **26**:709–712, 1975.

110. Margolin FG: Fetal alcohol syndrome: Report of a case. *J Am Osteopath Assoc* **77**:99–101, 1977.

111. Maroteaux P et al: Chondrodysplasie ponctuée et intoxication alcoolique maternelle. *Arch Fr Pédiatr* **41**:547–50, 1984.

112. Marsh DO: Occult neurologic effects of alcohol on brain and fetus. *NY State J Med* **83**:310–312, 1983.

113. Mau G, Netter P: Kaffee und Alkoholkonsum—Risikofaktoren in der Schwangerschaft? *Geburtshilfe Frauenheilkd* **34**:1018–1022, 1974.

114. McCarthy PA: Fetal alcohol syndrome and other alcohol-related birth defects. *Nurse Pract* **8**:33–34, 1983.

115. Miller HC: A model for studying the pathogenesis and incidence of low-birth-weight infants. *Am J Dis Child* **137**:323–327, 1983.

116. Mills JL, Graubard BI: Is moderate drinking during pregnancy associated with an increased risk for malformations? *Pediatrics* **80**:309–314, 1987.

117. Moller J et al: Hepatic dysfunction in patient with fetal alcohol syndrome. *Lancet* **1**:605–606, 1979.

118. Mukherjee SP: The foetal alcohol syndrome. *Br J Clin Prac* **38**:35, 1984.

119. Mulvihill JJ et al: Fetal alcohol syndrome: Seven new cases. *Am J Obstet Gynecol* **125**:937–941, 1976.

120. Mulvihill JJ, Yeager AM: Fetal alcohol syndrome. *Teratology* **13**:345–348, 1976.

121. Nakada T, Knight RT: Alcohol and the central nervous system. *Med Clin North Am* **68**:121–131, 1984.

122. Neidengard L et al: Klippel–Feil malformation complex in fetal alcohol syndrome. *Am J Dis Child* **132**:929–930, 1978.

122a. Neri G et al: Facial midline defect in the fetal alcohol syndrome: Embryogenetic considerations in two clinical cases. *Am J Med Genet* **29**:477–482, 1988.

123. Newman SL et al: Simultaneous occurrence of extrahepatic biliary atresia and fetal alcohol syndrome. *Am J Dis Child* **133**:101, 1979.

124. Noonan JA: Association of congenital heart disease with syndromes of other defects. *Pediatr Clin North Am* **25**:797–816, 1978.

125. Noonan JA: Congenital heart disease in the fetal alcohol syndrome. *Am J Cardiol* **37**:160, 1976.

126. O'Shea KS, Kaufman MH: The teratogenic effect of acetaldehyde. Implications for the study of fetal alcohol syndrome. *J Anat* **128**:65–76, 1979.

127. Ouellette EM: The fetal alcohol syndrome. *J Dent Child* **51**:222–224, 1984.

128. Ouellette EM: The fetal alcohol syndrome. *Bol Asoc Med P R* **76**:492–494, 1984.

129. Ouellette EM et al: Adverse effects on offspring of maternal alcohol abuse during pregnancy. *N Engl J Med* **297**:528–530, 1977.

130. Palmer RH et al: Congenital malformations in offspring of a chronic alcoholic mother. *Pediatrics* **53**:490–494, 1974.

131. Pauli RM, Feldman PF: Major limb malformations following intrauterine exposure to ethanol: Two additional cases and literature review. *Teratology* **33**:273–280, 1986.

132. Peiffer J et al: Alcohol embryo- and fetopathy. *J Neurol Sci* **41**:125–137, 1979.

133. Pocsy T, Balassa E: Alkoholos embryopathia. *Orv Hetil* **119**:209–211, 1978.

134. Poskitt EME: Foetal alcohol syndrome. *Alcohol Alcoholism* **19**:159–165, 1984.

135. Qazi QH, Masakawa A: Altered sex ratio in fetal alcohol syndrome. *Lancet* **1**:42, 1976.

136. Qazi QH et al: Renal anomalies in fetal alcohol syndrome. *Pediatrics* **63**:886–889, 1979.

137. Rice M: Fetal alcohol syndrome: A clinical study. *J Am Med Wom Assoc* **40**:23–27, 1985.

138. Riekman GA: Fetal alcohol. *J Can Dent Assoc* **11**:841–842, 1984.

139. Ronen G: Personal communication, 1987.

140. Root AW et al: Hypothalamic-pituitary function in the fetal alcohol syndrome. *J Pediatr* **87**:585–586, 1976.

141. Rosenbicht J et al: Fetal alcohol syndrome. *Oral Surg* **47**:8–10, 1979.

142. Rosett HL et al: Effects of maternal drinking on neonate state regulation. *Dev Med Child Neurol* **21**:464–473, 1979.

143. Rosett HL et al: Patterns of alcohol consumption and fetal development. *Obstet Gynecol* **61**:539–546, 1983.

144. Russell M: Intrauterine growth in infants born to women with alcohol-related psychiatric diagnoses. *Alcoholism Clin Exp Res* **1**:225–231, 1977.

145. Ryle PR, Thomson AD: Acetaldehyde and the fetal alcohol syndrome. *Lancet* **2**:219–220, 1983.

146. Sander LW et al: Effects of alcohol intake during pregnancy on newborn state regulation: A progress report. *Alcoholism Clin Exp Res* **1**:233–241, 1977.

147. Sandor GS et al: Cardiac malformations in the fetal alcohol syndrome. *J Pediatr* **98**:771–773, 1981.

148. Saule H: Fetales Alkohol-Syndrom: ein Fallbericht. *Klin Paediatr* **186**:452–455, 1974.

149. Scheiner AP et al: Fetal alcohol syndrome in child whose parents had stopped drinking. *Lancet* **1**:1077–1078, 1979.

150. Scheppe KJ: Alkohol-embryopathie: Zur Differentialdiagnose und postnatal retardierter Kinder. *Paediatr Praxis* **18**:207–218, 1977.

151. Scher MS et al: The effects of prenatal alcohol exposure on sleep cycling and arousal. *Pediatr Res* **20**:165A, 1986.

152. Schinzel AA: Cardiovascular defects associated with chromosomal aberrations and malformation syndromes. *Prog Med Genet* **5**:303–379, 1983.

153. Segal M: The 1990 prevention objectives for alcohol and drug misuse: Progress report. *Public Health Rep* **98**:426–435, 1983.

154. Shaywitz BA: Fetal alcohol syndrome: An ancient prroblem rediscovered. *Drug Thèr* **95**:108, 1978.

155. Shaywitz SE et al: Developmental language disability as a consequence of prenatal exposure to ethanol. *Pediatrics* **68**:850–855, 1981.

156. Shaywitz SE et al: Behavior and learning difficulties in children of normal intelligence born to alcoholic mothers. *J Pediatr* **96:**978–982, 1980.

157. Singh SP et al: Effects of ethanol ingestion on maternal and fetal glucose homeostasis. *J Lab Clin Med* **104**:176–184, 1984.

158. Smith DW: *Recognizable Patterns of Human Malformation.* Saunders, Philadelphia, 1976.

159. Smithells RW, Smith IJ: Alcohol and the fetus. *Arch Dis Childh* **59**:1113–1114, 1984.

160. Sparks SN: Speech and language in fetal alcohol syndrome. *ASHA* **26**:27–31, 1984.

161. Spiegel PG et al: The orthopedic aspects of the fetal alcohol syndrome. *Clin Orthop* **139**:58–63, 1979.

162. Spohr HL, Steinhausen HC: Der Verlauf der Alkoholembryopathie. *Monatsschr Kinderheilkd* **132**:844–849, 1984.

163. Staisey NL, Fried PA: Relationships between moderate maternal alcohol consumption during pregnancy and infant neurological development. *J Stud Alcohol* **44**:262–270, 1983.

164. Stanage WF: Fetal alcohol syndrome—intrauterine child abuse. *SD J Med* **36**:35, 1983.

165. Stein ZA, Susser M: Intrauterine growth retardation: Epidemiological issues and public health significance. *Semin Perinatol* **8**:5–14, 1984.

166. Steinhausen HC et al: Psychopathology in the offspring of alcoholic parents. *J Am Acad Child Psychiat* **4**:465–471, 1984.

167. Streissguth AP: Alcohol and pregnancy: An overview and an update. *Substance Alcohol Actions* **4**:149–173, 1983.

168. Streissguth, AP: The behavioral teratology of alcohol: Performance behavioral, and intellectual deficits in prenatally exposed children, in *Alcohol and Brain Development,* West J (ed), Oxford University Press, New York, 1986, pp. 3–44.

169. Streissguth AP: Maternal alcoholism and the outcome of pregnancy: A review of the fetal alcohol syndrome, in *Alcoholism Problems in Women and Children,* Greenblatt M, Schuckit MA (eds), Grune & Stratton, New York, 1976, pp. 251–274.

170. Streissguth AP: Maternal drinking and the outcome of pregnancy: Implications for child mental health. *Am J Orthopsychiatr* **47**:422–431, 1977.

171. Streissguth AP: Psychological handicaps in children with fetal alcohol syndrome. *Ann NY Acad Sci* **273**:140–145, 1976.

172. Streissguth AP et al: Attention, distraction and reaction time at age 7 years and prenatal alcohol exposure. *Neurobehav Toxicol Teratol* **8**:717–725, 1986.

173. Streissguth AP et al: Intelligence, behavior, and dysmorphogenesis in the fetal alcohol syndrome: A report on 20 patients. *J Pediatr* **92**:363–367, 1978.

174. Streissguth AP et al: Stability of intelligence in the fetal alcohol syndrome: A preliminary report. *Alcoholism Clin Exp Res* **2**:165–170, 1978.

174a. Strömland K: Ocular abnormalities in the fetal alcohol syndrome. *Acta Ophthalmol* **63**:*Suppl* **171**:1–50, 1985.

175. Sulik KK, Johnston MC: Embryonic origin of holoprosencephaly: Interrelationship of the developing brain and face. *Scan Electron Microsc* **1**:309–322, 1982.

176. Sulik KK et al: Fetal alcohol syndrome: Embryogenesis in a mouse model. *Science* **214**: 936–938, 1981.

177. Sulik KK et al: Fetal alcohol syndrome and Di George anomaly: Critical ethanol exposure periods for craniofacial malformations as illustrated in an animal model. *Am J Med Genet Suppl* **2**:97–112, 1986.

178. Sullivan WC: A note on the influence of maternal inebriety on the offspring. *J Ment Sci* **45**:489–503, 1899.

179. Tenbrinck MS, Buchin SY: Fetal alcohol syndrome. Report of a case. *JAMA* **232**:1144–1147, 1975.

180. Terrapon et al: Aortic arch interruption type *a* with aortopulmonary fenestration in an offspring of a chronic alcoholic mother. *Helv Paediatr Acta* **32**:141–148, 1977.

181. Ticha R et al: Fetal alcohol syndrome: Amino acid pattern. *Acta Paediatr Hung* **24**:143–148, 1983.

182. Tillner I, Majewski F: Furrows and dermal ridges of the hand in patients with alcohol embryopathy. *Humangenetik* **42**:307–314, 1978.

183. Tze WJ et al: Growth hormone response in fetal alcohol syndrome. *Arch Dis Childh* **51**:703–706, 1976.

184. Ulleland CN: The offspring of alcoholic mothers. *Ann NY Acad Sci* **197**:167–179, 1972.

185. van Biervliet JP: The fetal alcohol syndrome. *Acta Paediatr Belg* **30**:113–116, 1977.

186. Veghelyi PV: Fetal abnormality and maternal ethanol metabolism. *Lancet* **2**:53–54, 1983.

187. Veghelyi PV et al: The fetal alcohol syndrome: Symptoms and pathogenesis. *Acta Paediatr Acad Sci Hung* **19**:171–189, 1978.

188. Veghelyi PV et al: Maternal alcohol consumption and birth weight. *Br Med J* **2**:1365–1366, 1978.

189. Villermaulaz A: Syndrome de l'alcoolisme foetal. *Rev Méd Suisse Romande* **97**:613–619, 1977.

190. Vosanelli K: Schäden bei Kindern von Müttern mit chronischem Alkoholismus. *Wien Med Wochenschr* **127**:544–546, 1977.

191. Walsh AC: Damaged babies of alcoholic mothers: A rational explanation of the cause of the defects. *Lex Sci* **13**:206–209, 1977.

192. Wood RE: Fetal alcohol syndrome: Its implications for dentistry. *J Am Dent Assoc* **95**:596–599, 1977.

193. Wilsnack SC et al: Drinking and reproductive dysfunction among women in a 1981 national survey. *Alcoholism Clin Exp Res* **8**:451–458, 1984.

194. Wisniewski K et al: A clinical neuropathological study of the fetal alcohol syndrome. *Neuropediatrics* **14**:197–201, 1983.

195. Wood N, Turner JW: Fetal alcohol syndrome—a review. *J Dent Child* **48**:198–199, 1981.

196. Wood RE: Fetal alcohol syndrome: Its implications for dentistry. *J Am Dent Assoc* **95**:596–599, 1977.

197. Wright JT et al: Alcohol consumption, pregnancy, and low birthweight. *Lancet* **1**:663–665, 1983.

198. Wright JT et al: Alcohol and the fetus. *Br J Hosp Med* **29**:260, 262, 264 passim, 1983.

199. Yellin AM: The study of brain function impairment in fetal alcohol syndrome: Some fruitful directions for research. *Neurosci Behav Rev* **8**:1–4, 1984.

200. Zaunschirm A et al: Fetal alcohol syndrome and malignant disease. *Eur J Pediatr* **143**:160–161, 1984.

## Vitamin A congeners

Though information linking vitamin A to birth defects in animals has been available for over 30 years (23), the relevance of these observations to human beings remained uncertain until studies of isotretinoin in the early part of this decade revealed a characteristic pattern of birth defects associated with its use (1a,2,4,8,9,14,16–18). Subsequently, a growing body of both animal and human studies including anecdotal clinical observations has implicated both a growing group of vitamin A congeners and vitamin A itself as having a human teratogenic potential. Among the agents currently of concern are isotretinoin, etretinate, and "megadoses" of vitamin A (retinol) (5,8,15,19,20). There is little direct evidence about the frequency of children damaged by this class of agents. Currently less than 100 reports are available. However, the growing use of these agents in treating common skin disorders such as acne and psoriasis and large scale marketing of high dose vitamin A preparations have caused escalating concern (7,19).

Dosage and timing appear to be major factors with the highest risk period being between 2 and 5 weeks after conception (13,19). However, it should be noted that unlike isotretinoin, which has a half life of less than 1 day, both etretinate and retinol (vitamin A) have half lives of weeks to months, leading to both longer risk periods after discontinuation of use and to the possibility of substantial accumulation. Current data suggest that chronic use of 0.4–1.5 mg/kg/day of isotretinoin or etretinate or 0.2–1.5 mg/kg/day (greater than 15,000 U/day) of vitamin A during pregnancy may be associated with a significant increase in risk for serious congenital anomalies in the fetus (11,13,15,19). Unfortunately the latter is not rare. Recent Federal Drug Administration (FDA) surveys suggest that nearly one-third of the female population may be using vitamin A supplements with 1–2% exceeding this maximum allowable level (19).

A variety of studies in animal models and *in vitro* systems have explored pathogenetic mechanisms for vitamin A congener-related teratogenesis. The studies of Lammer et al (11,12) and Webster et al (22) suggest that an effect on neural crest development and migration and excessive cell death are of special importance. Little is known about interactions with other environmental agents or genetic differences in susceptibility.

**Clinical manifestations.** Current evidence suggests a relatively characteristic pattern of defects with striking similarities to the DiGeorge sequence (11). Craniofacial anomalies characteristically in-

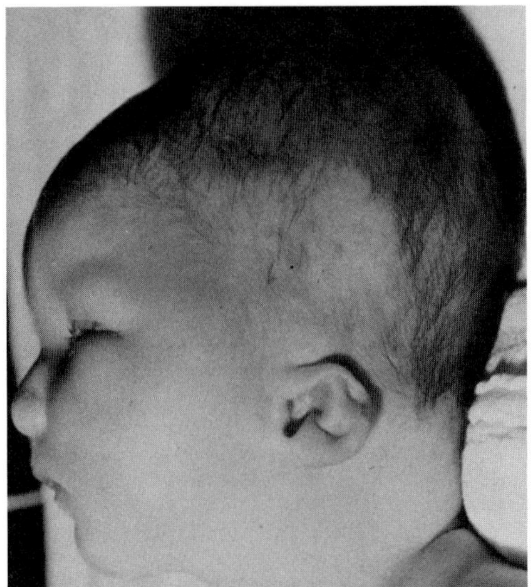

Fig. 2–2. *Isoretinoin embryopathy*. Note abnormal ears. (Courtesy of *P Fernhoff*, Atlanta).

clude microcephaly associated with serious developmental central nervous system anomalies often resulting in hydrocephaly and posterior fossa cysts (11,14). Cortical blindness and facial nerve plasies may result. Dysmorphic facial features also include facial asymmetry with midfacial hypoplasia, metopic synostosis, microphthalmia, oculomotor palsies, cleft palate and possibly cleft lip as well, microtia, and anomalies of the external auditory canal (Fig. 2–2) (11,16,19).

The most characteristic defects in other systems include cardiovascular anomalies such as hypoplastic aortic arch, VSD, transposition of the great vessels, and hypoplastic left ventricle (11). Thymic hypoplasia and genitourinary anomalies including hypoplastic kidneys and hydroureter are also common (18).

Other defects have been reported such as facial asymmetry, neural tube defects, holoprosencephaly, tracheoesophageal fistula, epibulbar lipodermoids, auricular pits, dimples, and creases, adrenal hypoplasia, sirenomelia, pterygium colli, hemivertebra, femoral hypoplasia, clubfoot, radial defects and other limb abnormalities, hepatic abnormalities, congenital cataracts, inguinal hernia, craniosynostosis, and syndactyly (3,8,13a,13b,19).

**Differential diagnosis.** Differential diagnosis includes *DiGeorge sequence, Bixler syndrome,* and *oculo-auriculo-vertebral spectrum.* In children with neural tube defects, other causes should be considered. Likewise, in children with VATER association, other conditions should be carefully sought and distinguished. A patient with some similarities to isotretinoin syndrome but without isotretinoin has been discussed (1,14a).

#### References (Vitamin A congeners)


1. Begleiter ML, Harris DJ: Letter to the editor: A phenocopy of the isotretinoin syndrome? *Am J Med Genet* **29**:225–226, 1988.

1a. Benke PJ: The isotretinoin teratogen syndrome. *JAMA* **251**:3267–3269, 1984.

2. Braun JT et al: Isotretinoin dysmorphic syndrome. *Lancet* **1**:506–507, 1984.

3. Cohen MM: *Craniosynostosis, Diagnosis, Evaluation, and Management.* Ravan Press, New York, 1986, pp. 67–69 and 577.

4. de la Cruz E et al: Multiple congenital malformations associated with maternal isotretinoin therapy. *Pediatrics* **74**:428–430, 1984.

5. Fabro S: The teratogenicity of retinoids. *Reprod Toxicol* **5**:5–8, 1986.

6. Hall JG: Vitamin A teratogenicity. *N Engl J Med* **311**:797–798, 1984.

7. Hall JG: Vitamin A: A newly recognized human teratogen. Harbinger of things to come? *J Pediatr* **105**:583–584, 1984.

8. Happle R et al: Teratogene Wirkung von Etretinat beim Menschen. *DMW* **109**:1476–1480, 1984.

9. Hill RM: Isotretinoin teratogenicity. *Lancet* **1**:1465, 1984.

10. Kassis I et al: Isotretinoin (Accutane) and pregnancy. *Teratology* **32**:145–146, 1985.

11. Lammer EJ: Retinoic acid embryopathy. *N Engl J Med* **313**:837–841, 1985.

12. Lammer EJ: Altered cell differentiation and an unusual pattern of external ear malformation in human retinoic acid embryopathy. *Pediatr Res* **20**:49A, 1986.

13. Lammer EJ et al: Risk for major malformation among human fetuses exposed to isotretinoin (13-cis-retinoic acid). *Teratology* **35**:68A, 1987.

13a. Lammer EJ et al: Auricular pits, dimples, and creases in retinoic acid embryopathy. *MOD Clin Genet Conf,* Minneapolis, Minnesota, July 19–22, 1987.

13b. Lammer EJ et al: Facial asymmetry in retinoic acid embryopathy. *MOD Clin Genet Conf,* Minneapolis, Minnesota, July 19–22, 1987.

14. Lott IT et al: Fetal hydrocephalus and ear anomalies associated with maternal use of isotretinoin. *J Pediatr* **105**:597–600, 1984.

14a. Lungarotti MS, Calabro A: Response to Drs. Lungarotti and Calabro. *Am J Med Genet* **29**:227–228, 1988.

15. Miller RK et al: Recommendation for vitamin A use during pregnancy. *Teratology* **35**:267–275, 1987.

16. Mounoud RL et al: À propos d'un cas de syndrome de Goldenhar: intoxication aigue à la vitamine à chez la mère pendant la grossesse. *J Génét Hum* **23**:135–154, 1975.

17. Orfanos CE: Teratogenität von Isotretinoin. *Hautarzt* **35**:503–505, 1984.

18. Pilotti G: Impervitaminosi A in gravidanza e malformazioni dell'apparato urinario nel feto. *Min Pediatr* **27**:682–684, 1975.

19. Rosa FW et al: Teratogen update: Vitamin A congeners. *Teratology* **33**:355–364, 1986.

20. Shepard TH et al: Megadoses of vitamin A. *Teratology* **34**:366, 1986.

21. Stern RS et al: Isotretinoin and pregnancy. *J Am Acad Dermatol* **10**:851–854, 1984.

22. Webster WS et al: Isotretinoin embryopathy and the cranial neural crest. An in vivo and in vitro study. *J Craniofac Genet Dev Biol* **6**:211–222, 1986.

23. Wilson J, Fraser FC (eds): *Handbook of Teratology,* Vol. 1, Plenum Press, New York, 1977.

24. Zarowny DP: Accutane Roche: Risk of teratogenic effects. *Can Med Assoc J* **131**:273, 1984.


### Folate antagonist chemotherapeutic agents

Folic acid antagonists have been recognized as having human teratogenic potential since the 1950s when studies of aminopterin (4-amino pteroylglutamic acid) use as an abortifacient revealed serious congenital malformations (10,15,21,35,36,42). Compounds in this category include aminopterin and methotrexate. Though other specific antagonists have not been studied, it seems reasonable to include them as likely to be hazardous to the fetus during pregnancy. Whether or not folic acid deficiency on a nutritional basis during pregnancy is teratogenic remains to be established (17).

Currently information is available on over 50 pregnancies in which such exposures have been documented (14,24,33,34,41,43). Little information is available on the frequency with which such exposures occur. However, concern is prompted not only by the use of such agents in the treatment of malignancy but also by the increasing use of such agents for the treatment of disorders such as psoriasis and rheumatoid arthritis (25).

The pathogenesis is incompletely understood. Practically no information is available about interactions with other environmental agents or genetic differences in drug metabolism. A similar pattern of anomalies has been reported in several children for whom no antifolate prenatal exposure was identified [4,11(case 2), 16a]. Patient 1 of Fraser et al (11) appears to have another condition, also described by Quintana-Castilla (personal communication, 1989).

Thirty years experience with folic acid antagonists has given a clear indication of a characteristic pattern of alterations of growth and development attributable to this class of agents. Though insufficient numbers and inadequate epidemiologic data presently exist to clearly establish the range and severity of effects attributable to this class of compounds, certain abnormalities emerge as being rather characteristic of the effects of these agents. Specific attack rates associated with

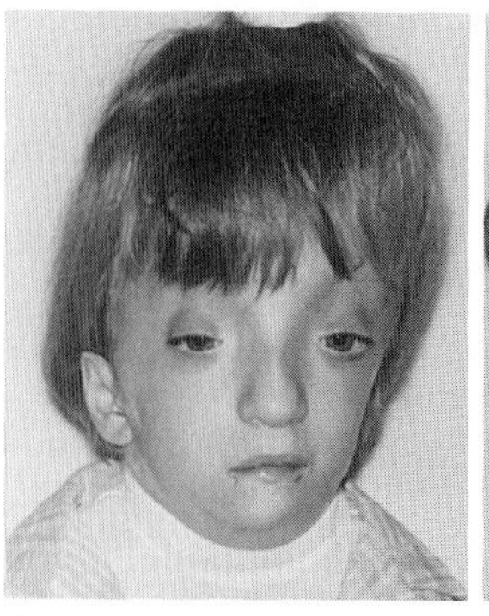
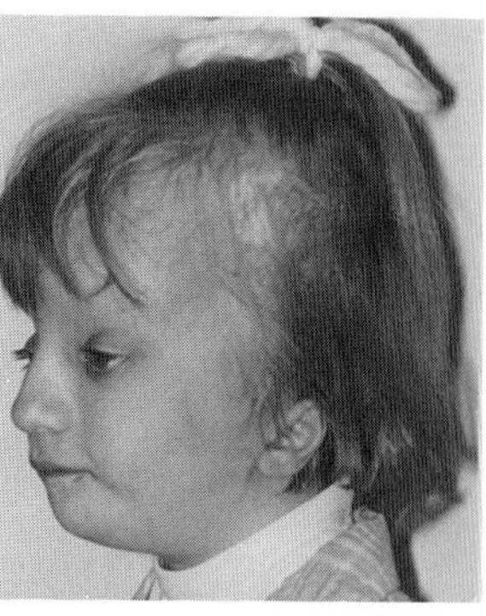

A

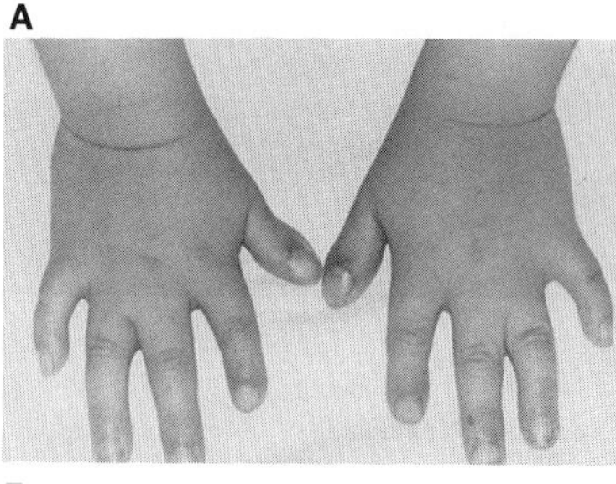
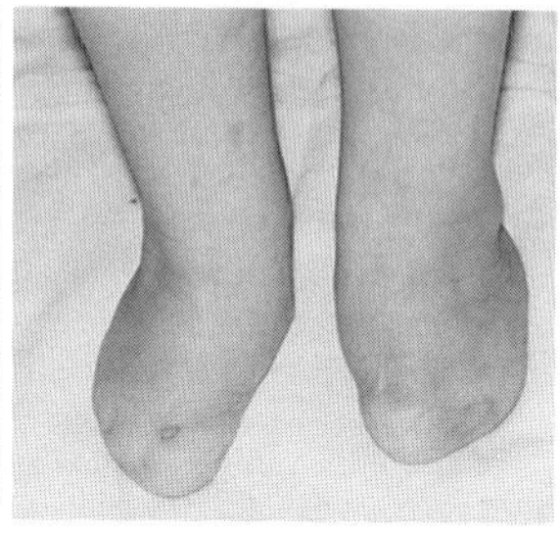

B

Fig. 2–3. (A) *A folic acid antagonist-induced syndrome* caused by methotrexate, a methyl derivative of aminopterin. (B) Brachydactyly and soft tissue syndactyly involving digits 3 and 4 of the hands together with absence of toes. (From *MM Cohen Jr,* Craniosynostosis: Diagnosis, Evaluation, and Management, Raven Press, New York, 1986, p 550.)

particular drug levels are not available. RJ Gorlin has calculated the susceptible period is during the 8–9th weeks of gestation (10,21, 22,28,30,42,44).

**Clinical manifestations.** Growth deficiency of prenatal onset may be present. Craniofacial anomalies (Fig. 2–3A) include abnormalities of central nervous system development with hydrocephalus and other defects, often resulting in mental retardation. Often abnormalities of the developing calvaria occur with delays in mineralization resulting in widely patent cranial sutures, craniolacunae, abnormal skull shape, and sometimes craniosynostosis (22). Less frequently, neural tube defects have been reported, including both anencephaly and spina bifida. Dysmorphic facial features include hypertelorism, auricular anomalies and striking micrognathia sometimes associated with cleft palate (5,13,18,22,28,41,43).

Other abnormalities include skeletal defects such as stenotic changes of the long bones, rib anomalies, and reduction defects of the distal extremities including absence or hypoplasia of digits (Fig. 2–3B). Syndactyly and talipes equinovarus have been reported (13,28,41).

**Differential diagnosis.** Differential diagnosis includes various genetic syndromes associated with *craniosynostosis* and other anomalies of the developing calvaria. Conditions causing micrognathia with cleft palate including the whole range of disorders associated with *Robin sequence* and skeletal defects should be considered. In cases with limb reduction defects and/or syndactyly, the *oromandibular limb–hypogenesis* group of disorders may be suggested.

### References (Folate antagonist chemotherapeutic agents)


1. Baranov VS: Characteristics of the teratogenic effect of aminopterin compared with that of other teratogenic agents. *Bull Exp Biol Med (USSR)* **61**:77–81, 1966.

2. Brandner M, Nusslé D: Foetopathie due a l'aminoptérine avec sténose congénitale de l'espace médullaire des os tubularies longs. *Ann Radiol* **12**:703–710, 1969.

3. Burnier AM: Acute myelomonocytic leukemia in pregnancy: Report of a case. *Am J Obstet Gynecol* **143**:41–43, 1982.

4. Centerwall W: Personal communication, 1987.

5. Char F: Denouement and discussion: Aminopterin embryopathy syndrome. *Am J Dis Child* **133**:1189–1190, 1979.

6. Dara P et al: Successful pregnancy during chemotherapy for acute leukemia. *Cancer* **47**:845–846, 1981.

7. deAlvarez RR: Discussion (to: An evaluation of aminopterin as an abortifacient). *Am J Obstet Gynecol* **83**:1476–1477, 1962.

8. Diniz EMA et al: [Effect, on the fetus, of methotrexate (amethopterin) administered to the mother. Presentation of a case]. *Rev Hosp Clin Fac Med São Paulo* **33**:286–290, 1978.

9. Dobbing J: Pregnancy and leukaemia. *Lancet* **1**:1155, 1977.

10. Emerson DJ: Congenital malformation due to attempted abortion with aminopterin. *Am J Obstet Gynecol* **84**:356–357, 1962.

11. Fraser FC et al: An aminopterin-like syndrome without aminopterin. *Clin Genet* **32**:28–34, 1987.

12. Gautier E: Demonstrations cliniques, embryopathie de l'aminoptérin, kwashiorkor, enfant maltraite, listeriose congénitale et saturnisme, maladie de Weil. *Schweiz Med Wochenschr* **99**:33–42, 1969.

13. Gellis SS, Feingold M: Aminopterin embryopathy syndrome, *Am J Dis Child* **133**:1189–1190, 1979.

14. Gilliland J, Weinstein L: The effects of cancer chemotherapeutic agents on developing fetus. *Cancer Rev* **38**:6–13, 1983.

15. Goetsch C: An evaluation of aminopterin as an abortifacient. *Am J Obstet Gynecol* **83**:1474–1477, 1962.

16. Grella P et al: Antineoplastici e gravidanza. *Minerva Med* **75**:1643–1649, 1984.

16a. Herrmann J, Opitz JM: An unusual form of acrocephalosyndactyly. *Birth Defects* **5**(3):39–42, 1969.

17. Hibbard BM: Folates and the fetus. *S Afr Med J* **49**:1223–1226, 1975.

18. Howard NJ, Rudd NL: The natural history of aminopterin-induced embryopathy. *Birth Defects* **13**(3C):85–93, 1977.

19. Jensen JK, Nyfors A: Cytogenetic effect of methotrexate on human cells in vivo. Comparison between results obtained by chromosome studies on bone-marrow cells and blood lymphocytes and by the micronucleus test. *Mutat Res* **64**:339–343, 1979.

20. McLain C: Leukemia in pregnancy. *Clin Obstet Gynecol* **17**:185–194, 1974.

21. Meltzer HJ: Congenital anomalies due to attempted abortion with 4-aminopteroglutamic acid. *JAMA* **161**:1253, 1956.

22. Milunsky A et al: Methotrexate-induced congenital malformations. *J Pediatr* **72**:790–795, 1968.

23. Mondello C et al: Chromosomal effects of methotrexate on cultured human lymphocytes. *Mutat Res* **139**:67–70, 1984.

24. Nicholson HO: Cytotoxic drugs in pregnancy: Review of reported cases. *J Obstet Gynaecol Br Commonw* **75**:307–312, 1968.

25. Perry W: Methotrexate and teratogenesis. *Arch Dermatol* **119**:874–875, 1983.

26. Pizzuto J et al: Treatment of acute leukemia during pregnancy: Presentation of nine cases. *Cancer Treat Rep* **64**:679–683, 1980.

27. Powell HR, Ekert H: Methotrexate-induced congenital malformations. *Med J Aust* **2**:1076–1077, 1971.

28. Reich EW et al: Recognition in adult patients of malformations induced by folic-acid antagonists. *Birth Defects* **14**(6B):139–160, 1977.

29. Schilsky RL et al: Gonadal dysfunction in patients receiving chemotherapy for cancer. *Ann Intern Med* **93**:109–114, 1980.

30. Shaw EB, Steinbach HL: Aminopterin-induced fetal malformation. Survival of an infant after attempted abortion. *Am J Dis Child* **115**:477–482, 1968.

31. Shaw EB: Fetal damage due to maternal aminopterin ingestion. Follow-up at 9 years of age. *Am J Dis Child* **124**:93–94, 1972.

32. Shaw EB, Reese EL: Fetal damage due to aminopterin ingestion: Follow-up at 17-½ years of age. *Am J Dis Child* **134**:1172–1173, 1980.

33. Sieber S, Adamson R: Toxicity of antineoplastic agents in man: Chromosomal aberrations, antifertility effects, congenital malformations, and carcinogenic potential. *Adv Cancer Res* **22**:57–155, 1975.

34. Sweet DL, Kinzie J: Consequences of radiotherapy and antineoplastic therapy for the fetus. *J Reprod Med* **17**:241–246, 1976.

35. Thiersch JB: Therapeutic abortions with a folic acid antagonist, 4-aminopteroylglutamic acid (4-amino P.G.A.) administered by the oral route. *Am J Obstet Gynecol* **63**:1298–1304, 1952.

36. Thiersch JB: Effect of 4-amino-pteroylglutamic acid (aminopterin) on early pregnancy. *Proc Soc Exp Biol Med* **74**:204–208, 1950.

37. Thiersch JB: The control of reproduction in rats with the aid of antimetabolites and early experiences as abortifacient agents in man. *Acta Endocrinol Suppl* **28**:37–45.

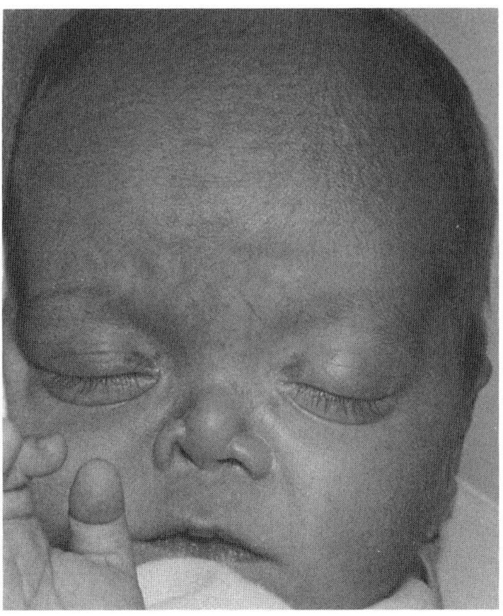
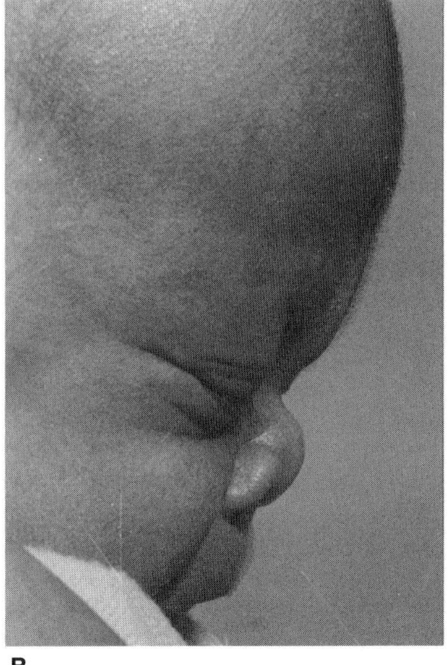
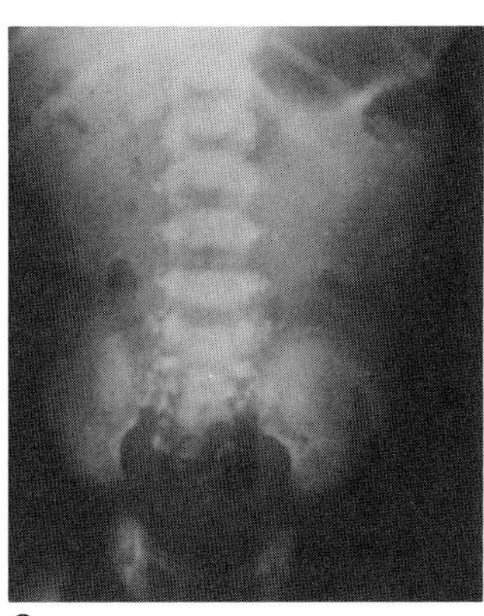

A B C

Fig. 2–4. *Fetal Warfarin syndrome.* (A,B) Hypoplastic nose. (C) Note stippling around vertebral column. (A,B courtesy of *JG Hall,* Vancouver, BC.)


38. Van Thiel DH et al: Pregnancies after chemotherapy or trophoblastic neoplasms. *Science* **169**:1326–1327, 1970.

39. Voorhees JJ et al: Cytogenetic evaluation of methotrexate-treated psoriatic patients. *Arch Dermatol* **100**:269–274, 1969.

40. Walden PAM, Bagshawe KD: Pregnancies after chemotherapy for gestational trophoblastic tumours. *Lancet* **2**:1241, 1979.

41. Warkany J: Aminopterein and methotrexate: Folic acid deficiency. *Teratology* **17**:353–358, 1978.

42. Warkany J et al: Attempted abortion with aminopterin (4-amino-pteroyl glutamic acid). *Am J Dis Child* **97**:274–281, 1959.

43. Warkany J: Teratogenicity of folic acid antagonists. *Cancer Bull* **33**:76–77, 1981.

44. Werthemann A: Allgemeine und spezielle Probleme bei der Analyse von Missbildungsursachen, in Sonderheit bei Thalidomid- und Aminopterinschäden. *Schweiz Med Wochenschr* **93**:223–227, 1963.


## Coumarin anticoagulants

Anticoagulant therapy during pregnancy is associated with a range of adverse maternal and fetal outcomes, many of which are attributable to maternal hemorrhage or the underlying maternal disease state. Maternal treatment with the coumarin group of anticoagulant agents is associated with a characteristic pattern of fetal effects. Whether other vitamin K antagonists may produce similar changes is unknown. However, recent reports of children with an apparent genetic disorder of vitamin K metabolism, affecting vitamin K-dependent coagulation factors, with a similar pattern of alterations of growth and development suggest that this might be the case (13a,15).

To date fewer than 50 cases of infants exposed to coumarin anticoagulants prenatally have been reported. A fairly wide range in severity of fetal effects is evident. Prothrombin levels in the range of 40–60% of normal have not been associated with obvious fetal damage (5). This suggests that concomitant exposure to vitamin K-containing compounds in the diet or from other sources might be a significant factor in predicting fetal outcome.

Approximately 35% of infants exposed to coumarin anticoagulants during pregnancy die prenatally or suffer serious teratogenic consequences. Studies of the timing of fetal exposure to this class of agents suggest that the period of greatest concern extends from approximately 6 to 9 weeks after conception (5,11,15). During this period of time, a characteristic pattern of alterations of craniofacial growth and other defects is often produced. Beyond 9 weeks, ocular defects and midline CNS malformations are more characteristic.

**Clinical manifestations.** Growth deficiency of prenatal onset, developmental delay, mental retardation, and other neurologic abnormalities may be present. Most characteristic are central nervous system malformations leading to microcephaly and occasionally hydrocephaly. Other reported defects include cerebral agenesis, occipital meningoencephalocele, agenesis of corpus callosum, and other midline anomalies. Seizures and hearing loss have also been reported. Striking facial features include severe midfacial hypoplasia with hypoplasia of nasal development (Fig. 2–4A,B) often associated with choanal atresia. There may also be optic atrophy, corneal opacities, cataract, and other ocular defects (8,11,14,16,17).

Defects in other systems characteristic of prenatal exposure to coumarin anticoagulants include cardiovascular anomalies (Table 2–2), widely spaced nipples, and abnormalities of skeletal development, particularly disproportionate short stature and stippled epiphyses reflecting alterations of calcification (Fig. 2–4C) (2,14,17).

**Differential diagnosis.** One must exclude *chondrodysplasia punctata* of various types as well as other causes of short stature and calcific stippling of growth centers, such as *Zellweger syndrome,* hypothyroidism, and genetic defects of vitamin K metabolism (13a,15). Some cases with neural tube defects may prompt consideration of chromosome anomalies that are also associated with this class of anomalies.

### References (Coumarin anticoagulants)


1. Barr M Jr, Burdi AR: Warfarin-associated embryopathy in a 17-week-old abortus. *Teratology* **14**:129–134, 1976.

2. Becker MH et al: Chondrodysplasia punctata. *Am J Dis Child* **129**:356–359, 1975.

3. Boefinger MK, Warkany J: Warfarin and fetal abnormality. *Lancet* **1**:911, 1976.

4. Carson M, Reid M: Warfarin and fetal abnormality. *Lancet* **2**:1127, 1976.

5. Chong MKB et al: Follow-up study of children whose mothers were treated with warfarin during pregnancy. *Br J Obstet Gynaecol* **91**:1070–1073, 1984.

6. Fourie DT, Hay IT: Warfarin as a possible teratogen. *S Afr Med J* **49**:2081–2083, 1975.

7. Hall JG: Warfarin and fetal abnormality. *Lancet* **1**:1127, 1976.

8. Hall JG: Embryopathy associated with oral anticoagulant therapy. *Birth Defects: Original Article Series* **12**(5):33–37, 1976.

9. Hanson JW, Smith DW: Teratogenicity of anticoagulants. *J Pediatr* **87**:838, 1975.

10. Holzgreve W et al: Warfarin-induced fetal abnormalities. *Lancet* **2**:914–915, 1976.

11. Kaplan LC: Congenital Dandy-Walker malformation associated with first trimester warfarin: A case report and literature review. *Teratology* **32**:333–337, 1985.

12. Kaplan LC: First trimester warfarin exposure and Dandy–Walker malformation without Warfarin embryopathy. *Proc Greenwood Genet Ctr* **5**:167, 1986.

13. Lamontagne JM et al: Warfarin embryopathy—a case report. *J Otolaryngol* **13**:127–129, 1984.

13a. Leonard CO: Vitamin K responsive bleeding disorder: A genocopy of the warfarin embryopathy. David W. Smith Conference on Malformations and Morphogenesis, Greenville, South Carolina, August, 1987.

14. Pauli RM et al: Warfarin therapy initiated during pregnancy and phenotypic chondrodysplasia punctata. *J Pediatr* **88**:506–508, 1976.

15. Pauli RM et al: Association of congenital deficiency of multiple vitamin K dependent coagulation factors and the phenotype of the warfarin embryopathy: Clues to the mechanism of teratogenicity of coumarin derivatives. *Am J Hum Genet* **41**:566–583, 1987.

16. Pettifor JM et al: Congenital malformations associated with the administration of oral anticoagulants during pregnancy. *J Pediatr* **86**:459–462, 1975.

17. Shaul WL et al: Chondrodysplasia punctata and maternal warfarin use during pregnancy. *Am J Dis Child* **129**:360–362, 1975.

18. Sherman S, Hall BD: Warfarin and fetal abnormalitiy. *Lancet* **1**:692, 1976.

## Anticonvulsant drugs

For over 30 years, debate has continued over the question of teratogenicity of anticonvulsant drugs (1–76). In spite of a substantial body of literature, problems such as confounding ascertainment bias, poor statistical power, biologically inappropriate or inconsistent case definition, and variations in seizure disorder management have plagued epidemiologic studies, rendering firm conclusions problematic (33). Nonetheless, it has become increasingly clear that infants born to mothers receiving anticonvulsant medications during pregnancy are at an increased risk for congenital defects. At least three major groups of anticonvulsant medications have teratogenic effects in some exposed fetuses that are not attributable to other concurrently existing prenatal risk factors. These include phenytoin and related hydantoin anticonvulsants, trimethadione and related oxazolidinediones, and valproic acid. Alleged effects of other anticonvulsant categories are presented in Table 2–3 at end of chapter.

### References (Anticonvulsant drugs)

1. Aase JM: Anticonvulsant drugs and congenital abnormalities. *Am J Dis Child* **127**:758, 1974.

2. Anderson RC: Cardiac defects in children of mothers receiving anticonvulsant therapy during pregnancy. *J Pediatr* **89**:318–319, 1976.

3. Annegers JF et al: Do anticonvulsants have a teratogenic effect? *Arch Neurol* **31**:364–373, 1974.

4. Annegers JR et al: Epilepsy, anticonvulsants, and congenital malformations. *Trans Am Neurol Assoc* **99**:184–186, 1974.

5. Annegers JF et al: Teratogenicity of anticonvulsant drugs, in *Epilepsy,* AA Ward Jr et al (eds), Raven Press, New York, 1983.

6. Barr M Jr et al: Digital hypoplasia and anticonvulsants during gestation: A teratogenic syndrome? *J Pediatr* **84**:254–256, 1974.

7. Barry JE, Danks DM: Anticonvulsants and congenital abnormalities. *Lancet* **2**:48–49, 1974.

8. Bethenod M, Frederich A: Les enfants des antiépileptiques. *Pédiatrie* **30**:227–248, 1975.

9. Biale Y et al: Congenital malformations due to anticonvulsive drugs. *Obstet Gynecol* **45**:439–442, 1975.

10. Biale Y et al: Congenital malformations and decreased blood level of folic acid induced by antiepileptic drugs. *Acta Obstet Gynecol Scand* **55**:187, 1976.

11. Bird AV: Anticonvulsant drugs and congenital abnormalities. *Lancet* **1**:311, 1979.

12. Bjerkedal T, Bahna SL: The course and outcome of pregnancy in women with epilepsy. *Acta Obstet Gynecol Scand* **52**:245–248, 1974.

13. Clericuzio CL: Fetal primidone effects. *Proc Greenwood Genet Ctr* **4**:93, 1985.

14. Clifford DB: Seizures and pregnancy. *Arch Fr Pédiatr* **29**:271–275, 1984.

15. Dalessio DJ: Seizure disorders and pregnancy. *N Engl J Med* **312**:559–563, 1985.

16. Danks DM et al: Digital hypoplasia and anticonvulsants during pregnancy. *J Pediatr* **85**:877–878, 1974.

17. Dieterich E: Antiepileptica-Embryopathien. *Ergeb Inn Med Kinderheilkd* **43**:93–107, 1979.

18. Espir MLE, Hytten FE: Pregnancy with epilepsy—the need for combined care? *Br J Obstet Gynaecol* **90**:1105–1106, 1983.

19. Fedrick J: Epilepsy and pregnancy: A report from the Oxford record linkage study. *Br Med J* **2**:442–448, 1973.

20. Finnell RH: Phenytoin-induced teratogenesis: A mouse model. *Science* **211**:483–484, 1981.

21. Friis ML: Antiepileptic drugs and teratogenesis. *Acta Neurol Scand* **94**:39–43, 1983.

21a. Gailey E et al: Minor anomalies in offspring of epileptic mothers. *J Pediatr* **112**:520–529, 1988.

22. Gustavson EE, Chen H: Goldenhar syndrome, anterior encephalocele, and aqueductal stenosis following fetal primidone exposure. *Teratology* **32**:13–17, 1985.

23. Higgins TA, Comerford JB: Epilepsy in pregnancy. *J Irish Med Assoc* **67**:317–320, 1974.

24. Hiilesmaa VK et al: Serum folate concentrations during pregnancy in women with epilepsy: Relation to antiepileptic drug concentrations, number of seizures, and fetal outcome. *Br Med J* **287**:577–579, 1983.

25. Hill RM: Fetal malformations and antiepileptic drugs. *Am J Dis Child* **130**:923–925, 1976.

26. Hill RM et al: Antiepileptic drugs and fetal well-being, in *Fetal Pharmacology,* L Boreus (ed), Raven Press, New York, 1973, pp 375–380.

27. Hill RM et al: Infants exposed in utero to antiepileptic drugs. *Am J Dis Child* **127**:645–653, 1974.

28. Ho NK, Loo DSC: Sympodia. *Am J Dis Child* **128**:391–393, 1974.

28a. Holmes LB: Teratogenic effects of anticonvulsant drugs. *J Pediatr* **112**:579–581, 1988.

29. Jager-Roman E et al: Somatic parameters, diseases, and psychomotor development in the offspring of epileptic parents, in *Epilepsy, Pregnancy and the Child,* D Janz et al (eds), Raven Press, New York, 1982.

30. Janz D: Schwangerschaft und Kindesentwicklung bie Epilepsie. *Geburtsh Frauenheilk* **44**:428–434, 1984.

31. Janz D: The teratogenic risk of antiepileptic drugs. *Epilepsia* **16**:159–169, 1975.

32. Janz D, Fuchs U: Are anti-epileptic drugs harmful when given during pregnancy? *Germ Med Mth* **9**:20–22, 1964.

33. Kelly TE: Teratogenicity of anticonvulsant drugs. I: Review of the literature. *Am J Med Genet* **19**:413–434, 1984.

34. Kelly TE et al: Teratogenicity of anticonvulsant drugs. II: A prospective study. *Am J Med Genet* **19**:435–443, 1984.

35. Kelly TE: Teratogenicity of anticonvulsant drugs. III: Radiographic hand analysis of children exposed in utero to diphenylhydantoin. *Am J Med Genet* **19**:445–450, 1984.

36. Kelly TE et al: Teratogenicity of anticonvulsant drugs. IV: The association of clefting and epilepsy. *Am J Med Genet* **19**:451–458, 1984.

37. Knox JDE: Teratogenic effect of anticonvulsants. *Lancet* **1**:198, 1973.

38. Koch S et al: Antiepileptika wahrend der Schwangerschaft. *DMW* **108**:250–257, 1983.

39. Koch S et al: Major malformations in children of epileptic parents—due to epilepsy or its therapy? in *Epilepsy, Pregnancy, and the Child,* D Janz et al (eds), Raven Press, New York, 1982, pp 313–315.

40. Kuhnz W et al: Carbamazepine and carbamazepine-10,11-epoxide during pregnancy and postnatal period in epileptic mothers and their nursed infants: Pharmacokinetics and clinical effects. *Pediatr Pharmacol* **3**:199–208, 1983.

41. Kuhnz W et al: Ethosuximide in epileptic women during pregnancy and lactation period. Placental transfer, serum concentrations in nursed infants and clinical status. *Br J Clin Pharmacol* **18**:671–677, 1984.

42. Landon MJ: Maturation of children on anticonvulsant therapy. *Lancet* **2**:1327, 1974.

43. Livingston S et al: Maternal epilepsy and abnormalities of the fetus and newborn. *Lancet* **2**:1265, 1973.

44. Lowe CR: Congenital malformations among infants born to epileptic women. *Lancet* **1**:9, 1973.

45. Macnaughton MC: Maternal epilepsy and abnormalities of the fetus and newborn. *Lancet* **2**:1088, 1972.

46. Magureanu S et al: O asociatie morbida particulara: oligofrenie cu dismorfie cranio-facial (inclusiv hipertelorism), absenta unghiilor si falangelor terminale la degetele IV–V si alte malformatii la o fetita de 7 ani. *Neurol Psichiatr Neurochir* **18**:255–262, 1973.

47. Majewski F et al: Zur Teratogenität von Antikonvulsiva. *Deutsche Med Wochenschrift* **105**:719–723, 1980.

48. Meadow SR: Anticonvulsant drugs and congenital abnormalities. *Lancet* **2**:1296, 1968.

49. Meadow SR: Congenital abnormalities and anticonvulsant drugs. *Proc Roy Soc Med* **63**:48–49, 1970.

50. Meinardi H: Het verband tussen het gebruik van anti-epileptica door de zwangere vrouw en het ontstaan van aangeboren afwijkingen bij haar kind. *New Tijdschr Geneeskd* **127**:2012–2016, 1983.

51. Milkovich L, Van Den Berg BJ: An evaluation of the teratogenicity of certain antinauseant drugs. *Am J Obstet Gynecol* **125**:244–248, 1976.

52. Morselli PL: Carbamazepine: Absorption, distribution, and excretion. *Adv Neurol* **11**:279–293, 1975.

53. Niermeijer MF: Gebruik van valproinezuur door de zwangere vrouw met epilepsie en de kans op aangeboren afwijkingen bij haar kind. *Ned Tijdschr Geneeskd* **128**:2460–2461, 1984.

54. Niswander JD, Wertelecki W: Congenital malformation among offspring of epileptic women. *Lancet* **1**:1062, 1973.

55. Norriss JW, Pratt RF: Folic acid deficiency and epilepsy. *Drugs* **8**:366–385, 1974.

56. Pearl KN et al: Functional palatal incompetence in the fetal anticonvulsant syndrome. *Arch Dis Childh* **59**:989–990, 1984.

57. Philbert A et al: Die epilepsiekranke Mutter und ihr Kind. *Wiener Klin Wochenschrift* **96**:786–790, 1984.

58. Pinilla ER et al: Antiepilepticos y malformaciones congenitas: estudio en una poblacion espanola (1976–1983). *Farmacoterapia* **2**:148–153, 1985.

59. Rating D et al: Minor anomalies in the offspring of epileptic parents, in *Epilepsy, Pregnancy, and the Child,* D Janz et al (eds), Raven Press, New York, 1982, pp 283–288.

60. Rating D et al: Antiepileptika in der Neugeborenenperiode. *Monatsschr Kinderheilkd* **131**:6–12, 1983.

61. Rating D et al: Teratogenic and pharmacokinetic studies of primidone during pregnancy and in the offspring of epileptic women. *Acta Paediatr Scand* **71**:301–311, 1982.

62. Robertson IG et al: Cranial nerve agenesis in a fetus exposed to carbamazepine. *Dev Med Child Neurol* **25**:540–541, 1983.

63. Rumeau-Rouquette C et al: Les medicaments du systeme nerveux sont-ils teratogenes? *Arch Fr Pédiatr* **33**:5–10, 1976.

64. Seip M: Growth retardation, dysmorphic facies and minor malformations following massive exposure to phenobarbitone in utero. *Acta Paediatr Scand* **65**:617–621, 1976.

65. Shapiro S et al: Anticonvulsants and parental epilepsy in the development of birth defects. *Lancet* **1**:272–275, 1976.

66. South J: Teratogenic effect of anticonvulsants. *Lancet* **2**:1154, 1972.

67. Speidel BD, Meadow SR: Epilepsy, anticonvulsants and congenital malformations. *Drugs* **8**:354–365, 1974.

68. Speidel BD, Meadow SR: Maternal epilepsy and abnormalities of the fetus. *Lancet* **2**:839–843, 1972.

69. Spellacy WN: Maternal epilepsy and abnormalities of the fetus and newborn. *Lancet* **2**:1196–1197, 1972.

70. Starreveld-Zimmerman AAE et al: Teratogenicity of antiepileptic drugs. *Clin Neurol Neurosurg* **2**:81–95, 1974.

71. Starreveld-Zimmerman AEE et al: Are anticonvulsants teratogenic? *Lancet* **2**:48–49, 1973.

72. Svigos JM: Epilepsy and pregnancy. *Aust NZ J Obstet Gynaecol* **24**:182–185, 1984.

73. Van Lang QC et al: Effect of in utero exposure to anticonvulsants on craniofacial development and growth. *J Craniofac Genet Dev Biol* **4**:115–133, 1984.

74. Visser GHA et al: Anticonvulsants and fetal malformations. *Lancet* **1**:970, 1976.

75. Watson JD, Spellacy WN: Neonatal effects of maternal treatment with the anticonvulsant drug diphenylhydantoin. *Obstet Gynecol* **37**:881–885, 1974.

76. Welch CE et al: Teratogenesis and antiepileptic drugs. *N Engl J Med* **289**:1089–1090, 1973.

**Hydantoins.** A full pattern of abnormalities sufficient to be recognizable as the fetal hydantoin syndrome is probably present in no more than 5–10% of exposed infants (21). Careful studies of more subtle indicators of altered growth and morphogenesis such as radiographic investigations of craniofacial or limb development or careful clinical investigation for other structural variations may identify subtle prenatal hydantoin effects in an additional 30% or more of cases (24,27,50). Several authors have suggested that manifestations are likely to be more severe or more frequent in infants exposed to multiple anticonvulsant agents (28,43,44,51). Developmental delay and neurological abnormalities have been suggested to be more frequent in infants whose mothers had seizures during pregnancy in spite of anticonvulsant therapy (37). An especially good review is that of Dieterich (13a).

The role of genetic factors in predicting fetal outcome remains a source of confusion (15). Though a number of authors have reported affected siblings (19,21), in some instances the effects among siblings have been dissimilar. Of particular interest is the observation of heteropaternal twins born to an epileptic mother treated with phenytoin during pregnancy, only one of whom displayed features of the fetal hydantoin syndrome (42). Subsequent metabolic studies have identified different epoxide hydratase activities, suggesting that this enzyme may be important in determining the manifestations of fetal hydantoin effects (7,7a,22). Strickler et al (49) correlated measurements of an apparent genetic defect in arene oxide detoxification with the presence of more serious malformations in infants exposed prenatally to phenytoin (17).

To date the pathogenetic mechanisms responsible for fetal hydantoin effects remain uncertain. The results or risks of short-term or acute exposure to these agents in pregnancy remain unknown. A role has been postulated for arene oxide metabolites. However, other metabolic factors, such as interference with folic acid metabolism, may be important as well. Goldman et al (18a) suggested that an elevated level of glucocorticoid receptors in circulating lymphocytes may be a marker for susceptibility to fetal hydantoin syndrome, but this finding needs confirmation.

Clinical manifestations. Fetal hydantoin syndrome consists of abnormalities of growth including mild prenatal onset growth deficiency and microcephaly, abnormalities of central nervous system function including dull mentality, and dysmorphic craniofacial and limb features (Fig. 2–5) (23). The most frequent and characteristic craniofacial abnormalities include a rather hypoplastic midface with short nose, a broad low nasal bridge and epicanthic folds, mild hypertelorism, ptosis, strabismus, and wide mouth with accentuated cupid's bow of the upper lip. In addition, the calvaria often has sutural ridging and there may be a short neck with mild webbing. Less commonly, clefts of the lip and/or palate are identified. Mild micrognathia is a variable feature.

Distal phalangeal hypoplasia with an increased frequency of low arch finger tip patterns and other limb defects, occasionally including finger-like thumbs, may also be seen (23). Concern has also been expressed about possible increased frequency of major genitourinary, central nervous system, and cardiac malformations (Table 2–2) (3,46), diaphragmatic hernia, and more severe limb reduction malformations (4,13). Phenytoin has a carcinogenic potential in adults. Also, tumors, particularly those of neural crest origin (such as neuroblastoma and melanotic neuroectodermal tumor), mesenchymoma, and Wilms tumor have now been reported in nine infants (1,2,5,6,9,34,41,47).

Differential diagnosis. Among the conditions that may cause differential diagnostic problems are *Coffin–Siris, Noonan, Williams,* and *Aarskog syndromes, neurofibromatosis,* and the effects of other teratogenic agents such as alcohol and maternal phenylketonuria. Furthermore, the craniofacial effects of other anticonvulsant drugs including barbiturates, primidone, and valproic acid may be superficially similar.

#### References (Hydantoins)

1. Allen RW, Jung AL: Fetal hydantoin syndrome and malignancy. *J Pediatr* **105**:681, 1984.

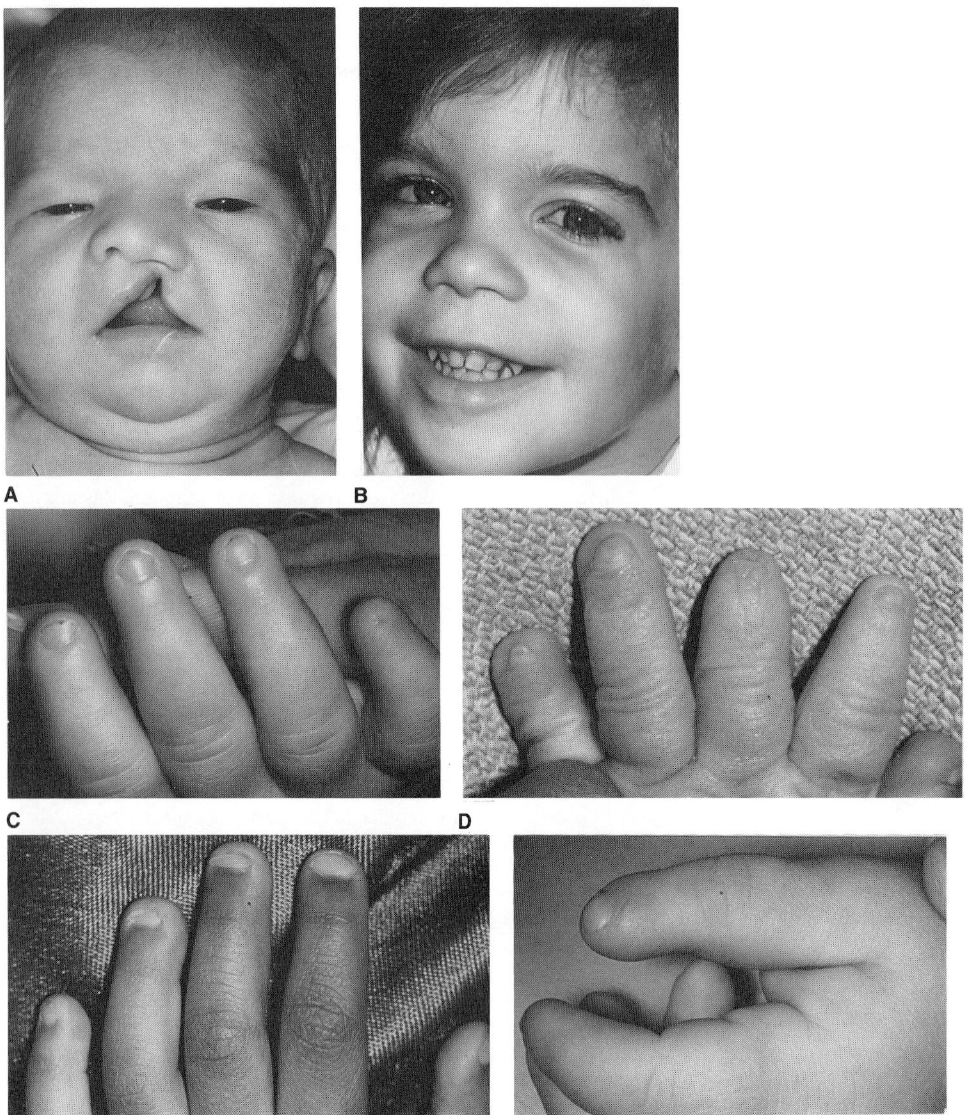


Fig. 2–5. *Fetal hydantoin syndrome.* (A) Cleft lip. (B) Small nose, low nasal bridge, hypertelorism, and strabismus. (C,D,E,F) Nail hypoplasia. (A,B,C,E,F courtesy of *JW Hanson,* Iowa City, Iowa).

2. Allen RW, Bentley FL: Fetal hydantoin syndrome, neuroblastoma, and hemorrhagic disease in a neonate. *JAMA* **244**:1464–1465, 1980.

3. Anderson RC: Cardiac defects in children of mothers receiving anticonvulsant therapy during pregnancy. *J Pediatr* **89**:318–319, 1976.

4. Barr M Jr et al: Digital hypoplasia and anticonvulsants during gestation: A teratogenic syndrome? *J Pediatr* **84**:254–256, 1974.

5. Blattner WA et al: Malignant mesenchymoma and birth defects: Prenatal exposure to phenytoin. *JAMA* **238**:334–335, 1977.

6. Bostrom B, Nesbit ME Jr: Hodgkin disease in a child with fetal alcohol-hydantoin syndrome. *J Pediatr* **103**:760–762, 1983.

7. Buehler BA, Delimont D: Epoxide hydralase activity: A direct assay for prediction of potential dilantin teratogenesis. *Proc Greenwood Genet Ctr* **4**:92, 1985.

7a. Buehler BA et al: "Stippled epiphyses": Clinical and biochemical associations. David W. Smith Meeting, Greenville, South Carolina, August, 1987.

8. Chen H et al: Fetal hydantoin syndrome associated with Turner's syndrome. *Birth Defects* **13**(3B):237–238, 1977.

9. Cohen MM Jr: Neoplasia and the fetal alcohol and hydantoin syndromes. *Neurobehav Toxicol Teratol* **3**:161–162, 1981.

10. Corcoran R, Rizk MW: VACTERL congenital malformation and phenytoin therapy? *Lancet* **2**:970, 1976.

11. Czeizel A: Diazepam, phenytoin, and etiology of cleft lip and/or cleft palate. *Lancet* **1**:810, 1976.

12. Dabee V et al: Teratogenic effects of diphenylhydantoin. *Can Med Assoc J* **112**:75–76, 1975.

13. Danks DM et al: Digital hypoplasia and anticonvulsants during pregnancy. *J Pediatr* **85**:877–878, 1974.

13a. Dieterich E: Antiepileptica–Embryopathien. *Ergeb Inn Med Kinderheilkd* **43**:93–107, 1979.

14. Escobar V, Bixler D: The fetal hydantoin syndrome. *Birth Defects* **13**(3D):275, 1977.

15. Finnell RH, Chernoff GF: Genetic background: The elusive component in the fetal hydantoin syndrome. *Am J Med Genet* **19**:459–462, 1984.

16. Finnell RH, Chernoff GF: Mouse fetal hydantoin syndrome: Effects of maternal seizures. *Epilepsia* **23**:423–429, 1982.

17. Finnell RH, DiLiberti JH: Hydantoin-induced teratogenesis: Are arene oxide intermediates really responsible? *Helv Paediat Acta* **38**:171–177, 1983.

18. Gadisseux JF et al: Pontoneocerebellar hypoplasia—a probable consequence of prenatal destruction of the pontine nuclei and a possible role of phenytoin intoxication. *Clin Neuropathol* **3**:160–167, 1984.

18a. Goldman AS et al: Elevated glucocorticoid receptor levels in lymphocytes of children with the fetal hydantoin syndrome (FHS). *Am J Med Genet* **28**:607–618, 1987.

19. Goodman RM et al: Congenital malformations in four siblings of a mother taking anticonvulsant drugs. *Am J Dis Child* **130**:884–887, 1976.

20. Hanson JW: Anticonvulsants and the fetus: The fetal hydantoin syndrome and related problems. *Proc Epilepsy Int Symp,* Vancouver, B.C., September 10–14, 1978.

21. Hanson JW: Teratogen update: Fetal hydantoin effects. *Teratology* **33**:349–353, 1986.

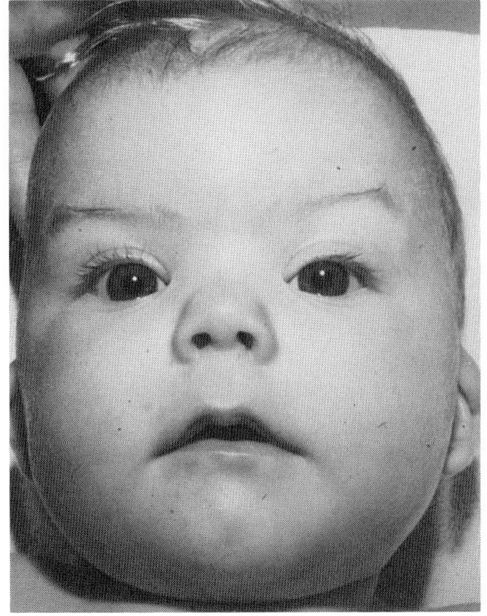
A

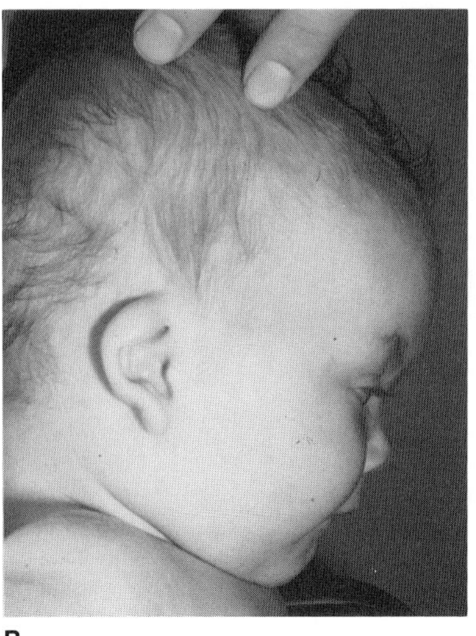
B

Fig. 2–6. *Fetal trimethadione syndrome.* (A) V-shaped eyebrows and (B) dysplastic ears. (Courtesy of *JW Hanson,* Iowa City, Iowa.)


22. Hanson JW, Buehler BA: Fetal hydantoin syndrome: Current status. *J Pediatr* **101**:816–818, 1982.

23. Hanson JW, Smith DW: The fetal hydantoin syndrome. *J Pediatr* **87**:285–290, 1976.

24. Hanson JW et al: Risks to the offspring of women treated with hydantoin anticonvulsants, with emphasis on the fetal hydantoin syndrome. *J Pediatr* **89**:662–668, 1976.

25. Hicks HE, Johnston MC: Biochemical changes in embryos after maternal administration of phenytoin: Possible clues for understanding the nature of an induced birth defect. *Proc Greenwood Genet Ctr* **4**:91, 1985.

26. Hirschberger M, Kleinberg F: Maternal phenytoin ingestion and congenital abnormalities: Report of a case. *Am J Dis Child* **129**:984, 1975.

27. Holmes LB: The effects of exposure to phenytoin in utero. *Pediatr Res.* **20**:92, 1986.

28. Jager-Roman E et al: Somatic parameters, diseases, and psychomotor development in the offspring of epileptic parents, in *Epilepsy, Pregnancy and the Child,* D Janz et al (eds), Raven Press, New York, 1982.

29. Johnson JP: Acquired craniofacial features associated with chronic phenytoin therapy. *Clin Pediatr* **23**:671–674, 1984.

30. Kogutt MS: Fetal hydantoin syndrome. *South Med J* **77**:657–658, 1984.

31. Kogutt MS: Radiological case of the month. *Am J Dis Child* **138**:405–406, 1984.

32. Leiber B, Hovels O: Embryopathisches Hydantoin-Syndrom. *Mschr Kinderheilkd* **124**:634–637, 1976.

33. Lewin PK: Phenytoin-associated congenital defects with Y-chromosome variant. *Lancet* **1**:559, 1973.

34. Lipson A, Bale P: Ependymoblastoma associated with prenatal exposure to diphenylhydantoin and methylphenobarbitone. *Cancer* **55**:1859–1862, 1985.

35. Livanainen M, Savolainen H: Side effects of phenobarbital and phenytoin during long-term treatment of epilepsy. *Acta Neurol Scand* **68**:49–67, 1983.

36. Loughnan PM et al: Phenytoin teratogenicity in man. *Lancet* **1**:70–72, 1973.

37. Majewski F et al: The teratogenicity of hydantoins and barbiturates in humans, with considerations on the etiology of malformations and cerebral disturbances in the children of epileptic parents. *Biol Res Pregnancy* **2**:37–45, 1981.

38. Mercier-Parot L, Tuchmann-Duplessis H: The dysmorphogenic potential of phenytoin: Experimental observations. *Drugs* **8**:340–353, 1974.

39. Monson RR et al: Diphenylhydantoin and selected congenital malformations. *N Engl J Med.* **289**:1049–1052, 1973.

40. Morin RA, Menolascino FJ: The fetal hydantoin syndrome: A case report and review. *Nebr Med J* **68**:51–53, 1983.

41. Pendergrass TW: Fetal hydantoin syndrome and neuroblastoma. *Lancet* **1**:150, 1976.

42. Phelan MC et al: Discordant expression of fetal hydantoin syndrome in heteropaternal dizygotic twins. *N Engl J Med* **307**:99–101, 1982.

43. Rating D et al: Minor anomalies in the offspring of epileptic parents, in *Epilepsy, Pregnancy, and the Child,* D Janz et al (eds), Raven Press, New York, 1982, pp. 283–288.

44. Rating D et al: Antiepileptika in der Neugeborenenperiode. *Monatsschr Kinderheilkd* **131**:6–12, 1983.

45. Robinow M: Fetal hydantoin syndrome characteristics. *Am J Dis Child* **138**:1154–1155, 1984.

46. Schardein JL: *Chemically Induced Birth Defects.* Dekker, New York, 1985.

47. Sherman S, Roizen N: Fetal hydantoin syndrome and neuroblastoma. *Lancet* **2**:517, 1976.

48. Smith DW: Fetal drug syndromes: Effects of ethanol and hydantoins. *Pediatr Rev* **1**:165–172, 1979.

49. Strickler SM et al: Genetic predisposition to phenytoin-induced birth defects. *Lancet* **2**:746–749, 1985.

50. Van Lang QC et al: Effect of in utero exposure to anticonvulsants on craniofacial development and growth. *J Craniofac Genet Dev Biol* **4**:115–133, 1984.

51. Waziri M et al: Teratogenic effect of anticonvulsant drugs. *Am J Dis Child* **130**:1022–1023, 1976.

52. Weiswasser WH et al: Coffin–Siris syndrome. *Am J Dis Child* **125**:838–840, 1973.

53. Zablen M, Brand N: Cleft lip and palate with the anticonvulsant ethotoin. *N Engl J Med* **297**:1404, 1977.


**Oxazolidinediones.** A unique pattern of abnormalities of growth and development in infants exposed prenatally to trimethadione and related anticonvulsant agents has been noted since 1970 (2). Among 55 pregnancies associated with prenatal trimethadione exposure, at least 47 have resulted in either fetal loss or birth of a child with a major congenital abnormality (1). Little is known of the pathogenesis of the disorder or of the contribution played by concomitant exposure to other teratogens or genetic factors.

Clinical manifestations. Craniofacial anomalies include microcephaly and characteristic facial appearence with flattening of the midface, short upturned nose, synophrys with V-shaped configuration of eyebrows, epicanthic folds, ocular defects including strabismus and myopia, and auricular anomalies, particularly overfolded, cup-like helices (Fig. 2–6). There is increased frequency of palatal defects, and malocclusion is common (4).

Children exposed to trimethadione prenatally show prenatal onset growth deficiency and are at increased risk for fetal death or miscarriage. Abnormalities in other body systems include cardiac malformations such as anomalous right subclavian and left common carotid arteries, patent ductus arteriosus, transposition of the great vessels, hypoplastic heart, valvular atresia, and tetralogy of Fallot (Table 2–2). Genitourinary anomalies include absent kidney and ureter, fetal renal lobulation, hypospadias, and inguinal hernia. Omphalocele and other gastrointestinal defects are occasionally described. Scoliosis, club hand, pterygium colli, meningomyelocele, and respiratory tract de-

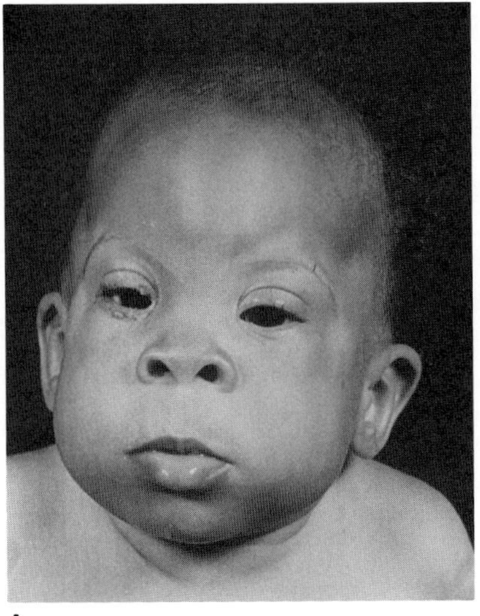

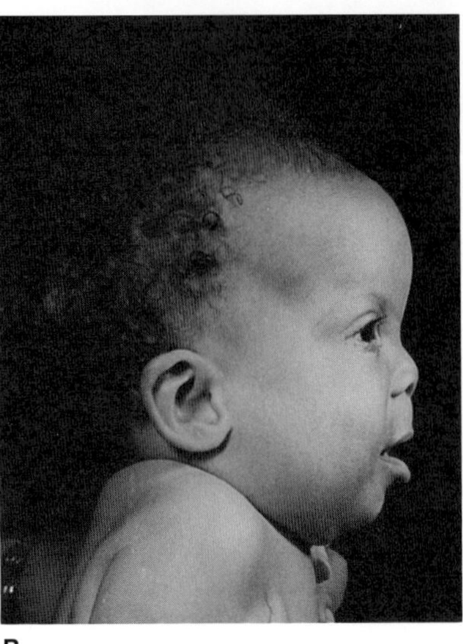

A B

Fig. 2–7. *Valproic acid embryopathy.* (A,B) Narrow bifrontal diameter, relative deficiency of outer orbital region, midface hypoplasia, short nose, broad flat nasal bridge, long flat philtrum, thin vermilion border of upper lip, and low-set posteriorly angulated ears. (From *HH Ardinger et al,* Am J Med Genet **29:**171, 1988.)

Fig. 2–8. *Toluene embryopathy.* (A,B) Short palpebral fissures, deeply set eyes, small midface, prominent nasal bridge, low-set ears, and micrognathia. (C,D) Compare with patient in A,B. (From *JH Hersh et al,* J Pediatr **106:**922, 1985.)

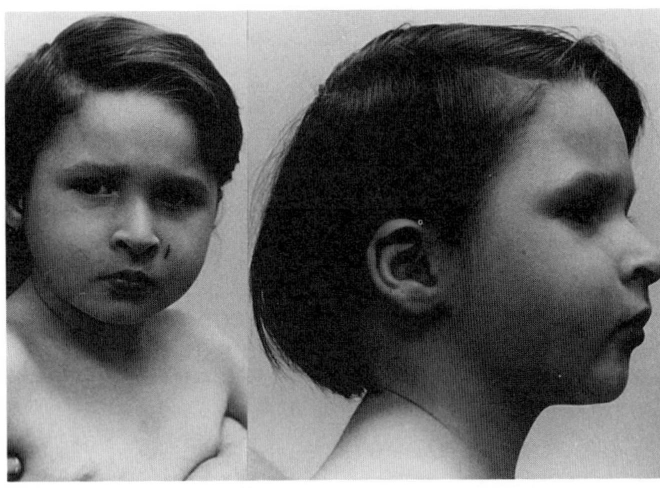

A B

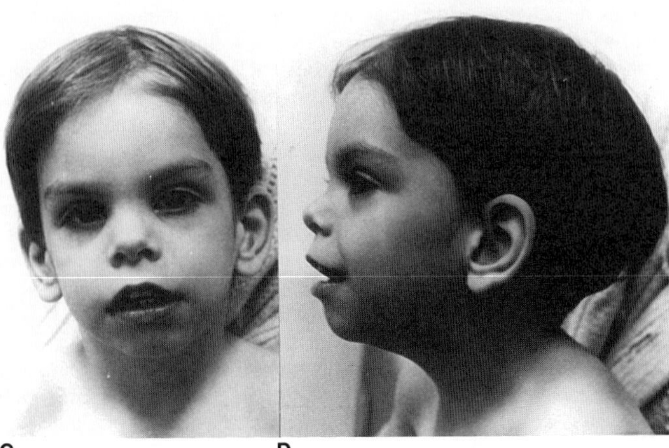

C D

fects of the trachea, larynx, and esophagus have been reported. Dermatoglyphic alterations have also been noted (2,3).

Differential diagnosis. Fetal oxazolidinedione effects may be reminiscent of other syndromes associated with striking synophrys, ear anomalies, and/or short stature such as *de Lange syndrome.* There is also some overlap between oxazolidinedione anticonvulsant effects and those of other anticonvulsant agents.

### References (Oxazolidinediones)

1. Feldman GL et al: Fetal trimethadione syndrome—report of an additional family and further delineation of this syndrome. *Am J Dis Child* **131**:1389–1402, 1977.

2. German J et al: Trimethadione and human teratogenesis. *Teratology* **3**:349–362, 1970.

3. Schroer RJ: Fetal trimethadione syndrome. *Proc Greenwood Genet Ctr* **4**:3–4, 1985.

4. Zackai EH et al: The fetal trimethadione syndrome. *J Pediatr* **87**:280–284, 1975.

**Valproic acid.** Valproic acid has been available for the treatment of seizure disorders in the United States for less than a decade. Numerous investigations in Europe support a causal relationship (6,14, 15,20,22–25). Little is known of the effects on the fetus attributable to interactions between this compound and other environmental agents nor is there available information on possible genetic–metabolic effects that might influence outcome. The observation that increased risk for spina bifida, often with hydrocephalus, is not apparently associated with increased risk for anencephaly suggests that the pathogenetic mechanism for birth defects produced by prenatal valproic acid exposure may not be related to neural tube closure but rather to factors that affect canalization posterior to the caudal neuropore (2). Whereas the risk for spina bifida subsequent to fetal exposure is thought to be in the range of 1–2%, the risk for a broader pattern of abnormalities in the fetus is not yet known (13a).

Clinical manifestations. Recent articles have suggested that there is a characteristic craniofacial appearance (Fig. 2–7) associated with prenatal valproic acid exposure (2,6a,9). Craniofacial features include abnormalities of the calvaria with metopic ridging, trigonocephaly, narrow bifrontal diameter, relative deficiency in the outer orbital re-

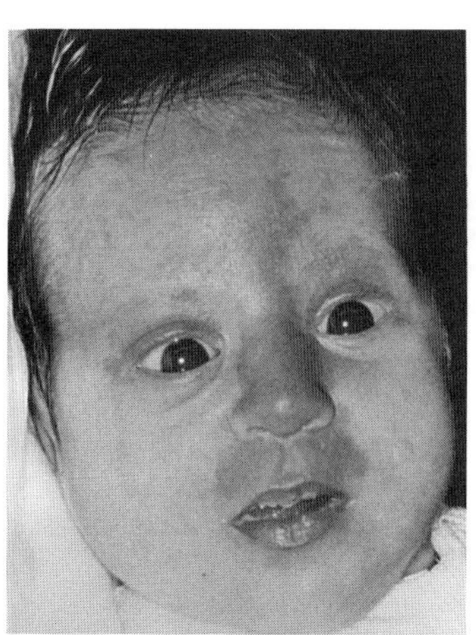
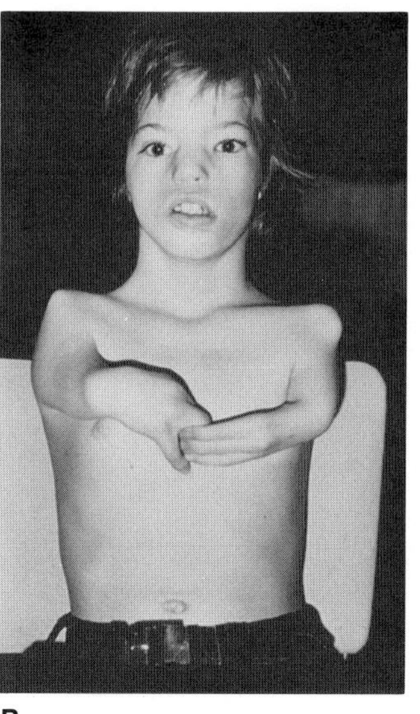
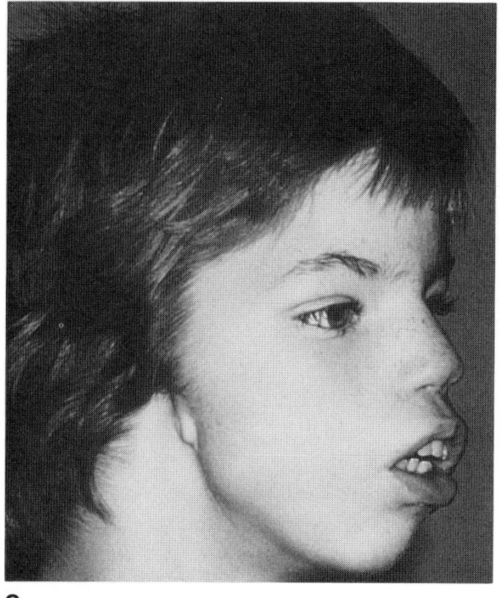

A B C

Fig. 2–9. *Thalidomide embryopathy.* (A) Midfacial hemangioma. (B,C) Phocomelia of both arms, absent ears. Patient also had ankylosis of temporomandibular joint, bifid uvula, and absence of one parotid gland. Mother had used thalidomide during the first three months of pregnancy. (A courtesy of *H-R Wiedemann,* Kiel, Germany. B,C courtesy of *PJW Stoelinga,* Arnhem, The Netherlands.)

gion, midfacial hypoplasia, short upturned nose with a broad flat bridge, and long flat philtrum with a thin vermilion border of the upper lip. Minor anomalies in ear position or shape are common as is micrognathia. Cleft lip and/or palate may be seen occasionally.

Abnormalities in other systems seen occasionally include tracheomalacia often leading to respiratory stridor, lower respiratory tract anomalies (abnormalities of lung lobulation), inguinal hernia, genital anomalies such as hypospadias, microscrotum, cryptorchidism and, less frequently, incomplete fusion of Müllerian duct structures in females, and cardiac anomalies. Among the specific types of heart defects are VSD, aortic coarctation, and PDA (Table 2–2). It has been suggested that such defects may be produced by a similar mechanism of cardiac dysmorphogenesis, that is, alterations in embryonic blood flow (1,7).

A number of low-frequency limb anomalies have included clinodactyly, arachnodactyly, camptodactyly, and distal phalangeal hypoplasia in patients who had been exposed to other anticonvulsants as well. Rare defects have included acromicria, dysplastic nails, talipes equinovarus, pes cavus, and polydactyly.

Low lumbosacral meningomyeloceles and other defects of vertebral segmentation have also been observed occasionally.

Differential diagnosis. The principal source of diagnostic confusion is with different anticonvulsant drugs used during pregnancy.

### References (Valproic acid)


1. Ardinger HH et al: Cardiac malformations associated with fetal valproic acid exposure. *Proc Greenwood Genet Ctr* **5**:162, 1986.

2. Ardinger HH et al: Verification of the fetal valproate syndrome phenotype. *Am J Med Genet* **29**:171–185, 1988.

3. Bailey CJ et al: Valproic acid and fetal abnormality. *Br Med J* **286**:190, 1983.

4. Bantz ME: Valproic acid and congenital malformations. *Clin Pediatr* **23**:352–353, 1984.

5. Blaw ME, Woody RC: Valproic acid embryopathy? *Neurology* **33**:255, 1983.

6. Castilla E: Valproic acid and spina bifida. *Lancet* **2**:683, 1983.

6a. Chitayat D et al: Congenital abnormalities in two sibs exposed to valproic acid in utero. *Am J Med Genet* **31**:369–373, 1988.

7. Clark EB: Mechanisms in the pathogenesis of congenital cardiac malformations, in *Genetics of Cardiovascular Disease,* Pierpont MEM, Moller JH (eds), Martinus Nijhoff, Boston, 1987, pp. 3–12.

8. Curran ACW: Spina bifida and sodium valproate. *Med J Aust* **1**:401, 1983.

9. DiLiberti JH et al: The fetal valproate syndrome. *Am J Med Genet* **19**:473–481, 1984.

10. Felding I, Rane A: Congenital liver damage after treatment of mother with valproic acid and phenytoin? *Acta Paediatr Scand* **73**:565–568, 1984.

11. Frew J: Valproate link to spina bifida. *Med J Aust* **1**:150, 1983.

12. Hurd RW: Valproate, birth defects, and zinc. *Lancet* **1**:181, 1983.

13. Koch S et al: Possible teratogenic effect of valproate during pregnancy. *J Pediatr* **103**:1007–1008, 1983.

13a. Lammer EJ et al: Teratogen update: Valproic acid. *Teratology* **35**:465–474, 1987.

14. Lindhout D, Meinardi H: Gebruik van valproinezuur gedurende de zwangerschap: een indicatie voor prenataal onderzoek op spina bifida. *Ned Tijdschr Geneeskd* **128**:2438–2440, 1984.

15. Mastroiacovo P et al: Maternal epilepsy, valproate exposure, and birth defects. *Lancet* **2**:1499, 1983.

16. Nau H et al: Valproic acid and its metabolites: Placental transfer, neonatal pharmacokinetics, transfer via mother's milk and clinical status in neonates of epileptic mothers. *J Pharmacol Exp Ther* **219**:768–777, 1981.

17. Pacifici GM et al: Valpromide/carbamazepine and risk of teratogenicity. *Lancet* **1**:397–398, 1985.

18. Philbert A, Pedersen B: Treatment of epilepsy in women of child-bearing age: Patients' opinion of teratogenic potential of valproate. *Acta Neurol Scand* **94**:35–38, 1983.

19. Pruitt AW et al: Valproate teratogenicity. *Pediatrics* **71**:980, 1983.

20. Robert E: Valproic acid in pregnancy—association with spina bifida: A preliminary report. *Clin Pediatr* **22**:336, 1983.

21. Robert E, Lofkvist E: Valproate and spina bifida. *Lancet* **2**:1392, 1984.

22. Robert E et al: L'acide valproique est-Il teratogene? *Rev Neurol* **139**:445–447, 1983.

23. Robert E: Valproate and birth defects. *Lancet* **2**:1142, 1983.

24. Ten Kate LP et al: Spina bifida door valproinezuur? *Ned Tijdschr Geneeskd* **127**:469–470, 1983.

25. Winter RM et al: Fetal valproate syndrome: Is there a recognizable phenotype? *J Med Genet* **24**:692–695, 1987.


## Toluene

In the past, evidence for human teratogenicity from *in utero* exposure to aromatic hydrocarbons has been inconclusive. In 1985 (1), three children with microcephaly, central nervous system dysfunction, minor craniofacial and limb anomalies, and variable growth deficiency were described who were born to women who inhaled large quantities

Table 2–3A. Other drug and chemical hazards to the fetus

| Agent | Congenital abnormalities |
|---|---|
| Environmental chemicals | |
| Organic mercurials | Microcephaly, cerebral palsy, mental retardation |
| Polychlorinated biphenyls (PCBs) | "Cola-colored" babies, growth deficiency, natal teeth, exophthalmos |
| Prescription drugs | |
| Anticancer agents | |
| Alkylating agents (e.g., chlorambucil, nitrogen mustards) | Fetal death, growth and mental deficiency, ocular, genitourinary, and limb malformations, cleft palate |
| Antibiotics | |
| Tetracyclines | Brownish yellow teeth, hypoplastic enamel |
| Streptomycin | Hearing loss |
| Kanamycin | Hearing loss[a] |
| Aminoglycosides (e.g., gentamicin) | Hearing loss[a] |
| Chloroquine | Ocular defects,[a] hearing loss[a] |
| Anticonvulsants | |
| Barbiturates (and primidone) | Possible syndrome, present evidence inconclusive[a] |
| Diazepam | Low risk for facial clefts[a] |
| Psychotropic agents | |
| Meprobamate[b] | Brain, heart, palate, and limb defects[a] |
| Benzodiazepines[b] | Cleft palate[a] |
| Thalidomide[c] | Facial asymmetry, cranial nerve palsies, cochlear hypoplasia with deafness, hypoplastic teeth, low nasal bridge with broad nasal tip, choanal atresia, colobomas, microphthalmia, cataracts, microtia, facial capillary hemangiomas, phocomelic limb defects (13) |
| Lithium | Ebstein anomaly |
| Methimazole | Aplasia cutis, choanal atresia, absent nipples (4a) |

[a]Proposed association; magnitude of risk uncertain; more information needed.
[b]These drugs have not been clearly associated with congenital defects, and available information is insufficient to ascribe specific risk estimates. In each case, if the risk is increased at all, it is estimated to be very low.
[c]Though thalidomide is of major historical interest, its clinical significance has waned somewhat, except for continuing legal problems and the health management issues among affected persons. Nonetheless, it is important to note that this drug continues to be used in some parts of the world for the treatment of epilepsy and major aphthae. Accordingly, new cases still appear sporadically.

Table 2–3C. Infectious teratogenic agents

| Agent | Congenital abnormalities |
|---|---|
| Viruses | |
| Rubella | Ocular and cardiovascular anomalies, deafness, microcephaly, immune and endocrine disturbances, delayed skeletal development, growth deficiency, mental deficiency |
| Cytomegalovirus | Ocular and cardiovascular anomalies, deafness, microcephaly, hydrocephaly, gastrointestinal tract anomalies,[a] growth deficiency,[a] mental deficiency |
| Herpes simplex | Ocular anomalies, microcephaly, patent ductus arteriosus, hypoplastic distal phalanges, growth deficiency,[a] mental deficiency |
| Varicella zoster | Ocular anomalies, microcephaly, hypoplastic limbs, neurogenic muscular atrophy, cutaneous defects, growth deficiency, mental deficiency |
| Bacteria | |
| *Treponema pallidum* | Ocular, dental and skeletal anomalies, hydrocephaly, microcephaly, cutaneous lesions, nerve palsies, nephrosis, growth deficiency, mental deficiency |
| Mycoplasmas | Neural tube defects,[a] aneuploidy,[a] growth deficiency,[a] mental deficiency[a] |
| Parasites | |
| *Toxoplasma gondii* | Ocular anomalies, microcephaly, hydrocephaly, deafness, growth deficiency, mental deficiency |
| Possibly hazardous agents | |
| Influenza viruses | Neural tube defects,[a] hydrocephaly[a] |
| Lymphocytic choriomeningitis virus | Hydrocephaly[a] |
| HTLV-III (AIDS virus)[b] | Growth deficiency, microcephaly, hypertelorism, prominent forehead, short nose with low nasal bridge, long slanting palpebral fissures, blue sclerae, triangular philtrum, patulous lips (11,21,25) |
| Pestiviruses | Microcephaly (16) |

[a]Proposed association; magnitude of risk uncertain; more information needed.
[b]Present evidence clearly indicates that HTLV-III can be passed to the fetus from an infected mother. However, the dysmorphic features reported in some infants with documented prenatal infection are still of uncertain significance. No data have yet established whether these reported findings are the consequence of prenatal disturbances of growth, or may represent postnatal changes; nor have susceptible prenatal periods been defined.

Table 2–3B. Hazardous maternal metabolic and genetic factors for the fetus

| Factor | Effects |
|---|---|
| Diabetes mellitus (insulin dependent) | Cardiovascular, renal, and neural tube defects, facial clefts, holoprosencephaly, caudal regression syndrome (6) |
| Phenylketonuria | Mental retardation, growth deficiency, microcephaly, heart defects,[a] vertebral defects,[a] fetal death[a] (14) |
| Myotonic dystrophy | Severe early-onset myotonic dystrophy, arthrogryposis, heart defects,[a] hernia,[a] hydrocephaly,[a] hydronephrosis,[a] facial clefts[a] (7) |

[a]Proposed association; magnitude of risk uncertain; more information needed.

Table 2–3D. Physical hazards to the fetus

| Agent | Effects |
|---|---|
| Radiation | |
| High levels | Microcephaly, ocular defects,[a] mutations, malignancy (20) |
| Heat or fever | Neural tube defects, other CNS anomalies,[a] mental retardation[a] (12) |
| Mechanical factors | |
| Oligohydramnios, multiple gestation, uterine malformation | Positional deformities, limb reduction defects,[a] mandibular hypoplasia[a] (23) |
| Amniotic bands | Syndactyly, prenatal limb amputations, atypical facial clefts, exencephaly, encephalocele, thoracogastroschisis, ocular defects |

[a]Proposed association; magnitude of risk uncertain; more information needed.

of toluene throughout pregnancy. Significant phenotypic abnormalities included a small midface, narrow bifrontal diameter, short palpebral fissures, deep set eyes, lowset ears, micrognathia (Fig. 2–8), blunted fingertips, and small fingernails. Two of the patients also had structural renal lesions. These manifestations were felt to be reminiscent of patterns of malformation following *in utero* exposure to other teratogens.

Two more cases of toluene embryopathy were reported in 1987 by the same group of investigators (2). Clinical findings included alterations of growth and development and phenotypic abnormalities similar to patients in the first report, although no renal defects were identified. Close monitoring of at-risk pregnancies should permit further delineation of the full clinical spectrum of toluene embryopathy and its frequency among offspring of toluene abusers.

### References (Toluene)

1. Hersh JH et al: Toluene embryopathy. *J Pediatr* **106**:922–927, 1985.

2. Hersh JH: Toluene embryopathy: Two new cases. *J Med Genet* **26**:333–337, 1989.

## Other teratogenic agents

A number of other agents have less dramatic or less well-characterized effects on craniofacial growth and development (Fig. 2–9). These are summarized in Table 2–3 and the reader is referred to references 1–26. It can be anticipated that this list may be expanded substantially as further information becomes available.

### References (General considerations)

1. Brent RL, Holmes LB: Chemical and basic science lessons from the thalidomide tragedy. What have we learned about the causes of limb defects? *Teratology* **38**:241–251, 1988.

1a. Briggs GG et al: *Drugs in Pregnancy and Lactation, a Reference Guide to Fetal and Neonatal Risk.* Williams & Wilkins, Baltimore, 1983.

2. Chung CS, Myrianthopoulos NC: *Factors Affecting Risks of Congenital Malformations.* Stratton Intercontinental Medical Book Corp, New York, 1975.

3. Cohen MM: *Craniosynostosis, Diagnosis, Evaluation, and Management.* Raven Press, New York, 1986, pp. 67–69 and 577.

4. Gal P, Sharpless MK: Fetal drug exposure—behavioral teratogenesis. *Drug Intell Clin Pharm* **18**:186–201, 1984.

4a. Greenberg F: Choanal atresia and athelia: Methimazole teratogenicity or a new syndrome? *Am J Med Genet.* **28**:931–934, 1987.

5. Hanshaw JB, Dudgeon JA: *Viral Diseases of the Fetus and Newborn.* Saunders, Philadelphia, 1978.

6. Hanson JW: Teratogenic agents, in *Principles and Practice of Medical Genetics,* AH Emory and DL Rimoin (eds), Churchill Livingstone, Edinburgh, 1983, pp. 127–151.

7. Harper P: *Myotonic Dystrophy.* Saunders, Philadelphia, 1979.

8. Heinonen OP et al: *Birth Defects and Drugs in Pregnancy.* Publishing Sciences Group, Littleton, MA, 1977.

9. Hill LM: Effects of drugs and chemicals on the fetus and newborn (second of two parts). *Mayo Clin Proc* **59**:755–765, 1984.

9a. Jones KL: *Smith's Recognizable Patterns of Human Malformations.* WB Saunders, Philadelphia, 1988.

10. Kalter H, Warkany J: Congenital malformations. *N Engl J Med* **308**:491–497, 1983.

10a. Laegreid L et al: Teratogenic effects of benzodiazepine use during pregnancy. *J Pediatr* **114**:126–131, 1989.

11. Marion RW et al: Human T-cell lymphotropic virus type III (HTLV-III) embryopathy. *Am J Dis Child* **140**:638–640, 1986.

12. Miller P et al: Hyperthermia as one possible etiology of anencephaly. *Lancet* **1**:519, 1978.

13. Newman CGH: Teratogen update: Clinical aspects of thalidomide embryopathy—a continuing preoccupation. *Teratology* **32**:133–144, 1985.

14. Nyhan WL: *Heritable Disorders of Amino Acid Metabolism.* Wiley, New York, 1974.

15. Pexider T: Teratogens, in *Genetics of Cardiovascular Disease,* Pierpont MEM, Moller JH (eds), Martinus Nijhoff, Boston, 1987, pp. 25–68.

16. Potts B et al: Studies of patients and controls for possible role of pestiviruses in the development of microcephaly. *Teratology* **35**:51A, 1987.

17. Remington JS, Klein JO: *Infectious Diseases of the Fetus and Newborn Infant.* Saunders, Philadelphia, 1976.

18. Schardein JL: *Chemically Induced Birth Defects.* Dekker, New York, 1985.

19. Schinzel AA: Cardiovascular defects associated with chromosomal aberrations and malformation syndromes. *Prog Med Genet* **5**:303–379, 1983.

20. Sever JL, Brent RL: *Teratogen Update, Environmentally Induced Birth Defect Risks.* Alan R. Liss, New York, 1986.

21. Sheikh T et al: Lack of evidence for cranio-facial dysmorphism in children with AIDS. *Pediatr Res* **21**:230A, 1987.

22. Shepard TH: *Catalog of Teratogenic Agents,* 5th ed. Johns Hopkins University Press, Baltimore, 1986.

23. Smith DW: *Recognizable Patterns of Human Malformation.* Saunders, Philadelphia, 1976.

24. Stephens TD: Proposed mechanisms of action in thalidomide embryopathy. *Teratology* **38**:229–239, 1988.

25. Stevenson RE: Prenatal AIDS: A cause of postnatal progressive dysmorphism. *Proc Greenwood Genet Ctr* **5**:22–25, 1986.

26. Wilson J, Fraser FC (eds): *Handbook of Teratology,* Vols. 1–4. Plenum Press, New York, 1977.

# Chapter 3
# Chromosomal Syndromes: Common and/or Well-Known Syndromes

## Trisomy 21 syndrome (Down syndrome)

Trisomy 21 is the most common and well known of all malformation syndromes. More than 100 different signs have been reported in several reviews (109,160). The reader is referred to the following sources for comprehensive coverage: Smith and Berg (160) and Benda (8,9) for general coverage; Apgar (5) and Shapiro (154) for the biology of trisomy 21 syndrome; Bersu (10), Stephens and Shepard (169), and Dunlap et al (48) for anatomic aspects; Hall (76) and Tolksdorf and Wiedemann (175) for pediatric aspects; Benda (9), Cowie (39), Smith and Berg (160), and Zellweger (187) for central nervous system manifestations; Cronk (41a) for growth charts; Kisling (98) for craniofacial aspects; Cohen and Cohen (35) for oral manifestations; Warkany (181) for the history of etiologic explanations; numerous authors for cytogenetic considerations and recurrence risk counseling (2,4,14a,23,51, 54,70,74,79,84–86,94,96,111,115,117,122,123,126,130,145,160,164–168,177,182); Lilienfeld and Benesch (110) for epidemiology; and Smith and Wilson (159) and Horrobin and Rynders (87) for presentations of the subject suitable for the layman, particularly for parents of an affected child.

In 1866, Langdon Down described a condition that he named ''mongolian idiocy.'' A description of the syndrome also appeared in the works of Séguin (150) who called the condition ''furfuraceous cretinism'' as early as 1846. Lejeune (108) demonstrated that the condition was associated with an extra G group chromosome in 1959, Polani et al (133) reported translocation type Down syndrome in 1960, and Clarke et al (28) observed mosaicism for an extra G group chromosome in 1961.

**Prevalence.** The birth prevalence of trisomy 21 syndrome is generally stated to be 1/650 live births, but it is known to vary in different populations from 1/600 to 1/2000 live births. Fifteen percent of patients institutionalized for mental retardation have trisomy 21 syndrome (77,110,160). It has been estimated that from 65 to 80% of trisomy 21 conceptions result in spontaneous abortions (160)

**Cytogenetics and recurrence risks.** Approximately 95% of all cases of Down syndrome result from nondisjunction. Although the syndrome occurs in offspring of mothers of all ages, the risk increases with increasing maternal age. The birth prevalence is 0.9/1000 when the mother is less than 33 years of age, 2.8/1000 when the mother is 35 to 38 years old, and 38/1000 when the mother is 44 years old or older (74). The association of advanced maternal age and nondisjunction has been discussed by several authors (86,94,177). Nondisjunction may occur during the first or the second meiotic division in either the female or the male parent (94,115,117) (Table 3–1). The possibility of paternal age effect has been discussed by several investigators (145,167). The occurrence of a genetic predisposition or nondisjunction gene has been debated (4) and although the risk of trisomy 21 in second and third degree relatives of an individual with nondisjunction-type Down syndrome has been shown to be no greater than the risk to the general population (2,54), the recurrence risk for free trisomy 21 in couples with one affected child is generally accepted to be 1% (168). Rarely, unusual recurrences of free trisomy 21 are reported, such as three affected first cousins (23). Approximately 1 in 200 patients with trisomy 21 syndrome has a double primary nondisjunction, the most frequent type being 48,XXY,+21 (78).

Approximately 4.8% of Down syndrome cases are caused by a translocation, either arising *de novo* or being transmitted from one of the parents (70). Information available about translocation-type Down syndrome appears in the following references and tables: the frequency of different types of translocations (74,111,160) (Table 3–2), the ratio of *de novo* to inherited translocations (111) (Table 3–3), the frequency of maternal and paternal origins of inherited translocations (77,94,111) (Table 3–1), and recurrence risks for translocation-type Down syndrome when one parent is a carrier (51,77,111,165,166) (Table 3–4).

Table 3–1. Origin of nondisjunctions and inherited translocations[a]

| | Maternal (%) | Paternal (%) |
|---|---|---|
| Overall origin of nondisjunction | 80 | 20 |
| Ratio of first-to-second meiotic nondisjunction | 80/20 | 60/40 |
| Origin of inherited D/G translocations | 93 | 7 |
| Origin of inherited G/G translocations | 50 | 50 |

[a]Based on Lister and Frota-Pessoa (111), Juberg and Mowrey (94), and Hamerton (77).

Table 3–2. Frequencies of different types of translocations[a]

| Type | Percentage |
|---|---|
| t(Dq21q) | 54 |
| t(14q21q) | (58) |
| t(13q21q) | (22) |
| t(15q21q) | (20) |
| t(21qGq) | 41 |
| t(21q21q) | (83) |
| t(21q22q) | (17) |
| t(21qOther[b]) | 5 |

[a]Based on de Grouchy and Turleau (74) and Lister and Frota-Pessoa (111)
[b]For example, chromosomes 1, 2, 6, 7, 12, and 19 (160).

Table 3–3. Translocations[a]

| | Type of translocation | |
|---|---|---|
| Characteristic | D/21 (%) | G/21 (%) |
| *De novo* | 53 | 91 |
| Inherited | 47 | 9 |

[a]Based on Lister and Frota-Pessoa (111).

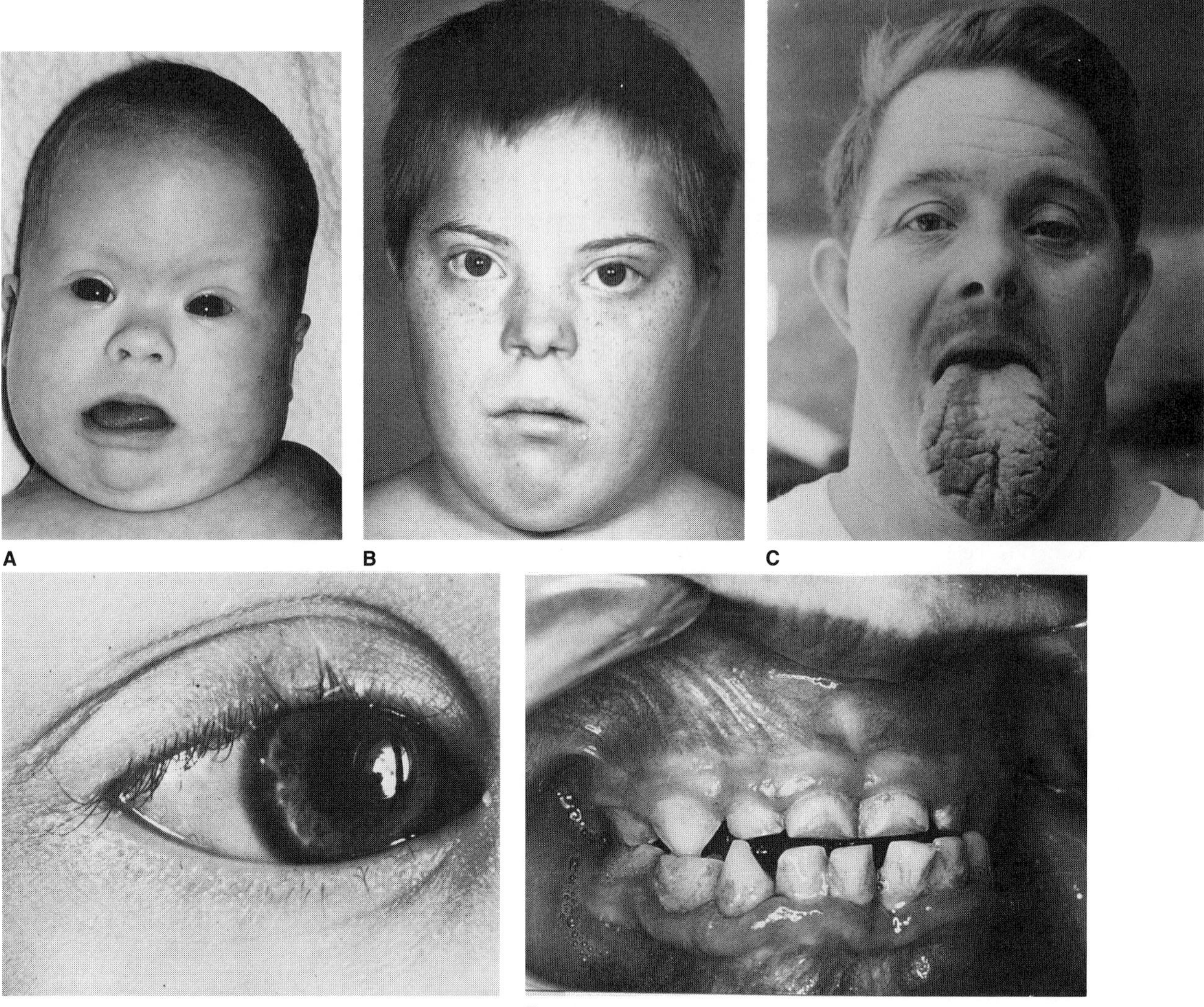

Fig. 3–1. *Trisomy 21 syndrome (Down syndrome).* (A) Typical facial appearance with upslanting palpebral fissures, low nasal bridge, button-like nose, tendency of the mouth to be open with the tongue protruding. Note also overfolding of helix and cutis marmorata. (B) Compare facies to A and C. (C) Severe macroglossia. (D) Brushfield spots. (E) Primary dentition showing enamel hypoplasia, severe gingival and periodontal involvement with mobility and drifting of lower incisors. (A from *MM Cohen Jr,* The Child with Multiple Birth Defects, Raven Press, New York, 1982, p 20. B,D from *B Russell,* Gentofte, Denmark. C courtesy of *J Červenka,* Minneapolis, Minnesota. E courtesy of *MM Cohen Sr,* Boston, Massachusetts.)

Detectable mosaicism is found in approximately 3% of trisomy 21 cases (79); the Down syndrome phenotype may not be fully expressed. Although it is frequently assumed that the extra chromosome 21 in mosaic Down syndrome arises from nondisjunction in a chromosomally normal zygote, evidence based on maternal ages suggests that a large proportion of such cases arise from meiotic nondisjunction followed by a "normalizing" mitotic error (126,140). Mosaic normal/trisomy 21 patients show a significantly decreasing percentage of trisomy 21 cells with age (130,182). If gonadal mosaicism occurs in a parent, it is possible to have a child with Down syndrome (77).

Fertility in female Down syndrome patients has been well documented. Not uncommonly a male relative is the parent. In about 50% of cases, the offspring has Down syndrome. Fertility in male Down syndrome patients is probably severely reduced but a documented example of fatherhood has been published (155a).

**Mortality.** In spite of advances in health care of the retarded and a gradual increase in the life span of those affected with Down syndrome, the average life expectancy is still 35 years (172). The periods of highest mortality risk are in infancy when congenital heart disease, leukemia, and respiratory diseases are factors and late adulthood when Alzheimer's disease and declining immunological function are significant factors. Estimates during the 1960s and 1970s vary from a 12-fold to a 124-fold increase in mortality from pneumonia and other infectious diseases. Mortality for congenital heart disease is highest during the first 2 years of life and only 40–60% of Down syndrome children with congenital heart disease live to age 10. Recent life expectancy figures have been provided by Baird and Sadovnick (6b). Neoplastic disease accounts for approximately twice the expected number of deaths, mostly attributable to leukemia, but lymphomas and other neoplasms such as testicular carcinoma, retinoblastoma, and central nervous system tumors have also been reported with Down syndrome (124,160,172).

**Common clinical diagnostic features.** Hall (76) noted 10 common signs in the newborn period: hypotonia, poor Moro reflex, hyperextensibility of joints, loose skin on the posterior part of the neck, flat facial profile, upslanting palpebral fissures, short ears with overhanging helices, dysplastic pelvis, clinodactyly of fifth fingers, and single palmar creases. At least four of these abnormalities were pres-

Table 3–4. Recurrence risks for translocation type trisomy 21 syndrome when one parent is a carrier[a]

| Type of translocation | Risk (%) |
|---|---|
| D/G maternal | 10 |
| D/G paternal | 2–5 |
| G/G (21/22) | 4 |
| G/G (21/21) | 100[c] |

[a]Based on Lister and Frota-Pessoa (111), Stene (165,166), Dutrillaux and Lejeune (51), and Hamerton (77).
[b]The theoretical risk is that one-third of the offspring of a translocation carrier will have Down syndrome, but the actual risk is much lower, being related, in part, to *in utero* loss of Down syndrome fetuses.
[c]t(21q21q) results in monosomic or trisomic zygotes. Since monosomy for chromosome 21 does not produce a viable zygote, all viable zygotes will have translocation type Down syndrome.

Table 3–5. Hall's ten cardinal features of trisomy 21 syndrome in the newborn[a,b]

| Feature | Percentage |
|---|---|
| Hypotonia | 80 |
| Poor Moro reflex | 85 |
| Hyperextensibility of joints | 80 |
| Excess skin on back of neck | 80 |
| Flat facial profile | 90 |
| Slanted palpebral fissures | 80 |
| Anomalous auricles | 60 |
| Dysplasia of pelvis | 70 |
| Brachymesophalangy of fifth finger | 60 |
| Single palmar crease | 45 |

[a]Based on B. Hall (76).
[b]100% have at least four features and 89% have six or more features.

ent in all of his cases, and six or more were present in 89% (Table 3–5). Other common signs include short stature, mental retardation, brachycephaly, flat occiput, epicanthic folds, Brushfield spots, fine lens opacities, open mouth with protruding tongue (Fig. 3–1A–D), short neck, short broad hands, distally placed axial triradii, wide gap between the first and second toes, and tibial arch pattern (109,160). Common physical findings in children and adults are listed together with their respective frequencies in Table 3–6, which is based on the combined data of Øster (129), Levinson et al (109), and Domino and Newman (47).

A number of diagnostic indices have been developed. Using 10 features, Jackson et al (88) were able to properly classify 158 of 169 Down syndrome individuals under 2 years of age. Rex and Preus (139) developed an index of eight phenotypic features capable of clinically diagnosing 95% of patients suspected of having Down syndrome. Various dermatoglyphic indices have also been developed (vide infra).

**Growth and skeletal abnormalities.** Both prenatal and postnatal growth deficiency are evident in Down syndrome. The average length is 2–3 cm less; average weight is 400 g less than normal infants. There is also a tendency toward premature birth, although prematurely born Down syndrome infants are smaller than their premature counterparts for gestational age. Postnatally, both height and weight are usually 2–4 SD below those of the general population. Bone age is normal to advanced at birth and thereafter slows down so that by 3 years of age, osseous maturation is significantly delayed. For final height attainment, males average approximately 151 cm and females average approximately 141 cm. Among the most characteristic skeletal findings are flaring of the iliac wings and brachymesophalangy of the fifth fingers (160).

**Central nervous system and performance.** Fetal brain growth is clearly delayed, so that infants commonly are microcephalic at birth (9,50). A decrease in total brain weight has been observed (127,161). Simplification of gyriform patterns is frequently mentioned, but such observations are difficult to evaluate because they are based on subjective evidence. Neuropathologic examination has demonstrated that the cerebellum and certain nuclei in the brain stem appear to be smaller than normal (41). Specific deficits have been documented in certain areas, such as auditory sequencing (12,46,137,149), color retention (158), short-term memory (114,137), articulation (13,46,55,102), visual-motor tasks (114,137), ability to differentiate between symbols (114,137,149), and language development.

Table 3–6. Physical findings in trisomy 21 children and adults[a]

| Findings | Percentages |
|---|---|
| Craniofacial | |
| Flat nasal bridge | 61 |
| Flat occiput | 76 |
| Eyes | |
| Upslanting palpebral fissures | 79 |
| Epicanthic folds | 48 |
| Brushfield spots | 53 |
| Strabismus | 22 |
| Nystagmus | 11 |
| Ears | |
| Dysplastic | 53 |
| Absent lobules | 70 |
| Mouth | |
| Open mouth | 61 |
| Fissured lips | 56 |
| Protruding tongue | 42 |
| Macroglossia | 43 |
| Furrowing of tongue | 61 |
| Narrow palate[b] | 67 |
| Irregular alignment of teeth | 71 |
| Neck | |
| Broad, short | 53 |
| Chest | |
| Flat nipples | 56 |
| Pectus excavatum | 10 |
| Pectus carinatum | 8 |
| Dorsolumbar kyphosis | 11 |
| Abdomen | |
| Diastasis recti | 82 |
| Umbilical hernia | 5 |
| Genitalia | |
| Small penis | 70 |
| Cryptorchidism | 21 |
| Small scrotum | 37 |
| Hands | |
| Short broad hands | 70 |
| Brachydactyly | 67 |
| Single palmar crease | 52 |
| Clinodactyly | 59 |
| Short fifth finger | 59 |
| Single flexion crease on fifth finger | 20 |
| Feet | |
| Gap between hallux and second toe | 50 |
| Plantar furrow | 31 |
| Joints | |
| Hyperflexibility | 62 |

[a]Based on combined data of Øster (129), Levinson et al (109), and Domino and Newman (47). Physical signs of children and adults but not newborns.
[b]In trisomy 21 syndrome, the palate appears high because it is narrow. Palatal height, however, is no higher than that observed in the general population.

Various other lesions of the brain have been observed, including anomalous structure of the peduncles of the cerebellar flocculi (89) and fibrous gliosis of the white matter (121). Many pathologic findings have been reported in the brain and spinal cord by Benda (8). Pi et al (132) and Urioste et al (177a) reported holoprosencephaly in association with trisomy 21 syndrome.

Pathologic observations of the brain in older patients demonstrate the atrophic changes characteristic of Alzheimer's disease. Senile plaques, neurofibrillary tangles, and granulovacuolar changes have been observed with both light and electron microscopy. In some patients, dementia follows slowly progressive intellectual and emotional deterioration; others seem less obviously affected (17,53,61,75,91, 101,116,128,161). Heston (81) and Heston and Mastri (82) reported that relatives of probands with classic Alzheimer's disease had an excessive frequency of trisomy 21 syndrome and myeloproliferative disorders. The gene for $\beta$-amyloid protein, which is found deposited in the brains of patients with Alzheimer's disease as well as in older individuals with Down syndrome, maps to chromosome 21. The overexpression of the gene appears to be responsible for the Alzheimer-like histopathologic features of Down syndrome (4a).

Neoplasms of the brain and retinoblastoma have been reported occasionally with Down syndrome, but far less frequently than leukemia (124).

The most extensive EEG investigation was carried out by Ellingson et al (52). They reported that 21% of Down syndrome cases had abnormalities including asymmetry and/or asynchrony, diffuse slow activity, and diffuse focal seizure activity. The frequency of epilepsy varies from less than 1% to nearly 10%. Epilepsy may be of the grand mal variety but other types, including myoclonic seizures and petit mal, may occur. In adults, epilepsy may sometimes be related to the cerebral changes of Alzheimer's disease (75,97,113,134,148,151, 195,180,185).

Mental retardation is considered to be a hallmark of nonmosaic trisomy 21 syndrome. The degree of hypotonia also is important as it affects not only motor ability but language as well (38,64,113,170). In the study of Domino and Newman (47), no systematic relationship between IQ and various physical stigmata was found. Typically, the IQ is reported to vary between 30 and 50 (49,59). The IQ range for home-reared patients varies from 27.4 to 62.4 compared to a range of 17.4 to 37.4 for institutionalized patients (45). Furthermore, developmental milestones such as walking and talking are achieved at a much earlier age in affected children reared in the home (24,27, 58,119,156,157,163,174). It is believed by some that the developmental potential of offspring with trisomy 21 syndrome is higher in families in which the parents have higher IQs (45,63). The study of Golden and Pashayan (71) shows that parental education is also a factor; the mean IQ of patients whose parents had graduated from high school was 50 (range 32–74), whereas the mean IQ was 35.6 (range 0–45) if the parents had completed grade school only.

The pattern of mental development usually demonstrates an early rise in IQ, which plateaus from ages 2 to 5 years, followed by a gradual decline (21,59,60,119). The social quotient tends to be substantially ahead of the mental age. The ability to make social adjustments sometimes makes Down syndrome individuals appear more intelligent than they are (37).

Investigation of the performance of individuals mosaic for trisomy 21 syndrome has shown that their mean performance level is above that of those with pure trisomy 21. Data suggest that the level of functioning tends to be higher in those with higher percentages of normal cells (59,60,62,140,144). Few mosaic individuals, however, test in the normal range. They often manifest milder physical stigmata than nonmosaic trisomy 21 persons (20,28,59,60,69). Kousseff (103) reported a Down syndrome patient with average intelligence. Chromosomal analysis of cultured peripheral lymphocytes showed trisomy 21 in all 62 mitoses counted.

Although in older patients dementia may develop as a consequence of Alzheimer's disease, psychiatric problems are encountered in some children. In the study of Menolascino (120), 11 of 86 children were found to be emotionally disturbed, and absence of speech development was common.

Newborn infants are frequently described as being good babies if they are not easily disturbed and they cause their mothers very little trouble. Such traits probably reflect reduced response to external stimuli and marked hypotonia. Later, children are often described as happy, cheerful, good tempered, and easily amused. They tend to mimic and may be mischievous. Langdon Down observed that they were humorous and had a lively sense of the ridiculous. Another common feature is obstinacy. Only a minority is judged to be aggressive or hostile or to display other varieties of maladaptive behavior (107,125,186).

The overwhelming majority have articulation defects; pronunciation is often slurred making speech incomprehensible. Sibilants and affricates are more severely compromised than other consonants. More than one-third also have dysrhythmic or explosive speech, which is commonly identified with stuttering or inexact rapid speech leading to distortion of sound and phrasing. The voice is often hoarse, raucous, and low pitched (13,95).

**Craniofacial manifestations.** Brachycephaly and flat occiput result in a cephalic index that is usually greater than 0.80 and may exceed 1.00 (normal, 0.75 to 0.80) (142). Fontanels are large, and closure is late (160). In the study of Chemke and Robinson (25), a "third fontanel" was noted in all affected patients. Persistent metopic suture is found in 67% of males (normal, 8.8%) and in 42% of females (normal, 12.3%) (142). Frontal and sphenoidal sinuses are absent and maxillary sinuses are hypoplastic in over 90% of cases (11,90,142,143,162). Bony midface hypoplasia produces ocular hypotelorism, a small nose with flattening of the nasal bridge, and relative mandibular prognathism (15,68,73). The nasion–sella–basion angle is increased (18).

Upward slanting of palpebral fissures and epicanthic folds are common. Other ocular findings include Brushfield spots, fine lens opacities, convergent strabismus, nystagmus, keratoconus, and cataract (160).

The ears tend to be small (1). Overlapping of the superior rim of the helix and small or absent lobes are common (76,160).

Odontoid abnormalities are found in 16% and atlantoaxial instability, a feature found in 10–20% and sometimes found in association with occipitoatlantal instability, result from generalized congenital laxity of ligaments, which is especially noticeable at the craniocervical joint (40).

The lips are broad, irregular, fissured, and dry (19). An open mouth with a protruding tongue is observed. The tongue appears relatively large because of the small oral cavity. That the "large tongue" is relative was demonstrated by Ardran et al (6). Occasionally, true macroglossia may be present. A fissured tongue is common. Lingual papillae have been noted to be large even during infancy (35).

The palate is narrower and shorter but not higher than average (155). Radiographically, palatal length averages about 25 mm (normal, $31 \pm 3$ mm) in the newborn. Cleft of the lip and/or palate is present in 0.5% (72).

Parotid salivary flow rate is decreased. A significant rise in pH, sodium, calcium, bicarbonate, uric acid, and nonspecific esterase in pure parotid salvia has been reported (35,44,183,184).

Periodontal disease has been observed in over 90% of cases. Severe involvement even below the age of 6 years is particularly common in the mandibular anterior and maxillary molar regions (Fig. 3–1E). Exfoliation of the lower central incisors from periodontal bone loss occurs frequently. However, calculus formation is neither common nor severe. Necrotizing ulcerative gingivitis has been reported to occur in about 30% of patients (15,30,92,99,147,171).

The prevalence of dental caries has been stated to be low by several authors (15,35), although these findings have been challenged (105).

Eruption of both deciduous and permanent teeth is delayed in 75% of cases. An irregular sequence of eruption is common, deciduous first molars sometimes preceding incisors (7,90,106,107a,141,171).

Missing teeth have been reported in 23–47% of patients. Third molars, second premolars, and lateral incisors are most frequently absent in the permanent dentition. In 12–17% of patients, deciduous lateral incisors are absent. Peg-shaped maxillary lateral incisors have been observed in 10%. Extreme hypodontia and anodontia have been noted occasionally (35,153,173).

Microdontia in permanent (67) and macrodontia in deciduous (3) dentitions have been found. Crown-size asymmetry (66) and a gradient of reduction along a mesial-to-distal axis (31) have been reported. Fusion of a deciduous mandibular lateral incisor with a canine or, less commonly, with a central incisor is a low-frequency finding (7,19,35).

Morphologic crown alterations have been reported (16,33,35,104, 135,176). Almost 50% of patients have three or more dental irregularities (104). Enamel hypoplasia (Fig. 3–1E) and enamel hypocalcification have also been noted (29,35). Jaspers (88a) indicated that taurodontism occurs with greater than expected frequency.

Irregular alignment of teeth is common. Posterior crossbite, mandibular overjet, mesiocclusion, anterior open bite, crowded teeth, and widely spaced teeth have been discussed by several authors (15,34,35,98).

**Cardiovascular system.** Cardiovascular anomalies occur in approximately 40% of cases and are a frequent cause (about 20%) of the recorded trisomy 21 syndrome deaths. Atrioventricular communis occurs in one-third of Down syndrome patients with congenital heart defects. In contrast, it is rare as an isolated defect in the general population. Another one-third of cardiac anomalies in Down syndrome are ventricular septal defects. Approximately one-fourth are either tetralogy of Fallot (7%), atrial septal defect (10%), or patent ductus arteriosus (3%). Transposition of the great vessels and coarctation of the aorta occur less frequently in Down syndrome than in the general population. Among patients with Eisenmenger complex, it has been noted that the proportion with trisomy 21 syndrome is higher than expected. It has also been reported that the highest mortality rate in Down syndrome patients with congenital heart defects occurred in those with right ventricular outflow tract obstruction or pulmonary vascular pressure at the systemic level, whereas the group with normal or slightly higher pulmonary vascular resistance did quite well (22,26,42, 56,57,152,160).

**Gastrointestinal system.** Gastrointestinal malformations occur in 10–18% of cases. Findings include tracheoesophageal fistula, pyloric stenosis, duodenal atresia, annular pancreas, Hirschsprung's disease, and imperforate anus. Knox and ten Bensel (100) reported that approximately 8% die from their gastrointestinal anomalies.

**Dermatoglyphics.** Many dermatoglyphic studies of trisomy 21 syndrome have been carried out, and several dermatoglyphic indices have been developed (14,43,83,112,131,136,138,160,179). Dermatoglyphic indices have been particularly useful as a diagnostic aid in trisomy 21 syndrome. The Walker index allows discrimination of approximately 70% of affected individuals from controls (179). Borgaonkar et al (14) developed a method by which 88% of trisomy 21 syndrome patients could be discriminated from the control population. This method was simplified by Reed et al (138) as a nomogram that discriminated 81% of patients with trisomy 21 syndrome. Lu (112) devised a discriminant analysis that permits the identification of 89% of those with trisomy 21 syndrome.

Significant findings in Down syndrome include a hallucal tibial arch or small distal loop, distally placed axial triradius on the palm, single palmar creases, single flexion crease on the fifth finger, and an increased number of ulnar loops on the fingertips. An unusual finding of a radial loop on the fourth and fifth fingers is also increased in trisomy 21 syndrome. In Table 3–7, significant dermatoglyphic features are summarized as percentages in the Down syndrome population compared with percentages of such findings in the general population (136).

**Other anomalies.** A variety of low-frequency anomalies have been reported and these are extensively reviewed by several authors (57,160).

Table 3–7. Dermatoglyphic features of trisomy 21 syndrome[a]

| Dermatoglyphic pattern | Trisomy 21 syndrome (%) | Controls (%) |
|---|---|---|
| Hallucal tibial arch | 72 | ≃0.5 |
| Hallucal small distal loop | 32 | 11 |
| Bilateral t″ | 82 | 3 |
| Single crease, digit 5 | 17 | <1 |
| Bilateral single palmar crease | 31 | 2 |
| 10 ulnar loops on fingers | 31 | 7 |
| Radial loop, digit 4 or 5 | 13 | 4 |
| Bilateral $I_3$ pattern | 46 | 26 |
| Thenar pattern | 4 | 11 |

[a]From Preus M, Fraser FC: *Am J Dis Child* **124**:933, 1972.

**Hematologic system.** Congenital hematologic disorders are common in Down syndrome and a benign natural history is common. Newborns frequently have polycythemia. A number of Down syndrome patients, usually newborns, have had transient, severe disorders of hematopoiesis simulating leukemia but with full recovery. Down syndrome studies have also shown a preponderance of younger forms of polymorphonuclear leukocytes that may be explained by an increased turnover of granulocytes. Acute lymphoblastic leukemia occurs with increased frequency in Down syndrome and the peak in the leukemia mortality rate occurs at an earlier age than in childhood leukemia without Down syndrome (160).

**Immune system.** Immunodeficiency in Down syndrome is related to an increased susceptibility to infection, an increased risk for developing neoplasia, particularly leukemia, an increased frequency of autoantibodies, and early aging. Autoantibodies against thyroid antigen are frequently found during early life (175). An increased prevalence of hypothyroidism and, less commonly, hyperthyroidism has been reported (146). The disturbed immunoglobulin balance increases with age; IgG and IgA rise after 5 years of age whereas IgM stays within normal limits. T cell deficiency is observed from birth onward (175).

**Differential diagnosis.** The Down syndrome phenotype is distinctive. Conditions that are sometimes clinically confused with it include hypothyroidism, *XXXXY syndrome, penta-X syndrome* (65), and *Zellweger syndrome.*

**Laboratory aids.** Chromosome study is necessary to confirm all cases. Amniocentesis or chorionic villus biopsy can be offered to all mothers with a previous history of having a Down syndrome child, older mothers, and translocation carriers. Pelvic radiograph is sometimes diagnostically useful in suspected cases. There are low levels of $\alpha$-fetoprotein in maternal serum and amniotic fluid in trisomy 21 and in other trisomies (6a,178a). There is, however, considerable debate about its use in routine prenatal maternal serum screening.

### References [Trisomy 21 syndrome (Down syndrome)]


1. Aase JM et al: Small ears in Down's syndrome: A helpful diagnostic aid. *J Pediatr* **82**:845–847, 1973.

2. Abuelo D et al: Risk for trisomy 21 in offspring of individuals who have relatives with trisomy 21. *Am J Med Genet* **25**:365–367, 1986.

3. Alexandersen V: *The Odontometrical Variation of the Deciduous and Permanent Teeth in Down's Syndrome,* Thesis, U Wisconsin, Madison, 1970.

4. Alfi OS et al: Evidence for genetic control of non-disjunction in man. *Am J Hum Genet* **32**:477–483, 1980.

4a. Anonymous: Alzheimer's disease, Down's syndrome, and chromosome 21. *Lancet* **1**:1011–1012, 1987.

5. Apgar V: Down's syndrome (mongolism). *Ann NY Acad Sci* **171**:303–688, 1970.

6. Ardran GM et al: Tongue size in Down's syndrome. *J Ment Defic Res* **16**:160–166, 1966.

6a. Ashwood ER et al: Maternal serum alpha-fetoprotein and fetal trisomy 21 in women 35 years and older: Implications for alpha-fetoprotein screening programs. *Am J Med Genet* **26**:531–540, 1987.

6b. Baird PA, Sadovnick AD: Life tables for Down syndrome. *Hum Genet* **82**:291–292, 1989.

7. Barkla DH: Eruption of permanent teeth in mongols. *J Ment Defic Res* **10**:190–197, 1966.

8. Benda CE: *The Child with Mongolism.* Grune & Stratton, New York, 1960.

9. Benda CE: *Down's Syndrome. Mongolism and Its Management.* Grune & Stratton, New York, 1969.

10. Bersu ET: Anatomical analysis of the developmental effects of aneuploidy in man: The Down syndrome. *Am J Med Genet* **5**:399–420, 1980.

11. Betlejewski S et al: Radiologische Untersuchungen der Entwicklung der Nasennebenhöhlen im Down-Syndrom. *Ann Paediatr (Basel)* **203**:355–362, 1964.

12. Bilovsky D, Share J: The ITPA and Down's syndrome: An explanatory study. *Am J Ment Defic* **70**:78–82, 1965.

13. Blanchard I: Speech pattern and etiology in mental retardation. *Am J Ment Defic* **68**:612–617, 1964.

14. Borgaonkar DS, et al: Evaluation of dermal patterns in Down's syndrome by predictive discrimination: 1: Preliminary analysis based on frequencies of patterns. *Johns Hopkins Med J* **128**:141–152, 1971.

14a. Bricarelli FD et al: Parental age and the origin of trisomy 21. *Hum Genet* **82**:20–26, 1989.

15. Brown RH, Cunningham WM: Some dental manifestations of mongolism. *Oral Surg* **14**:664–676, 1961.

16. Brown T, Townsend GC: Size and shape of mandibular first molars in Down syndrome. *Ann Hum Biol* **11**:281–290, 1984.

17. Burger PC, Vogel FS: The development of the pathologic changes of Alzheimer's disease and senile dementia in patients with Down's syndrome. *Am J Pathol* **73**:457, 1973.

18. Burwood RJ et al: The skull in mongolism. *Clin Radiol* **24**:475–480, 1973.

19. Butterworth T et al: Cheilitis of mongolism. *J Invest Dermatol* **35**:347–351, 1960.

20. Carlin ME et al: A comparison between a trisomy 21 child (probably mosaic) with usual intelligence and a mosaic Down syndrome population. *Birth Defects* **14**(6C):327–341, 1978.

21. Carr J: Mental and motor development in young mongol children. *J Ment Defic Res* **14**:205–220, 1970.

22. Carter CO: A life-table for mongols with the cause of death. *J Ment Defic Res* **2**:64–74, 1958.

23. Cavalli IJ et al: Letter to the editor: Down syndrome owing to simple trisomy 21 in three first cousins. *Am J Med Genet* **22**:831, 1985.

24. Centerwall SA, Centerwall WR: A study of children with mongolism reared in the home compared to those reared away from home. *Pediatrics* **25**:678–685, 1960.

25. Chemke J, Robinson A: The third fontanelle. *J Pediatr* **75**:617–622, 1969.

26. Chi TPL, Krovetz L: The pulmonary vascular bed in children with Down syndrome. *J Pediatr* **86**:533–538, 1975.

27. Cicchetti D, Sroufe LA: The relationship between affective and cognitive development in Down's syndrome infants. *Child Dev* **47**:920–929, 1976.

28. Clarke CM et al: 21-trisomy/normal mosaicism in an intelligent child with some mongoloid features. *Lancet* **1**:1028–1030, 1961.

29. Cohen MM, Winer RA: Dental and facial characteristics in Down's syndrome (mongolism). *J Dent Res* **44**:197–208, 1965.

30. Cohen MM et al: Oral aspects of mongolism. Periodontal disease in mongolism. *Oral Surg* **14**:92–107, 1961.

31. Cohen MM et al: Crown-size profile pattern in trisomy G. *J Dent Res* **49**:460, 1970.

32. Cohen MM et al: Occlusal disharmonies in trisomy G (Down's syndrome, mongolism). *Am J Orthodont* **58**:1386–1393, 1970.

33. Cohen MM et al: Abnormalities of the permanent dentition in trisomy G. *J Dent Res* **49**:1386–1393, 1970.

34. Cohen MM et al: Occlusal disharmonies in trisomy G (Down's syndrome, mongolism). *Am J Orthodont* **58**:367–372, 1970.

35. Cohen MM Sr, Cohen MM Jr: The oral manifestations of trisomy G (Down's syndrome). *Birth Defects* **7**(7):241–251, 1971.

36. Conen PE, Erkman B: Combined mongolism and leukemia. *Am J Dis Child* **112**:429–443, 1966.

37. Cornwall AC, Birch HG: Psychological and social development in home-reared children with Down's syndrome (mongolism). *Am J Ment Defic* **74**:341, 1969.

38. Cornwall AC: Development of language, abstraction and numerical concept formation in Down's syndrome children. *Am J Ment Defic* **79**:179–190, 1974.

39. Cowie VA: *A Study of the Early Development of Mongols.* Pergamon, Oxford, 1970.

40. Coria F et al: Craniocervical abnormalities in Down's syndrome. *Dev Med Child Neurol* **25**:252–255, 1983.

41. Crome L: The pathology of Down's disease, in *Mental Deficiency,* 2nd ed, Hilliard LT, Kirman BH (eds), Churchill, London, 1965.

41a. Cronk C et al: Growth charts for children with Down syndrome. *Pediatrics* **81**:102–110, 1988.

42. Cullum L, Liebman J: The association of congenital heart disease with Down's syndrome. *Am J Cardiol* **24**:354–357, 1969.

43. Cummins H: Dermatoglyphic stigmata in mongolism idiocy. *Anat Res* **73**:407–415, 1939.

44. Cutress TW: Composition, flow-rate and pH of mixed and parotid saliva from trisomic and other mentally retarded subjects. *Arch Oral Biol* **17**:1081–1094, 1972.

45. Dicks-Mireaux MJ: Mental development of infants with Down's syndrome. *Am J Ment Defic* **77**:26–32, 1972.

46. Dodd B: Recognition and reproduction of words by Down's syndrome and non-Down's syndrome retarded children. *Am J Ment Defic* **80**:306–311, 1975.

47. Domino G, Newman D: Relationship of physical stigmata to intellectual subnormality in mongoloids. *Am J Ment Defic* **69**:541–547, 1964.

48. Dunlap SS et al: Comparative anatomical analysis of human trisomies 13, 18, and 21: I. The forelimb. *Teratology* **33**:159–186, 1986.

49. Dunsdon MI et al: Upper end of range of intelligence in mongolism. *Lancet* **1**:565–568, 1960.

50. Durling D, Benda CE: Mental growth curve in untreated institutionalized mongoloid patients. *Am J Ment Defic* **56**:578–588, 1952.

51. Dutrillaux B, Lejeune J: Étude de la descendance des porteurs d'une translocation t(21qDq). *Ann Génét (Paris)* **12**:77–82, 1969.

52. Ellingson RJ et al: Clinical-EEG relationships in mongoloids confirmed by karyotype. *Am J Ment Defic* **74**:645–648, 1970.

53. Ellis WG et al: Presenile dementia in Down's syndrome: Ultrastructural identity with Alzheimer's disease. *Neurology (Minneap.)* **24**:101–106, 1974.

54. Eunpu DL: Trisomy 21: Rate in second-degree relatives. *Am J Med Genet* **25**:361–363, 1986.

55. Evans D, Hampson M: The language of mongols. *Br J Disord Commun* **3**:171–181, 1969.

56. Fabia J, Drolette M: Life tables up to age 10 for mongols with and without congenital heart defects. *J Ment Defic Res* **14**:235–242, 1970.

57. Fabia J, Drolette M: Malformations and leukemia in children with Down's syndrome. *Pediatrics* **45**:60–70, 1970.

58. Farrell MJ: Adverse effects of early institutionalization of mentally subnormal children. *Am J Dis Child* **91**:278–281, 1956.

59. Fishler K: Mental development in mosaic Down's syndrome as compared with trisomy 21, in *Down's Syndrome (Mongolism): Research, Prevention and Management,* Kock R, de la Cruz FF (eds), Brunner/Mazel, New York, 1975.

60. Fishler K et al: Comparison of mental development in individuals with mosaic and trisomy 21 Down's syndrome. *Pediatrics* **58**:744–748, 1976.

61. Fishman MA: Will the study of Down syndrome solve the riddle of Alzheimer disease? *J Pediatr* **8**:627–629, 1986.

62. Ford CE: *Mongolism,* Ciba Foundation Study Group No. 25, Wolstenholme GEW, Porter R (eds), Churchill, London, 1967, p 71.

63. Fraser FC, Sadovnick AD: Correlation of IQ in subjects with Down's syndrome and their parents and siblings. *J Ment Defic Res* **20**:179–182, 1976.

64. Frith U, Frith CD: Specific motor disabilities in Down's syndrome. *J Child Psychol Psychiat* **15**:293–301, 1974.

65. Gardner LI: Polysomy X masquerading as Down's syndrome. *Am J Dis Child* **133**:253, 1979.

66. Garn SM et al: Increased crown-size asymmetry in trisomy G. *J Dent Res* **49**:464, 1970.

67. Geciauskas MA, Cohen MM: Mesiodistal crown diameters of permanent teeth in Down's syndrome (mongolism). *Am J Ment Defic* **74**:563–567, 1970.

68. Gerald BE, Silverman FN: Normal and abnormal interorbital distances with special reference to mongolism. *Am J Roentgenol* **95**:154–161, 1965.

69. Gibson D, Gibbons RJ: The relation of mongolian stigmata to intellectual status. *Am J Ment Defic* **63**:345–353, 1958.

70. Giraud F, Mattei JF: Aspects epidémiólogiques de la trisomie 21. *Journées Européenes de Conseil Génétique. Medicine et Hygiène (Génève)* 1–30, 1975.

71. Golden W, Pashayan HM: The effect of parental education on the eventual mental development of non-institutionalized children with Down's syndrome. *J Pediatr* **89**:604–605, 1976.

72. Gorlin RJ et al: Facial clefting and its syndromes. *Birth Defects* **7**(7):3–49, 1971.

73. Gosman SD: Facial development in mongolism. *Am J Orthodont* **37**:332–349, 1951.

74. Grouchy J de, Turleau C: Autosomal disorders, in *Principles and Practice of Medical Genetics,* Emery EH, Rimoin DL (eds), Churchill Livingstone, Edinburgh, 1983, pp 170–192.

75. Haberland C: Alzheimer's disease in Down's syndrome: Clinical-neuropathological observations. *Acta Neurol Belg* **68**:369, 1969.

76. Hall B: Monglism in newborns: A clinical and cytogenetic study. *Acta Paediatr Scand (Suppl)* **154**:1–95, 1964.

77. Hamerton JL: *Human Cytogenetics,* Vol 2. Academic Press, New York, 1971.

78. Hamerton JL et al: Cytogenetics of Down's syndrome (mongolism). *Cytogenetics* **4**:171–185, 1965.

79. Harris DL et al: Parental trisomy 21 mosaicism. *Am J Hum Genet* **34**:125–133, 1982.

80. Hecht F et al: Nonrandomness of translocations in man. *Science* **161**:371–372, 1968.

81. Heston LL: Alzheimer's disease, trisomy 21, and myeloproliferative disorders: Associations suggesting a genetic diathesis. *Science* **196**:322–323, 1977.

82. Heston LL, Mastri AR: The genetics of Alzheimer's disease. *Arch Gen Psychiat* **34**:976–981, 1977.

83. Holt SB: Dermatoglyphics in mongolism. *Ann NY Acad Sci* **171**:602, 1970.

84. Hook EB: Differences between rates of trisomy 21 (Down syndrome) and other chromosomal abnormalities diagnosed in livebirths and in cells cultured after second-trimester amniocentesis—suggested explanations and implications for genetic counseling and program planning. *Birth Defects* **14**(6C):249–267, 1978.

85. Hook EB: Risk of trisomy 21 among relatives of trisomy 21 individuals. *Am J Med Genet* **22**:213–214, 1985.

86. Hook EB, Lindsjö A: Down syndrome in live births by single year maternal age interval in a Swedish study: Comparison with results from a New York State study. *Am J Hum Genet* **30**:19–27, 1978.

87. Horrobin JM, Rynders JE: *To Give an Edge: A Guide for New Parents of Down's Syndrome (Mongoloid) Children.* Colwell Press, Minneapolis, 1974.

88. Jackson JF et al: Clinical diagnosis of Down's syndrome. *Clin Genet* **9**:483–487, 1976.

88a. Jaspers MT: Taurodontism in the Down syndrome. *Oral Surg* **51**:632–636, 1981.

89. Jelgersma HC: On the tuberflocculi in mongolism idiocy. *Psychiat Neurol Neurochir* **66**:131, 1963.

90. Jensen GM: Dentoalveolar morphology and developmental changes in Down's syndrome. *Am J Orthodont* **64**:607–618, 1973.

91. Jervis GA: Early senile dementia in mongoloid idiocy. *Am J Psychiat* **105**:102, 1948.

92. Johnson NP, Young MA: Periodontal disease in mongols. *J Periodont* **34**:41–47, 1963.

93. Johnson NP et al: Tooth ring analysis in mongolism. *Aust Dent J* **10**:282–286, 1965.

94. Juberg RC, Mowrey PN: Origin of nondisjunction in trisomy 21 syndrome: All studies compiled, parental age analysis, and international comparisons. *Am J Med Genet* **16**:111–116, 1983.

95. Kendler-Zisk PK, Bialer I: Speech and language problems in mongolism: A review of the literature. *J Speech Dis* **32**:228–241, 1967.

96. Kikuchi Y et al: Translocation Down's syndrome in Japan: Its frequency, mutation rate of translocation and paternal age. *Jpn J Hum Genet* **14**:93–106, 1969.

97. Kirman BH: Epilepsy in mongolism. *Arch Dis Childh* **26**:501, 1951.

98. Kisling E: *Cranial Morphology in Down's Syndrome: A Roentgenocephalometric Study in Adult Males.* Munksgaard, Copenhagen, 1966.

99. Kisling E, Krebs G: Periodontal conditions in adult patients with mongolism (Down's syndrome). *Acta Odontol Scand* **21**:391–405, 1963.

100. Knox GE, ten Bensel RW: Gastrointestinal malformations in Down's syndrome. *Minn Med* **55**:542–544, 1972.

101. Kolata G: Down syndrome—Alzheimer's linked. *Science* **230**:1152–1153, 1985.

102. Kolstoe OP: Language training of low-grade mongoloid children. *Am J Ment Defic* **63**:17–30, 1958.

103. Kousseff BG: Trisomy 21 with average intelligence? *Birth Defects* **14**(6C):323–325, 1978.

104. Kraus BS et al: Mental retardation and abnormalities of the dentition. *Am J Ment Defic* **72**:905–917, 1968.

105. Kroll RG et al: Incidence of dental caries and periodontal disease in Down's syndrome. *NY State Dent J* **36**:151–156, 1970.

106. Kučera J: Age at walking, age at eruption of deciduous teeth and response to ephedrine in children with Down's syndrome. *J Ment Defic Res* **13**:143–148, 1969.

107. Langdon Down J: Observations on an ethnic classification of idiots. *Clin Lect Rep London Hospital* **3**:249, 1866.

107a. Le Clech G et al: La première dentition du trisomique 21: A propos de 114 enfants suivis régulierèment. *Ann Pédiatr* **33**:795–798, 1986.

108. Lejeune J: Le mongolisme. Premier exemple d'aberration autosomique humaine. *Ann Génét* **1**:41–49, 1959.

109. Levinson A et al: Variability of mongolism. *Pediatrics* **16**:43–54, 1955.

110. Lilienfeld AM, Benesch CH: *Epidemiology of Mongolism.* Johns Hopkins, Baltimore, 1969.

111. Lister TJ, Frota-Pessoa O: Recurrence risks for Down syndrome. *Hum Genet* **55**:203–208, 1980.

112. Lu KH: An informative and discriminate analysis of fingerprint patterns pertaining to identification of mongolism and mental retardation. *Am J Hum Genet* **20**:24–43, 1968.

113. Macgillivray RC: Epilepsy in Down's anomaly. *J Ment Defic Res* **11**:43, 1967.

114. Machay DN, McDonald G: The effects of varying digit message structures on their recall by mongols and non-mongol subnormals. *J Ment Defic Res* **20**:191–197, 1976.

115. Magenis RE et al: Parental origin of the extra chromosome in Down's syndrome. *Hum Genet* **37**:7–16, 1977.

116. Malamud N: Neuropathology of organic brain syndromes associated with aging, in *Aging and the Brain,* Gaits CM (ed), Plenum Press, New York, 1972.

117. Mattei JF et al: Origin of the extra chromosome in trisomy 21. *Hum Genet* **46**:107–110, 1979.

118. McMillan RS: Relation of human abnormalities of structure and function to abnormalities of the dentition. II. Mongolism. *J Am Dent Assoc* **63**:368–373, 1961.

119. Melyn M, White D: Mental and developmental milestones of non-institutionalized Down's syndrome children. *Pediatrics* **52**:542–545, 1973.

120. Menolascino FJ: Psychiatric aspects of mongolism. *Am J Ment Defic* **69**:653–660, 1965.

121. Meyer A, Jones J: Histological changes in the brain in mongolism. *J Ment Sci* **85**:206, 1939.

122. Mikkelsen M: Down's syndrome. *Humangenetik* **12**:1–28, 1971.

123. Mikkelsen M, Stene J: Genetic counselling on Down's syndrome. *Hum Hered* **20**:465–472, 1970.

124. Miller RW: Neoplasia and Down's syndrome, in *Down's Syndrome (Mongolism),* Apgar V (ed), *Ann NY Acad Sci* **171**:637–644, 1970.

125. Moore BC et al: Mongoloid and non-mongoloid retardates: A behavioral comparison. *Am J Ment Defic* **73**:433–436, 1968.

126. Niikawa N, Kajii T: The origin of mosaic Down syndrome: Four cases with chromosome markers. *Am J Hum Genet* **36**:123–130, 1984.

127. Norman RM: Malformations of the nervous system, birth injury and diseases of early life. *Neuropathology,* Edward Arnold, London, 1958.

128. Ohara PT: Electron microscopical study of the brain in Down's syndrome. *Brain* **92**:147, 1969.

129. Øster J: *Mongolism.* Danish Science Press, Copenhagen, 1953.

130. Parloir C et al: Down's syndrome in brother and sister without evident trisomy 21. *Hum Genet* **51**:227–230, 1979.

131. Penrose LS, Loesch D: Diagnosis with dermatoglyphic discriminants. *J Ment Defic Res* **15**:185–195, 1971.

132. Pi Sy et al: Brief clinical reports: Holoprosencephaly in a Down syndrome child. *Am J Med Genet* **5**:201–206, 1980.

133. Polani P et al: A mongol girl with 46 chromosomes. *Lancet* **1**:721–724, 1960.

134. Pollack MA et al: Infantile spasms in Down syndrome: A report of 5 cases and review of the literature. *Ann Neurol* **3**:406–408, 1978.

135. Prahl-Andersen B, Oerlemans J: Characteristics of permanent teeth in persons with trisomy G. *J Dent Res* **55**:633–638, 1976.

136. Preus M, Fraser FC: Dermatoglyphics and syndromes. *Am J Dis Child* **124**:933–943, 1972.

137. Ray AB, Shotick AL: Short-term and long-term recall of familiar objects by trainable and educable mentally retarded and normal individuals of comparable mental age. *J Ment Defic Res* **20**:183–190, 1976.

138. Reed TE et al: Dermatoglyphic nomogram for the diagnosis of Down's syndrome. *J Pediatr* **77**:1024–1032, 1970.

139. Rex AP, Preus M: A diagnostic index for Down syndrome. *J Pediatr* **100**:903–906, 1982.

140. Richards BW: Investigation of 142 mosaic mongols and mosaic parents of mongols; cytogenetic analysis and maternal age at birth. *J Ment Defic Res* **18**:199–208, 1974.

141. Roche AF, Barkla DH: The eruption of deciduous teeth in mongols. *J Ment Defic Res* **8**:54–65, 1964.

142. Roche AF et al: Growth changes in the mongoloid head. *Acta Paediatr Scand* **50**:133–140, 1961.

143. Roche AF et al: Non-metrical observations on cranial roentgenogram in mongolism. *Am J Roentgenol* **85**:659–662, 1961.

144. Rosecrans CJ: The relationship of normal/21 trisomy mosaicism and intellectual development. *Am J Ment Defic* **72**:562–566, 1968.

145. Roth M-P et al: Reexamination of paternal age effect in Down's syndrome. *Hum Genet* **63**:149–152, 1983.

146. Sare Z et al: Prevalence of thyroid disorder in Down syndrome. *Clin Genet* **14**:154–158, 1978.

147. Saxén L et al: Periodontal disease associated with Down's syndrome: An orthopantomographic evaluation. *J Periodont* **48**:337–340, 1977.

148. Schacter M: Les convulsions chez les mongoliens. *Méd Infant* **63**:5, 1956.

149. Scheffelin M: A comparison of four stimulus response channels in paired-associate learning. *Am J Ment Defic* **73**:303–307, 1968.

150. Séguin E: *Le traitement moral, l'hygiene et l'education des idiots.* J.B. Bailliere, Paris, 1846.

151. Seppäläinen AM, Kivalo E: EEG findings and epilepsy in Down's syndrome. *J Ment Defic Res* **11**:116, 1967.

152. Shaher RM et al: Clinical aspects of congenital heart disease in mongolism. *Am J Cardiol* **20**:497, 1972.

153. Shapiro BL: Prenatal dental anomalies in mongoloids. Comments on the basis and implications of variability. *Ann NY Acad Sci* **171**:562–577, 1970.

154. Shapiro BL: Down syndrome—A disruption of homeostasis. *Am J Med Genet* **14**:241–269, 1983.

155. Shapiro BL et al: The palate and Down's syndrome. *N Engl J Med* **276**:1460–1463, 1967.

155a. Sheridan R et al: Fertility in a male with trisomy 21. *J Med Genet* **26**:294–298, 1989.

156. Shipe D, Shotwell AM: Effect of out-of-home care on mongoloid children: A continuation study. *Am J Ment Defic* **69**:649–652, 1965.

157. Shotwell AM, Shipe D: Effect of out-of-home care on intellectual and social development of mongoloid children. *Am J Ment Defic* **68**:693–699, 1964.

158. Sinson JC, Wetherick NE: The nature of the colour retention deficit in Down's syndrome. *J Ment Defic Res* **19**:97–100, 1975.

159. Smith DW, Wilson AA: *The Child with Down's Syndrome (Mongolism).* Saunders, Philadelphia, 1973.

160. Smith GF, Berg JM: *Down's Anomaly.* Churchill Livingstone, Edinburgh, 1976.

161. Solitaire GB, Lamarche JB: Brain weight in the adult mongol. *J Ment Defic Res* **11**:79, 1967.

162. Spitzer R et al: A study of the abnormalities of the skull, teeth, and lenses in mongolism. *Can Med Assoc J* **84**:567–572, 1961.

163. Stedman DJ, Eichorn DH: A comparison of the growth and development of institutionalized and home-reared mongoloids during infancy and early childhood. *Am J Ment Defic* **69**:391–401, 1964.

164. Steinberg C et al: Recurrence rate for de novo 21q21q translocation Down syndrome: A study of 112 families. *Am J Med Genet* **17**:523–530, 1984.

165. Stene J: Statistical inference on segregation ratios for D/G translocation, when the families are ascertained in different ways. *Ann Hum Genet* **34**:93–115, 1970.

166. Stene J: A statistical segregation analysis of (21q22q) translocation. *Hum Hered* **20**:465–472, 1970.

167. Stene J et al: Paternal age and Down's syndrome: Data from prenatal diagnoses (DFG). *Hum Genet* **59**:119–124, 1981.

168. Stene J et al: Risk for chromosome abnormality at amniocentesis following a child with a non-inherited chromosome aberration. *Prenat Diag* **4**:81–95, 1984.

169. Stephens TD, Shepard TH: The Down syndrome in the fetus. *Teratology* **22**:37–41, 1980.

170. Strazzula M: Speech problems of the mongoloid child. *Q Rev Pediatr* **8**:268–273, 1953.

171. Swallow JN: Dental diseases in children with Down's syndrome. *J Ment Defic Res* **8**:102–118, 1964.

172. Thase ME: Longevity and mortality in Down's syndrome. *J Ment Defic Res* **26**:177–192, 1982.

173. Thomas DH: Anodontia in mongolism. *Br J Psychiatr* **85**:566–568, 1939.

174. Tizard J: Residential care of mentally handicapped children. *Br Med J* **1**:1041–1046, 1960.

175. Tolksdorf M, Wiedemann H-R: Clinical aspects of Down's syndrome from infancy to adult life, in *Trisomy 21: An International Symposium,* Burgio GR et al (eds), Springer-Verlag, Heidelberg, 1981, pp. 1–31.

176. Townsend GC, Brown RH: Tooth morphology in Down's syndrome: Evidence for retardation in growth. *J Ment Defic Res* **27**:159–169, 1983.

177. Trimble BK, Baird PA: Maternal age and Down syndrome: Age-specific incidence rates by single-year intervals. *Am J Med Genet* **2**:1–5, 1978.

177a. Urioste M et al: Holoprosencephaly and trisomy 21 in a child born to a nondiabetic mother. *Am J Med Genet* **30**:925–928, 1988.

178. Veall RM: The prevalence of epilepsy among mongols related to age. *J Ment Defic Res* **18**:99, 1974.

178a. Wald N, Cuckle H: AFP and age screening for Down syndrome. *Am J Med Genet* **31**:197–209, 1988.

179. Walker NF: The use of dermal configurations in the diagnosis of mongolism. *J Pediatr* **50**:19–26, 1957; *Pediatr Clin North Am* **5**:531–543, 1958.

180. Walter RD et al: Mongolism and convulsive seizures. *Arch Neurol Psychiat (Chic)* **74**:559, 1955.

181. Warkany J: Etiology of Down's syndrome, in *Down's Syndrome (Mongolism): Research, Prevention and Management,* Koch R, de la Cruz FF (eds), Brunner/Mazel, New York, 1975, pp. 9–15.

182. Wilson MG et al: Decreasing mosaicism in Down's syndrome. *Clin Genet* **17**:335–340, 1980.

183. Winer RA, Feller RP: Composition of parotid and submandibular saliva and serum in Down's syndrome. *J Dent Res* **51**:449–454, 1972.

184. Winer RA et al: Composition of human saliva, parotid gland secretory rate, and electrolyte concentration in mentally retarded persons. *J Dent Res* **44**:632–634, 1965.

185. Wolcott JG, Chun RWM: Myoclonic seizures in Down's syndrome. *Dev Med Child Neurol* **15**:805–808, 1973.

186. Wunsch WL: Some characteristics of mongoloids evaluated in a clinic for children with retarded mental development. *Am J Ment Defic* **62**:122–127, 1957.

187. Zellweger H: Down syndrome, in *Handbook of Clinical Neurology,* Vol 31, Vinken PJ, Bruyn GW (eds), North Holland, Amsterdam, 1977, pp 367–469.

## Trisomy 13 syndrome

In 1960, Patau et al (21) first identified trisomy 13 in the laboratory, although the clinical description of the syndrome may date back as early as 1657 in the writings of Bartholin (10). Trisomy 13 syndrome is characterized by microcephaly, scalp defects, frequent holoprosencephaly, microphthalmia, orofacial clefting, congenital heart defects, polydactyly, severe developmental retardation, and early demise (Fig. 3–2). Besides trisomy 13 syndrome, the condition has also been called Patau syndrome, $D_1$ trisomy, and 13–15 trisomy. The reader is referred to the following sources for extensive coverage: general review (1,7,8,10,14,26–30), epidemiology (6,15,17,30), cytogenetic aspects (7,8,10–11a,15,16,18), partial trisomy (24,25), anatomic studies (5,22), pathologic studies (1,19,19a,27), holoprosencephaly (3,4), neoplasia (9,12), and long-term survival (13,23).

Birth prevalence for free trisomy 13 is approximately 1/12,000, that of Robertsonian translocations, 1/56,000 to 1/80,000, and that of familial cases, 1/33,000 to 1/42,000. The rate of trisomy 13 spontaneous abortions is about 100-fold greater than the rate of live births. Trisomy 13 spontaneous abortions constitute approximately 1% of all recognized spontaneous abortions (15). A slightly greater number of females than males are affected. Free trisomy 13 occurs in approximately 75% of all cases and is associated with a maternal age effect (mean age, 31.6 years) (8). Variation in space or time of maternal age-specific live birth rates of trisomy 13 remains a possibility but, to date, is unproven. Methodological artifacts such as sampling variation and prenatal diagnosis must be taken into account in any such study (15).

Free trisomy originates in nondisjunction. Ishikiriyama and Niikawa (16) have shown that in trisomy 13, the ratio of nondisjunction in maternal and paternal meiosis was 14:3. For maternal meiosis, nondisjunction occurred more frequently during the first than during the second meiotic division (9:4) (11a). Approximately 20% result from translocations, mostly t(13q14). The overwhelming majority arise *de novo*. Approximately 5% of the translocation type are transmitted by one of the parents, with recurrence risk being 5% and the risk of spontaneous abortion being approximately 20%. In the unusual translocation

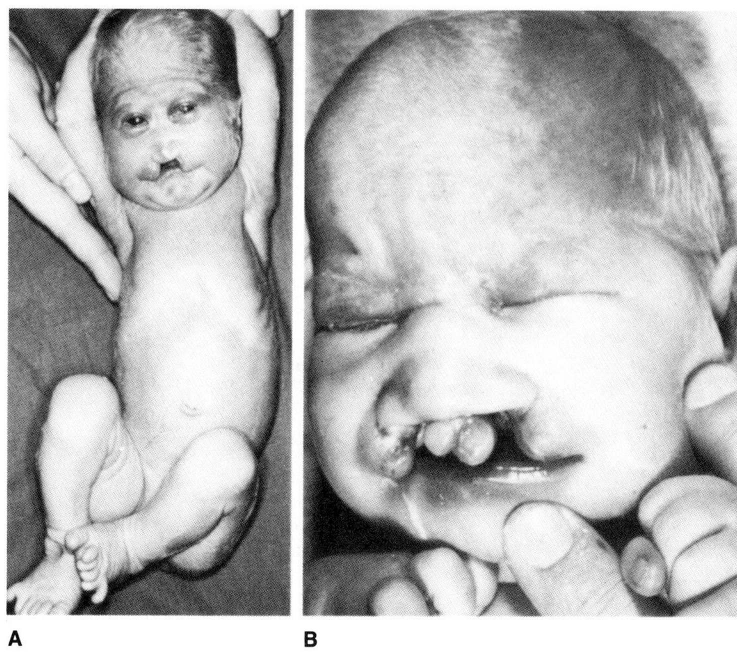

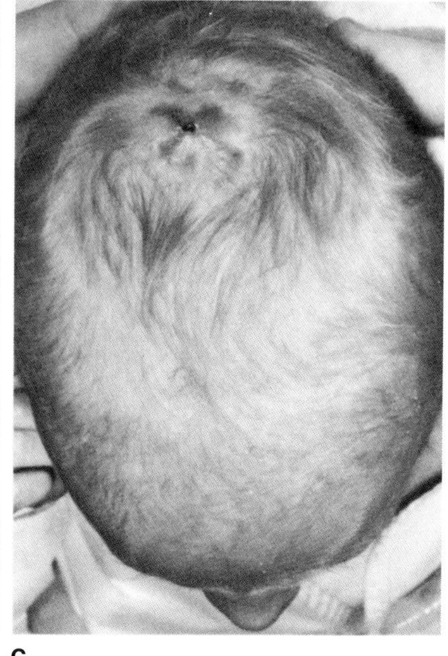



Fig. 3–2. *Trisomy 13 syndrome.* (A) Premaxillary agenesis type of holoprosencephaly with trisomy 13. Note hypotelorism, lack of nasal bones, and extra digits. (B) Bilateral cleft lip–palate, microphthalmia, ulnar hexadactyly, and superficial angioma over brow. (C) Skin defect at vertex of skull. (A from *PE Conan,* Am J Dis Child **111**:236, 1966.)

t(13q13q), the risk of recurrence or abortion is 100%. About 5% of all cases are mosaic for trisomy 13 (7,8). In such instances, the phenotype tends to be less severe (2). Distinct clinical syndromes involving a partial proximal or partial distal trisomy segment of chromosome 13 have been phenotypically defined (24,25). Different features of trisomy 13 syndrome are associated with these partial trisomic segments (details are provided in the next chapter). Double trisomy has been discussed by Mailhes et al (18).

Mean life expectancy is 130 days. Approximately 45% die during the first month, 70% during the first 6 months, and 86% during the first year (8). Survival beyond 3 years is exceptional. The oldest living patients to be reported are a 19-year-old female and an 11-year-old male. Both have recurrent chronic cellulitis, affecting the parotid region, axilla, groin, and abdominal wall, with sinus formation and resistance to antibiotic treatment. Both have wasting of the distal limb muscles and both are black (23). Survival appears to be better with translocation cases than with free trisomy (15). Females appear to live longer than males, but this difference has not been statistically significant (15). Better understanding of factors that predict long-term survival is needed for counseling (13).

Trisomy 13 syndrome and trisomy 18 syndrome share a number of features in common and these are listed in Table 3–8. Features more common to trisomy 13 syndrome and features more common to trisomy 18 syndrome are also listed. Extensive clinical and pathological findings of trisomy 13 syndrome are noted in Table 3–9 together with approximate percentages.

**Growth.** Mean birth weight is 2600 g. Feeding difficulties and failure to thrive are characteristic (7,27).

**Central nervous system.** Moderate microcephaly with sloping forehead and wide sagittal suture and fontanels are characteristic. Some degree of holoprosencephaly is common and is accompanied by apneic episodes and seizures. Severe developmental retardation is the rule and presumptive deafness is common. Hypotonia, hypertonia, and hydrocephaly have been encountered in some instances. Cerebellar hypoplasia and meningomyelocele are less commonly observed (3, 14,26,27,29).

**Craniofacial features.** Scalp ulceration is commonly observed at the vertex and may be variable in size. Sloping forehead and capillary hemangiomas, particulary in the glabellar region, are commonly observed. Accompanying variable degrees of holoprosencephaly are ocular hypotelorism and various associated features including, most commonly, lateral cleft lip, median cleft lip, and cebocephaly. Cyclopia occurs infrequently and ethmocephaly is rare. Ocular findings include microphthalmia, iris coloboma, and retinal dysplasia characterized by intraocular cartilage extending from the retrolental region to the sclera at the site of the iris coloboma. Cleft palate, micrognathia, and malformed ears, which may be low set in some instances, are also observed. Inner ear anomalies are of the Mondini or Scheibe types (3,4,10,14,26,27,29).

**Neck.** The neck is short and loose skin on the nape may be seen (27). Fetal cystic hygroma has been reported (20).

**Cardiovascular system.** Cardiovascular anomalies are common and autopsy studies have shown that more than one type may be present in the same individual. Most commonly observed are patent ductus arteriosus, infundibular ventricular septal defect, and atrial septal defect. Also reported are left superior vena cava, dextrocardia, aorta arising from the left ventricle, bicuspid aortic valve, bicuspid pulmonary valve, coarctation of the aorta, atretic pulmonary valve, hypoplastic pulmonary trunk, hypoplastic left atrium, hypoplastic left ventricle and abnormal semilunar valves (1,19a). Renal abnormalities are common and may include cystic dysplasia, renal hyperlobulation, hydronephrosis, hydroureters, double ureters, horseshoe kidney, and persistent nodular renal blastema (1,19a).

**Genitalia.** Cryptorchidism and scrotal anomalies are present in males. Bicornuate uterus is found in females. Other findings have been reported including hypertrophy of the clitoris, double vagina, hypoplastic ovaries, gonadal dysgenesis, and hypospadias (7,10,26,27,29).

**Limb anomalies.** Postaxial polydactyly, flexion of the fingers, sometimes with overlapping, and hyperconvex nails are common. Prominent calcaneus may be observed in some instances.

Dermatoglyphic alterations include single palmar creases, distal palmar axial triradius, and fibular S-shaped hallucal pattern or hallucal

Table 3–8. Comparison of trisomy 13 and 18 syndromes
*Features Common to Both Syndromes*[a]

| Features | Trisomy 13 syndrome (%) ($n=19$) | Trisomy 18 syndrome (%) ($n=29$) |
|---|---|---|
| Common to both syndromes | | |
| Ear anomalies | 87 | 100 |
| Cardiac defects | 94 | 93 |
| Micrognathia | 66 | 96 |
| Overlapping fingers | 73 | 100 |
| Microcephaly | 86 | 70 |
| Prominent heels | 62 | 87 |
| Highly arched palate | 72 | 87 |
| Microphthalmia | 88 | 82 |
| Hypertonia | 67 | 60 |
| Umbilical hernia | 83 | 67 |
| Cryptorchidism[a] | 100 | 43 |
| More common to trisomy 13 | | |
| Cleft palate | 63 | |
| Polydactyly | 78 | |
| Scalp defects | 75 | |
| Cleft lip | 50 | |
| Apneic episodes | 100 | |
| Capillary hemangiomas | 88 | |
| Dextrocardia | 100 | |
| Hypotelorism/hypertelorism | 83 | |
| Iris coloboma | 67 | |
| Sloping forehead | 100 | |
| Flat head | 75 | |
| Hypoplastic nipples | 100 | |
| Prominent nasal bridge | 100 | |
| Short neck | 100 | |
| More common to trisomy 18 | | |
| Prominent occiput | | 91 |
| Hip dislocation | | 82 |
| Hypoplastic nails | | 100 |
| Clubfoot | | 89 |
| Widely spaced nipples | | 90 |
| Hypertrophic clitoris[b] | | 89 |
| Hammertoes | | 89 |
| Narrow palpebral fissures | | 80 |
| Short sternum | | 100 |
| Small mouth | | 86 |
| Excess skin, neck | | 86 |
| Seizures | | 62 |
| Abnormal head | | 83 |
| Hypoplastic labia | | 100 |

[a]Adapted from ME Hodes et al, *J Med Genet* **15**:48–60, 1978.
[b]Sex adjusted percentage.

Table 3–9. Features of trisomy 13 syndrome[a]

| Feature | Percentage |
|---|---|
| Growth | |
| Failure to thrive | 87 |
| Central nervous system | |
| Microcephaly | 86 |
| Holoprosencephaly | 70 |
| Apneic episodes | 58 |
| Seizures | 25 |
| Hypotonia | 48 |
| Hypertonia | 26 |
| Severe developmental retardation | 100 |
| Presumptive deafness | 50 |
| Craniofacial features | |
| Scalp defects | 75 |
| Sloping forehead | 100 |
| Capillary hemangiomas | 72 |
| Ocular hypotelorism | 83 |
| Epicanthic folds | 56 |
| Microphthalmia | 76 |
| Iris coloboma | 33 |
| Other eye defects | 88 |
| Cleft lip | 58 |
| Cleft palate | 69 |
| Micrognathia | 84 |
| Malformed ears | 80 |
| Neck | |
| Short neck | 79 |
| Loose skin, nape | 59 |
| Cardiovascular anomalies | |
| Patent ductus arteriosus | (82)[b] |
| Ventricular septal defect | (73) |
| Atrial septal defect | (91) |
| Left superior vena cava | (14) |
| Dextrocardia | (24) |
| Aorta from left ventricle | (11) |
| Bicuspid aortic valve | (8) |
| Bicuspid pulmonary valve | (8) |
| Coarctation of aorta | (9) |
| Renal anomalies | 30–60 |
| Polycystic kidneys | (70)[c] |
| Renal hyperlobulation | (22) |
| Hydronephrosis | (25) |
| Hydroureters | (10) |
| Double ureters | (12) |
| Horseshoe kidney | (9) |
| Genitalia | |
| Cryptorchidism | 100[b] |
| Bicornuate uterus | 50[b] |
| Limb anomalies | |
| Polydactyly | 76 |
| Flexion of fingers, sometimes with overlapping | 68 |
| Hyperconvex nails | 68 |
| Single palmar crease | 64 |
| Distal axial triradius | 74 |
| Prominent calcaneus | 28 |
| Fibular S-shaped hallucal pattern | 39 |
| Other findings | |
| Inguinal/umbilical hernia | 40 |

[a]Adapted from AI Taylor, *J Med Genet* **5**:227–252, 1968; ME Hales et al, *J Med Genet* **15**:48–60, 1978; C Addor et al, *J Genet Hum* **23**:83–109, 1975; MM Cohen Jr, *Teratology,* **40:**211–235, 1989.
[b]Percentages in parentheses are from necropsy series that biases findings.
[c]Sex-adjusted percentage.

loop tibial. Also frequent are thenar exit of the A mainline (80%) and radial loops on other than the index finger (50%) (10,14,27).

**Other findings.** Also reported have been single umbilical artery, inguinal hernia, umbilical hernia, thin posterior ribs with or without a missing rib, hypoplastic pelvis with shallow acetabular angle, retroflexible thumb, ulnar deviation at the wrist, talipes equinovarus, as well as other abnormalities (7,10,14,26,27,29). Pancreatic dysplasia is a characteristic feature (19a).

**Hematologic findings.** Polymorphonuclear leukocytes often have nuclear projections (25–80%). Persistence in the newborn period of embryonic hemoglobin Gower-2 has been observed (7,10).

**Neoplasia.** Leukemia has been noted (12). The association of trisomy 13 syndrome and neuroblastoma in the same family has been reported (9).

**Anatomic features.** Autosomal trisomies are known to have rather specific constellations of muscle, peripheral nerve, and vascular variations that are not usually examined by either clinicians or pathologists. Trisomy 13 syndrome has the following characteristics: absence of the palmaris longus, palmaris brevis, plantaris and peroneus tertius, presence of a pectorodorsalis muscle and unusual muscles from the central tendon of the diaphragm to the pericardium, and variations in extensor indicis, extensors carpi radialis longus and brevus, biceps brachii, and suprahyoid muscles (5,22).

**Differential diagnosis.** Many syndromes have *holoprosencephaly* as one feature. *Pseudotrisomy 13 syndrome, Meckel syndrome* and even *Pallister–Hall syndrome* in instances can have holoprosencephaly and polydactyly concurrently (4). *Smith–Lemli–Opitz syndrome* and *hydrolethalus syndrome* have some individual features in common with trisomy 13, but the overall pattern of anomalies in each of these syndromes is distinctive.

**Laboratory aids.** Diagnosis is established by banded chromosome study.

### References (Trisomy 13 syndrome)

1. Addor C et al: Patau's syndrome: A pathological and cytogenetic study of two cases. *J Génét Hum* **23**:83–109, 1975.

2. Beer S et al: Trisomy 13 mosaic presenting as cleft lip and palate. *Hum Hered* **26**:321–323, 1976.

3. Cohen MM et al: Holoprosencephaly and facial dysmorphia: Nosology, etiology and pathogenesis. *Birth Defects* **7**(7):125–135, 1971.

4. Cohen MM Jr: Perspectives on holoprosencephaly. Part I. Epidemiology, genetics, and syndromology. *Teratology,* **40:**211–235, 1989.

5. Colacino SC, Pettersen JC: Analysis of the gross anatomical variations found in four cases of trisomy 13. *Am J Med Genet* **2**:31–50, 1978.

6. Conen PE, Erkman-Balis B: Frequency and occurrence of chromosomal syndromes. I. D-trisomy *Am J Hum Genet* **18**:374–398, 1966.

7. de Grouchy J, Turleau C: *Clinical Atlas of Human Chromosomes.* Wiley, New York, 1977.

8. de Grouchy J, Turleau C: Autosomal disorders, in *Principles and Practice of Medical Genetics,* Emery EH, Rimoin DL (eds), Churchill Livingstone, Edinburgh, 1983, Ch. 15, pp 170–192.

9. Feingold M et al: Familial neuroblastoma and trisomy 13. *Am J Dis Child* **121**:451, 1971.

10. Gorlin RJ: Classical chromosome disorders, in *New Chromosomal Syndromes,* Yunis JJ (ed), Academic Press, New York, 1977, pp 59–117.

11. Hamerton JL: *Human Cytogenetics,* Vol 2. Academic Press, New York, 1971.

11a. Hassold T et al: Cytogenetic and molecular studies of trisomy 13. *J Med Genet* **24**:625–732, 1987.

12. Hecht F: Letters to the editor: D1 trisomy syndrome, leukemia, and immunologic abnormalities. *J Pediatr* **73**:296, 1968.

13. Hecht F: Letter to the editor: Who will survive with trisomy 13 or 18? A call for cases 10 years old or above. *Am J Med Genet* **10**:417–418, 1981.

14. Hodes ME et al: Clinical experience with trisomies 18 and 13. *Am J Med Genet* **15**:48–60, 1978.

15. Hook EB: Rates of 47,+13 and 46 translocation D/13 Patau syndrome in live births and comparison with rates in fetal deaths and at amniocentesis. *Am J Hum Genet* **32**:849–858, 1980.

16. Ishikiriyama S, Niikawa N: Origin of extra chromosome in Patau syndrome. *Hum Genet* **68**:266–268, 1984.

17. Magenis RE et al: Trisomy 13 (D1) syndrome: Studies on parental age, sex ratio, and survival. *J Pediatr* **73**:222–228, 1968.

18. Mailhes JB et al: A case of double trisomy in a liveborn infant: 48,XXY,+13. *Clin Genet* **11**:147–150, 1977.

19. Marin-Padilla M: Structural organization of the cerebral cortex (motor area) in human chromosomal aberrations. A Golgi study. I. D1 (13–15) trisomy, Patau syndrome. *Brain Res* **66**:375–391, 1974.

19a. Moerman P et al: The pathology of trisomy 13 syndrome. A study of 12 cases. *Hum Genet* **80**:349–356, 1988.

20. Nakazato Y et al: Fetal cystic hygroma, web neck and trisomy 13 syndrome. *Br J Radiol* **58**:1011–1013, 1985.

21. Patau K et al: Multiple congenital anomalies caused by an extra autosome. *Lancet* **1**:790–793, 1960.

22. Pettersen JC et al: An examination of the spectrum of anatomic defects and variations found in eight cases of trisomy 13. *Am J Med Genet* **3**:183–210, 1979.

23. Redheendran R et al: Long survival in trisomy-13-syndrome: 21 cases including prolonged survival in two patients 11 and 19 years old. *Am J Med Genet* **8**:167–172, 1981.

24. Rogers JF: Clinical delineation of proximal and distal partial 13q trisomy. *Clin Genet* **25**:221–229, 1984.

25. Schinzel A et al: Further delineation of the clinical picture of trisomy for the distal segment of chromosome 13. *Hum Genet* **32**:1–12, 1976.

26. Smith DW: The 18 trisomy and the 13 trisomy syndromes. *Birth Defects* **5**(5):67–71, 1969.

27. Taylor AI: Autosomal trisomy syndromes: A detailed study of 27 cases of Edward's syndrome and 27 cases of Patau's syndrome. *J Med Genet* **5**:227–252, 1968.

28. Taylor MB et al: Chromosomal variability in the D1 trisomy syndrome. *Am J Dis Child* **120**:374–381, 1970.

29. Warkany J, Passarge E: Congenital malformations in autosomal trisomy syndromes. *Am J Dis Child* **112**:502–517, 1966.

30. Yu F-C et al: Trisomy 13 syndrome in Chinese infants. Clinical findings and incidence. *J Med Genet* **7**:132–137, 1970.

## Trisomy 18 syndrome

In 1960, Edwards et al (11) described a new syndrome associated with the presence of an extra chromosome in the E group that was subsequently shown to be chromosome 18. Features included growth deficiency, developmental retardation, prominent occiput, low-set malformed ears, micrognathia, short sternum, congenital heart defects, overlapped flexed fingers, dorsiflexed halluces, and prominent calcaneus (9,10,15,33) (Fig. 3–3). Before exact chromosome identification, trisomy 18 was known as E trisomy and/or Edwards syndrome. The reader is referred to the following sources for extensive coverage: general aspects (9,10,12,15,18,33,35,38,39), epidemiology (2a,7, 17,19,27,42), cytogenetics (16,22,24,28,29,40), anatomic aspects (3,37), pathology (18,22,36), central nervous system anomalies (21,25), ophthalmological aspects (4), cardiovascular defects (6,22,26), renal anomalies (22,26), long-term survival (5,14,20,23,31,32,34), prenatal diagnosis and counseling (1,25), and neoplasia (2,8,14,20,30). A diagnostic scoring system has been developed (216).

Large surveys have indicated a prevalence of approximately 1/5000 to 1/7000 (5,14,27). The possibility of a concentration of births at certain times during the year or in certain regions has been suggested (2a,27). Sex ratio shows an excess of affected females (4F:1M) (10). Trisomy 18 is associated with a maternal age effect, with the mean maternal age being 32.5 years (35).

The overwhelming majority of cases of trisomy 18 syndrome are due to *de novo* meiotic nondisjunction (10). Translocations may arise *de novo* or may be transmitted in a family. Mosaicism occurs in approximately 10% of cases (9,10). Classic trisomy 18 in two sibs was reported by Pauli et al (28). Translocation, isochromosome formation, partial trisomy 18, and mosaicism have been discussed by several authors (13,21a,24,28,29,40). Tetrasomy for 18p produces a distinct phenotype (29a).

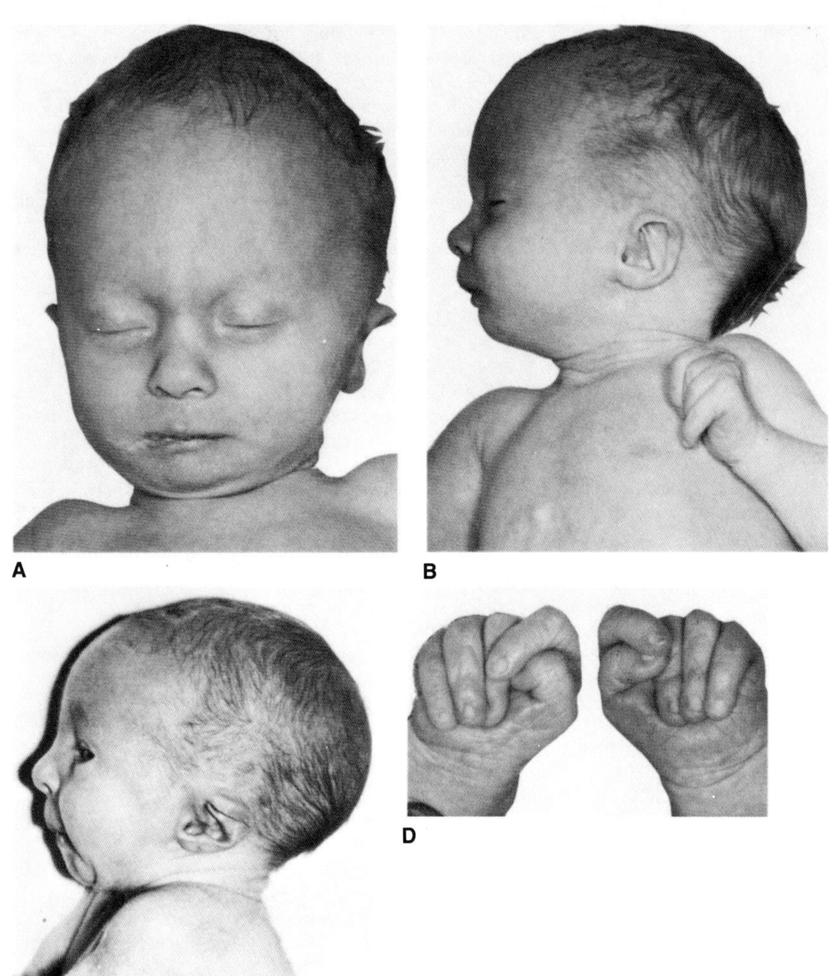

Fig. 3–3. *Trisomy 18 syndrome.* (A) Narrow bifrontal diameter and small lower jaw. (B) Prominent occiput, ptosis of the eyelid, low-set ears, and micrognathia. (C) Compare facies with patient in B. (D) Overlapping fingers. (A,B,C from *P Paerregård,* Acta Pathol Microbiol Scand **67:**479, 1966.)

The median life expectancy for liveborn infants with trisomy 18 is 5 days with a range of 1 hour to 18 months. Mean age at death is 48 days. In patients without accompanying cardiac defects or life-threatening gastrointestinal anomalies, median life expectancy is 40 days with a range of 4 hours to 18 months; this opts poorly for corrective surgery, even if this were indicated on other grounds. Although females survive longer than males, the difference is trivial (5). Approximately 30% fail to survive more than 1 month, 50% succumb by 2 months, and less than 10% live more than 1 year (10,15). There are reports of 6, 12, 13, and 15 year olds (23,31,32,34).

Average duration of pregnancy is 42 weeks with low fetal activity, polyhydramnios, and small placenta. A single umbilical artery is common. Altered gestational timing may occur with premature delivery in some instances (9,10,33).

Clinical findings are summarized in Table 3–10 and internal malformations are listed in Table 3–11. The features of trisomy 18 and trisomy 13 are compared in Table 3–8.

**Growth.** Mean birth weight is 2240 g. Postnatally, there is failure to thrive. Characteristically, hypoplasia of skeletal muscle, subcutaneous tissue, and adipose tissue is encountered. Some examples of trisomy 18 mosaicism have been associated with asymmetry of the body and/or face and, in addition, anomalies that lateralize, such as limb defects, may occur on one side but not on the other (9,29,33).

**Central nervous system.** Mental deficiency is severe. Hypotonia during the neonatal period is followed by hypertonia. The cry is weak and there is diminished response to sound (9,33). Variable degrees of holoprosencephaly associated with apneic episodes, seizures, and poikilothermia can be found on occasion (6,21). With apparently somewhat higher frequency, simple absence of the corpus callosum has been noted (38). Rarely, hydrocephaly, anencephaly, meningomyelocele, and facial palsy may be observed (25,33). Paucity of myelinization, microgyria, cerebellar hypoplasia, absent geniculate body, absent occipital lobes, and occipital lobe hemorrhage have been recorded.

Heterotopias of well-formed neurons either with or without undifferentiated neuroblastic cells have been found in the periventricular areas of the brain, located most commonly above or lateral to the head of the caudate nucleus and/or in the roof and lateral wall of the inferior and posterior horns, within fibers converging on the internal capsule and around the optic radiation. Heterotopias appear to result from excessive formation of embryonic neuroblasts with focally arrested migration in the periventricular white matter without resulting in deficient neuronal composition of the cerebral cortex (4,36).

**Craniofacial features.** The head is dolichocephalic in shape. The bifrontal diameter is narrow and the occiput is prominent. The ears are malformed and low set. The mouth is small, the palate is narrow, and micrognathia is evident (9,15,33).

Less frequently occurring craniofacial abnormalities include microcephaly, wide fontanels, hypoplasia of orbital ridges, Wormian bones, shallow elongated sella turcica (33), and various eye anomalies: globe (corneal opacities, corneal clouding, microphthalmia, iris colobomas, cataracts, persistent hyaloid artery, glaucoma, blue sclera, absent retinal pigment, persistent iridopupillary membrane), adnexa (slanted or narrow palpebral fissures, epicanthic folds, ptosis of the eyelids, abnormally thick eyelids, abnormally long or sparse eyelashes, inability to close the eyelids, blepharophimosis), orbit (hypertelorism, hypote-

Table 3–10. Features of trisomy 18 syndrome[a]

| Feature | Percentage |
|---|---|
| Growth | |
| Growth deficiency | 96 |
| Central nervous system | |
| Severe developmental retardation | 96 |
| Hypertonia | 60 |
| Craniofacial | |
| Microcephaly | 70 |
| Dolichocephaly | 93 |
| Prominent occiput | 91 |
| Narrow palpebral fissures | 80 |
| Small mouth | 86 |
| Micrognathia | 96 |
| Low-set, malformed ears | 88 |
| Neck | |
| Loose skin, nape | 56 |
| Thorax | |
| Short sternum | 68 |
| Widely spaced nipples | 90 |
| Cardiovascular anomalies | 85 |
| Abdominal wall | |
| Inguinal or umbilical hernia | 67 |
| Urogenital system | |
| Renal anomalies | 30 |
| Cryptorchidism[b] | 100 |
| Prominent clitoris[b] | 89 |
| Pelvis and hips | |
| Small pelvis, limited hip abduction | 68 |
| Limbs | |
| Overlapped, flexible fingers | 89 |
| Hypoplastic nails | 100 |
| Arch fingertip patterns | 96 |
| Clubfoot | 89 |
| Prominent calcaneus | 77 |
| Rockerbottom feet | 10–50 |
| Dorsiflexed hallux | 75 |
| Syndactyly, second and third toes | 10–50 |

[a]Adapted from ME Hodes et al, *J Med Genet* **15**:48–60, 1978 and AI Taylor, *J Med Genet* **5**:227–252, 1968.
[b]Sex-adjusted percentage.

lorism), and neuroophthalmology (strabismus, lateral gaze, asynergy of extraocular movement, decreased response to visual stimuli, nystagmus, anisocoria). Ocular histopathology has been reviewed extensively by Calderone et al (4).

Inner ear defects may include complete or partial absence of the auditory nerve, defects of the osseous spiral laminae and interscalar septa, shortened organ of Corti, abnormalities of the utriculoendolymphatic valve, atresia or absence of the semicircular ducts, and deformities of the organ of Corti (41).

**Limbs.** The hands are clenched with a tendency for overlapping of the second finger over the third finger and the fifth finger over the fourth. The distal crease on the fifth finger may be absent with less frequently occurring absence of the distal creases on the third and fourth fingers. The nails are hypoplastic, particulary the fifth finger and the toes (9,33).

Dermatoglyphic analysis shows an increased frequency of arched fingertip patterns on six or more fingers. Abnormalities of the feet include dorsiflexed halluces, prominent calcaneus, talipes equinovarus, rockerbottom feet, and syndactyly of the second and third toes.

Table 3–11. Internal malformations in trisomy 18 syndrome[a]

| Finding | Frequency |
|---|---|
| Central nervous system | |
| Hypoplasia or absence of corpus callosum | 3/15 |
| Absent septum pellucidum | 2/15 |
| Short frontal lobe | 2/15 |
| Abnormal gyri | 2/15 |
| Persistent paraventricular nerve cells | 3/15 |
| Cardiovascular system | |
| Polyvalvular disease | 15/15 |
| Ventricular septal defect | 13/15 |
| High takeoff of the right coronary ostium | 12/15 |
| Patent ductus arteriosus | 11/15 |
| Common brachiocephalic trunk | 7/15 |
| Coarctation of aorta | 3/15 |
| Mitral atresia with hypoplastic left ventricle | 1/15 |
| Patent foramen ovale | 14/15 |
| Urogenital system | |
| Cystic kidneys | 5/15 |
| Horseshoe kidney | 7/15 |
| Ovarian cyst | 1/15 |
| Double ureter | 2/15 |
| Alimentary tract | |
| Esophageal atresia (TEF) | 1/15 |
| Meckel's diverticulum | 8/15 |
| Incomplete fixation of the colon | 7/15 |
| Ectopic pancreas | 10/15 |
| Anal atresia | 2/15 |
| Other | |
| Thyroglossal duct cyst | 2/15 |
| Eventration of diaphragm | 2/15 |
| Skeletal anomalies | 5/15 |
| Single umbilical artery | 2/15 |
| Diaphragmatic hernia | 1/15 |

[a]Adapted from R Matsuoka et al, *Am J Med Genet* **14**:657–668, 1983, autopsy cases.

Low-frequency anomalies include syndactyly of the third and fourth fingers, polydactyly, short fifth metacarpals, and limb reduction malformations (6a,9,33).

**Cardiovascular system.** Cardiovascular anomalies occur in about 85% of cases and include polyvalvular disease, VSD, high takeoff of the right coronary ostium, PDA, and various other abnormalities (18,22,35). These are listed in Table 3–11.

**Urogenital system.** Urogenital anomalies are common (Table 3–10) and may include cryptorchidism, prominent clitoris, cystic kidneys, horseshoe kidneys, gonadal dysgenesis, and other abnormalities (18,22,35) (Table 3–11).

**Other anomalies.** A variety of other abnormalities have been reported, including tracheoesophageal fistula, Meckel's diverticulum, incomplete fixation of the colon, ectopic pancreas, anal atresia, thyroglossal duct cyst, eventration of the diaphragm, diaphragmatic hernia, and skeletal anomalies (22) (Table 3–11).

**Anatomical studies.** Bersu and Ramirez-Castro (3) dissected eight infants and found hypoplastic occipitofrontalis, auricular, and nasal muscles. Extensive fusion of muscles occurred around the corners of the mouth, a supernumerary muscle band extending from the corner of the mouth to the occipital attachment of the trapezius muscle. The otomandibular region showed a variable spectrum of muscular, skeletal, arterial, and salivary gland variations. Absence of muscles, supernumerary muscles, and variations in musculature of the upper and lower limbs have been described by Urban and Bersu (37).

**Neoplasia.** Various neoplasms have been reported on occasion, including Wilms tumor (14,20), hepatoblastoma (8), neurogenic tumor (30), and benign congenital papillary tumor of the bicuspid valve (2).

**Differential diagnosis.** Differential diagnosis includes *Pena–Shokeir syndrome I* and *trisomy 13 syndrome*.

**Laboratory aids.** Diagnosis is established by banded chromosome study. Cases may be referred for ultrasonographic evaluation of polyhydramnios around the thirtieth week of gestation.

### References (Trisomy 18 syndrome)

1. Adler B, Kushnick T: Genetic counseling in prenatally diagnosed trisomy 18 and 21: Psychosocial aspects. *Pediatrics* **69**:94–99, 1982.

2. Anderson KR et al: Congenital papillary tumor of tricuspid valve. An unusual cause of right ventricular outflow obstruction in a neonate with trisomy E. *Mayo Clin Proc* **52**:665–669, 1977.

2a. Barsel-Bowers GSM et al: A cluster of infants born with trisomy 18 syndrome. *N Engl J Med* **302**:120, 1980.

3. Bersu ET, Ramirez-Castro JL: Anatomical analysis of the developmental effects of aneuploidy in man—the 18-trisomy syndrome: I. Anomalies of the head and neck. *Am J Med Genet* **1**:173–193, 1977.

4. Calderone JP et al: Intraocular pathology of trisomy 18 (Edwards's syndrome): Report of a case and review of the literature. *Br J Ophthalmol* **67**:162–169, 1983.

5. Carter PE et al: Survival in trisomy 18, *Clin Genet* **27**:59–61, 1985.

6. Castillo JA et al: Aspects cardiologiques du syndrome d'Edwards. *Pédiatrie* **33**:277–286, 1978.

6a. Christianson AL, Nelson MM: Four cases of trisomy 18 syndrome with limb reduction malformations. *J Med Genet* **21**:293–297, 1984.

7. Conen PE, Erkman B: Frequency and occurrence of chromosomal syndromes II. E-trisomy. *Am J Hum Genet* **18**:374–398, 1966.

8. Dasouki M, Barr M Jr: Brief clinical report: Trisomy 18 and hepatic neoplasia. *Am J Med Genet* **27**:203–205, 1987.

9. de Grouchy J, Turleau C: *Clinical Atlas of Human Chromosomes.* Wiley, New York, 1977.

10. de Grouchy J, Turleau C: Autosomal disorders, in *Principles and Practice of Medical Genetics,* Emery EH, Rimoin DL (eds), Churchill Livingstone, Edinburgh, 1983, Ch. 15, pp. 170–192.

11. Edwards JH et al: A new trisomic syndrome. *Lancet* **1**:787–789, 1960.

12. Emanuel I et al: Trisomy 18 syndrome in Chinese infants. Clinical findings and incidence. *J Med Genet* **7**:138–141, 1970.

13. Geiser CF, Schindler AM: Long survival in a male with trisomy 18 syndrome and Wilms' tumor. *Pediatrics* **44**:111–115, 1969.

14. Goldstein H, Gjerum Nielsen K: Rates and survival of individuals with trisomy 18 and 18. *Clin Genet* **34**:366–372, 1988.

15. Gorlin RJ: Classical chromosome disorders, in *New Chromosomal Syndromes,* Yunis JJ (ed), Academic Press, New York, 1977, pp 59–117.

16. Hamerton JL: *Human Cytogenetics,* Vol 2. Academic Press, New York, 1971.

17. Hasold T et al: Effect of maternal age on autosomal trisomies. *Ann Hum Genet, Lond* **44**:29–36, 1980.

18. Hodes ME et al: Clinical experience with trisomies 18 and 13. *J Med Genet* **15**:48–60, 1978.

19. Hook EB et al: Rates of trisomy 18 in livebirths, stillbirths, and at amniocentesis. *Birth Defects* **15**(5C):81–93, 1979.

20. Karayalcin G et al: Wilms' tumor in a 13 year old girl with trisomy 18. *Am J Dis Child* **135**:665–667, 1981.

21. Lang AP et al: Trisomy 18 and cyclopia. *Teratology* **14**:195–204, 1976.

21a. Kohn G, Shohat M: Trisomy 18 mosaicism in an adult with normal intelligence. *Am J Med Genet* **26**:929–931, 1987.

21b. Marion RW et al: Trisomy 18 score: A reliable diagnostic test for trisomy 18. *J Pediatr* **113**:45–48, 1988.

22. Matsuoka R et al: Congenital heart anomalies in the trisomy 18 syndrome, with reference to congenital polyvalvular disease. *Am J Med Genet* **14**:657–668, 1983.

23. Mehta L et al: Trisomy 18 in a 13 year old girl. *J Med Genet* **23**:256–278, 1986.

24. Mendez HMM et al: Letter to the editor: Trisomy 18 mosaicism with few stigmata including macrogenitalia. *Am J Med Genet* **26**:229–230, 1987.

25. Merrild U et al: Anencephaly in trisomy 18 associated with elevated alpha-1-fetoprotein in amniotic fluid. *Hum Genet* **45**:85–88, 1978.

26. Moerman P et al: Spectrum of clinical and autopsy findings in trisomy 18 syndrome. *J Génét Hum* **30**:17–38, 1982.

27. Nielsen J et al: Prevalence of Edwards' syndrome. Clustering and seasonal variation? *Humangenetik* **26**:113–116, 1975.

28. Pauli R et al: Trisomy 18 in sibs and maternal chromosome 9 variant. *Birth Defects* **14**(6C):297–301, 1978.

29. Rao K et al: Asymmetric clinical and cytogenic findings in a 4-year-old girl with trisomy 18 mosaicism. *Birth Defects* **14**(6C):349–354, 1978.

29a. Rivera H et al: Tetrasomy 18p: A distinctive syndrome. *Ann Genet* **27**:187–189, 1984.

30. Robinson MG, McQuorquodale MM: Trisomy 18 and neurogenic neoplasia. *J Pediatr* **99**:428–429, 1981.

31. Roussey M et al: Trisomie 18 a survie prolongée. *Pédiatrie* **34**:819–826, 1979.

32. Smith A, Dulk GMD: Follow-up of case of advanced survival and trisomy 18. *J Ment Defic Res* **24**:157–158, 1980.

33. Smith DW: The 18 trisomy and the 13 trisomy syndromes. *Birth Defects* **5**:67–71, 1969.

34. Surana RB et al: 18-trisomy in a 15-year-old girl. *Am J Dis Child* **123**:75–77, 1972.

35. Taylor AI: Autosomal trisomy syndromes: A detailed study of 27 cases of Edwards' syndrome and 27 cases of Patau's syndrome. *J Med Genet* **5**:227–252, 1968.

36. Terplan KL et al: Histologic structural anomalies in the brain in trisomy 18 syndrome. *Am J Dis Child* **119**:228–235, 1970.

37. Urban B, Bersu ET: Chromosome 18 aneuploidy: Anatomical variations observed in cases of full and mosaic trisomy 18 and a case of deletion of the short arm of chromosome 18. *Am J Med Genet* **27**:425–434, 1987.

38. Warkany J et al: Congenital malformations in autosomal trisomy syndromes. *Am J Dis Child* **112**:502–517, 1966.

39. Weber FM, Sparkes RS: Trisomy E (18) syndrome: Clinical spectrum in 12 new cases, including chromosome autoradiography in 4. *J Med Genet* **7**:363–366, 1970.

40. Wiswell TE, Edwards RG: Presentation of the isochromosome trisomy 18 syndrome in an infant with the Robin anomalad. *Hawaii Med J* **45**:126–128, 1986.

41. Wright CG et al: Inner ear anomalies in two cases of trisomy 18. *Am J Otolaryngol* **6**:392–404, 1985.

42. Young ID et al: Changing demography of trisomy 18. *Arch Dis Childh* **61**:1035–1036, 1986.

## del(4p) syndrome (Wolf–Hirschhorn syndrome)

Although the defect is due to deletion of the 4p16 band (5,9), the deletion may be submicroscopic (3a). In most cases one-third to two-thirds of the short arm is deleted. Translocation is responsible for 10–15% of the cases (1,6a). The rest are largely *de novo.* Frequency appears to be about 1/50,000 births with a 2F:1M sex predilection. At least 35% die during the first 2 years of life, but some survive to adulthood. The phenotype is quite striking. In spite of normal gestation time, birth weight is usually reduced (~2000 g). Fetal activity is diminished and the child is characteristically hypotonic. In addition to very severe psychomotor and growth retardation, mild microcephaly, craniofacial asymmetry, high forehead, wide nasal bridge with prominent glabella, and nasal beaking, hypertelorism and epicanthal folds are virtually constant features (Fig. 3–4). About 10% have a midline scalp defect (4). The eyebrows are highly arched and somewhat sparse medially. About 50% exhibit ptosis, downward slanting palpebral fissures, facial angiomas, and divergent strabismus. About 35% have coloboma of the iris and corectopia (10). The ears are deep seated, poorly differentiated, and have lobeless pinnae (12). They have narrow external canals. Most have a preauricular dimple or skin tag. The philtrum is short and deep and the mouth usually has downturned corners. Cleft lip with cleft palate (Fig. 3–4A) (10%), cleft palate (40%), and micrognathia (50%) have been documented (7). Hypodontia has been reported (1a).

Congenital heart malformations, most often ASD or VSD, are noted in about 50%. Seizures occur in about 50%. Cryptorchidism and particularly hypospadias are found in almost all affected males and absent uterus and streak gonads have been described along with a wide spec-

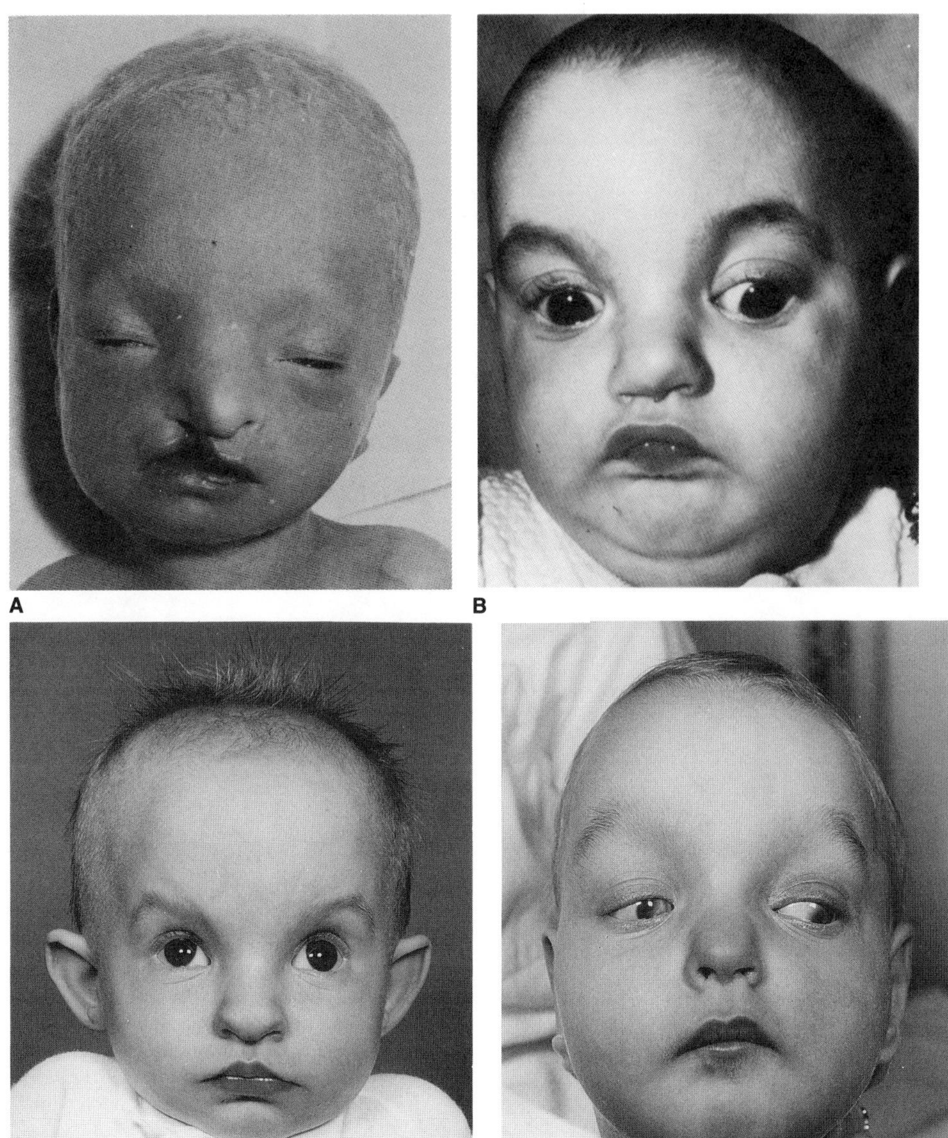


Fig. 3–4. *del(4p) syndrome (Wolf–Hirschhorn syndrome).* (A) Small head, hypertelorism, flat nose, cleft lip, short philtrum, down-turned mouth. (B) Microcephaly, cranial asymmetry, hypertelorism, strabismus, broad-based nose with asymmetric nares, short philtrum, and down-turned mouth. (C,D) Compare facies with patients A and B. (A from *AI Taylor,* J Med Genet **5:**227, 1968. B from *D Arias et al,* J Pediatr **76:**82, 1970. C,D from *A Schinzel* and *W Schmid,* Arch Genet **45:**88, 1972.)

trum of renal anomalies (2,3,5). Sacral dimple is an almost constant feature (11). The trunk is long and the limbs are thin. Talipes equinovarus is relatively common. Radiographically, proximal radioulnar synostosis, anterior fusion of vertebrae, fused ribs, and dislocated hips have been reported (6). The pelvis and carpal bones are late in ossification and pseudoepiphyses are seen at the base of each metacarpal.

### References [del(4p) syndrome (Wolf–Hirschhorn syndrome)]

1. Bauer K et al: Wolf–Hirschhorn syndrome owing to 1:3 segregation of a maternal 4;21 translocation. *Am J Med Genet* **21**:351–358, 1985.

1a. Burgersdijk R, Tan HL: Oral symptoms of the Wolf syndrome. *J Dent Child* **45**:488–489, 1978.

2. Fryns JP et al: The 4p− syndrome, with a report of two new cases. *Humangenetik* **19**:99–109, 1973.

3. Gonzalez CH et al: Pathologic findings in the Wolf–Hirschhorn (4p−) syndrome. *Am J Med Genet* **9**:183–187, 1981.

3a. Greenberg F et al: Molecular confirmation of Wolf–Hirschhorn syndrome with apparently normal chromosomes. *David W. Smith Workshop on Malformations and Morphogenesis,* Madrid, Spain, 23–29 May, 1989.

4. Johnson VP et al: The Wolf–Hirschhorn (4p−) syndrome. *Clin Genet* **10**:104–112, 1976.

5. Lurie IW et al: The Wolf–Hirschhorn syndrome. *Clin Genet* **17**:375–384, 1980; **18**:6–12, 1980.

6. Magill HL et al: 4p− (Wolf–Hirschhorn) syndrome. *Am J Roentgenol* **135**:283–288, 1980.

6a. Martsolf JT et al: Familial transmission of Wolf syndrome resulting from specific deletion 4p16 from t(4;8) (p16;p21) mat. *Clin Genet* **31**:366–369, 1987.

7. Morishita M et al: The oral manifestations of 4p− syndrome. *J Oral Maxillofac Surg* **41**:601–605, 1983.

8. Preus M et al: A taxonomic approach to the del(4p) phenotype. *Am J Med Genet* **21**:337–345, 1985.

9. Rivas F et al: On the deletion 4p16 Wolf–Hirschhorn syndrome. *Ann Génét* **22**:228–231, 1979.

10. Wilcox LM et al: Ophthalmic features of chromosome deletion 4p− (Wolf–Hirschhorn syndrome). *Am J Ophthalmol* **86**:834–839, 1978.

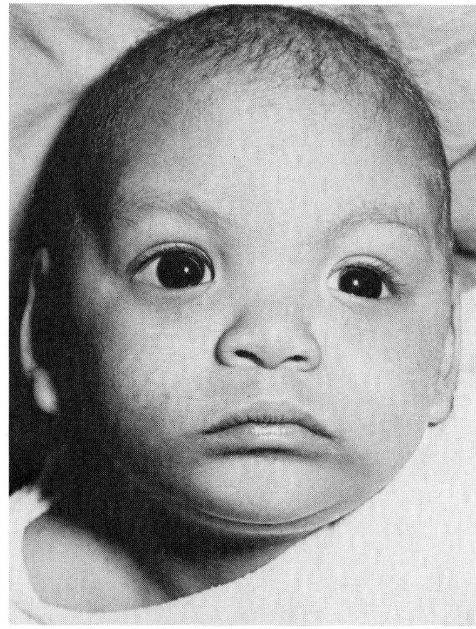
A

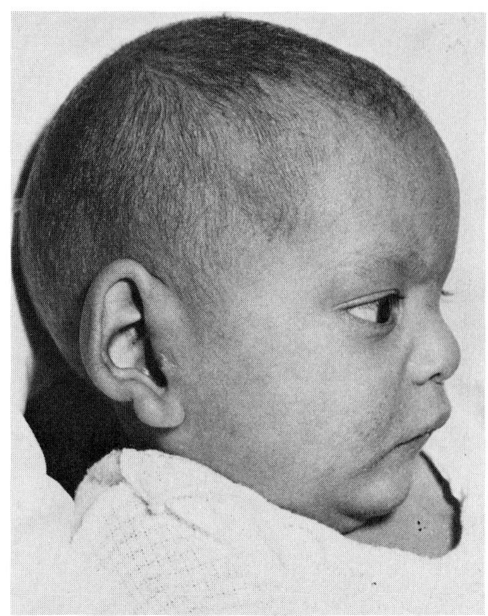
B

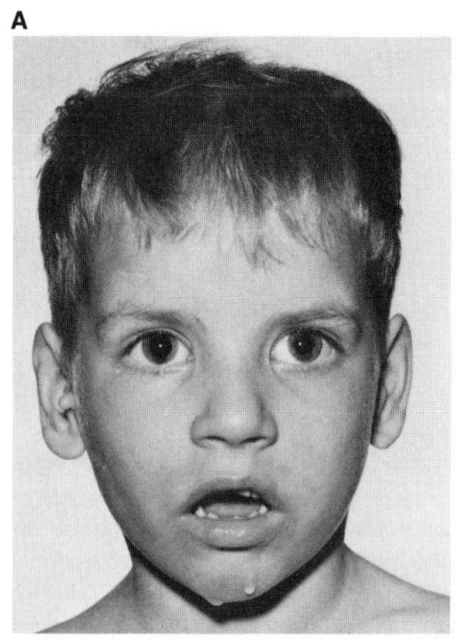
C

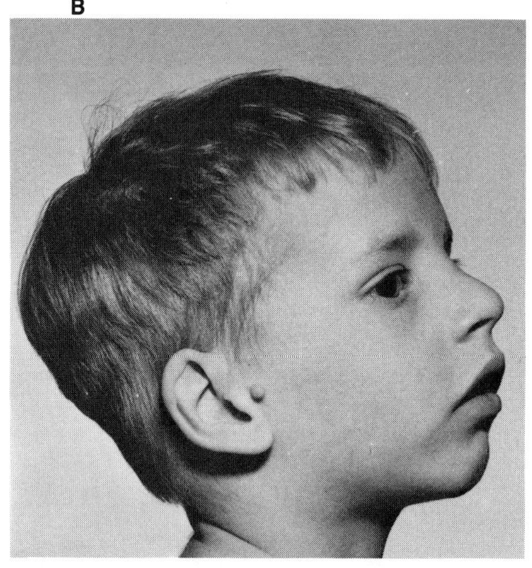
D

Fig. 3–5. *del(5p) syndrome (cri-du-chat syndrome).* (A,B) Microcephaly, round face, hypertelorism with broad nasal bridge, and malformed ears. (C,D) As child ages, round face disappears. Note preauricular tag. (A,B from *C Weinkove* and *R McDonald,* S Afr Med J **43:**218, 1969. C,D from *HB Dyggve* and *M Mikkelsen* Arch Dis Childh **40:**82, 1965.)

11. Wilson MG et al: Genetic and clinical studies in 13 patients with the Wolf–Hirschhorn syndrome (del(4p)). *Hum Genet* **59**:297–307, 1981.

12. Zellweger H et al: The short arm deletion syndrome of chromosome 4(4p− syndrome). *Arch Otolaryngol* **101**:29–32, 1975.

## del(5p) syndrome (cri-du-chat syndrome)

The basic defect is due to a partial deletion, either terminal or interstitial, of the short arm of chromosome 5 in the area of p14 to p15 (2,5). This may result from either a *de novo* deletion of the short arm (about 85%), or unbalanced translocation inherited from a carrier parent (about 15%) (1,3a,3b), the latter being more severely affected (9). The greater the deletion the lower the intelligence, height, and weight and the more severe the microcephaly (8). The frequency has been estimated at about 1/50,000 births. There is usually reduced life span. The syndrome is seen in approximately 1% of institutionalized mentally retarded patients.

The syndrome is characterized by a high shrill cry during infancy. However, the cry is neither pathognomonic nor is it present in all patients. It appears to be central, not laryngeal (4). In addition to severe somatic and mental retardation, the child exhibits microcephaly, increased inner canthal distance, and a round face (3). There are downward slanting palpebral fissures (60%), hypertelorism (75%), epicanthal folds, posteriorly rotated pinnae, preauricular tags (20%), and a broad nasal bridge with prominent nasal root and micrognathia (Fig. 3–5). With time, the face becomes asymmetric and the plumpness is lost (1a). Malocclusion is common, particularly overjet. The hair becomes prematurely gray. Cleft lip/palate occurs in 8–15% (6,9). The hands are smaller than normal with clinodactyly (7). Various congenital defects of the heart (30–50%) and frequent upper respiratory infections, otitis media, and feeding problems are common (8).

Musculoskeletal anomalies include talipes, dislocated hips, and inguinal hernia (9). Hypotonia, very marked in infancy, disappears and reflexes become hyperactive. The gait becomes shuffling. Malrotation of the bowel or megacolon is found in about 25% of the cases resulting from a parental translocation (9).

### References [del(5p) syndrome (cri-du-chat syndrome)]

1. Beemer FA et al: Familial partial monosomy 5p and trisomy 5q; three cases due to paternal pericentric inversion 5 (p151q333). *Clin Genet* **26**:209–215, 1984.

1a. Breg W et al: The cri-du-chat in adolescence and adults, the clinical findings in 13 older patients with partial deletion of the short arm of chromosome No. 5 (5p−). *J Pediatr* **77**:782–791, 1970.

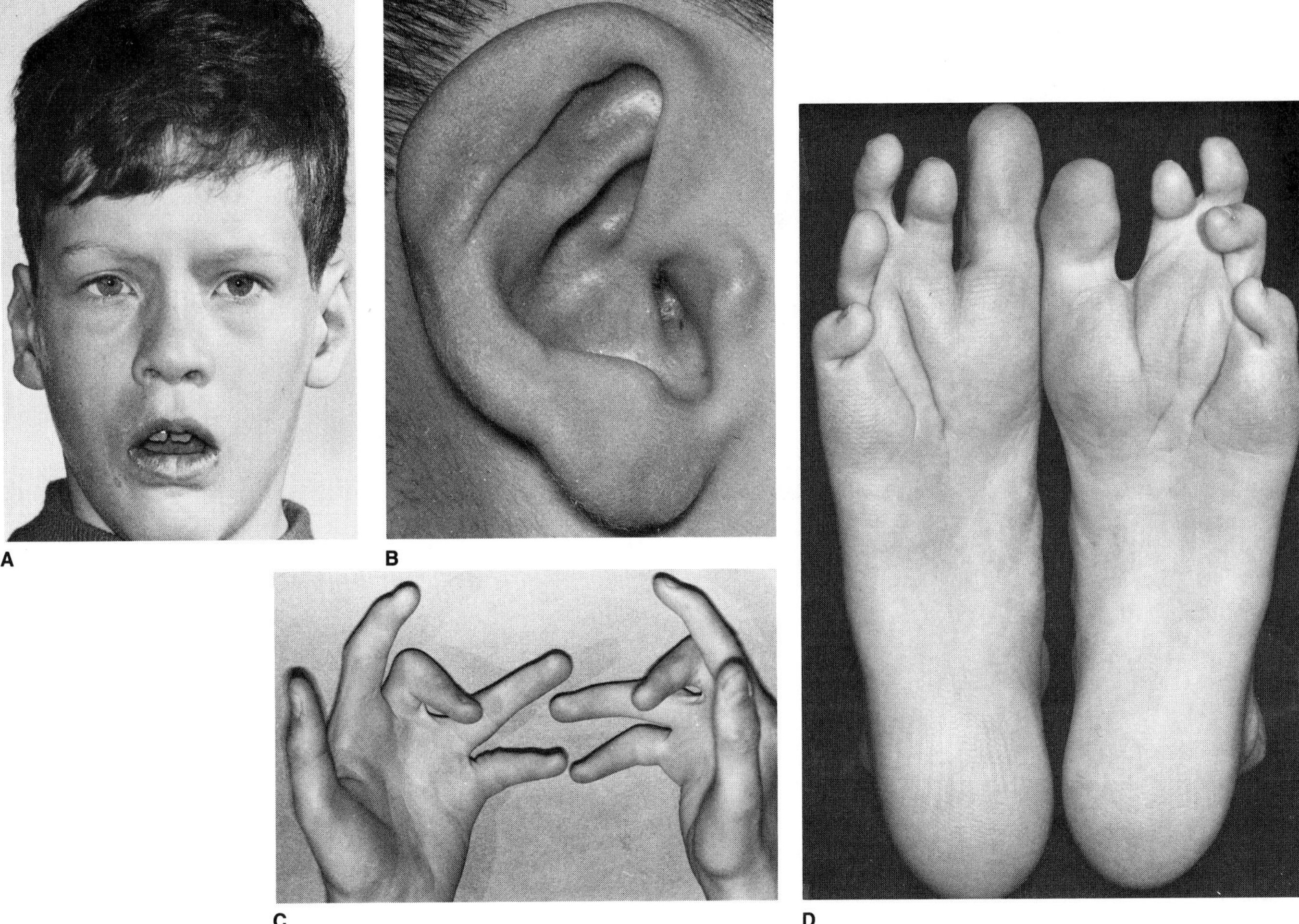


Fig. 3–6. *Trisomy 8 syndrome.* (A,B,C,D) Mentally retarded boy with joint contractures of fingers and toes, absent patellae, malformed ears, and vertical grooves on soles.

2. Fenger K, Niebuhr E: Measurements on band radiographs from 32 cri-du-chat probands. *Radiology* **129**:137–141, 1978.

3. Gordon RR, Cooke P: Facial appearance in the cri-du-chat syndrome. *Dev Med Child Neurol* **10**:69–76, 1968.

3a. Hashimoto T et al: Reciprocal translocation t(5;6) (p13q27) through three generations: Case report of cri du chat syndrome. *Hum Genet* **53**:145–147, 1980.

3b. Kushnick T et al: Familial 5p-syndrome. *Clin Genet* **26**:472–476, 1984.

4. Manning KP: The larynx in the cri-du-chat syndrome. *J Laryngol Otol* **91**:887–892, 1977.

5. Niebuhr E: Cytologic observations in 35 individuals with a 5p− karyotype. *Hum Genet* **42**:143–156, 1978.

6. Niebuhr E: The cri-du-chat syndrome. *Hum Genet* **42**:143–152, 1978; **44**:227–275, 1978.

7. Niebuhr E: Anthropometry in the cri-du-chat syndrome. *Clin Genet* **16**:82–95, 1979.

8. Wilkins LE et al: Psychomotor development in 65 home-reared children with cri-du-chat syndrome. *J Pediatr* **97**:401–405, 1980.

9. Wilkins LE et al: Clinical heterogeneity in 80 home-reared children with cri-du-chat syndrome. *J Pediatr* **102**:528–533, 1983.

## Trisomy 8 syndrome

Trisomy 8 (15%), or more frequently trisomy 8 mosaicism (85%) syndrome, is a relatively common autosomal chromosomal disorder. The abnormal cell line tends to disappear from the lymphocytes with age (8). The estimated frequency is about 1/25,000 to 50,000 children. There is at least a 5:1 male predilection. There is no significant difference in phenotype between so-called pure trisomy 8 and trisomy 8 mosaicism. Over 100 cases have been reported (1–10).

Life expectancy is essentially normal. Most infants have normal birth weight for gestational age. Mental retardation, usually of moderate degree (IQ 40–75), is a virtually constant feature. Agenesis of the corpus callosum is somewhat increased (10).

The forehead usually is high and prominent. The skull is often scaphocephalic. The face is elongated and the pinnae dysplastic. Mild hypertelorism and strabismus are evident in over 50% of the patients. Corneal opacities are frequent. The nose is broad based with an upturned tip in 60%. The mandible is small and retruded and the lower lip is commonly everted in about 40% (Fig. 3–6A, B). Cleft palate has been reported (1,6). Beemer et al (1a) reported a cephalometric study of three cases.

About 70% manifest contractures of fingers and toes, and generalized progressive joint restriction is common (Fig. 6C, D). About one-third of the patients have a long slender truck and slender pelvis and about two-thirds have a spinal deformity, most often scoliosis. Broad ribs, extra ribs, spina bifida occulta, and butterfly vertebrae are common. Absent or hypoplastic patellae are frequent.

Cardiac anomalies found in 25% include VSD, PDA, pulmonary stenosis, coarctation of the aorta, total anomalous pulmonary venous connection, and truncus arteriosus. Hydronephrosis and/or hydroureter are relatively common. Cryptorchidism has been noted in about 50%, and hydronephrosis, hydroureter, or ureteral obstruction in 40%. Deep

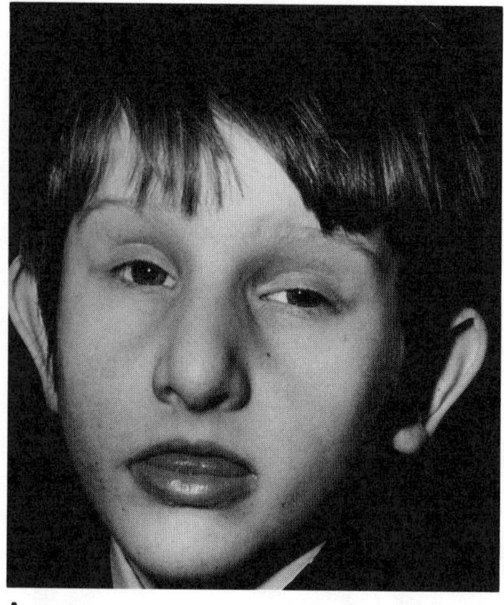

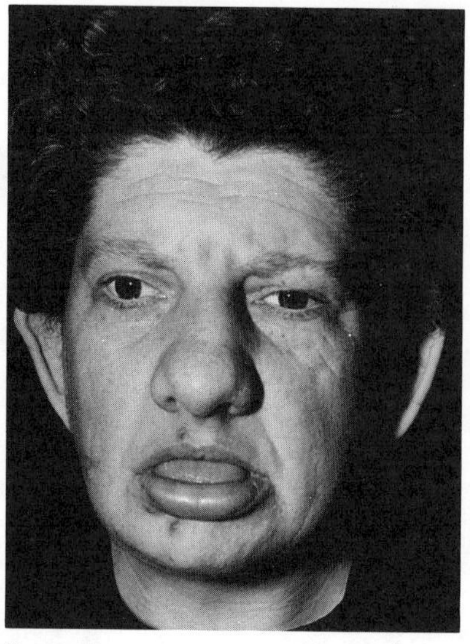

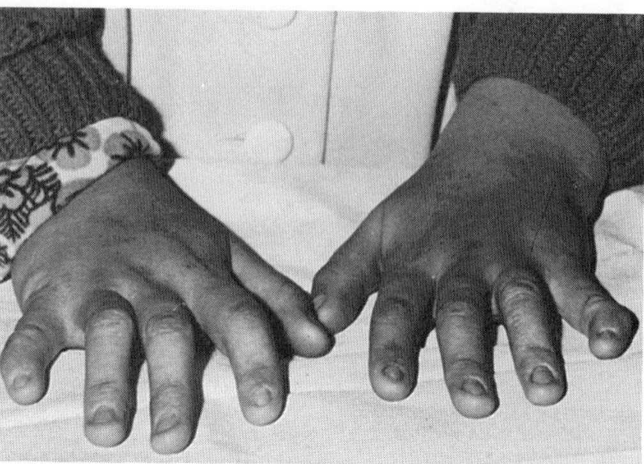

A B C

Fig. 3–7. *dup(9p) syndrome.* (A,B,C) Globular nasal tip, divergent strabismus, down-turned mouth, clinodactyly of fifth fingers, and small nails. (A from *CE Blank et al,* Clin Genet **7**:261, 1975. B,C from *P Balicek,* Hum Genet **27**:253, 1975.)

palmar and/or particularly plantar furrows are seen in about 75%. Other dermatoglyphic changes have been found (11).

### References (Trisomy 8 syndrome)


1. Annerén G et al: Trisomy 8 syndrome. *Helv Paediatr Acta* **36**:465–472, 1981.

1a. Beemer FA et al: Roentgencephalometric measurements in trisomy 8 mosaicism: Report of three cases. *J Craniofac Genet Dev Biol* **4**:233–241, 1984.

2. Bernsen AH et al: Trisomy 8 syndrome. *Acta Paediatr Scand* **66**:397–402, 1977.

3. Berry AC et al: Mosaicism and the trisomy 8 syndrome. *Clin Genet* **14**:105–114, 1978.

4. Frangoulis M, Taylor D: Corneal opacities—a diagnostic feature of the trisomy 8 mosaic syndrome. *Br J Ophthalmol* **67**:619–622, 1983.

5. Gagliardi ART et al: Trisomy 8 mosaicism. *J Med Genet* **15**:70–72, 1988.

6. Kakati S et al: An attempt to establish trisomy 8 syndrome. *Humangenetik* **19**:293–300, 1973.

7. Kosztolànyi G et al: Trisomy 8 mosacisism. *Eur J Pediatr* **123**:293–300, 1976.

8. Meisel-Stosiek M et al: Extremer Gewebe-Mosaizismus bei Trisomie 8-Syndrom. Trisomie in Fibroblasten bei normalem Karyotyp in Lymphozyten. *Klin Pädiatr* **195**:365–368, 1983.

9. Reyes PG et al: Trisomy 8 mosaicism syndrome: Report of monozygotic twins. *Clin Genet* **14**:90–97, 1978.

10. Riccardi VM: Trisomy 8: An international study of 70 patients. *Birth Defects* **13**(3C):171–184, 1977.

11. Rodewald A et al: Dermatoglyphic patterns in trisomy 8 syndrome. *Clin Genet* **12**:28–38, 1977.

12. Silengo MC et al: Radiologic features in trisomy 8. *Pediatr Radiol* **8**:116–118, 1979.


## dup(9p) syndrome

Although less striking than 13 or 21 trisomy, the phenotype of dup(9p) nevertheless has classic craniofacial stigmata. There is a 2:1 female sex predilection. Over 125 cases have been reported (11).

The classic features involve characteristic facies, variable mental retardation, and hypoplasia and/or dysplasia of the terminal phalanges, particularly those of the second and fifth fingers (Fig. 3–7). The facies is characterized by a high, broad forehead, large fontanel and open metopic suture in childhood, mild microbrachycephaly, flat occiput, enophthalmus, mild hypertelorism, and divergent strabismus. The eyes appear relatively small and deeply set with mild downward slanting palpebral fissures. The nose has a large globular tip, broad nasal root, and short philtrum. The mouth is large with the angles turned down and the lower lip everted. The pinnae are large and low set and outstanding with abnormal anthelix. The auditory canals are narrow. Cleft lip and/or palate has been found in 5%. The neck is short and webbed with a low hairline (1,3–6,8,9,11).

Intelligence quotients have varied between 30 and 65. There may be mild finger contractures, brachydactyly, and mild syndactyly of the third and fourth fingers. The clinodactylous fifth fingers often have a single flexion crease. The nails are frequently hypoplastic and most have transverse palmar creases. Congenital heart disease is found in 15–20%. Skeletal anomalies include hallux valgus, limited extension at the elbow, genua valga, kyphosis, and/or lumbar hyperlordosis.

Radiographically there is delayed bone age and retarded closure of anterior fontanel, thoracolumbar scoliosis, hypoplastic terminal and middle phalanges of fingers and toes, and proximal ossification centers of the metacarpals.

Among the several examples of tetrasomy for 9p, about one-third have cleft lip–palate, congenital heart disease, and hydrocephalus (2,7,10,11).

### References [dup(9p) syndrome]


1. Bacchichetti C et al: Partial trisomy 9: Clinical and cytogenetic correlations. *Ann Génét* **22**:199–204, 1979.

2. Balestrazzi P et al: Tetrasomy 9p confirmed by Galt. *J Med Genet* **20**:396–399, 1983.

3. Centerwall WR et al: Familial partial 9p trisomy. *J Med Genet* **13**:57–61, 1976.

4. Cuoco C et al: Duplication of the short arm of chromosome 9: Analysis of five cases. *Hum Genet* **61**:3–7, 1982.

5. Fryns JP et al: Partial duplication of the short arm of chromosome 9 (p13–p22) in a child with typical 9p trisomy phenotype. *Hum Genet* **46**:231–235, 1979.

6. Lewandowski RC et al: Trisomy for the distal one-half of the short arm of chromosome 9. *Am J Dis Child* **130**:663–667, 1976.

7. Moedjono SJ et al: Tetrasomy 9p: Confirmation by enzyme analysis. *J Med Genet* **17**:227–242, 1980.

8. Rodewald A et al: The dermatoglyphic pattern of the trisomy 9p syndrome. *Clin Genet* **16**:405–417, 1979.

9. Sutherland GR et al: Partial and complete trisomy 9: Delineation of a trisomy 9 syndrome. *Hum Genet* **32**:133–140, 1976.

10. Wisniewski L et al: Partial tetrasomy 9 in a liveborn infant. *Clin Genet* **14**:147–153, 1978.

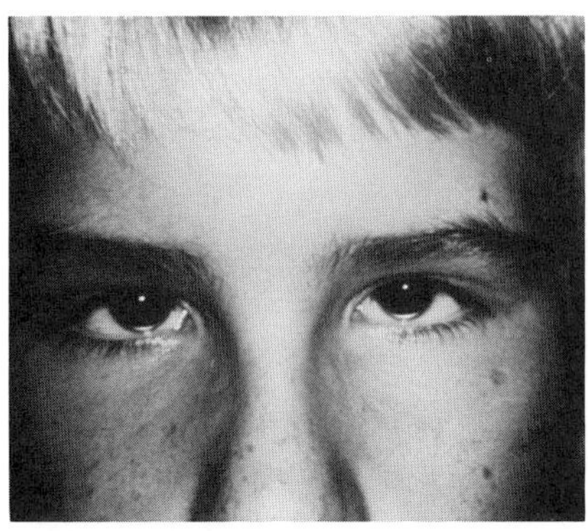
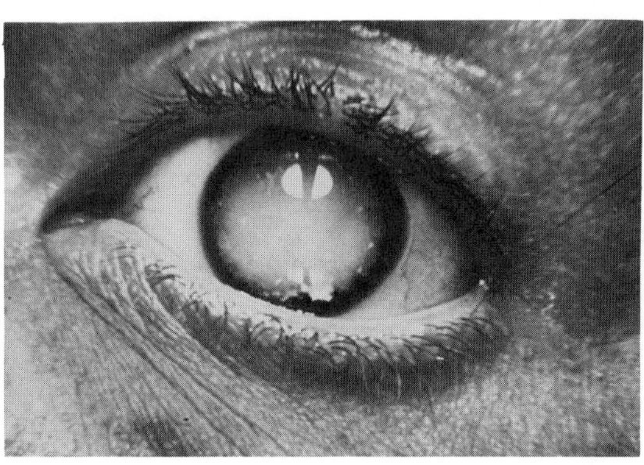
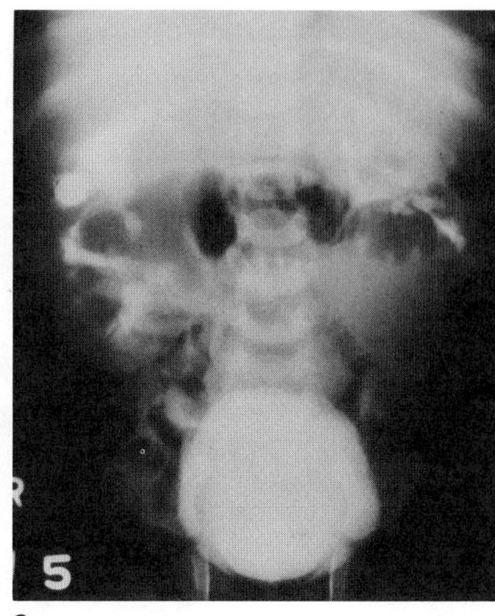


A B C

Fig. 3–8. *del(11)(p13) syndrome (aniridia—Wilms tumor syndrome).* (A) Bilateral aniridia. (B) Cataract developing in patient with aniridia. (C) Radiograph of bilateral Wilms tumor.

11. Young RS et al: The dermatoglyphic and clinical features of the 9p trisomy and partial 9p monosomy syndromes. *Hum Genet* **62**:31–39, 1982.

## del(11)(p13) syndrome (aniridia–Wilms tumor syndrome, AGR triad)

The syndrome of aniridia, mental retardation, and Wilms tumor (Fig. 3–8) is associated with deletion of the distal half of band 13 of the short arm of chromosome 11 (8,10). Maternal age appears to be advanced (9). The syndrome is generally sporadic. However, familial occurrence has been reported (6,12).

Craniofacial alterations include microcephaly with a prominent forehead, cranial asymmetry, long narrow face, high nasal root, ptosis of eyelids, and low-set poorly lobulated pinnae.

Mental retardation is nearly always a constant feature as well as hypospadias and cryptorchidism in males. Hyperkinesis is common. About one-half of the children exhibit growth retardation.

Among children with Wilms tumor, the syndrome is seen in about 1–2%. There is a definite male predilection, possibly as great as 3:1. In about 35% of the cases the Wilms tumor is bilateral, contrasting sharply with bilateral Wilms tumor *not* associated with the syndrome (2–4%). In the syndrome, the tumor appears at a somewhat younger age (2–3 years) than in those with isolated Wilms tumor (4–5 years).

Rarely gonadoblastoma is found instead of Wilms tumor (1,10). The syndrome has been concordant in monozygotic twins with a variable degree of expression (e.g., Wilms tumor in only one twin) indicating that the event is postzygotic and that the deletion is not predisposing to the Wilms tumor (2,4). Although the interstitial deletion of 11p has varied from case to case, all have in common del(11)(p13). The gene for catalase is located in the same band. This has been demonstrated by its dosage effect in normal individuals versus those who are trisomic and monosomic for this region. Thus, catalase levels allow for differentiation of either isolated aniridia or isolated Wilms tumor from the syndrome, particularly since the syndrome is variably expressed (3).

Wilms tumor may also be seen is association with *congenital hemihyperplasia (hypertrophy), Klippel–Trénaunay syndrome, Beckwith–Wiedemann syndrome,* and crossed renal ectopia (7).

### References [del(11)(p13) syndrome (aniridia–Wilms tumor syndrome, AGR triad)]

1. Andersen SR et al: Aniridia, cataract, and gonadoblastoma in a mentally retarded girl with deletion of chromosome 11. *Ophthalmologica* **176**:171–177, 1978.
2. Cotlier E et al: Aniridia, cataracts and Wilms tumor in monozygous twins. *Am J Ophthalmol* **86**:129–132, 1978.
3. Ferrell RE, Riccardi VM: Catalase levels in patients with aniridia and/or Wilms' tumor: Utility and limitations. *Cytogenet Cell Genet* **31**:120–123, 1981.
4. Maurer HS et al: The role of genetic factors in the etiology of Wilms' tumor. Two pairs of monozygous twins with congenital abnormalities (aniridia, hemihypertrophy) and discordance for Wilms' tumor. *Cancer* **43**:205–208, 1979.
5. Narahara N et al: Regional mapping of catalase in Wilms' tumor: Aniridia, genitourinary abnormalities and mental retardation triad loci to the chromosome segment 11p1305→p1306. *Hum Genet* **66**:181–185, 1984.
6. Rangecroft L, O'Donnell B: Wilms' tumor and associated congenital anomalies. *Ir J Med Sci* **149**:191–193, 1980.
7. Redman JF, Berry DL: Wilms' tumor in crossed fused renal ectopia. *J Pediatr Surg* **12**:601–603, 1977.
8. Riccardi VM et al: Chromosomal imbalance in the aniridia–Wilms' tumor association: 11p interstitial deletion. *Pediatrics* **61**:604–610, 1978.
9. Shannon RS et al: Wilms' tumor and aniridia: Clinical and cytogenetic features. *Arch Dis Childh* **57**:685–690, 1982.
10. Turleau C et al: Aniridia, male pseudohermaphoditism, gonadoblastoma, mental retardation and del 11p13. *Hum Genet* **57**:300–306, 1981.
11. Turleau C et al: Del 11p/aniridia complex. Report of three patients and review of 37 observations from the literature. *Clin Genet* **26**:356–362, 1984.
12. Yunis JJ, Ramsay NKC: Familial occurrence of the aniridia–Wilms' tumor syndrome with deletion of 11p13–14.1. *J Pediatr* **96**:1027–1030, 1980.

## del(13q) syndrome

Virtually constant features are reduced birth weight, psychomotor retardation, hypotonia, and microcephaly. About one-half of the patients exhibit trigonocephaly (1–10).

The facies is characterized by frontal bossing, broad prominent nasal root, protruding maxilla and incisors, and large prominent malrotated ears with deep helical sulcus. Eye anomalies consist of ptosis, epicanthal folds, microphthalmia, and colobomas. Retinoblastoma occurs in approximately 15%. Osteosarcoma and synovial sarcoma are found with increased frequency. The neck is short with redundant skin folds. Various congenital heart anomalies have been noted in approximately 35%. About 60% of the males have genital malformations including hypospadias, small or bifid scrotum, cryptorchidism, micropenis, and perineal fistula (Fig. 3–9). Anal atresia has been noted in about 20%. Hip dislocation and talipes equinovarus have increased frequency and the thumbs are hypoplastic or absent in about 30%.

Most of the 125 reported cases have involved loss of the distal two-thirds of the long arm. In some, this has resulted from deletion, in others translocation, and in others deletion and fusion, producing a ring chromosome. Deletion of bands q33→qter results in severe men-

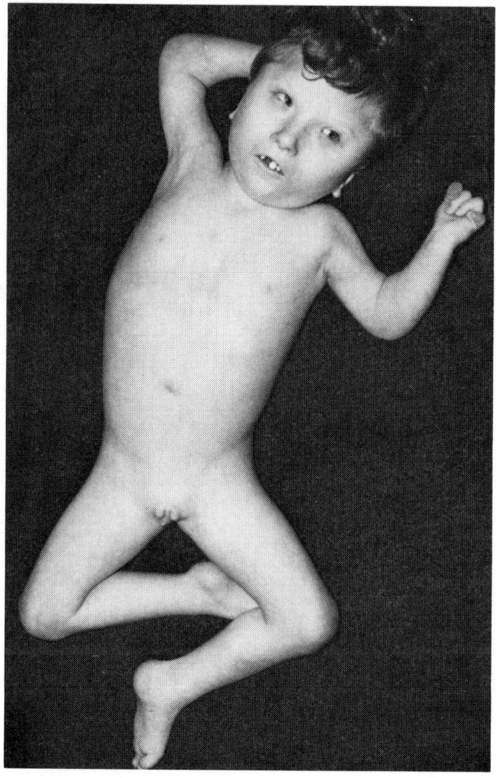

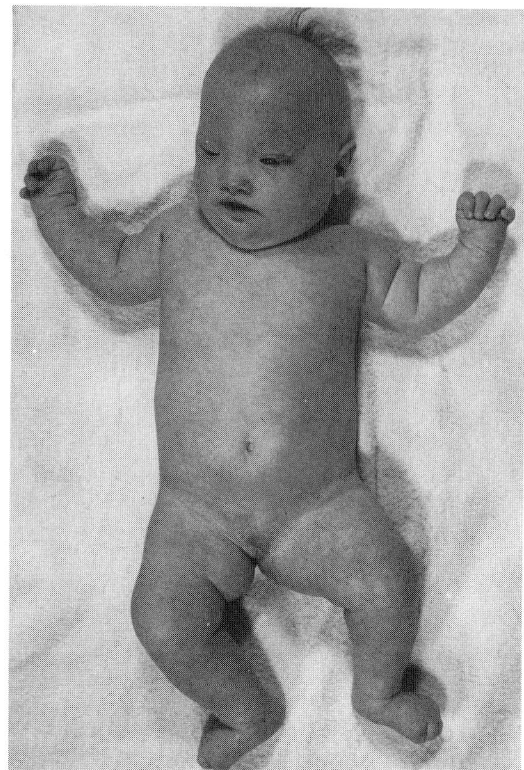

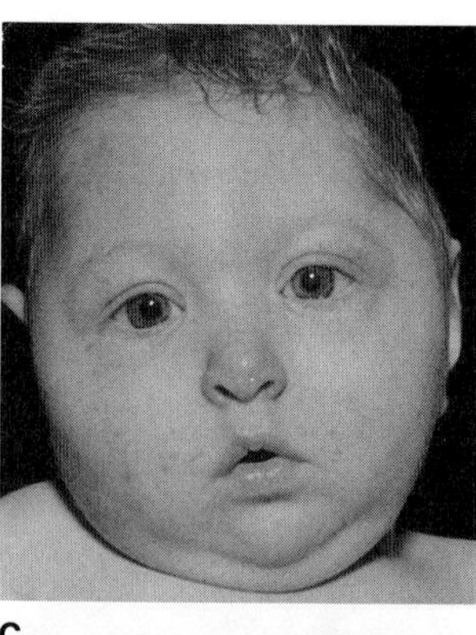

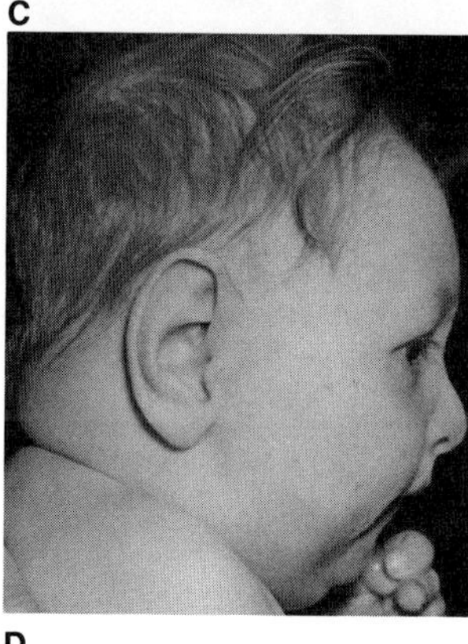

A B C D

Fig. 3–9. *del(13q) syndrome.* (A) Absence of thumbs, micropenis, cleft scrotum in severely retarded child. (B) Broad-based nose, ptosis of eyelids, and absent thumbs. (C,D) Microcephaly, trigonocephaly, apparent hypertelorism, prominent nasal bridge, and short neck. (A from *RS Sparkes et al,* Am J Hum Genet **19:**644, 1967. B from *RC Juberg et al,* J Med Genet **6:**314, 1969. C,D from *E Grace et al,* J Med Genet **8:**351, 1971.)

tal retardation, microcephaly, ocular hypertelorism, frontal bossing, protruding maxilla, and large ears. Those with additional deletion of bands q31 and q32 exhibit mental retardation, microcephaly, trigonocephaly, hypoplastic or absent thumbs and metacarpals, and male genital abnormalities. Retinoblastoma is associated with deletion of band q14, but only about 20% of those with that band deletion have the tumor. With large deletions holoprosencephaly may be found. The gene for esterase D is proximate; thus it can be used for prenatal diagnosis and prediction for occurrence of retinoblastoma (2).

### References [del(13q) syndrome]


1. Chemke J et al: Multiple skeletal anomalies in the "13q−" syndrome. *Eur J Pediatr* **128**:27–31, 1978.
2. Colwell JK et al: The need to screen all retinoblastoma patients for esterase D activity: Detection of submicroscopic chromosomal deletion. *Arch Dis Childh* **62**:8–11, 1987.
3. Cuschieri A et al: Partial deletion of the long arm of chromosome No. 13. *Hum Genet* **36**:341–344, 1977.
4. Genčik A et al: Retinoblastoma and chromosome 13 deletion. *Helv Paediatr Acta* **37**:457–464, 1982.
5. Martin NJ et al: The ring chromosome 13 syndrome. *Hum Genet* **61**:18–23, 1982.
6. Motegi T et al: Deletion (13)(q13q14.3) with retinoblastoma: Confirmation and extension of a recognisable pattern of clinical features in retinoblastoma patients with 13q deletion. *J Med Genet* **24**:696–712, 1987.
7. Noel B et al: Partial deletions and trisomies of chromosome 13: Mapping of bands associated with particular malformations. *Clin Genet* **9**:593–602, 1976.
8. Parcheta B: Clinical features in a case of ring chromosome 13. *Eur J Pediatr* **144**:409–412, 1985.
9. Wilson WG et al: Deletion (13)(q14.1q14.3) in two generations. Variability of ocular manifestations and definition of the phenotype. *Am J Med Genet* **28**:675–683, 1987.
10. Yunis J, Ramsay N: Retinoblastoma and subband deletion of chromosome 13. *Am J Dis Child* **132**:161–163, 1978.


## del(18p) syndrome

At least 120 cases of del(18p) syndrome have been reported (1–8). There is a 3F:2M predilection. Mean parental age is increased (6). At least 85% arise from *de novo* deletions. Translocation is responsible for about 10%. The extent of deletion of the short arm of chromosome 18 does not correlate with the clinical picture (4).

In addition to the mental (100%) and somatic (75%) retardation, the phenotype is not striking unless associated with holoprosencephaly. Microcephaly (50%), round face (40%), ocular hypertelorism (50%) ptosis (50%), strabismus (40%), epicanthus (40%), broad flat nose (80%), large outstanding ears (60%), carp mouth (65%), microretrognathia (45%), and short neck (45%) characterize the facies (Fig. 3–10) (3,5). Perhaps 25% exhibit pterygium colli and about 50% of the males have micropenis and/or cryptorchidism. About 5% have cleft lip and 10% have cleft palate (7,8). As noted, about 10% exhibit semilobar or alobar holoprosencephaly (2,7,8). Single central incisor has been reported (1). No more than 5% have congenital heart anomalies. These are heterogeneous (VSD, coarctation of the aorta, AV canal, PDA, and transposition of large vessels) (7,8).

### References [del(18p) syndrome]


1. Carratu A et al: Sindrome da monosomia 18p con oloprosencephalia. *Min Pediatr* **35**:1225–1228, 1983.

1a. Dolan LM et al: 18p− syndrome with single central maxillary incisor. *J Med Genet* **18**:396–397, 1981.

2. Faust J et al: The 18p− syndrome: Report of four cases. *Eur J Pediatr* **123**:59–66, 1976.
3. Faust J et al: Cranial morphology in the 18p− syndrome. *Eur J Pediatr* **129**:61–65, 1978.
4. de Grouchy J: The 18p−, 18q−, and 18r syndromes. *Birth Defects* **5(5)**:74–87, 1969.

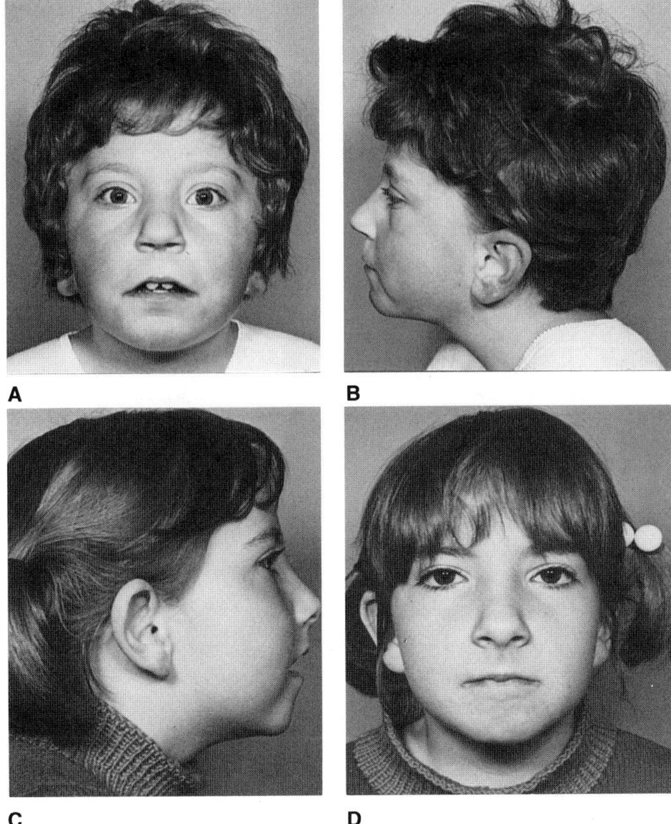


Fig. 3–10. *del(18p) syndrome.* Essentially normal facies but mild hypertelorism and wide mouth. (From *A Schinzel et al,* Arch Genet **47:**1, 1974.)


5. Habedank M, Trost-Brinkhues G: Monosomy 18p− and pure 18p− in a family with translocation (7;18). *J Med Genet* **20**:377–379, 1983.

6. Lurie IW, Lazjuk GI: Partial monosomies 18: Review of cytogenetical and phenotypical variants. *Humangenetik* **15**:203–222, 1972.

7. Schinzel A et al: The 18p− syndrome. *Arch Genet* **47**:1–15, 1974.

8. Schinzel A: Partielle Monosomie von Chromosom 18. *Arch Genet* **52**:125–147, 1979.


## del(18q) syndrome

In 1964, de Grouchy et al (5) reported partial deletion of the long arm of chromosome 18. The phenotype is distinctive, consisting of hypotonia, mental retardation, characteristic facial dysmorphism (Fig. 3–11), abnormal genitalia, and tapered fingers. More than 80 cases have been reported (1–6,8–17) and several extensive reviews are available (3,4,6,11–13). Wilson et al (16) indicated that the typical syndrome had deletion in band 18q21 and identified a different clinical syndrome in association with a more proximal deletion within band 18q12.

In 80%, the deletion occurs *de novo,* in 10%, the deletion results from parental pericentric inversion or translocation (1,9,14), and in 10%, the deletion is in the mosaic state, resulting in a less severe phenotype (4). One female patient had six pregnancies (14). There is a predilection for females with a 2:3 male to female sex ratio.

**Growth.** Mean birth weight is below 2700 g and, with growth and development, short stature below the fifth percentile is observed in 78% (13). Approximately 10% die within the first few months of life (4).

**Central nervous system.** Hypotonia is a constant feature and seizures are frequent. Mental retardation (100%) is profound, with few having an IQ over 30. The voice is often low pitched (6,12,13,16).

**Craniofacial features.** Microcephaly with head circumference below the second percentile is observed in 68%. In some instances, fontanel closure is delayed. The midface is retruded (85%) and relative mandibular prognathism becomes evident with age. The eyes are deeply set. Other features may include epicanthic folds (42%), strabismus (34%), nystagmus (80%), coloboma of the iris (7%), pale optic discs (84%), prominent anthelix or antitragus (84%), stenotic ear canals (50%), impaired hearing (61%), broad nasal bridge (81%), carp-like mouth (87%), cleft lip (9%), and cleft palate (29%) (4,6,7,10,12,13,16). A small subcutaneous nodule may be evident at the site of cheek dimples (6).

**Thorax and abdomen.** The nipples are widely spaced (79%) and bilateral subacromial dimples are observed. Umbilical hernia is a feature in 16%. Congenital heart defects occur in 35% (4,13,16), with atrial septal defect, pulmonary stenosis, patent ductus arteriosus, and ventricular septal defect occurring in decreasing order of frequency.

**Genitalia.** The genitalia are abnormal in both sexes. Findings include cryptorchidism (52%), hypospadias, micropenis, and hypoplasia of the labia minora (47%) (4,11,13). Inguinal hernia occurs in 13% (13).

**Extremities.** Dimples are present on the epitrochleal regions, over the knuckles, and on the lateral surfaces of the knees. The hands are long, thin, and tapered (90%). The thumbs are proximally placed in 90% and transverse palmar creases occur in 92%. An excess number of fingertip whorl patterns and a high total finger ridge count are characteristic. Other findings include abnormal implantation of the second toes (84%) and talipes equinovarus (21%) (4,13).

**Skeletal system.** Approximately half the cases are associated with osteoarticular anomalies including supernumerary or hypoplastic ribs, costal synostoses, spina bifida occulta, and coxa valga (4).

**Laboratory findings.** IgA deficiency is found in approximately 30% (4). Diagnosis is based on banded chromosome study.

### References [del(18q) syndrome]


1. Aarskog D: A familial 3/18 reciprocal translocation resulting in duplication-deficiency (3?+, 18q−). *Acta Paediatr Scand* **58**:397–406, 1969.

2. Aksu F et al: Numerische und strukturelle Aberrationen des Chromosoms Nr. 18. *Klin Pädiatr* **188**:220–232, 1976.

3. de Grouchy J: The 18p−, 18q−, and 18r syndromes. *Birth Defects* **5(5)**:74–87, 1969.

4. de Grouchy J, Turleau C: Partial 18q monosomy or 18q− syndrome. *Clinical Atlas of Human Chromosomes.* Wiley, New York, 1977, pp 170–176.

5. de Grouchy J et al: Deletion partielle des bras longs du chromosome 18. *Path Biol* **12**:579–581, 1964.

6. Gorlin RJ: 18q− syndrome, in *New Chromosomal Syndromes,* Yunis JJ (ed), Academic Press, New York, 1977, Ch. 3, pp 72–74.

7. Gorlin RJ et al: Facial clefting and its syndromes. *Birth Defects* **7(7)**:3–49, 1971.

8. Kunze J et al: Ring-Chromosom 18. *Humangenetik* **15**:289–318, 1972.

9. Law EM, Masterson JG: Familial 18q-syndromes. *Ann Génét* **12**:215–222, 1969.

10. Lurie I, Lazjuk G: Partial monosomies 18. *Humangenetik* **15**:203–222, 1972.

11. Parker CE et al: The syndrome associated with the partial deletion of the long arms of chromosome 18 (18q−). *Calif Med* **117**:65–71, 1972.

12. Rethoré MO: Deletions and ring chromosomes, in *Handbook of Clinical Neurology,* Vinken JP, Bruyn GW (eds), North-Holland Publishing Company, Amsterdam, Vol 26–27, *Congenital Malformations of the Brain and Skull,* 1977.

13. Schinzel A et al: Structural aberrations of chromosome 18. II. The 18q− syndrome. Report of three cases. *Humangenetik* **26**:123–132, 1975.

14. Subrt I, Pokorny J: Familial occurrence of 18q−. *Humangenetik* **10**:181–187, 1970.

15. Wertelecki W, Gerald PS: Clinical and chromosomal studies of the 18q− syndrome. *J Pediatr* **78**:44–52, 1971.

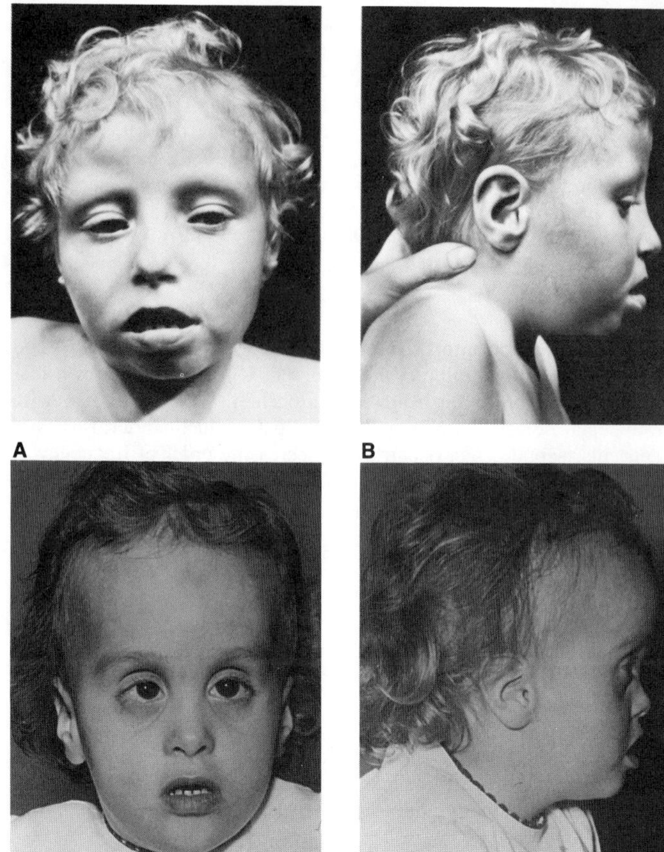


Fig. 3–11. *del(18q) syndrome.* (A,B) Midface hypoplasia, deeply set eyes, and prominent anthelix and antitragus in patient with 18r karyotype. (C,D) Compare facies of patient with that of patient in A,B. (A,B from *JD Mürken et al,* Z Kinderheilkd **109:**1, 1970.)

16. Wilson MG et al: Syndromes associated with deletion of the long arm of chromosome 18[del(18q)]. *Am J Med Genet* **3**:155–174, 1979.
17. Wolf U et al: Deletion on long arm of chromosome 18 (46,XX,18q−). *Humangenetik* **5**:70–71, 1967.

## Turner syndrome

In 1938, Turner (43) recognized the combination of sexual infantilism, webbed neck, and cubitus valgus as a distinct entity. In 1959, Ford et al (11) showed that patients with Turner syndrome were missing one sex chromosome (45,X). The syndrome is now known to consist of short stature, streak gonads, webbed neck, shield chest, peripheral lymphedema at birth, coarctation of the aorta, hypoplastic nails, short metacarpals, and multiple pigmented nevi (Fig. 3–12) (9,12). The reader is referred to the following sources for extensive coverage: general aspects (7,12,13), cytogenetics (4,8,9,13,29,37), growth and development (23,33), cognitive and psychosocial aspects (1,13,25), cardiovascular anomalies (20,27,28,32), craniofacial and oral aspects (10,14,15,26,40), renal anomalies (24,31), and neoplasia (39,45).

The prevalence is 1/2500 female births (9,17). Approximately 98–99% of Turner syndrome fetuses are spontaneously aborted; about 20% of all spontaneously aborted fetuses have Turner syndrome (5,6,12,17). Hall et al (13) noted that about one-third of patients were diagnosed in the newborn period, one-third during childhood, and one-third during the teenage years when failure to go through puberty became evident. Minimal diagnostic criterion is an abnormal karyotype in which all or part of one of the X chromosomes is absent. The overwhelming majority of patients have gonadal dysgenesis and short stature.

Approximately one-half of patients with Turner syndrome have 45,X karyotypes. The frequencies of different karyotypes are listed in Table 3–12 and the most common nonmosaic and mosaic karyotypes are listed in Table 3–13. Unusual mosaic karyotypes are also known to occur (30). Clinical diagnosis is based on the overall pattern of anomalies; there are no obligatory malformations and even phenotypic fea-

Table 3–12. Frequency of chromosomal constitutions seen in Turner syndrome[a]

| Chromosomal constitution | Approximate percentage |
|---|---|
| 45,X | 50 |
| Isochromosome X | 12–20 |
| Mosaicism | 30–40 |
| 45,X/46,XX | (10–15) |
| 45,X/46,XY | (2–5) |

[a]Adapted from JG Hall et al, *West J Med* **137**:32, 1982.

Table 3–13. Commonest karyotypes associated with Turner syndrome[a]

| |
|---|
| Commonest nonmosaic karyotypes |
| 45,X |
| 46,X,i(Xq) |
| 46,X,del(Xp) or 46,XXp− |
| 46,X,del(Xq) or 46,XXq− |
| 46,X,r(X) |
| 46,X,i(Xp) |
| 46,X,i(Yq) |
| 46,X,t(X;X) or 46,X,ter rea (X;X) |
| 46,X,t(X;any autosome) or 46,X,t(X;Y) |
| Commonest mosaic karyotypes |
| 45,X/46,XX |
| 45,X/47,XXX |
| 45,X/46,XX/47,XXX |
| 45,X/46,XY |
| 45,X/any karyotype with a structurally abnormal X or Y |

[a]Adapted from A de la Chapelle, Sex chromosome abnormalities, in *Principles and Practice of Medical Genetics,* Emery EH, Rimoin DL (eds), Churchill Livingstone, Edinburgh, 1983, Ch. 16, pp 193–215.

Table 3–14. Comparison of findings in pure Turner syndrome and Turner mosaicism[a]

| Finding | 45,X (%) (*n*=59) | Mosaic Turner (%) (*n*=41) |
|---|---|---|
| Head and neck | | |
| Epicanthic folds | 37 | 12 |
| Ptosis of the eyelids | 14 | 7 |
| Myopia | 14 | 15 |
| Abnormal ears | 56 | 37 |
| Low hairline | 66 | 37 |
| Webbed neck | 51 | 27 |
| Thorax | | |
| Shield chest | 66 | 39 |
| Coarctation of aorta | 12 | 5 |
| Heart murmur | 46 | 39 |
| Limbs | | |
| Lymphedema | 53 | 12 |
| Abnormal nails | 53 | 20 |
| Cubitus valgus | 56 | 54 |
| Renal abnormalities | 32 | 20 |

[a]Modified from JG Hall et al, *West J Med* **137**:32, 1982.

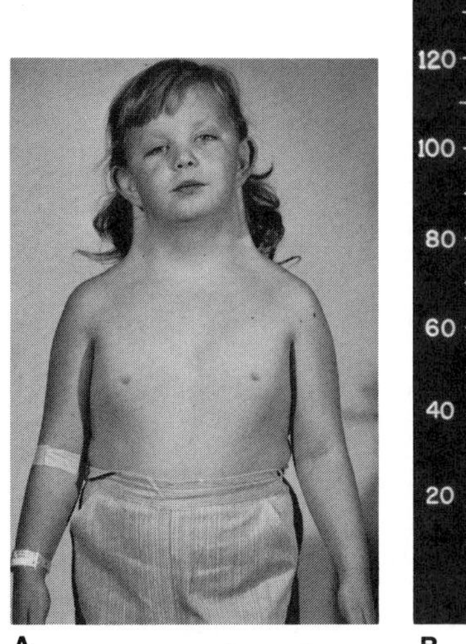
A

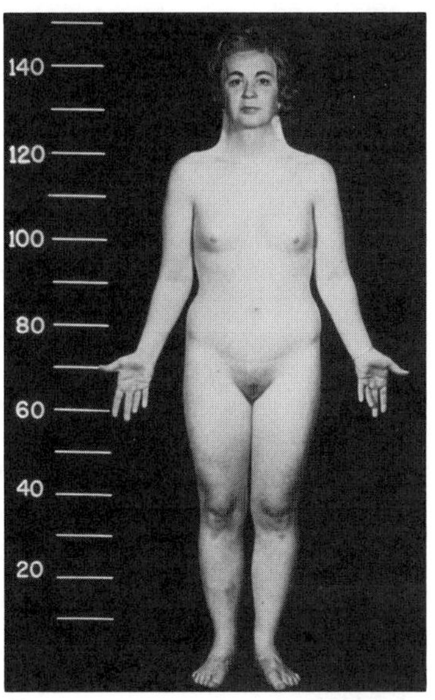

B

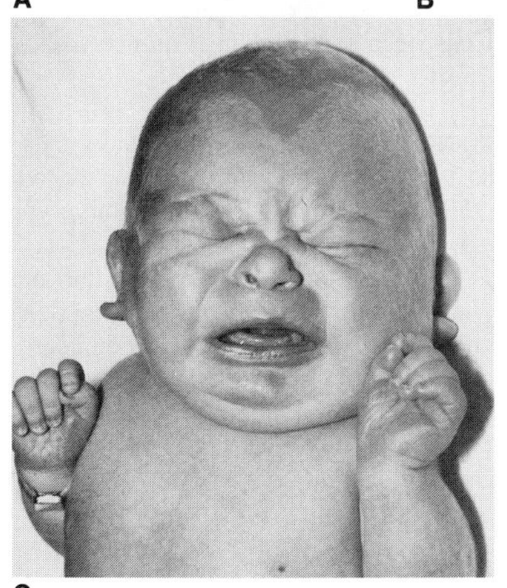
C

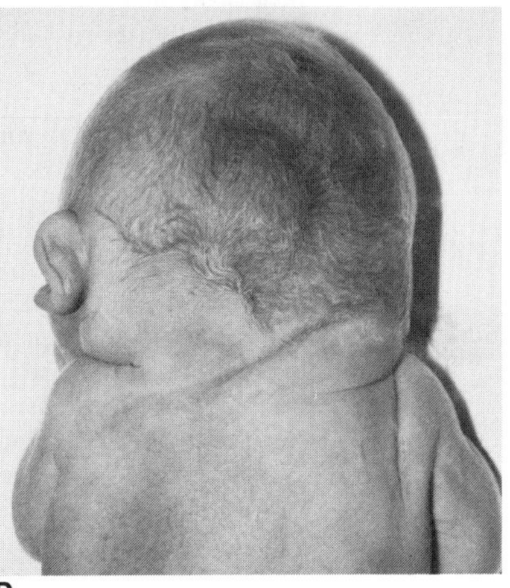
D

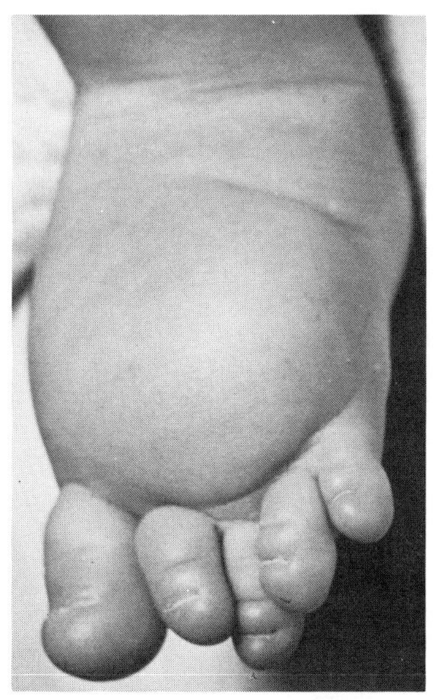
E

Fig. 3–12. *Turner syndrome.* (A) Pterygium colli, protruding ears, broad shield-like chest with small nipples. (B) Short stature, webbed neck, cubitus valgus, incomplete sexual development. (C,D) At birth, excess skin is present at the nape of the neck. Note protruding ears. (E) Lymphedema of foot with hypoplastic toenails. (C,D from *RR Gordon,* Br Med J **1**:483, 1969.)

tures such as amenorrhea and short stature may be absent (39). Patients mosaic for Turner syndrome tend to have fewer phenotypic features than patients with isochromosome of the long arm of the X chromosome or with pure 45,X Turner syndrome (Table 3–14) (13). The clinical severity of Turner syndrome mosaics increases with the relative increase in the abnormal cell line population (35). Patients with isochromosome formation of the short arm of the X chromosome are frequently normal phenotypically (39). Mosaicism of the 45,X/46,XY type exhibits variable findings. Approximately 15% have features of Turner syndrome, about 80% have ambiguous genitalia, and about 5% have a male phenotype with bilateral cryptorchidism (12). Phenotypic features of pure 45,X Turner syndrome are listed together with percentages in Table 3–15.

**Growth.** Average birth length is 47 cm (5) and birth weight is lower than normal, being 2933±467 in a Swedish series (21). Ranke et al (33) studied growth in 150 patients from three German American centers, observing that growth could be divided into four phases: (a) intrauterine growth retardation, (b) height development, which is normal up to a bone age of 2 years, (c) bone age of 2 to 11 years, when growth is markedly stunted, and (d) bone age after 11 years when the growth phase is prolonged but total height gain is below normal. Ranke et al (33) observed no difference in height between 45,X patients and other chromosomal variants. Lyon et al (23) provided growth data for four published series of European patients. Their results permit reasonable prediction of adult height in any patient with Turner syndrome. Lyon et al (23) also noted that estrogen treatment, although resulting in initial accelerated growth, had no significant effect on final height attainment. Rosenfeld at all (34a) reported significantly increased growth rates with human growth hormone and oxandrolone. Lemli and Smith (19) and Brook et al (2) observed that familial height played a role in determining final height attainment in patients with Turner syndrome, taller parents having taller daughters. Final height attainment is usually between 122 and 152 cm (9). Ranke et al (33) found a mean adult height attainment of 146.8 cm.

**Central nervous system.** Studies of performance in Turner syndrome in the past have been subject to methodological shortcomings including (a) faulty reporting of mental retardation in some patients when, in fact, reduced performance IQ in the presence of a normal verbal IQ results from specific deficit in spatial ability rather than global reduction in intelligence; (b) ascertainment bias of patients by

Table 3–15. Features of 45,X Turner syndrome[a]

| Feature | Percentage or finding |
|---|---|
| Growth | |
| Birth length | $\bar{X}=47$ cm |
| Birth weight | $\bar{X}=2933\pm467$ g |
| Final height attainment | 122–152 cm |
| Performance | |
| Cognitive | Deficits[b] |
| Intelligence | Normal |
| Psychiatric | Slightly increased risk for anorexia nervosa[c] |
| Head and neck | |
| Epicanthic folds | 25 |
| Highly arched palate | 36 |
| Visual abnormalities, usually strabismus | 22 |
| Auditory problems | 50 |
| Webbed neck | 46 |
| Short, broad neck, low hairline | 74 |
| Chest | |
| Shield chest | 53 |
| Cardiovascular | |
| Coarctation of the aorta, ventricular septal defect | 10–16 |
| Renal | 38 |
| Horseshoe kidney | — |
| Duplicated or otherwise anomalous ureters | — |
| Unilateral renal aplasia or hypoplasia | — |
| Gastrointestinal | |
| Telangiectases | — |
| Skin and lymphatics | |
| Pigmented nevi | 63 |
| Lymphedema of hands and feet | 38 |
| Nails | |
| Hypoplasia | 66 |
| Skeletal | |
| Cubitus valgus | 54 |
| Short metacarpals, metatarsals (usually 4) | 48 |
| Deformed medial tibial condyle | 65 |
| Osteoporosis | 50 |

[a] Adapted in part from JL Simpson, *Disorders of Sexual Differentiation*, Academic Press, New York, 1976 and A de la Chapelle, Sex chromosome abnormalities, in *Principles and Practice of Medical Genetics*, Emery EH, Rimoin DL (eds), Churchill Livingstone, Edinburgh, 1983, Ch. 16, pp 193–215.
[b] For specific deficits, see text and B Bender et al, *Pediatrics* **73**:175, 1984.
[c] For other psychiatric abnormalities, see text and JG Hall et al, *West J Med* **137**:32, 1982. For psychosocial adjustment, see E McCauley et al, *Clin Genet* **29**:284, 1986.

severe phenotypic stigmata instead of independently by karyotype; and (c) samples composed of pooled karyotypic subgroups instead of examination of each subgroup independently (1,13).

Visual–spatial deficit has been related to reduced functioning in the right cerebral hemisphere, specifically the right parietal lobe (1). Bender et al (1) found that 45,X Turner syndrome subjects were slightly delayed in walking, had a moderately decreased full-scale and performance IQ, and demonstrated striking deficit in perceptual organization and fine motor skills, but had average language skills. Linden et al (20a) noted that features such as slow speech development, hyperactivity, learning disabilities, neuromotor deficits, and short stature could contribute to poor self-image and lowered self-esteem during adolescence. Several psychiatric disturbances have been reported. A slight increase in the risk for anorexia nervosa has been mentioned (18,44). McCauley et al (25), studying 30 adult subjects, reported a significant subgroup with major psychiatric problems, especially depression with markedly low self-esteem.

**Head and neck abnormalities.** Epicanthic folds, ptosis of the eyelids, prominent ears, and micrognathia are common facial features. Visual abnormalities, particularly strabismus, are found in approximately 22% (12). Chronic suppurative otitis with resultant hearing loss occurs in some cases (18a,41).

In infants, excess skin on the nape of the neck is common. During embryonic life, neck blebs or cystic hygromas are common (3). With age, the excess skin on the neck metamorphoses into pterygium colli. The ears are prominent and the posterior hairline is low (12).

The palate is highly arched in approximately 36% and cleft palate may occur with a somewhat higher than normal frequency. The teeth may erupt prematurely, the first permanent molars appearing between 1.5 and 4 years of age (10,14–16,26,40).

**Chest.** The chest is broad with seemingly wide spaced, hypoplastic, at times, inverted nipples. Breast development is poor (12).

**Genitalia.** Gonadal dysgenesis or streak gonads are characteristic. The histologic pattern consists of long streaks of white wavy connective tissue stroma without follicles. However, follicles are present in fetal and infantile ovaries of patients with Turner syndrome. Patients have primary amenorrhea and sterility. Exogenous hormone replacement is essential for establishing secondary sexual characteristics. Fertility is a rare possibility and has been recorded in a number of instances (12,13).

**Cardiovascular abnormalities.** Coarctation of the aorta occurs in approximately 15% of patients with various Turner karyotypes, but occurs with higher frequency (50%) in patients with 45,X karyotype (13). Dissecting aortic aneurysm has also been reported (20). Coarctation of the aorta is the most common cause of hypertension and blood pressure often returns to normal following surgical repair. Adults have an increased frequency of hypertension (30%) even when coarctation of the aorta, renal parenchymal, and renal vascular disease have been eliminated as causes (13). Less common anomalies include ventricular septal defect, atrial septal defect, dextrocardia, bicuspid aortic valves, and hypoplastic left heart (13,28,32).

A variety of other vascular anomalies may be observed infrequently, including intestinal telangiectasias, hemangiomas, and lymphangiectasia. Gastrointestinal bleeding may indicate hemangiomas of the intestinal tract. Since the frequency of ulcerative colitis and Crohn's disease is higher in Turner syndrome than in the general population, gastrointestinal bleeding may also indicate primary bowel disease. Protein-losing enteropathy from gastrointestinal lymphangiomas has also been recorded. Turner syndrome patients also have a higher frequency (90%) of increased numbers of renal arteries (13).

Lymphedema occurs in approximately 80% of newborns. The hands and feet may appear puffy but lymphedema is usually transient, resolving in childhood (12,13). Recurrent lymphedema of the extremities may be observed in some patients and, rarely, severe lymphedema may be found in adulthood with chylous ascites (42). Lymphedema is almost always secondary to congenital hypoplasia of the lymphatic channels (13). Shepard et al (38) documented hypoalbuminemia in 45,X fetuses and suggested that lowered plasma albumin concentration could contribute to the edema by lowering osmotic pressure in the blood vessels.

**Renal anomalies.** Renal findings include horseshoe kidneys (20%), duplication of the collecting ducts (20%), and malrotation of the kidney (15%) (13,24,31). Potter sequence has been observed with Turner syndrome (22,46) and may result from cystic dysplasia, small kidneys, or obstructive uropathy (31). When the latter occurs, prune belly may be present (36).

**Skeletal abnormalities.** Bone age remains within normal limits until 12–14 years of age when the adolescent growth spurt fails to take place. Without hormone therapy, epiphyses usually fail to fuse until the 20's. Common skeletal abnormalities include cubitus valgus (approximately 75%), short fourth metacarpals (about 65%), deform-

Table 3–16. Common roentgenographic features of Turner syndrome[a]

| |
|---|
| Hand |
| Drumstick distal phalanges |
| Short fourth metacarpals |
| Carpal sign: change in angulation of carpal bones |
| Shortening of all hand bones |
| Madelung's deformity |
| Feet |
| Similar to hands |
| Pes cavus |
| Knees |
| Lateral dislocation of patellae |
| Hypoplastic patellae |
| Irregularity of tibial metaphysis and epiphysis |
| "Mushroom" projections, medial surface of proximal tibial metaphysis (medial tibial condyle) |
| Spine |
| Scoliosis |
| Lack of lumbar lordosis |
| Schmorl's nodes (abnormalities of cartilaginous endplates) |
| Hypoplasia of arch of atlas |
| Shortening of anteroposterior diameter of vertebral bodies |
| Ribs |
| Thin |
| Developmental abnormalities |
| Pelvis |
| Android configuration (50%) |
| Occasional widening of symphysis pubis |
| Skull |
| Midfacial hypoplasia |
| Deepening of posterior cranial fossa |
| Widely spaced mandibular rami |

[a]Adapted from CG Brook et al, *Ann Hum Biol* **4**:17, 1977; WD Risch et al, *Am J Roentgenol* **126**:1302, 1976; and JG Hall et al, *West J Med* **137**:32, 1982.

ity of the medial tibial condyle (about 65%), hypoplasia of the cervical vertebrae (about 80%), and small carpal angle (12,13,34). Some degree of osteoporosis is found in about 50% (12). However, bone radiolucency and coarse trabeculations can be observed in childhood (13). Common roentgenographic findings are listed in Table 3–16. Cohen (4a) reviewed five cases with craniosynostosis.

**Dermatologic features.** Hypoplastic, deeply set nails and multiple pigmented nevi are common (12). Seborrhea, xerosis, hirsutism, and keloid formation occur with increased frequency (13). Dermatoglyphic findings include an increased total finger ridge count. Redundant folds of skin resulting in pterygium colli and low nuchal hairline have already been discussed.

**Neoplasia.** Patients who are mosaic for 45,X/46,XY have an increased risk of gonadoblastoma. Such neoplasia develops in a high percentage during early childhood, but there is also an increase around puberty (39). Wertelecki et al (45) studied 289 patients with Turner syndrome and found nongonadal neoplasia in 2.8%. Three tumors were of neural origin, three were gastrointestinal, and one case of leukemia and one of carcinoma of the thyroid were noted. Wertelecki et al (45) listed all other cases of nongonadal neoplasia, including among others, two pituitary tumors, two adrenal tumors, and four brain tumors.

**Autoimmune disease.** Hypothyroidism, diabetes mellitus, and inflammatory bowel disease occur more frequently in the Turner syndrome population than in the general population. Acute Hashimoto's thyroiditis occurs infrequently, but hypothyroidism on an autoimmune basis occurs in approximately 20% of adult women with Turner syndrome (13). Papendieck et al (29a) reported that 55% had thyroid disturbances ($n=49$).

**Differential diagnosis.** Short stature can be observed with *Noonan syndrome,* familial short stature, dyschondrosteosis, type E brachydactyly, growth hormone deficiency, hypothyroidism, glucocorticoid excess, *multiple pterygium syndrome, Klippel–Feil anomaly,* and short stature due to chronic disease. Amenorrhea or failure to begin puberty occurs in pure gonadal dysgenesis, Stein–Leventhal syndrome, and primary or secondary amenorrhea. Lymphedema occurs as Milroy's disease, lymphedema with distichiasis, *Hennekam syndrome,* lymphedema with recurrent cholestasis, and lymphedema with intestinal angiectasia (13). Fetal cystic hygroma may be seen in *fetal alcohol syndrome, trisomy 21, trisomy 18, del(13q), del(18p), trisomy 22 mosaicism,* and *Noonan syndrome.*

**Laboratory aids.** Banded chromosome studies should be carried out when a clinical diagnosis of Turner syndrome is suspected. Fifty cells should be counted to rule out mosaicism. If leukocyte studies are normal but clinical suspicion of Turner syndrome is strong, fibroblast cultures should be carried out. Buccal smears should no longer be used because patients with isochromosome X often show Barr body material, resulting in misdiagnosis (13).

### References (Turner syndrome)


1. Bender B et al: Cognitive development of unselected girls with complete and partial X monosomy. *Pediatrics* **73**:175–182, 1984.

2. Brook CG et al: Height correlations between parents and mature offspring in normal subjects and in subjects with Turner's and Klinefelter's and other syndromes. *Ann Hum Biol* **4**:17–22, 1977.

3. Carr RF et al: Fetal cystic hygroma and Turner's syndrome. *Am J Dis Child* **140**:580–583, 1986.

4. Coco R, Bergada C: Cytogenetic findings in 125 patients with Turner's syndrome and abnormal karyotypes. *J Genet Hum* **25**:95–107, 1977.

4a. Cohen MM Jr: Craniosynostosis in the Turner syndrome. *Am J Med Genet,* in press.

5. De Grouchy J, Turleau C: *Clinical Atlas of Human Chromosomes.* Wiley, New York, 1977.

6. de la Chapelle A: Sex chromosome abnormalities, in *Principles and Practice of Medical Genetics,* Emery EH, Rimoin DL (eds), Churchill Livingstone, Edinburgh, 1983, Ch. 16, pp. 193–215.

7. Dickens JA: Concurrence of Turner's syndrome and anorexia nervosa. *Br J Psychiat* **117**:237, 1970.

8. Engel E, Forbes AP: Cytogenetic and clinical findings in 48 patients with congenitally defective or absent ovaries. *Medicine* **44**:135–164, 1965.

9. Ferguson-Smith MA: Karyotype–phenotype correlations in gonadal dysgenesis and their bearings on the pathogenesis of malformations. *J Med Genet* **2**:142–155, 1965.

10. Filipsson R et al: Time of eruption of the permanent teeth, cephalometric and tooth measurement and sulphation factor activity in 45 patients with Turner's syndrome with different types of X chromosome aberrations. *Acta Endocrinol (Kbh)* **48**:91–113, 1965.

11. Ford CE et al: A sex chromosomal anomaly in a case of gonadal dysgenesis (Turner's syndrome). *Lancet* **1**:711–713, 1959.

12. Gorlin RJ: Classical chromosome disorders, in *New Chromosomal Syndromes,* Yunis JJ (ed), Academic Press, New York, 1977, pp 59–117.

13. Hall JG et al: Turner's syndrome. *West J Med* **137**:32–44, 1982.

14. Horowitz SL, Morishima A: Palatal abnormalities in the syndrome of gonadal dysgenesis and its variants and in Noonan's syndrome. *Oral Surg* **38**:839–844, 1974.

15. Jensen BL: Craniofacial morphology in Turner syndrome. *J Craniofac Genet Dev Biol* **5**:327–340, 1985.

16. Johnson R, Baghdady VS: Maximum palatal height in patients with Turner's syndrome. *J Dent Res* **48**:472–476, 1969.

17. Kajii T et al: Anatomic and chromosomal anomalies in 639 spontaneous abortuses. *Hum Genet* **55**:87–98, 1980.

18. Kron L et al: Anorexia nervosa and gonadal dysgenesis—further evidence of a relationship. *Arch Gen Psychiat* **34**:332–335, 1977.

18a. Leheup BP et al: Otologic signs of Turner syndrome and early diagnosis of Turner syndrome. Reevaluation of 30 cases. *J Génét Hum* **36**:315–321, 1988.

19. Lemli L, Smith DW: The XO syndrome. A study of the differentiated phenotype in 25 patients. *J Pediatr* **63**:577–588, 1963.

20. Lin AE et al: Aortic dilation, dissection, and rupture in patients with Turner syndrome. *J Pediatr* **109**:820–826, 1986.

20a. Linden MG et al: A longitudinal study of Turner syndrome and Turner mosaicism in 14 females: Results, management, and genetic counseling. 39th

Annu Meeting Am Soc Hum Genet, New Orleans, October 12–15, 1988, Abstract No. 236.

21. Linsten J, Fraccaro M: Turner's syndrome, in *Genital Anomalies*, Rashad MN, Morton WRM (eds), Thomas, Springfield, 1969, Ch. 22, pp 396–456.

22. Lubinsky M et al: The association of "prune belly" with Turner's syndrome. *Am J Dis Child* **134**:1171–1172, 1980.

23. Lyon AJ et al: Growth curve for girls with Turner syndrome. *Arch Dis Childh* **60**:932–935, 1985.

24. Matthies F et al: Renal anomalies in Turner's syndrome—types and suggested embryogenesis. *Clin Pediatr (Phila)* **10**:561–565, 1971.

25. McCauley E et al: Psychosocial adjustment of adult women with Turner syndrome. *Clin Genet* **29**:284–290, 1986.

26. Melosky LC: A study of dental facial aspects in individuals with X-chromosome aberrations. Master of Science in Dentistry Thesis, University of Washington, 1966.

27. Miller MJ et al: Echocardiography reveals a high incidence of bicuspid aortic valve in Turner syndrome. *J Pediatr* **102**:47, 1983.

28. Nora JJ et al: The Ullrich–Noonan syndrome (Turner phenotype). *Am J Dis Child* **127**:48–55, 1974.

29. Palmer CG, Reichmann A: Chromosomal and clinical findings in 110 females with Turner syndrome. *Hum Genet* **35**:35, 1976.

29a. Papendieck LG de et al: High incidence of thyroid disturbances in 49 children with Turner syndrome. *J Pediatr* **111**:258, 1987.

30. Pincheira JV et al: 45 XO/49 XYYYY mosaicism in a male with stigmata of Turner's syndrome. *Clin Genet* **24**:384–388, 1983.

31. Rahal F et al: Gonadal dysgenesis associated with a multicystic kidney. *Am J Dis Child* **126**:505–506, 1973.

32. Rainier-Pope CR et al: Cardiovascular malformation in Turner's syndrome. *Pediatrics* **33**:919, 1964.

33. Ranke MB et al: Turner syndrome: Spontaneous growth in 150 cases and review of the literature. *Eur J Pediatr* **141**:81–88, 1983.

34. Risch WD et al: Bone mineral content in patients with gonadal dysgenesis. *Am J Radiol* **126**:1302–1309, 1976.

34a. Rosenfeld RG et al: Three-year results of a randomized prospective trial of methionyl human growth hormone and oxandrolone in Turner syndrome. *J Pediatr* **113**:393–400, 1988.

35. Sarkar R, Marimuthu KM: Association between the degree of mosaicism and the severity of syndrome in Turner mosaics and Klinefelter mosaics. *Clin Genet* **24**:420–428, 1983.

36. Savanelli A et al: Prune belly appearance in a Turner subject. *J Med Genet* **23**:92–93, 1986.

37. Schmid W et al: Cytogenetic findings in 89 cases of Turner's syndrome with abnormal karyotypes. *Hum Genet* **24**:93–104, 1974.

38. Shepard TH et al: Lowered plasma albumin concentration in fetal Turner syndrome. *J Pediatr* **108**:114–116, 1986.

39. Simpson JL: Gonadal dysgenesis and abnormalities of the human sex chromosomes: Current status of phenotypic–karyotypic correlations. *Birth Defects* **11**(4):23–59, 1975.

40. Spiegel RN et al: Cephalometric study of children with various endocrine diseases. *Am J Orthod* **59**:362–375, 1971.

41. Szpunar J, Rybak M: Middle ear disease in Turner's syndrome. *Arch Otolaryngol* **87**:34–40, 1968.

42. Treisman J, Collins FS: Adult Turner syndrome, associated with chylous ascites and vascular anomalies. *Clin Genet* **1**:218–233, 1987.

43. Turner HH: A syndrome of infantilism, congenital webbed neck and cubitus valgus. *Endocrinology* **23**:566–578, 1938.

44. Walinder J, Melibin G: Karyotyping of women with anorexia nervosa. *Br J Psychiat* **130**:48–49, 1977.

45. Wertelecki W et al: Nongonadal neoplasia in Turner's syndrome. *Cancer* **26**:485–488, 1970.

46. Wolf EL et al: Diagnosis of oligohydramnios-related pulmonary hypoplasia (Potter's syndrome): Value of portable voiding cystourethrography in newborns with respiratory distress. *Pediatr Radiol* **125**:769–773, 1977.

## Klinefelter syndrome and its variants

Klinefelter et al (21) reported postpubertal males with small testes and tubular hyalinization, normal numbers of Leydig cells, azospermia, gynecomastia, elevated urinary gonadotropins, and decreased urinary 17-ketosteroids. By 1956, such patients were shown to have Barr bodies (4,28). In 1959, Jacobs and Strong (19) described an XXY sex chromosome complement in chromatin-positive Klinefelter syndrome.

The birth prevalence of chromatin-positive males is approximately 2/1000 and is composed of several X-aneuploidy variants: 47,XXY, 48,XXYY, 46,XY/47,XXY, 48,XXXY, and 49,XXXXY. Approximately 80% are 47,XXY, about 10% represent mosaics, and the remainder have more unusual karyotypes (12). The prevalence of "classic" Klinefelter syndrome (47,XXY) is approximately 1.18/1000. At the time of conception, maternal age has been reported to be advanced (12). Mean maternal age has been reported as 31.3 years and mean paternal age as 35.5 years (7). Robinson et al (30), however, found no increase in parental age. Klinefelter syndrome occurs with a frequency of about 0.5–1.0% in males institutionalized for mental retardation, seizures, or mental illness and in about 10% of males who have sterility. Double aneuploidy of Klinefelter syndrome and Down syndrome occurs with greater frequency than expected by chance (12). Mosaicism has been discussed by several authors (8,9,11,20).

Except for screening, it is unusual for clinical diagnosis to be made during infancy or childhood, except for severe conditions such as 49,XXXXY. Classic Klinefelter syndrome is diagnosed most commonly at the time of puberty, although in some cases clinical cues may be evident in childhood (5).

With increasing X-aneuploidy (47,XXY to 49,XXXXY) several clinical trends are evident: decrease in stature, increase in mental retardation, decrease in total finger ridge count, increase in varicosities, increase in hypostatic leg ulcerations, increased frequency of radioulnar synostosis, increased frequency of mandibular prognathism, and increased frequency of taurodontism (2,6,12,22,27). It should be emphasized that these are statistical trends and need not apply to individual cases.

### 47,XXY

**Growth.** Mean birth weight is 3048 g, which tends to be lower than in normal brothers of Klinefelter infants (30). Until 3 years of age, height distribution in a large series of Klinefelter children is unremarkable, but after 3 years of age, the distribution is skewed, with significantly fewer boys below the 25th percentile than expected. Head circumference distribution during infancy is similar to that found in the general population. However, about 4 years of age, head circumference distribution tends to be skewed to the lower half of the normal curve (30). In adulthood, typical Klinefelter individuals are of average or somewhat above average height. In most human populations, Klinefelter individuals are 2–5 cm taller than average normal males. Mean height in northern Europe is 177.4 cm. Tall stature is primarily the result of an increase in leg length. Increased height from increased leg length is present before puberty, but is not particularly obvious. In approximately 60%, arm span exceeds height by 3 cm or more (8,12,30).

**Central nervous system and performance.** Delayed speech is found in 51%. Delays in emotional development are common (32%), and school maladjustment has been reported in 44%. Poor gross motor coordination is a feature in approximately 27%. During childhood, psychiatric problems do not occur more frequently than in normal individuals. Average IQ is approximately 90. Approximately 29% have IQs below 90. In general, Klinefelter individuals are usually neither highly intelligent nor severely retarded (8,13,30,31).

In adults, there may be disturbances of behavior, deviations in personality, as well as neurotic and psychotic reactions. Antisocial behavior, alcoholism, aggressiveness, depression, and periods of mania have been reported to occur commonly (8). Feelings of inadequacy and poor body image often accompany gynecomastia and testicular atrophy (23). However, most Klinefelter individuals tend to lead a quiet, passive type of existence (8).

Mechanisms by which these disturbances arise are not understood. Reports of cerebral dysfunction on the basis of electroencephalographic abnormalities suggest an organic basis. On the other hand, disturbances may arise from the combination of low normal intelligence and personality damage from abnormal sexual development (8).

Psychosexual orientation is male. Erection, coitus, and ejaculation occur but libido is often subnormal. Many Klinefelter individuals lead normal married lives (8).

Genitalia. Penile size is normal or slightly reduced. Small testes are a constant feature, with adult size measuring 1–2 cm compared with 3.5–4.5 cm in normal males. In nearly all cases, the testes are descended. The scrotum is normal in size and in pigmentation. Approximately 50% have a female pubic escutcheon. The prostate is smaller than normal (8,12,35).

The testes are small, soft, and often insensitive to pressure. Prepubertal testes are of normal size and microscopic appearance, but during adolescence they fail to enlarge. The seminiferous tubules are usually shrunken, hyalinized, and irregularly arranged. Tubules that are not sclerotic are immature and lined exclusively with Sertoli cells. Elastic fibers are absent around the tunica propria of the tubules. Leydig cells are clumped. Rarely, spermatogenesis can be demonstrated. Generally, Klinefelter syndrome is associated with sterility, but in a number of instances, indisputable evidence of paternity has been found (9,11,12,35).

Secondary sexual characteristics. Typically, Klinefelter syndrome patients do not have female fat distribution, high pitched voice, or notably scanty body hair. However, these features are present in many individuals. Facial hair is sparse in 60–90%. Typically untreated Klinefelter individuals shave only once or twice per week. Gynecomastia develops after puberty in approximately 50% (8,12).

Hormones. Leydig cells are defective; plasma testosterone is low in the presence of normal or high follicle-stimulating hormone (FSH) and leutinizing hormone (LH). Typically, patients have 50% or less of normal levels of plasma testosterone and a 4-fold increase in urinary excretion of pituitary gonadotropin (8).

Dermatologic findings. Varicose veins and hypostatic leg ulceration have been reported (2,6). There are no suggestive dermatoglyphic findings, although the total finger ridge count tends to be lower than average because of an increased number of arches (7).

Craniofacial features. Cephalometric investigation shows smaller calvarial size, smaller cranial base angle, and larger gonial angle than normal. Both maxillary and mandibular prognathism tend to occur (16). Alvesalo and Portin (1a) found permanent tooth crowns to be larger in 47,XXY males than in control males. Taurodontism has been reported in some instances (10,22).

Congenital malformations. Major malformations are found in approximately 18%, but no clear patterns have emerged. Findings may include cleft palate, inguinal hernia, cryptorchidism, unilateral renal aplasia, microcephaly, corneal opacity, aortic stenosis, mitral valve prolapse (10a), omphalocele, nerve deafness, hypospadias, pectus excavatum, or scoliosis. Minor anomalies are observed in 26%, clinodactyly being the most common (19%). Other findings may include ear anomalies, single palmar creases, strabismus, external rotation of the legs, "third" fontanel, micropenis, downslanting palpebral fissures, or genu recurvatum (30).

Other findings. Some evidence indicates an increase in pulmonary disorders (12) and, perhaps, lupus erythematosus (36). Diabetes mellitus is present in 8% of adults.

Neoplasia. An increased frequency of carcinoma of the breast has been noted, the risk being 66 times the risk in normal men and approaching the risk in normal women (18). Conversely, Klinefelter syndrome is found in 3.3% of men with breast cancer (14). Leukemia and malignant lymphoma, earlier thought to occur with increased risk, are now known not to be increased compared with the general population (34). Isurugi et al (17) reported seminoma of the testis.

**48,XXYY.** Birth prevalence is 0.04/1000 male births. The frequency is 50 to 100 times greater among males in mental institutions and prisons (7). Most likely, the disorder is caused by nondisjunction in both the first and second meiotic divisions during spermatogenesis, with the production of an XYY sperm. The less likely possibility of nondisjunction at the second meiotic division in both parents must be rare. Advanced parental age has not been observed (12).

In general, individuals with 48,XXYY karyotype tend to be approximately 4 cm taller, more aggressive, and more mentally retarded than those with 47,XXY karyotype. Characteristic features include small testes, eunuchoid habitus, sparse body hair, gynecomastia, and elevated gonadotropins (12). Dermatoglyphic studies have shown that fingertip arch patterns are more common in 48,XXYY individuals. Thus, the total finger ridge count is low (7).

**48,XXXY.** The condition is rare (8) and arises from successive nondisjunction in either maternal or paternal meiotic divisions (12).

Mental retardation is a constant finding. The penis is hypoplastic in 50% and gynecomastia is observed in approximately 35%. Other findings include facial asymmetry, epicanthic folds (25%), ocular hypertelorism, protruding lips, mandibular prognathism, short neck, radioulnar synostosis (10%), clinodactyly of the fifth finger (30%), coxa valga, and other abnormalities (12).

**49,XXXXY.** Over 100 cases of 49,XXXXY syndrome have been reported (3a). Postzygotic nondisjunction in an XXY zygote appears to be the cause for the XXXXY state, all the X chromosomes being of maternal origin. Parental age does not appear to be advanced (12). Sarto et al (32) suggested that two hypotheses seemed to explain best the abnormal phenotype that accompanies 49,XXXXY karyotype. First, the number of always active regions (tip of $X_p$) and of the possibly always active regions (Q dark regions on both sides of the centromere) is increased from one to four. Second, the replication pattern of the late-replicating X chromosomes is highly asynchronous, which might affect the phenotype.

Average birth weight is approximately 2500 g. Height is often below the third percentile and bone age is delayed in 89%. However, Borghgraef et al (3a) reported small for dates infants who had significant catchup growth, being at the 50th to 75th percentiles after age 4 years. Severe mental retardation is found, with IQs usually ranging from 20 to 60. Affected individuals may be extremely shy and timid (3a). Hypotonia, joint laxity, or both are found in approximately 33% (7, 12, 24).

The phenotype is distinctive (Fig. 3–13). Hypogonadism is severe, with pea-sized testes, micropenis, and pronounced infantilism of secondary sex characteristics. The scrotum is usually hypoplastic (79%) and the testes may be cryptorchid (24%). Histologically, Leydig cells are hypoplastic and germ cells are absent (3,7,8,12,29).

Clinical features (Table 3–17) include mild microcephaly, ocular hypertelorism (30%), upslanting palpebral fissures (71%), epicanthic folds (85%), strabismus (57%), myopia (25%), low nasal bridge sometimes with upturned nasal tip (96%), poorly modeled ears (78%), and short neck, sometimes with webbing (7,12,15,24,26).

During infancy, the face is often rounded. With growth, the midface appears retruded with relative mandibular prognathism (47%) (12). Taurodontism is a common finding (22).

Skeletal anomalies (Table 3–17) occur in over half of the cases and include sclerotic cranial sutures (57%), thick sternum, radioulnar synostosis, cubitus valgus, elongation of the distal ulna and proximal radius, wide proximal ulna, pseudoepiphyses of metacarpals and metatarsals, hypoplasia of the middle phalanx of the fifth digit with clinodactyly (93%), coxa valga (84%), genua valga (13%), and pes planus (54%). Other findings may include malformed cervical vertebrae, thoracic kyphosis, and scoliosis (12,24,25,26,33).

Congenital heart defects, particularly patent ductus arteriosus, are present in approximately 18% (12,24). Gynecomastia is not a feature of 49,XXXXY individuals. Gonadotropins are not elevated (12). A low total finger ridge count with an increased number of fingertip arch patterns may be observed (7).

**46,XX males.** Suggested birth prevalence is 1/25,000 newborn males. Phenotypic features are very similar to 47,XXY Klinefelter syndrome with two major differences. First, mean height attainment (168.2 cm)

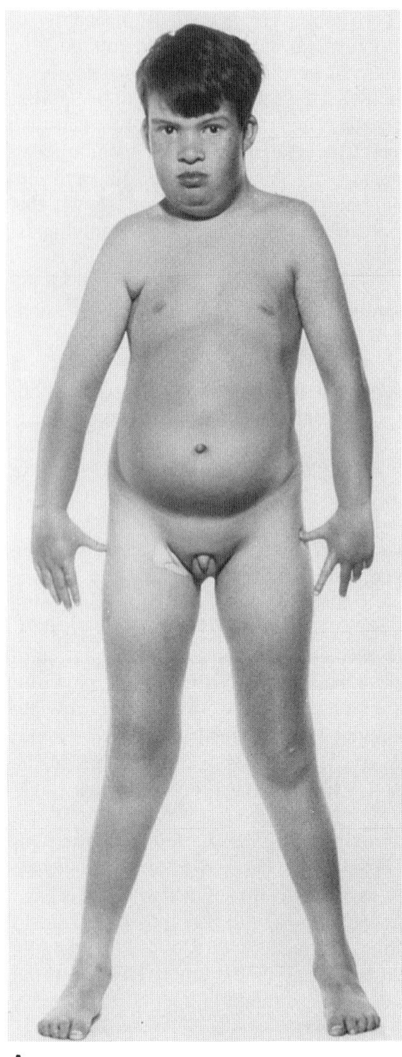
A

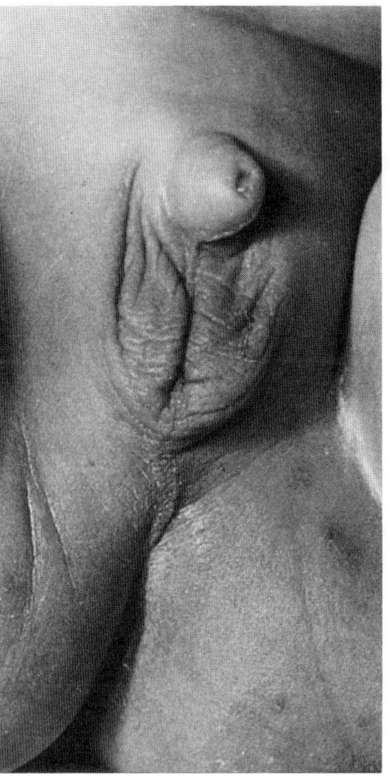
B

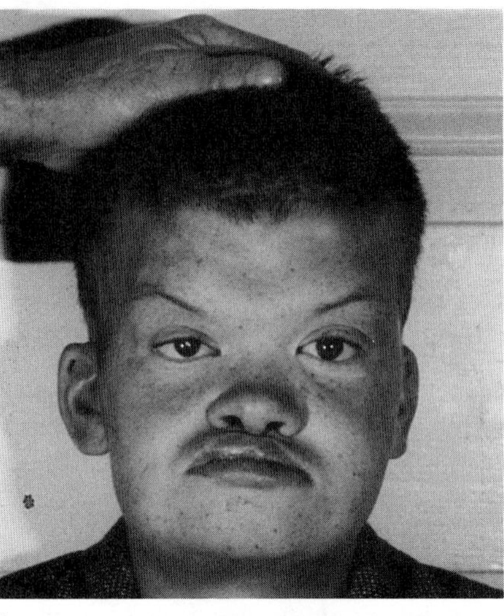
C

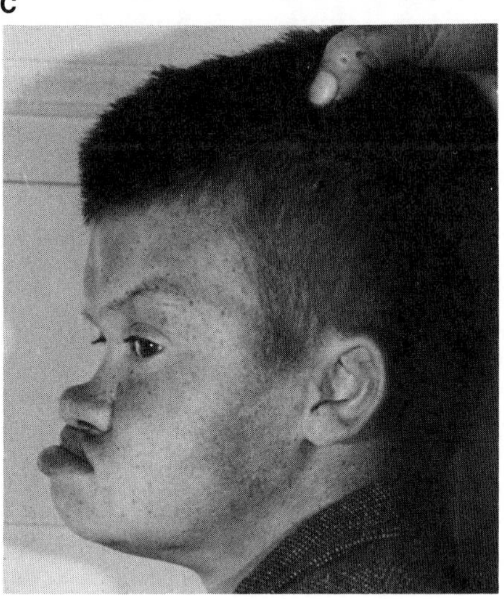
D

Fig. 3–13. *49,XXXXY Klinefelter syndrome.* (A) Upslanting palpebral fissures, small ears, hypogenitalism, and cubitus valgus. (B) Hypogenitalism. (C,D) Marked mandibular prognathism. (A from *MC Joseph et al,* J Med Genet **1**:95, 1964. B from *RA Pfeiffer,* Z Kinderheilkd **87**:356, 1962. C,D courtesy of *H Schade,* Münster, Germany.)

is below that of 47,XXY subjects (177.4 cm). Second, disproportion between trunk and limbs found in 47,XXY is not found in 46,XX males. Other differences are less striking. Gynecomastia occurs slightly less frequently than in classic Klinefelter syndrome (7,8). Measurements of permanent tooth size indicate that teeth in 46,XX males are smaller than those of normal males, being similar in size to those of normal females (1).

At present, no single hypothesis explains the 46,XX male condition, which may turn out to be etiologically heterogeneous. One possibility is undetected mosaicism involving a Y chromosome-containing cell line. A second possibility involves translocation of the testis-determining gene from the short arm of the Y chromosome to the X chromosome or to an autosome during paternal meiosis. Most cases are due to Y–X interchange. This hypothesis is supported by observing an increased length in the short arm of one X chromosome in some 46,XX males, although short arm length of X chromosomes is normal in others. Finally, testes and maleness in 46,XX individuals might result from a gene mutation, suggested by the small number of families with more than one 46,XX male (8).

**Differential diagnosis.** Differential diagnosis includes eunuchoidism, *homocystinuria,* and *47,XYY syndrome*. Patients with 49,XXXXY syndrome are distinctive clinically, but sometimes suggest *trisomy 21 syndrome* in fetuses (29) and newborns, although the overall pattern of anomalies is at variance with the latter.

Clinical clues leading to the detection of classic 47,XXY Klinefelter syndrome during childhood include dull mentality, school or behavioral problems, altered body habitus with relatively long legs and slim build, and small testes and small or inadequately developed phallus (5).

**Laboratory aids.** Banded chromosome study is essential when a clinical diagnosis of Klinefelter syndrome is suspected. Fifty cells should be counted to rule out mosaicism.

#### References (Klinefelter syndrome and its variants)


1. Alvesalo L, de la Chapelle A: Permanent tooth sizes in 46,XX-males. *Ann Hum Genet* **43**:97–102, 1979.

1a. Alvesalo L, Portin P: 47,XXY males: Sex chromosomes and tooth size. *Am J Hum Genet* **32**:955–959, 1980.

2. Andersen KE: Sex chromosomal anomalies: A possible association with leg ulcers. *Clin Exp Dermatol* **4**:223–226, 1979.

3. Autio-Harmainen H et al: Fetal gonadal histology in XXXXY, XYY and XXX syndromes. *Clin Genet* **18**:1–5, 1980.

3a. Borghgraef M et al: The 49,XXXXY syndrome. Clinical and psychological follow-up data. *Clin Genet* **33**:429–434, 1988.

Table 3–17. Clinical and radiological features of 49,XXXXY syndrome[a]

| Feature | Percentage |
|---|---|
| Craniofacial | |
| Ocular hypertelorism | 30 |
| Upslanting palpebral fissures | 71 |
| Epicanthic folds | 85 |
| Strabismus | 57 |
| Broad flat nose | 96 |
| Mandibular prognathism | 47 |
| Malformed ears | 78 |
| Central nervous system | |
| Mental retardation | 100 |
| Cardiac abnormalities | 18 |
| Genitalia | |
| Hypogonadism | 91 |
| Small penis | 79 |
| Abnormal scrotum | 79 |
| Cryptorchidism | 24 |
| Limb anomalies | |
| Limitation of elbow movement | 89 |
| Radioulnar synostosis | 32 |
| Clinodactyly, fifth finger | 93 |
| Coxa valga | 84 |
| Genua valga | 13 |
| Gap between hallux and second toe | 55 |
| Pes planus | 54 |
| Other skeletal findings | |
| Retarded bone age | 89 |
| Sclerotic cranial sutures | 57 |
| Capitate defect | 83 |
| Thoracic kyphosis | 53 |
| Scoliosis | 35 |

[a]Modified from CL Levy et al, *J Med Genet* **15**:301–316, 1978.


4. Bradbury JT et al: Chromatin test in Klinefelter's syndrome. *J Clin Endocrinol Metab* **16**:689, 1956.

5. Caldwell PD, Smith DW: The XXY (Klinefelter's) syndrome in childhood: Detection and treatment. *J Pediatr* **80**:250–258, 1972.

6. Campbell WA et al: Hypostatic leg ulceration and Klinefelter's syndrome. *J Ment Defic Res* **24**:115–117, 1980.

7. De Grouchy J, Turleau C: *Clinical Atlas of Human Chromosomes.* Wiley, New York, 1977.

8. de la Chapelle A: Sex chromosome abnormalities, in *Principles and Practice of Medical Genetics,* Emory EH, Rimoin DL (eds), Churchill Livingstone, Edinburgh, 1983, Ch. 16, pp 193–215.

9. Donlan, MA et al: Brief clinical report: Trisomy Xq in a male: The isochromosome X Klinefelter's syndrome. *Am J Med Genet* **27**:189–194, 1987.

10. Feichtinger C, Rossiwall B: Taurodontism in human sex chromosome aneuploidy. *Arch Oral Biol* **22**:327–329, 1977.

10a. Fricke GR et al: Mitral prolapse in Klinefelter syndrome. *Lancet* **2**:1414, 1981.

11. Gordon DL et al: Pathologic testicular findings in Klinefelter's syndrome. *Arch Intern Med* **130**:726–729, 1972.

12. Gorlin RJ: Classical syndrome disorders, in *New Chromosomal Syndromes,* Yunis JJ (ed), Academic Press, New York, 1977, pp 59–117.

13. Haka-Ikse K et al: Early development of children with sex chromosome aberrations. *Pediatrics* **62**:761–766, 1978.

14. Harnden DG et al: Carcinoma of the breast and Klinefelter's syndrome. *J Med Genet* **8**:460–461, 1971.

15. Hecht F: Letter to the editor: Observations on the natural history of 49,XXXXY individuals. *Am J Med Genet* **13**:335–336, 1982.

16. Ingerslev CH, Kreiborg S: Craniofacial morphology in Klinefelter syndrome: A roentgencephalometric investigation. *Cleft Palate J* **15**:100–108, 1978.

17. Isurugi K et al: Seminoma in Klinefelter's syndrome with 47,XXY, 15s+ karyotype. *Cancer* **39**:2041–2047, 1977.

18. Jackson AW et al: Carcinoma of male breast in association with the Klinefelter's syndrome. *Br Med J* **1**:223–225, 1965.

19. Jacobs PA, Strong JA: A case of human intersexuality having a possible XXY sex-determining mechanism. *Nature (London)* **183**:302–303, 1959.

20. Kardon NB et al: 47,XXY/48,XXXY/49,XXXXY mosaicism in a 4-year-old child. *Am J Dis Child* **122**:160–162, 1971.

21. Klinefelter HF Jr et al: Gynecomastia, aspermatogenesis without aleydigism and increased excretion of follicle-stimulating hormone. *J Clin Endocrinol* **2**:615–627, 1942.

22. Komatz Y et al: Taurodontism and Klinefelter's syndrome. *J Med Genet* **15**:452–454, 1978.

23. Kvale JN, Fishman JR: The psychosocial aspects of Klinefelter's syndrome. *JAMA* **193**:567–572, 1965.

24. Levy CL et al: Chromosome banding studies in two patients with XXXXY syndrome. *J Med Genet* **15**:301–316, 1978.

25. Ohsawa T et al: Roentgenographic manifestations of Klinefelter's syndrome. *Am J Roentgenol* **112**:178–184, 1971.

26. Pallister PD: Letter to the editor: 49,XXXXY syndrome. *Am J Med Genet* **13**:337–339, 1982.

27. Peterson WC et al: Cutaneous aspects of the XXYY genotype. *Arch Dermatol* **94**:695–698, 1966.

28. Plunkett ER, Barr ML: Testicular dysgenesis affecting the seminiferous tubules principally, with chromatin positive males. *Lancet* **2**:853–857, 1956.

29. Rehder H et al: The fetal pathology of the XXXXY-syndrome. *Clin Genet* **30**:213–218, 1986.

30. Robinson A et al: Summary of clinical findings: Profiles of children with 47,XXY, 47,XXX and 47,XYY karyotypes. *Birth Defects* **15**(1):261–266, 1979.

31. Salbenblatt JA et al: Gross and fine motor development in 47,XXY and 47,XYY males. *Pediatrics* **80**:240–244, 1987.

32. Sarto GE et al: What causes the abnormal phenotype in a 49,XXXXY male? *Hum Genet* **76**:1–4, 1987.

33. Schmidt R et al: Epiphysial dysplasia: A constant finding in the XXXXY syndrome. *J Med Genet* **15**:282–287, 1978.

34. Sohn K-Y, Boggs DR: Klinefelter's syndrome, LSD usage and acute lymphoblastic leukemia. *Clin Genet* **6**:20–22, 1974.

35. Topper E et al: Puberty in 24 patients with Klinefelter syndrome. *Eur J Pediatr* **139**:8–12, 1982.

36. Tsung SH, Heckman MG: Klinefelter syndrome, immunological disorders, and malignant neoplasm: Report of a case. *Arch Pathol* **98**:351–354, 1974.


## 47,XYY males

The 47,XYY chromosomal constitution was first observed by Sandberg et al (27) in 1961; the subject was a man of normal intelligence who had, with two different women, numerous progeny, including an amenorrheic female, twins (one with trisomy 21, the other a "blue baby"), and two spontaneous abortions (13). The 47,XYY karyotype has aroused more public attention than any other chromosomal abnormality because of the many reports of criminal behavior, most of which have a known ascertainment bias favoring behavioral disability (3,5,8,14,16,21,23,24,33). Birth prevalence is approximately 1/1000 male births (10). Sumner et al (31), utilizing a fluorescent technique, showed that over 1% of sperm from normal males contained two Y chromosomes, implying marked selection against such sperm.

**Growth.** Tall stature in adult life is characteristic. Mean final height attainment varies from 180 to 186 cm (11). Leg length and trunk length are increased, but the leg/trunk ratio is normal (17). Birth weights and birth lengths are normal (26). In a study of 43 affected boys (26), overall distribution of growth percentiles was not significantly different from normal, although one subset from Edinburgh (25) suggested a growth spurt between ages 2 and 6 years in approximately one-third of the cases. By 5 years, all boys were above the 50th percentile for height and approximately 38% were above the 90th percentile.

**Central nervous system and performance.** Approximately one-third of affected individuals have delayed speech or language development and there is some evidence of fine motor problems (26). Muscle weakness and poor coordination are commonly noted (9). In a study of 43 affected subjects (26), IQ scores ranged from 78 to 145. Approximately 38% showed IQs from 70 to 89. Thus, average IQ is lower than in normal males and verbal IQs tend to be more affected than performance IQs.

Males with 47,XYY chromosomal constitution are common and most probably blend into the general population as normal individuals. The most frequently observed stigmatizing features are excessive height for age, excessively impulsive behavior, and excessive temper tantrums in childhood. Stigmatization does not correlate with socioeconomic class (23). Psychological studies have shown that infantilism, lack of emotional control, increased impulsiveness after emotional stimulation, and weak sense of self are so characteristic of 47,XYY men that they can be recognized by psychological tests alone (24).

Even correcting for earlier biased ascertainment from mental institutions and prisons, it is now well established that among 47,XYY males, an excess degree of criminal behavior exists compared to 46,XY males. For example, Daly and Harley (8) karyotyped 3011 males from five Wisconsin state correctional institutions and found a 47,XYY frequency of 1%, which is five times greater than the newborn prevalence for 47,XYY males. Crimes are similar to those committed by 46,XY men. The preponderance of serious aggressive behavior against people has not been substantiated. Rather, theft, arson, and burglary are the crimes most often cited. The cause of antisocial behavior leading to conflict with the law is not entirely resolved and still much debated. None of the suggested causes, which include impaired intellectual function, abnormal electroencephalographic findings in a few instances, low socioeconomic status, or very tall stature, either separately or together, explains the increased risk of antisocial behavior (11).

**Gonadal status and fertility.** Gonadal development, testicular size, and testicular histology are normal (7,10,11). Many 47,XYY males have fathered offspring who are chromosomally normal. Pregnancies resulting from XYY individuals have ended in miscarriage or perinatal deaths, or have produced offspring with various chromosomal abnormalities (12a). In rare instances, subjects have procreated 47,XYY sons (22,32). In a few instances, however, small testes, decreased spermatogenesis, subfertility, and sterility have occurred. Cryptorchidism, micropenis, or hypospadias are rarely observed (10).

**Congenital malformations.** In a study of 43 affected infants, Robinson et al (26) found the overwhelming majority to be completely normal in appearance. Major malformations were not increased, although approximately 20% had minor anomalies. Although no clear pattern of minor anomalies emerged, clinodactyly with single fifth finger crease, inguinal hernia, and abnormal ears were observed in two instances each. Other minor anomalies and subtle phenotypic alterations have been noted, including mild facial asymmetry, mild pectus carinatum or excavatum, mild winging of the scapula, glabellar mounding, long ears, highly arched palate, and bony chin point (4,9,26). A number of major abnormalities have also been noted including urinary tract malformations (20) and radioulnar synostosis (6).

**Dermatologic findings.** Nodulocystic acne involving the face, chest, and back has been reported in association with 47,XYY subjects (36).

**Dentition.** Careful measurement of teeth has indicated that tooth size in 47,XYY males is larger than normal in both the deciduous and permanent dentitions (1,2). In a study of shovel-shaped maxillary incisors, affected central incisors were similar to normal controls, but lateral incisors were more shoveled and showed deeper lingual fossae in 47,XYY subjects than in control subjects (18).

**Differential diagnosis.** Differential diagnosis includes *Klinefelter syndrome* and *Marfan syndrome* (12). Males with *48,XXYY karyotype* have features that combine Klinefelter syndrome with above-normal stature and frequent mental deficiency. Several poly-Y karyotypes have been reported, including 48,XYY (15,28,34) and 49,XYYYY (29,30,35). Townes et al (34) described a 48,XYYY patient with mild mental deficiency, inguinal hernia, cryptorchidism, valvular pulmonic stenosis, and single palmar creases. Schoepflin and Centerwall (28) described a 48,XYY patient with mental retardation, impulsive aggressive behavior, single palmar creases, clinodactyly, delayed bone age, pseudoepiphyses at the bases of the metacarpals and metatarsals, and lack of patellar epiphyseal calcification. Hunter and Quaife (15) noted a 48,XYYY individual with no stigmata other than sterility. Sirota et al (29,30) reported a 49,XYYYY patient with trigonocephaly, upslanting palpebral fissures, epicanthic folds, highly arched palate, micrognathia, low-set ears, limitation of motion at the elbows and knees, and an IQ of 50. Mosaic 45,X/49,XYYYY was documented by van den Berghe et al (35). Findings included mental retardation, facial asymmetry, cataracts, and clinodactyly.

**Laboratory findings.** Karyotypic findings are usually unrelated to the reason for testing. Whether discovery occurs prenatally, during childhood, or later, it is a formidable challenge to counselors and patients alike. The ethical controversies of XYY screening are discussed elsewhere (19).

### References (47,XYY males)


1. Alvesalo L et al: The 47,XYY male, Y chromosome, and tooth size. *Am J Hum Genet* **27**:53–61, 1975.
2. Alvesalo L, Kari M: Size and deciduous teeth in 47,XYY males. *Am J Hum Genet* **29**:486–489, 1977.
3. Bartlett DJ et al: Chromosomes of male patients in a security prison. *Nature (London)* **219**:351–354, 1968.
4. Baughman FA, Mann JD: Ascertainment of seven YY males in a private neurology practice. *JAMA* **222**:446–448, 1972.
5. Borgaonkar DA, Shah SA: The XYY chromosome male—or syndrome? *Progr Med Genet* **10,** 135–222, 1974.
6. Cleveland WW et al: Radioulnar synostosis, behavioral disturbance, and XYY chromosomes. *J Pediatr* **74**:103–106, 1969.
7. Court Brown WM: Males with an XYY sex chromosome complement. *J Med Genet* **5**:341–359, 1968.
8. Daly R, Harley JP: Frequency of XYY males in Wisconsin state correctional institutions. *Clin Genet* **18**:116–122, 1980.
9. Daly RF et al: The XYY condition in childhood: Clinical observations. *Pediatrics* **43**:852–857, 1969.
10. De Grouchy J, Turleau C: *Clinical Atlas of Human Chromosomes.* Wiley, New York, 1977.
11. de la Chapelle A: Sex chromosome abnormalities, in *Principles and Practice of Medical Genetics,* Emory EH, and Rimoin DL (eds), Churchill Livingstone, Edinburgh, 1983, Ch. 16, pp 193–215.
12. Dignan P StJ et al: Arachnodactyly (Marfan's syndrome) with XYY karyotype. *Am J Dis Child* **124**:266–270, 1972.
12a. Grass F et al: Reproduction in XYY males: Two new cases and implications for genetic counseling. *Am J Med Genet* **19**:553–560, 1984.
13. Hauschka JS et al: An XYY man with progeny indicating familial tendency to non-disjunction. *Am J Hum Genet* **14**:22–30, 1962.
14. Hook EB, Kim D-S: Prevalence of XYY and XXY karyotypes in 337 nonretarded young offenders. *N Engl J Med* **283**:410–411, 1970.
15. Hunter H, Quaife R: A 48XYYY male: A somatic and psychiatric description. *J Med Genet* **10**:80–82, 1973.
16. Jacobs PA et al: Chromosome studies on men in a maximum security hospital. *Ann Hum Genet* **31**:339–347, 1968.
17. Keutel J, Dauner I: XYY Status bei Kindern. *Z Kinderheilkd* **106**:314–332, 1969.
18. Kirveskari P, Alvesalo L: Shovel shape of maxillary incisors in 47,XYY males. *Proc Finn Dent Soc* **77**:79–81, 1981.
19. Kopelman L: Ethical controversies in medical research: The case of XYY screening. *Persp Biol Med* **21**:196–204, 1978.
20. Machin GA: Urinary tract malformation in the XYY male. *Clin Genet* **14**:370–372, 1978.
21. Marinello MJ et al: A study of the XYY syndrome in tall men and juvenile delinquents. *JAMA* **208**:321–325, 1969.
22. Melnyck J et al: Failure of transmission of the extra chromosome in subjects with 47,XYY karyotype. *Lancet* **2**:797–798, 1969.
23. Money J et al: XYY syndrome, stigmatization, social class, and aggression. Study of 15 cases. *South Med J* **68**:1536–1542, 1975.
24. Noel B et al: The XYY syndrome: Reality or myth? *Clin Genet* **5**:387–394, 1974.
25. Ratcliffe SG et al: The Edinburgh study of growth and development in children with sex chromosome abnormalities. *Birth Defects* **15**(1):243–260, 1979.
26. Robinson A et al: Summary of clinical findings: Profiles of children with 47,XXY, 47,XXX and 47,XYY karyotypes. *Birth Defects* **15**(1):261–266, 1979.

27. Sandberg AA et al: An XYY human male. *Lancet* **2**:488–489, 1961.
28. Schoepflin C, Centerwall WR: 48,XYYY, a new syndrome. *J Med Genet* **9**:356–360, 1972.
29. Sirota L et al: 49,XYYYY—a case report. *Clin Genet* **19**:87–93, 1981.
30. Sirota L et al: Neurodevelopmental and psychological aspects in a child with 49,XYYYY karyotype. *Clin Genet* **30**:471–474, 1986.
31. Sumner AT et al: Distinguishing between X, Y, and YY-bearing human spermatozoa by fluorescence and DNA content. *Nature (London)* **229**:231–233, 1971.
32. Sundequist U, Hellstrom E: Transmission of 47,XYY karyotype. *Lancet* **2**:1367, 1969.
33. Telfer MA et al: Incidence of gross chromosomal errors among tall criminal American males. *Science* **159**:1249–1250, 1968.
34. Townes PL et al: A patient with 48,XYYY. *Lancet* **1**:1041–1044, 1965.
35. van den Berghe H et al: A male with 4Y chromosomes. *J Clin Endocrinol Metab* **28**:1370–1372, 1968.
36. Voorhees JJ et al: Nodulocystic acne as a phenotypic feature of the XYY genotype. *Arch Dermatol* **105**:913–919, 1972.

## 47,XXX, 48,XXXX and 49,XXXXX syndromes

**47,XXX.** No specific phenotype is characteristic and no increase in major malformations has been observed. In a study of 43 females with 47,XXX (15), the overwhelming majority of infants appeared normal at birth and no unusual karyotype would have been suspected. Minor anomalies have been noted in some instances, and an increased frequency of epicanthic folds and clinodactyly was reported by Robinson et al (15). Judisch and Patil (10) described a case with bilateral retinoblastoma and pinealoma.

In general, birthweights tend to be low but within normal limits. Height percentiles increase with age (15,17), most being above the 50th percentile by age 6. In contrast, head circumference values tend to be skewed to the left with most below the 50th percentile. Frank microcephaly has been noted in an occasional instance (15).

Delay in both receptive and expressive language is common (8,9,11a,15). Full scale IQs tend to be significantly lower than normal (15). Fryns et al (5) found mental deficiency as the presenting symptom in over 25%, but noted that his sample was biased. Deficit in gross motor skills has been reported (8). Also encountered have been emotional immaturity, social problems, learning disorders, and psychosis (5,8,9,11a).

Sexual development is normal, as a rule. However, fertility may range from normal to complete infertility with streak gonads (5). Some triple-X patients suffer from recurrent urinary tract infections (5).

**48,XXXX.** More than 20 cases have been recorded to date (7,18). No specific phenotype is associated with tetra-X females, although mental deficiency appears to be a constant feature, IQs averaging 55 with a range of 30–80. Speech and behavioral problems are frequently encountered. No increased frequency in congenital heart defects has been observed (11). Facial anomalies have included midface hypoplasia, mild hypertelorism, epicanthic folds, and mild micrognathia (7,18). On occasion, the face has been suggestive of trisomy 21 syndrome. A patient reported by Fryns et al (5) had features of Turner syndrome, although final height attainment was normal. Occasionally, clinodactyly of the fifth finger and radioulnar synostosis have been observed. The shoulder girdle may be narrow. Development of secondary sexual characteristics is incomplete with small breasts, scanty axillary and pubic hair, and frequently hypoplastic external genitalia. Menarche may occur spontaneously, but disturbances in the menstrual cycle have been noted. Ovarian tissue may be normal in some instances, although bilateral gonadal agenesis has also been reported (5).

**49,XXXXX.** Approximately 20 cases (at least 3 mosaic) of the penta-X syndrome have been described to date (1–6,12–14,16,19,20). The phenotype is characterized by short stature, severe mental deficiency, upslanting palpebral fissures, and radioulnar synostosis. Some patients are initially thought to have Down syndrome. Several reviews of clinical findings are available (6,19). Features of the penta-X syndrome appear in Table 3–18.

Table 3–18. Features of the penta-X syndrome[a]

| Feature | Percentage ($n = 20$)[b] |
|---|---|
| Growth | |
| Low birthweight | 55 |
| Failure to thrive | 35 |
| Short stature | 65 |
| Delayed bone age | 30 |
| Small head circumference | 55 |
| Performance | |
| Mental or psychomotor retardation | 80 |
| Craniofacial | |
| Upslanting palpebral fissures | 60 |
| Flat nasal bridge | 55 |
| Ears abnormal in position or structure | 65 |
| Lips everted and furrowed/thick | 15 |
| Dental abnormalities | 50 |
| Cleft palate | 10 |
| Micrognathia | 25 |
| Short neck | 45 |
| Low hairline | 20 |
| Limbs | |
| Radioulnar synostosis/abnormal elbows | 45 |
| Joint hyperflexion or dislocations | 35 |
| Micromelia | 30 |
| Camptodactyly/clinodactyly | 75 |
| Finger dermal ridge hypoplasia | 40 |
| Genua valga | 20 |
| Talipes | 25 |
| Metatarsus varus | 5 |
| Cardiac | |
| Congenital heart defect | 40 |
| Genitourinary | |
| Small uterus/abnormal ovaries | 25 |
| Renal hypoplasia | 10 |

[a]Adapted from SJ Funderburk et al, *Am J Med Genet* **8**:27, 1981.
[b]Features considered present only when specifically mentioned by authors.

Growth deficiency of prenatal onset with failure to thrive and short stature are common (6). Delayed bone age has been reported. The head circumference is small in about 55% and mental or psychomotor retardation occurs in approximately 80%. Mental deficiency is moderate to severe (6,13). Hypotonia has been reported (5,14).

The craniofacial appearance can be striking with upslanting palpebral fissures, flat nasal bridge, abnormal ears, and short neck (6). The hairline may be low. Ocular findings have included hypertelorism, epicanthic folds, ptosis of the eyelids, and iris coloboma (5,6,13,16). Small ears have been described and several reports mention ear tags (5,13). Hearing loss has been noted (13). Other features have been recorded in some patients, such as everted and furrowed or thick lips, cleft palate, and micrognathia. Dental abnormalities include malocclusion, hypodontia, and taurodontism (1,6,20).

The phenotypic appearance may be striking with narrow shoulders and genua valga (Fig. 3–14) (1,2,16). Radioulnar synostosis occurs in about 45% and joint hyperflexion or dislocations are found in approximately 35%. The patient reported by Dryer et al (4) had multiple dislocations suggestive of Larsen syndrome. Camptodactyly and/or clinodactyly are common, occurring in about 75%. Other reported findings have included micromelia, low total finger ridge count, clubfoot, metatarsus varus, and malposed toes (5,6).

Congenital heart defects occur in about 40%, PDA and VSD being particularly common (11,13,20). Small uterus and ovarian agenesis have been recorded in some cases. Delayed puberty has been noted (6,13,19). The kidneys have been reported to be hypoplastic in some instances (6,19). In one patient, pyelonephritis and renal failure were described (6). Moedjono et al (12) reported ketotic hypoglycemia.

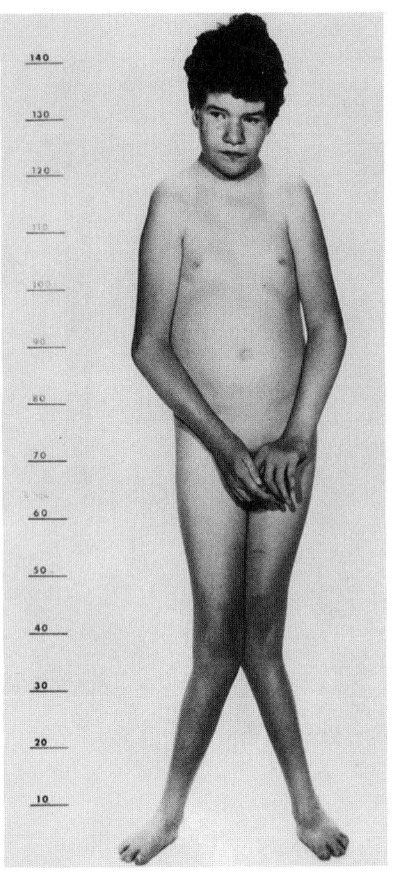

Fig. 3–14. *49,XXXXX syndrome.* Narrow shoulders and genua valga.

### References (47,XXX, 48,XXXX, and 49,XXXXX syndromes)


1. Archidiacono N et al: X Pentasomy: A case and review. *Hum Genet* **52**:69–77, 1979.
2. Berger R et al: Syndrome 49,XXXXX. *Ann Pédiatr* **20**:965–967, 1973.
3. Brody J et al: A female child with five X chromosomes. *J Pediatr* **70**:105–109, 1967.
4. Dryer RF et al: Pentasomy X with multiple dislocations. *Am J Med Genet* **4**:313–321, 1979.
5. Fryns JP et al: X-chromosome polysomy in the female: Personal experience and review of the literature. *Clin Genet* **23**:341–349, 1983.
6. Funderburk SJ et al: : Pentasomy X: Report of patient and studies of X-inactivation. *Am J Med Genet* **8**:27–33, 1981.
7. Gardner RJM et al: XXXX syndrome: Case report, and a note on genetic counselling and fertility. *Humangenetik* **17**:323–330, 1973.
8. Haka-Ikse K et al: Early development of children with sex chromosome aberrations. *Pediatrics* **62**:761–765, 1978.
9. Hier DB et al: Learning disorders and sex chromosome aberrations. *J Ment Defic Res* **24**:17–26, 1980.
10. Judisch GF, Patil SR: Concurrent heritable retinoblastoma, pinealoma, and trisomy X. *Arch Ophthalmol* **99**:1767–1769, 1981.

Keane JF et al: : Congenital heart disease in a tetra-X woman. *Chest* **66**:726–729, 1974.

11a. Linden MG et al: 47,XXX: What is the prognosis? *Pediatrics* **82**:619–630, 1988.

12. Moedjono S et al: : Ketotic hypoglycemia in the penta X and other chromosome imbalance syndromes. *Clin Genet* **14**:367–369, 1978.
13. Monheit A et al: The penta-X syndrome. *J Med Genet* **17**:392–396, 1980.
14. Mulcahy MT, Stevens JB: Pentasomy of X chromosome. *Lancet* **II**:1213–1214, 1975.
15. Robinson A et al: Summary of clinical findings: Profiles of children with 47,XXY, 47,XXX and 47,XYY karyotypes. *Birth Defects* **15**(1):261–266, 1979.
16. Sergovich F et al: The 49,XXXXX chromosome constitution: Similarities to the 49,XXXXY condition. *J Pediatr* **78**:285–290, 1971.
17. Stewart DA et al: Growth and development of children with X and Y chromosome aneuploidy: A prospective study. *Birth Defects* **15**(1):75–114, 1979.
18. Telfer MA et al: Divergent phenotypes among 48,XXXX and 47,XXX females. *Am J Hum Genet* **22**:326–335, 1970.
19. Toussi T et al: Brief clinical report: Renal hypodysplasia and unilateral ovarian agenesis in the penta-X syndrome. *Am J Med Genet* **6**:153–162, 1980.
20. Varrela J, Alvesalo L: Taurodontism in females with extra X chromosomes. *J Craniofac Genet Dev Biol* **9**:129–133, 1989.


## Triploidy syndrome

Triploidy is a frequent cause of fetal wastage (20%) prior to and during the second trimester of intrauterine development (1a,2). Most newborns have died within the first few days of life, but some survive into early infancy. It has been estimated that true triploidy occurs in approximately 1/2500 births (2,9), about 2% of all recognized conceptions. About 60% are 69,XXY with the rest mostly 69,XXX. The 69,XYY state is rare, constituting no more than 3%.

Over 35 well-documented cases of liveborn true triploidy and 20 cases of diploid/triploid mosaicism have been reported to date (2). In general, the latter condition is less severe, more compatible with life, and harder to diagnose clinically (2). Peripheral lymphocytes may be normal. Fibroblast culture may be required to show the chromosomal changes. The triploid cell population appears to be largely paternally derived, either from dispermy (65%) or from faulty meiotic division in the male (25%). Far less often, errors in maternal meiosis (10%) or mitotic errors in germ cell precursors are responsible (7,13,14,16,19,26,28).

In true triploidy, polyhydramnios is common. The placenta is usually large with hydatidiform degeneration in 65%. Mosaic triploid pregnancy complications are unusual. Although the placenta may be large, there is no hydatidiform degeneration. About 70% exhibit birth weight below 1900 g. Pronounced asymmetry has been observed in about 50% with either pure or mosaic triploidy. This may lead to the misdiagnosis of *Silver–Russell syndrome* (17a). The degree of asymmetry does not reflect the degree of mosaicism. Hypotonia has been documented in about 60%. Syndactyly of the third and fourth fingers (Fig. 3–15) and variable syndactyly of toes, transverse palmar creases, and clubfoot have been noted in at least 60% of both true and mosaic triploids (2). Camptodactyly of the fifth finger and an increased number of digital whorls have been reported (9,12).

Mental deficiency has been noted in most true examples. Hydrocephalus (20%), absent corpus callosum (15%), dilated ventricles (35%), large posterior fontanel (90%) (30), lumbosacral meningomyelocele (20%) and, frequently, Arnold–Chiari malformation and hydranencephaly have been reported. Holoprosencephaly has been noted (8,23,29).

The male genitalia in those with true 69,XXY are often abnormal. Hypospadias (40%), cryptorchidism (85%), micropenis (75%), and ambiguous genitalia (40%) are common. Scrotal abnormalities (60%) include bifidity and hypoplasia or agenesis of scrotum. Leydig cell hyperplasia has been documented in 20% (2). Females (69,XXX) usually manifest gonadal dysgenesis. Kidney anomalies (cystic dysplasia, glomerulosclerosis, hydronephrosis), adrenal hypoplasia, and congenital heart defects, particularly ventricular septal defect, patent ductus arteriosus, and atrial septal defect, occur in 50% of those with complete triploidy but do not occur in the mosaic form (1,2,4,6,15,20,27,29). In true triploidy, abdominal wall defects such as omphalocele, gastroschisis, umbilical hernia, and diastasis recti are seen in 50% (2). Agenesis of the gallbladder has been reported.

Radiographic changes include harlequin orbits, small anterior fontanel, gracile ribs, upswept clavicles, diaphyseal overtubulation of long bones, vertical ilia, and proximal radioulnar synostosis (25a).

**Craniofacial abnormalities.** The posterior fontanel is always large at birth. Asymmetry of the occipitoparietal calvaria occurs in 50%. Microphthalmia, iris and choroid colobomas, and epicanthal folds are each seen in 25% of true triploidy (2). The ears are dysplastic and/or low set in 35–50% (Fig. 3–15).

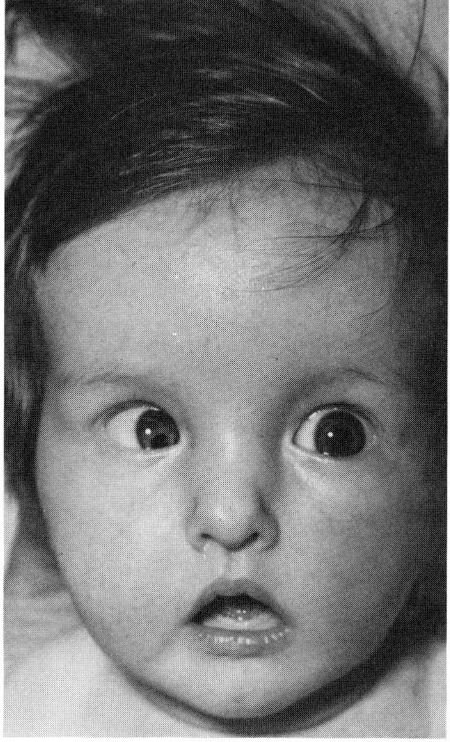

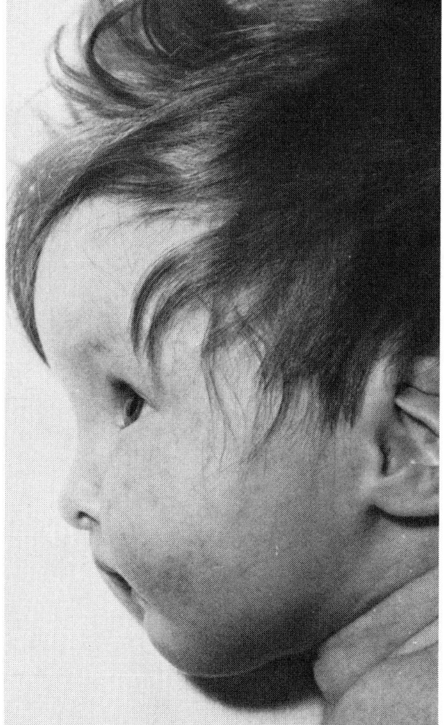

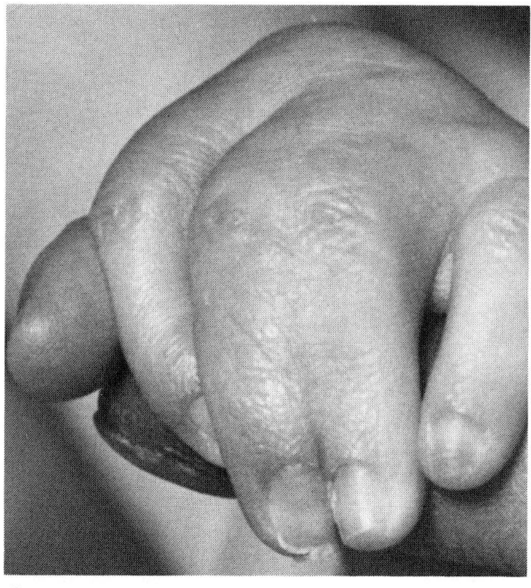

A B C

Fig. 3–15. *Triploidy syndrome.* (A,B) Frontal bossing, coloboma of iris, strabismus, and malformed ear. (C) Soft tissue syndactyly of third and fourth fingers. (From *W Schmid* and *D Vischer,* Cytogenetics **6:**145, 1967.)

Micrognathia is common. Isolated cleft palate (1,6,8,15,17,22,24,27) or cleft lip with or without cleft palate has been noted (1,20,24–27) in about 35%. Oral asymmetry, macroglossia, and smaller teeth on one side have been described (5,10,18,28).

### References (Triploidy syndrome)


1. Al Saadi A et al: Triploidy syndrome. *Clin Genet* **9**:43–50, 1976.

1a. Bendon RW et al: Prenatal detection of triploidy. *J Pediatr* **112**:149–153, 1988.

2. Blackburn WR et al: Comparative studies of infants with mosaic and complete triploidy, an analysis of 55 cases. *Birth Defects* **18**(3B):251–274, 1982.

3. Cassidy AB et al: Five month extrauterine survival in a female triploid (69,XXX) child. *Ann Génét* **20**:277–279, 1977.

4. David M et al: La triploidie chez l'enfant. *Pédiatrie* **30**:281–298, 371–390, 1975.

5. Dudakow E et al: Triploidy in man: A clearly recognizable syndrome? *Isr J Med Sci* **13**:493–499, 1977.

6. Finley WH et al: Triploidy in a liveborn male infant. *J Pediatr* **81**:855–856, 1972.

7. Fryns JP et al: Unusually long survival in a case of full triploidy of maternal origin. *Hum Genet* **38**:147–155, 1977.

8. Fulton AB et al: Ocular findings in triploidy. *Am J Ophthalmol* **84**:859–867, 1977.

9. Gosden CM et al: Clinical details, cytogenetics studies, and cellular physiology of a 69,XXX fetus, with comments on the biological effect of triploidy in man. *J Med Genet* **13**:371–380, 1976.

10. Graham JM Jr et al: Diploid–triploid mixoploidy: Clinical cytogenetic aspects. *Pediatrics* **68**:23–28, 1981.

11. Halbrecht I et al: Triploidy 69,XXX in a stillborn girl. *Clin Genet* **4**:210–212, 1973.

12. Harris MJ et al: Triploidy in 40 human spontaneous abortuses: Assessment of phenotypic embryos. *Obstet Gynecol* **57**:600–606, 1981.

13. Jacobs PA et al: The origin of human triploids. *Ann Hum Genet* **42**:49–57, 1978.

14. Jacobs PA et al: Late replicating X chromosomes in human triploidy. *Am J Hum Genet* **31**:447–457, 1979.

15. Keutel J et al: Triploidie (69,XXY) bei einem lebend geborenen Kind. *Z Kinderheilkd* **109**:104–117, 1970.

16. Lauritsen JG et al: Origin of triploidy in spontaneous abortuses. *Am Hum Genet* **43**:1–5, 1979.

17. Lejeune J et al: Chimére 46, XX/69, XXY. *Ann Génét* **10**:188–192, 1967.

17a. Meinecke P, Engelbrecht R: Fehlbildungs-Retardierungs-Syndrom infolge inkompletter Triploidie. *Monstschr Kinderheilkd* **135**:206–209, 1988.

18. Nicolas G: Myxodiploidia in a young child—its effect on orodontal anatomy. *Rev Fr Odontostomatol* **16**:97–108, 1969.

19. Niebuhr E et al: Triploidy in man: Cytogenetical and clinical aspects. *Humangenetik* **21**:103–126, 1974.

20. Paterson WG et al: Two cases of hydatidiform degeneration of the placenta with fetal abnormality and triploidy chromosome constitution. *J Obstet Gynaecol Br Commonw* **78**:136–142, 1971.

21. Proctor SE et al: Triploidy, partial mole and dispermy. An investigation of 12 cases. *Clin Genet* **26**:46–51, 1984.

22. Sacrez R et al: La triploidie chez l'enfant. *Pédiatrie* **22**:267–275, 1967.

23. Schinzel A et al: Triploidie als Ursache von Schwangerschaftsgestosse im 2. Trimenor. *Arch Gynäkol* **218**:113–123, 1975.

24. Shepard TH et al: Chromosomal aberrations in two embryos from the same mother. *Am J Obstet Gynecol* **102**:48–52, 1968.

25. Sherard J et al: Long term survival in a 69,XXY triploid male. *Am J Med Genet* **25**:307–312, 1986.

25a. Silverthorn KG et al: Radiographic findings in liveborn triploidy. *Pediatr Radiol* **19**:237–241, 1989.

26. Tarkkanen A: Ocular pathology in triploidy (69,XXY). *Ophthalmologica* **163**:90–97, 1971.

27. Walker S et al: Three further cases of triploidy in man surviving to birth. *J Med Genet* **10**:135–141, 1973.

28. Wertelecki W et al: The clinical syndrome of triploidy. *Obstet Gynecol* **47**:69–76, 1976.

29. Zergollern L et al: A liveborn infant with triploidy (69,XXX). *Z Kinderheilkd* **112**:293–300, 1972.

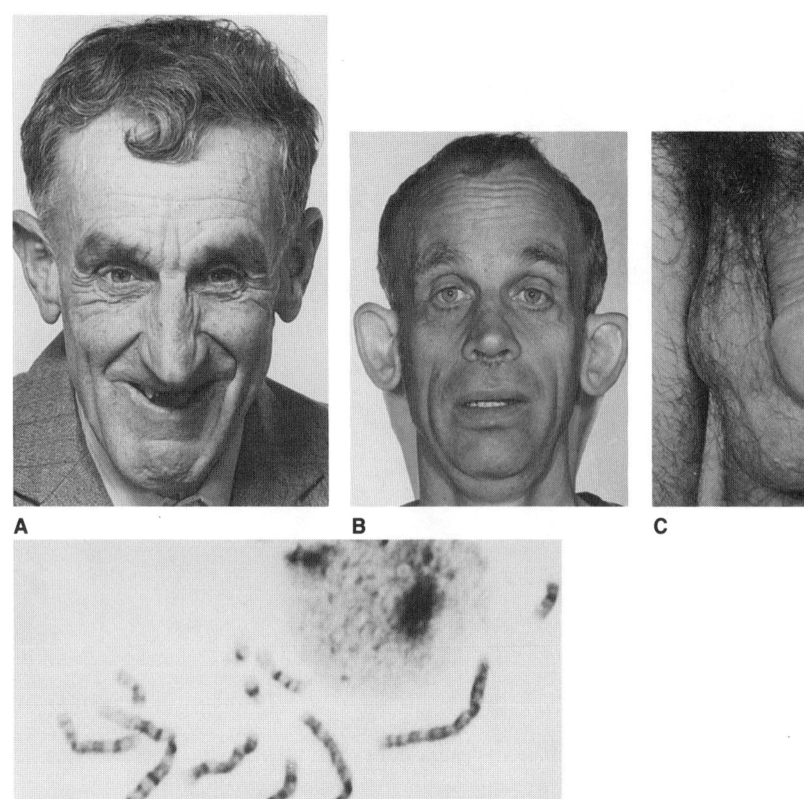

Fig. 3–16. *Fragile X syndrome.* (A) Long narrow face with large quadrangular forehead, somewhat underdeveloped midface, broad based nose, prominent chin, and prominent ears. (B) Compare facies to patient in A. (C) Macro-orchidism. (D) G-banded fragile X site. (A from *A McDermott et al,* J Med Genet **20:**169, 1983. D courtesy of *J Červenka,* Minneapolis, Minnesota.)

## Fragile X syndrome (Martin–Bell syndrome, macro-orchidism-marker X syndrome)

X-linked mental retardation appears to be second only to Down syndrome in its frequency. It represents a heterogeneity and has been classified into several groups. The only form that we shall consider here is the macro-orchidism/fragile X syndrome, which constitutes about 50% of cases of X-linked mental retardation. The chromosome defect was originally reported by Lubs (24) in 1969. Good reviews for history of the syndrome are those of Turner and Opitz (42), Opitz and Sutherland (28), and Schinzel and Largo (33). Several hundred cases have been reported to date. Its prevalence among mentally retarded males has been estimated to be about 2–6% (1). There have been three international conferences on the fragile X syndrome as of this writing (28,29).

The syndrome has variable expression in the hemizygote and especially in the female heterozygote (22). A significant number of unaffected males have transmitted the fragile X chromosome (6,13,16,45). None of these males has exhibited the abnormal chromosome themselves. This has been explained by an autosomal gene that, if homozygous, can suppress the condition in males and, if heterozygous, can suppress the disorder in female heterozygotes (14). Its frequency appears to be about 1/1200 males. About 35% are due to new mutation. There probably is genetic heterogeneity (2). Close linkage has been established with G6PD (4) and hemophilia B (2). A method of recurrence risk calculation for the relatives of isolated cases has been published (37a).

Birth weight may be elevated (1). Adult height is somewhat decreased (23). Head circumference is slightly increased and there is dolichocephaly. Hand and foot lengths are slightly reduced (27). In the postpubertal male, only about 60% exhibit the triad of typical facies, mental retardation, and macro-orchidism (3,6). Occasionally, hemizygotes appear entirely normal (16,45).

**Facies.** A long narrow face is present in roughly 60% of postpubertal hemizygotes. The forehead is large and quadrangular with prominent supraorbital ridges. Palpebral length is increased and inter-inner canthal distance is decreased (29). The cheeks feel somewhat thickened. The mandible becomes prominent during adolescence and there is some degree of midface retraction. The nose is usually broad-based. The pinnae tend to be very large (>7 cm), somewhat soft, and outstanding or cupped with simple helices and absent lobes (9a) (Fig. 3–16A,B). The palate is usually highly arched (1,5,6,19,25,27,30,34,43). Perhaps 25–40% of carrier females, especially the more retarded ones, have the typical facies (long face, prognathism, large everted pinnae) 7,21a,34,44,47). In the newborn, relative macrocephaly and large fontanel may be noted, but not the long face.

**Central nervous system.** Cognitive variability has been discussed by Theobald et al (41) and Chudley et al (4). Mental retardation in the hemizygote is moderately severe with IQs ranging from 25 to 69. However, in over 75%, IQs are less than 39. Retardation seems to increase with age. Speech delay is constant. Those with higher IQ exhibit cluttered speech (dysfluencies, stuttering). Often there is characteristic rhythmic intonation (litany speech). Those with low IQs are less verbal with short bursts of repetitive phrases. Patients exhibit retardation of motor development, clumsiness, hyperactivity, mild hypotonia, increased deep tendon reflexes, emotional instability, and automutilation (especially handbiting). Approximately 15% exhibit autism (avoidance of eye contact, hand flapping or other stereotypic movements, echolalia) (1,6,25,34,37b,46). About 50% manifest aggression. Seizures, noted in 15%, occur principally in those with the lowest intelligence. However, at least 50% of the adult males have abnormal EEG recordings. About 35% of obligate heterozygotes have borderline to subnormal intelligence (7,20,37) and another 15% have learning disability. Intelligence in female carriers appears to be inversely correlated with the expression of the fragile site (13,15,26). Some female heterozygotes are psychotic (7). Loesch and Hay (21a) suggested that 85% of female carriers have less than an 85 IQ.

**Connective tissues.** Joint laxity, especially of the fingers, knees, and ankles, may be marked. The feet are flat in 40%. The skin feels velvety soft and is somewhat lax. Mitral valve prolapse has been found in 80% of hemizygotes over 18 years, with mild aortic dilatation in about 15% (9,21).

**Genitourinary.** Macro-orchidism (Fig. 3–16C), unilateral or bilateral, noted in only 40% prior to puberty, is found in about 75% of adult hemizygotes (3). The testes tend to be softer than normal. Hyperpigmentation of the scrotum and enlargement of the penis have been noted in over 50%. Carrier females exhibit high fertility, a higher frequency of twinning (7), and an increased rate of miscarriages (21a).

**Oral findings.** The palate is high and narrow. In one study, cleft palate was found in about 8% (30), but this has not been mentioned in other surveys and it has not been our experience. Crossbite and openbite are relatively common (36). Tooth crown diameter asymmetry is frequent (31). The lateral palatine ridges are prominent in 60% (27).

**Differential diagnosis.** In the absence of a family history, one must exclude other forms of nonspecific mental retardation, especially Renpenning syndrome, which is characterized by relatively severe mental retardation without other central nervous system involvement, small head circumference, short stature, normal facies, normal ears, and normal or small testes. It should be emphasized, however, that fragile X syndrome, because of its variable expression, may be difficult to diagnose, especially before puberty. Several patients were initially thought to have *cerebral gigantism* (3). About 7–15% of autistic males are positive for the fragile X chromosome (3).

R-banded chromosomes can be used to rule out a fragile site near 6qter, which is present in about 10% of patients. It should be emphasized that macro-orchidism is not an invariable finding in the fragile X syndrome. Conversely, one may find macro-orchidism in otherwise normal males.

**Laboratory aids.** The fragile site, from which the name of the syndrome is derived, is evident only in chromosomes grown for a prolonged period (up to 96 hours) in a folic acid-deficient medium (such as TC 199) at elevated pH, supplemented with only 5% serum. The fragile site of Xq27-28 is seen in 5–50% (median, 20%) of lymphocyte cultures in the hemizygote (Fig. 3–16D). Scanning electron microscopy has established the break point at Xq27.3. Various agents such as bromodeoxyuridine, methotrexate, aminopterin, methionine, fluorodeoxyuridine, or trimethoprim are added for the last 24 hours of culture to enhance the expression of the fragile site (26). Colcemid is not used. Air drying of the preparation helps to increase the number of fragile sites manifested. Usually 100 cells are analyzed (29). It is important that the original blood sample not be stored for more than a few days. The female carrier exhibits the fragile site in 0.0–15.0% (median, 3%) of cells, and only about 50% of obligate carriers express the fra(X). A minimum of 4.0 and 1.5% of cells must show the fra(X) before a secure diagnosis can be made of the hemizygote and heterozygote, respectively. There is good evidence to indicate that fragility decreases with age. Hence, diagnosis of the heterozygote over 25 years of age may be extremely difficult (1). DNA marker analysis has been successfully employed (29).

Testicular volume is usually measured by the formula $\pi/6 \times \text{length} \times \text{width}^2$. Testicular volume in the postpubescent male is significantly increased (25–127 ml) above normal values (15–25 ml) (49).

Prenatal diagnosis has been carried out on amniocytes, but with variable success (11,17,37,39,48). Chorionic villus sampling, induction with at least two cytogenetic methods, and RFLP analysis have been advocated (35).

Dermatoglyphic features of the hemizygote include increased radial loops, whorls, and arches on fingertips (especially the second and third fingers), abnormal palmar and plantar creases, absence of c-triradii on palms, dysplasia of papillary ridges, and low frequency of true patterns on soles. Female heterozygotes show some of the same changes (32,38).

A metacarpal phalangeal index has been established for the syndrome (2a).

### Reference [Fragile X syndrome(Martin–Bell syndrome, macro-orchidism-marker X syndrome)]


1. Brondum-Nielsen K: Diagnosis of the fragile X syndrome (Martin–Bell syndrome): Clinical findings in 27 males with the fragile site at Xq28. *J Ment Defic Res* **27**:211–226, 1983.

2. Brown WT et al: Further evidence for genetic heterogeneity in the fragile X-syndrome. *Hum Genet* **75**:311–321, 1987.

2a. Butler MG et al: Metacarpophalangeal pattern profile analysis in fragile X syndrome. *Am J Med Genet* **31**:767–773, 1988.

3. Chudley AE, Hagerman RJ: Fragile X syndrome. *J Pediatr* **110**:821–831, 1987.

4. Chudley AE et al: Invited editorial comment: Cognitive variability in the fragile X syndrome. *Am J Med Genet* **28**:13–15, 1987.

5. De Arce MA, Kearns A: The fragile X syndrome: The patients and their chromosomes. *J Med Genet* **21**:89–91, 1984.

6. Fryns JP: The fragile X syndrome: A study of 83 families. *Clin Genet* **26**:497–528, 1984.

7. Fryns JP: The female and the fragile X: A study of 144 obligate female carriers. *Am J Med Genet* **23**:157–169, 1986.

8. Fryns JP et al: The psychological profile of the fragile X syndrome. *Clin Genet* **25**:131–134, 1984.

9. Hagerman RJ et al: Consideration of connective tissue dysfunction in the fragile X syndrome. *Am J Med Genet* **17**:111–121, 1984.

9a. Hagerman RJ et al: Institutional screening for the fragile X syndrome. *Am J Dis Child* **142**:1216–1221, 1988.

10. Harrison CJ et al: The fragile X: A scanning electron microscopic study. *J Med Genet* **20**:280–285, 1985.

11. Hecht F et al: Fragile X chromosome: Current methods. *Am J Med Genet* **11**:489–495, 1982.

12. Herbst DE et al: Further delineation of X-linked mental retardation. *Hum Genet* **58**:366–372, 1981.

13. Howard-Peebles PN, Friedman JM: Unaffected carrier males in families with fragile X syndrome. *Am J Hum Genet* **37**:956–964, 1985.

14. Israel MH: Autosomal suppressor gene for fragile X: An hypothesis. *Am J Med Genet* **26**:19–31, 1987.

14a. Jacky PB, Dill FJ: Expression in fibroblast culture of the satellited-X chromosomes associated with familial sex-linked mental retardation. *Hum Genet* **53**:267–269, 1980.

15. Jacobs PA et al: X-linked mental retardation: A study of 7 families. *Am J Med Genet* **7**::471–489, 1980.

16. Jacobs PA et al: A cytogenetic study of a population of mentally retarded males with special reference to the marker (X) syndrome. *Hum Genet* **63**:139–147, 1983.

17. Jenkins EC et al: Experience with prenatal fragile X detection. *Am J Med Genet* **17**:215–239, 1984.

18. Jennings M et al: Significance of phenotype and chromosomal abnormalities in X-linked mental retardation (Martin-Bell or Renpenning syndrome). *Am J Med Genet* **7**:417–432, 1980.

19. Kähkönen M et al: Marker X-associated mental retardation: A study of 150 retarded males. *Clin Genet* **23**:397–404, 1983.

20. Knoll JH et al: Frequency and replication pattern of fragile (X) (q28) in heterozygotes. *Am J Med Genet* **36**:640–645, 1984.

21. Loehr JP et al: Aortic root dilatation and mitral valve prolapse in the fragile X syndrome. *Am J Med Genet* **23**:189–194, 1986.

21a. Loesch DZ, Hay DA: Clinical features and reproductive patterns in fragile X female heterozygotes. *J Med Genet* **25**:407–414, 1988.

22. Loesch DZ et al: Phenotypic variation in male-transmitted fragile X: Genetic inferences. *Am J Med Genet* **27**:401–418, 1987.

23. Loesch DZ et al: Anthropometry in Martin–Bell syndrome. *Am J Med Genet* **30**:149–164, 1988.

24. Lubs HA: A marker X chromosome. *Am J Hum Genet* **21**:231–244, 1969.

25. Mattei JF et al: X-linked mental retardation with the fragile X: A study of 15 families. *Hum Genet* **59**:281–289, 1981.

26. McDermott A et al: Fragile X chromosome: Clinical and cytogenetic studies from seven families. *J Med Genet* **20**:169–171, 1983.

27. Meryash DL et al: An anthropometric study of males with the fragile X syndrome. *Am J Med Genet* **17**:159–174, 1984.

28. Opitz JM, Sutherland GR: International workshop on the fragile X and X-linked mental retardation. *Am J Med Genet* **17**:5–94, 1984.

29. Opitz JM et al: X-linked mental retardation 3. *Am J Med Genet* **30**:1–702, 1988.

30. Partington MW: The fragile X-syndrome: Preliminary data on growth and development in males. *Am J Med Genet* **17**:175–194, 1984.

31. Peretz B et al: Crown size asymmetry in males with fra(X) or Martin–Bell syndrome. *Am J Med Genet* **30**:185–190, 1988.

32. Rodewald A et al: Dermatoglyphic peculiarities in families with X-linked mental retardation and fragile Xq27: A collaborative study. *Clin Genet* **30**:1–13, 1986.

33. Schinzel A, Largo RH: The fragile X syndrome (Martin–Bell syndrome). *Helv Paediatr Acta* **40**:133–152, 1985.

34. Schmidt A: Fragile site Xq27 and mental retardation: Clinical and cytogenetic manifestation in heterozygotes and hemizygotes of five kindreds. *Hum Genet* **60**:322–327, 1982.

35. Shapiro LR et al: Experience with multiple approaches to the prenatal diagnosis of the fragile X syndrome: Amniotic fluid, chorionic villi, fetal blood and molecular methods. *Am J Med Genet* **30**:347–354, 1988.

36. Shellhart WC et al: Oral findings in fragile X syndrome. *Am J Med Genet* **23**:179–187, 1986.

37. Sherman SL et al: The marker (X) syndrome: A cytogenetic and genetic analysis. *Ann Hum Genet* **48**:21–37, 1984.

37a. Sherman SL et al: Recurrence risks of relatives in families with an isolated case of the fragile X syndrome. *Am J Med Genet* **31**:753–765, 1988.

37b. Simko A et al: Fragile X syndrome: Recognition in younger children. *Pediatrics* **83:**547–552, 1989.

38. Simpson NE: Dermatoglyphic indices of males with the fragile X syndrome and of the female heterozgotes. *Am J Med Genet* **23**:171–178, 1986.

39. Tehada I et al: Prenatal diagnosis of fragile X chromosome in amniotic fluid cells. *Ann Génét* **26**:247–250, 1983.

40. Thake A et al: Is it possible to make a clinical diagnosis of the fragile X syndrome in a boy? *Arch Dis Childh* **60**:1001–1007, 1985.

41. Theobald TM et al: Individual variation and specific cognitive deficits in the fra (X) syndrome. *Am J Med Genet* **28**:1–11, 1987.

42. Turner G, Opitz JM: X-linked mental retardation. *Am J Med Genet* **7**:407–415, 1980.

43. Turner G et al: X-linked mental retardation, macro-orchidism and the Xq27 fragile site. *J Pediatr* **96**:837–841, 1980.

44. Turner G et al: Heterozygous expression of X-linked mental retardation and X-chromosomes marker for (X)(q27). *N Engl J Med* **303**:662–664, 1980.

45. Van Roy BC et al: Fragile X trait in a large kindred: Transmission also through normal males. *J Med Genet* **20**:286–289, 1983.

46. Veenema H et al: The fragile X syndrome in a large family. II. Psychological investigations. *J Med Genet* **24**:32–48, 1988.

47. Webb GC et al: Fragile (X) (q27) sites in a pedigree with female carriers showing mild to severe mental retardation. *J Med Genet* **19**:44–48, 1982.

48. Webb T et al: Prenatal diagnosis of X-linked mental retardation with fragile (X) using fetoscopy and fetal blood sampling. *Prenatal Diag* **3**:131–137, 1983.

49. Zachmann M et al: Testicular volume during adolescence. *Helv Paediatr Acta* **29**:61–72, 1974.

# Chapter 4
# Chromosomal Syndromes: Unusual Variants

## del(1q) syndrome

There are about 30 reports of patients with variable deletions of the long arm of chromosome number 1. Most have been *de novo* and have involved q24→q32. Some have ring chromosome 1 (11a), others del (1)(q42) (13a). Somatic and mental retardation has been marked. Nearly all have seizures and hypotonia. Some have a high-pitched cry. The head is microbrachycephalic. The hair is sparse or fine. The face is round and flat with a prominent metopic suture. There is usually upward slanting palpebral fissures, apparent hypertelorism, epicanthal folds, strabismus, heavy cheeks, and somewhat short broad bulbous nose. The pinnae are malformed and posteriorly rotated without lobes. The philtrum is long and prominent. The mouth is often downturned, the vermilion thin, the mandible small, and the neck short (Fig. 4–1A). Skeletal anomalies have been variable. Approximately 40% of the patients have congenital heart disease (VSD, pseudotruncus arteriosus). Hypospadias and cryptorchidism are common features among male patients (1–17).

Several have cleft lip and/or palate or missing or bifid uvula (1,5,5a,10,16,18).

### References [del(1q) syndrome]


1. Andrle M et al. Terminal deletion of (1)(q42) and its phenotypic manifestations. *Hum Genet* **41**:115–120, 1978.

2. Dignan P, Soukup S: Terminal long-arm deletion of chromosome 1 in a male infant. *Hum Genet* **48**:151–156, 1979.

3. Johnson VP et al: Deletion of the distal long arm of chromosome 1: A definable syndrome. *Am J Med Genet* **22**:685–694, 1985.

4. Juberg RC et al: New deletion syndrome 1q43. *Am J Hum Genet* **33**:455–463, 1981.

5. Kessel E et al: Terminal deletion of the long arm of chromosome 1 in a malformed newborn. *Hum Genet* **42**:333–337, 1978.

5a. Kjesseler B et al: Apparently non-deleted ring-1 chromosome and extreme growth failure in a mentally retarded girl. *Clin Genet* **14**:8–15, 1978.

6. Koivisto M et al: A primary hypothyroidism, growth hormone deficiency and congenital malformations in a child with karyotype 46,XY,del(1)(q25q32). *Acta Pediatr Scand* **65**:513–518, 1976.

7. Manouvrier-Hanu S et al: A new case of distal deletion of the long arm of chromosome 1. *Am J Med Genet* **25**:599–600, 1986.

8. Meinecke P, Vögtel D: A specific syndrome due to deletion of the distal long arm of chromosome 1. *Am J Med Genet* **28**:371–376, 1987.

9. Menks Ribiero MC: Terminal deletion 1q43 in a newborn with hydrocephalus. *Ann Génét* **30**:126–128, 1987.

10. Montero MR et al: Terminal deletion of (1)(q42) in a newborn. *Ann Génét* **27**:178–179, 1984.

11. Neu RL et al: A 1q42 deletion in a Vietnamese infant. *Ann Génét* **25**:154–155, 1982.

11a. Patel SV, Verma RS: Genetic consequences of ring chromosome 1 in humans. *Dysmorphol Clin Genet* **1**:148–151, 1988.

12. Schinzel A, Schmid W: Interstitial deletion of the long arm of chromosome 1, del(1)(q21→q25) in a profoundly retarded 8-year-old girl with multiple anomalies. *Clin Genet* **18**:305–313, 1980.

13. Steinbach P et al: Multiple congenital anomalies/mental retardation (MCA/MR) syndrome due to interstitial deletion 1q. *Am J Med Genet* **19**:131–136, 1984.

13a. Tolkendorf E et al: A new case of deletion 1q42 syndrome. *Clin Genet* **24**:289–292, 1989.

14. Turleau C et al: Distal 1q monosomy: Two new observations and syndrome's description. *Ann Génét* **26**:161–164, 1983.

15. Turleau C et al: La monosomie 1q distal. *Ann Génét* **26**:161–164, 1983.

16. Watson MS et al: Chromosome deletion 1q42–43. *Am J Med Genet* **24**:1–6, 1986.

17. Wright LL et al: An unusual ocular finding associated with chromosome 1q deletion syndrome. *Pediatrics* **77**:786, 1986.

18. Zabel BU, Baumann WA: 1q deletion syndrome. *Clin Genet* **19**:544–545, 1981.


## dup(1q) syndrome

In addition to somatic and mental retardation, those with trisomy for about two-thirds of the long arm of chromosome 1 (1q2→qter) exhibit low birth weight, neonatal death, and abnormal facies characterized by wide forehead, midface hypoplasia, low-set pinnae, small downslanting palpebral fissures, broad nasal bridge, long beaked nose, and microretrognathia (1–10) (Fig. 4–1B). Several have exhibited cleft lip and/or cleft palate (3,5,9,11). Variable cardiac malformations and thymic hypoplasia or aplasia have been present in all.

Cranial anomalies are more often present in trisomic insertions and in trisomy for the distal one-third of the long arm (q42→qter): macrocephaly with large fontanel and wide sutures, frontal bossing, large low-set slanting pinnae, microphthalmia, downslanting short palpebral fissures, broad flat nasal bridge, long nose, small mandible, and short neck. Other anomalies in this group include long tapering overlapping fingers, anomalous implantation of toes, cardiac malformations (VSD, tetralogy of Fallot), cryptorchidism, and short chorda penis (1,2,4, 5,6,10).

Ocular, genitourinary, intestinal (stenosis), and thymic abnormalities are not associated with interstitial trisomy, but coloboma of the iris is common in trisomy for the terminal segment (5).

### References [dup(1q) syndrome]


1. Chen H et al: Omphalocele and partial trisomy 1q syndrome. *Hum Genet* **53**:1–4, 1979.

2. Chia NL et al: Trisomy (2)(q42→qter): Confirmation of a syndrome. *Clin Genet* **34**:224–229, 1988.

3. Garrett JH et al: Fetal loss and familial chromosome translocation. *Clin Genet* **8**:341–348, 1975.

4. Hustinx TWJ et al: Partial trisomy of chromosome 1. *Am J Med Genet* **3**:353–358, 1979.

5. Leisti J, Aula P: Partial trisomy 1 (q42→ter). *Clin Genet* **18**:371–378, 1980.

6. Liberfarb RM et al: Multiple congenital anomalies/mental retardation (MCA/MR) syndrome due to partial 1q duplication and possible 18p deletion: A study of four individuals in two families. *Am J Med Genet* **4**:27–37, 1979.

7. Lungarotti MS et al: De novo duplication 1q32→42: Variability of phenotypic features in partial 1q trisomies. *J Med Genet* **17**:398–402, 1980.

8. Michels VV et al: Duplication of part of chromosome 1q. *Am J Med Genet* **18**:125–134, 1984.

9. Norwood TH, Hoehn H: Trisomy for the long arm of human chromosome 1. *Humangenetik* **25**:79–82, 1974.

10. Rehder H, Friedrich U: Partial trisomy 1q syndrome. *Clin Genet* **15**:534–540, 1979.

11. Rosenthal J et al: Clinical variability of partial duplication 1q: A clinical report and literature review. *Am J Med Genet* **27**:787–792, 1987.

12. Schinzel A: Partial trisomy 1q25→1q32 in a malformed girl with a de novo insertion in 1q. *Hum Genet* **49**:167–173, 1979.

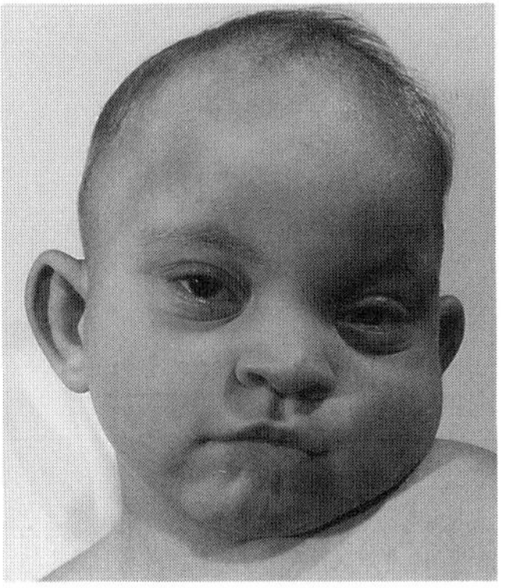

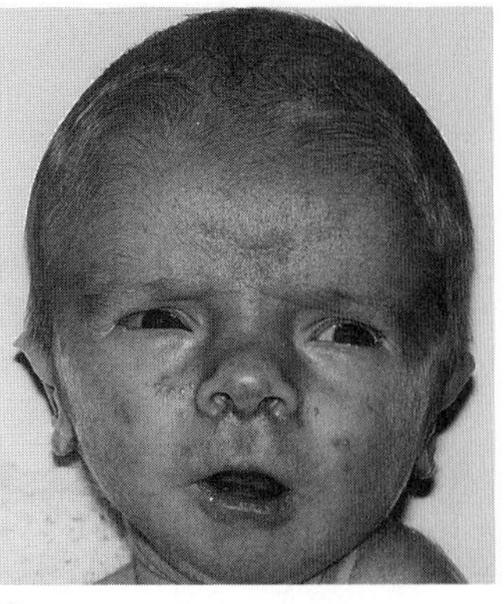

A B

Fig. 4–1. (A) *del(1q)syndrome.* Microbrachycephaly, round flat face, upslanting palpebral fissures with apparent hypertelorism, epicanthal folds, heavy cheeks, malformed pinnae, short neck. (B) *dup(1q) syndrome.* Five-week-old infant presenting macrocephaly, craniofacial asymmetry, apparent hypertelorism, and anteverted nostrils. (A from *M Andrle et al,* Hum Genet **41:**115, 1978; B from *NL Chia et al,* Clin Gent **34:**224, 1988.)

## dup(2p) syndrome

Approximately 25 cases have been described involving trisomy of the 2p21 or p23→pter region (1–10). In addition to being severely retarded, all patients show microcephaly. There is high prominent forehead with frontal upsweep of hair, hypertelorism, strabismus, dysmorphic pinnae, short nose with prominent tip, maxillary hypoplasia, broad nasal bridge, small mouth, broad alveolar ridges, and micrognathia (Fig. 4–2).

There is pre- and postnatal growth retardation. Body build is slender with hypotonia. The toes are widely spaced (40%). Skeletal anomalies include long tapering hyperflexible fingers (60%), scoliosis (15%), pectus excavatum (33%), and limited hip movement. Congenital heart anomalies (35%) are inconsistent in type. Micropenis (60%), shawl scrotum (15%), hypospadias (15%), and cryptorchidism (10%) are found.

Fig. 4–2. *dup(2p) syndrome.* Slender body, thin extremities, wide spaced toes, high prominent forehead, wide flat nasal bridge, ptosis, pointed chin. (From *U Francke* and *KL Jones,* Am J Dis Child **130:**1244, 1976.)

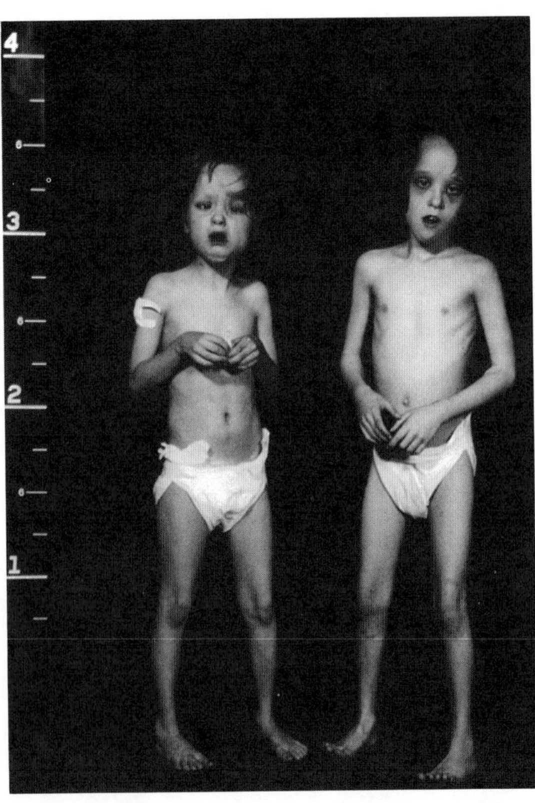


### References [dup(2p) syndrome]

1. Armendares S, Salamanca-Gómez F: Partial 2p trisomy (p21→pter) in two siblings of a family with a 2p−:15q+ translocation. *Clin Genet* **13**:17–24, 1978.

2. Fineman RM et al: Variable phenotype associated with duplication of different regions of 2p. *Am J Med Genet* **15**:451–456, 1983.

3. Francke U, Jones KL: The 2p partial trisomy syndrome. *Am J Dis Child* **130**:1244–1249, 1976.

4. Mayer U et al: Trisomie partielle 2p par translocation familiale 2-6. *Ann Génét* **23**:172–176, 1978.

5. Monteleone PE et al: De novo partial 2p duplication with postmortem description. *Am J Med Genet* **10**:451–456, 1983.

6. Neu RL et al: Partial 2p trisomy in a 46,XY,der(5), t(2;5) (p23;p15) pat infant; autopsy findings. *Ann Génét* **22**:33–34, 1979.

7. Pueschel SM et al: Partial trisomy 2p. *J Ment Defic Res* **31**:293–298, 1987.

8. Rosenfeld W et al: Partial duplication for the short arm of chromosome 2: The 2p23→pter syndrome. *Ann Génét* **25**:28–31, 1982.

9. Sekhon GS et al: Partial trisomy for the short arm of chromosome 2 due to familial balanced translocation. *Hum Genet* **44**:99–103, 1978.

10. Wakita Y et al: Duplication of 2p 25: Confirmation of the assignment of soluble acid phosphate ($HCP_1$) locus to 2p25. *Hum Genet* **71**:259–260, 1985.

## del(2q) syndrome

Too few cases with too many different deletions have been reported to form definitive patterns. For at least two deletions, however, there seem to be reasonably consistent findings. The first involves deletion q21–q31. These children have somatic and mental retardation, relative macrocephaly, small nose, cataracts, microphthalmia, ptosis, micrognathia, flexion deformity of fingers, and congenital heart disease (Fig. 4–3). All of these children have cleft palate. There is some clinical similarity to trisomy 18 (5,6).

The other syndrome involves deletion 2q31→q33. In addition to mental and somatic retardation, the children also exhibit the eye anomalies and finger deformities seen in the former syndrome. However, the nose tends to be prominent. At least three of these patients had cleft lip and/or cleft palate (1–4). There is again some clinical resemblance to trisomy 18 (2,6,6a).

### References [del(2q) syndrome]

1. Al-Awadi SA et al: Interstitial deletion of the long arm of chromosome 2:del(2)(q31q33). *J Med Genet* **20**:464–465, 1983.

2. Benson K et al: Interstitial deletion of the long arm of chromosome 2 in a malformed infant with karyotype 46,XX,del(2)(q31q33). *Am J Med Genet* **25**:405–412, 1986.

3. Buchanan PD et al: Interstitial deletion 2q31→q33. *Am J Med Genet* **15**:121–126, 1983.

4. Franceschini P et al: Interstitial deletion of the long arm of chromosome 2(q31→q33) in a girl with multiple anomalies and mental retardation. *Hum Genet* **64**:98, 1983.

5. Fryns JP et al: Interstitial deletion of the long arm of chromosome 2 in a polymalformed newborn; karyotype 46,XX,del(2)(q21q24). *Hum Genet* **39**:233–238, 1977.

6. McConnell TS et al: Partial deletion of chromosome 2 mimicking a phenotype of trisomy 18. *Hum Pathol* **11**:202–205, 1980.

6a. Ramer JC et al: A review of phenotype-karyotype correlations in individuals with interstitial deletions of the long arm of chromosome 2. *Am J Med Genet* **32**:359–363, 1989.

7. Shabtai F et al: Partial monosomy of chromosome 2; delineation syndrome. *Ann Génét* **25**:156–158, 1982.

8. Young RS: Deletion 2q; two new cases with karyotype 46,XY,del(2)(q31→q33) and 46,XX,del(2)(q36). *J Med Genet* **20**:199–202, 1983.

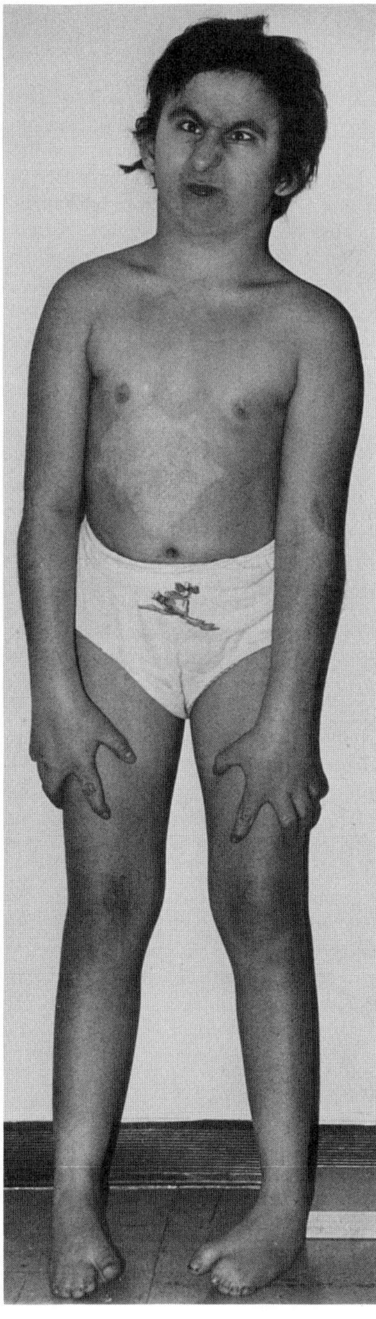

Fig. 4–3. *del(2q) syndrome.* Microphthalmia, strabismus, prominent nose, micrognathia, camptodactyly of fingers 3–5, hyperextensible index fingers, syndactyly of toes 2–5 in patient with interstitial deletion of 2(q31q33). Cleft palate was also present. (From *P Franceschini et al,* Hum Genet **64:**98, 1983.)

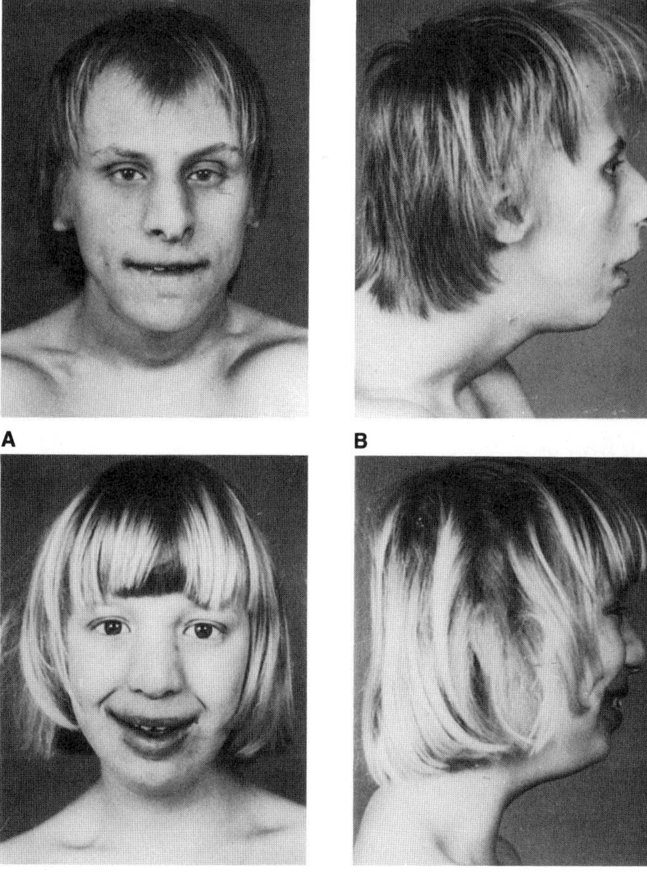

Fig. 4–4. *dup(2q) syndrome.* (A–D) Microbrachycephaly, beaked nose, wide mouth in brother and sister. (From *M Kyllerman et al,* Helv Paediatr Acta **39:**499, 1984.)

## dup(2q) syndrome

Approximately 25 examples of various degrees of partial trisomy 2q have been analyzed by Kyllerman et al (3), Yu and Chen (8), and Siffroi et al (5). Phenotype seems to vary little with the degree of deletion q31→qter to q34→qter (2,3). Birth weight and length are usually normal. In addition to the constant mental retardation, frequent findings include frontal bossing, microbrachycephaly with temporal retraction, hypertelorism, short beaked nose, elongated philtrum, and abnormal pinnae (Fig. 4–4A–D). Other eye findings include reduced vision, myopia, exotropia, glaucoma, nystagmus, iris defects, and fundus lesions (9). Some patients have thoracic kyphosis (1,6) and clinodactyly of the fifth finger. Visceral abnormalities have been extremely rare.

Cleft palate has been reported in several cases (1,4,7).

### References [dup(2q) syndrome]

1. Cotlier et al: The eye in the partial trisomy 2q syndrome. *Am J Ophthalmol* **84**:251–258, 1977.

2. Howard Peebles PN, Goldsmith JP: Duplication of region 2q31→2qter in a family with 2;9 translocation. *Hum Hered* **30**:84–88, 1980.

3. Kyllerman M et al: Delineation of a characteristic phenotype in distal trisomy 2q. *Helv Paediatr Acta* **39**:499–508, 1984.

4. Rosenthal IM et al: Trisomy of the distal portion of chromosome 2, a new familial syndrome associated with mental retardation and characteristic facies. *Am J Hum Genet* **26**:73A, 1974.

5. Siffroi JP et al: Trisomie partielle pour le bras long du chromosome 2 par malsegregation, d'une translocation t(2;7) (q321;p22) maternelle. *Ann Génét* **27**:241–244, 1984.

6. Strömland K: Eye findings in partial trisomy 2q. *Ophthalmol Paediatr Genet* **5**:145–150, 1985 (same case as Ref. 3).

7. Wisniewski L et al: Partial trisomy 2q and familial translocation t(2;18)(q31;p11). *Hum Genet* **45**:225–228, 1978.

8. Yu CW, Chen H: De novo tandem duplication of the long arm of chromosome 2 (q34–q37). *Birth Defects* **18**(3B):311–320, 1982.

9. Zankl M et al: Distal 2q duplication: report of two familial cases and an attempt to define a syndrome. *Am J Med Genet* **4**:5–16, 1979.

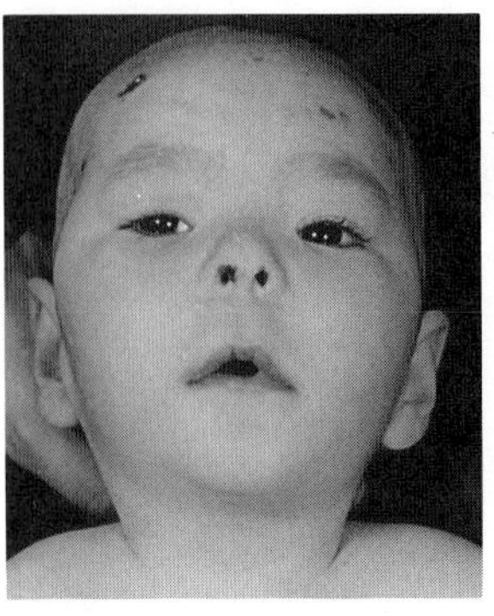
A

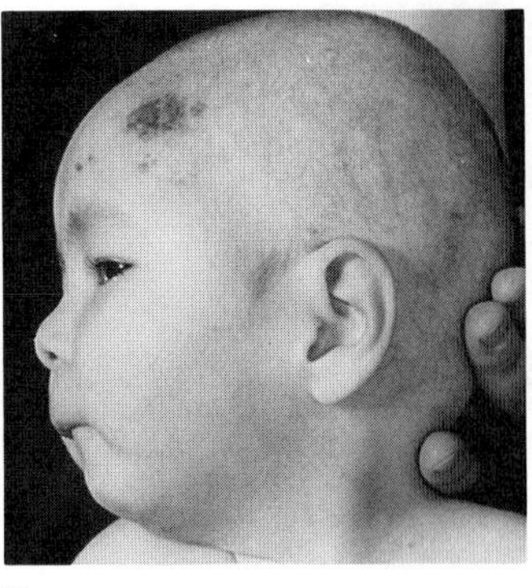
B

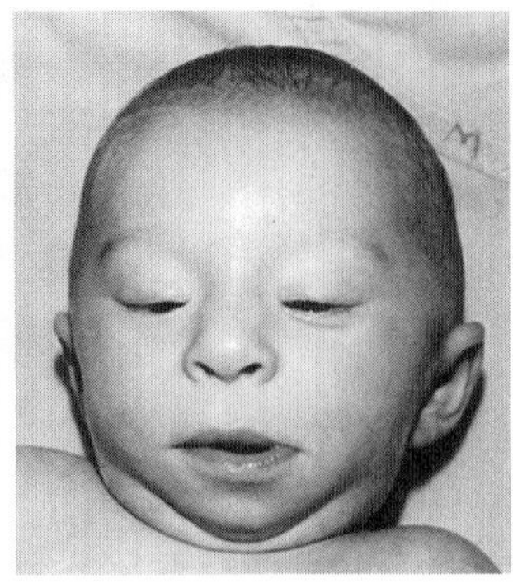
C

Fig. 4–5. *del(3p) syndrome.* (A,B) Microcephaly, dolichocephaly, somewhat triangular face with high forehead, ptosis, thin lips with long philtrum and small mandible. (C) Microcephaly, ptosis, inverse epicanthus, small nose with prominent bridge, narrow vermilion, and small mandible. (A,B from *DR Witt et al,* Clin Genet **27**:402, 1985; C from *MC Higginbottom et al,* J Med Genet **19**:71, 1982.)

## del(3p) syndrome

About a dozen cases of del(3p) syndrome have been reported. The most common features include low birthweight, severe postnatal growth retardation, severe mental retardation, microcephaly, brachycephaly, unusual facies, and developmental delay with severe psychomotor retardation (1–9).

The face is somewhat triangular with a high forehead. There are arched eyebrows, prominent nasal bridge, upslanting palpebral fissures, synophrys, ptosis, and low and dysmorphic pinnae. The lips are thin, the philtrum long, the angles downturned, and the mandible small (Fig. 4–5A–C).

Supernumerary postaxial digits of hands and feet and minor renal and cardiovascular anomalies have been reported in about 40%.

### References [del(3p) syndrome]


1. Beneck D et al: Deletion of the short arm of chromosome 3: A case report with necropsy findings. *J Med Genet* **21**:307–310, 1984.
2. Garcia-Sagredo JM et al: The phenotype of partial monosomy 3 (p25→pter) observed in two unrelated patients. *Clin Genet* **20**:387, 1981.
3. Gonzales J et al: Deletion partielle du bras court du chromosome 3. *Ann Génét* **23**:119–122, 1980.
4. Higginbottom MC et al: A second patient with partial deletion of the short arm of chromosome 3; karyotype 46,XY,del(3)(p25). *J Med Genet* **19**:71–73, 1982.
5. Merrild U et al: Partial deletion of the short arm of chromosome 3. *Eur J Pediatr* **136**:211–216, 1987.
6. Reifen RM et al: Partial deletion of the short arm of chromosome 3: Further delineation of the 3p25→3pter syndrome. *Clin Genet* **30**:127–130, 1986.
7. Schwyzer U et al: Terminal deletion of the short arm of chromosome 3, del(3pter-p25): A recognizable syndrome. *Helv Paediatr Acta* **42**:309–315, 1987.
8. Tolmie JL et al: Partial deletion of the short arm of chromosome 3. *Clin Genet* **29**:538–542, 1986.
9. Witt DR et al: Partial deletion of the short arm of chromosome 3 (3p25→3pter). *Clin Genet* **27**:402–407, 1985.


Fig. 4–6. *dup(3p) syndrome.* (A,B) Brachycephaly, frontal bossing, full cheeks, prominent philtrum, short neck. (From *JA Reiss et al,* Clin Genet **30**:50, 1986.)

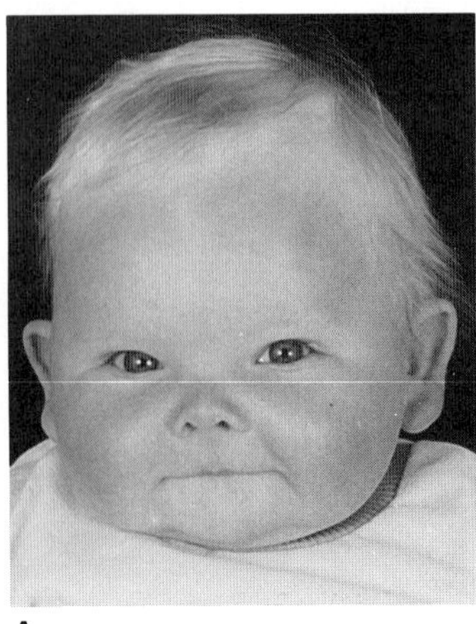
A

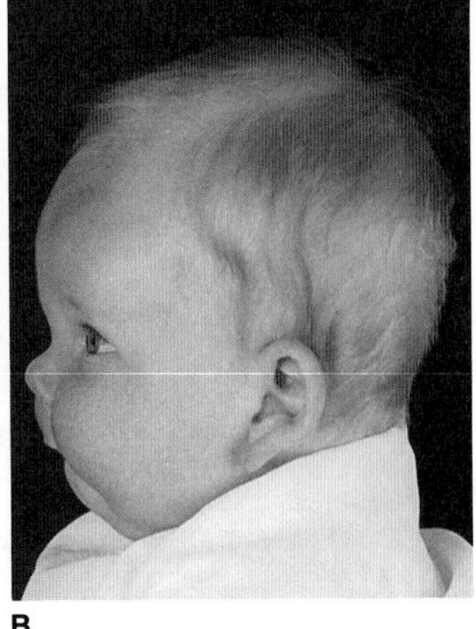
B

## dup(3p) syndrome

Approximately 25 patients have been reported with partial trisomy for the short arm of chromosome 3. These have been reviewed by Reiss et al (8). The size of the trisomic segment seems to have little effect on the phenotype (11).

Prenatal growth is retarded in only 15% and slowed postnatal growth in about 25%. Over 40% have died within the first 2 years of life. All exhibit psychomotor retardation. Hypospadias, micropenis, or cryptorchidism has been found in 75% of males. Congenital heart defect has been documented in 85%. Excessive fingertip whorls were noted in almost 90%.

Facial features include brachycephaly, frontal bossing, temporal indentation, hypertelorism, epicanthal folds, and square-shaped face with full cheeks, prominent philtrum, large mouth, micrognathia, and short neck. About 30% have cleft lip with or without cleft palate (Fig. 4–6A, B). The facies become less distinctive with age (1–12). Possibly 10% are associated with holoprosencephaly (4–6).

### References [dup(3p) syndrome]


1. Charrow JM et al: Duplication of 3p syndrome: Report of a new case and review of the literature. *Am J Med Genet* **8**:431–436, 1981.
2. Francke U: Clinical syndromes associated with partial duplications of chromosomes 2 and 3: dup(2p), dup(2q), dup(3p), dup(3q). *Birth Defects* **14**(6C):191–217, 1978.
3. Gimelli G et al: Dup(3)(p2→pter) in two families, including one infant with cyclopia. *Am J Med Genet* **20**:341–348, 1985.
4. Kurtzman DN et al. Duplication 3p21→3pter and cyclopia. *Am J Med Genet* **27**:33–37, 1987.
5. Lurie IW et al: Trisomy for the distal part of the short arm of chromosome 3. *Helv Paediatr Acta* **41**:509–573, 1986.
6. Martin NJ, Steinberg B: The dup(3)(p25→pter) syndrome: A case with holoprosencephaly. *Am J Med Genet* **14**:767–772, 1983.
7. Parloir C et al: Partial trisomy of the short arm of chromosome 3 (3p25→3pter). A distinct clinical entity. *Hum Genet* **47**:(3):239–244, 1979.
8. Reiss JA et al: Partial trisomy 3p syndrome. *Clin Genet* **30**:50–58, 1986.
9. Say B et al: Familial translocation (3p15p) with partial trisomy for the upper arm of chromosome 3 in two sibs. *J Pediatr* **88**:447–450, 1976.
10. Schinzel A et al: Trisomy 3(p23→pter) resulting from maternal translocation t(3;4)(p23q35). *Ann Génét* **21**(2):168–171, 1978.
11. Van Regemorter NE et al: Partial trisomy 3p in two siblings: Clinical and pathological findings. *Eur J Pediatr* **141**:53–56, 1983.
12. Yunis JJ: Trisomy for the distal end of the short arm of chromosome 3. A syndrome. *Am J Dis Child* **132**:30–33, 1978.


## dup(3q) syndrome

Falek et al (2) first described the clinical features of dup(3q), reporting the condition as "familial de Lange syndrome." At least 40 examples

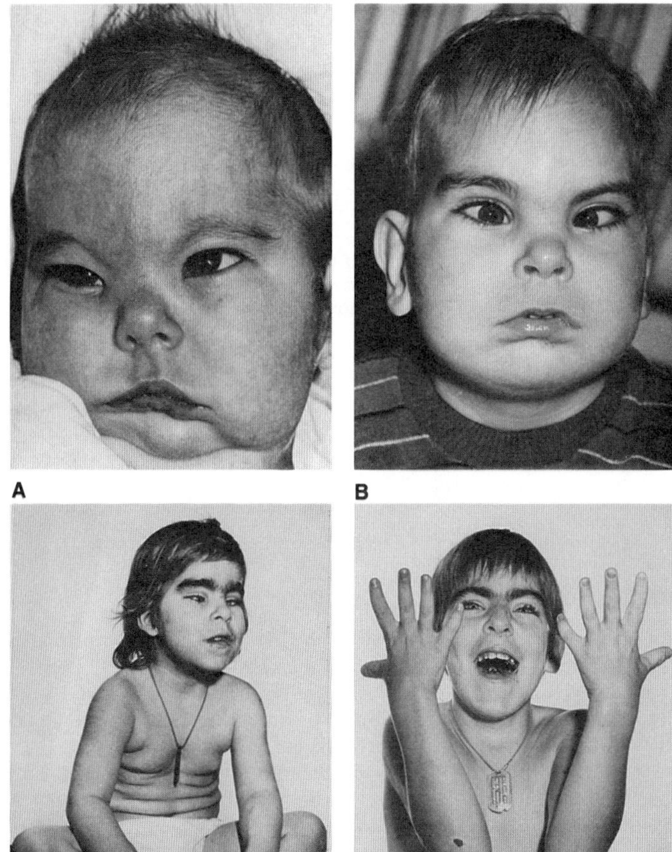


Fig. 4–7. *dup(3q) syndrome.* (A,B) Bushy eyebrows, synophrys, upslanting palpebral fissures, broad nasal root, anteverted nostrils, long philtrum, downturned corners of mouth. (C,D) Similar phenotype in older children. (A from *MT Mulcahy et al,* Ann Genet **22:**217, 1979; B from *L Tranebjaerg et al,* Clin Genet **32:**137, 1987; C,D from *A Falek et al,* Pediatrics **37:**92, 1966.)

have been reported, about 60% resulting from a familial translocation (8). Although the degree of region duplication has been rather variable, the phenotype has been remarkably constant. At least one-third have died before the end of the first year of life as a result of heart malformation and infections (8). Mental retardation is severe, and underlying brain anomalies (polymicrogyria, hypoplastic olfactory bulbs) and seizures have been noted in at least 85%. Craniofacial anomalies are remarkably constant. Hypertrichosis and synophrys are marked and most infants have abnormal head shape due to craniosynostosis. Palpebral fissures are often upslanting. There is a broad nasal root with anteverted nostrils. The upper lip is long and the maxilla is prominent. The corners of the mouth are downturned. The pinnae are malformed and approximately 80% have cleft palate. The neck is short and occasionally webbed (1–10) (Fig. 4–7A–D).

Hand anomalies include clinodactyly, camptodactyly, single palmar crease, and nail hypoplasia. In extreme cases, the hand may resemble that seen in trisomy 18. Talipes is present in at least 65%.

Omphalocele occurs in about 25%. Various urogenital abnormalities (micropenis, hypospadias, cryptorchidism, ambiguous genitalia, absent scrotal folds) have been found in approximately 50% and cardiac defects (ASD, double outlet right ventricle, subaortic stenosis, aberrant right subclavian artery) were noted in at least 75%.

The clinician should have little difficulty in differentiating dup(3q) syndrome from the *Brachmann–de Lange syndrome.* The latter far more frequently has severe intrauterine growth retardation, prominent philtrum, oligodactyly, proximally placed thumbs, and syndactyly of toes 2–3. Far more frequent in dup(3q) are craniosynostosis, cleft palate, and urinary tract anomalies.

Approximately 75% of the cases are derived from parental rearrangements involving a pericentric inversion of chromosome 3 or balanced translocation. Duplication of the 3q25→qter region is sufficient to generate the characteristic face, although a slightly more severe phenotype is produced by complete duplication of 3q.

#### References [dup(3q) syndrome]


1. Annerén G, Gustavson KH: Partial trisomy 3q (3q25→qter) syndrome in two siblings. *Acta Paediatr Scand* **73**:281–284, 1984.
2. Falek A et al: Familial de Lange syndrome with chromosomal abnormalities. *Pediatrics* **37**:92–101, 1966.
3. Francke U: Clinical syndromes associated with partial duplications of chromosomes 2 and 3: dup(2p), dup(2q), dup(3p), dup(3q). *Birth Defects* **14**(6C):191–217, 1978.
4. Francke U, Opitz JM: Chromosome 3q duplication and the Brachmann–de Lange syndrome. *J Pediatr* **95**:161–162, 1979.
5. Rosenfeld W et al: Duplication 3q: Severe manifestations in an infant with duplication of a short segment of 3q. *Am J Med Genet* **10**:187–192, 1981.
6. Steinbach P et al: The dup(3q) syndrome: Report of 8 cases and review of the literature. *Am J Med Genet* **10**:159–177, 1981.
7. Stengel-Rutkowski S et al: Partial trisomy 3q. *Eur J Pediatr* **130**:111–125, 1979.
8. Tranebjaerg L et al: Partial trisomy 3q syndrome inherited from familial t(3;9)(q26.1;23). *Clin Genet* **32**:137–143, 1987.
9. Wilson GN et al: The association of chromosome 3 duplication and the Cornelia de Lange syndrome. *J Pediatr* **93**:783–788, 1978.
10. Wilson GN et al: Further delineation of the dup(3q) syndrome. *Am J Med Genet* **22**:117–123, 1985.


## dup(4p) syndrome

In most cases the partial trisomy has been for most or all of the short arm. There does not appear to be good correlation between the amount of short arm replication and clinical stigmata. Over 50 cases have been reported. There is no sex predilection (3). About 75% have arisen from a balanced parental translocation, with most involving an acrocentric chromosome. Death has occurred prior to the age of 2 years in about 25%. Intelligence quotients have ranged from 20 to 65 (1–11).

In addition to the constant finding of psychomotor retardation, noted with a frequency of 50% or greater, are postnatal growth retardation, microcephaly, prominent glabella or supraorbital ridge, deep-set eyes, hypertelorism, short (usually large) nose with rounded tip, broad and low nasal bridge, apparent macrostomia, pointed or prominent chin, malformed, large posteriorly angulated pinnae, and short neck with low hairline (Figs. 4–8A,B, 4–9, and 4–10). Puberty is often delayed and micropenis is frequent. Somewhat less common (25–49%) are neonatal feeding problems, seizures, strabismus that resolves in 6 months, hypertonia, hypotonia, flexion contractures, scoliosis, clinodactyly of fifth fingers, camptodactyly, abnormally long fingers, hallux valgus, prominent heels, widely spaced nipples, cryptorchidism, hypoplastic scrotum, and hypospadias. Additional eye anomalies, seen in 15–20%, include microphthalmia, nystagmus, downward slanting palpebral fissures, asymmetry in size of eyes or pupils, synophrys, and long bushy eyebrows. Inguinal hernia and congenital heart disease of various types are found in about 15% (2–4, 6).

There is an increased frequency of fingertip whorls (7).

#### References [dup(4p) syndrome]


1. Andrle M et al: Two cases of trisomy 4p with translocation t(4p−,7q+) in several members of one family. *Hum Genet* **33**:155–160, 1976.
2. Crane J et al: 4p trisomy syndrome: Report of 4 additional cases and segregation analysis of 21 families with different translocations. *Am J Med Genet* **4**:219–229, 1979.

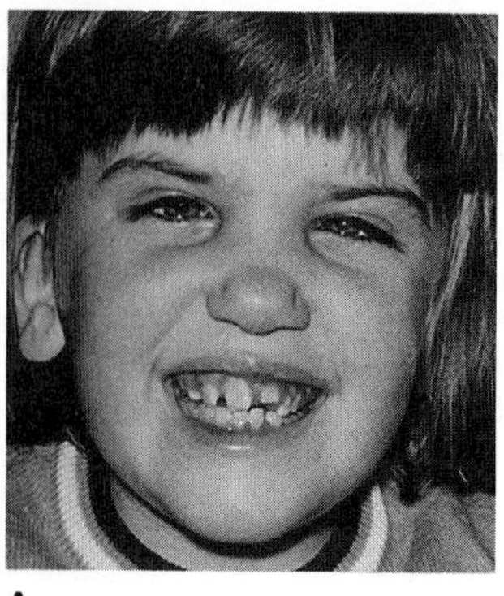

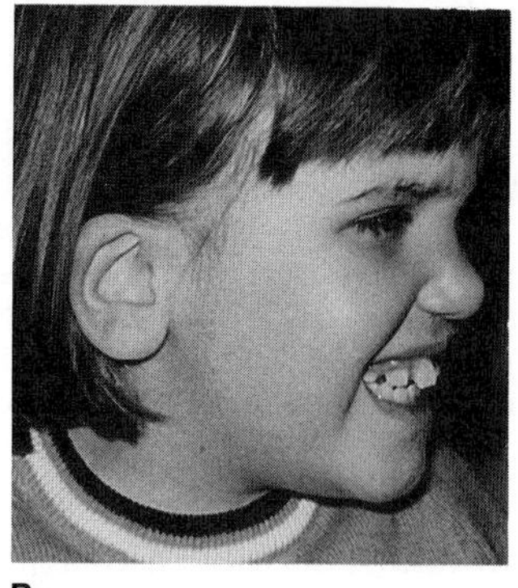

A B

Fig. 4–8. *dup(4p) syndrome.* (A,B) Prominent glabella, narrow palpebral fissures, apparent hypertelorism, bulbous nasal tip, posteriorly rotated pinnae with prominent anthelix. (From *TWJ Hustinx et al,* Ann Génét **18:**13, 1975.)

Fig. 4–9. *dup(4p) syndrome.* Round asymmetric face with chubby cheeks, somewhat bulbous nose, unilateral ptosis of eyelid, epicanthal folds, enophthalmos, malformed pinnae, short neck. (From *JG Mortimer et al,* Hum Hered **30:**58, 1980.)

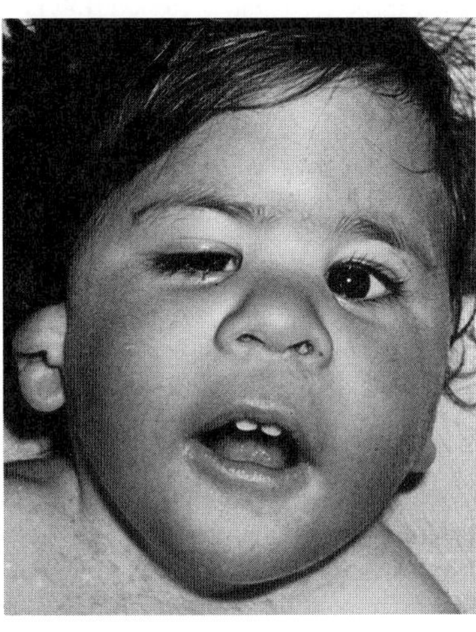

Fig. 4–10. *dup(4p) syndrome.* Prominent forehead and glabella, downslanting palpebral fissures, telecanthus, strabismus, flat nasal bridge, bulbous nasal tip, long philtrum, thin upper lip, short neck. (From *RC Rogers et al,* Proc Greenwood Genet Ctr **5:**29, 1986.)

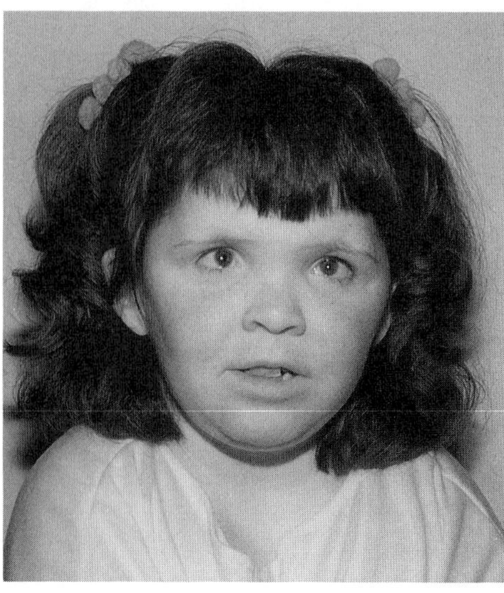


3. Dallapiccola B et al: Trisomy 4p: Five new observations and overview. *Clin Genet* **12**:344–356, 1977.
4. Gonzalez CH et al: The trisomy 4p syndrome: Case report and review. *Am J Med Genet* **1**:137–156, 1977.
5. Keren G et al: The trisomy 4p syndrome. *Eur J Pediatr* **138**:273–274, 1982.
6. Kessel E et al: Der Phänotyp der Trisomie des kurzen Arms eines Chromosoms Nr. 4. *Klin Pädiatr* **188**:215–219, 1976.
7. Mastroiacovo P et al: Hand dermatoglyphics in trisomy 4p. *Hum Genet* **34**:271–276, 1976.
8. Mortimer JG et al: Trisomy 4p and deletion 4p− in a family living a translocation t(4p−;12p+). *Hum Hered* **28**:132–140, 1978.
9. Qazi QH et al: Partial chromosome 4 trisomy. *Clin Genet* **20**:179–184, 1981.
10. Reynolds JF et al: Trisomy 4p in four relatives: Variability and lack of distinctive features in phenotypic expression. *Clin Genet* **24**:365–374, 1983.
11. Rogers RC et al: Partial trisomy of 4p. *Proc Greenwood Genet Ctr* **5**:29–38, 1986.


## del(4q) syndrome

In about 10% of the cases, one of the parents is a balanced translocation carrier. The critical deletion segment is 4q31→qter. Over 30 patients have been described. About 70% die within the first 2 years of life from congenital heart disease or from aspiration pneumonia (1–10). A most comprehensive survey is that of Lin et al (4).

Although birth weight is normal, mild postnatal growth retardation and mild to moderately severe mental retardation are common features. Craniofacial anomalies include microcephaly, low-set posteriorly angulated pinnae with malformed helices, short nose having low nasal bridge and anteverted nostrils, and micrognathia. Cleft palate with or without cleft lip is common (75%) as are hypertelorism, epicanthal folds, and laterally displaced inner canthi (7) (Figs. 4–11A,B and 4–12A,B). Small upward slanting palpebral fissures, oropharyngeal hypotonia, cardiac defects (mostly ASD and VSD), limited extension at the elbows, clinodactyly and tapering of the fifth finger with a pointed nail, transverse palmar crease, and overlapping toes occur (Fig. 4–12C). Anomalies having a frequency of 25% or less include low-set posteriorly angulated pinnae and various skeletal anomalies (proximately implanted thumbs, camptodactyly, short distal phalanx of the fifth finger, absence of the fourth metacarpal, dislocated hips, and talipes equinovarus).

### References [del(4q) syndrome]


1. Davis JM et al: The del(4)(q31) syndrome—a recognizable disorder with atypical Robin malformation sequence. *Am J Med Genet* **9**:113–117, 1981.
2. Frappaz D et al: Le syndrome de deletion terminale du bras long du chromosome 4. *Pédiatrie* **38**:261–270, 1983.
3. Frias JL et al: Deletion of the long arm of chromosome 4: A clinically identifiable syndrome? *Birth Defects* **14**(6C):355–358, 1978.
4. Lin AE et al: Interstitial and terminal deletions of the long arm of chromosome 4: Further delineation of phenotypes. *Am J Med Genet* **31**:533–548, 1988.
5. Lipson A et al: Partial deletion of the long arm of chromosome 4: A clinical syndrome. *J Med Genet* **19**:155–157, 1982.
6. Mitchell JA et al: Deletions of different segments of the long arm of chromosome 4. *Am J Med Genet* **8**:73–89, 1981.
7. Sandig KR et al: The partial 4q monosomy. *Eur J Pediatr* **138**:254–257, 1982.
8. Stamberg J et al: Terminal deletion (4)(q33) in a male infant. *Clin Genet* **21**:125–129, 1982.
9. Tomkins DJ et al: Two children with deletion of the long arm of chromosome 4 with breakpoint at band q33. *Clin Genet* **22**:348–355, 1982.
10. Townes PL et al: The 4q− syndrome. *Am J Dis Child* **133**:383–385, 1979.

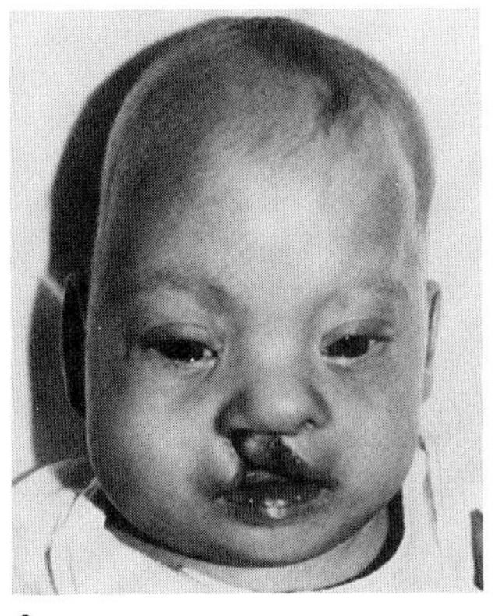

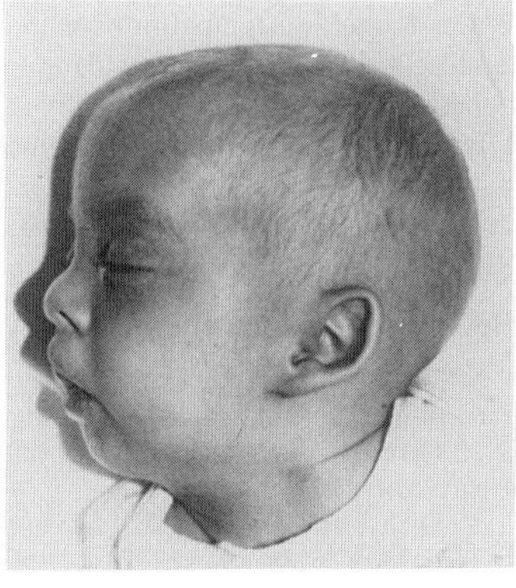

A B

Fig. 4–11. *del(4q) syndrome.* (A,B) Asymmetric forehead, upslanting palpebral fissures, pointed helix, epicanthal folds, anteverted nostrils, cleft lip–palate. [From *JL Frias et al,* Birth Defects **14**(6C):355, 1978.]

Fig. 4–12. *del(4q) syndrome.* (A,B) Upslanting palpebral fissures, short nose with anteverted nostrils, depressed nasal bridge, and micrognathia. Child had cleft palate. (C) Tapering and abbreviation of fifth finger with pointed nail. (A,B courtesy of *SK Clarren,* Seattle, Washington. C from *AE Lin et al,* Am J Med Genet **31**:533, 1988.)

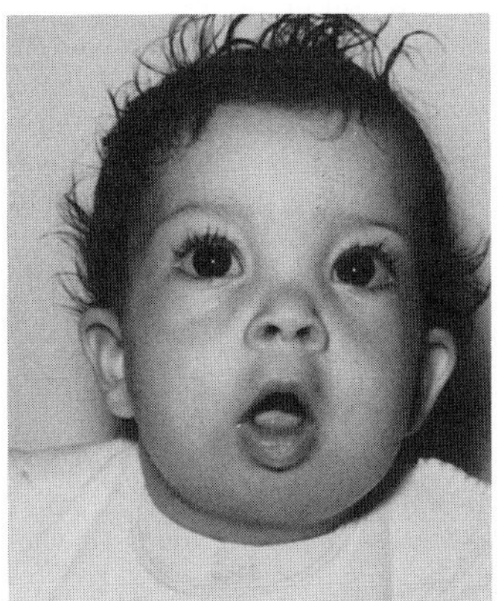

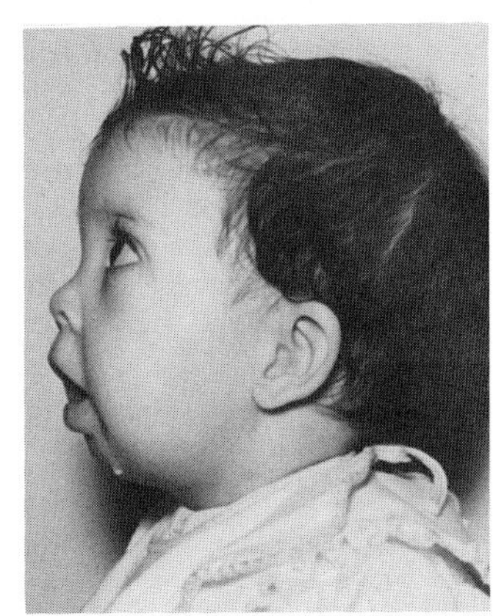

A B

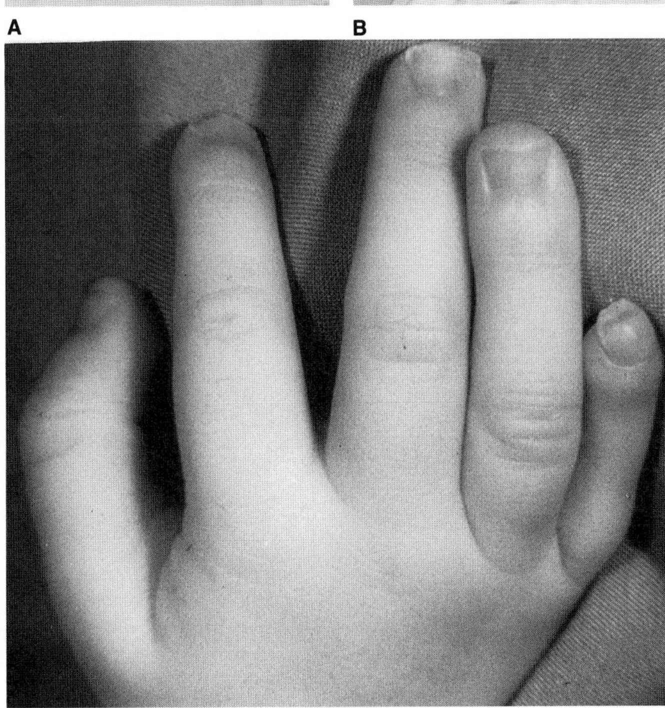

C

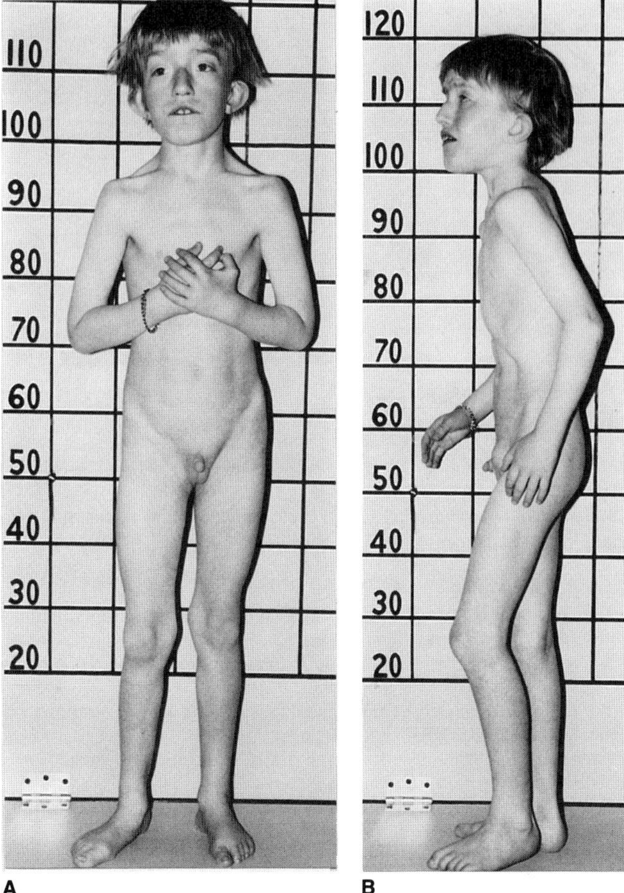


A B

Fig. 4–13. *dup(4q) syndrome.* (A,B) Broad forehead, downslanting palpebral fissures, large outstanding pinnae, pectus carinatum, hypoplastic genitalia. (From *RS Sparkes et al,* Ann Génét **20**:31, 1977.)

## dup(4q) syndrome

In 90% of the cases one of the parents has a balanced translocation. These have ranged from 4q21→qter to 4q32→qter. There is poor correlation with clinical findings (1–7).

Prognosis is poor because of marked psychomotor retardation. Neonatal morality is about 30%, largely in those with more severe cardiac or renal anomalies.

In addition to severe psychomotor retardation, cardiac and genitourinary anomalies are frequent. Birth weight is low in about half the cases. The craniofacies is often characterized by microcephaly with sloping forehead, epicanthal folds, bushy eyebrows, hypertelorism, strabismus, narrow downslanting palpebral fissures, large or prominent nasal bridge, straight nasofrontal angle, short philtrum with protruding lateral margins, downturned corners of mouth, horizontal dimple beneath the lower lip, low-set posteriorly angulated or malformed pinnae with prominent anthelix and hypoplastic tragus, pointed chin, micrognathia, and short neck (Figs. 4–13A,B and 4–14A,B). Hyper- or hypotonia has been noted in over 60% and umbilical or inguinal hernia in about 30%. About half the patients have cardiovascular anomalies, including tetralogy of Fallot and various venous-return anomalies, and about half exhibit genitourinary abnormalities (horseshoe kidney, renal hypoplasia, urethrovesicular reflux with or without hydronephrosis). In males, cryptorchidism is a constant feature. The hands may have unusual form, and rockerbottom feet are common.

### References [dup(4q) syndrome]

1. Andrle M et al: Partial trisomy 4q in two unrelated cases. *Hum Genet* **49**:179–183, 1979.

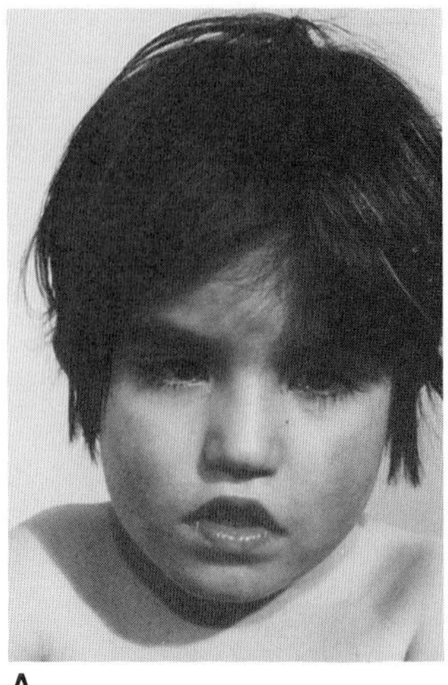

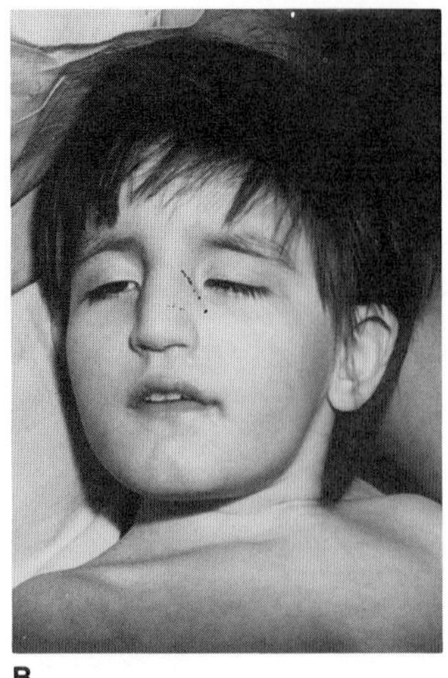

Fig. 4–14. *dup(4q) syndrome.* (A,B) Note similar facial features in two unrelated patients. (From *M Andrle,* Hum Genet **49:**179, 1979.)


2. Bonfante A et al: Partial trisomy 4q: Two cases resulting from a familial translocation t(4;18)(q27;p11). *Hum Genet* **52**:85–90, 1979.

3. Cervenka J et al: Partial trisomy 4q: Case report and review. *Hum Genet* **34**:1–7, 1976.

4. Fryns JP, van den Berghe H: Partial duplication of the long arm of chromosome 4. *Ann Génét* **23**:52–53, 1980.

5. Nielsen J et al: A family with a high risk of segregation for an autosomal unbalanced reciprocal translocation. *Hum Genet* **32**:343–348, 1976.

6. Sparkes RS et al: Partial 4q duplication due to inherited der (20), t(4;20)(q25;q13) mat. *Ann Génét* **20**:31–35, 1977.

7. Stella M et al: Partial trisomy 4q: Two cases with a familial translocation t(4;18)(q27;q23). *Hum Genet* **47**:245–251, 1979.


## dup(5p) syndrome

In addition to severe psychomotor retardation, the main features of this disorder are postnatal growth retardation, hypotonia, seizures, slender extremities with long fingers, short first toes, and club feet. Craniofacial changes include macrodolichocephaly, hypertelorism, upward slanting palpebral fissures, narrow palpebral fissures, epicanthic folds, low nasal bridge, bulbous nose, lowset pinnae, jowly appearance, long philtrum, full lips, and macroglossia (Fig. 4–15A–D). Dermatoglyphic changes include excess ulnar loops and arch tibial patterns in the hallucal area (2,3). An extensive review is that of Kleczkowska et al (4). About 40 cases have been reported (1–7).

### References [dup(5p) syndrome]


1. Brimblecombe FS et al: Complete 5p trisomy: 1 case and 19 translocation carriers in 6 generations. *J Med Genet* **14**:271–275, 1977.

2. Carnevale A et al: A clinical syndrome associated with dup (5p). *Am J Med Genet* **13**:279–283, 1982.

3. Di Liberti JH et al: Trisomy 5p: Delineation of clinical features. *Birth Defects* **13**(3C):185–194, 1977.

4. Kleczkowska A et al: Trisomy of the short arm of chromosome 5: Autopsy data in a malformed newborn with inv dup(5)(13.1→p15.3). *Clin Genet* **32**:49–56, 1987.

5. Leschot NJ, Lim KS: Complete trisomy 5p: De novo translocation 6(2;5)(q36;p11) with isochromosome 5p. *Hum Genet* **46**:271–278, 1979.

6. Opitz JM, Patau K: A partial trisomy 5p syndrome. *Birth Defects* **11**(5):191–200, 1975.

7. Vowles M et al: Trisomy 5p; A second case occurring in a previously described kindred. *J Med Genet* **21**:144–156, 1984.


## del(5q) syndrome

The effect of various deletions of the long arm of chromosome 5 has been studied (2,4,6). Mental retardation is severe but growth retardation has not been evident.

Brachycephaly, abundant coarse scalp hair, narrow forehead, epicanthal folds, anteverted nostrils, long deep philtrum, microretrognathia, and short neck characterize the craniofacies (Fig. 4–16A–D). Cleft palate has been noted in a few cases (1,3,5).

Bilateral pes adductus and inguinal hernias have been frequently observed.

### References [del(5q) syndrome]


1. Fielding I, Kristoffersson U: A child with interstitial deletion of chromosome No. 5. *Hereditas* **93**:337–339, 1980.

2. Harprecht-Beato W et al: Interstitial deletion in the long arm of chromosome no. 5. *Clin Genet* **23**:167–171, 1983.


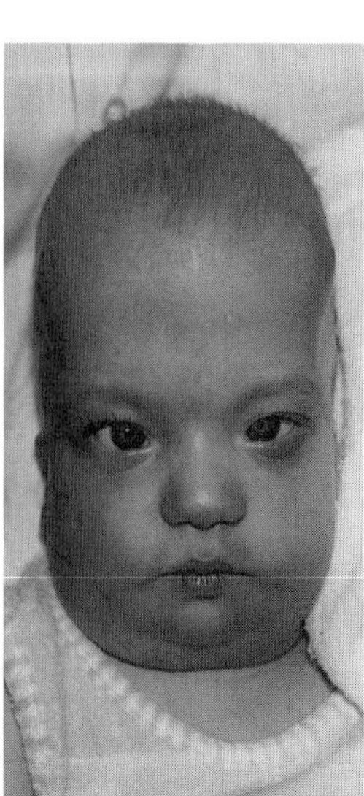

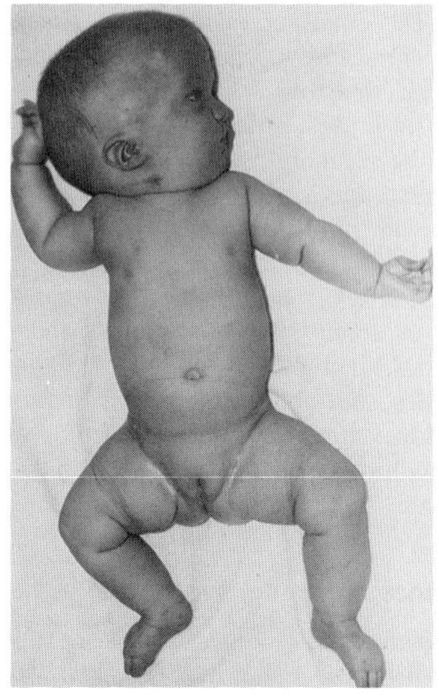

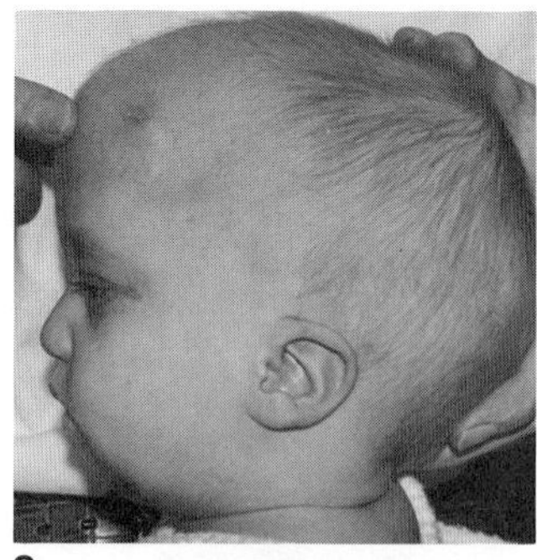

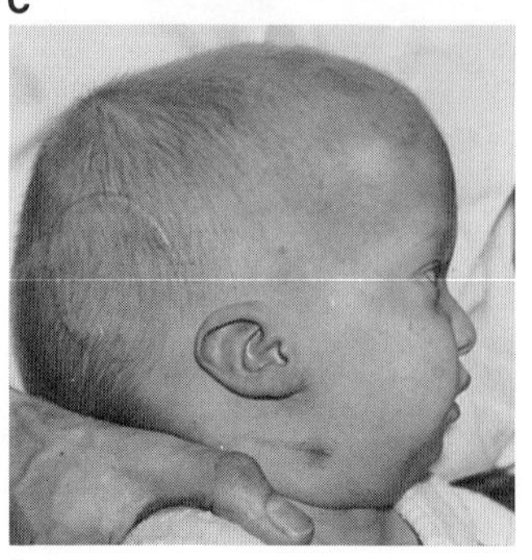

Fig. 4–15. *dup(5p) syndrome.* (A–D) Dolichocephaly, frontal bossing, strabismus, upslanting palpebral fissures, hypertelorism, unusually modeled pinnae, hypotonic posture. [From *JM Opitz* and *K Patau,* Birth Defects **11**(5):191, 1975.]

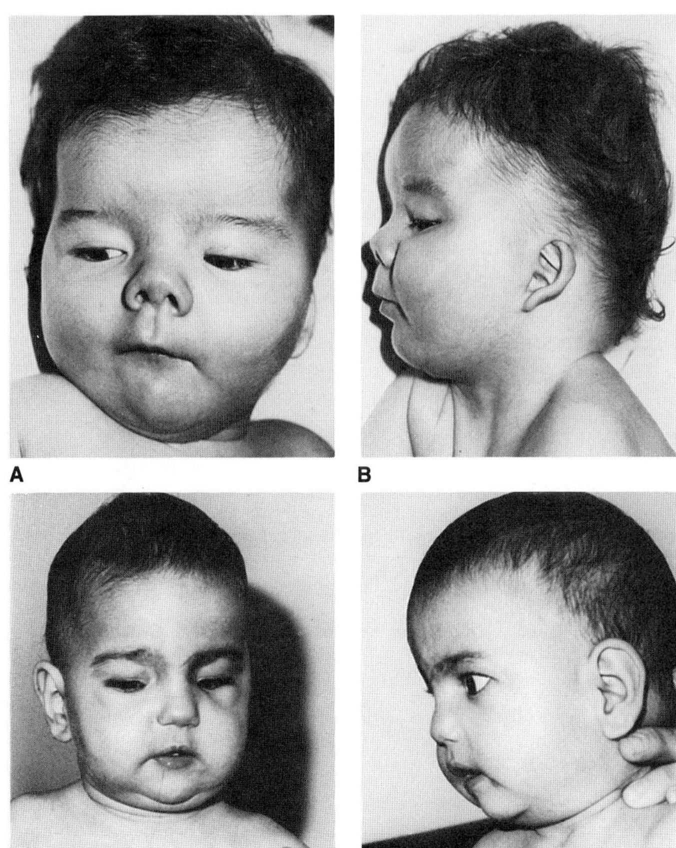


Fig. 4–16. *del(5q) syndrome.* (A,B) Brachycephaly, short nose with anteverted nostrils, long deep philtrum, microretrognathia. (C,D) Note similarity of phenotypes. (A,B from *W Harprecht-Beato et al,* Clin Genet **23:**167, 1983; C,D from *C Stoll et al,* J Med Genet **17:**486, 1980.)

3. Ohdo S et al: Interstitial deletion of the long arm of chromosome 5: 46,XX,del(5)(q13q22). *J Med Genet* **19**:479, 1982.

4. Rodewald A et al: Interstitial de novo deletion of the long arm of chromosome 5: Mapping of 5q bands associated with particular malformations. *Clin Genet* **22**:226–230, 1982.

5. Silengo MC et al: Interstitial deletion of the long arm of chromosome No. 5 in two unrelated children with congenital anomalies and mental retardation. *Clin Genet* **19**:174–180, 1981.

6. Stoll C et al: Interstitial deletion of the long arm of chromosome 5 in a deformed boy: 46,XY,del(5)(q13q15). *J Med Genet* **17**:486–487, 1980.

## dup(5q) syndrome

Trisomy for 5q31→5qter results in severe psychomotor retardation, low birth weight, microcephaly, high forehead, hypertelorism, downward slanting palpebral fissures, epicanthal folds, strabismus, large outstanding pinnae, long philtrum, large upper lip, and micrognathia. Musculoskeletal anomalies include brachydactyly, clinodactyly, preaxial polydactyly, and hernia (2–8). Various congenital heart defects, mostly VSD and ASD, are found in about one-half the cases.

In trisomy 5q13→q22, the children have high forehead, bulbous nose, short philtrum, large protruding pinnae, and micrognathia (3).

A smaller deletion 5q34→qter results only in failure to thrive and strabismus (1,3).

### References [dup(5q) syndrome]

1. Curry CJ et al: Partial trisomy for the distal long arm of chromosome 5 (region q34→qter): A new clinically unrecognized syndrome. *Clin Genet* **15**:454–461, 1979.

2. Elias-Jones A: The trisomy (5)(q31→qter) syndrome: Study of a family with a 5(5;14) translocation. *Arch Dis Childh* **63**:427–431, 1988.

3. Gilgenkrantz S et al: Partial proximal trisomy of the long arm of chromosome 5 (q13→q22) resulting from maternal insertions der ins (10,5). *J Med Genet* **18**:465–469, 1981.

4. Jones LA et al: Partial duplication of the long arm of chromosome 5. *Hum Genet* **51**:37–42, 1979.

5. Lazjuk GI et al: Partial trisomy 5q and partial monosomy 5q within the same family. *Clin Genet* **28**:122–129, 1985.

6. Passarge E et al: Fetal manifestation of a chromosomal disorder: Partial duplication of the long arm of chromosome 5(5q33→qter). *Teratology* **25**:221–225, 1982.

7. Rodewald A et al: Partial trisomy 5q: Three different phenotypes depending on different duplication segments. *Hum Genet* **55**:191–198, 1980.

8. Zabel B et al: Partial trisomy for short and long arms of chromosome No. 5. *J Med Genet* **15**:143–147, 1978.

## dup(6p) syndrome

About a dozen cases of this partial trisomy have resulted from parental translocation or inversion duplication. There is low birth weight. Most have died during infancy from respiratory or feeding problems (1–7).

The facies is characterized by high prominent forehand, craniosynostosis, flat occiput, wide fontanel, ptosis, blepharophimosis, cataracts, microcornea, strabismus, flat nasal root, very short nose, long philtrum, thin lips, small mouth with cupid's bow, and large simple low-set pinnae with poorly developed lobes and small pointed chin.

Congenital cardiac abnormalities (ASD, VSD, PDA) are found in 65%. Renal anomalies have included hydronephrosis and hypoplastic kidney. Musculoskeletal abnormalities have involved umbilical or inguinal hernia and talipes equinovarus.

### References [dup(6p) syndrome]

1. Bernheim A et al: Partial trisomy 6p. *Hum Genet* **48**:13–16, 1979.

2. Côte GB et al: Partial trisomy 6p with karyotype 46,XY, der (22), t(6;22)(p22; q13)mat. *J Med Genet* **15**:479–481, 1978.

3. Pagano L et al: Hereditary 3;6 translocation: Three cases of multiple malformations with partial trisomy 6p21→pter. *Ann Génét* **23**:173–175, 1980.

4. Phelan MC et al: Trisomy 6p due to a tandem duplication. *Proc Greenwood Genet Ctr* **5**:39–43, 1986.

5. Rosi G et al: Trisomy 6p22→6pter due to familial t(6;13)(p22;q34 or 33) translocation. *Hum Genet* **51**:67–72, 1979.

6. Smith BS, Petterson JC: An anatomical study of duplication 6p band on two sibs. *Am J Med Genet* **20**:649–652, 1985.

7. Turleau C et al: La trisome 6p partielle. *Ann Génét* **21**:88–91, 1978.

## del(6q) syndrome

Characteristics are mental retardation, microcephaly, facial asymmetry, upslanting palpebral fissures, hypertelorism, microphthalmia, strabismus, epicanthal folds, broad nasal bridge, prominent nose, long philtrum, large low-set dysmorphic pinnae, thin upper lip, micrognathia, cleft palate, and short neck (1,2,4,9).

Hand anomalies occur in about one-half the cases: clinodactyly of V, single palmar creases, fingernails on dorsal and volar aspect of fifth finger (3,5,6,7,10).

Similar nail anomalies and cleft palate were reported by Roger et al (8) in a retarded child but no karyotype was published.

Various forms of congenital heart disease have been found.

### References [del(6q) syndrome]

1. Bartoshesky L et al: Developmental abnormalities associated with long arm deletion of chromosome No. 6. *Clin Genet* **13**:68–71, 1978.

2. Goldberg R et al: Deletion of a portion of the long arm of chromosome 6. *Am J Med Genet* **5**:73–80, 1980.

3. Kalisman M et al: Dorsal skin and fingernails on the volar aspect of the hand: An unusual autosomal deformity. *Plast Reconstr Surg* **69**:694–696, 1982.

4. Kueppers F et al: Exclusion of the HLA locus for a large portion of the long arm of chromosome 6. *Hum Hered* **27**:242–246, 1977.

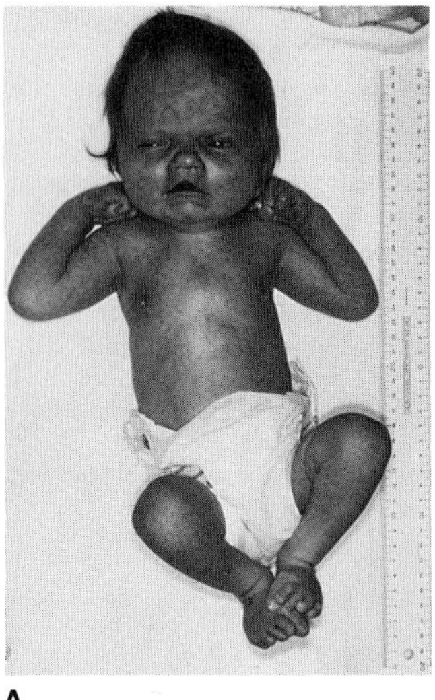

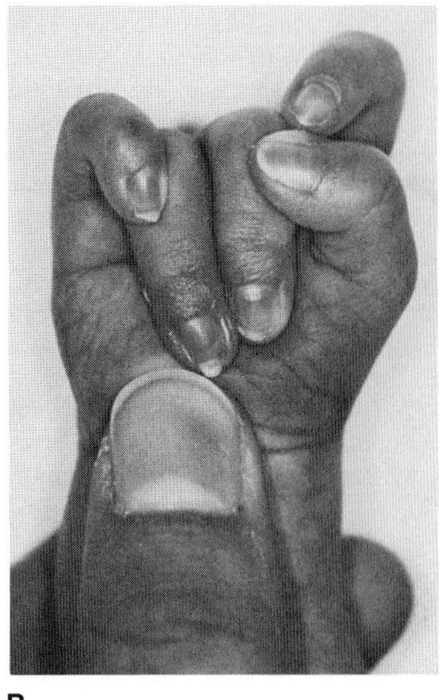

A B

Fig. 4–17. *dup(6q) syndrome.* (A) High prominent forehead, round flat face, large fontanel, wide metopic suture, bow-shaped mouth, thin lips, short philtrum, micrognathia, short neck, wide-spread nipples, and flexion contractures of elbows. (B) Flexion contracture of fingers with camptodactyly and overriding digits similar to those in trisomy 18. (From *A Sommer et al,* Proc Greenwood Genet Ctr **3**:22, 1983.)

5. Liberfarb RM et al: Chromosome 6q and associated malformations. *Ann Génét* **21**:223–225, 1978.

6. McNeal RM et al: Congenital anomalies including the VATER association in a patient with a del(6)q deletion. *J Pediatr* **91**:957–960, 1977.

7. Milosevic J, Kalicanin P: Long arm deletion of chromosome No. 6 in mentally retarded boy with multiple physical malformations. *J Ment Defic Res* **19**:139–144, 1975.

8. Roger H et al: Onychohétérotopie avec polynychie associée à un syndrome de Pierre Robin: à propos d'une nouvelle observation. *Ann Dermatol Venereol* **113**:235–242, 1986.

9. Yamamoto Y et al: Deletion of proximal 6q: A clinical report and review of the literature. *Am J Med Genet* **25**:467–471, 1986.

10. Young RS et al: Deletions of the long arm of chromosome 6: Two new cases and review of the literature. *Am J Med Genet* **20**:21–29, 1985.

## dup(6q) syndrome

About 20 patients are known, all arising from balanced parental translocation or inversion (1–10).

The facies is characterized by microbrachycephaly or turricephaly, prominent forehead, flat face and occiput, almond-shaped eyes, microphthalmia, hypertelorism, downslanting palpebral fissures, broad flat nasal bridge, bow-shaped mouth with thin lips, short nose, short philtrum, micrognathia, short webbed neck, and low-set ears with thick earlobes. The lower lip may have a median indentation (Figs. 4–17A,B and 4–18A–C). Cleft lip–palate or bifid uvula has been reported (4,5).

In addition to severe somatic and mental retardation, there are joint contractures, flexed or deviated fingers and wrists, club feet, and scoliosis. In the absence of congenital heart defects, survival is normal. Genital anomalies include hypoplastic labia, penis, and/or scrotum.

#### References [dup(6q) syndrome]

1. Clark CE et al: Trisomy 6q25→6qter in two sisters resulting from maternal 6;11 translocation. *Am J Med Genet* **5**:171–178, 1980.

2. Duca D et al: Familial partial trisomy: 6q25→6qter. *J Génét Hum* **28**:31–37, 1980.

3. Fitch N: Partial trisomy 6. *Clin Genet* **14**:181–185, 1978.

4. Neu RL et al: An infant with trisomy 6q21→6qter. *Ann Génét* **24**:167–169, 1981.

5. Pierpont MEM et al: Partial trisomy 6q and bilateral retinal detachment. *Ophthalmol Paediatr Genet* **7**:175–180, 1986.

6. Robertson KP et al: Acrocephalosyndactyly and partial trisomy 6. *Birth Defects* **11**(5):267–271, 1975.

7. Schmid W et al: Trisomy 6q25→6qter with a severely retarded 7-year-old boy with turricephaly, bow-shaped mouth, hypogenitalism and club feet. *Hum Genet* **46**:279–284, 1979.

8. Taysi K et al: Trisomy 6q22→6qter due to maternal 6;21 translocation. *Ann Génét* **26**:243–246, 1983.

9. Tipton RE et al: Duplication 6q syndrome. *Am J Med Genet* **3**:325–330, 1979.

10. Turleau C, de Grouchy J: Trisomy 6qter. *Clin Genet* **19**:202–206, 1981.

## del(7p) syndrome

The syndrome is characterized by variable craniosynostosis (turricephaly, microcephaly, flat occiput, prominent forehead, craniosynostosis (40%), trigonocephaly, cranial asymmetry, etc.) (2,4–6). Intelligence has varied from severe retardation to normal. Facial nevus flammeus, hypotelorism, downward slanting palpebral fissures, ptosis, epicanthal folds, saddle nose, small low-set dysplastic pinnae, and transverse palmar creases are frequent (1–10). Congenital heart disease and genital malformations are found in 50% (9). Cleft palate has been described (2,6,7) (Fig. 4–19A,B).

#### References [del(7p) syndrome]

1. Crawfurd M d'A et al: Partial monosomy 7 with interstitial deletions in two infants with differing congenital abnormalities. *J Med Genet* **16**:453–460, 1979.

2. Dhadial RK, Smith MF: Terminal 7p deletion and 1;7 translocation associated with craniosynostosis. *Hum Genet* **50**:285–289, 1979.

3. Friedrich U et al: A girl with karyotype 46,XX,del(7)(qter→p15). *Humangenetik* **26**:161–165, 1975.

4. Fryns JP et al: De novo partial 2q3 trisomy/distal 7p22 monosomy in a malformed new born with 7p deletion phenotype and craniosynostosis. *Ann Génét* **28**:45–48, 1985.

5. Garcia-Esquivel L et al: De novo del (7)(pter→p21.2::p15.2→qter) and craniosynostosis. *Ann Génét* **29**:36–38, 1986.

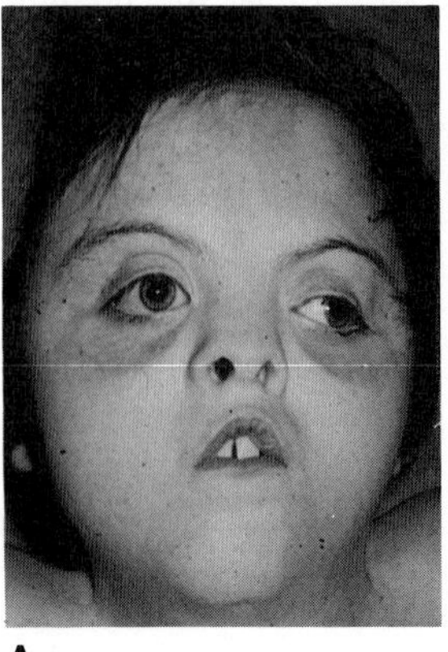

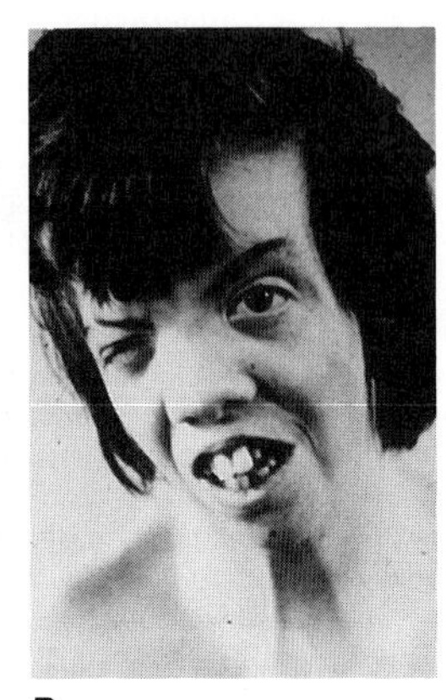

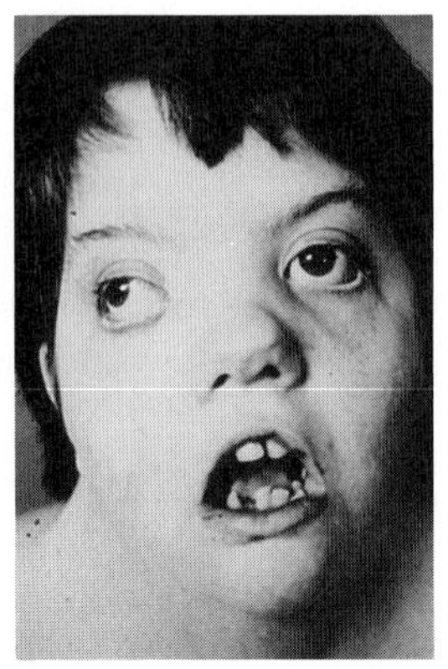

A B C

Fig. 4–18. *dup(6q) syndrome.* (A–C) Microcephaly, acrocephaly, prominent forehead, almond-shaped eyes, hypertelorism, downslanting palpebral fissures, flat facial profile with depressed nasal bridge and malar areas, carp mouth, micrognathia, and extremely short webbed neck. B and C are sisters. (A courtesy of *RE Tipton,* Memphis, Tennessee; B,C courtesy of *C Clark,* Wilmington, Delaware.)

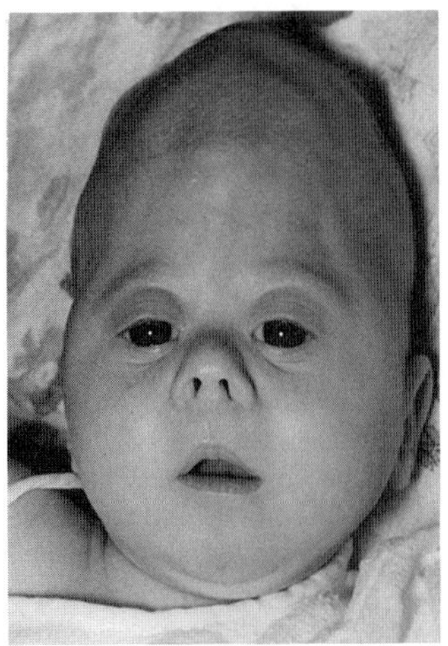
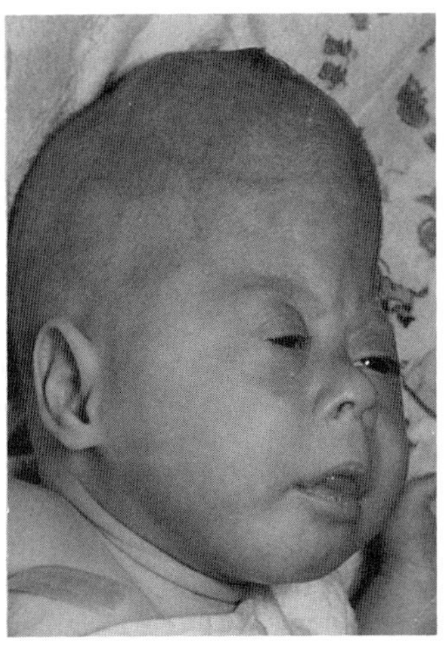

A B

Fig. 4–19. *del(7p) syndrome.* (A,B) Turribrachycephaly with prominent metopic and coronal areas, hypotelorism, short palpebral fissures, epicanthal folds, prominent eyeglobes, short nose with anteverted nostrils. (Courtesy of *JG Hall,* Vancouver, British Columbia, Canada.)

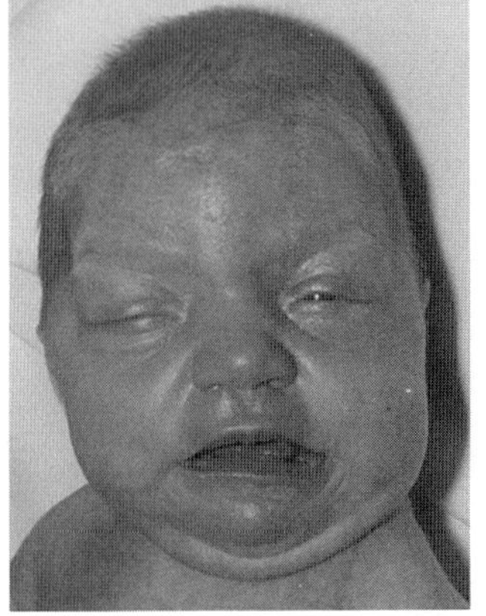
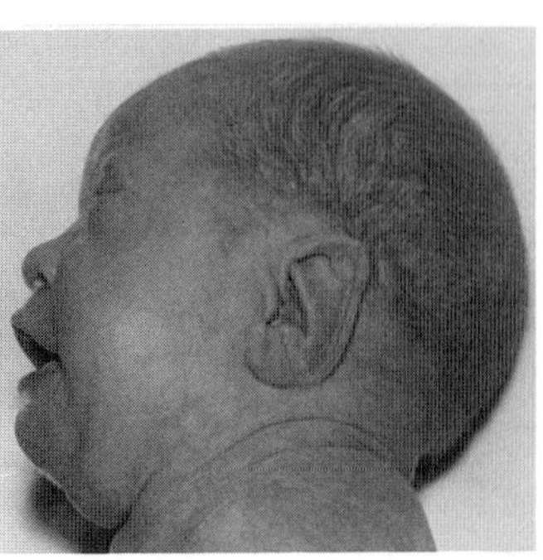

A B

Fig. 4–21. *del(7q) syndrome.* (A,B) Prominent forehead with broad nasal bridge and bulbous tip. Large dysplastic pinna, large mouth, micrognathia, and short neck. (Courtesy of *G Schwanitz,* Erlangen, Nürnberg, West Germany.)

5a. Hinkel GK et al: 7p− Deletions-Syndrom. *Mschr Kinderheilkd* **136**:824–827, 1988.

6. McPherson E et al: Chromosome 7 short arm deletion and craniosynostosis: A 7p− syndrome. *Hum Genet* **35**:117–123, 1976.

7. Miller M et al: Familial balanced insertional translocation of chromosome 7 leading to offspring with deletion and duplication of the inserted segment, 7p15→7p21. *Am J Med Genet* **4**:323–332, 1979.

8. Nakano S, Miyamoto N: A ring C7 chromosome in a mentally and physically retarded male with various somatic abnormalities. *Jpn J Hum Genet* **22**:33–41, 1977.

9. Schömig-Spingler M et al: Chromosome 7 short arm deletion, 7p21→pter. *Hum Genet* **74**:323–325, 1986.

10. Zackai EH, Breg WR: Ring chromosome 7 with variable phenotype expression. *Cytogenet Cell Genet* **12**:40–48, 1973.

## dup(7p) syndrome

Patients, severely retarded, rarely survive infancy. Dolichocephaly, wide fontanels, hypertelorism, full cheeks, short beaked nose, and micrognathia characterize the face. Cleft palate has been described (3,5). Willner (6) found craniosynostosis, choanal atresia, arachnodactyly, congenital hip dislocation, and club foot (Fig. 4–20A,B).

Fig. 4–20. *dup(7p) syndrome.* (A,B) Large fontanel, hypertelorism, upslanting palpebral fissures, full cheeks, short beaked nose, dysplastic pinna, downturned corners of mouth, micrognathia, short neck. (Courtesy of *G Schwanitz,* Erlangen, Nürnberg, West Germany.)

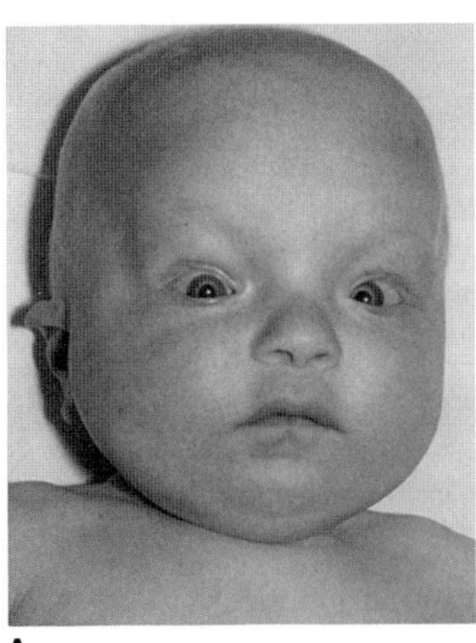
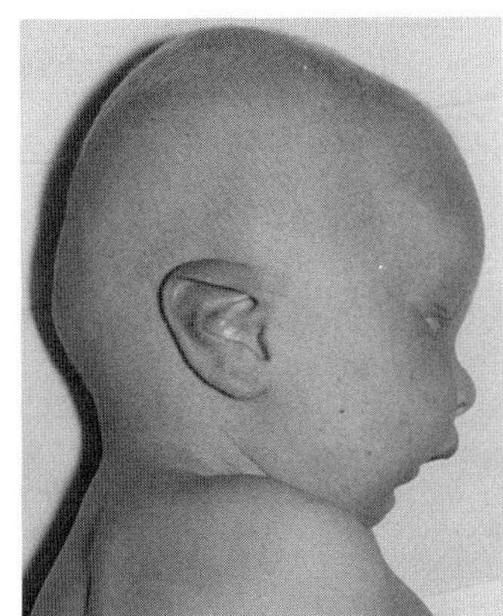

A B

Skeletal anomalies are frequent but of no recurrent type.

Various congenital cardiac anomalies have been reported. Transverse palmar crease is frequent.

### References [dup(7p) syndrome]

1. Berry AC et al: Two children with partial trisomy for 7p. *J Med Genet* **16**:320–321, 1979.

2. Carnevale A et al: Partial trisomy of the short arm of chromosome 7 due to a familial translocation rcp(7;14)(p11;11). *Clin Genet* **14**:202–206, 1978.

3. Larson LM et al: Partial trisomy 7p associated with familial 7p;22q translocation. *J Med Genet* **14**:258–261, 1977.

4. Miller M et al: Familial balanced insertional translocation of chromosome 7 leading to offspring with deletion and duplication of the inserted segment, 7p15→7p21. *Am J Med Genet* **4**:323–332, 1979.

5. Moore CM et al: Partial trisomy 7p in two families resulting from different balanced translocations. *Clin Genet* **21**:112–121, 1982.

6. Willner J: Personal communication, 1979.

## del(7q) syndrome

Proximal interstitial deletions involving 7q21→7q32 do not result in a recognizable clinical syndrome (7,8). However, deletion of 7q32→7qter results in pre- and postnatal growth retardation, feeding problems, severe mental retardation, hypotonia, microcephaly with prominent forehead, broad nasal bridge with bulbous tip, various eye anomalies, large dysplastic pinnae, large mouth, micrognathia, and short neck (4,5). Cleft lip with or without cleft palate has been noted in 25% (1–3) (Fig. 4–21A,B and 4–22).

Various congenital heart defects, distal limb anomalies, hypospadias, small penis, and abnormal palmar creases are frequent.

Holoprosencephaly has also been described (6).

### References [del(7q) syndrome]

1. Bernstein R et al: Two unrelated children with distal long arm deletion of chromosome 7: Clinical features, cytogenetic and gene marker studies. *Clin Genet* **17**:228–237, 1980.

2. Harris EL et al: 7q deletion syndrome (7q32→7qter). *Clin Genet* **12**:233–238, 1977.

3. Klep-dePater JM et al: Two cases with different deletions of the long arm of chromosome 7. *J Med Genet* **16**:151–154, 1979.

4. Kousseff BG et al: A partial long arm deletion of chromosome 7:46,XY, del (7)(q32). *J Med Genet* **14**:144–147, 1977.

5. Reynolds JD et al: Ocular abnormalities in terminal deletion of the long arm of chromosome seven. *J Pediatr Ophthalmol Strab* **21**:28–32, 1984.

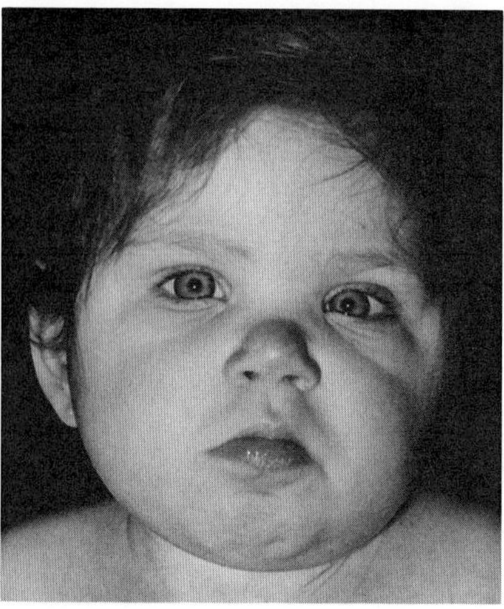

Fig. 4–22. *del(7q) syndrome.* Facies in older child similar to that seen in Fig. 4–21. (Courtesy of *B Biederman,* Edmonton, Alberta, Canada.)

6. Schwartz S et al: Cebocephaly-holoprosencephaly in a newborn girl with a terminal 7q deletion (46,XX,del(7)(pter→q32)). *Am J Med Genet* **15**:141–144, 1983.

7. Stallard R, Juberg RC: Partial monosomy 7q syndrome due to distal interstitial deletion. *Hum Genet* **57**:210–213, 1981.

8. Young RS et al: Terminal and interstitial deletions of the long arm of chromosome 7. *Am J Med Genet* **17**:437–480, 1984.

## dup(7q) syndrome

About 25 patients have been documented. An excellent analysis has been carried out by Johnson et al (4). For purposes of simpler presentation, two rather than six groups will be briefly described. All exhibited mental and somatic retardation with apparently low-set abnormal pinnae, and apparent hypertelorism.

Patients with dup(7q31→7qter) have large fontanels, square prominent forehead, short downslanting palpebral fissures, long eyelashes, short nose, long philtrum, thin vermilion, downcurved upper lip, and micrognathia. All had cleft palate (Fig. 4–23A–C). Anomalies seen in a few cases include ventricular dilatation and severe hypospadias. Death is early (1–4,9).

Those with dup(7q32–7qter) differ by failing to have cleft palate. In addition they exhibit hypotonia, epicanthal folds, scoliosis, congenital hip dislocation, and strabismus (3a,4–6,8–10).

Total dup(7p) has been described in only a few cases (11), but full trisomy 7 has been associated with Potter sequence (7,11).

### References [dup(7q) syndrome]

1. Alfi OS et al: Partial trisomy of the long arm of chromosome No. 7. *J Med Genet* **10**:187–189, 1973.

2. Al Saadi A, Moghadam HA: Partial trisomy of the long arm of chromosome 7. *Clin Genet* **9**:250–254, 1976.

3. Berger R et al: Les trisomies partielles du bras long du chromosome 7. *Nouv Presse Méd* **3**:1801–1804, 1976.

3a. Couzin DA et al: Partial trisomy 7(q32–qter) syndrome in two children. *J Med Genet* **27**:461–465, 1986.

4. Johnson DD et al: Duplication of 7q31.2→7qter and deficiency of 18qter. Report of two patients and literature review. *Am J Med Genet* **25**:477–488, 1986.

5. Klasen M et al: Partial trisomy 7q in two siblings. *Ann Génét* **26**:100–102, 1983.

6. Novales MA et al: Partial trisomy for the long arm of chromosome 7. *Hum Genet* **62**:378–381, 1982.

7. Pflueger SM et al: Trisomy 7 and Potter syndrome. *Clin Genet* **25**:543, 1984.

8. Schmid M et al: Partial trisomy for the long arm of chromosome 7 due to familial balanced translocation. *Hum Genet* **49**:283–289, 1979.

9. Vogel W et al: Partial trisomy 7q. *Ann Génét* **16**:277–280, 1973.

10. Winsor EJ et al: Meiotic analysis of a pericentric inversion, inv (7) (p22q32), in the father of a child with a duplication—deletion of chromosome 7. *Cytogenet Cell Genet* **20**:169–184, 1978.

11. Yunis E et al: Full trisomy 7 and Potter syndrome. *Hum Genet* **54**:13–18, 1980.

## del(8p) syndrome

At least a dozen patients have been described with spontaneous deletion of the distal portion of the short arm of chromosome 8. Intrauterine growth retardation, abnormal facial appearance, congenital heart defects, and genital anomalies in males have been observed in most patients. Later, postnatal growth deficiency, mental retardation, developmental delay, microcephaly, and some lessening of facial changes become evident.

The most prominent facial alterations during the first year of life include a wide low nasal bridge, epicanthal folds, short broad nose, malformed or malpositioned ears, prominent alveolar ridges, and micrognathia. Virtually all facial changes except epicanthal folds disappear with increasing age. Minor hand anomalies are present in about 50%: single transverse palmar crease, prominent palmar creases, and clinodactyly (1–7).

A variety of congenital heart anomalies have been seen in 65% [VSD, PS, ASD, transposition of great arteries] (4–7).

Genitourinary abnormalities have included cryptorchidism, hypospadias, and hypogonadism.

### References [del(8p) syndrome]

1. Bresson JL et al: Délétion partielle du bras court du chromosome 8. *Ann Génét* **20**:70–72, 1977.

2. Dobyns WB et al: Deficiency of chromosome 8p21.1→8pter: Case report and review of the literature. *Am J Med Genet* **22**:125–134, 1985.

3. Leisti J, Aula P: A case of deletion of short arm of chromosome 8. *Birth Defects* **13**(3B):187–194, 1977.

4. Orye E, Craen M: A new chromosome deletion syndrome: Report of a patient with a 46,XY,8p− chromosome constitution. *Clin Genet* **9**:289–301, 1976.

5. Patil SR, Hanson JW: Partial 8p− syndrome. *J Génét Hum* **28**:123–129, 1980.

6. Reiss JA et al: The 8p− syndrome. *Hum Genet* **47**:135–140, 1979.

7. Rodewald A et al: Partial monosomy 8p. An attempt to establish a new chromosome deletion syndrome. *Eur J Pediatr* **125**:45–47, 1977.

## dup(8p) syndrome

Approximately 30 cases have been reported, either spontaneous or resulting from a parental translocation.

dup(8p) results in high forehead with frontal and parietal bossing, but with temple retraction, full cheeks, round face, low nasal bridge, anteverted nostrils, wide mouth, cleft palate and/or bifid uvula, everted lower lip with carp mouth, large earlobes, and a short neck with redundant skin folds (Fig. 4–23D). Severe mental retardation and absence of the corpus callosum has been demonstrated. Various congenital heart malformations have been reported (1–7). An excellent phenotype–segment duplication correlation study is that of Walker and Bocian (7).

Trunk and extremities are long and contractures may restrict movement. Micropenis and cryptorchidism are frequent.

### References [dup(8p) syndrome]

1. Allen EF, Hodgkin WE: Trisomy for 8p21→pter owing to a familial translocation. *J Med Genet* **20**:68–69, 1983.

2. Clark CE et al: A case of partial trisomy 8p resulting from a maternal balanced translocation. *Am J Med Genet* **7**:21–25, 1980.

3. Fineman RM et al: Complete and partial trisomy of different segments of chromosome 8: Case report and review. *Clin Genet* **16**:390–396, 1979.

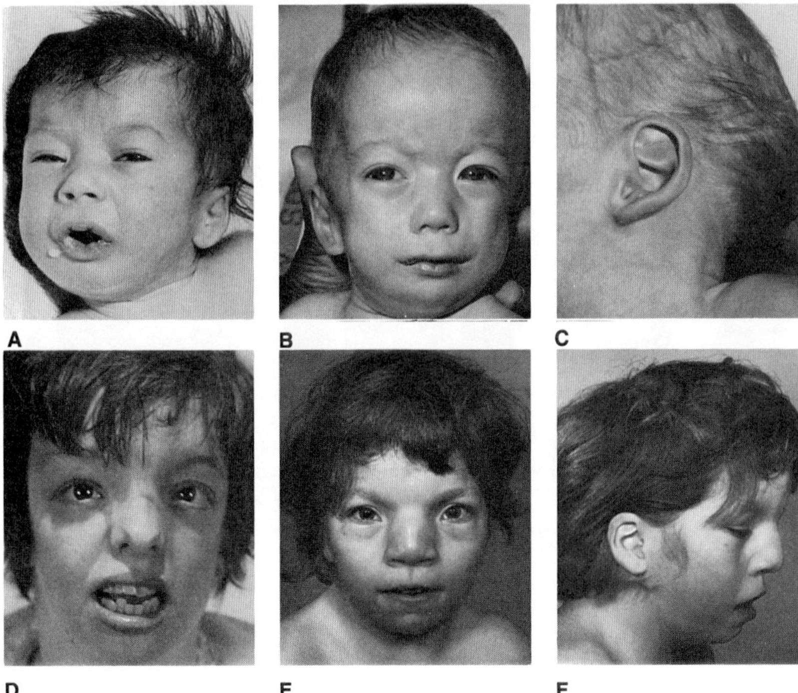


Fig. 4–23. *dup(7q) syndrome.* (A–C) Mild frontal bossing, prominent occiput, small palpebral fissures, low-set pinnae with cupping of outer helices, and micrognathia. *dup(8p) syndrome.* (D) Prominent forehead, strabismus, hypertelorism, narrow palpebral fissures, bulbous nose, everted lower lip, micrognathia. *dup(8q) syndrome.* (E,F) short protruding forehead, hypertelorism, upslanting palpebral fissures, broad flat nose with short septum, short upper lip, and small pinnae. (A from *W Vogel et al,* Ann Génét **16:**277, 1973. B,C from *OS Alfi et al,* J Med Genet **10:**187, 1973. D from *CE Clark et al,* Am J Med Genet **7:**21, 1980. E,F from *A Schinzel,* Hum Genet **37:**17, 1977.)

4. Funderburk SJ et al: Report of a trisomy 8p infant with carrier father. *Ann Génét* **21**:219–222, 1978.

5. Lazjuk GI et al: Trisomy 8p due to the 3:1 segregation of the balance translocation 5(8;15) mat. *Hum Genet* **46**:335–339, 1979.

6. Mattei JF et al: Clinical, enzyme, and cytogenetic investigations in three new cases of trisomy 8p. *Hum Genet* **53**:315–321, 1980.

7. Walker AP, Bocian M: Partial duplication 8q12→q21.1 in two sibs with maternally derived insertional and reciprocal translocations: Case reports and review of partial duplications of chromosome 8. *Am J Med Genet* **27**:3–22, 1987.

## dup(8q) syndrome

Approximately 50 cases of dup(8q) have been reported (7). Apart from mental retardation (IQ 20–70), phenotypic changes have been very inconstant: low birth weight (35%), prominent forehead and flat occiput (25%), hypertelorism (35%), upward slanting palpebral fissures (25%), short nose with broad base (40%), beaked nose (25%), thin upper lip and drooping lower lip (40%), and low-set pinnae (65%) (1–8). Cleft palate has been reported (1,4,8) (Fig. 4–23E,F). An excellent phenotype–segment duplication analysis is that of Walker and Bocian (8).

Skeletal anomalies present in about 40% are pectus excavatum or carinatum and kyphoscoliosis.

Congenital heart defects have been variable.

### References [dup(8q) syndrome]

1. Bowen P et al: Duplication 8q syndrome due to familial chromosome ins(10;8)(q21; q212q22). *Am J Med Genet* **14**:635–646, 1983.

2. Giorgi PL et al: Partial trisomy 8q resulting from maternal translocation t(2;8)(q373;q23). *Acta Genet Med Gemellol* **27**:75–79, 1978.

3. Rethoré MD et al: Chromosome 8: Trisomie complete et trisomies segmentaires. *Ann Génét* **20**:5–11, 1977.

4. Sachs ES, van Waveren G: Phenotype of partial 8 (q21→qter) in two unrelated patients with de novo translocation. *J Med Genet* **18**:204–208, 1981.

5. Schinzel A: Partial trisomy 9q in half-sisters with distinct dysmorphic patterns not similar to the trisomy 9 mosaicism syndrome. *Hum Genet* **37**:17–26, 1977.

6. Schinzel A: Karyotype-phenotype correlation: Mosaic trisomy 8 and partial trisomies of different segments of chromosome 8. *Hum Genet* **41**:363–367, 1978.

7. Townes PL, White MR: Trisomy 8q. *Am J Dis Child* **132**:498–501, 1978.

8. Walker AP, Bocian M: Partial duplication 8q12→q21.2 in two sibs with maternally derived insertional and reciprocal translocations. *Am J Med Genet* **27**:3–22, 1987.

## del(9p) syndrome

Approximately 80 cases of del(9p) have been described (1–12). Several r(9) examples have been documented, most having the phenotype of del(9p) (7). In 20%, the disorder results from a parental translocation. Alterations include mental and somatic retardation but normal birthweight (100%), trigonocephaly (90%), flat occiput (70%), hypertelorism and upward slanting small palpebral fissures (75%), epicanthal folds, broad flat nasal bridge (90%), anteverted nostrils (100%), low-set malformed pinnae with abnormal lobules (85%), low hair line, long philtrum (95%), small mouth with protruding lips (75%), micrognathia (85%), short somewhat webbed neck (95%), pterygium colli (75%), wide-set nipples (100%), congenital heart disease (VSD, PDA, PS) (65%), omphalocele or umbilical hernia (65%), inguinal hernia, long fingers or toes because of dolichomesophalangy, and relative shortness of the metacarpals, square hyperconvex nails, and increased number of fingertip whorls (90%) (2,12). Many of these findings change with age (4). Among 30 cases, three had cleft palate or bifid uvula (10,12). Some patients developed leukemia or lymphoma (1,3) (Figs. 4–24A,B, 4–25, 4–26A,B).

At birth, because of the upward slanting palpebral fissures, a child may be mistaken for having *Down syndrome*.

### References [del(9p) syndrome]

1. Bigner SH et al: 9p− in a girl with acute lymphocytic leukemia and sickle cell disease. *Cancer Genet Cytogenet* **21**:267–269, 1986.

2. Chaves-Carballo E et al: Neurologic aspects of the 9p− syndrome. *Pediatr Neurol* **1**:57–59, 1985.

3. Chilcote RR et al: Lymphocytic leukemia with lymphomatous features associated with abnormalities of the short arm of chromosome 9. *N Engl J Med* **313**:286–291, 1985.

4. Fryns JP et al: Deletion of the short arm of chromosome 9: A clinically recognizable entity. *Eur J Pediatr* **134**:201–204, 1980.

5. Funderburk et al: The 9p− syndrome. *J Med Genet* **16**:75–79, 1979.

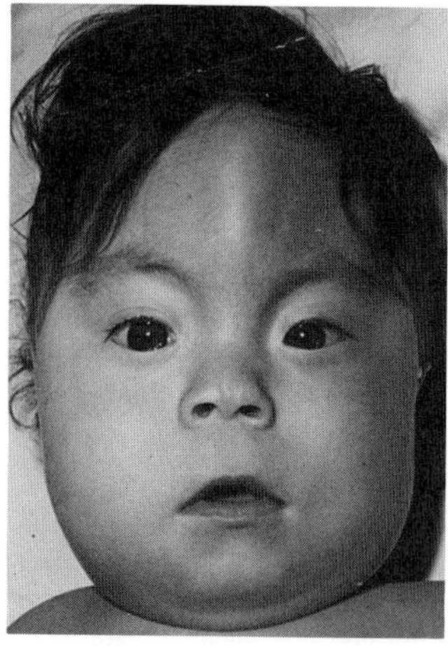

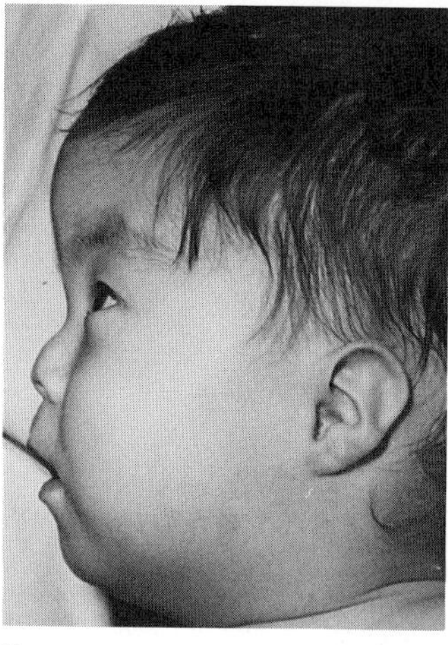

A B

Fig. 4–24. *del(9p) syndrome.* (A,B) Prominent forehead with trigonocephaly, mild upslanting palpebral fissures, apparent hypertelorism, wide nasal bridge, anteverted nostrils, long philtrum, short neck. (From *OS Alfi,* Ann Génét **16:**17, 1973.)

5a. Huret JL et al: Eleven new cases of del(9p) and features from 80 cases. *J Med Genet* **25**:741–749, 1988.

6. Hoo JJ et al: Complex de novo rearrangement of chromosome 9 with clinical features of monosomy 9p syndrome. *Clin Genet* **16**:151–155, 1979.

7. Leung AKC, Rudd NL: A case of ring (9)/del(9p) mosaicism associated with gastroesophageal reflux. *Am J Med Genet* **29**:43–48, 1988.

8. Rutten FJ et al: A case of partial 9p monosomy with some unusual clinical features. *Ann Génét* **21**:51–55, 1978.

9. Syzmánska J et al: 9p− syndrome: Two new observations. *Klin Pädiatr* **196**:121, 1984.

10. Wisniewski L et al: Two new cases of 9p− syndrome. *Klin Pädiatr* **192**:270–274, 1980.

11. Young RS et al: The dermatoglyphic and clinical features of the 9p trisomy and partial monosomy 9p syndromes. *Hum Genet* **62**:31–39, 1982.

12. Young RS et al: Two children with *de novo* del(9p). *Am J Med Genet* **14**:751–757, 1983.

## Tetrasomy (9p) syndrome

Tetrasomy (9p) is extremely rare, with less than a dozen reports having been written on this condition (1–8).

Fig. 4–25. *del(9p) syndrome.* Trigonocephaly, upslanting palpebral fissures, synophrys, small nose. [From *MT Mulcahy,* Ann Génét **21**(1), 47, 1978.]

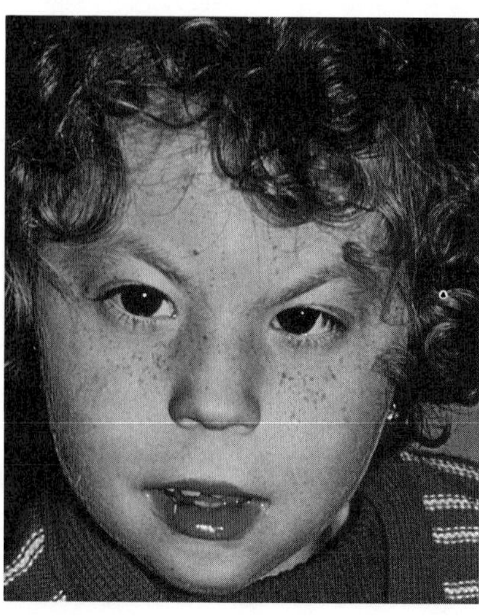

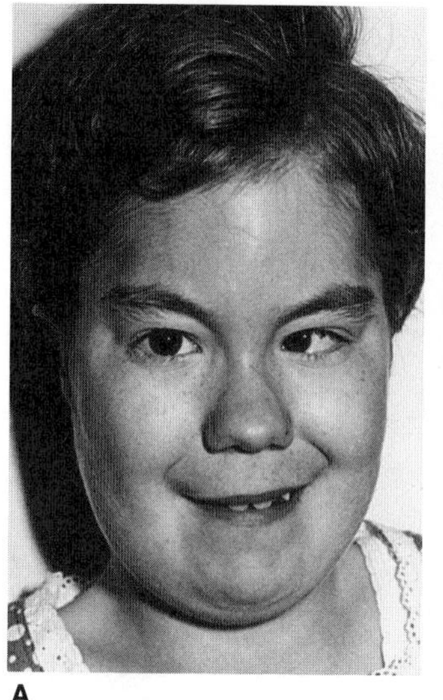

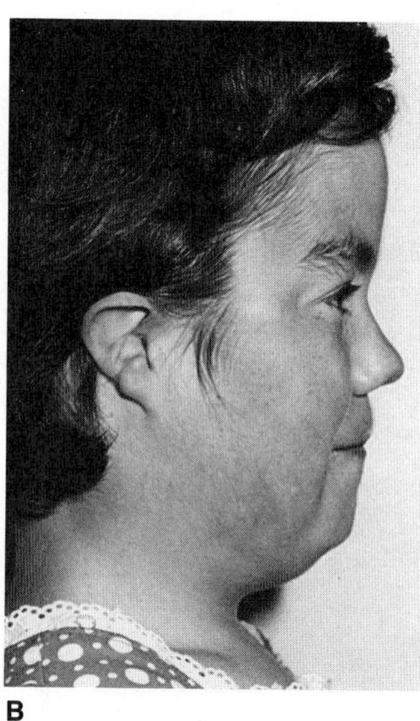

A B

Fig. 4–26. *del(9p) syndrome.* (A,B) Trigonocephaly, upslanting palpebral fissures, strabismus, broad flat nasal bridge, malformed pinna, long philtrum, short neck, (From *FJ Rutten et al,* Ann Génét **21:**51, 1978.)

In addition to psychomotor retardation, the following have been found in 50% or more: hypotonia, microcephaly, hydrocephaly, wide sutures and fontanels, hypertelorism, enophthalmos, epicanthal folds, strabismus, bulbous-beaked nose, low-set malformed pinnae, downslanting mouth, retromicrognathia, and short neck. Approximately half the patients have cleft lip/palate. Various congenital heart and urogenital anomalies have been reported (Fig. 4–27A–C).

### References [Tetrasomy (9p) syndrome]

1. Abe T et al: Partial tetrasomy 9 (9pter→9q2101) due to an extra isodicentric chromosome. *Ann Génét* **20**:111–114, 1977.

2. Balestrazzi P et al: Tetrasomy 9p confirmed by GALT. *J Med Genet* **20**:396–399, 1983.

3. Calvalcanti DP et al: Tetrasomy 9p caused by idic(9)(pter→q13→pter). *Am J Med Genet* **27**:497–503, 1987.

4. Garcia-Cruz D et al: Tetrasomy 9p: Clinical aspects and enzyme gene dosage expression. *Ann Génét* **25**:237–242, 1982.

5. Ghymers D et al: Tétrasomie partielle du chromosome 9, à l'état de mosaïque, chez un enfant porteur de malformations multiples. *Humangenetik* **20**:273–282, 1973.

6. Moedjono SJ et al: Tetrasomy 9p: Confirmation by enzyme analysis. *J Med Genet* **17**:227–230, 1980.

7. Rutten FJ et al: A presumptive tetrasomy for the short arm of chromosome 9. *Humangenetik* **25**:163–170, 1974.

8. Wisniewski L et al: Partial tetrasomy 9 in a live-born infant. *Clin Genet* **14**:147–153, 1978.

## dup(9q) syndrome

Approximately 20 patients are known to have tetrasomy for the long arm of chromosome 9 (1–9). Mental retardation is severe. The relatively constant features are microdolichocephaly, deep-set eyes, prominent beaked nose, relatively large pinnae, and microretrognathia (1–4,6) (Fig. 4–28).

Approximately 35% of those with tetrasomy for 9pter→q21-32 have cleft lip/palate (4,9).

The fingers and toes tend to be long. Flexion of fingers and/or limitation of joint mobility, dislocated hips, talipes equinovarus, hypoplastic genitalia, and congenital heart disease have also been found.

### References [dup(9q) syndrome]

1. Aftimos SF et al: Partial trisomy 9q due to maternal 9/17 translocation. *Am J Dis Child* **134**:848–850, 1980.

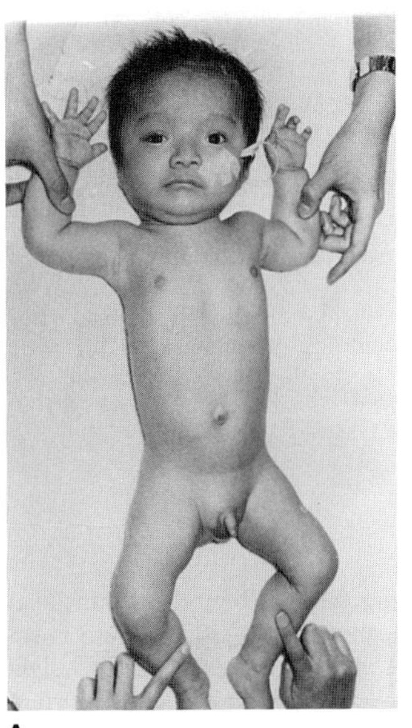

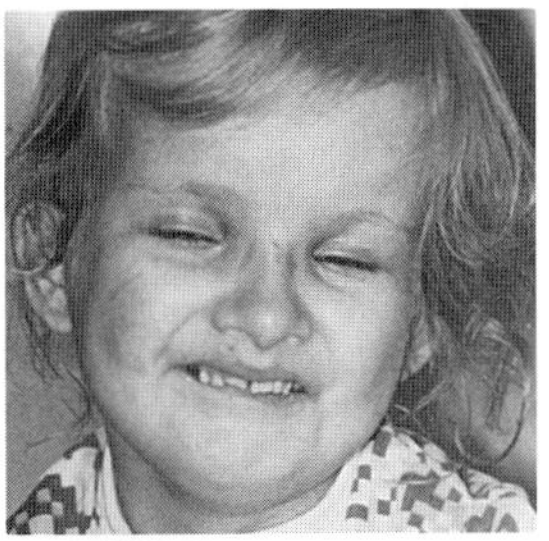

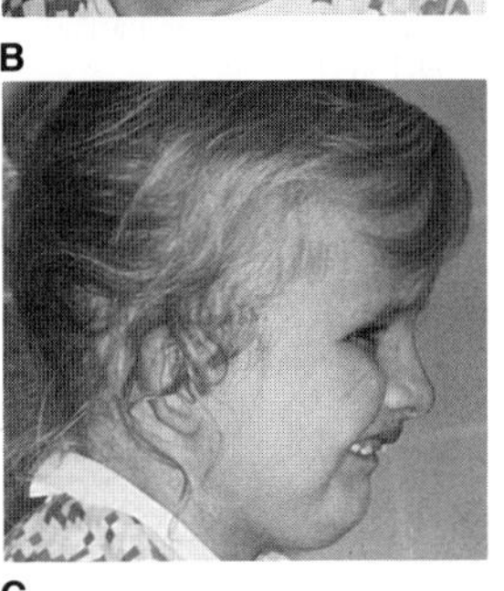

A B C

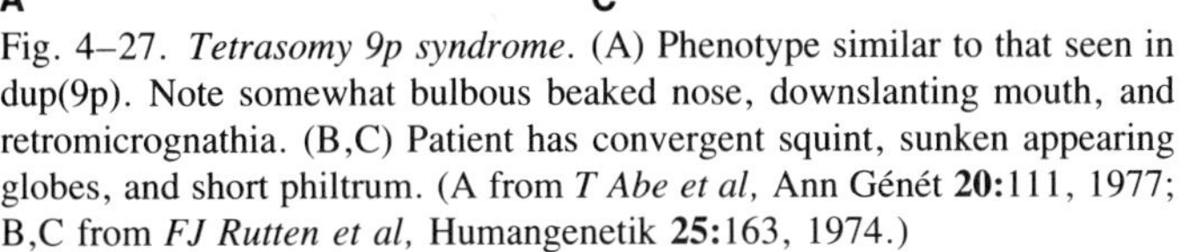

Fig. 4–27. *Tetrasomy 9p syndrome.* (A) Phenotype similar to that seen in dup(9p). Note somewhat bulbous beaked nose, downslanting mouth, and retromicrognathia. (B,C) Patient has convergent squint, sunken appearing globes, and short philtrum. (A from *T Abe et al,* Ann Génét **20:**111, 1977; B,C from *FJ Rutten et al,* Humangenetik **25:**163, 1974.)

2. Baccichetti C et al: Partial trisomy 9: Clinical and cytogenetic correlations. *Ann Génét* **22**:199–204, 1979.

3. Nakahori Y, Nakagomi Y: A malformed girl with duplication of chromosome 9q. *J Med Genet* **21**:387–388, 1984.

4. Schwanitz G et al: Partial trisomy 9 in the case of familial translocation 8/9 mat. *Ann Génét* **17**:163–166, 1974.

5. Soltan HC et al: Partial trisomy 9q resulting from a familial translocation t(9;16)(q32;q24). *Clin Genet* **25**:449–454, 1984.

6. Šubrt I et al. Partial trisomy 9q chromosomal syndrome. *Hum Genet* **34**:151–154, 1976.

Fig. 4–28. *dup(9q) syndrome.* Microdolichocephaly, upslanting small palpebral fissures, prominent beaked nose, short philtrum, microretrognathia. (From *SF Aftimos et al,* Am J Dis Child **134:**848, 1980.)

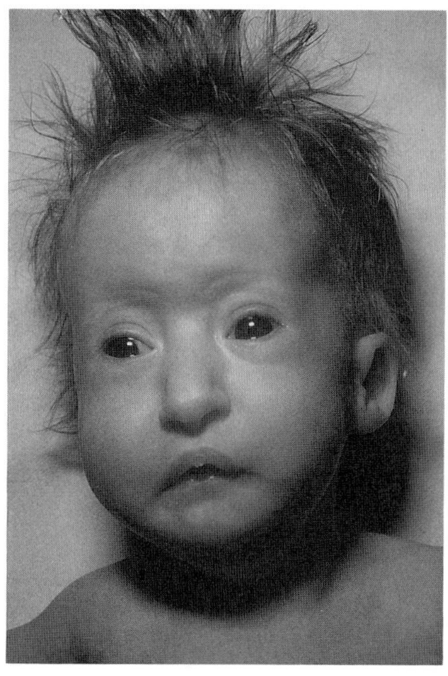

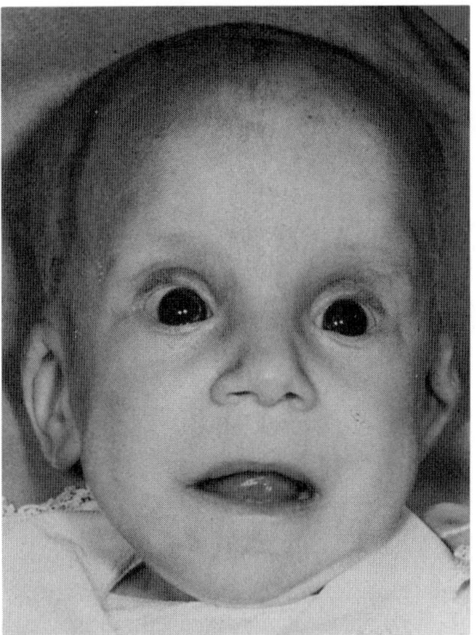

Fig. 4–29. *Trisomy 9 syndrome.* Microcephaly, dolichocephaly, and high forehead. Eyes are deeply sunken into the sockets. Palpebral fissures are upslanting. Pinnae are dysmorphic.

7. Ten SK et al: Three cases of partial trisomy 9q in one generation due to maternal reciprocal t(6;8;9) translocation. *Clin Genet* **31**:359–365, 1987.

8. Turleau C et al: Partial trisomy 9q: A new syndrome. *Hum Genet* **29**:233–242, 1975.

9. Wilson GN et al: The phenotype and cytogenetic spectrum of partial trisomy 9. *Am J Med Genet* **20**:277–282, 1985.

## Trisomy 9 syndrome

Trisomy 9 and trisomy 9 mosaicism do not appreciably differ in phenotype. All patients have psychomotor retardation. Most have low birth weight and/or failure to thrive and neurologic impairment, and most die before the fourth month of life (1–14). About 25 cases have been reported (5a).

Approximately one-third exhibit microcephaly. The palpebral fissures tend to be short with an upward slant. There are also microphthalmia and deeply-set eyes (5). The nose has a bulbous tip. Cleft lip/palate has been seen in 25% (6–9, 11–14). The ears tend to be low set and/or malformed, the chin small, and the neck short or webbed (Fig. 4–29).

Musculoskeletal abnormalities are common: congenital dislocation of the hips or other joints, joint limitations or dysplastic hands and/or feet, rockerbottom feet, and talipes. Congenital heart anomalies are frequent: VSD, PDA, double outlet right ventricle, and persistent left superior vena cava (3,5,7).

Urogenital abnormalities are also frequent: hydronephrosis, duplication of collecting system, cryptorchidism, micropenis, and hypospadias. The external genitalia of females are normal. Deep palmar and/or plantar creases are frequent.

### References (Trisomy 9 syndrome)

1. Annerén G, Sedin G: Trisomy 9 syndrome. *Acta Paediatr Scand* **70**:125–128, 1981.

2. Delicado A et al: Complete trisomy 9: Two additional cases. *Ann Génét* **28**:63–66, 1985.

3. Frohlich G: Delineation of trisomy 9. *J Med Genet* **19**:316–317, 1982.

4. Ginsberg J et al: Pathologic features of the eye in trisomy 9. *J Pediatr Ophthalmol Strab* **19**:37–41, 1982.

5. Katayama PK et al: Clinical delineation of trisomy 9 syndrome. *Obstet Gynecol* **56**:665–668, 1980.

5a. Levy I et al: Gastrointestinal abnormalities in the syndrome of mosaic trisomy 9. *J Med Genet* **26**:280–281, 1989.

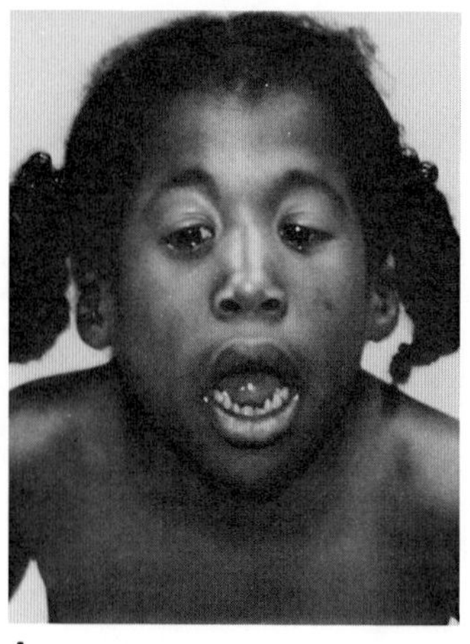
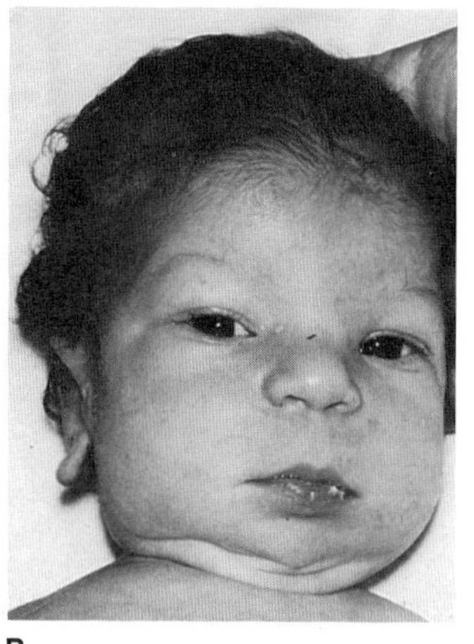

A B

Fig. 4–30. *del(10p) syndrome.* (A) Narrow facies, low nasal root, anteverted nostrils, dysmorphic pinnae. *dup(10p) syndrome.* (B) Dolichocephaly with high bulky forehead, wide open anterior fontanel, mild upslanting palpebral fissures, prominent nasal bridge, small triangular mouth, micrognathia. [A from *U Francke et al,* Birth Defects **11**(5):207, 1975. B from *E Orye et al,* J Génét Hum **33:**63, 1985.]

6. Mace SE et al: The trisomy 9 syndrome: Multiple congenital anomalies and unusual pathologic findings. *J Pediatr* **92**:446–448, 1978.

7. Mantagos S et al: Complete trisomy 9 in two live born infants. *J Med Genet* **18**:377–382, 1981.

8. Pfeiffer RA, Müller R: Der Phänotyp der Trisomie 9. *Mschr Kinderheilk* **132**:797–800, 1984.

9. Romain DR, Sullivan J: Delineation of trisomy 9 syndrome. *J Med Genet* **20**:156–157, 1983.

10. Sanchez JM et al: Report of a new case and clinical delineation of mosaic trisomy 9 syndrome. *J Med Genet* **19**:384–389, 1982.

11. Schinzel A: Mosaic trisomy and pericentric inversion of chromosome 9 in a malformed boy. *Humangenetik* **25**:171–177, 1974.

12. Seabright M et al: Trisomy 9 associated with enlarged 9qh segment in a liveborn. *Hum Genet* **34**:323–325, 1976.

13. Southerland GR et al: Partial and complete trisomy 9: Delineation of a trisomy 9 syndrome. *Hum Genet* **32**:133–140, 1976.

14. Williams T et al: Complex cardiac malformation in a case of trisomy 9. *J Med Genet* **22**:230–232, 1985.

## del(10p) syndrome

Less than 20 cases are known, most being male (1–10). There have been several break points yielding various phenotypes. Mental retardation is severe in all. Postnatal growth retardation is frequent (2). About 50% have died in infancy. The more constant facial features include microcephaly, frontal bossing, hypertelorism, short downslanting palpebral fissures, epicanthus, ptosis, strabismus, low nasal root with anteverted nostrils, small low-set pinnae, micrognathia, and short neck. Cleft lip–palate has been reported in a few cases (8,10) (Fig. 4–30A).

The nipples are widely spaced. Various congenital heart anomalies have been found in about 40%. Cryptorchidism, hernia, and renal abnormalities are frequent. Hypoplasia or aplasia of olfactory bulb and tracts have been found in those with more proximal deletions (5).

### References [del(10p) syndrome]

1. Berger R et al: Deletion of the short arm of chromosome No. 10. *Acta Paediatr Scand* **66**:659–662, 1977.

1a. Danesino G et al: Deficiency 10p. *Ann Génét* **27**:162–166, 1984.

2. Elstner CL et al: Further delineation of the 10p deletion syndrome. *Pediatrics* **73**:670–675, 1984.

3. Fryns JP et al: Distal 10p deletion syndrome. *Ann Génét* **24**:189–190, 1981.

4. Gentik A et al: Partial monosomy of chromosome 10 short arms. *J Med Genet* **20**:107–111, 1983.

5. Greenberg F et al: Hypoparathyroidism and T-cell immune defect in a patient with 10p deletion syndrome. *J Pediatr* **109**:489–492, 1986.

6. Klep-de Pater JM et al: Partial monosomy 10p syndrome. *Eur J Pediatr* **137**:243–246, 1981.

7. Koenig R et al: Partial monosomy 10p syndrome. *Ann Génét* **28**:173–176, 1985.

8. Shokeir MHK et al: Deletion of the short arm of chromosome no. 10. *J Med Genet* **12**:99–103, 1975.

9. Slinde S, Hansteen IL: Two chromosomal syndromes in the same family: Monosomy and trisomy for part of the short arm of chromosome 10. *Eur J Pediatr* **139**:153–157, 1982.

10. Sucui S, Nanulescu M: A case of 10p− syndrome. *Ann Génét* **26**:109–111, 1983.

## dup(10p) syndrome

Most examples of dup(10p) result from familial reciprocal translocation. Approximately 30 cases have been reported (1–7). There is severe mental and motor retardation with little or no speech. Birth weight is usually normal. Infants are often dolichocephalic with a high prominent forehead and wide open sutures and anterior fontanel. Palpebral fissures have a slight downward slant. The eyebrows are highly arched. The maxilla tends to protrude. The nasal root, initially broad, becomes prominent in older patients and the mouth triangular with a thin inverted upper lip (Fig. 4–30B). Cleft lip/palate has been noted in over 40% (2–7). The ears are often low set and somewhat angulated. Elbows, wrists, and fingers are often hyperextensible. Flexion deformities of fingers and toes and club feet are common. The kidneys tend to be cystic. For various other anomalies, see Stengel-Rutkowski et al (6) and Lurie et al (4).

### References [dup(10p) syndrome]

1. Gonzalez C et al: Duplication 10p in a girl due to a maternal translocation t(10;14) (p11;p12). *Am J Med Genet* **14**:159–167, 1983.

2. Hustinx TWJ et al: Trisomy for the short arm of chromosome No. 10. *Clin Genet* 408–415, 1974.

3. Johnson G et al: Partial trisomy 10p and familial translocation t(7;10)(p22;p12). *Hum Genet* **35**:353–356, 1977.

4. Lurie IW et al: Partial trisomy 10p in two generations. *Hum Genet* **41**:235–241, 1978.

5. Schleiermacher E et al: Brother and sister with trisomy 10p: A new syndrome. *Humangenetik* **23**:163–172, 1974.

6. Stengel-Rutkowski S et al: Trisomy 10p. *Eur J Pediatr* **126**:109–125, 1977.

7. Yunis E et al: Trisomy 10p. *Ann Génét* **19**:57–60, 1976.

## del(10q) syndrome

There have been about 18 cases reported (1–10). There is a possible excess of females. In addition to severe mental and growth retardation, there are a few relatively nonspecific facial features: microcephaly, prominent beaked nose, apparent hypertelorism, strabismus, malformed pinnae, and short neck. Cleft lip has been noted (5).

### References [del(10q) syndrome]

1. Chieri P, Iölster N: Monosomy 10qter due to a balanced maternal translocation: t(10;8)(q23;p23). *Clin Genet* **24**:147–150, 1983.

2. Curtis H et al: Terminal deletion of the long arm of chromosome 10. *J Med Genet* **23**:478–480, 1986.

3. Evans-Jones G et al: A further case of monosomy 10qter. *Clin Genet* **24**:216–219, 1983.

4. Gorinati M et al: Terminal deletion of the long arm of chromosome 10. *Am J Med Genet* **33:**502–504, 1989.

5. Mulcahy MT et al: Is there a monosomy 10qter syndrome? *Clin Genet* **21**:33–35, 1982.

6. Shapiro SD et al: Deletions of the long arm of chromosome 10. *Am J Med Genet* **20**:181–196, 1985.

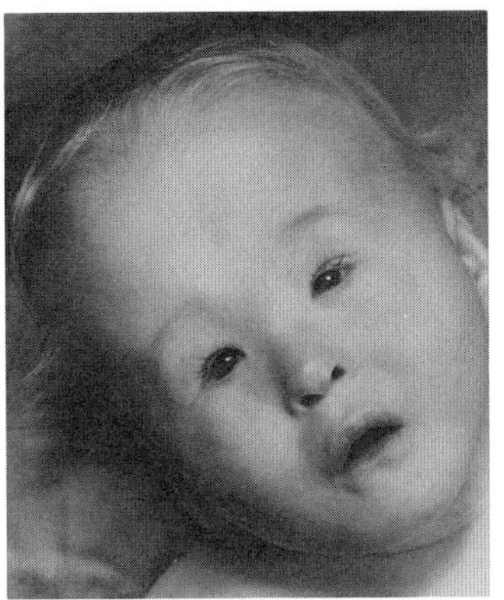
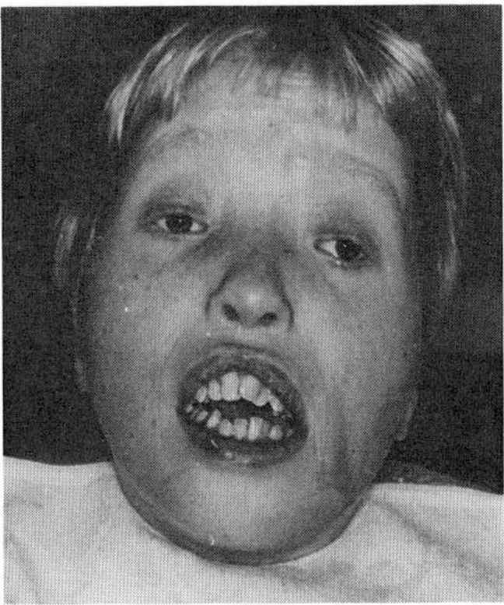

A B

Fig. 4–31. *dup(10q) syndrome.* (A,B) Large forehead, round face, arched eyebrows, small palpebral fissures, hypertelorism, and bow-shaped mouth. (A from *JJ Yunis* and *O Sanchez,* J Pediatr **84:**567, 1974. B from *S Kröyer* and *E Niebuhr,* Ann Génét **18:**50, 1975.)


7. Turleau C et al: Monosomy 10qter. *Hum Genet* **47**:233–237, 1979.

8. Wegner RD et al: Monosomy 10qter due to a balanced familial translocation: t(10;16)(q25.2; q24). *Clin Genet* **19**:130–133, 1981.

9. Wulfsberg EA et al: Chromosome 10qter deletion syndrome: A review and report of three new cases. *Am J Med Genet* **32**:364–367, 1989.

10. Zatterale A et al: Clinical features of monosomy 10qter. *Ann Génét* **26**:106–108, 1983.


## dup(10q) syndrome

Over 90% of cases have resulted from a balanced translocation in a parent. A significant majority are male. The prognosis is poor. Death has occurred prior to the age of 4 years in approximately half the cases, largely because of cardiac, renal, or respiratory complications. Those that survive are severely retarded (7).

Trisomy (10q25→qter) produces a rather distinctive clinical picture: severe mental retardation, pre- and postnatal growth retardation, marked hypotonia, microcephaly, large high forehead, somewhat flattened round face, fine arched eyebrows, narrow palpebral fissures, epicanthus, microphthalmia, flat nasal bridge, small short nose with anteverted nostrils, prominent cheek bones, bow-shaped mouth with prominent upper lip, small mandible, and low-set malformed pinnae. About 35 cases have been reported (1–6) (Fig. 4–31A,B).

About 50% exhibit ptosis, cleft palate, and long philtrum. Bilateral epicanthal folds create an illusion of hypertelorism, whereas reduced corneal diameter may simulate microphthalmia.

Anomalies of the hands and feet include camptodactyly, proximally implanted thumbs and/or great toes with a wide space between the hallux and second toe, overlapping and/or fusiform fingers, and rockerbottom feet. Deep plantar furrows have been noted in over one-third of the patients. At least half the males exhibit cryptorchidism. Various congenital heart defects have been found. Delayed bone age, scoliosis, and thin ribs are noted in about one-third of patients. Various kidney abnormalities include hypoplasia, cystic alterations, hydronephrosis, and hydroureter.

### References [dup(10q) syndrome]


1. Berger R et al: Trisomie 10q partielle de novo. *J Génét Hum* **24**:261–269, 1976.

2. Fryns JP et al: Partial trisomy of the distal portion of the long arm of chromosome number 10 (10q24→10qter): A clinical entity. *Acta Paediatr Belg* **32**:141–143, 1979.

3. Klep-dePater J et al: Partial trisomy 10q: A recognizable syndrome. *Hum Genet* **46**:29–40, 1979.

4. Prieur M et al: La trisomie 10q24→qter. *Ann Génét* **18**:217–222, 1975.

5. Stoll C, Roth MP: Observation familiale de trisomie pour la partie terminale du chromosome 10. *Arch Fr Pédiatr* **38**:273–274, 1981.

6. Taysi K et al: Partial trisomy 10q in three unrelated patients. *Ann Génét* **26**:79–85, 1983.

7. Yunis JJ, Sanchez O: A new syndrome resulting from partial trisomy for the distal third of the long arm of chromosome 10. *J Pediatr* **84**:567–570, 1974.


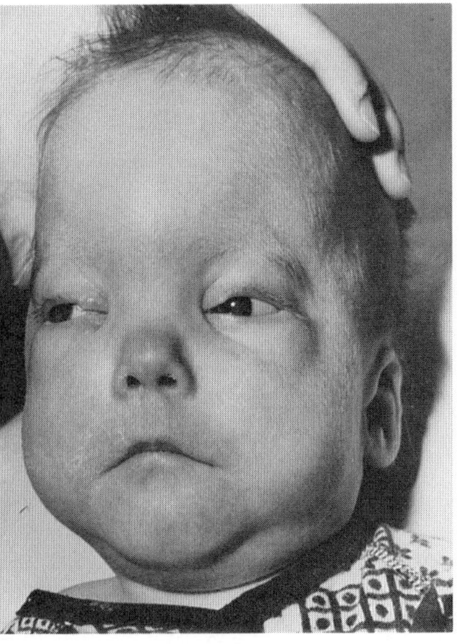

Fig. 4–32. *del(11q) syndrome.* Trigonocephaly, flat broad nasal bridge, carp mouth, epicanthal folds, mildly dysmorphic pinnae, hypertelorism, mild colobomas of eyelids, and micrognathia. (From *I Felding* and *F Mitelman,* Acta Paediatr Scand **68:**635, 1979.)

## del(11q) syndrome

All patients who have del(11)(q23→qter) or terminal deletion have exhibited mental retardation. The critical band is 11q24.1 (6). Most are *de novo* deletions. At least 35 cases have been reported. Only about one-half have been small for gestational age. About 60% have exhibited postnatal somatic retardation. Over 80% are female. Most have manifest frequent respiratory infections. About 25% die before 2 years of age, usually from congenital heart disease (1–15).

The facies is characterized by trigonocephaly (85%), hypertelorism (80%), microcephaly (50%), flat occiput (40%), broad nasal bridge with short upturned nose (95%), abnormally modeled pinnae (95%), carp mouth (95%), and micrognathia (90%). Eye anomalies have included coloboma of upper lid, downslanting palpebral fissures, ptosis, epicanthus, strabismus, and coloboma of iris (Fig. 4–32).

Various congenital heart anomalies (VSD, single ventricle, ASD) (65%), joint contractures and/or minor digital anomalies (70%), and single flexion creases (65%) have been found. Systemic anomalies are unusual (15). About 50% have exhibited thrombocytopenia.

There is clinical similarity with *Opitz trigonocephaly (C) syndrome.*

### References [del(11q) syndrome]


1. Cassidy SB et al: Trigonocephaly and the 11q− syndrome. *Ann Génét* **20**:67–69, 1977.

2. Cousineau AJ et al: Ring-11 chromosome: Phenotype–karyotype correlation with deletions of 11q. *Am J Med Genet* **14**:29–35, 1983.

3. Dörr V: Das klinische Erscheinungsbild der partiellen Monosomie von Chromosomen 11q. *Mschr Kinderheilkd* **134**:808–811, 1986.

4. Faust J et al: A case with 46,XX,del(11)(q21). *Clin Genet* **6**:90–97, 1974.

5. Felding I, Mitelman F: Deletion of the long arm of chromosome 11. *Acta Paediatr Scand* **68**:635–638, 1979.

6. Ferry AP et al: Ocular abnormalities in deletion of the long arm of chromosome 11. *Ann Ophthalmol* **13**:1373–1377, 1981.

7. Fryns JP et al: Distal 11q monosomy: The typical 11q monosomy syndrome is due to deletion of subband 11q24.1. *Clin Genet* **30**:255–260, 1986.

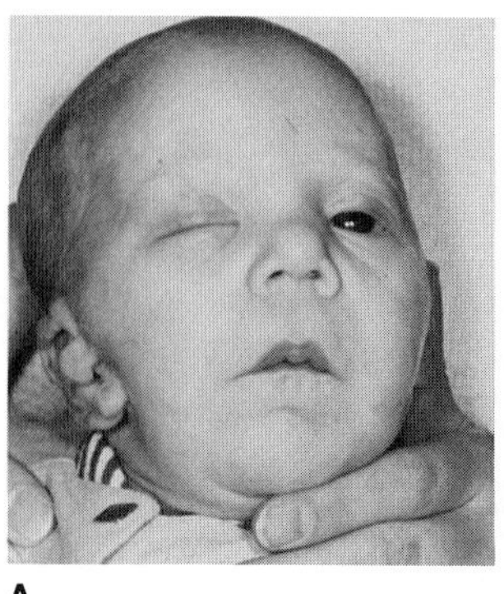
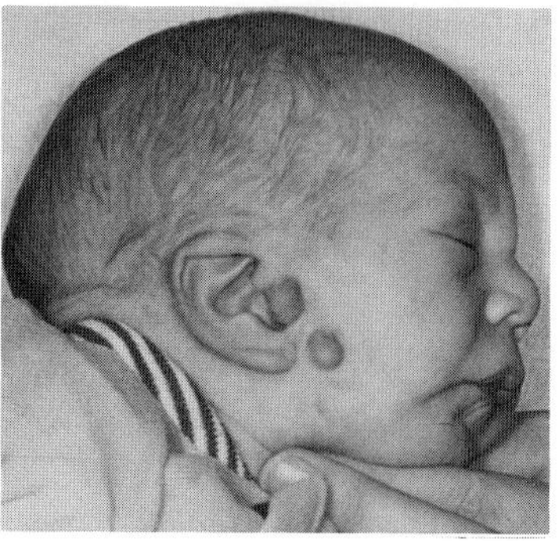

A B

Fig. 4–33. *dup(11q) syndrome.* (A,B) Facial asymmetry, epicanthal folds, ear tags, retracted lower lip.

7a. Helmuth RA et al: Holoprosencephaly, ear anomalies, congenital heart defect, and microphallus in a patient with 11q− mosaicism. *Am J Med Genet* **32**:178–181, 1989.

8. Küster W et al: Report of a deletion 11 (qter→q23.3) and short review of the literature. *Eur J Pediatr* **144**:286–288, 1985.

9. Larson SA et al: Deletion of 11q: Report of two cases and a review. *Birth Defects* **12**(5):125–130, 1976.

10. Leonard C et al: Monosomie partielle par deletion du bras long du chromosome 11: del(11)(q23). *Ann Génét* **22**:115–120, 1979.

11. McPherson E, Meissner L: 11q− syndrome: Review and report of two cases. *Birth Defects* **18**(3B):295–300, 1982.

12. Niikawa N et al: Ring chromosome 11 associated with clinical features of the 11q− syndrome. *Ann Génét* **24**:172–175, 1981.

13. O'Hare AE et al: Deletion of the long arm of chromosome 11 [46,XX,del(11)(q24.1→qter)]. *Clin Genet* **25**:373–377, 1984.

14. Schinzel AP et al: Partial deletion of long arm of chromosome 11 [del(11)(q23)]. *J Med Genet* **14**:438–444, 1977.

15. Sirota L et al: New anomalies found in the 11q− syndrome. *Clin Genet* **26**:569–573, 1984.

## dup(11q) syndrome

In most cases, dup(11q) is associated with 11q/22q translocation, inherited in nearly all cases through the carrier mother and having a recurrent risk of 2–6% (3,7). The frequency of this disorder appears to be far higher than supposed. There is also a group of cases representing dup(11q), most often 11q23→11qter (1,2,4–6,8,9). Although the first group may include trisomy for a small portion of 22q, the phenotypic overlap with the second group is marked.

Reduced birth weight and postnatal growth are nearly constant features. Mental retardation, ranging from low normal to moderate, is present in over 85%. Hypertonia is found in about 50%.

Craniofacial asymmetry and microcephaly are present in 35%. Inconstant features include hypertelorism, epicanthal folds, downslanting palpebral fissures, and strabismus. The nose is large and beaked in 25%. About 85% have preauricular (usually bilateral) ear tags or pits. The lower lip is usually retracted and the mandible small. Cleft palate has been found in 60% (2–5) (Figs. 4–33A,B and 4–34).

Genitourinary anomalies include imperforate anus or anal atresia (15%), associated with a fistula into the bladder, vestibule, or perineum. Cryptorchidism occurs in 35% with micropenis being an occasional finding.

Musculoskeletal anomalies include bipartite clavicular defect and dislocation or dysplasia of the hips (30%).

Congenital heart anomalies, found in 40% are variable and nonspecific.

Fig. 4–34. *dup(11q) syndrome.* Hypertelorism, broad flat nasal bridge, small mouth, retracted lower lip, micrognathia, and short neck. (From *HD Rott et al,* Humangenetik **14:**300, 1972.)

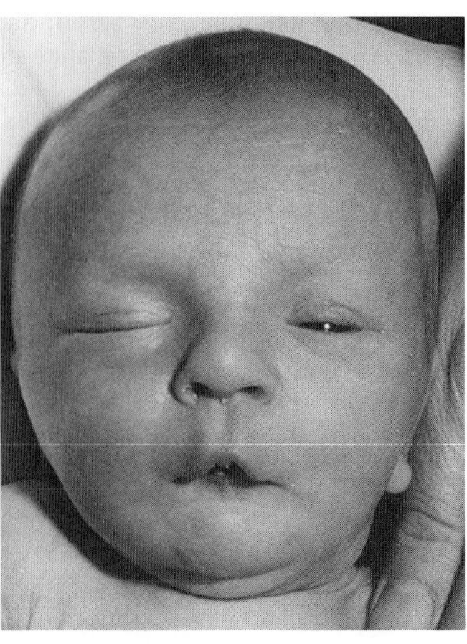

#### References [dup(11q) syndrome]

1. Aurias A, Laurent C: Trisomie 11q. Individualisation d'un nouveau syndrome. *Ann Génét* **18**:189–191, 1975.

2. DeFrance HF et al: Partial trisomy 11q due to paternal t(11q;18p); further delineation of the clinical picture. *Clin Genet* **25**:295–299, 1984.

3. Fraccaro M et al: The 11q;22q translocation: A European collaborative analysis of 43 cases. *Hum Genet* **56**:21–51, 1980.

4. Francke U et al: Duplication 11(q21 to 23→qter) syndrome. *Birth Defects* **13**(3B):167–187, 1977.

5. Giraud F et al: Trisomie partielle 11q et translocation familiale 11–12. *Humangenetik* **28**:343–347, 1975.

6. Greig F et al: Duplication 11(q22→qtr) in an infant: A case report with review. *Ann Génét* **28**:185–188, 1985.

7. Iselius L et al: The 11q;22q translocation: A collaborative study of 20 new cases and analysis of 110 families. *Hum Genet* **64**:343–355, 1983.

8. Lurie IV et al: Partial trisomy 11q as the result of sporadic translocation. *Hum Genet* **51**:63–66, 1979.

9. Pihko H et al: Partial 11q trisomy syndrome. *Hum Genet* **58**:129–123, 1981.

## Pallister–Killian syndrome (mosaic tetrasomy 12p, isochromosome 12p syndrome)

In 1981, Teschler-Nicola and Killian (20) described a child with mental retardation, postnatal growth deficiency, slow growing hair, and a distinctive facies. Numerous additional examples have been reported (1–25). It is caused by mosaic tetrasomy for 12p. The syndrome is the same as the Pallister mosaic syndrome, a disorder reported by Pallister et al (13) in 1977 and those described as mosaic tetrasomy 21 (2,8) . The identity of the chromosome disorder has been amply documented (9,10,14,21). Although peripheral lymphocytes may be normal, cultured fibroblasts and direct bone marrow analysis (23) will show the chromosome marker, but it may be lost in long-term culture (14) and with age (15,23). A phenocopy has been reported (17). Mothers of affected infants tend to be older (24).

In infancy, the hair is sparse and fine, particularly in the frontal and temporal area extending to the vertex. With age, hair distribution becomes normal. The facies is coarse. Hypopigmented areas of the upper facial skin have been noted. The forehead is high and the metopic skin and eyelids are puffy. Eyebrows and eyelashes are sparse medially. The pinnae are fleshy. Hearing loss has been documented in nearly all patients. There is ptosis, strabismus, and hypertelorism. The nose is short with a flat nasal bridge and anteverted nostrils. The philtrum is long and prominent. The mouth is downturned. The upper lip is thin and has a cupid-bow shape. The lower lip protrudes. The mandible is small and dental eruption is delayed. The neck is short (Figs. 4–35A,B and 4–36A–D). The adult phenotype is one of severe retardation, epilepsy, coarse facies and macroglossia (15).

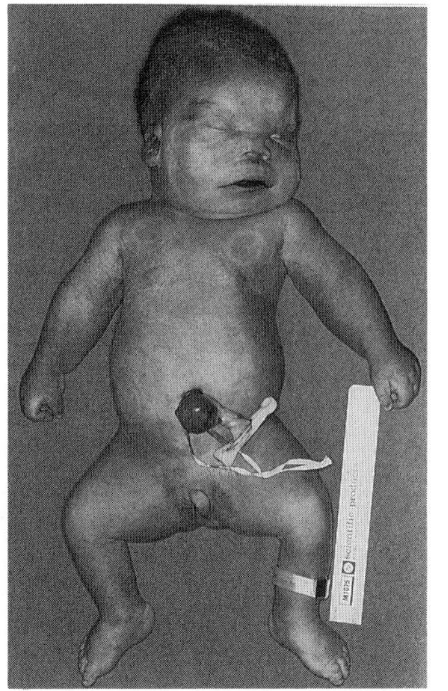

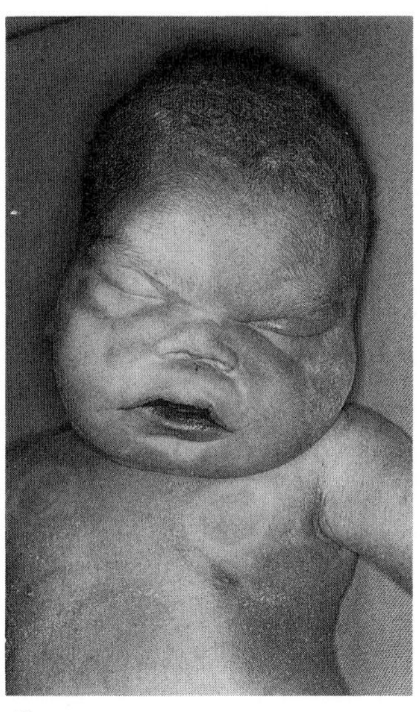

A B

Fig. 4–35. *Pallister–Killian syndrome. Tetrasomy i(12p).* (A) Note altered body proportions and multiple depigmented areas. (B) Short depressed nasal bridge, prominent premaxilla, and large mouth. (From *L Shivashankar et al,* Prenatal Diag **8:**85, 1988.)

Fig. 4–36. *Pallister–Killian syndrome.* (A,B) Fine sparse hair, particularly on frontal and temporal areas. Facies coarse. Note hypopigmented area of scalp, high forehead, eyebrows, medially sparse eyelashes, short nose with flat bridge and anteverted nostrils, downturned mouth, and thin upper lip. (C,D) Sparse hair, hypopigmented areas of forehead, telecanthus, lop ears with prominent lobes, prominent maxilla, and everted lip. [A,B from *W Killian et al,* J Clin Dysmorphol **1**(3):6, 1983. C,D from *AGW Hunter,* Clin Genet **28:**47, 1985.]

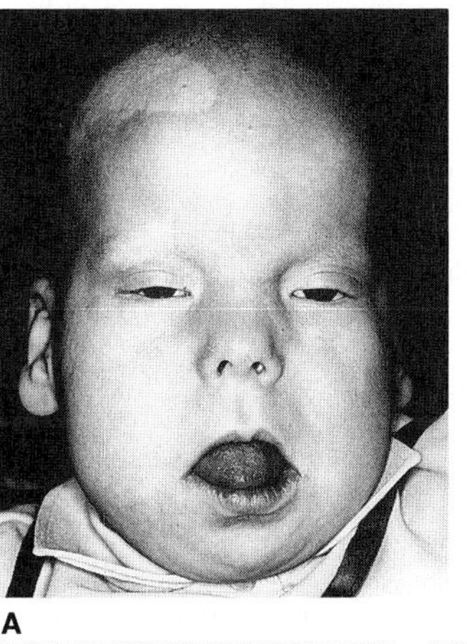

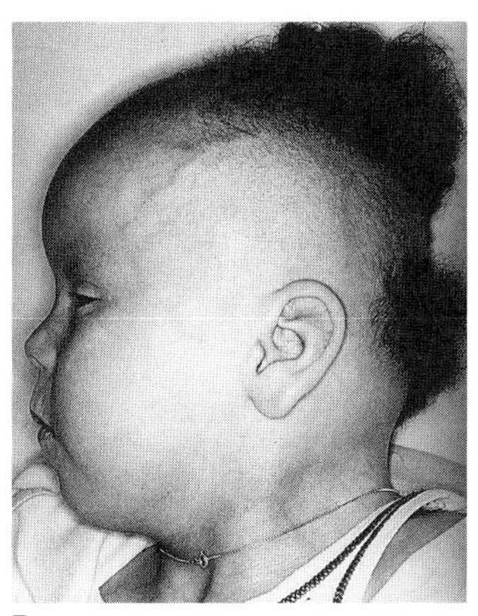

A B

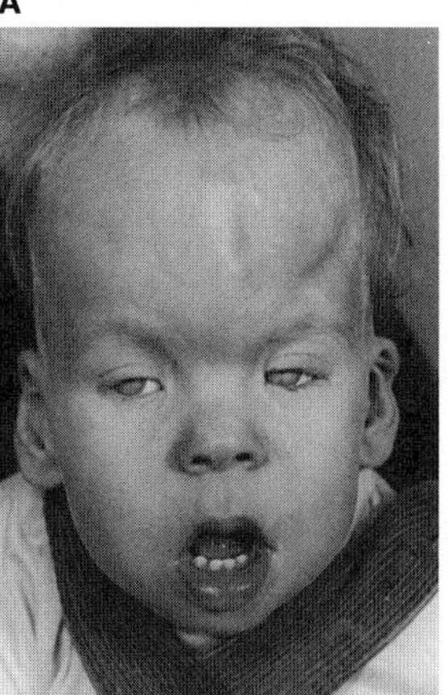

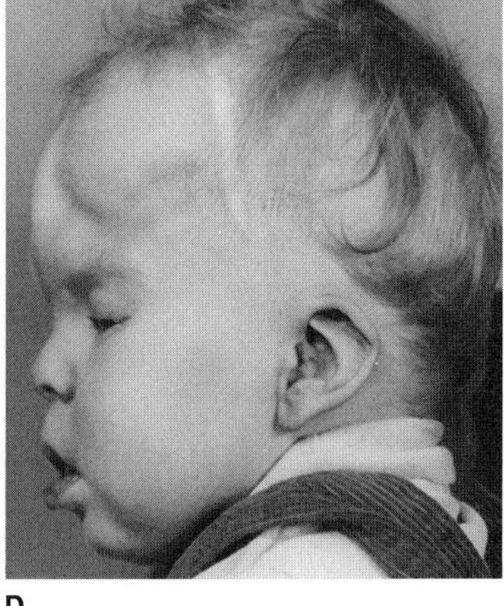

C D

Developmental retardation is profound. Speech is not attained. Neonatal asphyxia is frequent. Bone age is retarded. Postnatal growth deficiency and natal gigantism have been reported (16). Marked hypotonia, hypermobile joints, and areflexia have been noted in infancy, but these improve with age. Supernumerary nipples are a common finding. Several children have unexplained fever. Increased height of the vertebral bodies, mild thoracic scoliosis and atlanto-occipital fusion have been documented. Diaphragmatic hernia has been reported. The limbs and fingers tend to be short and there is often a sacral dimple. The thumb may be proximally inserted and the hallux is large (6a). Vine et al (21) noted elevated LDH-B in a patient. The gene for this enzyme has been mapped to the short arm of chromosome 12. *In situ* hybridization has confirmed the identity of the chromosome (19).

Direct bone marrow analysis is best employed, since the isochromosome tends to disappear with age (23).

### References [Pallister–Killian syndrome (mosaic tetrasomy 12p, isochromosome 12p syndrome)]


1. Buyse ML, Korf BR: Killian syndrome, Pallister mosaic syndrome, or mosaic tetrasomy 12p? An analysis. *J Clin Dysmorphol* **1**(3):2–3, 1983.

2. Fryns JP et al: Mosaic tetrasomy 21 in severe mental handicap. *Eur J Pediatr* **139**:87–89, 1982.

3. Gilgenkrantz S et al: Mosaic tetrasomy 12p. *Clin Genet* **28**:495–502, 1985.

4. Hall BD: Teschler-Nicola/Killian syndrome: A sporadic case in an 11-year-old male. *J Clin Dysmorphol* **1**(3):14–17, 1983.

5. Hersh JH et al: Teschler-Nicola/Killian: A case report. *J Clin Dysmorphol* **1**(3):20–24, 1983.

6. Hunter AGW et al: The characteristic physiognomy and tissue specific karyotype distribution in the Pallister–Killian syndrome. *Clin Genet* **28**:47–53, 1985.

6a. Kawashima H: Skeletal anomalies in a patient with the Pallister/Teschler–Nicola/Killian syndrome. *Am J Med Genet* **27**:285–289, 1987.

7. Killian W et al: Abnormal hair, craniofacial dysmorphism, and severe mental retardation—a new syndrome? *J Clin Dysmorphol* **1**(3):6–13, 1983.

8. Kwee ML et al: Mosaic tetrasomy 21 in a male child. *Clin Genet* **26**:150–155, 1984.

9. Lubinsky M: A case report of Teschler-Nicola/Killian syndrome. *J Clin Dysmorphol* **1**(3):25–27, 1983.

10. Lubinsky M, Sanger W: ''Killian syndrome'' and mosaic tetrasomy 12p. *J Clin Dysmorphol* **2**(1):5, 1984.

11. Pagon RA: Teschler-Nicola/Killian syndrome. *J Clin Dysmorphol* **1**(3):18–19, 1983.

12. Pauli RM et al: Mosaic isochromosome 12p. *Am J Med Genet* **27**:291–294, 1987.

13. Pallister PD et al: The Pallister mosaic syndrome. *Birth Defects* **13**(3B):103–110, 1977.

14. Peltomäki P et al: Pallister–Killian syndrome: Cytogenetic and molecular studies. *Clin Genet* **31**:399–405, 1987.

15. Quarrell OWJ et al: Pallister–Killian mosaic syndrome with emphasis on the adult phenotype. *Am J Med Genet* **31**:841–844, 1988.

15a. Raffel LJ et al: Chromosomal mosaicism in the Killian/Teschler-Nicola syndrome. *Am J Med Genet* **24**:607–611,. 1986.

16. Reynolds JF et al: Isochromosome 12p mosaicism (Pallister mosaic aneuploidy or Pallister–Killian syndrome): Report of 11 cases. *Am J Med Genet* **27**:257–274, 1987.

17. Russell LJ et al: Pallister–Killian phenotype in the absence of 12p aneuploidy. *Proc Greenwood Genet Ctr* **7**:177, 1988.

18. Schroer RJ, Stevenson RE: Further clinical delineation of the syndrome of unusual facial appearance, abnormal hair, and mental retardation reported by Teschler–Nicola and Killian. *Proc Greenwood Genet Ctr* **2**:3–5, 1983.

19. Shivashankar L et al: Prenatal diagnosis of tetrasomy 47,XY,+i(12p) confirmed by in situ hybridization. *Prenatal Diag* **8**:85–91, 1988.

20. Teschler–Nicola M, Killian W: Mental retardation, unusual facial appearance, abnormal hair. *Synd Ident* **7**(1):6–7, 1981.

21. Vine DT et al: Teschler–Nicola/Killian syndrome and mosaic tetrasomy 12p. *J Clin Dysmorphol* **2**(1):7, 1984.

22. Warburton D et al: Mosaic tetrasomy 12p: Four new cases and confirmation of the chromosomal origin of the supernumerary chromosome in one of the original Pallister-mosaic syndrome cases. *Am J Med Genet* **27**:275–283, 1987.

23. Ward BE et al: Isochromosome 12p mosaicism (Pallister–Killian syndrome): Newborn diagnosis by direct bone marrow analysis. *Am J Med Genet* **31**:835–840, 1988.

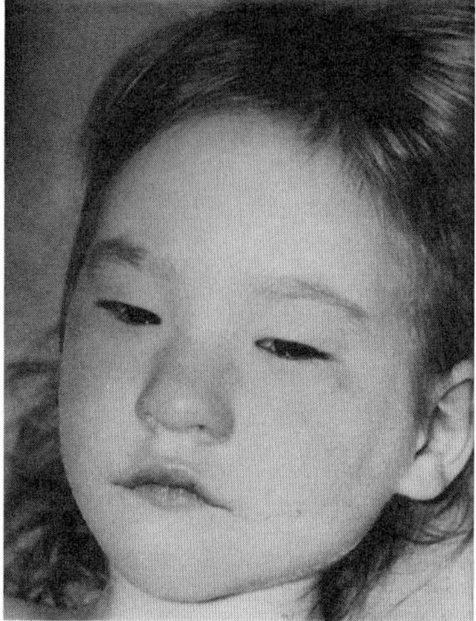

Fig. 4–37. *del(14q) syndrome.* Microcephaly, short narrow palpebral fissures, flat nasal bridge. (From *MO Rethoré et al,* Ann Génét **27**:91, 1984.)

24. Wenger SL et al: Risk effect of maternal age in Pallister i(12p) syndrome. *Clin Genet* **34**:181–184, 1988.
25. Wyatt PR: Pallister–Killian syndrome—an update of a clinical case. *Am J Med Genet* **29**:229, 1988.

## del(14q) syndrome

Most cases of del(14q) have ring chromosomes. The main features include microcephaly with flat occiput, epicanthal folds, downward slanting palpebral fissures, narrow elongated face, short palpebral fissures, flat nasal bridge, large low-set pinnae, micrognathia, and short neck. Retinal dystrophy may be specific (hyperpigmentation and yellow-white spots of the macula) (1–7) (Fig. 4–37).

Mental retardation, hypotonia, and seizures are severe. The lateral ventricles are moderately enlarged. Ring 14 syndrome is compatible with extended survival but recurrent respiratory infections are common.

### References [del(14q) syndrome]

1. Caille B et al: Deux nouvelles observations de chromosome 14 en anneau. *Ann Pédiatr* **32**:441–446, 1985.
2. Fryns JP et al: Ring chromosome 14: A distinct clinical entity. *J Génét Hum* **31**:367–375, 1983.
3. Gilgenkrantz S et al: Le syndrome r(14). *Ann Génét* **27**:73–78, 1984.
4. Hreidarsson SJ, Stamberg J: Distal monosomy 14 not associated with ring formation. *J Med Genet* **20**:147–149, 1983.
5. Jalbert P et al: Chromosome 14 en anneau chez des jumelles monozygotes. *Ann Génét* **20**:59–62, 1977.
6. Rethoré MO et al: Chromosome 14 en anneau. II. Une observation de r(14) en mosaïque le phenotype r(14). *Ann Génét* **27**:91–95, 1984.
7. Schmidt R et al: Ring chromosome 14: A distinct clinical entity. *J Med Genet* **18**:304–307, 1981.

## dup(14q) syndrome

In addition to mental retardation and pre- and postnatal growth retardation, there is a large face, chubby cheeks, facial asymmetry, hypertelorism, broad nose, short prominent philtrum, carp mouth, posteriorly rotated pinnae with prominent antitragus, and micrognathia (1–10). Cleft palate has been found in at least 50% (1,3,4,9) (Figs. 4–38A,B and 4-39A,B).

The nipples are high and widely spaced. Males exhibit hypogenitalism. Brain, lung, and congenital heart defects are common.

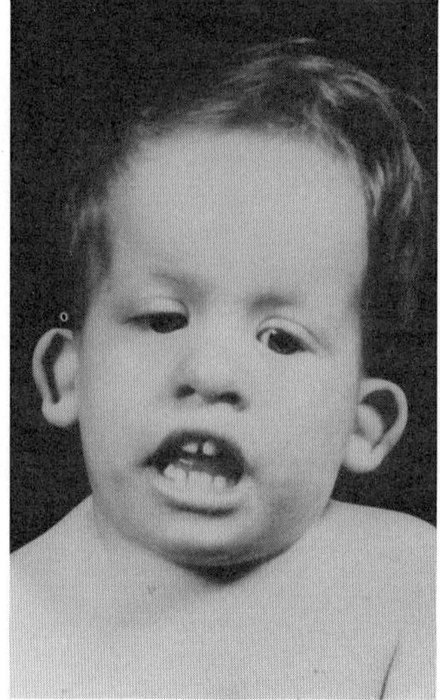

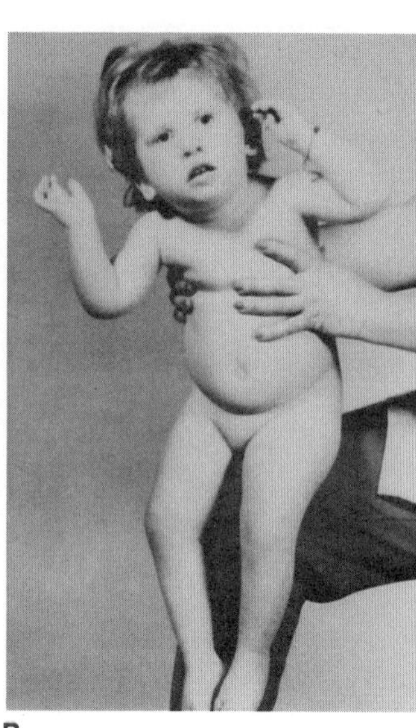

A B

Fig. 4–38. *dup(14q) syndrome.* (A,B) Microcephaly, broad flat nose with bulbous tip, downslanting palpebral fissures, short philtrum with arched upper lip and downturned mouth. (From *JQ Miller et al,* J Med Genet **16**:60, 1979.)

### References [dup(14q) syndrome]

1. Cottrall K et al: A case of proximal 14 trisomy with pathological findings. *J Ment Defic Res* **25**:1–6, 1981.
2. Johnson VP et al: Trisomy 14 mosaicism: Case report and review. *Am J Med Genet* **3**:331–339, 1979.
3. Martin AO et al: 46,XX/47,XX, + 14 mosaicism in a liveborn infant. *J Med Genet* **14**:214–218, 1977.
4. Miller JQ et al: Familial partial 14 trisomy. *J Med Genet* **16**:60–65, 1979.
5. Nikolis J et al: Tandem duplication of chromosome 14 (q24→q32) in a male newborn with congenital malformations. *Clin Genet* **23**:321–324, 1983.
6. Orye E et al: Distal trisomy 14q due to tandem duplication. *Ann Génét* **26**:238–239, 1983.
7. Pajaris IL et al: Partial trisomy 14q. *Hum Genet* **46**:243–247, 1979.
8. Pfeiffer RA, Kessel E: Balanced and unbalanced pericentric inversion of chromosome 14. *Hum Genet* **43**:103–106, 1978.
9. Raoul O et al: Trisomie 14q partielle. *Ann Génét* **18**:35–39, 1975.
10. Turleau C et al: La trisomie 14q distale. *Ann Génét* **26**:165–170, 1983.

## Trisomy 14 mosaicism syndrome

Although trisomy 14 has been reported in abortuses, most living examples have been mosaic. Polyhydramnios is frequent. Although birth

Fig. 4–39. *dup(14q) syndrome.* (A,B) Similar facies. (From *C Turleau et al,* Ann Génét **26**:166, 1983.)

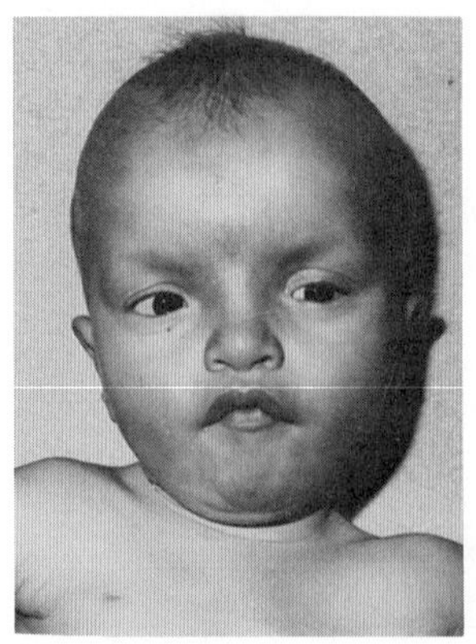

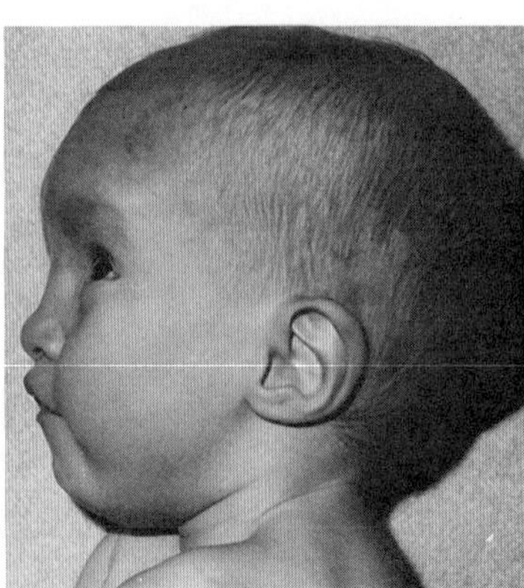

A B

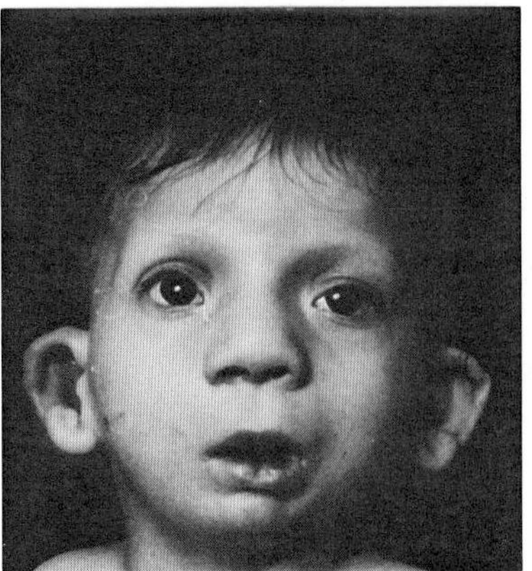

Fig. 4–40. *Trisomy 14 mosaicism syndrome.* Hypertelorism, broad saddle nose, asymmetric palpebral fissures, outstanding dysmorphic pinnae, long philtrum, micrognathia. (From *JD Murken et al,* Humangenetik **10:**254, 1970.)

weight is normal, postnatal growth and psychomotor development are severely retarded.

The forehead is prominent. The eyes may be deeply set and hypertelоric with an evanescent translucent film over the eyes. Wide nasal bridge, prominent maxilla, long philtrum, large mouth, thick lips, dysmorphic low-set pinnae, micrognathia, and short neck are characteristic. A number of patients have cleft palate (2,4–6,8) (Fig. 4–40). Several patients have exhibited reticular hyperpigmentation of the skin that resembles that seen in *incontinentia pigmenti.* Body and facial asymmetry has been seen in about one-third of the patients. Congenital heart anomaly, particularly tetralogy of Fallot, is common (1,3,7,9).

### References (Trisomy 14 mosaicism syndrome)

1. Fujimoto A et al: Trisomy 14 mosaicism with t(14;15)(q11p11) in offspring of a balanced translocation carrier mother. *Am J Med Genet* **22**:333–342, 1985.

2. Jenkins MV et al: Trisomy 14 mosaicism in a translocation 14q15q carrier: Probable dissociation and isochromosome formation. *J Med Genet* **18**:68–71, 1981.

3. Kaplan L et al: Trisomy 14 mosaicism in a liveborn male: Clinical report and review of the literature. *Am J Med Genet* **23**:925–930, 1986.

4. Murken JD et al: Trisomie D2 bei einem 2½ jährigen Mädchen (47,XX,+14). *Humangenetik* **10**:254–268, 1970.

5. Raoul O et al: Trisomie 14q partielle. *Ann Génét* **18**:35–39, 1975.

6. Reiss JA et al: Mosaicism with translocation: Autoradiographic and fluorescent studies of an inherited reciprocal translocation t(2q+;14q−). *J Med Genet* **9**:280–286, 1972.

7. Rethoré MO et al: Trisomie 14 en mosaïque chez une enfant multimalformée. *Ann Génét* **18**:71–74, 1975.

8. Simpson J, Zellweger H: Partial trisomy 14q− and parental translocation of No. 14 chromosome. *J Med Genet* **14**:124–127, 1977.

9. Turleau C et al: Trisomie 14 en mosaïque par isochromosome dicentrique. *Ann Génét* **23**:238–240, 1980.

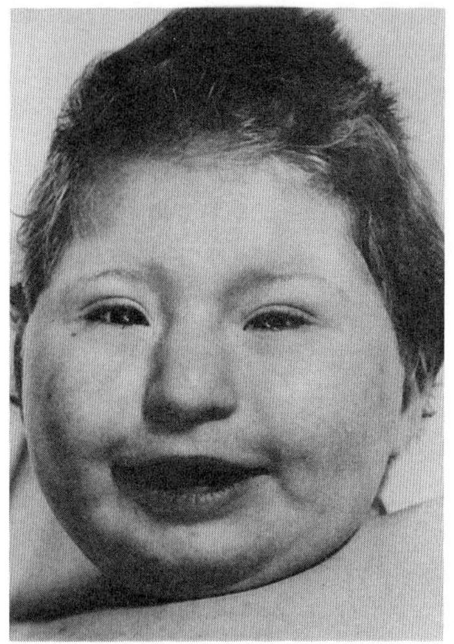

Fig. 4–41. *del(15q) syndrome.* Interstitial deletion of chromosome 15. Child has microcephaly, upslanting palpebral fissures, sloping forehead, high nasal bridge. (From *MY Yip et al,* J Med Genet **24:**709, 1987.)

## del(15q) syndrome

Deletion of the long arm of chromosome 15 usually results in a ring. At least 25 patients have been described (1–11).

In the young patient, the most characteristic findings include prenatal growth retardation (100%), variable mental retardation (95%), microcephaly (85%), hypertelorism (45%), a triangular face somewhat resembling that of Silver–Russell syndrome (40%) (8), and limb anomalies including delayed bone age (75%), brachymesophalangy (45%), clinodactyly of the fifth fingers, and thumb hypoplasia. About 30% have congenital heart anomalies. Café-au-lait spots are found in 30%.

During adult years, severe mental and somatic retardation become evident. The forehead is bossed (35%), the face triangular (40%), the pinnae anomalous (30%), and the nose has a broad high bridge. Males are hypogonadal (Figs. 4–41 and 4–42A–C).

### References [del(15q) syndrome]

1. Butler MG et al: Two patients with ring chromosome 15 syndrome. *Am J Med Genet* **29**:149–154, 1988.

2. Fryns JP et al: Ring chromosome 15 syndrome. *Hum Genet* **51**:43–48, 1979.

3. Fryns JP et al: Ring chromosome 15 syndrome. *Ann Génét* **29**:47–48, 1986.

4. Kousseff BG: Ring chromosome 15 and failure to thrive. *Am J Dis Child* **134**:798–799, 1980.

5. Meinecke P, Koske-Westphal T: Ring chromosome 15 in a male adult with radial defects. Evaluation of the phenotype. *Clin Genet* **18**:428–433, 1980.

6. Moreau N, Teyssier M: Ring chromosome 15: Report of a case in an infertile man. *Clin Genet* **21**:273–279, 1982.

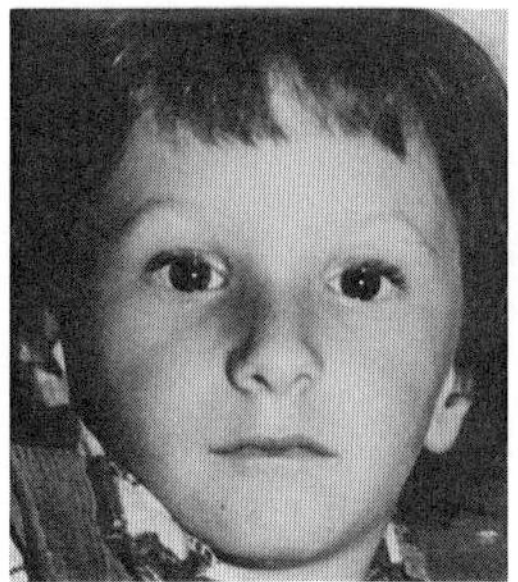
A

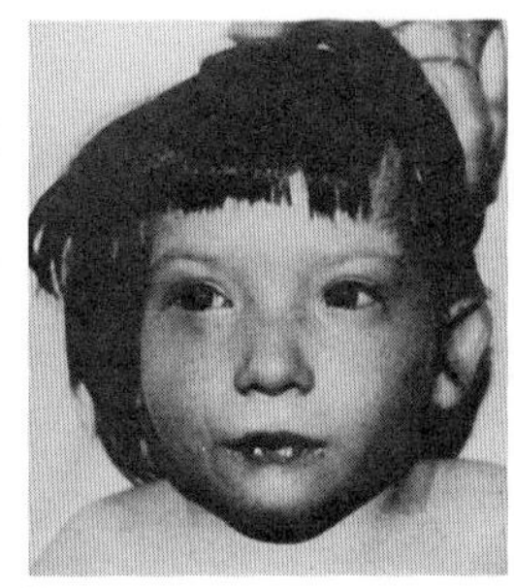
B

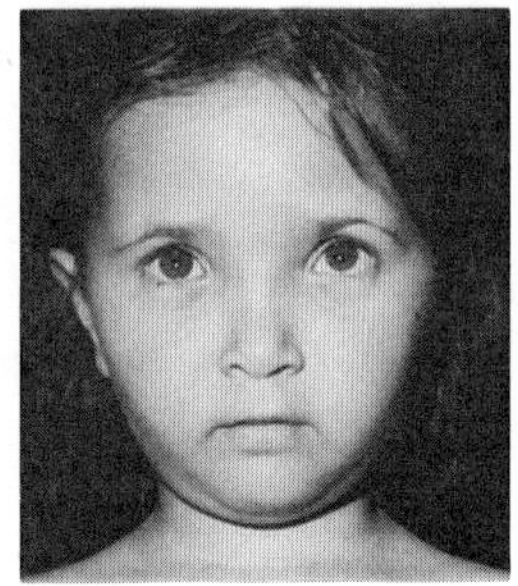
C

Fig. 4–42. *del(15q) syndrome.* (A) Patient having triangular face somewhat resembling that of Russell–Silver syndrome. (B,C) Note similar facies. (A from *MG Butler et al,* Am J Med Genet **29:**149, 1988; B from *E Yunis et al,* Hum Genet **57:**207, 1981; C from *E Ferrante et al,* Min Pediatr **29:**2163, 1977.)

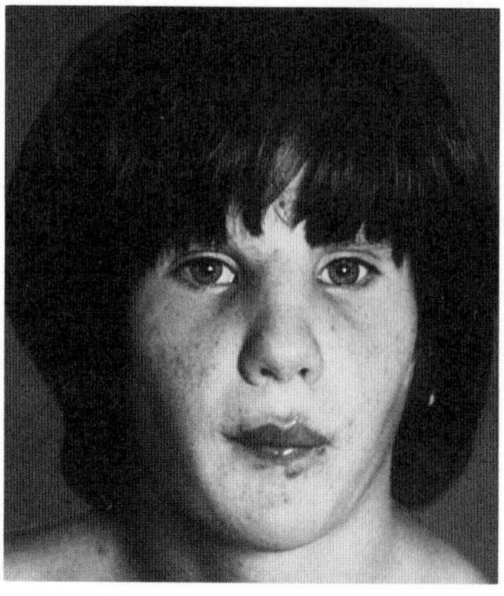

Fig. 4–43. *dup(15q) syndrome. Duplication of proximal chromosome 15.* Patient has mental retardation, microcephaly, flat nasal bridge, cleft palate, small mouth, and micrognathia. (From *G Annerén* and *KH Gustavson,* Clin Genet **22:**16, 1982.)

7. Otto J et al: Dysplastic features, growth retardation, malrotation of the gut and fatal VSD in a 4-month-old girl with r(15). *Eur J Pediatr* **142**:229–231, 1984.

8. Wilson GN et al: Phenotypic delineation of ring chromosome 15 and Russell–Silver syndrome. *J Med Genet* **22**:233–236, 1985.

9. Wisniewski L et al: The child with chromosome ring 15. *Klin Pädiatr* **191**:429–432, 1979.

10. Yip MY et al: Deletion 15q21.1→q22.1 resulting from paternal insertion into chromosome 5. *J Med Genet* **24**:709–712, 1987.

11. Yunis E et al: Ring (15) chromosome. *Hum Genet* **57**:207–209, 1981.

## dup(15q) syndrome

Most examples have resulted from unbalanced translocations (1–11). About 30 cases have been described. Microdolichocephaly, occasionally hydrocephaly, narrow or short and downward slanting palpebral fissures, bulbouse nose, malformed pinnae, long philtrum, cleft palate, and micrognathia characterize the facies of trisomy for the proximal long arm of chromosome 15 (1,6,7,11) (Fig. 4–43).

Those with distal 15q trisomy exhibit microcephaly with sloping forehead, facial asymmetry, downslanting and short or narrow palpebral fissures, ptosis, prominent nose with broad bridge, long philtrum, downturned mouth, highly arched palate, midline crease of the lower lip, puffy cheeks, micrognathia, and short neck (8,9). Postnatal growth deficiency, scoliosis, pectus excavatum, cryptorchidism, arachnodactyly, camptodactyly, hyperextensible thumbs, and cardiovascular anomalies are relatively common. Severe mental retardation is constant and hypotonia is frequent. Seizures are noted in 30% (Fig. 4–44A–H).

An example of complete trisomy 15 has been described (3).

### References [dup(15q) syndrome]

1. Annerén G, Gustavson KH: A boy with proximal trisomy 15 and a male foetus with distal trisomy 15 due to a familial 13p;15q translocation. *Clin Genet* **22**:16–21, 1982.

2. Coco R, Penchaszadeh VB: Inherited partial duplication deficiency of chromosome 15 (p12q22) *J Génét Hum* **26**:203–210, 1978.

3. Coldwell S et al: A case of trisomy of chromosome 15. *J Med Genet* **18**:146–148, 1981.

4. Gregoire MJ et al: Duplication 15q22→15ter and its phenotypic expression. *Hum Genet* **59**:429–433, 1981.

5. Lacro RV et al: Duplication of distal 15q. *Am J Med Genet* **26**:719–728, 1987.

6. Mankinen CB et al: Partial trisomy 15 in a young girl. *Clin Genet* **10**:27–32, 1976.

7. Pfeiffer RA, Kessel E: Partial trisomy 15q1. *Hum Genet* **33**:77–83, 1976.

8. Schnatterly P et al: Distal 15q trisomy: Phenotype comparison of nine cases in an extended family. *Am J Hum Genet* **36**:444–451, 1984.

9. Tzancheva M et al: Two familial cases with trisomy 15q distal due to a rep(5;15)(p14; q21). *Hum Genet* **56**:275–277, 1981.

10. Yip MY et al: A de novo tandem duplication 15(q21→qter) mosaic. *Clin Genet* **22**:1–6, 1982.

11. Zabel B, Baumann W: Trisomie partielle pour la partie distale du bras long du chromosome 15 par translocation X/15 maternelle. *Ann Génét* **20**:285–289, 1977.

## dup(16p) syndrome

Microcephaly, hypertelorism, round low-set pinnae, prominent maxilla, and micrognathia were found. About 50% died in infancy. Mental retardation was severe. Aplasia or hypoplasia of the thumb, ASD, and tetralogy of Fallot were noted (1,2).

### References [dup(16p) syndrome]

1. Leschot NJ et al: Five familial cases with a trisomy 16p syndrome due to translocation. *Clin Genet* **16**:205–214, 1979.

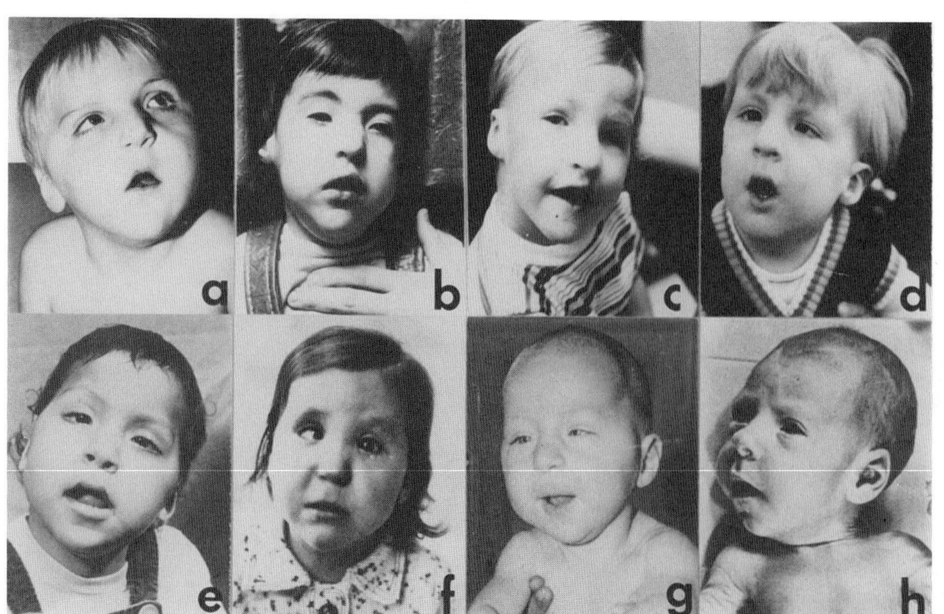


Fig. 4–44. *dup(15q) syndrome.* (A–H) *Duplication of distal chromosome 15.* Note facial asymmetry, narrow downslanting palpebral fissures, ptosis, prominent nose, long philtrum, downturned mouth, and puffy cheeks in eight reported patients. (From *P Schnatterly et al,* Am J Hum Genet **36:**444, 1984.)

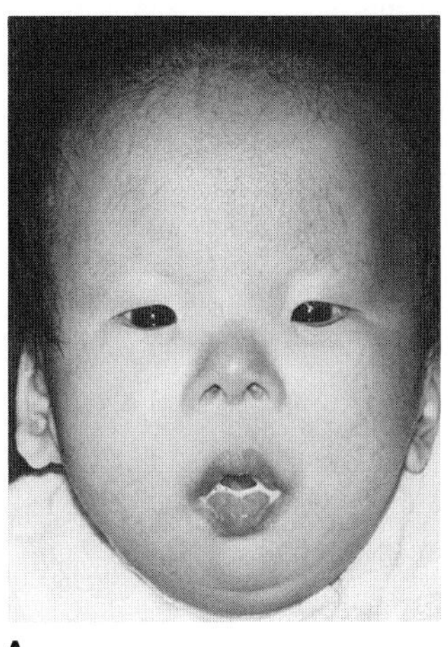
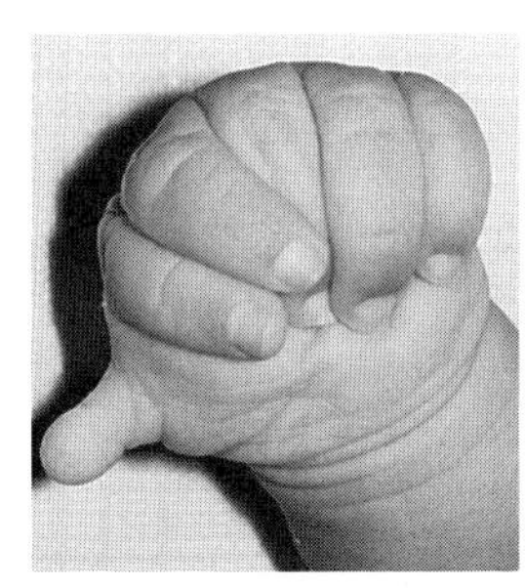
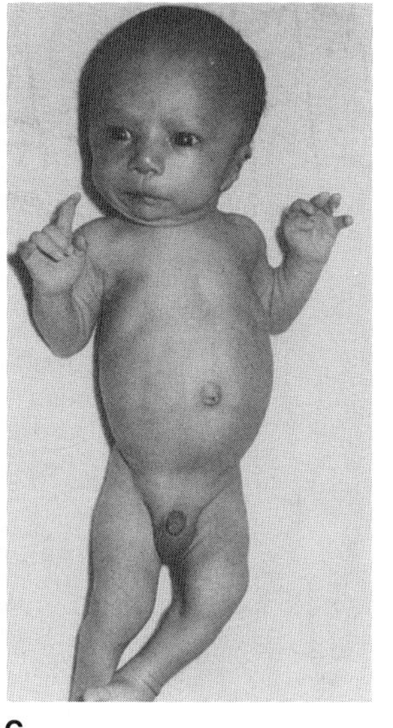

A B C

Fig. 4–45. *del(16q) syndrome.* (A) High forehead, large anterior fontanel, narrow palpebral fissures, small upturned nose, malformed pinnae, and short neck. Child had cleft palate. (B) Polydactyly and overlapping flexed fingers. (C) Relatively large skull with widely patent anterior fontanel, brachycephaly, high forehead, prominent metopic suture, triangular facies, small nose, long philtrum, micrognathia, and short neck. (A,B from *CC Lin et al,* Hum Genet **65:**134, 1985; C courtesy of *JM Cantú,* Guadalajara, Mexico.)

2. Roberts SH, Duckett DP: Trisomy 16p in a live born infant and a review of partial and full trisomy 16. *J Med Genet* **15**:375–381, 1978.

## del(16q) syndrome

Both interstitial and terminal deletions have been reported (1–9).

The phenotype is characterized by low birth weight, delayed growth and development, feeble suck, hypotonia, and distinct craniofacies: microcephaly, high forehead, prominent metopic suture, large anterior fontanel with or without wide cranial sutures, narrow palpebral fissures, small upturned nose, low-set folded helices, micrognathia, and short neck. Cleft palate has been described (6) as well as natal teeth (2) (Fig. 4–45A).

Diverse musculoskeletal anomalies (narrow thorax, small hands and feet, talipes, umbilical hernia, polydactyly, flexed fingers, broad halluces) have been reported (Fig. 4–45B,C).

Cardiac (coarctation of aorta, VSD), renal (cystic dysplasia), and intestinal (ectopic anus, malrotation) anomalies have been described.

### References [del(16q) syndrome]

1. Brenholz P et al: Fragile site on chromosome 16 and 16q− syndrome. *Am J Hum Genet* **34**:119A, 1982.
2. Elder FFB et al: Identical twins with deletion 16q: Evidence that 16q12.2−q13 is the critical band region. *Hum Genet* **67**:233–236, 1984.
3. Fryns JP et al: Partial monosomy of the long arm of chromosome 16 in a malformed newborn: Karyotype 46,XX,del(16)(q12). *Hum Genet* **38**:343–346, 1977.
4. Fryns JP et al: Partial monosomy of the long arm of chromosome 16: A distinct clinical entity? *Hum Genet* **46**:115–120, 1979.
5. Fryns JP et al: Interstitial 16q deletion with typical dysmorphic syndrome. *Ann Génét* **24**:124–125, 1981.
6. Lin CC et al: Interstitial deletion for a region of the long arm of chromosome 16. *Hum Genet* **65**:134–138, 1983.
7. Rivera H et al: Monosomy 16q: A distinct syndrome. *Clin Genet* **28**:84–86, 1985.
8. Taysi K et al: A terminal deletion long arm deletion of chromosome 16 in a dysmorphic infant: 46,XY,del(16)(q22). *Birth Defects* **14**:(6C):343–347, 1978.
9. Yunis E et al: Partial monosomy 16q−. *Hum Genet* **38**:347–350, 1977.

## dup(16q) syndrome

Trisomy 16 is the most common trisomy among spontaneous abortions during the first trimester, but there are only a few known liveborn examples (2,4,8,9).

In addition to low birth weight and severe psychomotor and mental retardation, common features included round face, frontal bossing, prominent glabella, hypertelorism, depressed nasal bridge, round nasal tip with anteverted nostrils, low-set pinnae, and prominent maxilla.

Common features include low birth weight, psychomotor retardation, low-set small pinnae, asymmetric skull, high forehead, small palpebral fissures, hypertelorism, flat nasal bridge, malformed long philtrum, hypertrichosis, flexion contractures, cryptorchidism, VSD, and foot deformities (1,3,5–7). Early death is common.

### References [dup(16q) syndrome]

1. Balestrazzi P et al: Partial trisomy 16q resulting from maternal translocation. *Hum Genet* **49**:229–235, 1979.
2. Dallapiccola B et al: De novo trisomy 16q11→pter. *Hum Genet* **49**:1–6, 1979.
3. Francke U: Quinacrine mustard fluorescence of human chromosomes: Characterizations of unusual translocations. *Am J Med Genet* **24**:189–213, 1972.
4. Fryns JP et al: Partial monosomy of the long arm of chromosome 16: A distinct clinical entity? *Hum Genet* **46**:115–120, 1979.
5. Garau A et al: Trisomy 16q21→qter. *Hum Genet* **53**:165–167, 1980.
6. Ridler MAC, McKeown JA: Trisomy 16q arising from a maternal 15p;16q translocation. *J Med Genet* **16**:317–319, 1979.
7. Schmickel R et al: 16q trisomy in a family with a balanced 15–16 translocation. *Birth Defects* **11**:(5):229–236, 1975.
8. Taysi K et al: A terminal long arm deletion of chromosome 16 in a dysmorphic infant: 46,XY,del(16)(q22). *Birth Defects* **14**(6C):343–347, 1978.
9. Yunis E et al: Partial trisomy 16q−. *Hum Genet* **38**:347–350, 1977.

## del(17p) syndrome

Less than a dozen patients have been described with del(17p). In addition to developmental delay, these patients have brachycephaly, midface hypoplasia with broad nasal bridge, highly arched palate, malformed and/or malpositioned pinnae, hearing deficit, relative mandibular prognathism, and short broad hands (2–6).

There have been several patients with ring 17 chromosome but a clinical pattern has not yet emerged (1).

### References [del(17p) syndrome]

1. Bridge J et al: Partial deletion of distal 17q. *Am J Med Genet* **21**:225–229, 1985.
2. Patil SR, Bartley JA: Interstitial deletion of the short arm of chromosome No. 17. *Hum Genet* **67**:237–238, 1984.

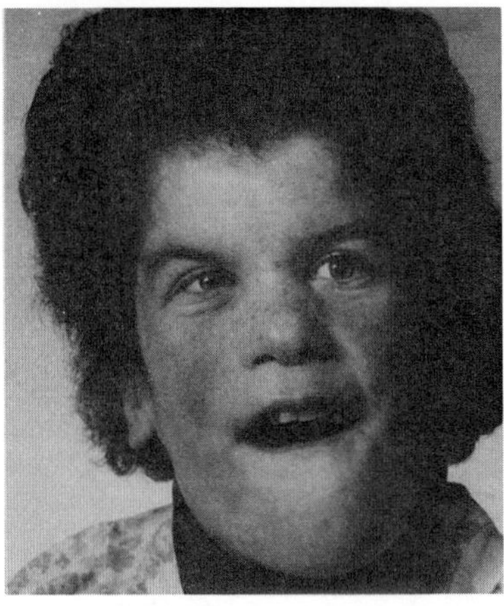

Fig. 4–46. *dup(17q) syndrome.* Narrow bifrontal diameter, temporal hair, widow's peak, and thin upper lip. [From *M Berberich et al,* Birth Defects **14**(6C):287, 1978.]

3. Popp DW et al: An additional case of deletion 17p11.2. *Am J Med Genet* **26**:493–495, 1987.

4. Smith ACM et al: Deletion of the 17 short arm in 2 patients with facial clefts. *Am J Hum Genet* **34**:410A, 1982.

5. Smith ACM et al: Interstitial deletion of (17)(p11.2p11.2) in 9 patients. *Am J Med Genet* **24**:393–414, 1986.

6. Stallard R et al: Monosomy of 17p 11.2 in 2 unrelated infants with developmental delay. *Am J Hum Genet* **36**:115S, 1984.

## dup(17p) syndrome

Severe mental and somatic retardation, microcephaly, narrow palpebral fissures, hypertelorism, broad nasal bridge, dysplastic low-set ears, chronic open mouth, micrognathia, and short webbed neck are characteristic (1–9).

Flexion abnormalities of the first four digits with extension of the fifth finger is common. The fingers tend to be long and tapered. Inguinal hernia has been noted (2,3,6). A transverse palmar crease has been found in most patients. The male genitalia are hypoplastic.

### References [dup(17p) syndrome]

1. Bartsch-Sandhoff M, Hieronimi G: Partial duplication of 17p. *Hum Genet* **49**:123–127, 1979.

2. Feldman GM et al: The dup(17p) syndrome. *Am J Med Genet* **11**:299–304, 1982.

3. Jinno Y et al: Trisomy 17p due to a t(5;17)(p15;p11) pat translocation. *Ann Génét* **25**:123–125, 1982.

4. Latta E, Hoo JJ: Trisomy of the short arm of chromosome 17. *Hum Genet* **23**:213–217, 1974.

5. Palutke W et al: An extra small metacentric chromosome identified as a deleted chromosome No. 17. *Clin Genet* **9**:454–458, 1976.

6. Rethoré MO et al: La trisomie 17p. *Ann Génét* **26**:17–20, 1983.

7. Salamanca-Gomez F, Armendares S: Identification of isochromosome 17 in a girl with mental retardation and congenital malformations. *Ann Génét* **18**:235–238, 1975.

8. Shabtai F et al: Pure trisomy 17p in 60% of cells. *Hum Genet* **52**:263–268, 1979.

9. Yamamoto Y et al: A case of partial trisomy 17 resulting from an X-autosomal translocation. *J Med Genet* **16**:395–399, 1979.

## dup(17q) syndrome

Duplication of the distal portion of 17q is quite rare. Most cases have involved bands q21, q22, or q23→qter (1–6).

The phenotypic alterations appear to be remarkably similar. Stature is short and psychomotor retardation is profound. Most patients exhibit microcephaly, plagiocephaly, frontal bossing, temporal retraction, facial asymmetry, and widow's peak. Several patients manifest downward slanting palpebral fissures, hypertelorism, and epicanthal folds with a flat nasal bridge. The mouth is often wide with a thin upper lip and downturned corners. Over half have either cleft lip or cleft palate (1,2,6). The pinnae are often low-set, posteriorly angulated, and malformed. The neck is short and broad and may occasionally be webbed with low posterior hairline (Fig. 4–46). Postaxial polydactyly of the hands and/or feet has been reported in several patients and there has been hyperlaxity of limb joints. Serious congenital heart anomalies have been present in about 50%. Central nervous system abnormalities have been a constant feature. Renal anomalies (hydronephrosis, cystic kidneys) are also common. All males have exhibited cryptorchidism.

### References [dup(17q) syndrome]

1. Berberich M et al: Duplication (partial trisomy) of the distal long arm of chromosome 17: A new clinically recognizable chromosome disorder. *Birth Defects* **14**(6C):287–295, 1978.

2. Bridge J et al: Partial duplication of distal 17q. *Am J Med Genet* **22**:229–235, 1985.

3. Fryns J et al: Partial trisomy 17q. *Hum Genet* **49**:361–364, 1979.

4. Gallien JU et al: An infant with duplication of 17q21→17qter. *Am J Med Genet* **8**:111–115, 1981.

5. Naccache NF et al: Duplication of distal 17q. *Am J Med Genet* **17**:633–639, 1984.

6. Turleau C et al: Distal trisomy 17q. *Clin Genet* **16**:54–57, 1979.

## dup(19q) syndrome

Less than a dozen cases of trisomy for the distal third of the long arm of chromosome 19 are known (1–3). In addition to mental retardation, the phenotype includes low birth weight and length, marked postnatal growth retardation, microbrachycephaly, widely open sutures, downward slanting palpebral fissures, hypertelorism, ptosis, short nose, short philtrum, downturned mouth, cleft palate, and short neck with redundant skin. The thorax is barrel shaped. Musculoskeletal disorders include hypotonia, diastasis recti, kyphosis, duplicated thumb, valgus deformity of feet, and laterally curved halluces (1–3)(Figs. 4–47A–D and 4–48).

### References [dup(19q) syndrome]

1. Chen H et al: Mosaic trisomy 19 syndrome. *Ann Génét* **24**:32–33, 1981.

2. Lange M, Alfi OS: Trisomy 19q. *Ann Génét* **19**:17–21, 1976.

3. Schmid W: Trisomy for the distal third of the long arm of chromosome 19 in brother and sister. *Hum Genet* **46**:263–270, 1979.

## dup(20p) syndrome

The clinical picture of patients trisomic for the short arm of chromosome 20 is quite variable. Approximately 25 cases have been analyzed, about 90% of which have originated from parental translocation, more often in the female parent (1–8, 10).

The few examples of so-called trisomy 20, based on banding pattern, do not form a homogeneous phenotype (9).

Clinical features include mild to severe, but usually moderate, psychomotor retardation, poor motor coordination, and poor, delayed speech. There is flat occiput, coarse hair, round face, full cheeks, sloping forehead, deep widely-set eyes with upward slanting palpebral fissures, hypertelorism, strabismus, short upturned nose with large nostrils, low-set large and posteriorly angulated pinnae, microretrognathia, and positional abnormalities of the feet, fingers, and toes. Various vertebral abnormalities resulting in scoliosis have been noted (Fig. 4–49A–D).

Cardiac anomalies have been found in about 35%, most frequently VSD and tetralogy of Fallot.

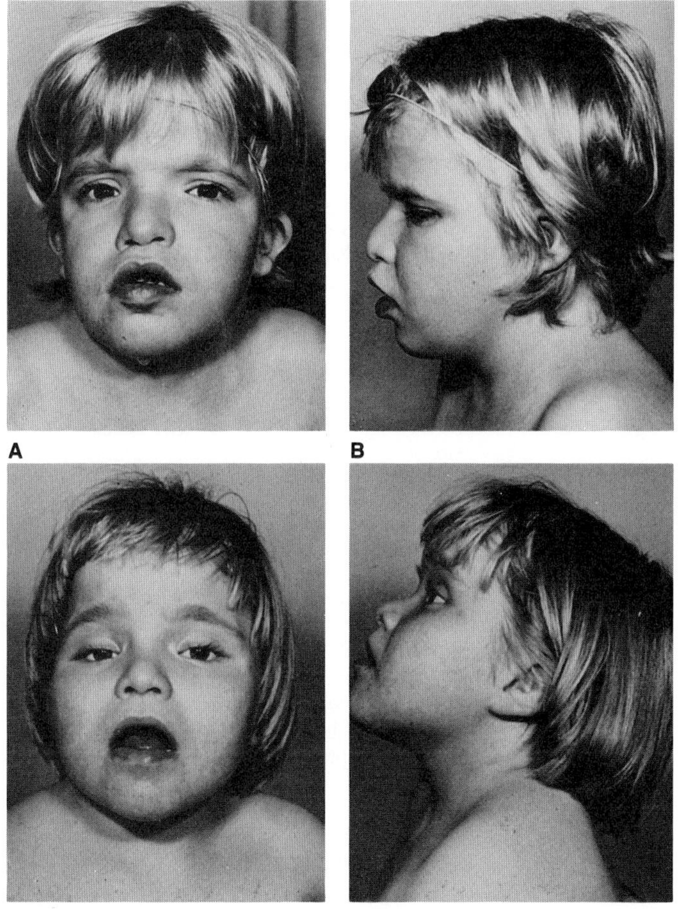

Fig. 4–47. *dup(19q) syndrome.* (A–D) Sisters with microbrachycephaly, ptosis, short philtrum, downturned mouth, short neck. (From *W Schmid,* Hum Genet **46:**263, 1979.)

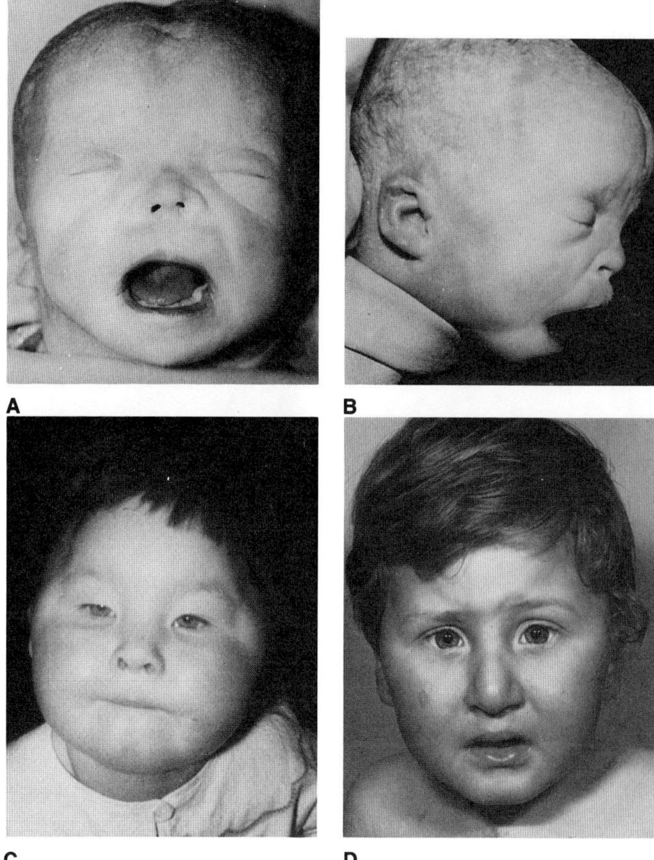

Fig. 4–49. *dup(20p) syndrome.* (A,B) Front and side views showing widened sagittal suture, hypertelorism, posteriorly rotated pinnae, and micrognathia. (C) Craniosynostosis, upslanting palpebral fissures, prominent cheeks, broad nasal bridge. (D) Beaked nose, small mouth, and full cheeks. (A,B from *SF Pan et al,* Clin Genet **9:**449, 1976; C from *IW Lurie et al,* J Génét Hum **33:**67, 1985; D from *A Schinzel,* Hum Genet **53:**169, 1980.)

Trisomy 20 mosaicism, a not uncommon finding in amniotic fluid cell cultures, has no clinical significance.

### References [dup(20p) syndrome]


1. Archidiacono N et al: Trisomy 20p from maternal t(3;20) translocation. *J Med Genet* **16**:229–232, 1979.
2. Balestrazzi P et al: De novo trisomy 20p with macro-orchidism in a prepubertal boy. *Ann Génét* **27**:58–59, 1984.
3. Centerwall W, Francke U: Familial trisomy 20p. Five cases and two carriers in three generations: A review. *Ann Génét* **20**:77–83, 1977.
4. Chen H et al: Partial trisomy 20p syndrome and maternal mosaicism. *Ann Génét* **26**:21–25, 1983.
5. Cohen MM et al: A familial F/G translocation (t(20p – ;22q + )) observed in three generations. *Clin Genet* **7**:120–127, 1975.
6. Delicado A et al: Partial trisomy 20. *Ann Génét* **24**:54–56, 1981.
7. Funderburk SJ et al: Trisomy 20p due to a paternal reciprocal translocation. *Ann Génét* **26**:94–97, 1983.
8. Lurie IW et al: Trisomy 20p. *J Génét Hum* **33**:67–75, 1985.
9. Pan SF et al: Trisomy of chromosome 20. *Clin Genet* **9**:449–453, 1976.
10. Schinzel A: Trisomy 20 pter →q11 in a malformed boy from a t(13;20)(p11; q11) translocation-carrier mother. *Hum Genet* **53**:169–172, 1980.


Fig. 4–48. *Mosaic trisomy 19 syndrome.* Microcephaly, widened fontanel, hydrops, hypertelorism, flat nasal bridge, short nose, small mouth, malformed pinnae, short neck and chest, and talipes. (From *H Chen et al,* Ann Génét **24:**32, 1981.)

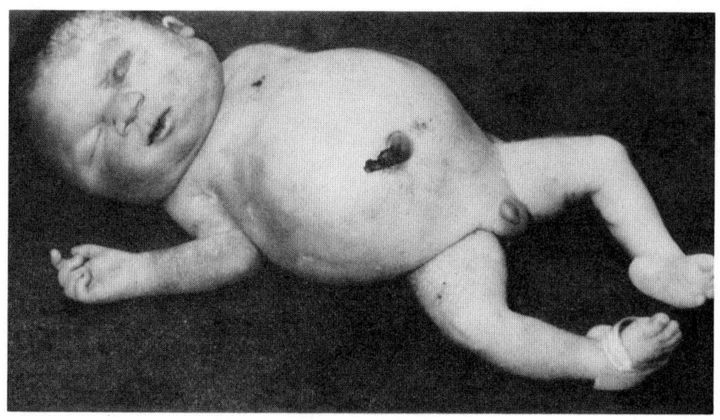

## Monosomy 21 syndrome

Because of the difficulty in differentiating chromosomes 21 and 22 prior to chromosome banding, we have carried out critical analysis of cases published only since 1973. (1–12).

The infant is small at birth and fails to thrive, the child often succumbing within the first year of life. Head circumference is markedly reduced, measuring between the 3rd and 10th percentile. The occiput is prominent. Most infants have been hypertonic.

The facies is characterized by downward slanting palpebral fissures, broad nasal base, large nose, somewhat low-set dysmorphic large pinnae, carp mouth, micrognathia, and short neck. Most have cleft lip

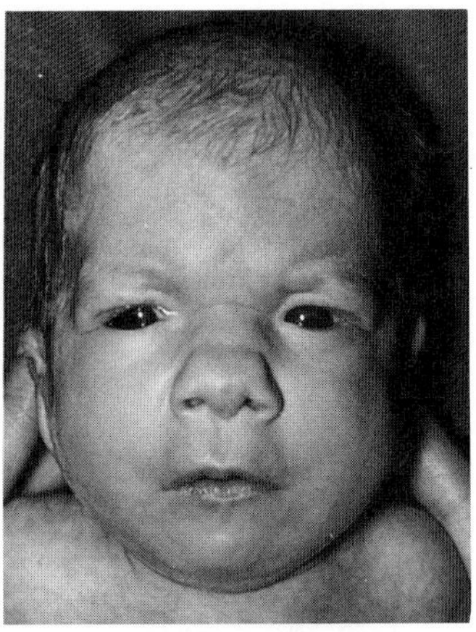

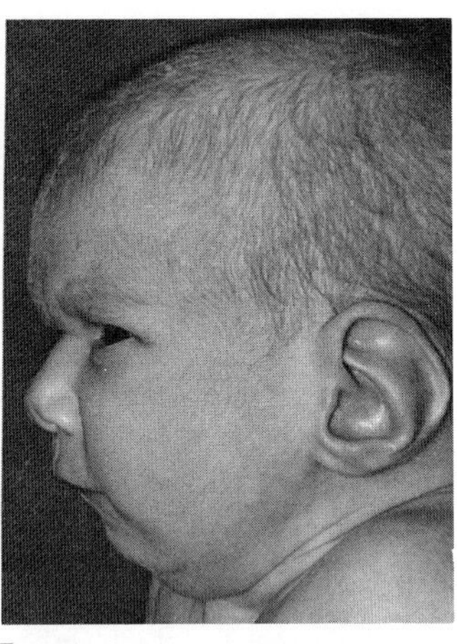

A B

Fig. 4–50. *Monosomy 21 syndrome.* (A,B) Downslanting palpebral fissures, broad nasal base, large nose, large dysmorphic pinna, micrognathia, and short neck. (From *M Mikkelsen* and *S Vestermark,* J Med Genet **11:**389, 1974.)

and/or cleft palate (2,4,6) (Fig. 4–50A,B). Skeletal anomalies include overlapping and/or flexed fingers and toes, kyphoscoliosis, short thorax, and narrow pelvis (7). Inconstant features include ambiguous genitalia or micropenis, cryptorchidism, and imperforate anus (10).

Cardiac anomalies have been relatively rare, the most common being preductal coarctation and patent ductus arteriosus.

Thrombocytopenia has been described in about 20%.

### References (Monosomy 21 syndrome)

1. Abeliovich D et al: Monosomy 21: A possible stepwise evolution of the karyotype. *Am J Med Genet* **4**:279–286, 1979.

1a. Carpenter NJ et al: Partial deletion 21: Case report with biochemical studies and review. *J Med Genet* **24**:706–708, 1987.

2. Davis JG et al: A child with presumptive monosomy 21 (45,XY,-21) in a family in which some members are Gq−. *Cytogenet Cell Genet* **17**:65–77, 1976.

3. Dziuba P et al: A female infant with monosomy 21. *Hum Genet* **31**:351–353, 1976.

4. Fryns JP et al: Full monosomy 21: A clinically recognized syndrome? *Hum Genet* **37**:155–159, 1977.

5. Halloran KH et al: 21 monosomy in a retarded female infant. *J Med Genet* **11**:386–389, 1974.

6. Herva R et al: 21-Monosomy in a liveborn male infant. *Eur J Pediatr* **140**:57–59, 1983.

7. Houston CS, Chudley AE: Separating monosomy 21 from the "arthrogryposis basket." *J Can Assoc Radiol* **32**:220–223, 1981.

8. Mikkelsen M, Vestermark S: Karyotype 45,XX,-21/46,XX,21q− in an infant with symptoms of G− deletion syndrome I. *J Med Genet* **11**:389–393, 1974.

9. Palmer CG et al: Four new cases of ring 21 and 22 including familial transmission of ring 21. *J Med Genet* **14**:54–60, 1977.

10. Philip N et al: Three new cases of partial monosomy 21 resulting from one ring 21 chromosome and two unbalanced reciprocal translocations. *Eur J Pediatr* **142**:61–64, 1984.

11. Richmond HG et al: A "G" deletion syndrome: Antimongolism. *Acta Paediatr Scand* **62**:216–220, 1973.

12. Wisniewski K et al: Monosomy 21 syndrome: Further delineation including clinically neuropathological, cytogenetic, and biochemical studies. *Clin Genet* **23**:102–110, 1983.

## del(22q) syndrome

Loss of the distal portion of the long arm of chromosome 22 usually results in a ring. Approximately 25 patients have been described (1,6,8,12). The phenotype is not striking. Head circumference has been decreased in about 65%. The face tends to be round in young children. There are horizontal and wide almond-shaped palpebral fissures. Epicanthal folds are frequent. Ptosis is present in approximately 35%. The eyebrows tend to be low set, the nose has a bulbous tip in infancy, and the pinnae are often large. Mental retardation is usually severe and becomes even more pronounced with age. Hypotonia is essentially a constant feature along with poor motor coordination (Figs. 4–51A,B).

Deletion of 22q11 has been found in several infants with *DiGeorge sequence* (thymic aplasia, congenital heart anomaly, hypoparathyroidism) (2,4,5,7,9,10,15,16). In some cases the chromosome abnormality has resulted from *de novo* rearrangements and in others from unbalanced transmission of familial translocations.

DiGeorge sequence, a developmental field defect, involves failure of the thymus and parathyroids to develop, resulting from disturbances of the third and fourth pharyngeal pouches (2a,13). The sequence is usually isolated but may be seen in *CHARGE association.* Occasionally DiGeorge sequence seems to have autosomal recessive or dominant inheritance (9a,11,14). Some of these examples may represent chromosome translocation. However, in several cases, no chromosome change has been detected with high-resolution cytogenetics (14).

Congenital heart anomalies, mostly conotruncal abnormalities, vary

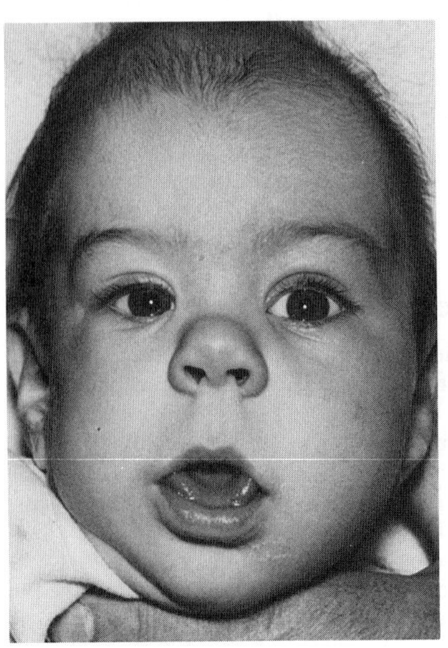

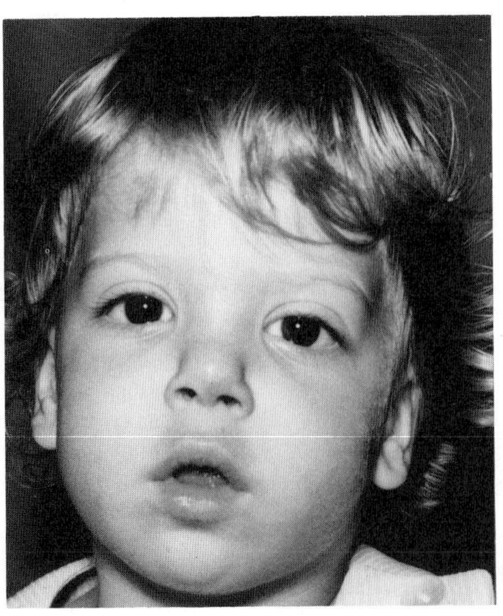

A B

Fig. 4–51. *del(22q) syndrome.* (A) Epicanthal folds, ptosis, broad upturned nose and long philtrum. (B) Compare facies with that of older child. (A courtesy of *D Hoefnagel,* Dartmouth, New Hampshire. B courtesy of *R Warren,* Miami Florida.)

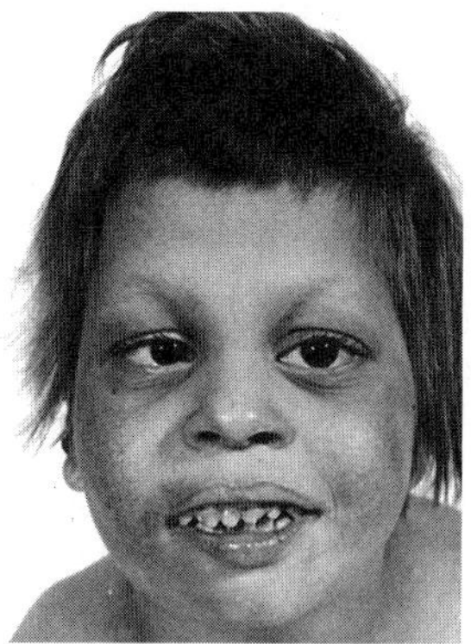
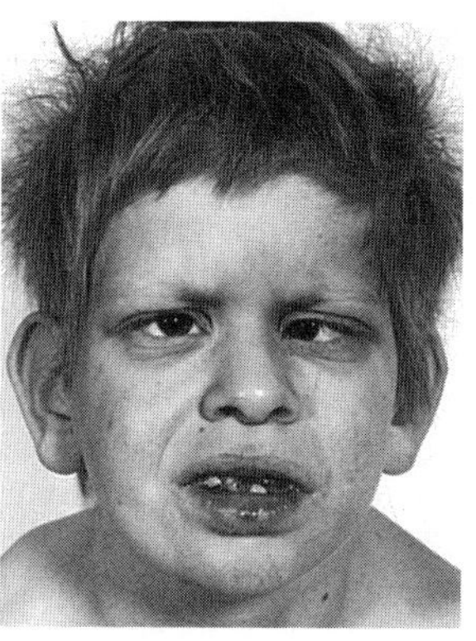

A B

Fig. 4–52. *Trisomy 22 syndrome.* (A,B) Esotropia, broad flat nose, prominent upper lip, everted lower lip. Female had preauricular tag on right, severe microtia on left. (A,B from *A Schinzel et al,* Hum Genet **56**:269, 1981.)

from VSD (most common) to interrupted aortic arch, right aortic arch, truncus arteriosus, tetralogy of Fallot, PDA, and single ventricle (3,16).

Cleft lip and/or palate have been noted (5,10,16). The pinnae may be malformed or posteriorly rotated (5,7,9). Hypertelorism, short palpebral fissures, prominent nose, short philtrum, and micrognathia are frequent.

#### References [del(22q) syndrome]


1. Aller V et al: An r(22) (p11→q13) in a mildly mentally retarded girl. *Hum Genet* **51**:157–162, 1979.

2. Augusseau S et al: DiGeorge syndrome and 22q11 rearrangements. *Hum Genet* **74**:206, 1986.

2a. Belohradsky BH: Thymusaplasie und -hypoplasie mit Hypoparathyreoidismus, Herz- und Gefässmissbildungen (DiGeorge-Syndrom). *Ergeb Inn Med Kinderheilk* **54**:36–106, 1985.

3. Conley ME et al: The spectrum of the DiGeorge syndrome. *J Pediatr* **94**:883–890, 1979.

4. Couly G et al: Le syndrome de DiGeorge neurocristopathie rhombencéphalique exemplaire. *Rev Stomatol Chir Maxillofac* **84**:103–108, 1983.

5. DelaChapelle A et al: A deletion in chromosome 22 can cause DiGeorge syndrome. *Hum Genet* **57**:253–256, 1981.

6. Fowler G et al: The use of sequential silver and quinacrine staining to determine the parental origin and break points of a ring-22 human chromosome. *Clin Genet* **18**:274–279, 1977.

7. Greenberg F et al: Familial DiGeorge syndrome and associated partial monosomy of chromosome 22. *Hum Genet* **65**:317–319, 1984.

8. Hunter AGW et al: Phenotypic correlations in patients with ring chromosome 22. *Clin Genet* **12**:239–249, 1977.

9. Kelley RI et al: The association of the DiGeorge anomalad with partial monosomy of chromosome 22. *J Pediatr* 101:197–200, 1982.

9a. Keppen LD et al: Confirmation of autosomal dominant transmission of the DiGeorge malformation complex. *J Pediatr* **13**:506–508, 1988.

10. Pong AJE et al: DiGeorge syndrome: Long term survival complicated by Graves disease. *J Pediatr* **106**:619–620, 1985.

11. Raatika M et al: Familial third and fourth pharyngeal pouch syndrome with truncus arteriosus: DiGeorge syndrome. *Pediatrics* **67**:173–175, 1981.

12. Rethoré MO et al: Le syndrome r(22). A propos de quatre nouvelles observations. *Ann Génét* **111**:117, 1976.

13. Robinson HB Jr: DiGeorge's or the III–IV pharyngeal pouch syndrome: Pathology and a theory of pathogenesis. *Perspect Pediatr Pathol* **2**:173–206, 1975.

14. Rohn RD: Familial third–fourth pharyngeal pouch syndrome with apparent autosomal dominant transmission. *J Pediatr* **105**:47–51, 1984.

15. Rosenthal IM et al: Multiple anomalies including thymic aplasia associated with monosomy 22. *Pediatr Res* **6**:358–363, 1972.

16. Schwanitz G, Zerres K: Partial monosomy 22 as a result of an X/22 translocation in a newborn with DiGeorge syndrome. *Ann Génét* **30**:80–84, 1984.


## Trisomy 22 syndrome

There has been extensive debate regarding the existence of trisomy 22. Yet there are more than 25 individuals with what appears to be full trisomy 22 documented by chromosomal banding (1–15). Schinzel (11) has argued that "22 trisomy" is, in fact, trisomy for the terminal bands of the long arm of chromosome 11, which has resulted from 3:1 segregation of the reciprocal translocation (11;22) as the phenotypes overlap to a marked degree. Thus, what we have to say should be understood in that context.

Probably all have exhibited some degree of somatic and mental retardation. However, birth weight has not been remarkably abnormal. The head is often microcephalic. The facies is not extremely dysmorphic. The nasal tip is often flat as a result of a short septum. The philtrum tends to be long (12). Craniofacial asymmetry has been cited (10,12).

Most patients have preauricular tags or sinuses. The pinnae are usually large and/or low set with prominent anthelix. Convergent strabismus is frequent. Cleft palate is common and the mandible tends to be small. Many of the infants have been classified as having *Robin sequence.* Rarely, there is cleft lip (6) (Fig. 4–52A,B).

The fingers are long and slender and the thumb is often fingerlike. About 40% exhibit hip dysplasia and/or luxation. Hypotonia has been noted in about 50%.

Abnormal external genitalia in males consist of micropenis, bifid scrotum, and cryptorchidism (3,5,10).

Congenital heart anomalies, present in 75%, have included a diverse group of abnormalities without obvious pattern: PS (10), VSD (10,12), ASD of secundum type, PDA (6–8), tricuspid valve atresia (6,11), hypoplastic right ventricle (6,11), aberrant subclavian artery (13), coarctation of aorta (5,8), left superior vena cava (5,11), and bicuspid aortic valve (8). An excellent survey of congenital heart disease in the syndrome is that of Lin et al (9).

#### References (Trisomy 22 syndrome)


1. Alfi S et al: Trisomy 22: A clinically identifiable syndrome. *Birth Defects* **11**(5):241–245, 1975.

2. Bass HN et al: Probable trisomy 22, identified by fluorescent and trypsin-Giemsa banding. *Ann Génét* **16**:189–192, 1973.

3. Begleiter ML et al: Confirmation of the trisomy 22 by trypsin-Giemsa. *J Med Genet* **13**:517–520, 1975.

4. Cervenka J et al: Trisomy 22 with "cat eye" anomaly. *J Med Genet* **14**:288–290, 1977.

5. Emanuel BS et al: Abnormal chromosome 22 and recurrence of trisomy 22 syndrome. *J Med Genet* **13**:50–56, 1976.

6. Hirschhorn K et al: Precise identification of various chromosomal abnormalities. *Ann Hum Genet* **36**:375–379, 1973.

7. Hsu LYF et al: Trisomy 22: A clinical entity. *J Pediatr* **79**:12–19, 1971.

8. Iselius L, Faxelius G: Trisomy 22 in a newborn girl with multiple malformations. *Hereditas* **12**:193–199, 1975.

9. Lin AE et al: Congenital heart disease in supernumerary der(22),t(11;22) syndrome. *Clin Genet* **29**:269–275, 1986.

10. Penchaszadeh VB, Coco R: Trisomy 22: Two new cases and delineation of the phenotype. *J Med Genet* **12**:193–199, 1975.

10a. Petersen MB et al: Full trisomy 22 in a newborn infant. *Ann Génét* **30**:101–104, 1987.

11. Schinzel A: Mosaic—trisomy 22 and the problem of full trisomy 22. *Hum Genet* **56**:269–273, 1981.

12. Shokeir MHK: Complete trisomy 22. *Clin Genet* **14**:139–146, 1978.

13. Uchida IA, Brynes EM: Confirmation of trisomy 22 with fluorescent banding. *Am J Hum Genet* **28**:189–190, 1976.

14. Vianello MG, Boniolo E: Trisomy 22. *J Génét Hum* **23**:239–250, 1975.

15. Welter DA et al: Trisomy 22 in a 20-year-old female. *Hum Genet* **43**:347–351, 1978.


## Cat eye syndrome (trisomy or tetrasomy 22pter→q11)

The so-called "cat eye syndrome" has been recognized for over 100 years, but credit should be given to Schachenmann et al (6) for defin-

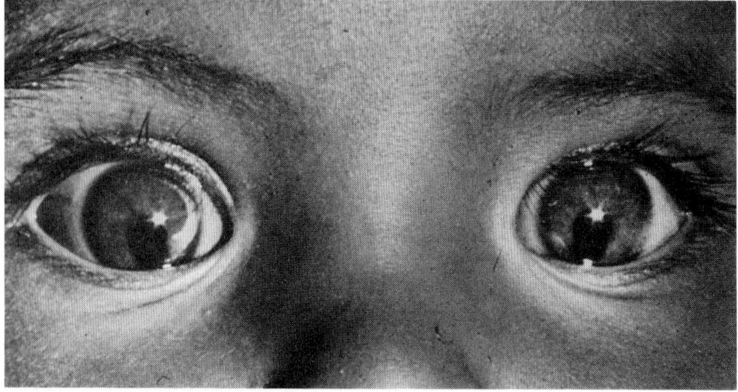

Fig. 4–53. *Cat eye syndrome.* Colobomata at six o'clock. (From *RA Peterson,* Arch Ophthalmol **90:**287, 1973.)

Fig. 4–54. *Tetraploidy syndrome.* (A,B) Microcephaly with narrow bifrontal diameter, sparse blond hair, lowset simplified pinnae, pretragal tags, anophthalmia and micrognathia, short philtrum, and beaked nose. (C) Newborn female with tetraploidy. Note microphthalmia, beaked nose, and overlapping fingers. (D) Note facies similar to that seen in (C). (A,B from *M Golbus et al,* J Med Genet **13:**329, 1976. C from *H Shiono et al,* Am J Med Genet **29:**543, 1988. D from *C Lafer et al,* Am J Med Genet **31:**375, 1988.)

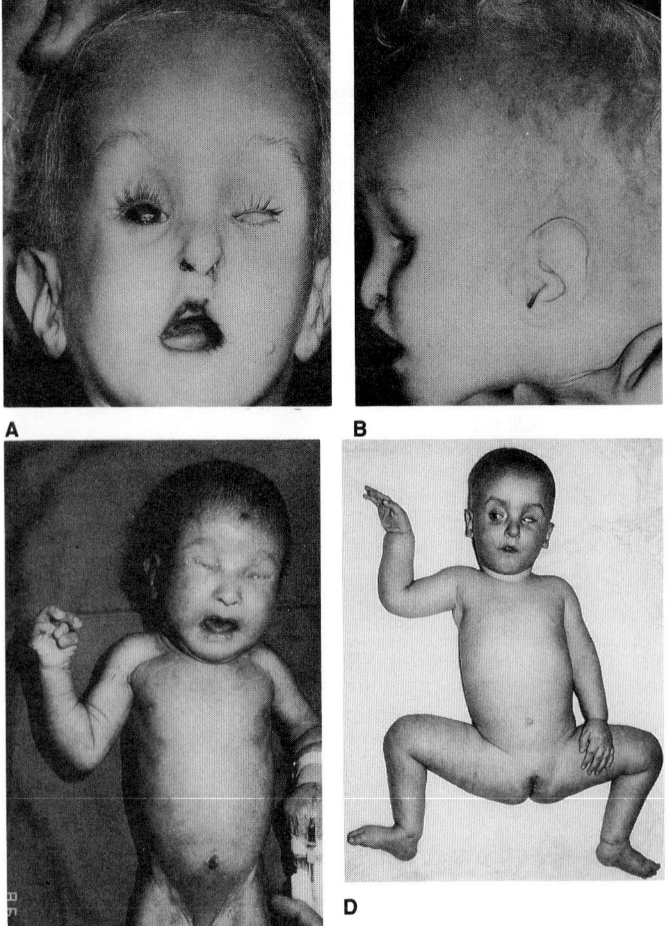


ing the syndrome and noting its association with a small G-like chromosome fragment. The identity of the fragment was demonstrated by Schinzel et al (7) to be trisomy or tetrasomy of the entire short arm of chromosome 22 to band 11 on the long arm (22pter→q11). The marker chromosome is transmitted from one generation to another, but expression of the syndrome being quite variable, only those at the extreme end of the spectrum are identified.

The term ''cat eye syndrome'' is derived from coloboma of the eye, frequently bilateral, which is present in about 70% (Fig. 4–53). The coloboma may involve the iris, choroid, and/or retina. Hypertelorism and downslanting palpebral fissures are found in 60% and epicanthus in 25%. About 20% exhibit unilateral microphthalmus and 65% have preauricular, usually bilateral, ear tags or pits (1,3,5,8). The auditory canals are occasionally (20%) atretic, resulting in mild conductive loss.

Mental retardation ranging from low normal to moderate is present in over 80%. Reduced somatic growth is a constant feature. Cleft palate occurs in 25% (2,4,9). Renal malformations, seen in about 50%, consist of unilateral aplasia, unilateral or bilateral hypoplasia, and cystic dysplasia.

A high imperforate anus or anal atresia occurs in 70%. It is associated with a fistula into the bladder, vestibule, or perineum. Inguinal and/or umbilical hernia and mobile cecum have been noted in several patients. Scoliosis is seen in about 30%.

Cardiac anomalies, present in about 40%, include total anomalous pulmonary venous return to the innominate vein (35%), tetralogy of Fallot (20%), and, less frequently, VSD, persistence of the left superior vena cava, absence of the inferior vena cava, tricuspid atresia, and Eisenmenger complex.

Ocular colobomata, preauricular tags, cardiac and renal anomalies, and anal atresia may be seen in a variety of disorders including the *VATER* and *CHARGE associations.* These disorders are not usually associated with an extra small chromosome.

#### References [Cat eye syndrome (trisomy or tetrasomy 22pter→q11)]

1. Balci S et al: The cat eye syndrome with unusual skeletal malformations. *Acta Pediatr Scand* **63**:623–626, 1974.

2. Gerald PS et al: Syndromal associations of imperforate anus. *Birth Defects* **8**(2):79–84, 1972.

3. Ginsburg J et al: Ocular abnormalities associated with extra small autosome. *Am J Ophthalmol* **65**:740–746, 1968.

4. Krmpotic E et al: Secondary dysjunction in partial trisomy 13. *Obstet Gynecol* **37**:381–390, 1971.

5. Kunze J et al: Cat-eye Syndrom. *Humangenetik* **26**:271–289, 1975.

6. Schachenmann G et al: Chromosome in coloboma and anal atresia. *Lancet* **2**:290, 1965.

7. Schinzel A et al: The ''cat eye syndrome'': Dicentric small marker chromosome probably derived from a No. 22 (tetrasomy pter→q11) associated with a characteristic phenotype. Report of 11 patients and delineation of the clinical picture. *Hum Genet* **57**:148–158, 1981.

8. Wilson GN et al: Cat eye syndrome owing to tetrasomy 22pter→q11. *J Med Genet* **21**:60–63, 1984.

9. Zellweger H et al: Two cases of multiple malformations with an autosomal chromosome aberration: Partial trisomy D? *Helv Paediatr Acta* **17**:290–300, 1962.

## Tetraploidy syndrome

Although relatively common (2–3%) in embryos lost spontaneously during the first trimester, tetraploidy or tetraploid–diploid mosaicism in liveborn infants is so rare that few generalizations can be made (3a). Some have been 92,XXYY and others 92,XXXX. Tetraploidy is not uncommon as an artifact in amniotic cell cultures (4).

All living examples have exhibited severe mental retardation, low birthweight, and microcephaly. The forehead is narrow and prominent, the nose beaked, and the philtrum short. The pinnae are low set and often lack cartilage (1,5,8). In some, there is microphthalmia or anophthalmia and short palpebral fissures (1,8,9) whereas others have corneal opacity, aphakia, and retinal detachment (3). The mouth appears small (Fig. 4–54A–D).

Anomalies of the extremities are common: absent phalanges of fingers and/or toes (2,3,10), arachnodactyly (5,8,9), syndactyly of toes (3,8), talipes (5,8), and single palmar creases (1–3,5–8).

## References (Tetraploidy syndrome)

1. Golbus MS et al: Tetraploidy in a liveborn infant. *J Med Genet* **13**:329–332, 1976.

2. Kelly TE, Rary JM: Mosaic tetraploid in a two-year-old female. *Clin Genet* **6**:221–224, 1976.

3. Kohn G et al: Tetraploidy–diploid mosaicism in a surviving infant. *Pediatr Res* **1**:461–469, 1967.

3a. Lafer CZ, Neu RL: A liveborn infant with tetraploidy. *Am J Med Genet* **31**:375–378, 1988.

4. Milunsky A et al: Polyploidy in prenatal genetic diagnosis. *J Pediatr* **79**:303–305, 1971.

5. Pitt D et al: Tetraploidy in a liveborn infant with spina bifida and other anomalies. *J Med Genet* **18**:309–311, 1981.

6. Quiroz E et al: Diploid–tetraploid mosaicism in a malformed boy. *Clin Genet* **27**:183–186, 1985.

7. Reddy CM et al: Diploid/tetraploid mosaicism in the offspring of a 46,XX/47,XXX mosaic mother. *J Natl Med Assoc* **69**:563–564, 1977.

8. Scarborough PR et al: Tetraploidy: A report of three live-born infants. *Am J Med Genet* **19**:29–37, 1984.

9. Shiono H et al: Tetraploidy in a 25-month-old girl. *Am J Med Genet* **29**:543–546, 1988.

10. Veenema H et al: Mosaic tetraploid in a male neonate. *Clin Genet* **22**:295–298, 1982.

# Chapter 5
# Metabolic Disorders

## Mucopolysaccharidoses

The mucopolysaccharidoses (MPS) are a family of inherited metabolic diseases that result from the deficiency of different lysosomal enzymes involved in the degradation of the mucopolysaccharides [now known as glycosaminoglycans (GAGs)], dermatan sulfate (DS), heparan sulfate (HS), and keratan sulfate (KS) (2,8,9,11–15,17,21). Most of the disorders exhibit leukocyte inclusions (10).

Based on clinical and biochemical studies, these disorders have been designated MPS I through MPS VII (1; Table 5–1). An extensive review of inborn errors of complex carbohydrate metabolism was published by Spranger (17a). A deficiency of a specific lysosomal enzyme has been demonstrated in each of the MPS disorders. Each has autosomal recessive inheritance with the exception of Hunter disease (MPS II), which is X-linked recessive. The genes encoding five of these enzymes have been mapped to specific chromosomes (Table 5–2), and recently, the gene encoding one of these enzymes ($\beta$-glucuronidase; MPS VII) has been cloned (4).

Although the clinical phenotypes of these disorders are unusually heterogeneous, certain findings are common and permit provisional clinical diagnoses. For example, all of the disorders have characteristic skeletal abnormalities (termed ''dysostosis multiplex''), with the exception of Morquio disease (MPS IV) (18). Marked short stature is observed [except in Scheie disease (MPS I-S)] and joints are often stiff. In addition, unusual hair, corneal clouding, hepatosplenomegaly, deafness, arteriosclerosis, and stiffening of the thoracic cage are common findings (19,20). However, the clinical heterogeneity within each of these disease entities can be quite remarkable. For example, patients with MPS I-H, MPS I-S, and MPS I-H/S have in common deficiency of $\alpha$-L-iduronidase, an enzyme activity required to degrade DS and HS (2).The genes appear to be allelic (14a,15a). Yet, patients with MPS I-H have severe dysostosis multiplex, short stature, and mental retardation, and expire in childhood, whereas patients with MPS I-S have mild skeletal abnormalities, normal height, and normal intelligence, and survive into adulthood. As might be expected, patients with MPS I-H/S have an intermediate phenotype. Further heterogeneity has been discussed by Spranger (17a). The suspected clinical diagnosis of patients with these disorders is confirmed by the demonstration of the specific enzymatic defect in isolated leukocytes, cultured fibroblasts, or lymphoid cells. Progressive deterioration generally leads to death in childhood with the exception of MPS I-S, MPS IV, and the mild form of MPS VI, in which patients typically survive into adulthood. Animal models have been described for MPS I (5,16), MPS VI (6), and MPS VII (7), which should facilitate the development and evaluation of various therapeutic strategies including bone marrow transplantation, enzyme replacement, and, perhaps, gene therapy. Prenatal diagnosis has ben accomplished in all of these disorders, most recently using cultured chorionic villi cells (3).

### References (General)


1. Astrin KH, Desnick RJ: Chromosomal localization of the structural genes encoding the human lysosomal hydrolases and their activator and stabilizer proteins, in *Molecular Basis of Lysosomal Storage Disorders,* Barranger JA, Brady, RO (eds), Academic Press, New York, 1984, pp 325–386.

2. Bach G et al: The defect in the Hurler and Scheie syndrome: Deficiency of $\alpha$-L-iduronidase. *Proc Natl Acad Sci USA* **69**:2048–2051, 1972.

3. Desnick RJ et al: Antenatal metabolic diagnoses: A compendium, in *Human Prenatal Diagnosis,* Filkins K, Kaminetzky H (eds), Dekker, New York, 1984, pp 59–108.

3a. Eggli KD, Dorst JP: The mucopolysaccharidoses and related conditions. *Sem Roentgenol* **21**:275–294, 1986.

4. Guise KS et al: Isolation and expression and *Escherichia coli* of a cDNA clone encoding human $\beta$-glucuronidase. *Gene* **34**:105–110, 1985.

5. Haskins ME et al: $\alpha$-L-Iduronidase deficiency in a cat: A model of mucopolysaccharidosis I. *Pediatr Res* **13**:1294–1297, 1982.

6. Haskins ME et al: Mucopolysaccharidosis in a domestic short-haired cat—a disease distinct from that seen in the Siamese cat. *J Am Vet Med Assoc* **175**:384, 1979.

7. Haskins ME et al: $\beta$-Glucuronidase deficiency in a dog: A model of mucopolysaccharidosis VII. *Pediatr Res* **18**:980–984, 1984.

8. Kenyon KR et al: The systemic mucopolysaccharidoses. *Am J Ophthalmol* **73**:811–833, 1972.

9. Kirchner M et al: Die Mukopolysaccharid-Speicherkrankheiten. *Kinderärztl Prax* **48**:19–32, 455–470, 1980.

10. Markesbery WR: Mucopolysaccharidoses: Ultrastructure of leukocyte inclusions. *Ann Neurol* **8**:332–336, 1980.

11. McKusick VA, Neufeld EF: The mucopolysaccharide storage diseases, in *The Metabolic Basis of Inherited Disorders,* Stanbury JB, Wyngaarden JB, Fredrickson DS, Goldstein, JL, Brown, MS (eds), McGraw-Hill, New York, 1983, pp 751–777.

12. McKusick VA: *Heritable Disorders of Connective Tissue,* 4th ed. Mosby, St. Louis, 1972.

13. McKusick VA et al: The genetic mucopolysaccharidoses. *Medicine (Baltimore)* **44**:445–484, 1965.

14. McKusick VA: Genetic heterogeneity and allelic variation in the mucopolysaccharidoses. *Johns Hopkins Med J* **146**:71–72, 1980.

14a. Mueller OT et al: Apparent allelism of the Hurler, Scheie, and Hurler/Scheie syndromes. *Am J Med Genet* **18**:547–556, 1984.

14b. Neufeld EF, Muenzer J: The mucopolysaccharidoses, in *The Metabolic Basis of Inherited Disease,* 6th ed, Scriver CR et al (eds), McGraw–Hill, New York, 1989.

15. Pennock CA, Barnes IC: The mucopolysaccharidoses. *J Med Genet* **13**:169–181, 1976.

15a. Roubicek M et al: The clinical spectrum of $\alpha$-L-iduronidase deficiency. *Am J Med Genet* **20**:471–481, 1985.

16. Shull RM et al: Canine $\alpha$-L-iduronidase deficiency: A model of mucopolysaccharidosis I. *Am J Pathol* **109**:244–248, 1982.

17. Spranger J: Lysosomale Enzymdefekte. *Mschr Kinderheilk* **131**:311–317, 1983.

17a. Spranger J: Mini review: Inborn errors of complex carbohydrate metabolism. *Am J Med Genet* **28**:489–499, 1987.

18. Spranger JW et al: *Bone Dysplasias.* Saunders, Philadelphia, 1984, pp 144–170.

19. Sugar J: Corneal manifestations of the systemic mucopolysaccharidoses. *Ann Ophthalmol* **11**:531–535, 1979.

20. Teschler-Nicola M, Killian W: Observations on hair shaft morphology in mucopolysaccharidoses. *J Ment Defic Res* **26**:193–202, 1982.

21. Watts RWE, Gibbs DA: *Lysosomal Storage Diseases: Biochemical and Clinical Aspects.* Taylor and Francis, Philadelphia, 1986.


**Mucopolysaccharidosis I-H (Hurler syndrome).** MPS I-H was first described in 1919 by Hurler (20) at the suggestion of Pfaundler. It is the classic prototype of the mucopolysaccharidoses, having the following cardinal features: growth failure after infancy, marked mental retardation, characteristic craniofacial dysmorphism and physical habitus, dysostosis multiplex, corneal clouding, histochemical and

Table 5–1. Classification of the mucopolysaccharidoses

| Type | Synonym | Clinical dysmorphism | Skeletal dysplasia | Corneal opacities | Mental retardation | Excessive urinary AMPS | Defective enzyme | Genetic transmission | Type |
|---|---|---|---|---|---|---|---|---|---|
| I-H | Hurler | Severe | Severe | Yes | Yes | DS and HS | α-L-Iduronidase | AR | I-H |
| I-H/S | Hurler/Scheie | Intermediate | Moderate | Yes | No | DS and HS | α-L-Iduronidase | AR | I-H/S |
| I-S[a] | Scheie | Mild | Mild | Yes | No | DS and HS | α-L-Iduronidase | AR | I-S |
| II-A | Hunter A | Late (moderate) | Moderate | No | No | HS and DS | Iduronate sulfate-sulfatase | XR | II-A |
| II-B | Hunter B | Early (moderate) | Moderate | No | Yes | DS and HS | Iduronate sulfate-sulfatase | XR | II-B |
| III-A | Sanfilippo A | Mild | Minimal | No | Yes | HS | Heparan-*N*-sulfatase | AR | III-A |
| III-B | Sanfilippo B | Mild | Minimal | No | Yes | HS | α-*N*-Acetyl glucosaminidase | AR | III-B |
| III-C | Sanfilippo C | Mild | Minimal | No | Yes | HS | Acetyl CoA:α-glucosaminide-*N*-acetyltransferase | AR | III-C |
| III-D | Sanfilippo D | Mild | Minimal | No | Yes | HS | α-*N*-Acetylgluco-saminide-6-sulfatase | AR | III-D |
| IV-A | Morquio A | Severe | Severe | Yes | No | KS | Galactosamine-6-sulfate-sulfatase | AR | IV-A |
| IV-B | Morquio B | Severe | Severe | Yes | No | KS | β-Galactosidase | AR | IV-B |
| VI-A | Maroteaux–Lamy A | Mild to Moderate | Moderate | Yes | No | DS | Arylsulfatase B | AR | VI-A |
| VI-B | Maroteaux–Lamy B | Severe | Severe | Yes | Mild | DS | Arylsulfatase B | AR | VI-B |
| VII | Sly | Severe | Severe | None | Late | DS and HS | β-Glucuronidase | AR | VII |

[a]Formerly classified as Type V.

Table 5–2. Chromosomal assignment of the structural genes for the lysosomal enzymes that degrade glycosaminoglycans

| Chromosome | Region | Enzyme | Abbreviation | EC number | Mucopoly-saccharidosis |
|---|---|---|---|---|---|
| 3 | p21–q21 | β-Galactosidase | GLB1 | 3.2.1.23 | MPS IV |
| 5 | p11–q13 | Arylsulfatase B | ARSB | 3.1.6.1 | MPS VI |
| 7 | cen–q22 | β-Glucuronidase | GUSB | 3.2.1.31 | MPS VII |
| 22 | pter–q11 | α-L-Iduronidase | IDA | 3.2.1.76 | MPS I |
| X | q26–q27 | α-L-Iduronide sulfatase | SIDS | — | MPS II |

biochemical evidence of intracellular lysosomal storage of GAGs, and excessive urinary excretion of DS and HS. Danes (10) described a variant that was neither Hurler nor Scheie syndrome.

In the first months of life there are a few relatively nonspecific findings, such as hernias, macrocephaly, limited hip abduction, and recurrent respiratory infections. The full clinical picture usually develops in the second year of life (Figs. 5–1 to 5–5). Death usually occurs before 10 years of age from pneumonia and cardiac failure.

The frequency of mucopolysaccharidosis I-H is approximately 1/100,000 births (23). The specific enzymatic defect is the deficiency of α-L-iduronidase activity (1,24), lack of which precludes intralysosomal degradation of the α-L-iduronide-containing GAGs, DS and HS. Subsequent intracellular accumulation of undegraded or partially degraded GAGs interferes with normal function of the affected cells and leads to the characteristic clinical signs and symptoms.

**Facies.** Slight coarsening of facial features at 3–6 months of age is usually the first abnormality noted. The head is large, and the frontal bones bulge. Premature closure of the sagittal and metopic sutures and hyperostosis in this area frequently lead to scaphocephaly. Hirsutism is represented on the face only as synophrys. The nasal bridge is depressed, the tip of the nose is broad, and the nostrils are wide and anteverted. The interpupillary distance is greater than normal. Corneal clouding appears during the third year of life, rarely earlier. Rarely, glaucoma is an early complication (28a). The lower eyelids and nasolabial folds are prominent, and the cheeks are full. The earlobes are thick. The lips are enlarged and patulous, and the mouth is usually held open, particularly after the age of 3 years. Chronic nasal discharge is usually marked even between the frequent bouts of upper respiratory infection (Fig. 5–1A–D). Nasal congestion with stertorous breathing through the mouth is severe, being related to hyperplastic adenoid tissue and a deep cranial fossa that narrows the airway between the sphenoid bone and the hard palate. Gag and swallowing reflexes become progressively diminished (22).

**Musculoskeletal system.** Length at birth is not decreased below the norm. The majority of patients with MPS I-H studied between the ages of 6 and 12 months are at or above the 87th percentile for total body length. Some remain among the tallest infants until 18 months, but growth ceases in all patients before 2 years of age. By age 3 years, all MPS I-H patients are below the 3rd percentile for stature. The neck is short. A serious complication is subluxation of $C_1$,$C_2$ (5). Both pectus carinatum and pectus excavatum occur, and usually there is lumbodorsal kyphosis or gibbus (Fig. 5–2). Range of motion is lim-

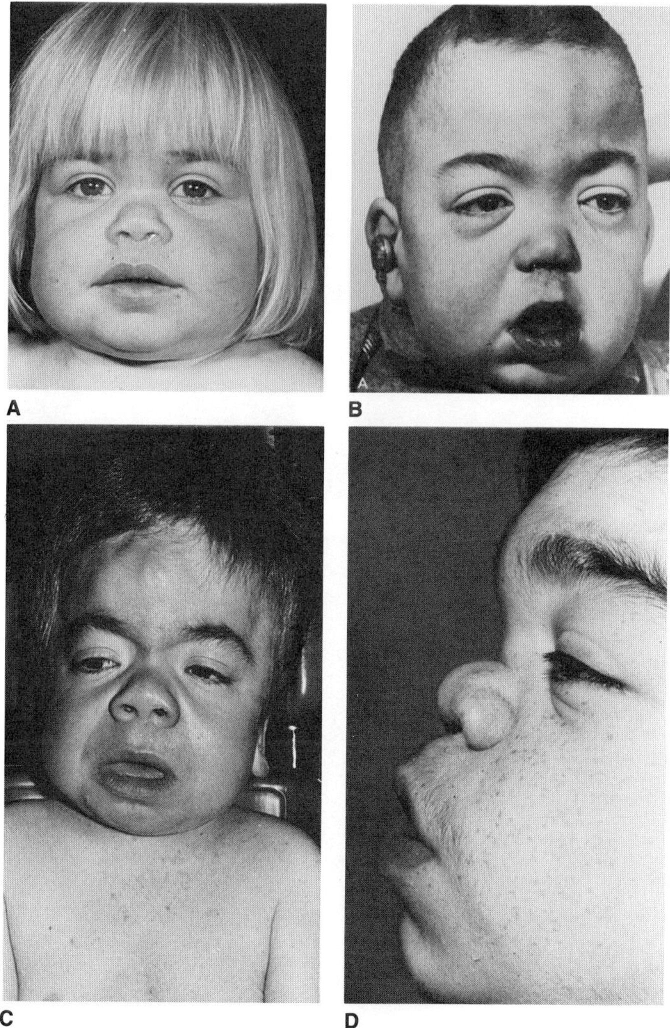


Fig. 5–1. *Hurler syndrome.* (A–D) Characteristic facies. Large head, prominent forehead, coarse features, hypertelorism, heavy lids, low nasal bridge, snub nose, long philtrum, and open mouth. (A,D courtesy of *RJ Desnick,* New York, New York. C courtesy of *E Passarge,* Hamburg, Germany.)

Fig. 5–2. *Hurler syndrome.* Two-year-old girl. Note characteristic facies and gibbus.

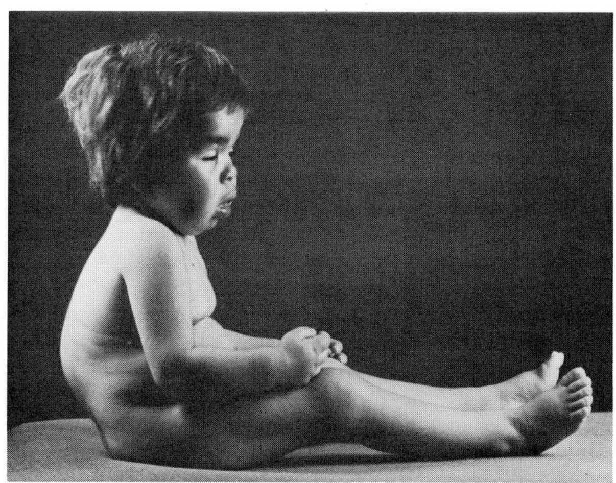

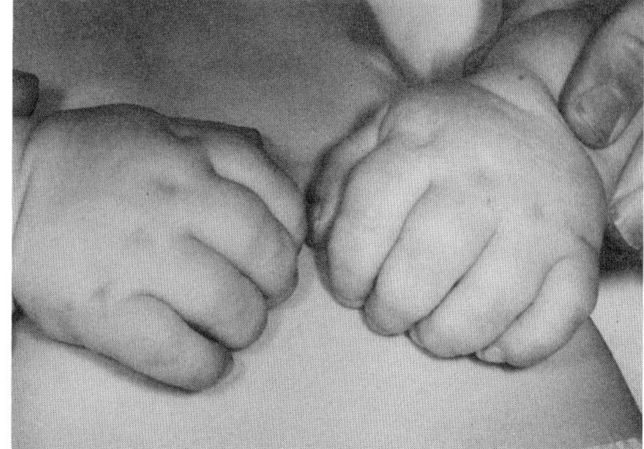

Fig. 5–3. *Hurler syndrome.* Clawhand deformity.

ited in all joints; in the hands, the so-called clawhand deformity (Fig. 5–3) results (25).

Radiographically, in infancy, bone trabeculation is coarse. In late infancy and early childhood, a pattern of skeletal changes called "dysostosis multiplex" emerges: the skull becomes large and deformed, the sphenoidal plane is depressed, and the sella is J-shaped, possibly from arachnoid cysts (Fig. 5–4A). The sagittal and lambdoidal sutures close prematurely. The cranial base and orbital roofs are particularly thick and dense. The orbits are shallow. Communicating hydrocephalus is often present. The stylohyoid ligment is almost always calcified (normal, 25%) and thicker than normal (29). The ribs are wide in their lateral and ventral portions, with overconstriction at their paravertebral ends (Fig. 5–4B). The vertebral bodies are dysplastic, with biconvex end plates and hook-shaped configuration of the lower thoracic and upper lumbar bodies, after 12–18 months of age (Fig. 5–4C). The basilar portions of the ilia are underdeveloped, with flaring of the iliac wings (Fig. 5–4D). The long tubular bones show marked diaphyseal widening and distortion, with small and deformed epiphyses. The shafts of the short tubular bones are underconstricted, with bullet-shaped phalanges and proximal pointing of the second to fifth metacarpals (37) (Fig. 5–4E).

The abdomen protrudes because of hepatic and splenic enlargement, deformity of the chest, shortness of the spine, and laxity of the abdominal wall. These changes are noted during the second year of life. Hepatomegaly may be detected as early as 1 month of age.

Inguinal hernia, present at birth or developing with the first 3 months of life, is a constant feature in boys. The hernias tend not to recur following surgery; they are a classic part of the patients' history before diagnosis is established but are not common during the subsequent course of the disorder. Umbilical hernias, usually small at birth in both sexes, gradually reach major proportions.

**Other findings.** The skin is pale, coarse, and dry and is covered by fine, lanugo-like fuzz, particularly on the back and extremities. Mental retardation is conspicuous and progressive. Moderate cardiomegaly, as a result of deposition of AMPS in the myocardium and valves, is usually present. Echocardiographic mitral valve deformity has been demonstrated (21). Hypertension is a frequent finding in older patients (38).

**Oral manifestations.** Oral changes (Fig. 5–5A–C) have been reviewed by Gardner (14). The lips are enlarged and patulous, with flattened philtrum, and the upper lip is particularly long. The mouth is usually held open, with protruding tongue, from about 3 years of age. Lip and tongue enlargement becomes marked after the age of 5 years (22).

The teeth are widely spaced, often exhibiting severe attrition. The incisors may exhibit some degree of conical crown form but are otherwise normal structurally. Because of macroglossia, there may be anterior open bite. Eruption is probably delayed in at least half the patients (7), particularly in areas of bone destruction. The second pri-

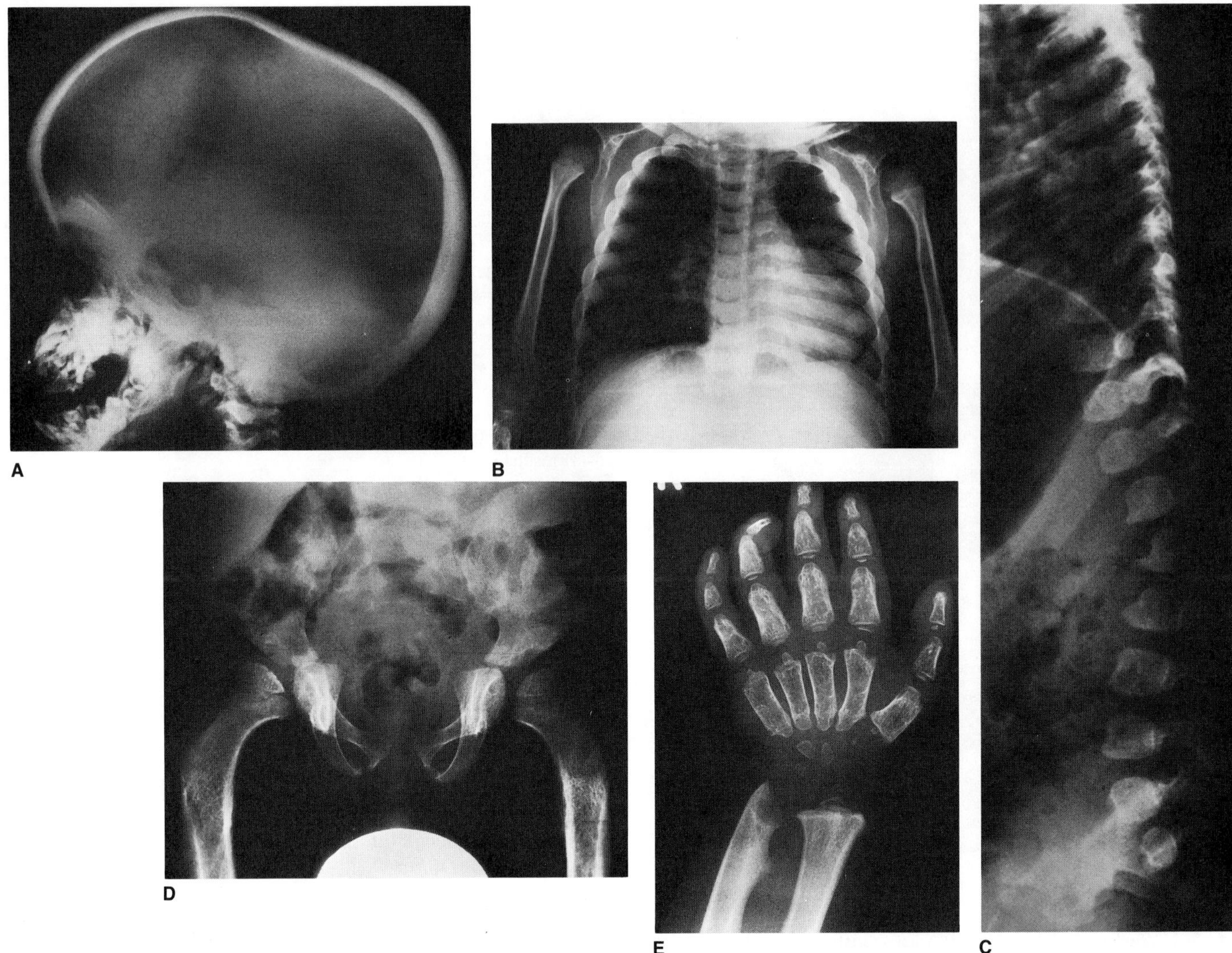


Fig. 5–4. *Hurler syndrome*. Roentgenograms. (A) Skull, 8-year-old patient. Macrocephaly, dolichocephaly, thickened calvaria, wide sella. (B) Chest of same patient. Wide, oar-shaped ribs, expanded medial portion of clavicle, valgus deformity of humeral neck, submetaphyseal overconstriction of proximal humerus, expansion of distal shaft of humerus. (C) Spine of 4-year-old patient. Flattening and biconvex endplates of vertebral bodies, hooked-shaped configuration of first two-thirds of the lumbar vertebrae. (D) Pelvis of 8-year-old patient. Hypoplasia of basilar portion of ilia, flared iliac crest, long femoral neck, coxa valga. (E) Hand of 6-year-old patient. Expanded shafts of short tubular bones, bullet-shaped phalanges, proximal pointing of second to fifth metacarpals, small carpal bones, tilted distal ends of radius and ulna.

mary molars or first and second permanent molars are often distoangularly positioned, with the distal surface of the crown being situated more deeply than the mesial. In some cases, there is dilaceration of the distal roots (41). These changes occur more frequently in the mandible.

Extremely common are localized areas of bone destruction that have been designated as "dentigerous cysts." These are often present by 3 years of age and more often involve the second primary molars and first and second permanent mandibular molars. The margins of the radiolucencies are usually smooth and clearly defined (7,14,19). We have believed that the "cysts" represent pooling of dermatan sulfate in hyperplastic dental follicles, since they also occur in MPS I-S and MPS VI but not in MPS III. However, the material in the follicles in MPS VI has been shown to be hyaluronic acid (vide infra).

The alveolar ridges are nearly always hyperplastic, resulting in spacing of the teeth. Some patients exhibit hyperplastic gingivitis, because of poor oral hygiene and mouth breathing. Rarely, there is true hyperplastic gingiva with eruption cysts (14). Histochemical study of the gingiva has demonstrated metachromatic cells (13).

The mandible is short and broad, with wide bigonial distance (41). The rami are short and narrow, and the condyle is replaced by a flat inclined surface or cup-shaped excavation. The mandibular notch is irregular or cleft. The temporomandibular joint may exhibit limited motion.

Airway obstruction and sleep apnea have been reported (33a).

**Diagnosis.** The earliest diagnostic tests for the MPS syndromes were based on the urinary excretion of GAGs. Although the reported ratios vary significantly, there is generally more DS found than HS in MPS I-H. Classic "cell culture correction" analyses performed by Neufeld and co-workers (3,27,28,34) to discriminate the various MPS disorders have been replaced by direct enzyme assays (15,17,24,40). For MPS I-H, the deficiency of $\alpha$-L-iduronidase can readily be detected using artificial (fluorogenic, colorigenic, or radioactive) substrates in isolated leukocytes or cultured skin fibroblasts. Heterozygote testing has been complicated by the marked overlap of normal and carrier groups (35,39). MPS I-H may be readily detected in second trimester pregnancies by examining the $\alpha$-L-iduronidase levels in cultured amniotic cells (12). Prenatal diagnosis in the first trimester has been accomplished using cultured chorionic villi cells (15).

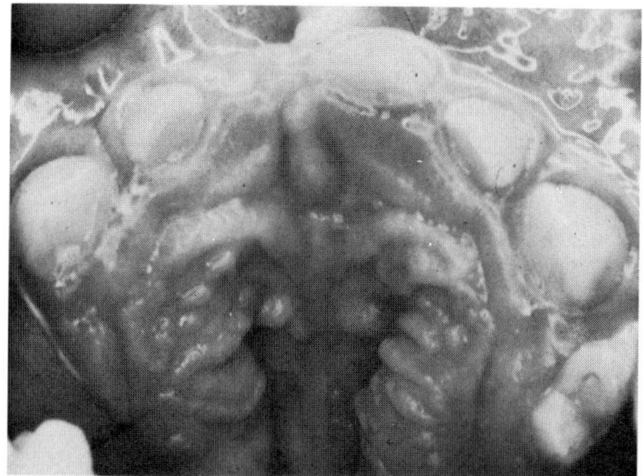
A

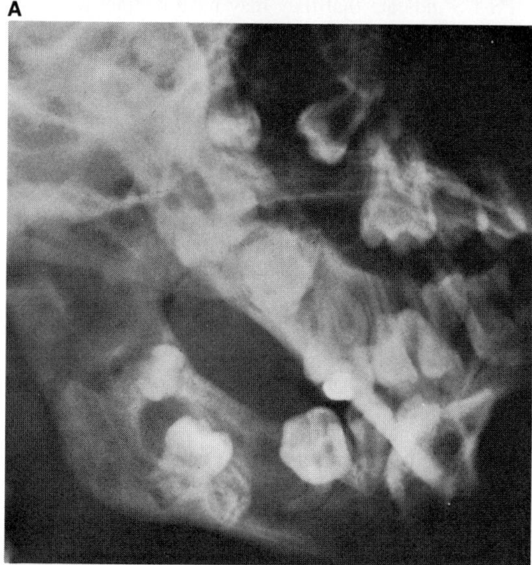
B

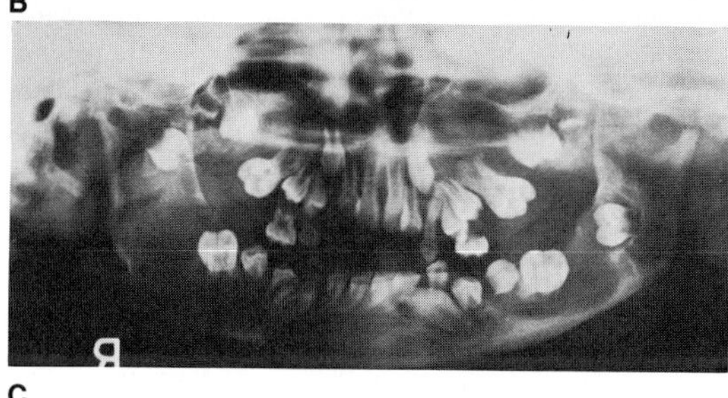
C

Fig. 5–5. *Hurler syndrome.* (A) Widened alveolar processes. (B) Note distoangular mandibular first molar with dilaceration of distal root, cystic bone destruction around first and second mandibular molars, partially inverted mandibular third molars in ramus, and clearly discernible cleft in mandibular notch. Condyle is hypoplastic. (C) More extensive alterations in 9-year-old male with mucopolysaccharidosis. (A courtesy of *Maj J Fay,* US Army. B from *HM Worth,* Oral Surg **22:**21, 1966. C from *MA Germann,* Dtsch Zahn Mund Kieferheilkd **59:**59, 1972.)

Biochemical findings. $\alpha$-L-Iduronidase, the enzyme deficient in the MPS I disorders, has been purified to homogeneity from a number of human sources and extensively characterized (8,9,30,31). The precursor protein has a molecular weight of about 70,000 and is glycosylated (26). The enzyme is a monomer that is proteolytically processed within the lysosome to a number of unique peptide species (26), which have been shown to form catalytically active aggregates (8). The structural gene encoding this enzyme has been mapped to chromosome 22 [Table 5–2 (32)]. Individuals with MPS I-H have from 0.0 to about 1% of normal enzymatic activity depending on the substrate and assay conditions used (18). All individuals studied to date with MPS I-H have been shown to have cross-reactive immunologic material (CRIM). Animal models of $\alpha$-L-iduronidase deficiency have been described and breeding colonies established (16,36). As indicated earlier, the MPS I-H mutation appears to be allelic with those of MPS I-S and MPS I-H/S (11).

Laboratory aids. Histochemical evidence of GAG storage is found in peripheral leukocytes, bone marrow cells, and cultured fibroblasts (2,4). Abnormal amounts of GAGs are excreted in the urine, most notably DS and HS. Patients with MPS I-H excrete more DS than HS. Several rapid screening tests have been devised, the most popular of which are the toluidine blue spot test and the gross albumin turbidity test, the latter being far more accurate (6). More discriminating electrophoretic methods for GAG analysis also have been described (e.g., 33).

## References [Mucopolysaccharidosis I-H (Hurler syndrome)]


1. Bach G et al: The defect in the Hurler and Scheie syndrome: Deficiency of $\alpha$-L-iduronidase. *Proc Natl Acad Sci USA* **69**:2048–2051, 1972.
2. Bartman J, Blanc W: Fibroblast cultures in Hurler's and Hunter's syndromes. *Arch Pathol* **89**:279–285, 1970.
3. Barton RW, Neufeld EF: The Hurler corrective factor. *J Biol Chem* **246**:7773–7779, 1971.
4. Belcher R: Ultrastructure and cytochemistry of lymphocytes in the genetic mucopolysaccharidoses. *Arch Pathol* **93**:1–7, 1972.
5. Brill CB et al: Spastic quadriparesis due to $C_1$–$C_2$ subluxation in Hurler syndrome. *J Pediatr* **92**:441–443, 1978.
6. Carter CH et al: Commonly used tests in the detection of Hurler's syndrome. *J Pediatr* **73**:217–221, 1968.
7. Cawson RA: The oral changes in gargoylism. *Proc R Soc Med* **55**:1066–1070, 1962.
8. Clements PR et al: Human $\alpha$-L-iduronidase. 1. Purification, monoclonal antibody production, native and subunit molecular mass. *Eur J Biochem* **152**:21–28, 1985.
9. Clements PR et al: Human $\alpha$-L-iduronidase. 2. Catalytic properties. *Eur J Biochem* **152**:29–34, 1985.
10. Danes BS: Variant of iduronidase deficient mucopolysaccharidoses. *J Med Genet* **14**:346–351, 1977.
11. Fortuin JJH, Kleijer WF: Hybridization studies of fibroblasts from Hurler, Scheie and Hurler–Scheie compound patients: Support for the hypothesis of allelic mutants. *Hum Genet* **53**:155–159, 1980.
12. Fratantoni FC et al: Intrauterine diagnosis of the Hurler and Hunter syndromes. *N Engl J Med* **280**:686–688, 1969.
13. Gardner DG: Metachromatic cells in the gingiva in Hurler's syndrome. *Oral Surg* **26**:782–789, 1968.
14. Gardner DG: The oral manifestations of Hurler's syndrome. *Oral Surg* **32**:46–57, 1971.
15. Hall CW, Neufeld EF: $\alpha$-L-Iduronidase activity in cultured skin fibroblasts and amniotic fluid cells. *Arch Biochem Biophys* **158**:817–821, 1973.
16. Haskins ME et al: $\alpha$-L-Iduronidase deficiency in a cat: A model of mucopolysaccharidosis I. *Pediatr Res* **13**:1294–1297, 1979.
17. Hopwood JJ: $\alpha$-L-Iduronidase, $\alpha$-D-glucuronidase and 2-sulfo-L-iduronate-2-sulfatase: Preparation and characterization of radioactive substrates from heparin. *Carbohydrate Res* **69**:203–207, 1979.
18. Hopwood JJ, Muller V: Biochemical discrimination of Hurler and Scheie syndromes. *Clin Sci* **57**:265–272, 1979.
19. Horrigan WD, Baker DH: Gargoylism: A review of the roentgen skull changes with description of a new finding. *Am J Roentgenol* **86**:473–477, 1961.
20. Hurler G: Über einen Typus multipler Abartungen, vorwiegend am Skelettsystem. *Z Kinderheilkd* **24**:220–234, 1919.
21. Johnson GL et al: Echocardiographic mitral valve deformity in the mucopolysaccharidoses. *Pediatrics* **67**:401–406, 1981.
22. Leroy JG, Crocker AC: Clinical definition of Hurler–Hunter phenotypes. *Am J Dis Child* **112**:518–530, 1966.
23. Lowry RD, Renwick DHG: Relative frequency of the Hurler and Hunter syndromes. *N Engl J Med* **284**:221–222, 1971.
24. Matalon R, Deanching M: The enzymic basis for the phenotypic variation of Hurler and Scheie syndrome. *Pediatr Res* **11**:519, 1977.
25. McKusick VA: *Heritable Disorders of Connective Tissue,* 4th ed. Mosby, St. Louis, 1972.
26. Myerowitz R, Neufeld EF: Maturation of $\alpha$-L-iduronidase in cultured human fibroblasts. *J Biol Chem* **256**:3044–3048, 1981.

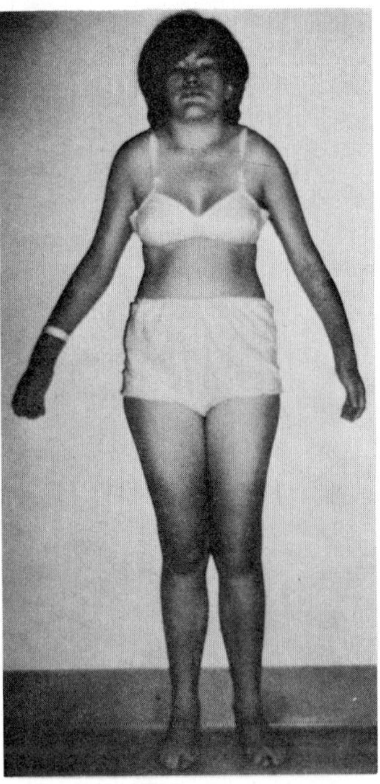

Fig. 5–6. *Scheie syndrome*. Fifteen-year-old patient. Round face, downturned mouth, relatively coarse facial features. (Courtesy of *RL Summitt*, Nashville, Tennessee.)

27. Neufeld EF, Cantz MJ: Corrective factors for inborn errors of mucopolysaccharide metabolism. *Ann NY Acad Sci* **179**:580–587, 1971.

28. Neufeld EF, Fratantoni JC: Inborn errors of mucopolysaccharide metabolism. *Science* **169**:141–145, 1970.

28a. Novacyzk MJ et al: Glaucoma as an early complication in Hurler's disease. *Arch Dis Childh* **63**:1091–1093, 1988.

29. Oestreich AE: The stylohyoid ligament in Hurler syndrome and related conditions: Comparison with normal children. *Radiology* **154**:665–666, 1985.

30. Schuchman EH et al: Human α-L-iduronidase. I. Purification and properties of the high uptake (higher molecular weight) and the low uptake (processed) forms. *J Biol Chem* **259**:3132–3138, 1984.

31. Schuchman EH et al: Human α-L-iduronidase. II. Comparative biochemical and immunologic properties of the purified low and high uptake forms. *Enzyme* **31**:166–175, 1984.

32. Schuchman EH et al: Regional assignment of the structural gene for human α-L-iduronidase. *Proc Natl Acad Sci USA* **81**:1169–1171, 1984.

33. Schuchman EH, Desnick RJ: A new, continuous monodimensional electrophoretic system for the separation and quantitation of individual glycosaminoglycans. *Anal Biochem* **117**:419–423, 1981.

33a. Shapiro J et al: Airway obstruction and sleep apnea in Hurler and Hunter syndromes. *Ann Otol Rhinol Laryngol* **94**:458–461, 1985.

34. Shapiro, LJ et al: The relationship of α-L-iduronidase and Hurler corrective factor. *Arch Biochem Biophys* **172**:156–161, 1976.

35. Shapiro LJ: Current status and future direction for carrier detection in lysosomal storage diseases, in *Lysosomes and Lysosomal Storage Diseases*, Callahan JW, Lowden JA (eds), Raven Press, New York, 1981, p 343.

36. Shull RM et al: Canine α-L-iduronidase deficiency: A model of mucopolysaccharidosis I. *Am J Pathol* **109**:244–248, 1982.

37. Spranger, J: The systemic mucopolysaccharidoses. *Ergeb Inn Med Kinderheilkd* **32**:165–265, 1972.

38. Taylor J et al: Nephrotic syndromes and hypertension in two children with Hurler syndrome. *J Pediatr* **108**:726–729, 1986.

39. Wappner RS, Brandt IK: Hurler syndrome: α-L-Iduronidase activity in leukocytes as a method for heterozygote detection. *Pediatr Res* **10**:629–632, 1976.

40. Weissmann B: Synthetic substrates for α-L-iduronidase. *Methods Enzymol* **50**:141–142, 1978.

41. Worth HM: Hurler's syndrome: A study of radiologic appearances in the jaws. *Oral Surg* **22**:21–35, 1966.

**Mucopolysaccharidosis I-S (Scheie syndrome).** Sibs with mucopolysaccharidosis I-S were described in 1962 by Scheie et al (6). Patients referred to in the older European literature as having "late Hurler disease" may have had the same condition (7,8).

The disease is rarely recognized in childhood. It is characterized by normal stature, corneal opacities, deformity of the hands (clawhands), involvement of the aortic valve, normal intelligence, and biochemical evidence of lysosomal storage of HS and DS and their excessive urinary excretion.

The disorder has autosomal recessive inheritance. Its frequency has been estimated at 1/500,000 births (3). Abnormal intracellular accumulation of DS and HS results from the deficient activity of α-L-iduronidase, the same enzyme deficiency responsible for MPS I-H (1). It is now clear the MPS I-H and I-S are caused by allelic mutations (5). Although some reports of residual α-L-iduronidase activities in MPS I-H and MPS I-S indicate that they may have distinct biochemical characteristics (4), it is not possible to clearly distinguish the different underlying α-L-iduronidase lesions in these disorders, and investigators currently are turning to molecular genetic techniques to differentiate the allelic mutations that result in MPS I-H, MPS I-S, and MPS I-H/S (see following).

**Facies.** No major abnormalities are noted in early childhood; symptoms usually become apparent by 5–15 years of age. In adults, the face is somewhat coarse, but not Hurler-like. It is broad with increased midfacial height and with mandibular prognathism. In most cases, the corners of the mouth are turned downward. Macroglossia may be present. Occasionally, the nose is broad and the nares are wide. Corneal clouding starts in early life, being initially peripheral, but by the third or fourth decade the corneal dystrophy can severely curtail vision (Fig. 5–6).

**Skeletal system.** Patients are normal or near normal in height. The neck may be short. In some, the trunk is relatively shorter than the extremities. Hands and feet are broad and short, and fingers and toes are fixed in a clawlike position (Fig. 5–7). The range of mobility is limited in all joints. Genua valga and pes cavus are common.

The most prominent roentgenographic changes are small carpal bones and the claw deformity of the fingers. Cystic changes are frequent in the carpals and metacarpals. Bone cysts of the femoral head have been reported (2). The carpal tunnel syndrome, complicated by median nerve entrapment, is common. In addition, there are widened ribs and sometimes mild hypoplasia of the basilar portion of the iliac bones.

**Other findings.** Intelligence is usually normal. Liver and spleen may be enlarged. Inguinal and/or umbilical hernias are frequently present. Most adult patients show signs of aortic stenosis and/or regurgitation; the murmurs are detected in childhood, but are not clinically significant until maturity. Life span is normal or may be reduced because of cardiac disease.

**Oral manifestations.** Study of the jaws and teeth has been made in only two sibs (1a). The changes are similar to those seen in MPS I-H, MPS II and MPS VI, i.e., cystic changes around unerupted first permanent molars. The mandibular condyles are underdeveloped.

**Laboratory aids.** As is true in MPS I-H, diagnosis may be readily accomplished by direct assay for α-L-iduronidase activity in isolated leukocytes or cultured skin fibroblasts (1,4). In childhood and late adolescence, pseudo-Hurler polydystrophy should be excluded, since the body habitus of young patients with this condition may be similar to that in MPS I-S. However, in pseudo-Hurler polydystrophy the GAG excretion is normal, the corneae are usually clear, and there is mental retardation.

### References [Mucopolysaccharidosis I-S (Scheie syndrome)]

1. Bach G et al: The defect in the Hurler and Scheie syndrome: Deficiency of α-L-iduronidase. *Proc Natl Acad Sci USA* **69**:2048–2051, 1972.

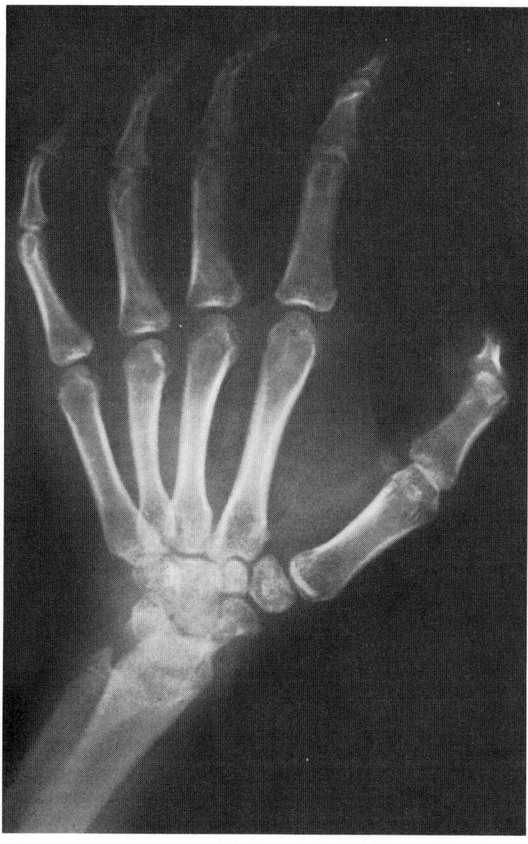

Fig. 5–7. *Scheie syndrome*. Same patient as in Fig. 5–6. Clawing deformities of fingers, small carpal bones with reduced carpal space. (Courtesy of *RL Summitt*, Nashville, Tennessee.)


1a. Keith O, Scully C: Orofacial features of Scheie's syndrome. *Oral Surg Oral Med Oral Pathol* (in press).

2. Lamon JM et al: Bone cysts in mucopolysaccharidosis I-S (Scheie syndrome). *Johns Hopkins Med J* **146**:73–74, 180.

3. McKusick VA: *Heritable Disorders of Connective Tissue,* 4th ed. Mosby, St. Louis, 1972.

4. McKusick VA, Neufeld EF: The mucopolysaccharide storage diseases, in *The Metabolic Basis of Inherited Diseases,* Stanbury JB, Wyngaarden JB, Fredrickson DS, Goldstein JL, Brown MS (eds), 5th ed, McGraw-Hill, New York, 1983, pp 751–777.

5. Mueller OT et al: Apparent allelism of the Hurler, Scheie, and Hurler/Scheie syndromes. *Am J Med Genet* **18**:547–556, 1984.

6. Scheie HG et al: A newly recognized forme fruste of Hurler's disease (gargoylism). *Am J Ophthalmol* **53**:733–769, 1962.

7. Schinz HR, Furtwängler A: Zur Kenntnis einer hereditären Osteoarthropathie mit rezessivem Erbgang. *Dtsch Z Chir* **207**:309–416, 1928.

8. Spranger J: The systemic mucopolysaccharidoses. *Ergeb Inn Med Kinderheilkd* **32**:165–265, 1972.


**Mucopolysaccharidosis I-H/S (Hurler–Scheie syndrome).** McKusick (7) described a third disorder having a phenotype intermediate between Hurler and Scheie diseases, designated Hurler–Scheie disease (MPS IH/S). This disorder also is caused by a deficiency of $\alpha$-L-iduronidase that results in lysosomal accumulation of DS and HS. The fact that MPS I-H/S patients have milder phenotype than MPS I-H individuals and more severe manifestations than MPS I-S patients suggests that they represented the genetic compound of MPS I-H and I-S alleles (4,7–9). However, reports of MPS I-H/S patients from consanguineous parents indicate that at least some, if not the majority, of these patients represent the expression of other mutant alleles at the $\alpha$-L-iduronidase locus (4). About 35 patients have been reported (1).

Clinical manifestations of the disease are usually evident by 2 years of age (1,5,7,10,11). Affected individuals have certain features in common with MPS I-H patients, including marked short stature, dysostosis multiplex, hepatosplenomegaly, umbilical and/or inguinal hernias, and corneal clouding (Fig. 5–8). (1,2,5,6,10,12,15). However, the symptoms present later and are milder than those seen in MPS I-H patients of the same age. The disease is progressive, particularly with regard to cardiopulmonary disease, and death usually occurs by age 25 years (13). Most notably, these patients have later onset (2–5 years) and milder course (12). Nearly all have normal mentation (1). Psychotic symptoms have been observed in older patients (3). Other anomalies include hearing loss (12%), thickened skin (25%), hirsutism (20%), and carpal tunnel syndrome (10%) (1).

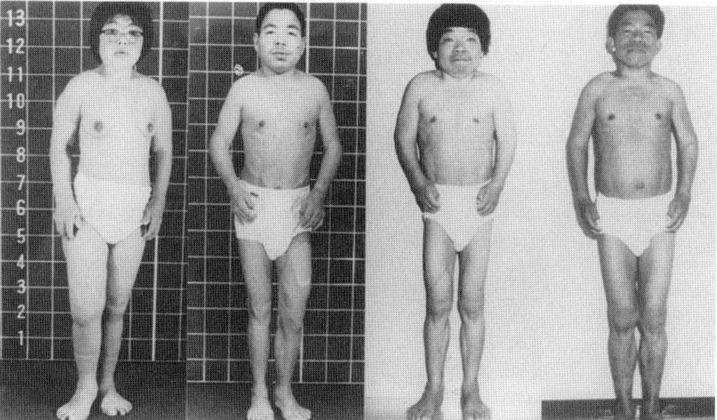


Fig. 5–8. *Hurler–Scheie syndrome*. Four patients from two families. Note similarity of phenotype. (From *N Kaibara et al,* Hum Genet **53**:37, 1979.)

Oral findings. Oral findings are similar to those found in MPS I-H patients. McKusick and Neufeld (8) noted that some patients with MPS I-H/S have micrognathia as well as symptomatic subarachnoid cysts of the sella turcica. Particular care must be taken in manipulating the oral cavity for anesthesia (11). Airway obstruction and sleep apnea have been reported, as in both Hurler and Hunter syndromes.

Laboratory aids. The diagnosis of MPS I-H/S can be confirmed by demonstration of deficient $\alpha$-L-iduronidase activity in isolated leukocytes or cultured fibroblasts, or by the massive urinary excretion of DS and HS. Residual $\alpha$-L-iduronidase activity has been detected in all three MPS I subtypes, but the activity levels are too low to permit discrimination of different mutations using commercially available substrates. Until the $\alpha$-L-iduronidase gene is cloned and the allelic mutations characterized at the molecular level, classification of patients with MPS I-H, I-S, and I-H/S must be made on the basis of the clinical findings. Most likely, a variety of different mutations at the $\alpha$-L-iduronidase locus produce the spectrum of disease severity, which has arbitrarily been categorized as severe (MPS I-H), intermediate (MPS I-H/S), and mild (MPS I-S). There is even a nonpathological allele (14). Finally, it is anticipated that the explanation for the presence of severe neurologic manifestations in MPS I-H and I-H/S and their absence in MPS I-S patients will provide further insight into the pathogenesis of $\alpha$-L-iduronidase deficiency.

### References [Mucopolysaccharidosis I-H/S (Hurler–Scheie syndrome)]


1. Colavita N et al: A further contribution to the knowledge of mucopolysaccharidosis I H/S compound. Presentation of two cases and review of the literature. *Australas Radiol* **30**:142–149, 1986.

2. Chijiwa T et al: Ocular manifestations of Hurler/Scheie phenotype in two sibs. *Jpn J Ophthalmol* **27**:54–62, 1983.

3. Dugas M et al: Psychotic symptoms during the evolution of dementia in mucopolysaccharidosis of Hurler–Scheie phenotype. *Arch Fr Pédiatr* **42**:373–375, 1985.

4. Fortuin JJH, Kleijer WF: Hybridization studies of fibroblasts from Hurler, Scheie and Hurler–Scheie compound patents: Support for the hypothesis of allelic mutants. *Hum Genet* **53**:155–159, 1980.
5. Kaibara N et al: Hurler–Scheie phenotype: A report of two pairs of inbred sibs. *Hum Genet* **53**:37–41, 1979.
6. Kaibara N et al: Hurler–Scheie phenotype with parental consanguinity. Report of an additional case supporting the concept of genetic heterogeneity. *Clin Orthop* **175**:233–236, 1983.
7. McKusick VA et al: Allelism, non-allelism and genetic compounds among the mucopolysaccharidoses. *Trans Assoc Am Phys* **85**:15–25, 1972.
8. McKusick VA, Neufeld EF: The mucopolysaccharide storage diseases, in *The Metabolic Basis of Inherited Diseases,* Stanbury JB, Wyngaarden JB, Fredrickson DS, Goldstein JL, Brown MS (eds), 5th ed, McGraw-Hill, New York, 1983, pp 751–777.
9. Mueller OT et al: Apparent allelism of the Hurler, Scheie and Hurler/Scheie syndromes. *Am J Med Genet* **18**:547–556, 1984.
10. Roubicek M et al: The clinical spectrum of alpha-L-iduronidase deficiency. *Am J Med Genet* **20**:471–481, 1985.
11. Sjögren P, Pedersen T: Anaesthetic problems in Hurler–Scheie syndrome. Report of two cases. *Acta Anaesthesiol Scand* **30**:484–486, 1986.
12. Stevenson R et al: The iduronidase deficiency mucopolysaccharidoses—clinical and roentgenographic studies. *Pediatrics* **57**:111–122, 1976.
13. Wassman ER et al: Postmortem findings in the Hurler–Scheie syndrome (mucopolysaccharidosis I-H/S). *Birth Defects* **18(3B)**:13–18, 1982.
14. Whitley CB et al: A non-pathologic allele (I-W) for low alpha-L-iduronidase enzyme activity vis-à-vis prenatal diagnosis of Hurler syndrome. *Am J Med Genet* **28**:233–243, 1987.
15. Winters PR et al: α-L-Iduronidase deficiency and possible Hurler–Scheie genetic compound. Clinical, pathologic, and biochemical findings. *Neurology* **26**:1003–1007, 1976.


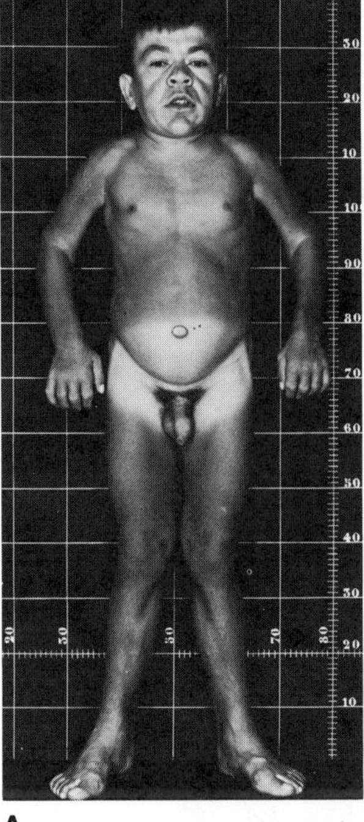
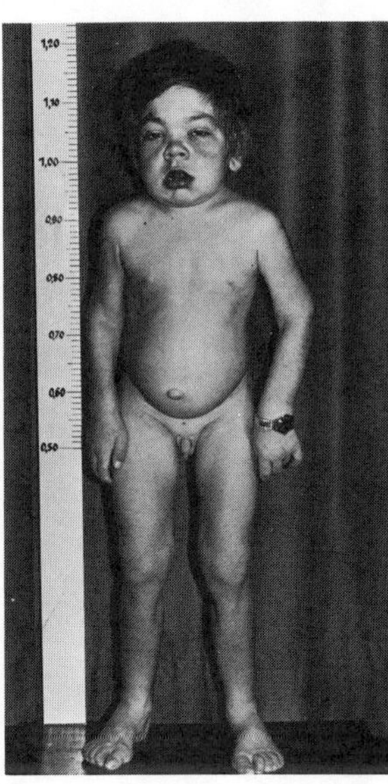

A B

Fig. 5–9. *Hunter syndrome.* (A) Sixteen-year-old male with mild MPS II. Mild coarseness of facies with broad face, low nasal bridge, joint contractures, genua valga. He has mixed hearing loss and normal intelligence. (B) Thirteen-year-old patient with severe MPS II. Growth retardation is evident. Abundant and coarse scalp hair, coarse facies, low nasal bridge, open mouth. (A from *UN Wiesmann* and *S Rampini,* Helv Paediatr Acta **29:**73, 1974).

**Mucopolysaccharidosis II (Hunter syndrome).** The first patients manifesting the X-linked form of MPS were reported by Hunter in 1917 (8). These two brothers were mildly affected at ages 8 and 10 years, respectively, with clear corneas and apparently normal intelligence. Although they died at 11 and 16 years, they probably suffered from the mild form of the disease (MPS II-mild), which is compatible with survival to adulthood (25,32,34). In the severe form (MPS II-severe), which is three to four times more frequent than the mild form, rapid psychomotor deterioration and progression of physical deformities may be observed after the third year of life, with death usually occurring between 4 and 14 years of age (27,37).

Both forms of MPS II result from allelic mutations (16,32,34) at the MPS II locus on the X chromosome, which has been localized close to the region Xq25-q27 (3,20a,30). Based on linkage analysis, the MPS II locus is closely linked to the arbitrary DNA probe, DSX13 (4). Both forms of MPS II result from deficient activity of the lysosomal hydrolase, iduronate sulfatase (3,5,6,24), resulting in progressive accumulation of DS and HS. To date, studies of residual enzyme activity have not distinguished between the mild and severe forms (32,34). In spite of earlier estimates, the frequency of MPS II is probably about 1/65,000 births (13,16), higher than that of MPS I-H (21). However, an estimate of 1/132,000 was made by Young and Harper (36). It may be more common among Jews (21). The occurrence of a wide range of severity suggests that other allelic forms of MPS II may exist (29). About 25% represent new mutations (4).

Interestingly, full expression of the MPS II phenotype has been observed in two females, one with an X:5 translocation [breakpoint presumably involving the region q26(17)], the other with a small X chromosome deletion involving band Xq25 (3). In the latter case, the affected girls' mother had a half-normal iduronate sulfatase activity, consistent with heterozygosity for MPS II. The affected girl had less than 1% of enzymatic activity and massive urinary excretion of DS and HS, suggesting that she expressed the mutant gene since the normal locus was involved in the chromosomal deletion.

Facies. Macrocephaly is present, all measurements being enlarged. The facies, although similar to that observed in MPS I-H, is sufficiently different as to be distinguishable. Although all MPS I-H facies are very similar to one another, the facies of any MPS II patient bears a coarse resemblance to the facies of his family members (10). The facies of patients with the more common MPS II-severe form usually is more striking than that of the milder form (25,32,34–37) (Figs. 5–9 and 5–10).

Skeletal system. The neck is short, the chest is broad, and the abdomen protrudes with umbilical hernias. Moderate thoracolumbar kyphosis may be present. The trunk is relatively shorter than the extremities. Joint mobility is restricted, with clawlike deformities of the fingers. The gait is stiff, with the trunk bent forward. Typically, there is shortness of stature only from about 3 years of age; however, some patients are not dwarfed until 5 or 6 years of age. MPS II-severe patients grow as fast or faster than normal children during the first 2–3 years of life. Adults with the mild type reach a height ranging between 120 and 140 cm (32,34), whereas patients with the severe form reach a height between 105 and 115 cm.

Roentgenographic changes are qualitatively similar to, but quantitatively less pronounced than those observed in MPS I-H when compared at identical ages (13,25).

Other findings. Intelligence is only slightly impaired in the mild form. In the severe type, there is progressive loss of intellectual function after the age of 2–3 years. In retrospect it can be seen that intellectual function of these patients never was normal. At 3 years of age, patients may be brought to a physician because of lack of speech. Patients become restless, hyperactive, and destructive (27,37). Diarrhea is common (37). Later, muscle tone increases, muscle reflexes become hyperactive, and, within a few years, the patient is bedridden, with flexion contractures and loss of environmental contact. Terminal convulsions are common (37).

Corneal clouding does not occur in either the mild or severe forms with rare exception (26). Vision may be impaired because of retinitis pigmentosa or chronic papilledema associated with chronic hydrocephalus. In both forms of the disease, progressive hearing loss occurs

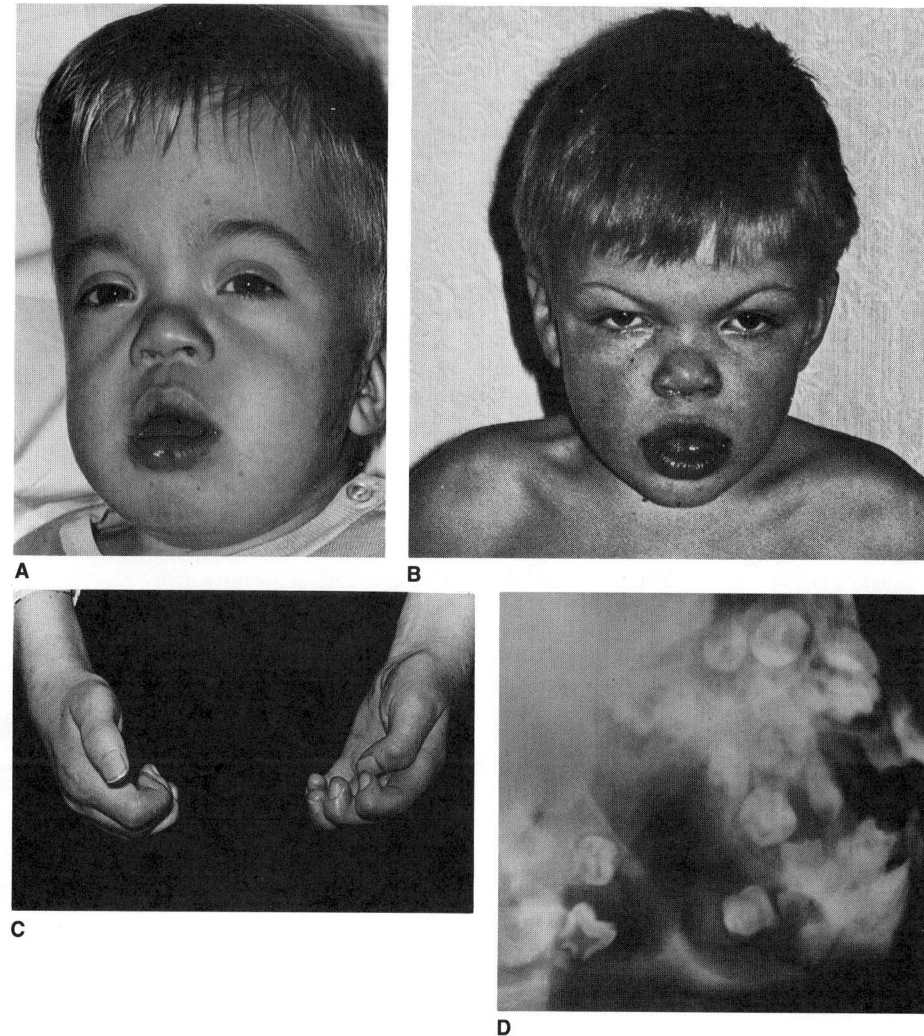

Fig. 5–10. *Hunter syndrome.* (A) Coarse facies, low nasal bridge, open mouth. (B) Compare facies with patient in A. (C) Clawhands. (D) Roentgenogram showing cysts around crowns of malaligned teeth. (B,C from *E Passarge et al,* Dtsch Med Wochenschr **99:**144, 1974. D from *HM Worth,* Oral Surg **22:**21, 1966.)

in most patients (20). It is often of the mixed type but may be predominantly conductive or sensorineural (22). Hepatosplenomegaly, inguinal and/or umbilical hernia, and cardiovascular defects are often found. Skin changes, present in a minority of patients, consist of hard, nontender, irregularly shaped papules varying in size from a few millimeters to a centimeter in diameter located over the scapulae and deltoid areas.

Death in patients with the severe form usually occurs by 15 years of age, preceded by complications of progressive neurologic deterioration. Cardiopulmonary disease, primarily resulting from infiltrative valvular disease, is a common cause of death. In contrast, patients with the mild form typically live into the sixth and seventh decades of life. Disease complications in these adult patients include carpal tunnel syndrome with medial nerve entrapment, degenerative disease of the hips, compromised respiratory function, and anesthetic risk (33).

Oral manifestations. The teeth are widely spaced. The tongue is enlarged, particularly after 5 years of age in patients with the severe form. The same bony alterations (condylar deformities, cystic alterations) found in patients with MPS I-H are also noted in the severe form of MPS II (4). Airway obstruction and sleep apnea have been reported (23).

Laboratory findings. Most patients with MPS II excrete approximately equal amounts of DS and HS in their urine, which is readily determined by routine screening tests. Alder–Reilly granulations may be present in peripheral granulocytes and bone marrow cells. Lymphocytes show metachromatic granules within vacuoles following toluidine blue staining. Ultrastructural studies show single membrane-bound structures. They are even evident on conjunctival biopsy (15).

Hemizygotes with both forms of MPS II can be enzymatically diagnosed by the demonstration of iduronate sulfatase deficiency in serum, isolated leukocytes, and cultured fibroblasts (2,5,11,32). Carrier detection is more difficult because of random X inactivation, but has been accomplished by enzyme assay using plasma, cultured cells, and hair roots (1,4,18,28,38,39). Prenatal diagnosis of MPS II can be accomplished by the demonstration of a male fetus with deficient iduronate sulfatase activity in chorionic villi (7,19) or in amniotic fluid and/or amniotic cells (12). An adjunct to the prenatal diagnosis of this disorder is the demonstration of increased iduronate sulfatase in the serum of pregnant heterozygotes (40). Prenatal diagnosis of a female heterozygous fetus has been carried out (9). Iduronate sulfatase, the lysosomal enzyme deficient in this disorder, has been purified from human plasma and is a monomer with a molecular weight of about 80,000–90,000 (31).

## References [Mucopolysaccharidosis II (Hunter syndrome)]


1. Archer IM et al: Carrier detection in Hunter syndrome. *Am J Med Genet* **16**:61–69, 1983.
2. Bach GH et al: The defect in the Hunter syndrome: Deficiency of sulfoiduronate sulfatase. *Proc Natl Acad Sci USA* **70**:2134–2138, 1973.
3. Broadhead PM et al: Full expression of Hunter's disease in a female with an X-chromosome deletion leading to non-random inactivation. *Clin Genet* **30**:392–398, 1986.
4. Chase DS et al: Genetics of Hunter syndrome: Carrier detection, new mutations segregation and linkage analysis. *Ann Hum Genet* **50**:349–360, 1986.

5. Dean MF: The iduronate sulphatase activities of cells and tissue fluids from patients with Hunter syndrome and normal controls. *J Inherit Metab Dis* **6**:108–111, 1983.

6. Fratantoni JC et al: The defect in Hurler and Hunter syndromes. II. Deficiency of specific factors involved in mucopolysaccharide degradation. *Proc Natl Acad Sci USA* **64**:350–366, 1969.

7. Harper PS et al: Chorion biopsy for prenatal testing in Hunter's syndrome (letter). *Lancet* **2**:812–813, 1984.

8. Hunter C: A rare disease in two brothers. *Proc R Soc Med* **10**(pt.1):104–116, 1917.

9. Kleijer WJ et al: Prenatal monitoring for the Hunter syndrome: The heterozygous female fetus. *Clin Genet* **15**:113–117, 1979.

10. Leroy JG, Crocker AC: Clinical definition of the Hurler–Hunter phenotypes. *Am J Dis Child* **112**:518–530, 1966.

11. Liebaers I, Neufeld EF: Iduronate sulfatase activity in serum, lymphocytes, and fibroblasts—simplified diagnosis of the Hunter syndrome. *Pediatr Res* **10**:733–736, 1976.

12. Liebaers I et al: Iduronate sulfatase in amniotic fluid: An aid in the prenatal diagnosis of the Hunter syndrome. *J Pediatr* **90**:423–425, 1977.

13. Lowry RB, Renwick DHG: Relative frequency of the Hurler and Hunter syndromes. *N Engl J Med* **284**:221–222, 1971.

14. Lustmann J et al: Dentigerous cysts and radiolucent lesions of the jaw associated with Hunter's syndrome. *J Oral Surg* **33**:679–685, 1975.

15. McDonnell JM et al: Ocular histopathology of systemic mucopolysaccharidosis, type II-A (Hunter syndrome, severe). *Ophthalmology* **92**:1772–1779, 1985.

16. McKusick VA: The relative frequency of the Hurler and Hunter syndromes. *N Engl J Med* **283**:853–854, 1970.

17. Mossman J et al: Hunter's disease in a girl: Association with X:5 chromosomal translocation disrupting the Hunter gene. *Arch Dis Childh* **58**:911–915, 1983.

18. Nwokoro N, Neufeld EF: Detection of Hunter heterozygotes by enzymatic analysis of hair roots. *Am J Hum Genet* **31**:42–49, 1979.

19. Pannone N et al: Prenatal diagnosis of Hunter syndrome using chorionic villi. *Prenatal Diag* **6**:207–210, 1986.

20. Peck JE: Hearing loss in Hunter's syndrome—mucopolysaccharidosis II. *Ear Hear* **5**:243–246, 1984.

20a. Roberts SH et al: Further evidence localising the gene for Hunter's syndrome to the distal region of the X chromosome long arm. *J Med Genet* **26**:309–313, 1989.

21. Schaap T, Bach G: Incidence of mucopolysaccharidosis in Israel: Is Hunter disease a "Jewish disease"? *Hum Genet* **56**:221–226, 1980.

22. Schachern PA et al: Mucopolysaccharidosis I-H (Hurler's syndrome) and human temporal bone histopathology. *Ann Otol Rhinol Laryngol* **93**:65–69, 1984.

23. Shapiro J et al: Airway obstruction and sleep apnea in Hurler and Hunter syndromes. *Ann Otol Rhinol Laryngol* **94**:458–461, 1985.

24. Sjoberg I et al: Hunter's syndrome: A deficiency of L-idurono-sulfate sulfatase. *Biochem Biophys Res Commun* **54**:1125, 1973.

25. Spranger J: The systemic mucopolysaccharidoses. *Ergeb Inn Med Kinderheilkd* **32**:165–265, 1972.

26. Spranger J et al: Mucopolysaccharidosis II (Hunter disease) with corneal opacities. *Eur J Pediatr* **129**:11–16, 1978.

27. Thurmon TF et al: Clinical heterogeneity in mucopolysaccharidosis II: Evidence for epistasis. *Birth Defects* **10**(8):125–127, 1974.

28. Tonnesen T: The use of fructose 1-phosphate to detect Hunter heterozygotes in fibroblast cultures from high-risk carriers. *Hum Genet* **66**:212–216, 1984.

29. Tsuzaki S et al: An unusually mild variant of Hunter's syndrome in a 14-year-old boy: Normal growth and development. *Acta Paediatr Scand* **76**:844–846, 1987.

30. Upadhyaya M et al: Localization of the gene for Hunter syndrome on the long arm of X chromosome. *Hum Genet* **74**:391–398, 1986.

31. Wasteson A, Neufeld EF: Iduronate sulfatase from human plasma. *Methods Enzymol* **83**:573–578, 1982.

32. Wiesmann UN, Rampini S: Mild form of the Hunter syndrome: Identity of the biochemical defect with the severe type. *Helv Paediatr Acta* **29**:73–78, 1974.

33. Young ID, Harper PS: Long-term complications in Hunter's syndrome. *Clin Genet* **16**:125–132, 1979.

34. Young ID, Harper PS: Mild form of Hunter's syndrome: Clinical delineation based on 31 cases. *Arch Dis Childh* **57**:828–836, 1982.

35. Young ID, Harper PS: A clinical and genetic study of Hunter's syndrome. *J Med Genet* **19**:401–407, 408–411, 1982.

36. Young ID, Harper PS: Incidence of Hunter's syndrome. *Hum Genet* **60**:391–392, 1982.

37. Young ID, Harper PS: The natural history of the severe form of Hunter's syndrome: A study based on 52 cases. *Dev Med Child Neurol* **25**:481–489, 1983.

38. Yutaka T et al: Iduronate sulfatase analysis of hair roots for identification of Hunter syndrome heterozygotes. *Am J Hum Genet* **30**:575, 1978.

39. Zlotogora J, Bach G: Heterozygote detection in Hunter syndrome. *Am J Med Genet* **17**:661–665, 1984.

40. Zlotogora J, Bach G: Hunter syndrome: Prenatal diagnosis in maternal serum. *Am J Hum Genet* **38**:253–260, 1986.

**Mucopolysaccharidoses III (Sanfilippo A, B, C, and D syndrome).** MPS III was first recognized in 1958 by Meyer et al (18) and later described by Meyer et al (19) and Sanfilippo et al (24). The disease is characterized by severe mental and neurologic degeneration associated with relatively mild MPS features resulting from the progressive lysosomal accumulation of HS (4).

Four enzymatic steps are required for the normal degradation of HS, and deficient activity of each enzyme in the pathway has been found in patients with MPS III (see Table 5–1). MPS IIIA results from deficient heparan-N-sulfate activity (12,13,17), MPS IIIB from defective $\alpha$-*N*-acetylglucosaminidase activity (21,32), MPS IIIC from deficient *N*-acetyltransferase activity (1,11), and MPS IIID from deficient $\alpha$-*N*-acetylglucosamine-6-sulfate sulfatase activity (2,7,9a,14). Most notably, all four enzymatic defects result in the lysosomal accumulation of HS and cannot be differentiated phenotypically. All four subtypes have autosomal recessive inheritance. Taken together, the MPS III subtypes represent the most common type of the mucopolysaccharide disorders, their frequency estimated to be about 1/24,000 in the Netherlands (30). Because of their mild somatic phenotype, the frequency estimate may be lower than the actual incidence of the disease. MPS IIIA is found most often in the United States and the U.K. while MPS IIIB is more common in Greece (1a,27a).

Facies. No abnormalities are noted in the young child. Older children may develop mild facial coarsening resembling patients with the MPS I-H phenotype, but the facial features never become as strikingly abnormal as in that disorder. About 80% of children with MPS III have a dull appearance with a slightly sunken nasal bridge and abundant, coarse scalp hair (23). The latter finding is the most consistent clinical feature even in children without other morphologic alterations. Hairs have been described as triangular (3) or as having marked variability in shape and pigment (26a) in cross section. Corneal clouding is absent (Fig 5–11).

Skeletal system. Height may be slightly reduced or normal. Joint mobility may be mildly restricted in elbows and knees. Roentgenographically, thickening of the posterior calvaria, sclerotic mastoids, ovoid-shaped vertebral bodies, and minimal hypoplasia of the supraacetabular portions of the ilia are the most consistent abnormalities (6,15,25,34)(Fig. 5–12). Otherwise, dysostosis multiplex develops slowly and is very mild. For example, the hands in MPS III are normal.

Other findings. Early development is usually normal. In the second to fifth year of life, development ceases, and behavioral problems, such as restlessness, aggressiveness, diminished attention span, and sleep disturbances, become manifest. Aggressive hyperactivity is frequently the parents' reason for seeking medical help. Subsequently, there is progressive loss of mental and motor skills. Loss of environmental contact is evident prior to a "vegetative" state, with spastic diplegia and death occurring between the ages of 10 and 20 years (20a). A comparison of the clinical course of patients with MPS IIIA, B, and C indicates that patients with MPS IIIA are generally more severely affected clinically than those with B or C subtypes (30,31a). The disorder in type A has earlier onset, is more marked, and results in earlier death. In addition, intrafamilial variation in the clinical expression of each of the subtypes was noted (29,30), consistent with different allelic mutations causing genetic heterogeneity in each subtype.

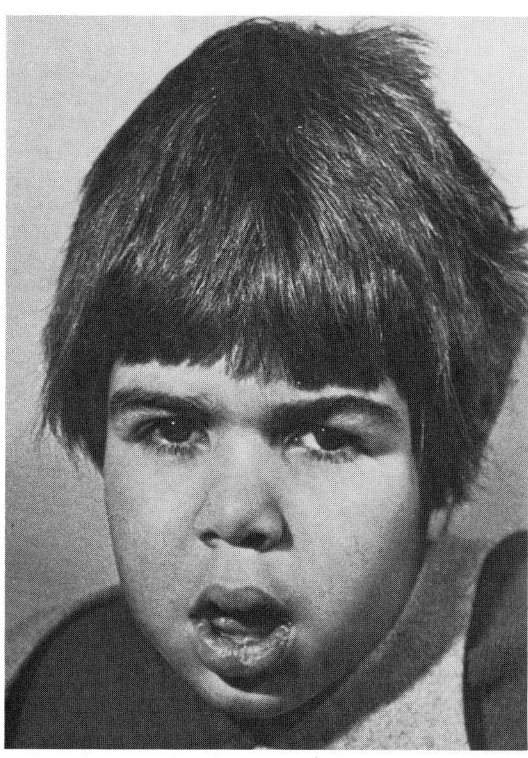

Fig. 5–11. *Sanfilippo syndrome.* Seven-year-old girl with MPS III. Note coarse features and thick abundant scalp hair. (From *E Passarge et al,* Dtsch Med Wochenschr **99:**144, 1974.)

Pathologic study of the brain has revealed marked deposition of HS, as well as ceramide polyhexoside and $G_{M1}$ ganglioside in the cerebrum (48). Hearing loss is frequently suspected but difficult to prove. Hepatosplenomegaly is present in more than 80% of patients. Histochemical and ultrastructural studies have been reported (5,9,26).

Oral manifestations. There seem to be no remarkable oral manifestations, but tooth abscesses are of major concern during the final stage. Obliteration of pulp chambers has been reported in one child. (35). The significance, if any, needs to be explored. The tongue is not large but may protrude late in the disorder.

Fig. 5–12. *Sanfilippo syndrome.* Thickening of parietal and occipital portions of cranial vault in 11-year-old girl with MPS III.

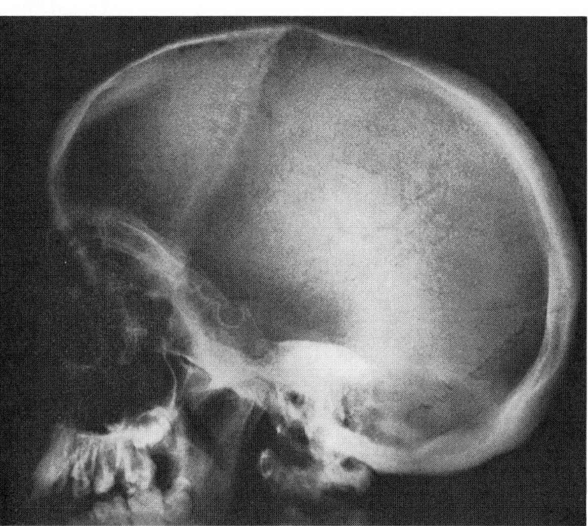

Laboratory findings. Clusters of coarse granulations are seen in the cytoplasm of about 35% of peripheral lymphocytes and in plasmacellular and reticulohistiocytic cells of the bone marrow. The granulations are frequently surrounded by areas of diminished stainability. The inclusions stain metachromatically with toluidine blue. Excessive amounts of HS (in the absence of DS) are excreted in the urine. It should be noted that the diagnosis of this disorder by demonstration of increased urinary excretion of HS may be missed by certain procedures, such as the toluidine blue filter paper test.

Definitive diagnosis of MPS III and carrier detection can be made by the demonstration of the deficient enzymatic activity in each of the subtypes using appropriate sources (e.g. serum, cultured fibroblasts and leukocytes) (13,16,22,27,27a,31,33). Prenatal diagnosis of all four MPS III subtypes is feasible by demonstration of the deficient enzymatic activity in chorionic villi or cultured amniocytes (5a,10,20,28).

## References [Mucopolysaccharidosis III (Sanfilippo A, B, C, and D syndrome)]


1. Bartsocas CS et al: Sanfilippo type C disease: Clinical findings in four patients with a new variant of mucopolysaccharidosis III. *Eur J Pediatr* **130**:251–258, 1979.

1a. Beratis NG et al: Sanfilippo disease in Greece. *Clin Genet* **29**:129–132, 1986.

2. Coppa GV et al: Clinical heterogeneity in Sanfilippo disease (mucopolysaccharidosis type D): Presentation of two new cases. *Eur J Pediatr* **140**:130–132, 1983.

3. Crump IA, Danks DM: Simple method for cutting transverse section of hair: Comments on shape of hair in Hurler and Sanfilippo syndrome. *Arch Dis Childh* **46**:383–386, 1971.

4. Danks DM et al: The Sanfilippo syndrome: Clinical, biochemical radiological, haematological, and pathological features of nine cases. *Aust Paediatr J* **8**:174–186, 1972.

5. DelMonte MA et al: Histopathology of Sanfilippo's syndrome. *Arch Ophthalmol* **101**:1255–1262, 1983.

5a. DiNatale P et al: First-trimester prenatal diagnosis of Sanfilippo C disease. *Prenatal Diag* **7**:603–605, 1987.

6. Farriaux JP et al: Etude comparative des aspects cliniques, radiologiques, biochemiques et génétiques de la maladie de Sanfilippo de type A et type B. *Helv Paediatr Acta* **29**:349–370, 1974.

7. Gatti R et al: Sanfilippo type D disease: Clinical findings in two patients with a new variant of mucopolysaccharidosis III. *Eur J Pediatr* **138**:168–171, 1982.

8. Ghatak NR et al: Neuropathology of Sanfilippo syndrome. *Ann Neurol* **2**:161–166, 1977.

9. Haust MD, Gordon BA: Ultrastructural and biochemical aspects of the Sanfilippo syndrome, type III genetic mucopolysaccharidosis. *Connect Tissue Res* **15**:57–64, 1986.

9a. Kaplan P, Wolfe LS: Sanfilippo syndrome type D. *J Paediatr* **110**:267–271, 1987.

10. Kleijer WJ et al: First trimester diagnosis of mucopolysaccharidosis IIIA (Sanfilippo A disease). *N Engl J Med* **314**:185–186, 1986.

11. Klein U et al: Sanfilippo syndrome type C: Assay for acetyl-CoA:$\alpha$-glucosaminide N-acetyltransferase in leukocytes for detection of homozygous and heterozygous individuals. *Clin Genet* **20**:55–59, 1981.

12. Kresse H et al: Biochemical heterogeneity of the Sanfilippo syndrome. *Biochem Biophys Res Commun* **42**:892–989, 1971.

13. Kresse H: Mucopolysaccharidosis IIIA (Sanfilippo A disease): Deficiency of heparin sulfamidase in skin fibroblasts and leucocytes. *Biochem Biophys Res Commun* **54**:111, 1973.

14. Kresse H et al: Sanfilippo disease type D: Deficiency of N-acetylglucosamine-6-sulfate sulfatase required for heparan sulfate degradation. *Proc Natl Acad Sci USA* **77**:6822–6826, 1980.

15. Langer LO: The radiographic manifestations of the HS-mucopolysaccharidosis of Sanfilippo. *Ann Radiol* **7**:315–325, 1964.

16. Marsh J, Fensom AH: 4-Methylumbelliferyl alpha-N-acetylglucosaminidase activity for diagnosis of Sanfilippo B disease. *Clin Genet* **27**:258–262, 1985.

17. Matalon R, Dorfman A: Sanfilippo A syndrome: Sulfamidase deficiency in cultured skin fibroblasts and liver. *J Clin Invest* **54**:907–912, 1974.

18. Meyer K et al: Excretion of sulfated mucopolysaccharides in gargoylism (Hurler's syndrome). *Proc Soc Exp Biol Med* **97**:275–279, 1958.

19. Meyer K et al: Sulfated mucopolysaccharides of urine and organs in gargoylism. *Proc Soc Exp Biol Med* **102**:587–590, 1959.

20. Mossman J et al: Prenatal tests for Sanfilippo disease type B in four pregnancies. *Prenatal Diag* **3**:347–350, 1983.

20a. Nidiffer FD, Kelly TE: Developmental and degenerative patterns associated with cognitive, behavioral and motor difficulties in the Sanfilippo syndrome: An epidemiological study. *J Ment Defic Res* **27**:185–203, 1983.

21. Nobili B et al: Sanfilippo's disease of type B. Study of the enzymatic deficiency in a family. *Pediatria (Napoli)* **91**:295–301, 1983.

22. Pallini R et al: Sanfilippo type C diagnosis: Assay of acetyl-CoA:alpha-glucosaminide N-acetyltransferase using [$^{14}$C]glucoseamine as substrate and leukocytes as enzyme source. *Pediatr Res* **18**:543–545, 1984.

23. Rampini S: Das Sanfilippo–Syndrom. *Helv Paediatr Acta* **24**:55–91, 1969.

24. Sanfilippo SJ et al: Mental retardation associated with acid mucopolysacchariduria (heparitin sulfate type). *J Pediatr* **63**:837–838, 1963.

25. Spranger J et al: Die HS-Mucopolysaccharidose von Sanfilippo (polydystrophe Oligophrenie). Bericht über 10 Patienten. *Z Kinderheilkd* **101**:71–84, 1967.

26. Tamagawa K et al: Neuropathological study and chemico-pathological correlation in sibling cases of Sanfilippo syndrome type B. *Brain Dev* **7**:599–609, 1985.

26a. Teschler-Nicola M, Killian W: Observations on hair shaft morphology in mucopolysaccharidoses. *J Ment Defic Res* **26**:193–202, 1982.

27. Thompson JN et al: Oligosaccharide substrates for heparin sulfamidase. *Anal Biochem* **152**:412–422, 1986.

27a. Toone JR, Applegarth DA: Carrier detection in Sanfilippo A syndrome. *Clin Genet* **33**:401–403, 1988.

28. Torok O et al: Prenatal diagnosis of Sanfilippo A disease (mucopolysaccharidosis III A). *Orv Hetil* **127**:2385–2387, 1986.

29. Uvebrant P: Sanfilippo type C syndrome in two sisters. *Acta Paediatr Scand* **74**:137–139, 1985.

30. Van de Kamp JJP et al: Genetic heterogeneity and clinical variability in the Sanfilippo syndrome (types A, B and C). *Clin Genet* **20**:152–160, 1981.

31. Vance JM et al: Carrier detection in Sanfilippo syndrome type B: Report of six families. *Clin Genet* **20**:135–140, 1981.

31a. van Schrojenstein-de Valk HMJ, van de Kamp JJP: Follow-up on seven adult patients with mild Sanfilippo B-disease. *Am J Med Genet* **28**:125–129, 1987.

32. von Figura K, Kresse H: The Sanfilippo B corrective factor: An N-acetyl-α-D-glucosaminidase. *Biochem Biophys Res Commun* **48**:262, 1972.

33. von Figura K et al: Sanfilippo B disease: Serum assays for detection of homozygous and heterozygous individuals in three families. *J Pediatr* **83**:607–611, 1973.

34. Wallace BJ et al: Mucopolysaccharidosis type III. *Arch Pathol* **82**:462–473, 1966.

35. Webman MS et al: Obliterated pulp cavities in the Sanfilippo syndrome (mucopolysaccharidosis III). *Oral Surg* **43**:734–738, 1977.

**Mucopolysaccharidoses, Types IVA and IVB (Morquio syndrome).** In 1929, Morquio (31) and Brailsford (5) independently reported cases of a disorder characterized by short trunk dwarfism, progressive spinal deformity, short neck, pectus carinatum, genua valga, pes planus, odontoid hypoplasia, and normal intelligence (Figs. 5–13 to 5–15). Clinical, and later, biochemical heterogeneity was demonstrated. Dale (6), as early as 1931, and others (3,16) described patients with milder clinical forms. About 1960, the disorder was recognized to be a mucopolysaccharidosis, caused by the lysosomal accumulation and urinary excretion of the mucopolysaccharide, KS.

Matalon et al (25) discovered the more common severe form (MPS IVA) to be caused by a deficiency of galactosamine-6-sulfate sulfatase, an enzyme that degrades KS (20,37). Arbisser et al (1) described a patient with normal GalNAc-6-S-sulfatase but deficient lysosomal β-galactosidase. This generally milder condition, known as MPS IVB, has been reported in several patients (1,6a,13,14,17,18,32,33,42–44). However, not all examples have been mild (17,44). MPS IVA has been subsequently shown to have severe, intermediate and even mild (8,15) forms, implying different alleles (2,11,35). Difference in severity may also be explained by the associated finding of reduced neuraminidase activity in some cases (12). All forms of Morquio syndrome have autosomal recessive inheritance. The frequency of MPS IVA has been estimated to be about 1/40,000 births (2). The frequency of MPS IVB has not been estimated, but is rarer than MPS IVA.

A number of milder phenotypic variants have been described (2). Some result from different allelic mutations of the two enzymes. In addition, there may be a third subtype, MPS IVC, resulting from β-N-acetylhexosaminidase B, another enzyme in the degradative pathway for keratan sulfate (26). In one patient known to us (DEC Cole et al, personal communication, 1988), epiphyseal changes, first noted at 5 years, were most severe in the hips and led to complete loss of hip motion by 13 years. Platyspondyly was progressive through adolescence, but punctate corneal opacities were not observed until adulthood. Cultured fibroblast homogenates had normal levels of β-galactosidase and GalNAc-6-sulfatase activity but were deficient for the β-N-acetylhexosaminidase B isozyme assessed by heat stability and specific substrate tests. Profiling of hexosaminidase fractions by DEAE-cellulose chromotography revealed a normal acidic (hexosaminidase A) peak but the basic (hexosaminidase B) isozyme was absent. Molecular cloning of the genes for the α and β subunits of the hexosaminidase A and B isozymes should permit more precise delineation of this new form of Morquio syndrome.

MPS IVA. Type IVA has three grades of severity, and dental findings are present in all grades (32a).

*Facies.* The facies is not specific, but the lower half of the face is often outstanding because of shortness and hyperextension of the neck.

*Musculoskeletal system.* There is reduced height because of shortened neck and trunk and, to a lesser extent, shortened extremities. Adult height rarely exceeds 100 cm (range 80–120 cm). The head is essentially normal, but mild scaphocephaly may be present because of premature closure of sutures. The head seems to rest directly on the shoulders. The neck is greatly shortened, with exaggerated cervical curvature and restricted movement (7). The thorax, after the second year of life, exhibits marked kyphosis or kyphoscoliosis, with general flattening of vertebrae and a characteristic pectus carinatum, the sternum extending almost horizontally from its clavicular junction, then angling downward in midsection. The lumbar region of the spine frequently exhibits a gibbuslike kyphosis or, less often, lordosis in the region of the first lumbar vertebra (Fig. 5–13A,B). Spinal cord compression may occur in the upper cervical segment as a complication from either atlantoaxial dislocation or subluxation at the thoracolumbar gibbus (4), which may result in death.

Extremities appear disproportionately long. There may be excessive joint mobility, and the wrists are usually enlarged. Genua valga, thickened knee joints, and pes planus are nearly constant findings (Fig. 5–13A,B). The stance is semicrouching. Usually there is prominent potbelly.

Radiographically, generalized platyspondyly with hypoplasia of the last thoracic and first lumbar vertebrae, coxa valga, flared ilia, and progressive femoral head flattening and fragmentation are found. In the young child, the vertebral bodies are ovoid and the superior acetabulae are deficiently ossified. The odontoid process is hypoplastic or absent. The bases of the second to fifth metacarpals are conical, but their shafts are normally constricted (23)(Fig. 5–14A–C). The distal ends of the radius and ulna are inclined toward each other. All the bones become markedly osteoporotic.

*Other findings.* The corneas become slowly but diffusely opacified in the form of a filmy haze. This is rarely obvious to the unaided eye before the tenth year of life (38,45). Progressive deafness usually begins in adolescence.

Intelligence is nearly always normal. Aortic regurgitation has been reported (29,35).

MPS IVB. In contrast to patients with MPS IVA, affected individuals with MPS IVB have a significantly milder phenotype. They have normal intelligence, milder dysostosis multiplex, mild pectus carinatum, corneal clouding, odontoid hypoplasia, moderate lumbar kyphosis, and minimal genua valga. Radiographs show platyspondyly and tongue-like protrusions of the lumbar spine. $C_2$–$C_3$ subluxation has been noted. Typically, the patients have normal hearing and no cardiac murmurs. Corneal clouding may be obvious or a slit-lamp examination may be required to observe fine corneal deposits. Often

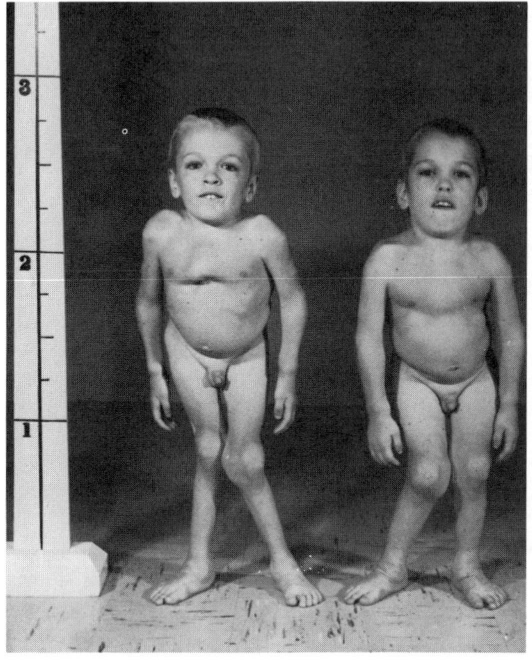


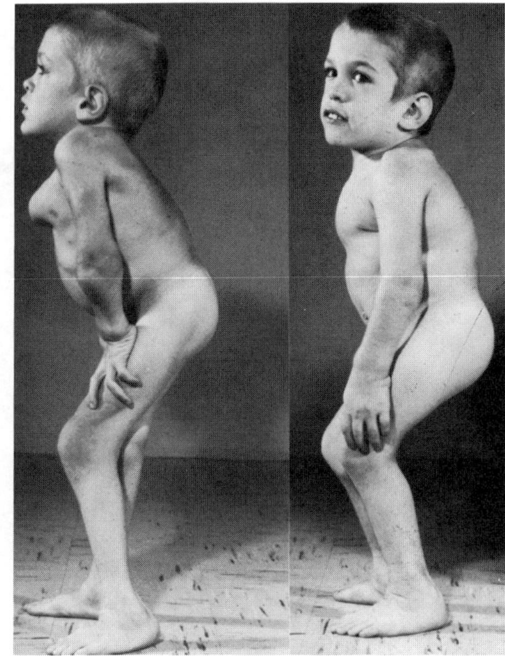

A B

Fig. 5–13. *Morquio syndrome.* (A,B) Brothers, ages 10½ and 9½ with dwarfism, kyphosis, genua valga, flexed knees, abnormally short neck, sternal protrusion, and large joints. (From *H Zellweger et al,* J Pediatr **59:**549, 1961.)

the clinical onset of the disease is characterized by an unstable, waddling gait.

*Oral manifestations.* In patients with MPS IVA, both the deciduous and permanent teeth have dull, gray crowns with pitted enamel that is very thin and has a tendency to flake off, causing small diastemas between the teeth. The cusps are small, flattened, and poorly formed, and caries is frequent (9,10,24,36)(Fig. 5–15A–C). The mandibular condyles may be flat or concave. In contrast, in MPS IVB, the enamel appears normal and provides a means to clinically distinguish the two types. Patients with MPS IVB tend to have a wide palate and widely spaced teeth.

Differential diagnosis. Practically all types of short-spine dwarfism have been confused with Morquio syndrome. In contrast to Morquio syndrome, *achondroplasia* is usually apparent at birth, with skeletal changes entirely different from those in MPS IVA. MPS I-H has radiographic similarities to Morquio syndrome during the first few years of life, but mentality is reduced and the gross physical appearance is strikingly characteristic. Although corneal clouding and deafness were thought to be distinguishing factors, they are also features of Morquio syndrome (45). Multiple epiphyseal dysplasia, a dominantly inherited disorder, may also simulate Morquio syndrome, but spinal involvement, if present, is of lesser degree.

*Diastrophic dysplasia* is characterized by progressive kyphoscoliosis, but normal vertebral body height, micromelia, clubfoot, widening of metaphyses and epiphyses of long bones, some limitation of joint motion, and thickened pinnae easily distinguish the conditions.

Children with metatropic dysplasia at first exhibit a long-trunked and later a short-trunked dwarfism. The disorder is characterized by progressive kyphoscoliosis and anisospondyly without hypoplasia of the last thoracic and first lumbar vertebral bodies.

*Spondyloepiphyseal dysplasia congenita,* inherited as a dominant trait, is a short-trunk dwarfism. There is platyspondyly but little or no involvement of the hands and feet, no corneal clouding, and no keratansulfaturia. Myopia is often severe.

Rickets and *hypophosphatasia* should also be considered.

Individuals with *Dyggve–Melchior–Clausen syndrome* somewhat resemble those with Morquio disease in skeletal alterations but do not have corneal clouding, do not excrete KS in the urine, and are mentally retarded. They do not have enamel deficiency. Pointing of the proximal metacarpals does not occur. There is neither hypoplasia of the odontoid process nor of the inferior thoracic or lumbar vertebrae. The iliac crest has a lacy border. The disorder has autosomal recessive inheritance. Since several patients have had normal intelligence, this disorder may have genetic heterogeneity.

Congenital dysplasia of the odontoid process with atlantoaxial dislocation can be seen in a number of disorders: Morquio syndrome, *Aarskog syndrome, Dyggve–Melchior–Clausen syndrome,* pseudoachondroplasia, *cartilage-hair hypoplasia, spondyloepiphyseal dysplasia congenita,* and spondylometaphyseal dysplasia.

Laboratory findings. Marked excretion of KS in the urine in childhood is a constant feature but excretion diminishes markedly by teenage years in MPS IVA. Some younger patients with MPS IVB do not excrete KS (21,34). *N*-Acetylgalactosamine-6-sulfatase may be determined in cultured fibroblasts, leukocytes, cultured amniotic cells, and chorionic villi (46). $\beta$-Galactosidase may be assayed using a *p*-nitrophenyl or 4-methylumbelliferyl $\beta$-galactosides (1).

Marked excretion of KS in the urine, in childhood, is a constant feature of MPS type IVA (11). It slowly decreases, reaching normal levels in adults. Unusually coarse granular inclusions may be found in peripheral neutrophilic granulocytes and fibroblasts exhibit metachromasia. Ultrastructural studies of epiphyseal plates (21) and brain (22) have been described.

### References [Mucopolysaccharidosis Types IVA and IVB (Morquio syndrome)]

1. Arbisser AI et al: Morquio-like syndrome with beta-galactosidase deficiency and normal hexosamine sulfatase activity: Mucopolysaccharidosis IVB. *Am J Med Genet* **1**:195–205, 1977.

2. Beck M et al: Heterogeneity of Morquio disease. *Clin Genet* **29**:325–331, 1986.

3. Bracher M: Chondrodystrophia congenita tarda. *Z Orthop Chir* **58**:503–518, 1933.

4. Blaw ME, Langer LO: Spinal cord compression in Morquio–Brailsford's disease. *J Pediatr* **74**:593–600, 1969.

5. Brailsford JF: Chondro-osteo-dystrophy. *Am J Surg* **7**:404–409, 1929.

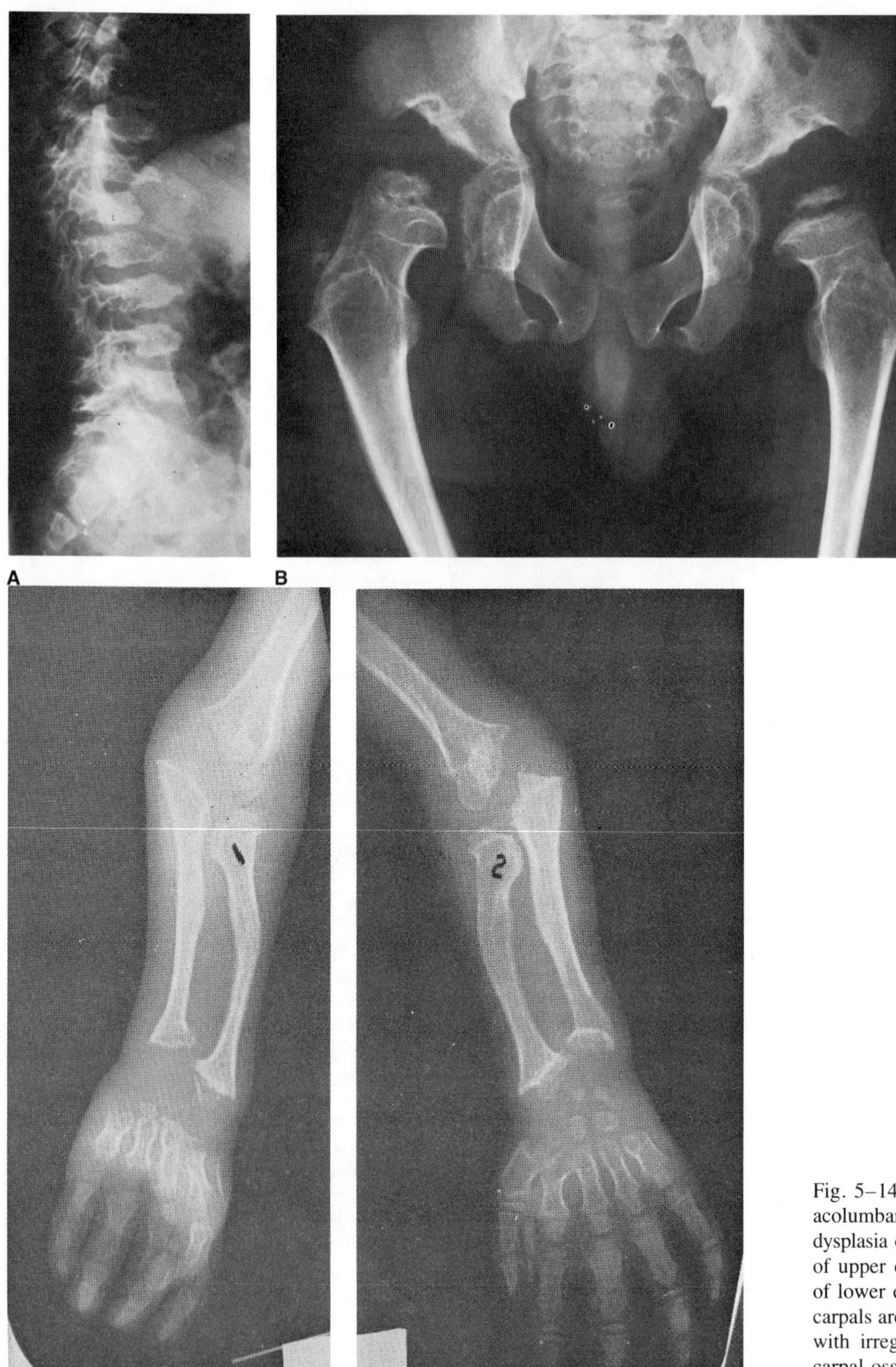


Fig. 5–14. *Morquio syndrome.* (A) Platyspondyly of thoracolumbar vertebrae. (B) Deficiently ossified acetabula, dysplasia of femoral heads, and coxa valga. (C) Long bones of upper extremity are more severely involved than those of lower extremity. The humerus, radius, ulna, and metacarpals are short, coarse, curved, and irregularly tubulated, with irregular epiphyseal plates. Also note deficiency of carpal ossification centers. (From *H Zellweger et al,* J Pediatr **59:**549, 1961.)

6. Dale F: Unusual forms of familial osteochondrodystrophy. *Acta Radiol* **12**:337–358, 1931.

6a. DiFerrante N et al: Deficiencies of glucosamine-6-sulfate or galactosamine-6-sulfate sulfatases are responsible for different mucopolysaccharidoses. *Science* **199**:79, 1978.

7. Edwards MK et al: CT metrizamide myelography of the cervical spine in Morquio syndrome. *AJNR* **3**:666–669, 1982.

8. Fujimoto A, Horowitz AL: Biochemical defect of non-keratan-sulfate excreting Morquio syndrome. *Am J Med Genet* **15**:265–273, 1983.

9. Gardner DG: The dental manifestations of the Morquio syndrome (MPS type IV): A diagnostic aid. *Am J Dis Child* **129**:1445–1448, 1975.

10. Garn SM, Hurme VO: Dental defects in three sibings afflicted with Morquio's disease. *Br Dent J* **93**:210–212, 1952.

11. Glössl J et al: Different properties of residual N-acetylgalactosamine-6-sulfate sulfatase in fibroblasts from mild and severe forms of Morquio disease type A. *Pediatr Res* **15**:976–978, 1981.

12. Glössl J et al: Partial deficiency of glycoprotein neuraminidase in some patients with Morquio disease type A. *Pediatr Res* **18**:302–305, 1984.

13. Groebe H et al: Morquio syndrome (mucopolysaccharidosis IVB) associated with beta-galactosidase deficiency. Report of two cases. *Am J Hum Genet* **32**:258–272, 1980.

14. Guibaud P et al: Morquio syndrome moderated by beta galactosidase

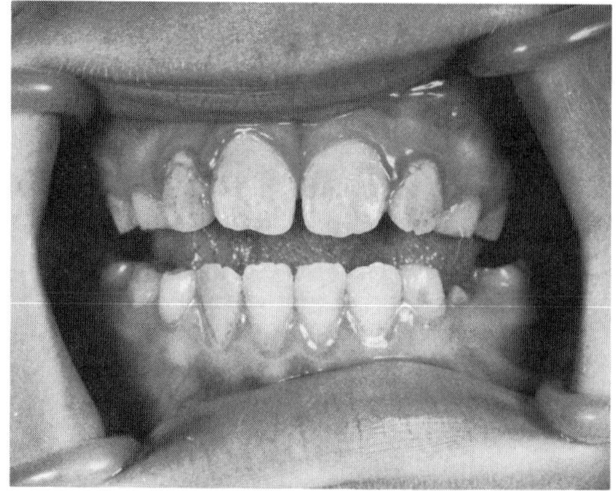

A

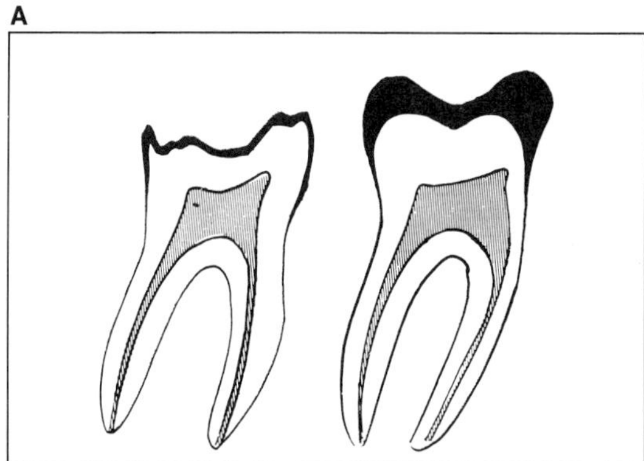

B

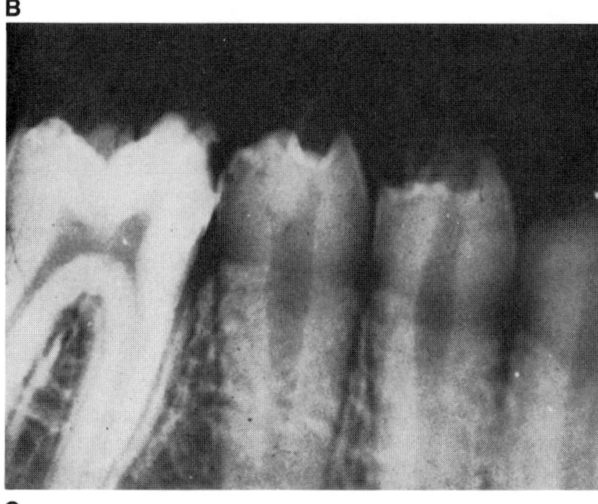

C

Fig. 5–15. *Morquio syndrome.* (A) Hypoplasia of enamel. (B) Diagram of lower molar showing reduced thickness of enamel layer in affected tooth (left) compared with corresponding tooth in normal sibling. (C) Dental roentgenogram showing reduced enamel thickness. (B,C from *SM Garn* and *VO Hurme,* Br Dent J **93:**210, 1952.)

deficiency. Mucopolysaccharidosis (type IV B) or oligosaccharidosis. *Ann Pédiatr* **30**:681–686, 1983.

15. Hecht JT et al: Mild manifestations of the Morquio syndrome. *Am J Med Genet* **18**:369–371, 1984.

16. Hochheim W et al: Beitrag zu den polytopen, erblichen, enchondralen Dysostosen. *Arch Orthop Unfallchir* **47**:463–480, 1955.

17. Holzgreve W et al: Morquio syndrome: Clinical findings in 11 patients with MPS IVA and two patients with MPS IVB. *Hum Genet* **57**:360–365, 1981.

18. Hoogeveen AT et al: Processing of human beta galactosidase in $G_{M1}$-gangliosidosis and Morquio B syndrome. *J Biol Chem* **259**:1974–1977, 1984.

19. Hopwood JJ, Elliott H: Detection of Morquio A syndrome using radiolabelled substrates derived from keratan sulphate for the estimation of galactose 6-sulphate sulphatase. *Clin Sci* **65**:325–332, 1983.

20. Horwitz AL, Dorfman A: The enzymatic defect in Morquio's disease: The specificity of N-acetylhexosamine sulfatases. *Biochem Biophys Res Commun* **80**:819, 1978.

21. Jenkins P et al: Morquio–Brailsford disease. A report of four affected sisters with absence of excessive keratan sulfate in the urine. *Br J Radiol* **46**:668–675, 1973.

22. Koto A et al: The Morquio syndrome: Neuropathology and biochemistry. *Ann Neurol* **4**:26–56, 1978.

23. Langer LO, Carey LS: The roentgenographic features of the KS mucopolysaccharidosis of Morquio disease. *Am J Roentgenol* **97**:1–20, 1966.

24. Levin LS et al: Oral findings in the Morquio syndrome (MPS IV). *Oral Surg* **39**:390–395, 1975.

25. Matalon R et al: Morquio's syndrome: Deficiency of a chondroitin sulfate N-acetylhexosamine sulfate sulfatase. *Biochem Biophys Res Commun* **61**:759–765, 1974.

26. Maroteaux P et al: Heterogeneite des formes frustes de la maladie de Morquio. *Arch Fr Pédiatr* **39**:761–765, 1982.

27. Maynard JA et al: Morquio's disease (mucopolysaccharidosis type IV), ultrastructure of epiphyseal plates. *Lab Invest* **28**:194–205, 1973.

28. McClure J et al: The histological and ultrastructural features of the epiphyseal plate in Morquio type A syndrome (mucopolysaccharidosis type IVA). *Pathology* **18**:217–221, 1986.

29. McKusick VA: *Heritable Disorders of Connective Tissue,* 4th ed. Mosby St. Louis, 1972.

30. Michalski JC et al: The structure of six urinary oligosaccharides that are characteristic for a patient with Morquio syndrome type B. *Carbohydrate Res* **1**:351–363, 1982.

31. Morquio L: Sur une forme de dystrophie osseuse familiale. *Arch Méd Enf* **32**:129–140, 1929.

32. Mutoh T et al: Atypical adult $G_{M1}$-gangliosidosis: Biochemical comparison with other forms of primary beta-galactosidase deficiency. *Neurology* **36**:1237–1241, 1986.

32a. Nelson J et al: Clinical findings in 12 patients with MPS IVA (Morquio disease). *Clin Genet* **33**:110–130, 1988.

33. O'Brien JS et al: Spondyloepiphyseal dysplasia, corneal clouding, normal intelligence, and acid β-galactosidase deficiency. *Clin Genet 9*:495–504, 1976.

34. Norman ME: Two brothers with nonkeratan sulfate-excreting Morquio syndrome. *Birth Defects* **10**(12):467–469, 1974.

35. Orii T et al: Late onset N-acetylgalactosamine-6-sulfate sulfatase deficiency in two brothers. *Conn Tissue* **13**:169–175, 1981.

36. Sela M et al: Oral manifestations of Morquio's syndrome. *Oral Surg* **39**:583–589, 1975.

37. Singh J et al: N-Acetylgalactosamine 6-sulfate sulfatase in man: Absence of the enzyme in Morquio disease. *J Clin Invest* **57**:1036–1040, 1976.

38. Spranger J: The systemic mucopolysaccharidoses. *Ergeb Inn Med Kinderheilkd* **32**:166–265, 1972.

39. Spranger J: Beta-galactosidase and the Morquio syndrome. *Am J Med Genet* **1**:207–209, 1977.

40. Stevenson RE et al: β-Galactosidase deficiency: Prolonged survival in three patients following central nervous system deterioration. *Clin Genet* **13**:305–313, 1978.

41. Taylor HA et al: Beta-galactosidase deficiency: Studies of two patients with prolonged survival. *Am J Med Genet* **5**:235–245, 1980.

42. Trojak JE et al: Morquio-like syndrome (MPS IVB) associated with deficiency of a β-galactosidase. *Johns Hopkins Med J* **146**:75–79, 1980.

43. Van der Horst GT et al: Morquio B syndrome: A primary defect in beta-galactosidase. *Am J Med Genet* **16**:261–275, 1983.

44. van Gemund JJ et al: Morquio B disease, spondyloepiphyseal dysplasia associated with acid β-galactosidase deficiency. Report of three cases in one family. *Hum Genet* **64**:50–54, 1983.

45. Von Noorden GK et al: Ocular findings in the Morquio–Ullrich's disease. *Arch Ophthalmol* **64**:581–591, 1960.

46. Yven M, Fensom AH: Diagnosis of classical Morquio's disease: N-acetylgalactosamine 6-sulfate sulfatase activity in cultured fibroblasts, leukocytes, amniotic cells and chorionic villi. *J Inherit Metab Dis* **8**:80–86, 1985.

**Mucopolysaccharidosis VI (Maroteaux–Lamy syndrome).** In 1963, Maroteaux et al (18) described a patient with a moderately severe Hurler-like phenotype but normal intelligence and high urinary excretion of DS.

Three forms of mucopolysaccharidosis VI exist: a mild type (25,40), an intermediate type, and a severe type (21,27,28,33,34). There may even be a very mild form (31). Children with the mild type develop reasonably well until about 6 years of age when short stature, corneal

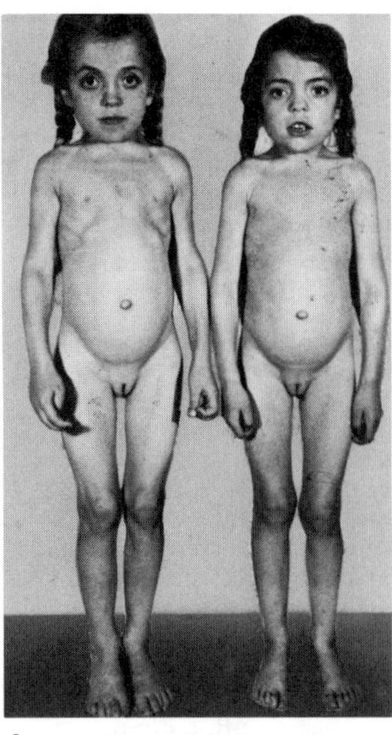

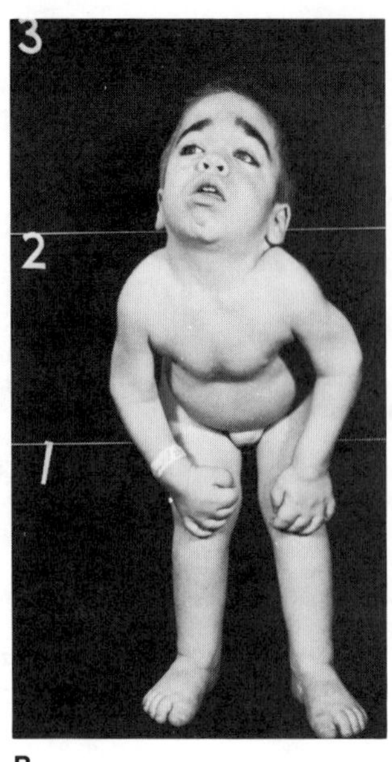


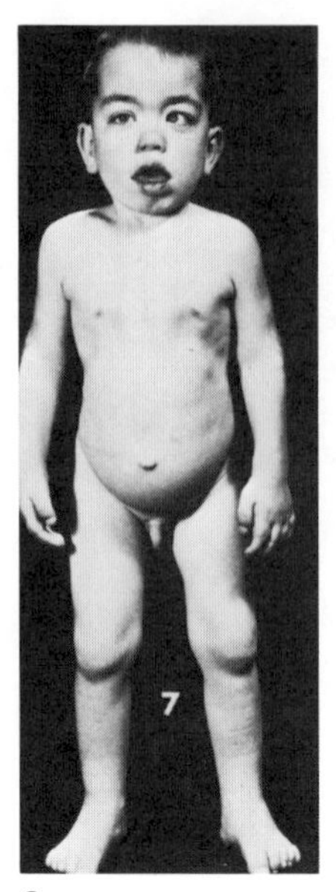


A B C

Fig. 5–16. *Maroteaux–Lamy syndrome.* (A) Seven- and 6-year-old patients with mild form of MPS V. (B) Seven-year-old male with severe form. (C) Seven-year-old patient with severe form. (A courtesy of *H-R Wiedemann,* Kiel, Germany. B courtesy of *LO Langer,* Minneapolis, Minnesota. C from *DA Stumpf et al,* Am J Dis Child **126:**747, 1973).

Fig. 5–17. *Maroteaux–Lamy syndrome.* (A–D) Note coarse facies, large head, hypertelorism, flat nasal bridge with arched nostrils, fissured tongue, anterior sternal protrusion, genua valga, protruding abdomen, and growth retardation. (Courtesy of *VA McKusick,* Baltimore, Maryland.)

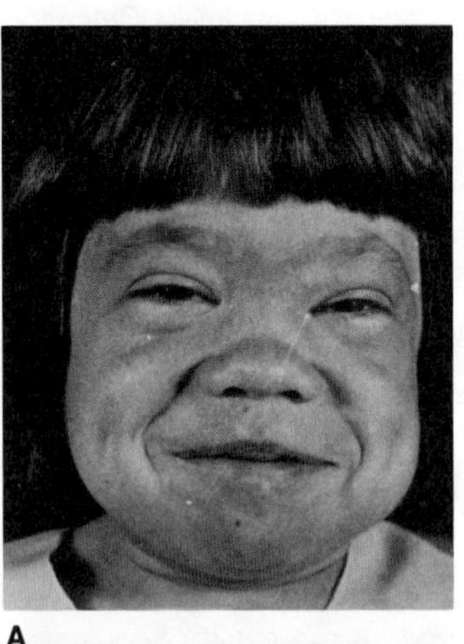
A

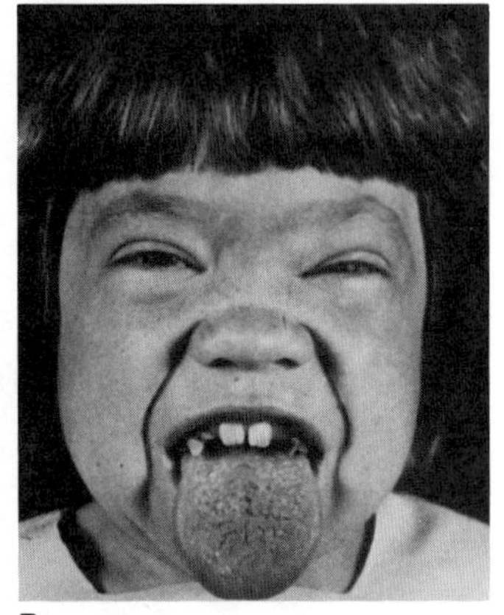
B

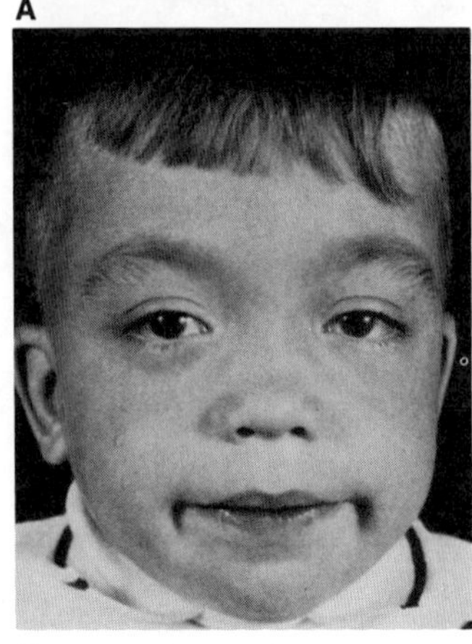
C

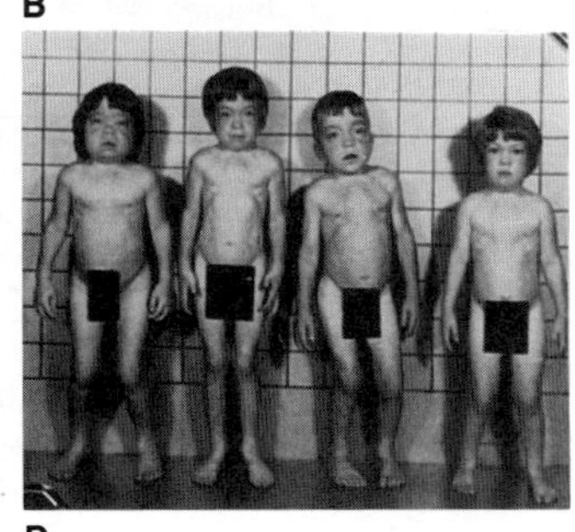
D

clouding and spinal deformities are noted. Legg–Perthes-like disease of the hips and aortic stenosis become apparent. The patients usually survive to adulthood (22,26). In patients with the severe type, morphologic changes are noted in early childhood and the disease progresses more rapidly to a state of severe disability with strikingly short stature, coarse facial appearance, hyperextended head, musculoskeletal abnormalities, severe corneal clouding, markedly reduced hearing, and prominent cardiac defects that frequently lead to death in adolescence (Figs. 5–16–5–19).

MPS VI has autosomal recessive inheritance, the three types being the result of allelic mutations in the structural gene encoding arylsulfatase B. The gene has been mapped to chromosome 5 (6). An animal model of MPS VI in Siamese and other cats has been extensively described and characterized (9, 10, 12, 19). The clinical and biochemical features are caused by abnormal intracellular accumulation of DS in mesenchymal cells and, secondarily, in parenchymal cells of internal organs, such as the liver. Although *N*-acetylgalactosamine-4-sulfate residues are also present on chondroitin-4-sulfate, there is no evidence for the accumulation of this GAG in MPS VI individuals. There are no estimates on the frequency of MPS VI in any population.

Facies. A prominent forehead may be noted at birth. The facies, similar to that in Hurler syndrome, with apparent hypertelorism, depressed nasal bridge, full cheeks and lips, relatively broad jaws, large cranium, and abundant eyebrows and scalp hair, becomes evident about the sixth year of life, occasionally earlier (36). Marked corneal opacity is regularly present (8, 21, 34, 35) (Figs. 5-16–5-19).

Musculoskeletal system. Adult height is usually 110–140 cm but those having the mild form may near 168 cm (24). The chest is deformed, with a prominent sternum. Multiple joint contractures and clawhand deformity begin after the first year of life.

Genua valga, lumbar kyphosis, and sternal protrusion are common. The radiographic changes in the severe type are similar to those of MPS I-H. Ossification of the superior portion of the femoral capital epiphysis may be markedly defective. In the mild type, there are cra-

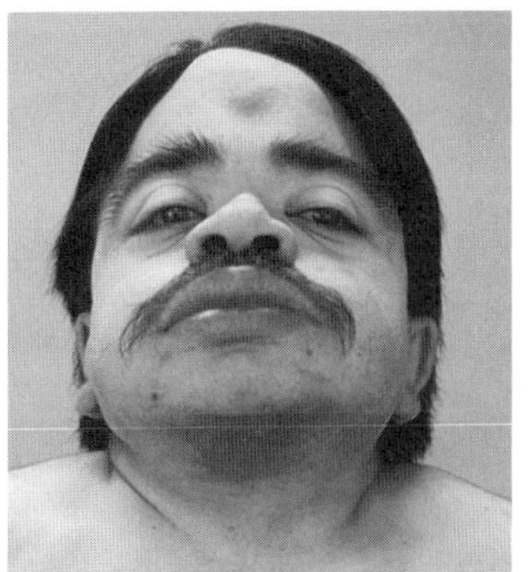
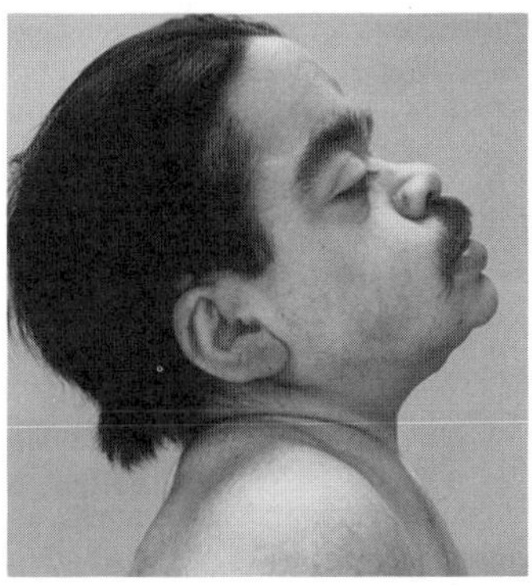

A B

Fig. 5–18. *Maroteaux–Lamy syndrome.* (A,B) Severe form. Note head hyperextension. (Courtesy of *S Rampini,* Zurich, Switzerland.)

nial changes, wide ribs, and pelvic dysplasia but few changes in spine and tubular bones (33) (Fig. 5-19).

Hernias are common in the severe form.

Other findings. Hepatomegaly is almost invariably present in the severe form. The spleen is enlarged in about half the cases. Cardiovascular involvement is common with aortic stenosis, mitral valve regurgitation, and narrowing of the coronary and other arteries (3,32,33,39).

Hearing defects, both conductive and sensorineural, may be detected audiometrically. Mentation is nearly always normal, with rare exception (38). However, impaired vision and hearing, restricted mobility, and secondary psychologic reaction may impede intellectual performance (33–35). Neurologic deficits most frequently include hydrocephalus, peripheral nerve compression (e.g., carpal tunnel syndrome), or hypoplasia of odontoid process associated with atlantoaxial subluxation. Myelopathy and radiculopathy have also been reported (13,23,24,40). Marked tracheal stenosis has been documented (13a,27).

Oral manifestations. The tongue becomes large as soon as the clinical picture is fully developed. The teeth are frequently widely spaced. Eruption of permanent molar teeth is retarded. Some are deeply buried, angulated in the mandible, and surrounded by radiolucent bony defects that we have assumed represented the accumulation of dermatan sulfate in hyperplastic follicles as in MPS I-H and in MPS II-B. Study of the follicular fluid in MPS-VI has shown it to be composed of hyaluronic acid (29).

Differential diagnosis. See Table 5-1.

Laboratory findings. There is an abundance of coarse, dense inclusions in granulocytes, monocytes, and of a large proportion of lymphoctyes in peripheral blood smears. Bone marrow preparations exhibit coarse inclusions in reticulohistiocytes, granulocytes, and their precursors. Biopsy may be easily done on the conjunctiva (17). Electron microscopic studies have shown numerous electrolucent vacuoles in cells from brain, liver, lung, and skin (15,16). Large quantities of DS are excreted in the urine, but the level decreases with age. Abnormal amounts of $^{35}$S-labeled GAGs are stored in cultured fibroblasts (1). Liver, kidney, spleen, brain, and fibroblast cultures from homozygotes and heterozygotes exhibit about 7 and 40% enzyme activity for arylsulfatase B, respectively (2,35). Heterozygote testing has been difficult, but a method has been reported that uses ratios of arylsulfatase B to another lysosomal sulfatase, arylsulfatase A, to minimize the overlap between normal and heterozygote populations.

The condition has been diagnosed prenatally (37) by assay of arylsulfatase B activity and [$^{35}$S]sulfate incorporation in cultured amniotic fluid cells. Radiolabeled oligosaccharides may be used to determine arylsulfatase B activity (11). Liquid chromatographic analysis has also been employed (14). Diagnosis from chorionic villi may not be accurate (30).

### References [Mucopolysaccharidosis VI (Maroteaux–Lamy syndrome)]


1. Barton RW, Neufeld EF: A distinct biochemical deficit in the Maroteaux–Lamy syndrome (mucopolysaccharidosis VI). *J Pediatr* **80**:114–116, 1972.
2. Beratis NG et al: Arylsulfatase B deficiency in Maroteaux–Lamy syndrome: Cellular studies and carrier identification. *Pediatr Res* **9**:475–480, 1975.
3. Betremieux P et al: Insuffisance cardiaque aiguë chez une fillette attiente de maladie de Maroteaux–Lamy. *Ann Pédiatr* **32**:639–641, 1985.
4. Black SH et al: Maroteaux–Lamy syndrome in a large consanguineous kindred: Biochemical and immunological studies. *Am J Med Genet* **25**:273–279, 1986.
5. DiFerrante N et al: Mucopolysaccharidosis VI (Maroteaux–Lamy disease). Clinical and biochemical study of a mild variant case. *Johns Hopkins Med J* **135**:42–54, 1974.
6. Fidzianska E et al: Assignment of the gene for human arylsulfatase B, ARSB, to chromosome region 5p11–5qter. *Cytogenet Cell Genet* **38**:150–151, 1984.
7. Fluharty AL: The mucopolysaccharidoses: A synergism between clinical and basic investigation. *J Invest Dermatol* **79**(Suppl):38s–48s, 1982.
8. Goldberg MF et al: Hydrocephalus and papilledema in the Maroteaux–Lamy syndrome. *Am J Ophthalmol* **69**:969–974, 1970.
9. Haskins ME et al: Mucopolysaccharidosis in a domestic short-haired cat—


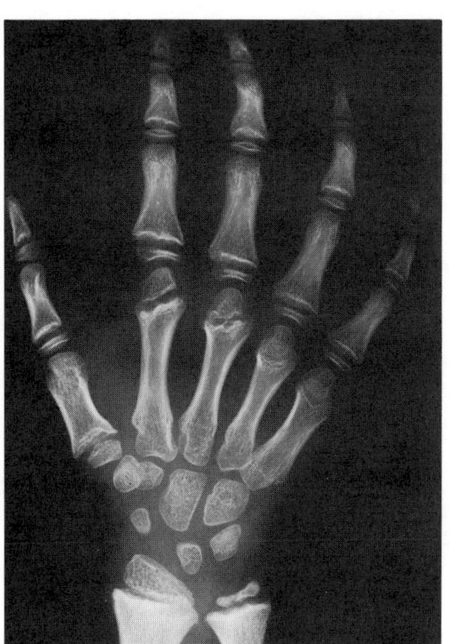
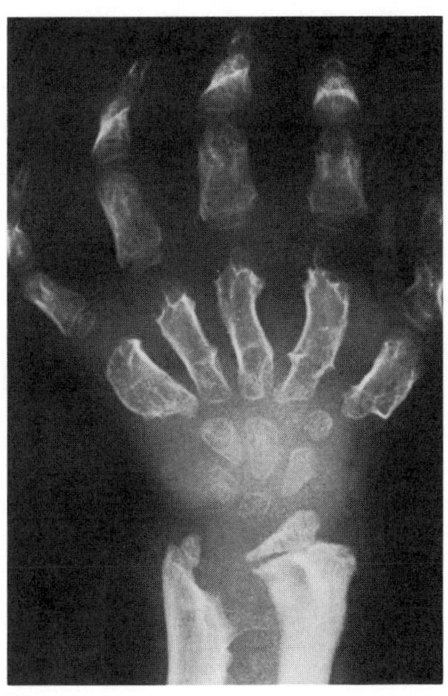

A B

Fig. 5–19. *Maroteaux–Lamy syndrome.* Roentgenograms. (A) Mild involvement of short tubular bones in 11-year-old. (B) Severe shortening and distortion of short tubular bones with marked epiphyseal and metaphyseal dysplasia in 14-year-old.

a disease distinct from that seen the Siamese cat. *J Am Vet Med Assoc* **175**:384, 1979.

10. Haskins ME et al: Mucopolysaccharide storage disease in three families of cats with arylsulfatase B deficiency: Leukocyte studies and carrier identification. *Pediatr Res* **13**:1203–1210, 1979.

11. Hopwood JJ et al: Diagnosis of Maroteaux–Lamy syndrome by the use of radiolabeled oligosaccharides as substrates for the determination of arylsulphatase B activity. *Biochem J* **234**:507–514, 1986.

12. Jezyk PF et al: Mucopolysaccharidosis in a cat with arylsulfatase B deficiency: A model of Maroteaux–Lamy syndrome. *Science* **198**:834–836, 1977.

13. Kaufman HH et al: Cervical myelopathy due to dural compression in mucopolysaccharidosis. *Surg Neurol* **17**:404–410, 1982.

13a. Keller C et al: Mukopolysaccharidose Typ VI-A (Morbus Maroteaux–Lamy): Korrelation der klinischen und pathologisch—anatomischen Befunde bei einem 27 jährigen Patienten. *Helv Paediatr Acta* **42**:317–333, 1987.

14. Kodama C et al: Liquid chromatographic determination of urinary glycosaminoglycans for differential diagnosis of genetic mucopolysaccharidoses. *Clin Chem* **32**:30–34, 1986.

15. Levy LA et al: Ultrastructure of Reilly bodies (metachromatic granules) in the Maroteaux–Lamy syndrome (mucopolysaccharidosis VI): A histochemical study. *Am J Clin Pathol* **73**:416–422, 1980.

16. Liberti J: Diagnosis of lysosomal storage diseases by the ultrastructural study of conjunctival biopsies. *Pathol Ann* **15**(Part 1):37–66, 1980.

17. Markesbery WR et al: Mucopolysaccharidosis: Ultrastructure of leukocyte inclusions. *Ann Neurol* **8**:332–335, 1980.

18. Maroteaux M et al: Une nouvelle dysostose avec élimination urinaire de chondroitine-sulfate B. *Presse Méd* **71**:1849–1852, 1963.

19. McGovern MM et al: Animal model studies of allelism: Characterization of arylsulfatase B mutations in homoallelic and heteroallelic (genetic compound) homozygotes with feline mucopolysaccharidosis VI. *Genetics* **110**:733–749, 1985.

20. McGovern MM et al: Bone marrow transplantation in Maroteaux–Lamy syndrome (MPS type VI): Status 40 months after BMT. *Birth Defects* **22**(1):41–53, 1986.

21. McKusick VA, Neufeld EF: The mucopolysaccharide storage diseases, in *The Metabolic Basis of Inherited Diseases,* Stanbury JB, Wyngaarden JB, Fredrickson DS, Goldstein JL, Brown MS (eds), 5th ed, McGraw-Hill, New York, 1982, pp 751–777.

22. Paterson DE et al: Maroteaux–Lamy syndrome mild form—MPS VIb. *Br J Radiol* **55**:805–812, 1982.

23. Peterson DI et al: Myelopathy associated with Maroteaux–Lamy syndrome. *Arch Neurol* **32**:127–129, 1975.

24. Pilz H et al: Deficiency of arysulfatase B in two brothers aged 40 and 38 years (Maroteaux–Lamy syndrome type B). *Ann Neurol* **6**:315–325, 1979.

25. Poser C et al: MGH CPC case 44-1983. *N Engl J Med* **309**:1109–1117, 1983.

26. Quigley HA, Kenyon KR: Ultrastructural and histochemical studies of a newly recognized form of systemic mucopolysaccharidosis (Maroteaux–Lamy syndrome, mild phenotype). *Am J Ophthalmol* **77**:809–818, 1974.

27. Rampini S et al: Mukopolysaccharidose VI-A (Morbus Maroteaux–Lamy, schwere Form). *Helv Paediatr Acta* **41**:515–530, 1986.

28. Riggio S et al: Radiologic aspects of a severe form of Maroteaux–Lamy syndrome. *Radiol Med (Torino)* **70**:629–630, 1984.

29. Roberts MW et al: Occurrence of multiple dentigerous cysts in a patient with the Maroteaux–Lamy syndrome (mucopolysaccharidosis, type VI). *Oral Surg* **58**:169–175, 1984.

30. Sanguinetti N et al: The arylsulfatases of chorionic villi: Potential problems in the first trimester diagnosis of metachromatic leucodystrophy and Maroteaux–Lamy disease. *Clin Genet* **30**:302–308, 1986.

31. Saul RA et al: Atypical presentation with normal stature in Maroteaux–Lamy syndrome (MPS VI). *Proc Greenwood Genet Ctr* **3**:49–52, 1984.

32. Schieken RM et al: Cardiac manifestations of the mucopolysaccharidoses. *Circulation* **52**:700–705, 1975.

33. Spranger J et al: Mucopolysaccharidosis VI (Maroteaux–Lamy disease). *Helv Paediatr Acta* **25**:337–362, 1970.

34. Spranger J: The systemic mucopolysaccharidoses. *Ergeb Inn Med Kinderheilkd* **32**:165–265, 1972.

35. Stürmer J: Mucopolysaccharidose type VI-A (Morbus Maroteaux–Lamy syndrome). *Klin Mbl Augenheilk* **194**:273–281, 1989.

36. Van Biervliet JP et al: Un cas de maladie de Maroteaux–Lamy decouvert precocement. *Arch Fr Pédiatr* **34**:362–370, 1977.

37. Van Dyke DL et al: Prenatal diagnosis of Maroteaux–Lamy syndrome. *Am J Med Genet* **8**:235–242, 1981.

38. Vestermark S et al: Mental retardation in a patient with Maroteaux–Lamy. *Clin Genet* **31**:114–117, 1987.

39. Wilson CS et al: Aortic stenosis and mucopolysaccharidosis. *Ann Intern Med* **92**:496–498, 1980.

40. Young R et al: Compressive myelopathy in Maroteaux–Lamy syndrome: Clinical and pathologic findings. *Ann Neurol* **8**:336–340, 1980.

**Mucopolysaccharidosis VII (Sly syndrome, β-glucuronidase deficiency).** MPS VII, resulting from β-glucuronidase deficiency, was first described by Sly et al (19) in 1973. The disease is characterized by short stature, hepatosplenomegaly, progressive dysostosis multiplex, and progressive mild mental retardation after the age of 2 years. Since the original description, about a dozen examples have been reported (1–9,11–18,20,21). These appear to constitute a clinical heterogeneity (10) ranging from rather severely affected patients (Type I) (1–3,12,16,17), to those moderately affected (Type II) (5,8,18,19), to those mildly affected (Type III) (1,3,6). There may also be a mild adult form. Type I has its onset at birth, Type II at 2 to 3 years of age, and Type III at adolescence (Figs. 5-20 and 5-21). Inheritance is autosomal recessive for each type. The gene for MPS VII is at 7q11.2–7q22 (4,21). A canine model has been described (10). Deficiency of lysosomal β-glucuronidase activity in fibroblasts, leukocytes, and most tissues leads to an inability to degrade DS and HS, the predominant GAGs containing β-linked glucuronic acid residues, and results in their lysosomal accumulation and urinary excretion.

Facies. Rarely, in the severe form, the disorder may present a nonimmune hydrops fetalis (12,16). In Type I and, to some degree, in Type II, there are moderate Hurler-like changes, with hypertelorism, depressed nasal bridge, prominent alveolar processes, and anteverted nostrils (11,20). The corneas appear cloudy in Type I, but clear in Types II and III.

Skeletal system. In Type I and, to some degree in Type II, short stature becomes apparent in the second year of life, height falling below the 3rd percentile. The head is large, with frontal prominence and premature closure of the sagittal and lambdoidal sutures. Pectus excavatum or carinatum and thoracolumbar gibbus, already noted in infancy, increase with age in Types I and II. Talipes and hernia also occur in Types I and II patients. Type III patients may present with mild kyphosis or scoliosis at puberty.

Radiographically, in Types I and II, there are moderately severe changes of dysostosis multiplex with premature closure of cranial sutures, J-shaped sella, oar-shaped ribs, hook-like deformities of the lower thoracic and upper lumbar vertebrae, underdevelopment of the basilar portions of the ilia, aseptic necrosis of femoral heads, shortening of tubular bones, and proximal pointing of metacarpals II–V. There is no dysostosis multiplex in Type III patients.

Other findings. Hepatosplenomegaly, inguinal and/or umbilical hernia, and developmental retardation are present after the age of 2 years in Type I and II patients, but psychomotor retardation is evident only in Type I patients. Recurrent pulmonary infections are common.

*Oral manifestations.* Widened alveolar ridges have been described.

Differential diagnosis. See Table 5-10.

Laboratory findings. Coarse metachromatic inclusions are present in peripheral granulocytes, granulocyte precursors in bone marrow, and in cultured fibroblasts (9). Ultrastructurally, there are clear vacuoles and granular inclusions in nearly all granulocytes and mononuclear cells in MPS VII (15,16a).

The definitive diagnosis can be made by demonstration of markedly deficient β-glucuronidase activity in serum, leukocytes, and cultured skin fibroblasts. Intermediate activity is round in heterozygotes (7,9).

Prenatal diagnosis can be accomplished by demonstrating deficiency of β-glucuronidase deficiency in chorionic villi or cultured amniocytes (14).

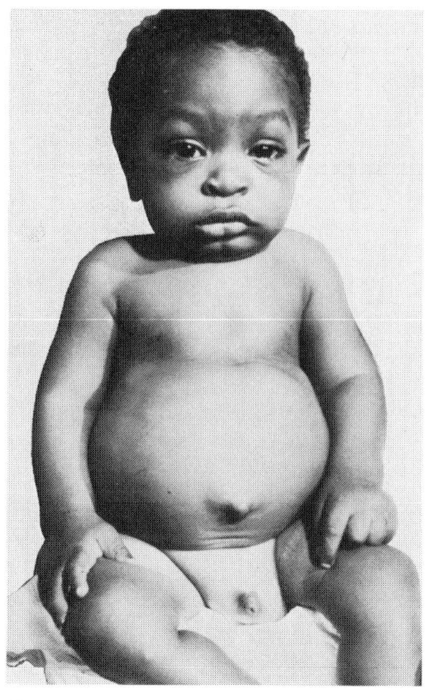

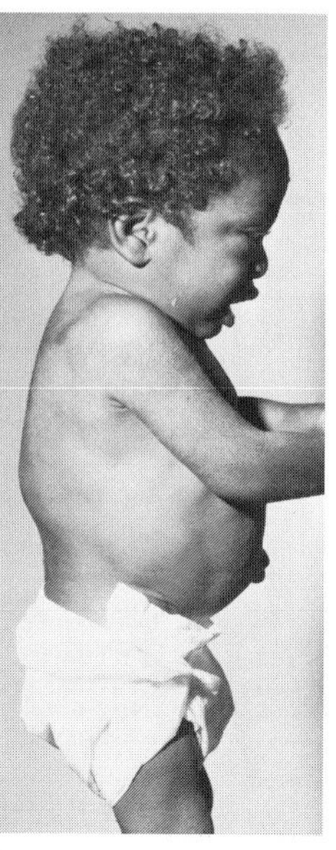

A B
Fig. 5–20. *Sly syndrome, β-glucuronidase deficiency.* (A,B) Coarse facies, pot belly, gibbus. (From *WS Sly et al,* J Pediatr **82**:249, 1973.)

### References [Mucopolysaccharidosis VII (Sly syndrome, β-glucuronidase deficiency)]


1. Beaudet AL et al: Variation in the phenotypic expression of β-glucuronidase deficiency. *J Pediatr* **86**:388–394, 1975.
2. Capdeville R et al: A new case of mucopolysaccharidosis type VII with major skeletal abnormalities. *Ann Pédiatr* **30**:689–692, 1983.


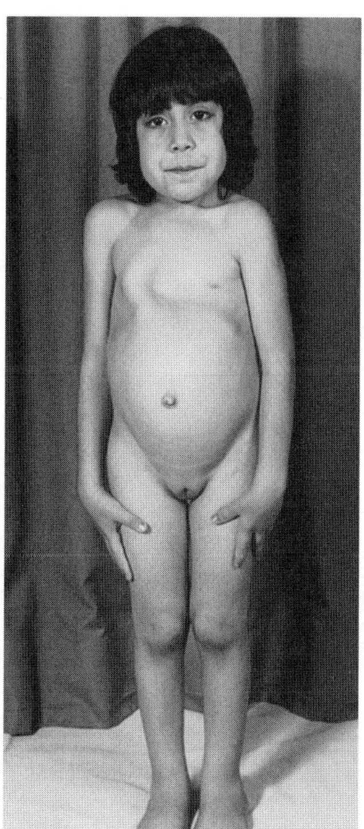

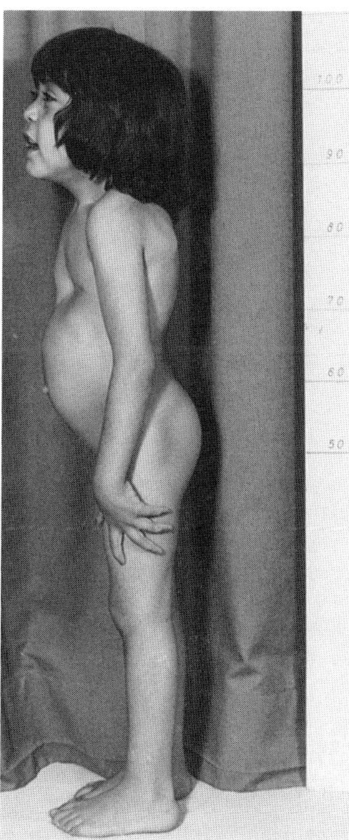

A B
Fig. 5–21. *Sly syndrome, β-glucuronidase deficiency.* (A,B) Six-year-old showing disproportionate growth retardation. There is right-sided sternal protrusion with resultant pectus excavatum, mild kyphoscoliosis. (From *AC Sewell et al,* Clin Genet **21**:366, 1982).


3. Danes BS, Degnan M: Different clinical and biochemical phenotypes associated with β-glucuronidase deficiency. *Birth Defects* **10**:(12):251–257, 1974.
4. Francke U: The human gene for β-glucuronidase is on chromosome 7. *Am J Hum Genet* **28**:357–362, 1976.
5. Gehler J et al: Mucopolysaccharidosis VII: β-Glucuronidase deficiency. *Humangenetik* **23**:149–158, 1974.
6. Gitzelmann R et al: Unusually mild course of β-glucuronidase deficiency in two brothers (mucopolysaccharidosis VII). *Helv Paediatr Acta* **33**:413–428, 1978.
7. Glaser JH et al: β-Glucuronidase deficiency mucopolysaccharidosis: Methods for enzymatic diagnosis. *J Lab Clin Med* **82**:969–977, 1973.
8. Guibaud P et al: Mucopolysaccharidose type VII: par déficit en β-glucuronidase: etude d'une famille. *J Génét Hum* **27**:29–43, 1979.
9. Hall CW et al: A β-glucuronidase deficiency mucopolysaccharidosis: Studies in cultured fibroblasts. *Arch Biochem Biophys* **155**:32–38, 1973.
10. Haskins ME et al: Beta glucuronidase deficiency in a dog: A model of human mucopolysaccharidosis VII. *Pediatr Res* **18**:980–984, 1984.
11. Hoyme HE et al: Presentation of mucopolysaccharidosis VII (beta-glucuronidase deficiency) in infancy. *J Med Genet* **18**:237–239, 1981.
12. Irani D et al: Postmortem observations on β-glucuronidase deficiency presenting as hydrops fetalis. *Ann Neurol* **14**:486–490, 1983.
13. Lee JES et al: β-Glucuronidase deficiency: A heterogeneous mucopolysaccharide disorder. *Am J Dis Child* **139**:57–59, 1985.
14. Maire I et al: β-Glucuronidase deficiency: Enzyme studies in an affected family and prenatal diagnosis. *J Inherit Metab Dis* **2**:29–34, 1980.
15. Markesbery WR et al: Mucopolysaccharidosis: Ultrastructure of leukocyte inclusions. *Ann Neurol* **8**:332–335, 1980.
16. Nelson A et al: Mucopolysaccharidosis VII (β-glucuronidase deficiency) presenting as a nonimmune hydrops fetalis. *J Pediatr* **101**:574–576, 1982.
16a. Peterson L et al: Mucopolysaccharidosis VII: A morphologic, cytochemical, and ultrastructural study of the blood and bone marrow. *Am J Clin Pathol* **78**:544–548, 1982.
17. Pfeiffer RA et al: Beta-glucuronidase deficiency in a girl with unusual clinical features. *Eur J Pediatr* **126**:155–161, 1977.
18. Sewell AC et al: Mucopolysaccharidosis type VII (β-glucuronidase deficiency): A report of a new case and a survey of those in the literature. *Clin Genet* **21**:366–373, 1982.
19. Sly WS et al: Beta-glucuronidase deficiency: Report of clinial, radiologic, and biochemical features of a new mucopolysaccharidosis. *J Pediatr* **82**:249–257, 1973.
20. Teyssier G et al: Mucopolysaccharidose de type VII a révélation néonatale. *Arch Fr Pédiatr* **38**:603–608, 1981.
21. Ward JC et al: Regional gene mapping of human β-glucuronidase (GUSB) by dosage analysis: Assignment to region 7q11.23-7q21. *Am J Hum Genet* **35**:56A, 1983.


## Oligosaccharidoses and related disorders

This section deals with various oligosaccharidoses and related conditions. Each disorder together with its storage substance and enzymatic defect is listed in Table 5-3.

**$G_{M1}$ gangliosidosis, Type I.** $G_{M1}$ gangliosidosis is an oligosaccharidosis caused by the lysosomal accumulation of $G_{M1}$ ganglioside, asialo-$G_{M1}$ ganglioside, and galactose-containing oligosaccharides (5,6). The disease was first recognized in 1964 by Landing et al (3).

There are three major forms: infantile, juvenile, and adult. However, since only the infantile form is dysmorphic, the latter two types will not be discussed.

The infantile form is manifest shortly after birth. In its full expression, it is characterized by progressive cerebral deterioration, with death usually occurring between 6 months and 2 years of age from bronchopneumonia. The other clinical and radiographic features resemble either I-cell disease or MPS I-H.

$G_{M1}$ gangliosidosis of all types is the consequence of the homozygous state of the mutant gene that produces a functionally deficient acid β-D-galactosidase. Another disorder, Morquio syndrome, Type B (MPS IVB), shares the same enzyme defect.

Inheritance is autosomal recessive (1). The gene has been located at 3p21-cen (10).

Facies. The facial features are coarse at birth, in contrast to the face in MPS I-H, which is normal for the first 6 months of life. Mild

Table 5–3. The oligosaccharidoses and related disorders[a]

| Group designation type | Primary metabolic deficiency | Secondary metabolic features | Chromosomal location |
|---|---|---|---|
| Oligosaccharidoses | | | |
| $G_{M1}$ gangliosidosis | | | |
| Infantile (Type 1) $G_{M1}$ gangliosidosis<br>Juvenile (Type 2) $G_{M1}$ gangliosidosis<br>Adult (Type 3) $G_{M1}$ gangliosidosis<br>Morquio syndrome Type B | β-D-Galactosidase | | 3p21–cen |
| Fucosidosis | α-L-Fucosidase | — | 1p34 |
| Aspartylglucosaminuria | 1-Aspartamido-β-N-GlcNAc-amino hydrolase | Aspartyl-glucosamine | 4q21-qter |
| Mannosidosis | Acid α-D-mannosidase | — | 19p13.3–q12 |
| Sialidosis | | | |
| Neonatal hydropic sialidosis<br>Infantile sialidosis (nephrosialidosis)<br>Childhood dysmorphic sialidosis (Spranger syndrome)<br>Juvenile (adult) normosomatic sialidosis (cherry-red spot myoclonus syndrome) | Glycoprotein sialidase | — | 6p21.3 |
| Galactosialidosis | | | |
| Neonatal hydropic galactosialidosis<br>Late infantile galactosialidosis<br>Juvenile (adult) galactosialidosis | 32-kDa protective glycoprotein | Glycoprotein sialidase and β-D-galactosidase | 20 |
| Recognition marker phosphotransferase deficiencies | | | |
| I-cell disease<br>Pseudo-Hurler polydystrophy | GlcNAc-1 phosphotransferase | Non-membrane-bound hydrolases | |
| Winchester syndrome[b] | ? | Special trisaccharide | ? |
| Related disorders[c] | | | |
| Sialic acid storage disease | | | |
| Infantile type<br>Salla disease (late infantile type) | ? | Free sialic acid | ? |
| Berman syndrome | Ganglioside sialidase? | Ganglioside sialidase | ? |
| Mucosulfatidosis (multiple sulfatase deficiency, Austin syndrome) | Protective factor common sulfatases | Multiple sulfatases | ? |

[a]Modified from J. Leroy, David W. Smith Meeting, Mills College, Oakland, CA, 3–7 August 1988.
[b]Dealt with in Chapter 9.
[c]Not true oligosaccharidoses according to the strictly chemical definition.

macrocephaly with frontal bossing is found in about 60%. The nasal bridge is depressed, the philtrum is prominent, the cheeks are full, and the eyelids puffy. In mild cases, these facial changes are present only in later infancy. Corneal opacities are rarely found, but cherry-red macular spots are detected in most (Fig. 5-22).

**Skeletal system.** Kyphoscoliosis is an early finding. The hands are short and stubby. There are multiple flexion contractures of the joints.

The roentgenographic changes are those of dysostosis multiplex. They appear earlier and are more severe than in MPS I-H. The ribs are wide, and the vertebral bodies are short in their anteroposterior diameter, with convex end plates and hook-shaped deformities at the thoracolumbar junction.

The basilar portions of the ilia are hypoplastic. In young infants, there is periosteal cloaking of the shafts of the long tubular bones. This is not observed in MPS I-H, but is a well-known early finding in patients with I-cell disease. In older infants and in young children, the shafts of the long bones are overtubulated, with irregular contours. The short tubular bones appear swollen, with proximal pointing of the second to fourth metacarpals. Bone trabeculation is coarse (2).

**Central nervous system.** The infants are hypotonic. They suck and swallow poorly, fail to thrive, and never learn to crawl or sit. In addition to gross motor delay, they exhibit seizures, blindness, deafness, and often spastic quadriplegia.

**Other findings.** Hepatosplenomegaly is present; hydrocele is frequent (1).

**Oral manifestations.** The tongue and alveolar processes are enlarged (Fig. 5-23) (3,7,8). There is inadequate documentation but some evidence of accumulation of storage material about unerupted first permanent molar teeth (J. Dorst, personal communication, 1972).

**Differential diagnosis.** Cherry-red macular spots can also be found in Tay–Sachs disease, Sandhoff disease, metachromatic leukodystro-

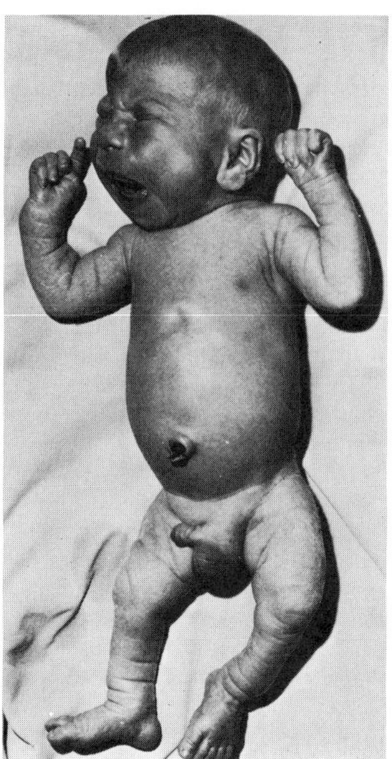

Fig. 5–22. *$G_{M1}$ gangliosidosis, Type I.* Coarse facial features, widened alveolar processes, large ears. Wrist and ankle deformities were apparent at 2 weeks of age. (From *CR Scott et al,* J Pediatr **71:**357, 1967).

phy, infantile Niemann–Pick disease, the sialidoses, and Farber lipogranulomatosis.

Marked dysostosis multiplex-like skeletal anomalies in an infant are more compatible with $G_{M1}$ gangliosidosis or I-cell disease than with a mucopolysaccharidosis.

The differentiation of $G_{M1}$ gangliosidosis Type I from other related disorders is summarized in Table 5-3.

Laboratory findings. Between 10 and 80% of peripheral lymphocytes are vacuolated. Foam cells are found in the bone marrow and in the viscera. Glomerular epithelial cells are vacuolated. The neurons of the central nervous system and retina have a granular appearance because of $G_{M1}$ ganglioside-loaded lysosomes.

Under electron microscopy these appear as whorled and striped "zebra bodies," identical to those seen in Tay–Sachs disease (9). Urinary excretion of MPS is usually normal, although galactose-containing oligosaccharides are markedly elevated. The activity of acid $\beta$-galactosidase is deficient in tissues and body fluids, including lymphocytes, cultured fibroblasts, and urine. Prenatal and heterozygote detection are possible. The gene has been located on the short arm of chromosome 3.

Fig. 5–23. *$G_{M1}$ gangliosidosis, Type I.* Gingival enlargement. (Courtesy of *CR Scott,* Seattle, Washington.)

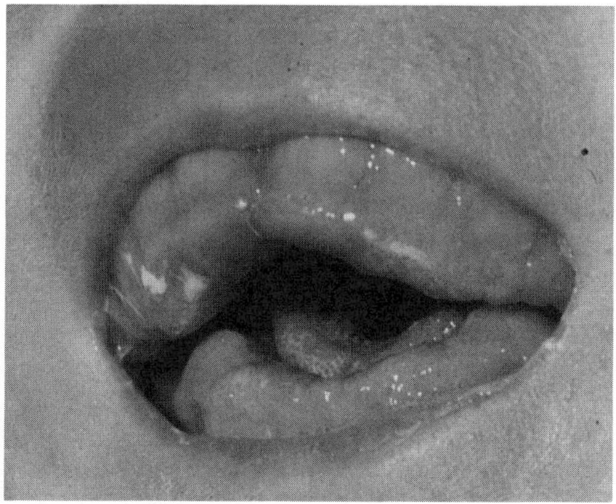

### References ($G_{M1}$ gangliosidosis, Type I)


1. Giugliani R et al: $G_{M1}$ gangliosidosis: Clinical and laboratory findings in eight families. *Hum Genet* **70**:347–354, 1985.
2. Grossman H, Danes BS: Neurovisceral storage disease: Roentgenologic features and mode of inheritance. *Am J Roentgenol* **103**:149–153, 1968.
3. Landing BH et al: Familial neurovisceral lipidosis. *Am J Dis Child* **108**:503–522, 1964.
4. O'Brien JS: Generalized gangliosidosis. *J Pediatr* **75**:167–186, 1969.
5. O'Brien JS: Five gangliosidoses. *Lancet* **2**:805, 1969.
6. O'Brien J: The gangliosidoses, in *The Metabolic Basis of Inherited Disease,* 5th ed, Stanbury JB et al (eds), McGraw-Hill, New York, 1983, pp. 945–969.
7. Sandhoff K, Christomanou H: Biochemistry and genetics of gangliosidoses. *Hum Genet* **50**:107–143, 1979.
8. Scott CR et al: Familial neuro-visceral lipidosis. *J Pediatr* **71**:357–366, 1967.
9. Severi F et al: Infantile $G_{M1}$ gangliosidoses: Histochemical, ultrastructural and biochemical studies. *Helv Paediatr Acta* **26**:192–209, 1971.
10. Sips HJ et al: The chromosomal localization of human $\beta$-galactosidase revisited: A locus for $\beta$-galactosidase on human chromosome 3 and for its protective protein on human chromosome 22. *Hum Genet* **69**:340–344, 1985.


**Fucosidosis.** Fucosidosis is an oligosaccharidosis caused by the abnormal intracellular accumulation of fucose-containing glycolipids and oligosaccharides due to lack of $\alpha$-L-fucosidase. It was first described by Durand et al in 1966 (4).

Fucosidosis has autosomal recessive inheritance. It appears to be more common in southern Italians, New Mexican Spanish-Americans, and Navajo (3,5,13a).

The gene is located at chromosome 1p34 (3a,7). Originally thought to be heterogeneous (1,6,8–10,14,16,18), there is such variability of expression within the same family that Clericuzio (3) and Willems (19) suggested that heterogeneity was dubious.

Facies. The facial features become progressively coarse with large lips, periorbital puffiness, and frontal bossing. There is some resemblance to patients with MPS III. The hair, however, is not as coarse. The tongue may be large. The corneas are clear and the fundi unremarkable.

Skeletal system. Patients exhibit small stature and mild dysostosis multiplex. The vertebral bodies are initially ovoid with subsequent flattening, marginal irregularity, and beaking in lateral projection. There is mild hypoplasia of the supraacetabular portions of the ilia. Bone trabeculation is coarse (2,11,15).

Other findings. There may be hepatosplenomegaly and cardiomegaly. Tortuous conjunctival vessels and bull's eye retinopathy have been reported (17). Hernias may be present. Recurrent respiratory infections regularly occur.

Oral findings. Gingival and labial telangiectasia (Fig. 5-24) has been reported (10,13).

Laboratory findings. Peripheral lymphocytes are vacuolated. Skin biopsy shows deposits of homogeneous eosinophilic material between the dermis and epidermis. Electron microscopic studies reveal numerous membrane-bound vacuoles in all tissues. Urinary excretion of MPS is normal, but there is excess excretion of fucosyl-containing oligosaccharides. There is deficient activity of the lysosomal enzyme, $\alpha$-L-fucosidase, in tissues, in cultured fibroblasts, leukocytes, serum, and urine.

Sweat chlorides are increased from 2- to 5-fold (10).

Prenatal diagnosis has been accomplished (5,12).

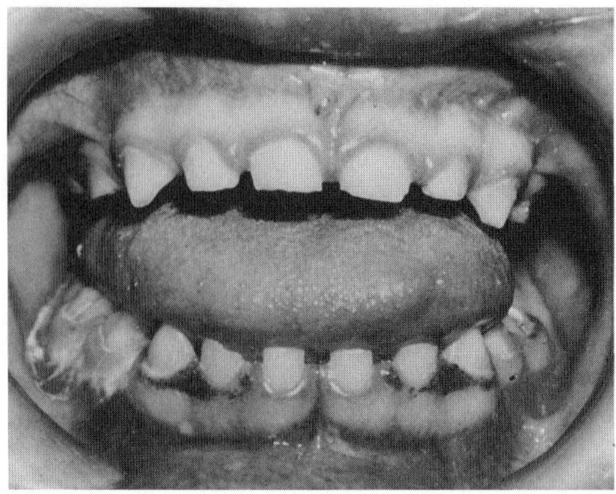

Fig. 5–24. *Fucosidosis.* Telangiectatic lesions of gingiva. (From *DE Prindiville* and *D Stern,* J Oral Surg **34:**603, 1967.)

### References (Fucosidosis)


1. Borrone C et al: Fucosidosis: Clinical, biochemical, immunologic and genetic studies in two new cases. *J Pediatr* **84**:727–780, 1974.
2. Brill PW et al: Roentgenographic findings in fucosidosis type 2. *Am J Roentgenol* **124**:75–82, 1975.
3. Clericuzio C: Phenotypic variability of fucosidosis: Review and six new cases. David W. Smith Workshop, Oakland, CA, August 3–10, 1988.

3a. Darby JK et al: Restriction analysis of the structural α-L-fucosidase gene and its linkage to fucosidosis. *Am J Hum Genet* **43**:749–755, 1988.

4. Durand P et al: New mucopolysaccharide lipid storage disease? *Lancet* **2**:1313–1314, 1966.
5. Durand P et al: Detection of carriers and prenatal diagnosis for fucosidosis in Calabria. *Hum Genet* **51**:195–201, 1979.
6. Epinette WW et al: Angiokeratoma corporis diffusum with α-L-fucosidase deficiency. *Arch Dermatol* **107**:754–757, 1973.
7. Fowler ML et al: Chromosome 1 localization of the human alpha-L-fucosidase structural gene with a homologous site on chromosome 2. *Cytogenet Cell Genet* **43**:103–108, 1986.
8. Honjoh M et al: Fucosidosis type 3 with angiokeratoma corporis diffusum. *J Dermatol* **12**:174–182, 1985.
9. Kornfeld M et al: Fucosidosis with angiokeratoma. *Arch Pathol Lab Med* **101**:478–485, 1977.
10. Kousseff BG et al: Fucosidosis type 2. *Pediatrics* **57**:205–213, 1976.
11. Lee FA et al: Radiographic features of fucosidosis. *Pediatr Radiol* **5**:204–208, 1977.
12. Poënaru L et al: Prenatal diagnosis of fucosidosis. *Clin Genet* **10**:260–264, 1976.
13. Prindiville DE, Stern D: Oral lesions in fucosidosis. *J Oral Surg* **34**:603–608, 1976.

13a. Sangiorgi S et al: Genetic and demographic characterization of a population with high incidence of fucosidosis. *Hum Hered* **31**:100–105, 1982.

14. Schoonderwaldt HC et al: Two patients with an unusual form of type II fucosidosis. *Clin Genet* **18**:348–354, 1980.
15. Shafer IA et al: Lysosomal bone disease. *Pediatr Res* **5**:391–392, 1971.
16. Smith EB et al: Fucosidosis. *Cutis* **19**:195–197, 1977.
17. Snodgrass MB: Ocular findings in a case of fucosidosis. *Br J Ophthalmol* **60**:508–511, 1976.
18. Sovik O et al: Fucosidosis: Severe phenotype with survival to adult age. *Eur J Pediatr* **135**:211–216, 1980.
19. Willems PJ et al: Intrafamilial variability in fucosidosis. *Clin Genet* **34**:7–14, 1988.


**Aspartylglucosaminuria.** Aspartylglucosaminuria, an oligosaccharidosis in which glycoprotein-derived aspartylglucosamine accumulates in various tissues, was first identified in England by Jenner and Pollitt (10). However, most of the patients have been identified in Finland (1–4). Autio et al (3,4) carried out detailed studies on 57 Finnish patients. A few cases have been reported in the United States (7–9,12) and in Finns living in Norway (5).

Inheritance is autosomal recessive. The basic defect is deficiency of 1-aspartamido-β-*N*-GlcNAc-amino hydrolase. Heterozygotes may be identified and prenatal detection is possible (1,2a). The gene is located at 4q21-qter (2).

Infancy and childhood are usually characterized by recurrent diarrhea and frequent respiratory infections. There is early onset of splenomegaly with abdominal protrusion.

Facies. Remarkable resemblance is noted among the affected. The skull is frequently asymmetric. The features gradually become coarse during childhood. the nasal bridge is broad and low. The nostrils are anteverted and the lips thickened. Mild hypertelorism, epicanthal folds, and crystal-like lens opacities have been present in about half the cases. The facial skin, particularly that of the eyelids and cheeks, has a tendency to sag with age (Figs. 5-25A,B and 5-26A,B). There is some degree of sun sensitivity. Acne, particularly of the face, has been noted in several patients.

Skeletal manifestations. Inguinal and/or, more often, umbilical hernia has been found in over one-third of patients before the age of 3 months. Muscular hypotonia has been present in about 20%. Genua valga has been noted in at least 75%. The long bones, metacarpals, and phalanges have thin cortices. Spondylolisthesis and spondylolysis are described (5a). Kyphosis or scoliosis and protuberant abdomen have also been frequently reported. Growth retardation is seen only after 15 years of age.

The calvaria is characteristically thickened and brachycephalic. The frontal sinuses are absent or poorly developed (Fig. 5-26C). Mild dysostosis multiplex in the spine is common. The ulna is somewhat shortened (12,13).

Nervous system. Progressive mental retardation to an IQ value of 40 or less in the second decade is a constant feature. It usually first becomes evident around 5 years of age. Speech is severely delayed. In about one-third of the patients, the voice becomes raspy in adult life. Periodic hyperactivity, hyperirritability, and/or aggressive reactions have been noted in about 50%.

Mild to moderate hearing loss has been found in about 20% of adults with this disorder. Clumsy gait and poor coordination of the hands are noted early in life.

Skin. Angiokeratomata have been reported (6).

Other findings. Macro-orchidism has been documented (5a). Amenorrhea and oligomenorrhea in females and scant beard and pubic hair have been noted (3).

Oral manifestations. The teeth have often been noted to be spaced. The gingiva and tongue were stated to be enlarged in about half the cases.

Laboratory findings. From 5 to 20% of the blood lymphocytes are vacuolated in 75% of the patients (Fig. 5-26D) and about one-half exhibit mild neutropenia and decreased prothrombin time.

The definitive diagnosis is made by finding markedly decreased aspartylglycosaminidase in the plasma, leukocytes, or cultured skin fibroblasts (8,9) or aspartylglucosaminuria by chromatographic or electrophoretic methods (4,11).

Prenatal diagnosis has been accomplished (2a).

### References (Aspartylglucosaminuria)


1. Aula P et al: Enzymatic diagnosis and carrier detection of aspartylglucosaminuria using blood samples. *Pediatr Res* **10**:625–629, 1976.
2. Aula P et al: Assignment of the structural gene encoding human aspartylglucosaminidase to the long arm of chromosome 4 (4q21→4qter). *Am J Hum Genet* **36**:1215–1224, 1984.

2a. Aula P et al: Prenatal diagnosis and fetal pathology of aspartylglycosaminuria. *Am J Med Genet* **19**:359–367, 1984.

3. Autio S: Aspartylglycosaminuria. Analysis of thirty-four patients. *J Ment Defic Res Monograph Ser* **1**:1–93, 1972.
4. Autio S et al: Aspartylglucosaminuria (AGU). Further aspects of its clin-

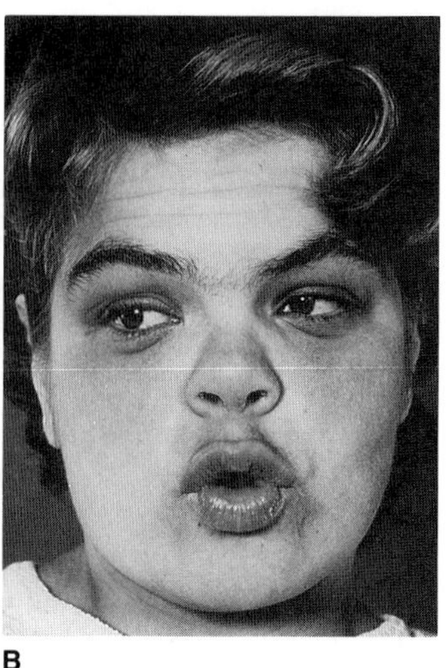

A B

Fig. 5–25. *Aspartylglucosaminuria.* (A,B) Remarkable resemblance between affected. Coarse facial features include furrowed brow, broad low nasal bridge, mild hypertelorism, epicanthal folds, sagging of facial skin, and thick lips. (Courtesy of *S Autio,* Helsinki, Finland.)


ical picture mode of inheritance and epidemiology based on a series of 57 patients. *Ann Clin Res.* **5**:149–155, 1973.

4a. Beaudet AL, Thomas GH: Disorders of glycoprotein degradation, in *The Metabolic Basis of Inherited Disease,* 6th ed, Scriver CR et al (eds), McGraw–Hill, New York, 1989.

5. Borud O et al: Aspartylglucosaminuria in Northern Norway in eight patients: Clinical heterogeneity and variations with the diet. *J Inherit Metab Dis* **1**:95–97, 1978.

5a. Chitayat D et al: Aspartylglucosaminuria in a Puerto Rican family: Additional features of a panethnic disorder. *Am J Med Genet* **31**:527–532, 1988.

6. Gehler J et al: Clinical and biochemical delineation of aspartylglycosaminuria as observed in two members of an Italian family. *Helv Paediatr Acta* **36**:179–189, 1981.

7. Hreidarsson S et al: Aspartylglycosaminuria in the United States. *Clin Genet* **23**:427–435, 1983.

8. Isenberg JN, Sharp HL: Aspartylglycosaminuria: Psychomotor retardation masquerading as a mucopolysaccharidosis. *J Pediatr* **86**:713–718, 1975.

9. Isenberg JN, Sharp HL: Aspartylglucosaminuria: Unique biochemical and ultrastructural characteristics. *Hum Pathol* **7**:469–481, 1976.

10. Jenner FA, Pollitt RJ: Large quantities of 2-acetamido-1-(beta-L-aspartamido)-1,2-dideoxyglucose in the urine of mentally retarded siblings. *Biochem J* **103**:48p–49p, 1967.

11. Maury CP, Palo J: N-Acetylglucosamine-asparagine levels in tissues of patients with aspartylglucosaminuria. *Clin Chem Acta* **108**:293–299, 1980.

12. Schmidt H et al: Skelettveränderungen bei zwei deutschen Kindern. *Roefo* **149**:143–146, 1988.

13. Stevenson RE et al: Aspartylglucosaminuria. *Proc Greenwood Genet Ctr* **1**:69–72, 1982.


Fig. 5–26. *Aspartylglucosaminuria.* (A) Radiograph showing thickened calvaria. (B) Vacuolated lymphoctyes in peripheral blood smear. (A from *RB Foundation,* Proc R Soc Med **67:**878, 1974. B from *JN Isenberg* and *HL Sharp,* J Pediatr **86:**713, 1975.)

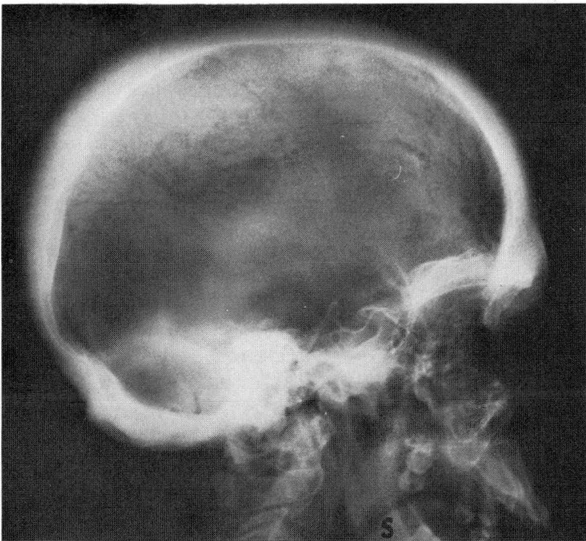

A

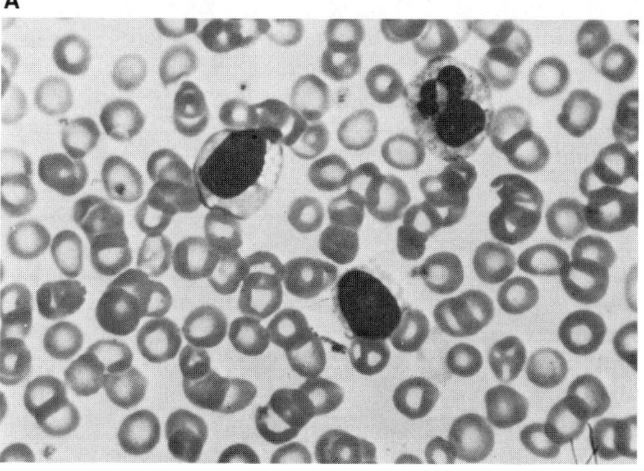

B

**Mannosidosis.** Mannosidosis, a type of oligosaccharidosis, was first characterized in 1967 by Öckerman (13). At least 50 cases have been reported to date (12). Inheritance is autosomal recessive with parental consanguinity in 25% (18). There is genetic heterogeneity, a more severe infantile form (Type I), and a milder juvenile–adult form (type II) (4). Nevertheless, there may be considerable variability among affected family members (11). The gene is located at chromosome 19p13.2→q12 (7).

The children are essentially normal for the first year of life but about 60% exhibit a propensity toward recurrent respiratory infections (3,6). Expression has varied from death in childhood to few clinical signs (5,8,12,14). There is stunted growth in adults (3).

Facies. The coarse facies (prominent supraorbital ridges, hypertelorism, broad based nose, prominent jaw) noted after the first few years of life becomes progressive but neither to the degree noted in MPS I-H nor as early as in I-cell disease (6). The nasal bridge tends to be depressed; the forehead and mandible are prominent. The neck is somewhat short (Fig. 5-27).

Central nervous system. There is delayed early motor development, which becomes manifest as clumsy motor function and ataxia. There is rapid progression of mental deterioration in Type I patients. Speech is delayed. Tendon reflexes are brisk.

Eyes and ears. Spoke-wheel posterior lenticular or superficial corneal opacities have been noted in 25% (1,2,9). Severe high-frequency sensorineural hearing loss is a common, if not constant, feature of Type II patients (2).

Musculoskeletal system. There is general mild hypotonia. The abdomen is protuberant. Umbilical hernia is found in 60%.

All have mild dysostosis multiplex, which becomes more severe in some and improves in others (17). The calvaria is thick with hypoplastic to absent paranasal sinuses in at least 60%. The long bones are osteoporotic. The ulna and radius are broad with curved diaphyses and a thin cortex. Joint mobility has been remarkaby variable in severity (5,18). The vertebrae are ovoid, flattened, and beaked in some cases with gibbus formation (6,10,16,19). Mild bony deformity at the hip is common.

Other findings. Hepatosplenomegaly has been noted in 50% (18) but may disappear in adulthood.

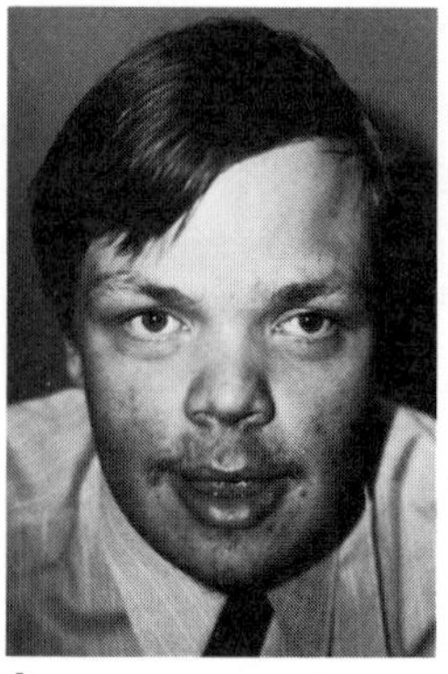

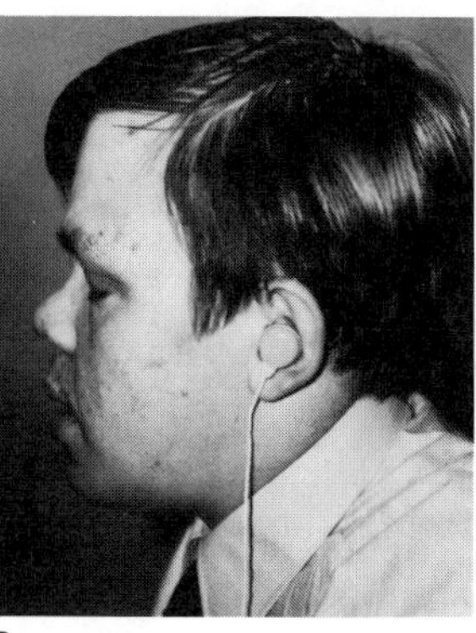

A B

Fig. 5–27. *Mannosidosis.* (A,B) Coarse features in adult. Note hearing aid. (Courtesy of *S Autio,* Helsinki, Finland.)

**Oral findings.** Macroglossia and widely spaced teeth have been noted (13).

**Laboratory aids.** The peripheral and bone marrow lymphocytes are vacuolated in possibly 90% of the cells counted (3). Coarse dark granules are present in the neutrophils. There is a defect in neutrophil chemotaxis. Pancytopenia has been found (16). Decreased serum IgG has been noted. The condition is diagnosed by finding markedly reduced acid $\alpha$-D-mannosidase in leukocytes and cultured fibroblasts. Mannose-rich oligosaccharides can be readily detected in the urine of affected individuals (2). Heterozygote detection and prenatal diagnosis can be accomplished by finding acid $\alpha$-D-mannosidase deficiency (15).

### References (Mannosidosis)

1. Arbisser AE et al: Ocular findings in mannosidosis. *Am J Ophthalmol* **82**:465–471, 1976.

2. Autio S et al: Mannosidosis: Clinical, fine-structural and biochemical findings in three cases. *Acta Paediatr Scand* **62**:555–565, 1973.

3. Autio S et al: The clinical course of mannosidosis. *Ann Clin Res* **14**:93–97, 1982.

4. Bach G et al: A new variant of mannosidosis with increased residual enzymatic activity and mild clinical manifestation. *Pediatr Res* **12**:1010–1015, 1978.

4a. Beaudet AL, Thomas GH: Disorders of glycoprotein degradation, in *The Metabolic Basis of Inherited Disease,* 6th ed, Scriver CR et al (eds), McGraw–Hill, New York, 1989.

5. Booth CW et al: Mannosidosis: Clinical and biochemical studies in a family of affected adolescents and adults. *J Pediatr* **88**:821–824, 1976.

6. Desnick RJ et al: Mannosidosis: Clinical, morphologic, immunologic and biochemical studies. *Pediatr Res* **10**:985–996, 1976.

7. Kaneda Y et al: Regional assignment of five genes on human chromosome 19. *Chromosoma* **95**:8–12, 1987.

8. Kistler JP et al: Mannosidosis—new clinical presentation, enzyme studies and carbohydrate analysis. *Arch Neurol* **34**:45–51, 1977.

9. Letson RD, Desnick, RJ: Punctate lenticular opacities in type II mannosidosis. *Am J Ophthalmol* **85**:218–224, 1978.

10. Milla PJ et al: Mannosidosis: Clinical and biochemical study. *Arch Dis Childh* **52**:937–942, 1977.

11. Mitchell ML et al: Mannosidosis: Two brothers with different degrees of disease severity. *Clin Genet* **20**:191–202, 1981.

12. Montgomery TR et al: Mannosidosis in an adult. *Johns Hopkins Med J* **151**:113–117, 1982.

13. Öckerman PA: A generalized storage disorder resembling Hurler's syndrome. *Lancet* **2**:239–241, 1967.

14. Patton MA et al: Mannosidosis in two brothers: Prolonged survival in the severe phenotype. *Clin Genet* **22**:284–289, 1982.

15. Poënaru L et al: Antenatal diagnosis in three pregnancies at risk for mannosidosis. *Clin Genet* **16**:428–432, 1979.

16. Press OW et al: Pancytopenia in mannosidosis. *Arch Intern Med* **143**:1266–1268, 1983.

17. Spranger J et al: The radiographic features of mannosidosis. *Radiology* **119**:401–407, 1976.

18. Vidgoff J et al: Mannosidosis in three brothers—a review of the literature. *Medicine* **56**:335–348, 1977.

19. Yunis JJ et al: Clinical manifestations of mannosidosis—a longitudinal study. *Am J Med* **61**:841–848, 1976.

Fig. 5–28. *Sialidosis.* Note prominent scalp veins, mildly coarse features with broad hypoplastic nasal bridge. There is abdominal distension and large hydroceles. (From *AS Aylesworth et al,* J Pediatr **96:**662, 1980.)

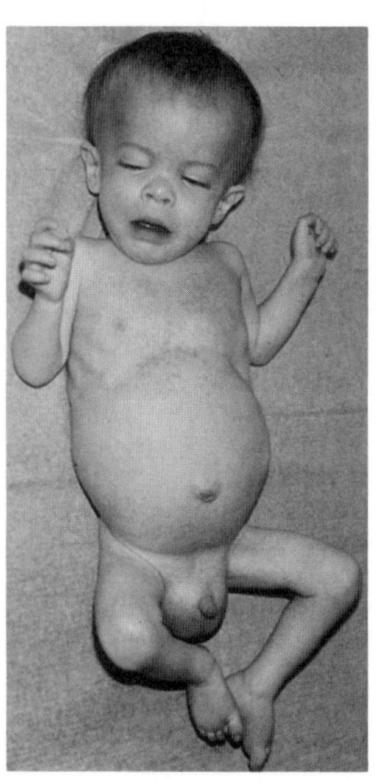

**Sialidosis (neuraminidase deficiency).** The sialidoses are a group of oligosaccharidoses that exhibits a primary deficiency of glycoprotein sialidase activity toward glycoproteins and oligosaccharides. One form, the juvenile or adult normosomatic sialidosis with normal stature and intelligence, exhibits macular cherry-red spots with visual loss and myoclonus. Since the facies is normal and there is no dysostosis multiplex, we will not discuss this type further. Good reviews are those of Spranger (16) and Young et al (22).

There are three forms of sialidosis with dysmorphic phenotype and coarse facies (Figs. 5-28–5-31): (1) congenital lethal variety presenting with hydrops fetalis, (2) infantile form (nephrosialidosis) initially presenting from birth to the end of the first year of life, and (3) childhood form (Spranger syndrome) usually having onset between 8 and 15 years and becoming progressively worse.

All types have autosomal recessive inheritance. The gene for glycoprotein sialidase is located at 6p21.3. The childhood subtype occurs predominantly in Japanese.

The childhood form (Spranger syndrome) has the mildest course, the life span usually being somewhat reduced, but most have survived to the fourth and fifth decades. There is variable coarsening of features ranging from thick lips, flat nasal bridge, and mild hypertelorism to a facies resembling MPS I-H, short trunk, relatively long limbs, and moderately severe dysostosis multiplex. Myoclonus, similar to that in the normosomatic type, is noted in 75%. Neuromuscular findings have commonly included ataxic gait, tremor, myoclonic jerks, generalized seizures, impaired hearing, hypotonia, wasting, and peripheral neuropathy. Mental function is essentially normal until adolescence, when it tapers off to an IQ of 60–70. Eye findings include progressive visual loss, cherry-red macular spots, punctate lenticular opacities, and occasional corneal opacities. Angiokeratomas similar to those found in Fabry disease and fucosidosis have been reported (12).

The infantile form (nephrosialidosis) is characterized by more severe dysostosis multiplex, hepatosplenomegaly, glomerular nephropathy, hearing loss, seizures, pyramidal tract signs, and severe mental retardation. Survival to the second decade is frequent. As in the child-

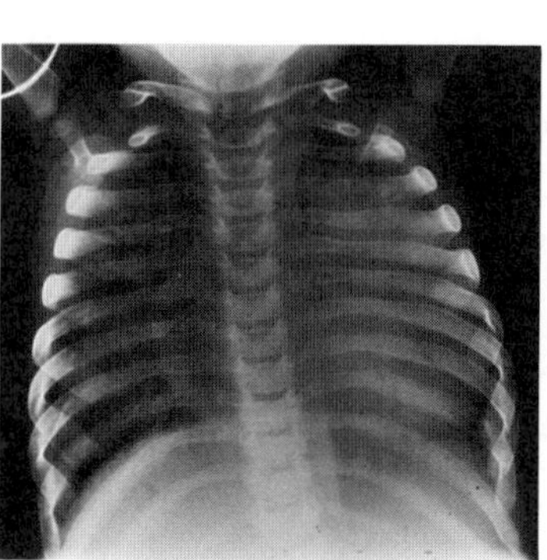
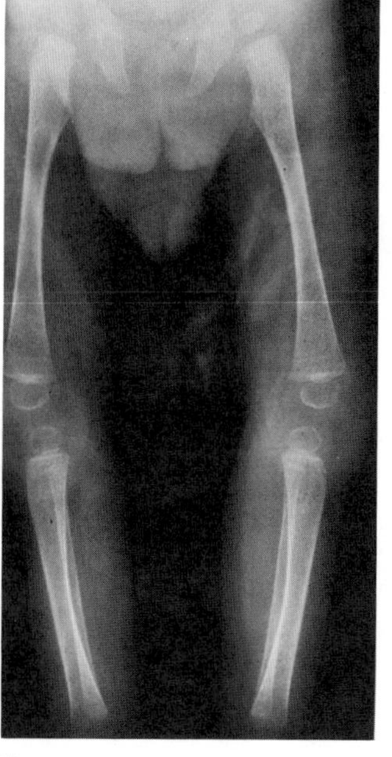
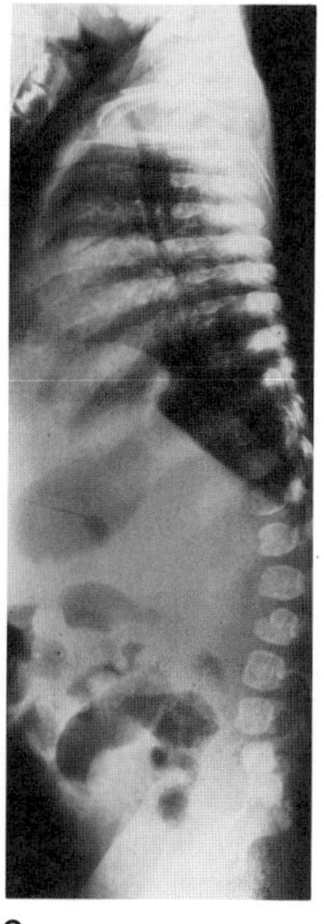

A B C

Fig. 5–29. *Sialidosis.* (A–C) Mild dysostosis multiplex at 15½ months showing broad irregular ribs, broad femoral and tibial diaphyses with poor modeling, broad femoral neck with coxa valga, and hook-shaped lumbar vertebral body. (From *AS Aylesworth et al,* J Pediatr **96:**662, 1980).

hood form, visual loss, cherry-red spots, myoclonus, and ataxia are seen in older children (3–5,8–11,14,17–22).

In the so-called congenital or fetal form, the infants are frequently stillborn with hydrops fetalis. Those not stillborn usually die within a few months. In addition to ascites with pericardial effusion, corneal opacities, hepatosplenomegaly, inguinal hernia, stippled epiphyses, and periosteal cloaking of long bones are seen. Renal involvement with proteinuria also occurs in the congenital form (1,6,7,15).

Differential diagnosis. Neuraminidase deficiency with β-galactosidase deficiency has been classified as a separate entity (see *Galactosialidosis*).

Laboratory findings. Vacuolated lymphocytes and bone marrow and placental foam cells are lacking in the normosomatic form but prominent in the other types. There is tissue storage of sialyloligosaccharides, increased urinary excretion of sialic acid containing oligosaccharides, and marked elevation of individual oligosaccharides. The defect is in glycoprotein sialidase (2,13).

Fig. 5–30. *Sialidosis.* Mild coarseness of facial features. (From *T Miyatake et al,* Ann Neurol **6:**232, 1979.)

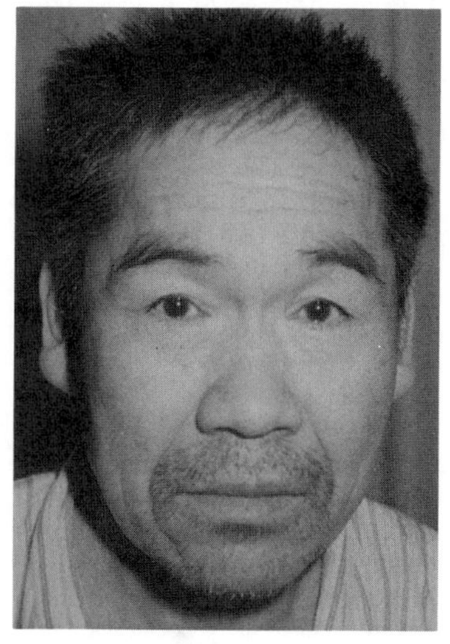

### References [Sialidosis (neuraminidase deficiency)]


1. Aylesworth AJ et al: A severe infantile sialidosis: Clinical, biochemical and microscopic features. *J Pediatr* **96**:662–668, 1980.
2. Cantz M. Messer H: Oligosaccharides and ganglioside neuraminidase activities of mucolipidosis I (sialidase) and mucolipidosis II (I-cell disease) fibroblasts. *Eur J Pediatr* **97**:113–118, 1979.
3. Dufier JL et al: Manifestation oculaires d'une forme particuliere de mucolipidose: la néphrosialidose. *J Fr Ophtalmol* **3**:247–256, 1980.
4. Goldberg MF et al: Macular cherry-red spot corneal clouding and β-galactosidase deficiency. *Arch Intern Med* **128**:387–398, 1971.
5. Kelly TE, Graetz G: Isolated acid neuraminidase deficiency: A distinct lysosomal storage disease. *Am J Med Genet* **1**:31–46, 1977.
6. Kelly TE et al: Mucolipidosis I (acid neuraminidase deficiency). Three cases and delineation of the variability of the phenotype. *Am J Dis Child* **135**:705–708, 1981.
7. Laver J et al: Infantile lethal neuraminidase deficiency (sialidosis). *Clin Genet* **23**:97–101, 1983.


Fig. 5–31. *Sialidosis.* Cherry-red spot. (From *T Miyatake et al,* Ann Neurol **6:**232, 1979.)

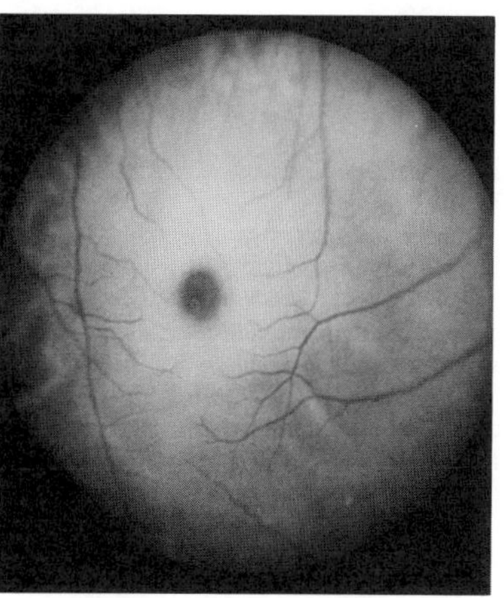

8. Louis JJ et al: Une observation de mucolipidose de type I par defect primaire en alpha D neuraminidase. *J Génét Hum* **31**:79–91, 1983.
9. Lowden JA, O'Brien JS: Sialidosis: A review of human neuraminidase deficiency. *Am J Hum Genet* **31**:1–18, 1979.
10. Maroteaux P et al: Sialidose par déficit en alpha (2-6) neuraminidase sans atteinte neurologique. Mucolipidose de type I? *Arch Fr Pédiatr* **35**:286–291, 1978.
11. Maroteaux P et al: Un nouveau type de sialidose avec atteinte renal: la nephrosialidose. *Arch Fr Pédiatr* **35**:819–844, 1978.
12. Miyataki T et al: Adult type neuronal storage disease with neuraminidase deficiency. *Ann Neurol* **6**:232–244, 1979.
13. Mueller OT, Wenger DA: Mucolipidosis I: Studies of sialidase activity and a prenatal diagnosis. *Clin Chim Acta* **109**:313–324, 1981.
14. O'Brien JS, Warner TG: Sialidosis: Delineation of subtypes by neuraminidase assay. *Clin Genet* **17**:35–38, 1980.
15. Riches WG, Schmuckler EA: A severe infantile mucolipidosis. *Arch Pathol Lab Med* **107**:147–152, 1983.
16. Spranger J: Mucolipidosis I: Phenotype and nosology. *Perspect Inherit Metab Dis* **4**:303–315, 1981.
17. Spranger J et al: Lipomucopolysaccharidosis. *Z Kinderheilkd* **103**:285–306, 1968.
18. Spranger J et al: Mucolipidosis I—a sialidosis. *Am J Med Genet* **1**:21–29, 1977.
19. Spranger, J, Cantz M: Mucolipidosis I, the cherry-red spot-myoclonus syndrome and neuraminidase deficiency. *Birth Defects* **14**(6B):105–112, 1978.
20. Thomas GH et al: Neuraminidase deficiency in the original patient with the Goldberg syndrome. *Clin Genet* **16**:323–330, 1979.
21. Winter RM et al: Sialidosis type 2 (acid neuraminidase deficiency): Clinical and biochemical features of a further case. *Clin Genet* **18**:203–210, 1980.
22. Young ID et al: Neuraminidase deficiency: Case report and review of the phenotype. *J Med Genet* **24**:283–290, 1987.

**Galactosialidosis.** Galactosialidosis, an oligosaccharidosis, occurs in three forms: (a) neonatal hydropic, (b) late infantile, and (c) juvenile or adult. The disorder has been reported on several occasions (1–16), probably more often in Japanese patients.

The disorder has autosomal recessive inheritance. There is evidence to suggest that the gene is located on chromosome 20 (11). The primary defect seems to be in a 32-kDa protective glycoprotein. A combined $\beta$-galactosidase and acid $\alpha$-neuraminidase deficiency is found. There is undoubtedly genetic heterogeneity (13).

Galactosialidosis is characterized by growth retardation, dysostosis multiplex, mental retardation, cerebellar ataxia, myoclonus, and seizures. There is coarse facies, corneal clouding, macular cherry-red spot, and hearing loss. Chitayat et al (3) did not find a cherry-red spot, however. Angiokeratomas apparently occur only in Japanese patients (5).

In contrast to many of the other MPS and oligosaccharide disorders, the viscera are not enlarged, there are no vacuolated blood cells, and no mucopolysacchariduria.

### References (Galactosialidosis)

1. Andria G et al: Infantile neuraminidase and $\beta$-galactosidase deficiencies (galactosialidosis) with mild clinical courses. *Perspect Inherit Metab Dis* **4**:379–395, 1985.
2. Berard-Badier M et al: Étude ultrastructurale du parenchyme hepatique dans les mucopolysaccharidoses. *Path Biol (Paris)* **18**:117–128, 1970 (Case 3).
3. Chitayat D et al: Juvenile galactosialidosis in a white male: A new variant. *Am J Med Genet* **31**:887–901, 1988.
4. Gravel RA et al: Infantile sialidosis: A phenocopy of type I $G_{M1}$ gangliosidosis distinguished by genetic complementation and urinary oligosaccharides. *Am J Hum Genet* **31**:669–679, 1979.
5. Ishibashi A et al: $\beta$-Galactosidase and neuraminidase deficiency associated with angiokeratoma corporis diffusum. *Arch Dermatol* **120**:1344–1346, 1984.
6. Kleijer WJ et al: Prenatal diagnosis of sialidosis with combined neuraminidase in $\beta$-galactosidase deficiency. *Clin Genet* **16**:60–61, 1979.
7. Kobayashi T et al: Adult type mucolipidosis with $\beta$-galactosidase and sialidase deficiency. *J Neurol* **221**:137–149, 1979.
8. Kuriyama M et al: Adult mucolipidosis with $\beta$-galactosidase and neuraminidase deficiencies. *J Neurol Sci* **46**:245–254, 1980.
9. Loonen MC et al: Combined sialidase (neuraminidase) and $\beta$-galactosidase deficiency: Clinical, morphological, and enzymological observations in a patient. *Clin Genet* **26**:139–149, 1984.
10. Maire I, Nivelon-Chevallier A: Combined deficiency of $\beta$-galactosidase and neuraminidase: Three affected siblings in a French family. *J Inherit Metab Dis* **4**:221–223, 1981.
11. Mueller OT et al: Sialidosis and galactosialidosis: Chromosomal assignment of two genes associated with neuraminidase-deficiency disorders. *Proc Natl Acad Sci USA* **83**:1817–1821, 1986.
11a. O'Brien JS: $\beta$-Galactosidase deficiency, in *The Metabolic Basis of Inherited Disease,* 6th ed, Scriver CR et al (eds), McGraw–Hill, New York, 1989.
12. Okada S et al: A case of neuraminidase deficiency associated with a partial $\beta$-galactosidase defect. *Eur J Pediatr* **130**:239–249, 1979.
13. Palmeri S et al: Galactosialidosis: Molecular heterogeneity among distinct clinical phenotypes. *Am J Hum Genet* **38**:137–148, 1986.
14. Suzuki Y et al: Macular cherry-red spots and $\beta$-galactosidase deficiency in an adult: An autopsy case with progressive cerebellar ataxia, myoclonus, thrombocytopathy and accumulation of polysaccharide in liver. *Arch Neurol* **34**:157–161, 1977.
15. Suzuki Y et al: Beta-galactosidase deficiency—juvenile and adult patients. *Hum Genet* **36**:219–229, 197.
16. Wenger DA et al: Macular cherry-red spots and myoclonus with dementia: Coexistent neuraminidase and $\beta$-galactosidase deficiencies. *Biochem Biophys Res Commun* **82**:589–595, 1978.

**I-Cell disease (mucolipidosis II).** I-cell disease, formerly known as mucolipidosis II, was originally described in 1967 by Leroy and Demars (3,7). Really an oligosaccharidosis, it is characterized by severe psychomotor retardation, marked shortness of stature, facial features reminiscent of MPS I-H, impressive gingival enlargement, a rapid deteriorating course, and death from heart failure (hypertrophic cardiomyopathy), bronchopneumonia, or pulmonary atelectasis, usually by the age of 5 years (6a,14a,25) (Figs. 5-32 to 5-34). This disorder, like pseudo-Hurler polydystrophy, is the result of a deficiency in recognition marker phosphotransferase. Inheritance is autosomal recessive. Consanguinity is high (16). The disease may be somewhat more frequent in Japan and there is probably genetic heterogeneity (5a,15). The gene locus has been identified at 4q21–q23.

Important negative signs and symptoms are absent splenomegaly, equivocal or absent corneal cloudiness, and normal urinary excretion of MPS (8,9).

It received the name of I-cell disease because of the numerous granular inclusions in the cytoplasm of cultured fibroblasts and amniotic fluid cells observed under phase contrast microscopy (1,3,7). The inclusions are large lysosomes containing heterogeneous undegraded material (4,12).

At birth, the infants are small-for-dates ($<$ 10th percentile), and have coarse facies, muscular hypotonia, dislocated hips, inguinal hernias in males, and sometimes tight and thickened skin that becomes more pliable with age (2,20,23). Hirsutism has been noted in about 60%. During the first year, there is a history of recurrent upper respiratory infections (rhinitis, otitis media), failure to thrive, and marked lack of psychomotor development. The full clinical picture is reached by 1 year of age. There is severe shortness of stature. Patients rarely become heavier than 15 kg and few have been taller than 80 cm. Most never reach the average height of a 1-year-old child and growth ceases by the third year. (Note difference with MPS I-H, where excessive growth from 6 to 18 months of age has been frequently documented.)

Motor retardation is more severe than mental retardation. Many patients do not accomplish unaided ambulation, but several older patients could walk without support. Most patients over 4 years can speak two-word sentences and are toilet trained.

Facies. Some patients have exhibited premature lightening of hair color. Head circumference remains normal with respect to stature. The facies is reminiscent of that in MPS I-H patients. There are small orbits, flat supraorbital ridges, puffy eyelids, a slight degree of exophthalmos, and a pattern of tortuous veins about the orbits and temporal areas (24). The cheeks are full and pink, partly because of mul-

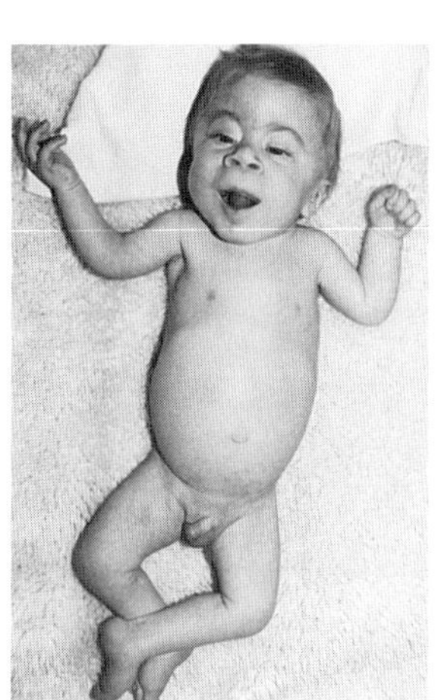

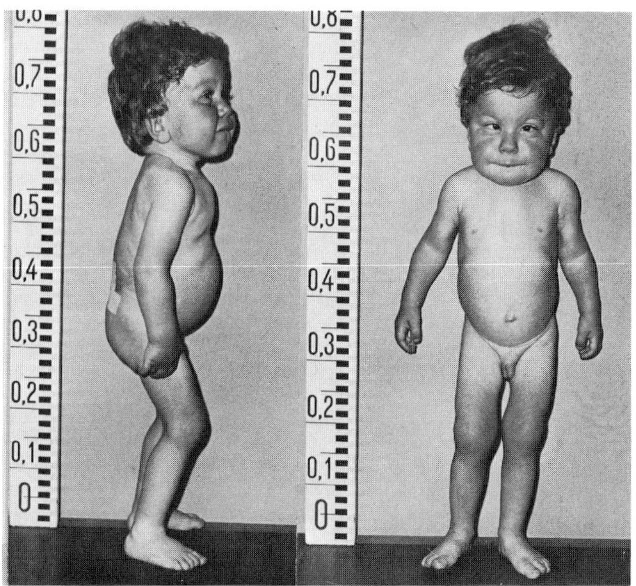


A B

Fig. 5–32. *I-cell disease.* (A,B) Hurleroid features are evident from early infancy. (A courtesy of *R Schutz,* Göttingen, West Germany. B from *U Wiesmann et al,* Acta Paediatr Scand **63:**9, 1974.)

tiple fine telangiectasias. The lower part of the face is fishlike when viewed in profile, mainly because of enlargement of the gingiva (Fig. 5-33A,B).

There is intermittent copious nasal discharge but to a lesser degree than in MPS I-H patients.

Mild corneal clouding as a late sign has been documented in about 40% but vision is not impaired. However, on slit lamp examination, all have some degree of corneal opacification. Glaucoma and megalocornea have been occasionally noted (10).

Musculoskeletal system. There is shortness of the neck and deformed thoracic cage. Umbilical and/or inguinal hernias are found in 50 and 75%, respectively. Considerable restriction of joint mobility, particularly in the shoulders and wrists, is evident. The hands and fingers are stubby and the wrists broadened. Restriction of motion is less impaired in the lower limbs, which appear hypotrophic. Thoracolumbar kyphosis with gibbus formation may be present, but is not observed in any patient who can stand upright. The costochondral junctions are knoblike. Pes valgus has been noted in 25% of the neonates.

Radiographically, generalized demineralization, coarse trabecular pattern, and extensive periosteal cloaking of all long bones are seen in early infancy (2,13). This phenomenon is also observed in newborns with $G_{M1}$ gangliosidosis Type I and the disorders cannot be differentiated radiographically at this stage. Periosteal new bone formation can be observed until 4–6 months of age. Subsequently, this overgrowth becomes confluent with the underlying cortex and disappears entirely between 8 and 12 months of age. From that point, dysostosis multiplex is observed as in MPS I-H, the bony abnormalities always being more severe in I-cell disease patients at comparable ages. Other differences are minor involvement of the calvaria in I-cell disease and minor to moderate diaphyseal widening in long bones, particularly of the lower limbs.

Stippled epiphyses, particularly of the calcaneus and knees, and pathologic fractures have been documented (6). Premature synostosis of skull sutures has been reported (18). No lamina dura is found around

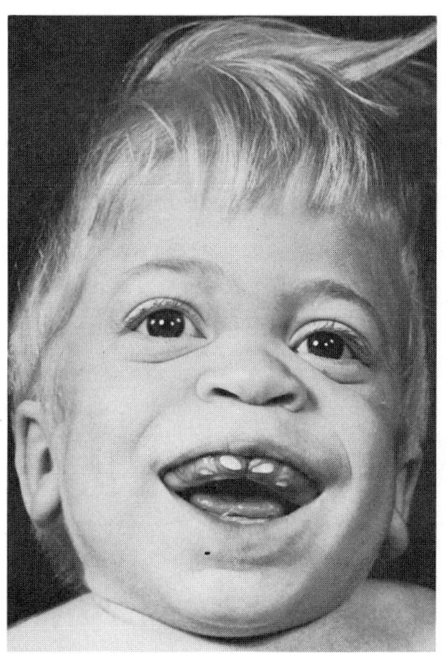

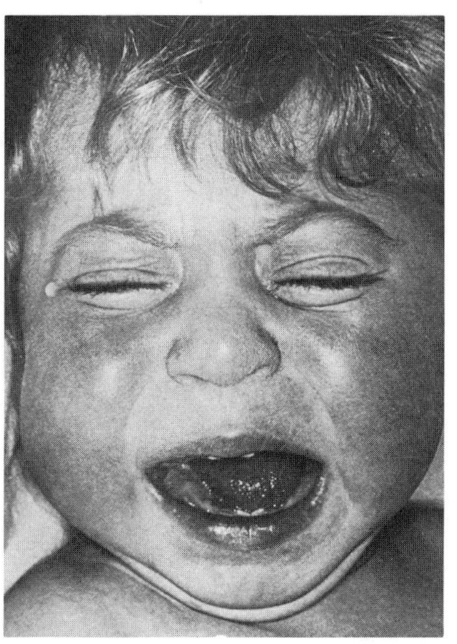

A B

Fig. 5–33. *I-cell disease.* (A,B) Characteristic facial appearance. (A courtesy of *CI Scott,* Wilmington, Delaware. B from *NS Gordon,* Postgrad Med J **49:**359, 1973.)

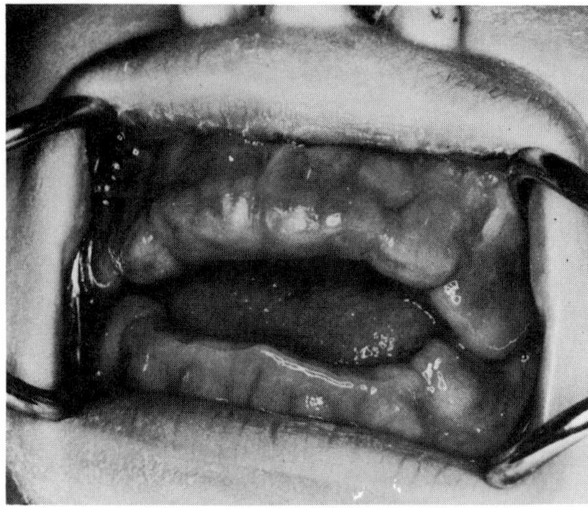

A

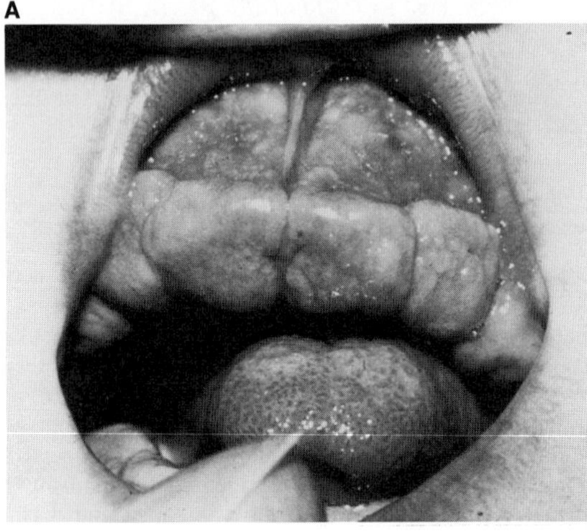

B

Fig. 5-34. *I-cell disease.* (A,B) Marked thickening of gingiva. Note anterior open bite in A. (A from *D Galili et al,* Oral Surg **37:**533, 1974. B from *DT Whelan et al,* Clin Genet **24:**90, 1983.)

the teeth. The metacarpals are proximally pointed, the distal phalanges being poorly modeled (21).

Other findings. Minimal to moderate hepatomegaly, evident at birth, occurs in 40%.

Oral manifestations. Enlargement of the gingiva and anterior alveolar process is present as early as 4 months of age. It is slowly progressive. In some patients, it reaches grotesque propotions (8,9) and together with a thick tongue prevents proper closure of the mouth (Fig. 5-34A,B). Usually the teeth are deeply buried in the hypertrophied tissue or do not erupt at all. Radiographic examination of the teeth reveals that the enamel is quite hypocalcified and that there is accumulation of storage material about the crowns of unerupted first molar teeth.

Differential diagnosis. See Table 5-3.

Laboratory findings. Peripheral lymphocytes contain large lysosomal cytoplasmic inclusions. The urinary excretion of GAGs is normal. All cultured fibroblasts contain an abundance of coarse cytoplasmic inclusions, hence I-cell disease, with a characteristic inclusion-free perinuclear zone (7).

Most lysosomal acid hydrolase activities are considerably increased (at least 20-fold for some) in the serum but are decreased (10–20% of normal) in cultured fibroblasts.

The enzymatic defect is *N*-acetylglucosamine-1-phosphotransferase, the same as in pseudo-Hurler polydystrophy (14). Heterozygotes can be identified by having intermediate levels (5,19,22). Measuring the activity of acid hydrolases in amniotic fluid and cultured amniotic cells is reliable for prenatal detection (17). This has also been accomplished by assay of *N*-acetylglucosamine-1-phosphotransferase assay of chorionic villi (1a).

### References [I-Cell disease (mucolipidosis II)]


1. Abe K et al: Ultrastructural studies in fetal I-cell disease. *Pediatr Res* **10**:669–676, 1976.

1a. Ben-Yoseph Y et al: First trimester prenatal evaluation for I-cell disease by N-acetyl-glucosamine 1-phosphoribosyltransferase assay. *Clin Genet* **33**:38–43, 1988.

2. Cipollani C et al: Neonatal mucolipidosis II (I-cell disease) clinical, radiological and biochemical studies in a case. *Helv Paediatr Acta* **35**:85–95, 1980.

3. DeMars, RI, Leroy JG: The remarkable cells cultured from a human with Hurler's syndrome: *In Vitro* **2**:107–118, 1967.

4. Galili D et al: Massive gingival hyperplasia preceding dental eruption in I-cell disease. *Oral Surg* **37**:533–539, 1974.

5. Hasilik A et al: Enzymatic phosphorylation of lysosomal enzymes in the presence of UDP-N-acetylglucosamine. Absence of activity in I-cell fibroblasts. *Biochem Biophys Res Commun* **98**:761–767, 1981.

5a. Honey NK et al: The mucolipidoses: Identification by abnormal electrophoretic patterns of lysosomal hydrolases. *Am J Med Genet* **9**:239–253, 1981.

6. Lemaitre L et al: Radiological signs of mucolipidosis II or I-cell disease. *Pediatr Radiol* **7**:97–105, 1978.

6a. Leroy JG: The oligosaccharidoses (formerly mucolipidoses), in *Principles and Practice of Medical Genetics,* Emery AE. Rimoin DL (eds), Churchill Livingstone, Edinburgh, 1983, pp. 1348–1365.

7. Leroy JG, DeMars RI: Mutant enzymatic and cytological phenotypes in cultured human fibroblasts. *Science* **157**:804–806, 1967.

8. Leroy JG et al: I-cell disease. *Birth Defects* **5**(4):174–185, 1969.

9. Leroy JF, Spranger JW: I-cell disease continued. *N Engl J Med* **283**:598–599, 1970.

10. Libert J et al: Ocular findings in I-cell disease (mucolipidosis type II). *Am J Ophthalmol* **83**:617–628, 1977.

11. Martin JJ et al: I-cell disease (mucolipidosis II): A report on its pathology. *Acta Neuropathol 33*:285–305, 1975.

12. Martin JJ et al: I-cell disease: A further report on its pathology. *Acta Neuropathol* **64**:234–242, 1984.

13. Michels VV et al: Mucolipidosis II: Unusual presentation with a congenital angulated fracture. *Clin Genet* **21**:225–227, 1982.

14. Mueller OT et al: Mucolipidosis II and III. The genetic relationships between two disorders of lysosomal enzyme biosynthesis. *J Clin Invest* **72**:1016–1023, 1983.

14a. Nolan CM, Sly WS: I-cell disease and pseudo-Hurler polydystrophy, in *The Metabolic Basis of Inherited Disease,* 6th ed, Scriver CR et al (eds), McGraw–Hill, New York, 1989.

15. Okada S et al: Heterogeneity in mucolipidosis II (I-cell disease). *Clin Genet* **23**:155–159, 1983.

16. Okada S et al: I-cell disease: Clinical studies of 21 Japanese cases. *Clin Genet* **28**:207–215, 1985.

17. Owada M et al: Prenatal diagnosis of I-cell disease by measuring altered α-mannosidase activity in amniotic fluid. *J Inherit Metab Dis* **3**:117–121, 1980.

18. Pazzaglia UE et al: Mucolipidosis II: Correlation between radiological features and histopathology of the bones. *Pediatr Radiol* **19:**406–413, 1989.

19. Reitman ML et al: Fibroblasts from patients with I-cell disease and pseudo-Hurler polydystrophy are deficient in uridine 5-diphosphate-N-acetylglucosamine: Glycoprotein N-acetyl glucosaminylphosphotransferase activity. *J Clin Invest* **67**:1574–1579, 1981.

20. Spritz RA et al: Neonatal presentation of I-cell disease. *J Pediatr* **93**:954–958, 1978.

21. Taber P et al: Roentgenographic manifestations of Leroy's I-cell disease. *Am J Roentgenol* **118**:213–221, 1973.

22. Varki A et al: Demonstration of the heterozygous state for I-cell disease and pseudo-Hurler polydystrophy by assay of N-acetylglucosaminylphosphotransferase in white cells and fibroblasts. *Am J Hum Genet* **34**:717–729, 1982.

23. Whelan DT et al: Mucolipidosis II: The clinical, radiological and biochemical features in three cases. *Clin Genet* **24**:90–96, 1983.

24. Wiesmann U et al: Mucolipidosis II (I-cell disease). A clinical and biochemical study. *Acta Paediatr Scand* **63**:9–16, 1974.

25. Wiesmann UN, Herschkowitz NN: Mucolipidosis II and III. The clinical pictures and the pathogenetic mechanisms. *Persp Inherit Metab Dis* **4**:437–451, 1981.

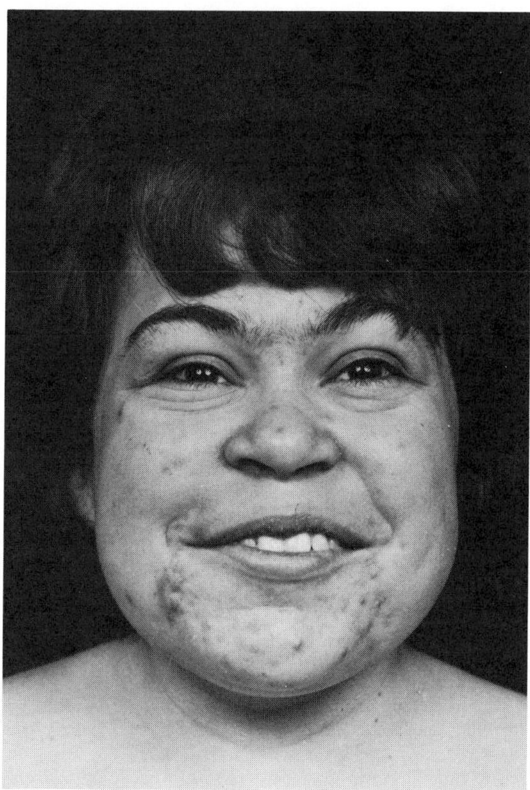

Fig. 5–35. *Pseudo-Hurler polydystrophy.* Mildly Hurleroid facies.

**Pseudo-Hurler polydystrophy (mucolipidosis III).** Reported initially by Maroteaux and Lamy (6) in 1966, pseudo-Hurler polydystrophy is characterized by mild mental retardation, early restriction of joint mobility, dysostosis multiplex, and normal mucopolysacchariduria. At least 70 patients have been reported (Fig. 5-35–5-37) (1–5,11).

Like I-cell disease, it results from a deficiency in recognition marker phosphotransferase (5a, 7a).

The disorder has autosomal recessive inheritance. The gene for GlcNAc-1-phosphotransferase has been located at 4q21–q23. There is genetic heterogeneity (4).

Typically, in the second or third year of life, restricted joint mobility, small stature, short neck, tight indurated skin, scoliosis, and hip dysplasia are noted. Mild nonprogressive mental retardation (IQ 65–85) has been present in most patients (9).

Fig. 5–36. *Pseudo-Hurler polydystrophy.* Fingers cannot be flexed or extended. (Courtesy of *J Sensenbrenner,* Baltimore, Maryland.)

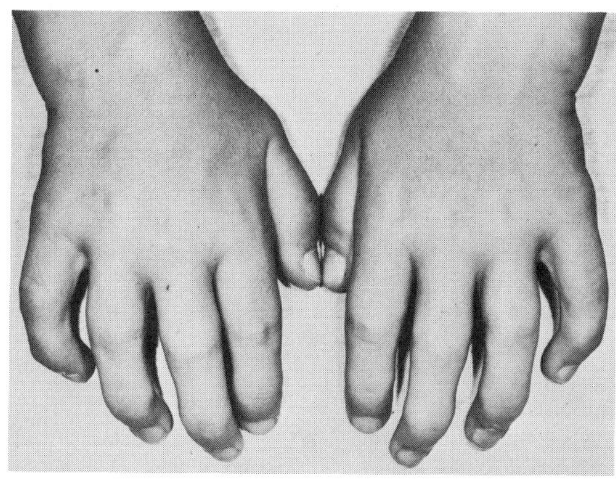

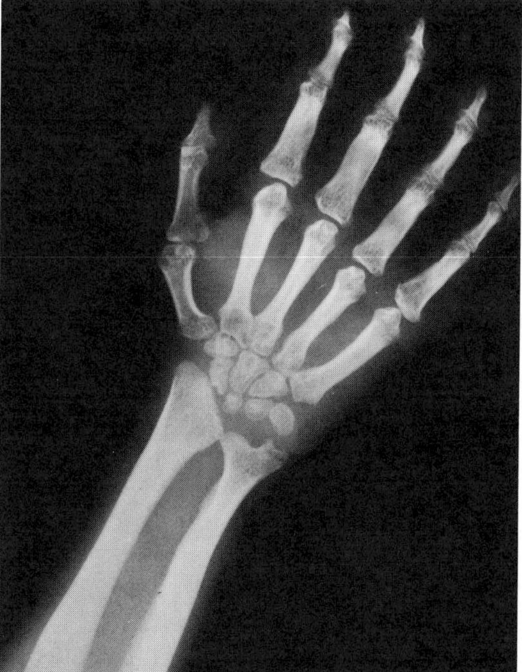

A

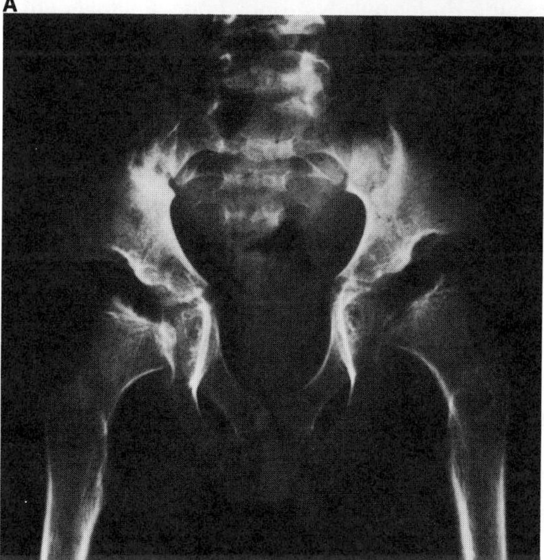

B

Fig. 5–37. *Pseudo-Hurler polydystrophy.* (A) Hand bones coarsely trabeculated. Small proximal carpal bones, irregular distal radius and ulna. (B) Hypoplastic iliac bodies and femoral heads and necks. (A courtesy of *M Robinow,* Charlottesville, Virginia. B courtesy of *J Dorst,* Baltimore, Maryland.)

Facies. Facies have been variable, but most patients exhibit some coarsening of features (Fig. 5-35). The mandible becomes progressively prognathic.

Eyes. Under slit-lamp examinations nearly all patients have corneal clouding that clinically is little more than a slight corneal haze (7).

Musculoskeletal system. Short stature, decreased upper-to-lower segment ratio, and shortened arm span are present in all patients. Joint stiffness, which begins about the age of 3 years, progresses slowly until puberty (Fig. 5-36) (7).

Premature closure of cranial sutures is frequent, but the skull is normal in shape. The foramen magnum is small. The clavicles are short and thick, with the midportions bowed superiorly. Vertebral body alterations are quite variable but generally are mild. They can be observed within the first year of life (8). There is flaring of the iliac wings with constriction of the iliac bones and prominent anterior superior iliac spines. The acetabula are shallow with oblique roofs. Progressive destruction of the capital femoral epiphyses is a striking feature in most patients (Fig. 5-37B). The metacarpals are pointed proximally. The carpal bones are small and irregular. Bone age is considerably retarded (Fig. 5-37A) (1–3,5,7,8).

Differential diagnosis. See Table 5-30.

Laboratory findings. Peripheral leukocytes are normal. Vacuolated plasma cells are often found in the bone marrow. Finely granulated intracytoplasmic material with staining characteristics of MPS has been found in bone marrow cells. Ultrastructural studies of cultured fibroblasts reveal lysosomal vacuoles containing abnormally accumulated lipids, glycoproteins, and MPS storage (10). Urinary excretion of MPS is normal but that of sialyloligosaccharides is excessive. The activities of most lysosomal enzymes are low in fibroblasts, but high levels (5–50 times normal) are found in serum, findings similar to those demonstrated in I-cell disease (8,12).

The enzymatic defect is *N*-acetylglucosaminylphosphotransferase, the same enzyme as that in I-cell disease (11). Heterozygotes can be identified by intermediate values of enzymatic activity (13).

### References [Pseudo-Hurler polydystrophy (mucolipidosis III)]

1. Aviad I et al: Roentgen findings of pseudo-Hurler polydystrophy in the adult with a note on cephalometric changes. *Am J Roentgenol* **122**:56–66, 1974.

2. Gericke GJ: Mucolipidosis III: Two patients displaying genetic pleiotropism. *S Afr Med J* **51**:140–144, 1977.

3. Herd JK et al: Mucolipidosis type III: *Am J Dis Child* **132**:1181–1186, 1978.

4. Honey NK et al: Mucolipidosis type III is genetically heterogeneous. *Proc Natl Acad Sci USA* **79**:7420–7424, 1982.

5. Kelly TE et al: Mucolipidosis III (pseudo-Hurler polydystrophy): Clinical and laboratory studies in a series of 12 patients. *Johns Hopkins Med J* **137**:156–175, 1975.

5a. Leroy JG: The oligosaccharidoses (formerly mucolipidoses), in *Principles and Practice of Medical Genetics,* Emery AE. Rimoin DL (eds), Churchill Livingstone, Edinburgh, 1983.

6. Maroteaux P, Lamy M: La pseudopolydystrophie de Hurler. Presse Méd **74**:2889–2892, 1966.

7. Melhem R et al: Roentgen findings in mucolipidosis III (pseudo-Hurler polydystrophy). *Radiology* **166**:153–160, 1973.

7a. Nolan CM, Sly WS: I-cell disease and pseudo-Hurler polydystrophy, in *The Metabolic Basis of Inherited Disease,* 6th ed, Scriber CR et al (eds), McGraw–Hill, New York, 1989.

8. Nolte K, Spranger J: Early skeletal changes in mucolipidosis III. *Ann Radiol* **19**:151–159, 1976.

9. Stern H et al: Pseudo-Hurler polydystrophy (mucolipidosis 3). A clinical, biochemical and ultrastructural study. *Isr J Med Sci* **10**:463–475, 1974.

10. Taylor HA: Mucolipidosis III (pseudo-Hurler polydystrophy): Cytological and ultrastructural observations of cultured fibroblast cells. *Clin Genet* **4**:388–397, 1973.

11. Reitman M et al: Fibroblasts from patients with I-cell disease and pseudo-Hurler polydystrophy and deficient in uridine 5-diphosphate N-acetyglucosamine: Glycoprotein N-acetylglucosaminylphosphotransferase activity. *J Clin Invest* **67**:1574–1579, 1981.

12. Sewell AC: Urinary oligosaccharide excretion in disorders of glycolipid, glycoprotein and glycogen metabolism. *Eur J Pediatr* **134**:183–194, 1980.

13. Varki A et al: Demonstration of the heterozygous state of I-cell disease and pseudo-Hurler polydystrophy by assay of N-acetylglucosaminylphosphotransferase in white blood cells and fibroblasts. *Am J Hum Genet* **34**:717–729, 1982.

**Sialic acid storage disease.** In addition to severe mental and somatic retardation, the main features of infantile sialic acid storage disease are sparse hair, coarse facies, and hepatosplenomegaly. Most patients have ascites and diarrhea (5–7).

The facies becomes progressively coarse with epicanthal folds, apparent hypertelorism, anteverted nostrils, prominent philtrum, and enlarged gingiva (5).

Vacuolated lymphocytes are demonstrable, and there is free sialic acid in the urine. Similar biochemical and ultrastructural findings of lysosomal storage (conjunctival biopsy) are noted in Salla disease. Inheritance is autosomal recessive.

Salla disease or late infantile sialic acid storage disease is characterized by mental retardation and clumsiness that become evident after the first year of life (1–4). Life span is normal. The disease is named for the geographic area in which the original Finnish kindred was found. It has been occasionally found outside the Finnish population (2). Patients also exhibit ataxia, rigidity, athetosis, spasticity, and impaired speech but do not manifest dysmorphic features or hepatosplenomegaly. Prenatal diagnosis has been accomplished (3).

### References (Sialic acid storage disease)

1. Aula P et al: "Salla disease": A new lysosomal storage disorder. *Arch Neurol* **36**:88–94, 1979.

2. Echenne B et al: Salla disease in one non-Finnish patient. *Eur J Pediatr* **145**:320–322, 1986.

3. Renlund M. Aula P: Prenatal detection of Salla disease based upon increased free sialic acid in amniocytes. *Am J Med Genet* **28**:377–384, 1987.

4. Renlund M et al: Salla disease: A new lysosomal storage disorder with disturbed sialic acid metabolism. *Neurology* **33**:57–66, 1983.

5. Stevenson RE et al: Sialic acid storage disease with sialuria: Clinical and biochemical features in the severe infantile type. *Pediatrics* **72**:441–449, 1983.

6. Thomas GH et al: Alterations in cultured fibroblasts of sibs with an infantile form of a free (unbound) sialic acid storage disorder. *Pediatr Res* **17**:307–312, 1983.

7. Tondeur R et al: Infantile form of sialic acid storage disorder: Clinical, ultrastructural and biochemical studies in two siblings. *Eur J Pediatr* **139**:142–147, 1982.

**Berman syndrome (mucolipidosis IV).** The Berman syndrome, first described in 1974 by Berman et al (4), is a rare lysosomal storage disease characterized by bilateral corneal opacities in infancy, full facial features, and progressive psychomotor retardation (7,16).

There is storage of intralysosomal phospholipids and glycoconjugates in tissues as a result of deficiency of soluble ganglioside sialidase (2,5). Technically it is not an oligosaccharidosis.

It has autosomal recessive inheritance, with about one-half the patients being Jews of Ashkenazi origin, probably from southern Poland (15). About 40 cases have been documented (7). There may be heterogeneity (10,18).

Facies. The face is full, but unlike that of other mucolipidosis, it is not truly coarse. Some patients have a somewhat bulbous nose and full lower lip (6,9,10a,13).

Eyes. Strabismus, amblyopia, myopia, photophobia, increased lacrimation, and bilateral moderate to severe corneal clouding are seen in infancy (8,11,12,15). Corneal involvement is congenital in about one-half the cases (1). The latter improves with time but tapetoretinal changes and marked pallor of the optic discs become enhanced in older children (15).

Organomegaly. There is no organomegaly.

Skeletal anomalies. No skeletal changes or restricted joint movement are observed.

Neurologic. Mild to severe mental and motor retardation appear during the first year of life. Initially, hypotonia with normal or decreased reflexes is followed in time by athetosis and spasticity with hyperreflexia and clonus (13,17).

Differential diagnosis. The severely affected corneal epithelium with an intact Bowman's membrane distinguishes Berman syndrome from MPS and $G_{M1}$ gangliosidosis. Lack of identifiable enzyme changes in

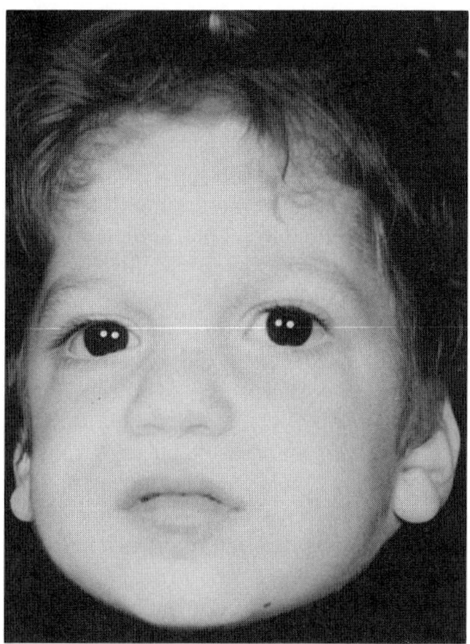

Fig. 5–38. *Mucosulfatidosis.* Coarse facial features. (Courtesy of *RD Burck,* New York, New York.)

serum and tissues rules out *sialidosis, I-cell disease,* and *pseudo-Hurler polydystrophy.*

Laboratory findings. There is no increase in urinary MPS. Phase contrast examination reveals numerous 1 × 2 μm inclusions in cultured fibroblasts similar to those in I-cell disease.

Diagnosis is confirmed by electron microscopic observation of characteristic storage bodies (single membrane, limited vesicles), filled with fibrillogranular material (MPS) and lamellar cytoplasmic bodies (phospholipids) in biopsied tissues (cornea, conjunctiva, skin, fibroblasts). Fibroblasts have reduced (but not absent) ganglioside α-neuraminidase levels (3). The disorder has been diagnosed prenatally by ultrastructural examination of cultured amniotic fluid cells and chorionic villi for inclusions (9,14).

#### References [Berman syndrome (mucolipidosis IV)]


1. Amir N et al: Mucolipidosis type IV: Clinical spectrum and natural history. *Pediatrics* **79**:953–959, 1987.
2. Bach G et al: Ganglioside sialidase deficiency. *Biochem Biophys Res Commun* **90**:1341–1347, 1979.
3. Ben-Yoseph Y et al: Catalytically defective ganglioside neuraminidase in mucolipidosis IV. *Clin Genet* **21**:374–381, 1982.
4. Berman ER et al: Congenital corneal clouding with abnormal systemic storage bodies: A new variant of mucolipidosis. *J Pediatr* **85**:519–526, 1974.
5. Caimi L et al: Mucolipidosis IV, a sialolipidosis due to ganglioside sialidase deficiency. *J Inherit Metab Dis* **5**:218–224, 1982.
6. Crandall BF et al: Mucolipidosis IV. *Am J Med Genet* **12**:301–308, 1982.
7. Goutières F et al: Mucolipidosis IV. *Neuropaediatrie* **10**:321–331, 1981.
8. Kenyon KR et al: Mucolipidosis IV: Histopathology of conjunctiva, cornea and skin. *Arch Ophthalmol* **97**:1106–1111, 1979.
9. Kohn G et al: Prenatal diagnosis of mucolipidosis IV by electron microscopy. *J Pediatr* **90**:62–66, 1977.
10. Lake BD et al: A mild variant of mucolipidosis type 4 (ML4). *Birth Defects* **18**(6):391–404, 1982.

10a. Livni N. Merin S: Mucolipidosis IV. *Arch Pathol Lab Med* **102**:600–604, 1978.

11. Merin S et al: Mucolipidosis IV: Ocular, systemic, and ultrastructural findings. *Invest Ophthalmol* **14**:437–448, 1975.
12. Merin S et al: The cornea in mucolipidosis IV. *J Pediatr Ophthalmol* **13**:289–295, 1976.
13. Newell FW et al: A new mucolipidosis with psychomotor retardation, corneal clouding and retinal degeneration. *Am J Ophthalmol* **80**:440–449, 1975.
14. Ornoy A et al: Early prenatal diagnosis of mucolipidosis IV. *Am J Med Genet* **27**:983–985, 1987.
15. Riedel KG et al: Ocular abnormalities in mucolipidosis IV. *Am J Ophthalmol* **99**:125–136, 1985.
16. Tellez-Nagel I et al: Mucolipidosis IV: Clinical, ultrastructural, histochemical and chemical studies of a case including brain biopsy. *Arch Neurol* **33**:828–835, 1976.
17. Zlotogora J et al: A muscle disorder as presenting symptom in a child with mucolipidosis IV. *Neuropediatrics* **14**:104–105, 1983.
18. Zwaan J, Kenyon KR: Two brothers with presumed mucolipidosis IV. *Birth Defects* **18**(6):381–390, 1982.


**Mucosulfatidosis (Austin syndrome, multiple sulfatase deficiency).** The clinical features of mucosulfatidosis, a disorder described by Austin (1) in 1973, are those found in steroid sulfate sulfatase deficiency (ichthyosis), mucopolysaccharidoses [dysostosis multiplex, psychomotor delay, coarse facial features (Fig. 5-38), hearing loss, hepatosplenomegaly], and late infantile metachromatic leukodystrophy (motor weakness, psychomotor delay, demyelinization, and gliosis of white matter of the brain) (1–11). An early example is that of Thieffrey et al (10).

Presenting during the first 2 years of life, the children gradually lapse into a vegetative state and succumb during the first decade.

The decrease in arylsulfatase A, B, and C activities is caused either by impaired enzyme production or excessive degradation. There are mucopolysacchariduria and sulfatiduria. Inheritance is autosomal recessive.

An increase in cerebrospinal fluid protein has been demonstrated. Alder–Reilly granules are found in leukocytes.

The disorder is most often confused with MPS II.

#### References [Mucosulfatidosis (Austin syndrome, multiple sulfatase deficiency)]


1. Austin J: Multiple sulfatase deficiency. *Arch Neurol* **28**:258–264, 1973.
2. Burch M et al: Multiple sulphatase deficiency presenting at birth. *Clin Genet* **30**:409–415, 1986.
3. Burk RD et al: Early manifestations of multiple sulfatase deficiency. *J Pediatr* **104**:574–578, 1984.
4. Couchot J et al: La mucosulfatidose: étude de trois cas familiaux. *Arch Fr Pédiatr* **31**:775–795, 1974.
5. Crawfurd M d'A: Genetics of steroid sulphatase deficiency and X-linked ichthyosis. *J Inherit Metab Dis* **5**:153–163, 1982.
6. Perlmutter-Cremer N et al: Unusual early manifestation of multiple sulfatase deficiency. *Ann Radiol* **24**:43–48, 1981.
7. Rampini S et al: Die Kombination von metachromatischer Leukodystrophie und Mukopolysaccharidose als selbstandiges Krankeitsbild (Mukosulfatidose). *Helv Paediatr Acta* **5**:436–461, 1970.
8. Rose FA: The mammalian sulfatases and placental sulfatase deficiency in man. *J Inherit Metab Dis* **5**:145–152, 1982.
9. Soong BW et al: Multiple sulfatase deficiency. *Neurology* **28**:1273–1275, 1988.
10. Thieffry S et al: Leukodystrophie métachromique (sulfatidose) et mucopolysaccharidose associées chez un meme malade. *Rev Neurol* **114**:193–200, 1966.
11. Vamos E et al: Multiple sulphatase deficiency with early onset. *J Inherit Metab Dis* **4**:103–104, 1981.


## Metabolic disorders with dysmorphic features

**Fabry syndrome (angiokeratoma corporis diffusum universale).** The clinical syndrome was described independently in 1898 by two dermatologists, Anderson (2) in England and Fabry (21) in Germany. The characteristic cutaneous lesions led Fabry to term this disease *angiokeratoma corporis diffusum universale.* The disorder results from the deficient activity of α-galactosidase A (9,26), a lysosomal enzyme encoded by a recently cloned gene (8) located on the long arm of the X chromosome (Xq22) (28a). The enzymatic defect leads to the systemic accumulation of the glycolipid globotriaosylceramide, particularly in the plasma (39) and lysosomes of the vascular endothelium and smooth muscle cells (25). The progressive endothelial glycolipid deposition in affected males results in ischemia and infarction, and leads to the major clinical manifestations (18). The disease is inherited as an X-linked recessive trait with complete penetrance and

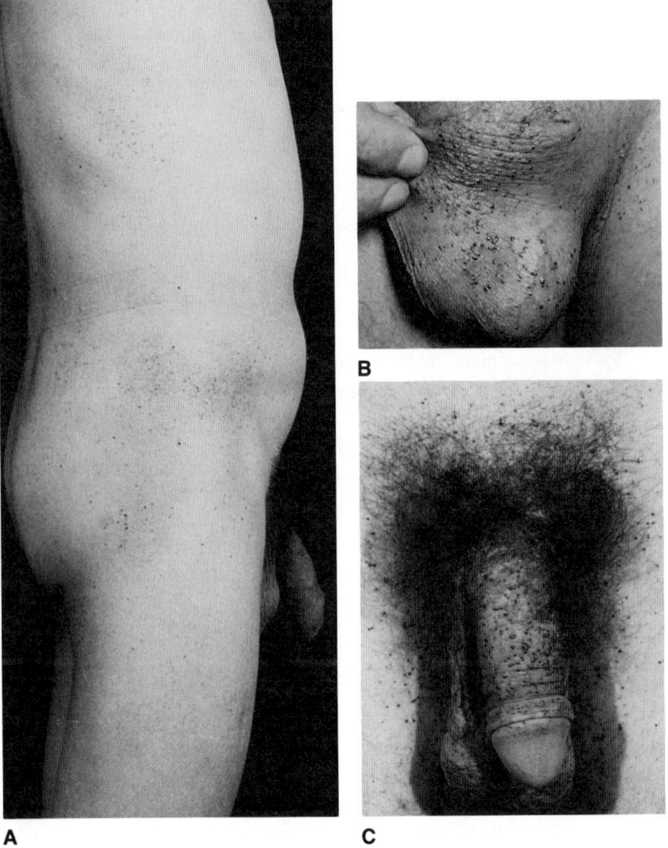


Fig. 5–39. *Fabry syndrome.* (A) Note distribution of skin lesions in 27-year-old man. (B,C) Small, somewhat raised, vascular lesions on scrotum and penis of patients. (A,B from *A Rahmen,* Trans Assoc Am Physician **75:**371, 1961.)

variable clinical expressivity in hemizygous males (32). Disease expression in most heterozygous females is limited to the keratopathy; however, a few have been described with manifestations as severe as those of affected males (18). Over 300 affected hemizygotes have been described. A recent, comprehensive review is available (19).

Onset of the disease in affected males usually occurs during childhood or adolescence. Early manifestations include periodic crises of severe pain in the extremities (acroparesthesias), the appearance of angiokeratomas, hypohidrosis, and the characteristic corneal and lenticular opacities. With advancing age, progressive vascular involvement causes ischemia and infarction leading to cardiac, cerebral, and renal vascular disease with death typically occurring the the fourth or fifth decade of life. Affected individuals who are blood group B or AB have a more severe course since the blood group B substance also accumulates due to the deficient $\alpha$-galactosidase A activity (18).

Facies. There is no pathognomonic facies; however, frontal bossing and prominent lower jaw and lips have been reported in several male patients. Many patients appear young for chronologic age (41).

Skin. The cutaneous vascular lesions (angiokeratomas) are telangiectases, which usually appear as clusters of individual punctate, dark-red to blue angiectases in the superficial layers of the skin. The lesions may be flat or slightly raised and do not blanch with pressure. There may be a slight hyperkeratosis over these lesions. They usually appear during childhood and, with age, progressively increase in size and number. Characteristically, these lesions are most dense between the umbilicus and the knees (over the iliosacral area, scrotum, posterior thorax, thighs, buttocks, and umbilicus) and have a tendency toward bilateral symmetry (18) (Fig. 5-39A–C). The face, with the exception of the submental area, may be involved.

Variants without the characteristic skin lesions have been described (1,3,4,7,12). Hypohidrosis is a common symptom, and atrophic or sparse sweat and sebaceous glands have been reported (34,37). Males shave infrequently, and body hair may be slight.

Eyes. Ocular manifestations include aneurysmal dilatation and tortuosity of conjunctival and retinal vessels as well as characteristic corneal and lenticular changes (35). The conjunctival and retinal vascular lesions are common and part of the diffuse systemic vascular involvement. The keratopathy is characterized by diffuse haziness and whorled streaks extending from a central vortex in the corneal epithelium (40). The corneal lesions resemble the changes seen in chloroquine intoxication and must be observed by slit lamp microscopy (18); they occur in all hemizygous males and in about 80% of heterozygous females. The lenticular changes include a granular anterior capsular or subcapsular deposit seen in about 30% of males and a unique linear opacity (termed the "Fabry cataract") in hemizygous males and some heterozygous females that is best observed by retroillumination. The opacity appears as a whitish, spoke-like deposit of fine granular material near the posterior capsule (35). These lesions do not impair vision.

Cardiac, cerebral, and renal vascular manifestations. With increasing age, the major morbid symptoms result from the progressive involvement of the vascular system. Early in the course of the disease, casts, red cells, and lipid inclusions with characteristic birefringent "Maltese crosses" appear in the urinary sediment. Proteinuria, isosthenuria, and gradual deterioration of renal function and development of azotemia occur in the second to fourth decades of life (18). Renal transplantation and chronic hemodialysis are lifesaving (14,18). Cardiovascular findings may include hypertension, left ventricular hypertrophy, anginal chest pain, myocardial ischemia or infarction, cardiomyopathy, and congestive heart failure (4,5,10,16,23,29). Mitral insufficiency is the most common valvular lesion (4,16,23,34a). Abnormal electrocardiographic and echocardiographic findings are common (4,5,23,29). Cerebrovascular manifestations result primarily from multifocal small vessel involvement. Death most often results from uremia or vascular disease of the heart or brain (18). The mean age of death for affected males who did not receive hemodialysis or renal transplantation was 41 years (13).

Microscopic examination of various tissues demonstrates the accumulation of the glycolipid, predominantly in endothelial, perithelial, and smooth muscle cells of blood vessels, epithelial cells of the cornea and of glomeruli and tubules of the kidney, muscle fibers of the heart, ganglion cells of the autonomic nervous system, and peripheral Schwann cells (36). Foamy, lipid-laden macrophages are seen in the bone marrow and lymph nodes.

Acroparesthesias. The single most debilitating symptom of this disease is pain. Typically, affected males experience episodic crises of excruciating burning pain in fingers and toes in childhood that may become more frequent and severe in adolescence. These painful acroparesthesias may last several days to a week and are associated with low-grade fever and elevation of the erythrocyte sedimentation rate; these symptoms have led to the misdiagnosis of rheumatic fever. During the second and third decades of life, these recurrent, painful episodes may occur only infrequently, usually associated with a fever. However, in few patients, they may become progressively more frequent and severe, radiate to proximal extremities, and occasionally persist for 1–2 weeks. Affected individuals may be incapacitated for prolonged periods of time with pain that is so unrelenting that suicide has been attempted. It has been suggested that the etiology of the acroparesthesias may be due to impaired autonomic function (11) and involvement of the peripheral nervous system (31). Their frequency and severity are decreased by use of phenylhydantoin and carbamazepine (27,28).

Other clinical findings. Nausea, vomiting, diarrhea, and abdominal or flank pain are common gastrointestinal symptoms (30,33). Other less frequent features include massive lymphedema of the legs (Fig.

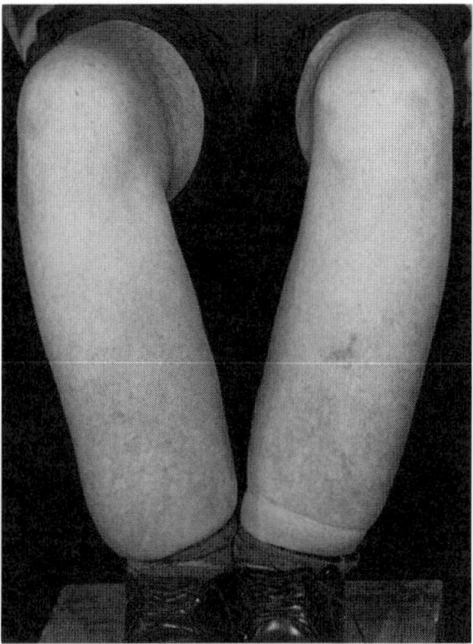

Fig. 5–40. *Fabry syndrome.* Extensively swollen lower legs.

5-40) and dyspnea. Musculoskeletal system findings have included a permanent deformity of the distal interphalangeal joints of the fingers (41) and avascular necrosis of the head of the femur and talus (22). Mild normochromic, normocytic anemia, presumably caused by decreased red cell survival, has been observed. Many hemizygotes appear to have growth retardation or delayed puberty.

Oral manifestations. The majority of patients have symmetric, pinpoint, macular, purplish spots (angiokeratomas) on the lips, particularly on the lower lip near the skin–mucosal junction, on either side of the midline (41) (Fig. 5-41). The lesions are smaller than those on the skin. The buccal mucosa appears to be involved to a lesser degree (18,41). The gingiva, soft palate, and uvula are only rarely involved (18,38,41). The tongue is not affected. Involvement of the nasal mucosa with resultant epistaxis has been reported (41). Glycosphingolipid accumulation has been demonstrated in dental pulp from hemizygous males (20).

Differential diagnosis. The condition can be diagnosed in hemizygous males by a history of acroparesthesias, the presence of characteristic skin lesions, and the observation of the corneal and/or lenticular changes. Heterozygotes usually are asymptomatic; however, about 80% have corneal lesions that can be observed by slit-lamp microscopy.

The skin lesions are so characteristic in distribution that need for differential diagnosis is extremely limited. The lesions of *hereditary hemorrhagic telangiectasia* are larger, do not involve the lower trunk and thighs, and are less numerous and more irregular. The Fordyce type of angiokeratoma is usually limited to the scrotum, and the Mibelli type forms warty lesions on the extremities or ears (18,41).

Fig. 5–41. *Fabry syndrome.* Pinpoint lesions on upper lip. Oral lesions are generally limited to the lips and do not involve the tongue.

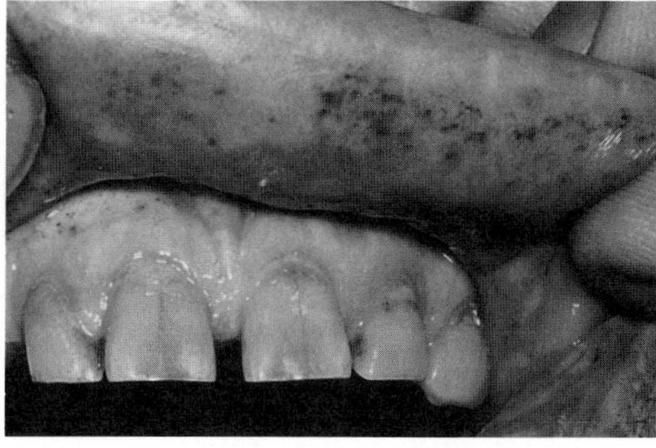

Laboratory aids. Histological and ultrastructural examination of biopsied skin, kidney, or other tissues will reveal birefringent lipid inclusions and lamellar inclusions in lysosomes, respectively (18). All suspect cases should be confirmed enzymatically by the demonstration of deficient $\alpha$-galactosidase A activity in plasma, isolated leukocytes, tears, or cultured fibroblasts or lymphoblasts (15, 18, 24, 26). Prenatal detection can be accomplished by the demonstration of deficient $\alpha$-galactosidase A activity in chorionic villi obtained in the first trimester or in cultured amniocytes obtained by amniocentesis in the second trimester of pregnancy.

The recent cloning of the cDNA encoding $\alpha$-galactosidase A (8) has permitted characterization of the molecular nature of the mutations that led to the enzymatic defect. Approximately 6% of the 75 unrelated families studied had gene rearrangements (partial gene deletions or duplications). In addition, restriction fragment length polymorphisms (RFLPs) have been identified in about 20% of Fabry families studied using the cDNA as a gene-specific probe (6). Thus, heterozygotes can be definitively diagnosed in informative families using the cloned $\alpha$-galactosidase A cDNA probe.

### Reference [Fabry syndrome (angiokeratoma corporis diffusum universale)]


1. Ainsworth SK, Smith RM: A case study of Fabry's disease occurring in a black kindred without peripheral neuropathy or skin lesions. *Lab Invest* **38**:373–379, 1978.

2. Anderson W: A case of "angiokeratoma." *Br J Dermatol* **10**:113–117, 1898.

3. Bach G et al: Pseudodeficiency of alpha-galactosidase A. *Clin Genet* **21**:59–64, 1982.

4. Bass JL et al: The M-mode echocardiagram in Fabry's disease. *Am Heart J* **100**:807–812, 1980.

5. Becker AE et al: Cardiac manifestations of Fabry's disease. Report of a case with mitral insufficiency and electrocardiographic evidence of myocardial infarction. *Am J Cardiol* **36**:829–835, 1975.

6. Bernstein HS et al: Fabry disease: Analysis of mutations in the human alpha-galactosidase A gene. *Am J Hum Genet* **39**:A188, 1986.

6a. Bird TD, Lagunoff D: Neurological manifestations of Fabry disease in female carriers. *Ann Neurol* **4**:537–540, 1978.

7. Bishop DF et al: Fabry disease: An asymptomatic hemizygote with significant residual alpha-galactosidase A activity. *Am J Hum Genet* **33**:71A, 1982.

8. Bishop DF et al: Human alpha-galactosidase A: Nucleotide sequence of a cDNA clone encoding the mature enzyme. *Proc Natl Acad Sci USA* **83**:4859–4863, 1986.

9. Brady RO et al: Enzymatic defect in Fabry's disease: Ceramide trihexosidase deficiency. *N Engl J Med* **275**:1163–1167, 1967.

10. Broadbent JC et al: Fabry cardiomyopathy in the female confirmed by endomyocardial biopsy. *Mayo Clin Proc* **56**:623–628, 1981.

11. Cable WJL et al: Fabry disease: Impaired autonomic function. *Neurology* **32**:498–502, 1982.

12. Clarke JT et al: Ceramide trihexosidosis (Fabry's disease) without skin lesions. *N Engl J Med* **284**:233–235, 1971.

13. Colombi A et al: Angiokeratoma corporis diffusum—Fabry's disease. *Helv Med Acta* **34**:67–72, 1967.

14. Desnick RJ et al: Treatment of Fabry's disease: Correction of the enzymatic deficiency by renal transplantation. *J Lab Clin Med* **78**:989–990, 1971.

15. Desnick RJ et al: Enzymatic diagnosis of hemizygotes and heterozygotes: alpha-Galactosidase activities in plasma, serum, urine, and leukocytes. *J Lab Clin Med* **81**:157–171, 1973.

16. Desnick RJ et al: Cardiac valvular anomalies in Fabry's disease: Clinical, morphologic and biochemical studies. *Circulation* **54**:818–825, 1976.

17. Desnick RJ et al: Enzyme therapy XII: Enzyme therapy in Fabry's disease: Differential enzyme and substrate clearance kinetics of plasma and splenic alpha-galactosidase A. *Proc Natl Acad Sci USA* **76**:5326–5330, 1979.

18. Desnick RJ, Bishop DF: Fabry disease: Alpha-galactosidase deficiency, in *The Metabolic Basis of Inherited Disease,* 6th ed, Scriver CR et al (eds), McGraw–Hill, New York, 1989, pp. 1751–1796.

19. Desnick RJ, Sweeley CC: Fabry disease: alpha-Galactosidase A deficiency (angiokeratoma corporis diffusum universale), in *Dermatology in General Medicine,* 3rd ed, Fitzpatrick TB, Eisen AZ, Wolff K, Freedberg IM, Austin KF (eds), McGraw-Hill, New York, 1987, pp 1739–1760.

20. Desnick SJ et al: Fabry's disease (ceramide trihexosidase deficiency): Diagnostic confirmation by analysis of dental pulp. *Arch Oral Biol* **17**:1473–1479, 1972.

21. Fabry J: Ein Beitrag zur Kenntnis der Purpura haemorrhagica nodularis (Purpura papulosa hemorrhagica Hebrae). *Arch Dermatol Syphilol (Berl)* **43**:187–200, 1898.

22. Fone DJ, King WE: Angiokeratoma corporis diffusum (Fabry's syndrome). *Aust Ann Med* **13**:339–348, 1964.

23. Goldman M et al: Echocardiographic abnormality and disease severity in Fabry disease. *J Am Coll Cardiol* **7**:1157–1161, 1986.

24. Johnson DL et al: Fabry disease: Diagnosis of hemizygotes and heterozygotes by alpha-galactosidase A activity in tears. *Clin Chim Acta* **63**:81–90, 1975.

25. Johnson DL, Desnick RJ: Molecular pathology of Fabry's disease: Physical and kinetic properties of alpha-galactosidase A in cultured human endothelial cells. *Biochim Biophys Acta* **538**:195–204, 1978.

26. Kint JA: Fabry's disease: α-Galactosidase deficiency. *Science* **167**:1268–1269, 1970.

27. Lenoir G et al: La maladie de Fabry. Traitement du syndrome acrodyniforme par la carbamazepine. *Arch Fr Pédiatr* **34**:704–709, 1977.

28. Lockman LA et al: Relief of pain of Fabry's disease by diphenylhydantoin. *Neurology* **23**:871–875, 1973.

28a. MacDermot KD et al: Anderson-Fabry disease: A close linkage with highly polymorphic DNA markers: DXS17 DXS87 and DXS88. *Hum Genet* **77**:263–266, 1987.

29. Mehta J et al: Electrocardiographic and vectocardiographic abnormalities in Fabry's disease. *Am Heart J* **93**:699–705, 1977.

30. O'Brien BD et al: Pathophysiologic and ultrastructural basis for intestinal symptoms in Fabry's disease. *Gastroenterology* **82**:957–960, 1982.

31. Ohnishi A, Dyck PJ: Loss of small peripheral sensory neurons in Fabry disease. *Arch Neurol* **31**:120–127, 1974.

32. Opitz JM et al: The genetics of angiokeratoma corporis diffusum (Fabry's disease), and its linkage with Xg(a) locus. *Am J Hum Genet* **17**:325–342, 1965.

33. Rowe JW et al: Intestinal manifestations of Fabry's disease. *Ann Intern Med* **81**:628–631, 1974.

34. Sagebiel RW, Parker F: Cutaneous lesions of Fabry's disease: Glycolipid lipidosis—light and electron microscopic findings. *J Invest Dermatol* **50**:208–213, 1968.

34a. Sakuraba H et al: Cardiovascular manifestations in Fabry's disease: A high incidence of mitral valve prolapse in hemizygotes and heterozygotes. *Clin Genet* **29**:276–283, 1986.

35. Sher NA et al: The ocular manifestations in Fabry's disease. *Arch Ophthalmol* **97**:671–676, 1979.

36. Sung JH: Autonomic neurons affected by lipid storage in the spinal cord of Fabry's disease: Distribution of autonomic neurons in the sacral cord. *J Neuropathol Exp Neurol* **38**:87–96, 1979.

37. Tarnowski WM, Hashimoto K: New light microscopic skin findings in Fabry's disease. *Acta Derm Venereol* **49**:386–393, 1969.

38. Uono M: Fabry's disease—from the standpoint of neuronal lipidosis. *Jpn J Clin Med* **25**:1587–1594, 1967.

39. Vance DE et al: Concentrations of glycosyl ceramides in plasma and red cells in Fabry's disease. *J Lipid Res* **10**:188–192, 1969.

40. Wallace HJ: Anderson–Fabry disease. *Br J Dermatol* **88**:1–24, 1973.

41. Wise D et al: Angiokeratoma corporis diffusum: A clinical study of eight affected families. *Q J Med* **31**:177–206, 1962.


**Homocystinuria (cystathionine synthase deficiency).** Homocystinuria (cystathionine β-synthase deficiency) is accompanied by a distinctive clinical syndrome that includes ectopia lentis, arteriovenous thromboembolic episodes, dolichostenomelia, mental retardation, and osteoporosis (Figs. 5-42–5-45). It was first clearly distinguished from Marfan syndrome by Carson and Neill (6) and by Gerritsen et al (15). Delineation of the biochemical defect followed soon after (16,17,27). More recently, our knowledge of the natural history of this condition has been expanded by the remarkable international survey of Mudd and others (30), compiling the principal clinical findings for 629 affected individuals. An excellent review is that of Przyrembel (31a).

Etiology and genetics. Cystathionine β-synthase is a heterodimeric enzyme whose size varies with the tissue of origin (37,38). Both subunits arise from the same parent polypeptide of molecular weight 63,000 but they are cleaved to fragments of 48,000 molecular weight in the process of enzyme activation. Thus, the size of the molecule and its

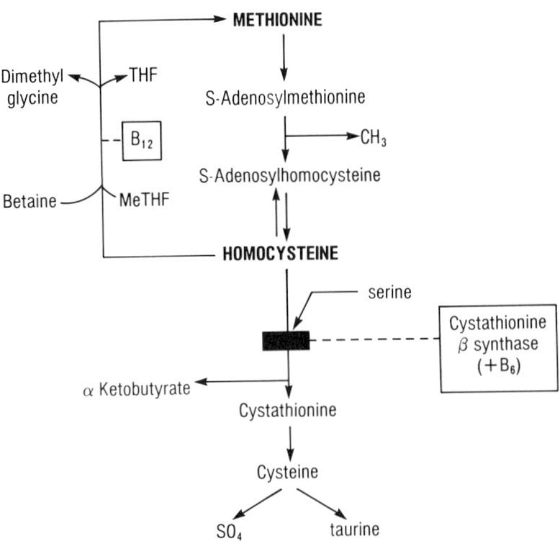


Fig. 5–42. *Homocystinuria.* Transsulfuration pathway. Cystathionine β-synthase is key enzyme for conversion of sulfur-containing enzyme methionine to cysteine.

activity are probably a function of tissue-specific proteolytic enzymes (38). Cystathionine β-synthase is a key enzyme in the transsulfuration pathway that is responsible for conversion of the sulfur-containing amino acid methionine to cysteine before it can be ultimately catabolized to its component parts—carbon dioxide, urea, and inorganic sulfate (Fig. 5-42). The enzyme normally requires pyridoxal-5′-phosphate, an active form of vitamin $B_6$, as a cofactor and catalyzes the replacement of the β-hydroxyl group of serine (another amino acid) by homocysteine to form cystathionine. In an important salvage pathway, homocysteine can also acquire a methyl group from an appropriate methyl donor, such as methylenetetrahydrofolate ($MeH_4F$) or betaine (trimethylglycine), to regenerate methionine. A form of vitamin $B_{12}$ known

Fig. 5–43. *Homocystinuria.* Marfanoid habitus and genua valga.

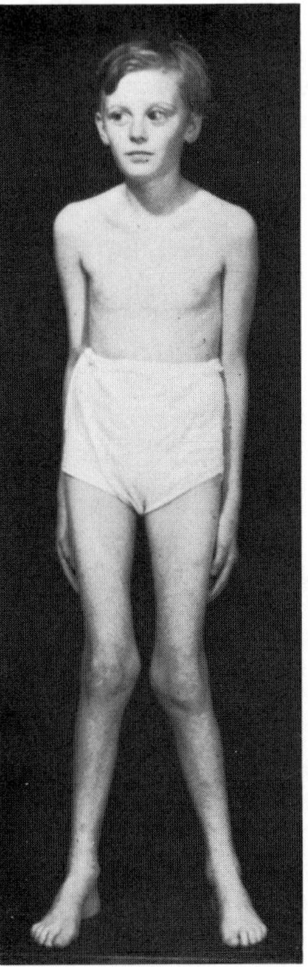

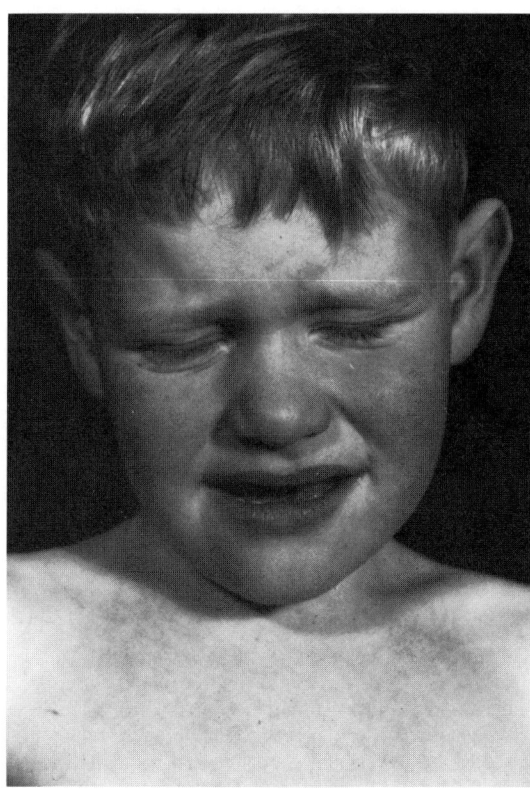

Fig. 5–44. *Homocystinuria.* Note malar flushing. (From *NAJ Carson* and *G Gaull,* New York.)

as methylcobalamin (MeCbl) is essential to this remethylation reaction (29).

Cystathionine β-synthase deficiency is an autosomal recessive disorder with an estimated birth prevalence of about 1/200,000 (12). These figures are partly based on data from newborn screening programs that probably fail to recognize all patients. The much higher prevalence estimates in certain local surveys [e.g., New South Wales (1/58,000); Northern Ireland (1/24,000)] further indicate an underestimate of the true frequency in populations with largely Irish ancestry (6,7,29,45).

In cystathionine β-synthase deficiency, both methionine and homocystine accumulate in various tissues, as well as in blood and urine. The deficiency has been studied in detail through the culture and biochemical or immunological characterization of mutant fibroblasts (3,35–40). These studies show an extremely high concordance between the severity of the clinical syndrome and the extent of the enzyme defect, suggesting that there is little heterogeneity of the genetic locus (40). This is supported by the evidence of somatic cell hybridization that provisionally assigns a single gene for the enzyme to human chromosome 21 (36). However, there appear to be many different mutant alleles, judging from the heterogeneity of clinical, enzymatic, and immunochemical findings (21,39,40,44). This allelic heterogeneity has clinical implications. Those individuals with no enzyme activity are essentially unresponsive to treatment with any amount of vitamin $B_6$; those with residual activity *in vitro* are, as a rule, $B_6$ responsive in proportion to that activity (35,42). The degree of residual activity is relatively constant among family members, as are the clinical manifestations. More recent studies have also revealed the existence of allelic compound heterozygotes—individuals with two different mutations for the cystathionine β-synthase gene—who may manifest a syndrome indistinguishable from that in apparently homozygous mutants. Studies of messenger RNA production for the enzyme in fibroblasts indicate that there are few deletional mutations that span the entire cystathionine β-synthase gene (40). The molecular details of smaller mutations will no doubt be reported in the near future. The pyridoxine effect is discussed elsewhere (14,28,29,35,39,40).

Fig. 5–45. *Homocystinuria.* Inferior dislocation of lens associated with acute glaucoma. (From *MC Carey,* Am J Med **45**:7, 1968.)

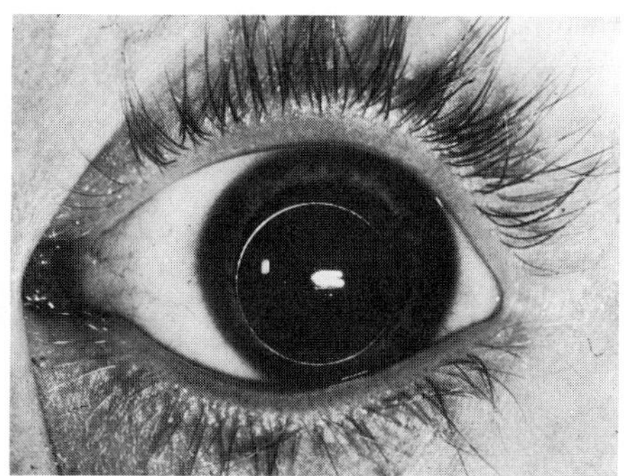

Pathophysiology. Homocysteine is an amino acid with an active sulfhydryl side chain and readily forms disulfide bonds in place of its physiologic analog, cysteine. The proteins most susceptible to these abnormal reactions appear to be the collagens, although other structural peptides such as elastin may also be affected. Used as a drug, penicillamine is another sulfhydryl compound that can produce changes similar to those associated with cystathionine β-synthase deficiency via the same mechanism. The end result is weakened fibrous tissue that leads to altered bone formation, rupture of the ocular ligaments of the lens, and abnormalities in joints, skin, and blood vessels (29).

Previously, the neurologic abnormalities have been thought to arise from deficiency of the enzyme product, cystathionine, which is a putative CNS neurotransmitter. More recently, the mental deficit has been attributed to high concentrations of a potentially toxic precursor, *S*-adenosylhomocysteine (Fig. 5-42). Adenosine itself is a known neurotoxin *in vivo* and *S*-adenosylmethionine accumulation leads to significant cytotoxic changes in cell culture. A third view—that the neurological changes are the result of multiple subclinical thromboembolic events—is not supported by autopsy examinations of brain (29).

The relationship of homocystinuria to thrombosis and atherosclerosis has fascinated investigators since the disorder was first described (23). Much attention has been focused on abnormalities of platelet function (24), but most recent investigations suggest that platelet adhesiveness, ultrastructural features, and survival time (19) are normal. There is also no convincing evidence for an abnormality in the coagulation cascade (29).

However, there are quite distinctive changes in the structure of the walls of large and small arteries, and vascular endothelial cell dysfunction has been described (11). Focal intimal and medial fibrosis together with perivascular connective tissue proliferation and widening of the internal elastic lamina have been linked to vascular damage initiated by high homocysteine concentrations. These lesions are also observed in animal studies with homocysteine loading and in homocystinuria as a result of 5,10-methylenetetrahydrofolate reductase (MTHFR) deficiency (Table 5-4) (29)

Clinical manifestations.

*General.* Individuals with cystathionine β-synthase deficiency are normal at birth. Although the disorder is usually identified in childhood, more than 90% of patients reach their sixteenth birthday and more than 75% survive to the fifth decade (30).

*Ocular.* Ectopia lentis (Fig. 5-45) is a hallmark, being mentioned in 86% of the 472 case reports that give an initial presenting sign (30). Dislocation results from disruption of the suspensory ligaments and is downward in the majority of cases but may occur in any direction. Common secondary features include iridodonesis (particularly evident when the head is moved), marked myopia, and astigmatism. Anterior dislocation may cause pupillary block, glaucoma, and acute ocular pain. Cataract is usualy the result of lens trauma (29).

Lens dislocation has occurred in more than 50% of patients by age 8 and in more than 90% by age 25 (30). $B_6$ responders are less likely

Table 5–4. Causes of homocystinuria

| |
|---|
| Cystathionine β-synthase deficiency |
| 5,10-Methylenetetrahydrofolate reductase (MTHFR) deficiency |
| Disorders of $B_{12}$ metabolism |
| Nutritional deficiency |
| Defective $B_{12}$ absorption |
| Defective methylcobalamin formation |
| Cobalamin C and D diseases (cytosolic cobalamin reductase deficiency) |
| Cobalamin E disease (methionine synthase and associated cobalamin reductase deficiency) |

to dislocate regardless of treatment, but early treatment appears to offer added protection to this group. Other reported ocular findings are optic atrophy, cystic degeneration of the retina, and retinal detachment (9,17,29).

*Skeletal system.* Affected individuals are tall and thin as children. True arachnodactyly and Marfanoid habitus with radiographic evidence of excessively long, thin bones (dolichostenomelia) are seen in one-third of the cases (2,29). Other somatic manifestations include joint contractures, joint laxity, pes planus, pes cavus, pectus carinatum, pectus excavatum, kyphoscoliosis, widened metaphyses and epiphyses (most recognizable at the knee), genua valga, and increased carrying angle at the elbow (Fig. 5-43). Associated radiographic features include tibial growth arrest lines, metaphyseal spicules, enlarged carpal bones, short fourth metacarpals, elongated talus, and retarded calcification of the lunate bone in the hand (4,26,29,33,41,42).

Of all radiographic features, osteoporosis is the most frequent finding (41). Although present in less than 5% of patients under 4 years of age, it affects 70% of individuals over 20 (30). It appears to be less severe in the $B_6$-responsive group. It is most often identified in the spine and is associated with biconcavity of the vertebral bodies. Some have suggested that the latter finding is not a feature of osteoporosis per se but is of vascular origin, as the biconcavity is posteriorly placed—a feature typical of the osteopenia associated with chronic hemolytic diseases such as sickle cell anemia (29). However, vertebral wedging, kyphoscoliosis, and pathological fractures that heal slowly are all directly attributable to loss of bone mass (29). The progression of osteoporosis is less rapid in the $B_6$ responders, but the effect of $B_6$ treatment is not clear (30).

*Thromboembolic events.* Thrombosis and embolism are life-threatening complications in this disorder. Large or small arteries and veins may be affected, and vascular occlusions can occur at any age (29). In the international survey of 629 patients, 253 events were reported in 29% of patients (30). Of these, 32% were cerebrovascular accidents, 51% involved peripheral veins (including 13% pulmonary embolism), 4% produced myocardial infarction, and 11% affected peripheral arteries or damaged major organs. In spite of a 4-year, event-free period in early infancy, there was a very sharp increase in thromboembolic events at puberty. By the age of 30, half of the reported patients had experienced at least one event. $B_6$ responders were only marginally better off than nonresponders and the beneficial effects of $B_6$ therapy were not clearly demonstrable. Surgery poses a significantly increased risk for thromboembolism (30).

*Central nervous system.* The degree of mental retardation associated with homocystinuria varies widely. Both profound retardation (IQ$<$30) and normal intelligence are described. In the international survey, the median IQ for $B_6$ responders was 86 but for nonresponders was 64, a highly significant difference (30).

Seizures are characteristic, being reported at an earlier age in $B_6$-nonresponsive patients. The great majority are of the grand mal type and affect 21% of all patients (30). Also reported are nonspecific EEG changes (10), psychiatric disturbances (including schizophenia), spasticity, and hyperreflexia (29). In some cases, these latter two signs are related to a cerebrovascular thromboembolic event, but in others, they have been attributed to the disorder itself (29–31). Abbott et al (1) found significant psychiatric disorders in 51% in their series.

*Miscellaneous.* Other connective tissue findings include bilateral malar flush (Fig. 5-44) (50%) and livedo reticularis. Hepatomegaly caused by fatty liver is observed occasionally, and electromyographic abnormalities have been found in those with clinical myopathy. There is an increased incidence of bilateral inguinal hernia, and omphalocele has been described (29).

*Oral manifestations.* The palate is often narrow and highly arched. The teeth have been reported to be crowded and irregularly aligned. Mandibular prognathism has been noted (29).

**Diagnosis and laboratory aids.** Mass neonatal screening programs detect some of the patients with homocystinuria (45). They are more likely to miss those with milder disease who, in turn, are more likely to respond effectively to pyridoxine (29). Infants and children at risk are usually screened by the cyanide–nitroprusside spot test for urine, but other disorders of sulfur amino metabolism will also produce a positive test (8,43). Moreover, false negatives have been known to occur. Quantitation of serum and urine amino acids by column chromatography reliably shows abnormal amounts of homocysteine in both fluids, but diagnostic proof, particularly in individuals without clinical findings, rests with the demonstration of enzyme deficiency in fibroblast culture (21,28,35–40,44). Prenatal diagnosis is possible using this assay on cultured amniocytes (3,13).

**Differential diagnosis.**

*Biochemical.* Other causes of homocystinuria are listed in Table 5-4. Inborn errors affecting the remethylation pathway from homocysteine to methionine (Fig. 5-42) constitute the majority of these.

Inherited deficiency of the MTHFR enzyme is an autosomal recessive condition characterized by mental retardation and sometimes early death (34). Enzyme activity is impaired if the $B_{12}$ cofactor is deficient, a situation that can arise if there is dietary $B_{12}$ deficiency, a $B_{12}$ malabsorption syndrome, or a defect in the conversion of $B_{12}$ to the methylcobalamin form used by the remethylation enzyme (18,25,31,32). Because they are unable to resynthesize methionine, these patients have low plasma methionine, clearly distinguishing them from patients with cystathionine β-synthase deficiency. They may also have other abnormalities associated with disturbed $B_{12}$ metabolism, such as methylmalonic aciduria and/or megaloblastic anemia, but they do not have a Marfanoid somatotype or associated clinical features (32).

*Clinical.* Lens subluxation is a feature of other disorders of sulfur amino acid metabolism as well as *Marfan syndrome* and related disorders (8,20). The Marfanoid habitus is found in *Marfan syndrome,* congenital contractural arachnodactyly, *XXY syndrome, XYY syndrome,* sickle cell anemia, *nevoid basal cell carcinoma syndrome,* and *multiple mucosal neuroma syndrome.*

### References [Homocystinuria (cystathionine synthase deficiency)]


1. Abbott MH et al: Psychiatric manifestations of homocystinuria due to cystathionine β-synthase deficiency: Prevalence, natural history, and relationship to neurologic impairment and vitamin-$B_6$-responsiveness. *Am J Med Genet* **26**:959–969, 1987.
2. Beals RK: Homocystinuria: A report of two cases and review of the literature. *J Bone Jt Surg* **51A**:1564–1572, 1969.
3. Bittles AH, Carson N: Tissue culture techniques as an aid to prenatal diagnosis and genetic counselling in homocystinuria. *J Med Genet* **10**:120–123, 1973.
4. Brill PW et al: Homocystinuria due to cystathionine synthase deficiency: Clinical roentgenologic correlations. *Am J Roentgenol* **121**:45–54, 1974.
5. Carey MC et al: Homocystinuria. *Am J Med* **45**:7–25, 1968.
6. Carson NAJ, Neill DW: Metabolic abnormalities detected in a survey of mentally backward individuals in Northern Ireland. *Arch Dis Childh* **37**:505–513, 1962.
7. Carson NAJ et al: Homocystinuria: Clinical and pathological review of ten cases. *J Pediatr* **66**:565–583, 1965.
8. Crawhall JC: A review of the clinical presentation and laboratory findings in two uncommon hereditary disorders of sulfur amino acid metabolism, β-mercaptolactate cysteine disulfideuria and sulfite oxidase deficiency. *Clin Biochem* **18**:139–142. 1985.
9. Cross HE, Jensen AD: Ocular manifestations in the Marfan syndrome and homocystinuria. *Am J Ophthalmol* **76**:405–419, 1973.

10. DelGiudice A et al: Electroencephalographic abnormalities in homocystinuria due to cystathionine synthase deficiency. *Am Neurol Neurosurg* **85**:164–168, 1983.

11. DeGroot PG et al: Endothelial cell dysfunction in homocystinuria. *Eur J Clin Invest* **13**:405–410, 1983.

12. Finkelstein JD et al: Homocystinuria due to cystathionine synthase deficiency: The model of inheritance. *Science* **146**:785–787, 1964.

13. Fleischer LD et al: Homocystinuria: Investigations of cystathionine synthase in cultured fetal cells and the prenatal determination of genetic status. *J Pediatr* **85**:677–680, 1974.

14. Frimpter GW: Homocystinuria: Vitamin $B_6$ dependent or not? *Ann Intern Med* **71**:209–211, 1969.

15. Gerritsen T et al: The identification of homocystine in the urine. *Biochem Biophys Res Commun* **9**:493–496, 1962.

16. Gerritsen T, Waisman HA: Homocystinuria: Absence of cystathionine in the brain. *Science* **145**:588, 1964.

17. Gerritsen T, Waisman HA: Homocystinuria, an error in the metabolism of methionine. *Pediatrics* **33**:413–420, 1964.

18. Higginbottom MC et al: A syndrome of methylmalonic aciduria, homocystinuria, megaloblastic anemia and neurologic abnormalities in a vitamin $B_{12}$-deficient breast-fed infant of a strict vegetarian. *N Engl J Med* **299**:317–356, 1978.

19. Hill-Zobel RL et al: Kinetics and distribution of $^{111}$indium-labelled platelets in patients with homocystinuria. *N Engl J Med* **307**:781–786, 1982.

20. Jansen AD: Heritable ectopia lentis, in *Genetic and Metabolic Eye Disease*, Little, Brown, Boston, 1974, p 325.

21. Lipson MH et al: Affinity of cystathionine β-synthase for pyridoxal 5′-phosphate in cultured cells. *J Clin Invest* **66**:188–193, 1980.

22. Matsui SM et al: The natural history of the inherited methylmalonic acidemias. *N Engl J Med* **308**:857–861, 1983.

23. McCully KS, Ragsdale BD: Production of arteriosclerosis by homocystinuria. *Am J Pathol* **61**:1–8, 1970.

24. McDonald L et al: Homocystinuria, thrombosis and the blood-platelets. *Lancet* **1**:745–746, 1964.

25. Mitchell GA et al: Clinical heterogeneity in cobalamin C variant of combined homocystinuria and methylmalonic aciduria. *J Pediatr* **108**:410–415, 1986.

26. Morreels SH et al: The roentgenographic features of homocystinuria. *Radiology* **90**:1150–1158, 1968.

27. Mudd SH et al: Homocystinuria: An enzymatic defect. *Science* **143**:1443–1445, 1964.

28. Mudd SH et al: Homocystinuria due to cystathionine synthase deficiency: The effect of pyridoxine. *J Clin Invest* **49**:1762–1773, 1970.

29. Mudd SH, Levy HL: Disorders of transsulfuration, in *The Metabolic Basis of Inherited Disease*, Stanbury JB et al (eds), McGraw-Hill, New York, 1983, p 522.

30. Mudd SH et al: The natural history of homocystinuria due to cystathionine β-synthase deficiency. *Am J Hum Genet* **37**:1–31, 1985.

31. Newman G, Mitchell JRA: Homocystinuria presenting as multiple arterial occlusions. *Q J Med* **210**:251–258, 1984.

31a. Przyrembel H: Homocystinuria. *Ergeb Inn Med Kinderheilkd* **49**:77–135, 1982.

32. Rosenblatt DS et al: Vitamin $B_{12}$ responsive homocystinuria and megaloblastic anemia: Heterogeneity in methylcobalamin deficiency. *Am J Med Genet* **26**:377–383, 1987.

33. Schedewie H et al: Skeletal findings in homocystinuria. *Pediatr Radiol* **1**:12–23, 1973.

34. Rowe PB: Inherited disorders of folate metabolism, in *The Metabolic Basis of Inherited Disease*, Stanbury JB et al (eds), McGraw-Hill, New York, 1983, p 516.

35. Skovby F et al: Immunochemical studies on cultured fibroblasts from patients with homocystinuria due to cystathionine β-synthase deficiency. *Am J Hum Genet* **34**:73–83, 1982.

36. Skovby F et al: Assignment of the gene for cystathionine β-synthase to human chromosome 21 in somatic cell hybrids. *Hum Genet* **65**:291–294, 1984.

37. Skovby F et al: Biosynthesis of human cystathionine β-synthase in cultured fibroblasts. *J Biol Chem* **259**:583–587, 1984.

38. Skovby F et al: Biosynthesis and proteolytic activation of cystathionine β-synthase in rat liver. *J Biol Chem* **259**:588–593, 1984.

39. Skovby F: Homocystinuria: Clinical, biochemical and genetic aspects of cystathionine β-synthase and its deficiency in man. *Acta Paediatr Scand (Suppl)* **321**:1–21, 1985.

40. Skovby F, Mudd SH: Clinical and biochemical studies of homocystinuria due to cystathionine β-synthase deficiency, in *Inherited Diseases of Amino Acid Metabolism*, Bvickel H, Wachtel U (eds), Georg Thieme Verlag, New York, 1985, p 238.

41. Smith SW: Roentgen findings in homocystinuria. *Am J Roentgenol* **100**:147–154, 1967.

42. Tamburrini O et al: Short fourth metacarpal in homocystinuria. *Pediatr Radiol* **15**:209–210, 1985.

43. Thomas GH, Howell RR: *Selected Screening Tests for Genetic Metabolic Disease*, Year Book Medical Publishers Inc., Chicago, 1974, pp 30–35.

44. Uhlendorf BW et al: Homocystinuria: Studies in tissue culture. *Pediatr Res* **7**:645–658, 1973.

45. Wilcken B, Turner G: Homocystinuria in New South Wales. *Arch Dis Childh* **53**:242–245, 1978.

46. Wilcken DEL et al: Homocystinuria—the effects of betaine in the treatment of patients not responsive to pyridoxine. *N Engl J Med* **309**:448–453, 1983.

47. Wilcken DEL et al: Homocystinuria due to cystathionine β-synthase deficiency—the effects of betaine treatment in pyridoxine-responsive patients. *Metabolism* **34**:1115–1121, 1985.

**Hypophosphatasia.** Hypophosphatasia is an inherited disorder of bone mineralization characterized by rachitic changes in childhood or osteomalacia in adult life, absence of dental cementum and premature loss of teeth, and decreased alkaline phosphatase enzyme activity (Figs. 5–46 to 5–50). Hypophosphatasia was first identified as a separate entity by Rathbun (25) in 1948, although there are earlier case reports (24). A wide range of presentations are described but all patients have some deficit in serum and tissue alkaline phosphatase. From a clinical perspective, most individuals can be classified as having either the neonatal, infantile, childhood, or adult form of the disorder (7,11,15,24). (Table 5-5). Scriver and Cameron (30) described a patient with clinical features of the childhood disorder but with normal alkaline phosphatase activity (6) and designated this variant pseudohypophosphatasia.

Genetics and biochemistry. The human alkaline phosphatases share the property of hydrolyzing artificial phosphoester substrates at alkaline pH. They constitute a system of multiple molecular forms that originate from at least three different genes and multiple tissue-specific posttranslational modifications (17,18). In hypophosphatasia, both the placental and intestinal isoenzymes, localized to chromosome 2 (13a), are largely intact, but the so-called "tissue nonspecific" or "liver/kidney/bone" group of isoenzymes is usually markedly deficient (18). A single gene localized to the short arm of chromosome 1 (4,34a,37) codes for the protein precursor that is then modified to produce at least three electrophoretically and biochemically distinct isoenzymes found in bone, liver, and kidney, respectively (13,31). These isoenzymes are antigenically similar but have different developmental sequences (17,18,31,32). The bone isoenzyme is believed to play an important role in calcification, and its production varies with the rate of bone mineralization (27). Possible functions may include the supply of inorganic phosphate through phosphoester hydrolysis for mineralization, the clearing of pyrophosphate, a calcification inhibitor, and the binding or storage of inorganic phosphate (22,24).

Although pathologically similar to D-deficiency rickets, hypophosphatasia is clearly differentiated as a disorder of bone matrix rather than bone mineral, as secondary mineralization is intact (11,24). Moreover, vitamin D metabolism is normal in this disorder (41). Acid phosphatase activity, a marker of osteoclastic function (18), is also normal (6,11).

The deficit in the tissue alkaline phosphatase is reflected in decreased total serum activity, a composite of the various isoenzyme fractions (18). In serum, too, it is the tissue nonspecific enzyme that is very low, but intestinal isoenzymes may also be depressed (19,24). Neutrophils and cultured skin fibroblasts are also deficient in alkaline phosphatase activity (4,13,33,40,42).

The birth prevalence of hypophosphatasia has been estimated to be 1/100,000 (12). However, this represents incomplete ascertainment because asymptomatic adults and those with milder forms of the disorder are excluded.

Both recessive and dominant modes of inheritance have been observed in different hypophosphatasia kindreds (23,24,38). The fre-

Table 5–5. Hypophosphatasia

| Signs | Neonatal (%) $n=47$ | Early infantile (%) $n=90$ | Childhood (%) $n=108$ | Adult (%) $n=17$ |
|---|---|---|---|---|
| Morbidity | 100 | 40 | 1 | — |
| Craniotabes | 91 | 37 | — | — |
| Bone deformities | 89 | 59 | 46 | 60 |
| Respiratory distress | 74 | 15 | — | — |
| Cyanosis | 59 | 9 | NA | — |
| Fractures | 30 | 23 | — | — |
| Rachitic signs | 30 | 69 | 4 | 60 |
| Rachitic rosary | 23 | 47 | 46 | 18 |
| Cranial sutures widened | 23 | 54 | 26 | — |
| Seizures | 22 | 24 | 2 | 6 |
| Pneumonia | 17 | 22 | 4 | — |
| Nephrocalcinosis | 11 | 27 | 4 | — |
| Shrill cry | 11 | 15 | NA | — |
| Hypercalcemia | 6 | 28 | 5 | 6 |
| Muscle hypotonia | 6 | 28 | 10 | — |
| Vomiting | 4 | 42 | 6 | — |
| Failure to thrive | NA | 70 | 15 | — |
| Craniosynostosis | NA | 40 | 26 | 12 |
| Fever | NA | 35 | — | — |
| Open fontanel | 2 | 34 | 1 | — |
| Early tooth loss | NA | 29 | 78 | 53 |
| Growth retardation | NA | 19 | 35 | 53 |
| Walking difficulties | NA | 10 | 28 | 47 |
| Constipation | NA | 10 | 1 | — |
| Blue sclerae | ? | 9 | — | — |
| Genua valga | 2 | 8 | 32 | 18 |
| Bone pain | NA | 6 | 14 | 65 |
| Pseudofractures | NA | — | — | 100 |
| Osteoporosis | — | — | — | 53 |

Data derived from HG Terheggen, A Wischermann, *Monatschr Kinderheilkd* **132:**512, 1984.

quency of consanguinity and the recurrence rates for infantile hypophosphatasia are clearly indicative of a recessive mode (23,), but the suggestion that hypophosphatasia is an autosomal dominant condition with homozygous lethality is in keeping with the data on the adult form of the disorder (8,9,38). Both suggest that if this is a single gene disorder, there is substantial allelic heterogeneity (11). A three allele system has been proposed by Igbokwe (136). The availability of cloned cDNA probes for the tissue nonspecific alkaline phosphatase enzyme should permit the molecular mapping of this disorder and speed delineation of the basic defect (37).

### Clinical features.

*Neonatal.* In this form of hypophosphatasia, markedly impaired mineralization occurs *in utero*. The extremities are shortened, the long bones are deformed, and the cranial vault fails to mineralize (craniotabes) (Figs. 5–46 and 5–47). Polyhydramnios has been observed more frequently in hypophosphatasia pregnancies. Premature stillbirth deliveries are not uncommon (20,24,46). Radiographs show small, sclerotic bones at the base of the skull and a membranous calvaria. The ribs are small, thin, and deformed. Sclerotic patches are also observed in the ribs and other tubular bones. In some infants, almost no bone is formed. In others, spikes may present at the elbows and knees (RJ Gorlin and P Bader, personal observations, 1989, 21, 31a). In live births, the outcome depends on the extent of pulmonary and neurological compromise, but demise usually occurs within a few days (7,14,15,20,24,35,46). (Table 5-5).

*Infantile.* Those with the infantile form commonly present some time after birth because of failure to thrive (11,23,24). Apparently difficult to recognize with radiographs, most newborns do well for a short period and then experience a wide variety of problems related to impaired bone growth. Hypercalcemia may be marked, explaining a history of irritability, poor feeding, anorexia, vomiting, hypotonia, polydipsia, polyuria, dehydration, and constipation. Episodes of unexplained fever, tender bones, and respiratory distress are also described. Renal function may be impaired by hypercalciuria and nephrocalcinosis. Traumatic fractures are frequently found (11,14,15,24,35). Mortality may be as high as 40% (35a) (Table 5-5).

The anterior fontanel is often enlarged and may bulge. The membranous cranial sutures are also frequently widened and some degree of ocular prominence resulting from shallow orbits may be apparent within the first few months of life. Head circumference also increases more slowly than expected as premature sutural fusion sets in. Radiographs show widespread demineralization and rachitic changes in the metaphyses (Fig. 5–48), but usually with less diaphyseal bowing than would be expected with severe metaphyseal disease (7,15). A rachitic "rosary" is common. In infants who survive, there is often spontaneous improvement in mineralization and remission of clinical problems (23,45) other than craniostenosis. Although the sutures appear widened and membranous, intense mineralization activity may be detectable by nuclear scintigraphy (34). Moderate short stature in adulthood and premature loss of deciduous teeth are also common (Table 5-5).

*Childhood.* Childhood hypophosphatasia is a milder condition that often presents as "rickets" in the second and third year of life (3,15). Signs of intracranial hypertension or failure to thrive are typical (11,15,24). Some long bone deformity is not unusual but tends to recede with time. The most serious treatable complication in this group is also craniosynostosis. All sutures appear to be involved but ocular prominence resulting from shallow orbits can be quite characteristic. Other ocular signs include keratopathy and conjunctival calcification caused by hypercalcemia and blue sclerae (2,12). Spontaneous remission of bone disease is common. In at least one case, this has been accompanied by an increase in serum alkaline phosphatase activity (45) (Table 5-5).

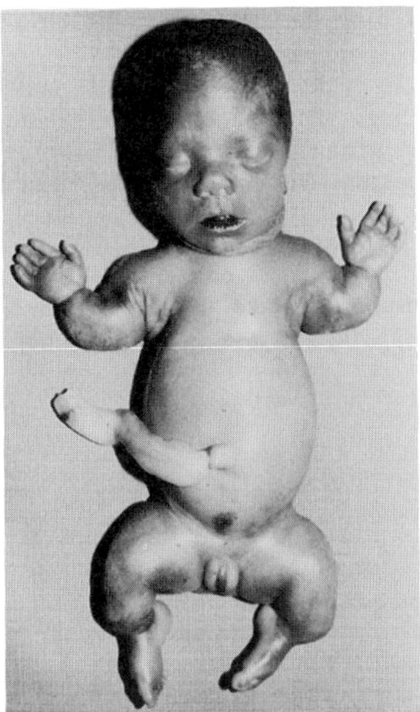

Fig. 5–46. *Hypophosphatasia*. Severe lethal hypophosphatasia. Note short limbs.

*Adult*. The adult form is mild but osteomalacia may produce significant pseudofractures, severe bone pain, and increased susceptibility to traumatic fracture (3,11,23,24,39). The proximal femur is a frequent site of pseudofractures that extend to complete transverse fractures and loss of mobility (5). In this group, a bone scan can be helpful in identifying and clarifying the sources of pain. There is also a predilection for chrondrocalcinosis and marked osteoarthropathy later in life (40) (Table 5-5).

Fig. 5–47. *Hypophosphatasia*. Note markedly impaired mineralization and short extremities.

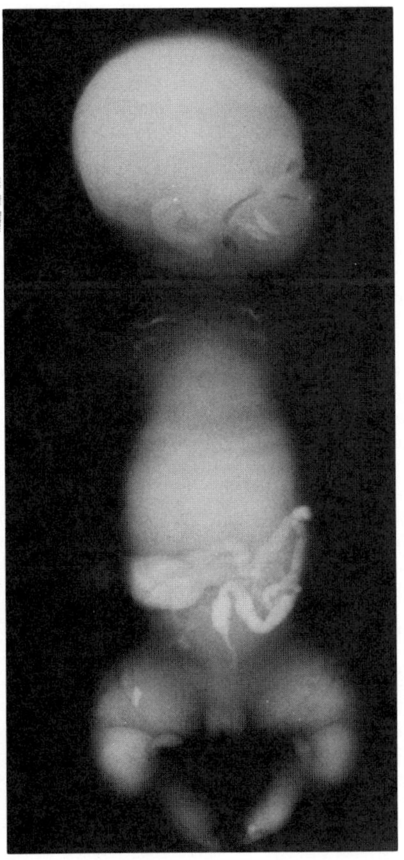

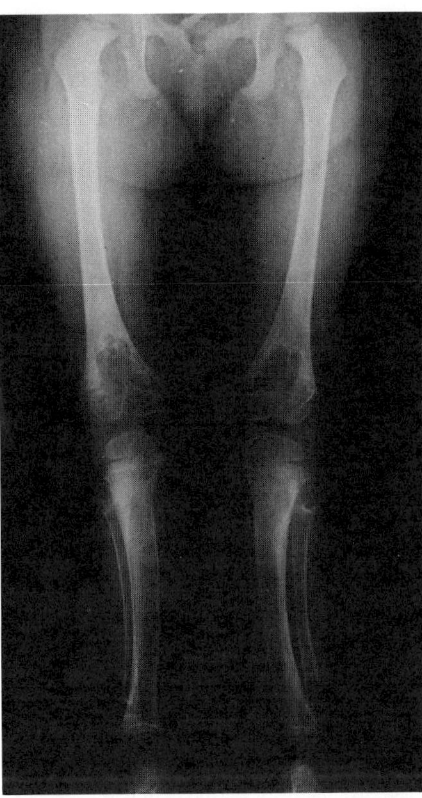

Fig. 5–48. *Hypophosphatasia*. Note large midmetaphyseal ossification defects and angulated tibias.

**Oral manifestations.** Delayed dentition, premature loss of deciduous teeth, and spontaneous loss of permanent teeth are characteristic of hypophosphatasia (Fig. 5–49) (1,3,15a,23). They may be the only clinical signs of disease, giving rise to the term odontohypophosphatasia for this variant condition (23). The anterior deciduous teeth are more likely to be affected and the most frequent loss involves the incisors (1,3). The process is that of relatively painless extrusion and does not invoke periodontal inflammation (1). Dental X-rays show reduced alveolar bone, enlarged pulp chambers and root canals, but normal enamel (1,3,23).

**Pathology.** In the infant, bone histomorphometry reveals a marked excess of osteoid volume and an osteomalacic pattern of tetracycline labeling in dynamic studies (11). Bone alkaline phosphatase is usually undetectable and electron microscopy shows otherwise normal subcellular architecture of osteoblasts and their associated matrix vesicles (11,22). Iliac crest biopsies in adults show less dramatic and more variable changes. The severity of the osteomalacia, as measured by relative osteoid volume, is inversely correlated with the amount of detectable alkaline phosphatase and with the concentrations of serum alkaline phosphatase activity (11). In shed teeth, marked deficiency or absence of cementum (Fig. 5–50) is a striking characteristic, accounting for the ready loss of teeth (1,3,23,24). This appears to be the result of aplasia since resorption of cementum has never been observed. Dentin formation is delayed with less being formed. Interglobular dentin and osteodentin have also been observed (1,3,23).

**Diagnosis and laboratory aids.** Hypophosphatasia can be diagnosed on clinical grounds alone, but low serum concentrations of alkaline phosphatase will often help to rule out other disorders with similar findings. In most cases, the detection of increased urinary phosphoethanolamine or pyrophosphate will serve to confirm the diagnosis (8,10,24,29). Measurements of the $B_6$ vitamin, pyridoxal-$PO_4$, in serum is a more sensitive test, as levels tend to be abnormally low in other bone diseases characterized by increased alkaline phosphatase

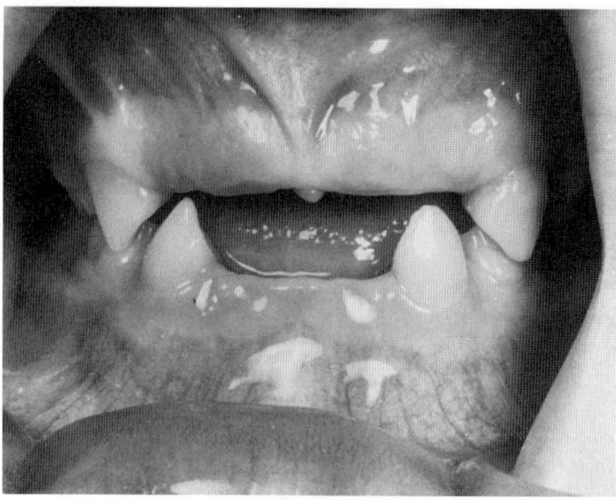

Fig. 5–49. *Hypophosphatasia.* Premature loss of deciduous teeth, most commonly affecting the anterior region.

activity (6,44). It may also be helpful in identifying mildly affected patients or detecting variants such as pseudohypophosphatasia (6,16,30). When preceded by an oral $B_6$ loading test, heterozygotes can sometimes be identified (MP Whyte et al, personal communication, 1987). Alkaline phosphatase is also characteristically low in cultured fibroblasts from hypophosphatasia (41,43), but shows poor correlation with the severity of the disorder (41).

In the severe neonatal and infantile forms, second trimester prenatal diagnosis has been achieved using ultrasonic detection of the bony malformations (20,28,36). Determination of amniotic fluid alkaline phosphatase activity discriminates poorly between normal, heterozygotic, and affected individuals (20,28), but monoclonal antibody testing of chorionic villus samples has been used successfully for first-trimester prenatal diagnosis (36). Diagnosis by determination of alkaline phosphatase isoenzyme activities in cultured amniocytes has also been described (20,26).

Fig. 5–50. *Hypophosphatasia.* Section of tooth in hypophosphatasia. Note absence of cementum.

Differential diagnosis. Neonatal hypophosphatasia can be distinguished from *osteogenesis imperfecta, achondrogenesis of various types, campomelic dysplasia,* and other congenital osteochondrodysplasias by radiographic and biochemical findings. Childhood and adult hypophosphatasia should be differentiated from treatable forms of rickets and osteomalacia and from other metaphyseal chondrodysplasias.

### References (Hypophosphatasia)


1. Beumer J et al: Childhood hypophosphatasia and the premature loss of teeth. A clinical and laboratory study of seven cases. *Oral Surg* **35**:631–640, 1973.

2. Brenner RL et al: Eye signs of hypophosphatasia. *Arch Ophthalmol* **81**:614–617, 1969.

3. Bruckner RJ et al: Hypophosphatasia with premature shedding of teeth and aplasia of cementum. *Oral Surg* **15**:1351–1369, 1962.

4. Chodirker BN et al: Infantile phosphatasia-linkage with the Rh locus. *Genomics* **1**:280–282, 1987.

5. Coe JD et al: Management of femoral fractures and pseudofractures in adult hypophosphatasia. *J Bone Joint Surg* **68A**:981–990, 1986.

6. Cole DEC et al: Increased serum pyridoxal-5′-phosphate in pseudohypophosphatasia (letter). *N Engl J Med* **314**:992–993, 1986.

7. Currarino G: Hypophosphatasia. *Progr Pediatr Radiol* **4**:469–494, 1973.

8. Eastman JR, Bixler D: Urinary phosphoethanolamine: Normal values by age (letter). *Clin Chem* **26**:1757–1758, 1980.

9. Eastman JR, Bixler D: Lethal and mild hypophosphatasia in half-sibs. *J Craniofac Genet Dev Biol* **2**:35–44, 1982.

10. Eastman JR, Bixler D: Clinical, laboratory, and genetic investigations of hypophosphatasia: Support for autosomal dominant inheritance with homozygous lethality. *J Craniofac Genet Dev Biol* **3**:213–234, 1983.

11. Fallon MD et al: Hypophosphatasia: Clinicopathologic comparison of the infantile childhood, and adult forms. *Medicine* **63**:12–24, 1984.

12. Fraser D: Hypophosphatasia. *Am J Med* **22**:730–745, 1957.

13. Gainer AL, Stinson RA: Evidence that alkaline phosphatase from human neutrophils is the same gene product as the liver/kidney/bone isoenzyme. *Clin Chim Acta* **123**:11–17, 1982.

13a. Griffin CA et al: Human placental and intestinal alkaline phosphatase genes map to 2q34-q37. *Am J Hum Genet* **41**:1025–1034, 1987.

13b. Igbokwe EC: Inheritance of hypophosphatasia. *Med Hypotheses* **18**:1–5, 1985.

14. Jelke H: Hypophosphatasia. *Acta Paediatr Scand* **49**:297–308, 1960.

15. Kozlowski K et al: Hypophosphatasia. *Pediatr Radiol* **5**:103–117, 1976.

15a. Macfarlane JD, Swart JGN: Dental aspects of hypophosphatasia: A case report, family study and literature review. *Oral Surg* **67**:521–526, 1989.

16. Mehes K et al: Hypophosphatasia: Screening and family investigations in an endogamous Hungarian village. *Clin Genet* **3**:60–66, 1972.

17. Moss DW: Alkaline phosphatase isoenzymes. *Clin Chem* **28**:2007–2016, 1982.

18. Moss DW: Multiple forms of acid and alkaline phosphatases: Genetics, expression and tissue-specific modification. *Clin Chim Acta* **161**:123–135, 1986.

19. Mueller HD et al: Isoenzymes of alkaline phosphatase in infantile hypophosphatasia. *J Lab Clin Med* **102**:24–29, 1983.

20. Mulivor RA et al: Prenatal diagnosis of hypophosphatasia: Genetic, biochemical, and clinical studies. *Am J Hum Genet* **30**:271–282, 1978.

21. Oestreich AE, Bofinger MK: Prominent transverse (Bowder) bone spurs as a diagnostic clue in a case of neonatal hypophosphatasia without metaphyseal irregularity. *Pediatr Radiol* **19**:341–342, 1989.

22. Ornoy A et al: Histologic and ultrastructural studies on the mineralization process in hypophosphatasia. *Am J Med Genet* **22**:743–758, 1985.

23. Pimstone B et al: Hypophosphatasia: Genetic and dental studies. *Ann Intern Med* **65**:722–729, 1966.

24. Rasmussen H: Hypophosphatasia, in *The Metabolic Basis of Inherited Disease,* Stanbury WB et al (eds), McGraw-Hill, New York, 1983, pp 1497–1506.

25. Rathbun JC: Hypophosphatasia, a new developmental anomaly. *Am J Dis Childh* **75**:822–831, 1948.

26. Rattenbury JM et al: Prenatal diagnosis of hypophosphatasia (letter). *Lancet* **1**:306, 1976.

27. Register TC et al: Roles of alkaline phosphatase and labile internal mineral in matrix vesicle-mediated calcification. *J Biol Chem* **261**:9354–9360, 1986.

28. Rudd N et al: Prenatal diagnosis of hypophosphatasia. *N Engl J Med* **295**:146–148, 1976.

29. Russell RGG: Metabolism of inorganic pyrophosphate ($PP_i$). *Arthritis Rheum* **19**:465–478, 1976.

30. Scriver C, Cameron D: Pseudohypophosphatasia. *N Engl J Med* **281**:604–606, 1969.

31. Seargent LZ, Stinson RA: Evidence that three structural genes code for human alkaline phosphatases. *Nature (London)* **281**:152–154, 1979.

31a. Spranger J: "Spur-limbed" dwarfism identified as hypophosphatasia. *Dysmorphol Clin Genet* **2**:123, 1988.

32. Stepan JJ et al: Age and sex dependency of the biochemical indices of bone remodelling. *Clin Chim* **151**:273–283, 1985.

33. Stinson RA et al: Neutrophil alkaline phosphatase in hypophosphatasia. *N Engl J Med* **312**:1642–1643, 1985.

34. Sty JR et al: Skull scintigraphy in infantile hypophosphatasia. *J Nucl Med* **20**:305–306, 1979.

34a. Swallow DW et al: The liver/bone/kidney isozyme of alkaline phosphatase (ALPL) is coded by a gene on chromosome 1. *Cytogenet Cell Genet* **40**:756, 1985.

35. Teree TM, Klein L: Hypophosphatasia: Clinical and metabolic studies. *J Pediatr* **72**:41–50, 1968.

35a. Terheggen HG, Wischermann A: Congenitale Hypophosphatasie. *Mschr Kinderheild* **132**:512–522, 1984.

36. Warren RC et al: First trimester diagnosis of hypophosphatasia with a monoclonal antibody to the liver/bone/kidney isoenzyme of alkaline phosphatase. *Lancet* **2**:856–858, 1985.

37. Weiss MJ et al: Isolation and characterization of a cDNA encoding a human liver/bone/kidney type alkaline phosphatase. *Proc Natl Acad Sci USA* **83**:7182–7186, 1986.

37a. Whyte MP: Hypophosphatasia. In *The Metabolic Basis of Inherited Disease,* 6th ed, Scriver CR et al (eds), McGraw–Hill, New York, 1989.

38. Whyte MP et al: Adult hypophosphatasia dominant inheritance in a large kindred. *Trans Assoc Am Phys* **91**:144–155, 1978.

39. Whyte MP et al: Adult hypophosphatasia: Clinical laboratory and genetic investigation of a large kindred with review of the literature. *Am J Med* **58**:329–347, 1979.

40. Whyte MP et al: Adult hypophosphatasia with chondrocalcinosis and arthropathy. *Am J Med* **72**:631–641, 1982.

41. Whyte MP et al: Adult hypophosphatasia: Generalized deficiency of alkaline phosphatase activity demonstrated with cultured skin fibroblasts. *Trans Assoc Am Phys* **95**:253–263, 1982.

42. Whyte MP, Seino Y: Circulating vitamin D metabolite levels in hypophosphatasia. *J Clin Endocrinol Metabol* **55**:178–180, 1982.

43. Whyte MP et al: Alkaline phosphatase deficiency in cultured skin fibroblasts from patients with hypophosphatasia: Comparison of the infantile, childhood, and adult forms. *J Clin Endocrinol Metabol* **57**:831–837, 1983.

44. Whyte MP et al: Markedly increased circulating pyridoxal-5′-phosphate levels in hypophosphatasia. *J Clin Invest* **76**:752–756, 1985.

45. Whyte MP et al: Infantile hypophosphatasia: Normalization of circulating bone alkaline phosphatase activity followed by skeletal remineralization. *J Pediatr* **108**:82–88, 1986.

46. Wolff C, Zabransky S: Hypophosphatasia congenita letalis. *Eur J Pediatr* **138**:197–199, 1982.

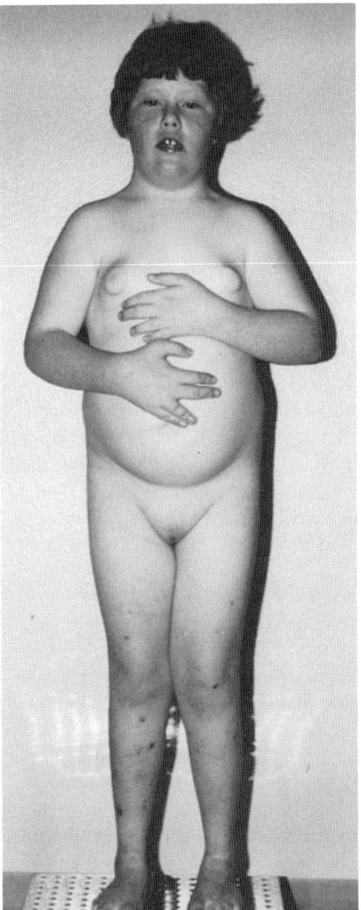

Fig. 5–51. *Pseudohypoparathyroidism.* Short stature and obesity.

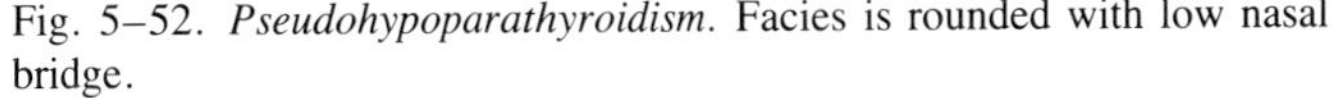

Fig. 5–52. *Pseudohypoparathyroidism.* Facies is rounded with low nasal bridge.

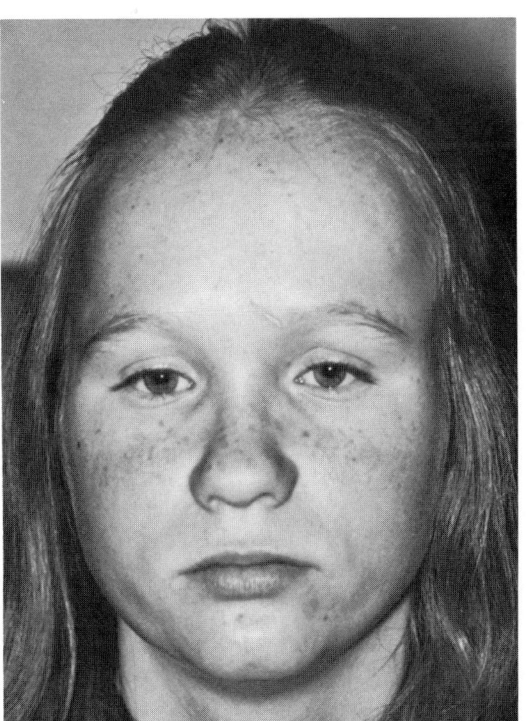

**Pseudohypoparathyroidism (Albright hereditary osteodystrophy, acrodysostosis).** Albright et al (2) first described pseudohypoparathyroidism (PHP), a hypocalcemic syndrome similar to hypoparathyroidism but with renal and skeletal resistance to parathyroid hormone (PTH). Ten years later, Albright and associates (3) defined a normocalcemic variant of PHP, calling it pseudopseudohypoparathyroidism (PPHP). Chase et al (9) delineated two different types of PHP on the basis of the biochemical response to exogenous PTH. Most PHP individuals have a Type I defect, in which the normally increased urinary excretion of both cAMP and phosphate in response to PTH is lacking. In others with Type II PHP, a cAMP response to PTH is observed, but there is no phosphate diuresis. A decade later, Farfel et al (16) and Levine et al (27) reported that a receptor–cyclase coupling protein was defective in the erythrocytes of some PHP Type I patients. This protein, now called stimulatory guanine nucleotide-binding protein (Gs), is responsible for coupling the cellular receptor that binds PTH—an external cell membrane event—with the formation and release of cAMP on the internal membrane surface. The defect involves many tissues but is manifest in only some PHP patients, now designated Type Ia (27). Other PHP Type I individuals without the coupling protein defect (Type Ib) may express abnormalities of the PTH molecule itself (34,39) or PTH bioactivity may be blocked (31,39). Biochemical heterogeneity is further complicated by significant clinical heterogeneity (Figs. 5–51 to 5–57). For example, the dissociation of renal and skeletal resistance to PTH is well recognized (6). Thus,

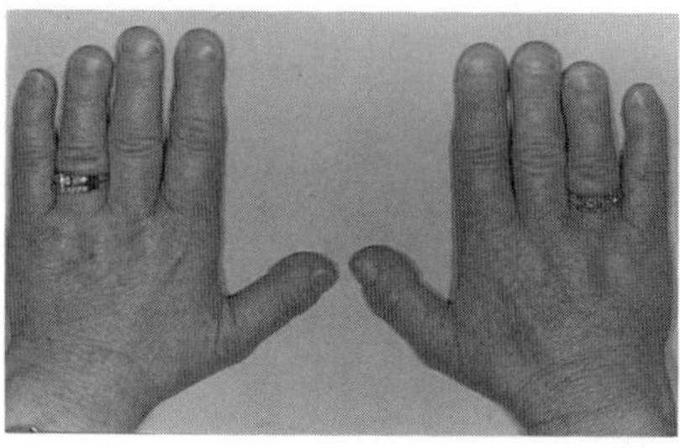

A

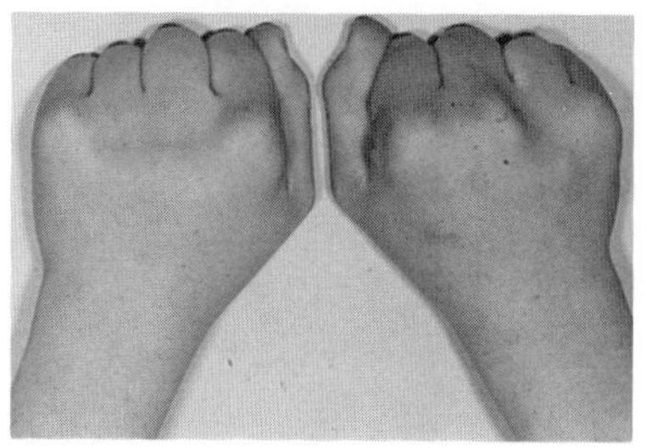

B

Fig. 5–53. *Pseudohypoparathyroidism.* (A,B) Short metacarpals manifest by absent knuckles when patient makes a fist. Also note short distal thumb phalanx associated with wide thumb nail.

some patients may display radiological signs of hyperparathyroidism in the presence of hypocalcemia and positive tests for renal resistance to PTH. (6,13). Finally, there are many examples of dissociation between the biochemical abnormalities and the somatic phenotype, commonly called Albright hereditary osteodystrophy (AHO) (19), although most individuals with AHO have Type Ia PHP with the receptor-coupling defect (16,18,19,30,39). A comprehensive review is that of Werder (50).

Genetics. The generally held concept of X-linked dominant inheritance for PHP or AHO is no longer valid (19). The syndrome is generally more severe and is twice as frequent in females, but male-to-male transmission has been described in a number of pedigrees (20,24,26,33,46,47). Deficient receptor-coupling activity shows a dominant pattern in most pedigrees, but recessive patterns have been reported (8,16–18,30). Male-to-male transmission of the biochemical trait has also been described (17,46). There may be different patterns of inheritance for different forms of the disorder, although most pedigrees of PHP with AHO and autosomal dominant inheritance frequently have pronounced clinical and age-related variability, particularly in relation to hypocalcemia (5,30). Thus, an individual might be normocalcemic in infancy (19), hypocalcemic and more dysmorphic in early childhood (35,43), and revert to normocalcemic or PPHP phenotype in adulthood (19). Others become hypocalcemic only later in life (19). Johnson (25) has attempted to explain PHP as an example of the metabolic interference hypothesis. The same hypothesis has been advanced to explain variable expression in the receptor-coupling defect (30).

Fig. 5–54. *Pseudohypoparathyroidism.* Severe shortening of toes from abbreviated metatarsals.

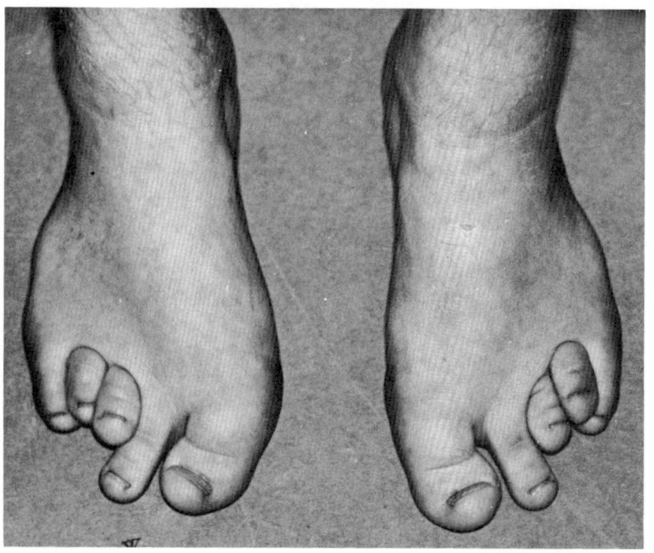

Growth. Newborns with PHP are of normal length, but growth usually lags in childhood and 62% are less than the third percentile for height at maturity (19). Birth weight may be slightly greater than normal but obesity (Fig. 5–51) characterizes 65% of patients under 18 years (19). Patients with PPHP tend to be taller and less obese (44).

Fig. 5–55. *Pseudohypoparathyroidism.* (A) All metacarpals shortened. Also observe cone-shaped epiphyses in index fingers. (B) Markedly shortened third and fourth metatarsals.

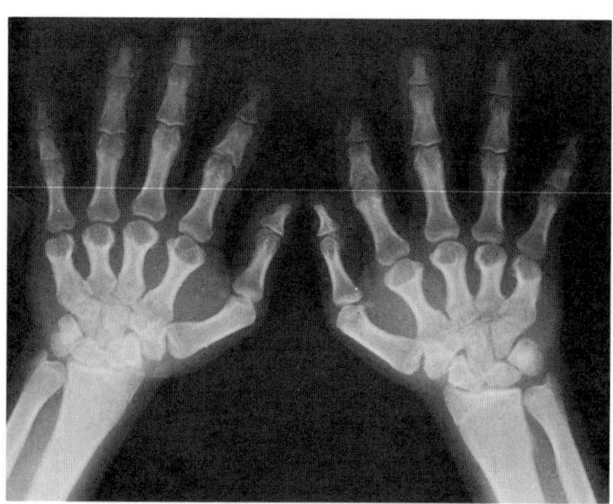

A

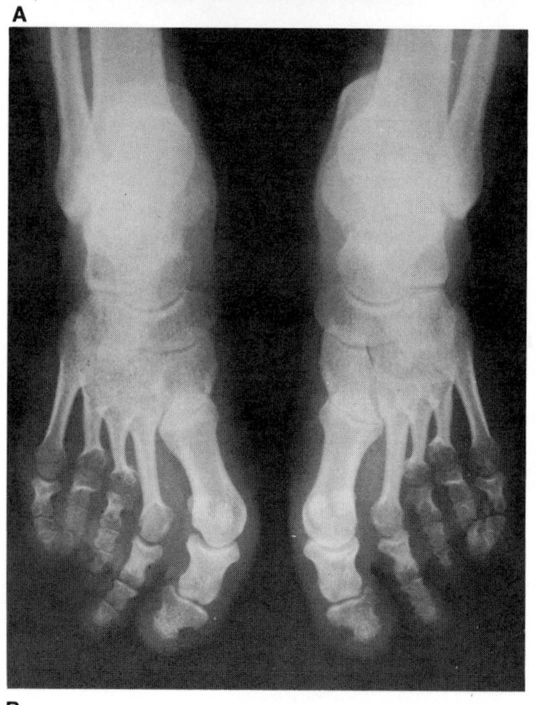

B

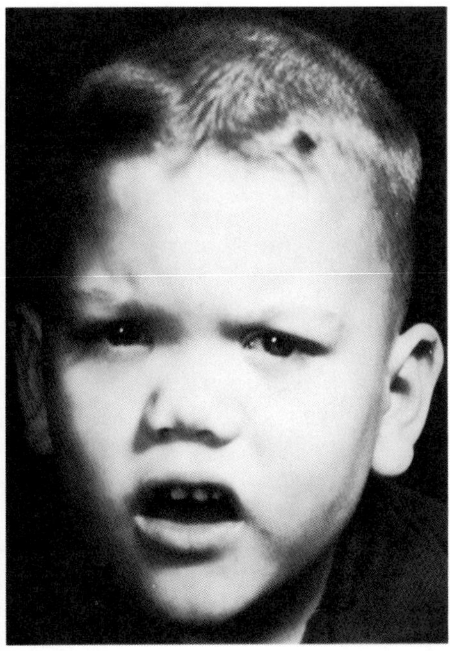

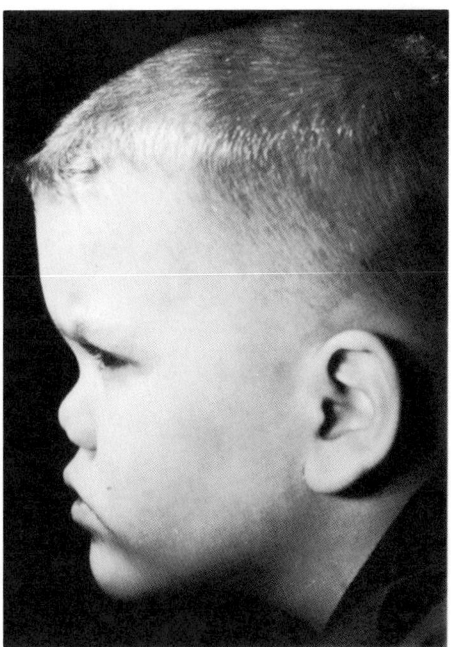

A B

Fig. 5–56. *Pseudohypoparathyroidism.* (A,B) Acrodysostosis. Marked hypoplasia of nose and midface. (From *M Robinow et al,* Am J Dis Child **121:**195, 1971.)

Central nervous system. Mental retardation is found in 70% of hypocalcemic PHP patients but in only 30% of normocalcemic PPHP patients (19,36,50). Farfel and Friedman (15a) found 65% of Type Ia and none with Type Ib were mentally retarded. Personalities are generally affable and pleasant. Seizures unrelated to hypocalcemia have been described. It is not clear whether seizures are secondary to earlier insult, such as hypothyroidism or intracranial calcifications, or whether they are primary.

Olfactory dysfunction is associated with the disorder (10,23,48). The receptor coupling protein, Gs, is an essential link in the transduction of the olfactory signal. Thus, anosmia is found in Type Ia patients with deficient Gs protein activity and other disturbances of cAMP-dependent pathways and in no patient with normal Gs activity (48).

Craniofacial features. The face is characteristically rounded with full cheeks, low nasal bridge, and short neck (Fig. 5–52). These features are more prevalent in PHP than in PPHP. With PHP, cataracts may be found in 25%, but with PPHP, cataracts occur in less than 10% (42). Enamel hypoplasia, widened root canals, and delayed eruption have been noted in over one-third of PHP patients (4, 11,12,32,40,45).

Musculoskeletal system. Most common of distinctive features associated with AHO are short metacarpals and metatarsals, particularly of the fourth and fifth digits (42). Short metacarpals are manifest by absent knuckles when the patient makes a fist (Figs. 5–53 to 5–55). Also characteristic is a short distal thumb phalanx associated with a wide thumbnail (19,21a,38). Such changes may not be evident until later childhood (35,43). Brachycephaly and premature suture closure have been reported. Hyperostosis of the cranial vault is seen in about one-third of patients. Long bone findings such as curvature of the radius, bowing of the tibia, and genua valga have also been noted (43,44).

Generalized osteopenia may also be observed, and trabecular coarsening is a feature of PHP with skeletal resistance. In PHP with bone responsiveness, manifestations of hyperparathyroidism may be seen, including subperiosteal bone resorption, bone cysts, and focal osteosclerosis (6).

Ectopic calcification. Propensity for soft-tissue calcification is well known. Subcutaneous deposits are found in the scalp and along the extremities, particularly the periarticular areas of the hands and feet. Calcifications may also be found in the brain, particularly the basal ganglia and choroid plexus (41a). As a rule, deposits are not present in muscle, viscera, or cartilage (4,14,19,44).

Endocrine findings. Disturbances of other endocrine systems are present in some but not all PHP patients (19,28,41). Hypogonadism and hypothyroidism are frequently found in PHP and tend to show high concordance with deficiencies in Gs protein activity (28). Other deficits include altered prolactin response to thyroid-releasing hormone (7,28) and a deficient metabolic response to glucagon (28). Such changes are predictable on the basis of the widespread abnormality in receptor-coupling protein, but syndromic variability and inconsistency in the biochemical findings in Type 1a patients have not been satisfactorily explained (18,30).

Differential diagnosis. In idiopathic hypoparathyroidism, tetanic and epileptiform convulsions, increased thickness of the skull, hypoplastic enamel, and cataracts may be present, but other features of AHO are absent. Hypoparathyroidism may occur with *Kenny syndrome* in which short stature is associated with internal cortical thickening and medullary stenosis of the tubular bones (15). Short metacarpals occur in 10% of the general population and in a variety of conditions (21,22,37) including Type E brachydactyly, peripheral dysostosis, *Turner syndrome,* and *nevoid basal cell carcinoma syndrome.* Acrodysostosis, originally defined as a separate entity on the basis of peripheral dysostosis, nasal hypoplasia, and mental retardation, is now known to be part of the spectrum of pseudohypoparathyroidism (Figs. 5–56 and 5–57) (1,19a).

Laboratory aids. In PHP, routine laboratory tests reveal hypocalcemia, hyperphosphatemia, and increased immunoreactive PTH. Type II PHP may be differentiated from Type I by observing a rise in urinary excretion of cAMP but not phosphate. Overlap often makes it necessary to examine the renal response to exogenous hormone (9). Measurement of Gs receptor-coupling protein can be used to detect

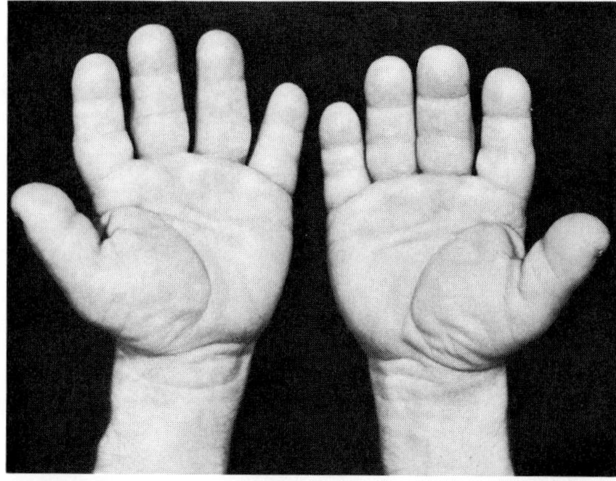

A

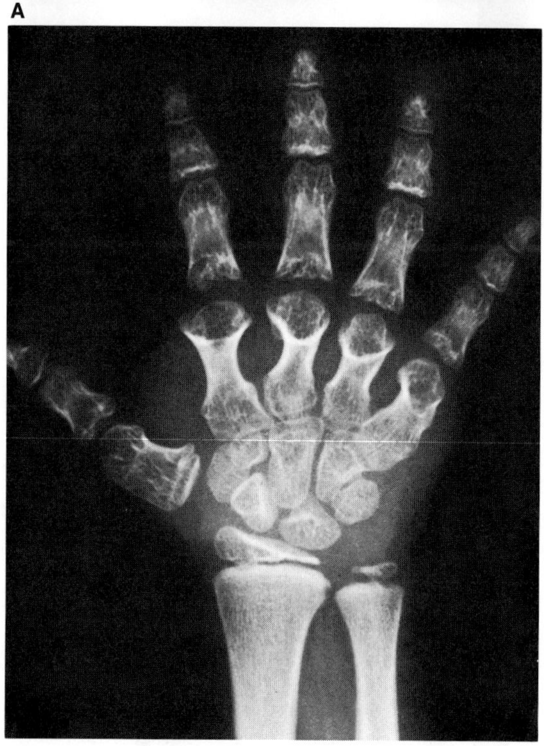

B

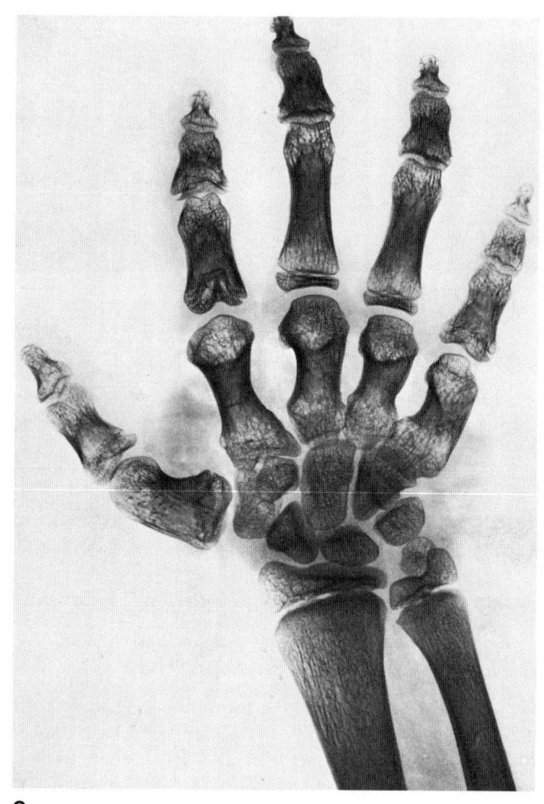

C

Fig. 5–57. *Pseudohypoparathyroidism. Acrodysostosis.* (A) Hands are short and stubby. Feet are similarly deformed. (B) Radiograph showing shortened metacarpals and phalanges. (C) Similar radiographic alterations in another patient. (From *A Giedion,* Fortschr Roentgenstr **110:**507, 1969.)

the Type Ia subgroup; this may be clinically significant as these PHP patients are at greatest risk for related endocrine disturbances. Infants with PHP should be carefully tested for hypothyroidism as this is a treatable cause of mental deficiency (29,49).

### References [Pseudohypoparathyroidism (Albright hereditary osteodystrophy, acrodysostosis)]

1. Ablow RC et al: Acrodysostosis coinciding with pseudohypoparathyroidism and pseudo-pseudohypoparathyroidism. *Am J Roentgenol* **128**:95–99, 1977.

2. Albright F et al: Pseudo-hypoparathyroidism—example of "Seabright Bantam" syndrome. *Endocrinology* **30**:922–932, 1942.

3. Albright F et al: Pseudo-pseudohypoparathyroidism. *Trans Assoc Am Physicians* **65**:337–350, 1952.

4. Armstein AR et al: Albright's hereditary osteodystrophy. *Am Intern Med* **64**:996–1008, 1966.

5. Boscherini B et al: Albright's hereditary osteodystrophy. *Acta Paediatr Scand* **69**:305–309, 1980.

6. Burnstein MI et al: Metabolic bone disease in pseudohypoparathyroidism: Radiological features. *Radiology* **155**:351–356, 1985.

7. Carlson HE et al: Prolactin deficiency in pseudohypoparathyroidism. *N Engl J Med* **296**:140–144, 1977.

8. Cederbaum SD, Lippe BM: Probable autosomal recessive inheritance in a family with Albright's hereditary osteodystrophy and an evaluation of the genetics of the disorder. *Am J Hum Genet* **2**:638–645, 1973.

9. Chase LR et al: Pseudohypoparathyroidism: Defective excretion of 3′, 5′-AMP in response to parathyroid hormone. *J Clin Invest* **48**:1832–1844, 1969.

10. Christian JC et al: Hypogonadotrophic hypogonadism with anosmia: The Kallmann syndrome. *Birth Defects* **7**(6):166–171, 1971.

11. Cranin AN, Katz HE: Pseudohypoparathyroidism. *J Am Dent Assoc* **74**:741–746, 1967.

12. Croft LK et al: Pseudohypoparathyroidism. *Oral Surg* **20**:758–770, 1965.

13. Dabbagh S et al: Renal-nonresponsive, bone-responsive pseudohypoparathyroidism. A case with normal vitamin D metabolite levels and clinical features with rickets. *Am J Dis Child* **138**:1030–1033, 1984.

14. Elrick H et al: Further studies on pseudohypoparathyroidism: Report of four new cases. Acta Endocrinol **5**:199–225, 1950.

15. Fanconi S et al: Kenny syndrome: Evidence for idiopathic hypoparathyroidism in two patients and for abnormal parathyroid hormone in one. *J Pediatr* **109**:469–475, 1986.

15a. Farfel Z, Friedman E: Mental deficiency in pseudohypoparathyroidism type I is associated with Ns-protein deficiency. *Ann Intern Med* **105**:197–199, 1986.

16. Farfel Z et al: Defect of receptor-cyclase coupling protein in pseudohypoparathyroidism. *N Engl J Med* **303**:237–242, 1980.

17. Farfel Z et al: Pseudohypoparathyroidism: Inheritance of deficient receptor-coupling activity. *Proc Natl Acad Sci USA* **78**:3098–3102, 1981.

18. Fischer JA et al: Pseudohypoparathyroidism: Inheritance and expression of deficient receptor-cyclase coupling protein activity. *Clin Endocrinol* **19**:747–754, 1983.

19. Fitch N: Albright's hereditary osteodystrophy: A review. *Am J Med Genet* **11**:11–29, 1982.

19a. Frey G et al: Die Akrodysostose-ein autosomal-dominant vererbte periphere Dysplasie. *Kinderärztl Prax* **50**:149–153.

20. Goeminne C: Albright's hereditary poly-osteochondrodystrophy (pseudopseudohypoparathyroidism with diabetes, hypertension, arteritis and polyarthrosis). *Acta Genet Med (Roma)* **14**:226–281, 1966.

21. Gorlin RJ, Sedano HO: Cryptodontic brachymetacarpalia. *Birth Defects* 7(7):200–203, 1971.

21a. Graudal N et al: The pattern of shortened hand and foot bones in D- and E-brachydactyly and pseudohypoparathyroidism/pseudopseudohypoparathyroidism. *Roefo* **148**:460–462, 1988.

22. Halal F et al: Differential diagnosis in young women with oligomenorrhea and the pseudo-pseudohypoparathyroidism variant of Albright's hereditary osteodystrophy. *Am J Med Genet* **21**:551–568, 1985.

23. Henkin RT: Impairment of olfaction and the tastes of sour and bitter in pseudohypoparathyroidism. *J Clin Endocrinol Metab* **28**:624–628, 1968.

24. Hermans PE et al: Pseudo-pseudohypoparathyroidism (Albright's hereditary osteodystrophy). A familial study. *Mayo Clin Proc* **39**:81–91, 1964.

25. Johnson WG: Metabolic interference and the +/− heterozygote. A hypothetical form of simple inheritance which is neither dominant nor recessive. *Am J Hum Genet* **32**:374–386, 1980.

26. Lee JB et al: Familial pseudohypoparathyroidism. *N Engl J Med* **276**:1179–1184, 1968.

27. Levine MA et al: Deficient activity of guanine nucleotide regulatory protein in erythrocytes from patients with pseudohypoparathyroidism. *Biochem Biophys Res Commun* **94**:1319–1324, 1980.

28. Levine MA et al: Resistance to multiple hormones in patients with pseudohypoparathyroidism. Association with deficient activity of guanine nucleotide regulatory protein. *Am J Med* **74**:545–555, 1983.

29. Levine MA et al: Infantile hypothyroidism in two sibs: An unusual presentation of pseudohypoparathyroidism type Ia. *J Pediatr* **107**:919–922, 1985.

30. Levine MA et al: Activity of the stimulatory guanine nucleotide-binding protein is reduced in erythrocytes from patients with pseudohypoparathyroidism and pseudopseudohypoparathyroidism: Biochemical, endocrine, and genetic analysis of Albright's hereditary osteodystrophy in six kindreds. *J Clin Endocrinol Metab* **62**:497–502, 1986.

30a. Levine MA et al: Genetic deficiency of the alpha subunit of the guanine nucleotide-binding protein G(s) as the molecular basis for Albright hereditary osteodystrophy. *Proc Natl Acad Sci USA* **85**:617–621, 1988.

31. Loveridge N et al: Inhibition of cytochemical bioactivity of parathyroid hormone by plasma in pseudohypoparathyroidism type I. *J Clin Endocrinol Metab* **54**:1274–1275, 1982.

32. Mackler H et al: Familial pseudohypoparathyroidism. *Calif Med* **77**:332–334, 1952.

33. Minozzi M et al: Su un caso di osteodistrofia ereditaria de Albright varieta normocalcemia con documenta transmisione da maschio a maschio. *Folia Endocrinol (Roma)* **16**:168–188, 1963.

34. Mitchell J, Goltzman D: Examination of circulating parathyroid hormone in pseudohypoparathyroidism. *J Clin Endocrinol Metab* **61**:328–334, 1985.

35. Monn E et al: Pseudohypoparathyroidism: A difficult diagnosis in early childhood. *Acta Paediatr Scand* **65**:487–493, 1976.

36. Papaioannoa AC, Matsas BE: Albright's hereditary osteodystrophy (without hypocalcemia). *Pediatrics* **31**:599–607, 1963.

37. Poznanski AK: *The Hand in Radiologic Diagnosis*. Saunders, Toronto, 1974.

38. Poznanski AK et al: The pattern of shortening of the bones of the hand in PHP and PPHP—a comparison with brachydactyly E, Turner syndrome, and acrodysostosis. *Radiology* **12**:707–716, 1977.

39. Radeke HH et al: Multiple pre- and postreceptor defects in pseudohypoparathyroidism (a multicenter study with twenty four patients). *J Clin Endocrinol Metab* **62**:393–402, 1986.

40. Ritchie GM: Dental manifestations of pseudohypoparathyroidism. *Arch Dis Childh* **40**:565–673, 1965.

41. Shapiro MS et al: Multiple abnormalities of anterior pituitary hormone secretion in association with pseudohypoparathyroidism. *J Clin Endocrinol Metab* **51**:483–487, 1980.

41a. Smit L et al: Intracerebral bilateral symmetrical calcifications, demonstrated in a patient with pseudohypoparathyroidism. *Clin Neurol Neurosurg* **90**:145–150, 1988.

42. Spranger JW: Skeletal dysplasias and the eye: Albright's hereditary osteodystrophy. *Birth Defects* 5(4):122–128, 1969.

43. Steinbach HL et al: Evolution of skeletal lesions in pseudohypoparathyroidism. *Radiology* **85**:670–676, 1965.

44. Steinbach HL, Young DA: The roentgen appearance of pseudohypoparathyroidism (PH) and pseudo-pseudohypoparathyroidism (PPH). *Am J Roentgenol* **97**:49–66, 1966.

45. Trevathan TH: Delayed eruption of teeth in pseudohypoparathyroidism. *N Z Dent J* **57**:20–23, 1961.

46. Van Dop C et al: Father to son transmission of decreased N(s) activity in pseudohypoparathyroidism type Ia. *J Clin Endocrinol Metab* **59**:825–828, 1984.

47. Weinberg AG, Stone RT: Autosomal dominant inheritance in Albright's hereditary osteodystrophy. *J Pediatr* **79**:996–999, 1971.

48. Weinstock RS et al: Olfactory dysfunction in humans with deficient guanine nucleotide-binding protein. *Nature (London)* **322**:635–636, 1986.

49. Weisman Y et al: Pseudohypoparathyroidism type Ia presenting as congenital hypothyroidism. *J Pediatr* **107**:413–415, 1985.

50. Werder EA: Pseudohypoparathyroidism. *Ergeb Inn Med Kinderheilkd* **42**:191–221, 1979.

**Williams syndrome.** The syndrome of characteristic facial appearance, mental retardation, growth deficiency, cardiovascular anomalies, and infantile hypercalcemia (Figs. 5–58 to 5–61) was described in a joint paper in 1952 by Fanconi and co-workers (16) in Zürich and in London. Important early papers are those of Williams (62) and Beuren (4–6). A good historical account has been presented by Myers and Willis (47). Several important clinical reviews are available (7,27,35,43,46a). Parental perspective on the disorder (1) is instructive for clinicians.

Naming the syndrome has been problematical. The disorder has been known in the past as the idiopathic hypercalcemia–supravalvular aortic stenosis syndrome in spite of the fact that both features are frequently absent (35). The term "elfin face" syndrome (14) presents the problem of naming a disorder after a mythical being. The currently accepted designation, Williams syndrome, or, less often, the Williams–Beuren syndrome, ignores the historical precedent mentioned earlier.

Delineating the phenotypic spectrum of abnormalities has also been difficult. Defining a syndrome on the basis of sporadic occurrence when the basic defect is unknown truncates the phenotype toward the severe end of the spectrum. Thus, reported frequencies of various findings are not particularly meaningful. The problem is further compounded by ascertainment on the basis of cardiovascular anomalies in some reports (5) and on the basis of facial features in others (35). It has been suggested that the syndrome may represent a spectrum that overlaps with hypercalcemia with or without mental retardation and supravalvular aortic stenosis with and without mental retardation (5).

Although several authors have discussed presumed familial examples of the Williams syndrome (10,28,61), to date all reported cases at the severe end of the phenotypic spectrum are apparently sporadic (26). We are unaware of examples of more than one full-blown case in the same family. On the other hand, reported relatives have had supravalvular aortic stenosis in some instances (28). Using the Jones–Smith criteria of facial–CNS–growth deficiency features without cardiovascular involvement (35), the affected cousins reported by White et al (61) qualify as a familial example.

The monozygotic twins noted by Wiltse et al (63) and Oorthuys (47a) are convincing; reported cases of presumed dizygotic twins (31,39) are unconvincing. Jones and Smith (35) and Martin et al (43) found no evidence for a paternal age effect, although Greenberg and Lewis (27) noted that paternal age appeared to be increased. A mildly affected father and a severely affected daughter have been observed by JG Hall (personal communication). Chromosomal rearrangements have been reported, but no consistent pattern has emerged and their relationship to the disorder remains unclear (23,33,43,46).

Natural history has been particularly well documented by Morris et al (46a) who tabulated data on 109 subjects. Features of infancy,

Table 5–6A. Medical problems in infants with Williams syndrome

| Problem | Percentage ($n=42$) |
|---|---|
| Early symptoms | |
| Feeding difficulty | 71 |
| Failure to thrive | 81 |
| Vomiting | 40 |
| Constipation | 43 |
| Colic | 67 |
| Chronic otitis media | 38 |
| Hypercalcemia | 4/6 |
| Birth defects | |
| Congenital heart defects | 79 |
| Umbilical hernia | 14 |
| Inguinal hernia | 38 |

Adapted from CA Morris et al, *J Pediatr* **113**:318–326, 1988.

childhood, and adulthood are listed together with their respective frequencies in Tables 5–6A, 5–6B, and 5–6C. Pagon et al (47a) studied physical, neurodevelopmental, and behavioral characteristics during late childhood and adolescence.

**Growth.** In most cases, there is mild to moderate growth deficiency of prenatal onset with more striking growth failure in the postnatal period (35). Mean birth weights are lower than average (27,43). In the series of 31 patients reported by Greenberg and Lewis (27), mean birthweight was 2830 g. In the study of Martin et al (43), mean male adult height was 159 cm, mean female adult height was 147 cm, mean male adult head circumference was 54.8 cm, and mean female adult head circumference was 52.8 cm. Only one patient has been reported with documented growth hormone deficiency who responded to therapy with human growth hormone (56).

**Central nervous system and performance.** Mild microcephaly, which is most striking in bifrontal diameter, is found in 65%. Other characteristic features include mental deficiency (95%), mild neurologic dysfunction (50%), and unusual personality (65%). Intelligence quotients have varied from 41 to 80 with an average IQ of 56 among 14 noninstitutionalized patients (35) and may correlate with severity of the condition as judged by age at diagnosis and extent of hypercalcemia (58a). Developmental delay and neurologic dysfunction are highly distinctive and can be diagnostic (3,44,45a,47b). In childhood, most patients have an echolalic, loquacious, friendly pattern of speech that

Table 5–6B. Medical problems in children with Williams syndrome

| Problem | Percentage ($n=42$) |
|---|---|
| Central nervous system | |
| Developmental delay with specific learning disability | 97 |
| Attention deficit disorder | Common |
| Ocular | |
| Esotropia | 50 |
| Hyperopia | 24 |
| Auditory | |
| Chronic otitis media | 43 |
| Dental | |
| Enamel hypoplasia | 48 |
| Microdontia | 55 |
| Malocclusion | 85 |
| Cardiovascular | |
| Congenital heart defects | 79 |
| Supravalvular aortic stenosis | 64 |
| Supravalvular pulmonic stenosis | 24 |
| Ventricular septal defect | 12 |
| Patent ductus arteriosus | 5 |
| Hypertension | 17 |
| Genitourinary | |
| Renal anomalies | ? |
| Enuresis | 52 |
| Gastrointestinal | |
| Constipation | 43 |
| Musculoskeletal | |
| Joint limitation | 50 |
| Kyphosis | 21 |
| Lordosis | 38 |
| Scoliosis | 12 |
| Awkward gait | 60 |
| Extra sacral crease | 52 |

Adapted from CA Morris et al, *J Pediatr* **113**:318–326, 1988.

Table 5–6C. Medical problems of adults with Williams syndrome

| Problem | Percentage ($n=17$) |
|---|---|
| Central nervous system | |
| Mental retardation (IQ<70) | 59 |
| Borderline intellectual functioning (IQ 70–85) | 41 |
| Hyperreflexia in lower extremities | 50 |
| Ocular | |
| Hyperopia | 18 |
| Cardiovascular | |
| Supravalvular aortic stenosis | 76 |
| Diminished peripheral pulse | 35 |
| Pulmonic artery stenosis | 35 |
| Hypertension | 47 |
| Diffuse aortic hypoplasia | 18 |
| Other documented arterial stenoses | 18 |
| Genitourinary | |
| Renal anomalies | ? |
| Urinary tract infections | 29 |
| Vesicoureter reflux | 3/4 |
| Bladder diverticuli | 3/4 |
| Nephrocalcinosis | 0[a] |
| Gastrointestinal | |
| Constipation | 41 |
| Peptic ulcer | 18 |
| Cholelithiasis | 12 |
| Diverticulitis | 12 |
| Diabetes mellitus | 12 |
| Obesity | 29 |
| Integumentary | |
| Prematurely gray hair | 60 |

Adapted from CA Morris et al, *J Pediatr* **113**:318–326, 1988.
[a]Reported as a cause of renal failure in other reports, however.

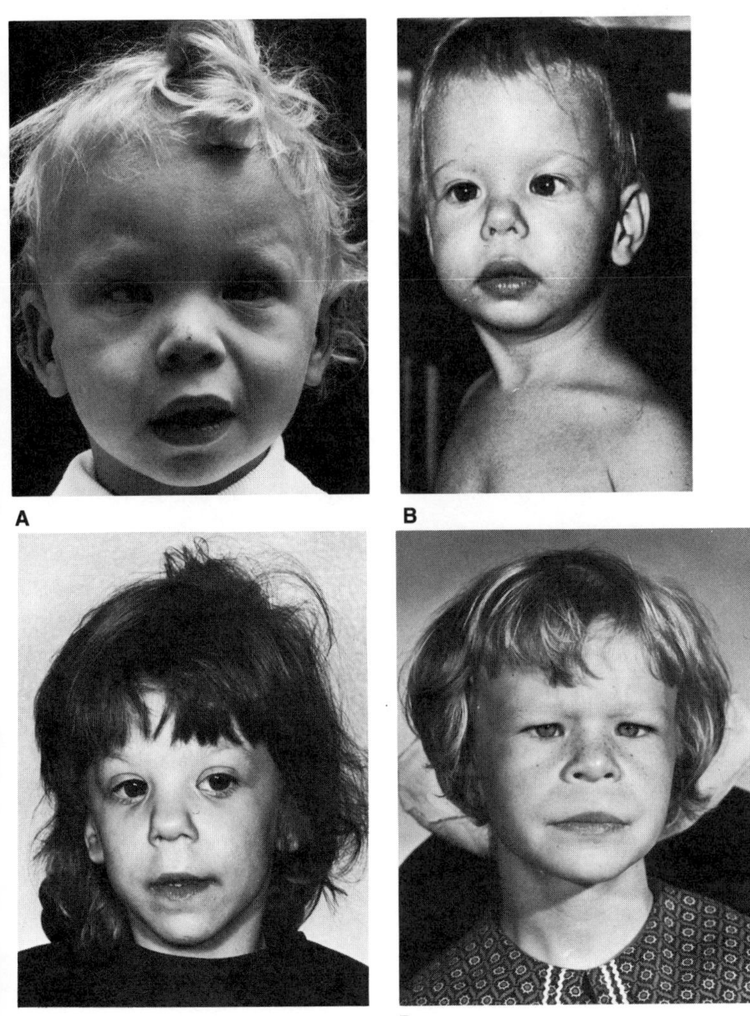


Fig. 5–58. *Williams syndrome.* (A–D) Compare facies for anteverted nostrils, long philtrum, mild midface hypoplasia, epicanthic folds, strabismus. (A from *A Dupont,* Dan Med Bull **17**:33, 1970. B from *F Ray,* J Pediatr Ophthalmol **8**:188, 1971.)

has been described as a "cocktail party manner" (35). In a positive sense, this can be related to a relatively better command of verbal and language skills (3,35,44). However, language performance is significantly below normal. Disordered syntax, overuse of cliches, and tangential reasoning have been noted. The ability to tell time or make change is relatively poor (1a). In a few, impulsivity and destructive behavior predominate (35). Frank autism has also been observed (51).

Major motor development is relatively spared (43,44), but visual–motor integration is usually impaired to an extent that it is out of proportion to the global deficit (44). The very frequent history of hyperacusis without physical findings (1,43) suggests that the impairment of perceptual–motor integration can include auditory and other modalities.

Facial features. Facial features are distinctive and become more striking with age. The combination of flat midface, depressed nasal bridge, anteverted nostrils, long philtrum, thick lips, wide intercommissural distance, and open mouth is characteristic (Fig. 5–58A–D). Ocular findings may include medial eyebrow flaring (80%), short palpebral fissures (50%), hypotelorism (50%), epicanthic folds (50%), periorbital fullness, strabismus (especially esotropia) (35%), and over 60% of blue- or hazel-eyed individuals with a stellate or lacy iris pattern (Fig. 5–59) (75%) (5,6,16,27,35,48,54,60). Uncommonly, corneal and/or lenticular opacities (54) and ptosis of eyelids (personal observation) may be observed. Hypermetrous discs (55%) and simple vertical branching of the central retinal vessels (70%) have been documented (27). In some cases, the ears may be prominent. The thyroid cartilage becomes more prominent with age (27).

Oral manifestations. Thick lips, wide intercommissural distance, and long philtrum are characteristic. The voice is often hoarse or brassy (5,6,35,40). The maxillary arch has been described as being too broad for the mandibular arch. Hypodontia, microdontia, small slender roots, and dens invaginatus have been reported (5,6,14,20,62). Hypoplastic bud-shaped maxillary deciduous second molars and mandibular permanent first molars (Fig. 5–60) have been noted by some authors (5,6). In a series of 17 patients studied metrically, microdontia was not observed nor were hypoplastic bud-shaped teeth present in any patient (MR Heinemann, personal communication). Mild micrognathia, widened mandibular angle, osteosclerotic changes in the lamina dura (particularly in the premolar–molar region), delayed mineralization of teeth, folding and thickening of the buccal mucous membranes, and prominent and accessory labial frenula have also been noted (5,6,14,38).

Cardiovascular system. Supravalvular aortic stenosis (Fig. 5–61) and pulmonary artery stenosis are the most common findings, although various other cardiovascular defects have been recorded. When

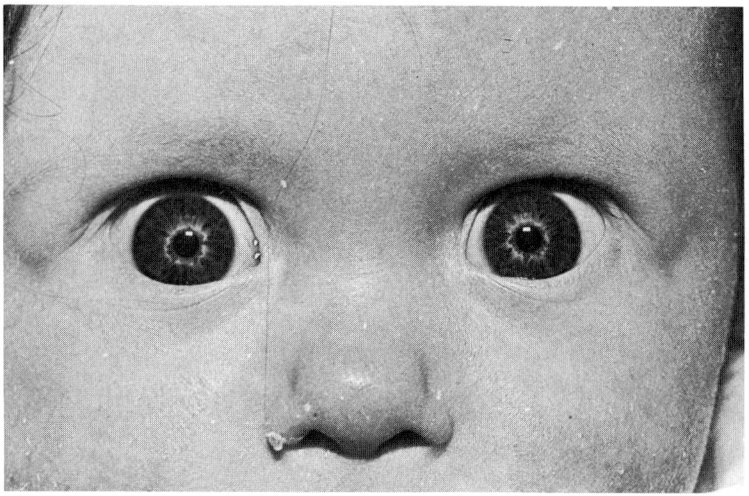

Fig. 5–59. *Williams syndrome*. Stellate iris pattern. (From *FO Jensen* and *AC Begg*, NZ Med J **68:**364, 1968.)

studied echocardiographically, all have supravalvular aortic narrowing (28a). These include valvular aortic stenosis, aortic hypoplasia, bicuspid aortic valves, coarctation of the aorta, coronary artery stenosis, renal artery stenosis, various peripheral artery stenoses (carotid, innominate, subclavian, celiac, and mesenteric), atrial septal defect, ventricular septal defect, anomalous pulmonary venous return, arteriovenous fistula (lung), interruption of the aortic arch, and aplasia of the portal vein (8,14–16,29,35,41,49,50,53). About 10–15% have mitral valve prolapse or bicuspid aortic valve (28a). About 60% of individuals over 22 years of age have been recorded to have hypertension (46a). Some Williams syndrome patients have no cardiovascular anomalies, but are easily diagnosed by characteristic facial appearance and performance (35).

Longitudinal studies indicate that supravalvular aortic stenosis tends to progress with age; pulmonary artery stenosis improves (32). Coronary artery stenosis can lead to myocardial infarction and sudden death and probably contributes to the increased risk of death at cardiac catheterization (55,58). Renal artery stenosis also contributes to the increased frequency (29%) (43) of systemic hypertension (13,55,63).

Genitourinary abnormalities. Micropenis is a relatively common feature (35). Information about reproductive function is not available. Multiple bladder diverticula associated with urinary incontinence are common (1b). Nephrocalcinosis can be a complication of hypercalcemia, but may not be visualized by standard radiographic technique (13,16,43). Ultrasonography or computerized tomography are likely to be more helpful. Renal calculi are also known to occur (43). Renal insufficiency and criteria for investigation have been discussed by Biesecker et al (6a).

Skeletal system. The hypercalcemic phase may result in widespread osteosclerotic changes that regress in time. Craniosynostosis (secondary to microcephaly) is uncommon. Increased density of the metaphyses, epiphyses, and skull base may be present in some in-

Fig. 5–60. *Williams syndrome*. (A) Upper dental arch completely overlaps lower arch. (B) Note expanded upper arch, hypoplastic upper second deciduous molar on right. (C) Detail showing bud-shaped maxillary teeth. (A through C from *AJ Beuren et al*, Am J Cardiol **13:**471, 1964.)

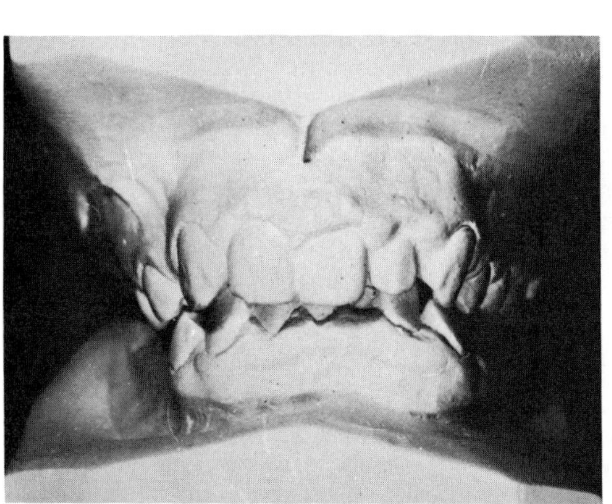

A

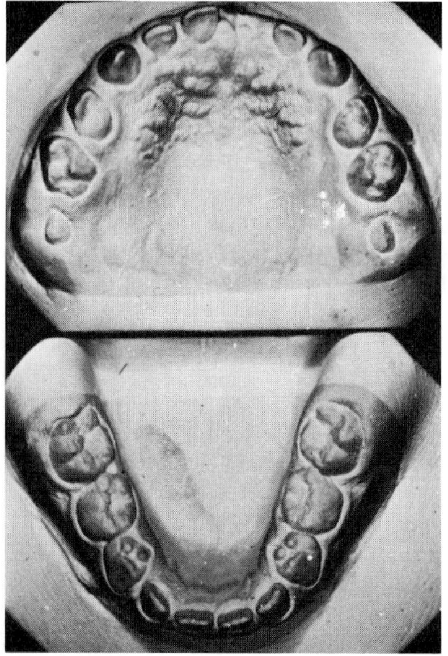

B

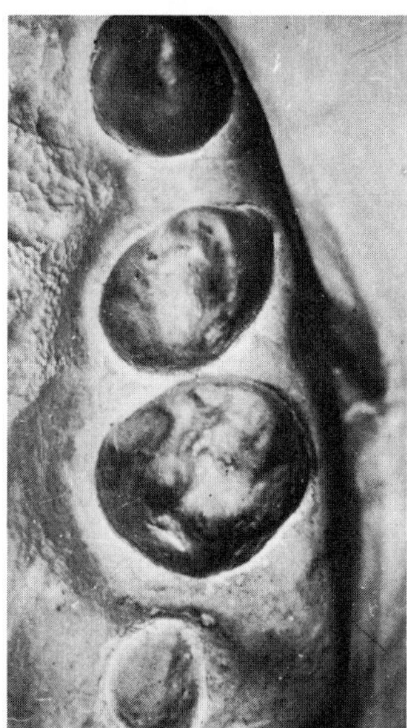

C

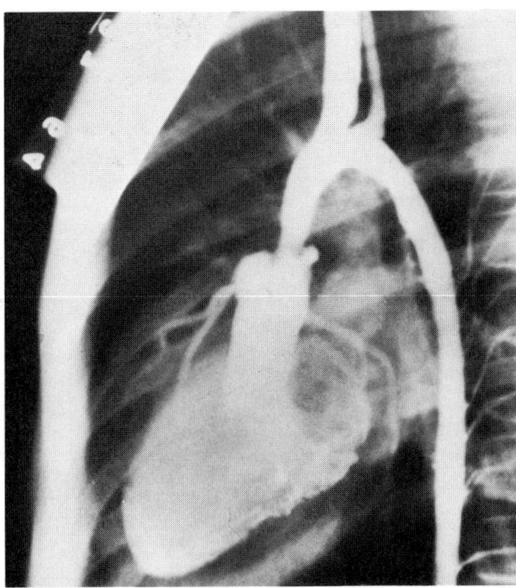

Figure 5–61. *Williams syndrome.* Note area of stenosis and hypoplasia of entire aorta distal to obstruction. (From *K Jue et al,* J Pediatr **67:**1130, 1965.)

stances (14,16,38). Other findings may include pectus excavatum (40%), kyphoscoliosis (19%), minor structural and postural curvatures (55%), hallux valgus (75%), and fifth finger clinodactyly (40%) (35,43).

Other findings. Hypoplastic, deeply set nails are found in 65% (35). Other findings include inguinal hernia (38%), umbilical hernia (14%), and rectal prolapse (12%) (35,43). Contractures (50%) and prematurely gray hair (60%) have been recorded (37a,46a).

Hypercalcemia. Hypercalcemia has not been documented in most cases (35), but has been a feature in some instances (37). When present, it usually disappears during the second year of life. Prolonged symptomatic hypercalcemia has been observed in some patients (52). Retrospective interviewing may reveal a history of failure to thrive, hypotonia, anorexia, constipation, or renal impairment (40).

From the earliest descriptions, a disorder of vitamin D has been repeatedly implicated in some way (2,21,22,24,31,39,57). Initially, vitamin D intoxication was considered the likely etiology in Britain because of reported excessive dietary intake during and after pregnancy, but reduction of vitamin D intake did not eliminate the condition. Forfar and associates (19) advanced the postulate that Williams syndrome infants are abnormally sensitive to vitamin D *in utero* or in early infancy. The induction of similar craniofacial, dental, and cardiovascular anomalies in newborn rabbits born to mothers who had received prodigious amounts of vitamin D (21,22) lends support to this hypothesis, but also suggests that abnormalities of vitamin D metabolism may be secondary (18). The regulation of 25-hydroxyvitamin D [25(OH)D] may be abnormal (57). Circulating levels of 25(OH)D, 1,25-dihydroxyvitamin D [1,25(OH)2D], and vitamin D-binding protein have not been elevated in most series (9,11,12,25) with few exceptions (24), irrespective of hypercalcemia. In a few reports, instances of decreased levels of 1,25(OH)D have been found, possibly signifying appropriate feedback suppression (25).

In contrast, there seems to be general agreement that the response of serum calcium to a calcium-loading test is abnormal, even in individuals with no history, clinical, or radiographic findings, or laboratory evidence of hypercalcemia (11). Although both control of endogenous parathyroid hormone and response to exogenous parathyroid hormone are not abnormal, endogenous response of calcitonin to a calcium load is strikingly deficient (11).

Differential diagnosis and laboratory aids. Isolated supravalvular aortic stenosis and peripheral pulmonary artery stenosis have been reported to follow autosomal dominant transmission (17,34,36,42,45,55), but occur most frequently on a sporadic basis. Peripheral pulmonary stenoses may also be found in *thalidomide embryopathy, rubella embryopathy, Down syndrome, and Alagille syndrome.* Other causes of hypercalcemia, such as familial hypocalciuric hypercalcemia (30), hyperparathyroidism, and vitamin D intoxication can usually be distinguished by history and routine laboratory tests. Martin et al (43) distinguished benign infantile hypercalcemia (Lightwood syndrome) from severe Fanconi-type hypercalcemia on the basis of absence of associated anomalies, normal development, and lack of hypercalcemia symptoms. Ultrasonography is a valuable tool for noninvasive detection of subtle cardiovascular and renal abnormalities. Cross-sectional echocardiography is helpful for diagnosing supravalvular aortic stenosis (59). The calcium loading test may be useful in cases in which there is some doubt about the somatic phenotype (13).

### References (Williams syndrome)


1. Anonymous: Case history of a child with Williams syndrome. *Pediatrics* **75**:962–968, 1985.

1a. Arnold R et al: The psychological characteristics of infantile hypercalcemia. *Dev Med Child Neurol* **27**:49–59, 1985.

1b. Babbitt DB et al: Multiple bladder diverticula in Williams "Elfin-Facies" syndrome. *Pediatr Radiol* **8**:29–31, 1979.

2. Becroft DMO, Chambers D: Supravalvular aortic stenosis—infantile hypercalcemia syndrome: In vitro hypersensitivity to vitamin D2 and calcium. *J Med Genet* **13**:223–228, 1976.

3. Bennett FC et al: The Williams elfin facies syndrome: The psychological profile as an aid in syndrome identification. *Pediatrics* **61**:303–306, 1978.

4. Beuren AJ: Supravalvular aortic stenosis: A complex syndrome with and without mental retardation. *Birth Defects* **8**(5):45–56, 1972.

5. Beuren AJ et al: Supravalvular aortic stenosis in association with mental retardation and a certain facial appearance. *Circulation* **26**:1235–1240, 1962.

6. Beuren AJ et al: The syndrome of supravalvular aortic stenosis, peripheral pulmonary stenosis, mental retardation and similar facial appearance. *Am J Cardiol* **13**:471–483, 1964.

6a. Biesecker LG et al: Renal insufficiency in Williams syndrome. *Am J Med Genet* **28**:131–135, 1987.

7. Burn J: Williams syndrome. *J Med Genet* **23**:389–395, 1986.

8. Char F: Williams facies with portal vein aplasia and mental retardation. *Birth Defects* **8**(5):262–263, 1972.

9. Chesney RW et al: Increased plasma 1,25-dihydroxyvitamin D in infants with hypercalcemia and elfin facies (letter). *N Engl J Med* **313**:889, 1985.

10. Cortada X et al: Familial Williams syndrome. *Clin Genet* **18**:173–176, 1980.

11. Culler FL, et al: Impaired calcitonin secretion in patients with Williams syndrome. *J Pediatr* **107**:720–723, 1985.

12. Daiger SP et al: Vitamin-D-binding protein in the Williams syndrome and idiopathic hypercalcemia. *N Engl J Med* 298:687–688, 1978.

13. Daniels SR et al: Systematic hypertension secondary to peripheral vascular anomalies in patients with Williams syndrome. *J Pediatr* **106**:249–251, 1985.

14. Dupont B et al: Idiopathic hypercalcemia of infancy: The elfin face syndrome. *Dan Med Bull* **17**:33–46, 1970.

15. Eie H et al: Localized supravalvular aortic stenosis combined with mental retardation and peculiar facial appearance. *Acta Med Scand* **191**:517–520, 1972.

16. Fanconi G et al: Chronische Hypercalcämie, kombiniert mit Osteosklerose, Hyperazotämie. Minderwuchs und kongenitalen Missbildungen. *Helv Paediatr Acta* **7**:314–334, 1952.

17. Feigl A et al: Supravalvular aortic and peripheral pulmonary arterial stenosis. A report of eight cases in two generations. *Isr J Med Sci* **16**:496–502, 1980.

18. Forbes GB: Vitamin D in pregnancy and the infantile hypercalcemia syndrome (letter). *Pediatr Res* **13**:1382, 1979.

19. Forfar JO et al: Idiopathic hypercalcemia of infancy. *Lancet* **i**:981, 1956.

20. Frank RM et al: Aspects odonto-stomatologiques de la stenose aortique susvalvulaire. *Rev Stomatol (Paris)* **67**:223–232, 1966.

21. Friedman WF: Vitamin D and the supravalvular aortic stenosis syndrome. *Adv Teratol* **3**:85–96, 1968.

22. Friedman WF, Mills LF: The relationship between vitamin D and the craniofacial and dental anomalies of the supravalvular aortic stenosis syndrome. *Pediatrics* **43**:12–18, 1969.

23. Fryns JP et al: The elfin face syndrome and the short arm of chromosome 15. *Ann Génét* **25**:181–182, 1982.

24. Garabedian M et al: Elevated plasma 1,25-dihydroxyvitamin D concentrations in infants with hypercalcemia and an elfin facies. *N Engl J Med* **312**:948–952, 1985.
25. Goodyer PR et al: Observations on the evolution and treatment of idiopathic hypercalcemia. *J Pediatr* **105**:771–773, 1984.
26. Gorlin RJ: Risk of recurrence in usually nongenetic malformation syndromes. *Birth Defects* **15**(5C):181–188, 1979.
27. Greenberg F, Lewis RA: The Williams syndrome. Spectrum and significance of ocular findings. *Ophthalmology* **95**:1608–1612, 1988.
28. Grimm T, Wesselhoeft H: The genetic aspects of Williams–Beuren syndrome and the isolated form of the supravalvular aortic stenosis. Investigation of 128 families. *Z Kardiol* **69**:168–172, 1980.
28a. Hallidie-Smith KA, Karas S: Cardiac anomalies in Williams–Beuren syndrome. *Arch Dis Childh* **63**:809–813, 1988.
29. Harris LL, Nghiem QX: Idiopathic hypercalcemia of infancy with interruption of the aortic arch. *J Pediatr* **73**:84–88, 1968.
30. Hooft C et al: Familial incidence of hypercalcemia. *Helv Paediatr Acta* **16**:199–210, 1961.
31. Illig R, Prader A: Kasuistische Beiträge zur idiopathischen Hypercalcemie und Vitamin D Intoxication. *Helv Paediatr Acta* **14**:618–646, 1959.
32. Ino T et al: Progressive vascular lesions in Williams syndrome. *J Pediatr* **107**:826, 1985.
33. Jefferson RD et al: A terminal deletion of the long arm of chromosome 4 {46,XX,del(4)(q33)} in an infant with phenotypic features of Williams syndrome. *J Med Genet* **23**:474–480, 1986.
34. Johnson LW et al: Familial supravalvular aortic stenosis. *Chest* **70**:494–560, 1976.
35. Jones KL, Smith PW: The Williams elfin facies syndrome: A new perspective. *J Pediatr* **86**:718–723, 1975.
36. Jorgensen G, Beuren AJ: Genetische Untersuchungen bei supravalvularen Aortenstenosen. *Humangenetik* **1**:497–515, 1965.
37. Joseph MC, Parrott D: Severe infantile hypercalcemia with special reference to the facies. *Arch Dis Childh* **33**:385–395, 1958.
37a. Kaplan P et al: Contractures in Williams syndrome. *Eighth Annual Workshop on Malformations and Morphogenesis,* Greenville, South Carolina, August 15–19, 1987.
38. Kelly JR, Barr ES: The elfin facies syndrome. *Oral Surg* **40**:205–218, 1975.
39. Kenny FM et al: Metabolic studies in a patient with idiopathic hypercalcemia of infancy. *J Pediatr* **62**:531–537, 1963.
40. Kivaol E et al: Mental retardation, typical facies and aortic stenosis syndrome. *Ann Med Intern Fenn* **54**:81–87, 1965.
41. Levy EP: Infantile hypercalcemia facies and mental retardation associated with atrial septal defect and anomalous pulmonary venous return. *Birth Defects* **8**(5):73–74, 1972.
42. Lewis AJ et al: Supravalvular aortic stenosis. Report of a family with peculiar somatic features and normal intelligence. *Dis Chest* **55**:372–379, 1969.
43. Martin NDT et al: Idiopathic infantile hypercalcemia—a continuing enigma. *Arch Dis Childh* **59**:605–613, 1984.
44. MacDonald WG, Roy DC: Williams syndrome: A developmental profile. *Clin Invest Med* **9**:A131, 1986.
45. McCue CM et al: Familial supravalvular aortic stenosis. *J Pediatr* **73**:889–895, 1968.
45a. Meyerson MD, Frank RA: Language, speech and hearing in Williams syndrome: Intervention approaches and research needs. Reprint from *Dev Med Child Neurol* **29**:258–262, 1987.
46. Miles JH, Michalski KA: Familial 15q12 duplication associated with Williams phenotype. *Am Soc Hum Genet* 34th meeting, Norfolk, Virginia, October 30–November 2, 1983, Abstr. #429.
46a. Morris CA et al: The natural history of Williams syndrome: Physical characteristics. *J Pediatr* **113**:318–326, 1988.
47. Myers AR, Willis PW: Clinical spectrum of supravalvular aortic stenosis. *Arch Intern Med* **118**:553–561, 1966.
47a. Oorthuys JWE: Monozygote tweeling met het Williams-Beuren of 'elfin-face' syndroom. *T Kindergeneesk* **52**:197–200. 1984.
47b. Pagon RA et al: Williams syndrome: Features in late childhood and adolescence. *Pediatrics* **80**:85–91, 1987.
48. Preus M: Iris pattern in patients with Williams syndrome. *J Pediatr* **87**:840, 1975.
49. Przybojewski JZ et al: Supravalvular aortic stenosis in the adult. *S Afr Med J* **59**:796–803, 1981.
50. Rashkind WJ et al: Cardiac findings in idiopathic hypercalcemia of infancy. *J Pediatr* **58**:464–469, 1961.
51. Reiss AL et al: Autism associated with Williams syndrome. *J Pediatr* **106**:247–249, 1985.
52. Robb M et al: Prolonged symptomatic hypercalcemia in Williams syndrome: An analysis of parathyroid function. *Proc Greenwood Genet Ctr* **3**:131–132, 1984.
53. Roberts NK, Moes CAF: Supravalvular pulmonary stenosis and unusual facial appearance. *Birth Defects* **8**(5):57–59, 1972.
54. Roy FH et al: Infantile hypercalcemia in supravalvular aortic stenosis. *J Pediatr Ophthalmol* **8**:188–193, 1971.
55. Schmidt RE et al: Generalized arterial fibromuscular dysplasia and myocardial infarction in familial supravalvular aortic stenosis syndrome. *J Pediatr* **74**:576–584, 1969.
56. Spadoni GL et al: Williams syndrome and growth hormone deficiency. *J Pediatr* **102**:640–641, 1983.
57. Taylor AB et al: Abnormal regulation of circulating 25-hydroxyvitamin D in the Williams syndrome. *N Engl J Med* **306**:972–975, 1982.
58. Terlune PE et al: Myocardial infarction associated with supravalvular aortic stenosis. *J Pediatr* **106**:251–254, 1985.
58a. Udwin O et al: Age at diagnosis and abilities in idiopathic hypercalcemia. *Arch Dis Childh* **61**:1164–1167, 1986.
59. Vogt J et al: Qualitative and quantitative investigations in supravalvular aortic stenosis by cross-sectional echocardiography. *Z Kardiol* **69**:70–73, 1980.
60. Wesselhoeft H et al: The spectrum of supravalvular aortic stenosis. Clinical findings of 150 patients with Williams–Beuren syndrome and the isolated lesion. *Z Kardiol* **69**:131–140, 1980.
61. White RA et al: Familial occurrence of the Williams syndrome. *J Pediatr* **91**:614–616, 1977.
62. Williams JCP et al: Supravalvular aortic stenosis. *Circulation* **24**:1311–1318, 1961.
63. Wiltse HE et al: Infantile hypercalcemia syndrome in twins. *N Engl J Med* **275**:1157–1160, 1966.

**Zellweger syndrome (cerebrohepatorenal syndrome).** DeLange (12a), in 1949, probably reported the first case of Zellweger syndrome. Bowen and associates (4) in 1964 and, independently, Smith and co-workers in 1965 (54) described a syndrome consisting of hypotonia, high forehead and other craniofacial anomalies, hepatomegaly and liver dysfunction, and renal cortical cysts (Figs. 5–62 to 5–65). In 1973, Goldfischer et al (18) reported that liver and kidney in affected individuals lacked peroxisomes. The significance of this finding has become apparent more recently as the biochemical consequences of deficient peroxisomal function have been defined. Other generalized disorders of peroxisomal function have been revealed and now include the neonatal form of adrenoleukodystrophy, the infantile form of Refsum's disease, hyperpipecolic acidemia, and rhizomelic chondrodysplasia punctata (Table 5–7). The nosology of these newly defined conditions and the criteria for distinguishing among variants are still being delineated (38,40). There are also new and less well-defined conditions such as pseudo-Zellweger syndrome, in which peroxisomal enzymes are diminished but peroxisomal structure and number are intact (20), and milder variants of Zellweger syndrome that can be identified biochemically but do not present with typical craniofacial features or cerebrohepatic disease in early infancy (2,3,71,72). Several reviews of peroxisomal disorders and their clinical manifestations have been published (35,36,52,53,56,71). An extensive review of older Zellweger syndrome references has been compiled (40) as well as a survey of over 100 cases (69a).

Genetics. Affected sibs (4,6,14–16,31,39,41,54,55,58) and parental consanguinity (6,21,39,47,54,70) indicate autosomal recessive inheritance. The frequency of the disorder has been estimated to be 1/100,000 live births (11), but identification of milder variants and atypical cases by biochemical means is likely to reveal a higher frequency of this disorder (38).

Cell complementation studies indicate that neonatal adrenoleukodystrophy and rhizomelic chondrodysplasia punctata are biochemically distinct entities (Table 5–7). Fibroblasts from patients with infantile Refsum's disease and hyperpipecolic acidemia do not complement those from patients with Zellweger syndrome, but it remains uncertain whether further biochemical and genetic heterogeneity may exist within this group (56).

Zellweger syndrome as a peroxisomal disorder. Biochemical and ultrastructural findings in Zellweger syndrome and in related peroxisomal disorders are summarized in Table 5–7.

Table 5–7. Biochemical and ultrastructural findings in Zellweger syndrome and related peroxisomal disorders

| Finding (sample source) | Peroxisomal defect | Group 2[a] | | | Group 3[a] | Group 1[a] |
|---|---|---|---|---|---|---|
| | | Zellweger syndrome | Hyper-pipecolic acidemia | Infantile Refsum's disease | Neonatal adrenoleuko-dystrophy | Rhizomelic chrondro-dysplasia punctata |
| Absent peroxisomes (liver, fibroblasts)[b] | ?Peroxisomal biogenesis | + | ± | + | + | ± |
| Pipecolic acidemia (urine, plasma, CSF) | ?D-Amino acid oxidase[c] | + | + | + | + | ? |
| Phytanic acid (serum, urine) | Phytanic acid oxidase | + | ? | + | + | + |
| Very-long-chain fatty acids (VLCFAs) (plasma, urine) | VLCFA $\beta$-oxidation[d] | + | + | + | + | – |
| Bile acid intermediates (bile, plasma, urine) | $C_{26}$ bile acid side chain cleavage[e] | + | ? | + | + | – |
| Plasmalogen deficiency (serum, fibroblasts) | DHAP-AT and alkyl-DHAP synthetase[f] | + | + | + | + | + |

[a]Complementation established after somatic cell fusion by restoration of activity of DHAP-AT (56).
[b]In fibroblasts, peroxisomes are reduced to <10% of normal.
[c]Tentatively localized to peroxisomes (52,61) but there are still questions about the pathway and its location (34).
[d]The entire complex of three enzymes—acyl-CoA oxidase, bifunctional protein (enoyl-CoA hydratase and 3-hydroxyacyl-CoA dehydrogenase), and 3-oxoacyl-CoA thiolase—is deficient (50).
[e]This defect, which involves a $\beta$-oxidation complex, results in the accumulation of dihydroxy- and trihydroxycoprostanic acids (25,46).
[f]Both enzymes are essential first steps in the synthesis of the other linkage that distinguishes plasmalogens from other phospholipid species (35,71).

*Ultrastructural.* Peroxisomes can be identified by electron microscopy but can be more reliably differentiated from other organelles by histochemical techniques or immunocytochemical procedures using antibodies against peroxisomal enzymes (19). In classic Zellweger syndrome, recognizable peroxisomes are essentially absent from liver and kidney and are greatly decreased in cultured skin fibroblasts (19,52).

*Biochemical.* The first biochemical abnormality to be described was increased urinary pipecolic acid. Earlier reported as an isolated inborn error of metabolism (15), its association with the syndrome was first noted by Danks and co-workers (11) and has since been confirmed as a frequent finding (77%) (21,23,32,33,52,71). Pipecolic acid, a minor product of lysine metabolism, can be detected by routine amino acid chromatography (11,23,30). Elevated pipecolic acid concentrations may be found in urine, plasma, and cerebrospinal fluid (CSF) (21,23,33) but normal urine and plasma values have been reported (10,21). In patients given lysine loads, there is decreased ability to dispose of the pipecolic acid metabolite (59). This abnormality is presumably related to a specific defect in a peroxisomal pathway (52,61), although this has been questioned (34). Because elevated pipecolic acid concentrations are found in other disorders of peroxisomal function (33,52) and in unrelated conditions (10,23,30), detection of this metabolite constitutes helpful information but is not diagnostic.

The first reports of abnormal bile acid synthesis suggested a mitochondrial origin (26,69) for the defect, but more recent information indicates that impaired side chain cleavage of the cholesterol precursor in the peroxisome is responsible for the accumulation of distinctive bile acid intermediates in the plasma, urine, and bile (25,69).

Similarities between adrenoleukodystrophy and Zellweger syndrome led Brown and co-workers (5) to the discovery that both share a common defect in the catabolism of very-long-chain fatty acids (VLCFAs) by a specific $\beta$-oxidation enzyme pathway (52). These distinctive lipids are readily quantitated in plasma, cultured fibroblasts, brain, amniocytes, amniotic fluid, or chorionic villus tissue and therefore serve as useful markers for pre- and postnatal diagnosis (1,24, 37,49,51,52).

Similarly, the peroxisomal localization of the biosynthetic pathway for plasmalogens, a minor species of either phospholipids, led to the identification of plasmalogen deficiency as another characteristic of the Zellweger syndrome (12). The first enzyme in the pathway, acyl-CoA:dihydroxyacetone-phosphate acyltransferase (DHAP-AT), is also readily assayed in a variety of tissues and has been utilized for pre- and postnatal diagnosis (27,33,49,51,52,71).

Phytanic acid accumulation also occurs in Zellweger syndrome, as it does in other peroxisomal disorders, notably infantile Refsum's disease (1,44,45).

Although the biochemical abnormalities of the Zellweger syndrome can be related to the lack of peroxisomal organelles, the connection is not a simple one. For example, catalase, the prototypical soluble peroxisomal enzyme, shows normal activity in Zellweger syndrome cells but the assay no longer requires detergent solubilization, suggesting that the enzyme is synthesized normally but remains directly accessible in the cytosol because there is no peroxisomal membrane to limit its diffusion (52,65). For the $\beta$-oxidation enzymes responsible for VLCFA catabolism, however, it appears that peroxisomal processing functions lead to impaired maturation and very rapid degradation of the enzymes, so that neither immunodetectable protein nor enzymatic activity is found in cultured cells (50).

It should also be noted that concomitant ultrastructural and biochemical changes in mitochondrial function remain unexplained (52,60) and that patients with normal complements of renal and hepatic peroxisomes have been reported (19,20,33).

**Somatic manifestations.** Gestation and delivery are usually uneventful (21), but the neonate is small and the complications of cerebral and hepatic disease commonly manifest shortly after birth (71). Failure to thrive and delayed development are characteristic (94%). Although demise in the first year of life is common, survival into later childhood has occurred (21); milder variants may not present in the newborn period (2,3,52,72).

*Craniofacial features.* The forehead is high and the skull may be pear shaped or show some degree of turribrachycephaly. The occiput is sometimes flattened (50%) (42) as is the face, particularly the supraorbital ridges (92%). Puffy eyelids, ocular hypertelorism, mild downward obliquity of the palpebral fissures, and epicanthic folds may be observed. The cheeks are full and the nostrils anteverted (Fig. 5–62A and 5–63). Micrognathia may be found and there is redundant

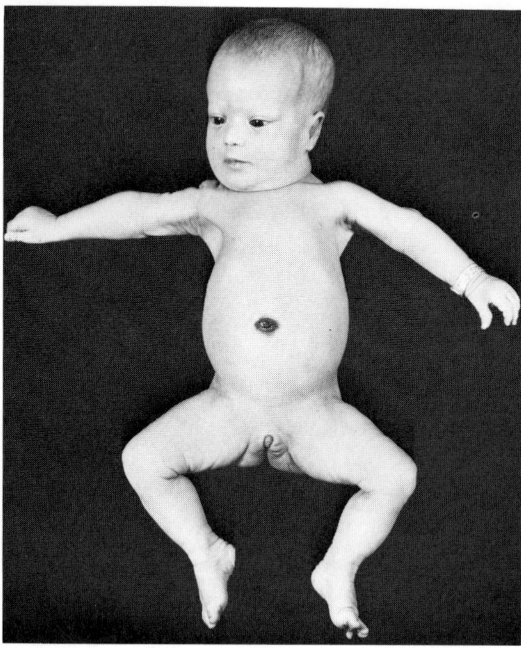

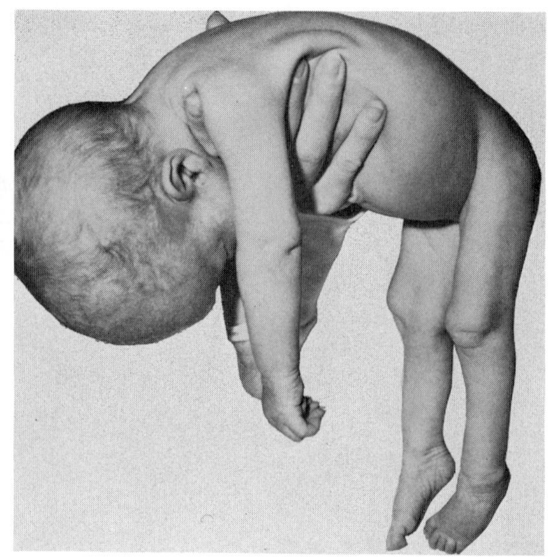

A B

Fig. 5–62. *Zellweger syndrome*. (A) High forehead, upslanting palpebral fissures in severely hypotonic female infant. (B) Extreme hypotonia. (A from *JE Jan et al*, Am J Dis Child **119:**274, 1970. B from *E Passarge* and *AJ McAdams*, J Pediatr **71:**691, 1967.)

skin on the neck. The ears may be posteriorly angulated and the helices may be abnormal (39,54). Narrow highly arched palate (72%) and protruding tongue have been noted (34,71).

Brushfield spots, cloudy corneas, and cataracts (71%) are common. The cortical cataracts do not appear to progress; a prominent lenticular Y suture may also be found (29,32). Funduscopic examination may reveal optic disc pallor (36%) or retinal pigment changes (38%) that resemble other forms of retinitis pigmentosa (13,44). Visual impairment is progressive and is accompanied by nystagmus and electroretinographic changes.

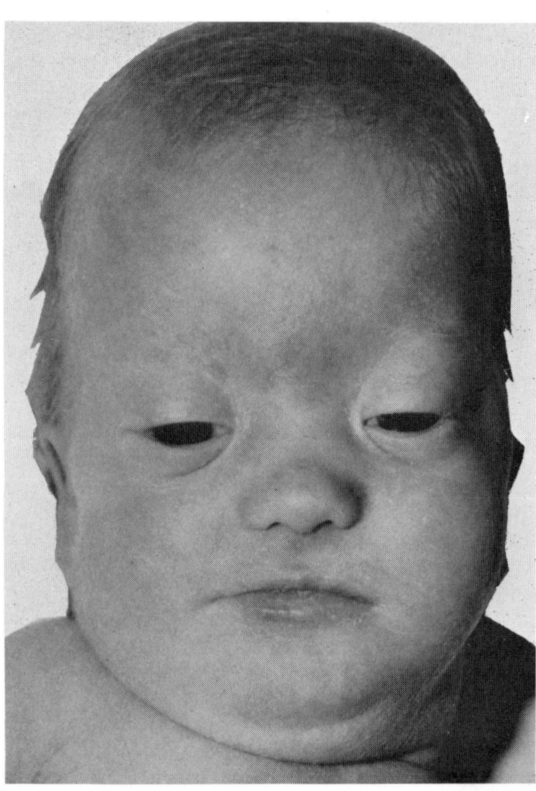

Fig. 5–63. *Zellweger syndrome*. Characteristic craniofacies. Compare with facies shown in Fig. 1A. (From *E Passarge* and *AJ McAdams*, J Pediatr **71:**691, 1967.)

*Central nervous system.* Macrocephaly and a large anterior fontanel (94%) are often noted. Profound hypotonia is almost universal (Fig. 5–62A,B) (99%); mental retardation and seizures (oculogyric fits) are also characteristic. Neonatal reflexes are sometimes absent but electromyography may be normal. Ventricular dilatation is frequently detected on CT scan and the EEG is abnormal (24). At autopsy, distinctive maldevelopmental and degenerative changes are found (Fig. 5–64A) including micropachygyria (67%), lissencephaly and hypoplastic or absent corpus callosum (20%), dysplasia of the olivary nucleus, olfactory tracts, and the cerebellum (27%), and heterotopias of the cerebrum and cerebellum (48%). Microscopic changes include sudanophilic leukoencephalomyelopathy (26%), increased glycogen storage, and gliosis (35%) (4,38,39,41,42,47,64,71).

*Liver.* An enlarged liver is rarely encountered at birth but is a common finding (69%) in the first year (21,71). Prolonged neonatal jaundice may be observed (60%) and abnormal liver function (85%) can be biochemically documented early in life. Histopathologic changes suggest a progression from hepatic giant cell transformation through periportal fibrosis to cirrhosis without widespread inflammatory disease or necrosis (8,17,33). In some, this sequence is completed within a few months time but in older children, hepatic liver function may stabilize (21) and hepatic fibrosis predominates (17). Increased hepatic iron content has been reported frequently (63%) (11,17,32,33,39,63) but the finding is not without exception (11,42). Hemosiderin deposits appear to be unrelated to progression of hepatic disease (17,32,33,42). Serum iron, iron saturation, and transferrin concentrations are also elevated in 62%, but values vary widely and are not necessarily correlated with the degree of tissue iron storage (11,17,21,32). Biliary dysgenesis and intrahepatic cholestasis are also described (17,32).

Related gastrointestinal findings include islet cell hyperplasia and hypoglycemia (17,42), pyloric hypertrophy, malrotation of the colon (4,39), and pancreatic fibrosis (11).

*Kidney.* The kidneys are variable in size and sometimes studded with multiple macroscopic cortical cysts (Fig. 5–64B). Glomerular and tubular microcystic disease are almost always found (93%). Horseshoe kidneys, foci of renal dysgenesis, and interstitial fibrosis

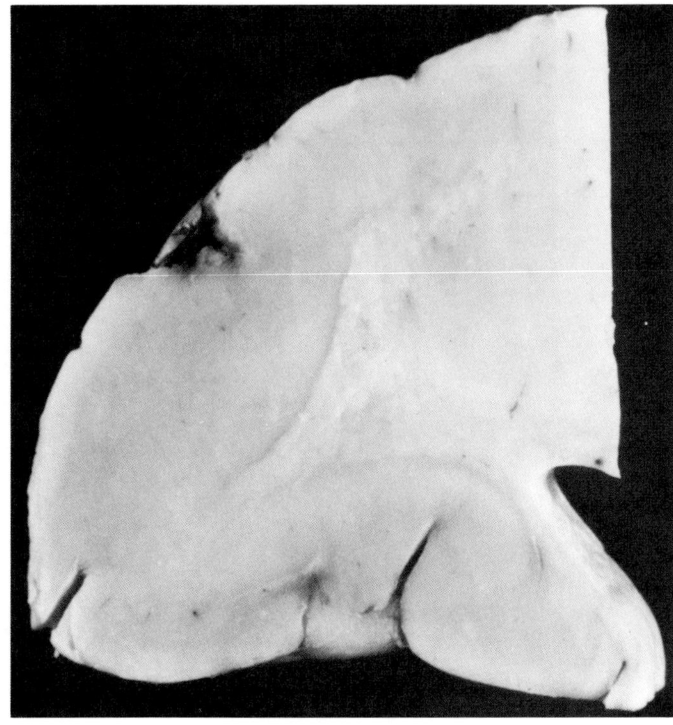

A

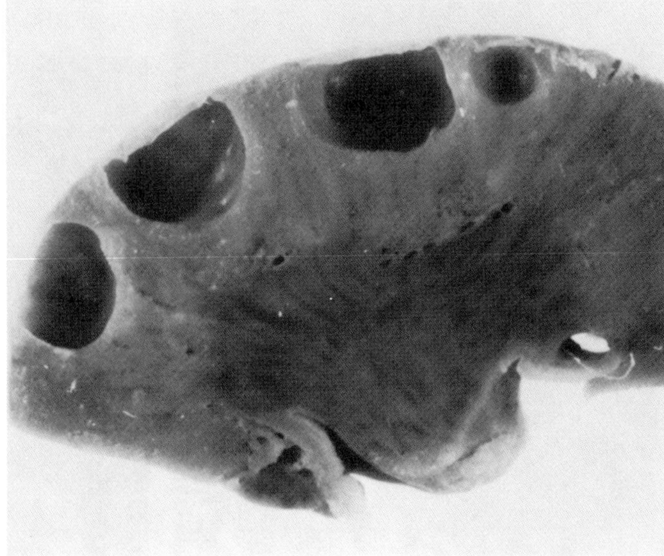

B

Fig. 5–64. *Zellweger syndrome*. Gross pathologic specimens. (A) Cross-section of cerebrum showing macrogyria, flattening of cortical surface, and thickened cortical gray matter. (B) Kidney showing large cortical cysts. (A,B from *A Poznanski et al,* Am J Roentgenol **109:**313, 1970.)

have also been described (17,33,39,62). Proteinuria and generalized aminoaciduria are often detected (58%) (11,33,39,47).

*Pulmonary and cardiovascular findings.* Focal pneumonia may be noted and pulmonary hypoplasia is frequently found.Cardiovascular malformations are also common. PDA, VSD, and aortic arch anomalies have been reported (17,32,71).

*Skeletal findings.* Bone age is often retarded and hypomineralization and Wormian bones have been noted (42,47,70). Calcific stippling (Fig. 5–65A,B) has been observed, particularly in the acetabular cartilages and along the inferior medial margin of the patellas. Stippled epiphyses are present in long bones in 69% of cases (42,71). Calcification of the hyoid bone and thyroid cartilage has also been noted. Metaphyseal radiolucencies have been described (11,31,42, 47,70).

*Other findings.* Other features include camptodactyly (10%), single palmar creases (79%), ulnar deviation of the hands (15%), cubitus valgus, flexion at the knees and hips, talipes equinovarus (56%), metatarsus adductus, rockerbottom feet, and dorsiflexion of the fourth toes. *DiGeorge sequence* has been reported (17). Other anomalies include widely spaced nipples, deep sacral dimple, hypoplastic dermal ridges, small penis, hypospadias, cryptorchidism (75%), prominent clitoris (56%), umbilical hernia, and diastasis recti (4,11,39,42,47, 54,71).

Differential diagnosis. Zellweger syndrome may be confused on occasion with *trisomy 21 syndrome*. Differential diagnosis also includes hyperpipecolic acidemia (33,38,52), neonatal adrenoleukodystrophy (33), infantile Refsum's disease (7,43–46,67), Leber disease (13), and *rhizomelic chondrodysplasia punctata* (28). Biochemical overlap and genetic distinctions between these entities are discussed elsewhere (56,57,68). Pseudo-Zellweger syndrome and milder variants of Zellweger syndrome have also been discussed (2,3,20,71,72). Beemer and coworkers (3a,3b) reported a new peroxisomal disorder in sibs characterized by lethality, hydrocephalus, unusual facies, dense bones, and sex reversal.

Laboratory aids. Elevated pipecolic acid concentrations in urine and plasma, detected by routine amino acid chromatography, are helpful but not diagnostic (21,23,33). VLCFAs can be quantified from plasma, cultured fibroblasts, amniocytes, and amniotic fluid (1,37,49,51,52). DHAP-AT is readily assayed from various tissues and can be used for pre- and postnatal diagnosis (27,33,49,51,52,71). Dihydroxycoprastanoic acid levels are decreased in amniotic fluid (55a).

### References [Zellweger syndrome (cerebrohepatorenal syndrome)]


1. Aubourg P et al: The cerebro-hepato-renal (Zellweger) syndrome: Lamellar lipid profiles in adrenocortical, hepatic mesenchymal, astrocyte cells and increased levels of very long chain fatty acids and phytanic acid in the plasma. *J Neurol Sci* **69**:9–25, 1985.

2. Barth PG et al: A milder variant of Zellweger syndrome. *Eur J Pediatr* **144**:338–342, 1985.

3. Barth PG et al: A sibship with a mild variant of Zellweger syndrome. *J Inher Metab Dis* **10**:253–259, 1987.

3a. Beemer Fa. Ertbruggen IV: Peculiar facial appearance, hydrocephalus, double-outlet right ventricle, genital anomalies and dense bones with lethal outcome. *Am J Med Genet* **19**:391–394, 1984.

3b. Beemer FA, Wanders RJA: A formerly new syndrome revisited: A biochemical basis. *David W. Smith Workshop on Malformations and Morphogenesis,* Madrid, Spain, 23–29, 1989.

4. Bowen P et al: A familial syndrome of multiple congenital defects. *Johns Hopkins Med J* **114**:402–414, 1964.

5. Brown FR et al: Cerebro-hepato-renal (Zellweger) syndrome and neonatal adrenoleukodystrophy: Similarities in phenotype and accumulation of very-long-chain fatty acids. *J Hopkins Med J* **151**:344–361, 1982.

6. Brun A et al: The Zellweger syndrome: Subcellular pathology, neuropathology, and the demonstration of *Pneumocystis carinii* pneumonitis in two siblings. *Eur J Pediatr* **127**:229–245, 1978.

7. Budden SS et al: Dysmorphic syndrome with phytanic acid oxidase deficiency, abnormal very long chain fatty acids, and pipecolic acidemia: Studies in four children. *J Pediatr* **108**:33–39, 1986.

8. Carlson BR, Weinberg AG: Giant cell transformation cerebrohepatorenal syndrome. *Arch Pathol Lab Med* **102**:596–599, 1978.

9. Clayton PT et al: Plasma bile acids in patients with peroxisomal dysfunction syndromes: Analysis by capillary gas chromatography-mass spectrometry. *Eur J Pediatr* **146**:166–173, 1987.

10. Dancis J, Hutzler J: The significance of hyperpipecolatemia in Zellweger syndrome. *Am J Hum Genet* **38**:707–711, 1986.

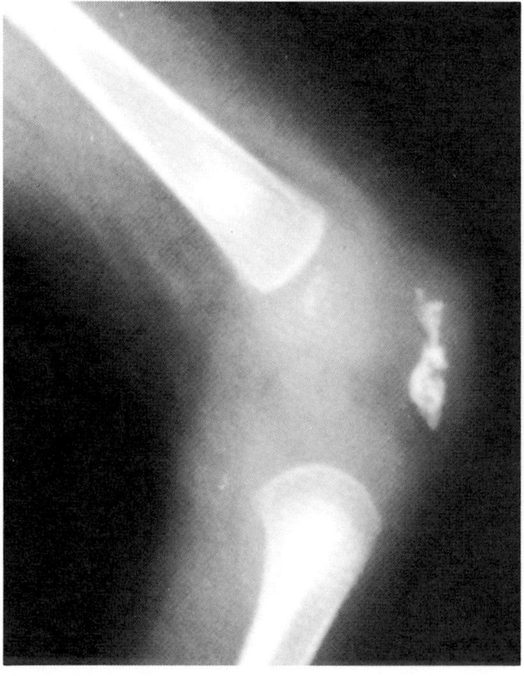

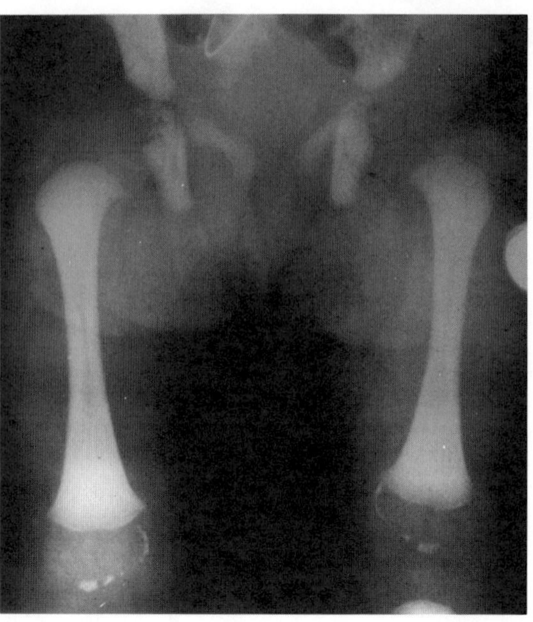

A B

Fig. 5–65. *Zellweger syndrome*. Roentgenograms. (A) Extensive calcification of patella. (B) Note calcification of hips and kneecaps. (A from *A Poznanski et al,* Am J Roentgenol **109:**313, 1970.)


11. Danks DM, et al: Cerebro-hepato-renal syndrome of Zellweger. *J Pediatr* **86**:382–387, 1975.

12. Datta NS et al: Deficiency of enzymes catalyzing the biosynthesis of glycerol-ether lipids in Zellweger syndrome. *N Engl J Med* **311**:1080–1083, 1984.

12a. DeLange C, Janssen T: Chondrodystrophia calcifikans congenita als onderdeel van een reeks aangeboren afwijkingen. *Maandschr Kindergeneesk* **17**:67–74, 1949.

13. Ek J et al: Peroxisomal dysfunction in a boy with neurologic symptoms and amaurosis (Leber disease): Clinical and biochemical findings similar to those observed in Zellweger syndrome. *J Pediatr* **108**:19–24, 1986.

14. Garzuly F et al: Neuronale Migrationsstörung bei Cerebro-hepato-renal Syndrom "Zellweger." *Neuropädiatrie* **5**:319–328, 1974.

15. Gatfield PD et al: Hyperpipecolatemia: A new metabolic disorder associated with neuropathy and hepatomegaly. *Can Med Assoc J* **99**:1215–1233, 1968.

16. Gilchrist KW et al: Immunodeficiency in cerebro-hepato-renal syndrome. *Lancet* **1**:164–165, 1974.

17. Gilchrist KW et al: Studies of malformation syndromes of man XIB: The cerebro-hepato-renal syndrome of Zellweger: Comparative pathology. *Eur J Pediatr* **121**:99–118, 1976.

18. Goldfischer S et al: Peroxisomal and mitochondrial defects in the cerebrohepatorenal syndrome. *Science* **182**:62–64, 1973.

19. Goldfischer S, Reddy JK: Peroxisomes (microbodies) in cell pathology. *Int Rev Exp Pathol* **26**:45–84, 1984.

20. Goldfischer S et al: Pseudo-Zellweger syndrome: Deficiencies in several peroxisomal oxidative activities. *J Pediatr* **108**:25–32, 1986.

21. Govaerts L et al: Cerebro-hepato-renal syndrome of Zellweger: Clinical symptoms and relevant laboratory findings in 16 patients. *Eur J Pediatr* **139**:125–128, 1982.

22. Govaerts L et al: Disturbed adrenocortical function in cerebro-hepatorenal syndrome of Zellweger. *Eur J Pediatr* **143**:10–12, 1984.

23. Govaerts L et al: Pipecolic acid levels in serum and urine from neonates and normal infants: Comparison with values reported in Zellweger syndrome. *J Inher Metab Dis* **8**:87–91, 1985.

24. Govaerts L et al: Disturbed very long chain (C24–C26) fatty acid pattern in fibroblasts of patients with Zellweger's syndrome. *J Inher Metab Dis* **8**:5–8, 1985.

25. Gustafsson J et al: Zellweger's cerebro-hepato-renal syndrome—variations in expressivity and in defects of bile acid synthesis. *Clin Genet* **24**:313–319, 1983.

26. Hanson RF et al: Defects of bile acid synthesis in Zellweger's syndrome. *Science* **203**:1107–1108, 1979.

27. Heymans SA et al: Deficiency of plasmalogens in cerebro-hepato-renal (Zellweger) syndrome. *J Pediatr* **142**:10–15, 1984.

28. Heymans HSA et al: Peroxisomal abnormalities in rhizomelic chondrodysplasia punctata. *J Inher Metab Dis* 9(Suppl 2):329–331, 1986.

29. Hittner HM, Kretzer FL: Lenticular opacities indicating carrier status and lens abnormalities characteristic of homozygotes. *Arch Ophthalmol* **99**:1977–1982, 1981.

30. Hutzler J, Dancis J: The determination of pipecolic acid: Method and results of hospital survey. *Clin Chim Acta* **128**:75–82, 1983.

31. Jan JE et al: Cerebro-hepato-renal syndrome of Zellweger. *Am J Dis Child* **119**:274–277, 1970.

32. Kelley RI: Review: The cerebrohepatorenal syndrome of Zellweger, morphologic and metabolic aspects. *Am J Med Genet* **16**:505–517, 1983.

33. Kelley RI et al: Neonatal adrenoleukodystrophy: New cases, biochemical studies, and differentiation from Zellweger and related peroxisomal polydystrophy syndromes. *Am J Med Genet* **23**:869–901, 1986.

34. Lam S et al: L-Pipecolaturia in Zellweger syndrome. *Biochim Biophys Acta* **882**:254–257, 1986.

35. Lazarow PB: The role of peroxisomes in mammalian cellular metabolism. *J Inher Metab Dis* **10**(Suppl 1):11–22, 1987.

36. Monnens L, Heymans H: Peroxisomal disorders: Clinical characterization. *J Inher Metab Dis* **10**(Suppl 1):23–32, 1987.

37. Moser AE et al: The cerebrohepatorenal (Zellweger) syndrome. Increased levels and impaired degradation of very-long-chain fatty acids and their use in prenatal diagnosis. *N Engl J Med* **310**:1141–1146, 1984.

38. Moser HW: Peroxisomal disorders (Editorial). *J Pediatr* **108**:89–91, 1986.

39. Opitz JM et al: The Zellweger syndrome. *Birth Defects* **5**(2):143–158, 1969.

40. Opitz JM: The Zellweger syndrome: Book review and bibliography. *Am J Med Genet* **22**:419–426, 1985.

41. Passarge E, McAdams AJ: Cerebro-hepato-renal syndrome. *J Pediatr* **71**:691–702, 1967.

42. Patton RG et al: Cerebro-hepato-renal syndrome of Zellweger: Two cases with islet cell hyperplasia, hypoglycemia, and thymic anomalies, and comments on iron metabolism. *Am J Dis Child* **124**:840–844, 1972.

43. Poll-Thé BT et al: Impaired plasmalogen metabolism in infantile Refsum's disease. *Eur J Pediatr* **144**:513–514, 1986.

44. Poulos A et al: Infantile Refsum's disease (phytanic acid storage disease): A variant of Zellweger's syndrome? *Clin Genet* **26**:579–586, 1984.

45. Poulos A et al: Cerebro-hepato-renal (Zellweger) syndrome, adrenoleukodystrophy, and Refsum's disease: Plasma changes and skin fibroblast phytanic acid oxidase. *Hum Genet* **70**:172–177, 1985.

46. Poulos A, Whiting MJ: Identification of 3 alpha, 7 alpha, 12 alpha-trihydroxy-5-beta-cholestan-26-oic acid, an intermediate in cholic acid synthesis, in the plasma of patients with infantile Refsum's disease. *J Inher Metab Dis* **8**:13–17, 1985.

47. Poznanski AK et al: The cerebro-hepato-renal syndrome (CHRS): Zellweger's syndrome. *Am J Roentgenol* **109**:313–322, 1970.

48. Punnett HH, Kirkpatrick JA: A syndrome of ocular abnormalities, calcification of cartilage and failure to thrive. *J Pediatr* **73**:602–606, 1968.

49. Roscher A et al: The cerebrohepatorenal (Zellweger) syndrome: An improved method for the biochemical diagnosis and its potential value for prenatal detection. *Pediatr Res* **19**:930–933, 1985.

50. Schram AW et al: Biosynthesis and maturation of peroxisomal beta-oxidation enzymes in fibroblasts in relation to Zellweger syndrome and infantile Refsum disease. *Proc Natl Acad Sci USA* **83**:6156–6158, 1986.

51. Schutgens RBH et al: The cerebro-hepato-renal (Zellweger) syndrome: Prenatal detection based on impaired biosynthesis of plasmalogens. *Prenatal Diag* **5**:337–344, 1985.

52. Schutgens RBH et al: Peroxisomal disorders: A newly recognized group of genetic diseases. *Eur J Pediatr* **144**:430–440, 1986.

53. Schutgens RBH et al: Zellweger syndrome: Biochemical procedures in diagnosis, prevention and treatment. *J Inher Metab Dis* **10**(Suppl 1):33–45, 1987.

54. Smith DW et al: A syndrome of multiple developmental defects including polycystic kidneys and intrahepatic biliary dysgenesis in 2 siblings. *J Pediatr* **67**:617–624, 1965.

55. Sommer A et al: The cerebro-hepato-renal syndrome (Zellweger's syndrome). *Biol Neonate* **25**:219–230, 1974.

55a. Stellaard F et al: Prenatal diagnosis of Zellweger syndrome by determination of trihydroxycoprostanic acid in amniotic fluid. *Eur J Pediatr* **148**:175–176, 1988.

56. Tager JM: Inborn errors of cellular organelles: An overview. *J Inher Metab Dis* **10**(Suppl 1):3–10, 1987.

57. Tager JM et al: More on Zellweger's syndrome, infantile Refsum's disease, and rhizomelic chondrodysplasia punctata. *N Engl J Med* **315**:766–767, 1986.

58. Taylor JC et al: A new case of the Zellweger syndrome. *Birth Defects* **5**(2):159–160, 1969.

59. Trijbels JMF et al: Biochemical studies in the cerebro-hepato-renal syndrome of Zellweger: A disturbance in the metabolism of pipecolic acid. *J Inher Metab Dis* **2**:39–42, 1979.

60. Trijbels JMF et al: Biochemical studies in the liver and muscle of patients with Zellweger syndrome. *Pediatr Res* **17**:514–517, 1983.

61. Trijbels JMF et al: Localization of pipecolic acid metabolism in rat liver peroxisomes: Probable explanation for hyperpipecolitemia in Zellweger syndrome. *J Inher Metab Dis* **10**:128–134, 1987.

62. Vincens A et al: A propos d'un cas de syndrome de Zellweger (syndrome hepato-cérebro-renal). *Ann Pédiatr* **20**:553–560, 1972.

63. Vitale L et al: Congenital and familial iron overload. *N Engl J Med* **280**:642–645, 1969.

64. Volpe JJ, Adams RD: Cerebro-hepato-renal syndrome of Zellweger. An inherited disorder of neuronal migration. *Acta Neuropathol* **20**:175–198, 1972.

65. Wanders RJA et al: Peroxisomal matrix enzymes in Zellweger syndrome: Activity and subcellular localization in liver. *J Inher Metab Dis* **8**(Suppl 2):151–152, 1985.

66. Wanders RJA et al: Pre- and postnatal diagnosis of the cerebro-hepato-renal (Zellweger) syndrome via a simple method directly demonstrating the presence or absence of peroxisomes in cultured skin fibroblasts, amniocytes or chorionic villi fibroblasts. *J Inher Metab Dis* **9**(Suppl 2):317–320, 1986.

67. Wanders RJA et al: Deficiency of dihydroxyacetonephosphate acyltransferase and catalase-containing particles in patients with infantile Refsum's disease. *J Inher Metab Dis* **9**(Suppl 2):325–328, 1986.

68. Wanders RJA et al: Genetic relation between the Zellweger syndrome, infantile Refsum's disease, and rhizomelic chondrodysplasia punctata. *N Engl J Med* **31**:787–788, 1986.

69. Wanders RJA et al: Impaired cholesterol side chain cleavage activity in liver from patients with the cerebro-hepato-renal (Zellweger) syndrome in relation to the accumulation of di- and trihydroxycoprostanoic acid in serum. *J Inher Metab Dis* **9**(Suppl 2):321–324, 1986.

69a. Wanders RJ et al: Peroxisomal disorders in neurology. *J. Neurol Sci* **88**:1–39, 1988.

70. Williams JP et al: Roentgenographic features of the cerebrohepatorenal syndrome of Zellweger. *Am J Roentgenol* **115**:607–610, 1972.

71. Wilson GN et al: Zellweger syndrome: Diagnostic assays, syndrome delineation, and potential therapy. *Am J Med Genet* **24**:69–82, 1986.

72. Wilson GN et al: Syndrome delineation and dysmorphogenetic mechanisms in peroxisomal disorders. *Eighth Annual Workshop on Malformations and Morphogenesis,* Greenville, South Carolina, August 15–19, 1987, p 52.

# Chapter 6
# Syndromes Affecting Bone: The Osteogenesis Imperfectas

## The osteogenesis imperfectas

Osteogenesis imperfecta is a heterogeneous group of heritable disorders of connective tissues characterized by bone fragility. Associated features in some affected individuals include blue sclerae, opalescent teeth with characteristic radiologic features, hearing loss, deformity of the long bones and spine, and joint hyperextensibility. Dominant as well as recessive pedigrees have been reported and many sporadic cases have been described. The first reported case was probably in an Egyptian whose remains, dating from 1000 BCE, are in the British Museum (44). Credit is usually given to Ekman (33) for performing, in 1788, the first comprehensive study of the syndrome and discussing its inheritance. Lobstein (62) and Vrolik (115), in the early nineteenth century, described the disease in the adult and the newborn, respectively. Spurway (105) and Eddowes (32) described blue sclerae, and van der Hoeve and de Kleijn (113) in the early 1900s mentioned deafness as part of the syndrome. Preiswerk (83) may have been the first to describe the dental abnormalities. The basis for the current clinical classification is the study by Sillence et al (97). A monograph on the disorder has been written by Smith et al (103).

The severity of the disorder varies widely between and even within families; some individuals have minimum involvement of the skeleton and may never have a fracture, whereas others have very severe involvement and many fractures. Clinical and genetic studies delineate at least four major syndrome groups (Table 6–1) (97), although all of these syndromes are likely heterogeneous at the clinical, radiographic, and molecular levels (16,32a,49a,56,57,96,98–101). An excellent discussion is that of van der Harten et al (112a).

Nomenclature classifying patients into congenita and tarda forms is not useful, as individuals with any osteogenesis imperfecta syndrome can be born with fractures; in addition, this feature cannot be consistently correlated with inheritance pattern, prognosis, or recurrence risk (96,97,104). Classification into broad-boned and thin-boned types is also unsatisfactory, as change from one shape to another can occur with age (98). There appears to be little benefit in classification based on severity of long bone deformity (4), as severe deformity can occur in many osteogenesis imperfecta syndromes; in addition, this method of classification does not take into account the pattern of inheritance or other clinical findings (98,104a,109).

Until recently, the phenotypes of patients with the syndrome, as well as the inheritance patterns, have not always been completely delineated. Thus, it is difficult to interpret much of the large body of literature on the disorder in view of the current classification.

The estimated prevalence of all types combined is about 0.5/10,000 births (107a).

### Type I

General features. This disorder is characterized by autosomal dominant inheritance (97) (Fig. 6–1). Carothers et al (20) analyzed a heterogeneous group of patients with osteogenesis imperfecta Type I and Type IV, presumed to be the result of new mutations. These investigators demonstrated that the mean paternal age at birth was significantly higher than that of controls and that paternal age effect in new mutation osteogenesis imperfecta Type IA, when analyzed alone, was also significantly increased. The effect of paternal age on the mutation rate, however, was smaller than in other dominant disorders, such as achondroplasia.

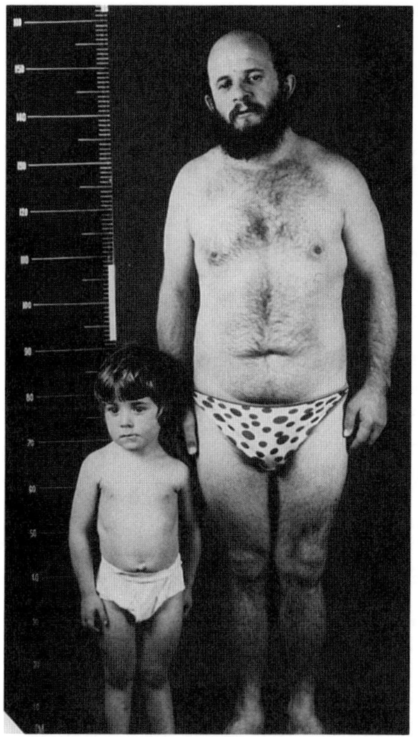

Fig. 6–1. *Osteogenesis imperfecta Type I.* Father and 4-year-old son with relatively mild skeletal deformity. The father had bilateral hearing aids. (From *DO Sillence,* Symposium on Heritable Disorders of Connective Tissue, CV Mosby Co., St. Louis, 1982.)

Facies. Triangular facies is frequently noted (72), as is temporal bulging (97). The maxilla may be hypoplastic with a relative mandibular prognathism (26).

Ophthalmologic abnormalities. A hallmark of this syndrome is blue sclerae. Scleral color appears consistent within families, although the degree of blueness varies from one family to another (LS Levin, unpublished). Scleral blueness is believed to arise from disordered molecular organization (102). Sclerae are of normal width, but there is evidence for increased noncollagenous matrix (34). Low ocular rigidity in a heterogeneous group of patients with osteogenesis imperfecta has been found (53); some had blue sclerae, and thus may have had osteogenesis imperfecta Type I. Lanting et al (55) found reduced optical scattering properties in the sclerae of two patients with this syndrome, findings they attributed, in part, to reduction of collagen fiber thickness and decreased variability of fiber diameter.

Embryotoxon has also been noted (36, LS Levin, unpublished). It is unknown if this abnormality is the result of chance association of

Table 6–1. Major osteogenesis imperfecta syndromes

| Type | Salient features | Inheritance pattern | Comments |
|---|---|---|---|
| IA | Mild to moderately severe bone fragility<br>Blue sclerae<br>Hearing loss<br>Normal teeth | Autosomal dominant | Additional heterogeneity in OI Type I is likely |
| IB | Mild to moderately severe bone fragility<br>Blue sclerae<br>Hearing loss<br>Opalescent teeth | Autosomal dominant | |
| IC | Mild to moderately severe bone fragility<br>Blue sclerae<br>Hearing loss<br>Dentition resembles dentin dysplasia Type II | Autosomal dominant | |
| IIA,B,C | Very severe bone fragility<br>Blue sclerae<br>Stillborn or death shortly after birth | Autosomal dominant<br>Autosomal recessive | Distinction based on clinical and radiologic features equivocal<br>Some may represent severe Type III<br>However, additional heterogeneity likely |
| III | Moderately severe to severe bone fragility<br>Blue sclerae in infancy that fades with age<br>Generally not lethal in infancy, but death in first decades of life not uncommon | Autosomal dominant<br>Autosomal recessive | Heterogeneous |
| IVA | Mild to moderately severe bone fragility<br>Normal sclerae, but may be pale blue in early childhood<br>Hearing loss<br>Normal teeth | Autosomal dominant | Distinction between OI Type IV and Type III may be difficult<br>Additional heterogeneity in Type IV is likely |
| IVB | Mild to moderately severe bone fragility<br>Normal sclerae, but may be pale blue in early childhood<br>Hearing loss<br>Opalescent teeth | Autosomal dominant | |

hypercholesterolemia or whether it is one of the pleiotropic effects of the syndrome, at least in some families.

Otolaryngologic abnormalities. Hearing loss, rarely detected before 10 years of age, usually begins with a conductive deficit in the late second or early third decade (88). With age, mixed and sensorineural hearing losses are observed (27,88,93). Riedner et al (88) noted that by the fifth decade, half of all patients had hearing losses, whereas by the seventh decade, all individuals had hearing losses, although the number of older individuals tested was small. Cox and Simmons (27) reported similar findings. Shapiro et al (95) reported audiologic abnormalities in a heterogeneous group of patients; half younger than 30 years of age, and 95% over 30, had hearing losses. Half of all patients examined had sensorineural losses. Conductive loss in this syndrome has been attributed to ossicular immobility at the stapes footplate (75,88). Fracture of the stapedial crura and atrophy of the stapes may also contribute to loss of hearing acuity (88).

Neurologic manifestations. Results of CT scans of the head have been normal in the few patients tested (112), and ventricles are normal in size. In the three-member family with osteogenesis imperfecta Type I and dental abnormalities reported by Pozo et al (82), advanced basilar impression resulted in ventricular dilatation, multiple neurological disturbances of the foramen magnum compression syndrome, and death from acute brain stem compression; these findings, however, are likely rare in other patients with this syndrome. These investigators noted that all patients reported with osteogenesis imperfecta and basilar impression had mild skeletal disease. Half the 56 patients with osteogenesis imperfecta studied by Reite and Solomons (87) had abnormal electroencephalograms; however, their patients were a heterogeneous group and were not classified.

Cardiovascular involvement. Pyeritz (85) reviewed the cardiovascular features of this syndrome. The frequency of symptomatic cardiovascular anomalies is low. Hortop et al (50) reported nonprogressive aortic root dilatation in about 12% of affected patients. In one study of dominant osteogenesis imperfecta, 9% of males and females had asymptomatic mitral valve prolapse; 24% of males but only 4% of females had asymptomatic aortic root dilatation (86). Aortic regurgitation has been observed in patients after the third decade, as has mitral regurgitation (1,86,106). Aortic aneurysm and dissection do not occur, although left atrial rupture has been reported in one patient (89). Mitral valve leaflets were thin in half the patients reported by White et al (118). Microscopic findings in the valves include myxoid degeneration and atrophy and, in the aorta, cystic medial necrosis (1).

Joint abnormalities. Joint hypermobility and dislocation are found (72).

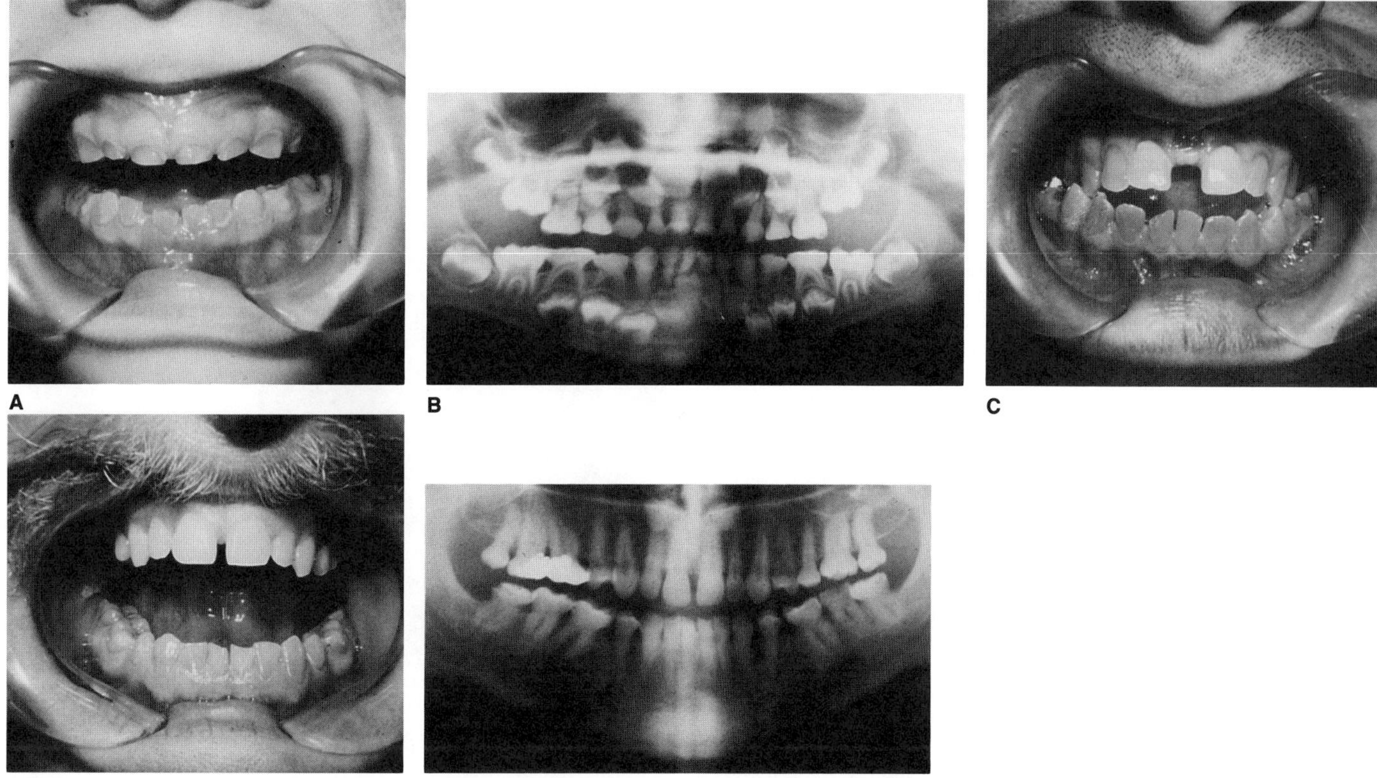

Fig. 6–2. *Type I.* (A) Opalescent deciduous teeth in a 5.5-year-old with osteogenesis imperfecta Type IB. Marked incisal wear is evident. (From *LS Levin,* Clin Orthop Rel Res **159:**64, 1981.) (B) Radiographs of deciduous and developing permanent dentitions in a 5.5-year-old with Type IB. Pulp chambers and root canals of the deciduous teeth are almost completely obliterated, and greater constriction than normal is present at the crown–root junctions of these teeth. Pulp chambers and root canals of the developing mandibular permanent first molars are wider than normal for this stage of development. (C) Permanent dentition in a patient with osteogenesis imperfecta Type IB. All teeth are clinically opalescent. (D) Permanent dentition in a patient with osteogenesis imperfecta Type IB demonstrating variability; the maxillary teeth are opalescent. (E) Radiographs of the permanent dentition in Type IB. Findings are similar to those found in the deciduous dentition. The root canals of some teeth in this patient, however, are patent.

Skeletal manifestations. Macrocephaly has been reported (97,112). Head size is usually large for height: the median of the distribution of head sizes is above the 50th percentile for normal (97). Wormian bones may be present (97). Platybasia and occipitalization of the upper cervical vertebrae may produce a ''tam o'shanter'' appearance (51,72).

Multiple fractures usually occur, although about 10% of patients may not have fractures (97). There is considerable variability within and between families in the age of onset and frequency of fractures. Reduction in fracture frequency at puberty has been noted, followed by an increase in fracture frequency in women after the menopause (73). Long bone deformity consists of bowing and angulation (97); however, it is not as severe in this osteogenesis imperfecta syndrome as it can be in others. About 20% of adults have kyphosis or scoliosis, which may be progressive (97). Trunk shortening has also been described (6). Osteopenia may be minimal and undetectable on skeletal radiographs.

Paterson et al (71) found a significantly higher fracture rate between ages 5 and 20 years in patients with Type I osteogenesis imperfecta and opalescent teeth (Type IB) than in those with Type I who had normal teeth (Type IA). They also found that individuals with Type IB were more likely to have had a fracture at birth, to have a higher fracture frequency, and to have a height below the second percentile. Patients with normal teeth were more likely than patients with opalescent teeth to have prolonged fracture-free periods during childhood. These two groups, however, were similar in the frequency of joint hyperextensibility, bruising, hearing impairment, and joint dislocations.

Birth weights and birth lengths are generally normal. Short stature is of postnatal onset and usually mild; by adulthood, half the affected patients are less than the third percentile for height (97). Individuals with the disorder whose height exceeds 6 feet have been noted (LS Levin, unpublished).

Gillerot et al (42) reported a family with osteogenesis imperfecta Type I in which a newborn infant had features identical to Type II. Heyes et al (48) reported a kindred with osteogenesis imperfecta Type I and opalescent teeth; a stillborn had sustained multiple fractures *in utero,* and had a membranous cranium, resembling osteogenesis imperfecta Type II; no radiographs were depicted.

Oral manifestations. Dental abnormalities have been described in many individuals with osteogenesis imperfecta; unfortunately, in many of these reports, the complete phenotype and inheritance pattern are not described, and interpretation in view of current classification is not possible. Heterogeneity based on the presence or absence of dental abnormalities has been noted (56,57,71,92). Paterson et al (71) recognized that two groups of families with osteogenesis imperfecta Type I can be distinguished: a group with normal teeth (Type IA) and a group with specific dental abnormalities (Type IB). In patients with dental abnormalities, deciduous and permanent teeth are opalescent, and amber or blue-grey on eruption (59) (Fig. 6–2). On radiographic examination, there is increased constriction at the coronal–radicular junctions, and pulps become obliterated with secondary dentin. However, pulps may be wider than normal during early development (59). Roots are thinner and shorter than normal (92). There is little variation in expression within the deciduous dentition, and these teeth may wear rapidly; on the other hand, there may be considerable variability in color within an individual's permanent dentition. Dental radiographs may be necessary to establish the diagnosis where tooth discoloration is mild. When dental abnormalities are present, they may be helpful in making the diagnosis of osteogenesis imperfecta when the presence of other abnormalities is equivocal.

Lukinmaa et al (63) and Paterson et al (72) noted that opalescent teeth were rarer in families with Type I osteogenesis imperfecta than in Type IV. Malocclusion has been described and may be more common in individuals with opalescent teeth than in those with normal teeth (92).

Lukinmaa et al (64) described "thistle-tube" or flame-shaped pulps in several patients with osteogenesis imperfecta, one of which may have had Type I, although scleral color was not stated. These dental findings are similar to those found in dentin dysplasia Type II (41). Levin et al (61a) found similar abnormalities in the permanent dentitions of five members of a family with osteogenesis imperfecta Type I; these teeth were not opalescent, and they lacked the other radiologic abnormalities seen in the dentition in Type IB. However, the deciduous teeth were opalescent, had increased constrictions at the coronal–radicular junctions, and had partially obliterated pulp chambers.

There are few systematic studies of the histopathology of the teeth in what is unequivocally osteogenesis imperfecta Type I. In general, however, on microscopic examination, the dentin–enamel junction has been reported by some investigators to be flatter than normal, and to lack normal scalloping (90,119). However, Levin et al (57), using scanning electron microscopy, and Lukinmaa et al (63), using light microscopy, described a normal junction. On light microscopic examination, laminated dentin, tubules of abnormal size and shape, and structures resembling entrapped blood vessels have been described (77,90,119). Enamel surfaces and prism organization were normal on scanning electron microscopy of multiple deciduous and permanent teeth from two families with osteogenesis imperfecta Type IB (57). However, dentin tubules were short, narrow, and tortuous compared with normal teeth from other osteogenesis imperfecta families and controls. Calcification fronts were composed of nodules of abnormal sizes and shapes (57). Gage et al (39) studied teeth from a group of patients with a variety of osteogenesis imperfecta syndromes. They concluded that the majority of teeth from these patients were biochemically abnormal even if dental abnormalities were not present; however, the criteria for determining whether clinical abnormalities of the dentition existed in the patients studied were not presented.

Other abnormalities. Easy bruisability has been found in about 75% of affected patients (72). Hernias and excessive sweating have also been reported (72).

### Type II

General features. This clinically and biochemically heterogeneous group of osteogenesis imperfecta syndromes is characterized by extreme bone fragility consistently leading to intrauterine or early infant death (17a,65a,97,99). Sillence et al (99) subclassified this syndrome into three, possibly different, disorders (Groups A, B, and C) distinguished from one another on the basis of roentgenographic features. This classification has been somewhat more sharply defined by van der Harten et al (112a).

Features of newborns in Group A have been well delineated (99). At birth, infants are small for gestational age and usually premature. Twenty percent are stillborn and the remainder die within hours or days of birth; 90% are no longer living by 4 weeks of age. Breech delivery occurs in 15%. General connective tissue fragility is present and the head or a limb may be torn off during delivery (97). Infants in Group B have a mean gestational age of 37.6 weeks; mean survival is about 14 hours (109). Length of gestation and survival of individuals in Group C have not been described in detail because few examples are reported; however, two cases reported by Thompson et al (109) were stillborn at 28 and 30 weeks, respectively. Since most, if not all, cases of osteogenesis imperfecta Type II have been ascertained through early death, the true average length of survival is unknown (17).

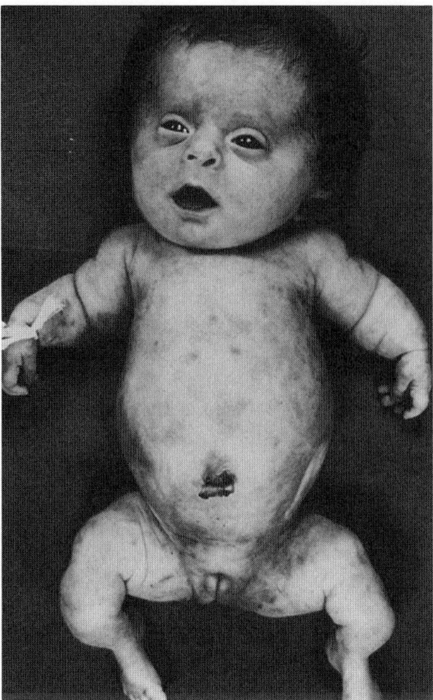

Fig. 6–3. *Type II.* Perinatal death with frontal and temporal bossing, limb shortening with external rotation, and abduction of the thighs and angulation of the legs. (From *DO Sillence,* Am J Med Genet **17:**407, 1984.)

Although instances of familial recurrence and parental consanguinity have been observed in Group A (99,109), the majority are sporadic cases. It was originally thought that most cases are inherited as autosomal recessives (69,78,99). However, based on a study of 30 cases, Young et al (120) concluded that most cases of osteogenesis imperfecta Type IIA were the result of new dominant mutations; no sib pairs were ascertained, 25 were born to nonconsanguineous couples, and increased paternal age effect was observed. Byers et al (17a,17b) and others have confirmed this. Thompson et al (109) calculated an empirical recurrence risk of 7.7% for Type IIB; no increase in parental age over the general population was found. Familial occurrence with phenotypically normal parents has also been observed in Type IIB (112a). Although too few families with Type IIC are available for analysis, autosomal recessive inheritance is suggested (109,112a).

Conflicting estimates of recurrence risk for osteogenesis imperfecta Type II likely reflect, in part, heterogeneity; a proportion of cases may be inherited as autosomal recessive and some as dominant disorders. Other modes of inheritance cannot be excluded. Cohn et al (23a) confirmed, at the molecular level, the occurrence of gonadal mosaicism for a mutation in the $\alpha 1$ gene of Type I collagen.

Facies. The face and cranium are molded and the cranium often appears disproportionately large for the face (Fig. 6–3). Commonly mild micrognathia and a small narrow nose are noted (99).

Ophthalmologic findings. Blue/black sclerae are present in virtually all affected individuals (99). On examination of the cornea of an infant who died at 17 days of age with presumed osteogenesis imperfecta Type II, collagen fiber diameter was reduced, normal cross-striations were not seen, and collagen fibers were more densely packed than normal (13).

Cardiovascular system. Information about the cardiovascular system in osteogenesis imperfecta Type II is derived from autopsy studies (85). Abnormalities include thickening of the valve leaflets, myxoid degeneration of the valves, calcification of the pulmonary, cerebral, or peripheral arteries, intimal proliferation and medial calcification of the pulmonary artery, thickening of the media and adventitia in small and medium-sized pulmonary arteries, and atherosclerotic changes in

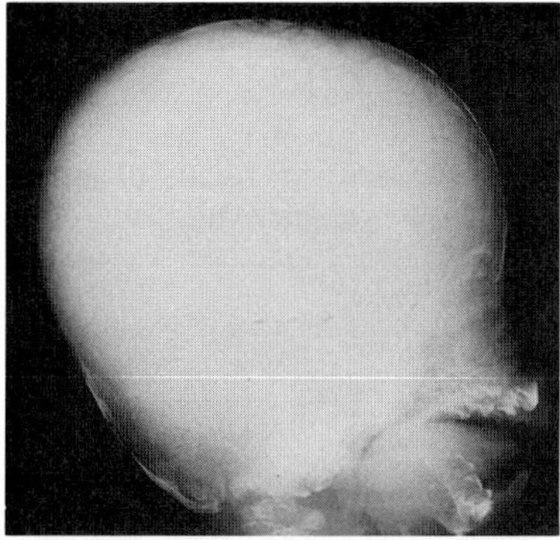
A

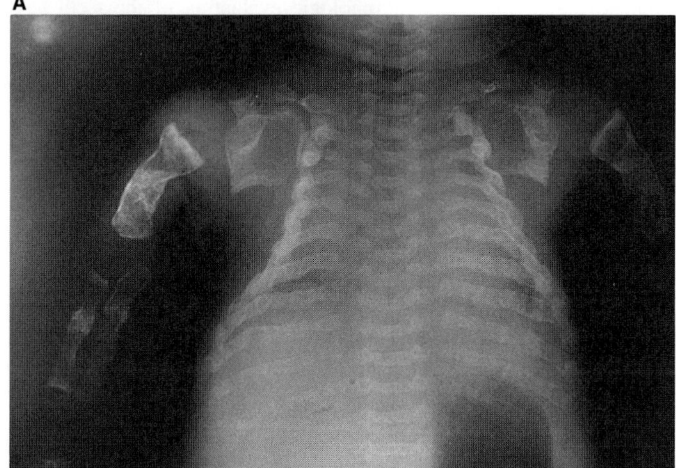
B

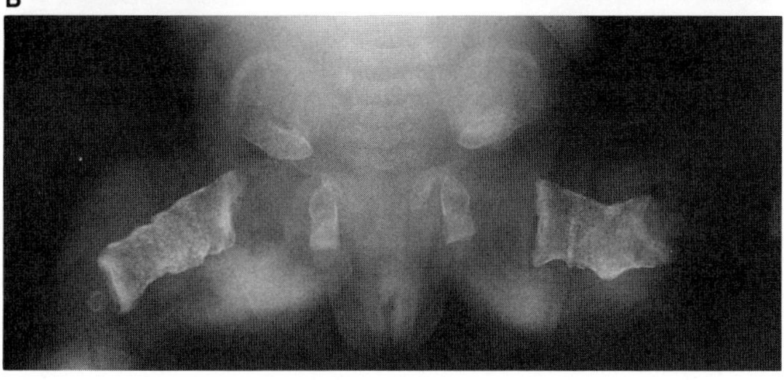
C

Fig. 6–4. *Type II.* (A) Lateral radiograph of skull and face showing extreme osteopenia and multiple bone islands (Wormian bones) in frontal bone and vertex. (B) Chest radiograph showing continuously beaded ribs and short broad dysplastic humeri. (C) Femora showing crumpled and dysplastic (concertina-like) appearance. Pelvis showing osteopenia and dysplastic changes. (A,B,C from *DO Sillence,* Am J Med Genet **17**:407, 1984.)

the aorta. Microscopic calcification throughout the aorta and endocardium has also been noted (99).

Skeletal manifestations. In Type IIA, there is marked reduction in ossification in the facial bones and in the cranial vault (Fig. 6–4). The chest is small. Ribs are slightly short, thick, and continuously beaded. The humeri and femora are crumpled (accordion shaped), short, and broad. Thighs are held in abduction at right angles to the body. Tibias are broad, accordion shaped, and angulated. Vertebrae are flattened, the ilia are broad and round, and the ischia and pubic bones are broad and without form (99,109).

Few cases of osteogenesis imperfecta Types IIB and IIC have been studied (65a,99,109,112a). In Type IIB, the skull exhibits spotty mineralization and the ribs are thin and wavy without beading, but ribs show only occasional beading and are not uniformly thick (99). Femurs are short, broad, and crumpled, and tibias are thickened and angulated. In Type IIB there are more well-modeled humeri with wide metaphyses, and more normal vertebral body height than Type IIA. In Type IIC, the face is not remarkably abnormal. The arms are longer than in Type IIA, but the legs are similarly bent inward. Marked underossification of the skull and slender but not so uniformly beaded ribs as in Type IIA have been noted (109). Long bones are slender and shafts are inadequately modeled. Angulation deformities of the shafts of all long bones, and multiple fractures are seen. Heights of the vertebral bodies are near normal (109).

Oral abnormalities. Dental abnormalities have been reported. Dean and Hiramato (29) noted argyrophilic fiber-like structures in the dentin, absence of predentin, an irregular pulpal–dentin junction, a paucity of argyrophilic granules in the odontoblast cytoplasm, an abundance of argyrophilic fibers in the coronal pulp, and dilated capillaries in the coronal pulp. No abnormalities were found in the enamel organ or in the morphology of developing teeth. Calonius et al (18) reported the dental findings in an infant who died 8 days after birth: the dentin was thinner than normal, interglobular dentin extended to the cemento–enamel junction, dentin tubules in the circumpulpal dentin were few and wide, and degenerated osteoblasts were found entrapped in the dentin. Mantle dentin was normal and irregular at the enamel–dentin interface; the dental papilla and pulp were normal. Similar findings have been reported elsewhere (3,5,10,52). Haebara et al (46), however, reported normal teeth in a newborn with presumed osteogenesis imperfecta Type II who died shortly after birth, suggesting heterogeneity. Although other studies have reported dental abnormalities in patients with lethal osteogenesis imperfecta, phenotypes have not been sufficiently characterized; thus, it cannot be determined whether they had osteogenesis imperfecta Type II or another lethal osteogenesis imperfecta syndrome (43,47,114). Levin et al (60) reported normal dentition in an infant with a lethal osteogenesis imperfecta syndrome of unknown type who survived for 10 months.

### Type III

General features. This osteogenesis imperfecta syndrome, characterized as progressively deforming with normal sclerae, is usually not lethal, at least in the newborn period (97). Beighton and Versfeld (8) and Viljoen and Beighton (114a) have reported a high prevalence of this disorder in South Africa.

The diagnosis in sporadic cases with an osteogenesis imperfecta Type III phenotype is confounded by clinical and radiographic heterogeneity, and clinical similarity to other patients with severe osteogenesis imperfecta (98,100,109). The extent of this heterogeneity is unknown (98), but Hanscom and Bloom (46a) recognize four subtypes. In a recent study (100), approximately one-third of patients survived long term, reflecting not only the severity of the disorder but also heterogeneity within the group. Of 17 patients with this disorder, 4 died in the first year and 5 in the second and third decades. Death usually results from complications of severe bone fragility, skeletal deformity including kyphoscoliosis, pulmonary hypertension, and cardiopulmonary failure. Without a family history of recurrence in siblings or parental consanguinity, a definite diagnosis of osteogenesis imperfecta Type III is difficult.

Thompson et al (109) calculated an empirical recurrence risk of 7% for families of children who have severe osteogenesis imperfecta but who survive the perinatal period. They concluded that about 75% were new autosomal dominant mutations, whereas 25% were autosomal recessives. Clinical, genetic, and radiologic findings have been reviewed by Sillence et al (97–101), Thompson et al (109) and Hanscom and Bloom (46a).

Ophthalmologic abnormalities. Although blue sclerae are observed in infancy, the blueness fades by 1 year of age, so that older patients have pale blue or white sclerae (97).

Otolaryngologic abnormalities. Although hearing loss has been said to occur infrequently (97), audiologic findings have not been well

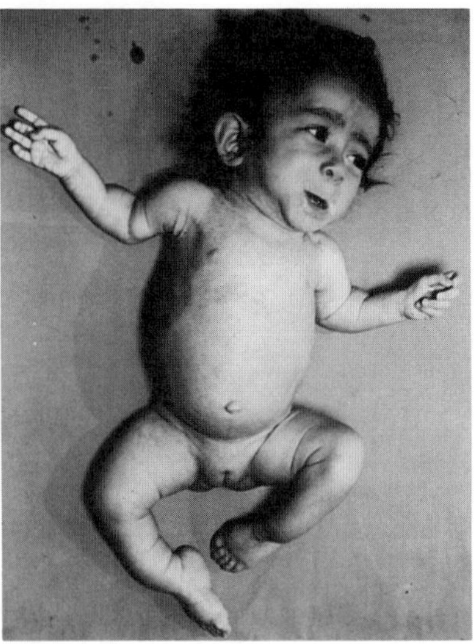

Fig. 6–5. *Type III.* Seven-month-old with many healing fractures and limb deformity but relatively normal sclerae. (From *DO Sillence,* Symposium on Heritable Disorders of Connective Tissue, CV Mosby Co., St. Louis, 1982.)

documented in patients with unequivocal osteogenesis imperfecta Type III.

Neurologic abnormalities. On head CT scans, diffuse ventricular dilatation and diffuse cortical atrophy have been described (112). No associated increase in intracranial pressure, cranial nerve dysfunction, or herniation syndrome was noted.

Cardiovascular system. In a recent review of cardiovascular abnormalities in the osteogenesis imperfecta syndromes, Pyeritz (85) noted that significant cardiac or vascular lesions had not been described. Asymptomatic mitral valve prolapse may be found.

Joint abnormalities. Ligamentous laxity is marked in children but is less severe in adults (97).

Skeletal findings. Infants with this syndrome are usually born at or near term with normal birth weight (97); birth length is usually normal, but some have marked deformity (97) (Fig. 6–5). With age, all fall well below the third percentile in height for age and sex (97,100).

Head size is disproportionately large compared to the rest of the body (97). The ossification defect in the skull is not as severe as in osteogenesis imperfecta Type II (98). Frontal and temporal bossing contributes to the triangular facies (100). The anterior fontanelle, sutures, and posterior fontanelle are wide (98). Wormian bones or bone islands may be palpable along the posterior sutures (97).

Fractures are present at birth in half or more of these infants; all have numerous fractures by 1 or 2 years of age (100). Thompson et al (109) suggested that the better the bone morphology at birth, the better the survival. Although multiple rib fractures occur, continuous beading as seen in osteogenesis imperfecta Type II is not found (98). Long bones are also subject to multiple fractures and bowing and, in some cases, metaphyseal flaring (98). The limbs are not as short or as deformed as in osteogenesis imperfecta Type II (98). Some femora are normal, whereas others are short and broad (97). Femurs and tibias may be markedly angulated (97). In some patients, neonatal radiographs demonstrate broad metaphyses with centrally overmodeled diaphyses and angulation deformities; during the first year of life, diaphyses broaden (100). Sillence et al (97), in a longitudinal study, demonstrated that many patients had marked thickening of the femoral shafts during the first few years of life; this morphology makes it difficult to distinguish these patients from those with osteogenesis imperfecta Type IIB. With time, however, progressive narrowing occurs, so that in older patients, femurs were thin.

In the first few years, metaphyses develop increasing density and irregularity, which progress so that by the end of the first decade, metaphyses and epiphyseal zones are replaced by whorls of radiodensity (97). Progressive and marked vertebral flattening with "codfish" changes are also observed (100). Trunk shortening is common (6). Severe kyphoscoliosis also develops (97,98).

Most patients become markedly handicapped (97,107b). Bowing and angulation deformities are likely to be progressive unless early intramedullary rods are placed (100).

Oral abnormalities. Dental abnormalities similar to those found in the other osteogenesis imperfecta syndromes may be present, but complete radiographic and morphologic evaluation of the teeth has been reported for few patients. The patient reported by Nicholls et al (67) and Pope et al (80) had normal teeth. LS Levin has seen families where some with this syndrome have normal teeth, whereas others have teeth that are opalescent.

### Type IV

General features. Paterson et al (72) have provided the most complete clinical description of this disorder, although Sillence et al (97) first proposed this condition as a separate syndrome. Segregation in more than two generations and male-to-male transmission have been reported (72,97). Evidence thus favors autosomal dominant inheritance. The condition may be less common than Type I (72,97). Beighton et al (6) have proposed that this disorder is heterogeneous.

Paterson et al (72) determined that, in addition to differences in scleral color compared to osteogenesis imperfecta Type I, patients with osteogenesis imperfecta Type IV more commonly have fractures at birth and opalescent teeth. Bruising and nosebleeds were found to be less common in osteogenesis imperfecta Type IV than in Type I.

Facies. Facial appearance is similar to osteogenesis imperfecta Type I (Fig. 6–6).

Ophthalmologic abnormalities. Sclerae are usually normal, although they may be pale blue in early childhood (72). Paterson et al (72) noted that no adolescents or adults with the condition had abnormal sclerae.

Otolaryngologic findings. In patients over 30, the frequency of hearing impairment (30%) is significantly less than that in osteogenesis imperfecta Type I (72).

Neurologic manifestations. Basilar impression resulted in neurologic signs and symptoms in the family with dominant osteogenesis imperfecta, opalescent teeth, and Wormian bones reported by Hurwitz and McSwiney (51); scleral color, however, was not mentioned.

Cardiovascular system. A few reports describe a tendency to aortic root dilatation (85).

Joint abnormalities. A frequency of joint hypermobility, joint dislocations, and hernias similar to that found in osteogenesis imperfecta Type I is found in Type IV.

Skeletal abnormalities. Just over one-quarter of patients have fractures at birth (72). There is wide variation among patients in total number of fractures. However, fracture frequency is maximal during childhood but decreases markedly after puberty. Some individuals with this syndrome may never have fractures (97). Sillence et al (97) proposed that some patients have a progressively deforming phenotype; thus, it might be difficult in some cases to distinguish this syndrome from osteogenesis imperfecta Type III.

Paterson et al (72) have studied skeletal radiographs of a large number of patients with this condition. Although radiographs taken at the

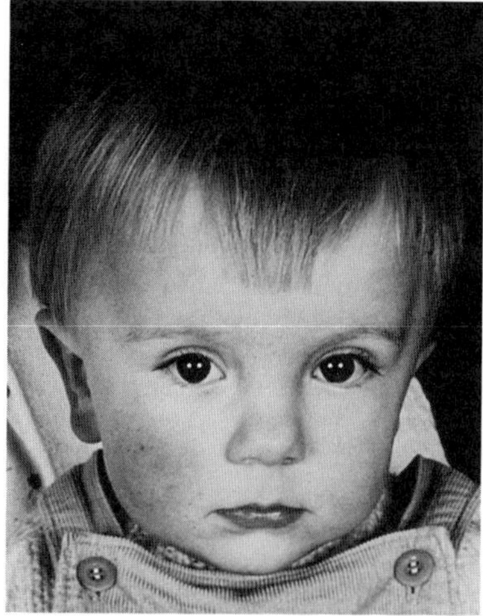

Fig. 6–6. *Type IV.* Boy age 18 months with macrocephaly and mild frontal and temporal bossing.

time of first fracture may show no osteoporosis, with repeated fractures, osteoporosis and cystic changes are found.

The frequency and average number of Wormian bones in the skull in this syndrome are unknown.

**Oral findings.** Dental findings are similar to those found in osteogenesis imperfecta Type I. Paterson et al (72) and Levin et al (56,57) noted that dentinogenesis imperfecta was either consistently present or consistently absent in each family. Dental abnormalities are significantly more common in Type IV than in Type I (72). It was also noted that these abnormalities, when present, were the one consistent marker within a family although other clinical features might vary (72). Thus, specific dental abnormalities, when present in a family, may be very helpful for diagnosis when other clinical abnormalities are mild.

**Other features.** A tendency to bruising, hernias, and excessive sweating has been noted (72).

**Differential diagnosis.** There are a relatively large number of syndromes classified as osteogenesis imperfecta but distinguishable from the major groups by their associated features (Table 6–2). Levin et al (58) described three kindreds in which multiple radiolucent–radiopaque lesions in the jaws, as well as fractures and other skeletal abnormalities, were found; teeth were normal and inheritance was autosomal dominant.

Cole and Carpenter (25) recognized a previously unreported form of osteogenesis imperfecta in which metaphyseal fractures were most prominent during the first 4 months of life followed by diaphyseal

Table 6–2. Other osteogenesis imperfecta syndromes

| Salient features | Inheritance pattern | Reference |
|---|---|---|
| Mild bone fragility<br>Normal sclerae<br>Multiple radiolucent–radiopaque jaw lesions<br>Normal teeth | Autosomal dominant | Levin et al (58) |
| Multiple fractures<br>Ocular proptosis<br>Craniosynostosis<br>Hydrocephaly<br>Distinctive facies | Unknown | Cole and Carpenter (25) |
| Multiple fractures<br>Osteoporosis<br>Wormian bones<br>Frontal bossing<br>Hyperextensible joints<br>Retinitis pigmentosum | Autosomal recessive | Pfeiffer et al (76a)<br>Heide (47a)<br>Blechmann and Crommer (12a) |
| Multiple fractures<br>Blue sclerae<br>Wormian bones<br>Generalized moderate osteoporosis<br>Dentinogenesis imperfecta | Autosomal dominant | Beighton (7)<br>?Crawfurd and Winter (28) |
| Blue sclerae<br>Wormian bones<br>Mandibular hypoplasia<br>Joint laxity<br>Easy bruisability<br>Normal dentition | Dominant | McLean et al (65) |
| Prenatal fractures<br>Congenital cataracts<br>Microcephaly<br>Thin calvaria<br>Blue sclerae | Recessive | Buyse and Bull (15) |
| Multiple bone fractures<br>Osteopenia<br>Microphthalmia<br>Hyperextensible joints | Autosomal recessive | Frontali et al (38)<br>Beighton et al (9) |
| Mild bone fragility<br>Generalized osteoporosis<br>Destructive generalized joint disease | Unknown | Pentinnen et al (76) |
| Bone fragility<br>Osteoporosis<br>Sclerotic skull lesions<br>Normal teeth? | Unknown | Colavita et al (24) |
| Bone fragility<br>Areas of rarefaction in pelvis and long bones<br>Normal sclerae<br>Normal hearing | Autosomal dominant | Keats and Anast (54) |
| Bone fragility<br>Marfanoid habitus<br>Slight basilar impression<br>Wormian bones<br>Other osseous abnormalities | Recessive? | Miegel et al (66) |
| Bone fragility<br>Joint hypermobility<br>Wrinkled skin<br>Blue sclerae<br>Other osseous abnormalities | Unknown | Biering and Ivesen (11) |
| No fractures<br>Blue sclerae<br>Keratoconus<br>Spondylolisthesis | Autosomal recessive | Greenfield et al (45) |

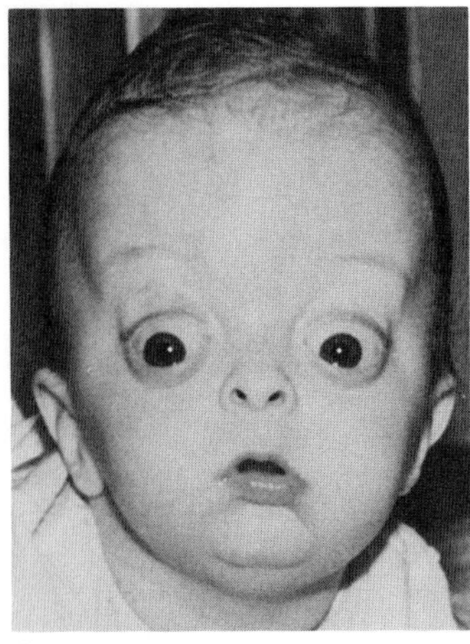

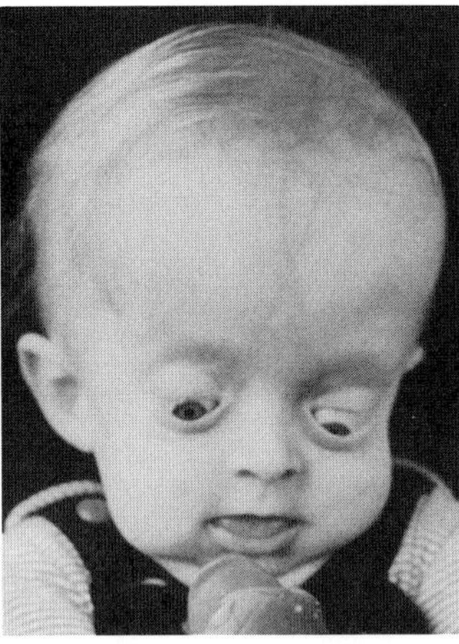

A B

Fig. 6–7. *Cole–Carpenter syndrome.* (A,B) Compare facies. Note bulging forehead, striking proptosis, midfacial deficiency, and micrognathia. (From *DEC Cole* and *TO Carpenter,* J Pediatr **110:**76, 1987.)

fractures and bowing deformities as weight bearing increased. Other distinctive features included craniosynostosis, proptosis, hydrocephaly, and distinctive facial appearance. Also noted were blue sclerae, micrognathia, and high-pitched voice. One patient had dentinogenesis imperfecta (Figs. 6–7 and 6–8). Both patients were males born to unrelated parents and both represented essentially sporadic occurrences in their respective families. Other as yet unreported cases are known to MM Cohen Jr.

Pfeiffer et al (76a) and Heide (47a) described three sibs with severe osteoporosis, fractures, dislocated hips, macrocephaly, Wormian bones, frontal bossing, persistent anterior fontanel, brachytelephalangy, hyperextensible joints, congenital retinitis pigmentosa with blindness, and severe mental retardation (Figs. 6–9 and 6–10). Bone age was retarded. There was severe dental malocclusion. A somewhat similar case was described by Blechmann and Crommer (12a).

Fig. 6–8. *Cole–Carpenter syndrome.* (A) At age 4 months, distinctive metaphyseal lucencies or compression fractures were apparent in all long bones. (B) By age 8 months, cortical thinning and demineralization had progressed. Diaphyseal fractures subsequently healed, but deformities and bowing were already well advanced. (A,B from *DEC Cole* and *TO Carpenter,* J Pediatr **110:**76, 1987.)

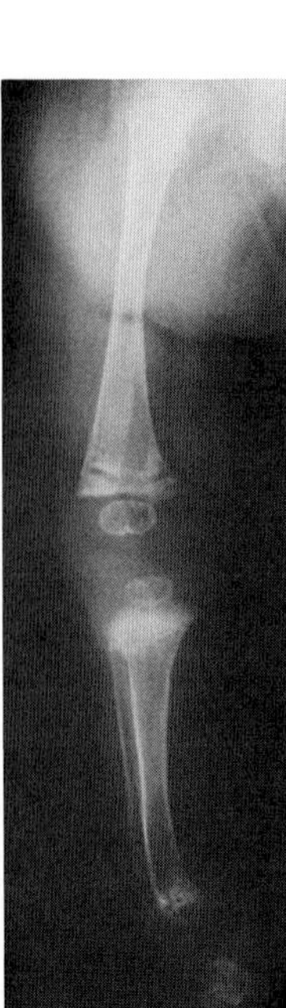

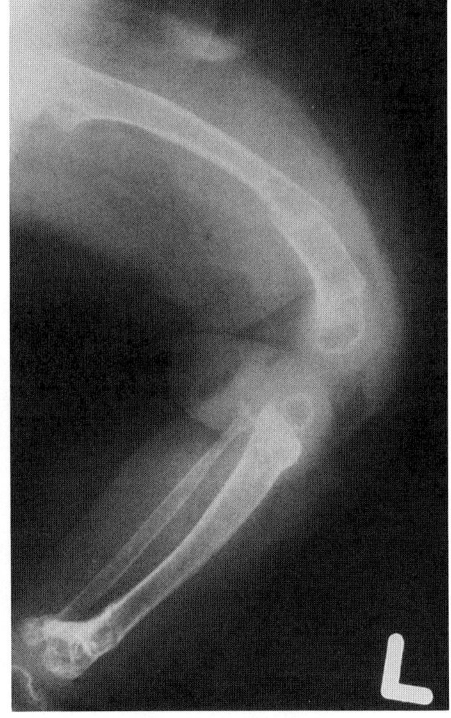


A B

Beighton (7) described a kindred with dentinogenesis imperfecta, blue sclerae, multiple Wormian bones, generalized moderate osteoporosis, mild femoral bowing, and mild vertebral flattening. One person had multiple fractures with mild trauma. Stature and intelligence were normal. Crawfurd and Winter (28) reported a patient, similar to the one reported by Beighton (7), who lacked femoral bowing and vertebral flattening, but who had frontal bossing, joint laxity, and easy bruisability. McLean et al (65) reported a father and daughter with Wormian bones, blue sclerae, mandibular hypoplasia, loose joints with dislocations, and femoral and tibial bowing; dentition was normal and there were no fractured bones.

Buyse and Bull (15) described three sibs with congenital cataracts, microcephaly, thin calvaria, prenatal fractures, and blue sclerae. The brain had a smooth cortex without convolutions, sulci, or gyri. These sibs were stillborn or died shortly after birth. Parents were normal. Frontali et al (38) and Beighton et al (9) reported an autosomal recessive disorder characterized by microphthalmia, vitreoretinal dysplasia or phthisis bulbi, blindness, hypotonia, hyperextensible joints, multiple bone fractures, osteopenia, and other bone abnormalities; this syndrome is better referred to as the osteoporosis-pseudoglioma syndrome. Pentinnen et al (76) described a 14-year-old girl who had two fractures, generalized osteoporosis, and destructive generalized joint disease resembling juvenile rheumatoid arthritis; on biochemical investigations, an increased ratio of Type III collagen to Type I collagen was found. *Campomelic dysplasia* should also be considered in the differential diagnosis.

Colavita et al (24) reported a 21-year-old male with fractures, bumps on the head, osteoporosis, and multiple sclerotic lesions of the skull; although the patient was said to have dentinogenesis imperfecta, radiographs showed normal teeth. Osteogenesis imperfecta has been found with congenital joint contractures as an autosomal recessive condition (Bruck syndrome) (114b).

Keats and Anast (54) reported a three-generation kindred with normal sclerae, normal hearing, multiple bone fractures, and circumscribed areas of rarefaction in the pelvis and long bones. It is also of note that a premature, stillborn infant with short, bowed lower extremities as well as fractures was born to this kindred. Miegel et al (66) described a 10-year-old male born to a consanguineous couple; the boy had multiple fractures, marfanoid habitus, narrow thorax, thin extremities and mild arachnodactyly, lumbar lordosis, genua valga, moderate muscular hypotonia, and mildly hyperextensible joints. Radiographs showed slight basilar impression, Wormian bones, flattened vertebral bodies, and bowing of the long bones; cystic irregularities in the bones were noted. Biering and Iveson (11) described a 5-month-old male with blue sclerae, wrinkled, loose, and pale skin with a conspicuous venous network, excessive joint laxity, hypotonia, bilateral hip dislocation, multiple spontaneous fractures of the vertebrae and femurs, thoracic kyphosis, severe generalized osteoporosis, and slender long bones.

Greenfield et al (45) reported a family with blue sclerae, keratoconus, childhood onset of otosclerotic-like hearing loss, and spondylolisthesis, but no fractures. Inheritance was autosomal recessive. Carey et al (19) described three children with bone fractures, blue sclerae, embryotoxon, gradual hearing loss beginning in the second decade of life, arachnodactyly, tall stature, dolichostenomelia, pectus excavatum, joint hypermobility, and scoliosis but no ectopia lentis. The proband had marked dilatation of the ascending aorta and severe aortic incompetence; he died suddenly at age 32. The father was a member of a large kindred with dominantly inherited osteogenesis imperfecta; the mother had scoliosis, pectus excavatum, clinodactyly and camptodactyly of the fifth fingers and toes, ligamentous laxity, and flat feet.

Wormian bones are found in *cleidocranial dysplasia, pycnodysostosis, progeria, mandibuloacral dysplasia, acroosteolysis,* and Menkes syndrome. Fractures can also be the result of child abuse (70,74).

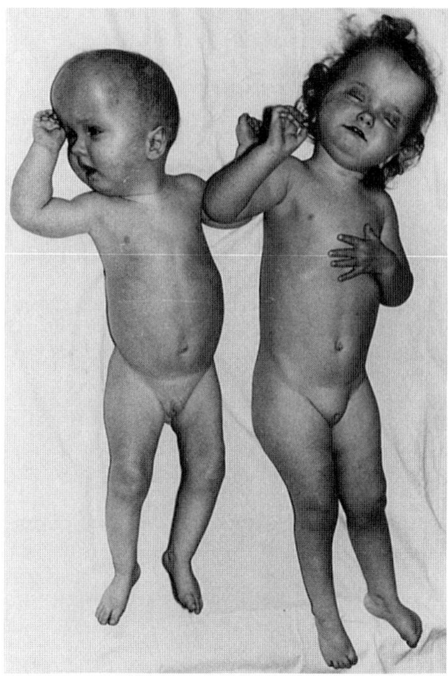

Fig. 6–9. *Osteoporosis, fractures, macrocephaly, blindness, and severe mental retardation.* Two sibs with macrocephaly, frontal bossing, and congenital blindness. (Courtesy of *W Lenz,* Münster, West Germany.)

Fig. 6–10. *Osteoporosis, fractures, macrocephaly, blindness, and severe mental retardation.* Roentgenogram showing severe osteoporosis and leg length disparity due to dislocated hips. (Courtesy of *W Lenz,* Münster, West Germany.)

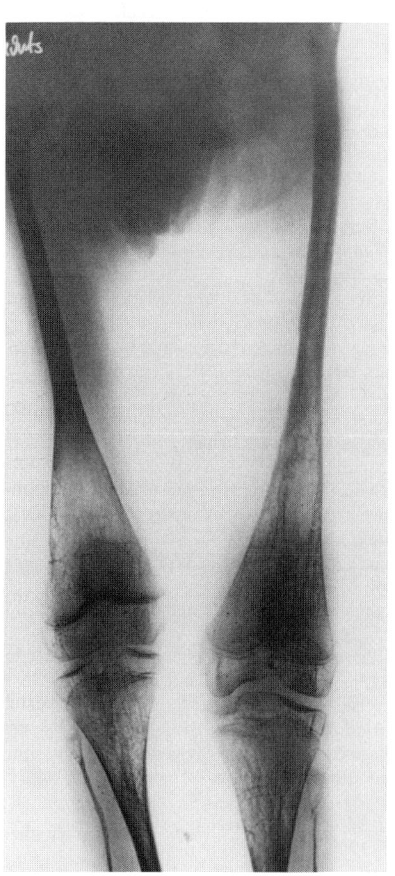

Opalescent teeth with radiologic abnormalities similar to those found in some patients with osteogenesis imperfecta are also seen in dentinogenesis imperfecta, a disorder affecting the dentition only. This disorder, sometimes called hereditary opalescent dentin, has autosomal dominant inheritance (12). Similar clinical and radiologic dental abnormalities are found in dentinogenesis imperfecta Type III in the Brandywine isolate (61) and in the deciduous dentition in patients with dentin dysplasia Type II (41). Devitalized teeth may also become darkened with time.

Laboratory aids. Sonography has been used as a method of prenatal diagnosis for individuals at risk for osteogenesis imperfecta Type I. Hobbins et al (49) reported a displaced femoral fracture in a fetus at risk for osteogenesis imperfecta Type I, using ultrasound at 19 weeks of gestation; femurs were not significantly shorter than normal. Chervenak et al (22), using ultrasound, reported a fetus that was normal at 20 weeks of gestation; however, at 24 weeks both femurs were markedly bowed, and at 32 and 38 weeks, long bone demineralization was suggested, but no fractures were found. At birth, there were no abnormalities, but by the ninth day of life, a midshaft fracture of the right femur had occurred. The mother had osteogenesis imperfecta Type I.

Elejalde and de Elejalde (35), using ultrasound at 17, 18, and 19 weeks in a pregnancy at risk for osteogenesis imperfecta Type II, documented short femurs and tibias, angulated long bones, diminished mineralization, and an abnormally shaped rib cage in the fetus; limited fetal movements were also noted. Shapiro et al (94) reported the results of prenatal diagnosis of osteogenesis imperfecta Type II using ultrasonography at 17.5 weeks of gestation, fetal radiographs at 21 weeks, analysis of procollagen synthesis in amniotic fluid cells, and ultrastructural studies of fetal dermis. Dinno et al (31), on ultrasonic examination at 19 weeks of gestation of a fetus at risk for osteogenesis imperfecta Type II, found a compressed soft cranium, limb distortion, and faint, irregular echoes of the spine and ribs. On radiographs, deficient ossification of the spine, ribs, limbs, and calvaria were noted. Using ultrasonic diagnosis at 16 weeks of pregnancy, Ghosh et al (40) found short femurs with irregular outlines in a fetus at risk for osteogenesis imperfecta Type II. Two weeks later, a repeat scan demonstrated evidence of fracture and deformity and shortening of other long bones. Patel et al (68) identified a fetus as having severe osteogenesis imperfecta (likely Type II), because of a dense, bent, and short femur at 17 weeks of gestation. The skull and spine were normal. Stephens et al (107) found abnormally shaped fetal femurs at 13.5 weeks of gestation on examination using ultrasound.

Aylsworth et al (2) reported results of ultrasonic examination on an affected fetus with osteogenesis imperfecta Type III at 19 weeks of gestation; both femurs were bowed. On radiographs, there was diminished ossification. Both parents were unaffected, but their first child had osteogenesis imperfecta with severe bone deformity, slightly blue sclerae, and normal dentition. Carpenter et al (21) found an affected fetus on sonography at 18 weeks of gestational age, based on thoracic–abdominal disproportion and lower limb shortening; the father had severe deforming osteogenesis imperfecta of unspecified type.

Using genetic linkage analysis, Tsipouras et al (111) have excluded inheritance of osteogenesis imperfecta Type IV in a fetus at risk; the child born was determined to be normal, since no fractures had been sustained at least during follow-up for 10 months.

A broad spectrum of molecular defects involving collagen has been incriminated in the etiology of osteogenesis imperfecta. These defects have been reviewed by Byers and Bonadio (16), Pope et al (79), and Prockop (84), and confirm and expand our knowledge of heterogeneity in this group of connective tissue disorders. Mutations that affect synthesis of Type I collagen result in osteogenesis imperfecta Type I. These mutations are heterozygous and lead to a reduction of approximately 50% in the amount of Type I collagen synthesized; secretion and posttranslational modification of these molecules as well as the rate of molecular assembly appear normal (16). The osteogenesis imperfecta Type II phenotype is produced by many different mutations that disrupt the formation of the collagen triple helix; at least four classes of mutations in this molecule have been identified (2a,16,23b,84). These mutations encode the pro $\alpha 1$(I) and pro $\alpha 2$(I) chains of Type I procollagen (17a,17b,109a,118a). Superti-Furga et al (107c) have shown that mutations in the triple helical region of $\alpha 2$(I) chains produce a milder phenotype than those in the $\alpha 1$(I) chains.

There have been a few studies of patients with osteogenesis imperfecta Type III (16). One patient had a molecular defect that prevents the incorporation of pro-$\alpha 2$(I) chains into Type I collagen so that the major collagen of bone and skin consists of $\alpha$(I) trimers (16,67,79). In another patient, the mutation resulted in overmodification of the collagen molecule. Byers and Bonadio (16) have suggested that the location at which the overmodification begins is related to the variability of the highly heterogeneous phenotype in this osteogenesis imperfecta syndrome.

Studies of patients with osteogenesis imperfecta Type IV suggest that the defect is the result of heterozygous mutations in the triple helical domain of the pro-$\alpha 2$(I) chain that affect modification, molecular stability, secretion, and perhaps molecular aggregation (16). Wenstrup et al (117) examined fibroblasts from two members of a large kindred in which the disorder was linked to the pro-$\alpha 2$(I) gene of Type I collagen; they found a deletion of about 10 amino acid residues that altered the triple helical structure of Type I collagen.

The use of linkage analysis using restriction fragment length polymorphisms for the pro-$\alpha 1$(I) and pro-$\alpha 2$(I) collagen genes has considerably advanced our knowledge of the segregation of different osteogenesis imperfecta syndromes (23b,37,110) and the types of collagen mutations producing the disorders. For example, it would appear that in the majority of families with osteogenesis imperfecta Type I studied, the syndrome segregates linked to the Type I collagen $\alpha 1$ gene but not to the $\alpha 2$ gene (16,110). On the other hand, in the majority of families studied with osteogenesis imperfecta Type IV, the disorder segregates linked to polymorphisms of the Type I collagen $\alpha 2$ gene (37).

Heterogeneity in osteogenesis imperfecta Type I has been demonstrated using DNA polymorphisms (116). Using restriction fragment length polymorphisms for human Type II collagen, Sykes et al (108) failed to demonstrate segregation of this gene with dominant Type I osteogenesis imperfecta in the three pedigrees examined.

Using cloned DNA probes, Chu et al (23) demonstrated an internal deletion in one allele for the pro-$\alpha 1$(I) chain in a patient with osteogenesis imperfecta Type II. Pope et al (81) demonstrated a 300-base pair deletion in an $\alpha 1$(I)-like collagen gene in six patients with osteogenesis imperfecta Type II. In each family, one of the two clinically unaffected parents of each infant carried an identical gene defect; these investigators suggested that the other parent had an unidentified collagen abnormality, and that the infants represented, in fact, genetic compounds. Skyes and Ogilvie (108a) however, suggested that the deleted gene was, in reality, a common allelic variant, and might not be relevant to perinatal lethal osteogenesis imperfecta.

An animal model of osteogenesis imperfecta with opalescent teeth has been described in Friesian calves (30).

### References (The osteogenesis imperfectas)


1. Acar J et al: Osteogénese imparfaite et insuffisance aortique. *Ann Méd Interne* **8**:514–518, 1980.

2. Aylsworth AS et al: Prenatal diagnosis of a severe deforming type of osteogenesis imperfecta. *Am J Med Genet* **19**:707–714, 1984.

2a. Bateman JF et al: Lethal perinatal osteogenesis imperfecta due to the substitution of arginine for glycine at residue 391 of the $\alpha 1$(I) chain of type I collagen. *J Biol Chem* **262**:7021–7027, 1987.

3. Bauer KH: Über Osteogenesis imperfecta. Zugleich ein Beitrag zur Frage einer allgemeinen Erkrankung sämtlicher Stützgewebe. *Deutsche Z Chir* **154**:166–213, 1920.

4. Bauze RJ et al: A new look at osteogenesis imperfecta: A clinical, radiological and biochemical study of forty-two patients. *J Bone Jt Surg* **57B**:2–12, 1975.

5. Becks H: Histologic study of tooth structure in osteogenesis imperfecta. *Dent Cosmos* **73**:437–454, 1931.

6. Beighton P et al: Skeletal complications in osteogenesis imperfecta. A review of 153 South African patients. *S Afr Med J* **64**:565–568, 1983.

7. Beighton P: Familial dentinogenesis imperfecta, blue sclerae, and Wormian bones without fractures: Another type of osteogenesis imperfecta? *J Med Genet* **88**:124–128, 1985.

8. Beighton P, Versfeld GA: On the paradoxically high relative prevalence of osteogenesis imperfecta type III in the black population of South Africa. *Clin Genet* **27**:398–401, 1985.

9. Beighton P et al: The ocular form of osteogenesis imperfecta: A new autosomal recessive syndrome. *Clin Genet* **28**:69–74, 1985.

10. Beibl M: Beitrag zur Frage der Osteogenesis imperfecta durch Untersuchungen am Zahnsystem. *Virchows Arch Pathol Anat* **255**:54–70, 1925.

11. Biering A, Ivesen T: Osteogenesis imperfecta associated with Ehlers-Danlos syndrome. *Acta Paediatr* **44**:279–286, 1955.

12. Bixler D et al: Dentinogenesis imperfecta: Genetic variations in a six-generation family. *J Dent Res* **49**:1196–1199, 1969.

12a. Blechmann G, Crommer J: Front olympien à la naissance et cécité. *Arch Fr Pédiatr* **5**:374, 1948.

13. Blümcke S et al: Histochemical and fine structural studies on the cornea in osteogenesis imperfecta congenita. *Virchows Arch (Zellpathol)* **11**:124–132, 1972.

14. Bonadio J, Byers PH: Subtle structural alterations in the chains of type I procollagen produce osteogenesis imperfecta type I. *Nature (London)* **316**:363–366, 1985.

15. Buyse M, Bull M: A syndrome of osteogenesis imperfecta, microcephaly, and cataracts. *Birth Defects* **14**(6B):95–98, 1978.

16. Byers PH: Disorders of collagen metabolism, in *Metabolic Basis of Inherited Disease,* 6th ed, Scriver CR, Beaudet AL, Sly WS, Valle D (eds). McGraw-Hill, New York, 1989.

17. Byers PH et al: Osteogenesis imperfecta—update and perspective. *Am J Med Genet* **17**:429–435, 1984.

17a. Byers PH et al: Perinatal lethal osteogenesis imperfecta (OI type II); a biochemically heterogeneous disorder usually due to new mutations in the gene for type I collagen. *Am J Hum Genet* **42**:237–248, 1988.

17b. Byers PH et al: A novel mutation causes a perinatal lethal form of osteogenesis imperfecta. An insertion of one $\alpha 1$(I) collagen allele (COL1A1). *J Biol Chem* **263**:7855–7861, 1988.

18. Calonius PEB et al: Tooth germ changes in osteogenesis imperfecta. A case report. *Suom Hammasläák Toim* **63**:220–226, 1967.

19. Carey MC et al: Osteogenesis imperfecta in twenty-three members of a kindred with heritable features contributed by a nonspecific skeletal disorder. *Q J Med* **37**:437–449, 1968.

20. Carothers AD et al: Risk of dominant muttion in older fathers: Evidence from osteogenesis imperfecta. *J Med Genet* **23**:227–230, 1986.

21. Carpenter MW et al: Midtrimester diagnosis of severe deforming osteogenesis imperfecta with autosomal dominant inheritance. *Am J Perinatol* **3**:80–83, 1986.

22. Chervenak FA et al: Antenatal sonographic findings of osteogenesis imperfecta. *Am J Obstet Gynecol* **143**:228–230, 1982.

23. Chu M-L et al: Internal deletion in a collagen gene in a perinatal lethal form of osteogenesis imperfecta. *Nature (London)* **304**:78–80, 1983.

23a. Cohn D et al: Molecular confirmation of gonadal mosaicism in osteogenesis imperfecta, type II. David W. Smith Workshop on Malformations and Morphogenesis, Oakland, California, August 1988.

23b. Cohn DH et al: Lethal osteogenesis imperfecta resulting from a single nucleotide change in one human pro $\alpha 1$(I) collagen allele. *Proc Natl Acad Sci USA* **83**:6045–6047, 1986.

24. Colavita N et al: Calvarial doughnut lesions with osteoporosis, multiple fractures, dentinogenesis imperfecta and tumorous changes in the jaws (report of a case). *Australas Radiol* **28**:226–231, 1984.

25. Cole DEC, Carpenter TO: Bone fragility, craniosynostosis, ocular proptosis, hydrocephalus, and distinctive facial features: A newly recognized type of osteogenesis imperfecta. *J Pediatr* **110**:76–80, 1987.

26. Cole NL et al: Surgical management of patients with osteogenesis imperfecta. *J Oral Maxillofac Surg* **40**:578–584, 1982.

27. Cox JR, Simmons CL: Osteogenesis imperfecta and associated hearing loss in five kindreds. *S Med J* **75**:1222–1226, 1982.

28. Crawfurd M d'A, Winter RM: A new type of osteogenesis imperfecta (letter). *J Med Genet* **19**:158, 1982.

29. Dean DH, Hiramato RN: Osteogenesis imperfecta congenita: Dental features of a rare disease. *J Oral Med* **39**:119–121, 1984.

30. Denholm LJ, Cole WG: Heritable bone fragility, joint laxity and dysplastic dentin in Friesian calves: A bovine syndrome of osteogenesis imperfecta. *Aust Vet J* **60**:9–17, 1983.

31. Dinno ND et al: Midtrimester diagnosis of osteogenesis imperfecta, type II. *Birth Defects* **18**(3A):125–132, 1982.

32. Eddowes A: Dark sclerotics and fragilitas ossium. *Br Med J* **2**:222, 1900.

32a. Edwards MJ et al: Clinical and biochemical correlation in osteogenesis imperfecta. David W. Smith Workshop on Malformations and Morphogenesis, Oakland, California, August, 1988.

33. Ekman OS: Descriptionem et casus aliquot osteomalaciae sistens. Dissertatio Medica, Uppsala, Sweden, 1788, cited in Weil UV: Osteogenesis imperfecta: Historical background. *Clin Orthop Rel Res* **159**:6–11, 1981.

34. Eichenholtz W, Mueller D: Electron microscopic findings on the cornea and sclera in osteogenesis imperfecta. *Klin Monatsbl Augenkheilkd* **161**:646–653, 1972.

35. Elejalde BR, de Elejalde MM: Prenatal diagnosis of perinatally lethal osteogenesis imperfecta. *Am J Med Genet* **14**:353–359, 1983.

36. Eustace P: Hypercholesterolaemia in osteogenesis imperfecta. *Br J Clin Pract* **27**:225–227, 1973.

37. Falk CT et al: Use of molecular haplotypes specific for the pro α2(I) collagen gene in linkage analysis of the mild autosomal dominant forms of osteogenesis imperfecta. *Am J Hum Genet* **38**:269–279, 1986.

38. Frontali M et al: Osteoporosis-pseudoglioma syndrome: Report of three affected sibs and an overview. *Am J Med Genet* **22**:35–47, 1985.

39. Gage JP et al: Dentine is biochemically abnormal in osteogenesis imperfecta. *Clin Sci* **70**:339–346, 1986.

40. Ghosh A et al: Simple ultrasonic diagosis of osteogenesis imperfecta type II in early second trimester. *Prenat Diag* **4**:235–240, 1984.

41. Giansanti JS, Allen JD: Dentin dysplasia, Type II, or dentin dysplasia, coronal type. *Oral Surg Oral Med Oral Pathol* **38**:911–917, 1974.

42. Gillerot Y et al: Lethal perinatal type II osteogenesis imperfecta in a family with a dominantly inherited type I. *Eur J Pediatr* **141**:119–122, 1983.

43. Godfrey JL: A histological study of dentin formation in osteogenesis imperfecta congenita. *J Oral Pathol* **2**:95–111, 1973.

44. Gray PHK: A case of osteogenesis imperfecta, associated with dentinogenesis imperfecta, dating from antiquity. *Clin Radiol* **20:**106–108, 1969.

45. Greenfield G et al: Blue sclerae and keratoconus: Key features of a distinct heritable disorder of connective tissue. *Clin Genet* **4**:8–16, 1973.

46. Haebara H et al: An autopsy case of osteogenesis imperfecta congenita-histochemical and electron microscopical studies. *Acta Pathol Jpn* **19**:377–394, 1969.

46a. Hanscom DA, Bloom BA: The spine in osteogenesis imperfecta. *Orthoped Clin N Am* **19**:449–458, 1988.

47. Haubach: Über abnormale und normale Dentin-bildung bei Osteogenesis imperfecta. *Vjschr Zahnheilk* **45**:268–271, 1929.

47a. Heide T: Ein Syndrom béstehend aus Osteogenesis imperfecta, Makrozephalus mit Schaltknochen und prominenten Stirnhöckern, Brachytelephalangie, Gelenküberstreckbarkeit, kongenitaler Amaurose und Oligophrenie bei drei Geschwistern. *Klin Pädiatr* **193**:334–340, 1981.

48. Heyes FM et al: Osteogenesis imperfecta and odontogenesis imperfecta: Clinical and genetic aspects in eighteen families. *J Pediatr* **56**:234–245, 1960.

49. Hobbins JC et al: Diagnosis of fetal skeletal dysplasias with ultrasound. *Am J Obstet Gynecol* **142**:306–312, 1982.

49a. Hollister DW: Molecular basis of osteogenesis imperfecta. *Curr Probl Dermtol* **17**:76–94, 1987.

50. Hortop J et al: Cardiovascular involvement in osteogenesis imperfecta. *Circulation* **73**:54–61, 1986.

51. Hurwitz LJ, McSwiney RR: Basilar impression and osteogenesis imperfecta in a family. *Brain* **83**:138–149, 1960.

52. Jeckeln E: Systemgebundene mesenchymale Erschöpfung. Eine neue Begriffsfassung der Osteogenesis imperfecta. *Virchow Arch Pathol Anat* **280:**351–373, 1931.

53. Kaiser-Kupfer KI et al: Low ocular rigidity in patients with osteogenesis imperfecta. *Inv Ophthalmol Vis Sci* **20**:807–809, 1981.

54. Keats TE, Anast GS: Circumscribed skeletal rarefactions in osteogenesis imperfecta. *AJR* **84**:492–498, 1960.

55. Lanting PJH et al: Decreased scattering coefficient of blue sclerae. *Clin Genet* **27**:187–190, 1985.

56. Levin LS et al: Classification of osteogenesis imperfecta by dental characteristics. *Lancet* **1**:332–333, 1978.

57. Levin LS et al: Scanning electron microscopy of teeth in autosomal dominant osteogenesis imperfecta: Support for genetic heterogeneity. *Am J Med Genet* **5**:189–199, 1980.

58. Levin LS et al: Osteogenesis imperfecta with unusual skeletal lesions. Report of three families. *Am J Med Genet* **21**:257–269, 1985.

59. Levin LS et al: The dentition in the osteogenesis imperfecta syndromes. *Clin Orthop Rel Res* **159**:64–74, 1981.

60. Levin LS et al: Osteogenesis imperfecta lethal in infancy: Case report and scanning electron microscopic studies of the deciduous teeth. *Am J Med Genet* **12**:359–368, 1982.

61. Levin LS et al: Dentinogenesis imperfecta in the Brandywine isolate (D.I. type III). Clinical, radiologic, and scanning electron microscopic studies of the dentition. *Oral Surg Oral Med Oral Pathol* **56**:267–274, 1983.

61a. Levin LS et al: Osteogenesis imperfecta type I with unusual dental abnormalities. *Am J Med Genet* **31**:921–932, 1988.

62. Lobstein JFGCM: De la fragilité des os, ou do l'osteopsathyrose, Traité de l'Anatomie Pathologique, Vol 2, FG Lerrault, Paris, 1833, pp 204–212. Cited in Weil UH: Osteogenesis imperfecta: Historical background. *Clin Orthop Rel Res* **159**:6–10, 1981.

63. Lukinmaa P-L et al: Dental findings in osteogenesis imperfecta: I. Occurrence and expression of type I dentinogenesis imperfecta. *J Craniofac Genet Dev Biol* **7**:115–125, 1987.

64. Lukinmaa P-L et al: Dental findings in osteogenesis imperfecta: II. Dysplastic and other developmental defects. *J Craniofac Genet Dev Biol* **7**:127–135, 1987.

65. McLean JR et al: The Grant syndrome. Persistent Wormian bones, blue sclerae, mandibular hypoplasia, shallow glenoid fossae and campomelia—an autosomal dominant trait. *Clin Genet* **29**:523–529, 1986.

65a. Maroteaux P et al: Les formes anténatal de l'ostéogenese imparfaite: Essay de classification. *Arch Fr Pédiatr* **43**:235–241, 1986.

66. Meigel WN et al: A constitutional disorder of connective tissue suggesting a defect in collagen biosynthesis. *Klin Wschr* **52**:906–912, 1974.

67. Nicholls AC et al: The clinical features of homozygous α2(I) collagen deficient osteogenesis imperfecta. *J Med Genet* **21**:257–262, 1984.

68. Patel ZM et al: Prenatal diagnosis of lethal osteogenesis imperfecta (OI) by ultrasonography. *Prenat Diag* **3**:261–263, 1983.

69. Paterson CR: Heterogeneity of osteogenesis imperfecta congenita. *Lancet* **1**:821, 1980.

70. Paterson CR: Unexplained fractures in childhood: Differential diagnosis of osteogenesis imperfecta and other disorders from non-accidental injury. *J Neurolog Orthopaed Med Surg* **7**:253–255, 1986.

71. Paterson CR et al: Osteogenesis imperfecta type I. *J Med Genet* **20**:203–205, 1983.

72. Paterson CR et al: Osteogenesis imperfecta with dominant inheritance and normal sclerae. *J Bone Jt Surg* **65B**:35–39, 1983.

73. Paterson CR et al: Osteogenesis imperfecta after the menopause. *N Engl J Med* **310**:1694–1696, 1984.

74. Paterson CR: Osteogenesis imperfecta in the differential diagnosis of child abuse. *Child Abuse Neglect* **1**:449–452, 1977.

75. Pedersen U: Osteogenesis imperfecta. Clinical features. Hearing loss and stapedectomy. *Acta Otolaryngol* Suppl **145**:1–36, 1985.

76. Pentinnen R et al: An arthropathic form of osteogenesis imperfecta. *Acta Paediatr Scand* **69**:263–267, 1980.

76a. Pfeiffer RA et al: Amaurosis congenita Leber. *Arch Kinderheilk* **182**:179–191, 1971.

77. Pindborg JJ: Dental aspects of osteogenesis imperfecta. *Acta Pathol Microbiol Scand* **24**:47–64, 1947.

78. Pope FM, Nicholls AC: Heterogeneity of osteogenesis imperfecta congenita (letter). *Lancet* **1**:820–821, 1980.

79. Pope FM et al: Collagen genes and proteins in osteogenesis imperfecta. *J Med Genet* **22**:466–478, 1985.

80. Pope FM et al: Clinical features of homozygous α2(I) collagen deficient osteogenesis imperfecta (letter). *J Med Genet* **23**:377, 1986.

81. Pope FM et al: Lethal osteogenesis imperfecta congenita and a 300 base pair gene deletion for an α1(I)-like collagen. *Br Med J* **288**:431–434, 1984.

82. Pozo JL et al: Basilar impression in osteogenesis imperfecta. A report of three cases in one family. *J Bone Jt Surg* **66B**:233–238, 1984.

83. Preiswerk R: Ein Beitrag zur Kenntnis der Osteogenesis imperfecta (Vrolik). *Jahrb Kinderheilk* **76**:40–57, 1912.

84. Prockop DJ: Osteogenesis imperfecta: Phenotypic heterogeneity, protein suicide, short and long collagen. *Am J Hum Genet* **36**:499–505, 1984.

85. Pyeritz RE: Heritable disorders of connective tissue, in *Genetics of Cardiovascular Disease,* Pierpont ME, Moller JH (eds), Marinus Nijhoff Publishing, Boston, 1986, pp 265–303.

86. Pyeritz RE, Levin LS: Aortic root dilatation and valvular dysfunction in osteogenesis imperfecta. *Circulation* **64**:1193A, 1981.

87. Reite M, Solomons C: The EEG in osteogenesis imperfecta. *Clin Electroencephalog* **11**:16–21, 1980.

88. Riedner ED et al: Hearing patterns in dominant osteogenesis imperfecta. *Arch Otolaryngol* **106**:737–740, 1980.

89. Rogerson ME et al: Left atrial rupture in osteogenesis imperfecta. *Br Heart J* **56**:187–189, 1986.

90. Rushton MA: Structures of the teeth in late cases of osteogenesis imperfecta. *J Pathol Bacteriol* **48**:591–603, 1939.

91. Saint-Martin J et al: Malformations osseuse complexes d'evolution letale. *Arch Fr Pédiatr* **36**:188–193, 1979.

92. Schwartz S, Tsipouras P: Oral findings in osteogenesis imperfecta. *Oral Surg Oral Med Oral Pathol* **57**:161–167, 1984.

93. Seedorf KS: Osteogenesis imperfecta. A study of clinical features and heredity based on 55 Danish families comprising 180 members. Universitetsforlaget I Århus, Copenhagen, 1949.

94. Shapiro JE et al: Prenatal diagnosis of lethal osteogenesis imperfecta. *J Pediatr* **100**:127–133, 1982.

95. Shapiro JR et al: Hearing and middle ear function in osteogenesis imperfecta. *JAMA* **247**:2120–2126, 1982.

96. Sillence DO: Osteogenesis imperfecta. An expanding panorama of variants. *Clin Orthoped Rel Res* **159**:11–25, 1981.

97. Sillence DO et al: Genetic heterogeneity in osteogenesis imperfecta. *J Med Genet* **16**:101–116, 1979.

98. Sillence DO: Osteogenesis imperfecta: Clinical variability and classification, in *Heritable Disorders of Connective Tissue,* Akeson WH, Bornstein P, Glimcher KJ (eds), CV Mosby, St Louis, 1982, pp 223–237.

99. Sillence DO et al: Osteogenesis imperfecta type II. Delineation of the phenotype with reference to genetic heterogeneity. *Am J Med Genet* **17**:407–423, 1984.

100. Sillence DO et al: Osteogenesis imperfecta type III. Delineation of the phenotype with reference to genetic heterogeneity. *Am J Med Genet* **23**:821–832, 1986.

101. Sillence DO et al: Clinical variability in osteogenesis imperfecta—variable expressivity or genetic heterogeneity. *Birth Defects* **15**(5B):113–129, 1979.

102. Smith R et al: The eye and collagen in osteogenesis imperfecta. *Birth Defects* **12**(3):563–566, 1976.

103. Smith R et al: *The Brittle Bone Syndrome. Osteogenesis Imperfecta.* Butterworths, London, 1983.

104. Spranger J: Osteogenesis imperfecta 1982, in *Skeletal Dysplasias,* Papadatos CJ, Bartsocas CS (eds), Alan R. Liss, New York, 1982, pp 223–233.

104a. Spranger J: Osteogenesis imperfecta: A pasture for splitters and lumpers. *Am J. Med Genet* **17**:425–428, 1984.

105. Spurway J: Hereditary tendency to fracture. *Br Med J* **2**:844, 1896.

106. Stein D, Kloster FE: Valvular heart disease in osteogenesis imperfecta. *Am Heart J* **94**:637–641, 1977.

107. Stephens JD et al: Prenatal diagnosis of osteogenesis imperfecta type II by real-time ultrasound. *Hum Genet* **64**:191–193, 1983.

107a. Stoll C et al: Birth prevalence rates of skeletal dysplasias. *Clin Genet* **35**:88–92, 1989.

107b. Stöss H et al: Heterogeneity of osteogenesis imperfecta. Biochemical and morphologic findings in a case of type III according to Sillence. *Eur J Pediatr* **145**:34–39, 1986.

107c. Superti-Furga A et al: Clinical variability of osteogenesis imperfecta linked to COL1A2 and associated with a structural defect in the type I collagen molecule. *J Med Genet* **26**:358–362, 1989.

108. Sykes B et al: Exclusion of the $\alpha$I(II) cartilage collagen gene as the mutant locus in type IA osteogenesis imperfecta. *J Med Genet* **22**:187–191, 1985.

108a. Sykes B, Ogilvie D: Lethal osteogenesis imperfecta and a gene deletion (letter). *Br Med J* **288**:1380–1381, 1984.

109. Thompson EM et al: Recurrence risks and prognosis in severe sporadic osteogenesis imperfecta. *J Med Genet* **24**:390–405, 1987.

109a. Thompson EM et al: Osteogenesis imperfecta type IIA: Evidence for dominant inheritance. *J Med Genet* **24**:386–389, 1987.

110. Tsipouras P et al: Molecular heterogeneity of the mild autosomal dominant forms of osteogenesis imperfecta. *Am J Hum Genet* **36**:1172–1179, 1984.

111. Tsipouras P et al: Prenatal prediction of osteogenesis imperfecta (OI Type IV): Exclusion of inheritance using a collagen gene probe. *J Med Genet* **24**:406–409, 1987.

112. Tsipouras P et al: Neurologic correlates of osteogenesis imperfecta. *Arch Neurol* **43**:150–152, 1986.

112a. van der Harten H et al: Perinatal lethal osteogenesis imperfecta. Radiologic and pathologic evaluation of seven prenatally diagnosed cases. *Pediatr Pathol* **8**:233–252, 1988.

113. van der Hoeve J, de Kleijn A: Blaue Sclerae, Knochenbrüchigkeit und Schwerhörigkeit. *Arch Ophthalmol* **95**:91–93, 1918.

114. Velley J: Étude clinique et génétique de la dentinogenèse imparfaite héréditaire. *Actual Odontostomatol (Paris)* **28**:519–532, 1974.

114a. Viljoen D, Beighton P: Osteogenesis imperfecta type III: An ancient mutation in Africa? *Am J Med Genet* **27**:907–912, 1987.

114b. Viljoen D et al: Osteogenesis imperfecta with congenital joint contractures (Bruck syndrome). *Clin Genet* **36**:122–126, 1989.

115. Vrolik W: Tabulae ad illustrandum embryogenesim hominis et mammalium, tam naturatem quam abnormen, Amsterdam, 1949. Cited in Weil UH: Osteogenesis imperfecta: Historical background. *Clin Orthop Rel Res* **159**:6–10, 1981.

116. Wallis G et al: Mutations linked to the pro $\alpha$2(I) collagen gene are responsible for several cases of osteogenesis imperfecta type I. *J Med Genet* **23**:411–416, 1986.

117. Wenstrup RJ et al: Osteogenesis imperfecta type IV. Biochemical confirmation of genetic linkage to the pro $\alpha$2(I) gene of type I collagen. *J Clin Invest* **78**:1449–1455, 1986.

118. White NJ et al: Cardiovascular abnormalities in osteogenesis imperfecta. *Am Heart J* **106**:1416–1420, 1983.

118a. Willing MC et al: Heterozygosity for a large deletion in the $\alpha$2(I) collagen gene (COL1A2) has a dramatic effect on type I collagen secretion and produces perinatal lethal osteogenesis imperfecta. *J Biol Chem* **263**:8398–8404, 1988.

119. Witkop CJ Jr, Rao S: Inherited defects in tooth structure. *Birth Defects* **7**(7):153–184, 1971.

120. Young ID et al: Osteogenesis imperfecta typa IIA: Evidence for dominant inheritance. *J Med Genet* **24**:386–389, 1987.

# Chapter 7
# Syndromes Affecting Bone: Chondrodysplasias and Chondrodystrophies

## Chondrodysplasias and chondrodystrophies

We cannot emphasize strongly enough the need for radiographic examination of lethal bone dysplasias. There surely must be in excess of 100 or more diseases that fall in this category. The minimum number of radiographic views which should be taken include AP and lateral views of the whole body (babygram), lateral skull, and hand and foot views.

At autopsy, trachea, vertebrae, costochondral junction, and the head of the humerus and femur should be obtained (1,2).

### References (Chondrodysplasias and chondrodystrophies)


1. Koslowski K: The radiographic clues in the diagnosis of bone dysplasia. *Pediatr Radiol* **15**:1–3, 1985.
2. Yang SS et al: Lethal short-limbed chondrodysplasia in early infancy. In *Perspectives in Pediatric Pathology,* Rosenberg HS, Bolande RP (eds), vol. 3, Yearbook Medical Publishers, 1976, Chicago, pp 1–40.


## Achondrogenesis

Achondrogenesis, separated in 1936 by Parenti (39), from achondroplasia, is a type of lethal chondrodysplasia characterized by short trunk, severe micromelia, normal size head, and specific skeletal changes. There are, however, earlier examples (12). The term achondrogenesis, first coined by Fraccaro (14) in 1952, represents a heterogeneity of at least four or more types* (Figs. 7–1 to 7–4) (27). There has been and still is considerable confusion regarding terminology. Camera et al (9) have suggested that Type I with dense bones be designated "pycnochondrogenesis," citing earlier examples (43). Borochowitz et al (7) indicated further heterogeneity of Type I with two distinct subgroups. We suggest that the reader peruse Whitley and Gorlin (47), Borochowitz et al (5), Yang et al (51), Chen et al (10), and Van der Harten et al (46) for discussion of nomenclature.

The three severe forms (Types I–III) of achondrogenesis have autosomal recessive inheritance. Where multiple sibs have been born, they have been of the same type within each of the several groups. All patients with hypochondrogenesis (Type IV) have been isolated examples (35, 46). They probably represent new autosomal dominant mutations (16,17). D Hollister (personal communication, 1989) indicates that type IV is due to a glycine to serine substitution at position 943 of the collagen molecule. Over 150 cases have been recorded in aggregate, and several large series have been analyzed. The frequency has been estimated at 1/40,000 births (4,44).

Achondrogenesis of the more severe types (Types I–III) is incompatible with life, half of the infants being stillborn and the rest succumbing within the first few hours. Mean weight for Type I patients is about 1200 g, whereas Type II infants average 2100 g. Infants with hypochondrogenesis (Type IV), the mildest form, have survived for as long as a few months, but no long-term survivors have been reported (5,20,34,35). There is a history of polyhydramnios in over 50% of Types I–III (42). Gestation is about 30 weeks in Type I and 35 weeks in Types II and III. Fetal hydrops is common in Types I and II.

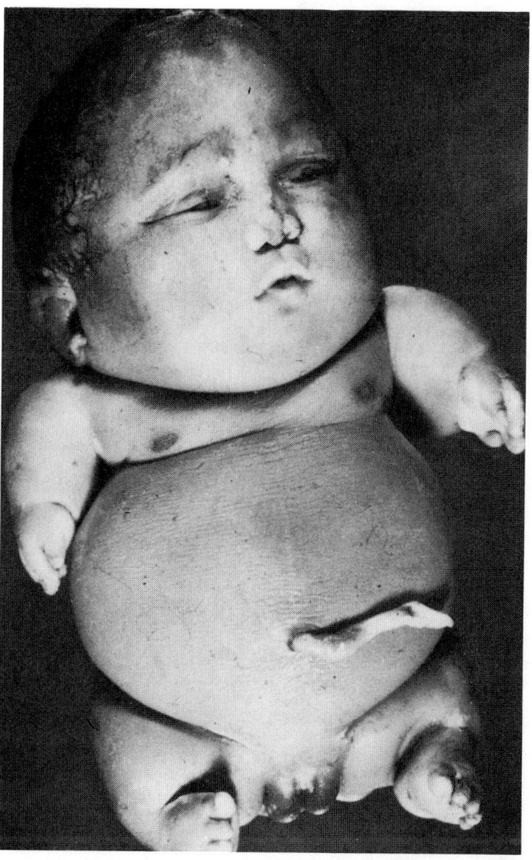

Fig. 7–1. *Achondrogenesis.* Phenotype of lethal dwarfism with disproportionately large head, disproportionately short limbs, and pot belly.

Facies. The head is disproportionately large relative to reduced neck, trunk, and limb length, causing the infant to be erroneously considered to have hydrocephaly. The head is often greater than 40% of body length and appears to sit on the chest. In Type I, the forehead slopes and the face appears puffy. The nose is small with anteverted nares and long philtrum, and there is retrognathia with double chin. Types II and III infants have a large prominent forehead, flat face, depressed nose with marked anteversion of nostrils, normal philtrum, and more normal chin. The neck is short in all types. The face in hypochondrogenesis is not remarkable.

*The terminology nomenclature we have used is different from that advocated by the International Nomenclature of Bone Dysplasias. Their terms Types IA and IB imply that these are variants of a single disorder. However, histopathologic and radiologic evidence clearly indicates otherwise. Therefore we have replaced their Type IA with Type I, their Type IB with Type II, and their Type II with Type III.

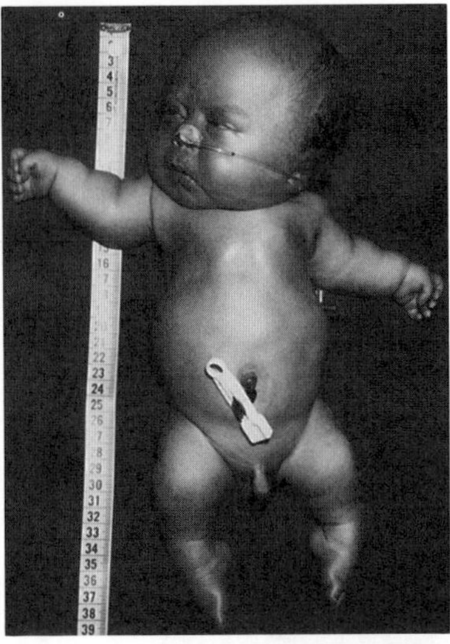

Fig. 7–2. *Achondrogenesis, Type IV.* Hypochondrogenesis. (From *G Hendrickx et al,* Eur J Pediatr **140**:278, 1983.)

Skeletal alterations. In Types I–III, the extremities are bowed, rarely exceeding 10 cm in length. Total body length seldom is greater than 36 cm at term (range 22–40 cm). The belly is greatly enlarged, partly from the short chest cavity and partly from hydrops.

General radiographic characteristics shared by the three severe types consist of marked underossification of vertebral bodies, sternum, ilia, ischia, pubic bones, talus, and calcaneus. The ribs are short and cupped with flared ends. Since failure of endochondral bone growth, manifest as reduced linear growth of long bones, appears to be a major defect in achondrogenesis, the "femoral cylinder index" (length/width) was proposed as means of distinguishing four types of achondrogenesis (46,37). We will discuss briefly the specific characteristic of each of the four major types.

**Type I (Houston–Harris).** In this form, also called Type IA, the calvaria is very poorly mineralized (Table 7–1). This form has marked limb shortening and striking multiple rib fractures. The long bones are very short, a femoral cylinder index of 1.0–2.8 being characteristic. The vertebral bodies are so inadequately mineralized that they may appear to be absent, particularly in the lower thoracic and lumbar spine (36,39,46,48,50). Multiple affected sibs have been reported (3,4,19,22,30,38,43,45,46).

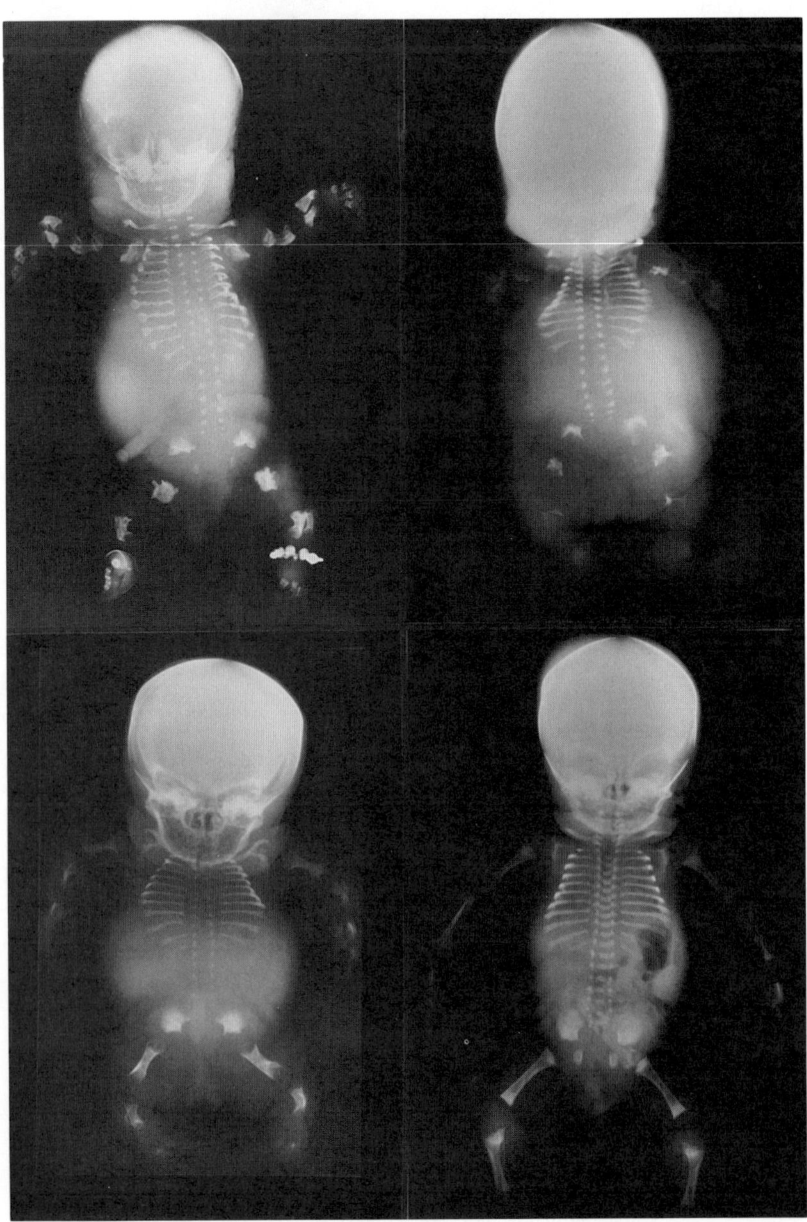

Fig. 7–3. *Roentgenograms of the four types of achondrogenesis.* Top left: Type I (Houston-Harris) with severe limb and multiple rib fractures. See also Fig. 7–4 top left. Top right: Type II (Fraccaro) with severe limb reduction without rib fractures. See also Fig. 7–4 top right. Bottom left: Type III (Langer–Saldino) with mushroom-stem femora, intermediate limb reduction, and halberd ilia. See also Fig. 7–4 bottom left. Bottom right: Type IV (hypochondrogenesis) with least severe limb reduction. See also Fig. 7–4 bottom right. (From *CB Whitley and RJ Gorlin,* Radiology **148:**693, 1983. Top left courtesy of *R Wapner and LG Jackson,* Philadelphia, Pennsylvania. Top right courtesy of *RM Saldino,* San Diego, California. Bottom left courtesy of *PE Andersen,* Odense, Denmark. Bottom right courtesy of *K Kozlowski,* Sydney, Australia.)

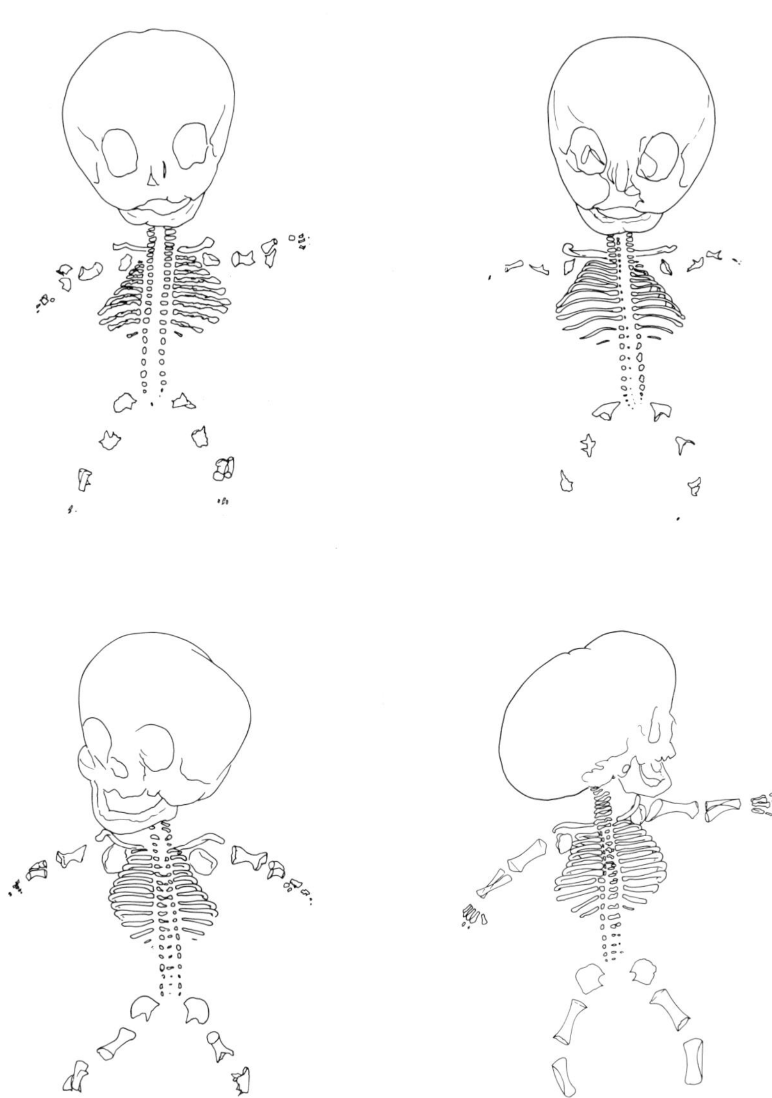

Fig. 7–4. *Diagrammatic sketches of* the *four types of achondrogenesis.* Top left: Type I (Houston–Harris). Top right: Type II (Fraccaro). Bottom left: Type III (Langer–Saldino). Bottom right: Type IV (hypochondrogenesis). (From *CB Whitley* and *RJ Gorlin,* Radiology **148:**693, 1983.)

**Type II (Fraccaro).** In this form, also called Type IB, there are crenated ilia and stellate long bones. The vertebral bodies exhibit minimal ossification; the pedicles are ossified. There are no rib fractures (8) (Table 7–1). Clinically, the limbs are shorter than those with Type I and the calvaria is ossified. A femoral cylinder index of 0.2–2.8 is observed in this group (46,47). Multiple affected sibs have been described (41,46,48).

**Type III (Langer–Saldino, severe type).** This form has also been called Type II. Infants have well ossified calvariae and exhibit better developed mushroom-stem femora and almost triangular halberd-shaped ilia (Table 7–1). Ossification of the spine is variable. A more mild defect in endochondral bone growth is reflected by a femoral cylinder index ranging from 2.0 to 4.9 (2,25,29,32,40,46,48,49). Ischial and pubic bones are unossified. Multiple affected sibs have been reported (11).

**Type IV (Hypochondrogenesis, Langer–Saldino, mild type).** By radiographic criteria, the same pattern of chondrodysplasia is observed, but to a relatively mild degree (Table 7–1). Thoracolumbar vertebral bodies are only slightly ossified and manifest as thin lamellae. Ossification is incomplete in the cervical, upper thoracic, and lower lumbar regions. The sternum is very short. The ilia are hypoplastic but more sculptured. Ischial ossification is variably present but pubic ossification is never observed. The sacrum is deficient. Limb bones are thick and short with metaphyseal widening and irregularity. There are no epiphyseal centers. Femoral cylinder index ranges from 4.9 to 8.0 (47) or as high as 10.7 (46). The distinction between Type III (Langer–Saldino) and Type IV (hypochondrogenesis) has been challenged on the basis of the apparent continuity of clinical, radiographic, and histopathologic changes in a large series of patients (5,35).

Pathologic findings. Histopathologic studies have demonstrated different patterns of morphologic abnormalities in the most severe types (5). In Type I the cartilage is hypercellular with clustered chondrocytes within a diffuse matrix. The resting chondrocytes contain PAS-positive, diastase-resistant, round to oval intracytoplasmic inclusions. The lacunae are dilated. There is defective column formation. Type II has more randomly dispersed chondrocytes. Their cytoplasm is vacuolated but there are no inclusions. The lacunae are not dilated. As in Type I, there is no columnization. In Types III and IV, the chondrocytes are densely packed with dilated lacunae (42a,46,52).

Studies (16,17,24,37) have implicated abnormal, poorly secreted Type II collagen in Type IV (hypochondrogenesis). Nonexpression of

Table 7–1. Achondrogenesis: Skeletal Changes

| | Type I (IA) Houston–Harris | Type II (IB) Fraccaro | Type III (II) Langer–Saldino, severe | Type IV[a] (II) Langer–Saldino, mild |
|---|---|---|---|---|
| Skull | Poorly ossified | Poorly ossified Occipital defect | Normally ossified Occipital defect | Normally ossified Occipital defect |
| Vertebral column | Bodies unossified Pedicles ossified only in cervical and upper thoracic region | Body minimally ossified Arches ossified | Bodies poorly ossified Arches ossified | Bodies flattened, but ossified |
| Ribs | Short with multiple fractures, cupped ends | Short without fractures, cupped ends, flared | Short, not flared | Short, not flared |
| Clavicles | Short, wide | Mildly elongated | Normal | Normal |
| Scapulae | Hypoplastic | Hypoplastic | Almost normal | Almost normal |
| Pelvis | Ilia crenated Ischia poorly ossified, widely spaced, low Pubic bones ossified | Ilia crenated Ischia and pubic bones unossified unossified | Ilia small and halberd-shaped Ischia and pubic bones unossified | Ilia better modeled, Ischia ossified Pubic bones unossified |
| Humerus | Short, stellate | Short, stellate | Short, flared, cupped | Short, rounded ends |
| Radius/ulna | Short, metaphyseal irregularities | Poorly ossified | Short, flared, cupped | Short, rounded ends |
| Femur | Proximal, metaphyseal spike, distal clubbing | Crenated, distal metaphyseal flare | Short, flared cupped | Short, rounded ends |
| Tibia–fibula | Short, wide, metaphyseal irregularities | Stellate | Short, flared at both ends, cupped | Short, rounded ends |
| Hands/feet | Unossified | Unossified | Short, talus and calcaneus unossified | Short, talus and calcaneus unossified |

[a]Mild cases indistinguishable radiographically and biochemically from hypochondrogenesis and SED congenita. International Nomenclature is in parentheses.

Type II collagen has been reported in Type II (Fraccaro) achondrogenesis (13).

**Other findings.** Other than hypoplastic lungs, the viscera are usually normal. PDA (11,30,41,48) and hydronephrosis (26,30), cryptorchidism (26,30,48), and inguinal hernia (25) have been noted.

**Oral manifestations.** Cleft palate has been observed in Type III cases (4,48,49), and in one-half of Type IV cases (1,5,35).

**Differential diagnosis.** The clinical appearance alone would allow the clinician to separate this disorder from *thanatophoric dysplasia, homozygous achondroplasia, atelosteogenesis Type II, hypophosphatasia,* and *fibrochondrogenesis.* Multiple rib fractures in Type II may suggest *osteogenesis imperfecta.* Differentiation from *Schneckenbecken dysplasia* is more difficult but can be made on the basis of the characteristic "snail back" iliac silhouette in the latter. Another unique disorder in the achondrogenesis spectrum has been identified (28) but is distinguished by more delicate and curved clavicles and other long bones. Hypochondrogenesis (Type IV) greatly resembles *spondyloepiphyseal dysplasia congenita.* There is recent evidence (16,17) that hypochondrogenesis shares with *SED congenita, Kniest dysplasia,* and *Stickler syndrome* defects in Type II collagen production.

**Laboratory aids.** Prenatal diagnosis of Types I and III has been made sonographically early in the second trimester by several investigators (3,11,15,18,22,41,43,46). Findings include polyhydramnios, hydrops, extremely short limbs, short trunk, prominent abdomen, cervical hygroma, and underossification of spine.

### References (Achondrogenesis)


1. Aylsworth AS et al: New observations in two types of lethal dwarfism: Achondrogenesis with cleft palate, and thanatophoric dysplasia with cloverleaf skull. *Proc Greenwood Genet Ctr* **5**:153, 1986.
2. Anderson PE: Achondrogenesis type II in twins. *Br J Radiol* **54**:61–65, 1981.
3. Anteby SO et al: Prenatal diagnosis of achondrogenesis. *Radiol Clin* **46**:109–114, 1977.
4. Bokesoy I et al: A case of achondrogenesis type I. *Hum Genet* **67**:349–350, 1984.
5. Borochowitz Z et al: Achondrogenesis II—hypochondrogenesis: Variability versus heterogeneity. *Am J Med Genet* **24**:273–288, 1986.
6. Borochowitz Z et al: Achondrogenesis type I—further heterogeneity. *Proc Greenwood Genet Ctr* **5**:157–186, 1986.
7. Borochowitz Z et al: Achondrogenesis, type I: Delineation of further heterogeneity and identification of two distinct subgroups. *J Pediatr* **112**:23–31, 1988.
8. Bremer K: Eine seltene Missbildung aus dem Formenkreis der Chondrodystrophie. *Z Geburtsch Gynäkol* **140**:198–202, 1953.
9. Camera G et al: Pycnochondrogenesis: An association of skeletal defects resembling achondrogenesis with generalized bone sclerosis: A new condition? *Clin Genet* **30**:335–337, 1986.
10. Chen H et al: Achondrogenesis: A review with special consideration of achondrogenesis type II (Langer–Saldino). *Am J Med Genet* **10**:379–394, 1981.
11. Curran JP et al: Lethal forms of chondrodysplastic dwarfism. *Pediatrics* **53**:76–85, 1974.
12. Donath J, Vogl A: Untersuchungen über den chondrophischen Zwergwuchs. *Wien Arch Intern Med* **10**:1–44, 1925 (Fig. 12).
13. Eyre DR et al: Nonexpression of cartilage type II collagen in a case of Langer–Saldino achondrogenesis. *Am J Hum Genet* **39**:52–67, 1986.
14. Fraccaro M: Contributo allo studio delle malattie del mesenchima osteopoietico l'achondrogenesi. *Folia Hered Path* **1**:190–207, 1952.

15. Garten KJ, Pulliam KP: Prenatal diagnosis of lethal short-limbed dwarfism. *J Diag Med Sonography* **1**:7–12, 1985.
16. Godfrey M, Hollister DW: Type II achondrogenesis—hypochondrogenesis: Identification of abnormal type II collagen. *Am J Hum Genet* **43**:904–913, 1988.
17. Godfrey M et al: Type I achondrogenesis—hypochondrogenesis: Morphologic and immunohistopathologic studies. *Am J Hum Genet* **43**:894–903, 1988.
18. Golbus MS et al: Prenatal diagnosis of achondrogenesis. *J Pediatr* **91**:464–466, 1977.
19. Harris R et al: Pseudo-achondrogenesis with fractures. *Clin Genet* **3**:435–441, 1972.
20. Hendrickx G et al: Hypochondrogenesis; an additional case. *Eur J Pediatr* **140**:278–281, 1983.
21. Horton WA et al: Achondrogenesis type II: Abnormalities of extracellular matrix. *Pediatr Res* **22**:324, 1987.
22. Houston S et al: Fatal neonatal dwarfism. *J Canad Assoc Radiol* **23**:45–61, 1976.
23. Hwong WH et al: The pathology of cartilage in chondrodysplasias. *J Pathol* **127**:11–18, 1979.
24. Keene GM et al: Type II achondrogenesis—hypochondrogenesis: Evidence for a molecular defect of Type II collagen. March of Dimes Clinical Genetics Conference, Baltimore, July 10–13, 1988.
25. Jiminez RP et al: Achondrogenesis. *Pediatrics* **51**:1087–1090, 1973.
26. Johnson VP et al: Midtrimester prenatal diagnosis of achondrogenesis. *J Ultrasound Med* **3**:223–226, 1984.
27. Kozlowski K et al: Neonatal death dwarfish (report of 17 cases). *Australas Radiol* **21**:164–183, 1977.
28. Kozlowski E et al: A new type of achondrogenesis. *Pediatr Radiol* **16**:430–342, 1986.
29. Langer LO Jr et al: Thanatophoric dwarfism: A condition confused with achondroplasia in the neonate with brief comments on achondrogenesis and homozygous achondroplasia. *Radiology* **92**:285–294, 1969.
30. Lauder J et al: Achondrogenesis type I, a familial subvariant. *Arch Dis Childh* **51**:550–557, 1976.
31. Laxova R et al: Family with probable achondrogenesis and lipid inclusions in fibroblasts. *Arch Dis Childh* **48**:212–216, 1973.
32. Legrand J: Un cliché de foetus achondroplastique in utero. *J Radiol Électrol* **37**:82–84, 1956.
33. MacHenry JC et al: Achondrogenesis. *Irish J Med Sci* **147**:404–406, 1978.
34. Macpherson RI, Wood BP: Spondylo-epiphyseal dysplasia congenita: A cause of lethal neonatal dwarfism. *Pediatr Radiol* **9**:217–224, 1980.
35. Maroteaux P et al: Hypochondrogenesis. *Eur J Pediatr* **141**:14–22, 1983.
36. Molz G, Spycher MA: Achondrogenesis type I: Light and electron-microscopic studies. *Eur J Pediatr* **134**:69–74, 1980.
37. Murray LW et al: Type II collagen defects in the chondrodysplasia. March of Dimes Clinical Genetics Conference, Baltimore, July 10–13, 1988.
38. Ornoy A et al: Achondrogenesis type I in three sibling fetuses. *Am J Pathol* **82**:71–84, 1976.
39. Parenti GC: La anosteogenesi: Una varieta della osteogenesi imperfetta. *Pathologica* **28**:447–462, 1936.
40. Rolland JC et al: L'achondrogénese: Aspects cliniques et radiologiques. *Ann Pédiatr* **20**:329–334, 1973.
41. Saldino RM: Lethal short-limbed dwarfism: Achondrogenesis and thanatophoric dwarfism. *Am J Roentgenol* **112**:185–197, 1971.
42. Schulte MJ et al: Lethale Achondrogenesis: eine Übersicht über 56 Fälle. *Klin Pädiatr* **190**:327–340, 1978.
42a. Sillence DO et al: Morphologic studies in the skeletal dysplasia. *Am J Pathol* **96**:813–870, 1979.
43. Smith WL et al: "In utero" diagnosis of achondrogenesis, type I. *Clin Genet* **19**:51–54, 1981.
44. Stoll C et al: Birth prevalence rates of skeletal dysplasia. *Clin Genet* **35**:88–92, 1989.
45. Urso FP, Urso MJ: Achondrogenesis in two sibs. *Birth Defects* **10**(12):11–18, 1974.
46. Van der Harten HJ et al: Achondrogenesis—hypochondrogenesis: The spectrum of chondrogenesis imperfecta. A radiological, ultrasonographic and histopathologic study of 23 cases. *Pediatr Pathol* **8**:571–597, 1988.
47. Whitley CB, Gorlin RJ: Achondrogenesis: New nosology with evidence of genetic heterogeneity. *Radiology* **148**:693–698, 1983.
48. Wiedemann H-R et al: Achondrogenesis within the scope of connately manifested generalized skeletal dysplasias. *Z Kinderheilkd* **116**:223–251, 1974.
49. Xanthakos UF, Riejent MM: Achondrogenesis: Case report and review of the literature. *J Pediatr* **82**:658–663, 1973.
50. Yang SS et al: Two types of heritable lethal achondrogenesis. *J Pediatr* **85**:796–801, 1974.
51. Yang SS et al: Proposed readjustment of eponyms for achondrogenesis. *J Pediatr* **88**:333–334, 1975.
52. Yang SS et al: Lethal short-limbed chondrodysplasia in early infancy. *Perspect Pediatr Pathol* **3**:1–40, 1976.

## Achondroplasia

The term achondroplasia was first used by Parrot (31) in 1878 to describe a rhizomelic form of short-limbed dwarfism associated with enlarged head, depressed nasal bridge, short stubby trident hands, lordotic lumbar spine, prominent buttocks, and protuberant abdomen (Figs. 7–5 to 7–7.). Achondroplasia is a misleading term because cartilage is, in fact, formed in the disorder; however, the term is well established. Until recently, a variety of chondrodysplasias were frequently confused with achondroplasia (19,30,45). The Egyptian gods, Bes and Ka, are depicted as having the disorder (43). The reader is referred to an excellent report on the 1986 International Symposium on Human Achondroplasia (28a).

More than 80% of recorded cases of achondroplasia are sporadic, representing new mutations. Increased paternal age at time of conception is associated with sporadic cases (28). Among the familial cases, autosomal dominant inheritance can be demonstrated (28).

The gene frequency of achondroplasia has been estimated as ranging between 1/16,000 and 1/35,000 (11,28,29,48a). Earlier ascertainments of frequency (25,48) were probably overestimates because other chondrodysplasias, in addition to achondroplasia, were undoubtedly included in these surveys (28,46).

Affected individuals are heterozygous for the achondroplasia gene. Presumed homozygosity has been reported in a few instances in which both parents were achondroplasts (32,42,57). Homozygous achondroplastic infants are more severely affected, clinically and radiologi-

Fig. 7–5. *Achondroplasia.* Note enlarged head with frontal bossing and low nasal bridge. Rhizomelia and short hands with trident deformity are evident.

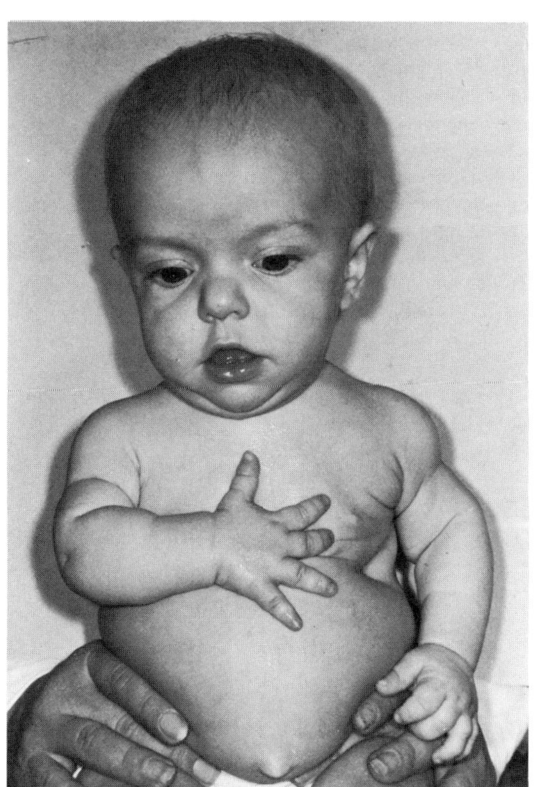

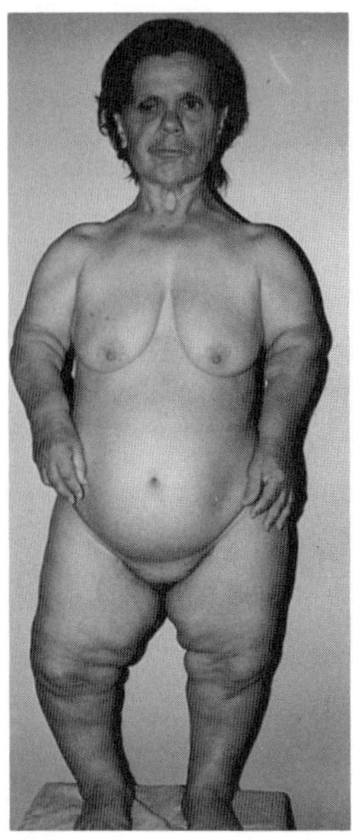
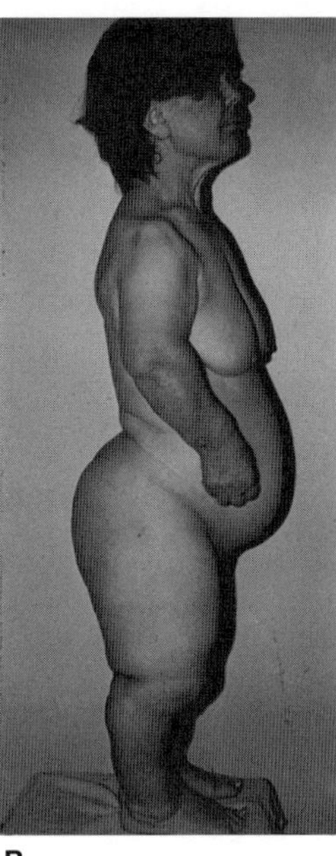

A B

Fig. 7–6. *Achondroplasia.* (A,B) Rhizomelia, lordotic lumbar spine, prominent buttocks, protuberant abdomen, and genua vara. (From *MM Cohen Jr,* The Child with Multiple Birth Defects, Raven Press, New York, 1982, p 103.)

cally, than are infants heterozygous for the disorder, and the condition is lethal during infancy (32), although aggressive respiratory and surgical measures kept one infant alive for several years (26a). Death appears to result from brain stem compression (16a). Deaths attributable to cardiovascular causes are increased in patients between 25 and 54 years of age (16b). The homozygous state resembles thanatophoric dysplasia in many respects but is still distinguishable.

Presumed examples of an autosomal recessive achondroplasia describe either a recessively inherited chondrodysplasia misdiagnosed as achondroplasia or are insufficiently documented to establish the diagnosis with certainty (23). Instances of affected sibs with normal parents can probably be explained by gonadal mosaicism (4,8,10,38). Cases of achondroplasia within the same kindred that seemingly do not show complete penetrance have been reported on rare occasions (30,37,53). Mosaic achondroplasia has been reported by Rimoin and McKusick (40). Achondroplasia and hypochondroplasia are allelic and an achondroplasia–hypochondroplasia compound has been reported (22,45a). The clinical and radiologic abnormalities were intermediate between those of heterozygous and homozygous achondroplasia and mental retardation was marked.

The basic defect is unknown. Achondroplasia has been shown not to be caused by a mutation in the gene for type II collagen (9a). Early histologic studies, which suggested gross disorganization of endochondral ossification, were misleading because they described patients with thanatophoric dysplasia, metatropic dysplasia, or achondrogenesis, rather than true achondroplasia (41). Rimoin and associates (41) found well-organized endochondral ossification with longitudinal columns of cartilage cells in chondroosseous rib junctions. Iliac crest cartilage was normal. These findings suggest that the abnormality in achondroplasia might be quantitative, affecting the rate of cartilage growth (41). However, Stanescu (47) found clusters of proliferative cells in biopsies of the tibial growth plate, rather than columns. The clusters were separated by wide septa of fibrous material. Ponseti and co-workers (24,36) also found abnormalities in the growth plates of the fibula, whereas the iliac crest cartilage and growth plate were nearly normal. Histologic, histochemical, ultrastructural, and biochemical studies of growth plates from different anatomic locations (including both weight-bearing and non-weight-bearing areas) in different age groups are necessary to resolve further the pathogenesis of achondroplasia.

Defective oxidative energy formation with decreased phosphorylation at the NADH dehydrogenase region of the terminal respiratory system has been demonstrated in achondroplastic muscle (21). Absent oxidative phosphorylation has been observed in homozygous achondroplasia (21a). A defect in peripheral glucose utilization has also been shown (7).

Mean birth lengths are 47.7 cm for males and 47.2 cm for females. Mean birth weights are 3500 g for males and 3150 g for females. Growth curves have been described by Horton et al (17). Final adult height is 130 cm for males and 123 cm for females. Mean adult weights are 55 kg for males and 46 kg for females. There is a tendency for obesity (16c).

Motor milestones are slow. Head control may not occur until 3 to 4 months and affected children may not walk until 24 to 36 months. Ultimately, however, development is normal.

Reproductive fitness is considerably reduced among achondroplastics because of social difficulties in finding mates and because of obstetrical problems of achondroplastic women (prematurity and the necessity for caesarean deliveries due to cephalopelvic disproportion (50).

**Facies and skull.** The head is enlarged, with frontal bossing and low nasal bridge (Fig. 7–8A,B). Occasionally, these features are not present at birth, but disproportionate growth of the head occurs during the first year of life and then parallels the normal curve (5,20,28a). Cephalometric analysis has been performed by Cohen et al (6) and Pederson (34).

**Central nervous system.** Intelligence is almost always normal, although acquisition of motor skills may be delayed because of the large head and short extremities (5,49). Mild ventricular dilatation has been reported by several authors (5,9,18,27,55). Gross mechanical block caused by obliteration of the basal cisterns, by obliteration at the level of the foramen magnum, or by kinking of the cerebral aqueduct has not been demonstrated in most cases (19). However, there is a relatively higher sudden unexpected death rate during infancy and early childhood (3,33). Cervicomedullary compression and its evaluation have been discussed by several authors (26a,36b,54a). Significant hydrocephaly (stepwise increase in the headgrowth slope) with neurologic signs and symptoms has occurred in a few instances (5,27,35,55) and is probably caused by cerebrospinal fluid obstruction at the level of the foramen magnum.

Most evidence to date seems to favor communicating hydrocephaly. James et al (18) demonstrated communicating hydrocephaly in two achondroplastic children with cisternography. Mueller et al (27) postulated two possibilities. First, early hydrocephaly may be caused by cerebrospinal fluid outlet obstruction resulting from a small posterior fossa that becomes compensated later in life secondary to bony structural maturation. Second, patients with achondroplasia may have obstruction of cerebrospinal fluid flow at the subarachnoid villi or in the venous sinuses secondary to retrograde pressure from marginal jugular veins. These could be compensated in size secondary to small jugular

Fig. 7–7. *Achondroplasia.* Small hands with short fingers. Note trident hand deformity.

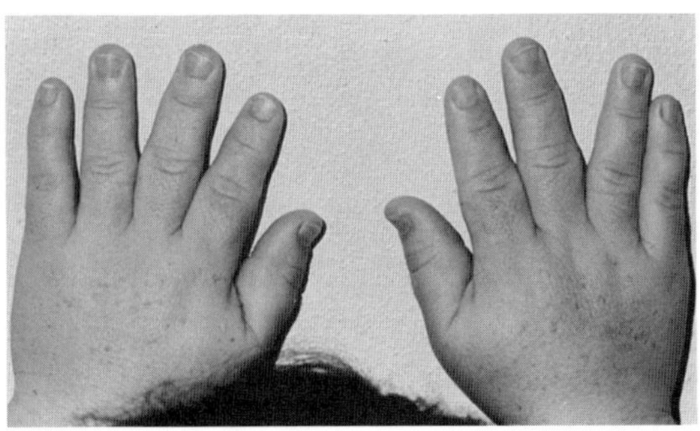

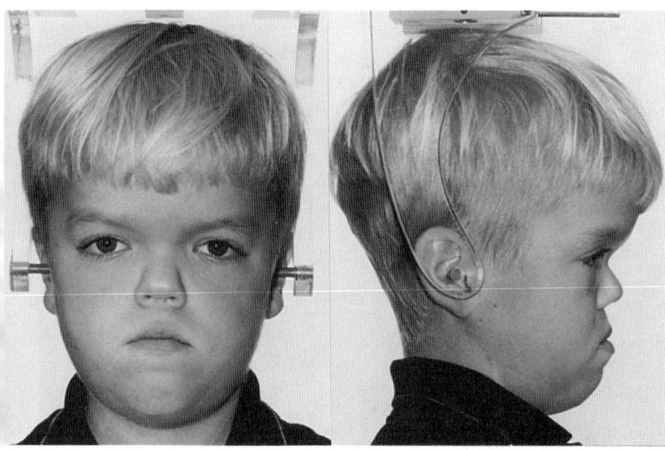


Fig. 7–8. *Achondroplasia.* (A,B) Craniofacial configuration showing enlarged calvaria with frontal bossing, low nasal bridge, and midface recession. [From *MM Cohen JR,* Mutations affecting craniofacial cartilage, in Cartilage: Biomedical Aspects, Vol. 3, Hall BK (ed), Academic Press, New York, 1983, p 191.]

foramina that resulted from faulty endochondral ossification in the posterior fossa.

Pierre-Kahn et al (35) studied hydrocephaly in 25 achondroplastic patients. They suggested that the hydrocephaly was related to constriction of the sigmoid sinus at the level of narrowed jugular foramina, resulting in a rise in intracranial venous pressure. They further noted that, in most instances, the hydrocephaly stabilized spontaneously in earlier life. Further studies are necessary to determine if the jugular foramina are small or if their emissary vein foramina are enlarged in achondroplasia.

The narrow spinal canal predisposes to neurologic complications with age. Compression of the spinal cord and nerve rootlets results from osteophytes, prolapsed intervertebral disks, or deformed vertebral bodies (16b,26,36a,51,52).

**Skeletal system.** Enlarged calvaria and basilar kyphosis are constant features. The anterior cranial base length is normal, but the posterior cranial base length is shorter than normal (6). The foramen magnum is small (16). The maxilla is hypoplastic resulting in midface deficiency and relative mandibular prognathism (6). The frontal and occipital bones and, in some cases, the temporal bones may be prominent (19,20). Partial occipitalization of the first cervical vertebra occurs in most cases.

The interpediculate distances progressively narrow from the upper to the lower lumbar spine, the pedicles are shortened in anteroposterior diameter, the posterior aspect of the vertebral bodies is concave, and the bony spinal canal diameters are decreased, particularly in the lumbar region (Fig. 7–9). Anterior wedging of the vertebral bodies (particularly in the region of the thoracolumbar junction) with resultant kyphosis may be prominent (19,20,44). Kyphosis occurs in about 10% and scoliosis in 7% (56). A thoracolumbar gibbus is more common in South African achondroplastic patients (2).

The lumbar spine appears to articulate low in relation to the crests of the iliac bone. The sacrum is narrow and horizontally oriented. The pelvis is broad and short. Narrowing of the pelvic inlet prevents vaginal delivery in pregnant achondroplastic females. The superior acetabular margins are oriented horizontally, and the sacrosciatic notch is acute (Figs. 7–9 and 7–10). The thoracic cage is relatively small in anteroposterior diameter (1,19,20). Legs are frequently bowed because of lax knee ligaments.

Limb bones are shortened in a rhizomelic pattern, which is more prominent in the upper extremities. There is incomplete extension at the elbows. The carpus is relatively large. The metacarpals and phalanges, although shortened, are disproportionately large in relation to

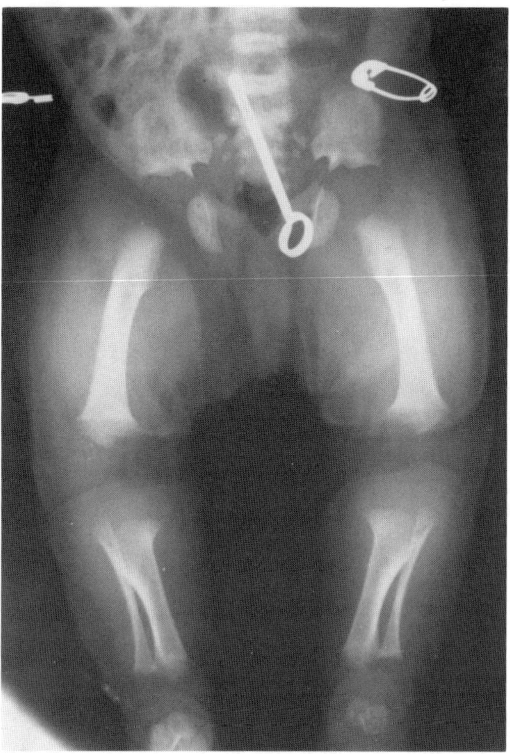

Fig. 7–9. *Achondroplasia.* Roentgenogram showing shortening of long bones, low articulation of lumbar spine in relation to iliac crest, short broad pelvis, and downward diminishing interpediculate distances in lumbar spine.

the humerus, radius, and ulna (19,20). Metacarpophalangeal relations have been described in detail (17b). Genua vara is found in 15% (56). The fibula is overlong at the ankle compared to the tibia, leading in some cases to varus foot deformity. There is limitation of elbow extension.

**Otolaryngologic findings.** Otitis media is likely common during the first 6 years of life, but its frequency, severity, and sequelae are not well documented. Hall (14) surveyed 150 achondroplasts over age 18; 75% had a history of ear infections and 11% indicated that they had a significant hearing loss. Among the 88 achondroplasts studied by Glass et al (12), 97% reported having had ear infection and/or hearing loss; on audiometric testing, 72% had a hearing loss of 22 dB or greater. In two studies of limited series of achondroplasts, most of whom were younger than 30, conductive hearing loss was noted in half; sensorineural and mixed hearing losses were also noted, but less frequently (12, 28a, R Young and T Davis, 1983, unpublished data). Progressive otosclerosis was documented by Carlin et al (4a).

**Differential diagnosis.** Achondroplasia should be distinguished from *achondrogenesis, thanatophoric dysplasia, Ellis–van Creveld syndrome,* metatropic dysplasia, *diastrophic dysplasia,* asphyxiating thoracic dystrophy, *hypochondroplasia,* pseudoachondroplasia, *Nance–Sweeney chondrodysplasia,* Schmid type metaphyseal dysplasia, various spondylo-epiphyseal and spondylo-metaphyseal dysplasias, and other types of short-limbed dwarfism (13,19,20,22,23,45,54) (Fig. 7–10B–D). The changes that distinguish heterozygous achondroplasia from homozygous achondroplasia and thanatophoric dysplasia reside largely in the vertebral column, the pelvis, and the limb bones (Fib. 7-10E–F). The differences have been elegantly discussed by Pauli et al (32).

**Laboratory aids.** Roentgenographic studies permit differentiation from other forms of dwarfism that simulate achondroplasia (46). Prenatal diagnosis by ultrasound has not been possible (9,15), although

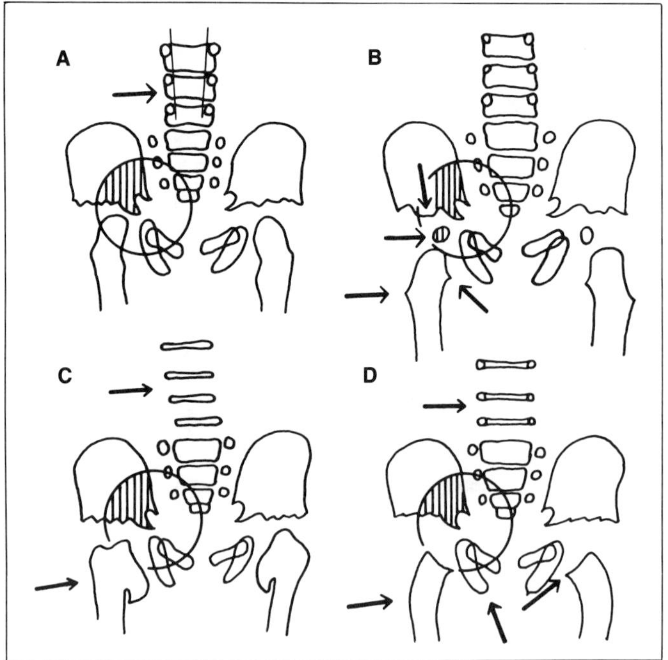


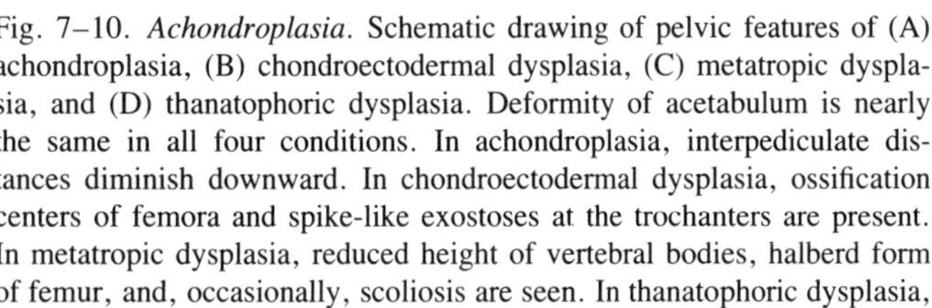


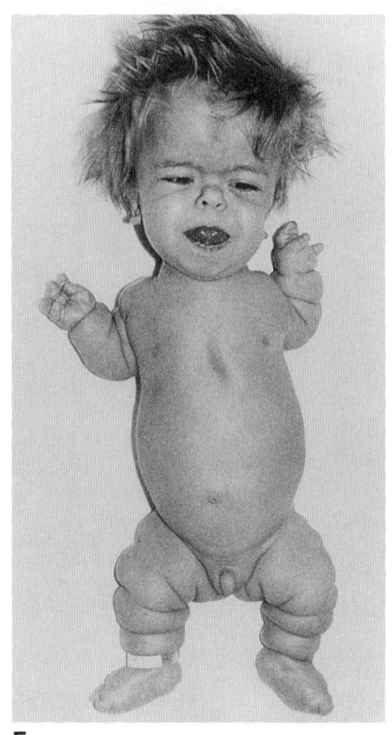


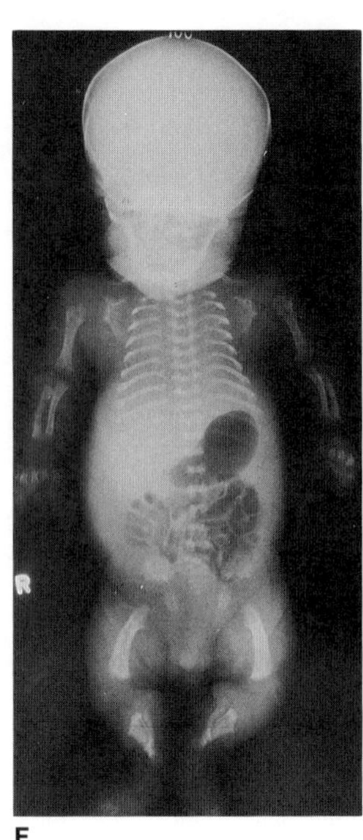


Fig. 7–10. *Achondroplasia.* Schematic drawing of pelvic features of (A) achondroplasia, (B) chondroectodermal dysplasia, (C) metatropic dysplasia, and (D) thanatophoric dysplasia. Deformity of acetabulum is nearly the same in all four conditions. In achondroplasia, interpediculate distances diminish downward. In chondroectodermal dysplasia, ossification centers of femora and spike-like exostoses at the trochanters are present. In metatropic dysplasia, reduced height of vertebral bodies, halberd form of femur, and, occasionally, scoliosis are seen. In thanatophoric dysplasia, vertebral bodies are flat, spike-like exostoses are present at the os pubis and at the femur, and the femur is bowed. (E,F) *Homozygous achondroplasia.* In E, note large head, small extremities with redundant skin folds, and narrow rib cage. Radiograph (F) showing small thoracic cage, marked platyspondyly, small iliac bones with scalloped lower margins, and short tubular bones, especially the humeri and femora. (A–D from *K Gefferth,* Prog Pediatr Radiol **4:**137, 1973. E from *RM Pauli et al,* Am J Med Genet **16:**459, 1983. F courtesy of *SS Yang,* Royal Oak, Michigan.)

second trimester ultrasound studies have been recorded on occasion. Hummel et al (17a) suspected at 18 weeks gestation and confirmed at 24 weeks gestation the diagnosis of homozygous achondroplasia by ultrasonography.

## References (Achondroplasia)


1. Bailey JA: Orthopaedic aspects of achondroplasia. *J Bone Jt Surg* **52A**:1285–1301, 1970.

2. Beighton P, Bathfield CA: Gibbal achondroplasia. *J Bone Jt Surg* **63B**:328–329, 1981.

3. Bland JD, Emery JL: Unexpected death of children with achondroplasia after the prenatal period. *Dev Med Child Neurol* **24**:489–492, 1982.

4. Bowen P: Achondroplasia in two sisters with normal parents. *Birth Defects* **10**(12):31–36, 1974.

4a. Carlin ME et al: Does achondroplasia predispose to otosclerosis: March of Dimes Clinical Genetics Conference, Baltimore, July 10–13, 1988.

5. Cohen ME et al: Neurological abnormalities in achondroplastic children. *J Pediatr* **71**:367–376, 1967.

6. Cohen MM Jr et al: A morphometric analysis of the craniofacial configuration in achondroplasia. *J Craniofac Genet Dev Biol (Suppl)* **1**:139–165, 1985.

7. Collipp PJ et al: Abnormal glucose tolerance in children with achondroplasia. *Am J Dis Child* **124**:682–689, 1972.

8. David TJ: Germinal mosaicism. *Clin Genet* **26**:79–80, 1984.

9. Elejalde BR et al: Prenatal diagnosis in two pregnancies of an achondroplastic woman. *Am J Med Genet* **15**:437–439, 1983.

9a. Francomano CA, Pyeritz RE: Achondroplasia is not caused by mutation in the gene for Type II collagen. *Am J Med Genet* **29**:955–961, 1988.

10. Fryns JP et al: Germinal mosaicism in achondroplasia: A family with 3 affected siblings of normal parents. *Clin Genet* **24**:156–158, 1983.

11. Gardner RJM: A new estimate of the achondroplasia mutation rate. *Clin Genet* **11**:31–38, 1977.

12. Glass L et al: Audiologic findings of patients with achondroplasia. *Int J Pediatr Otorhinolaryngol* **3**:129–135, 1981.

13. Hall BD, Spranger J: Hypochondroplasia: Clinical and radiological aspects in 39 cases. *Radiology* **133**:95–100, 1979.

14. Hall JG: Unpublished survey of 150 patients with achondroplasia. 1974. [Quoted by Glass L et al (see Ref. 12)].

15. Hall JG et al: Failure of early prenatal diagnosis in classic achondroplasia. *Am J Med Genet* **3**:371–375, 1979.

16. Hecht JT et al: Computerized tomography of the foramen magnum: Achondroplastic values compared to normal standards. *Am J Med Genet* **20**:355–360, 1985.

16a. Hecht JT et al: Foramen magnum stenosis in homozygous achondroplasia. *Eur J Pediatr* **145**:545–547, 1986.

16b. Hecht JT et al: Mortality in achondroplasia. *Am J Hum Genet* **41**:454–464, 1987.

16c. Hecht JT et al: Obesity and achondroplasia. *Am J Med Genet* **31**:597–602, 1988.

17. Horton WA et al: Standard growth curves for achondroplasia. *J Pediatr* **93**:435–438, 1978.

17a. Hummel M et al: Prenatal diagnosis of homozygous achondroplasia. March of Dimes Clinical Genetics Conference, Baltimore, July 10–13, 1988.

17b. Ingemarsson S, Fenger K: Metacarphophalangeal relations in 21 Danish patients with achondroplasia. *Dan Med Bull* **35**:104–107, 1988.

18. James AE et al: Hydrocephalus in achondroplasia studied by cisternography. *Pediatrics* **49**:46–49, 1972.

19. Langer LO et al: Achondroplasia. *Am J Roentgenol* **100**:12–26, 1967.

20. Langer LO et al: Achondroplasia. Clinical radiologic features with comment on genetic implications. *Clin Pediatr* **7**:474–485, 1968.

21. Mackler B et al: Oxidative energy deficiency. II. Human achondroplasia. *Arch Biochem Biophys* **159**:885–888, 1973.

21a. Mackler B et al: Studies of human achondroplasia: Oxidative metabolism in tissue culture cells. *Teratology* **33**:9–13, 1986.

22. McKusick VA et al: Observations suggesting allelism of the achondroplasia and hypochondroplasia genes. *J Med Genet* **10**:11–16, 1973.

23. Maroteaux P, Lamy M: Achondroplasia in man and animals. *Clin Orthop* **33**:91–103, 1964.

24. Maynard JA et al: Histochemistry and ultrastructure of the growth plate in achondroplasia. *J Bone Jt Surg* **63A**:969–979, 1981.

25. Mørch ET: *Chondrodystrophic Dwarfs in Denmark.* Munksgaard, Copenhagen, 1941.
26. Morgan DF, Young RF: Spinal neurological complications of achondroplasia: Results of surgical treatment. *J Neurosurg* **52**:463–472, 1980.
26a. Moskowitz N et al: Foramen magnum decompression in an infant with homozygous achondroplasia. *J Neurosurg* **70**:126–128, 1989.
27. Mueller SM et al: Achondroplasia and hydrocephalus. *Neurology* **27**:430–434, 1977.
28. Murdock JL et al: Achondroplasia—a genetic and statistical survey. *Ann Hum Genet.* **33**:227–244, 1970.
28a. Nicoletti B et al (eds): Human achondroplasia. A multidisciplinary approach. First International Symposium on Human Achondroplasia, 1986, Rome, Italy, 1988. Plenum Press, New York and London.
29. Oberland F et al: Achondroplasia and hypochondroplasia: Comments on frequency, mutation rate, and radiological features in skull and spine. *J Med Genet* **16**:140–146, 1979.
30. Opitz JM: "Unstable premutation" in achondroplasia: Penetrance vs. phenotrance. *Am J Med Genet* **19**:251–254, 1984.
31. Parrot JM: Sur les malformations achondroplasiques et le dieu Ptah. *Bull Soc Anthropol (Paris)* **1**:296, 1878.
32. Pauli RM et al: Homozygous achondroplasia with survival beyond infancy. *Am J Med Genet* **16**:459–474, 1983.
33. Pauli RM et al: Apnea and sudden unexpected death in infants with achondroplasia. *J Pediatr* **104**:342–348, 1984.
34. Pedersen PV: *A Roentgenographic Cephalometric Survey of Cranial and Facial Structures in the Human Achondroplastic Dwarf.* Master of Science Thesis, University of Washington, 1970.
35. Pierre-Kahn A et al: Hydrocephalus and achondroplasia: A study of 24 observations. *Child's Brain* **7**:205–219, 1980.
36. Ponseti IV: Skeletal growth in achondroplasia. *J Bone Jt Surg* **52A**:701–716, 1970.
36a. Pyeritz RE et al: Thoracolumbosacral laminectomy in achondroplasia: Long-term results in 22 patients. *Am J Med Genet* **28**:433–444, 1987.
36b. Reid CS et al: Cervicomedullary compression in young patients with achondroplasia: Value of comprehensive neurologic and respiratory evaluation. *J Pediatr* **110**:522–530, 1987.
37. Reiser CA: *Search for Premutation in Achondroplasia and Hypochondroplasia.* MS Thesis, University of Wisconsin, 1980.
38. Reiser CA et al: Achondroplasia: Unexpected familial recurrence. *Am J Med Genet* **19**:245–250, 1984.
39. Rimoin DL: Histopathology and ultrastructure of cartilage in the chondrodystrophies. *Birth Defects* **10**(9):1–18, 1974.
40. Rimoin DL, McKusick VA: Somatic mosaicism in an achondroplastic dwarf. *Birth Defects* **5**(4):17–19, 1969.
41. Rimoin DL et al: Endochondral ossification in achondroplastic dwarfism. *N Engl J Med.* **283**:728–735, 1970.
42. Rogovits N et al: Homozygote Achondroplasie und thanatophorer Zwergwuchs-pränatal diagnostizerbare Skelettstörungen. *Geburtsch Frauenheilk* **32**:184–191, 1972.
43. Scott CI: Achondroplastic and hypochondroplastic dwarfism. *Clin Orthoped Rel Res* **114**:18–30, 1976.
44. Siebens AA et al: Curves of the achondroplastic spine: A new hypothesis. *Johns Hopkins Med J* **142**:205–210, 1978.
45. Silverman FN: A differential diagnosis of achondroplasia. *Radiol Clin North Am* **6**:223–237, 1968.
45a. Sommer A et al: Achondroplasia–hypochondroplasia complex, *Am J Med Genet* **26**:949–958, 1987.
46. Spranger J et al: *Bone Dysplasias.* Gustav Fischer-Verlag, Stuttgart, 1974.
47. Stanescu V: Study of bone growth. *N Engl J Med* **284**:110–111, 1971 (letter).
48. Stevenson AC: Achondroplasia: An account of the condition in Northern Ireland. *Am J Hum Genet* **9**:81–91, 1957.
48a. Stoll C et al: Birth prevalence rates of skeletal dysplasias. *Clin Genet* **35**:88–92, 1989.
49. Todorov AB et al: Developmental screening tests in achondroplastic children. *Am J Med Genet* **9**:19–23, 1981.
50. Tyson JE et al: Obstetric and gynecologic considerations of dwarfism. *Am J Obstet Gynecol* **108**:688–704, 1970.
51. Vogl A: The fate of the achondroplastic dwarf (neurological complications of achondroplasia). *Exp Med Surg* **20**:108–117, 1962.
52. Vogl A, Osborne RL: Lesions of the spinal cord (transverse myelopathy) in achondroplasia. *Arch Neurol Psychiatr (Chic)* **61**:644–662, 1949.
53. Wadia R: Achondroplasia in two first cousins. *Birth Defects* **5**(4):227–229, 1969.
54. Walker BA et al: Hypochondroplasia. *Am J Dis Child* **122**:95–104, 1971.
54a. Wassman ER Jr, Rimoin DL: Cervicomedullary compression with achondroplasia. *J Pediatr* **113**:411, 1988.
55. Wise BL et al: Achondroplasia and hydrocephalus. *Neuropädiatrie* **3**:106–113, 1971.
56. Wynne-Davies R et al: Achondroplasia and hypochondroplasia: Clinical variation and spinal stenosis. *J Bone Jt Surg* **63B**:508–515, 1981.
57. Yang SS et al: Upper cervical myelopathy in achondroplasia. *Am J Clin Pathol* **68**:68–72, 1977.

## Hypochondroplasia

Hypochondroplasia is a common form of disproportionate short stature (Fig. 7–11A,B) with relatively few clinical manifestations, but with radiographic features similar to those found in achondroplasia, although milder in degree. Over a hundred cases have been recorded to date (1,3–8,10,12–15,18,19) and several excellent reviews are available (6,18). Hypochondroplasia and achondroplasia are allelic (9,11,16). Inheritance is autosomal dominant. Most instances represent new mutations, although familial instances have been encountered (18).

Diagnosis is difficult and is often by exclusion. It is not commonly made in the newborn period, although Hall and Spranger (6) noted that 66% are macrocephalic at birth. Short stature is not usually recognized until approximately 22 months of age. Final height attainment varies between 132 and 147 cm (1,6,18). Clinical and radiographic features are summarized in Table 7–2.

The skull may be rectangular in shape with a slightly prominent forehead. The facial appearance is normal. Macrocephaly is present in 57% of all cases and in two-thirds of newborns. Mental retardation is found in approximately 10% (6).

Bowlegs appear in early childhood, but tend to straighten spontaneously with age. The limbs are disproportionately short (100%), elbow extension is limited (100%), and brachydactyly is mild to moderate (97%). Lumbar lordosis is observed in about 34%. Mild joint pain on exercise is often observed in adults (1,18).

Radiographically, the interpediculate distances are moderately narrowed, the pedicles are shortened anteroposteriorly, and the vertebral

Fig. 7–11. *Hypochondroplasia.* (A,B) Disproportionately short limbs. (A courtesy of *DL Rimoin,* Los Angeles, California. B from *MM Cohen Jr,* The Child With Multiple Birth Defects, Raven Press, New York, 1982, p 103.)

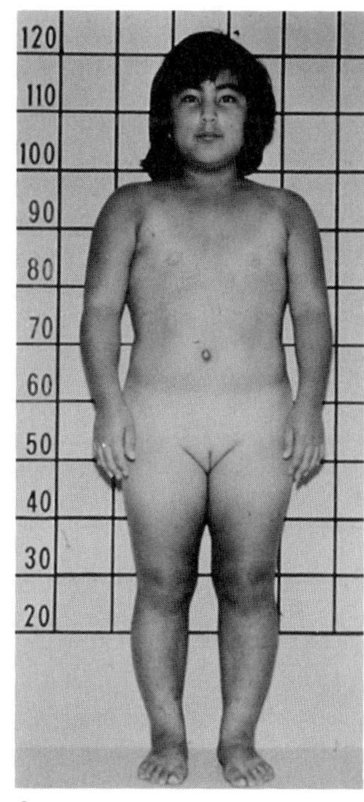


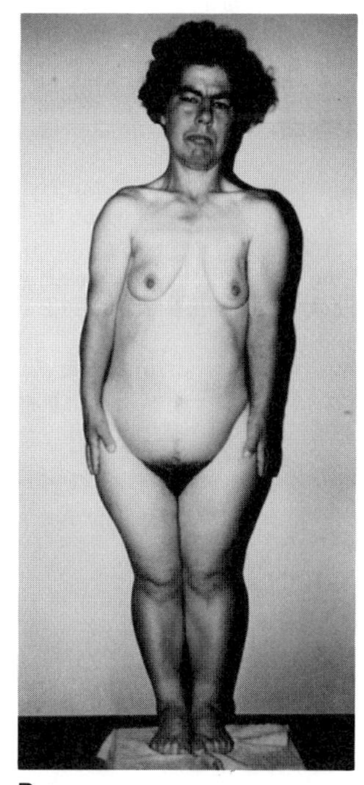

A B

Table 7–2. Clinical and radiographic features of hypochondroplasia

| Striking features | Approximate percentage |
|---|---|
| Clinical | |
| Macrocephaly | 57 |
| Mental retardation | 10 |
| Disproportionately short limbs | 100 |
| Limited elbow extension | 100 |
| Mild-to-moderate brachydactyly | 97 |
| Lumbar lordosis | 34 |
| Tubular bones | |
| Short broad femoral neck | 92 |
| Short long bones with mild metaphyseal flare | 100 |
| Long distal portion of fibula | 92 |
| Long distal portion of ulna | 73 |
| Long ulnar styloid[a] | 68 |
| Lumbar spine | |
| Narrow or unchanged interpedicular distance | 80 |
| AP shortening of lumbar pedicles (lateral view) | 89 |
| Dorsal concavity (lateral view) | 81 |
| High vertebrae (lateral view) | 33 |
| Platyspondyly (lateral view) | 37 |
| Pelvis | |
| Squared shortened ilia | 100 |

Adapted from BD Hall and J Spranger, *Radiology* **133**:95–100, 1979 and NG Heselson et al, *Clin Radiol* **30**:79–85, 1979.
[a]Presence related to age.

bodies of the lumbar spine have an increased dorsal concavity. The tubular bones are shortened and relatively squared. The femoral neck is broad and short. The distal end of the fibula is elongated in relation to the tibia. The ilia are squared and shortened (1,6,10,12,18).

For pregnancy, Caesarean section may be necessary. Prenatal diagnosis has been reported in a fetus at risk for hypochondroplasia (17). Ultrasound examination at 22 weeks showed decreased length of limb bones by measurement.

Differential diagnosis includes *achondroplasia,* which is similar but much more severe both clinically and radiographically. In achondroplasia, both the craniofacial appearance and the pelvic configuration are very distinctive in contrast to hypochondroplasia. In achondroplasia–hypochondroplasia compound, clinical and radiographic findings are different from either achondroplasia or hypochondroplasia (16).

Long bone changes in hypochondroplasia may be similar to those in metaphyseal chondrodysplasia, Schmidt type, although the vertebral abnormalities of hypochondroplasia are not present in the Schmidt type.

Desch and Horton (2) reported an autosomal recessive bone dysplasia resembling hypochondroplasia. In the former condition, birth length was much less than normal, the interpediculate distances were normal, and the humeri were shortened with no significant shortening of the tibiae and the ulnae, no significant brachydactyly, and normal head circumference.

Finally, hypochondroplasia may be confused with familial short stature at the lower end of the normal curve in the general population.

#### References (Hypochondroplasia)


1. Beals RK: Hypochondroplasia: A report of five kindreds. *J Bone Jt Surg* **51A**:728–736, 1969.
2. Desch LW, Horton WA: An autosomal recessive bone dysplasia syndrome resembling hypochondroplasia. *Pediatrics* **75**:786–789, 1985.
3. Dorst JP: Hypochondroplasia. *Birth Defects* **5**(4):260–261, 1969.
4. Frydman M et al: The genetic entity of hypochondroplasia. *Clin Genet* **5**:223–229, 1974.
5. Glasgow JFT et al: Hypochondroplasia. *Arch Dis Childh* **53**:868–872, 1978.
6. Hall BD, Spranger J: Hypochondroplasia: Clinical and radiological aspects in 39 cases. *Radiology* **133**:95–100, 1979.
7. Hall JG: Hypochondroplasia. *Birth Defects* **5**(4):262–272, 1969.
8. Heselson NG et al: The radiographic manifestations of hypochondroplasia. *Clin Radiol* **30**:79–85, 1979.
9. Kelly TE: Probable case of achondroplasia–hypochondroplasia compound. *Birth Defects* **10**(12):360, 1974.
10. Kozlowski: K: Hypochondroplasia, in *Progress in Pediatric Radiology. Intrinsic Diseases of Bones,* Karger, Basel, 1973, Vol 4, pp 238–249.
11. McKusick VA et al: Observations suggesting allelism of the achondroplasia and hypochondroplasia genes. *J Med Genet* **10**:11–16, 1973.
12. Murdoch JL: Hypochondroplasia. *Birth Defects* **5**(4):273–276, 1969.
13. Newman DE, Dunbar JC: Hypochondroplasia. *J Can Assoc Radiol* **26**:95–103, 1975.
14. Oberklaid F et al: Achondroplasia and hypochondroplasia. *J Med Genet* **16**:140–16, 1979.
15. Remy J et al: L'hypochondroplasie. A propos de cinq observations. *Ann Radiol (Paris)* **16**:481–493, 1973.
16. Sommer A et al: Achondroplasia–hypochondroplasia complex. *Am J Med Genet* **26**:949–957, 1987.
17. Stoll C et al: Prenatal diagnosis of hypochondroplasia. *Prenatal Diagnosis* **5**:423–426, 1985.
18. Walker BA et al: Hypochondroplasia. *Am J Dis Child* **122**:95–104, 1971.
19. Wynne-Davies R et al: Achondroplasia and hypochondroplasia: Clinical variation and spinal stenosis. *J Bone Jt Surg* **63B**:508–515, 1981.


## Acromesomelic dysplasia

Maroteaux et al (12), in 1971, first defined a rare short-limbed dwarfism that they termed acromesomelic dysplasia. There were earlier recorded cases (4,10). An 11,000-year-old acromesomelic skeleton was exhumed in southern Italy (5).

Approximately 30 examples of the disorder have been documented (1–3,7,8,13–15).

Diagnosis is usually not made until about 2 years of age. Inheritance is probably autosomal recessive. There has been parental consanguinity (1,12) and sibs have been affected (2,7,10,12,14).

The disorder is characterized by disproportionate short stature. Adult height ranges from 94 to 123 cm. Puberty may be delayed. The fingers and toes are particularly abbreviated. The forearms are relatively shorter than the lower legs. The arms are often bowed whereas the legs are straight. The facies is characterized by frontal bossing, low nasal root, and slightly flattened midface (Figs. 7–12 and 7–13). Low thoracic kyphosis and lumbar hyperlordosis are common. Hyperlaxity of the joints, particularly of the hands and feet, is marked.

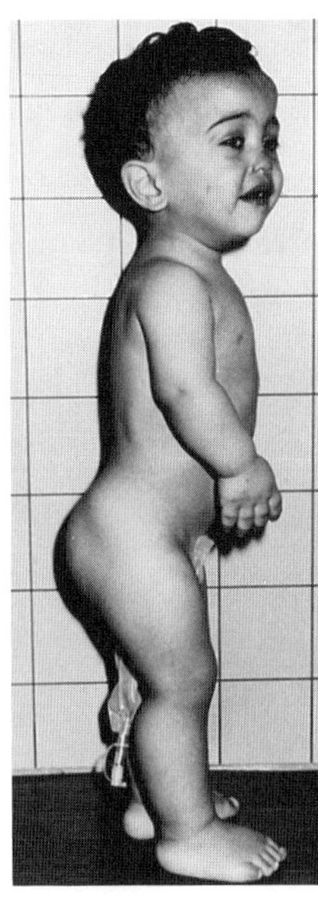

Fig. 7–12. *Acromesomelic dysplasia.* Relatively large head, frontal bossing, low nasal bridge, short forearms and hands, lumbar lordosis. (From *M Raes et al,* Helv Paediat Acta **40:**415, 1985.)

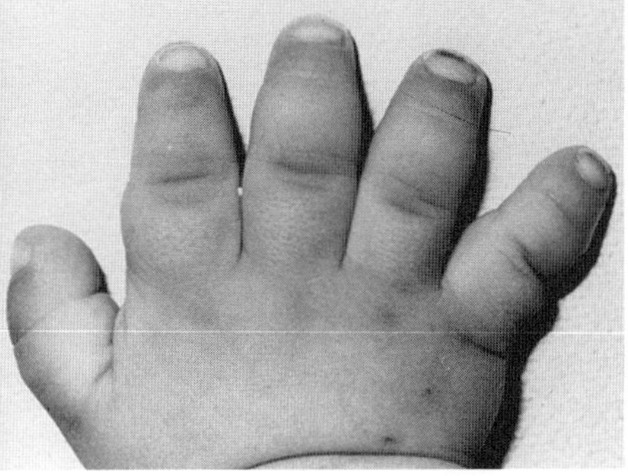

Fig. 7–13. *Acromesomelic dysplasia.* Broad hand with short stubby fingers. (From *M Raes et al,* Helv Paediatr Acta **40:**415, 1985.)

Fig. 7–14. *Acromesomelic dysplasia.* Short bent radius and ulna, increased radioulnar distance. (From *M Raes et al,* Helv Paediatr Acta **40:**415, 1985.)

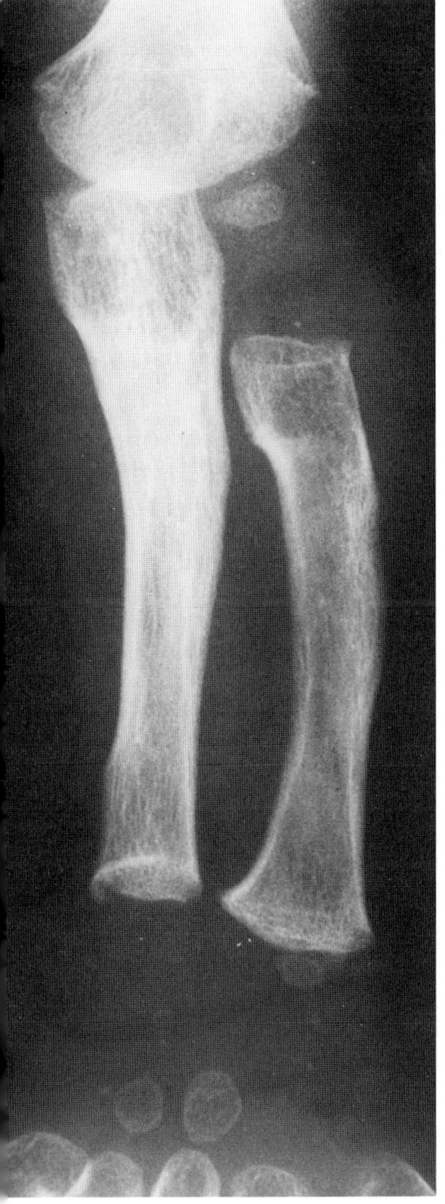

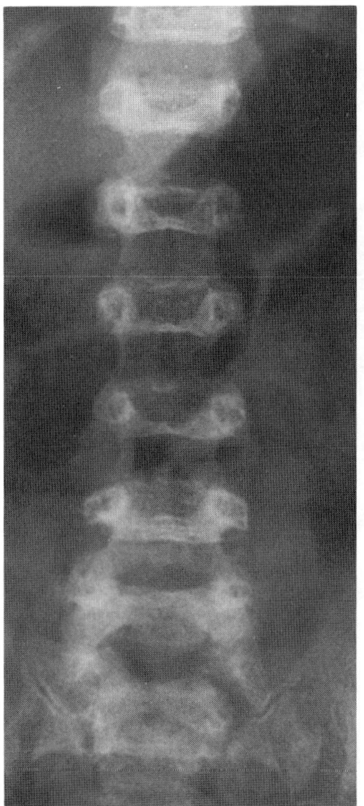

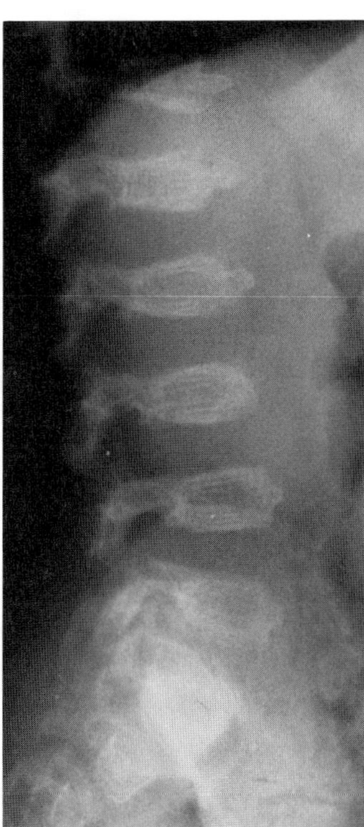

A B

Fig. 7–15. *Acromesomelic dysplasia.* (A,B) Vertebral bodies showing reduced vertical height and anterior beaking. (From *M Raes et al,* Helv Paediatr Acta **40:**415, 1985.)

Radiologic changes include frontal bossing and marked occipital prominence and generalized shortening of the long bones with metaphyseal flaring. The radii are bowed, and there is an increased distance between the ulnae and carpal bones. The metacarpals, metatarsals, and phalanges are extremely short, the latter being cone shaped. There is premature fusion between the epiphyses and metaphyses. The vertebral height is reduced, particularly in the thoracolumbar region with anterior beaking (Figs. 7–14 and 7–15A,B).

Differential diagnosis includes *pseudohypoparathyroidism* and pseudoachondroplasia. In the former, the vertebral changes seen in acromesomelic dysplasia are absent. The nose is smaller and mental retardation is common.

Several other unusual disorders with acromesomelia have been described. Israel and Vasan (6) noted a condition with acromesomelic dysplasia; bony abnormalities of the cervical spine, pelvis, ribs, and long bones; ASD, webbed esophagus; and stenosed larynx. The case was sporadic and chromosomes were normal. Langer et al (9) reported a severe form of acromesomelic dysplasia in which abnormalities were restricted to the limbs, the lower being most severely affected. Dislocations involved the hips and knees. They named the condition Hunter–Thompson type acromesomelic dysplasia. Inheritance is likely autosomal recessive. Leroy (11) observed two unrelated infants with acromesomelic dysplasia, growth and psychomotor retardation, synophrys, small nose, and flat face, giving a mild ''de Lange-like'' appearance.

### References (Acromesomelic dysplasia)


1. Borelli P et al: Acromesomelic dwarfism in a child with an interesting family history. *Pediatr Radiol* **13**:165–168, 1983.
2. Goodman RM et al: Peripheral dysostosis: An autosomal recessive form. *Birth Defects* **10**(12):137–146, 1974.
3. Hall CM et al: Acromesomelic dwarfism. *Br J Radiol* **53**:999–1003, 1980.
4. Hobaek A: *Problems in Hereditary Chondrodysplasias.* Oslo University Press, Oslo, 1961, pp 116–119.

5. Horgan J: Paleolithic compassion: Did tender loving care help a Stone Age dwarf to survive? *Sci Am* February 1988, 17–18.

6. Israel JN, Vasan U: A new syndrome of acromesomelic skeletal dysplasia. March of Dimes Clinical Genetics Conference, Baltimore, July 10–13, 1988.

7. Langer LO Jr, Garrett RT: Acromesomelic dysplasia. *Radiology* **137**:349–355, 1980.

8. Langer LO Jr et al: Acromesomelic dwarfism: Manifestations in childhood. *Am J Med Genet* **1**:87–100, 1977.

9. Langer LO Jr et al: A severe acromesomelic dysplasia, the Hunter–Thompson type and comparison with the Grebe type. Ninth Annual David W. Smith Workshop on Malformations and Morphogenesis, Oakland, August 3–7, 1988.

10. Lannois M: Deux cas de nanisme achondroplastique chez le frere et la soeur. *Lyon Méd* **98**:893–900, 1902.

11. Leroy JG: Growth and psychomotor retardation, synophrys and acromesomelic dysplasia: A syndrome in need of nosological definition. Ninth Annual David W. Smith Workshop on Malformations and Morphogenesis, Oakland, August 3–7, 1988.

12. Maroteaux P et al: Le nanisme acromésomelique. *Presse Méd* **79**:1839–1842, 1971.

13. Pallister PD: A 59 year old multiparous woman with acromesomelic dwarfism. *Am J Med Genet* **1**:343–346, 1978.

14. Pfeiffer RA: Akromesomeleter Zwergwuchs. *Roefo* **125**:171–173, 1976.

15. Raes M et al: A boy with acromesomelic dysplasia. *Helv Paediatr Acta* **40**:415–420, 1985.

## Atelosteogenesis type I

Atelosteogenesis, a term derived from the Greek referring to incomplete bone formation and coined by Maroteaux et al (3), is a rare neonatally lethal chondrodysplasia characterized by marked rhizomelic shortening of the limbs. A number of cases have been reported, all isolated examples (1–6). Hence, we do not know the heredity of the disorder. Each may represent a new autosomal dominant mutation or there may be an insufficient number of cases for affected sibs to have been born. Recently heterogeneity has been demonstrated, but several cases of Maroteaux et al (3) are hard to classify and there may be variable expression. The term *atelosteogenesis I* will here refer to the so-called classic disorder (Figs. 7–16 and 7–17). The disorder has also been called giant cell chondrodysplasia and spondylohumerofemoral hypoplasia (5,6).

Fig. 7–16. *Atelosteogenesis Type I.* Note low nasal bridge, rhizomelic shortness, and severe talipes equinovarus. (From *SS Yang et al,* Am J Med Genet **15:**615, 1983.)

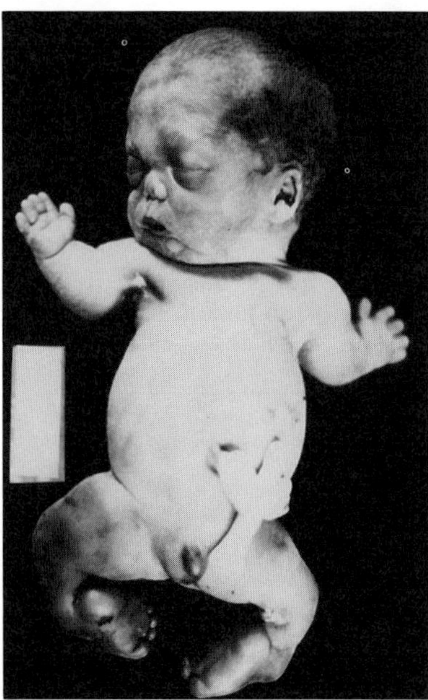

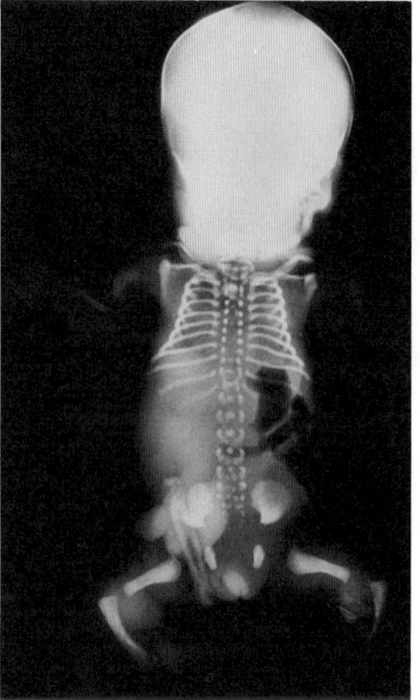

Fig. 7–17. *Atelosteogenesis Type I.* Severe ossification deficiency of vertebral bodies, eleven pairs of ribs, agenesis of fibulae. Humeri and femora are shortened with rounded proximal and tapered distal ends. (From *SS Yang et al,* Am J Med Genet **15:**615, 1983.)

The facies is characterized by frontal bossing, prominent globes, edematous eyelids, depressed nasal bridge, hypoplastic nose, micrognathia, and short neck (Fig. 7–16). All have cleft palate. Laryngeal stenosis has been described (6). The extremities are shortened rhizomelically and talipes is usually severe. Male infants have cryptorchidism. In some cases, there has been polyhydramnios.

Radiographically there is hypoplasia of the distal humeri and often of the distal femora. The fibulae may be hypoplastic or absent. In some examples, the forearm bones may be hypoplastic. The vertebral centra are hypoplastic with coronal clefts, and there is uneven ossification of most of the proximal and middle phalanges of the hands and feet. The pubic bones may be small but otherwise the pelvis is essentially normal (Fig. 7–17).

Histopathologic changes include clusters of chondrocytes surrounded by fibrous capsules and degenerated zones containing degenerated chondrocytes and copious amounts of metachromatic material in the epiphyses and basal zone of the growth plate (3). Multinucleated giants cells may be scattered throughout the resting cartilage (5,6). However, they are not specific for the disorder (4).

#### References (Atelosteogenesis type I)

1. Chervenak FA et al: Antenatal diagnosis of frontal cephalocele in a fetus with atelosteogenesis. *J Ultrasound Med* **5**:111–113, 1986.

1a. Kozlowski K, Bateson EM: Atelosteogenesis. *Roefo* **140**:224–225, 1984.

2. Kozlowski K et al: New forms of neonatal death dwarfism. Report of three cases. *Pediatr Radiol* **10**:155–160, 1981 (Case 3).

3. Maroteaux P et al: Atelosteogenesis. *Am J Med Genet* **13**:15–25, 1982.

4. Sillence D, Kozlowski K: "Giant cell" chondrodysplasia. *Am J Med Genet* **15**:637, 1983.

5. Sillence DO et al: Spondylohumerofemoral hypoplasia (giant cell chondrodysplasia): A neonatally lethal short-limb skeletal dysplasia. *Am J Med Genet* **13**:7–14, 1982.

6. Yang SS et al: Two lethal chondrodysplasias with giant chondrocytes (Case 1). *Am J Med Genet* **15**:615–625, 1983.

## Atelosteogenesis type II

McAlister et al (2), Sillence et al (3), and Whitley et al (4) reported sibs who died at birth with a new short-limbed dwarfism. The infant

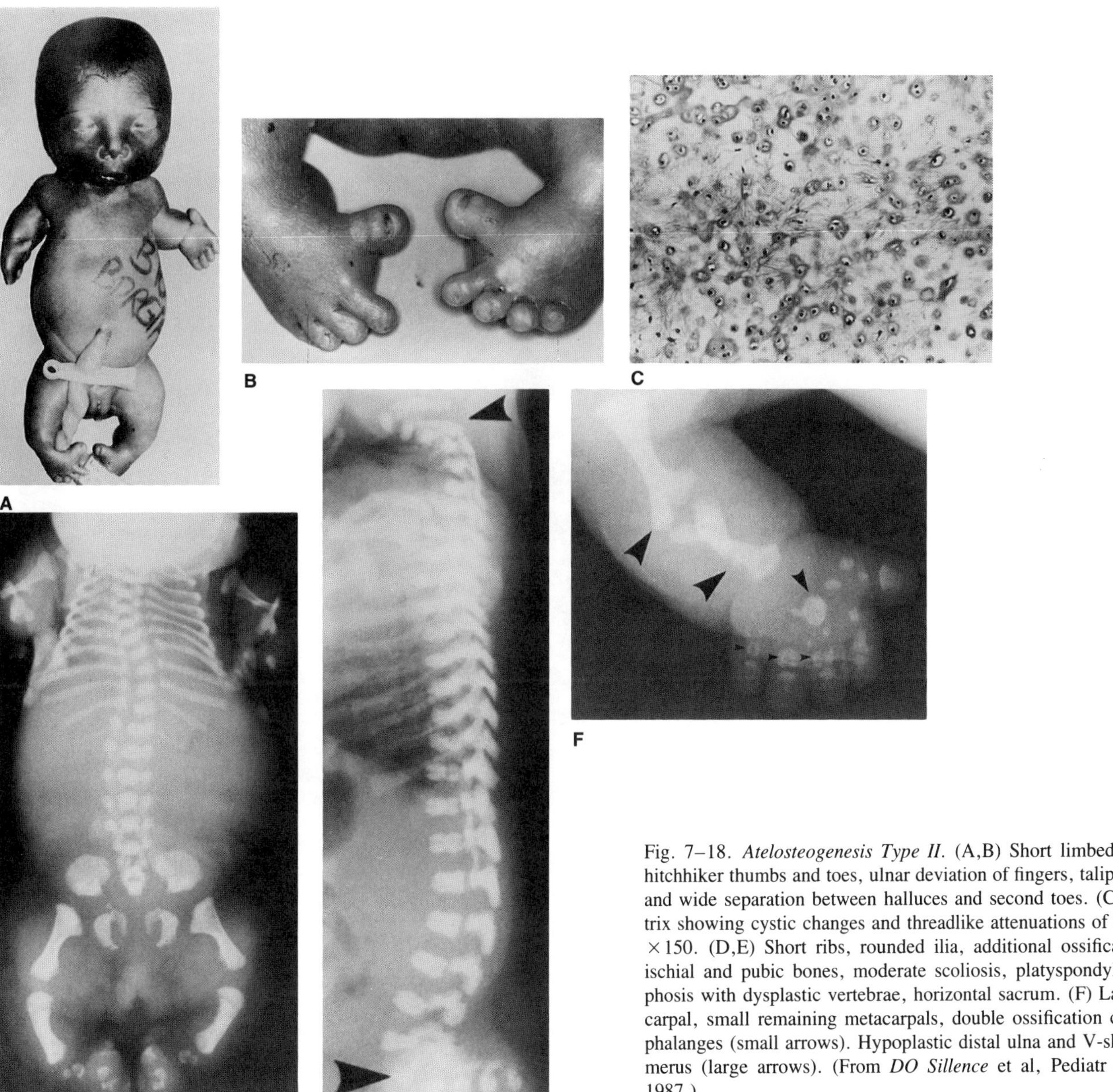


Fig. 7–18. *Atelosteogenesis Type II.* (A,B) Short limbed dysplasia with hitchhiker thumbs and toes, ulnar deviation of fingers, talipes equinovarus, and wide separation between halluces and second toes. (C) Cartilage matrix showing cystic changes and threadlike attenuations of matrix. H & E, ×150. (D,E) Short ribs, rounded ilia, additional ossification centers at ischial and pubic bones, moderate scoliosis, platyspondyly, cervical kyphosis with dysplastic vertebrae, horizontal sacrum. (F) Large third metacarpal, small remaining metacarpals, double ossification centers of some phalanges (small arrows). Hypoplastic distal ulna and V-shaped distal humerus (large arrows). (From *DO Sillence* et al, Pediatr Radiol **17**:112, 1987.)

described by Herzberg et al (1) probably has a different disorder. Sillence et al (3) suggested the term *atelosteogenesis II*. Inheritance is autosomal recessive.

The facies was characterized by frontal bossing, flat nasal bridge, and short neck. About 50% had cleft palate. The trunk and limbs were markedly shortened and incurved. The abdomen was protuberant. The thumbs were radially deviated and the halluces and second toes were widely separated. Feet were held in equinovarus position (Fig. 7–18A,B).

Radiographically, the vertebral bodies were moderately flattened. Some infants had mild scoliosis, but there was marked kyphosis of the cervical spine with hypoplastic/dysplastic changes and a horizontal sacrum. The iliac bones were rounded with shortened sacrosciatic notches and flat acetabular roofs. The ischial and pubic bones were well formed with additional ossification centers. The limb bones were short, particularly proximally, with metaphyseal flaring. The end of the distal humerus manifested a U- or V-shaped depression and the distal femur was rounded. The radius and tibia were particularly bowed. The distal ulna was hypoplastic and the proximal ulna was broad. The second and/or third metacarpals and the first and second metatarsals were larger than the remaining bones of the hand or foot. Some of the middle phalanges had double ossification centers (Fig. 7–18D–F). Bronchial rings were irregular in contour and, in some areas, there was increased perichondral fibrous tissue. The reserve zone of the cartilage was attenuated with many cystic areas containing only radiating threads of matrix (Fig. 7–18C) (2).

Because of the "hitchhiker thumb and hallux," *diastrophic dysplasia* must be excluded. *Atelosteogenesis Type I* can be excluded on histopathologic and radiologic grounds. In contrast, *Atelosteogenesis Type II* on lateral view shows cervical kyphosis and lumbosacral hyperlordosis. The pelvic shape in Type I is more nearly normal than in Type II. An unusual variant was reported by Herzberg et al (1).

## References (Atelosteogenesis type II)

1. Herzberg AJ et al: Brief clinical report: Variant of atelosteogenesis?: Report of a 20-week fetus. *Am J Med Genet* **29**:883–890, 1988.

2. McAlister WH et al: A new neonatal short limbed dwarfism. *Skel Radiol* **13**:271–275, 1985.

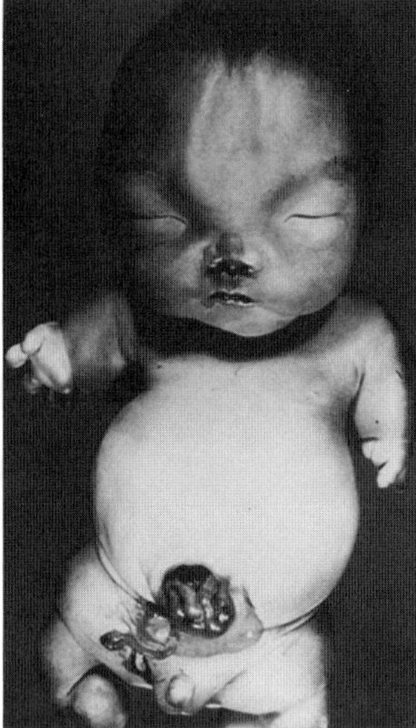

Fig. 7–19. *Boomerang dyslasia.* Facies similar to that seen in frontonasal malformation. In addition to short extremities, note four digits on each hand. (Courtesy of *Y Sugiura,* Nagoya, Japan.)

3. Sillence DO et al: Atelosteogenesis: Evidence for heterogeneity. *Pediatr Radiol* **17**:112–188, 1987.
4. Whitley CB et al: De la Chapelle dysplasia. *Am J Med Genet* **25**:29–35, 1986.

## Boomerang dysplasia

The term *boomerang dysplasia* was coined by Tenconi et al (3) to describe a lethal short-limbed dwarfism. The name is reflective of the shape of the femur, ulna, and tibia (Figs. 7–19 to 7–22). Another example was reported earlier (1), a third example was reported in 1985 (2), and a fourth by P Beighton (personal communication, 1989). All were isolated cases.

The head circumference was large, the eyes hyperteloric, and the forehead was full. The nasal root was broad, the nose small, the nostrils anteverted. The nasal septum and lateral cartilages were severely hypoplastic. The philtrum was prominent. The palate was cleft in one of the infants. Malar hypoplasia and micrognathia were present. The trunk was short, the abdomen prominent. The limbs, particularly the upper, exhibit rhizo- and mesomelic shortening with limitation of joint movement. The lower limbs were bowed anteriorly. The feet were in a calcaneovalgus position. There were soft tissue syndactyly of the third and fourth fingers, hypoplasia of the thumb nails, and partial duplication of the terminal phalanx of the index fingers.

Radiographically, the radii and fibulae were missing, the humeral ossification centers hypoplastic or missing, and the ulnae, femora, and tibiae were boomerang in shape. The metapodial bones were abnormal in form and the phalanges were not well ossified.

The bodies of the ilia were poorly formed and the pubic bones were absent. There were 13 pairs of ribs.

Giant chondrocytes were described (2).

Fig. 7–20. *Boomerang dysplasia.* Observe four toes on each foot. (Courtesy of *Y Sugiura,* Nagoya, Japan.)

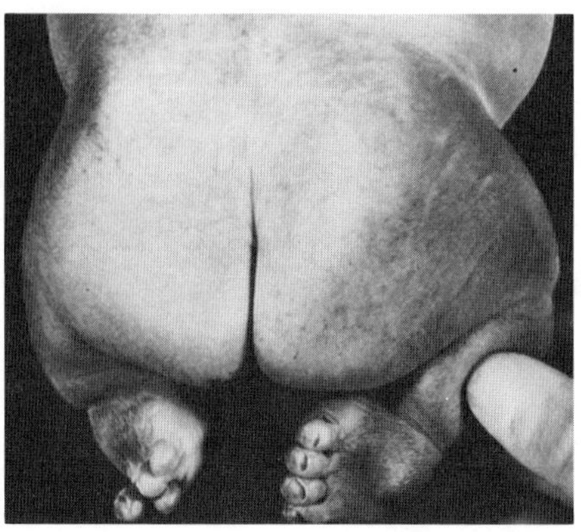

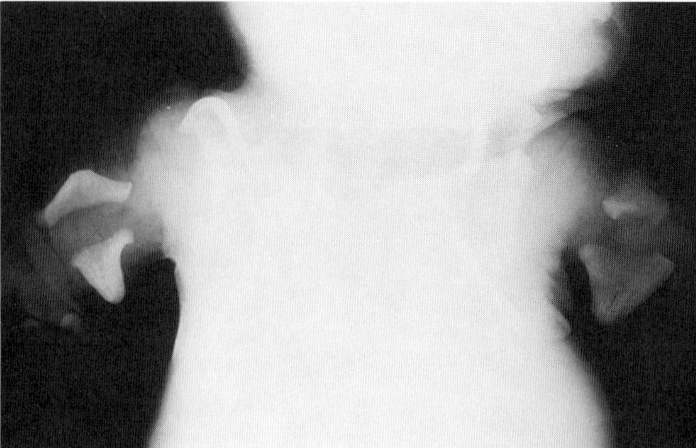

A

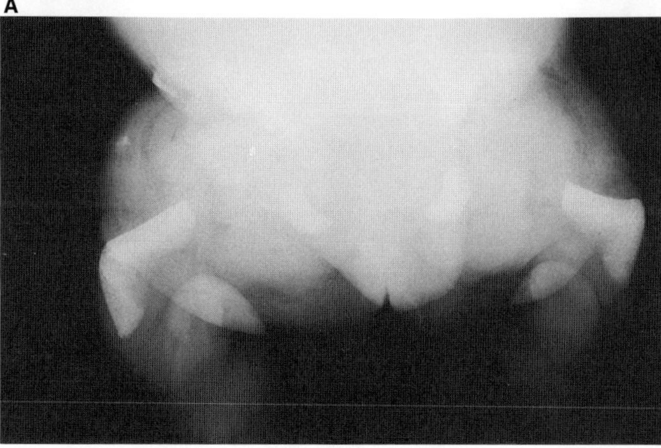

B

Fig. 7–21. *Boomerang dysplasia.* (A,B) Absent radii and fibulae with ulnae, femora, and tibiae boomerang in shape. Body of ilia poorly formed and absent pubic bones. (Courtesy of *Y Sugiura,* Nagoya, Japan.)

Fig. 7–22. *Boomerang dysplasia.* (A,B) Abnormal metapodial bones and poorly ossified phalanges. (From *K Kozlowski et al,* Br J Radiol **58**:369 1985.)

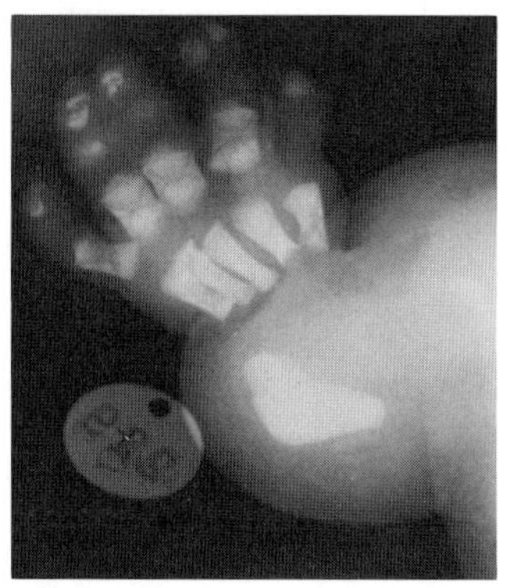

A

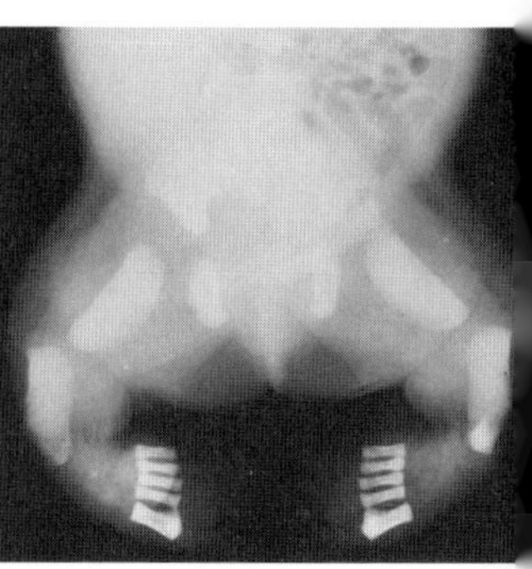

B

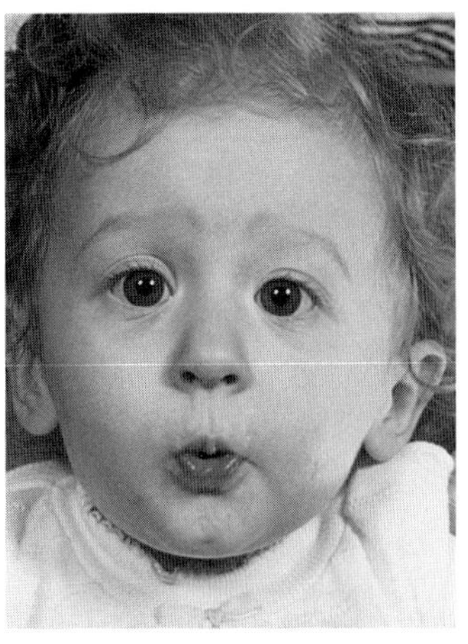
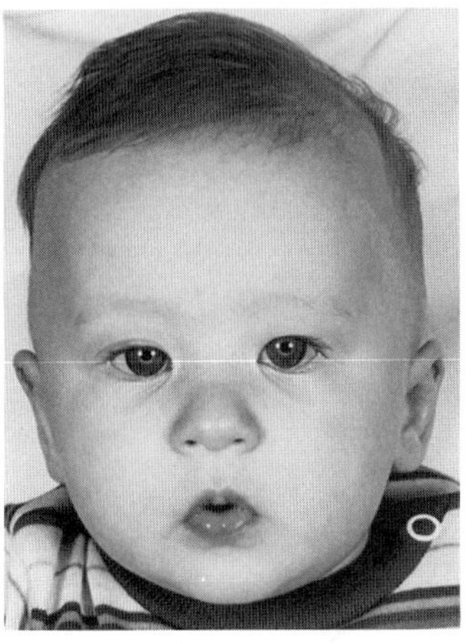

A B
Fig. 7–23. *Burton syndrome*. (A,B) Note pursed lips. (From *BK Burton et al*, J Pediatr **109:**642, 1986.)

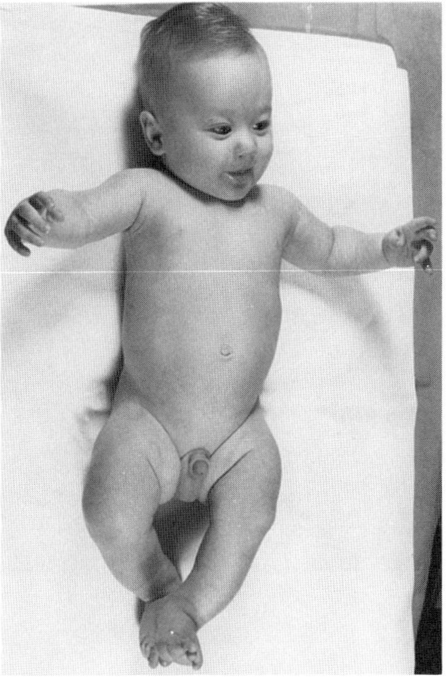

Fig. 7–25. *Burton syndrome*. Note short limbs. (From *BK Burton et al*, J Pediatr **109:**642, 1986.)

#### References (Boomerang dysplasia)

1. Kozlowski K et al: New forms of neonatal death dwarfism. *Pediatr Radiol* **10**:155–160, 1981.
2. Kozlowski K et al: Boomerang dysplasia. *Br J Radiol* **58**:369–371, 1985.
3. Tenconi R et al: Boomerang dysplasia: A new form of neonatal death dwarfism. *Roefo* **138**:378–380, 1983.

## Burton syndrome

In 1986 Burton et al (1) reported male and female sibs with a Kniest-like skeletal dysplasia. In addition, there were microstomia, pursed lips, and dislocated lenses (Figs. 7–23 to 7–28).

Growth, which was retarded, was at the 5th percentile. The face, other than having a small mouth with pursed lips, was not distinctive but looked somewhat like the *Schwartz–Jampel syndrome* facies. In both sibs, bilateral downward subluxation of the lenses was noted at about the age of 2 years.

The limbs were short and bowed, and the joints were stiff and enlarged.

Radiographically, the long bones were somewhat dumbbell-shaped with flared metaphyses and mildly shortened diaphyses. The thoracic and lumbar vertebral bodies were somewhat flattened. There was also cervical kyphosis and increased lumbosacral angle. The chest was rather bell shaped and the pelvis exhibited somewhat narrowed sacrosciatic notches with a notched lateral margin of the acetabula and wide ilia.

The cartilage showed scattered, dense patches within the matrix with disturbed column formation. The chondrocytes appeared large and mature rather than degenerated and typically hypertrophic. Within the scattered dense patches there were collagen bundles that were 10–30 times broader than normal.

Fig. 7–24. *Burton syndrome*. Downward subluxation of lens. (From *BK Burton et al*, J Pediatr **109:**642, 1986.)

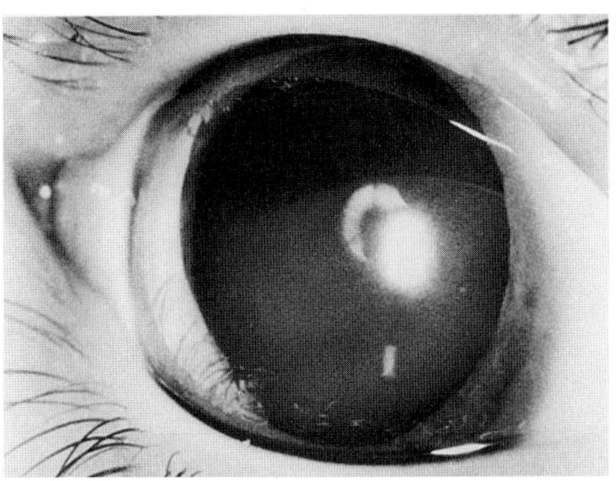

#### Reference (Burton syndrome)

1. Burton BK et al: A new skeletal dysplasia: Clinical, radiologic, and pathologic findings. *J Pediatr* **109**:642–648, 1986.

## Campomelic syndrome

Although described earlier by a number of authors (2,5,20), campomelic syndrome (Figs. 7–29 to 7–32A,B) was first recognized as an entity by Spranger et al (27) in 1970 and Maroteaux et al (18) in 1971. Excellent reviews of over 100 published cases have been done by Hall and Spranger (10), Beluffi and Fraccaro (4), and Houston et al (13). The disorder has been diagnosed prenatally (3,9,32). Prevalence has been estimated as 1/100,000 births (27a).

The inheritance pattern is possibly autosomal recessive (7,8,10). Consanguinity is somewhat increased and affected sibs have been described (6,7,19,26,29). The disorder has also been seen in monozygotic twins (21). The phenotypic sex ratio among 100 cases was 3F:1M, but karyotype analysis of 43 phenotypic females showed about half were sex-reversed 46,XY examples. The sex-reversed infants were H-Y antigen negative (6,12,14,22,23).

In about 50% of the cases, the child is either born dead or dies within a few hours. Nearly all have succumbed by 10 months of age. A few have lived for many years (13). Hearing loss and mental retardation are evident in nearly all those that survive (13). At least 85% exhibit respiratory distress as a result of small thoracic cage, narrow larynx, hypoplastic trachea, and, possibly, CNS-based hypotonia. Polyhydramnios, beginning at about 32 weeks, is common.

Frequent craniofacial features are macrocephaly, dolichocephaly, large anterior fontanel and sutures, disproportionately small face, short narrow palpebral fissures, apparent hypertelorism, flat nasal bridge, low-set cartilage-poor pinnae, small nose with anteverted nostrils, long philtrum, small mouth, retroglossia, micrognathia, and short neck with redundant skin. Cleft palate is present in at least 80% (4).

The long bones of the lower extremities are bent, but to varied

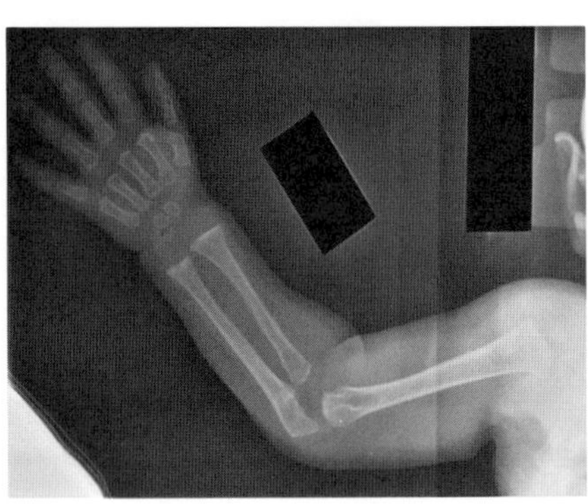

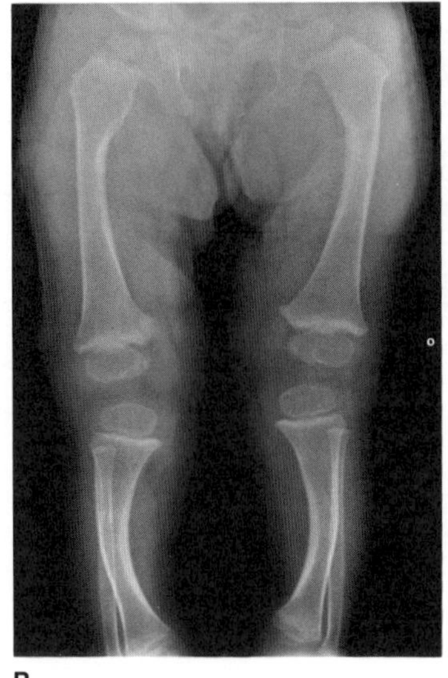

A B

Fig. 7–26. *Burton syndrome.* (A,B) Long bones somewhat dumbbell shaped with flared metaphyses and mildly shortened diaphyses. (From *BK Burton et al,* J Pediatr **109:**642, 1986.)

degrees. The genesis of bowing and shortening of the lower limbs has been discussed by Lazjuk et al (16a) and Pazzaglia and Beluffi (22a). The bones of the upper extremities are mildly bowed in about 25%. The elbows are occasionally dislocated. Pretibial skin dimples over the most convex site are found in about 90%. Talipes equinovarus is a common feature. There is often a wide space between the hallux and the second toe.

Radiographic changes include tall narrow orbits (90%), hypoplastic bladeless scapulae (90%), small bell-shaped chest (80%), nonmineralized sternum (80%), slender ribs (85%), 11 ribs (70%), slender trachea (70%), flattened and/or hypoplastic vertebral bodies (particularly cervical) with nonmineralized pedicles (80%), and kyphoscoliosis (70%). Bowed shortened tibias and femurs, hypoplastic fibulas, narrow iliac wings with increased acetabular angles, late developing pubic bones, vertical and widely spaced ischia, and dislocated hips are constant findings. The proximal tibial and distal femoral epiphyses are absent in 85%. The talus is nonmineralized in 75%. The hands exhibit clinodactyly, brachydactyly, and small middle phalanges in 70% (4,10).

Autopsy findings in about 60 cases showed absence or hypoplasia of olfactory tracts or bulbs (25%), hydrocephalus (25%), variable congenital heart anomalies [VSD, ASD, PDA, tetralogy of Fallot, stenosis of aortic isthmus (30%)], deficiency of laryngeal and tracheobronchial cartilages (40%), and hydroureter and hydronephrosis (30%). Renal hypoplasia is also found. Those with sex reversal are not always completely reversed; some have ambiguous genitalia (22). The inner ears showed no cartilage cells in the otic capsule and the ossicles were malformed (30).

Fig. 7–27. *Burton syndrome.* (A,B) Somewhat flattened thoracic and lumbar vertebral bodies. Chest is rather bell shaped and pelvis exhibits somewhat narrowed sacrosciatic notches with notched lateral margin of acetabula and wide ilia. Note increased lumbosacral angle. (From *BK Burton et al,* J Pediatr **109:**642, 1986.)

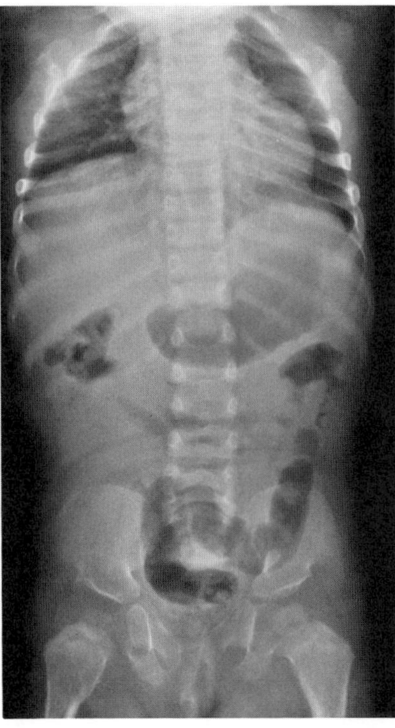

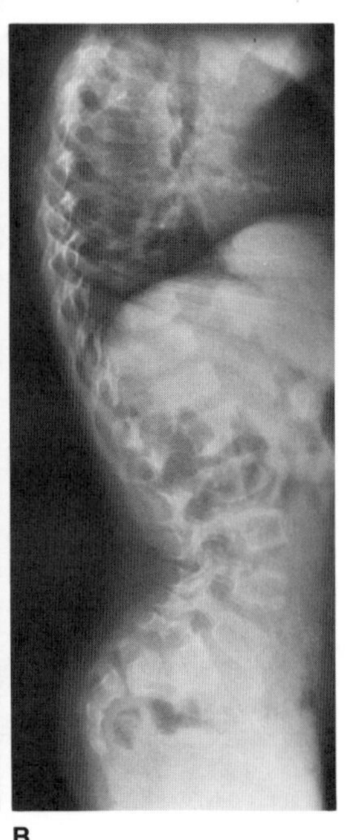

A B

Fig. 7–28. *Burton syndrome.* Cervical kyphosis. (From *BK Burton et al,* J Pediatr **109:**642, 1986.)

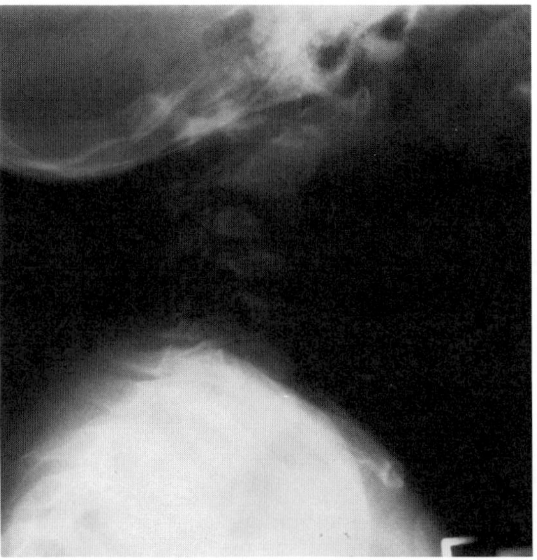

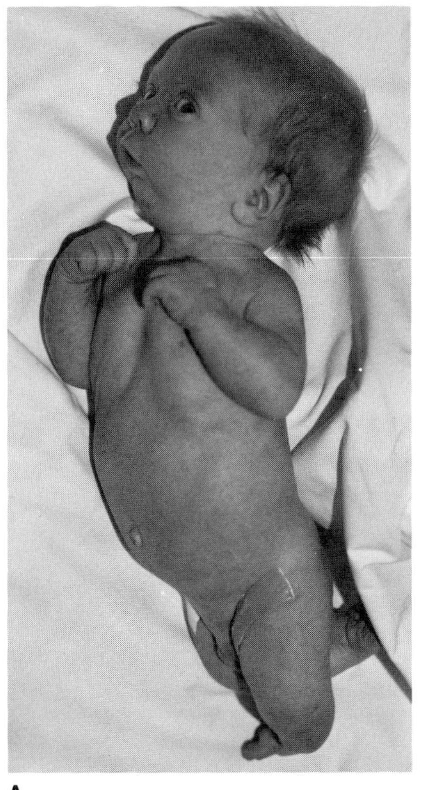
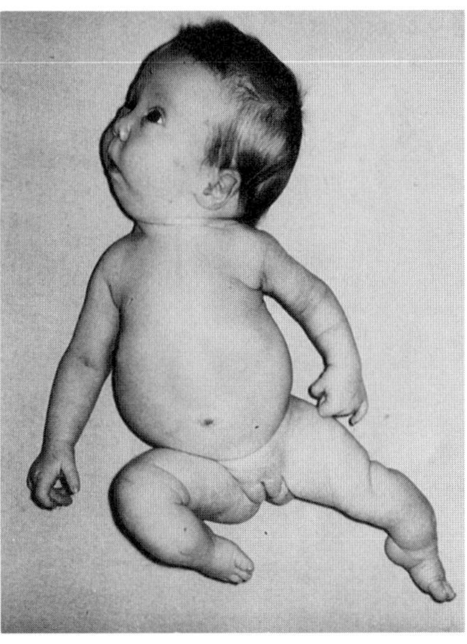
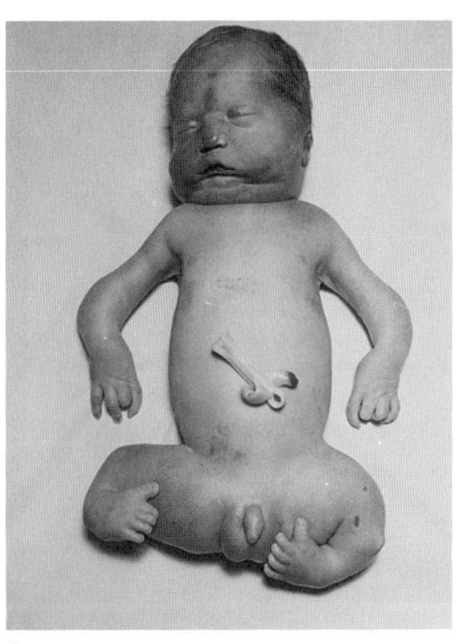

A B C

Fig. 7–29. *Campomelic syndrome*. (A–C) Note bent femora and tibiae, micrognathia. Note curved upper limbs and gap between great toe and second toe in C. (A courtesy of *JM Opitz*, Helena, Montana. B courtesy of *J Lindsten*, Stockholm, Sweden. C courtesy of *RE Stevenson*, Greenwood, South Carolina.)

**Differential diagnosis.** Kozlowski et al (16) described a few disorders with bent bones, but Hall and Spranger (11) listed almost 30 conditions having congenital bowing of long bones.

Kyphomelic dysplasia (Fig. 7–32C,D) is an autosomal recessively inherited disorder characterized by short stature with both short trunk and short limbs, small chest, narrow shoulders, pectus carinatum, limited joint movements, and skin dimples over the greater trochanters. Radiographically, the humeri, tibiae, and radii are mildly bowed with wide irregular rachitic-like metaphyses. The femora are very short and severely bowed. The ribs are short with flared ends and there may be 11 pair of ribs. Platyspondyly, increased acetabular angles, and underossification of proximal tibial epiphyses are less frequently encountered. These changes disappear by five years of age (9–11,15,17,24,28a, 31). The mandible remains small. Cleft palate is not seen. Intelligence is normal.

A somewhat heterogeneous group of disorders is characterized by short, thick, and bent bones, generalized osteopenia, brown-green teeth, normal sclerae, and minimal fractures. In two cases, there was polyhydramnios. The changes resolve with time (11,25). Still a different, probably autosomal recessively inherited disorder was described by Stüve and Wiedemann (28). The feet were abnormally positioned. There was fatal respiratory distress. Finally there is another group of disorders that has scaphomacrocephaly, impaired CNS function, thin bones, bending limited to the femora, talipes, and, frequently, cleft palate (11). This form most closely resembles campomelic syndrome.

Fig. 7–30. *Campomelic syndrome*. Note short palpebral fissures, apparent hypertelorism, flat nasal bridge, severe micrognathia, and short neck with redundant skin. (Courtesy of *RE Stevenson*, Greenwood, South Carolina.)

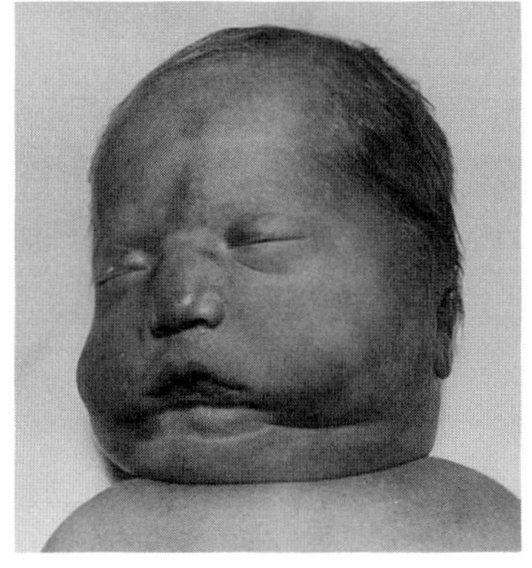

### References (Campomelic syndrome)


1. Austin GE et al: Long-limbed campomelic dwarfism. *Am J Dis Child* **134**:1035–1042, 1980.
2. Bain AD, Barrett HS: Congenital bowing of the long bones. *Arch Dis Childh* **34**:516–524, 1959.
3. Balcar I, Bieber FR: Sonographic and radiologic findings in campomelic dysplasia. *Am J Roentgenol* **141**:481–482, 1983.
4. Beluffi G, Fraccaro M: Congenital and clinical aspects of campomelic dysplasia. *Prog Clin Biol Res* **104**:53–65, 1982.
5. Bound JP et al: Congenital anterior angulation of the tibia. *Arch Dis Childh* **27**:179–184, 1952.
6. Bricarelli F et al: Sex reversed XY females with campomelic dysplasia are H-Y negative. *Hum Genet* **57**:15–22, 1981.
7. Cremin BJ et al: Autosomal recessive inheritance in camptomelic dwarfism. *Lancet* **1**:488–489, 1973.
8. Fontaine P et al: Le conseil génétique dans dysplasie campomelique. *J Génét Hum* **82**:267–279, 1980.
9. Fryns JP et al: Congenital bowing of the long bones: An example of a campomelic syndrome of the short-limbed normocephalic subtype. *Acta Paediatr Scand* **72**:789–791, 1983.
10. Hall BD, Spranger JW: Familial congenital bowing with short bones. *Radiology* **132**:611–614, 1979.

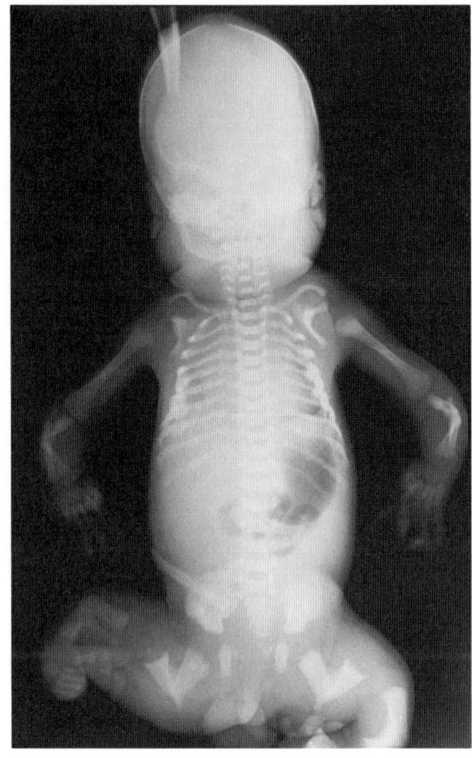
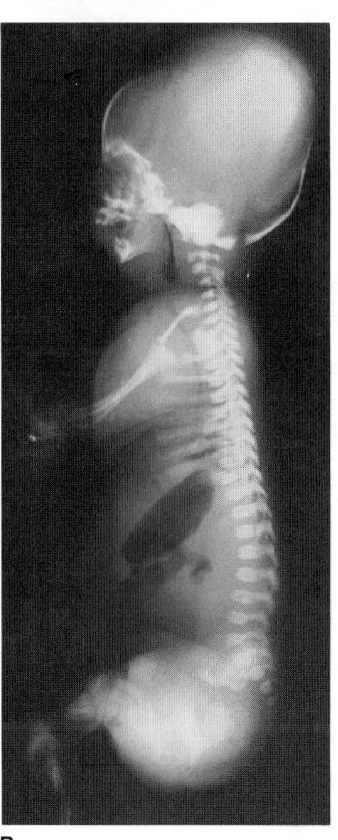

A B

Fig. 7–31. *Campomelic syndrome.* (A,B) Note dolichocephaly, bowing of long bones, hypoplastic scapulae, small bell-shaped chest, flattened vertebral bodies, narrow iliac wings with increased acetabular angles, talipes equinovarus. (Courtesy of *RE Stevenson,* Greenwood, South Carolina.)

11. Hall BD, Spranger JW: Congenital bowing of the long bones: A review and phenotype analysis of 13 undiagnosed cases. *Eur J Pediatr* **133**:131–138, 1980.

12. Hoefnagel D et al: Camptomelic dwarfism associated with XY-gonadal dysgenesis and chromosomal anomalies. *Clin Genet* **13**:489–499, 1978.

13. Houston CS et al: The campomelic syndrome: A review, report of 17 cases and follow-up on the currently 17-year-old boy first reported by Maroteaux in 1971. *Am J Med Genet* **15**:3–28, 1983.

14. Hovmöller ML et al: Campomelic dwarfism: A genetically determined mesenchymal disorder combined with sex reversal. *Hereditas* **86**:51–62, 1977.

15. Khajavi A et al: Heterogeneity in the campomelic syndrome. Long and short bone varieties. *Radiology* **120**:641–647, 1976.

16. Kozlowski K et al: Syndromes of congenital bowing of the long bones. *Pediatr Radiol* **7**:40–48, 1979.

16a. Lazjuk GI et al: Campomelic syndromes: Concepts of the bowing and shortening in the lower limbs. *Teratology* **35**:1–8, 1987.

17. Maclean RN et al: Skeletal dysplasia with short angulated femora (kyphomelic dysplasia). *Am J Med Genet* **14**:373–380, 1983.

18. Maroteaux P et al: Le syndrome campomelique. *Presse Méd* **22**:1157–1162, 1971.

19. Mellows HJ et al: The camptomelic syndrome in two females siblings. *Clin Genet* **18**:137–141, 1980.

20. Middleton PS: Studies on prenatal lesions of striated muscle as a cause of congenital deformation (Case 3). *Edinb Med J* **41**:401–442, 1934.

21. Moedjano SJ et al: The campomelic syndrome in a singleton and monozygotic twins. *Clin Genet* **18**:397–401, 1980.

22. Pauli RM, Pagon RA: Abnormalities of sexual differentiation in campomelic dwarfs. *Clin Genet* **18**:223–225, 1980.

22a. Pazzaglia UE, Beluffi G: Radiology and histopathology of the bent limbs in campomelic dysplasia. Implications in the aetiology of the disease and review of theories. *Pediatr Radiol* **17**:50–55, 1987.

23. Puck SM et al: Absence of H-Y antigen in an XY female with campomelic dysplasia. *Hum Genet* **57**:23–27, 1981.

24. Rezza E et al: Familial congenital bowing with thick bones and metaphyseal changes, a distinct entity. *Pediatr Radiol* **14**:323–327, 1984.

25. Rogers JG et al: A variant of campomelia. *Birth Defects* **11**(6):119–125, 1975.

26. Shafai T, Schwartz L: Camptomelic dwarfism in siblings. *J Pediatr* **89**:512–513, 1976.

27. Spranger JW et al: Increasing frequency of a syndrome of multiple osseous defects. *Lancet* **2**:716, 1970.

27a. Stoll C et al: Birth prevalence rates of skeletal dysplasia. *Clin Genet* **35**:88–92, 1989.

28. Stüve A, Wiedemann H-R: Angeborene Verbiegungen langer Röhrenknochen-eine Geschwisterbeobachtung. *Z Kinderheilkd* **111**:184–192, 1971.

28a. Temple IK et al: Kyphomelic dysplasia. *J Med Genet* **26**:457–468, 1989.

29. Thurmon TF et al: Familial campomelic dwarfism. *J Pediatr* **83**:841–843, 1973.

30. Tokita N et al: The campomelic syndrome. Temporal bone histopathologic features and otolaryngologic manifestations. *Arch Otolaryngol* **105**:449–454, 1979.

31. Viljoen D, Beighton P: Kyphomelic dysplasia: Further delineation of the phenotype. *Dysmorphol Clin Genet* **1**(4):136–141, 1988.

32. Winter R et al: Prenatal diagnosis of campomelic dysplasia by ultrasonography. *Prenat Diagn* **5**:1–8, 1985.

## Cartilage–hair hypoplasia (metaphyseal chondrodysplasia, type McKusick)

McKusick et al (29,30) in 1964–1965, first described a syndrome in the Old-Order Amish characterized by short-limbed dwarfism and fine, sparse, light-colored hair. Subsequently, non-Amish patients with the disorder have been reported (1,12,25–27). An increased birth prevalence has been reported in Finland (24,33,45).

Autosomal recessive inheritance has been established (29,30). However, the observed number of affected individuals is significantly less than would be expected on the basis of a recessive hypothesis, even when infant deaths are excluded (30). Considerable variability in expression has been noted, even within Amish kindreds (30).

**Growth.** Perheentupa and Kaitila (33a) studied growth in 88 patients. Mean male birth length was $45.9 \pm 2.3$ cm and mean female birth length was $44.9 \pm 2.6$ cm. Birth weight was $3430 \pm 430$ g for males and $3220 \pm 440$ g for females. Final height attainment was $133 \pm 10.3$ cm for males and $123 \pm 11.3$ cm for females. Mean adult weights were $57.1 \pm 13.9$ kg for males and $45.4 \pm 7.7$ kg for females.

**Facies and hair.** The head is of normal size. The hair is blond, brittle, fine, silky, and sparse on the scalp and elsewhere on the body (29). Eyebrows, eyelashes, and beard are also sparse (29) (Fig. 7–33). The diameter of the hair shaft is 50–65% of normal and there is no central pigment core (2,10–12,22,26,29,30,47), although the latter

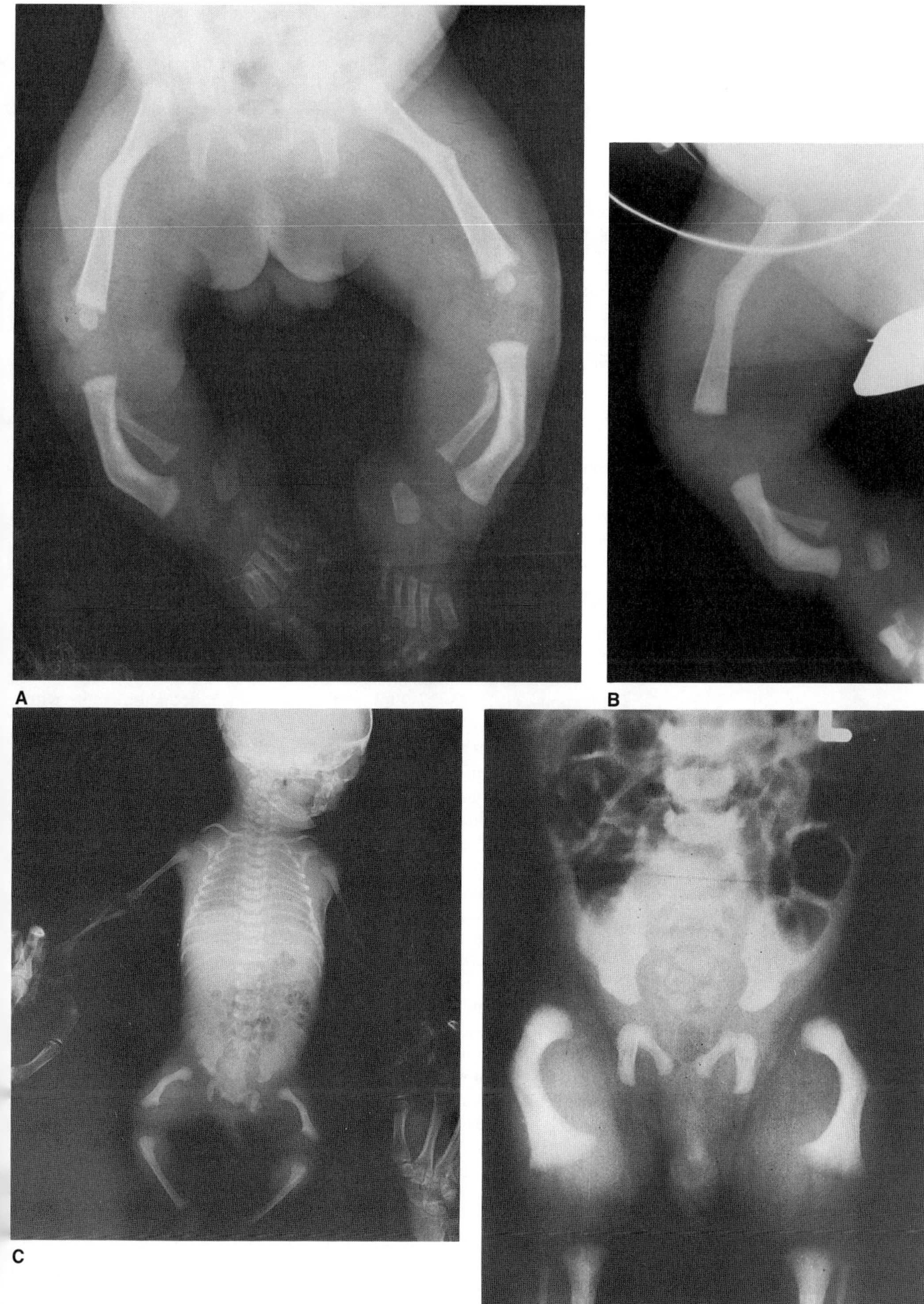


Fig. 7–32 (A,B). *Campomelic syndrome.* Roentgenograms showing campomelia (A courtesy of *JM Opitz,* Helena, Montana. B courtesy of *GE Austin,* Decatur, Georgia.) (C,D). *Kyphomelic dysplasia.* Roentgenograms showing small chest, short ribs with flared ends, short and severely bowed femora, underossification of proximal tibiae, increased acetabular angles. (C,D from *D Viljoen* and *P Beighton,* Dysmorphol Clin Genet **1**(4):136, 1988.)

finding is not a constant feature (46). Blackston and Brown (2) reported wide spacing between overlapping cuticular scales as well as increased copper hair content.

**Skeletal alterations.** The legs are relatively short, and the femurs are mildly bowed (Fig. 7–34). The hands are short and pudgy, and the fingernails and toenails are small (27) (Fig. 7–35). Some patients have marked hyperextensibility of joints, particularly of the hands, wrists, and feet. However, most are unable to fully extend their elbows. Several patients were noted to have been "floppy" babies (30).

Radiologically, irregularly scalloped metaphyses with sclerotic margins are noted (Fig. 7–36A). Small, cloudy, cystic radiolucencies may be scattered throughout the metaphyses, particularly at the distal femoral metaphyses (Fig. 7–36B) (42). Epiphyses tend to be flattened.

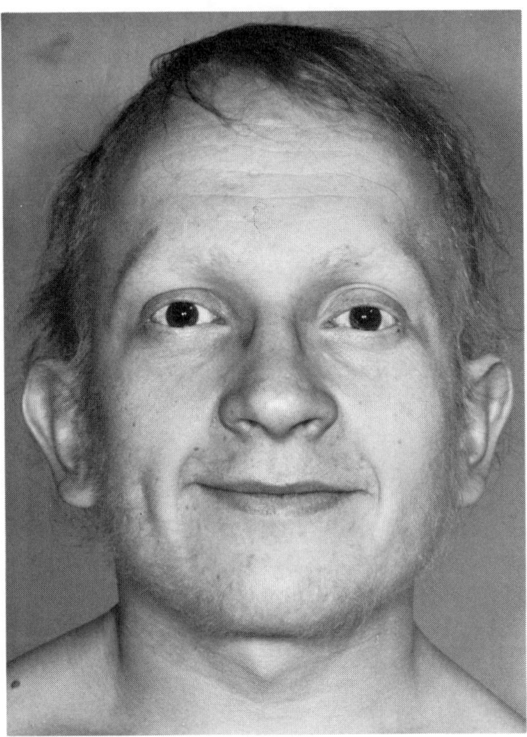

Fig. 7–33. *Cartilage–hair hypoplasia.* Note sparse blond hair and deficiency of eyebrows and eyelashes.

Fig. 7–34. *Cartilage–hair hypoplasia.* Marked shortening of extremities.

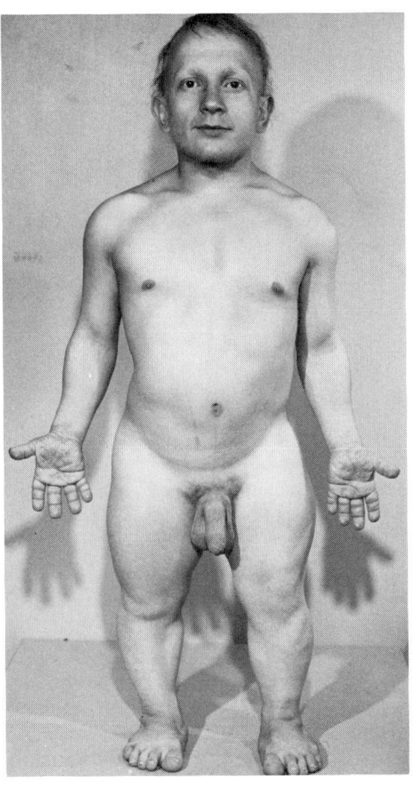

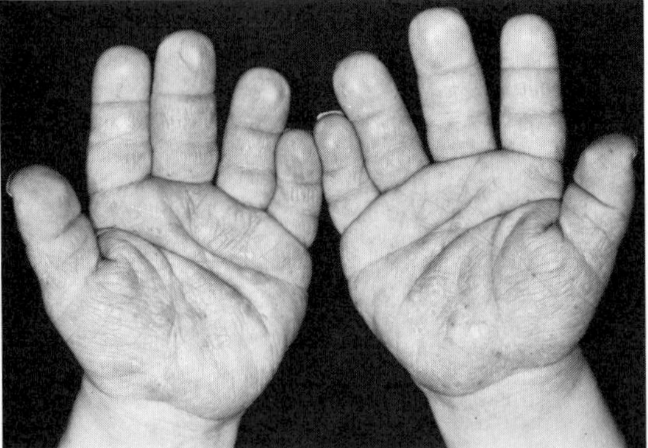

Fig. 7–35. *Cartilage–hair hypoplasia.* Short hands.

There is often narrowing of interpediculate distances, and about one-third of patients have mild odontoid hypoplasia. There is mild flaring of the lower rib cage. The sternum is prominent proximally. Vertebral height may be increased and mild lumbar lordosis may be found. The tibia is characteristically shorter than the fibula (Fig. 7–36B). Cephalometric studies have not demonstrated striking abnormalities (39).

On microscopic examination of the costochondral junction, few cartilage cells are found; those present do not form orderly columns (30).

**Infections.** An important clinical feature of severe cartilage–hair hypoplasia is unexplained susceptibility to severe varicella and other infections (1,18,23–27,29,30). Chronic noncyclic neutropenia with maturation arrest has also been reported (1,7,26,30,48). Immunologic investigation in two children revealed persistent lymphopenia, diminished skin hypersensitivity, diminished responsiveness of their lymphocytes to phytohemagglutinin *in vitro,* and, in one, delayed rejection of a skin allograft. Serum immunoglobulin levels were normal or elevated (26). Patients have not been able to synthesize antibodies to a variety of viral and bacterial antigens. It has been suggested that these persons have a distinct form of cellular immune defect that is responsible for their unusual susceptibility to varicella infection. Trojak et al (45), using mixed lymphocyte culture studies and mitogen-induced stimulation studies, found significant defects in several T-lymphocyte functions in Old-Order Amish patients with the disorder. Similar abnormalities have been found in Finnish patients (36). These investigators concluded that an impairment of cellular immunity was an integral part of the syndrome. Polmar and Pierce (35) found a defect in T-lymphocyte proliferation and concluded that it was not caused by excess suppressor-cell activity or impaired accessory-cell function. Defective proliferation was also found in B cells and fibroblasts. They also noted that individuals with the disorder had marked impairment of proliferation-dependent cytotoxic mechanisms, whereas proliferation-independent natural killer cell activity was normal or even above normal.

Smallpox vaccination and live polio vaccine should be avoided (25,41).

**Malignancy.** Among 110 patients, about 6% had Hodgkin's disease, lymphomas, or leukemia (13a). RJ Gorlin has a patient with Hodgkin's disease.

**Other findings.** Caesarean section is necessary for childbirth. Malabsorption and megacolon have been noted in a few patients (3,6,13,18,25,29,30,38).

**Differential diagnosis.** Fine blond hair may be seen in *hypohidrotic ectodermal dysplasia* (as well as in many other ectodermal dysplasia syndromes).

Three other metaphyseal chondrodysplasias must be distinguished from cartilage–hair hypoplasia. The Jansen type is associated with mental and motor retardation, decreased muscle mass, abnormally shaped skull,

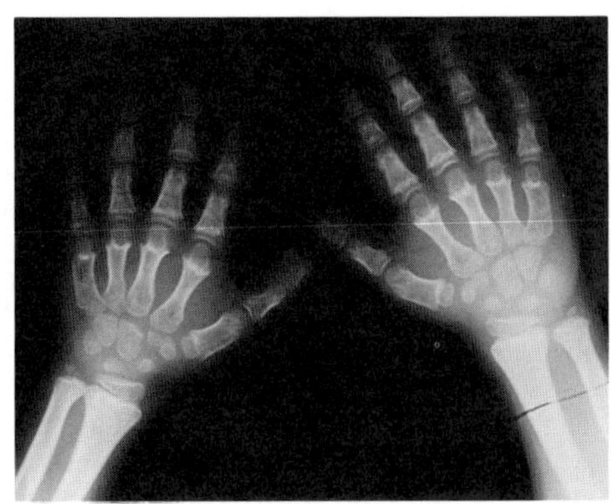
A

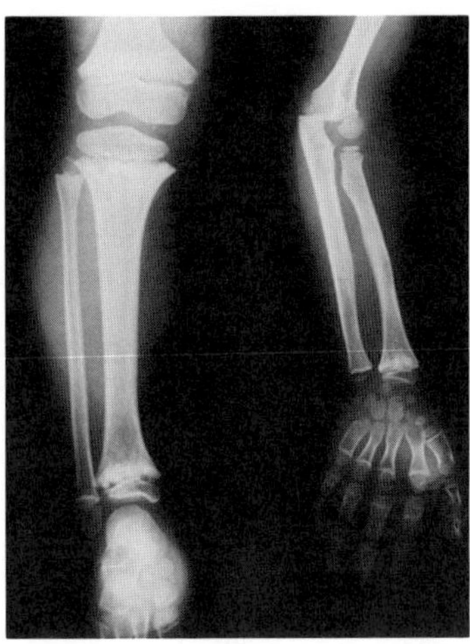
B

Fig. 7–36. *Cartilage–hair hypoplasia.* (A) Roentgenogram of hands showing scalloped metaphyses with sclerotic margins and flattened epiphyses. (B) Roentgenogram of long bones. Note metaphyseal irregularities. Tibia is characteristically shorter than fibula.

hypertelorism, flexion deformities of many joints, beading at the costochondral junction, and gross enlargement of metaphyses (16). Inheritance is possibly autosomal dominant. The Schmidt type of metaphyseal chondrodysplasia is characterized by mild growth retardation, normal intelligence, normal facies, and autosomal dominant inheritance (40). Transient abnormalities are seen at the growth plate, mostly slight widening and irregularity. The Spahr type is similar to the Schmidt type, except that it apparently has autosomal recessive inheritance. Other metaphyseal chondrodysplasias have been described by Spranger (43).

Disorders in which chondroosseous dysplasia is associated with abnormalities of immune function may be distinguished from cartilage–hair hypoplasia by radiologic examination. One of these disorders is the syndrome of chondroosseous dysplasia, adenosine deaminase deficiency, and severe combined immunodeficiency (8,31). In cartilage–hair hypoplasia, adenosine deaminase levels are normal (21). Gatti and associates (15) described a syndrome of lymphopenic agammaglobulinemia, short-limbed dwarfism, cutis laxa, alopecia of the scalp, ichthyosiform dermatosis, erythroderma, and absence of hair and eyebrows. The disorder probably has autosomal recessive inheritance. Patients reported by McKusick and Cross (28) in an Old-Order Amish family had a skeletal dysplasia, ataxia telangiectasia, and Swiss-type agammaglobulinemia. The Schwachmann–Diamond syndrome (43) is characterized by metaphyseal chondrodysplasia, malabsorption, pancreatic insufficiency, and neutropenia. Blackfan–Diamond anemia has been reported in cartilage–hair hypoplasia (17). Nezelof (32) described patients with T-cell deficiency and little or no abnormality of $\gamma$-globulin; the defect may be limited to the thymus. Some patients with this disorder (12) had metaphyseal dysostosis.

*Achondroplasia* and *hypophosphatasia* may be easily distinguished from cartilage–hair hypoplasia on skeletal radiographs. In addition, patients with hypophosphatasia prematurely exfoliate deciduous teeth, mainly the incisors and canines, during the first few years of life (5). Metaphyseal irregularities in cartilage–hair hypoplasia are sharp, in contrast to the frayed and indistinct metaphyses in vitamin D-resistant rickets.

LS Levin evaluated sibs and an isolated patient with a cartilage–hair hypoplasia syndrome distinguishable from the McKusick type by having dental abnormalities. The permanent incisors and premolars were smaller than normal; each permanent incisor had a notch centrally on the incisal edge. The lingual cusps of the lower premolars were bifurcated on their occlusal surfaces.

Bellini and Bardare (4) and Jequier et al (19) reported a metaphyseal chondrodysplasia with cone-shaped epiphyses and cupped metaphyses in the tubular bones and in the hands and feet, abnormally shaped vertebral bodies, odontoid hypoplasia, and alopecia. There was early growth but premature closure of the epiphyses.

A checklist of conditions associated with retarded longitudinal growth has been developed (23,37).

**Laboratory aids.** Steffensen and Østergaard (44) and others (14,34,36,45,46,48) demonstrated selective dysfunction of cell-mediated immunity and inverted ratio of T-helper/T-suppressor cells.

### References [Cartilage–hair hypoplasia (metaphyseal chondrodysplasia, type McKusick)]


1. Ammann AJ et al: Antibody-mediated immunodeficiency in short-limbed dwarfism. *J Pediatr* **84**:200–203, 1974.

2. Blackston RD, Brown AC: Cartilage-hair hypoplasia, in *Hair, Trace Elements and Human Illness,* Brown AC, Crounse R (eds), Prager Publishers, New York, 1980, pp 257–272.

3. Beals RK: Cartilage–hair hypoplasia: A case report. *J Bone Jt Surg* **50A**:1245–1249, 1968.

4. Bellini F, Bardare M: Su un caso di disostosi periferica. *Minerva Pediatr* **18**:105–110, 1966.

5. Beumer J III et al: Childhood hypophosphatasia and the premature loss of teeth. A clinical and laboratory study of seven cases. *Oral Surg Oral Med Oral Pathol* **35**:631–640, 1973.

6. Boothby CB, Bower BC: Cartilage-hair hypoplasia. *Arch Dis Childh* **48**:919–921, 1973.

7. Burke V et al: Association of pancreatic insufficiency and chronic neutropenia in childhood, *Arch Dis Childh* **42**:147–157, 1967.

8. Cederbaum SD et al: The chondro-osseous dysplasia of adenosine deaminase deficiency with severe combined immunodeficiency. *J Pediatr* **89**:737–742, 1976.

9. Clinicopathological conference. A case of Swiss-type agammaglobulinemia and achondroplasia, demonstrated at the Royal Postgraduate Medical School. *Br Med J* **2**:1371–1374, 1966.

10. Coupe RL, Lowry RB: Abnormality of the hair in cartilage-hair hypoplasia. *Dermatologica* **141**:329–334, 1970.

11. D'Apuzzo V, Joss E: Metaphysäre Dysostose und Hypoplasie der Haare: Knorpel-Haar-Hypoplasie. *Helv Paediatr Acta* **27**:241–252, 1972.

12. Fauchier C: Nanisme diastrophique ou "dysostose metaphysaire." *Ann Pédiatr (Paris)* **17**:876–881, 1970.

13. Fauchier C: Nanisme diastrophique familial avec maladie de Hirschprung. *Ann Pédiatr (Paris)* **16**:496–502, 1969.

13a. Francomano CA et al: Cartilage hair hypoplasia in the Amish. Increased susceptibility to malignancy. *Am J Hum Genet* **35**:98A, 1983.

14. Fulginiti VA et al: Agammaglobulinemia and achondroplasia (letter). *Br Med J* **2**:242, 1967.

15. Gatti RA et al: Hereditary lymphopenic agammaglobulinemia associated with a distinctive form of short-limb dwarfism and ectodermal dysplasia, *J Pediatr* **75**:675–684, 1969.

16. Gram PB et al: Metaphyseal chondrodysplasia of Jansen. *J Bone Jt Surg* **41A**:951–959, 1959.

17. Harris RF et al: Cartilage-hair hypoplasia, defective T-cell function, and

Diamond–Blackfan anemia in an Amish child. *Am J Med Genet* **8**:291–297, 1981.

18. Irwin GA: Cartilage-hair hypoplasia (CHH), variant of familial metaphyseal dysostosis. *Radiology* **86**:920–928, 1966.
19. Jequier S et al: Metaphyseal chondrodysplasia with ectodermal dysplasia. *Skel Radiol* **7**:107–112, 1981.
20. Kaitila I, Perheentupa J: Cartilage-hair hypoplasia (CHH), in *Population Structure and Genetic Disorders*, Ericsson AW et al (eds), Academic Press, New York, 1980, pp 588–591.
21. Kaitila IJ et al: Normal red cell adenosine deaminase activity in cartilage-hair hypoplasia. *J Pediatr* **87**:153–154, 1975.
22. Kelling C et al: Biophysical and biochemical studies of the hair in cartilage-hair hypoplasia. *Clin Genet* **4**:500–506, 1973.
23. Langer LO: Short stature. Check list of conditions associated with retarded longitudinal growth. *Clin Pediatr* **8**:142–153, 1969.
24. Lodin H, Sjögren I: Chondro-ectodermal dysplasia, *Acta Paediatr Scand* **53**:583–590, 1964.
25. Lowry RB et al: Cartilage-hair hypoplasia: A rare and recessive cause of dwarfism. *Clin Pediatr* **9**:44–46, 1970.
26. Lux SE et al: Chronic neutropenia and abnormal cellular immunity in cartilage-hair hypoplasia. *N Engl J Med* **282**:231–236, 1970.
27. McKusick VA: *Heritable Disorders of Connective Tissue*, 4th ed., CV Mosby, St Louis, 1972, p. 789.
28. McKusick VA, Cross HE: Ataxia-telangiectasia and Swiss-type agammaglobulinemia: Two genetic disorders of the immune mechanism in related Amish sibships. *J Am Med Assoc* **195**:739–745, 1966.
29. McKusick VA et al: Dwarfism in the Amish, *Trans Assoc Am Physicians* **77**:151–168, 1964.
30. McKusick VA et al: Dwarfism in the Amish. II. Cartilage-hair hypoplasia. *Bull Johns Hopkins Hosp* **116**:285–326, 1965.
31. Meuwissen HJ et al: Combined immunodeficiency disease associated with adenosine deaminase deficiency, *J Pediatr* **86**:169–181, 1975.
32. Nezelof C: Thymic dysplasia with normal immunoglobulins and immunologic deficiency: Pure alymphocytosis, in *Immunologic Deficiency Diseases*, Good RA (ed), The National Foundation, New York, 1968, pp 104–115.
33. Norio R et al: Hereditary diseases in Finland, *Ann Clin Res* **5**:109–141, 1973.

33a. Perheentupa MO, Kaitila I: Growth of patients with cartilage-hair hypoplasia. March of Dimes Clinical Genetics Conference, Baltimore, July 10–13, 1988.

34. Pierce GF, Polmar SH: Lymphocyte dysfunction in cartilage-hair hypoplasia: Evidence for an intrinsic defect in cellular proliferation. *J Immunol* **129**:570–575, 1982.
35. Polmar SH, Pierce GF: Cartilage hair hypoplasia: Immunological aspects and their clinical implications. *Clin Immunol Immunopathol* **40**:87–93, 1986.
36. Ranki A et al: In vitro T- and B-cell reactivity in cartilage-hair hypoplasia. *Clin Exp Immunol* **32**:352–360, 1978.
37. Ray HC, Dorst JP: Cartilage-hair hypoplasia. *Prog Pediatr Radiol* **4**:270–298, 1973.
38. Roberts PAL et al: Hirschsprung's disease associated with a variant form of achondroplasia in sister and brother. *Proc R Soc Med* **62**:329, 1969.
39. Rönning O et al: Craniofacial and dental characteristics of cartilage-hair hypoplasia. *Cleft Palate J* **15**:49–55, 1978.
40. Rosenbloom AL, Smith WD: The natural history of metaphyseal dysostosis. *J Pediatr* **66**:857–868, 1965.
41. Saulsbury FT et al: Combined immunodeficiency and vaccine-related poliomyelitis in a child with cartilage-hair hypoplasia. *J Pediatr* **86**:868–872, 1975.
42. Seige M: Metaphysäre Chondrodysplasie vom typ McKusick (Knorpel-Haar Hypoplasie). *Mschr Kinderheilkd* **128**:157–159, 1980.
43. Spranger J: Metaphyseal chondrodysplasias. *Birth Defects* **12**(6):33–46, 1976.
44. Steffensen O, Østergaard PA: An inverted ratio of T-helper/T-suppressor cells and selected deficiency of cell-mediated immunity in a girl with cartilage-hair hypoplasia. *Eur J Pediatr* **135**:55–58, 1980.
45. Trojak JE et al: Immunologic studies of cartilage-hair hypoplasia in the Amish. *Johns Hopkins Med J* **148**:157–164, 1981.
46. Virolainen M et al: Cellular and humoral immunity in cartilage-hair hypoplasia. *Pediatr Res* **12**:961–966, 1978.
47. Wiedemann HR et al: Knorpel-Haar Hypoplasie. *Arch Kinderheilkd* **176**:74–85, 1967.
48. Wilson WG et al: Cartilage-hair hypoplasia (metaphyseal chondrodysplasia, type McKusick) with combined immune deficiency. Variable expression and development of immunologic functions in sibs. *Birth Defects* **14**(6A):117–129, 1978.

## Chondrodysplasia punctata (general)

The chondrodysplasia punctata disorders are a heterogeneous group of skeletal dysplasias whose common feature is the radiographic appearance of punctate or stippled calcifications at long-bone epiphyses during infancy. The punctate calcifications of epiphyseal and growth plate cartilage are presumed to represent intrauterine injury of cartilage with subsequent healing by fibrosis, calcification, and ossification. Thus, "chondrodysplasia punctata" is not a specific disease designation but is, more properly, a radiographic sign evoking consideration of a spectrum of genetic and teratogen-induced disorders.

First described by Conradi (1), in 1914, chondrodysplasia punctata was later split into two categories by Spranger (4) who differentiated the relatively mild and presumably autosomal dominant *Conradi–Hünermann disease* from lethal *rhizomelic chondrodysplasia punctata*. Happle (3) further distinguished an X-linked dominant type that accounts for all cases of Conradi–Hünermann disease. Most recently, Curry et al (2) have proposed an *X-linked recessive chondrodysplasia punctata* that may actually represent a chromosome microdeletion syndrome.

Thus, three major clinical entities have been distinguished on the basis of degree of skeletal aberration, presence of cataracts, and skin involvement (Table 7–3). Although these attempts to differentiate clinical phenotypes appear justified, there is uncertainty as to whether these phenotypic differences represent true genetic heterogeneity or are merely gradations in the variable expression of a single X-linked gene or gene complex. Specifically, the lack of male-to-male transmission casts doubt on the existence of an autosomal dominant form of chondrodysplasia; alternately, this mild phenotype might represent the forme fruste of the X-linked dominant type. Likewise, the apparent X-linked recessive type might result from homozygosity for mutations of the locus (or loci) responsible for X-linked dominant Conradi–Hünermann disease. The heritable phenotypes must be further differentiated from the teratogenic effects of anticoagulant therapy with vitamin K antagonists (e.g., coumarin embryopathy) and a number of other conditions in which transient epiphyseal stippling may occur.

#### References [Chondrodysplasia punctata (general)]

1. Conradi E: Vorzeitiges Auftreten von Knochen und eigenartigen Verkalkungskernen bei Chondrodystrophia foetalis hypoplastica: Histologische und Roentgenuntersuchungen. *Jb Kinderheilkd* **80**:86–97, 1914.
2. Curry CJR et al: Inherited chondrodysplasia punctata due to a deletion of the terminal short arm of an X chromosome. *N Engl J Med* **311**:1010–1015, 1984.
3. Happle R et al: Sex-linked chondrodysplasia punctata? *Clin Genet* **11**:73–76, 1977.
4. Spranger JW et al: Heterogeneity of chondrodysplasia punctata. *Hum Genet* **11**:190–212, 1971.

**Conradi–Hünermann disease: X-linked dominant type.** The X-linked dominant pattern (Figs. 7–37 and 7–38) was first recognized by Happle et al (11–14) to be limited to females, being lethal in hemizygous males. The mother of an affected daughter may have mild manifestations such as short stature or skin or eye changes (1a,6,11,16–18,21).

It has been proposed that this form accounts for at least 25% (20), but our review suggests that all reported cases of Conradi–Hünermann syndrome, including the "autosomal dominant" cases, are actually the X-linked dominant condition. The disorder is intermediate in severity with a good prognosis. It has been suggested that peroxisome dysfunction is present (14a,14b). The disorder occurs exclusively in females because the underlying gene defect is lethal in hemizygous males (7a). A stillborn male infant has been noted (2). Familial observations have been reported (7,10,20,24). Intelligence appears to be normal, although performance on specific tests may be impaired by poor vision and skeletal limitations on motor skills (20).

Facies. Frontal bossing is common with macrocephaly noted in several patients (5,8–10,13–15). The nasal root is flat and broad (3,5,8).

Table 7–3. Chondrodysplasia punctata: various forms

| Feature | X-Linked dominant type chondrodysplasia punctata | X-Linked recessive chondrodysplasia punctata | Rhizomelic chondrodysplasia punctata |
|---|---|---|---|
| Head circumference | Normal for age | Mild microcephaly | Small for age |
| Nose | Hypoplastic | Hypoplastic | Hypoplastic |
| Cataracts | 65%, often unilateral and asymmetric | 2 of 4 cases, bilateral | 65%, bilateral and symmetric |
| Skin changes | Congenital ichthyosiform erythroderma<br>Ichthyosis in older child<br>Systematic atrophoderma, pseudopelade, coarse twisted scalp hair | Mild ichthyosis as neonate, sparse unruly hair | Dry scaly rash in 25% |
| Contractures | Mild in 25% | Mild | Frequent, severe |
| Skeletal changes | Asymmetric shortening of limbs, usually femur and humerus; scoliosis after first year | Distal phalangeal hypoplasia<br>Carriers have broad wrists and short arms | Severe bilateral shortening of femur and/or humerus with severe metaphyseal changes |
| Stippling | Asymmetric involvement of long bones; paravertebral, laryngeal, tracheal | Bilateral symmetric involvement of long bones; paravertebral, laryngeal, tracheal | Proximal and distal humeri and femora, knee. No paravertebral stippling |
| Mental development | Normal to mildly retarded | Developmental delay, hyperactivity, behavioral problems | Severely retarded |
| Prognosis | Good; lethal in hemizygous male | Relatively good | Lethal, usually in the first year of life |
| Inheritance | X-linked dominant | X-linked recessive | Autosomal recessive |

The face is often asymmetric because of hypoplasia of one side (9,14,21,24). The neck may be short (7,13,14,25).

**Eyes.** Congenital diffuse cataracts have been observed in about 65% and may be unilateral (35%) or bilateral, often (65%) with asymmetric intensity (12). Happle (12) pointed out that this exception to the general rule that hereditary cataracts tend to be bilateral and symmetrical is probably caused by lyonization. Microphthalmia and microcornea have also been reported (16,24).

**Skin.** Newborn patients exhibit ichthyosiform erythroderma with thick, adherent scales arrayed in a linear, blotchy pattern presumably reflecting lyonization. The hyperkeratotic eruption is followed by systematized atrophoderma distributed in a mosaic pattern leaving patchy linear areas of alopecia on the scalp. A linear pattern of hyperpigmentation, which is not congruent with the pattern of hairlessness, has been observed and compared to incontinentia pigmenti (7). The nail plates are often flattened (8). Microscopic sections of the skin show a thin stratum granulosum. These skin changes can be distinguished from other forms of ichthyosis on the basis of histologic and ultrastructural criteria (4).

**Musculoskeletal alterations.** In infancy, punctate calcifications are scattered throughout the spinal column, costal cartilages, sternum,

Fig. 7–37. *Chondrodysplasia punctata.* X-Linked dominant type. (A) Height is reduced. Note disproportionate limb length and scoliosis. (B) Sparse hair, flat midface, anteverted nostrils, and cataract. (C) Patchy alopecia on scalp. (D) Skin changes. (From *D Comings et al,* J Pediatr **72:**63, 1968.)

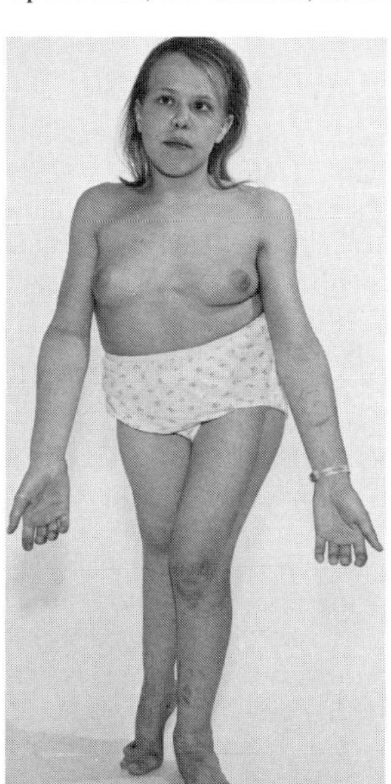
A

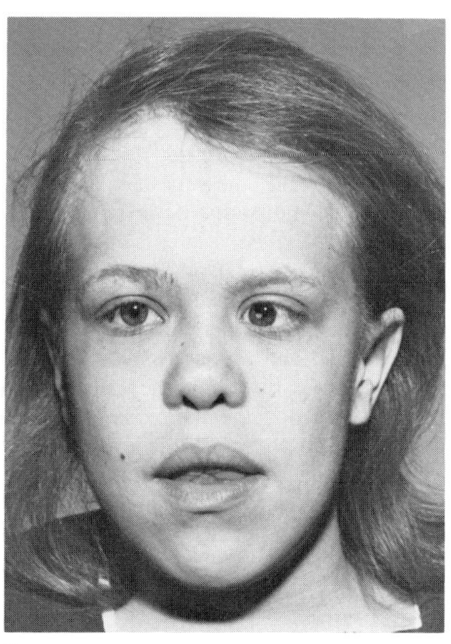
B

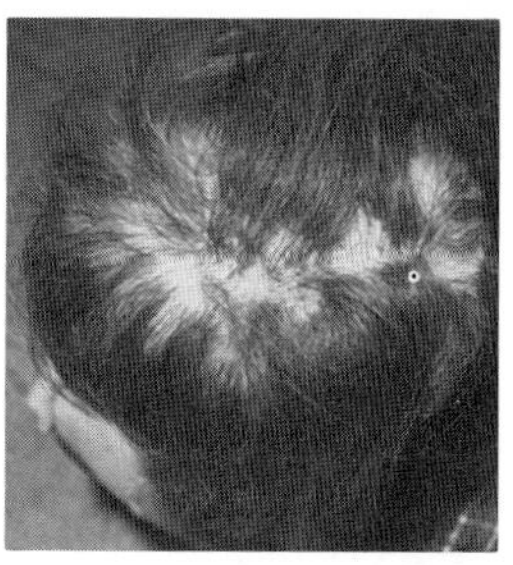
C

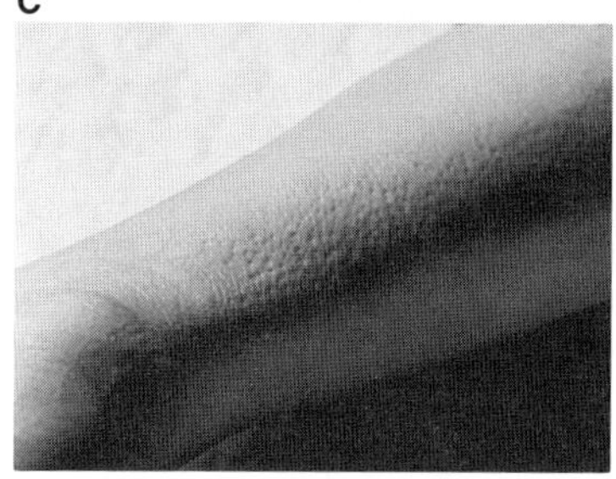
D

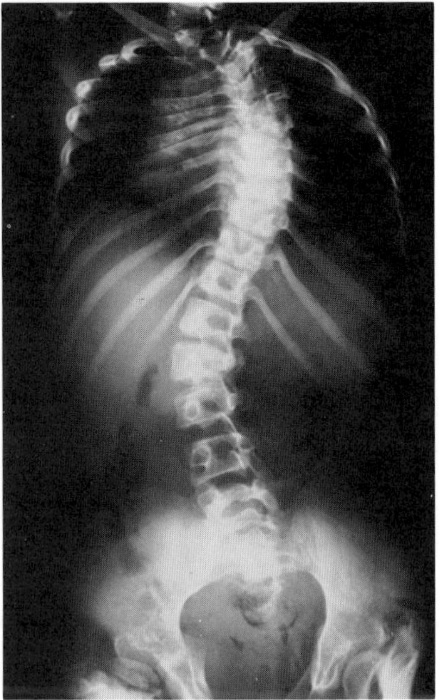

Fig. 7–38. *Chondrodysplasia punctata.* X-Linked dominant type with severe scoliosis.

clavicle, scapulae, and in the epiphyseal centers of the extremities. Loss of the characteristic stippled epiphyses with time makes the diagnosis difficult. In some individuals there is true dysplasia of vertebral bodies (5,7,18). Stature is usually reduced (1,13,25). Scoliosis or kyphoscoliosis with asymmetric shortening of limb bones, particularly the femur and humerus, is common. Shortening of the limbs is asymmetrical. Flexion contractures involving the hip or knee joints (10,13), elbow (16,22), or fingers (7,11,21), and hip dysplasia (1,6,7,15,21,24) have been reported. Metapodial bones may also be involved (12,16,18). Some patients exhibited postaxial supernumerary digits of the hands (12,16,20,22–24). Talipes is not uncommon (9,10,13,16). Intrauterine radiographic examination of a pregnancy at 16 weeks failed to show diagnostic features, although the characteristic extensive calcifications in sites of endochondral bone formation were apparent by radiographic examination of the abortus.

**Other manifestations.** Mental retardation has been described (7) but is rare. Congenital paraplegia resulting from uncharacteristic maldevelopment of the spinal cord has been reported (6). Cardiovascular disorder has been rarely mentioned (15).

### References (Conradi–Hünermann disease: X-linked dominant type)

1. Allansmith M, Senz E: Chondrodystophia congenita punctata. *Am J Dis Child* **100**:109–116, 1960.

1a. Andersen PE Jr, Justesen P: Chondrodysplasia punctata. *Skeletal Radiol* **16**:233–236, 1987.

2. Bergstrom K et al: Chondrodysplasia calcificans congenita (Conradi's disease) in a mother and her child. *Clin Genet* **3**:158–161, 1972.

3. Bodian EL: Skin manifestations of Conradi's disease (chondrodystrophia congenita punctata). *Arch Dermatol* **94**:743–748, 1966.

4. Colde G, Happle RP: Histologic and ultrastructural features of the ichthyotic skin in X-linked dominant chondrodysplasia punctata. *Acta Derm Venerol* **64**:389–394, 1985.

5. Comings DE et al: Conradi's disease. *J Pediatr* **72**:63–69, 1968.

6. Curless RG: Dominant chondrodysplasia punctata with neurologic symptoms. *Neurology* **33**:1095–1097, 1983.

7. Curth HO: Follicular atrophoderma and pseudopelade associated with chondrodysplasia calcificans congenita. *J Invest Dermatol* **13**:233–247, 1949.

7a. De Raeve L et al: Lethal course of X-linked dominant chondrodysplasia punctata in male newborn. *Dermatologica* **178**:162–170, 1989.

8. Edidin DV et al: Chondrodysplasia punctata (Conradi–Hünermann syndrome). *Arch Dermatol* **113**:1431–1434, 1977.

9. Finkel JJ, McKusick VA: Case report R-chondrodystrophia calcificans congenita. *Birth Defects* **5**(4):322–325, 1969.

10. Goerttler E: Chondrodysplasia punctata Typ Conradi–Hünermann. *Z Hautkr* **54**:676–677, 1979.

11. Happle R: Homologous genes for X-linked chondrodysplasia punctata in man and mouse. *Hum Genet* **63**:24–27, 1983.

12. Happle R: Cataracts as a marker of genetic heterogeneity in chondrodysplasia punctata. *Clin Genet* **19**:64–66, 1981.

13. Happle R: X-gekoppelt dominate Chondrodysplasia punctata? *Mschr Kinderheilkd* **128**:203–207, 1980.

14. Happle R et al: Sex-linked chondrodysplasia punctata? *Clin Genet* **11**:73–76, 1977.

14a. Heymans HSA et al: Rhizomelic chondrodysplasia punctata: Another peroxisomal disorder. *N Engl J Med* **313**:187–188, 1985.

14b. Holmes RD et al: Peroxisomal enzyme deficiency in the Conradi–Hünermann form of chondrodysplasia punctata. *N Engl J Med* **316**:1608, 1987.

15. Jeune M et al: Les maladie congénitale des epiphyses pointillees du calcinose foetale epiphysaire chondrodystrophiante. *Arch Fr Pédiatr* **10**:914–943, 1953.

16. Joosten R, Habedanke M: Sex-linked chondrodysplasia punctata due to a new mutation. *Acta Paediatr Belg* **32**:275–278, 1979.

17. Kaser H: Chondrodysplasia calcificans congenita. *Schweiz Med Wschr* **87**:157–172, 1957.

18. Laugier J: Maladie des epiphyses ponctuées. *Pédiatrie* **24**:723–732, 1969.

19. Maleville J et al: Atrophodermie folliculaire, pseudopelade, keratose pilaire des sourcils et etat ichthyosique. *Bull Soc Fr Dermatol Syph* **76**:85–86, 1969.

20. Manzke H et al: Dominant sex-linked inherited chondrodysplasia punctata: A different type of chondrodysplasia punctata. *Clin Genet* **17**:97–107, 1980.

21. Miescher G: Atypische Chondrodystrophie, Typus Morquio, kombiniert mit follikularer Atrophodermie. *Dermatologica* **89**:38–40, 1944.

21a. Mueller RF et al: X-linked dominant chondrodysplasia punctata. *Am J Med Genet* **20**:137–144, 1985.

22. Norum RE et al: Chondrodysplasia punctata, dominant type, with peripheral cataracts. *Birth Defects* **13**(3C):244–245, 1977.

23. Scott C: Addendum. Heterogeneity of chondrodysplasia punctata. *Hum Genet* **11**:212, 1971.

24. Silengo MC et al: Clinical and genetic aspects of Conradi–Hünermann disease. *J Pediatr* **97**:911–917, 1980.

25. Spranger JW et al: Heterogeneity of chondrodysplasia punctata. *Hum Genet* **11**:190–212, 1971.

26. Tasker WG et al: Chondrodystrophia calcificans congenita. *Am J Dis Child* **119**:122–127, 1970.

27. Thamdrup E, Zachau-Christiansen B: Dysplasia epiphysialis punctata. *Acta Paediatr Scand* **51**:589–593, 1962.

28. Thiel HJ et al: Katarakt bei Chondrodystrophia calcificans connata. *Klin Mbl Augenheilkd* **154**:536–545, 1969.

29. Weber A: Zur Frage der Chondrodysplasia calcificans congenita. *Helv Paediatr Acta* **13**:228–238, 1958.

**Rhizomelic chondrodysplasia punctata.** This type, having autosomal recessive inheritance, is the most severe form of chondrodysplasia punctata, leading to death usually before the second year of life (Figs. 7–39 and 7–40). Familial occurrence has been reported (6,11) and both sexes are equally affected. Parental consanguinity has been estimated at 8–10%. Biochemical studies have recently indicated that this type results from a disorder of peroxisomal metabolism (9). Patients exhibit profound deficiency of plasmalogen (ether lipid) synthesis, the presence of unprocessed peroxisomal thiolase in the liver, reduced alkyldehydroxyacetone phosphate synthase activity in fibroblasts, and elevated phytanic acid levels in the plasma and liver, indicating reduction of phytanic acid oxidation to 1–5% of controls (9b,13a). Other peroxisomal functions are normal. Somatic cell hybridization studies have led to the suggestion that rhizomelic chondrodysplasia punctata represents one of the disorders with a specific impairment of peroxisomal function that can be differentiated from at least two other complementation groups because of more generalized defects of peroxisomal functions (group 2: Zellweger syndrome, hyperpipecolic acidemia; group 3: neonatal adrenoleukodystrophy) (9b,14,16).

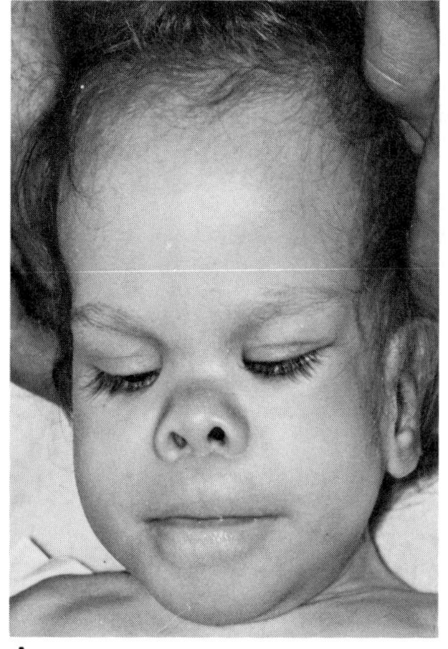
A

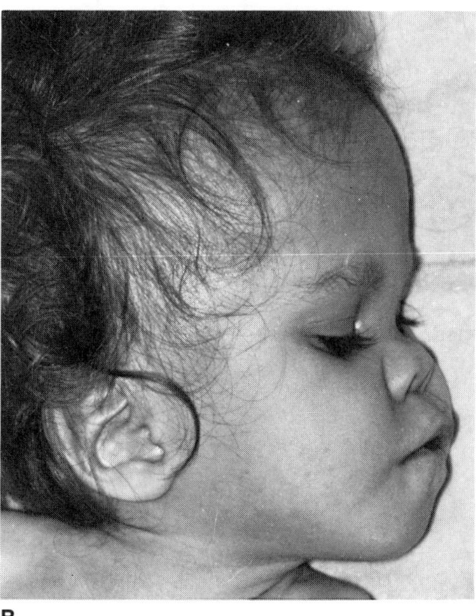
B

Fig. 7–39. *Chondrodysplasia punctata, rhizomelic form.* Autosomal recessive form. (A) Flat midface and small upturned nose. (B) Deficient midface and posteriorly angulated ears.

Fig. 7–40. *Chondrodysplasia punctata, rhizomelic form.* Note abbreviation of humeri and femora with stippling at proximal and distal ends. (From *JM Connor et al,* Am J Med Genet **22:**243, 1985.)

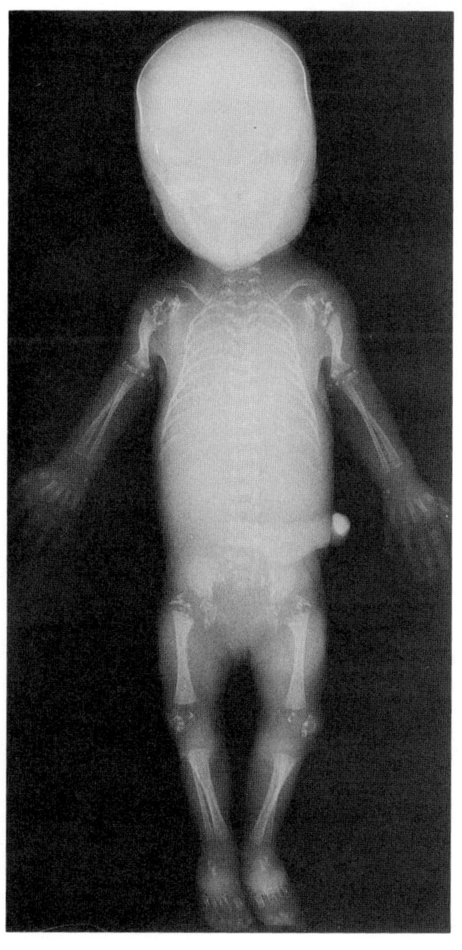

Facies. The face is symmetric but notable for frontal bossing, flat nasal bridge, and small nares.

Eyes. Cataracts observed in about two-thirds of the cases are usually bilateral and of equal density (14).

Skin. At least 25% of patients have ichthyosis, which develops shortly after birth.

Musculoskeletal alterations. In contrast to the other types, there is severe congenital rhizomelic shortening of the extremities. Small head circumference tends to be present at birth and is a constant finding in older infants and children. Contractures have been noted in over 60% and foot deformities in about 10%.

Radiographic skeletal abnormalities include severe shortening, metaphyseal cupping, splaying, and disturbed ossification of the humerus and/or femur. Epiphyseal and extraepiphyseal calcifications are usually severe (Fig. 7–41). Lateral views of the spine show a coronal cleft of the vertebral bodies (14). There are marked degenerative changes in resting chondrocytes (13a).

Oral manifestations. Cleft palate (2–8,12,13) and submucous palatal cleft (1) have been noted in the rhizomelic type.

### References (Rhizomelic chondrodysplasia punctata)


1. Armaly MF: Ocular involvement in chondrodystrophia calcificans congenita punctata. *Arch Ophthalmol* **57**:491–502, 1957.
2. Brogdon BG, Crow NE: Chondrodystrophia calcificans congenita. *Am J Roentgenol* **80**:443–448, 1958.
3. Condron CJ: Conradi's disease: A case without cutaneous manifestations. *Birth Defects* **7**(8):214–215, 1971.
4. Côté PE: Observations sur la chondrodysplasie epiphysaire. *Laval Méd* **20**:481–489, 1955.
5. Coughlin EJ et al: Chondrodystrophia calcificans congenita. *J Bone Jt Surg* **32A**:938–942, 1950.
6. Fraser FC, Scriver JB: A hereditary factor in chondrodystrophia calcificans congenita. *N Engl J Med* **250**:272–277,1954.
7. Gilbert EF et al: Chondrodysplasia punctata–rhizomelic form. *Eur J Pediatr* **123**:89–109, 1976.
8. Hewitt HL, Bochove W: Chondrodystrophia calcificans congenita (Case 1). *Radiol Clin Biol* **40**:175–183, 1971.
9. Heymans HSA et al: Rhizomelic chondrodysplasia punctata: Another peroxisomal disorder (letter). *N Engl J Med* **313**:187–188, 1985.

9a. Hoefler S et al: Prenatal diagnosis of rhizomelic chondrodysplasia punctata. *Prenat Diag* **8**:571–576, 1988.

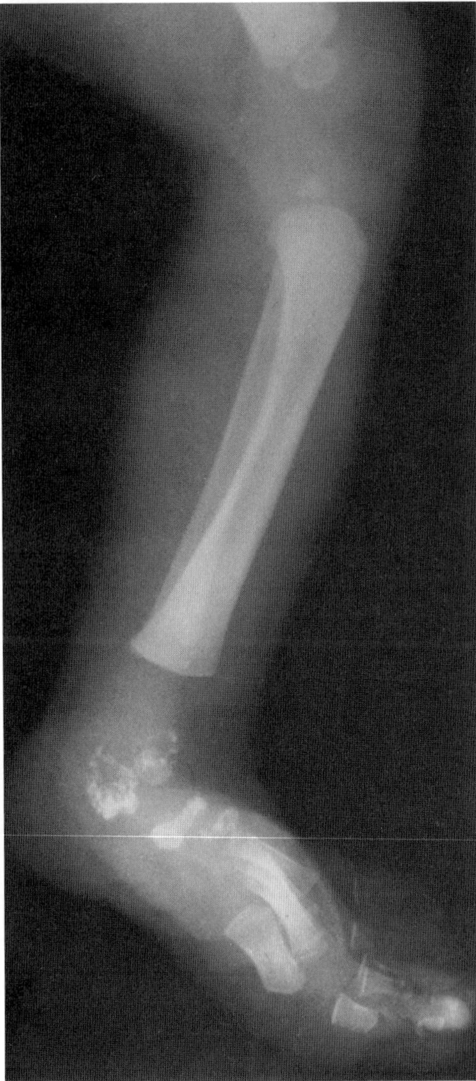

Fig. 7–41. *Chondrodysplasia punctata, rhizomelic form.* Note stippling of heel.


9b. Hoefler G et al: Biochemical abnormalities in rhizomelic chondrodysplasia punctata. *J Pediatr* **112**:726–733, 1988.

10. Koischwitz D, Anders G: Die Chondroplasia punctata. *Roefo* **132**:689–694, 1980.

11. Maitland DG: Punctate epiphyseal dysplasia occurring in two members of the same family. *Br J Radiol* **12**:91–93, 1939.

12. Louvar RD et al: Conradi–Hünermann syndrome. *Clin Pediatr* **13**:680–685, 1974.

13. Phillips LI: Chondrodystrophia calcificans congenita (Case 1). *NZ Med J* **56**:22–27, 1957.

13a. Paulos A et al: Rhizomelic chondrodysplasia punctata. Clinical, pathologic and biochemical findings in two patients. *J Pediatr* **113**:685–690, 1988.

14. Schutgens RBH et al: Peroxisomal disorders: A newly recognized group of genetic diseases. *Eur J Pediatr* **144**:430–440, 1986.

15. Spranger JW et al: Heterogeneity of chondrodysplasia punctata. *Humangenetik* **11**:190–212, 1971.

16. Tager JM et al: To the editor. *N Engl J Med* **315**:767, 1986.


**X-linked recessive chondrodysplasia punctata.** An X-linked recessive disorder with chondrodysplasia punctata, ichthyosis, and mental retardation was reported by Curry et al (1). Cytogenetic studies revealed a small deletion of the distal short arm of the X-chromosome (Xp22.32) in four affected males, and in 11 of 25 apparently normal related females. The presence of ichthyosis in affected males prompted biochemical studies that demonstrated functional deletion of three genes previously mapped to this region of the X chromosome corresponding to the observed deletion (steroid sulfatase; $Xg^a$; and the M1C2X locus for expression of 12E7 antigen). It has been suggested that extensive calcification of blood vessels may play an important role in the pathology of this form (2). Heterozygous female carriers are clinically normal but, as a group, are slightly shorter, with broad wrists and short arms, in comparison to noncarrier relatives (1).

Facies. Nasal hypoplasia was present in all children with some tendency toward improvement in later childhood.

Eyes. Bilateral cataracts were observed in some patients.

Skin. All had sparse and unruly hair with mild ichthyosis particularly over the chest, back of legs, neck, and axillae.

Musculoskeletal alterations. Bilaterally symmetric punctate stippling of multiple epiphyseal centers is characteristic, including the paravertebral epiphyses and those of the larynx, trachea, and long bones; these disappeared relatively rapidly with age (3,4).

#### References (X-linked recessive chondrodysplasia punctata)


1. Curry CJR et al: Inherited chondrodysplasia punctata due to a deletion of the terminal short arm of an X chromosome. *N Engl J Med* **311**:1010–1015, 1984.

2. Kurczynski TW et al: Lethal chondrodysplasia punctata with extensive calcifications. *Am J Hum Genet* **37**:63A, 1985.

3. Raap G: Chondrodystrophia calcificans congenita. *Am J Roentgenol* **49**:77–82, 1943.

4. Silverman FN: Discussion on the relation between stippled epiphyses and the multiplex form of epiphyseal dysplasia. *Birth Defects* **5**(4):68–70, 1969.


**Other disorders with stippled epiphyses.** Epiphyseal stippling in infancy is a nonspecific sign and may also be seen in *Zellweger syndrome,* multiple epiphyseal dysplasia, *alcohol embryopathy* (1), *hydantoin embryopathy, coumarin embryopathy* (2,10,14,15), chondritis secondary to bacteremia, $G_{M1}$-*gangliosidosis, Smith–Lemli–Opitz syndrome, trisomies 18 and 21,* congenital hypothyroidism, and other disorders (6–8,13). Burck (3,4) described a form of mesomelic dysplasia characterized by mild punctate epiphyseal calcification, relatively short distal ulnae and long proximal fibulae, dislocated radial heads and patellae, and a distinct facies marked by prominent forehead, flat midface, flat nasal bridge, and micrognathia. Two other cases were reported by Rogers et al (11). Stippled epiphyses may be found in single digits (9) and in the CHILD (congenital hemidysplasia, ichthyosiform erythroderma, limb defects) syndrome (5). A mild variety, undoubtedly heterogeneous, was reported by Sheffield et al (12).

**Laboratory aids.** Prenatal diagnosis of the rhizomelic type is possible by demonstrating lack of phytanic acid oxidation and defects in plasmalogen synthesis (9a).

#### References (Other disorders with stippled epiphyses)


1. Badois C et al: Punctate epiphyses disease associated with alcoholic foetopathy: First three reported cases. *Ann Radiol* **26**:244–250, 1983.

2. Becker MH et al: Chondrodysplasia punctata: Is maternal warfarin therapy a factor? *Am J Dis Child* **129**:356–359, 1975.

3. Burck U: Mesomelic dysplasia with punctate epiphyseal calcifications—a new entity of chondrodysplasia punctata? *Eur J Pediatr* **138**:67–72, 1982.

4. Burck U et al: Mesomelic dysplasia with short ulna, long fibula, brachymetacarpy, and micrognathia. Clinical and radiological differential diagnostic features. *Pediatr Radiol* **9**:161–165, 1980.

5. Happle R et al: The CHILD syndrome: Congenital hemidysplasia with ichthyosiform erythroderma and limb defects. *Eur J Pediatr* **134**:27–33, 1980.

6. Haynes ER, Wagner, WF: Chondroangiopathia calcarea seu punctata. *Radiology* **57**:547–550, 1951.

7. Hunter AGW et al: Chondrodysplasia punctata in an infant with duplication 16p due to a 7:16 translocation. *Am J Med Genet* **21**:581–589, 1985.

8. Lawrence JJ et al: Unusual manifestations of chondrodysplasia punctata. *Skel Radiol* **18**:15–19, 1989.
9. Mason RC, Kozlowski K: Chondrodysplasia punctata. A report of 10 cases. *Radiology* **109**:145–150, 1973.
10. Pauli RM et al: Warfarin therapy initiated during pregnancy and phenotypic chondrodysplasia punctata. *J Pediatr* 88:506–508, 1976.
11. Rogers JG et al: Mesomelic dysplasia—confirmation of a new clinical type of chondrodysplasia punctata. March of Dimes Clinical Genetics Conference, Baltimore, July 10–13, 1988.
12. Sheffeld LJ et al: Chondrodysplasia punctata—23 cases of a mild and relatively common variety. *J Pediatr* **89**:916–923, 1976.
13. Spranger JW et al: Heterogeneity of chondrodysplasia punctata. *Hum Genet* **11**:190–212, 1971.
14. Strüwe FE et al: Coumarin-Embryopathie. *Radiologue* **24**:68–71, 1984.
15. Tamburrini O et al: Chondrodysplasia punctata after Warfarin. *Pediatr Radiol* **17**:323–324, 1987.

## Diastrophic dysplasia

Lamy and Maroteaux (15), in 1960, used the term "diastrophic dwarfism" to describe a syndrome consisting of micromelic dwarfism, progressive scoliosis, bilateral talipes equinovarus, and various deformities of the digits, hip dysplasia, characteristic external ear deformities, and, frequently, cleft palate (Figs. 7–42 to 7–45). Over 300 cases have been described with over 100 of these coming from Finland (21). Prior to its delineation, cases in infants were generally classified as examples of "achondroplasia with clubbed feet," and in adults, as Morquio syndrome (4,13). Early examples are those of Schenk (22) in 1910 and Duken (4) in 1921.

The syndrome has autosomal recessive inheritance (1). The so-called "diastrophic variant" presently is considered to be a mild form of diastrophic dysplasia, reflecting the wide variability in phenotypic expression of this condition (8,14). Gustavson et al (5) suggested that there were lethal and nonlethal forms with the lethal form exhibiting lower birthweight, overlapping joints, dislocation of cervical spine, and, frequently, congenital heart disease. Fertility seems to be reduced (27).

The basic defect may be a metabolic abnormality of the chondrocyte that induces cell death and/or a defect in the synthesis of collagen and/or proteoglycan (9). Stanescu et al (24) have demonstrated that the basic defect may be in collagen Type II that presents an abnormal pattern of the segment long spacing at the level of band 42 and 43, strongly suggesting alterations in the CB 10.5 peptide.

About 25% die in infancy of aspiration pneumonia or respiratory distress (glossoptosis, tracheomalacia). The somewhat hoarse cry, noted in several infants, may be related to abnormal laryngeal cartilages that may, in turn, reflect poor prognosis (27). Death later in life results from progressive cervical kyphosis with medullary compression (13a) (Fig. 7–45F). Intelligence is normal.

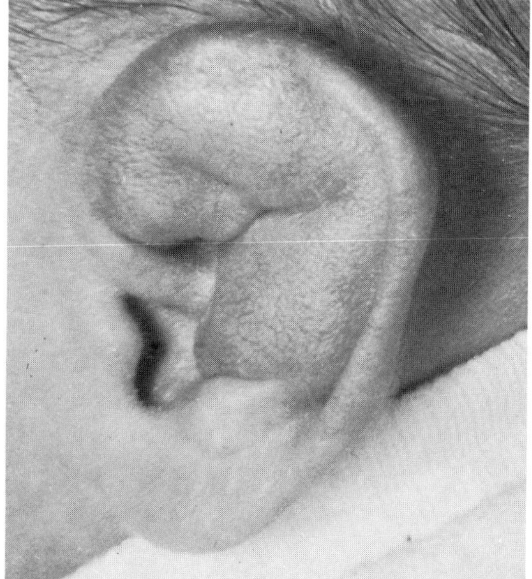

Fig. 7–43. *Diastrophic dysplasia.* Cystic ear during hemorrhagic phase, which resolves, leaving pinna calcified and distorted in form.

**Facies.** The face tends to be square, with a narrow nasal bridge, broad midnose, flared nostrils, and circumoral fullness (18,27). Bilateral, but at times asymmetric, deformity of the pinnae has been noted in over 80% (27). It may be evident within the first few days or weeks of life as a cystic swelling from which serosanguineous fluid may be

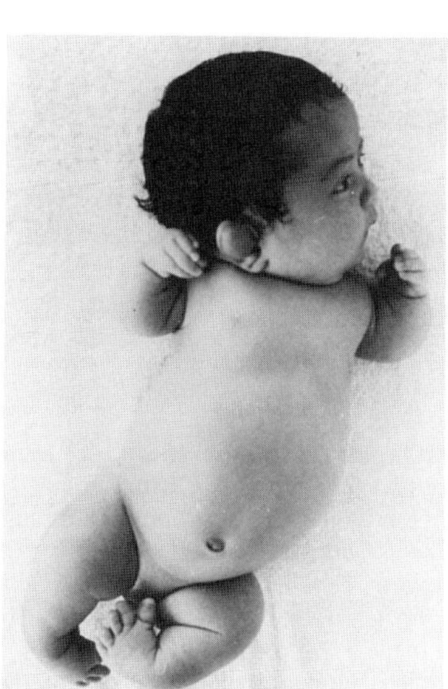

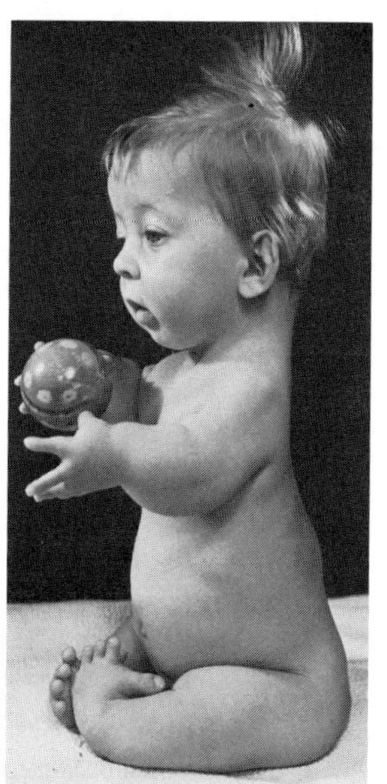

A B

Fig. 7–42. *Diastrophic dysplasia.* (A) Micromelia, talipes equinovarus, cystic ear. (B) Micromelic dwarfism, scoliosis, bilateral talipes equinovarus, deviated thumbs and halluces. (A courtesy of *C Gonzales,* São Paulo, Brazil, in MM Cohen Jr, The Child with Multiple Birth Defects, Raven Press, New York, 1982, p 104. B from *J Spranger* and *H Gerken,* Z Kinderheilkd **98:**227, 1967.)

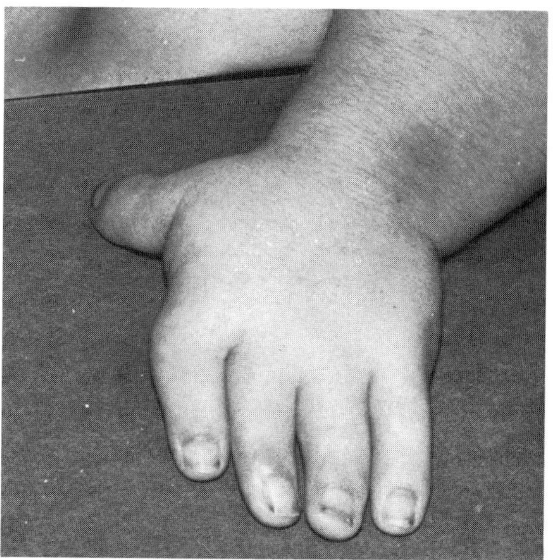

Fig. 7–44. *Diastrophic dysplasia.* Hand and fingers are short, with thumb proximally placed.

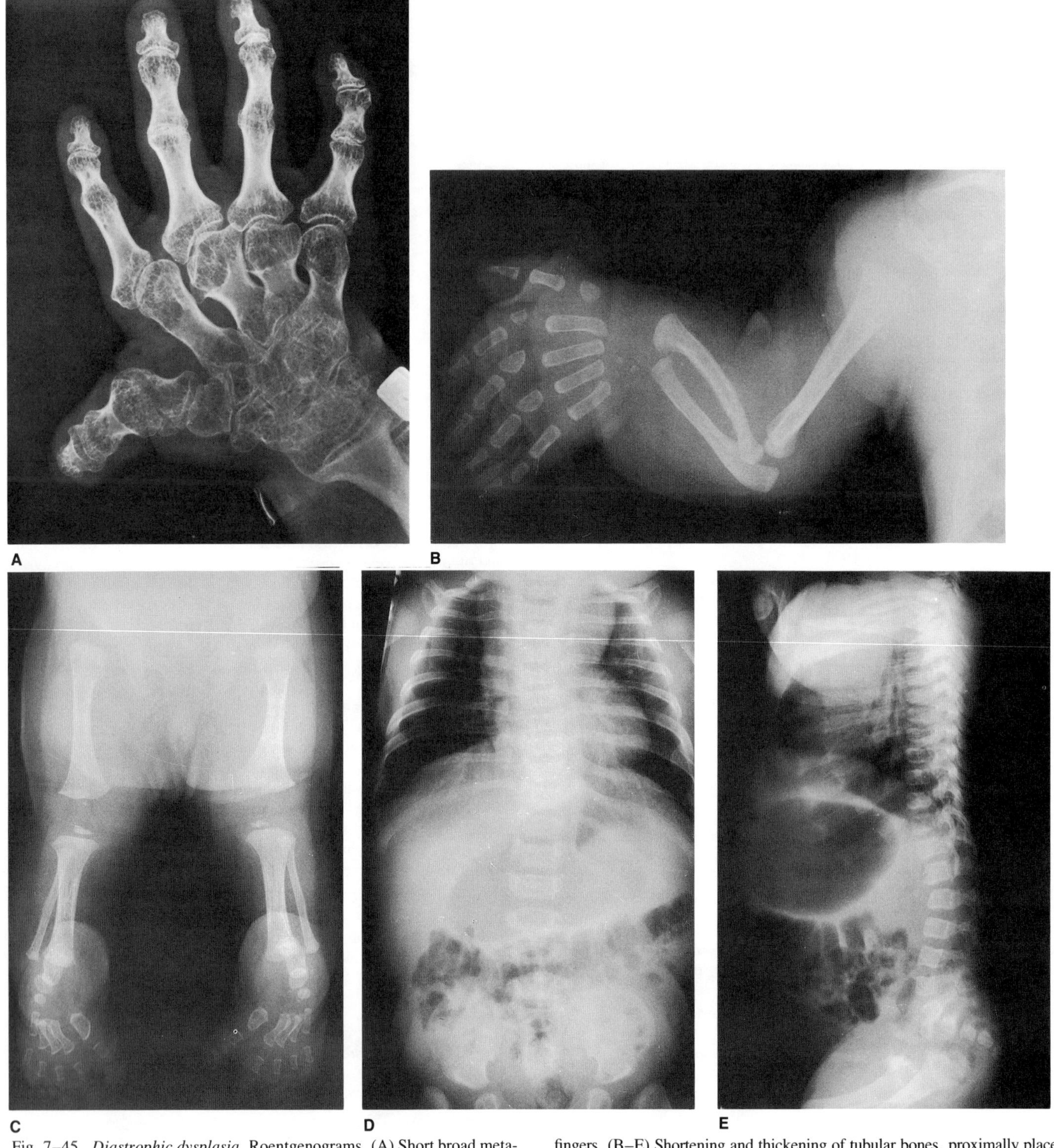


Fig. 7–45. *Diastrophic dysplasia.* Roentgenograms. (A) Short broad metacarpals with widened metaphyses. First metacarpal is particularly shortened. Fusion is evident at proximal interphalangeal joint of second to fourth fingers. (B–E) Shortening and thickening of tubular bones, proximally placed thumb, talipes, narrow interpediculate distance below third lumbar vertebra.

extracted (13,27) (Fig. 7–43). This resolves within a month, but the architecture of the pinna becomes distorted with calcification of the cartilage. The external auditory canals may be narrowed.

**Musculoskeletal alterations.** Mesomelic dwarfism is a constant feature. Mean birth length is about 42 cm. Horton et al (10), studying 72 patients, established growth curves. Mean adult height is about 118 cm with males ranging from 86 to 127 cm and females from 104 to 122 cm (10,27). In a series of 100 patients, Kaitila et al (10a) found a mean adult height of $141.3 \pm 12.1$ cm for males with a range of 129–160 cm and a mean adult height of $124.0 \pm 12.2$ cm with a range of 90–143 cm for females. There is shortening of all limbs (Fig. 7–42A,B). Bilateral talipes equinovarus is severe, becomes worse with age, is resistant to treatment, and tends to recur after therapy. The patients bear their weight on their toes; thus, walking is limited.

The thumbs are proximally inserted, hypermobile, and laterally dis-

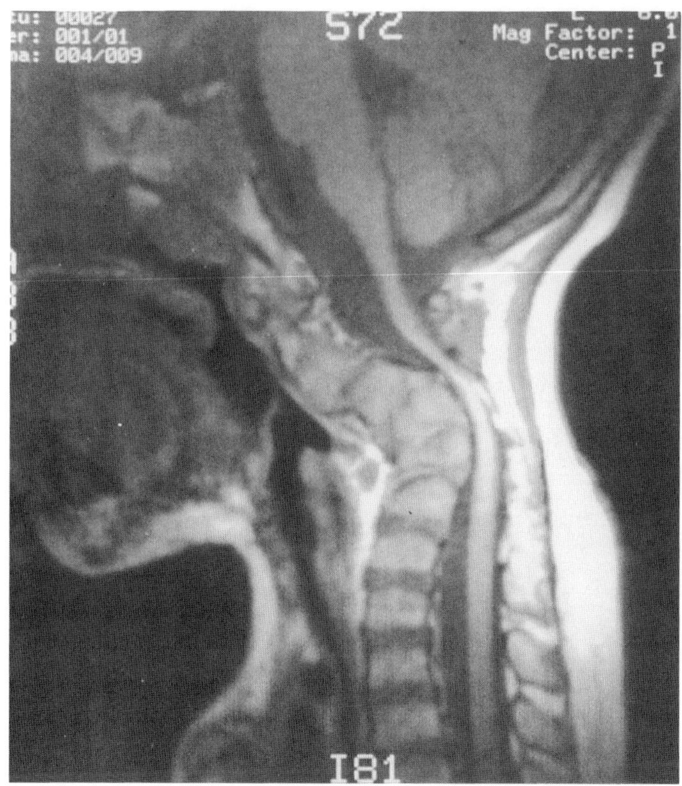


F

Fig. 7–45. (*cont.*) (F) MR showing compression of spinal cord at level of C4. (A from *J Spranger* and *H Gerken,* Z Kinderheilkd **98**:227, 1967. B–E courtesy of *G Aicardi,* Genoa. F from *J Krecak* and *RJ Starshak,* Pediatr Radiol **17**:321, 1987.)

placed (''hitchhiker's thumb'') (7). The broad hands and shortened fingers often exhibit ulnar deviation. Frequent webbing, contractures, and fixation of the interphalangeal finger joints occur (Fig. 7–44). The metacarpophalangeal profile pattern has been discussed by Butler et al (2a).

Scoliosis is not present at birth but is progressive, particularly in preadolescent years. It tends to become severe and rigid with a large rotary component (2,3,6,7,11). Kyphosis is occasionally associated. Lordosis is frequent and is not progressive (5). All patients reported by Herring (6) and Bethem et al (2) presented spina bifida occulta in the cervical spine. Flexion contracture and/or subluxation or dislocation, particularly of the hips and, to a lesser extent, of the knees and shoulders, are common and progressive, further reducing height (27). Inguinal hernia has also been described (26,27).

Radiographic changes include shortening and thickening of nearly all tubular bones. The epiphyses have delayed appearance and are flattened and distorted. With time, the metaphyses become widened, irregular, and deformed. The humerus is less shortened than the radius and ulna. The thumb is proximally placed, and the first metacarpal is small and rounded (Fig. 7–45A–C). Synostosis of the proximal interphalangeal joints is a constant feature. Carpal development is accelerated, but secondary centers for the metacarpals, metatarsals, and phalanges are retarded in appearance. The metacarpals are broader at the distal end than at the proximal end. The first metatarsal is broader and wider than the others (1,3,16,25,28).

Dislocation or subluxation of the hips and coxa vara are associated with flattening of the acetabular roof and delayed appearance and poor development of the capital femoral epiphysis. The patella is often subluxated. Progressive kyphosis of the cervical region of the spine, with subluxation of the second and third cervical vertebrae, is frequent and may result in spinal cord compression (2,6,13a,16). The interpediculate distance tends to narrow below the third lumbar vertebra (2) (Fig. 7–45C–F).

Precocious ossification of costal cartilages and calcification of pinnal cartilage occur in at least 80%. Intracranial calcification has also been described (27).

Histopathologic study has shown a generalized degenerative disorder of cartilage, with focal death of cells followed by matrix dissolution, cyst formation, fibrovascular scarring, and dystrophic ossification (20). Histopathologic characteristic findings include degeneration of chondrocytes, abnormal distribution of collagen in resting cartilage, and large cystic lesions in the resting cartilage which exhibits intracartilaginous ossification (9).

**Oral manifestations.** The mouth is full and broad, with the lower lip slightly larger than the upper. Cleft palate has been found in 25–50% (1,16,23,26). Rintala et al (21) reported 43% with cleft palate and 32% with submucous palatal cleft. Oddly, this latter group did not have nasal speech. They further estimated that 10% had glossoptosis due to micrognathia. Most patients have some degree of micrognathia.

**Differential diagnosis.** The disorder may be confused with a plethora of chondrodysplatic and other disorders, such as arthrogryposis, *achondroplasia,* and *cartilage–hair hypoplasia* (16,27). *Pseudodiastrophic dysplasia,* a rare recessive bone dysplasia, has a somewhat similar phenotype, but has interphalangeal joint dislocations and platyspondyly. The characteristic degeneration of cartilage is not evident.

Pinnal calcification can rarely be seen in other disorders such as ochronosis, Addison's disease, acromegaly, systemic chondromalacia, familial cold hypersensitivity (27), *Nance–Sweeney chondrodysplasia* (17), and *Keutel syndrome* (12).

**Laboratory aids.** Prenatal diagnosis has been made by ultrasonography (4a,4b,19).

### References (Diastrophic dysplasia)


1. Amuso SJ: Diastrophic dwarfism. *J Bone Jt Surg* **50A**:113–118, 1968.

2. Bethem D et al: Disorders of the spine in diastrophic dwarfism. *J Bone Jt Surg* **62A**:529–536, 1980.

2a. Butler MG et al: Metacarpophalangeal pattern profile analysis in diastrophic dysplasia. *Am J Med Genet* **28**:685–689, 1987.

3. Dorn U et al: Diastrophic dwarfism. *Z Orthopäd* **118**:359–366, 1980.

4. Duken J: Zur Frage der mechanischen Entstehung der Chondrodystrophie. *Mschr Kinderheilkd* **22**:348–355, 1921.

4a. Gembruch U et al: Diastrophic dysplasia: A specific prenatal diagnosis by ultrasound. *Prenat Diagn* **8**:539–545, 1988.

4b. Gollop TR, Eigier A: Brief clinical report: Prenatal ultrasound diagnosis of diastrophic dysplasia at 16 weeks. *Am J Med Genet* **27**:321–324, 1987.

5. Gustavson KH et al: Lethal and nonlethal diastrophic dysplasia. *Clin Genet* **28**:321–334, 1985.

6. Herring JA: The spinal disorders in diastrophic dwarfism. *J Bone Jt Surg* **60A**:177–182, 1978.

7. Hollister DW, Lachman RS: Diastrophic dwarfism. *Clin Orthoped* **114**:11–69, 1976.

8. Horton W et al: The phenotypic variability of diastrophic dysplasia. *J Pediatr* **93**:609–613, 1978.

9. Horton W et al: Diastrophic dwarfism: A histochemical and ultrastructural study of the endochondral growth plate. *Pediatr Res* **13**:904–909, 1979.

10. Horton W et al: Growth curves for height for diastrophic dysplasia, spondyloepiphyseal dysplasia congenita and pseudoachondroplasia. *Am J Dis Child* **136**:316–319, 1982.

10a. Kaitila I et al: Clinical expression and course of diastrophic dysplasia. March of Dimes Clinical Genetics Conference, Baltimore, July 10–13, 1988.

11. Kash IJ et al: Cervical cord compression in diastrophic dwarfism. *J Pediatr* **84**:862–864, 1974.

12. Keutel J et al: A new autosomal recessive syndrome: Peripheral pulmonary stenosis, brachytelephalangism, neural hearing loss and abnormal cartilage calcification/ossification. *Birth Defects* **8**(5):60–68, 1972.

13. Kratz RC: Congenital perichondritis associated with achondroplasia. *Laryngoscope* **66**:93–97, 1956.

13a. Krecak J, Starshak RJ: Cervical kyphosis in diastrophic dwarfism: CT and MR findings. *Pediatr Radiol* **17**:321–322, 1987.

14. Lachman R et al: Diastrophic dysplasia: The death of a variant. *Radiology* **140**:79–86, 1981.

15. Lamy M, Maroteaux P: Le nanisme diastrophique. *Presse Méd* **68**:1977–1980, 1960.
16. Langer LO: Diastrophic dwarfism in early infancy. *Am J Roentgenol* **93**:399–404, 1965.
17. Nance WE, Sweeney A: A recessively inherited chondrodystrophy. *Birth Defects* **6**(4):25–27, 1970.
18. Naselli A et al: La displasia diastrofica. *Min Pediatr* **35**:891–897, 1983.
19. O'Brien GD et al: Early prenatal diagnosis of diastrophic dwarfism by ultrasound. *Br Med J* **280**:1300, 1980.
20. Rimoin D: Histopathology and ultrastructure of cartilage in the chondrodystrophies. *Birth Defects* **10**(9):1–18, 1974.
21. Rintala A et al: Cleft palate in diastrophic dysplasia. *Scand J Plast Surg* **20**:45–49, 1986.
22. Schenk AK: L'achondroplasie chez l'homme. Thesis, St. Petersbourg Military Academy, 1910 (cited by Ref 15).
23. Spranger J, Gerken H: Diastrophischer Zwergwuchs. *Z Kinderheilkd* **98**:227–234, 1967.
24. Stanescu V et al: Pathogenesis of pseudoachondroplasia and diastrophic dysplasia. *Prog Clin Biol Res* **104**:385–394, 1982.
25. Taybi H: Diastrophic dwarfism. *Radiology* **80**:1–10, 1963.
26. Vazquez AM, Lee FA: Diastrophic dwarfism. *J Pediatr* **72**:234–242, 1968.
27. Walker BA et al: Diastrophic dwarfism. *Medicine* **51**:41–59, 1972.
28. Wilson DW et al: Diastrophic dwarfism. *Arch Dis Childh* **44**:48–59, 1969.

## Pseudodiastrophic dysplasia

Burgio et al (1) in 1974 were the first to describe an autosomal recessive condition that they called pseudodiastrophic dwarfism. Five additional patients have been reported (2,3). Although the clinical findings are somewhat similar to those of diastrophic dysplasia, distinct radiologic and chondroosseous histologic differences are evident. There are rhizomelic shortening of the limbs and severe club foot deformity, but no ''hitchhiker's thumb'' (Figs. 7–46 and 7–47). The patients described by Burgio et al (1) and Canki et al (2) died within the first year of life. However, Eteson et al (3) reported two children over the age of 4.

Among the seven cases described to date, three had cleft palate, one submucous cleft palate, and one bifid uvula. At birth, the cranium is enlarged with midface hypoplasia. The sclerae appear somewhat bluish. There is mild hypertelorism, flat nasal bridge, large malformed earlobes, and folded superior helices, dislocated elbows, interphalangeal dislocations, and club feet, long clavicles, slightly short ribs with anterior flaring, and platyspondyly with scoliosis. The pelvis appears normal with horizontal acetabular roofs and medial and lateral acetabular spikes. Rhizomelic shortening of limbs, elbow dislocations, and interphalangeal joint dislocations of the hands and club foot are characteristic. Later in infancy and early childhood, the vertebral bodies appear hypoplastic and platyspondylic scoliosis progressively worsens. The sacrosciatic notches become smaller and there is improvement of the hand and foot abnormalities with physical therapy.

Radiographic features that distinguish the syndrome from that of *diastrophic dysplasia* are platyspondyly and elbow and proximal interphalangeal joint dislocations. The degeneration of cartilage, characteristic of diastrophic dysplasia, is not seen in pseudodiastrophic dysplasia. In pseudodiastrophic dysplasia there is no cystic enlargement of the pinnae.

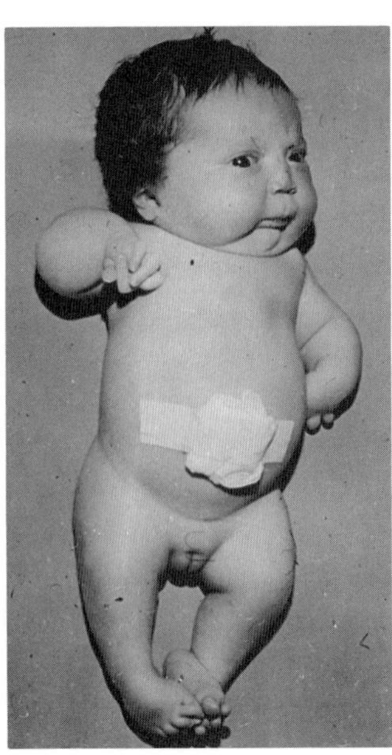
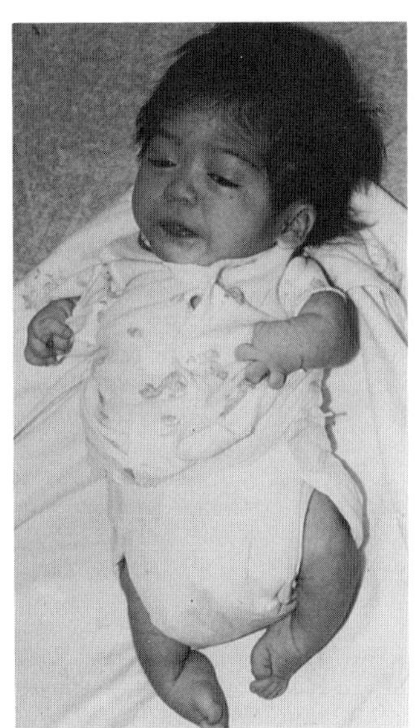

Fig. 7–46. *Pseudodiastrophic dysplasia.* (A,B) Flat nasal bridge, short limbs, finger contractures, clubfeet. (A from *G-R Burgio et al,* Arch Fr Pédiatr **31:**681, 1974. B from *DE Eteson et al,* J Pediatr **109:**635, 1986.)

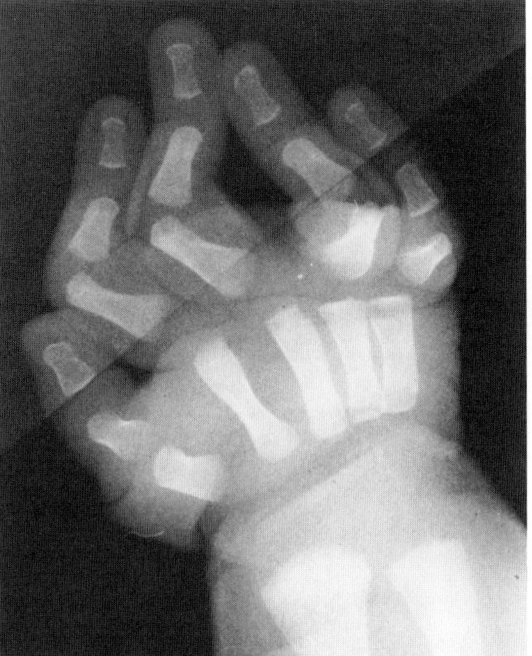

Fig. 7–47. *Pseudodiastrophic dysplasia.* Roentgenogram of hand showing multiple interphalangeal and metacarpophalangeal joint dislocations with resultant ulnar and radial deviation of digits. (From *DE Eteson et al,* J Pediatr **109:**635, 1986.)

### References (Pseudodiastrophic dysplasia)

1. Burgio GR et al: Nanisme pseudodiastrophique: étude de deux soeurs nouveau-nées. *Arch Fr Pédiatr* **31**:681–696, 1974.
2. Canki N et al: Le nanisme pseudodiastrophique: à propos d'une observation. *J Génét Hum* **27**:247–252, 1979.
3. Eteson DJ et al: Pseudodiastrophic dysplasia: A distinct newborn skeletal dysplasia. *J Pediatr* **109**:635–641, 1986.

## Dyggve–Melchior–Clausen syndrome

The syndrome was probably first defined by Dyggve et al (2) in 1962, although earlier examples appear to have been published (5). It is a disorder of short-trunk dwarfism and mental retardation and has most often been mistaken for Morquio syndrome (Figs. 7–48 to 7–50). About 35 cases have been reported.

There is genetic heterogeneity, both autosomal recessive (1–3,8,11,16,17) and X-linked recessive (Smith–McCort syndrome) (6,9,10,14,16,18) forms having been recognized. Among those with autosomal recessive inheritance, several are of Lebanese origin (1).

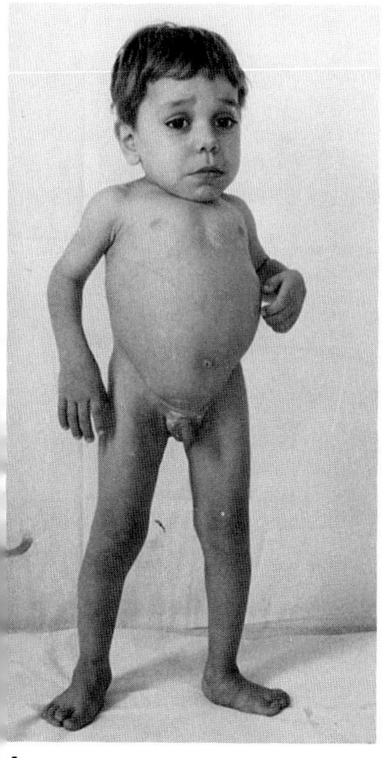
A

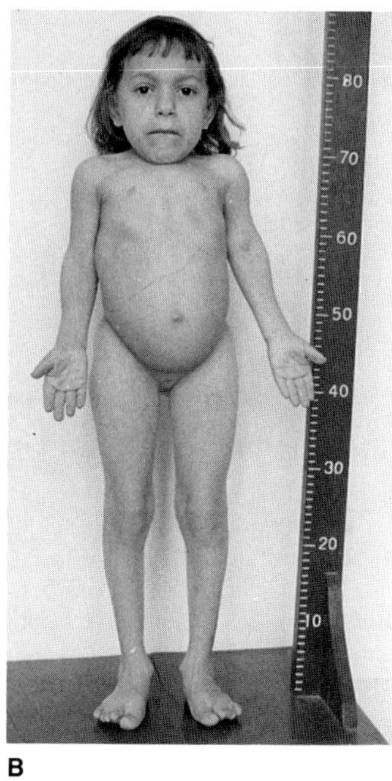

B

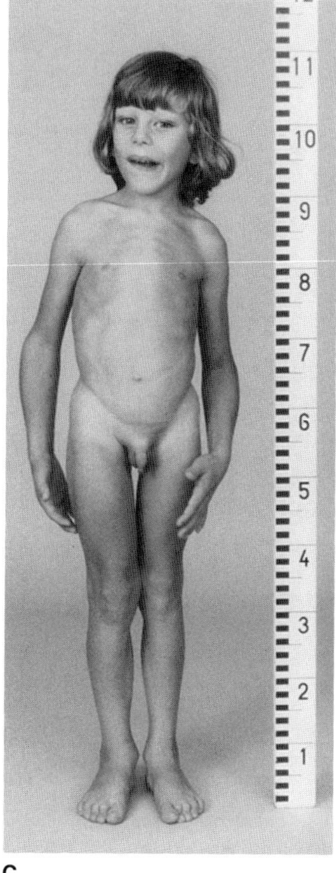

C

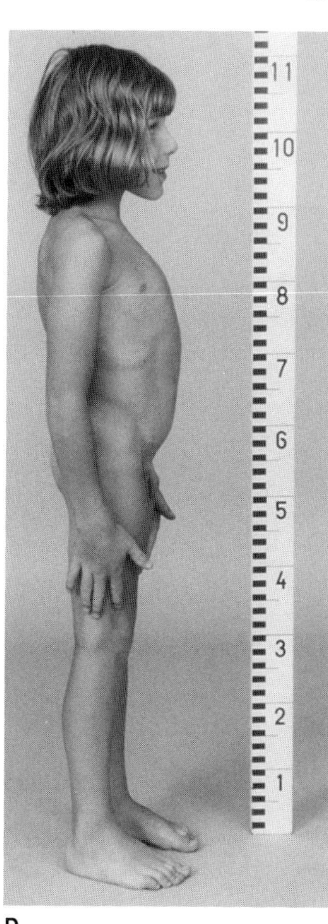

D

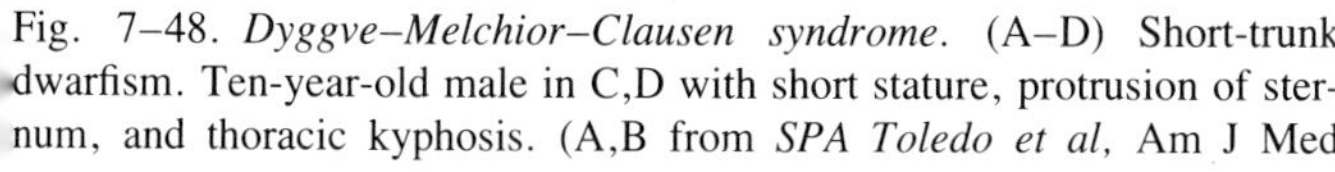

Fig. 7–48. *Dyggve–Melchior–Clausen syndrome.* (A–D) Short-trunk dwarfism. Ten-year-old male in C,D with short stature, protrusion of sternum, and thoracic kyphosis. (A,B from *SPA Toledo et al,* Am J Med Genet **4**:255, 1979. C,D from *R Schlaepfer et al,* Helv Paediatr Acta **36:**543, 1981.)

Neonatal measurements are normal. Life span, apart from the risk of atlantoaxial instability, is normal (vide infra).

**Facies.** The face is coarse with a prominent mandible.

**Central nervous system.** In the autosomal recessive form, milestones are delayed and progressive mental deficiency is mild to moderate (IQ 35–65); there is also microcephaly, with the adult head circumference rarely exceeding 47 cm. In the X-linked form, head circumference is only mildly reduced (53–54 cm) and intelligence is normal.

**Skeletal changes.** In the autosomal recessive form, atlantoaxial instability predisposes to spinal cord compression (11). Adult height ranges between 115 and 127 cm and in the X-linked form between 130 and 150 cm. By the age of 5–8 years, disproportionate stature becomes evident. The trunk and neck are short, the chest is barrel shaped with protruding sternum, and there are thoracic kyphosis and lumbar lordosis (Fig. 7–48). Joint mobility is somewhat restricted. There is rhizomelic limb shortening and genua valga. Most assume a crouching stance and walk with a waddling gait. The hands and feet are broad. Radiographic changes (Figs. 7–49 and 7–50) include disproportionately large facial bones with hyperpneumatization of paranasal sinuses and calvarial thickening in the parietal and occipital regions. The vertebrae are flattened with irregular end plates and notching of the anterosuperior bodies in the lumbar region. The ilia are small and irregular with a wide pubic symphysis. The iliac crests have a lacy outline during childhood. The acetabula are sloping and dysplastic. The femoral heads are small, flattened, and usually dislocated. The shoulder joints are similarly affected. The humoral diaphysis is short and curved with distal metaphyseal flaring. The radial heads and olecranons are dysplastic and the metacarpals shortened. The proximal row of carpal bones is small. Otherwise the hand bones are essentially normal. The bones of the lower extremities are less severely affected (1,12,13,15).

On light microscopy, resting cartilage cells show lacunae containing clusters of five or more chondrocytes in some areas. The cartilage matrix is very fibrous. Horton and Scott (7) described clusters of degenerating chondrocytes. Electron microscopic studies revealed that the chondrocytes exhibit widened cisternae of rough endoplasmic reticulum and vesicles coated with a single smooth single-layered membrane (4). Biochemical analysis of cartilage showed increased glucosaminoglycans (4).

**Differential diagnosis.** In contrast to *Morquio syndrome,* there is no mucopolysacchariduria and no corneal clouding. There are normal hearing and normal teeth, microcephaly, and mental retardation. The skeletal changes are truly different in spite of somewhat similar appearance (e.g., the lacy iliac crest, the eventually normal bones of the hands).

**Laboratory aids.** Some investigators have found elevated chondroitin sulfate-*N*-acetylgalactosamine 6-sulfate sulfatase and reduced aryl sulfatase glycoprotein-AMP metabolism (3).

### References (Dyggve–Melchior–Clausen syndrome)

1. Bonafede RP, Beighton P: The Dyggve–Melchoir–Clausen syndrome in adult siblings. *Clin Genet* **14**:24–30, 1978.

2. Dyggve H et al: Morquio-Ullrich's disease: An inborn error of metabolism? *Arch Dis Childh* **37**:525–534, 1962.

3. Dyggve H et al: The Dyggve–Melchior–Clausen (DMC) syndrome: A 15 year follow-up and a survey of the present clinical and chemical findings. *Neuropädiatrie* **8**:429–442, 1977.

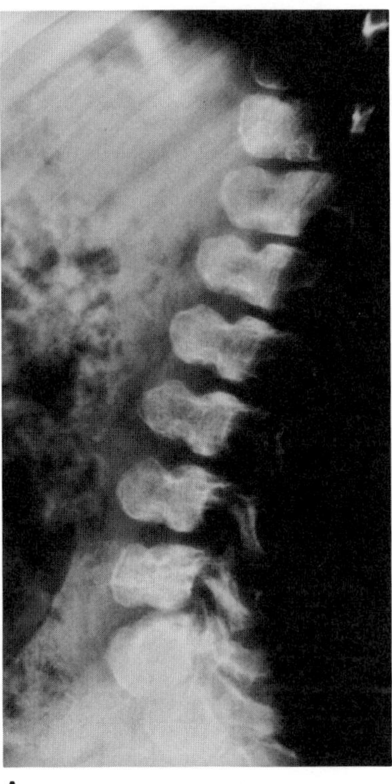

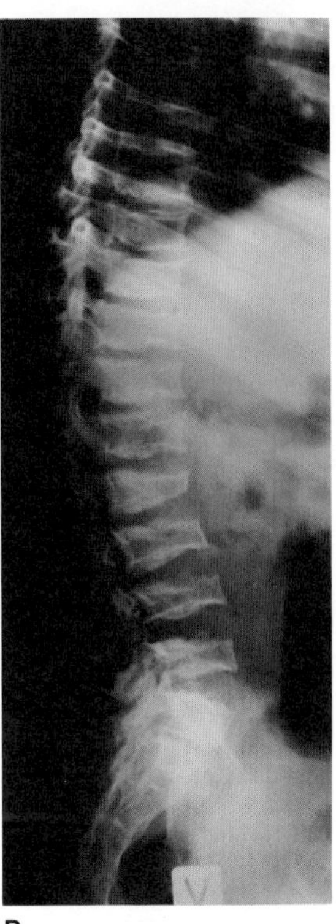

A B

Fig. 7–49. *Dyggve–Melchior–Clausen syndrome.* (A) Platyspondyly with central indentation and ventral pointing. (B) Marked platyspondyly in older individual. (A from *R Schlaepfer et al,* Helv Paediatr Acta **36:**543, 1981. B from *HV Dyggve et al,* Neuropädiatrie **8:**429, 1977.)

4. Engfeldt B et al: Dyggve–Melchior–Clausen dysplasia. *Acta Paediatr Scand* **72**:269–274, 1983.

5. Farrell MJ et al: Morquio's disease with mental defect. *Arch Neurol Psychiatr* **48**:456–464, 1942.

6. Gwinn JL, Barnes GR: Morquio–Brailsford's disease. *Am J Dis Child* **115**:347–348, 1968.

7. Horton WA, Scott CI: Dyggve–Melchior–Clausen syndrome: A histochemical study of the growth plate. *J Bone Jt Surg* **64A**:408–414, 1982.

Fig. 7–50. *Dyggve–Melchior–Clausen syndrome.* Short and broad iliac wings with lacy crests. Ischial and pubic bones somewhat plump with flattened, poorly demarcated acetabula; hypoplasia and shortening of femoral necks. (From *HV Dyggve et al,* Neuropädiatrie **8:**429, 1977.)

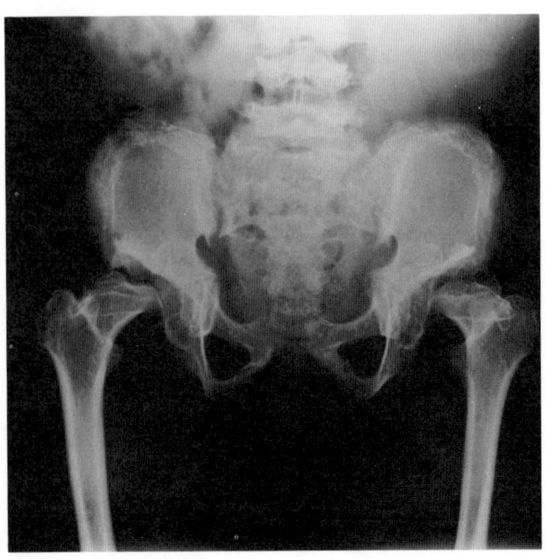

8. Kaufman RL et al: The Dyggve–Melchior–Clausen syndrome. *Birth Defects* **7**(1):144–149, 1971.

9. Koppers B: Smith–McCort syndrome. *Roefo* **130**:213–222, 1979.

10. Linker A et al: Morquio's disease and mucopolysaccharide excretion. *J Pediatr* **77**:1039–1047, 1970.

11. Naffah J: The Dyggve–Melchior–Clausen syndrome. *Am J Hum Genet* **28**:607–614, 1976.

12. Schlaepfer R et al: Das Dyggve–Melchior–Clausen-Syndrom. *Helv Paediatr Acta* **36**:543–559, 1981.

13. Schorr R et al: The Dyggve–Melchior–Clausen syndrome. *Am J Roentgenol* **128**:107–113, 1977.

14. Smith R, McCort J: Osteochondrodystrophy (Morquio–Brailsford type). *Calif Med* **88**:55–59, 1958.

15. Spranger J et al: The Dyggve–Melchior–Clausen syndrome. *Radiology* **114**:415–421, 1975.

16. Spranger J et al: Heterogeneity of Dyggve–Melchior–Clausen dwarfism. *Hum Genet* **33**:279–287, 1976.

17. Toledo SPA et al: Dyggve–Melchior–Clausen syndrome: Genetic studies and report of affected sibs. *Am J Med Genet* **4**:255–261, 1979.

18. Yunis E et al: X-linked Dyggve–Melchior–Clausen syndrome. *Clin Genet* **18**:284–290, 1980.

## Dyssegmental dysplasia, type Silverman–Handmaker

Apparently the first illustrated report of this bone dysplasia, which is lethal at birth, was that of Simmonds (12) who, in 1900–1901, was trying to illustrate the value of radiographic diagnosis.

Almost 70 years were to pass before the second case was described by Silverman (11). There have been more than a dozen examples published under various terms (1–13), a possible example being that of Maisonneuve et al (9). Gorlin and Langer (4) described the early history of the disorder.

Inheritance is autosomal recessive (1–3,8).

Polyhydramnios has been noted in about 25%. Birth weight is generally normal but birth length is reduced as a result of short spine and curved lower limbs. The infants are either stillborn (40%) or live only a day or two. Hydrocephaly and occipital exencephalocele have been reported (1,7). In a few cases there was a defect in the occipital bone (1,5). Polymicrogyria has been reported (5) as well as Dandy–Walker malformation and cerebellar hypoplasia (1).

Fig. 7–51. *Dyssegmental dysplasia, type Silverman–Handmaker.* Abbreviated limbs, overlapping fingers with cortical thumbs, flat nose, hirsutism of lower legs. [From *RJ Gorlin* and *LO Langer Jr,* Birth Defects **14**(6B):193, 1978.]

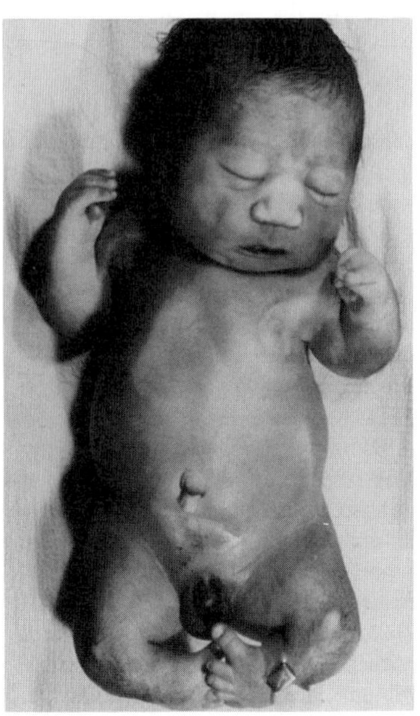

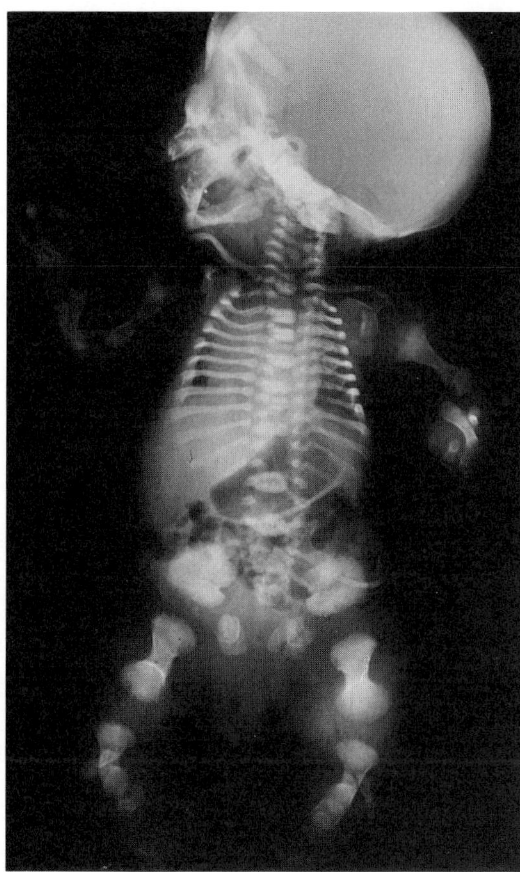

Fig. 7–52. *Dyssegmental dysplasia, type Silverman–Handmaker.* Roentgenogram showing severe symmetric shortening of all tubular bones with marked metaphyseal flaring and cupping. The ulnae appear to be somewhat folded on themselves. The trunk is short and the thorax narrow. Vertebral bodies are of variable sizes, thickness, and width. The iliac bones have decreased vertical dimensions with hypoplasia of the horizontal and inferior margins giving them a rounded shape. Acetabula are small in comparison to the wide proximal femora. The sacrosciatic notches are narrow and deep.

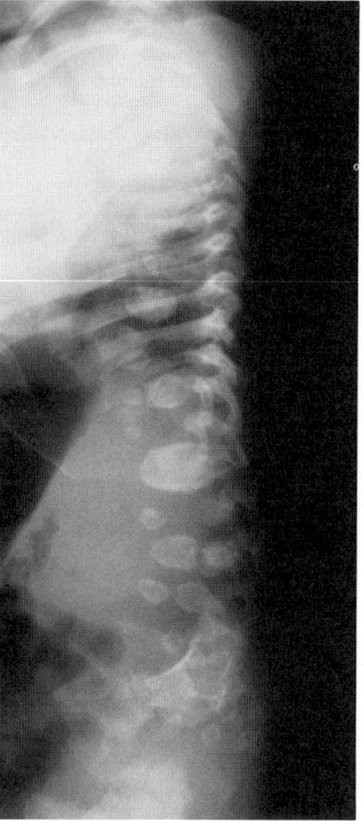

Fig. 7–53. *Dyssegmental dysplasia, type Silverman–Handmaker.* Roentgenogram showing abnormal vertebral bodies with anisospondyly.

The facies is somewhat unusual. There is mild blepharophimosis, flat nasal bridge, and hypoplastic supraorbital ridges. The mandible is small and the neck remarkably short (Fig. 7–51). Cleft palate is reported in one-half of the cases (1,7,10,11).

The chest is narrow. Reduced joint mobility is a constant feature. The limbs are short and curved (microcampomelia) with talipes equinovarus (Fig. 7–51). The fingers may wedge and the thumbs may be adducted. Lumbosacral kyphosis may be marked. Hirsutism has been present in about 90% of the cases. Inguinal and/or umbilical hernia have been reported (7). In the few cases that have come to autopsy, hydroureter and hydronephrosis have been observed in the more severe cases (1,3,6,7,10,13). Patent ductus has been found in several patients (1,3,6,7).

Radiographically, there are short trunk and narrow thorax. The vertebral bodies are of variable size, thickness, and width (anisospondyly). Most vertebral bodies consist of two or more ossified masses separated by vertical radiolucent clefts in lateral view. The iliac bones have decreased vertical dimension with hypoplasia of the horizontal and inferior margins giving them a rounded shape. The acetabula are small in comparison to the wide proximal femora. The sacrosciatic notches are narrow and deep. There is severe symmetric shortening of all tubular bones with marked metaphyseal flaring and cupping. The long bones may appear to be somewhat folded upon themselves, particularly the ulnae (Figs. 7–52 and 7–53). The scapulae are small and rounded in comparison with the wide humeral metaphyses. The radius is disproportionately short. The first metacarpals are particularly abbreviated. Prenatal diagnosis by ultrasound has been accomplished (7)

Microscopically, there is a disturbance in enchondral ossification: lack of columnization, ballooning of cartilage cells, mucoid degeneration of resting cartilage, and prominent large unfused calcospherites in the growth plate and calcifying zones. There is accumulation of acid mucopolysaccharides and no increase in collagen fibers (Fig. 7–54) (5,6). The disorder may have a disturbance in formation of $\alpha$1-collagen chains (13).

Differential diagnosis includes *Kniest syndrome* and *Burton syndrome*. More disturbed chondrocyte columnization is seen in Kniest syndrome but the same "Swiss cheese" cystic changes are observed. Prenatal diagnosis by ultrasound has been accomplished (1a).

Fig. 7–54. *Dyssegmental dysplasia, type Silverman–Handmaker.* Puddle-like spaces in cartilage matrix.

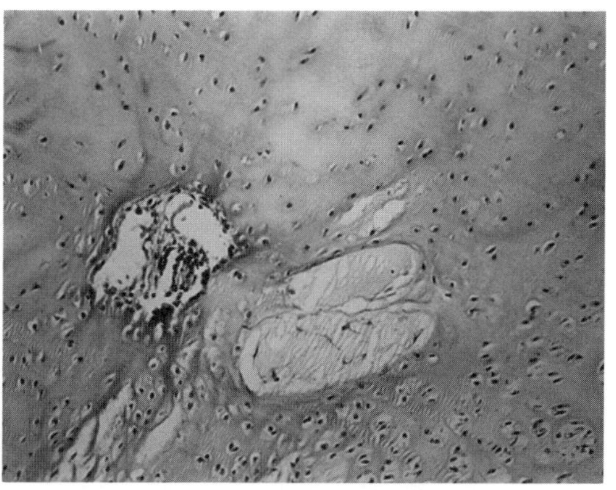

### References (Dyssegmental dysplasia, type Silverman–Handmaker)

1. Aleck KA et al: Dyssegmental dysplasias: Clinical, radiographic, and morphologic evidence of heterogeneity. *Am J Med Genet* **27**:295–312, 1987.

1a. Andersen PE Jr et al: Dyssegmental dysplasia in siblings: Prenatal ultrasonic diagnosis. *Skel Radiol* **17**:29–31, 1988.

2. Fasanelli S et al: Dyssegmental dysplasia (report of two cases with review of the literature). *Skel Radiol* **14**:173–177, 1985.

3. Goodlin RC, Lowe EW: Unexplained hydramnios associated with a thanatophoric dwarf. *Am J Obstet Gynecol* **118**:873–875, 1974.

4. Gorlin RJ, Langer LO Jr: Dyssegmental dwarfism: Lethal anisospondylic camptomicromelic dwarfism. *Birth Defects* **14**(6B):193–197, 1978.

5. Greco MA et al: Dyssegmental dwarfism: A histologic study of osseous and nonosseous cartilage. *Hum Pathol* **15**:490–493, 1984.

6. Gruhn JG et al: Dyssegmental dwarfism: A lethal anisospondylic camptomicromelic dwarfism. *Am J Dis Child* **132**:382–386,1978.

7. Handmaker SD et al: Dyssegmental dwarfism: A new syndrome of lethal dwarfism. *Birth Defects* **13**(3D):79–90, 1978.

8. Kim HJ et al: Prenatal diagnosis of dyssegmental dwarfism. *Prenat Diagn* **6**:143–150, 1986.

9. Maisonneuve J et al: Nanisme dyssegmentaire. *Pédiatrie* **39**:273–277, 1984.

10. Miething R et al: Dyssegmentaler Zwergwuchs: Bericht über zwei Fälle. *Radiologe* **21**:190–194, 1981.

11. Silverman FN: Forms of dysostotic dwarfism of uncertain classification. *Ann Radiol* **12**:1005–1008, 1969 (Figs. 76A,B).

12. Simmonds M: Untersuchungen von Missbildungen mit Hilfe des Röntgenverfahrens (Case 2) *Fortschr Geb Röntgenstr* **4**:197–211, 1900–1901.

13. Svejcar J: Biochemical abnormalities on connective tissue of osteodysplasty of Melnick–Needles and dyssegmental dwarfism. *Clin Genet* **23**:369–375, 1983.

## Dyssegmental dysplasia, type Rolland–Desbuquois

Rolland et al (5), in 1972, reported an infant with a short-limbed chondrodysplasia that resembled Kniest syndrome (Figs. 7–55 to 7–58). About a dozen examples have been described by other authors (1–4). It is considered to be a mild form of dyssegmental dysplasia (1–2). The disorder is also lethal but many survive past infancy.

Fig. 7–55. *Dyssegmental dysplasia, type Rolland–Desbuquois.* (A,B) Infant with hydrocephaly, occipital encephalocele, short limbs. (From *N Dinno et al,* Eur J Pediatr **123:**39, 1976.)

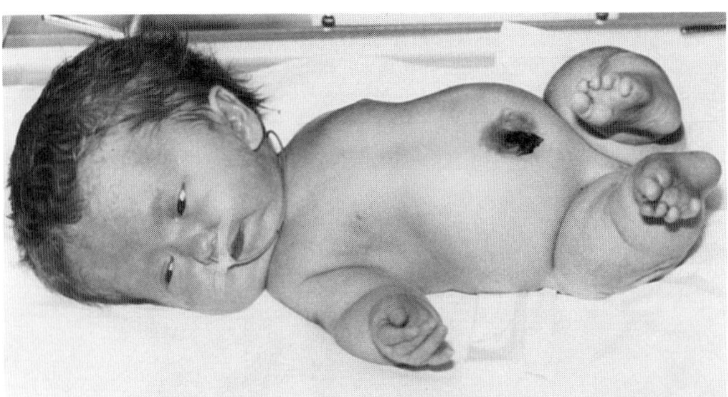

A

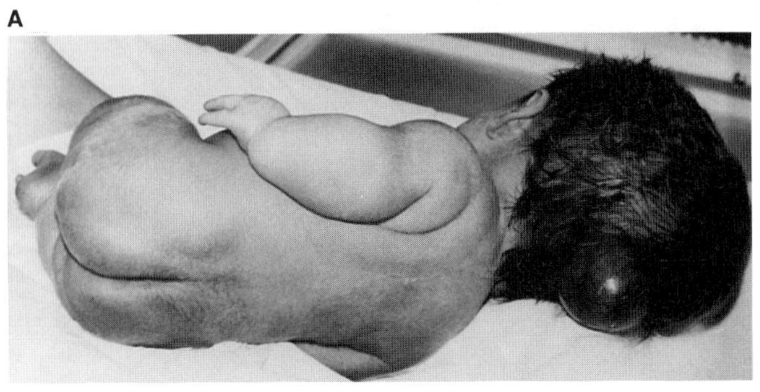

B

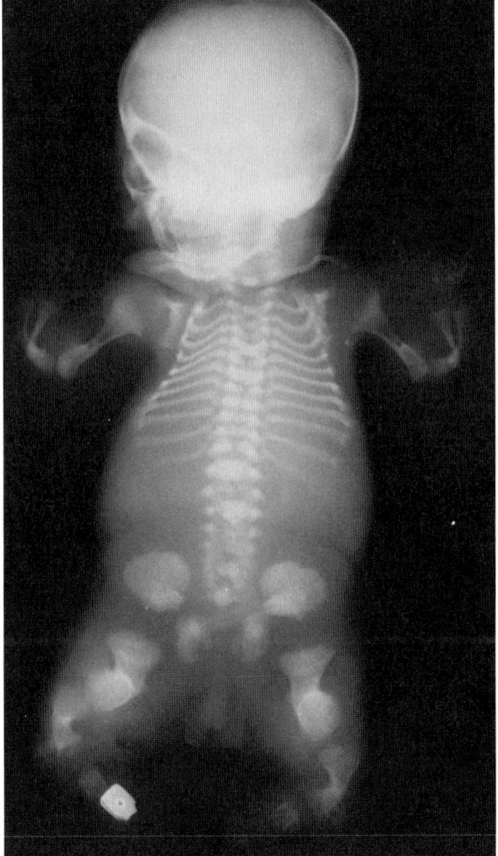

Fig. 7–56. *Dyssegmental dysplasia, type Rolland–Desbuquois.* Roentgenogram showing short broad long bones with metaphyseal widening. The chest is narrow and the ribs are short and flared. (Courtesy of *LO Langer Jr,* Minneapolis, Minnesota.)

Inheritance is autosomal recessive (1,2,5). All exhibit neonatal distress.

The facies is similar to that seen in dyssegmental dysplasia, type Silverman–Handmaker. The orbits are shallow. The face is round and flat with relative micrognathia. One infant manifested an occipital encephalocele (3), another an occipital defect (1), and two had hydrocephalus (3,4). Another had dislocated lenses (4). Cleft palate has been found in most cases (1–5).

As in the Silverman–Handmaker type, there are microcampomelia, decreased joint mobility, narrow chest, and hirsutism. Hernia is noted in a few infants (1,4,5), but the radiologic changes are less severe. The long bones are short, broad, and bowed with some degree of metaphyseal widening. The pelvis appears essentially normal. Lateral spinal views demonstrate variability in size and shape of the vertebral bodies and prominent coronal clefting, but less than that seen in the Silverman–Handmaker type. There is accelerated carpal bone maturation (4). The ribs are short and flared.

Chondroosseous morphology is characterized by patches of broad collagen fibers. The growth palate is normal but foamy Kniest-like changes may be observed in the resting cartilage cells.

### References (Dyssegmental dysplasia, type Rolland–Desbuquois)

1. Aleck KA et al: Dyssegmental dysplasias: Clinical, radiographic, and morphologic evidence for heterogeneity. *Am J Med Genet* **27**:295–312, 1987.

2. Bueno M et al: Dysplasie dyssegmentaire: à propos de 2 cas familiaux d'evolution letale. *Arch Fr Pédiatr* **41**:269–271, 1984.

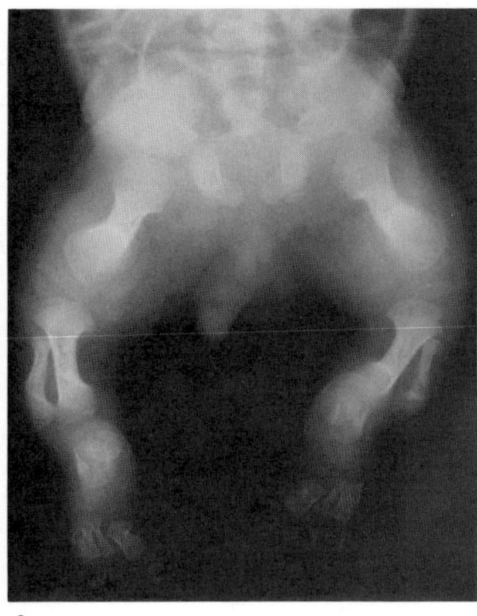
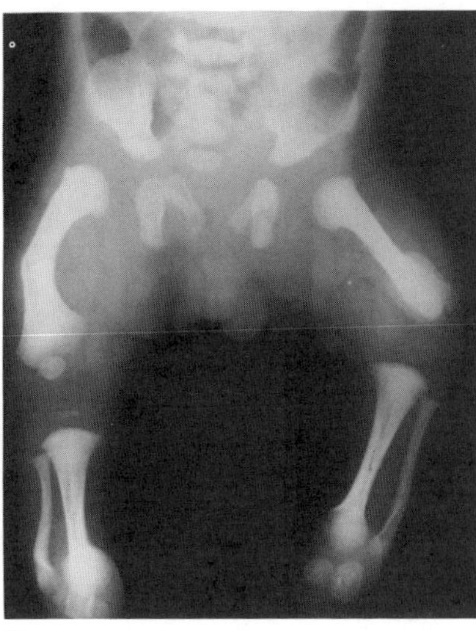

A B

Fig. 7–57. *Dyssegmental dysplasia, type Rolland–Desbuquois.* (A,B) Roentgenograms of lower limbs at different ages. Long bones are short and broad with some metaphyseal widening. Note bowing in B. (Courtesy of *LO Langer Jr,* Minneapolis, Minnesota.)

3. Dinno ND et al: Chondrodysplastic dwarfism, cleft palate and micrognathia in a neonate, a new syndrome? *Eur J Pediatr* **123**:39–42, 1976.

4. Langer LO et al: A severe infantile micromelic chondrodysplasia which resembles Kniest disease. *Eur J Pediatr* **123**:29–38, 1976.

5. Rolland JC et al: Nanisme chondrodystrophique et division palatine chez un nouveau-né. *Ann Pédiatr* **19**:139–143, 1972.

Fig. 7–58. *Dyssegmental dysplasia, type Rolland–Desbuqouis.* Lateral spine roentgenogram showing variability in size and shape of vertebral bodies and prominent coronal clefting. (Courtesy of *LO Langer Jr,* Minneapolis, Minnesota.)

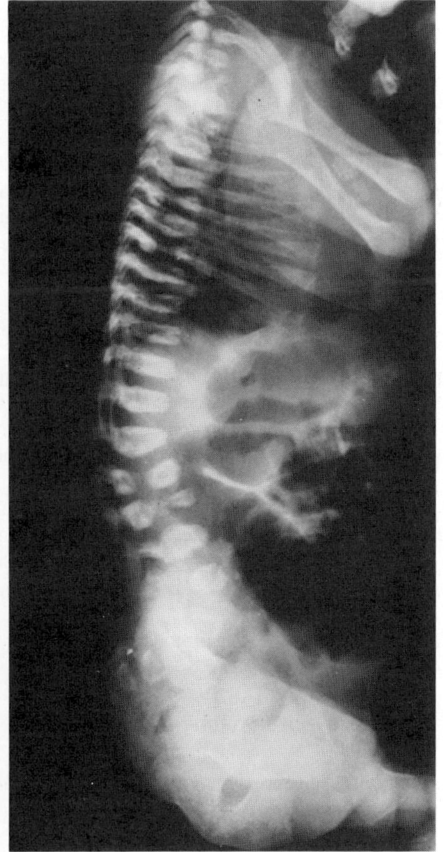

## Ellis–van Creveld syndrome (chondroectodermal dysplasia)

Credit is given to Ellis and van Creveld (10) for describing the complete syndrome and calling it chondroectodermal dysplasia. The disorder, partially described earlier in several reports (1,4,21,24,38), consists of bilateral postaxial polydactyly of the hands, chondrodysplasia of long bones resulting in acromesomelic dwarfism, ectodermal dysplasia affecting nails and teeth, and less often congenital heart malformations (Figs. 7–59 to 7–62).

Fig. 7–59. *Ellis–van Creveld syndrome.* (A) Long appearing thorax with pectus carinatum; mesomelia of lower extremities. (B) Compare phenotype with A. (A from *GB Winter* and *M Geddes,* Br Dent J **122:**103, 1967. B from *HO Bützler et al,* Fortschr Roentgenstr **118:**537, 1973.)

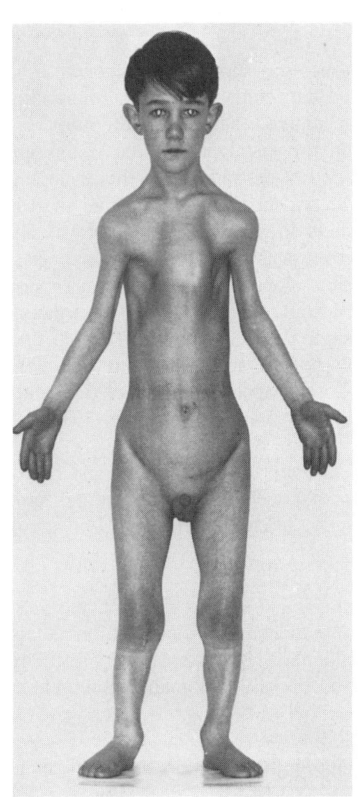
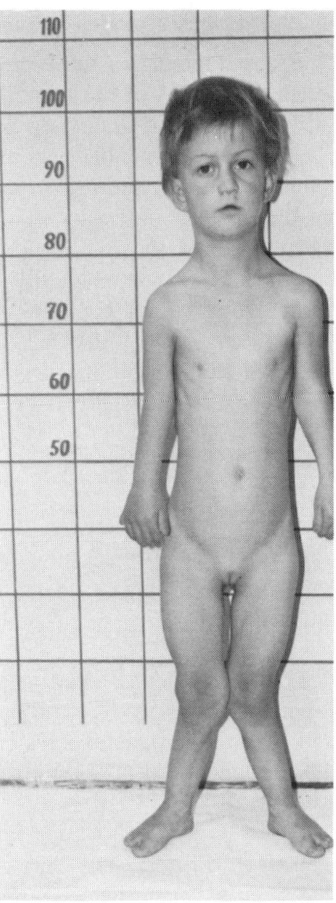


A B

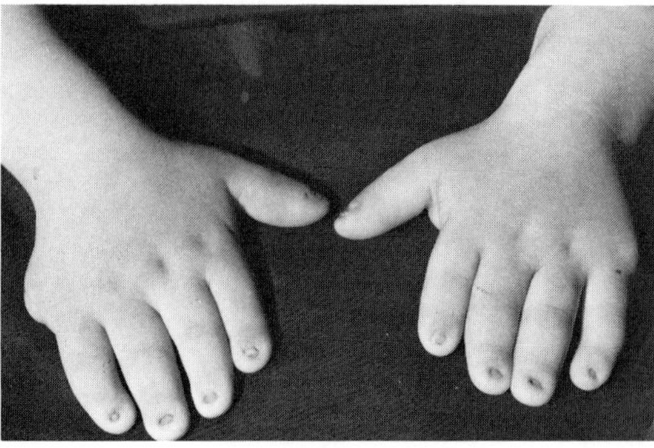

Fig. 7–60. *Ellis–van Creveld syndrome.* Digits abbreviated, with hypoplastic nails. Note that digits have been amputated on ulnar side. (Courtesy of *DH Altman,* Miami, Florida.)

The syndrome has autosomal recessive inheritance (19,20,26); parental consanguinity has been confirmed in about 30% (8,10,11, 19,22,26). Ellis–van Creveld syndrome is the most common type of dwarfism among the Amish. McKusick et al (19) found 52 cases in 50 sibships among the Amish isolate of Lancaster County, Pennsylvania. The disorder has been reported in non-Amish populations as well (8,17). Birth prevalence has been estimated at 7/1,000,000 (29a). There are about 250 reported cases (40).

**Facies.** The facies is not especially characteristic except for a mild defect in the middle of the upper lip, which, although often present, is usually not striking.

**Skeletal anomalies.** The extremities are often plump and markedly shortened progressively distalward, that is, from the trunk to the phalanges (Fig. 7–59). Bilateral postaxial hexadactyly is frequent (Fig. 7–60) and heptadactyly has also been noted (9,14). Frequently, the patient cannot make a tight fist. Only rarely are there extra toes (11,14,19,20), and a wide space is often present between the hallux and the other toes (29). Genua valga (6,22), curvature of the humerus, talipes equinovarus (33,35), talipes calcaneovalgus (8), and pectus carinatum with thoracic constriction (9,19,29) have also been reported.

Radiographically, the tubular bones are short and thickened. The diaphyseal ends of the humerus and the femur are plump. Shortening of the radius and ulna is even more marked than that of the humerus. The proximal end of the ulna and the distal end of the radius are unusually large, and the proximal end of the radius and the distal end of the ulna are unusually small. The widened end of the tibial shaft is irregular, and the ossification centers in the proximal epiphysis are hypoplastic (Fig. 7–61A,B). There is peaking of the proximal tibia, with a long lateral and a short medial slope, resulting in genua valga after the age of 6 years (6,19). The fibula is most severely shortened, being only about 50% of normal length (9). Syncarpalism (hamate and capitate), synmetacarpalism, and polymetacarpalism are frequent (5,14,19,22,23) (Fig. 7–61C,D). Cone-shaped epiphyses of the hands (type 37 of Giedion) are pathognomonic for the syndrome (13) (Fig. 7–61A).

In infancy, the pelvis is dysplastic with low iliac wings and hook-like downward projection of the medial acetabulum. The capital femoral epiphysis may ossify prematurely. In childhood, the pelvic shape normalizes (6).

**Heart.** Congenital heart defects are found in 50–60% (17), the most frequent being single atrium (40%) and endocardial cushion defect (11,17). Some patients have cor triloculare (22) or even cor biloculare (29). Lynch et al (17) have reviewed the heart anomalies comprehensively.

**Hair and nails.** The hair, particularly the eyebrows and pubic hair, has been stated to be thin and sparse (10,14,19,22,36). However, RJ Gorlin has not been impressed by this feature. Nearly all patients have severe dystrophy of the fingernails, which are markedly hypoplastic, thin, and often wrinkled or spoon-shaped (Fig. 7–60).

**Eyes.** The eyes are usually normal, but esotropia (22) and congenital cataract (14,20) have been observed.

**Genitourinary system.** About one-third of male patients have genital anomalies. These anomalies have included cryptorchidism (10), mild epispadias (20), and hypospadias (19,34) Miscellaneous abnormalities of the urinary system have been reviewed by Rosemberg et al (26).

**Central nervous system.** Some patients are mentally retarded (34), but McKusick (19) suggests that retardation is not an integral part of the disorder. Hydrocephaly has been noted in several instances (3,12). Other central nervous system anomalies have been reviewed by Rosemberg et al (26). Dandy–Walker malformation has been described (6a,40).

**Oral manifestations.** The most striking and constant finding is fusion of the middle portion of the upper lip to the maxillary gingival margin so that no mucobuccal fold or sulcus is present anteriorly (2,8,10–14,20,22,25,36). The middle portion of the upper lip appears to have a notch (2) (Fig. 7–62A–D).

Natal teeth have been observed in at least 25% and may be more frequent than reported. Congenitally missing teeth are also a constant finding, particularly in the mandibular anterior region (2,7–11,20, 21,25,27,34) where the alveolar ridge is often serrated (2,8,10,29, 33,40) (Fig. 7–62C). Notching of the lower alveolar process may represent continuation of the normal serrated condition of the gingiva from the third to the seventh month *in utero* (37). Erupted teeth are usually small (2,20), have conical crowns (2,9,10,21,39), and are irregularly spaced (2,10,11,21). Teeth that are not conical are somewhat bicuspid in form, with accentuated cuspal height and deep fissures (Fig. 7–62D). Supernumerary teeth have also been noted on occasion (25,27,39).

**Differential diagnosis.** It may be almost impossible to differentiate radiographically the Ellis–van Creveld syndrome from asphyxiating thoracic dystrophy (16,23). Patients may have identical changes in hands, pelvis, and long bones. Differential diagnosis is based on the following clinical changes present in the Ellis–van Creveld syndrome: cardiac anomalies, nail hypoplasia, fusion of upper lip and gingiva, and, when present, neonatal teeth. Later in life, genua valga in Ellis–van Creveld syndrome and renal failure with hypertension in asphyxiating thoracic dystrophy help distinguish the two disorders.

The Ellis–van Creveld syndrome is differentiated from other chondrodystrophies such as *achondroplasia, chondrodysplasia punctata, Morquio syndrome,* and *cartilage–hair hypoplasia* by its distinctive radiographic features.

Polydactyly and hypodontia or other dental anomalies have been seen in several generations without other apparent stigmata (32) and in association with *acrodental dysostosis (Weyers)* and *trisomy 13.* Postaxial polydactyly is also seen as a component of Bardet–Biedl syndrome (28).

Partial fusion of the upper lip as a result of hyperplastic frenula is seen in the *oro-facial–digital syndromes.* Natal teeth are observed in *pachyonychia congenita* and *Hallermann–Streiff syndrome.* Natal teeth may also occur alone and may, in some instances, be familial (15,30).

**Laboratory aids.** The disorder has been diagnosed prenatally by ultrasound and fetoscopy (12,18).

### References [Ellis–van Creveld syndrome (chondroectodermal dysplasia)]

1. Baisch A: Anonychia congenita, kombiniert mit Polydaktylie und verzögertem abnormen Zahndurchbruch. *Dtsch Z Chir* **232**:450–457, 1931.

2. Biggerstaff RH, Mazaheri M: Oral manifestations of the Ellis–van Creveld syndrome. *J Am Dent Assoc* **77**:1090–1095, 1968.

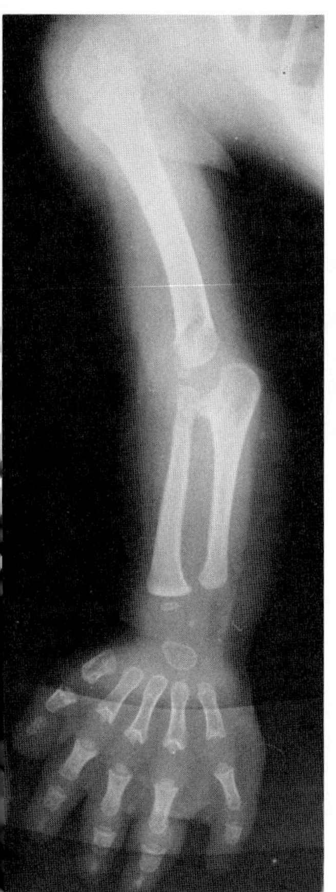

A

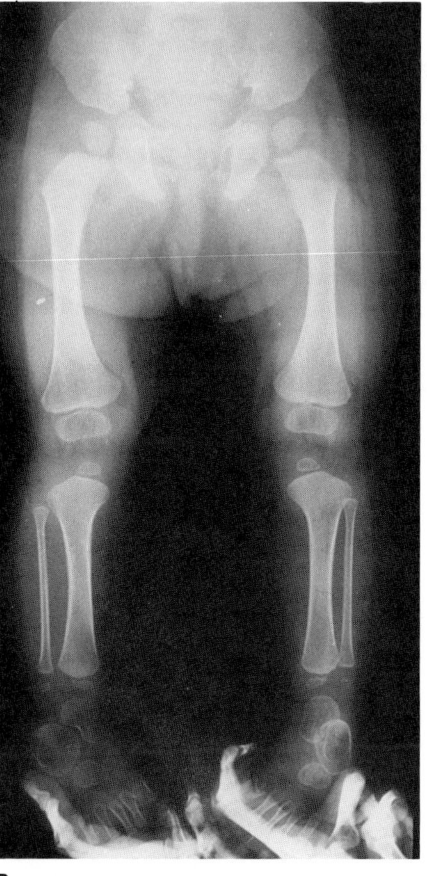

B

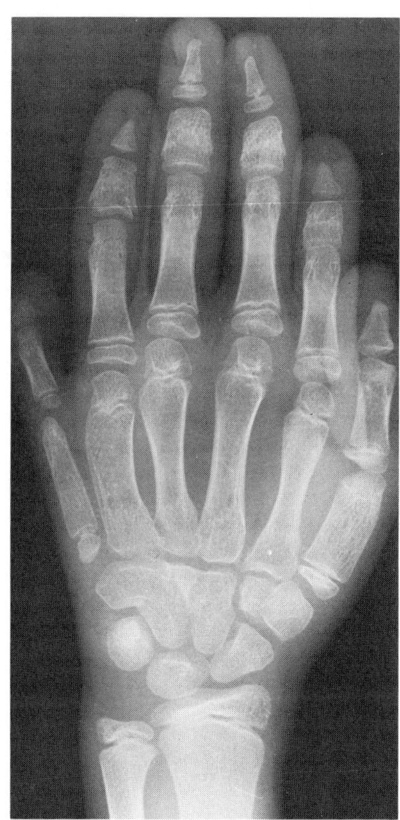

C

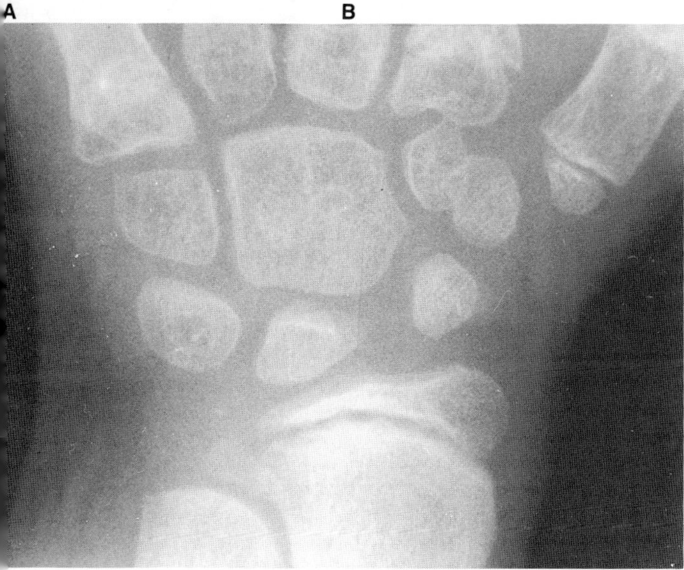

D

Fig. 7–61. *Ellis–van Creveld syndrome.* (A,B) Radiographs demonstrating unusually large proximal end of ulna and distal end of radius, and unusually small proximal end of radius, as well as progressive shortening. Extra digit had been amputated shortly after birth. Also note peaking of proximal tibia. (C) Malformed middle phalanges with cone-shaped epiphysis of middle phalanx of fifth finger. Note extra finger on ulnar side, malformed fifth metacarpal, capitate–hamate fusion. (D) Syncarpalism. (A,B courtesy of *D Gutman* and *A Jungmann,* Hadera, Israel. D courtesy of *A Poznanski,* Chicago, Illinois.)

3. Blackburn MG, Belliveau RE: Ellis–van Creveld syndrome: A report of previously undescribed anomalies in two siblings. *Am J Dis Child* **122**:267–270, 1971.

4. Bode: *Entwicklung des Zahnsystems gekoppelt mit Polydaktylie und Anonychia congenita.* Thesis, Göttingen, 1935.

5. Böhm N et al: Chondroectodermal dysplasia (Ellis–van Creveld syndrome) with dysplasia of renal medulla and bile ducts. *Histopathology* **2**:267–281, 1978.

6. Bützler HO et al: Die Röntgendiagnose der Skelettveränderungen des Ellis–van Creveld-Syndroms in Wachstumsalter. *Fortschr Roentgenstr* **118**:538–552, 1973.

6a. Christian JC et al: A family with three recessive traits and homozygosity for a long 9qh+ chromosome agent. *Am J Med Genet* **6**:301–306, 1980.

7. Da Silva EO et al: Ellis–van Creveld syndrome: Report of 15 cases in an inbred kindred. *J Med Genet* **17**:349–356, 1980.

8. Douglas WF et al: Chondroectodermal dysplasia (Ellis–van Creveld syndrome): Report of two cases in a sibship and review of literature. *Am J Dis Child* **97**:473–478, 1959.

9. Ellis RWB, Andrew JD: Chondroectodermal dysplasia. *J Bone Jt Surg* **44B**:626–636, 1962.

10. Ellis RWB, van Creveld S: A syndrome characterized by ectodermal dysplasia, polydactyly, chondro-dysplasia and congenital morbus cordis. *Arch Dis Childh* **15**:65–84, 1940.

11. Engle MA, Ehlers KH: Ellis–van Creveld syndrome with asymmetric polydactyly and successful surgical correction of common atrium. *Birth Defects* **5**(4): 65–67, 1969.

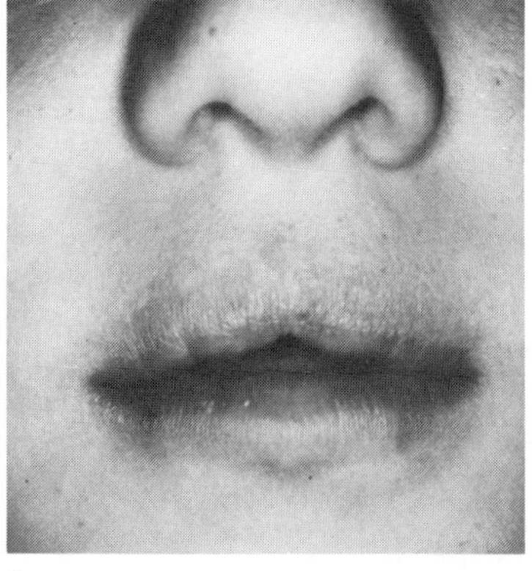
A

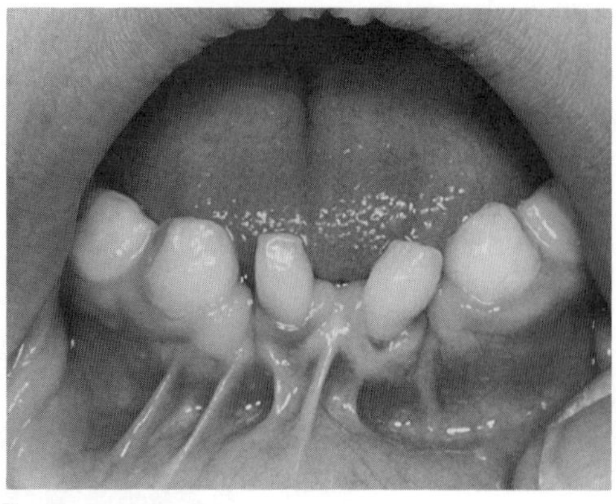
B

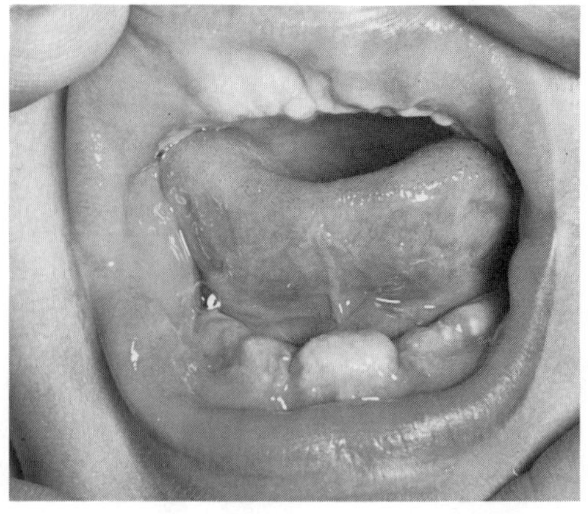
C

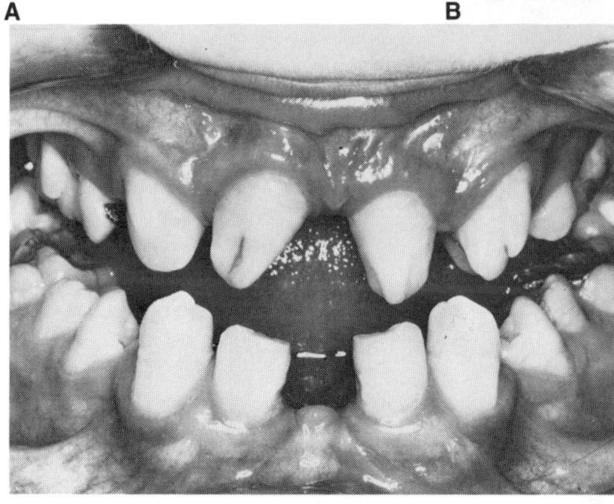
D

Fig. 7–62. *Ellis–van Creveld syndrome.* (A) Mild midline defect of upper lip. (B) Multiple frenula connect lip to lower alveolar ridge. (C) In infant, note absence of superior mucobuccal fold, serrated lower anterior alveolar process. (D) Malformed and absent incisors. (A from *RH Biggerstaff* and *M Mazaheri*, J Am Dent Assoc **77:**1090, 1968. D from *GB Winter* and *M Geddes*, Br Dent J **122:**103, 1967.)


12. Filly RA et al: Short-limbed dwarfism: Ultrasonographic diagnosis by mensuration of fetal femoral length. *Radiology* **138**:653–656, 1981.

13. Giedion A: Cone-shaped epiphyses of the hands and their diagnostic value: The trich-rhino-phalangeal syndrome. *Ann Radiol* **10**:322–329, 1967.

14. Hartwein L: Zur Kasuistik des Ellis–van Creveld Syndroms. *Kinderärztl Prax* **27**:229–233, 1959.

15. Kates GA et al: Natal and neonatal teeth: A clinical study. *J Am Dent Assoc* **109**:441–443, 1984.

16. Langer LO Jr: The thoracic-pelvic-phalangeal dystrophy. *Birth Defects* **5**(4):55–64, 1969.

17. Lynch JI et al: Congenital heart disease and chondroectodermal dysplasia. Report of two cases, one in a Negro. *Am J Dis Child* **115**:80–87, 1968.

18. Mahoney MJ, Hobbins JC: Prenatal diagnosis of chondroectodermal dysplasia (Ellis–van Creveld syndrome) with fetoscopy and ultrasound. *N Engl J Med* **297**:258–260, 1977.

19. McKusick VA et al: Dwarfism in the Amish. I. The Ellis–van Creveld syndrome. *Bull Johns Hopkins Hosp* **115**:306–336, 1964.

20. Metrakos JD, Fraser FC: Evidence for a hereditary factor in chondroectodermal dysplasia (Ellis–van Creveld syndrome). *Am J Hum Genet* **6**:260–269, 1954.

21. Miller HA: Dental abnormalities in a patient with achondroplasia. *Int J Orthodont* **23**:296–299, 1937.

22. Mitchell FN, Waddell WW Jr: Ellis–van Creveld syndrome: Report of two cases in siblings. *Acta Paediatr* **47**:142–151, 1958.

23. Neiman N et al: Syndrome d'Ellis–van Creveld et dystrophique thoracique asphyxiante. *Pédiatrie* **28**:253–263, 1973.

24. Pires de Lima JA: Dents à la naissance, *Bull Mêm Soç Anthropol Paris* **4**:71–74, 1923.

25. Prabhu SR, Dholakia HM: Chondroectodermal dysplasia (Ellis–van Creveld syndrome): Report of two cases. *J Oral Surg* **36**:631–637, 1978.

26. Rosemberg S et al: Chondroectodermal dysplasia (Ellis–van Creveld) with anomalies of CNS and urinary tract. *Am J Med Genet* **15**:291–295, 1983.

27. Sarnat H et al: Developmental dental anomalies in chondroectodermal dysplasia (Ellis–van Creveld syndrome). *J Dent Child* **47**:28–31, 1980.

28. Schachat AP, Maumenee IH: The Bardet–Biedl syndrome and related disorders. *Arch Ophthalmol* **100**:285–288, 1982.

29. Smith HL, Hand AM: Chondroectodermal dysplasia (Ellis–van Creveld syndrome). *Pediatrics* **21**:298–307, 1958.

29a. Stoll C et al: Birth prevalence rates of skeletal dysplasias. *Clin Genet* **35**:88–92, 1989.

30. Tarlow MJ: Erupted teeth in the newborn. 6 members in a family. *Arch Dis Childh* **39**:492–493, 1974.

31. Taylor GA et al: Polycarpy and other abnormalities of the wrist in chondroectodermal dysplasia: The Ellis–van Creveld syndrome. *Radiology* **151**:393–396, 1984.

32. Thomas: Über einen Fall von hereditären Polydaktylie mit Anomalien der Zähne. *Dtsch Mschr Zahnheilkd* **6**:407–408, 1888.

33. Turner EK: The Ellis–van Creveld syndrome: Report of a case. *Med J Aust* **1**:366–367, 1956.

34. Walls WT et al: Chondroectodermal dysplasia (Ellis–van Creveld syndrome): Report of a case and review of the literature. *Am J Dis Child* **98**:242–248, 1959.

35. Weiss H, Crosett AD Jr: Chondroectodermal dysplasia: Report of a case and review of the literature. *J Pediatr* **46**:268–275, 1955.

36. Weller SDV: Chondroectodermal dysplasia (Ellis–van Creveld syndrome). *Proc R Soc Med* **44**:731–732, 1951.

37. West CM: The development of the gums and their relationship to the deciduous teeth in the human foetus. *Carnegie Inst Contrib Embryol* **16**:23–46, 1925.

38. Willner H: Ektodermale Missbildungen. Kasuistischer Beitrag zur Unterzahl von Zähnen. *Dtsch Zahn Mund Kieferheilkd* **3**:279–285, 1936.

39. Winter GB, Geddes M: Oral manifestations of chondroectodermal dysplasia (Ellis–van Creveld syndrome). *Br Dent J* **122**:103–107, 1967.

40. Zangwill KM et al: Dandy–Walker malformation in Ellis–van Creveld syndrome. *J Med Genet* **31**:121–129, 1988.


## Fibrochondrogenesis

Lazzaroni–Fossati (3,4), in 1978–1979, described a neonatal lethal form of short-limb dwarfism, abnormally developed vertebrae, and abnormal facial appearance (Figs. 7–63 to 7–65). Several additional examples have been reported (1,2,5). Inheritance is clearly autosomal recessive (3,5).

The face is round and flat with prominent eyes. There is a narrow

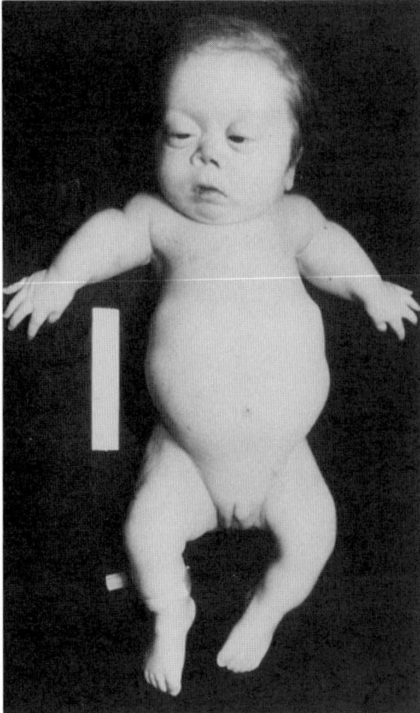

Fig. 7–63. *Fibrochondrogenesis.* Note round face, protuberant eyes, flat nasal root, anteverted nares, short neck, rhizomelic shortening of limbs, proportionate head and trunk length. (From *CB Whitley et al,* Am J Med Genet **19**:265, 1984.)

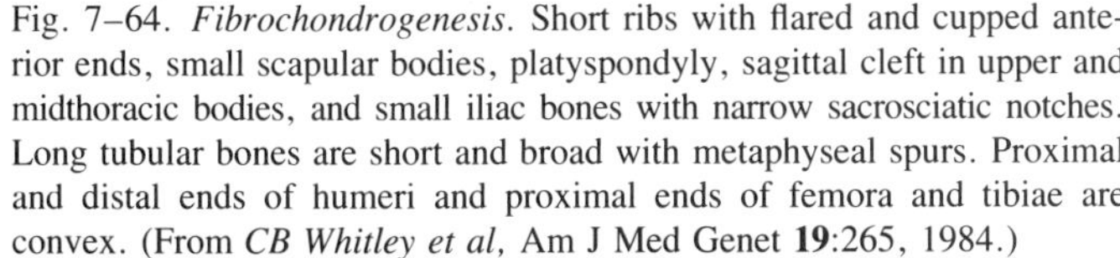

Fig. 7–64. *Fibrochondrogenesis.* Short ribs with flared and cupped anterior ends, small scapular bodies, platyspondyly, sagittal cleft in upper and midthoracic bodies, and small iliac bones with narrow sacrosciatic notches. Long tubular bones are short and broad with metaphyseal spurs. Proximal and distal ends of humeri and proximal ends of femora and tibiae are convex. (From *CB Whitley et al,* Am J Med Genet **19**:265, 1984.)

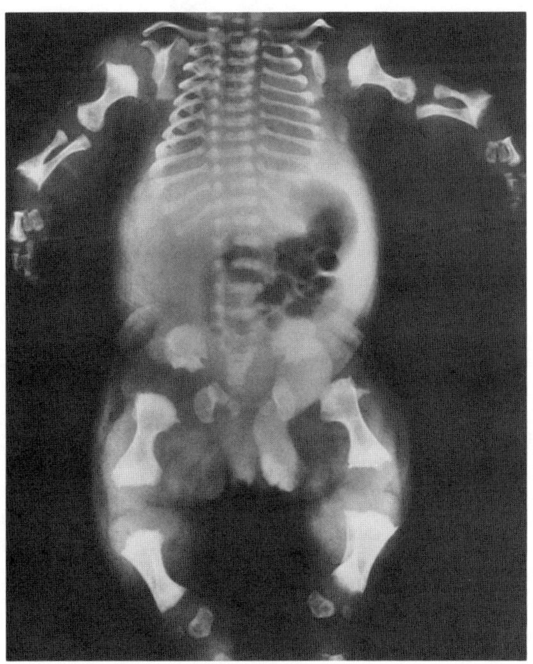

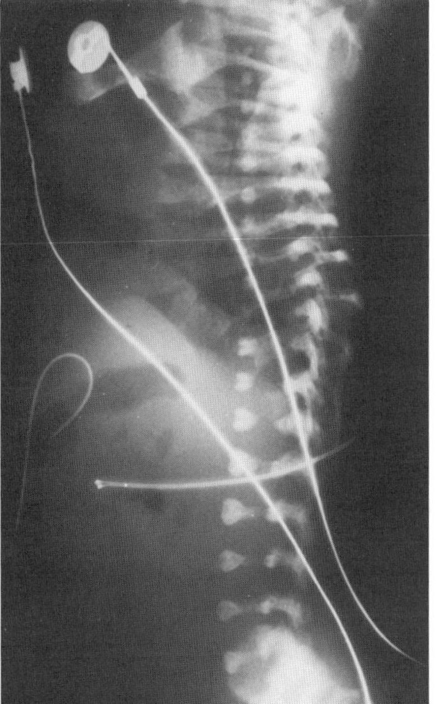

Fig. 7–65. *Fibrochondrogenesis.* Lateral spine showing pear-shaped silhouette due to ossification of only upper and midthoracic bodies. (From *CB Whitley et al,* Am J Med Genet **19:**265, 1984.)

chest, moderately severe micromelia, and markedly enlarged joints. The head and trunk are proportionate.

Radiographically, the clavicles are long and thin. The ribs are short with wide, cupped anterior ends. The tubular bones are short and dumbbell shaped with metaphyseal flare. The vertebral bodies are flattened and project a diagnostic pear-shaped silhouette in lateral view.

The ilia are small with narrow sacrosciatic notches and a medial acetabular spike. The ischia and pubic bones are short and relatively broad and the fibulae are short.

Cartilage histopathology is distinctive (Fig 7–66). The growth plate is disorganized with fibroblastic dysplasia of chondrocytes that are often clustered, two to four cells per lacuna. There are characteristic extracellular densely fibrous collagenic septa.

Fig. 7–66. *Fibrochondrogenesis.* Columnar alignment of hypertrophic cells is distorted. Resting cartilage composed of hypercellular, often spindle-shaped cells with hyperchromatic nuclei. (From *CB Whitley et al,* Am J Med Genet **19:**265, 1984.)

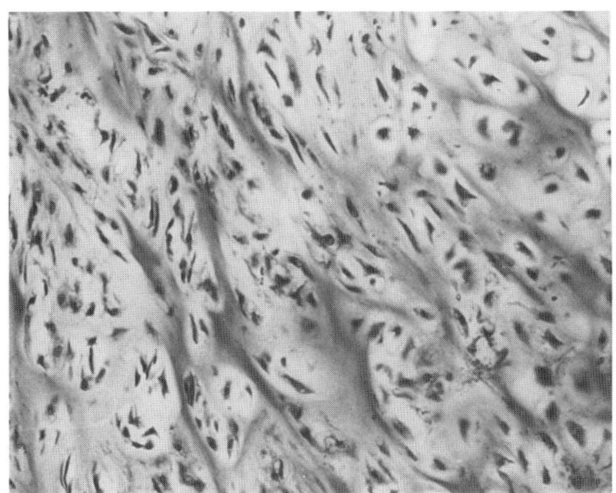

Ultrastructural studies show a paucity of endoplasmic reticulum in chondrocytes.

Cleft palate has been present in several patients (3–5).

### References (Fibrochondrogenesis)


1. Colavita N, Kozlowski K: Neonatal death dwarfism—a new form. *Pediatr Radiol* **14**:451–452, 1984 (same case as Ref. 2).

2. Eteson DJ et al: Fibrochondrogenesis: Radiologic and histologic studies. *Am J Med Genet* **19**:277–290, 1984.

3. Lazzaroni–Fossati F: La fibrodiscondrogenesi. *Minerva Pediatr* **31**:1273–1280, 1979.

4. Lazzaroni–Fossati F et al: La fibrochondrogenese. *Arch Fr Pédiatr* **35**:1096–1104, 1978.

5. Whitley CB et al: Fibrochondrogenesis: Lethal, autosomal recessive chondrodysplasia with distinctive cartilage histopathology. *Am J Med Genet* **19**:265–276, 1984.


## Geleophysic dysplasia

Spranger et al (3), in 1971, reported a skeletal disorder characterized by a happy natured (geleophysic) face, short stature, and short hands and feet (Figs. 7–67 to 7–69). Subsequently other cases have been reported (1,6). A child with similar findings but with an unhappy face has also been noted (5). Inheritance is autosomal recessive (1,4). Shohat et al (2) found lysosomal vacuoles in skin epithelial cells and postulated a lysosomal storage disease.

The facies is indeed pleasant. The palpebral fissures are upward slanting, the cheeks full, the philtrum long, and there is thin vermilion of the upper lip that, in some cases, may be inverted.

Height is below the 3rd percentile. The arms are somewhat short. There is joint limitation at the elbows, hips, and knees and decreased finger extension. All manifest a tip-toe gait as a result of pes equinovarus.

Radiographically, tubular bones are somewhat short and plump, capital femoral epiphyses are short and irregular, and shafts of the first and fifth metacarpals and proximal and middle phalanges are widened.

Psychomotor development is usually delayed. Late in infancy, recurrent respiratory and middle ear infections become more frequent. The liver is enlarged and there are umbilical and inguinal hernias (4). Death may result from infiltration of the mitral and aortic valves with an abnormal glycoprotein (4).

Acromicric dysplasia should be excluded. In geleophysic dysplasia, the facial appearance is different. The liver is usually enlarged with some type of storage material, there is progressive thickening of heart valves with a poor prognosis, and inheritance is autosomal recessive (1a).

### References (Geleophysic dysplasia)


1. Koiffmann CP et al: Familial recurrence of geleophysic dysplasia. *Am J Med Genet* **19**:483–486, 1984.

1a. Maroteaux P et al: Acromicric dysplasia. *Am J Med Genet* **24**:447–459, 1986.

2. Shohat M et al: Geleophysic dysplasia: A storage disorder involving the skin, liver, heart, and trachea. March of Dimes Clinical Genetics Conference, Baltimore, July 10–13, 1988.

3. Spranger JW et al: Geleophysic dwarfism—a "focal" mucopolysaccharidosis? *Lancet* **2**:97–98, 1971.

4. Spranger JW et al: Geleophysic dysplasia. *Am J Med Genet* **19**:489–499, 1984.

5. Spranger J et al: Acrofacial dysplasia resembling geleophysic dysplasia. *Am J Med Genet* **19**:501–506, 1984.

6. Vanace PW et al: Mitral stenosis in an atypical case of gargoylism: A case with pathologic and histochemical studies of the cardiac tissues. *Circulation* **21**:80–89, 1960.


## Kniest dysplasia (metatropic dysplasia, type II)

The disorder described by Kniest (7) in 1952 is a form of generalized spondyloepimetaphyseal bone dysplasia with disproportionate dwarfism. Most patients have been isolated examples but several authors

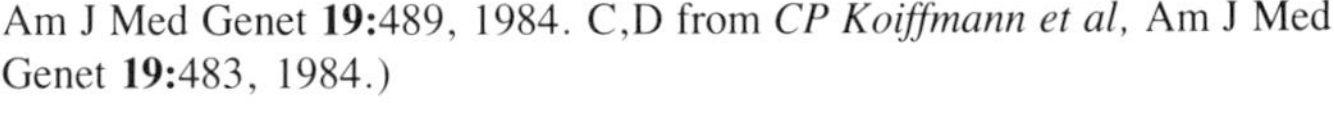

Fig. 7–67. *Geleophysic dysplasia.* (A,B) Short stature, short arms, joint limitation, and tip-toe gait due to pes equinovarus. (C,D) Compare phenotype with patient shown in A and B. (A and B from *JW Spranger et al,* Am J Med Genet **19:**489, 1984. C,D from *CP Koiffmann et al,* Am J Med Genet **19:**483, 1984.)

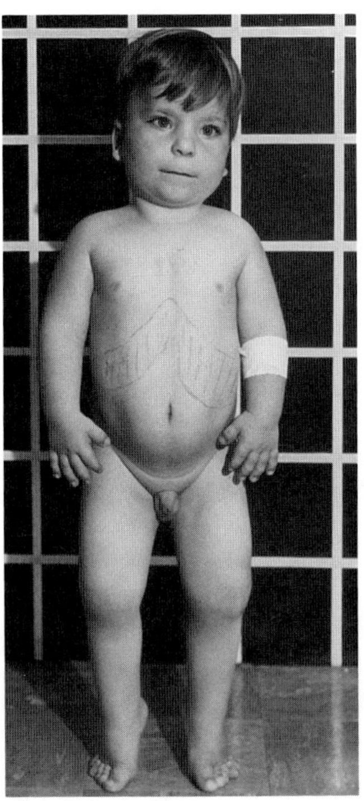

A

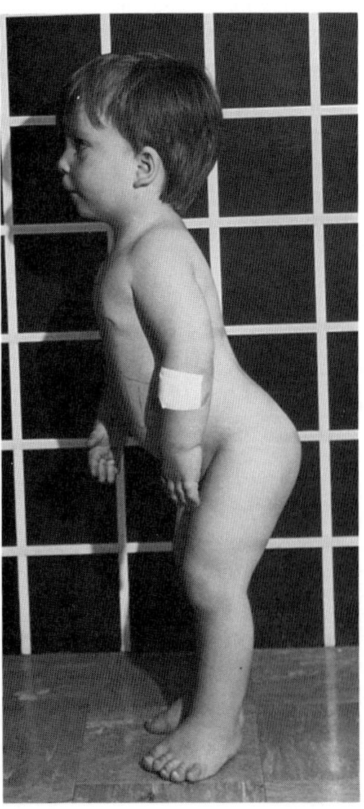

B

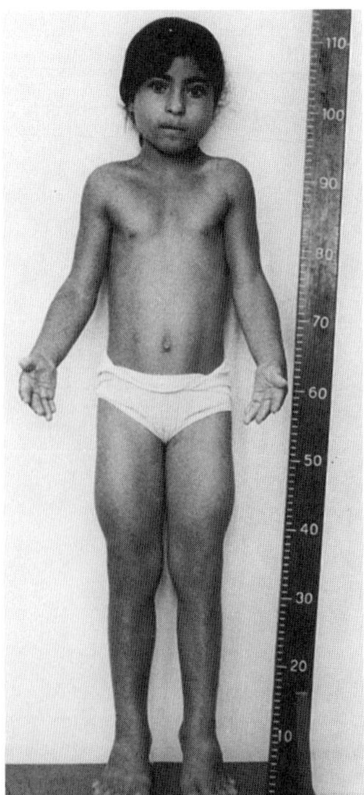


C

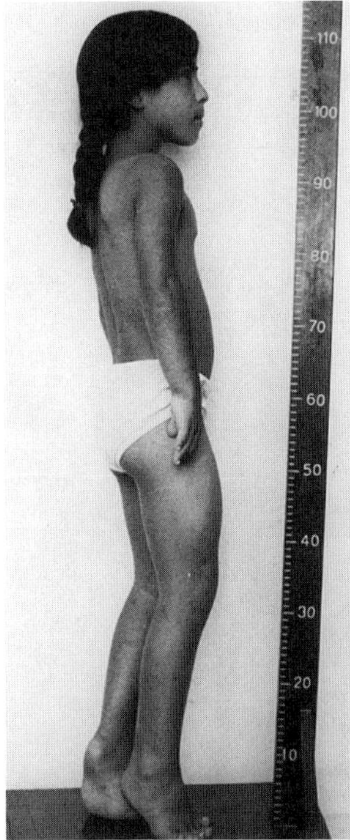


D

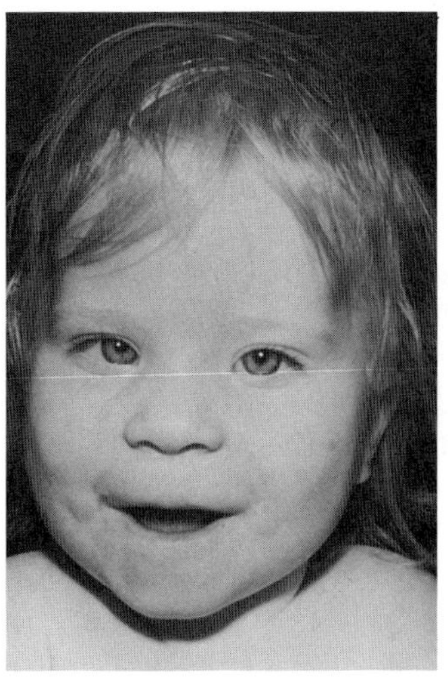
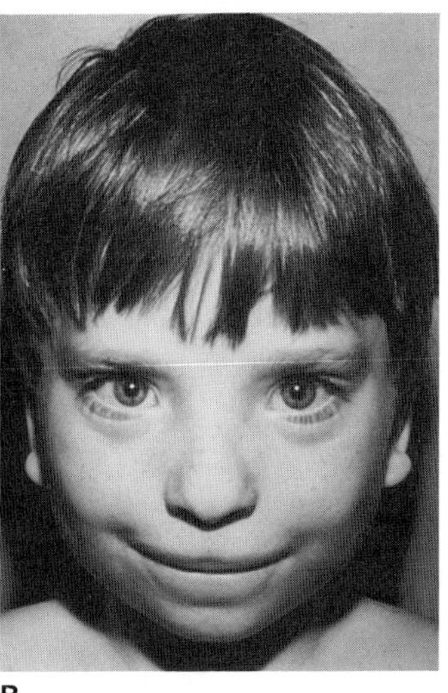

A B

Fig. 7–68. *Geleophysic dysplasia.* (A,B) Upslanting palpebral fissures, full cheeks, long philtrum, and thin vermilion of upper lip resulting in pleasant facies. (A,B from *JW Spranger et al,* Am J Med Genet **19:**489, 1984.)

Fig. 7–69. *Geleophysic dysplasia.* (A) Brachydactyly. Note decreased finger extension. (B) Roentgenogram showing somewhat short and plump tubular bones. Shafts of first and fifth metacarpals and proximal and middle phalanges widened. (A from *CP Koiffmann et al,* Am J Med Genet **19:**483, 1984. B from *JW Spranger et al,* Am J Med Genet **19:**489, 1984.)

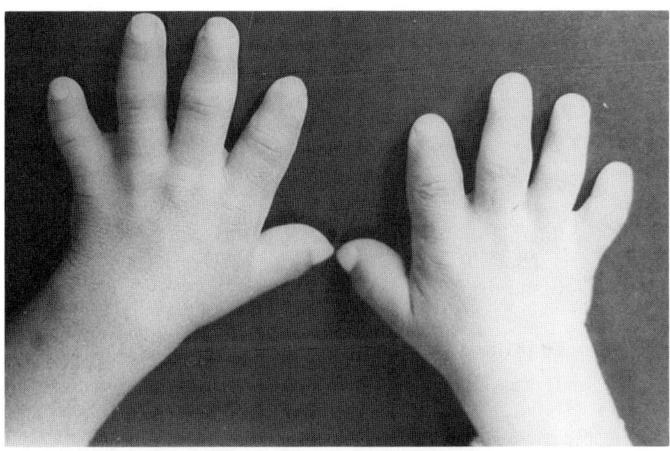

A

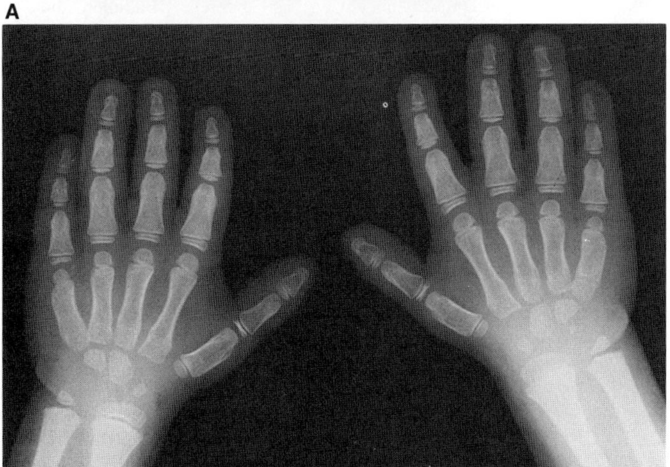

B

(4,6,12,13) have noted the disorder in two generations. Identical twins have been described (20). An autosomal recessive Kniest-like syndrome has also been reported (19). Poole et al (15a) demonstrated that the C-propeptide is missing in Type II collagen. The genetic heterogeneity can be explained by a defect in the C-peptide cleavage enzyme (AR) and amino acid substitution at cleavage site, making Type II collagen resistant to cleavage (AD). Several authors (11,12) have confused the condition with metatropic dysplasia, Type I.

The face is round, with the midface flat and the nasal bridge depressed, giving the eyes a somewhat exophthalmic appearance. The neck is usually short. The head appears to sit on the thorax (Fig. 7–70A,B). At birth, cleft palate, clubfoot, and prominent knees may be noted (1,3,15,21,22). Lordosis and/or dorsal kyphosis and tibial bowing usually develop within the first few years of life. The child may not sit and walk until 2 and 3, respectively. By that time, most joints become progressively enlarged and stiff. The gait is waddling. Movement at the metacarpophalangeal joint is normal, but the child cannot make a fist. The fifth fingers are generally not involved. The palms may have a violaceous hue. The elbows, wrists, and knees become particularly enlarged, and flexion and extension of most joints become progressively reduced. The feet are flat and outturned. Hernia is frequent. Adult height ranges between 105 and 145 cm.

Severe myopia and lattice degeneration with or without retinal detachment and/or cataract formation have been present in about 40%, as has cleft palate (1,6,18,21). A large series was studied by Maumenee and Traboulsi (14). Conduction and/or sensorineural hearing loss, a frequent finding, may develop before puberty. Recurrent otitis media and respiratory infections are common. Dobin and Daniel (1a) reported a case with aortic stenosis, ASD, strabismus, choroidal mass, and tortuous vascular pattern.

Radiographically, the neurocranium is large in comparison with the facial skeleton. The anterior fontanel is late to close. The cranial base angle is flattened and the sella turcica is anteriorly displaced. The odontoid is short and wide (3,6,20). Platyspondyly, particularly of the upper thoracic part of the spine, is severe. The vertebrae exhibit vertical clefts. The long bones are somewhat short, slightly bowed, and have flared metaphyses. The epiphyses are large, irregular, and punctate. The hands show epiphyseal and carpal retardation with generalized osteopenia. Later, the carpal bones assume bizarre shapes and sizes. The iliac bones are small, particularly in relation to the large capital femoral epiphysis and proximal femoral metaphysis. The pubic rami are poorly ossified. The femoral capital epiphysis forms late, the neck is wide and short with a poorly ossified central area, and there

Fig. 7–70. *Kniest dysplasia.* (A,B) Five-year-old girl with flat midface, myopia, cleft palate. At birth, legs were noted to be abnormally short and hips were stiff. Umbilical and inguinal hernia were repaired. During early childhood, joints became enlarged, painful, and stiff. Note flexion deformities.

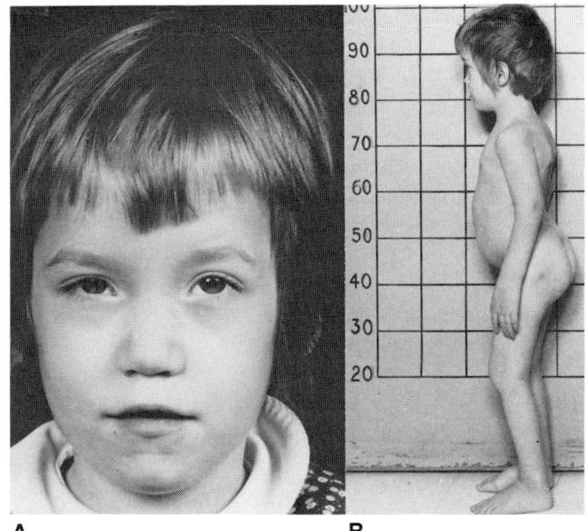


A B

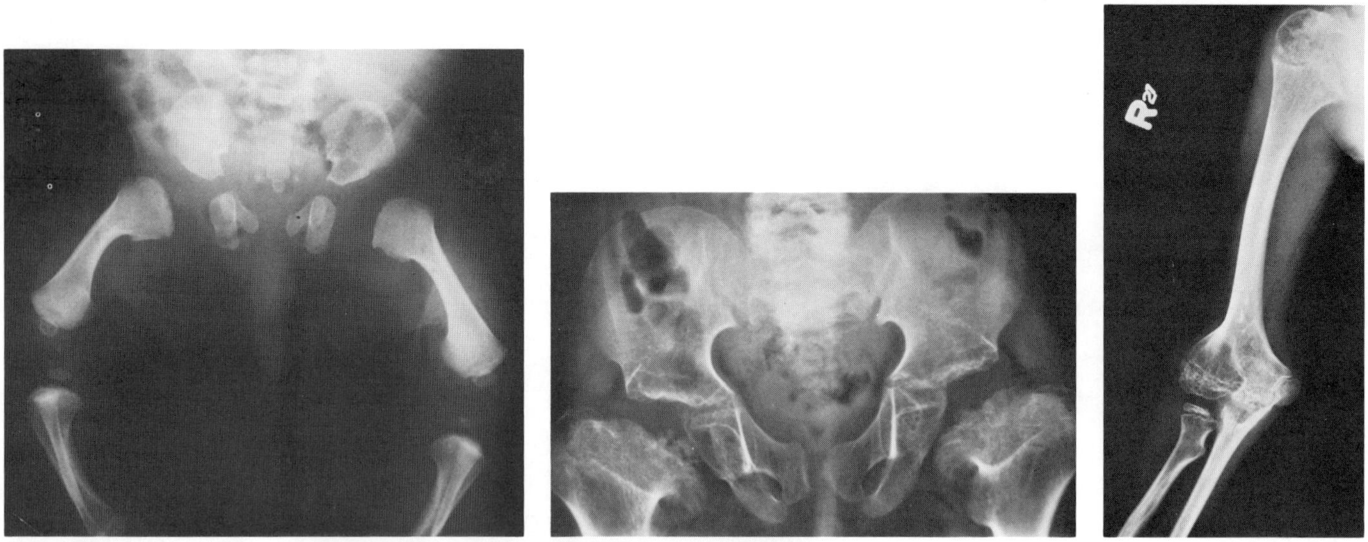

Fig. 7–71. *Kniest dysplasia.* Roentgenograms. (A) Dumbbell femora. (B) Coxa vara with irregular mineralization of femoral capital epiphyses, wide femoral heads, and trochanters. Irregularity of acetabular roofs. (C) Irregularity of epiphyses and flared metaphyses of shortened humerus. (D) Platyspondyly with vertical clefts in the lumbar spine. (E) Age 12 years. Bulbous enlargement of both ends of the bones and diffuse osteoporosis. Note short tuft of thumb. (A,D,E from *RS Lachman et al,* Am J Roentgenol **123:**805, 1975.)

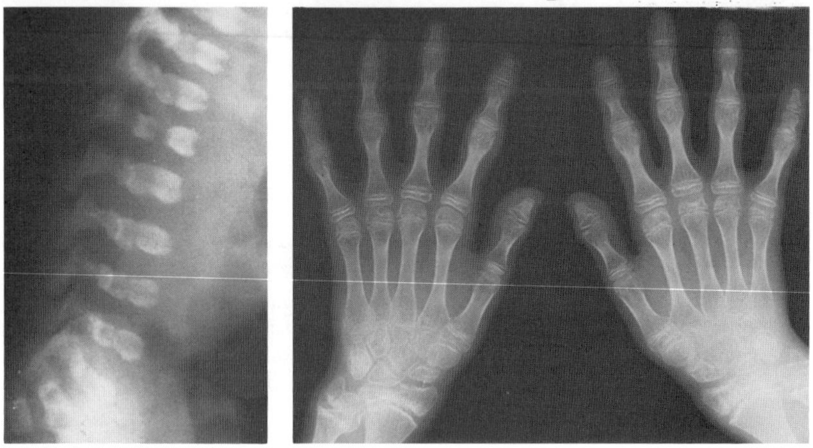

may be coxa vara. The trochanter is prominent (9,10) (Fig. 7–71A–E).

Histopathologic examination of the bones has shown that the cartilage contains large chondrocytes that lie in a very loosely woven matrix containing numerous empty spaces (''Swiss cheese cartilage'') (Fig. 7–72) (1,2,16,17). Ultrastructural studies of cartilage cells have shown dilated cisternae of endoplastic reticulum, a finding noted in other skeletal dysplasias. There is vacuolar degeneration of extralacunar matrix in the area of the resting cartilage adjacent to the growth plate (5). Keratansulfaturia has been found (3,6,15).

Fig. 7–72. *Kneist dysplasia.* ''Swiss-cheese'' cartilage. (From *RS Lachman et al,* Am J Roentgenol **123:**805, 1975.)

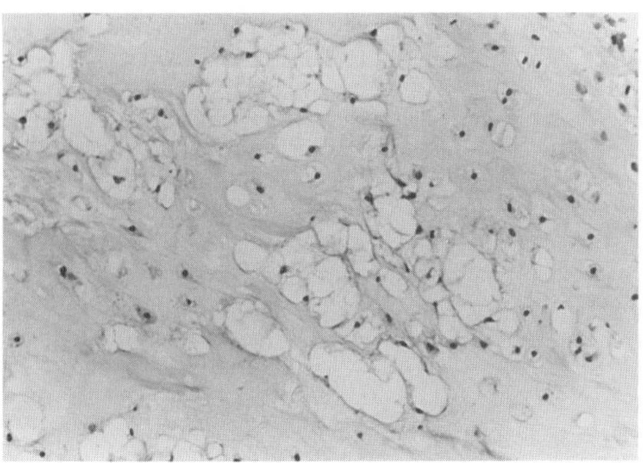

In differential diagnosis *dyssegmental dysplasia, type Rolland–Desbuquois* and the *Burton syndrome* should be excluded.

## References [Kniest dysplasia (metatropic dysplasia, type II)]


1. Chen H et al: Kniest dysplasia: Neonatal death with necropsy. *Am J Med Genet* **6**:171–178, 1980.

1a. Dobin SM, Daniel CA: Further delineation of the natural history of Kniest syndrome. March of Dimes Clinical Genetics Conference, Baltimore, July 10–13, 1988.

2. Frayha R et al: The Kniest (Swiss cheese cartilage) syndrome. *Arthr Rheum* **22**:286–289, 1979.

3. Friede H et al: Craniofacial and mucopolysaccharide abnormalities in Kniest dysplasia. *J Craniofac Genet Dev Biol* **5**:267–276, 1985.

4. Gnamey D et al: La maladie de Kniest: Une observation familiale. *Arch Fr Pédiatr* **33**:143–151, 1976.

5. Horton WA, Rimoin DL: Kniest dysplasia: A histochemical study of the growth plate. *Pediatr Res* **13**:1266–1270, 1970.

6. Kim HJ et al: Kniest syndrome with dominant inheritance and mucopolysacchariduria. *Am J Med Genet* **27**:755–764, 1975.

7. Kniest W: Zur Abgrenzung der Dysostosis enchondralis von der Chondrodystrophie. *Z Kinderheilk* **70**:633–640, 1952.

8. Kniest W, Leiber B: Kniest-Syndrom. *Mschr Kinderheilk* **125**:970–973, 1977.

9. Kozlowski K et al: Metatropic dwarfism and its variants. *Australas Radiol* **20**:367–385, 1976.

10. Lachman RS et al: The Kniest syndrome. *Am J Roentgenol* **123**:805–814, 1975.

11. Larose JH, Gary RB: Metatropic dwarfism. *Am J Roentgenol* **106**:156–161, 1969.

12. Marchese GS: Una nuova entita morbosa: Il nanismo metatropico. *Minerva Pediatr* **19**:649–653, 1967.

13. Maroteaux P, Spranger J: La maladie de Kniest. *Arch Fr Pédiatr* **30**:735–750, 1973.

14. Maumenee I, Traboulsi EI: The ocular findings in Kniest dysplasia. *Am J Ophthalmol* **100**:155–160, 1985.

15. Pennock CA et al: Keratan sulfate excretion in a patient with Kniest dysplasia. *J Inherited Metab Dis* **2**:75–78, 1979.

15a. Poole AR et al: Kniest dysplasia is characterized by an apparent abnormal processing of the C-propeptide of type II cartilage collagen resulting in imperfect fibril assembly. *J Clin Invest* **81**:579–589, 1988.

16. Rimoin D: Histopathology and ultrastructure of cartilage in the chondrodystrophies. *Birth Defects* **10**(9):1–18, 1974.

17. Rimoin DL et al: Chondro-osseous pathology in the chondrodystrophies. *Clin Orthoped* **114**:137–152, 1976.

18. Roaf R et al: A childhood syndrome of bone dysplasia, retinal detachment and deafness (Case 2). *Dev Med Child Neurol* **9**:463–473, 1967.

19. Sconyers SM et al: A distinct chondrodysplasia resembling Kniest dysplasia: Clinical, roentgenographic, histologic and ultrastructural findings. *J Pediatr* **103**:898–904, 1983.

20. Siggers DC et al: The Kniest syndrome. *Birth Defects* **10**(9):193–208, 1974.

21. Silengo MC et al: Kniest disease with Pierre Robin syndrome and hydrocephalus. *Pediatr Radiol* **13**:106–109, 1983.

22. Spranger J, Maroteaux P: Kniest disease. *Birth Defects* **10**(12):50–56, 1974.

## Lethal short-limbed dysplasias

This chapter describes 10 types of rare lethal short-limbed dysplasias.

**Type Astley–Kendall.** Astley and Kendall (1), in 1980, Colavita and Kozlowski (4), in 1984, and Nairn and Chapman (10a), in 1989, reported a lethal bone dysplasia characterized by barely calcified calvaria, normal skull base, and extreme shortening of the extremities with hypoplastic or dysplastic changes in long bones and flaring of metaphyses (Fig. 7–73A). There were spicule-like projections at the ends of the shafts and the long bones had transverse defects. The spine and extremities had ectopic ossification centers. The ribs were short and thin with flaring ends. There was a space between the ribs and the abnormally thin vertebral bodies. A peculiar spicule at the anterior end of the mandible was noted in one patient (4). The ilia were remarkably short and crescentic in shape. Multiple ossification centers were found in the hands, feet, scapulae, and ischia. The carpus, tarsus, and epiphyses were dense.

Inheritance is autosomal recessive (10a).

**Type Beemer–Winter.** Beemer et al (2) and Winter (13) described a lethal short limbed dysplasia characterized by macrocephaly, fat face, midline cleft lip, cleft palate, ascites, omphalocele, narrow thorax, short ribs, highly placed clavicles, small scapulae, small ilia, short tubular bones, and bowing of the radius and the ulna (Fig. 7 73B–D). No polydactyly was present. In one instance reported by Beemer et al (2), parents were consanguineous and an affected sib of their Case 1 was diagnosed prenatally by ultrasound. Thus, inheritance is probably autosomal recessive. The condition appears to be different from any of the known *short rib-polydactyly syndromes*.

**Type Blomstrand.** Blomstrand et al (3) described a lethal dysplastic infant born to consanguineous parents, suggesting the possibility of autosomal recessive inheritance. Hydramnios occurred at the twenty-eighth week. Findings (Fig. 7–74A,B) included normal facies, macroglossia, short limbs, normal sized hands and feet, general edema, and preductal aortic coarctation. Skeletal anomalies included relatively small facial skeleton, short skull base, low nasal bridge, hypoplastic mandible, and accelerated osseous maturation. The thorax was long and narrow and showed uniform widening of the anterior ends of short horizontal ribs. The sacral and coccygeal vertebrae were ossified and all long tubular bones were extremely short with marked metaphyseal flaring and cupping. The carpal, metacarpal, tarsal, and metatarsal bones were wide and short with advanced ossification. The hyoid bone was mineralized as well as the laryngeal cartilages. The phalanges and metacarpals were calcified by the twenty-ninth week. The condition should be distinguished from *achondrogenesis* and *thanatophoric dysplasia*.

**Type Greenberg.** Greenberg et al (6), in 1988, reported two sibs, the offspring of consanguineous parents, with a unique, lethal severe short-limb dwarfism. One sib presented with hydrops fetalis (Fig. 7–74C,D). Acute polyhydramnios was noted during the first pregnancy. The second sib was diagnosed by serial sonographc examinations that showed markedly short fetal limbs. The pregnancy was terminated at 20 weeks. RJ Gorlin has seen an identical case.

The thorax was narrow, the limbs markedly rhizomelic, and the hands broad with short fingers. Radiographically, there was extremely poor ossification of the membranous bones of the skull, but a dense skull base. Midface hypoplasia was severe (Fig. 7–74,E). The cervical vertebrae appeared moth eaten, and there were calcifications near the larynx and trachea. The midthoracic and lower thoracic lumbar spine exhibited platyspondyly with multiple ossification centers of the vertebral bodies and dense bones in the spinous processes with poor pedicular and laminar development. The ribs exhibited unusual ossification gaps with an anterior ossified tail-like extension. The iliac wings appeared moth eaten and a lace-like quality was noted in the pubis and ischium. Most long bones appeared moth eaten (Fig. 7–74F). The clavicles were elongated.

Microscopically, chondroosseous changes were characterized by marked disorganization of tissue with interspersed masses of cartilage, bone, and mesenchymal tissues.

**Type Holmgren–Connor.** Holmgren et al (7), Connor et al (5), and Maroteaux et al (10) described an autosomal recessive severe micromelic dysplasia with slightly incurved limbs, hyperlaxity of extremities, disproportionately large skull, slightly flattened nasal bridge, and, in some cases, congenital heart defects. Most of the patients succumbed from respiratory distress within hours, days, or during the first weeks of life.

Radiographs show short long bones with incurved diaphyses, especially the femora, enlarged and slightly irregular metaphyses, striking well-developed and rounded lower femoral epiphysis, short square iliac wings with a wide iliac angle, and short ribs with wide and slightly irregular anterior ends (Fig. 7–74G,H).

Histologically, the upper tibial growth cartilage has an irregular arrangement of cells, demasked fine fibers in the matrix, short primary trabeculae, and closure of some of the vascular lacunae by bony bridging.

**Type Koide.** In 1983, Koide et al (8) reported a newly recognized form of lethal short-limbed dwarfism consisting of midface hypoplasia, prominent nasal bones, lack of mandibular angle, poor ossification of the occipital bone, severe platyspondyly, diminished ossification in the sacral region with some punctate calcifications, marked lordosis, lack of ossification of the cervical spine, markedly trident pelvis on its inferior aspect near the acetabular roofs with narrowing of the sacrosciatic notches and extreme hypoplasia of the iliac wings, widened ischia, short ribs, and micromelic hypoplasia (Fig. 7–75A–C). Changes in the cartilage matrix were striking, with dense collagenous material in the matrix surrounding the chondrocytes but little such material among the chondrocytes. Instead, fragmented large bands of collagen fibrils were noted. Since the patient reported by Koide et al (8) was the product of a consanguineous mating, autosomal recessive inheritance is possible. The condition should be distinguished from various types of *achondrogenesis*.

**Type Kozlowski.** Kozlowski et al (9) reported a new type of lethal short-limbed dysplasia resembling the achondrogenesis syndromes clinically, but presenting distinctive radiographic and microscopic fea-

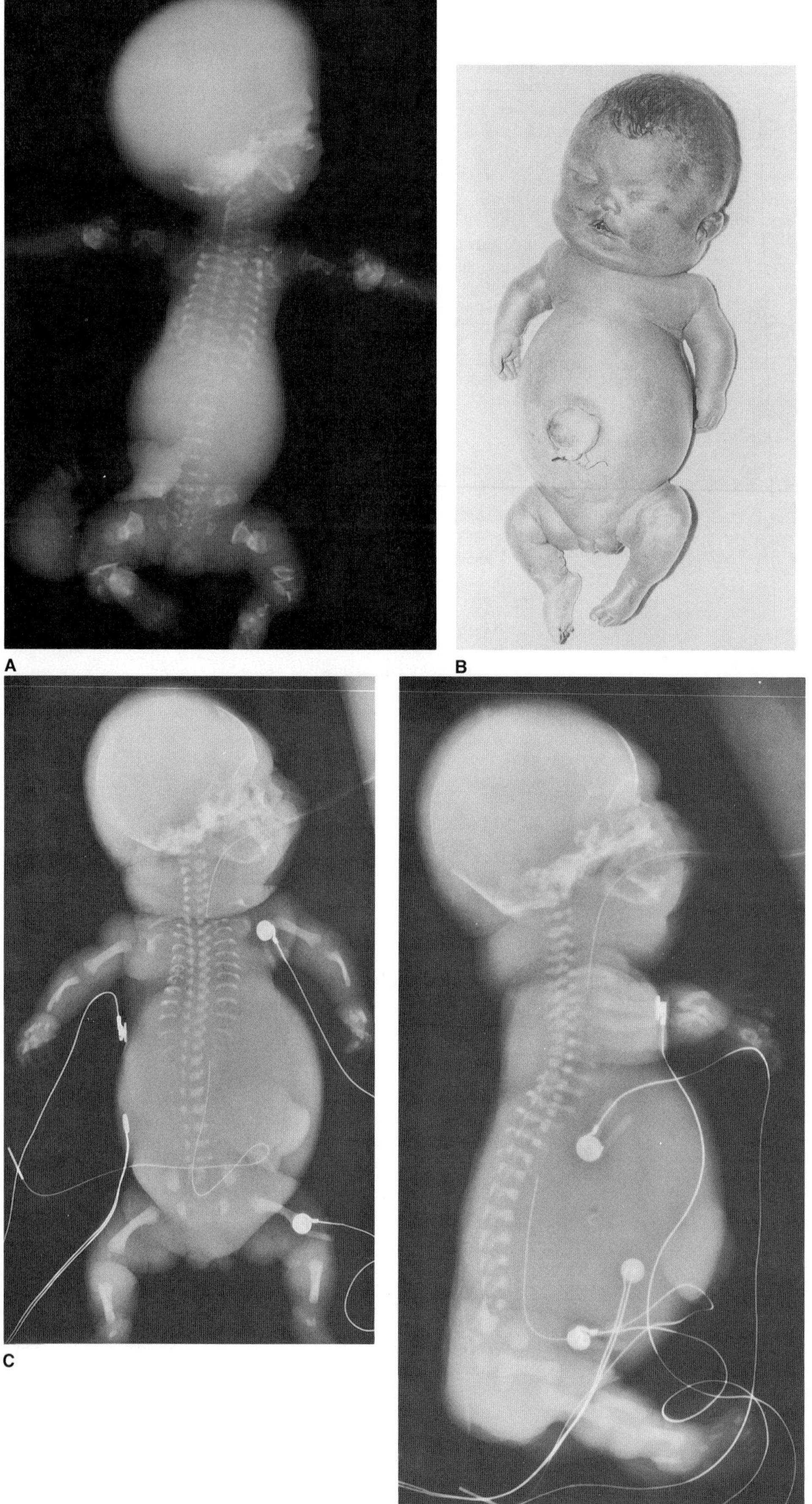


Fig. 7–73 (A) *Lethal short-limbed dysplasia, type Astley–Kendall.* Note barely calcified calvaria, shortened extremities with transverse defects of humeri, femora, tibiae with flared metaphyses, short ribs, space between ribs and vertebral bodies, platyspondyly, crescent-shaped ilia. (From *R Astley* and *AC Kendall,* Ann Radiol **23:**121, 1980.) (B–D) *Lethal short-limbed dysplasia, type Beemer–Winter.* (B) Generalized edema, large head with flat face, broad nasal base, median cleft of upper lip, small edematous pinnae, short narrow thorax, and distended abdomen with omphalocele. (C,D) Short horizontal ribs, high clavicles, relatively normal ilia, and metaphyses of abbreviated long bones. No polydactyly of hands or feet. (From *FA Beemer et al,* Am J Med Genet **14:**115, 1983.)

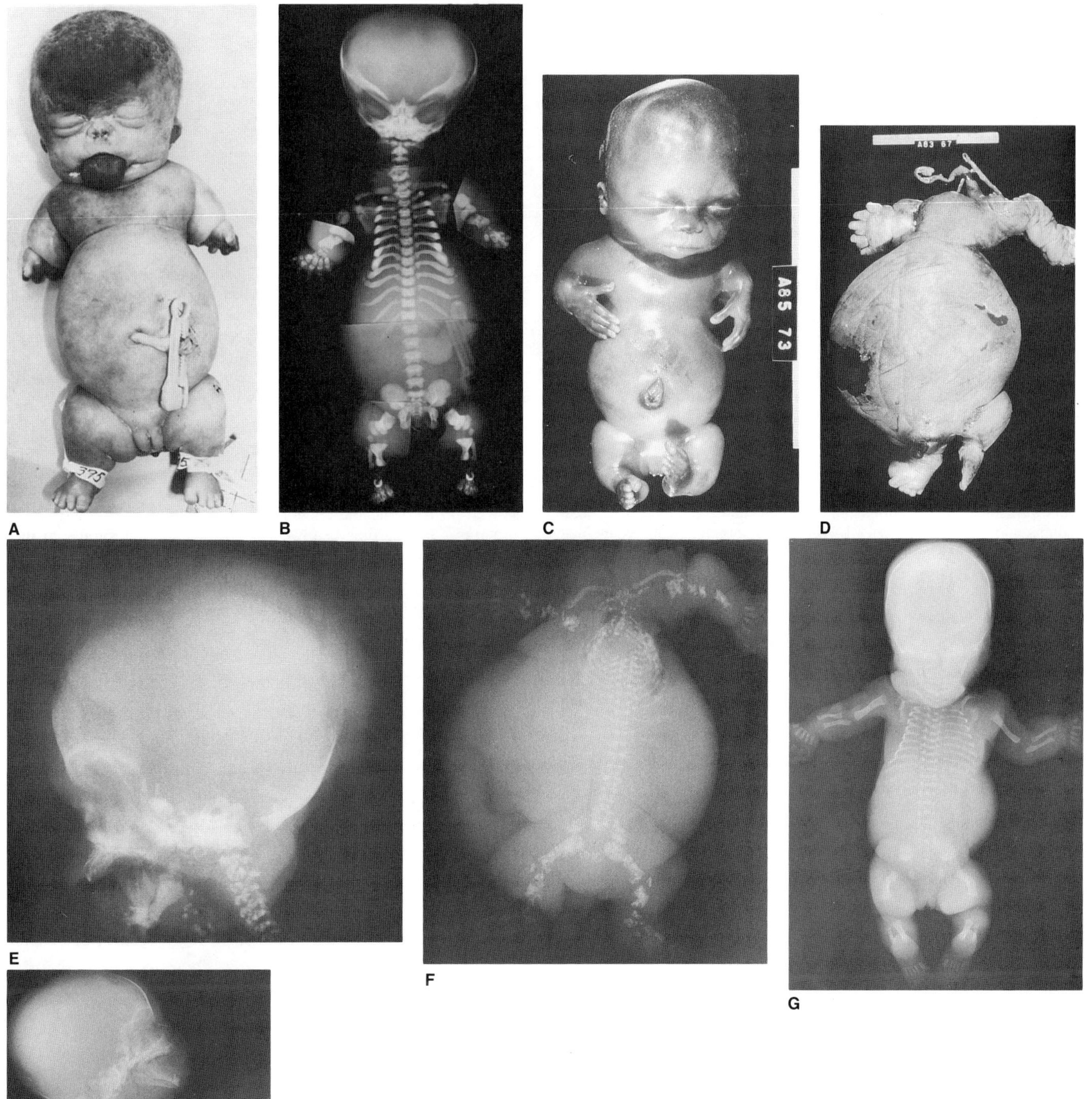


Fig. 7–74 (A,B) *Lethal short-limbed dysplasia, type Blomstrand.* (A) Note low nasal bridge, macroglossia, short limbs, narrow chest with large abdomen. (B) Radiograph showing long narrow thorax, widened anterior ribs, severely abbreviated long bones with flared cupped metaphyses. Note extremely advanced bone age. (A,B from *S Blomstrand et al,* Pediatr Radiol **15:**141, 1985.) (C–F) *Lethal short-limbed dysplasia, type Greenberg.* (C,D) Two sibs with severe dysplasia. Note hydrops fetalis in (D). (E) Radiograph showing extremely poor ossification of membranous bones, dense cranial base, and midface hypoplasia. (F) Radiograph showing moth-eaten appearance of bone. (From *CR Greenberg et al,* Am J Med Genet **29:**623, 1988.) (G,H) *Lethal short-limbed dysplasia, type Holmgren–Connor.* Shortening of all bones with femoral and humeral bowing, short skull base, micrognathia, mild platyspondyly, hypoplasia of iliac, pubic, and ischial bones. (From *JM Connor et al,* Am J Med Genet **22:**23, 1985.)

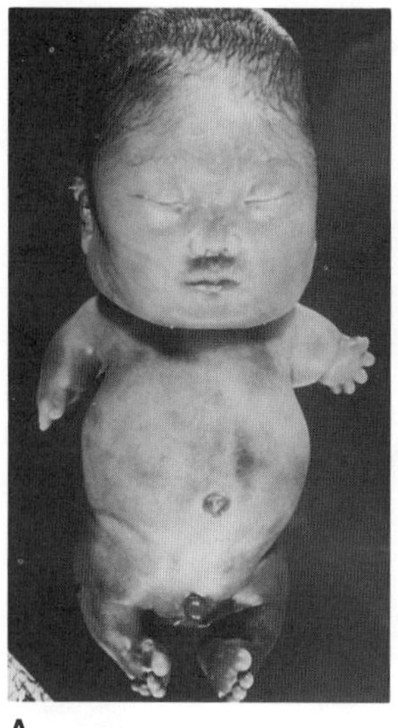
A

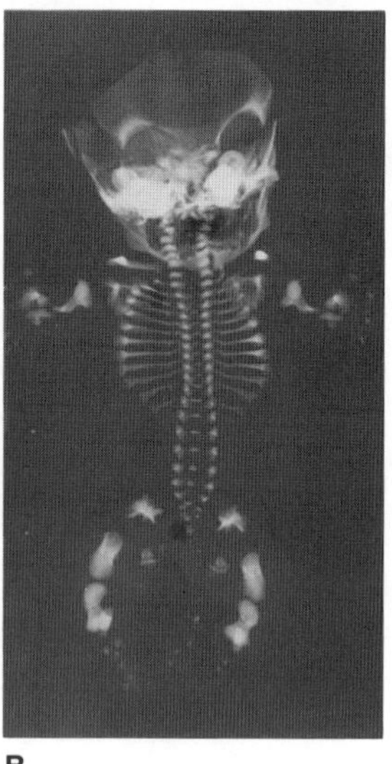
B

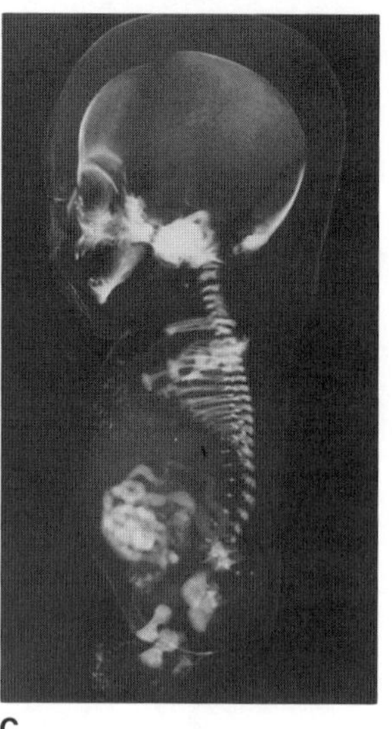
C

Fig. 7–75. *Lethal short-limbed dysplasia, type Koide.* (A) Micromelia, large head, flat facies, large abdomen. (B,C) Radiographs showing short ribs, platyspondyly, crenated ilia, less severely involved long bones. Note distally tapered humeri. (From *T Koide,* Pediatr Radiol **13:**102, 1983.)

tures. Micromelia was observed together with a large head, edema of the scalp and scrotum, protuberant abdomen, midface hypoplasia, small mouth, and fleshy ears (Fig. 7–76A). Radiographic features include long distally hooked clavicles, thin ribs, platyspondyly, and severely affected rhizomelic segments with relatively minor changes in the mesomelic segments as well as in the pelvis and head (Fig. 7–76B). Histologic study showed loss of smooth perichondrium of the proximal femoral epiphysis with intermingling of fibrous perichondrium and fibrous ingrowth into epiphyseal cartilage. There was gross disruption of chondrocyte organization in all zones of the growth plate. Uneven distribution of reserve zone chondrocytes was noted, many of which were large and appeared to be degenerating. No zone of proliferation was detected. In the chondroosseous zone of transformation, many randomly oriented degenerating chondrocytes were observed. Primary trabeculae were distorted and calcified cartilage cores were observed in primary trabeculae deep in the metaphyses. Diaphyseal bone margins were irregular and scalloped.

**Type Piepkorn.** Piepkorn et al (11) described a severely dysplastic infant with marked lack of ossification of all bones except the clavicle. There was polyhydramnios. The arms and legs were flipperlike. The palate was cleft. There was no evident ossification in the limbs. Persistent left superior vena cava and urogenital abnormalities were evident (Fig. 7–76C–F).

**Type Tsang.** Tsang et al (12) described a sporadic instance of a condition characterized by rhizomesoacromelic shortening of the upper limbs, long clavicles, markedly short ribs, hypoplastic vertebral bodies with well-formed pedicles and posterior processes, hypoplastic iliac bones with a "snail-like" configuration, marked shortening of both femora with metaphyseal flaring and irregularity, and no evidence of ossification of the talus or calcaneus.

Histologic findings included hypercellularity of resting cartilage in the femoral epiphysis, round chondrocytes with large round central nuclei, and absence of lacunar space around chondrocytes. The irregular growth plate was distinctly different from that of *Schneckenbecken dysplasia.*

The patient of Tsang et al (12) also had frontal bossing and cleft palate.

**Type Carty.** Carty et al (3a) reported a condition they termed *dappled diaphyseal dysplasia.* The fetus, a product of nonconsanguineous parents, was hydropic and short limbed. Radiographs showed rudimentary calcification of the cranium and multifocal ossification of all long bones. The epiphyses and diaphyses could not be distinguished. The ribs, because of the dappled ossification, appeared fractured. Platyspondyly was marked (Fig. 7–76G,H). Microscopic examination showed no recognizable chondroosseous transformation or trabecular formation. The cartilage showed changes somewhat similar to those of *Kniest dysplasia* and *diastrophic dysplasia.*

### References (Lethal short-limbed dysplasias)


1. Astley R, Kendall AC: A bone dysplasia for diagnosis. *Ann Radiol* **23**:121–123, 1980.
2. Beemer FA et al: A new short rib syndrome: Report of two cases. *Am J Med Genet* **14**:115–123, 1983.
3. Blomstrand S et al: A case of lethal congenital dwarfism with accelerated skeletal maturation. *Pediatr Radiol* **15**:141–143, 1985.

3a. Carty H et al: Dappled diaphyseal dysplasia. *Roefo* **150**:228–229, 1989.

4. Colavita N, Kozlowski K: Neonatal death dwarfism—a new form. *Pediatr Radiol* **14**:451–452, 1984.
5. Connor JM et al: Lethal neonatal chondrodysplasias in the West of Scotland with a description of thanatophoric dysplasialike, autosomal recessive disorder, Glasgow variant. *Am J Med Genet* **22**:243–253, 1985.
6. Greenberg CR et al: A new autosomal recessive lethal chondrodystrophy with congenital hydrops. *Am J Med Genet* **29**:623–632, 1988.
7. Holgren G et al: Semilethal bone dysplasia in three sibs: A new genetic disorder. *Clin Genet* **26**:249–251, 1984.
8. Koide T et al: A case of new chondrodystrophy. *Pediatr Radiol* **13**:102–105, 1983.
9. Kozlowski K et al: A new type of achondrogenesis. *Pediatr Radiol* **16**:430–432, 1986.
10. Maroteaux P et al: Recessive lethal chondrodysplasia, "round femoral inferior epiphysis type." *Eur J Pediatr* **147**:408–411, 1988.

10a. Nairn ER, Chapman S: A new type of lethal short-limbed dwarfism. *Pediatr Radiol* **19**:253–257, 1989.

11. Piepkorn M et al: A lethal neonatal dwarfing condition with short ribs, polysyndactyly, cranial synostosis, cleft palate, cardiovascular and urogenital anomalies and severe ossification defect. *Teratology* **16**:345–358, 1977.
12. Tsang WC et al: A new lethal short limbed dwarfism—case report. Third Manchester Birth Defects Conference, Manchester, October 25–28, 1988.
13. Winter RM: A lethal short rib syndrome without polydactyly. *J Med Genet* **25**:349–357, 1988.


## Megepiphyseal dysplasia

Gorlin et al (1973) described a male with cleft palate, dislocated lenses, deafness, somatic and mental retardation, epicanthal folds, and snub nose. Striking were enlarged joints (shoulders, elbows, hips, knees, ankles). Radiographic study showed marked shortening of long bones, with flared metaphyses and extremely large proximal and distal epiphyses. The carpal bones were large (Fig. 7–77A–E).

Homocystinuria, found on biochemical study, would account for

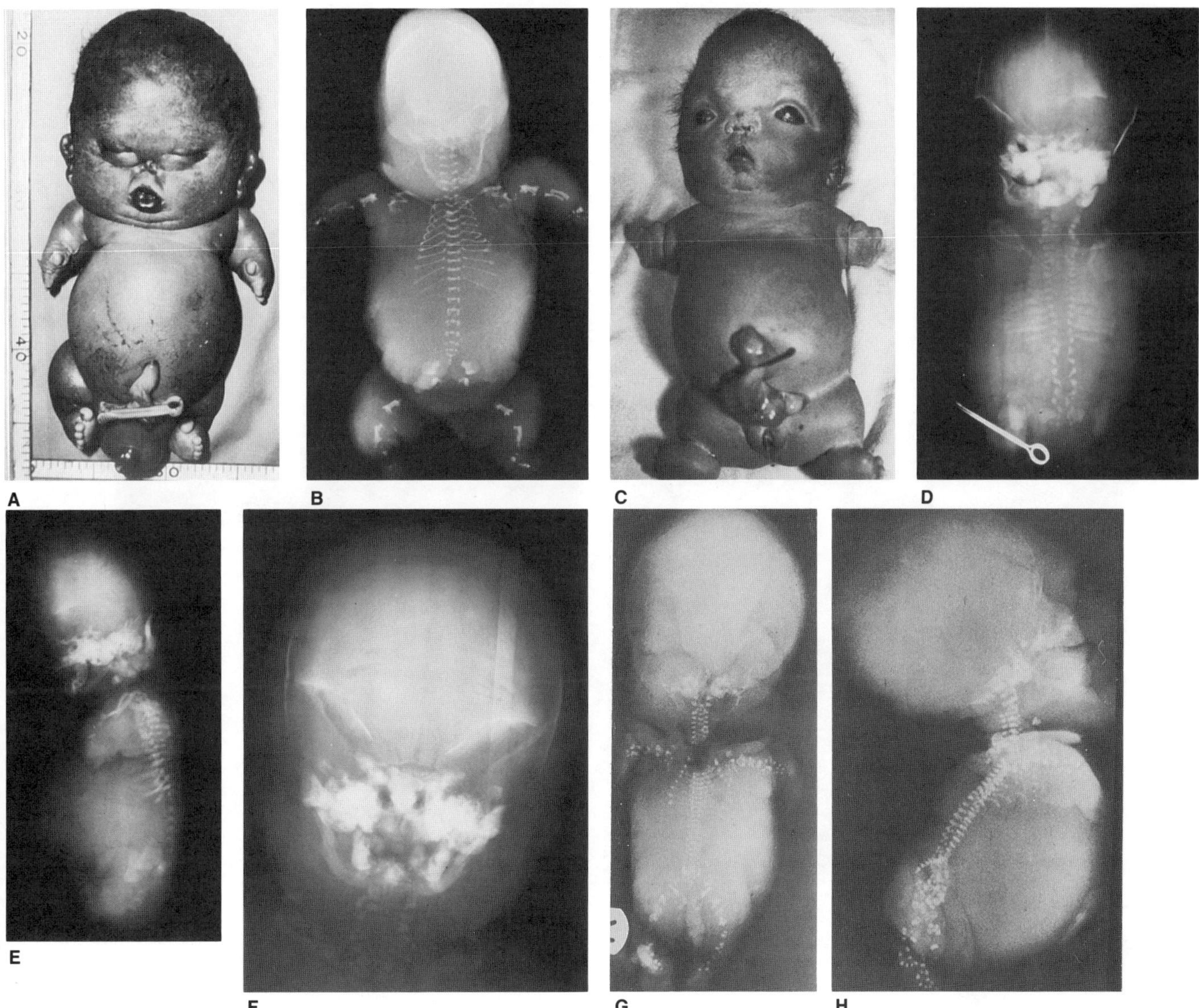

Fig. 7–76 (A,B) *Lethal short-limbed dysplasia, type Kozlowski.* (A) Stillborn, less than 30 cm in length. Note micromelia, hydrops with marked swelling of scrotum, midface hypoplasia, small mouth, fleshy ears. (B) Roentgenogram showing long, distally hooked clavicles of equal width, thin ribs, platyspondyly, relatively minor changes in the head, pelvis, and mesomelic segments of the extremities with severely affected rhizomelic segments. (From *K Kozlowski et al,* Pediatr Radiol **16:**430, 1986.) (C–F) *Lethal short-limbed dysplasia, type Piepkorn.* Note flipper-like arms and legs, facial anomalies, marked lack of ossification of all bones except clavicles. (From *M Piepkorn et al,* Teratology **16:**345, 1977.) (G,H) *Lethal short-limbed dwarfism, type Carty (dappled diaphyseal dysplasia).* Long bones and pelvis replaced by multiple small ossified foci. Cranium is very poorly calcified. Note narrow rib cage and playtspondyly. Ribs, because of similar changes, appear fractured. (From *H Carty et al,* Roefo, **150:**228, 1989.)

the dislocated lenses and the mental retardation, but the skeletal alterations were unique. The child was a product of father–daughter incest. The other disorder, made homozygous by the incest, is possibly the same as *short stature, low nasal bridge, cleft palate, and sensorineural hearing loss.*

### Reference (Megepiphyseal dysplasia)

1. Gorlin RJ et al: Megepiphyseal dwarfism. *J Pediatr* **83**:633–635, 1973.

## Nance–Sweeney chondrodysplasia

Nance and Sweeney (1) noted a unique chondrodysplasia having autosomal recessive inheritance. Cousins were similarly affected and there was parental consanguinity. Clinically, the cases superficially resembled achondroplasia. In addition, thick leathery skin, soft-tissue calcifications, and dysplastic ears were noted. Cleft palate was present in at least two of four affected sibs (Fig. 7–78A–D).

### Reference (Nance–Sweeney chondrodysplasia)

1. Nance WE, Sweeney AA: Recessively inherited chondrodysplasia. *Birth Defects* **6**(4):25–27, 1970.

## Opsismodysplasia

Maroteaux et al (1), in 1984, employed the term opsismodysplasia (Gr. delayed maturation) to describe a rhizomicromelic dwarfism orig-

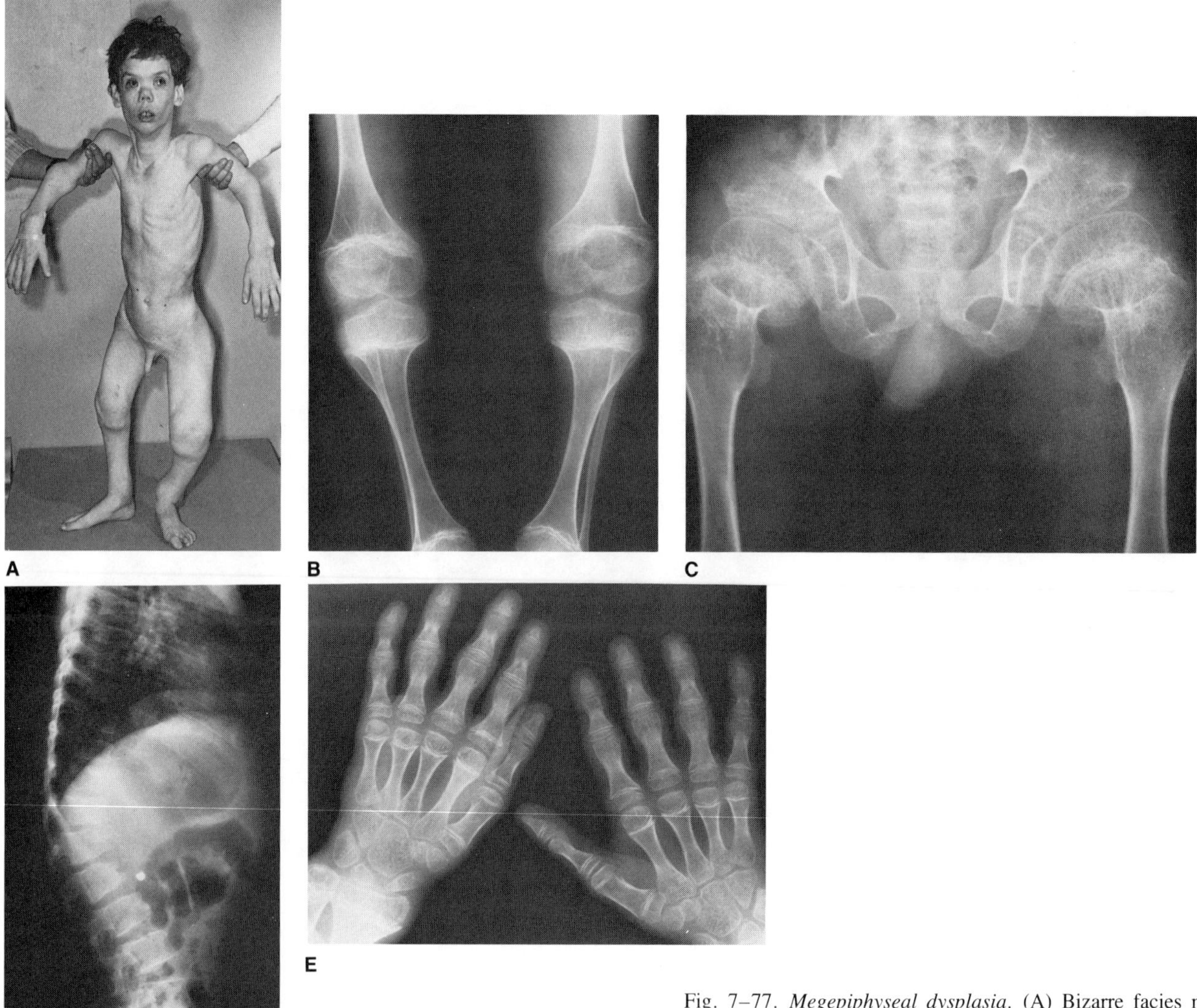

Fig. 7–77. *Megepiphyseal dysplasia.* (A) Bizarre facies marked by circumferential staphyloma, retrousse nose with anteverted nostrils. Note enlarged joints. (B) Huge epiphyses, widened metaphyses of shortened long bones. (C) Markedly enlarged femoral heads, trochanters. (D) Anterior wedging of several lumbar vertebrae. (E) Note flattened epiphyses of metacarpals, large carpal bones.

inally reported by Zonana et al (2) in 1977. The disorder, recognizable at birth, is characterized by short stature, hypotonia, and short hands with digits of equal length. The thumbs and halluces tend to be wide. The forehead is high and bossed with large anterior and posterior fontanels. The nose is small with anteverted nostrils. The philtrum is long (Figs. 7–79 and 7–80).

Radiographic changes include severe bone retardation (delayed appearance of distal femoral and proximal tibial epiphyses), shortness of long bones with short, thick diaphyses, and irregular cup-shaped metaphyses, those of the metacarpals and metatarsals being especially concave (Fig. 7–81A–C). The pelvis somewhat resembles that of asphyxiating thoracic dysplasia. The vertebral bodies are thin and lamellar.

Inheritance is autosomal recessive.

#### References (Opsismodysplasia)

1. Maroteaux P et al: Opsismodysplasia: A new type of chondrodysplasia with predominant involvement of the bones of the hand and the vertebrae. *Am J Med Genet* **19**:171–182, 1984.

2. Zonana J et al: A unique chondrodysplasia secondary to a deficit in chondroosseous transformation. *Birth Defects* **13**(3D):155–163, 1977.

### Osteoglophonic dysplasia

The term osteoglophonic dysplasia, derived from Greek and meaning hollowed-out bone, was introduced by Spranger (see 1) to refer to a distinctive skeletal dysplasia characterized by disproportionate short stature, craniosynostosis, unerupted teeth, multiple lucent metaphyseal defects, and anterior beaking of the vertebral bodies (Figs. 7–82 to 7–85). First described by Fairbank (2) in 1958 as a case report of "acrocephaly with abnormalities of the extremities," several other instances have been recorded since that time (1,3–6). Patients of both sexes, as well as father–son transmission in one instance (4), suggest autosomal dominant inheritance.

Craniosynostosis has occurred in all cases to date, and multiple sutures are usually involved. Head shape may be brachycephalic, oxycephalic, or cloverleaf (4). Hydrocephaly occurred in one instance (4).

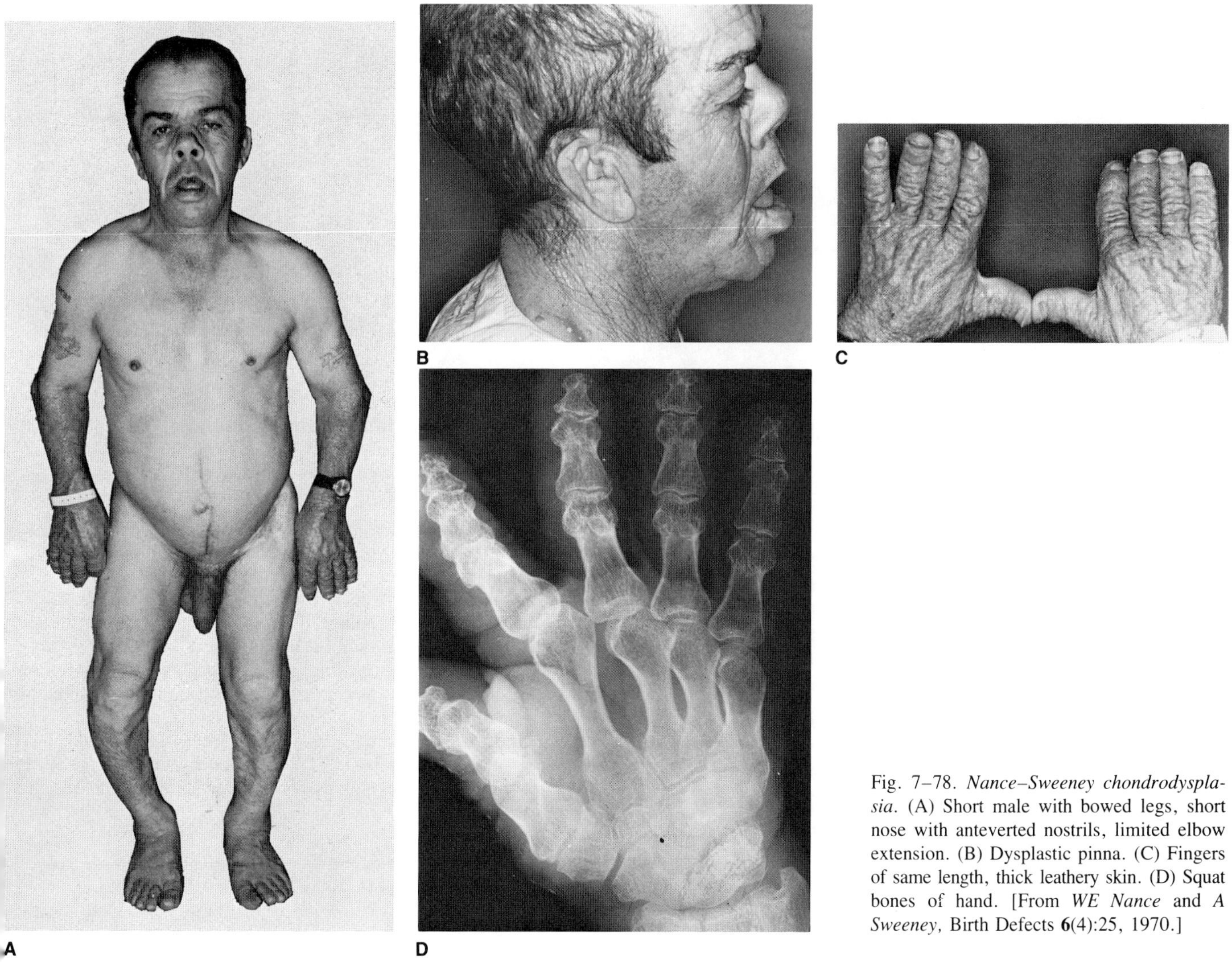

Fig. 7–78. *Nance–Sweeney chondrodysplasia.* (A) Short male with bowed legs, short nose with anteverted nostrils, limited elbow extension. (B) Dysplastic pinna. (C) Fingers of same length, thick leathery skin. (D) Squat bones of hand. [From *WE Nance* and *A Sweeney,* Birth Defects **6**(4):25, 1970.]

Fig. 7–79. *Opsismodysplasia.* High forehead with frontal bossing, small nose with anteverted nostrils, long philtrum. (From *P Maroteaux et al,* Am J Med Genet **19:**171, 1984.)

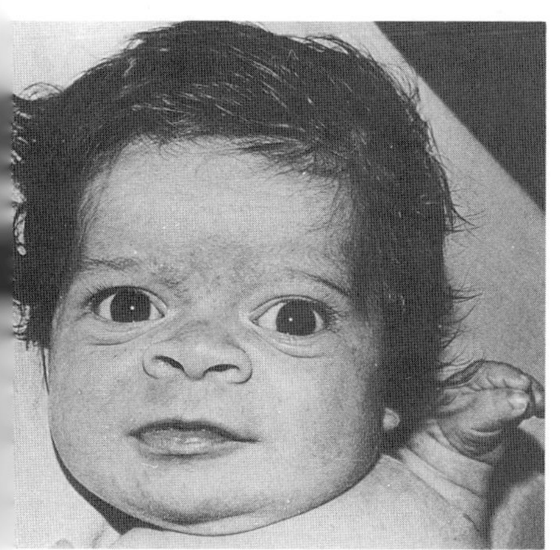

Fig. 7–80. *Opsismodysplasia.* Frontal bossing more clearly evident in this view. (From *P Maroteaux et al,* Am J Med Genet **19:**171, 1984.)

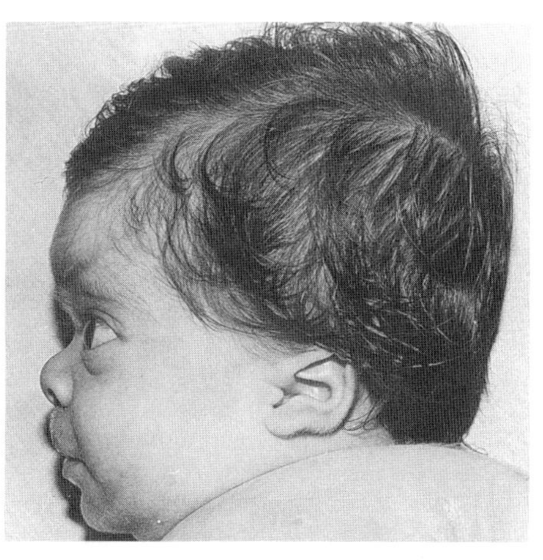

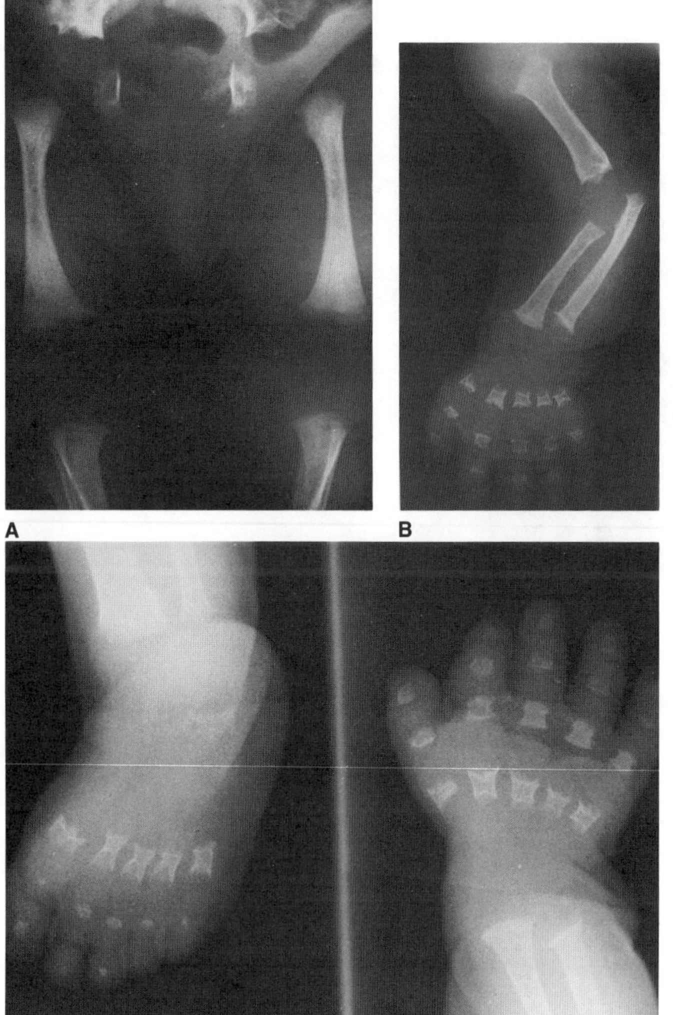

Fig. 7–81. *Opsismodysplasia.* Roentgenograms. (A) Delay of bone maturation with square appearance of iliac bone. Horizontal roof of the acetabulum. Infant of 1 day. (B) Shortness of long bones. Thick diaphyses and irregularly cup-shaped metaphyses, those of the metacarpals being especially concave. Infant of 15 months. (C) Shortness of metacarpal, metatarsal, and phalangeal bones. Note especially irregular concave metacarpals and metatarsals. Infant of 15 months. (From *P Maroteaux et al,* Am J Med Genet **19:**171, 1984.)

Facial features include frontal bossing, shallow orbits with maxillary hypoplasia, proptosis, hypertelorism, low nasal bridge, anteverted nares, and mandibular prognathism. Radiographically, there are multiple unerupted teeth.

Shortening of the extremities may be noted at birth (4). The long bones are undermodeled with generalized osteoporosis, cortical thickening, and loss of normal trabecular paterning. Lucent fibrous dysplasia-like changes are present throughout the metaphyses and are most prominent in the distal femurs and proximal tibias. The tubular bones of the extremities are short and broad; markedly dysplastic changes are observed in the epiphyseal ossification centers. In the spine, platyspondyly with anterior projection of the vertebral bodies is a prominent feature. In affected adults, resolved lucent defects result in grossly distorted, somewhat widened, and somewhat cystic-appearing metaphyses in many long bones. Other features include symmetrical flattening and lateral migration of the femoral heads, dysplastic and dislocated humeral heads, Type-B brachydactly, hypoplastic or absent middle phalanges in the feet, and thoracolumbar scoliosis (4).

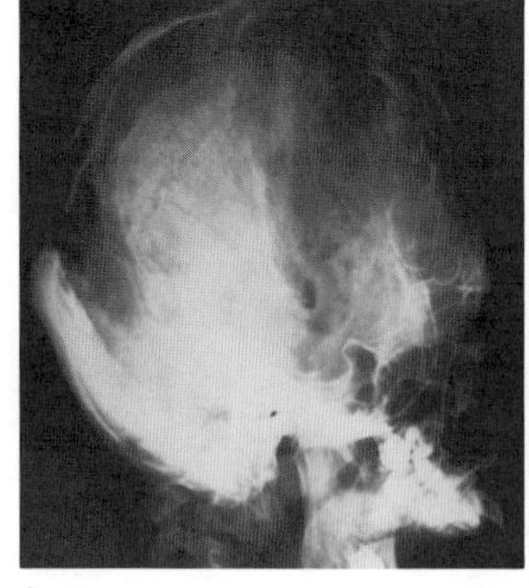

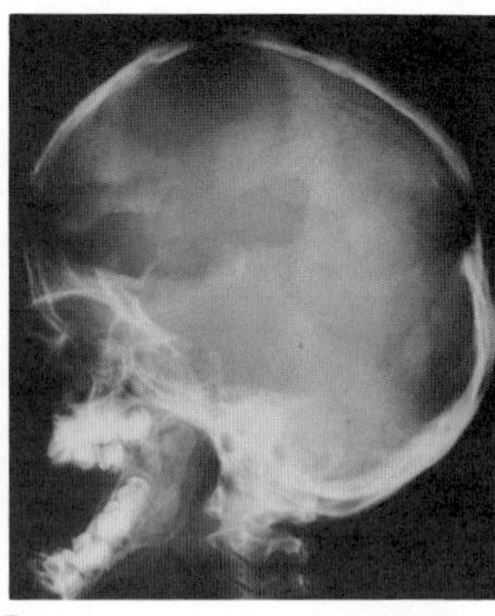

Fig. 7–82. *Osteoglophonic dysplasia.* (A,B) Oxycephalic skulls in patients of different ages. Note unerupted teeth in adult in A. Note hypoplastic maxilla and relative mandibular prognathism in B. Calvarial defects secondary to surgery for craniosynostosis. (A from *MM Cohen Jr,* Craniosynostosis, Raven Press, New York, 1986, p 488.)

Instances of cryptorchidism and inguinal hernia have been noted (4). A patient with classic findings of osteoglophonic dysplasia studied by MM Cohen had, in addition, documented renal phosphate wasting and osteomalacia. Serum calcium, serum phosphorus, and alkaline phosphatase concentrations have been within normal limits in all cases except for one instance with renal phosphate wasting.

Although osteoglophonic dysplasia is still in the early stages of its delineation, preliminary aspects of the natural history deserve mention. First, an early history of psychomotor retardation and feeding

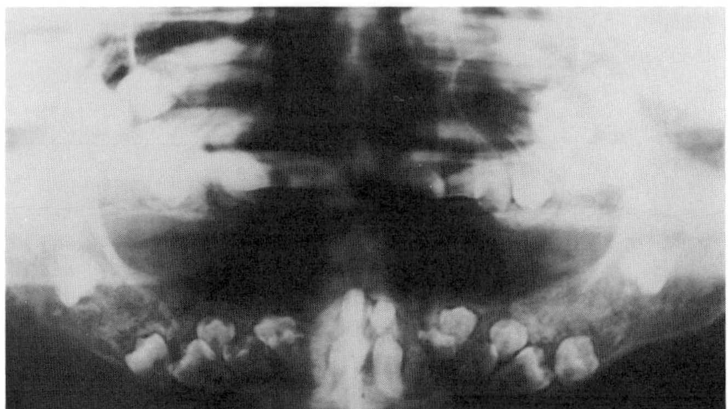

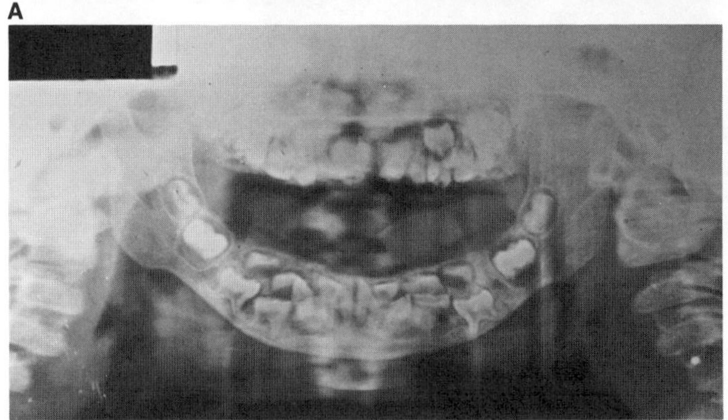

Fig. 7–83. *Osteoglophonic dysplasia.* (A,B) Unerupted teeth in patients of different ages. (A from *MM Cohen Jr,* Craniosynostosis, Raven Press, New York, 1986, p 488. B courtesy of *JF Reynolds,* Helena, Montana.)

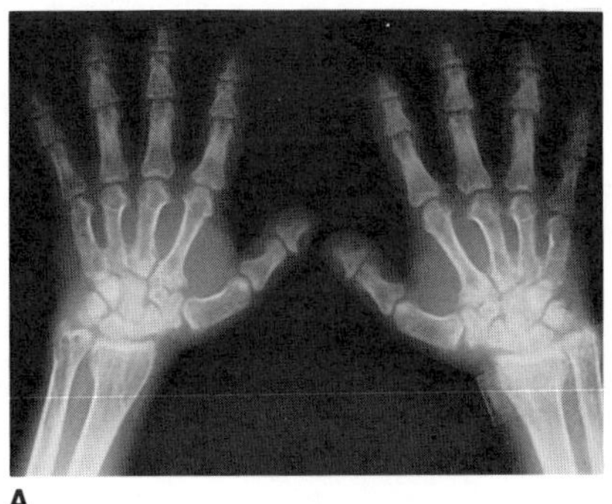


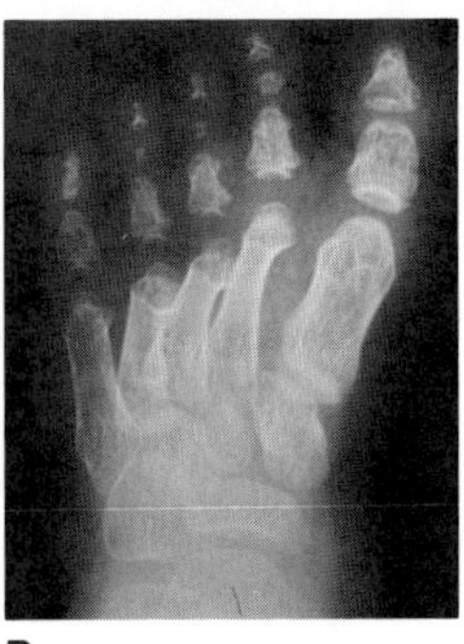


Fig. 7–84. *Osteoglophonic dysplasia.* (A,B) Brachydactyly in patients of different ages. Note abnormal epiphyseal ossification centers in B. (A from *MM Cohen Jr,* Craniosynostosis, Raven Press, New York, 1986, p 488.)

difficulties is commonly observed with later normal intelligence and nutrition. The lytic metaphyseal defects, histologically composed of moderately cellular fibrous tissue in a slightly whorled pattern (3), resolve with age. In Fairbank's case (2), the osteolytic defects were present at age 10, but had resolved by radiographic examination at age 24. Disproportionate dwarfism has resulted in an average adult height of 4 ft ($N = 3$). One patient died at age 27 of pneumonia.

### References (Osteoglophonic dysplasia)


1. Beighton P et al: Osteoglophonic dwarfism. *Pediatr Radiol* **10**:46–50, 1980.
2. Fairbank T: *An Atlas of General Affections of the Skeleton,* Case 84, ES Livingstone, Edinburgh/London, 1951.
3. Keats TE et al: Craniofacial dysostosis with fibrous metaphyseal defects. *Am J Roentgenol* **124**:271–275, 1975.
4. Kelley RI et al: Osteoglophonic dwarfism in two generations. *J Med Genet* **20**:436–440, 1983.
5. Reynolds JF et al: Osteoglophonic dysplasia: Natural history and confirmation of autosomal dominant inheritance. *Proc Greenwood Genet Ctr* **7**:167–168, 1988.
6. Santos H et al: Osteoglophonic dysplasia: A new case. *Eur J Pediatr* **147**:547–549, 1988.


Fig. 7–85. *Osteoglophonic dysplasia.* Lower femoral and upper tibial metaphyses are expanded and contain multiple defects; patellae are dislocated. (From *P Beighton et al,* Pediatr Radiol **10:**46, 1980.)

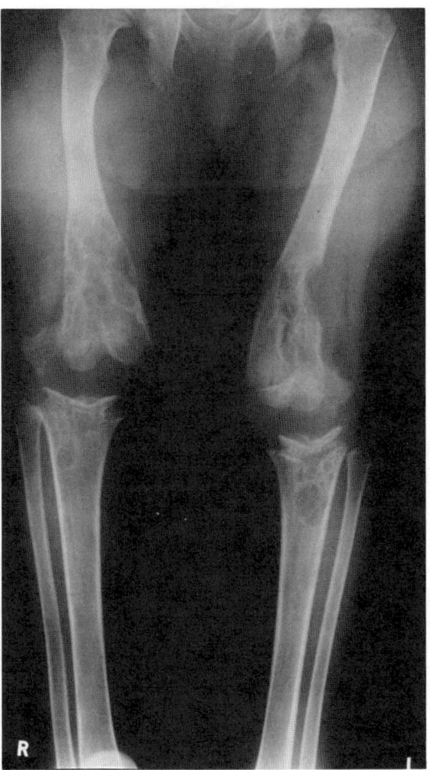


## Schneckenbecken dysplasia

Graff et al (2), in 1972, and Laxova et al (4), in 1973, reported a lethal neonatal disorder described as thanatophoric dysplasia. A similar condition was described in 1986 by Borochowitz et al (1) in 10 infants from three families. Inheritance is autosomal recessive. Affected sibs and parental consanguinity have been documented (1,2).

The infants were edematous, and hydramnios was an almost constant feature. The head was large, the neck was short, and the face exhibited midfacial flattening (1–4). One child had cleft palate (1).

The long bones were short with thick dumbbell-like metaphyses. The clavicles had a handlebar form and the scapulae were hypoplastic. The vertebral bodies were flat and the ribs were short and splayed. The ischia were short, vertical, and precociously ossified. There was a snail-like appearance to the ilia, hence Schneckenbecken from the German. The pubic bones were hypoplastic (Figs. 7–86 and 7–87). The ankle bones were precociously ossified with three ossification centers. Brachydactyly was evident. Microscopically, the cartilage exhibited increased cellular density and hypervascularity with each chondrocyte containing a large round central nucleus. Horseshoe kidneys, hydroureters, and hydronephrosis were found (2).

Fig. 7–86. *Schneckenbecken dysplasia.* Note short long bones with enlarged epiphyses, handle-bar shaped clavicles, hypoplastic scapulae, flattened vertebral bodies, short splayed ribs, abnormal ilia.

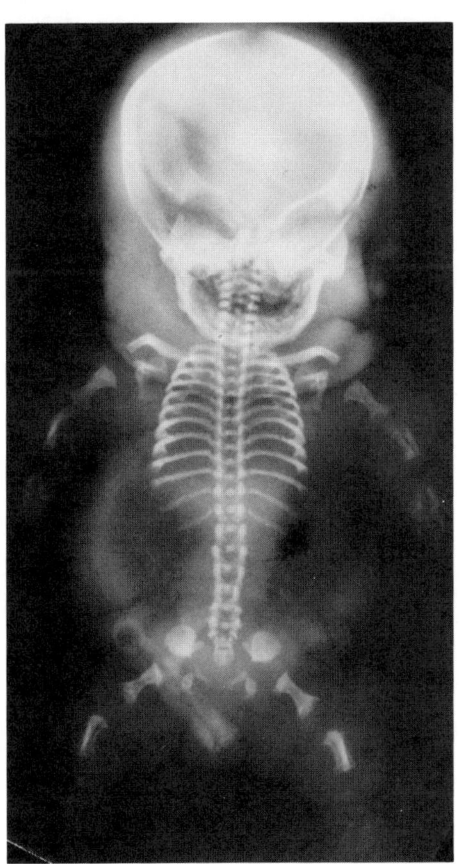

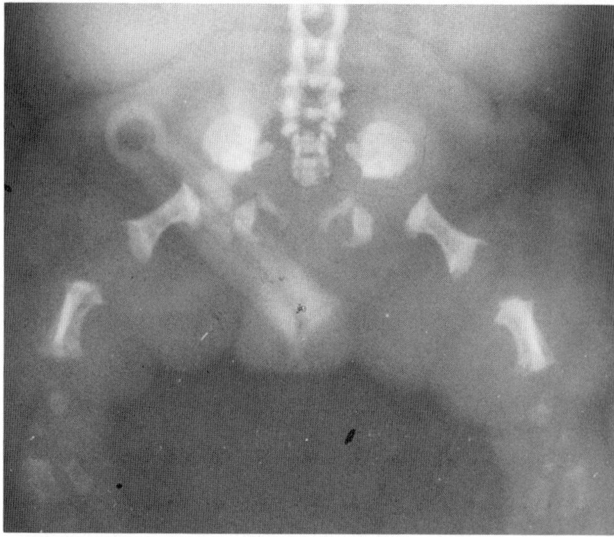

Fig 7–87. *Schneckenbecken dysplasia.* Enlarged radiograph showing snail-like ilia (Schneckenbecken). (From *Z Borochowitz et al,* Am J Med Genet **25:**47, 1986.)

### References (Schneckenbecken dysplasia)

1. Borochowitz Z et al: A distinct lethal neonatal chondrodysplasia with snail-like pelvis: Schneckenbecken dysplasia. *Am J Med Genet* **25**:47–60, 1986.

2. Graff G et al F: Familial recurring thanatophoric dwarfism. *Obstet Gynecol* **39**:515–520, 1972.

3. Knowles S et al: A new category of lethal short-limbed dwarfism. *Am J Med Genet* **25**:41–46, 1986 (same patients as Ref. 4).

4. Laxova R et al: Family with probable achondrogenesis and lipid inclusions in fibroblasts. *Arch Dis Childh* **48**:212–216, 1973.

## Short rib–polydactyly syndromes, type I (Saldino–Noonan) and type III (Naumoff)

Saldino and Noonan (15), in 1972, first reported a lethal syndrome characterized by micromelia, narrow chest, protuberant abdomen, postaxial polydactyly, remarkably short ribs, abnormally contoured ilia, peg-shaped or pointed femora, notched tibiae, general underossification of short tubular bones with poor corticomedullary demarcation, and various visceral anomalies. Additional patients with very similar radiologic findings were reported by several authors (4,12,14,18). This is called Type I short rib–polydactyly syndrome (SRP I).

Inheritance is clearly autosomal recessive, but all severely affected cases are clinically female. This in part is the result of failure of secondary sexual differentiation in genetic and gonadal males. Thus, sex-reversed and constitutional females exhibit more severe manifestations (17a). Affected sibs have been described (4,12,14) and there is marked concordance in radiologic features, particularly with regard to the ilia and femora. The condition can easily be separated from short rib–polydactyly (SRP) Type II Majewski, which is discussed in the next section.

However, Verma et al (21) and Naumoff et al (11) suggested that there is another lethal autosomal recessive inherited short rib–-polydactylous dwarfism with less severe changes in the ilia and long bones. This is called Type III short rib–polydactyly syndrome (SRP III). The metaphyses are widened with longitudinal spurring at the margin of the metaphyses of the distal humeri and proximal and distal radii and femora, clear corticomedullary demarcation, short cranial base, and horizontal trident-shaped acetabular margins. The tibiae are broad with constriction at the lower and middle third. All hand bones are hypoplastic. Similar cases were subsequently recorded by several authors (1–3,5,7–10,13). Doubt has been strongly expressed regarding separation of the two types (2a,3,16) but Yang et al (22) have rejected the fusion.

The clinical phenotypes of SRP Types I and III are strikingly similar (Figs. 7–88 and 7–89). Prenatally there is oligohydramnios or polyhydramnios and hydrops. The forehead is prominent and the nose wide with a depressed bridge and anteverted nostrils. The pinnae are usually malformed. Cleft lip or cleft palate had been reported (4,14). The tongue may be cleft and alveolar ridges furrowed (14). There may be hyperplastic frenula. The facies may resemble that seen in *Potter sequence*. The extremities are very short, the chest narrow, and the abdomen protuberant. The polydactyly (six or seven digits are frequent) is postaxial, usually asymmetric, and more often involves the hands. There is also variable syndactyly on the postaxial side. There may be minimal ossification of the short bones of the hands (9).

Visceral anomalies include hypoplastic lungs and atresia of the esophagus, duodenum, and particularly the anus in SRP Type I. Both forms may exhibit short or malrotated intestines. Various congenital heart anomalies (transposition of great vessels, VSD, double outlet left ventricle, endocardial cushion defect, persistence of left superior vena cava) have been documented in SRP Type I but not in SRP Type III. Coarctation of the aorta, hypoplastic left or right heart, and/or cor biloculare have occurred in SRP Type III but not in SRP Type I. In both types the kidneys may be absent, dysplastic, and/or cystic, and the ureters and bladder hypoplastic. In Type I SRP there is a small or absent, rarely septate, urethrovaginal opening, and absent and double uterus have been reported. Several phenotypic males with 46,XY karyotype have had ambiguous genitalia (2a,18), and several clinical females have been shown to have 46,XY karyotype (2a).

Various miscellaneous anomalies include agenesis of the gallbladder and fibrosis of the pancreas (2a,5,11,12,14).

Both disorders have irregular chondrocytic proliferation with clumping or absence of flattened chondrocytes at the growth plate. Hypertrophic cells are irregularly dispersed. The primary trabecular bone is distorted and bridged (17).

Yang et al (22) described PAS-positive chondrocytic cytoplasmic inclusions in SRP Type III but not in SRP Type I. However, Sillence (16) has not borne out this observation. Chondrocytic inclusions may be seen in *Type I achondrogenesis, spondyloepiphyseal dysplasia (SED) congenita, Kniest dysplasia,* and in several types of pseudoachondroplastic SED. *Beemer–Winter lethal short-limbed dysplasia* does not exhibit polydactyly.

Short–rib polydactyly syndromes have been diagnosed prenatally (6,19).

### References [Short rib–polydactyly syndromes, type I (Saldino–Noonan) and type III (Naumoff)]

1. Barrio AL, Martinez MCM: A proposito de un case de sindrome de Saldino–Noonan. *Radiologia (Madrid)* **21**:373–376, 1979.

2. Belloni C, Beluffi G: Short rib–polydactyly syndrome, type Verma–Naumoff. *Roefo* **134**:431–435, 1981.

2a. Bernstein R et al: Short rib–polydactyly syndrome: A single or heterogeneous entity? A reevaluation prompted by four new cases. *J Med Genet* **22**:46–53, 1985.

3. Cherstvoy ED et al: Difficulties in classification of the short rib-polydactyly syndromes. *Eur J Pediatr* **133**:57–61, 1980.

4. Gordon IRS, Brown NJ: The syndrome of micromelic dwarfism and multiple anomalies. *Ann Radiol* **19**:161–165, 1976.

5. Grote W et al: Prenatal diagnosis of a short rib–polydactylia syndrome type Saldino–Noonan at 17 weeks' gestation. *Eur J Pediatr* **140**:63–66, 1983.

6. Johnson VP et al: Midtrimester prenatal diagnosis of short-limb dwarfism (Saldino–Noonan syndrome). *Birth Defects* **18**(3A):133–141, 1982.

7. Kaibara N et al: Short rib–polydactyly syndrome type I, Saldino–Noonan. *Eur J Pediatr* **133**:63–65, 1980.

8. Kozlowski K et al: New forms of neonatal death dwarfism. *Pediatr Radiol* **10**:155–160, 1981.

9. Krepler R et al: Nicht lebensfähiger, mikromeler Zwergwuchs: Thoraxdystrophie-Polydaktylie-Syndrom Typ Saldino–Noonan. *Mschr Kinderheilkd* **124**:167–173, 1976.

10. Lowry RB, Wignalle N: Saldino–Noonan short rib–polydactyly dwarfism syndrome. *Pediatrics* **56**:121–123, 1975.

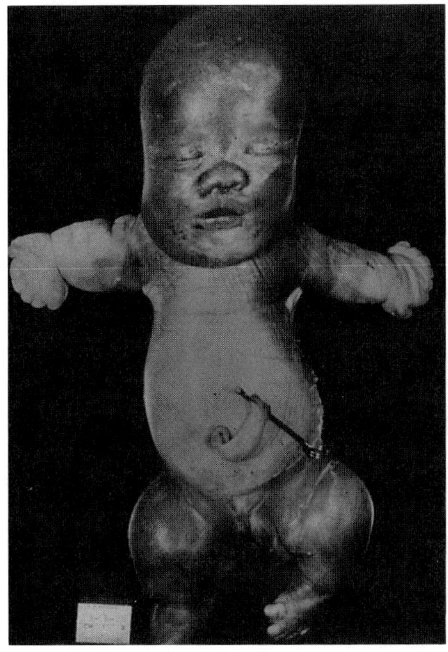
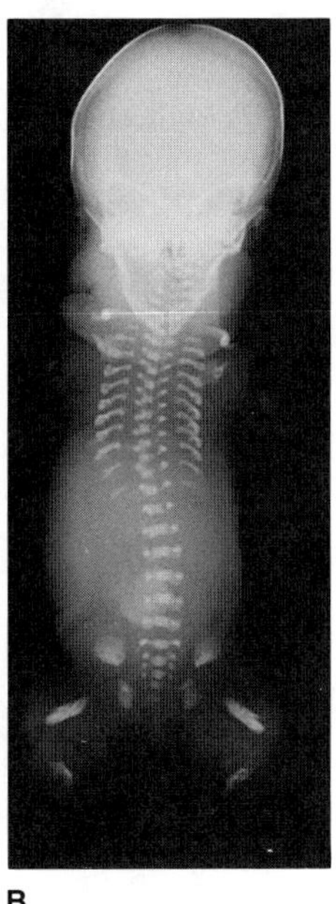
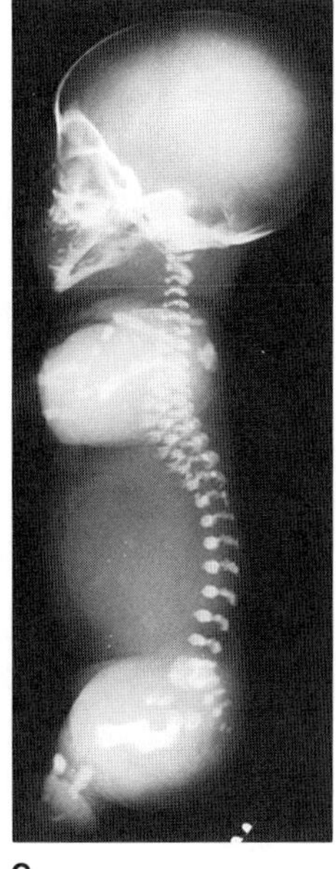

A B C

Fig. 7–88. *Short rib–polydactyly syndrome, type I (Saldino–Noonan).* (A) Note micromelia, narrow chest, protuberant abdomen, postaxial polydactyly. Forehead is prominent, nose wide with low nasal bridge; nostrils are anteverted. (B,C) The ribs are remarkably short, the ilia abnormally contoured, the femora peg shaped, the tibiae notched, the vertebral bodies small and flattened. (A from *M Richardson et al,* J Pediatr **91:**467, 1977. B,C courtesy of *AL Baudet,* Houston, Texas.)


11. Naumoff P et al: Short rib–polydactyly syndrome type 3. *Radiology* **122**:443–447, 1977.

12. Richardson MM et al: Prenatal diagnosis of recurrence of Saldino–Noonan dwarfism. *J Pediatr* **91**:467–471, 1977.

13. Rosani R, Bertoli G: La sindrome di Saldino–Noonan. *Minerva Pediatr* **31**:1351–1356, 1979.

14. Rupprecht E, Gurski A: Kurzrippen-Polydaktylie-Syndrom Typ Saldino–Noonan. *Helv Paediat Acta* **37**:161–169, 1982.

15. Saldino RM, Noonan CD: Severe thoracic dystrophy with striking micromelia, abnormal osseous development, including the spine and multiple visceral anomalies. *Am J Roentgenol* **114**:257–263, 1972.


Fig. 7–89. *Short rib–polydactyly syndrome, type III (Naumoff).* (A) Clinical phenotype strikingly similar to Type I (Majewski). (B) Changes in ilia and long bones less severe than in Type I. (From *R Bernstein et al,* J Med Genet **22:**46, 1985.)

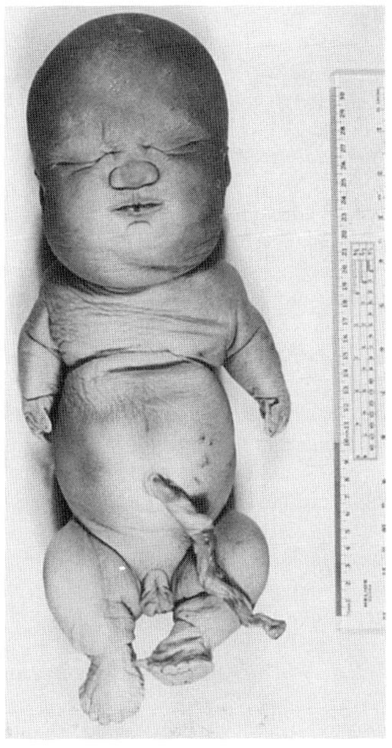
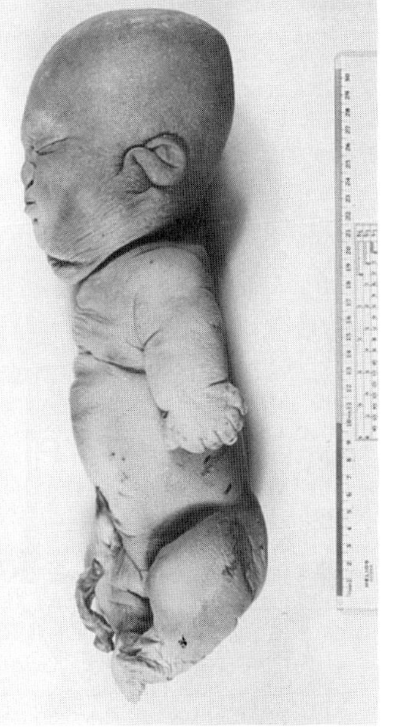

A B


16. Sillence DO: Non-Majewski short rib–polydactyly syndrome. *Am J Med Genet* **7**:223–229, 1980.

17. Sillence DO et al: Morphological studies in the skeletal dysplasias. *Am J Pathol* **96**:811–870, 1979.

17a. Sillence D et al: Perinatally lethal short rib–polydactyly syndromes. 1. Variability in known syndromes. *Pediatr Radiol* **17**:474–480, 1987.

18. Spranger J et al: Short rib–polydactyly (SRP) syndromes, types Majewski and Saldino–Noonan. *Z Kinderheilk* **116**:73–94, 1974.

19. Toftager-Larsen K, Benzie RJ: Fetoscopy in prenatal diagnosis of the Majewski and the Saldino–Noonan types of the short rib–polydactyly syndromes. *Clin Genet* **26**:56–60, 1984.

20. Vecchi R: Morphologic features in a case of Saldino–Noonan syndrome. *Folia Hered Pathol* **30**:39–49, 1981.

21. Verma IC et al: An autosomal recessive form of lethal chondrodystrophy with severe thoracic narrowing, rhizoacromelic type of micromelia, polydactyly and genital anomalies. *Birth Defects* **11**(6):167–174, 1975.

22. Yang SS et al: Short rib–polydactyly syndrome, type 3 with chondrocytic inclusions. *Am J Med Genet* **7**:205–213, 1980.


## Short rib–polydactyly syndrome, type II (Majewski)

There are several autosomal recessively inherited lethal forms of short-limbed dwarfism characterized by short ribs and polydactyly. In the Majewski type, the infant is hydropic and there is polyhydramnios. The thorax is short and narrow. The abdomen is protuberant. The extremities are abbreviated (mesomelic brachymelia), particularly the lower. There is pre- and postaxial polysyndactyly with up to nine digits per extremity (Figs. 7–90 to 7–93). The degree of polysyndactyly is extremely variable. There have been a few cases in which the hands have exhibited only postaxial polysyndactyly and the feet only preaxial polysyndactyly. We cannot classify the infant with a Majewski-like syndrome and cebocephaly (14).

On postmortem examination, there is hypoplasia of the epiglottis, larynx, and lungs. About 50% exhibit imperforate anus and intestinal malrotation and/or short bowel and fibrocystic pancreas. Cardiac anomalies include persistent superior vena cava and, occasionally, VSD and coarctation of the aorta. Genitourinary anomalies observed are small or absent urogenital opening, hypospadias, micropenis, cryptorchidism, septate uterus, and septate or rudimentary vagina. The kidneys are hypoplastic with multiple glomerular cysts and focal cystic dilatation of the distal tubules. Often there is cystic and/or hypoplastic uterus.

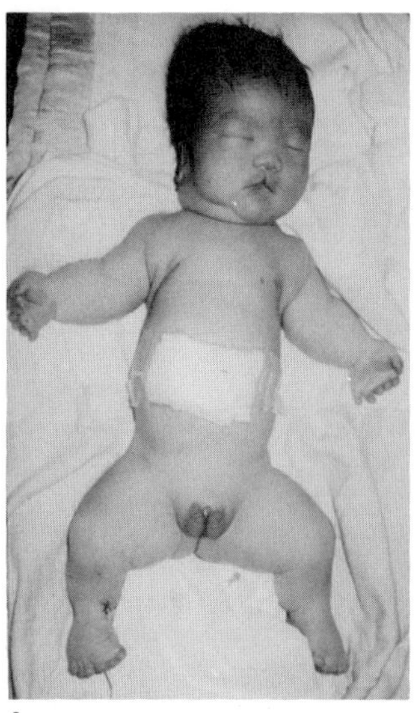
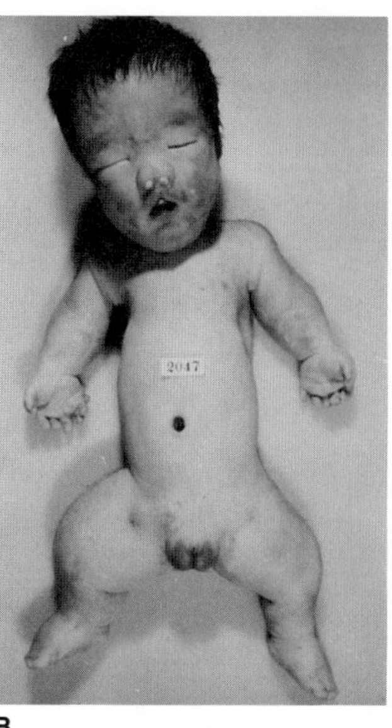

A B

Fig. 7–90. *Short rib–polydactyly syndrome, type II Majewski.* (A,B) Affected sibs. Note midline cleft of upper lip, short limbs, narrow chest, polysyndactyly. (From *T Motegi et al,* Hum Genet **49:**269, 1979.)

Radiographic changes include very short horizontal ribs, pre- and postaxial polydactyly, and short rounded tibiae. The ilia and vertebral bodies appear essentially normal. There is premature ossification of the proximal epiphyses of the humeri and femora. The middle and distal phalanges are poorly ossified.

Oral–facial anomalies include somewhat malformed pinnae, median cleft lip and cleft palate (2,4–6,8,12,18–22), bifid or trifid tongue (2,12), notched gingiva (7), ankyloglossia (8,12,18) and natal teeth (5,7).

In *Beemer–Winter lethal short-limbed dysplasia,* there is no polydactyly (1,9,16). An infant described by Bidot-López et al (3) had only mild shortness of ribs and tibiae, preaxial polydactyly of the feet only, and normal internal organs and genitalia.

Prenatal diagnosis has been accomplished by fetoscopy (21) and ultrasound (10,17).

### References [Short rib–polydactyly syndrome, type II (Majewski)]

1. Beemer FA et al: A new short rib syndrome: Report of two cases. *Am J Med Genet* **14**:115–123, 1983.

Fig. 7–91. *Short rib–polydactyly syndrome, type II Majewski.* (A,B) Median cleft lip and somewhat malformed pinnae. (Courtesy of *JS Fitzsimmons,* J Med Genet **19:**141, 1982.)

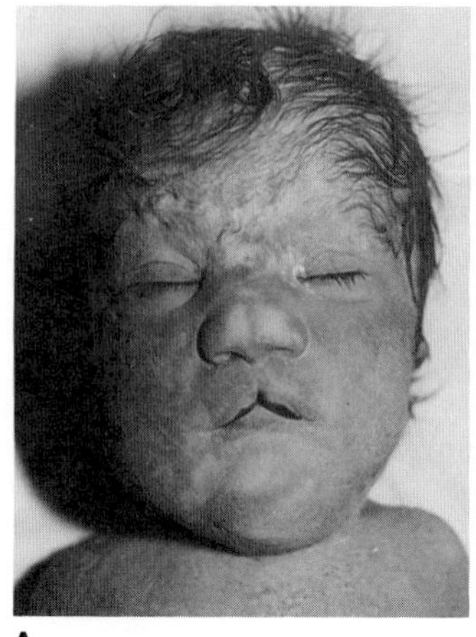
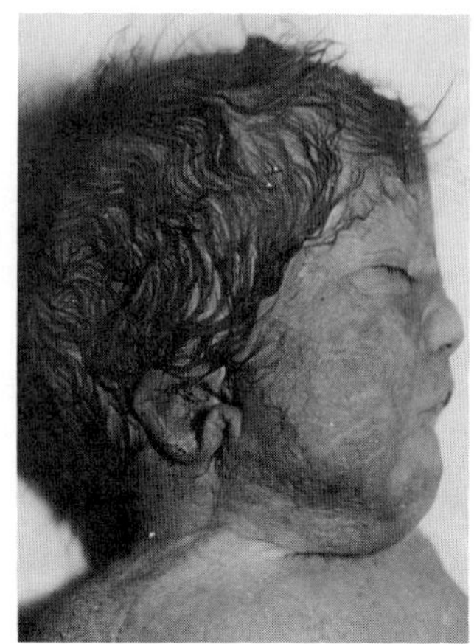

A 
B

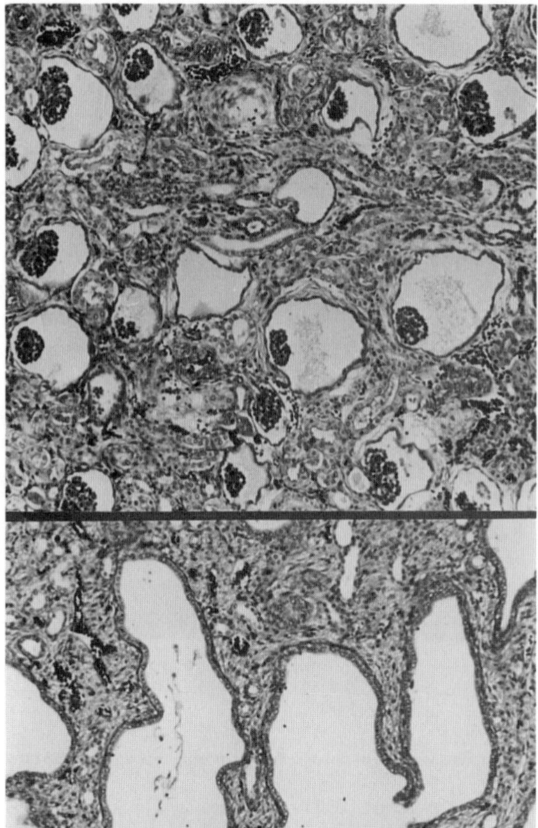

Fig. 7–92. *Short rib–polydactyly syndrome, type II Majewski.* Cortex (top) and medulla (bottom) of kidney showing multiple glomerular cysts and focal cystic dilatation of distal tubules. (From *T Motegi et al,* Hum Genet **49:**269, 1979.)

2. Bergström K et al: A case of Majewski syndrome with pathoanatomic examination. *Skel Radiol* **4**:134–140, 1979.

3. Bidot-López et al: A case of short rib polydactyly. *Pediatrics* **61**:427–432, 1978.

4. Black IL et al: Parental consanguinity and the Majewski syndrome. *J Med Genet* **19**:141–143, 1982.

5. Casper JL: Ein Missgeburt seltenster Art. *Berl Klin Wochenschr* **1**:9–10, 1864.

6. Chen H et al: Short rib polydactyly syndrome, Majewski type. *Am J Med Genet* **7**:215–222, 1980.

7. Cooper CP, Hall CM: Lethal short rib–polydactyly syndrome of the Majewski type: A report of three cases. *Radiology* **144**:513–517, 1982.

8. Dreibholz E: *Beschriebung einer sogenannten Phokomelie.* Thesis, Berlin.

9. Garcia H et al: Short rib–polydactyly syndromes. *Klin Pädiatr* **200**:141–144, 1988.

10. Gembruch U et al: Early prenatal diagnosis of short–rib polydactyly (SRP) syndrome type I (Majewski) by ultrasound in a case at risk. *Prenat Diagn* **5**:357–362, 1985.

11. Kozlowski K et al: Neonatal death dwarfism (report of 14 cases). *Australas Radiol* **21**:164–183, 1977 (Case 1).

12. Majewski F et al: Polysyndaktylie, verkürzte Gliedmassen und Genitalfehlbildungen: Kennzeichen eines selbständigen Syndroms? *Z Kinderheilkd* **111**:118–138, 1971.

13. Motegi T et al: Short rib–polydactyly, Majewski type, in two male siblings. *Hum Genet* **49**:269–275, 1979.

14. Nivelon-Chevallier A et al: Chondrodysplasie letale à côtes courtes type Majewski: Diagnostic in utero. *Pédiatrie* **37**:453–460, 1982.

15. Otto AW: *Seltene Beobachtung zur Anatomie, Physiologie und Pathologie gehörig.* Breslau, Holaufer, 1816.

16. Passarge E: Familial occurrence of a short rib syndrome with hydrops fetalis but without polydactyly. *Am J Med Genet* **14**:403–405, 1983.

17. Pauli RM et al: Short rib polydactyly, type Majewski: Prenatal diagno-

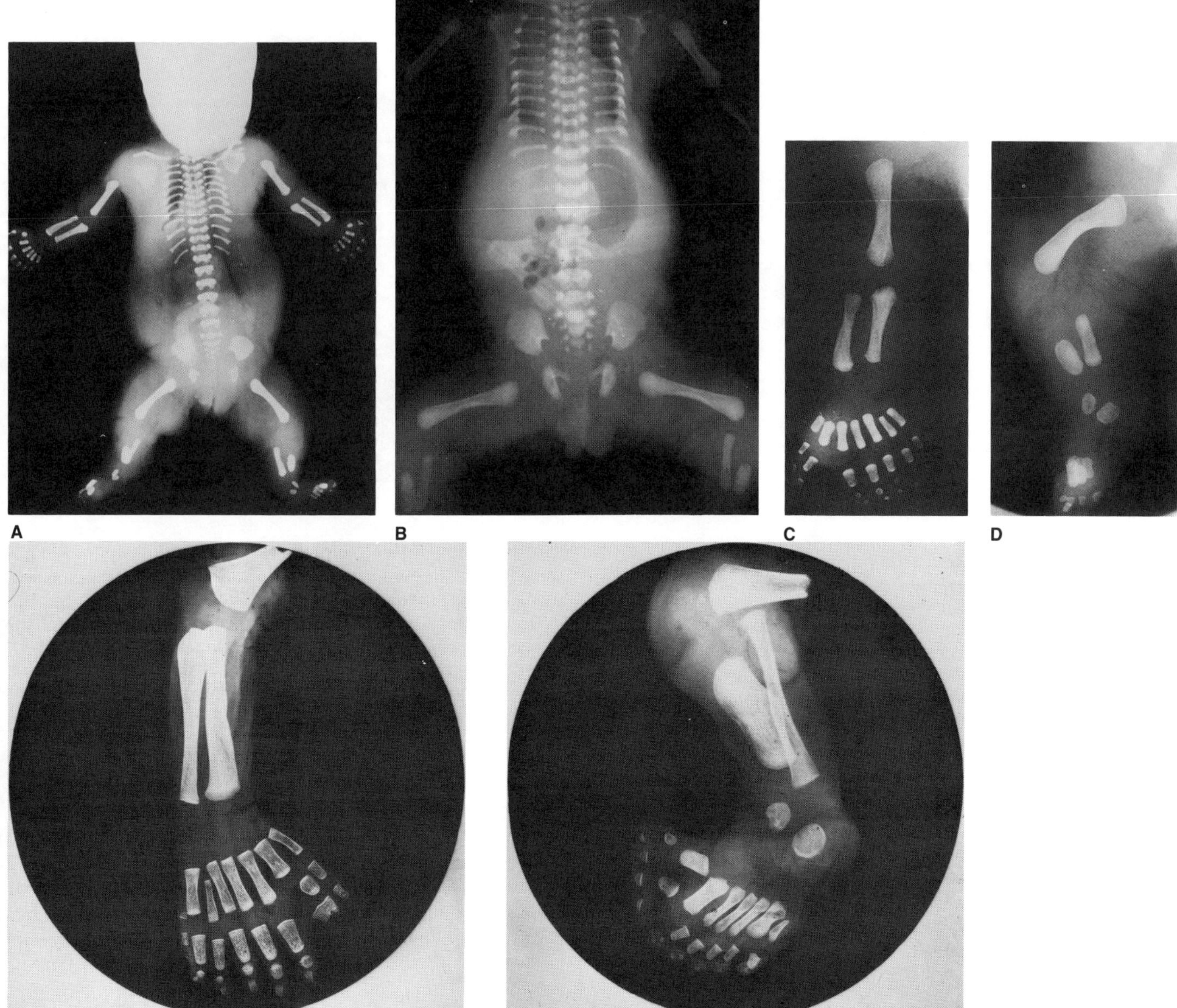


Fig. 7–93. *Short rib–polydactyly syndrome, type II Majewski.* (A) Micromelia. (B) Short ribs, normal pelvis, very abbreviated tibiae. (C) Shortened humerus, radius and ulna, pre- and postaxial polydactyly. (D) Shortened femur and lower leg bones, particularly the tibia. (E,F) Compare with C and D. (B,C,D, courtesy of *H Jorulf,* Uppsala, Sweden. E,F from *E Pitschi,* Thesis, Zürich, 1904–1905.)

sis, and possible clue to chromosomal localization of the Majewski gene. Eighth David W. Smith Workshop on Malformations and Morphogenesis, Greenville, South Carolina, August 15–19, 1987.

18. Pitschi E: *Zur Kasuistik der Poly- und Syndaktylie aller Extremitäten nebst beiderseitigem partiellem Tibiadefekt und anderen Missbildungen (doppelte Anlage des Unterkiefers).* Thesis, Zürich, 1904–1905.

19. Spranger J et al: Short rib–polydactyly (SRP), syndromes, types Majewski and Saldino–Noonan. *Z Kinderheilk* **116**:73–94, 1974.

20. Thomson GSM et al: Antenatal detection of recurrence of Majewski dwarfs (short rib–polydactyly syndrome type 2 Majewski). *Clin Radiol* **33**:509–517, 1982.

21. Toftager-Larsen K, Benzie RJ: Fetoscopy in prenatal diagnosis of the Majewski and the Saldino–Noonan type of short rib–polydactyly syndromes. *Clin Genet* **26**:56–60, 1984.

22. Walley VM et al: Short rib–polydactyly syndrome, Majewski type. *Am J Med Genet* **14**:445–452, 1983.

## Spondyloepiphyseal dysplasia congenita

Spondyloepiphyseal dysplasia (SED) congenita (Figs. 7–94 to 7–98) was first described by Spranger and Wiedemann in 1966 (15). The disorder is rare, the prevalence being approximately 1/100,000 (16a,18). The classic form of the disorder has autosomal dominant inheritance (5,12,15). Significant heterogeneity is evident (see below). SED congenita is associated with congenitally reduced stature (adult height 84–128 cm) resulting largely from disproportionate shortness of the neck and trunk and coxa vara. The head appears to sit upon the trunk and is often held in retroflexion. Horton et al (6) established growth curves for the disorder. The extremities are proportionately shortened but the hands and feet are normal. There is a small bell-shaped chest and protuberant abdomen (Fig. 7–94A,B). Stiffness, limitation at the hips,

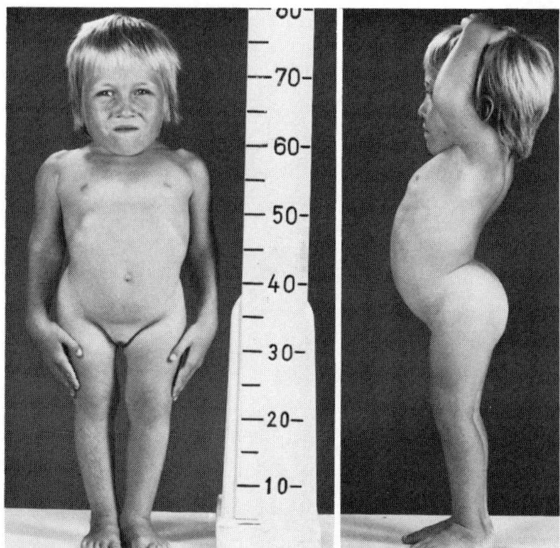


Fig. 7–94. *Spondyloepiphyseal dysplasia congenita.* Markedly reduced stature, shortened neck and trunk, severe myopia, and retinal detachment. Note marked lumbar lordosis. (Courtesy of *J Spranger,* Kiel, Germany.)

and waddling gate are evident. Most patients exhibit pectus carinatum, moderate thoracic kyphoscoliosis, and, in particular, lumbar lordosis. Talipes varus occurs in about 10–15%. Nonprogressive myopia of 5 diopters or greater has been documented in about half of the children. In those with high myopia, vitreoretinal degeneration is encountered and vitreous syneresis is apparently present in all patients. Retinal detachment, in spite of earlier reports (12,15), is apparently rare (3,18). The disorder apparently results from a defect in Type II collagen (11a,11b).

Radiographically, in the infant, the vertebral bodies appear ovoid in lateral view. The odontoid is usually hypoplastic and may dislocate. As the child matures, there is platyspondyly with posterior wedging of the vertebral bodies. Mild to moderate metaphyseal alterations are noted in the long bones of infants. There is retardation in ossification of the sternum, pubic bones, distal femoral and proximal tibial epiphyses, talus, and calcaneus. The iliac bones are hypoplastic. The upper femoral epiphyses are small and deformed, late in development, and in coxa vara position (Figs. 7–95–7–98).

The frequency of cleft palate is difficult to determine since few authors have tabulated these data. Our impression is that 15–20% exhibit cleft palate (2,5,7,9,11,12,17,18).

There appears to be genetic heterogeneity. A lethal variant exhibits

Fig. 7–95. *Spondyloepiphyseal dysplasia congenita.* Note retardation in ossification of pubic bones. [From *W Holthusen,* Ann Radiol (Paris) **15:**253, 1972.]

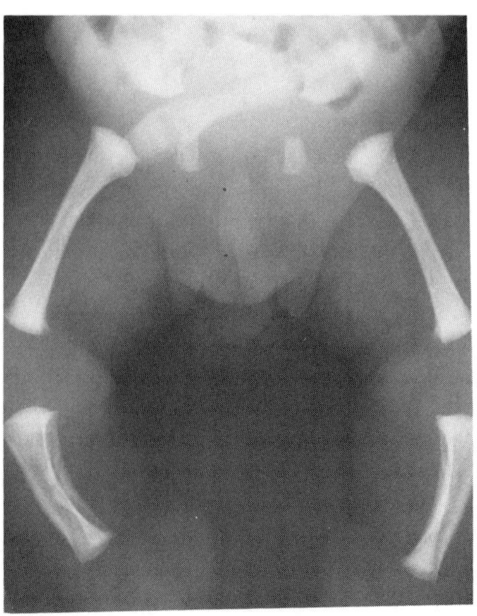

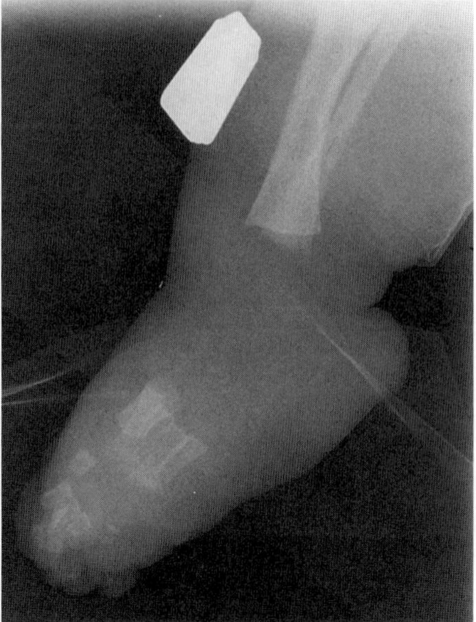

Fig. 7–96. *Spondyloepiphyseal dysplasia congenita.* Failure of calcification of talus and calcaneus.

Fig. 7–97. *Spondyloepiphyseal dysplasia congenita.* Ovoid vertebral bodies.

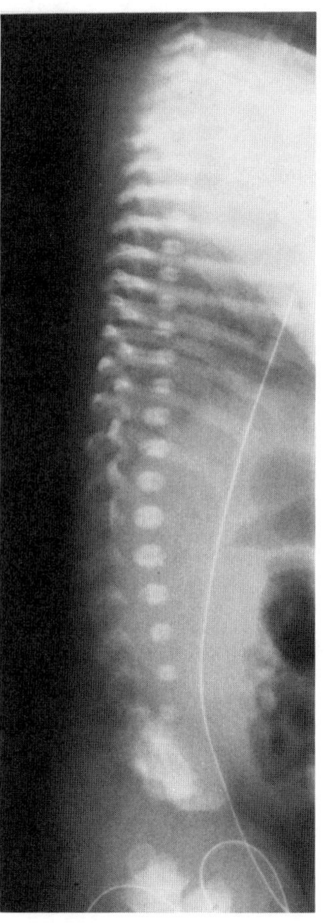

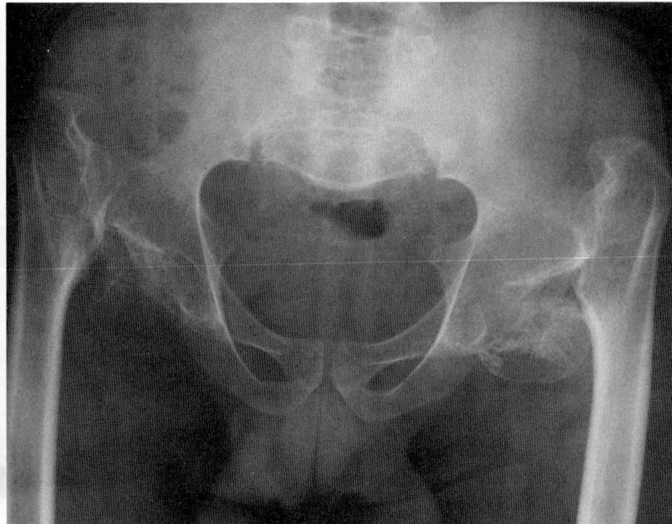

Fig. 7–98. *Spondyloepiphyseal dysplasia congenita.* Dislocated hips with femoral heads in the acetabula.

radiographic changes virtually identically to those in the classic form; but there is somewhat more rhizomelic shortening of long bones, and the clavicles appear longer and the scapulae more square. There are also histochemical and ultrastructural differences (16,19). Wynne-Davies and Hall (18) noted two forms within classic SED congenita. In the first type, comprising about 65% of cases, patients have very short stature as adults (104–127 cm) and grossly disorganized hips with severe coxa vara. The other 35% have a stature of 130–145 cm with only mild coxa vara. The two forms cannot be separated radiographically until after the fourth year of life. Kozlowski et al (8) suggested that there is heterogeneity based on mild and severe degrees of metaphyseal involvement. Those with more marked metaphyseal involvement had more severe scoliosis and less myopia. A possible autosomal recessive form in affected sibs with normal parents has been described (4,13,17). Whether this represents the early manifestation of spondylometepiphyseal dysplasia, Strudwick type, in which cleft palate can also be found or whether it is a separate entity is not currently known (1,9,10,14).

Huson et al (6a) reported a newly recognized type of spondyloepiphyseal dysplasia with dysmorphic features. Three of four children of a first cousin marriage were affected. Features included blue sclerae, broad palate, short philtrum, prominent vermilion borders, broad nasal root, broad flat chest, mild pectus excavatum, long fingers and toes, and marked lumbar lordosis. Radiographic survey showed a unique form of spondyloepiphyseal dysplasia.

#### References (Spondyloepiphyseal dysplasia congenita)


1. Anderson CE et al: Spondylometepiphyseal dysplasia, Strudwick type. *Am J Med Genet* **13**:243–256, 1982.

2. Bach C et al: Dysplasie spondylo-epiphysaire congénitale avec anomalies multiples. *Arch Fr Pédiatr* **24**:23–33, 1967.

3. Hamidi-Toosi S, Maumenee IH: Vitreoretinal degeneration in spondyloepiphyseal dysplasia congenita. *Int Orthoped* **2**:47–51, 1978.

4. Harrod MJE et al: Genetic heterogeneity in spondyloepiphyseal dysplasia congenita. *Am J Med Genet* **18**:311–320, 1984.

5. Holthusen W: Dysplasia spondyloepiphysaria congenita. *Radiologe* **16**:286–287, 1976.

6. Horton WA et al: Growth curves for height for diastrophic dysplasia, spondyloepiphyseal dysplasia congenita, and pseudoachondroplasia. *Am J Dis Child* **136**:316–319, 1982.

6a. Huson S et al: A previously unrecognized spondyloepiphyseal dysplasia with associated dysmorphic features. Third Manchester Birth Defects Conference, Manchester, October 25–28, 1988.

7. Kozlowski K et al: Spondylo-epiphyseal dysplasia congenita. *Ann Radiol* **11**:367–375, 1968.

8. Kozlowski K et al: Dysplasia spondylo-epiphysealis congenita Spranger–Wiedemann: A critical analysis. *Australas Radiol* **21**:260–280, 1977.

9. Ludthardt T: Dysplasia spondylo-epiphysaria congenita. *Klin Pädiatr* **187**:538–545, 1975.

10. Maroteaux P et al: Spondylo-epiphyseal dysplasia congenita. *Pediatr Radiol* **10**:250, 1981.

11. Michaelis E et al: Dysplasia spondyloepiphyseal congenita. *Fortschr Roentgenstr* **119**:429–438, 1973.

11a. Murray LW, Rimoin DL: Type II collagen abnormalities in the spondyloepi- and spondyloepimetaphyseal dysplasias. *Am J Hum Genet* **37**:13A, 1985.

11b. Murray TG et al: Spondyloepiphyseal dysplasia congenita: Light and electron microscopic studies of the eye. *Arch Ophthalmol* **103**:407–411, 1985.

12. Spranger JW, Langer LO Jr: Spondyloepiphyseal dysplasia congenita. *Radiology* **94**:313–322, 1970.

13. Spranger JW, Langer LO Jr: Spondylo-epiphyseal dysplasias. *Birth Defects* **10**(9):19–61, 1974.

14. Spranger JW, Maroteaux P: Genetic heterogeneity of spondyloepiphyseal dysplasia congenita? *Am J Med Genet* **14**:601–602, 1983.

15. Spranger J, Wiedemann H-R: Dysplasia spondyloepiphysaria congenita. *Helv Paediatr Acta* **21**:598–611, 1966.

16. Stanescu R et al: La dysplasie spondyloépiphysaire congénitale et son hétérogénéité. *Arch Fr Pédiatr* **37**:527–530, 1980.

16a. Stoll C et al: Birth prevalence rates of skeletal dysplasias. *Clin Genet* **35**:88–92, 1989.

17. Sugiura Y et al: Spondylo-epiphyseal dysplasia congenita. *Int Orthoped* **2**:47–51, 1978.

18. Wynne-Davies R, Hall C: Two clinical variants of spondylo-epiphyseal dysplasia congenita. *J Bone Jt Surg* **64B**:435–441, 1982.

19. Yang SS et al: Spondyloepiphyseal dysplasia congenita: A comparative study of chondrocyte inclusions. *Arch Pathol Lab Med* **104**:208–211, 1980.


### Spondyloepimetaphyseal dysplasia with joint laxity

Beighton and Kozlowski (1), in 1980, first defined this entity in Afrikaners. Beighton and co-workers (2–4) subsequently reported 18 patients with the disorder. Farag et al (5) noted an affected sibship with consanguinity. Inheritance is autosomal recessive (3).

The syndrome, evident at birth, is constantly manifested by dwarfism, articular hypermobility, spinal malalignment, thoracic asymmetry, dislocation of radial heads, and talipes equinovarus (Fig. 7–99A–E). The hips are dislocated in about 30% and there is genua valga in 80%. The terminal phalanges, particularly of the thumbs, are spatulate. Spine malalignment is progressive (6). Paraplegia or death from cardiorespiratory failure is common. Congenital heart anomalies (VSD, ASD) were found in 30%.

The face tends to be oval with protuberant eyes, blue sclerae, and a long philtrum. Cleft palate is present in 30%. The skin is somewhat hyperelastic with a doughy consistency.

#### References (Spondyloepimetaphyseal dysplasia with joint laxity)


1. Beighton P, Kozlowski K: Spondylo-epi-metaphyseal dysplasia with joint laxity and severe, progressive kyphoscoliosis. *Skel Radiol* **5**:205–212, 1980.

2. Beighton P et al: Spondylo-epimetaphyseal dysplasia with joint laxity and severe progressive kyphoscoliosis. *S Afr Med J* **64**:772–775, 1983.

3. Beighton P et al: The manifestations and natural history of spondylo-epi-metaphyseal dysplasia with joint laxity. *Clin Genet* **26**:308–317, 1984.

4. Beighton P, Kozlowski K: Spondylo-epi-metaphyseal dysplasia with joint laxity and severe progressive kyphoscoliosis. Clinical Genetics Conference, Baltimore, July 10–13, 1988.

5. Farag TI et al: A family with spondyloepimetaphyseal dwarfism: A "new" dysplasia or Kniest disease with autosomal recessive inheritance? *J Med Genet* **24**:597–601, 1987.

6. Kozlowski K, Beighton P: Radiographic features of spondyloepimetaphyseal dysplasia with joint laxity and progressive kyphoscoliosis. *Roefo* **141**:337–341, 1984.


### Thanatophoric dysplasia

The word *thanatophoros,* from Greek, means *death bringing,* and thanatophoric dysplasia, first described by Maroteaux et al (49,50), is

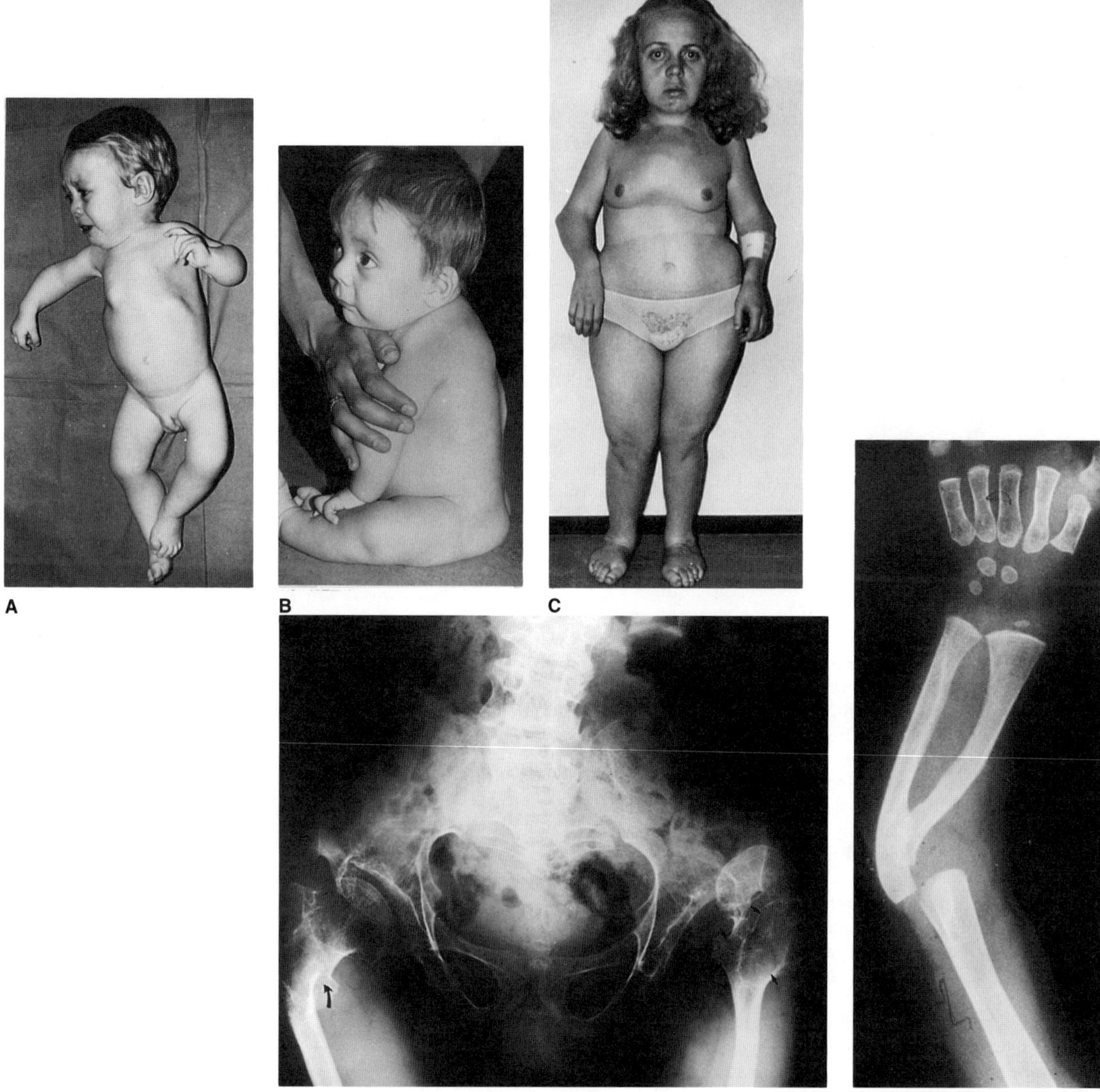

Fig. 7–99. *Spondyloepimetaphyseal dysplasia with joint laxity*. (A) Short limbs, thoracic asymmetry, elbow deformity. (B) Oval face, protuberant eyes, malaligned spine. (C) Age 20. Note short stature, thoracic asymmetry, genua valga, oval face, prominent eyes, long upper lip. (D) Dislocated left hip, osteoporotic ends of narrow femoral shafts show highly abnormal trabecular pattern with cyst formation. Note decreased interpediculate distances. (E) Proximal shortening of radius and ulna with subluxation of elbow joint. (A, B from *P Beighton et al,* S Afr Med J **64:**772, 1983. C from *P Beighton et al,* Clin Genet **26:**308, 1984. D,E from *P Beighton* and *K Kozlowski,* Skel Radiol **5**:205, 1980.)

most always incompatible with life. The condition is characterized by marked shortening of the extremities with numerous skin folds, relatively normal trunk length, narrow thorax, and either disproportionately large head with frontal bossing, protruding eyes, and low nasal bridge, or, less commonly, cloverleaf skull. Many articles on the subject have appeared (2–11,13,14,16–18,20,21,24,25,27–50,52–61,63–75) dealing with a variety of different topics including frequency (7,12), affected sibs (17a,54), monozygotic twins (29,61a,64), sibs misdiagnosed as having thanatophoric dysplasia (23,26,62), differential diagnosis (1,11,15,19,22,42,51,59,63), separate types of thanatophoric dysplasia (43), radiologic features (43), histopathology of long bones (53,59,60,73), cloverleaf skull histopathology (5,38), neuropathology (27,72), cloverleaf skull heterogeneity (51,56,161), and prenatal diagnosis (6,7,13,17).

Although usually lethal at birth, long-term (150–170 days) survival has been reported (43,47,52,68). More than 100 cases have been described to date. The overwhelming majority are sporadic. An early case with cloverleaf skull is that of Vrolik (70) in 1849.

**Epidemiology.** Connor et al (12) found a birth prevalence of 1/42,221 and suggested a mutation rate of $11.8 \pm 4.1 \times 10^{-6}$ mutations per gene per generation. Stoll et al (68a) reported a birth prevalence of 1/25,000. Hall (25) noted a propensity of births during the summer and early fall.

**Histology and pathogenesis.** Endochondral ossification is severely disturbed in thanatophoric dysplasia and is manifested by abnormal columnization, maturation, and hypertrophy of chondrocytes in the growth plates (74,75). Cultured fibroblasts from achondroplasts can be distinguished from normal fibroblasts on the basis of total intracellular mucopolysaccharide content and the relative proportion of dermatan sulfate (14). Ornoy et al (53) studied light microscopic, transmission, and scanning electron microscopic findings in 13 cases. In growth plates, areas with less abnormal cartilage and bone alternated with areas of severely abnormal cartilage and bone. They suggested that the pathogenesis of the skeletal abnormalities is based on focal replacement of the growth plate and periosteum by persisting abnormal, mesenchymal-like tissue from which the abnormal bone originates.

**Nosology and genetics.** Langer et al (43) distinguished two types of thanatophoric dysplasia (Figs. 7–100 to 7–104). In Type 1, the long tubular bones, particularly the femora, are curved and the vertebral bodies are very flat (Figs. 7–100A,B and 7–102). In Type 2, the femora are straight, the vertebral bodies are not as flat as in Type 1, and cloverleaf skull is virtually always present (Figs. 7–100C, 7–102, and 7–104). In contrast, Type 1 occurs without cloverleaf skull in the vast majority of instances, and, when occasionally present, it is mild. Although Langer et al (43) noted some overlap in the diagnostic criteria of Type 1 and Type 2, they suggested that, on the basis of the evidence available to date, thanatophoric dysplasia represents two closely related but separate entities rather than variable expression of a single entity. They further suggested that both types are probably lethal autosomal dominant traits. This is supported by the finding of advanced paternal age (50a) and its occurrence in identical twins (74). Type 2 is less common than Type 1 but is more often associated with cloverleaf skull (1a,4,5,32,33,38a,39). The affected sibs with Type 2 thanatophoric dysplasia reported by Partington et al (54) could be explained on the basis of gonadal mosaicism. Horton et al (29) demonstrated discordance for the cloverleaf skull anomaly in affected monozygotic twins. Cloverleafing was mild in one twin and absent in the other; this may represent variability within Type 1 thanatophoric dysplasia. Although several claims have been made about affected sibs with presumed thanatophoric dysplasia, subsequent radiographic and histologic studies have demonstrated that these infants had other types of lethal short-limbed dwarfism (29,59,60).

**Natural history.** Thanatophoric infants are either stillborn or survive for a few days, and, on occasion, several months, usually succumbing to respiratory distress that can be explained in most instances by the narrow thorax, muscular hypotonia, and alterations in the bronchial cartilages. Survival beyond 1 year has rarely been observed (47,52,68). Langer et al (43) noted a patient who survived to 19 years of age. In approximately 70% of cases, there is a history of hydramnios. At least one-third of thanatophoric infants are premature and born by breech presentation (59).

Fig. 7–100. *Thanatophoric dysplasia (curved bone type).* (A,B) Xeroradiograph showing midfacial hypoplasia, micrognathia, narrow thorax, very short ribs, hypoplastic flattened vertebral bodies, abbreviated long bones, curved humeri and femora, short metapodial bones, and phalanges. (C) *Thanatophoric dysplasia with cloverleaf skull.* Note cloverleaf skull, abnormal clavicles and ribs, hypoplastic vertebrae, small pelvic bones, abbreviated long bones, and small bones of extremities. In contrast to classic thanatophoric dysplasia, the femora and humeri are straight with irregular metaphyseal plates. (From *R Elejalde,* Am J Med Genet **22:**669, 1985.)

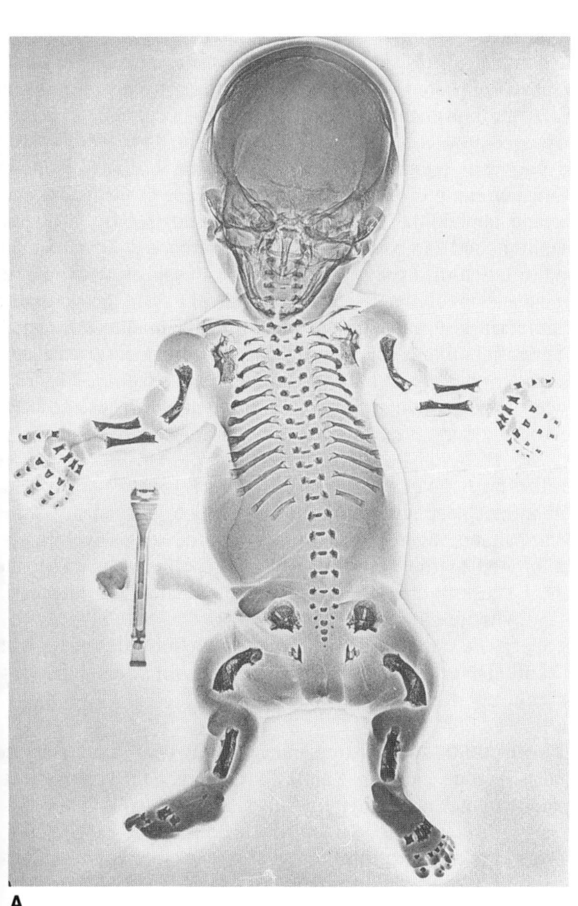

A

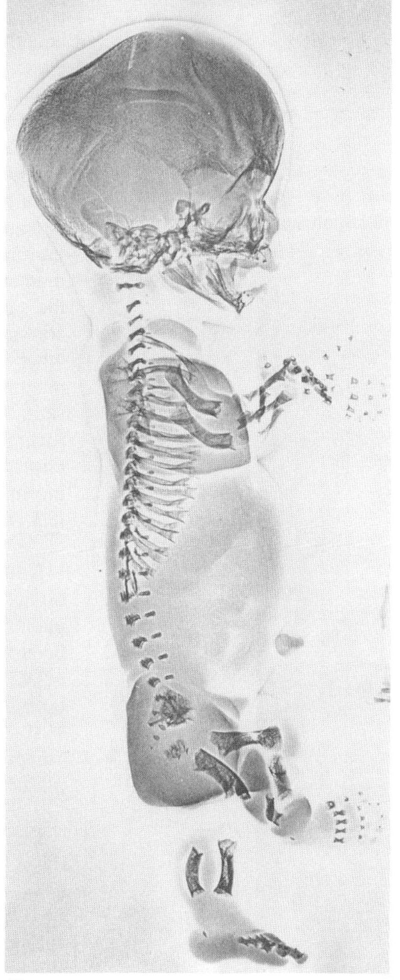

B

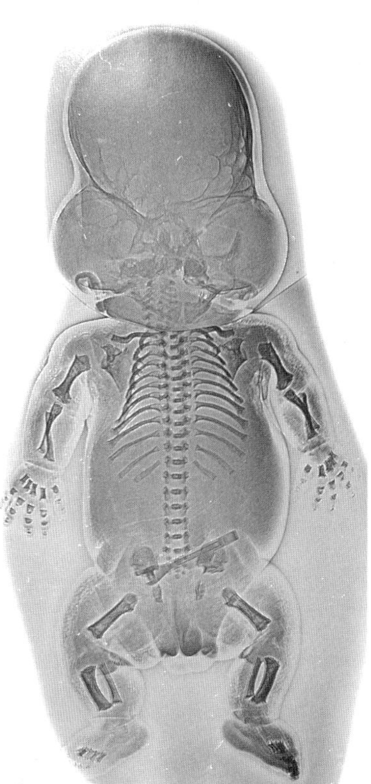

C

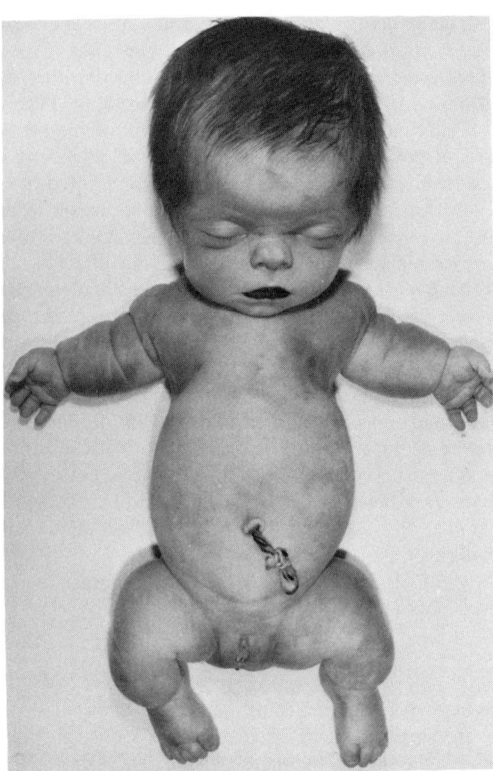

Fig. 7–101. *Thanatophoric dysplasia (curved bone type).* Short neck, short bowed extremities with shortened digits, constricted upper part of chest, large abdomen.

Fig. 7–102. *Thanatophoric dysplasia (curved bone type).* Enlarged head circumference with frontal bossing, hypertelorism, proptosis. (From *A Giedion,* Helv Paediatr Acta **23:**175, 1968.)

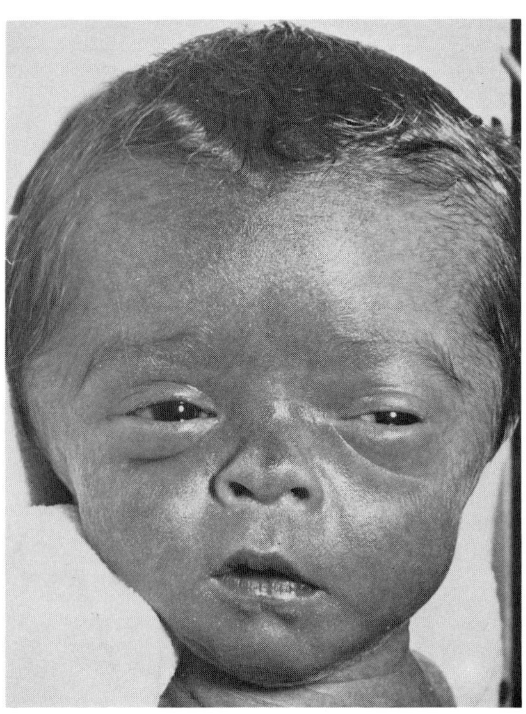

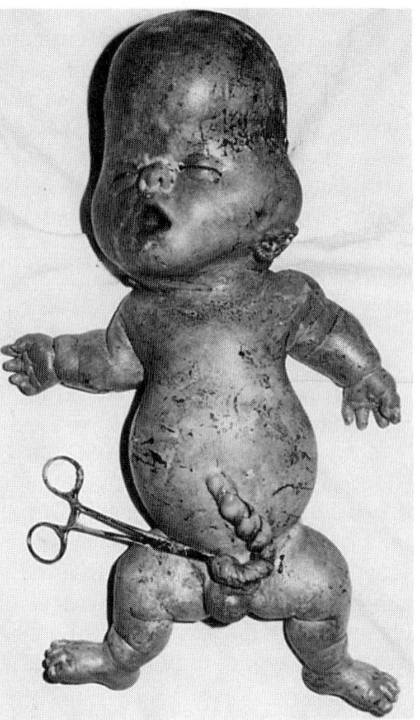

Fig. 7–103. *Thanatophoric dysplasia (straight bone type).* Trilobular skull, depressed ears parallel to shoulders, short neck, small thoracic cage, protuberant abdomen, micromelia, and relatively normal trunk length. (From *MW Partington et al,* Arch Dis Childh **46:**656, 1971.)

**Craniofacial features.** In Type 1 thanatophoric dysplasia, the head is disproportionately large, with head circumference as large as 40 cm. Characteristic features usually include frontal bossing, protruding eyes, and low nasal bridge (Fig. 7–102). Cloverleaf skull anomaly is found in Type 2 (Fig. 7–103) and such cases have been reviewed elsewhere (11,40). Histologic study of the cranial base demonstrates disarrayed chondrocytes (5). In a detailed anatomic and histologic study, Kokich et al (38) showed that the synchondroses usually present at birth between the presphenoid, basisphenoid, basioccipital, exoccipital, and supraoccipital components of the skeletal floor were absent; the cranial base was completely united as one bone. This results in dramatic foreshortening of the cranial base. In some cloverleaf skulls, the sagittal and lambdoidal sutures are synostosed (5,54). In others, coronal, sagittal, and lambdoidal sutures are involved (38). On the other hand, in cases of Type 1 thanatophoric dysplasia without premature synostosis involving the cranial vault (11,38), the trilobular skull configuration is usually absent or minimally manifested. Thus, severe cloverleaf skull anomaly is probably not dependent on the dysplastic changes in the cranial base per se, but results from the restricting influence of the particular pattern of prematurely synostosed cranial sutures on the normal expansion of the brain.

**Neuropathology.** Neuropathologic study of several cases demonstrated abnormal deep sulci in the temporal lobe, dysgenesis of the parahippocampal area, agenesis of Ammon's horn, and periventricular heterotopia limited to the temporal lobe with polymicrogyria in the adjacent area. These findings were present in thanatophoric dysplasia both with and without cloverleaf skull (27,72).

Hydrocephaly is a common finding in thanatophoric dysplasia with cloverleaf skull. Diverticle formation in areas of focal temporal dysplasia has been reported in one instance (27).

**Other abnormalities.** Low-frequency findings in thanatophoric dysplasia may include cleft lip–palate (24), congenital heart defect (33), and adenomyosis of the pyloric muscular layer (67).

**Recurrence risk and prenatal diagnosis.** Overall recurrence risk for thanatophoric dysplasis is very low (29,64). Prenatal diagnosis of

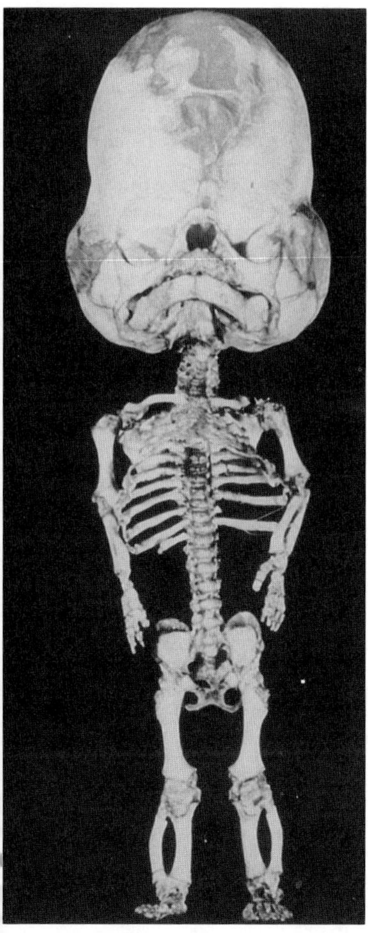

Fig. 7–104. *Thanatophoric dysplasia (straight bone type).* Skeleton of infant with cloverleaf skull. Skull is large in comparison with rest of skeleton. Note normal clavicles, narrow thoracic cage, short extremities, straight femora, and relatively normal length of spine. (Courtesy of Pathology Museum of St. Bartholomew's Hospital, London.)

the basis of ultrasonic and radiographic findings has been reported in several instances (6,7,13,17). Since almost all cases of thanatophoric dysplasia are sporadic, it has not been possible to single out couples at risk. Thus, early prenatal diagnosis, which is theoretically possible, has not been the rule.

**Differential diagnosis.** Differential diagnosis of thanatophoric dysplasia from heterozygous *achondroplasia,* homozygous *achondroplasia,* different types of *achondrogenesis,* severe *hypophosphatasia,* and different *short rib–polydactyly syndromes* is discussed by Rimoin (59). The *cloverleaf skull* malformation is known to be both etiologically and pathogenetically heterogeneous.

## References (Thanatophoric dysplasia)

1. Amman J et al: Antibody-mediated immunodeficiency in short-limbed dwarfism. *J Pediatr* **84**:200–203, 1974.

1a. Andersen PE Jr, Kock K: Micromelic bone dysplasia with cloverleaf skull. *Skel Radiol* **17**:551–555, 1989.

2. Aylsworth AS et al: New observations in two types of lethal dwarfism. Achondrogenesis with cleft palate, and thanatophoric dysplasia with cloverleaf skull. *Proc Greenwood Genet Ctr* **5**:153, 1986.

3. Beaudoing A et al: Thanatophoric dwarfism. *Pédiatrie* **24**:459–462, 1969.

4. Bloomfield JA: Cloverleaf skull and thanatophoric dwarfism. *Aust Radiol* **14**:429–434, 1970.

5. Bonucci E, Nardi F: The cloverleaf skull syndrome: Histological, histochemical and ultrastructural findings. *Virchows Arch Pathol Anat* **357**:199–212, 1972.

6. Burrows PE et al: Early antenatal sonographic recognition of thanatophoric dysplasia with cloverleaf skull deformity. *AJR* **143**:841–843, 1984.

7. Camera G et al: Prenatal diagnosis of thanatophoric dysplasia at 24 weeks. *Am J Med Genet* **18**:39–43, 1984.

8. Campbell RE: Thanatophoric dwarfism *in utero*. *Am J Roentgenol* **112**:198–200, 1971.

9. Canton E: Sobre tres fetos acondroplásticos y sus radiografiás respectivos (Cases 1,3). *Sem Med (B Aires)* **10**:489–505, 1903.

10. Centa A, Camera G: Il considetto "nanismo tanatoforo." *Minerva Pediatr* **21**:447–453, 1969.

11. Cohen MM Jr: *Craniosynostosis: Diagnosis, Evaluation, and Management.* Raven Press, New York, 1986.

12. Connor JM et al: Lethal neonatal chondrodysplasias in the west of Scotland 1970–1983 with a description of a thanatophoric dysplasialike, autosomal recessive disorder, Glasgow variant. *Am J Med Genet* **22**:243–253, 1985.

13. Cronenberg NE: A case of chondrodystrophia foetalis. Discovered by X-ray examination before delivery. *Acta Obstet Gynecol Scand* **13**:275–282, 1933.

14. Danes BS: Achondroplasia and thanatophoric dwarfism. A study of cell culture. *Birth Defects* **10**(12):37–42, 1974.

15. Davis JA: A case of Swiss-type agammaglobulinemia and achondroplasia. *Br Med J* **2**:1371–1374, 1966.

16. Edwards JA et al: Lethal short-limbed dwarfism. *Birth Defects* **10**(12):18–20, 1974.

17. Elejalde BR, de Elejalde MM: Thanatophoric dysplasia: Fetal manifestations and prenatal diagnosis. *Am J Med Genet* **22**:669–683, 1985.

17a. Frahm R: Thanatophorer Zwergwuchs. *Radiologe* **26**:598–601, 1986.

18. Franceschini P et al: Le nanisme thanatophore dans le cadres des nanismes pseudo-achondroplastiques. *Ann Radiol* **13**:399–404, 1970.

19. Gatti RA et al: Hereditary lymphopenic agammaglobulinemia associated with a distinctive form of short-limbed dwarfism and ectodermal dysplasia. *J Pediatr* **75**:675–684, 1969.

20. Giedion A: Thanatophoric dwarfism. *Helv Paediatr Acta* **23**:175–183, 1968.

21. Goard KE, Kozlowski K: Thanatophoric dwarfism II. *Pediatr Radiol* **1**:8–11, 1973.

22. Gotoff SP et al: Granulomatous reaction in an infant with combined immunodeficiency disease and short-limbed dwarfism. *J Pediatr* **80**:1010–1017, 1972.

23. Graff G et al: Familial recurring thanatophoric dwarfism. *Obstet Gynecol* **39**:515–520, 1972.

24. Gwinn JL et al: Thanatophoric dwarfism. *Am J Dis Child* **120**:141–142, 1970.

25. Hall JG: Thanatophoric dwarfism may be genetic but not polygenic. *Pediatrics* **52**:469–470, 1973.

26. Harris R, Patton JT: Achondroplasia and thanatophoric dwarfism in the newborn. *Clin Genet* **2**:61–72, 1971.

27. Hori A et al: Ventricular diverticles with localized dysgenesis of the temporal lobe in cloverleaf skull anomaly. *Acta Neuropathol (Berlin)* **60**:132–136, 1983.

28. Horton WA et al: Abnormal chondrocyte differentiation in thanatophoric dysplasia. *Proc Greenwood Genet Ctr* **5**:152, 1985.

29. Horton WA et al: Discordance for the Kleeblattschädel anomaly in monozygotic twins with thanatophoric dysplasia. *Am J Med Genet* **15**:97–101, 1983.

30. Horton WA et al: Further heterogeneity within lethal neonatal short-limbed dwarfism: The platyspondylic types. *J Pediatr* **94**:736–742, 1979.

31. Huguenin M et al: Two different mutations within the same sibship. *Helv Paediatr Acta* **24**:239–245, 1969.

32. Iannaccone G, Gerlini G: The so-called "cloverleaf skull syndrome." *Pediatr Radiol* **2**:175–184, 1974.

33. Isaacson G et al: Thanatophoric dysplasia with cloverleaf skull. *Am J Dis Child* **137**:876–898, 1983.

34. Jeannin C, Surun: Foetus achondroplastique (présentation) de pièces. *Bull Soc Obstét Gynécol Paris* **13**:181–184, 1910.

35. Jurczok F, Schollmeyer R: Zur Frage des gehäuften Auftretens von Extremitätenmissbildungen bei Neugeborenen (Case 4). *Geburtshilfe Frauenheilkd* **22**:400–421, 1962.

36. Kaufman HJ: "New" skeletal dysplasias in the newborn: New X-ray findings. *Birth Defects* **10**(12):1–9, 1974.

37. Keats TE et al: Thanatophoric dwarfism. *Am J Roentgenol* **108**:473–480, 1970.

38. Kokich VG et al: The cloverleaf skull anomaly. An anatomic and histologic study of two specimens. *Cleft Palate J* **19**:89–99, 1982.

38a. Kozlowski K et al: Cloverleaf skull and bone dysplasias. *Australas Radiol* **31**:309–314, 1987 (Cases 3,4).

39. Kozlowski K et al: Cloverleaf skull with generalized bone dysplasia. Report of a case with short review of the literature. *Pediatr Radiol* **15**:412–414, 1985.

40. Kremens, B et al: Thanatophoric dysplasia with cloverleaf-skull: Case report and review of the literature. *Eur J Pediatr* **139**:298–303, 1982.

41. Langenbach E: Ein Fall von Chondrodystrophia foetalis mit Asymmetrie des Schädels. *Virchows Arch Pathol Anat* **189**:12–17, 1907.

42. Langer, LO Jr et al: Thanatophoric dwarfism: A condition confused with achondroplasia in the neonate, with brief comments on achondrogenesis and homzygous achondroplasia. *Radiology* **92**:285–294, 1969.

43. Langer LO Jr et al: Thanatophoric dysplasia and cloverleaf skull. A report of nine new cases and review of the literature. *Am J Med Genet Suppl* **3**:167–179, 1987.

44. Lenz W et al: Thanatophoric Zwergwuchs. *Z Kinderheilkd* **111**:162–174, 1971.

45. Leroy JG et al: Fatal neonatal dwarfism: Examples of thanatophoric dwarfism and hypophosphatasia. *Birth Defects* **10**(12):21–30, 1974.

46. Levi L, Bouchacourt L: Radiographies de foetus achondroplases (Case 1). *Rev Hyg Med Inf* **3**:517–528, 1904.

47. MacDonald IM et al: Prolonged survival in two cases of thanatophoric dysplasia. *Proc Greenwood Genet Ctr* **5**:151–152, 1986.

48. Magrier C: Présentation d'un foetus achondroplastique. *Bull Soc Obstet Gynécol Paris* **1**:248–256, 1898.

49. Maroteaux P et al: Le nanisme, thanatophore. *Presse Méd* **75**:2519–2524, 1967.

50. Maroteaux P, Lamy M: Le diagnostic des nanismes chondrodystrophiques chez les nouveau-nés. *Arch Fr Pédiatr* **25**:241–262, 1967.

50a. Martinez-Frias ML et al: Thanatophoric dysplasia: An autosomal dominant condition? *Am J Med Genet* **31**:815–820, 1988.

51. McKusick VA, Cross HE: Ataxia-telangiectasia and Swiss-type agammaglobulinemia. *JAMA* **195**:739–745, 1966.

52. Moir DH, Kozlowski K: Long term survival in thanatophoric dwarfism. *Pediatr Radiol* **5**:123–125, 1976.

53. Ornoy A et al: The role of mesenchyme-like tissue in the pathogenesis of thanatophoric dysplasia. *Am J Med Genet* **21**:613–630, 1985.

54. Partington MW et al: Cloverleaf skull and thanatophoric dwarfism. Report of four cases, two in the same sibship. *Arch Dis Childh* **46**:656–664, 1971.

55. Pena SDJ, Goodman HO: The genetics of thanatophoric dwarfism. *Pediatrics* **51**:104–109, 1973.

56. Porak C, Durante G: Les micromélies congénitales. Achondroplasie vraie et dystophie périostale. *Nouv Iconogr Salpet* **18**:481–538, 1905.

57. Raffel L et al: Thanatophoric dysplasia with and without Kleeblattschädel: Variability rather than heterogeneity. *Am Soc Hum Genet* 34th, Norfolk, Oct. 30–Nov. 2, 1983.

58. Raffele F: Di l'achondroplasia nel feto. Considerazione cliniche e anatomopatologische. *Chir Organi Mov* **5**:467–502, 1921.

59. Rimoin DL: The chondrodystrophies. *Adv Hum Genet* **5**:1–118, 1975.

60. Rimoin DL et al: Histologic studies in the chondrodystrophies. *Birth Defects* **10**(12):274–295, 1974.

61. Rischbieth H, Barrington A: Dwarfism (Fig. 31–2). *Treas Hum Inherit* **7–8:** 559, 1921.

61a. Rupprecht E: Der Wert postmortales Röntgendiagnostik am Beispiel frühletaler Skelettdysplasien. *Kinderärzt Prax* **53**:133–140, 1985.

62. Sabry A: Thanatophoric dwarfism in triplets. *Lancet* **2**:533, 1974.

63. Saldino RM: Lethal short-limbed dwarfism: Achondrogenesis and thanatophoric dwarfism. *Am J Roentgenol* **112**:185–197, 1971.

64. Serville F et al: Thanatophoric dysplasia of identical twins. *Am J Med Genet* **17**:703–706, 1984.

65. Shah K: Thanatophoric dwarfism. *J Med Genet* **10**:243–252, 1973.

66. Southion CL: Thanatophoric dwarfism. *Austral Radiol* **16**:316–319, 1972.

67. Sršeň Š et al: Thanatophorer Zwergwuchs. *Nanismus Thanatophorus* bei vier Neugeborenen. *Pädiatr Pädol* **9**:336–343, 1974.

68. Stensvold K et al: An infant with thanatophoric dwarfism surviving 169 days. *Clin Genet* **29**:157–159, 1986.

68a. Stoll C et al: Birth prevalence rates of skeletal dysplasias. *Clin Genet* **35**:88–92, 1989.

69. Thompson BH, Parmley TH: Obstetric features in thanatophoric dwarfism. *Am J Obstet Gynecol* **109**:396–401, 1971.

70. Vrolik W: *Tabulae ad illustrandam embryogenes in Hominis et Mamalium.* G.M.P., London, 1849.

71. Widdig K et al: Beitrag zum Kleeblattschädel-Syndrom. *Zentralbl Allg Pathol* **118**:358–366, 1974.

72. Wongmongkolrit T et al: Neuropathological findings in thanatophoric dysplasia. *Arch Pathol Lab Med* **107**:132–135, 1983.

73. Yang SS et al: Histopathologic examination in osteochondrodysplasia. *Arch Pathol Lab Med* **110**:10–12, 1986.

74. Young ID et al: Thanatophoric dysplasia in identical twins. *J Med Genet* **26**:276–279, 1989.

75. Young RS et al: Thanatophoric dwarfism and cloverleaf skull. *Radiology* **106**:401–405, 1973.

# Chapter 8
# Syndromes Affecting Bone: Craniotubular Bone Disorders

## General considerations

There has been much confusion attendant to genetic disorders of bone characterized by modeling errors of tubular and cranial bones. Gorlin et al (3) divided craniotubular dysplasias into Pyle disease, craniometaphyseal dysplasia, craniodiaphyseal dysplasia, frontometaphyseal dysplasia, Schwarz–Lélek syndrome, dysosteosclerosis, and oculodentoosseous dysplasia. The craniotubular hyperostoses consists of Van Buchem disease, sclerosteosis, congenital hyperphosphatasia, autosomal dominant osteosclerosis, and Camurati–Engelmann disease. More extensive discussion may be found in Konigsmark and Gorlin (4), Beighton and Cremin (2), and Beighton (1).

### References (General considerations)


1. Beighton P: *Inherited Disorders of the Skeleton,* (ed 2), Churchill–Livingstone, Edinburgh and New York, 1988.

2. Beighton P, Cremin BJ: *Sclerosing Bone Dysplasias.* Springer Verlag, New York, 1980.

3. Gorlin RJ et al: Genetic craniotubular bone dysplasias and hyperostoses: A critical analysis. *Birth Defects* **5**(4):79–95, 1969.

4. Konigsmark B, Gorlin RJ: *Genetic and Metabolic Deafness.* WB Saunders, Philadelphia, 1976.


## Craniometaphyseal dysplasia, autosomal dominant and recessive forms

This disorder, often erroneously reported as Pyle disease, is characterized by unusual facies. Usually within the first year of life, the root of the nose begins to broaden and an elevated wing of bone gradually extends bilaterally over the nasal bridge to the zygomas.

Increasing bony sclerosis narrows the nasal lumen, leading to obstruction, with resultant open mouth (Fig. 8–1 A,B). Bony alterations in the temporal bone and pyramid produce mixed hearing loss that becomes evident in childhood in about one-half the cases. It is slowly progressive until there is moderate to severe (30–90 dB) loss by the fourth decade. In about 30% there is peripheral facial nerve paralysis, headache, or vertigo (1,3).

Hypertelorism is a constant feature. Nystagmus is common. Rarely, there is visual loss due to optic atrophy (7). This suggests bony encroachment on the optic foramina. The alveolar ridges may be thickened. Occasionally there is delayed eruption of permanent teeth.

Radiographically, hyperostosis and sclerosis involve the frontal and occipital portions of the calvaria, the base of the skull, and, less often, the mandible. There is increased bone deposit on the walls of the paranasal sinuses and underpneumatization of mastoid cells. Most marked is frontonasal hyperostosis (Fig. 8–1C). The long bones have a club-shaped metaphyseal flare that is far milder than that seen in Pyle disease and may be minimal during the first years of life. Cortical hyperostosis of diaphyses is noted in the young, but disappears with age. The short tubular bones exhibit the same changes as those noted in long bones (Fig. 8–1D,E). Reported favorable responses to calcitonin and to calcitriol in single patients (4a,7a) have been critically discussed by Cole and Cohen (3b).

Craniometaphyseal dysplasia may have autosomal dominant (1–14a) or recessive (15–24) transmission. Extreme variability in the dominant form does not allow for differentiation from the recessive form. The latter appears to be somewhat less variable. Although some recessive cases appear to be more severe than dominant examples (Fig. 8–2A–C), in a sporadic case, it is not possible to clinically distinguish between the two forms (18). We cannot classify the case of Girdwood et al (14a) into either the dominant or recessive form.

### References (Craniometaphyseal dysplasia–autosomal dominant form)


1. Beighton P et al: Craniometaphyseal dysplasia: Variability of expression within a large family. *Clin Genet* **15**:252–258, 1979.

2. Carlson DH, Harris GBC: Craniometaphyseal dysplasia: A family with three documented cases. *Radiology* **103**:147–151, 1972.

3. Carnevale A et al: Autosomal dominant craniometaphyseal dysplasia: Clinical variability. *Clin Genet* **23**:17–22, 1983.

3a. Colavita N et al: Cranio-metaphyseal dysplasia. *Australas Radiol* **32**:257–262, 1988.

3b. Cole DEC, Cohen MM Jr: A new look at craniometaphyseal dysplasia. *J Pediatr* **112**:577–579, 1988.

4. Cooper JC: Craniometaphyseal dysplasia: A case report and review of the literature. *Br J Oral Surg* **12**:196–204, 1974.

4a. Fanconi S et al: Craniometaphyseal dysplasia with increased bone turnover and secondary hyperparathyroidism: Therapeutic effect of calcitonin. *J Pédiatr* **112**:587–590, 1988.

5. Guibaud P et al: La dysplasie cranio-métaphysaire. *Pédiatrie* **28**:149–161, 1973.

6. Holt JF: The evolution of cranio-metaphyseal dysplasia. *Ann Radiol* **9**:209–224, 1966.

7. Jend HH et al: Cranio-metaphyseal stratiform dysplasia—conventional radiography and CT findings. *Eur J Radiol* **1**:261–265, 1981.

7a. Key LL et al: Treatment of craniometaphyseal dysplasia with high-dose calcitriol. *J Pediatr* **112**:583–586, 1988.

8. Kietzer G, Paparella MM: Otolaryngological disorders in craniometaphyseal dysplasia. *Laryngoscope* **79**:921–941, 1969.

9. Martin FW: Cranio-metaphyseal dysplasia. *J Laryngol Otol* **91**:159–169, 1977.

10. Puliafito CA et al: Optic atrophy and visual loss in craniometaphyseal dysplasia. *Am J Ophthalmol* **92**:696–701, 1981.

11. Rimoin DL et al: Cranio-metaphyseal dysplasia (Pyle's disease): Autosomal dominant inheritance in a large kindred. *Birth Defects* **5**(4):96–104, 1969.

12. Shea J et al: Cranio-metaphyseal dysplasia: The first successful surgical treatment for associated hearing loss. *Laryngoscope* **91**:1369–1374, 1981.

13. Spiro PC et al: Radiology of the autosomal dominant form of craniometaphyseal dysplasia. *South Afr Med J* **49**:839–842, 1975.

14. Spitzer W, Steinhauser EW: Die kraniometaphysäre Dysplasie. *Dtsch Zahnärtztl Z* **36**:96–100, 1981.

14a. Taylor DB, Sprague P: Dominant craniometaphyseal dysplasia–family studies over four generations. *Australas Radiol* **33**:87–89, 1989.


### References (Craniometaphyseal dysplasia—autosomal recessive form)


15. Girdwood TG et al: Craniometaphyseal dysplasia congenita—Pyle's disease in a young child. *Br J Radiol* **42**:299–303, 1969.

15a. Graf K: Die Bedeutung des Pyle-Syndroms (Leontiasis ossea) für die Oto-Rhino-Laryngologie. *Z Laryngol Rhinol* **44**:438–445, 1965.

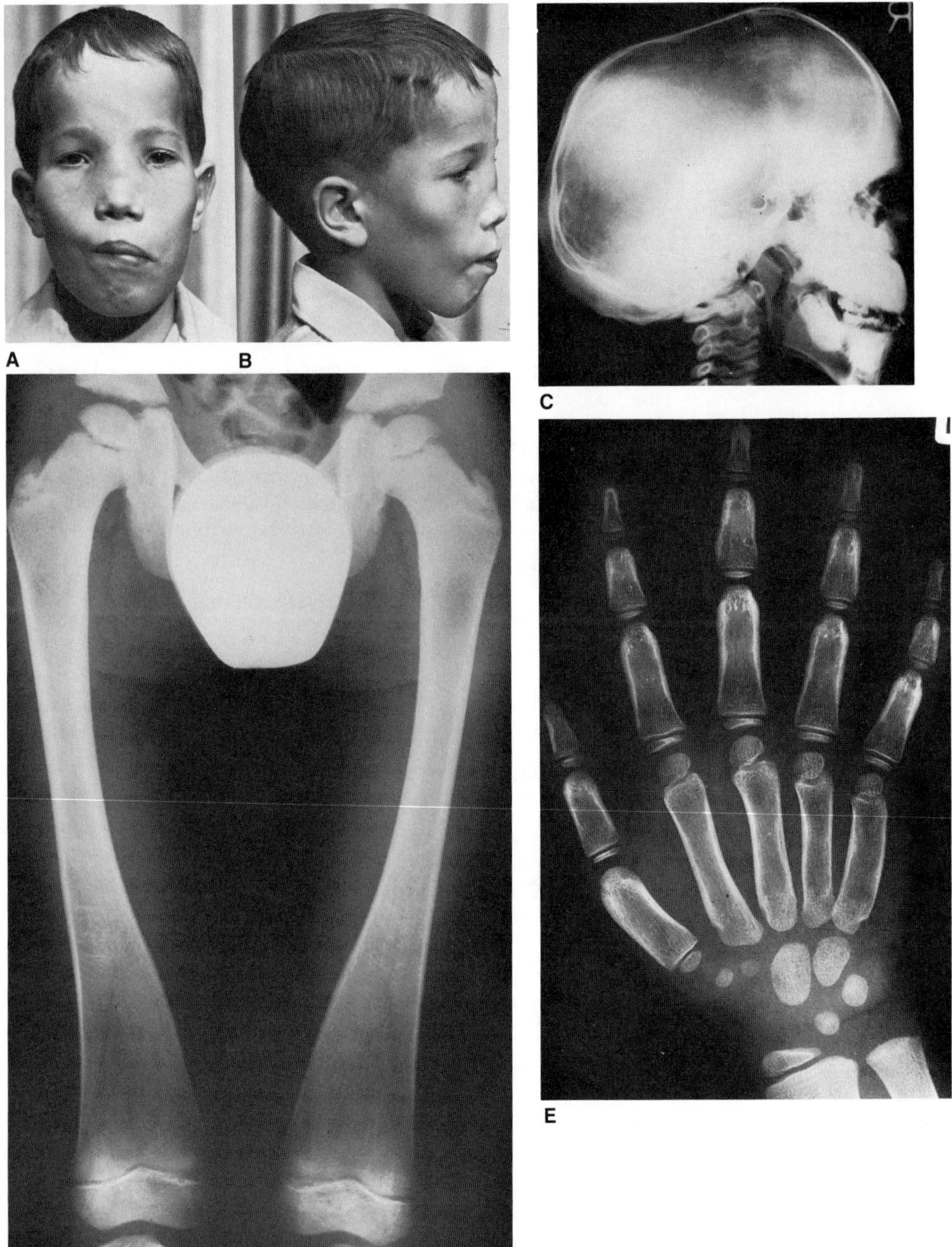


Fig. 8–1. *Craniometaphyseal dysplasia—dominant form.* (A,B) Eleven-year-old boy with progressive hearing loss. Note widened nasal bridge, left facial palsy. Mother and sister were similarly affected. (C) Skull of 6-year-old male showing frontooccipital hyperostosis, sclerosis of skull base and facial bones, underpneumatization of sinuses and mastoids, and dolichocephaly with postcoronal depression of parietal bones. (D) Femora of same child showing club-shaped metaphyseal flare, minimal diaphyseal sclerosis. (E) Hands of same child exhibiting undermodeling of short tubular bones, with distal cortical sclerosis of phalanges.

16. Jackson WP et al: Metaphyseal dysplasia, epiphyseal dysplasia, diaphyseal dysplasia and related constitutions. *Arch Intern Med* **94**:871–885, 1957.

17. Lehmann ECH: Familial osteodystrophy of the skull and face. *J Bone Jt Surg* **39B**:313–315, 1957.

18. Lièvre JA, Fischgold H: Leontiasis ossea chez l'enfant (osteopetrose partielle probable). *Presse Méd* **64**:763–765, 1956.

19. Millard DR et al: Craniofacial surgery in craniometaphyseal dysplasia. *Am J Surg* **113**:615–621, 1967.

20. Nicolo A, Briani S: La displasia cranio-metafisaria. *Ann Radiol Diagn* **39**:185–202, 1966.

21. Penchaszadeh VB et al: Autosomal recessive craniometaphyseal dysplasia. *Am J Med Genet* **5**:43–55, 1980.

22. Ross MW, Altman DH: Familial metaphyseal dysplasia: Review of the clinical and radiological features of Pyle's disease. *Clin Pediatr* **6**:143–149, 1967.

23. Sommer F: Eine besondere Form einer generalisierten Hyperostose mit Leontiasis ossea faciei et cranii. *Radiol Clin (Basel)* **23**:65–75, 1954.

24. Wemmer V, Böttger E: Die kraniometaphysäre Dysplasie (Jackson). *Roefo* **128**:66–69, 1978.

## Craniodiaphyseal dysplasia

Joseph et al (5) first used the term craniodiaphyseal dysplasia to designate a severe bone disorder characterized by massive generalized

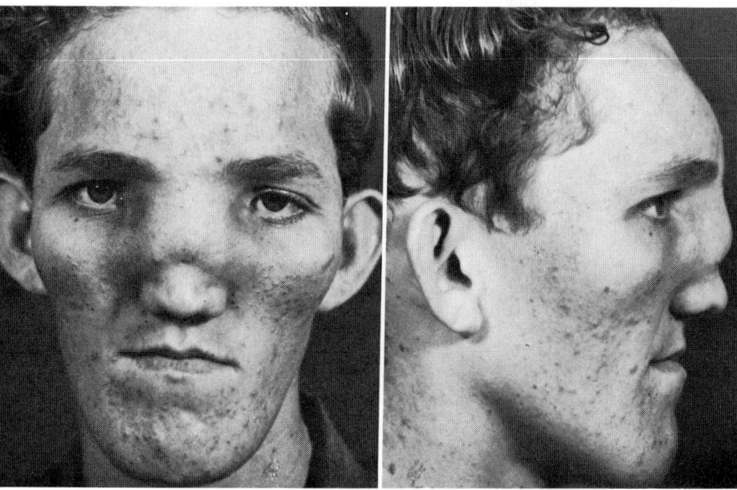

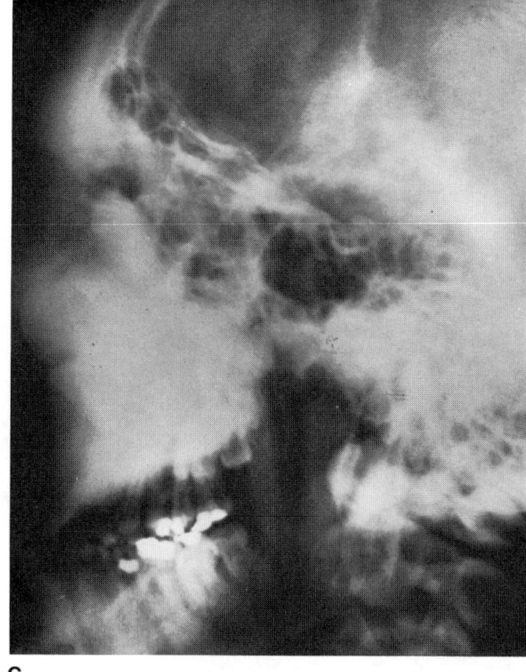

A B C

Fig. 8–2. *Craniometaphyseal dysplasia—recessive form.* (A,B) Seventeen-year-old boy exhibiting bony overgrowth of frontal, nasal, and maxillary bones, hypertelorism, and mild mandibular prognathism. Sister was more severely affected. (C) Marked bony overgrowth, particularly of maxillary, nasal, and frontal areas. (From *DR Millard et al,* Am J Surg **113:**615, 1967.)

hyperostosis and sclerosis, involving in particular the skull and facial bones (1,2,4–9,11) (Fig. 8–3A–D). The patient described by Gemmell (3) possibly had a mild form of the disease and the patient described by Schaefer et al (10) really had craniometaphyseal dysplasia of the dominant type.

Facial and cranial thickening, distortion, and enlargement are severe. Nasal obstruction and recurrent upper respiratory infection appear within the first few years or even first few months of life. Marked bony thickening, hypertelorism, nasal flattening, and severe dental malocclusion generally follow. Bilateral choanal stenosis can be demonstrated within the first few years. Growth is retarded and early death is common.

All patients have severe hypertelorism, lacrimal duct obstruction resulting from bony overgrowth, and diminshed visual acuity or blindness as a result of optic atrophy.

Developmental milestones are delayed. Compression of cranial nerves results from bony overgrowth. This relentless process is associated with headache, progressive mental retardation, and seizures. Often there is lack of sexual maturity. Hearing loss, generally mixed, has been described in all cases. Radiographically, the skull and facial bones as well as the mandible are severely sclerotic and hyperostotic. The paranasal sinuses and mastoids do not develop. There is moderate thickening and marked sclerosis of the ribs and clavicles. The long tubular bones do not exhibit metaphyseal flare, but rather have a policeman's nightstick shape and show diaphyseal endostosis.

The short tubular bones of the hands and feet, particularly the first metapodial, exhibit cylinderization. A few investigators have found elevated levels of serum alkaline phosphatase, but normal levels of calcium and phosphorus.

Inheritance is likely autosomal recessive.

### References (Craniodiaphyseal dysplasia)


1. de Souza O: Leontiasis ossea, *Porto Allegre (Brazil) Faculdade de Med Dos Cursos* **13**:47–54, 1927.
2. Fosmoe RJ et al: Van Buchem's disease (hyperostosis corticalis generalistata familiaris). *Radiology* **90**:771–774, 1968.
3. Gemmell JH: Leontiasis ossea: A clinical and roentgenographical entity, *Radiology* **25**:723–729, 1935.
4. Halliday, J: A rare case of bone dysplasia. *Br J Surg* **37**:52–63, 1949–1950.
5. Joseph R et al: Dysplasie cranio-diaphysaire progressive: Ses relations avec la dysplasie diaphysaire progressive de Camurati-Engelmann. *Ann Radiol* **1**:477–490, 1958.
6. Kaitila I et al: Craniodiaphyseal dysplasia. *Birth Defects* **11**(6):359–361, 1975.
7. Kirkpatrick DB et al: The craniotubular bone modeling disorders: A neurosurgical introduction to rare skeletal dysplasias with cranial nerve compression. *Surg Neurol* **7**:221–232, 1977 (same as Ref. 6).

7a. Levy MH, Kozlowski K: Cranio-diaphyseal dysplasia. *Australas Radiol* **31**:431–435, 1987.

8. Macpherson RI: Craniodiaphyseal dysplasia, a disease or group of diseases? (Case 1). *J Can Assoc Radiol* **25**:22–23, 1974.
9. Scarfo GB et al: Idrocephalo associato a displasia cranio-diafisaria. *Radiol Med* **65**:249–252, 1979.
10. Schaefer B et al: Dominantly inherited craniodiaphyseal dysplasia. A new craniotubular dysplasia. *Clin Genet* **30**:381–391, 1986.
11. Stransky E et al: On Paget's disease with leontiasis ossea and hypothyreosis starting in early childhood. *Ann Paediatr* **199**:393–408, 1962.


## Schwarz–Lélek syndrome

The patient described by Schwarz (2) appears to have a distinct syndrome comprising severe genua vara and marked frontal bossing (Fig. 8–4A). Affected sibs were reported by Williams et al (3). The patient described by Lélek (1) possibly has a different disorder (L Langer, personal communication, 1989).

Radiographically, the changes in the humerus, hands, clavicles, and ribs were similar to those in Pyle disease. However, there was massive internal bowing of the femur, with radiolucent splaying of the metaphyseal area (Fig. 8–4B). In infancy, there was large anterior fontanel as well as thinning of the calvaria with wormian bones and natal teeth. Hyperostosis and mild sclerosis of the skull, particularly in the frontal and occipital areas, as well as of the maxilla and man-

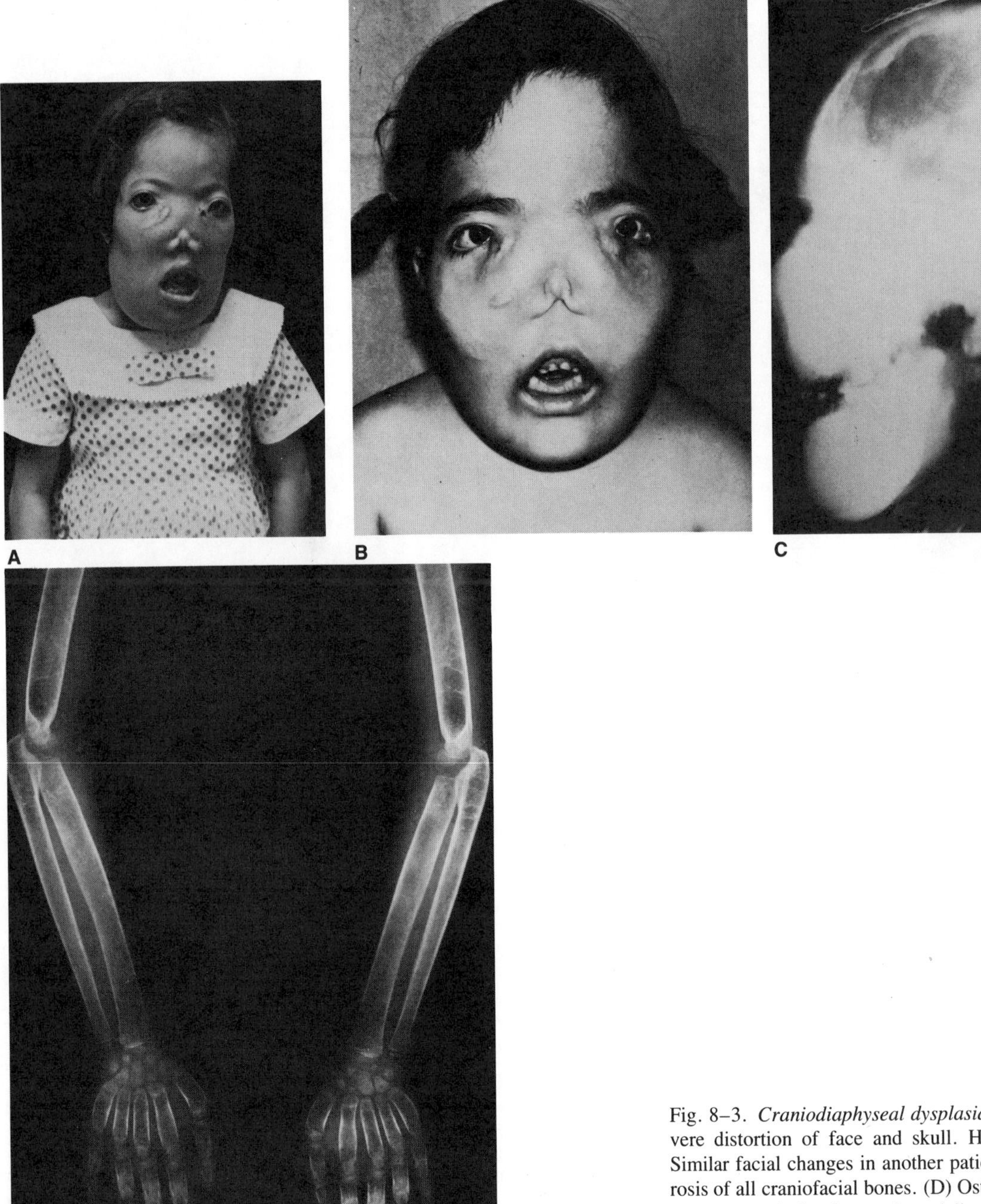


Fig. 8–3. *Craniodiaphyseal dysplasia.* (A) Fourteen-year-old girl with severe distortion of face and skull. Head circumference was 57 cm. (B) Similar facial changes in another patient. (C) Marked thickening and sclerosis of all craniofacial bones. (D) Osteoporosis, "policeman's nightstick" appearance, lack of normal modeling. (A,B from *RI Macpherson,* J Can Assoc Radiol **25**:22, 1974. C,D from *E Stransky et al,* Ann Paediatr **199**:393, 1962.)

dible, were marked. The paranasal sinuses were obliterated. Occipital horns were noted (2,3).

Inheritance is probably autosomal recessive (3).

### References (Schwarz–Lélek syndrome)

1. Lélek I: Camurati-Engelmann'sche Erkrankung. *Fortschr Röntgenstr* **94**:393–408, 1962.

2. Schwarz E: Craniometaphyseal dysplasia. *Am J Roentgenol* **84**:461–466, 1960.

3. Williams CA et al: A new syndrome of craniotubular dysplasia having diaphyseal and metaphyseal modeling defects and increased susceptibility to fracture. March of Dimes Birth Defects Conference, Baltimore, July, 1988.

## Osteopetrosis

Osteopetrosis is characterized by failure of resorption of the primary spongiosa by osteoclasts, resulting in increased osseous density in which cortical and cancellous bone cannot be distinguished radiographically. There is an increased number of osteoclasts histologically (6,18,21).

Osteopetrosis has been traditionally divided into two groups: the so-called congenital or malignant autosomal recessive type and the adult or benign autosomal dominant form. Actually, there is considerable genetic heterogeneity. We shall review four distinct forms.

**Severe autosomal recessive osteopetrosis (Albers–Schönberg disease).** This disorder is characterized by increased density of

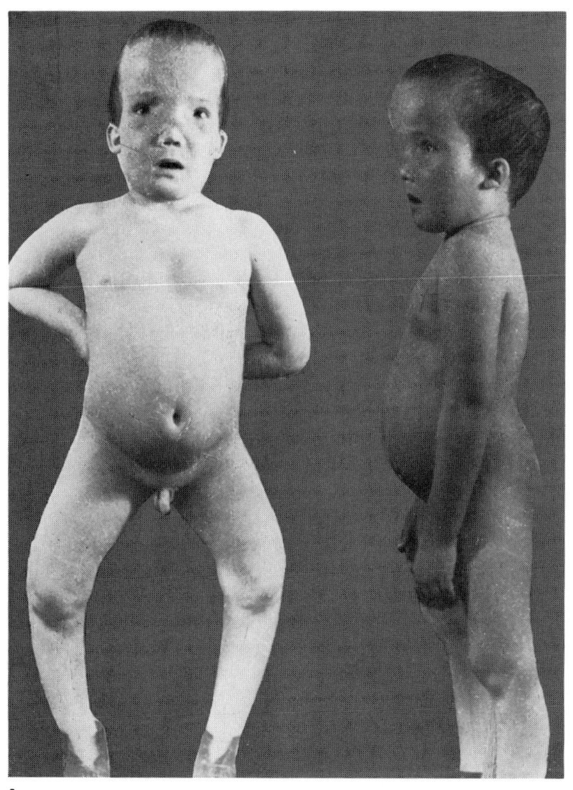

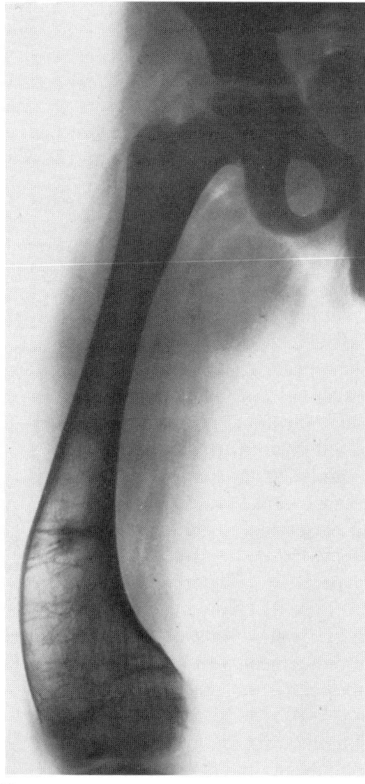

Fig. 8–4. *Schwarz–Lélek syndrome.* (A) Nine-year-old male, 122 cm tall. Note large skull, frontal bossing, severe genua vara. (B) Femur is symmetrically broadened and bent inward in distal portion. Proximal femur is sclerotic, distal femur thinned, genua vara. (From *I Lélek,* Fortschr Rontgenstr **94:**702, 1965.)

nearly all bones and the complications that occur from failure of resorption of the primary spongiosa and its resultant persistence: anemia, hepatosplenomegaly, blindness, deafness, facial paralysis, and osteomyelitis. The involved bones are expanded, splayed, and dense, with the epiphysis, metaphysis, and diaphysis being involved to a similar degree. The cortical and cancellous bones are indistinguishable radiographically (Fig. 8–5A,B). Pathogenesis is extensively discussed by Reeves et al (20).

The frequent occurrence in sibs and consanguinity have been demonstrated by a large number of investigators (17). Over 400 cases have been reported. A particularly severe form was reported in two sibs by El Khazen et al (7a). The infants had *in utero* fractures, hip dislocation, hydrocephaly, and hypoplasia of the cerebellum. No osteoclasts were found.

This form may even be recognized *in utero* or at birth. Severe anemia, jaundice, hepatosplenomegaly, and failure to thrive characterize the neonatal form. The infant may be stillborn or survive only a few months (30).

All tubular bones may be involved, but growth is usually normal. The skull is thickened and dense, mainly at its base, but the calvaria is involved as well, without the recognizable diploë. The mastoid bones and paranasal sinuses are poorly aerated, and the facial bones appear denser than normal.

Defective vision and nystagmus are extremely common. Optic atrophy eventuating from pressure of bone on optic veins is a relatively common complication. Facial paralysis results from the pressure of dense bone on the foramen of the seventh cranial nerve (16).

Osteomyelitis of the jaws seems to be a significant complication of dental extraction, presumably the result of a deficient blood supply (7,23). It may lead to extraoral fistulas. Primary molars and all permanent teeth are greatly distorted and remain totally or partially embedded in basal bone (3). Ankylosis of cementum to bone has been described (30). The teeth appear to be secondarily affected by failure of bone resorption and/or osteomyelitis. Many authors have remarked on the high incidence of dental caries.

**Mild autosomal recessive osteopetrosis.** This form is rare. It is characterized by short stature, increased upper/lower segment ratio, mandibular prognathism, fractures following minimal trauma, mild to moderate anemia with extramedullary hematopoiesis, unerupted teeth, and osteomyelitis.

The occurrence in sibs and parental consanguinity indicate autosomal recessive inheritance. Kahler et al (14) and others (2,5,10,11,26) have nicely summarized the published cases. This group is probably heterogeneous. Several cases with mental retardation cannot be accurately classified (8,12) but they may possibly represent the disorder discussed below. Horton et al (11) found a decreased number of osteoclasts.

Fig. 8–5. *Osteopetrosis—severe autosomal recessive form.* (A,B) Skull is thickened and dense at cranial base; calvaria is involved as well without recognizable diploë; facial bones appear denser than normal. Defective vision and facial paralysis become evident.

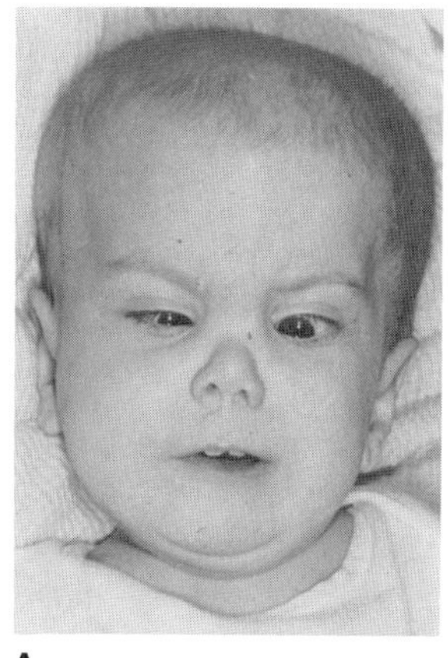

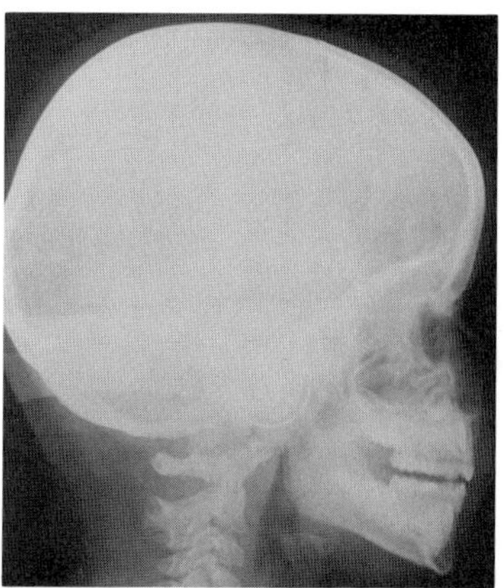

**Autosomal recessive osteopetrosis with renal tubular acidosis.** There have been a few families in which severe osteopetrosis has been found in combination with short stature, dull mentality, visual impairment, renal tubular acidosis, extramedullary hematopoiesis, basal ganglion calcification, hepatosplenomegaly, and pancytopenia (4,9,16a,19,19a,22,27,29). A deficiency of carbonic anhydrase II has been demonstrated in erythrocytes (22). Heterozygotes have one-half the enzyme levels. An improved assay has been developed (23a).

**Benign autosomal dominant osteopetrosis.** Benign autosomal dominant osteopetrosis usually appears somewhat later in life than the autosomal recessive types (3a,3b). It rarely is associated with fractures following minor trauma (13,14a,15,24,31). It appears to be heterogeneous (1). In one form (Type I) there are sclerosis and thickening of the calvaria, no endplate sclerosis of vertebral bodies, and no "bone within bone" appearance in the ilia. In the second form (Type II), the base of the skull is thickened, and there are endplate thickening of vertebral bodies and iliac "bone within bone" appearance (3a,3b,3c). In Type I there is often involvement of V with narrowing of the internal auditory meatus. In Type II, cranial nerve VII may be affected. Conductive hearing loss is associated with Type I (3c). There is no narrowing of the internal auditory meatus in Type II. Not uncommonly, the condition is discovered upon routine X-ray films of the chest or upon radiographic survey of the family of a patient with known benign dominant osteopetrosis. It is more common than the recessive form. There may be a severe dominant form (31).

The condition has been shown to appear silently within the first few years of life, being manifest by increased radiopacity of the skull. Density is most marked at the diaphyseal ends of long bones, gradually extending to the epiphyses and to the marrow cavity (13).

In contrast to the autosomal recessive forms, the autosomal dominant form is *not* associated with anemia, hepatosplenomegaly, blindness, deafness, or mental retardation. In exception to this is the mentally retarded, blind patient reported by Thomson (25). Rarely (10–20%), patients have cranial nerve palsy (13,28).

The times of initiation of calcification and closure of the epiphyses are not altered. There is usually some lack of remodeling, particularly in the femur and tibia. Nearly all bones are ultimately involved. In the more common form the vertebrae assume a "sandwich" form early in the course of the disease as a result of calcification of the upper and lower surfaces, and usually there is a crescentric band of either increased or decreased density parallel to the iliac crest.

In the skull, there is thickening of the base with clubbing of the anterior and posterior clinoid processes. The sinuses become involved and ultimately disappear. There is increased density of the calvaria with disappearance of the diploë, but no enlargement of the head.

About 10% have osteomyelitis of the mandible (13). Dyson (7) also described this complication. Elevated serum acid phosphatase has been noted in all forms (13,15,21).

### References (Osteopetrosis)


1. Anderson PE, Bollerslev J: Heterogeneity of autosomal dominant osteopetrosis. *Radiology* **164**:223–224, 1987.

2. Beighton P et al: Osteopetrosis in South Africa. The benign, lethal and intermediate forms. *S Afr Med J* **55**:659–665, 1979.

3. Bjorvatn K et al: Oral aspects of osteopetrosis. *Scand J Dent Res* **87**:245–252, 1979.

3a. Bollerslev J: Osteopetrosis. A genetic and epidemiological study. *Clin Genet* **31**:86–90, 1987.

3b. Bollerslev J et al: Autosomal dominant osteopetrosis. *J Laryngol Otol* **101**:1088–1091, 1987.

3c. Bollerslev J et al: Autosomal dominant osteopetrosis. An otoneurological investigation of the two radiological types. *Laryngoscope* **98**:411–413, 1988.

4. Bourke E et al: Renal tubular acidosis and osteopetrosis. *Nephron* **28**:268–272, 1981.

5. Boyko A, Smylski PT: Osteopetrosis. *J Oral Surg* **32**:859–863, 1974.

6. Brown DM, Dent PB: Pathogenesis of osteopetrosis: A comparison of human and animal species. *Pediatr Res* **5**:181–191, 1971.

7. Dyson DP: Osteomyelitis of the jaws in Albers-Schönberg disease. *Br J Oral Surg* **7**:178–187, 1970.

7a. El Khazen N et al: Lethal osteopetrosis with multiple fractures in utero. *Am J Med Genet* **23**:811–819, 1986.

8. Funderburk SJ: Osteopetrosis in two brothers with severe mental retardation. *Birth Defects* **11**(6):91–98, 1975.

9. Guibaud P et al: Ostéopetrose et acidose rénale tubulaire: deux cas de cette association dans une fratrie. *Arch Fr Pédiatr* **29**:269–286, 1972.

10. Hasenhuttl K: Osteopetrosis: Review of the literature and comparative studies in a case with a twenty year follow-up. *J Bone Jt Surg* **44A**:359–370, 1962.

11. Horton WA et al: Osteopetrosis: Further heterogeneity. *J Pediatr* **97**:580–585, 1980.

12. Hunter AGW, Macpherson RI: Mental retardation and osteosclerosis. *Am J Med Genet* **2**:267–273, 1978.

13. Johnston CC et al: Osteopetrosis: A clinical, genetic, metabolic and morphologic study of the dominantly inherited benign form. *Medicine* **47**:149–167, 1968.

14. Kahler SG et al: A mild autosomal recessive form of osteopetrosis. *Am J Med Genet* **17**:451–464, 1984.

14a. Kuhlencordt F et al: Die Osteopetrosis Albers-Schönberg. *Ergeb Inn Med Kinderheilk* **39**:135–160, 1977.

15. Kukla LF: Dominant osteopetrosis. *Clin Pediatr* **16**:846–847, 1977.

16. Lehman RAW et al: Neurological complications of infantile osteopetrosis. *Ann Neurol* **2**:378–384, 1977.

16a. Leone G: Osteopetrosi recessiva con calcificazioni cerebrale. Studio de 3 soggetti adulti in due famiglie consanguinee. *Radiol Med* **68**:373–378, 1982.

17. Loria-Cortés R et al: Osteopetrosis in children. *J Pediatr* **91**:43–47, 1977.

18. Milgram JW, Jasty M: Osteopetrosis. A morphological study of twenty-one cases. *J Bone Jt Surg* **64A**:912–929, 1982.

19. Ohlsson A: Marble bone disease: Recessive osteopetrosis, renal tubular acidosis and cerebral calcifications in three Saudi Arabian families. *Dev Med Child Neurol* **22**:72–84, 1980.

19a. Rajeh SA et al: The syndrome of osteopetrosis, renal acidosis and cerebral calcification in two sisters. *Neuropediatrics* **19**:162–165, 1988.

20. Reeves J et al: The pathogenesis of infantile malignant osteopetrosis. Bone mineral metabolism and complications in five infants. *Metab Bone Dis Res* **3**:135–142, 1981.

21. Shapiro F et al: Human osteopetrosis. A histologic, ultrastructural and biochemical study. *J Bone Jt Surg* **62A**:384–399, 1980.

22. Sly WS et al: Carbonic anhydrase II deficiency in 12 families with the autosomal recessive syndrome of osteopetrosis with renal tubular acidosis and cerebral calcification. *N Engl J Med* **313**:139–144, 1985.

23. Steiner M et al: Osteomyelitis of the mandible associated with osteopetrosis. *J Oral Maxillofac Surg* **41**:395–405, 1983.

23a. Sudaram V et al: Carbonic anhydrase deficiency: Diagnosis and carrier detection using differential enzyme inhibition and inactivation. *Am J Hum Genet* **38**:125–136, 1986.

24. Svoboda PJ et al: Albers-Schönberg disease complicated with periodontal disease. *J Periodontol* **54**:592–597, 1983.

25. Thomson J: Osteopetrosis in successive generations. *Arch Dis Childh* **24**:143–148, 1949.

26. Trias A, Fery A: Osteopetrosis in adults. *Rev Chir Orthop* **60**:593–606, 1974.

27. Vainsel M et al: Osteopetrosis associated with proximal and distal renal tubular acidosis. *Acta Paediatr Scand* **61**:429–434, 1972.

28. Welford NT: Facial paralysis associated with osteopetrosis (marble bones): Report of a case of the syndrome occurring in five generations of the same family. *J Pediatr* **55**:67–72, 1959.

29. Whyte MP et al: Osteopetrosis, renal tubular acidosis and basal ganglia calcification in three sisters. *Am J Med* **69**:64–74, 1980.

30. Younai F et al: Osteopetrosis: A case report including gross and microscopic findings in the mandible at autopsy. *Oral Surg* **65**:214–221, 1988.

31. Yu JS et al: Osteopetrosis. *Arch Dis Childh* **46**:257–263, 1971.


## Oculodentoosseous dysplasia (oculodentodigital syndrome)

Although reported as early as 1920 by Lohmann (19), and independently by several other investigators (1,21,25), Meyer-Schwickerath et al (20), in 1957, were the first to fully describe a syndrome characterized by narrow nose with hypoplastic alae and thin nostrils, microcornea with iris anomalies, syndactyly and/or camptodactyly of postaxial fingers, hypoplasia or aplasia of the middle phalanx of the fifth fingers and toes, and enamel hypoplasia (Fig. 8–6A–J). About 80 cases have been reported to date.

The syndrome has autosomal dominant inheritance (5,24,27). New mutations represent approximately 50% of cases (27). Jones et al (15) found an advanced age in fathers of isolated cases of the disorder. Although affected sibs with normal parents have been described (6,10), these cases can be attributed to the variable expressivity of the disorder. There may, however, be genetic heterogeneity (vide infra).

Short narrow palpebral fissures, epicanthal folds, and a long thin nose with prominent nasal bridge and hypoplastic alae nasi produce a characteristic physiognomy. Head circumference may be somewhat reduced (18,27,28). Hyperostosis of the skull has been reported (4,24,27).

The most striking eye changes consist of short narrow palpebral apertures, microcornea (6–10 mm in diameter), and epicanthal folds in childhood (5,8,9,17). The oft-quoted findings of hypertelorism and microphthalmia are spurious (8,9). The pupil may be eccentric. The iris may consist of fine porous spongy tissue. Between the frill and the pupillary rim are crypts and lacunae, and the iris frill may overlie the pupillary rim. Remnants of the pupillary membrane may be present along the iris margin rather than across the pupil (4,6,11,20). A number of patients have exhibited strabismus or secondary glaucoma (8,16,25,29). There may be an increase in the number of disc vessels (16). Persistent hyperplastic primary vitreous has been noted on two occasions (12,31). It has been suggested that this is characteristic of the recessive form of the disorder (31). Radiographically, orbital hypotelorism has been demonstrated in 40% (9). Dry, lusterless hair that fails to grow to normal length has been described in about 30% (11,18,20,24,28). One author noted microscopic changes of monilethrix and pili annuli (30). The pinnae may be somewhat abnormally modeled and/or outstanding. Conduction hearing loss has been described in a number of cases (7,10,15,27,30), in part because of recurrent otitis media.

Most patients have had a gracile build. Camptodactyly of the fifth or, less often, of the fourth fingers is a common finding. Clinically, the fifth finger appears to be shortened. Bilateral syndactyly of the fourth and fifth fingers (rarely the third) with ulnar clinodactyly and syndactyly of the third and fourth toes are often present (25,27).

Radiographic examination reveals the middle phalanx of the fifth finger to be cuboid or deltoid or occasionally absent (30). The feet, clinically normal, on radiographic examination exhibit aplasia or hypoplasia of the middle phalanx of one or more toes. In at least one case there was postaxial hexadactyly of toes (17). Lack of modeling of the metaphyseal area of the long bones is relatively common (4,5,7,11,18,26).

Generalized enamel hypoplasia has been noted by a number of investigators (6,7,27,28). The alveolar ridge of the mandible may be wider than normal (3,11,16,26,32). Cleft lip–palate has been seen by several authors (7,10,22,32) and by RJ Gorlin. A number of authors have observed microdontia (8,13,32).

A child with many similar stigmata, but with rudimentary dry and brittle nails, was described by Whitwell (33). Sibs reported by Beighton et al (2) exhibited marked cranial hyperostosis, massive mandibular overgrowth, gross clavicular widening, blindness, microphthalmia, calcification of the basal ganglia, cataracts, cleft lip–palate, and spastic quadriplegia. Perhaps this represents an autosomal recessive form. Spastic paraplegia was also noted by Nivelon-Chevallier (22).

Although the eye anomalies appear to be similar to those observed in *Rieger syndrome,* there is neither microcornea nor enamel hypoplasia in the latter although tooth formation is suppressed. Microcornea in combination with glaucoma, epicanthal folds, absent frontal sinuses, and hyperkeratosis of the palms may exhibit autosomal dominant inheritance (14).

### References [Oculodentoosseous dysplasia (oculodentodigital syndrome)]


1. Bauer KH: Homoiotransplantation von Epidermis bei eineiigenen Zwillin-gen. *Bruns Beitr Klin Chir* **141**:442–447, 1927.
2. Beighton P et al: Oculo-dento-osseous dysplasia: Heterogeneity or variable expression? *Clin Genet* **16**:169–177, 1979.
3. Cowan A: Leontiasis ossea. *Oral Surg* **12**:983–989, 1959.
4. David JEA, Palmer PES: Familial metaphyseal dysplasia. *J Bone Jt Surg* **40B**:87–93, 1958.
5. Dudgeon J, Chisolm JA: Oculo-dento-digital dysplasia. *Trans Ophthalmol Soc UK* **94**:203–210, 1974.
6. Eidelman E et al: Orodigitofacial dysostosis and oculodentodigital dysplasia. *Oral Surg* **23**:311–319, 1967.
7. Fára M et al: Oculodentodigital dysplasia. *Acta Chir Plast* **19**:110–122, 1977.
8. Fára M, Gorlin RJ: The question of hypertelorism in oculodentoosseous dysplasia. *Am J Med Genet* **10**:101–102, 1981.
9. Farman AG et al: Oculodentodigital dysplasia. *Br Dent J* **142**:405–408, 1977.
10. Gillespie FD: Hereditary dysplasia oculodentodigitalis. *Arch Ophthalmol* **71**:187–192, 1964.
11. Gorlin RJ et al: Oculodentodigital dysplasia. *J Pediatr* **63**:69–75, 1963.
12. Gutierrez Diaz A et al: Oculodentodigital dysplasia. *Ophthalmol Paediatr Genet* **1**:227–232, 1982.
13. Haines JO, Rogers SC: Oculodento-digital dysplasia: A rare syndrome. *Br J Radiol* **48**:932–936, 1975.
14. Holmes LB, Walton DS: Hereditary microcornea, glaucoma and absent frontal sinuses. *J Pediatr* **74**:968–972, 1969.
15. Jones KL et al: Older paternal age and fresh gene mutation: Data on additional disorders. *J Pediatr* **86**:84–88, 1975.
16. Judisch GF et al: Oculodentodigital dysplasia. *Arch Ophthalmol* **97**:878–884, 1979.
17. Kadrnka-Lorrenčić M et al: Die oculo-dento-digitale Dysplasie (Das Meyer-Schwickerath-Syndrom). *Mschr Kinderheilkd* **121**:595–599, 1973.
18. Kurlander GJ et al: Roentgen differentiation of the oculodentodigital syndrome and the Hallermann–Streiff syndrome of infancy. *Radiology* **86**:77–85, 1966.
19. Lohmann W: Beitrag zur Kenntnis des reinen Mikrophthalmus. *Arch Augenheilkd* **86**:136–141, 1920.
20. Meyer-Schwickerath G et al: Mikrophthalmussyndrome. *Klin Monatsbl Augenheilkd* **131**:18–30, 1957.
21. Mohr OL: Dominant acrocephalosyndactyly. *Hereditas* **25**:193–203, 1939.
22. Nivelon-Chevallier A et al: Dysplasie oculo-dento-digitale. *J Génét Hum* **29**:171–179, 1981.
23. Patton MA: Oculodentoosseous syndrome. *J Med Genet* **22**:386–389, 1985.
24. Pfeiffer RA et al: Oculo-dento-digitale Dysplasie. *Klin Monatsbl Augenheilkd* **152**:247–262, 1968.
25. Pitter J, Svejda J: Über den Einfluss der Röntgenstrahlen auf die Entstehung von Missbildungen der menschlichen Frucht. *Ophthalmologia* **123**:386–393, 1952.
26. Rajic DS, de Veber LL: Hereditary oculodentoosseous dysplasia. *Ann Radiol* **9**:224–231, 1966.
27. Reisner SH et al: Oculodentodigital dysplasia syndrome. *Am J Dis Child* **118**:600–607, 1969.
28. Sugar HS et al: The oculo-dento-digital dysplasia syndrome. *Am J Ophthalmol* **61**:1448–1451, 1966.
29. Sugar HS: Oculodentodigital dysplasia syndrome with angle-closure glaucoma. *Am J Ophthalmol* **86**:36–38, 1978.
30. Thodén CJ et al: Oculodentodigital dysplasia syndrome. *Acta Paediatr Scand* **66**:635–638, 1977.
31. Traboulsi EI: Persistent hyperplastic primary vitreous and recessive oculo-dento-osseous dysplasia. *Am J Med Genet* **24**:95–100, 1986.
32. Weintraub DM et al: A family with oculodentodigital dysplasia. *Cleft Palate J* **12**:323–329, 1975.
33. Whitwell GPB: A case of ectodermal defect associated with hypertelorism. *Br J Dermatol* **43**:648–652, 1931.


## Frontometaphyseal dysplasia

Gorlin and Cohen (6), in 1969, separated frontometaphyseal dysplasia from other craniotubular dysplasias. The condition consists of pronounced bony supraorbital ridge, mixed hearing loss, and generalized skeletal dysplasia. About two dozen cases were subsequently described by numerous authors (1–5,7–14,16a–22). A probable earlier example is that of Lischi (15).

Inheritance appears to be X-linked with variable expression in carrier females (2,7,10). Although some authors have suggested autoso-

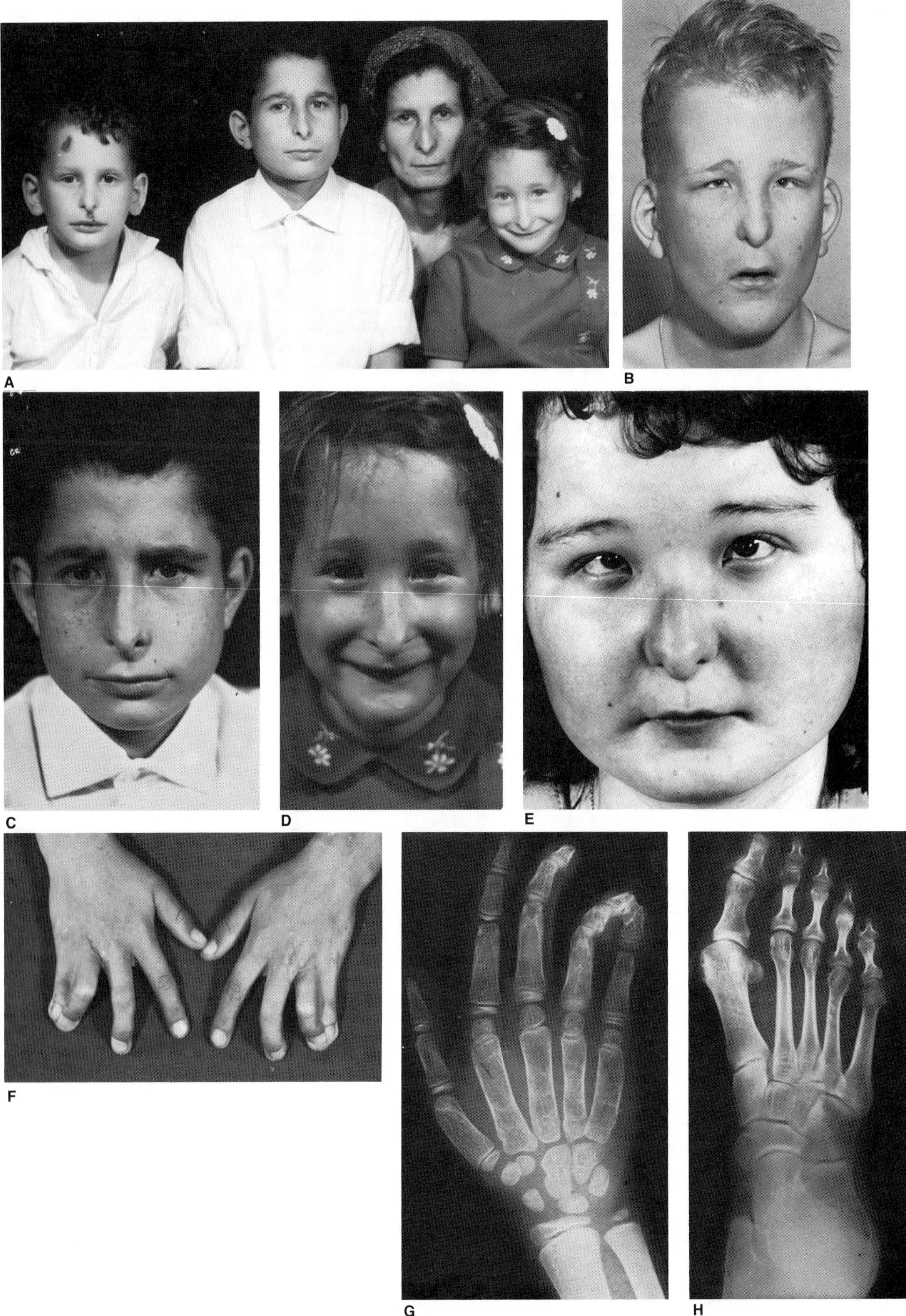
A B C D E F G H

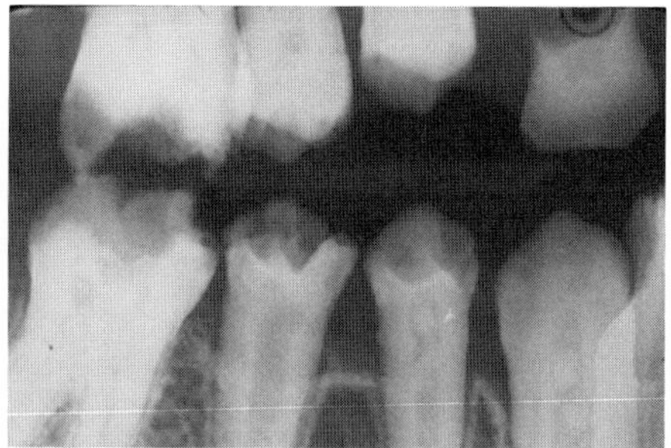

I

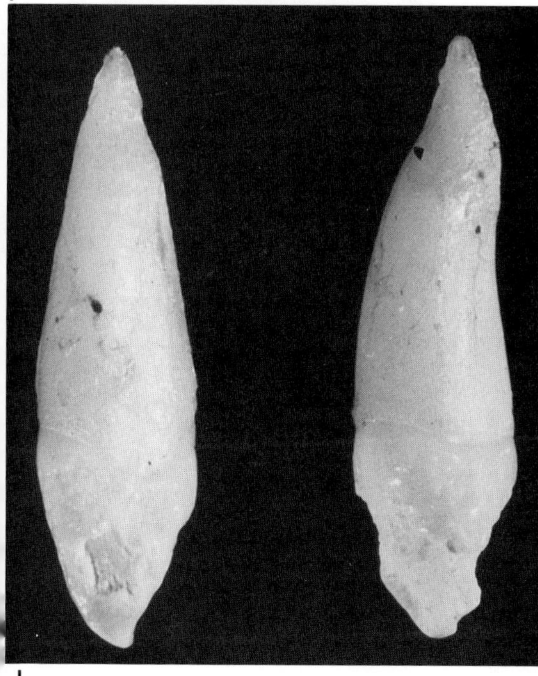

J

Fig. 8–6. *Oculodentoosseous dysplasia.* (A) Affected family. (B–E) Note similar facies. Microphthalmia, thin nose without alar flare. (F) Bilateral 4–5 syndactyly with ulnar deviation. (G,H) Note poor modeling of long bones, bony fusion of terminal phalanges of fourth and fifth fingers, absence of middle phalanges of toes. (I,J) Marked enamel hypoplasia. (A,C,D,F,G,H from *SH Reisner et al,* Am J Dis Child **118:**600, 1969. B from *RJ Gorlin et al,* J Pediatr **63:**69, 1963. E from *H Reich,* Hautarzt **31:**515, 1980.)

mal dominant inheritance (3,13,22), there has been no male-to-male transmission.

The marked supraorbital ridge, wide nasal bridge, downward slanting palpebral fissues, and small pointed chin give the patient a striking appearance. Enlargement of the supraorbital ridge becomes evident before puberty (4) (Fig. 8–7A–H).

There is both primary and secondary wasting of muscles of the hands. Dorsiflexion of the wrist and extension of the elbows are reduced, with pronation and supination being extremely limited. Flexion deformities of the fingers and ulnar deviation of the wrist are progressive. Finger mobility is essentially limited to the metacarpophalangeal joints. The thumbs tend to be broad. Hammertoes have been also noted.

Radiographic findings include a thick torus-like frontal ridge, absence of frontal sinuses, "Hershey kiss" or "top of the mosque" defects of supraorbital rims, arched superior border of maxillary sinuses, short maxilla, elongated cranial base, and antegonial notching of the mandible with marked hypoplasia of the angle and condyloid process (1,6,9,12,14,19).

The foramen magnum is greatly enlarged, and numerous vertebral anomalies have been noted; for example, the odontoid process is located too far anteriorly, the atlas has no posterior arch, and the lumbar vertebrae are flattened. There are fusion of the second and third cervical vertebrae, and subluxation of the third and fourth vertebrae. The shoulders may be highly positioned. Scoliosis is usually marked with resultant shortening of the trunk. The long bones manifest increased density in the diaphyseal region, with lack of modeling in the metaphyseal area producing an Erlenmeyer flask deformity. The legs may be laterally bowed. Marked flaring of the iliac bones and coxa valga are noted, as well as fused and eroded carpal bones, wide elongated middle phalanges, and increased interpediculate distances in the lumbar region of the spine (4,6,10,14,19,22). The ribs and vertebrae are irregularly contoured (9) and the lower ribs are "coat hanger" in form. Poznanski (18) and Jend-Rossmann et al (10) suggested a characteristic metacarpophalangeal profile.

Progressive mixed hearing loss has been reported (1,6,19,21,22). Hirsutism of the buttocks and thighs is common.

Urinary tract anomalies (hydroureter and hydronephrosis) (5,12,13,19) and obstructive airway disease (1,5,6) are probably relatively common complications.

Missing permanent teeth and retained deciduous teeth (1,4,6) have been found. Most patients have malocclusion.

Mitral valve prolapse has been reported (17) as well as bands of soft tissue extending from the medial edge of the scapula to the vertebral column (20).

### References (Frontometaphyseal dysplasia)


1. Arenberg JK et al: Otolaryngologic manifestations of frontometaphyseal dysplasia. The Gorlin–Holt syndrome. *Arch Otolaryngol* **99**:52–58, 1974.
2. Balestrazzi P: Hérédite liée au sexe dans la dysplasie fronto-metaphysaire. *J Génét Hum* **33**:419–425, 1985.
3. Beighton P, Hamersma H: Frontometaphyseal dysplasia: Autosomal dominant or X-linked. *J Med Genet 17*:53–56, 1980.
4. Danks DM et al: Fronto-metaphyseal dysplasia. *Am J Dis Child* **123**:254–258, 1972.
5. Fitzsimmons JS et al: Fronto-metaphyseal dysplasia: Further delineation of the clinical syndrome. *Clin Genet* **22**:1955–205, 1982.
6. Gorlin RJ, Cohen MM Jr: Frontometaphyseal dysplasia: A new syndrome. *Am J Dis Child* **118**:487–494, 1969.
7. Gorlin RJ, Winter RB: Frontometaphyseal dysplasia—evidence for X-linked inheritance. *Am J Med Genet* **5**:81–84, 1980.
8. de Haas WH et al: Metaphyseal dysostosis. *J Bone Jt Surg* **51B**:290–299, 1969.
9. Holt JF et al: Frontometaphyseal dysplasia. *Radiol Clin N Am* **10**:225–243, 1972.
10. Jend-Rossmann I et al: Frontometaphyseal dysplasia: Symptoms and possible mode of inheritance. *J Oral Maxillofac Surg* **42**:743–748, 1984.
11. Jervis GA, Jenkins EC: Frontometaphyseal dysplasia. *Syndrome Ident* **3**(1):18–19, 1975.
12. Kanemura T et al: Frontometaphyseal dysplasia with congenital urinary tract malformations. *Clin Genet* **16**:399–404, 1979.
13. Kassner EG et al: Frontometaphyseal dysplasia: Evidence for autosomal dominant inheritance. *Am J Roentgenol* **127**:927–933, 1976.
14. Kleinsorge H, Böttger E: Das Gorlin–Cohen-Syndrom (fronto-metaphysäre Dysplasie). *Roefo* **127**:451–458, 1977.
15. Lischi G: Le torus supraorbitalis: Variation craniênne rare. *J Radiol Électrol* **48**:463–466, 1967.
16. Medlar RC, Crawford AH: Frontometaphyseal dysplasia presenting as scoliosis: A report of a family with four cases. *J Bone Jt Surg* **60A**:392–394, 1978.

16a. Mersten A et al: Cranio-metaphyseal dysplasia. *Radiol Diag* **21**:70–74, 1980.

17. Park JM et al: Mitral valve prolapse in a patient with frontometaphyseal dysplasia. *Clin Pediatr* **25**:469–471, 1986.
18. Poznanski AK: *The Hand in Radiologic Diagnosis: With Gamuts and Pattern Profiles,* 2nd ed. W. B. Saunders, Philadelphia, 1984.
19. Sauvegrain J et al: Dysplasie fronto-metaphysaire. *Ann Radiol* **18**:155–162, 1975.
20. Ullrich E et al: Frontometaphyseal dysplasia: Report of two familial cases. *Australas Radiol* **23**:265–271, 1979.
21. Walker BA: A craniodiaphyseal dysplasia or craniometaphyseal dysplasia, ? type. *Birth Defects* **5**(4):298–300, 1969.
22. Weiss L et al: Frontometaphyseal dysplasia—evidence for dominant inheritance. *Birth Defects* **11**(5):55–56, 1975.

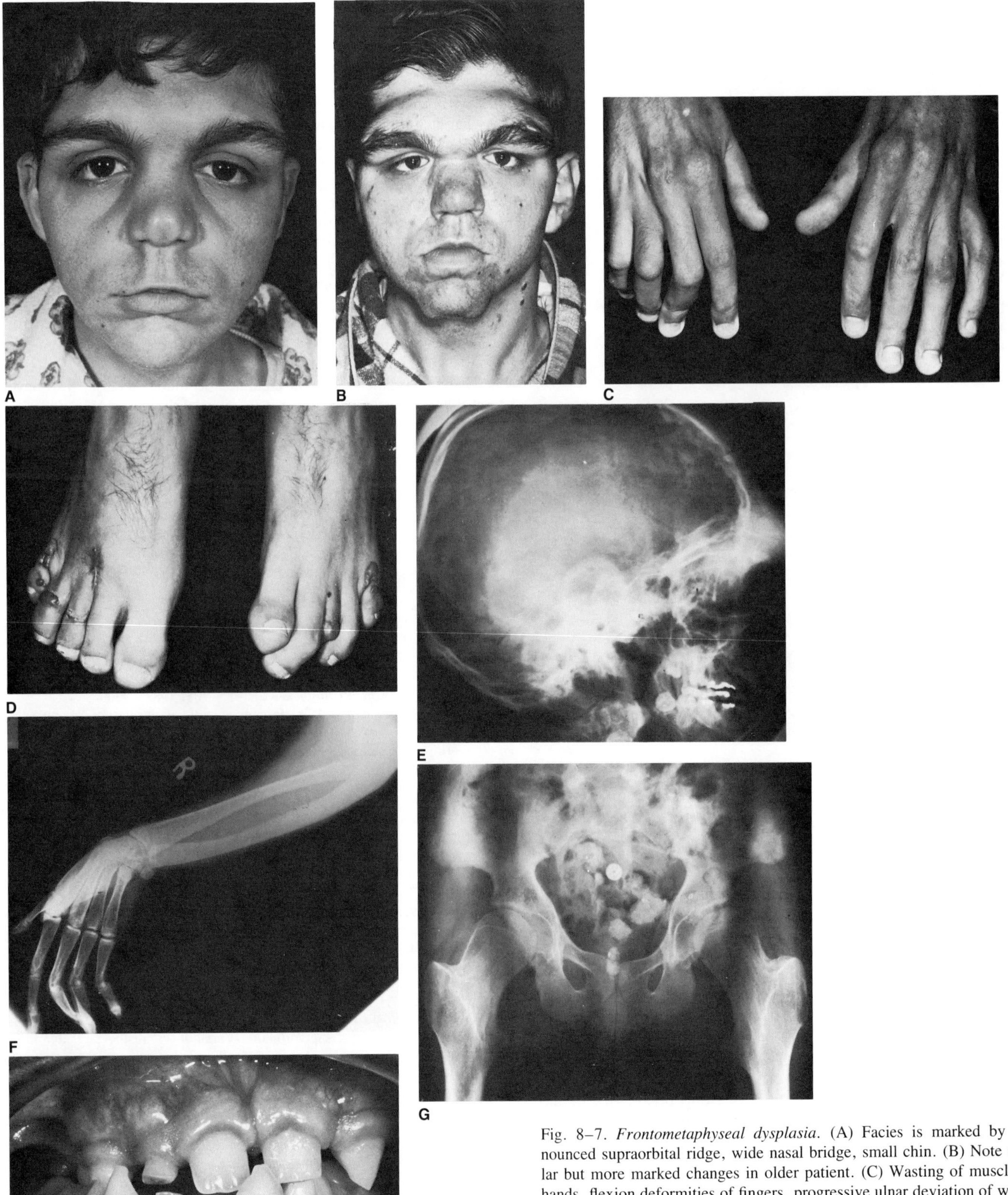


Fig. 8–7. *Frontometaphyseal dysplasia.* (A) Facies is marked by pronounced supraorbital ridge, wide nasal bridge, small chin. (B) Note similar but more marked changes in older patient. (C) Wasting of muscles of hands, flexion deformities of fingers, progressive ulnar deviation of wrists. (D) Hammertoes and keloid formation. (E) Supraorbital torus, hypoplastic mandible with irregular lower border. (F) Relative absence of metaphyseal modeling. (G) Marked flaring of iliac bones and coxa valga. (H) Missing permanent teeth, retained deciduous teeth, teeth without conical crown form in 18-year-old patient. (A from *D Danks et al,* Am J Dis Child **123**:254, 1972. B–G from *RJ Gorlin and MM Cohen Jr,* Am J Dis Child **118**:287, 1969.)

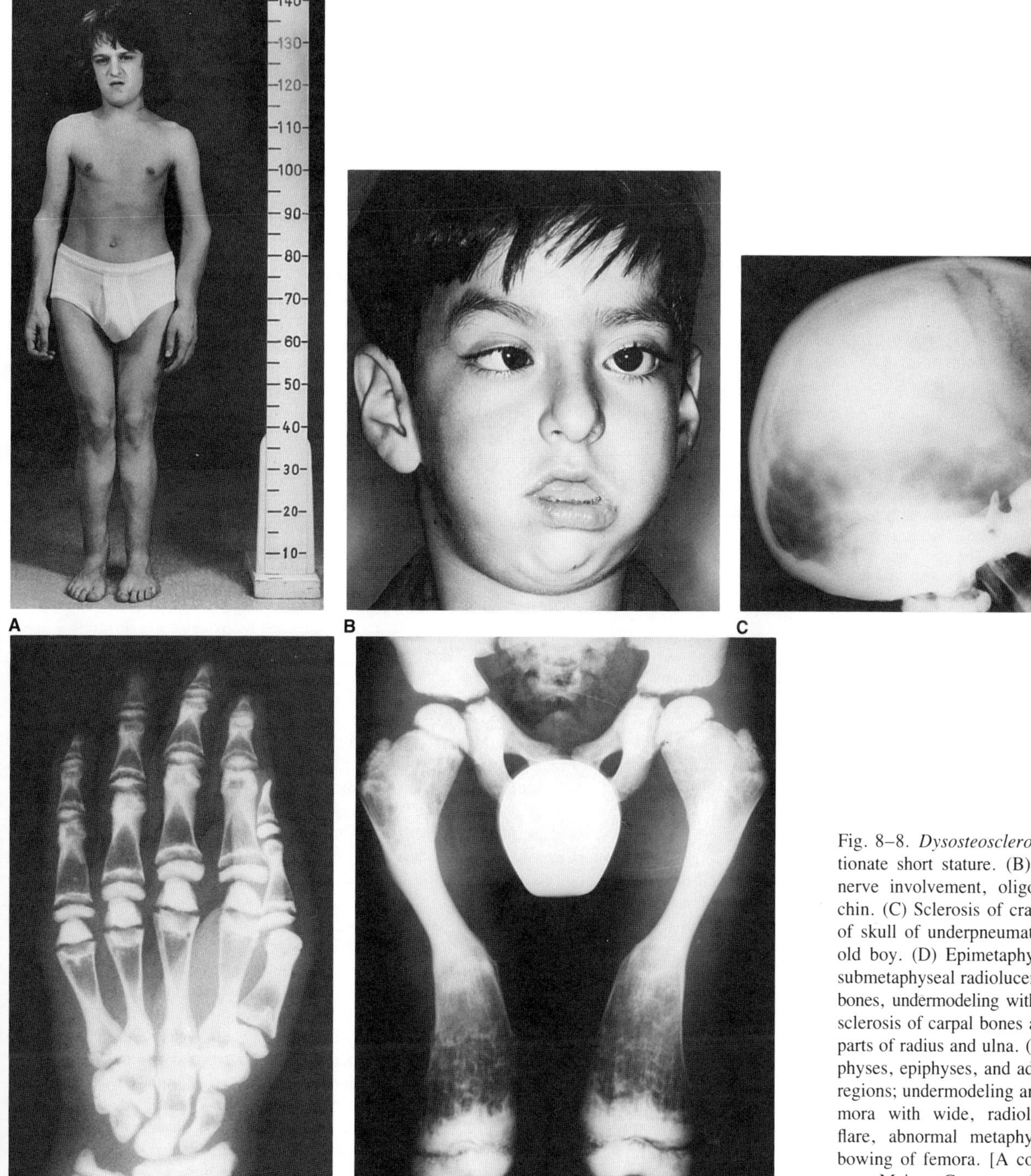


Fig. 8–8. *Dysosteosclerosis.* (A) Disproportionate short stature. (B) Strabismus, facial nerve involvement, oligodontia, and small chin. (C) Sclerosis of cranial vault and base of skull of underpneumatization in 10-year-old boy. (D) Epimetaphyseal sclerosis with submetaphyseal radiolucency of short tubular bones, undermodeling with metaphyseal flare; sclerosis of carpal bones and epimetaphyseal parts of radius and ulna. (E) Sclerosis of diaphyses, epiphyses, and adjacent metaphyseal regions; undermodeling and shortening of femora with wide, radiolucent metaphyseal flare, abnormal metaphyseal trabeculation; bowing of femora. [A courtesy of *J Spranger,* Mainz, Germany. B courtesy of *DL Rimoin,* Los Angeles, California. C–E from *RJ Gorlin et al,* Birth Defects **5**(4):79, 1969.]

## Dysosteosclerosis

Dysosteosclerosis, a term first employed by Spranger (12) in 1968, had been described as early as 1933 by Ellis (1). Additional examples have been described (2–4,6,7,9,11,13,14,16) (Fig. 8–8A–E).

Affected sibs (1,2,8,13) and parental consanguinity (1,2,4,11,16) indicate autosomal recessive inheritance. However, there appears to be an X-linked recessive form (8). The patient reported by Ventruto et al (16) surely has dysosteosclerosis.

The anterior fontanel tends to remain open. There is frontal and biparietal bossing and narrow chin. Oligodontia and poorly calcified teeth with late eruption have been described (4,8,15). Natal teeth have also been noted (3).

The patient is short and there is a tendency to bone fractures (6,12,13). The limbs are disproportionately shortened in comparison to the trunk and somewhat bowed. Pectus carinatum has been noted in several patients.

During early childhood, there may be cranial nerve involvement such as optic atrophy, abducens palsy, and facial paralysis. Some degree of spasticity and exaggerated reflexes have been evident (1,2,6,11).

Others have manifested progressive mental retardation. Some have exhibited progressive otosclerosis (6). Macular atrophy of the skin has been found in several cases (1,2,10,11,14).

Radiographically, the calvaria and skull base are thickened. There is sclerosis of the orbital roofs, absent paranasal sinuses, and constriction of the foramens. The clavicles, scapulae, and ribs are sclerotic. The vertebral bodies are flattened and irregularly dense. Long tubular bones are bent in the region of the shortened, thickened diaphyses. The metaphyses are bottle shaped. The epiphyses and metaphyses are sclerotic, but the submetaphyseal areas are clear and their trabecular structure is coarse and irregular. Short tubular bones exhibit similar changes. Iliac bones are hypoplastic and sclerotic.

Microscopic study of the growth plates suggests that the metaphyses are filled with irregular trabeculae consisting mainly of cartilaginous matrix with few chondrocytes (5).

#### References (Dysosteosclerosis)


1. Ellis RWB: Osteopetrosis. *Proc R Soc Med* **27**:1563–1571, 1933–4.
2. Field CE: Albers–Schönberg disease: An atypical case. *Proc R Soc Med* **32**:320–324, 1938–9.
3. Fryns JP et al: Dysosteosclerosis in a mentally retarded boy. *Acta Paediatr Belg* **33**:53–56, 1980.
4. Houston CS et al: Dysosteosclerosis. *Am J Roentgenol* **130**:988–991, 1978.
5. Kaitila I, Rimoin DL: Histologic heterogeneity in the hyperostotic bone dysplasias. *Birth Defects* **12**(6):71–79, 1976.
6. Kirkpatrick DB et al: The craniotubular bone modeling disorders: A neurological introduction to rare skeletal dysplasias with cranial nerve compression. *Surg Neurol* **7**:221–232, 1977.
7. Leisti J et al: Dysosteosclerosis. *Birth Defects* **11**(6):349–351, 1975.
8. Nema HV: Craniometaphyseal dysplasia. *Br J Ophthalmol* **58**:107–109, 1974.
9. Parascandolo S et al: Su un caso clinico di displasia cranio-metaphisaria. *Min Stomatol* **34**:671–675, 1985.
10. Pascual-Castroviejo I et al: X-Linked dysosteosclerosis. *Eur J Pediatr* **126**:127–138, 1977.
11. Roy C et al: Un nouveau syndrome osseux avec anomalies cutanées et troubles neurologiques. *Arch Fr Pédiatr* **25**:893–905, 1968.
12. Spranger J et al: Die Dysosteosclerose—eine Sonderform der generalisierten Osteosklerose. *Fortschr Röntgenstr* **109**:504–512, 1968.
13. Stehr L: Pathogenese und Klinik der Osteosklerosen. *Arch Orthop Unfall-Chir* **41**:156–182, 1942.
14. Temtamy SA et al: Metaphyseal dysplasia, anetoderma and optic atrophy: An autosomal recessive syndrome. *Birth Defects* **10**(12):61–71, 1974.
15. Utz W: Manifestation der Dysosteosklerose im Kieferbereich. *Deutsch Zahnärztl Z* **25**:48–50, 1970.
16. Ventruto V et al: A case of autosomal recessive form of cranio-metaphyseal dysplasia with unusual features and with bone fragility. *Australas Radiol* **31**:79–81, 1987.


## Van Buchem disease and pseudo-Van Buchem disease

Van Buchem disease (generalized cortical hyperostosis) is characterized by osteosclerosis of the skull, mandible, clavicles, and ribs and by hyperplasia of the diaphyseal cortex of the long and short bones (Fig. 8–9A–E).

The disorder exhibits autosomal recessive inheritance (2–9). It has been alleged to be a mild form of sclerosteosis (see below). The family described by Dixon et al (1) has a different disorder. We have seen the same condition in affected sibs. We have termed this *pseudo-Van Buchem disease* (Fig. 8–9F–H).

Facial changes develop slowly, but usually become apparent before the second decade. A most striking finding is a wide and thickened mandible, suggesting acromegaly. Rarely, skull circumference is enlarged. Occasionally there is mild exophthalmos.

Patients experience headache, unilateral or rarely bilateral facial paralysis, optic atrophy, and progressive sensorineural or mixed hearing loss.

Radiographic changes include thickening of the calvaria and increased density of the skull base. The body of the mandible is greatly enlarged in all measurements; the angle is obtuse. The long tubular bones exhibit diaphyseal thickening and are rough textured. The cortical hyperostosis is predominantly endosteal in character. In severe cases, the medullary cavity is occluded. The transverse diameter of the diaphysis is normal or increased.

Elevated serum alkaline phosphatase level has been noted in most cases.

RJ Gorlin has observed a sporadic case of huge mandible, sclerosed long tubular bones especially in the metaphyseal areas, and similar involvement of the scapula and tarsal bones. The ribs were broad, but not sclerosed. The distal portions of the limbs and vertebral column were normal. Alkaline phosphatase levels were extremely high.

#### References (Van Buchem disease and pseudo-Van Buchem disease)


1. Dixon JM et al: Two cases of Van Buchem's disease. *J Neurol Neurosurg Psychiatr* **45**:913–918, 1982.
2. Garland LH: Generalized leontasis ossea. *Am J Roentgenol* **55**:37–43, 1946.
3. Jacobs P: Van Buchem disease. *Postgrad Med J* **53**:497–505, 1977.
4. Sala O: ORL importance of some aspects of osteopetrosis. *Boll Mal Orecch* **71**:577–592, 1953.
5. Toledo F et al: Hiperostosis endosteal (enfermedad de Van Buchem). *Radiologia* **24**:49–52, 1982.
6. Van Buchem FSP et al: Hyperostosis corticalis generalisata familiaris. *Acta Radiol* **44**:109–114, 1955.
7. Van Buchem FSP et al: Hyperostosis corticalis generalisata: Report of seven cases. *Am J Med* **33**:387–397, 1962.
8. Van Buchem FSP et al: *Hyperostosis Corticalis Generalisata Familiaris (Van Buchem's Disease)*. American Elsevier, New York, 1976.
9. Van der Wouden A: Deafness caused by hyperostosis corticalis generalisata. *Pract Otorhinolaryngol* **30**:91–92, 1968.


## Sclerosteosis

The disorder was described as early as 1929 by Hirsch (10). Several other reports of the disorder (7,9,11–13,15,18,19) antedate Hansen's (8) definition of sclerosteosis. The disorder is characterized by generalized osteosclerosis with hyperostosis of the calvaria, mandible, clavicles, and pelvis rather different from that observed in Van Buchem disease. Usually there are syndactyly and other abnormalities of the digits (Fig. 8–10A–C). The disorder appears to be one of osteoblast hyperactivity (16).

Sclerosteosis has autosomal recessive inheritance. Most patients have been South African of Dutch ancestry (1,4,5,7,12). Its frequency there has been estimated to be about 1/60,000 Afrikaners (1). It has been seen among Japanese (17). Heterozygotes exhibit increased calvarial width and density (P Beighton, personal communication, 1989).

The typical facies, evident by the age of 5 years, is characterized by frontal prominence, hypertelorism, and broad flat nasal root. The mandible is prognathic, broadened and squared, and dental malocclusion is frequent. The face may be distorted with relative midfacial hypoplasia. Head circumference is enlarged. In most cases, mixed hearing loss appears in childhood (14). Facial nerve paralysis, transient in infancy, is common in adulthood. Characteristically, it is unilateral for many years. There is increased intracranial pressure in 80% (2,5,16). Ataxia has been reported (12,16). Exophthalmos, optic atrophy, reduced visual fields, convergent strabismus, nystagmus, chronic headache, and decreased sensory function of the trigeminal nerve have been described in adults. Visual loss occurs in 30%. Only rarely, however, is there total blindness. Several patients have died suddenly from impaction of the medulla in the foramen magnum (2,6).

Body height is over 180 cm in 70%. In about 75% there is asymmetric partial or complete cutaneous syndactyly of the index and middle fingers. There may be radial deviation of the distal phalanx of the index fingers. The nails on the involved digits are hypoplastic in 80%. Possibly height is correlated with syndactyly.

Radiographically, the calvaria becomes thickened and sclerotic in infancy. This gradually increases to about the age of 30 years. The base is dense and the foramina obliterated. The mandible is massive, prognathic, often asymmetric, and with an obtuse angle. The clavicles

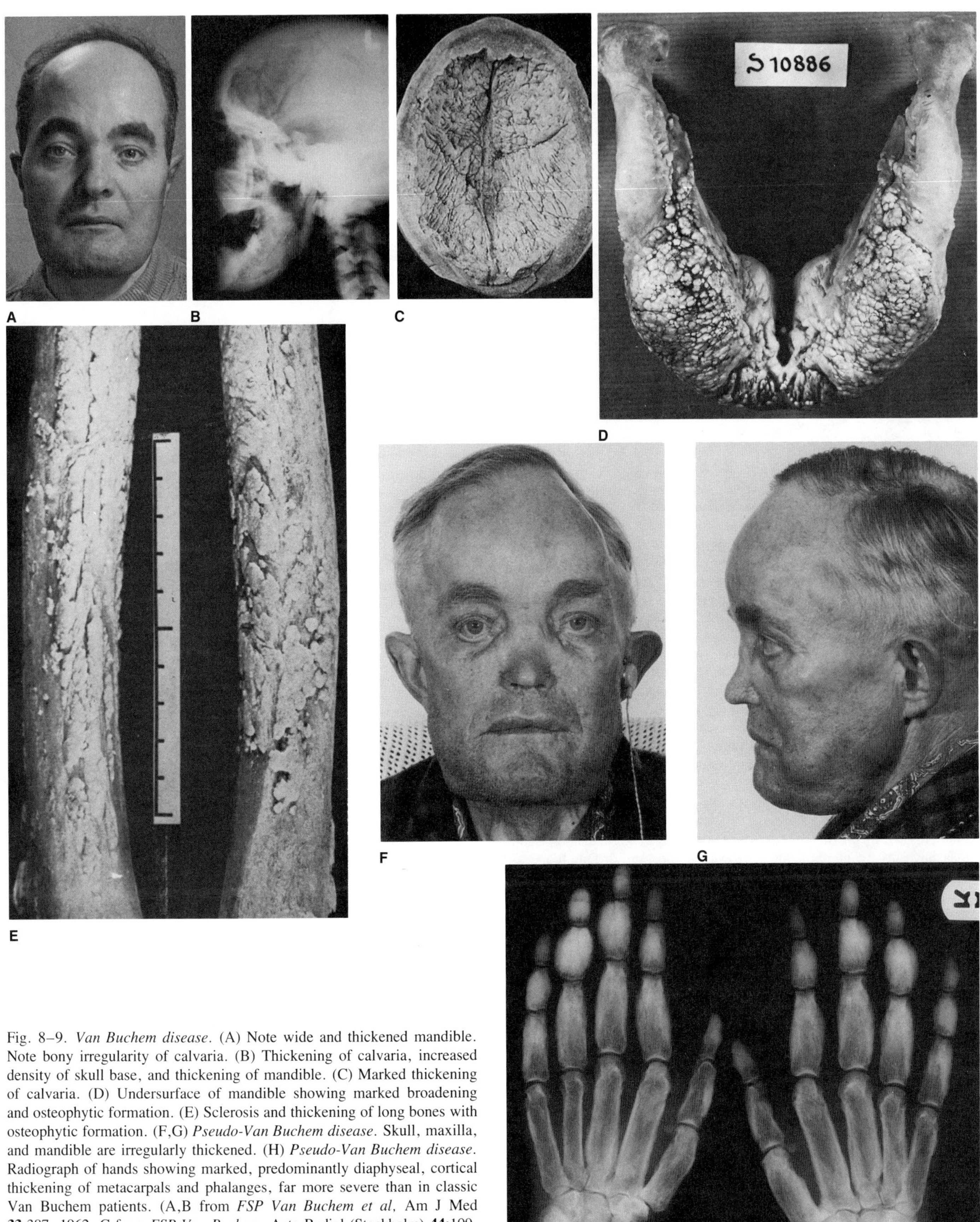


A B C D E F G H

Fig. 8–9. *Van Buchem disease.* (A) Note wide and thickened mandible. Note bony irregularity of calvaria. (B) Thickening of calvaria, increased density of skull base, and thickening of mandible. (C) Marked thickening of calvaria. (D) Undersurface of mandible showing marked broadening and osteophytic formation. (E) Sclerosis and thickening of long bones with osteophytic formation. (F,G) *Pseudo-Van Buchem disease.* Skull, maxilla, and mandible are irregularly thickened. (H) *Pseudo-Van Buchem disease.* Radiograph of hands showing marked, predominantly diaphyseal, cortical thickening of metacarpals and phalanges, far more severe than in classic Van Buchem patients. (A,B from *FSP Van Buchem et al,* Am J Med **33:**387, 1962. C from *FSP Van Buchem,* Acta Radiol (Stockholm) **44:**109, 1955. D from *FSP Van Buchem* and *HN Hadders,* Schweiz Med Wochenschr **87:**231, 1957. E courtesy of *FSP Van Buchem,* Haarlem, The Netherlands. F–H from *JM Dixon et al,* J Neurol Neurosurg Psychiat **45:**913, 1982.)

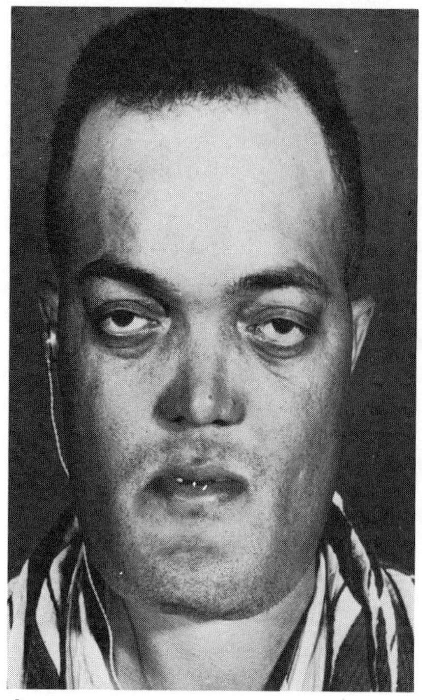

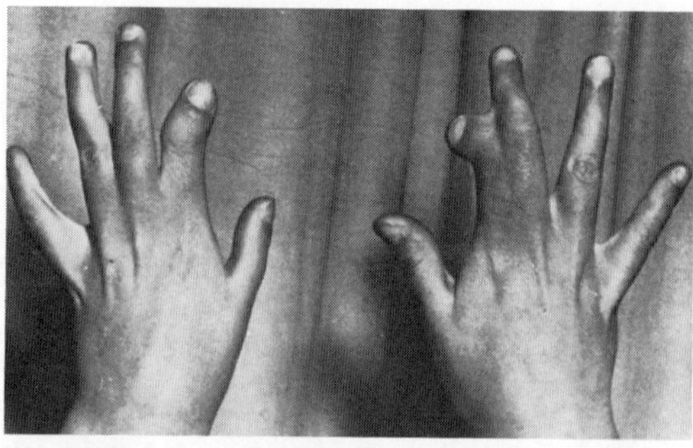

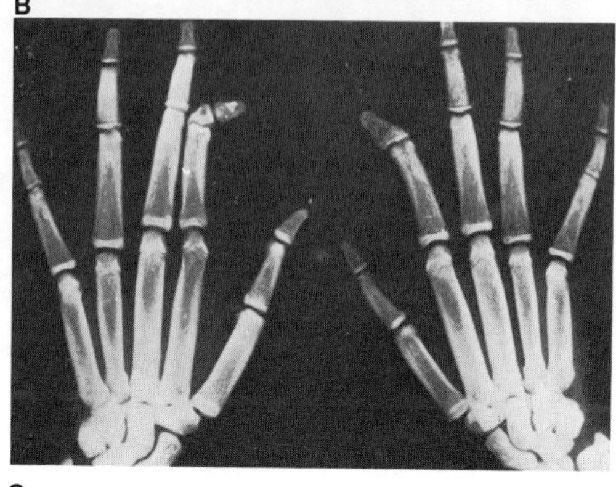

A B C

Fig. 8–10. *Sclerosteosis.* (A) Square appearance of mandible, resulting in part from flattening of mandibular plane. Patient was deaf and had facial palsy. (B) Soft-tissue syndactyly of second and third fingers bilaterally. Third and fourth fingers partly fused unilaterally. Radial clinodactyly of index fingers. (C) Roentgenogram showing absent middle phalanx of one index finger and delta phalanx of other index finger, as well as general lack of modeling of tubular bones. (A courtesy of *CJ Witkop Jr*, Minneapolis, Minnesota. B,C from *AS Truswell,* J Bone Jt Surg **40B:**208, 1958.)

and ribs are broadened and dense because of cortical thickening. The scapulae, pelvis, and vertebral endplates and pedicles are uniformly sclerotic. The tubular bones, in addition to increased density, exhibit a lack of diaphyseal modeling. The index finger may have no middle phalanx or only a small triangular bone (delta phalanx) producing radial deviation. Bony syndactyly may involve the second and third fingers (3).

Microscopic study of the bone reveals only increased density with osteoblastic hyperactivity (16).

Patients with *Van Buchem disease* tend to be of normal height and never have involvement of digits. However, most are of Dutch ancestry. Sclerosteosis tends to be more severe in its manifestations. Hearing loss (90%) and raised intracranial pressure (80%) are more common than in those with Van Buchem disease. Beighton et al (5), having examined 80 Afrikaners with sclerosteosis in South Africa and 15 patients with Van Buchem disease in Holland, have posited that they are the same disorder, the difference being the occurrence of an additional epistatic gene in South Africa.

#### References (Sclerosteosis)

1. Beighton P, Hamersma H: Sclerosteosis in South Africa. *S Afr Med J* **55**:783–788, 1979.

2. Beighton P et al: The clinical features of sclerosteosis: A review of the manifestations in twenty-five affected individuls. *Ann Intern Med* **84**:393–397, 1976.

3. Beighton P et al: The radiology of sclerosteosis. *Br J Radiol* **49**:934–939, 1976.

4. Beighton P et al: Sclerosteosis—an autosomal recessive disorder. *Clin Genet 11*:1–7, 1977.

5. Beighton P et al: The syndromic status of sclerosteosis and Van Buchem disease. *Clin Genet* **25**:175–181, 1984.

6. Epstein S et al: Endocrine function in sclerosteosis. *S Afr J Med* **55**:1105–1110, 1979.

7. Falconer AW, Ryrie BJ: Report on a familial type of generalized osteosclerosis. *Presse Méd* **195**:12–14, 1937.

8. Hansen HG: Sklerosteose, in *Handbuch der Kinderheilkunde,* Vol 6, Opitz H, Schmid F (eds), Springer-Verlag, Berlin, Göttingen, Heidelberg, New York, 1967, pp 351–355.

9. Higinbotham NL, Alexander SF: Osteopetrosis: Four cases in one family. *Am J Surg* **53**:444–454, 1941.

10. Hirsch IS: Generalized osteitis fibrosa. *Radiology* **13**:44–84, 1929.

11. Kelley CH, Lawlah JW: Albers–Schönberg disease: A familial survey. *Radiology* **47**:507–513, 1946.

12. Klintworth GK: Neurologic manifestations of osteopetrosis (Albers–Schönberg's disease). *Neurology* **13**:512–519, 1963.

13. Kretzmar JH, Roberts RA: Case of Albers–Schönberg disease. *Br Med J* **1**:837–838, 1936.

14. Nager GT, Hamersma H: Sclerosteosis involving the temporal bone: Clinical and radiologic aspects. *Am J Otolaryngol* **7**:1–16, 1986.

15. Pietruschka G: Weitere Mitteilungen über die Marmorknochenkrankheit (Albers–Schönbergsche Krankheit) nebst Bemerkungen zur Differential-diagnose. *Klin Mbl Augenheilk* **132**:509–525, 1958.

16. Stein SA et al: Sclerosteosis: Neurogenic and pathophysiologic analysis of an American kinship. *Neurology* **33**:267–277, 1983.

17. Sugiura Y, Yasuhara T: Sclerosteosis. *J Bone Jt Surg* **57A**:273–276, 1975.

18. Truswell AS: Osteopetrosis with syndactyly. A morphological variant of Albers–Schönberg disease. *J Bone Jt Surg* **40B**:208–218, 1958.

19. Witkop CJ Jr: Genetic disease of the oral cavity, in *Oral Pathology,* Tiecke RW (ed), McGraw-Hill, New York, 1965. (Same kindred as in Ref. 11.)

### Autosomal dominant osteosclerosis (endosteal hyperostosis, Worth type)

Clinical manifestation is essentially limited to square jaw, that is, widened and deepened mandible with increased gonial angle (1–14,16–18) (Fig. 8–11A–C). We are not certain how to classify the case of Scott and Gautby (15). The changes begin at puberty and plateau with cessation of growth. Possibly, torus palatinus is more frequently associated with the condition (2,9,13,17,18).

Radiographically, there is endosteal sclerosis of the neurocranium

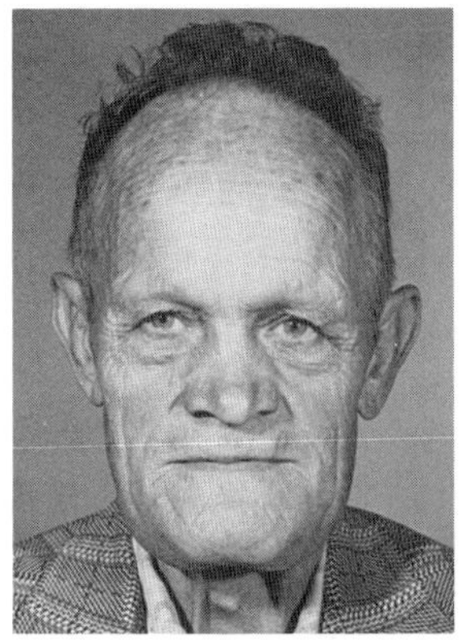
A

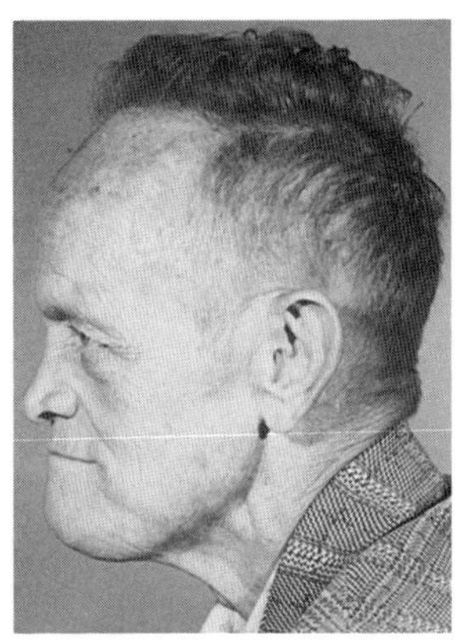
B

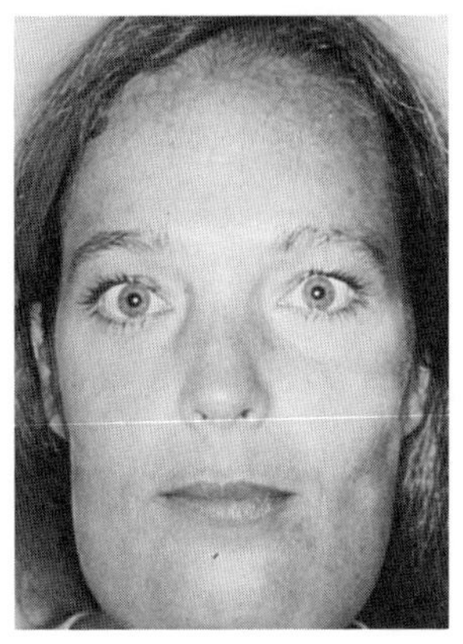
C

Fig. 8–11. *Autosomal dominant osteosclerosis.* (A–C) Prominent frontal area and squared jaw. Torus palatinus is often associated. (A,B from *RK Beals,* J Bone Jt Surg **58A:** 1172, 1976. C from *EW Ruckert et al,* J Oral Maxillofacial Surg **43:**801, 1985.)

with loss of the diploë, osteosclerosis, and hyperostosis of the mandible with absence of the normal antegonial notches, endosteal sclerosis of the diaphyses of long bones (including metacarpals and metatarsals), and osteosclerosis of the pelvis. The vertebral bodies, ribs, and clavicles are involved to a minor degree.

In contrast to *Van Buchem disease,* there is usually no compression of cranial nerves as a result of foraminal encroachment or elevation of serum alkaline phosphatase with rare exception (2,4,8,10,12) and there is autosomal dominant inheritance (6). *Autosomal dominant osteopetrosis* is not associated with enlarged mandible and, furthermore, skeletal involvement is generalized.

### References [Autosomal dominant osteosclerosis (endosteal hyperostosis, Worth type)]


1. Beals RK: Endosteal hyperostosis. *J Bone Jt Surg* **58A**:1172–1173, 1976.
2. Demonchy A et al: Hyperostose corticale généralisée. *Nouv Presse Méd* **7**:2849–2851, 1978.
3. Dyson DP: Van Buchem's disease (hyperostosis corticalis generalisata familiaris). *Br J Oral Surg* **9**:237–245, 1972 (same case as Ref. 11).
4. Eastman JR, Bixler D: Generalized cortical hyperostosis (Van Buchem disease): Nosologic considerations. *Radiology* **125**:297–304, 1977.
5. Gelman MI: Autosomal dominant osteosclerosis. *Radiology* **125**:289–296, 1977.
6. Gorlin RJ, Glass L: Autosomal dominant osteosclerosis. *Radiology* **125**:547–548, 1977.
7. Jacobs P: Van Buchem disease. *Postgrad Med J* **53**:497–505, 1977.
8. Lapresle J et al: Hyperostose corticalis généralisée dominante avec atteinte multiple des nerfs crâniens. *Nouv Presse Méd* **5**:2703–2706, 1975.
9. Maroteaux P et al: L'hyperostose corticle généralisée a transmission dominante. *Arch Fr Pédiatr* **28**:685–698, 1971.
10. Moretti C et al: Iperostosi endostale a transmissione dominante. Descrizione de 8 casi in 3 generazioni dello stesso nucleo familiare. *Radiol Med (Torino)* **68**:151–158, 1982.
10a. Nakamura T et al: Autosomal dominant type of endosteal hyperostosis with unusual manifestations of sclerosis of the jaw bones. *Skeletal Radiol* **16**:48–51, 1987.
11. Owen RJ: Van Buchem's disease (hyperostosis corticalis generalisata). *Br J Radiol* **49**:126–132, 1976.
12. Perez-Vicente JA et al: Autosomal dominant endosteal hyperostosis. Report of a Spanish family with neurological involvement. *Clin Genet* (in press).
13. Ruckert WE et al: Surgical treatment of Van Buchem's disease. *J Oral Maxillofac Surg* **43**:801–805, 1985.
14. Russell WJ et al: Idiopathic osteosclerosis: A report of six related cases. *Radiology* **90**:70–76, 1968.
15. Scott WC, Gautby THT: Hyperostosis corticalis generalisata familiaris. *Br J Radiol* **47**:500–503, 1974.
16. Segond P et al: Le retrécissement du canal médullaire des os a transmission dominante. *Nouv Presse Méd* **2**:2728–2732, 1973.
17. Vayssairat M et al: Nouveaux cas familiaux d'hyperostose cortiale généralisée a transmission dominante (type Worth). *J Radiol Electrol* **57**:719–724, 1976.
18. Worth HM, Wollin DG: Hyperostosis corticalis generalisata congenita. *J Can Assoc Radiol* **17**:67–74, 1966.


## Progressive diaphyseal dysplasia (Camurati–Engelmann disease)

Progressive diaphyseal dysplasia is a sclerotic and hyperostotic disorder of bone. Originally reported by Cockayne (3) in 1920, it was defined by Camurati (2) in 1922 and Engelmann (6) in 1929. More than 100 cases have been reported (12).

Inheritance is autosomal dominant with considerable variation in expression (12). New mutations account for about 50%.

The most common clinical findings include delayed ambulation, bone pain, generalized neuromuscular weakness, flat feet, broad based waddling gait, thin musculature with disproportionately long limbs and bowed tibiae and can be manifest as early as the third to fifth year of life, although the mean age is about 15–20 years (1,7,8,19). There may be genua vara, genua valga, lumbar lordosis, or scoliosis. Less often there is hepatosplenomegaly (5) (Fig. 8–12A–H).

Secondary sexual development is poor. Some patients exhibit frontal bossing, exophthalmos, papilledema, epiphora, optic atrophy, and headache (9,11,12,16). Mixed hearing loss has been noted in 5–7% (4,5,11,15,17,18).

Radiographically there is symmetric irregular spindle-shaped, sclerotic cortical thickening of the middiaphyses of the long tubular bones and narrowing of the medullary cavities. With age, the process extends proximally and distally toward the metaphyses which are rarely involved. The epiphyses are not affected. The base of the skull and calvaria are sclerotic in 70%. The cervical vertebrae, clavicles, pelvic bones, hand and foot bones, and ribs are affected in about 20%. The mandible is sclerotic in 25% and occasionally is significantly enlarged (4,11,13).

Serum alkaline phosphatase, urinary hydroxyproline, and erythrocyte sedimentation rate may be elevated (14). Anemia is relatively frequent (6a). The scintigraphic changes are striking and not always correlated with radiographic changes (10).

### References [Progressive diaphyseal dysplasia (Camurati–Engelmann disease)]


1. Beighton P, Cremin BJ: *Sclerosing Bone Dysplasias.* Springer-Verlag, New York, 1980.
2. Camurati M: Di un raro di osteite simmetrica ereditaria degli arti inferiori. *Clin Organi Mov* **6**:622–665, 1922.
3. Cockayne EA: Case for diagnosis. *Proc R Soc Med* **13**:132–136, 1920.
4. Cohen J, States JD: Progressive diaphyseal dysplasia. *Lab Invest* **5**:492–508, 1956.
5. Crisp AJ, Brenton DP: Engelmann's disease of bone—a systemic disorder? *Ann Rheum Dis* **41**:183–188, 1982.
6. Engelmann G: Ein Fall von Osteopathia hyperostotica (sclerosis) multiplex infantiles. *Fortschr Roentgenol* **39**:1011–1116, 1929.
6a. Ghosal SP et al: Diaphyseal dysplasia associated with anemia. *J Pediatr* **113**:49–57, 1988.
7. Hundley JD, Wilson FC: Progressive diaphyseal dysplasia. *J Bone Jt Surg* **55A**:461–474, 1973.

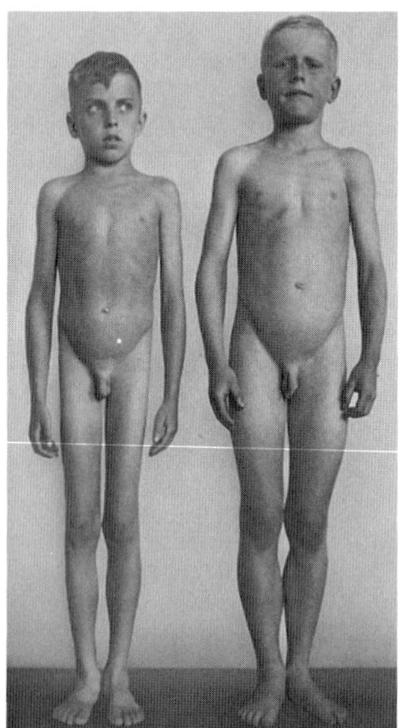

A

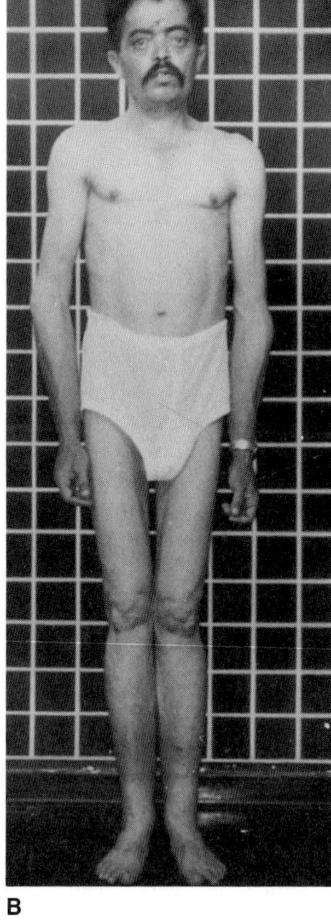

B

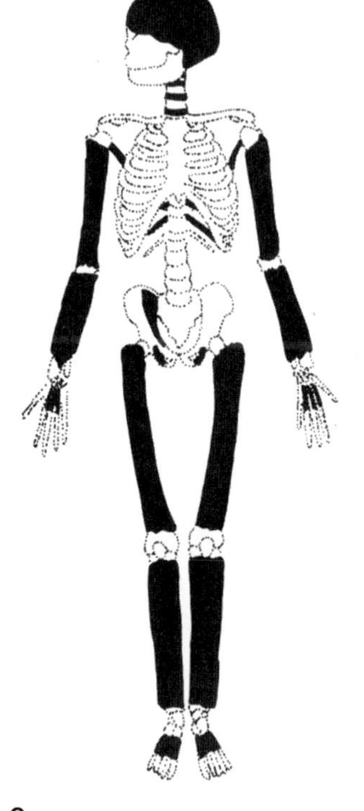

C

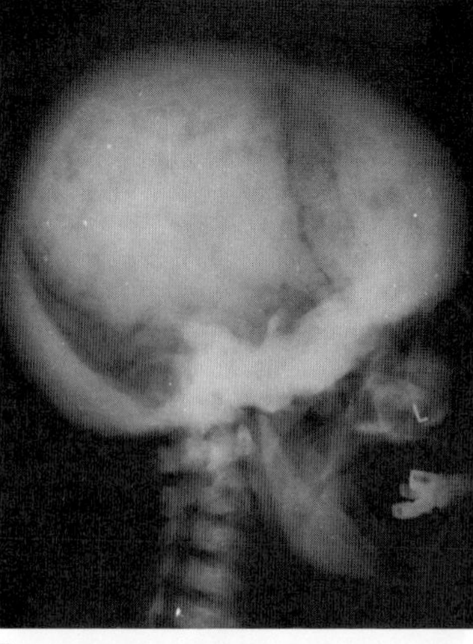

D

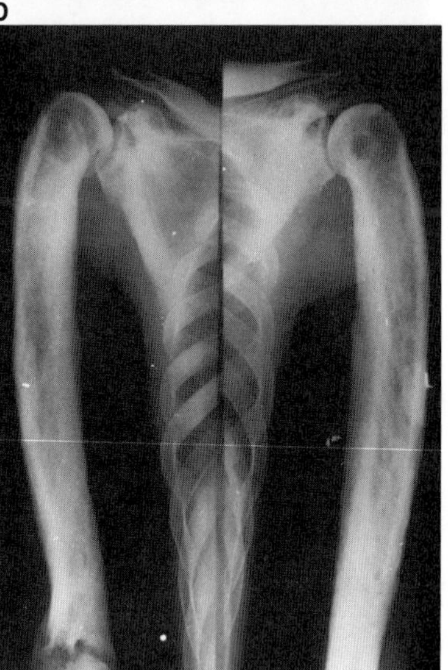

E

8. Kaitila I et al: Craniodiaphyseal dysplasia. *Birth Defects* **11**(6):359–362, 1975.

9. Kuhlencordt F et al: Diaphysäre Dysplasie (Camurati–Engelmann-Syndrom) mit fortschreitendem Visusverlust. *Dtsch Med Wschr* **106**:617–621, 1981.

10. Kumar B et al: Progressive diaphyseal dysplasia (Engelmann's disease): Scintigraphic radiographic-clinical correlations. *Radiology* **140**:87–92, 1981.

11. Morse PH et al: Ocular findings in hereditary diaphyseal dysplasia (Engelmann's disease). *Am J Ophthalmol* **68**:100–104, 1969.

12. Naveh Y et al: Progressive diaphyseal dysplasia: Genetics and clinical and radiologic manifestations. *Pediatrics* **74**:399–405, 1984.

13. Ramon Y, Buchner A: Camurati–Engelmann's disease affecting the jaws. *Oral Surg* **22**:592–599, 1966.

14. Smith R et al: Clinical and biochemical studies in Engelmann's disease (progressive diaphyseal dysplasia). *Quart J Med* **46**:273–294, 1977.

15. Sparkes RS, Graham CB: Camurati–Engelmann disease. *J Med Genet* **9**:73–85, 1972.

16. Tucker AS et al: Craniodiaphyseal dysplasia: Evolution over a five-year period. *Skel Radiol* **1**:47–55, 1976.

17. Trunk G et al: Progressive and hereditary diaphyseal dysplasia. *Arch Int Med* **123**:417–422, 1969.

18. Van Dalsem VF et al: Progressive diaphyseal dysplasia. *J Bone Jt Surg* **61A**:596–598, 1979.

19. Yoshioka H et al: Muscular changes in Engelmann's disease. *Arch Dis Childh* **55**:716–719, 1980.

## Osteopathia striata with cranial sclerosis

Approximately 30 individuals have been described with the autosomal dominantly inherited syndrome of cranial sclerosis with osteopathia striata. Oddly, most patients are females. The syndrome has been well reviewed by Winter et al (15), in 1980, and de Keyser et al (9), in 1983.

The cranium is usually biparietally enlarged. This is evident at birth but is frequently mildly progressive so that adult head circumference is often 60–65 cm. The facies appears somehat squared. Nasal obstruction may be evident in infancy (2,6,7,11). There is frontal bossing, the nasal bridge is broad, and the eyes appear wide set (11b). Hearing loss is usually mixed, quite variable in degree, and often involves the low frequencies. Some patients have facial palsy (6,9a) whereas others have been mildly mentally retarded (1,2,6–8,11,13,14). A few have scoliosis (7,12,14). Some manifest transient cardiac murmurs in childhood (13,14) (Fig. 8–13A–G).

Cleft palate or bifid uvula has been noted in several cases (1,2,4–6,8,9a,10,11a,12–14).

Cataracts and abbreviated tooth roots or unerupted teeth have been noted by RJ Gorlin and by others (2,6,14).

Radiographically, there is scoliosis and hyperostosis of the cranial vault and marked increase in density of the cranial base. The sinuses

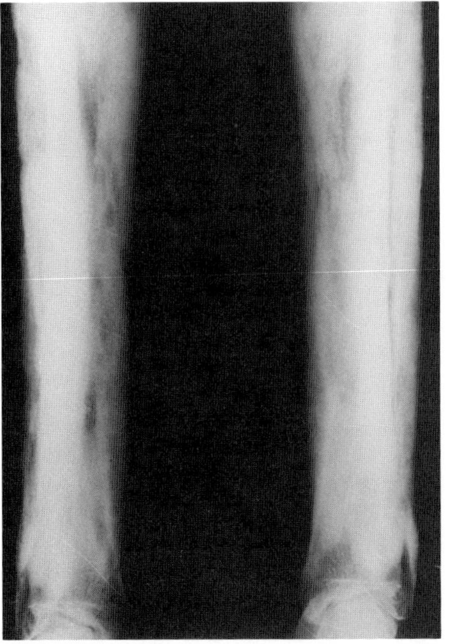
F

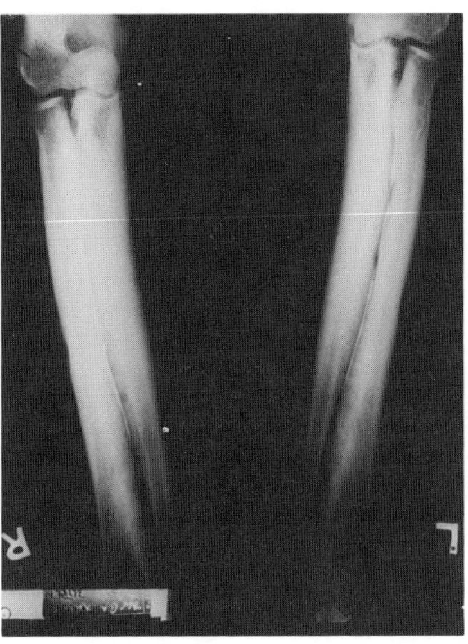

G

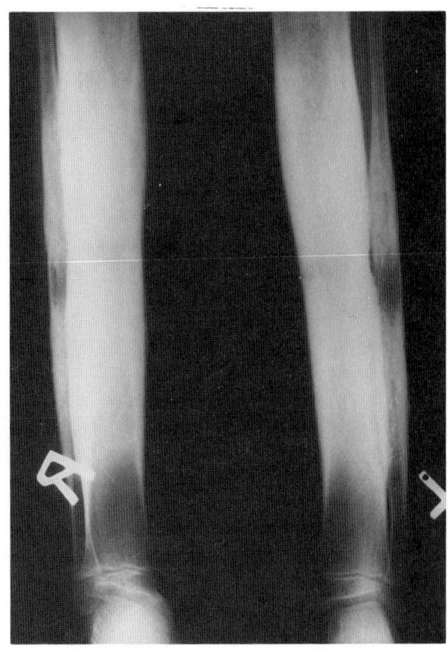

H

Fig. 8–12. *Progressive diaphyseal dysplasia (Camurati–Engelmann disease).* (A) Note reduced muscle mass of child on left with that of age-matched control. (B) Compare with A. (C) Diagrammatic sketch of skeleton indicating sclerotic and hyperostotic nature of disorder. (D) Lateral view of skull showing markedly severe sclerosis of cranial bones and cranial base. Mandible is also involved. (E) Anteroposterior roentgenogram of arms showing sclerosis, bowing, and thickening of humeri. Diaphyses are thicker than proximal epiphyses. (F) Anteroposterior roentgenogram of legs showing extreme sclerosis and thickening of tibiae and fibulae, with sparing of periarticular space. (G) Anteroposterior roentgenogram of forearms showing sclerosis and thickening of radii and ulnae. Diaphyses and proximal metaphyses are involved. Joint space is spared. (H) Anteroposterior view of both legs showing symmetrical fusiform enlargement and cortical thickening in diaphyses with narrowing of medullary cavities. (A from *H-R Wiedemann,* Z Kinderheilkd **65:**346, 1948. C adapted from *P Rubin,* Dynamic Classification of Bone Dysplasias, Yearbook Medical Publishers, Chicago, 1964. B,D,E,F,G from *Y Naveh et al,* Pediatrics **74:**399, 1984. H from *Y Naveh et al,* Pediatrics **76:**944, 1985.)

may be obscured and the mastoid air cells diminished. The anterior fontanel closes late (5). The long bones and iliac wings appear combed, hence the name *osteopathia striata*. Some have general increased bone density. Spina bifida occulta in the lumbar region is common (7,14).

### References (Osteopathia striata with cranial sclerosis)

1. Bass HN et al: Osteopathia striata syndrome: Clinical, genetic and radiologic considerations. *Clin Pediatr* **19**:369–373, 1980.

2. Bloor DV: A case of osteopathia striata. *J Bone Jt Surg* **36B**:261–265, 1954.

3. de Boer SM, van Gool AV: Schedel-en gebitsafwijkingen bij een patiënte met osteopathia striata. *Ned T Geneesk* **118**:1373–1380, 1974.

4. Clément A et al: Une affection osseuse rare, me a ne pas meconnaitre: L'osteopathie striée. *J Radiol* **63**:673–676, 1982.

5. Cortina H et al: Familial osteopathia striata with cranial condensation. *Pediatr Radiol* **11**:87–90, 1981.

6. Franklyn PP, Wilkinson D: Two cases of osteopathia striata, deafness and cranial osteopetrosis. *Ann Radiol* **21**:91–93, 1978.

7. Horan FT, Beighton PH: Osteopathia striata with cranial sclerosis: An autosomal dominant entity. *Clin Genet* **13**:201–206, 1978 (same as Ref. 2).

8. Jones MD, Mulcahy ND: Osteopathia striata, osteopetrosis and impaired hearing. *Arch Otolaryngol* **87**:116–118, 1968.

9. de Keyser J et al: Osteopathia striata with cranial sclerosis. *Clin Neurol* **85**:41–48, 1983.

9a. Kornreich L et al: Osteopathia striata, cranial sclerosis with cleft palate and facial nerve palsy. *Eur J Pediatr* **147**:101–103, 1988.

10. Nakamura T et al: Osteopathia striata with cranial sclerosis affecting three family members. *Skel Radiol* **14**:267–269, 1985.

11. Paling MR et al: Osteopathia striata with sclerosis and thickening of the skull. *Br J Radiol* **54**:344–348, 1981.

11a. Piechowiak H et al: Cranial sclerosis with striated bone disease (osteopathia striata). *Klin Pädiatr* **198**:418–424, 1986.

11b. Robinow M, Unger F: Syndrome of osteopathia striata, macrocephaly, and cranial sclerosis. *Am J Dis Child* **138**:821–823, 1984.

12. Sevaux G, Galmiche P: Sur un cas d'osteopathie striée. *Rev Rhum* **3**:248–252, 1970.

13. Taybi H, Nurock AB: Discussion of osteopathia striata. *Birth Defects* **5**(4):105–108, 1969.

14. Walker BA: Osteopathia striata with cataracts and deafness. *Birth Defects* **5**(4):295–297, 1969.

15. Winter RM et al: Osteopathia striata with cranial sclerosis: Highly variable expression within a family including cleft palate in two neonatal cases. *Clin Genet* **18**:462–474, 1980.

## Hyperphosphatasemia

Variously described under designations such as juvenile Paget's disease, osteochalasia desmalis familiaris, familial osteoectasia with macrocranium, fragile bones with macrocranium, and hereditary chronic hyperphosphatasemia, the disorder was first described by Sorrel and LeGrand-Lambling (17) in 1938 and Choremis et al (5) in 1958. It has been well reviewed by Fanconi et al (11).

The condition has autosomal recessive inheritance (3,9,10,13,19,20). About half of the patients have been of Puerto Rican origin (1, 2,10,18,20).

The syndrome is characterized by fever, bone pain, and swelling of the extremities during the first year of life (7). Later, enlargement of the calvaria and often numerous fractures and bending of the bones of the extremities occur, particularly anterior bowing of the legs and general broadening of the diaphyseal areas of tubular bones (Fig. 8–14A–C). However, healing is normal. Headache and hypertension are frequent (14,15). Cardiomegaly has been described (13). The sclerae may

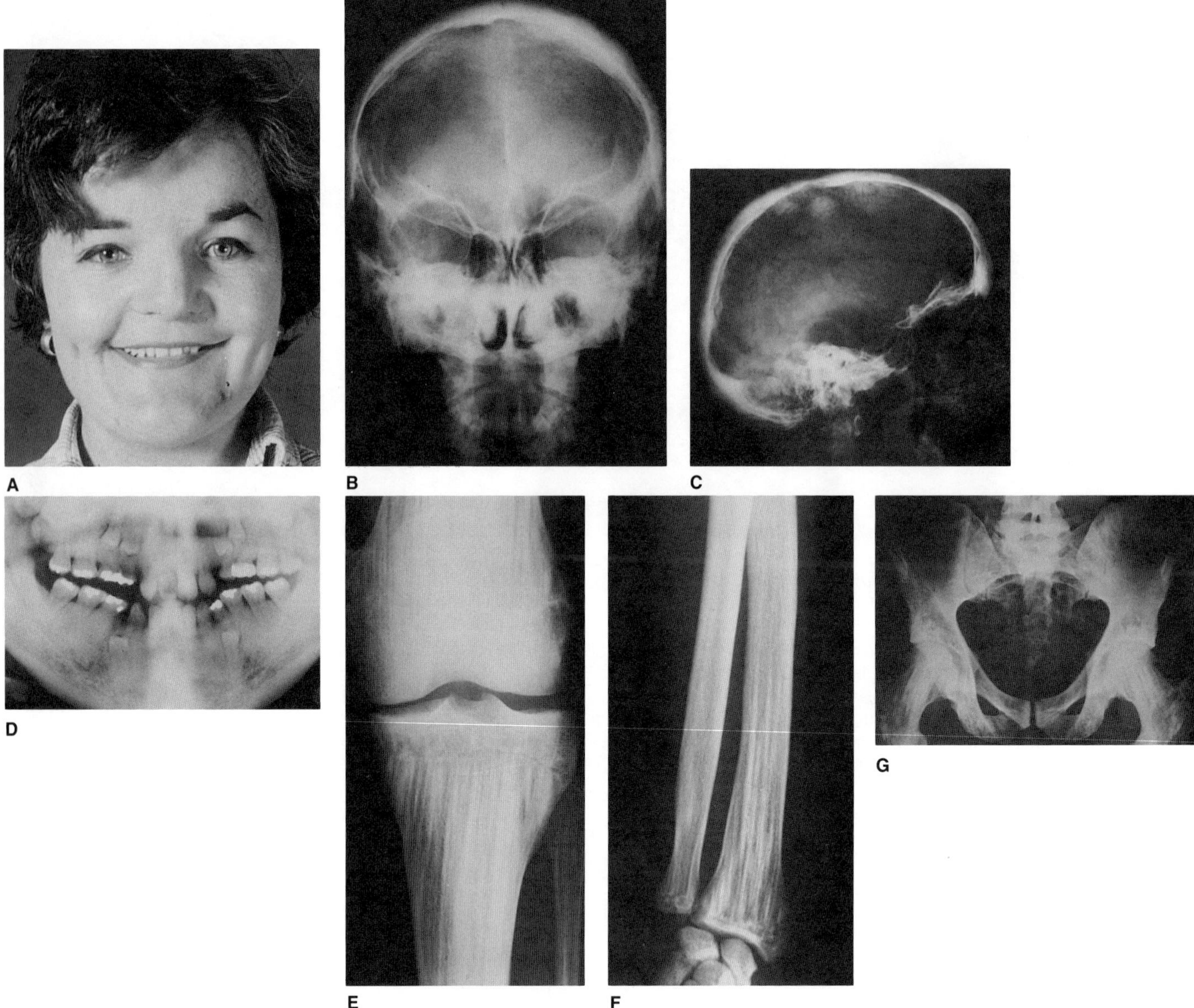


Fig. 8–13. *Osteopathia striata with cranial sclerosis.* (A) Facies appears somewhat squared with frontal bossing, broad nasal bridge, and the appearance of wideset eyes. (B,C) Markedly thickened sclerotic areas in cranial vault and base with particularly dense petrous portion of temporal bone. Note rudimentary paranasal sinuses and mastoids. (D) Abbreviated tooth roots. (E) Dense linear striations parallel to long axis are present in femur, tibia, fibula. (F) Note combed appearance of radius and ulna. (G) Characteristic striations present in proximal end of each femur and numerous areas of bone condensation are present in acetabula and ischia. (B,C,E,F,G from *PA Schnyder,* Skeletal Radiol **5**:19, 1980.)

be blue (10). Intelligence is normal. Hearing is commonly diminished and angioid streaking of the retina has been reported (10,15,20). The skin exhibits pseudoxanthoma elasticum (6a,8,12,15–16).

Histologically, there is intensive metaplastic fibrous bone formation as well as increased osteoblastic and osteoclastic activity, very similar to that seen in Paget's disease but without typical mosaic or regression lines.

Since chondral ossification is not markedly disturbed (epiphyses are normally formed and the joints are not involved), growth is not seriously dimished. Muscle weakness is frequent, which retards walking, running, and jumping.

Radiographic examination of the skull reveals changes (''cotton ball patches'') remarkably like those seen in Paget's disease. There is flattening of vertebral bodies. Long bones exhibit bending, overcylindricalization, and generalized cortical widening. The bone trabeculation is coarse and bone density diminished. Short bones are involved to a lesser degree, mostly on the endosteal side. The facial bones, except in the patient reported by Marshall (14), have not been involved. Scintigraphic changes are striking (13). Teeth are shed early due to root resorption (10).

The blood picture is generally normal, although anemia was described in Swoboda's (19) patients. Serum alkaline phosphatase (normal < 25) may exceed 500 King–Armstrong units (KAU) (15). Serum acid phosphatase is also elevated (normal 1.5–3.5 KAU) as well as urinary hydroxyproline and leucine aminopeptidase (6).

### References (Hyperphosphatasemia)

1. Bakwin H, Eiger MS: Fragile bones and macrocranium. *J Pediatr* **49**:558–564, 1956.

2. Bakwin H et al: Familial osteoectasia with macrocranium. *Am J Roentgenol* **91**:609–617, 1964.

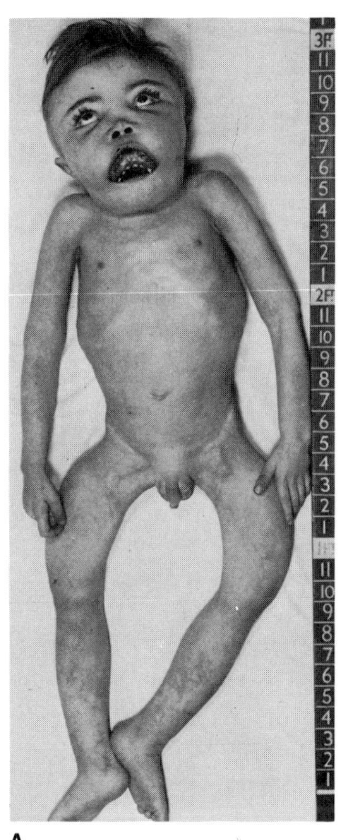
A

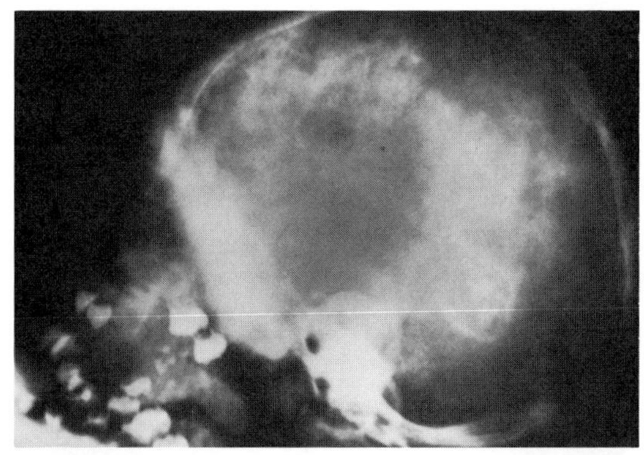
B

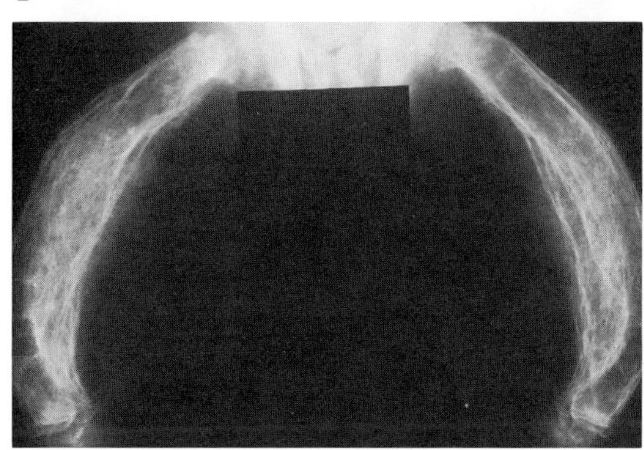
C

Fig. 8–14. *Hyperphosphatasemia.* (A) Four-year-old child exhibiting enlargement of skull, obstruction of upper air passages, bowing of lower extremities, and extensive enlargement of maxilla. (B) Note changes similar to those observed in Paget's disease of bone. (C) Marked thickening and bending of femora with diaphyseal widening resulting from periosteal new bone formation. (Courtesy of *WC Marshall,* London, England.)


3. Blanco O et al: Familial idiopathic hyperphosphatasia. *J Bone Jt Surg* **59B**:421–427, 1977.

4. Caffey J: *Caffey's Pediatric X-ray Diagnosis,* 8th ed. Year Book Medical Publishers, Chicago, 1985, p 651.

5. Choremis C et al: Osteitis deformans (Paget's disease) in an 11-year-old boy. *Helv Paediatr Acta* **13**:185–188, 1958.

6. Desai MP et al: Chronic idiopathic hyperphosphatasia in an Indian child. *Am J Dis Child* **126**:626–628, 1973.

6a. Dohler JR et al: Idiopathic hyperphosphatasia with dermal pigmentation—a 20 year follow-up. *J Bone Jt Surg* **68B**:305–310, 1986.

7. Dunn V et al: Familial hyperphosphatasemia: Diagnosis in early infancy and response to human thyrocalcitonin therapy. *AJR* **132**:541–545, 1979.

8. Eng AM, Bryant J: Clinical pathologic observations in pseudoxanthoma elasticum. *Int J Dermatol* **14**:585–605, 1975.

9. Eroglu M, Taneli NN: Congenital hyperphosphatasia (juvenile Paget's disease)—eleven years follow-up of three sisters. *Ann Radiol (Paris)* **20**:145–150, 1977.

10. Eyring EJ, Eisenberg E: Congenital hyperphosphatasia. *J Bone Jt Surg* **50A**:1099–1117, 1968.

11. Fanconi G et al: Osteochalasia desmalis familiaris. *Helv Paediatr Acta* **19**:279–295, 1964.

12. Fretzin DF: Pseudoxanthoma elasticum in hyperphosphatasia. *Arch Dermatol* **111**:271–272, 1975.

13. Iancu TC et al: Chronic familial hyperphosphatasemia. *Radiology* **129**:669–676, 1978.

14. Marshall WC: A case of progressive osteopathy with hyperphosphatasia. *Proc R Soc Med* **55**:238–239, 1962.

15. Mitsudo SM: Chronic idiopathic hyperphosphatasia associated with pseudoxanthoma elasticum. *J Bone Jt Surg* **53A**:303–314, 1971.

15a. Saraf SK, Gupta SK: Juvenile Paget's disease. *Australas Radiol* **33**:189–191, 1989.

16. Saxe N, Beighton P: Cutaneous manifestations of osteoectasia. *Clin Exp Dermatol* **7**:605–609, 1982.

17. Sorrel E, LeGrand-Lambling: Dystrophie osseuse généralisée. *Bull Soç Pédiatr Paris* **36**:89–92, 1938.

18. Stemmermann GN: A histologic and histochemical study of familial osteoectasia. *Am J Pathol* **48**:641–651, 1966.

19. Swoboda W: Hyperostosis corticalis deformans juvenilis: Ungewöhnliche generalistierte Osteopathie bei zwei Geschwistern. *Helv Paediatr Acta* **13**:292–312, 1958.

20. Thompson RC et al: Hereditary hyperphosphatasia: Studies in three siblings. *Am J Med* **47**:209–219, 1969.

21. Whalen JP et al: Calcitonin treatment in hereditary bone dysplasia with hyperphosphatasemia: A radiographic and histologic study of the bone. *Am J Roentgenol* **129**:29–35, 1977.

22. Woodhouse N et al: Paget's disease in a 5-year-old: Acute response to human calcitonin. *Br Med J* **4**:267–268, 1972.

# Chapter 9
# Syndromes Affecting Bone: Other Skeletal Disorders

## Calvarial doughnut lesions, osteoporosis, and dentigerous cysts

In 1969, Keats and Holt (4) reported multiple round calvarial radiolucencies surrounded by dense rings of sclerotic bone (Fig. 9–1A,B). Bartlett and Kishore (1) described a father and two sons with lumpy skulls and similar radiographic skull changes. Royen and Ozonoff (6) and Colavita et al (2) reported isolated males with a history of multiple long bone fractures and calvarial doughnut lesions and, in the latter case, there were cysts surrounding the roots of several teeth. Histologically, the lesions represent a fibrous nidus surrounded by sclerotic bone, essentially a fibrous dysplasia. Serum alkaline phosphatase has been elevated (1,2,6). The disorder should be differentiated from *calvarial hyperostosis* (5).

### References (Calvarial doughnut lesions, osteoporosis, and dentigerous cysts)


1. Bartlett JE, Kishore PRS: Familial "doughnut" lesions of the skull. *Radiology* **119**:385–387, 1976.
2. Colavita N et al: Calvarial doughnut lesions with osteoporosis, multiple fractures, dentinogenesis imperfecta and tumourous changes in the jaws. *Australas Radiol* **28**:226–231, 1984.
3. Ford KB et al: Case report 290. *Skel Radiol* **12**:232–233, 1984 and **13**:68–71, 1985.
4. Keats TE, Holt JF: The calvarial "donut lesion": A previously undescribed entity. *Am J Roentgenol* **105**:314–318, 1969.
5. Pagon RB: Calvarial hyperostosis: A benign X-linked recessive disorder. *Clin Genet* **29**:73–78, 1986.
6. Royen PM, Ozonoff MB: Multiple calvarial "doughnut lesions." *Am J Roentgenol* **121**:121–123, 1974.


Fig. 9–1. *Calvarial doughnut lesions, osteoporosis, and dentigerous cysts.* (A,B) Note multiple round calvarial radiolucencies surrounded by dense rings of sclerotic bone.

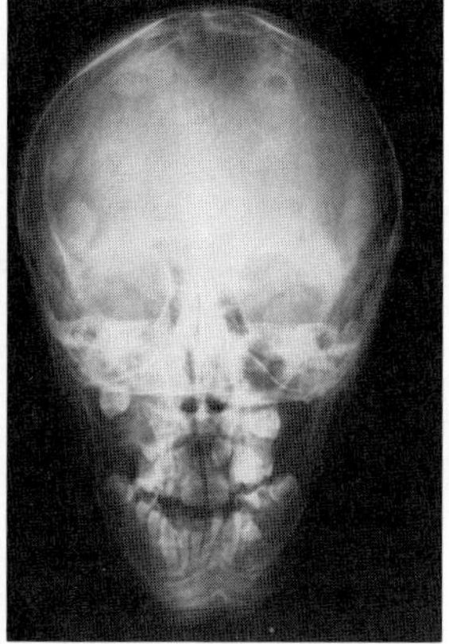

A

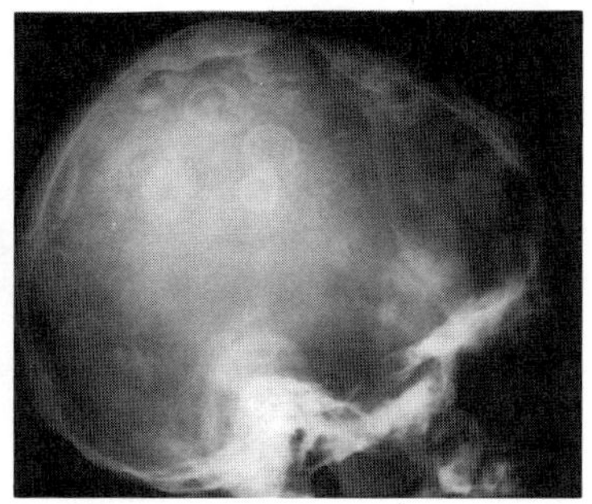

B

## Cleidocranial dysplasia

Over 700 cases of cleidocranial dysplasia have been documented since the reports of Martin (29) in 1765 and Meckel (30) in 1760. It possibly was first manifest in Neanderthal man (16). Scheuthauer (38) accurately described the syndrome in 1871. Marie and Sainton (28), in 1897, reported the combination of aplasia or hypoplasia of one or both clavicles, exaggerated development of the transverse diameter of the cranium, and delayed ossification of the fontanels. They named the syndrome "cleidocranial dysostosis."

Since then, many extensive reviews and analyses of the syndrome have been carried out and over 100 associated anomalies have been recorded (6–8, 31). A classic anatomic study is that of Hultkrantz (24), a good clinical review is that of Schuch and Fleischer-Peters (39), and an excellent historical review was carried out by Soule (43).

The syndrome has autosomal dominant inheritance (2,7,8,31,39). It has been suggested that 20–40% represent new mutations (7,39).

**Facies and general appearance.** The appearance is generally pathognomonic (Figs. 9–2 and 9–3). Affected individuals are usually short, with males averaging 156.6 cm and females 144.6 cm. The skull is brachycephalic, with pronounced frontal and parietal bossing, and the maxilla and zygomas are hypoplastic; thus, the face appears small. The nose is broad at the base, with the bridge depressed. There is hypertelorism. The neck appears long, and the shoulders are narrow and droop markedly.

Fig. 9–2. *Cleidocranial dysplasia.* Frontal and parietal bossing, glabellar groove in 13-year-old girl attempting to approximate shoulders. [From *M Føns,* Acta Otolaryngol (Stockholm) **67**:483, 1969.]

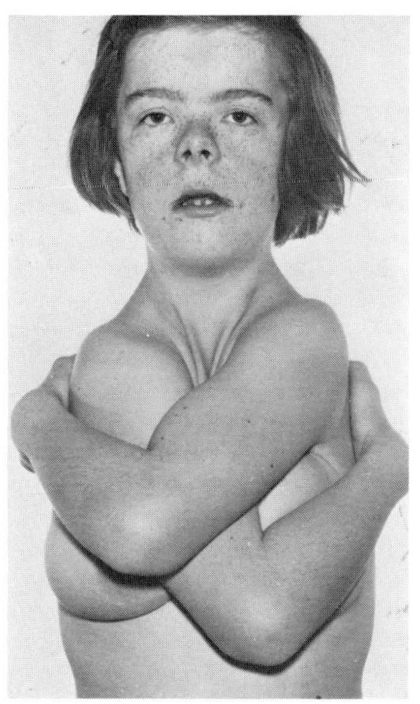

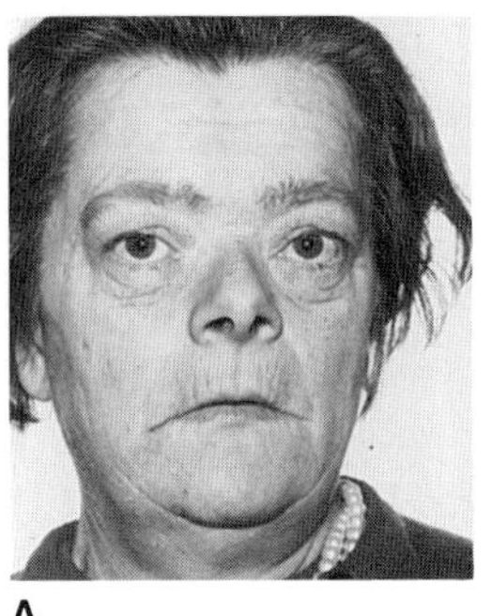
A

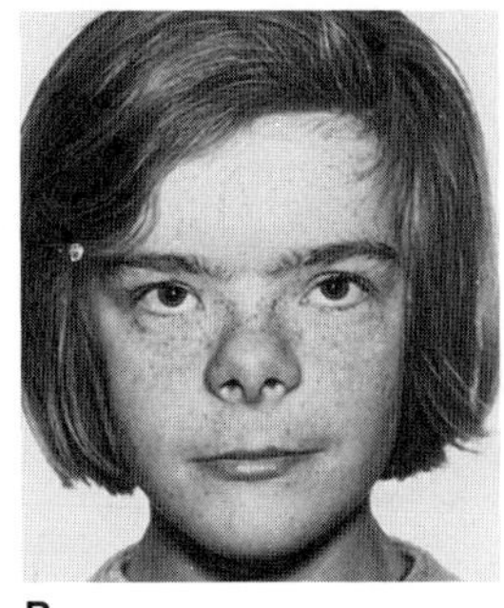
B

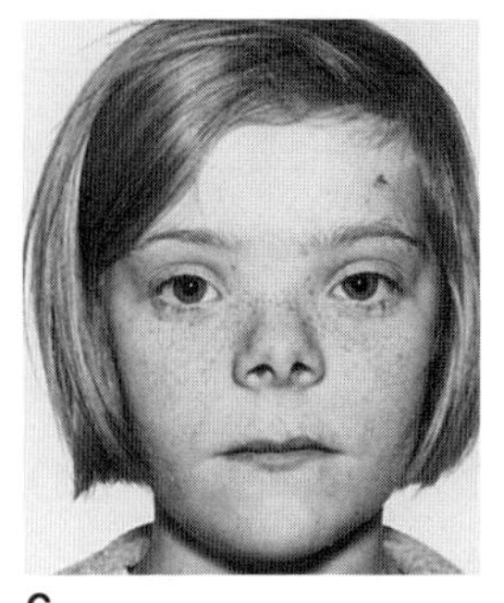
C

Fig. 9–3. *Cleidocranial dysplasia.* (A–C) Note brachycephalic skull, frontal and parietal bossing with the appearance of small face. [From *M Føns,* Acta Otolaryngol (Stockholm) **67**:483, 1969.]

Fig. 9–4. *Cleidocranial dysplasia.* (A) Numerous Wormian bones found in lambdoidal sutures, delayed cranial bone formation. (B) Wormian bones in lambdoidal sutures. (A courtesy of *F Silverman,* Cincinnati, Ohio.)

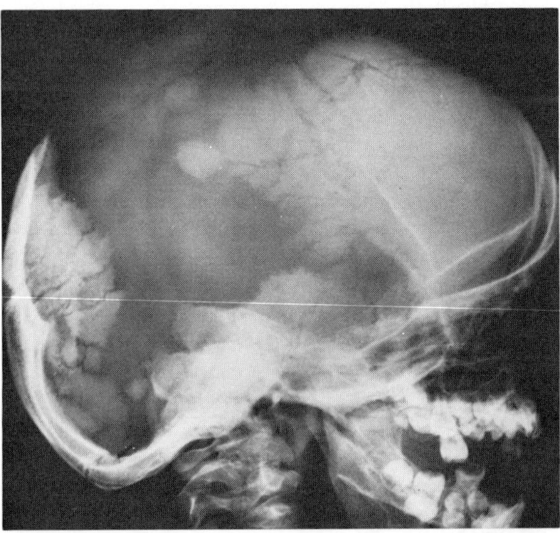
A

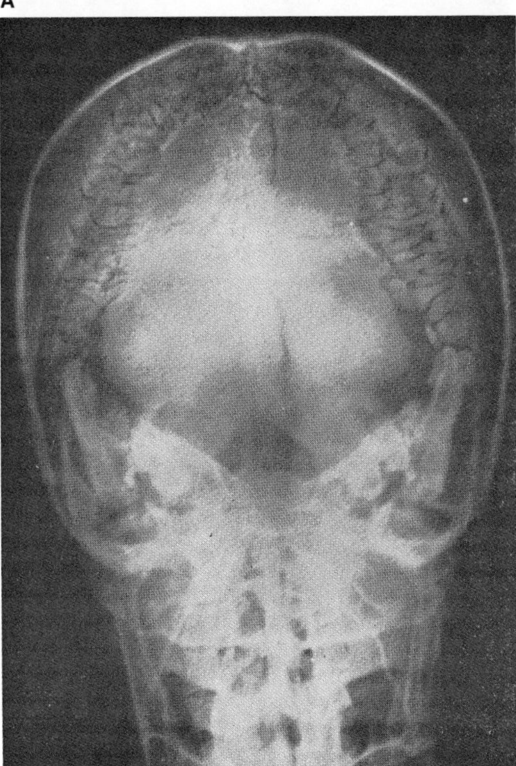
B

**Cranium.** The skull is large and short; the cephalic index is usually in excess of 80. In most patients, a groove, overlying the metopic suture, extends from the nasion to the sagittal suture. Closure of the anterior fontanel and sagittal and metopic sutures is delayed, often for life (45). There is segmental calvarial thickening in the supraorbital portion of the frontal bone, the squama of the temporal bones, and the occipital bone above the inion. Secondary centers of ossification appear in the suture lines, and many Wormian bones are formed (45,47) (Fig. 9–4A,B). In extreme cases, the parietal bones are not present at birth (44). The cranial base has short sagittal diameter. The foramen magnum, which is large, often exhibits defects in the posterior wall (43). Paranasal sinuses (2,12) and mastoids (47) are often underdeveloped or absent.

**Clavicle.** Clavicles are absent unilaterally or bilaterally in about 10% (42); more frequently, they are defective at the acromial end (25). The clavicles of some patients have a central gap (pseudoarthrosis), with bone replacement by fibrous connective tissue (8) (Fig. 9–5A–C). When the defect is unilateral, it is more frequently on the right side (25a).

Deficiency of the clavicle is responsible for the long appearance of the neck and the narrow shoulders. The range of shoulder movements permitted by this bony defect is often remarkable, frequently allowing the individual to approximate the shoulders in front of the chest. This ability is not always recognized by the patient, nor are the parents of an affected child necessarily aware of it (Fig. 9–2).

In this syndrome, there are variations in size, origin, and insertion of muscles related to the clavicles, especially the sternocleidomastoid, trapezius, deltoid, and pectoralis major, yet function is remarkably good (7).

**Other skeletal deformities.** Although cleidocranial dysplasia was originally believed to involve only bones of membraneous origin, involvement of endochondral bones has been recognized since Fitchet's (7) report. The most frequent abnormalities include delayed closure of the pubic symphysis, coxa vara, or (less often) coxa valga with lateral notching of the capital femoral epiphysis (25a), cone-shaped thorax, spina bifida occulta of the cervical, thoracic, or lumbar regions of the spine, lumbar spondylolysis, pseudoepiphyses at the base of one or more metacarpals, abnormally pointed terminal phalangeal tufts of the hands and feet, and cone-shaped epiphyses of the distal phalanges. The reduced pelvic diameter requires caesarean section in about 35% of affected women (Fig. 9–5D,E) (25a).

**Other findings.** Conduction hearing impairment (10,11,20) has been described.

**Oral manifestations.** The palate is highly arched. Submucous cleft palate and complete cleft of the hard and soft palates have been described (47). Delayed union at the mandibular symphysis is characteristic. Development of the premaxilla is poor, and since growth of the mandible is usually normal, relative prognathism results (42).

The literature presents a rather chaotic picture of the dentition including multiple supernumerary teeth (Fig. 9–6A), multiple crown and root abnormalities, crypt formation around impacted teeth, ectopic lo-

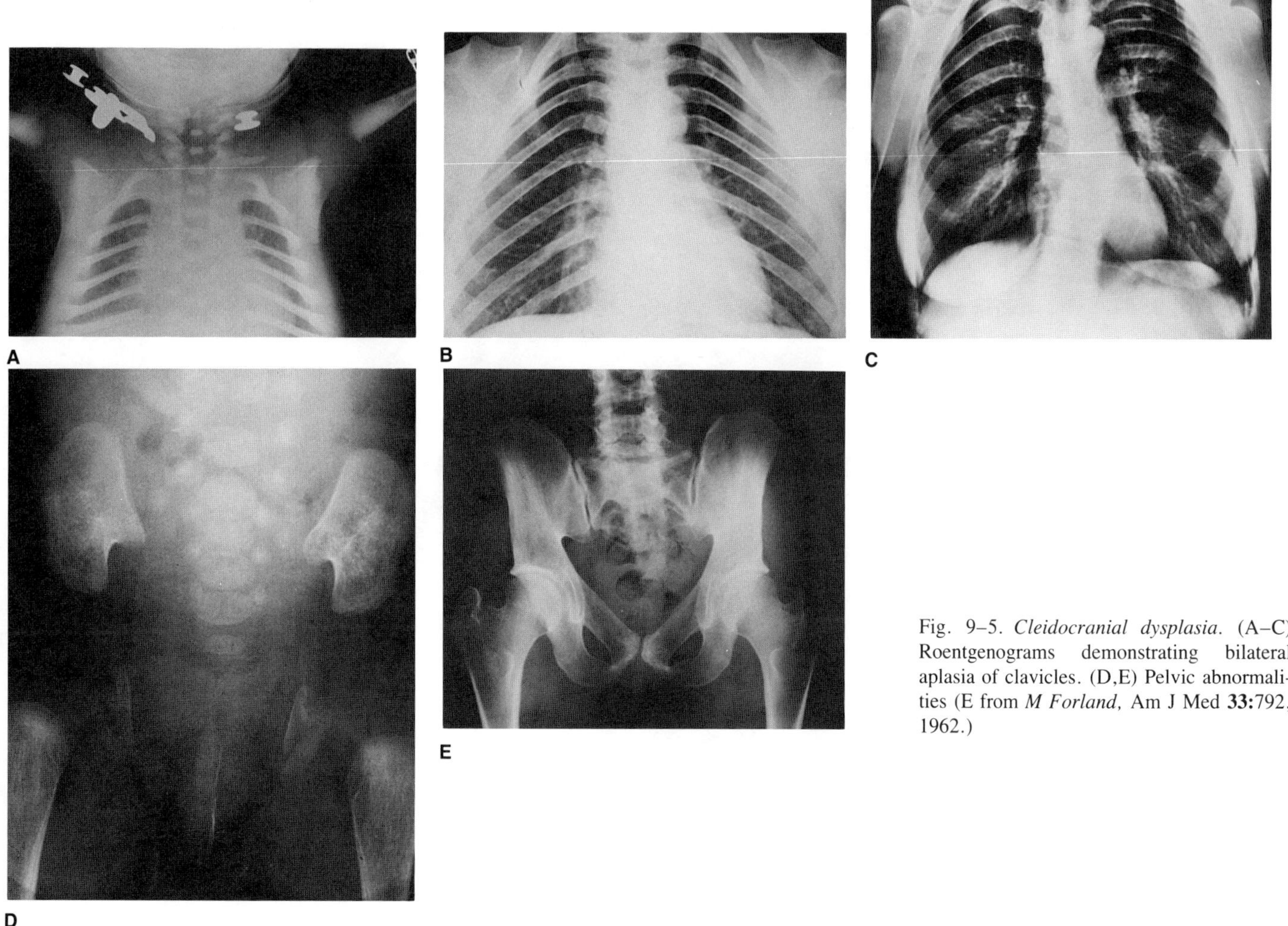


Fig. 9–5. *Cleidocranial dysplasia.* (A–C) Roentgenograms demonstrating bilateral aplasia of clavicles. (D,E) Pelvic abnormalities (E from *M Forland,* Am J Med **33:**792, 1962.)

calization of teeth, and lack of tooth eruption (12,19,21,32,34–36,45,47). It is known that extraction of deciduous teeth does not promote eruption of permanent teeth (47). Rushton (36,37) and others (1,17,22,42,48) studied teeth microscopically and observed that roots lacked a layer of cellular cementum (Fig. 9–6B,C). Fleischer–Peters (9) suggested that the greater than normal bone density of the jaws might inhibit tooth eruption. Rushton (36) and Hitchin and Fairley (23) attributed noneruption of teeth to failure of bone to resorb.

Recently, Jensen and Kreiborg (26) carried out a remarkable study of 20 patients (most followed longitudinally), utilizing cephalograms, panorex and intraoral x-rays, intraoral photographs, and surgically removed teeth. They found that the dental lamina produced normal primary teeth and permanent tooth crowns. The primary teeth erupted, but not the permanent teeth, except for first molars and occasionally other teeth. Following normal development of the permanent crowns, the dental lamina was reactivated to form supernumerary teeth. These occurred with highest frequency in the maxillary central incisor and canine regions and areas. The bizarre supernumerary crown and root morphology appeared to be related to spacial crowding. Deciduous root resorption is extremely delayed or arrested, and can probably be explained by diminished bone resorption. Abnormalities of root morphology in the permanent dentition appear secondary to arrested eruption.

**Differential diagnosis.** Although the facies as well as body habitus are characteristic, brachycephaly and frontal bossing may suggest rickets, prenatal syphilis, *achondroplasia,* hydrocephaly, *osteogenesis imperfecta,* and *pycnodysostosis.* A deficient premaxilla may also be seen in *Apert syndrome* and in *Crouzon syndrome.* Depressed nasal bridge is seen in *hypohidrotic ectodermal dysplasia, Stickler syndrome,* and prenatal syphilis. The appearance of the shoulders is similar to that seen in intrauterine or natal fracture. True widening of the pubic symphysis is seen most often with extrophy of the bladder and epispadias (33). *Mandibuloacral dysplasia* is characterized by mild cranial and clavicular dysplasia, but lacks the dental abnormalities seen in cleidocranial dysplasia. In *Hajdu–Cheney syndrome (acroosteolysis)* clavicular demineralization, cranial dysplasia, and severe acroosteolysis occur, but the dental abnormalities found in cleidocranial dysplasia are absent.

Goodman et al (15) reported a disorder resembling cleidocranial dysplasia in three male offspring of two consanguineous matings. These patients had short stature, brachycephaly, open fontanels, bilateral absence of clavicles, Wormian bones, nail dysplasia, and radiologic abnormalities of the spine and pelvis. Inheritance was compatible with an autosomal or X-linked recessive disorder.

The *Yunis-Varón syndrome* is characterized by absent clavicles, macrocrania, diastasis of sutures, micrognathia, absent thumbs and distal phalanges of fingers, hypoplasia of the proximal phalanx and absence of the distal phalanx of great toes, pelvic dysplasia, bilateral hip dislocation, and retracted and poorly delineated lips. Also, patients wth enlarged parietal foramina, macrocephaly, occipital hair tuft, and lateral clavicular aplasia have been described by several investi-

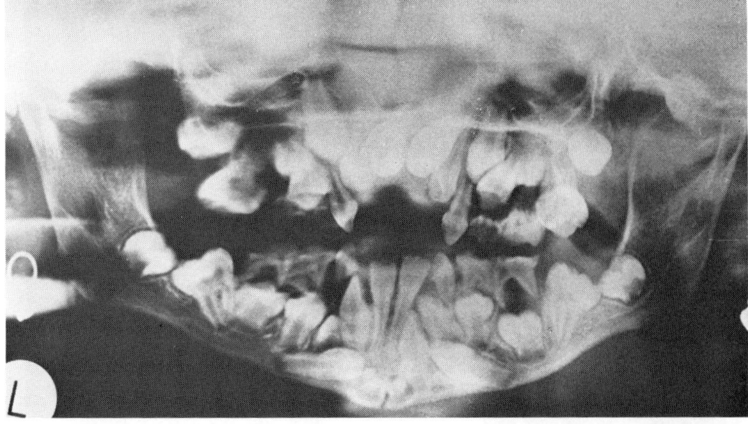


A

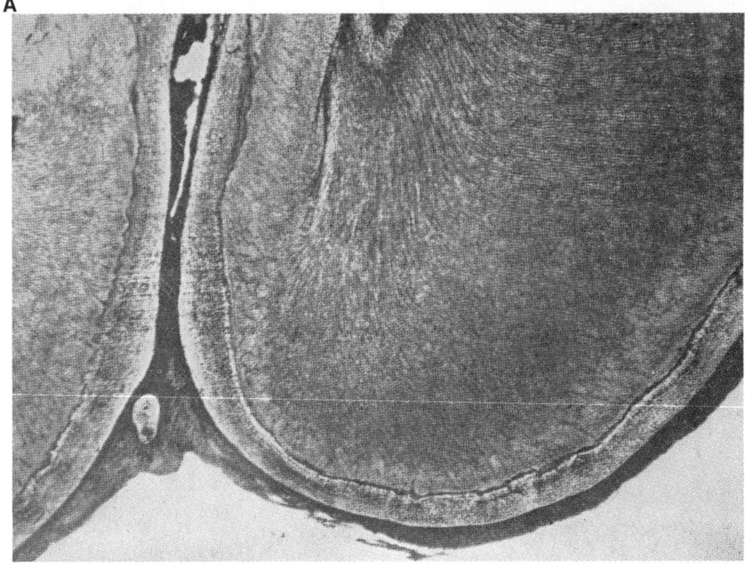

B

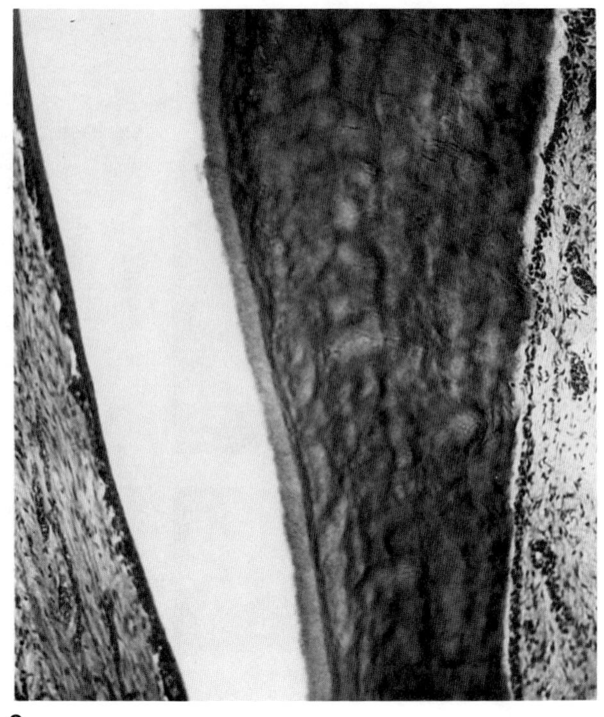

C

Fig. 9–6. *Cleidocranial dysplasia.* (A) Panorex showing multiple supernumerary, unerupted teeth. (B) Tooth roots, exhibiting complete absence of cellular cementum. (C) Absent cementum. [A from *A Fleischer-Peters,* Stoma (Heidelb) **23:**212, 1970. B from *M Rushton,* Br Dent J **100:**81, 1956.]

gators (4,14,18); this condition may have autosomal dominant inheritance.

Postnatal clavicular hypoplasia or dysgenesis occurs in *progeria,* disorders of *acroosteolysis,* and possibly *pycnodysostosis.* Congenital clavicular dysplasia may occur in many disorders reviewed by Hall (18), including *focal dermal hypoplasia, dup(11q) syndrome, Floating-Harbor syndrome,* and *Antley–Bixler syndrome.* Isolated pseudoarthrosis of the right clavicle as an isolated phenomenon is largely limited to females. Crane and Heise (4) reported a lethal, autosomal recessive condition characterized by poorly mineralized calvaria, cleft lip and palate, micrognathia, upturned nares, ocular hypertelorism, depressed nasal bridge, and hypoplastic, posteriorly angulated and low set helices. In addition, absent cervical vertebrae and clavicles, talipes equinovarus, soft tissue syndactyly of fingers and toes, short penis, and undescended testes were noted.

Wallis et al (45a) reported autosomal dominant inheritance of lateral clavicular defects and rhizomelic short stature (cleidorhizomelic syndrome).

Absence of clavicles, anal atresia, and psoriasis-like lesions were reported in sibs by Fukuda et al (13). The affected boy also had hypospadias and urethrorectal fistula. His sister had rectovaginal fistula. The boy's teeth were described as cone shaped with enamel hypoplasia. The parents were normal.

A peculiar variant, in which there were dolichocephaly, severe lordosis, generalized joint hypermobility, and dystrophic toenails, was described by Winkler (46). RJ Gorlin has seen another patient with the disorder. Patients with cleidofacial dysplasia (27) lack clavicles bilaterally, are microbrachycephalic, have exophthalmos with hypoplasia of lids, and are mentally retarded; inheritance is autosomal recessive.

Silverman and Reiley (41) described spondylomegepiphyseal metaphyseal dysplasia, an autosomal recessive condition similar to cleidocranial dysplasia, but without its cranial and clavicular features.

Delayed ossification of the pubic ramus can also occur in a number of conditions: prematurity, *del(4p) syndrome,* Sjögren–Larsson syndrome, *campomelic dysplasia,* and *spondyloepiphyseal dysplasia congenita* (3).

RJ Gorlin and MM Cohen Jr have seen a girl with congenital absence of both clavicles and sternum and contractures of the fingers.

Failure of eruption of most permanent teeth has been reported as an autosomal dominant trait (40). It also occurs in *GAPO syndrome.*

Wormian bones are seen in a number of disorders, among them *pycnodysostosis, osteogenesis imperfecta,* hypothyroidism, and *Hajdu–Cheney syndrome.*

## References (Cleidocranial dysplasia)


1. Chapman LA, Main JHP: Cementum in cleidocranial dysplasia. *J Can Dent Assoc* **42**:139–142, 1976.
2. Chemin JP et al: Une nouvelle observation de dysostose cleidocranienne. *Rev Stomatol (Paris)* **61**:900–905, 1960.
3. Cortina H et al: The non-ossified pubis. *Pediatr Radiol* **8**:87–92, 1979.
4. Crane JP, Heise RL: New syndrome in three affected siblings. *Pediatrics* **68**:235–237, 1981.

5. Eckstein HB, Hoare RD: Congenital parietal "foramina" associated with faulty ossification of the clavicles. *Br J Radiol* **36**:220–221, 1963.
6. Eldridge WW et al: Cleidocranial dysostosis. *Am J Roentgenol* **34**:41–49, 1935.
7. Fitchet SM: Cleidocranial dysostosis: Hereditary and familial. *J Bone Jt Surg* **11**:838–866, 1929.
8. Fitzwilliams DCL: Hereditary cranio-cleido-dysostosis; with a review of all the published cases of this disease: Theories of the development of the clavicle suggested by this condition. *Lancet* **2**:1466–1475, 1910.
9. Fleischer-Peters A: Zur Pathohistologie des Alveolarknochens bei Dysostosis cleidocranialis. *Stoma (Heidelb)* **23**:212–215, 1970.
10. Føns M: Ear malformations in cleidocranial dysostosis. *Acta Oto-laryngol (Stockholm)* **67**:483–489, 1969.
11. Forland M: Cleidocranial dysostosis. *Am J Med* **33**:792–799, 1962.
12. Frøhlich E: Die Erblichkeit der Dysostosis cleidocranialis. *Dtsch Zahn Mund Kieferheilkd* **41**:157–168, 1937.
13. Fukuda K et al: Two siblings with cleidocranial dysplasia associated with atresia ani and psoriasis-like lesions: A new syndrome? *Eur J Pediatr* **136**:109–111, 1981.
14. Golabi M et al: Parietal foramina-clavicular hypoplasia: An autosomal dominant syndrome. *Am J Dis Child* **138**:596–599, 1984.
15. Goodman RM et al: Evidence for an autosomal recessive form of cleidocranial dysostosis. *Clin Genet* **8**:20–29, 1975.
16. Greig DM: A Neanderthaloid skull presenting features of cleidocranial dysostosis and other peculiarities. *Edinburgh Med J* **40**:497–557, 1933.
17. Gundlach KKH, Breurman R: Dysplasia cleidocranialis-Histologische Befunde am Zahnzement. *Dtsch Zahnärtzl Z* **33**:574–578, 1978.
18. Hall BD: Syndromes and situations associated with congenital clavicular hypoplasia or agenesis. *Prog Clin Biol Res* **104**:279–288, 1982.
19. Hall RK, Hyland AL: Combined surgical and orthodontic management of the oral abnormalities in children with cleidocranial dysplasia. *Int J Oral Surg* **7**:267–273, 1978.
20. Hawkins HB et al: The association of cleidocranial dysostosis witth hearing loss. *Am J Roentgenol* **125**:944–947, 1975.
21. Hesse G: Dysostosis cleidocranialis unter besonderer Berücksichtigung des Gebisses. *Vjschr Zahnheilkd* **41**:162–177, 1925.
22. Hitchin AD: Cementum and other root abnormalities of permanent teeth in cleidocranial dysostosis. *Br Dent J* **139**:313–318, 1975.
23. Hitchin AD, Fairley JM: Dental management in cleidocranial dysostosis. *Br J Oral Surg* **12**:46–55, 1974.
24. Hultkrantz JW: Über Dysostosis cleidocranialis. *Z Morphol Anthropol* **11**:385–528, 1908.
25. Jackson WPU: Osteo-dental dysplasia (cleido-cranial dysostosis). *Acta Med Scand* **139**:292–303, 1951.
25a. Jarvis JL, Keats TE: Cleidocranial dysostosis: A review of 40 new cases. *Am J Roentgenol* **121**:5–16, 1974.
26. Jensen BL, Kreiborg S: Development of the dentition in cleidocranial dysplasia. *J Oral Pathol* (in press).
27. Kozlowski K et al: Dysplasia cleido-facialis. *Z Kinderheilkd* **108**:331–338, 1970.
28. Marie P, Sainton P: Observation d'hydrocéphalie héréditaire (père et fils) par vice de développment du crâne et du cerveau. *Bull Soc Méd Hôp Paris* **14**:706–712, 1897.
29. Martin: Sur un déplacement natural de la clavicule. *J Méd Chir Pharmacol* **23**:456–460, 1765.
30. Meckel (1760), cited by Siggers CD: Cleidocranial dysostosis. *Dev Med Child Neurol* **17**:522–524, 1975.
31. Miles PW: Cleidocranial dysostosis: A survey of six new cases and 126 from the literature. *J Kansas Med Soc* **41**:462–468, 1940.
32. Miller R et al: Cleidocranial dysostosis: A multidisciplinary approach to treatment. *J Am Dent Assoc* **96**:296–300, 1978.
33. Muecke EC, Currarino G: Congenital widening of the pubic symphysis. *Am J Roentgenol* **103**:178–185, 1968.
34. Oatis GW Jr et al: Cleidocranial dysostosis with mandibular cyst. Report of a case. *Oral Surg* **40**:62–67, 1975.
35. Rushton MA: The dental condition in cleidocranial dysostosis. *Guy's Hosp Rep* **87**:354–361, 1937.
36. Rushton MA: The failure of eruption in cleidocranial dysostosis. *Br Dent J* **63**:641–645, 1937.
37. Rushton MA: An anomaly of cementum in cleidocranial dysostosis. *Br Dent J* **100**:81–83, 1956.
38. Scheuthauer G: Kombination rudimentärer Schlüsselbeine mit Anomalien des Schädels beim erwachsenen Menschen. *Allg Wien Med Ztg* **16**:293–295, 1871.
39. Schuch P, Fleischer-Peters A: Zur Klinik der Dysostosis cleidocranialis. *Z Kinderheilkd* **98**:107–132, 1967.
40. Shokeir MHK: Complete failure of eruption of all permanent teeth: An autosomal dominant disorder. *Clin Genet* **5**:322–326, 1974.
41. Silverman FN, Reiley MA: Spondylo-megaepiphyseal metaphyseal dysplasia: A new bone dysplasia resembling cleidocranial dysplasia. *Radiology* **156**:365–371, 1986.
42. Smith NH: A histologic study of cementum in a case of cleidocranial dysostosis. *Oral Surg* **25**:470–478, 1968.
43. Soule AB Jr: Mutational dysostosis. *J Bone Jt Surg* **28**:81–102, 1946.
44. Tan KL, Tan LKA: Cleidocranial dysostosis in infancy. *Pediatr Radiol* **11**:114–115, 1981.
45. Todorov AB et al: Cleidocranial dysplasia. *Birth Defects* **10**(4):355–359, 1974.
45a. Wallis C et al: Newly recognized autosomal dominant syndrome of rhizomelic shortness with clavicular defect. *Am J Med Genet* **31**:881–886, 1988.
46. Winkler H: Ein eigenartiger Fall von Dysostosis cleidocranialis bei einem siebenjährigen Kinde. *Jb Kinderheilkd* **149**:238–260, 1937.
47. Winter GR: Dental conditions in cleidocranial dysostosis. *Am J Orthodont* **29**:61–89, 1943.
48. Yamamoto H et al: Cleidocranial dysplasia: A light microscope, electron microscope, and crystallographic study. *Oral Surg Oral Med Oral Pathol* **68**:195–200, 1989.

## Dysplastic clavicles, sparse hair, and digital anomalies (Yunis–Varón syndrome)

In 1980, Yunis and Varón (4) described five children with prenatal and postnatal deficiency, hypoplastic clavicles, absence of thumbs and first metatarsals, and distal aphalangia. An additional patient was reported by Hennekam and Vermeulen-Meiners (1). Another possible far more mildly affected child was described by Hughes and Parting-

Fig. 9–7. *Dysplastic clavicles, sparse hair, and digital anomalies (Yunis–Varón syndrome).* (A–D) Dolichocephaly with sparse hair, eyebrows, and lashes, outstanding pinnae with wide concha, thin lips, and micrognathia. (A,B from *E Yunis* and *H Varón,* Am J Dis Child **134:**649, 1980. C,D from *HE Hughes* and *MW Partington,* Am J Med Genet **14:**539, 1983.)

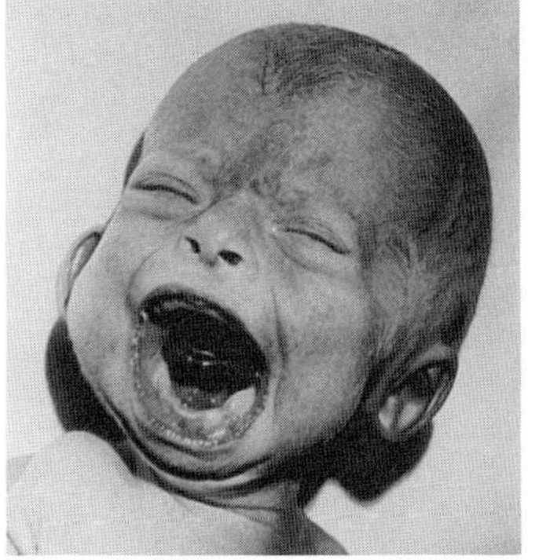
A

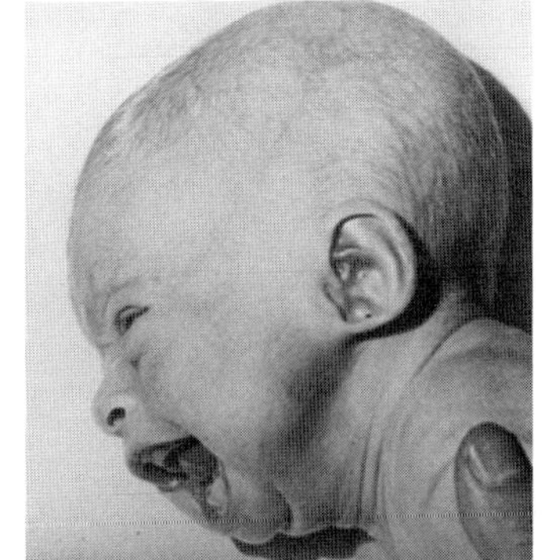
B

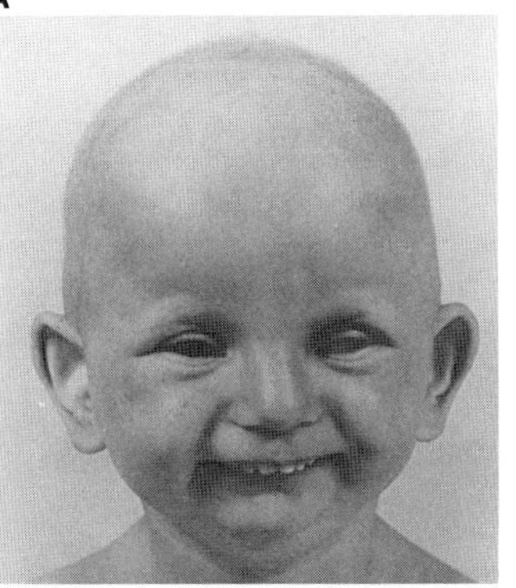
C

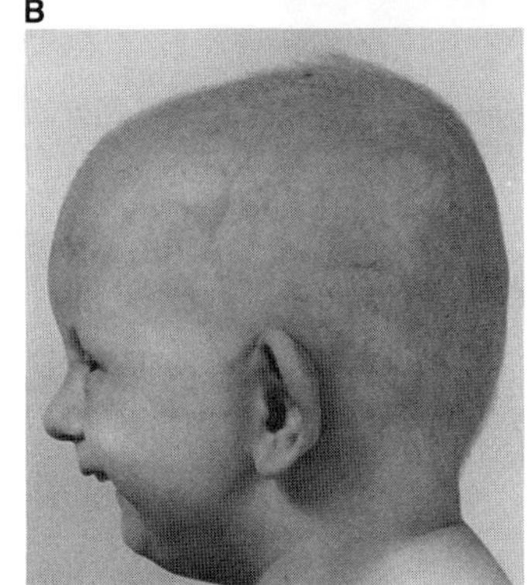
D

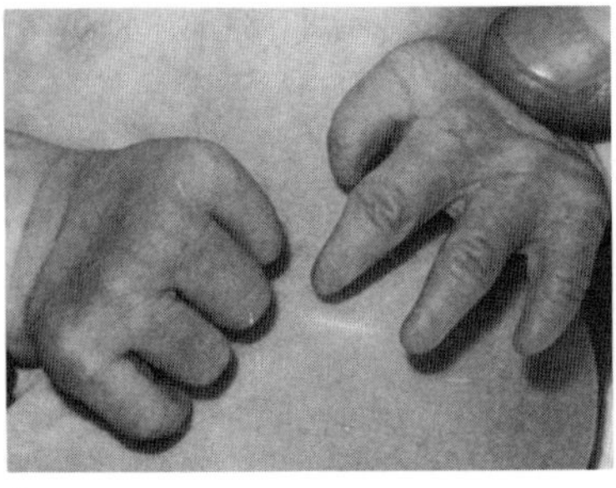

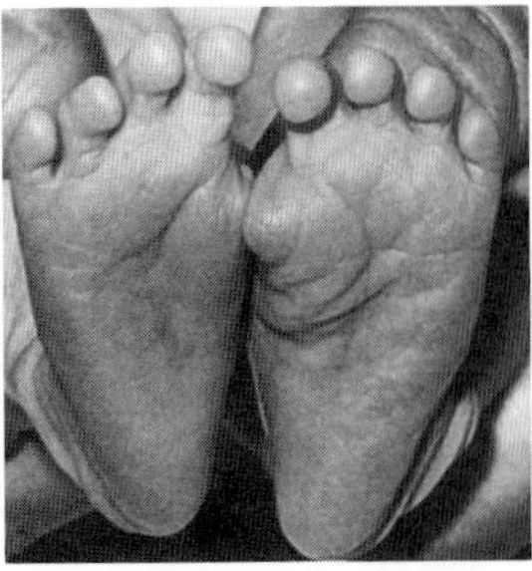

A B

Fig. 9–8. *Dysplastic clavicles, sparse hair, and digital anomalies (Yunis–Varón syndrome).* (A,B) Absent terminal phalanges, agenesis of thumbs and halluces. (From *E Yunis* and *H Varón,* Am J Dis Child **134**:649, 1980.)

ton (2). The disorder is markedly lethal in the neonatal period (Figs. 9–7 to 9–10B).

Parental consanguinity has been demonstrated (1,4). Inheritance is probably autosomal recessive.

The facies appeared distinctive: microdolichocephaly, sparse hair, eyebrows, and lashes, small nose with anteverted nostrils, outstanding pinnae with wide concha, absent tragus, antitragus, and earlobe, short philtrum, thin lips, narrow palate, and micrognathia (Fig. 9–7). Some exhibited small eyes and proptosis. There is variable soft tissue syndactyly of fingers and toes, absence of thumbs, abbreviated halluces, and hypoplasia or agenesis of nails and terminal phalanges (Fig. 9–8). Premature loss of deciduous teeth has been noted (3).

Radiographic changes include agenesis or hypoplasia of one or both clavicles, absent sternal ossification, hip dislocation, agenesis or hypoplasia of thumbs, hypoplasia of first metatarsals, agenesis of middle phalanges of second fingers, hypoplasia or agenesis of phalanges of halluces, and variable distal aphalangia of the second to fifth fingers or toes (Fig. 9–7 to 9–10B). Craniofacial changes involve unusual head form, skull dysostosis, wide fontanels and separated sutures, hypoplastic facial bones, micrognathia, and cystic dental follicles. Cardiomyopathy has been reported (3).

Fig. 9–9. *Dysplastic clavicles, sparse hair, and digital anomalies (Yunis–Varón syndrome).* Agenesis of right clavicle. (From *E Yunis* and *H Varón,* Am J Dis Child **134**:649, 1980.)

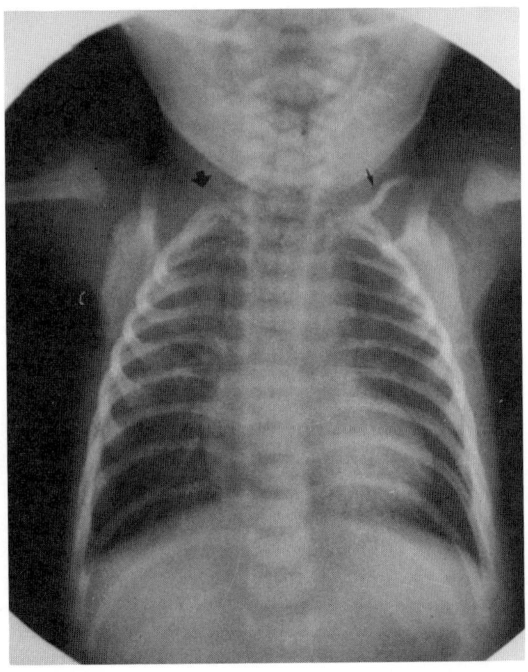

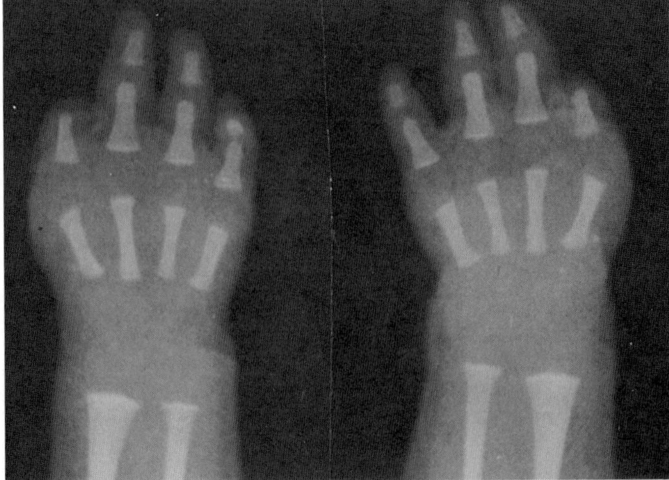

A

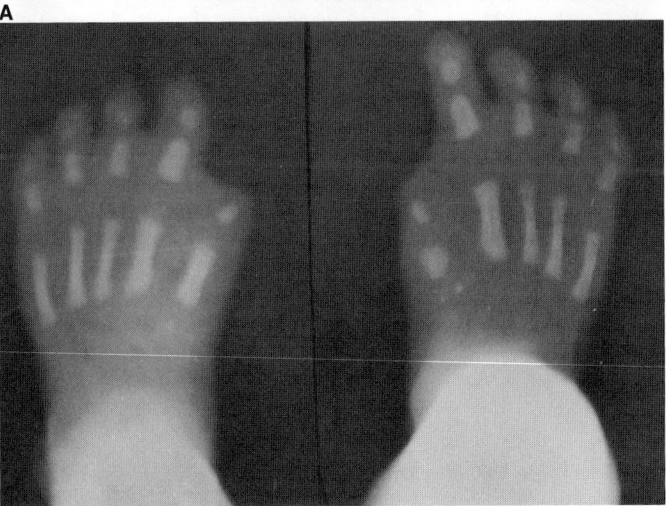

B

Fig. 9–10. *Dysplastic clavicles, sparse hair, and digital anomalies (Yunis–Varón syndrome).* (A) Absence of entire first ray and distal phalanges, bilateral absence of middle phalanges in second finger. (B) Hypoplasia of phalanges in halluces and of first metatarsals. (From *E Yunis* and *H Varón,* Am J Dis Child **134**:649, 1980.)

#### References [Dysplastic clavicles, sparse hair, and digital anomalies (Yunis–Varón syndrome)]

1. Hennekam RCM, Vermeulen-Meiners C: Further delineation of the Yunis–Varón syndrome. *J Med Genet* **26**:55–58, 1989.
2. Hughes HE, Partington MW: The syndrome of Yunis and Varón: Report of a further case. *Am J Med Genet* **14**:539–544, 1983.
3. Partington MW: Cardiomyopathy added to the Yunis–Varón syndrome. *Proc Greenwood Genet Ctr* **7**:224–225, 1988.
4. Yunis E, Varón H: Cleidocranial dysostosis, severe micrognathia, bilateral absence of thumbs and first metatarsal bone, and distal aphalangia. *Am J Dis Child* **134**:649–653, 1980.

## Fibrodysplasia ossificans progressiva

Fibrodysplasia ossificans progressiva (FOP) is a rare autosomal dominant condition with progressive ectopic ossification and characteristic skeletal malformations. Features include malformations of the hallux, reduction defects of all digits, hearing loss, and baldness. Progressive disability due to ectopic calcification is erratic, but severe restriction of movement eventually occurs and is especially evident in the shoulders and spine by the age of 10 years (Fig. 9–11A–G) (9,20). The condition was first recorded by Guy Patin in 1692 (22). The term

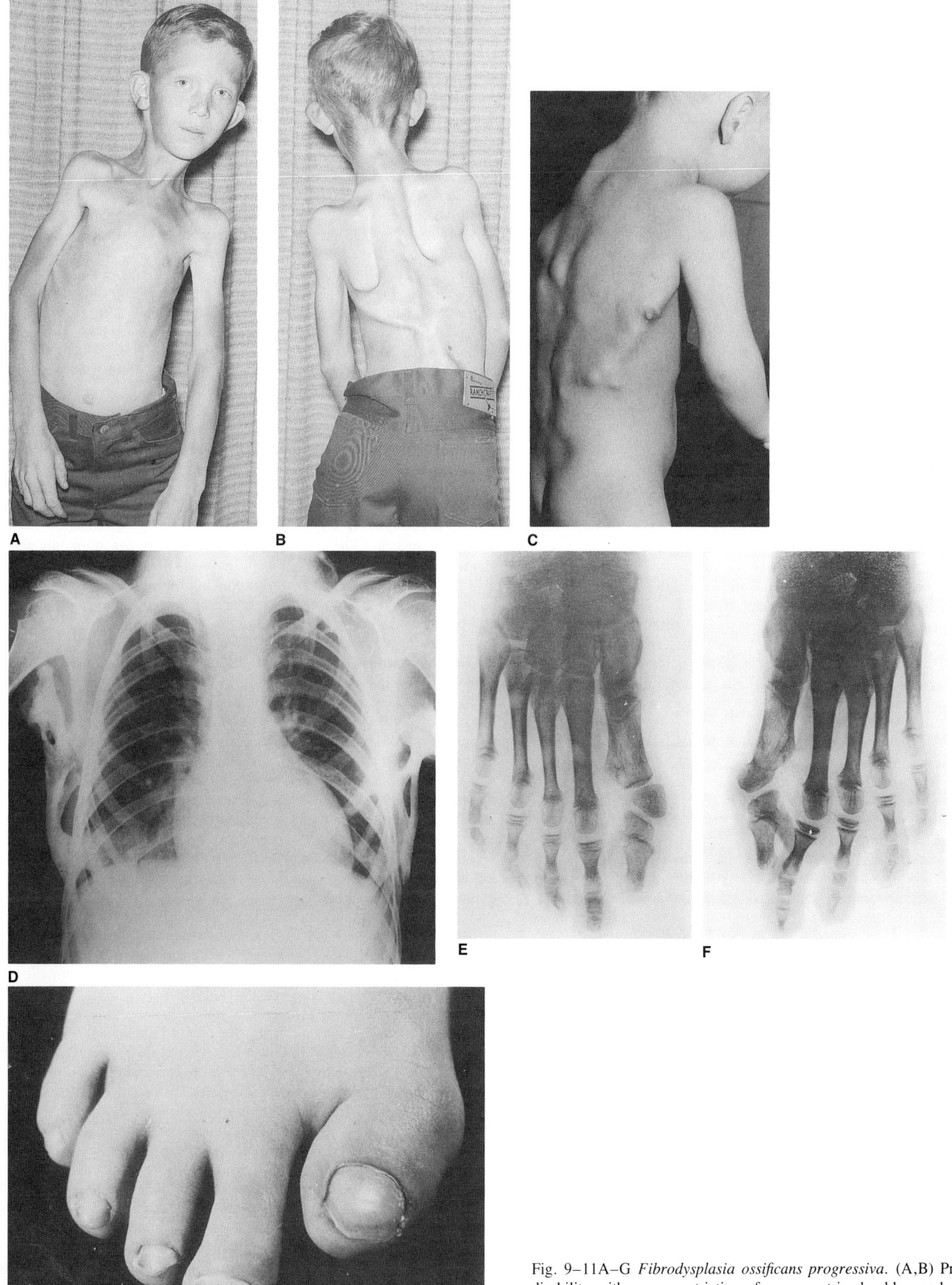


Fig. 9–11A–G *Fibrodysplasia ossificans progressiva.* (A,B) Progressive disability with severe restriction of movement in shoulders and spine. (C) Lumpy areas of ossification. (D) Note calcifications. (E–G) Hallucal abnormalities.

myositis ossificans progressiva is said to have been assigned by von Dusch in 1868 (20). Since connective tissue is primarily affected, especially aponeuroses, fasciae, and tendons, the term "myositis" is no longer appropriate. In 1918, Rosenstirn (28) analyzed 119 cases and added one of his own. By 1982, in excess of 550 instances had been noted (8). Although many single cases appear in the literature, a number of large series have been followed (9,20,27,40). Whites are reported most commonly, but a number of black patients have also been observed (6,33).

**Genetics.** The overwhelming majority of cases are sporadic (20,40). Autosomal dominant inheritance is based on several instances of parent-to-child and male-to-male transmission (4,15) and on concordant monozygotic twins (14,42). Thus, most cases arise as new mutations. Genetic fitness appears to be close to zero, physical disability probably being the main reason, although infertility has also been suggested (40).

Evidence from published reports to date indicates that in families in which the FOP gene is transmitted through two generations, penetrance is complete. However, variable expressivity is observed with respect to skeletal malformations and the extent of ectopic ossification. On occasion, a parent has exhibited only the characteristic skeletal malformations, with the child expressing the full FOP phenotype with ectopic ossification (38). There is no evidence for genetic heterogeneity (8).

With autosomal dominant inheritance, the expected 1:1 sex ratio has not been demonstrated. Some studies show that males are more frequently affected, with ratios ranging from 4:1 to 3:2, although other studies show a strong female predilection as high as 67% of the patients in one series (20). The reasons for this are not entirely clear, but it seems obvious that problems of sample size in some studies and ascertainment bias in others play a major role.

A significant paternal age effect has been demonstrated for new mutations in three studies (8,26,40). A similar paternal age effect has also been found with other autosomal dominant mutations (e.g., Apert syndrome, Marfan syndrome, achondroplasia) and indicates that, with increasing age, the chance of fertilization by a sperm that has a single gene replication error increases.

Connor and Evans (8), attempting complete ascertainment of FOP in the United Kingdom, indicated a point prevalence of $0.61 \times 10^{-6}$. They also calculated a direct estimate of the mutation rate: 1.8 $(SE \pm 1.04) \times 10^{-6}$ mutations per gene per generation, a rate compatible with other known human mutations.

**Pathogenesis.** FOP is a distinctive histopathologic entity that can be differentiated from other soft tissue lesions that ossify, such as myositis ossificans, extraosseous osteosarcoma, and osseous metaplasia. Early FOP is characterized by multifocal, interconnecting nodules of spindle-shaped, fibroblast-like cells in a distinctive connective tissue matrix with bone spicules occupying the central area. Foci of chondroid differentiation may sometimes be observed. Lesions evolve to become mature lamellar bone with adipose and hematapoietic tissue in the cancellous spaces; the rim of fibroblast-like cells is no longer evident. Such pathologic features suggest that the spindle-shaped cells, like periosteum, are precursors of the osseous tissue found in FOP lesions (11).

Recent attempts to understand pathogenesis have viewed FOP lesions as a reaction to dystrophic calcification (18,30,35,36). Metaplastic bone formation has been observed with various lesions of the skin and subcutaneous tissue that are known to be associated with dystrophic calcification concurrently (29). Studies of periosteal grafts also support the notion that aberrant periosteal differentiation may be the fundamental disturbance in FOP. Transplanted periosteum initially produces woven bone and later produces mature lamellar bone with adipose and hematopoietic marrow elements (34,41); poorly vascularized and nonimmobilized periosteal grafts can make cartilage. Thus, ectopically placed periosteum can produce the full range of histologic patterns observed in FOP.

Evidence to date strongly suggests that the disturbance is not systemic, but multifocal. Biochemical studies have shown normal calcium and phosphate balance, normal levels of parathyroid hormone, normal tubular handling of phosphate, and normal responsiveness to parathyroid stimulation (6,23). High levels of alkaline phosphatase have been demonstrated in surrounding connective tissue (13) and conflicting results have been obtained by other investigators (3,16,21). However, these studies considered only single patients and none used a fluorometric assay system. Connor and Evans (7) investigated *in vitro* skin fibroblasts from 6 FOP patients and 27 normal controls; using a fluorometric assay system, they concluded that alkaline phosphatase has normal qualitative and quantitative regulation.

Clinical evidence indicates that lesions tend to be located near flat bones that are formed primarily by membranous ossification. Lack of involvement of the abdominal wall, perineum, and internal viscera strongly suggests multifocal rather than truly systemic disturbance.

Associated hypoplasia of normally located osseous structures, such as the hallucal anomalies, is consistent with reduction in a precursor cell pool derived from apparent migration; shape and location are maintained by the end-product, but size is reduced. Ultrastructural and histochemical studies also support the idea that fibroblast-like cells show evidence of osteogenic differentiation (19). Fibroblast-like cells are not all biologically the same (31); their functions can be modified by perturbations in the cellular environment (32). Fibroblasts are also known to respond to different connective tissue peptides in various ways (5). Interaction of fibroblasts with collagen and other connective tissue molecules is mediated by cell surface proteins that modulate the behavior of fibroblasts (43). It should be noted that bone, ligament, fascia, and tendon all share Type I collagen as their dominant connective tissue protein. The fact that fibroblast activity can be modulated pharmacologically may be rationally exploited some day to treat patients with FOP (24). The key to unlocking the mystery of FOP may be the discovery of what factors regulate or induce osteogenic activity by fibroblast-like cells in normal and diseased tissues (25).

**Clinical features and natural history.** Clinical features include four possible types of malformation of the hallux, reduction defects of the digits, hearing loss, baldness, and rarely mental deficiency. Findings in FOP (Fig. 9–11A–G) together with their frequencies are presented in Table 9–1. All patients have skeletal abnormalities of the hallux (Fig. 9–11E–G) that are present at birth. Type I is the commonest; the halluces are short, lack a skin crease, and possess a single phalanx. In Type II, the halluces are of normal length but are stiff from early childhood and show progressive osseous fusion with age. In Type III, two phalanges are present that become rigid during the second decade due to osteophytic lipping. Variable reduction defects are observed in Type IV. Other skeletal abnormalities include short thumbs due to short first metacarpals; clinodactyly of the fifth fingers; short broad femoral necks; abnormal cervical vertebrae with small bodies, large pedicles, and large spinous processes; progressive bony ankylosis of the cervical spine; and, occasionally, exostoses of the proximal tibiae (9,20,27). The radiographic spectrum of abnormalities has been reviewed elsewhere (12,39).

Ectopic ossification is progressive and begins in early childhood. The site of onset is most commonly the neck or paraspinal region and less commonly the head or limbs. When new lumps appear, reddening of the overlying skin may occur and pain may sometimes be present. Certain areas within the connective tissue are prone to ossification, especially the paraspinal muscles, limb girdle muscles, and the muscles of mastication. Involvement of joint capsules, ligaments, and plantar fasciae is common. In FOP patients, various factors are known to precipitate ectopic ossification, such as muscle trauma, biopsy, surgical procedures to excise ectopic bone, intramuscular injections, careless venipuncture, and dental treatment (9,20,27). All patients eventually develop restriction of movement and physical handicap. Episodes of ossification and subsequent disability are characteristically erratic. The disorder is known to have long periods of inactivity. Although ectopic ossification is most marked prior to puberty, new lumps may occur during the sixth and seventh decades. Ectopic calcification has the most severe effect on axial connective tissues, and limb involvement

Table 9–1. Features of fibrodysplasia ossificans progressiva[a]

| Feature | Percentage |
|---|---|
| Abnormal hallux | 100 |
| Type I (one phalanx) | 79 |
| Type II (two phalanges with progressive bony fusion) | 9 |
| Type III (two phalanges with osteophytic lipping) | 6 |
| Type IV (variable reduction defects) | 6 |
| Short thumbs secondary to short first metacarpals | 59 |
| Clinodactyly, fifth finger | 44 |
| Short broad femoral necks | 55 |
| Ectopic calcification, site of onset | |
| Neck | 38 |
| Paraspinal region | 32 |
| Head | 9 |
| Limbs | 12 |
| Joint involvement | |
| Spine | 100 |
| Shoulder | 100 |
| Elbow | 55 |
| Wrist | 7 |
| Hip | 59 |
| Knee | 38 |
| Ankle | 32 |
| Temporomandibular | 71 |
| Hearing loss[b] | 24 |
| Diffuse thinning of hair[c] | 24 |
| Mental deficiency | 6 |

[a]Adapted from data in JM Connor and DAP Evans, *J. Bone Jt Surg* **64B**:76–83, 1982.
[b]Of these 63% were conductive.
[c]Of these 75% were female.

is most marked proximally (9). Chest wall fixation may lead to diminished pulmonary reserve, and most patients eventually die from respiratory failure (10).

Other abnormalities such as hearing loss and baldness occur in approximately 25% of all patients. The hearing loss may be conductive or sensorineural. The diffuse type of baldness, when present, becomes evident in middle age and the majority of those affected are female. It appears to be a primary feature of FOP, although it might conceivably represent a secondary effect of nutritional deficiency based on inability to open the jaws. Mental deficiency is found only as a low-frequency abnormality (9,20).

**Differential diagnosis.** Delayed diagnosis is commonplace in patients with FOP even though they have characteristic skeletal malformations. Common misdiagnoses are hallus valgus, diaphyseal aclasia, *Klippel–Feil anomaly,* and various forms of arthrogryposis. Swellings, depending upon their site, may be mistaken for lymphadenopathy, sarcoma, or even mumps (6).

A number of entities have similarities to FOP. Osseous metaplasia may occur in various organs and tissues in association with inflammatory, neoplastic, and vascular disorders. It may be found in the deep soft tissues of paraplegic patients beneath the level of the spinal cord lesion. Metaplastic bone formation in the skin and subcutaneous tissue is known to occur with pilomatrixoma (12).

Myositis ossificans has histologic similarities to FOP. However, soft tissue lesions are solitary, grow rapidly, and are usually located in a limb. A previous history of trauma is often present (1).

Extraskeletally occurring osteosarcoma is rare. Although it resembles both FOP and myositis ossificans, most cases occur in adults and consist of large solitary masses of the extremities (2). FOP lesions are small and multinodular in character; extraskeletal osteosarcoma tends to occur as a large single tumor mass. Both myositis ossificans and extraosseous osteosarcoma resemble FOP in having the capacity to express fibrous, chondroid, and osseous tissue patterns. However, lesions of extraskeletal osteosarcoma are more cellular, have more atypia and more frequent mitoses, and produce malignant osteoid; these features are not present in FOP (2,37).

**Laboratory aids.** Diagnosis depends on clinical and radiographic demonstration of characteristic skeletal malformations. Routine laboratory tests are usually normal (36); biopsy of lumps should be avoided (27), as should careless venipuncture, intramuscular injections, and surgery to excise ectopic bone. Dental treatment should be carried out cautiously and anesthesiologists should be aware of the possibility of atlantoaxial subluxation and also that intubation may be technically difficult because of fixation of the jaws (9).

### References (Fibrodysplasia ossificans progressiva)

1. Ackerman LV: Extra-osseous localized non-neoplastic bone and cartilage formation (so-called myositis ossificans). *J Bone Jt Surg* **40A**:279–298, 1958.

2. Allan CJ, Soule EH: Osteogenic sarcoma of the somatic soft tissues—clinicopathologic study of 26 cases and review of the literature. *Cancer* **27**:1121–1132, 1971.

3. Beratis NG et al: Alkaline phosphatase activity in cultured fibroblasts from fibrodysplasia ossificans progressiva. *J Med Genet* **13**:307–309, 1976.

4. Burton-Fanning FW, Vaughan AL: A case of myositis ossificans. *Lancet* **2**:849–850, 1901.

5. Chiang TM et al: Binding of chemotactic collagen derived peptides to fibroblasts—the relationship to fibroblast chemotaxis. *J Clin Invest* **62**:916–922, 1978.

6. Connor JR, Beighton P: Fibrodysplasia ossificans progressiva in South Africa: Case reports. S Afr Med J **61**:404–406, 1982.

7. Connor JM, Evans DAP: Quantitative and qualitative studies on skin fibroblast alkaline phosphatase in fibrodysplasia ossificans progressiva. *Clin Chim Acta* **117**:355–360, 1981.

8. Connor JM, Evans DAP: Genetic aspects of fibrodysplasia ossificans progressiva. *J Med Genet* **19**:35–39, 1982.

9. Connor JM, Evans DAP: Fibrodysplasia ossificans progressiva: The clinical features and natural history of 34 patients. *J Bone Jt Surg* **64B**:76–83, 1982.

10. Connor JM et al: Cardiopulmonary function in fibrodysplasia ossificans progressiva. *Thorax* **36**:419–423, 1981.

11. Cramer SF et al: Fibrodysplasia ossificans progressiva: A distinctive bone-forming lesion of the soft tissue. *Cancer* **48**:1016–1021, 1981.

12. Cremin B et al, The radiological spectrum of fibrodysplasia ossificans progressiva. *Clin Radiol* **33**:499–508, 1982.

13. Dixon TF et al: Myositis ossificans progressiva: Report of a case in which ACTH and cortisone failed to prevent reossification after excision of ectopic bone. *J Bone Jt Surg* **36B**:445–449, 1954.

14. Eaton WL et al: Early myositis ossificans progressiva occurring in homozygotic twins. A clinical and pathologic study. *J Pediatr* **50**:591–598, 1957.

15. Gaster A: Discussion in meeting in West London Medico-Chirurgical Society, 7 October 1904. *W London Med J* **10**:37, 1905.

16. Herrmann J et al: Fibrodysplasia ossificans progressiva and the XXXY syndrome in the same sibship. *Birth Defects* **19**(5):43–49, 1969.

17. Kewalramani LS: Ectopic ossification. *Am J Phys Med* **59**:99–121, 1977.

18. Lutwak L: Myositis ossificans progressiva—mineral, metabolic and radioactive calcium studies of the effects of hormones. *Am J Med* **37**:269–293, 1964.

19. Maxwell WA et al: Histochemical and ultrastructural studies in fibrodysplasia ossificans progressiva (myositis ossificans progressiva). *Am J Pathol* **87**:483–498, 1977.

20. McKusick VA: Fibrodysplasia ossificans progressiva, in *Heritable Disorders of Connective Tissue,* 4th ed. C.V. Mosby, St. Louis, 1972, pp 400–415.

21. Miller RL et al: Studies on alkaline phosphatase activity in cultured cells from a patient with fibrodysplasia ossificans progressiva. *Lab Invest* **37**:254–259, 1977.

22. Patin G: Lettres choisies de feu Monsieur Guy Patin. Letter of 27 August 1648 written to AF. Cologne: Laurens, **1**:28, 1692.

23. Pitt P, Hamilton EBD: Myositis ossificans progressiva. *J Royal Soc Med* **77**:68–70, 1984.

24. Prockop DJ et al: The biosynthesis of collagen and its disorders. *N Engl J Med* **301**:13–23, 77–85, 1979.

25. Reddi AH et al: Transitions in collagen types during matrix induced cartilage, bone and bone marrow formation. *Proc Natl Acad Sci USA* **74**:5589, 1977.

26. Rogers JG, Chase GA: Paternal age effect in fibrodysplasia ossificans progressiva. *J Med Genet* **16**:147–148, 1979.

27. Rogers JG, Geho WB: Fibrodysplasia ossificans progressiva. *J Bone Jt Surg* **61A**:909–914, 1979.

28. Rosenstirn J: A contribution to the study of myositis ossificans progressiva. *Ann Surg* **68**:485–520, 591–637, 1918.

29. Roth SI et al: Cutaneous ossification. *Arch Pathol* **76**:44–54, 1963.

30. Ruderman RJ et al: A possible etiologic mechanism for fibrodysplasia ossificans progressiva. *Birth Defects* **10**(12):299, 1974.

31. Schneider EL et al: Tissue specific differences in cultured human diploid fibroblasts. *Exp Cell Res* **108**:1–6, 1977.

32. Schwartz RI, Bissell MJ: Dependence of the differentiated state on the cellular environment. Modulation of collagen synthesis in tendon cells. *Proc Natl Acad Sci USA* **74**:4453, 1977.

33. Shipton EA et al: Anaesthesia in myositis ossificans progressiva: A case report and clinical review. *S Afr Med J* **67**:26–28, 1985.

34. Skoog T: The use of periosteum and surgical for bone restoration in congenital clefts of the maxilla. A clinical report and experimental investigation. *Scand J Plast Reconstr Surg* **1**:113–130, 1967.

35. Smith DA et al: Myositis ossificans progressiva—Case report with metabolic and histochemical studies. *Metabolism* **15**:521–528, 1966.

36. Smith R: Myositis ossificans progressiva. *Sem Arth Rheum* **4**:369–380, 1975.

37. Stout AP, Lattes R: Tumors of the soft tissues, in *Atlas of Tumor Pathology*. Second Series, Fascicle I. Armed Forces Institute of Pathology, Bethesda, 1967, p 162.

38. Sympson T: Case of myositis ossificans. *Br Med J* **ii**:1026–1027, 1986.

39. Thickman D et al: Fibrodysplasia ossificans progressiva *Am J Roentgenol* **139**:935–941, 1982.

40. Tünte W et al: Zur Genetik der Myositis ossificans progressiva. *Humangenetik* **4**:320–351, 1967.

41. Uddstromer L, Ritsila V: Osteogenic capacity of periosteal grafts. A qualitative and quantitative study of membranous and tubular bone periosteum in young rabbits. *Scand J Plast Reconstr Surg* **12**:207–214, 1978.

42. Vastine JA et al: Myositis ossificans progressiva in homozygotic twins. *Am J Roentgenol* **59**:204–212, 1948.

43. Zetter BR et al: Role of the high-molecular weight glycoprotein in cellular morphology, adhesion, and differentiation. *Ann NY Acad Sci* **312**:299–316, 1978.

## Hajdu–Cheney syndrome (acroosteolysis)

The Hajdu–Cheney syndrome consists of dissolution of the terminal phalanges, bizarrely shaped skull, premature loss of teeth, and short stature (Figs. 9–12 to 9–16). It was first described by Hajdu and Kauntze (12) in 1948. Cheney (5) reported familial occurrence in 1965.

The syndrome has autosomal dominant inheritance with very variable expressivity (5,8,24,32), but the vast majority have been isolated examples. About 35 patients have been described (1–5,7,8,11, 12,14,15,17,24,26–28,28,30–35,37–39). A few examples are less certain (23,25). We cannot identify some presumed cases (29).

Patients have been generally healthy except for recurrent upper respiratory infections or asthma (1,7,30,34).

**Facies.** The face is characteristic (Figs. 9–12 and 9–13). The head appears disproportionately large. The scalp hair and eyebrows are thick and coarse with synophrys (13,15,34,35). The hair is low on the forehead and nape. The outer supraorbital ridges are often enlarged. There may be mild exophthalmos and hypertelorism. The midface is somewhat hypoplastic and the philtrum is long. The lower third of the face is shortened, probably in large part due to premature loss of teeth. The mouth tends to be small and the chin usually recedes. The neck is often short.

**Ears.** Conduction (12,30,32,34,35,37) and sensorineural (14,33) hearing loss has been noted in a number of cases.

**Eyes.** Myopia, epicanthal folds, nystagmus, reduced visual fields, abducens palsy, disc pallor, and optic atrophy have been found (1,12, 15,34,37).

**Skin.** Generalized hirsutism is relatively frequent (30,32,34,35,38). The skin may be somewhat more elastic than normal (15,25). The nails are often wider than they are long and may become coarse and curved (1,8,25).

**Central nervous system.** A serious complication can result from impaction of the cerebellum into the foramen magnum (12,15,18,24,27, 30,37). This can cause occipital headache, hydrocephaly, and progressive neurologic deterioration with involvement of the lower cranial nerves (gruff or low-pitched voice, paralyzed palate, anesthesia of pharynx) and cerebellar dysfunction (1,14,15,18,34,35,37). Seventh nerve palsy has also been noted (1).

**Musculoskeletal alterations.** Progressive basilar invagination, dolichocephaly, and unusual protuberance of the squamous portion of the occipital bone (bathrocephaly) are striking. Widening of the metopic, coronal, and lambdoidal sutures with multiple Wormian bones and depression at the anterior fontanel are evident in most patients

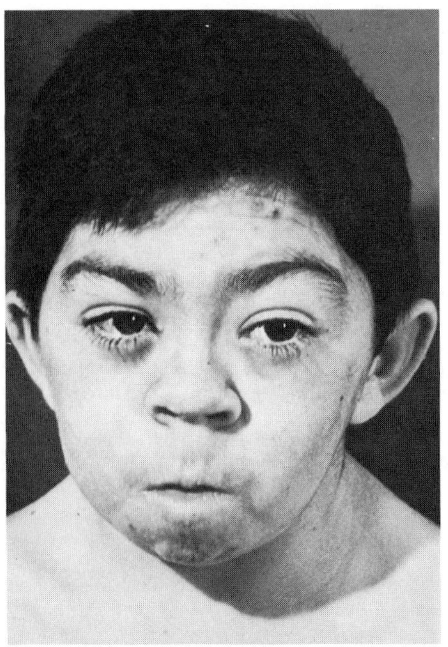
A

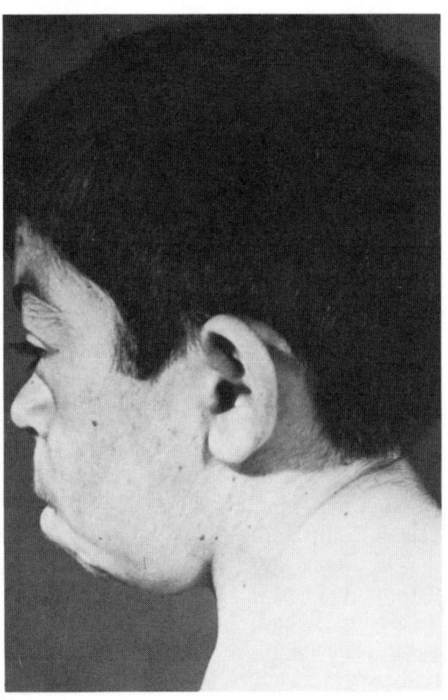
B

Fig. 9–12. *Hajdu–Cheney syndrome.* (A,B) Unusual facies marked by midface hypoplasia, fullness of outer supraorbital ridges, mild hypertelorism, small mouth, and receding chin. (A,B from *A Matisonn* and *F Zaidy,* S Afr Med J **47:**2060, 1973.)

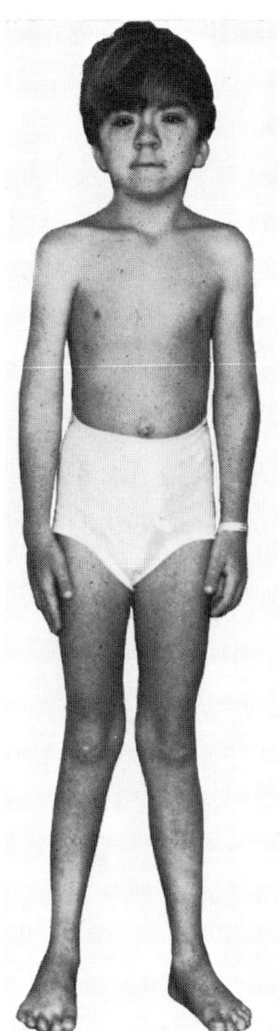

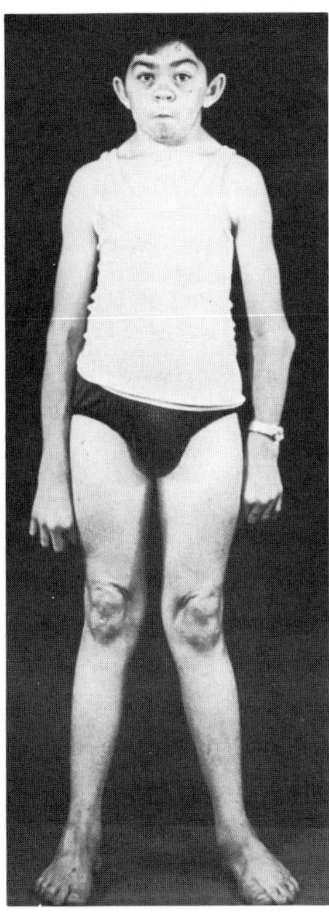

A B

Fig. 9–13. *Hajdu–Cheney syndrome.* (A) Ten-year-old boy exhibiting shortness of stature, hirsutism, genua valga, unusual facial appearance. (B) Compare with 18-year-old boy. Note similar facies, genua valga, subluxation of radial head. (A from *J Herrmann et al,* Z Kinderheilkd **114:**93, 1973. B from *A Madison* and *F Zaidy,* S Afr Med J **47:**2060, 1973.)

(Fig. 9–14) (3–5,12,32). The frontal sinuses are absent and the maxillary antra underdeveloped. The sella turcica is enlarged, elongated (J-shaped), and wide open with slender clinoids (Fig. 9–14). The anterior nasal spine resorbs (32). The mandibular condyles are positioned anterior to the glenoid fossae and there may be resorption of the condylar heads (1) or mandibular rami (10,11). The mandibular chin button is often missing.

Adult height has ranged from 140 to 157 cm but decreases with age (8). This is in part due to progressive kyphosis and/or scoliosis, marked osteoporosis, and compression of thoracic vertebrae. There is associated pain (4,5,12) due, in part, to compression fractures of the spine. Extension and flexion of the neck are often limited. The superior and inferior surfaces of the vertebrae are concave, assuming a so-called "fish-bone" shape (37). The cervical spine is often straighter than normal (3,31). Intervertebral disks may appear denser than the vertebral bodies. Shortening and clubbing due to resorption of the distal portion of fingers and toes, primarily the former, begin around the third or fourth year of life. In severe cases, the middle phalanges can be involved (5,7,8,28). The terminal portion of the thumb is especially abbreviated (5,26,32) (Fig. 9–15A,B). All joints are somewhat hyperflexible, especially the interphalangeal joints (2,3,4,11,15, 27,28,30,31,34,38). Hip dislocation has been reported (2). Pain or paresthesia may be experienced in the joints, especially on motion (30). Genua valga is frequent. Long bones, metacarpals, and metatarsals often fracture (3,5,7,26–28). The metaphyseal area of metapodial bones tends to undergo dissolution. Narrowing of the metacarpophalangeal and/or metatarsophalangeal spaces (5,11,12,28) and osteolysis of the radial head (12,33) have been documented. The tibiae and fibulae may be somewhat curved and some patients have a proximal anterior tibial groove. Some patients have exhibited fusion of the dorsal processes of the third to fifth cervical vertebrae. Radial dislocation has also been noted (24,38). Club feet (1,25,28,34) and umbilical and/or inguinal hernia (30,34) have been reported in a few cases.

Fig. 9–14. *Hajdu–Cheney syndrome.* Bathrocephaly with many Wormian bones in suture lines, depressed anterior fontanel, and basilar impression. (From *WD Cheney,* Am J Roentgenol **94:**595, 1965.)

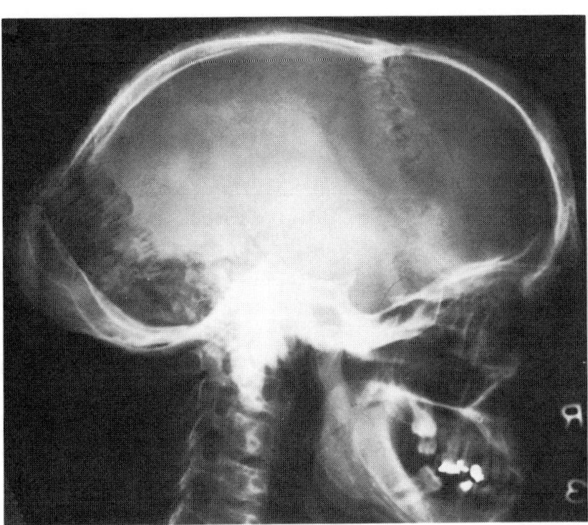

**Genitourinary.** Anomalies have included renal cortical cysts (1,27), glomerulonephritis (27), urinary reflux (35), hypogonadism (1,24), cryptorchidism (15), and hypospadias (38).

**Other findings.** VSD (30,32) and malrotation of the bowel (32) have been reported.

Fig. 9–15. *Hajdu–Cheney syndrome.* (A) Observe abbreviation of terminal phalanges. (B) Radiograph showing lysis of terminal phalanges. (A from *A Madison* and *F Zaidy,* S Afr Med J **47:**2060, 1973. B from *WD Cheney,* Am J Roentgenol **94:**595, 1965.)

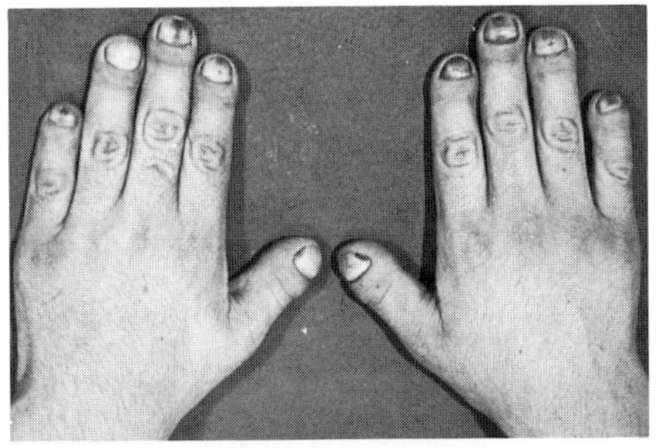

A

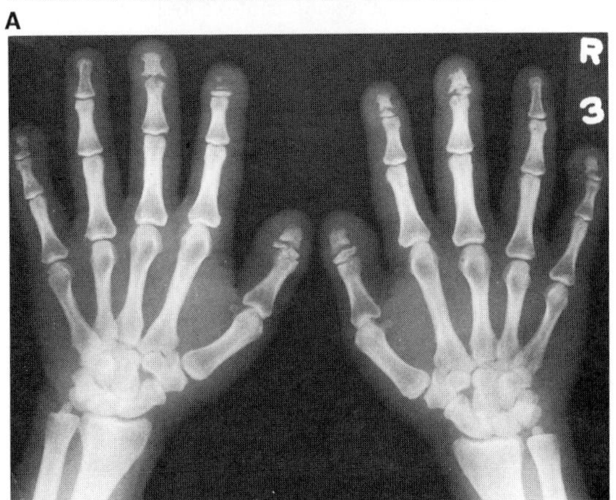


B

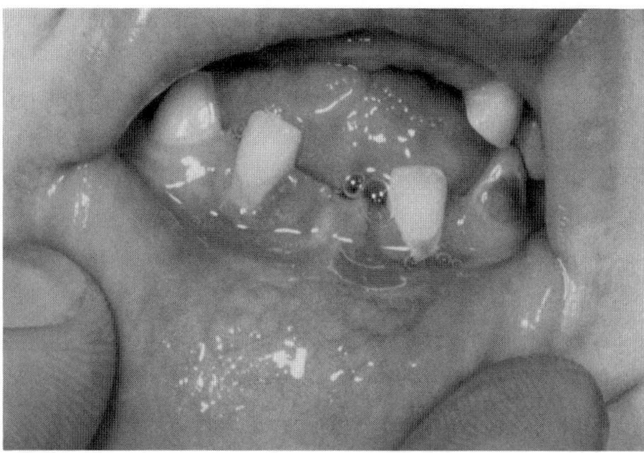

Fig. 9–16. *Hajdu–Cheney syndrome.* Early loss of teeth due to periodontal disease.

**Oral manifestations.** Early loss of teeth due to periodontal disease with marked resorption of the alveolar ridges within 6 months after loss of teeth is a constant feature (Fig. 9–16). Permanent teeth are often impacted (1,4,5,11,27,32,33,37). Malocclusion is constant. Molar roots may be resorbed (1,7,26). Cleft palate, cleft uvula, and velopharyngeal incompetence have been mentioned (3,18,27,30,34). The mandible, small in childhood, may become relatively prognathic with age (34).

**Differential diagnosis.** The term ''acroosteolysis'' is used nonspecifically to refer to dissolution of the terminal phalanges of the hands and feet. Acroosteolysis may also be seen in *pycnodysostosis, progeria, mandibuloacral dysplasia, epidermolysis bullosa, Murray–Puretić–Drescher syndrome, Winchester syndrome,* Gorham disease, scleroderma, syringomyelia, leprosy, syphilis, psoriasis, trauma, dominant acroosteolysis, neurogenic ulcerative acropathy, manual exposure to polyvinyl chloride intoxication, and a host of other disorders (6,8,9,13,16,17,19–22,29,36).

**Laboratory aids.** Several investigators (3,15,24,33) described somewhat elevated serum alkaline phosphatase levels but others have noted normal values (8,34). Vaněk (33) found elevated acid phosphatase. Brown et al (3) noted elevated plasma $\beta$-glucuronidase activity.

### References [Hajdu–Cheney syndrome (acroosteolysis)]


1. Allen CM et al: The acro-osteolysis (Hajdu–Cheney) syndrome: Review of the literature and report of a case. *J Periodontol* **55**:224–229, 1984.
2. Blery M et al: Acro-ostéolyse d'Hajdu–Cheney et anéurysme calcifié sur canal artériel redux. *Ann Radiol* **27**:27–30, 1984.
3. Brown DM et al: The acro-osteolysis syndrome: Morphologic and biochemical studies. *J Pediatr* **88**:573–580, 1976.
4. Chawla S: Cranio-skeletal dysplasia with acro-osteolysis. *Br J Radiol* **37**:702–705, 1964.
5. Cheney WD: Acro-osteolysis *Am Roentgenol* **94**:595–607, 1965.
6. Crecelius W: Ein Beitrag zum Krankheitsbild der Osteopathia dysplastica familiaris. *Fortschr Roentgenstrahl* **76**:196–202, 1952.
7. Dorst JP, McKusick VA: Acroosteolysis (Cheney syndrome). *Birth Defects* **5**(3):215–217, 1969.
8. Elias AN et al: Hereditary osteodysplasia with acro-osteolysis (the Hajdu–Cheney syndrome). *Am J Med* **65**:627–636, 1978.
9. Giaccai L: Familial and sporadic neurogenic acro-osteolysis. *Acta Radiol* **38**:17–29, 1952.
10. Gilula LA et al: Idiopathic nonfamilial acro-osteolysis with cortical defects and mandibular ramus osteolysis. *Radiology* **121**:63–68, 1976.
11. Greenberg BD, Street D: Idiopathic nonfamilial acro-osteolysis. *Radiology* **69**:259, 1957.
12. Hajdu N, Kauntze R: Cranio-skeletal dysplasia. *Br J Radiol* **21**:42–48, 1948.
13. Harris DK, Adams WGF: Acro-osteolysis occurring in men engaged in the polymerisation of vinyl chloride. *Br Med J* **3**:712–714, 1967.
14. Herrmann J et al: Arthro-dento-osteo-dysplasia (Hajdu–Cheney syndrome). *Z Kinderheilkd* **114**:93–110, 1973.
15. Iwaya T et al: Hajdu–Cheney syndrome. *Arch Orthop Traumat Surg* **95**:293–302, 1979.
16. Jänner M et al: Zum Krankheitsbild der familiärer Akroosteolyse. *Z Haut Geschlechtskr* **34**:65–73, 1963.
17. Kaur S et al: Acro-osteolysis (report of two cases and brief review of the literature). *Clin Neurol Neurosurg* **82**:45–56, 1980.
18. Kawamura J et al: Hajdu–Cheney syndrome: Report of a non-familial case. *Neuroradiology* **21**:295–301, 1981.
19. Kleinsorge H: Acroosteolystische Erscheinungen der Osteomalacie. *Fortschr Roentgenstr* **73**:471–475, 1950.
20. Kozlowski K et al: Neurogene ulcerierende Akropathie. Akroosteolyse-Syndrom. *Mschr Kinderheilkd* **119**:169–175, 1971.
20a. Kozlowski K et al: Acroosteolysis: Problems of diagnosis—report of four cases. *Pediatr Radiol* **8**:79–86, 1979.
21. Krikler DM: The case of the vanishing toes. *S Afr Med J* **29**:1050–1052, 1955.
22. Lamy M, Maroteaux P: Acro-osteolyse dominante. *Arch Fr Pédiatr* **18**:693–702, 1961.
23. Lièvre JA, Gama G: L'acroosteolyse. *Bull Soç Med Hôp Paris* **73**:109–120, 1957.
24. Matisonn A, Zaidy F: Familial acro-osteolysis. *S Afr Med J* **47**:2060–2063, 1973.
25. Newton TH, Carpenter ME: Ehler–Danlos syndrome with acro-osteolysis. *Br J Radiol* **32**:739–743, 1959.
26. Papavasiliou CG et al: Idiopathic nonfamilial acro-osteolysis associated with other bone abnormalities. *Am J Roentgenol* **83**:687–691, 1960.
27. Rosenmann E et al: Sporadic idiopathic acro-osteolysis with cranio-skeletal dysplasia, polycystic kidneys and glomerulonephritis: A case of the Hajdu–Cheney syndrome. *Pediatr Radiol* **6**:116–120, 1977.
28. Shaw DG: Acro-osteolysis and bone fragility. *Br J Radiol* **42**:934–936, 1969.
29. Silver J: Acro-osteolysis and internal lymphangiectaia. *Proc R Soc Med* **65**:723–724, 1972.
30. Silverman FN et al: Acro-osteolysis (Hajdu–Cheney syndrome). *Birth Defects* **10**(12):106–123, 1974. (Case 1 same as Ref. 14 and Ref. 39; Case 2 same as Ref. 12.)
31. Toglia JV: Hereditary dysostosis. *Tex State J Med* **62**:36–41, 1966.
32. Van den Houten BR et al: The Hajdu–Cheney syndrome: A review of the literature and report of 3 cases. *Int J Oral Surg* **14**:113–125, 1985.
33. Vaněk J: Idiopathische Osteolyse von Hajdu–Cheney. *Roefo* **128**:75–79, 1978.
34. Weleber RG, Beals RK: The Hajdu–Cheney syndrome: Report of two cases and review of the literature. *J Pediatr* **88**:243–249, 1976.
35. Wendel U, Kemperdick H: Idiopathische Osteolyse vom Typ Hajdu–Cheney. *Mschr Kinderheilkd* **127**:581–584, 1979.
36. Wieland H: Ein Beitrag zur Kenntnis der Akroosteolyse. *Fortschr Roentgenstr* **77**:193–198, 1952.
37. Williams B: Foramen magnum impaction in a case of acro-osteolysis. *Br J Surg* **64**:70–73, 1977.
38. Zahran M et al: Arthro-osteo-renal dysplasia. *Acta Radiol Diagn* **25**:39–43, 1984.
39. Zugibe FT et al: Arthrodentoosteodysplasia: A genetic acroosteolysis syndrome. *Birth Defects* **10**(5):145–152, 1974.


## Infantile cortical hyperostosis (Caffey–Silverman syndrome)

Although infantile cortical hyperostosis was originally described by Röske (23) in 1930, it was not until 1945–1946 that the clinical and roentgenographic studies of Caffey and Silverman (7) and of Smyth et al (27) called attention to it. The disorder affects infants under 6 months of age. Its most constant features are bilateral swelling over the mandible or other bones, roentgenographic evidence of new bone formation in the area, hyperirritability, and mild fever. At least 250 cases have been recorded. A number of large series have been published (10,13–15,17,24) and autosomal dominant inheritance has been repeatedly demonstrated (1,10,12,14,15,21a,24,29). Saul et al (24) and MacLaughlan et al (20a) indicated that familial infantile cortical hyperostosis differed in several respects from sporadic instances, the former having an earlier onset of disease (24% at birth), less frequent mandibular, ulnar, and clavicular involvement, no involvement of the ribs and scapulae, and more frequent lower extremity involvement. There is reason to believe that the sporadic form is disappearing (20a). The onset in the sporadic form is 10 weeks and in the familial form is 7 weeks (20a). It has been observed roentgenographically as early as 5 weeks prenatally and has been found 20 months after birth (1,2,3,7,14,15,19,20). Blank (4) indicated that some unexplained episodes of pain and cortical thickening in older children may represent recurrences in patients in whom the infantile phase of the disease has not been severe enough to call attention to it.

Cayler and Peterson (8) estimated that the disorder is found in 3 of every 1000 registered patients under 6 months of age. A low mortality rate has been described. It has been suggested that pathogenesis is based on congenital abnormality of the vessels supplying the periosteum of the involved bones, hypoxia effecting focal necrosis of the overlying soft tissue, resulting in new subperiosteal bone formation (25,26).

We will consider the sporadic and autosomal dominant cases collectively.

**Facies.** Because of the swelling, the facies is so striking that the condition may be diagnosed with considerable assurance even prior to confirmatory X-ray evidence. The swelling is symmetric and located over the body and ramus of the mandible (Fig. 9–17). Pallor is often observed as well.

**Soft tissues.** The condition is initiated by tender, soft-tissue swelling over the face, around the orbits, thorax, or extremities; this swelling often undergoes remission and exacerbation (21). It is firm, brawny, and often so painful as to cause pseudoparalysis of an extremity. It is not accompanied by redness or increased heat.

**Fever and irritability.** Pain, fever of mild degree, and hyperirritability are seen in at least two-thirds of the patients (6,26). These signs commonly precede the appearance of the swelling and bone involvement. One or all may, however, be absent, Anemia, leukocytosis, and elevation of the sedimentation rate occur in more than half the patients (17).

Fig. 9–17. *Infantile cortical hyperostosis.* Note bilateral swellings over ramus of mandible.

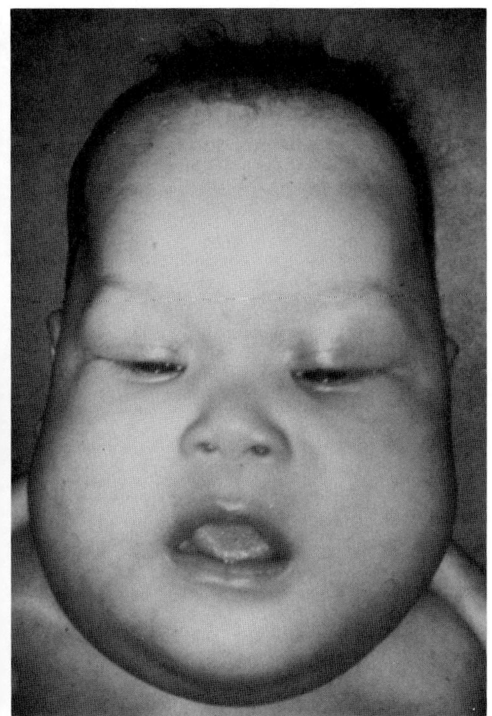

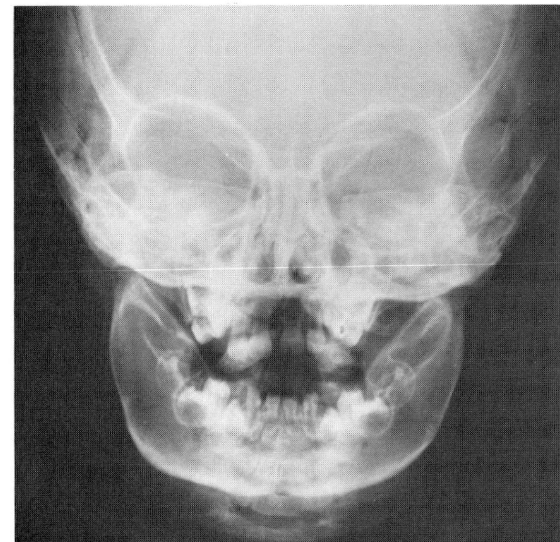

Fig. 9–18. *Infantile cortical hyperostosis.* Roentgenogram of jaws showing symmetric mandibular enlargement 6 months after onset of infantile cortical hyperostosis. (From *PM Burbank et al,* Oral Surg **11:**1126, 1958.)

**Skeletal system.** The most frequently affected bone is the mandible, at least three of every four patients experiencing mandibular enlargement (Fig. 9–18). Less commonly involved are the clavicle, tibia, ulna, femur, rib, humerus, maxilla, and fibula (6,16–18, 26,28). Usually several bones are affected at the same time (Fig. 9–19A,B). As noted above, in the autosomal dominant form, the mandible, ulna, and clavicle are less often involved than in the sporadic form and the ribs and scapulae almost never (20a).

New periosteal bone formation, appearing most often during the ninth week, undergoes resolution slowly. Though complete clinical resolution takes place within 3 to 30 months (average, 9 months), roentgenographic evidence may persist for many years (7,22). Bone bridges between the radius and ulna and between ribs have been described (7). Forward bowing of the tibia is common (15,19,31). Pleural effusion has been reported in cases in which there has been rib involvement (7). Later recurrences of the disorder, although uncommon, have been described (4).

**Oral manifestations.** The oral findings have been discussed, in part, above. Involvement of the mandible was formerly thought to be necessary for diagnosis of the condition, but analysis of large series of cases has revealed that this is not so (26). Nevertheless, swelling of the jaws is the most common presenting sign. In a follow-up survey of 11 cases, Burbank et al (6) demonstrated that in six cases the mandible was the only bone involved. Follow-up showed that the fever had no effect on the enamel or on the eruption sequence. However, 8 of the 11 patients had roentgenographic evidence of residual bony asymmetry of the mandible at the angle and ramus, and some had severe malocclusion.

**Pathology.** Several microscopic studies have been performed (5), the most comprehensive being that of Eversole et al (11). In the early stages, foci of polymorphonuclear neutrophilic leukocytes are seen within the periosteum. The periosteum is swollen and mucoid in appearance, losing its well-defined limits and blending into the muscle, fascia, and tendons. At this stage, there is some resemblance to osteosarcoma, and erroneous diagnosis and treatment may result. The small arteries of the periosteum and overlying soft tissue show intimal proliferation. In the later stages, poorly vascularized and incompletely structured new bone is laid down. Neither hemorrhage nor inflammation is seen at this stage (17).

**Differential diagnosis.** Bocian et al (4a) reported a newly recognized skeletal dysplasia that simulates infantile cortical hyperostosis.

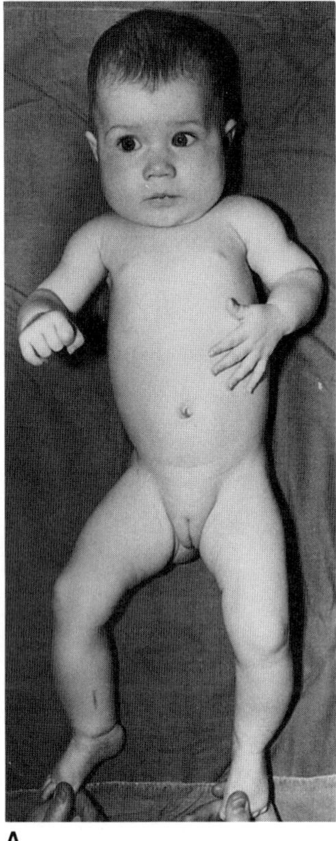

A

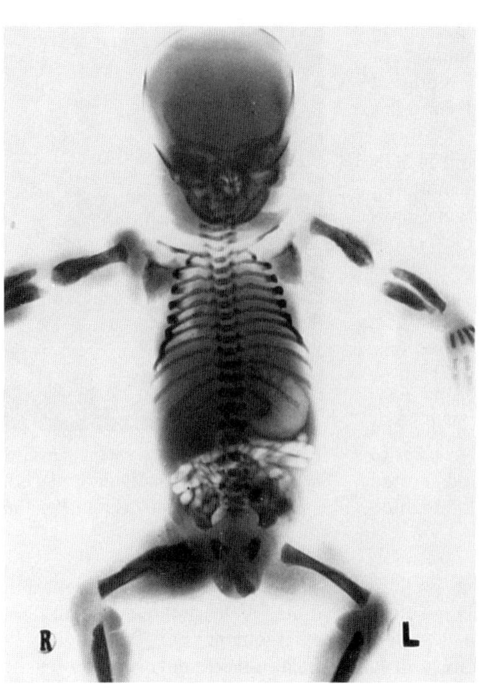


B

Fig. 9–19. *Infantile cortical hyperostosis.* (A) Involvement of mandible and lower extremities. (B) Roentgenogram showing thickening of periosteum and new bone formation in long bones. (A from *J W Gerrard et al,* J Pediatr **59:**543, 1961.)

Inheritance is probably autosomal recessive. Also to be considered are epidemic parotitis (mumps), vaccinial and pyogenic osetomyelitis, parotid tumor, rickets, congenital syphilis, subperiosteal hematoma, scurvy, and vitamin A intoxication (9,19).

Koslowski and Tsuruta (19a), in 1989, described a new form of lethal neonatal dwarfism which they called *dysplastic cortical hyperostosis.* It was characterized by generalized symmetric cortical thickening and sclerosis of the long bones. The ribs appeared somewhat wavy. To be excluded are *infantile cortical hyperostosis (Caffey-Silverman) syndrome, I-cell disease,* and $G_{M1}$ *gangliosidosis.*

**Laboratory aids.** Roentgenographic study not only of the mandible but of the chest and long bones confirms the clinical impression. Serum alkaline phosphatase level is elevated in cases with marked bone deposition (17). In over 80%, the sedimentation rate is elevated. Anemia and leukocytosis are common. Elevated IgM level and thrombocytopenia have also been noted (30).

### References [Infantile cortical hyperostosis (Caffey–Silverman syndrome)]

1. Ball MJ, Feingold M: Autosomal dominant inheritance of Caffey's disease. *Birth Defects* **10**(9):139–146, 1974.

2. Barba WP, Freriks DJ: The familial occurrence of infantile cortical hyperostosis in utero. *J Pediatr* **42**:141–150, 1953.

3. Bennett HS, Nelson TR: Prenatal cortical hyperostosis. *Br J Radiol* **26**:47–49, 1953.

4. Blank E: Recurrent Caffey's cortical hyperostosis and persistent deformity. *Pediatrics* **55**:856–860, 1975.

4a. Bocian M et al: A probable new recessive dysplasia simulating Caffey disease. Eighth David W. Smith Annual Workshop on Malformations and Morphogenesis, Greenville, South Carolina, August 15–19, 1987.

5. Brooksaler F, Miller JE: Infantile cortical hyperostosis. *J Pediatr* **48**:739–753, 1956.

6. Burbank PM et al: The dental aspects of infantile cortical hyperostosis. *Oral Surg* **11**:1126–1137, 1958.

7. Caffey J, Silverman WA: Infantile cortical hyperostosis. *Am J Roentgenol* **54**:1–16, 1945.

8. Cayler GC, Peterson CA: Infantile cortical hyperostosis: Report of 17 cases. *Am J Dis Child* **91**:119–125, 1956.

9. Cochran W: Infantile cortical hyperostosis: A review with illustrative case report. *Acta Paediatr Scand* **51**:442–453, 1962.

10. Emmery L et al: Familial infantile cortical hyperostosis. *Eur J Pediatr* **141**:56–58, 1983.

11. Eversole SL Jr et al: Hitherto undescribed characteristics of the pathology of infantile cortical hyperostosis (Caffey's disease). *Bull Johns Hopkins Hosp* **101**:80–100, 1957.

12. Faul R: Familiäre Auftreten der infantilen kortikalen Hyperostose. *Arch Kinderheilkd* **164**:271–276, 1961.

13. Finsterbush A, Rang M: Infantile cortical hyperostosis. *Acta Orthop Scand* **46**:727–736, 1975.

14. Fráňa L, Sekanina M: Infantile cortical hyperostosis. *Arch Dis Childh* **51**:589–595, 1976.

15. Fried K, et al: Autosomal dominant inheritance with incomplete penetrance of Caffey disease (infantile cortical hyperostosis). *Clin Genet* **19**:271–274, 1981.

16. Galyean J, Robertson WO: Caffey's syndrome: Some unusual ocular manifestations. *Pediatrics* **45**:122–125, 1970.

17. Holman GH: Infantile cortical hyperostosis: A review. *Q Rev Pediatr* **17**:24–31, 1962.

18. Jackson DR, Lyne DE: Infantile cortical hyperostosis. *J Bone Jt Surg* **61**(A):770–772, 1979.

19. Kaser H: Das Krankheitsbild der infantilen corticalen Hyperostose. *Helv Paediatr Acta* **17**:153–184, 1962.

19a. Koslowski K, Tsuruta T: Dysplastic cortical hyperostosis: A new form of neonatal dwarfism. *Br J Radiol* **62**:376–378, 1989.

20. Kühl J et al: Ein Beitrag zur Krankheitsbild der infantilen kortikalen Hyperostose. *Arch Kinderheilkd* **179**:209–229, 1969.

20a. MacLachlan AK et al: Familial infantile cortical hyperostosis in a large Canadian family. *Can Med Assoc J* **130**:1172–1174, 1984.

21. Minton LR, Elliott J: Ocular manifestations of infantile cortical hyperostosis. *Am J Ophthalmol* **64**:902–907, 1967.

21a. Newberg AH, Tampas JP: Familial infantile cortical hyperostosis: An update. *Am J Roentgenol* **137**:93–96, 1981.

22. Pajewski M, Vure E: Late manifestations of infantile cortical hyperostosis (Caffey's disease). *Br J Radiol* **40**:90–95, 1967.

23. Röske G: Eine eigenartige Knochenerkrankung im Säuglingsalter. *Monatsschr Kinderheilkd* **47**:385–400, 1930.

24. Saul RA et al: Caffey's disease revisited. *Am J Dis Child* **136**:56–60, 1982.

25. Sidbury JB: Infantile cortical hyperostosis. *Postgrad Med* **22**:211–215, 1957.

26. Sidbury JB Jr, Sidbury JB: Infantile cortical hyperostosis: Inquiry into the etiology and pathogenesis. *N Engl J Med* **250**:304–314, 1954.

27. Smyth FS et al: Periosteal reaction, fever and irritability in young infants: A new syndrome? *Am J Dis Child* **71**:333–350, 1946.

28. Taillefer R et al: Aspect scintigraphique de l'hyperostose corticale infantile (maladie de Caffey). *J Assoc Can Radiol* **34**:12–15, 1983.

29. Tampas JP et al: Infantile cortical hyperostosis. *JAMA* **175**:491–493, 1961.

30. Temperley IJ et al: Raised immunologlobulin levels and thrombocytopenia in infantile cortical hyperostosis. *Arch Dis Childh* **47**:982–983, 1972.

31. Van Zeben W: Infantile cortical hyperostoses. *Acta Paediatr Scand* **35**:10–20, 1948.

## Kenny syndrome (tubular stenosis)

Kenny and Linarelli (5) and Caffey (2) first described the syndrome of proportional growth retardation with macrocephaly, low birthweight, and episodic hypocalcemia with hyperphosphatemia leading to tetany. About 20 cases have been reported. Inheritance is autosomal dominant (5,8,9).

Adult height has ranged from 121 to 149 cm. (7). The shafts of the long bones are narrow, with stenosis of the medullary cavities. Bone age is delayed in about 65%. There is mild brachymetacarpalia. Delayed closure of the anterior fontanel (90%) and wide metopic suture with absent diploic spaces are seen. We cannot agree that the patient of Wilson et al (11) has the disorder. The patients of Majewski et al (8) lacked medullary stenosis. Mental development is retarded in 15% (3).

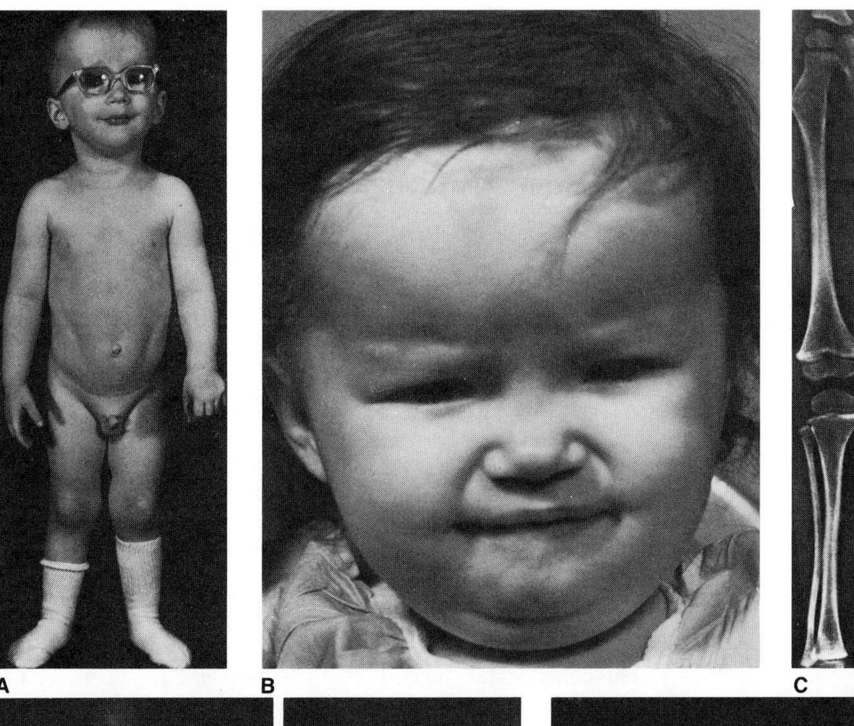


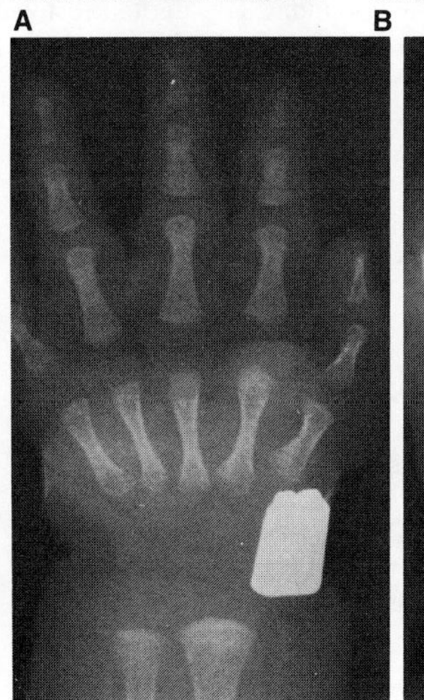
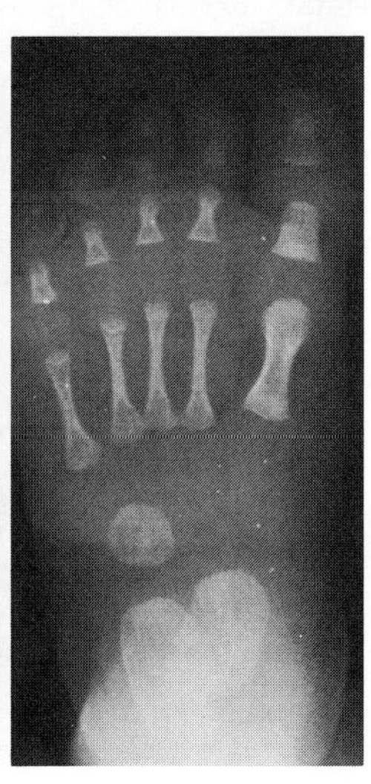


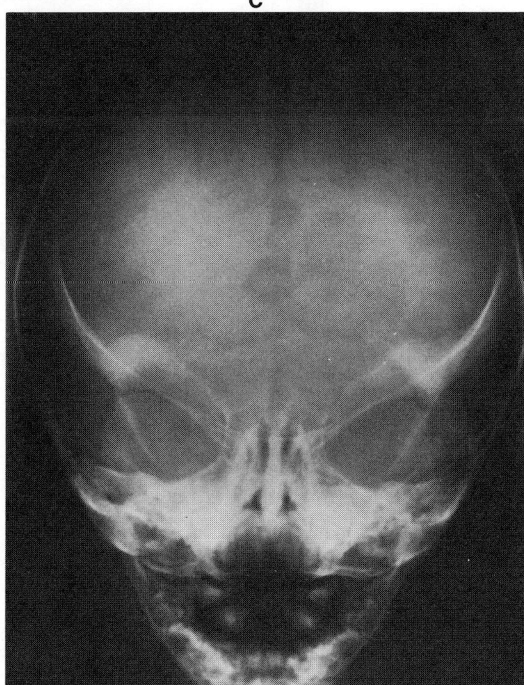


Fig. 9–20. *Kenny syndrome.* (A) Proportional dwarfism, myopia. (B) Frontal bossing, high hairline, diminished eyebrows and lashes. (C) Inner cortical thickening of thin tubular bones. (D) Similar appearance in bones of hands and feet. (E) Wide fontanel and metopic suture, harlequin configuration of orbital roofs. (A and C from *FM Kenny* and *L Linarelli,* Am J Dis Child **111:**201, 1967. D and E from *R Frech* and *W McAlister,* Radiology **91:**457, 1968.)

The facies is only mildly dysmorphic. There are frontal bossing, a high hairline, and diminished eyebrows and lashes (Fig. 9–20A–E). Ocular findings found in 80% have ranged from uncomplicated microphthalmia and hyperopia to extreme pseudopapilledema, vascular tortuosity, and macular clouding. Corneal and retinal calcification has been seen on autopsy (1a,7).

Episodic hypocalcemic tetany has been noted in about 65% (3). Anemia has been documented in about 30%. Idiopathic hypoparathyroidism has been noted (3).

### References [Kenny syndrome (tubular stenosis)]


1. Abdel-al Y et al: Kenny–Caffey syndrome. *Clin Pediatr* **28**:175–179, 1989.

1a. Boynton JR et al: Ocular findings in Kenny syndrome. *Arch Ophthalmol* **97**:896–900, 1979.

2. Caffey J: Congenital stenosis of medullary spaces in tubular bones and calvaria in two proportional dwarfs—mother and son coupled with transitory hypocalcemic tetany. *Am J Roentgenol* **100**:1–11, 1967.

3. Fanconi S et al: Kenny syndrome. Evidence for idiopathic hypothyroidism in two patients and for abnormal parathyroid hormone in one. *J Pediatr* **109**:469–475, 1986.

4. Frech RS, McAlister WH: Medullary stenosis of the tubular bones associated with hypocalcemic convulsions and short stature. *Radiology* **91**:457–461, 1968.

5. Kenny FM, Linarelli L: Dwarfism and cortical thickening of the tubular bones: Transient hypocalcemia in a mother and son. *Am J Dis Child* **111**:201–207, 1966.

6. Larsen JL et al: Unusual cause of short stature. *Am J Med* **78**:1025–1032, 1985.

7. Lee WK et al: The Kenny–Caffey syndrome: Growth retardation and hypocalcemia in a young boy. *Am J Med Genet* **14**:773–782, 1983.

8. Majewski F et al: The Kenny syndrome, a rare type of growth deficiency with tubular stenosis, transient hypoparathyroidism and anomalies of rarefaction. *Eur J Pediatr* **136**:21–30, 1981.

9. Sarria A et al: Estenosis tubular diafisaria (sindrome de Kenny–Caffey). *An Esp Pediatr* **13**:373–379, 1980.

10. Weiland P et al: Severe dwarfism associated with hypocalcemia and unusual parathyroid hormone findings. *Pediatr Res* **15**:119A, 1981.

11. Wilson MG et al: Dwarfism and congenital medullary stenosis (Kenny syndrome). *Birth Defects* **10**(12):128–132, 1974.


Fig. 9–21. *Lenz–Majewski syndrome.* (A,B) Note disproportionately large head contrasting with reduced trunk and limbs. Note prominent veins and syndactyly in A. (From *M Robinow,* J Pediatr **91:**417, 1977.)

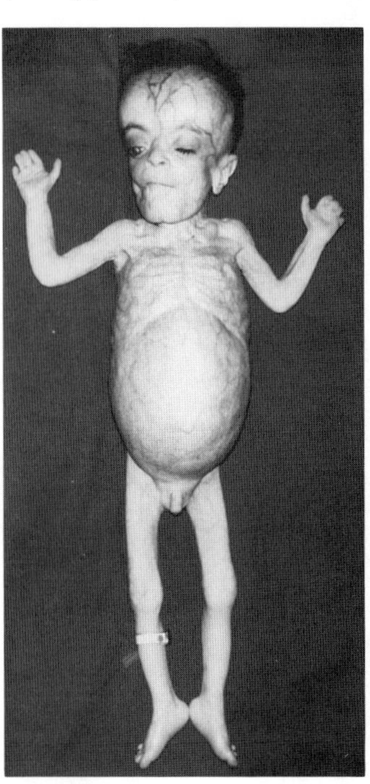

A

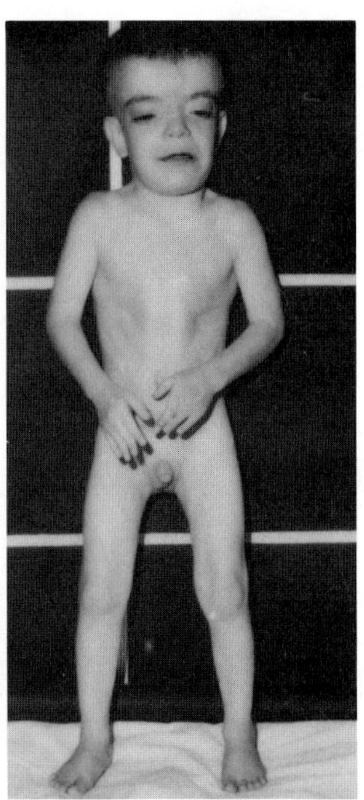

B

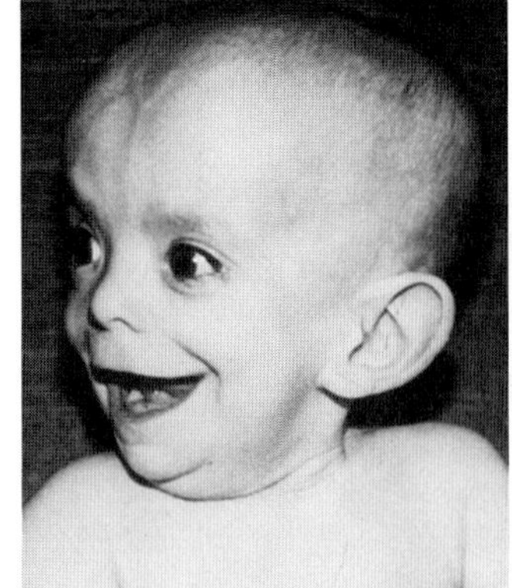

A

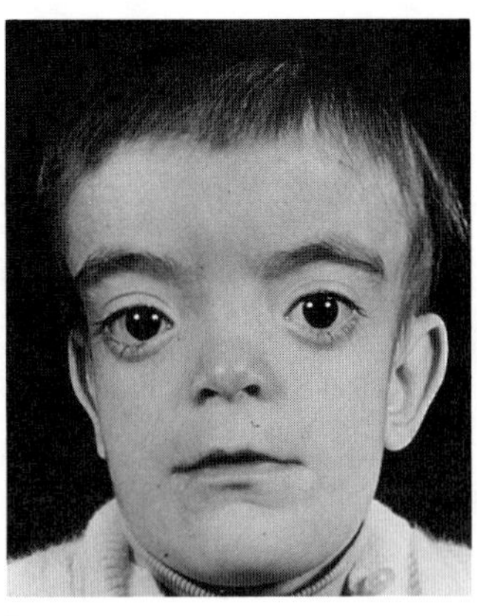

B

Fig. 9–22. *Lenz–Majewski syndrome.* Disproportionately large craniofacies. (A) Age 8 months. Note thin atrophic skin with prominent veins. (B) Note hypertelorism. (B from *M Robinow,* J Pediatr **91:**417, 1977.)

## Lenz–Majewski syndrome

The syndrome characterized by large head, characteristic facies, loose skin, mental retardation, and skeletal findings (Figs. 9–21 to 9–24) was first reported by Braham (1) in 1969. This report went largely ignored until the entity was rediscovered by Lenz and Majewski (6) in 1974. Eleven reported examples were reviewed in 1983 by Gorlin and Whitley (3). Additional examples are those of Elefant et al (2) and Chrzanowska et al (1a).

All cases have been isolated. Paternal age appears advanced (3). Chromosome studies have been normal.

**Facies.** The head appears disproportionately great with large fontanels and widely separated sutures that close late. The size of the head contrasts sharply with the reduced trunk and limbs (Figs. 9–21 and 9–22). Prominent veins, especially in the scalp, are evident. The ears are very large and floppy. Commonly there are choanal atresia or stenosis and nasolacrimal duct obstruction. Hypertelorism is evident.

**Musculoskeletal.** Inguinal hernia is common. The digits are hyperflexible and there may be generalized hypotonia (4). Radiographic features include progressive sclerosis of the skull (especially at the base), facial bones, and vertebrae. The clavicles and ribs are broad. The middle phalanges are short or absent. The long bones exhibit diaphyseal undermodeling and midshaft cortical thickening. However, there is marked hypostosis of the metaphyses and epiphyses. In general, skeletal maturation is retarded (Fig. 9–24A–G).

Fig. 9–23. *Lenz–Majewski syndrome.* Loose and wrinkled atrophic skin of hands with short digits and partial syndactyly.

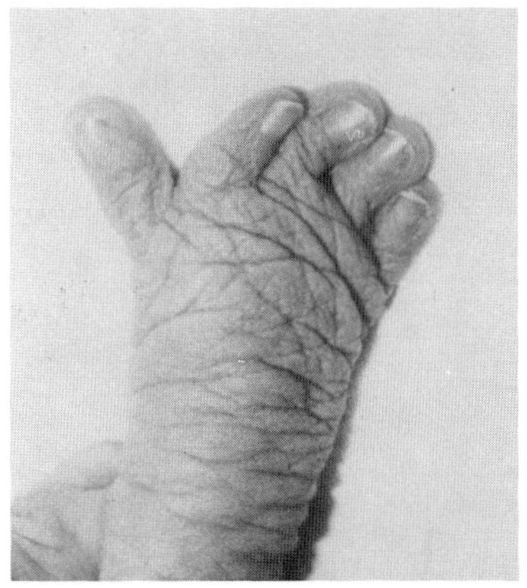

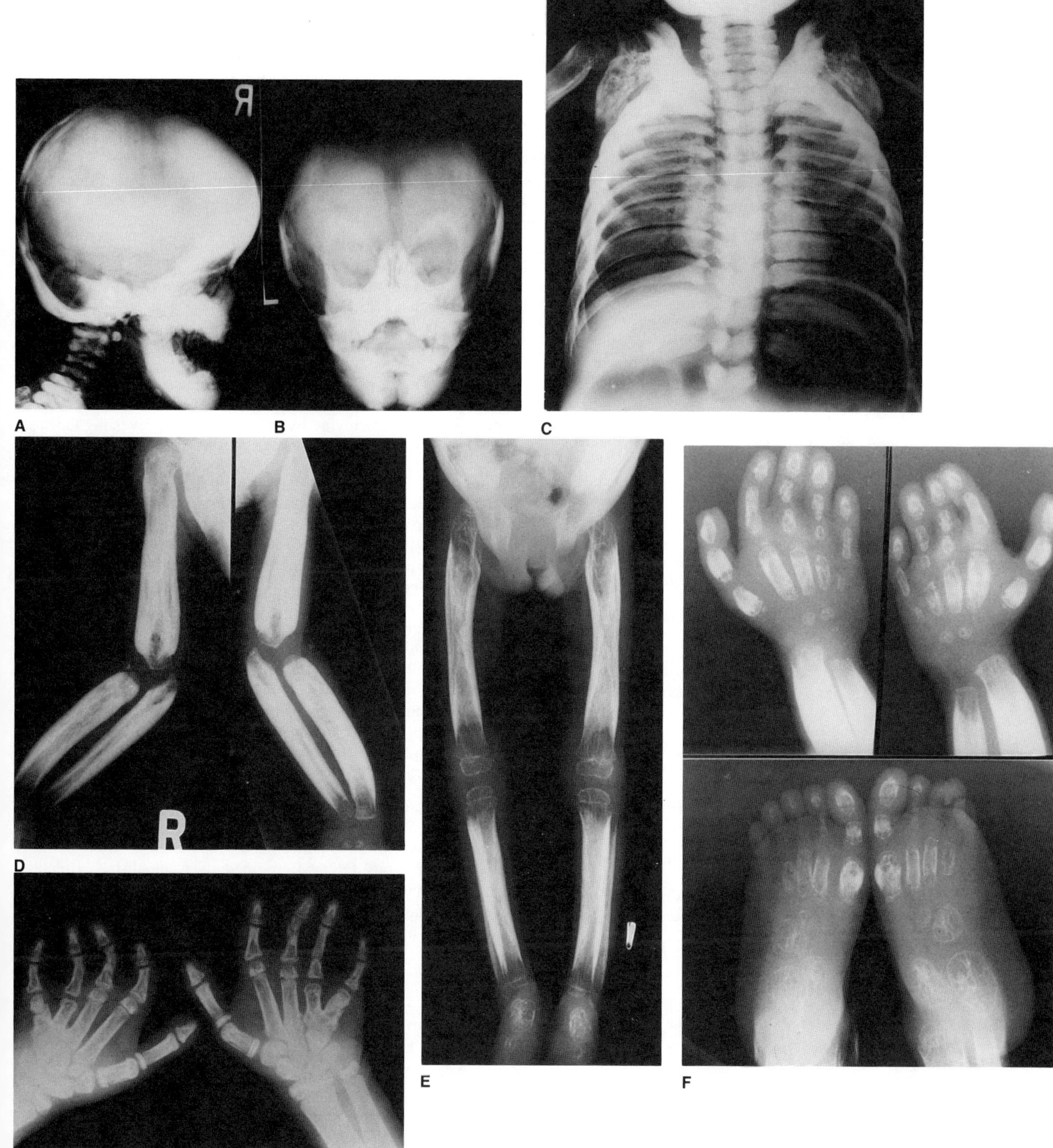


Fig. 9–24. *Lenz–Majewski syndrome.* Radiographs. (A,B) Hyperostosis of facial bones, cranial base, mandible, and proximal cervical spine; paranasal sinus and mastoids obliterated. (C) Hyperostosis of ribs, clavicles, scapulae. (D,E) Diaphyseal hyperostosis of long bones. (F) Hands and feet showing hypoplasia of bones on medial aspects of hands and lateral aspects of feet as well as diaphyseal hyperostosis. (G) Note abbreviated digits, absent middle phalanges, and clinodactyly. (A–F from *RI Macpherson,* J Can Assoc Radiol **25**:22, 1974.)

**Genitourinary.** Cryptorchidism has been a uniform finding in affected males. Hypospadias and/or chordee have been noted (5,8). The anus may be anteriorly displaced.

**Central nervous system.** All children with disorder have been mentally retarded, intelligence quotients ranging from 20 to 40. Sensorineural hearing loss is frequent.

**Skin.** The skin is thin, loose, wrinkled, and atrophic (Fig. 9–23). Veins, especially in the scalp, are prominent and cutis marmorata is evident (4). Proximal interdigital webbing of the fingers is frequent (3).

**Oral manifestations.** Tooth enamel has been defective in all patients. No microscopic studies have been carried out.

**Differential diagnosis.** Radiographically the disorders most often mistaken for Lenz–Majewski syndrome are *craniometaphyseal dysplasia* and *craniodiaphyseal dysplasia.* In contrast to those disorders, there does not appear to be any impairment of cranial nerves. One example was thought to represent *Camurati–Engelmann syndrome* (1).

**Laboratory aids.** While alkaline phosphatase levels have been elevated in some cases (3,8), the significance is not known.

### References (Lenz–Majewski syndrome)


1. Braham RL: Multiple congenital abnormalities with diaphyseal dysplasia (Camurati–Engelmann's syndrome). *J Oral Surg* **27**:20–26, 1969.

1a. Chrzanowska KH et al: Skeletal dysplasia syndrome with progeroid appearance, characteristic facial and limb anomalies, multiple synostoses, and distinct skeletal changes: A variant example of the Lenz-Majewski syndrome. *Am J Med Genet* **32**:470–474, 1989.

2. Elefant E et al: Acrogeria: A case report. *Ann Paediatr* **204**:273–280, 1965.

3. Gorlin RJ, Whitley CB: Lenz–Majewski syndrome. *Radiology* **149**:129–131, 1983.

4. Hood OJ et al: Cutis laxa with craniofacial, limb, genital and brain defects. *J Clin Dysmorphol* **2**(4):23–26, 1984.

5. Kay CI et al: Cutis laxa, skeletal anomalies, and ambiguous genitalia. *Am J Dis Child* **127**:115–117, 1974.

6. Lenz WD, Majewski F: A generalized disorder of the connective tissues with progeria, choanal atresia, symphalangism, hypoplasia of dentin and craniodiaphyseal hypostosis. *Birth Defects* **10**(12):133–136, 1974.

7. Macpherson RI: Craniodiaphyseal dysplasia, a disease or group of diseases. *J Can Assoc Radiol* **25**:22–33, 1974.

8. Robinow et al: The Lenz–Majewski hyperostotic dwarfism: A syndrome of multiple congenital anomalies, mental retardation, and progressive skeletal sclerosis. *J Pediatr* **91**:417–421, 1977.


## Mandibuloacral dysplasia

Cavallazzi et al (1), in 1960, were probably the first to report a syndrome characterized by mandibular hypoplasia, delayed cranial suture closure, dysplastic clavicles, abbreviated club-shaped terminal phalanges, acroosteolysis, and atrophy of skin over the hands and feet (Figs. 9–25 to 9–27). They did not recognize the condition as a new syndrome. Others reported cases but did not associate the disorder with that in Cavallazzi's patient or mistakenly classified their patients as examples of either Werner syndrome or acrogeria (2,4,7). Young et al (11) coined the name for this disorder. Danks et al (3) employed the term "craniomandibular dermatodysostosis" to describe what we believe is the same condition. The occurrence of the disorder in sibs strongly suggests autosomal recessive inheritance (7–10,12). It has an apparently high frequency in Italians (9).

The shoulders are narrow, changes usually becoming evident by the sixth or seventh year. Fat deposits are marked over the abdomen (5,6,10).

**Facies.** The eyes appear exophthalmic. The nose is sharp or pointed. With age, the face becomes pinched. The scalp hair becomes sparse and lustreless without graying by the third decade, revealing the scalp veins. The mandible becomes progressively smaller.

**Musculoskeletal.** Stature is usually reduced by 3 SD (3,11). There appears to be progressive stiffness of the joints (3,11,12) and swelling and redness may occur. The hands and feet appear small. The terminal phalanges are short and club-shaped which, together with the cutaneous atrophy, cause them to appear spatulate.

Radiographic changes include slow osteolysis of the mandibular body and ramus, resulting in micrognathia, delayed cranial suture closure, numerous Wormian bones, and acroosteodysplasia of fingers and toes. Clavicular osteolysis is progressive, eventuating in severe hypoplasia or aplasia.

Variable findings include abnormal modeling of femur, humerus, and tibia with patchy thickening of the cortex, hypoplasia of the first through fourth ribs and coxa valga.

**Skin.** The skin over the hands and feet becomes atrophic and mottled giving a poikilodermatous appearance at about two years of age. A mottled brown skin rash may extend over the trunk and extremities (5–11). Subcutaneous fat is diminished over the distal extremities causing the veins and tendons to become evident. Calcium deposits frequently extrude from the scalp, ears, elbows and fingertips. Plantar hyperkeratoses have been noted (7). The nails become brittle.

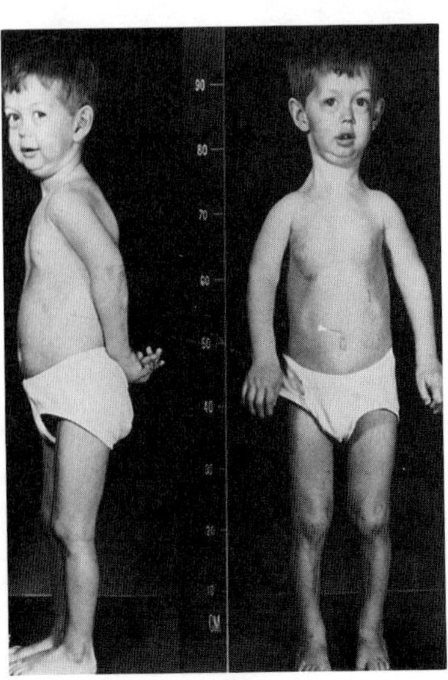


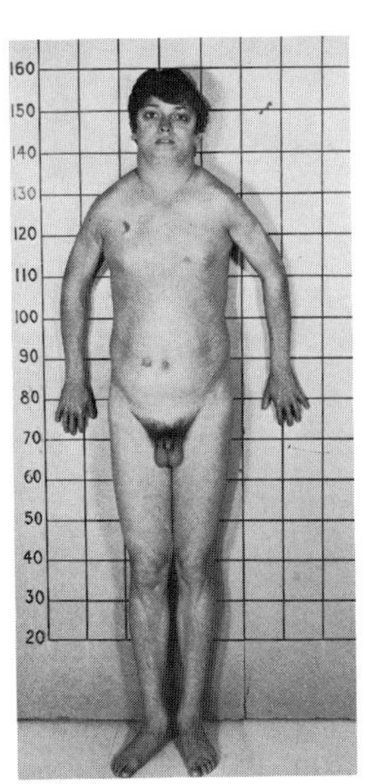


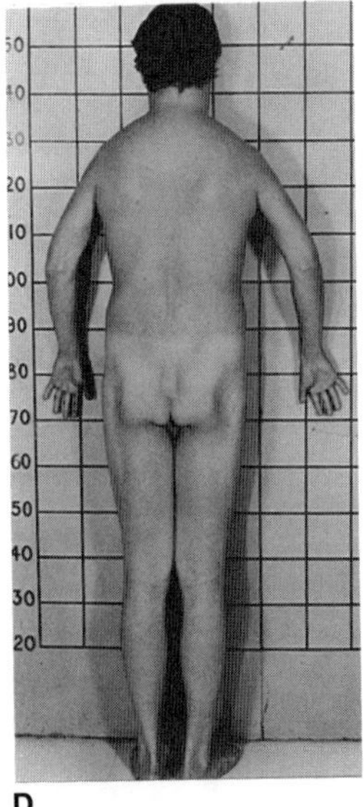


Fig. 9–25. *Mandibuloacral dysplasia.* (A–D) Short stature, sloping shoulders, thick trunk and neck, micrognathia. [A,B from *DM Danks et al,* Birth Defects **10**(12):99, 1974. C,D from *VA McKusick et al,* Birth Defects **7**(7):291, 1971.]

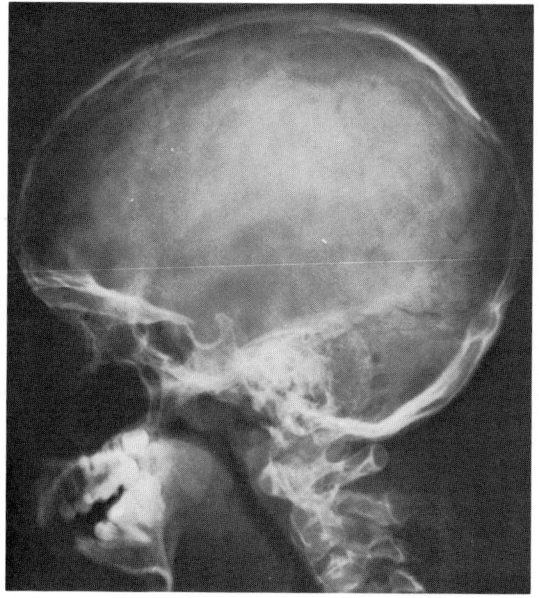

A

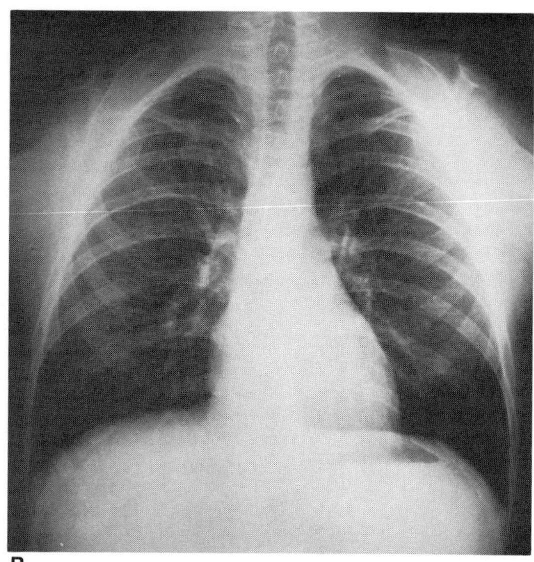

B

Fig. 9–26. *Mandibuloacral dysplasia.* Roentgenograms. (A) Wide sagittal sutures and hypoplastic body and ramus of mandible. (B) Hypoplastic clavicles. [A,B from *VA McKusick et al,* Birth Defects **7**(7):291, 1971.]

Fig. 9–27. *Mandibuloacral dysplasia.* (A) Short terminal phalanges and stiffness of interphalangeal joints. (B) Similar acroosteolytic alterations in toes. [A,B from *VA McKusick et al,* Birth Defects **7**(7):291, 1971.]

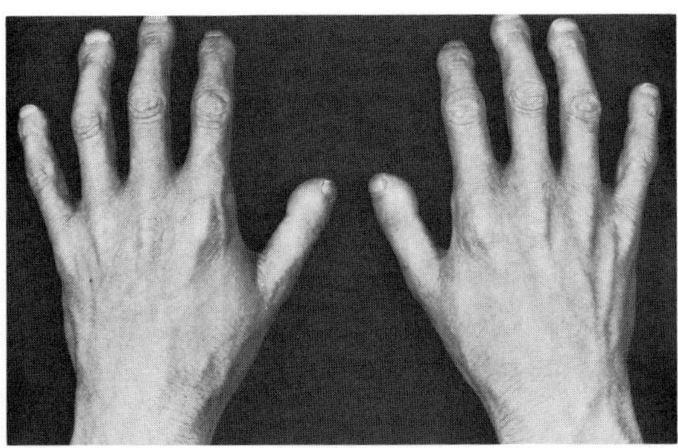

A

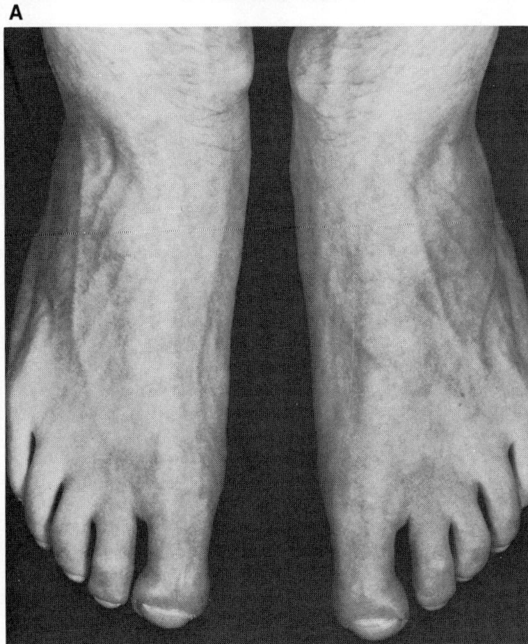

B

**Oral manifestation.** Mandibular hypoplasia with inability to open the mouth widely has been a common feature. Crowding of the mandibular teeth with absence of cellular cementum has been demonstrated (3). The roots are hypoplastic. The teeth are lost during adolescence (12).

**Differential diagnosis.** Similar cutaneous changes are seen in *progeria* and *Werner syndrome*. *Cleidocranial dysplasia* should easily be excluded.

**Laboratory findings.** Diabetes mellitus has been described (7).

### References (Mandibuloacral dysplasia)

1. Cavallazzi C et al: Su di un caso di disostosi cleidocranica. *Riv Clin Pediatr* **65**:312–326, 1960.
2. Cohen LK et al: Werner's syndrome. *Cutis* **12**(1):76–80, 1973.
3. Danks DM et al: Craniomandibular dermatodysostosis. *Birth Defects* **10**(12):99–105, 1974.
4. Levi L et al: L'acrogeria di Gottron: Descrizione di un caso. *G Ital Dermatol* **105**:645–651, 1970.
5. McKusick VA: *Heritable Disorders of Connective Tissue,* 4th ed. C.V. Mosby, St. Louis, 1972 pp. 822–823.
6. McKusick VA et al: *Medical Genetics 1961–1963*. Pergamon Press, New York, 1966, p. 447.
7. Mensing H et al: Werner-Syndrom-artige Erkrankung bei drei Brüdern. *Hautarzt* **33**:542–547, 1982.
8. Pallotta R, Morgese G: Mandibuloacral dysplasia: A rare progeroid syndrome: two brothers confirming autosomal recessive inheritance. *Clin Genet* **26**:133–138, 1984.
9. Tenconi R et al: Another Italian family with mandibuloacral dysplasia: Why does it seem more frequent in Italy? *Am J Med Genet* **24**:357–364, 1986.
10. Welsh O: Study of a family with a new progeroid syndrome. *Birth Defects* **11**(5):25–38, 1975 and *Mod Prob Paediatr* **17**:44–58, 1975.
11. Young LW et al: A new syndrome manifested by mandibular hypoplasia, acroosteolysis, stiff joints and cutaneous atrophy (mandibuloacral dysplasia) in two unrelated boys. *Birth Defects* **7**(7):291–297, 1971.
12. Zina AM et al: Familial mandibuloacral dysplasia. *Br J Dermatol* **105**:719–723, 1981.

## Marfan syndrome

The main features of the Marfan syndrome include disproportionate skeletal growth with dolichostenomelia and arachnodactyly, ectopia lentis, and fusiform and dissecting aneurysms of the aorta. Marfan syndrome was first mentioned in the fifth century in the Babylonian Talmud (28). It has been suggested that Abraham Lincoln had Marfan syndrome (29,48) but because Lincoln was known to have a Marfanoid habitus and his son, Tad, had a flat face with cleft palate, the possibility of Stickler syndrome has also been raised (34). A case has been made for Marfan's original patient having cystathionine synthase deficiency (homocystinuria) (7), but the most convincing argument has been made for congenital contractural arachnodactyly (6). The famous violin virtuoso, Nicolo Paganini, had long fingers and hyperextensible joints that were said to be attributable to Marfan syndrome (67). In 1986, the volleyball star and U.S. Olympic Team member Flo Hyman died from a ruptured aortic aneurysm associated with Marfan syndrome. This tragic event focused widespread public attention on the syndrome for the first time (58).

Marfan syndrome follows an autosomal dominant mode of transmission with a high degree of penetrance and variable expressivity. The homozygous state is lethal (13,68). Approximately 85% of cases are familial with the rest arising as new mutations (3,44,48,53,60). Advanced paternal age at the time of conception is associated with sporadic cases (44,48). The prevalence has been estimated to vary from 1.5 to 10/100,000 in the general population, depending on the mode of ascertainment (23,27,58,60).

Marfan syndrome is thought to result from a structural defect in one of the connective tissue proteins. Various abnormalities in collagen metabolism have been observed; the best defined is an increase in hydroxyproline excretion, which is presumed to result from increased collagen turnover (8,48). The initial report of an association between a 38 base pair insertion in the pro-$\alpha$2(I) collagen gene and Marfan syndrome (11) suggested that the basic cause of the disorder might be identified, but further investigations have shown that this DNA insertion is a common polymorphism (21). Linkage analyses using various collagen gene probes have been uniformly uninformative (PH Byers, personal communication, 1987). Ogilvie et al (55) published evidence against mutations in the major fibrillar collagen genes being involved.

Other studies focusing on elastin-related proteins appear more promising. In vascular tissue, the histopathologic changes are restricted to the elastic layers of the media (1,57). Although a defect in the elastin protein itself has been suggested (1), significant differences in hyaluronic acid metabolism (4) and in the elastin-associated microfibrillar fiber array (36,37) have also been demonstrated in fibroblast cultures.

**General features.** Characteristic skeletal, cardiovascular, and ocular changes may be present from birth, or may be appreciated in childhood, adolescence, or only in later adulthood because of an affected relative (Table 9–2). Normal growth is the rule, although affected individuals are taller and thinner than average. Head circumference is normal and intelligence is not affected. An extensive long term study of the natural history was reported by Marsalese et al (46).

**Musculoskeletal system.** Dolichostenomelia, defined by an upper segment-to-lower segment ratio (US/LS) of at least 2 SD below the mean, is found in 76%. Arachnodactyly occurs in 88% (Figs. 9–28 and 9–29). Other features include pectus deformity (excavatum more often than carinatum) (68%), scoliosis (44%), and weakness of joint capsules manifested by pes planus (44%) and hyperextensibility of joints (56%) with recurrent dislocation (hips, patellae, clavicles). Enlarged vertebrae and widened spinal cord have also been noted (60,79,83).

**Ocular changes.** Ectopia lentis, resulting from lax suspensory ligaments, is found in 50–80% and is almost always bilateral and symmetrical (14,19,44). The direction of dislocation is upward in about 75% (47,54,60). The zonular fibers remain intact, permitting normal accommodation. Some degree of backward dislocation is frequently found, but most cases of glaucoma are the result of surgical extirpation rather than acute blockage of the anterior chamber. An enlarged axial diameter is responsible for the increased tendency to myopia

Table 9–2. Characteristics of Marfan syndrome[a] ($n = 50$)

| Clinical feature | Percentage |
|---|---|
| Ocular | 70 |
| Ectopia lentis | 60 |
| Myopia | 34 |
| Cardiovascular | 98 |
| Midsystolic click only | 30 |
| Midsystolic click and late-systolic murmur | 18 |
| Aortic regurgitant murmur | 10 |
| Mitral regurgitant murmur only | 6 |
| Prosthetic aortive valve | 10 |
| Abnormal echocardiogram | 96 |
| Aortic enlargement | 84 |
| Mitral valve prolapse | 58 |
| Prosthetic aortic valve | 10 |
| Musculoskeletal | 100 |
| Arachnodactyly | 88 |
| Upper segment/lower segment at least 2 SD below mean for age | 77 |
| Pectus deformity | 68 |
| High, narrow palate | 60 |
| Height greater than 95th percentile for age | 58 |
| Hyperextensible joints | 56 |
| Vertebral column deformity | 44 |
| Pes planus | 44 |
| Striae distensae | 24 |
| Inguinal hernia | 22 |
| Family history | 85 |
| Additional documented cases of syndrome | 85 |
| Sporadic cases (new mutations) | 15 |
| Unclear or unknown pedigree | 6 |

[a] From RE Pyeritz and VA McKusick, *N Engl J Med* **300**:772, 1979.

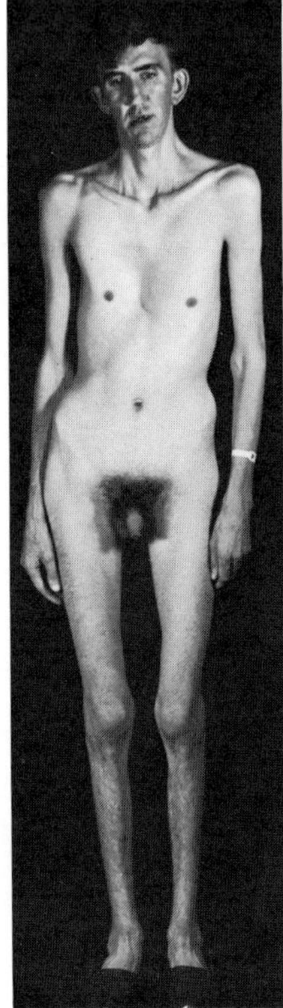

Fig. 9–28. *Marfan syndrome.* Patient has dolichostenomelia, arachnodactyly, scoliosis, pectus excavatum, and bilaterally subluxated lenses. (From *N Tuna* and *AP Thal*, Circulation **24:**1154, 1961.)

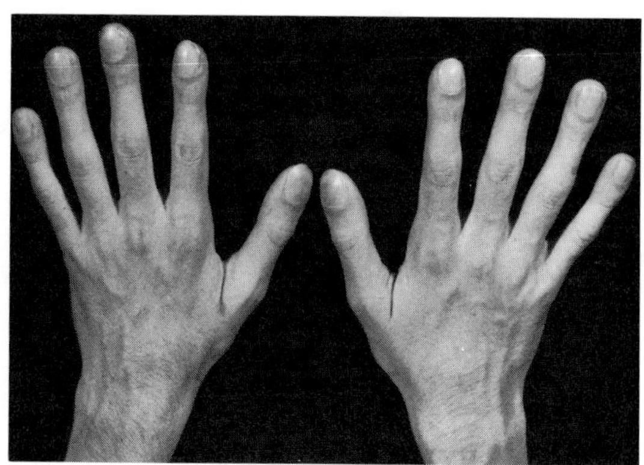
A

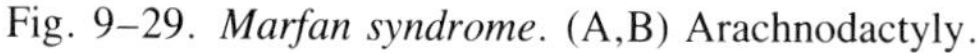

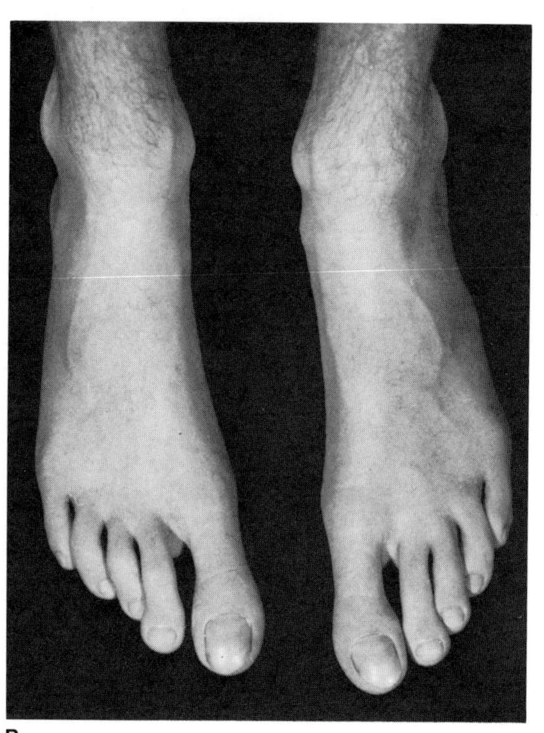
B

Fig. 9–29. *Marfan syndrome.* (A,B) Arachnodactyly.

(34%) that occurs most often in the first two decades. Of 300 Marfan syndrome patients, 6.4% developed retinal detachment (49).

An increased corneal diameter gives the appearance of megalocornea, but corneal flattening is a consistent finding. Hypopigmentation of the posterior iris epithelial layer permits transillumination of the iris in 10% (63). Increased size of the eyeglobe accounts for choroid thinning, apparent microphakia, and blue sclerae. Ocular changes are also found in the anterior chamber angle, ciliary body, and pupil (2). The most frequent cause of visual loss is amblyopia from delayed refraction or inadequate correction of myopia; the visual loss is usually reversible. More than 80% of patients have normal vision, even with dislocated lenses. Rarely, Rieger anomaly has been noted (32).

**Cardiovascular abnormalities.** Cystic necrosis of the vascular media leads to diffuse, progressive dilatation of the ascending aorta (65). The resulting aneurysm can be associated with increasing aortic regurgitation and congestive heart failure or dissection and rupture. Together, these constitute the leading causes of death in affected individuals. Before the development of an effective procedure for presymptomatic, elective repair of the aorta, the average age of death was 32 years (9,30,52,82). Echocardiographic evidence of mitral valve prolapse is found in 68% (61) (Table 9–3).

Valvular abnormalities may be present at birth and aortic dilatation is detectable by echosonography from early infancy (10,70). Mitral valve prolapse is found in the majority of children (67–100%). Approximately 69% develop progressive mitral valvular dysfunction and approximately 33% develop potentially malignant ventricular dysrhythmias (14,70).

In adults, significant auscultatory signs are found in only one-third to one-half of cases, whereas abnormal echocardiograms are found in the majority (80–95%). Characteristically, the aortic root diameter is above the upper limit of normal (approximately 50 mm) and there is mitral valve prolapse and left atrial dilatation. Symptomatic aortic regurgitation, which is found in less than 25% of patients, and aneurysmal dissection are both associated with dilatation of the aortic root beyond 60 mm (18,60,61). De Sanctis et al (22) published an excellent review of aortic dissection. Serious cardiovascular complications during pregnancy are also more likely with dilatation greater than 60 mm (24,59). Other reported cardiovascular complications include aneurysms of the descending aorta or pulmonary artery (48). As with

Table 9–3. Mitral valve prolapse syndromes

| Name | Connective tissue findings | Other findings | Lab findings | Inheritance |
|---|---|---|---|---|
| Cutis laxa | Loose, sagging skin | | Abnormal elastic fibers | AD |
| Ehlers–Danlos syndromes | | | | |
| I | Easy bruising, prominent scars, joint hypermobility, soft, hyperextensible skin, herniae, varicose veins | | Large, irregular collagen fibrils | AD |
| II | Easy brusising, joint hypermobility, soft skin with moderate hyperextensibility, hernias, varicose veins | | Large, irregular collagen fibrils | AD |
| III | Soft skin, marked joint hypermobility, varicose veins, hernias | | Large, irregular collagen fibrils | AD |

Table 9–3. Mitral valve prolapse syndromes (*continued*)

| Name | Connective tissue findings | Other findings | Lab findings | Inheritance |
|---|---|---|---|---|
| VI | Scoliosis, ocular fragility, soft, hyperextensible skin, hypermobile joints | | Lysyl hydroxylase deficiency | AR |
| VIII[a] | Skin fragility with abnormal scars, moderate joint laxity, periodontitis | | Small collagen fibrils in dermis | AD |
| X | Soft, mildly hyperextensible skin, mild joint laxity, easy bruising | | Fibronectin defect | AR |
| Fragile X syndrome | Hyperelastic skin, hypermobile joints, pectus, highly arched palate | Mental retardation, autistic-like behavior | Fragile site on X | XL |
| Osteogenesis imperfecta | | | | |
| I | Frequent fractures, hearing loss, blue sclerae | | Decreased pro $\alpha 1$(I) chains | AD |
| III | Scleral and bone fragility, bone deformities | | Pro $\alpha 2$(I) chain mutation | ?AR, AD |
| IV | Frequent fractures, hearing loss | | Col 1A2 mutation | AD |
| Pseudoxanthoma elasticum | Skin lesions, angioid streaks in eye | | Abnormal elastic fibers | AD |
| MPS III | Hernias, deafness | Mental retardation, hepatosplenomegaly | Excess heparan sulfate excretion | AR |
| Contractural arachnodactyly | Clubfoot, contractures, scoliosis, myopia, crumpled ear | | | AD |
| Marfan syndrome | Striae, myopia, retinal detachment, highly arched palate, pectus, scoliosis, hernias, hypermobile joints, arachnodactyly | | | AD |
| Marfan hypermobility syndrome | Striae, slight skin hyperextensibility, myopia, highly arched palate, dental crowding, pectus, scoliosis, joint laxity, joint dislocations | | | AD |
| Stickler syndrome | Prominent joints, myopia, highly arched palate | Midface hypoplasia, cleft palate, sensorineural deafness | | AD |
| Mitral valve prolapse syndrome | Myopia, highly arched palate, scoliosis, absent thoracic kyphosis, pectus | | | AD |
| Forney syndrome | Highly arched palate, deafness, dental crowding, joint fusions | Freckling, iris pigmentary changes, exotropia, short stature | | AD |
| Pfeiffer–Palm–Teller syndrome | Highly arched palate, cup-shaped ears, limited joint mobility | Enamel hypoplasia, highly pitched voice, brachydactyly, short stature, hypertonia | AR | |
| Tamminga syndrome | Myopia, limited knee and elbow movement | Cerebral atrophy, low-set ears, deep-set eyes, iris dysplasia, optic coloboma, clinodactyly, mental retardation, hypotonia | | ? |
| Furlong syndrome | Myopia, highly arched palate, scoliosis, hernias, joint contractures, arachnodactyly | Craniosynostosis, proptosis, hypertelorism, hypospadias | | ? |
| Shprintzen–Goldberg syndrome | Pliable ears, pectus, hernias, pectus, contractures, arachnodactyly | Craniosynostosis, strabismus, exophthalmos, maxillary and mandibular hypoplasia, hypertelorism, optic atrophy, lowset ears, cryptorchidism, mental retardation | | ? |
| Sugarman–Vogel syndrome[b] | Hyperextensible skin, highly arched palate, pectus, hernias, joint hypermobility, arachnodactyly, genu recurvatum | Hypertelorism, strabismus, exophthalmos, micrognathia, low-set ears, cryptorchidism, clinodactyly, hypotonia | | ? |
| Jaffer–Beighton syndrome[b] | Highly arched palate, dental crowding, pectus, joint hypermobility, arachnodactyly, soft skin | Brachycephaly | | ? |

[a] Dubious existence.

[b] Mitral valve prolapse has not been described in this syndrome but should be looked for in future cases.

all patients with valvular disease, there is an increased risk of bacterial endocarditis.

**Pulmonary pathology.** Because of the connective tissue defect, the tall asthenic habitus, and the frequency of thoracic cage deformities, affected individuals are at increased risk for spontaneous pneumothorax. Although the frequency is low (4.4%), recurrence is likely unless the underlying lesion, usually an apical bulla, is resected. Pulmonary infections and chronic emphysematous changes occur with increased frequency. The pulmonary vital capacity is generally reduced, adding significantly to the anesthetic risk (33,40,76,81).

**Skin and integument.** Connective tissue involvement leads to an increased frequency of striae distensae (24%) and inguinal herniae (22%). Miescher elastomata, especially on the neck, may also be found (42,48,60).

**Craniofacial features.** Dolichocephaly usually occurs with prominent supraorbital ridges. Frontal bossing is common, and decreased retrobulbar tissue is associated with enophthalmos (54).

Although earlier authors estimated the prevalence of a high palatal vault to be between 15 and 40% (48), some have observed this condition in all their patients (Fig. 9–30) (44). Cleft palate or bifid uvula has been reported in several instances (44,48,80). The teeth have been noted to be long and narrow and frequently maloccluded. Mandibular prognathism is common and temporomandibular joint disease is found more frequently than expected. Large maxillary sinuses may be noted radiographically (5,42,48).

**Miscellaneous findings.** Abnormalities of the central nervous system, presumably of connective tissue origin, include dural ectasia, sacral meningocele, and dilated cisterna magna, but neurological manifestations are rare (15,35,58).

**Diagnosis.** Early diagnosis is important for genetic counseling and for preventing or delaying the complications of cardiac lesions (12). The US/LS ratio (48) and the metacarpal index (39,56) are useful anthropometric measurements but must be related to age- and sex-matched normal values. The Walker–Murdock wrist sign (48) and the Steinberg thumb sign (71) are subject to interpretation and are of limited value in differentiating Marfan syndrome from other hypermobility conditions (60,64,78). Growth data and anthropometric measurements in the Marfan syndrome have been especially well analyzed by Pyeritz et al (62). Detailed ophthalmological evaluation and echocardiography will most often provide the objective signs necessary to substantiate the diagnosis (58,60).

During the third trimester, diagnosis can be strongly suspected on the basis of ultrasonographic analysis of limb lengths. Although positive determination is too late to allow the option of termination, some parents may benefit from emotional preparation for the birth of an affected child (66).

Fig. 9–30. *Marfan syndrome.* High palatal vault.

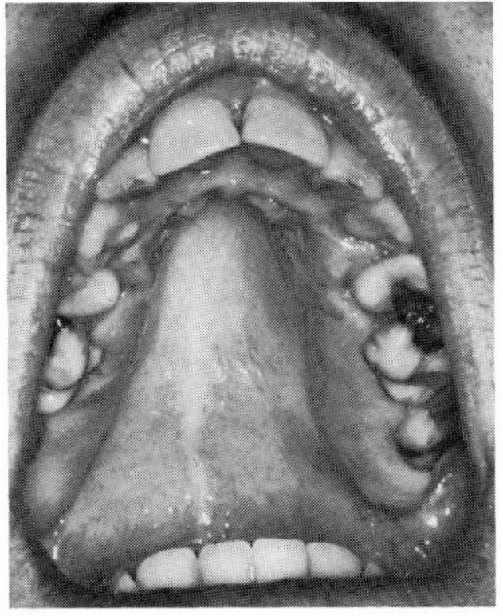

Graham et al (21) reported experience in diagnosing Marfan syndrome during infancy in 22 patients. Serious cardiac pathology and congenital contractures of the hands were frequent findings. Impressive also was the association with megalocornea, particularly in the more severely affected infants.

**Differential diagnosis.** A Marfanoid habitus may be observed in *cystathionine synthase deficiency (homocystinuria),* congenital contractural arachnodactyly (6), Marfanoid hypermobility syndrome (77), eunuchoidism, *Klinefelter syndrome, XXY syndrome* (23), sickle cell anemia, *multiple endocrine adenomatosis, Type IIb,* and occasionally in *nevoid basal cell carcinoma syndrome.* Marfanoid habitus may also be a feature in some cases of *Stickler syndrome* (34) and, surprisingly, has been observed with multiple endocrine neoplasia, Type I, in a patient with mitral valve prolapse, mental retardation, and bilateral optic atrophy (45).

Marfanoid features may also be seen in other disorders. In the *Shprintzen–Goldberg syndrome* (69,72), arachnodactyly is associated with craniosynostosis, mental deficiency, ocular proptosis, micrognathia, and other features. Furlong et al (27) reported a patient with the typical characteristics of Marfan syndrome, but additionally had craniosynostosis, hypospadias, spondylolisthesis, and absence of ectopia lentis. Cohen (17) reported a patient with Marfanoid features, sagittal synostosis, preauricular tags, and other anomalies. Clunie and Mason (16) described three sibs, products of a consanguineous union; a Marfanoid habitus and multiple inguinal and femoral hernias were described. Jaffer and Beighton (38) noted a patient with arachnodactyly, pectus carinatum, spondylolisthesis, joint laxity, and mental deficiency. Lujan et al (43) and Fryns and Buttiens (26) reported an X-linked disorder characterized by a Marfanoid habitus, large head, mental retardation, long narrow face, highly arched palate, small mandible, atrial septal defect, and other anomalies. Tamminga et al (73) reported an infant with a Marfanoid phenotype, congenital contractures, microspherophakia, optic nerve colobomas, prolapse of the mitral and tricuspid valves, cerebral white matter hypoplasia, and spinal axonopathy. Currarino and Friedman (20) reported a severe form of congenital contractural arachnodactyly associated with unusual histopathologic changes in the metaphyses and epiphyses together with other anomalies such as ankyloblepharon, esophageal atresia with tracheoesophageal fistula, duodenal atresia, and vertebral malformations. They suggested that congenital contractural arachnodactyly may be etiologically heterogeneous.

Fragoso and Cantú (25) described four sibs with psychomotor retardation, flat and coarse facies, dolichocephaly, low posterior hairline, synophrys, hypertelorism, broad nose with bifid columella, malar hypoplasia, small mouth, and large ears (Fig. 9–31A,B). Musculoskeletal alterations included pectus excavatum, muscle hypoplasia, dolichostenomelia, osteopenia, and thin metapodial bones, phalanges, and ribs. Inheritance is probably autosomal recessive.

Thieffry–Kohler syndrome (40,75), a dubious entity, has been characterized by Marfanoid appearance, frontal bossing, micrognathia, mild scoliosis, pes cavus, overlapping toes, and plantar cysts. Bone destruction of the wrists and ankles, beginning with the carpal and tarsal bones and spreading to involve adjacent bones, occurred during childhood and progressed, with painless, at times asymmetric, osteolysis. Serum alkaline phosphatase and hydroxyproline levels were elevated.

The Mirhosseini syndrome is characterized by microcephaly, severe mental deficiency, pigmentary retinal degeneration, cataracts, hyperextensible joints, arachnodactyly, and mild scoliosis (50,51). Inheritance is autosomal recessive.

A decreased US/LS ratio is a normal finding in Nilotic blacks (48). Ectopia lentis may occur in Weil–Marchesani syndrome (48), *Ehlers–Danlos syndromes, homocystinuria, osteogenesis imperfecta,* and as an isolated autosomal recessive trait (48). In homocystinuria, the lens tends to dislocate, nasally, inferonasally, or inferiorly (19). Aortic dilatation is also seen in Erdheim's cystic medical necrosis and in tertiary syphilis (48). Joint hypermobility is seen in a number of disorders, including *osteogenesis imperfecta,* the *Ehlers–Danlos syn-*

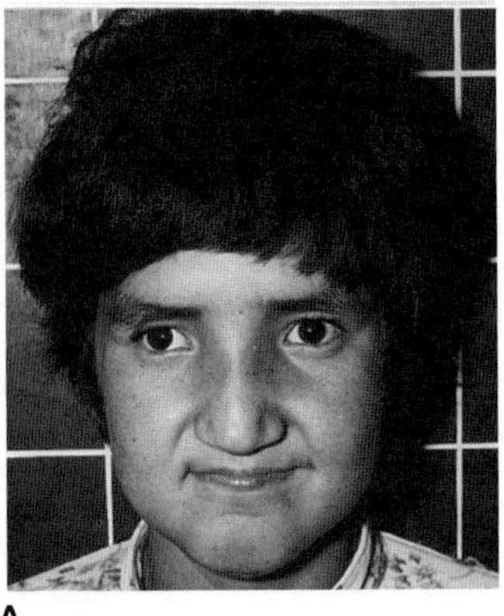
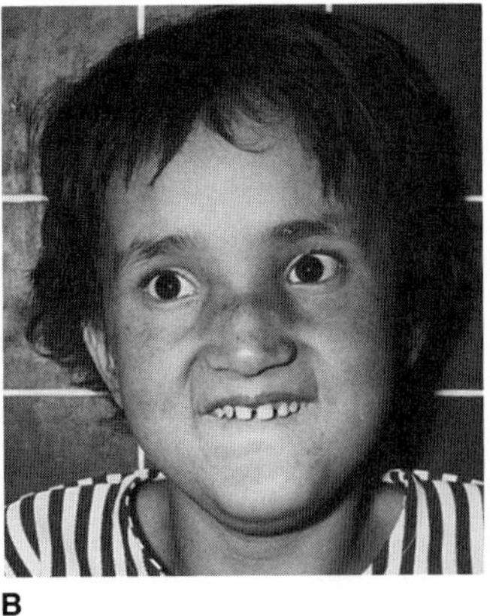

A B

Fig. 9–31. *Coarse facies, Marfanoid habitus, and mental retardation.* (A,B) Facies of two of four affected sibs. In addition to mental retardation, all have flat coarsened facies, synophrys, hypertelorism, esotropia, broad nose, short philtrum, somewhat small mouth. (From *R Fragoso* and *JM Cantu*, Clin Genet **25:**187, 1984.)

*dromes, homocystinuria, Stickler syndrome,* and Marfanoid hypermobility syndrome (34,48,77).

Temtamy et al (74) compared ultrastructural findings of the dental pulp and gingiva in Marfan syndrome and homocystinuria, noting the accumulation of finely granular and fibrillar material between elastic and collagen fibers and in nerve endings of the latter.

### References (Marfan syndrome)

1. Abraham PA et al: Marfan syndrome: Demonstration of abnormal elastin in aorta. *J Clin Invest* **70**:1245–1252, 1982.

2. Allen RA et aL: Ocular manifestation of the Marfan syndrome. *Trans Am Acad Ophthalmol Otolaryngol* **71**:18–38, 1967.

3. Ambani LM et al: Variable expression of Marfan syndrome in monozygotic twins. *Clin Genet* **8**:358–363, 1975.

4. Appel A et al: Cell-free synthesis of hyaluronic acid in Marfan syndrome. *J Biol Chem* **254**:12199–12203, 1979.

5. Barr M: Letter to the Editor. Marfan's syndrome. *N Engl J Med* **301**:273, 1979.

6. Beals RK, Hecht F: Congenital contractural arachnodactyly. *J Bone Jt Surg* **53A**:987–993, 1971.

7. Bianchine J: The Marfan syndrome revisited. *J Pediatr* **79**:717–719, 1971.

8. Boucek RJ et al: The Marfan syndrome: A deficiency in chemically stable collagen cross-links. *N Engl J Med* **305**:988–1012, 1981.

9. Bruno L et al: Cardiac, skeletal, and ocular abnormalities in patients with Marfan's syndrome and in their relatives: Comparison with the cardiac abnormalities in patients with kyphoscoliosis. *Br Heart J* **51**:220–230, 1984.

10. Buchanan R, Wyatt GP: Marfan's syndrome presenting as an intrapartum death. *Arch Dis Childh* **60**:1074–1076, 1985.

11. Byers PH et al: Marfan syndrome: Abnormal $\alpha 2$ chain in type I collagen. *Proc Natl Acad Sci US* **78**:7745–7749, 1981.

12. Chan K-L et al: Marfan syndrome diagnosed in patients 32 years of age or older. *Mayo Clin Proc* **62**:589–594, 1987.

13. Chemke J et al: Homozygosity for autosomal dominant Marfan's syndrome. *J Med Genet* **21**:173–177, 1984.

14. Chen S-C et al: Ventricular dysrhythmias in children with Marfan's syndrome. *Am J Dis Child* **139**:273–276, 1985.

15. Chu NS: Marfan's syndrome and epilepsy: Report of two cases and review of the literature. *Epilepsia* **24**:49–55, 1983.

16. Clunie GJA, Mason JM: Visceral diverticula and the Marfan syndrome. *Br J Surg* **50**:51–52, 1962.

17. Cohen MM Jr: Genetic perspectives on craniosynostosis and syndromes with craniosynostosis. *J Neurosurg* **47**:886–898, 1977.

18. Come PC: Echocardiographic assessment of cardiovascular abnormalities in the Marfan syndrome: Comparison with clinical findings and with roentgenographic estimation of aortic root size. *Am J Med* **74**:465–474, 1983.

19. Cross HE, Jensen AD: Ocular manifestations in the Marfan syndrome and homocystinuria. *Am J Ophthalmol* **75**:405–420, 1973.

20. Currarino G, Friedman JM: A severe form of congenital contractural arachnodactyly in two newborn infants. *Am J Med Genet* **25**:763–773, 1986.

21. Dalgleish R et al: Length polymorphism in the pro $\alpha 2(I)$ collagen gene: An alternative explanation in a case of Marfan syndrome. *Hum Genet* **73**:91–92, 1986.

22. DeSanctis RW et al: Aortic dissection. *N Engl J Med* **317**:1060–1066, 1987.

23. Dignan PStJ et al: Arachnodactyly (Marfan's syndrome) with XYY karyotype. *Am J Dis Child* **124**:266–277, 1972.

24. Ferguson JE et al: Marfan's syndrome: Acute aortic dissection during labor, resulting in fetal distress and cesarean section, followed by successful surgical repair. *Am J Obstet Gynecol* **147**:759–762, 1983.

25. Fragoso R, Cantú JM: A new psychomotor retardation syndrome with peculiar facies and Marfanoid habitus. *Clin Genet* **25**:187–190, 1984.

26. Fryns J-P, Buttiens M: X-linked mental retardation with Marfanoid habitus. *Am J Med Genet* **28**:267–274, 1987.

27. Furlong J et al: New Marfanoid syndrome with craniosynostosis. *Am J Med Genet* **26**:599–604, 1987.

28. Goodman RM et al: The Marfan syndrome in Israel, First International Symposium on the Marfan syndrome, Baltimore, July 8–10, 1988.

29. Gordon AM: Abraham Lincoln—a medical appraisal. *J Kentucky Med Assoc* **60**:249–253, 1962.

30. Gott VL et al: Surgical treatment of aneurysms of the ascending aorta in the Marfan syndrome: Results of composite-graft repair in 50 patients. *N Engl J Med* **314**:1070–1074, 1986.

31. Graham JM et al: Infantile Marfan syndrome. Ninth David W. Smith Workshop on Malformations and Morphogenesis, Oakland, California, August 3–7, 1988.

32. Grin TR, Nelson LB: Rieger's anomaly associated with Marfan's syndrome. *Ann Ophthalmol* **19**:380–384, 1987.

33. Hall JR et al: Pneumothorax in the Marfan syndrome: Prevalence and therapy. *Ann Thorac Surg* **37**:500–504, 1984.

34. Herrmann J et al: The Stickler syndrome (hereditary arthroophthalmopathy). *Birth Defects* **11**(2):77–103, 1975.

35. Hofman KJ et al: Marfan syndrome: Neuropsychological aspects. *Am J Med Genet* **31**:331–338, 1988.

36. Hollister DW: Marfan syndrome: Abnormality of the microfibrillar fiber array. *Clin Res* **35**:211A, 1987.

37. Hollister DW et al: Marfan syndrome: Abnormalities of the microfibrillar fiber array detected by immunohistopathologic studies. First International Symposium on the Marfan Syndrome, Baltimore, July 8–10, 1988.

38. Jaffer Z, Beighton P: Syndrome identification case report 98—arachnodactyly, joint laxity, and spondylolisthesis. *J Clin Dysmorph* **1**(2):14–18, 1983.

39. Joseph MC, Meadow SR: The metacarpal index in infants. *Arch Dis Childh* **44**:515–516, 1969.

40. Keech MK et al: Familial studies of the Marfan syndrome. *J Chron Dis* **19**:57–83, 1966.

41. Kohler E et al: Hereditary osteolysis. *Radiology* **108**:99–105, 1973.

42. Loveman AB et al: Marfan's syndrome: Some cutaneous aspects. *Arch Dermatol* **87**:428–435, 1963.

43. Lujan JE et al: A form of X-linked mental retardation with Marfanoid habitus. *Am J Med Genet* **17**:311–322, 1984.

44. Lynas MA: Marfan's syndrome in Northern Ireland. *Ann Hum Genet* **22**:289–301, 1958.

45. Manning GS et al: Multiple endocrine neoplasia, Type I: Association with Marfanoid habitus, optic atrophy, and other abnormalities. *Arch Intern Med* **143**:2315–2316, 1983.

46. Marsalese DL et al: Marfan's syndrome: Natural history and long term follow-up: The Cleveland Clinic Experience. First International Symposium on the Marfan Syndrome, Baltimore, July 8–10, 1988.

47. Maumenee IH: The eye in the Marfan syndrome. *Birth Defects* **18**(6):515–524, 1982.

48. McKusick VA: *Heritable Disorders of Connective Tissue.* 4th ed. C.V. Mosby, St. Louis, 1972.

49. McWilliams WG, Maumenee IH: Retinal detachment in the Marfan syndrome. First International Symposium on the Marfan Syndrome, Baltimore, July 8–10, 1988.

50. Mendez HMM et al: The syndrome of retinal pigmentary degeneration, microcephaly, and severe mental retardation (Mirhosseini–Holmes–Walton syndrome): Report of two patients. *Am J Med Genet* **22**:223–228, 1985.

51. Mirhosseini SA et al: Syndrome of pigmentary retinal degeneration, cataract, microcephaly, and severe mental retardation. *J Med Genet* **9**:193–196, 1972.

52. Murdoch JL et al: Life expectancy and causes of death in the Marfan syndrome. *N Engl J Med* **286**:804–808, 1972.

53. Murdoch JL et al: Parental age effects on the occurrence of new mutations for the Marfan syndrome. *Ann Hum Genet* **35**:331–336, 1972.

54. Nelson LB, Maumenee IH: Ectopia lentis. *Surv Ophthalmol* **27**:143–160, 1982.

55. Ogilvie DJ et al: Segregation of all four major fibrillar collagen genes in the Marfan syndrome. *Am J Hum Genet* **41**:1071–1082, 1987.

56. Parish JG: Skeletal hand charts in inherited connective tissue diseases. *J Med Genet* **4**:227–238, 1967.

57. Perejda AJ et al: Marfan's syndrome: Structural, biochemical, and mechanical studies of the aortic media. *J Lab Clin Med.* **106**:376–383, 1985.

58. Pyeritz RE: The Marfan syndrome. *Am Fam Physician* **34**:83–94, 1986.

59. Pyeritz RE: Maternal and fetal complications of pregnancy in the Marfan syndrome. *Am J Med* **71**:784–790, 1981.

60. Pyeritz RE, McKusick VA: The Marfan syndrome: Diagnosis and management. *N Engl J Med* **300**:772–776, 1979.

61. Pyeritz RE, Wappel MA: Mitral valve dysfunction in the Marfan syndrome: Clinical and echocardiographic study of prevalence and natural history. *Am J Med* **74**:797–807, 1983.

62. Pyeritz RE et al: Growth and anthropometrics in the Marfan syndrome. *Prog Clin Biol Res* **200**:355–366, 1985.

63. Ramsey MS et al: The Marfan syndrome: A histopathologic study of ocular findings. *Am J Ophthalmol* **76**:102–116, 1973.

64. Rand TC et al: The metacarpal index in normal children. *Pediatr Radiol* **9**:31–32, 1980.

65. Roberts WC, Honig HS: The spectrum of cardiovascular disease in the Marfan syndrome. A clinico-morphologic study of 18 necropsy patients and comparison to 151 previously reported necropsy patients. *Am Heart J* **104**:115–125, 1982.

66. Sanders RC et al: Sonography of the Marfan syndrome *in utero*. First International Symposium on the Marfan Syndrome, Baltimore, July 8–10, 1988.

67. Schoenfeld MR: Nicolo Paganini: Musical magician and Marfan mutant? *JAMA* **239**:40–42, 1978.

68. Schollin J et al: Probable homozygotic form of the Marfan syndrome in a newborn child. *Acta Paediatr Scand* **77**:452–456, 1988.

69. Shprintzen RJ, Goldberg RB: A recurrent pattern syndrome of craniosynostosis associated with arachnodactyly and abdominal hernias. *J Craniofac Genet Dev Biol* **2**:65–74, 1982.

70. Sisk HE et al: The Marfan syndrome in early childhood: Analysis of 15 patients diagnosed at less than 4 years of age. *Am J Cardiol* **52**:353–358, 1983.

71. Steinberg I: A simple screening test for the Marfan syndrome. *Am J Roentgenol* **97**:118–124, 1966.

72. Sugarman G, Vogel MW: Craniofacial and musculoskeletal abnormalities—a questionable connective tissue disease. *Synd Ident* **7**(1):16–17, 1981.

73. Tamminga P et al: An infant with Marfanoid phenotype and congenital contractures associated with ocular and cardiovascular anomalies, cerebral white matter hypoplasia and spinal axonopathy. *Eur J Pediatr* **143**:228–231, 1985.

74. Temtamy SA et al: Comparative ultrastructural studies of gingiva and pulp in the Marfan syndrome and homocystinuria. First International Symposium on the Marfan Syndrome, Baltimore, July 8–10, 1988.

75. Thieffry S, Sorrell-Dejerine J: Forme spéciale d'ostéolysis essentielle héréditaire et familiale à stabilization spontanée, survenant dans l'enface. *Presse Méd* **66**:1858–1861, 1958.

76. Verghese C: Forum: Anaesthesia in Marfan's syndrome. *Anaesthesia* **39**:917–922, 1984.

77. Walker BA et al: The Marfanoid hypermobility syndrome. *Ann Intern Med* **71**:349–352, 1969.

78. Walker TM: The normal metacarpal index. *Br J Radiol* **52**:787–791, 1979.

79. Wilner HI, Finby N: Skeletal manifestations in the Marfan syndrome. *JAMA* **187**:490–495, 1964.

80. Wilson R: Marfan's syndrome: Description of a family. *Am J Med* **23**:434–444, 1957.

81. Wood JR et al: Pulmonary disease in patients with Marfan syndrome. *Thorax* **39**:780–784, 1984.

82. Young D: Familial dissecting aneurysm complicating Marfan's syndrome. *Am Heart J* **78**:577–578, 1969.

83. Zimprich H: Zur Genetik des Marfan-Syndroms. *Helv Paediatr Acta* **19**:483–489, 1964.

## McCune–Albright syndrome

The McCune–Albright syndrome is characterized by polyostotic fibrous dysplasia, multiple areas of cutaneous light brown pigmentation or café-au-lait spots, and autonomous hyperfunction of one or more of the endocrine glands, especially the gonads and the thyroid (Figs. 9–32 to 9–34). The syndrome was delineated by McCune (33,34) in 1936 and 1937 and by Albright et al (1,2) in 1937 and 1938, although it was described as early as 1922 by Weil (42). It has been suggested that cases 5, 6, and 7 of von Recklinghausen may possibly represent examples of this syndrome rather than neurofibromatosis (14). Because of multiple endocrine gland dysfunctions, the syndrome has been included as one of the multiple endocrine adenomatoses (16). The most extensive review is that of Danon and Crawford (14).

The etiology is unknown. Virtually all cases described to date have been isolated examples. Hibbs and Rush (27) reported a woman with skin pigmentation, possible sexual precocity, and bone lesions consistent with fibrous dysplasia; a daughter had multiple cystic bone lesions, but no endocrine or skin manifestations. Firat and Stutzman (18) described a mother and daughter with hyperparathyroidism, cystic lesions of bone, but no skin or other endocrine abnormalities; bone biopsy in the mother was consistent with fibrous dysplasia. The syndrome has also been observed in one of monozygotic (30) and in one of dizygotic (11) twins.

Happle (25) postulated an autosomal dominant lethal gene as the cause, resulting in loss of the zygote *in utero*. Mutated cells are thought to survive only when occurring together with normal cells in the mosaic state. The disorder possibly results from a gametic half chromatid mutation or from an early somatic mutation. Lethal gene survival by mosaicism permits explanation of asymmetric bone lesions and endocrinopathies of central or peripheral origin by random distribution of mutant cells.

The pathogenesis remains an enigma. Neurologic disturbance with widespread trophic consequences and embryonic defects affecting multiple systems were two possibilities originally proposed by Albright et al (1). Another theory considered the bone lesions primary with endocrine manifestations arising secondarily from sclerotic cranial base involvement stimulating the hypothalamus. However, sclerotic changes in the skull base are absent in about half the cases, making this hypothesis untenable. Autonomous hyperfunction of the peripheral target glands seems most likely, although whether hyperfunction is truly autonomous by enhanced end organ sensitivity or by trophic substances not measured by the usual assays is still uncertain (14). The possibility of a basic defect opposite in nature to those of pseudohypoparathyroidism deserves investigation. Functional failure of receptor complexes might underlie various manifestations of the syndrome (14).

Features of the syndrome occur with variable frequency. Diagnosis is made with certainty if all three principal features are present, but should also be considered with two features because each has special characteristics seldom encountered in other disorders. The presence of any one of the three features should prompt a search for one or both of the other features of the syndrome. In the series of Harris et al (26), about 38% had bone involvement alone, 40% had two features, and 22% had the full-blown syndrome. Pritchard (35) found cutaneous pigmentation in 43% ($n=181$). Approximately 45% of affected females exhibit sexual precocity, but only 6% of affected males have this feature. Of patients referred for sexual precocity, only 8% have McCune–Albright syndrome (36) and only 3% of patients with polyostotic fibrous dysplasia have McCune–Albright syndrome (14).

**Skeletal manifestations.** Although any bone may be involved, the long bones are most frequently affected, especially the upper end of the femur. Bowing resembling a hockey stick may be produced, resulting in leg-length discrepancy. Limp, leg pain, or fracture is the presenting complaint in about 70% (26). Other bones affected, in descending order of frequency, are the tibia, fibula, pelvis, humerus, radius, and ulna. Bilateral involvement occurs in about half the cases (8,17) (Fig. 9–33A,B). Occasionally, a single bone is involved. Incipient bowing of the legs may be seen as early as the first year of life and nearly always appears before the end of the first decade (33,34). The process may be asymptomatic or accompanied by pain and fracture. Fractures may be multiple and recurrent. At least 85% have one fracture, and over 40% have three or more (26). Occasionally, bones on only one side of the body may be involved (35).

Bone is replaced by a yellowish to red-brown fibrous tissue, its

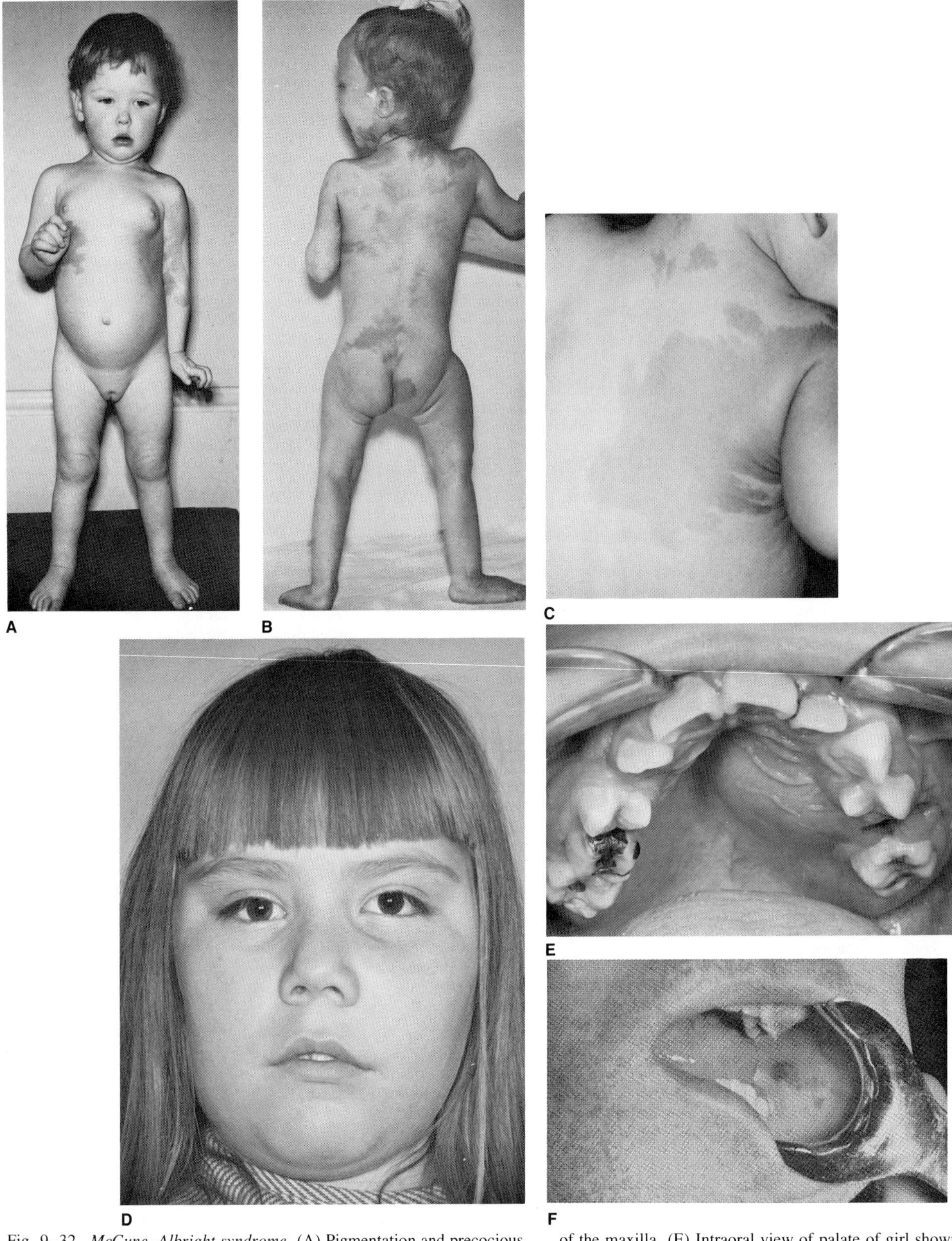

Fig. 9–32. *McCune–Albright syndrome.* (A) Pigmentation and precocious puberty. (B) Irregular cutaneous pigmentation and hockey-stick deformity. (C) Close-up showing irregular cutaneous pigmentation in another patient. (D) Note facial asymmetry, fullness of left cheek due to fibrous dysplasia of the maxilla. (E) Intraoral view of palate of girl shown in D. (F) Buccal and palatal melanotic mucosal pigmentation. (A from *S Agarwala* and *JB Heycock,* Br J Clin Pract **22:**339, 1969. B courtesy of *H Pande,* Oslo, Norway. E,F from *RJ Gorlin* and *AP Chaudhry,* Oral Surg **10:**857, 1957.)

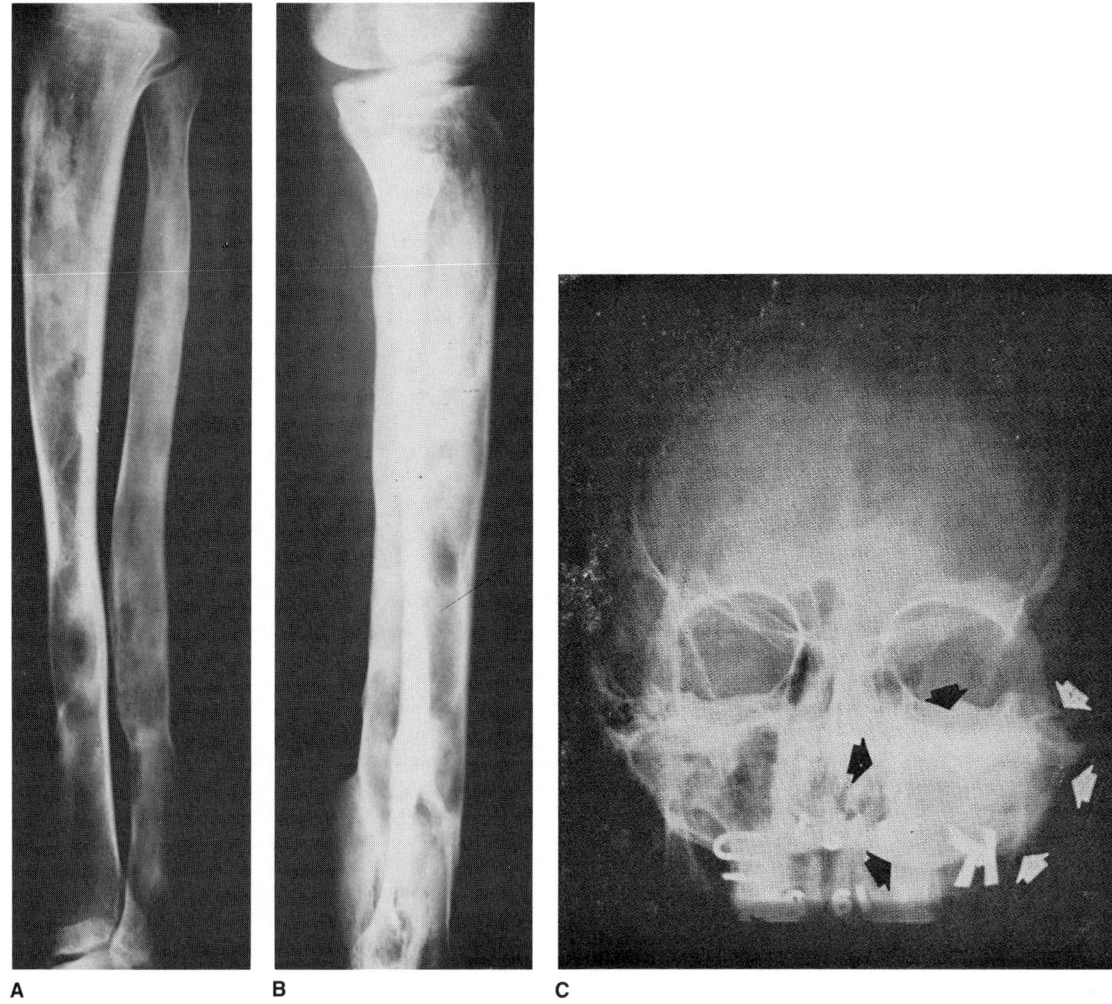


Fig. 9–33. *McCune–Albright syndrome.* Roentgenograms. (A,B) Thickening and pseudocystic involvement of tibia and fibula. (C) Hyperostotic involvement of maxillary sinus by fibrous dysplasia.

composition varying greatly in different parts of the body. It may be rich or poor in cells. The stroma may vary from a finely fibrillar one with a loose whorled arrangement to one that is densely collagenous. Some areas appear edematous, with numerous small cystic spaces. Foci of hemorrhage and multinucleated giant cells may be observed. The trabeculae are irregular in form, and occasionally a few fragments of cartilage are present (Fig. 9–34).

Facial asymmetry occurs in about 25% (Fig. 9–32D,E) and may be accompanied by protrusion of the eye with associated visual disturbances in some instances (17,35). The bony lesions of the skull and facial skeleton, in contrast to the cystic lesions of long bones, are hyperostotic (Fig. 9–33C). The skull base becomes thickened and dense, bulging upward into the cranial cavity. The calvaria may also become thickened, with marked occipital and frontal bulging. Bossing may be asymmetric, with unilateral, and occasionally bilateral, obliteration of the sinuses and nasal passages. The overgrowth of bone around foramina may result in deafness and/or blindness.

The jaws may be enlarged, expanded, and distorted. Roentgenographic examination may show a dense mass, especially in the maxilla, extending into and obliterating the sinuses and expanding the buccal plate in the tuberosity areas, or there may be a radiolucent area, more common in the mandible, similar to that seen in long bones. Often there is loss of trabeculae and a "ground-glass" appearance on roentgenographic examination (14,26,35,44).

**Cutaneous manifestations.** Pigmentation is of the café-au-lait type. Well-defined, generally unilateral, irregular macular spots are scattered over the forehead, nuchal area, and buttocks. Only rarely are the face, lips, or mucosa affected (9,21,37) (Fig. 9–32A–C,F).

There appears to be a correlation between the amount of pigmentation and the degree of bone involvement (7). It has been stated that pigmentation is more frequent on the side of unilateral bone involvement, although this has been denied (28). The pigment appears from the fourth month to the second year of life. In a few patients, it has become evident a few weeks postnatally (5).

**Endocrine manifestations.** Endocrine abnormalities, reviewed elsewhere (1–3,5,6,8,12,14,15,17,18,20,23,24,28,29,31–38,40,41,45), are characterized by autonomous hyperfunction and, to date, no instances of extraglandular trophic influences mediating hypersecretion have been identified. In affected patients, any given endocrine gland may function normally. At autopsy, characteristic endocrine nodular hyperplasia may be found that did not give rise to clinical signs of hyperfunction during life. In fact, endocrine organs, in spite of autonomy, may function so nearly normally that their independence is not recognized until appropriate tests have been performed.

Sexual precocity is the most common endocrine manifestation, especially in females (Fig. 9–32A). It is generally manifest earlier in girls than in boys. Menarche is reached between 1 and 5 years of age

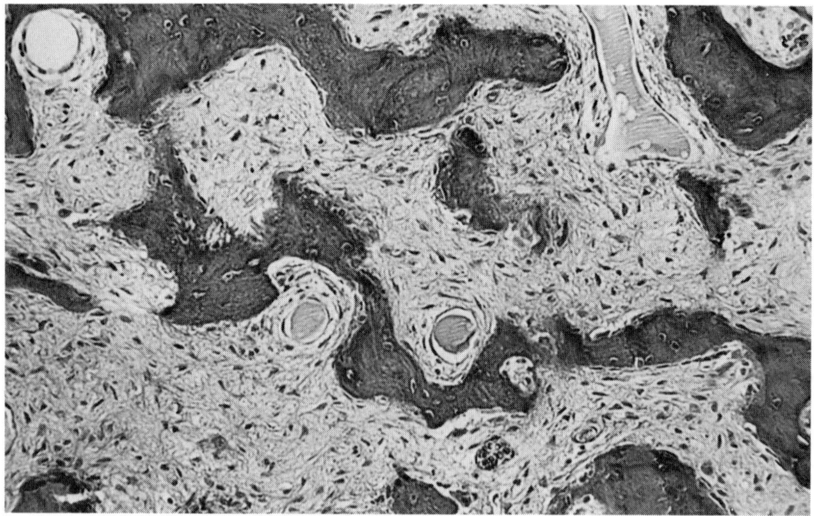

Fig. 9–34. *McCune–Albright syndrome*. C-shaped trabeculae composed of metaplastic woven or fiber bone in fibrous connective tissue stroma.

in 50% and between 6 and 10 years in another 33% (8). However, vaginal bleeding may occur within the first few months or even the first few days of life (1,2,5,23). It is usually irregular, lasts from 2 to 4 days, and may, on occasion, be profuse. Breast development and pubic and axillary hair appear after the menarche, usually from the fifth to the tenth year, but may be manifest as early as birth (33,34). Hypertrophy of the external genitalia has also been noted (1,2). It is intriguing that ovaries associated with autonomous estrogen secretion and sexual precocity become subservient to normal trophic influences in adulthood, permitting women to be fertile (14,23). In males with sexual precocity, enlargement of the penis and testes is accompanied by growth of public hair, suggesting hyperfunction of both spermatic tubules and Leydig cells (14,23). Precocious puberty may be accompanied by gynecomastia (17,35).

Hyperthyroidism is present in about 20%, occurring at an early age (2,34,35,37,38,45). The sex distribution is approximately equal. Thyrotoxic manifestations such as irritability, poor weight gain, and growth failure have been described during infancy, as early as 3 months in one instance (38). With untreated sexual precocity or hyperthyroidism, skeletal maturation is often rapid and premature closure of the epiphyses may result, producing short stature in adulthood.

Various other endocrine disorders have also been noted including Cushing's syndrome, hypersomatotropinism, hyperprolactinemia, hyperparathyroidism, and hypophosphatemic vitamin D-resistant rickets or osteomalacia without hypercalcemia or elevated parathyroid hormone levels (12,14,17,18,20,24,31,32,36).

**Central nervous system.** Although the overwhelming majority of patients are of normal intelligence, mental deficiency has been reported in a few instances (1,12,14). It may be secondary to factors such as prematurity, hypercorticalism, or grossly malformed skull. The significance of mental retardation in others remains unclear. In the original description of Albright et al (1), an accessory mammillary body was found at autopsy. However, no other patient has exhibited this finding.

**Neoplasms.** Malignancies are rarely described. The most unusual malignancy recorded to date has been carcinoma of the breast in an 11-year-old girl. Endometrial carcinoma has also been recorded (14). Instances of sarcoma arising in areas of fibrous dysplasia have been secondary to radiation therapy (26). Other neoplasms have been observed in patients with isolated polyostotic fibrous dysplasia, but could conceivably occur as components of the McCune–Albright syndrome. Multiple intramuscular myxomas have been noted in a number of such patients (43). Reticuloendothelial hyperplasia with lymphoid and myeloid metaplasia has been described (38). Leukemia (16) and osteoma of the skin (39) have also been noted.

**Other findings.** At autopsy, the thymus and spleen are frequently hyperplastic, but the significance of this finding is unknown (14).

**Differential diagnosis.** Fibrous dysplasia may occur without the McCune–Albright syndrome. The overwhelming majority of cases are monostotic; polyostotic involvement occurs much less frequently and, of these, only 3% have McCune–Albright syndrome (10,14,18,26,41). Radiographically, bone lesions should be distinguished from those of hyperparathyroidism, histiocytosis X, multiple myeloma, Paget's disease of bone, *neurofibromatosis*, and giant cell tumor. It is unfortunate that some authors have referred to cherubism, an autosomal dominant trait, as ''familial fibrous dysplasia of the jaws,'' a designation it does not merit. Jaffe–Campanacci syndrome consists of café-au-lait macules with nonossifying fibromas of long bones and giant cell granulomas of the jaws (34a,39a).

Alvarez-Arratia et al (4) reported a probable monogenic form of polyostotic fibrous dysplasia. Cole et al (13) described a congenital disorder of bone with unusual facial appearance, bone fragility, hyperphosphatasemia, and hypophosphatemia—a condition they called panostotic fibrous dysplasia.

Skin pigmentation is also seen in neurofibromatosis. ''Coast of California'' contour is typical of *neurofibromatosis* and a markedly irregular outline (''Coast of Maine'') usually occurs with McCune–Albright syndrome; exceptions have been noted (22). Giant pigment granules, characteristically seen in malpighian cells or melanocytes in neurofibromatosis, are very rare in McCune–Albright syndrome (7,19). Precocious puberty occurs in the adrenogenital syndrome, with ovarian granulosa cell tumor, and occasionally in *Peutz–Jeghers syndrome*.

**Laboratory aids.** Radiographic studies are useful for fibrous dysplastic lesions, advanced bone age, and for occasional findings such as rickets, osteomalacia, or osteoporosis. Serum alkaline phosphatase is elevated in about 50% (35). Appropriate endocrine investigations should be carried out.

### References (McCune–Albright syndrome)

1. Albright F et al: Syndrome characterized by osteitis fibrosa disseminata; areas of pigmentation and endocrine dysfunction with precocious puberty in females. *N Engl J Med* **216**:727–736, 1937.

2. Albright F et al: Syndrome characterized by osteitis fibrosa disseminata, areas of pigmentation and a gonadal dysfunction. *Endocrinology* **22**:411–421, 1938.

3. Alexander FW: Polyostotic fibrous dysplasia with raised steroid excretion. *Arch Dis Childh* **46**:91–94, 1971.

4. Alvarez-Arratia MC et al: A probable monogenic form of polyostotic fibrous dysplasia. *Clin Genet* **24**:132–139, 1983.

5. Arlien-Søborg U, Iversen T: Albright's syndrome: A brief survey and report of a case in a seven-year-old girl. *Acta Paediatr Scand* **45**:558–568, 1956.

6. Benedict PH: Sex precocity and polyostotic fibrous dysplasia. *Am J Dis Child* **111**:426–429, 1966.

7. Benedict PH et al: Melanotic macules in Albright's syndrome and in neurofibromatosis. *JAMA* **205**:618–626, 1968.

8. Boenheim F, McGavack TH: Polyostotische fibröse Dysplasie. *Ergeb Inn Med Kinderheilkd* **3**:157–184, 1952.

9. Bowerman JE: Polyostotic fibrous dysplasia with melanotic pigmentation. *Br J Oral Surg* **6**:188–191, 1969.

10. Brunt P, McKusick VA: Fibrous dysplasia: A report of genetic and clinical study with a review of the literature. *Medicine* **49**:343–374, 1970.

11. Caldwell GA, Broderick TG: Polyostotic fibrous dysplasia in one of negro twin girls. *Ann Intern Med* **27**:114–126, 1947.

12. Chung KF et al: Acromegaly and hyperprolactinemia in McCune–Albright syndrome: Evidence of hypothalamic dysfunction. *Am J Dis Child* **137**:134–136, 1983.

13. Cole DEC et al: Panostotic fibrous dysplasia: A congenital disorder of bone with unusual facial appearance, bone fragility, hyperphosphatasemia, and hypophosphatemia. *Am J Med Genet* **14**:725–735, 1983.

14. Danon M, Crawford JD: The McCune–Albright syndrome. *Ergeb Inn Med Kinderheilkd* **55**:82–115, 1987.

15. D'Armiento M et al: McCune–Albright syndrome: Evidence for autonomous multiendocrine hyperfunction. *J Pediatr* **102**:584–586, 1983.

16. DiGeorge AM: Albright syndrome: Is it coming of age? *J Pediatr* **87**:1018–1020, 1975.

17. Falconer MA et al: Fibrous dysplasia of bone with endocrine disorders and cutaneous pigmentation (Albright's disease). *Q J Med* **11**:121–154, 1942.

18. Firat D, Stutzman L: Fibrous dysplasia of the bone: Review of twenty-four cases. *Am J Med* **44**:421–429, 1968.

19. Frenk E: Etude ultrastructurale des taches pigmentaires du syndrome d'Albright. *Dermatologica* **143**:12–20, 1971.

20. Giovannelli G et al: McCune–Albright syndrome in a male child: A clinical and endocrinologic enigma. *J Pediatr* **92**:220–226, 1978.

21. Gorlin RJ, Chaudhry AP: Oral melanotic pigmentation in polyostotic fibrous dysplasia: Albright's syndrome. *Oral Surg* **10**:857–862, 1957.

22. Grant DB, Martinez L: The McCune–Albright syndrome without typical skin pigmentation. *Acta Paediatr Scand* **72**:467–478, 1983.

23. Hackett LJ Jr, Christopherson WM: Polyostotic fibrous dysplasia. *J Pediatr* **35**:767–771, 1949.

24. Hall R, Warrick C: Hypersecretion of hypothalamic releasing hormones; a possible explanation of the endocrine manifestations of polyostotic fibrous dysplasia (Albright's syndrome). *Lancet* **1**:1313–1316, 1972.

25. Happle R: The McCune–Albright syndrome: A lethal gene surviving by mosaicism. *Clin Genet* **29**:321–324, 1986.

26. Harris WH et al: The natural history of fibrous dysplasia: An orthopedic, pathological and roentgenographic study. *J Bone Jt Surg* **44A**:207–233, 1962.

27. Hibbs RE, Rush HP: Albright's syndrome. *Ann Intern Med* **37**:587–593, 1952.

28. Husband P, Snodgrass G: McCune–Albright syndrome with endocrinological investigations. *Am J Dis Child* **119**:164–167, 1970.

29. Lee PA et al: McCune–Albright syndrome: Long-term follow-up. *JAMA* **256**:2980–2984, 1986.

30. Lemli L: Fibrous dysplasia of bone: Report of female monozygotic twins with and without McCune–Albright syndrome. *J Pediatr* **91**:947–949, 1977.

31. Lightner ES et al: Growth hormone excess and sexual precocity in polyostotic fibrous dysplasia (McCune–Albright syndrome): Evidence for abnormal hypothalamic function. *J Pediatr* **87**:922–927, 1975.

32. Lipson A, Hsu T-H: The Albright syndrome associated with acromegaly: Report of a case and review of the literature. *Johns Hopkins Med J* **149**:10–14, 1981.

32a. Mauras N, Blizzard RM: The McCune–Albright syndrome. *Acta Endocrinol (Suppl)* **113**:207–217, 1986.

33. McCune DJ: Osteitis fibrosa cystica: The case of a nine year old girl who also exhibits precocious puberty, multiple pigmentation of the skin and hyperthyroidism. *Am J Dis Child* **52**:743–747, 1936.

34. McCune DJ, Bruch H: Osteodystrophia fibrosa. *Am J Dis Child* **54**:806–848, 1937.

34a. Mirra JM et al: Disseminated nonossifying fibromas in association with café-au-lait spots (Jaffe–Campanacci syndrome). *Clin Orthopaed* **168**:192–205, 1982.

35. Pritchard JE: Fibrous dysplasia of the bones. *Am J Med Sci* **222**:313–332, 1951.

36. Rieth KG et al: Pituitary and ovarian abnormalities demonstrated by CT and ultrasound in children with features of the McCune–Albright syndrome. *Radiology* **153**:389–393, 1984.

37. Robinson M: Polyostotic fibrous dysplasia of bone. *J Am Dent Assoc* **42**:47–57, 1951.

38. Samuel S et al: Hyperthyroidism in an infant with McCune–Albright syndrome: Report of a case with myeloid metaplasia. *J Pediatr* **80**:275–278, 1972.

39. Shelley WB: Alopecia with fibrous dysplasia and osteomas of the skin. A sign of polyostotic fibrous dysplasia. *Arch Dermatol* **112**:715–719, 1976.

39a. Steinmetz JC et al: Campanacci syndrome. *J Pediatr Orthoped* **8**:602–604, 1988.

40. Verghese A: Albright's syndrome in the male with situs inversus. *Proc R Soc Med* **55**:357–358, 1962.

41. Warrick CK: Some aspects of polyostotic fibrous dysplasia: Possible hypothesis to account for the associated endocrine changes. *Clin Radiol* **24**:125–138, 1973.

42. Weil: 9-jähriges Mädchen mit Pubertas praecox und Knochenbrüchigkeit. *Klin Wochenstr* **1**:2114–2115, 1922.

43. Wirth WA et al: Multiple intramuscular myxomas. Another extraskeletal manifestation of fibrous dysplasia. *Cancer.* **27**:1167–1173, 1971.

44. Zachariades N et al: Albright syndrome. *Int J. Oral Surg* **13**:53–58, 1984.

45. Zangeneh F et al: McCune–Albright syndrome with hyperthyroidism. *Am J Dis Child* **111**:644–648, 1966.

## Melnick–Needles syndrome (osteodysplasty)

In 1966, Melnick and Needles (21), described a syndrome characterized by generalized bone dysplasia and abnormal facies (Figs. 9–35 to 9–37). About 35 cases have been reported to date (2,3,5,7,8,12–14,16,18–24,28,29,31,33,34). In all but three examples, the patients were females with about one-half the cases representing new mutations. In only three kindreds was there transmission from one generation to another (9,21,31). Male patients, originally reported to be affected by Melnick and Needles (21), on follow-up were found to be normal (9). Three males born to normal parents had the same degree of expression as that of affected females (12,16,29); and four affected females had male stillborn infants with a distinct embryopathy (7a,16,28,33,35) characterized by exophthalmos, omphalocele and/or malrotation of the gut, and skeletal anomalies (thin calvaria with stellate ossification pattern, cervical lordosis, cervicothoracic kyphosis, thoracolumbar lordosis, thin irregular ribs, curved long bones with sheathing, and hypoplasia or absence of thumbs and halluces (Fig. 9–38). A less certain example is that of Deleporte et al (6). Several cases (1,26,32) cannot be accepted as examples of Melnick–Needles syndrome on either clinical or radiographic grounds or due to inadequate documentation. Others (4,25) clearly have frontometaphyseal dysplasia. The disorder reported by ter Haar et al (27), although having somewhat similar clinical, but far less severe radiographic, features, surely is not classic Melnick–Needles syndrome and has autosomal recessive inheritance. The lethal condition reported in sibs by Kozlowski et al (15) has autosomal recessive inheritance and is still a different entity.

Gorlin and Knier (9) suggested that Melnick–Needles syndrome has X-linked dominant inheritance lethal in the male.

Early childhood is marked by recurrent respiratory and ear infections (8,12,14,21,23,29,33). Height and weight are usually below the 10th percentile. The patients tend to remain thin. The breasts are very small and sexual hair is sparse (24).

**Facies.** The facies is characterized by high somewhat hirsute forehead, somewhat prominent supraorbital ridges, prominent eyes, full cheeks, large pinnae, and marked micrognathia (Fig. 9–36A,B). Strabismus (7,22,34) and blue sclerae (2) have also been reported.

**Skeletal alterations.** The neck is long, the chest is narrow, and the upper arms and terminal digits are often short. There is valgus deformity at the elbows and extension is often diminished. Mild pectus excavatum has been noted in over 50% of the cases (14,22,34). About 35% exhibit delay in motor development and abnormal gait.

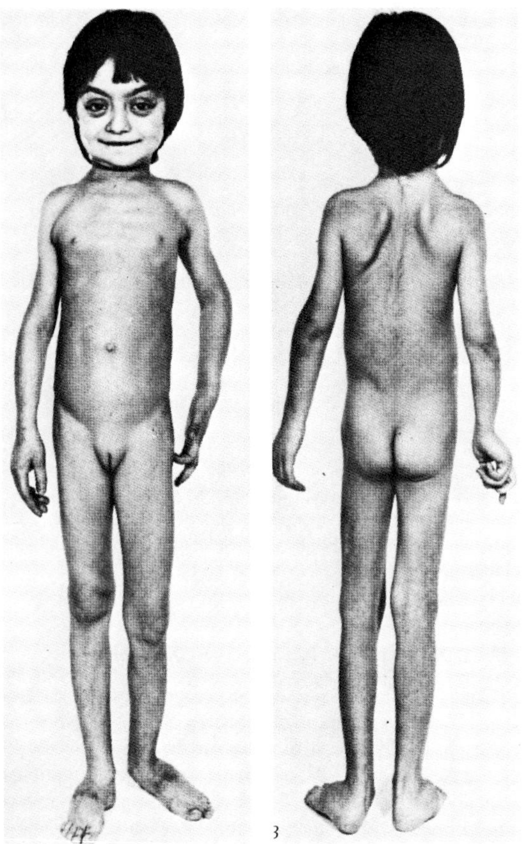

Fig. 9–35. *Melnick–Needles syndrome.* Dysplastic habitus, foot dysplasia. (From *N Moelter* and *A Walther,* Mschr Kinderheilkd **123:**178, 1975.)

Dorsal kyphosis and/or scoliosis is relatively common. Most patients exhibit genua valga and about one-third have pes planus or valgus. Some have dislocated hips.

Radiographically, the calvaria often has digital markings. There is delayed closure of the anterior fontanel. The skull base and mastoid processes are sclerosed. There is mildly increased interorbital distance. The paranasal sinuses tend to remain underdeveloped. The mandible is small with scalloped rami and lack of coronoid processes (10,11).

All vertebral bodies are unusually tall, especially those of the axis, atlas, and occipital condyles. The thoracic vertebrae exhibit an anterior concavity with double beaking. Decreased disc space has been found in the lumbar area. The clavicles have cortical irregularity with flaring. Sternal ossification is delayed. The scapulae are hypoplastic.

Most striking are the changes in long bones. Bowing of the radius and tibia produces an S-shaped appearance. The metaphyses at the proximal and distal ends of the humerus, fibula, and tibia are flared. The terminal phalanges are short and thick, especially those of the thumbs. Cone-shaped epiphyses are found in the middle phalanx of the abbreviated fifth fingers and in the terminal phalanx of the thumbs (8,13,18,29). Coxa valga is marked. The iliac bones are flared at the crest and constricted in the supraacetabular area, whereas the ischial bones are tapered. The ribs are ribbon-like, with cortical irregularity (Fig. 9–38A–E).

**Other findings.** Ureterovesicular obstruction has been documented in a few cases (2,7,14,21,29). Fryns et al (8a) reported hyperlaxity in affected males.

**Oral manifestations.** Micrognathia and marked malocclusion are constant features.

**Differential diagnosis.** The radiographic features are so distinctive as to differentiate this syndrome from all other disorders in which there is delayed closure of the anterior fontanel. We see little similarity between Melnick–Needles syndrome and the so-called autosomal recessive lethal type of "precocious osteodysplasty" reported by Kozlowski et al (15). A skeletal disorder characterized by overly long

Fig. 9–36. *Melnick–Needles syndrome.* Characteristic facial appearance. (A,B) Note mild prominence of eyes, long neck, small mandible. (C,D) Note exophthalmos, hypertelorism, outstanding nose, receding chin. (A courtesy of *FH Stelling* and *P Meunier,* Greenville, South Carolina. C,D from *N Moelter* and *A Walther,* Mschr Kinderheilkd **123:**178, 1975.)

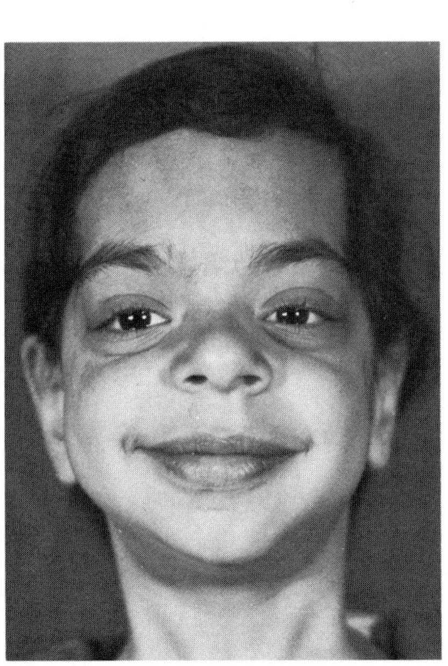

A

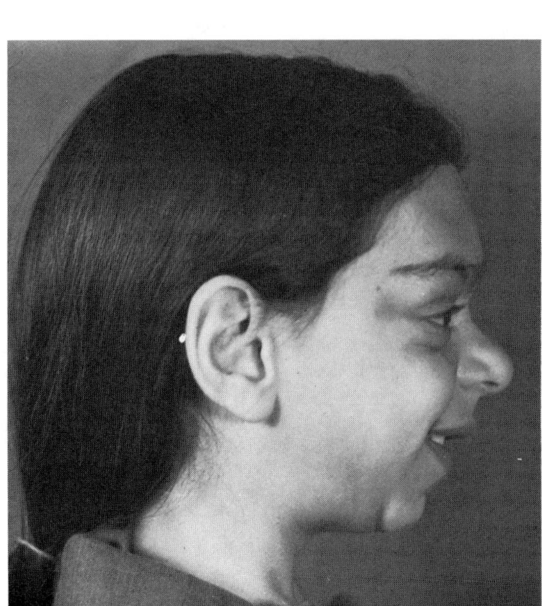

B

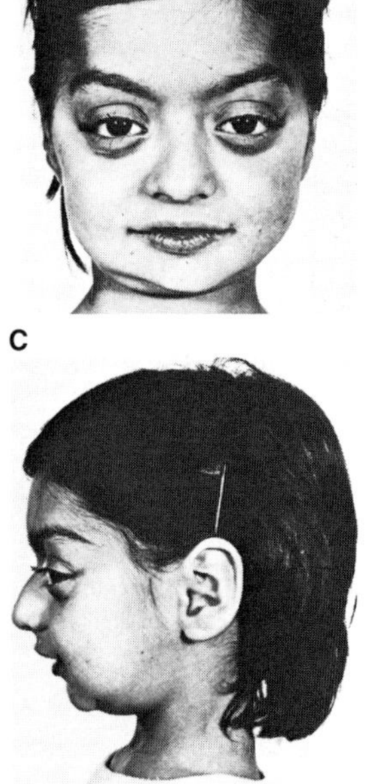

C

D

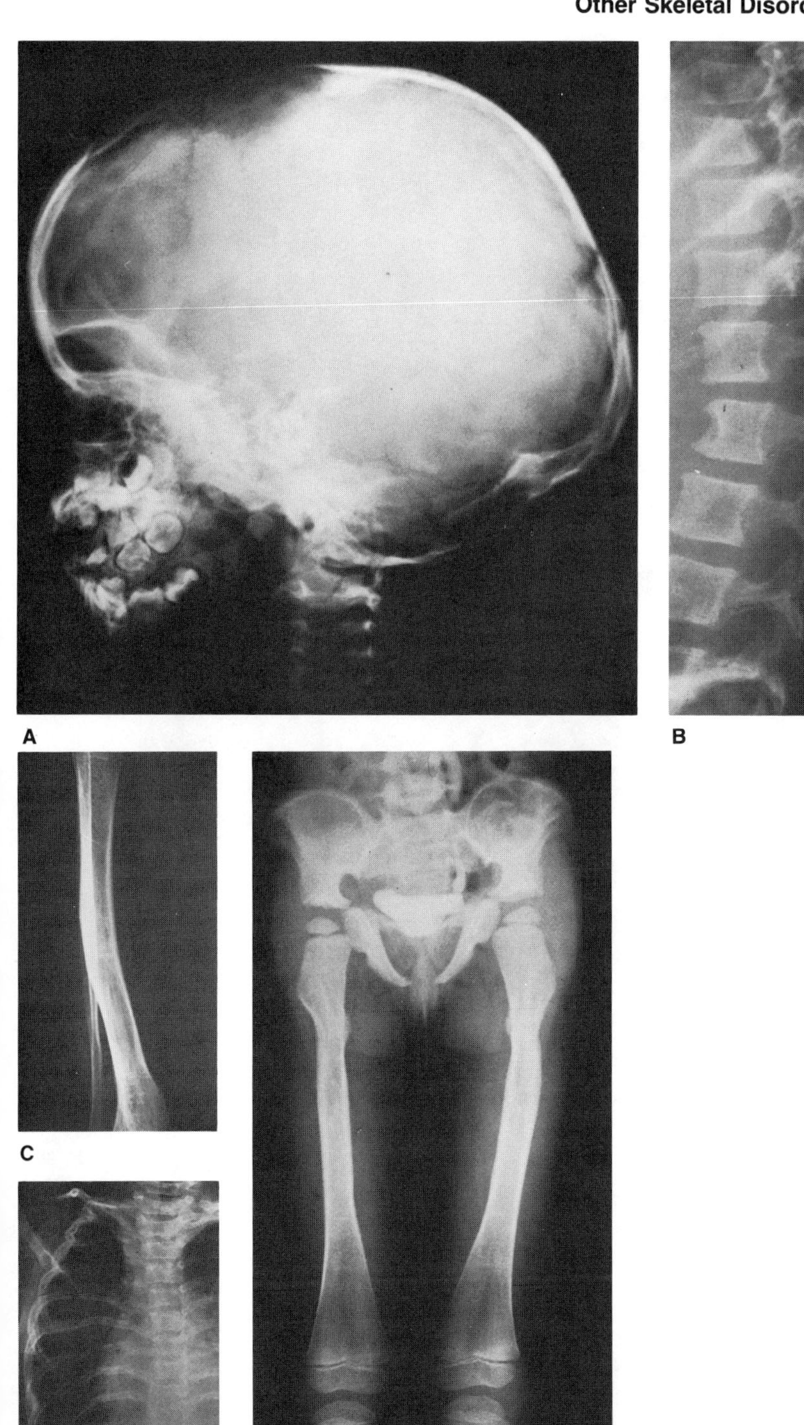


Fig. 9–37. *Melnick–Needles syndrome.* Roentgenograms. (A) Delayed closure of anterior fontanel; paranasal sinuses underdeveloped. (B) Disproportionately tall vertebral bodies having anterior concavity. (C) S-shaped bowing of tibia. (D) Flared ilia, flat acetabulae, tapered ischial bones, coxa valga, metaphyseal flare. (E) Ribbon-like ribs with cortical irregularities. (A,D from *J Melnick* and *C Needles,* Am J Roentgenol **97:**39, 1966 and *FH Stelling* and *P Meunier,* Greenville, South Carolina. B,C,E courtesy of *B Leiber,* Frankfurt, Germany.)

curved fibulas but without the skeletal abnormalities of Melnick–Needles syndrome was reported by Dereymaeker et al (6a, case 1) and Exner (7b). Possibly one of the patients had acroosteolysis (6a).

**Laboratory findings.** Prenatal diagnosis has been carried out by ultrasonography (17).

### References [Melnick–Needles syndrome (osteodysplasty)]

1. Barrio AL, Martinez MCM: Osteodisplasia. Sindrome de Melnick y Needles. *Radiologia (Madrid)* **21**:53–56, 1979.

2. Bartolozzi P et al: Melnick–Needles syndrome: Osteodysplasty with kyphoscoliosis. *J Pediatr Orthoped* **3**:387–391, 1983.

3. Bérard V et al: Un nouveau cas d'ostéodysplastie ou syndrome de Melnick et Needles. *Pédiatrie* **5**:425–430, 1980.

4. Böttger E et al: Differentialdiagnose der metaphaysären Dysplasien und der Osteodysplastie (Melnick–Needles-Syndrom). *Z Orthopäd* **116**:810–819, 1978. (Case 4.)

5. Coste F et al: Osteodysplasty (Melnick–Needles syndrome). *Ann Rheum Dis* **27**:360–366, 1968. (Same as Ref 18.)

6. Deleporte B et al: L'ostéodysplastie ou syndrome de Melnick et Needles (à propos d'une nouvelle observation). *J Génét Hum* **33**:13–20, 1985.

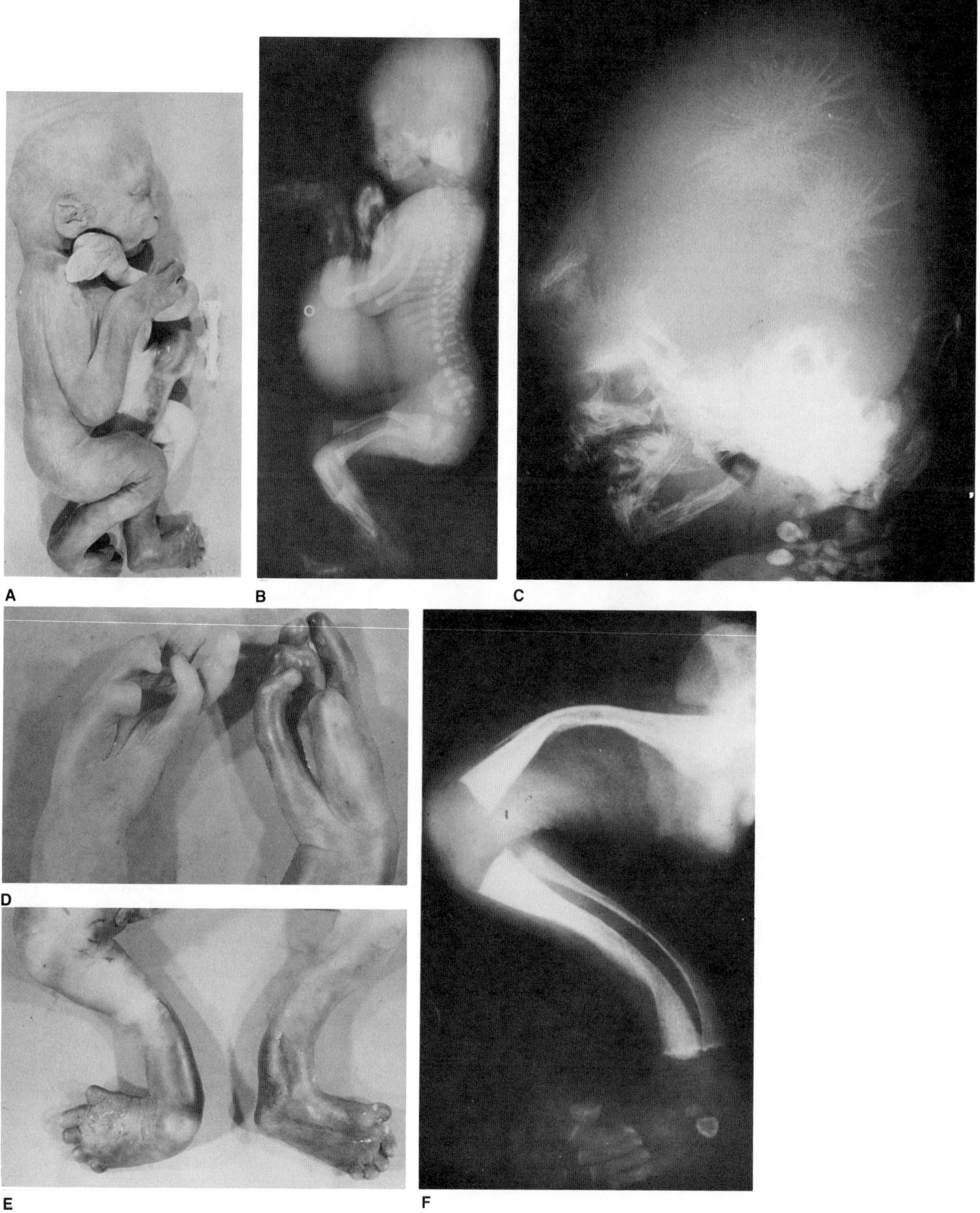


Fig. 9–38. *Melnick–Needles syndrome*. Male stillborn with distinct embryopathy, the product of an affected female. (A–C) Exophthalmos, omphalocele, thin calvaria with stellate ossification pattern, cervical lordosis, cervicothoracic kyphosis, thoracolumbar lordosis, and thin irregular ribs. (D–F) Curved long bones with sheathing, hypoplasia or absence of thumbs and halluces. (A courtesy of *M Barr,* Ann Arbor, Michigan. B,C from *G Theander* and *O Ekberg,* Acta Radiol Diag **22**:369, 1981.)

6a. Dereymaeker AM et al: Melnick–Needles syndrome (osteodysplasty). *Helv Paediatr Acta* **41**:339–352, 1986. (Case 1.)

7. de Toni T et al: La sindrome di Melnick–Needles. *Min Pediatr* **35**:447–454, 1983.

7a. Donnenfeld AE et al: Melnick–Needles syndrome in males: A lethal multiple congenital anomalies syndrome. *Am J Med Genet* **27**:159–173, 1987.

7b. Exner GU: Serpentine fibula-polycystic kidney syndrome: A variant of Melnick–Needles syndrome or a distinct entity? *Eur J Pediatr* **147**:554–564, 1988.

8. Fryns JP et al: Osteodysplasty: A rare skeletal disorder. *Acta Paediatr Belg* **32**:65–68, 1979.

8a. Fryns JP et al: Hyperlaxity in males with Melnick–Needles syndrome. *Am J Med Genet* **29**:607–611, 1988.

9. Gorlin RJ, Knier J: X-linked or autosomal dominant, lethal in the male, inheritance of the Melnick–Needles (osteodysplasty) syndrome? A reappraisal. *Am J Med Genet* **13**:465–467, 1982.

10. Gorlin RJ, Langer LO: Melnick–Needles syndrome: Radiographic alterations in the mandible. *Radiology* **128**:351–353, 1978.

11. Gorlin RJ, Leonard M: The nature of the mandibular lesions in Melnick–Needles syndrome. *Radiology* **150**:844–850, 1984.

12. Grosse KP, Böwing B: Osteodysplastie, in *Klinische Genetik in der Pädiatrie*, Spranger J, Tolksdorf M (eds), Vol I, Thieme Verlag, Stuttgart, 1980, pp 80–86.

13. Grumbach Y et al: Étude radiologique d'un cas d'ostéodysplastie. *J Radiol Électrol Med Nucl* **55**:129–131, 1974.

14. Klint RB et al: Melnick–Needles osteodysplasia associated with pulmonary hypertension, obstructive uropathy and marrow hypoplasia. *Pediatr Radiol* **6**:49–51, 1977.

15. Kozlowski K et al: Precocious type of osteodysplasia: A new autosomal recessive form. *Acta Radiol Diag* **14**:171–176, 1973.

16. Krajewska-Walasek M et al: Melnick–Needles syndrome in males. *Am J Med Genet* **27**:153–158, 1987.

17. Lachman R, Hall JG: The radiological prenatal diagnosis of the generalized bone dysplasia and other skeletal abnormalities. *Birth Defects* **15**(5A):3–24, 1979.

18. Leiber B, Hövels O: Melnick–Needles-Syndrom. *Mschr Kinderheilkd* **123**:178–182, 1975.

19. Maroteaux P et al: L'ostéodysplastie (syndrome de Melnick et de Needles). *Presse Méd* **76**:715–718, 1968.

20. Martin C et al: Un cas d'ostéodysplastie (syndrome de Melnick et de Needles). *Arch Fr Pédiatr* **28**:446–447, 1971.

21. Melnick JC, Needles CF: An undiagnosed bone dysplasia. A two family study of four generations and three generations. *Am J Roentgenol* **97**:39–48, 1966.

22. Moadel E, Bryk D: Osteodysplasty (Melnick–Needles syndrome). *Radiology* **123**:154, 1977.

23. Perry LO et al: Melnick–Needles syndrome. *J Pediatr Ophthalmol* **15**:226–230, 1978.

23a. Santinelli R et al: Osservazioni clinico–radiologiche su di un caso di sindrome di Melnick–Needles. *Pediatria (Napoli)* **91**:441–448, 1983.

24. Scheeper JH: A patient with skeletal abnormalities due to dysplasia of the muscular system. *Radiol Clin Biol* **37**:339–348, 1967. (Same case as Ref. 30.)

25. Sellars SL, Beighton P: Deafness in osteodysplasty of Melnick and Needles. *Arch Otolaryngol* **104**:225–227, 1978.

26. Stoll C et al: L'ostéodysplastie. *Pédiatrie* **31**:195–199, 1976.

27. ter Haar B et al: Melnick–Needles syndrome: Indication for an autosomal recessive form. *Am J Med Genet* **13**:469–477, 1982.

28. Theander G et al: Congenital malformations associated with maternal osteodysplasty: A new malformation syndrome. *Acta Radiol Diagn* **22**:369–377, 1981.

29. Theodorou SD et al: Osteodysplasty (Melnick–Needles syndrome) in a male, in *Skeletal Dysplasias*, Alan R. Liss, New York, 1982, pp 139–142.

30. Vaadrager JM: Craniotubular dysplasia. *Br J Plast Surg* **30**:127–133, 1977. (Case 3) (Same as Ref. 23.)

31. Vanek J et al: Osteodysplastie. *Acta Chir Orthop Traum* **43**:30–52, 1976.

32. Verger R et al: L'ostéodysplastie syndrome de Melnick et Needles. A propos d'un observation nouvelle. *Rev Pédiatr* **11**:131–138, 1975.

33. Von Oeyen P et al: Omphalocele and multiple severe congenital anomalies associated with osteodysplasty (Melnick–Needles syndrome). *Am J Med Genet* **13**:453–463, 1982.

34. Wendler H, Kellerer K: Osteodysplasie-Syndrom (Melnick–Needles). *Roefo* **122**:309–313, 1975.

35. Zackai EH et al: The male Melnick–Needles syndrome phenotype. *Am J Hum Genet* **39**:88A, 1986.

## Osteochondrodysplasia with hypertrichosis

Cantú et al (1) described two sibs and two unrelated patients with coarse facies, epicanthal folds, long curly eyelashes, anteverted nostrils, long philtrum, prominent mouth, congenital generalized hypertrichosis, macrosomy at birth, short neck, narrow thorax, wide ribs, platyspondyly, hypoplastic ischiopubic rami, coxa valga, long bones with Erlenmeyer flask shape, and generalized osteopenia (Figs. 9–39 and 9–40). All exhibited cardiomegaly. Inheritance may be autosomal recessive.

### Reference (Osteochondrodysplasia with hypertrichosis)

1. Cantú JM et al: A distinct osteochondrodysplasia with hypertrichosis—individualization of a probable autosomal recessive entity. *Hum Genet* **60**:36–41, 1982.

## Pachydermoperiostosis [Touraine–Solente–Golé syndrome, primary (idiopathic) hypertrophic osteoarthropathy]

Pachydermoperiostosis is characterized by coarsening of facial features with thickening and furrowing of the face, forehead, and scalp, and clubbing of the digits with periosteal new bone formation (Fig. 9–41 to 9–43). Touraine et al (39), in 1935, first recognized pachydermoperiostosis as a distinct clinical entity, although examples were noted as early as 1868 by Friedreich (10). Other early cases are those of Unna (40) and Grönberg (13).

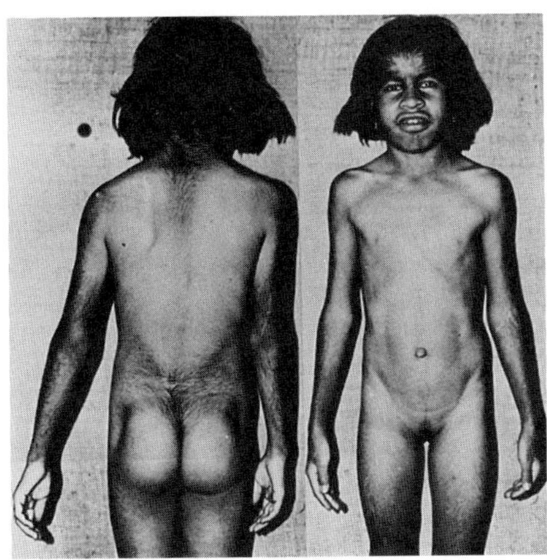

A B C

Fig. 9–39. *Osteochondrodysplasia with hypertrichosis.* (A–C) Coarse facial features, abundant facial and body hair. (From *J M Cantú et al,* Hum Genet **60**:36, 1982.)

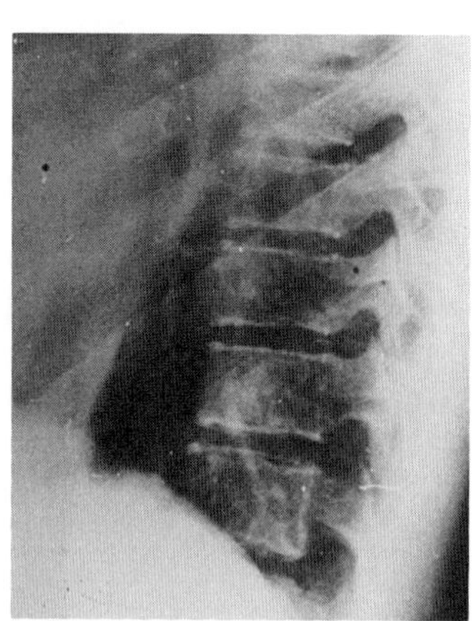

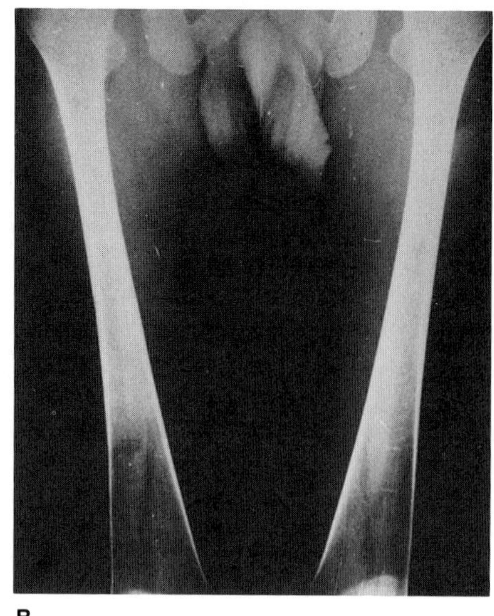

A B

Fig. 9–40. *Osteochondrodysplasia with hypertrichosis.* (A) Platyspondyly, irregular surfaces. (B) Erlenmeyer flask femora. (From *JM Cantú et al,* Hum Genet **60:**36, 1982.)

Most reports have been isolated cases of the condition. However, kindred have been reported in which the disorder appears to be autosomal dominant with variable expressivity (20,22,29,31a,41,43) but only about 15% of the affected are females (31a). The disorder usually appears in the second decade of life (4,11,14,17,18,20,26,31,32,38).

**Facies, skin, and skin appendages.** Thickening of the skin occurs over the face, forehead, scalp, hands, and feet. The face is drawn into thick folds, producing creasing or furrowing that causes the patient to look worried or angry, as well as prematurely aged (1,4,9,15,24,31,33,42). The nasolabial folds become deep (4,42) (Fig. 9–41A,B). Thickening of the scalp tends to produce a corrugated surface or cutis verticis gyrata (13,40), although patients without this characteristic have been described (14,37).

The skin has been reported to be greasy or oily (4,13,18,26,32,36). The dilated sebaceous pores are filled with plugs of sebum that can be easily expressed (13) (Fig. 9–41C). Acne may be marked (20). Pseudoptosis, caused by thickening of the eyelids, may be so severe as to impair vision (4,14,36). Skin biopsy shows sebaceous gland hyperplasia, thickening of the stratum corneum, and perivascular round-cell infiltrates (13,29,42). The nails may be thick and curved. Rimoin (29) and Kerber and Vogl (21) round that peripheral blood flow was reduced. Hyperhidrosis of the hands and feet (1,14,15,18,20) is common. Multiple basal cell carcinomas were found in one case (37).

**Skeletal alterations.** Bone changes affect primarily the long tubular bones, metacarpals, metatarsals, and proximal phalanges (13, 15,20,23,37,43) (Fig. 9–43A,B). Enlargement may be painful (5, 20,22,23). The ends of the fingers and toes become thickened during the second decade of life (13,15) and may become very bulbous (10,14,32,37,43) (Fig. 9–42). This clubbing is produced by soft-tissue hyperplasia, which stops abruptly at the distal interphalangeal joints. A diffuse irregular periosteal ossification increases the circumference of the affected bones and results in loss of normal tubulation of long bones. Acroosteolysis in the hands and feet has been described (14).

Joint effusions of the knees (15,20,23,26) and ossification of ligaments and tendons (14) lead to ankylosis of joints (14). The clavicles, patellas, and pubis may also be affected, but the carpal and tarsal bones, sella turcica, and articular surfaces are spared. The skull is usually not affected, but the posterior half of the calvaria has been described as thickened (32). Enlargement of the wrists and knees has been noted (5,14,15).

**Differential diagnosis.** All clinical aspects of pachydermoperiostosis have been described in secondary (pulmonary) hypertrophic osteoarthropathy (16). Patients with pulmonary hypertrophic osteoarthropathy lack a family history of the disorder but have a primary neoplasm, usually bronchogenic carcinoma. There appears to be increased blood flow in pulmonary osteoarthropathy (12), in contrast to reduced flow in pachydermoperiostosis (29).

In acromegaly, there is enlargement of the hands and feet, and thickening of the skin, particularly of the face (33,42). The mandible, nose, sella turcica, supraorbital ridges, and tongue are also enlarged, findings not seen in pachydermoperiostosis.

Thyroid acropachy may follow medical or surgical treatment of hyperthyroidism. As in pachydermoperiostosis, the distal parts of the limbs may be come enlarged, and clubbing of the fingers and toes may occur (27). There may be subperiosteal new bone formation in the hands. Severe exophthalmos and pretibial myxedema may be present, together with high levels of long-acting thyroid stimulator in the serum of such patients.

Simple hereditary clubbing (acropathy) has been described (8). Rimoin (29) has suggested that hereditary acropathy may be an incomplete form of pachydermoperiostosis.

Rosenthal and Kloepfer (30) and Harbison and Nice (17) described a combination of cutis verticis gyrata, corneal leukoma, marked supraorbital ridging, and enlarged frontal and sphenoid sinuses as an autosomal dominant syndrome in a large kindred; the cranium was thickened in some patients but no periosteal reaction was seen along the shafts of long bones.

*Beare–Stevenson cutis gyratum syndrome* (2) should also be excluded.

A patient reported by Vogl and Goldfischer (43) had oily and thickened facial skin, deep folds on the forehead, deep nasolabial folds, and clubbing of fingers and toes. Onset of these changes was noted by the patient at about age 40. No evidence of periosteal proliferation was found.

Familial idiopathic osteoarthropathy of childhood (6,7) is an autosomal recessive disorder characterized by eczema, clubbing of fingers, large hands and feet, thick arms and legs due to periosteal new bone formation, and persistent anterior fontanel. Onset has been reported during thc sccond year of life.

The patient reported by Herbert and Fessel (19) had large, unusually shaped extremities since birth, which increased in thickness during adolescence, periosteal new bone formation affecting the distal ends of some of the long bones and clubbing of all fingers and toes, thickening of the skin of the lower legs, and acroosteolysis of the distal phalanges of the fingers and toes; there was no hyperhidrosis or cutis verticis gyrata, and facial features were normal.

Members of the kindred reported by Boylen and Blackard (3) as

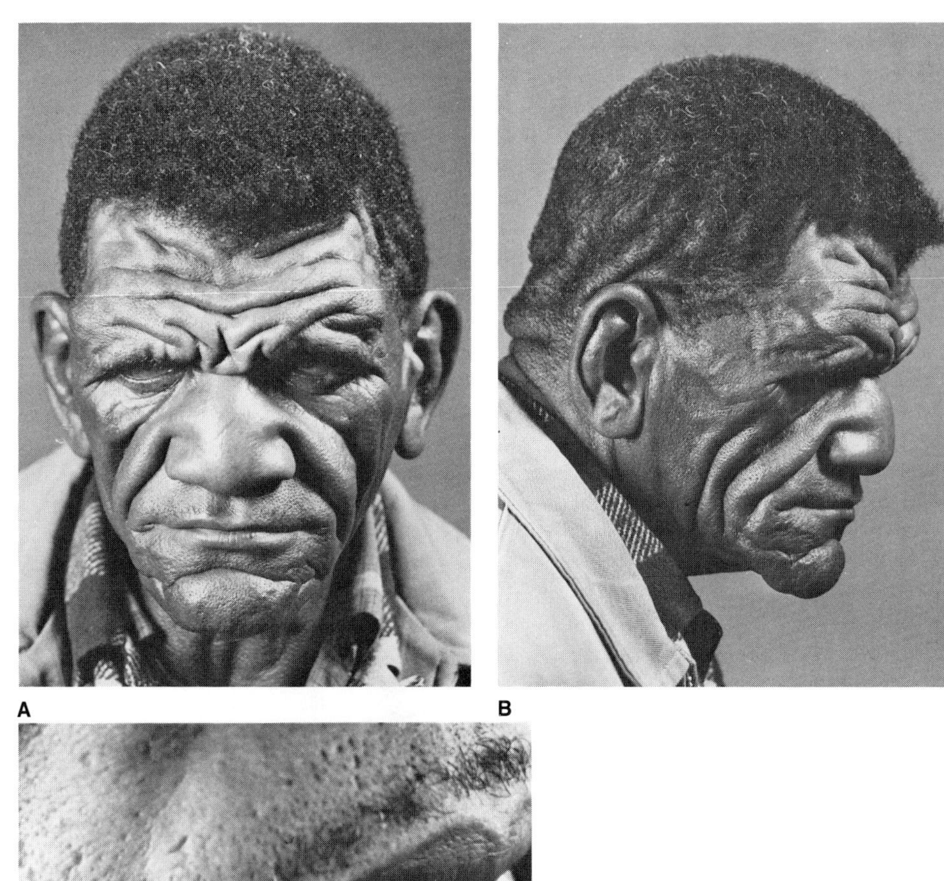

A B

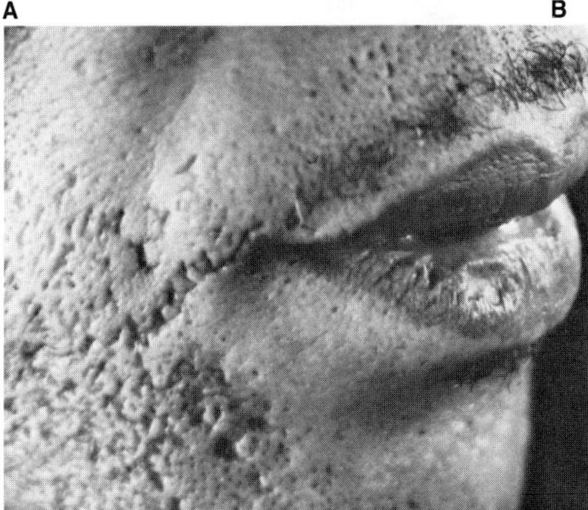

C

Fig. 9–41. *Pachydermoperiostitis.* (A,B) Coarse features, cutis gyrata. (C) Oily thickened skin with large sebaceous pores. No history of acne. (A,B courtesy of *HW Kloepfer,* New Orleans, Louisiana. C courtesy of *A Susmano,* Chicago, Illinois.)

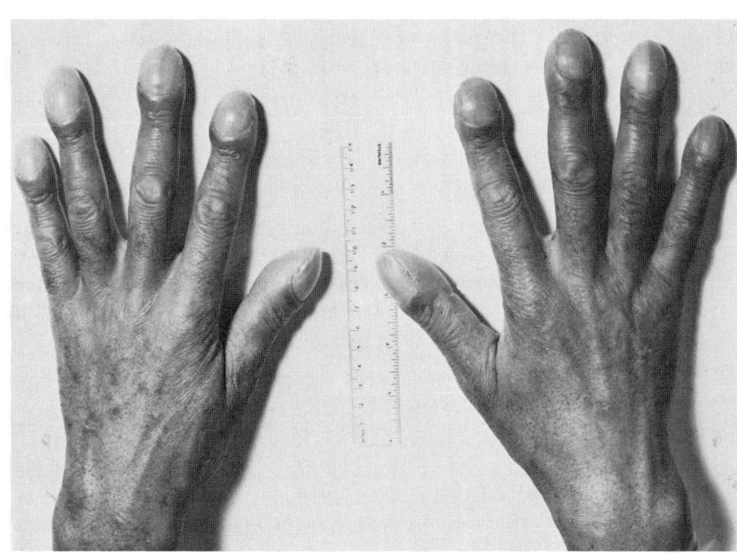

Fig. 9–42. *Pachydermoperiostitis.* Clubbing of terminal phalanges. (Courtesy of *A Susmano,* Chicago, Illinois.)

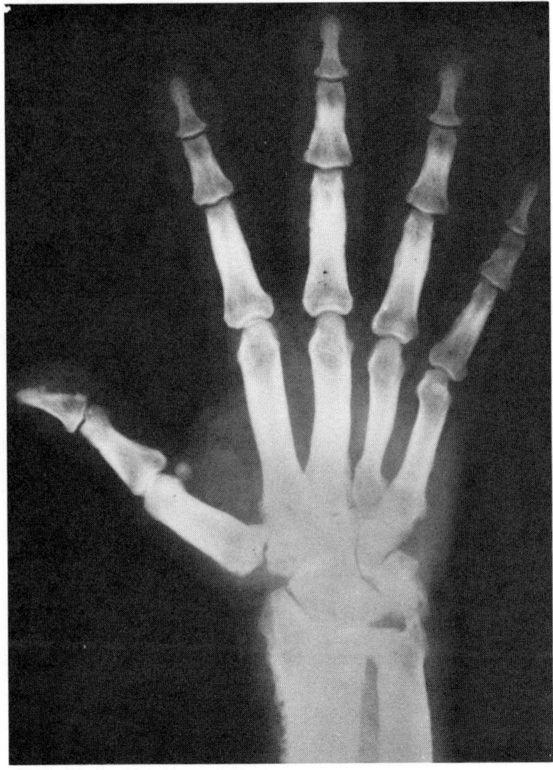
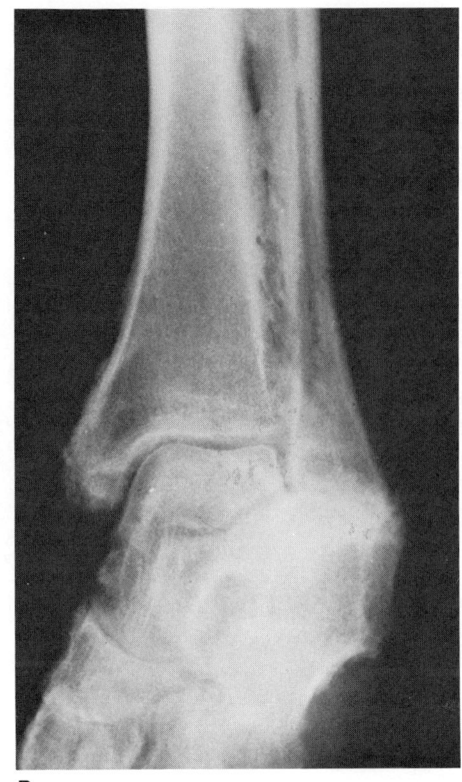

A B

Fig. 9–43. *Pachydermoperiostitis.* (A) Note thickening and increased density of proximal phalanges and metacarpals; periosteal thickening of radius. (B) Marked periosteal proliferation along entire length of tibia and fibula. (A from *G Pietruschka et al,* Klin Monatsbl Augenheilkd, **154:**525, 1969. B courtesy of *A Susmano,* Chicago, Illinois.)

pachydermoperiostosis had hyperthyroidism. They had coarse prominent facial features, frontal bossing, furrowing of the skin over the forehead, and large hands. Corneal leukomas were also found. The proband had no significant clubbing of the hands, but other members of the family exhibited this characteristic. There were no periosteal changes in the long bones, but radiographs of the proband's hands showed several cystic lesions. Inheritance was autosomal dominant.

Leibowitz and Kalk (25) reported a patient with coarsening of facial features beginning at puberty, cutis verticis gyrata, acneiform eruption on the face, unilateral ptosis, and prominent nasolabial folds; there was no clubbing or periostosis. The patient had depressed urinary excretion of 17-hydroxy- and corticosteroids, depressed response to thyrotropin-releasing hormone, and aminoaciduria.

Sirinavin et al (34) reported a 12-year-old girl, born to a consanguineous couple, with clubbing of fingers and toes since 7 months of age, hyperhidrosis of the hands and feet, slightly tender and swollen distal phalanges, dystrophic nails, thickening of soft tissues of knees and ankles, generalized osteoposis, joint effusion in the knees, and acroosteolysis. Although other skeletal abnormalities were present, subperiosteal ossification was not found.

Thomas (35) described a 22-year-old male with marked clubbing of fingers and toes of unknown onset and mild periosteal thickening of the tibias and forearms; the remainder of the physical examination was unremarkable. His father had similar hand and foot abnormalities.

### References [Pachydermoperiostosis (Touraine–Solente–Golé syndrome, primary (idiopathic) hypertrophic osteoarthropathy)]

1. Angel JH: Pachydermo-periostosis (idiopathic osteoarthropathy). *Br Med J* **2**:789–792, 1957.

2. Beare JM et al: Cutis gyratum, acanthosis nigricans and other congenital anomalies. A new syndrome. *Br J Dermatol* **81**:241–247, 1969.

3. Boylen CT, Blackard WG: Pachydermoperiostosis: A review and presentation of a case with associated Graves' disease. *S Med J* **64**:317–320, 1971.

4. Brugsch HG: Acropachyderma with pachydermoperiostosis. Report of a case. *Arch Intern Med* **68**:687–700, 1941.

5. Camp JD, Scanlan RL: Chronic idiopathic hypertrophic osteo-arthropathy. *Radiology* **50**:581–594, 1948.

6. Chamberlain DS et al: Idiopathic osteoarthropathy and cranial defects in children. *Am J Roentgenol* **93**:408–415, 1965.

7. Cremin BJ: Familial idioapathic osteoarthropathy of children: A case report and progress. *Br J Radiol* **43**:568–570, 1970.

8. Curth HO et al: Familial clubbed fingers. *Arch Dermatol* **83**:828–836, 1961.

9. Findlay GH, Oosthuizen WJ: Pachydermoperiostosis: The syndrome of Touraine, Solente and Golé. *S Afr Med J* **25**:747–752, 1951.

10. Friedreich N: Hyperostose des gesammten Skelettes. *Virchows Arch Pathol Anat* **43**:83–87, 1868.

11. Freund E: Idiopathic familial generalized osteophytosis. *Am J Roentgenol* **39**:216–217, 1938.

12. Ginsburg J: Observations on peripheral circulation in hypertrophic pulmonary osteoarthropathy. *Q J Med* **27**:335–352, 1958.

13. Grönberg A: Is cutis verticis gyrata a symptom in endocrine syndrome which has so far received little attention? *Acta Med Scand* **67**:24–42, 1927.

14. Guyer PB et al: Pachydermoperiostosis with acro-osteolysis. A report of five cases. *J Bone Jt Surg* **60B**:219–223, 1978.

15. Hambrick GW, Carter DM: Pachydermoperiostosis. *Arch Dermatol* **94**:594–608, 1966.

16. Hammarsten JF, O'Leary J: The features and significance of hypertrophic osteoarthropathy. *Arch Intern Med* **99**:431–441, 1957.

17. Harbison JB, Nice CM: Familial pachydermoperiostosis presenting as an acromegaly-like syndrome. *Am J Roentgenol* **112**:532–536, 1971.

18. Hedayati H et al: Acrolysis in pachydermoperiostosis. Primary or idiopathic hypertrophic osteoarthropathy. *Arch Intern Med* **140**:1087–1088, 1980.

19. Herbert DA, Fessel WJ: Idiopathic hypertrophic osteoarthropathy (pachydermoperiostosis). *West J Med* **134**:354–357, 1981.

20. Herman MA et al: Pachydermoperiostosis-clinical spectrum. *Arch Intern Med* **116**:918–923, 1965.

20a. Hochmuth WP et al: Touraine–Solente–Golé–Syndrome. *Med Klin* **70**:146–150, 1975.

21. Kerbert RE, Vogl A: Pachydermoperiostosis. Peripheral circulatory studies. *Arch Intern Med* **132**:245–248, 1973.

22. Langston HH: Bone dystrophy of unknown aetiology (presented for diagnosis). *Proc R Soc Med* **43**:299–303, 1950.

23. Lauter SA et al: Pachydermoperiostosis: Studies of the synovium. *J Rheumatol* **5**:85–95, 1978.

23a. Lazarus JH, Galloway JK: Pachydermoperiostosis: An unusual cause of finger clubbing. *Am J Roentgenol* **118**:308–313, 1973.

24. Lehman MA et al: Idiopathic hypertrophic osteoarthropathy (acropachyderma with pachydermoperiostosis). *Bull Hosp Joint Dis* **24**:56–67, 1963.

25. Leibowitz MR, Kalk WJ: Cutis verticis gyrata with metabolic abnormalities. *Dermatologica* **166**:146–150, 1983.

26. Nieman HL: Pachydermoperiostosis with bone marrow failure and gross extramedullary hematopoiesis. *Radiology* **110**:553–554, 1974.

27. Dixon DW, Samols E: Acral changes associated with thyroid diseases. *J Am Med Assoc* **212**:1175–1181, 1970.

28. Pietruschka G et al: Ein Beitrag zur Pachydermoperiostose. *Klin Monatsbl Augenheilkd* **154**:525–536, 1969.

29. Rimoin DL: Pachydermoperiostosis (idiopathic clubbing and periostosis). Genetic and physiologic considerations. *N Engl J Med* **272**:923–931, 1965.

30. Rosenthal JW, Kloepfer HW: An acromegaloid, cutis verticis gyrata, corneal leukoma syndrome: A new medical entity. *Arch Ophthalmol* **68**:722–726, 1962.

31. Roy JN: Hypertrophy of the palpebral tarsus, the facial integument and the extremities of the limbs associated with widespread osteo-periostosis: A new syndrome. *Can Med Assoc J* **34**:615–622, 1936.

31a. Salfeld K, Spalckhaver I: Zur Kenntnis der Pachydermoperiostosis. *Derm Wochenschr* **152**:497–511, 1966.

31b. Schneider I et al: Pachydermoperiostose (Touraine–Solente–Golé–Syndrom). *Hautarzt* **33**:221–223, 1982.

32. Schuster MM et al: Facial deformity in pachydermoperiostosis. Idiopathic hypertrophic osteoarthropathy. *Plast Reconstr Surg* **35**:666–674, 1965.

33. Shawarby K, Ibrahim MS: Pachydermoperiostosis. A review of literature and report on four cases. *Br Med J* **1**:763–766, 1962.

34. Sirinavin C et al: Digital clubbing, hyperhidrosis, acro-osteolysis and osteoporosis. A case resembling pachydermoperiostosis. *Clin Genet* **22**:83–89, 1983.

35. Thomas HB: Agnogenic congenital clubbing of the fingers and toes. *Am J Med Sci* **203**:241–246, 1942.

36. Sisson RJ: Cutis verticis gyrata. *J Am Med Assoc* **86**:1126–1127, 1926.

37. Thomas RHM, Kirby JDT: Pachydermoperiostosis with multiple basal cell carcinomata. *J R Soc Med* **78**:335–337, 1985.

38. Törnblum N et al: Osteodermatopathia hypertrophicans. *Acta Med Scand* **164**:325–339, 1959.

39. Touraine A et al: Un syndrome ostéo-dermatopathique: La pachydermie plicaturée avec pachypériostose des extrémités, *Presse Méd* **43**:1820–1824, 1935.

40. Unna PG: Cutis verticis gyrata. *Monatsschr Prakt Dermatol* **45**:227–233, 1907.

41. Ursing B: Pachydermoperiostosis. *Acta Med Scand* **188**:157–160, 1970.

42. Venecie PY et al: Pachydermoperiostosis with gastric hypertrophy, anemia and increased serum bone gla-protein levels. *Ann Dermatovenereol* **124**:1831–1834, 1988.

43. Vogl A, Goldfischer S: Pachydermoperiostosis. Primary or idiopathic hypertrophic osteoarthropathy. *Am J Med* **33**:166–187, 1962.

Fig. 9–44. *Pycnodysostosis.* (A,B) Eight-year-old boy with short stature. Note mild exophthalmos and small mandible. (From *A Giedion* and *M Zachmann,* Helv Paediatr Acta **21**:412, 1966.)

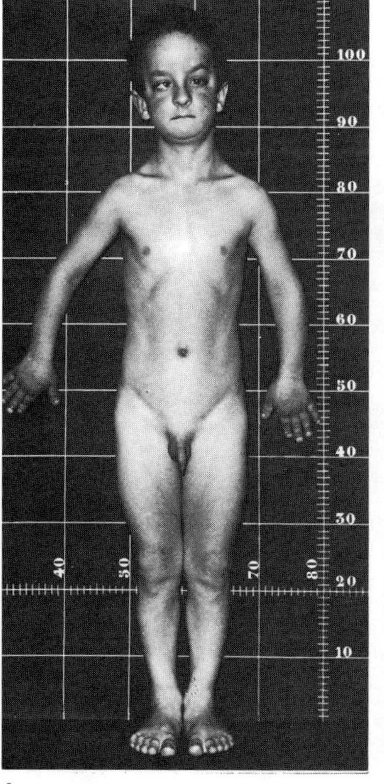
A

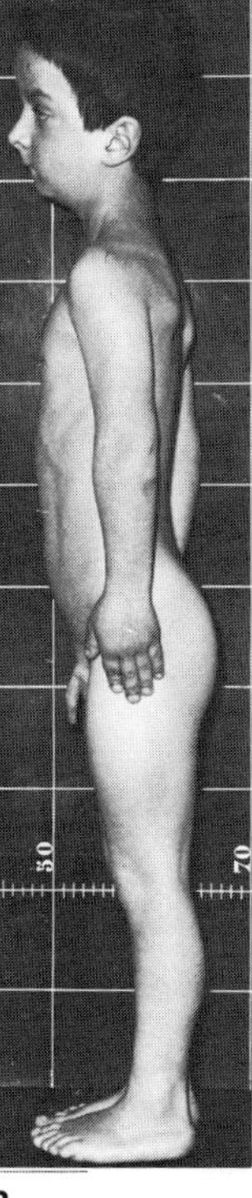
B

## Pycnodysostosis

Maroteaux and Lamy (19) and Andrén et al (2), in 1962, defined pycnodysostosis as a syndrome consisting of dwarfism, osteopetrosis, abbreviated terminal phalanges, cranial anomalies, such as persistence of fontanels and failure of closure of cranial sutures, and hypoplasia of the mandibular angle. Several cases of the syndrome were published as examples of other disorders, chiefly osteopetrosis and cleidocranial dysplasia. The first documented case is that of Montanari (22) in 1923. About 80 cases have been reported (14,32).

Affected sibs have been noted in many instances (1,10, 22,25,27,30,33,36,37). The syndrome has also been seen in identical twins (2). Parental consanguinity has been noted in more than 30% (28). Autosomal recessive inheritance is indicated. There is good evidence to suggest that Toulouse-Lautrec had the syndrome (20). The patient reported by Roth (26) does not have pycnodysostosis.

**Facies.** The head appears large because of occipital bulging. Parrot-like nose with mild exophthalmos and micrognathia are characteristic (Fig. 9–44A,B).

**Skeletal alterations.** Because of shortness of the extremities, adult height is reduced to 134–152 cm (53–60 in.). The trunk is not shortened but often exhibits marked pectus excavatum, with underdeveloped and widened terminal phalanges that often present a drum-stick appearance (Fig. 9–45A–C). The nails may be thin and hypoplastic.

The acromial end of the clavicle is usually somewhat hypoplastic (2,29,32,36). Bilateral genua valga is frequent. Partial disappearance of the hyoid bone has been reported (35).

On radiographic examination, the skull is dolichocephalic with frontal and occipital bossing (1,12,18,33). Most cranial sutures and fontanels are open, especially the parietooccipital. The bones of the calvaria are thin, dense, and without diploic markings. Wormian bones are commonly observed (2,27,28,30). Craniosynostosis has been reported rarely (6). The frontal sinuses are consistently absent, and other paranasal sinuses are hypoplastic or missing. The mastoid air cells are often not pneumatized (2,27–28,36) (Fig. 9–46A,B).

There is increased radiopacity of all bones, but especially of the long bones, spine, and cranial base. Bone fragility is increased, over 70% having multiple fractures during their lifetime (5,34). The terminal phalanges of the fingers and toes exhibit fragmentation of the heads with preservation of the bases, osteolysis of the unguiculate portions, or narrowing of the ends of otherwise normal terminal phalanges (4). Brachymesophalangy of the fifth fingers, less often of the index finger, is a common finding. The fourth metatarsal is occasionally abbreviated (32).

Microscopic studies of the involved bones carried out by Shuler (29), Taracena del Piñal (33), and Soto et al (30) have shown reduction in osteoclastic and osteoblastic activity, with reduced rates of bone formation and resorption. The bone has been found to be markedly sclerotic. Electron microscopic studies have demonstrated that osteoclasts have diminished or and even inactive secretory functions (21).

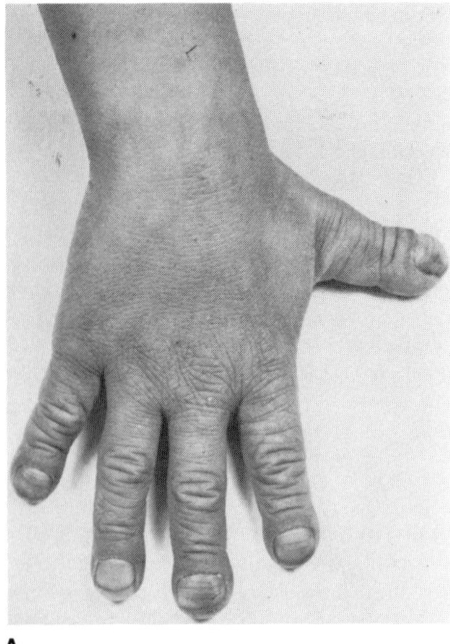

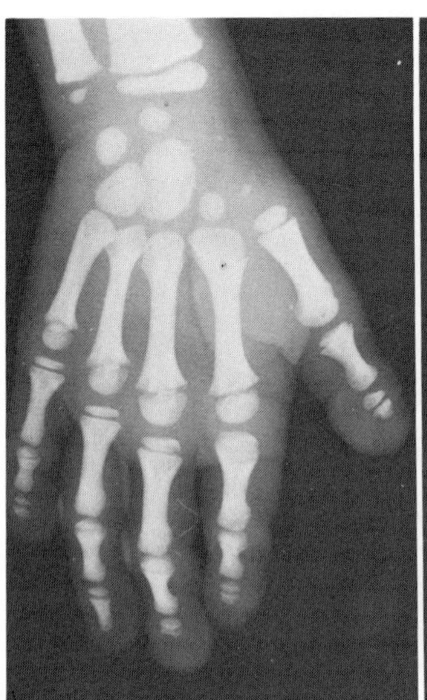

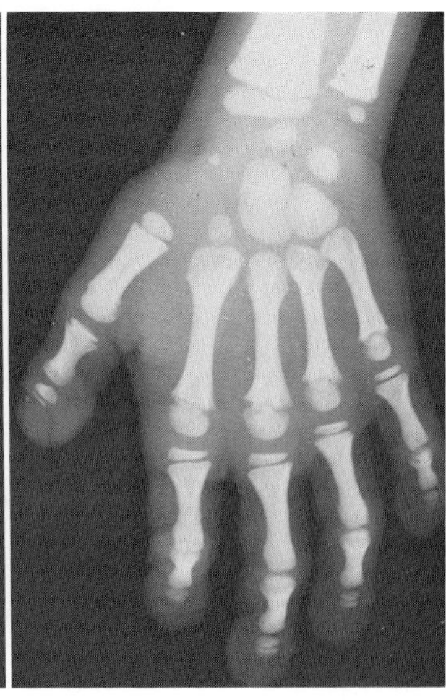

A B

Fig. 9–45. *Pycnodysostosis.* (A,B) Terminal digits of fingers reduced and widened. Nails often overlap ends of fingers. Note increased bone density, acroosteolysis of terminal phalanges, also pointed terminal phalanx. (From *SE Shuler,* Arch Dis Childh **38:**620, 1963.)

**Eyes.** The eyes may be somewhat exophthalmic (1,32,33,36) with blue sclerae (26,37).

**Oral manifestations.** Obtuse mandibular angle is a constant feature. Facial bones are often underdeveloped, with relative mandibular prognathism (2,12,13,36,38). Oral and dental anomalies include premature or delayed eruption (1,11,12,32,38), enamel hypoplasia (25,33), malposed teeth (7,11,32,33), and grooved palate (12,13,32,33). The soft palate tends to be long (23,40). Lacey et al (16) described a patient with short and blunted tooth roots and multiple congenitally missing permanent tooth germs.

**Differential diagnosis.** Differential diagnosis includes *osteopetrosis, acroosteolysis,* and *mandibuloacral dysplasia.* The open cranial fontanels and sutures may suggest *cleidocranial dysplasia.*

*Acroosteolysis* is associated with progressive reduced height, kyphosis, bathrocephaly, basilar impression, numerous Wormian bones, absence of frontal sinuses, and fusion of the spinous processes of the cervical vertebrae. The terminal phalanges are shortened and often exhibit tenderness, pain, and paresthesia. The alveolar process often is markedly atrophic, but the angle is not missing as in pycnodysostosis. It has autosomal dominant inheritance. So-called industrial acroosteolysis has been described in workers synthesizing polyvinyl

Fig. 9–46. *Pycnodysostosis.* Roentgenograms. (A,B) Absence of fusion of sutures and closure of fontanels. Increased bone density and absence of mandibular angle. (From *SE Shuler,* Arch Dis Childh **38:**620, 1963.)

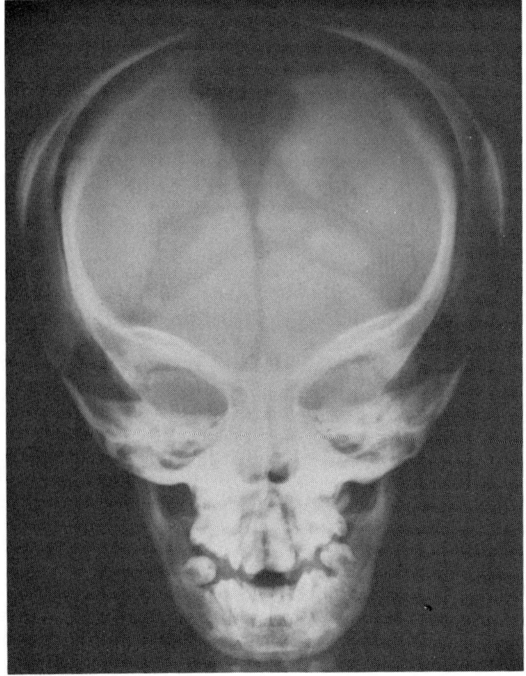

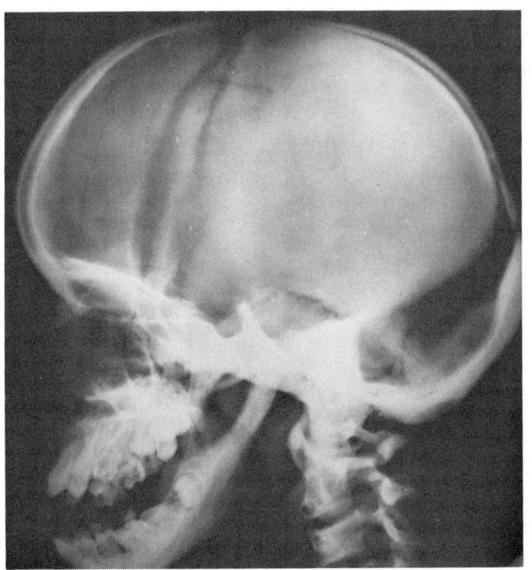

A B

chloride (18). Lamy and Maroteaux (17) described isolated autosomal dominant acroosteolysis.

*Stanescu osteosclerosis syndrome,* a rare form of craniofacial dysostosis, inherited as an autosomal dominant trait, is characterized by small skull, thin cranial bone, depressions over the frontoparietal and occipitoparietal sutures, poorly developed mandible with obtuse angle, exophthalmos, and very short limbs with massive thick cortices (31).

*Mandibuloacral dysplasia* resembles pycnodysostosis in delayed closure of skull sutures, Wormian bones, and hypoplasia of terminal phalanges, but there is no increase in bone density or aplasia of the mandibular angle. Instead there is antegonial mandibular notching, stiff joints, and cutaneous atrophy (9,39).

**Laboratory aids.** Reduced serum alkaline phosphatase has been found in a few cases (12,33,37). Baker et al (3) described intermittently high plasma calcitonin levels. Hepatosplenomegaly and iron deficiency anemia also have been found (4,15,21,24). Cabrejas et al (8) described increased intestinal calcium absorption and diminished exchangeable calcium pool turnover as well as diminished bone accretion rate.

### References (Pycnodysostosis)


1. Abboud MS et al: Albers-Schönberg disease with report of two cases in Egyptian family. *Arch Pediatr* **71**:131–138, 1954.
2. Andrén L et al: Osteopetrosis acro-osteolytica: A syndrome of osteopetrosis, acro-osteolysis and open sutures of the skull. *Acta Chir Scand* **124**:496–507, 1962.
3. Baker RK et al: Plasma calcitonin in pycnodysostosis. *J Clin Endocrinol Metab* **37**:46–55, 1973.
4. Balthazar E et al: Pycnodysostosis: An unusual case. *Br J Radiol* **45**:304–307, 1972.
5. Bennani-Smires C et al: La pycnodysostose. Aspects classiques et inhabituels. A propos de 7 cas. *J Radiol* **65**:689–695, 1984.
6. Bernard R et al: Pycnodysostose et craniostenose. *Ann Pédiatr* **27**:383–385, 1980.
7. Braun JP, Peterschmitt J: L'aspect radiologique de la pycnodysostose. *J Radiol Électrol* **46**:509–512, 1965.
8. Cabrejas ML et al: Pycnodysostosis: Some aspects concerning kinetics of calcium metabolism and bone pathology. *Am J Med Sci* **271**:215–220, 1976.
9. Cavallazzi C et al: Su di un caso di disostosi cleidocrania. *Riv Clin Pediatr* **65**:312–326, 1960.
10. Dusenberry JF, Kane JJ: Pycnodysostosis: Report of three new cases. *Am J Roentgenol* **99**:717–723, 1967.
11. Elmore SM et al: Pycnodysostosis with a familial chromosome anomaly. *Am J Med* **40**:273–282, 1966.
12. Giedion A, Zachmann M: Pyknodysostose. *Helv Paediatr Acta* **21**:612–621, 1966.
13. Janečka J, Bruna J: Pyknodysostosis in Röntgenbild. *Fortschr Röntgenstr* **118**:298–305, 1973.
14. Kemperdick H, Lehr HJ: Die Pyknodysostose. *Monatschr Kinderheilkd* **123**:52–57, 1975.
15. Kozlowski K, Yu JS: Pycnodysostosis, a variant with visceral manifestations. *Arch Dis Childh* **47**:804–807, 1972.
16. Lacey SH et al: Pycnodysostosis. A case report of a child with associated trisomy X. *J Pediatr* **77**:1033–1038, 1970.
17. Lamy M, Maroteaux P: Acro-ostéolyse dominant. *Arch Fr Pédiatr* **18**:693–702, 1961.
18. Markowitz SS et al: Occupational acroosteolysis. *Arch Dermatol* **106**:219–223, 1972.
19. Maroteaux P, Lamy M: La pycnodysostose. *Presse Méd* **70**:999–1002, 1962.
20. Maroteaux P, Lamy M: The malady of Toulouse-Lautrec. *J Am Med Assoc* **191**:715–717, 1965.
21. Meredith SC et al: Pycnodysostosis. *J Bone Jt Surg* **60**:1122–1127, 1978.
22. Montanari U: Acondroplasia e disostosi cleidocrania digitale. *Chir Organi Mov* **7**:379–391, 1923.
23. Nielsen EL: Pycnodysostosis. *Acta Paediatr Scand* **63**:437–442, 1974.
24. Norman CH, Dubowy J: Pycnodysostosis with splenomegaly and anemia. *NY J Med* **71**:2419–2421, 1971.
25. Perez-Cuadra E, Galvez-Galan F: Enfermedad de Albers-Schönberg: Dos casos en la misma familia. *Bol Inst Pat Med (Madrid)* **15**:322–332, 1960.
26. Roth VG: Pycnodysostosis presenting with bilateral subtrochanteric fractures. *Clin Orthoped Rel Res* **117**:247–253, 1976.
27. Sandomenico C, Del Vecchio E: La picnodisostosi. Inquadramento classificativo e patogenico. *Radiol Med (Torino)* **53**:160–181, 1967.
28. Sedano HO et al: Pycnodysostosis: Clinical and genetic considerations. *Am J Dis Child* **116**:70–77, 1968.
29. Shuler SE: Pycnodysostosis. *Arch Dis Childh* **38**:620–625, 1963.
30. Soto RJ et al: Pycnodysostosis: Metabolic and histologic studies. *Birth Defects* **5**(4):109–116, 1969.
31. Stanescu V et al: Syndrome héréditaire dominant, reunissant une dysostose cranio-faciale de type particulier, une insuffisance de croissance diaspect chondrodystrophique et une epaississement massif de la corticale des os longs. *Rev Fr Endocrinol Clin* **4**:221–231, 1963.
32. Sugiura Y et al: Pycnodysostosis in Japan: Report of six cases and a review of Japanese literature. *Birth Defects* **10**(12):78–98, 1974.
33. Taracena del Piñal B: Cuatro casos de una enfermedad nueva, la picnodisostosis. *Ann Med Quirg Cruz Roja Esp* **6**:5–24, 1964.
34. Taylor MM et al: Pycnodysostosis. *J Bone Jt Surg* **60A**:1128–1130, 1978.
35. Theander G: Partial disappearance of the hyoid bone in pyknodysostosis. *Acta Radiol Diag* **19**:237–242, 1978.
36. Thoms J: Cleido-cranial dysostosis: Report of two cases with special characteristics. *Acta Radiol* (Stockholm) **50**:514–520, 1958.
37. Weismann-Netter R, Lorch P: Nanisme familial avec densification generalisée du squelette et dysgenesies polytopiques. *Sem Hôp Paris* **32**:2713–2719, 1956.
38. Wiedemann H-R: Pyknodysostose. *Fortschr Roentgenstr* **103**:590–597, 1965.
39. Young LW et al: Mandibular hypoplasia, acroosteolysis, stiff joints and cutaneous atrophy. *Birth Defects* **7**(7):291–297, 1971.
40. Yousefzadeh DK et al: Radiographic studies of upper airway obstruction with cor pulmonale in a patient with pycnodysostosis. *Pediatr Radiol* **8**:45–47, 1979.


## Stanescu osteosclerosis syndrome

Striking features of Stanescu type osteosclerosis include short stature, brachycephaly, hypoplastic midface, ocular proptosis, micrognathia, brachydactyly, and dense cortices of long bones (Figs. 9–47 to 9–49). Other findings have included depression of the frontoparietal and occipitoparietal sutures, lack of pneumatization of frontal and sphenoidal bones, small crowded teeth with enamel hypoplasia, sacralization of $S_1$, and exostoses. Patients appear to have craniosynostosis judging from photographic and radiographic documentation of craniofacial shape (3), although to date no specific mention has been made of this in the texts of the three articles on the subject (1,3,4). Further documentation is needed in future reports of the disorder.

In the family reported by Stanescu et al (4), 11 affected individuals were reported. More recently, Maximilian et al (3) reexamined the family, reporting three new individuals with the disorder. Dipierri and Guzman (1) described an affected mother and daughter. Inheritance is autosomal dominant. Hall (2) noted a sporadic instance of a patient with some similarities, although differences included lack of brachydactyly and severe involvement of the spine and thorax with pectus excavatum and kyphoscoliosis. Hall's patient may represent a different entity.

### References (Stanescu osteosclerosis syndrome)


1. Dipierri JE, Guzman JD: A second family with autosomal dominant osteosclerosis-type Stanescu. *Am J Med Genet* **18**:13–18, 1984.
2. Hall JG: Craniofacial dysostosis—either Stanescu dysostosis or a new entity. *Birth Defects* **10**(12):521–523, 1974.3
3. Maximilian C et al: Syndrome de dysostose cranio-faciale avec hyperplasie diaphysaire. *J Génét Hum* **29**:129–139, 1981.
4. Stanescu V et al: Syndrome héréditaire dominant, réuinissant une dysostose craniofaciale de type particulier, une insuffisance de croissance d'aspect chondrodystrophique et un épaississement massif de la corticale des os longs. *Rev Fr Endocrinol Clin* **4**:219–231, 1963.

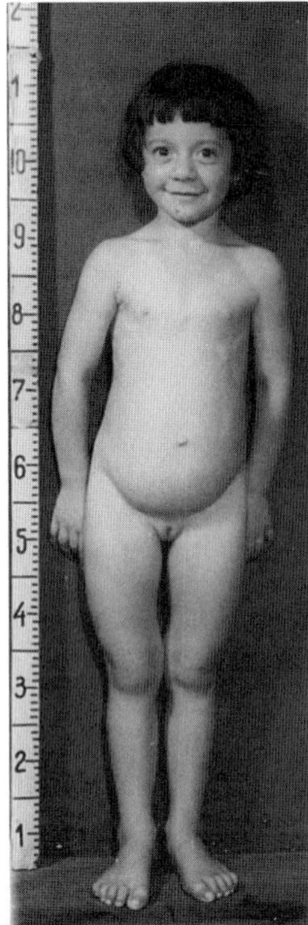
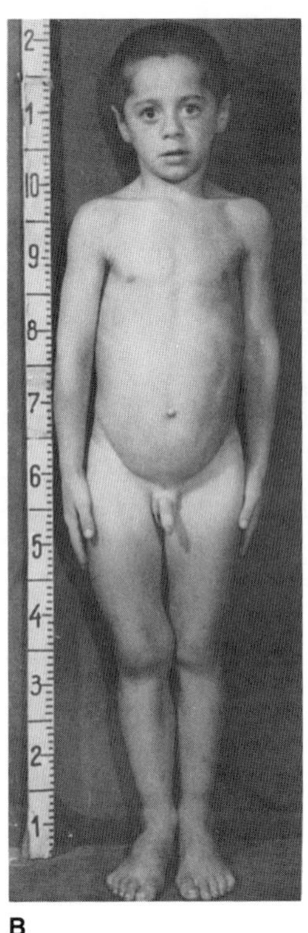
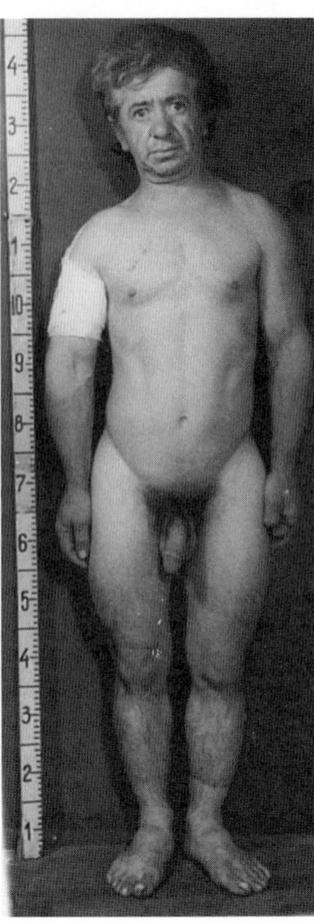

A B C

Fig. 9–47. *Stanescu osteosclerosis syndrome.* (A–C) Short stature, brachycephaly, ocular proptosis. (From *C Maximilian et al,* J Génét Hum **29:**129, 1981.)

## Stickler syndrome (Marshall–Stickler syndrome, Wagner–Stickler syndrome, hereditary arthroophthalmopathy)

The syndrome of flat midface, cleft palate, high myopia with retinal detachment and cataracts, hearing loss, and arthropathy with generally mild spondyloepiphyseal dysplasia (Fig. 9–50) has autosomal dominant inheritance with markedly variable age-dependent expressivity. Stickler and co-workers (34,35), in 1965–1967, reported a combination of eye findings, hearing loss, cleft palate, Marfanoid build, and bone changes. The binary combination of eye changes and cleft palate was described both earlier and later by a large number of authors (4,6–10,16,17,36). This has been reviewed by Cohen et al (4) and Herrmann et al (15). Although neither retinal detachment nor other features of Stickler syndrome were originally described in a large Swiss kindred reported by Wagner (autosomal dominant vitreoretinal degeneration), follow-up studies have disclosed some of these findings (3,11,37). However, vide infra. The radiographic presentation during infancy has been called the Weissenbacher–Zweymüller syndrome (14,19,31a,40). Ayme and Preus (1) believe that the "Marshall syndrome" (11a,18,23,25a,41) represents a separate entity while others view it as severe expression of Stickler syndrome (2,5). It has been suggested that 30% of infants with *Robin sequence* eventuate with Stickler syndrome (21,26,29,31,32). We believe that the Walden syndrome (38) is Stickler syndrome.

Francomano et al (9b) found that the ophthalmological condition described by Wagner is not caused by a mutation in COL2A1, that genetic heterogeneity exists between families clinically diagnosed as having Stickler syndrome, and that the mutation causing Wagner syndrome is genetically distinct from that causing Stickler syndrome in some families. They further indicate that the gene for Stickler syndrome is on chromosome 12, near the structural gene for type II collagen (9a).

Myopia, 8–18 diopters, is found in 75–80% of patients as early as 5 years of age. Before the twentieth year, vitreous and chorioretinal degeneration with broad zones of retinal detachment are observed in 70%, often bilaterally. If untreated, this leads to blindness. Associated eye findings are astigmatism (60%), cataracts (45%), strabismus (30%), and glaucoma (10%). Eye findings have been extensively discussed by a number of authors (20,22,24,40).

The craniofacial spectrum ranges from an essentially normal face (15–25%) to midfacial flattening due to short maxilla, prominent eyes, epicanthal folds, depressed nasal bridge, long philtrum, and small chin. Cleft palate, submucous cleft palate, and abnormal palatal mobility have been reported in 20% (12,13). Progressive sensorineural high tone hearing loss has been found in 80%.

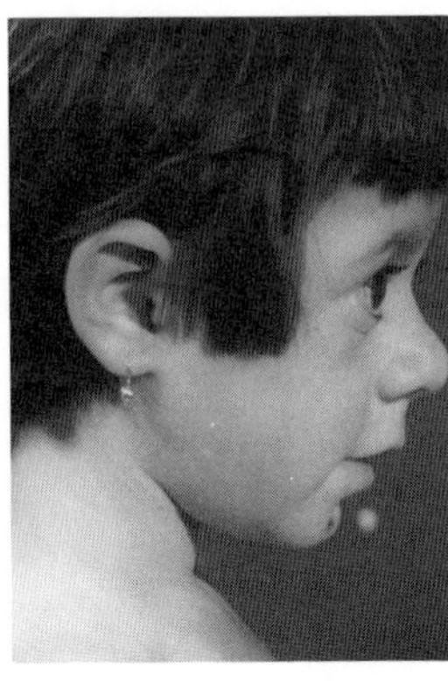
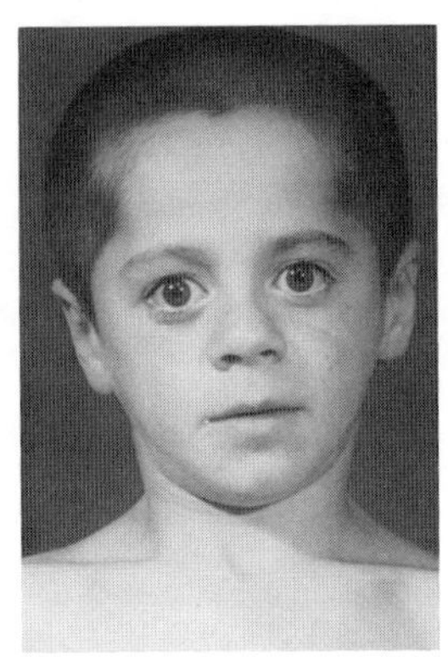
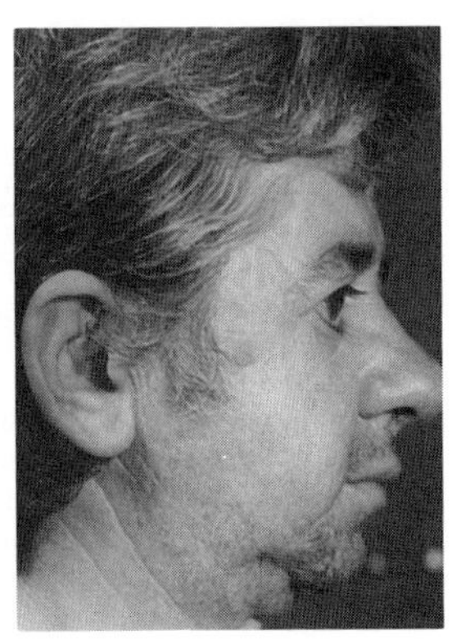

A B C

Fig. 9–48. *Stanescu osteosclerosis syndrome.* (A–C) Note brachycephalic skull in B, mild midface deficiency with ocular proptosis and micrognathia in A and C. (From *C Maximilian et al,* J Génét Hum **29:**129, 1981.)

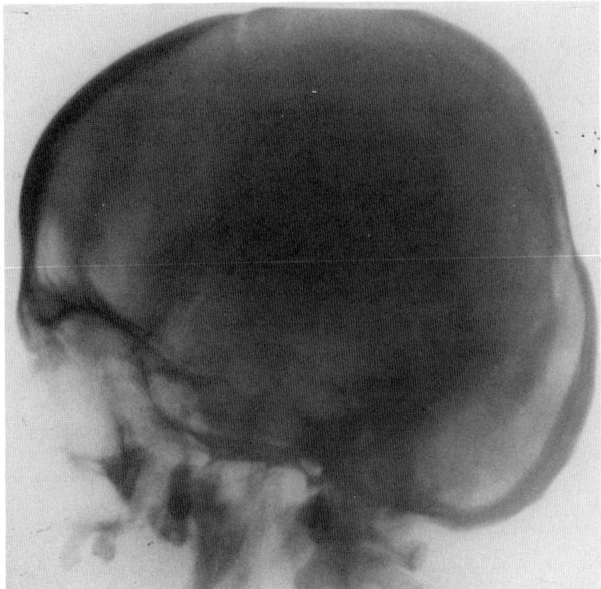

Fig. 9–49. *Stanescu osteosclerosis syndrome.* Radiograph of skull showing dense cranium, shallow sella, and obtuse mandibular angle. (From *V Stanescu et al,* Rev Fr Endocrinol Clin **4:**219, 1963.)

Some patients have a Marfanoid body habitus, but at least 25% are below the 3rd percentile in height. In childhood, joint hypermobility is common. The joints may be enlarged, often hyperextensible (35%), and sometimes painful and warm with use, becoming stiff with rest. Talipes equinovarus may occur.

Radiographically, in infancy there is rhizomelic shortening of the limbs, metaphyseal widening, and vertebral coronal clefts. During childhood, mild spondyloepiphyseal dysplasia (multiple epiphyseal ossification disturbances, moderate flattening of vertebral bodies) and diminution of the width of the shaft of tubular bones are noted. Scoliosis has been evident in 10%. The pelvic bones are hypoplastic, the femoral neck being poorly modeled and plump. There is progressively early joint degeneration in 30%, but rarely in individuals less than 30 years old. The skeletal features observed radiographically and the clinical joint involvement are not always present in Stickler syndrome (15,27,30). Short cranial base and hypoplastic midface have been borne out by cephalometric study (29).

All patients with *Robin sequence,* especially with an autosomal dominant history, should be examined periodically for severe myopia to prevent ocular complications of the Stickler syndrome. Other disorders with some degree of overlap include *Kniest dysplasia, SED congenita,* and *short stature, low nasal bridge, cleft palate and sensorineural hearing loss.*

Mitral valve prolapse was found in almost 50% (21a).

### References [Stickler syndrome (Marshall–Stickler syndrome, Wagner–Stickler syndrome, hereditary arthroophthalmopathy)]


1. Ayme S, Preus M: The Marshall and Stickler syndromes: Objective rejection of lumping. *J Med Genet* **21**:34–38, 1984.

2. Baraitser M: Marshall/Stickler syndrome. *J Med Genet* **19**:139–140, 1982.

3. Blair NP et al: Hereditary progressive arthro-ophthalmopathy of Stickler. *Am J Ophthalmol* **88**:876–888, 1979.

4. Cohen MM et al: A dominantly inherited syndrome of hyaloideoretinal degeneration, cleft palate and maxillary hypoplasia (Cervenka syndrome). *Birth Defects* **7**(7):83–86, 1971.

5. Cohen MM Jr: The demise of the Marshall syndrome. *J Pediatr* **85**:878, 1974.

6. Cotlier E, Reinglass H: Marfan-like syndrome with lens involvement. *Arch Ophthalmol* **93**:43–106, 1975.

7. Daniel R et al: Hyalo-retinopathy in the clefting syndrome. *Br J Ophthalmol* **58**:96–102, 1974.

8. Delaney WV et al: Inherited retinal detachment. *Arch Ophthalmol* **69**:44–50, 1963.

9. Falger ELF et al: Hereditary hyaloideo-retinal degeneration and palatoschisis. *Ophthalmologica* **160**:384, 1970.

9a. Francomano CA et al: The Stickler syndrome: Evidence for close linkage to the structural gene for type II collagen. *Genomics* **1**:293–296, 1987.

9b. Francomano CA et al: The Stickler and Wagner syndromes: Evidence for genetic heterogeneity. 39th Annual Meeting Am. Soc. Hum. Genet., New Orleans, October, 1988.

10. Frandsen E: Hereditary hyaloideo-retinal degeneration (Wagner) in a Danish family. *Acta Ophthalmol (Kbh)* **44**:223–232, 1966.

11. Godel V et al: The Wagner–Stickler syndrome complex. *Doc Ophthalmol* **52**:179–188, 1981.

11a. Günzel H et al: Marshall-Syndrom. Klinish-genetische Untersuchungen über eine Familie mit 8 Merkmalträgern. *Kinderärztl Prax* **56**: 25–31, 1988.

12. Hall J: Stickler syndrome presenting as a syndrome of cleft palate, myopia and blindness inherited as a dominant trait. *Birth Defects* **10**(8):157–171, 1974.

13. Hall JC, Herrod H: The Stickler syndrome presenting as a dominantly inherited cleft palate and blindness. *J Med Genet* **12**:397–404, 1975.

14. Haller JO et al: The Weissenbacher–Zweymüller syndrome of micrognathia and rhizomelic chondrodysplasia at birth with subsequent normal growth. *Am J Roentgenol* **125**:936–943, 1975.

15. Herrmann J et al: The Stickler syndrome (hereditary ophthalmopathy). *Birth Defects* **11**(2):76–103, 1975.

16. Hirose T et al: Wagner's hereditary vitreoretinal degeneration and retinal detachment. *Arch Ophthalmol* **89**:176–185, 1973.

17. Jansen LM: Degeneratio hyaloideo-retinalis hereditaria. *Ophthalmologia* **144**:458–464, 1962.

18. Keith CG et al: Abnormal facies, myopia and short stature. *Arch Dis Childh* **47**:787–793, 1973.

19. Kelly TE et al: The Weissenbacher–Zweymüller syndrome: Possible neonatal expression of the Stickler syndrome. *Am J Med Genet* **11**:113–119, 1982.

20. Knobloch WH: Inherited hyaloideo-retinopathy and skeletal dysplasia. *Trans Am Ophthalmol Soc* **73**:417–429, 1975.

21. Kreidler JF et al: Robin-Syndrom mit Oberkiefer- und Nasenhypoplasie—eine erbliche Missbildungs-kombination? *Fortschr Kiefer Gesichtschir* **21**:262–266, 1976.

21a. Liberfarb RM, Goldblatt A: Prevalence of mitral valve prolapse in the Stickler syndrome. *Am J Med Genet* **24**:387–392, 1986.

22. Liberfarb RM et al: The Wagner–Stickler syndrome: A study of 22 families. *J Pediatr* **99**:394–399, 1981.

23. Marshall D: Ectodermal dysplasia. Report of a kindred with ocular abnormalities and hearing defects. *Am J Ophthalmol (Suppl)* **45**:143–156, 1958.

24. Nielsen CE: Stickler's syndrome. *Acta Ophthalmol* **59**:286–295, 1981.

25. O'Donnell JJ et al: Generalized osseous abnormalities in the Marshall syndrome. *Birth Defects* **12**(5):299–314, 1976.

26. Perkins J: Pierre Robin syndrome. *Trans Ophthalmol Soc UK* 40:179–180, 1970.

27. Popkin JS, Polomeno RC: Stickler's syndrome (hereditary progressive arthro-ophthalmopathy). *Can Med Assoc J* **111**:1071–1076, 1974.

28. Ruppert ES et al: Hereditary hearing loss with saddle nose and myopia. *Arch Otolaryngol* **92**:95–98, 1970.

29. Saksena SS et al: Stickler syndrome: A cephalometric study of the face. *J Craniofac Genet Dev Biol* **3**:19–28, 1983.

30. Say B et al: The Stickler syndrome (hereditary arthro-ophthalmopathy). *Clin Genet* **12**:179–182, 1977.

31. Schreiner RL et al: Stickler syndrome in a pedigree of Pierre Robin syndrome. *Am J Dis Child* **126**:86–90, 1973.

31a. Scribanu N et al: The Weissenbacher–Zweymüller phenotype in the neonatal period as an expression in the continuum of manifestations of the hereditary arthroophthalmopathies. *Ophthalmol Paediatr Genet* **8**:159–162, 1987.

32. Smith WK: Pierre Robin syndrome in brothers. *Birth Defects* **5**(2):220–221, 1969.

33. Spranger J: Arthro-ophthalmopathia hereditaria. *Ann Radiol (Paris)* **11**:359–364, 1968.

34. Stickler GB et al: Hereditary progressive arthroophthalmopathy. *Mayo Clin Proc* **40**:433–455, 1968.

35. Stickler GB, Pugh DG: Hereditary progressive arthro-ophthalmopathy. II. Additional observation on vertebral anomalies, a hearing defect and a report of a similar case. *Mayo Clin Proc* **42**:495–500, 1967.

35a. Temple IK: Stickler's syndrome. *J Med Genet* **26**:119–126, 1989.

36. Van Balen ATM, Falger ELF: Hereditary hyaloideoretinal degeneration and palatoschisis. *Arch Ophthalmol* **83**:152–162, 1970.

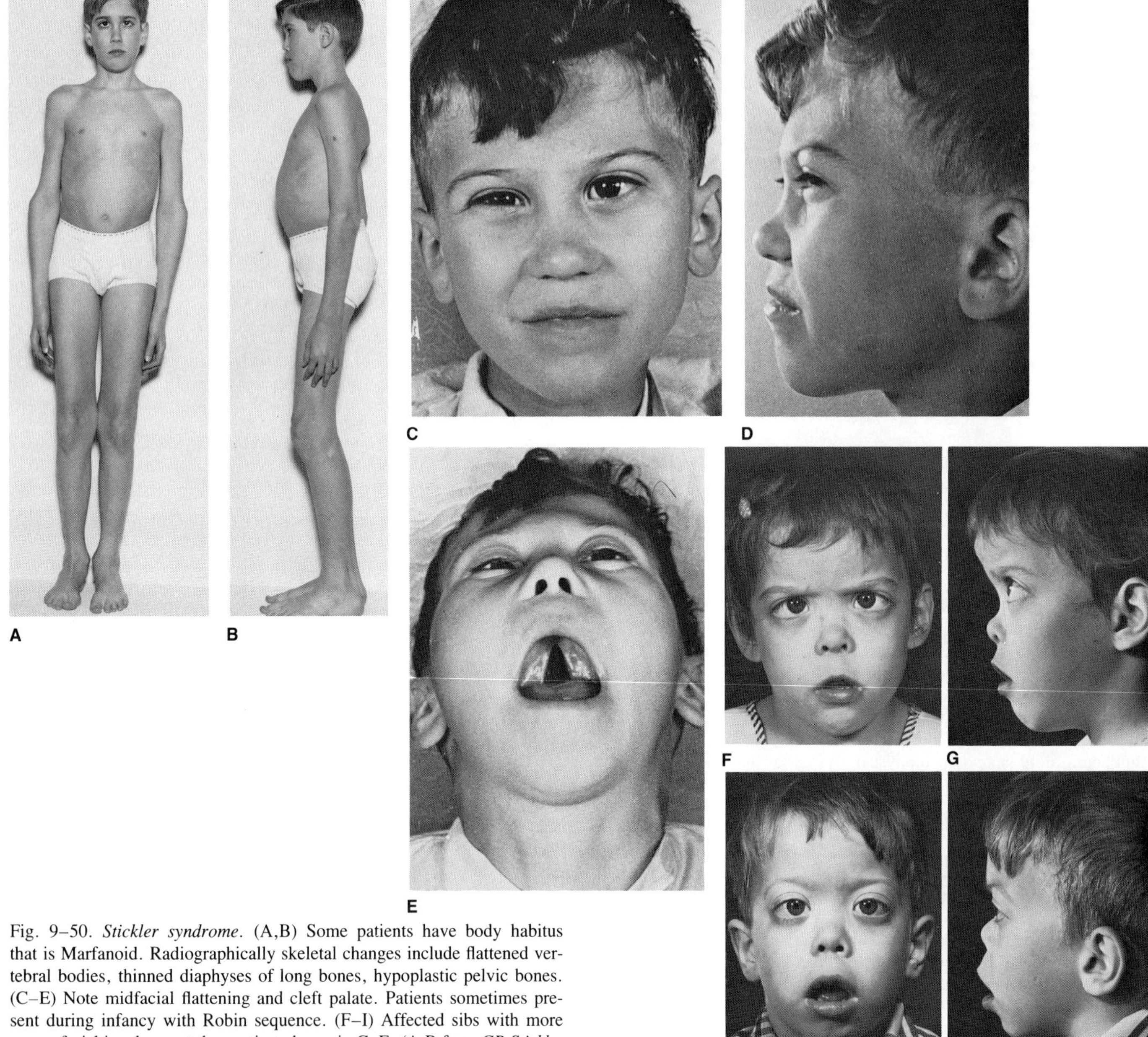


Fig. 9–50. *Stickler syndrome.* (A,B) Some patients have body habitus that is Marfanoid. Radiographically skeletal changes include flattened vertebral bodies, thinned diaphyses of long bones, hypoplastic pelvic bones. (C–E) Note midfacial flattening and cleft palate. Patients sometimes present during infancy with Robin sequence. (F–I) Affected sibs with more severe facial involvement than patient shown in C–E. (A,B from *GB Stickler* and *DG Pugh,* Mayo Clin Proc 42:495, 1967. F–I courtesy of *H Zellweger et al,* Iowa City, Iowa.)

37. Von Bohringer HR et al: Zur Klinik and Pathologie der Degeneration and hyaloideo-retinalis hereditaria (Wagner). *Ophthalmologica* **39**:330–338, 1960.
38. Walden RH et al: Pierre Robin syndrome in association with combined lengthening and shortening of long bones. *Plast Reconstr Surg* **48**:80–82, 1971.
39. Winter RM et al: The Weissenbacher–Zweymüller, Stickler and Marshall syndromes. Further evidence for their identity. *Am J Med Genet* **16**:189–200, 1983.
40. Young NJA et al: Stickler's syndrome and neovascular glaucoma. *Br J Ophthalmol* **63**:826–873, 1979.
41. Zellweger H et al: The Marshall syndrome: Report of a new family. *J Pediatr* **84**:868–871, 1974.

## Short stature, low nasal bridge, cleft palate, and sensorineural hearing loss

Several authors (1,3,4,6,7) have described a condition that simulates Stickler syndrome, but appears to have autosomal recessive inheritance because of occurrence in sibs and/or parental consanguinity (3,4,6). The cases described by Nance and Sweeney (5) and by Giedion (2) may represent a different condition.

Feeding difficulties were noted during the neonatal period, and infancy was characterized by recurrent respiratory problems (bronchitis,

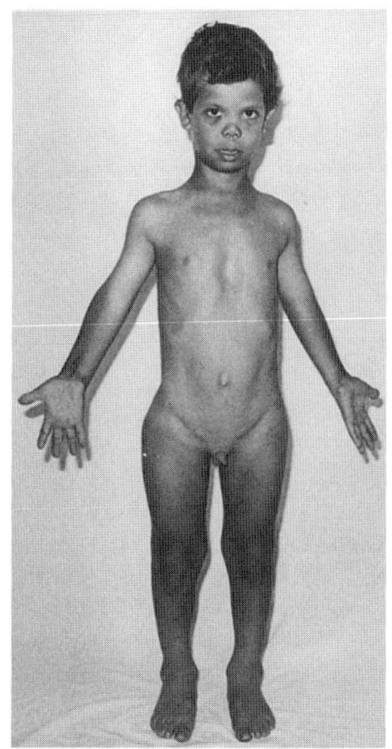
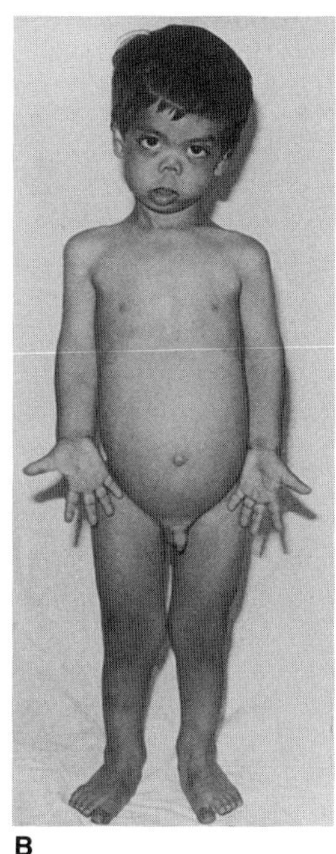
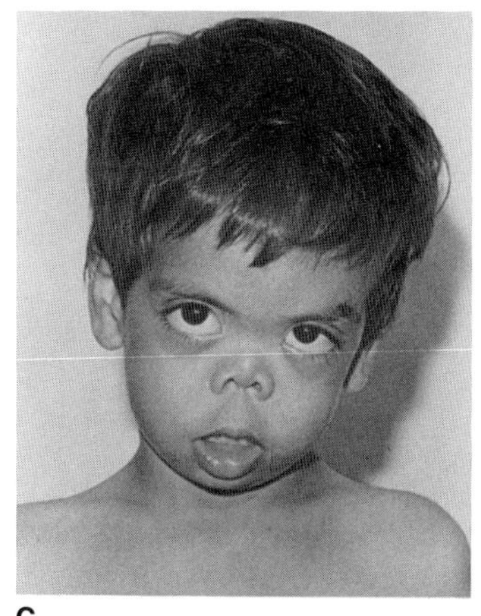
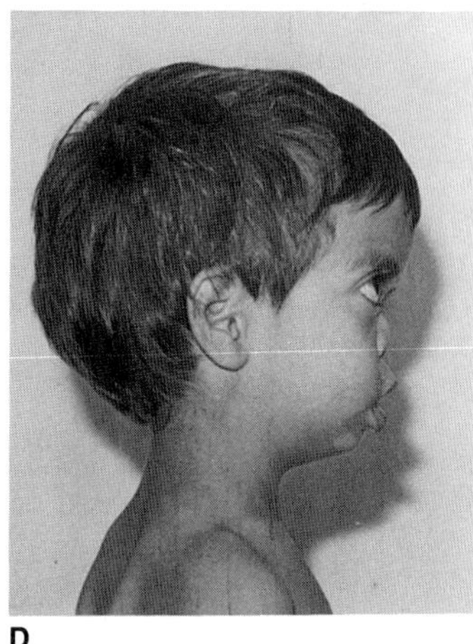

Fig. 9–51. A,B *Short stature, low nasal bridge, cleft palate, and sensorineural hearing loss.* (A,B) Two affected brothers. (C,D) Close-up of younger brother showing severe nasal saddling and anteverted nares. (From *C Salinas et al,* Am J Med Genet (in press).

pneumonia, etc.) and enteritis. The limbs are short. The metacarpophalangeal joints have somewhat reduced mobility. Progressive kyphosis and lumbar lordosis were noted.

The nose is very small with anteverted nostrils and the nasal bridge is severely depressed (Fig. 9–51). Cleft palate (2–4,6,7) has been observed in most cases. There may be a midfacial angioma (4,7). Moderate to severe sensorineural, possibly mixed, hearing loss has been documented in most cases.

Radiographically the leg bones are relatively short and broad with mild metaphyseal flaring. The epiphyses are enlarged. Kyphosis and lumbar lordosis are evident with defects of the upper anterior angles of the vertebral bodies at the thoracolumbar junction.

### References (Short stature, low nasal bridge, cleft palate, and sensorineural hearing loss)


1. Fanconi S et al: The SPONASTRISME dysplasia: Familial short-limbed dwarfism with saddle nose, spinal alterations and metaphyseal striation. *Helv Paediatr Acta* **38**:267–280, 1983.
2. Giedion A et al: Oto-spondylo-megaepiphyseal dysplasia (OSMED). *Helv Paediatr Acta* **37**:361–380, 1982 (Patient A).
3. Insley J, Astley R: A bone dysplasia with deafness. *Br J Radiol* **47**:244–251, 1974.
4. Miny P, Lenz W: Autosomal recessive deafness with skeletal dysplasia and facial appearance of Marshall syndrome. *Am J Med Genet* **21**:317–324, 1985.
5. Nance WE, Sweeney A: A recessively inherited chondrodystrophy. *Birth Defects* **6**(4):25–27, 1970.
6. Salinas C et al: Bone dysplasia, deafness and cleft palate syndrome. *Am J Med Genet* (in press).
7. Winter RM et al: The Weissenbacher–Zweymüller, Stickler, and Marshall syndromes: Further evidence for their identity. *Am J Med Genet* **16**:189–199, 1983 (Patient 2).


## Winchester syndrome

Winchester et al (9), in 1969, reported sibs with a connective tissue disorder of short stature, joint contractures with claw hands, peripheral corneal opacities, osteoporosis, carpal-tarsal osteolysis, stiff joints with severe rheumatoid-like joint destruction, skin lesions, and coarse facies (Figs. 9–52 to 9–55). Mentation was normal. Brown and Kuwabara (1) described the same sibs. Other patients have been recorded by several authors (2–6,8). Inheritance is autosomal recessive. Dunger et al (2) noted excessive collagen turnover in skin and gingival biopsies and identified an abnormal oligosaccharide in the urine of two patients. The trisaccharide contained one fucose and two galactose residues.

McKeown (7) reported two Indian sibs with progressive arthritis, mucosal hypertrophy, and skin pigmentation over the joints. Postmortem examination in one case showed findings consistent with infantile hyalinosis. Although both sibs resembled Winchester syndrome, biochemical abnormalities were not confirmed.

Fig. 9–52. *Winchester syndrome.* (A) Coarse facies. (B) Peripheral corneal opacity. (From *DW Hollister et al,* J Pediatr **84:**701, 1974.)

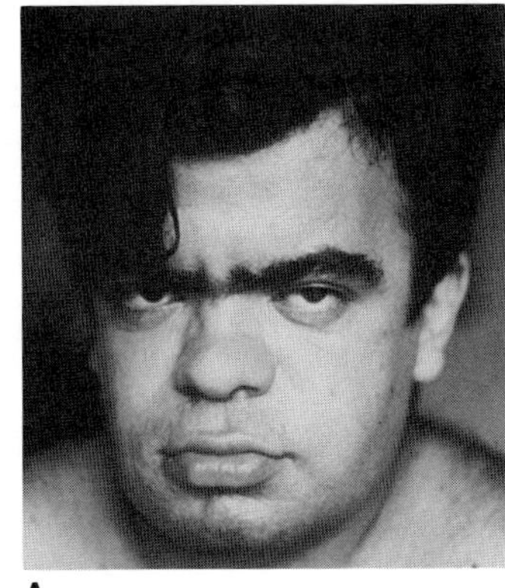
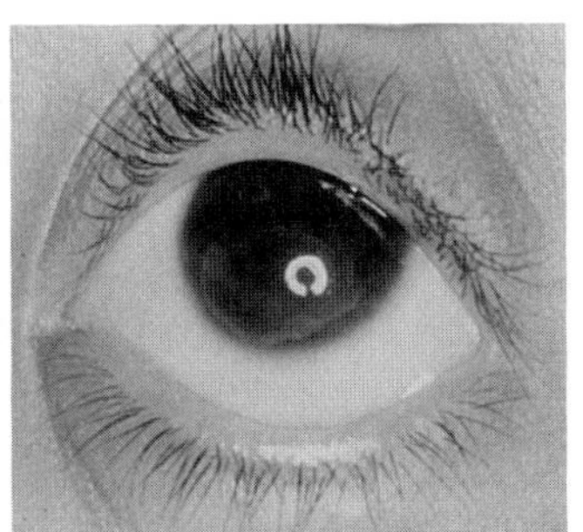

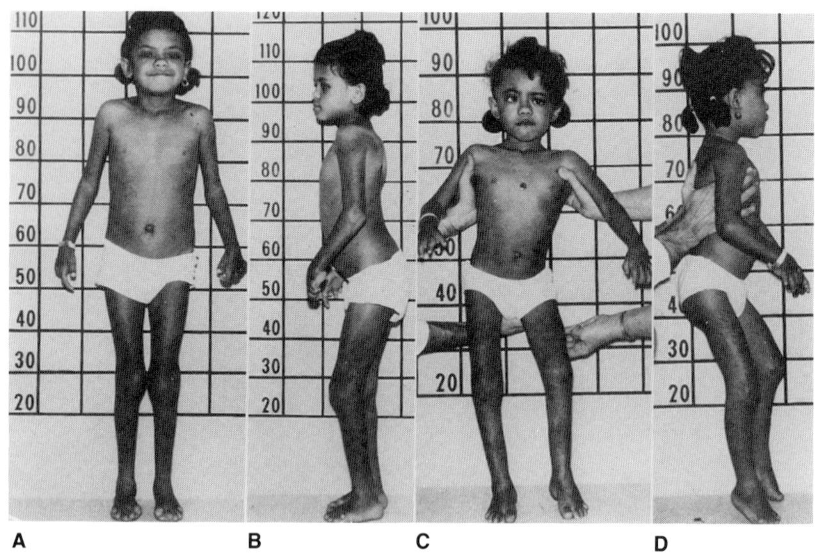

Fig. 9–53. *Winchester syndrome.* (A–D) Short stature and contractures in two sisters. (From *DW Hollister et al,* J Pediatr **84:**701, 1974.)

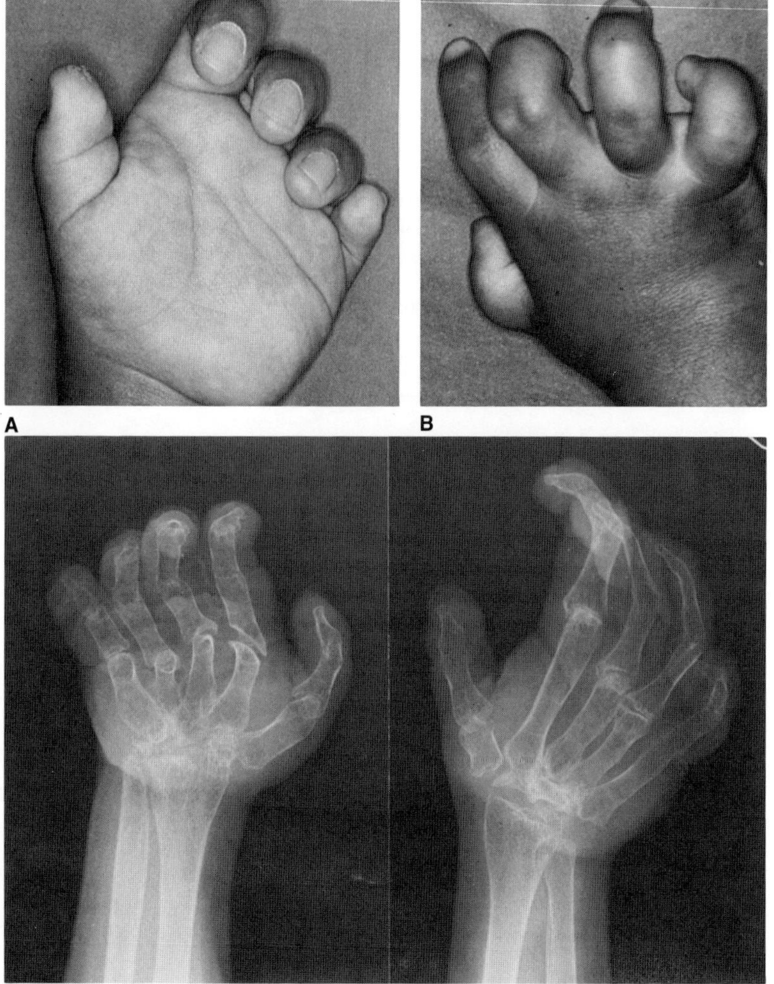

Fig. 9–54. *Winchester syndrome.* (A,B) Claw hands. (C) Roentgenogram showing osteoporosis, carpal osteolysis, and severe joint destruction. (From *DW Hollister et al,* J Pediatr **84:**701, 1974.)

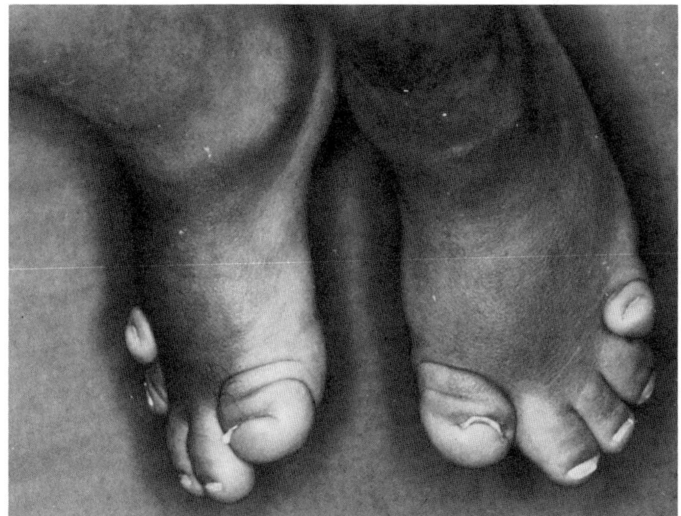

A

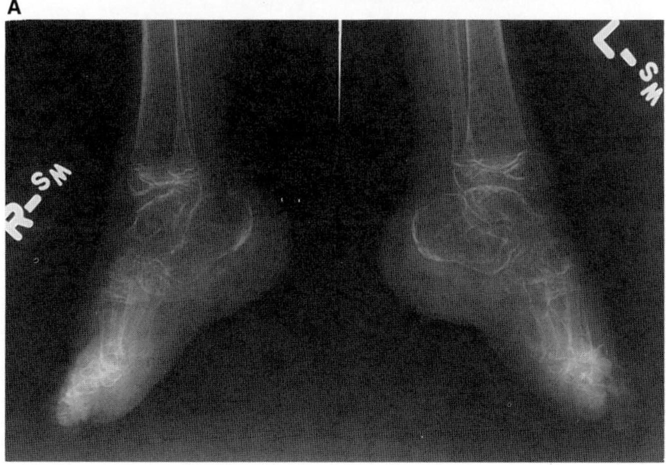


B

Fig. 9–55. *Winchester syndrome.* (A,B) Similar changes in feet. (From *DW Hollister et al,* J Pediatr **84:**701, 1974.)

## References (Winchester syndrome)

1. Brown SI, Kuwabara T: Peripheral corneal opacities and skeletal deformities: A newly recognized acid mucopolysaccharidosis simulating rheumatoid arthritis. *Arch Ophthalmol* **83**:667–677, 1970.

2. Dunger DB et al: Two case of Winchester syndrome: With increased urinary oligosaccharide excretion. *Eur J Pediatr* **146**:615–619, 1987.

3. Gheysens E, Durero S: Le syndrome de Winchester: deux cas familiaux. Thèse Docteur Médicine, Nice, 1984.

4. Hollister DW et al: The Winchester syndrome: A nonlysosomal connective tissue disease. *J Pediatr* **84**:701–709, 1974.

5. Irani A et al: The Winchester syndrome. A case report. *Indian Pediatr* **15**:861–863, 1978.

6. Landing BH, Nadorra R: Infantile systemic hyalinosis. *Paediatr Pathol* **6**:55–79, 1986.

7. McKeown CME: Winchester syndrome or systemic infantile hyalinosis? 3rd Manchester Birth Defects Conference, Manchester, England, October 25–28, 1988.

8. Vouzellaud B: Le syndrome de Winchester: Contribution à partir trois cas familiaux. M.D. Thèse, Université Claude Bernard, June 23, 1988.

9. Winchester P et al: A new acid mucopolysaccharidosis with skeletal deformities simulating rheumatoid arthritis. *Am J Roentgenol* **106**:121–128, 1969.

# Chapter 10
# Proportionate Short Stature Syndromes

## Aarskog syndrome (facial–digital–genital syndrome)

Credit is usually given to Aarskog (1) for describing the disorder in 1970. However, earlier descriptions were made (3,21). The condition is characterized by unusual facies, short stature, abnormalities of the hands and feet, and genital anomalies. In most families, inheritance is compatible with an X-linked recessive disorder (5,8,9–14,16–18,21, 23,24,26,33,34,36,38); female heterozygotes frequently exhibit minor stigmata (1,9,12,13,15b,16,18,23,24,26,28–30,34,36,38). Bawle et al (8) and Tyrkus et al (35) reported women with full expression of the syndrome, each of whom had an affected son. Bawle et al (8) found a balanced X-autosome translocation in both mother and son and suggested that Xq13 was the likely site of the Aarskog locus. Tyrkus et al (35) found a reciprocal 8pXq translocation and concluded that the gene locus was on the long arm of the X chromosome.

Autosomal dominant inheritance with limited expression in females, although a less likely mechanism (15), has also been suggested. Several pedigrees demonstrating male-to-male transmission have been reported (19,22,37). Genetic heterogeneity is likely.

**Facies.** The forehead is broad with prominent ridging of the metopic suture (9,11,14,33), and the face is usually round. There is a widow's peak in about 50% (9,11,14,15,17,18,33). The corneal diameter may be enlarged (25). Hypertelorism in 85%, ptosis of the upper eyelids in 50% (11,16,21), and a short, broad, stubby nose with anteverted nostrils in 85% characterize the face. The philtrum is usually long (14,15,18,19). A variety of malformations of the external ear have been described, most commonly thickened or fleshy earlobes (4,8,9,11,14,17,19,26,28,33,34). The maxilla is often hypoplastic (19,29,34). There may be a linear curved depression below the lower lip (9,11,26) (Fig. 10–1A,B).

**Musculoskeletal system.** Although birth size has been normal in most reports, growth retardation usually becomes evident during the first few years of life. Most patients are below the 3rd percentile in height, with adults rarely exceeding 160 cm (19). Hands and feet are short as are the fifth fingers (14,35), and there is mild soft tissue webbing between the fingers in over 60% (4,11,19,21,23). Hyperextension of the proximal interphalangeal joints and flexion of the distal interphalangeal joints are found in 80% (9,16,17,19,30,34) (Fig. 10–2). In about 65%, there is fifth finger clinodactyly (19). The feet are small, broad, and flat with splayed bulbous toes (11,23,28). Occasionally, there is metatarsus adductus (9,16,23). In over 50%, there is pectus excavatum (1,8,9,11,17–19,21,23,38). Inguinal hernia has been noted in over 60% (19,23). The umbilicus is prominent with a protruding buttonlike central area surrounded by a deep ovoid depression (10,17).

Radiographically, striking changes are generally limited to the cervical spine and hands (9,27,34). Spina bifida occulta or cervical vertebral defects have been found in over 50% (19). Scott (33) described hypoplasia of the first cervical vertebra with an unfused posterior arch; on extension, it entered the foramen magnum. Odontoid hypoplasia has also been described (6). With flexion, there was subluxation of the first and second cervical vertebras. RJ Gorlin has observed a patient with fusion of the second and third cervical vertebras. The terminal phalanges of the fingers and the middle phalanx of the fifth finger are hypoplastic in over 60% (19). Hanley et al (21) noted osteochondritis dissecans. Bone age is often retarded (19).

**Genital anomalies.** The scrotum appears bifid, with the scrotal fold extended ventrally around the base of the penis, somewhat resembling a shawl about the neck (Fig. 10–3). Presumably this genital anomaly results from failure of caudal shift of the fused labioscrotal folds. Commonly, one or both testes are undescended (19).

**Other findings.** Bilateral single palmar creases (7,11,19,23) and a single crease in the fifth fingers are frequent (9,11,23,33,38); distally displaced axial triradii have also been noted (10,19,23,30). Most patients are of normal or low normal intelligence (15). Ophthalmoplegia has been described in several patients (16,21,29,34). Growth hormone was deficient in one patient (26), but normal in several others (1,18,30,33).

**Oral manifestations.** Enamel hypoplasia and a "col" deformity of the anterior mandible have also been noted in one patient (29). Halse et al (20), in examining 10 patients in three families, found delayed eruption in 3 of 4 patients aged 7–11. Congenitally missing teeth and short roots were found in the family originally reported by Aarskog (1,2). Cleft lip and palate have been described in a few affected males (13,23,34,38); a carrier female (23) had cleft lip.

**Differential diagnosis.** The *Noonan syndrome* and the *Leopard syndrome* share features with this disorder such as short stature, hypertelorism, ptosis of upper eyelids and hypogonadism. The *Robinow syndrome* also resembles the Aarskog syndrome, as does *pseudohypoparathyroidism.*

Teebi et al (34a) reported a newly recognized autosomal recessive disorder with some similarities to Aarskog syndrome including short stature, hypertelorism, short stubby nose, anteverted nostrils, ear anomalies, small broad hands, and shawl scrotum. However, the eyes did not downslant nor was there ptosis. The hair was coarse, dry, and hypopigmented. Of five affected sibs, three were male and two were female.

### References [Aarskog syndrome (facial–digital–genital syndrome)]


1. Aarskog D: A familial syndrome of short stature associated with facial dysplasia and genital anomalies. *J Pediatr* **77**:856–861, 1970.
2. Aarskog D: A familial syndrome of short stature associated with facial dysplasia and genital anomalies. *Birth Defects* **7**(6):235–239, 1971.
3. Ainley RG: Hypertelorism (Greig's syndrome). A case report. *J Pediatr Ophthalmol* **5**:148–150, 1968.
4. Archibald RM, German J: The Aarskog–Scott syndrome in four brothers. *Birth Defects* **11**(2):25–29, 1975.
5. Andrassy RJ et al: The Aarskog syndrome: Significance for the surgeon. *J Pediatr Surg* **14**:462–464, 1979.
6. Baldellou AT et al: Risk of medullary damage in Aarskog–Scott syndrome. *Clin Genet* **26**:225, 1983 (Abstract).
7. Bartsocas CS, Dimitriou JK: Aarskog–Scott syndrome of unusual facies, joint hypermobility, genital anomaly and short stature. *Birth Defects* **11**(2):453–455, 1975.
8. Bawle E et al: Aarskog syndrome: Full male and female expression associated with an X-autosome translocation. *Am J Med Genet* **17**:595–602, 1984.

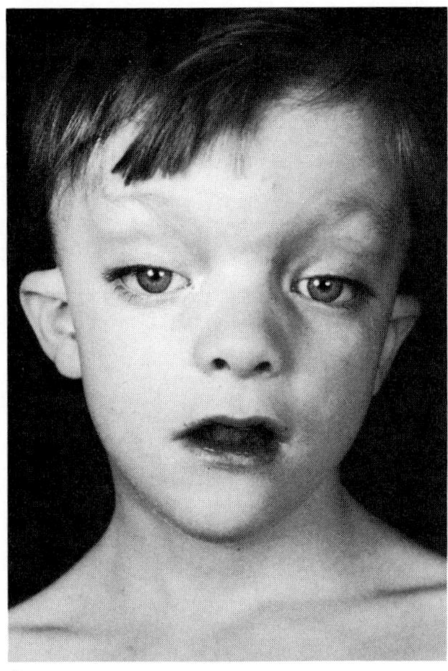
A

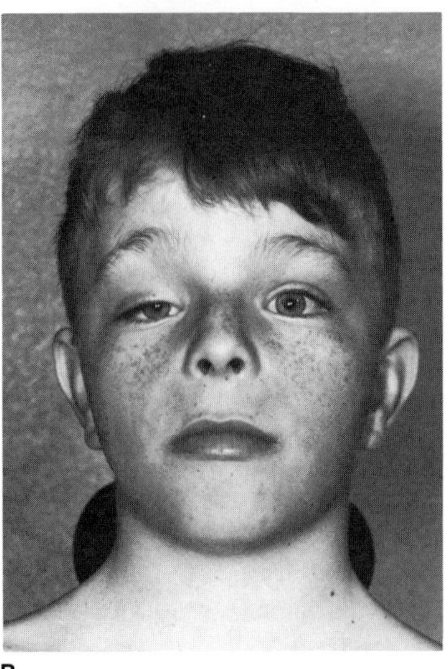
B

Fig. 10–1. *Aarskog syndrome*. (A) Facies is characterized by broad forehead, hypertelorism, bilateral ptosis of upper eyelids, and low-set ears. (B) Similar facies in patient from a different family. Note unilateral ptosis. [From *CI Scott Jr,* Birth Defects 7(6):240, 1971.]

9. Berman P et al: Inheritance of the Aarskog syndrome. *Birth Defects* **10**(7):151–159, 1974.
10. Berman P et al: The inheritance of the Aarskog facial-digital-genital syndrome. *J Pediatr* **86**:885–891, 1975.
11. Berry C et al: Aarskog's syndrome. *Arch Dis Childh* **55**:706–710, 1980.
12. Debrand J et al: Le syndrome facio-digito-génital ou syndrome d'Aarskog. *Pédiatrie* **32**:65–72, 1977.
13. De Saxe M et al: The Aarskog (facio-digital-genital) syndrome in South Africa. A report of three families. *S Afr Med J* **65**:299–303, 1984.
14. Duncan P et al: Additional features of the Aarskog syndrome. *J Pediatr* **91**:769–770, 1977.
15. Escobar V, Weaver DD: Aarskog syndrome, new findings and genetic analysis. *JAMA* **240**:2638–2641, 1978.
15a. Friedman JM: Umbilical dysmorphology: The importance of contemplating the belly button. *Clin Genet* **28**:343–347, 1985.
15b. Fryns JP et al: The Aarskog syndrome. *Hum Genet* **42**:129–135, 1978.
16. Funderburk SJ, Crandall BF: The Aarskog syndrome in three brothers. *Clin Genet* **6**:119–124, 1974.
17. Furukawa CT et al: The Aarskog syndrome. *J Pediatr* **81**:1117–1122, 1972.
18. Fryns JP et al: The Aarskog syndrome. *Hum Genet* **42**:129–135, 1978.
19. Grier RE et al: Autosomal dominant inheritance of the Aarskog syndrome. *Am J Med Genet* **15**:39–46, 1983.
20. Halse A et al: Dental findings in patients with Aarskog syndrome. *Scand J Dent Res* **87**:253–259, 1979.

Fig. 10–2. *Aarskog syndrome*. The hands are short and wide with frequent webbing of the fingers, hypermobility, and subluxation of the proximal interphalangeal joints. (Courtesy of *P Berman,* Montreal, Canada.)

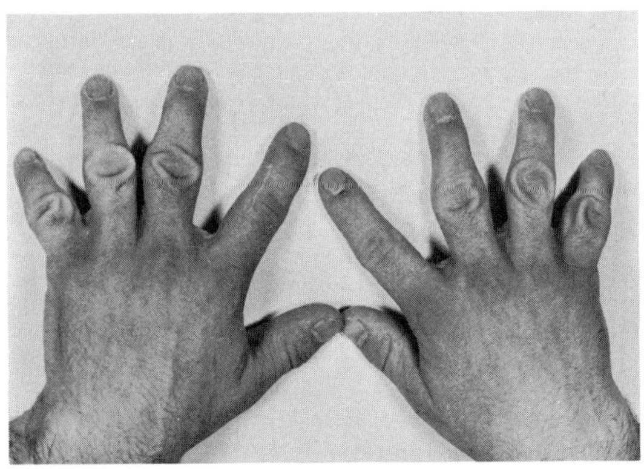

21. Hanley WB et al: Osteochondritis dissecans and associated malformation in two brothers. *J Bone Jt Surg* **49A**:925–937, 1967.
22. Harris DJ: Male-to-male transmission of the Aarskog syndrome. Excerpta Medica No. 426 (Fifth International Conference on Birth Defects, Montreal, 21–27 August 1977), Littlefield JW, Ebling FJG, Henderson IW (eds), p 95 (Abstract).
23. Hoo JJ: The Aarskog (facio-digito-genital) syndrome. *Clin Genet* **16**:269–276, 1979. .
24. Hurst DL: Metatarsus adductus in two brothers with Aarskog syndrome. *J Med Genet* **20**:477, 1983.
25. Kirkham TH et al: Ophthalmic manifestations of Aarskog (facial-digital-genital) syndrome. *Am J Ophthalmol* **79**:441–445, 1975.
26. Kodama M et al: Aarskog syndrome with isolated growth hormone deficiency. *Eur J Pediatr* **135**:273–276, 1981.
27. Kunze J, Spranger J: Aarskog-Syndrom. *Klin Pediätr* **185**:490–494, 1973.

Fig. 10–3. *Aarskog syndrome*. Abnormal penoscrotal configuration due to scrotal folds joining ventrally over the base of the penis. Note left inguinal hernia. [From *CI Scott Jr,* Birth Defects 7(6):240, 1971.]

28. Marcel B et al: Aarskog's syndrome. *Acta Med Auxol* **9**:217–226, 1977.
29. Melnick M, Shields ED: Aarskog syndrome: New oral-facial findings. *Clin Genet* **9**:20–24, 1976.
30. Oberiter V et al: The Aarskog syndrome. *Acta Paediatr Scand* **69**:567–570, 1980.
31. Pederson JC et al: The Aarskog syndrome. *Ann Génét* **23**:108–110, 1980.
32. Ricotti GC et al: La sindrome di Aarskog: Descrizione ed evoluzione de due casi, uno dei quali con anomalia cromosomica. *Acta Med Auxol* **10**:135–150, 1978.
33. Scott CI Jr: Unusual facies, joint hypermobility, genital anomaly and short stature: A new dysmorphic syndrome. *Birth Defects* **7**(6):240–246, 1971.
34. Sugarman GI et al: The facial-digital-genital (Aarskog) syndrome. *Am J Dis Child* **126**:248–252, 1973.
34a. Teebi AS et al: New autosomal recessive faciodigitogenital syndrome. *J Med Genet* **25**:400–406, 1988.
35. Tyrkus M et al: Aarskog–Scott syndrome inherited as an X-linked dominant with full male-female expression. *Am J Hum Genet* **32**:134A, 1980 (Abstract).
36. van den Bergh P et al: Anomalous cerebral venous drainage in Aarskog syndrome. *Clin Genet* **25**:288–294, 1984.
37. Van de Vooren MJ et al: The Aarskog syndrome in a large family, suggestive for autosomal dominant inheritance. *Clin Genet* **24**:439–445, 1983.
38. Welch JP: Elucidation of a 'new' pleiotropic connective tissue disorder. *Birth Defects* **10**(10):138–146, 1974.

## Bloom syndrome

Bloom syndrome consists of intrauterine growth retardation, sunlight sensitivity leading to telangiectatic erythema (Figs. 10–4 and 10–5), a tendency to chromosomal breakage with a high frequency of sister chromatid exchanges (Figs. 10–6 and 10–7), immunologic deficiency, hypogonadism and infertility in males, and an increased risk of neoplasia. The condition was first recognized by Bloom (5–7), a genetically oriented dermatologist, in 1954. Another case was reported by Torre and Cramer (43) as a case of discoid lupus erythematosis with primordial dwarfism. German (17,21) first recognized chromosomal breakage and emphasized the predilection for neoplasia. The Bloom Syndrome Registry currently has 130 patients on record (19,22,23). The reader is referred to the following sources for detailed coverage: extensive reviews (12a,14,15,44), neoplastic aspects (8,9,16,18,19), tumor surveillance (19,22,23,25), chromosomal breakage and sister chromatid exchange (3,10,12a,13,16–21,24,33,41), immunologic deficiency (40,44), endocrine aspects (1,30,44), dermatologic aspects (14,15), growth (14,15,44), and genetic aspects (15,26,28).

**Genetics.** The syndrome has autosomal recessive inheritance. Both sexes are affected, although a slight female deficit remains unexplained. Heterozygous carriers are normal, the only unusual finding to date being a tendency for low concentration of some blood immunoglobulin. The gene occurs with high frequency among Ashkenazi Jews who originated from Eastern Europe. The gene frequency is 0.0042 in Israeli Ashkenazim and more than 1 in 120 is a heterozygous carrier. The gene probably expanded to high frequency through the founder effect and random genetic drift in a relatively isolated population. Parental consanguinity is not increased in affected Jewish families. The gene does occur with very low frequency in other populations and, in contrast, parental consanguinity in non-Jewish families is very high (15,26), for example, in the Japanese population (15a).

**Cytogenetics.** Chromosomal breakage and sister chromatid exchanges are characteristic (20). The unusually high rate of about 60–120 sister chromatid exchanges per cell, representing a 5- to 10-fold increase, is pathognomonic (3,10,12a). Chromosome changes in leukemic clones are nonrandom (41a). No increased chromosomal breakage has been detected in heterozygous carriers (33). Phenotypic dimorphism occurs in patients with cells that are normal and cells that have a 10-fold increase in sister chromatid exchanges (24,44). Retarded DNA chain growth has also been documented, possibly resulting from a slow rate of replication fork movement. It has further been suggested that mitotic activity is low in fibroblast cultures (27). Emerit et al (13) noted that fibroblasts release a low-molecular-weight clastogenetic factor that breaks chromosomes and induces sister chromatid exchanges. Vijayalaxmi et al (46) indicated that about eight times the normal number of 6-thioguanine-resistant lymphocytes were detected in blood, the basis for the increase being unknown, with genomic instability from chromosomal aberrations as one possible explanation. Recently, Willis and Lindahl (48) and Chan et al (12) reported DNA ligase I deficiency in lymphoid cells.

**Growth.** Growth deficiency of prenatal onset is a prominent clinical feature, birthweight rarely exceeding 2300 g at term. Mean birth length is 44 cm. Final height attainment rarely exceeds 145 cm in males and 130 cm in females (14,15,32). Body habitus is normal except for its fragile appearance. Affected infants and children eat much less than normal and subcutaneous fat is sparse during infancy and childhood, although vigor and strength are generally normal (15). There is disproportionate microcephaly, accentuated by delicacy and narrowness of the face. The head is dolichocephalic. Some patients have

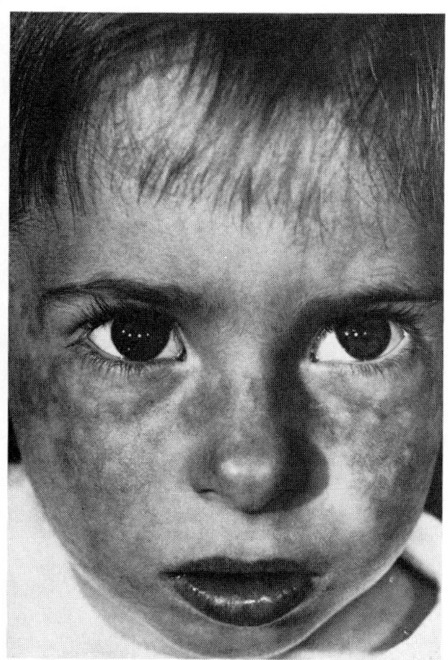

A

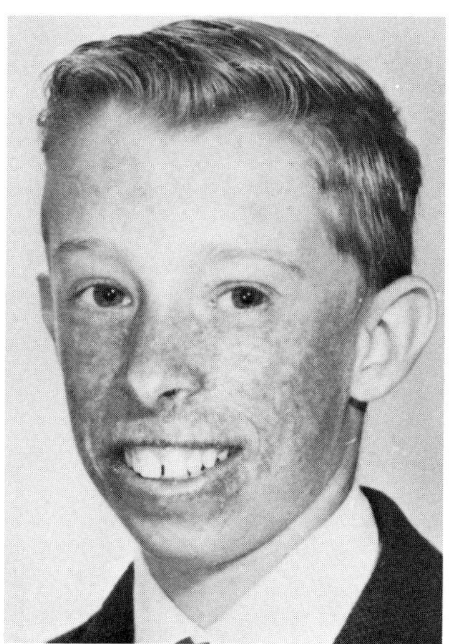

B

Fig. 10–4. *Bloom syndrome.* (A) Telangiectatic erythema of face. (B) Extensive facial telangiectasia in 14-year-old male. (From *J German,* Am J Hum Genet **21:**196, 1969.)

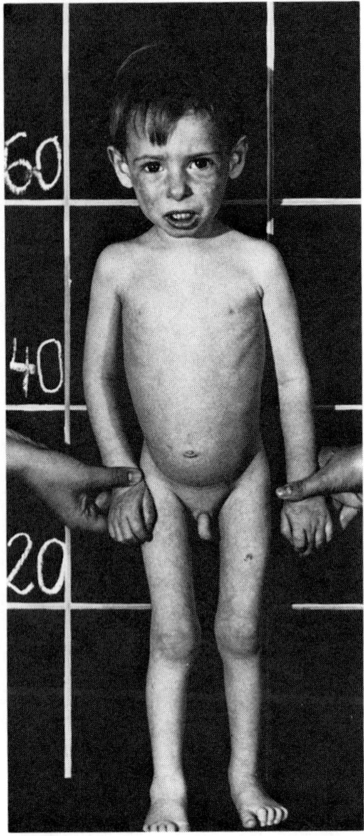


Fig. 10–5. *Bloom syndrome.* Four-year-old boy far below third percentile in height, with telangiectatic erythema of face. (From *J Keutel et al,* Z Kinderheilkd **101:**165, 1967.)

prominent nose and ears, with slightly receding mandible (Figs. 10–4 and 10–5) (7,9,15,29,32,34,42).

**Performance.** Intelligence is normal (15), although mild mental deficiency has been observed in some patients (8,14,32,42). Psychological problems are common and may result from short stature and unsightly facial lesions; interference with school progress may occur and simulate low intelligence (15). A squeaky, high-pitched voice is characteristic and occurs in both sexes (7,15,29,32).

**Skin.** Light sensitivity is noticed early in infancy and leads to development of telangiectatic erythema, which may appear anytime be-

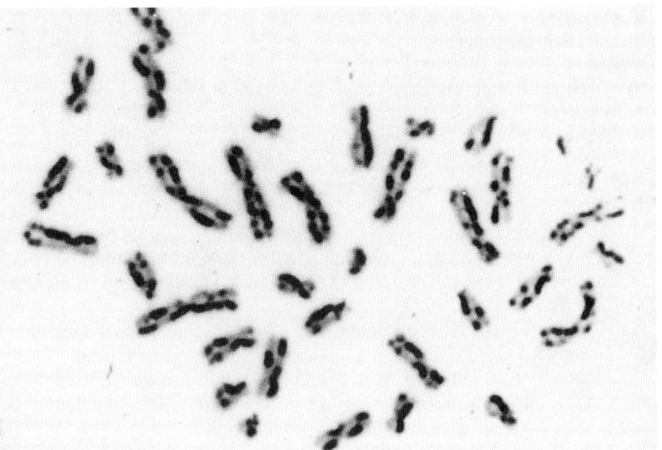

Fig. 10–7. *Bloom syndrome.* Note markedly increased number of sister chromatid exchanges (partial mitosis). (Courtesy of *J Cervenka,* Minneapolis, Minnesota.)

tween early infancy and 2 years of age. Erythema involves light-exposed areas of the face; superficially it resembles lupus erythematosus because of the butterfly distribution across the nose. Severe lesions also may occur on the lower eyelids, lips, ears, and neck. A chronic fissure or ulcer of the lower lip is a bothersome complication and chronic cheilitis is a prominent feature. The eyelashes may be lost. The forearms and dorsa of the hands may become involved, but rarely does the erythema extend to the trunk. Exposure to sunlight may cause bullae and vesicles. Milia may be seen in scarred areas (5,7,8,29,34, 39,42). German (15) noted that skin lesions appeared to be less severe in females than in males. Heavily pigmented individuals such as Mexicans (38) and Japanese (2,31) exhibit less telangiectasia and sun sensitivity than lightly pigmented individuals.

Hyperpigmented areas, irregular in outline and shape and varying from one to many centimeters in greatest dimension, are found mainly on the trunk, but also on various parts of the extremities. The spots are café-au-lait in color, although they are dark brown, almost black, in affected black patients. Less often observed are hypopigmented areas that are irregular in outline and smaller in size. In some instances, both hyper- and hypopigmented spots may be observed in the same patient (15). Keratosis follicularis is found in about 20%. Acanthosis nigricans was noted in one patient who developed diabetes mellitus during puberty (7,14).

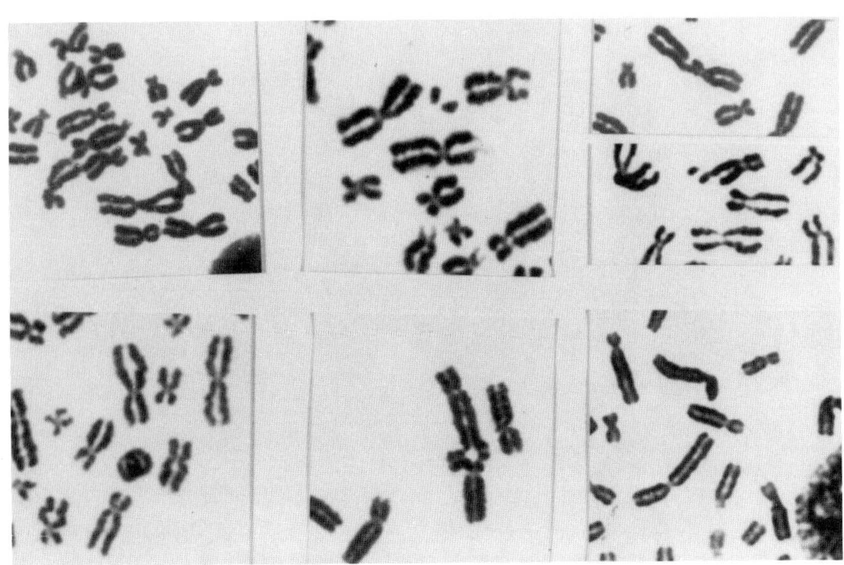

Fig. 10–6. *Bloom syndrome.* Various chromosome changes including dicentrics, acentric fragments, ring, breaks, and tetraradial figure. (Courtesy of *J German,* New York, New York.)

**Immunologic features.** Decreased immunoglobulin levels are characteristic and manifest as reduced IgA, IgG, or IgM. Although the immunologic disturbance has variable expression, a specific deficiency is similar in affected members of the same family (14,29,32,34,40,45,46). A single complementation defines patients of diverse ethnic background (47).

**Endocrine findings.** Hypogonadism of moderate degree and infertility are characteristic of affected males. Involvement of the tubular elements of the testes explains the occurrence of sterility and, although the Leydig cells also appear to be affected, patients have normal secondary sexual characteristics except for testicular development. Testes are small to diminutive in size, and cryptorchidism is common. Menstruation occurs in postpubertal females, but periods may be irregular and infrequent in some (14,15,30,44).

Normal growth hormone response has been observed following pituitary stimulation in two instances, although deficient release of growth hormone was documented in another case (1,44). Growth velocity was noted to increase during administration of exogenous human growth hormone in one instance (44).

**Neoplasia.** Commonly observed malignancies include leukemia, lymphoma, adenocarcinoma, and squamous cell carcinoma (4,8,9,22,23,25). Wilms tumor has been recorded three times (11). Of 61 reported neoplasms, 60 have been malignant (Table 10–1). The mean age of detection has been 24.8 years, the earliest occurring at 4 years of age and the latest at age 44. Three patients have each had two malignancies: adenocarcinoma of the sigmoid at age 37 and adenocarcinoma of the gastroesophageal junction at age 44 in one patient; adenocarcinoma of the sigmoid and squamous cell carcinoma of the esophagus, both detected at age 39, in another patient; and disseminated lymphoma and squamous cell carcinoma of the epiglottis, both detected at age 30, in still another patient (22,23,25).

There is an overall risk of at least 44% for developing a neoplasm ($n = 132$). This must be considered a minimal estimate because other malignancies will probably be detected in this same population with time.

Table 10–1. Neoplasms in 132 patients with Bloom syndrome[a]

| Type of neoplasm | Number of cases identified |
|---|---|
| Leukemia | |
| Acute lymphatic | 2 |
| Acute lymphyocytic | 4 |
| Acute myelomonocytic | 2 |
| Acute | 7 |
| Malignant lymphoma | |
| Lymphosarcoma | 2 |
| Pleomorphic reticulum cell sarcoma | 1 |
| Lymphoma | 5 |
| Hodgkin's | 2 |
| Adenocarcinoma, gastrointestinal tract | 7 |
| Squamous cell carcinoma[b] | 11 |
| Basal cell carcinoma | 6 |
| Metastatic carcinoma, primary site unidentified | 1 |
| Carcinoma of the breast | 4 |
| Wilms tumor | 3 |
| Meningioma | 1 |
| Osteogenic sarcoma | 1 |
| Carcinoma of cervix | 2 |
| Total neoplasms in 57 patients | 61 |

[a]Adapted from J German et al: *Clin Genet* **12**:162, 1977; *Clin Genet* **15**:361, 1979; *Clin Genet* **25**:166, 1984; *Clin Genet* **35**:57, 1989. AEL Cairney et al: *J Pediatr* 111:414–416, 1987.

[b]Tongue, esophagus, epiglottis, larynx, skin, lung.

**Other findings.** Conjunctivitis and telangiectasis of the conjunctival vessels may be observed. Colloid body-like spots in Bruch's membrane have been noted (35,42).

A variety of low-frequency abnormalities have been recorded including congenital heart defect, unequal leg lengths, absent toe, syndactyly, supernumerary digits, clinodactyly, hip dislocation, pes equinus, hypodontia, large protruding ears, single palmar crease, sacral dimple, and urethral or meatal narrowing (7,14,15,29,34).

**Differential diagnosis.** Bloom syndrome has been grouped with *ataxia telangiectasia, Cockayne syndrome, Rothmund–Thomson syndrome, Werner syndrome, dyskeratosis congenita,* Fanconi anemia, and *xeroderma pigmentosum* (14,34,37). Nearly all these disorders are characterized by growth retardation, increased neoplasia (both lymphoreticular and carcinomatous), immunologic deficiency, late-onset diabetes, and premature senility of the skin and conjunctiva (37).

Sunlight sensitivity is also a feature of Cockayne syndrome, Rothmund–Thomson syndrome, and xeroderma pigmentosum. Pigmentary skin changes are seen in sun-exposed areas of patients with ataxia-telangiectasia, Werner syndrome, dyskeratosis congenita, and Fanconi anemia.

Increased chromosomal breaks have been reported in Fanconi anemia, Cockayne syndrome, ataxia-telangiectasia, and dyskeratosis congenita (14,34). However, they are unlike the changes found in Bloom syndrome (41), and in no other disorder investigated to date are sister chromatid exchanges as strikingly increased in frequency as in Bloom syndrome (15).

**Laboratory aids.** A characteristically increased number of chromosome breaks and sister chromatid exchanges is demonstrable in blood lymphocytes, freshly aspirated bone marrow cells, skin fibroblasts and long-term culture, and some lymphoblastoid cell lines (14,29,32,34, 40,45,46). Serum IgA, IgG, and IgM are reduced and there is in vitro impaired response to pokeweed mitogen (45). Prenatal diagnosis may be possible using serial sonographic estimates of fetal size and growth rate to detect severe retardation (15).

### References (Bloom syndrome)

1. Ahmad U et al: Endocrine abnormalities and myopathy in Bloom's syndrome. *J Med Genet* **14**:418–421, 1977.

2. Arase S et al: Bloom's syndrome in a Japanese boy with lymphoma. *Clin Genet* **18**:123–127, 1980.

3. Bartram CR et al: Chromatid exchanges in ataxia-telangiectasia, Bloom syndrome, Werner syndrome, and xeroderma pigmentosum. *Ann Hum Genet* **40**:79–86, 1976.

4. Berkower AS, Biller HF: Head and neck cancer associated with Bloom's syndrome. *Laryngoscope* **98**:746–748, 1988.

5. Bloom D: Congenital telangiectatic erythema in a Levi-Lorain dwarf. *Arch Dermatol* **69**:526, 1954.

6. Bloom D: Congenital telangiectatic erythema resembling lupus erythematosus in dwarfs: Probably a syndrome entity. *Am J Dis Child* **88**:754–758, 1954.

7. Bloom D: The syndrome of congenital telangiectatic erythema and stunted growth. *J Pediatr* **68**:103–113, 1966.

8. Braun-Falco O, Marghescu S: Kongenitales telangiektatisches Erythem (Bloom-Syndrom) mit Diabetes insipidus. *Hautarzt* **17**:155–161, 1966.

9. Braun-Falco O, Marghescu S: Bloom-Syndrom. Eine Krankheit mit relativ hoher Leukämie-Morbiditat. *Münch Med Wochenschr* **111**:65–69, 1969.

10. Bryant EM et al: Normalisation of sister chromatid exchange frequencies in Bloom's syndrome by euploid cell hybridisation. *Nature (London)* **279**:795–796, 1979.

11. Cairney AEL et al: Wilms tumor in three patients with Bloom syndrome. *J Pediatr* **111**:414–416, 1987.

12. Chan JYH et al: Altered DNA ligase activity in Bloom's syndrome cells. *Nature (London)* **325**:357–359, 1987.

12a. Cohen MM, Levy HP: Chromosome instability syndromes. *Adv Hum Genet* **18**:43–149, 1989.

13. Emerit I et al: Chromosome breakage factor in the plasma of two Bloom's syndrome patients. *Hum Genet* **61**:65–67, 1982.

14. German J: Bloom's syndrome. I. Genetical and clinical observations in the first twenty-seven patients. *Am J Hum Genet* **21**:196–227, 1969.

15. German J: Bloom's syndrome. VIII. Review of clinical and genetic aspects, in *Genetic Diseases Among Ashkenazi Jews*. Goodman RM, Motulsky AG (eds), New York, Raven Press, 1979, pp 121–139.

16. German J: Bloom's syndrome. X. The cancer proneness points to chromosome mutation as a crucial event in human neoplasia, in *Chromosome Mutation and Neoplasia*. German J (ed), Alan R. Liss, New York, 1983, pp 347–357.

17. German J: Cytological evidence for crossing-over in vitro in human lymphoid cells. *Science* **144**:298–301, 1964.

18. German J: Patterns of neoplasia associated with the chromosome-breakage syndromes, in *Chromosome Mutation and Neoplasia*, German J (ed), Alan R. Liss, New York, 1983, pp 97–134.

19. German J, Passarge E: Bloom's syndrome. XII. Report for the register for 1987. *Clin Genet* **35**:57–69, 1989.

19a. German J, Takebe H: Bloom's syndrome. XIV. The disorder in Japan. *Clin Genet*, **35**:93–110, 1989.

20. German J et al: Bloom's syndrome. III. Analysis of the chromosome aberration characteristic of this disorder. *Chromosoma (Berlin)* 48:361–366, 1974.

21. German J et al: Chromosome breakage in a rare and probably genetically determined syndrome of man. *Science* **148**:506–507, 1965.

22. German J et al: Bloom's syndrome. VII. Progress report for 1978. *Clin Genet* **15**:361–367, 1979.

23. German J et al: Bloom's syndrome XI. Progress report for 1983. *Clin Genet* **25**:166–174, 1984.

24. German J et al: Bloom's syndrome. IV. Sister-chromatid exchanges in lymphocytes. *Am J Hum Genet* **29**:248–255, 1977.

25. German J et al: Bloom's syndrome. V. Surveillance for cancer in affected families. *Clin Genet* **12**:162–168, 1977.

26. German J et al: Bloom's syndrome. VI. The disorder in Israel and an estimation of the gene frequency in the Ashkenazim. *Am J Hum Genet* **29**:553–562. 1977.

27. Hand R, German J: Bloom's syndrome. DNA replication in cultured fibroblasts and lymphocytes. *Hum Genet* **38**:297–306, 1977.

28. Hustinx TWJ et al: Bloom's syndrome in two Dutch families. *Clin Genet* **12**:85–96, 1977.

29. Katzenellenbogen I, Laron Z: A contribution to Bloom's syndrome. *Arch Dermatol* **82**:609–616, 1960.

30. Kauli R et al: Gonadal function in Bloom's syndrome. *Clin Endocrinol* **6**:285–289, 1977.

31. Kawashima H et al: Bloom's syndrome in a Japanese girl. *Clin Genet* **17**:143–148, 1980.

32. Keutel J et al: Bloom-Syndrom. *Z Kinderheilkd* **101**:165–180, 1967.

33. Kuhn EM, Therman E: No increased chromosome breakage in three Bloom's syndrome heterozygotes. *J Med Genet* **16**:219–222, 1979.

34. Landau JW et al: Bloom's syndrome. *Arch Dermatol* **94**:687–694, 1966.

35. Landau J et al: Eye findings in congenital telangiectatic erythema and growth retardation. *Am J Ophthalmol* **62**:753–754, 1966.

36. Rauh JL, Soukup SW: Case report. Bloom's syndrome. *Am J Dis Child* **116**:409–413, 1968.

37. Reed WB et al: Cutaneous manifestations of ataxia-telangiectasia. *JAMA* **195**:746–753, 1966.

38. Rivera H et al: Bloom's syndrome in a Mexican mestizo girl. *Ann Génét* **29**:39–41, 1966.

39. Sawitsky A et al: Chromosomal breakage and acute leukemia in congenital telangiectatic erythema and stunted growth. *Ann Intern Med* **65**:487–495, 1966.

40. Schoen EJ, Shearn MA: Immunoglobulin deficiency in Bloom's syndrome. *Am J Dis Child* **113**:594–596, 1967.

41. Schroeder TM, German J: Bloom's syndrome and Fanconi's anemia. Demonstration of two distinctive patterns of chromosome disruption and rearrangement. *Humangenetik* **25**:299–306, 1974.

41a. Shabtai F et al: Non-random chromosome aberrations in a complex leukaemic clone of Bloom's syndrome patient. *Hum Genet* **80**:311–314, 1988.

42. Szalay GC: Dwarfism with skin manifestations. *J Pediatr* **62**:686–695, 1963.

43. Torre DP, Cramer J: Primordial dwarfism. Discoid lupus erythematosus. *Arch Dermatol* **69**:511–513, 1954.

44. Vanderschueren-Lodeweyckx M et al: Bloom's syndrome. *Am J Dis Child* **138**:812–816, 1984.

45. Van Kerckhove CW et al: Bloom's syndrome. Clinical features and immunologic abnormalities of four patients. *Am J Dis Child* **142**:1089–1093, 1988.

46. Vijayalaxmi HJ et al: Bloom's syndrome. Evidence for an increased mutation frequency in vivo. *Science* **221**:851–853, 1983.

47. Weksberg R et al: Bloom's syndrome: A single complementation defines patients of diverse ethnic origin. *Am J Hum Genet* **42**:816–824, 1988.

48. Willis AE, Lindahl T: DNA ligase I deficiency in Bloom's syndrome. *Nature (London)* **325**:355–357, 1987.

## de Lange syndrome (Brachmann-de Lange syndrome)

The syndrome of primordial growth deficiency, severe mental retardation, anomalies of the extremities, and characteristic facial appearance (Figs. 10–8 to 10–13), was independently described by Brachmann (7) in 1916 and by de Lange (22) in 1933. The condition has been known as de Lange syndrome (6,9,10,12,25), Cornelia de Lange syndrome (4,23,31,33,39), Brachmann–de Lange syndrome (27,28), and typus degenerativus Amstelodamensis (22,37,42). The reader is referred to the following sources for special coverage: extensive surveys (6,16,33), references (28), numerical taxonomy (32), epidemiology (2), chromosomal findings (4,8), genetic studies, familial cases, and twins (3,5,9,11,19,27,29,35,41), evolution of the phenotype (31), phenotypic overriding of racial characteristics (17), radiographic findings (15,20), neurological, psychometric, and behavioral aspects (1a,13,14,18,38), limb anomalies (30), cardiac defects (6,12,34), gastrointestinal abnormalities (21), dermatoglyphics (6,40), and pregnancy (26).

By 1971, at least 250 cases had been reported (5,6). At present, over 400 cases are known (4,16,32). Epidemiologic estimates vary. Beck (2) reported a population prevalence of 0.6/100,000 in Denmark. Huang et al (17) gave a birth prevalence of 1/16,744 in Taipei. Opitz (28) suggested a birth prevalence of 1/10,000 in USA.

Most cases are sporadic. Affected sibs have been reported on a number of occasions (3,5,13a,24,29). Several families with vertical transmission have been observed (1,19,23a,35). Pregnancy in a woman with de Lange syndrome has been described; a clinically normal female infant was delivered (26). Instances of consanguinity (4,32), concordant monozygotic twins (27, see also 28,43), discordant monozygotic twins (9), and discordant dizygotic twins (41) have also been noted. Chromosomal studies have been normal in most instances, although occasional, inconsistent chromosomal anomalies have been recorded (4,6,8). Recurrence risk has been estimated to be 2–5% (29).

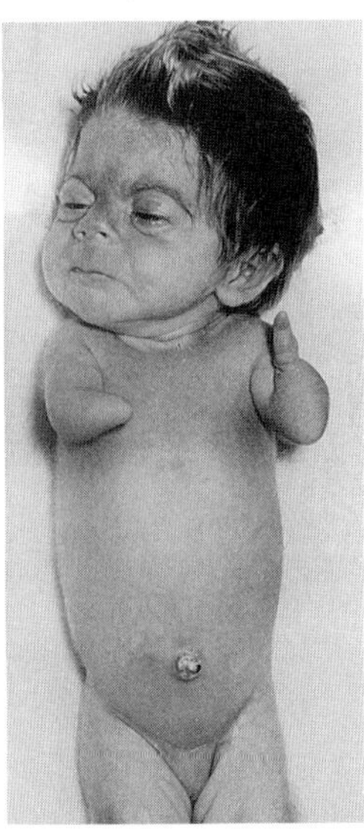

A

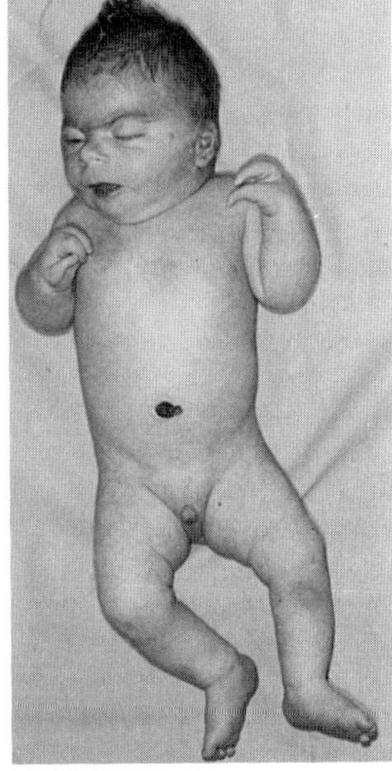

B

Fig. 10–8. *de Lange syndrome.* (A) Typical craniofacies and oligodactyly of upper limbs. Hirsutism of thighs. (B) Characteristic craniofacial appearance with microcephaly, small hands and feet, small external genitalia. (A courtesy of *RW Smithells*, Leeds, England. B courtesy of *M Silverman*, Atlanta, Georgia.)

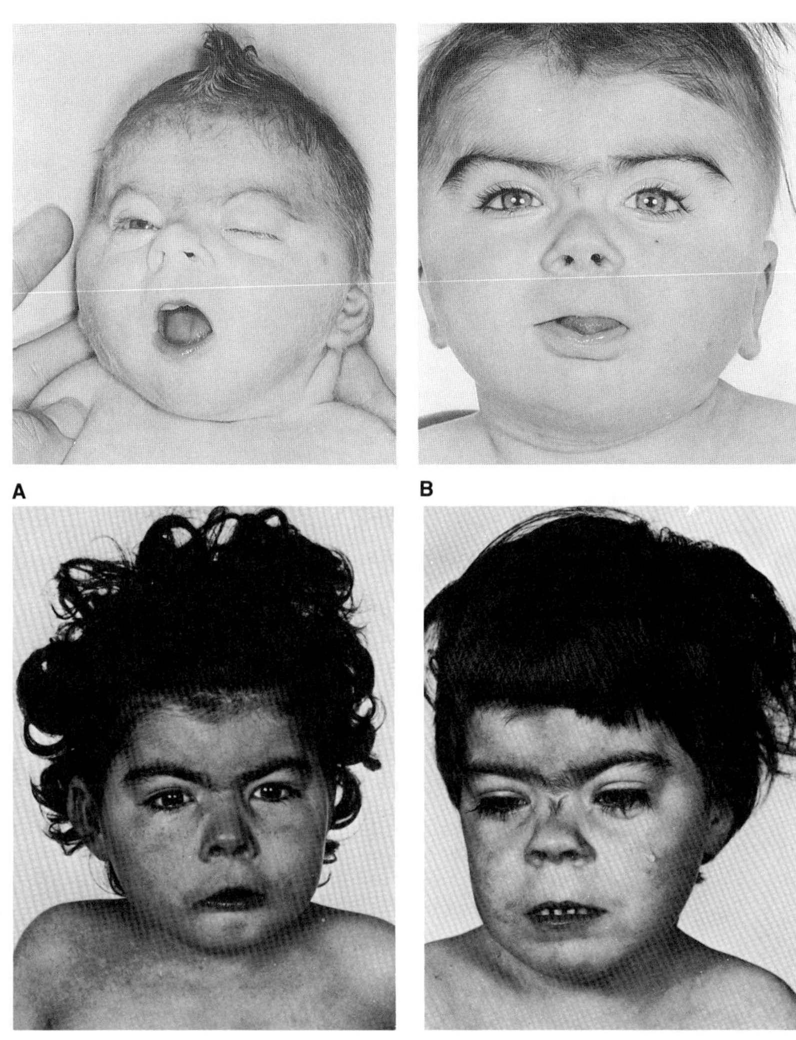

Fig. 10–9. *de Lange syndrome*. (A–D) Compare characteristic facies. Note microcephaly, low hairline, synophrys, small nose with anteverted nostrils, thin lips. (A,B courtesy of *M Silverman,* Atlanta, Georgia. C,D from *B Schlesinger et al,* Arch Dis Childh **38:**349, 1970.)

To date, no specific microdeletion or duplication has been found with high resolution banding.

**Clinical variability and diagnosis.** Lower birth weights are correlated with more severe phenotypic features, including more severe upper limb anomalies and greater psychomotor retardation. A significant excess of females is also found in the lower birth weight group (1a,16). Variability is marked. We believe that many with mild phenotypic changes are being missed (14a).

Using the technique of numerical taxonomy, Preus and Rex (32) constructed a diagnostic index of 30 characters that could discriminate de Lange from non-de Lange in 99% of the cases, leaving a 1% zone of doubt. The possibility of mildly affected individuals with minimal mental deficiency or borderline IQ, normal head circumference, normal height, and/or normal limbs has been discussed by Opitz (28). Findings in the de Lange syndrome are summarized in Table 10–2.

**Growth.** Birth weight is 2500 g or less in 72% in spite of normal duration of pregnancy. Postnatal growth is severe in 96%, both height and weight usually remaining far below the third percentile for age. Bone age is delayed, and, not uncommonly, the sequence of development of various centers of ossification may be disturbed (5). Recurrent respiratory infections and an early demise before 6 years of age are common (6).

**Central nervous system.** Microbrachycephaly is common, reported frequencies varying from 50 to 90% (2,6,8). Mental deficiency (75–100%) (8), abnormal speech development (75–100%) (8), behavioral problems (57%) (16), seizures (14–20%) (1a,16), hypertonia (less than 25%) (8), and, rarely, hypotonia (8) have been reported.

The cry is usually low pitched and growling (6). Patients frequently do not exhibit facial expression of emotion and commonly display stereotypic movements. Vestibular stimulation or vigorous movement tends to elicit pleasurable responses (18). Younger patients are dysphonic, with a frequency below normal pitch register, and older subjects are hoarse (13). In the study of Hawley et al (16), behavior problems included regurgitation, projectile vomiting, chewing and swallowing difficulties, lack of interest in food, and excessive screaming, biting, and hitting as well as frequent temper tantrums. Self mutilation has been discussed by Shear et al (38). Hearing loss occurs in 24% (16). IQ scores range from 4 to 85, and 80% of the patients reported by Barr et al (1a) were severely or profoundly retarded. An IQ of 78 was noted by Gadoth et al (14).

**Craniofacial features.** Patients resemble one another to a remarkable degree (Fig. 10–9A–D) and Huang et al (17) noted that the facial phenotype overrides racial characteristics. The typical facial appearance may not be evident during the first year of life (29), and Passarge et al (31) have discussed and illustrated evolution of the facial phenotype with age.

The skull is microbrachycephalic. The temporal and scalp veins may be conspicuous. The eyes are confluent (synophrys), the eyelashes long and curly, and the hairline is low. The nose is small with a flat nasal bridge. The nostrils are anteverted, and the philtrum is long. A

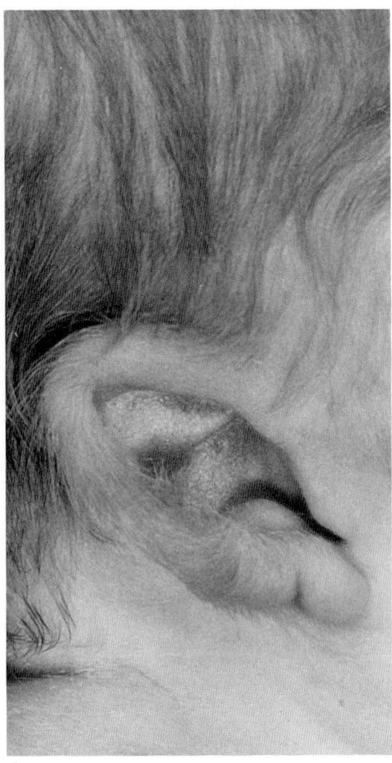


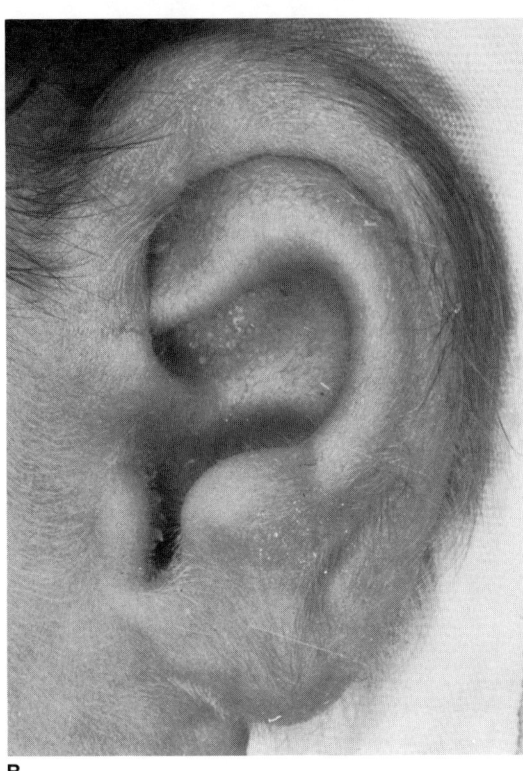


Fig. 10–10. *de Lange syndrome*. (A,B) Small malformed ears. Note hirsutism. (A courtesy of *M Silverman,* Atlanta, Georgia.)

bluish hue is often observed about the eyes, nose, and mouth. The ears may be malformed, small, apparently low-set, and hirsute (6) (Fig. 10–10A,B). The neck is short and thick.

Micrognathia is common and prominent mental spur may be seen. The lips are thin, with the corners of the mouth downturned. Cleft palate occurs in about 20% (5,6,11–12,25,29,33,42). Delayed tooth eruption and microdontia have been noted (5,25). Late eruption and widely spaced teeth were found in 93% of 64 patients reported by Hawley et al (16). A broad acellular zone with gracile fibers around blood vessels has been observed in the gingiva (36).

Fig. 10–11. *de Lange syndrome*. Examples of various types of malformed upper limbs found in de Lange syndrome. (From *JM Berg et al,* The de Lange Syndrome, Pergamon, Oxford, 1970.)

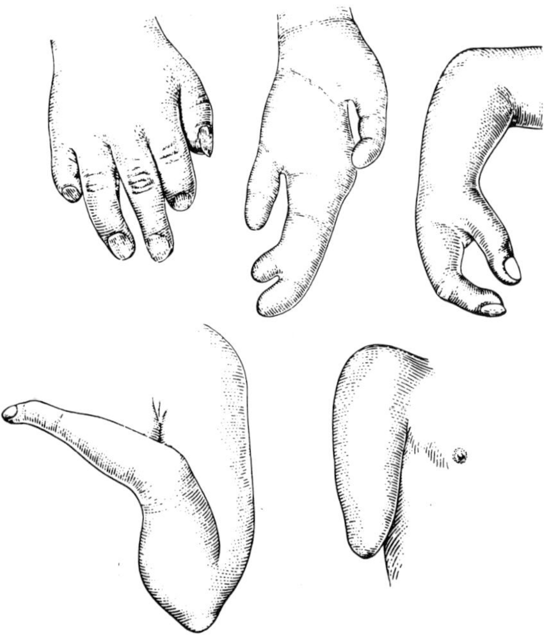

**Limbs.** Generally, the hands and feet are small. The fingers are often short and tapering with clinodactylous fifth digits that have only a single flexion crease. The thumbs are proximally placed in about 80% and the thenar muscles are hypoplastic. Flexion contractures of the elbow are present in approximately 80%. Soft tissue syndactyly of the second and third toes is common. About 20% exhibit severely malformed upper limbs, varying from oligodactyly to more severe phocomelia. Limb involvement may be unilateral or bilateral and, when bilateral, is not necessarily identical or even closely similar (Figs. 10–8A,B,10–11, and 10–12) (6,30).

Roentgenographically, the humerus, radius, and ulna are shortened. Hypoplasia and dorsal dislocation of the radial head are observed in 85%. Often the neck of the humerus is elongated. The semilunar notch of the ulna is shallow. In some cases, the forearm bones are absent. The first metacarpal and middle phalanx of the index and fifth fingers are often hypoplastic as are the third, fourth or fifth metatarsals (11a). The acetabular angle is low, especially when the child is less than 1 year or age. Coxa valga is a constant feature. The sternum is short with a reduced number of ossification centers, and the ribs are rather thin (6,20,22).

Dermatoglyphic findings include hypoplastic ridge patterns, single palmar creases, and increased *atd* angle. There is also an increase in radial loops on the third and fourth fingertips (40) and *c-d* interdigital triradius (6).

**Skin.** Hirsutism is often generalized, with hair whorls over the shoulders, lower back, and extremities (Fig. 10–13A,B). The nipples and umbilicus are frequently hypoplastic. Cutis marmorata is present in at least 50% (6,39). Pigmented nevi are relatively common.

**Genitourinary system.** The kidneys are often hypoplastic, dysplastic, or cystic (12). Cryptorchidism and/or hypospadias occur in 94% (16). Females commonly have bicornuate or septate uterus and long narrow ovaries (42).

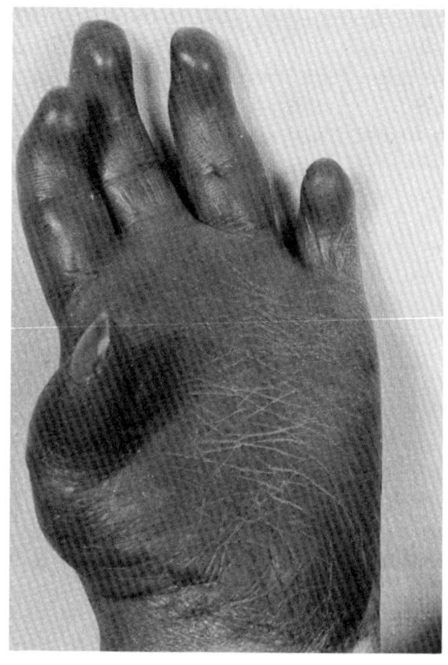
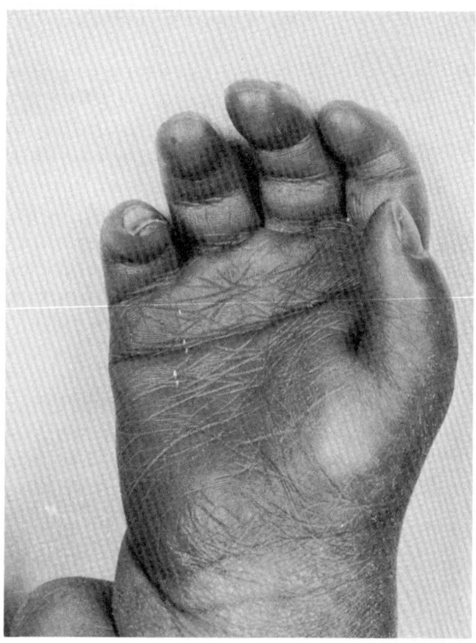

A B

Fig. 10–12. *de Lange syndrome*. (A,B) Micromelia with proximally placed thumbs, small fifth fingers with single flexion crease, finger contractures, right single palmar crease, and lack of palmar creases on left. (From *B Schlesinger et al*, Arch Dis Childh **38:**349, 1970.)

**Gastrointestinal abnormalities.** A variety of gastrointestinal abnormalities have been reported including malrotation, annular pancreas, pyloric stenosis and/or duodenal obstruction, hiatus hernia, inguinal hernia, umbilical hernia, colon duplication, Meckel's diverticulum, and gastric ulcer perforation (21,45).

**Cardiovascular system.** Berg et al (6) indicated that by 1970 congenital heart defects had been reported 42 times. Recorded defects included VSD, ASD, PDA, hypoplasia of the leaflets of the aortic valve, rudimentary left ventricle, anomalous venous drainage, overriding aorta, and ventricular fibroelastosis (12,34).

Fig. 10–13. *de Lange syndrome*. (A,B) Hirsutism of back, with whorling. (From *O Noe*, Clin Pediatr **3:**541, 1964.)

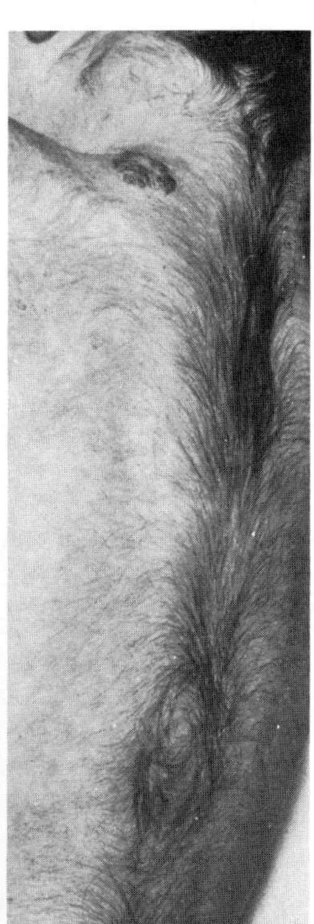
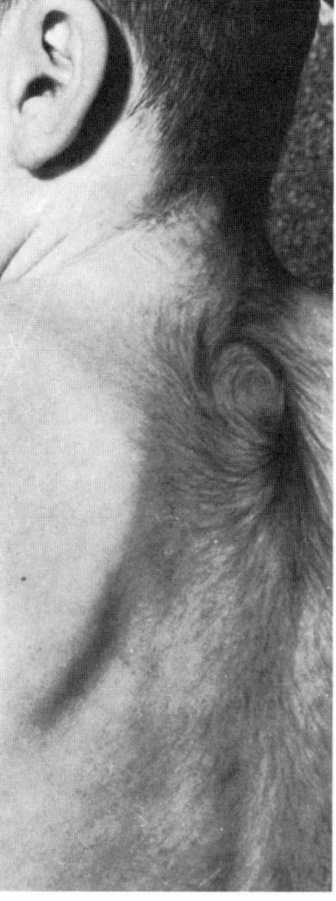

A B

**Other findings.** A variety of miscellaneous anomalies have been discussed by Berg et al (6) and others (24a,45). Endocrinological aspects have been discussed elsewhere (6,37). Laboratory studies have not revealed any consistent abnormalities (6). Daniel and Higgins (10), for example, reported elevated serum $\alpha$-ketoglutarate and glutamate levels.

**Differential diagnosis.** There is considerable phenotypic overlap with *dup(3q) syndrome*. Clinical features of the *fetal alcohol syndrome* may occasionally be confused with de Lange syndrome. MM Cohen and RJ Gorlin have observed a child with *Gorlin–Chaudhry–Moss syndrome* who bore facial resemblance to de Lange syndrome. It should also be noted that de Lange (23), in 1934, described another syndrome with congenital muscular hypertrophy, hypertonia, and developmental retardation. This condition, too, has sometimes been called de Lange syndrome, but it has nothing to do with the syndrome discussed here.

**Laboratory aids.** Pregnancy associated plasma protein A has been suggested as a marker in prenatal diagnosis (44).

### References [de Lange syndrome (Brachmann-de Lange syndrome)]


1. Bankier A et al: Familial occurrence of Brachmann–de Lange syndrome. *Am J Med Genet* **25**:163–165, 1986.

1a. Barr AN et al: Neurologic and psychometric findings in the Brachmann-de Lange syndrome. *Neuropädiatrie* **3**:46–66, 1971.

2. Beck B: Epidemiology of Cornelia de Lange's syndrome. *Acta Paediatr Scand* **65**:631–638, 1976.

3. Beck B: Familial occurrence of Cornelia de Lange's syndrome. *Acta Paediatr Scand* **63**:225–231, 1974.

4. Beck B, Mikkelsen M: Chromosomes in the Cornelia de Lange syndrome. *Hum Genet* **59**:271–276, 1981.

5. Beratis NG et al: Familial de Lange syndrome: Report of three cases in a sibship. *Clin Genet* **2**:170–176, 1971.

6. Berg JM et al: *The de Lange Syndrome*, Pergamon, New York, 1970.

7. Brachmann E: Ein Fall von symmetrischer Monodaktylie durch Ulnadefekt. *Jb Kinderheilk* **84**:224–235, 1916.

8. Breslau EJ et al: Prometaphase chromosomes in five patients with the Brachmann–de Lange syndrome. *Am J Med Genet* **10**:179–186, 1981.

9. Carakushansky G, Berthier C: The de Lange syndrome in one of twins. *J Med Genet* **13**:404–406, 1976.

10. Daniel WL, Higgins JV: Biochemical and genetic investigation of the de Lange syndrome. *Am J Dis Child* **121**:401–405, 1971.

11. Falek A et al: Familial de Lange syndrome with chromosome abnormalities. *Pediatrics* **37**:92–101, 1986.

Table 10–2. Findings in de Lange syndrome[a]

| Feature | Frequency[b] |
|---|---|
| Growth | |
| Low birth weight | + + + + |
| Prenatal growth retardation | + + + + |
| Performance | |
| Mental deficiency | + + + + |
| Abnormal speech development | + + + + |
| Seizures | +/− |
| Hypertonia | + |
| Hypotonia | +/− |
| Craniofacial | |
| Microbrachycephaly | + + + |
| Low hairline | + + + |
| Hypertelorism | +/− |
| Synophrys | + + + + |
| Long eyelashes | + + + |
| Eye abnormalities | + |
| Abnormal nasal bridge | + + + |
| Anteverted nostrils | + + + |
| Long philtrum | + + + |
| Downturned corners of mouth | + + |
| Micrognathia | + + + |
| Highly arched palate | + |
| Cleft palate/bifid uvula | + |
| Apparently low-set ears | + + + |
| Short neck | + + |
| Limbs | |
| Small hands and feet | + + + + |
| Proximally placed thumbs | + + + |
| Clinodactyly | + + + |
| Hypoplastic phalanges, fifth fingers | + |
| Single palmar crease | + + |
| Cutaneous syndactyly, 2–3 toes | + + + |
| Talipes | + |
| Skin | |
| Hirsutism | + + + |
| Cutis marmorata | + + + |
| Other | |
| Cardiac anomalies | + |
| Renal anomalies | +/− |
| Vertebral anomalies | + |

[a]Adapted from EJ Breslau et al: *Am J Med Genet* **10**:179, 1981.
[b]Key: + + + +, 75–100%; + + +, 50–75%; + +, 25–50%; +, 25% or less; +/−, rare.


11a. Filippi G: The de Lange syndrome; Report of 15 cases. *Clin Genet* **35**:343–363, 1989.

12. France NE et al: Pathological features in the de Lange syndrome. *Acta Paediatr Scand* **58**:470–480, 1969.

13. Fraser WI, Campbell BM: A study of six cases of de Lange Amsterdam dwarf syndrome, with special attention to voice, speech and language characteristics. *Dev Med Child Neurol* **20**:189–198, 1978.

13a. Fryns JP et al: The Brachmann–de Lange syndrome in two siblings of normal parents. *Clin Genet* **31**:413–415, 1987.

14. Gadoth N et al: Normal intelligence in the Cornelia de Lange syndrome. *Johns Hopkins Med J* **150**:70–72, 1982.

14a. Greenberg F, Robinson LK: Mild Brachmann–de Lange syndrome: Changes of phenotype with age. *Am J Med Genet* **32**:90–92, 1989.

15. Halal F, Preus M: The hand profile in de Lange syndrome: Diagnostic criteria. *Am J Med Genet* **3**:317–323, 1979.

16. Hawley PP et al: Sixty-four patients with Brachmann–de Lange syndrome: A survey. *Am J Med Genet* **20**:453–459, 1985.

17. Huang C-C et al: Two cases of the de Lange syndrome in Chinese infants. *J Pediatr* **71**:251–254, 1967.

18. Johnson HG et al: A behavioral phenotype in the de Lange syndrome. *Pediatr Res* **10**:843–850, 1976.

19. Kumar D et al: Cornelia de Lange syndrome in several members of the same family. *J Med Genet* **22**:296–300, 1985.

20. Kurlander GJ, DeMyer W: Roentgenology of the Brachmann–de Lange syndrome. *Radiology* **88**:101–110, 1967.

21. Lachman R et al: Gastrointestinal abnormalities in the Cornelia de Lange syndrome. *Mt Sinai J Med* **48**:236–240, 1981.

22. Lange C de: Sur un typ nouveau de dégéneration (typus Amstelodamensis). *Arch Méd Enf* **36**:713–718, 1933.

23. Lange C de: Congenital hypertrophy of the muscles, extrapyramidal motor disturbances and mental deficiency. *Am J Dis Child* **48**:243–268, 1934.

23a. Leavitt A et al: Cornelia de Lange syndrome in a mother and daughter. *Clin Genet* **28**:157–161, 1985.

24. Lieber E et al: Brachmann–de Lange syndrome. *Am J Dis Child* **125**:717–718, 1973.

24a. Maruiwa M et al: Cornelia de Lange syndrome associated with Wilms tumour and infantile hemangioendothelioma of the liver. Report of two autopsy cases. *Virchows Arch* [A] *(Pathol Anat)* **413**:463–468, 1988.

25. McArthur RG, Edwards IH: de Lange syndrome: Report of 20 cases. *Can Med Assoc J* **96**:1185–1198, 1967.

26. Mosher GA et al: Brief clinical report: Pregnancy in a woman with the Brachmann–de Lange syndrome. *Am J Med Genet* **22**:103–107, 1985.

27. Motl ML, Opitz JM: Phenotypic and genetic studies of the Brachmann–de Lange syndrome. *Hum Hered* **21**:1–16, 1971.

28. Opitz JM: Editorial comment: The Brachmann–de Lange syndrome. *Am J Med Genet* **22**:89–102, 1985.

29. Pashayan HM: Variability of the de Lange syndrome: Report of 3 cases and genetic analysis of 54 families. *J Pediatr* **75**:853–858, 1969.

30. Pashayan HM et al: Variable limb malformations in the Brachmann–Cornelia de Lange syndrome. *Birth Defects* **11**(5):147–156, 1975.

31. Passarge E et al: Cornelia de Lange syndrome: Evolution of the phenotype. *Pediatrics* **48**:833–836, 1971.

32. Preus M, Rex AP: Definition and diagnosis of the Brachmann–de Lange syndrome. *Am J Med Genet* **16**:301–312, 1983.

33. Ptacek LJ et al: The Cornelia de Lange syndrome. *J Pediatr* **63**:1000–1020, 1963.

34. Rao PS: Congenital heart disease in the de Lange syndrome. *J Pediatr* **79**:674–677, 1971.

35. Robinson LK et al: Brachmann–de Lange syndrome: Evidence for autosomal dominant inheritance. *Am J Med Genet* **22**:109–115, 1985.

36. Russell BG: The Cornelia de Lange syndrome. Typus degenerativus Amstelodamensis: Histologic studies of the marginal gingiva. *Scand J Dent Res* **78**:369–373, 1970.

37. Schlesinger B et al: Typus degenerativus Amstelodamensis. *Arch Dis Childh* **38**:349–357, 1970.

38. Shear CS et al: Self mutilative behavior as a feature of the de Lange syndrome. *J Pediatr* **78**:506–507, 1971.

39. Shuster DS, Johnson S: Cutaneous manifestations of the Cornelia de Lange syndrome. *Arch Dermatol* **93**:702–707, 1966.

40. Smith GF: A study of the dermatoglyphics in the de Lange syndrome. *J Ment Defic Res* **10**:241–247, 1966.

41. Stevenson RE, Scott CI: Discordance for Cornelia de Lange syndrome in twins. *J Med Genet* **13**:402–403, 1976.

42. Vischer D: Typus degenerativus Amstelodamensis (Cornelia de Lange syndrom). *Helv Paediatr Acta* **20**:415–445, 1965.

43. Watson A: Cornelia de Lange syndrome: Occurrence in twins. *Aust J Dermatol* **20**:7–9, 1979.

44. Westergaard JG et al: Pregnancy-associated plasma protein A: A possible marker in the classification and prenatal diagnosis of Cornelia de Lange syndrome. *Prenat Diag* **3**:225–232, 1983.

45. Wick MR et al: Duodenal obstruction, annular pancreas and horseshoe kidney in an infant with Cornelia de Lange syndrome. *Minn Med* **65**:539–541, 1982.


## Dubowitz syndrome

In 1965, Dubowitz (4) reported sibs with intrauterine growth retardation, primordial short stature, microcephaly, mental retardation, high-pitched voice, and characteristic facial appearance (Figs. 10–14 and 10–15). He assumed that the children had Bloom syndrome. In 1971, Grosse et al (7) suggested that the disorder was a distinct entity. Approximately 42 cases have been published to date (3,5,8–10,12,13,16–19,21,22). Autosomal recessive inheritance has been demonstrated because of multiple sib involvement and parental consanguinity (4,9,13,16,22,23). It has also been documented in identical twins (22).

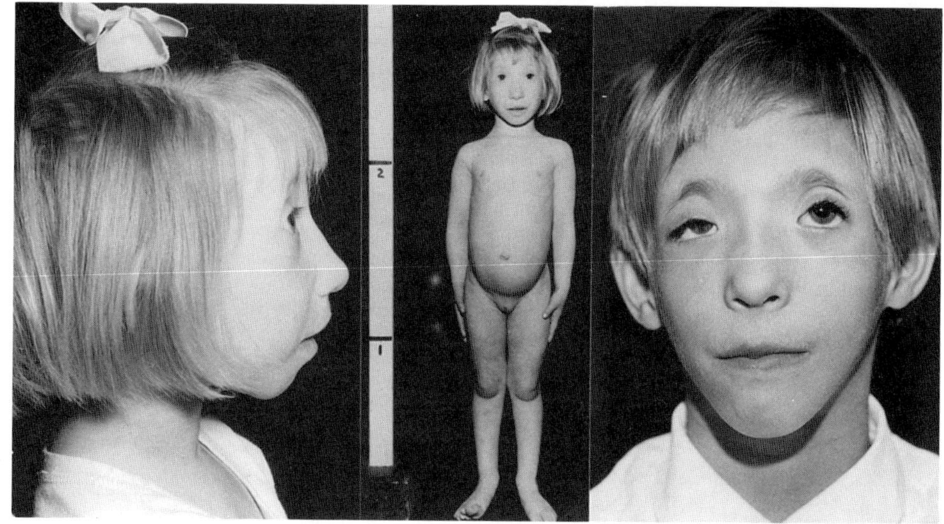

Fig. 10–14. *Dubowitz syndrome.* Primordial shortness of stature, blepharophimosis, ptosis of eyelids, and micrognathia. (From *F Grosse et al,* Z Kinderheilkd **110:**175, 1971.)

Moller and Gorlin (11) and Belohradsky et al (1) surveyed the disorder.

**Growth.** Birth weight is about 2.3 kg and birth length approximately 45 cm. Prenatal growth failure has been evident in about 70%. Postnatal growth is retarded in 90%. Head circumference at birth averages 30 cm. Delayed bone age has been reported in over 65%.

**Facies.** In nearly all cases there is microcephaly, the degree of which is not correlated with mental retardation. The small head circumference seen at birth remains small. High sloping forehead with sparse frontal hair and flat supraorbital ridges are seen in approximately 90%. Facial asymmetry has been noted occasionally. A broad based nose and nasal tip are seen in over 50%. Telecanthus has been reported in 90%, ptosis and/or blepharophimosis, often asymmetric, in 65%, and epicanthic folds in 75%. Somewhat dysmorphic pinnae and micrognathia have been noted in about 95%. The face elongates with age (Figs. 10–14 and 10–15). About 85% have microretrognathia (1).

**Skin.** An eczematous skin eruption, especially of the face and extremities, has been noted in approximately 60%, usually from birth. Variable minor soft tissue syndactyly of the second and third toes has been documented in 40%. The scalp hair is usually sparse. Hypoplasia of the eyebrows has been observed in 20%.

**Central nervous system.** Motor milestones are reached at normal times. Significant mental retardation has been evident in about 35%. Most children have been estimated to be in the dull normal range. Hyperactivity has been manifested by approximately 40% (15). Speech is delayed in 60%.

**Other findings.** High-pitched voice has been noted in 50%. About 65% exhibit poor feeding, frequent vomiting, and chronic diarrhea during infancy. About 70% of affected males manifest hypospadias and/or cryptorchidism.

Of considerable interest are reports that suggest an association with leukemia (6), lymphoma (16), neuroblastoma (16), and aplastic anemia (2,20). Although there have been some suggestions that these patients have an immune defect, its precise nature has not been defined (7,10,13,16). Sauer and Spelger (16) noted hypogammaglobulinemia in one sib and IgA deficiency in the other. Both of these children had malignancies.

**Oral manifestations.** Cleft palate and/or bifid uvula and submucous cleft palate have been found in about 35% (1,3,7,13,22).

**Differential diagnosis.** The Dubowitz syndrome may be confused with the *Bloom syndrome* or *fetal alcohol syndrome.*

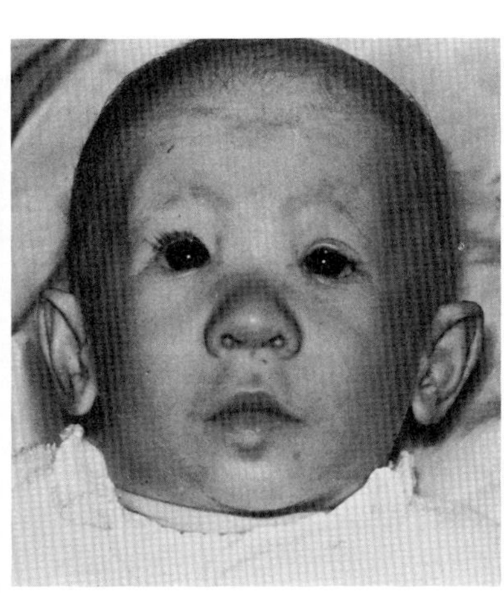

A

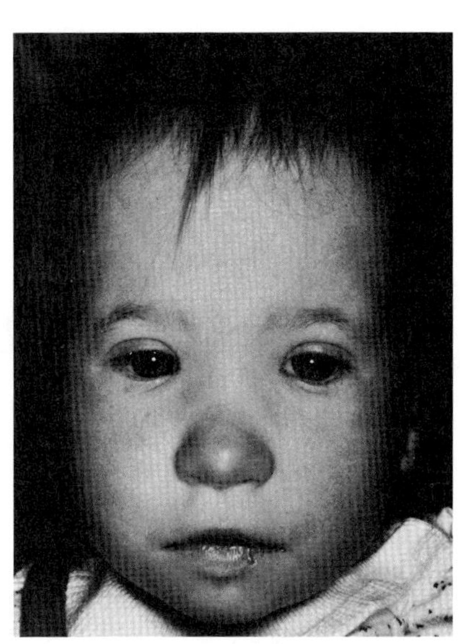

B

Fig. 10–15. *Dubowitz syndrome.* (A) Note high forehead with flat supraorbital ridges, widely spaced eyes, blepharophimosis, and small mandible. (B) Sister of patient shown in A. Note similar facies.

### References (Dubowitz syndrome)


1. Belohradsky BH et al: Das Dubowitz–Syndrom. *Ergeb Inn Med Kinderheilkd* **57**:146–184, 1988.
2. Berthold F et al: Fatal aplastic anaemia in a child with features of Dubowitz syndrome. *Eur J Pediatr* **146**:605–607, 1987.
3. Borkenstein M, Falk W: Zur Klinik des Dubowitz-Syndroms. *Wien Med Wochenschr* **128**:14–17, 1978.
4. Dubowitz V: Familial loss birth weight dwarfism with an unusual facies and a skin eruption. *J Med Genet* **2**:12–17, 1965.
5. Fryns JP et al: The Dubowitz syndrome in a teenager. *Am J Med Genet* **4**:345–347, 1979.
6. Gröbe H: Dubowitz-Syndrom und akute lymphatische Leukämie. *Mschr Kinderheilkd* **131**:467–468, 1983.
7. Grosse F et al: The Dubowitz syndrome. *Z Kinderheilkd* **110**:175–187, 1971.
8. Kondo I et al: A Japanese patient with Dubowitz syndrome. *Clin Genet* **31**:389–392, 1987.
9. Küster W, Majewski F: The Dubowitz syndrome. *Eur J Pediatr* **144**:574–578, 1986.
10. Majewski F et al: A rare type of low birth weight dwarfism: The Dubowitz syndrome. *Z Kinderheilkd* **120**:283–292, 1975.
11. Moller KT, Gorlin RJ: The Dubowitz syndrome: A retrospective. *J Craniofac Genet Dev Biol Suppl* **1**:283–286, 1985.
12. Müller W et al: Seckel-Syndrom. *Mschr Kinderheilkd* **126**:454–456, 1978.
13. Opitz JM et al: The Dubowitz syndrome. *Z Kinderheilkd* **115**:1–12, 1973.
14. Orrison WW et al: The Dubowitz syndrome: Further observations. *Am J Med Genet* **7**:155–170, 1980.
15. Parrish JM, Wilroy RS Jr: The Dubowitz syndrome: The psychological status of ten cases at follow-up. *Am J Med Genet* **6**:3–8, 1980.
16. Sauer O, Spelger G: Dubowitz-Syndrom mit Immunodefizienz und malignem Neoplasma bei zwei Geschwistern. *Mschr Kinderheilkd* **125**:885–887, 1977.
17. Shuper A et al: The diagnosis of Dubowitz in the neonatal period. *Eur J Pediatr* **145**:151–152, 1986.
18. Stoll C et al: Le syndrome de Dubowitz. *Pédiatrie* **35**:149–152, 1985.
19. Walter L: Das Dubowitz-Syndrom. *Z Ärtzl Fortbild* **74**:748–755, 1980.
20. Walters TR, Desposito F: Aplastic anemia in Dubowitz syndrome. *J Pediatr* **106**:622–623, 1985.
21. Wilhelm OL, Méhes K: Dubowitz syndrome. *Acta Paediatr Hung* **27**:67–75, 1986.
22. Wilroy RS Jr et al: The Dubowitz syndrome. *Am J Med Genet* **2**:275–284, 1978.
23. Winter RM: Dubowitz syndrome. *J Med Genet* **23**:11–13, 1986.


## Hallermann–Streiff syndrome

The Hallermann–Streiff syndrome is characterized by dyscephaly, hypotrichosis, microphthalmia, cataracts, beaked nose, micrognathia, and proportionate short stature. The condition was probably first described by Aubry (1) in 1893, although he did not report the complete syndrome. Hallermann (19), in 1948, and especially Streiff (43), in 1950, separated the syndrome from progeria and mandibulofacial dysostosis. An important early paper as that of François (14) who, in 1958, presented two personally-studied cases, reviewed 22 cases from the literature, and further delineated the syndrome. The condition has also been called François syndrome (2,4), François dyscephalic syndrome (6,13), and oculomandibulodyscephaly (39). Over 150 cases have been recorded to date. Several extensive reviews (2,5,13) and surveys (42,45) are available.

Virtually all cases have been sporadic. There is no sex predilection. The syndrome has been described as concordant (48) and discordant (37) in monozygotic twins, and an affected female has had two normal children (34). We cannot accept any of the familial cases known to us to date. For example, Guyard et al (17) described a father and daughter and Koliopoulos and Palimeris (26) reported what they called "atypical Hallermann–Streiff–François syndrome" in three successive generations. Neither of these papers describes Hallermann–Streiff syndrome, however. François (13) cited several familial cases as personal communication that have never been published. He further cited two affected sibs reported by Hall et al (18) as examples, although Hall et al specifically recognized the condition they reported as an entity different from Hallermann–Streiff syndrome; they named their disorder pseudoprogeria/Hallermann–Streiff (PHS) syndrome.

The affected cousins reported by Schanzlin et al (36) do not represent familial examples either. Although a second cousin clearly had Hallermann–Streiff syndrome, the proband had dup(10q) with buphthalmos, sclerocornea, total aniridia, nasal abnormality, micrognathia, and bifid halluces. Chromosomal studies have been normal in almost all instances (2,4,12,15,24,28), with few exceptions (20,25).

The sporadic cases reported by Scoppetta et al (38) and Steel et al (41) are not examples of the syndrome. Consanguinity was reported by Berbich et al (4), but in their sibship of eight children, only one was affected. François (13) cited a number of other instances of consanguinity. However, families that can be verified had only one affected individual per family.

**Natural history.** Clinical findings are summarized in Table 10–3. Narrow upper air passages may make feeding difficult during infancy (20). Pneumonia and/or severe feeding difficulties have led to the demise of affected infants in several instances. Anesthetic risks have been discussed by Ravindran and Stoops (35) and Sataloff and Roberts (35a). Sleep apnea has also been noted (14a).

Birth weight is normal in about 64%, prematurity and/or low birth weight occurring in the other 36% (45). Short stature is seen in 45–68% (13). Growth is diminished proportionately, being at least 2–5 SD below the mean. Final height attainment for females is about 152.4 cm, with males being 2.5–5.0 cm taller (42).

**Craniofacial features.** The face is small, with a long, thin, tapering, "pinched" nose, and receding chin (37a). An odd-shaped, bulging skull with brachycephaly is often accompanied by frontal or parietal bossing (Fig. 10–16A–F). Mild microcephaly and malar bone hypoplasia also occur but are not constant features (13,28). Gaping or dehiscence of sutures, as well as delayed closure of fontanels, has been described by nearly all authors. The nose is thin, pointed, and often curved and may have a tendency to septal deviation (3,34). Hypotrichosis, especially of the scalp, brows, and lashes, occurs in about 80–82% (2,5,15,15a,45). Scanning electron microscopy has shown absent or abnormal cuticle development in the hair shafts (15). Alopecia is most prominent about the frontal and occipital areas, but is especially marked along suture lines (44). Cutaneous atrophy, present in about 68–70% (2,5), is largely limited to the scalp and nose. Scalp skin is thin and taut, and scalp veins are prominent. Similar changes, often focal, are observed on the nose.

Ocular findings are summarized in Table 10–4. Previously undiagnosed patients tend to visit ophthalmologists because of visual impairment from congenital cataract. Microphthalmia of variable severity (78–83%) and bilateral congenital cataracts (81–90%) occur with high frequency (2,5,45). Cataracts consist of milky white liquefied lens masses that often resorb spontaneously (13,50). Blue scleras have been described in about 22–31%, nystagmus in 32–45%, strabismus in 33–37%, glaucoma in 7–11%, pupillary membrane persistence in 5%, and downslanting palpebral fissures in 12–13% (2,5,21,28,44–46,48). Donders (11) published a histopathologic study of the eye from an autopsy case. Anomalies of the fundus, conjunctiva, and

Table 10–3. Principal features of Hallermann–Streiff syndrome[a]

| Abnormality | Percentage |
|---|---|
| Dyscephaly | 98–99 |
| Cataract | 81–90 |
| Microphthalmia | 78–83 |
| Dental abnormalities | 80–85 |
| Hypotrichosis | 80–82 |
| Skin atrophy | 68–70 |
| Proportionate short stature | 45–68 |

[a]From D Barrucand et al: *Rev Oto-Neuro-Ophthal* **50**:305, 1978 and B Carles-Mermet et al: Thesis, Lyon, 1979.

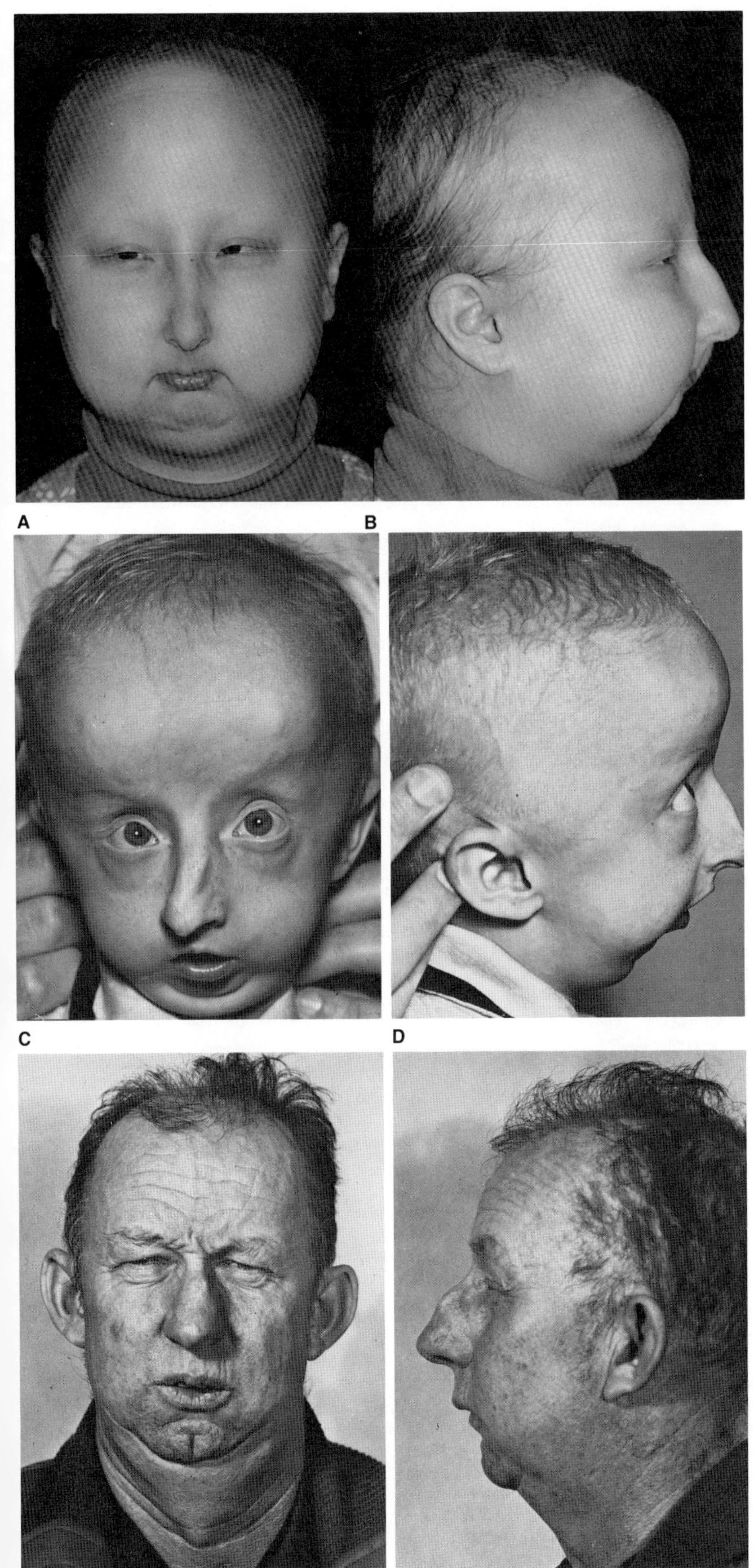


Fig. 10–16. *Hallermann–Streiff syndrome*. (A–F) Characteristic facial appearance with brachycephaly, hypotrichosis, thin pointed nose, and micrognathia. Note posteriorly angulated ears in patient shown in D. Note also chin groove in patient shown in E. [A,B from *MM Cohen Jr*, Malformation syndromes, in Surgical Correction of Dentofacial Deformities, *WH Bell et al* (eds), W.B. Saunders, Philadelphia, 1980, p 35. C,D from *D Hoefnagel* and *K Benirschke*, Arch Dis Childh **40:**57, 1965.]

Table 10–4. Ocular findings in Hallermann–Streiff syndrome[a]

| Abnormality | Percentage |
|---|---|
| Congenital cataract | 81–90 |
| Microphthalmia | 78–83 |
| Nystagmus | 32–45 |
| Strabismus | 33–37 |
| Blue scleras | 22–31 |
| Sparse eyelashes and eyebrows | 29 |
| Fundus anomalies | 18–22 |
| Conjunctival defects | 11 |
| Corneal abnormalities | 9–14 |
| Downslanting palpebral fissures | 12–13 |
| Intraocular hypertension | 7–11 |
| Iris atrophy | 10–14 |
| Vitreous degeneration | 8 |
| Eyelid anomalies | 6 |
| Iris coloboma | 5 |
| Pupillary membrane persistence | 5 |
| Enophthalmos | 2.5–4 |
| Hypotelorism | 2.5 |
| Epicanthic folds | 2–4 |
| Disc coloboma | 1 |
| Choroidal coloboma | 1 |
| Ptosis of the eyelids | 1–3 |
| Hypoplasia of the lacrimal puncta | 2 |
| Epibulbar tumor | 1 |

[a]From D Barrucand et al: *Rev Oto-Neuro-Ophthal* **50**:305, 1978 and B Carles-Mermet, Thesis, Lyon, 1979.

cornea as well as miscellaneous ocular findings (Table 10–4) are well documented in several accounts (2,3,5,12–14,16,19,28,30,32,44,47–50).

The mandible is hypoplastic with double chin and central clefting or dimpling (3,7,14,16,19,43). The ascending ramus is usually short, and the condyle may be missing or the fossa hypoplastic (14,25,27). Roentgenographic examination reveals a characteristic temporomandibular joint displacement of approximately 2 cm forward from its normal position that is located just in front of the external auditory meatus (48). The palate is high and narrow and the paranasal sinuses are diminished in size (12). Microstomia is present in about 10% (25,48). Cephalometric and anthropometric studies have been carried out (14a,17a).

Dental anomalies are common (80–85%) and may include absence of teeth, persistence of deciduous teeth, malocclusion and open bite, malformed teeth, severe and premature caries, supernumerary teeth, and natal teeth (1–3,5,7,12–14,19,20,23,30,32,33,37,40,42,43). A well-documented histopathologic study of dentoalveolar abnormalities observed at autopsy is available (39).

**Central nervous system.** Mental deficiency has been noted in about 15% (25,45); and although an estimate as high as 31% has been indicated elsewhere (2,5,13), Crevits et al (9) has warned that the psychometric aspects of the syndrome have been treated very subjectively, with only rare instances of psychometric testing. Hyperactivity, choreoathetosis, and generalized tonic–clonic seizures have been noted occasionally (9,12,25).

**Other findings.** Hypogenitalism has been reported in 10–12% (2,5,20,40). Also documented have been cryptorchidism, hypospadias, clitoral enlargement, breast asymmetry, and breast atrophy (2,13). Axillary and pubic hair may be scant (12,14).

Dinwiddie et al (10) recorded various cardiac anomalies including pulmonic stenosis, ASD, VSD, PDA, and tetralogy of Fallot. They indicated a frequency of about 4.8%.

Chandra et al (8) reported a case with deficiency of humoral immunity and hypoparathyroidism. Although immunodeficiency, hypoparathyroidism, and cardiac anomalies have now been observed in Hallermann–Streiff syndrome, true DiGeorge sequence has not occurred. The patient reported by Chandra et al (8) had normal cell-mediated immunity with reduced serum IgG and plasma opsonic function.

A great many other abnormalities have been observed including occipital bossing, hydrocephaly, calcification of the falx cerebri (2,13,29,51), osteoporosis (32), syndactyly (48), lordosis and/or scoliosis (30,37), spina bifida, winging of the scapula (25,34,45,48), pectus carinatum, and pectus excavatum (13,48). Vitiligo and livedo have also been noted (14,28,51).

**Differential diagnosis.** *Progeria* differs from the Hallermann–Streiff syndrome by having premature arteriosclerosis, nail dystrophy, acromicria, and chronic deforming arthritis. In addition, the eyes are normal.

*Mandibulofacial dysostosis* shares in common with the Hallermann–Streiff syndrome micrognathia, high palatal vault, and malar hypoplasia, but usually has lower eyelid colobomas and associated ear anomalies.

The pseudoprogeria/Hallermann–Streiff (PHS syndrome) (18) has similarities to the Hallermann–Streiff syndrome but has, in addition, severe spastic quadriplegia. Furthermore, appearance at birth is normal except for absence of eyebrows and eyelashes. The disorder is progressive, with limb spasticity and psychomotor delay evident by 6 months of age.

### References (Hallermann–Streiff syndrome)


1. Aubry M: Variété singulière d'alopécie congénitale; alopécie suturale. *Ann Dermatol Syphiligr (Paris)* **4**:899–900, 1893.

2. Barrucand D et al: Syndrome de François. À propos de deux cas. *Rev Oto-Neuro-Ophthal* **50**:305, 1978.

3. Blodi FC: Development anomalies of the skull affecting the eye. *Arch Ophthalmol* **57**:593–610, 1957.

4. Berbich A et al: Syndrome de François. *Arch Ophtalmol (Paris)* **37**:723–730, 1977.

5. Carles-Mermet B: Dyscéphalie à tête d'oiseau (signes oculaires et etiologie). À propos de 4 observations et revue de la literature. Thesis, Lyon, 1979.

6. Carones AV: Syndrome dyscéphalique de François. *Ophthalmologica* **142**:510–518, 1961.

7. Caspersen I, Warburg M: Hallermann–Streiff syndrome. *Acta Ophthalmol (Kbh)* **46**:385–390, 1968.

8. Chandra RK et al: Deficiency of humoral immunity and hypoparathyroidism associated with the Hallermann–Streiff syndrome. *J Pediatr* **93**:892–893, 1978.

9. Crevits L et al: Oculomandibular dyscephaly (Hallermann–Streiff–François syndrome) associated with epilepsy. *J Neurol* **215**:225–230, 1977.

10. Dinwiddie R et al: Cardiac defects in the Hallermann–Streiff syndrome. *J Pediatr* **92**:77, 1978.

11. Donders PC: Hallermann–Streiff syndrome. *Documenta Ophthalmol* **44**:161–166, 1977.

12. Falls HF, Schull WJ: Hallermann–Streiff syndrome: A dyscephaly with congenital cataracts and hypotrichosis. *Arch Ophthalmol* **63**:409–420, 1960.

13. François J: François' dyscephalic syndrome. *Birth Defects* **18**(6):595–619, 1982.

14. François MJ: A new syndrome: Dyscephalia with bird face and dental anomalies, nanism, hypotrichosis, cutaneous atrophy, microphthalmia and congenital cataract. *Arch Ophthalmol* **60**:842–862, 1958.

14a. Friede H et al: Cardiorespiratory disease with Hallermann–Streiff syndrome: Analysis of craniofacial morphology by cephalometric roentgenograms. *J Craniofac Genet Develop Biol (Suppl)* **1**:185–198, 1985.

15. Golomb RS, Porter PS: A distinct hair shaft abnormality in the Hallermann–Streiff syndrome. *Cutis* **16**:122–128, 1975.

15a. Gratton CEH et al: Atrophic alopecia in the Hallermann–Streiff syndrome. *Clin Esp Dermatol* **14**:250–252, 1989.

16. Gregersen E: Ocular abnormalities in progeria. *Acta Ophthalmol (Kbh)* **34**:347–354, 1956.

17. Guyard M et al: Sur deux cas de syndrome dyscéphalique à tête d'oiseau. *Bull Soc Fr Ophthal* **62**:443–447, 1962.

17a. Haberman H, Clement PAP: The value of anthropometrical measurements in a case of Hallermann–Streiff syndrome. *Rhinology* **17**:179–194, 1979.

18. Hall BD et al: Pseudoprogeria/Hallermann–Streiff (PHS) syndrome. *Birth Defects* **10**(7):137–146, 1974.

19. Hallermann W: Vogelgesicht und Cataracta congenita. *Klin Monatsbl Augenheilkd* **113**:315–318, 1948.

20. Hoefnagel D, Benirschke K: Dyscephalia mandibulo-oculo-facialis (Hallermann–Streiff syndrome). *Arch Dis Childh* **40**:57–61, 1965.

21. Hopkins DJ, Horan EC: Glaucoma in the Hallermann–Streiff syndrome. *Br J Ophthalmol* **54**:416–422, 1970.

22. Hudwinkle JB et al: An animal model of the Hallermann–Streiff syndrome in the Australian wombat. *Aust J Paediatr* **31**:950–952, 1978.

23. Hutchinson D: Oral manifestations of oculomandibulodyscephaly with hypotrichosis (Hallermann–Streiff syndrome). *Oral Surg* **31**:234–247, 1971.

24. Imamura S et al: Hallermann–Streiff syndrome: Case report. *Dermatologica* **160**:354–357, 1980.

25. Judge C, Chakanovskis JE: The Hallermann–Streiff syndrome. *J Ment Defic Res* **15**:115–120, 1971.

26. Koliopoulos J, Palimeris G: Atypical Hallermann–Streiff–François syndrome in three successive generations. *J Pediatr Ophthalmol* **12**:235–239, 1975.

27. Kurlander GJ et al: Roentgen differentiation of the oculodentodigital syndrome and the Hallermann–Streiff syndrome in infancy. *Radiology* **86**:77–85, 1966.

28. Lamy M et al: La dyscéphalie (syndrome de Hallermann–Streiff–François). *Arch Fr Pédiatr* **22**:929–938, 1965.

29. Larmande A et al Dyscéphalie à tête d'oiseau (Syndrome d'Hallermann–Streiff–François). *Pédiatrie* **49**:313, 1962.

30. Ludwig A, Korting G: Vogt-Koyanagi-ähnliches Syndrom und mandibulofaciale Dysostosis (Franceschetti-Zwahlen). *Arch Dermatol Syph (Berlin)* **190**:307–319, 1950.

31. Marchesani O: Über Beziehungen zwischen Wachstum und Nervensegmenten. *Dtsch Ophthalmol Gesell* **55**:34–48, 1949.

32. Moehlig RC: Progeria with nanism and congenital cataracts in a five-year-old child. *JAMA* **132**:640–642, 1946.

33. Patterson GT et al: Surgical correction of the dentofacial abnormality in Hallermann–Streiff syndrome. *J Oral Max-Fac Surg* **40**:380–384, 1982.

34. Ponte F: Further contributions to the study of the syndrome of Hallermann and Streiff. *Ophthalmologica* **143**:399–408, 1962.

35. Ravindran R, Stoops CM: Anesthetic management of a patient with Hallermann–Streiff syndrome. *Anesth Analg* **58**:254–255, 1979.

35a. Sataloff RT, Roberts BR: Airway management in Hallermann–Streiff syndrome. *Am J Otolaryngol* **5**:64–67, 1984.

36. Schanzlin DJ et al: Hallermann–Streiff syndrome associated with sclerocornea, aniridia, and a chromosomal abnormality. *Am J Ophthalmol* **90**:411–415, 1980.

37. Schondel A: Two cases of progeria complicated by microphthalmus. *Acta Paediatr* **30**:286–304, 1943.

37a. Sclaroff A, Eppely BL: Evaluation and surgical correction of the facial skeletal deformity in Hallermann–Streiff syndrome. *Int J Oral Maxillofac Surg* **16**:738–744, 1987.

38. Scoppetta C et al: Oligophrenia with the Hallermann–Streiff syndrome. *J Neurol* **220**:211–214, 1979.

39. Slootweg PJ, Huber J: Dento-alveolar abnormalities in oculomandibulodyscephaly (Hallermann–Streiff syndrome). *J Oral Pathol* **13**:147–154, 1984.

40. Srivastava S et al: Mandibulo-oculo-facial dyscephaly. *Br J Ophthalmol* **58**:543–549, 1966.

41. Steel HH et al: Bilateral dislocation of the hip in Hallermann–Streiff syndrome: A case report. *J Bone Jt Surg* **57A**:1002–1005, 1975.

42. Steele RW, Bass JW: Hallermann–Streiff syndrome: Clinical and prognostic considerations. *Am J Dis Child* **120**:462–465, 1970.

43. Streiff EB: Dysmorphie mandibulo-faciale (tête d'oiseau) et alteration oculaires. *Ophthalmologica* **120**:79–83, 1950.

44. Sugar A et al: Hallermann–Streiff–François syndrome. *J Pediatr Ophthalmol* **8**:234–238, 1971.

45. Suzuki Y et al: Hallermann–Streiff syndrome. *Dev Med Child Neurol* **12**:496–506, 1970.

46. Teuscher M: Beitrag zum Ullrich-Fremerey-Dohna oder François-Syndrom. *Acta Ophthalmol* **52**:534–540, 1974.

47. Ullrich O, Fremerey-Dohna H: Dyskephalie mit Cataracta congenita und Hypotrichose als typischer Merkmalskomplex. *Ophthalmologica* **125**:73–90, 144–154, 1953.

48. Van Balen ATM: Dyscephaly with microphthalmos, cataract and hypoplasia of the mandible. *Ophthalmologica* **141**:53–63, 1961.

49. Walbaum R et al: Le syndrome de Hallermann–Streiff–François. *Pédiatrie* **23**:789–794, 1968.

50. Wolter JR, Jones DH: Spontaneous cataract absorption in Hallermann–Streiff syndrome. *Ophthalmologica* **150**:401–408, 1965.

51. Zolog N et al: Syndrome dyscéphalique de François avec tetralogie de Fallot. *Rev Oto-Neuro-Ophthal* **39**:65, 1967.

## Rubinstein–Taybi syndrome

In 1963, Rubinstein and Taybi (47) observed the constellation of broad thumbs and halluces, characteristic facial dysmorphism, growth retardation, and mental deficiency (Figs. 10–17 to 10–19). However, the syndrome had been described as early as 1957 in the French literature by Michail et al (34). Rubinstein (45) reviewed 114 patients in 1969 and had catalogued over 225 patients by 1972 (see 10). At least 570 cases have been recorded to date (2–7,9–28, 30,31,33–38,40–56,59–62,66). Some of the European literature on the subject has been cited by Behrens-Baumann (3). The syndrome has been observed in whites, blacks, and orientals (46).

Frequency estimates have ranged from 1 in 300 to 1 in 720 mentally deficient institutionalized individuals over 5 years old (45). These data have been used to estimate the prevalence in the general population which is said to range from 1/300,000 to 1/720,000 (54). Hennekam et al (14a) estimated a birth prevalence of 1/275,000–300,000 in the Netherlands. Simpson and Brissenden (54) estimated a recurrence risk of approximately 1% for families with one affected child.

The etiology is uncertain. The overwhelming majority of reported cases are sporadic. High resolution banding of eight patients showed normal G-banded prometaphase chromosomes (68). Occasionally, inconsistent chromosomal abnormalities have been reported (43). Eleven MZ concordant and two MZ discordant twins have been recorded (1,5,18,20,22,28,51,59, MM Cohen, personal observation). Affected sibs were noted by Johnson (26), Pfeiffer (38), and Halamek et al (13a). The parents in one instance (16) were first cousins. Another pair was reported by Der Kaloustian (10), but broad thumbs and halluces were not present. Consanguinity was noted by Der Kaloustian (10) and by Padfield et al (37). Several family histories are noteworthy. One patient reported by Rubinstein and Taybi (47) had a father and five paternal relatives with broad thumbs. By 1969, Rubinstein (45) noted seven other patients with familial evidence of broad thumbs. Spencer (57) reported a single instance of presumed Rubinstein–Taybi syndrome without broad thumbs and halluces. Hennekam et al (20) recorded a mother and son affected with classic Rubinstein–Taybi syndrome. Cotsirilos et al (8) reported a mother and two sibs affected with a Rubinstein–Taybi-like disorder. Six other relatives of the mother were reported to have broad thumbs and halluces. They suggested that the lack of data indicating vertical transmission of Rubinstein–Taybi syndrome might be explained by reduced genetic fitness resulting from severe mental retardation. They further suggested that sporadic instances might possibly be considered new dominant mutations. The family with broad thumbs and mental deficiency reported by Robinow (39) probably represents a separate entity.

**Systemic manifestations.** The natural history has been particularly well documented by Stevens et al (58). Various findings are listed in Table 10–5. Polyhydramnios has been reported in 40% (58).

**Growth.** Length, weight, head circumference at birth are between the 25th and 50th centiles. Average birth length is 49.3 cm with a range of 43.9–53.3 cm. Average birth weight is 3.09 kg with a range of 2.05–4.28 kg. Mean head circumference is 33.4 cm with a range of 29–38 cm (58). Often there is poor weight gain during infancy (58).

Values for final height attainment are not available to date (58a).

**Performance and central nervous system.** Global mental deficiency is characteristic with the most severe delay in expressive speech. The average IQ is 51 (range 33–72) ($n=37$). There is no apparent behavioral phenotype. However, affected individuals tend to be loving

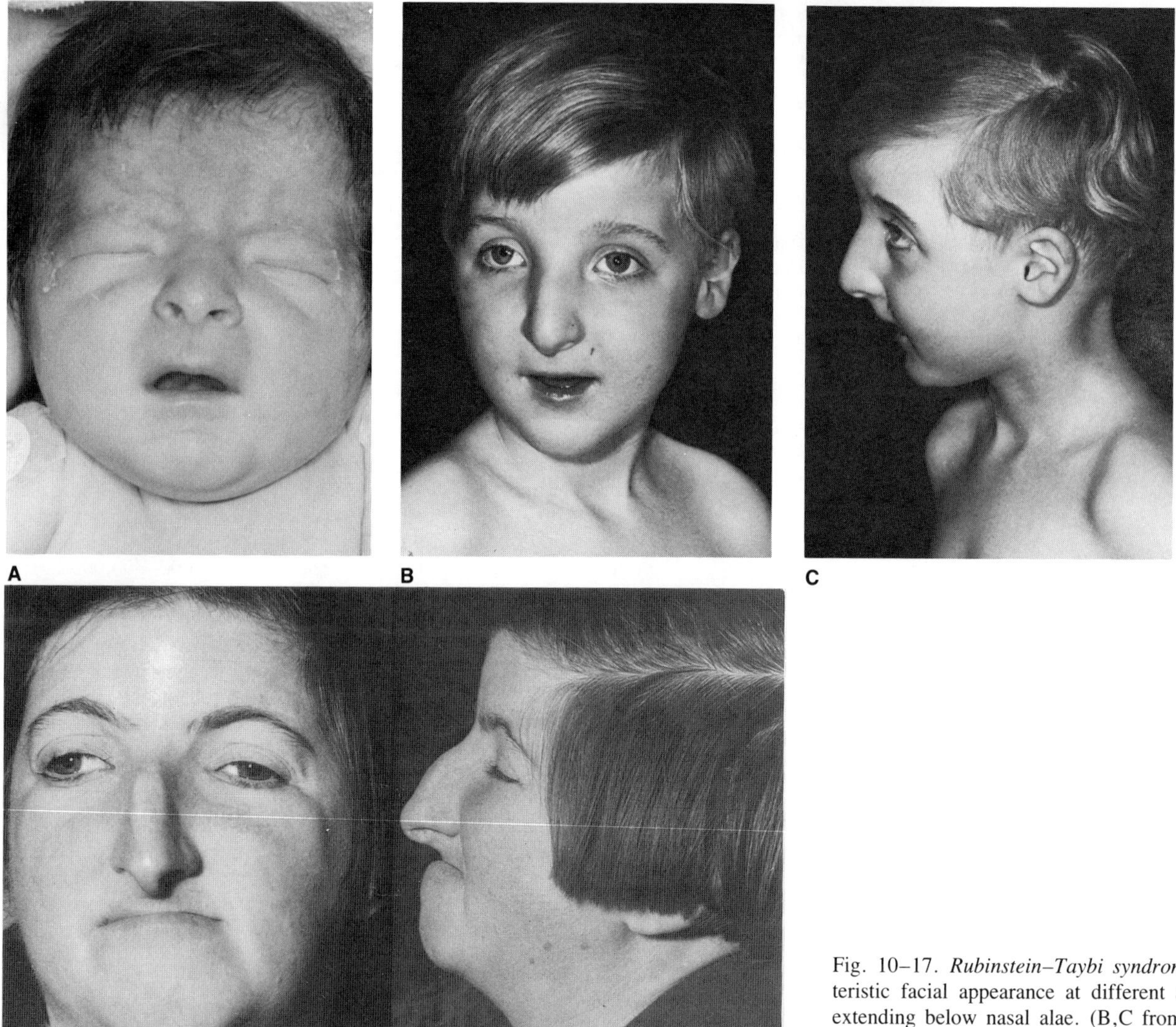


Fig. 10–17. *Rubinstein–Taybi syndrome.* (A–E) Characteristic facial appearance at different ages. Note septum extending below nasal alae. (B,C from *R Weiland,* Arch Kinderheilkd **179:**78, 1969. D,E from *JM Berg et al,* J Ment Defic Res **10:**204, 1966.)

and friendly, although maladaptive behavior has also been noted. Attention span is short (58).

Electroencephalographic abnormalities, seizures, absence of the corpus callosum, and hyperactive deep-tendon reflexes have been noted (30,36,45–47,58).

**Craniofacial features.** The facial appearance is striking, with microcephaly, prominent forehead, downslanting palpebral fissures, epicanthal folds, strabismus, broad nasal bridge, beaked nose with the nasal septum extending below the alae, highly arched palate, and mild micrognathia (Fig. 10–17A–E). The features are recognizable in the newborn (11a). Facial changes with time have been studied by Allanson (1). Grimacing or unusual smile has been observed frequently. Other findings may include long eyelashes, nasolacrimal duct obstruction, ptosis of eyelids, juvenile glaucoma, refractive error, and minor abnormalities in the shape, position, and degree of rotation of the ears (3–5,7,12,31,32,43,46,47). Low-frequency abnormalities have included bifid uvula, submucous palatal cleft, bifid tongue, macroglossia, short lingual frenum, natal teeth, and thin upper lip (7,24,45,50,58). Talon cusps (markedly enlarged cingulum on maxillary incisor teeth) have been observed in over 90% (14,20a,28a,62). (Fig. 10–20).

**Hands and feet.** Broad thumbs and great toes have been present in almost all reported cases. In most instances, the terminal phalanges of the fingers are also broad. Clinodactyly of the fifth fingers and overlapping of the toes are present in over half the cases. Angulation deformities of the thumbs and halluces, together with abnormally shaped proximal phalanges, occur in about 35% (58,67). Abnormally shaped first metatarsals and duplication of the proximal or distal phalanx of the halluces have also been reported. Rarely, hexadactyly of the feet, partial cutaneous syndactyly involving the toes, and absence of the distal phalanx of the hallux have been noted (5–7,12,23,44–47,58,60,61). (Figs. 10–18 and 10–19).

Alterations in the frequency of various fingerprint patterns have been observed, but findings have been inconsistent. Increased frequencies of loops, whorls, or arches have been reported. Significant dermatoglyphic findings have included an increased frequency of thenar, interdigital, and hypothenar patterns. Single palmar creases have been observed in many cases. Thumb tip triradius, thumb double pattern, distally placed axial triradius, deep plantar crease, and large hallucal loop with laterally displaced *f* triradius with or without associated *e* triradius have also been reported (9,12,15,21,40,44–47,54,55).

**Skeletal system.** Growth retardation and delayed bone age are common. Large anterior fontanel or delay in its closure, large foramen magnum, and parietal foramens have been reported in some cases. Other skeletal anomalies have included pectus excavatum, other sternal abnormalities, rib defects, scoliosis, kyphosis, lordosis, spina bi-

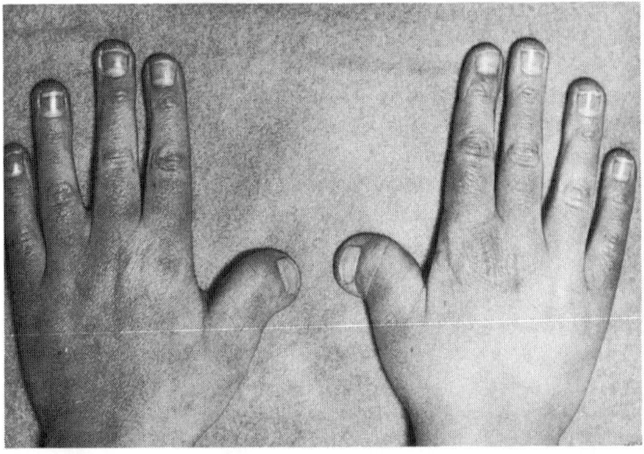

A

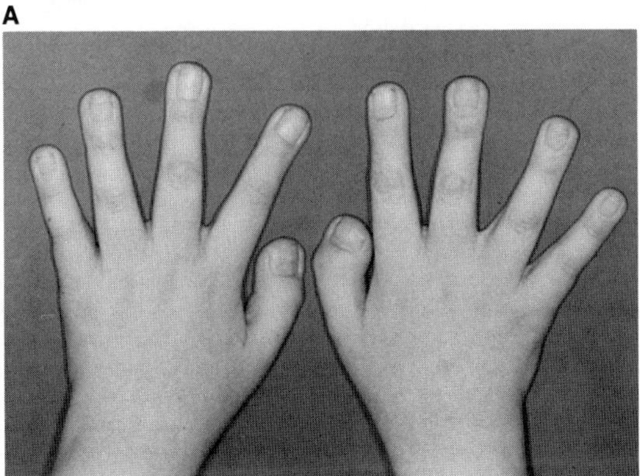

B

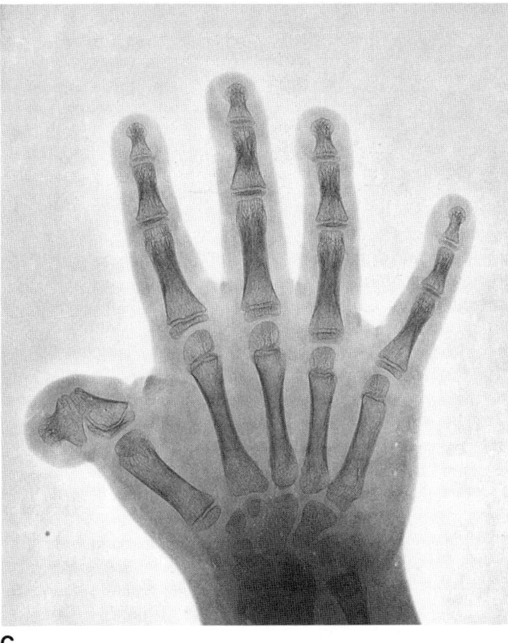

C

Fig. 10–18. *Rubinstein–Taybi syndrome.* (A) Broad and radially deviated thumbs. (B) Compare with A. (C) Roentgenogram showing broad short terminal phalanges and triangular proximal phalanges of thumbs (A from *JM Berg et al,* J Ment Defic Res **10:**204, 1966. C from *A Neuhold et al,* Roefo 150:49, 1989.)

fida, flat acetabular angles, flaring of ilia, and notched ischia (45,46). Various other low-frequency anomalies have been discussed by Rubinstein (45) and Robson et al (41).

The gait is commonly stiff. Hypotonia, lax ligaments, and hyperextensible joints have also been noted (45,46).

**Genitourinary system.** Incomplete or delayed descent of the testes has been reported in probably all males. Anomalies of the urinary tract, including duplication of the kidney and ureter, renal agenesis, and other abnormalities, have been recorded in a number of cases. Urinary tract infections occur in about 25% (58). Rarely, angulated penis and hypospadias have been noted (45).

Table 10–5. Medical problems of Rubinstein–Taybi syndrome[a]

| Feature | Percentage ($n = 50$) |
|---|---|
| Birth history | |
| Gestational length | 39.8 weeks (range 32–44 weeks) |
| Polyhydramnios | 39 |
| Infancy history | |
| Respiratory problems | 38 |
| Feeding problems | 88 |
| Constipation | 74 |
| Poor weight gain | 80 |
| Medical history | |
| Visual problems | 84 |
| Strabismus | (48) |
| Refractive error | (38) |
| Astigmatism | (18) |
| Other | (4) |
| Cataracts | 6 |
| Glaucoma | 2 |
| Coloboma | 6 |
| Tear duct obstruction | 30 |
| Ptosis | 41 |
| Hearing loss | 24 |
| Frequent middle ear infections | 52 |
| Congenital heart defects | 38 |
| PDA | (18) |
| VSD | (16) |
| ASD | (14) |
| Coarctation | (4) |
| Pulmonic stenosis | (4) |
| Urinary tract infection | 24 |
| Keloids or hypertrophic scarring | 28 |
| Severe constipation | 48 |

[a]Based on CA Stevens et al (58).

**Other findings.** A variety of congenital heart defects have been found in about 35% (58). Abnormal lung lobulation, supernumerary nipples, nevus flammeus of the forehead, nape, or back, hirsutism, and other abnormalities have been reported (5,6,17,26,45,58). Nasopharyngeal rhabdomyosarcoma (56), intraspinal neurilemoma (49), angioblastic meningioma (65), and acute leukemia (27) have been recorded. Keloids or hypertrophic scarring has been described in several cases (15a,29a,42,51a,58).

**Differential diagnosis.** Although many components of the syndrome may occur as isolated findings or as features of various other syndromes, the overall pattern of anomalies is sufficiently distinctive to permit diagnosis in most instances. The authors have seen numerous patients who have exhibited only some of the stigmata of the Rubinstein–Taybi syndrome. Some of these may represent incomplete forms. Differential diagnosis can be a problem in the newborn period. Occasionally, some cases have been confused with *de Lange syndrome* (29) or with *trisomy 13* (63). Broad thumbs may be observed in *Apert syndrome* and *Pfeiffer syndrome,* and short thumbs and fingers are seen in Type D brachydactyly.

### References (Rubinstein–Taybi syndrome)

1. Allanson J: The changing face: Rubinstein–Taybi syndrome. Ninth Annual David W. Smith Workshop on Malformations and Morphogenesis, Oakland, August 7–10, 1988.

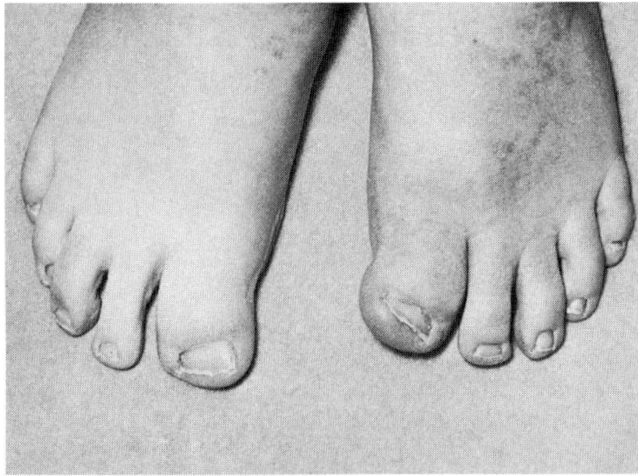

A

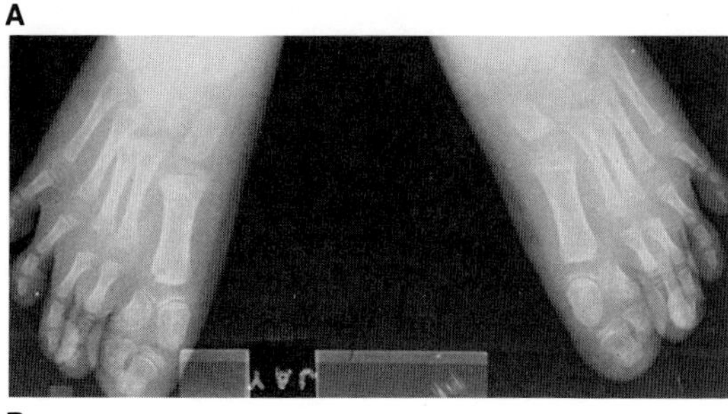

B

Fig. 10–19. *Rubinstein–Taybi syndrome.* (A) Broad and tibially deviated halluces. (B) Roentgenogram showing broad great toes with duplication of halluces (A from *JM Berg et al,* J Ment Defic Res **10:**204, 1966.)


2. Baraitser M, Preece MA: The Rubinstein–Taybi syndrome: Occurrence in two sets of identical twins. *Clin Genet* **23**:318–320, 1983.

3. Behrens-Baumann W: Augensymptome beim Rubinstein–Taybi-Syndrom. *Klin Mbl Augenheilk* **171**:126–135, 1977.

4. Berg JM et al: On the association of broad thumbs and first toes with other physical peculiarities and mental retardation. *J Ment Defic Res* **10**:204–220, 1966.

5. Buchinger G, Ströder J: Rubinstein–Taybi-Syndrom bei wahrscheinlich eineiigen Zwillingen und drei weiteren Kindern. *Klin Pädiatr* **185**:296–307, 1973.

6. Coffin GS: Brachydactyly, peculiar facies, and mental retardation. *Am J Dis Child* **108**:351–359, 1964.

7. Coffin GS: Three retarded children with unusual face and hands: A variant of the wide-thumbs syndrome? Symposium 10: Rubinstein–Taybi syndrome, in *Proceedings, First Congress of the International Association for the Scientific Study of Mental Deficiency,* Richards BW (ed), Montpellier, France, 1967, pp 600–605.


Fig. 10–20. *Rubinstein–Taybi syndrome.* Talon cusps of four maxillary permanent incisors, marked distal marginal ridge on left lateral incisor. (From *DG Gardner* and *SS Girgis,* Oral Surg **47:**519, 1979.)

8. Cotsirilos P et al: Dominant inheritance of a syndrome similar to Rubinstein–Taybi. *Am J Med Genet* **26**:85–93, 1987.

9. Davison BCC et al: Mental retardation with facial abnormalities, broad thumbs and toes and unusual dermatoglyphics. *Dev Med Child Neurol* **9**:588–593, 1967.

10. Der Kaloustian VM: The Rubinstein–Taybi syndrome: A clinical and muscle electron microscopic study. *Am J Dis Child* **124**:897–902, 1972.

11. De Toni T et al: La sindrome de Rubinstein–Taybi, presentazion di due nuovi casi. *Min Pediatr* **34**:765–770, 1982.

12. Filippi G: The Rubinstein–Taybi syndrome: Report of 7 cases. *Clin Genet* **3**:303–319, 1972.

13. Gambon R, Amato M: Rubinstein–Taybi syndrome in the neonate. *Helv Paediatr Acta* **39**:279–283, 1984.

14. Gardner DG, Girgis SS: Talon cusps: A dental anomaly in the Rubinstein–Taybi syndrome. *Oral Surg* **47**:519–521, 1979.

15. Giroux J, Miller JR: Dermatoglyphics of the broad thumb and great toe syndrome. *Am J Dis Child* **113**:207–209, 1967.

15a. Goodfellow A et al: Rubinstein–Taybi syndrome and spontaneous keloids. *Clin Exp Dermatol* **5**:369–370, 1980.

16. Halamek LP et al: Rubinstein–Taybi syndrome in two siblings of a first cousin marriage. Ninth Annual David W. Smith Workshop on Malformations and Morphogenesis, Oakland, August 7–10, 1988.

17. Hall BD: Keloids in Rubinstein–Taybi syndrome. Ninth Annual David W. Smith Workshop on Malformations and Morphogenesis, Oakland, August 7–10, 1988.

18. Hayem F et al: Le syndrome de Rubinstein–Taybi, discussion des formes incompletes et familiales. *Pédiatrie* **25**:89–102, 1970.

19. Hennekam RCM et al: Rubinstein–Taybi syndrome in the Netherlands. Ninth Annual David W. Smith Workshop on Malformations and Morphogenesis, Oakland, August 7–10, 1988.

20. Hennekam RCM et al: Rubinstein–Taybi syndrome in a mother and son. *Eur J Pediatr* **148**:439–441, 1989.

20a. Hennekam RCM, Van Doorne JM: Oral aspects of Rubinstein–Taybi syndrome. *Am J Med Genet* (in press).

21. Herrmann J, Opitz JM: Dermatoglyphic studies in a Rubinstein–Taybi patient, her unaffected dizygous twin sister and other relatives. *Birth Defects* **5**(2):22–24, 1969.

22. Holthusen W, Panteliadis C: Rubinstein–Taybi-Syndrom bei frühgeborenen (wahrscheinlich eineiigen) Zwillingen. *Mschr Kinderheilk* **119**:523–527, 1971.

23. Jancar J: Rubinstein–Taybi's syndrome. *J Ment Defic Res* **9**:265–270, 1965.

24. Jeliu G, Saint-Rome G: Le syndrome de Rubinstein–Taybi: A propos d'une observation. *Union Méd Can* **96**:22–29, 1967.

25. Job JC et al: Etudes sur les nanismes constitutionnels. II. Le syndrome de Rubinstein et Taybi. *Ann Pédiatr* **11**:646–650, 1964.

26. Johnson CF: Broad thumbs and broad great toes with facial abnormalities and mental retardation. *J Pediatr* **68**:942–951, 1966.

27. Jonas DM et al: Rubinstein–Taybi syndrome and acute leukemia. *J Pediatr* **92**:851–852, 1978.

28. Kajii T et al: Monozygotic twins discordant for Rubinstein–Taybi syndrome. *J Med Genet* **18**:312–314, 1981.

28a. Kinirons MJ: Oral aspects of Rubinstein–Taybi syndrome. *Br Dent J* **154**:46–47, 1983.

29. Kroth H von: Cornelia de Lange-Syndrom I bei Zwillingen. *Arch Kinderheilkd* **173**:273–283, 1966.

29a. Kurwa AR: Rubinstein–Taybi syndrome and spontaneous keloids. *Clin Exp Dermatol* **4**:251–254, 1978.

30. Kushnick T: Brachydactyly, facial abnormalities and mental retardation. Rubinstein–Taybi syndrome. *Am J Dis Child* **111**:96–98, 1966.

31. Levy NS: Juvenile glaucoma in the Rubinstein–Taybi syndrome. *J Pediatr Ophthalmol* **13**:141–143, 1976.

32. Lowry RB: Rubinstein–Taybi syndrome—A 14 year followup of a case with normal intelligence. Ninth Annual David W. Smith Workshop on Malformations and Morphogenesis, Oakland, August 7–10, 1988.

33. McArthur RG: Rubinstein–Taybi syndrome: Broad thumbs and great toes, facial abnormalities and mental retardation. A presentation of three cases. *Can Med Assoc J* **96**:462–466, 1967.

34. Michael J et al: Pouce bot arque en forte abduction-extension et autres symptomes concomitants. *Rev Chir Orthop* **43**:142–146, 1957.

35. Naveh Y, Friedman A: A case of Rubinstein–Taybi syndrome. *Clin Pediatr* **15**:779–783, 1976.

36. Neuhäuser G: Pneumoencephalographic findings in the Rubinstein–Taybi syndrome, Symposium 10: Rubinstein–Taybi syndrome, in *Proceedings, First Congress of the International Association for the Scientifc Study of Mental Deficiency,* Richards BW (ed), Montpellier, France, 1967, pp 615–617.

37. Padfield CJ et al: The Rubinstein–Taybi syndrome. *Arch Dis Childh* **43**:94–106, 1967.

38. Pfeiffer RA: Rubinstein–Taybi syndrom bei wahrscheinlich eineiigen Zwillingen. *Humangenetik* **6**:84–87, 1968.

39. Robinow M: A familial syndrome of mental deficiency and broad thumbs. *Birth Defects* **5**(2):42, 1969.

40. Robinson GC et al: Broad thumbs and toes and mental retardation: Unusual dermatoglyphic observations in two individuals. *Am J Dis Child* **111**:287–290, 1966.

41. Robson MJ et al: Cervical spondylolisthesis and other skeletal abnormalities in Rubinstein–Taybi syndrome. *J Bone Jt Surg* **62**(B):297–299, 1980.

42. Rohlfing B et al: Rubinstein–Taybi syndrome. *Am J Dis Child* **121**:71–74, 1971.

43. Roy FH et al: Ocular manifestations of the Rubinstein–Taybi syndrome: Case report and review of the literature. *Arch Ophthalmol* **79**:272–278, 1968.

44. Rubinstein JH: A syndrome of broad thumbs and first toes, mental retardation, and characteristic facial features—a follow-up report, Symposium 10: Rubinstein–Taybi syndrome, in *Proceedings, First Congress of the International Association for the Scientific Study of Mental Deficiency,* Richards BW (ed), Montpellier, France, 1967, pp 589–595.

45. Rubinstein JH: The broad thumbs syndrome—progress report 1968. *Birth Defects* **5**(2):25–41, 1969.

46. Rubinstein JH: Broad thumb-hallux syndrome, In *Birth Defects Compendium and Atlas,* Bergsma D (ed), National Foundation, Williams & Wilkins, 1973, pp 218–219.

47. Rubinstein JH, Taybi H: Broad thumbs and toes and facial abnormalities. *Am J Dis Child* **105**:588–603, 1963.

48. Rubinstein JH, Taybi H: The historical aspects of Rubinstein–Taybi syndrome. Ninth Annual David W. Smith workshop on Malformations and Morphogenesis, Oakland, August 7–10, 1988.

49. Russell NA et al: Intraspinal neurilemoma in association with the Rubinstein–Taybi syndrome. *Pediatrics* **47**:444–446, 1971.

50. Salmon MA: The Rubinstein–Taybi syndrome: A report of two cases. *Arch Dis Childh* **43**:102–106, 1968.

51. Schinzel A et al: Monozygotic twinning and structural defects. *J Pediatr* **95**:921–930, 1979.

51a. Selmanowitz VJ, Stiller MJ: Rubinstein–Taybi syndrome: Cutaneous manifestations and colossal keloids. *Arch Dermatol* **117**:504–506, 1981.

52. Simpson NE: The Rubinstein–Taybi syndrome: Familial and dermatoglyphic data. *Am J Hum Genet* **25**:225–229, 1973.

53. Simpson NE: The Rubinstein–Taybi syndrome: Chromosomal studies. *Am J Hum Genet* **25**:230–236, 1973.

54. Simpson NE, Brissenden JE: The Rubinstein–Taybi syndrome. *Am J Hum Genet* **25**:225–229, 1973.

55. Smith GF, Berg JM: Dermatoglyphics in Rubinstein–Taybi syndrome, in *Proceedings, First Congress of the International Association for the Scientific Study of Mental Deficiency,* Richards BW (ed), Montpellier, France, 1967, pp 606–612.

56. Sobel RA, Woerner S: Rubinstein–Taybi syndrome and nasopharyngeal rhabdomyosarcoma. *J Pediatr* **99**:1000–1001, 1981.

57. Spencer DA: Partial Rubinstein–Taybi syndrome. *Lancet* **2**:713–714, 1971.

58. Stevens CA et al: Rubinstein–Taybi syndrome. A natural history study. Ninth Annual David W. Smith Workshop on Malformations and Morphogenesis, Oakland, August 7–10, 1988.

58a. Stevens CA et al: Growth in the Rubinstein–Taybi syndrome. *Am J Med Genet* (in press).

59. Takeuchi M: Rubinstein's syndrome in two siblings. *Gunmma J Med Sci* **15**:17–22, 1966.

60. Taybi H, Rubinstein JH: Broad thumbs and toes, and unusual facial features: A probable mental retardation syndrome. *Am J Roentgenol* **93**:362–366, 1965.

61. Taybi H: Broad thumbs and great toes, facial abnormalities, and mental retardation syndrome, Symposium 10: Rubinstein–Taybi syndrome, in *Proceedings, First Congress of the International Association for the Scientific Study of Mental Deficiency,* Richards BW (ed), Montpellier, France, 1967, pp 596–599.

62. Tomich CE: Talon cusps: A dental anomaly in the Rubinstein–Taybi syndrome. *Oral Surg* **47**:519–521, 1979.

63. True CW, Rubenstein JH: Pathological findings in a case of the Rubinstein–Taybi syndrome, Symposium 10: Rubinstein–Taybi syndrome, in *Proceedings, First Congress of the International Association for the Scientific Study of Mental Deficiency,* Richards BW (ed), Montpellier, France, 1967, pp 613–614.

64. van Gelderen HH et al: Trisomy G/normal mosaics in non-mongoloid mentally deficient children. *Acta Paediatr Scand* **56**:517–525, 1967.

65. Wilson GN et al: Intracranial angioblastic meningioma and an aged appearance in an adult female with Rubinstein–Taybi syndrome. Ninth Annual David W. Smith Workshop on Malformations and Morphogenesis, Oakland, August 7–10, 1988.

66. Wilson MG: Rubinstein–Taybi and D trisomy syndromes. *J Pediatr* **73**:404–408, 1968.

67. Wood VE, Rubinstein JH: Surgical treatment of the thumb in the Rubinstein–Taybi syndrome. *J Hand Surg* **12B**:166–172, 1987.

68. Wulfsberg, EA et al: High resolution chromosome banding in the Rubinstein–Taybi syndrome. *Clin Genet* **23**:35–37, 1983.

## Seckel syndrome and other microcephalic primordial dwarfisms

Seckel syndrome is characterized by marked intrauterine and postnatal growth retardation, microcephaly, mental retardation, and typical facial appearance with beaklike protrusion of the midface (Figs. 10–21 and 10–22). In the past, the condition was overdiagnosed, and both Majewski and Goecke (22) and Thompson and Pembrey (42) have called attention to this. Of Seckel's two original patients (37), only case 1 fits the proper diagnostic criteria, case 2 being different enough to constitute some other condition.

Of the well over 60 cases of Seckel syndrome reported to date, two-thirds do not meet the proper diagnostic criteria; only 20 cases from the literature qualify as examples [1,2(case 1),5,7,10,15,17(case 2), 18,22,26,30(case 2),34,37(case 1),40,41,42(cases 1–3)] In addition, Schönenberg (35) reported a patient with only mild growth retardation who, on reexamination at approximately 7 years of age (22), exhibited classic features of the syndrome.

The male-to-female sex ratio is 9:11, and both sexes are equally severely affected. All parents have been normal and consanguinity has not been established in any instance to date. Affected sibs have been reported by six authors (1,5,8,15,34,42). The syndrome has autosomal recessive inheritance.

A great many cases from the literature do not qualify as examples of Seckel syndrome [4(cases 1,2), 5(case 3), 7,9,11,12,13(cases 1,2),

Fig. 10–21. *Seckel syndrome.* Proportionate short stature, microcephaly, curved nose, and small mandible in two affected sibs. (From *B Boscherini et al,* Eur J Pediatr **137**:237, 1981.)

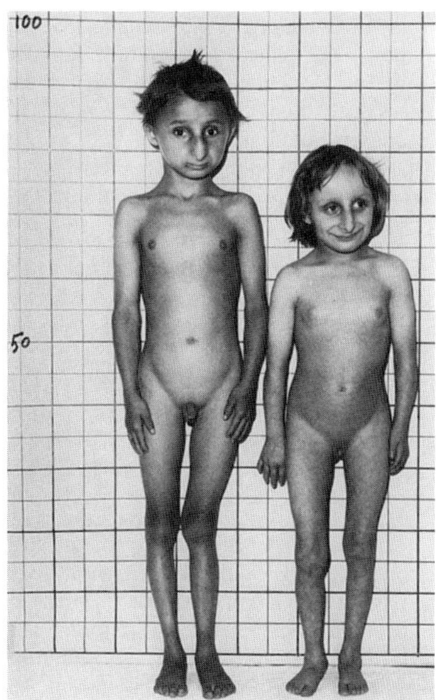

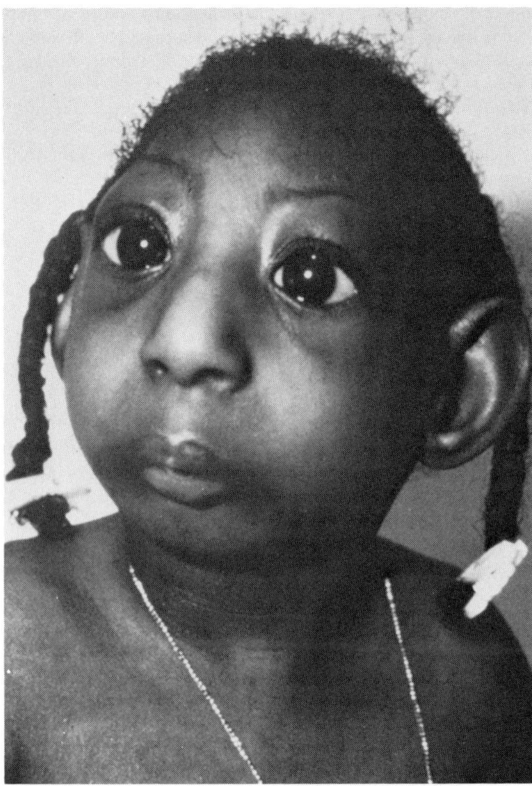

Fig. 10–22. *Seckel syndrome.* Characteristic facies, with small head circumference and micrognathia, lending prominence to the midface. (Courtesy of *JJ Sauk,* Baltimore, Maryland.)

Talbe 10–6. Major manifestations of Seckel syndrome[a]

| Striking features | Frequency |
|---|---|
| Growth | |
| Prenatal-onset growth deficiency | 17/17 |
| Postnatal growth deficiency | 17/17 |
| Delayed osseous maturation | 12/13 |
| Performance | |
| Mental deficiency | 17/17 |
| Severe mental deficiency | 6/16 |
| Craniofacial | |
| Microcephaly | 17/17 |
| Craniosynostosis | 7/13 |
| Relatively large eyes | 10/16 |
| Downslanting palpebral fissures | 7/11 |
| Malformed ears | 10/12 |
| Prominent curved nose | 17/17 |
| Micrognathia | 17/17 |
| Highly arched palate | 4/10 |
| Cleft palate | 2/17 |
| Enamel hypoplasia | 6/10 |
| Limbs | |
| Clinodactyly, fifth finger | 8/8 |
| Dislocation of radial head | 3/6 |
| Hip dysplasia | 5/9 |
| Genitalia | |
| Cryptorchidism | 3/4 |
| Clitoromegaly | 3/7 |
| Skin | |
| Hirsutism | 3/10 |

[a]Based on an analysis of 17 cases of Majewski and Goecke (22).

14,16,20(cases 1–3),21(cases 1–3),27(cases 1,2),29,30(cases 1,3),31, 32,36,37(cases 2–4,6–15),38,39,42(cases 4,5),43,47].

**Growth and development.** Striking features of Seckel syndrome are listed in Table 10–6. The mean birth weight of affected newborns at term is approximately 1543 g (range 1000–2055 g). Postnatal growth deficiency is, on the average, 7.1 SD below the mean, with a range of −5.1 to −13.5 SD. Head circumference is as retarded as height in half the cases; in the remaining instances, head circumference is the singularly most retarded parameter. All patients are mentally retarded with half having an IQ below 50. Moderate mental deficiency has been observed in some instances (22).

In addition to delayed osseous maturation, the phalanges exhibit ivory epiphyses and cone-shaped epiphyses in the proximal phalanges. The carpal bones are relatively small and disharmony occurs in maturation between the carpals and phalanges (33).

**Craniofacial features.** Craniofacial features are striking with severe microcephaly, receding forehead, and micrognathia, lending prominence to the midface and curved nose (Fig. 10–22). Synostosis of cranial sutures occurs in approximately 50%. Other craniofacial features may include facial asymmetry, relatively large eyes, downslanting palpebral fissures, lobeless ears, highly arched or cleft palate, enamel hypoplasia, crowded teeth, and class II malocclusion (15,22,42).

**Other findings.** Limb anomalies include clinodactyly of the fifth finger, abnormal finger flexion creases, dislocation of the radial head, and hip dysplasia. Other findings may sometimes include cryptorchidism, clitoromegaly, and hirsutism (22). Butler et al (7a) reported pancytopenia in one patient and chromosome instability in two. Findings not previously emphasized have been discussed by Thompson and Pembrey (42).

**Differential diagnosis.** The craniofacial features of Seckel syndrome allow differentiation from other syndromes of growth deficiency with microcephaly such as *Dubowitz syndrome, fetal alcohol syndrome, trisomy 18 syndrome, de Lange syndrome, Bloom syndrome,* and Fanconi syndrome.

Seckel syndrome should also be distinguished from three defined types of osteodysplastic primordial dwarfism. Type I or brachymelic primordial dwarfism (2,23) has low birth weight, microcephaly, prominent nose, micrognathia, sparse scalp hair, small anterior fontanel, short neck, low-set ears, and short bowed humeri and femora (Fig. 10–23A–D). Type II (2,7,24,45) has pre- and postnatal growth failure, microcephaly, premature suture closure, curved nose, receding chin, large eyes, dysplastic ears, disproportionately short forearms and legs, brachymesophalangy, brachymetacarpy I, V-shaped flaring of the distal femoral metaphyses, triangularly shaped distal femoral epiphyses, high narrow pelvis, proximal femoral epiphysiolysis, and coxa vara (Fig. 10–24A–C). Type III (25) has intrauterine growth retardation, alopecia, microcephaly, receding forehead and chin, large eyes, large prominent nose, elongated clavicles, cleft cervical vertebral arches, lumbar platyspondyly, hypoplastic iliac wings and acetabulae, and horizontal distal margin of the os ilium (Fig. 10–25A,B). Winter et al (46) and Haan et al (14a) reported patients with features of both Types I and III, and suggested that the osteodysplastic types of primordial dwarfism needed to be further delineated.

Verloes et al (44a) reported a Type II-like microcephalic osteodysplasia in sibs, although no alterations of pelvic shape or in the acetabula were observed. Majoor-Krakauer et al (25a) reported a Seckel-like variant with severe microcephaly, extreme micrognathia, hydrocephaly, abnormal gyral pattern, and delayed cortical neuronal migration. Toriello et al (42a) described two sibs with microcephalic primordial dwarfism, mental deficiency, cataracts, enamel hypoplasia, immune deficiency, and generalized delay in ossifications.

Patients reported as examples of Seckel syndrome, but with unusual manifestations [22 (Table 2)], also suggest further unsorted heteroge-

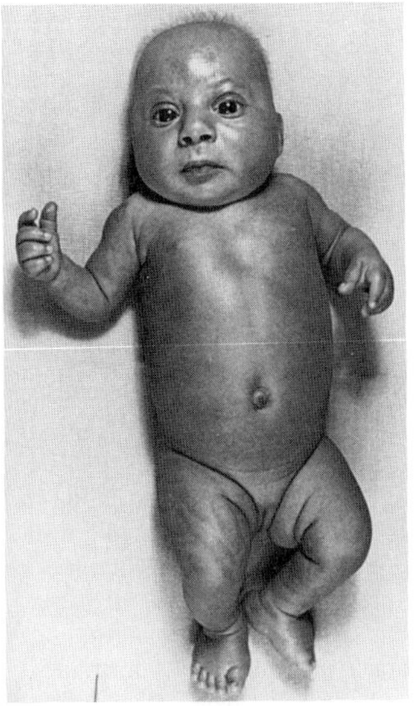

A

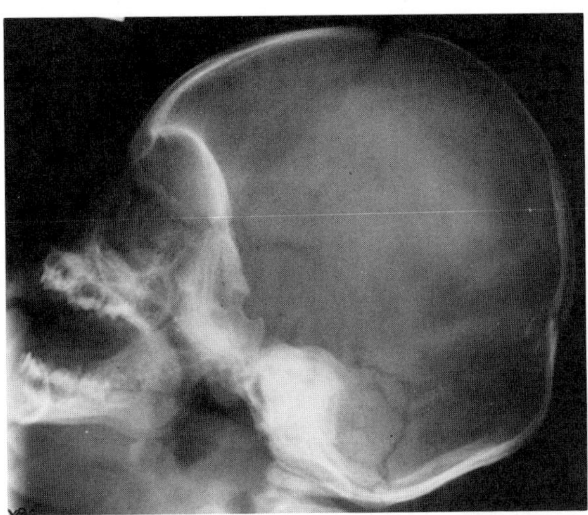

B

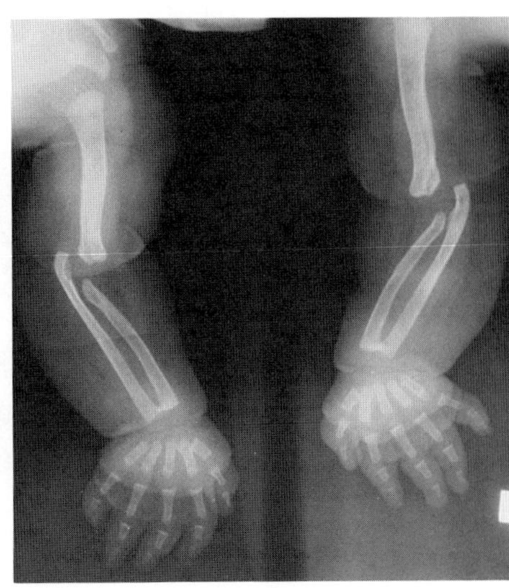

C

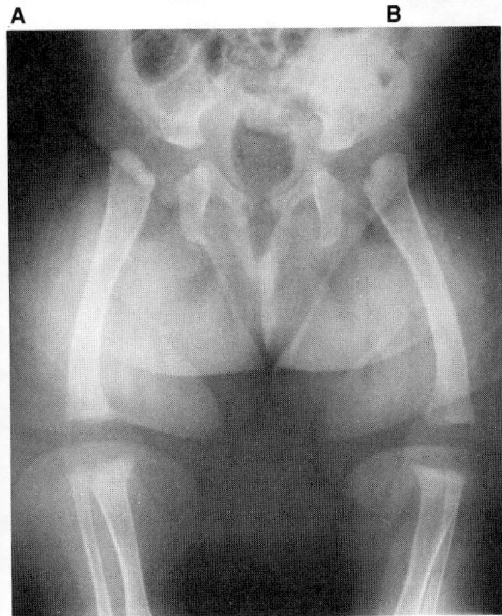

D

Fig. 10–23. *Microcephalic primordial dwarfism (osteodysplastic Type I).* (A) Microcephaly, hypotrichosis, prominent eyes, large fleshy nose, small dysplastic ears, and micrognathia. (B) Radiograph showing microcephaly, steep skull base. (C) Radiograph showing shortened humeri, bilateral dislocation at elbows, short metacarpals, absence of secondary ossification centers in humeral heads, carpals, and capitellum. (D) Hypoplasia of lower ilia, widened iliac angle, lateral bending of femoras, absent epiphyses at knees. [A courtesy of *F Majewski* and *J Spranger,* Mainz, Germany. B courtesy of *HN Bass* and *RS Sparkes,* Los Angeles, California. C,D from *HN Bass,* Syndrome Ident **3**(2):12, 1975.]

neity. Fitch et al (12) reported a child with microcephaly, premature graying, loss of scalp hair, eyelid ptosis, beaklike nose, micrognathia, low-set ears, wrinkled palmar skin, and cryptorchidism. However, growth deficiency was of postnatal onset. McKusick (personal communication, 1974) examined similarly affected sibs and suggested autosomal recessive inheritance.

The so-called Taybi–Lindner cephaloskeletal dysplasia may be the same as Type III (18a,41a).

### References (Seckel syndrome and other microcephalic primordial dwarfisms)


1. Aarons PH: Vogelkopdwergen. *Maandschr Kindergeneesk* **32**:384–394, 1964.

2. Anoussakis CH et al: Les nanismes congénitaux avec dysmorphie. II. Le nanisme congénital à tête d'oiseau (type Virchow-Seckel). *Pédiatrie* **29**:261–267, 1974.

3. Bass HN et al: Case report 33. *Syndrome Ident* **3**:12–14, 1975.

4. Bixler D, Antley RM: Microcephalic dwarfism in sisters. *Birth Defects* **10**(7):161–165, 1974.

5. Black J: Low birth weight dwarfism. *Arch Dis Childh* **36**:633–644, 1961.

6. Boscherini B et al: Intrauterine growth retardation: A report of two cases with bird-headed appearance, skeletal changes and peripheral GH resistance. *Eur J Pediatr* **137**:237–242, 1981.

7. Brizard J et al: Sur un cas de nanisme extrême à début intra-utérin vraisemblablement du type Seckel. *Ann Pédiatr* **20**:655–660, 1973.

7a. Butler MG et al: Do some patients with Seckel syndrome have hematological problems and/or chromosome breakage? *Am J Med Genet* **27**:645–649, 1987.

8. Cervenka J et al: Seckel's dwarfism: Analysis of chromosome breakage and sister chromatid exchanges. *Am J Dis Child* **133**:555–556, 1979.

9. Cordier J et al: Anomalies oculaires au cours du syndrome de Seckel (nanisme congénital à tête d'oiseau). *Bull Soc Ophtalmol Fr* **72**:119–122, 1972.

10. de la Cruz F: Bird headed dwarf: A case report. *Am J Ment Defic* **68**:54–62, 1963.

11. Dutau G et al: Syndrome malformativ associant un nanisme à des anomalies de la face et des extremités: forme de transition entre le syndrome de Franceschetti et le nanisme de Seckel. *Rev. Méd Toulouse* **7**:457–466, 1971.

12. Fitch N et al: A form of birdheaded dwarfism with features of premature senility. *Am J Dis Child* **120**:260–264, 1970.

13. Frijns JP, van den Berghe H: Familial bird headed dwarfism. *Acta Paediatr Belg* **29**:121–122, 1976.

14. Gellis SS et al: Picture of the month: Bird headed dwarfs (Seckel dwarfism). *Am J Dis Child* **114**:583–584, 1967.

14a. Haan EA et al: Osteodysplastic primordial dwarfism: Report of a fur-

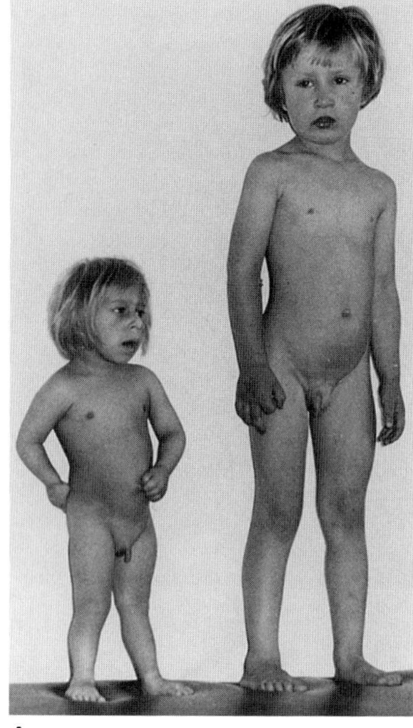
A

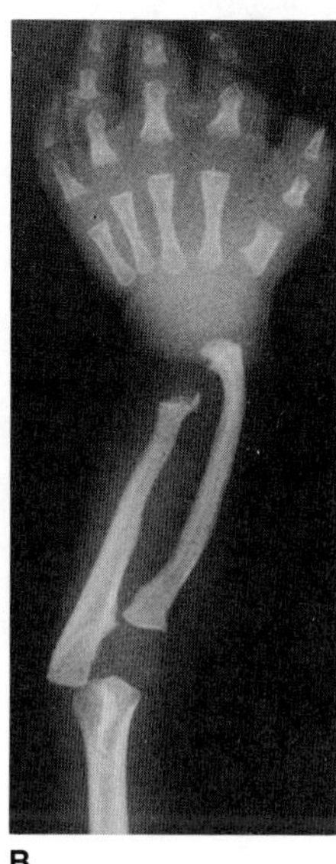
B

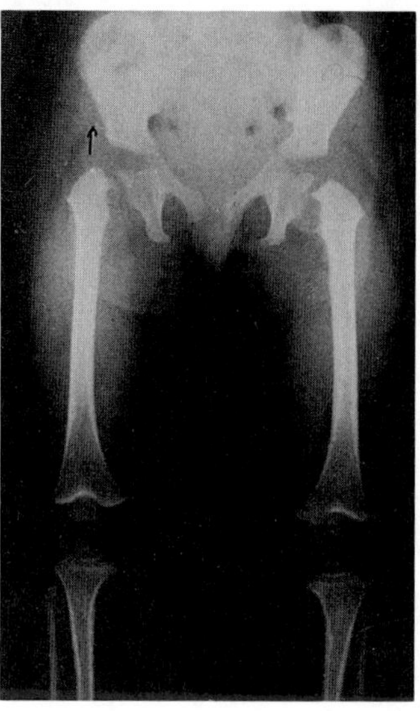
C

Fig. 10–24. *Microcephalic primordial dwarfism (osteodysplastic Type II).* (A) Patient next to age-matched control child. In addition to mental retardation, marked growth retardation, microcephaly, small forehead, moderately prominent nose, and micrognathia. (B) Note disproportionate shortness of forearms, brachymesophalangy, brachmetacarpy I. (C) Observe V-shaped flare of distal femoral metaphyses, triangular distal femoral epiphyses, proximal femoral epiphysiolysis, and coxa vara. (From *F Majewski et al,* Am J Med Genet **12:**23, 1982.)


ther case with manifestations similar to those of types I and III. *Am J Med Genet* **33**:224–227, 1989.

15. Harper RG et al: Bird headed dwarfs (Seckel's syndrome). *J Pediatr* **70**:699–704, 1967.

16. Heinisch HM: Zur Differentialdiagnose Vogelkopfzwerg und Rubinstein-Taybi-Syndrom. *Radiologe* **7**:387–390, 1967.

17. Lambotte C et al: Seckel syndrome-birdheaded dwarfism. *Acta Paediatr Belg* **29**:79–82, 1976.

18. de Lange C: Nanosomia vera. *Jb Kinderheilkd* **89**:264–268, 1919.

18a. Lavollay B et al: Nanisme familial congénital avec dysplasie céphalo-skelettique (syndrome de Taybi–Lindner). *Arch Fr Pédiatr* **41**:57–60, 1984.

19. Lilleyman JS: Constitutional hypoplastic anemia associated with familial "bird headed" dwarfism—Seckel syndrome. *Am J Pediatr Hematol Oncol* **6**:207–209, 1984.

20. Lim KH, Wong HB: Ocular anomalies in Seckel's syndrome. *Aust NZ J Med* **3**:520–522, 1973.

21. Maitinsky SP: Vogelköpfiger Zwergwuchs. Bericht über drei neue Fälle. Thesis, München, 1964.

22. Majewski F, Goecke T: Studies of microcephalic primordial dwarfism. I: Approach to a delineation of the Seckel syndrome. *Am J Med Genet* **12**:7–21, 1982.

23. Majewski F, Spranger J: Über einen neuen Typ des primordialen Minderwuchses: Der brachymele primordiale Minderwuchs. *Monatsschr Kinderheilkd* **124**:499–503, 1976.

24. Majewski F et al: Studies of microcephalic primordial dwarfism. II: *Am J Med Genet* **12**:23–35, 1982.

25. Majewski F et al: Studies of microcephalic primordial dwarfism. III: An intrauterine dwarf with platyspondyly and anomalies of pelvis and clavicles—osteodysplastic primordial dwarfism type III. *Am J Med Genet* **12**:37–42, 1982.

25a. Majoor-Krakauer DF et al: Microcephaly, micrognathia and bird-headed dwarfism: Prenatal diagnosis of a Seckel-like syndrome. *Am J Med Genet* **27**:183–188, 1987.

26. Mann TB, Russell A: A study of microcephalic midget of extreme type. *Proc R Soc Med* **52**:1024–1027, 1959.

27. McKuskick VA et al: Seckel's bird headed dwarfism. *N Engl J Med* **277**:279–286, 1967.

28. Mercier J et al: Le syndrome de Seckel (ou nanisme à tête d'oiseau). *Rev Stomatol Chir Maxillofac* **84**:264–268, 1983.

29. Müller W et al: Seckel-Syndrom. *Monatsschr Kinderheilkd* **126**:454–456, 1978.

30. Nuñez E et al: Sindroma de Seckel. *Rev Chil Pediatr* **42**:117–119, 1971.

31. Payet G: Nanisme et hyperlaxité, dysmorphie facial et luxations multiples. Syndrome de Larsen? *Arch Fr Pédiatr* **3**:601–602, 1975.

32. Pichler E et al: Eine Patientin mit Seckel-Syndrom. *Monatsschr Kinderheilkd* **121**:689–691, 1973.

33. Poznanski AK et al: Radiologic findings in the hand in Seckel syndrome (bird headed dwarfism). *Pediatr Radiol* **13**:19–24, 1983.

34. Sauk JJ et al: Familial bird-headed dwarfism (Seckel's syndrome). *J Med Genet* **10**:196–198, 1973.

35. Schönenberg H: Seckel-Syndrom. *Klin Paediatr* **188**:449–454, 1976.

36. Schulz IS et al: Sindrom "cabeça de passaro" de Seckel. *Arq Neuropsiquiatr* **33**:286–292, 1975.

37. Seckel HPG: *Bird Headed Dwarfs,* CC Thomas, Springfield, Illinois, 1960.

38. Singh SD et al: Bird headed dwarf (Seckel's syndrome). *Indian Pediatr* **12**:935–937, 1975.

39. Spennati GF, Persichetti B: Nanismo de Seckel. *Minerva Pediatr* **26**:851–855, 1974.

40. Szalay GC: Intrauterine growth retardation versus Silver's syndrome. *J Pediatr* **64**:234–240, 1964.

41. Szalay GC: Seckel syndrome. *J Med Genet* **11**:216, 1974.

41a. Thomas PS, Nevin NC: Congenital familial dwarfism with cephaloskeletal dysplasia (Taybi–Lindner syndrome). *Ann Radiol* **19**:187–192, 1976.

42. Thompson E, Pembrey M: Seckel syndrome: An overdiagnosed syndrome. *J Med Genet* **22**:192–201, 1985.

42a. Toriello HV et al: An apparently new syndrome of microcephalic primordial dwarfism and cataracts. *Am J Med Genet* **25**:1–8, 1986.

Toudic L et al: Nanisme intra-utérin majeur avec dysmophies et encéphalopathie profunde du type nanisme à tête d'oiseau (Virchow-Seckel). *Ann Pédiatr* **24**:653–656, 1977.

44. Tschiya H et al: Analysis of the dentition and orofacial skeleton in Seckel's bird-headed dwarfism. *J Maxillofac Surg* **9**:170–175, 1981. (Same cases as Ref. 6.)

44a. Verloes A et al: Microcephalic osteodysplasic dwarfism (type II-like) in siblings. *Clin Genet* **32**:88–94, 1987.

45. Willems PJ et al: A new case of the osteodysplastic primordial dwarfism type II. *Am J Med Genet* **26**:819–824, 1987.

46. Winter RM et al: Osteodysplastic primordial dwarfism: Report of a further patient with manifestations similar to those seen in patients with types I and II. *Am J Med Genet* **21**:569–574, 1985.

47. Zamboni G, Bernardi F: A familial case of Seckel's syndrome with abnormal karyotype. *Acta Med Auxol* **12**:33–38, 1980.


## Silver–Russell syndrome (Russell–Silver syndrome)

The syndrome of short stature of prenatal onset, triangular facies, body asymmetry, variation in the pattern of sexual development, and other abnormalities including café-au-lait pigmentation and clinodactyly of

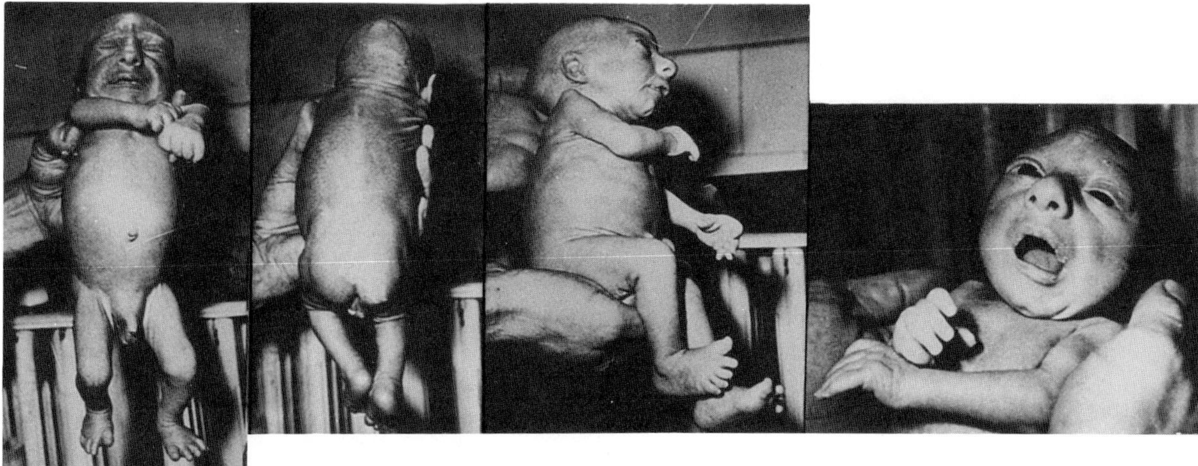

A

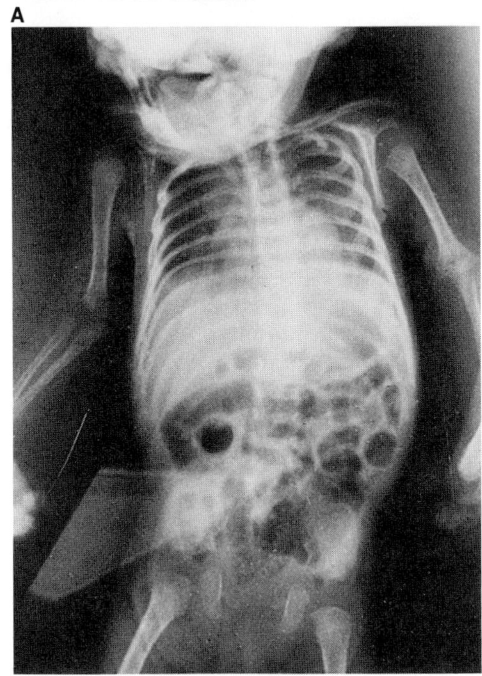

B

Fig. 10–25. *Microcephalic primordial dwarfism (osteodysplastic Type III).* (A) Alopecia, receding forehead and chin, large eyes, prominent nose. (B) Note elongated clavicles, platyspondyly, enlargement of proximal femora. (From *F Majewski et al,* Am J Med Genet **12:**37, 1982.)

the fifth fingers (Figs. 10-26 to 10-28) was independently described by Silver (37) in 1953 and by Russell (32) in 1954. Most authors regard the Silver–Russell syndrome as a single entity (13,18,24,31,40). There has been a recent tendency to reverse the order of the names in the syndrome to acknowledge priority.

The etiology is unknown and is probably heterogeneous. Syndrome heterogeneity has been discussed by several authors (27,33,39). More than 150 cases have been reported (13,23,24,31). Almost all examples have been sporadic, although a few familial instances have been noted. In several instances, the mother was stated to be short (13,23). Autosomal dominant inheritance with most cases representing fresh mutations is possible. Concordant monozygotic twins (30) and dizygotic twins (36) have been documented as well as discordant monozygotic twins (34). Cases have been tabulated and analyzed by Escobar et al (13). Fuleihan et al (16) described three affected sibs whose parents were consanguineous. They appear to have 3-M syndrome. The sibs reported by Schwingshackl et al (35) probably had Bloom syndrome. Our collective experience suggests that the diagnosis is overused.

**Growth.** Birth weight is usually less than 2200 g at full term and birth length is about 44 ± 2.5 cm (1). The placenta is small (2). Short stature is maintained throughout childhood, height usually being below the third percentile. Adult height averages 149.5 cm in males and 138 cm in females or about −3.6 SD (1,10,18,28,30,40). Females seen to gain some subcutaneous fat after puberty (10).

**Facies.** The facies is characterized by pseudohydrocephaly due to relative smallness of the face (Figs. 10–26 and 10–27). The calvaria, while appearing large, is really somewhat smaller than normal (1,14). The forehead is prominent or bossed and the face triangular with the chin small and pointed in about 65% (13). The sclerae may be bluish in infancy (23). The eyes seem large (28). The mouth appears wide and the corners are often turned downward. The upper lip vermilion is thin. The pinnae may protrude. Appearance becomes markedly less striking with age.

**Musculoskeletal system.** Congenital asymmetry has been noted in 65–80% (2,13) (Fig. 10-26A). Although occasionally total, it may involve only the head, trunk, or limbs. Rarely, the asymmetry becomes evident only with growth (36). Poor muscular development and delay in early gross motor performance are common.

Delayed closure of the anterior fontanel is found in 20% (13,18). Occasionally, there is hip or elbow dislocation. The fifth fingers are

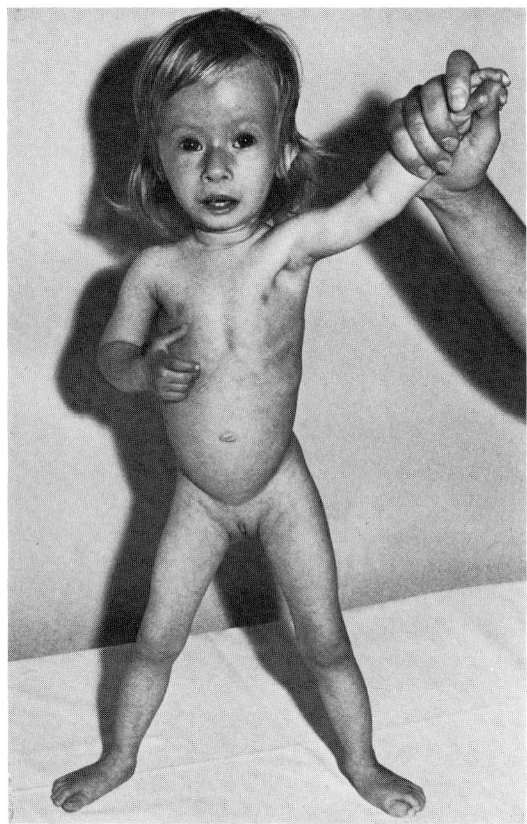

A

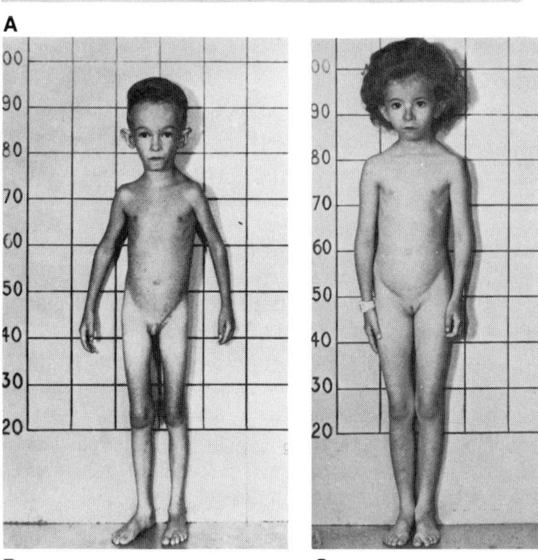


B C

Fig. 10–26. *Silver–Russell syndrome.* (A–C) Proportionate short stature. Note leg asymmetry and similar facies. (A from *G Schumacher* and *H Niederhoff,* Helv Paediatr Acta **22:**404, 1967. B,C from *RH Haslam et al,* Pediatrics **51:**216, 1973.)

abbreviated and exhibit clinodactyly in over 75% (2,32,40) (Fig. 10–28).

Radiographically, bone age is retarded in relation to both sexual development and chronologic age until puberty in 50% (18,21,40). The long bones tend to be slender. The humerus is somewhat shortened in 20% (13). Hypoplasia of the middle phalanges of the fifth fingers is evident in 80%. Pseudoepiphyses are found more often at the base of the second metacarpal than in the normal population (40) and frequently there are distal phalangeal ivory epiphyses (21). Soft-tissue syndactyly between the second and third toes is seen in about 20% (13). The total and posterior cranial base are reduced in length.

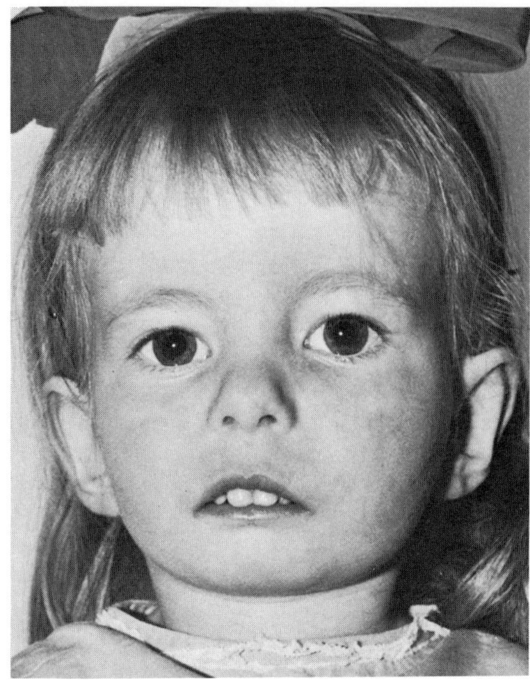

Fig. 10–27. *Silver–Russell syndrome.* Normal-sized cranium with disproportionately small facial bones. Note downturned angles of mouth. (Courtesy of *A Russell,* London, England.)

**Urogenital anomalies.** Variation in sexual development has been found in over 30% (36), but there is usually normal puberty (40). Cryptorchidism and/or hypospadias are present in over 35% of males (1,20,23,29,32,36,41). Females have exhibited premature estrogenation of the urethral or vaginal mucosa in about 25% (13). Ambiguous genitalia has also been reported (17,19,23). Renal and/or ureteral anomalies such as hydronephrosis, ureteropelvic obstruction, pyelonephritis with reflux, and enlarged kidneys have been found (2,16,20,38).

**Other findings.** Café-au-lait spots have been noted in about 25% (13). Hyperhidrosis and tachypnea are frequent findings in the neonatal period due to hypoglycemia (14,18,20c,27). Dermatoglyphics are not specific (3). Mild developmental delay or mental retardation has been reported in about 35% (13,33). Testicular cancer and craniopharyngioma have been reported (11,41).

**Oral manifestations.** Down-turned corners of the mouth have been observed in over 60% (36). The maxilla and mandible are small, the palate is high and narrow, and the teeth are crowded.

**Differential diagnosis.** Silver–Russell syndrome is one of a large group of conditions categorized as ''intrauterine growth retardation''

Fig. 10–28. *Silver–Russell syndrome.* Clinodactyly of fifth fingers. (Courtesy of *A Russell,* London, England.)

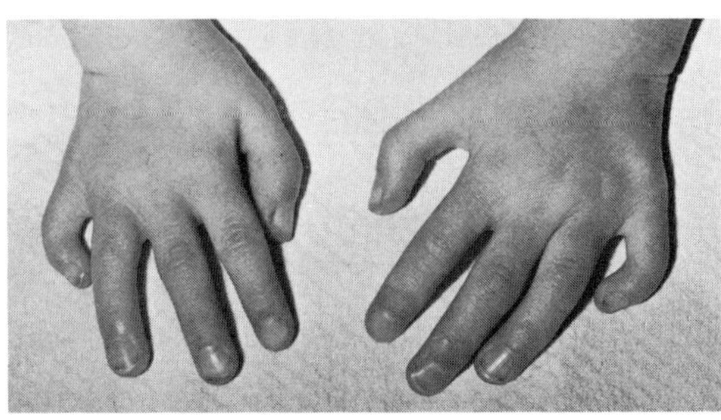

or "low-birthweight dwarfism." Differential diagnosis includes a plethora of conditions with short stature and precocious sexual development (36). To be excluded are *mulibrey nanism,* an autosomal recessive disorder associated with pericardial constriction and characteristic eye findings, and *3-M syndrome,* another autosomal recessive condition. An X-linked disorder of short stature with skin pigmentation has also been described (27). The authors suggested that other examples may be of this type (5,13,16).

Café-au-lait pigmentation and/or body asymmetry may be seen with *neurofibromatosis, hemihyperplasia (hemihypertrophy), McCune–Albright syndrome, Klippel–Trénaunay–Weber syndrome, Proteus syndrome,* and *diploid–triploid mosaicism, trisomy 18 mosaicism* (7,20,22), and *del(18p) syndrome* (8). *Ring 15 chromosome* with loss of IGF-I receptor gene results in a Silver–Russell-like phenotype (4,15).

**Laboratory findings.** Urinary gonadotropin levels have been elevated in about 10% (1,2,9,36,38). Hypoglycemia following short periods of fasting has been described (13,18,19), and growth hormone deficiency has been reported in a few cases (11,12,19,25). Cassidy et al (6) and Okabe and Ueda (26) have discussed hypopituitarism in the syndrome.

### References [Silver–Russell syndrome (Russell–Silver syndrome)]


1. Angehrn V et al: Silver–Russell syndrome: Observations in 20 patients. *Helv Paediatr Acta* **34**:297–308, 1979.
2. Anoussakis C et al: Le nanisme congénitale avec dysmorphie crânio-faciale et asymétrie corporelle (type Silver–Russell). *Pédiatrie* **29**:249–259, 1974.
3. Brehme H, Schröter R: Hautleistenbefunde von 15 Patienten mit Russell–Silver-Syndrom. *Humangenetik* **5**:28–35, 1967.
4. Butler MD et al: Two patients with ring chromosome 15 syndrome. *Am J Med Genet* **29**:149–154, 1988.
5. Callaghan KA: Asymmetrical dwarfism, or Silver's syndrome, in two male siblings. *Med J Austral* **2**:789–792, 1970.
6. Cassidy SB et al: Russell–Silver syndrome and hypopituitarism. *Am J Dis Child* **140**:155–159, 1986.
7. Chauvel PJ et al: Trisomy 18 mosaicism with features of Russell–Silver syndrome. *Dev Med Child Neurol* **17**:220–243, 1975.
8. Christensen, MF, Nielsen J: Deletion short arm 18 and Russell–Silver syndrome. *Acta Paediatr Scand* **67**:101–103, 1978.
9. Curi JFJ et al: Elevated serum gonadotrophins in Silver's syndrome. *Am J Dis Child* **144**:658–661, 1967.
10. Davies PSW et al: Adolescent growth and puberty progression in the Silver–Russell syndrome. *Arch Dis Childh* **63**:130–135, 1988.
11. Draznin MD et al: Russell–Silver syndrome and craniopharyngioma. *J Pediatr* **96**:887–889, 1980.
12. Eeckels R et al: Plasma growth hormone determination in the Silver–Russell syndrome. *Helv Paediatr Acta* **25**:363–369, 1970.
13. Escobar V et al: Phenotypic and genetic analysis of the Silver–Russell syndrome. *Clin Genet* **13**:278–288, 1978.
14. Fitch N, Pinsky L: The lateral facial profile of the Russell–Silver dwarf. *J Pediatr* **80**:827–829, 1972.
15. Francke U et al: Loss of IGF-I receptor gene in patients with ring chromosome 15 is related to Russell–Silver-like phenotype. Ninth Annual David W. Smith Workshop on Malformations and Morphogenesis, Oakland, August 7–10, 1988.
16. Fuleihan DS et al: The Russell–Silver syndrome: Report of three siblings. *J Pediatr* **78**:654–657, 1971.
17. Gardner LI et al: Abnormal genitalia in males with Silver–Russell (SR) dwarfism: A not infrequent complication. *Pediatr Res* **15**:643, 1981.
18. Gareis FJ et al: The Russell–Silver syndrome without asymmetry. *J Pediatr* **79**:775–781, 1971.
19. Hall JG: Microphallus, growth hormone deficiency and hypoglycemia in Russell–Silver syndrome. *Am J Dis Child* **132**:1149, 1978.
20. Haslam RH et al: Renal abnormalities in the Russell–Silver syndrome. *Pediatrics* **51**:216–222, 1973.
21. Herman TE et al: Hand radiographs in Russell–Silver syndrome. *Pedatrics* **79**:743–744, 1987.
22. Hook EB, Yunis JJ: Congenital asymmetry associated with trisomy 18 mosaicism. *Am J Dis Child* **110**:551–555, 1965.
23. Marks LJ, Bergeson PS: The Silver–Russell syndrome, a case with sexual ambiguity and a review of the literature. *Am J Dis Child* **131**:447–451, 1977.
24. Moss SH, Switzer HE: Congenital hypoplastic thumb in the Silver syndrome—a case report and review of the upper extremity anomalies in the world literature. *J Hand Surg* **8**:480–486, 1983.
25. O'Brien JE et al: Growth hormone deficiency in a patient with Silver–Russell syndrome. *J Pediatr* **93**:152–153, 1979.
26. Okabe Y, Ueda K: Russell–Silver syndrome and pituitary gonadal axis dysfunction. *Am J Dis Child* **141**:123–124, 1987.
27. Partington MW: X-linked short stature with skin pigmentation: Evidence for heterogeneity of the Russell–Silver syndrome. *Clin Genet* **29**:151–156, 1986.
28. Patton MA: Russell–Silver syndrome. *J Med Genet* **25**:557–560, 1988.
29. Reister HC, Scherz RG: Silver syndrome. *Am J Dis Child* **107**:410–416, 1964.
30. Rimoin DL: The Silver syndrome in twins. *Birth Defects* **5**(2):183–187, 1969.
31. Robichaux V et al: Silver–Russell syndrome: A family with symmetric and asymmetric siblings. *Arch Pathol Lab Med* **105**:157–159, 1981.
32. Russell A: A syndrome of "intrauterine" dwarfism recognizable at birth with cranio-facial dysostosis, disproportionately short arms, and other anomalies (5 examples). *Proc R Soc Med* **47**:1040–1044, 1954.
33. Saal HM et al: Reevaluation of Russell–Silver syndrome. *J Pediatr* **107**:733–737, 1985.
34. Samn M et al: Monozygotic twins discordant for Russell–Silver syndrome. Ninth Annual David W. Smith Workshop on Malformations and Morphogenesis, Oakland, August 7–10, 1988.
35. Schwingshackl A et al: Familiäres Russell-Syndrom. *Pädiatr Pädol* **9**:130–137, 1974.
36. Silver HK: Asymmetry, short stature and variations in sexual development: A syndrome of congenital malformations. *Am J Dis Child* **107**:495–515, 1964.
37. Silver HK et al: Syndrome of congenital hemihypertrophy, shortness of stature, and elevated urinary gonadotrophins. *Pediatrics* **12**:368–376, 1953.
38. Spirer Z et al: Renal abnormalities in the Russell–Silver syndrome. *Pediatrics* **54**:120, 1974.
39. Szalay GC: Russell–Silver dwarfism. *J Pediatr* **108**:1037, 1985.
40. Tanner JM et al: The natural history of the Silver–Russell syndrome: A longitudinal study of thirty-nine cases. *Pediatr Res* **9**:611–623, 1975.
41. Weiss GR, Garnick MB: Testicular cancer in a Russell–Silver dwarf. *J Urol* **126**:836–837, 1981.


## 3-M syndrome

Although first described by Fuhrmann et al (3) in 1972, Miller et al (7), in 1975, rediscovered this distinct type of low-birthweight proportionate dwarfism. They employed the term "3-M syndrome" to refer to the initials of the first three authors' last names. Additional cases have been reported (2,4,9–11). Hennekam et al (6) reported three affected sibs and reviewed the findings in 19 previously recorded cases. Inheritance is autosomal recessive (2,4,12). Presumably heterozygotes have thinner bones and a prominent talus (5).

Low birthweight at full term and proportionate small size characterize the disorder. Dolichocephaly and triangular facies are evident (Figs. 10–29 and 10–30). The malar region is flat, the ears outstanding, the lips patulous, and the chin pointed and prominent. The dental arch is V-shaped with malocclusion and anterior crowding. Frontal bossing and short broad neck with prominent trapezius muscles are evident. The shoulders are square and high and the thorax short. Pectus carinatum or excavatum is common and there may be transverse grooves above the costal margins. The scapulae are winged, the joints are hyperextensible, and the fifth fingers are short. Hypospadias has been seen in a few cases (12).

Radiographically, the bones are slender, especially the diaphyses. This becomes more prominent with age. The vertebral bodies are tall (2,8). The pubic and ischial bones are small and the iliac wings flared (Figs. 10–31 to 10–33). Bone age is slightly delayed.

Several patients with 3-M syndrome have been erroneously labeled as having *Silver–Russell syndrome* (4). Body asymmetry is not seen in the 3-M syndrome. Furthermore, there is no altered pattern of sexual development, craniofacial disproportion (pseudohydrocephaly), or delayed closure of the anterior fontanel. Height is shorter than in those with Silver–Russell syndrome (9). The facies somewhat resembles that seen in *Bloom syndrome* and the two conditions have been confused

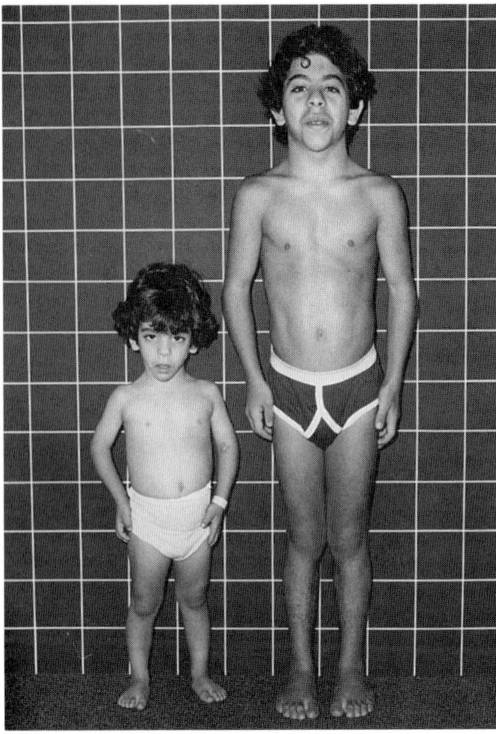

Fig. 10–29. *3-M syndrome.* Low-birthweight proportionate dwarfism in 4- and 15-year-old sibs. Note triangular face, outstanding pinnae, patulous lips, pointed outstanding chin. (From *J Spranger,* Eur J Pediatr **123:**115, 1976.)

(8). We cannot identify the disorder described by Callaghan (1). Also to be excluded is *mulibrey nanism.*

### References (3-M syndrome)

1. Callaghan KA: Asymmetrical dwarfism, or Silver syndrome in two male siblings. *Med J Aust* **2**:789–792, 1970.

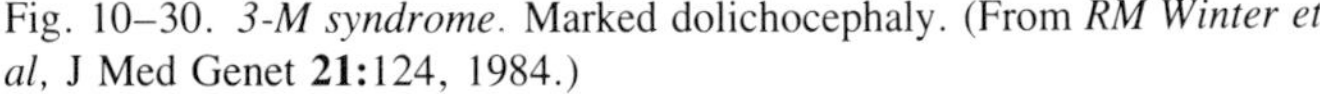

Fig. 10–30. *3-M syndrome.* Marked dolichocephaly. (From *RM Winter et al,* J Med Genet **21:**124, 1984.)

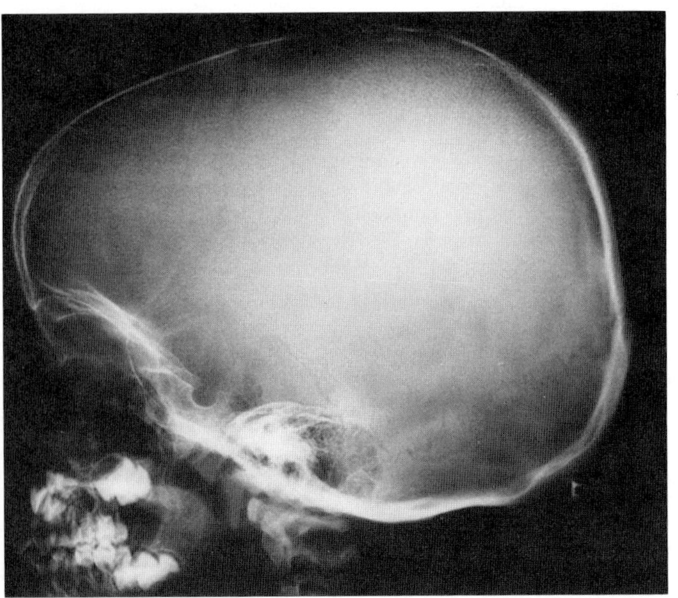

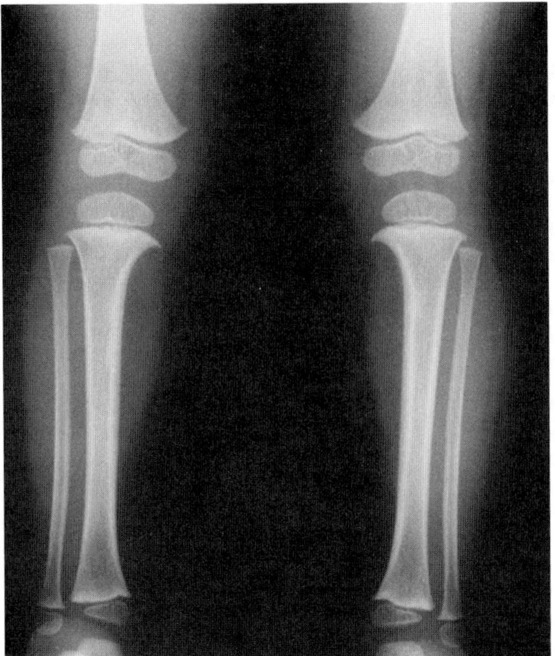

Fig. 10–31. *3-M syndrome.* Narrow long bones with overconstriction of diaphyses and metaphyseal flaring. (From *RM Winter et al,* J Med Genet **21:**124, 1984.

2. Cantú JM et al: 3-M slender-boned nanism. *Am J Dis Child* **135**:905–908, 1981.

2a. Feldmann M et al: 3M dwarfism. *J Med Genet* **26**:583–585, 1989.

3. Fuhrmann W et al: Familiärer Minderwuchs mit unproportionert hohen Wirbeln. *Humangenetik* **16**:271–282, 1972.

4. Fuleihan DS et al: The Russell–Silver syndrome: Report of three siblings. *J Pediatr* **78**:654–657, 1971.

5. Garcia-Cruz D, Cantú JM: Heterozygous expression in 3M slender-boned nanism. *Hum Genet* **52**:221–226, 1979.

6. Hennekam RCM et al: Further delineation of the 3-M syndrome with review of the literature. *Am J Med Genet* **28**:195–209, 1987.

Fig. 10–32. *3-M syndrome.* Tall vertebral bodies. (From *RM Winter et al,* J Med Genet **21:**124, 1984.)

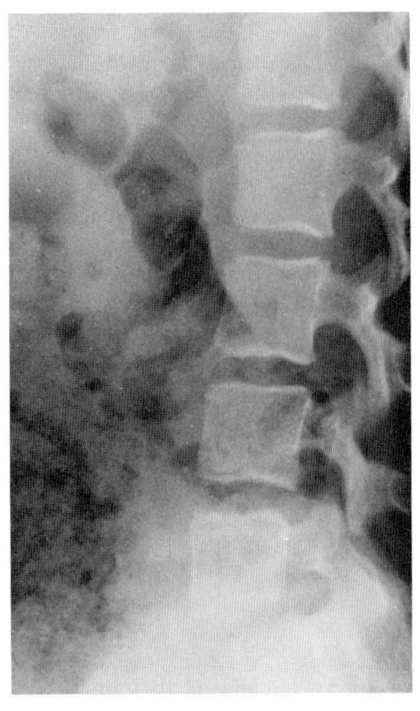

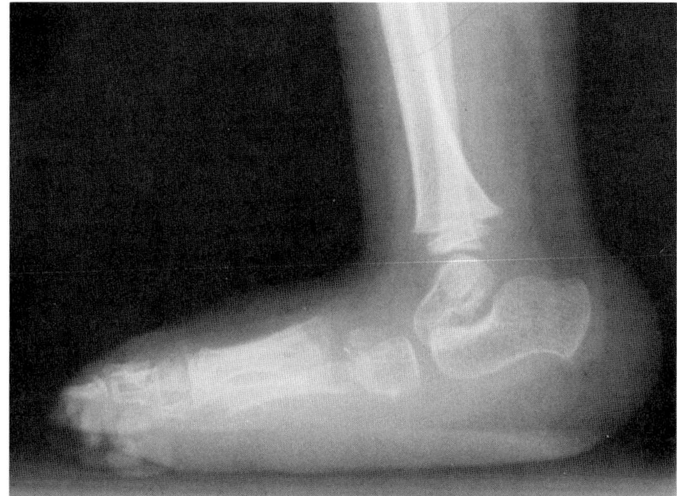

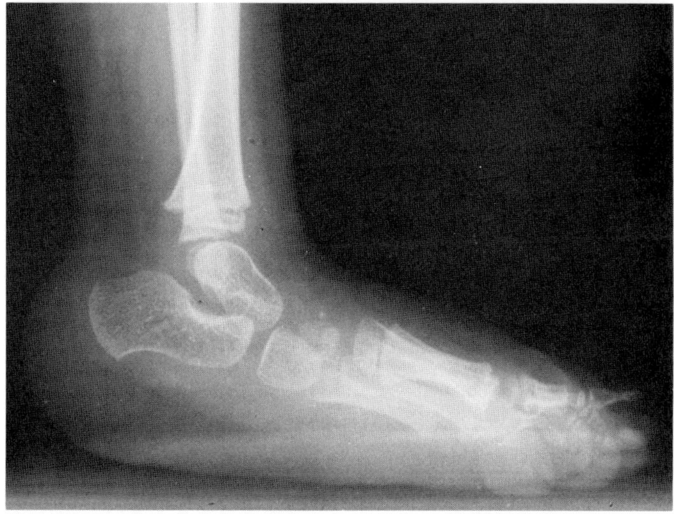

Fig. 10–33. *3-M syndrome.* Vertical talus with prominent calcaneus. (From *RM Winter et al,* J Med Genet **21:**124, 1984.)

7. Miller JD et al: The 3-M syndrome: A heritable low birthweight dwarfism. *Birth Defects* **11**(5):39–47, 1975.
8. Schwingshackl A et al: Familiäres Russell-Syndrom. *Pädiat Pädol* **9**:130–137, 1974.
9. Spranger J: "New" dwarfing syndrome. *Birth Defects* **13**(3B):11–29, 1977.
10. Spranger J et al: A new familial intrauterine growth retardation syndrome in the "3-M syndrome." *Eur J Pediatr* **123**:115–124, 1976.
11. Van Goethem H, Malvaux P: The 3-M syndrome: A heritable low birthweight dwarfism. *Helv Paediatr Acta* **42**:159–165, 1987.
12. Winter RM et al: The 3-M syndrome. *J Med Genet* **21**:124–128, 1984.

## Mulibrey nanism

Perheentupa et al (3), in 1973, described a form of prenatal growth retardation associated with anomalies of muscle, liver, brain, and eye, employing the mnemonic "mulibrey" (Figs. 10–34 to 10–36). At least 30 patients have been reported, most stemming from Finland (1–9). The disorder has autosomal recessive inheritance.

Birth, weight, and length are usually 1.5 to 2 SD below the mean. With time, growth becomes progressively retarded so that height is 3 SD below the mean. Adult males vary from 136 to 161 cm; adult females range from 126 to 151 cm. Some die in childhood from cardiac involvement but most survive to adulthood without incapacity.

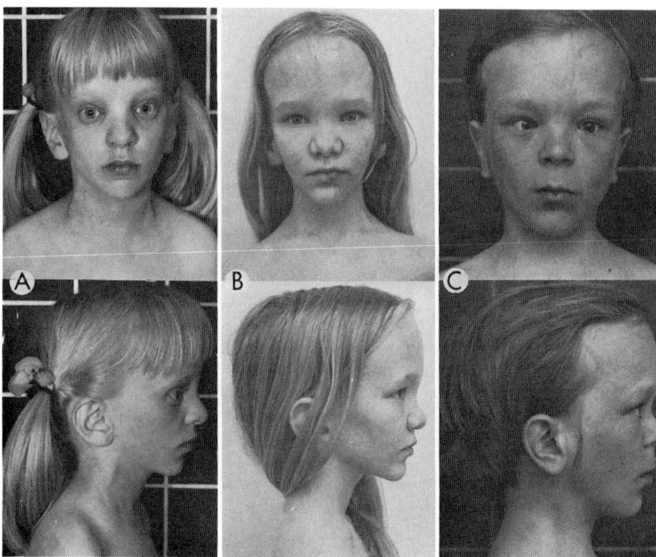


Fig. 10–34. *Mulibrey nanism.* (A–C) Three unrelated adolescent patients exhibiting similar facies. (From *J Perheentupa et al,* Lancet **2:**351, 1973.)

**Facies.** The face is triangular and the forehead is prominent and high. The nasal bridge is deep and broad (Fig. 10–34A–C).

**Eyes.** There is mild hypertelorism. Aggregation and dispersion of pigment in the midperiphery and more peripheral areas of the fundus with yellow dots are characteristic. The choroid is hypoplastic. The optic discs and maculas are normal (Fig. 10–35). Atrophy of the corneal epithelium and thickening of Bowman's membrane have been demonstrated (7).

**Heart.** Thickened and adherent pericardium results in cardiac constriction. Radiographically, this gives the heart a globular shape. There is elevated venous pressure, prominent left atrium and/or right ventricle, dilated neck veins, and, in some cases, ascites with congestive heart failure (10).

**Liver.** In the neonatal period, the liver is often enlarged about 2 cm or so below the costal margin. This is probably due to passive venous congestion, secondary to the constrictive pericarditis.

**Central nervous system.** Intelligence is mildly retarded. Pneumoencephalography has shown abnormally large cerebral ventricles and cisternae.

**Skin.** About 65% manifest cutaneous nevus flammeus.

Fig. 10–35. *Mulibrey nanism.* Fluorescein angiogram of fundus showing conglomerate of dots and severe hypoplasia of choroid. (From *J Perheentupa et al,* Lancet **2:**351, 1973.)

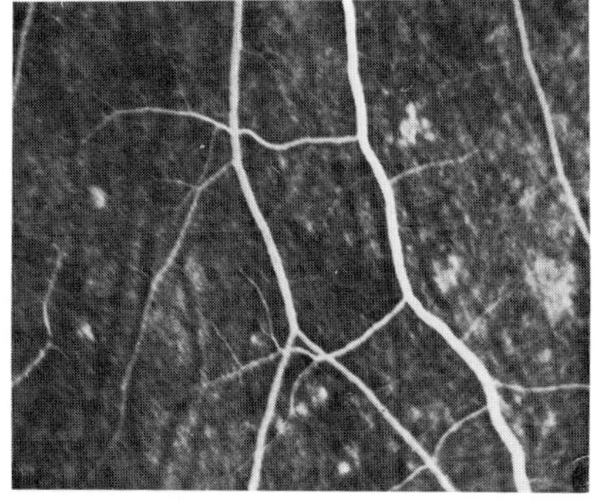

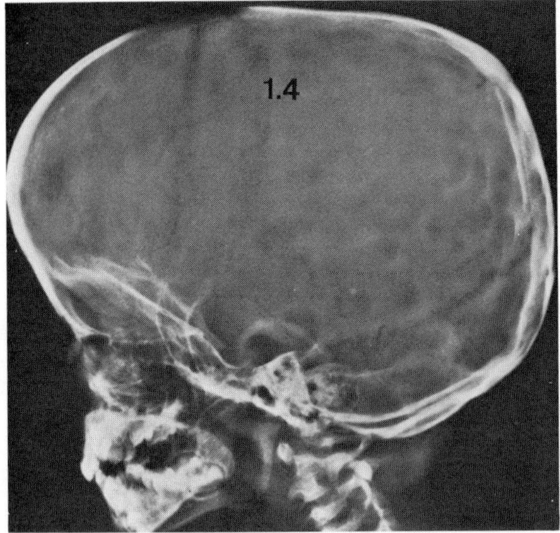


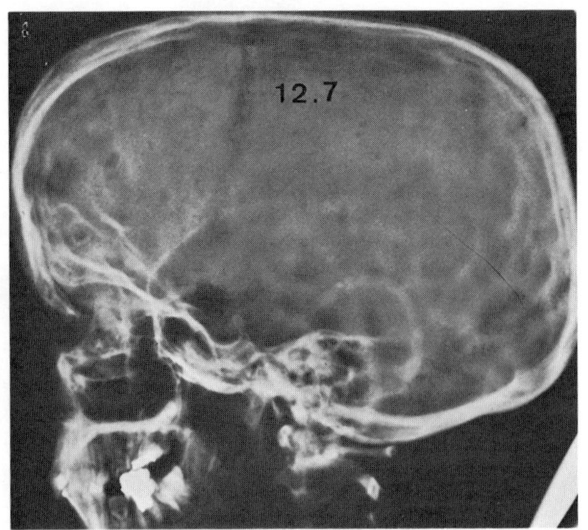


A B

Fig. 10–36. *Mulibrey nanism.* (A,B) Typical skull radiographs at 1.4 and 12.7 years exhibiting abnormal cranial form, J-shaped sella, and increased basilar angle. (From *J Perheentupa et al,* Lancet **2:**351, 1973.)

**Musculoskeletal systems.** The extremities are thin and short. Most patients are mildly hypotonic. Radiographically, bone age is normal. There is frontal and occipital bossing. The sella turcica is long and shallow. The frontal and sphenoidal sinuses are absent or hypoplastic (Fig. 10–36A,B). Fibrous dysplastic lesions of the tibia have been found in about 30%.

**Oral manifestations.** Dental malocclusion has been noted in about 50% and hypodontia in about 25%. The tongue appears small (2). For cephalometric findings, see (2).

**Differential diagnosis.** There is considerable overlap with *Silver–Russell syndrome*.

### References (Mulibrey nanism)

1. Cumming GR et al: Constrictive pericarditis with dwarfism in two siblings. *J Pediatr* **88**:569–572, 1976.

2. Myllärniemi S et al: Craniofacial and dental study of mulibrey nanism. *Cleft Palate J* **15**:369–377, 1978.

3. Perheentupa J et al: Mulibrey nanism, an autosomal recessive syndrome with pericardial constriction. *Lancet* **2**:351–355, 1973.

4. Perheentupa J et al: Mulibrey nanism: Review of 23 cases of a new autosomal recessive syndrome. *Birth Defects* **11**(2):3–17, 1975.

5. Raitta C, Perheentupa J: Mulibrey nanism: An inherited dysmorphic syndrome with characteristic ocular findings. *Acta Ophthalmol* **52**:162–171, 1974.

6. Similä S et al: A case of mulibrey nanism with associated Wilms' tumor. *Clin Genet* **17**:29–30, 1980.

7. Tarkkanen A et al: Mulibrey nanism, an autosomal recessive syndrome with ocular involvement. *Acta Ophthalmol* **60**:628–633, 1982.

8. Thorén L: The so-called mulibrey nanism with pericardial constriction. *Lancet* **2**:731, 1973.

9. Tuuteri L et al: The cardiopathy of mulibrey nanism, a new inherited syndrome. *Chest* **65**:628–631, 1974.

10. Voorhess ML et al: Growth failure with pericardial constriction. *Am J Dis Child* **130**:1146–1148, 1976.

# Chapter 11
# Overgrowth Syndromes and Postnatal Onset Obesity Syndromes

## Beckwith–Wiedemann syndrome [EMG (exomphalos–macroglossia–gigantism) syndrome]

In 1963, Beckwith (3) reported three cases of a newly recognized syndrome consisting of macroglossia, omphalocele, cytomegaly of the adrenal cortex, hyperplasia of gonadal interstitial cells, renal medullary dysplasia, and hyperplastic visceromegaly (Figs. 11–1 to 11–5). Subsequently, Beckwith (4) enlarged his series of patients, noting postnatal somatic gigantism, mild microcephaly, and severe hypoglycemia. In 1964, Wiedemann (97) independently reported the syndrome in three sibs and observed a further component—a dome-shaped defect of the diaphragm. Other important contributions have been made by many investigators (1,2,5,7,8,11,12,15–17,20,21,26,29,32–34,36–40,41–43,45–47,56–58,60,63,68,70,71,80–83,88,89,93,98–102,104). Over 400 cases have been reported. Its frequency has been estimated at 1/17,000 births (27). Exhaustive reviews have been published by Irving (39), Filippi and McKusick (28), Cohen (18,19), Niikawa et al (60), Pettenati et al (67), and Engström et al (27). Early diagnosis of this striking condition alerts the clinician to the dual threat of hypoglycemia and possible neoplasia. Clinical and laboratory findings in the Beckwith–Wiedemann syndrome (also known as Wiedemann–Beckwith syndrome) are listed in Table 11–1 together with frequencies based on Sotelo-Avila et al (84). These frequencies are at variance with those of Pettenati et al (67) and the reader may wish to compare them.

**Etiology.** Most cases of the Beckwith–Wiedemann syndrome are sporadic. In an investigation of the parental ages of 80 sporadic cases, Schultz and Sherman (75) found no evidence for increased maternal or paternal ages. Discordant monozygotic twins have been reported (6,7,50,52,61). Lunt et al (52) indicated that at least 15 twin pairs have been reported; they observed concordance in only one pair of monozygotic twins. Twenty-four familial occurrences have been recorded to date (60,67). Various etiologic hypotheses have been proposed including autosomal recessive inheritance (3,6,15,28,39,97), autosomal dominant inheritance with variable expressivity (5,8,42,60,81,82,89), autosomal dominant, sex-dependent inheritance (51), autosomal dominant inheritance with delayed mutation of an unstable, premutated allele (36,43), multifactorial inheritance (7,29,99), and chromosomal abnormalities (34,67,69,73,90,94,95).

In 1986, Niikawa et al (60) reported five families with 22 affected individuals. Their analysis of all 24 familial occurrences in the literature strongly suggested dominant inheritance, incomplete penetrance, and variable expressivity. They noted that clinical manifestations become less distinct with increasing age; growth rate slows, macroglossia tends to regress relative to the size of the oral cavity, and mild umbilical anomalies become less evident. Nevus flammeus lesions fade after the first few years and midface hypoplasia becomes less evident with age. This may explain why earlier generations in affected families appear to have less than the expected number of cases compared with sibships of the probands. The syndrome is transmitted from mother to offspring two to three times more frequently than from the father. Decreased reproductive fitness among affected fathers, who have an increased frequency of hypospadias and/or cryptorchidism, appears to be a reasonable explanation for this sex-influenced transmission. The present authors accept the genetic hypothesis and interpretation of Niikawa et al (60). Based on the autosomal dominant hypothesis, Pettenati et al (67) indicated that penetrance was about 30–40% for males and approximately 50–60% for females.

Chromosomal anomalies especially involving dup(11)(p15.5) have been reported in association with a number of Beckwith–Wiedemann syndrome patients (35,44a,90,94,95) and eight of these cases have been tabulated by Pettenati et al (67). However, most reported cases are cytogenetically normal. For example, high-resolution banding studies of 7 patients by Niikawa et al (60) and 19 patients by Pettanati et al (67) failed to demonstrate any dup(11p) abnormalities. Henry et al (35a) suggested that the 11p15.5 region is involved in those that have associated adrenocortical carcinoma.

Saal et al (74), using molecular hybridization with DNA probes, studied the dose of the insulin gene and the c-Ha-*ras*-1 oncogene, both of which are located on the short arm of chromosome 11. They found no evidence of an increased dose of either gene. Spritz et al (86) also provided data suggesting that the Beckwith–Wiedemann syndrome is not frequently associated with small duplications of 11p15 material that embed the insulin and insulin-like growth factor II genes. Little et al (49) reported loss of the calcitonin (11p13–15) and insulin (11p15–15.1) genes in a hepatoblastoma from a child with Beckwith–Wiedemann syndrome.

**Pathogenesis.** Growth hormone production has been normal in patients with the Beckwith–Wiedemann syndrome (85,101). In one case, Spencer and his associates (85) found increased circulatory somatomedin activity with normal growth hormone, suggesting a defect in the control of somatomedin production that might result in gigantism, macroglossia, and visceromegaly. Barlow (2) measured urinary excretion of polyamines in seven children with the Beckwith–Wiedemann syndrome. The findings of raised putrescine and low spermidine ratios were consistent with a disturbed metabolic pathway under growth hormone-like regulation.

Wiedemann et al (101) noted that features such as genital hypertrophy and impaired glucose tolerance placed the syndrome in close relationship to progressive generalized lipodystrophy and lipoatrophic diabetes with or without congenital muscular hypertrophy. They regarded congenital generalized lipodystrophy, Beckwith–Wiedemann syndrome, Sotos syndrome, and Silver–Russell syndrome as examples of "congenital diencephalic syndromes of childhood."

In one patient with the Beckwith–Wiedemann syndrome, Lazarus and associates (45) observed no glucagon-like immunoreactivity in peripheral serum during the basal condition or after oral glucose administration. They suggested that lack of glucagon secretion may be the cause of hypoglycemia in the disorder. However, this is not possible, as Carnelutti (13) has pointed out, since oral glucose inhibits glucagon secreted from the pancreas. Furthermore, intestinal immunoreactive glucagon secreted during glucose loading is a large molecule that has no hyperglycemic or glycogenolytic activity. Hypoglycemia in the Beckwith–Wiedemann syndrome is based on hyperplastic pancreatomegaly involving the islets.

Because of the endocrine cytomegaly, Beckwith (4) suggested that the fetal adrenal cortex is either overactive or underactive with excess

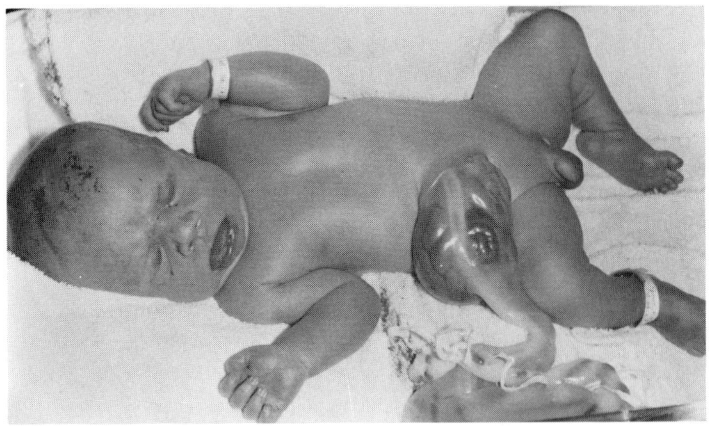

Fig. 11–1. *Beckwith–Wiedemann syndrome.* Note large tongue and omphalocele. (From *MW Moncrieff et al,* Postgrad Med J **46:**162, 1970.)

stimulation caused by a feedback mechanism similar to that found in the adrenogenital syndrome. He further noted that the abnormalities observed in the hypophysis, gonads, islets of Langerhans, and paraganglia should be considered in evaluating the abnormal growth present in the syndrome and that altered placental endocrine physiology could conceivably play a role in producing many of the features found during the neonatal period.

The visceromegaly of the syndrome may possibly result in omphalocele and diaphragmatic eventration. Against this view is Beckwith's observation (personal communication, 1969) in affected premature cadavers that profound visceromegaly is absent as well as the absence of abdominal and diaphragmatic defects in other conditions with early, increased abdominal pressure, such as infantile polycystic disease, universal nephroblastomatosis, and low intestinal atresia. Pathogenetic aspects of the Beckwith–Wiedemann syndrome have been analyzed at the mitotic, nuclear, tissue, organ, and body levels by Herrmann et al (37).

**Infant mortality.** The infant death rate is approximately 21%, usually resulting from either congestive heart failure or severe malformations associated with the syndrome (67).

**Growth.** Average length for males at birth is above the 95th percentile and growth throughout adolescence parallels the normal growth curve at or above the 95th percentile. Average birth length for females is at the 75th percentile and length increases to the 95th percentile by 18 months. Female height then remains at or above the 95th percentile throughout adolescence again paralleling the normal growth curve (67). Gigantism is not always present at birth. Growth may even be subnormal for a few months, although somatic gigantism eventually results in most cases (4,28). Advanced bone age is usually present and there may be widening of the metaphyses and cortical thickening of the long bones. Hemihyperplasia (hemihypertrophy) has been a feature in approximately 13% of the cases (4,20,28,38,39,84,100).

Recently, Sippell et al (79) studied the growth of seven patients

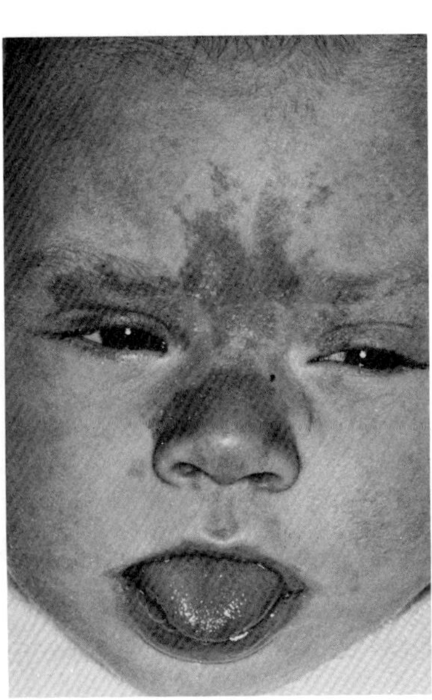

A

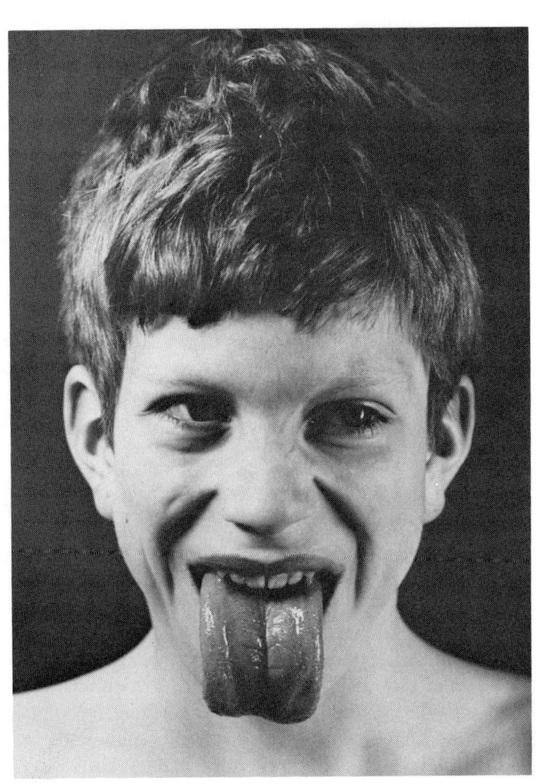

B

Fig. 11–2. *Beckwith–Wiedemann syndrome.* (A) Glabellar nevus flammeus and macroglossia. (B) Macroglossia with remarkable ability to extend tongue. (A from *H-R Wiedemann,* Z Kinderheilkd **106:**171, 1969. B courtesy of *H-R Wiedemann,* Kiel, Germany.)

Table 11–1. Estimated frequencies of clinical and laboratory findings in the Beckwith–Wiedemann syndrome[a]

| Finding | Percentages |
|---|---|
| Growth/skeletal | |
| Increased birth weight | 39 |
| Postnatal gigantism | 33 |
| Accelerated osseous maturation | 21 |
| Asymmetry | 13 |
| Skeletal anomalies | 14 |
| Performance | |
| Seizures, apnea, cyanosis | 22 |
| Mental deficiency | 12 |
| Craniofacial | |
| Macroglossia | 82 |
| Ear lobe grooves | 38 |
| Flame nevus | 32 |
| Craniofacial dysmorphism[b] | 39 |
| Mild microcephaly | 14 |
| Abdominal/genitourinary | |
| Omphalocele, umbilical hernia | 75 |
| Gastrointestinal anomalies | 13 |
| Hepatomegaly | 32 |
| Splenomegaly | 14 |
| Nephromegaly | 23 |
| Genitourinary anomalies | 24 |
| Cardiac anomalies | 16 |
| Diaphragmatic anomalies | 7 |
| Inguinal hernia | 6 |
| Laboratory findings | |
| Hypoglycemia | 30 |
| Polycythemia | 20 |
| Hypocalcemia | 5 |
| Hypercholesterolemia, hyperlipidemia | 2 |

[a]Frequencies estimated from 174 cases from the literature tabulated by C Sotelo-Avila et al, *J Pediatr* **96**:47–50, 1980. See MJ Pettenati et al, *Hum Genet* **74**:143–154, 1986 for different percentages. There is an obvious ascertainment bias in the frequencies of various findings from reported cases. On the one hand, the more severe cases are most likely to be reported. On the other hand, the presence or absence of various findings may be omitted from some reports.

[b]Includes maxillary hypoplasia, prominent occiput, flat nasal bridge, highly arched palate, frontal ridge, downslanting palpebral fissures, etc.

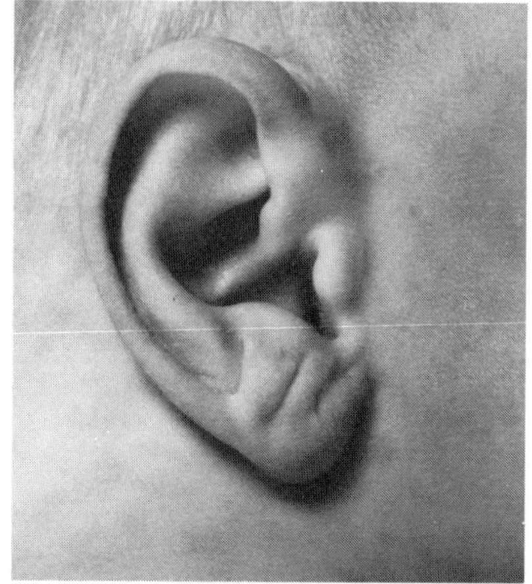

A

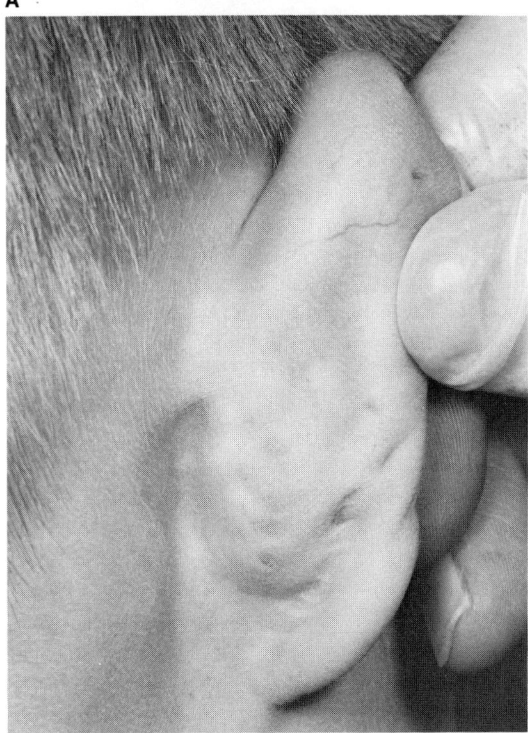

B

Fig. 11–3. *Beckwith–Wiedemann syndrome.* (A) Linear grooves on earlobe. (B) Punched-out depressions of posterior pinna. (A from *H-R Wiedemann,* Z Kinderheilkd **106:**171, 1969.)

longitudinally. They found that growth velocity remained above the 90th centile up to 4–6 years of age. After puberty it was +2.5 SD or about 13.2 cm greater than their parents' height. Bone age was markedly advanced, especially during the first 4 years. Weight was above the 90–97th centile during infancy and early childhood and remained there, usually being appropriate or slightly subnormal for height, until adulthood. Three girls, however, reached and maintained the 50th centile during or after puberty. Spontaneous pubertal development occurred within the normal limits of time.

**Performance and central nervous system.** Intelligence is usually normal (28), although mild to moderate retardation was a regular feature in Beckwith's series (4). Mild microcephaly has been an associated feature in some cases (21). In other cases, mental deficiency may be due, in part, to undetected hypoglycemic episodes during infancy. Seizures have been a feature in some instances. Rarely, hydrocephaly (38) and subdural hematomas (20) have been noted.

**Craniofacial features.** Macroglossia (Figs. 11–1 and 11–2A,B) is very common at birth (58), but is not an obligatory feature of the syndrome. Chronic alveolar hypoventilation has been reported secondary to macroglossia on occasion (80). Tongue biopsies have been normal. In some cases, macroglossia tends to regress with gradual accommodation of the tongue to the oral cavity. At present, it is not known whether this is caused by enlargement of the oral cavity relative to the tongue, shrinkage of the tongue relative to the oral cavity, or a combination of both processes. Persistent macroglossia leads to anterior open-bite, and requires surgical intervention (1,16,28a). Patients with the syndrome have also been observed to be prognathic. Is macroglossia directly responsible for the prognathic mandible? Several authors have noted cessation of prognathic growth following partial glossectomy (76,101). However, careful long-term studies of many patients are needed. It is conceivable that prognathism may reflect the generalized somatic gigantism that occurs in the syndrome. Further, is the prognathism a true mandibular prognathism, or is it relative to hypoplastic changes in the midface (38)?

Facial nevus flammeus (Fig. 11–2A), another frequent feature of the syndrome, tends to become less prominent during the first year of life. Ear lobe grooves (Fig. 11–3A) are very distinctive, consisting of slit-like linear indentations (38). Indented ear lesions on the posterior

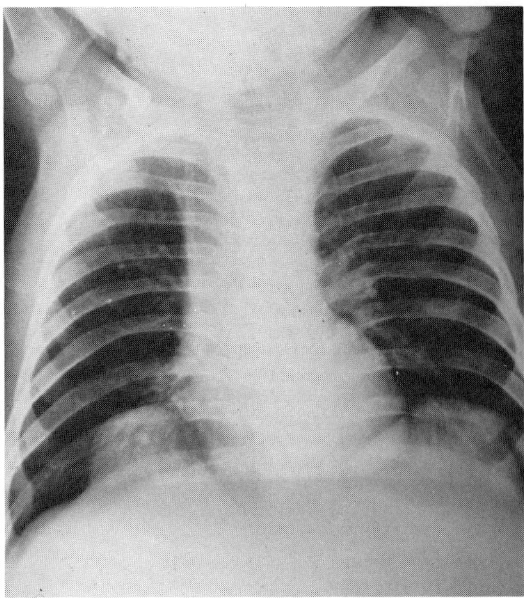

Fig. 11–4. *Beckwith–Wiedemann syndrome.* Bilateral diaphragmatic eventration (From *I Irving,* J Pediatr Surg **2:**499, 1967.)

rim of the helix or concha (Fig. 11–3B) may also be observed (42). Because of the rarity of these anomalies in the general population, they serve as valuable diagnostic signs of the Beckwith–Wiedemann syndrome when they are present. Mild microcephaly has been a feature of some cases (4,21). Cleft palate or submucous cleft palate has been noted occasionally (86a). Various other low-frequency anomalies of the head and neck are listed under Other Abnormalities.

Fig. 11–5. *Beckwith–Wiedemann syndrome.* Adrenocytomegaly. (From *K Bech,* Acta Pathol Microbiol Scand **79A:**279, 1971.)

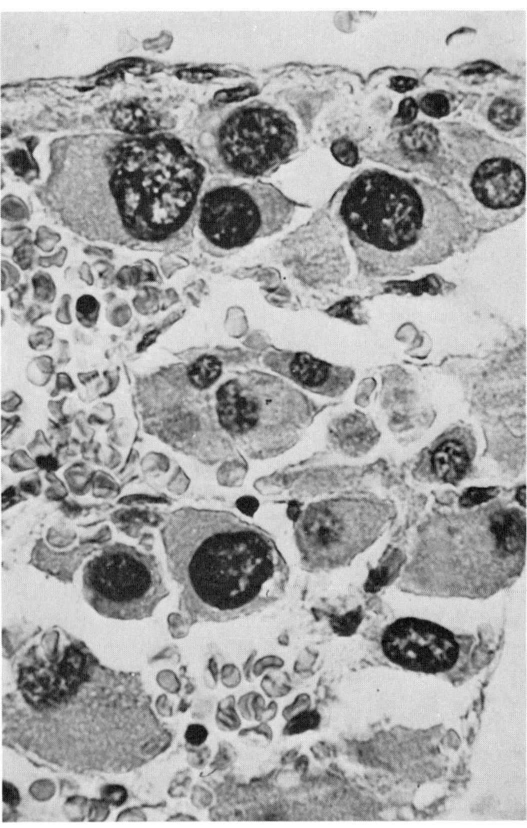

**Visceromegaly.** Omphalocele (Fig. 11–1) is common, although umbilical hernia or diastasis recti may be observed in some instances. Malrotation anomalies may be a feature. In considering the visceromegaly noted in the syndrome, hepatomegaly and nephromegaly occur frequently, but cardiomegaly is found in no more than 15%. Pancreatomegaly has been a feature of autopsied cases. Medullary sponge kidney has been observed on occasion. Genital overgrowth has been noted in some instances. Hyperplastic bladder, uterus, and thymus have also been reported (4,20,38,39,57,93,97,100a). A dome-shaped defect of the diaphragm (Fig. 11–4) has been seen in a number of cases. In this defect, the posterior part of the diaphragmatic leaf is elevated (73). Thorburn and her co-workers (88) have observed a case with diaphragmatic hernia.

**Histopathology.** The histopathologic features of the syndrome reflect the visceromegaly observed grossly. In the pancreas, hyperplastic changes are evident in the acini, islets, and ducts. In the kidney, the lobular arrangement is disordered. Each lobule is capped by a wide, persistent nephrogenic activity zone. Medullary dysplasia is evident with most pyramids showing an increased amount of stroma. Cytomegaly of the fetal adrenal cortex (Fig. 11–5) is prominent, the cells containing sudanophilic droplets. The adrenal cortex is cystic and the medulla is hyperplastic. In the pitutary gland, cells resembling amphophiles are increased in number. Gonadal interstitial cells are hyperplastic. The paraganglia are also hyperplastic (4). In some cases, both macroglossia and omphalocele are absent, but the visceral histologic lesions are florid (103).

**Cardiovascular anomalies.** Greenwood and associates (33) reported cardiovascular abnormalities in 12 of 13 patients. Seven had congenital heart defects including ASD, VSD, PDA, hypoplastic left heart, and tetralogy of Fallot; the other five had idiopathic cardiomegaly. VSD has also been observed in three patients by Kosseff et al (43) and in five patients by Niikawa et al (60). ASD has been noted (60,70) as has coarctation of the aorta (8,39). Pulmonary artery stenosis was reported by Raine et al (70).

**Other abnormalities.** Polyhydramnios has been found in about 50% (50a). Various low-frequency anomalies have been reported including, among others, persistent anterior fontanel, prominent metopic ridge, meningomyelocele, malformed cerebellum, preauricular pits, cleft palate, conductive hearing loss from fixation of the stapes, accessory nipples, pectus excavatum, pectus carinatum, congenital hip dislocation, clinodactyly, postaxial polydactyly, pyloric senosis, ileal stenosis, atresia of the colon, imperforate anus, pyelocalyceal diverticula, other renal and ureteral malformations, prune belly, inguinal hernia, hypospadias, cryptorchidism, unicornuate uterus, and bicornuate uterus (6,8,10,21,23,38–40,43,54,60,62,81,83,87). Wales et al (94) reported a premature Beckwith–Wiedemann infant who developed the bronze baby syndrome when exposed to phototherapy.

**Neoplasms.** Wiedemann (100) noted that of 388 reported and personally observed cases, 29 developed a total of 32 neoplasms. Of these, 26 were intraabdominal, 5 were extraabdominal, and 1 was a malignant lymphoma, giving a total reported tumor percentage of 7.5%. It is important to emphasize that reported cases of the Beckwith–Wiedemann syndrome with neoplasia have an ascertainment bias that favors the reporting of cases with tumors; the true frequency of neoplasia may be lower. Although approximately 13% of patients have hemihyperplasia (hemihypertrophy) (84,100), 40% with neoplasia have hemihyperplasia (100). Thus, Beckwith–Wiedemann patients with hemihyperplasia have an increased risk for developing tumors.

Nephroblastoma is reported most commonly followed by adrenal cortical carcinoma, followed by hepatoblastoma. Other reported neoplasms have included malignant lymphoma, neuroblastoma, glioblastoma, rhabdomyosarcoma, pancreatoblastoma, teratoma, adenoma of the adrenal cortex, fibroma of the heart, umbilical myxoma, retroperitoneal ganglioneuroma, carcinoid tumor of the appendix, and fibroadenoma of the breast (9,25a,26,49,67,70,71,79a,84,100,104). Tumors

Table 11–2. A comparison of reported neoplasms associated with the Beckwith–Wiedemann syndrome, hemihyperplasia (hemihypertrophy), and Sotos syndrome[a]

| | Beckwith–Wiedemann syndrome[b] | Hemihyperplasia (hemihypertrophy)[b] | Sotos syndrome[c] |
|---|---|---|---|
| Tumor frequency | 7.5% | 3.8% | 3.9% |
| Method of study | Retrospective literature review | Prospective mail survey | Retrospective literature review |
| Bias | Possible overestimate | Possibly none | Possible overestimate |
| Malignant tumors | Nephroblastoma[d]<br>Adrenal cortical carcinoma[d]<br>Hepatoblastoma[d]<br>Hepatocellular carcinoma<br>Glioblastoma<br>Neuroblastoma<br>Rhabdomyosarcoma<br>Malignant lymphoma<br>Pancreatoblastoma<br>Teratoma | Nephroblastoma[d]<br>Adrenal cortical carcinoma[d]<br>Hepatoblastoma[d]<br>Neuroblastoma<br>Pheochromocytoma<br>Testicular carcinoma<br>Undifferentiated sarcoma | Nephroblastoma[d]<br>Hepatocellular carcinoma[d]<br>Epidermoid carcinoma |
| Benign tumors | Adrenal adenoma<br>Carcinoid tumor<br>Fibroadenoma<br>Fibrous hamartoma<br>Ganglioneuroma<br>Myxoma | Adrenal adenoma | Cavernous hemangioma<br>Hairy pigmented nevus<br>Osteochondroma |

[a]See text and references provided. [From Cohen (19).]
[b]Not all cases of hemihypertrophy are part of the Beckwith–Wiedemann spectrum.
[c]Tumors occur with lower frequency in Sotos syndrome than with either Beckwith–Wiedemann syndrome or hemihyperplasia (hemihypertrophy).
[d]Most commonly observed neoplasms in descending order of frequency.

found with the syndrome are compared with tumors associated with hemihyperplasia (hemihypertrophy) and with Sotos syndrome in Table 11–2. Perhaps there is a common mechanism involving development of Beckwith–Wiedemann syndrome and embryonal tumors that represents somatic development of homozygosity for a mutant allele at a locus on chromosome 11 (44).

**Laboratory findings.** Approximately 30% of reported cases have had symptomatic hypoglycemia (84). The condition is transitory, being responsive to medical therapy with spontaneous regression usually occurring during the first 4 months of life. It is not possible to determine the overall frequency of neonatal hypoglycemia in the syndrome, however, because undetected hypoglycemia may occur in some cases and blood glucose levels have not been given in many reports. Polycythemia has been noted in 20% of the cases. Also reported in a few instances are hypocalcemia, hypercholesterolemia, and hyperlipidemia (84).

High serum $\alpha$-fetoprotein levels have been recorded in some instances and may indicate the presence of nephroblastoma (12), gigantism with visceromegaly (63), or possibly Beckwith–Wiedemann syndrome with associated hypothyroidism. On occasion, hypothyroidism (56), thyroxine-binding globulin deficiency (3), or low uptake with normal levels of protein-bound iodine (92) have been noted. Beckwith–Wiedemann syndrome patients with low serum thyroxine levels should be further evaluated for frank hypothyroidism or simply thyroxine-binding globulin deficiency (46).

**Prenatal diagnosis.** Ultrasound monitoring of fetal size, organ size, and abdominal wall contour may be useful for families with a positive history of the syndrome. Weinstein and Anderson (96) noted increased amniotic fluid, bilateral cystic kidneys, and a larger than expected fetus at 20 weeks. Winter et al (102), using serial ultrasound monitoring, showed enlarged abdominal circumference and an omphalocele at 18 weeks. Omphalocele has also been detected by $\alpha$-fetoprotein (19). MacMillin et al (54) detected omphalocele and polyhydramnios by ultrasonography. Shapiro et al (78) detected a large placenta in a Beckwith–Wiedemann fetus by ultrasound.

**Differential diagnosis.** The facial features and large tongue may, at times, suggest a *mucopolysaccharidosis,* an *oligosaccharidosis,* or hypothyroidism. The combination of macroglossia and umbilical hernia can be seen in both *trisomy 21 syndrome* and hypothyroidism. The syndrome of macroglossia, intrauterine growth retardation, and transient neonatal diabetes mellitus has been discussed by Dacou-Voutetakis et al (22). Isolated muscular macroglossia has been discussed by Bronstein et al (11). Autosomal dominant inheritance of macroglossia has been observed (72). Omphalocele may occur alone or with a variety of associated anomalies and syndromes (18). Familial omphalocele (25,53) and familial gastroschisis and abdominal hernia (91) are also known to occur.

In 1934, de Lange (24) described three children with a distinctive condition that bears some superficial resemblance to the Beckwith–Wiedemann syndrome. Findings include congenital muscular hypertrophy, hypertrophy, hypertonia, and developmental retardation. Macroglossia and large ears were observed in two of the three patients, and overgrowth at birth was a feature in one instance. Autopsy findings in one case demonstrated polymicrogyria and a widespread porencephalic process.

Infants of diabetic mothers sometimes resemble infants with the Beckwith–Wiedemann syndrome in being gigantic and hypoglycemic. Combs et al (21) have pointed out that newborn infants of diabetic mothers are overweight for their length, whereas Beckwith–Wiedemann syndrome patients are not. They further noted that the hypoglycemia seen in the Beckwith–Wiedemann syndrome may persist beyond the immediate neonatal period, whereas the hypoglycemia found in infants of diabetic mothers occurs very early and is of brief duration. Infant giants represent a rarely encountered disorder in which gross macrosomia occurs without associated malformations. Overgrowth is based on prenatal hyperinsulinism that persists into postnatal life. A history of maternal diabetes is always negative and maternal glucose tolerance tests are always normal (18).

Perlman and others (30,31,48,59,64,66,68) have reported a clinicopathologic complex consisting of visceromegaly, macrosomia, renal hamartomas, nephroblastomatosis, and Wilms tumor in the absence of macroglossia, omphalocele, and hemihyperplasia (hemihypertrophy).

In our opinion, these patients belong to the same spectrum of abnormalities as the Beckwith–Wiedemann syndrome and do not constitute a separate entity. Some cases of the syndrome have neither macroglossia nor omphalocele (103); other cases have nephroblastomatosis (14). Wilms tumor may occur in combination with abnormal renal function, and abnormal sexual differentiation leading to various gonadal neoplasms (Drash syndrome) (55).

Sotelo-Avila and his co-workers (84) have noted that the Beckwith–Wiedemann syndrome and hemihyperplasia (hemihypertrophy) are probably at either end of the same spectrum, intermediate forms being the connecting links. In our opinion, *hemihyperplasia (hemihypertrophy)* is etiologically heterogeneous and although some cases represent the Beckwtih–Wiedemann spectrum, other cases do not.

**Laboratory aids.** Early monitoring of blood glucose levels is indicated. High-resolution banding studies should be carried out. Laboratory tests and specialized procedures for the diagnosis of specific congenital heart defects may be indicated. Shah (77) recommends abdominal and renal ultrasound scans at 3-month intervals up to age 5, or more frequently if necessary, and then scans at 6-month intervals until adolescence. Daugbjerg and Everberg (23) recommend hearing tests at intervals from early childhood.

### References [Beckwith–Wiedemann syndrome (EMG [exomphalos–macroglossia–giantism] syndrome)]


1. Arons M et al: The macroglossia of Beckwith's syndrome. *Plast Reconstr Surg* **45**:341–344, 1970.

2. Barlow GB: Excretion of polyamines by children with Beckwith's syndrome. *Arch Dis Childh* **55**:40–42, 1980.

3. Beckwith JB: Extreme cytomegaly of the adrenal fetal cortex, omphalocele, hyperplasia of the kidneys and pancreas, and Leydig-cell hyperplasia: Another syndrome? Presented at the Annual Meeting of the Western Society for Pediatric Research, Los Angeles, Nov 11, 1963.

4. Beckwith JB: Macroglossia, omphalocele, adrenal cytomegaly, gigantism, and hyperplastic visceromegaly. *Birth Defects* **5**(2):188–196, 1969.

5. Ben-Galim E: Beckwith–Wiedemann syndrome in a mother and her son. *Am J Dis Child* **131**:801–803, 1977.

6. Benke PJ: Familial Beckwith–Wiedemann syndrome. March of Dimes Birth Defects Meeting, San Francisco, June, 1978.

7. Berry AC et al: Monozygotic twins discordant for Wiedemann–Beckwith syndrome and the implications for genetic counseling. *J Med Genet* **17**:136–138, 1980.

8. Best LG, Hoekstra RE: Wiedemann–Beckwith syndrome: Autosomal dominant inheritance in a family. *Am J Med Genet* **9**:291–299, 1981.

9. Borenstein TC et al: Congenital gastric teratoma in Beckwith–Wiedemann syndrome. Society for Pediatric Research, Washington, DC, May 1–5, 1989.

10. Bronk JB, Parker BR: Pyelocalyceal diverticula in the Beckwith–Wiedemann syndrome. *Pediatr Radiol* **17**:80–81, 1987.

11. Bronstein IP et al: Macroglossia in children. *Am J Dis Child* **54**:1328–1343, 1937.

12. Brown NJ, Goldie DJ: Beckwith's syndrome with renal neoplasia and alpha-fetoprotein secretion. *Arch Dis Childh* **53**:435, 1978.

13. Carnelutti M: E.M.G. syndrome and carbohydrate metabolism. *Lancet* **1**:374, 1969.

14. Chadarévian JP et al: Massive infantile nephro-blastomatosis. *Cancer* **39**:2294–2304, 1977.

15. Chemke J: Familial macroglossia-omphalocele syndrome. *J Génét Hum* **24**:271–279, 1969.

16. Cohen MM Jr: Comments on the macroglossia-omphalocele syndrome. *Birth Defects* **5**(2):197, 1969.

17. Cohen MM Jr: Macroglossia, omphalocele, visceromegaly, cytomegaly of the adrenal cortex and neonatal hypoglycemia. *Birth Defects* **7**(7):226–232, 1971.

18. Cohen MM Jr: Overgrowth syndrome, in *Associated Congenital Malformations,* El-Shafie M, Klippel CH, (eds), Williams & Wilkins, Baltimore, 1981, pp 71–104.

19. Cohen MM Jr: A comprehensive and critical assessment of overgrowth and overgrowth syndromes. *Adv Hum Genet* **18**:181–303, 1989.

20. Cohen MM Jr et al: Beckwith–Wiedemann syndrome: Seven new cases. *Am J Dis Child* **122**:515–519, 1971.

21. Combs JT et al: New syndrome of neonatal hypoglycemia. *N Eng J Med* **275**:236–243, 1966.

22. Dacou-Voutetakis C et al: Macroglossia, transient neonatal diabetes mellitus and intrauterine growth failure: A new distinct entity? *Pediatrics* **55**:127–131, 1975.

23. Daugberg P, Everberg G: A case of Beckwith–Wiedemann syndrome with conductive hearing loss. *Acta Paediatr Scand* **73**:408–410, 1984.

24. de Lange C: Congenital hypertrophy of the muscles, extrapyramidal motor disturbances and mental deficiency. *Am J Dis Child* **48**:243–268, 1934.

25. DiLiberti JH: Familial omphalocele: Analysis of risk factors and case report. *Am J Med Genet* **13**:263–268, 1982.

25a. Drut R, Jones MC: Congenital pancreatoblastoma in Beckwith–Wiedemann syndrome: An emerging association. *Pediatr Pathol* **8**:331–339, 1988.

26. Emery LG et al: Neuroblastoma associated with Beckwith–Wiedemann syndrome. *Cancer* **52**:176–179, 1983.

27. Engström W et al: Wiedemann–Beckwith syndrome. *Eur J Pediatr* **147**:450–457, 1988.

28. Filippi G, McKusick VA: Beckwith–Wiedemann syndrome (exomphalos-macroglossia-giantism syndrome): Report of two cases and review of the literature. *Medicine* **49**:279–298, 1970.

28a. Friede H, Figueroa AA: The Beckwith–Wiedemann syndrome: A longitudinal study of the macroglossia and dentofacial complex. *J Craniofac Genet Dev Biol* **1**:179–187, 1985.

29. Gardener JI: Pseudo-Beckwith–Wiedemann syndrome: Interaction with diabetes mellitus. *Lancet* **2**:911–912, 1973.

30. Greenberg F et al: The Perlman familial nephroblastomatosis syndrome. *Am J Med Genet* **24**:101–110, 1986.

31. Greenberg F et al: Expanding the spectrum of the Perlman syndrome. *Am J Med Genet* **29**:773–776, 1988.

32. Greene RJ et al: Immunodeficiency associated with exomphalos-macroglossia-gigantism syndrome. *J Pediatr* **82**:814–820, 1973.

33. Greenwood RD et al: Cardiovascular abnormalities in the Beckwith–Wiedemann syndrome. *Am J Dis Child* **131**:293–294, 1977.

34. Gustavson KH et al: A 4-5/21-22 chromosomal translocation associated with multiple congenital anomalies. *Acta Paediatr Scand* **53**:172–181, 1964.

35. Haas OA et al: Das Wiedemann–Beckwith-Syndrom: Klinische Characteristik konstitutionelle Chromosomeanomalien and Tumorinzidenz. *Klin Pädiatr* **199**:283–291, 1987.

35a. Henry I et al: Molecular definition of the 11p15.5 region involved in the Beckwith–Wiedemann and probably in predisposition to adrenocortical carcinoma. *Hum Genet* **81**:273–274, 1989.

36. Herrmann J, Opitz JM: Delayed mutation as a cause of genetic disease in man: Achondroplasia and the Wiedemann–Beckwith syndrome, in *Regulation of Cell Proliferation and Differentiation,* Nichols WW, Murphy DG (eds), Plenum Press, New York, 1977.

37. Herrmann J et al: Dysplasia, malformations and cancer, especially with respect to the Wiedemann–Beckwith syndrome, in *Regulation of Cell Proliferation and Differentiation,* Nichols WW, Murphy DG (eds), Plenum Press, New York, 1977.

38. Irving I: Exomphalos with macroglossia: A study of 11 cases. *J Pediatr Surg* **2**:499–507, 1967.

39. Irving I: The EMG syndrome (exomphalos, macroglossia, gigantism), in *Progress in Pediatrics,* Vol. 1, Rickham RP, Hacker WC, Prevolt J (eds), Urban and Schwarzenberg, Munich, 1970, pp 1–61.

40. Knight JA et al: Association of the Beckwith–Wiedemann and prune belly syndromes. *Clin Pediatr* **19**:485–488, 1980.

41. Koh THHG et al: Pancreatoblastoma in a neonate with Wiedemann–Beckwith syndrome. *Eur J Pediatr* **145**:435–438, 1986.

42. Kosseff AL et al: The Wiedemann–Beckwith syndrome: Genetic considerations and a diagnostic sign. *Lancet* **1**:844, 1972.

43. Kosseff AL et al: The Wiedemann–Beckwith syndrome: Clinical, genetic and pathogenetic studies of 12 cases. *Eur J Pediatr* **123**:139–166, 1976.

44. Koufos A et al: Loss of heterozygosity in three embryonal tumors suggests a common pathogenetic mechanism. *Nature (London)* **316**:330–334, 1985.

44a. Koufos A et al: Familial Wiedemann–Beckwith syndrome with a second Wilms tumor both mapped to 11p15.5 *Am J Hum Genet* **44**:711–719. 1989.

45. Lazarus L: EMG syndrome and carbohydrate metabolism. *Lancet* **2**:1347–1348, 1968.

46. Leung AKC: Wiedemann–Beckwith syndrome and hypothyroidism. *Eur J Pediatr* **144**:295, 1985.

47. Leung AKC et al: Thyroxine-binding globulin deficiency in Beckwith syndrome. *J Pediatr* **95**:752–753, 1979.

48. Liban E, Kozenitzky FL: Metanephric hamartomas and nephroblastomatosis in siblings. *Cancer* **25**:885–888, 1970.

49. Little MH et al: Loss of aileles on the short arm of chromosome 11 in a hepatoblastoma from a child with Beckwith–Wiedemann syndrome. *Hum Genet* **79**:186–189, 1988.

50. Litz CE et al: Absence of detectable chromosomal and molecular abnormalities in monozygotic twins discordant for the Wiedemann–Beckwith syndrome. *Am J Med Genet* **30**:821–833, 1988.
51. Lubinsky M et al: Autosomal dominant sex-dependent transmission of the Wiedemann–Beckwith syndrome. *Lancet* **1**:932, 1974.
52. Lunt PW et al: Concordant and discordant MZ twins in Beckwith syndrome—related but different aetiologies? 3rd Manchester Birth Defects Conference, Manchester, UK, October 25–28, 1988.
53. Lurie IW, Ilyina HG: Familial omphalocele and recurrence risk. *Am J Med Genet* **17**:541–543, 1984.
54. MacMillin MD et al: Prenatal diagnosis of Beckwith–Wiedemann syndrome by ultrasound in three pregnancies. 39th Annual Meeting American Society of Human Genetics, New Orleans, Lousiana, October 12–15, 1988, Abstract No. 953.
55. Manivel JC et al: Complete and incomplete Drash syndrome. *Hum Pathol* **18**:80–89, 1987.
56. Martínez R et al: The Wiedemann–Beckwith syndrome in four sibs including one with associated congenital hypothyroidism. *Eur J Pediatr* **143**:233–235, 1985.
57. McCarten KM et al: Renal ultrasound in Beckwith–Wiedemann syndrome. *Pediatr Radiol* **11**:46–48, 1981.
58. McManamny DS, Barnett JS: Macroglossia as a presentation of the Beckwith–Wiedemann syndrome. *Plast Reconstr Surg* **75**:170–176, 1985.
59. Neri G et al: The Perlman syndrome: Familial renal dysplasia with Wilms tumor, fetal gigantism, and multiple congenital anomalies. *Am J Med Genet* **19**:195–207, 1984.
60. Niikawa N et al: The Wiedemann–Beckwith syndrome: Pedigree studies on five families with evidence for autosomal dominant inheritance with variable expressivity. *Am J Med Genet* **24**:41–56, 1986.
61. Olney AH et al: Wiedemann–Beckwith syndrome in apparently discordant monozygotic twins. *Am J Med Genet* **29**:491–499, 1988.
62. Patterson GT et al: Macroglossia and ankyloglossia in Wiedemann–Beckwith syndrome. *Oral Surg* **65**:29–31, 1988.
63. Pavanello L et al: α-Fetoprotein in Wiedemann–Beckwith syndrome. *J Pediatr* **109**:391–392, 1986.
64. Perlman M: Letter to the editor: Perlman syndrome: Familial renal dysplasia with Wilms tumor, fetal gigantism, and multiple congenital anomalies. *Am J Med Genet* **25**:793–795, 1986.
65. Perlman M et al: Renal hamartomas, nephroblastomatosis, and fetal gigantism. *J Pediatr* **83**:414–418, 1973.
66. Perlman M et al: Syndrome of fetal gigantism, renal hamartomas, and nephroblastomatosis with Wilms' tumor. *Cancer* **34**:1212–1217, 1975.
67. Pettenati MJ et al: Wiedemann–Beckwith syndrome: Presentation of clinical and cytogenetic data on 22 new cases and review of the literature. *Hum Genet* **74**:143–154, 1986.
68. Piussan CH et al: Syndrome de Wiedemann et Beckwith: Une nouvelle observation familiale. *J Génét Hum* **28**:281–291, 1980.
69. Punnett HH et al: An (X;1) translocation, balanced, 46 chromosomes: Repository identification No. GM-97. *Cytogenet Cell Genet* **13**:406–407, 1974.
70. Raine PA et al: Breast fibroadenoma and cardiac anomaly associated with EMG (Beckwith–Wiedemann) syndrome. *J Pediatr* **94**:633–634, 1979.
71. Reddy JK et al: The Beckwith–Wiedemann syndrome. Wilms' tumor, cardiac hamartoma, persistent visceromegaly and glomeruloneogenesis in a 2-year-old boy. *Arch Pathol* **94**:523–531, 1972.
72. Reynoso MC et al: Autosomal dominant macroglossia in two unrelated families. *Hum Genet* **74**:200–202, 1986.
73. Ruffié J et al: Remaniements chromosomiques complexes portant sur les autosomes s'accompagnant d'anomalies crânio-faciales et d'un omphalocele. *CR Seances Acad Sci (III)* **252**:386–389, 1966.
74. Saal H et al: High resolution cytogenetics and molecular hybridization of the Beckwith–Wiedemann syndrome. *Am J Med Genet (Suppl)* **36**:110S, 1984.
75. Schulz CJ, Sherman SL: The Beckwith–Wiedemann syndrome: Investigation of parental ages of isolated cases. 39th Annual Meeting American Society of Human Genetics, New Orleans, Louisiana, October 12–15, 1988, Abstract No. 275.
76. Shafer AD: Primary macroglossia. *Clin Pediatr* **7**:353–363, 1968.
77. Shah KJ: Beckwith–Wiedemann syndrome: Role of ultrasound in its management. *Pediatr Radiol* **34**:313–319, 1983.
78. Shapiro LR et al: The placenta in familial Beckwith–Wiedemann syndrome. *Birth Defects* **18**(3B):203–206, 1982.
79. Sippell WG et al: Growth, bone maturation and pubertal development in children with the EMG-syndrome. *Clin Genet* **35**:20–28, 1989.
79a. Sirinelli D et al: Beckwith–Wiedemann syndrome and neural crest tumors. *Pediatr Radiol* **19**:242–245, 1989.
80. Smith DF et al: Chronic alveolar hypoventilation secondary to macroglossia in the Beckwith–Wiedemann syndrome. *Pediatrics* **70**:695–697, 1982.
81. Sommer A et al: Familial occurrence of Wiedemann–Beckwith syndrome and persistent fontanel. *Am J Med Genet* **1**:59–63, 1977.
82. Sommer A et al: Familial occurrence of the Wiedemann–Beckwith syndrome. *Birth Defects* **14**(6B):178–179, 1978.
83. Sotelo-Avila C, Singer DB: Syndrome of hyperplastic fetal visceromegaly and neonatal hypoglycemia (Beckwith's syndrome: A report of seven cases). *Pediatrics* **46**:240–251, 1970.
84. Sotelo-Avila C et al: Complete and incomplete forms of the Beckwith–Wiedemann syndrome: Their oncogenic potential. *J Pediatr* **96**:47–50, 1980.
85. Spencer GSG et al: Raised somatomedin associated with normal growth hormone. *Arch Dis Childh* **55**:151–153, 1980.
86. Spritz RA et al: Brief communication: Normal dosage of the insulin and insulin-like growth factor II genes in patients with the Beckwith–Wiedemann syndrome. *Am J Hum Genet* **39**:265–273, 1986.
86a. Takato T et al: Cleft palate in the Beckwith–Wiedemann syndrome. *Ann Plast Surg* **22**:347–349, 1989.
87. Taylor WN: Urological implications of the Beckwith–Wiedemann syndrome. *J Urol* **125**:439–441, 1981.
88. Thorburn MJ et al: Exomphalos-macroglossia-gigantism syndrome in Jamaican infants. *Am J Dis Child* **119**:316–321, 1970.
89. Tovar JA et al: L'hérédite du syndrome de Wiedemann–Beckwith. *Chir Pédiatr* **20**:187–189, 1979.
90. Turleau C et al: Trisomy 11p15 and Beckwith–Wiedemann syndrome. A report of two cases. *Hum Genet* **67**:219–221, 1984.
91. Ventruto V et al: Gastroschisis in two sibs with abdominal hernia in maternal grandfather and greatgrandfather. *Am J Med Genet* **21**:405–407, 1985.
92. Vidailhet M et al: Syndrome de Beckwith–Wiedemann. Étude de quatre observations, dont une avec néphroblastome. *Arch Fr Pédiatr* **28**:1011–1012, 1971.
93. Virdis R et al: Hypertension and medullary sponge kidneys in an adolescent with Beckwith–Wiedemann syndrome. *Pediatrics* **91**:761–763, 1977.
94. Wales JKH et al: Bronze baby syndrome, biliary hypoplasia, incomplete Beckwith–Wiedemann syndrome and partial trisomy 11. *Eur J Pediatr* **145**:141–143, 1986.
95. Waziri M et al: Abnormality of chromosome 11 in patients with features of Beckwith–Wiedemann syndrome. *J Pediatr* **102**:873–876, 1983.
96. Weinstein L, Anderson C: In utero diagnosis of Beckwith–Wiedemann syndrome by ultrasound. *Radiology* **134**:474, 1980.
97. Wiedemann H-R: Complexe malformatif familial avec hernia ombilicale et macroglossie—un syndrome nouveau? *J Génét Hum* **13**:223–232, 1964.
98. Wiedemann H-R: Das EMG Syndrom: Exomphalos, Makroglossie, Gigantismus aund Kohlenhydratstoffwechselstörung. *Z Kinderheilkd* **102**:1–36, 1968.
99. Wiedemann H-R: Exomphalos-Makroglossie-Gigantismus-Syndrom, Berardinelli-Seip-Syndrom und Sotos-Syndrom—ein vergleichende Betrachtung unter ausgewählten Aspekten. *Z Kinderheilkd* **115**:95–110, 1973.
100. Wiedemann H-R: Tumours and hemihypertrophy associated with Wiedemann–Beckwith syndrome. *Eur J Pediatr* **141**:129, 1983.
101. Wiedemann H-R et al: Über das Syndrom Exomphalos-Makroglossie-Gigantismus, über generalisierte Muskelhypertrophie, progressive Lipodystrophie und Miescher-Syndrom in Sinne diencephaler Syndrome. *Z Kinderheilkd* **102**:1–36, 1968.
101a. Wiedemann H-R: Genital overgrowth in the EMG syndrome. *Am J Med Genet* **32**:255–256, 2989.
102. Winter SC et al: Prenatal diagnosis of the Beckwith–Wiedemann syndrome. *Am J Med Genet* **24**:137–142, 1986.
103. Wöckel W et al: A variant of the Wiedemann–Beckwith syndrome. *Eur J Pediatr* **135**:319–324, 1981.
104. Wojciechowski AH, Pritchard JH: Beckwith–Wiedemann (exomphalos-macroglossia-gigantism-EMG) syndrome and malignant lymphoma. *Eur J Pediatr* **137**:317–321, 1981.

## Hemihyperplasia (hemihypertrophy)

Hemihyperplasia (hemihypertrophy), described in 1822 by Meckel (31) and in 1839 by Wagner (47), gained widespread recognition after the studies of Gesell (14) and Lenstrup (27) during the 1920s. Comprehensive reviews are those of Wakefield and Hines (48), Ward and Lerner (49), Ringrose et al (38), Parker and Skalko (34), Gorlin and Meskin (17), Cohen (4,6), Bell and McTigue (2a), and Pollock et al (35a). Contributions have been made by many authors (1,3,8,11, 12,16,18–20,37).

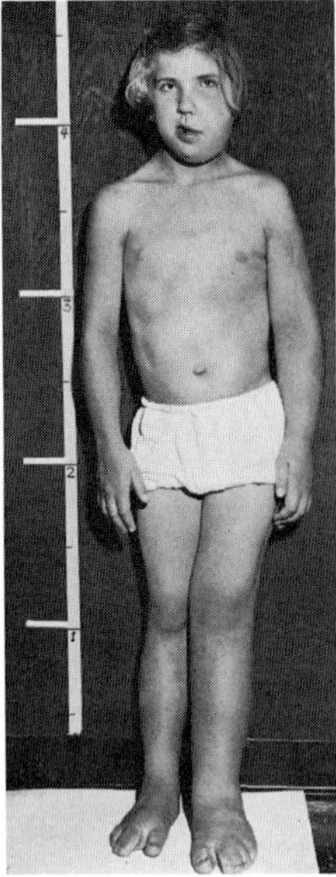

Fig. 11–6. *Hemihyperplasia.* Note asymmetry of body, with complete left-sided hemihyperplasia. Note also syndactyly of toes.

Although the term hemihypertrophy has been used conventionally and frequently in the medical literature, it is inappropriate, as the condition so obviously refers to hemihyperplasia. We will use the term hemihyperplasia with the term hemihypertrophy in parentheses as an interim measure. The differences between common asymmetry, hemihyperplasia, hypertrophy, hemiatrophy, and preferential laterality have been discussed by Cohen (5). In hemihyperplasia, the enlarged area may vary from a single digit, a single limb, or unilateral facial enlargement to involvement of half the body (16,17,38,45,46). Hemihyperplasia may be segmental, unilateral, or crossed. About 2% are crossed (35a). In some cases, the defect is limited to a single system, for example, muscular, vascular, skeletal, or nervous system, but it may frequently involve multiple systems (Figs. 11–6 to 11–9) (46).

Fig. 11–7. *Hemihyperplasia.* Unilateral hyperplasia of face and tongue.

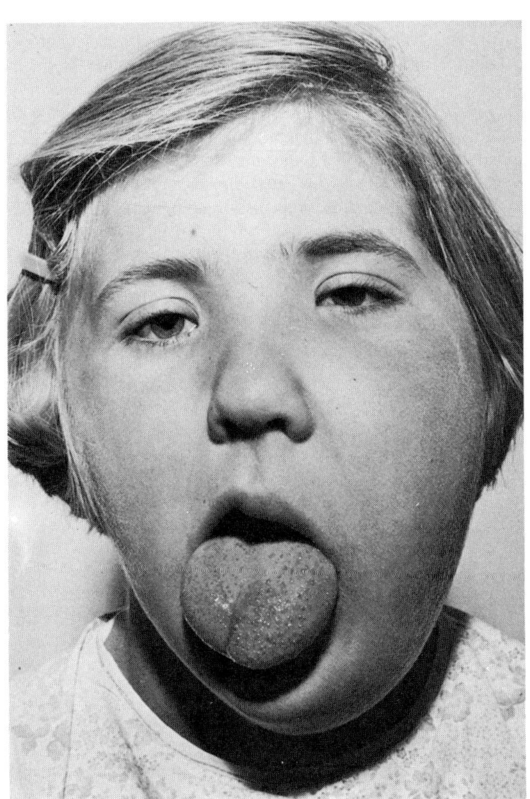

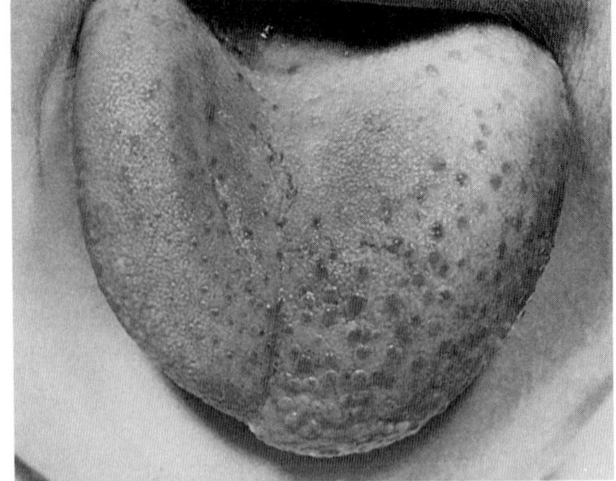

A

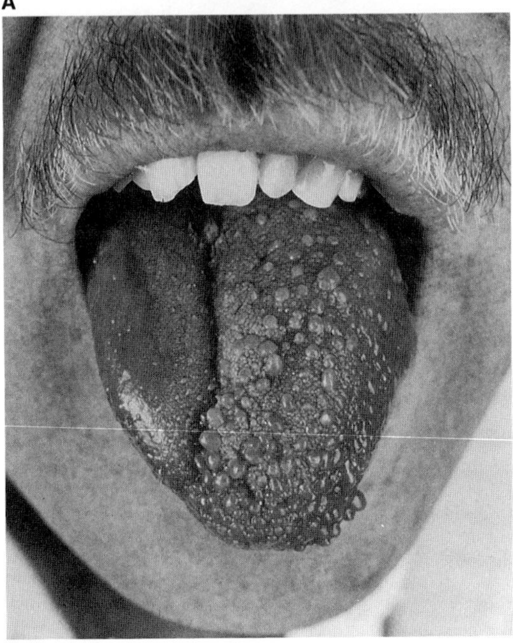

B

Fig. 11–8. *Hemihyperplasia.* (A,B) Marked hemihyperplasia of tongue with enlargement of fungiform papillae. Note sharp demarcation at midline. (B courtesy of *BB Horswell,* Farmington, Connecticut.)

Almost all cases are sporadic if incomplete forms of the Beckwith–Wiedemann syndrome and neurofibromatosis are excluded. Hoyme et al (21) found an approximately 2:1 right-sided predominance and a 2:1 female predilection in 104 cases.

The etiology and pathogenesis are poorly understood. A tendency toward dizygotic twinning has been observed in some cases (14). Chromosomal anomalies, including diploid/triploid mosaicism, trisomy 18 mosaicism, translocation 13q;7p, partial G and B monosomy with B/G translocation, and abnormally large chromosome number 3 have been reported (12,23,29). Many other theories have been advanced to explain hemihyperplasia, including anatomic and functional vascular or lymphatic abnormalities, lesions of the central nervous system leading to altered neurotrophic action, endocrine abnormalities, asymmetric cell division and deviation of the twinning process, fusion of two eggs following fertilization leading to unequal regulative ability in the two halves, mitochondrial damage to an overripened egg leading to overregeneration and unilateral enlargement of the neural tube, and proliferation of neural crest cells (33,35a). The range and variability of clinical abnormalities, together with the large number of sporadic cases, suggest etiologic heterogeneity. None of the proposed

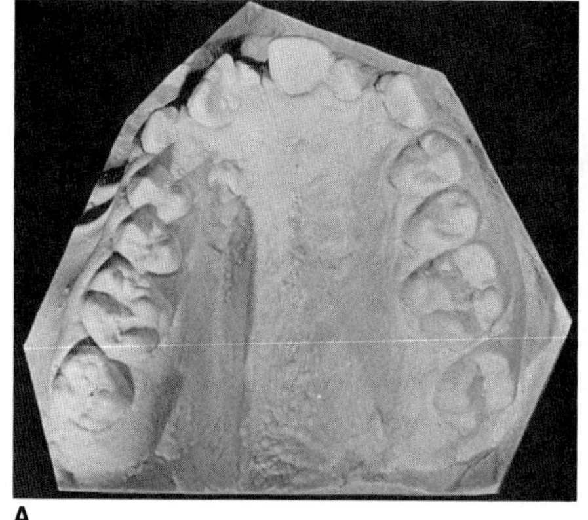


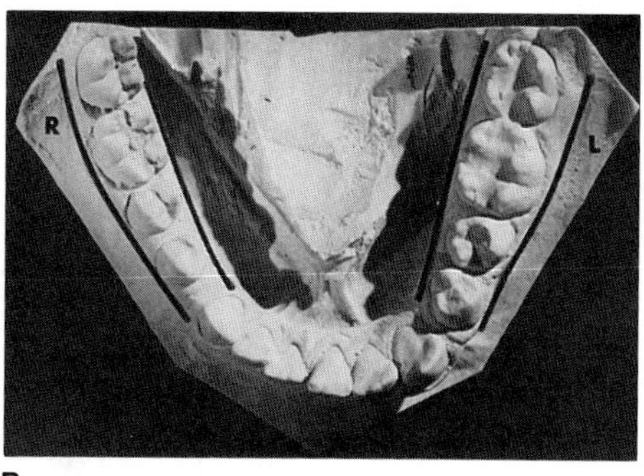


Fig. 11–9. *Hemihyperplasia.* (A,B) Casts of jaws. Note differences in width of bone and size of teeth on affected and normal sides. (From *RJ Gorlin* and *L Meskin,* J Pediatr **61:**870, 1962.)

theories, some of which are quite fanciful, explains adequately all cases of hemihyperplasia.

**Clinical manifestations.** Asymmetry is usually evident at birth and may become accentuated with age, especially at puberty. Occasionally, asymmetry has been stated not to be present at birth, but to develop later (3). However, such observations are valid only when measurements are taken at birth. The bones have been found to be unilaterally enlarged and increased bone age on the affected side has been reported (34). A variety of nonneoplastic abnormalities have been observed to affect the limbs, teeth, skin, central nervous sytem, cardiovascular system, liver, kidneys, and genitalia (Figs. 11–6 to 11–9). Many miscellaneous anomalies have also been noted. Table 11–3 lists the many observed abnormalities by system with appropriate references. The most extensive reviews of abnormalities by system and miscellaneous anomalies are those of Ringrose et al (38), Parker and Skalko (34), Gorlin and Meskin (17), and Bell and McTigue (2a).

Table 11–3. Miscellaneous abnormalities associated with hemihyperplasia (hemihypertrophy)[a]

| | |
|---|---|
| Skin | Central nervous system |
| Nevi (38) | Hemimegalencephaly (14, 38) |
| Pigmentation (28b,38) | Cerebral hemiatrophy (50) |
| Cutis marmorata (38) | Macrocephaly (43) |
| Telangiectasis (38,46) | Cyst of septum pellucidum (34) |
| Nevus flammeus (14,38,43,49) | Mental deficiency (32,46) |
| Coarse skin on affected side (38) | Seizures (34,37,43,46) |
| Ichthyosis on affected side (7) | Cardiovascular |
| Hirsutism (38) | Congenital heart defects (38,48) |
| Hypertrichosis (22) | Liver |
| Thicker hair on affected side (38) | Cyst (36) |
| Excessive secretions of sebaceous and sweat glands (40) | Focal nodular hyperplasia (10) |
| Increased skin temperature on affected side (38) | Kidney |
| Limb | Medullary sponge kidney (9,18,19b,28,41a,43a,45) |
| Macrodactyly (16,38) | Unilateral nephromegaly (24) |
| Polydactyly (38) | Abnormal collecting system (34) |
| Syndactyly (38) | Polycystic kidneys (39) |
| Clubfoot (38) | Genitalia |
| Skeletal | Hypospadias (34,38) |
| Increased bone age on affected side (14) | Cryptorchidism (34,38) |
| Compensatory scoliosis (38) | Macropenis (38) |
| Hip dysplasia (38) | Enlarged testis on affected side (34) |
| Dental, oral | Clitoromegaly (38) |
| Enlarged teeth on affected side (17,40) | Other |
| Early eruption of teeth on affected side (17) | Strabismus (38) |
| Abnormal tooth roots (40,41) | Tracheoesophageal fistula (34) |
| Enlarged alveolar ridge on affected side (17) | Supernumerary nipples (38) |
| Enlarged tongue on affected side (17) | Umbilical hernia (38) |
| | Inguinal hernia (34,38) |
| | Short stature (38) |

[a]See especially the reviews of Ringrose et al (38), Parker and Skalko (34), Gorlin and Meskin (17), and Bell and McTigue (2a) for many other anomalies.

**Neoplasms.** Various neoplasms have been reported in association with hemihyperplasia including, most commonly, Wilms tumor, adrenal cortical carcinoma, and hepatoblastoma, in that order. Other tumors have also been noted: neuroblastoma, pheochromocytoma, testicular carcinoma, undifferentiated sarcoma of the lung, and adrenal adenoma (3,11,12,17,20,21,28a,34,35,42–44). In the study of Hoyme et al (21) of 104 cases of isolated hemihyperplasia, 3.8% developed tumors. All known instances of neoplasia associated with nonsyndromic hemihyperplasia have been sporadic, except for the unusual family reported by Meadows (30): a mother with hemihyperplasia gave birth to three children each of whom developed Wilms tumor but each of whom had no evidence of hemihyperplasia. Neoplasms associated with isolated, nonsyndromic hemihyperplasia are compared with those that occur with the Beckwith–Wiedemann syndrome and Sotos syndrome in Table 11–2.

Since tumors that occur with hemihyperplasia have embryonal origin, study of their relationship to the teratogenic aspects of hemihyperplasia may lead to more specific information on oncogenic mechanisms. There is no relationship between laterality of hemihyperplasia and any of the solid tumors reported. When the oncogenic stimulus does not lateralize to the enlarged side, it is possible that such cases may represent "occult" crossed hemihyperplasia, affecting the internal organs of the contralateral side.

**Differential diagnosis.** So-called familial instances of hemihyperplasia are frequently incompletely documented and may represent other disorders, particularly *neurofibromatosis.* There is no question that the spectrum of the *Beckwith–Wiedemann syndrome* shades over into hemihyperplasia. However, not all cases of hemihyperplasia represent the Beckwith–Wiedemann syndrome, particularly those sporadically occurring cases with no other signs and symptoms of the latter (4). Hemihyperplasia may also be observed with the *linear verrucous epidermal nevus syndrome* (15) and with the syndrome of hemihypesthesia, hemiareflexia, and scoliosis (2). The HIPO syndrome consists of hemihyperplasia, intestinal webs, preauricular tags, and cloudy cornea (19a).

Asymmetry may also occur with arteriovenous aneurysm, congenital lymphedema, *Silver–Russell syndrome, McCune–Albright syndrome, Klippel–Trénaunay–Weber syndrome, Proteus syndrome,* multiple exostoses, Ollier syndrome, *Maffucci syndrome, Langer–*

*Giedion syndrome,* and facial tumors of childhood (13). Hemiatrophy may occur secondarily to early central nervous system insult and may also occur in *Romberg syndrome, Sturge–Weber angiomatosis,* and unilateral ichthyosiform erythroderma (7). *Bencze syndrome* (26) is an autosomal dominantly inherited form of facial asymmetry associated with esotropia, amblyopia, and submucous cleft palate.

**Laboratory aids.** Patients with isolated, nonsyndromic hemihyperplasia should be screened with ultrasound for tumors, especially Wilms tumor, every 6 months for the first 4 years of life and probably less frequently thereafter until age 7. Chromosome studies may, on occasion, reveal mosaicism of various types (12,23,29).

### References [Hemihyperplasia (hemihypertrophy)]

1. Arnold EB: Case of hemiacromegaly. *Int J Orthodont* **22**:1228–1233, 1933.

2. Andermann E et al: A syndrome of hemihypertrophy, hemihypaesthesia, hemiareflexia and scoliosis in 3 unrelated girls. *Fifth Int Conf Birth Defects,* Montreal, Quebec, August 21–27, 1977.

2a. Bell RA, McTigue DJ: Complex congenital hemihypertrophy: A case report and literature review. *J Pedodont* **8**:300–313, 1984.

3. Boxer LA, Smith DL: Wilms' tumor prior to onset of hemihypertrophy. *Am J Dis Child* **120**:564–565, 1970.

4. Cohen MM Jr.: Overgrowth syndromes, in *Associated Congenital Malformations,* El-Shafie M, Klippel CH (eds), Williams & Wilkins, Baltimore, 1981, pp 71–104.

5. Cohen MM Jr: *The Child with Multiple Birth Defects,* Raven Press, New York, 1982, pp 111–114.

6. Cohen MM Jr: A comprehensive and critical assessment of overgrowth and overgrowth syndromes. *Adv Hum Genet* **18**:181–303, 1989.

7. Cullen SI et al: Congenital unilateral ichthyosiform erythroderma. *Arch Dermatol* **99**:724–729, 1969.

8. Curtis F: Kongenitaler partieller Riesenwuchs mit endokrinen Störungen. *Dtsch Arch Klin Med* **147**:310–319, 1925.

9. Eisenberg RL, Pfister RC: Medullary sponge kidney associated with congenital hemihypertrophy (asymmetry). *Am J Roentgenol* **116**:773–777, 1972.

10. Everson RB et al: Focal nodular hyperplasia of the liver in a child with hemihypertrophy. *J Pediatr* **88**:985–987, 1976.

11. Fraumeni JF, Miller RW: Adrenocortical neoplasms with hemihypertrophy, brain tumors, and other disorders. *J Pediatr* **70**:129–138, 1967.

12. Fraumeni JF et al: Wilms' tumor and congenital hemihypertrophy: Report of five new cases and review of literature. *Pediatrics* **40**:886–899, 1967.

13. Furnas DW et al: Congenital hemihypertrophy of the face: Impersonator of childhood facial tumors. *J Pediatr Surg* **5**:344–348, 1970.

14. Gesell A: Hemihypertrophy and twinning: Further study of the nature of hemihypertrophy with report of a new case. *Am J Med Sci* **173**:542–555, 1927.

15. Goldschmidt H et al: Hemihypertrophie, Naevus sebaceus, multiple Knochenzysten und zerebroretinale Angiomatose: eine Komplexe Phakomatose. *Helv Paediatr Acta* **31**:487–498, 1976.

16. Gonzalez-Crussi F et al: The pathology of congenital localized gigantism. *Plast Reconstr Surg* **59**:411–417, 1977.

17. Gorlin RJ, Meskin LH: Congenital hemihypertrophy. *J Pediatr* **61**:870–879, 1962.

18. Groff DB, Buchino JJ: A child with hemihypertrophy and a right flank mass. *J Pediatr* **100**:500–504, 1982.

19. Haicken BN: Congenital hemihypertrophy. *Am J Dis Child* **120**:372–373, 1970.

19a. Hanley TB, Simon JW: Congenital HIPO syndrome. *Ann Ophthalmol* **16**:342–344, 1984.

19b. Harris RE et al: Medullary sponge kidney and congenital hemihypertrophy: Case report and literature review. *Urology* **126**:676–678, 1981.

20. Hennessy WT et al: Congenital hemihypertrophy and associated abdominal lesions. *Urology* **18**:576–579, 1981.

20a. Hirano T, Seeler RA: Congenital virilizing adrenal hyperplasia, seg mental hypertrophy, macrodactyly and cystic lymphangioma. *J Pediatr* **93**:326–327, 1978.

21. Hoyme HE et al: The incidence of neoplasia in children with isolated congenital hemihypertrophy. David Smith Meeting on Malformations and Morphogenesis, Burlington, Vermont, Aug 11–14, 1986.

22. Hurwitz S, Klaus SN: Congenital hemihypertrophy with hypertrichosis. *Arch Dermatol* **103**:98–100, 1971.

23. Johnston AW, Penrose LS: Congenital asymmetry. *J Med Genet* **3**:77–85, 1966.

24. Kirks DR, Schackeford GD: Idiopathic hemihypertrophy with associated ipsilateral benign nephromegaly. *Radiology* **115**:145–148, 1975.

25. Kogon SL et al: Hemifacial hypertrophy affecting the maxillary dentition. *Oral Surg Oral Med Oral Pathol* **58**:549–553, 1984.

26. Kurnit D et al: An autosomal dominantly inherited syndrome of facial asymmetry, esotropia, amblyopia and submucous cleft palate (Bencze syndrome). *Clin Genet* **16**:301–304, 1979.

27. Lenstrup E: Eight cases of hemihypertrophy. *Acta Paediatr Scand* **6**:205–213, 1926.

28. Levy M et al: Hemihypertrophy and medullary sponge kidney. *Can Med Assoc J* **96**:1322–1326, 1967.

28a. Lewis D, Geschickter CF: Tumors of the sympathetic nervous system. Neuroblastoma, paraganglioma, ganglioneuroma. *Arch Surg* **28**:16–22, 1934.

28b. Loh HS: Congenital hemifacial hypertrophy. *Br Dent J* **153**:111–112, 1982.

29. Marçallo FA et al: Hemihypotrophy in a girl with a translocation t(13q;7p). *Eur J Pediatr* **124**:167–171, 1977.

30. Meadows AT: Wilms' tumor in three children of a woman with hemihypertrophy. *N Eng J Med* **291**:23, 1974.

31. Meckel JF: Über die seitliche Asymmetrie im tierischen Körper, anatomische physiologische Beobachtungen und Untersuchungen. Renger, Haile, 1822, p 147.

32. Morris JV, MacGillivray RC: Mental deficiency and hemihypertrophy. *Am J Ment Defic* **59**:644–651, 1955.

33. Noé O, Berman HH: The etiology of congenital hemihypertrophy and one case report. *Arch Pediatr* **79**:278–288, 1962.

34. Parker DA, Skalko RG: Congenital asymmetry: Report of 10 cases with associated developmental abnormalities. *Pediatrics* **44**:584–589, 1969.

35. Parra G et al: Congenital total hemihypertrophy and carcinoma of undescended testicle. *J Urol* **113**:343–344, 1977.

35a. Pollock RA et al: Congenital hemifacial hyperplasia: An embryologic hypothesis and case report. *Cleft Palate J* **22**:173–184, 1985.

36. Porter FN: Hemihypertrophy associated with a benign congenital liver cyst. *Eur J Pediatr* **139**:89–90, 1982.

37. Reed EA: Congenital total hemihypertrophy. *Arch Neurol Psychiatr* **14**:824–827, 1925.

38. Ringrose RE et al: Hemihypertrophy. *Pediatrics* **36**:434–448, 1965.

39. Ritter R: Hemihypertrophy in a boy with renal polycystic disease. *Pediatr Radiol* **5**:98–102, 1976.

40. Rudolph CD, Norvold RW: Congenital partial hemihypertrophy involving marked malocclusion. *J Dent Res* **23**:133–139, 1944.

41. Rushton MA: A dental abnormality of size and race. *Proc R Soc Med* **41**:490–496, 1948.

41a. Saypol DC, Laudone VP: Congenital hemihypertrophy with adrenal carcinoma and medullary sponge kidney. *Urology* **21**:510–511, 1983.

42. Schnakenburg K et al: Congenital hemihypertrophy and malignant giant pheochromocytoma—a previously undescribed coincidence. *Eur J Pediatr* **122**:263–273, 1976.

43. Stephen MJ et al: Macrocephaly in association with unusual cutaneous angiomatosis. *J Pediatr* **87**:353–359, 1975.

43a. Thompson IA: Medullary sponge kidney and congenital hemihypertrophy. *South Med J* **80**:1455–1456, 1987.

44. Tommasi AF, Jitomirski F: Crossed congenital hemifacial hemihyperplasia. *Oral Surg Oral Med Oral Pathol* **67**:190–192, 1989.

45. Tomooka Y et al: Congenital hemihypertrophy with adrenal adenoma and medullary sponge kideny. *Br J Radiol* **61**:851–852, 1988.

46. Viljoen D et al: Manifestations and natural history of idiopathic hemihypertrophy: A review of eleven cases. *Clin Genet* **26**:81–86, 1984.

47. Wagner R, see Kottmeier HL: Über Hemihypertrophia und Hemiatrophia corporis totalis nebst spontane Extremitätengangräne bei Sauglingen im Anschluss zu einem ungewöhnlichen Fall. *Acta Paediatr Scand* **20**:543, 1938.

48. Wakefield EG, Hines EA Jr: Congenital hemihypertrophy: A report of eight cases. *Am J Med Sci* **195**:493–500, 1933.

49. Ward J, Lerner HH: A review of the subject of congenital hemihypertrophy and a complete case report. *J Pediatr* **31**:403–414, 1947.

50. Wyler AR, Ward AA: Cranial asymmetry secondary to unilateral hemispheric damage during late childhood. *J Neurosurg* **52**:423–425, 1980.

## Sotos syndrome (cerebral gigantism)

Sotos syndrome, known also as cerebral gigantism, is an overgrowth syndrome characterized by increased birth weight, excessive growth during the first 4 years of life, advanced bone age, and distinctive facial features including macrodolichocephaly, hypertelorism, and

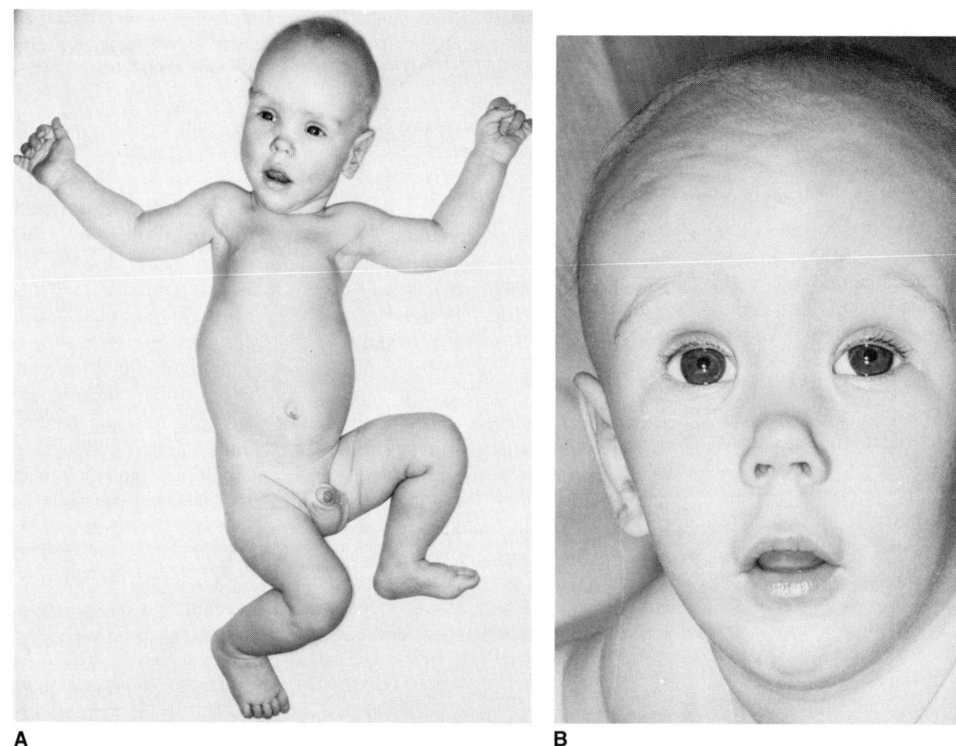

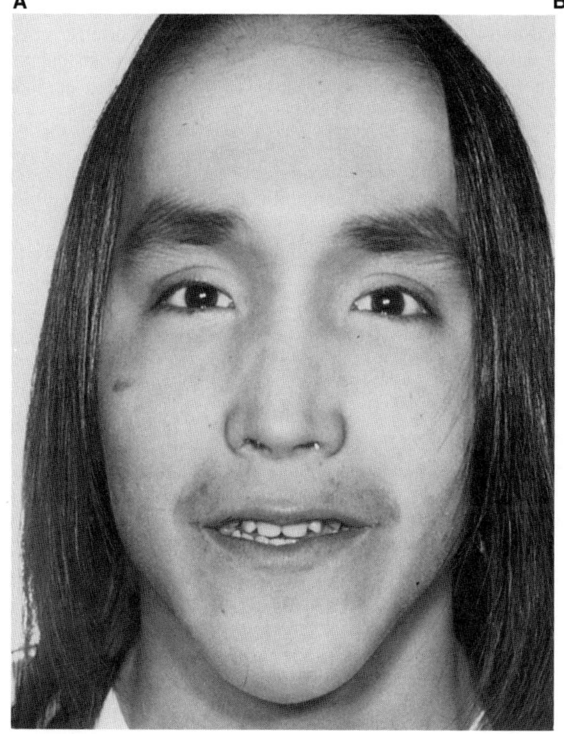

Fig. 11–10. *Sotos syndrome*. (A) Eighteen-month-old male, showing typical facies. Note receding hairline and small nose. (B) Six-month-old female with characteristic facies. (C) Seventeen-year-old American Indian boy with Sotos syndrome. (A courtesy of *MM Steiner*, Chicago, Illinois.)

prominent mandible (Fig. 11–10A–C). First reported by Sotos et al (48), in 1964, over 200 cases have been documented to date. Extensive reviews are those of Jaeken et al (27), Maes et al (32), Sakano et al (44), Sotos et al (49), Cohen (14,15), Dodge et al (17), and Wit et al (59). Growth and development are extensively dealt with by Wit et al (59). Several publications on the subject actually represent other entities (19,22,23,38,42).

Gigantism in Sotos syndrome is assumed to be of hypothalamic origin. The syndrome is frequently discussed in the context of other "diencephalic syndromes of childhood," such as Beckwith–Wiedemann syndrome, lipodystrophy, and the diencephalic emaciation syndrome (41,56).

The etiology is unknown. It must be emphasized that the overwhelming majority of cases are sporadic. Instances of concordant monozygotic twins have been reported (26,31). A number of familial cases have been consistent with autosomal dominant inheritance (4,8,21,24,34,47,55,58,60). Zonana et al (60) indicated that some sporadic cases may, in fact, have an affected parent who remains undiagnosed. The diagnosis of Sotos syndrome in adults can be difficult because advanced bone age is no longer a diagnostic aid, final height attainment may not be strikingly increased, intellectual deficits may be mild, and dysmorphic features may not be particularly striking. There also seems to be a lack of awareness of Sotos syndrome by most internists (13).

Possible evidence for autosomal recessive inheritance is less convincing. Boman and Nilsson (9) reported two affected sibs from a possibly consanguineous mating, but parental gonadal mosaicism for a new autosomal dominant mutation cannot be ruled out. Townes and Scheiner in an abstract (54) and a letter to the editor (53) noted the occurrence of three affected sibs—monozygotic twin girls and their brother. However, these cases have never been documented.

More complex family relationships have been reported. Bejar and co-workers (7) observed two affected brothers whose maternal grandfather and maternal uncle had macrocephaly as an isolated defect. Hooft et al (25) described affected first cousins and Krauel et al (29) reported affected third cousins. Wit et al (59) noted clinical signs in a brother and sister of the proband's grandfather; the family stemmed from a genetic isolate.

Maternal and paternal ages at the time of conception are normal (59) and contradict an earlier suggestion (17) that parental ages are slightly higher than normal.

The clinical findings of Sotos syndrome are listed in Table 11–4 with percentages derived from the combined 102 cases reported by Jaeken et al (27) and Wit et al (59).

**Growth and skeletal findings.** Overgrowth is commonly evident in the newborn, birth weight averaging 4200 g in males and 4000 g in females. Birth lengths average 55.3 cm in males; they are also increased in females, but small sample size does not permit a meaningful estimate (58). Excessive growth is especially pronounced during the first 4 years of life. Bone age is advanced in all patients and gradually increases until the fifth year of life after which the difference between bone age and chronological age stabilizes at 2–2.8 years (49,58). Dysharmonic maturation with abnormal sequences of appearances of carpal bones is found in some affected individuals. A characteristic metacarpophalangeal pattern has been reported (11,11a,16,59). From 4 years onward, growth curves usually remain above the 97th percentile. Mid-parental height of affected individuals is similar to normal mean mid-parental height. A small percentage of affected individuals has more extreme final height attainment (59). Mean adult arm span is 208 cm (17). The hands and feet are large (Fig. 11–10) (49). Mean hand and foot lengths in adults are 23 and 35 cm, respectively (17). Hemihyperplasia (hemihypertrophy) has been documented in several cases (45,56,59).

Table 11–4. Clinical findings in Sotos syndrome[a]

| | Total number of patients | Percentage with finding |
|---|---|---|
| Growth | | |
| Large birth weight | 19 | 84 |
| Excessive growth | 102 | 97 |
| Accelerated osseous maturation | 102 | 79 |
| Large hands and feet | 80 | 83 |
| Performance | | |
| Developmental retardation | 100 | 84 |
| Lack of fine motor control | 92 | 72 |
| Neonatal adaptation and/or feeding difficulties | 80 | 44 |
| Craniofacial | | |
| Macrocrania | 20 | 90 |
| Dolichocephaly | 100 | 85 |
| Receding hairline | 18 | 94 |
| Prominent forehead | 101 | 96 |
| Ocular hypertelorism | 98 | 92 |
| Downslanting palpebral fissures | 20 | 65 |
| Pointed chin | 101 | 79 |
| Highly arched palate | 97 | 96 |
| Premature eruption of teeth | 80 | 57 |

[a]Based on 80 cases reviewed by Jaeken et al (27), and 22 cases reviewed by Wit et al (59). Because the findings are based on sporadic cases, there is an obvious ascertainment bias toward the severe end of the phenotypic spectrum.

**Performance and central nervous system abnormalities.** Most patients have nonprogressive neurologic dysfunction manifested by unusual clumsiness. Dull intelligence is found in 85%, the remainder having normal intelligence. Delay in expressive language and motor development during infancy is especially common and in some instances may be followed by attainment of normal or near normal intelligence (17,26,39,60). Delay in walking until after 15 months of age and speech delay until after 2.5 years are usual. Seizures and respiratory and feeding problems have been noted in some patients (27). Often drooling is observed (1,48). Attention deficit may also be a component (27,39). Occasionally, other neurologic signs have been reported including nystagmus, strabismus, increased deep tendon reflexes, hypotonia, and muscle weakness (5,20,48,57). At times, behavior may be aggressive (1,28,36).

Dilatation of the cerebral ventricles is common. Other documented abnormalities include absent corpus callosum, prominent cortical sulci, cavum septum pellucidum, and cavum velum interpositi (20,40,60).

**Craniofacial features.** Dolichocephaly and marked frontal bossing are almost always present. The head circumference is usually well above the 97th percentile. Receded frontal hairline, bony hypertelorism, elongated anterior cranial base, downslanting palpebral fissures, increased mandibular length with pointed chin, highly arched palate, and premature eruption of deciduous teeth are observed in over 50% (Fig. 11–11A,B) (15a,27,36,37,40,49,55a,59).

**Other findings.** Many low-frequency abnormalities have been reported including, among others, congenital heart anomalies, juvenile macular degeneration, anterior fontanel bones, vertebra plana, kyphoscoliosis, brittle nails, syndactyly, functional megacolon, and autonomic failure with persistent fever (2,12,20,27a,35,40,48,57,59,60).

**Neoplasms.** A variety of neoplasms, both malignant and benign, have been described in association with Sotos syndrome. These include Wilms tumor (33), neuroblastoma (M Nance, personal communication, 1989), hepatocellular carcinoma (51), vaginal epidermoid carcinoma (46), osteochondroma (26), cavernous hemangioma (1), and hairy pigmented nevus (26). Keeping in mind the ascertainment bias favoring publication of associated tumors, less than 5% of reported cases have associated malignant or benign neoplasms. Tumors found in Beckwith–Wiedemann syndrome, isolated hemihypertrophy, and Sotos syndrome are compared in Table 11–2. It is intriguing to note that tumors of the kidney, tumors of the liver, and hemihyperplasia occur in both Beckwith–Wiedemann and Sotos syndrome, albeit with considerably lower frequency in the latter condition.

**Laboratory findings.** A 14% frequency of glucose intolerance has been demonstrated in Sotos syndrome by numerous investigators (27). Nineteen percent of families include members with diabetes mellitus (49). Growth hormone studies have been normal except for one patient who had a paradoxical rise in growth hormone level in response to hyperglycemia (26), which might suggest hypothalamic disregulation. Normal peripheral sensitivity to growth hormone is suggested by the demonstration of a normal rise in nonesterified fatty acid levels after the administration of growth hormone (40). Earlier studies of somatomedin were inconclusive (18,24,28,30,31,41,43,44). An in-depth study of 22 patients (58), two of whom were reported earlier (5,6), showed that somatomedin activity dropped from high to normal values during the first year, to below normal from 1–5 years, and thereafter returned to the lower half of normal or below the normal range. Hyperthyroidism has been observed in several patients (10,57) and one patient has had Kocher–Debré–Semelaigne syndrome, characterized by hypothyroidism and muscular hypertrophy (50).

The plasma concentration of branched chain essential amino acids was found to be considerably higher in two patients with Sotos syndrome than in control subjects. The ratios of some essential/nonessential amino acids were markedly different, the glycine/valine ratio being especially altered (7). In two other patients with Sotos syndrome, diminished cystine and elevated glutamic acid levels were found (31), but amino acid patterns are far from constant (3). In still another study of four patients, no amino acid abnormalities were found (27).

**Differential diagnosis.** Enlarged head circumference may be seen in several other conditions: hydrocephalus, *neurofibromatosis, achondroplasia,* autosomal dominant macrocephaly, etc. Although some syndromes occur with overgrowth and others are associated with macrocephaly, Sotos syndrome is a distinctive condition. Several syndromes have been confused with Sotos syndrome. *Nevo syndrome* (38), in addition to Sotosoid features, has generalized edema at birth, severe muscular hypotonia, contractures of the feet, wrist drop, and clinodactyly. The condition has autosomal recessive inheritance. *Weaver syndrome* has a distinctive facial appearance, widened distal long bones, and more accelerated osseous maturation. The *Bannayan–Riley–Ruvalcaba* syndrome (22,23,42) has some Sotosoid features but has distinctive pigmentary spotting of the penis and intestinal polyposis, especially of the colon. Inheritance is autosomal dominant. Some patients clinically suspected of having Sotos syndrome have also been reported with *fragile X syndrome* (54a). Goldstein et al (20a) observed two patients with overgrowth, congenital hypotonia, nystagmus, strabismus, and mental deficiency; they were thought to bear some resemblance to those with Sotos syndrome.

### References [Sotos syndrome (cerebral gigantism)]


1. Abraham JM, Snodgrass G: Sotos syndrome of cerebral gigantism. *Arch Dis Childh* **44**:203–210, 1969.

2. Appenzeller O, Snyder RD: Autonomic failure with persistent fever in cerebral gigantism. *J Neurol Neurosurg Psychiat* **32**:123–128, 1969.

3. Baerlocher K, Raggi M: Zur Frage von aminosäuren Veränderungen in Plasma beim zerebralen Gigantismus (Sotos-Syndrom). *Pädiatr Pädol* **16**:121–132, 1981.

4. Bale AE et al: Familial Sotos syndrome (cerebral gigantism): Craniofacial and psychological characteristics. *Am J Med Genet* **20**:613–624, 1985.

5. Barth PG et al: Unilateral delayed opercularization in a case of Sotos's syndrome (cerebral gigantism). *Neuroradiology* **20**:49–52, 1980.

6. Beemer FA: Cerebraal gigantisme (Sotos Syndroom). *T Kindergeneesk* **50**:168–171, 1982.

7. Bejar RL et al: Cerebral gigantism: Concentrations of amino acids in plasma and muscle. *J Pediatr* **76**:105–111, 1970.

8. Benke PJ: Dominantly inherited cerebral gigantism. March of Dimes Birth Defects Meeting, San Francisco, 1978.

9. Boman H, Nilsson D: Sotos syndrome in two brothers. *Clin Genet* **18**:420–427, 1980.

10. Bonvini E, Valente G: Gigantismo cerebrale: Descrizione di un caso. *Min Pediatr* **28**:1643–1649, 1976.

11. Butler MG et al: Metacarpophalangeal pattern profile analysis in Sotos syndrome. *Am J Med Genet* **20**:625–629, 1985.

11a. Butler MG et al: Metacarpophalangeal pattern profile analysis in Sotos syndrome: A follow-up report on 34 subjects. *Am J Med Genet* **29**:143–147, 1987.

12. Caffey J: Comment on Brown WH, Anterior fontanel bone: Report of a case, in *Yearbook of Pediatrics,* SS Gellis (ed), Yearbook Medical Publishers, Chicago, 1963, pp 401–402.

13. Cohen MM Jr: Diagnostic problems in cerebral gigantism. *J Med Genet* **13**:80, 1976.

14. Cohen MM Jr: A comprehensive and critical assessment of overgrowth and overgrowth syndromes. *Adv Hum Genet* **18**:181–303, 1989.

15. Cohen MM Jr: The large-for-gestational-age (LGA) infant in dysmorphic perspective, in *Clinical Genetics: Problems in Diagnosis and Counseling,* Wiley AM, Carter TP, Kelly S, Porter IH (eds), Academic Press, New York, 1982, pp 153–169.

15a. Crosher R: Advanced dental development in cerebral gigantism. *Br Dent J* **161**:514, 1986.

16. Dijkstra PF: Cerebral gigantism (Sotos' syndrome). Metacarpophalangeal pattern profiles. *Roefo* **143**:183–185, 1985.

17. Dodge PR et al: Cerebral gigantism. *Dev Med Child Neurol* **25**:248–252, 1983.

18. Du Caju MVL, Van der Brande JL: Plasma somatomedin levels in growth disturbances. *Acta Paediatr Scand* **62**:96, 1973 (Abstract).

19. Evans PR: Sotos syndrome (cerebral gigantism) with peripheral dysostosis. *Arch Dis Childh* **46**:199–202, 1971.

20. Ferrier PE et al: Cerebral gigantism (Sotos syndrome) with juvenile macular degeneration. *Helv Paediatr Acta* **35**:97–102, 1980.

20a. Goldstein DJ et al: Overgrowth, congenital hypotonia, nystagmus, strabismus, and mental retardation: Variant of dominantly inherited Sotos sequence? *Am J Med Genet* **29**:783–792, 1988.

21. Goumy P et al: Gigantisme cérébrale familial. Une nouvelle observation a transmission autosomique dominant. *Pédiatrie* **34**:249–256, 1979.

22. Halal F: Male-to-male transmission of cerebral gigantism. *Am J Med Genet* **12**:411–419, 1982.

23. Halal F: Letter to the editor: Cerebral gigantism, intestinal polyposis, and pigmentary spotting of the genitalia. *Am J Med Genet* **15**:161, 1983.

24. Hansen FJ, Friis B: Familial occurrence of cerebral gigantism, Sotos' syndrome. *Acta Paediatr Scand* **65**:387–389, 1976.

25. Hooft C et al: Gigantisme cérébral familial. *Acta Paediatr Belg* **22**:173–186, 1968.

26. Hook EB, Reynolds JW: Cerebral gigantism. *J Pediatr* **70**:900–914, 1967.

27. Jaeken J et al: Cerebral gigantism syndrome. *Z Kinderheilkd* **112**:332–346, 1972.

27a. Kaneko H et al: Congenital heart defects in Sotos sequence. *Am J Med Genet* **2**:569–576, 1987.

28. Kjellman B: Cerebral gigantism. *Acta Paediatr Scand* **54**:603–609, 1965.

29. Krauel X et al: Gigantisme cérébral: Deux cas familiaux. *J Génét Hum* **25**:205–214, 1977.

30. Lecornu M: Le facteur serique de sulfation (somatomédine) dans les retards de croissance, le gigantisme cérébral et l'acromegaly. *Arch Fr Pédiatr* **30**:595–608, 1973.

31. Lecornu M et al: Gigantisme cérébral chez des jumeaux. *Arch Fr Pédiatr* **33**:277–286, 1976.

32. Maes B et al: Gigantisme cérébral. Revue de la litterature à propos d'un cas. *Sem Hôp Paris* **52**:1537–1542, 1976.

33. Maldonado V et al: Cerebral gigantism associated with Wilms' tumor. *Am J Dis Child* **138**:486–488, 1984.

34. McKusick VA: Cerebral gigantism (Sotos syndrome), in *Mendelian Inheritance in Man,* 4th ed, Johns Hopkins University Press, Baltimore, 1975, p 385.

35. Mikulowski W et al: Gigantyzm i akromegalia u chlopca b-letniego. *Endokrynol Pol* **13**:407–412, 1962.

36. Milunsky A et al: Cerebral gigantism in childhood. *Pediatrics* **40**:395–402, 1967.

37. Motohashi N et al: Roentgencephalometric analysis of cerebral gigantism. Report of four patients. *J Craniofac Genet Dev Biol* **1**:73–94, 1981.

38. Nevo S et al: Evidence for autosomal recessive inheritance in cerebral gigantism. *J Med Genet* **11**:158–165, 1974.

39. Ott JE, Robinson A: Cerebral gigantism. *Am J Dis Child* **117**:357–368, 1969.

40. Poznanski AK, Stephenson JM: Radiologic finding in hypothalamic acceleration of growth associated with cerebral atrophy and mental retardation (cerebral gigantism). *Radiology* **88**:446–456, 1967.

41. Ranke MB, Bierich JR: Cerebral gigantism of hypothalamic origin. *Eur J Pediatr* **140**:109–111, 1983.

42. Ruvalcaba RHA et al: Sotos syndrome with intestinal polyposis and pigmentary changes of the genitalia. *Clin Genet* **18**:413–416, 1980.

43. Saenger P et al: Somatomedin in cerebral gigantism. *J Pediatr* **88**:155–156, 1976.

44. Sakano T et al: Cerebral gigantism: A report of two cases with elevated serum somatomedin A levels and a review of the Japanese literature. *Hiroshima J Med Sci* **26**:311–319, 1977.

45. See G et al: Association d'un gigantisme cérébral type Sotos et d'une hemihypertrophic corporelle totale congénitale. *Sem Hôp Paris* **50**:223–227, 1974.

46. Seyedabadi S et al: Epidermoid carcinoma of the vagina in a patient with cerebral gigantism. *J Arkansas Med Soc* **78**:123–127, 1981.

47. Smith A et al: Dominant Sotos's syndrome. *Arch Dis Childh* **55**:579, 1980.

48. Sotos JF et al: Cerebral gigantism in childhood. *N Engl J Med* **271**:109–116, 1964.

49. Sotos JF et al: Cerebral gigantism. *Am J Dis Child* **131**:625–627, 1977.

50. Sotos JF et al: Cerebral gigantism and primary hypothyroidism: Pleiotrophy or incidental concurrence *Am J Med Genet* **2**:201–205, 1978.

51. Sugarman GI et al: A case of cerebral gigantism and hepatocarcinoma. *Am J Dis Child* **131**:631–633, 1977.

52. Torgersen J: Hereditary factor in sutural pattern of the skull. *Acta Radiol* **36**:374–382, 1951.

53. Townes PL: Cerebral gigantism. *J Med Genet* **13**:80, 1976.

54. Townes PL, Scheiner AP: Cerebral gigantism (Sotos syndrome): Evidence for recessive inheritance. *Pediatr Res* **7**:349, 1973 (Abstract).

54a. Verloes A et al: Sotos syndrome and fragile X chromosomes. *Lancet* **2**:329, 1987.

55. Weber B et al: Cerebral gigantism: Hormonal studies and dermatoglyphic patterns. *Acta Paediatr Scand* **58**:662, 1969.

55a. Welbury RR, Fletcher HJ: Cerebral gigantism (Sotos syndrome)-two case reports. *J Paediatr Dent* **4**:41–44, 1988.

56. Wiedemann H-R: Exomphalos-Makroglossie-Gigantismus-Syndrom. Berardinelli-Seip-Syndrom und Sotos-Syndrom; eine vergleichende Betrachtung unter ausgewählten Aspekten. *Z Kinderheilkd* **115**:193–207, 1973.

57. Wilson TA et al: Cerebral gigantism and thyrotoxicosis. *J Pediatr* **96**:685–687, 1980.

58. Winship IM: Sotos syndrome—autosomal inheritance substantiated. *Clin Genet* **28**:243–246, 1985.

59. Wit JM et al: Cerebral gigantism (Sotos syndrome). Compiled data of 22 cases. *Eur J Pediatr* **144**:131–140, 1985.

60. Zonana J et al: Dominant inheritance of cerebral gigantism. *J Pediatr* **91**:251–256, 1977.

## Nevo syndrome

The syndrome was observed by Nevo et al (3) in 1974 in three sibs among a large inbred family from Israel. No other cases have been reported to date. Whereas Nevo and co-workers (3) described their patients as having cerebral gigantism, Cohen pointed out that the condition was clearly at variance with cerebral gigantism (1) and named the condition Nevo syndrome (2).

Fig. 11–11. *Nevo syndrome.* (A) Large infant. (B) Second affected infant. Note contractures and puffy hands. (From *S Nevo et al,* J Med Genet **11:**158, 1974.)

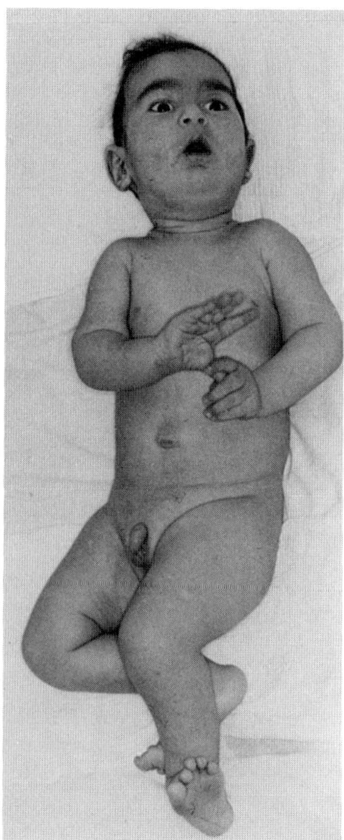

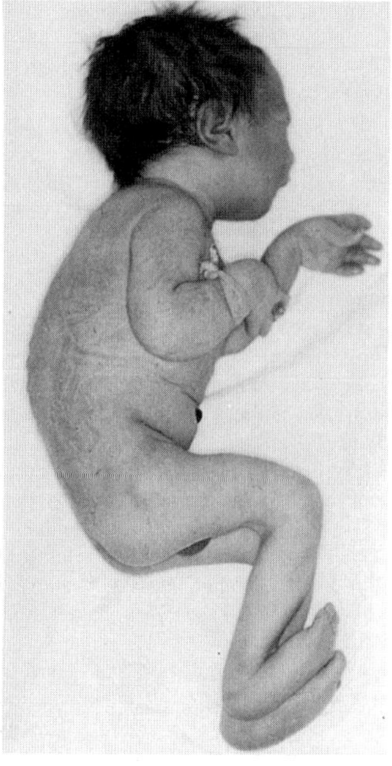

A B

Features of Nevo syndrome similar to those found in cerebral gigantism include intrauterine overgrowth, accelerated osseous maturation, dolichocephaly, large extremities, clumsiness, and retarded motor and speech development. Nevo syndrome has, in addition, generalized edema at birth, severe muscular hypotonia, contractures of the feet, wrist drop, and clinodactyly (Fig. 11–11A,B). Large, lowset, malformed ears and cryptorchidism were observed in some patients (3). Autosomal recessive inheritance was clearly established.

### References (Nevo syndrome)

1. Cohen MM Jr: Diagnostic problems in cerebral gigantism. *J Med Genet* **13**:80, 1976.

2. Cohen MM Jr: Overgrowth syndromes, in *Associated Congenital Malformations,* El-Shafie M, Klippel CH (eds), Williams & Wilkins, Baltimore, 1981, pp 71–104.

3. Nevo S et al: Evidence for autosomal recessive inheritance in cerebral gigantism. *J Med Genet* **11**:158–165, 1974.

## Bannayan–Riley–Ruvalcaba syndrome (Bannayan–Zonana syndrome, Ruvalcaba–Myhre syndrome, Riley–Smith syndrome)

Ruvalcaba et al (12) reported two males with macrocephaly, intestinal polyposis, and pigmented spotting of the penis (Figs. 11–12 and 11–13). Other cases were soon described (6,7). Cohen (2) noted that the condition was distinct from Sotos syndrome and named it Ruvalcaba–Myhre syndrome. Since then, additional examples have been recorded (3,4), using the name Ruvalcaba–Myhre–Smith syndrome.

Bannayan (1) described the combination of macrocephaly with multiple subcutaneous and visceral lipomas and hemangiomas. Zonana et al (18) also reported this combination. Earlier, Riley and Smith (11) had observed the association of macrocephaly, pseudopapilledema, and multiple hemangiomas.

Until recently, many medical geneticists have regarded these conditions as three separate entities, each having autosomal dominant inheritance. By 1986, Saul and Stevenson (15) had questioned whether Bannayan syndrome and Ruvalcaba–Myhre syndrome actually represented discrete entities. Dvir et al (5) recognized that all three conditions represented different combinations of the same syndrome. The process of lumping three earlier recognized syndromes as one and the same etiologic entity raises the problem of what to name the condition. We suggest combining the names of the first authors of the three original reports—Bannayan–Riley–Ruvalcaba syndrome. In that way, the combining of the three earlier recognized "syndromes" is reflected jn the name (2a).

A number of other cases of the Bannayan–Riley–Ruvalcaba syndrome have been recorded (5a,8,10,16). As already indicated, inheritance is autosomal dominant. To date, more males than females have been reported, perhaps because females are less severely affected and the disorder is less penetrant in females (4).

At least 60% of patients have had a myopathic process in the proximal muscles. Muscle biopsy has shown neutral fat accumulation, predominantly in enlarged type I muscle fibers. Type II fibers are smaller than normal and contain less fat (4).

**Growth.** Birth weight is usually in excess of 4000 g and birth length is above the 97th percentile. Postnatal growth decelerates, both reported older children and adults being well within the normal range (3,4,12).

**Performance.** Hypotonia, gross motor delay, and mild-to-severe mental deficiency has been reported in approximately half the patients. About 25% exhibit seizures (3). DiLiberti and Budden (4a) reported five children with asymmetric motor development. These minor motor asymmetries were transient and improved with age.

**Craniofacial features.** Head circumference is at least 4.5 SD above the mean. A few patients have exhibited delayed closure of the ante-

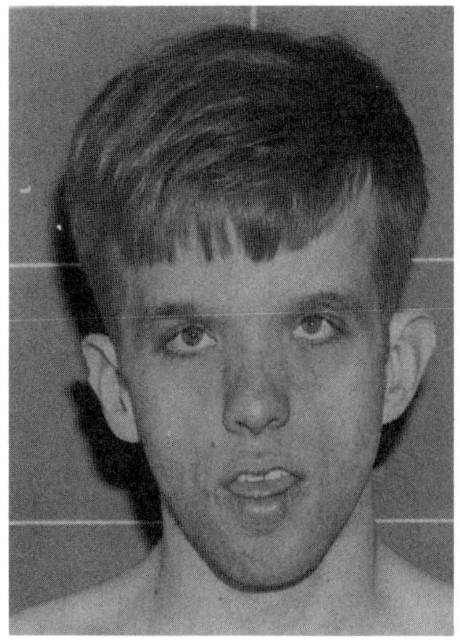

A

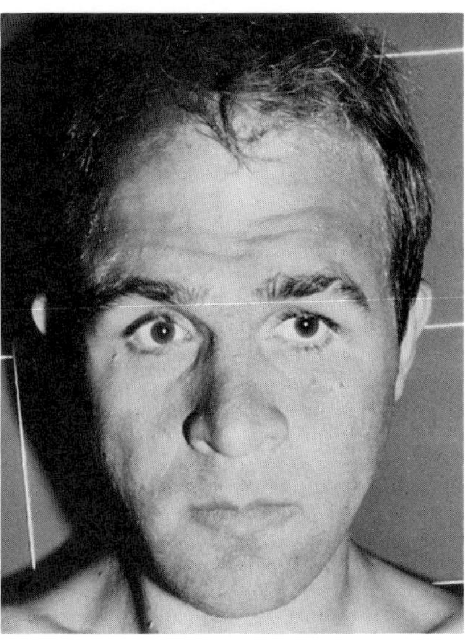

B

Fig. 11–12. *Bannayan–Riley–Ruvalcaba syndrome.* (A,B) Macrocephaly, prominent mandible. (From *R Ruvalcaba et al,* Clin Genet **18:**413, 1980.)

rior fontanel. Ocular hypertelorism has been noted in some instances (Fig. 11–12A,B). Downslanting palpebral fissures and strabismus or amblyopia are frequently observed. Examination of the eyes under slit lamp has demonstrated prominent Schwalbe lines and clearly visible corneal nerves in approximately 35% of patients (3,5a). Pseudopapilledema has been found in some cases (5,11).

**Skin.** Pigmented macules are found on the penile shaft in most affected males (Fig. 11–13A,B). This spotting is often subtle and will be missed if not specifically looked for. Cutaneous angiolipomas have been observed in over half the patients. They vary in number, size, and location (3). A few patients have had a small number of café-au-lait spots on the trunk and lower extremities. One patient had acanthosis nigricans-like lesions of the face. Another patient had accessory nipples.

**Gastrointestinal system.** Hamartomatous polyps, usually multiple and limited to the distal ileum and colon, may be associated with intussusception and/or rectal bleeding. They have been found in 45% of all patients. While some became manifested in childhood, others have not become evident until middle age (3). Protein-losing enteropathy has been observed (9; RJ Gorlin, personal observation). The infants described by Ruymann (13), Sachatello et al (14), and Le Luyer et al (9) had diffuse gastrointestinal polyposis, macrocephaly, alopecia, nail dystrophy, clubbing of fingers and toes, hypotonia, hepatosplenomegaly, anemia, and protein-losing enteropathy. These infants probably had Bannayan–Riley–Ruvalcaba syndrome.

Fig. 11–13. *Bannayan–Riley–Ruvalcaba syndrome.* (A,B) Pigmented macules on penile shaft. (A from *J DiLiberti,* Am J Med Genet **15:**491, 1983. B from *R Ruvalcaba et al,* Clin Genet **18:**413, 1980.)

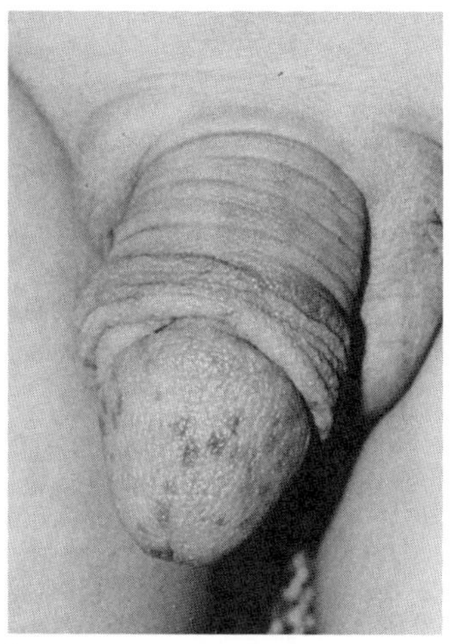

A

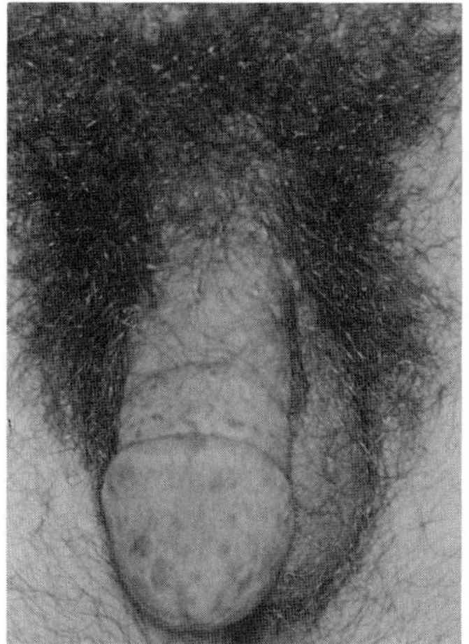

B

**Neoplasms.** Mesodermal hamartomatous masses are usually discrete lipomas (75%), less often hemangiomas (40%), and uncommonly lymphangiomas (10%). In 20% of patients, the tumors are mixed in type. Most of the hamartomas are subcutaenous, but may be intracranial (20%) or intrabony (10%). Some of the lipomas may be aggressive and can cause serious complications (10). Hamartomatous polyposis has already been discussed in the section on the gastrointestinal system.

Malignant tumors have been noted in some patients. One adult male had a thyroid tumor and two other affected males had thyroidectomies at a young age. One affected female has had rapid onset of bilateral invasive breast cancer at age 34 (DiLiberti, personal communication, 1986).

**Skeletal system.** Metacarpophalangeal profile patterns show accelerated growth of the first metacarpal and first and second middle phalanges (6). Joint hyperextensibility, pectus excavatum, and scoliosis have been reported in some patients (3).

**Other abnormalities.** Two adult males had enlarged testes (3).

**Differential diagnosis.** Although Bannayan–Riley–Ruvalcaba syndrome has some features suggestive of *Sotos syndrome* and other features suggestive of *Peutz–Jeghers syndrome,* the condition is easily distinguishable from both of these two syndromes. Lipomas, hemangiomas, and lymphangiomas may also occur in *Proteus syndrome.* The combination of macrocephaly, angiomatosis, and limb asymmetry in sporadic cases has been discussed by Stephen et al (17).

**Laboratory aids.** Electromyography and muscle biopsy should be performed on all infants with macrocephaly, normal CT scans, and hypotonia. Patients with Bannayan–Riley–Ruvalcaba syndrome should be monitored for gastrointestinal polyposis and thyroid neoplasms.

#### References [Bannayan–Riley–Ruvalcaba syndrome (Bannayan–Zonana syndrome, Ruvalcaba–Myhre syndrome, Riley–Smith syndrome)]

1. Bannayan GA: Lipomatosis, angiomatosis and macroencephaly: A previously undescribed congenital syndrome. *Arch Pathol* **92**:1–5, 1971.

2. Cohen MM Jr: The large-for-gestational-age (LGA) infant in dysmorphic perspective. *Clinical Genetics: Problems in Diagnosis and Counseling.* AM Willey, TP Carter, S Kelly, and IH Porter, eds, Academic Press, New York, 1982, 153–169.

2a. Cohen MM Jr: Bannayan–Riley–Ruvalcaba syndrome. Renaming a condition previously thought to represent three separate entities. *Am J Med Genet,* in press.

3. DiLiberti JH et al: Ruvalcaba–Myhre–Smith syndrome: Case report with probable autosomal-dominant inheritance and additional manifestations. *Am J Med Genet* **15**:491–495, 1983.

4. DiLiberti JH et al: A new lipid storage myopathy observed in individuals with the Ruvalcaba–Myhre–Smith syndrome. *Am J Med Genet* **18**:163–167, 1984.

4a. DiLiberti JH, Budden S: Transient motor asymmetry in young children with the Ruvalcaba–Myhre–Smith syndrome. Ninth Annual David W. Smith Workshop and Malformations and Morphogenesis, Oakland, California, August 3–7, 1988.

5. Dvir M et al: Heredofamilial syndrome of mesodermal hamartomas, macrocephaly and pseudopapilledema. *Pediatrics* **81**:287–290, 1988.

5a. Grezula JC et al: Ruvalcaba–Myhre–Smith syndrome. *Pediatr Dermatol* **5**:28–32, 1988.

6. Halal F: Male-to-male transmission of cerebral gigantism. *Am J Med Genet* **12**:411–419, 1982.

7. Halal F: Letter to the editor: Cerebral gigantism, intestinal polyposis, and pigmentary spotting of the genitalia. *Am J Med Genet* **15**:161, 1983.

8. Higginbottom MC, Schultz P: The Bannayan syndrome: An autosomal dominant disorder consisting of macrocephaly, lipomas, hemangiomas, and risk for intracranial tumors. *Pediatrics* **69**:632–634, 1982.

9. Le Luyer B et al: Generalized juvenile polyposis in an infant: Report of a case and successful management by endoscopy. *J Pediatr Gastroenterol Nutr* **4**:128–134, 1985.

10. Miles JH: Macrocephaly with hamartomas: Bannayan–Zonana syndrome. *Am J Med Genet* **19**:225–234, 1984.

11. Riley HD, Smith WR: Macrocephaly, pseudopapilledema and multiple hemangiomata. *Pediatrics* **26**:293–300, 1960.

12. Ruvalcaba RHA et al: Sotos syndrome with intestinal polyposis and pigmentary changes of the genitalia. *Clin Genet* **18**:413–416, 1980.

13. Ruymann FB: Juvenile polyps with cachexia. *Gastroenterology* **47**:431–438, 1969.

14. Sachatello CR et al: Juvenile gastrointestinal polyposis in a female infant: Report of a case and review of the literature of a recently recognized syndrome. *Surgery* **75**:107–114, 1974.

15. Saul RA, Stevenson RE: Are Bannayan syndrome and Ruvalcaba–Myhre–Smith syndrome discrete entities? *Proc Greenwood Genet Ctr* **5**:3–7, 1986.

16. Saul RA et al: Mental retardation in the Bannayan syndrome. *Pediatrics* **69**:642–644, 1982.

17. Stephan MJ et al: Macrocephaly in association with unusual cutaneous angiomatosis. *J Pediatr* **87**:353–359, 1975.

18. Zonana J et al: Macrocephaly with multiple lipomas and hemangiomas. *J Pediatr* **89**:600–603, 1976.

## Weaver syndrome

In 1974, Weaver et al (20) reported a syndrome of persistent overgrowth of prenatal onset, accelerated osseous maturation, distinctive craniofacial appearance, developmental delay, widened distal long bones, and camptodactyly (Fig. 11–14 and 11–15). Approximately 25 cases have been recorded (1–4,7,9,10,12,14–21), most representing isolated examples with a 3:1 male-to-female ratio (2). Females have been noted to be more mildly affected (2,17,18,18a).

Etiology is unclear. The syndrome has possibly been described in sibs twice (10,17). Positive family histories with a partially affected parent have also been reported on two occasions (2,12). In both instances, the mother of the proband had similar facial characteristics and a similar growth pattern, but normal intelligence. Two adults with classic Weaver syndrome have been reported to date (2,9a,9b). The mean paternal age at the time of conception in 18 cases is 31.3 years (2). Further studies are necessary to delineate the etiology, the phenotypic spectrum of anomalies, and the natural history.

Fitch (8, 8a) published extensive analyses of the syndrome and Ardinger et al (2) reported seven cases and reviewed the literature. Features of Weaver syndrome are summarized in Table 11–5.

**Growth.** Increased birth weight (4–6 kg) and length (57 cm) are significant, although, on occasion, growth may not be greatly accelerated for a few months. Bone age is remarkably advanced (Fig. 11–15); the growth parameters of head circumference, length, and weight proceed at two to three times the expected rate. Carpal maturation is accelerated over that of phalanges and metacarpals. Endocrine studies have yielded normal results in those tested. The patient reported by Shimura et al (18) did not have overgrowth at birth or by 19 months of age. A patient described by Weaver (20a) increased her weight dramatically from the 20th percentile to greater than the 97th percentile from 13–15 months.

Other skeletal findings include widened or splayed long bone metaphyses, especially of the femurs, and somewhat mottled epiphyses. The iliac wings may be broad and low (2,4,12,21).

**Performance and central nervous system.** Mild hypertonia or hypotonia is common, and motor development is mildly to moderately retarded. The cry is low-pitched and hoarse. Although the appetite is

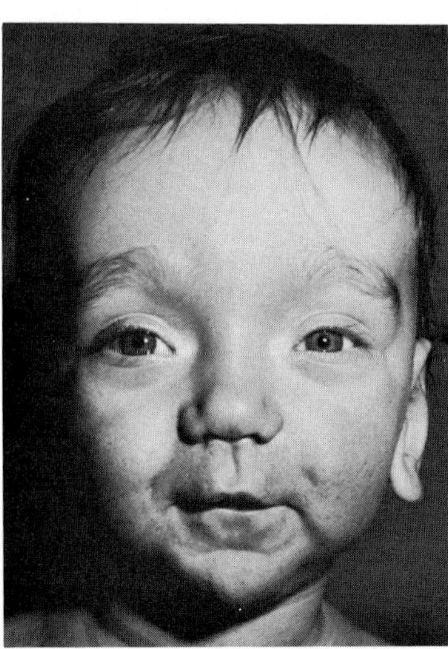

A

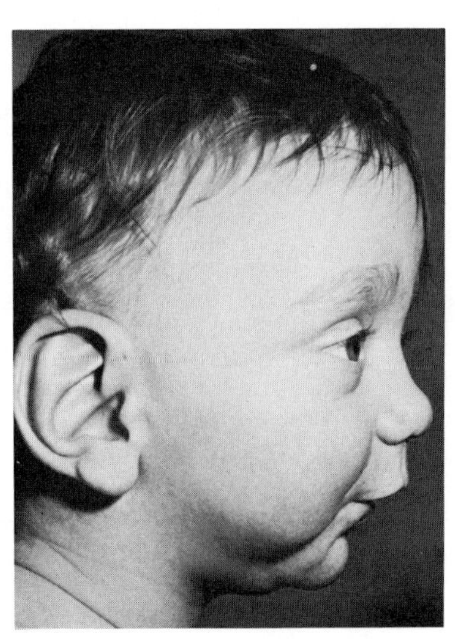

B

Fig. 11–14. *Weaver syndrome.* (A,B) Note broad forehead, hypertelorism, large ears, long philtrum, and micrognathia. (From *DD Weaver et al,* J Pediatr **84:**547, 1974.)

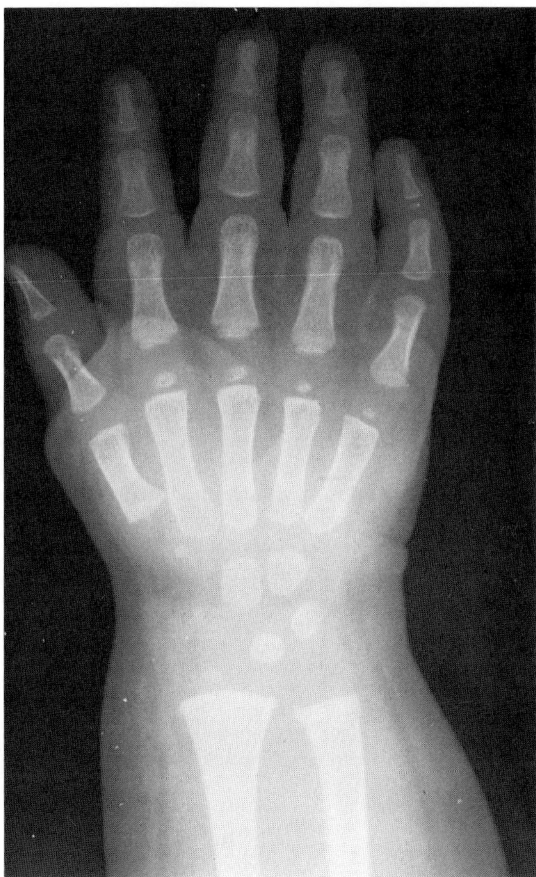

Fig. 11–15. *Weaver syndrome.* Hand-wrist roentgenogram at 11 months showing carpal bone age of a 4-year-old. (Courtesy of *DD Weaver,* Indianapolis, Indiana.)

Table 11–5. Features of Weaver syndrome

| Findings | Frequencies |
|---|---|
| Growth | |
| Prenatal growth excess | 16/20 |
| Postnatal growth excess | 19/20 |
| Accelerated osseous maturation | 19/19 |
| Performance | |
| Hypertonia | 10/18 |
| Hypotonia | 5/18 |
| Developmental delay | 19/19 |
| Hoarse, low-pitched cry | 14/17 |
| Craniofacial | |
| Broad forehead | 18/19 |
| Flat occiput | 6/11 |
| Large ears | 15/17 |
| Ocular hypertelorism | 19/19 |
| Prominent or long philtrum | 10/14 |
| Relative micrognathia | 17/19 |
| Limbs | |
| Camptodactyly | 11/16 |
| Prominent fingerpads | 6/9 |
| Thin, deeply set nails | 9/10 |
| Broad thumbs | 4/6 |
| Clinodactyly, toes | 4/5 |
| Limited elbow or knee extension | 10/13 |
| Widened distal long bones | 16/18 |
| Foot deformities[a] | 7/10 |
| Other | |
| Excess loose skin | 11/12 |
| Umbilical hernia or diastasis recti | 12/17 |
| Inguinal hernia | 4/14 |
| Inverted nipples | 3/4 |

[a]Talipes equinovarus, talipes calcaneovalgus, metatarsus adductus.

voracious, hypothalamic dysregulation has not been demonstrated (2). Difficulty in swallowing or breathing has been noted in several cases (4,21). Other findings have included small cysts of the septum pellucidum in two instances (2,21), dilation of the ventricles, basal cisterns, sylvian cistern, and interhemispheric fissure, consistent with nonspecific cerebral atrophy (17), and enlarged vessels and hypervascularization in the areas of the middle and left posterior cerebral arteries (12).

**Craniofacial features.** Macrocephaly, broad forehead, and flattened occiput are characteristic. Scalp hair is moderately thin. The ears are large and may be mildly dysmorphic or low set. Other features include hypertelorism, long prominent philtrum, relative micrognathia, and redundant nuchal skin folds (Fig. 11–14A,B). Low-frequency findings have been noted such as mild craniofacial asymmetry, upslanting or downslanting palpebral fissures, small palpebral fissures, ptosis, strabismus, and highly arched palate (2,8,12).

**Limbs.** Common findings include camptodactyly, prominent finger pads, thin deeply set nails, broad thumbs, clinodactyly of toes, and limited extension at elbows and knees. Foot deformities have been noted such as talipes equinovarus, talipes calcaneovalgus, metatarsus adductus, pes adductus, and pes cavus. Deep creases may be observed on the palmar and plantar surfaces (2,7,12,20).

**Other findings.** Excessive loose skin, umbilical hernia or diastasis recti, inguinal hernia, and inverted nipples have been reported (2).

**Neoplasms.** Although many overgrowth syndromes are associated with neoplasia, none of the 20 cases of Weaver syndrome now reported has shown this association. A presumed case of Weaver syndrome possibly associated with neuroblastoma has been communicated to MM Cohen, but no diagnostic confirmation is available. Because both overgrowth and neoplasia represent an increase in the number of cell divisions, it seems reasonable to recommend that Weaver syndrome patients should be monitored for possible neoplasia.

**Differential diagnosis.** Accelerated osseous maturation may be observed in a number of syndromes and these have been listed by Weisswichert et al (21, Table 1). Differential diagnosis includes *Marshall–Smith syndrome, Sotos syndrome,* and *Beckwith–Wiedemann syndrome;* similarities and differences among these disorders have been contrasted by Amir et al (1, Table 1).

Previous reports (4,18) have suggested that Weaver syndrome is simply a variant of Marshall–Smith syndrome. We (5,6) and others (2,8,9,20,21) believe that they represent two separate entities. Both share dysharmonic osseous maturation, early acceleration of linear growth, and developmental delay, but differ in craniofacial dysmorphism. Features of Marshall–Smith syndrome not found in Weaver syndrome include widened middle and proximal phalanges, failure to thrive in terms of weight, problems with respiratory secretions, and early demise.

Kelly (11) reported a Weaver-like syndrome with hyperprogesteronemia and maternal luteoma. Tsukahara et al (19) reported a Weaver-like syndrome with cleft lip, accessory nipples, pectus excavatum, bifid xyphoid process, irregularly shaped vertebral bodies, and inflexible right thumb.

### References (Weaver syndrome)

1. Amir N et al: Weaver–Smith syndrome: A case study with long-term follow-up. *Am J Dis Child* **138**:1113–1115, 1984.

2. Ardinger HH et al: Further delineation of Weaver syndrome. *J Pediatr* **108**:228–235, 1986.

3. Bailey-Wilson JE et al: An unusual case of early overgrowth and congenital anomalies (Abstract). *Am Soc Hum Genet 34th Annual Meeting,* Norfolk, 1983, p 75A.

4. Bosch-Banyeras JM et al: Acceleration du development postnatal, hypertonie, elargissement des phalanges médianes et des metaphyses distales du femur, facies particuliar: s'agit-il d'un syndrome de Weaver? *Arch Fr Pédiatr* **35**:177–183, 1978.

5. Cohen MM Jr: Overgrowth syndromes, in *Associated Congenital Anomalies,* El Shafie M, Klippel CH (eds), Williams & Wilkins, Baltimore, 1981, pp 71–104.

6. Cohen MM Jr: A comprehensive and critical assessment of overgrowth and overgrowth syndromes. *Adv Hum Genet* **18**:181–303, 1989.

7. Farrell SA, Hughes HE: Brief clinical report: Weaver syndrome with pes cavus. *Am J Med Genet* **21**:737–739, 1985.

8. Fitch N: The syndromes of Marshall and Weaver. *J Med Genet* **17**:174, 1980.

8a. Fitch N: Update on the Marshall–Smith–Weaver controversy. *Am J Med Genet* **20**:559–562, 1985.

9. Gemme G et al: The Weaver–Smith syndrome. *J Pediatr* **97**:962–964, 1980.

9a. Greenberg F et al: Weaver syndrome: The changing phenotype in an adult. *Am J Med Genet* **33**:127–129, 1989.

9b. Hoyme HE et al: The Weaver syndrome: Natural history through adulthood. *Proc Greenwood Genet Ctr* **3**:84, 1984.

10. Jalaguier J et al: Avance de la maturation osseuse et syndrome dysmorphique chez deux germains (syndrome de Marshall-Weaver). *J Génét Hum* **31**:385–395, 1983.

11. Kelly TE: Weaver-like syndrome with hyperprogesteronemia and maternal luteoma. *Proc Greenwood Genet Ctr* **2**:128–129, 1983.

12. Majewski F et al: The Weaver syndrome: A rare type of primordial overgrowth, *Eur J Pediatr* **137**:277–282, 1981.

13. Marshall RE et al: Syndrome of accelerated skeletal maturation and relative failure to thrive: A newly recognized clinical growth disorder. *J Pediatr* **78**:95–101, 1971.

14. Meinecke P et al: The Weaver syndrome in a girl. *Eur J Pediatr* **141**:58–59, 1983.

15. Moreno H, Kirkland R: Another candidate for the overgrowth syndrome (letter). *J Pediatr* **85**:583, 1974.

16. Moreno HD et al: Case report 18. *Syndrome Ident* **2**(1):22–25, 1974.

17. Roussounis MB, Crawford JJ: Siblings with Weaver syndrome. *J Pediatr* **102**:595–597, 1983.

18. Shimura T et al: Marshall–Smith syndrome with large bifrontal diameter, broad distal femora, camptodactyly, and without broad middle phalanges. *J Pediatr* **94**:93–95, 1979.

18a. Thomson EM et al: A girl with the Weaver syndrome. *J Med Genet* **24**:232–234, 1987.

19. Tsukahara M et al: A Weaver-like syndrome in a Japanese boy. *Clin Genet* **25**:73–78, 1984.

20. Weaver DD et al: A new overgrowth syndrome with accelerated skeletal maturation, unusual facies, and camptodactyy. *J Pediatr* **84**:547–552, 1974.

20a. Weaver DD et al: Delayed onset of the Weaver syndrome. *David W. Smith Workshop on Malformations and Morphogenesis.* Greenville, South Carolina, August 15–19, 1987.

21. Weisswichert PH et al: Accelerated bone maturation syndrome of the Weaver type. *Eur J Pediatr* **137**:329–333, 1981.

## Marshall–Smith syndrome

In 1971, Marshall et al (10) first reported a disorder characterized by accelerated skeletal maturation, mental and somatic retardation, failure to thrive with chronic respiratory distress and early death, characteristic facial appearance, and remarkable skeletal alterations (Figs. 11–16 to 11–18). At least 18 cases have been reported (2,3,5–9,11–14,16–18) and all have been isolated examples. Findings have been summarized by Hoyme and Bull (6a) and appear in Table 11–6.

**Growth, performance, and natural history.** Moderate to severe respiratory distress accompanied by noisy stridor may necessitate intubation. There is failure to thrive, death almost always being associated with pneumonia within months. Only a few children have survived for up to 3 years. The child will often lie with the head extended (10,17). Speech is poor and frequently the child is hypotonic. Psychomotor delay is common.

**Craniofacial features.** The forehead is prominent with frontal ridging while the supraorbital ridges are flattened. The eyes bulge or are proptosed and there is megalocornea. The sclerae are blue in 50%. The eyebrows are bushy with synophrys. The nose is small with a low nasal bridge and anteverted nostrils. Unilateral or bilateral choanal stenosis or atresia has been reported (5,7,14,16,17) in 30%. The helices of the pinnae are often folded with hypoplastic cartilages. The ears may be low set. The mandible is small (Figs. 11–16 to 11–18A). The mouth is small and the palate may be highly arched. The stridor appears to be related both to glossoptosis and to laryngomalacia (6,8,17).

**Skeletal system.** The hands and feet and/or digits are often stated to be long. Radiographically, carpal and tarsal bone age is markedly advanced (7,11,14,16). At birth, the phalangeal epiphyses, femoral heads, and patellas are usually ossified, suggesting a bone age of 4 years or more (9,11,12). Long bones tend to be thin (2,6,8–13). The proximal and middle phalanges of the hands are remarkably thick. The former are rectangular; the latter are bullet shaped. The terminal phalanges are greatly reduced in size. The metacarpals are also widened

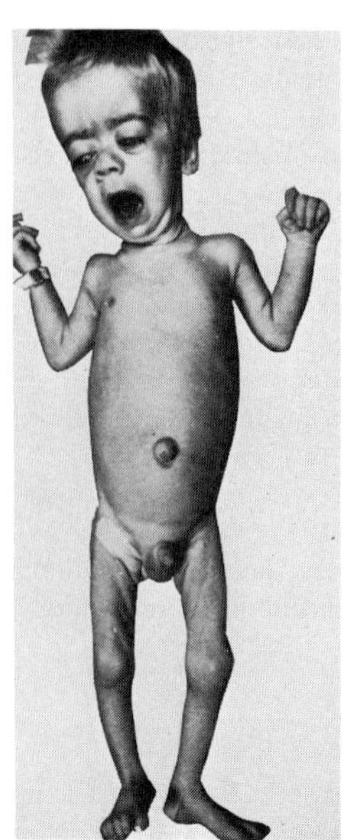
A

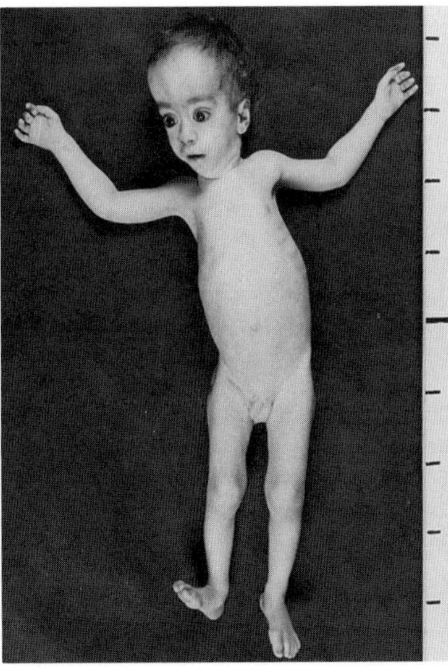
B

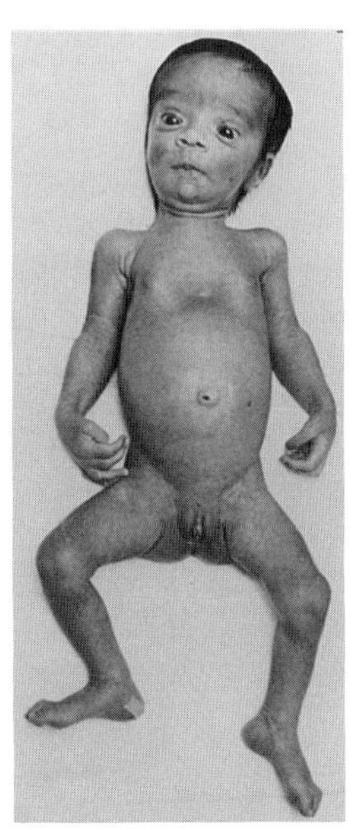
C

Fig. 11–16. *Marshall–Smith syndrome.* (A–C) Prominent forehead, bushy eyebrows, bulging eyes in patients with failure to thrive. (A from *RE Marshall et al,* J Pediatr **78:**95, 1971. B from *E de Toni et al,* Minerva Pediatr **28:**1499, 1976. C from *S Flatz* and *J Natzschka,* Klin Pädiatr **190:**592, 1978.)

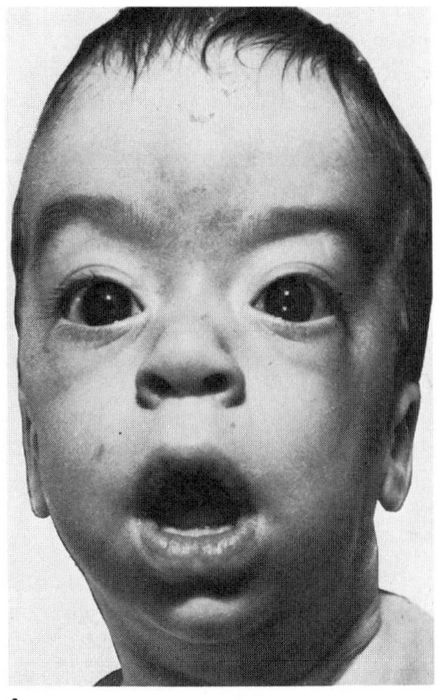
A

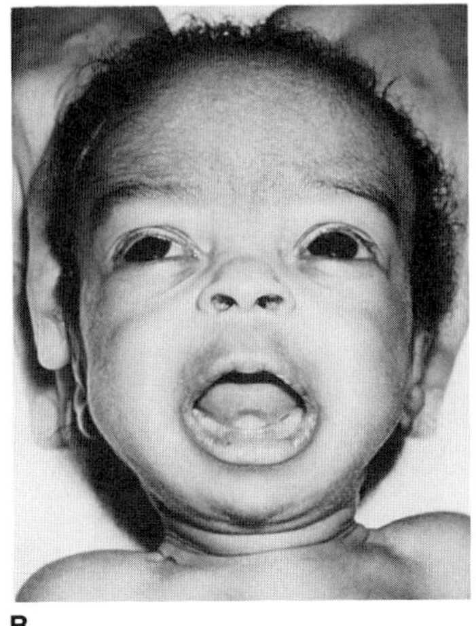
B

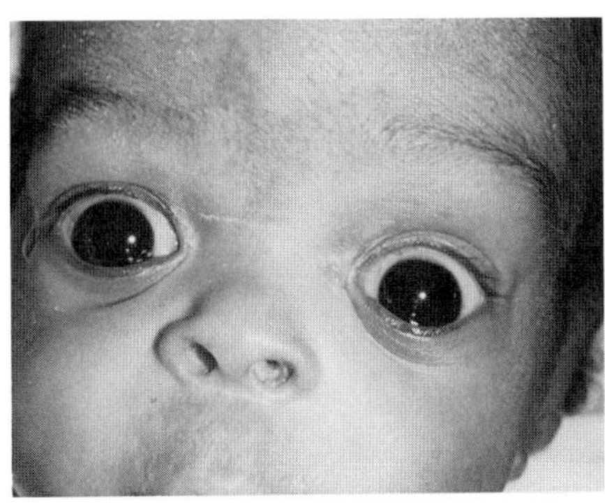
C

Fig. 11–17. *Marshall–Smith syndrome*. (A–C) Bulging eyes, coarse eyebrows, low nasal bridge, micrognathia. (A from *RE Marshall et al,* J Pediatr **78:**95, 1971. B,C from *JCS Perrin et al,* Birth Defects **12**(5):209, 1976.)

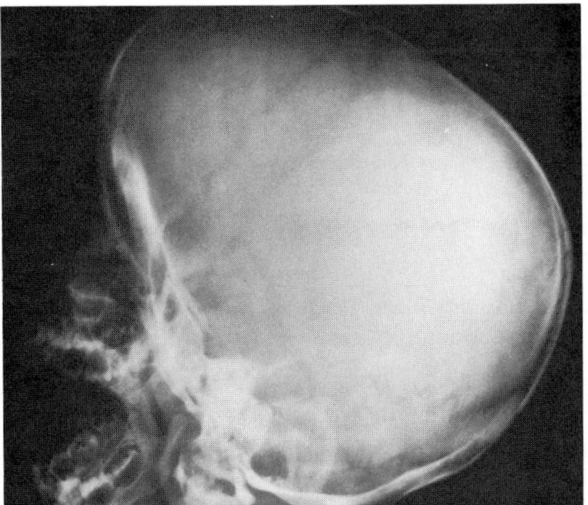
A

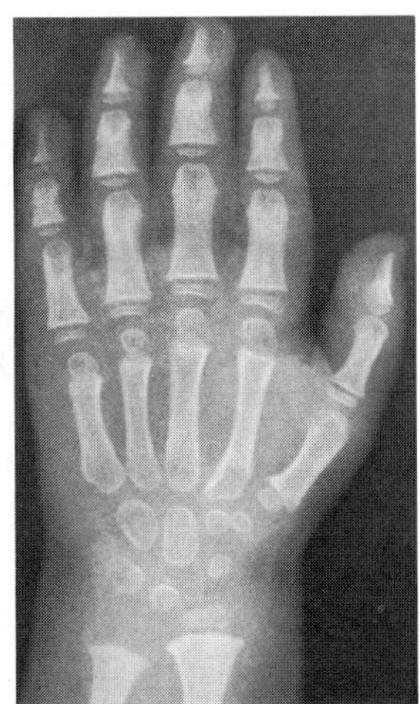
B

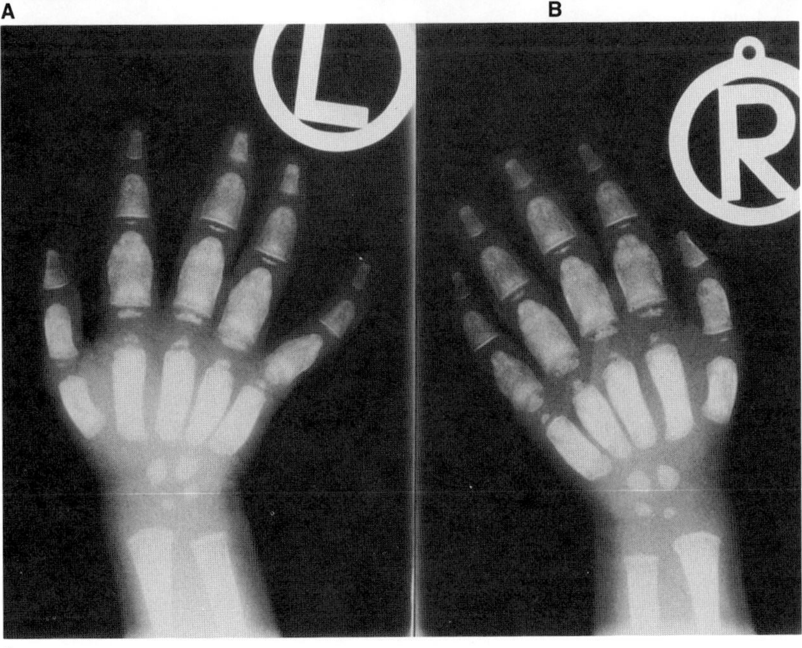

C

Fig. 11–18. *Marshall–Smith syndrome*. Roentgenograms. (A) Prominent calvaria, small facial bones, and hypoplastic mandibular rami. (B) Bone age of 6–7 years at chronologic age of 10 months. (C) Radiograph of hand at 22 days. Note enlarged bullet-shaped proximal and middle phalanges, very small distal phalanges, phalangeal epiphyses, curved first metacarpal, and advanced bone age. (A,B from *RE Marshall et al,* J Pediatr **78:**95, 1971. C from *M Hassan et al,* Pediatr Radiol **5:**53, 1976.)

Table 11–6. Features of the Marshall–Smith syndrome[a]

| Findings | Frequencies |
|---|---|
| Growth, skeletal | |
| Accelerated linear growth | 2/18 |
| Accelerated osseous maturation | 18/18 |
| Broad phalanges | 17/17 |
| Failure to thrive | 11/13 |
| Performance | |
| Neurodevelopmental abnormalities | 13/13 |
| Structural brain anomalies | 7/14 |
| Respiratory | |
| Respiratory tract abnormalities | 15/18 |
| Recurrent pneumonia | 14/18 |
| Pulmonary hypertension | 4/18 |
| Death in early infancy | 10/18 |
| Craniofacial | |
| Prominent forehead | 15/18 |
| Small face | 12/18 |
| Prominent eyes | 17/18 |
| Blue sclerae | 11/18 |
| Flat nasal bridge | 17/18 |
| Anteverted nares | 15/16 |
| Micrognathia | 14/18 |
| Glossoptosis | 6/17 |
| Choanal atresia/stenosis | 3/17 |
| Other | |
| Hypertrichosis | 7/18 |
| Umbilical hernia | 6/18 |

[a]Adapted from HE Hoyme and MJ Bull: *Eighth David W. Smith Workshop on Malformations and Morphogenesis,* Greenville, South Carolina, August 15–19, 1987.

distally and are poorly modeled (Fig. 11–18B,C). Scoliosis has been noted in a few cases.

The frontal bone is prominent and the calvaria thickened and prominent. The orbits are shallow and the facial bones small or hypoplastic. The mandibular rami are underdeveloped with absence of the angle (Fig. 11–18A) (3,5,6,10,16).

**Other abnormalities.** Umbilical hernia or omphalocele is present in approximately 60% (8,10,17). Generalized hypertrichosis occurs in over half the cases (8). Pachygyria of the occipital and temporal areas of the brain has been documented in some instances (10,16). Only a few children have had cardiovascular defects and these have included PDA, ASD, and hypertrophy of the pulmonary arteries (5).

**Differential diagnosis.** Comparison is sometimes made with *Weaver syndrome,* but except for accelerated osseous maturation, the phenotypes are distinct. Differences have been well reviewed by Fitch (4,4a). In *Sotos syndrome,* phalangeal age is ahead of carpal age, but in Marshall–Smith syndrome the opposite occurs. The case reported by Shimura et al (15) as an example of overlap between Marshall–Smith syndrome and Weaver syndrome represents simply Weaver syndrome, in our opinion.

### References (Marshall–Smith syndrome)

1. Chopard A: Les syndromes d'accélération de la maturation osseuse. *Pédiatrie* **33**:706–708, 1978.

2. de Toni E et al: Una rara sindrome con accelerazione della maturazione scheletrica (Sindrome de Marshall). *Minerva Pediatr* **28**:1499–1509, 1976.

3. Ferran JL et al: Acceleration de la maturation osseuse du nouveau-né avec dysmorphie faciale syndrome de Marshall-Smith. *J Radiol Electrol Méd Nuc* **59**:579–583, 1978.

4. Fitch N: The syndromes of Marshall and Weaver. *J Med Genet* **17**:174–178, 1980.

4a. Fitch N: Update on the Marshall–Smith–Weaver controversy. *Am J Med Genet* **20**:559–562, 1985.

5. Flatz S, Natzschka J: Syndrom der akzelarierten Skelettreifung vom Typ Marshall. Kasuistik und Überblick. *Klin Pädiatr* **190**:592–598, 1978.

6. Hassan M et al: The syndrome of accelerated bone maturation in the newborn infant with dysmorphism and congenital malformations. *Pediatr Radiol* **5**:53–57, 1976.

6a. Hoyme HE, Bull MJ: The Marshall–Smith syndrome: Natural history beyond infancy. Eighth David W. Smith Workshop on Malformations and Morphogenesis, Greenville, South Carolina, August 15–19, 1987.

7. Iafusco F et al: Su di un caso di accelerata maturazione scheletrica (sindrome di Marshall). *Pediatria (Napoli)* **85**:487–496, 1977.

8. Johnson JP et al: Marshall–Smith syndrome: Two case reports and a review of pulmonary manifestations. *Pediatrics* **71**:219–223, 1983.

9. LaPenna R, Folger GM Jr: Extreme upper airway obstruction with the Marshall syndrome. *Clin Pediatr* **21**:507–509, 1982.

10. Marshall RE et al: Syndrome of accelerated skeletal maturation and relative failure to thrive: A newly recognized clinical growth disorder. *J Pediatr* **78**:95–101, 1971.

11. Menguy C et al: Le syndrome de Marshall. *Ann Pédiatr* **33**:339–343, 1986.

12. Nábrády J, Bozalyi I: Accelerált csontérés és somatomentalis retardatio újabb esete. *Orv Hetil* **114**:2782–2785, 1973.

13. Nair P, Sabarinathan K: Syndrome of accelerated skeletal maturation and relative failure to thrive. *Indian Pediatr* **19**:1036–1039, 1982.

14. Perrin JCS et al: Accelerated skeletal maturation syndrome with pulmonary hypertension. *Birth Defects* **12**(5):209–217, 1976.

15. Shimura T et al: Marshall–Smith syndrome with large bifrontal diameter, broad distal femora, camptodactyly and without broad middle phalanges. *J Pediatr* **94**:93–95, 1979.

16. Tipton RE et al: Accelerated skeletal maturation in infancy syndrome. Report of a third case. *J Pediatr* **83**:829–832, 1973.

17. Viveshwara N et al: Syndrome of accelerated skeletal maturation in infancy, peculiar facies and multiple congenital anomalies. *J Pediatr* **84**:553–556, 1974.

18. Yoder CC et al: Marshall–Smith syndrome. Further delineation. *South Med J* **81**:1297–1300, 1988.

## Elejalde syndrome (acrocephalopolydactylous dysplasia)

In 1977, Elejalde and co-workers (1) described a spectacular overgrowth syndrome. Birthweights in two patients were 7500 and 4300 g.

Fig. 11–19. *Elejalde syndrome.* Fetus of 34 weeks delivered by Cesarean section weighing 4300 g. Note short limbs, acrocephalic skull, facial anomalies, excess subcutaneous tissue of neck, trunk, and limb, and small omphalocele. [From *BR Elejalde,* Birth Defects **13**(3B):53, 1977.]

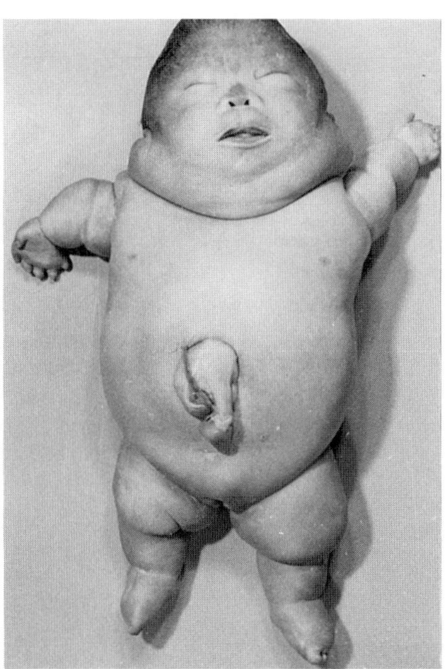

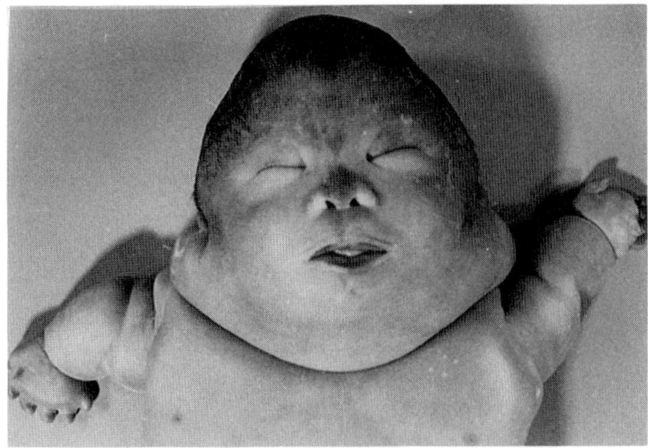

A

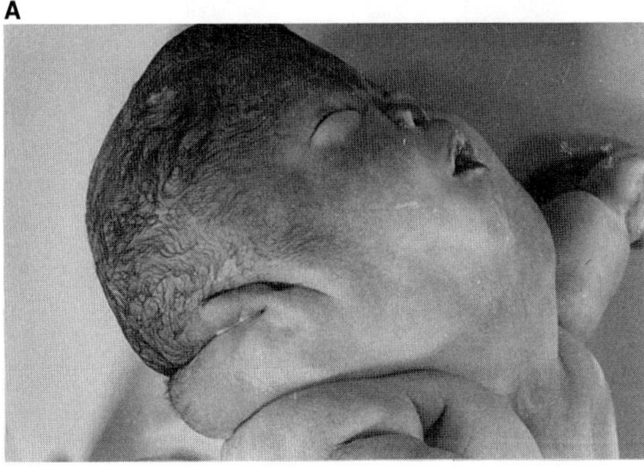

B

Fig. 11–20. *Elejalde syndrome.* (A,B) Complete fusion of all cranial sutures, closure of fontanels, epicanthic folds, downslanting palpebral fissures, hypoplastic nose, rudimentary auricles, and redundant tissue around neck. [From *BR Elejalde,* Birth Defects **13**(3B):53, 1977.]

Table 11–7. Features of Elejalde syndrome[a]

| Findings | Patient 1 | Patient 2 |
|---|---|---|
| Growth | | |
| Gigantism | + | + |
| Abdominal/genitourinary | | |
| Omphalocele | + | + |
| Accessory spleen | – | + |
| Megaureter | + | + |
| Megabladder | + | + |
| Megavagina | – | + |
| Redundant connective tissue | + | + |
| Proliferation of perivascular nerve fibers | + | + |
| Craniofacial | | |
| Craniosynostosis | + | + |
| Ocular hypertelorism | + | + |
| Epicanthic folds | + | + |
| Downslanting palpebral fissures | + | + |
| Hypoplastic nose | + | + |
| Rudimentary auricles | + | + |
| Limbs | | |
| Short limbs | + | + |
| Polydactyly | – | + |
| Other | | |
| Redundant neck skin | + | + |
| Hypoplastic lungs | + | + |

[a]Based on data from BR Elejalde et al: *Birth Defects* **13**(3B):53, 1977.

Features included a swollen globular body, omphalocele, short limbs, redundant neck skin, craniosynostosis, hypoplastic nose, and rudimentary auricles (Fig. 11–19 and 11–20). The pertinent findings of the Elejalde syndrome are summarized in Table 11–7. Three affected individuals in two sibships were recorded for which consanguinity was established. Thus, the Elejalde syndrome has autosomal recessive inheritance.

Autopsy findings showed enlarged kidneys with a thick fibrous capsule, lack of clear-cut differentiation between the cortex and the medulla, and multiple cysts. Excessive connective tissue was found throughout the body, except for the central nervous system. It was most prominent subcutaneously in the media of blood vessels, in the walls of the viscera, and interstitially in organs such as the pancreas and kidneys. Cell kinetic studies showed that fibroblasts from Elejalde syndrome patients completed the whole cell cycle in 63% of the normal cell cycle time. Perivascular proliferation of nerve fibers was found in many viscera, especially the spleen, thymus, colon, heart, and adrenal glands.

#### Reference [Elejalde syndrome (acrocephalopolydactylous dysplasia)]

1. Elejalde BR et al: Acrocephalopolydactylous dysplasia. *Birth Defects* **13**(3B): 53–67, 1977.

### Simpson–Golabi–Behmel syndrome (Golabi–Rosen syndrome)

The syndrome is an X-linked disorder characterized by prenatal and postnatal growth deficiency, intellectual impairment, characteristic facial appearance, and a variety of other anomalies (Figs. 11–21 to 11–23). In 1984, Golabi and Rosen (3) reported the syndrome in four affected males with partial manifestations in a female obligate carrier sometimes including characteristic facial changes. Striking features of

Fig. 11–21. *Simpson–Golabi–Behmel syndrome.* Note hypertelorism, broad flat nasal bridge, short upturned nose, large mouth. Weight greater than 90th percentile at birth. (From *M Golabi* and *L Rosen,* Am J Med Genet **17**:345, 1984.)

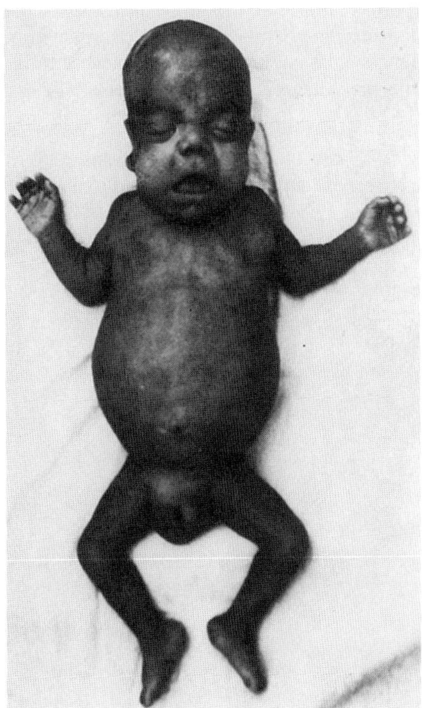

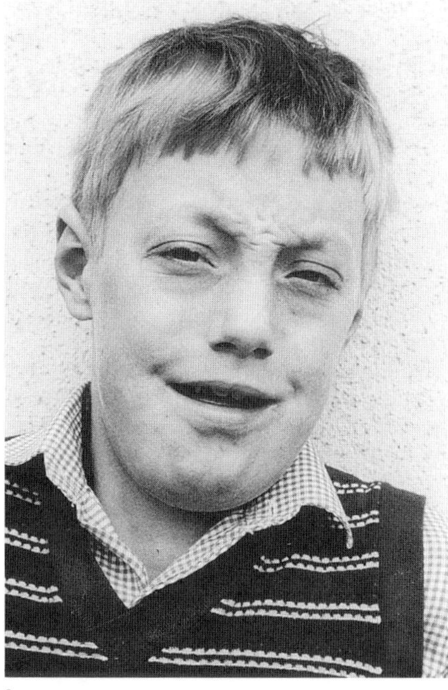
A

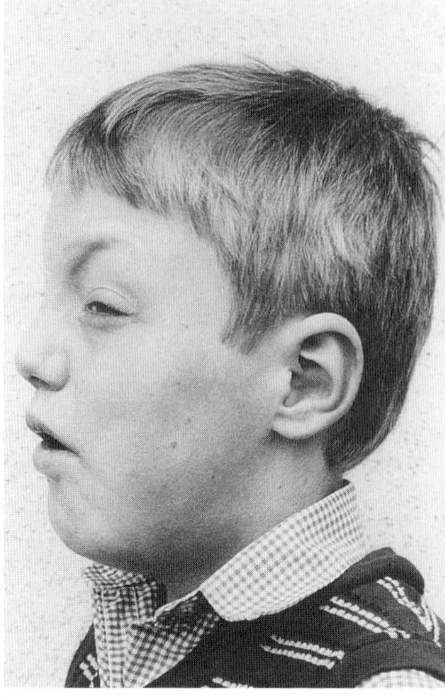
B

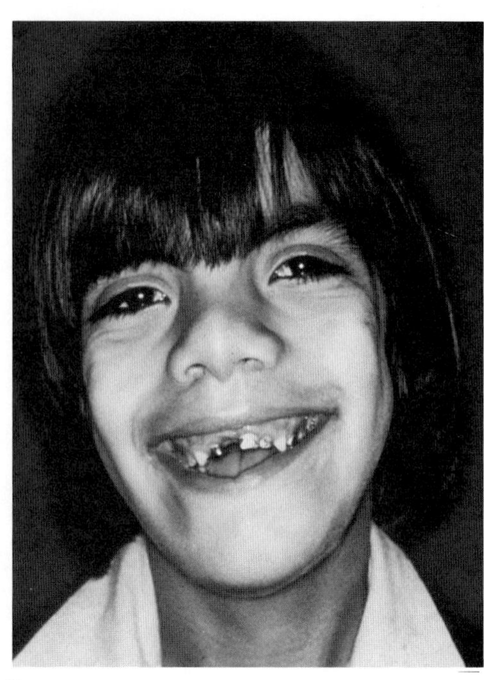
C

Fig. 11–22. *Simpson–Golabi–Behmel syndrome.* (A,B) Coarse facies, wide nasal bridge, wide mouth. (C) Note hypertelorism, short broad upturned nose, and large mouth. (A,B from *A Behmel et al,* Hum Genet **67**:409, 1984. C from *M Golabi* and *L Rosen,* Am J Med Genet **17**:345, 1984.)

the syndrome consisted of macrosomia, mental deficiency, broad nose, wide mouth, cleft palate, large cystic kidneys, and musculoskeletal and limb abnormalities. Opitz (6) reported what he considered three instances in another affected family and named the condition Golabi–Rosen syndrome. Behmel et al (1) reported an affected family and called attention to an earlier description by Simpson et al (8). They questioned the inclusion of Opitz's patients (6) because they lacked overgrowth and had severely impaired central nervous system function. Tsukahara et al (9) described what they called a Weaver-like syndrome in a Japanese boy. Later, Kajii and Tsukahara (4) thought their findings were more consistent with Golabi–Rosen syndrome. We are reluctant to consider their patient an example of either syndrome. In 1988, Neri et al (5) reported an affected family and reviewed all previous cases. Consideration of Opitz's patients (6,7) led to the conclusion that they had the same syndrome, albeit in a very severe form. Neri et al (5) noted that the precedent set by Simpson et al (8) and

Fig. 11–23. *Simpson–Golabi–Behmel* syndrome. Postaxial hexadactyly, mild soft tissue syndactyly between second and third digits. (From *M Golabi* and *L Rosen,* Am J Med Genet **17**:345, 1984.)

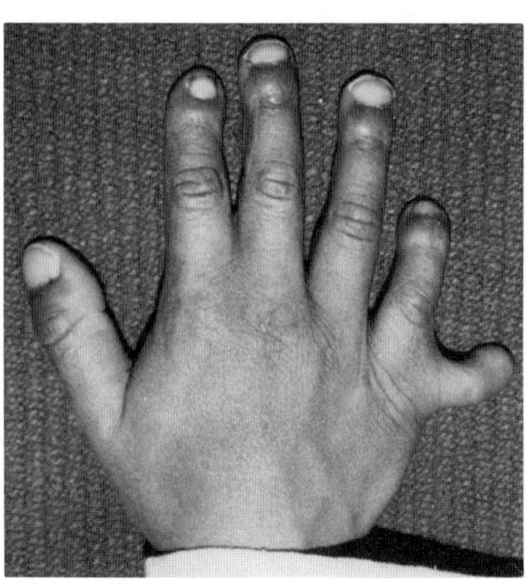

the contribution made by Behmel et al (1,2) necessitated a change in eponymic designation from Golabi–Rosen syndrome to Simpson–Golabi–Behmel syndrome. Perhaps as many as 26 cases have been described (5).

A family with presumed Simpson–Golabi–Behmel syndrome was noted by Garganta et al (2a). Because the proband had ear creases and other manifestations compatible with a diagnosis of Beckwith-Wiedemann syndrome, it is not known at present whether this family represents an example of the Beckwith–Wiedemann syndrome or of the Simpson–Golabi–Behmel syndrome. Publication of this family should aid in resolving the issue.

Three of four affected sibs reported by Golabi and Rosen (3) had an early demise within the first 4 months of life. The infant mortality rate is approximately 50% and may possibly result from respiratory/obstructive cor pulmonale, heart failure due to congenital heart defect and/or conduction defect, severe congenital hypotonia with its pulmonary consequences, or severe hypoglycemia from excessive islets of Langerhans.

**Growth** Impressive overgrowth is present at birth and continues postnatally. Birth weights from 4000 g to above 5000 g have been recorded (3,5). Birth weight may possibly correlate with the severity of the condition, the more severely affected infants having normal birth weights (5–7).

**Central nervous system.** Mild mental deficiency has been present in some, with moderate retardation in others (3,5–7). All of Opitz's patients (6,7) were severely retarded. Hypotonia of variable degree is common and its secondary effects may include visceroptosis of the liver, pectus excavatum, and neonatal respiratory distress syndrome. Strabismus and/or nystagmus have been recorded. Seizures are uncommon. Some patients had a stiff, wide-based gait, with clumsiness, calf-wasting, and winging of the scapulae (5).

**Craniofacial abnormalities.** The head is large, skull sutures may be prominent, and the facial appearance may be coarse. Other anomalies include hypertelorism, midface deficiency, broad nose, wide mouth, macroglossia, malocclusion with anterior open bite, and a narrow, highly arched palate. Low-frequency anomalies include coloboma of the optic disc, preauricular tags and pits, cleft lip, cleft palate, submucous cleft palate, and grooved lower lip.

**Limb anomalies.** The hands are large and square. Anomalies include, variably, broad halluces and thumbs, postaxial hexadactyly, partial soft tissue syndactyly of the second and third fingers and toes, foot deformities, hypoplastic nails, most frequently involving the index fingers, and single palmar creases (3,5).

**Skeletal system.** Vertebral segmentation defects are especially common, including anomalies such as fusion involving C2–3, cervical ribs, six lumbar vertebrae, and sacral or coccygeal defects associated with sacral dimpling or sinus, and eversion of the tip of the coccyx. Bone age is advanced. Mild scoliosis and pectus excavatum have been recorded. Clefting of the xiphisternum has been noted (3,5).

**Cardiovascular anomalies.** Conduction defects with partial AV block and incomplete right bundle branch block occur most commonly, but VSD and PDA have also been documented (5).

**Genitourinary system.** Large kidneys, lobulated kidneys, cystic kidneys, duplication of the renal pelvis, mild hydronephrosis, cysts involving both glomeruli and tubules, and cryptorchidism have been reported (5).

**Gastrointestinal tract.** Reported abnormalities have included intestinal malrotation, pyloric ring, and Meckel's diverticulum (5).

**Skin.** Supernumerary nipples have been noted. The skin may be thickened and brown in color. Spotty perioral and palatal pigmentation has been described (3,5).

**Other findings.** Inguinal hernia, umbilical hernia, and polycythemia have been noted (3,5).

**Differential diagnosis.** Other overgrowth syndromes, especially *Beckwith–Wiedemann syndrome* and *Weaver syndrome*, should be considered.

### References [Simpson–Golabi–Behmel syndrome (Golabi–Rosen syndrome)]


1. Behmel A et al: A new X-linked dysplasia gigantism syndrome: Identical with Simpson dysplasia syndrome? *Hum Genet* **67**:409–413, 1984.

2. Behmel A et al: A new X-linked dysplasia gigantism syndrome: Follow-up in the first family and a second Austrian family. *Am J Med Genet* **30**:275–285, 1988.

2a. Garganta CL et al: A third family with Golabi–Rosen syndrome. 39th Annual American Society of Human Genetics Meeting, New Orleans, Louisiana, October 12–15, 1988, Abstract No. 199.

3. Golabi M, Rosen L: A new X-linked mental retardation-overgrowth syndrome. *Am J Med Genet* **17**:345–358, 1984.

4. Kajii T, Tsukahara M: Letter to the editor: The Golabi–Rosen syndrome. *Am J Med Genet* **19**:819, 1984.

5. Neri G et al: Simpson–Golabi–Behmel syndrome: An X-linked encephalo-tropho-schisis syndrome. *Am J Med Genet* **30**:287–299, 1988.

6. Opitz JM: The Golabi–Rosen syndrome—report of a second family. *Am J Med Genet* **17**:359–366, 1984.

7. Opitz JM et al: Simpson–Golabi–Behmel syndrome: Follow-up of the Michigan family. *Am J Med Genet* **30**:301–308, 1988.

8. Simpson JL et al: A previously unrecognized X-linked syndrome of dysmorphia. *Birth Defects* **11**(2):18–24, 1973.

9. Tsukahara M et al: A Weaver-like syndrome in a Japanese boy. *Clin Genet* **25**:73–78, 1984.


## Prader–Willi syndrome

Prader–Willi syndrome was initially reported in 1956 by Prader, Labhart, and Willi (71). They described 14 patients with obesity, hypogonadism, cryptorchidism, mental retardation, and hypotonia (Figs. 11-24 to 11-27). No subsequent large series appeared until 1968 when Dunn (23) reported 9 cases and Zellweger and Schneider (86) published 12 cases. This was followed-up in 1972 by Hall and Smith (35) who described 32 cases. Other authors have reported a large number of additional patients (24,42c,53,54,76,80). The Prader–Willi syndrome has a prevalence of 1/25,000 (87). To date, over 500 cases have been documented (10). The disorder has been reported among whites, blacks, hispanics, orientals, and American indians (3,30,33). Excellent recent reviews are those of Bray (2,3), Cassidy (12,14), and Greenswag (33). A book on the Prader–Willi syndrome was published in 1981 (42) and a volume of the *American Journal of Medical Genetics* was devoted to the topic (1,6,11,16,31,32,39,48,51,57,58,63,70,82,84).

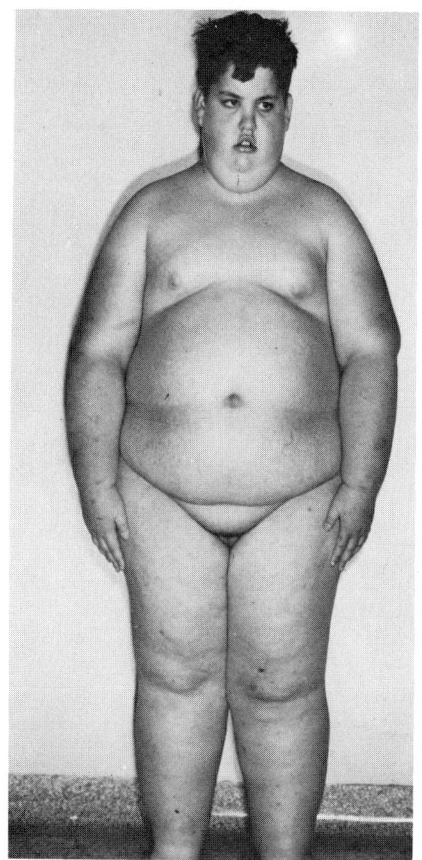

Fig. 11–24. *Prader–Willi syndrome.* Marked obesity, small hands, and hypoplastic genitalia. (From *MM Cohen Jr* and *RJ Gorlin,* Am J Dis Child **117**:213, 1969.)

**Etiology.** Originally, the Prader–Willi syndrome was thought to be potentially monogenic since rare recurrences have been noted. However, no particular pattern was apparent (3,4,17,24,35). The etiology was not established until 1981 when Ledbetter and colleagues (55), utilizing high-resolution banding techniques, identified a small chromosome deletion at the 15q11 to 15q13 region in 50% of their patients. The proximal 15q region had been suspect since previous reports had noted the Prader–Willi phenotype in some individuals who had translocations involving chromosome 15 (55). Follow-up studies confirmed the chromosome deletion in 50–92% of patients with classic Prader–Willi phenotype (10,28,50,56,64,77). Parental origin of del(15q) was inexplicably paternal in almost all cases although parental chromosomes were normal (9,10,47). Why some patients with aberrations such as deletions, duplications, and balanced/unbalanced states involving the 15q11 to 15q13 region do or do not have the full, partial, or absent Prader–Willi phenotype (21,26,27,45,61,68,72,77a) remains to be explained. In a 1984 report by Hasegawa and colleagues

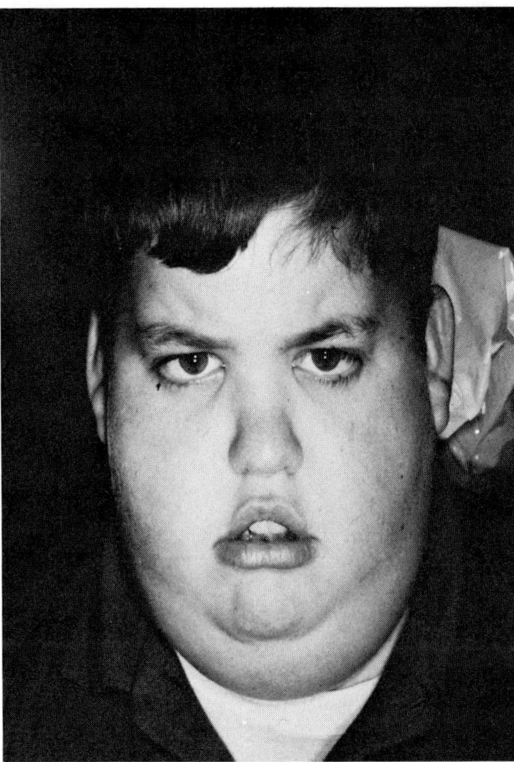

Fig. 11–25. *Prader–Willi syndrome.* Facial obesity, narrow bifrontal diameter, almond-shaped eyes, characteristic mouth. (From *MM Cohen Jr* and *RJ Gorlin,* Am J Dis Child **117:**213, 1969.)

(37), two maternal cousins with Prader–Willi phenotype had unbalanced translocations [46,XY or XX, −15, +der(14),rcp(14;15)(q11.2,q13)] through balanced reciprocal translocation [rcp(14;15)(q11.2,q13)] mothers. There is no difference in chromosome breakage between those that exhibit the deletion and those that do not (11a). We believe that some presumed familial cases of Prader–Willi syndrome (5,59) do not, in fact, represent the disorder. Other familial reports are incompletely documented (25). The two sibs reported by Ishikawa et al (42b) had normal 15 chromosomes; one sib had 46,X,del(X)(pter→q26.1:). Based on data from 1500 Prader–Willi families, Cassidy (15) concluded that a recurrence risk of less than 1/1000 seemed reasonable and that such recurrence is unlikely to occur when the proband has del(15q).

**Natural history.** Individuals with Prader–Willi syndrome have a predictable life scenario. About 7% are premature and 40% are breech

Fig. 11–26. *Prader–Willi syndrome.* Disproportionately small hands. (From *MM Cohen Jr* and *RJ Gorlin,* Am J Dis Child **117:**213, 1969.)

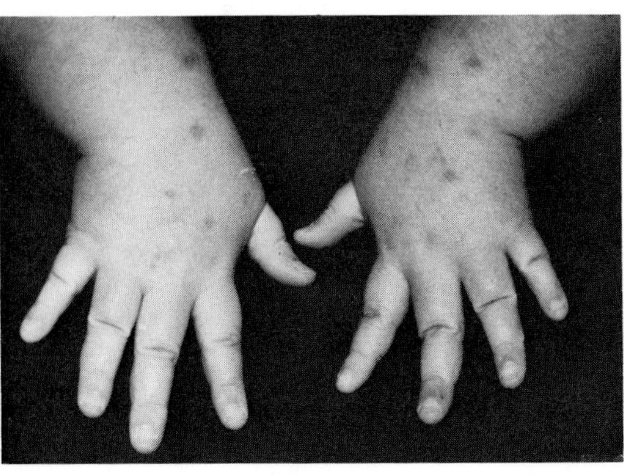

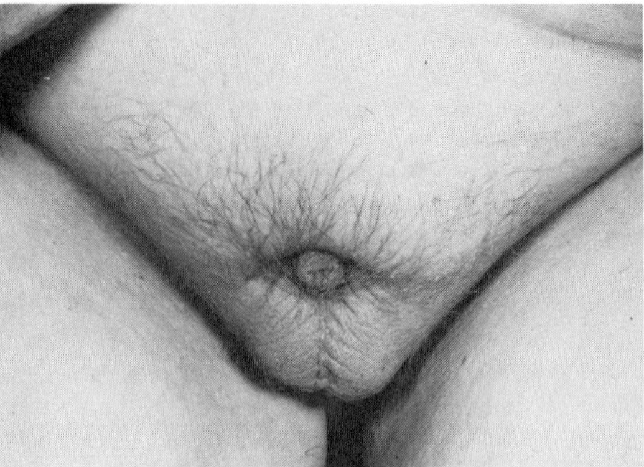

Fig. 11–27. *Prader–Willi syndrome.* Hypoplastic genitalia with penis buried in fat. (From *MM Cohen Jr* and *RJ Gorlin,* Am J Dis Child **117:**213, 1969.)

deliveries. They have low normal birth weight, severe hypotonia, and severe feeding difficulties in the newborn period (35). Cryptorchidism and hypogenitalism are noted in males. Otherwise, no significant dysmorphology is found (35). The severe hypotonia requires prolonged and special feeding techniques (12,14,18,23,35,61,86). Failure to thrive may be a problem during the first 6–12 months (12,14,40,65). Gradual improvement in muscle tone during the last half of the first year heralds an improved rate of weight gain (2,14,35,86). By 3 years, significant weight gain has occurred and delayed milestones are usually recognized (35). Unusual facial appearance, poor linear growth, and small hands and feet are noted at this time (35). Between 3 and 5 years, obesity and mild behavioral problems begin (35). Between midchildhood and adolescence, obesity becomes more prominent as do behavioral problems such as stubbornness, temper tantrums, and violent reactions (35,86). No signs or few signs of puberty exist (33). During teenage years and young adulthood, severe obesity often causes cardiorespiratory problems (33) and an increased frequency of diabetes mellitus (33,86). The teenager and/or adult with Prader–Willi syndrome may become unmanageable, causing much despair in the family (14,33). Diagnosis at any time, but particularly before 5 years of age, can potentially lead to effective treatment by weight reduction and control with an ameliorating effect on many problems (12, 14,20,33,35).

**Growth.** Low normal birthweight and failure to thrive are characteristic (35). Proportionate short stature by adulthood is usually in the mild to moderate range (8,33,40). Adult males have an average height of 61 in. and adult females, 59 in. (3). No consistent hormonal basis for short stature has been identified (3,67,73).

A number of anthropometric studies have been carried out (1a,8,16a,42a). Hudgins and Cassidy (42a) and Butler and Meaney (8) had large sample size. Foot length tends to be smaller than hand length and both hand and foot lengths frequently fall within the normal range before 10 years of age. As the discrepancy between height age and chronological age increases, hand and foot lengths become proportionately smaller. No anthropometric differences have been found between chromosome 15 deletion cases and nondeletion cases.

**Central nervous system and performance.** Severe congenital hypotonia is a constant feature during the neonatal period and early infancy. Improvement usually begins during the latter half of the first year (35). Hypotonia is clearly associated with poor fetal activity (74–85%), breech presentation (22–40%), and severe feeding difficulties (3,12,35). Early delays in developmental milestones (i.e., sitting up at 1 year, walking alone at 24–30 months) are also probably related (12,33,35,71,86). Various studies, including muscle biopsies, nerve conduction times, and electromyelograms, have either been normal, inconsistent, or nonspecific, which suggests that the hypotonia is probably central in origin (3,12,14).

Sleep disorders are common: daytime somnolence (50%), snoring (45%), restless sleep (40%), cataplexy (15%) (15b). Mental retardation is usually mild (63%) to moderate (31%), but in 3–8%, intelligence appears to be normal (33, Prader–Willi Syndrome Parent Association). Avoidance of obesity through early diagnosis and treatment may improve ultimate intelligence (20). Seizures occur in 16–20%, but are usually not a chronic problem (3,35). Speech difficulties are found in almost all children (12,62,86).

It is ironic that Prader–Willi children are known for their severe behavior problems when they are frequently loving, placid, and pleasant to be around (12,35). However, it is the behavior problems that cause families the most concern (33). Large, out-of-control, and not especially hypotonic adolescents and adults can be very destructive and dangerous to themselves (33,86).

Decreased pain sensitivity, particularly of the skin, may explain why ordinary cuts and bruises can be picked at constantly by Prader–Willi individuals without apparent pain (12,14,33,35). Chronic sores and scarred skin lesions are found in all children beyond midchildhood. Nasal bleeding secondary to chronic picking can be a difficult problem (33).

Some degree of oculocutaneous albinoidism occurs in approximately 50% (19,38,85). Misrouting of retinal–ganglion fibers at the optic chiasm, typically seen in albinos, may be responsible for the strabismus found in Prader–Willi individuals (19).

**Obesity.** Excessive weight is the major complicating factor in Prader–Willi syndrome. Once severe hypotonia remits, the child begins to gain weight at an excessive rate (35,65). Even before clinical obesity is evident, skinfold thickness indicates excessive adipose tissue (8). Obesity is usually obvious between 3 and 5 years (35) and has a generalized distribution, sparing only the distal finger tips and upper cranium (Figs. 11–24 and 11–26). Voracious appetite or binge eating can be dramatic, but weight gain can occur even on low daily caloric intake (41). Obesity can reach gigantic proportions causing severe cardiorespiratory distress as seen in Pickwickian syndrome (35,54,65,86). Frequent nodding-off or sleeping during normal waking hours and normal activities is typical for older obese individuals with Prader–Willi syndrome (15b,33,54,66,81). Diabetes mellitus occurs in 4.5–19% of late teenage and adult individuals with Prader–Willi syndrome (3,33,54) and seems to be related directly to age and degree of obesity. Insulin is required in about 66%, but with significant weight loss the need for insulin disappears (33,35). There is some evidence that weight reduction or avoidance of obesity in young Prader–Willi children stabilizes intelligence quotients (20). Some theorize, and common sense dictates, that effective weight control prolongs life expectancy in individuals with Prader–Willi syndrome (33). Anesthetic risk shows an increased but surprisingly low (2.5%) complication rate considering the obese state of many affected individuals (83).

**Sexual development.** Cassidy et al (15a) surveyed genital abnormalities and hypogonadism in 105 patients. In general, individuals with Prader–Willi syndrome have few or no signs of pubescence (33). Males have high frequencies of scrotal hypoplasia, cryptorchidism, and micropenis (Fig. 11–27) (33,35,79). When males have any signs of puberty, minimal development of axillary and pubic hair occurs with a slightly higher frequency of demonstrable but scant beard (33,36). Hypogonadism, hypogonadotropism, and hypergonadotropic-hypogonadism have been ascribed as causes of faulty sexual development (3,36,43,44,75). Testicular abnormalities and oligospermia have also been documented (3,36,46). Females have similar problems although usually there is some breast development, but menstrual periods, if they do occur, are usually delayed in onset, irregular, and scant (33). Occasionally, male and female patients have precocious puberty; however, it is almost always incomplete and not true precocious puberty (47,60,78). Neither males nor females are fertile (33).

**Craniofacial features.** Head circumference is usually normal, but the bifrontal diameter is narrow (8,35). The eyes are almond shaped (35). Because the medial upper lid hooks sharply downward, the eyes often have a slit-like appearance and an upward slant during infancy (76). Strabismus occurs in 40–95% (3,35). An open, inverted, V-shaped mouth is usually present in the young child (Fig. 11–25) (14,76). Enamel hypoplasia, dental caries, and malocclusion occur with increased frequency (18,22,35,42,50,86). Reduced salivation has been reported (3,14).

**Limbs.** The hands and feet are stated to be small (3,8,35) (Fig. 11–26). Although microsomic hands and feet are thought to be present congenitally, they are not usually recognized as such during the first 2 years (35,42a). The fingers are short and narrow, but can look wide proximally and tapered distally when obesity becomes significant (14,22,35,53,61,71,86). However, Chitayat et al (16a) found the hands and feet to be of normal size.

**Skeletal system.** Orthopedic and/or osseous problems are common. At birth, talipes (6.4%) and congenital hip dislocation (9.4%) occur with increased frequency (35,86). Later, scoliosis, osteoporosis, and a slight increase in the fracture rate are often noted (12,13,33,35). Delayed bone age can be seen (8,35). Metacarpophalangeal pattern profiles have been used to separate Prader–Willi syndrome from obese non-Prader–Willi controls (7). Skeletal roentgenograms show abnormalities that are both subtle and consistent, but that lack enough specificity to be useful as a major diagnostic tool (50).

**Hematologic findings.** Three instances of leukemia have been reported (33,34) and one case has been associated with factor XI deficiency (29).

**Differential diagnosis.** Prader–Willi syndrome should be distinguished from other postnatal onset obesity syndromes. In the neonatal period, differential diagnosis should include syndromes with severe hypotonia such as *Zellweger syndrome, trisomy 21 syndrome,* Werdnig–Hoffmann disease, *congenital myotonic dystrophy,* congenital myopathies, and *fetal akinesia* syndromes.

In Urban–Rogers–Meyer syndrome there is a Prader–Willi habitus, with osteopenia, camptodactyly, and mental retardation. The disorder has been described only in males, two of them brothers (66a). *Cohen syndrome* should also be excluded.

**Laboratory aids.** Karyotyping may render a neonatal diagnosis more secure in those with the 15q12 deletion.

### References (Prader–Willi syndrome)


1. Alexander RC et al: Rumination and vomiting in Prader–Willi syndrome. *Am J Med Genet* **28**:889–895, 1987.

1a. Aughton DJ, Cassidy SB: Physical features of Prader–Willi syndrome in the neonatal period. Ninth Annual David W. Smith Workshop on Malformations and Morphogenesis, Oakland, California, August 3–7, 1988.

2. Bray GA, Wilson WG: Prader–Labhart–Willi syndrome: An overview. *Growth Genet Hormones* **2**:1–5, 1986.

3. Bray GA et al: The Prader–Willi syndrome: A study of 40 patients and a review of the literature. *Medicine* **62**:59–80, 1983.

4. Brissenden JE, Levy EP: Prader–Willi syndrome in infant monozygous twins. *Am J Dis Child* **126**:110–112, 1973.

5. Burke MG et al: Clinical and cytogenetic survey of 39 individuals with Prader–Labhart–Willi syndrome. *Am J Med Genet* **23**:793–809, 1986.

6. Butler MG, Jenkins BB: Sister chromatid exchange analysis in the Prader–Labhart–Willi syndrome. *Am J Med Genet* **28**:821–827, 1987.

7. Butler MG, Meaney FJ: Metacarpophalangeal pattern profile analysis in Prader–Willi syndrome. A follow-up report on 38 cases. *Clin Genet* **28**:27–30, 1985.

8. Butler MG, Meaney FJ: An anthropometric study of 38 individuals with Prader–Labhart–Willi syndrome. *Am J Med Genet* **26**:445–455, 1987.

9. Butler MG, Palmer CG: Parental origin of chromosome 15 deletion in Prader–Willi syndrome. *Lancet* **1**:1285–1286, 1983.

10. Butler MG et al: Clinical and cytogenetic survey of 39 individuals with Prader–Labhart–Willi syndrome. *Am J Med Genet* **23**:793–809, 1986.

11. Butler MG et al: Plasma immunoreactive $\beta$-melanocyte stimulating hormone (lipotropin) levels in individuals with Prader–Labhart–Willi syndrome. *Am J Med Genet* **28**:839–844, 1987.

11a. Butler MG, Jenkins BB: Analysis of chromosome breakage in the Prader–Labhart–Willi syndrome. *Am J Med Genet* **32**:514–519, 1989.

12. Cassidy SB: Prader–Willi syndrome. *Curr Probl Pediatr* **14**:1–55, 1984.

13. Cassidy SB: Osteoporosis in Prader–Willi syndrome. *Proc Greenwood Genet Ctr* **5**:120, 1985.

14. Cassidy SB: Prader–Willi syndrome. Characteristics, management, and etiology. *Ala J Med Sci* **24**:169–175, 1987.

15. Cassidy SB: Letter to the editor: Recurrence risk in Prader–Willi syndrome. *Am J Med Genet* **28**:59–60, 1987.

15a. Cassidy SB et al: Genital abnormalities and hypogonadism in 105 patients with Prader–Willi syndrome. Eighth David W. Smith Workshop on Malformations and Morphogenesis, Greenville, South Carolina, August 15–19, 1987.

15b. Cassidy SB et al: Sleep disorders in Prader–Willi syndrome. David W. Smith Workshop on Malformations and Morphogenesis, Madrid, Spain, May 23–29, 1989.

16. Chasalow FI et al: Steroid metabolic disturbances in Prader–Willi syndrome. *Am J Med Genet* **28**:857–864, 1987.

16a. Chitayat D et al: Perinatal and first year follow up of patients with Prader–Willi syndrome—normal size of hands and feet. 9th Annual David W. Smith Workshop on Malformations and Morphogenesis, Oakland, California, August 3–7, 1988.

17. Clarren SK, Smith DW: Prader–Willi syndrome. Variable severity and recurrence risk. *Am J Dis Child* **131**:798–800, 1977.

18. Cohen MM Jr, Gorlin RJ: The Prader–Willi syndrome. *Am J Child* **117**:213–218, 1969.

19. Creel DJ et al: Abnormalities of the central visual pathways in Prader–Willi syndrome associated with hypopigmentation. *New Engl J Med* **314**:1606–1609, 1986.

20. Crnic KA et al: Preventing mental retardation associated with gross obesity in the Prader–Willi syndrome. *Pediatrics* **66**:787–789, 1980.

21. DeFraites EB et al: Familial Prader–Willi syndrome. *Birth Defects* **11**(4):123–126, 1975.

22. Duckett DP et al: Unbalanced reciprocal translocations in cases of Prader–Willi. *Hum Genet* **67**:156–161, 1984.

23. Dunn HG: The Prader–Labhart–Willi syndrome: Review of the literature and report of nine cases. *Acta Paediatr Scand Suppl* **186**:1–38, 1968.

24. Dunn HG et al: Clinical experience with 23 cases of Prader–Willi syndrome, in *The Prader–Willi Syndrome,* Holm VA et al (eds), University Park Press, Baltimore, 1981, pp 69–88.

25. Fernandez F et al: Prader–Willi syndrome in siblings, due to unbalanced translocation between chromosomes 15 and 22. *Arch Dis Childh* **62**:841–843, 1987.

26. Fraccaro M et al: Deficiency, transposition and duplication of one 15q region may be alternatively associated with Prader–Willi (or a similar) syndrome. Analysis of seven cases after varying ascertainment. *Hum Genet* **64**:388–394, 1983.

27. Fuhrmann-Rieger A et al: Duplication or insertion in 15q11–13 associated with mental retardation—short stature and obesity—Prader–Willi or Cohen syndrome? *Clin Genet* **25**:347–352, 1984.

28. Fukushima Y et al: The Prader–Willi syndrome and interstitial deletion of chromosome 15: High-resolution chromosome analysis of 14 patients with the Prader–Willi syndrome and of 5 suspected patients. *Jpn J Hum Genet* **29**:1–6, 1984.

29. Futterweit W et al: Coexistence of Prader–Willi syndrome, congenital ectropion uveae with glaucoma and factor X1 deficiency. *JAMA* **255**:3280–3282, 1986.

30. Golden WL et al: Prader–Willi in black females. *Clin Genet* **26**:161–163, 1984.

31. Greenberg F, Ledbetter DH: Deletions of proximal 15q without Prader–Willi syndrome. *Am J Med Genet* **28**:813–820, 1987.

32. Greenberg F et al: Neonatal diagnosis of Prader–Willi syndrome and its implications. *Am J Med Genet* **28**:845–856, 1987.

33. Greenswag LR: Adults with Prader–Willi syndrome: A survey of 232 cases. *Dev Med Child Neurol* **29**:145–152, 1987.

34. Hall BD: Leukemia and the Prader–Willi syndrome. *Lancet* **1**:46, 1985.

35. Hall BD, Smith DW: Prader–Willi syndrome. A resume of 32 cases including an instance of affected first cousins, one of whom is of normal stature and intelligence. *J Pediatr* **81**:286–293, 1972.

36. Hamilton CR et al: Hypogonadotropinism in Prader–Willi syndrome. Induction of puberty and spermatogenesis by clomiphene citrate. *Am J Med* **52**:322–329, 1972.

37. Hasegawa T et al: Cytogenetic studies of familial Prader–Willi syndrome. *Hum Genet* **65**:325–330, 1984.

38. Hittner HM et al: Oculocutaneous albinoidism as a manifestation of reduced neural crest derivatives in the Prader–Willi syndrome. *Am J Ophthalmol* **94**:328–337, 1982.

39. Holm VA: Cytogenetics in Prader–Willi syndrome: A questionnaire study. *Am J Med Genet* **28**:915–924, 1987.

40. Holm VA, Nugent JK: Growth in the Prader–Willi syndrome. *Birth Defects* **18**(3B):93–100, 1982.

41. Holm VA, Pipes PL: Food and children with Prader–Willi syndrome. *Am J Dis Child* **130**:1063–1067, 1976.

42. Holm VA et al: *Prader–Willi Syndrome.* University Park Press, Baltimore, 1981.

42a. Hudgins L, Cassidy SB: Hand and foot length in Prader–Willi syndrome. 9th Annual David W. Smith Workshop on Malformations and Morphogenesis, Oakland, California, August 3–7, 1988.

42b. Ishikawa T et al: Prader–Willi syndrome in two siblings: One with normal karyotype, one with a terminal deletion of distal Xq. *Clin Genet* **32**:295–299, 1987.

42c. Jancer J: Prader–Willi syndrome (hypotonia, obesity, hypogonadism growth and mental retardation). *J Ment Defic Res* **15**:20–29, 1971.

43. Jaskulsky SR, Stone NH: Hypogonadism in Prader–Willi syndrome. *Urology* **29**:207–208, 1987.

44. Jeffcoate WJ et al: Endocrine function in Prader–Willi syndrome. *Clin Endocrinol* **12**:81–89, 1980.

45. Kaplan LC et al: Clinical heterogeneity associated with deletion in the long arm of chromosome 15: Report of 3 new cases and their possible genetic significance. *Am J Med Genet* **28**:45–53, 1987.

46. Katcher M et al: Absence of spermatogonia in the Prader–Willi syndrome. *Eur J Pediat* **124**:257–260, 1977.

47. Kauli R et al: Pubertal development in the Prader–Labhart–Willi syndrome. *Acta Paediatr Scand* **67**:763–767, 1978.

48. Kousseff BG et al: Unique mosaicism in Prader–Labhart–Willi syndrome—A contiguous gene or aneuploidy syndrome? *Am J Med Genet* **28**:803–811, 1987.

49. Kyriakides M et al: Effect of naloxone on hyperphagia in Prader–Willi syndrome. *Lancet* **1**:876–877, 1980.

50. Labidi F, Cassidy SB: A blind prometaphase study of Prader–Willi syndrome. Frequency and consistency in interpretation of del15q. *Am J Hum Genet* **39**:452–460, 1986.

51. Lamb AS, Johnson WM: Premature coronary artery atherosclerosis in a patient with Prader–Willi syndrome. *Am J Med Genet* **28**:873–880, 1987.

52. Landwirth J et al: Prader–Willi syndrome. *Am J Dis Child* **116**:211–217, 1968.

53. Laurance BM: Hypotonia, mental retardation, obesity, and cryptorchidism associated with dwarfism and diabetes in children. *Arch Dis Childh* **42**:126–139, 1967.

54. Laurance BM et al: The Prader–Willi syndrome after the age of 15 years. *Arch Dis Childh* **56**:181–186, 1981.

55. Ledbetter DH et al: Deletions of chromosome 15 as a cause of the Prader–Willi syndrome. *N Engl J Med* **304**:325–329, 1981.

56. Ledbetter DH et al: Chromosome 15 abnormalities and the Prader–Willi syndrome: A follow-up report of 40 cases. *Am J Hum Genet* **34**:278–285, 1982.

57. Ledbetter DH et al: Conference report: Second annual Prader–Willi syndrome scientific conference. *Am J Med Genet* **28**:779–790, 1987.

58. Lee PDK et al: Linear growth response to exogenous growth hormone in Prader–Willi syndrome. *Am J Med Genet* **28**:865–871, 1987.

59. Lubinsky M et al: Familial Prader–Willi syndrome with apparently normal chromosomes. *Am J Med Genet* **28**:37–43, 1987.

60. MacMillan DR et al: Syndrome of growth resistance, obesity, and intellectual impairment with precocious puberty. *Arch Dis Childh* **47**:119–121, 1972.

61. Mattei MG et al: Chromosome 15 anomalies and the Prader–Willi syndrome: Cytogenetic analysis. *Hum Genet* **66**:313–334, 1984.

62. Meyerson MD: The effect on syndrome diagnosis on speech remediation. *Birth Defects* **21**(2):47–68, 1985.

63. Mitchell W, Cook KV: Social skills training of Prader–Willi staff. *Am J Med Genet* **28**:907–913, 1987.

64. Niikawa N, Ishikiriyama S: Clinical and cytogenetic studies of the Prader–Willi syndrome: Evidence of phenotype-karotype correlation. *Hum Genet* **69**:22–27, 1985.

65. Nugent JK, Holm VA: Physical growth in Prader–Willi syndrome, in *The Prader–Willi Syndrome,* Holm VA et al (eds), University Park Press, Baltimore, 1981, pp 269–280.

66. Orenstein DM et al: The obesity hypoventilation syndrome in children with the Prader–Willi syndrome: A possible role for familial decreased response to carbon dioxide. *J Pediatr* **97**:765–767, 1980.

66a. Pagnan NAB, Gollop TR: Prader–Willi habitus, osteopenia, and camptodactyly (Urban–Rogers–Meyer syndrome): A probable second report. *Am J Med Genet* **31**:787–792, 1988.
67. Parre A et al: Immunoreactive insulin and growth hormone response in patients with Prader–Willi syndrome. *J Pediatr* **83**:587–593, 1973.
68. Pauli RM et al: 'Expanded' Prader–Willi syndrome in a boy with an unusual 15q chromosome deletion. *Am J Dis Child* **137**:1087–1089, 1983.
69. Pearson KD et al: Roentgenographic manifestations of the Prader–Willi syndrome. *Radiology* **10**:369–377, 1971.
70. Pettigrew AL et al: Duplication of proximal 15q as a cause of Prader–Willi syndrome. *Am J Med Genet* **28**:791–802, 1987.
71. Prader A et al: Ein Syndrom von Adipositas, Kleinwuchs, Kryptorchismus, und Oligophrenie nach Myotonisartigem Zustand im Neugeborenenalter. *Schweiz Med Wschr* **86**:1260–1261, 1956.
72. Reynolds JF et al: Brief clinical report: Atypical phenotype associated with deletion (15)(pter→q11::q13→qter). *Am J Med Genet* **28**:55–58, 1987.
73. Rudd BT et al: Adrenal response to ACTH in patients with Prader–Willi syndrome, simple obesity, and constitutional dwarfism. *Arch Dis Childh* **44**:244–247, 1969.
74. Schwartz S et al: Deletions of proximal 15q and nonclassical Prader–Willi syndrome phenotypes. *Am J Med Genet* **20**:255–263, 1985.
75. Seyler LE et al: Hypergonadotropic-hypogonadism in the Prader–Labhart–Willi syndrome. *J Pediatr* **94**:435–437, 1979.
76. Stephenson JBP: Prader–Willi syndrome: Neonatal presentation and later development. *Dev Med Child Neurol* **22**:792–795, 1980.
77. Takano T et al: High-resolution cytogenetic studies in patients with Prader–Willi syndrome. *Clin Genet* **30**:241–248, 1986.
77a. Tantravahi U et al: Quantitative calibration and use of DNA probes for investigating chromosome abnormalities in the Prader–Willi syndrome. *Am J Med Genet* **33**:78–87, 1989.
78. Vanelli M et al: Precocious puberty in a male with Prader–Labhart–Willi syndrome. *Helv Paediat Acta* **39**:373–377, 1984.
79. Vehling D: Cryptorchidism in the Prader–Willi syndrome. *Urology* **124**:103–104, 1980.
80. Vischer VD et al: Das Prader–Labhart–Willi-Syndrom (myotonischer Diabetes), in *Handbuch des diabetes Pathophysiologie,* Vol. 2, Pfeiffer EF (ed), JF Lehmanns Verlag, München, 1971, pp 631–648.
81. Walsh JK et al: Excessive daytime somnolence in the Prader–Willi syndrome (PWS). Abstr 1915, *Ped Res* **20**(4):479A, 1986.
82. Wenger SL et al: Clinical comparison of 59 Prader–Willi patients with and without the 15(q12) deletion. *Am J Med Genet* **28**:881–887, 1987.
83. Wett RJ: Study on surgery and anesthesia in the person with Prader–Willi syndrome. *The Gathered View* (newsletter of Prader–Willi Syndrome Parent Association) **XI**(2):8, 1985.
84. Whitman BY, Accardo P: Emotional symptoms in Prader–Willi syndrome adolescents. *Am J Med Genet* **28**:897–905, 1987.
85. Wiesner GL et al: Hypopigmentation in the Prader–Willi sydnrome. *Am J Hum Genet* **40**:431–442, 1987.
86. Zellweger H, Schneider HJ: Syndrome of hypotonia-hypopigmentia-hypogonadism-obesity (HHHO) or Prader–Willi syndrome. *Am J Dis Child* **5**:588–598, 1968.
87. Zellweger H, Soper RT: The Prader–Willi syndrome. *Med Hygiene* **37**:3338–3345, 1979.

## Cohen syndrome

Cohen syndrome was first reported in 1973 as an obesity/hypotonia syndrome associated with mental retardation, narrow hands and feet, and characteristic facial appearance consisting of downslanting palpebral fissures, short philtrum, open mouth, prominent upper central incisors, maxillary hypoplasia, and mild micrognathia (Figs. 11-28 to 11-30) (4). This original report identified male and female sibs and one sporadic instance. In 1978, a report of four additional cases confirmed the Cohen syndrome as a distinct entity (3). To date, approximately 85 cases have been recorded under the designation of Cohen syndrome (2–11,13,15–17,19–24) while an additional five cases have been reported as "Cohen-like" or resembling Cohen syndrome (1,12,14).

A significant number of cases reported as Cohen syndrome do not appear to be secure diagnoses (2,7,11,24). Most articles reporting three or more cases have some suspect diagnoses, particularly among their sporadically occurring cases (3,8,10,13,17,19,21). In addition to the original patients reported by Cohen et al (4), patients representing the most consistent and identifiable phenotype of the Cohen syndrome include the patients of Carey and Hall (cases 3 and 4) (3), Goecke et al (cases 1 and 2) (13), and Norio et al (cases 1, 2, 3, and 4) (17). Not surprisingly, they represent occurrences in sibs. Goecke et al (13) recognized the difficulty in diagnosing sporadic cases because of the wide variability of features among reported cases.

Morris et al (16) attempted to divide Cohen syndrome into two separate groups: one with chorioretinal dystrophy, leukopenia, and lack

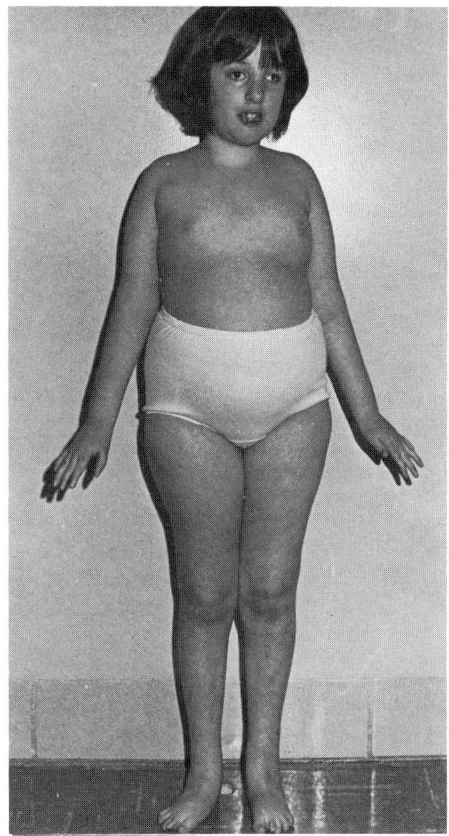

Fig. 11–28. *Cohen syndrome.* Obesity of mid-childhood onset, tapering hands and feet. (From *MM Cohen Jr et al,* J Pediatr **83:**280, 1973.)

Fig. 11–29. *Cohen syndrome.* (A,B) Characteristic facial appearance with mild microcephaly, downslanting palpebral fissures, short philtrum, and open mouth. (From *MM Cohen Jr et al,* J Pediatr **83:**280, 1973.)

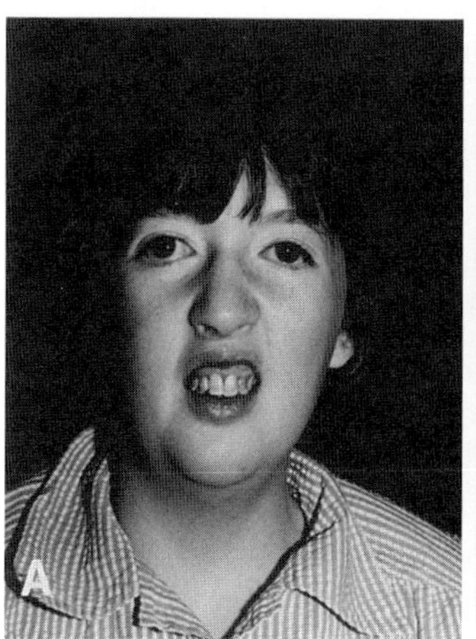



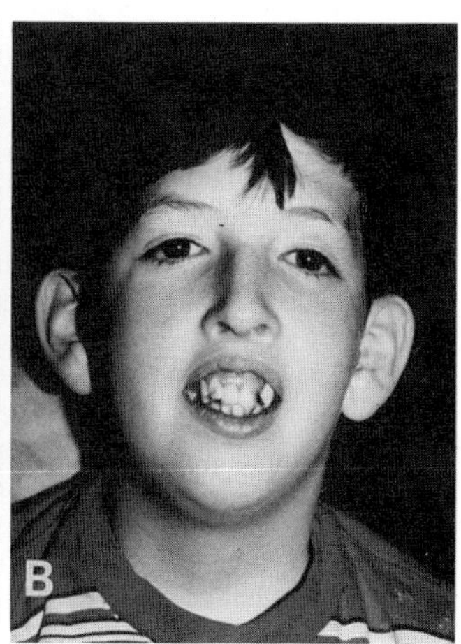

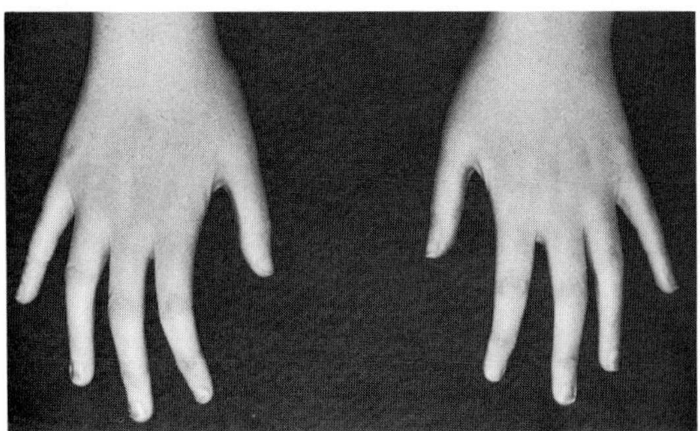

Fig. 11–30. *Cohen syndrome*. Narrow hands and fingers. (From *MM Cohen Jr et al,* J Pediatr **83:**280, 1973.)

of obesity (Norio syndrome) and the other with "classic" Cohen syndrome. Norio and Raitta (18) saw no need for such a division and, in fact, using the criteria of Morris et al (16), case 3 of Cohen et al (4) could not be classified in either category. The above diagnostic confusion is symptomatic of a distinct syndrome that may be heterogeneous, but is more apt to be related to features that are subject to loose interpretation and create pseudophenocopies that are hard to refute from history and photographs alone.

The inheritance pattern of Cohen syndrome is almost surely autosomal recessive. Eighteen sibships of two (3,4,7,9,10,13,16,17,19,21) and one sibship of four (15) have been reported. Consanguinity has been noted in four instances (17,21). Escobar et al (8) suggested autosomal dominant inheritance based on a family they feel shows vertical transmission (8). However, Sack and Friedman (21) reported 32 families (mostly Ashkenazi Jews) with at least one affected child and found no instance of parents or more distant relatives having any features of Cohen syndrome. The ratio of affected males to affected females is 1:1 (21).

**Growth.** Newborns have low–normal birthweights, averaging between the 10th and 25th percentiles (17,21). About 61–68% ultimately have short stature (17,19), although a 13% short stature rate and a 20% tall stature rate have also been documented (21).

Truncal obesity usually becomes evident between 5 and 12 years of age (Fig. 11–28) (3,4). Severe obesity or weight above the 97th percentile does not occur, as a rule, and not all patients develop truncal obesity (10,21).

Delayed puberty was first noted by Carey and Hall in 1978 (3). Delayed or absent signs of puberty occur in approximately 80% (19). Thirty-one percent of males are cryptorchid (21).

**Central nervous system and performance.** Hypotonia may be present at birth and can be of variable severity. It is usually noticeable in infancy (92%) and may persist into late childhood or early adolescence (3,19). The hypotonia may explain the high frequency of kyphosis and kyphoscoliosis (54%) (21).

Microcephaly is present in 50–60% (17,19), usually of mild to moderate degree, and may be prenatal or postnatal in onset (13,17). Two sibs reported by Friedman and Sack (10) had macrocephaly, but diagnostic accuracy is in question in such cases. Mental retardation is almost constant with intelligence quotients ranging from 30 to 80 (17,21). About 6% have minor, nonchronic seizures (19). The demeanor has been noted to be happy, pleasant, and affectionate (4), but no studies have yet been carried out on a large group of patients.

**Facial features.** Open mouth, exposed upper gingiva, and prominent upper central incisors are characteristic (4). The philtrum is short; the upper lip is arched and everted (4,17). Maxillary hypoplasia and mild micrognathia combined with high nasal bridge give the upper two-thirds of the midface a narrow prominence (Fig. 11–29A,B) (4,17,21). The palate is usually high and narrow (97%) (17). The eyes are commonly downslanted (3,4,17,21). Occasionally, iris and/or retinal colobomas may occur with or without microphthalmia (3,4,19). Myopia (46%) and strabismus (52%) are frequent eye findings (19). Retinal pigmentary abnormalities (or chorioretinopathy) are also common (3–5,10,17,19,20,21,21a). The helical rims may be hypoplastic (17).

**Limbs.** Cubitus valgus and hyperextensible joints are found in about 50% (19). Mild syndactyly of the fingers, sometimes of the second and third fingers, occurs in 30–35% (17,19). Hands are narrow and fingers are thin, nontapering, and long appearing (Fig. 11–30) (3,4,17,19,21). Feet and toes are similarly affected, but less dramatically. Arms and legs are slender, probably secondary to decreased muscle mass (17).

**Cardiovascular system.** Heart defects have been noted in 10% (21) and may include floppy mitral valve, mitral valve prolapse, pseudotruncus with VSD, and isolated VSD (15a,20,21). Norio et al (17) noted significant systolic murmurs in five of six of their cases, but the murmurs disappeared with age and no cardiac studies were performed.

**Laboratory findings.** Nonsymptomatic leukopenia was reported by Norio et al (17) and Warburg et al (21a) and has not been seen in any other cases of Cohen syndrome, but generally has not been looked for. ERG has shown chorioretinopathy (17). Hormonal studies for gonadal function have been nonspecific (3).

**Differential diagnosis.** Cohen syndrome is very distinctive; it is usually easy to separate from other postnatal onset obesity syndromes. Because a significant number of postnatal onset obesity syndromes of unknown genesis need to be further delineated, some of these can be confused with Cohen syndrome. Additional families with several affected siblings with Cohen syndrome need to be identified and closely followed.

Most cases of Cohen syndrome are not correctly diagnosed until midchildhood to early adolescence (21). It is important to note that open mouth and prominent upper central incisors often occur in mouth breathing (or adenoid) facies independently of Cohen syndrome. Furthermore, upper central incisors are prominent in all children from midchildhood to the onset of adolescence because of the naturally occurring contrast between permanent and deciduous teeth in the mixed dentition. Every patient with mouth breathing facies and either mental retardation or obesity does not have Cohen syndrome per se, nor does every child with obesity and mental retardation who cannot be diagnosed as having one of the other well delineated postnatal onset obesity syndromes.

### References (Cohen syndrome)

1. Bader PI: A new MR syndrome resembling Cohen syndrome. *Proc Greenwood Genet Ctr* **6**:124, 1987.

2. Balestrazzi P. et al: The Cohen syndrome: Clinical and endocrinological studies of two new cases. *J Med Genet* **17**:430–432, 1980.

3. Carey JC, Hall BD: Confirmation of the Cohen syndrome. *J Pediatr* **93**:239–244, 1978.

4. Cohen MM Jr et al: A new syndrome with hypotonia, obesity, mental deficiency, and facial, oral, ocular, and limb anomalies. *J Pediatr* **83**:280–284, 1973.

5. de Toni T, Caflero V: Sexual development in a girl with Cohen syndrome. *J Pediatr* **100**:1001–1002, 1982.

6. de Toni T et al: The Cohen syndrome: Presentation of the first Italian case. *Minerva Pediatr* **34**:261–266, 1982.

7. Doyard P, Mattei JF: Syndrome de Cohen chez deux soeurs. *Sem Hôp Paris* **60**:1143–1147, 1984.

8. Escobar V et al: A study of Cohen syndrome's inheritance. In press.

9. Ferre P et al: Le syndrome de Cohen, une affection autosomique recessive? *Arch Fr Pédiatr* **39**:159–160, 1982.

10. Friedman E, Sack J: The Cohen syndrome: Report of five new cases and a review of the literature. *J Craniofac Genet Dev Biol* **2**:193–200, 1982.

11. Fryns JP, Van Den Berghe H: The Cohen syndrome. *J Génét Hum* **29**:449–453, 1981.

12. Fuhrmann-Rieger A. et al: Duplication or insertion in 15q11–13 associated with mental retardation—short stature and obesity—Prader–Willi or Cohen syndrome? *Clin Genet* **25**:347–352, 1984.

13. Goecke T. et al: Mental retardation, hypotonia, obesity, ocular, facial, dental, and limb abnormalities (Cohen syndrome). Report of three patients. *Eur J Pediatr* **138**:338–340, 1982.

14. Grix A Jr, Nelson S: New obesity, mental retardation syndrome resembling Cohen syndrome. *Proc Greenwood Genet Ctr* **6**:123, 1987.

15. Kousseff BG: Cohen syndrome: Further delineation and inheritance. *Am J Med Genet* **17**:317–319, 1980.

15a. Méhes K et al: Cohen syndrome: A connective tissue disorder? *Am J Med Genet* **31**:131–133, 1988.

16. Morris CA et al: Defining the Cohen syndrome. *Proc Greenwood Genet Ctr* **6**:123, 1987.

17. Norio R et al: Further delineation of the Cohen syndrome; report on chorioretinal dystrophy, leukopenia and consanguinity. *Clin Genet* **25**:1–14, 1984.

18. Norio R, Raitta C: Are the Mirohosseini–Holmes–Walton syndrome and the Cohen syndrome identical? *Am J Med Genet* **25**:397–398, 1986.

19. North C. et al: The clinical features of the Cohen syndrome: Further case reports. *J Med Genet* **22**:131–134, 1985.

20. Sack J, Friedman E: Cardiac involvement in the Cohen syndrome: A case report. *Clin Genet* **17**:317–319, 1980.

21. Sack J, Friedman E: The Cohen syndrome in Israel. *Israel J Med Sci* **22**:766–770, 1986.

21a. Warburg M et al: The Cohen syndrome: Retinal lesions and granulocytopenia. *Ophthalmol Paediatr Genet* (in press).

22. Wilson S et al: Cohen syndrome: Case report. *Pediatr Dent* **7**:326–328, 1985.

22a. Young ID, Moore JR: Intrafamilial variation in Cohen syndrome. *J Med Genet* **24**:488–492, 1987.

23. Zeitler S et al: Cohen-Syndrom bei zwei Brüdern. *Klin Pädiatr* **199**:55–57, 1987.

24. Zeller WP et al: Cohen syndrome. A consideration in evaluation of short stature. *Ill Med J* **171**:31–33, 1987.

## Börjeson–Forssman–Lehmann syndrome

Börjeson et al (3), in 1962, reported an X-linked syndrome of characteristic facies (Fig. 11–31A–D), mental retardation, obesity, hypotonia, and hypogonadism (Fig. 11–31E). Additional cases clearly identify the disorder as having X-linked inheritance with variable expression in female heterozygotes (1,5,7,8,10). An additional example is insufficiently documented (2) and another possibly erroneously identified (9). Approximately a dozen cases have been reported (1,3–8,10). Over 80% of affected males are short and at least 60% obese. These findings may be secondary to hypogonadism.

All affected males have a coarse facial appearance characterized by microcephaly, prominent supraorbital ridges, deep-set eyes, nystagmus, ptosis, and large ears (Fig. 11–31A–D) (1,3,8,10).

Radiographic studies have shown thickened calvaria, narrow cervical spinal canal, Scheuermann-like vertebral changes, mild epiphyseal dysplasia, delayed closure of radial and ulnar epiphyses, and short distal phalanges (1).

Hemizygotes have moderate to severe mental retardation. About 50% exhibit seizures and all have hypotonia. Electroencephalographic study has shown a paucity of alpha rhythms (1). CT scans show mildly dilated lateral ventricles. Female heterozygotes may have somewhat dull intelligence.

Hemizygotes have small atrophic or nonpalpable testes that descend late, small penis, and hypoplastic prostate. Secondary sexual development is poor and puberty is delayed until late in the second decade (1,8). Gynecomastia is frequent (Fig. 11–31E). Robinson et al (8) noted twin female heterozygotes with ovarian dysfunction.

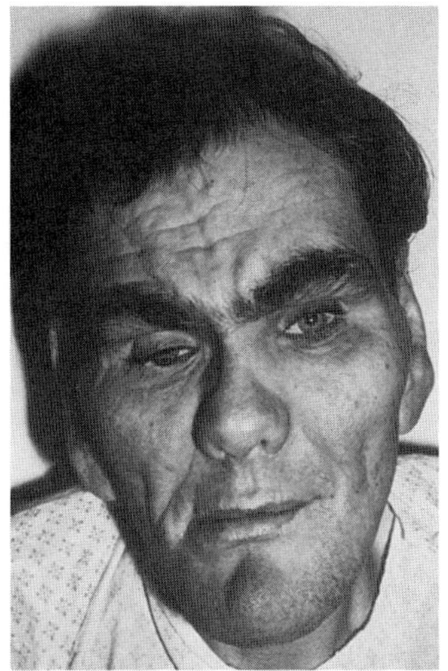
A

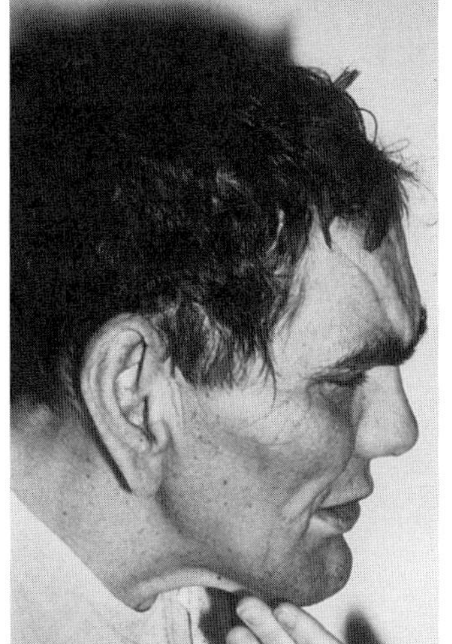
B

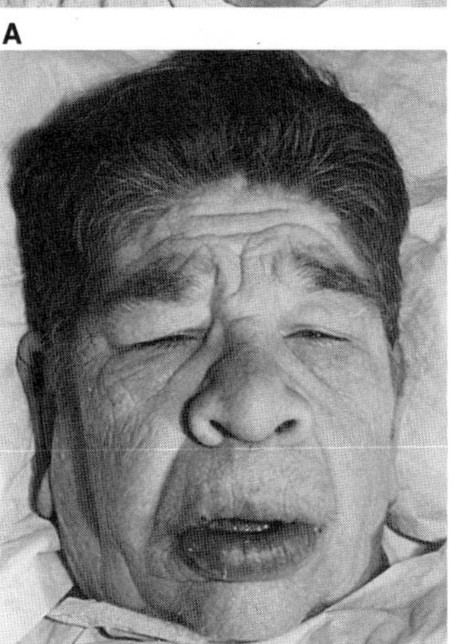
C

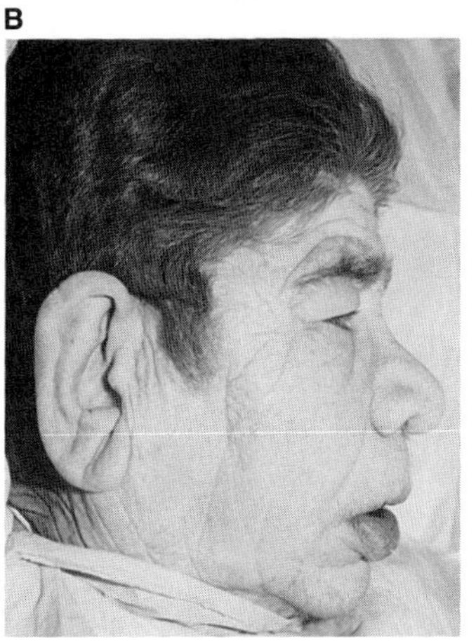
D

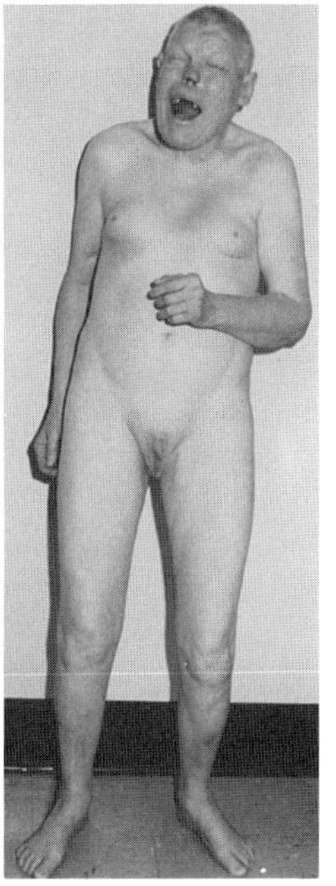
E

Fig. 11–31. *Börjeson–Forssman–Lehmann syndrome.* (A–D) Characteristic coarse facial appearance with microcephaly, prominent supraorbital ridges, deeply set eyes, nystagmus, ptosis, and large ears. (E) Characteristic facies, gynecomastia, lack of secondary sexual development. (A,B from *HH Ardinger,* Am J Med Genet **19:**653, 1984.)

Dermatoglyphic patterns of an affected male included bilateral $I_4$ whorls, double loop hypothenar patterns, Sydney lines, and fibular arch pattern (7).

The gene has been localized to Xq26-27 (8a).

### References (Börjeson–Forssman–Lehmann syndrome)


1. Ardinger HH et al: Börjeson–Forssman–Lehmann syndrome: Further delineation in five cases. *Am J Med Genet* **19**:653–664, 1984.

2. Baar HS, Galindo J: The Börjeson–Forssman–Lehmann syndrome. *J Ment Defic Res* **9**:125–130, 1965.

3. Börjeson M et al: An X-linked recessively inherited syndrome characterized by grave mental deficiency, epilepsy and endocrine disorder. *Acta Med Scand* **171**:12–21, 1962.

4. Börjeson et al: Combination of idiocy, epilepsy, hypogonadism, dwarfism, hypometabolism, and morphologic peculiarities inherited as an X-linked recessive syndrome. *Proc 2nd Int Cong Ment Retard, Vienna Part* **I**:188–192, 1963.

5. Brun A et al: An inherited syndrome with mental deficiency and endocrine disorder. A patho-anatomical study. *J Ment Defic Res* **18**:317–325, 1974.

6. Dereymaeker AM et al: The Börjeson–Forssman–Lehmann syndrome. A family study. *Clin Genet* **29**:317–320, 1986.

7. Flannery DB et al: Letter to the editor: Dermatoglyphics in Börjeson–Forssman–Lehmann syndrome. *Am J Med Genet* **21**:401–404, 1985.

8. Robinson LK et al: The Börjeson–Forssman–Lehmann syndrome. *Am J Med Genet* **15**:457–468, 1983.

8a. Turner G et al: Börjeson–Forssman–Lehmann syndrome: Clinical manifestations and gene localization to Xq26-27. *Am J Med Genet* **34**:463–469, 1989.

9. Veall RM et al: The Börjeson–Forssman–Lehmann syndrome: A new case. *J Ment Defic Res* **23**:231–242, 1979.

10. Weber FT et al: Primary hypogonadism in the Börjeson–Forssman–Lehmann syndrome. *J Med Genet* **15**:63–66, 1978.

# Chapter 12
# Hamartoneoplastic Syndromes

## Acanthosis nigricans

Although Pollitzer (49) and Janovsky (37) independently described acanthosis nigricans (AN) in 1890, it was Pollitzer (50), in 1909, who made the first extensive study of cases previously reported and who emphasized the relationship of the skin disease to abdominal malignancy. During the next 75 years, over 1000 cases have been reported, most of which have been associated with adenocarcinoma, principally of the stomach. Several excellent surveys (9,45) have been carried out, probably the greatest contributions being the numerous and extensive studies of Curth (15–23).

Throughout the years, AN had been classified in various ways with all classifications being based on the clinical presentation and association of AN with other conditions. The traditional classification divides AN into benign and malignant forms. It should be emphasized that this division is improper because the skin changes in both the so-called benign and malignant forms are histologically identical, with no cellular evidence of malignancy. The division was established to stress the association of AN with internal malignancy, especially gastrointestinal adenocarcinoma. Presently there is a tendency to divide AN into only two groups, the first associated with an internal malignant neoplasm and the second group composed of several entities that have their association with insulin resistance in common (26,33).

We believe that a division of AN according to clinical presentation can aid in the proper identification of the underlying associated condition(s). In reviewing the literature, the following associations of AN have been identified:

A. Neoplastic association
B. Nonneoplastic associations
   1. Insulin-resistant types
      a. Type A syndrome (insulin receptor; includes leprechaunism) (HAIR-AN syndrome)
      b. Type B syndrome (autoantibodies)
      c. Type C syndrome (postreceptor level)
      d. Obesity
      e. Other endocrinopathies (includes Prader–Willi syndrome)
   2. Congenital syndromes (Crouzon, Beare–Stevenson cutis gyratum syndrome)
   3. Autosomal dominant type
   4. Drug-induced type
   5. Miscellaneous types (idiopathic or isolated reports of association with other diseases or syndromes.)

The nonneoplastic association types occur with greater frequency than the neoplastic association type. The discussion that follows will present first, the clinical characteristics of AN and second, a summary of each of the associated conditions.

**Skin.** The typical findings are dark-brown, smooth, hyperkeratotic papules, ranging from slight discoloration to cases in which the entire skin can be affected. In approximate order of frequency of pigmentation and papillomatous changes are the axillae, neck, genitalia, groin and inner thighs, umbilicus, perianal area, other flexural surfaces, and areolas (5,44). In addition to these changes, there is an exaggeration of normal skin markings. The axillae or neck usually become pigmented before other areas are involved (Fig. 12–1A,B).

The clinical manifestations also include florid cutaneous papillomatosis and the sign of Leser–Trélat (multiple, rapidly growing seborrheic keratoses). These manifestations may occur individually or in association with each other (4,59).

The major histologic changes observed are marked deposition of surface keratin with abnormal stratum corneum, acanthosis of the spinous cell layer intermixed with atrophy (44,52), and marked increase in the number of melanocytes with increased deposition of melanin in the basal cell layer (26).

The actual mode of pathogenesis of AN is unknown. It has been suggested that a peptide or group of peptides is responsible for the production of the papillomas, pigmentation, and keratinization. It is assumed that these peptides may be produced by the adenocarcinomas and might be present in the endocrinopathies (6,38).

**Oral manifestations.** Possibly the earliest descriptions of oral lesions in AN was made by Pollitzer (50) in 1909. Masson and Montgomery (40) and Fladung and Heite (25) suggested that at least 50% of patients with the neoplastic association form have oral lesions. On the other hand, the present authors, on the basis of a survey of over 200 cases of the neoplastic association form, think that a truer value is probably about 30–40%. A similar figure was reported by Brown and Winkelmann (9). Unfortunately, the oral mucosa is seldom thoroughly inspected in the course of a general examination.

Of all oral tissues, the tongue and lips are involved most frequently and to the greatest degree. The dorsum of the tongue, or at times the lateral border, exhibits hypertrophy and elongation of papillae. These give the tongue, marked by deep fissures or furrows, a shaggy or prickly appearance. In addition, one may see papillomatous growths studding its surface (2,15,31,35,40–42,50,58,60). In contrast to the skin lesions, these growths are rarely pigmented (3,9).

The lips, especially the upper, may be markedly enlarged and covered by filiform or papillomatous growths. These are especially marked at the angles of the mouth (2,19,34,40,41,50) (Fig. 12–2A–E).

The buccal mucosa is usually less severely involved (2,4,44,66). There is generally a diffuse unevenness of its surface and a velvety white appearance. Occasionally, single fungiform growths are observed (13,19,40,58,60). The palate may be similarly affected (41,50). The gingiva, especially the interdental papillae, may become so much enlarged as almost to cover the teeth, resembling idiopathic fibromatosis (2,4,44,46,66). The connective tissue in these areas is well vascularized and with elongated papillae. Oral lesions are characterized by the absence of melanin deposition.

There is insufficient evidence to suggest the frequency of oral involvement in the nonneoplastic associations of the disease, but it would not appear to be great (9,31,39,65), Fladung and Heite (25) estimating 15%.

**Malignant neoplasia.** Curth and co-workers (19,22) and Rigel and Jacobs (54) presented impressive evidence that about 75% of the associated tumors are abdominal adenocarcinomas, of which almost 60% arise in the stomach. Other adenocarcinomas such as those of the uterus (5), pancreas, intestines (66), and, in a smaller percentage, bladder (2), lung (54), and breast (35,46) can be associated with AN. These carcinomas have a high degree of malignancy and prognosis is very poor.

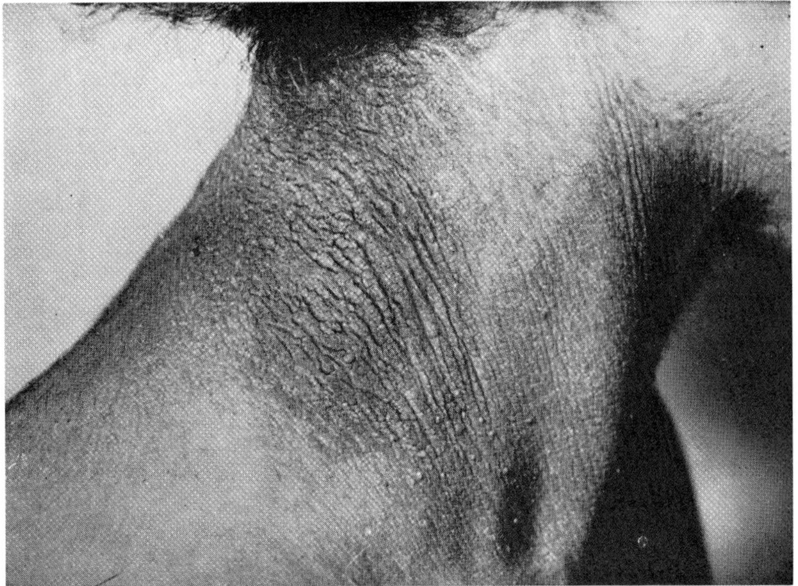


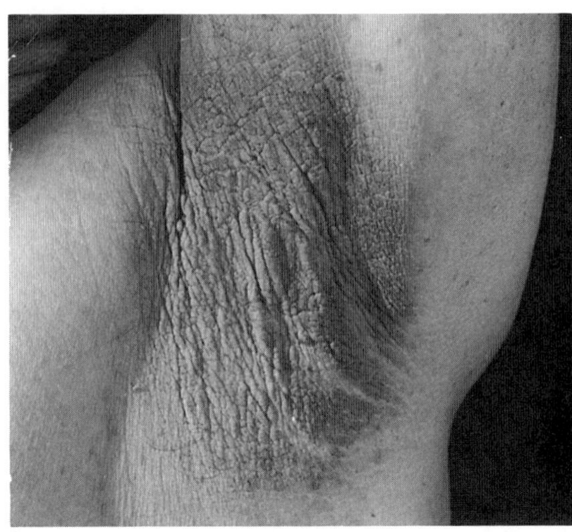


Fig. 12–1. *Acanthosis nigricans.* (A) Pigmentations and papillomatosis of cervical region associated with gastric adenocarcinoma. (B) Pigmentation and papillomatosis of axillary area associated with gastric adenocarcinoma. (A from *HO Curth,* New York, New York.)

The mortality rate is 100%, the average survival period after discovery being less than 2 years. Brown and Winkelmann (9) and Ackerman and Lantis (1) argue that although adenocarcinomas predominate, other tumors such as lymphomas may be part of the neoplasia-associated form (47). Garrott (29) suggested that when AN is associated with osteosarcoma, the distribution of skin lesions is different (extensor surfaces) and it affects young individuals.

Involvement of the skin may precede, accompany, or follow the detection of the cancer. In about 20%, the AN precedes the appearance of the malignancy by up to 16 years. It parallels the cancer in proportion to the degree of spread; it may regress with radiation therapy or surgical removal of the tumor and may reflourish with recurrence of the adenocarcinoma (19,22,47a). Generalized skin hyperpigmentation and pruritus occur in about 40% of neoplastic-associated type cases (2,9,30,44).

Palmar and plantar hyperkeratosis accompanies the neoplastic associated type in about 25% (8,9). More than 80% of affected persons are over 40 years at onset (40).

The vaginal and conjunctival mucous membranes may also be the

The vaginal and conjunctival mucous membranes may also be the
site of verrucous lesions (57). Other mucosae such as vaginal, esophageal, and pharyngeal can also be involved with papillary lesions (2,66).

**Insulin-resistant types.** This group is composed of conditions associated with AN and characterized by tissue resistance to the action of insulin. Patients with insulin resistance have been classified into *Type A* (HAIR-AN syndrome). This condition, observed mostly in adolescent or young females, is characterized by hyperandrogenism, insulin resistance, and AN (6). Virilization, polycystic ovaries, and accelerated growth are part of the clinical findings. Five percent of female patients with hyperandrogenism assessed for hyperinsulinemia when fasting and after oral glucose administration have been found to present this syndrome (27). This variety is due to genetic defects in insulin receptor pathways (26). Leprechaunism also falls into this type. *Type B* is observed in older females with autoimmune disorders in which the insulin resistance is due to circulating autoantibodies directed against the insulin receptor. Recently, a third variety has been described, *Type C.* The clinical findings are similar to those found in Type A but in Type C the insulin resistance seems to be located at postreceptor levels. The latter has been occasionally observed in males (28). Furthermore, Ritchie et al (55) have reported the association of AN, insulin resistance, hypogonadotrophic hypogonadism, hyperprolactinemia, and multiple organ-specific antibodies in a brother and sister. And excellent review is that of Redon et al (53).

**Obesity.** This variety is almost always observed in young female patients where receptor and postreceptor mechanisms are responsible for the insulin-resistant state (27).

**Other endocrinopathies.** AN occasionally associated with a variety of other endocrinopathies may reflect the insulin resistance frequently observed in those disorders. The endocrinopathies have included *lipodystrophic diabetes* (11), adrenogenital syndrome (67), Cushing's syndrome, acromegaly, Addison's disease, Stein–Leventhal syndrome, *Prader–Willi syndrome* (53), and hyper- or hypothyroidism (9,10,21,26,53). Diabetic patients with insulin resistance may also present AN (61). This type also can be seen in youngsters (juvenile AN) (12).

Mendenhall (43) described two sisters and a brother derived from first cousin parents with unusual facies with marked prognathism, premature aging, precocious eruption of both primary and secondary dentitions with dysplastic teeth, thickened nails, hirsutism, AN, abdominal protuberance, insulin resistance, diabetes mellitus, and phallic enlargement in males. Pineal hyperplasia was found at necropsy (51). Similar findings were reported by West and co-workers (68,69) in a brother and sister. Brown and Winkelmann (9) described two patients with malignant pinealoma and AN.

**Congenital syndromes.** AN has been observed in patients with *Bloom syndrome* and *Crouzon syndrome,* among others (20,21,53).

The association of AN and Crouzon syndrome was first briefly pointed out by Curth (20) in a dermatology congress without actual case presentation. However, a few cases have been published in which this association is fully documented (8a,52,63). *Beare–Stevenson cutis gyratum syndrome* contains AN with cutis gyratum, craniosynostosis, cleft palate, hypodontia, and other defects (7,62)

**Autosomal dominant type.** Genetic studies have revealed that in some families the nonneoplastic type of AN exhibits irregular autosomal dominant inheritance (15,25,32,64). This variety manifests as

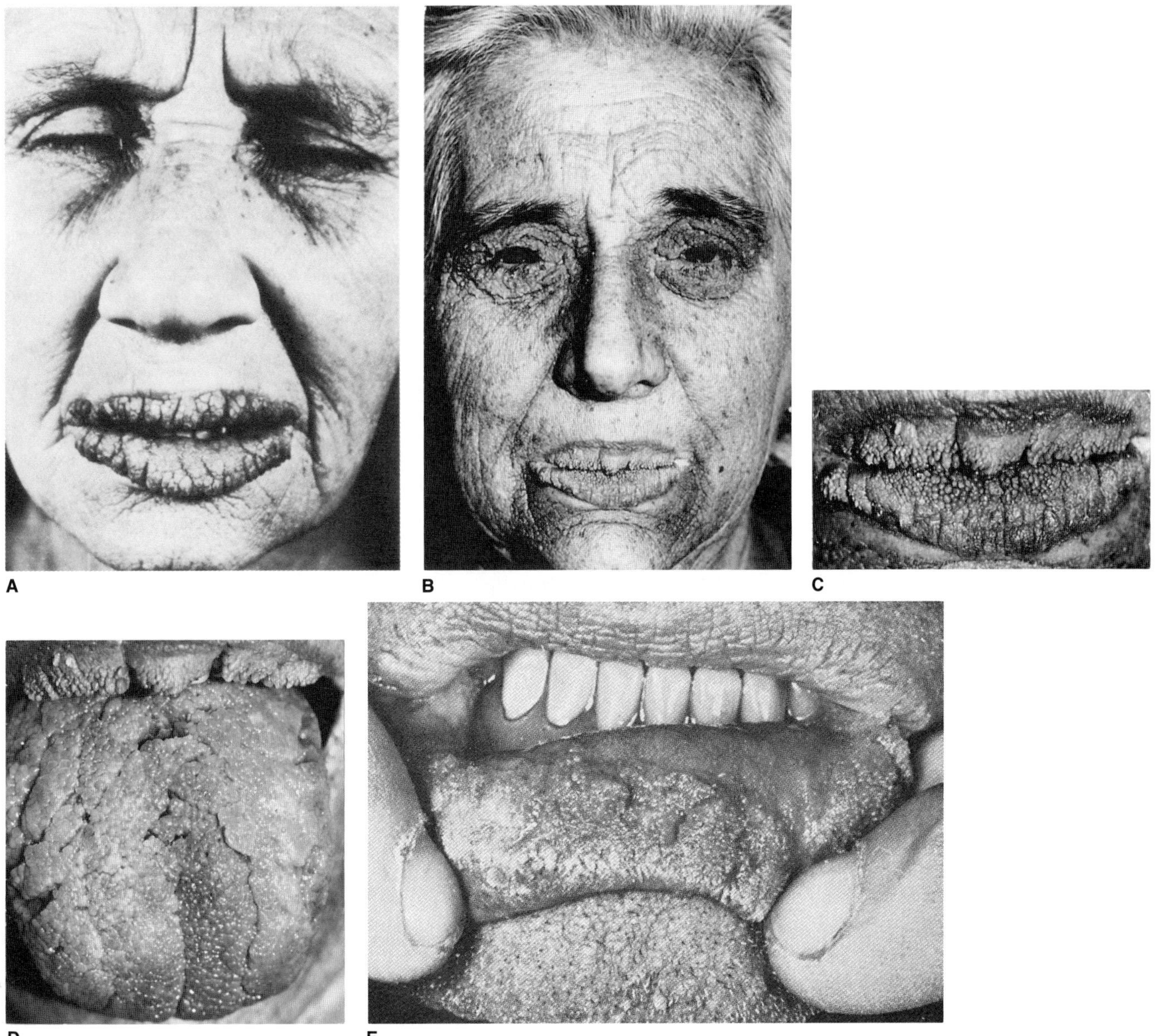

Fig. 12–2. *Acanthosis nigricans.* (A) Note papillomatosis of lips. (B) Extensive papillomatosis around eyes and lips. Patient had advanced gastric carcinoma. (C) Marked labial involvement. (D) Matlike thickening of dorsum of tongue. (E) Thickening of labial mucosa and papillomatosis. Pigmentation of oral mucosa is characteristically absent. (A courtesy of *N Danbolt,* Oslo, Norway. B–D courtesy of *K Wolff,* Vienna, Austria. E courtesy of *A Proppe,* Kiel, Germany.)

a genodermatosis resembling ichthyosis hystrix, which may be present at birth or may begin later, either in childhood or, more often, at puberty, at which time it becomes more active.

**Drug-induced type.** The ingestion of diethylstilbestrol or nicotinic acid, or even topical application of nicotinic acid, may occasionally induce the development of AN (24,48).

**Miscellaneous types.** A variety of conditions have been sporadically reported associated with AN such as lupoid hepatitis, pemphigus vulgaris (14), and *hyperphosphatasemia (juvenile Paget's disease of bone)* (56).

**Differential diagnosis.** Regarding endogenous pigmentation, one should consider Addison's disease, arsenic poisoning, and hemochromatosis, but in none of these conditions is there an associated papillomatosis. Ichthyosis hystrix, *Cowden syndrome,* bromoderma, pemphigus vegetans, *hyalinosis cutis et mucosae,* condyloma acuminatum, and hairy tongue must all be excluded. In Afro-Americans, dermatosis papulosa nigra should be ruled out. Hirschowitz et al (36) described AN in a syndrome of nerve deafness, absent gastric motility, small intestine diverticulitis, and progressive sensory neuropathy. Autosomal recessive inheritance appeared likely. A generalized cutaneous papillomatosis similar to AN, including oral mucosa and palmoplantar surfaces, with no associated internal malignancy and starting at age 63, has been described in an Italian male (71).

**Laboratory aids.** Abundant deposits of glycosaminoglycan (GAG) consisting mainly of hyaluronic acid have been found in the papillary dermis of lesions of AN in patients with polycystic ovaries and insulin resistance. Normal amounts of GAGs were found in nonaffected skin of the same patients (70).

### References (Acanthosis nigricans)


1. Ackerman AB, Lantis LR: Acanthosis nigricans with Hodgkin's disease. *Arch Dermatol* **95**:202–205, 1967.

2. Aloi FG et al: Acanthosis nigricans della mucosa orale associata a carcinoma vesicale. *G Ital Derm Venereol* **118**:95–97, 1983.

3. Andreev VC et al: Acanthosis nigricans of oral mucosa. *Dermatologica* **126**:25–29, 1963.

4. Andreev VC et al: Generalized acanthosis nigricans. *Dermatologica* **163**:19–24, 1981.

5. Azizi E et al: Generalized malignant acanthosis nigricans and primary fibrinolysis. *Arch Dermatol* **118**:955–956, 1982.

6. Barbieri RL, Ryan KJ: Hyperandrogenism, insulin resistance, and acanthosis nigricans syndrome: A common endocrinopathy with distinct pathophysiologic features. *Am J Obstet Gynecol* **147**:90–101, 1983.

7. Beare JM et al: Cutis gyratum, acanthosis nigricans and other congenital anomalies. A new syndrome. *Br J Dermatol* **81**:241–247, 1969.

8. Breathnach SM, Wells GC: Acanthosis palmaris: Tripe palms. A distinctive pattern of palmar keratoderma frequently associated with internal malignancies. *Clin Exp Dermatol* **5**:181–189, 1980.

8a. Breitbart AS et al: Crouzon's syndrome associated with acanthosis nigricans: Ramifications for the craniofacial surgeon. *Ann Plast Surg* **22**:310–315, 1989.

9. Brown J, Winkelmann RK: Acanthosis nigricans: A study of 90 cases. *Medicine* **47**:33–51, 1968.

10. Brown J et al: Acanthosis nigricans and pituitary tumors. *JAMA* **198**:619–623, 1966.

11. Brubaker MM et al: Acanthosis nigricans and congenital lipodystrophy. *Arch Dermatol* **91**:320–325, 1965.

12. Chaussain JL et al: Insulin-specific binding to erythrocytes in 6 girls with acanthosis nigricans. *Horm Res* **20**:186–191, 1984.

13. Cochrane T, Alexander J O'D: Acanthosis nigricans. *Br J Dermatol* **63**:225–230, 1951.

14. Coverton RW, Armstrong RB: Acanthosis nigricans developing in resolving lesions of pemphigus vulgaris. *Arch Dermatol* **118**:115–116, 1982.

15. Curth HO: Benign type of acanthosis nigricans. *Arch Dermatol Syph* **34**:353–366, 1936.

16. Curth HO: Cancer associated with acanthosis nigricans: Review of literature and report of a case of acanthosis nigricans with cancer of the breast. *Arch Surg* **47**:517–552, 1943.

17. Curth HO: Acanthosis nigricans and its association with cancer. *Arch Dermatol Syph* **57**:158–170, 1948.

18. Curth HO: Significance of acanthosis nigricans. *Arch Dermatol Syph* **66**:80–100, 1952.

19. Curth HO: Pigmentary changes of the skin associated with internal disease. *Postgrad Med* **41**:439–444, 1967.

20. Curth HO: The necessity of distinguishing four types of acanthosis nigricans. XIII Congr Int Dermatol, 1968, Springer, Munich, pp 558–577.

21. Curth HO: Classification of acanthosis nigricans. *Int J Dermatol* **15**:592–593, 1976.

22. Curth HO, Aschner BM: Genetic studies on acanthosis nigricans. *Arch Dermatol Syph* **79**:55–66, 1959.

23. Curth HO et al: The site and histology of the cancer associated with malignant acanthosis nigricans. *Cancer* **15**:364–382, 1962.

24. Elgart ML: Acanthosis nigricans and nicotinic acid. *J Am Acad Dermatol* **5**:709–710, 1981.

25. Fladung G, Heite HJ: Häufigkeitsanalytische Untersuchungen zur Frage der symptomologischen Abgrenzung verschiedener Formen der Acanthosis nigricans. *Arch Klin Exp Dermatol* **205**:282–311, 1957.

26. Flier JS: Metabolic importance of acanthosis nigricans. *Arch Dermatol* **121**:193–194, 1985.

27. Flier JS et al: Acanthosis nigricans in obese women with hyperandrogenism. Characterization of an insulin-resistant state distinct from the type A and B syndromes. *Diabetes* **34**:101–107, 1985.

28. Fukushima N et al: A case of insulin resistance associated with acanthosis nigricans. *Tohuku J Exp Med* **144**:129–138, 1984.

29. Garrott TC: Malignant acanthosis nigricans associated with osteogenic sarcoma. *Arch Dermatol* **106**:384–385, 1972.

30. Greengood R, Tring FC: Treatment of acanthosis nigricans with cyproheptadine. *Br J Dermatol* **106**:697–698, 1982.

31. Hellerström S: Zur Kenntnis der Acanthosis nigricans. *Acta Derm Venereol* **14**:86–98, 1933.

32. Hermann H: Zur Erbpathologie der Acanthosis nigricans. *Z Menschl Vererb Konstit Lehre* **33**:193–202, 1955.

33. Hernandez-Perez E: On the classification of acanthosis nigricans. *Int J Dermatol* **23**:605–606, 1984.

34. Herold WC et al: Acanthosis nigricans: Its occurrence in association wth gastric carcinoma in a seventeen-year-old girl. *Arch Dermatol Syph* **44**:789–799, 1941.

35. Heymans G et al: Acanthosis nigricans et patologies associeés. *Acta Clin Belg* **37**:1411–147, 1982.

36. Hirschowitz BI et al: Hereditary nerve deafness in three sisters with absent gastric mobility, small bowel diverticulitis and ulceration and progressive sensory neuropathy. *Birth Defects* **8**(2):27–41, 1972.

37. Janovsky V, in PG Unna et al (eds.), *Internationaler Atlas seltener Hautkrankheiten*. Leopold Voss, Leipzig, 1890, plate II.

38. Kahn CR et al: The syndromes of insulin resistance and acanthosis nigricans: Insulin receptor disorders in man. *N Engl J Med* **294**:739–745, 1976.

39. Lang E, Schmidt R: Beziehungen zwischen Acanthosis nigricans benigna und Pseudoacanthosis nigricans. *Hautarzt* **33**:640–645, 1982.

40. Masson JC, Montgomery H: Relationship of acanthosis nigricans to abdominal malignancy. *Am J Obstet Gynecol* **32**:717–726, 1936.

41. Matras A: Acanthosis nigricans mit Schleimhautveränderungen. *Zbl Haut Geschl-Kr* **63**:410, 1940.

42. McKenna RMB, Roxburgh I: Acanthosis nigricans. *Br J Dermatol* **61**:251–252, 1949.

43. Mendenhall EN: Tumor of the pineal body with high insulin resistance. *J Indiana Med Assoc* **43**:32–36, 1950.

44. Misch KJ et al: Warts as a presenting sign in acanthosis nigricans. *Clin Exp Dermatol* **8**:651–656, 1983.

45. Moncorps C: Acanthosis nigricans, in *Handbuch der Haut-und Geschlechtskrankheiten,* Jadassohn J (ed), Springer-Verlag, Berlin, 1931.

46. Mostofi RS et al: Oral malignant acanthosis nigricans. *Oral Surg* **56**:372–374, 1983.

47. Neill SM et al: Mycosis fungoides associated with acanthosis nigricans. *J R Soc Med* **78**:79–81, 1985.

47a. Kazuchio K et al: Improvement of oral lesions associated with malignant acanthosis nigricans after treatment of lung cancer. *Oral Surg Oral Med Oral Pathol* **68**:74–79, 1989.

48. Pascal J et al: Acanthosis nigricans induit par acide nicotinique local. *Ann Dermatol Venereol* **111**:739–740, 1984.

49. Pollitzer S, in *Internationaler Atlas seltener Hautkrankheiten,* Unna PG et al (eds), Leopold Voss, Leipzig, 1890, plate 10.

50. Pollitzer S: Acanthosis nigricans: A symptom of a disorder of the abdominal sympathetic. *JAMA* **53**:1369–1373, 1909.

51. Rabson SM, Mendenhall EN: Familial hypertrophy of pineal body, hyperplasia of adrenal cortex and diabetes mellitus. *Am J Clin Pathol* **26**:283–290, 1956.

52. Reddy BSN et al: An unusual association of acanthosis nigricans and Crouzon's disease. A case report. *J Dermatol* **12**:85–90, 1985.

53. Redon MI et al: Acanthosis nigricans: A cutaneous marker of tissue resistance to insulin. *J Am Acad Dermatol* **21**:461–469, 1989.

54. Rigel DS, Jacobs MI: Malignant acanthosis nigricans: A review. *J Dermatol Surg Oncol* **6**:923–927, 1980.

55. Ritchie CM et al: Unusual presentations of acanthosis nigricans and insulin resistance in a brother and sister. *Br Med J* **290**:1320, 1985.

56. Saxe N, Beighton P: Cutaneous manifestations of osteoectasia. *Clin Exp Dermatol* **7**:605–609, 1982.

57. Scheer M: Acanthosis nigricans. *Arch Dermatol Syph* **28**:118–119, 1933.

58. Schwartz JH, Miller EC: Acanthosis nigricans. *Arch Dermatol Syph* **18**:534–538, 1928.

59. Schwartz RA: Acanthosis nigricans, florid cutaneous papillomatosis and the sign of Leser-Trélat. *Cutis* **28**:319–322, 1981.

60. Scott OLS: Acanthosis nigricans. *Br J Dermatol* **64**:461–462, 1952.

61. Serrano Rios M et al: Pancreatic A and B cell hyperfunction in the Mendenhall syndrome. *Diabetologia* **25**:8–12, 1983.

62. Stevenson RE et al: Cutis gyratum and acanthosis nigricans with other anomalies: A distinctive syndrome. *J Pediatr* **92**:950–952, 1978.

63. Suslak L et al: Crouzon syndrome with periapical cemental dysplasia and acanthosis nigricans: The pleiotropic effect of a single gene? *Birth Defects* **21**(B):127–134, 1985.

64. Tasjian D, Jarrat M: Familial acanthosis nigricans. *Arch Dermatol* **120**:1351–1354, 1984.

65. Tolmach JA: Acanthosis nigricans: Juvenile type. *Arch Dermatol Syph* **40**:819–820, 1939.

66. Walton S et al: A case of acanthosis nigricans—an investigative approach. *Clin Exp Dermatol* **9**:58–63, 1984.

67. Weissman-Katzenelson V et al: Acanthosis nigricans in association with

the adrenogenital syndrome, hyperlipoproteinemia type IV, and congenital malformation of the urinary tract. *Arch Dermatol* **119**:953–954, 1983.

68. West RJ et al: Familial insulin-resistant diabetes, multiple somatic anomalies and pineal hyperplasia. *Arch Dis Childh* **50**:703–708, 1975.

69. West RJ, Leonard JV: Familial insulin resistance with pineal hyperplasia: Metabolic studies and effect of hypophysectomy. *Arch Dis Childh* **55**:619–621, 1980.

70. Wortsman J et al: Glycosaminoglycan deposition in the acanthosis nigricans lesions of the polycystic ovary syndrome. *Arch Intern Med* **143**:1145–1148, 1983.

71. Zina G et al: Black Sardinian wild piglet. *Int J Dermatol* **22**:35–36, 1983.

## Cowden syndrome (multiple hamartoma syndrome)

This syndrome was first described in detail by Lloyd and Dennis (36) in 1963 and its name is derived from an affected individual. However, Costello (15), in 1941, described the skin lesions in a patient who Brownstein (8) later documented with the disorder. Another early example is that of Witten and Kopf (59). Many cases of the syndrome have been published, emphasizing its hamartomatous character and delineating its principal involvement of the skin, gastrointestinal tract, breasts, and thyroid (Figs. 12–3 to 12–5). Inheritance is autosomal dominant (20,35,43,51). There are several excellent reviews (36,43a,47,48,51). The frequencies of various abnormalities found in Cowden syndrome are listed in Table 12–1.

**Skin.** Mucocutaneous abnormalities are present in 99% of the cases (51). The age of onset of the dermatologic lesions is not well documented; however, they are usually noted by the patient at the end of the second or third decade (48). Typical lesions are lichenoid or papillomatous papules and small nodules, primarily on and around the eyelids, alae nasi, nasolabial folds, mouth, pinnas, lateral neck, glabella, and the dorsa of hands and forearms (4a,20,26,36,38,58) (Fig. 12–3A,B). They are numerous, some individuals having 50 or more. Punctate keratoses are present on the palms and soles (5,20,26,41,48,54) (Fig. 12–3C).

The microscopic features of the skin lesions have been reported in detail by several authors (8,49,50,52). They noted features of hair follicle hamartomas, with most lesions showing the pattern of tricholemmomas, that is, composed of large, pale glycogen-rich epithelial cells centrally surrounded by a single layer of smaller palisaded cells. A few displayed features intermediate between those of tricholemmomas and inverted follicular keratoses, whereas others had characteristics intermediate between tricholemmoma and tumor of the follicular infundibulum. Some specimens showed nonspecific verrucous acanthomas or were not diagnostic. Dermal fibromas were also found, characterized by interwoven collagen bundles with a laminated or tortuous appearance, embedded in abundant mucin (51). Dames and Mahrle (16) found, in one patient, a tricholemmoma, inverted follicular keratoses, trichofolliculomas, trichoepitheliomas, and intermediate forms with characteristics of two different hair follicle tumors. Immunoperoxidase studies of facial and extrafacial skin lesions using antiserum to bovine papillomavirus Type I were negative (25,35,50). Electron microscopic studies (4,29) have also been negative for viral particles.

Squamous cell carcinoma of the skin of the nose (5) and basal cell carcinomas of the skin of the face (4) and perianal skin (11) have been reported.

Table 12–1. Abnormalities associated with Cowden syndrome[a] ($n=100$; M = 37; F = 63)

| Abnormalities | Percentages |
|---|---|
| Mucocutaneous | |
| Multiple facial papules | 85 |
| Acral keratoses | 73 |
| Palmoplantar keratoses | 54 |
| Multiple oral papillomas | 85 |
| Dermal fibromas | 24 |
| Multiple skin tags | 16 |
| Oral fibromas | 86 |
| Scrotal tongue | 20 |
| Lipomas | 31 |
| Vascular malformations | 18 |
| Cutaneous and oral malignancies | 8 |
| Thyroid gland | |
| Goiter, adenoma | 68 |
| Hyperthyroidism | 2 |
| Hypothyroidism | 3 |
| Thyroiditis | 3 |
| Thyroglossal duct cyst | 2 |
| Follicular adenocarcinoma | 3 |
| Female breast | |
| Fibrocystic disease | 52 |
| Anatomic abnormalities | 8 |
| Virginal hypertrophy | 6 |
| Ductal adenocarcinoma | 28 |
| Ductal papilloma | 14 |
| Male breast | |
| Benign gynecomastia | 7 |
| Female genitourinary system | |
| Menstrual irregularities | 20 |
| Ovarian abnormalities (mainly cysts) | 19 |
| Leiomyomas | 5 |
| Vaginal and vulvar cysts | 6 |
| Adenocarcinoma of uterus | 6 |
| Carcinoma of uterine cervix | 3 |
| Carcinoma of ovary | 2 |
| Transitional cell carcinoma of renal pelvis | 2 |
| Male genitourinary system | |
| Hydrocele, varicocele | 3 |
| Transitional cell carcinoma of bladder | 3 |
| Gastrointestinal tract | |
| Polyps of upper GI tract | 22 |
| Polyps of colon and rectosigmoid | 29 |
| Diverticula of colon and sigmoid | 2 |
| Ganglioneuromas and neuromas | 4 |
| Epithelioid leiomyoma of rectosigmoid | 1 |
| Hepatic hamartoma | 1 |
| Adenocarcinoma of cecum | 2 |
| Adenocarcinoma of colon | 1 |
| Facial dysmorphism and skeletal abnormalities | |
| High head circumference | 21 |
| Adenoid facies | 8 |
| Highly arched palate | 14 |
| Kyphosis, kyphoscoliosis | 14 |
| Hand and foot abnormalities | 6 |
| Pectus excavatum | 6 |
| Bone cysts | 4 |
| Nervous system | |
| Neuromas of cutaneous nerves | 5 |
| Neurofibroma | 3 |
| Meningioma | 3 |
| Hearing loss | 2 |
| Eye | |
| Cataracts | 3 |
| Angioid streaks | 2 |
| Congenital blood vessel anomaly | 1 |
| Myopia | 3 |

[a] Adapted from TM Starink et al, *Clin Genet* **29**:222–233, 1986.

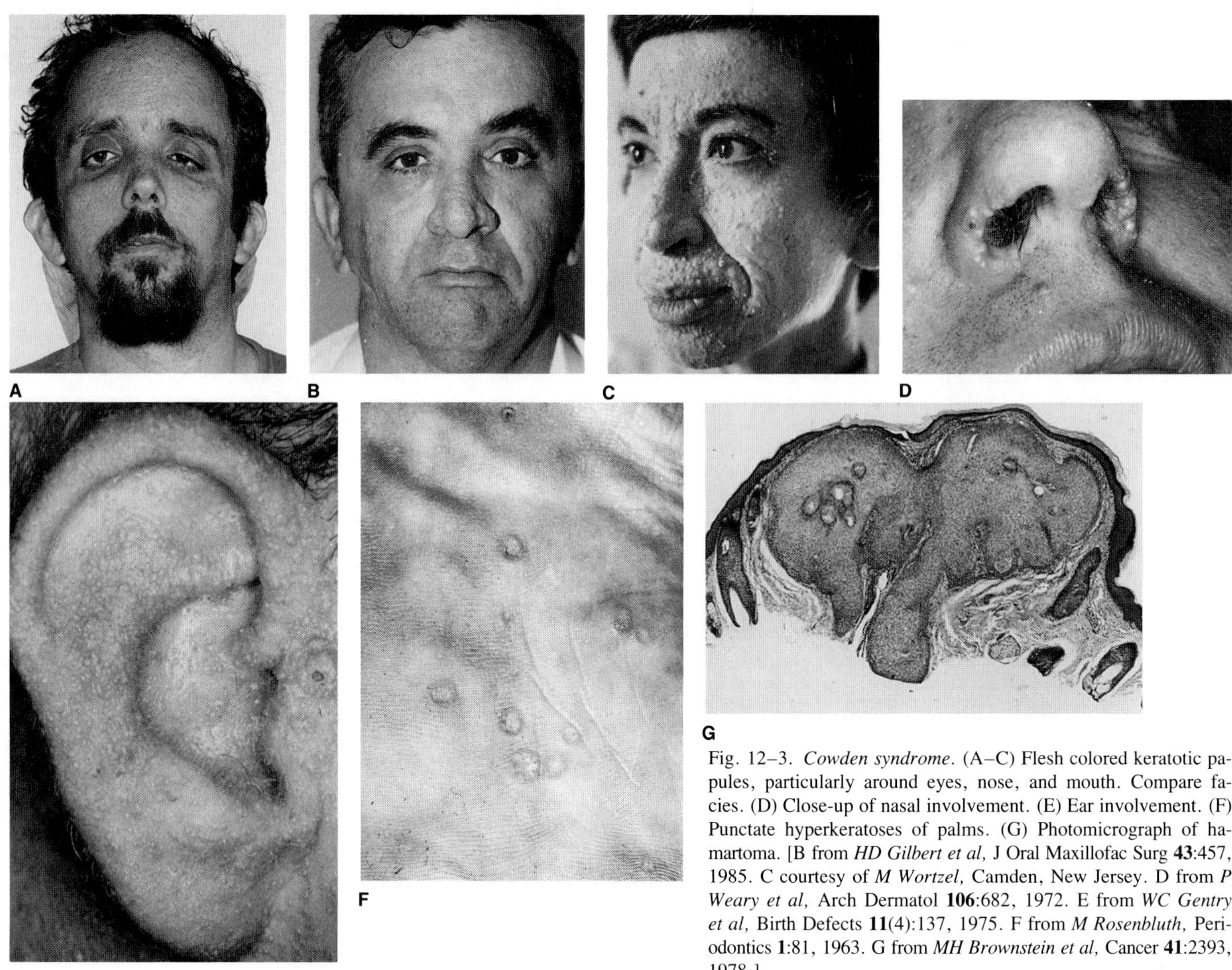

Fig. 12–3. *Cowden syndrome.* (A–C) Flesh colored keratotic papules, particularly around eyes, nose, and mouth. Compare facies. (D) Close-up of nasal involvement. (E) Ear involvement. (F) Punctate hyperkeratoses of palms. (G) Photomicrograph of hamartoma. [B from *HD Gilbert et al,* J Oral Maxillofac Surg **43**:457, 1985. C courtesy of *M Wortzel,* Camden, New Jersey. D from *P Weary et al,* Arch Dermatol **106**:682, 1972. E from *WC Gentry et al,* Birth Defects **11**(4):137, 1975. F from *M Rosenbluth,* Periodontics **1**:81, 1963. G from *MH Brownstein et al,* Cancer **41**:2393, 1978.]

**Breasts.** Seventy percent of females with the disorder have breast lesions (51): fibroadenomas (9), virginal hypertrophy (36,48), fibrocystic disease (4,36,44,48,54), ductal papillomas (48) and adenocarcinoma (7,9,20,34,36,38,48,53,56,57). The adenocarcinomas are found in 25% (51) and may be bilateral (7,9,25,51) and 65% have nodal metastases. Adenocarcinoma of the breast in males has not been reported but gynecomastia has been noted.

**Thyroid.** Thyroid gland abnormalities occur in about 68% (48), and include "goiter" (9,17,43,46,48,57,58), adenoma (4,11,23,26,28, 36,44,46,50), and follicular adenocarcinoma (9,54,57) (Fig. 12–4). Thyroid tumors have not developed in males (26).

**Gastrointestinal tract.** The frequency of GI lesions is not known since all reported patients have not received complete gastrointestinal examinations. However, Carlson et al (12) noted that 13 of 17 patients reported in the literature who had undergone lower gastrointestinal examinations had colonic polyposis. Polyps have been reported as early as 5 years of age (57) and have been found from one end of the gastrointestinal tract to the other: esophagus (24,26,30,31,33,40,48, 57,58), stomach (23,29,39,40,58), small intestine (23,29,39,40,58), and large intestine (23,25,35,51,58). On microscopic examination, polyps may be ganglioneuromatous (33), hyperplastic (25,30), benign lymphoid (30,40,48), hamartomatous (9,12,25,32,44,54), lipomatous (30,32), leiomyomatous (12), juvenile (30), and adenomatous (33,57). Esophageal polyps have been diagnosed on microscopic examination as exhibiting glycogenic acanthosis (33).

Gorensek et al (25) reported one patient with many hyperplastic polyps and adenocarcinoma of the descending colon. Burnett et al (9) reported adenocarcinoma of the cecum and Walton et al (56) described metastatic colon carcinoma; neither report noted preceding polyposis. Otherwise, no other cases of colonic carcinoma in patients with this syndrome have been documented. Thus, it has not been determined whether malignant degeneration of colonic polyps is characteristic of the syndrome.

**Other neoplasms.** Many other neoplasms have been noted: cerebral gangliocytoma (46), meningioma (19,26,57), dural arteriovenous malformation (48), hemangiomas (5,7,30,48,56,57), neurofibroma (17), granular cell tumor (48), xanthomas (6), neuroma of Auerbach's plexus (9), lipomas (9,12,20,30,34,38,39,41,45,46,48,54,57), angiolipomas (20,58), angiomyomas of the extremities (20,34), liposarcoma (7,48), acute myelogenous leukemia (42), non-Hodgkin's lymphoma (16a), melanoma (12,27,45,54), endometrial carcinoma (4), transitional cell carcinoma of the bladder (30,35), and transitional cell carcinoma of the renal pelvis (38).

**Skeletal abnormalities.** Lordosis (5), kyphosis (48), scoliosis (9), and kyphoscoliosis (36) have been documented. A large head circumference was found in 80% by Starink et al (51).

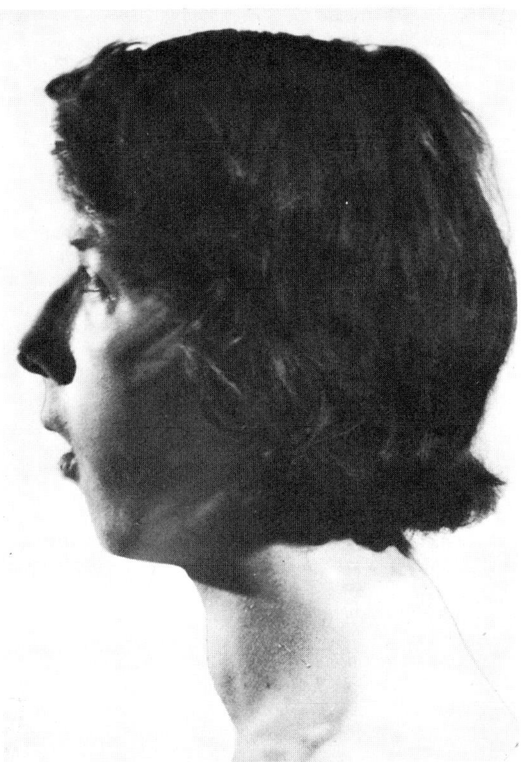

Fig. 12–4. *Cowden syndrome.* Note thyroid enlargement. (From *KM Lloyd* and *M Dennis,* Ann Intern Med **58**:136, 1963.)

**Other findings.** Hydronephrosis has been described (34,41); however, in one case (34) it was thought to be secondary to a huge retroperitoneal lipoma. Angioid streaks of the retina have been reported (2,3). Ovarian cysts have occurred (7,9,54,55,56). Laryngeal polyps which interfere with phonation have also been noted (25,57,58).

**Oral manifestations.** Papular and verrucous lesions of the lips, tongue, gingiva, edentulous alveolar ridges, buccal mucosa, palate, and tonsillar fossa have been seen in most patients (1,2,5,10,14,17, 18,20–22,26,28,35,41,53) (Fig. 12–5A–G). These lesions may coalesce and produce a cobblestone appearance (28,43,58). The light microscopic appearance of oral lesions is consistent with the diagnosis of fibroma according to some investigators (7–9,22,29,57). Others have described epithelial hyperplasia and papillomatosis (36,49,53). Electron microscopic studies of oral lesions have not shown virus particles (4,45). Squamous cell carcinoma of the tongue has been reported (11).

**Differential diagnosis.** Patients reported by Ackerman (1) and Byars and Jurkiewicz (10) exhibited giant fibroadenomas of the breast, secondary kyphosis, hypertrichosis, and gingival fibromatosis. This condition (Byars–Jurkiewicz syndrome) may represent a disorder distinct from Cowden syndrome (Fig. 12–6).

On clinical examination, cutaneous lesions in Cowden syndrome resemble those of acrokeratosis, epidermodysplasia verruciformis, Darier's disease, Torre–Muir syndrome, *lipoid proteinosis,* and *tuberous sclerosis.* However, the microscopic features and associated abnormalities serve to distinguish between these disorders. The intraoral lesions of focal epithelial hyperplasia and *multiple endocrine neoplasia syndrome, Type 2B* are similar to those of Cowden syndrome, but

Fig. 12–5. *Cowden syndrome.* Oral involvement. (A–C) Papillomatous lesions of lips. Compare. (D) Close-up of lower lip involvement. (E) Papillomatous tongue. (F, G) Gingival involvement. (A from *KM Lloyd* and *M Dennis,* Ann Intern Med **58**:136, 1963. C from *BS Allen et al,* J Am Acad Dermatol **2**:303, 1980. E from *P Fritsch et al,* Hautarzt **32**:285, 1981. G from *M Rosenbluth,* Periodontics **1**:81, 1963.)

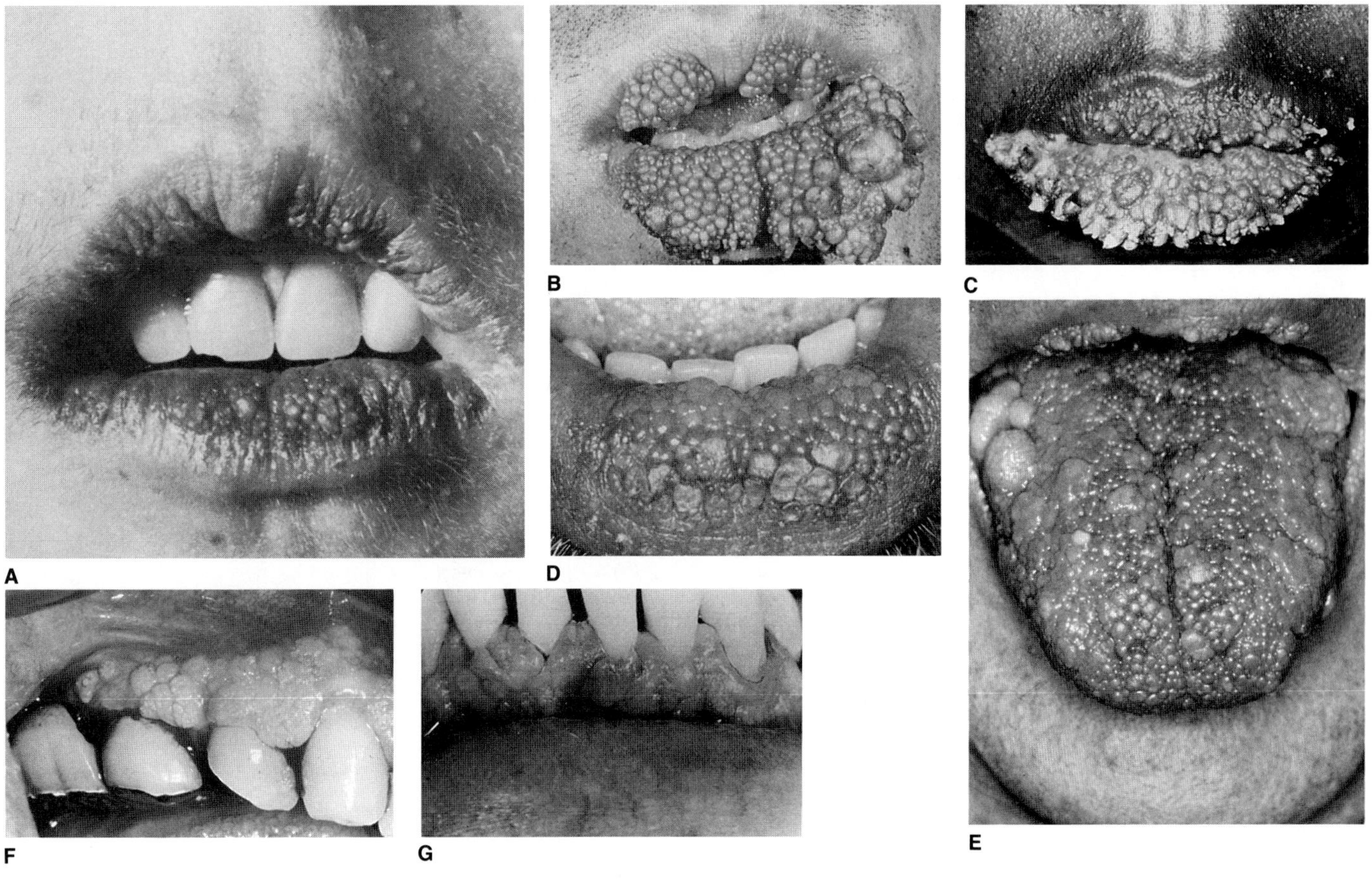

A B C D E F G

Fig. 12–6. *Byars–Jurkiewicz syndrome.* Giant fibroadenomas of the breast, secondary kyphosis, hypertrichosis, and gingival fibromatosis. (Courtesy of *L Ackerman,* St. Louis, Missouri.)

differ on microscopic examination. The lesions at the oral commissures simulate those of *acanthosis nigricans.*

Polyps of the colon are found in a number of disorders (see *Gardner syndrome*). Ganglioneuromatous proliferation of the bowel may also be associated with *multiple endocrine neoplasia type 2b syndrome, neurofibromatosis,* and with juvenile polyposis of the colon (40a).

**Laboratory aids.** Biopsy of cutaneous lesions should aid in differential diagnosis. T lymphocyte deficiency (29,42) has been reported but the significance is moot. The results of epidermal growth factor studies have been normal (13).

### References [Cowden syndrome (multiple hamartoma syndrome)]


1. Ackerman L: Personal communication, 1969.

2. Allen BS et al: Multiple hamartoma syndrome. A report of a new case with associated carcinoma of the uterine cervix and angioid streaks of the eyes. *J Am Acad Dermatol* **2**:303–308, 1980.

3. Aram H, Zidenbaum M: Multiple hamartoma syndrome (Cowden's disease). *J Am Acad Dermatol* **9**:774–776, 1983 (letter).

4. Aylesworth R, Vance JC: Multiple hamartoma syndrome with endometrial carcinoma and the sign of Leser–Trélat. *Arch Dermatol* **118**:136–138, 1982.

4a. Bardenstein DS et al: Cowden's disease. *Ophthalmology* **95**:1038–1041, 1988.

5. Bart RS, Kopf AW: Cowden's disease (multiple hamartoma syndrome). *J Dermatol Surg Oncol* **7**:378–380, 1981.

6. Brownstein MH et al: Trichilemmomas in Cowden's disease. *JAMA* **238**:26, 1977 (letter).

7. Brownstein MH et al: Cowden's disease. A cutaneous marker of breast cancer. *Cancer* **41**:2393–2398, 1978.

8. Brownstein MH et al: The dermatopathology of Cowden's syndrome. *Br J Dermatol* **100**:667–673, 1979.

9. Burnett JW et al: Cowden's disease. *Br J Dermatol* **93**:329–336, 1975.

10. Byars LT, Jurkiewicz M: Congenital macrogingivae and hypertrichosis with subsequent giant fibroadenomas of the breasts. *Plast Reconstr Surg* **27**:608–612, 1961.

11. Camisa C et al: Cowden's disease. Association with squamous cell carcinoma of the tongue and perianal basal cell carcinoma. *Arch Dermatol* **120**:676–678, 1984.

12. Carlson GJ et al: Colorectal polyps in Cowden's disease (multiple hamartoma syndrome). *Am J Surg Pathol* **8**:763–770, 1984.

13. Carlson HE et al: Cowden disease: Gene marker studies and measurement of epidermal growth factor. *Am J Hum Genet* **38**:908–917, 1986.

14. Civatte J et al: Maladie de Cowden. *Ann Dermatol Vénéreol* **104**:645–647, 1977.

15. Costello MJ: A case for diagnosis (keratosis follicularis?). *Arch Dermatol Syph* **44**:109–110, 1941.

16. Dames K, Mahrle G: Hair follicle tumors in a case of multiple hamartoma syndrome (MHS), in *Hair Research. Status and Future Aspects,* Orfanos CE, Montagna W, Stüttgen G (eds), Springer-Verlag, New York, 1981, pp 371–374.

16a. Elston DM et al: Multiple hamartoma syndrome (Cowden's disease) associated with non-Hodgkin's lymphoma. *Arch Dermatol* **122**:572–575, 1986.

17. Elton R et al: Cowden's syndrome. *Int J Dermatol* **20**:617–618, 1981.

18. Flageat J et al: La maladie de Cowden. *J Radiol* **65**:701–704, 1984.

19. Fritsch P et al: Das Multiple-Hamartome-Syndrom (Cowden-Syndrom). *Hautarzt* **32**:285–291, 1981.

20. Gentry WC et al: Multiple hamartoma syndrome (Cowden disease). *Arch Dermatol* **109**:521–530, 1974.

21. Gentry WC et al: Cowden syndrome. *Birth Defects* **11**(4):137–141, 1975.

22. Gertzman GBR et al: Multiple hamartoma and neoplasia syndrome (Cowden's syndrome). *Oral Surg Oral Med Oral Pathol* **49**:314–316, 1980.

23. Gilbert HD et al: Cowden's disease (multiple hamartoma syndrome). *J Oral Maxillofac Surg* **43**:457–460, 1985.

24. Gold BM et al: Radiologic manifestations of Cowden disease. *Am J Roentgenol* **135**:385–387, 1980.

25. Gorensek M et al: Disseminated hereditary gastrointestinal polyposis with orocutaneous hamartomatosis (Cowden's disease). *Endoscopy* **16**:59–63, 1984.

26. Graham RM, Emmerson RW: Multiple hamartoma and neoplasia syndrome. *Clin Exp Dermatol* **10**:262–268, 1985.

27. Greene SL et al: Cowden's disease with associated malignant melanoma. *Int J Dermatol* **23**:466–467, 1984.

28. Greer RO et al: Cowden's disease (multiple hamartoma syndrome). Report of a limited mucocutaneous form. *J Periodontol* **47**:531–534, 1976.

29. Halevy S et al: Cowden's disease in three siblings: Electron-microscope and immunological studies. *Acta Derm Venereol* **65**:126–131, 1985.

30. Hauser H et al: Radiological findings in multiple hamartoma syndrome (Cowden disease). A report of three cases. *Radiology* **137**:317–323, 1980.

31. König F et al: Morbus Cowden—"multiples Hamartome-Syndrom"—mit gastrointestinalen Manifestationen. *Radiologe* **23**:324–326, 1983.

32. Kuffer R et al: Maladie de Cowden. *Rev Stomatol Chir Maxillofac* **80**:246–256, 1979.

33. Lashner BA et al: Ganglioneuromatosis of the colon and extensive glycogenic acanthosis in Cowden's disease. *Dig Dis Sci* **31**:213–216, 1986.

34. Lattes R (New York NY), Gellmani S, Zuflacht J (Valley Stream NY): Personal communication, 1974.

35. Laugier P et al: Maladie de Cowden. *Ann Derm Venereol* **106**:453–463, 1979.

36. Lloyd KM II, Dennis M: Cowden's disease: A possible new symptom complex with multiple system involvement. *Ann Intern Med* **58**:136–142, 1963.

37. Monnier G et al: La maladie de Cowden. *Ann Dermatol Venereol* **112**:169–177, 1985.

38. Mulvihill JJ: Personal communication, 1975.

39. Nuss PD et al: Multiple hamartoma syndrome (Cowden's disease). *Arch Dermatol* **114**:743–746, 1978.

40. Ortonne JD et al: Involvement of the digestive tract in Cowden's disease. *Int J Dermatol* **19**:570–576, 1980.

40a. Pham BN, Villanueva RP: Ganglioneuromatous proliferation associated with juvenile polyposis coli. *Arch Pathol Lab Med* **113**:91–94, 1989.

41. Rosenbluth M: Multiple noduli cutanei: An unusual case of multiple noduli with gingival manifestations. *Periodontics* **1**:81–83, 1963.

42. Ruschak PJ et al: Cowden's disease associated with immunodeficiency. *Arch Dermatol* **117**:573–575, 1981.

43. Salem OH, Steck WD: Cowden's disease (multiple hamartoma and neoplasia syndrome). *J Am Acad Dermatol* **8**:686–696, 1983.

43a. Scully RE et al: Weekly clinicopathological exercises; Case 24-1987. *N Engl J Med* **316**:1531–1540, 1987.

44. Siegel JM: Tuberous sclerosis (forme fruste). Cowden syndrome. *Arch Dermatol* **110**:476–477, 1974.

45. Siegel JM: Cowden's disease: Report of a case with malignant melanoma. *Cutis* **16**:255–258, 1975.

46. Sogol PG et al: Cowden's disease: Familial goiter and skin hamartomas. A report of three cases, *West J Med* **139**:324–328, 1983.

47. Starink TM: *The Cowden Syndrome and Other Familial Multiple Hair Follicle Tumors* (sic) *Syndromes*. Free University Press, Amsterdam, 1986.

48. Starink RM: Cowden's disease: Analysis of fourteen new cases. *J Am Acad Dermatol* **11**:1127–1141, 1984.

49. Starink RM, Hausman R: The cutaneous pathology of facial lesions in Cowden's disease. *J Cutan Pathol* **11**:331–337, 1984.

50. Starink TM, Hausman R: The cutaneous pathology of extrafacial lesions in Cowden's disease. *J Cutan Pathol* **11**:338–344, 1984.

51. Starink TM et al: The Cowden syndrome: A clinical and genetic study in 21 patients. *Clin Genet* **29**:222–233, 1986.

52. Starink TM et al: The cutaneous pathology of Cowden's disease: New findings. *J Cutan Pathol* **12**:83–93, 1985.

53. Swart JGN et al: Oral manifestations in Cowden's syndrome. Report of four cases. *Oral Surg Oral Med Oral Pathol* **59**:264–268, 1985.

54. Thyresson NH, Doyle JA: Cowden's disease (multiple hamartoma syndrome). *Mayo Clin Proc* **56**:179–184, 1981.

55. Wade TR, Kopf AW: Cowden's disease: A case report and review of the literature. *J Dermatol Surg Oncol* **4**:459–464, 1978.

56. Walton BJ et al: Cowden's disease: A further indication for prophylactic mastectomy. *Surgery* **99**:82–86, 1986.

57. Weary P et al: The multiple hamartoma syndrome (Cowden's disease). *Arch Dermatol* **106**:682–690, 1972.

58. Weinstock JV, Kawanishi H: Gastrointestinal polyposis with orocutaneous hamartomas (Cowden's disease). *Gastroenterol* **74**:890–895, 1978.

58a. Witten VH, Kopf AW: Case for diagnosis: Verrucae planae? Adenoma sebaceum? Epidermodysplasia verucciformis? *Arch Dermatol* **76**:799–800, 1957.

## Encephalocraniocutaneous lipomatosis

Prominent features of encephalocraniocutaneous lipomatosis (ECCL) include seizures beginning in infancy, mental deficiency, and unilateral cutaneous and ophthalmologic lesions together with ipsilateral cerebral malformations (Figs. 12–7 and 12–8). The condition was first described by Haberland and Perou (4). Several cases have been recorded to date (1–6) and both males and females have been affected.

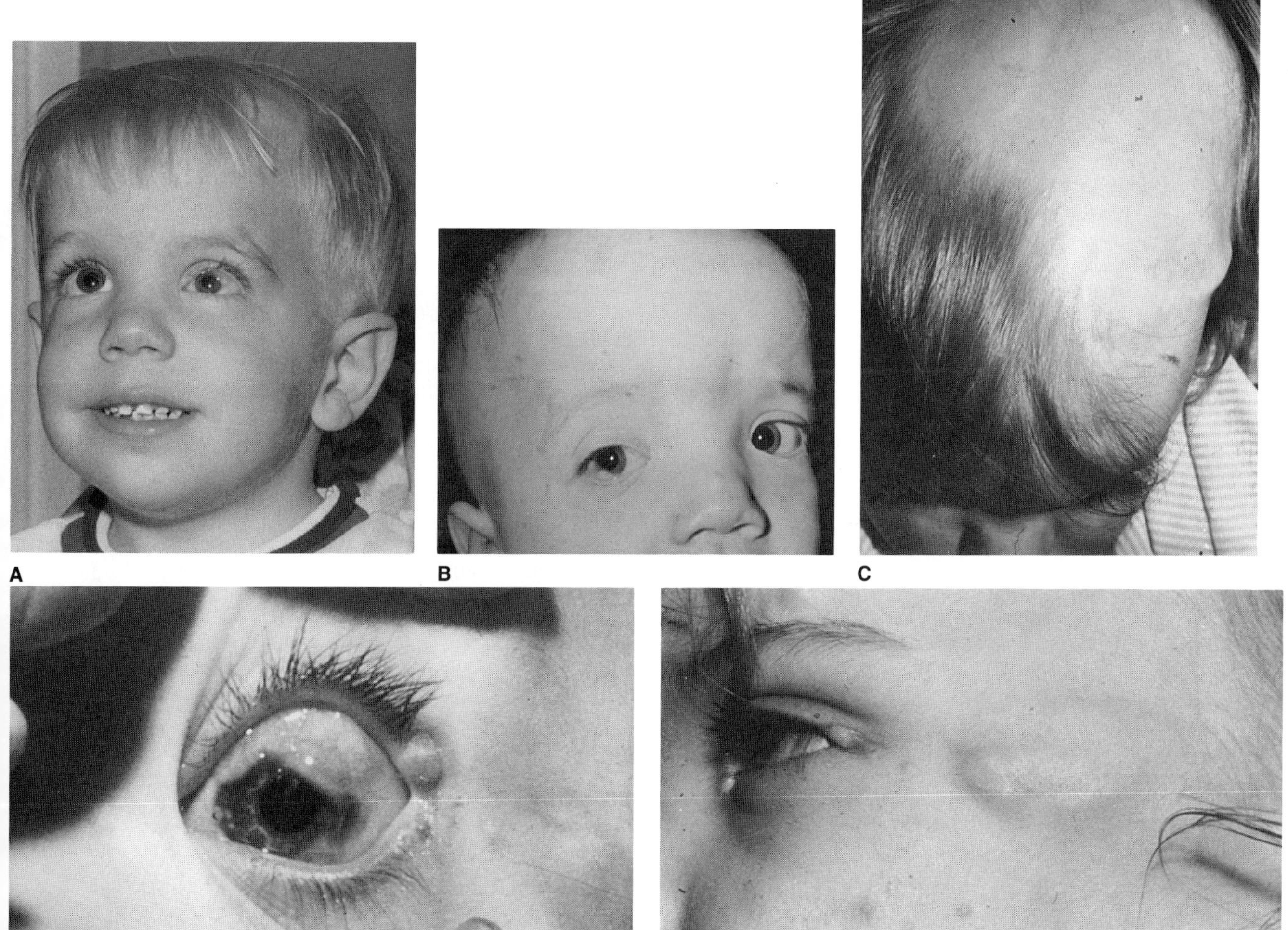


Fig. 12–7. *Encephalocraniocutaneous lipomatosis.* (A) Patient at 2 years. Note lesion of left sclera and patch of alopecia in left parietal area. (B) Another patient at 9 months. Note lesion of right sclera and extensive alopecia of right frontoparietal region. Small skin tag on right upper eyelid and on nose. (C) Hairless plaque involving most of left and part of right side of scalp. (D) Fleshy soft plaque on left bulbar conjunctiva obscuring outline of iris. Note soft papule on left upper lid margin, representing an angiofibroma. (E) Cutaneous lipomatosis and angiomatosis. Note soft colorless plaque and scattered papules. (A,B from *MA Fishman et al,* Pediatrics **61**:580, 1978; C,D from *NP Sanchez et al,* Br J Dermatol **104**:89, 1981; E courtesy of *AR Rhodes,* Boston, Massachusetts.)

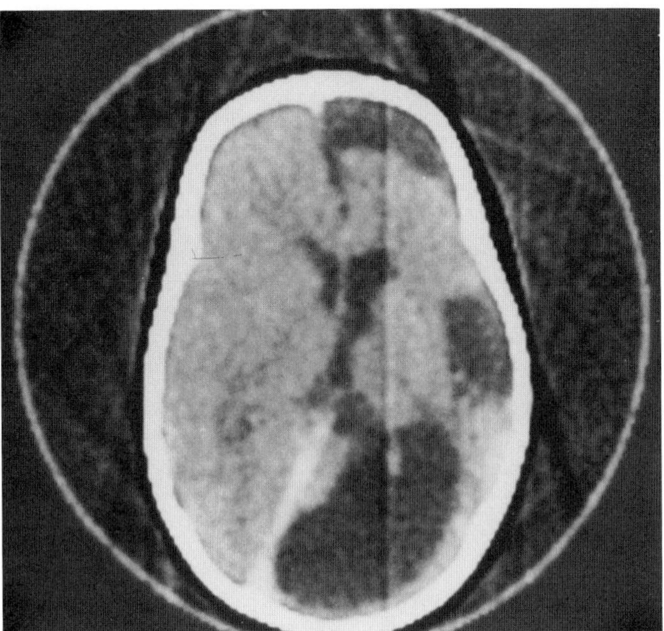

Fig. 12–8. *Encephalocraniocutaneous lipomatosis.* Computerized tomogram of cranium of patient shown in Fig. 12–7B. Note right hemiatrophy with enlarged subarachnoid spaces and porencephalic cyst. (From *MA Fishman et al,* Pediatrics **61**:580, 1978.)

All cases have been sporadic. Sanchez et al (5) noted that his patient's mother's sister had died at 2 weeks of age with hydrocephaly.

Although most instances have had unilateral involvement, bilateral involvement has been observed in one instance (5). Anatomic distribution of lesions, involving the face, eye, and leptomeninges, closely parallels the involvement of Sturge–Weber angiomatosis; however, mesodermal dysgenesis is usually limited to the vascular system. In ECCL, cutaneous and meningeal lipomatosis occur together with multiple lipomas of the nervous system (4,5). Because some cases include hyperostoses of the skull, (3,4,6), cutaneous lipomas outside the skull (4), and visceral lipomatosis (4), Wiedemann and Burgio (7) suggested that ECCL might possibly represent a more localized form of Proteus syndrome. If true, the relationship might be somewhat analagous to segmental neurofibromatosis and neurofibromatosis proper (NF I).

**Craniofacial skin and ocular lesions.** Superficial lipomas, lipofibromas, and connective tissue nevi have been observed. Angiofibromas have also been recorded (4) and isolated neurofibroma was noted in one instance (3). Soft tissue papules, tumors, and plaques 2 mm to 3 cm in diameter are found about the cheek, eyelid, bulbar conjunctiva, forehead, and neck. Smooth plaques occur over the frontoparietal area in association with focal alopecia (1–6). Bony protuberances of the skull have been described in some instances (3,4,6).

**Central nervous system.** Abnormalities consist of unilateral cerebral atrophy, dilated ventricular system, porencephaly, and lipomas of the leptomeninges (3,5). Haberland and Perou (4) described a lipoangiomatous meningeal malformation, polymicrogyria, cerebral calcification, and lipomas of the middle cranial fossa in their case. Bamforth et al (1) noted hydrocephaly and subarachnoid calcification in a gyral pattern.

Seizures begin during infancy. Mental retardation may be mild (3,5) or severe (4).

**Differential diagnosis.** Encephalocraniocutaneous lipomatosis is easily differentiated from *Sturge–Weber angiomatosis, focal dermal hypoplasia syndrome,* and *oculocerebrocutaneous syndrome. Epidermal nevus syndrome* may superficially resemble ECCL because the former is often associated with craniofacial skin lesions and neurologic abnormalities. However, epidermal nevi are clinically and histologically distinguishable. As indicated, Wiedemann and Burgio (7) suggested that ECCL might possibly represent a more localized form of *Proteus syndrome.*

**Laboratory aids.** CT scanning, EEG, and contrast studies are useful.

#### References (Encephalocraniocutaneous lipomatosis)

1. Bamforth JS et al: Abnormal unilateral development of brain and mesodermally derived tissue in encephalocraniocutaneous lipomatosis. 8th Annual Workshop on Malformations and Morphogenesis, August 15–19, 1987, Furman University, Greenville, South Carolina, p 26.
2. Bitoun P et al: Picture of the Month. Encephalocraniocutaneous lipomatosis. *Am J Dis Child* **136**:1085–1086, 1982.
3. Fishman MA et al: Encephalocraniocutaneous lipomatosis. *Pediatrics* **61**:580–582, 1978.
4. Haberland C, Perou M: Encephalocraniocutaneous lipomatosis. *Arch Neurol* **22**:144–155, 1970.
5. Sanchez NP et al: Encephalocraniocutaneous lipomatosis: A new neurocutaneous syndrome. *Br J Dermatol* **104**:89–96, 1981.
6. Schlack HG, Skopnik H: Encephalocraniocutane Lipomatose und lineärer Naevus sebaceus. *Mschr Kinderheilk* **133**:235–237, 1985.
7. Wiedemann H-R, Burgio GR: Encephalocraniocutaneous lipomatosis and Proteus syndrome. *Am J Med Genet* **25**:403–404, 1986.
8. Zuntová A: Asymetrická osteokutánní lipomatóza u tříletého chlapce provázená chybrým vývojem mozku. *Čs Pediatr* **38**:724–727, 1983.

## Epidermal nevus syndrome

The association of epidermal nevus, central nervous system abnormalities including mental deficiency and seizures, skeletal defects, and other anomalies was described in two patients by Feuerstein and Mims (12) in 1962, although isolated examples were recorded earlier by Schimmelpenning (33) in 1957 and Berg and Crome (3) in 1960. Other patients were reported in the 1960s (24,29), but it was Solomon and his co-workers (37) who had the major role in delineating the syndrome. The condition has been called the epidermal nevus syndrome (37), linear sebaceous nevus syndrome (20), Feuerstein–Mims syndrome (31), Schimmelpenning–Feuerstein syndrome (1,40), and Solomon syndrome (22). Over 100 instances have been recorded to date (9,11a,36,44). Several cases are less certain because of incomplete documentation (15,19). Comprehensive reviews are those of Larrègue et al (22) and Solomon and Esterly (36). The reader is referred to the following sources for extensive coverage: cutaneous manifestations (22,36), central nervous system features (9,17,25,40), ocular abnormalities (7,16,17a,34,42), skeletal defects (36), and neoplasms (11,20,23,27,36). Features of the syndrome are summarized in Table 12–2.

The overwhelming majority of cases are sporadic. Three instances of father-to-son transmission, one instance of mother-to-daughter transmission, and two affected sibs born to normal parents are known (36). Two families have been reported in which various relatives of the proband had either seizures or mental deficiency, but no other features of the syndrome (3,4). MM Cohen has also seen a family in which several relatives of the proband had isolated mental retardation and/or seizures. The syndrome has been reported in one of monozygotic (10) and in one of dizygotic (21) twins. Chromosomes have been normal (17,28,37). Solomon and Esterly (36) noted that six of their patients' mothers were nurses and two were radiology technicians ($n=60$). However, no formal epidemiologic investigation has been carried out to date.

**Cutaneous manifestations.** We concur with Solomon et al (37) that the name ''epidermal nevus'' is more appropriate than ''linear sebaceous nevus'' since sebaceous elements may be absent from cutaneous lesions. A spectrum of epidermal nevi may occur in the syndrome. Nomenclature is complex and has been discussed elsewhere (36).

Table 12–2. Features of the epidermal nevus syndrome[a]

| Findings | Percentage ($n = 60$) |
|---|---|
| Cutaneous | 100 |
| Epidermal nevi | |
| Nevus unius lateris | (60) |
| Ichthyosis hystrix | (20) |
| Linear nevus sebaceus | (10) |
| Localized acanthosis nigricans | (20) |
| Other verrucous or mixed lesions | (33) |
| Hemangiomas | 37 |
| Dermatomegaly | 15 |
| Hypopigmentation | 10 |
| Café-au-lait spots | 10 |
| Many melanocytic nevi | 10 |
| Skeletal | 70 |
| Kyphosis/scoliosis | 28 |
| Ankle and foot deformities | 15 |
| Genua valga | 5 |
| Hemihypertrophy (hemihyperplasia)[b] | 15 |
| Short limbs | 10 |
| Finger and toe deformities | 10 |
| Cranial and facial bone deformities | 15 |
| Bone cysts | 3 |
| Vitamin D-resistant rickets | 2 |
| Neurologic | 50 |
| Mental deficiency | 40 |
| Seizures | 33 |
| Hemiparesis | 14 |
| Cranial nerve involvement[c] | 8 |
| Ocular | 33 |
| Extension of nevus to lid | 19 |
| Lipodermoid | 15 |
| Nystagmus | 2 |
| Coloboma and cortical blindness | 2 |

[a]Adapted from LM Solomon and NB Esterly: *Curr Prob Pediatr* **6**(1):1–56, 1975.
[b]Or localized gigantism in some instances. Involves both bone and soft tissue.
[c]VI, VII, and VIII.

Epidermal nevi are present at birth or may develop during the first few months of life (Figs. 12–9A,B and 12–10A–I). Rarely, they may appear as late as puberty. Once the lesion has reached stability in late adolescence, further progression is not likely to occur. Approximately 60% have long warty streaks that are identical to nevus unius lateris but which are rarely completely unilateral. Generally, the more distally a lesion occurs or the closer it occurs to an articular joint, the more verrucous it tends to be. Limbs are commonly affected with streaking down a dorsal surface to involve one or more fingernails. When epidermal nevi involve flexural areas such as the neck or axilla, they appear dark, velvety, or mossy. Another type of lesion, known as ichthyosis hystrix and occurring in about 20%, is found mostly on the trunk. Lesions tend to be somewhat scaly and erythematous or light brown and have a whorled pattern. Still another type of lesion, found in about 10%, is the linear nevus sebaceus, a characteristic orange-tan (black in blacks), unilateral, verrucous plaque on the scalp that projects onto the cheek, ear, nose, eyelid, and/or oral mucous membrane. The lesion tends to stop at the midline. The acanthotic type of lesion, affecting approximately 20%, is a uniform, slightly scaly, dirty-looking discoloration of the skin. Its borders are indistinct and stop at the ventral midline. About 2% of lesions form a string or row of pigmented papillomas. Finally, mixtures of several different types can be found in about 33% of patients (36).

Nevi may extend to involve the mucous membranes of the mouth, anus, vagina, or penis. Oral and anal complications are especially bothersome since resultant friable papillomatous masses are subject to episodes of bleeding (5,6,31,36,37,41).

Histologically, lesions overlying the skull show sebaceous hypertrophy whereas lesions involving the fingers tend to show epithelial hyperplasia. Histologic variability of such organoid nevi indicates that linear nevus sebaceus, nevus unius lateris, and ichthyosis hystrix are morphologic, topographic, or temporal variants of the same pathologic process (36).

Other cutaneous lesions may occur. Approximately 37% have some form of cutaneous hemangioma other than commonplace nuchal nevus flammeus. Large areas of hypopigmentation, affecting primarily the thorax but sometimes involving the scalp, can be found in about 10%. Café-au-lait spots and small pigmented hairy nevi have also been observed on occasion. Visible scalp hair appears to be absent over scalp lesions, but sparse rudimentary follicles are present histologically. In-

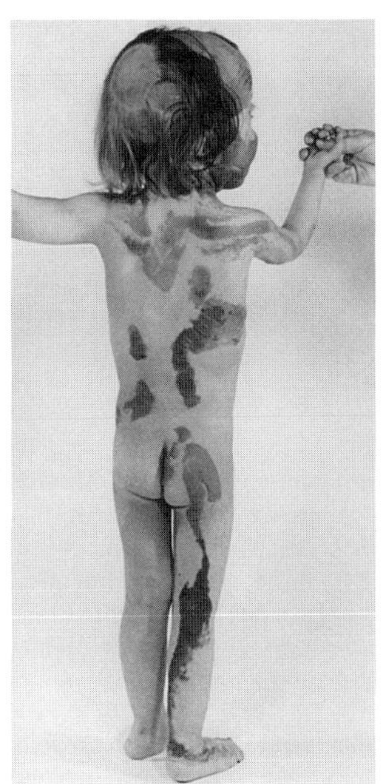
A

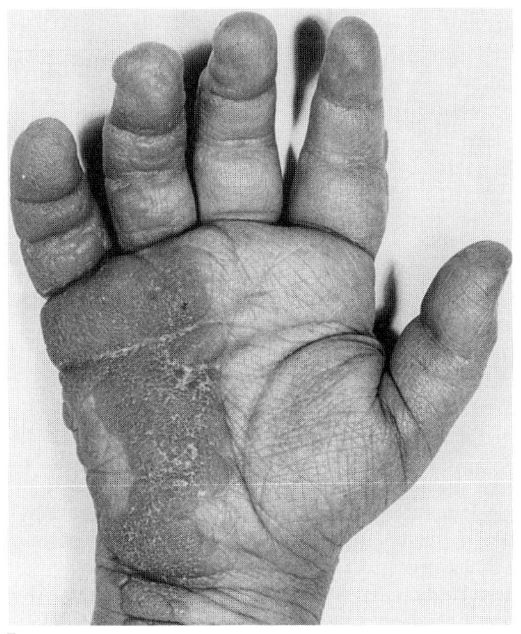
B

Fig. 12–9. *Epidermal nevus syndrome.* (A) Involvement of hair, face, neck, back, buttocks, and legs of 4.5-year-old child. (B) Hand lesion of patient shown in A. (A,B from *B Leiber,* Mschr Kinderheilkd **137**:585, 1979.)

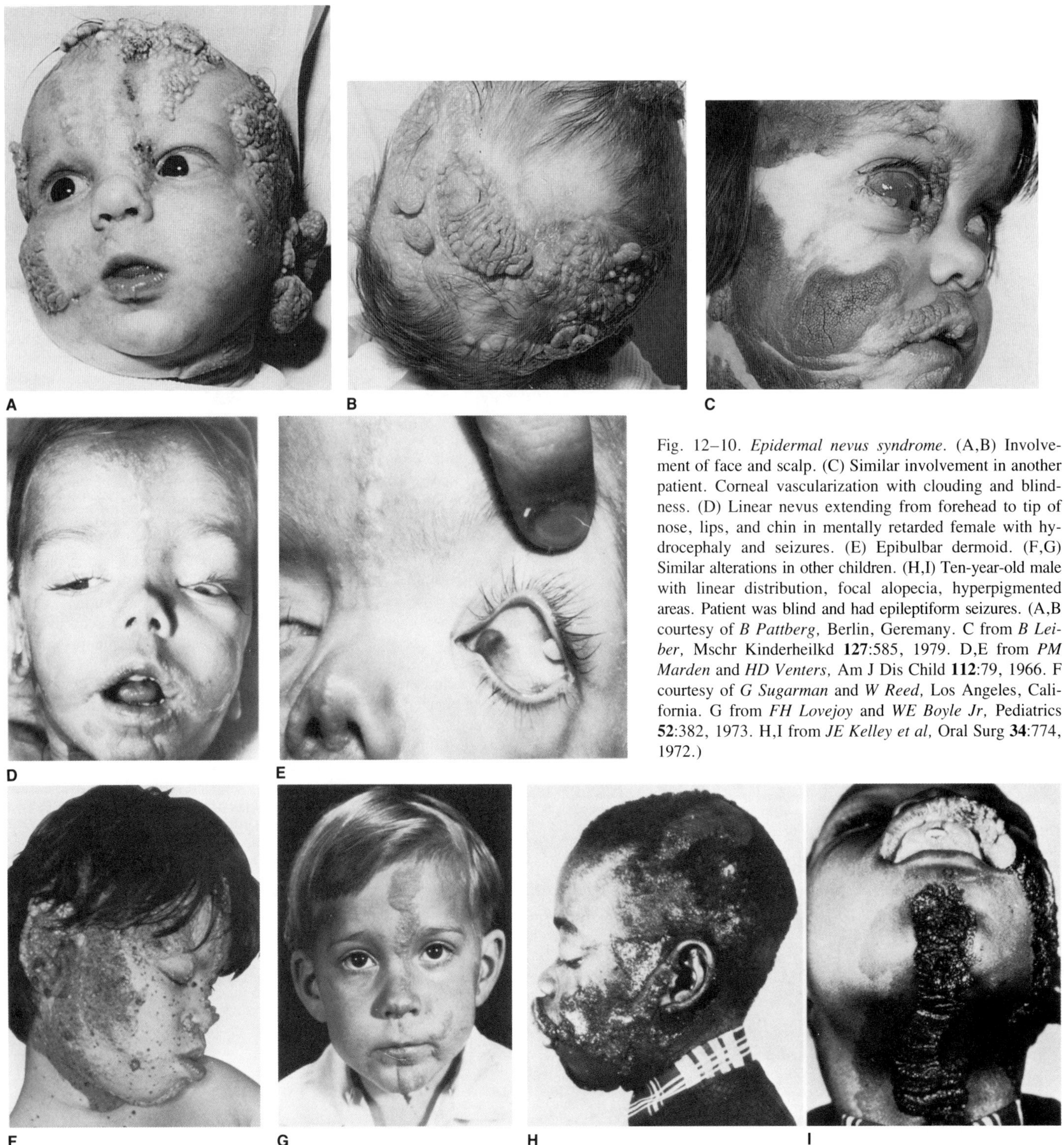

Fig. 12–10. *Epidermal nevus syndrome.* (A,B) Involvement of face and scalp. (C) Similar involvement in another patient. Corneal vascularization with clouding and blindness. (D) Linear nevus extending from forehead to tip of nose, lips, and chin in mentally retarded female with hydrocephaly and seizures. (E) Epibulbar dermoid. (F,G) Similar alterations in other children. (H,I) Ten-year-old male with linear distribution, focal alopecia, hyperpigmented areas. Patient was blind and had epileptiform seizures. (A,B courtesy of *B Pattberg,* Berlin, Geremany. C from *B Leiber,* Mschr Kinderheilkd **127**:585, 1979. D,E from *PM Marden* and *HD Venters,* Am J Dis Child **112**:79, 1966. F courtesy of *G Sugarman* and *W Reed,* Los Angeles, California. G from *FH Lovejoy* and *WE Boyle Jr,* Pediatrics **52**:382, 1973. H,I from *JE Kelley et al,* Oral Surg **34**:774, 1972.)

creased skin thickness, increased hairiness, and warmth are found in patients with hemihyperplasia (hemihypertrophy) or localized gigantism (36).

**Skeletal abnormalities.** Defects have included incomplete development of various bones, vertebral defects, incomplete development of the talus, camptodactyly, clinodactyly, brachydactyly, ischial and pubic hypoplasia, incomplete rib formation, abnormal clavicles, hypoplastic nasal and orbital bones, shortening of limb bones, kyphoscoliosis, posterior luxation of the ankle, pes equinovarus, and genua valga (17,21,22,22a,30,36,37). Some patients have exhibited frontal bossing, asymmetry of the skull, and/or premature closure of the sphenofrontal suture (29,30,38). Vitamin D-resistant rickets has been reported in several patients (1a,7a,29a,34a,36,38,44). Uncommonly, cystic changes have been observed in the long bones or mandible (36). Hemihyperplasia (hemihypertrophy) or localized gigantism involving bone and soft tissue may be observed in about 15% (36).

**Central nervous system.** About 50% have moderate-to-severe neurologic involvement. Mental deficiency occurs in approximately

40% and seizures, found in about 33%, may be focal or generalized. Spastic hemiparesis or paralysis with concomitant atrophy of the corresponding musculature and contractures is observed in about 14%. Cranial nerve involvement, a feature of approximately 8%, affects the oculomotor nerves, and especially VI, VII, and VIII. Cerebral hemangiomas or vascular malformations have also been recorded. Cortical atrophy, hydrocephaly, hemimegalencephaly, and porencephaly, have also been noted (8,9,17,18,22b,25,28,36,37,40).

**Ocular defects.** Approximately 33% of patients have ocular manifestations, the most common being extension of the epidermal nevus to involve the eyelid, eyelid margin, and conjunctiva (Fig. 12–10C,E). Other abnormalities have included coloboma of the lid and/or iris and retina, epibulbar lipodermoid, corneal opacity and pannus formation, oculomotor dysfunction, nystagmus, optic nerve hypoplasia, and cortical blindness (12,16,17a,18,18a,24,28,30,34,36,38,39,42).

**Dental findings.** Teeth are frequently hypoplastic and several patients have been reported with odontodysplasia (Fig. 12–11) (4,24,29,35).

**Other findings.** A number of other findings have been reported including cleft palate (3), bifid uvula (18), coarctation of the aorta, bizarre origin of the subclavian artery, PDA, horseshoe kidney, and intrahepatic cystic biliary adenomas (24,28,29,36).

**Neoplasms.** In the series of patients reported by Solomon and Esterly (36), syringocystadenoma papilliferum (18) was observed in eight patients and keratoacanthoma in four. Hemangiomas of various types involving the skin and central nervous sytem (36) have already been discussed. Other reported tumors have included Wilms tumor, rhabdomyosarcoma, salivary gland adenocarcinoma, carcinoma of the esophagus and stomach, adenocarcinoma of the breast, metastatic squamous cell carcinoma, ameloblastoma, astrocystoma and verruciform xanthoma (2,11,20,23,27,26).

**Differential diagnosis.** Isolated epidermal nevi may occur without any other features of the syndrome and two large series have been recorded (26,43). Café-au-lait spots are found in *neurofibromatosis, McCune–Albright sysndrome,* and Jaffe-Campanacci syndrome. *Encephalocraniocutaneous lipomatosis* (13,14,32) consists of unilateral cutaneous and ophthalmologic lesions together with ipsilateral cerebral malformations, seizures, and variable degrees of mental retardation. Epidermal nevi, epibulbar dermoids, and hyperostosis of the skull with craniosynostosis and mental deficiency are features of *Thanos syndrome.* In *Proteus syndrome,* skull hyperostoses, asymmetry of limbs, partial gigantism of the hands and/or feet, and verrucous epidermal nevi may occur, sometimes with mental deficiency. *Sturge–Weber angiomatosis* and *Klippel–Trénaunay–Weber syndrome* are easily distinguished.

**Laboratory aids.** Biopsies of epidermal nevi show variable but diagnostic histopathology. Other appropriate aids should include skull X rays, EEG, roentgenographic bone survey, serum calcium, and serum phosphate.

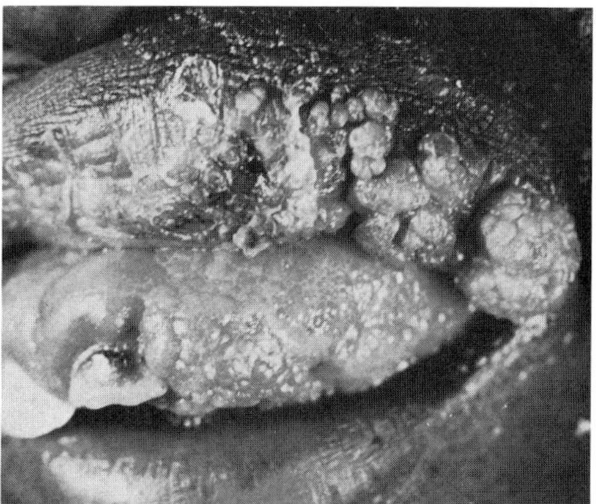
A

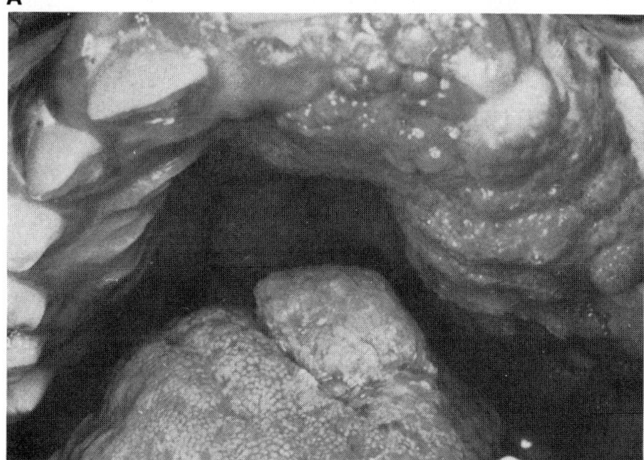
B

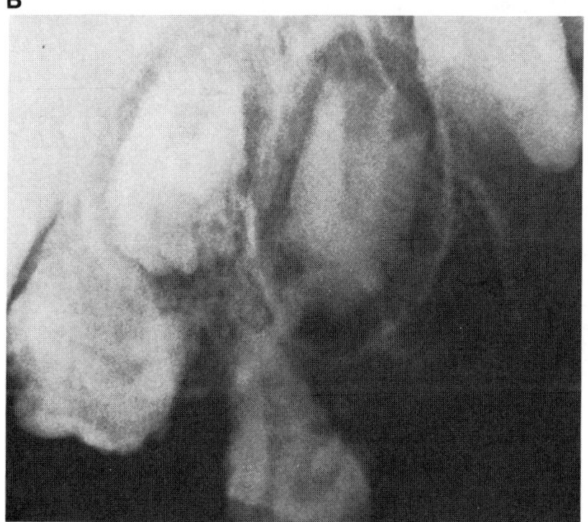
C

Fig. 12–11. *Epidermal nevus syndrome.* (A) Labial and gingival involvement. (B) Unilateral involvement of maxillary alveolar ridge, palate, and tongue. (C) Roentgenogram of teeth showing hypoplasia of enamel and dentin reminiscent of odontodysplasia. (A–C from *JE Kelley et al,* Oral Surg **34**:774, 1972.)

### References (Epidermal nevus syndrome)


1. Albrecht-Nebe H et al: Kasuistischer Beitrag zum Schimmelpenning-Feuerstein-Mims-Syndrom. *Dermatol Monatschr* **174**:257–266, 1988.

1a. Aschinberg LC et al: Vitamin D-resistant rickets associated with epidermal nevus syndrome: Demonstration of a phosphaturic substance in the dermal lesions. *J Pediatr* **91**:56–60, 1977.

2. Barr RJ, Plank CJ: Verruciform xanthoma of the skin. *J Cutan Pathol* **7**:422–428, 1980.

3. Berg JM, Crome J: A possible case of atypical tuberous sclerosis. *J Ment Defic Res* **4**:24–31, 1960.

4. Bianchine J: The nevus sebaceus of Jadassohn: A neurocutaneous syndrome and a potentially premalignant lesion. *Am J Dis Child* **120**:223–228, 1970.

5. Bitter K: Über die Erstbeobachtung eines angeborenen Naevus sebaceus im Trigeminusbereich mit Gehirnmissbildungen und Riesenzellgeschwulsten des Ober-und Unterkiefers. *Dtsch Zahn Mund Kieferheilkd* **56**:17–24, 1971.

6. Brown HM, Gorlin RJ: Oral mucosal involvement in nevus unius lateris (ichthyosis hystrix). *Arch Dermatol* **81**:509–515, 1960.

7. Burch JV et al: Ichthyosis hystrix (epidermal nevus syndrome) and Coats' disease. *Am J Ophthalmol* **89**:25–30, 1980.

7a. Carey DE et al: Hypophosphatemic rickets/osteomalacia in linear sebaceous nevus syndrome: A variant of tumor-induced osteomalacia. *J Pediatr* **109**:994–1000, 1986.

8. Chalhub EG et al: Linear nevus sebaceus syndrome associated with porencephaly and nonfunctioning major cerebral venous sinuses. *Neurology* **25**:857–860, 1975.

9. Clancy RR et al: Neurologic manifestations of the organoid nevus syndrome. *Arch Neurol* **42**:236–240, 1985.

10. Cockayne EA: *Inherited Abnormalities of the Skin and Its Appendages*, Oxford University Press, London, 1933, p 298.

11. Dimond RL, Amon RB: Epidermal nevus and rhabdomyosarcoma. *Arch Dermatol* **112**:1424–1426, 1976.

11a. Fahnenstick H et al: Organoides Naevussyndrom (Schimmelpenning–Feuerstein–Mims–Syndrom): Kasuistik und Literatur. *Klin Pädiatr* **201:**54–57, 1989.

12. Feuerstein RC, Mims LC: Linear nevus sebaceus with convulsions and mental retardation. *Am J Dis Child* **104**:675–679, 1962.

13. Fishman MA et al: Encephalocraniocutaneous lipomatosis. *Pediatrics* **61**:580–582, 1978.

14. Gellis SS, Feingold M: Linear nevus sebaceus syndrome. *Am J Dis Child* **120**:139–140, 1970.

14a. Goldberg LH et al: The epidermal nevus syndrome: Case report and review. *Pediatr Dermatol* **4**:27–33, 1987.

15. Gördüren S: Aberrant lacrimal gland associated with other congenital abnormalities. *Br J Ophthalmol* **46**:277–280, 1962.

16. Haslan RHA, Wirtschafter JD: Unilateral external oculomotor nerve palsy and nevus linearis sebaceus. *Neurology* **22**:879–887, 1972.

17. Holden KR et al: Neurological involvement in nevus unius lateris and nevus linearis sebaceus. *Neurology (Minneap)* **22**:879–897, 1972.

17a. Insler MS, Davlin L: Ocular findings in linear sebaceous naevus syndrome. *Br J Ophthalmol* **71**:268–272, 1987.

18. Jancar J: Naevus syringocystadenomatosus papilliferus with skull and brain lesions, hemiparesis, epilepsy and mental retardation. *Br J Dermatol* **82**:402–406, 1970.

18a. Katz B et al: Optic nerve hypoplasia in the syndrome of nevus sebaceus of Jadassohn. *Ophthalmology* **94:**1570–1576, 1987.

19. Lall K: Teratoma of conjunctiva (associated with nevus systematicus and epilepsia symptomatica). *Acta Ophthalmol (Kbh)* **40**:555–558, 1962.

20. Lansky LL et al: Linear sebaceous nevus syndrome. *Am J Dis Child* **128**:587–590, 1972.

21. Lantis S et al: Nevus sebaceus of Jadassohn: Part of new neurocutaneous syndrome. *Arch Dermatol* **98**:117–123, 1968.

22. Larrègue M et al: Le syndrome du naevus épidermique de Solomon. *Ann Dermat Syphiligraph (Paris)* **101**:45–55, 1974.

22a. Leonidas JC et al: Radiologic features of the linear sebaceous syndrome. *Am J Roentgenol* **132**:277–279, 1979.

22b. Levin S et al: Computed tomography appearances in the linear sebaceus naevus syndrome. *Neuroradiology* **26**:469–472, 1984.

23. Lovejoy FH, Boyle WE: Linear nevus sebaceus syndrome. *Pediatrics* **52**:382–387, 1973.

24. Marden PM, Venters HD: A new neurocutaneous syndrome. *Am J Dis Child* **112**:79–81, 1966.

25. McAuley DL et al: Neurological involvement in the epidermal naevus syndrome. *J Neurol Neurosurg Psychiatr* **41**:466–469, 1978.

26. Mehregan H, Pinkus H: Life history of organoid nevi. *Arch Dermatol* **91**:564–588, 1965.

27. Meyerson LB: Nevus unius lateralis, brain tumor, and diencephalic syndrome. *Arch Dermatol* **95**:501–504, 1967.

28. Mollica F et al: Linear sebaceous nevus in a newborn. *Am J Dis Child* **128**:868–871, 1974.

29. Monahan RH et al: Multiple choristomas, convulsion and mental retardation as a new neurocutaneous syndrome. *Am J Ophthalmol* **64**:529–532, 1967.

29a. Moorjani R, Shaw DG: Feuerstein and Mims syndrome with resistant rickets. *Pediatr Radiol* **5**:120–122, 1976.

30. Moynahan EJ, Wolff OH: A new neurocutaneous syndrome (skin, eye, brain) consisting of linear nevus, bilateral lipodermoids of the conjunctiva, cranial thickening, cerebral cortical atrophy, and mental retardation. *Br J Dermatol* **79**:651–652, 1967.

31. Reichart PA et al: Gingival manifestation in linear nevus sebaceus syndrome. *Int J Oral Surg* **12**:437–443, 1983.

32. Sanchez NP et al: Encephalocraniocutaneous lipomatosis: A new neurocutaneous syndrome. *Br J Dermatol* **104**:89–96, 1981.

33. Schimmelpenning GW: Klinischer Beitrag zur Symptomatologie der Phakomatosen. *Fortschr Geb Roentgenstr Nuklearmed* **87**:716–720, 1957.

34. Shochot Y et al: Eye findings in the linear sebaceous nevus syndrome: A possib. clue to the pathogenesis. *J Craniofac Genet Dev Biol* **2**:289–294, 1982.

34a. Skovby F et al: Hypophosphatemic rickets in linear sebaceous nevus sequence. *J Pediatr* **111**:855–857, 1987.

35. Slootweg PJ, Meuwissen PRM: Regional odontodysplasia in epidermal nevus syndrome. *J Oral Pathol* **14**:256–262, 1985.

36. Solomon LM, Esterly NB: Epidermal and other congenital organoid nevi. *Curr Prob Paediatr* **6**:3–56, 1975.

37. Solomon LM et al: The epidermal nevus syndrome. *Arch Dermatol* **97**:273–285, 1968.

38. Sugarman GI, Reed WB: Two unusual neurocutaneous disorders with facial cutaneous signs (case 1). *Arch Neurol* **21**:242–247, 1969.

39. Tripp JH: A new "neurocutaneous" syndrome (skin, eye, brain, and heart) syndrome. *Proc R Soc Med* **64**:23–24, 1971.

40. Vles JSH et al: Neuroradiological findings in Jadassohn nevus phakomatosis: A report of four cases. *Eur J Pediatr* **144**:290–294, 1985.

41. Wauschkuhn J, Rohde B: Systematisierte Talgdrüsen-, Pigment- und epitheliale Naevi mit neurologischer Symptomatik; Feuerstein-Mimssches Neuroektodermales-Syndrom. *Hautarzt* **22**:10–13, 1971.

42. Wilkes SR et al: Ocular malformation in association with ipsilateral facial nevus of Jadassohn. *Am J Ophthalmol* **92**:344–352, 1981.

43. Wilson Jones E, Heyl T: Naevus sebaceus: A report of 140 cases with special regard to development of secondary malignant tumours. *Br J Dermatol* **82**:99–117, 1970.

44. Zaremba J: Jadassohn's naevus phakomatosis: Study based on a review of 37 cases. *J Ment Defic Res* **22**:103–123, 1978.

## Gardner syndrome

During 1950–1953, Gardner and co-workers (35,36) recognized and reported a syndrome of multiple adenomatous polyposis of the large intestines, multiple osteomas of the facial bones, cutaneous epidermoid cysts, and desmoid tumors and fibrous hyperplasia of the skin and mesentery. Similar cases had been documented earlier (14,27,33,34), but Gardner was the first to recognize the hereditary pattern.

Initially, Gardner syndrome (GS) was thought to be an entity separate from familial colorectal polyposis (FCP). However, with the accumulation of data, many investigators have demonstrated that it represents one tail of the spectrum of that disorder (26,54,94,104,105). Cohen (19) found that about 50% of those with FCP had no extracolonic manifestations.

Inheritance is autosomal dominant with complete penetrance, and markedly variable expressivity (13,35,84,95). The FCP-GS gene has been localized to a small region on the long arm of chromosome 5 (10,72,11). GS has been reported in 28-year-old monozygous twins. One had complete expression of the disorder; the other had only osteomas of the jaws (38). The frequency of FCP-GS has been estimated to be between 1 in 12,000 (123) and 1 in 1400 (95). The mutation rate of $1.3 \times 10^{-5}$ mutations per gene per generation is on the same order of magnitude as other human mutations (95).

An inherited susceptibility of FCP-GS cells has been found to retrovirus-induced transformation and chromosomal aneuploidy (97). Chromosomal alterations in fibroblasts (increased tetraploidy, chromosomal instability, heteromorphism of 2q) of patients with GS have been reported (24–26,37).

**Osteomas.** The osteomas associated with FCP and particularly those associated with GS are of various sizes and have limited growth potential. The radiopacity and microscopy are similar to those of mature compact bone and they have well developed Haversian systems (4,51,82,90). The osteomas can be found in any bone, but most often are seen throughout the calvaria, frontal and ethmoidal sinuses and facial skeleton (Fig. 12–12A–C) (17,55,98). The tooth-bearing areas of the mandible and even of the maxilla are commonly involved (51). Those in the mandible commonly become confluent (4,51,121). In a Japanese study, 50% had skeletal involvement (119), and 46% had three or more osteomas of the jaws (51) whereas 81–93% had one or more (51,119,121). The osteomas may protrude (exostoses), but in most cases they appear as enostoses without palpable swelling (17,51,121). Bülow et al (11) studied individuals at risk for FCP and

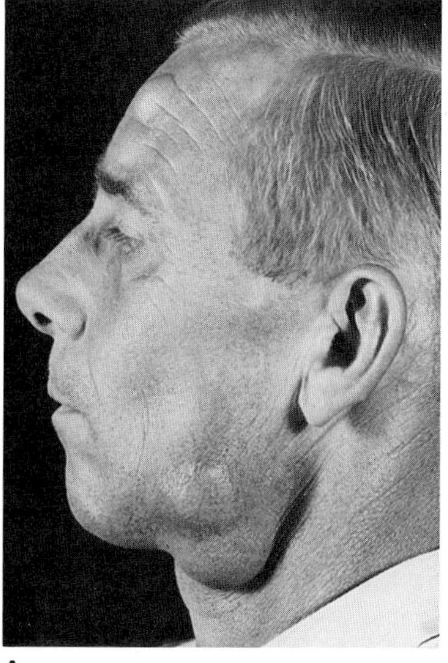
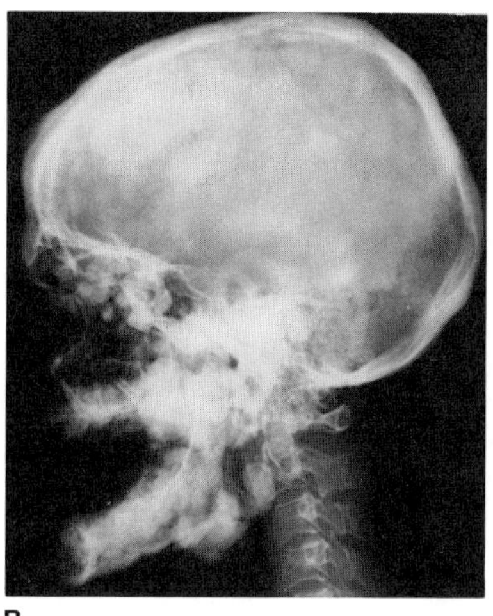
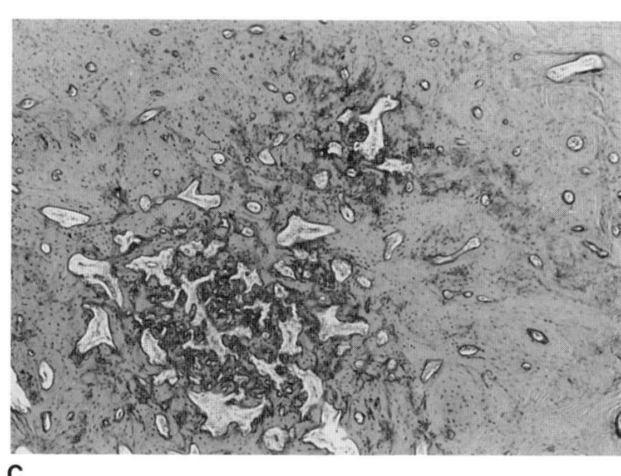

A B C

Fig. 12–12. *Gardner syndrome.* (A) Osteomas of forehead, zygomatic area, and mandible. (B) Roentgenogram showing numerous osteomas scattered throughout jaws and skull. (C) Histologic appearance of bony mass of mandible. (A courtesy of *EL Jones,* Washington, DC. C from *K Ooya et al,* J Oral Pathol **5**:305, 1976.)

found that 76% had mandibular densities, contrasting sharply with 4% in the normal population. Long tubular bones, most often the radius, ulna, and metacarpals, may be sites of small osteomas, but involvement—in contrast to that of the facial skeleton—is minimal, and usually manifests as rather diffuse subperiosteal cortical thickening. In a few cases, however, these osteomas presented as small, well-defined exostoses (17,55). Osteomas may be found in members of families with FCP who do not have manifest polyposis (39) and may precede the appearance of intestinal symptoms (82,98,119).

**Epidermoid cysts.** Cysts of the skin occur in about 50–60% of all cases of GS, but in some families have been present in nearly all of those affected (69,125). In one study of 74 patients with the syndrome, over 50% had cysts and all those with cysts had or subsequently developed colonic polyposis (69). The number of cysts has ranged from 1 to 20 with an average of 4 per patient. They occur most often on the legs, face, scalp, and arms, the trunk being seldomly affected (Figs. 12–13A) (69). Although the cysts may appear at any time from birth to 35 years, they most often become manifest around puberty, prior to the appearance of colorectal polyposis (35,39,69,71). New cysts appear periodically (39,125). The early appearance of the cysts can be used as a guide to indicate which members of a family are at risk for developing polyposis (21,29).

The histologic picture of the cyst shows the characteristic features of epidermoid cysts (Fig. 12–13B) (69), although they are often referred to as sebaceous cysts. It has been pointed out that about 50% of the cysts have shadow cells, resembling pilomatrixoma (22,71).

**Gastrointestinal system.** Multiple intestinal polyposis of the colon and rectum with a marked tendency to malignant degeneration is characteristic (Fig. 12–14A,B). Although the polyps may appear before puberty (21,29,95), the chance for malignant transformation at this age is less than 5%. The mean age for diagnosis of the polyps is 23–31 years (70). The mean age for diagnosis of the intestinal cancer is 37 years (95). By 30 years, about 50% of the patients exhibit malignant degeneration, but the frequency of malignancy with advanced age is probably near 100%. The origin of the polyps is multiclonal (50).

Polyposis of the stomach and small intestines has rarely been reported (39,41,44,47,59,120,122,131) and malignant degeneration of small bowel polyps appears to be relatively low (2–3%) (12). The reader is referred to a good summary of these and other cases (104).

The association of periampullary adenocarcinoma (carcinoma of the ampulla of Vater, duodenum or pancreas) has been established (14,56,79). It may develop in the absence of other extracolonic adenomatous polyps. The frequency has been estimated as high as 12%

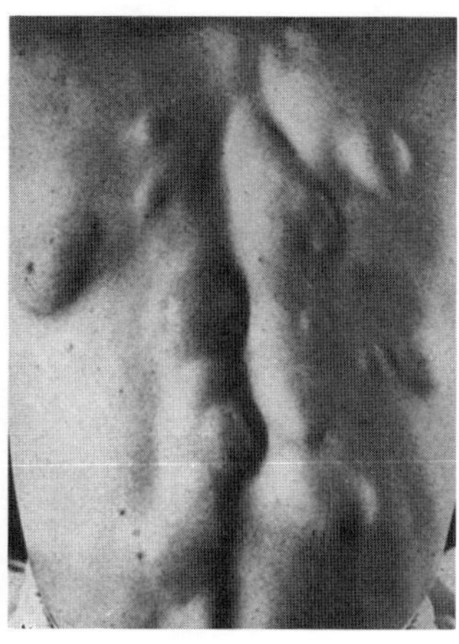
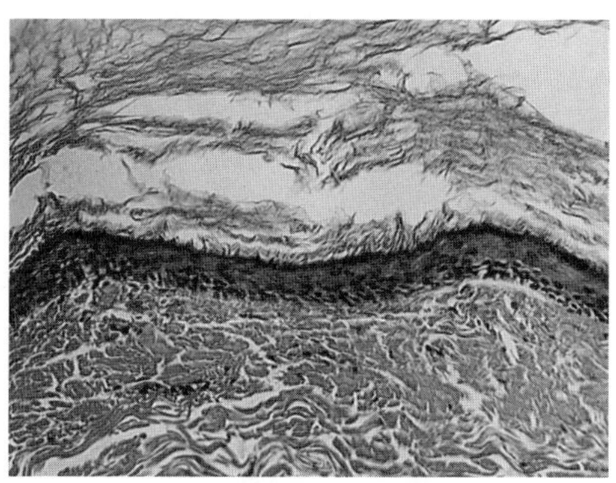

A B

Fig. 12–13. *Gardner syndrome.* (A) Multiple epidermoid inclusion cysts over dorsal region. (B) Histologic section of epidermoid inclusion cyst. (A from *MC Oldfields,* Br J Surg **41**:534, 1954).

A

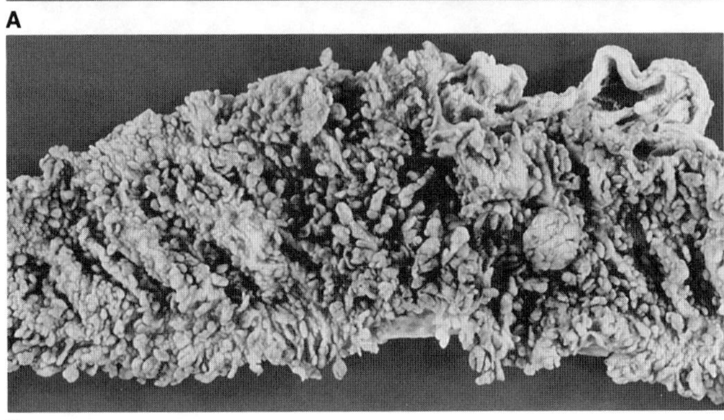

B

Fig. 12–14. *Gardner syndrome.* (A) Multiple polyps of colon undergoing malignant transformation to adenocarcinoma. (B) Close-up of multiple colonic polyps. (B from *K Ooya et al,* J Oral Pathol **5**:305, 1976.)

among patients with FCP but in only 2–3% of patients with GS (93). Lymphoid hyperplasia of the terminal ileum has also been noted (104).

Both abdominal and extraabdominal desmoids exhibiting diffuse fibrous infiltration have been seen in 15–30% of patients with GS (16b,55,87,106), but in only 6% of those with FCP (62,109). The tumors usually arise 1–3 years following abdominal surgery. Their removal may be followed by further desmoid formation, although desmoids, fibromas, and fibrosarcomas of the skin have been noted without prior surgery (87,88,109). It should be noted that among patients who exhibit both FCP and desmoids, about half will show other stigmata of GS (81,107).

Mesenteric and retroperitoneal fibromatosis has been reported in patients with no stigmata other than FCP (39,81,107). Occasionally, the fibroblastic tumors may precede the discovery of the colonic polyposis (81,104).

**Other neoplasms.** There appears to be a distinct proclivity for those with FCP to develop a variety of neoplasms. The extracolonic neoplasms may be diagnosed a long time prior to the symptoms of intestinal polyposis (7,33,88). Various tumors of the central nervous system (glioma, medulloblastoma) have been reported (3,15,16, 19,23,73,92,109,111,130). This has created confusion with Turcot syndrome, which some have alleged is the same as GS (see Differential Diagnosis). Papillary carcinoma of the thyroid has been frequently associated with FCP and GS (3,15,19,20,23,59,67,86,110,111). Most of these patients are female.

A wide variety of other neoplasms has been noted including adrenal adenoma (27,79,85), adrenal adenocarcinoma (80), hepatocellular carcinoma (132), hepatoblastoma (61,64,77), retroperitoneal leiomyoma (19,41,88), neurofibroma (8), osteosarcoma (46,124), osteochondroma (42), chondrosarcoma (40), lipoma (19,65), fibroma of the breast (42,107), basal cell carcinoma (19,76,83), and a host of other tumors (19,45,104,105,124).

**Eye findings.** Three or more patches of congenital hypertrophic retinal pigment epithelium have been reported in over 85% of patients. The ocular fundal lesions are usually bilateral (78%) (5,9,64, 74,75,113,116,117).

**Miscellaneous findings.** Increased skin pigmentation has been noted (127).

**Oral manifestations.** In 1943, Fitzgerald (33) appears to be the first to have described multiple odontomas in the syndrome. In 1962, Fader et al (31) added supernumerary teeth (Fig. 12–15A,B). These findings have been supported by many other investigators (17, 55,99,108).

In 97 patients with FCP, 17% had dental abnormalities; 11% had supernumerary teeth and/or osteomas, and 9% had impacted permanent teeth (112). Utsunomiya and Nakamura (121) found impacted teeth in about 35% of patients with intestinal polyposis. Söndergaard et al (112) noted supernumerary teeth in 10% and impacted teeth in 9%. Ida et al (51), in a study of 52 patients with polyposis coli, reported embedded teeth (27%), supernumerary teeth (21%), and compound odontomas (11%). One or more osteomas of the jaws were found in over 80%. Occult radiopaque lesions of the jaws are common (16a,89,91,128).

Various incidental findings have included hypercementosis (133), root resorption (82), ankylosis (4), and persistent primary teeth (1).

**Differential diagnosis.** Multiple polyps of the intestine have been described in a number of disorders (Table 12–3). The reader is referred to several excellent reviews (13,26,28,30,43,103,126).

Fig. 12–15. *Gardner syndrome.* (A,B) Panorex of jaws showing osteomas and odontomas. (A from *J Wolf et al,* Br J Oral Maxillofac Surg **24**:410, 1986. B From *F Sitzman* and *H Bruning,* Dtsch Zahnärztl Z **32:**781, 1977.

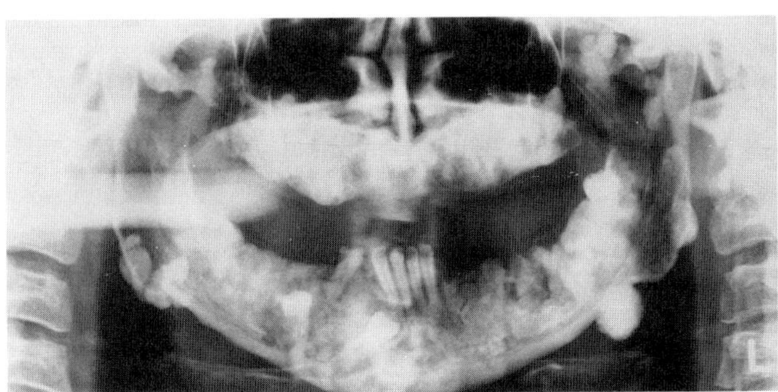

A

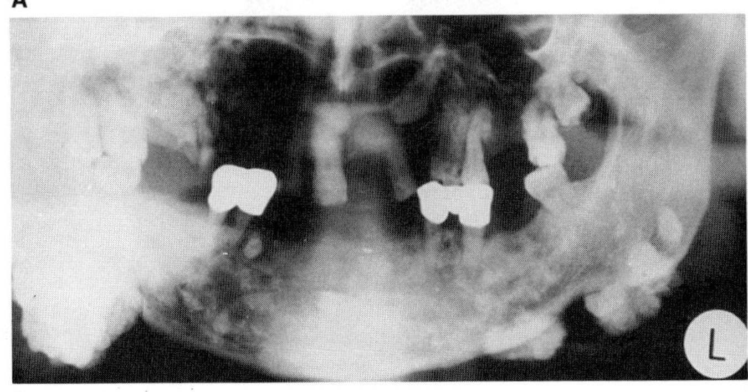


B

Table 12–3. Syndromes with polyposis

| Syndrome | Characteristics | References or page numbers in text |
|---|---|---|
| Cowden syndrome | Facial papules, acral keratoses, oral papillomas, thyroid adenoma, fibrocystic disease of the breast, menstrual irregularities, gastrointestinal polyposis, ductal adenocarcinoma of the breast, autosomal dominant | See pages 357–361 |
| Familial colorectal polyposis (FCP)/Gardner syndrome (GS) | FCP and GS represent a spectrum; FCP-GS gene localized to long arm of chromosome 5; autosomal dominant; osteomas of frontal bone, maxilla, and mandible; epidermoid cysts; intestinal polyposis with high predisposition to malignancy, desmoids, odontomas, supernumerary teeth | 10,26,35,51,54,55,72, 81,86,91,94,104, 105,125,128 |
| Juvenile polyposis of the colon | Hamartomatous polyps (nonprecancerous), autosomal dominant | 114 |
| Peutz–Jeghers syndrome | Gastrointestinal polyposis, macular pigmentation of face and lips, predisposition to malignancies (gastrointestinal, nongastrointestinal, ovarian), autosomal dominant | See pages 399–403 |
| Turcot syndrome[a] | Multiple intestinal polyposis; glioblastoma, medulloblastoma, or astrocytoma; suggested autosomal recessive inheritance | 6,53,58,83,115,118 |
| Cronkhite–Canada syndrome | Generalized gastrointestinal polyposis, edema, malabsorption, protein-losing enteropathy, generalized alopecia, nail dystrophy, middle-to-old age, sporadic occurrence | 25a,54a,57,60 |
| Infantile Cronkhite–Canada syndrome | Macrocephaly, gastrointestinal polyposis, hypotonia, hepatosplenomegaly, anemia, protein-losing enteropathy, alopecia, nail dystrophy, clubbing of fingers and toes, sporadic occurrence | 68,101,102 |
| Torre–Muir syndrome | Multiple sebaceous tumors; keratoacanthomas; adenocarcinomas of colon, endometrium, and ovary; sporadic occurrence | 32,49,129 |
| Perifollicular fibromas and intestinal polyposis | Perifollicular fibromas; skin tags of the face, neck, and trunk; adenomatous colorectal polyposis; autosomal dominant | 48,66,99a |
| Bannayan–Riley–Ruvalcaba syndrome | Macrocephaly, intestinal polyposis of distal ileum and colon, pigmentary spotting of the penis, hypotonia, mild-to-moderate mental deficiency, infantile overgrowth | See pages 336–338 |

[a]We have considerable skepticism re the existence of this as a separate syndrome.

**Juvenile polyposis of the colon.** Juvenile polyps of the colon may have autosomal dominant inheritance (114). The polyps are hamartomatous and not precancerous. They are composed of an excess of lamina propria in which epithelium-lined tubules are embedded with or without cystic dilatation and secondary inflammation.

**Turcot syndrome.** In 1959, Turcot et al (118) reported sibs with multiple intestinal polyposis. One had glioblastoma, the other medulloblastoma. Several authors (6,58,83,115) reported sibs with glioblastoma multiforme and multiple polyposis. Astrocytoma has also been reported (53,115). Multiple café-au-lait spots and axillary freckling, similar to that seen in neurofibromatosis, may be seen. In two of the kindred there was parental consanguinity. Autosomal recessive inheritance has been suggested. However, there have been both numerous examples of nonfamilial occurrence of brain tumors with colonic polyposis and several examples of brain tumors in FCP (vide supra). These have been reviewed elsewhere (53,63,63a,96,100,115). Colonic polyps in Turcot syndrome are presumably smaller in number, larger in size, and earlier to undergo malignant degeneration than in FCP. We doubt the separate existence of Turcot syndrome.

**Peutz–Jeghers syndrome.** This autosomal dominant syndrome, characterized by generalized gastrointestinal hamartomatous polyps and macular pigmentation of the face, lips, and oral mucosa, is described in detail later in this chapter. The typical polyp contains muscularis mucosa, and the epithelial element is related to the smooth muscle in the same manner as in normal mucous membrane.

**Cronkhite–Canada syndrome.** Generalized gastrointestinal polyposis in middle-aged to elderly individuals may be associated with edema, malabsorption, diarrhea, protein-losing enteropathy, generalized alopecia, and nail dystrophy. Brownish skin pigmentation may be diffuse over the face, neck, and hands, including palmar creases (25a,54a,57,60). The disorder is not hereditary. Most cases have been reported from Japan.

**Torre–Muir syndrome.** This syndrome, apparently nongenetic, includes multiple sebaceous neoplasms (generally benign), keratoacanthomas, and adenocarcinomas, most often of the colon, endometrium, and ovary (32,49,129).

**Perifollicular fibromas and intestinal polyposis.** Hornstein (48) reported autosomal dominant perifollicular fibromas and skin tags of the face, neck, and trunk associated with adenomatous colorectal polyposis as an entity separate from Gardner syndrome. The association of skin tags and intestinal polyps has been discussed by several authors (2,18,66,99a). Ishii et al (52) reported sibs with congenital atrichia, pigmented and papular cutaneous lesions in the 20s, and gastrointestinal polyposis in the 30s.

**Other multiple intestinal syndromes.** Polyps of the large intestine have been seen in *Cowden syndrome*. There are several inherited examples of associated polyposis of stomach and colon, familial polyposis of the entire gastrointestinal tract, and several cases of solitary polyps of the colon and rectum apparently inherited as an autosomal dominant trait (47). However, it is possible that these are merely examples of the variability of familial colonic polyposis (13,26,28, 103,126).

*Bannayan–Riley–Ruvalcaba syndrome* consists of macrocephaly, intestinal polyposis, and pigmentary spotting of the genitalia; inheritance is autosomal dominant. A condition that, for lack of a better term, we shall call *infantile Cronkhite–Canada syndrome* consists of diffuse gastrointestinal polyposis, macrocephaly, alopecia, nail dystrophy, clubbing of fingers and toes, hypotonia, hepatosplenomegaly, anemia, and protein-losing enteropathy (68,101,102). Some examples undoubtedly represent Bannayan–Riley–Ruvalcaba syndrome.

**Diagnosis.** The presence of multiple epidermoid inclusion cysts, desmoids, or bony growths, especially of the facial skeleton, should lead to a complete search for the intestinal component. Recording of these incidental findings should be just as mandatory as radiographic search for additional intestinal polyps if a single rectal polyp is detected. The discovery of multiple polyps, or of any other component, also places the onus of responsibility upon the investigator to search other relatives thoroughly for stigmata. Since a negative report does not mean that polyps or other components will not appear in future years, periodic reexamination of all persons with a parent or sib who had one or more signs is necessary.

**Laboratory aids.** Radiographic survey, especially of the facial skeleton, and barium studies of the large and small intestines are mandatory. Ornithine decarboxylase has been reported elevated in patients with familial polyposis (78).

### References (Gardner syndrome)


1. Akuamoa-Boateng E et al: Klinik und Diagnostik des Gardner-Syndroms. *Dtsch Zahnärztl Z* **37**:367–370, 1982.
2. Alcalay J et al: Skin tags and colonic polyps. *J Am Acad Dermatol* **16**:402–403, 1987.
3. Alm T, Licznerski G: The intestinal polyposes. *Clin Gastroenterol* **2**:577–602, 1973.
4. Amato AE, Small EW: Oral manifestations of Gardner's syndrome: Report of a case. *J Oral Surg* **28**:458–460, 1970.
5. Baker RH et al: Hyperpigmented lesions of the retinal pigment epithelium in familial adenomatous polyposis. *Am J Med Genet* **31**:427–435, 1988.
6. Baughman FA et al: The glioma-polyposis syndrome. *N Engl J Med* **281**:1345–1346, 1969.
7. Bessler W et al: Case report 253. Gardner syndrome with aggressive fibromatosis. *Skel Radiol* **11**:56–59, 1984.
8. Bochetto JF et al: Multiple polyposis, exostoses and soft tissue tumors. *Surg Gynecol Obstet* **117**:489–494, 1963.
9. Blair NP, Trempe CL: Hypertrophy of the retinal pigment epithelium associated with Gardner's syndrome. *Am J Ophthalmol* **90**:661–667, 1980.
10. Bodmer WF et al: Localization of the gene for familial adenomatous polyposis on chromosome 5. *Nature (London)* **328**:614–616, 1987.
11. Bülow S et al: Mandibular osteomas in familial polyposis coli. *Dis Colon Rect* **27**:105–108, 1984.
12. Burt RW et al: Upper gastrointestinal polyps in Gardner's syndrome. *Gastroenterology* **86**:295–301, 1984.
13. Bussey HJR et al: Genetics of gastrointestinal polyposis. *Gastroenterology* **74**:1325–1330, 1978.
14. Cabot RC (ed): Case records of the Massachusetts General Hospital. Case 21061. *N Engl J Med* **212**:263–267, 1935.
15. Camiel MR et al: Association of thyroid carcinoma with Gardner's syndrome in siblings. *N Engl J Med* **278**:1056–1058, 1968.
16. Capps WF et al: Carcinoma of the colon, ampulla of Vater and urinary bladder with familial multiple polyposis. *Dis Colon Rect* **10**:298–305, 1968.

16a. Carl W, Sullivan MA: Dental abnormalities and bone lesions associated with familial adenomatous polyposis. *JADA* **119**:137–139, 1989.

16b. Carr LY et al: Gardner's syndrome and mesenteric desmoids. *Am J Gastroenterol* **80**:310–312, 1985.

17. Chang CH et al: Bone abnormalities in Gardner's syndrome. *Am J Roentgenol* **102**:645–652, 1968.
18. Chobanian SJ: Skin tags and colonic polyps: A gastroenterologist's perspective. *J Am Acad Derm* **16**:407–409, 1987.
19. Cohen SB: Familial polyposis coli and its extracolonic manifestations. *J Med Genet* **19**:193–203, 1982.
20. Cole CW, Meban S: Familial polyposis coli: An unusual family. *Can J Surg* **26**:374–376, 1983.
21. Coli RD et al: Gardner's syndrome: A revisit to a previously described family. *Am J Dig Dis* **15**:551–568, 1970.
22. Cooper PH, Fechner RE: Pilomatricoma-like changes in the epidermal cysts of Gardner's syndrome. *J Am Acad Dermatol* **8**:639–644, 1983.
23. Crail HW: Multiple primary malignancies arising in the rectum, brain and thyroid. *US Naval Med Bull* **49**:123–128, 1949.
24. Danes BS: The Gardner syndrome: Increased tetraploidy in cultured skin fibroblasts. *J Med Genet* **13**:52–56, 1976.
25. Danes BS: Increased *in vitro* tetraploidy: Tissue specific within the heritable colorectal cancer syndromes with polyposis coli. *Cancer* **41**:2330–2334, 1978.

25a. Daniel ES et al: The Cronkhite-Canada syndrome. An analysis of clinical and pathologic features and therapy. *Medicine (Baltimore)* **61**:293–309, 1982.

26. DeCosse JT et al: Familial polyposis. *Cancer* **39**:267–273, 1977.
27. Devic A, Bussy MM: Un cas de polypose adenomateuse generalisée à tout l'intestin. *Arch Mal Appar Dig* **6**:278–289, 1912.
28. Dodd GD: Genetics and cancer of the gastrointestinal system. *Radiology* **123**:263–275, 1977.
29. Duncan BR et al: The Gardner syndrome: Need for early diagnosis. *J Pediatr* **72**:479–505, 1968.
30. Erbe RW: Inherited gastrointestinal polyposis syndromes. *N Engl J Med* **294**:1101–1104, 1976.
31. Fader M et al: Gardner's syndrome (intestinal polyposis, osteomas, sebaceous cysts) and a new dental discovery. *Oral Surg* **15**:153–172, 1962.
32. Fahmy A et al: Muir-Torre syndrome. *Cancer* **49**:1898–1903, 1982.
33. Fitzgerald GM: Multiple composite odontomas coincidental with other tumorous conditions: Report of a case. *J Am Dent Assoc* **30**:1408–1417, 1943.
34. Frangenheim P: Familiäre Hyperostosen der Kiefer. *Bruns Beitr Klin Chir* **90**:139–151, 1914.
35. Gardner EJ: Follow-up study of a family group exhibiting dominant inheritance for a syndrome including intestinal polyps, osteomas, fibromas and epidermal cysts. *Am J Hum Genet* **14**:376–390, 1962.
36. Gardner EJ: Discovery of the Gardner syndrome. *Birth Defects* **8**(2):48–51, 1972.
37. Gardner EJ et al: Numerical and structural chromosome aberrations in cultured lymphocytes and cutaneous fibroblasts of patients with multiple adenomas of the colorectum. *Cancer* **49**:1413–1419, 1982.
38. Gorlin RJ: Gardner's syndrome without polyposis (letter to the editor). *Humangenetik* **6**:380, 1968.
39. Gorlin RJ, Chaudhry AP: Multiple osteomatosis, fibromas, lipomas and fibrosarcomas of the skin and mesentery, epidermoid inclusion cysts of the skin, leiomyomas and multiple intestinal polyposis. *N Engl J Med* **263**:1151–1158, 1960.
40. Greer JA et al: Gardner's syndrome and chondrosarcoma of the hyoid bone. *Arch Otolaryngol* **103**:425–427, 1977.
41. Gumpel RC, Carballo JD: New concept of familial adenomatosis. *Ann Intern Med* **45**:1045–1058, 1956.
42. Haggitt RC, Booth JL: Bilateral fibromatosis of the breast in Gardner's syndrome. *Cancer* **25**:161–166, 1970.
43. Haggitt RC, Reid BJ: Hereditary gastrointestinal polyposis syndromes. *Am J Surg Pathol* **10**:871–887, 1986.
44. Hamilton SR, Bussey HJR: Ileal adenomas after colectomy in nine patients with adenomatous polyposis coli/Gardner's syndrome. *Gastroenterology* **77**:1252–1257, 1979.
45. Helson L: Pre-Gardner's syndrome thyroglossal cysts and undifferentiated tumor of neural crest origin. *Anticancer Res* **4**:247–250, 1984.
46. Hoffmann DC, Brooke BN: Familial sarcoma of bone in a polyposis coli family. *Dis Colon Rectum* **13**:119–120, 1970.
47. Hoffmann DC, Goligher JC: Polyposis of the stomach and small intestine in association with familial polyposis coli. *Br J Surg* **58**:126–128, 1971.
48. Hornstein OP: Generalized dermal perifollicular fibromas wth polyps of the colon. *Hum Genet* **33**:193–197, 1976.
49. Householder MS, Zeligman I: Sebaceous neoplasms associated with visceral carcinomas. *Arch Dermatol* **116**:61–64, 1980.
50. Hsu SH et al: Multiclonal origin of polyps in Gardner syndrome. *Science* **221**:951–953, 1983.
51. Ida M et al: Osteomatous changes and tooth abnormalities found in the jaws of patients with adenomatosis coli. *Oral Surg* **52**:2–11, 1981.
52. Ishii Y et al: Atrichia with papular lesions associated with gastrointestinal polyposis. *J Dermatol* **6**:111–116, 1979.
53. Itoh H et al: Turcot's syndrome and its mode of inheritance. *Gut* **20**:414–419, 1979.
54. Jarvinen HJ et al: Gardner's stigmata in patients with familial polyposis coli. *Br J Surg* **69**:718–721, 1982.

54a. Jenkins D et al: The Cronkhite–Canada syndrome. An ultrastructural study of pathogenesis. *J Clin Pathol* **38**:271–276, 1985.

55. Jones EL, Cornell WP: Gardner's syndrome. Review of the literature and report of a family. *Arch Surg* **92**:287–300, 1966.

56. Jones TR, Nance FC: Periampullary malignancy in Gardner's syndrome. *Ann Surg* **185**:565–571, 1977.

57. Katayama Y et al: Cronkhite-Canada syndrome associated with a rectal cancer and adenomatous changes in colonic polyps. *Am J Surg Pathol* **9**:65–71, 1985.

58. Kawanami K et al: Turcot's syndrome, report of an autopsy case. *Stomach Intestine* **11**:1075–1082, 1976.

59. Keshgegian AA, Enterline HT: Gardner's syndrome with duodenal adenomas, gastric adenomyoma and thyroid papillary-follicular adenocarcinoma. *Dis Colon Rectum* **21**:255–260, 1978.

60. Kinderblom LG et al: Cronkhite-Canada syndrome. *Cancer* **39**:2667–2673, 1977.

61. Kingston JE et al: Association between hepatoblastoma and polyposis coli. *Arch Dis Childh* **48**:959–962, 1983.

62. Klemmer S et al: Occurrence of desmoids in patients with familial adenomatous polyposis of the colon. *Am J Med Genet* **28**:385–392, 1987.

63. Krakowicz P: The Turcot syndrome. *Acta Chir Scand* **145**:113–115, 1979.

63a. Kropilak M et al: Brain tumors in familial adenomatous polyposis. *Dis Colon Rectum* **32**:778–782, 1989.

64. Krush AJ et al: Hepatoblastoma, pigmented ocular fundus lesions and jaw lesions in Gardner's syndrome. *Am J Med Genet* **29**:323–332, 1988.

65. Laberge MY et al: Soft tissue tumors associated with familial polyposis: Report of case. *Mayo Clin Proc* **32**:749–752, 1957.

66. Leavitt J et al: Skin tags: A cutaneous marker for colonic polyps. *Ann Intern Med* **98**:928–930, 1983.

67. Lee FI, MacKinnon MD: Papillary thyroid carcinoma associated with polyposis coli. A case of Gardner's syndrome. *Am J Gastroenterol* **76**:138–140, 1981.

68. Le Luyer B et al: Generalized juvenile polyposis in an infant: Report of case and successful management by endoscopy. *J Pediatr Gastroenterol Nutr* **4**:128 134, 1985.

69. Leppard B: Epidermoid cysts and polyposis coli. *Proc R Soc Med* **67**:1036–1037, 1974.

70. Leppard B, Bussey HJR: Epidermoid cysts, polyposis coli, and Gardner's syndrome. *Br J Surg* **62**:387–393, 1975.

71. Leppard B, Bussey HJR: Gardner's syndrome with epidermoid cysts showing features of pilomatrixomas. *Clin Exp Dermatol* **1**:75–82, 1976.

72. Leppert M et al: The gene for familial polyposis coli maps to the long arm of chromosome 5. *Science* **238**:1411–1413, 1987.

73. Lewis JH et al: Turcot's syndrome. Evidence for autosomal dominant inheritance. *Cancer* **51**:524–528, 1983.

74. Lewis RA, Strong LC: Congenital hypertrophy of the retinal pigment epithelium: A marker for Gardner syndrome and a clue to Turcot syndrome. 9th Annual David W. Smith Workshop for Malformations and Morphogenesis, Oakland, California, August 3–7, 1988.

75. Lewis RA et al: The Gardner syndrome. Significance of ocular features. *Ophthalmology* **91**:916–925, 1984.

76. Lewis RJ, Mitchell JC: Basal cell carcinoma in Gardner's syndrome. *Acta Dermatovenereol* **51**:67–68, 1971.

77. Li FP et al: Hepatoblastoma in families with polyposis coli. *JAMA* **257**:2475–2477, 1987.

78. Luk GD, Gaylin SB: Ornithine decarboxylase as a biologic marker in familial polyposis. *N Engl J Med* **311**:80–83, 1984.

79. MacDonald JM et al: Gardner's syndrome and periampullary malignancy. *Am J Surg* **113**:425–430, 1967.

80. Marshall WH et al: Gardner's syndrome with adrenal carcinoma. *Australas Ann Med* **16**:242–244, 1967.

81. McAdam WAF, Goligher JC: The occurrence of desmoids in patients with familial polyposis coli. *Br J Surg* **57**:618–631, 1970.

82. McFarland PH et al: Gardner's syndrome: Report of two families. *J Oral Surg* **26**:632–638, 1968.

83. Michels VV, Stevens JC: Basal cell carcinoma in a patient with intestinal polyposis. *Clin Genet* **22**:80–82, 1982.

84. Naylor EW, Gardner EJ: Penetrance and expressivity of the gene responsible for the Gardner syndrome. *Clin Genet* **11**:381–393, 1977.

85. Naylor EW, Gardner EJ: Adrenal adenomas in a patient with Gardner's syndrome. *Clin Genet* **20**:67–73, 1981.

86. Naylor EW, Lebenthal E: Gardner's syndrome: Recent developments in research and management. *Dig Dis Sci* **25**:945–959, 1980.

87. Naylor EW et al: Desmoid tumors and mesenteric fibromatosis in Gardner's syndrome. *Arch Surg* **114**:1181–1185, 1979.

88. O'Brien JP, Wells P: The synchronous occurrence of benign fibrous tissue neoplasia in hereditary adenosis of the colon and rectum. *NY State J Med* **55**:1877–1880, 1955.

89. Offerhaus GJA et al: Occult radiopaque for lesions in familial adenomatous polyposis coli and hereditary non-polyposis colorectal cancer. *Gastroenterology* **93**:490–497, 1987.

90. Ooya K et al: Sclerotic masses in the mandible of the patient with familial polyposis of the colon. *J Oral Pathol* **5**:305–311, 1976.

91. Palmer TH: Gardner's syndrome: Six generations. *Am J Surg* **143**:405–408, 1982.

92. Parks TG et al: Familial polyposis coli associated with extracolonic abnormalities. *Gut* **11**:323–329, 1970.

93. Pauli RM et al: Gardner syndrome and periampullary malignancy. *Am J Med Genet* **6**:205–219, 1980.

94. Peters PE et al: Röntgendiagnostik und Klinik des Gardner-Syndroms. *Roefo* **136**:133–137, 1982.

95. Pierce E et al: Gardner's syndrome: Formal genetics and statistical analysis of a large Canadian kindred. *Clin Genet* **1**:65–80, 1970.

96. Radin DR et al: Turcot syndrome: A case with spinal cord and colonic neoplasm. *Am J Roentgenol* **142**:475–476, 1984.

97. Rasheed S et al: Inherited susceptibility to retrovirus-induced transformation of Gardner syndrome cells. *Am J Hum Genet* **35**:919–931, 1983.

98. Rayne J: Gardner's syndrome. *Br J Oral Surg* **6**:11–17, 1968.

99. Redding SW et al: Gardner's syndrome: Report of case. *J Oral Surg* **39**:50–52, 1981.

99a. Rongioletti F et al: Fibrofolliculomas, trichodiscomas, and acrochordrons (Birt-Hogg-Dubé) associated with intestinal polyposis. *Clin Exp Dermatol* **14**:72–74, 1989.

100. Rothman D, Kendall AB: Dilemma in a case of Turcot's (glioma-polyposis) syndrome: Report of a case. *Dis Colon Rectum* **18**:514–515, 1975.

101. Ruymann FB: Juvenile polyps with cachexia. *Gastroenterology* **47**:431–438, 1969.

102. Sachatello CR et al: Juvenile gastrointestinal polyposis in a female infant: Report of a case and review of the literature of a recently recognized syndrome. *Surgery* **75**:107–114, 1974.

103. Schimke RN: Genetic syndromes with gastrointestinal cancer, in *The Genetics and Heterogeneity of Common Gastrointestinal Disorders,* Rotter JI et al (eds), Academic Press, New York, 1980.

104. Schuchardt WA Jr, Ponsky JL: Familial polyposis and Gardner's syndrome. *Surg Gynecol Obstet* **148**:97–103, 1979.

105. Sener SF et al: The spectrum of polyposis. *Surg Gynecol Obstet* **159**:525–532, 1984.

106. Shiffman MA: Familial multiple polyposis associated with soft-tissue and hard-tissue tumors. *JAMA* **179**:514–522, 1962.

107. Simpson RD et al: Mesenteric fibromatosis in familial polyposis: A variant of Gardner's syndrome. *Cancer* **17**:526–534, 1964.

108. Singer R: Ein Beitrag zum Gardner-Syndrom. *Dtsch Zahn Mund Kieferheilk* **62**:18–31, 1974.

109. Smith WG: Multiple polyposis, Gardner's syndrome and desmoid tumors. *Dis Colon Rectum* **1**:323–332, 1958.

110. Smith WG: Familial multiple polyposis: Research tool for investigating the etiology of carcinoma of the colon. *Dis Colon Rectum* **11**:17–31, 1968.

111. Smith WG, Kern BB: The nature of the mutation in familial multiple polyposis. Papillary carcinoma of the thyroid, brain tumors and familial multiple polyposis. *Dis Colon Rectum* **16**:264–271, 1973.

111a. Solomon E et al: Chromosome 5 allele loss in human colorectal carcinomas. *Nature* (London) **328**:616–619, 1987.

112. Søndergaard JO et al: Dental anomalies in familial adenomatous polyposis coli. *Acta Odontol Scand* **45**:61–63, 1987.

113. Stein EA, Brady KD: Ophthalmologic and electro-oculographic findings in Gardner's syndrome. *Am J Ophthalmol* **106**:326–331, 1988.

114. Stemper TJ et al: Juvenile polyposis and gastrointestinal carcinoma: A study of a kindred. *Ann Intern Med* **83**:639–646, 1975.

115. Todd DW et al: A family affected with intestinal polyposis and gliomas. *Ann Neurol* **10**:390–392, 1981.

116. Traboulsi EI et al: Prevalence and importance of pigmented ocular fundus lesions in Gardner's syndrome. *N Engl J Med* **316**:661–667, 1987.

117. Traboulsi EI et al: Pigmented ocular fundus lesions in the inherited gastrointestinal polyposis syndromes and in hereditary nonpolyposis colorectal cancer. *Ophthalmology* **95**:964–969, 1988.

118. Turcot J et al: Malignant tumors of the central nervous system associated with familial polyposis of the colon: Report of two cases. *Dis Colon Rectum* **2**:465–468, 1959.

119. Ushio K et al: Lesions associated with familial polyposis coli: Studies of lesions of the stomach, duodenum, bones and teeth. *Gastrointest Radiol* **1**:67–80, 1976.

120. Utsunomiya J et al: Gastric lesion of familial polyposis coli. *Cancer* **34**:745–754, 1974.

121. Utsunomiya J, Nakamura T: The occult osteomatous changes in the mandible in patients with familial polyposis coli. *Br J Surg* **62**:45–51, 1975.

122. Watanabe H et al: Gastric lesions in familial adenomatosis coli. Their incidence and histologic analysis. *Hum Pathol* **9**:269–283, 1978.
123. Watne AL: Syndromes of polyposis coli and cancer. *Curr Prob Cancer* **7**:3–31, 1982.
124. Watne AL et al: Gardner's syndrome. *Surg Gynecol Obstet* **141**:53–56, 1975.
125. Weary PE: Gardner's syndrome. *Arch Dermatol* **90**:20–30, 1964.
126. Wennstrom J et al: Hereditary benign and malignant lesions of the large bowel. *Cancer* **34**:850–857, 1974.
127. Weston SD, Wiener M: Familial polyposis associated with a new type of soft-tissue lesion (skin pigmentation): Report of three cases and a review of the literature. *Dis Colon Rectum* **10**:311–321, 1967.
128. Wolf J et al: Gardner's dento-maxillary stigmas in patients with familial adenomatous coli. *Br J Oral Maxillofac Surg* **24**:410–416, 1986.
129. Worret WI et al: Torre-Muir-Syndrom. Talgdrüsenneoplasien, Keratoacanthome, multiple interne Karzinome und Vererbung. *Hautarzt* **32**:519–524, 1981.
130. Yaffee HS: Gastric polyposis and soft tissue tumors. (A variant of Gardner's syndrome). *Arch Dermatol* **89**:806–808, 1964.
131. Yao T et al: Duodenal lesions in familial polyposis of the colon. *Gastroenterology* **73**:1086–1092, 1977.
132. Zeze F et al: Hepatocellular carcinoma associated with familial polyposis of the colon. *Dis Colon Rectum* **26**:465–468, 1983.
133. Ziter FMH: Roentgenographic findings in Gardner's syndrome. *JAMA* **192**:1000–1002, 1965.

## Gorlin syndrome (nevoid basal cell carcinoma syndrome)

The syndrome consists principally of nevoid basal cell carcinomas, odontogenic keratocysts, skeletal anomalies, and intracranial calcification (Figs. 12–16 to 12–20). It was independently described by Jarisch (58) and White (121) in 1894, but probably existed during early Egyptian times (104). Early American reports are those of Binkley and Johnson (6), Howell and Caro (54), and Gorlin and Goltz (35). There have been several good historic reviews (52,53). Excellent systematic surveys have been provided by Gorlin and his co-workers (34–39). The reader is encouraged to review the 1987 publication of Gorlin (34).

The syndrome has been designated by a variety of different terms including basal cell nevus syndrome, nevoid basal cell carcinoma syndrome, epitheliomatose multiple generalisée (type Ferrari), syndrome of jaw cysts, basal cell tumors, and skeletal anomalies, polycystoma, fünfte Phakomatose, hereditary cutaneomandibular polyoncosis, Gorlin syndrome, and Gorlin–Goltz syndrome (14,17,37).

Although nevoid basal cell carcinoma syndrome is personally preferred (34), not all affected adults have basal cell carcinomas. Furthermore, there is probably a strong ascertainment bias favoring syndrome patients who manifest skin tumors. In addition, a dermatologic designation may not be completely appropriate since many other systems besides skin are involved. On the other hand, the condition is known throughout Europe as Gorlin syndrome (e.g., 12,17,19,41, 47,69,79,80,93,96), because of Gorlin's contributions to the understanding of the condition (34–39).

The disorder is a complex hamartoneoplastic/malformation syndrome with over 100 different signs and symptoms primarily involving the skin, central nervous system, and skeletal system. Major features are listed in Table 12–4. Inheritance is autosomal dominant with complete penetrance and extremely variable expressivity (2,116). Approximately 40% of cases represent new mutations. A paternal age effect has been demonstrated in such instances (60). The syndrome occurs in approximately 1 in 200 patients with basal cell carcinomas. Approximately 500 cases have been reported to date. The syndrome seems to support Knudson's double hit theory, the first hit being the autosomal dominant gene and the second being radiation, actinic or X-ray (34,51).

**Craniofacial features.** A characteristic facies is present in about 70%. This is due in part to increased size of the calvaria (occipitofrontal circumference, 60 cm or more in adults) and in part to frontal and biparietal bulging, well-developed supraorbital ridges that may give the eyes a sunken appearance, heavy and often fused eyebrows, broadened nasal root, low position of the occiput, mild hypertelorism, exotropia, and exaggerated length of the mandible associated with pouting of the lower lip (Fig. 12–16A,B) (36,113). Several patients have been noted to have mildly slanted auricles, although this feature is frequently overlooked (17).

Table 12–4. Features of Gorlin syndrome[a]

| |
|---|
| 50% or greater frequency |
| Enlarged occipitofrontal circumference |
| Mild ocular hypertelorism |
| Multiple basal cell carcinomas |
| Odontogenic keratocysts of jaws |
| Epidermal cysts of skin |
| Palmar and/or plantar pits |
| Calcified ovarian fibromas (probably overestimated frequency) |
| Calcified falx cerebri |
| Rib anomalies (splayed, fused, partially missing, bifid, etc.) |
| Spina bifida occulta of cervical or thoracic vertebrae |
| Calcified diaphragma sellae (bridged sella, fused clinoids) |
| Hyperpneumatization of paranasal sinuses |
| 49 to 15% frequency |
| Calcification of tentorium cerebelli and petroclinoid ligament |
| Short fourth metacarpals |
| Kyphoscoliosis or other vertebral anomalies |
| Lumbarization of sacrum |
| Pectus excavatum or carinatum |
| Pseudocystic lytic lesion of bones (hamartomas) |
| Strabismus (exotropia) |
| 14% or less but not random |
| Medulloblastoma (true frequency not known) |
| Inguinal hernia (?) |
| Meningioma |
| Lymphomesenteric cysts |
| Cardiac fibroma |
| Fetal rhabdomyoma |
| Ovarian fibrosarcoma |
| Marfanoid build |
| Agenesis of corpus callosum |
| Cyst of septum pellucidum |
| Cleft lip and/or palate |
| Polydactyly, postaxial—hands or feet |
| Sprengel deformity of scapula |
| Congenital cataract, glaucoma, coloboma of iris, retina, optic nerve, medullated retinal nerve fibers |
| Subcutaneous calcifications of skin (possibly underestimated frequency) |
| Minor kidney malformations |
| Hypogonadism in males |
| Mental retardation |

[a]From RJ Gorlin, *Medicine* **66**:96, 1987.

A cephalometric study has been carried out by Dahl et al (20). Although "congenital communicating hydrocephaly" has been reported on several occasions, it is difficult to know how many examples represented merely benign macrocephaly, that is increase in head circumference (34).

Congenital blindness due to corneal opacity, congenital or precocious cataract, or glaucoma and/or coloboma of the iris, choroid, and optic nerve, coupled with convergent or divergent strabismus and nystagmus, has been reported in 10–15% (1–2,6,18,29,35,36,41,65,68). Retinitis pigmentosa, falciform folds of the retina with detachment, retinal hamartomas, and medullated retinal nerve fibers have also been noted (34).

Cleft lip and/or palate occur in about 5%. Seventeen cases with this manifestation were tabulated by van Dijk and Neering (118) in 1980, and other examples have been noted as well (11,18,41,65,88,102,107).

**Skin.** Basal cell carcinomas differ in several respects from classical basal cell carcinomas. First, lesions tend to be multiple rather than

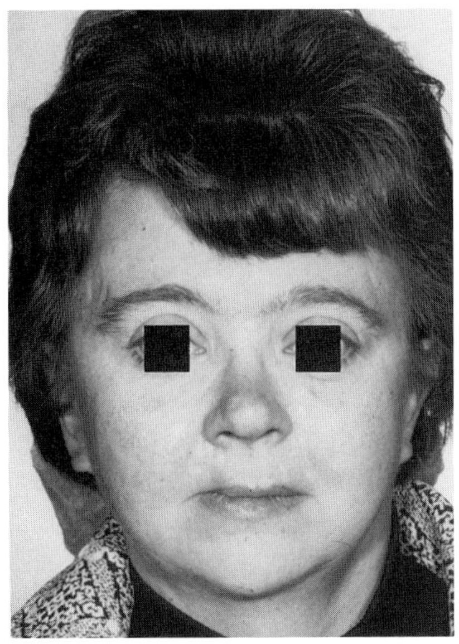
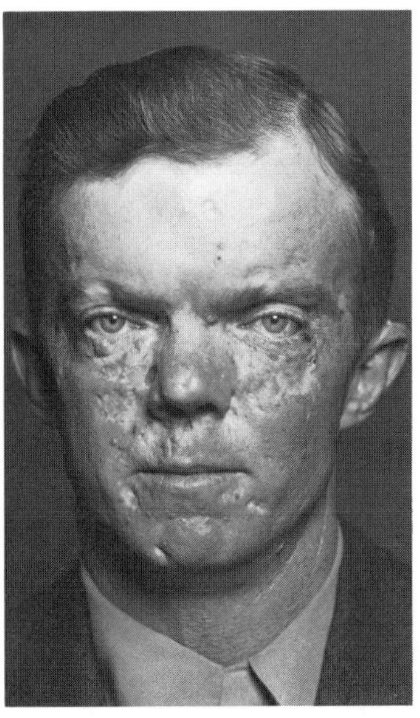
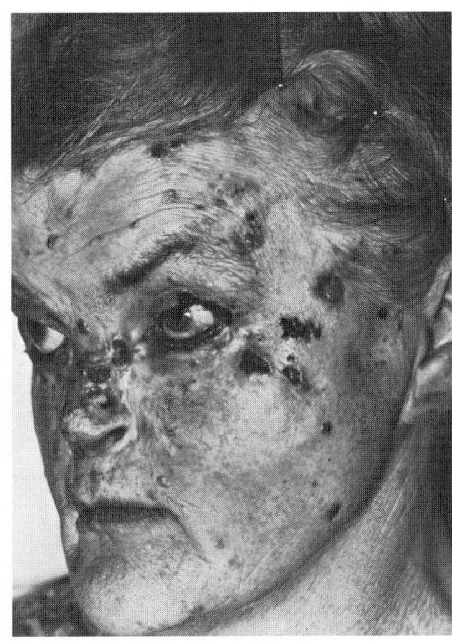

A B C

Fig. 12–16. *Gorlin syndrome.* (A) Typical facies of patient with nevoid basal cell carcinoma syndrome. Note increased head circumference, mild hypertelorism, and several scars from removed basal cell carcinomas. (B) Frontal and temporoparietal bossing, numerous basal cell carcinomas. (C) Extensive basal cell carcinomas. (B from *WD Maddox,* Thesis, University of Minnesota, 1963. C from *U Berendes,* Hautarzt **22**:261, 1971.)

single. Second, they may occur on nonexposed as well as sun-exposed areas of the skin. Third, they may occur at an earlier age. Fourth, lesions are more commonly associated with melanin pigmentation and foci of calcification. Such foci and, rarely, metaplastic bone formation may also occur in normal appearing skin of patients with the syndrome. Fifth, biologic behavior may differ since many lesions are quiescent and few become aggressive (17,34,67,82).

Nevoid basal cell carcinomas appear largely between puberty and 35 years of age, although they have been reported in young children, including the nape of the neck at birth (RJ Gorlin, personal communication, 1987). In only about 15% are they manifest before puberty (96). About 10% of patients over the age of 30 years have none. Conversely, however, while about 2% of patients under 45 years of age have the syndrome, this rises to 22% for those less than 19 years of age (108). Some individuals have only a few; others have literally thousands that arise in any region of the skin, but especially on the face, neck, and upper trunk (1,1a,67). The periorbital areas, eyelids, nose, malar region, and upper lip are the facial sites most often affected. Rarely are the abdomen, lower trunk, and extremities involved. Unilateral involvement has been described (42,106). Small groups may resemble moles, skin tags, ordinary nevus cell nevi, or hemangiomas. They are pearly, flesh colored, or reddish brown; they may be isolated or grouped (Figs. 12–16B,C and 12–17A,C) (2,13,37,39,92). They may grow rapidly for a few days to a few weeks, but most remain static. Prior to puberty, the lesions are harmless even when large numbers are present. Only a few become aggressively malignant, and then only after adolescence when they may be locally invasive, behaving like ordinary basal cell carcinomas. Evidence of aggressive transformation of an individual lesion is heralded by an increase in size, ulceration, bleeding, and crusting. Death has resulted in a few instances from invasion of the brain, lung, or peritoneum (34,111,122). Blacks with the syndrome tend to have fewer skin lesions than white, probably due to protective skin pigmentation (34).

Histologically, nevoid basal cell carcinomas cannot be differentiated from ordinary basal cell carcinomas. They are composed of nests, islands, or sheets of cells with large deeply staining nuclei with indistinct cell membranes and variable number of mitotic figures (Fig. 12–17D). In view of the pluripotentiality of the basal cells, a full spectrum of basal cell carcinoma may develop in a patient, including superficial, multicentric, solid, cystic, adenoid, and lattice-like. About one-third of the patients have two or more types of basal cell carcinoma patterns (67,71,73).

Milia (small keratin-filled cysts) are found intermixed with the basal cell cancers on the face in about 30%. Larger, often multiple, epidermoid cysts (1–2 cm) occur on the limbs and trunk in over 50% (4,57,66). Occasionally they occur on the palms (86). Several authors have described associated chalazion and comedones (2,36,37).

Asymmetric, small (1–2 mm) palmar and/or plantar pits are present in about 65% (Fig. 12–17E,F). They rarely occur on the sides or dorsa of the fingers or toes and are more common on the hands than on the feet. They may be age related since they are rarely noted in children. The pits are more obvious in patients who perform manual labor. Basal cell carcinomas have arisen in the base of these pits (46,55,56,107,114). Light microscopy of the pits show focal absence of the stratum corneum, thinning of the stratum granulosum, vacuolization of the spinous layer, and irregular rete ridges. Ultrastructurally, there are poorly developed tonofibrils, small keratohyaline granules, decreased desmosomal attachments, an increase in discharged cementosomes, and premature desquamation of horny cells (34,36,46,117).

**Musculoskeletal and radiographic features.** Patients may be very tall. Some exhibit a Marfanoid habitus. Radiographic findings (Fig. 12–18A–E) include a large calvaria with parietal and biparietal bossing, low occiput, and mildly increased interorbital distance. The frontal sinuses are enlarged in 60%. Platybasia is relatively frequent. Bridging of the sella turcica (calcification of the diaphragma sellae) is seen in at least 60–80%, a finding noted in only about 4% of the normal population (26,79). Lamellar calcification of the falx cerebri, which appears relatively early in life, is seen in at least 85% (normal is 5%). There can also be calcification of the tentorium cerebelli (40%), petroclinoid ligament (20%), dura, pia, and choroid plexus (26,85).

About 60% have anteriorly splayed, fused, partially missing, hypoplastic, or bifid ribs. Kyphoscoliosis with or without associated pectus excavatum or carinatum is present to some degree in about 30–40% and spina bifida occulta of the cervical or thoracic vertebrae is found in 60%. Cervical or upper thoracic vertebral fusion or lack of segmentation has been documented in about 40% (69). Lumbarization of the sacrum occurs in about 40%. Sprengel deformity is found in 5–10% and, in some surveys, as high as 25% (96). Medial hooking or dysplasia of the lower scapular borders has been noted in several patients (22,26,75).

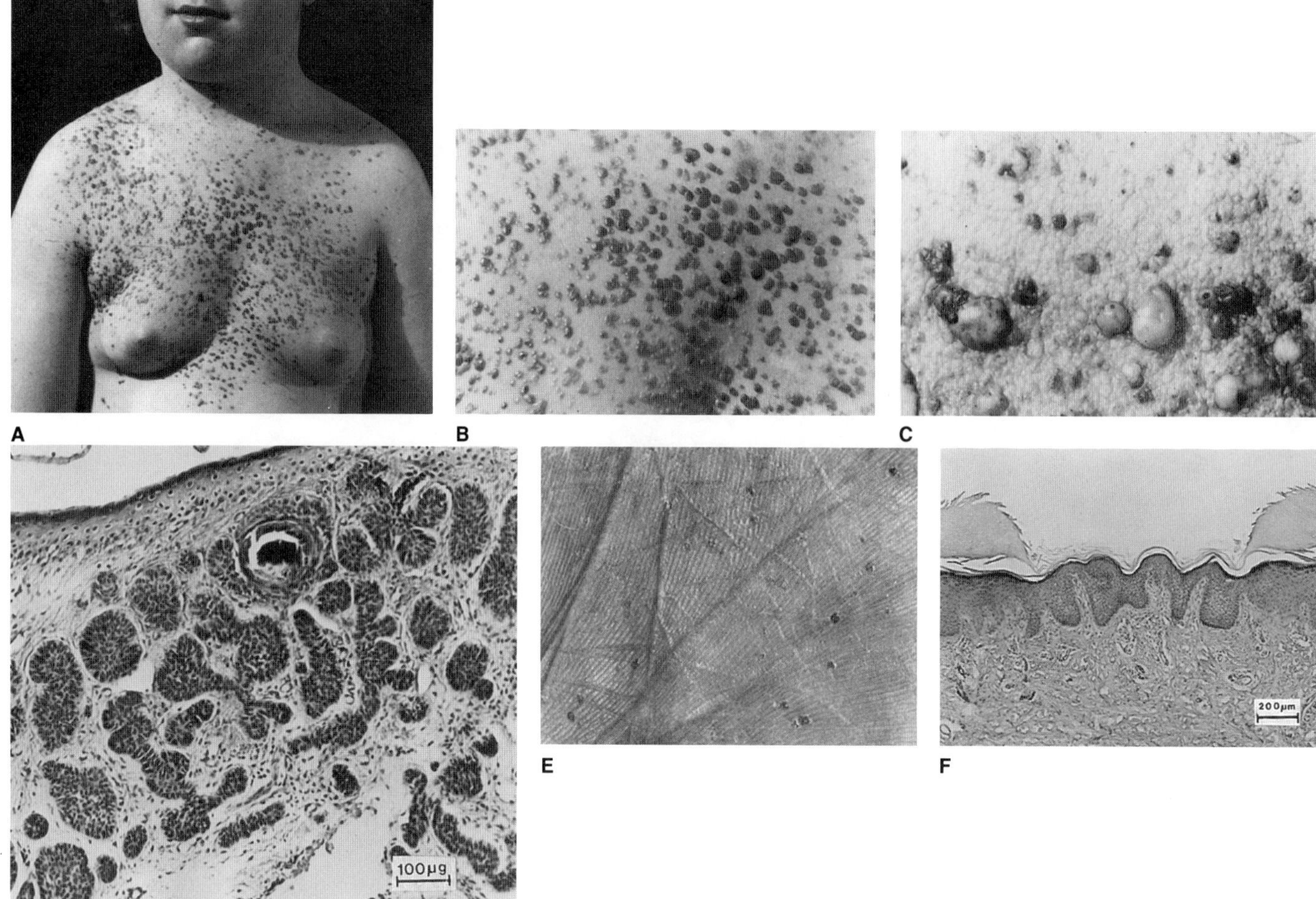


Fig. 12–17. *Gorlin syndrome.* (A) Multiple nevoid basal cell carcinomas scattered over chest. (B) Close-up of lesions shown in A. (C) Multiple basal cell carcinomas. Note differences in size and pigmentation. (D) Photomicrograph of nevoid basal cell carcinoma. Note ectopic calcification (100×, H & E). (E) Palmar pits. Pits are usually a few millimeters deep, more evident in those performing manual labor; relatively rare in children. (F) Photomicrograph of palmar pit. Note mild acantholysis of epithelium at bottom of pit. (A from *JB Howell* and *MR Caro,* Arch Dermatol **79**:67, 1959. B,F courtesy of *JB Howell,* Dallas, Texas. C courtesy of *RA Cawson,* London, England.)

Small pseudocystic lytic bone lesions are noted in at least 35%, most often in the phalanges and metapodial bones. However, the long bones, pelvis, and calvaria may also be affected (7,26,87). These bone radiolucencies represent hamartomas composed of fibrous connective tissue, nerves, and blood vessels (78). Probably some cases of ''metastatic medulloblastoma'' really represent these hamartomatous changes (47). Spotted sclerotic osteopoikolytic lesions have also been documented (7). Subcutaneous calcification of fingers and scalp has been documented by several authors (36).

Ovarian fibromas, which are bilateral and calcified, may be present in 50%. Radiographically, they may overlap as a single calcified mass. If one finds bilateral calcified pelvic masses in a prepubertal female, the syndrome should be considered. One or more enlarged chylous or lymphatic cysts of the mesentery may be calcified (34).

The fourth metacarpal is short in perhaps 20% but this ''metacarpal sign'' is of poor diagnostic help since several studies have shown that about 10% of the normal population have one or both short fourth metacarpal bones. Occasional findings include pes planus, defective medial portion of the clavicle, pre- or postaxial polydactyly, hallux valgus, and syndactyly of the second and third fingers (34,36,57,107).

**Odontogenic keratocysts.** Cysts of the jaws, aptly named odontogenic keratocysts (Fig. 12–19A–C), develop during the first decade of life (usually after the seventh year) to peak during the second or third decades (74,96). This is approximately a decade earlier than the much more common, isolated odontogenic keratocysts, not associated with the syndrome. About 15% of patients do not have radiographically demonstrable cysts by age 40. There are reports of individuals whose first jaw cysts occurred in the sixth decade (96). The cysts are continuous in their development. They are often very large before they effect expansion of the jaws. In spite of widespread extension throughout the jaws, they almost never cause symptoms unless secondarily infected following surgery. Most often they are detected on routine dental checkups. Rarely do they cause pathologic fracture (66,107). However, they may perforate the cortex and extend to soft tissues where they cause swelling. Adjacent teeth may be occasionally loosened. In the maxilla, the sinuses may be invaded.

Odontogenic keratocysts are found in over 80%, about three times as often in the mandible as in the maxilla (8,87). They may be relatively small, single, or multiple, but more often are large, bilateral, unilocular, or multilocular, and asymmetric, involving both jaws. In young patients they can cause displacement of developing permanent teeth. The cysts tend to occur mainly in the canine to premolar area, in the mandibular retromolar-ramus area, and in the region of the maxillary second molar. Cysts in the molar-ramus area may extend into the coronoid process, whereas those in the maxilla may involve the maxillary antra. The cysts may cross the midline in either jaw. Computerized tomography has been employed in estimating the size of jaw cysts (70). The recurrence rate of the cysts is high, estimates varying from 30 to 60% (25).

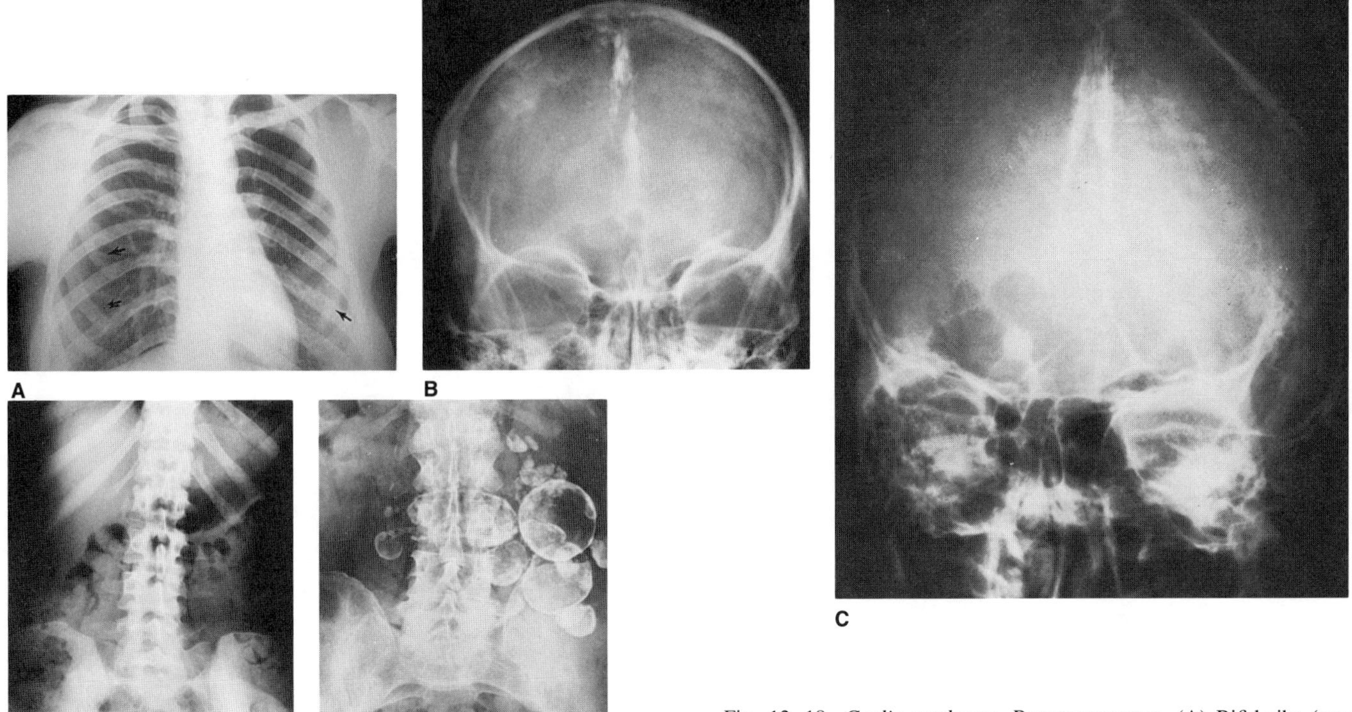

Fig. 12–18. *Gorlin syndrome.* Roentgenograms. (A) Bifid ribs (arrows) and scoliosis. (B) Calcified falx cerebri. (C) Lamellar calcification of falx cerebri. (D) Calcified ovarian fibromas. Most often mistaken for calcified uterine myomas. Usually bilateral, they often overlap in the midline. (E) Multiple lymphomesenteric (chylous) cysts.

Odontogenic keratocysts present as multilocular or invaginated cysts (along with microdaughter cysts or epithelial rests in 25–50% of the cases), with a parakeratinized, or rarely, an orthokeratinized (4%), stratified squamous epithelium consisting of five to eight rows of cells having a regularly oriented, well-defined basal epithelial cell layer, palisaded nuclei, but no rete ridges (Fig. 12–19D) (34). The continuous parakeratinized epithelial lining acts as a more efficient semipermeable membrane than the epithelium of other odontogenic cysts. Hence, the osmolality, rate of growth, and aggressiveness are greater in keratocysts than in other odontogenic cysts. About 70% of cysts associated with an unerupted tooth are not true dentigerous cysts when examined histologically. Usually a layer of fibrous tissue separates the crown from the adjacent cyst cavity. Some of the larger keratocysts expand in size to include tooth follicles (10,17).

In some cases, the epithelial rests proliferate to produce a picture like that of squamous odontogenic tumor (50). The mitotic index is comparable to that of the dental lamina. Budding of the epithelium into the connective tissue and suprabasilar splitting are noted in at least 50% (9). Inflammatory cells rarely are found in the underlying connective tissue. The cyst capsule is thin. Some cysts exhibit foci of calcification in the walls (18).

Woolgar et al (125) and Dominguez and Keszler (24) found significant differences between syndrome keratocysts and single keratocysts. Syndrome keratocysts were found to have a markedly increased number of satellite cysts, solid islands of epithelial proliferation, odontogenic rests within the capsule, and mitotic figures in the epithelial lining of the main cavity. An index of activity derived from these parameters suggested a greater growth potential in syndrome cysts. Elsewhere, Woolgar et al (126) noted that syndrome keratocysts tend to occur at a much earlier age than single keratocysts.

Most authors believe that odontogenic keratocysts arise from the dental lamina (10,125). Stoelinga et al (110), however, found that when keratocysts and overlying epithelium were surgically removed together, the cysts were adherent to the overlying soft tissue through a perforation in the bone. Serial sectioning of the overlying soft tissue showed a much greater number of epithelial islands, cords, and microcysts than elsewhere around keratocysts. Furthermore, a connection could be established in some instances between a microkeratocyst and the basal cell layer of the overlying oral epithelium. The smaller number of islands, cords, and microcysts observed elsewhere around keratocysts could be explained by displacement. Corroborating evidence that suggests the soft tissue origin of keratocysts is provided by reports of keratocyst formation within bone grafted from the rib or the iliac crest to the mandible (27). Other such examples have been reviewed by Stoelinga et al (110).

If keratocysts arise from the oral epithelium, how do they come to lie within bone? Inductive bone resorption is probably the major factor. Resorption of bone caused by mechanical pressure is most likely a contributing factor only. Some keratocysts may arise directly from budding of the keratocyst lining itself or from odontogenic rests. Both are, of course, derivatives of the oral epithelium (110).

**Central nervous system.** Congenital communicating hydrocephaly has been noted by a number of authors (36). In many instances, however, this simply represents an abnormality of cranial shape. Cysts of the choroid plexus of the third and lateral ventricles and glial nodules projecting into the walls of the lateral ventricles have been described (115). Agenesis of the corpus callosum has been found in a few patients (6,36,84,96). Mental deficiency, reported in about 3%, is probably more common (57,113).

**Other findings.** In males, the syndrome may be associated with hypogonadotrophic hypogonadism, anosmia, cryptorchidism, female pubic escutcheon, gynecomastia, and/or scanty facial or body hair (94,96,113). Lymphatic or chylous cysts or cystic lymphangiomas of

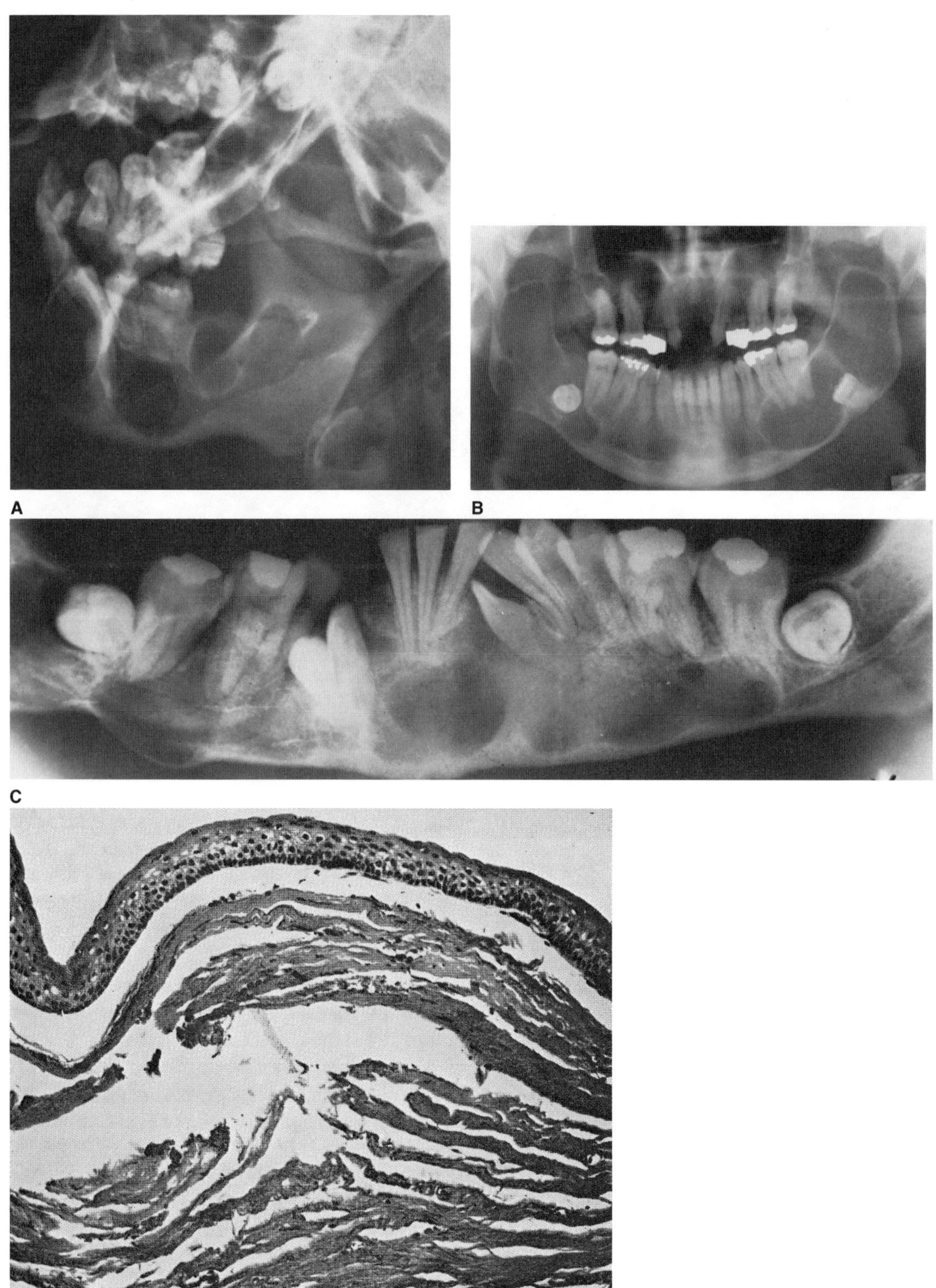

Fig. 12–19. *Gorlin syndrome.* (A) Multiple cysts scattered throughout both jaws. (B) Panorex showing multiple odontogenic keratocysts. Note displaced molars. (C) Extensive involvement of mandible with multiple cysts. (D) Photomicrograph of odontogenic keratocysts from 40-year-old female. (A from *RJ Gorlin et al,* Cancer **18**:89, 1965. C from *J Mills* and *J Foulkes,* Br J Radiol **40**:366, 1967.)

the mesentery, some of which are calcified, have been documented (16,36,48,79,97,101).

Various minor kidney malformations (horseshoe kidney, L-shaped kidney, unilateral renal agenesis, duplicaton of renal pelvis and ureter, and renal cysts) have been described (34), but data collection has been desultory.

**Neoplasms.** The commonly observed basal cell carcinomas have previously been discussed in the section on Skin. Various low-frequency neoplasms are also part of the syndrome. Medulloblastoma developing within the first 2 years of life has been described in several patients, in their sibs, in their offspring, or in more distant relatives (1a,19,29,34,36,63,66,85,94,96,107). These tumors may lead to the

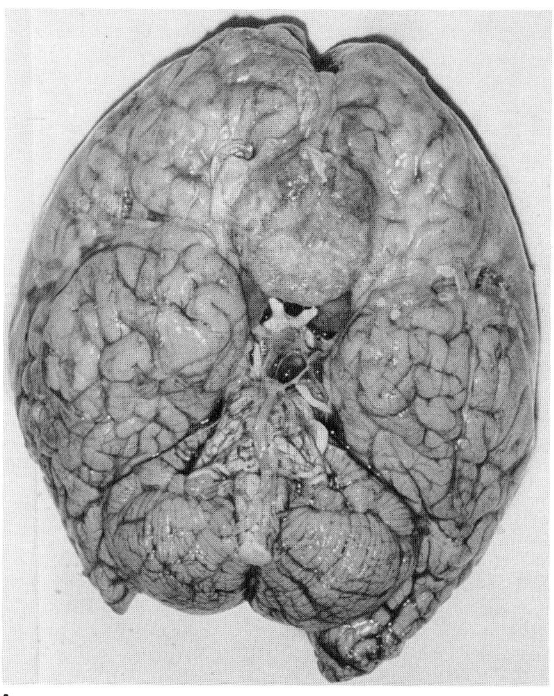

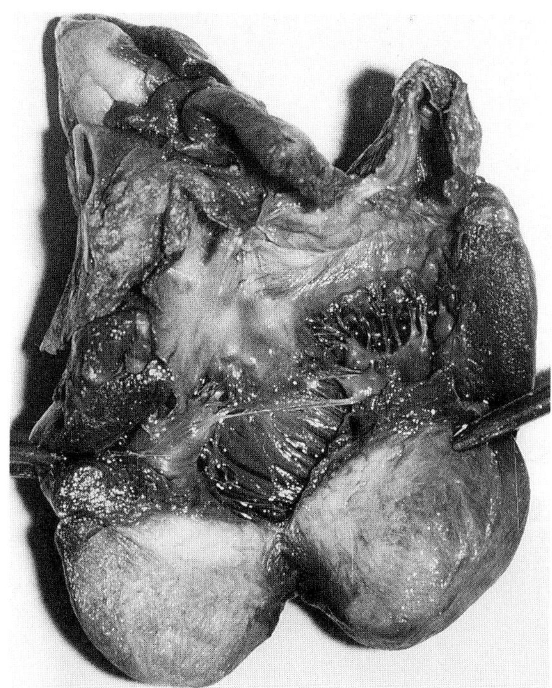

A B
Fig. 12–20. *Gorlin syndrome.* (A) Note meningioma. (B) Fibroma of ventricle of heart. (A from *PJW Stoelinga et al,* Oral Surg **36**:686, 1973. B courtesy of *C Reiter,* Vienna, Austria.)

death of the patient before the more usual features of the syndrome appear. The tumor's onset seems to be earlier and behavior seems to be less lethal than with isolated medulloblastoma (85). Meningioma (Fig. 12–20A). (81,107,111,112) and craniopharyngioma (112) have also been recorded.

Ovarian fibromas (12,93), usually bilateral and calcified or overlapping medially as a single calcified mass, are often not discovered unless they twist on their pedicles (7a,59a,108). They do not seem to reduce fertility. A number of ovarian fibrosarcomas (57,63,103) have been recorded. Cardiac fibroma has been observed in at least 11 instances (1a,11,49,59,98) (Fig. 12–20B).

Ameloblastoma (16,43,59,71) and squamous cell carcinoma (45a,80,94) have been noted to arise from keratocysts. Fibrosarcomas of the jaws (6,97,112) appear to be secondary to radiation therapy.

Examples of isolated neurofibroma (44) and schwannoma (62) have been recorded and MM Cohen has seen classic Gorlin syndome with a (nonplexiform) neurofibroma of the tongue. Other neoplasms have been noted including renal fibroma (34), melanoma (34,41), leiomyoma (48,97,101), fetal rhabdomyoma (21,62,105), rhabdomyosarcoma (5), benign mesenchymoma (101,124), adenoid cystic carcinoma (23), pleomorphic adenoma (48), adrenal cortical adenoma (115), fibroepithelial tongue polyp (96), hamartomatous bronchogenic cyst (116), theca cell tumor (17), seminoma (34), fibroadenoma of the breast (41), odontogenic myxoma (41), lipomas (41), thyroid adenoma (41,62), carcinoma of the bladder (41), Hodgkin's disease (91), and chronic leukemia (RJ Gorlin and MM Cohen, personal observation, 1975). Obviously, some of these latter tumors concur by chance. For other tumors, see Gorlin (34) and Stieler et al (109).

**Laboratory findings.** It has been shown that malignant transformation of cells *in vitro* with acetylaminofluorene occurs at a faster rate in cells from patients with the Gorlin syndrome than in cells from normal individuals (28). This test—known as the Elejalde test—needs confirmation. It may possibly be useful as a diagnostic aid in suspected sporadic instances of the disorder, in which basal cell carcinomas and/or jaw cysts have not yet developed (17). Ringborg et al (99) found a 25% decreased level of maximum DNA repair synthesis of ultraviolet light-damaged leukocytes. Featherstone et al (30) found that fibroblasts from patients with the syndrome were not unusually sensitive to potentially lethal damage following exposure to $\gamma$-irradiation although lymphocytes were. On the other hand, Nagasawa et al (83), studying survival curves of fibroblasts exposed to X rays, ultraviolet light, and mitomycin C, found that strains of fibroblasts from nevoid basal cell carcinoma patients were slightly hypersensitive to all three of these DNA-damaging agents. Chan and Little (15) and Artlett and Priestly (3) also found that fibroblasts taken from several patients were defective in their repair of potentially lethal damage, relating this to the increased frequency of basal cell cancers observed in exposed fields following irradiation of such individuals. Sensitivity to ultraviolet and X-irradiation has also been discussed by Frentz et al (31).

Happle and Hoehn (44) found an increased frequency of spontaneous chromosome breaks in fibroblasts from uninvolved skin. Gao et al (32) found no abnormalities in chromosome number or structure but did find an increase in the number of sister chromatid exchange frequency in those with aggressive basal cell carcinomas and suggested that this may serve as an indicator of aggressive transformation. Römke et al (100) reported no increase in chromosome breaks. Prometaphase G- and C-banding has disclosed no abnormalities (33).

It has been suggested that shortened fourth metacarpals and calcifications in various parts of the body might indicate abnormal calcium and phosphorus metabolism. Using the Ellsworth–Howard test, an almost complete lack of end-organ responsiveness to parathormone has been noted. However, interpretation of the Ellsworth–Howard test is fraught with danger since phosphaturia following the administration of parathormone can be minimal even in some normal individuals. Since calcemic and phosphaturic responses to parathormone are mediated by cyclic AMP, and since cyclic AMP is a more sensitive measure of responsiveness than phosphaturia, it follows that testing with cyclic AMP is preferable to the use of the Ellsworth–Howard test. On this basis, normal responsiveness to parathormone has been reported in patients with the Gorlin syndrome (61,82).

Odontogenic keratocysts contain a remarkably low content of protein (less than 4 g/dl) and numerous keratinized squamous cells in the cyst fluid in contrast to other dental cysts (10,64). Ultrastructural studies of the epithelium suggest a collagenolytic activity (89). This may reflect the high leucine aminopeptidase activity noted in the cyst capsule by Magnusson (72).

**Differential diagnosis.** Unilateral linear nevoid basal cell carcinomas (ULNBCC) with comedones may well represent a somatic mutational clonal manifestation of the syndrome (13,123). ULNBCC has been associated with diffuse osteoma cutis and unilateral anodontia (1). Also to be excluded are basaloid follicular hamartomas which are associated with myasthenia gravis (76). Bazex syndrome consists of multiple basal cell carcinomas (especially of the face), follicular atrophoderma (especially of tissue of hands and feet and elbows), hypotrichosis,and generalized hypohidrosis or anhidrosis of the face and head (40,90,120). It probably has X-linked dominant inheritance. The so-called autosomal dominant Rombo syndrome (77) has many features of Bazex syndrome but the skin is normal and sweating is normal. Rasmussen (95) described a syndrome of trichoepitheliomas, milia, and cylindromas. There is superficial resemblance to multiple seborrheic keratoses of the trunk and extremities that may arise in a patient with adenocarcinoma (Leser–Trélat sign) (119). Jaw cysts, if few in number, may be mistaken for conventional dentigerous cysts or isolated keratocysts. The calcification of several organs (falx cerebri, skin, ovaries) and short fourth metacarpals may suggest *pseudohypoparathyroidism*.

**Laboratory aids.** Radiographic examination of the chest, skull, jaws, and other areas as indicated is useful to establish diagnosis. Biopsy of skin lesions confirms the clinical impression of nevoid basal cell carcinomas. Odontogenic keratocysts can be distinguished histologically from the many other known types of jaw cysts.

### References [Gorlin syndrome (nevoid basal cell carcinoma syndrome)]


1. Aloi FG et al: Unilateral linear basal cell nevus associated with diffuse osteoma cutis, unilateral anodontia, and abnormal bone mineralization. J Am Acad Dermatol **20**:973–978, 1989.

1a. Anderson DE, Cook WA: Jaw cysts and the basal cell nevus syndrome. *J Oral Surg* **24**:15–26, 1966.

2. Anderson DE et al: The nevoid basal cell carcinoma syndrome. *Am J Hum Genet* **19**:12–22, 1967.

3. Artlett CF, Priestly A: Deficient recovery from potentially lethal damage in some gamma-irradiated human fibroblast strains. *Br J Cancer* **49**(*Suppl* **6**):227–232, 1984.

4. Barr RJ et al: Cutaneous keratin cysts of nevoid basal cell carcinoma syndrome. *J Am Acad Dermatol* **14**:572–576, 1986.

5. Beddis IR et al: Nasopharyngeal rhabdomyosarcoma and Gorlin's nevoid basal cell carcinoma syndrome. *Med Pediatr Oncol* **11**:178–179, 1983.

6. Binkley GW, Johnson HH: Epithelioma adenoides cysticum: Basal cell nevi, agenesis of corpus callosum and dental cysts. *Arch Dermatol Syph* **63**:73–84, 1951.

7. Blinder G et al: Widespread osteolytic lesions of the long bones in basal cell nevus syndrome. *Skel Radiol* **12**:195–198, 1984.

7a. Bosch-Banyeras JM et al: Calcified ovarian fibromas in prepubertal girls. *Eur J Pediatr* **148**:749–750, 1989.

8. Brannon RB: The odontogenic keratocyst: A clinicopathologic study of 312 cases. Part I. Clinical features. *Oral Surg* **42**:54–72, 1976.

9. Brannon RB: The odontogenic keratocyst: A clinicopathologic study of 312 cases. Part II. Histologic features. *Oral Surg* **43**:233–255, 1977.

10. Browne RM: The odontogenic keratocyst. Clinical aspects. *Br Dent J* **128**:225–231, 1970.

11. Bunting PD, Remensnyder JP: Basal cell nevus syndrome. *Plast Reconstr Surg* **60**:895–901, 1977.

12. Burket RL, Rauh JL: Gorlin's syndrome: Ovarian fibromas at adolescence. *Obstet Gynecol* **47**(*Suppl*):43–46, 1976.

13. Carney RG: Linear unilateral basal-cell nevus with comedones. *Arch Dermatol* **65**:471–476, 1952.

14. Case 10-1986. *N Engl J Med* **314**:700–706, 1986.

15. Chan GL, Little JB: Cultured diploid fibroblasts from patients with the nevoid basal cell carcinoma syndrome are hypersensitive to killing by ionizing radiation. *Am J Pathol* **111**:50–55, 1983.

16. Clendenning WE et al: Basal cell nevus syndrome. *Arch Dermatol* **90**:38–53, 1964.

17. Cohen MM Jr: In *Dysmorphic Syndromes with Craniofacial Manifestations in Oral Facial Genetics,* Stewart E, Prescott G (eds), C.V. Mosby, St. Louis, 1976, pp 596–606.

18. Cotton S et al: Nevoid basal cell carcinoma syndrome. *J Oral Med* **37**:69–73, 1982.

19. Cutler TP et al: Multiple naevoid basal cell carcinoma syndrome (Gorlin's syndrome). *Clin Exp Dermatol* **4**:373–379, 1979.

20. Dahl E et al: Craniofacial morphology in the nevoid basal cell carcinoma syndrome. *Int J Oral Surg* **5**:300–310, 1976.

21. Dahl I et al: Foetal rhabdomyoma. *Acta Pathol Microbiol Scand (Sect A)* **84**:107–112, 1976.

22. DeBoer EM, Bruynzeel DP: Basal cell nevus syndrome and webbed neck. *Dermatologica* **173**:245–247, 1986.

23. de la Plaza R et al: Two cases of nevoid basal cell carcinoma syndrome. *Plast Reconstr Surg* **71**:114–119, 1983.

24. Dominguez FV, Keszler A: Comparative study of keratocysts associated and non-associated with nevoid basal cell carcinoma syndrome. *J Oral Pathol* **17**:39–42, 1988.

25. Donatsky O, Hjorting-Hansen E: Recurrence of the odontogenic keratocyst in 13 patients with the nevoid basal cell carcinoma syndrome—a 6-year follow-up. *Int J Oral Surg* **9**:173–179, 1980.

26. Dunnick NR et al: Nevoid basal carcinoma syndrome: Radiographic manifestations including cystlike lesions of the phalanges. *Radiology* **127**:331–334, 1978.

27. Edwards JL, McMillan MD: Jaw cysts and the basal-cell naevus syndrome: Cyst recurrence in a bone graft. *NZ Dent J* **68**:229–237, 1972.

28. Elejalde BR: In-vitro transformation of cells from patients with naevoid basal-cell carcinoma syndrome. *Lancet* **2**:1199–1200, 1976.

29. Esser R, Bohnert B: Neurologic symptoms of basal cell nevus syndrome. *Eur Neurol* **19**:335–338, 1980.

30. Featherstone T et al: Studies on the radiosensitivity of cells from patients with basal cell naevus syndrome. *Am J Hum Genet* **35**:58–66, 1983.

31. Frentz G et al: The nevoid basal cell carcinoma syndrome: Sensitivity to ultraviolet and X-irradiation. *J Am Acad Dermatol* **17**:637–643, 1987.

32. Gao J et al: Studies on the genetics of basal cell nevus syndrome in one family. *Chinese Med J* **98**:538–542, 1985.

33. Gibbs PM et al: The multiple basal cell nevus syndrome: A cytogenetic study of six cases. *Cancer Genet Cytogenet* **20**:369–370, 1986.

34. Gorlin RJ: Nevoid basal cell carcinoma syndrome. *Medicine* **66**:96–113, 1987.

35. Gorlin RJ, Goltz RW: Multiple nevoid basal-cell epithelioma, jaw cysts and bifid rib syndromes. *N Engl J Med* **262**:908–912, 1960.

36. Gorlin RJ, Sedano HO: The multiple nevoid basal cell carcinoma syndrome revisited. *Birth Defects* **7**(8):140–148, 1971.

37. Gorlin RJ, Sedano HO: Multiple nevoid basal cell carcinoma syndrome, in *Handbook of Clinical Neurology,* Vol. 14, Vinken PJ, Bruyn GW (eds), North-Holland Publishing Company, Amsterdam, 1972, pp 455–473.

38. Gorlin RJ et al: Multiple nevoid basal cell carcinoma, odontogenic keratocysts and skeletal anomalies. *Acta Dermato-Venereol* **43**:39–55, 1963.

39. Gorlin RJ et al: The multiple basal-cell nevi syndrome. An analysis of a syndrome consisting of multiple nevoid basal-cell carcinoma, jaw cysts, skeletal anomalies, medulloblastoma, and hyporesponsiveness to parathormone. *Cancer* **18**:89–103, 1965.

40. Gould DJ, Barker DJ: Follicular atrophoderma and multiple basal cell carcionomas (Bazex). *Br J Dermatol* **99**:431–435, 1978.

41. Gundlach KKH, Kiehn M: Multiple basal cell carcinomas and keratocysts—the Gorlin and Goltz syndrome. *J Maxillofac Surg* **7**:299–307, 1979.

42. Gutierrez MM, Mora GG: Nevoid basal cell carcinoma syndrome: A review and case report of a patient with unilateral basal cell nevus syndrome. *J Am Acad Dermatol* **15**:1023–1030, 1986.

43. Happle R: Naevobasalom und Ameloblastom. *Hautarzt* **24**:290–294, 1973.

44. Happle R: Neurofibromas and nevoid basal cell carcinomas. *Arch Dermatol* **108**:582–583, 1973.

45. Happle R, Hoehn H: Cytogenetic studies on cultured fibroblast-like cells derived from basal cell carcinoma tissue. *Clin Genet* **4**:17–24, 1973.

45a. Hasegawa K et al: Basal cell nevus syndrome with squamous cell carcinoma of the maxilla. *J Oral Maxillofac Surg* 47:629–632, 1989.

46. Hashimoto K et al: Electromicroscopic studies of palmar and plantar pits of nevoid basal cell epithelioma. *J Invest Dermatol* **59**:380–393, 1972.

47. Hawkins JC et al: Multiple nevoid basal cell carcinoma syndrome (Gorlin's syndrome): Possible confusion with metastatic medulloblastoma. *J Neurosurg* **50**:100–102, 1979.

48. Helseth A, Mylius EA: Gorlins syndrom. *Tidsskr Norsk Laegef* **106**:2851, 1986.

49. Hess J, Bink-Boelkens MTE: Fibroma cordis bij een zuigeling met het basocellulaire naevussyndroom. *Ned Tijdschr Geneeskd* **120**:1796–1799, 1976.

50. Hodgkinson DJ et al: Keratocysts of the jaw: Clinicopathologic study of 79 patients. *Cancer* **41**:803–813, 1978.

51. Howell JB: Nevoid basal cell carcinoma syndrome: Profile of genetic and environmental factors in oncogenesis. *J Am Acad Dermatol* **11**:98–104, 1984.

52. Howell JB: The roots of naevoid basal cell carcinoma syndrome. *Clin Exp Dermatol* **5**:339–348, 1980.

53. Howell JB, Anderson DE: Transformation of epithelioma adenoides cysticum into multiple rodent ulcers: Fact or fallacy. *Br J Dermatol* **95**:233–242, 1976.

54. Howell JB, Caro MR: Basal cell nevus: Its relationship to multiple cutaneous cancers and associated anomalies of development. *Arch Dermatol* **79**:67–80, 1959.

55. Howell JB, Freeman RG: Structure and significance of the pits with their tumors in the nevoid basal cell carcinoma syndrome. *J Am Acad Dermatol* **2**:224–238, 1980.

56. Howell JB, Mehregan AH: Story of the pits. *Arch Dermatol* **102**:583–597, 1970.

57. Jackson R, Gardere S: Nevoid basal cell carcinoma syndrome. *Can Med Assoc J* **105**:850–859, 1971.

58. Jarisch W: Zur Lehre von den Hautgeschwülsten. *Arch Dermatol Syph (Berlin)* **28**:162–222, 1894.

59. Jeanmougin M et al: Naevomatose basocellulaire et améloblastome. *Ann Dermatol Venereol (Paris)* **106**:691–693, 1979.

59a. Johnson AD et al: Nevoid basal cell carcinoma syndrome: Bilateral ovarian fibromas in a 3½-year-old girl. *J Am Acad Dermatol* **14**:371–374, 1986.

60. Jones KL et al: Older paternal age and fresh gene mutation: Data on additional disorders. *J Pediatr* **86**:84–88, 1975.

61. Kaufman, RL, Chase LR: Normal stimulation of cyclic 3′, 5′-AMP by parathormone. *Birth Defects* **7**(8):149–155, 1971.

62. Klijanienko J et al: Naevomatose basocellulaire associée à un rhabdomyome foetal multifocal. *Presse Méd* **17**:2247–2250, 1988.

63. Kraemer BB et al: Fibrosarcoma of the ovary: A new component in the nevoid basal cell carcinoma syndrome. *Am J Surg Pathol* **8**:231–236, 1984.

64. Kramer IRH, Toller PA: The use of exfoliative cytology and protein estimations in preoperative diagnosis of odontogenic keratocysts. *Int J Oral Surg* **2**:143–151, 1973.

65. Lahti A et al: Polycystic jaws in patients with cleft lip and palate. *Ann Chir Gynecol Fenn* **62**:161–165, 1973.

66. Leppard BJ: Skin cysts in the basal cell naevus syndrome. *Clin Exp Dermatol* **8**:603–612, 1983.

67. Lindeberg H, Jepsen FL: The nevoid basal cell carcinoma syndrome: Histopathology of the tumors. *J Cutan Pathol* **10**:68–73, 1983.

68. Lindeberg H et al: The nevoid basal cell carcinoma syndrome. Otoneurological aspects. *J Laryngol Otol* **100**:1181–1185, 1986.

69. Littler BO: Gorlin's syndrome and the heart. *Br J Oral Surg* **17**:135–146, 1979.

70. MacKenzie GD et al: Computerized tomography in the diagnosis of an odontogenic keratocyst. *Oral Surg* **59**:302–305, 1985.

71. Maddox WD et al: Multiple nevoid basal cell epitheliomas, jaw cysts and skeletal defects. *JAMA* **188**:106–111, 1964.

72. Magnusson BC: Odontogenic keratocysts: A clinical and histochemical study with special reference to enzyme histochemistry. *J Oral Pathol* **7**:8–18, 1978.

73. Mason JK et al: Pathology of the nevoid basal cell carcinoma syndrome. *Arch Pathol* **79**:401–408, 1965.

74. McClatchey K et al: Odontogenic keratocysts and nevoid basal cell carcinoma syndrome. *Arch Otolaryngol* **101**:613–616, 1975.

75. McEvoy BF, Gatzek H: Multiple nevoid basal cell carcinoma syndrome: Radiological manifestations. *Br J Radiol* **42**:24–28, 1969.

76. Mehregan AH, Baker S: Basaloid follicular hamartoma: Three cases with localized and systematized unilateral lesions. *J Cutan Pathol* **12**:55–65, 1985.

77. Michaelson G et al: The Rombo syndrome: A familial disorder with vermiculate atrophoderma, milia, hypotrichosis, trichoepitheliomas, basal cell carcinomas and peripheral vasodilation with cyanosis. *Acta Dermatovener (Stockholm)* **61**:497–503, 1981.

78. Miller RF, Cooper RR: Nevoid basal cell carcinoma syndrome. Histogenesis of skeletal lesions. *Clin Orthoped Rel Res* **89**:246–252, 1972.

79. Mills JJ, Foulkes J: Gorlin's syndrome: A radiological and cytogenetic study of 9 cases. *Br J Radiol* **40**:366–371, 1967.

80. Moos KF, Rennie JS: Squamous cell carcinoma arising in a mandibular keratocyst in a patient with Gorlin's syndrome. *Br J Oral Max Surg* **25**:280–284, 1987.

81. Mortimer PS et al: Basal cell naevus syndrome and intracranial meningioma. *J Neurol Neurosurg Psychiatr* **47**:210–212, 1984.

82. Murphy KJ: Subcutaneous bone formation in the naevoid basal cell carcinoma syndrome: Normal urinary cyclic AMP response to parathyroid hormone infusion. *Clin Radiol* **26**:37–39, 1974.

83. Nagasawa H et al: Study of basal cell nevus syndrome fibroblasts after treatment with DNA damaging agents. *Basic Life Sci* **29B**:775–785, 1984.

84. Naguib MG et al: Central nervous system involvement in the nevoid basal cell carcinoma syndrome: Case report and review of the literature. *Neurosurgery* **11**:52–56, 1982.

85. Neblett CR et al: Neurological involvement in the nevoid basal cell carcinoma syndrome. *J Neurosurg* **35**:577–584, 1971.

86. Nicolai JP: The basal naevus syndrome and palmar cysts. *Hand* **11**:98–102, 1979.

87. Novak KD, Bloss N: Rontgenologische Aspekte des Basalzell-Naevus Syndroms (Gorlin–Goltz-Syndrom). *Roefo* **124**:11–16, 1976.

88. Olson RAJ et al: Nevoid basal cell carcinoma syndrome. *J Oral Surg* **39**:308–312, 1981.

89. Philipsen HP et al: Ultra-structure of epithelial lining of keratocysts in nevoid basal cell carcinoma syndrome. *Int J Oral Surg* **5**:71–81, 1973.

90. Plosila M et al: The Bazex syndrome: Follicular atrophoderma with multiple basal cell carcinomas, hypotrichosis and hypohidrosis. *Clin Exp Dermatol* **6**:31–41, 1981.

91. Potaznik D, Steinherz P: Multiple nevoid basal cell carcinoma syndrome and Hodgkin's disease. *Cancer* **53**:2713–2715, 1984.

92. Pratt MD, Jackson R: Nevoid basal cell carcinoma syndrome. *J Am Acad Dermatol* **16**:964–970, 1987.

93. Raggio M et al: Recurrent ovarian fibromas with basal cell nevus syndrome (Gorlin syndrome). *Obstet Gynecol* **61** *(Suppl)*:95–96, 1983.

94. Ramsden RT, Barrett A: Gorlin's syndrome. *J Laryngol Otol* **89**:615–621, 1978.

95. Rasmussen JE: A syndrome of trichoepitheliomas, milia and cylindromas. *Arch Dermatol* **111**:610–614, 1971.

96. Rayner CRW et al: What is Gorlin's syndrome? The diagnosis and management of basal cell naevus syndrome based on a study of thirty-seven patients. *Br J Plast Surg* **30**:62–67, 1977.

97. Reed JC: Nevoid basal cell carcinoma syndrome with associated fibrosarcoma of the maxilla. *Arch Dermatol* **97**:304–306, 1968.

98. Reiter C et al: Cardiac fibroma in familial Gorlin syndrome with virilism. *Wien Klin Wochenschr* **94**:430–434, 1982.

99. Ringborg V et al: Decreased UV-induced DNA repair synthesis in peripheral leukocytes from patients with the nevoid basal cell carcinoma syndrome. *J Invest Dermatol* **76**:268–270, 1981.

100. Römke C et al: Investigations of chromosomal stability in Gorlin–Goltz syndrome. *Arch Dermatol Res* **277**:370–372, 1985.

101. Rossi R et al: Neurocutaneous syndromes and retroperitoneal tumors. *Urology* **13**:292–294, 1979.

102. Ruprecht A et al: Cleft lip and palate, seldom seen features of the Gorlin-Goltz syndrome. *Dentomaxillofac Radiol* **16**:99–103, 1987.

103. Ryan DE, Burkes EJ: The multiple basal cell nevus syndrome in a Negro famly. *Oral Surg* **36**:831–840, 1973.

104. Satinoff MI, Wells C: Multiple basal cell naevus syndrome in ancient Egypt. *Med Hist* **13**:294–297, 1969.

105. Schweisguth O et al: Naevomatose baso-cellulaire; association à un rhabdomyosarcome congénital. *Arch Fr Pédiatr* **25**:1083–1093, 1968.

106. Shelley WB: Quadrant distribution of basal cell nevi. *Arch Dermatol* **100**:741–743, 1969.

107. Southwick GJ, Schwartz RA: The basal cell nevus syndrome. Disasters occurring among a series of 36 patients. *Cancer* **44**:2294–2305, 1979.

108. Springate JE: The nevoid basal cell carcinoma syndrome. *J Pediatr Surg* **21**:908–910, 1986.

109. Stieler W et al: Basalzellnävus-Syndrom mit Plattenepithelkarzinom des Larynx. *Z Hautkr* **63**:113–120, 1988.

110. Stoelinga PJW et al: The origin of keratocysts in the basal cell nevus syndrome. *J Oral Surg* **33**:659–663, 1975.

111. Stoelinga PJW et al: Some new findings in the basal cell carcinoma syndrome. *Oral Surg* **36**:686–692, 1973.

112. Tamoney HJ Jr: Basal cell nevoid syndrome. *Am Surg* **35**:279–283, 1969.

113. Tasanen A: Skeletal anomalies and keratocysts in the basal cell nevus syndrome. *Int J Oral Surg* **4**:225–235, 1975.

114. Taylor WB, Wilkins JW: Nevoid basal cell carcinomas of the palm. *Arch Dermatol* **102**:654–655, 1970.

115. Taylor WB et al: The nevoid basal cell carcinoma syndrome. *Arch Dermatol* **98**:612–614, 1968.

116. Totten JR: The multiple nevoid basal cell carcinoma syndrome: Report of its occurrence in four generations of a family. *Cancer* **46**:1456–1462, 1980.

117. Ullman S et al: Ultrastructure of palmar and plantar pits in basal cell nevus syndrome. *Acta Dermatonen (Stockholm)* **52**:329–336, 1972.

118. van Dijk E, Neering H: The association of cleft lip and palate with basal cell nevus syndrome. *Oral Surg* **50**:214–216, 1980.

119. Venencie PY, Perry HO: Sign of Leser-Trélat: Report of two cases and review of the literature. *J Am Acad Dermatol* **10**:83–88, 1984.

120. Viknins P, Berlin A: Follicular atrophoderma and basal cell carcinoma: The Bazex syndrome. *Arch Dermatol* **113**:948–951, 1977.

121. White JC: Multiple benign cystic epitheliomas. *J Cutan Genitourin Dis* **12**:477–484, 1894.

122. Winkler PA, Guyuron B: Multiple metastases from basal cell naevus syndrome. *Br J Plast Surg* **40**:528–531, 1987.

123. Wirth H, Tilgen W: Linearer unilateraler Basalzellnävus. *Hautarzt* **34**:620–624, 1983.

124. Wolthers OD, Stellfeld M: Benign mesenchymoma in the trachea of a patient with the nevoid basal cell carcinoma syndrome. *J Laryngol Otol* **101**:522–526, 1987.

125. Woolgar JA et al: A comparative histological study of odontogenic keratocysts in basal cell naevus syndrome and control patients. *J Oral Pathol* **16**:75–80, 1987.

126. Woolgar JA et al: The odontogenic keratocysts and its occurrence in the nevoid basal cell carcinoma syndrome. *Oral Surg* **64**:727–730, 1987.

## Klippel–Trénaunay–Weber syndrome

The combination of cutaneous angiomatosis, varicose veins, and enlargement of bone and soft tissue was first described by Klippel and Trénaunay (23) in 1900. Parkes-Weber (37), in 1907, added hemodynamically significant arteriovenous communications. Although the presence of arteriovenous fistulas worsens the prognosis by heralding occasional high output cardiac failure, we cannot accept division into two separate syndromes, as recommended by some authors (5). Pathologic evaluation of the vascular lesions has shown no consistent pattern (21,27). Venous abnormalities such as phlebectasia of superficial and deep venous systems predominate, but arterial defects are not unusual. Because of the variability of the vascular anomalies, this syndrome remains a clinical diagnosis, not a pathologic one.

Although early descriptions emphasized unilateral leg hypertrophy with cutaneous and subcutaneous hemangiomas, varicosities, phlebectasia, and, occasionally, arteriovenous fistula, subsequent reports expanded these findings to include almost every conceivable body area. Many additional abnormalities have also been recognized, including visceral hemangiomas, lymphangiomatous anomalies, macrodactyly, syndactyly, polydactyly, and oligodactyly.

Over 1000 cases have been recorded to date (4,44,47) and several recent, extensive reviews are available (5,44,50). Servelle (44) noted elongation of the involved limb with edema in 84%, varicose veins in 36%, and flat hemangiomas in 32%; venography and surgical exploration demonstrated malformation of the deep veins involving the popliteal vein in 51%, superficial femoral vein in 16%, popliteal and superficial femoral veins in 29%, iliac veins in 3%, and inferior vena cava in 1%.

Most patients are of normal mentality, exceptions occurring when vascular abnormalities involve the craniofacial area (15,19,27). The majority of these patients have Sturge–Weber angiomatosis with Klippel–Trénaunay–Weber syndrome (6,38,51). There is ample evidence to consider the two disorders as differing only in location of involvement (12,15).

Baskerville et al (5) suggested that a mesodermal defect acting primarily on angiogenesis could conceivably explain features of the syndrome. Persistence of the embryonic vascular network, which usually regresses in the developing limb bud, could produce increased skin blood flow and temperature and an increase in size and number of veins. This might result in increased bone growth and histologic changes of intimal thickening, elastosis, and ectasia of superficial veins. Early mesodermal involvement could also explain deep vein atresia, absence of valves in the deep veins, reduplication of axial veins, and nonvascular anomalies such as syndactyly and polydactyly. Finally, persistence of embryonic vascular structures is compatible with those cases associated with Sturge–Weber angiomatosis.

Almost all cases are sporadic. Koch (24) cited a number of familial cases of the Klippel–Trénaunay–Weber syndrome from the literature. Close scrutiny of these cases reveals inadequate documentation of vascular involvement; relatives of the proband with minor findings, such as isolated varicosities or birthmarks of the posterior neck, are included as well as other disorders, particularly neurofibromatosis. In a review of 135 cases (18), only Lindenauer's example of brother–sister involvement (28) is convincing.

**Extremities.** Unilateral leg hypertrophy is the most frequent finding (28,29). However, hypertrophy can symmetrically or asymmetrically involve any or all limbs (Fig. 12–21A,B and 12–22). Hypertrophy is usually noted at birth, but may occur at any age and progressively increase in degree (24). Circumference and/or length of extremities are usually increased but, rarely, only the proximal or distal limb segment may be disproportionately large (28). Infrequently, atrophy of a limb may occur or an enlarged limb may become atrophic over many years (48). Ultimate height is rarely excessive or stunted except in severe cases.

In most instances, a visible vascular abnormality is present in the hypertrophied area (25,42,46); however, this is not always the case (16). Hemangiomatous lesions of almost every variety have been noted, but varicosities, phlebectasia, nevus flammeus (portwine mark), and vascular masses predominate (1,28,29). These lesions may minimally involve the extremity, but they may also cause severe distortion. The larger vascular masses may sometimes result in a generalized bleeding diathesis of the Kasabach–Merritt variety (20). The occurrence of arteriovenous fistulas is not uncommon (28). Lymphectasia in association with the vascular abnormalities can result in marked limb swelling with recurrent cellulitis. Viljoen et al (52) reported aseptic cellulitis as a potentially severe complication in 40% of their series.

Macrodactyly may involve one or many digits, some assuming gigantic proportions (8,27,33,34). (Fig. 12–23). In those instances in which the digits are not enlarged, they are often of unequal length. There also may be clinodactyly at the most involved joints and/or relative brachydactyly. Cutaneous syndactyly is frequent, but rarely involves more than two digits on any one limb. Polydactyly and oligodactyly are relatively uncommon, usually being found in patients with more severe and grotesque limb abnormalities (8,26,27).

Roentgenographic studies of the extremities usually show enlargement of subcutaneous tissue, muscle, and bone; however, only one of these tissues may be hypertrophied (40). Phlebectasis, phleboliths, arteriovenous fistulas, hyperostosis, bony sclerosis, and bone atrophy can readily be demonstrated when present (1,19,28,29,48).

**Skin.** Typically, vascular lesions are distributed over the lower limb (28), with frequent extension to the buttocks and less often with involvement of the lower part of the back, flank, lateral area of the chest, and axillary area (24). The upper extremities are sometimes involved, as are the abdomen, chest, neck, and face (7,12). Telangiectatic spots have been reported (1). Hyperpigmented streaks and spots have been noted particularly in the area of the vascular lesion (21). Rarely, pigmentation of areas has been preceded by a vesicular rash (38). Some patients have had skin ulcers at birth or at a later age (22,27).

**Craniofacial features.** Craniofacial involvement, when present, is quite similar to that seen in Sturge–Weber angiomatosis, both in degree of variability and distribution of occurrence (12,32). Hemangiomatous involvement of the craniofacial area (Fig. 12–24A) may place the patient at increased risk for neurologic complications, including mental retardation. Double contoured tramline calcifications may be evident roentgenographically (32). Macrocephaly has been observed in some instances (49).

Eye abnormalities have included enophthalmos, conjunctival telangiectasis, heterochromia iridis, iris coloboma, oculosympathetic palsy, glaucoma, scleral pigmentation, retinal varicosities, choroidal

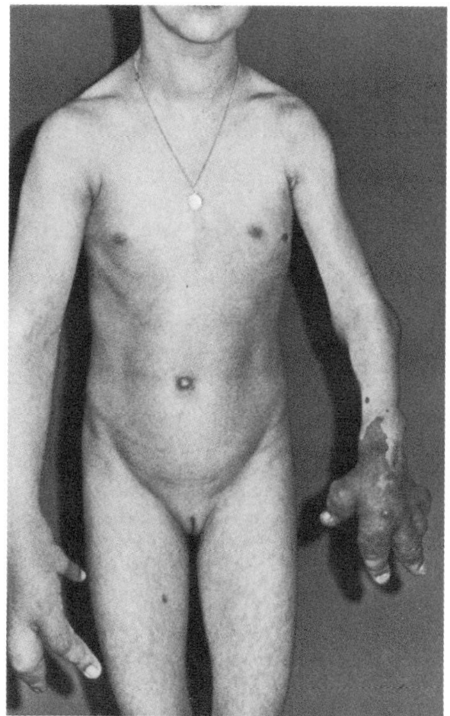

A

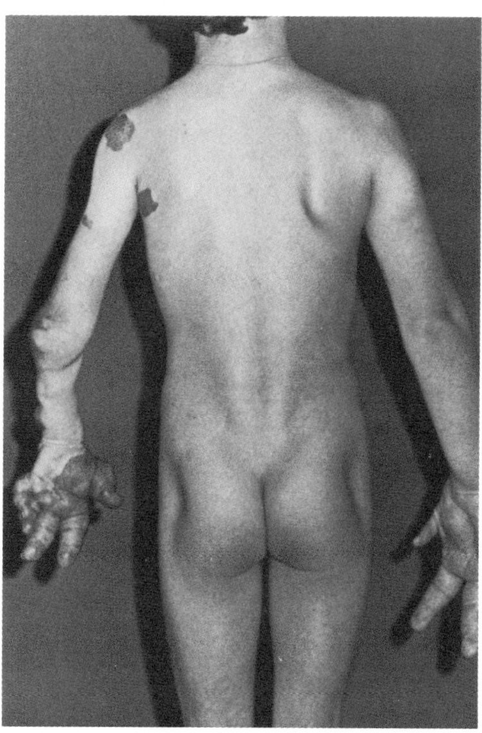

B

Fig. 12–21. *Klippel–Trénaunay–Weber syndrome.* (A,B) Extreme enlargement of hands, disparity in size of arms, fixation of left elbow joint. (From *E Nöh* and *R Steckenmesser*, Z Orthopäd **112**:243, 1974.)

angiomas, Marcus Gunn pupil, strabismus, orbital varix, and disc anomalies (1,12,34,35).

Oral manifestations (Fig. 12–24B) have been described by a number of authors (25,31,43,47,48) and may include angiomatous involvement of the lips, buccal mucosa, tongue, palate, gingiva, and/or oropharynx, enlarged maxilla, premature eruption of teeth on the affected side, displacement of teeth, and malocclusion.

Fig. 12–22. *Klippel–Trénaunay–Weber syndrome.* Asymmetry of legs. Patient had angiomatous lesions on right lateral aspect of trunk.

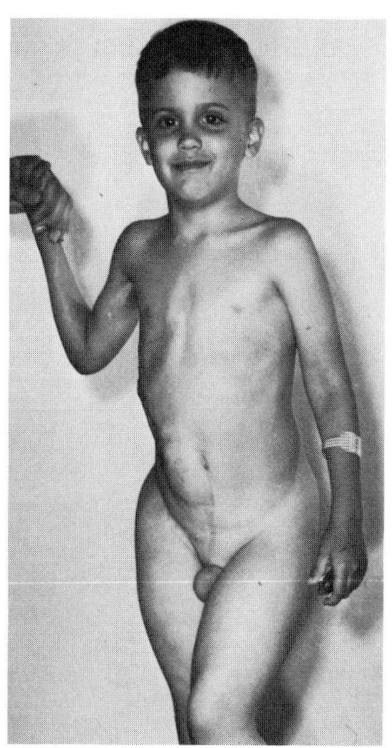

**Viscera.** Involvement of the viscera is not rare. The major manifestations are hemangiomatous lesions of the gastrointestinal tract (13,26,45), urinary system (14), visceral organs (21), mesentery, and pleura (26). The types of vascular lesions are as varied as those found on the extremities. Pulmonary vein varicosity has been reported (36).

Generalized nonspecific visceromegaly has been noted (9), particularly involving the kidney (12,19). However, it is important to note that no patient with Klippel–Trénaunay–Weber syndrome has had associated Wilms tumor, although a patient with nephroblastomatosis has been described (30). Abdominal lymphectasia and protein-losing enteropathy secondary to lymphectasia have been reported (10).

**Other findings.** Enlargement of the genitalia with ambiguity secondary to direct hemangiomatous involvement has been observed. Li-

Fig. 12–23. *Klippel–Trénaunay–Weber syndrome.* Gigantiform toes. One toe has been surgically removed from the right foot.

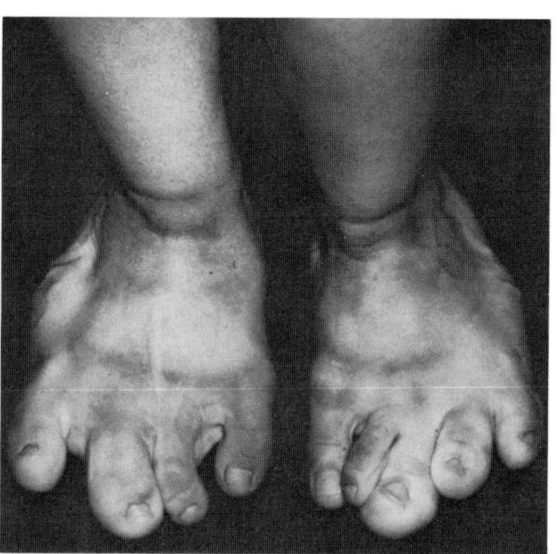

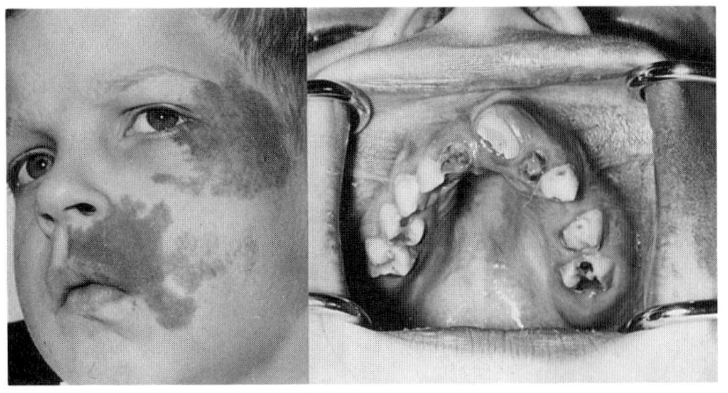

A B

Fig. 12–24. *Klippel–Trénaunay–Weber syndrome.* (A) Flat hemangioma of left maxillary-zygomatic region. (B) Enlargement of left upper jaw accompanied by premature eruption of upper left central incisor. (A,B from *R Stellmach,* Fortschr Kiefer Gesichtschir **4**:54, 1958.)

podystrophy involving the upper extremities has occurred in a few severe examples (8,26). Scoliosis secondary to unequal leg length has been noted (40). Paraplegia due to compression by a vertebral and epidural cavernous hemangioma has been reported (17). Hypospadias has been described (5).

**Differential diagnosis.** *Neurofibromatosis* must be excluded, since limb hypertrophy and skin hemangiomas may be associated with the disorder (2). A relative lack of multiple discrete café-au-lait spots occurs in Klippel–Trénaunay–Weber syndrome. *Beckwith–Wiedemann syndrome* may have associated hemihyperplasia and skin hemangiomas. Cutis marmorata telangiectatica congenita when associated with discrete vascular skin lesions and aberrations of limb size may be difficult to differentiate from Klippel–Trénaunay–Weber syndrome (11,53). *Maffucci syndrome,* with its frequent limb hypertrophy and vascular skin lesions, may cause confusion, but the usual late onset of the vascular lesions and the presence of enchondromatosis clearly separate the two disorders. *Sturge–Weber angiomatosis,* which may occur with some instances of Klippel–Trénaunay–Weber syndrome, is discussed in detail later in this chapter.

*Hemihyperplasia (hemihypertrophy)* can be associated with vascular abnormalities of the skin (41). Some cases reported under this designation have been examples of the Klippel–Trénaunay–Weber syndrome (39). Large isolated hemangiomas of a limb can cause hypertrophy. Single limb hypertrophy or isolated macrodactyly without hemangiomatous involvement should not be considered as constituting the Klippel–Trénaunay–Weber syndrome. However, such limbs and digits should be watched closely for possible development of vascular or pigmented skin lesions.

Other syndromes with hemangiomatous involvement include *Bannayan–Riley–Ruvalcaba syndrome* (3,14) and *Proteus syndrome.* Both of these syndromes also have lipomatous involvement. Stephan et al (49) reported the association of macrocephaly and angiomatosis.

### References (Klippel–Trénaunay–Weber syndrome)


1. Arrighi F: Hamartose ecto-mesodérmique: Un cas d'angiomatose diffuse avec fusion de maladie de Sturge-Weber-Krabbe et de maladie de Parkes Weber. *Bull Soc Fr Dermatol Syphiligr* **67**:562–563, 1960.
2. Arrighi F: Hamartose ecto-mesodérmique: Un cas de fusion de maladie de Recklinghausen (avec éléphantiasis nevromateux de Virchow) et de maladie de Klippel-Trénaunay-Parkes Weber. *Bull Soc Fr Dermatol Syphiligr* **67**:564, 1960.
3. Banhayan GA: Lipomatosis, angiomatosis and macroencephalia: A previously undescribed congenital syndrome. *Arch Pathol* **92**:1–5, 1971.
4. Barek L et al: The Klippel–Trénaunay syndrome: A case report and review of the literature. *Mt Sinai J Med* **49**:66–70, 1982.
5. Baskerville PA et al: The etiology of the Klippel–Trénaunay syndrome. *Ann Surg* **202**:624–627, 1985.
6. Bonse G: Röntgenbefunde bei einer Phakomatose (Sturge–Weber kombiniert mit Klippel–Trénaunay). *Fortschr Roentgenstr* **74**:727–729, 1951.
7. Brooksaler F: The angioosteohypertrophy syndrome (Klippel–Trénaunay–Weber syndrome). *Am J Dis Child* **112**:161–164, 1966.
8. Buchanec J, Galanda V: Polymalformačny-Klippelov-Trenaunayov-Weberov syndróms progresivnou lipodistrofiou u 6-ročného chlapca. *Čs Pediatr* **24**:228–232, 1969.
9. Cagiati L: Klinischer und pathologischer Beitrag zum Studium der halbseitigen Hypertrophie. *Dtsch Z Nervenheilkd* **32**:282–293, 1907.
10. Caplan DB et al: Angioosteohypertrophy syndrome with protein-losing enteropathy. *J Pediatr* **74**:119–123, 1969.
11. Fahrig H: Zur Cutis marmorata teleangiectatica congenita (Phlebectasia congenita) und ihren Beziehungen zu fakultativ mit Naevi teleangiectatici kombinierten Missbildungen. *Z Kinderchir* **12**:101–110, 1973.
12. Furukawa T et al: Sturge–Weber and Klippel–Trénaunay syndrome with nevus of Ota and Ito. *Arch Dermatol* **102**:640–645, 1970.
13. Ghahremani GG et al: Diffuse cavernous hemangioma of the colon in the Klippel–Trénaunay syndrome. *Ped Radiol* **118**:673–678, 1976.
14. Geley L et al: Isolierte Gliedmassen-monohypertrophie mit Lipomatosis derselben Körperhälfte-ein neues Syndrom? *Z Kinderchir* **12**:101–110, 1973.
15. Gottron HA, Schnyder UW: Vererbung von Hautkrankheiten, In *Handbuch der Haut- und Geschlechtskrankheiten,* Vol. 7, Springer-Verlag, Berlin, 1966, p 715.
16. Gougerot H, Filliol P: Naevus variquex ostéo-hypertrophique de Klippel ou hémiangiectasie hypertrophique de Parkes Weber. *Arch Dermatol Syph (Paris)* **1**:404–411, 1929.
17. Gourie-Devi M, Prakash B: Vertebral and epidural hemangioma with paraplegia in Klippel–Trénaunay–Weber syndrome. *J Neurosurg* **48**:814–817, 1978.
18. Hall BD: Bladder hemangiomas in Klippel–Trénaunay–Weber syndrome. *N Engl J Med* **285**:1032–1033, 1971.
19. Hall R: A case of melorheostosis with cutaneous haemangioma and lymphatic vesicles. *J Bone Jt Surg* **43B**:335–337, 1961.
20. Inceman S, Tangun Y: Chronic defibrination syndrome due to a giant hemangioma associated with microangiopathic hemolytic anemia. *Am J Med* **46**:997–1002, 1969.
21. Inui M et al: An autopsy case of Klippel–Trénaunay–Weber's disease. *Acta Pathol Jpn* **19**:251–263, 1969.
22. Ippen H: Systematisierte Angiektasie mit Gliedmassenatrophie (Ein Beitrag zum "Klippel–Trénaunay-Syndrom"). *Hautarzt* **8**:317–320, 1959.
23. Klippel M, Trénaunay P: Du naevus variqueux ostéo-hypertrophique. *Arch Gen Méd* **185**:641–672, 1900.
24. Koch G: Zur Klinik, Symptomatologie, Pathogenese und Erbpathologie des Klippel–Trénaunay–Weber'schen Syndroms. *Acta Genet Med (Roma)* **5**:326–370, 1956.
25. Kontras SB: The Klippel–Trénaunay–Weber syndrome. *Birth Defects* **10**(7):177–188, 1974.
26. Kuffer FR et al: Klippel–Trénaunay–Weber syndrome, visceral angiomatosis and thrombocytopenia. *J Pediatr Surg* **3**:65–72, 1968.
27. Lamar LM et al: Klippel–Trénaunay syndrome. *Arch Dermatol* **91**:58–59, 1965.
28. Lindenauer SM: The Klippel-Trénaunay syndrome: Varicosity, hypertrophy and hemangioma with no arteriovenous fistula. *Ann Surg* **162**:303–314, 1965.
29. Lindenauer SM: Congenital arteriovenous fistula and the Klippel–Trénaunay syndrome. *Ann Surg* **174**:246–263, 1971.
30. Mankad VN et al: Bilateral nephroblastomatosis and Klippel–Trénaunay syndrome. *Cancer* **33**:1462–1467, 1974.
31. Miescher G: Über plane Angiome (Naevi hyperaemici). *Dermatologica* **106**:176–183, 1958.
32. Nellhaus G et al: Sturge-Weber disease with bilateral intracranial calcifications at birth and unusual pathologic findings. *Acta Neurol Scand* **43**:314–347, 1967.
33. Nöh E, Steckenmesser R: Der angeborene Riesenwuchs, klinische und arteriographische Befunde an Hand und Arm beim Klippel–Trénaunay-Syndrom. *Z Orthopäd* **112**:243–252, 1974.
34. Noriega-Sanchez A et al: Oculocutaneous melanosis associated with the Sturge–Weber syndrome. *Neurology (Minneap)* **22**:256–268, 1972.
35. O'Connor PS, Smith JL: Optic nerve variant in the Klippel–Trénaunay–Weber syndrome. *Ann Ophthalmol* **10**:131–137, 1978.
36. Owens DW et al: Klippel–Trénaunay–Weber syndrome with pulmonary vein varicosity. *Arch Dermatol* **108**:111–113, 1973.
37. Parkes-Weber F: Angioma formation in connection with hypertrophy of limbs and hemi-hypertrophy. *Br J Dermatol* **19**:231–235, 1907.
38. Rademacher R: Über einen Fall einer Kombination von Sturge–Weber

und Klippel–Trénaunay-Syndrom mit konstitutioneller Neurodermitis. *Dermatol Wochenschr* **143**:381–386, 1961.
39. Ringrose RE et al: Hemihypertrophy. *Pediatrics* **36**:434–448, 1965.
40. Rose LM: Hypertrophy of the lower limbs with cutaneous naevus and varicose veins. *Arch Dis Childh* **25**:162–169, 1950.
41. Sabanas AO, Chatterton CC: Crossed congenital hemihypertrophy. *J Bone Jt Surg* **37A**:871–874, 1955.
42. Schönenberg H, Redemann M: Klippel–Trénaunay–Weber-Syndrom. *Klin Pädiatr* **184**:449–460, 1972.
43. Sciubba JJ, Brown AM: Oral-facial manifestations of Klippel–Trénaunay–Weber syndrome: Report of two cases. *Oral Surg* **43**:227–232, 1977.
44. Servelle M: Klippel and Trénaunay's syndrome. *Ann Surg* **201**:365–373, 1985.
45. Sheperd JA: Angiomatous conditions of the gastro-intestinal tract. *Br J Surg* **40**:409–421, 1953.
46. Silva M da, Neves H: Über einen Fall von Klippel–Trénaunay-schem Symptomenkomplex, der erfolgreich mit Röntgenstrahlen behandelt wurde. *Fortschr Roentgenstr* **90**:475–482, 1959.
47. Steiner M et al: Klippel–Trénaunay–Weber syndrome. *Oral Surg* **63**:208–215, 1987.
48. Stellmach R: Zwei Beobachtungen von partiellem Riesenwuchs kindlicher Kiefer beim sogenannten planen Hämangiom des Geischtshaut. *Fortschr Kiefer Gesichtschir* **4**:54–57, 1958.
49. Stephan MJ et al: Macrocephaly in association with unusual cutaneous angiomatosis. *J Pediatr* **87**:353–359, 1975.
50. Telander RL et al: Prognosis and management of lesions of the trunk in children with Klippel–Trénaunay syndrome. *J Pediatr Surg* **19**:417–422, 1984.
51. Teller H, Lindner B: Über Mischformen der phakomatösen Syndrome von Sturge–Weber und Klippel–Trénaunay. *Z Haut Geschlechtskr* **13**:113–120, 1952.
52. Viljoen D et al: The cutaneous manifestations of the Klippel–Trénaunay–Weber syndrome. *Clin Exp Dermatol* **12**:12–17, 1987.
53. Way BH et al: Cutis marmorata telangiectatica congenita. *J Cutan Pathol* **1**:10–25, 1974.

## Maffucci syndrome (enchondromatosis and hemangiomatosis)

This syndrome is characterized by enchondromatosis, bone deformities, hemangiomas, and phlebolithiasis. The condition was described by Maffucci (36) in 1881 and by Kast and von Recklinghausen (27) in 1889. Carleton et al (13), Bean (9,10), Anderson (2), and Lewis and Ketcham (32) have reviewed the disorder. About 150 cases have been described.

The etiology is unknown. All reported cases have been sporadic and there is neither racial nor sex predilection (2,32). Hall (21a) suggested that perhaps the disorder results from somatic mutation, lethal in the nonmosaic state. Patients with the disorder may differ in the severity of their clinical manifestations; thus, heterogeneity cannot be excluded.

**Skeletal system.** Enchondromas are usually diagnosed between 1 and 5 years of age (34,39). These cartilaginous tumors are most numerous in the phalanges of the hands and feet, but may involve any bone preformed in cartilage (14,15,32,37b,45). They have been unilateral in approximately 40% (2), but are frequently bilateral, although asymmetric. Skeletal abnormalities may become so severe as to produce gross bone deformity (Fig. 12–25 and 12–26) (11,15,22,23,32,37b,45). Limb length discrepancy (2,5,16,17,19,26,30,37,38,50), scoliosis (5,10,26,30,31,48), and bowing of limbs (5,24,25,30,34,37) may also occur. The facies may be asymmetric (32). Short stature is not uncommon (26,30,39). Fractures have been reported (5,10,14,21–23,25,29,30,37,45,48). Chondrosarcoma may develop in the enchondromas (8,17,18,21,25,32,45,46). Although malignant change has been reported in about 20% (2,9,19,21,25,32), this figure is suspect; malignant transformation is probably less frequent because individuals who develop chondrosarcoma may be more likely to be reported than those who do not.

**Hemangiomas.** Hemangiomas may be present at birth (3,12,35,38,40,50) or develop as late as the second decade of life (1,45). They occur most commonly on the skin and are superficial or deep. They may be localized or generalized, unilateral or bilateral but asymmetrical. Although one of the most common sites of occurrence is the hand, they can occur anywhere on the skin. On biopsy, most have been described as cavernous rather than capillary (2,14,15,17–19,32,42). Hemangiomas of the meninges (18), pharynx (33,34,48), esophagus, ileum, and anal mucosa have been noted (21,23,42). Phlebectasia has been reported in approximately 25% (2,42) and phlebolithiasis is common (3,7,8,11,19,24,25,30,32,37,39,46). Lymphangiomas also occur (5,30,33,37,47). One report describes capillary intraosseous hemangioma involving areas of bone rarefaction on angiography (3).

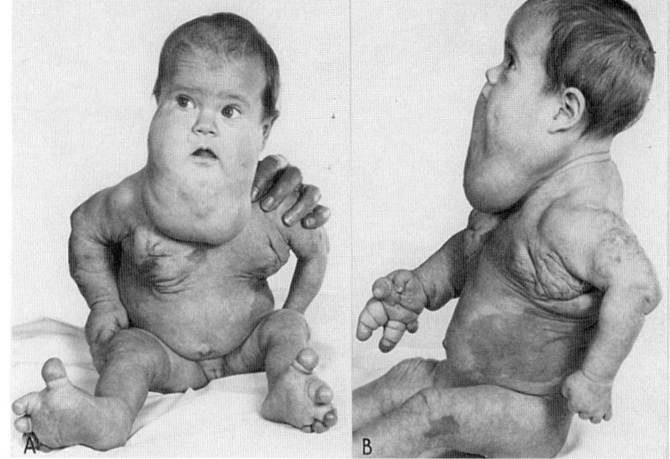

Fig. 12–25. *Maffucci syndrome.* (A,B) Gross distortion of body due to multiple enchondromas of the hands and feet, hemangiomas, lymphangioma. (A,B from *D Matthews,* Br J Plast Surg **17**:366, 1964.)

**Other neoplasms.** Carotid body tumor (4), angiosarcoma (9), fibrosarcoma (25), pancreatic carcinoma (25,46), hepatic adenocarcinoma (46), ovarian teratoma (30), ovarian cystadenocarcinoma (34), malignant ovarian tumor of mesenchymal origin, otherwise unspecified (32), glioma (13,18), astrocytoma (18), pituitary adenoma (37a, also see 43), and unspecified brain tumor (46) have been reported.

**Oral manifestations.** Oral hemangiomas have been described. In some reports, the oral site has not been specified (3,26). They have been most frequently noted on the tongue (Fig. 12–27) (28–31,33,35,37), although the buccal mucosa (6,30,33), lips (20), and palate (3,22,35,42,45,49) have been documented sites.

**Differential diagnosis.** The syndrome should not be confused with Ollier disease (enchondromatosis without hemangiomas). Occasionally, difficulty may be encountered in distinguishing the Maffucci syndrome from the blue rubber bleb nevus syndrome in which hemangiomas are found on the skin, but multiple enchondromas are not present. Sakurane et al (42) described a patient with Maffucci syndrome and the blue rubber bleb nevus syndrome who probably really had the Maffucci syndrome. The patient described by Richardson et al (41) had hemangiomas without enchondromas, and therefore is not an example of Maffucci syndrome. Schnall and Genuth (43) reported a patient with enchondromas, hemangiomas, pituitary adenoma, parathyroid adenoma, and neurilemoma of the nerve roots at the level of the sixth thoracic vertebra who may represent a disorder different from Maffucci syndrome. Patients with the *Klippel–Trénaunay–Weber syndrome* also have hemangiomas of the skin but lack enchondromatosis. The *Proteus syndrome* should be excluded. Spranger et al (44) and Zwerina (51) discuss various forms of enchondromatosis.

**Laboratory aids.** Hemogram and studies for occult blood in the feces may be indicated in some instances.

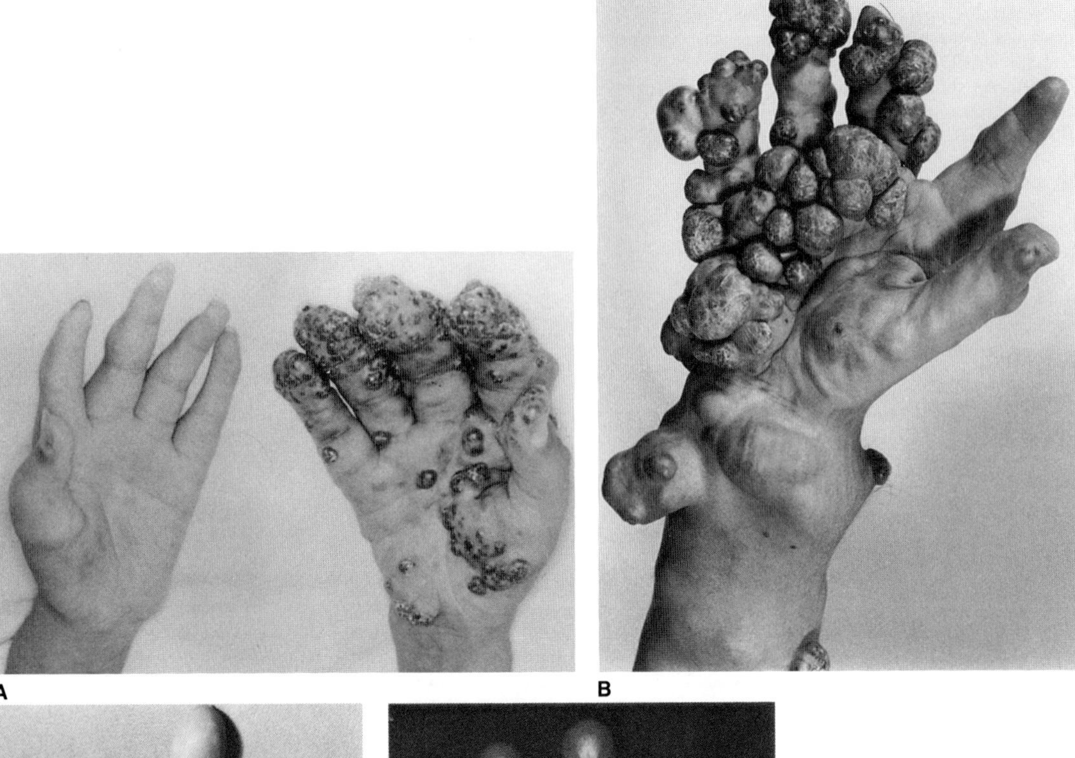

A B

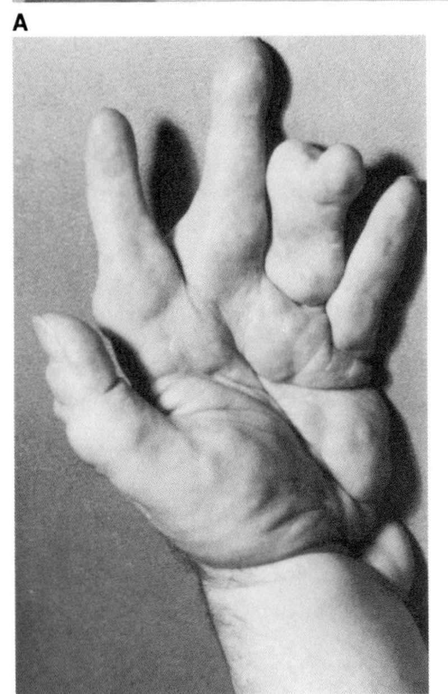

C

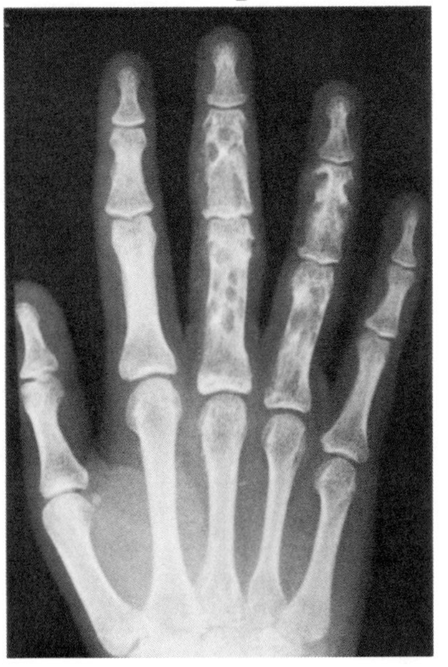

D

Fig. 12–26. *Maffucci syndrome.* (A) Hemangiomatous and enchondromatous involvement. (B) More severe involvement. (C) Distortion from multiple enchondromas. (D) Radiograph showing enchondromas of several phalanges. (A from *GFY Ma* and *PC Leung,* Br J Plast Surg **37**:615, 1984. B from *DA Tilsley* and *PW Burden,* Br J Dermatol **105**:331, 1981. C from *WB Bean,* Arch Intern Med **95**:767, 1955. D from *WG Cauble* and *HS Bowman,* Arch Surg **97**:678, 1968.)

#### References [Maffucci syndrome (enchondromatosis and hemangiomatosis)]


1. Allen BR: Maffucci's syndrome. *Br J Dermatol* **99**(Suppl **16**):31–33, 1978.
2. Anderson IF: Maffucci's syndrome: Report of a case with a review of the literature. *S Afr Med J* **39**:1066–1070, 1965.
3. Andrén L et al: Maffucci's syndrome. Report of four cases. *Acta Chir Scand* **126**:397–405, 1963.
4. Armstrong EA et al: Maffucci's syndrome complicated by an intracranial chondrosarcoma and a carotid body tumor. Case report. *J Neurosurg* **55**:479–483, 1981.
5. Ashenhurst EM: Dyschondroplasia with hemangiomata (Maffucci's syndrome). *Arch Neurol* **2**:552–555, 1960.
6. Bachert C: Ein besonderer Fall von multiplen Hämangiomen im Kopf-Hals-Bereich: Variation des Maffucci-Syndroms? *HNO* **33**:472–474, 1985.
7. Bahk YW: Dyschondroplasia with hemangiomata (Maffucci's syndrome). *Radiology* **82**:407–409, 1964.
8. Banna J, Parwani GS: Multiple sarcomas in Maffucci's syndrome. *Br J Radiol* **42**:304–307, 1969.
9. Bean WB: Dyschondroplasia and hemangiomata (Maffucci's syndrome). *Arch Intern Med* **95**:767–778, 1955.
10. Bean WB: Dyschondroplasia and hemangiomata (Maffucci's syndrome). II. *Arch Intern Med* **102**:544–550, 1958.
11. Beranbaum SL, Tzamouranis G: Maffucci's syndrome. Dyschondroplasia with hemangiomas. Report of a case. *Am J Roentgenol* **80**:479–481, 1958.
12. Berlin R: Maffucci's syndrome. Dyschondroplasia with vascular hamartomas. *Acta Med Scand* **177**:299–307, 1965.
13. Carleton A et al: Maffucci's syndrome (dyschondroplasia with haemangeomata). *Q J Med* **11**:203–228, 1942.
14. Cameron JM: Maffucci's syndrome. *Br J Surg* **44**:596–598, 1957.

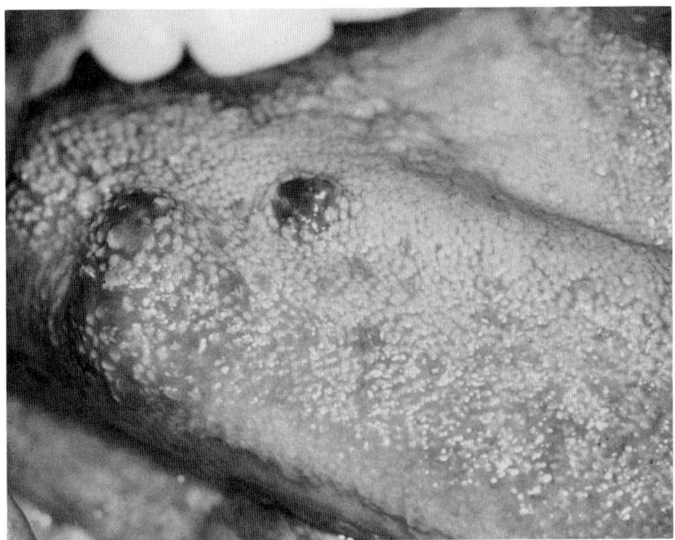

Fig. 12–27. *Maffucci syndrome.* Angiomas of tongue. (From *G Laskaris* and *C Skouteris,* Oral Surg **57**:263, 1984.)


15. Cauble WG, Bowman HS: Dyschondroplasia and hemangiomas (Maffucci's syndrome). Presentation of one case. *Arch Surg* **97**:678–681, 1968.

16. Chen VT, Harrison DA: Maffucci's syndrome. *Hand* **10**:292–298, 1978.

17. Cook PL, Evans PG: Chondrosarcoma of the skull in Maffucci's syndrome. *Br J Radiol* **50**:833–836, 1977.

18. Cremer H et al: The Maffucci–Kast syndrome. Dyschondroplasia with hemangiomas and frontal lobe astrocytoma. *J Cancer Res Clin Oncol* **101**:231–237, 1981.

19. Elmore SM, Cantrell WC: Maffucci's syndrome. Case report with a normal karyotype. *J Bone Jt Surg* **48A**:1607–1613, 1966.

20. Gutman E et al: Enchondromatosis with hemangiomas (Maffucci's syndrome). *South Med J* **71**:466–467, 1978.

21. Hall BD: Intestinal hemangiomas and Maffucci's syndrome. *Arch Dermatol* **105**:608, 1972.

21a. Hall JG: Somatic mosaicism: Observations related to clinical genetics. *Am J Hum Genet* **43**:355–363, 1988.

22. Halper H, Wedlick L: Maffucci's syndrome: With a report of a case. *Med J Aust* **1**:936–939, 1951.

23. Ikram-ul-Haq et al: Maffucci's syndrome. *J Int Coll Surg* **43**:133–140, 1965.

24. Indra KG et al: Dyschondroplasia with multiple haemangiomata—Maffucci's syndrome. *Br J Radiol* **36**:697–698, 1963.

25. Johnson JL et al: Maffucci's syndrome (dyschondroplasia with hemangiomas). *Am J Med* **28**:864–866, 1960.

26. Kaibara N et al: Generalised enchondromatosis with the unusual complications of soft tissue calcifications and haemangiomas. Follow-up for a twelve year period. *Skeletal Radiol* **8**:43–46, 1982.

27. Kast A, von Recklinghausen F: Ein Fall von Enchondrom mit ungewöhnlicher Multiplikation. *Virchows Arch Pathol Anat* **118**:1–18, 1889.

28. Kennedy JG: Dyschondroplasia and haemangiomata (Maffucci's syndrome). Report of a case with oral and intracranial lesions. *Br Dent J* **135**:18–21, 1973.

29. Krause GR: Dyschondroplasia with hemangioma (Maffucci's syndrome). Case report, *Am J Roentgenol* **52**:620–623, 1944.

30. Kuzma JF, King JM: Dyschondroplasia with hemangiomatosis (Maffucci's syndrome) and teratoid tumor of the ovary. *Arch Pathol* **46**:74–82, 1948.

31. Laskaris G, Skouteris C: Maffucci's syndrome. Report of a case with oral hemangiomas. *Oral Surg Oral Med Oral Pathol* **57**:263–266, 1984.

32. Lewis RJ, Ketcham AS: Maffucci's syndrome: Functional and neoplastic significance. Case report and review of the literature. *J Bone Jt Surg* **55A**:1465–1479, 1973.

33. Loewinger RJ et al: Maffucci's syndrome: A mesenchymal dysplasia and multiple tumour syndrome. *Br J Dermatol* **96**:317–322, 1977.

34. Lowell SH, Mathog RH: Head and neck manifestations of Maffucci's syndrome. *Arch Otolaryngol* **105**:427–430, 1979.

35. Ma GFY, Leung PC: The management of the soft-tissue haemangiomatous manifestations of Maffucci's syndrome. *Br J Plast Surg* **37**:615–618, 1984.

36. Maffucci A: Di un caso di encondroma ed angioma multiple. Contribuzione alla genesi embrionale dei tumor. *Mov Med Chir* **3**:399–412, 565–575, 1881.

37. Marberg K et al: Dyschondroplasia with multiple hemangiomata (Maffucci's syndrome). *Ann Intern Med* **49**:1216–1228, 1958.

37a. Marymount JV et al: Maffucci's syndrome complicated by carcinoma of the breast, pituitary adenoma, and mediastinal hemangioma. *South Med J* **80**:1429–1431, 1987.

37b. Matthews D: The congenitally deformed hand. *Br J Plast Surg* **17**:366–375, 1964.

38. Mullins JF, Livingood CS: Maffucci's syndrome (dyschondroplasia with hemangiomas). A case with early osseous changes. *Arch Dermatol Syph* **63**:478–482, 1951.

39. Niechajev IA, Hansson LI: Maffucci's syndrome. Case report. *Scand J Plast Reconstr Surg* **16**:215–219, 1982.

40. Phelan EMD et al: Generalised enchondromatosis associated with haemangiomas, soft-tissue calcifications and hemihypertrophy. *Br J Radiol* **59**:69–74, 1986.

41. Richardson JA et al: Maffucci's syndrome. *Arch Intern Med* **109**:186–191, 1962.

42. Sakurane HF et al: The association of blue rubber bleb nevus and Maffucci's syndrome. *Arch Dermatol* **95**:28–36, 1967.

43. Schnall AM, Genuth SM: Multiple endocrine adenomas in a patient with the Maffucci syndrome. *Am J Med* **61**:952–956, 1976.

44. Spranger J et al: Two peculiar types of enchondromatosis. *Pediatr Radiol* **7**:215–219, 1978.

45. Strang C, Rannie I: Dyschondroplasia with haemangiomata (Maffucci's syndrome). Report of a case complicated by intracranial chondrosarcoma. *J Bone Jt Surg* **32B**:376–383, 1950.

46. Sun T-C et al: Chondrosarcoma in Maffucci's syndrome. *J Bone Jt Surg* **67A**:1214–1219, 1985.

47. Suringa DWR, Ackerman AB: Cutaneous lymphangiomas with dyschondroplasia (Maffucci's syndrome). A unique variant of an unusual syndrome. *Arch Dermatol* **101**:472–474, 1970.

48. Tilsley DA, Burden PW: A case of Maffucci's syndrome. *Br J Dermatol* **105**:331–336, 1981.

49. Torri O: Angiome ed encondromi multiple nello istesso individuo. *Clin Chir (Milano)* **10**:81–105, 1902.

50. Umansky AL: Dyschondroplasia with hemangiomata (Maffucci's syndrome). Report of an early case with mild osseous manifestations. *Bull Hosp Jt Dis* **7**:59–68, 1946.

51. Zwerina H: Zur Nomenklatur der Dyschondroplasie, *Z Orthopäd* **110**:659–662, 1972.


## Multiple endocrine neoplasia, type 2B (multiple mucosal neuroma syndrome)

Initially described in part by Wagenmann (76) and by Froboese (32) in 1922–1923, the syndrome of multiple mucosal neuromas, pheochromocytoma, medullary carcinoma of the thyroid, and asthenic body build with muscle wasting of the extremities was delineated in more detail by Williams and Pollock (82), Gorlin et al (34–36), Schimke (63), Levy et al (49), Khairi et al (45), Carney (14–21), and Fryns and Chrzanowska (32a). About 150 patients have been described.

The syndrome has autosomal dominant inheritance. About 50% represent new mutations (18). Detailed analyses of cases have been reported (18,35,45). Most aspects of the syndrome can be explained by hyperplasia and/or neoplasia of neural crest derivatives (8,73,80,82), but some aspects, such as asthenic or Marfanoid habitus, cannot.

Babu et al (6,6a) carried out high-resolution G-banded chromosome studies on 12 multiple endocrine neoplasia (MEN) 2A and 7 MEN 2B families. Their findings suggest that the dominant mutation in many of these families is a visible deletion within band 20p12.2. However, recent independent reports place the gene for MEN 2A on chromosome 10 near the centromere (50a,65b). Possibly MEN 2B represents a microdeletion involving flanking genes.

**Facies.** In infancy, there is often a history of profound difficulty in feeding with failure to thrive (42). The distinct facies is elongated and characterized by a wide-eyed expression, broad-based nose, and large nodular lips with submucosal nodules on the vermilion border. The tarsal plates are thickened resulting in eversion of the upper eyelids. The lower face appears long (Fig. 12–28). The characteristic

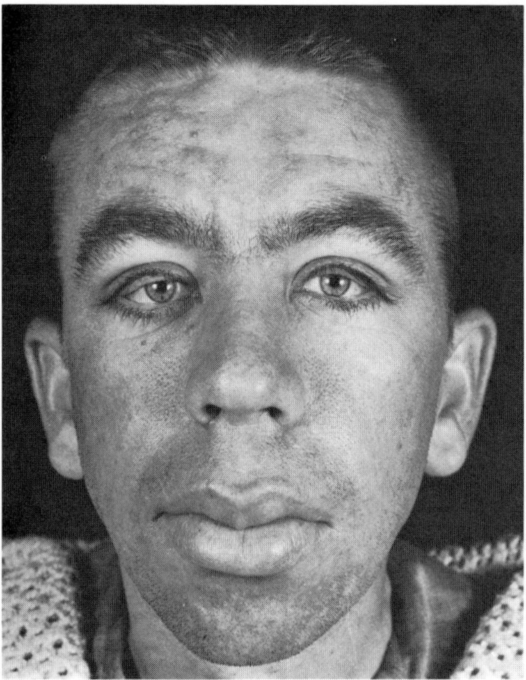

Fig. 12–28. *Multiple endocrine neoplasia, type 2B.* Facies is characterized by coarse features, large nodular lips, eversion of upper eyelids. (From *RN Schimke et al,* N Engl J Med **279**:1, 1968.)

facies is not always present (65a). Circumoral and/or midfacial lentiginosis has been seen (8,10,13a,24).

**Ophthalmologic manifestations.** The upper eyelid margins are thickened and often everted (13,22,37,45,50). Pedunculated neuromas are present on the palpebral conjunctiva, eyelid margins, or rarely the cornea in about 60%.

The cornea is the site of thickened white medullated nerve fibers (2,7,12,13,32,45,53,64,66,67a,77) that can easily be seen under slit lamp examination. They extend into the pupillary area, where they anastomose (Fig. 12–29).

Fig. 12–29. *Multiple endocrine neoplasia, type 2B.* White medullated corneal nerve fibers which anastomose in pupillary area. [From *DL Knox,* Birth Defects **7**(3):161, 1971.]

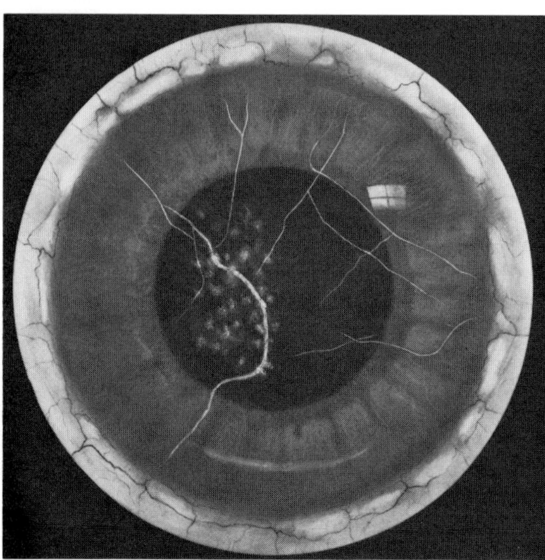

**Otolaryngologic manifestations.** Nasal, laryngeal, and bronchial mucosa may also be the site of neuromas (7,13,35,45,46,53,66). Perinasal skin may rarely be affected (65,72). Hyperplasia of neurenteric ganglion cells has been found in the bronchi.

**Thyroid gland.** Medullary carcinoma of the thyroid has been found in more than 90% (45,58). Most patients have been between 18 and 25 years of age at the time of initial diagnosis of their tumors but about 75% have metastasis at the time of diagnosis (45). Rarely has the diagnosis been made before age 12. However, patients in the first decade of life, including one as young as 23 months (70), have been reported to have already developed this neoplasm (31,44,51,54,55,58). Prospective surveillance should begin at birth and continue to the age of about 40 years. The carcinoma occurs with equal frequency in both sexes and may be present without symptoms or palpable nodules in the thyroid gland (30,70). Several tumors often develop within the same or within different lobes of the thyroid gland (11,26,30,31,53). This neoplasm tends to spread through lymphatic vessels to the cervical lymph nodes and mediastinum. Metastasis to regional lymph nodes is frequently found at initial thyroid surgery (Fig. 12–30A) (22,23,35,44,58,66,81). The average age at death from this neoplasm is 21 years (62a).

Medullary carcinoma of the thyroid arises from parafollicular cells (C cells) that have their origin in the embryonic ultimobranchial body, which in turn is derived at least in part from the neural crest (35,49). It is the only thyroid tumor that contains amyloid (Fig. 12–30B) which is found interspersed among the cells and fibrous septa. The C cells on electron microscopy exhibit characteristic secretory granules (Fig. 12–30C). C-cell hyperplasia is frequently found adjacent to areas of carcinoma; occasionally, in young patients, only C-cell hyperplasia is present, indicating that it is a precursor of medullary carcinoma. In addition to elaborating calcitonin, medullary thyroid carcinoma may also produce amyloid, serotonin, carcinoembryonic antigen, 5-hydroxyindole acetic acid, histaminase, bradykinin, various prostaglandins, DOPA decarboxylase, somatostatin, and an ACTH-like peptide (4,35,75).

**Adrenal gland.** Pheochromocytoma has been diagnosed in about 50%. It increases in frequency with age, probably being present in 90% of older patients. It most often becomes evident during the second and third decades of life, especially when provoked during investigation of the thyroid neoplasm (74). About 10% of patients die from a cardiovascular crisis just before or after surgery; the average age at death is 21 years (62a). Although rare, extraadrenal lesions have been reported (51). The presence of pheochromocytomas due to catecholamine secretion is often heralded by paroxysmal weakness, flushing, pounding headache, nausea, hypertension, dyspnea, palpitation, flatulence, paresthesia, blanching of the extremities, profuse sweating, and intractable diarrhea. Abdominal discomfort or cramping is frequently experienced.

Pheochromocytomas arise from cells derived from the neural crest. The tumors are multiple in 50% and very frequently bilateral when associated with medullary carcinoma of the thyroid (Fig. 12–31A) (7,11,12,51,56,58,61,75,78,82,83). When bilateral, one of the tumors may precede the other by decades. They range in size from a few millimeters to several centimeters in diameter. Malignant change is uncommon before the tumor exceeds 4 cm in diameter (67). Adrenal medullary hyperplasia is a precursor to pheochromocytoma (18,78).

**Gastrointestinal tract.** Gastrointestinal abnormalities may be present shortly after birth (55). Abdominal distension is common. At least 30% have some intestinal complaint: megacolon (Fig. 12–32A), diverticulosis, and chronic constipation alternating with watery diarrhea (2,14,17,25,27,53,57). Achalasia has also been reported (24). These abnormalities are in part related to a vasoactive intestinal peptide (76). Histopathologic study has revealed intestinal ganglioneuromatosis (Fig. 12–32B), which may involve the entire gut as well as the liver, gallbladder, and pancreas (14,17,56,82,83). Rectal carcinoid has also been found (28). Adenomatous polyposis coli has been described (59).

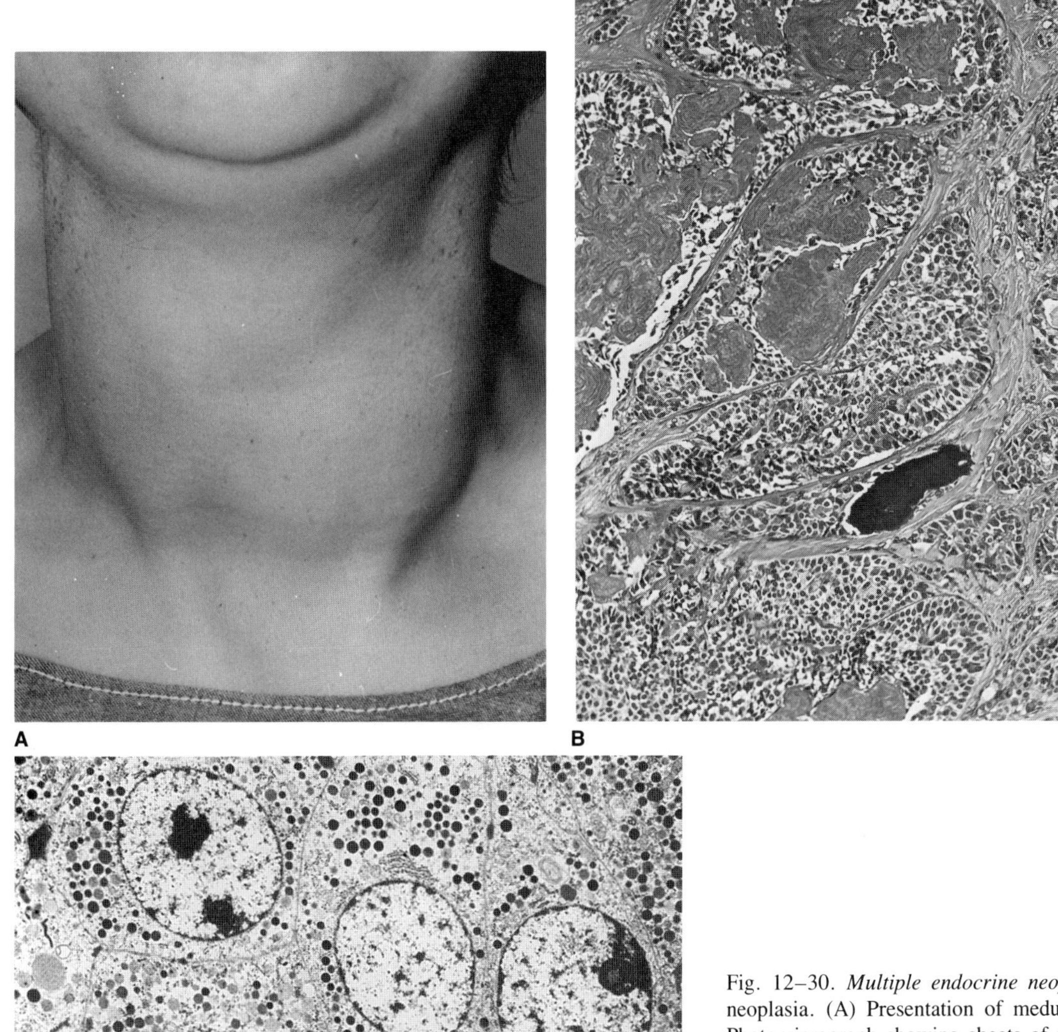


Fig. 12–30. *Multiple endocrine neoplasia, type 2B*. Multiple endocrine neoplasia. (A) Presentation of medullary carcinoma of the thyroid. (B) Photomicrograph showing sheets of round to spindle-shaped cells among which are masses of amyloid. (C) Medullary carcinoma of thyroid (original mag. ×6500). The C cells containing multiple secretory granules are easily visualized. (B from *GH Friedell,* Cancer **15**:241, 1962. C courtesy of *GM Dodd* and *B Mackay,* MD Anderson Hospital and Turner Institute, Houston, Texas.)

**Musculoskeletal alterations.** At least 75% have an asthenic or Marfanoid habitus with severe muscular wasting and weakness especially of the proximal extremities, simulating a myopathic state (Fig. 12–33A,B) (13,18,21,23,27,31,35,50,53,56,57a,66). Many skeletal alterations have been noted: pectus excavatum, talipes equinovarus, pes cavus (25%), slipped femoral epiphysis (10%), aseptic necrosis of the lumbar spine, kyphoscoliosis (25%), lordosis, and increased joint laxity (12%). These findings have been extensively reviewed elsewhere (30,35,57a). Mandibular prognathism has been noted in several patients (27,45).

**Other findings.** Melanotic skin pigmentation has been reported, possibly reflecting elaboration of an MSH-like peptide (23,64,69). Hypertrophy of peripheral nerves has been found (43,84). Dyck et al (29) studied the neurologic features of patients with the disorder and found involvement of autonomic nerves as well as of somatic motor and sensory neurons; these investigators postulated that neurologic symptoms in this disease could be attributed to neuroma formation. The cutaneous nerves are enlarged (20). Pubertal delay has been described (49,63,80).

**Oral manifestations.** Oral and labial involvement is the first component of the syndrome to appear, almost always in the first decade of life (43). In possibly 50%, these lesions are congenital or noticed in early infancy (14,17,35–37,43,53,81). The oral lesions consist of mucosal neuromas that principally involve the lips and tongue, although the buccal mucosa, gingiva, and palatal mucosa may be affected (18,35–37,48) (Fig. 12–34A,B). Both lips are extensively enlarged and nodular and have been described as blubbery. The lingual lesions are most commonly found on the anterior dorsal surface of the tongue and appear as pink pedunculated nodules; they have also been reported on the ventral tongue surface (48).

On light microscopy, the mucosal nodules are plexiform neuromas (16,22,26,31,54,84), that is, unencapsulated masses of convoluted myelinated and unmyelinated nerves (Fig. 12–34C) which elaborate calcitonin (75). Histochemical investigation demonstrates absence of

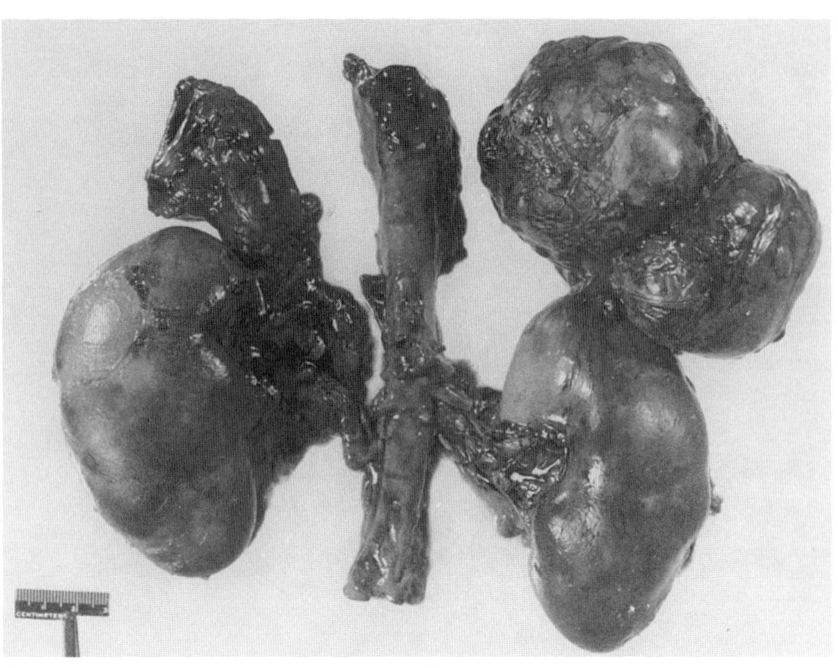

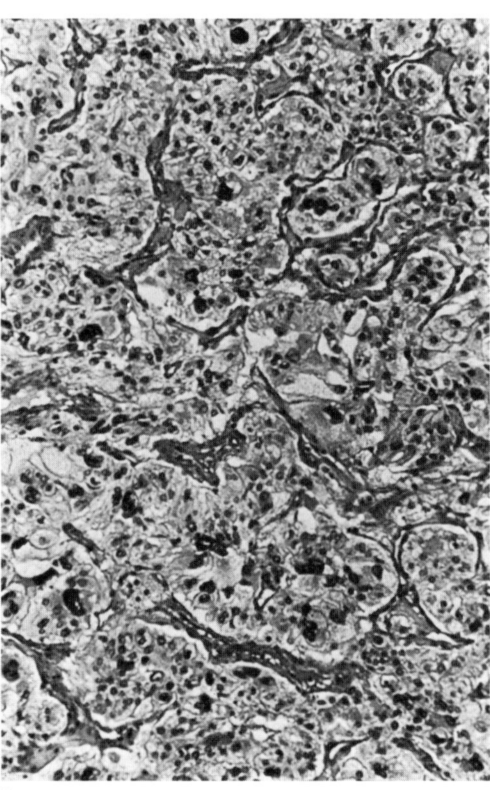

A B

Fig. 12–31. *Multiple endocrine neoplasia, type 2B*. (A) Bilateral pheochromocytomas. One on left is much larger than that on right. (B) Photomicrograph showing polyhedral cells separated by thin connective tissue septa, rich in blood vessels.

Fig. 12–32. *Multiple endocrine neoplasia, type 2B*. (A) Barium demonstrates megacolon. (B) Hyperplastic neuromatous infiltrate with ganglion cells in intestinal walls between muscle layers.

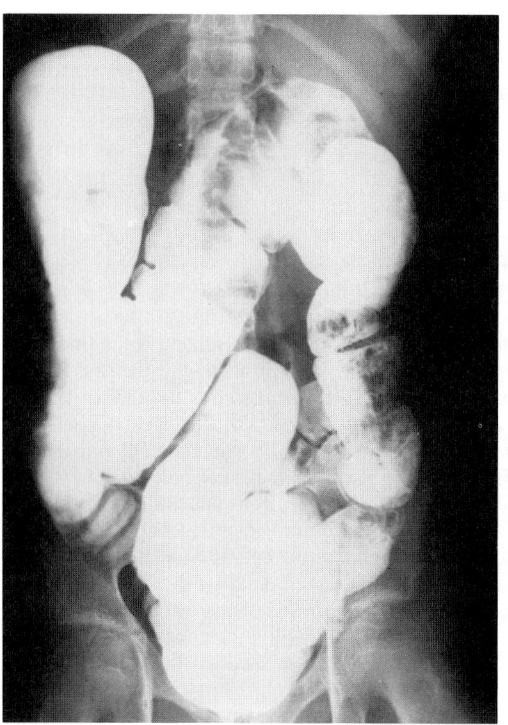

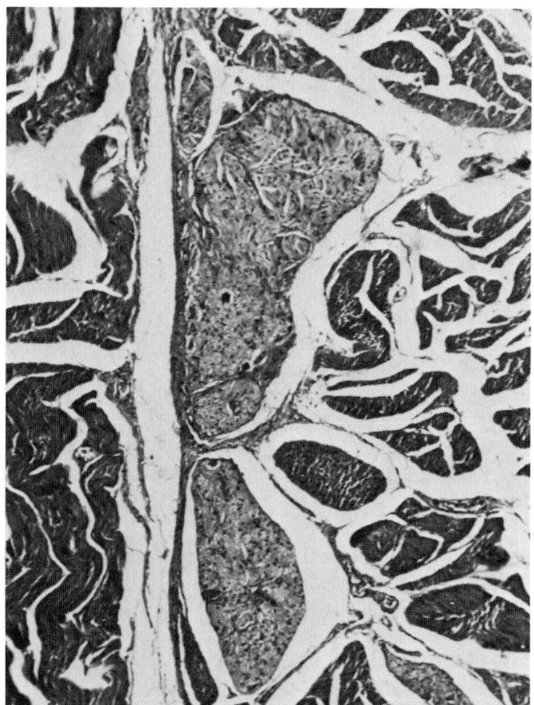

A B

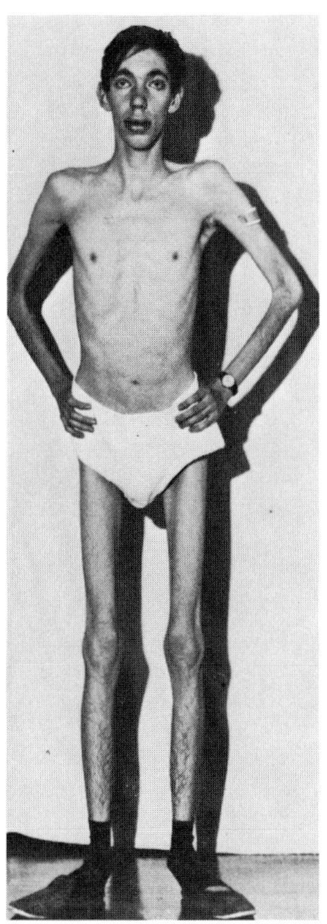
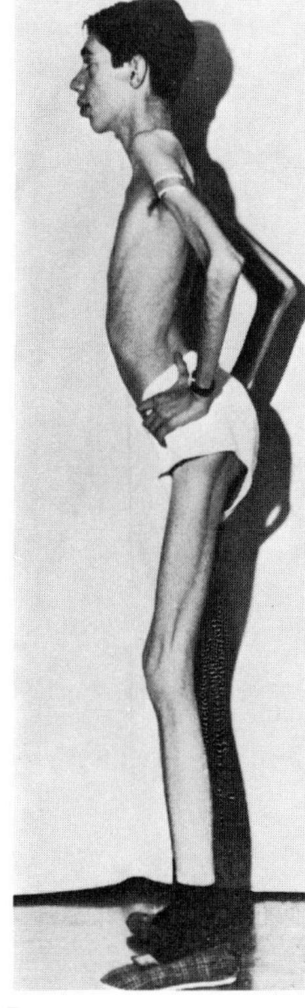


Fig. 12–33. *Multiple endocrine neoplasia, type 2B*. (A,B) Asthenic habitus with muscle wasting and lumbar lordosis. (A,B from *M Levy et al*, Arch Fr Pédiatr **27**:561, 1970.)

both specific and nonspecific cholinesterase activity, in contrast to neurofibroma, which rarely contains axons but is cholinesterase positive (2,36). Electron microscopic studies of the lingual mucosal neuromas reveal myelinated and nonmyelinated nerve fibers (36).

Cephalometric studies report high palatal vault, posterior crossbite, steep mandibular plane, and mandibular retrognathism (60). The mandibular canal and mental foramen are widened (Fig. 12–34D) (3).

**Differential diagnosis.** Pheochromocytoma may occur as an isolated tumor or as an autosomal dominant trait (71). It is also found in about 1% of cases of *neurofibromatosis*. The tumor may also be seen with von Hippel–Lindau syndrome (5) and with brain tumors including cerebellar hemangioblastoma, ependymoma, astrocytoma, meningioma, and spongioblastoma. Medullary carcinoma of the thyroid has been reported to occur without other abnormalities or as an autosomal dominant trait (40). Pheochromocytoma and medullary carcinoma of the thyroid (MEN 2A, MEN 2, Sipple syndrome) is also an autosomal dominant disorder and is about six times as common as MEN 2B. Some patients with MEN 2B have hyperparathyroidism secondary to hyperplasia or adenomas of the parathyroids due to excess calcitonin (69). Hyperparathyroidism may rarely have autosomal dominant inheritance or exhibit central giant cell granulomas or fibroosseous lesions of the jaws (61a,76a). Carney et al (19) reported three families in which pancreatic islet cell tumors and pheochromocytomas were inherited as an autosomal dominant disorder. Griffiths et al (40) reported the association of pheochromocytoma, neurofibromatosis, and duodenal carcinoid. Ganglioneuromatous polyps and juvenile or adenomatous polyposis of the large bowel have been reported. Intestinal ganglioneuromatosis as an isolated finding may have autosomal dominant inheritance (63). Multiple systematized neuromas of the skin and mucosa have been reported in the absence of the syndrome (1).

**Laboratory aids.** Elevated calcitonin levels may be expected if medullary carcinoma of the thyroid is present (9) even if the tumor is not clinically apparent. Several measurements are useful in detection of recurrence, metastases, or growth of residual tumor. Both serum and urinary assays are available (9). The pentagastrin stimulation test with or without calcium infusion is an effective technique for producing maximum calcitonin secretion (19,69,75). Intradermal injection of 1:1000 histamine produces a wheal but no flare in patients with medullary carcinoma of the thyroid, a finding that, in our opinion, is not related to its high histaminase content (7,35) as normal flare response does not return following tumor removal. It appears to be a type of dysautonomia. Irregular calcification of the tumor and pulmonary (57a), liver (41a), and bone (41a) metastases may be demonstrated in about 10% (28) by new imaging (scintigraphic) techniques (41a,57a). Fine needle aspiration biopsy has been used effectively in patients with palpable tumor (75). Fluorescamine can be employed for cytology on smears pretreated with formaldehyde gas (75).

Increased vanilmandelic acid and altered epinephrine/norepinephrine ratios occur with pheochromocytoma (42). Radiographically, the kidney may be displaced downward or egg-shell calcification of the upper pole may be found (38). Preferred methods of imaging are CT and ultrasound. Scintigraphy may aid in diagnosis of the tumor or its metastases (11a,31a,47a,69). Intestinal ganglioneuromatosis has a characteristic appearance on barium enema such as alternating areas of spasm and dilatation (2,28).

### References [Multiple endocrine neoplasia, type 2B (multiple mucosal neuroma syndrome)]

1. Altmeyer P, Merkel KH: Multiple systematisierte Neuroma der Haut und der Schleimhaut. *Hautarzt* **32**:240–244, 1981.
2. Anderson TE et al: Roentgen findings in intestinal ganglioneuromatosis. Its association with medullary thyroid carcinoma and pheochromocytoma. *Radiology* **101**:93–96, 1971.
3. Anneroth G, Hermdahl A: Syndrome of multiple mucosal neurofibromas, pheochromocytoma and medullary thyroid carcinoma. *Int J Oral Surg* **7**:126–131, 1978.
4. Arkins FL et al: Dopa decarboxylase in medullary carcinoma of the thyroid. *N Engl J Med* **289**:545–548, 1973.
5. Atuk NO et al: Familial pheochromocytoma, hypercalcemia, and von Hippel–Lindau disease. A ten year study of a large family. *Medicine* **58**:209–218, 1979.
6. Babu VR et al: Chromosome 20 deletion in human multiple endocrine neoplasia types 2A and 2B: A double-blind study. *Proc Natl Acad Sci USA* **81**:2525–2528, 1984.
6a. Babu VR et al: Chromosome 20 deletion in multiple endocrine neoplasia type 2: Expanded double-blind studies. *Am J Med Genet* **27**:739–748, 1987.
7. Baum JL: Abnormal intradermal histamine reaction in the syndrome of pheochromocytoma, medullary carcinoma of the thyroid gland and multiple mucosal neuromas. *N Engl J Med* **284**:963–964, 1971.
8. Baylin SB: The multiple endocrine neoplasia syndromes: Implications for the study of inherited tumors. *Sem Oncol* **5**:35–45, 1978.
9. Baylin SB et al: Elevated histaminase activity in medullary carcinoma of the thyroid gland. *N Engl J Med* **283**:1239–1244, 1970.
10. Bazex A et al: Syndrome de Sipple héréditaire. *Ann Dermatol Venereol* **104**:103–114, 1977.
11. Block MB et al: Multiple endocrine adenomatosis Type IIB. *JAMA* **234**:710–714, 1975.
11a. Bomanji J et al: Imaging neural crest tumors with $^{123}$I-metaiodobenzylguanidine and X-ray computed tomography. A comparative study. *Clin Radiol* **39**:502–506, 1988.
12. Braley AE: Medullated corneal nerves and plexiform neuromas associated with pheochromocytoma. *Trans Am Ophthalmol Soc* **52**:189–197, 1954.
13. Brown RS et al: The syndrome of multiple mucosal neuromas and medullary carcinoma of the thyroid in childhood. Importance of recognition of the phenotype for the early detection of malignancy. *J Pediatr* **86**:77–83, 1975.
13a. Calmettes L et al: Manifestations oculopalpébrales des neuromes myéliniques muquex. *Arch Ophtal* **19**:257–269, 1959.
14. Carney JA, Hayles AB: Alimentary tract manifestations of muliple endocrine neoplasia, type 2b, *Mayo Clin Proc* **52**:543–548, 1977.
15. Carney JA et al: Bilateral adrenal medullary hyperplasia in multiple endocrine neoplasia type 2: Precursor of bilateral pheochromocytoma. *Mayo Clin Proc* **50**:3–10, 1975.

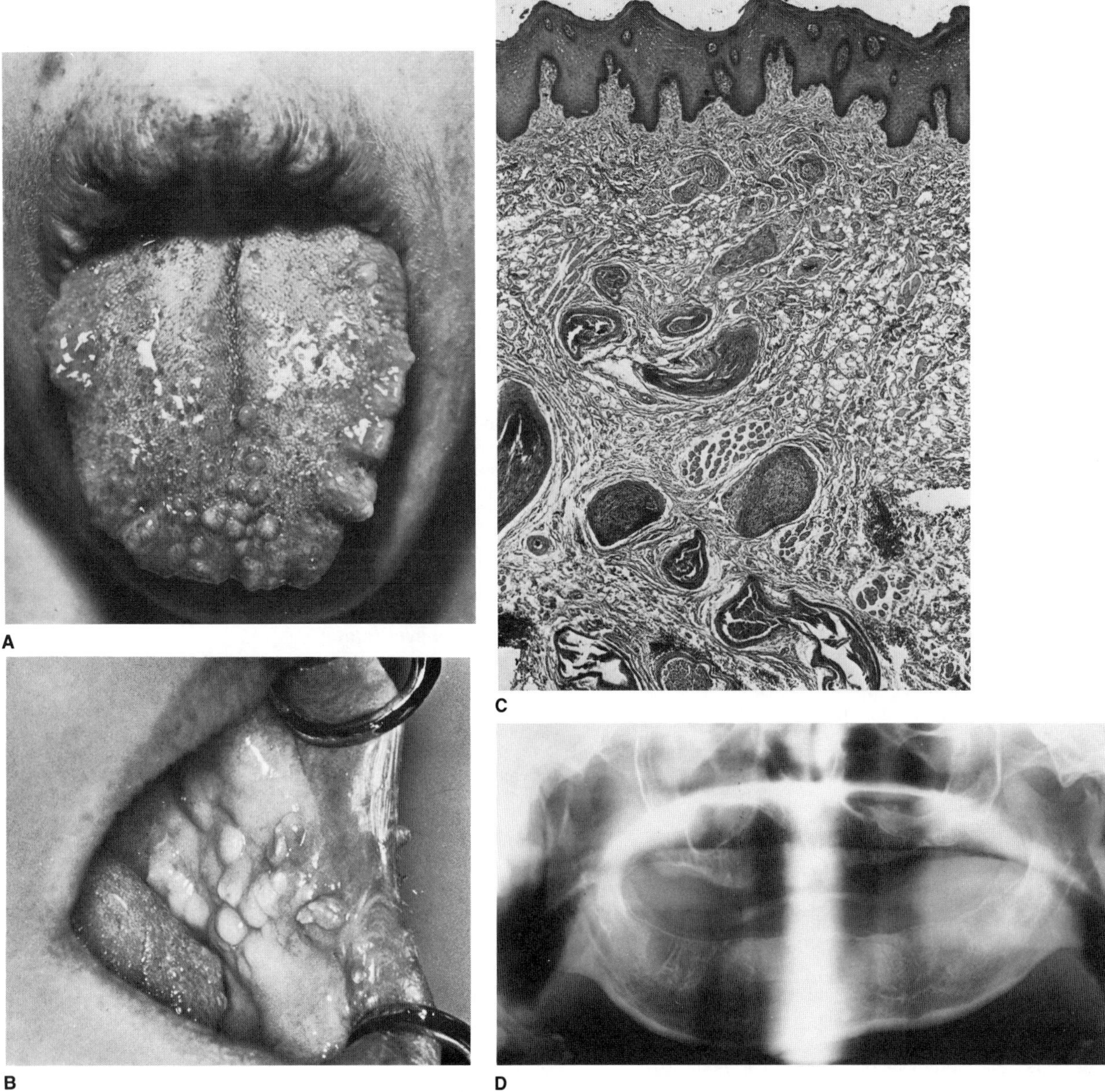

Fig. 12–34. *Multiple endocrine neoplasia, type 2B.* (A) Mucosal neuromas of lips and anterior and lateral dorsum of tongue. (B) Neuromas of buccal mucosa. (C) Congeries of axons of nerves. Histochemically, these are cholinesterase negative. (D) Widened mandibular canal. (A from *KW Bruce,* Oral Surg **7**:1150, 1954. B from *HE Simpson,* Oral Surg **19**:228, 1965.)

16. Carney JA et al: Mucosal ganglioneuromatosis, medullary thyroid carcinoma and pheochromocytoma: Multiple endocrine neoplasia, type 2b. *Oral Surg Oral Med Oral Pathol* **41**:739–752, 1976.

17. Carney JA et al: Alimentary tract ganglioneuromatosis: A major component of the syndrome of multiple endocrine neoplasia, type 2b. *N Engl J Med* **295**:1287–1291, 1976.

18. Carney JA et al: Multiple endocrine neoplasia, type 2B. *Pathobiol Annual* **8**:105–153, 1978.

19. Carney JA et al: Familial pheochromocytoma and islet cell tumor of the pancreas. *Am J Med* **68**:515–521, 1980.

20. Carney JA et al: Abnormal cutaneous innervation in multiple endocrine neoplasia, type 2b. *Ann Intern Med* **94**:362–363, 1981.

21. Carney JA et al: Multiple endocrine neoplasia with skeletal manifestations. *J Bone Jt Surg* **63A**:405–410, 1981.

22. Casino AJ et al: Oral-facial manifestations of the multiple endocrine neoplasia syndrome. *Oral Surg Oral Med Oral Pathol* **51**:517–523, 1981.

23. Cunliffe WJ et al: A calcitonin-secreting medullary thyroid carcinoma associated with mucosal neuromas, marfanoid features, myopathy and pigmentation. *Am J Med* **48**:120–126, 1970.

24. Cuthbert JA et al: Colonic and esophageal disturbance in a patient with multiple endocrine neoplasia, type 2B. *Aust NZ J Med* **8**:518–520, 1978.

25. Demos TC et al: Multiple endocrine neoplasia (MEN) syndrome type 2 B: Gastrointestinal manifestations. *Am J Roentgenol* **140**:73–78, 1983.

26. Dent CE et al: Medullary carcinoma of the thyroid gland in a girl aged 10 years. *Arch Dis Childh* **51**:223–226, 1976.

27. Dodd GD: The radiologic features of multiple endocrine neoplasia types IIA and IIB. *Sem Roentgenol* **20**:64–90, 1985.

28. Dunn EL et al: Medullary carcinoma of the thyroid gland. *Surgery* **73**:848–858, 1973.

29. Dyck PJ et al: Multiple endocrine neoplasia, type 2b: Phenotype recognition; neurological features and their pathological basis. *Ann Neurol* **6**:302–314, 1979.

30. Frank K et al: Importance of early diagnosis and follow-up in multiple endocrine neoplasia MEN IIB. *Eur J Pediatr* **143**:112–116, 1984.

31. Flensborg EW, Schiøtz PO: Abnormal intradermal histamine reaction and elevated serum calcitonin in the syndrome of: Medullary carcinoma of the thyroid gland, pheochromocytoma and multiple mucosal neuromas. A survey and a case in a child with symptoms from early infancy: *Dan Med Bull* **21**:20–26, 1974.

31a. Francis IR et al: Complementary roles of CT and $^{131}$I-MIBG scintigraphy in diagnosing pheochromocytomas. *Am J Roentgenol* **141**:719–725, 1983.

32. Froboese C: Das aus markhaltigen Nervenfascern bestehende, gangliezellenlose echte Neurom in Rankenform-zugleich ein Beitrag zu den nervösen Geschwülsten der Zunge und des Augenlides, *Virchows Arch Pathol Anat* **240**:312–327, 1923.

32a. Fryns J, Chrzanowska K: Mucosal neuromata syndrome [MEN type IIb(III)]. *J Med Genet* **25**:703–706, 1988.

33. Gagel RF et al: Age-related probability of development of hereditary medullary thyroid carcinoma. *J Pediatr* **101**:941–946, 1982.

34. Gorlin RJ: Skin test for medullary thyroid carcinoma, *N Engl J Med* **284**:983–984, 1971.

35. Gorlin RF, Mirkin BL: Multiple mucosal neuromas, pheochromocytoma, medullary carcinoma of the thyroid and Marfanoid body build with muscle wasting. Syndrome of hyperplasia and neoplasia of neural crest derivatives—an unitarian concept. *Z Kinderheilkd* **113**:313–325, 1972.

36. Gorlin RJ, Vickers RA: Multiple mucosal neuromas, pheochromocytoma, medullary carcinoma of the thyroid and Marfanoid body build with muscle wasting: Reexamination of a syndrome of neural crest malmigration. *Birth Defects* **7**(6):69–72, 1971.

37. Gorlin RJ et al: Multiple mucosal neuromas, pheochromocytoma and medullary carcinoma of the thyroid. A syndrome. *Cancer* **22**:293–299, 1968.

38. Grainger RG et al: Egg-shell calcification: A sign of pheochromocytoma. *Clin Radiol* **18**:282–286, 1967.

39. Graze K et al: Natural history of familial medullary thyroid carcinoma. *N Engl J Med* **299**:980–985, 1978.

40. Griffiths DFR et al: Multiple endocrine neoplasia associated with von Recklinghausen's disease. *Br Med J* **287**:1341–1342, 1983.

40a. Grün R, Eberle F: Multiple endocrine neoplasia, type II (MEN II). *Ergeb Inn Med Kinderheilkd* **46**:152–201, 1981.

41. Hamilton BP et al: Measurement of urinary epinephrine in screening for pheochromocytoma in multiple endocrine neoplasia type II. *Am J Med* **65**:1027–1032, 1978.

41a. Johnson DG et al: Bone and liver images in medullary carcinoma of the thyroid. *J Nuclear Med* **25**:419–422, 1984.

42. Jones BA, Sisson JC: Early diagnosis and thyroidectomy in multiple endocrine neoplasia, type 2B. *J Pediatr* **102**:219–223, 1983.

43. Joosten E et al: Hypertrophy of peripheral nerves in the syndrome of multiple mucosal neuromas, endocrine tumors and marfanoid habitus. Autonomic distribution of sural nerve findings. *Acta Neuropathol (Berlin)* **30**:251–261, 1974.

44. Kaufman FR et al: Metastatic medullary thyroid carcinoma in young children with mucosal neuroma syndrome. *Pediatrics* **70**:263–267, 1982.

45. Khairi MRA et al: Mucosal neuroma, pheochromocytoma and medullary thyroid carcinoma: Multiple endocrine neoplasia, type 3. *Medicine* **54**:89–112, 1975.

46. Kilp H, Walzer P: Ein Neurolemmom der Bindehaut und der Lider mit Veränderungen der Zunge und des Nasenraumes. *Klin Monatsbl Augenheilkd* **162**:251–254, 1973.

47. Koppang HS et al: Multiple endocrine neoplasia, type III. *Int J Oral Maxillofac Surg* **15**:483–488, 1986.

47a. Kumar R et al: Adrenal scintigraphy. *Sem Roentgenol* **23**:243–249, 1988.

48. Le Dourin N, Le Lièvre C: Démonstration de l'origine neurale des cellules à calcitonine du corps ultimobranchial chez l'embryon de poulet. *CR Acad Sci [D] (Paris)* **270**:2857–2860, 1970.

49. Levy M et al: Neuromatose et épithélioma à stroma amyloïde de la thyroïde chez l'enfant. *Arch Fr Pédiat* **27**:561–583, 1970.

50. Marks AD, Channick BJ: Extra-adrenal pheochromocytoma and medullary thyroid carcinoma with pheochromocytoma. *Arch Intern Med* **134**:1106–1112, 1974.

50a. Mathew CGP et al: A linked genetic marker for multiple endocrine neoplasia type 2A on chromosome 10. *Nature (London)* **328**:527–528, 1987.

51. Mendelsohn G, Diamond MP: Familial ganglioneuromatous polyposis of the large bowel. *Am J Surg Pathol* **8**:515–520, 1984.

52. Mielke JE et al: Diverticulitis of the colon in a young man with Marfan's syndrome. Associated with carcinoma of the thyroid gland and neurofibromas of the tongue and lips. *Gastroenterology* **48**:379–382, 1965.

53. Miller RL et al: The ultrastructure of oral neuromas in multiple mucosal neuromas, pheochromocytoma, medullary thyroid carcinoma syndrome. *J Oral Pathol* **6**:253–263, 1977.

54. Moyes CD, Alexander FW: Mucosal neuroma syndrome presenting in a neonate. *Dev Med Child Neurol* **19**:518–534, 1977.

55. Netzloff ML et al: Medullary carcinoma of the thyroid in the multiple mucosal neuromas syndrome. *Ann Clin Lab Sci* **9**:368–373, 1979.

56. Normann T, Otnes B: Intestinal ganglioneuromatosis, diarrhoea and medullary thyroid carcinoma. *Scand J Gastroenterol* **4**:553–559, 1969.

57. Norton JA et al: Multiple endocrine neoplasia type 2B. The most aggressive form of medullary thyroid carcinoma. *Surg Clin North Am* **59**:109–118, 1979.

57a. Ohta H et al: A new imaging agent for medullary carcinoma of the thyroid. *J Nuclear Med* **25**:323–325, 1984.

58. Perkins JT et al: Adenomatous polyposis coli and multiple endocrine neoplasia type 2B: A pathogenetic relationship. *Cancer* **55**:375–380, 1985.

59. Pryse JC et al: A dento-craniofacial study of the multiple endocrine neoplasia (MEN) type 3 syndrome. *J Oral Med* **34**:85–88, 1979.

60. Raue F et al: Pheochromocytoma in multiple endocrine neoplasia. *Cardiology* **72** (Suppl **1**):147–149, 1985.

61. Robertson DM et al: Thickened corneal nerves as a manifestation of multiple endocrine neoplasia. *Trans Am Acad Ophthalmol Otol* **79**:772–787, 1975.

61a. Rosen IB, Palmer JA: Fibro-osseous tumors of the facial skeleton in association with primary hyperparathyroidism: An endocrine syndrome or a coincidence? *Am J Surg* **142:**494–498, 1981.

62. Roy AD et al: Idiopathic intestinal pseudo-obstruction: A familial visceral neuropathy. *Clin Genet* **18**:291–297, 1980.

62a. Saltzman CL et al: Thick lips, bumpy tongue, and slipped capital femoral epiphyses, a deadly combination. *J Pediatr Orthoped* **8**:219–222, 1988.

63. Schimke RN: Multiple endocrine adenomatosis syndromes. *Adv Intern Med* **21**:249–266, 1974.

64. Schnitzler L et al: Neuromes cutanés et muqueux avec étude histopathologique et ultrastructurale. *Ann Dermatol Syph* **100**:241–260, 1973.

65. Schweitzer NMJ, Van der Pol BAE: Multiple mucosal neuroma (MMN) or multiple endocrine neoplasia (MEN) type 3 syndrome. Ocular manifestations: A case report. *Doc Ophthalmol* **44**:151–159, 1977.

65a. Sciubba JJ et al: The occurrence of multiple endocrine neoplasia, type IIb in two children of an affected mother. *J Oral Pathol* **16**:310–316, 1987.

65b. Simpson NE et al: Assignment of multiple endocrine neoplasia type 2a to chromosome 10 by linkage. *Nature (London)* **328**:528–530, 1987.

66. Sisson JC et al: Scintigraphy with I-131 MIBG as an aid in the treatment of pheochromocytomas in patients with the multiple endocrine neoplasia type 2 syndromes. *Henry Ford Hosp Med J* **32**:254–261, 1985.

67. Sizemore GW, Go VLW: Stimulation tests for diagnosis of medullary thyroid carcinoma. *Mayo Clin Proc* **50**:53–56, 1975.

67a. Spector B et al: Histologic study of the ocular lesions in multiple endocrine neoplasia type IIb. *Am J Ophthalmol* **91**:204–215, 1981.

68. Steiner AL et al: Study of a kindred with pheochromocytoma, medullary thyroid carcinoma, hyperparathyroidism and Cushing's disease: Multiple endocrine neoplasia, type 2. *Medicine* **47**:371–410, 1968.

69. Stjernholm MR et al: Medullary carcinoma of the thyroid before age 2 years. *J Clin Endocrinol Metab* **51**:252–253, 1980.

70. Swinton NW et al: Hypercalcemia and familial pheochromocytoma. Correction after adrenalectomy. *Ann Intern Med* **76**:455–457, 1972.

71. Thies W: Multiple echte fibrilläre Neurome (Rankenneurome) der Haut und Schleimhaut. *Arch Klin Exp Dermatol* **218**:561–573, 1964.

72. Temple WJ et al: The APUD system and its apudomas. *Int Adv Surg Oncol* **4**:255–276, 1981.

73. Telenius-Berg M et al: Impact of screening on prognosis in the multiple endocrine neoplasia type 2 syndromes: Natural history and treatment results in 105 patients. *Henry Ford Hosp Med J* **32**:225–232, 1984.

74. Voelkel EF et al: Concentrations of calcitonin and catecholamines in pheochromocytomas, a mucosal neuroma and medullary thyroid carcinoma. *J Clin Endocrinol Metab* **37**:297–307, 1973.

75. Voorhess ML: Functioning tumors. *Am J Dis Child* **134**:14–15, 1980.

76. Wagenmann A: Multiple Neurome des Auges und der Zunge. *Ber Dtsch Ophthalmol Ges* **43**:282–285, 1922.

76a. Warnakulasuriya S et al: Familial hyperparathyroidism associated with cementifying fibroma of the jaws in two siblings. *Oral Surg* **59**:269–274, 1985.

77. Webb TA et al: Differences between sporadic pheochromocytoma and pheochromocytoma in multiple endocrine neoplasia, type 2. *Am J Surg Pathol* **4**:121–126, 1980.

78. Weidner N et al: Mucosal ganglioneuromatosis associated with multiple colonic polyps. *Am J Surg Pathol* **8**:779–786, 1984.

79. Welbourn RB: Current status of the apudomas. *Ann Surg* **185**:1–12, 1977.

80. White MP et al: Mucosal neuroma syndrome—a phenotype for malignancy. *Arch Dis Childh* **60**:870–871, 1985.

81. Whittle TS Jr, Goodwin MN Jr: Intestinal ganglioneuromatosis with the mucosal neuroma-medullary thyroid carcinoma-pheochromocytoma syndrome. A case report and review of the literature. *Am J Gastroenterol* **65**:249–257, 1976.

82. Williams ED, Pollock DJ: Multiple mucosal neuromata with endocrine tumours: A syndrome allied to von Recklinghausen's disease. *J Pathol Bacteriol* **91**:71–80, 1966.

83. Winkelmann RK, Carney JA: Cutaneous neuropathology in multiple endocrine neoplasia, type 2B. *J Invest Dermatol* **79**:307–312, 1982.

84. Wright BA, Wysocki GP: Traumatic neuroma and multiple endocrine neoplasia type III. *Oral Surg Oral Med Oral Pathol* **51**:527–530, 1981.

## The neurofibromatoses (NF I Recklinghausen type, NF II acoustic type, other types)

In 1849, Robert Smith (86a), the First Professor of Surgery at Dublin Medical School, reported the clinical and necropsy findings in two cases of neurofibromatosis and cited 75 references from the earlier medical literature. He did not recognize that the tumors contain neural elements, however, and it was von Recklinghausen's (93) publication in 1882 that convinced the medical world that neurofibromatosis (NF) was a distinct entity.

Today neurofibromatosis is known to be etiologically heterogeneous and two distinct types have been well delineated; other clinical types are known that need to be further defined. There were earlier indications that neurofibromatosis might be etiologically heterogeneous. The mutation rate had been calculated to be $1 \times 10^{-4}$ mutations per gamete per generation (16), the highest in man. Cohen and Hayden (12b), who first differentiated *Proteus syndrome* from neurofibromatosis in 1979, suggested that the mutation rate might conceivably reflect the pooling of several heterogeneous entities.

The reader is referred to the following sources for more detailed coverage: general aspects (69,70,72,96), genetics (5,8,19,23,41,60, 71,73,78,83,84,92,100), NF I Recklinghausen type (2,5,16,40,70,83), NF II acoustic type (40,56a,70,78,84), NF V segmental type (16,45a, 45b,60a,70,77a,78a,81a), NF IX neurofibromatosis/Noonan type (1,1c,12a,46,58,65,81), tumors and neoplasms (12,20,32,38,49,54, 63,95,99), cutaneous manifestations (6,15,16,42, 45,57,101,105,107), skeletal findings (36,37,39,44,48,85), central nervous system manifestations (17,56a,76), ophthalmologic features (25,27,29,54,68,108), endocrine findings (80), pediatric aspects (9,13,14,21,24,36,80,105), and animal models (34,82,see 70).

**Types of neurofibromatosis.** Known and suggested types of neurofibromatosis appear in Table 12–5. Suggested types are heuristic in nature. Types I through VII have been proposed elsewhere (70). We have added Types VIII and IX.

NF I Recklinghausen type. This classic form of neurofibromatosis accounts for about 90% of all cases. Major features include six or more café-au-lait spots, cutaneous neurofibromas, and Lisch nodules. Axillary freckling develops in approximately 66% of all patients. Inheritance is autosomal dominant with about 50% of cases representing new mutations. The gene responsible is located near the centromere of chromosome 17 (2,5,19,70,83).

NF II acoustic type. The hallmark consists of bilateral acoustic neuromas. Symptoms are usually caused by pressure on the vestibulocochlear and facial nerve complex, the first symptom usually being hearing loss that often begins during the teenage years or early twenties, but occasionally may occur as early as the first or as late as the seventh decade of life. Café-au-lait spots and cutaneous neurofibromas are also present, but occur less commonly than in NF I. Neurofibromas are easy to overlook; they are generally less than 2 cm in diameter, minimally raised, and often have a roughened surface that may have more prominent hairs than the surrounding skin. Axillary freckling is uncommon. On occasion, deep plexiform neurofibromas may result in palpable masses. Tumors of the central nervous system are especially common, Schwann cell tumors occurring most frequently. However, multiple tumors of meningeal or glial origin may also occur. Other Schwann cell tumors may develop along the cranial nerves or spinal roots. Presenile lens opacities or subcapsular cataracts occur in about 50%. Lisch nodules are absent. NF II has autosomal dominant inheritance with penetrance of over 95%. The gene responsible has been mapped to chromosome 22 (40,56a,70,78,84).

Because NF I and NF II may both have central and peripheral manifestations, the previously used terms ''peripheral neurofibromatosis'' for NF I and ''central neurofibromatosis'' for NF II have been discarded as misleading and confusing (56a).

NF III mixed type. This type of neurofibromatosis has features of NF I and NF II. Café-au-lait spots are usually pale, few in number, and may be large. Cutaneous neurofibromas are especially common on the palms. A multiplicity of brain tumors occurs including acoustic neuromas, meningiomas, and spinal/paraspinal neurofibromas. Optic gliomas do not occur. Lisch nodules are absent. Brain tumors are usually of early onset with a rapid course that precludes procreation. To date, all cases have been sporadic (70).

NF IV variant type. This is a residual category of variant phenotypes for patients that do not fit neatly into any other known type (70).

NF V segmental type. In this form, neurofibromas and café-au-lait spots are restricted to one area of the body. To be classified as the segmental type, the café-au-lait spots and/or axillary freckling must be ipsilateral to tumor involvement with no crossing of the midline (70). This localized form of neurofibromatosis is probably not rare. For example, Crowe et al (16) reported four instances, Miller and Sparkes (60a) noted two cases, and Roth et al (77a) summarized 23 cases including four of their own and proposed a subclassification of NF-V. Jung (45a) added several more examples. Segmental neurofibromatosis probably arises as a somatic mutation that affects a single cell early during embryonic development, producing limited distribution in patients. Most cases of the segmental type have been sporadic (16,60a,70), but familial occurrence has been reported (78a).

NF VI café-au-lait type. In this form, only café-au-lait spots are found. Neurofibromas and Lisch nodules are absent. In some instances, however, nonspecific features, such as pectus excavatum, may occur. Inheritance is autosomal dominant (70).

NF VII late onset type. Neurofibromas do not become apparent until the end of the third decade or later. Café-au-lait spots and Lisch nodules are absent. To date, all cases have been sporadic (70).

NF VIII gastrointestinal type. In this form, neurofibromatous involvement is limited to the gastrointestinal tract. Onset of symptoms is delayed until adulthood and some carriers are asymptomatic until their middle or late adult years. Increased risk of intestinal problems include those of bleeding, intussusception, and obstruction (33a,54a). Inheritance is dominant and most likely autosomal, although no male-to-male transmission has been recorded as of this writing (33a). Reciprocal translocation between chromosomes 12 and 14 has been described in one family and may be due to chance, or due to linakge of the gene for intestinal neurofibromatosis to one of the breakpoints in chromosome bands 12q13 and 14q13 (92).

NF IX neurofibromatosis/Noonan type. Patients have features of both neurofibromatosis and Noonan syndrome. A number of interpretations are possible: (1) chance concurrence of Noonan syndrome and neurofibromatosis, (2) neurofibromatosis/Noonan syndrome as an unusual variant of Noonan syndrome, (3) neurofibromatosis/Noonan syndrome as an unusual type of neurofibromatosis, and (4) neurofibromatosis/Noonan syndrome as a newly recognized entity. Most cases have been sporadic to date (1b,46,58,65,81), although vertical transmission has been noted (1). Clayton-Smith and Donnai (12a) reported a family in which a male child and his mother had Noonan syndrome

Table 12–5. Types of neurofibromatosis

| Type | Characteristic features | Comments | References |
|---|---|---|---|
| NF I von Recklinghausen type | Six or more café-au-lait spots, cutaneous neurofibromas, Lisch nodules, axillary freckling | Accounts for 90% of all cases. Autosomal dominant. Gene located near centromere of chromosome 17 | 2,5,16,40,70,83 |
| NF II acoustic type | Bilateral acoustic neuromas, café-au-lait spots, and cutaneous neurofibromas less commonly than in NF I. Schwann cell tumors of cranial nerves or spinal roots, meningiomas, gliomas, subcapsular cataracts. No Lisch nodules | Autosomal dominant. Gene maps to chromosome 22 | 40,56a,70,78,84,100 |
| NF III mixed type | Café-au-lait spots that are pale, few in number, and may be large; cutaneous neurofibromas especially of the palms; multiplicity of brain tumors (acoustic neuromas, meningiomas, spinal/paraspinal neurofibromas); absence of optic gliomas and Lisch nodules | All sporadic to date. Multiple brain tumors of early onset with rapid course precludes procreation | 70 |
| NF IV variant type | Variant phenotypes | Describes patients that do not fit neatly into any other type | 70 |
| NF V segmental type | Neurofibromas and café-au-lait spots in restricted area of the body | Differentiating characteristic is that café-au-lait spots and/or axillary freckling be ipsilateral to tumors with no crossing of midline. Most cases sporadic, occasional familial instance | 16,45a,60a,70,77a, 78a,81a |
| NF VI café-au-lait type | Café-au-lait spots only, nonspecific features, e.g., pectus excavatum in some cases, absence of neurofibromas and Lisch nodules | Autosomal dominant | 70 |
| NF VII late onset type | Neurofibromas not apparent until end of third decade or later. Absence of café-au-lait spots and Lisch nodules | All cases sporadic to date | 70 |
| NF VIII gastrointestinal type | Neurofibromas limited to gastrointestinal tract | Dominant inheritance, most likely autosomal. Gene may possibly be linked to one of the breakpoints in chromosome bands 12q13 and 14q13 | 33a,54a,92 |
| NF IX neurofibromatosis/ Noonan type | Features of neurofibromatosis and Noonan syndrome combined | Most cases sporadic to date. Familial cases noted on occasion | 1,1c,12a,46,58,65, 81 |

with NF, a sister had NF alone, and a brother had Noonan syndrome with no signs of NF. We also know of the association of neurofibromatosis and cherubism, possibly due to contiguous gene deletion.

**Classic neurofibromatosis (NF I).**

Etiology and pathogenesis. The classic form of neurofibromatosis, as described by von Recklinghausen (93), appears with a frequency of one case per 2500 to 3000 births and occurs approximately once in 200 individuals with mental deficiency (16). About 80,000 people in the United States are affected (70). Inheritance is autosomal dominant with almost 100% penetrance (70). Although it has been suggested that 25% of all cases represent new mutations (60), an earlier study (16) and a recent, large study (70) show that 50% represent new mutations, and an increase in paternal age at the time of conception has been found as an associated feature (73). Some authors (60) have suggested that maternal neurofibromatosis increases the overall severity of the disorder in affected offspring of the mothers, but in a large, well-controlled study (71), no significant differences in overall severity were found between maternal affected, paternal affected, and sporadically occurring cases.

In our opinion, it is incorrect to classify neurofibromatosis as a neurocutaneous syndrome because not all defects are attributable to neural crest derivatives. Many of the features such as café-au-lait spots, neurofibromas, sphenoid bone dysplasia, and pheochromocytoma are, in fact, of neural crest origin (70,72,96). Other findings, however, such as cerebral cortical heterotopias and optic gliomas, appear to be derived from the neural tube itself. Still other findings such as pseudoarthrosis, coarctation of aorta, renal artery stenosis, rhabdomyosarcoma, and leukemia appear to be of mesodermal origin (70,72). Finally, findings such as generalized short stature (70) defy explana-

ml—mesodermal layer
SM—SOMATIC MESODERM
—kyphosis
—scoliosis
—hemihypertrophy
—bowed tibia
—pseudarthrosis
—ribbon-like ribs
—pelvis deformities
—coarctation of aorta
—renal artery stenosis
—microaneurysms
—fibroma molluscum
—fibrosing alveolitis (?)

ng — neural groove
+
np — neural plate
NT—NEURAL TUBE:
—cortical cerebral heteropias
—optic glioma
—brain tumors

nj—neuroectodermal junction
NC—NEURAL CREST:
meningiomas
sphenoid bone dysplasia
pheochromocytoma
café au lait spots
neurofibroma
schwannoma
giant pigmented nevi

ml ng np nj nj
ml nj ng np
NT NC ml
NT NC SM

A

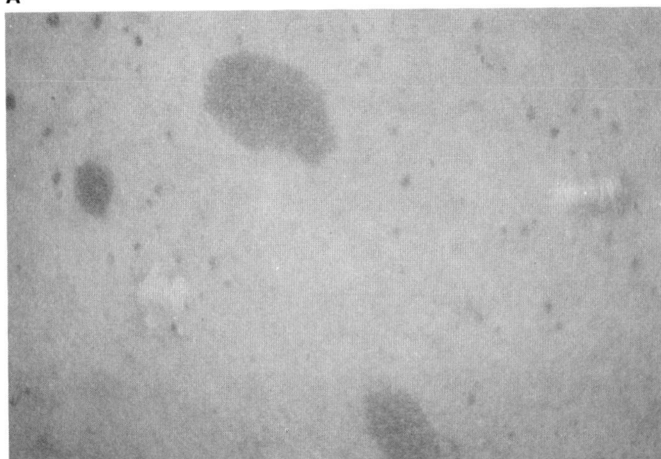

B

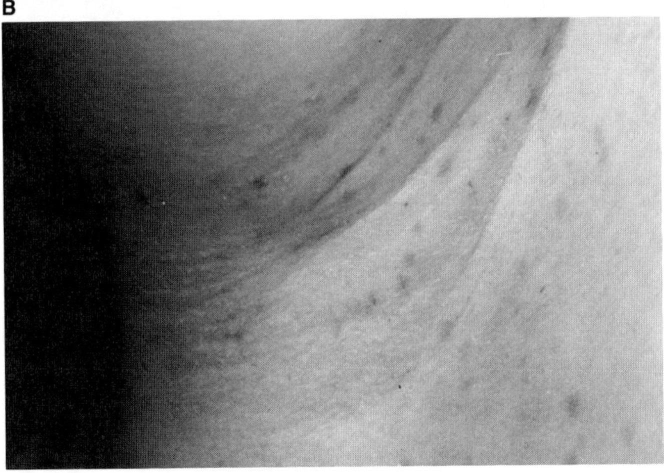

C

Fig. 12–35. *Neurofibromatosis.* (A) Embryonic involvement of somatic mesoderm, neural tube, and neural crest. (B) Cáfe-au-lait spots. (C) Axillary freckling. (A from *JV Wander* and *TK Das Gupta*, Curr Prob Surg **14**:3, 1977.)

tion. Although most evidence points to neurofibromatosis as a disorder of neural crest derivation, some controversy remains about whether the neural and mesenchymal components are interrelated or arise independently of each other. Figure 12–35A illustrates embryonic involvement of somatic mesoderm, neural tube, and neural crest (96). At present, it is not known whether features should be explained on the basis of the same gene expressed in different tissues or whether only neural crest cells are primarily affected followed by interaction between crest cells and other types of cells. We favor the former interpretation in which neural crest abnormalities are considered the major pleiotropic effect.

Natural history. Natural history has been discussed by a number of authors (4a,70,86b). Over 40% of patients have some manifestations at birth, and over 60% by the second year of life (24). Café-au-lait spots usually develop first, with multiple lesions present within the first year of life. In about 50%, axillary freckling appears later (79). Cutaneous neurofibromas appear around the onset of puberty and increase in number throughout life (Figs. 12–35B,C to 12–38). Lisch nodules, best observed in slit lamp examination, begin to appear in early childhood and have been observed in almost all affected adults (40). Average height is reduced and 16% are below the third centile (70).

About 33% of all patients develop one or more complications. Plexiform neurofibromas occur in 30%. About 6% of those over 18 develop some form of malignancy. Other important complications include neurological problems in 10% (including epilepsy, aqueductal stenosis, and spinal neurofibromas), scoliosis in 5%, pseudoarthrosis in 3%, gastrointestinal neurofibromas in 2%, endocrine neoplasms in 2%, and renal artery stenosis in 2% (40). Approximately 8–9% have mental retardation, but learning disabilities of various kinds affect 25% (70).

Neoplasia. The most distinctive and common skin neoplasm is the neurofibroma, especially the plexiform variety (32). Fialkow et al (23) analyzed neurofibromas from glucose-6-phosphate dehydrogenase A-B heterozygotes and concluded that each tumor had multiple cell origin, with tumorogenesis minimally involving 150 cells.

Neoplasms may be present at birth or appear during childhood or even later. They vary greatly in size with localized enlargement of many nerve trunks in larger neurofibromas. They are most striking on the skin, with some patients manifesting few, hundreds, or even thousands of individual neurofibromas and others having large unilateral pendulous masses (Fig. 12–36C,D). Many organs may be involved, including stomach, intestines, kidney, bladder, larynx, and heart (7,9,10,18,24,30,75). In the head and neck region, the most commonly affected sites are the scalp, cheek, neck, and oral cavity (28,50,55). Neurofibrosarcomatous transformation has been reported in 3–12% (24,37,40,49). Schwannomas, meningiomas, astrocytomas (especially optic gliomas), ependymomas, and rarely medulloblastomas have been observed (17,38,42,70,76).

A variety of other tumors and neoplasms have been recorded including cutaneous angiomas in 53% (99), subcutaneous leiomyomas (22a), carcinoid tumor (22a), xanthogranulomas with excessive frequency in young patients (63,70), subcutaneous neurofibromas in and about the cervical spinal cord in about 2–5% (70), pheochromocytoma in about 1% or less (38,70), neuroblastoma (38), rhabdomyosarcoma (38), Wilms tumor (38,95), leukemia with a striking excess of nonlymphocytic forms especially juvenile chronic myelogenous leukemia (12,38,70), adenocarcinoma of the pancreas (70), lipoma (24), liposarcoma (20), and virilizing adrenal carcinoma (24).

Skin. In addition to nodular tumors of the skin, café-au-lait spots (Fig. 12–35B) are found in over 99%. The smooth-edged pigmented macules are usually present at birth, but may take months, or even a year, to appear. They increase in size during the first decade and vary in size from 1 to 2 mm to over 15 cm. Their distribution is random over the body except for a disproportionately small number on the face (69). The color varies from yellowish to chocolate-brown. The

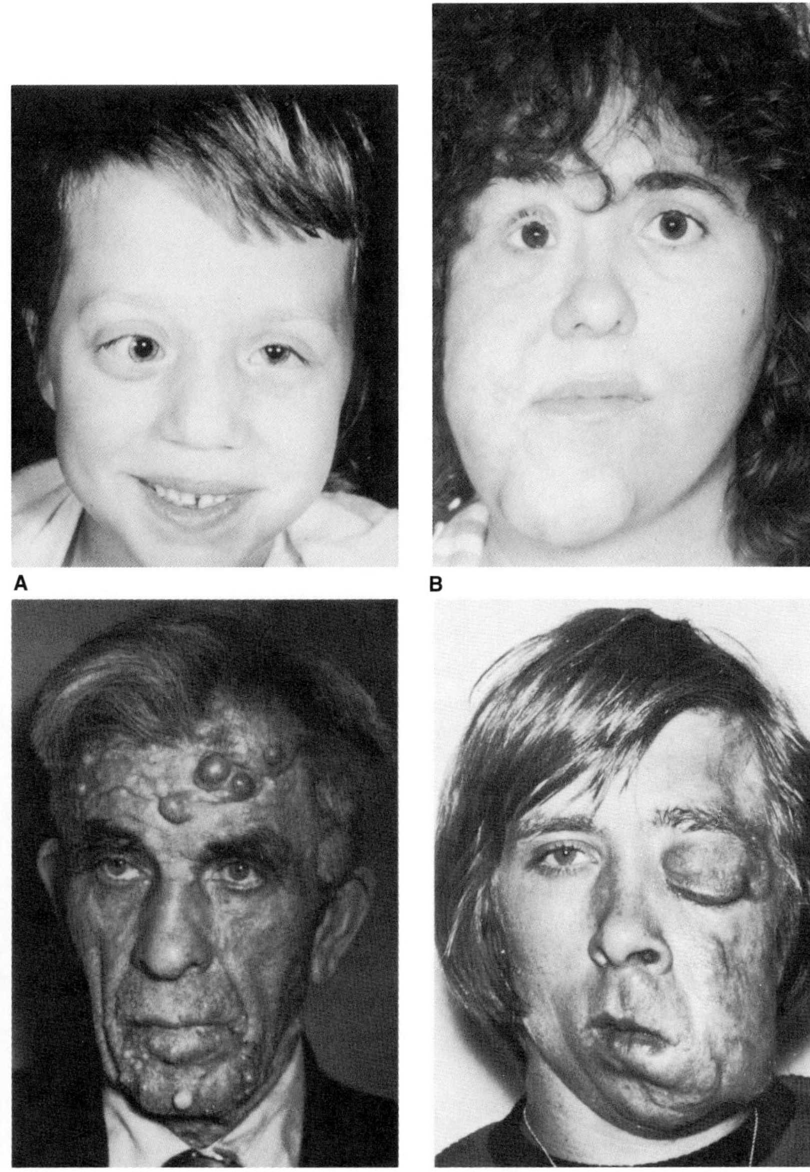

Fig. 12–36. *Neurofibromatosis.* Various craniofacial features of neurofibromatosis. (A) Unilateral exophthalmos due to bony defect of the posterosuperior orbital wall. See Fig. 12–37. (B) Facial asymmetry. (C) Patient with hundreds of cutaneous neurofibromas. (D) Severe involvement with unilateral neurofibromatosis. (D from *I Coblin* and *B Reil,* J Maxillofac Surg **3**:23, 1975.)

presence of six or more café-au-lait spots greater than 1.5 cm in diameter has come to represent the criterion for diagnosing neurofibromatosis, although fewer are present in some instances (15,16,69,77, 105). Axillary freckling (Fig. 12–35C) is present in approximately 50% (79) and, if present, is a significant diagnostic clue. Pigmented hairy nevi may also be noted (32,88). Cutaneous blue-red and pseudoatrophic macules and palmar melanotic macules have been reported as additional cutaneous signs (101,107). Dermatoglyphic findings include an excess of digital central pocket patterns (94).

Central nervous system. Mental deficiency with an IQ under 70 occurs in about 8–9%. Learning disabilities, present in 25%, include easy distractability, impulsiveness, deficient visual–motor coordination, excessive scatter of scores from one set of test items to another, and language and vocabulary deficits (70). Seizures are seen in about 5% (70), and frank hydrocephaly with aqueductal stenosis (38a) as well as asymptomatic ventricular dilation (38a,70) have been recorded. Distortion of cortical architecture from glial proliferation and neuronal heterotopias deep in the cerebral white matter have been described (17,32,76). Rosman and Pearce (76) suggested that neuronal heterotopias, best explained by cortical neuron migration arrest during brain development, might be linked to mental retardation. Ventriculomegaly with Chiari type 1 malformations has been observed in NF-V (1a). Riviello et al (74a) reported two cases of aqueductal stenosis and reviewed 25 cases in the literature, estimating a frequency of about 1%. They also found macrocephaly in about 30%.

Eyes. Any part of the eye may be involved. Neurofibromas of the eyelids have been occasionally observed. Intraorbital lesions may produce proptosis and muscle palsies. Sphenoid bone dysplasia may produce pulsating exophthalmos (Figs. 12–36A and 12–37A,B). Phakoma, congenital glaucoma, corneal opacity, detached retina, optic atrophy, and congenital ptosis of the eyelids have also been reported (25,27,29,55,68). Lisch nodules of various sizes may be found anywhere in the iris (Fig. 12–37C). These lesions are melanocytic hamartomas and are seen only in NF patients. Lisch nodules have a direct relationship to increasing patient's age and severity of skin lesions (108).

Skeletal system. Bony abnormalities have been particularly well discussed by Crawford (13) and Crawford and Bagamery (14). Scoliosis, the most common skeletal defect observed in NF (14), ranges

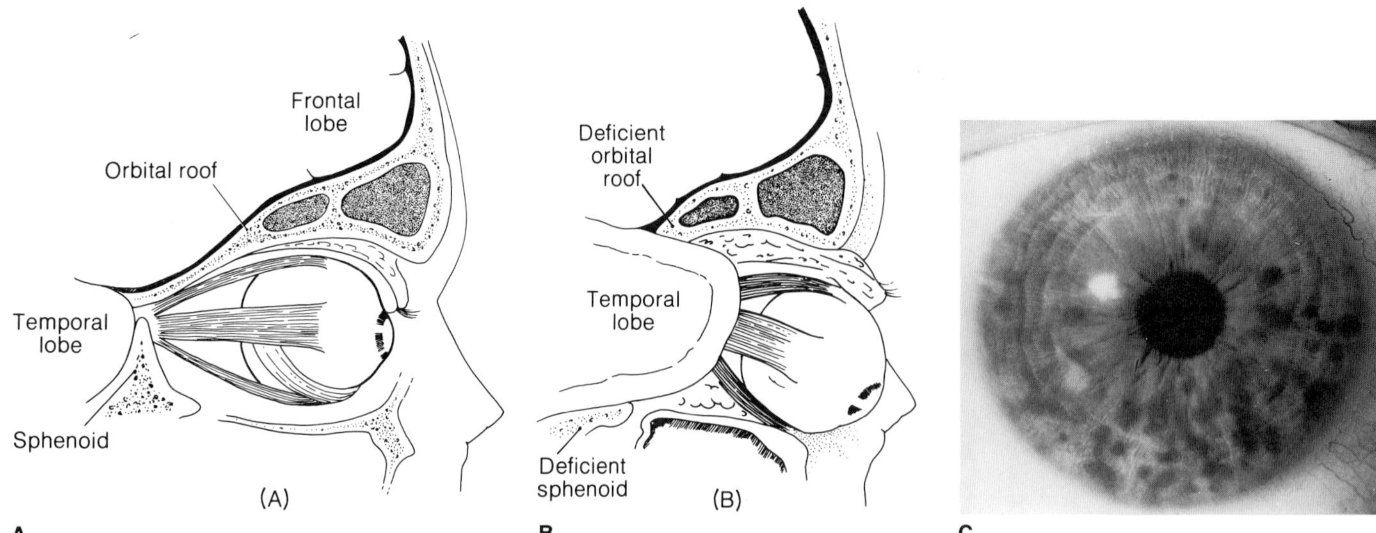


Fig. 12–37. *Neurofibromatosis.* (A) Normal anatomy. (B) Deficient orbital roof and sphenoid wing allowing prolapse of temporal lobe with exophthalmos. See Fig. 12–36A. (C) Lisch nodules. (A,B adapted from *A Bruwer,* Arch Ophthalmol **53**:3, 1955. C courtesy of *CG Summers,* Minneapolis, Minnesota.)

from mild to severe curvature; the etiology is thought to be secondary to a localized neurofibroma eroding and infiltrating bone. Other spinal defects include kyphosis, cervical spine abnormalities, and spondylolisthesis. Commonly observed are subperiosteal erosive changes caused by pressure from proliferating neurofibromatous tissue in the periosteum and overlying soft parts. Central ''cystic'' lesions of bone result from expansive growth of neurofibromas within the medullary cavity in some patients. In other cases, no cause for central lesions can be found (44).

Pseudoarthroses (commonly with bowing of the tibia and fibula) are also common (35). Bony defects of the skull, especially of the posterosuperior orbital wall (Fig. 12–37A,B), overgrowth of cranial bones, and craniofacial asymmetry (Fig. 12–36B,D) have been reported. Macrocephaly is a well known association (70). Hypertelorism has been found in 24% in the series reported by Westerhof et al (102). A variety of other anomalies may be observed, including hemihyperplasia of a limb or digit, spina bifida, absent patella, elevated scapulas, congenital dislocations (especially of the hip, radius, and ulna),

Fig. 12–38. *Neurofibromatosis.* Oral involvement. (A) Neurofibromatosis tumor of tongue. (B) Photomicrograph of plexiform neurofibroma of tongue seen in A. (C) Involvement of maxilla of patient seen in Fig. 12–36D. (D) Roentgenogram of patient shown in Fig. 12–36D. Note changes in mandibular ramus and angle. (C,D from *I Coblin* and *B Reil,* J Maxillofac Surg **3**:23, 1975.)

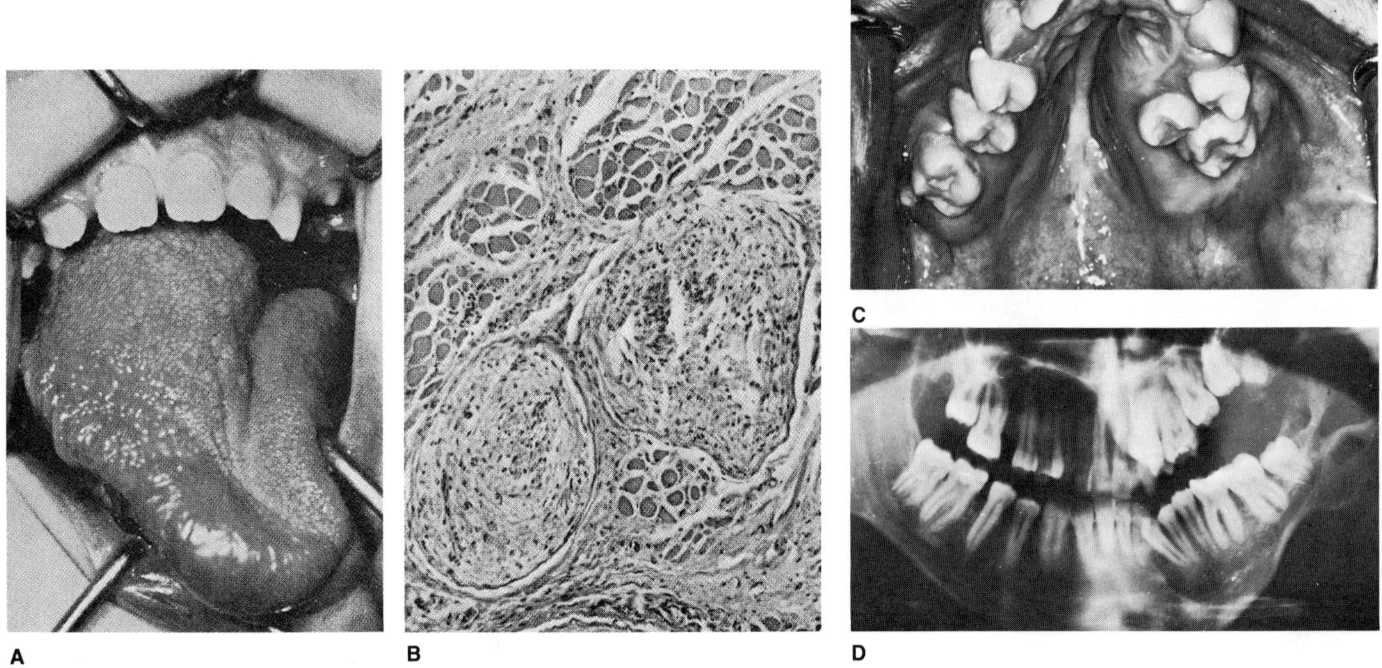

clubfoot, syndactyly, and complete or partial absence of limb bones (14,16,24,32,37,39,44,64,74). Postaxial polydactyly was noted in one affected family by Merlob et al (59).

Endocrine system. Findings have been well reviewed by Saxena (80). In childhood, the most common endocrine abnormality is sexual precocity (52). Other findings have included hypopituitarism, hypogonadism, gigantism, acromegaly, delayed sexual development, obesity, hypoglycemia, diabetes insipidus, goiter, myxedema, and hyperparathyroidism.

Cardiovascular system. Cardiovascular anomalies are low-frequency findings: pulmonic valvular stenosis, supravalvular aortic stenosis, coarctation of the aorta, ASD, congenital heart block, stenotic renal arteries, and other defects (18,32,47,75,86). Vascular malformations of the parotid gland have been reported (22).

Oral manifestations. The frequency of oral lesions in the past was stated to be 4–7%, but recent studies of a large series of patients by Shapiro et al (85) ascertained by clinical and complete radiographic studies suggest a higher frequency. D'Ambrosio et al (16a) reported an excellent study of jaw and skull changes, noting intraoral manifestations in 66% and skeletal involvement of the maxilla and mandible in 58%. Their overall sample had involvement of some kind in 92%. Although the age range spanned 6–66 years, we assume that children made up a small percentage of the sample. Most common are oral neurofibromas, enlarged fungiform papillae, intrabony lesions, wide inferior alveolar canals, and enlarged mandibular foramina (4,85). Tumors may involve any oral soft tissue, although there is some predilection for the tongue (3,26,31,43,53,62,87,90,98,104,106). Tongue lesions, consisting mostly of macroglossia and/or single or multiple tumors, have a sex ratio of almost 2M:1F (4) (Fig. 12–38A,B). Lesions of the bony maxilla and mandible are uncommon (74,106) (Fig. 12–38C,D), but with marked involvement of the face there may be combined maxillo-zygomatico-temporomandibular hypoplasia not from pressure atrophy but from maldevelopment of these bones (48). Hyperplasia of the soft oral tissues has been described associated with underlying bony hypoplasia (103). Teeth are consequently in malposition with total or partial retention. No enamel or dentin defects have been reported (103).

Other findings. Fibrosing alveolitis has been an associated finding (67,96). Velopharyngeal insufficiency with no cause identified after extensive workup was recorded in seven patients (66). The reader is referred to several sources for other low-frequency abnormalities (9,16,21,24,70).

**Differential diagnosis.** Neurofibromatosis should be distinguished from *multiple mucosal neuroma syndrome (multiple endocrine neoplasia, type 2B), Klippel–Trénaunay–Weber syndrome,* multiple lipomatosis, *LEOPARD syndrome,* and *hemihyperplasia (hemihypertrophy).* The café-au-lait spots of *McCune–Albright syndrome* tend to be more markedly scalloped (coast of Maine appearance) in contrast to the smooth-edged lesions (coast of California appearance) of neurofibromatosis (77,105). Café-au-lait spots may also be observed in association with pulmonic stenosis and mental deficiency in Watson syndrome (1a,51,97). Allanson et al (1d) have demonstrated that the genes for Watson syndrome and NF I are either allelic or tightly linked. Stephan et al (89) described isolated cases of cutaneous angiomatosis and macrocephaly in association with the Klippel–Trénaunay–Weber syndrome. The *Bannayan–Riley–Ruvalcaba syndrome* (109) characterized by macrocephaly, multiple lipomas, and hemangiomas with dominant inheritance should be excluded as should *Proteus syndrome* (91). Stein et al (88) reported a case of neurofibromatosis that presented as *epidermal nevus syndrome.* The reader should consult the chapter on *Noonan syndrome* for features of the disorder sometimes found with NF IX neurofibromatosis/Noonan type.

Café-au-lait macules have been associated with disseminated nonossifying fibromas of the long bones and jaw under the name of Jaffe–Campanacci syndrome (7a,61,88a). The jaw lesions are histologically similar to giant cell granulomas. No other signs of neurofibromatosis have been described in this syndrome.

It should be carefully noted that single or even multiple neurofibromas or schwannomas without neurofibromatosis is frequently observed by surgeons and pathologists. Cutaneous and central nervous system schwannomatosis may be seen with meningiomas, gliomas, and astrocytomas (67a). Familial glioblastoma multiforme was reported without neurofibromatosis by Chemke et al (11).

**Laboratory aids.** Biopsy of individual lesions is useful for establishing the diagnosis in questionable cases. The café-au-lait macules of NF I tend to have more large pigmented granules than those in McCune–Albright syndrome (6). Johnson and Chorneco (45) reported more dopa-positive melanocytes/$mm^2$ in the café-au-lait macules of neurofibromatosis than in those of normal individuals. The density of melanin macroglobules is significantly higher in biopsies of café-au-lait macules of patients with NF I than in patients with NF II or normal controls (57). Soft tissue neurofibromas have been found to take up $^{99m}$Tc DTPA. The localization of this radioisotope in soft tissue benign neoplasms of NF facilitates their identification by scintigraphy (56).

### References [The neurofibromatoses (NF I Recklinghausen type, NF II acoustic type, other types )]


1. Abuelo DN, Meryash DL: Neurofibromatosis with fully expressed Noonan syndrome. *Am J Med Genet* **29**:937–941, 1988.

1a. Afifi AK et al: Ventriculomegaly in neurofibromatosis—1. *Neurofibromatosis* **1**:299–305, 1988.

1b. Allanson JE, Watson GH: Watson syndrome—nineteen years on. *Proc Greenwood Genet Ctr* **6**:173, 1987.

1c. Allanson JE et al: Noonan phenotype associated with neurofibromatosis. *Am J Med Genet* **21**:457–462, 1985.

1d. Allanson JE et al: Watson syndrome: Is it a subtype of neurofibromatosis? David W. Smith Workshop on Malformations and Morphogenesis, Madrid, Spain, May, 1989.

2. Anonymous: Neurofibromatosis. *Lancet* **1**:663–664, 1987.

3. Baden E et al: Multiple neurofibromatosis with oral lesions: Review of the literature and report of a case. *Oral Surg* **8**:268–280, 1955.

4. Baden E et al: Neurofibromatosis of the tongue: A light and electronmicroscopic study with review of the literature from 1849 to 1981. *J Oral Med* **39**:157–164, 1984.

4a. Bader JL: Neurofibromatosis and cancer. *Ann NY Acad Sci* **486**:57–65, 1986.

5. Barker D et al: Gene for von Recklinghausen neurofibromatosis is in the pericentric region of chromosome 17. *Science* **236**:1100–1102, 1987.

6. Benedict PH et al: Melanotic macules in Albright's syndrome and in neurofibromatosis. *JAMA* **205**:618–626, 1968.

7. Buntin PT, Fitzgerald JF: Gastrointestinal neurofibromatosis. *Am J Dis Child* **119**:521–523, 1970.

7a. Campanacci M et al: Multiple non-ossifying fibromata with extraskeletal anomalies: A new syndrome? *J Bone Jt Surg* **65B**:627–632, 1983.

8. Carey JC et al: Penetrance and variability in neurofibromatosis: A genetic study of 60 families. *Birth Defects* **15**(5B):271–281, 1979.

9. Chao DH-C: Congenital neurocutaneous syndromes in childhood. I. Neurofibromatosis. *J Pediatr* **55**:189–199, 1959.

10. Charron JW, Gariepy G: Neurofibromatosis of the bladder: Case report and review of the literature. *Can J Surg* **13**:303–306, 1970.

11. Chemke J et al: Familial glioblastoma multiforme without neurofibromatosis. *Am J Med Genet* **21**:731–735, 1985.

12. Clark RD, Hutter JJ Jr: Familial neurofibromatosis and juvenile chronic myelogenous leukemia. *Hum Genet* **60**:230–232, 1982.

12a. Clayton-Smith J, Donnai D: Neurofibromatosis–Noonan syndrome—Independent segregation of the two conditions within a family. Third Manchester Birth Defects Conference, Manchester, U.K., October 25–28, 1988.

12b. Cohen MM Jr, Hayden PW: A newly recognized hamartomatous syndrome. *Birth Defects* **15**(5B):291–296, 1979.

13. Crawford AH: Neurofibromatosis in children. *Acta Orthoped Scand* **57**:1–60, 1986.

14. Crawford AH: Bagamery N: Osseous manifestations of neurofibromatosis in childhood. *J Pediat Orthoped* **6**:72–88, 1986.

15. Crowe FW, Schull WJ: Diagnostic importance of café-au-lait spot in neurofibromatosis. *Arch Intern Med* **91**:758–766, 1953.

16. Crowe FW et al: *Multiple neurofibromatosis,* Charles C Thomas, Springfield, Ill., 1956.

16a. D'Ambrosio JA et al: Jaw and skull changes in neurofibromatosis. *Oral Surg* **66**:391–396, 1988.

17. Davidson KC: Cranial and intracranial lesions in neurofibromatosis. *Am J Roentgenol* **98**:550–556, 1966.

18. Diekmann L et al: Ungewöhnliche Erscheinungsformen der Neurofibromatose (von Recklinghausensche Krankheit) im Kindesalter. *Z Kinderheilkd* **101**:191–222, 1967.

19. Diehl SR et al: A refined genetic map of the region of chromosome 17 surrounding the von Recklinghausen neurofibromatosis (NF1) gene. *Am J Hum Genet* **44**:33–37, 1989.

20. Dreyfuss U et al: Liposarcoma—a rare complication in neurofibromatosis. Case report. *Plast Reconstr Surg* **61**:287–290, 1978.

21. Dunn DW: Neurofibromatosis in childhood, in *Current Problems in Pediatrics,* Vol. 17, Lockhart JD (ed), Year Book Medical Publishers, Chicago, August, 1987.

22. Farag MZ: Vascular malformation of the parotid gland in von Recklinghausen's disease. *J Laryngol Otol* **97**:571–574, 1983.

22a. Fernández MT et al: Von Recklinghausen neurofibromatosis with carcinoid tumors and subcutaneous leiomyomas of the duodenum. *Neurofibromatosis* **1**:294–298, 1988.

23. Fialkow PJ et al: Multiple cell origin of hereditary neurofibromas. *N Engl J Med* **284**:298–300, 1971.

24. Fienman NL, Yakovac W: Neurofibromatosis in childhood. *J Pediatr* **76**:339–346, 1970.

25. Freeman AG: Proptosis and neurofibromatosis. *Lancet* **1**:1032–1033, 1987.

26. Freeman MJ, Standish SM: Facial and oral manifestations of familial disseminated neurofibromatosis. *Oral Surg* **19**:52–59, 1965.

27. Grant WM, Walton DS: Distinctive gonioscopic findings in glaucoma due to neurofibromatosis. *Arch Ophthalmol* **79**:127–134, 1968.

28. Griffith BH et al: Neurofibromas of the head and neck. *Surg Gynecol Obstet* **160**:534–538, 1985.

29. Hall BD: Letter to the editor: Congenital lid ptosis associated with neurofibromatosis. *Am J Med Genet* **25**:595–597, 1986.

30. Halpern M, Currarino G: Vascular lesions causing hypertension in neurofibromatosis. *N Engl J Med* **273**:248–252, 1965.

31. Hankey GT: Von Recklinghausen's disease with local tumors of the palate. *Proc R Soc Med* **26**:959–961, 1933.

32. Harkin JC, Reed RJ: *Tumors of the Peripheral Nervous System,* 2d ser., fasc. 3, Armed Forces Institute of Pathology, Washington, D.C., 1969.

33. Hayes DM et al: Von Recklinghausen's disease with massive intraabdominal tumor and spontaneous hypoglycemia: Metabolic studies before and after perfusion of abdominal cavity with nitrogen mustard. *Metabolism* **10**:183–199, 1961.

33a. Heimann R et al: Hereditary intestinal neurofibromatosis. I. A distinctive genetic disease. *Neurofibromatosis* **1**:26–32, 1988.

34. Hinrichs SH et al: A transgenic mouse model for human neurofibromatosis. *Science* **237**:1340–1343, 1987.

35. Hofmann P, Galanski M: Kongenitale Unterschenkelverbiegung bei Neurofibromatose von Recklinghausen. *Roefo* **125**:417–421, 1976.

36. Holt JF: Neurofibromatosis in children. *Am J Roentgenol* **130**:615–639, 1978.

37. Holt JF, Wright EM: The radiologic features of neurofibromatosis. *Radiology* **51**:647–663, 1948.

38. Hope DG, Mulvihill JJ: Malignancy in neurofibromatosis, in *Advances in Neurology, Vol. 29, Neurofibromatosis (von Recklinghausen Disease),* Riccardi VM, Mulvihill JJ (eds), Raven Press, New York, 1981, pp 33–56.

38a. Horwich A et al: Brief clinical report: Aqueductal stenosis leading to hydrocephalus—an unusual manifestation of neurofibromatosis. *J Med Genet* **14**:577–581, 1983.

39. Hunt JC, Pugh DG: Skeletal lesions in neurofibromatosis. *Radiology* **76**:1–20, 1961.

40. Huson SM: The different forms of neurofibromatosis. *Br Med J* **294**:1113–1114, 1987.

41. Huson SM et al: Linkage analysis of peripheral neurofibromatosis (von Recklinghausen disease) and chromosome 19 markers linked to myotonic dystrophy. *J Med Genet* **23**:55–57, 1986.

42. Izumi AK et al: Von Recklinghausen's disease associated with multiple neurilemomas. *Arch Dermatol* **104**:172–176, 1971.

43. Jacobs MH: Oral manifestations in von Recklinghausen's disease (neurofibromatosis). *Am J Orthodont (Oral Surg)* **32**:28–33, 1946.

44. Jaffe HL: *Tumors and Tumorous Conditions of the Bones and Joints,* Lea & Febiger, Philadelphia, 1958, pp 242–255.

45. Johnson BL, Chorneco DR: Café-au-lait spot in neurofibromatosis and in normal individuals. *Arch Dermatol* **102**:442–446, 1970.

45a. Jung EG: Segmental neurofibromatosis (NF-5). *Neurofibromatosis* **1**:306–311, 1988.

45b. Kaplan DL, Pestana A: Cutaneous segmental neurofibromatosis. *South Med J* **82**:516–517, 1989.

46. Kaplan P, Rosenblatt B: A distinctive facial appearance in neurofibromatosis von Recklinghausen. *Am J Med Genet* **21**:463–470, 1985.

47. Kaufman RL et al: Family studies in congenital heart disease. IV. Congenital heart disease associated with neurofibromatosis. *Birth Defects* **8**(5):92–95, 1972.

48. Koblin I, Reil B: Changes in the facial skeleton in cases of neurofibromatosis. *J Maxillofac Surg* **3**:23–27, 1975.

49. Knight WA et al: Neurofibromatosis associated with malignant neurofibromas. *Arch Dermatol* **107**:747–750, 1973.

50. Kragh LV et al: Neurofibromatosis of the head and neck. *Plast Reconstr Surg* **25**:565–573, 1960.

51. Kumar BB: Watson's syndrome. *Am J Dis Child* **123**:612, 1972.

52. Laue L et al: Precocious puberty associated with neurofibromatosis and optic gliomas. *Am J Dis Child* **139**:1097–1100, 1985.

53. LeClerc G, Pont J: Un cas de maladie de Recklinghausen avec tumeur majeure siégeant dans l'espace maxillopharyngien et provoquant des troubles dans le domaine du sympathique cervical. *Rev Chir (Paris)* **70**:735–737, 1932.

54. Lewis RA et al: Von Recklinghausen neurofibromatosis. II. Incidence of optic gliomata. *Ophthalmology* **91**:929–935, 1984.

54a. Lipton S, Zuckerbrod M: Familial enteric neurofibromatosis. *Med Times* **94**:544–548, 1966.

55. Maceri DS, Saxon KG: Neurofibromatosis of the head and neck. *Head Neck Surg* **6**:842–850, 1984.

56. Mandell GA et al: Neurofibromas: Location by scanning with Tc-99 in DTPA. *Radiology* **157**:803–806, 1985.

56a. Martuza RL, Eldridge R: Neurofibromatosis 2 (bilateral acoustic neurofibromatosis). *N Engl J Med* **318**:684–688, 1988.

57. Martuza RL et al: Melanin macroglobules as a cellular marker of neurofibromatosis: A quantitative study. *J Invest Dermatol* **85**:347–350, 1985.

58. Mendez HMM: The neurofibromatosis-Noonan syndrome. *Am J Med Genet* **21**:471–476, 1985.

59. Merlob P et al: Postaxial polydactyly in association with neurofibromatosis. *Clin Genet* **32**:202–205, 1987.

60. Miller M, Hall JG: Possible maternal effect of severity of neurofibromatosis. *Lancet* **2**:1071–1074, 1978.

60a. Miller RM, Sparkes RS: Segmental neurofibromatosis. *Arch Dermatol* **113**:837–838, 1977.

61. Mirra JM et al: Disseminated nonossifying fibromas in association with café-au-lait spots (Jaffe–Campanacci syndrome). *Clin Orthopaed* **168**:192–205, 1982.

62. Muller H: Makroglossia neurofibromatosa congenita. *Zbl Allg Pathol* **57**:55–56, 1933.

63. Newell GB et al: Juvenile xanthogranuloma and neurofibromatosis. *Arch Dermatol* **107**:262, 1973.

64. Norman ME: Neurofibromatosis in a family. *Am J Dis Child* **123**:159–160, 1972.

65. Opitz JM, Weaver DD: The neurofibromatosis—Noonan syndrome. *Am J Med Genet* **21**:477–490, 1985.

66. Pollack MA, Shprintzen RJ: Velopharyngeal insufficiency in neurofibromatosis. *Int J Pediatr Otorhino-laryngol* **3**:257–262, 1981.

67. Porterfield JK et al: Brief clinical report: Pulmonary hypertension and interstitial fibrosis in von Recklinghausen neurofibromatosis. *Am J Med Genet* **25**:531–535, 1986.

67a. Purcell SM, Dixon SL: Schwannomatosis. An usual variant of neurofibromatosis or a distinct clinical entity. *Arch Dermatol* **125**:390–393, 1989.

68. Reese AB: *Tumors of the Eye,* 2d ed, Hoeber, New York, 1963.

69. Riccardi VM: Von Recklinghausen neurofibromatosis. *N Engl J Med* **305**:1617–1627, 1981.

70. Riccardi VM, Eichner JE: *Neurofibromatosis. Phenotype, Natural History and Pathogenesis,* Johns Hopkins Univ. Press, Baltimore, 1986.

71. Riccardi VM, Wald JS: Discounting an adverse maternal effect on severity of neurofibromatosis. *Pediatrics* **79**:386–393, 1987.

72. Riccardi VM et al: *Neurofibromatosis (von Recklinghausen Disease), Advances in Neurology,* Vol. 29, Raven Press, New York, 1981.

73. Riccardi VM et al: The pathophysiology of neurofibromatosis: IX. Paternal age as a factor in the origin of new mutations. *Am J Med Genet* **18**:169–176, 1984.

74. Rittersma J et al: Neurofibromatosis with mandibular deformities. *Oral Surg* **33**:718–727, 1972.

74a. Riviello JJ Jr et al: Aqueductal stenosis in neurofibromatosis. *Neurofibromatosis* **1**:312–317, 1988.
75. Rosenquist GC et al: Acquired right ventricular outflow obstruction in a child with neurofibromatosis. *Am Heart J* **79**:103–108, 1970.
76. Rosman NP, Pearce J: The brain in neurofibromatosis. *Brain* **90**:829–838, 1970.
77. Ross DE: Skin manifestations of von Recklinghausen's disease and associated tumors (neurofibromatosis). *Am Surg* **31**:729–740, 1965.
77a. Roth RR et al: Segmental neurofibromatosis. *Arch Dermatol* **123**:917–920, 1987.
78. Rouleau GA et al: Genetic linkage of bilateral acoustic neurofibromatosis to a DNA marker on chromosome 22. *Nature (London)* **329**:246–248, 1987.
78a. Rubinstein AE et al: Familial transmission of segmental neurofibromatosis. *Neurology* **33**:76, 1983.
79. Samuelsson B, Axelsson R: Neurofibromatosis. *Acta Dermatovenereol Suppl* **95**:67–71, 1981.
80. Saxena K. Endocrine manifestations of neurofibromatosis in children. *Am J Dis Child* **120**:265–271, 1970.
81. Saul RA: Letter to the editor: Noonan syndrome in a patient with hyperplasia of the myenteric plexuses and neurofibromatosis. *Am J Med Genet* **21**:491–492, 1985.
81a. Saul RA, Stevenson RE: Segmental neurofibromatosis: A distinct type of neurofibromatosis? *Proc Greenwood Genet Ctr* **3**:3–6, 1984.
82. Schlumberger HG: Limbus tumors as a manifestation of von Recklinghausen's neurofibromatosis in goldfish. *Am J Ophthalmol* **34**:415–422, 1951.
83. Schmidt MA et al: Cases of neurofibromatosis with rearrangements of chromosome 17 involving band 17q11.2. *Am J Med Genet* **28**:771–777, 1987.
84. Seizinger BR et al: Loss of genes on chromosomes 22 in tumorigenesis of human acoustic neuroma. *Nature (London)* **322**:644–647, 1986.
85. Shapiro SD et al: Neurofibromatosis: Oral and radiographic manifestations. *Oral Surg* **58**:493–498, 1984.
86. Smith CJ et al: Renal artery dysplasia as a cause of hypertension in neurofibromatosis. *Arch Intern Med* **125**:1022–1026, 1970.
86a. Smith RW: *Treatise on the Pathology, Diagnosis and Treatment of Neuroma,* Hodges and Smith, Dublin, 1849.
86b. Sørensen SA et al: On the natural history of von Recklinghausen neurofibromatosis. *Ann NY Acad Sci* **486**:30–44, 1986.
87. Spencer WG, Shattock SG: A case of macroglossia neurofibromatosa. *Proc R Soc Med* **1**:8, 1908 (Path. Sect.).
88. Stein KM et al: Neurofibromatosis presenting as the epidermal nevus syndrome. *Arch Dermatol* **105**:229–232, 1972.
88a. Steinmetz JC et al: Campanacci syndrome. *J Pediatr Orthopaed* **8**:602–604, 1988.
89. Stephan MJ et al: Macrocephaly in association with unusual cutaneous angiomatosis. *J Pediatr* **87**:353–359, 1975.
90. Stillman FS: Neurofibromatosis. *J Oral Surg* **10**:112–117, 1952.
91. Tibbles JAR, Cohen MM Jr: The Proteus syndrome: The Elephant Man diagnosed. *Br Med J* **293**:683–685, 1986.
92. Verhest A et al: Hereditary intestinal neurofibromatosis. II. Translocation between chromosomes 12 and 14. *Neurofibromatosis* **1**:33–36, 1988.
93. von Recklinghausen F: *Über die multiplen Fibroma der Haut und ihre Beziehung zu den multiplen Neuromen,* A. Hirschwald, Berlin, 1882.
94. Vormittag W et al: Dermatoglyphics and creases in patients with neurofibromatosis von Recklinghausen. *Am J Med Genet* **25**:389–395, 1986.
95. Walden PAM et al: Wilms' tumour and neurofibromatosis. *Br Med J* **1**:813, 1977.
96. Wander JV, Das Gupta TK: Neurofibromatosis, in *Current Problems in Surgery,* Vol. XIV, Year Book Medical Publishers, Chicago, February 1977.
97. Watson GH: Pulmonary stenosis, café-au-lait spots and dull intelligence. *Arch Dis Childh* **42**:303–307, 1967.
98. Weber FP: Neurofibromatosis of the tongue in a child, together with a note on the classification of incomplete and anomalous cases of Recklinghausen's disease. *Br J Child Dis* **7**:13–16, 1910.
99. Wertelecki W et al: Multiple cutaneous angiomas in neurofibromatosis. *Proc Greenwood Genet Ctr* **6**:105–106, 1987.
100. Wertelecki W et al: Neurofibromatosis: Clinical and DNA linkage studies of a large kindred. *N Engl J Med* **319**:278–283, 1988.
101. Westerhof W, Konrad K: Blue-red macules and pseudoatrophic macules: Additional cutaneous signs in neurofibromatosis. *Arch Dermatol* **118**:577–581, 1982.
102. Westerhof W et al: Neurofibromatosis and hypertelorism. *Arch Dermatol* **120**:1579–1581, 1984.
103. Westphal D, Koblin I: Zahn-und Kieferbefunde bei Neurofibromatose im Kiefer-und Gesichtsbereich. *Deut Zahnärztl Z* **32**:418–420, 1977.
104. Whitfield A: Cutaneous neurofibromatosis in which newly formed nerve fibers were found in the tumors. *Lancet* **1**:1230–1232, 1903.
105. Whitehouse D: Diagnostic value of the café-au-lait spot in children. *Arch Dis Childh* **41**:416–419, 1966.
106. Winters SE et al: Neurofibromatosis (von Recklinghausen's disease) with involvement of the mandible. *Oral Surg* **13**:76–79, 1960.
107. Yesudian P et al: Palmar melanotic macules. A sign of neurofibromatosis. *Int J Dermatol* **23**:468–471, 1984.
108. Zehavi C et al: Iris (Lisch) nodules in neurofibromatosis. *Clin Genet* **29**:51–55, 1986.
109. Zonana J et al: Macrocephaly with multiple lipomas and hemangiomas. *J Pediatr* **89**:600–603, 1976.

## Peutz–Jeghers syndrome

The syndrome of mucocutaneous melanotic pigmentation associated with intestinal polyposis was probably first described in 1896 by Sir Jonathan Hutchinson (27), although he was not aware of the presence of polyposis at the time. Follow-up of one of his patients revealed the cause of death to be intussusception (66). Credit for pointing out the relationship between these ostensibly unrelated conditions goes to Peutz (47), a Dutch physician, who, in 1921, described the syndrome in three generations. However, knowledge of the disorder did not become widespread until Jeghers et al (29) in 1949 published a comprehensive account of 10 cases. Dozois et al (15) were the first to point out the increased rate of ovarian tumors in female patients. Dormandy (13) and Klostermann (30,31) provided excellent summaries of the syndrome. Giardiello et al (20a) showed that the overall risk of cancer is considerably higher than previously supposed.

The syndrome has autosomal dominant inheritance with a high degree of penetrance (13,29,31,63). Almost 35% of cases represent new mutations (5). Bartholomew and Dahlin (5) indicated, in their survey of 117 case, that 43% had a family history of both polyps and pigmentation and 13% of pigmentation alone.

**Gastrointestinal system.** Polyposis of the gastrointestinal tract is the clinically most important component of the syndrome. The polyps are hamartomatous in origin (4,5,14). Bartholomew and Dahlin (5) suggested that the following sites are involved, with the annotated frequency; jejunum, 65%; ileum, 55%; large intestine and rectum, 36% each; stomach, 23%; duodenum, 15%. A higher frequency of gastric and colonic polyps was found by Utsunomiya et al (63).

Thus, polyps may be found anywhere in the mucus-secreting portion of the gastrointestinal tract and may make themselves apparent by producing intussusception. Usually the intussusception is self-resolving, but it may lead to serious intestinal obstruction and death. The age of onset of symptoms varies from a few weeks to 82 years (average, 29 years), and may present somewhat earlier in males. However, about 70% experience some type of gastrointestinal symptoms prior to diagnosis: intermittent colicky pain (85%) and melena or rectal bleeding (35%) (5). Rectal prolapse occurs in about 7%. In the majority of patients, the period before diagnosis is about 5 years. Hypochromic anemia due to intestinal bleeding has been found in 15–25%.

The polyps are usually described as benign hamartomas, varying in size from 0.5 to 7.0 cm in diameter (18,67). Dormandy (13) suggested that these growths arise from primitive adenomatous vesicles embedded in the intestinal wall (Fig. 12–39A–C). This view has been supported by other investigators (4,5,52,67) and adenomatous epithelium may be found in the hamartomas (46). Several authors (9, 12,20a,23,25,33,35,38,41,43,52,53,60,63,67a) have described malignant degeneration of a hamartomatous polyp. Perzin and Bridge (46) reviewed the literature and found adenomatous and carcinomatous changes within a single polyp. Since the glands of the polyps are proliferative in nature, it is not surprising that there are areas of glandular dysplasia and cancerous change (68). In a review of cases with malignant transformation, Dozois et al (14) found no parallelism between location of the malignant tumor and the site of polyps, that is,

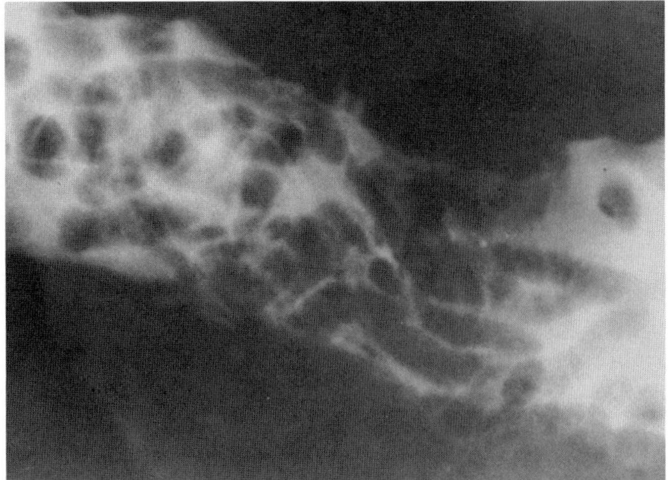

A

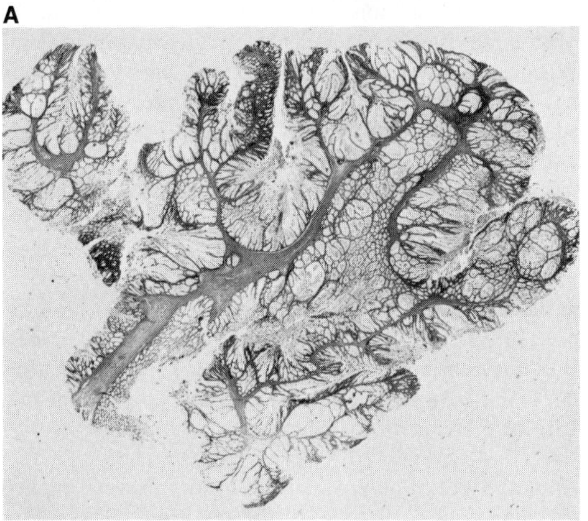

B

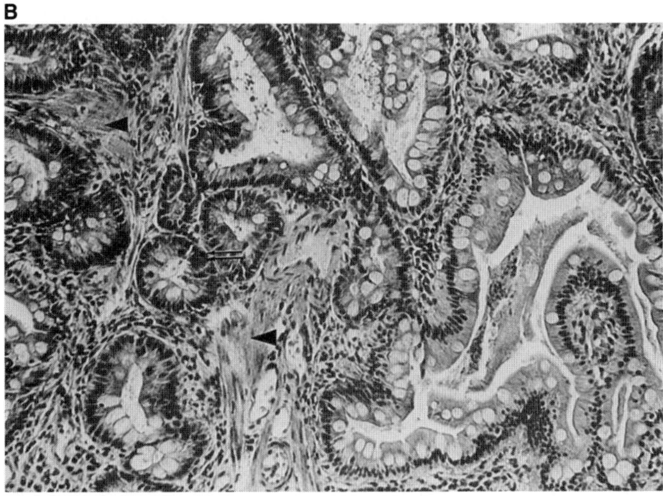

C

Fig. 12–39. *Peutz–Jeghers syndrome.* (A) Roentgenogram of small intestine demonstrating multiple polyposis on barium swallow. (B) Low-power view of polyp from large intestine. Note arborization of nonstriated muscle. (C) Photomicrograph of adenomatous polyp. (A,B from *JD Reid,* Cancer **18**:970, 1965. C from *KJ Walecki et al,* Pediatr Radiol **14**:62, 1984.)

the most common locations for the adenocarcinomas (stomach and colon) are the least likely to be sites of polyps.

Malignancy has been discussed by a number of authors (9,36,48,52,63). It was generally believed that malignant tumors developed in 10–20%. However, in the extensive long-term study of Giardiello et al (20a), cancer developed in 48%. There was an increased risk for neoplasms to develop at both gastrointestinal and nongastrointestinal sites. The mean interval between diagnosis of the syndrome and diagnosis of cancer was $25 \pm 20$ years (range, 1–64 years). Neoplasms occurred in 8 of 13 families and in a few pedigrees were not clustered. Nongastrointestinal cancers included a 100-fold excess of pancreatic carcinoma and others such as ductal carcinoma of the breast, adenocarcinoma of the lung, and multiple myeloma in addition to colonic malignancy and ovarian tumors. Within the gastrointestinal tract, only adenomatous polyps became malignant, never hamartomatous polyps. Reasons for the lower malignancy rates found by other investigators (14,36,63) are well analyzed by Giardiello et al (20a).

Microscopically, gastrointestinal hamartomas represent focal overgrowths in improper proportions of tissues indigenous to that part of the gastrointestinal tract. A branching-tree arrangement of smooth muscle may be seen scattered throughout the growths. Frequent mitotic figures are characteristic. The growths may extend to the serosal surface.

Enteritis cystica profunda characterized by mucosal glands and mucinous cysts that penetrate the tunica muscularis of the small and large bowel has been documented (1,34). The condition, in our opinion, represents extreme hamartomatosis or pseudoinvasion.

Polyps of other organs include the nose (13,22,31,47,72), choanae and antrum (28,37), uterus (31,48,72), ureter and bladder (13,22,31, 47,57), gallbladder (19), bile duct (55), and esophagus (2,44). Bronchial adenosis has also been noted (13,22,47).

**Skin.** In about 50%, numerous, usually discrete, brown to bluish-black macules are present on the skin, especially about the body orifices—perioral, periorbital, perianal, and perigenital (61a) (Fig. 12–40A–E). Though some patients exhibit only a few pigmented macules, others are markedly pigmented. Pigmented spots occur on the extremities in about 65% (5,58,63), fingers, toes, (69,72), sometimes on the palms or soles, or, occasionally, in other areas, such as the umbilicus, axilla, or shoulder (20,72). Pigmentation of the nails has also been described (32,64). The pigmentation usually appears in infancy and seems to fade somewhat at puberty. It may, however, appear as late as the eighth decade.

**Ovarian cysts and tumors and uterine adenocarcinoma.** Ovarian tumors occur in about 10–14% of female patients (10,14). They may be found even in very young children (14). Granulosa cell tumors (6,8,11,14,15,22,29,48) have been associated with precocious puberty (11,56). Brenner tumor, dysgerminoma, (15,62) and cystadenoma (15) have also been found. Associated cervical adenocarcinoma is of considerable interest (1,39,40,70).

Scully (54) suggests that a distinctive ovarian neoplasm, which he called "sex cord tumor with annular tubules" (SCTAT), may occur in the syndrome. The tumors tend to be bilateral, often of microscopic size, frequently calcified, and nonmalignant. The neoplasm, probably derived from granulosa cells, produces endometrial hyperplasia (Fig. 12–41). Christian (11) reviewed the literature and reported 15 patients with ovarian tumors among 125 female patients with the syndrome and questioned whether SCTAT is a tumor or a hamartoma. Steenstrup (59), too, proposed that SCTAT is a hamartoma. Hart et al (24) classified SCTAT as a distinctive annular and membranous variant of the granulosa cell tumor in view of the morphologic similarities and clinical behavior. In a review of 74 cases of SCTAT, Young et al (70) found that 27 patients had the syndrome and four of these also had cervical adenomas. Young et al (71) found three different gynecologic neoplasms: SCTAT, well-differentiated mucinous adenocarcinoma of the cervix, and a distinctive sex cord-stromal tumor. Rodu and Martinez (51) summarized the relationship between Peutz–Jeghers syndrome and cancer.

**Other findings.** Breast carcinoma (7,21,35,49,61) and pancreatic adenocarcinoma (20a) have also been reported. Cantú et al (10) reported a 6-year-old boy with feminizing Sertoli cell tumor, and Dubois et al (16) reported gynecomastia and testis enlargement in a 3-year-old boy, with SCTAT developing in a normal testis.

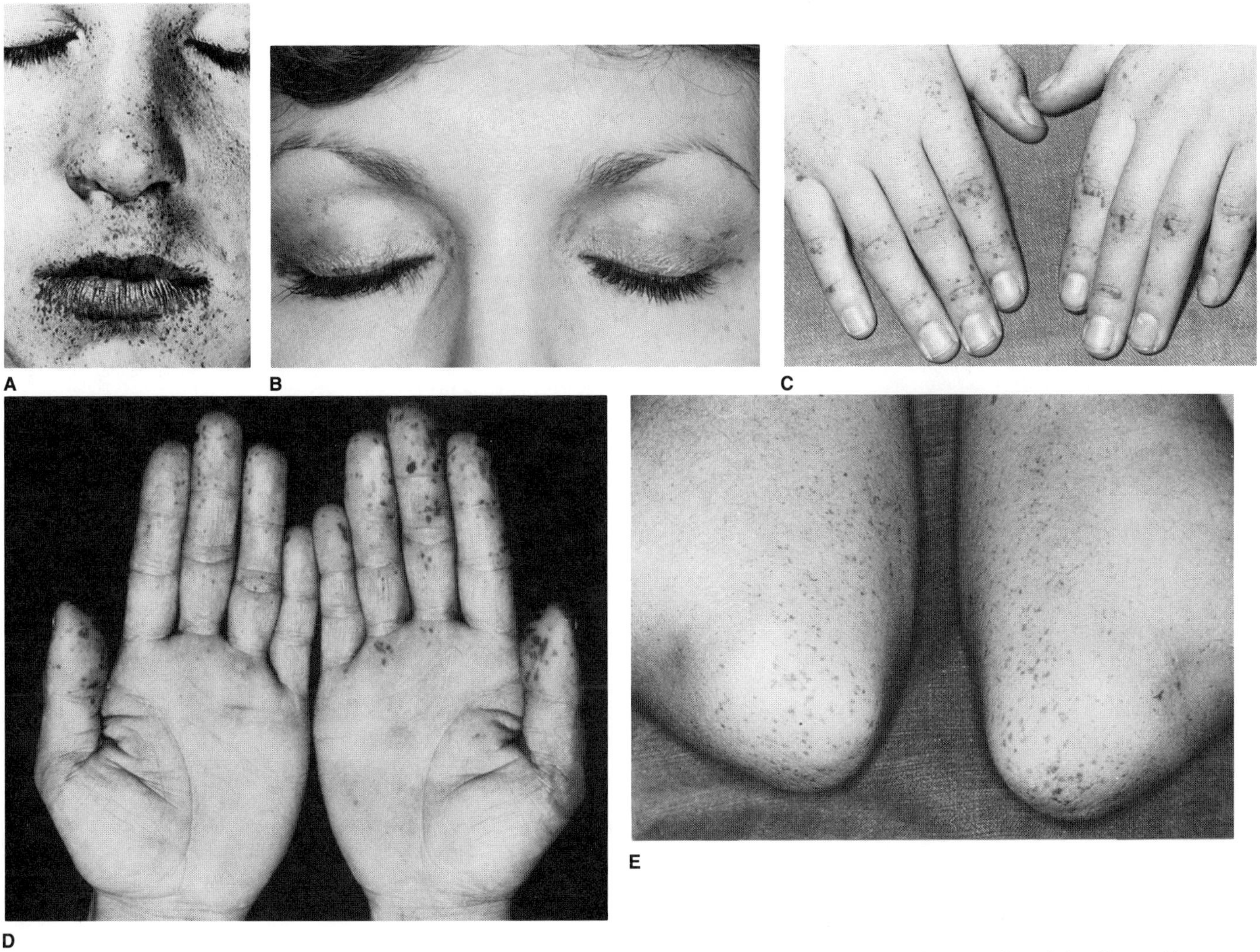

Fig. 12–40. *Peutz–Jeghers syndrome.* (A) Note extensive melanotic pigmentation of lips, numerous very small freckle-like spots about mouth, nose, and eyes. (B) Compare mild faded involvement of eyelids with patient shown in A. (C) Melanotic pigmentation on fingers. (D) Pigmented macules on hands. (E) Spotting in the region of the elbows. (A courtesy of *J Calnan,* London, England. C,E courtesy of *G Klostermann,* G Thieme Verlag, Stuttgart, Germany. D from *K Yamada et al,* J Dermatol (Tokyo) **8**:367, 1981).

Fig. 12–41. *Peutz–Jeghers syndrome.* Sex cord tumor. Note simple and complex annular tubules with external calcification. (From *RE Scully,* Cancer **25**:1107, 1970.)

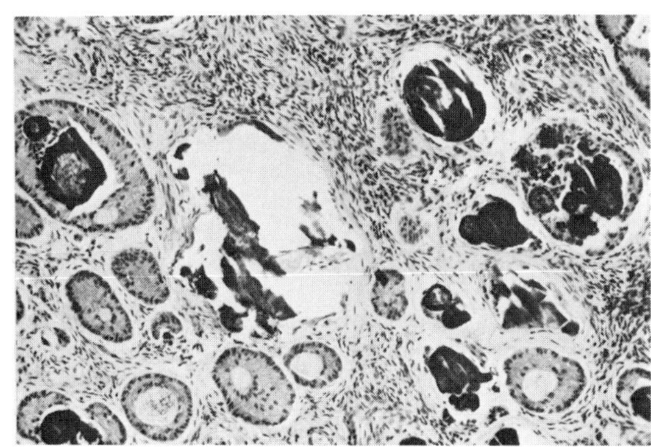

**Oral manifestations.** On the lips, especially the lower, and on the oral mucosa, round, oval, or irregular, rarely confluent macules of bluish-grey pigment of variable intensity may be seen (73). They vary in size from 1 to 12 mm and are usually somewhat larger than those on the skin. About 98% of 117 patients had pigmentation of the lips, and 88% had involvement of the buccal mucosa (5). Less frequently pigmented are the palate and gingiva. Only rarely are the tongue and the oral floor involved (31,47,58). There does not appear to be any relationship between the amount of oral pigmentation and the degree or distribution of the visceral polyposis (Fig. 12–42A–C).

Several investigators have pointed out that labial pigment also tends to fade, and that the pigment on the buccal mucosa fades to a lesser degree (29,31,63), thus being helpful in diagnosis (4,5). Rarely, pigmentation may be present without polyposis (13,29,72).

Pigmentation of other mucosal surfaces, viz. conjunctival (3,42,61a), nasal (13,37), and anal (45), may also be seen. Pigmented oral papillomatosis (37) and oral polyposis with a polyp of the tonsillar pillar (26) have been reported.

**Differential diagnosis.** Oral mucosal pigmentation may be seen normally in over 90% of blacks or members of other dark-skinned races and in 5% of whites (17,73). In Addison's disease, the cutaneous pigmentation is generalized and often increased along body folds

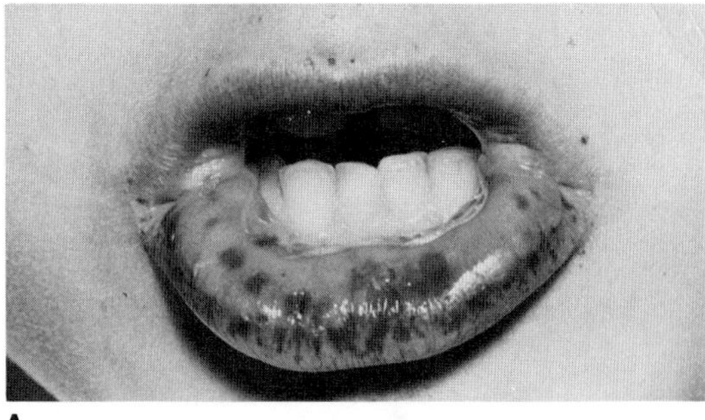

A

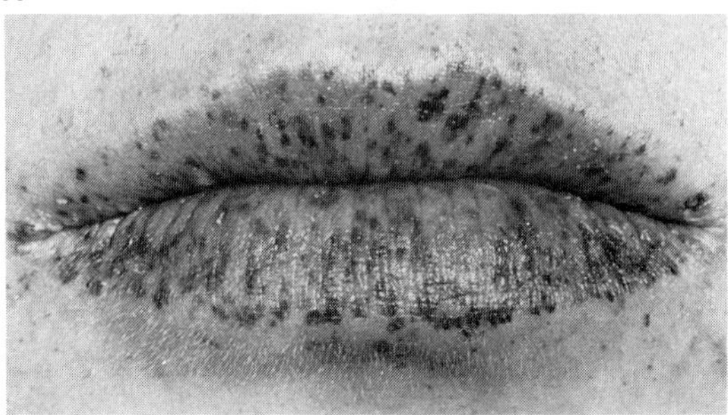

B

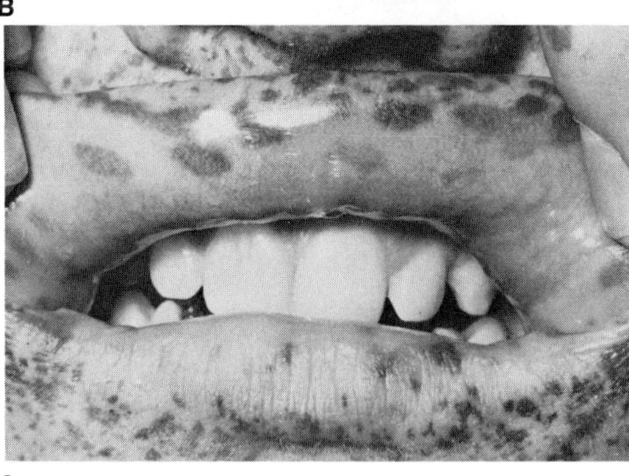

C

Fig. 12–42. *Peutz–Jeghers syndrome.* (A–C) Melanotic pigmentation of the lips. Note involvement of mucosal surface in C. (A from *LG Bartholomew et al.* Gastroenterology **32**:434, 1957. B from *M Zingsheim,* Hautarzt **17**:85, 1966. C from *G Klostermann,* G Thieme Verlag, Stuttgart, Germany.)

and scars. The distribution of freckles does not ordinarily occur periorally or labially and intraorally is more generalized. Lentiginosis profusa *(LEOPARD syndrome)* is generalized over the skin but does not involve mucosal surfaces. Lentiginosis also occurs in *Carney syndrome.* Oral pigmentation may rarely occur in *McCune–Albright syndrome.* Polyposis and pigmentation have been described in association with alopecia and nail dystrophy, that is, Cronkhite–Canada syndrome (see *Gardner syndrome*).

Other forms of polyposis of the intestinal tract are usually limited to the colon. These include familial polyposis coli, *Gardner syndrome,* juvenile polyposis, and disseminated polyposis of the colon and rectum. These and other types are discussed in the section on Gardner syndrome.

Sex cord tumor of the ovary may occur in women without Peutz–Jeghers syndrome. In contrast to those with the syndrome, the tumors tend to be unilateral, large, rarely calcified, and mixed with Sertoli and/or granulosa cells; they are malignant in about 15%.

**Laboratory aids.** The presence of labial and/or oral pigmentation should suggest a thorough history and examination of the gastrointestinal tract by proctoscopic and radiographic means. Diagnosis of early intussusception may be possible by ultrasonography (65).

### References (Peutz–Jeghers syndrome)


1. Anderson NJ et al: Peutz–Jeghers syndrome with cervical adenocarcinoma and enteritis cystica profunda. *West J Med* **141**:242–244, 1984.
2. André R et al: Syndrome de Peutz–Jeghers avec polypose oesophagienne. *Bull Soç Méd Hôp Paris* **117**:505–510, 1966.
3. Andrew R: Generalized intestinal polyposis with melanosis. *Gastroenterology* **23**:495–499, 1953.
4. Bartholomew LG et al: Intestinal polyposis associated with mucocutaneous melanin pigmentation: Peutz–Jeghers syndrome. Review of literature and report of six cases with special reference to pathologic findings. *Gastroenterology* **32**:434–451, 1957.
5. Bartholomew LG, Dahlin DC: Intestinal polyposis and mucocutaneous pigmentation. *Minn Med* **41**:848–852, 1958.
6. Berkowitz SB et al: Syndrome of intestinal polyposis with melanosis of the lips and buccal mucosa: Study of incidence and location of malignancy with three new case reports. *Ann Surg* **141**:129–133, 1955.
7. Besson A et al: Syndrome de Peutz–Jeghers. Invaginations et cancers associés. *Chirurgie* **104**:117–130, 1978.
8. Burdick D et al: Peutz–Jeghers syndrome. *Cancer* **16**:854–867, 1963.
9. Burdick D, Prior JT: Peutz–Jeghers syndrome. A clinicopathologic study of a large family with a 27 year follow-up. *Cancer* **50**:2139–2146, 1982.
10. Cantú JM et al: Peutz–Jeghers syndrome with feminizing Sertoli cell tumor. *Cancer* **46**:223–228, 1980.
11. Christian CD: Ovarian tumors—an extension of the Peutz–Jeghers syndrome. *Am J Obstet Gynecol* **111**:529–534, 1971.
12. Cochet B et al: Peutz–Jeghers syndrome associated with gastrointestinal carcinoma: Report of two cases in a family. *Gut* **20**:169–175, 1979.
13. Dormandy TL: Gastrointestinal polyposis with mucocutaneous pigmentation: Peutz–Jeghers syndrome. *N Engl J Med* **256**:1093–1102, 1141–1146, 1186–1190, 1957.
14. Dozois RR et al:The Peutz–Jeghers syndrome. *Arch Surg* **98**:509–517, 1969.
15. Dozois RR et al: Ovarian tumors associated with the Peutz–Jeghers syndrome. *Ann Surg* **172**:233–238, 1970.
16. Dubois RS et al: Feminizing sex cord tumor with annular tubules in a boy with Peutz–Jeghers syndrome. *J Pediatr* **101**:568–571, 1982.
17. Dummett CO, Barens G: Pigmentation of the oral tissues: A review of the literature. *J Periodontol* **38**:369–378, 1967.
18. Estrada R: Hamartomatous polyps in Peutz–Jeghers syndrome: A light, histochemical, and electron-microscopical study. *Am J Surg Pathol* **7**:747–754, 1983.
19. Foster DR, Foster DB: Gall bladder polyps in Peutz–Jeghers syndrome. *Postgrad Med J* **56**:373–376, 1980.
20. Fristche W, Fleischhauer G: Chirurgischen Beitrag zur hereditären Dünndarm-polyposis (Peutz–Jeghers-Syndrom). *Chirurg* **28**:266–269, 1957.

20a. Giardiello FM et al: Increased risk of cancer in the Peutz–Jeghers syndrome. *N Engl J Med* **316**:1511–1514, 1987.

21. Gloor E: Un cas de syndrome de Peutz–Jeghers associé à un carcinome mammaire bilateral, à un adenocarcinome du col utérin e a des tumeurs des cordons sexuels à tubules annelés bilaterales dans les ovaries. *Schweiz Med Wochenschr* **108**:717–721, 1978.
22. Hafter E: Gastrointestinale Polypose mit Melanose der Lippen und Mundschleimhaut (Peutz–Jeghers'sche Syndrom). *Gastroenterologia (Basel)* **84**:341–348, 1955.
23. Halbert RE: Peutz–Jeghers syndrome with metastasizing gastric adenocarcinoma. *Arch Pathol Lab Med* **106**:517–520, 1982.
24. Hart WR et al: Ovarian neoplasms resembling sex cord tumors with annular tubules. *Cancer* **45**:2352–2363, 1980.
25. Hsu SD: Peutz–Jeghers syndrome with intestinal carcinoma: Report of the association in one family. *Cancer* **44**:1527–1532, 1979.
26. Humphries AL et al: Peutz–Jeghers syndrome with colonic adenocarcinoma and ovarian tumor. *JAMA* **197**:296–298, 1966.
27. Hutchinson J: Pigmentation of lips and mouth. *Arch Surg* **7**:290, 1896.
28. Janku J: Peutz–Jeghers syndrome: Involvement of gastrointestinal and upper respiratory tracts. *Am J Gastroenterol* **56**:545–549, 1971.
29. Jeghers H et al: Generalized intestinal polyposis and melanin spots of the oral mucosa, lips, and digits. *N Engl J Med* **241**:993–1005, 1031–1036, 1949.
30. Klostermann GF: *Pigmentfleckenpolypose. Klinische, histologische und erbbiologische Studien am sogenannten Peutz-Syndrom,* Thieme, Stuttgart, 1960.
31. Klostermann GF: Zur Kenntnis der Pigmentfleckenpolypose. Bemerkungen zu Diagnostik, Verlauf und Erbbiologie des sogenannten Peutz–Jeghers—Syndroms auf Grund katamnestischer Daten. *Arch Klin Exp Dermatol* **226**:182–189, 1966.
32. Kyle J: Peutz–Jeghers syndrome. *Scot Med J* **6**:361–367, 1961.

33. Kyle J: Gastric carcinoma in Peutz–Jeghers syndrome. *Scot Med J* **29**:187–190, 1984.
34. Kyriakos M, Condon SC: Enteritis cystica profunda. *Am J Clin Pathol* **69**:77–85, 1978.
35. Lehur PA et al: Peutz–Jeghers syndrome. Association of duodenal and bilateral breast cancers in the same patient. *Dig Dis Sci* **29**:178–182, 1984.
36. Linos DA et al: Does Peutz–Jeghers syndrome predispose to gastrointestinal malignancy? A later look. *Arch Surg* **116**:1182–1184, 1981.
37. Lowe NJ: Peutz–Jeghers syndrome with pigmented oral papillomas. *Arch Dermatol* **111**:503–505, 1975.
38. Matuchansky C et al: Peutz–Jeghers syndrome with metastasizing carcinoma arising from a jejunal hamartoma. *Gastroenterology* **77**:1311–1315, 1979.
39. McGowan L et al: Peutz–Jeghers syndrome with "adenoma malignum" of the cervix. A report of two cases. *Gynecol Oncol* **10**:125–133, 1980.
40. McKelvey JL, Goodlin RR: Adenoma malignum of the cervix. *Cancer* **16**:549–557, 1963.
41. Miller LF et al: Adenocarcinoma of the rectum arising in a hamartomatous polyp in a patient with Peutz–Jeghers syndrome. *Dig Dis Sci* **28**:1047–1051, 1983.
42. Oshiro T et al: Treatment of the pigmentation of the lips and oral mucosa in Peutz–Jeghers syndrome using ruby and argon lasers. *Br J Plast Surg* **33**:346–349, 1980.
43. Papaioannou A, Criteselis A: Malignant changes in Peutz–Jeghers syndrome. *N Engl J Med* **289**:694, 1973.
44. Parker MC, Knight M: Peutz–Jeghers syndrome causing obstructive jaundice due to a polyp in common bile duct. *J R Soc Med* **76**:701–703, 1983.
45. Pastinsky I, Vankos J: Beiträge zur Pathologie der Lentiginosis periorificialis cum Polyposi intestinali hereditaria (Peutz–Touraine–Jeghers-Syndrom). *Hautarzt* **8**:276–279, 1957.
46. Perzin KH, Bridge MF: Adenomatous and carcinomatous changes in hamartomatous polyps of the small intestine (Peutz–Jeghers syndrome): Report of a case and review of the literature. *Cancer* **49**:971–983, 1982.
47. Peutz JL: Over een zeer merkwaardige, gecombineerde familiaire polyposis van de slijmvliezen van den tractus intestinis met die van de neuskeelholte en gepaard met eigenaardige pigmentaties van huid-en slijmvliezen. *Nederl Maandschr Geneesk* **10**:134–146, 1921.
48. Reid JD: Duodenal carcinoma in the Peutz–Jeghers syndrome. *Cancer* **18**:970–977, 1965.
49. Riley E, Swift M: A family with Peutz–Jeghers syndrome and bilateral breast cancer. *Cancer* **46**:815–817, 1980.
50. Rintala A: The histological appearance of gastrointestinal polyps in Peutz–Jeghers syndrome. *Acta Chir Scand* **117**:366–373, 1959.
51. Rodu B, Martinez MG: Peutz–Jeghers syndrome and cancer. *Oral Surg* **58**:584–588, 1984.
52. Ryo UY et al: Extensive metastases in Peutz–Jeghers syndrome. *JAMA* **239**:2268–2269, 1978.
53. Schneider A, Jakobs W: Ein Fall von Peutz–Jeghers-Syndrom mit Adenokarzinom an der Flexura duodenojejunalis. *Dtsch Z Verdau Stoffwechselkr* **42**:210–216, 1982.
54. Scully RE: Sex cord tumor with annular tubules: A distinctive ovarian tumor of the Peutz–Jeghers syndrome. *Cancer* **25**:1107–1121, 1970.
55. Shibata HR, Phillips MJ: Peutz–Jeghers syndrome with jejunal and colonic adenocarcinomas. *Can Med Assoc J* **103**:285–289, 1970.
56. Solh HM et al: Peutz–Jeghers syndrome associated with precocious puberty. *J Pediatr* **103**:593–595, 1983.
57. Sommerhaug RG, Mason T: Peutz–Jeghers syndrome and ureteral polyposis. *JAMA* **211**:120–122, 1970.
58. Staley CJ, Schwartz H: Gastrointestinal polyposis and pigmentation of the oral mucosa (Peutz–Jeghers syndrome). *Int Abstr Surg* **105**:1–15, 1957.
59. Steenstrup EK: Ovarian tumors and Peutz–Jeghers syndrome. *Acta Obstet Gynecol Scand* **51**:237–240, 1972.
60. Stockdale AD et al: Gastrointestinal malignancy in association with Peutz–Jeghers syndrome—three further cases. *Clin Oncol* **10**:299–301, 1984.
61. Trau H et al: Peutz–Jeghers syndrome and bilateral breast carcinoma. *Cancer* **50**:788–792, 1982.
61a. Traboulsi EI, Maumenee IH: Periorbital pigmentation in the Peutz–Jeghers syndrome. *Am J Ophthalmol* **102**:126–127, 1986.
62. Tseng HL, Braunstein L: Polyposis of small intestine. *Gastroenterology* **27**:426–430, 1954.
63. Utsunomiya J et al: Peutz–Jeghers syndrome: Its natural course and management. *Johns Hopkins Med J* **136**:71–82, 1975.
64. Valero A, Sherf K: Pigmented nails in Peutz–Jeghers syndrome. *Am J Gastroenterol* **43**:56–58, 1965.
65. Walecki KJ et al: Ultrasound contribution to diagnosis of Peutz–Jeghers syndrome. *Pediatr Radiol* **14**:62–64, 1984.
66. Weber FP: Patches of deep pigmentation of oral mucous membrane not connected with Addison's disease. *Q J Med* **12**:404, 1919.
67. Weller RO, McColl I: Electron microscope appearance of juvenile and Peutz–Jeghers polyps. *Gut* **7**:265–270, 1966.
67a. Williams JP, Knudsen A: Peutz–Jeghers syndrome with metastasizing duodenal carcinoma. *Gut* **6**:179–184, 1965.
68. Yaguchi T et al: Peutz–Jeghers polyp with several foci of glandular dysplasia. *Dis Colon Rectum* **25**:592–596, 1982.
69. Yamada K et al: Ultrastructural studies on pigment macules of Peutz–Jeghers syndrome. *J Dermatol (Tokyo)* **8**:367–377, 1981.
70. Young RH et al: Ovarian sex cord tumor with annular tubules: Review of 74 cases including 27 with Peutz–Jeghers syndrome and four with adenoma malignum of the cervix. *Cancer* **50**:1384–1402, 1982.
71. Young RH et al: A distinctive ovarian sex cord-stromal tumor causing sexual precocity in the Peutz–Jeghers syndrome. *Am J Surg Pathol* **7**:233–243, 1983.
72. Zegarelli EV et al: Melanin spots of the oral mucosa and skin associated with polyps: Report of a case of peculiar pigmentation of the lips and mouth. *Oral Surg* **7**:972–978, 1954.
73. Zegarelli EV et al: Atlas of oral lesions observed in the syndrome of oral melanosis with associated intestinal polyposis (Peutz–Jeghers syndrome). *Am J Dig Dis* **4**:479–489, 1959.

## Proteus syndrome

Protean manifestations of the disorder include partial gigantism of the hands and/or feet, asymmetry of the limbs, plantar hyperplasia, hemangiomas, lipomas, lymphangiomas, varicosities, verrucous epidermal nevi, macrocephaly, cranial hyperostoses, and long bone overgrowth (Figs. 12–43 to 12–45). The syndrome was first delineated in 1979 by Cohen and Hayden (10) who reported two patients under the title of "a newly recognized hamartomatous syndrome" and proposed the disorder as a recurrent pattern syndrome distinct from neurofibromatosis and Klippel–Trénaunay–Weber syndrome. Wiedemann et al (39), unaware of Cohen and Hayden's earlier paper (10), reported four cases in 1983, further delineated the syndrome, named it Proteus syn-

Fig. 12–43. *Proteus syndrome.* Unusual craniofacial appearance, kyphoscoliosis, and protuberant abdomen. (From *JT Fay* and *SR Schow,* J Oral Surg **26**:739, 1968.)

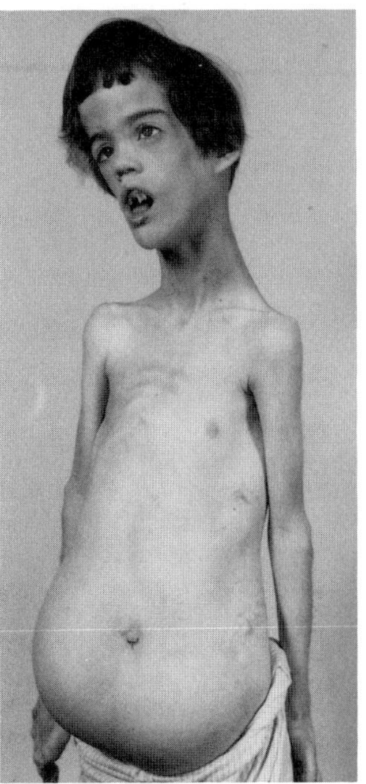

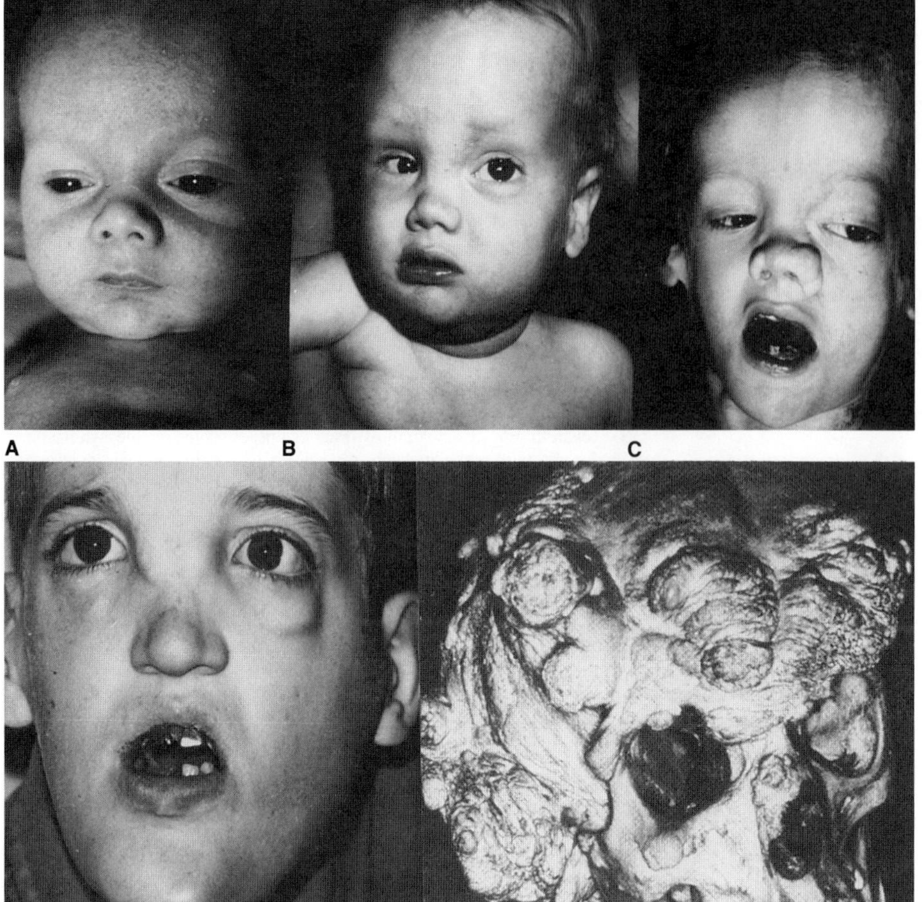


Fig. 12–44. *Proteus syndrome*. Craniofacial features. (A–C) Evolution of features. (A) Patient at 1 month. (B) Same patient at 5 months. (C) Same patient at 5.5 years. (D) Boy developing hyperostoses of nasal bridge, left infraorbital region and mandible. (E) Hyperostoses in Joseph Merrick's skull. [A–C from *JAR Tibbles* and *MM Cohen Jr,* Br Med J **293**:683, 1986. D from *MM Cohen Jr* and *PW Hayden,* Birth Defects **15**(5B):291, 1979. E from *M Howell* and *P Ford,* The True History of the Elephant Man, Allison and Busby, London, 1980.]

drome after the Greek god Proteus (the polymorphous, who could change his shape at will to avoid capture), and called it to the attention of pediatricians. Since then, many significant reviews have appeared (4,6,9,11,15,19,25,26,29,36,41). An exhaustive review is that of Cohen (9). At least 45 examples have been published to date, including some early case reports (1–5,13,14,17,20–24,27,28,30–33,38,40). Less certain are the cases of Graetz (16) and Temtamy and Rogers (34).

No sex predilection and no advanced paternal age have been found (36). All instances to date have been sporadic (15,25). The patient reported by Horton (20, Case 2) had a normal child. Happle (18) and Hall (17a) suggested that the disorder may result from somatic mutation, lethal in the nonmosaic state. Viljoen et al (36) suggested that pathogenesis may be due to triggering of various growth factors.

Elsewhere, Cohen (7–9) and Tibbles and Cohen (35) have presented evidence that Joseph Merrick, the famous Elephant Man (whom Sir Frederick Treve called ''John'') had Proteus syndrome, not neurofibromatosis. Merrick had no family history of neurofibromatosis, no evidence of café-au-lait spots in adulthood (which surely would have been recognized and described by the dermatologist HR Crocker), and no evidence of neurofibromas. What is clear is that Merrick had no congenital abnormalities, but went on to develop bizarre manifestations that were more severe than those encountered with neurofibromatosis, and this is characteristic of Proteus syndrome (Table 12–6). In particular, a plaster cast of Merrick's foot shows the typical moccasin-type of plantar hyperplasia characteristic of Proteus syndrome (Fig. 12–45F). The fact that Merrick is now known to have Proteus syndrome, which is uncommon, should help undo the unnecessary psychosocial burden placed on some families with neurofibromatosis, a common disorder.

**Growth, overgrowth, and craniofacial features.** Birth weight may be increased, and a number of newborns have weighed as much as 4000 g or more (6,39). Increased growth velocity occurs during the first few years of life (10,39) and may be generalized involving the whole body, unilateral involving one limb, focal resulting in macrodactyly, or some combination of these. The limbs may be asymmetric with bony overgrowth; and partial gigantism of the hands and/or feet is common. Gyriform hyperplasia of the hands and feet, including moccasin-type plantar hyperplasia, is striking (Fig. 12–45A–F). Development of rounded hyperostoses leads to progressive enlargement, asymmetry, and disfigurement of the skull (Fig. 12–44D,E). Exostoses have also occurred in the external auditory canal, on the nasal bridge, and on the alveolar ridges. Frontal bossing is common. The craniofacial appearance may become progressively bizarre with age (Fig. 12–44A–C). Other abnormalities of the musculoskeletal system are especially common including genua valga, hip dislocation, coxa valga, scoliosis, kyphosis, valgus deformities of the halluces and feet, clinodactyly, bony protuberances, and dysplastic vertebrae. Final height attainment is normal (1,10,12,14,15,19,20,26,27a,41).

**Tumors and neoplasms.** Subcutaneous hamartomas may be noted soon after birth. Some may grow rapidly; others may remain unchanged or regress. Hemangiomas, lipomas, lymphangiomas, and various compound tumors have been recorded. The thorax and upper abdomen are most commonly involved, and abdominal and pelvic lipomatosis has been noted. Varicose veins have been recorded with increased frequency (6,10,11,29,36).

Because the syndrome is incompletely delineated at present, patients should be carefully monitored for possible development of other

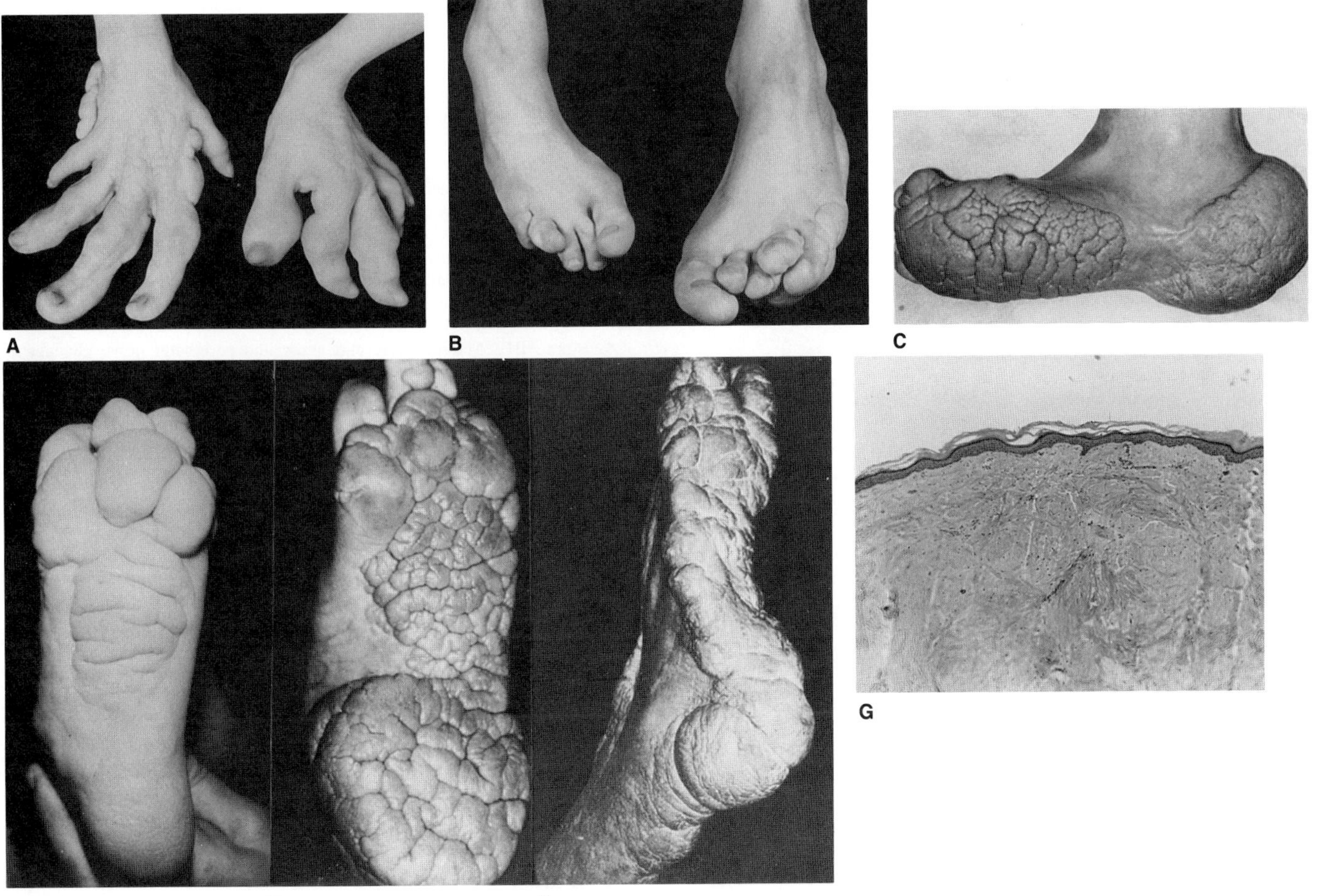


Fig. 12–45. *Proteus syndrome*. (A–F) Enlargement of hands and/or feet associated with gigantic fingers and/or toes, sometimes with fibrous overgrowth. Note "moccasin" lesion in C. (D–F) Evolution of "moccasin" lesion typical of Proteus syndrome. (D) Patient at 5.5 years. (E) Second patient at 13 years. (F) Third patient at 29 years (plaster cast of the Elephant Man's foot). (G) Photomicrograph of fibrous lesion from foot. Note highly collagenized fibrous connective tissue and no evidence of neurofibroma. (A,B courtesy of *N O'Doherty*, Dublin, Ireland.) C,E from *MM Cohen Jr* and *PW Hayden*, Birth Defects **15**(5B):291, 1979. D from *JAR Tibbles* and *MM Cohen Jr*, Br Med J **193**:683, 1986. F from *M Howell* and *P Ford*, The True History of the Elephant Man, Allison and Busby, London, 1980.]

neoplasms since Proteus syndrome is so obviously a hamartomatous disorder. In one of our patients (10), a monomorphic adenoma of the parotid gland developed and several years later, fibroadenomas of the breast appeared. A yolk sac tumor of the testis and a papillary adenoma of the epididymis have also been noted in other patients (19,36).

**Skin.** Pigmented intradermal nevi, hyperpigmented and nonpigmented diffuse areas, linear streaks, or linear verrucous epidermal nevi may be found on the neck, trunk, and extremities, frequently evident at birth or soon thereafter (10,11,37,39,41).

**Performance.** Moderate mental deficiency has been evident in about 55% and seizures have been documented in 13% (4,6,10,11,19,26).

**Other findings.** Goiter, cyst-like alterations of the lungs, enlarged penis, macroorchidism, strabismus, epibulbar tumor, prominent globes, ptosis, myopia, anisocoria, heterochromia iridis, unilateral microphthalmia, retinal detachment, cataract, chorioretinitis, rarely blindness, submucous cleft palate, enlarged incisor teeth, enlarged alveolar process, and a variety of other low frequency abnormalities have been observed (10,11,14,19,25,26,36–40). Nephrogenic diabetes insipidus has been noted in one case (21).

**Differential diagnosis.** To be ruled out are *neurofibromatosis* and *Klippel–Trénaunay–Weber syndrome*. Ollier-type enchondromatosis and *Maffucci syndrome* can be excluded since Proteus syndrome has no enchondromas. *Bannayan–Riley–Ruvalcaba syndrome* consists of macrocephaly, multiple lipomatosis, angiomatosis, and other abnormalities.

## References (Proteus syndrome)


1. Azouz EM et al: Radiologic findings in the Proteus syndrome. *Pediatr Radiol* **17**:481–485, 1987.
2. Bender BL, Yunis E: Fibrocartilagenous lesions of bone and hemangiomas and lipomas of soft tissues resembling Maffucci's syndrome. *J Bone Jt Surg* **61A**:1104–1108, 1979.
3. Bialer MG et al: Proteus syndrome versus Bannayan–Zonana syndrome: A problem in differential diagnosis. *Eur J Pediatr* **148**:122–125, 1988.
4. Burgio GR, Wiedemann H-R: Further and new details on the Proteus syndrome. *Eur J Pediatr* **143**:71–73, 1984.
5. Chandler FA: Local overgrowth. *JAMA* **109**:1411–1414, 1937.
6. Clark RD et al: Proteus syndrome: An expanded phenotype. *Am J Med Genet* **27**:99–118, 1987.
7. Cohen MM Jr: The Elephant Man did not have neurofibromatosis. *Proc Greenwood Genet Ctr* **6**:187–192, 1987.
8. Cohen MM Jr: Invited historical comment: Further diagnostic thoughts about the Elephant Man. *Am J Med Genet* **29**:777–782, 1988.

Table 12–6. Clinical features of Proteus syndrome[a]

| Features | Previously reported cases | Joseph Merrick |
|---|---|---|
| Growth | | |
| Asymmetry of limbs | 38/38 | + |
| Partial gigantism of hands and/or feet | 31/38 | + |
| Skeletal | | |
| Bone hypertrophy | 31/38 | + |
| Scoliosis/kyphois/abn vert | 20/38 | + |
| Joint limitation/angulation | 17/38 | + |
| Skull hyperostoses | 20/38 | + |
| Skin | | |
| Verrucous epidermal nevi | 26/38 | +[b] |
| Thickening of skin and subcutaneous tissue | 23/38 | + |
| Lipomas[c] | 26/38 | + |
| Vascular anomalies[d] | 20/38 | − |
| Dilated veins | 15/38 | − |
| Unspecified subcutaneous masses | 31/38 | + |
| Craniofacial | | |
| Macrocephaly and/or frontal bossing | 7/38 | + |
| Epibulbar tumor | 5/38 | − |
| Enlarged eye | 6/38 | − |
| Strabismus | 9/38 | − |
| Prognathism | 3/11 | − |
| Performance | | |
| Mental deficiency[e] | 11/37 | − |
| Motor delay | 6/38 | − |
| Seizures | 2/13 | − |

[a]From MM Cohen Jr (7) and MG Bialer (3).
[b]Treves writes of "papillomatous" skin, which in one area merges into "a mere roughening of the skin." The latter may represent a verrucous epidermal nevus.
[c]Pelvic lipomatosis has been found in three instances, thus increasing the total number of patients affected with lipomas.
[d]Vascular anomalies include hemangiomas, lymphangiomas, "venous dilation," "purplish discoloration," and varicosities.
[e]Mental deficiency is a variable feature, so Merrick's normal mentation is fully compatible with Proteus syndrome.


9. Cohen MM Jr: Understanding Proteus syndrome, unmasking the Elephant Man, and stemming elephant fever. *Neurofibromatosis* **1**:260–280, 1988.
10. Cohen MM Jr, Hayden PW: A new recognized hamartomatous syndrome. *Birth Defects* **15**(5B):291–296, 1979.
11. Costa T et al: The Proteus syndrome: Report of two cases of pelvic lipomatosis. *Pediatrics* **76**:984–989, 1985.
12. Cremin BJ et al: The Proteus syndrome: The magnetic resonance and radiologic features. *Pediatr Radiol* **17**:486–488, 1987.
13. Fauchet R, Stagnara P: La mégaspondylodysplasia: Aspects orthopédiques. *Rev Chir Orthoped* **67**:647–653, 1981.
14. Fay JT, Schow SR: A possible case of Maffucci's syndrome. *J Oral Surg* **26**:739–744, 1968.
15. Gorlin RJ: Proteus syndrome. *J Clin Dysmorphol* **2**(1):8–9, 1984.
16. Graetz I: Über einen Fall von sogenannter "totaler halbseitiger Korperhypertrophie." *Z Kinderheilkd* **45**:381–403, 1928.
17. Granges Y: A propos d'un cas inédit de syndrome polymalformatif avec gigantisme partiel des extrémities. Etude comparative des 100 cas de syndrome de Maffucci. Thesis. Lausanne, 1973.
17a. Hall JG: Somatic mosaicism: Observations related to clinical genetics. *Am J Hum Genet* **43**:355–363, 1988.
18. Happle R: Cutaneous manifestation of lethal genes. *Hum Genet* **72**:280, 1986.
19. Hornstein L et al: Linear nevi, hemihypertrophy, connective tissue hamartomas and unusual neoplasms in children. *J Pediatr* **110**:404–408, 1987.
20. Horton WA: Klippel–Trénaunay–Weber syndrome. *Birth Defects* **7**(8):316–318, 1971 (Case 2).
21. Hotamisligil GS, Ertogan F: The Proteus syndrome: Report of a case with nephrogenic diabetes insipidus. *Clin Genet* (in press).
22. Kontras SB: Case report #19 *Synd Ident* **2**(2):1–3, 1974.
23. Lacassie Y et al: Proteus syndrome: Further delineation. Personal communication, 1988.
24. Langsteiner F, Steifler G: Über die kongenitalen Hypertrophien (Hyperplasien). *Dtsch Z Nervenheilk* **138**:274–304, 1935.
25. Lezama DB, Buyse ML: The Proteus syndrome. *J Clin Dysmorphol* **2**(1):10–13, 1984.
26. Malamitsi-Puchner A et al: Severe Proteus syndrome in an 18-month-old boy. *Am J Med Genet* **27**:119–126, 1987.
27. Marcaillou A: Contribution á l'ètude des déformations vertébrales dans les dysplasies osseuses. (Dysplasies spondylo-épiphysaires essentiellement). A propos de 49 observations dont 47 au Centre de Massues. Thesis, Lyon, 1975.
27a. Mayatepek E et al: Expanding the phenotype of the Proteus syndrome: A severely affected patient with new findings. *Am J Med Genet* **32**:402–406, 1989.
28. Moore BH: Macrodactyly and associated peripheral nerve changes. *J Bone Jt Surg* **24**:617–631, 1942.
29. Mücke J et al: Variability in the Proteus syndrome: Report of an affected child with progressive lipomatosis. *Eur J Pediatr* **143**:320–323, 1985.
30. Pawlaczyk B, Sioda T: Hypertrophied lumbar and muscular atrophy. *Synd Ident* **4**(1):3–4, 1976.
31. Romestan-Boutin E: La mégaspondylodysplasie. A propos de 14 observations dont 6 au Centre de Massues. Thesis, Lyon, 1982.
31a. Samlaska CP et al: Proteus syndrome. *Arch Dermatol* **125**:1109–1114, 1989.
32. Schnake C et al: Sindrome Proteus: Contribucion a su dilineacion clinica. *Rev Child Pediatr* **57**:585–594, 1986.
33. Stevenson RE, Saul RA: Limb overgrowth associated with nevi and lipomas. *Proc Greenwood Genet Ctr* **1**:29–33, 1982.
34. Temtamy SA, Rogers JG: Macrodactyly, hemihypertrophy, and connective tissue nevi: Report of a new syndrome and review of the literature. *J Pediatr* **89**:600–603, 1976.
35. Tibbles JAR, Cohen MM Jr: Proteus syndrome: The Elephant Man diagnosed. *Br Med J* **293**:683–685, 1986.
36. Viljoen DL et al: Proteus syndrome in Southern Africa: Natural history and clinical manifestations in six individuals *Am J Med Genet* **27**:87–98, 1987.
37. Viljoen DL et al: Cutaneous manifestation of the Proteus syndrome. *Pediatr Dermatol* **5**:14–21, 1988.
38. Werthemann A: Die Entwicklungsstörungen der Extremitäten. *Handbuch speziellen pathol Anat Histol* **9**(6):377–393, 1952.
39. Wiedemann H-R et al: The Proteus syndrome. Partial gigantism of the hands and/or feet, nevi, hemihypertrophy, subcutaneous tumors, macrocephaly, skull anomalies and possible accelerated growth and visceral affections. *Eur J Pediatr* **140**:5–12, 1983.
40. Wieland E: Zur Pathologie der dystrophischen Form des angeborenen partiellen Riesenwuchses. *Jahrb Kinderheilk* **65**:519–594, 1907.
41. Zuntová A: Asymmetrical osteocutaneous lipomatosis in a three-year-old boy associated with defective development of the brain. *Čs Pediatr* **38**:724–727, 1983.


## Sturge–Weber angiomatosis

Sturge–Weber angiomatosis is characterized by venous angiomatosis of the leptomeninges. Other findings most often include ipsilateral facial angiomatosis, ipsilateral gyriform calcifications of the cerebral cortex, seizures, mental deficiency, hemiplegia, and ocular defects (Figs. 12–46 to 12–49). Many eponyms have been used to denote the disorder, and priority of discovery has been debated. A representative but by no means exhaustive list includes Sturge–Weber syndrome, Sturge–Weber disease, encephalotrigeminal angiomatosis, Sturge–Weber–Krabbe syndrome, Sturge–Weber–Dimitri syndrome, Sturge–Kalischer–Weber syndrome, and meningofacial angiomatosis (9–11,22,28,29,31,32,35,37,46). Several references are of historical interest (13,20,22,23,38,43,49). The most extensive study of Sturge–Weber angiomatosis was provided by Alexander and Norman (2).

All cases to date have been sporadic. Although a few histories have suggested incomplete forms in first degree relatives who had single features such as seizures, mental retardation, or vascular nevi, no instance of full-blown Sturge–Weber angiomatosis has ever been recorded in more than one individual within the same family. Neither sex preponderance nor predilection for left- or right-sided involvement has been found (2). Hall (17a) suggested that the syndrome may rep-

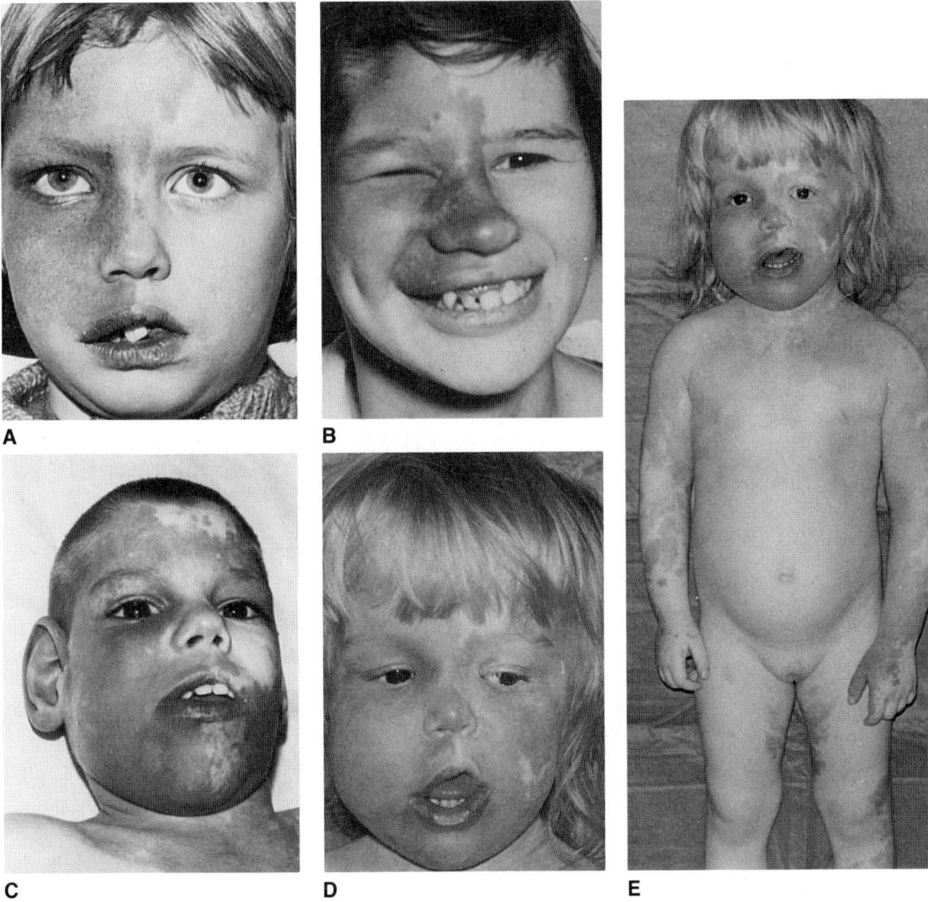


Fig. 12–46. *Sturge–Weber angiomatosis.* (A,B) Note facial distribution of angiomatosis. (C,D) Note bilateral involvement. (E) Same patient as observed in D. Note angiomatosis on body and enlargement of left arm and hand. This patient represents an example of Sturge–Weber angiomatosis as part of Klippel–Trénaunay–Weber syndrome. [C–E from *MM Cohen Jr,* Dysmorphic syndromes with craniofacial manifestations, in Orofacial Genetics, Stewart RE, Prescott GH (eds), Mosby, St. Louis, 1976.]

resent a lethal dominant mutation that is maintained because of somatic mosaicism.

Sturge–Weber angiomatosis can be explained on the basis of an embryonic anomaly with secondary consequences. During the sixth week of normal intrauterine development, a vascular plexus develops around the cephalic portion of the neural tube and under the ectoderm destined to become facial skin. Normally, this vascular plexus regresses during the ninth week, but in Sturge–Weber angiomatosis, it persists, resulting in angiomatosis of the leptomeninges overlying the cerebral cortex (Fig. 12–48), together with facial angiomatosis on the ipsilateral side. Variation in the degree of persistence or regression of the vascular plexus accounts for cases of bilateral involvement and also for cases with unilateral occurrence in which angioma of the leptomeninges occurs in the absence of facial involvement (4).

Other features of the association seem to be secondary to the leptomeningeal lesion. A poorly understood alteration in the vascular dynamics of the angioma results in precipitation of calcium deposits in the cerebral cortex underlying the angioma. Seizures and mental de-

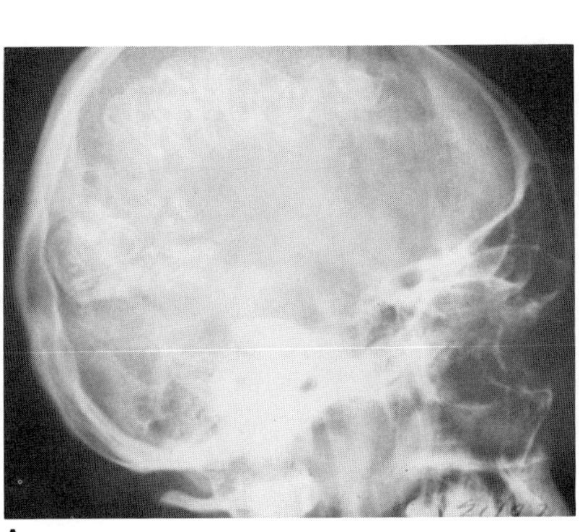


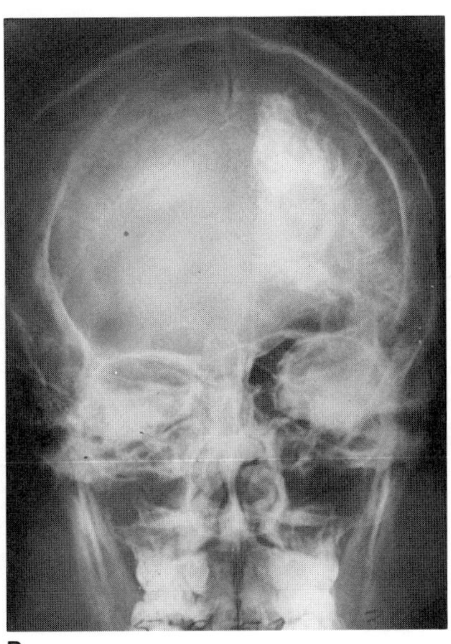


Fig. 12–47. *Sturge–Weber angiomatosis.* Roentgenograms. (A,B) Typical double contoured calcification. Note unilateral distribution.

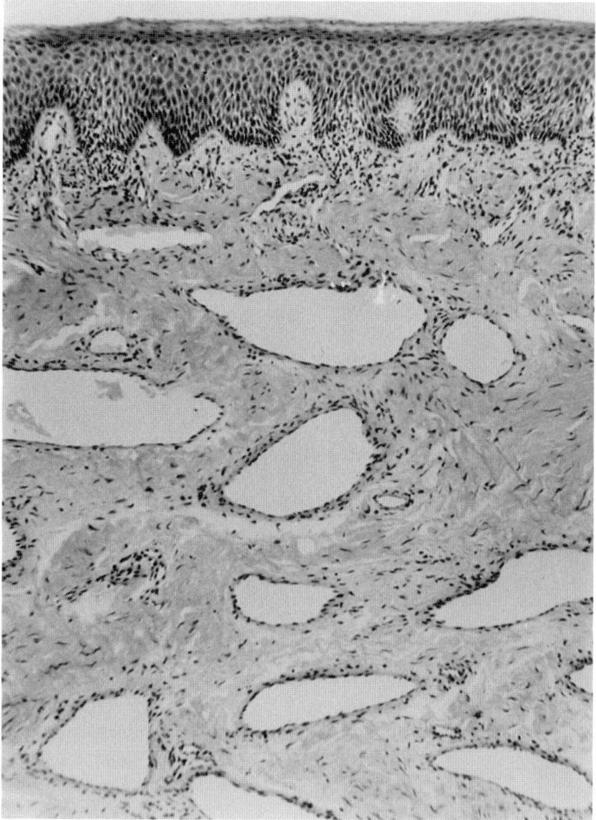

Fig. 12–48. *Sturge–Weber angiomatosis.* Histology of angiomatous lesion.

ficiency are probably secondary to this process. Thus, it is unnecessary to insist on multiple criteria for Sturge–Weber angiomatosis. A single initiating defect—an angioma of the leptomeninges—can be said to constitute the minimum criterion for the disorder.

Because angiomas of the skin and, in some cases, of the viscera may be observed in the Klippel–Trénaunay–Weber syndrome, it is not surprising that Sturge–Weber angiomatosis occurs in association with this syndrome. In our literature review, over 20 such cases were noted (e.g., 16,18,45) and we have personally observed this association in several patients (Fig. 12–46E). Since angiomas may occur anywhere in the Klippel–Trénaunay–Weber syndrome, it is possible that this syndrome and Sturge–Weber angiomatosis represent the same basic disorder, the latter occurring whenever an angioma happens to be present in the leptomeninges.

**Face.** A nevus flammeus lesion occurs on the ipsilateral side of the face in approximately 90% of patients (32) (Fig. 12–46A,B). In some instances, the nevus may extend onto the neck, chest, and back. In other instances, only a smaller periocularly circumscribed lesion or just involvement of the upper eyelid may occur. Bilateral facial nevus or lack of facial nevus has also been observed (Fig. 12–46C–E) (2,4). The color varies from pink to purplish-red and may decrease in intensity with age. The lesion is sharply demarcated and usually flat but on occasion can be raised or even saccular.

Cushing (12) noted a correlation between the distribution of the nevus flammeus and the course of the trigeminal nerve. However, Alexander and Norman (2) found the trigeminal relationship to be secondary and fortuitous. They assumed that the distribution of the facial nevus was determined, in part, by the position of the processes and fissures in the developing face. As the critical positioning of the port-wine nevus is associated with the ophthalmic ($V_1$) division of the trigeminal nerve, we view the syndrome as a dysmorphogenesis of cephalic neuroectoderm.

Ocular and intraoral involvement may occur. Choroidal angioma is common, and buphthalmos, glaucoma, and hemianopsia have been reported (2,26). Intraoral angiomatosis (Fig. 12–49A,B) occurs most frequently on the buccal mucosa and lips, macrocheilia occurring when the lips are involved (1,2,9,48). The palate is less frequently affected. Tongue involvement may be accompanied by hemihypertrophy (5,32,37). Gingival lesions, when present, may range from slight vascular hyperplasia to monstrous overgrowth, making closure of the mouth impossible (6,14,15,17,34). Vascular gingival hyperplasia, which blanches on pressure, should be distinguished from fibrous hyperplasia, which may accompany medication with diphenylhydantoin. Multiple pyogenic granulomas of the gingiva were reported in one case by Thoma (46). Both unilateral hypertrophy and hypotrophy of the alveolar process have been reported (2,19,28). Ipsilateral premature eruption of permanent teeth (28,39), ipsilateral delayed eruption (15), and ipsilateral normal eruption (48) have been noted. Unilateral premature eruption causes irregular positioning of teeth, leading to malocclusion. The size of the teeth may vary in the affected area, macrodontia being most frequently observed (2,10).

Fig. 12–49. *Sturge–Weber angiomatosis.* (A) Gingival enlargement and angiomatosis. (B) Intraoral hemangioma in 8-year-old girl. Eruption of teeth on affected side is more advanced than on unaffected side. (A from *HE Royle,* Oral Surg **22**:490, 1966.)

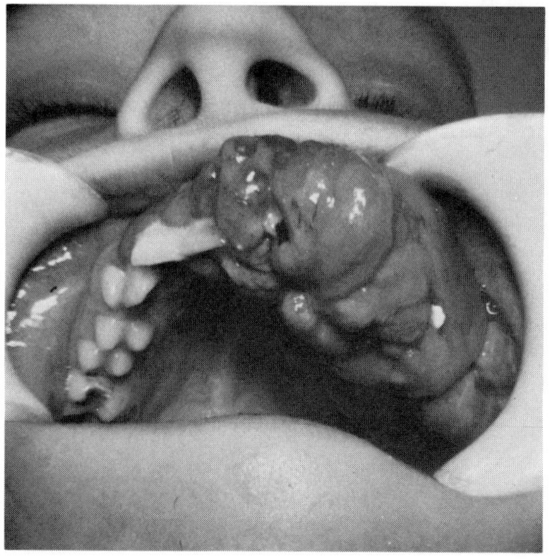

A

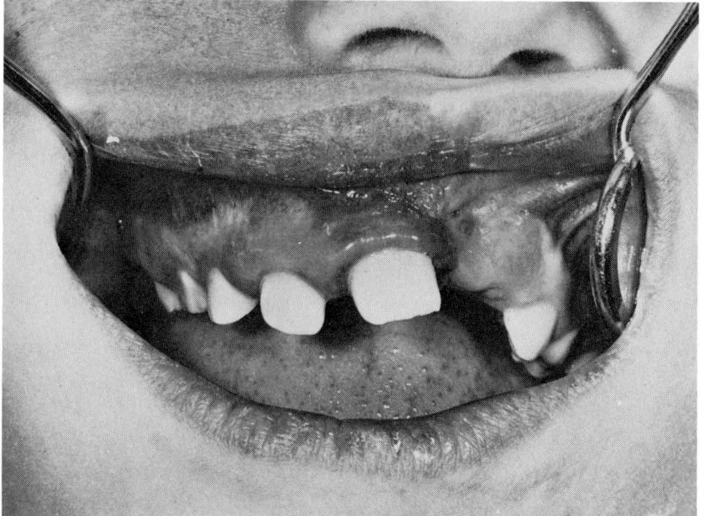

B

**Central nervous system.** The characteristic lesion consists of a unilateral, thin-walled angioma in the leptomeninges (Fig. 12–48) overlying the posterior temporal, posterior parietal, and occipital areas. Occasionally, bilateral involvement may be present (2,24,27,29,47,50). Abnormalities of the cerebral vascular system occur in about 46% (33). There may be lack of superficial cortical veins and associated nonfilling of the superior sagittal sinus, enlargement and tortuosity of the deep subependymal and deep medullary veins, and occasionally bizarre courses of cerebral veins (7). Arteriovenous malformations, arterial thromboses, abnormalities of external carotid orientation, subdural hematoma, and cerebral hypoplasia or atrophy have also been recorded (33).

Intracranial calcification is visible radiographically in about 57% (33). The smallest and possibly earliest calcification occurs in perithelial cells. Norman and Schoene (31) have hypothesized that calcification is caused by anoxic injury to endothelial, perithelial, and possibly glial mitochondria due to stasis and abnormal vessel permeability in the cerebral vessels composing the angioma. Gyriform, double-contoured lines of calcification develop in the underlying cerebral cortex (Fig. 12–47A,B) and the roentgenographic appearance is pathognomonic for the disorder when present (2,11). Calcification is usually visible after the second year of life, but may rarely be present at birth. On occasion, rapid development of massive calcification occurs. In general, cases with radiographically visible calcification become stationary by the second decade. Increasing age and the presence of intracranial calcification decrease the chances of demonstrating vascular abnormalities by cerebral angiography (2,11,33). Macrocephaly (42) and asymmetry of the skull (19) have been observed in some instances.

Seizures were observed in 90% of cases in one series (32). Symptoms appear during infancy, seizures occurring contralaterally to the angiomatosis. Most often they are focal, but generalized convulsions may occur. Hemiparesis is less frequent, and the paretic limb is sometimes hypotrophic (2). At least 30% exhibit mental deficiency. With extensive cerebral changes, retardation may be pronounced (2,32).

**Other findings.** Angiomatosis has also been observed on occasion in the choroid plexus (8,50), thyroid (8,50), pituitary gland, lung, gastrointestinal tract, pancreas, ovaries (50), and thymus (8). Melanosis has been reported in the leptomeninges (8,29), and oculocutaneous involvement has also been observed (30). Nevus of Ota (blue nevus around area of trigeminal innervation) and nevus of Ito (blue nevus around supraclavicular, scapular, and deltoid regions) have been recorded (16). Sturge–Weber angiomatosis is not known to be associated with neoplasia (11,32). An exceptional case is multiple myeloma noted in one instance (21). Any of the various findings associated with Klippel–Trénaunay–Weber syndrome may be associated with Sturge–Weber angiomatosis since the two syndromes may occur together in the same patient (16,18,45).

**Differential diagnosis.** The relationship between the Sturge–Weber angiomatosis and *Klippel–Trénaunay–Weber syndrome* has already been discussed. Transitory nevus flammeus lesions during the neonatal period are extremely common (44). However, in many cases, they may not be as intense in color as the Sturge–Weber lesion, which tends to be darker and most frequently unilateral in distribution. Although the facial nevus in Sturge–Weber angiomatosis varies considerably in extent, dark, supraocular involvement should arouse suspicion (2) as the port-wine stain in Sturge–Weber syndrome has an association only with the ophthalmic division of the trigeminal nerve (14a).

The association of macrocephaly and angiomatosis may occur in disseminated hemangiomatosis, *neurofibromatosis, Bannayan–Riley–Ruvalcaba syndrome, Beckwith–Wiedemann syndrome, Klippel–Trénaunay–Weber syndrome,* and cutis marmorata telangiectatica congenita (42).

In Coats disease (telangiectatic retinal vessels that subsequently cause retinal detachment), facial involvement may occur. For example, Allen and Parlette (3) reported a patient with maxillary and cervical nevus flammeus lesions. However, the central nervous system is never involved. In *Bannyan–Riley–Ruvalcaba syndrome,* hemangiomas and lipomas occur with macrocephaly and mild neurologic dysfunction. Inheritance is autosomal dominant (25,51). Reinke et al (36) reported the combination of multiple hemangiomas, nevus of Ota, and Takayasu arteritis. Selmanowitz (40) reported a family with nevus flammeus of the forehead inherited as an autosomal dominant trait.

**Diagnostic aids.** Because severe brain damage can exist, it is important to establish the presence of leptomeningeal angiomatosis as soon as possible by means of angiography. However, once cerebral calcification occurs, the chances of obtaining a positive angiogram are considerably diminished (2,33). Roentgenograms after 2 years of age frequently reveal the pathognomonic, gyriform, double-contoured calcification lines. Electroencephalographic studies are abnormal, usually consisting of unilateral depression of cortical activity (2,3,5). Elevated spinal fluid protein has been reported (41).

### References (Sturge–Weber angiomatosis)


1. Achslogh J: Hémisphérectomie dans un cas de maladie de Sturge–Weber. *Acta Neurol Belg* **58**:837–847, 1958.
2. Alexander GL, Norman RM: *The Sturge–Weber Syndrome,* John Wright and Sons, Bristol, 1960.
3. Allen HB, Parlette HL: Coats disease. *Arch Dermatol* **108**:413–415, 1973.
4. Andriola M, Stolfi J: Sturge–Weber syndrome. *Am J Dis Child* **123**:507–510, 1972.
5. Arendt A: Sturge–Weber'sche Erkrankung. *Dtsch Gesundheitsw* **8**:137–141, 1953.
6. Baer PN et al: Gingival hemangioma associated with Sturge–Weber syndrome. *Oral Surg* **14**:1383–1390, 1961.
7. Bentson JR et al: Cerebral venous drainage pattern of the Sturge–Weber syndrome. *Am J Roentgenol* **101**:111–118, 1971.
8. Bentz M et al: Sturge–Weber syndrome. *Arch Pathol Lab Med* **106**:75–78, 1982.
9. Born E: Über die Sturge–Weber Angiomatose. *Zentralbl Allg Pathol* **97**:569–576, 1957–1958.
10. Brushfield T, Wyatt W: Sturge–Weber disease. *Br J Child Dis* **24**:98–106, 209–213, 1927; **25**:96–101, 1928.
11. Chao DH-C: Congenital neurocutaneous syndromes of childhood. III. Sturge–Weber disease. *J Pediatr* **55**:635–649, 1959.
12. Cushing H: Cases of spontaneous intracranial hemorrhage associated with trigeminal nevi. *JAMA* **47**:178–183, 1906.
13. Dimitri V: Tumor cerebral congenital (angioma cavernoso). *Rev Med Argent* **36**:1029, 1923.
14. El Mostehy MR, Stallard RE: The Sturge–Weber syndrome: Its periodontal significance. *J Periodontol* **40**:243–246, 1969.

14a. Enjolras O et al: Facial port-wine stains and Sturge–Weber syndrome. *Pediatrics* **76**:48–51, 1985.

15. Falk W: Beitrag zur Aetiologie und Klinik den Sturge–Weberschen Krankheit. *Oest Z Kinderheilkd* **5**:175–185, 1950.
16. Furukawa T et al: Sturge–Weber and Klippel–Trénaunay syndrome with nevus of Ota and Ito. *Arch Dermatol* **102**:640–645, 1970.
17. Gyarmati I: Oral change in Sturge–Weber's disease. *Oral Surg* **13**:795–801, 1960.

17a. Hall JG: Somatic mosaicism: Observations related to clinical genetics. *Am J Hum Genet* **43**:355–363, 1988.

18. Heuser M: De l'entite nosologique des angiomatoses neuro-cutanées (Sturge–Weber et Klippel–Trénaunay). *Revue Neurol (Paris)* **124**:213–228, 1971.
19. Höring H: Zur Lokalisation mesenchymaler Dysplasien bei Sturge–Weber Krankheit. *Arch Klin Exp Dermatol* **209**:615–624, 1960.
20. Kalischer S: Demonstration des Gehirns eines Kindes mit Telangiectasie der linksseitigen Gesichts-Kopfhaut und Hirnoberfläche. *Berl Klin Wochenschr* **34**:1059, 1897.
21. Kiely JM, Perry HO: Sturge–Weber syndrome associated with multiple myeloma. *Arch Dermatol* **100**:63–65, 1969.
22. Krabbe KH: Facial and meningeal angiomatosis associated with calcifications of brain cortex: Clinical and anatomo-pathologic contribution. *Arch Neurol Psychiatr* **32**:737–755, 1934.
23. Krabbe KH, Wissing O: Calcifications de la piemère du cerveau (d'origine angiomateuse) demontrée par la radiographie. *Acta Radiol* **10**:523–532, 1929.
24. Lagenstein I, Gruttner R: Sturge–Weber-Syndrom mit isolierter kortikaler Gefässmissbildung. *Klin Pädiatr* **190**:572–575, 1978.
25. Miles JH et al: Macrocephaly with hamartomas: Bannayan–Zonana syndrome. *Am J Med Genet* **19**:225–234, 1984.

26. Miller SJN: Ophthalmic aspects of the Sturge–Weber syndrome. *Proc R Soc Med* **56**:419–421, 1965.

27. Morgan G: Pathology of the Sturge–Weber syndrome. *Proc R Soc Med* **56**:422–423, 1963.

28. Myle G: Sémiologie de l'angiomatose encéphalotrigéminée ou encéphalo-cranio-faciale. *Acta Neurol Belg* **50**:713–785, 1950.

29. Nellhaus G et al: Sturge–Weber disease with bilateral intracranial calcifications at birth and unusual pathologic findings. *Acta Neurol Scand* **43**:314–347, 1967.

30. Noriega-Sanchez A et al: Oculocutaneous melanosis associated with the Sturge–Weber syndrome. *Neurology* **22**:256–262, 1972.

31. Norman MG, Schoene WC: The ultrastructure of Sturge–Weber disease. *Acta Neuropath (Berlin)* **37**:199–205, 1977.

32. Peterman AF et al: Encephalotrigeminal angiomatosis (Sturge–Weber disease): Clinical study of thirty-five cases. *JAMA* **167**:2169–2176, 1958.

33. Poser CM, Taveras JM: Cerebral angiography in encephalotrigeminal angiomatosis. *Radiology* **68**:237–336, 1957.

34. Protzel MS: Sturge–Weber syndrome. *Oral Surg* **10**:388–399, 1957.

35. Radermecker J: L'électroencéphalographie dans l'angiomatose encéphalotrigéminée de Sturge–Weber–Krabbe. *Acta Neurol Belg* **51**:427–451, 1951.

36. Reinke RT et al: Ota nevus, multiple hemangiomas, and Takayasu arteritis. *Arch Dermatol* **110**:447–449, 1974.

37. Roizin L et al: Congenital vascular anomalies and their histopathology in Sturge–Weber–Dimitri syndrome (naevus flammeus with angiomatosis and encephalosis calcificans). *J Neuropathol Exp Neurol* **18**:75–97, 1959.

38. Schirmer R: Ein Fall von Telangiektasie. *Albrecht von Graefes Arch Ophthalmol* **7**:119–121, 1860.

39. Schuermann H: *Krankheiten der Mundschleimhaut und der Lippen,* 2nd ed., Urban and Schwarzenberg, Berlin, 1958.

40. Selmanowitz VJ: Nevus flammeus of the forehead. *J Pediatr* **73**:755–757, 1968.

41. Skoglund RR et al: Elevated spinal-fluid protein in Sturge–Weber syndrome. *Dev Med Child Neurol* **20**:99–102, 1978.

42. Stephan MJ et al: Macrocephaly in association with unusual cutaneous angiomatosis. *J Pediatr* **87**:353–359, 1975.

43. Sturge WA: Case of partial epilepsy apparently due to lesion of one of vasomotor centres of brain. *Trans Clin Soc London* **12**:162–167, 1879.

44. Tan KL: Nevus flammeus of the nape, glabella and eyelids. *Clin Pediatr* **11**:112–118, 1972.

45. Teller H, Lindner B: Über Mischformen der phakomatösen Syndrome von Sturge–Weber und Klippel–Trénaunay. *Z Haut Geschlechtskr* **13**:113–120, 1952.

46. Thoma KH: Sturge–Kalischer–Weber syndrome with pregnancy tumors. *Oral Surg* **5**:1124–1131, 1952.

47. Tönnis W, Friedmann G: Roentgenographische und klinische Befunde bei 23 Patienten mit Sturge–Weber Erkrankung. *Zentralbl Neurochir* **25**:1–10, 1964.

48. Wannenmacher MF, Forck G: Mundschleimhautveränderungen beim Sturge–Weber-Syndrom. *Dtsch Zahnärztl Z* **25**:1030–1035, 1970.

49. Weber FP: Right-sided hemi-hypotrophy resulting from right-sided congenital spastic hemiplegia, with morbid condition of left side of brain, revealed by radiograms. *J Neurol Psychopathol* **3**:134–139, 1922.

50. Wohlwill FJ, Yakovlev PI: Histopathology of meningofacial angiomatosis (Sturge–Weber's disease): Report of four cases. *J Neuropathol Exp Neurol* **16**:341–364, 1957.

51. Zonana J et al: Macrocephaly with multiple lipomas and hemangiomas. *J Pediatr* **89**:600–603, 1976.

## Tuberous sclerosis

In tuberous sclerosis, the classic triad consists of seizures, mental deficiency, and angiofibromas. Bourneville (6), in 1880, gave the first detailed description of the neurologic symptoms and cerebral pathology in three patients and coined the term tuberous sclerosis. Other papers coauthored with Brissaud were published in 1881 (7) and 1900 (8). However, as early as 1862, von Recklinghausen (74) described cardiac myomas and sclerotic lesions of the brain in an infant. In 1890, Pringle (57) reported a 25-year-old mildly retarded woman with facial lesions that he called "adenoma sebaceum," a term coined earlier by Balzer and Ménétrier (1). Historical aspects of the disorder have been especially well reviewed by Gomez (23) and Morgan and Wolfort (50). Tuberous sclerosis is also known as Bourneville–Pringle syndrome, epiloia, and adenoma sebaceum syndrome. The last term, so long used, should be abandoned because facial lesions are angiofibromas, not sebaceous adenomas. Several important reviews are available (10,20,23,34,55,78).

Tuberous sclerosis is characterized by a potential for hamartomatous growth in multiple organs and has a broad range of expression (Figs. 12–50 to 12–55). The classic triad of seizures, mental retardation, and angiofibromas (Fig. 12–50A–E) represents one end of the spectrum of severity; mild cases can be difficult to detect. Autosomal dominant inheritance with variable expressivity has been demonstrated (10,23,40,78). Earlier studies indicated that approximately 70% were isolated examples, representing new mutations, but this is an overestimate (10). Lack of penetrance exists but appears to be relatively uncommon (1a). Up to 44% of cases may be familial (20). Linkage studies support the assignment of the gene to the distal long arm of chromosome 9 (11a,21).

Published estimates of the prevalence of clinically ascertained tuberous sclerosis in institutions for the mentally retarded range from 0.32 to 1.1% (23). A recent prevalence estimate from the general population is 1/23,000 (78), although actual clinically ascertained population studies of specific geographical areas vary from 1/100,000 to 1/170,000 in the general population (23,70). These represent serious underestimates because patients with subtle or no clinical manifestations were not identified. Surveys of autopsy data indicate a frequency of approximately 2.5–2.8% (23,78).

The studies of Cassidy et al (10) and Gomez (23) clearly show that minimal diagnostic criteria for tuberous sclerosis can be used only if family members at risk are thoroughly evaluated. Detailed skin examination was the most sensitive diagnostic test (96%). However, as skin findings may not be present in infancy and early childhood, other tests can be useful, including cranial CT (67%), skull roentgenograms (46%), hand–foot roentgenograms (39%), funduscopic examination through dilated pupils (33%), and renal ultrasound (30%) (10). Using such criteria, Gomez (23) has listed the findings for primary and secondary diagnosis of tuberous sclerosis (Table 12–7).

**Skin.** Hypomelanotic macules are the earliest and most commonly observed skin lesions (Fig. 12–51A,B) (90%). Often they are present at birth, but sometimes they can be observed only with a Wood's lamp. Most hypomelanotic muscles are leaf-shaped or round on one end and tapered on the other. In about 20%, depigmented hair patches are found. They may be the first sign of the disorder (45). Facial angiofibromas occur in 70%. Other findings include raised leathery plaques in the lumbosacral area (shagreen patches) (36%), periungual or subungual fibromas (Fig. 12–52A,B) usually appearing at puberty (18%), pigmented lesions sometimes called "café-au-lait" or "brown" spots (16%) [Bell and Macdonald (4) have demonstrated that café-au-lait spots are found no more frequently than in the normal population], and pedunculated cutaneous nodules or skin tags (molluscum fibrosum pendulum) (14%) (19,23,35,62). Fryer et al (22) has discussed forehead plaque as a presenting sign in tuberous sclerosis. When present,

Table 12–7. Diagnostic criteria for tuberous sclerosis[a]

| Primary criteria (only one required for diagnosis) | Secondary criteria (two required for diagnosis) |
|---|---|
| Facial angiofibromas | Infantile spasms |
| Ungual fibromas | Hypopigmented macules |
| Cortical tuber (at necropsy) | Shagreen patch |
| Subependymal hamartomas (at necropsy or on CT) | Single retinal hamartoma |
| | Bilateral renal angiomyolipomas or cysts |
| Multiple retinal hamartomas | Cardiac rhabdomyoma |
| Fibrous plaque on forehead | First-degree relative with a primary diagnosis of tuberous sclerosis |

[a]Adapted from MR Gomez, *Tuberous Sclerosis,* Raven Press, New York, 1979.

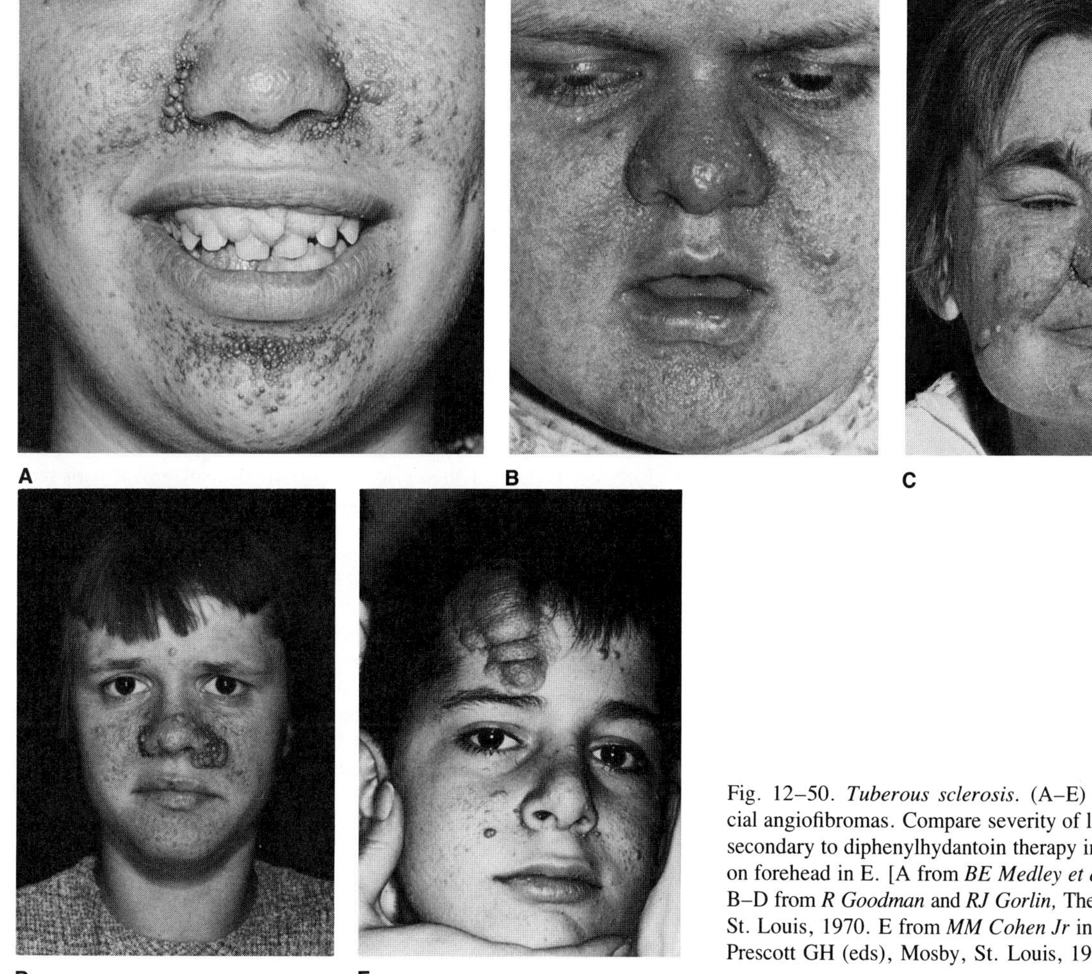


Fig. 12–50. *Tuberous sclerosis.* (A–E) Characteristic distribution of facial angiofibromas. Compare severity of lesions. Note gingival hyperplasia secondary to diphenylhydantoin therapy in A. Note large fibromatous mass on forehead in E. [A from *BE Medley et al,* Sem Roentgenol **11**:35, 1976. B–D from *R Goodman* and *RJ Gorlin,* The Face in Genetic Disease, Mosby, St. Louis, 1970. E from *MM Cohen Jr* in Orofacial Genetics, Stewart RE, Prescott GH (eds), Mosby, St. Louis, 1976.]

they are smooth patches of slightly raised skin with a reddish or yellowish discoloration.

**Performance and central nervous system.** CT reveals most gross pathological processes, which include hamartomatous foci, dilated ventricles, and subependymal giant cell astrocytomas. Hamartomatous mineralization, if not present at birth, may develop as early as 5 months of age and progress throughout infancy and childhood. Hamartomatous foci consist of subependymal nodules and cerebral or cerebellar cortical tubers (Fig. 12–54A) (33,39,46).

Seizures occur in 88%. Approximately 29% have focal seizures and 15% have both generalized and partial seizures. The frequency of various types of seizures are as follows: tonic–clonic seizures (41%), infantile spasms (30%), myoclonic seizures (16%), atypical absence (7%), tonic seizures (6%), and akinetic seizures (4%). Partial seizures break down as follows: motor seizures (16%), complex symptoms (10%), and unknown types (2%) (24,34,43,48,53,76,77).

Approximately 60% have mental deficiency. A close correlation has been established between seizures and mental retardation, 89% of all individuals having seizures being mentally subnormal. In addition, the age of seizure onset and the severity of mental subnormality are directly related (24,76). Fryer et al (22a) reported an affected family through five generations with no history of seizures or mental retardation.

Approximately 56% have intracranial calcifications (Fig. 12–55A,B) (48). Most occur in the cerebrum, 63% of lesions being bilateral, the remaining 37% being unilateral. Cerebellar calcifications occur in approximately 12% (33,39,46). The calcification is progressive: in children below 1 year—15%, below 5 years—35%, but by 14 years 50–60% (48).

Increased intracranial pressure caused by tumor obstruction of cerebral spinal fluid circulation occurs infrequently, and less than 4% of patients have separated cranial sutures. Mid-ventricular dilatation occurs in about 50% (48). Space occupying lesions, usually subependymal giant cell astrocytomas, frequently cause obstruction of the foramen of Monro (48). Convincing examples of other types of central nervous system tumors are extremely rare (9,12,36,59,63).

Childhood schizophrenia and autism and dementia in adults are almost always associated with the presence of seizures or growth of a tumor, which frequently causes obstruction with increased intracranial pressure (24).

**Skeletal system.** In approximately half the cases, the skull exhibits thickened calvaria with an irregular outer table and exostoses are often observed on the inner table of the frontal bone. Areas of increased density appear throughout the skull, especially in the parietal region after puberty (26,32,33). Approximately 65% have cyst-like areas in the phalanges and irregular periosteal new bone formation along the shafts of the metacarpals and metatarsals (Fig. 12–55C). Less frequently, the long bones are involved (33,65). Gigantism of a single digit has been reported on several occasions (36a,78).

**Ocular findings.** Retinal hamartomas are found in approximately 50%. Two basic types of retinal lesions occur. The most common type is smooth, semitransparent, relatively flat, circular, or oval-shaped with distinct boundaries, being grayish or yellowish in color and being found

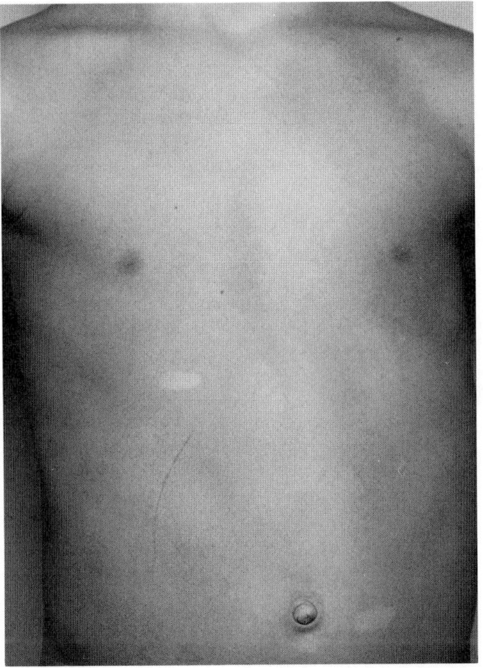

A

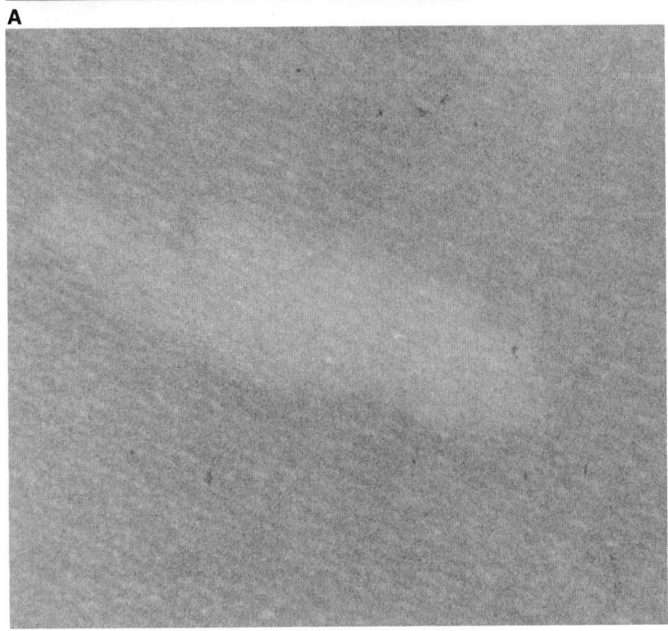

B

Fig. 12–51. *Tuberous sclerosis.* (A) Note hypomelanotic macules. (B) Achromic patch of skin. (A courtesy of *MR Gomez,* Rochester, Minnesota. B from *BE Medley et al,* Sem Roentgenol **11**:35, 1976.)

in 55% of patients with retinal lesions. The opaque and nodular type occurs in approximately 45% of patients with retinal hamartomas. The lesions are elevated and multinodular, resembling the grains of tapioca, salmon eggs, or mulberries. They frequently occur along the disc margin or halfway toward the periphery of the retina. Some patients have retinal hamartomas with the morphologic characteristics of both types. This accounts for 15% of patients with hamartomas (27,29,37,42,51,61,69).

**Renal abnormalities.** Characteristic renal abnormalities in tuberous sclerosis are angiomyolipomas (Fig. 12–54B) (60) and polycystic kidneys (60,75). Approximately 60–65% of autopsied cases have angiomyolipomas (72a). Most commonly these are multiple and bilateral. Their prevalence increases with age. They are seldom troublesome, although exceptions have been noted (2,3,5,17,23,28,43,47,49,52, 56,60). Renal angiomyolipoma by itself may be a forme fruste of tuberous sclerosis (72a).

Histologically disturbing changes occur in the spindled smooth muscle cells that form disorganized sheets, sinuous bands, or small clusters between the fat cells and frequently spread centrifugally from vessel walls. The frequent findings of hyperchromatic nuclei, pleomorphism, and occasional mitotic activity are atypical histologic features, but should not be taken as evidence of malignancy (5,17,60). An instance of Wilms tumor has been recorded (26a).

**Cardiac manifestations.** Approximately 30% have cardiac rhabdomyomas and the same proportion of patients with cardiac rhabdomyomas have tuberous sclerosis. In the largest series of patients with cardiac rhabdomyomas (18), 78% died during the first year of life and 17% of these were either stillborn or expired within 24 hours. Most rhabdomyomas occur intramurally and only a small percentage represent intracavitary tumors. Cardiac signs and symptoms occur on the basis of obstruction of blood flow, myocardial involvement with secondary deterioration of ventricular function, or disturbance of cardiac rhythm. Most lesions are multiple rather than single and their most frequent occurrence in infants suggests that the entity is a hamartoma rather than a neoplasm (13,16,18,23,38,41,58,68,72).

**Other findings.** Cystic disease of the lungs, pulmonary lymphangiomyomatosis (honeycomb lung), occurs most often in females (11F:1M), presenting in the third decade (48). Hemangiomas of the spleen and racemose angioma of the liver have been reported. Hamartomas have also been noted in the thyroid, pancreas, and testes (11,15,23). Various endocrinopathies have been recorded (31,64) including precocious puberty (14).

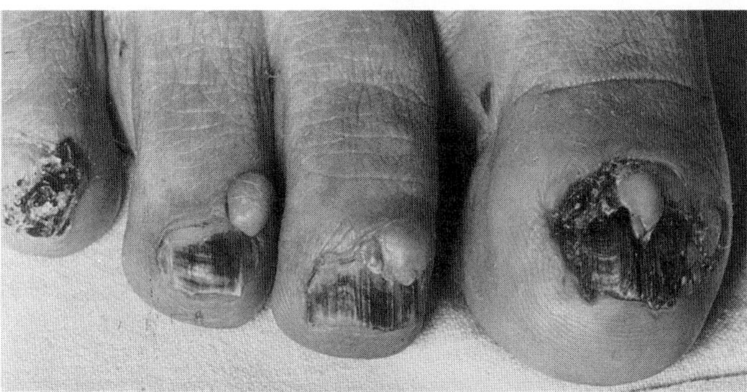

A

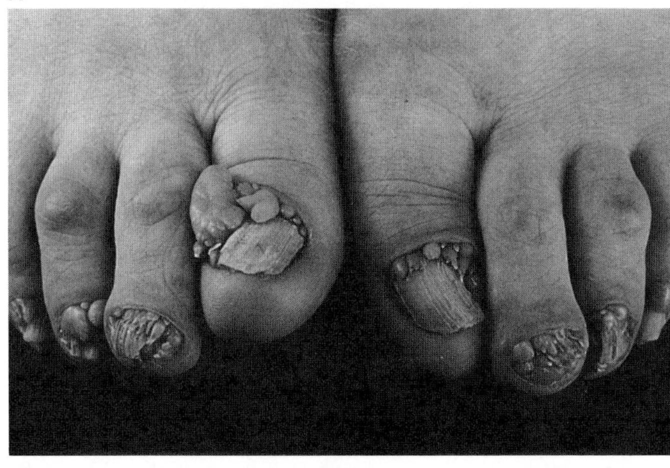

B

Fig. 12–52 *Tuberous sclerosis.* (A,B) Subungual fibromas. (A courtesy of *A Lodin,* Stockholm, Sweden. B courtesy of *M Klingbeil,* Berlin, Germany.)

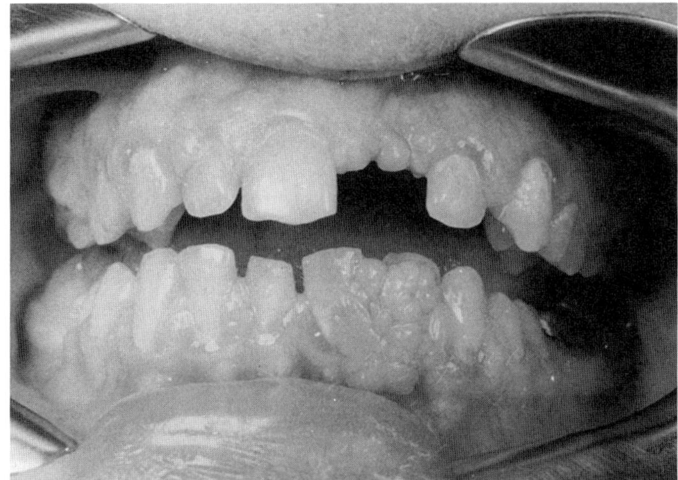
A

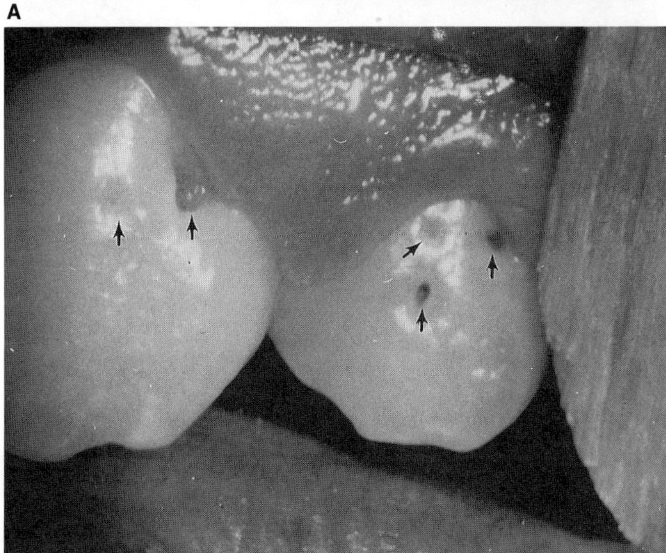
B

Fig. 12–53. *Tuberous sclerosis.* Oral lesions. (A) Gingival fibromas. (B) Pit-shaped enamel defects. (B from *F Vikilzadeh* and *R Happle,* Hautarzt **31**:336, 1980.)

**Oral manifestations.** The oral mucosa may be the site of fibrous growths (Fig. 12–53A). Approximately 11% of patients have such lesions. Most frequently they occur on the anterior gingiva but may be found on the oral mucosa (25,54,67). Enamel pits (Fig. 12–53B) have been observed (30,73). In some series, the frequency has been very high—on the order of 70% (40a,74a) and 90% (G Mlynarczyk, personal communication, 1988).

**Differential diagnosis.** Most frequently mistaken for facial angiofibromas are multiple trichoepitheliomas, syringocystadenomas, colloid milia, atypical xanthomas, or, occasionally, intradermal nevi.

Roentgenographic differential diagnosis of the skull should include *Sturge–Weber angiomatosis,* calcifying subdural hematoma and calcifying neoplasms, cytomegalic inclusion disease, toxoplasmosis, rubella, HTLV III intrauterine infections, and *hyalinosis cutis et mucosae.* Long bone changes in tuberous sclerosis should be differentiated from those of *neurofibromatosis,* fibrous dysplasia, enchondromatosis, gout, sarcoid, and hypertrophic arthritis. Shagreen patches may also be seen in association with osteopoikilosis and *Proteus syndrome.* Rarely, *Klippel–Trénaunay–Weber syndrome* has been observed with tuberous sclerosis (23,71).

**Laboratory aids.** Laboratory aids include Wood's lamp examination of the skin, cranial computed tomography, renal ultrasound, and roentgenograms of the hands, feet, and skull (48,66). Funduscopic examination through dilated pupi' preferably with fluorescein angiography is essential. Disclosing solution can be used to demonstrate enamel pits. All parents of patients with tuberous sclerosis should be thoroughly examined before recurrence risk counseling is given. Electron microscopy in combination with the Dopa reaction may be helpful in recognizing incomplete forms by identifying white leaf-shaped macules (29a).

Prenatal diagnosis has been accomplished through detection of cardiac rhabdomyoma by ultrasound and echocardiography (35a,54a).

### References (Tuberous sclerosis)


1. Balzer F, Ménétrier P: Étude sur un cas d'adénomes sébacés de la face et du cuir chevelu. *Arch Physiol North Pathol (Serie III)* **61**:564–576, 1885.

1a. Baraitser M, Patton MA: Reduced penetrance in tuberous sclerosis. *J Med Genet* **22**:29–31, 1985.

2. Becker JA et al: Angiomyolipoma (hamartoma) of the kidney: An angiographic review. *Acta Radiol (Diagn) (Stockholm)* **14**:561–568, 1973.

3. Beh WP et al: A renal cause for massive retroperitoneal hemorrhage-renal angiomyolipoma. *J Urol* **116**:372–374, 1976.

4. Bell SD, Macdonald DM: The prevalence of café-au-lait patches in tuberous sclerosis. *Clin Exper Dermatol* **10**:562–565, 1985.

5. Bernstein J et al: The renal lesion in syndromes of multiple congenital malformations. Cerebrohepatorenal syndrome; Jeune asphyxiating thoracic dystrophy; tuberous sclerosis; Meckel syndrome. *Birth Defects* **10**(4):35–43, 1974.

6. Bourneville DM: Sclérose tubéreuse des circonvolutions cérébrales: idiotie et épilepsie hemiplégique. *Arch Neurol (Paris)* **1**:81–91, 1880.

7. Bourneville DM, Brissaud E: Encéphalite ou sclérose tubéreuse des circonvolutions cérébrales. *Arch Neurol (Paris)* **1**:390–412, 1881.

8. Bourneville DM, Brissaud E: Idiotie et épilepsie symptomatiques de sclérose tubéreuse ou hypertrophique. *Arch Neurol (Paris)* **10**:29–39, 1900.

9. Brown J: Tuberous sclerosis with malignant astrocytoma. *Med J Aust* **1**:811–814, 1975.

10. Cassidy SB et al: Family studies in tuberous sclerosis. *JAMA* **249**:1302–1304, 1983.

11. Comptom WR et al: The abdominal angiographic spectrum of tuberous sclerosis. *Am J Roentgenol* **126**:807–813, 1976.

11a. Connor JM et al: Linkage of the tuberous sclerosis locus to a DNA polymorphism detected by v-abl. *J Med Genet* **24**:544–546, 1987.

12. Cooper JR: Brain tumors in hereditary multiple system hamartomatosis (tuberous sclerosis). *J Neurosurg* **34**:194–202, 1971.

13. Crawford DC et al: Cardiac rhabdomyomata as a marker for the antenatal detection of tuberous sclerosis. *J Med Genet* **20**:303–312, 1983.

14. Cummings JL et al: Tuberous sclerosis. *Am J Dis Child* **132**:1215–1216, 1978.

15. Darden JW et al: Hamartoma of the spleen: A manifestation of tuberous sclerosis. *Am Surg* **41**:564–566, 1975.

16. Farooki ZQ et al: Ultrasonic pattern of ventricular rhabdomyoma in two infants. *Am J Cardiol* **34**:842–844, 1974.

17. Farrow GM et al: Renal angiomyolipoma: A clinicopathologic study of 32 cases. *Cancer* **22**:564–570, 1968.

18. Fenoglio JJ Jr et al: Cardiac rhabdomyoma: A clinicopathologic and electron microscopic study. *Am J Cardiol* **38**:241–251, 1976.

19. Fitzpatrick TB et al: White leaf-shaped macules: Earliest visible sign of tuberous sclerosis. *Arch Dermatol* **98**:1–6, 1968.

20. Fleury P et al: The incidence of sporadic cases versus familial cases. *Brain Dev* **2**:107–117, 1979.

21. Fryer AE et al: Evidence that the gene for tuberous sclerosis is on chromosome 9. *Lancet* **1**:659–660, 1987.

22. Fryer AE et al: Forehead plaque: A presenting skin sign in tuberous sclerosis. *Arch Dis Childh* **62**:292–304, 1987.

22a. Fryer AE et al: Tuberous sclerosis: A large family with no history of seizures or mental retardation. *J Med Genet* **24**:547–548, 1987.

23. Gomez MR: *Tuberous Sclerosis,* Raven Press, New York, 1979.

24. Gomez MR: Neurologic and psychiatric symptoms, in *Tuberous Sclerosi.* Gomez MR (ed), Raven Press, New York, 1979, pp 85–93.

25. Gorlin RJ et al: Oral manifestations of the Fitzgerald–Gardner, Pringle–Bourneville, Robin, adrenogenital and Hurler–Pfaundler syndromes. *Oral Surg* **13**:1236–1244, 1960.

26. Green GJ: The radiology of tuberous sclerosis. *Clin Radiol* **19**:135–147, 1968.

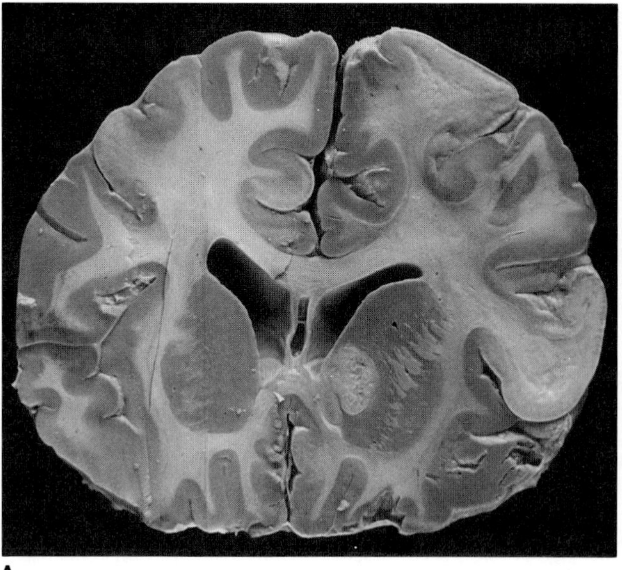

A

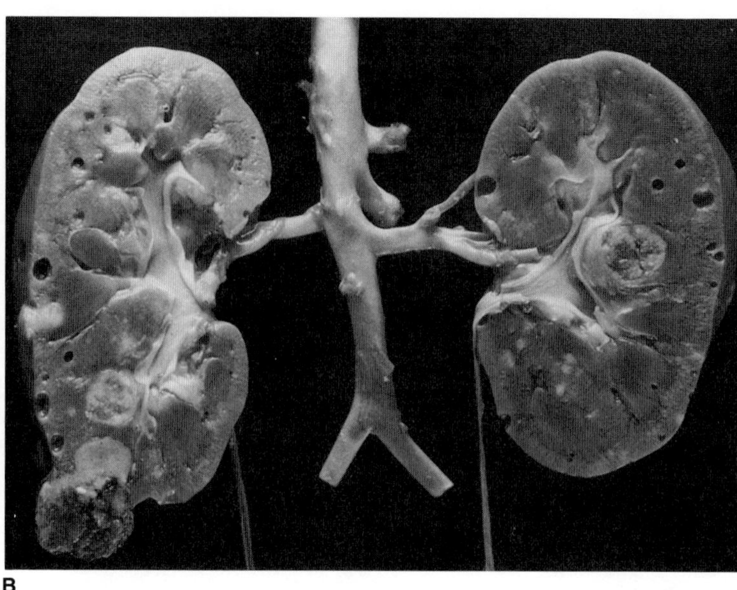

B

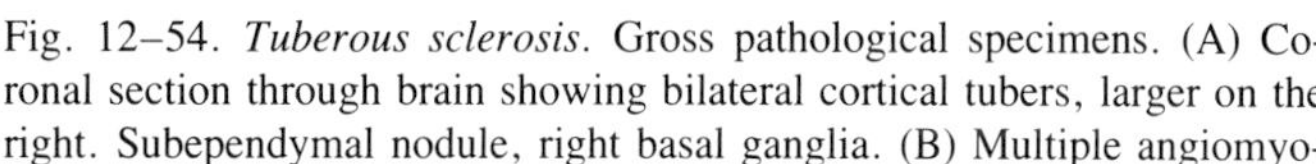

Fig. 12–54. *Tuberous sclerosis.* Gross pathological specimens. (A) Coronal section through brain showing bilateral cortical tubers, larger on the right. Subependymal nodule, right basal ganglia. (B) Multiple angiomyolipomas of kidneys. (A,B from *BE Medley et al,* Sem Roentgenol **11**:35, 1976.)

Fig. 12–55. *Tuberous sclerosis.* Roentgenograms. (A) Intracranial calcifications. (B) Calcification in large posterior frontoparietal cortical tuber. (C) Lytic lesions of hand bones. (A courtesy of *A Lodin,* Stockholm, Sweden. B from *BE Medley et al,* Sem Roengenol **11**:35, 1976. C from *TD Hawkins,* Br J Radiol **32**:157, 1959).

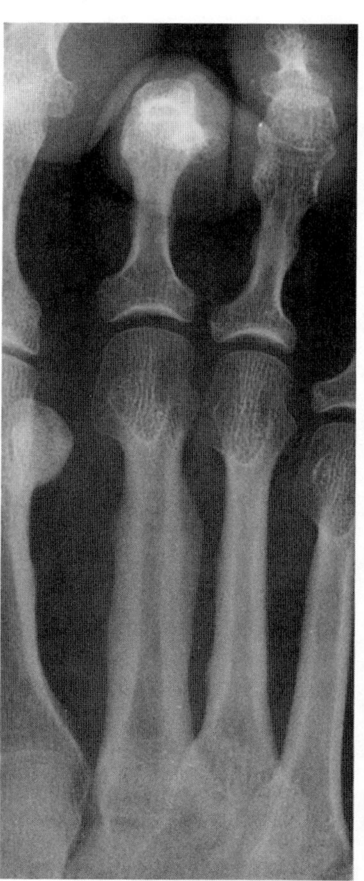

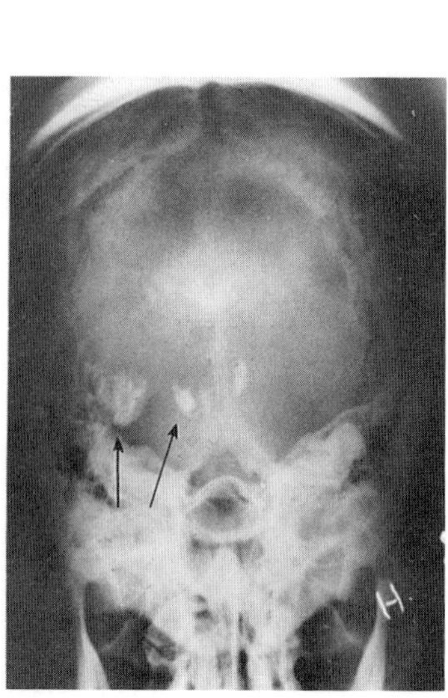

A

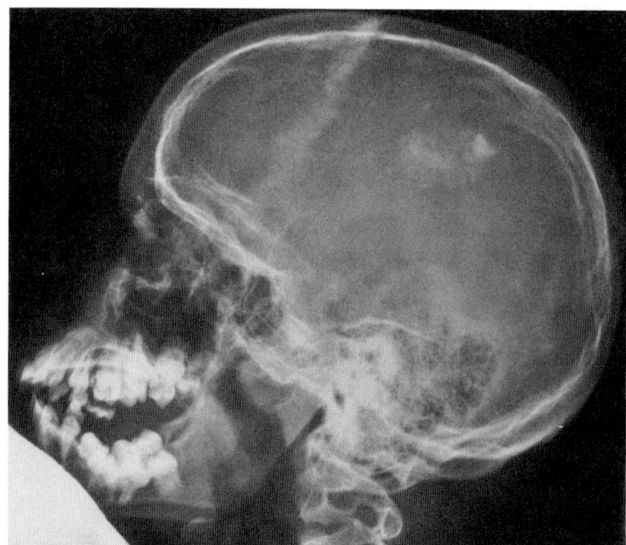

B

C

26a. Grether P et al: Wilms' tumor in an infant with tuberous sclerosis. *Ann Genet* **30**:183–185, 1987.

27. Grover WD, Harley RD: Early recognition of tuberous sclerosis by funduscopic examination. *J Pediatr* **75**:991–995, 1969.

28. Hadju SI, Fotte FW Jr: Angiomyolipoma of the kidney. Report of 27 cases and review of the literature. *J Urol* **102**:396–401, 1969.

29. Hall GS: The ocular manifestations of tuberous sclerosis. *Q J Med* **15**:209–220, 1946.

29a. Hausser I, Anton-Lamprecht I: Electron microscopy as a means for carrier detection and genetic counselling in families at risk of tuberous sclerosis. *Hum Genet* **76**:73–80, 1987.

30. Hoff M et al: Enamel defects associated with tuberous sclerosis. *Oral Surg* **40**:261–269, 1975.

31. Hoffman WH et al: Acromegalic gigantism and tuberous sclerosis. *J Pediatr* **93**:478–480, 1978.

32. Holt JF, Dickerson WW: The osseous lesions of tuberous sclerosis. *Radiology* **58**:1–7, 1952.

33. Houser WO, McLeod RA: Roentgenographic experience at the Mayo Clinic, in *Tuberous Sclerosis,* Gomez MR (ed), Raven Press, New York, 1979, pp 27–53.

34. Hunt A: Tuberous sclerosis: A survey of 97 cases. I: Seizures, pertussis immunization and handicap. *Dev Med Child Neurol* **25**:346–349, 1983.

35. Hurwitz S, Braverman IM: White spots in tuberous sclerosis. *J Pediatr* **77**:587–594, 1970.

35a. Journel H et al: Prenatal diagnosis of familial tuberous sclerosis following detection of cardiac rhabdomyoma by ultrasound. *Prenat Diag* **6**:283–290, 1986.

36. Kapp JP et al: Brain tumors with tuberous sclerosis. *J Neurosurg* **26**:191–202, 1967.

36a. Kousseff BG: Tuberous macrodactyly. *Dysmorphol Clin Genet* **3**:5–7, 1989.

37. Kranias G, Romano PE: Depigmented iris sector in tuberous sclerosis. *Am J Ophthalmol* **83**:758, 1977.

38. Kuehl KS et al: Left ventricular rhabdomyoma: A rare cause of subaortic stenosis in the newborn infant. *Pediatrics* **46**:464–468, 1970.

39. Lee BCP, Gawler J: Tuberous sclerosis: Comparison of computed tomography and convention neuroradiology. *Radiology* **127**:403–407, 1978.

40. Lowry RB et al: Inheritance of tuberous sclerosis. *Lancet* **1**:53, 1979.

40a. Lygadakis NA, Lindenbaum RH: Pitted enamel hypoplasia in tuberous sclerosis patients and first degree relatives. *Clin Genet* **32**:216–221, 1987.

41. Mair DD: Cardiac manifestations, in *Tuberous Sclerosis,* Gomez MR (ed), Raven Press, New York, 1979, pp 155–169.

42. Martyn L: Tuberous sclerosis of Bourneville, in *Retinal Diseases in Children,* Tasman W (ed), Harper & Row, New York, 1971, pp 98–103.

43. Maruyama K et al: Tuberous sclerosis, a clinical study of 38 cases in an epilepsy clinic. *Brain Develop* **8**:16–27, 1976.

44. McCullogh DL et al: Renal angiomyolipoma (hamartoma). Review of the literature and report of seven cases. *J Urol* **105**:32–44, 1971.

45. McWilliam RC, Stephenson JBP: Depigmented hair, the earliest sign of tuberous sclerosis. *Arch Dis Childh* **53**:961–963, 1978.

46. Medley BE et al: Tuberous sclerosis. *Sem Roentgenol* **11**:35–54, 1976.

47. Moorhead JD et al: Management of hemorrhage secondary to renal angiomyolipoma with selective arterial embolization. *J Urol* **117**:122–123, 1977.

48. Monaghan HP et al: Tuberous sclerosis complex in children. *Am J Dis Child* **135**:912–917, 1981.

49. Monteforte WJ, Kohnen PW: Angiomyolipomas in case of lymphangiomatosis syndrome: Relationships to tuberous sclerosis. *Cancer* **34**:317–321, 1974.

50. Morgan JE, Wolfort F: The early history of tuberous sclerosis. *Arch Dermatol* **115**:1317–1319, 1979.

51. Nyboer JG et al: Retinal lesions in tuberous sclerosis. *Arch Ophthalmol* **94**:1277–1280, 1976.

52. O'Callaghan TJ et al: Tuberous sclerosis with striking renal involvement in a family. *Arch Intern Med* **135**:1082–1087, 1975.

53. Pampligione G, Moynahan EJ: The tuberous sclerosis syndrome: Clinical and EEG studies in 100 children. *J Neurol Neurosurg Psychiatr* **39**:666–673, 1976.

54. Papanayotou P, Vezirtzi E: Tuberous sclerosis with gingival lesions. *Oral Surg* **39**:578–582, 1975.

54a. Platt LD et al: Prenatal diagnosis of tuberous sclerosis: The use of fetal echocardiography. *Prenat Diag* **7**:407–412, 1987.

55. Ponsot G, Lyon G: La sclérose tubéreuse de Bourneville. Etude clinique de 59 observations chez l'enfant. *Arch Fr Pédiatr* **34**:9–22, 1977.

56. Price EB Jr, Mostofi K: Symptomatic angiomyolipoma of the kidney. *Cancer* **18**:761–774, 1965.

57. Pringle JJ: A case of congenital adenoma sebaceum. *Br J Dermatol* **2**:1–14, 1890.

58. Probst A, Ohnacker H: Sclérose tubéreuse de Bourneville chez un premature. Ultrastructure des cellules atypiques: Presence de microvillosities. *Acta Neuropathol (Berlin)* **40**:157–161, 1977.

59. Reagan TJ: Neuropathology, in *Tuberous Sclerosis,* Gomez MR (ed), Raven Press, New York, 1979, pp 69–83.

60. Robbins TO, Bernstein J: Renal involvement, in *Tuberous Sclerosis,* Gomez MR (ed), Raven Press, New York, 1979, pp 143–153.

61. Robertson DM: Ophthalmic findings, in *Tuberous Sclerosis,* Gomez MR (ed), Raven Press, New York, 1979, pp 121–141.

62. Rogers RS III: Dermatologic manifestations, in *Tuberous Sclerosis,* Gomez MR (ed), Raven Press, New York, 1979, pp 95–119.

63. Russell DE, Rubinstein LJ: *Pathology of Tumors of the Nervous System,* 3rd ed, Williams & Wilkins, Baltimore, 1971, pp 33–34, 124–126, 227.

64. Sareen CK et al: Tuberous sclerosis. *Am J Dis Child* **123**:34–39, 1972.

65. Schöner et al: Morbus Pringle mit Skelettveränderungen. *Hautarzt* **31**:334–335, 1980.

66. Scott LN et al: The value of CT in genetic counseling in tuberous sclerosis. *Pediatr Radiol* **9**:1–4, 1980.

67. Scully C: Oral mucosal lesions in association with epilepsy and cutaneous lesions. The Pringle–Bourneville syndrome. *Int J Oral Surg* **10**:68–72, 1981.

68. Shaher RM et al: Clinical presentation of rhabdomyoma of the heart in infancy and childhood. *Am J Cardiol* **30**:95–103, 1972.

69. Shelton RW: The incidence of ocular lesions in tuberous sclerosis. *Ann Ophthalmol* **7**:771–773, 1975.

70. Singer K: Genetic aspects of tuberous sclerosis in a Chinese population. *Am J Hum Genet* **23**:33–40, 1971.

71. Troost BT et al: Tuberous sclerosis and Klippel–Trénaunay–Weber syndromes. Association of two complete phakomatoses in a single individual. *J Neurol Neurosurg Psychiatr* **38**:500–504, 1975.

72. Tsakraklides V et al: Rhabdomyomas of heart. *Am J Dis Child* **128**:639–646, 1974.

72a. Van Baal JG et al: Tuberous sclerosis and the relationship with renal angiomyolipoma. A genetic study and the clinical aspects. *Clin Genet* **35**:167–173, 1989.

73. Vikilzadeh F, Happle R: Grübchenformige Schmelzdefekte bei tuberöse Sklerose. *Hautarzt* **31**:336–337, 1980.

74. Von Recklinghausen F: Ein Herz von einen Neugeborenen welches mehrere theils nach aussen, theils nach den Höhlen prominirende Tumoren (Myomen) trug, Verh Ges Geburtsch 25 März. *Monatschr Geburtsk* **20**:1–2, 1862.

74a. Weits-Binnerts JJ et al: Dental pits in deciduous teeth—an early sign in tuberous sclerosis. *Lancet* **2**:1344–1345, 1982.

75. Wenzl JE et al: Tuberous sclerosis presenting as polycystic kidneys and seizures in an infant. *J Pediatr* **77**:673–676, 1970.

76. Westmoreland BF: Electroencephalographic experience at the Mayo Clinic, in *Tuberous Sclerosis,* Gomez MR (ed), Raven Press, New York, 1979, pp 55–67.

77. Yakolev PI, Guthrie RH: Congenital ectodermoses (neurocutaneous syndromes) in epileptic patients. Bourneville tuberous sclerosis (Epiloia). *Arch Neurol Psychiatr (Chic)* **26**:1145–1194, 1931.

78. Zaremba J: Tuberous sclerosis: A clinical and genetical investigation, in *The Etiology of Inherited Disorders,* MSS Information Corp, New York, 1974.

## Multiple odontoma–esophageal stenosis syndrome

Herrmann (3) reported a young male with huge tumors of the maxilla and mandible containing 1200 and 900 teeth, respectively, in various stages of development, including geminated and invaginated teeth. In 1973, Schmidseder and Hausamen (5) studied the same patient, noting that his two sons (one dying from pneumonia soon after birth) also manifested multiple odontomas of both jaws in infancy. The surviving infant was found to have a liver disorder and pulmonary stenosis. Subsequently, a daughter was born who again manifested odontomas. The boy experienced recurrences of odontomas that exhibited a higher degree of differentiation. He also suffered esophageal stenosis, as did his father (7).

Bader (1) reported multiple odontomas of both jaws in a female infant with calcified aortic stenosis, congenital cylindric bronchiectasis, leiomyomatosis of the esophagus with stenosis, hyperplasia of the myenteric plexus, and chronic interstitial cirrhosis of the liver. GL Barnes (personal communication, 1974) observed sibs with odontomas in four quadrants, malrotation and stenosis of the bowel, and iris colobomas.

Multiple odontomas were documented in male sibs by Schmitz and Witzel (6), but associated anomalies were not mentioned. Beisser (2) and Malik and Khalid (4) also reported multiple bilateral odontomas in both jaws.

In view of the occurrence of what appears to be a syndrome of multiple odontomas, chronic interstitial cirrhosis of the liver, and esophageal stenosis in two generations, it must be assumed that the disorder is inherited as an autosomal dominant trait. In case of Schmitz and Witzel, the parents of the male sibs were normal.

### References (Multiple odontoma–esophageal stenosis syndrome)

1. Bader G: Odontomatosis (multiple odontomas). *Oral Surg* **23**:770–773, 1967.

2. Beisser V: Ein seltene Fall von selbständigem, multiplen Odontom beiderseits im Ober- und Unterkiefer und ein Literaturstudium über diese Geschwülste des Zahn-, Mund- und Kieferbereiches. Thesis, Düsseldorf. 1964.

3. Herrmann M: Über von Zahnsystem ausgehende Tumoren bei Kindern. *Fortschr Kiefer Gesichtschir* **4**:226–229, 1958.

4. Malik SA, Khalid M: Odontomatosis—a case report. *Br J Oral Surg* **11**:262–264, 1974.

5. Schmidseder R, Hausamen JR: Multiple odontogenic tumors and other anomalies. *Oral Surg* **39**:249–258, 1975.

6. Schmitz, Witzel A: Neubildung von Zähnen und zahnähnlichen Gebilden. *Dtsch Mschr Zahnheilkd* **19**:126–130, 1901.

7. Schönberger W: Angeborene multiple Odontome und Dysphagie bei Vater und Sohn—eine syndromhafte Verknüpfung? *Z Kinderheilk* **117**:101–108, 1974.

# Chapter 13
# Syndromes Affecting the Skin and Mucosa

## Aplasia cutis congenita

Aplasia cutis congenita is localized or widespread absence of skin at birth. First described by Cordon in 1767 (16), the defect became more widely known with Campbell's 1826 report (13) of two sibs with involvement of the scalp vertex. Over 500 cases have been reported to date. Clinical associations, classification, and etiology have been discussed in several noteworthy reviews (19,24,59a,61).

Lesions commonly present as ulcerated or membranous defects of variable size near the midline of the scalp vertex (Figs. 13–1 and 13–2). Small lesions are usually located at or in proximity to the parietal hair whorl (59). Absent hair is a constant feature. The defect is single (75%) in most cases, but double, triple, or multiple vertex defects may occur (19). Although most single defects (66%) are oval-shaped and small (0.5–3 cm), more extensive defects over 10 cm in diameter have been reported (19). Defects may involve the epidermis only or may extend to full skin thickness. Osseous involvement of the calvaria, found in 20–30%, is detected roentgenographically by demonstrating an area of bone deficiency underlying the cutaneous defect (19).

Small, superficial midline defects tend to heal gradually without complications during the initial weeks after birth, leaving a residual atrophic or hypertrophic scar with alopecia. Healing of a calvarial defect is generally complete during the first year of life (19). Extensive scalp defects, which may extend to involve the dura (19,39), pose significant risk for infection, potential meningitis, venous thrombosis (42), and sagittal sinus hemorrhage (8,13,19,26,56).

Although most cases are sporadic, many familial instances have been recorded. Autosomal dominant inheritance with incomplete penetrance and variable expressivity occurs most frequently (47,61,69a), but autosomal recessive inheritance is implicated occasionally, usually in offspring of consanguineous parents (47). Concordant monozygotic (33) and discordant monozygotic (71) twins have been documented. Aplasia cutis congenita involving the trunk and/or limbs is etiologically heterogeneous, and familial examples have been noted in some instances (22,55).

Aplasia cutis congenita is a feature of many syndromes and also occurs on a disruptive, infectious, and teratogenic basis. Association with many isolated malformations has been documented as well. Conditions associated with aplasia cutis congenita are summarized in Table 13–1, together with pertinent references and page numbers in the text.

**Differential diagnosis.** Congenital scalp defects should be considered separately from aplasia cutis congenita involving the trunk and extremities. Differential diagnosis includes localized scalp infection, congenital dermoid cyst, small meningocele, heterotopic brain or glial tissue (35,51), and traumatic lesions secondary to fetal scalp monitoring (10). As the child grows and scarring of the cutis aplasia lesions

Fig. 13–1. *Aplasia cutis congenita.* Healed oval lesion of aplasia cutis congenita at site of parietal hair whorl. (From *MJ Stephan et al,* J Pediatr **101**:850, 1982.)

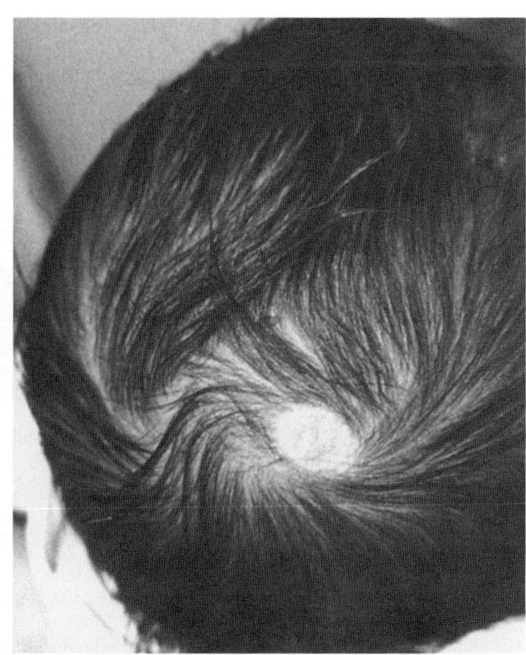

Fig. 13–2. *Aplasia cutis congenita.* Extensive ulcerative lesion in trisomy 13 syndrome.

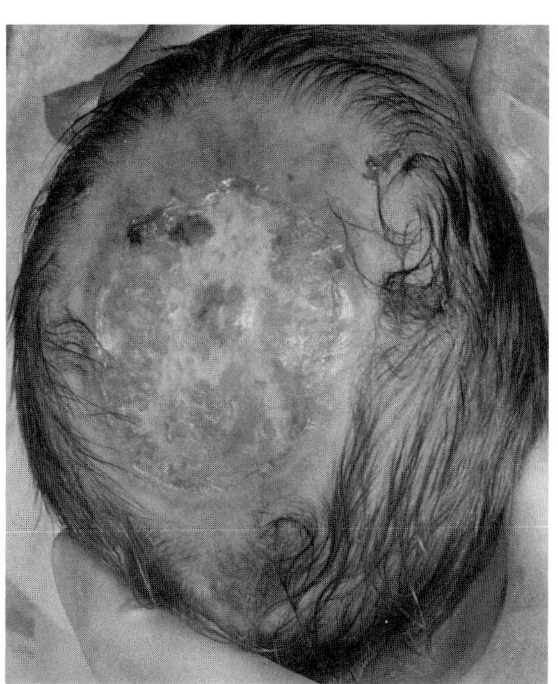

Table 13–1. Conditions with aplasia cutis congenita

| Conditions | References | (pages in text) |
|---|---|---|
| Syndromal | | |
| Chromosomal | | |
| Trisomy 13 syndrome | 1, 13 | (40) |
| del(4p) syndrome | 29,32 | (46) |
| Monogenic | | |
| Adams–Oliver syndrome | 2,7,25,38,61,63 | |
| Postaxial polydactyly | 12,19a | |
| Ear anomalies and rudimentary nipples | 20 | |
| Unilateral facial paresis, dermal sinuses, and ear malformations | 4 | (419) |
| Johanson–Blizzard syndrome | 36 | (812) |
| Setleis syndrome | 45 | (514) |
| Familial 46,XY gonadal dysgenesis with anomalies of ectodermal and mesodermal structures | 9 | |
| Ectodermal dysplasia-abnormal facies | 43a | |
| Unknown genesis | | |
| Focal dermal hypoplasia | 27 | (472) |
| Ectodermal dysplasias (various types of the tricho-odonto-onychial subgroup) | 23,53,64,65 | |
| Epidermal nevus syndrome | 41 | (362) |
| Multiple hamartomas, giant pigmented nevocellular nevus, and central nervous system malformation | 49 | |
| Oculo-cerebro-cutaneous syndrome | 18 | (511) |
| Sakati syndrome | 15 | (558) |
| Spear–Mickle syndrome | 15,57 | (548) |
| Disruptive | | |
| Amnionic rupture sequence | 30 | (11) |
| Infectious | | |
| Congenital herpes simplex | 60,62 | |
| Congenital varicella | 3,58,65a | |
| Teratogenic | | |
| Methimazole | 37,48,50,66 | |
| Malformational | | |
| Central nervous system | | |
| Hydrocephaly | 13,34 | |
| Holoprosencephaly | 39,40 | |
| Meningocele | 34 | |
| Congenital midline porencephaly | 72 | |
| Early telencephalic defects | 21 | |
| Occult spinal dysraphism | 31 | |
| Craniospinal rachischisis | 28 | |
| Leptomeningeal angiomatosis and aneurysm of distal posterior cerebral artery | 54 | |
| Cardiovascular system | | |
| Arteriovenous malformation | 67 | |
| Congenital heart defects | 17 | |
| Gastrointestinal system | | |
| Tracheoesophageal fistula | 17 | |
| Omphalocele | 68 | |
| Intestinal lymphangiectasia | 8 | |
| Cleft lip–palate | 34,39 | |
| Pyloric atresia | 14 | |
| Other | | |
| Piebaldism | 11 | |

occurs, sebaceous nevus, epidermal nevus, and alopecia of pseudopelade type may be confused with it. Particular care must be taken to distinguish linear or stellate lesions of aplasia cutis congenita from linear alopecia seen in "en coup de sabre" morphea.

Isolation of herpes simplex type 2 virus has been reported from a newborn with three occipitoparietal ulcerated scalp defects, with serologic confirmation of elevated antiherpes complex IgM (62). These lesions healed spontaneously with localized scarring.

For extensive aplasia cutis congenita involving the trunk and extremities, it is important to exclude various types of *epidermolysis bullosa* (5,6,24,70). Clinically helpful clues for epidermolysis bullosa include frequent oral mucosal involvement and development of further lesions after birth. If doubt remains, ultrastructural examination of a skin biopsy from the margin of a lesion will distinguish epidermolysis bullosa and define the genetic type. Nonfamilial truncal and extremity aplasia cutis congenita have sometimes been attributed to thromboplastic emboli derived from a fetus papyraceus resulting from death of a co-twin early in the second trimester of pregnancy (44,46). In other cases, when no fetus papyraceus is discovered, the emboli may originate from the placenta itself (43). *Restrictive dermopathy* should be excluded.

**Laboratory aids.** Histologic sections from skin biopsy will confirm the diagnosis in uncertain cases. The defective area is devoid of elastic fibers, dermal papillae, and normal blood vessels. No well-demarcated border between normal skin and aplasia cutis congenita is observed; rather, there exists an intermediate zone of skin within which dermal appendages such as hair papillae, sebaceous glands, and sweat glands are found to be small to rudimentary and to decrease in size and number centripetally with gradual transition from normal skin to the ulcer of aplasia cutis (69).

### References (Aplasia cutis congenita)


1. Abuelo D, Feingold M: Scalp defects in trisomy 13. *Clin Pediatr* **8**:416–417, 1969.
2. Adams FH, Oliver CP: Hereditary deformities in man: Due to arrested development. *J Hered* **36**:2–7, 1945.
3. Alkalay AL et al: Fetal varicella syndrome. *J Pediatr* **111**:320–323, 1987.
4. Anderson CE et al: Autosomal dominantly inherited cutis aplasia congenita, ear malformations, right-sided facial paresis, and dermal sinuses. *Birth Defects* **15**(5B):265–270, 1979.
5. Bart BJ et al: Congenital localized absence of skin and associated abnormalities resembling epidermolysis bullosa: A new syndrome. *Arch Dermatol* **93**:296–304, 1966.
6. Bart BJ: Epidermolysis bullosa and congenital localized absence of skin. *Arch Dermatol* **101**:78–81, 1970.
7. Bonafede RP, Beighton P: Autosomal dominant inheritance of scalp defects with ectrodactyly. *Am J Med Genet* **3**:35–41, 1979.
8. Bronspiegel N et al: Aplasia cutis congenita and intestinal lymphangiectasia. *Am J Dis Child* **139**:509–513, 1985.
9. Brosnan PG et al: A new-familial syndrome of 46,XY gonadal dysgenesis with anomalies of ectodermal and mesodermal structures. *J Pediatr* **97**:586–590, 1980.
10. Brown ZA et al: Aplasia cutis congenita and fetal scalp electrode. *Am J Obstet Gynecol* **129**:351–352, 1977.
11. Bull M: Cutis aplasia with piebaldism. David W. Smith Workshop on Malformations and Morphogenesis, Greenville, SC, August, 1987.
12. Buttiens M et al: Scalp defect associated with postaxial polydactyly: Confirmation of a distinct entity with autosomal dominant inheritance. *Hum Genet* **71**:86–88, 1985.
13. Campbell W: Case of congenital ulcer on the cranium of a fetus, terminating in fatal hemorrhage on the 18th day after birth. *Edinb J Med Sci* **2**:82–83, 1826.
14. Carmi R et al: Aplasia cutis congenita in two sibs discordant for pyloric atresia. *Am J Med Genet* **11**:319–328, 1981.
15. Cohen MM Jr.: *Craniosynostosis: Diagnosis, Evaluation, and Management,* Raven Press, New York, 1986.
16. Cordon M: Extrait d'une lettre au sujet de trois enfants de la même mère nés avec partie des extrémites dénuéé de peau. *J Méd Chir Pharm* **26**:556–557, 1767.
17. Deeken JH, Caplan RM: Aplasia cutis congenita. *Arch Dermatol* **102**:386–389, 1970.
18. Delleman JW, Oorthuys JWE: Orbital cyst in addition to congenital

cerebral and focal dermal malformations: A new entity? *Clin Genet* **19**:191–198. 1981.

19. Demmel U: Clinical aspects of congenital skin defects. *Eur J Pediatr* **121**:21–50, 1975.

19a. Dumont M, Fischer L: Les aplasies congénitales du cuir chevelu chez le nouveau-né. *Lyon Méd* **9**:373–377, 1962.

20. Finlay AY, Marks R: An hereditary syndrome of lumpy scalp, odd ears, and rudimentary nipples. *Br J Dermatol* **99**:423–430, 1978.

21. Fowler CW, Dumars, KW: Cutis aplasia and cerebral malformation. *Pediatrics* **52**:861–864, 1973.

22. Freire-Maia N et al: Recessive aplasia cutis congenita of the limbs. *J Med Genet* **17**:123–126, 1980.

23. Freire-Maia N, Pinheiro M: *Ectodermal Dysplasias: A Clinical and Genetic Study*, Alan R. Liss, New York, 1984.

24. Frieden IJ: Aplasia cutis congenita: A clinical review and proposal for classification. *J Am Acad Dermatol* **14**:646–660, 1986.

25. Fryns JP et al: Congenital scalp defect with distal limb reduction anomalies. *Eur J Pediatr* **126**:289–295, 1977.

26. Glasson DW, Duncan GM: Aplasia cutis congenita of the scalp: Delayed closure complicated by massive hemorrhage. *Plast Reconstr Surg* **75**:423–425, 1985.

27. Goltz RW et al: Focal dermal hypoplasia syndrome. *Arch Dermatol* **101**:1–11, 1970.

28. Grieg DM: Localized congenital defects of the scalp. *Edinb Med J* **38**:341–358, 1931.

29. Guthrie RD et al: The 4p− syndrome: A clinically recognizable chromosomal deletion syndrome. *Am J Dis Child* **122**:421–425, 1971.

30. Higginbottom MC et al: The amniotic band disruption complex: Timing of amniotic rupture and variable spectra of consequent defects. *J Pediatr* **95**:544–549, 1979.

31. Higginbottom MC et al: Aplasia cutis congenita: A cutaneous marker of occult spinal dysraphism. *J Pediatr* **96**:687–689, 1980.

32. Hirschhorn K et al: Deletion of short arms of chromosome 4–5 in a child with defects of midline fusion. *Humangenetik* **1**:479–482, 1965.

33. Hodgman JE et al: Congenital scalp defects in twin sisters. *Am J Dis Child* **110**:293–295, 1965.

34. Ingalls NW: Congenital defects of the scalp. Studies in the pathology of development. *Am J Obstet Gynecol* **25**:861–873, 1933.

35. Jackson FE, Moore BS: Ectopic glial tissue in the occipital scalp. *Arch Dis Childh* **44**:428–430, 1969.

36. Johanson A, Blizzard R: Syndrome of congenital aplasia of the alae nasi, deafness, hypothyroidism, dwarfism, absent permanent teeth and malabsorption. *J Pediatr* **79**:982–987, 1971.

37. Kalb RE, Grossman ME: The association of aplasia cutis congenita with therapy of maternal thyroid disease. *Pediatr Dermatol* **3**:327–330, 1986.

38. Koiffman CP et al: Congenital scalp skull defects with distal limb anomalies (Adams–Oliver syndrome-McKusick 10030): Further suggestion of autosomal recessive inheritance. *Am J Med Genet* **29**:263–268, 1988.

39. Kosnik EJ, Sayers MP: Congenital scalp defects: Aplasia cutis congenita. *J Neurosurg* **42**:32–36, 1975.

39a. Küster W, Traupe H: Klinik und Genetik angeborener Hautdefekte. *Hautarzt* **39**:553–563, 1988.

40. Lacro RV et al: Aplasia cutis congenita and other cutaneous lesions at the vertex of the scalp: Cutaneous markers for defects in neural tube closure. Abstract 29, David W. Smith Workshop on Malformation and Morphogenesis, Greenville, South Carolina, August 15–19, 1987.

41. Lantis S et al: Nevus sebaceus of Jadassohn: Part of a new neurocutaneous syndrome? *Arch Dermatol* **98**:117–123, 1968.

42. Lavine D et al: Congenital scalp defect with thrombosis of the sagittal sinus. *Plast Reconstr Surg* **61**:599–602, 1978.

43. Levin DL et al: Congenital absence of skin. *J Am Acad Dermatol* **2**:203–206, 1980.

43a. Leichtman LG et al: Ectodermal dysplasia, cutis aplasia, and facial dysmorphism in mother and daughter: A recognizable autosomal dominant syndrome. David W. Smith Workshop on Malformations and Morphogenesis, Madrid, Spain, May, 1989.

44. Mannino FL et al: Congenital skin defects and fetus papyraceus. *J Pediatr* **91**:559–564, 1977.

45. Marion RW et al: Autosomal recessive inheritance in the Setleis bitemporal "forceps marks" syndrome. *Am J Dis Child* **141**:895–897, 1987.

46. Markman L et al: Association of aplasia cutis congenita and fetus papyraceus in a triplet pregnancy. *Aust Paediatr J* **18**:294–296, 1982.

47. McMurray BR et al: Hereditary aplasia cutis congenita and associated defects. *Clin Pediatr* **16**:610–614, 1977.

48. Milham S Jr, Elledge W: Maternal methimazole and congenital defects in children. *Teratology* **5**:125, 1972.

49. Mimouni F et al: Multiple hamartomas associated with intracranial malformation. *Pediatr Derm* **3**:219–225, 1986.

50. Mujtaba Q, Burrow GN: Treatment of hyperthyroidism in pregnancy with propylthiouracil and methimazole. *Obstet Gynecol* **46**:282–286, 1975.

51. Orkin M, Fisher I: Heterotopic brain tissue (heterotopic neural rest). *Arch Dermatol* **94**:699–707, 1966.

52. Paltzik RL, Aiello AM: Aplasia cutis congenita associated with valvular heart disease. *Cutis* **36**:57–58, 1985.

53. Pinheiro M et al: A previously undescribed condition: tricho-odonto-onycho-dermal syndrome: A review of the tricho-odonto-onychial subgroup of ectodermal dysplasias. *Br J Dermatol* **105**:371–382, 1981.

54. Pozzati E et al: Leptomeningeal angiomatosis and aplasia congenita of the scalp. *J Neurosurg* **58**:937–940, 1983.

55. Rauschkolb RR, Enriquez SI: Aplasia cutis congenita. *Arch Dermatol* **86**:54–57, 1962.

56. Schneider BM et al: Aplasia cutis congenita complicated by sagittal sinus hemorrhage. *Pediatrics* **66**:948–950, 1980.

57. Spear SL, Mickle JP: Simultaneous cutis aplasia congenita of the scalp and cranial stenosis. *Plast Reconstr Surg* **71**:413–417, 1983.

58. Srabstein JC et al: Is there a congenital varicella syndrome? *J Pediatr* **84**:239–243, 1974.

59. Stephan MJ et al: Origin of scalp vertex aplasia cutis. *J Pediatr* **101**:850–853, 1982.

59a. Stevenson RE, DeLoache WR: Aplasia cutis congenita of the scalp. *Proc Greenwood Genet Ctr* **7**:14–18, 1988.

60. Strawn EY, Scrimenti RJ: Intrauterine herpes simplex infection. *Am J Obstet Gynecol* **115**:581–582, 1973.

61. Sybert VP: Aplasia cutis congenita: A report of 12 new families and review of the literature. *Pediatr Dermatol* **3**:1–14, 1985.

62. Tomer A, Harel A: Congenital absence of scalp skin and herpes simplex virus. *Isr J Med Sci* **19**:950–951, 1983.

63. Toriello HV et al: Scalp and limb defects with cutis marmorata telangiectatica congenita: Adams–Oliver syndrome? *Am J Med Genet* **29**:269–276, 1988.

64. Tsakalakos N et al: A previously undescribed ectodermal dysplasia of the tricho-odonto-onychial subgroup in a family. *Arch Dermatol* **122**:1047–1053, 1986.

65. Tuffli GA, Laxova R: New autosomal dominant form of ectodermal dysplasia. *Am J Med Genet* **14**:381–384, 1983.

65a. Unger-Köppel J et al: Varizellenfetopathie. *Helv Paediatr Acta* **40**:399–404, 1985.

66. Van Dijke CP et al: Methimazole, carbimazole, and congenital skin defects. *Ann Intern Med* **106**:60–61, 1987.

67. Vasconez LO: Congenital defect of the skull and scalp due to an arteriovenous malformation. *Plast Reconstr Surg* **51**:692–695, 1973.

68. Vinocur CD et al: Surgical management of aplasia cutis congenita. *Arch Surg* **111**:1160–1164, 1976.

69. Walker JC et al: Congenital absence of skin (aplasia cutis congenita). *Plast Reconstr Surg* **26**:209–218, 1960.

69a. Weippl G, Ader H: Kongenitaler Skalp-Defekt in vier Generationen. *Klin Pädiatr* **187**:84–86, 1975.

70. Wojnarowska FT et al: Dystrophic epidermolysis bullosa presenting with congenital localized absence of skin: Report of four cases. *Br J Dermatol* **108**:477–483, 1983.

71. Yagupsky P et al: Aplasia cutis congenita in one of monozygotic twins. *Pediatr Dermatol* **3**:403–405, 1986.

72. Yokota A, Matsukado Y: Congenital midline porencephaly: A new brain malformation associated with scalp anomaly. *Childs Brain* **5**:380–397, 1979.

## Aplasia cutis congenita, ear malformations, facial paresis, and dermal sinuses

In 1979, Anderson et al (2) reported a Mexican-American family in which four generations were affected with aplasia cutis congenita of the scalp, ear defects (uni- or bilateral lop ear, conductive and possibly mixed hearing loss), dermal sinuses (both pretragal and parasternal), and unilateral facial palsy. Hypoplastic facial canals were verified by tomography in those with facial palsy. Finlay and Marks (3) and Aase and Wilroy (1) reported somewhat similarly involved families with the SEN syndrome, an autosomal dominant disorder of *s*calp defects (aplasia cutis congenita), *e*ar abnormalities (small or rudimentary tragus, antitragus, and lobule), and rudimentary or absent *n*ipples (Figs. 13–3 to 13–5). Tuffli and Laxova (4) described autosomal dom-

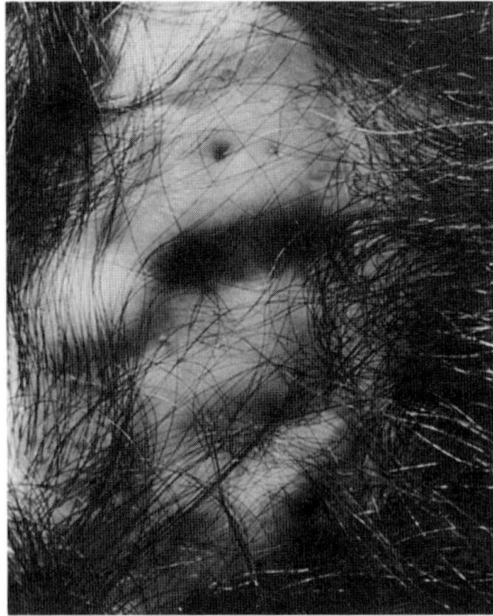

Fig. 13–3. *Aplasia cutis congenita, ear malformations, facial paresis, and dermal sinuses.* In and adjacent to midline, over posterior scalp, are firm raised nodules, not covered by hair. (From *AY Finlay* and *R Marks,* Br J Dermatol **99**:423, 1978.)

inant inheritance of aplasia cutis verticis, hypohidrosis, nipple/breast hypoplasia, nail dysplasia, and delayed tooth eruption. Differential diagnosis would include *branchio-oto-renal syndrome.*

#### References (Aplasia cutis congenita, ear malformations, facial paresis, and dermal sinuses)

1. Aase JM, Wilroy SR: The Finlay–Marks (SEN) syndrome: Report of a new case and review of the literature. *Proc Greenwood Genet Ctr* **7**:177–178, 1988.
2. Anderson CE et al: Autosomal dominantly inherited cutis aplasia congenita, ear malformations, right-sided facial paresis and dermal sinuses. *Birth Defects* **15**(5B):265–270, 1979.
3. Finlay AY, Marks R: An hereditary syndrome of lumpy scalp, odd ears and rudimentary nipples. *Br J Dermatol* **99**:423–430, 1978.
4. Tuffli GA, Laxova R: New autosomal dominant form of ectodermal dysplasia. *Am J Med Genet* **14**:381–384, 1983.

Fig. 13–4. *Aplasia cutis congenita, ear malformations, facial paresis, and dermal sinuses.* Superior helix is largely absent but rudiment is downturned. (From *AY Finlay* and *R Marks,* Br J Dermatol **99**:423, 1978.)

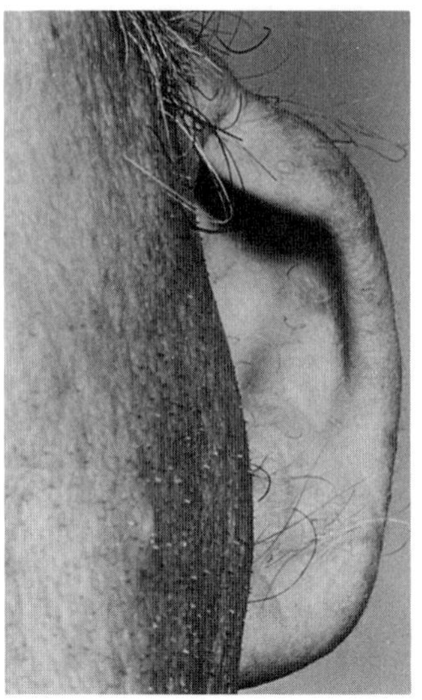

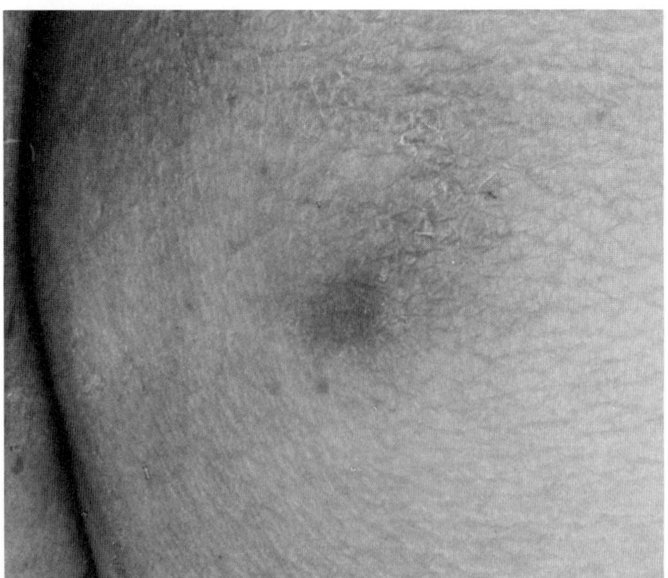

Fig. 13–5. *Aplasia cutis congenita, ear malformations, facial paresis, and dermal sinuses.* Rudimentary nipple manifest by brown flat area only. (From *AY Finlay* and *R Marks,* Br J Dermatol **99**:423, 1978.)

## Ascher syndrome

In 1920 and 1922, Ascher brought to the attention of ophthalmologists the syndrome of blepharochalasis, double lip, and nontoxic thyroid enlargement, although the combination of blepharochalasis and double lip was described in 1909 by Laffer (16). Subsequently, additional examples of the complete syndrome have been reported (8,9,13–15,20,22). Examination of other reports (3–7,10–12,17–19,21–28) indicates that blepharochalasis may occur as an isolated abnormality as may double lip. Blepharochalasis was reported as an isolated finding as early as the nineteenth century (12). In Ascher syndrome autosomal dominant transmission has been suggested (3,11,19,24), many sporadic cases apparently representing fresh mutations.

**Eyes.** Sagging eyelids are striking. The upper lids and less commonly the lower are characterized by relaxation of the tarsal fold, which allows the tissue between the eyebrow and the edge of the lid to hang slack over the palpebral fissure. The lid skin is markedly thin and atrophic. Atrophy and drooping of the lid often follow repeated angioneurotic edema-like episodes. In several cases, swelling of the lids has appeared during the first 8 years of life. In others, its occurrence was noted at about the time of puberty (15,22). The swelling of the lids and the enlargement of the lips may occur simultaneously (8,20) (Figs. 13–6 to 13–8).

Gross surgical examination of tissue removed from the relaxed skin of the lid has shown prolapsed orbital fat or, more frequently, hyperplastic lacrimal gland tissue. The lids, on microscopic examination, have exhibited an increased number of blood vessels, but no unanimity of opinion exists about changes in the elastic fibers (18).

**Lips.** The lip, almost always the upper, is the site of a horizontally running duplication located between the inner (pars villosa) and outer (pars glabrosa) parts of the lip (18). The fold cannot be seen when the lips are closed, only when the patient is smiling or talking. The enlargement of the lip may exist from childhood (3,5,11). Rarely, the lower lip is also enlarged (7,11) (Figs. 13–6 to 13–8).

Microscopic examination of the excessive labial tissue usually reveals loose areolar tissue and hyperplastic mucous glands, numerous blood-filled capillaries, and perivascular infiltration with plasma cells and lymphocytes (3,5,7,10,14).

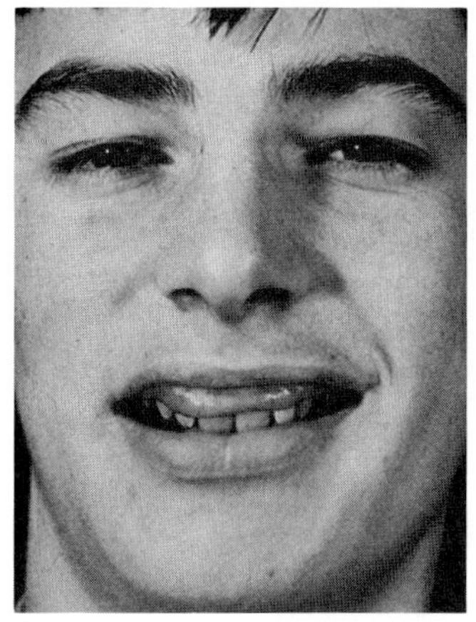
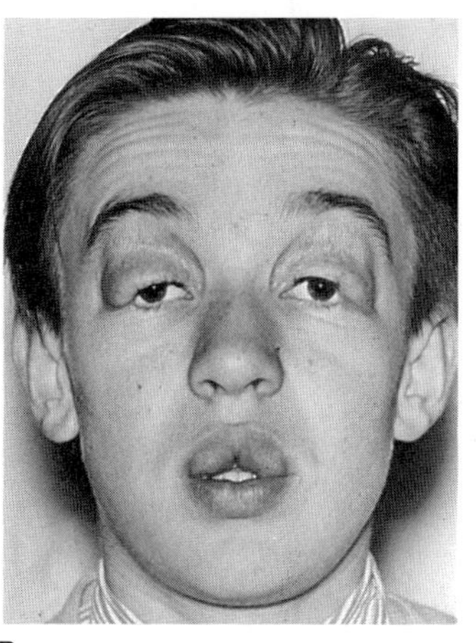

A B

Fig. 13–6. *Ascher syndrome.* (A) Note transverse crease with "doubling" of upper lip and sagging of lateral portion of upper eyelids. (B) Patient with appearance similar to that of patient in A. (A from *MC Oldfield,* Br J Surg **47**:58, 1959. B from *K Stehr,* Dtsch Med Wochenschr **87**:1148, 1962.)

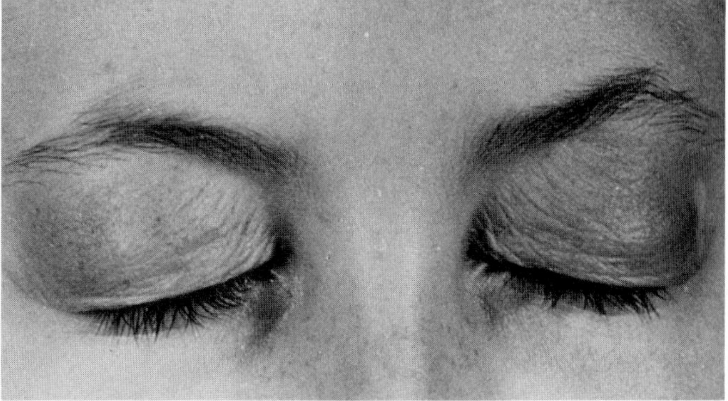

Fig. 13–7. *Ascher syndrome.* Blepharochalasis. (From *R Michalowski,* Derm Wschr **149**:232, 1964.)

**Thyroid.** Thyroid gland enlargement is variable and not usually associated with toxic symptoms. Barnett et al (3), however, reported a patient with hypothyroidism and myxedema. Enlargement may appear several years after eyelid involvement (2), but usually appears during the second decade. It may be evident only on scanning with radioactive iodine (20).

**Differential diagnosis.** Blepharochalasis and double lip may each occur as isolated anomalies. Most instances of so-called double lower lip are actually lip sinuses of the transverse furrow type (21). Floppy eyelid syndrome, the combination of blepharochalasis and chronic papillary conjunctivitis, has been discussed by Goldberg et al (12). The *Melkersson–Rosenthal syndrome* (cheilitis glandularis, facial paralysis, and fissured tongue) and vascular neoplasms (hemangioma, lymphangioma) should be considered. The syndrome of *acromegaloid features and thickened oral mucosa* should be excluded.

### References (Ascher syndrome)


1. Ascher KW: Blepharochalasis mit Struma und Doppellippe. *Klin Monatsbl Augenheilkd* **65**:86–97, 1920.
2. Ascher KW: Das Syndrom Blepharochalasis, Struma und Doppellippe. *Klin Wochenschr* **1**:2287–2288, 1922.
3. Barnett ML et al: Double lip and double lip with blepharochalasis (Ascher's syndrome). *Oral Surg* **34**:727–733, 1972.
4. Brazin SA et al: Unilateral blepharochalasis. *Arch Dermatol* **115**:479–481, 1979.
5. Calnan J: Congenital double lip: Record of a case with a note on the embryology. *Br J Plast Surg* **5**:197–202, 1952–1953.
6. Delaire J: Les doubles lèvres. *Actual Odontostomatol (Paris)* **87**:365–380, 1969.
7. Dingham RO, Billman HR: Double lip. *J Oral Surg* **5**:146–148, 1947.
8. Eigel W: Blepharochalasis und Doppellippe, ein thyreotoxisches Oedem? *Dtsch Med Wochenschr* **51**:1947–1949, 1925.
9. Eisenstodt MD. Blepharochalasis and double lip. *Am J Ophthalmol* **32**:128–130, 1949.
10. Findlay GH: Idiopathic enlargements of the lips: Cheilitis granulomatosa, Ascher's syndrome and double lip. *Br J Dermatol* **66**:129–138, 1954.
11. Franceschetti A: Manifestation de blépharochalasis chez le père associé à des doubles lèvres apparaissant également chez sa filette âgée d'un mois. *J Génét Hum* **4**:181–182, 1955.
12. Goldberg R et al: Floppy eyelid syndrome and blepharochalasis. *Am J Ophthalmol* **102**:376–381, 1986.
13. Hartmann K: Blepharochalasis mit Struma und Doppellippe. *Klin Monatsbl Augenheilkd* **89**:376–380, 1932.
14. Hausamen JE et al: Klinischer Beitrag zum Ascher-Syndrom. *Dtsch Zahnärztl Z* **24**:983–987, 1969.
15. Klemens F: Blepharochalasis, Struma und Doppellippe. *Klin Monatsbl Augenheilkd* **105**:474–482, 1940.
16. Laffer WB: Blepharochalasis: Report of a case of this trophoneurosis involving also the upper lip. *Cleveland Med J* **8**:131–135, 1909.
17. Lebuisson D et al: Le syndrome de Laffer-Ascher. *J Fr Ophtal* **12**:750–752, 1978.
18. Neurstaetter O: Uber den Lippensaum beim Menschen: seinen Bau, seine Entwicklung und seine Bedeutung. *Jena Z Med Naturw* **29**:345–390, 1895.
19. Panneton P: La blepharo-chalazis. *Arch Ophtalmol (Paris)* **53**:729–755, 1936.
20. Papanayatou PH, Hatziotis JC: Ascher's syndrome. *Oral Surg* **35**:467–471, 1973.
21. Rintala AE: Congenital double lip and Ascher syndrome: II. Relationship to the lower lip sinus syndrome. *Br J Plast Surg* **34**:31–34, 1981.
22. Schimpf A: Das Ascher-Syndrom. *Dermatol Wochenschr* **132**:1077–1086, 1955.


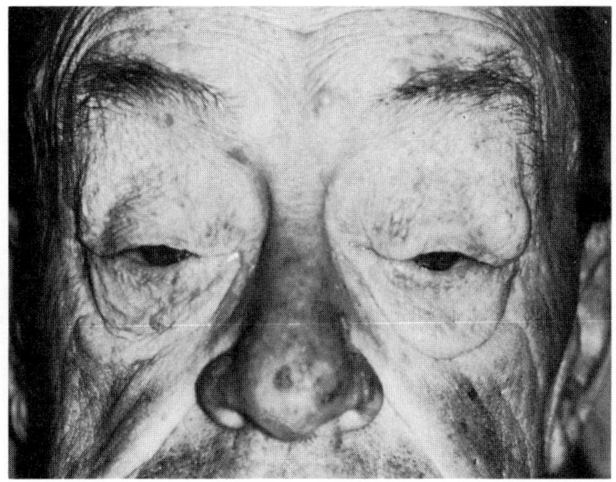
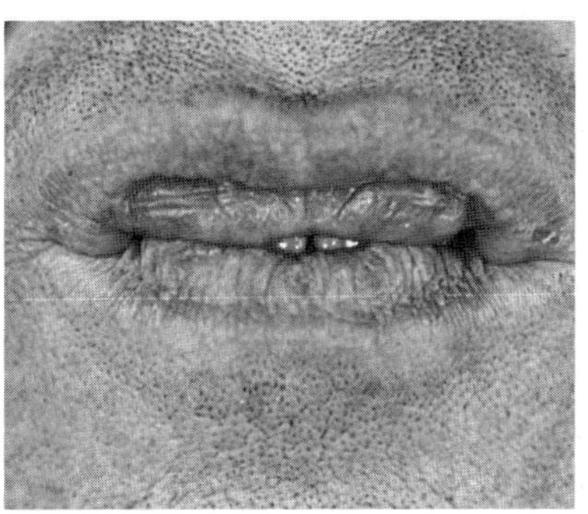

A B

Fig. 13–8. *Ascher syndrome.* (A,B) Compare similar alterations in older patients (A courtesy of *AF Morgan,* Seattle, Washington.)

23. Segal P, Jablonska S: Le syndrome d'Ascher. *Ann Ocul* **194**:511–526, 1961.
24. Stehr K et al: Pathogenese und Therapie des Ascher-Syndroms. *Dtsch Med Wochenschr* **87**:1148–1154, 1962.
25. Suter L, Vakilzadeh F: Ascher-Syndrom. *Hautarzt* **28**:257–259, 1977.
26. Swerdloff G: Double lip. *Oral Surg* **13**:627–629, 1960.
27. Tapasztó I et al: Some data on the pathogenesis of blepharochalasis (case I). *Acta Ophthalmol (Kbh)* **41**:167–175, 1963.
28. Wirths M: Beiderseitige Lidgeschwulst, kombiniert mit Geschwulstbildung der Oberlippe. *Z Augenheilkd* **44**:176–178, 1920.

## Acromegaloid features and thickened oral mucosa (Hughes syndrome)

Hughes et al (1) described a four-generation kindred in which 13 individuals were affected with a progressively coarse acromegaloid facial appearance and thickening of the lips and oral mucosa. Inheritance is autosomal dominant.

The most striking features are thickened lips without true doubling, overgrowth of the oral mucosa resulting in exaggerated rugae and frenula, and thickened upper eyelids leading to narrow palpebral fissures. Both facial appearance and large doughy hands remind the observer of acromegaly (Figs. 13–9 and 13–10).

In differential diagnosis *pachydermoperiostosis, Ascher syndrome,* and *multiple mucosal neuroma syndrome* should be excluded. In the syndrome described by Hughes et al (1) the skin was not furrowed. The nose becomes rather significantly bulbous, far more than that seen in pachydermoperiostosis. Although the oral mucosa becomes thickened in Ascher syndrome, it is not to the degree found in Hughes syndrome, and although the hand joints are hyperextensible (Fig. 13–11), there is no clubbing as there is in pachydermoperiostosis.

#### Reference [Acromegaloid features and thickened oral mucosa (Hughes syndrome)]

1. Hughes HE et al: An autosomal dominant syndrome with "acromegaloid" features and thickened oral mucosa. *J Med Genet* **22**:119–125, 1985.

Fig. 13–9. *Acromegaloid features and thickened oral mucosa.* Thickened lips, fullness of upper eyelids leading to narrowed palpebral fissures. (From *HE Hughes,* Cardiff, Wales.)

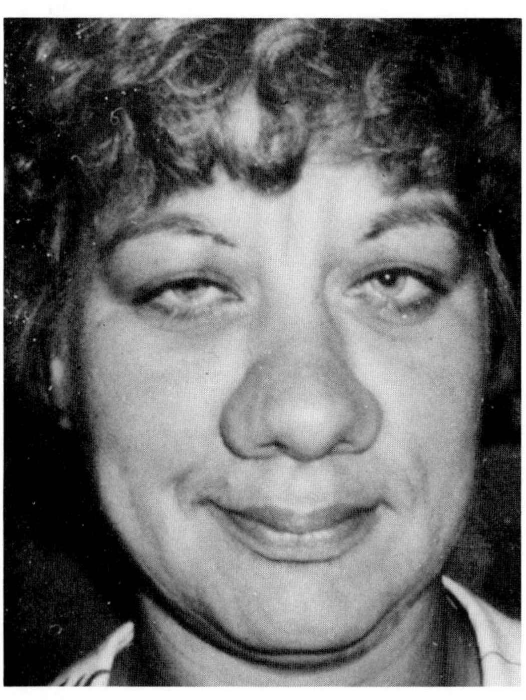

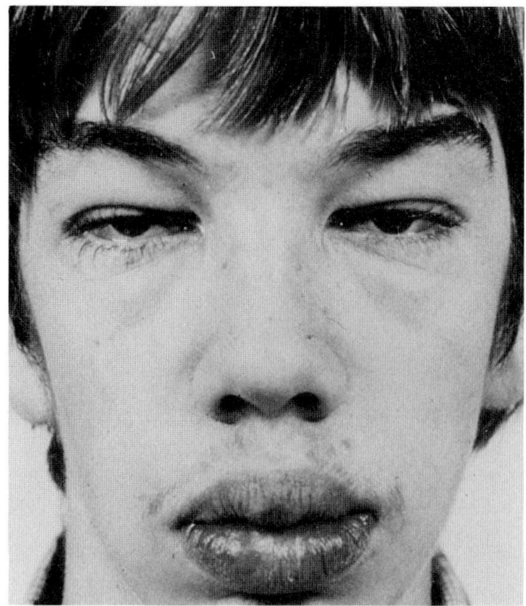

Fig. 13–10. *Acromegaloid features and thickened oral mucosa.* Similar coarse features and bulbous nose. (From *HE Hughes et al,* J Med Genet **22**:119, 1985.)

## Cutis laxa syndromes

Cutis laxa is characterized by skin that hangs in loose folds. There are many forms, some genetic and others acquired. Confusion existed in the early literature since the Ehlers–Danlos syndromes (EDS) also had that designation. Although the term *cutis laxa* was used as early as 1833 by Alibert (2), the generalized form was probably first described by Graf (24) in 1836. A good general review is that of Mensing et al (43).

**Autosomal recessive form.** This is the most common congenital genetic type of cutis laxa. Birth weight is normal but often there is

Fig. 13–11. *Acromegaloid features and thickened oral mucosa.* Hyperextensible joints. (From *HE Hughes et al,* J Med Genet **22**:119, 1985.)

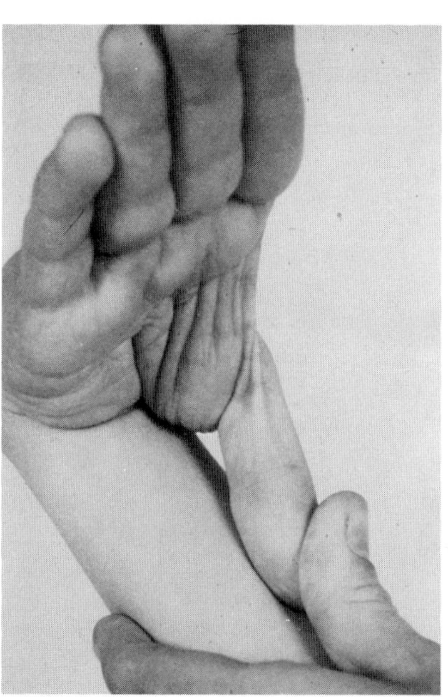

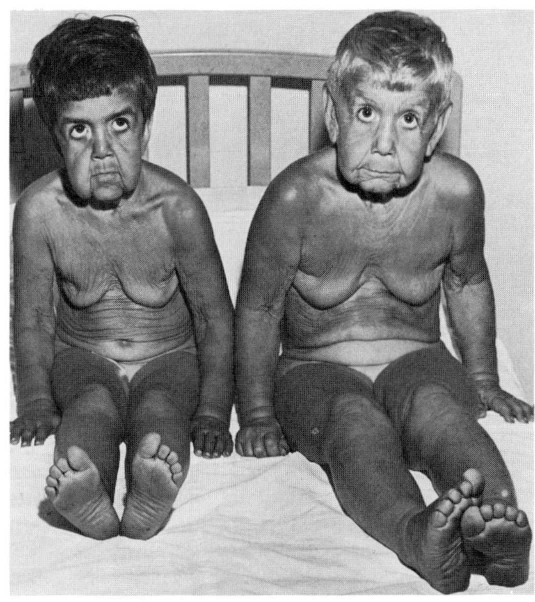

Fig. 13–12. *Cutis laxa.* Affected sibs look much older than their chronologic ages of 5 and 6 years. (Courtesy of *FA Balboni,* Garden City, New York.)

mild postnatal growth deficiency. Most of these children die by the third year of life.

Affected sibs (23,41,42,60) and parental consanguinity (1,16,32, 42,59,67) indicate autosomal recessive inheritance. We believe that the syndrome reported by Dallaire et al (13) does not represent a separate disorder as they suggest.

Facies and skin. Due to accentuation of skin folds, the infant appears aged. The upper lip is long and the columella is short. Blepharochalasis and, at times, ectropion add to the aged appearance (23,26,57). The skin of the entire body appears too large. There is no hyperelasticity, fragility, or difficulty in healing (Fig. 13–12).

Fig. 13–13. *Cutis laxa.* Multiple herniae of bladder.

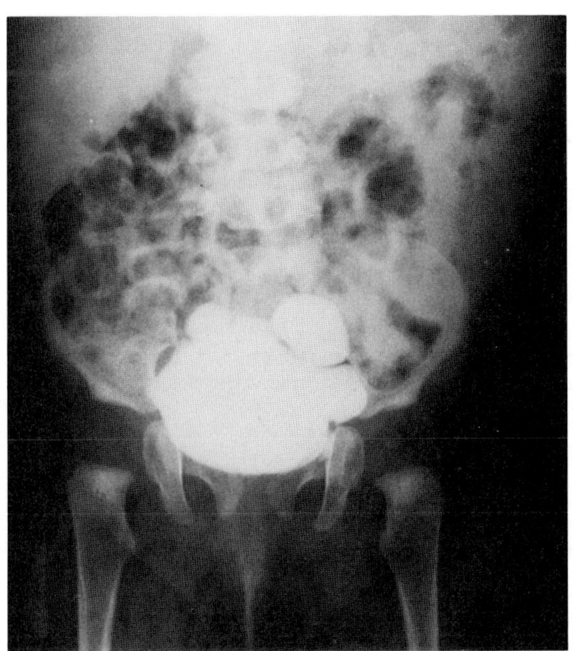

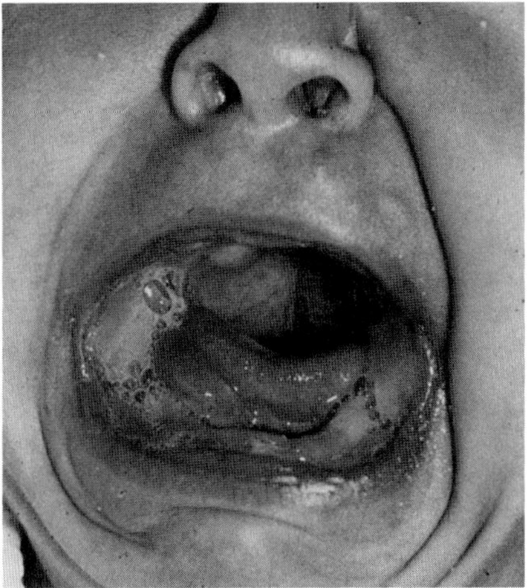

Fig. 13–14. *Cutis laxa.* Oral mucosa hangs loosely. (Courtesy of *AH Mehregan,* Monroe, Michigan.)

Cardiorespiratory. Tachypnea, pneumonitis, and airway obstruction associated with emphysema are common features. This results in severe hypoxemia, respiratory failure, cor pulmonale, and right ventricular enlargement. These findings together with diaphragmatic hernia lead to early demise (23,26,57,60). The emphysema results from generalized elastolysis. Bundle branch block (23), pulmonary artery stenosis (42,67), and dilated and tortuous carotid, vertebral and pulmonary arteries, and dilated aortic root (23,65,67) have been noted.

Musculoskeletal. Hernia is common, including inguinal (1,10,23, 41,57,60), diaphragmatic (1,11,23,59), and ventral and/or umbilical (23,60). Loose joints are rarely noted (1).

Gastrointestinal. Diverticula of the gastrointestinal tract involving the pharynx (32), esophagus (23,57), and rectum (23,41,42,60) have been documented.

Genitourinary. Bladder diverticula have been described by several authors (1,22,23,32,57,60) (Fig. 13–13). Vaginal prolapse has also been noted.

Oral. The voice is deep and resonant in quality due to laxity of the vocal cords (5,23,69). There is also looseness of the oral and pharyngeal mucosa (23) (Fig. 13–14).

**Autosomal dominant form.** This form is less common than the autosomal recessive type and runs a quite benign course (4,5,

Fig. 13–15. *Cutis laxa.* Autosomal dominant type of cutis laxa in 15-year-old girl who looks older than her middle-aged mother. (From *P Beighton,* Br J Plast Surg **23**:285, 1970.)

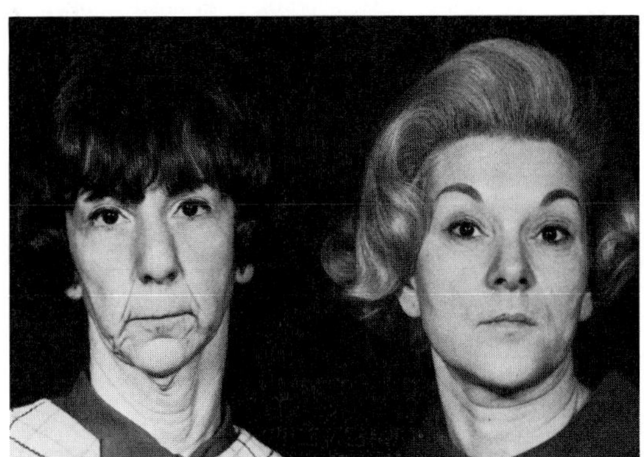

9,29,37,49,57,59,62,65,68). Perhaps the earliest published example is that of Rossbach (52) in 1884. Onset is usually somewhat later than the recessive form. Drooping of eyelids and sagging facial skin together with accentuation of the nasolabial and other facial folds produce an aged appearance. The nose is often hooked, the nostrils everted, and the philtrum long. An affected pubertal child will often look older than the unaffected parent! (Fig. 13–15). Complications are few and lifespan is normal.

Rarely associated abnormalities have included hoarseness (4,9,29), pulmonary artery stenosis (29,65), mitral valve prolapse (9), hernia (59), bronchiectasis (5), joint dislocations (9), tortuosity and dilatation of carotid arteries and aorta (15,29), dilatation of sinuses of Valsalva (9), and coarctation of aorta (4).

**Cutis laxa, late closure of fontanels, intrauterine growth retardation, and hyperlaxity of joints.** Debré et al (14), in 1937, were possibly the first to describe this form of cutis laxa. Several other authors have reported examples (1,3,7,17,18,21,30,33–35,46,47, 50,51,54,63). It should be noted that the patient reported by Theopold and Wildhack (63) is a cousin of the patients reported by Fittke (17). Possibly the patient described by Schirren et al (56) had this condition. Inheritance is probably autosomal recessive, but most patients are female (33,46,51,54). The rate of parental consanguinity is high.

Birth weight and length are under the 10th percentile. Delayed postnatal growth, mild gross motor retardation, poor feeding, mild mental retardation, frontal bossing, sagging jowls, reversed V-shaped eyebrows, epicanthic folds, sunken nasal bridge and apparent mild hypertelorism are common. In addition to congenital cutis laxa, which is especially severe over the abdomen, hands, and feet, there is late closure of the fontanels, especially the anterior one. The skull sutures are markedly separated. The hips, either one or more often both, are congenitally dislocated in 85%. All joints, especially those of the hands, exhibit generalized laxity, which leads to unusual positioning. All appear to have fifth finger clinodactyly. Talipes equinovarus has been noted in a few cases. The subcutaneous veins of the abdomen are markedly dilated and the abdominal musculature is lax. Large inguinal hernias have been found. Less constant findings include downslanting palpebral fissures, macular coloboma, myopia, iris hypoplasia, single transverse palmar crease, and hydronephrosis (3,56). Cleft lip has been reported (46).

**Differential diagnosis.** X-linked cutis laxa formerly called *Ehlers–Danlos syndrome, type IX* is termed *occipital horn syndrome.*

Acquired cutis laxa may follow nonspecific cutaneous inflammation, persistent urticaria, or nephrotic syndrome (64). Occasionally there is no history of prior skin disorder (6). Some patients have only cutaneous involvement whereas others have internal manifestations: pulmonary emphysema with cardiovascular complications, femoral, inguinal, ventral, or diaphragmatic hernias, diverticula of the intestinal tract, cystocele, rectocele, and uterine prolapse (6,31,32,36). Ruptured patellar tendons and associated multiple myeloma have been reported (12,58). The reader is referred to several thorough reviews of acquired cutis laxa (27,40,48,66,69).

Braun-Falco et al (8) were dealing with a case of prune belly. Tracheobronchiomegaly (Mounier–Kuhn syndrome) is characterized by dilatation and diverticulas of the trachea and bronchi (19) with blepharochalasis and redundancy of oral mucosa *(Ascher syndrome).* The "wrinkly" skin syndrome, a recessively inherited disorder characterized by congenital wrinkling of the skin of the chest, abdomen, and dorsa of the hands and feet, must be excluded. The palms and soles have increased numbers of wrinkles. There is generalized hypotonia with winging of the scapulae (20).

Cutis laxa of the facial skin has also been found in association with a dominantly inherited syndrome of systemic amyloidosis with corneal lattice dystrophy, cranial nerve palsy, hair loss, nephropathy, and, occasionally, cardiopathy (Meretoja syndrome) (44,50a). It has also been seen with severe osteoporosis (53). Cutis laxa is seen in *gerodermia osteodysplastica, De Barsy syndrome,* and *Lenz–Majewski syndrome.* A congenital generalized cutis laxa may result from chronic maternal penicillamine therapy during pregnancy. These infants may have severe pulmonary complications (38,61,66). Tsukahara et al (64a) described a disorder intermediate between cutis laxa and *Ehlers-Danlos syndrome* in a mother and child.

Bladder diverticula may also be seen in Menkes syndrome, prune belly sequence, *Ehlers–Danlos syndromes,* and *Williams syndrome.*

**Pathology.** With orcein or Weigert elastic fiber stain, absence or marked diminution of elastic fibers has been observed, especially in the papillary layer of the dermis. Degenerative changes are marked by fragmentation of elastic fibers. Some fibers are short and thin.

Ultrastructural changes include normal microfibril formation, deficiency of elastin, and uneven distribution of dense amorphous or granular substances that produce dense bundles (28,39,45,55). The various types of cutis laxa cannot be differentiated by light or electron microscopic study.

**Laboratory aids.** The mechanism by which there is insufficient elastin is not known. Whether it is based on deficient formation or accelerated destruction has not been shown. While low serum copper and increased elastase and/or decreased antielastase levels (23,25,26,28) have been reported, others have found normal or even elevated values (28,29). Acquired cutis laxa is associated with normal enzyme levels (29).

### References (Cutis laxa syndromes)

1. Agha A et al: Two forms of cutis laxa presenting in the newborn period. *Acta Paediatr Scand* **67**:775–780, 1978.

2. Alibert JL: Histoire d'un berger des environs de Gisore (dermatose hypermorphe). *Monogr Dermatol* **2**:719, 1833.

3. Allanson J et al: Congenital cutis laxa with retardation of growth and motor development: A recessive disorder of connective tissue with male lethality. *Clin Genet* **29**:133–136, 1986.

4. Balboni FA: Cutis laxa and multiple vascular anomalies, including coarctation of the aorta. *Bull St Francis Hosp (Roslyn)* **19**:26–34, 1963.

5. Beighton P et al: The dominant and recessive forms of cutis laxa. *J Med Genet* **9**:216–221, 1972.

6. Bettman AG: Excessively relaxed skin and the pituitary gland. *Plast Reconstr Surg* **15**:489–501, 1955.

7. Bittel-Dobrzynska N, Siniecki B: Cutis laxa (Ehlers–Danlos syndrome) with congenital dislocation of the hips. *Endokr Pol* **15**:469–479, 1964.

8. Braun-Falco O et al: Angeborene Dermatochalasis als Leitsymptom eines Symptomenkomplexes. *Arch Klin Exp Dermatol* **220**:166–182, 1964.

9. Brown FR et al: Cutis laxa. *Johns Hopkins Med J* **150**:148–153, 1982.

10. Chadfield HW, North JF: Cutis laxa. *Trans St Johns Hosp Dermatol Soc* **57**:181–189, 1971.

11. Christiaens L et al: Emphyseme congénital et cutis laxa. *Presse Méd* **62**:1799–1801, 1954.

12. Cho SY, Maguire RF: Multiple myeloma associated with acquired cutis laxa. *Cutis* **26**:209–211, 1980.

13. Dallaire L et al: A syndrome of generalized elastic fiber deficiency with leprechaunoid features: A distinct genetic disorder with an autosomal recessive mode of inheritance. *Clin Genet* **10**:1–11, 1976.

14. Debré R et al: "Cutis laxa" avec dystrophies osséuses. *Bull Soç Méd Hôp Paris* **53**:1038–1039, 1927.

15. Dingman RO et al: Cutis laxa congenita—generalized elastosis. *Plast Reconstr Surg* **44**:431–435, 1969.

16. Duperrat B et al: Cutis laxa et anétodermie. *Ann Chir Plast* **22**:11–16, 1977.

17. Fittke H: Über eine ungewöhnliche Form "multiplier Erbabartung" (Chalodermie und Dysostose). *Z Kinderheilkd* **63**:510–523, 1942.

18. Fitzsimmons JS et al: Variable clinical presentation of cutis laxa. *Clin Genet* **28**:284–295, 1985.

19. Gay S, Dee P: Tracheobronchomegaly—the Mounier–Kuhn syndrome. *Br J Radiol* **57**:640–644, 1984.

20. Gazit E et al: The wrinkly skin syndrome: A new heritable disorder of connective tissue. *Clin Genet* **4**:186–192, 1973.

21. Goldblatt J et al: Cutis laxa, retarded development and joint hypermobility syndrome. *Dysmorphol Clin Genet* **1**:142–144, 1988.

22. Goltz RW, Hult AM: Generalized elastolysis (cutis laxa) and Ehlers-Danlos syndrome (cutis hyperelastica). *South Med J* **58**:848–854, 1965.

23. Goltz RW et al: Cutis laxa, a manifestation of generalized elastolysis. *Arch Dermatol* **92**:373–387, 1965.

24. Graf NI: Ortliche erbliche Erschlaffung der Haut. *Wschr Ges Heilkd* 225, 1836.

25. Grahame R, Beighton P: The physical properties of skin in cutis laxa. *Br J Dermatol* **84**:326–329, 1971.

26. Hajjar BA, Joyner EN: Congenital cutis laxa with advanced cardiopulmonary disease. *J Pediatr* **73**:116–119, 1968.

27. Harris RB et al: Generalized elastolysis (cutis laxa). *Am J Med* **65**:815–822, 1978.

28. Hashimoto K, Kanzaki T: Cutis laxa: Ultrastructural and biochemical studies. *Arch Dermatol* **111**:861–873, 1975.

29. Hayden JG et al: Cutis laxa associated with pulmonary artery stenosis. *J Pediatr* **72**:506–509, 1968.

30. Hinkel GK, Rupprecht E: Zur Einordnung multiplier mesenchymaler Fehlbildungen (Eine Beobachtungen von Dysostosis cleidocranialis mit Dermatochalasis). *Arch Kinderheilk* **175**:292–302, 1967.

31. Jablonska S: Inflammatorische Hautveränderungen die einer erworbenen Cutis laxa vorausgehen. *Hautarzt* **17**:341–346, 1966.

32. Janik JS et al: Cutis laxa and hollow viscus diverticula. *J Pediatr Surg* **17**:318–320, 1982.

33. Karrar ZA: Cutis laxa, ligamentous laxity and delayed development. *Pediatrics* **74**:903–904, 1984.

34. Karrar ZA et al: Cutis laxa, intrauterine growth retardation, and bilateral dislocation of the hips: A report of five cases. *Prog Clin Res* **104**:215–221, 1982.

35. Khaldi F et al: Cutis laxa. *Ann Pédiatr* **34**:165–168, 1987 (Case 2).

36. Lally JF et al: The roentgenographic manifestations of cutis laxa (generalized elastolysis). *Radiology* **113**:605–606, 1974.

37. Lewis E: Cutis laxa. *Proc R Soc Med* **41**:864–865, 1948.

38. Linares A et al: Reversible cutis laxa due to maternal *D*-penicillamine treatment. *Lancet* **2**:43, 1979.

39. Marchase P et al: A familial cutis laxa syndrome with ultrastructural analysis of collagen and elastin. *J Invest Derm* **75**:399–403, 1980.

40. Marshall J et al: Post-inflammatory elastolysis and cutis laxa. *S Afr Med J* **40**:1016–1022, 1966.

41. Mehregan AH et al: Cutis laxa (generalized elastolysis): A report of four cases with autopsy findings. *J Cutan Pathol* **5**:116–126, 1978.

42. Mendoza HR: Cutis laxa. *Rev Esp Pediatr* **24**:735–740, 1968.

43. Mensing H et al: Cutis laxa: Klassifikation, Klinik and molekulare Defekte. *Hautarzt* **35**:506–511, 1984.

44. Meretoja J: Genetic aspects of familial amyloidosis with corneal lattice dystrophy and cranial neuropathy. *Clin Genet* **4**:173–185, 1973.

45. Nanko H et al: Acquired cutis laxa (generalized elastolysis): Light and electron microscopic studies. *Acta Dermatovenereol* **59**:315–324, 1979.

46. Patton M et al: Congenital cutis laxa with retardation of growth and development. *J Med Genet* **24**:556–561, 1987.

47. Philip AGS: Cutis laxa with intrauterine growth retardation and hip dislocation in a male. *J Pediatr* **93**:150–151, 1978.

48. Reed WB et al: Acquired cutis laxa. *Arch Dermatol* **103**:661–669, 1971.

49. Reidy JP: Cutis hyperelastica (Ehlers–Danlos) and cutis laxa. *Br J Plast Surg* **16**:84–91, 1963 (Fig. 10, 11).

50. Reisner SH et al: Cutis laxa associated with severe intrauterine growth retardation and congenital dislocation of the hip. *Acta Paediatr Scand* **60**:357–360, 1971.

50a. Rintala AE et al: Primary hereditary systemic amyloidosis (Meretoja's syndrome): Clinical features and treatment by plastic surgery. *Scand J Plast Reconstr Surg* **22**:141–145, 1988.

51. Rogers JG, Danks DM: Cutis laxa with delayed development. *Aust Paediatr J* **21**:283–285, 1985.

52. Rossbach MJ: Ein merkwürdiger Fall von greisenhafter Veränderung der allgemeinen Körperdecke bei einem achtzehnjährigen Jungling. *Dtsch Arch Klin Med* **36**:197–203, 1884.

53. Sakati NO, Nyhan WL: Congenital cutis laxa and osteoporosis. *Am J Dis Child* **137**:452–454, 1983.

54. Sakati NO et al: Syndrome of cutis laxa, ligamentous laxity and delayed development. *Pediatrics* **72**:850–856, 1983.

55. Sayers CP et al: Pulmonary elastic tissue in generalized elastolysis (cutis laxa) and Marfan's syndrome: A light and electron microscopic study. *J Invest Dermatol* **65**:451–457, 1975.

56. Schirren C et al: Ektodermale Dysplasie mit Hypohidrosis, Hypotrichosis und Hypodontie. *Hautarzt* **11**:70–75, 1960.

57. Schreiber MM, Tilley NJ: Cutis laxa. *Arch Dermatol* **84**:266–272, 1961.

58. Scott MA et al: Acquired cutis laxa associated with multiple myeloma. *Arch Dermatol* **112**:853–855, 1976.

59. Sestak Z: Ehlers–Danlos syndrome and cutis laxa: An account of families in the Oxford area. *Ann Hum Genet* **25**:313–321, 1962.

60. Siegmund L: Über das sogennante Oedema lymphangiectaticum. *Zbl Allg Pathol* **70**:243, 1938.

61. Solomon L et al: Neonatal abnormalities associated with D-penicillamine treatment during pregnancy. *N Engl J Med* **296**:54–55, 1977.

62. Talbot FB: Metabolism study of a case simulating premature senility. *Mschr Kinderheilkd* **25**:643–646, 1923.

63. Theopold W, Wildhack R: Dermatochalasis in Rahmen multipler Abartungen. *Mschr Kinderheilkd* **99**:213–218, 1951.

64. Tsuji T et al: Acquired cutis laxa concomitant with nephrotic syndrome. *Arch Dermatol* **123**:1211–1216, 1987.

64a. Tsukahara M et al: A disease with features of cutis laxa and Ehlers–Danlos syndrome. *Hum Genet* **78**:9–12, 1988.

64b. Van Maldergem L et al: Facial anomalies in congenital cutis laxa with retarded growth and skeletal dysplasia. *Am J Med Genet* **32**:265, 1989.

65. Wagstaff LA et al: Vascular abnormalities in congenital generalized elastolysis (cutis laxa). *S Afr Med J* **44**:1125–1127, 1970.

66. Walshe JM: Congenital cutis laxa and maternal D-penicillamine. *Lancet* **2**:144–145, 1979.

67. Weir EK et al: Cardiovascular abnormalities in cutis laxa. *Eur J Cardiol* **5**:255–261, 1977.

68. Wiener K: Gummihaut (Cutis laxa) mit dominanter Vererbung. *Arch Dermatol Syph* **148**:599–601, 1925.

69. Wilsch L et al: Spätmanifeste Dermatochalasis. *Dtsch Med Wschr* **102**:1451–1454, 1977.

## Occipital horn syndrome (Ehlers–Danlos syndrome, type IX)

Formerly called Ehlers–Danlos syndrome (EDS) type IX and first reported by Lazoff et al (6) in 1975, the occipital horn syndrome is characterized by occipital exostoses, soft easily bruisable skin, hyperextensible joints, and widening and bowing of multiple long bones. A number of kindreds have been reported (1–5,7,9,10). The disorder is due to a defect in copper metabolism leading to lysyl oxidase deficiency and abnormal collagen (1,2,3–5). Serum ceruloplasmin and copper levels are low. High levels of copper are found in cultured fibroblasts with low levels in hair. The disorder may be allelic with Menkes syndrome (3,5,8). Inheritance is X-linked.

The face, neck, and trunk are long and thin (1,2) (Fig. 13–16A). The hyperelastic, soft, and bruisable skin with resultant atrophic scars and the hyperelastic joints are not as marked as in EDS type I or as mild as EDS type II. Varicose veins have been noted in 70% (9,10) (Fig. 13–16B).

Patients exhibit limited extension of the narrow shoulders, elbows, and knees, hypermobility of finger joints, pes planus, and genua valga (Fig. 13–16A). Pectus excavatum or carinatum has been found in 40% and saber skins in 30%. Various hernias (hiatal, femoral, inguinal) have been noted in about 35%. Bladder diverticula with bladder neck obstruction were found in 60–70%. Chronic diarrhea has been noted in 40% (1,2,4,6,9). Borderline IQ has been documented in 40%.

Radiologic findings include occipital exostoses (horns) symmetrically located on each side of the foramen magnum (Fig. 13–17A). These may be palpated, are a constant feature, and may become large with age. The horns may represent ectopic bone formation within the trapezius and sternocleidomastoid aponeuroses. The clavicles are very short with a widened medullary cavity and hammer-shaped distal extremities (Fig. 13–17B). The femora exhibit focal hyperostosis at sites of tendon and ligament insertion (Fig. 13–18A). Carpal fusion involving capitate–hamate and trapezium–trapezoid coalescence has been noted in over 50% (9,10) (Fig. 13–18B). Less specific are deformation of the humerus, radius, ulna, tibia, and fibula (90%), osteoporosis (70%), narrowing of rib cage (65%), dislocation of radial head (40%), mild platyspondyly, coxa valga, and flattening of acetabular roofs.

### References [Occipital horn syndrome (Ehlers–Danlos syndrome, type IX)]

1. Blackston RD et al: Ehlers–Danlos syndrome (EDS), type IX: Biochemical evidence for X-linkage. Abst 140. *Am J Hum Genet* **41**(3):A49, 1987.

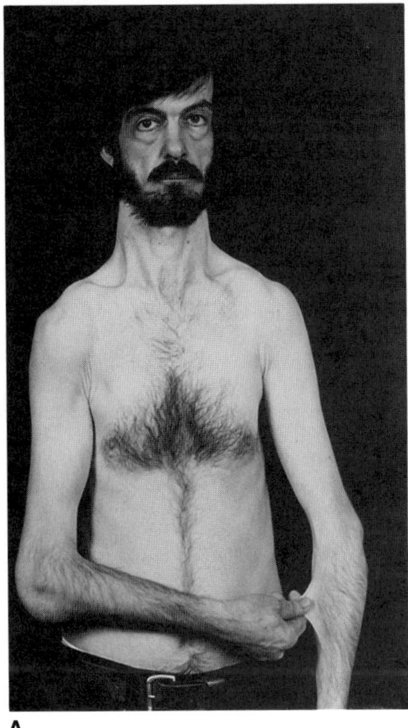
A

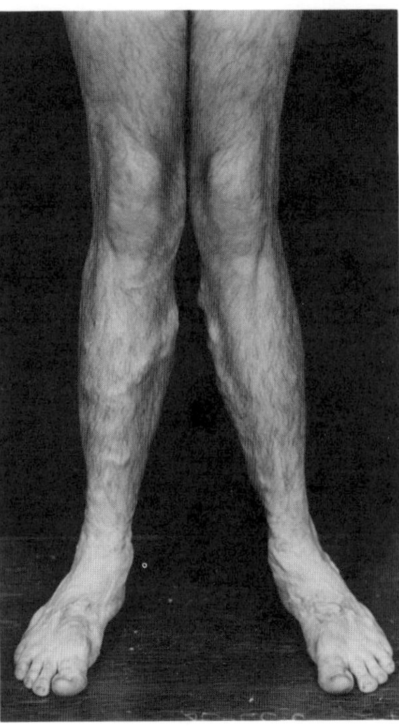
B

Fig. 13–16. *Occipital horn syndrome.* (A) Patient showing long neck, narrow sloping shoulders, long thorax, hyperelasticity of skin. (B) Genua valga, varicosities, pes planus. (Courtesy of *DD Weaver,* Indianapolis, Indiana.)

1a. Byers PH et al: An X-linked form of cutis laxa due to deficiency of lysyl oxidase. *Birth Defects* **12**(5):293–298, 1976.

2. Byers PH et al: X-linked cutis laxa: Defective collagen cross link formation due to decreased lysyl oxidase activity. *N Engl J Med* **303**:61–65, 1980.

3. Kaitila I et al: A skeletal and connective tissue disorder associated with lysyl oxidase deficiency and abnormal copper metabolism, in *Skeletal Dysplasias,* Papadatos CJ, Bartsocas CS (eds), Alan R. Liss, New York, 1982, pp 307–316.

4. Kuivaniemi H et al: Type IX Ehlers–Danlos syndrome and Menkes syndrome: The decrease in lysyl oxidase activity is associated with a corresponding deficiency in the enzyme protein *Am J Hum Genet* **37**:798–808, 1985.

5. Kuivaniemi H et al: Abnormal copper metabolism and deficient lysyl oxidase activity in a heritable connective tissue disorder. *J Clin Invest* **69**:730–733, 1982.

6. Lazoff SG et al: Skeletal dysplasia, occipital horns, intestinal malabsorption, and obstructive uropathy—a new hereditary syndrome. *Birth Defects* **11**(5):71–74, 1975.

7. MacFarlane JD et al: A new Ehlers–Danlos syndrome with skeletal dysplasia. *Am J Hum Genet* **32**:118A, 1980.

8. Peltonen L et al: Alterations in copper and collagen metabolism in the Menkes syndrome and a new subtype of the Ehlers–Danlos syndrome. *Biochemistry* **22**:6146–6163, 1983.

9. Sartoris DJ et al: Type IX Ehlers–Danlos syndrome. A new variant with pathognomonic radiographic features. *Radiology* **152**:665–670, 1984.

10. Weaver D: Personal communication, 1987.

Fig. 13–17. *Occipital horn syndrome.* (A) Arrows point to occipital exostoses. Note long neck. (B) Club-shaped distal clavicles. (A, B from *DJ Sartoris,* Radiology **152**:665, 1984.)

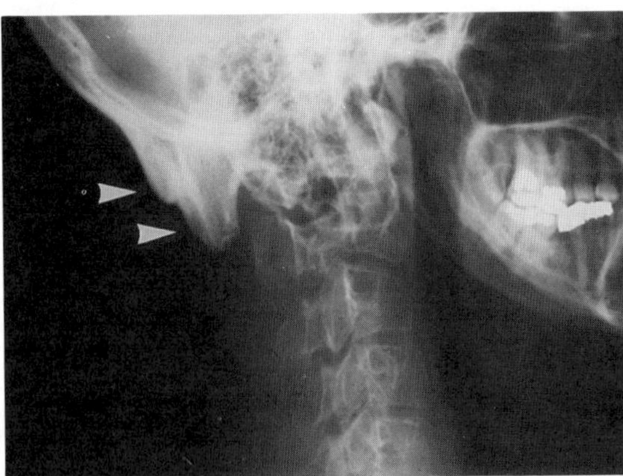
A

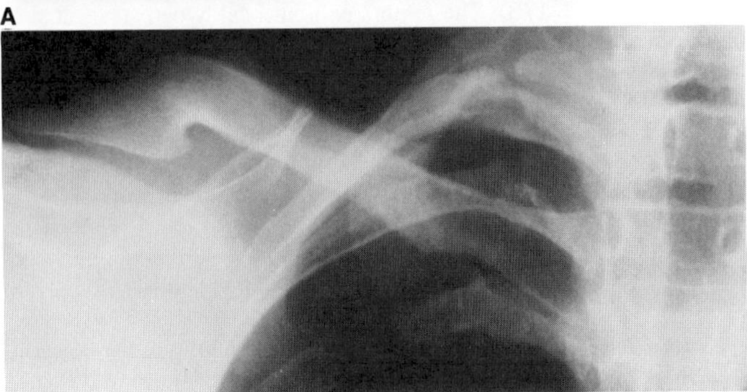
B

## De Barsy syndrome

De Barsy et al (3) and others (2,4–8) described a syndrome of intrauterine growth retardation, wrinkled atrophic skin, open fontanels and sutures, somatic and mental retardation, brisk deep tendon reflexes, athetoid posturing, hypermobility of small joints, and muscular hypotonia. The facies is characterized by frontal bossing in early childhood but postnatal changes include microcephaly, prematurely aged appearance, corneal clouding, large dysplastic ears, hypertelorism, and thin lips (Figs. 13–19 to 13–23). Some patients have cataracts. We are not convinced that the child noted by Bartsocas (1) has the syndrome. Inheritance is autosomal recessive (6,8). Possibly the child reported by Wiedemann (9) in 1969 has De Barsy syndrome or *gerodermia osteodysplastica.*

Elastic fibers are frayed and reduced in number. Chemotactic migration of cultured fibroblasts is reduced and there is impaired granulocyte function (5,7). The disorder must be differentiated from *cutis laxa syndromes* and those of premature aging (acrogeria, metageria, *gerodermia osteodysplastica, Wiedemann–Rautenstrauch syndrome*).

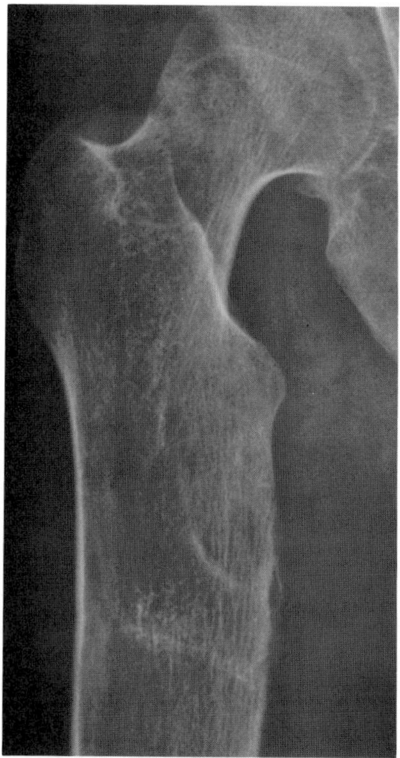

A

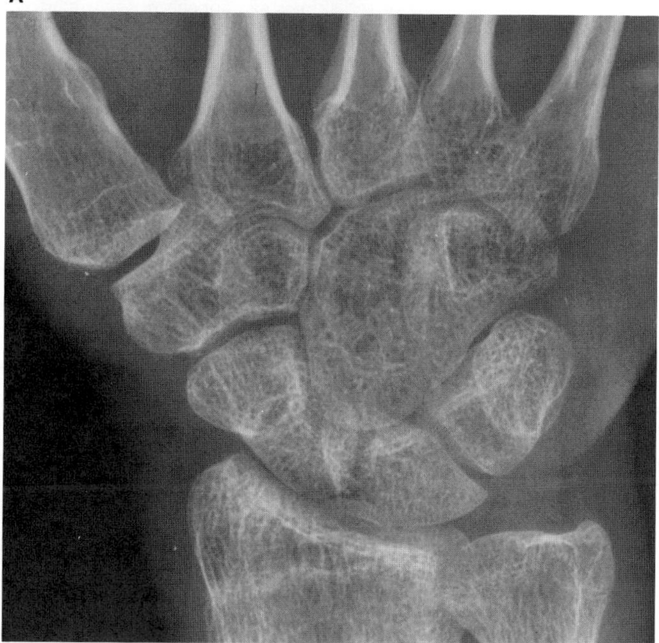

B

Fig. 13–18. *Occipital horn syndrome.* (A) Femur shows focal hyperostosis at site of tendon insertion. (B) Capitate–hamate and trapezium–trapezoid coalescence. (A,B from *DJ Sartoris,* Radiology **152**:665, 1984.)

## References (De Barsy syndrome)


1. Bartsocas C: De Barsy syndrome, in *Skeletal Dysplasias* Papadatos C, Bartsocas C (eds), Alan R. Liss, New York, 1982, pp. 157–160.
2. Burck U: DeBarsy-Syndrom—ein weitere Beobachtung. *Klin Pädiatr* **186**:441–444, 1974.
3. De Barsy AM et al: Dwarfism, oligophrenia, and degeneration of the elastic tissue in skin and cornea. A new syndrome? *Helv Paediatr Acta* **23**:305–313, 1968.
4. Goecke T et al: Cutis laxa, Hornhautstrübung und Retardierung-Das de Barsy-Syndrom. *Klin Genet-Pädiatr* **2**:139–143, 1980.


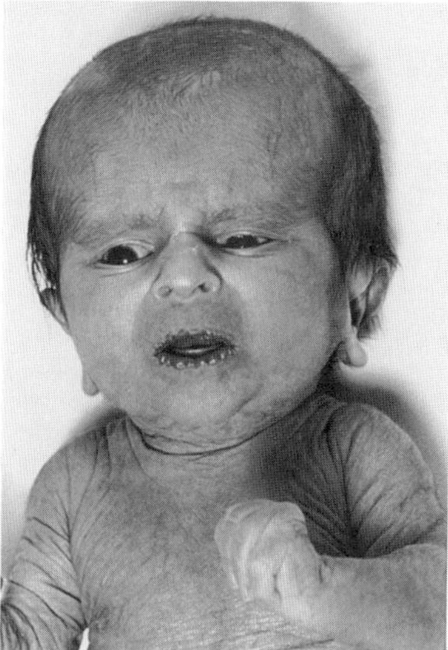

Fig. 13–19. *De Barsy syndrome.* Wizened face of 10-day-old. Note loose wrinkled skin. (From *BF Pontz et al,* Eur J Pediatr **145**:428, 1986.)


5. Hoefnagel D et al: Congenital athetosis, mental deficiency, dwarfism and laxity of skin and ligaments. *Helv Paediatr Acta* **26**:397–402, 1971.
6. Kunze J et al: De Barsy syndrome—an autosomal recessive progeroid syndrome. *Eur J Pediatr* **144**:348–354, 1985.
7. Pontz BF et al: Biochemical, morphological and immunological findings in a patient with a cutis laxa-associated inborn disorder (De Barsy syndrome). *Eur J Pediatr* **145**:428–434, 1986.
8. Riebel T: De Barsy–Moens–Dierckx-Syndrom: Beobachtung bei Geschwistern. *Mschr Kinderheilk* **124**:96–98, 1976.
9. Wiedemann H-R: Über einige progeroide Krankheitsbilder und deren diagnostische Einordnung. *Z Kinderheilkd* **107**:91–106, 1969.


## Gerodermia osteodysplastica (Bamatter syndrome)

The family first reported by Bamatter et al (1), in 1949, was reviewed by the same group of investigators over a 20-year period (2,3,8). The disorder manifests a characteristic facies, hyperlaxity of the skin and joints, and growth retardation. Other families have been reported (7,12). Others are probable examples. In the interest of linguistic purity, we have changed the name from geroderma osteodysplastica to gerodermia osteodysplastica (13).

Fig. 13–20. *De Barsy syndrome.* Microcephaly, large pinnae with poor modeling, downslanting palpebral fissures, strabismus, and right-sided cataract. (From *J Kunze et al,* Eur J Pediatr **144**:348, 1985.)

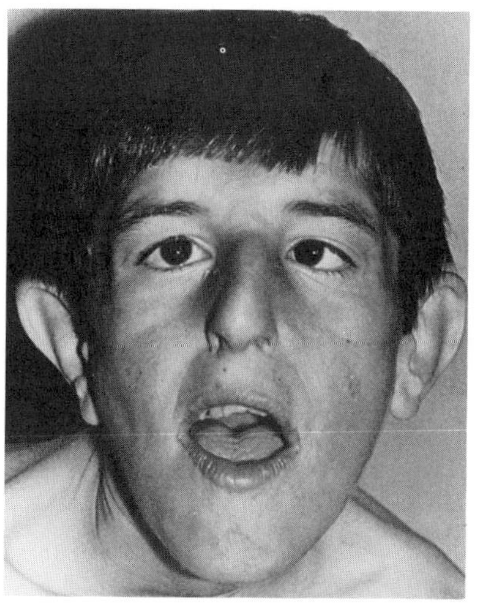

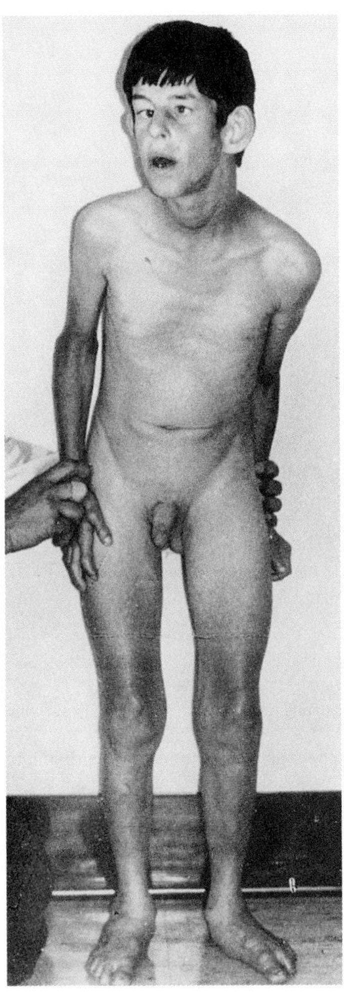

Fig. 13–21. *De Barsy syndrome.* Progeroid aspect, microcephaly, reduced subcutaneous fat, pectus excavatum, and growth retardation. (From *J Kunze et al,* Eur J Pediatr **144**:348, 1985.)

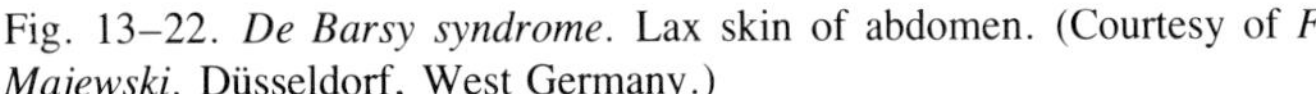

Fig. 13–22. *De Barsy syndrome.* Lax skin of abdomen. (Courtesy of *F Majewski,* Düsseldorf, West Germany.)

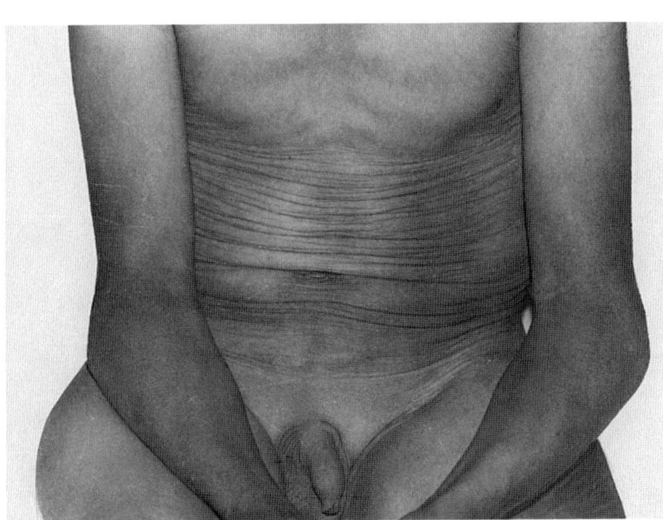

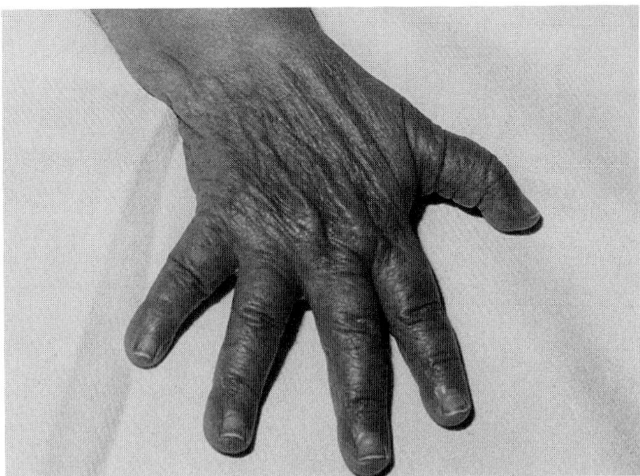

Fig. 13–23. *De Barsy syndrome.* Lax thin skin of hand. (Courtesy of *F Majewski,* Düsseldorf, West Germany.)

Originally thought to be X-linked (8), recent evidence suggests autosomal recessive inheritance (7,9). No doubt the disorder is underreported.

Growth retardation is below the third percentile in perhaps a third of the children. Frequently, span is greater than height.

**Facies.** Due to looseness of the skin, a sad appearance is imparted. The eyelids and cheeks droop. The head tends to be brachycephalic. The forehead is prominent. The nose is often jutting and fleshy. The lower lip is downturned and the midface is frequently mildly hypoplastic with relative mandibular prognathism. About half the patients exhibit malar flush (Figs. 13–24A, B and 13–25).

**Skin.** The skin is thin and creased with recoil. It is especially without turgor over the hands and feet.

**Musculoskeletal.** Constant features are marked muscular hypotonia and hyperlaxity of joints (hands, feet, knees, hips) with associated hernias, flat feet, and dislocated hips. Inguinal hernia has been found in approximately 35%. A virtually constant features is osteoporosis with resultant platyspondyly, anterior wedging, and occasionally biconcave vertebral bodies (4). Compression fractures may be observed within the first few years of life. Long bones tend to fracture with minimal trauma. Mild scoliosis and sternal anomalies are present in about 65%. Radiographic changes include multiple Wormian bones in the lambdoidal suture.

**Oral manifestations.** At least 70% have downturned lower lip. Almost constant features are malocclusion and highly arched palate. The alveolar bone is decreased in amount.

**Differential diagnosis.** Gerodermia osteodysplastica is most likely to be confused with the *cutis laxa syndromes.* Acrogeria is characterized by wrinkled and aged appearing skin, most marked over the dorsa of the hands and feet, flat feet, and dislocated hips. Many of these cases of acrogeria represent EDS IV. However, in gerodermia osteodysplastica, the skin is not atrophic and the superficial veins are not highly visible. Commonly associated with acrogeria are wide cranial sutures, blue sclerae, diaphyseal narrowing of long bones, and nail dysplasia. Furthermore, individuals who have acrogeria are not jowly and do not exhibit the joint hyperextensibility and the osteoporotic bony changes seen here.

The syndrome should be differentiated from wrinkly skin syndrome reported by Gazit et al (5). This autosomal recessively inherited disorder is characterized by wrinkled skin of the anterior chest and/or dorsal surfaces of the hands, hypotonia, kyphosis, winged scapulas,

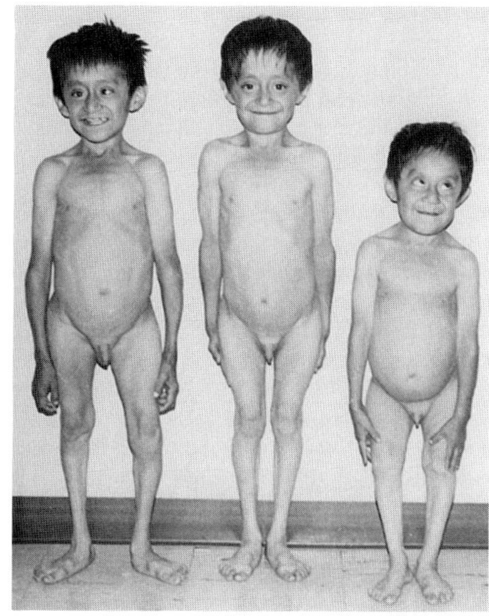

A

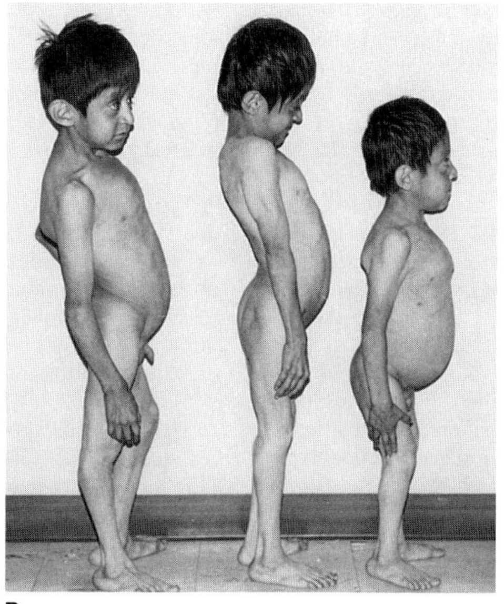

B

Fig. 13–24. *Gerodermia osteodysplastica.* (A,B) Six-, seven-, and eight-year-old sibs with saggy cheeks, premature wrinkling of skin of face, abdomen, dorsum of hands and feet, stooped posture, winged scapulae, and flat feet. (From *R Lisker et al,* Am J Med Genet **3**:389, 1979.)

and prominant cutaneous venous pattern. There are also microcephaly, severe myopia, and chorioretinitis.

The *De Barsy syndrome* characterized by congenital athetosis, mental retardation, severe growth retardation, and laxity of skin and ligaments must also be excluded. *Lenz–Majewski syndrome* and *Patterson syndrome (pseudoleprechaunism)* have wrinkly skin as one of their components but the phenotypes of those disorders are so distinct that discussion is not relevant.

### References [Gerodermia osteodysplastica (Bamatter syndrome)]

1. Bamatter F et al: Gérodermie ostéodysplastique héréditaire. Un nouveau biotype de la ''progeria''. *Confin Neurol* **9**:397, 1949.

Fig. 13–25. *Gerodermia osteodysplastica.* Facies showing premature aging. (Courtesy of *D Klein,* Geneva, Switzerland.)

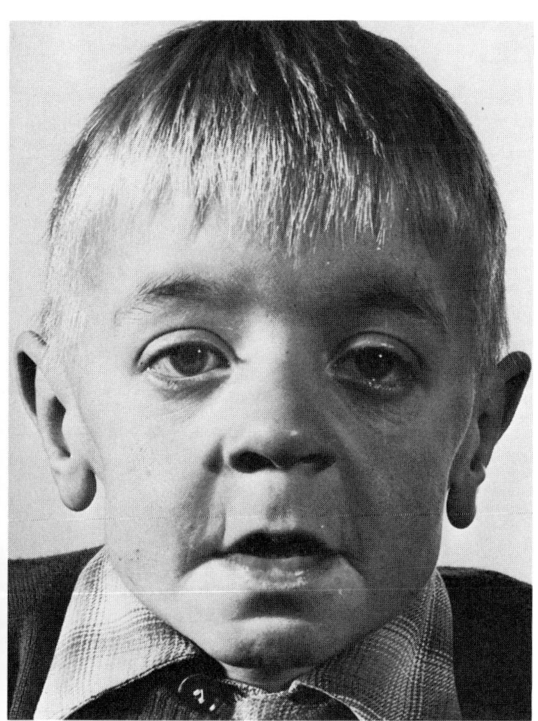

2. Bamatter F et al: Gérodermie ostéodysplastique héréditaire. *Ann Paediatr* **174**:126–127, 1950.

3. Boreaux G: La gérodermie ostéodysplastique a hérédite liée au sexe nouvelle entité clinique et génétique. *J Génét Hum* **17**:137–178, 1969.

4. Brocher JEW et al: Roentgenologische Befunde bei Geroderma ostéodysplastica héréditaria. *Fortschr Roentgenstr* **109**:185–198, 1968.

5. Gazit E et al: The wrinkly skin syndrome: A new heritable disorder of connective tissue. *Clin Genet* **4**:186–192, 1973.

6. Hunter AGW: Is geroderma osteodysplastica underdiagnosed? *J Med Genet* **25**:843–846, 1988.

7. Hunter AGW et al: Geroderma osteodysplastica: Report of two affected families. *Hum Genet* **40**:311–324, 1978.

8. Klein D et al: Une affection liée au sexe: La gérodermie ostéodysplastique héréditaire (20 ans d'observation). *Rev Otoneurophthalmol* **40**:415–421, 1968.

9. Lisker R et al: Gerodermia osteodysplastica hereditaria: Report of three affected brothers and literature review. *Am J Med Genet* **3**:389–395, 1979.

10. Patton MA et al: Congenital cutis laxa with retardation of growth and development. *J Med Genet* **24**:556–561, 1987.

11. Sakati NO, Nyhan WL: Congenital cutis laxa and osteoporosis. *Am J Dis Child* **137**:452–454, 1983.

12. Suter H et al: Geroderma osteodysplastica hereditaria (GOH) in a girl, in *Skeletal Dysplasias,* Papadatos CJ and Bartsocas CS (eds), Alan R. Liss, New York, 1982, pp 327–329.

13. Wiedemann H-R: Geroderma osteodysplastica—what would Virchow have thought about it? *Hum Genet* **43**:245, 1978.

## Ehlers–Danlos syndromes

**Introduction.*** The Ehlers–Danlos syndromes (EDS) constitute a heterogeneous group of generalized connective tissue disorders with skin fragility, skin hyperextensibility, and joint hypermobility. The earliest recognized description was of a young Spaniard, George Albes, who was exhibited to the Academy of Leyden in Holland in 1657 by Van Meek'ren (1,22). The first scientific description was by Chernogubov, a Russian dermatologist, who presented an affected 17-year-old boy to the Moscow Venereology and Dermatology Society meeting in 1892. He attributed the multisystem features to a generalized defect in embryonic formation of connective tissue. Throughout the Soviet medical literature, the condition is known as Chernogubov syndrome (10).

Ehlers (11), in 1901, reported the association of hyperelastic skin, skin hemorrhages, and loose jointedness. Danlos (9), in 1908, added cutaneous pseudotumors and fragility, and Weber and Aitken (32) included subcutaneous spherules. Terminology became especially complex during the early part of the twentieth century. Among descriptive

*References for Introduction appear on p. 440.

features used to name the condition were dermatorrhexis, dermatolysis, Gummihaut, cutis pendula, cutis hyperelastica, and chalasoderma. The eponym Ehlers–Danlos syndrome finally emerged as the name used most commonly in the West.

More than 750 affected individuals had been reported in over 450 articles by 1989. Comprehensive surveys are those of McKusick (22), Beighton et al (1–3), and Hollister (16,17). The reader is referred to the following references for detailed coverage: molecular defects and connective tissue metabolism (5–7,20,24,25,27,28), ultrastructural studies of the skin (4,15,19), histologic studies of the skin (29), collagen studies (13,26,31), cardiovascular abnormalities (8,21), and pregnancy (30).

During the last several years, genetic and biochemical studies have defined more than 10 types of Ehlers–Danlos syndrome, and the molecular basis of several of them has been identified. It is important to specify the type of Ehlers–Danlos syndrome a patient has because the natural history and mode of inheritance differ depending on type. Clinical features, inheritance, and the biochemical defect of various forms are summarized in Table 13–2. Besides types I–X, several other types have been identified but are not numbered as types. Approximately 80% of affected individuals have Type I or Type II. Approximately 10% have Type III, about 4% have Type IV, and about 6% have various other types (3). Type VIII probably cannot be separated from Type IV; and Type IX, now called the *occipital horn syndrome*, is a disorder of copper dysmetabolism.

Many patients with Ehlers–Danlos-like clinical findings do not fit the general classification, and as more biochemical studies are completed, it is likely that the classification will expand. From the clinical perspective, important considerations are the mode of inheritance and whether the natural history can be predicted from family or biochemical studies. It is vital to distinguish the recessively inherited form of Type VI because prenatal diagnosis is potentially available. It is also important to identify patients with Type IV so that clear discussion of pregnancy risks and problems of surgery can take place in a nonemergent setting. The initial evaluation of each patient should include careful physical examination and detailed family history. Biochemical studies of collagens synthesized by fibroblasts cultured from dermal punch biopsies can identify patients with Type IV, Type VI, and some patients with Type VII. Defects in collagen processing or of the structure or synthesis of other matrix macromolecules will probably be identified in patients with Types I and II in the future, and research in these areas is underway. Several laboratories in the United States, Great Britain, Europe, and Australia can provide the laboratory investigations needed to characterize the various types of Ehlers–Danlos syndrome.

**Types I (severe form) and II (mild form).** Types I (severe form) and II (mild form) have autosomal dominant inheritance. Only rarely are they characterized by major complications. The clinical findings in both types may be dramatic with markedly soft, velvety, hyperextensible skin, impressive joint hypermobility, easy bruising, and thin, atrophic, "cigarette-paper" scars. Varicose veins are common. Birth prevalence of these two disorders is estimated to be approximately 1/10,000–20,000 (4,24).

The biochemical bases of Types I and II are not known. Early studies suggested that the solubility of dermal collagens was increased (17) and that skin tensile strength, contributed largely by collagen, was decreased (4). More recently, morphologic studies have shown that the architecture of collagen fibrils in skin is unusual. Collagen fibril diameter is larger than normal and there are multiple composite fibrils (29,35). Shinkai et al (30) suggested that the cellular processing of Type I collagen was not normal. The studies of PH Byers (personal communication, 1987) suggest that cleavage of the COOH-terminal propeptide extension may be far slower than normal in cell culture. These findings are compatible with the suggestion that Type I procollagen, which retains COOH-terminal propeptide extensions, is unable to form normal fibril structures (12). To date, linkage studies have not located the genes for Types I and II but the genes are probably allelic. A family with reduced amounts of Type III collagen has been documented (9b). A form even milder than the mitis type is seen in 10% of the population (20a).

**General features.** Prematurity, due to early rupture of fetal membranes, is observed in as many as half of Type I cases (1,31–33) and to a lesser degree in Type II cases. In addition, there may be postpartum hemorrhage, hemorrhage following episiotomy or lacerations, and uterine or bladder prolapse (32). Children with Types I or II EDS are often thought to have joint hypermobility as newborns, but may not

Table 13–2. Clinical features, modes of inheritance, and biochemical defects in Ehlers–Danlos syndromes

| Type | Clinical features | Inheritance | Biochemical defect |
|---|---|---|---|
| I: Gravis | Soft velvety, hyperextensible skin, easy bruising, "cigarette paper" scars, hypermobile joints, varicose veins, prematurity | AD | Not known |
| II: Mitis | Similar to Ehlers–Danlos Type I, but less severe | AD | Not known |
| III: Familial hypermobility | Soft skin, no scarring, marked large and small joint hypermobility | AD | Not known |
| IV: Arterial | Thin, translucent skin with visible veins, marked bruising, skin and joints have normal extensibility, arterial, bowel, and uterine rupture | AD, AR | Abnormal Type III collagen synthesis, secretion, or structure |
| V: X linked | Similar to Ehlers–Danlos Type II | XLR | Not known |
| VI: Ocular | Soft, velvety, hyperextensible skin, hypermobile joints, scoliosis, ocular fragility and keratoconus | AR | Lysyl hydroxylase deficiency |
| VII: Arthrochalasis multiplex congenita | Congenital hip dislocation, joint hypermobility, soft skin with normal scarring | AD | Structural mutation at amino-terminal cleavage site in pro$\alpha$1(I) and pro$\alpha$2(I); defective aminopeptidase |
| VIII: Periodontal[a] | Generalized periodontitis, skin similar to Ehlers–Danlos Type II | AD | Not known |
| IX: Occipital horn syndrome[b] | Soft extensible, lax skin, bladder diverticulae and rupture, short arms, limited pronation, and supination, broad clavicles, occipital horns | XLR | Abnormal copper utilization with defect in lysyl oxidase |
| X: Fibronectin | Similar to Ehlers–Danlos Type II | AR | Defect in fibronectin |

[a]Probably does not exist as form separate from Type IV.

[b]Now separated as form of copper dysmetabolism, see pp. 425–426.

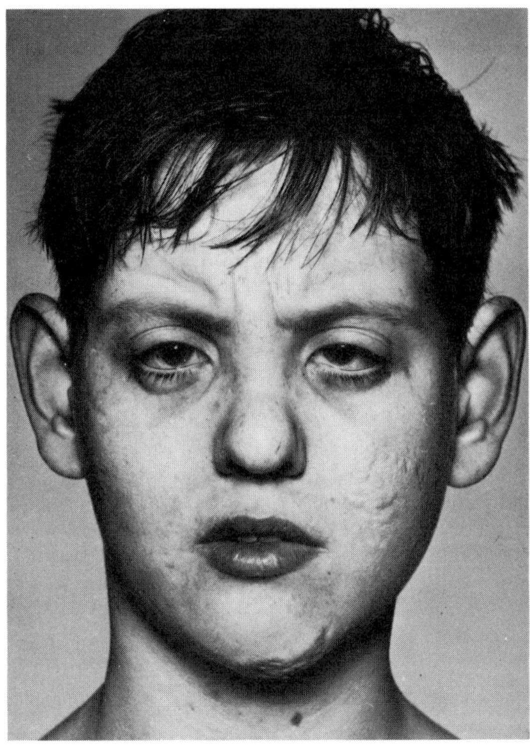

Fig. 13–26. *Ehlers–Danlos syndrome. Type I.* Numerous "cigarette paper" scars of face, epicanthal folds, flat nasal bridge. (From *GM Barabas* and *AP Barabas,* Br Dent J **123**:473, 1967.)

Fig. 13–27. *Ehlers–Danlos syndrome, Type I.* Hyperelastic skin returns to normal position after being stretched. (From *P Beighton,* The Ehlers–Danlos Syndrome, Heinemann, London, 1970.)

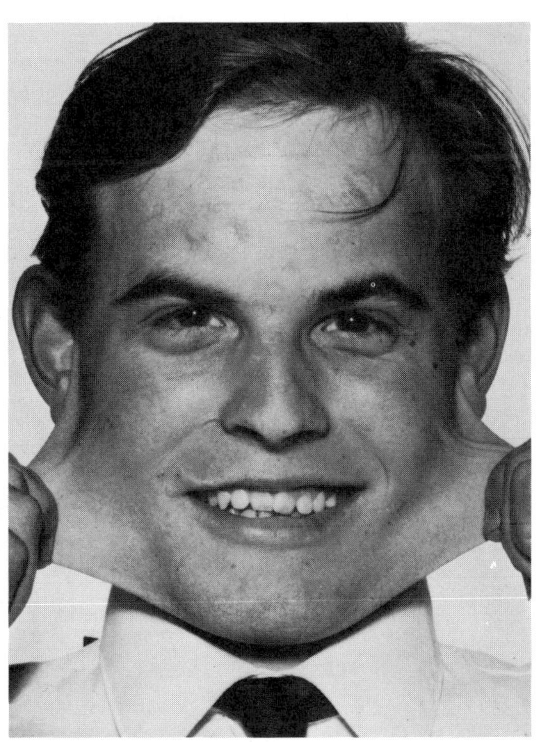

come to medical attention until about 1 year when they are noted to be slow to walk because of poor joint stability. Susceptibility to skin tearing, forming of thin scars, and easy bruising become apparent. Referral for evaluation for child abuse is consequently not uncommon (25). Evidence of mitral valve prolapse by clinical criteria or by echocardiography may be present in childhood and certainly by adolescence.

Early onset of joint symptoms occurs, apparently resulting in premature osteoarthritis from mechanical stress on joints produced by excessive mobility and mild joint instability. There is often concern that routine surgery is hazardous, but although tissue is slightly more friable than normal, careful attention to detail during surgical procedures makes the likelihood of complications minimal. Sutures are generally left in place two or three times longer than usual to ensure adequate healing. Life expectancy for individuals with Type I or Type II Ehlers–Danlos syndrome is normal (4,19,20,24).

Skin. The skin has a velvety feel and is hyperelastic, especially over the major joints. After being stretched, it returns to its normal position. The skin is thinner than normal, brittle, and fragile (15). Minimal trauma may produce gaping wounds. After healing, pigmented papyraceous scars result. These can usually be detected over the forehead, chin, knees, shins, and/or elbows (Figs. 13–26, 13–27, and 13–29). Bruising is variable in degree. Goodman et al (14) demonstrated secondary skin creases over the palms.

Molluscoid pseudotumors over the heels and major joints are not uncommon and often the skin is redundant on the hands and feet (9,21) (Fig. 13–29). Calcified, cyst-like structures, 2–10 mm in diameter, may be found subcutaneously, especially over bony prominences of the forearms and shins in about 30% (7). Varicose veins are common and acrocyanosis has been noted with increased frequency.

Musculoskeletal system. Hyperextensibility of joints, weak hand clasp, and pes planus are usually present (6,10,36) (Fig. 13–28). Genu recurvatum has been noted in about 25%, and there may be recurrent joint dislocations. Talipes has been found in approximately 5%. The ulnar styloid process may be elongated. Kyphoscoliosis and thoracic asymmetry are observed in 15–20% but usually are not severe (7,10). Hernia, either inguinal or umbilical, has been found in 10–20% (8).

Cardiovascular defects. Leier et al (23) studied 16 patients with Type I. Roentgenograms, electrocardiograms, and echocardiograms were done. Many underwent cardiac catheterization. Most patients had mi-

Fig. 13–28. *Ehlers–Danlos syndrome, Type I.* Extreme joint hypermobility. (From *GM Barabas* and *AP Barabas,* Br Dent J **123**:473, 1967.)

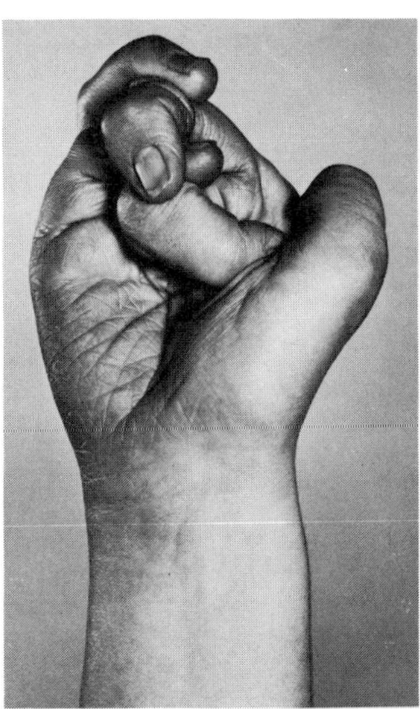

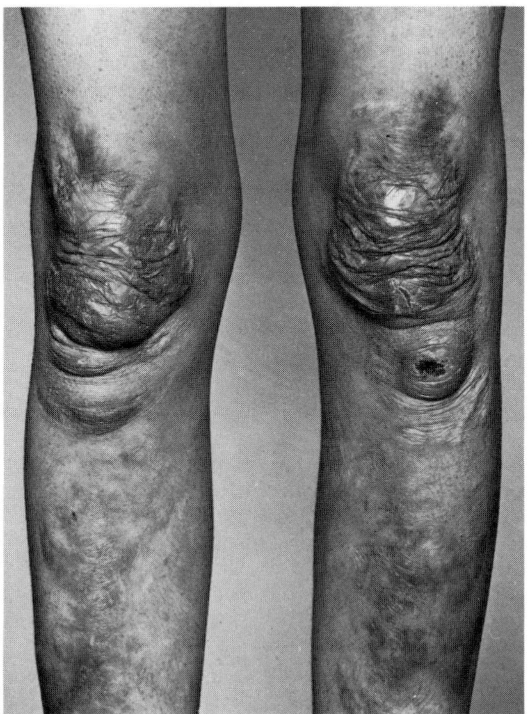

Fig. 13–29. *Ehlers–Danlos syndrome, Type I.* Papyraceous scars of knees with pseudotumor below left knee. (From *GM Barabas* and *AP Barabas,* Br Dent J **123**:473, 1967.)

Fig. 13–30. *Ehlers–Danlos syndrome, Type I.* Easy eversion of upper lids (Méténier sign). (From *P Beighton,* The Ehlers–Danlos Syndrome, Heinemann, London, 1970.)

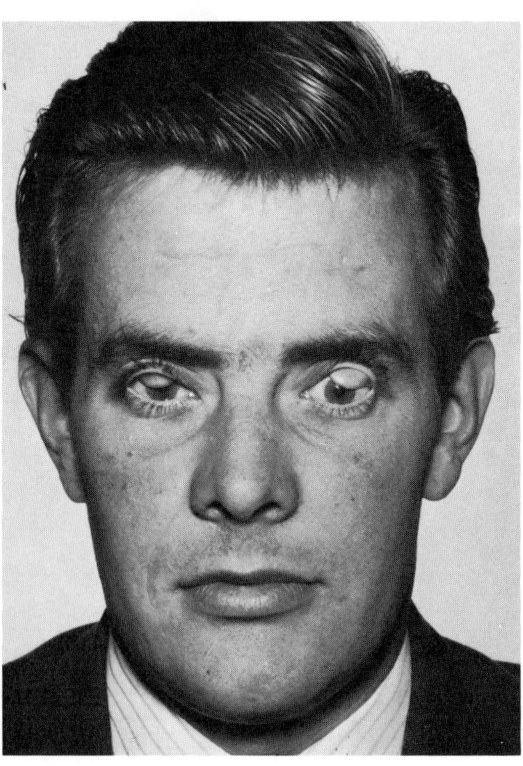

tral valve prolapse and six had tricuspid valve prolapse. Dilatation of the aortic root or ectasia of the sinuses of Valsalva, or both, also occurred in six. Congenital heart defects included two cases of bicuspid aortic valve, one instance of pulmonic valvular stenosis, two VSDs, and one ASD.

Craniofacial features. Epicanthic folds are seen in about 25% (5,9,21,24). Blue sclerae are noted in less than 10% (9,10,21,22). Myopia and strabismus have been observed with far greater frequency than blue sclerae. Also noted have been microcornea, retinal detachment, keratoconus, and angioid streaks (16,26). Hyperextensibility of skin allows for easy eversion of the upper eyelids (Méténier's sign) (4) (Fig. 13–30). The pinnae may project outward (11).

Approximately 50% can touch the nose with the tip of the tongue (Gorlin's sign), an ability found in only 8–10% of the general population (Fig. 13–31C). The oral mucosa is fragile and easily bruised. Oral healing may be slightly retarded, since the edges of a wound draw apart, but there is no evidence of abnormal healing or excessive scar formation in the mouth (3). The gingiva are more liable to injury, and periodontal disease has been reported at an earlier age (9,21). However, it is possible that one of these patients might have had Type IV EDS (9) whereas the other child might merely have exhibited gingivitis (21). Recurrent subluxation of the temporomandibular joint has been reported (6,13,34).

Barabas and Barabas (3) found that premolar and molar teeth had high cusps and deep occlusal fissures. Roentgenographically, teeth may have stunted and deformed roots and large pulp stones in the coronal part of the pulp chamber (2,3,14a,18,27,28) (Fig. 13–31A,B). Barabas (2) found hypoplastic areas in enamel, irregularities of amelodentinal and cementodentinal junctions, and formation of pathologic dentin, more frequently in the root than in the crown containing vascular inclusions, abnormal dentinal tubules, and many denticles. Carr and Green (9a) found multiple odontogenic keratocysts in Type II.

### References [Types I (severe form) and II (mild form)]


1. Barabas AP: Ehlers–Danlos syndrome associated with prematurity and premature rupture of foetal membranes; possible increase in incidence. *Br Med J* **2**:682–684, 1966.

2. Barabas GM: The Ehlers–Danlos syndrome: Abnormalities of the enamel, dentine, cementum and the dental pulp: A histological examination of 24 teeth from 6 patients. *Br Dent J* **126**:509–515, 1969.

3. Barabas GM, Barabas AP: The Ehlers–Danlos syndrome: A report of the oral and haematological findings in nine cases. *Br Dent J* **123**:473–479, 1967.

4. Beighton P: *The Ehlers–Danlos Syndrome,* Heinemann, London, 1970.

5. Beighton P: Serious ophthalmological complications in the Ehlers–Danlos syndrome. *Br J Ophthalmol* **54**:263–268, 1970.

6. Beighton P, Horan F: Orthopaedic aspects of the Ehlers–Danlos syndrome. *J Bone Jt Surg* **51B**:444–453, 1969.

7. Beighton P, Thomas ML: The radiology of the Ehlers–Danlos syndrome. *Clin Radiol* **20**:354–361, 1969.

8. Beighton P et al: Gastrointestinal complications of the Ehlers–Danlos syndrome. *Gut* **10**:1004–1008, 1969.

9. Benjamin B, Weiner H: Syndrome of cutaneous fragility and hyperelasticity and articular hyperlaxity. *Am J Dis Child* **65**:246–257, 1943.

9a. Carr RJ, Green DM: Multiple odontogenic keratocysts in a patient with type II (mitis) Ehlers–Danlos syndrome. *Br J Oral Maxillofac Surg* **26**:205–214, 1988.

9b. DePaepe A et al: Ehlers–Danlos syndrome type I: A clinical and ultrastructural study of a family with reduced amounts of collagen type III. *Br J Dermatol* **117**:89–97, 1987.

10. Durham DG: Cutis hyperelastica (Ehlers–Danlos syndrome) with blue scleras, microcornea and glaucoma. *Arch Ophthalmol* **49**:220–221, 1953.

11. Ellis FE, Bundick WR: Cutaneous elasticity and hyperelasticity. *Arch Dermatol* **74**:22–32, 1956.

12. Fleischmajer R et al: Collagen fibrillogenesis in human skin. *Ann NY Acad Sci* **460**:246–257, 1985.

13. Goodman RM, Allison ML: Chronic temporomandibular joint subluxation in Ehlers–Danlos syndrome. *J Oral Surg* **27**:659–661, 1969.

14. Goodman RM et al: Evolution of palmar skin creases in the Ehlers–Danlos syndrome. *Clin Genet* **3**:67–72, 1972.

14a. Gosney MBE: Unusual presentation of a case of Ehlers–Danlos syndrome. *Br Dent J* **163**:54–56, 1987.

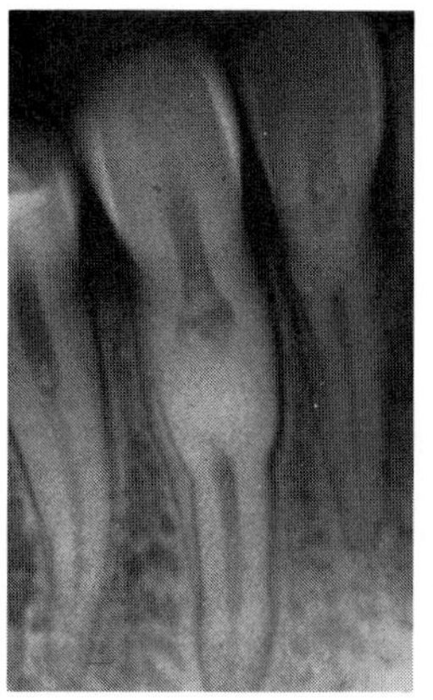
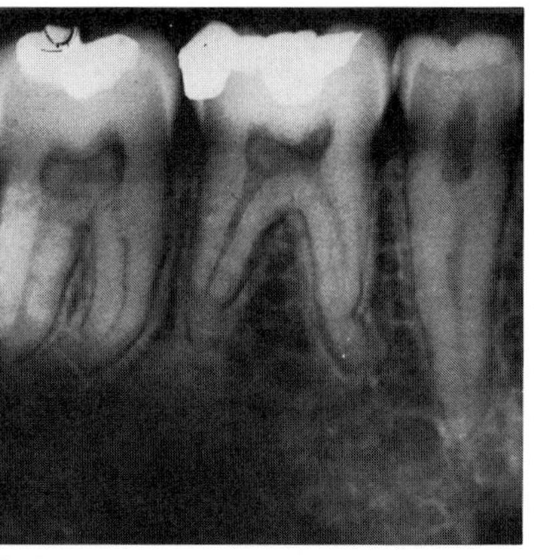
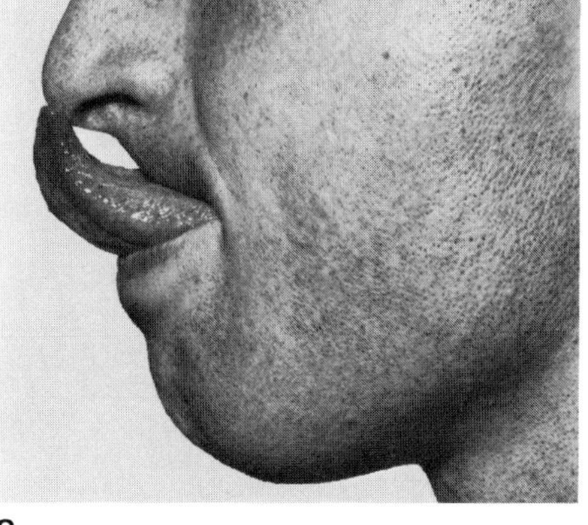

A B C

Fig. 13–31. *Ehlers–Danlos syndrome, Type I.* (A,B) Stunted and deformed roots and large pulp stones in coronal portion of pulp chamber. (C) Ability to touch nose with tongue tip, present in 50% of those with Ehlers–Danlos syndrome, occurs in less than 10% of normal persons. (A,B from *JJ Pindborg,* Pathology of the Hard Dental Tissues, W.B. Saunders, Philadelphia, 1970; C from *P Beighton,* The Ehlers–Danlos Syndrome. Heinemann, London, 1970.)


15. Grahame R, Beighton P: Physical properties of the skin in the Ehlers–Danlos syndrome. *Ann Rheum Dis* **28**:246–251, 1969.

16. Green WR et al: Angioid streaks in Ehlers–Danlos syndrome. *Arch Ophthalmol* **76**:197–204, 1966.

17. Harris ED, Sjoerdsma A: Collagen profile in various conditions. *Lancet* **2**:707–709, 1966.

18. Hoff M: Dental manifestations in Ehlers–Danlos syndrome. *Oral Surg* **44**:864–871, 1977.

19. Hollister DW: Heritable disorders of connective tissues: Ehlers–Danlos syndrome. *Ped Clin N Am* **25**:575–591, 1978.

20. Hollister DW et al: Genetic disorders of collagen metabolism. *Adv Hum Genet* **12**:1–87, 1982.

20a. Holzberg M et al: The Ehlers–Danlos syndrome: Recognition, characterization, and importance of a milder variant of the classic form. *J Am Acad Dermatol* **19**:656–666, 1988.

21. Johnson SAM, Falls HF: Ehlers–Danlos syndrome: A clinical and genetic study. *Arch Dermatol Syph (Chic)* **60**:82–104, 1949.

22. Kanof A: Ehlers–Danlos syndrome: Report of a case with suggestion of a possible causal mechanism. *Am J Dis Child* **83**:197–202, 1951.

23. Leier CV et al: The spectrum of cardiac defects in the Ehlers–Danlos syndrome, types I and III. *Ann Intern Med* **92:** 171–178, 1980.

24. McKusick VA: *Heritable Disorders of Connective Tissue,* 4th ed., C.V. Mosby, St. Louis, 1972.

25. Owen SM, Durst RD: Ehlers–Danlos syndrome simulating child abuse. *Arch Dermatol* **120**:97–101, 1984.

26. Pemberton JW et al: Familial retinal detachment and the Ehlers–Danlos syndrome. *Arch Ophthalmol* **76**:817–824, 1966.

27. Pindborg JJ: *Pathology of the Dental Hard Tissues,* Saunders, Philadelphia, 1970.

28. Selliseth NE: Odontologische Befunde bei einer Patientin mit Ehlers–Danlos-Syndrom. *Acta Odontol Scand* **23**:91–101, 1965.

29. Sevenich M et al: Ehlers–Danlos syndrome: A disease of fibroblasts and collagen fibrils. *Arch Dermatol Res* **267**:237–251, 1965.

30. Shinkai H et al: Connective tissue metabolism in cultured fibroblasts of a patient with Ehlers–Danlos syndrome type I. *Arch Dermatol Res* **257**:113–122, 1976.

31. Smith CV, Phelan JP: Pregnancy and the Ehlers–Danlos syndrome. *J Reproduct Med* **27**:757–760, 1982.

32. Snyder RR et al: Ehlers–Danlos and pregnancy. *Obstet Gynecol* **61**:649–650, 1983.

33. Taylor DJ et al: EDS syndrome during pregnancy. *Obstet Gynecol Surg* **36**:277–281, 1981.

34. Thexton A: A case of Ehlers–Danlos syndrome presenting with recurrent dislocation of the temporomandibular joint. *Br J Oral Surg* **2**:190–193, 1965.

35. Vogel A et al: Abnormal collagen fibril structure in the gravis form (type I) of the Ehlers–Danlos syndrome. *Lab Invest* **40**:201–206, 1979.

36. Welbury RR: Ehlers-Danlos syndrome: Historical review, report of two cases in one family and treatment needs. *J Dent Child* **56**:220–224, 1989.


**Type III (familial hypermobility).** Type III (familial hypermobility) is an autosomal dominantly inherited disorder characterized by marked joint laxity and soft, but minimally hyperextensible or fragile, skin and minimal bruising. In the newborn, joint laxity is often striking, although dislocation is rare. Infants may be slow to walk because of joint laxity. Major complications are recurrent dislocations, which may require surgery for stabilization, and early onset degenerative joint disease. Pregnancy and delivery are usually uncomplicated (13). Approximately 10% of Ehlers–Danlos syndrome patients have Type III. The basic defect is unknown (1–3,12).

Formerly, Type III was known as the benign hypermobile variety of Ehlers–Danlos syndrome (6,7). However, there appears to be an increased risk of mitral valve prolapse as an associated complication (5,11). Type III is most commonly confused with normal variant hypermobility found in approximately 4–7% of the general population (4,9,10). Horton et al (8) reported an autosomal dominant joint instability syndrome characterized by dislocation of the hip, patella, and elbows. Type VII and Type Viljoen also have marked joint hypermobility.

#### References [Type III (familial hypermobility)]


1. Beighton P: *The Ehlers–Danlos syndrome,* Heinemann, London, 1970.

2. Beighton P et al: Variants of the Ehlers–Danlos syndrome: Clinical, biochemical, haematological, and chromosomal features of 100 patients. *Ann Rheum Dis* **28**:228–245, 1969.

3. Beighton P et al: Dominant inheritance in familial generalized articular hypermobility. *J Bone Jt Surg* **52B**:145–147, 1970.

4. Biro F et al: The hypermobility syndrome. *Pediatrics* **72**:701–706, 1983.

5. Cabeen WR Jr et al: Mitral valve prolapse and conduction defects in Ehlers–Danlos syndrome. *Arch Intern Med* **137**:1227–1231, 1977.

6. Hollister DW: Heritable disorders of connective tissue: Ehlers–Danlos syndrome. *Ped Clin N Am* **25**:575–591, 1978.

7. Hollister DW et al: Genetic disorders of collagen metabolism. *Adv Hum Genet* **12**:1–87, 1982.

8. Horton WA et al: Familial joint instability syndrome. *Am J Med Genet* **6**:221–228, 1980.

9. Jessee EF et al: The benign hypermobile joint syndrome. *Arthritis Rheum* **23**:1053–1056, 1980.

10. Kirk JA et al: The hypermobility syndrome. *Ann Rheum Dis* **26**:419–425, 1967.

11. Leier CV et al: The spectrum of cardiac defects in the Ehlers–Danlos syndrome, types I and III. *Ann Intern Med* **92**:171–178, 1980.

12. McKusick VA: *Heritable Disorders of Connective Tissue,* 4th ed., C.V. Mosby, St. Louis, 1972.

13. Skovgaard N, Pelle J: Pregnancy and delivery in a patient with Ehlers–Danlos syndrome, type III. *Ugeskr Laegr* **148**:2230, 1986.


**Type IV (ecchymotic, arterial, or Sack–Barabas).** Type IV, also known as the ecchymotic, arterial, or Sack–Barabas variety of Ehlers–Danlos syndrome, has the most ominous prognosis because of the predilection for catastrophic bleeding from major arteries and spontaneous perforation of the gastrointestinal tract. There is low birthweight and prematurity (16a). About 50% die before reaching 40 years of age (6). Joint hypermobility is usually limited to the digits, and skin hyperextensibility may be minimal or absent in adults, but congenitally dislocated hips are not uncommon. The skin is characteristically thin and translucent with an easily observed prominent venous network. Minor trauma leads to extensive ecchymosis, resulting in scarring over bony prominences. In childhood, "battering" may be suspected. Musculoskeletal and orthopedic problems do not occur in Type IV. Hypermobility is restricted to the small joints. The condition is etiologically heterogeneous. Prevalence estimates range from 1/100,000 to less than one in a million (2,5,11). About 35 patients have been reported (6).

Biochemical heterogeneity has been demonstrated and most likely results from different mutations in the same gene rather than in different genes. Type IV results from gene mutations for Type III procollagen (4a,16b,21). Mutations have been found that (1) interfere with secretion of assembled molecules, (2) alter expression of one allele, or (3) lead to increased intracellular degradation of Type III collagen molecules. Superti-Furga et al (22a,22b) described multiple exon deletion resulting in heterozygosity at the COL3A1 locus. Various skin changes have been observed, depending on the nature of the mutation. Small collagen fibrils and a mixed population of fibril sizes may be present. In patients with defects in secretion, marked dilatation of the rough endoplasmic reticulum is found in fibroblasts (5,7,14,15,21,25). Radioimmunoassay has revealed low levels of procollagen Type III aminopropeptide in a subgroup of patients with EDS Type IV (20a).

Both autosomal dominant (1,2,6,16,25) and autosomal recessive (1,15,16,22) forms have been observed. Consanguinity was recorded in a family of two affected sibs reported by Sulh et al (22). In 1979, Byers et al (4) proposed a preliminary classification of Type IV with subtypes A, B, C, and D. Subtype IVA has autosomal dominant inheritance and a life expectancy of 30–50 years (1). Further heterogeneity is known; Tsipouras et al (26) demonstrated different mutations affecting the stability and secretion of pro$\alpha$1 (III) chains of Type III procollagen in two separate families, both having autosomal dominant transmission. Subtype IVB has autosomal recessive inheritance (1,15) and a life expectancy of 15–30 years. Subtype IVC is represented by a sporadic case in one family; extreme dilatation of the rough endoplasmic reticulum of dermal fibroblasts was noted (4). Subtype IVD (4), now thought to have autosomal recessive inheritance (22), is associated with collagen fibrils that are small in diameter but vary in size. As Byers et al (4) recognized in 1979, the classification is heuristic, and it should be anticipated that further studies will result in significant changes. Hartsfield et al (6a) and Beighton (2) found periodontitis in a family with Type IV, raising the question of the validity of Type VIII.

Temple et al (25), using antihuman Type III collagen antibodies, demonstrated the collagen deficit by immunofluorescence.

General features. The skin is thin, inextensible, and translucent (30%), showing a venous pattern over the trunk, abdomen, and extremities, and early aging. Minimal joint hypermobility is evident in 30%, being limited to the small joints of the hands and feet. Bruising is marked, and extensive ecchymoses may follow even minor trauma in 65% (6). Bony prominences are covered with thin, darkly pigmented scars which differ from those observed in Types I and II (Fig. 13–32C). The facies is characterized by large appearing eyes, a somewhat pinched nose, and thin lips, the so-called "acrogeric facies" (25) (Fig. 13–32A). Often, the skin over the face has a parchment-like appearance. Large varicose veins may be present (Fig. 13–32B). Patients are at greatly increased risk for arterial aneurysm or dissection (80%), arteriovenous fistulas (20%), spontaneous rupture of the colon (10%), and rupture of the gravid uterus (6,18). Recurrent abdominal pain without major findings is common and may result from mural

Fig. 13–32. *Ehlers–Danlos syndrome, Type IV.* (A) Note prominent eyes and outstanding ears. (B) Varicosity of veins. (C) Ecchymotic type showing severe bruising following minor trauma over knees and shins. Elbows also commonly involved. Skin and joint hyperextensibility are less marked than in severe type (see Table 13–2). (A,B from *HMB Sulh et al,* Clin Genet **25**:278, 1984; C from *P Beighton,* The Ehlers–Danlos Syndrome, Heinemann, London, 1970.)

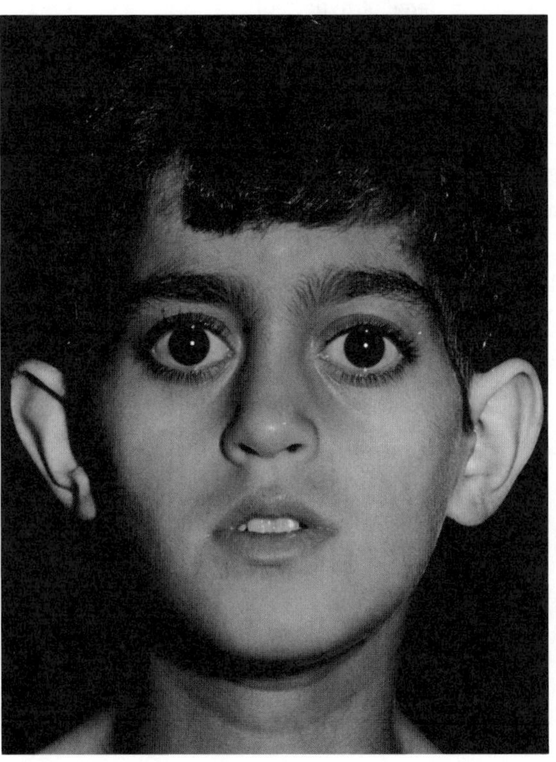
A

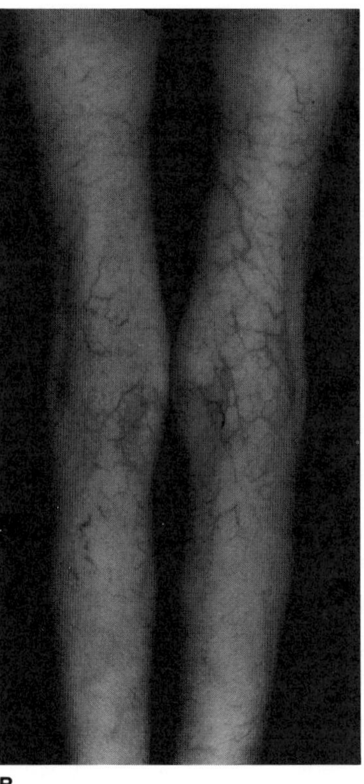
B

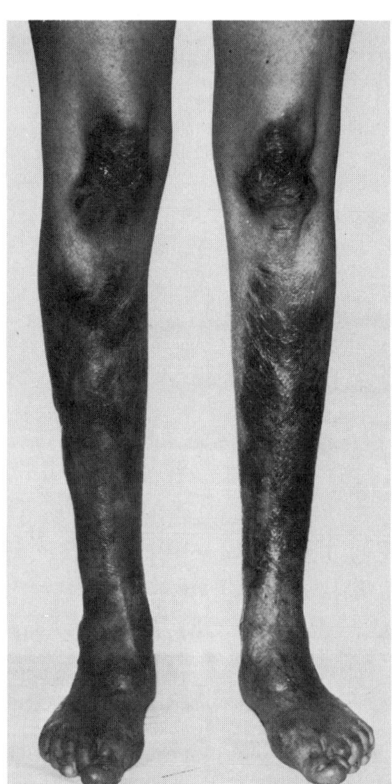
C

hemorrhage in the small bowel. The location of arterial hemorrhage determines the presenting symptoms in some patients (stroke, abdominal bleeding, limb compartmental syndrome), whereas bowel rupture is the first complication in others. Removal of the distal two-thirds of the colon, the most common site of rupture, may decrease the likelihood of recurrence. Over 40% die prior to surgery (6). Surgical attempts at vascular repair have been especially unsuccessful due to the friability of vessel walls and other tissues. Usually following prolonged intraoperative attempts to achieve hemostasis, arteries are simply ligated. Conservative treatment, whenever possible, is recommended. Because of the extreme vascular fragility, angiographic studies should be carried out only in the most dire circumstances since complications occur in about 65% (6). The dramatic, deceptive, and deadly aspects of Type IV have been well discussed in the surgical literature. The extreme importance of proper diagnosis of Type IV is especially evident when dealing with an undiagnosed patient. In such instances, rupture of an appendix or Meckel's diverticulum, perforation of a bowel by a foreign body, or pelvic inflammatory disease is suspected during surgery for lower abdominal pain. Hemorrhagic tendencies lead to unnecessary and expensive coagulopathy studies. The friability of tissues, development of hematomas, excessive bleeding encountered during surgery, and partial dehiscence of sutured wounds are perplexing to the surgeon (2,3,8,9,11,20,23).

Life span is generally shortened. Death has been reported from 12 to 60 years, most commonly during the 20s and 30s. Causes of death include exsanguination from major artery rupture, sepsis from bowel rupture, and shock from uterine rupture (2,4,11).

The complications of pregnancy, in addition to vascular rupture and uterine rupture, include tearing of vaginal tissues during delivery (1,12,13,18,24). Life-threatening complications may occur in 15–20%. In undiagnosed children, easy bruising with minimal or no trauma has resulted in suspicion of child abuse (17).

Other findings. Other findings have been noted on occasion including infantile polycystic kidneys (10), keratoconus, and periodontal disease (2,6a). The reader is referred to EDS Type VIII for a discussion of periodontal disease in EDS.

### References [Type IV (ecchymotic, arterial, or Sack–Barabas)]


1. Barabas AP: Heterogeneity of the Ehlers–Danlos syndrome: Description of three clinical types and a hypothesis to explain the basic defects. *Br Med J* **2**:612–613, 1967.

2. Beighton P: *The Ehlers–Danlos syndrome,* Heinemann, London, 1970.

3. Burnett HF et al: Abdominal aortic aneurysmectomy in a 17-year-old patient with Ehlers–Danlos syndrome: Case report and review of the literature. *Surgery* **74**:617–620, 1973.

4. Byers PH et al: Clinical and ultrastructural heterogeneity of type IV Ehlers–Danlos syndrome. *Hum Genet* **47**:141–150, 1979.

4a. Byers PH et al: Altered secretion of type III collagen in a form of type IV Ehlers–Danlos syndrome. *Lab Invest* **44**:336–341, 1981.

5. Byers PH et al: Type IV Ehlers–Danlos syndrome, in *Proceedings of the Workshop on Heritable Disorders of Connective Tissue,* Akeson W, Bornstein P, Glimcher MJ (eds), C.V. Mosby, St. Louis, 1982.

6. Cikrit DF et al: Spontaneous arterial perforation: The Ehlers–Danlos specter. *J Vasc Surg* **5**:248–255, 1987.

6a. Hartsfield JK et al: Periodontitis in Ehlers–Danlos syndrome type IV. *Am J Hum Genet* **41**(3):A67, 1987.

7. Holbrook KA, Byers, PH: Ultrastructural characteristics of the skin in a form of Ehlers–Danlos syndrome type IV: Storage in the rough endoplasmic reticulum. *Lab Invest* **44**:336–341, 1981.

8. Imahori S et al: Ehlers–Danlos syndrome with multiple arterial lesions. *Am J Med* **47**:967–977, 1969.

9. Lach B et al: Spontaneous carotid-cavernous fistula and multiple arterial dissections in type IV Ehlers–Danlos syndrome. *J Neurosurg* **66**:462–467, 1987.

10. Mauseth R et al: Infantile polycystic disease of the kidneys and Ehlers–Danlos syndrome in an 11-year-old patient. *J Pediatr* **90**:81–83, 1977.

11. McKusick VA: *Heritable Disorders of Connective Tissue,* 4th ed., C.V. Mosby, St. Louis, 1972.

12. Peaceman AM, Cruikshank DP: Ehlers–Danlos syndrome and pregnancy: Association of type IV disease with maternal death. *Obstet Gynecol* **69**:428–431, 1987.

13. Pearl W, Spicer M: Ehlers–Danlos syndrome. *South Med J* **74**:80–81, 1981.

14. Pope FM et al: Patients with Ehlers–Danlos syndrome type IV lack type III collagen. *Proc Natl Acad Sci USA* **72**:1314–1316, 1975.

15. Pope FM et al: Inheritance of Ehlers–Danlos type IV syndrome. *J Med Genet* **14**:200–204, 1977.

16. Pope FM et al: EDS IV (acrogeria): New autosomal dominant and recessive types. *J R Soc Med* **73**:180–186, 1980.

16a. Pope FM et al: Clinical presentations of Ehlers–Danlos syndrome, type IV. *Arch Dis Childh* **63**:1016–1025, 1988.

16b. Pyeritz RE et al: Ehlers–Danlos syndrome IV due to a novel defect in type III procollagen. *Am J Med Genet* **19**:607–622, 1984.

17. Roberts DLL et al: Ehlers–Danlos syndrome type IV mimicking nonaccidental injury in a child. *Br J Dermatol* **111**:341–345, 1984.

18. Rudd NL et al: Pregnancy complications in type IV Ehlers–Danlos syndrome. *Lancet* **1**:50–53, 1983.

19. Snyder RR et al: Ehlers–Danlos syndrome and pregnancy. *Obstet Gynecol* **61**:649–650, 1983.

20. Sparkman RS: Ehlers–Danlos syndrome type IV: Dramatic, deceptive and deadly. *Am J Surg* **147**:703–704, 1984.

20a. Steinmann B et al: Ehlers–Danlos syndrome type IV: A subset of patients distinguished by low serum levels of the amino-terminal propeptide of type III procollagen. *Am J Med Genet* **34**:68–71, 1989.

21. Stolle CA et al: Synthesis of an altered type III procollagen in a patient with type IV Ehlers–Danlos syndrome. *J Biol Chem* **260**:1927–1944, 1985.

22. Sulh HMB et al: Ehlers–Danlos syndrome type IVD: An autosomal recessive disorder. *Clin Genet* **25**:278–287, 1984.

22a. Superti-Furga A et al: Ehlers–Danlos syndrome type IV: A multiple exon deletion is one of the two COL1A3 alleles affecting structure, stability, and processing of type III procollagen. *J Biol Chem* **263**:6226–6232, 1988.

22b. Superti-Furga A et al: Molecular defects of type III procollagen in Ehlers-Danlos syndrome type IV. *Hum Genet* **82**:104–108, 1989.

23. Sykes EM Jr,: Colon perforation in Ehlers–Danlos syndrome. Report of two cases and review of the literature. *Am J Surg* **147**:410–413, 1984.

24. Taylor DJ et al: Ehlers–Danlos syndrome during pregnancy: A case report and review of the literature. *Obstet Gynecol Surv* **36**:277–281, 1981.

25. Temple AS et al: Detection of type III collagen in skin fibroblasts from patients with Ehlers–Danlos syndrome type IV by immunofluorescence. *Br J Dermatol* **118**:17–26, 1988.

26. Tsipouras P et al: Ehlers–Danlos syndrome type IV: Cosegregation of the phenotype to a COL2A1 allele of type III procollagen. *Hum Genet* **74**:41–46, 1986.


**Type V (X-linked).** In 1968, Beighton (1) first reported this entity in two English families. They were reexamined in 1985 (3). Skin hyperextensibility was striking, but joint hypermobility, most evident in the fingers, was mild. Cutaneous fragility, bruising, and scarring tended to be moderate. Both spheroid and molluscoid pseudotumors were found. There was generalized musculoskeletal weakness (dorsal kyphosis, hernia, pes planus, genua valga). Clinical features were similar to those affected with Type II.

X-linked recessive inheritance was established (1–9). Biochemical studies failed to identify any abnormality in the activity of lysyl oxidase, an enzyme involved in cross-linking of collagen (9), and no other biochemical defect has yet been identified (6). Life span is probably normal and female carriers are asymptomatic (2). Intestinal obstruction and brain abscess were noted in two male sibs. All affected members of the English families had red hair (3).

### References [Type V (X-linked)]


1. Beighton P: X-linked recessive inheritance in the Ehlers–Danlos syndrome. *Br Med J* **3**:409–411, 1968.

2. Beighton P: *The Ehlers–Danlos Syndrome,* Heinemann, London, 1970.

3. Beighton P, Curtis D: X-linked Ehlers–Danlos syndrome type V: The next generation. *Clin Genet* **27**:472–478, 1985.

4. Beighton P et al: Variants of the Ehlers–Danlos syndrome: Clinical, biochemical, haematological, and chromosomal features of 100 patients. *Ann Rheum Dis* **28**:228–245, 1969.

5. Hollister DW: Heritable disorders of connective tissue: Ehlers–Danlos syndrome. *Ped Clin N Am* **25**:575–591, 1978.

6. Hollister DW et al: Genetic disorders of collagen metabolism. *Adv Hum Genet* **12**:1–87, 1982.

7. Kobayasi T et al: Dermal changes in the Ehlers–Danlos syndrome. *Clin Genet* **25**:477–484, 1984.

8. McKusick VA: *Heritable Disorders of Connective Tissue,* 4th ed., C.V. Mosby, St. Louis, 1972.

9. Siegel RC et al: Cross-linking of collagen in the X-linked Ehlers–Danlos type V. *Biochem Biophys Res Commun* **88**:281–287, 1979.

**Type VI (ocular).** Among the various forms of EDS, Type VI has been the most confusing for us to evaluate. It appears to be extremely heterogeneous. Pinnell et al (24), in 1972, described a form of EDS characterized by low skin hydroxylysine and lysyl hydroxylase deficiency. In addition to the classic features of EDS Type I, there were severe scoliosis and fragility of ocular tissues. However, during the past 15 years, a large number of examples have been reported that do not quite fit the definition. Ihme et al (15) suggested three forms of EDS VI: Type A—severe with absent hydroxylysine in skin collagen and low lysyl hydroxylase activity in skin, Type B—clinically similar but with normal skin hydroxylysine and low activity of lysyl hydroxylase in skin fibroblasts, and Type C—principally ocular form with normal biochemical findings. It may be convincingly argued that these groups represent at least three different diseases and thus should not be grouped. However, in most reported cases of Type C, biochemical studies have not been carried out. All types have autosomal recessive inheritance. Intermediate values of lysyl hydroxylase have been found in heterozygotes and prenatal diagnosis is possible (9).

For heuristic purposes we will attempt to define the subtypes as we presently view them.

Both Type A (9,11,17,18,24,30,33) and Type B (8,16,25,27,29) patients exhibit newborn hypotonia, loose jointedness with dislocation, excessively stretchable soft velvety, bruisable, and fragile skin, moderate to severe kyphoscoliosis, and rupture of the globes and/or retinal detachment (Fig. 13–33 to 13–36). Type B patients have macrocephaly (8). We cannot classify the case of Kuming and Joffe (19). It would be appear to be EDS Type IV, but the patient had keratoconus. One family could be either Type A or B (21). Patients may exhibit arterial rupture (33).

Fig. 13–33. *Ehlers–Danlos syndrome, Type VI.* Note scoliosis, arachnodactyly, scarred skin, splayed feet. (From *B Steinmann et al,* Helv Paediatr Acta **30**:255, 1975.)

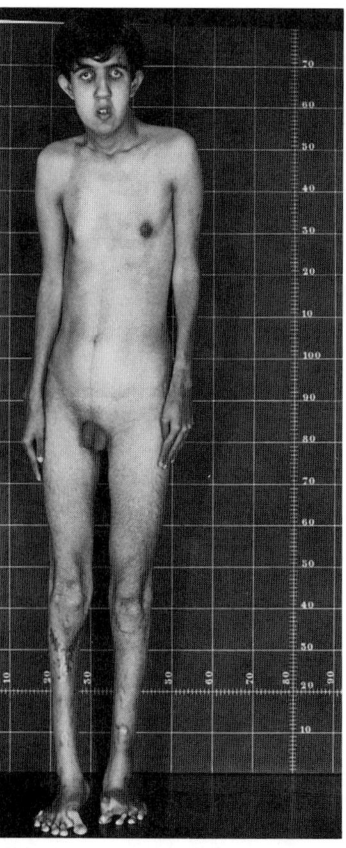

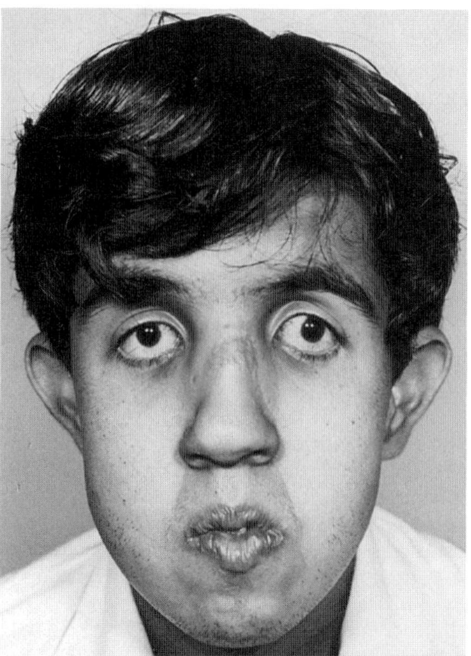

Fig. 13–34. *Ehlers–Danlos syndrome, Type VI.* Microcorneae, ectropion, downward-slanting palpebral fissures, puckered mouth. (From *B Steinmann et al,* Helv Paediatr Acta **30**:255, 1975.)

Often there is delay in motor development. Height is originally normal but scoliosis develops during adolescence and is progressive (33).

Epicanthal folds, blue sclerae, keratoconus, keratoglobus, and angioid streaks are common. Glaucoma (30) manifests in the third decade with retinal detachment in the fourth. The cornea or globes may rupture following minimal trauma. Corneal diameter may be somewhat reduced (29). Moderate myopia is common, as is hearing loss. Joint laxity can be found in a large number of conditions: *Marfan syndrome, osteogenesis imperfecta, Larsen syndrome,* pseudoachondroplasia, and *cartilage–hair hypoplasia.* The osteoporosis–pseudoglioma syndrome has radiologic changes similar to those in osteogenesis imperfecta as well as mild mental retardation, retinal detachment, and ligamentous laxity (20). Possibly the child reported by Biering and Iverson (6) actually had a form of cutis laxa.

There are numerous reports of Type C but again we may be dealing with heterogeneity. Brittle cornea, blue sclera, keratoglobus, keratocornea, and microcornea have been reported as an autosomal recessive disorder in various combination with or without joint laxity, fractures, macrocephaly, hearing loss, and mild mental retardation (1–5,7,10,12–14,22,28,32) (Fig. 13–37A). Red hair has been noted in a few kindred (14,28). Robertson (26) discovered that 50% of his patients with keratoconus had joint hypermobility (Fig. 13–37B). Retinal detachment may rarely be seen in EDS Type I (23).

A possibly new type with associated polyneuropathy has been reported (11a).

### References [Type VI (ocular)]

1. Arkin W: Blue scleras with keratoglobus. *Am J Ophthalmol* **58**:678–682, 1964.

2. Babel J, Houber J: Kératocône et sclérotiques bleues dans une anomalies congénitale du tissu conjonctif. *J Génét Hum* **17**:241–246, 1969.

3. Badtke G: Über einer eigenartigen Fall von Keratokonus und blauen Skleren bei Geschwistern. *Klin Mbl Augenheilkd* **106**:585–592, 1941.

4. Behrens-Baumann W et al: Blaue Sklera-Syndrom und Keratoglobus (okulären Typ des Ehlers–Danlos-Syndroms). *Albrecht Graefes Arch Klin Ophthalmol* **204**:235–246, 1977.

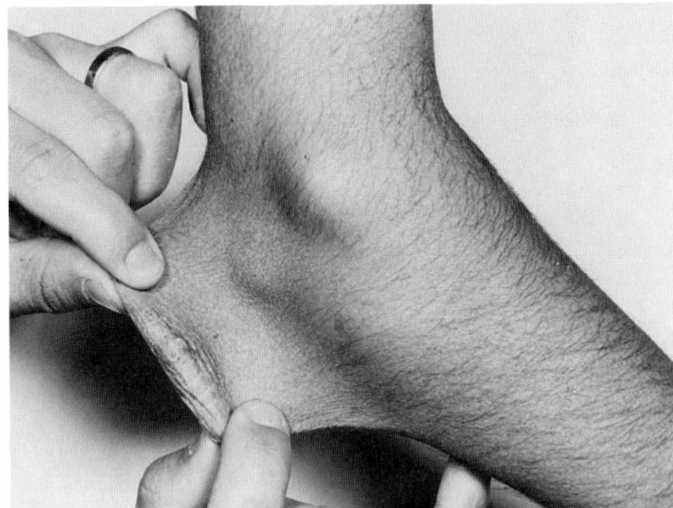
A

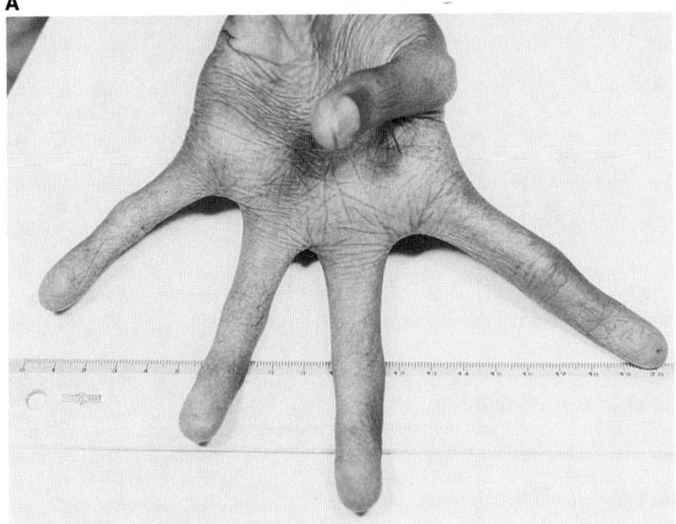
B

Fig. 13-35. *Ehlers–Danlos syndrome, Type VI.* (A,B) Hyperelasticity of skin, arachnodactyly, numerous skin markings. (From *B Steinmann et al,* Helv Paediatr Acta **30**:255, 1975.)

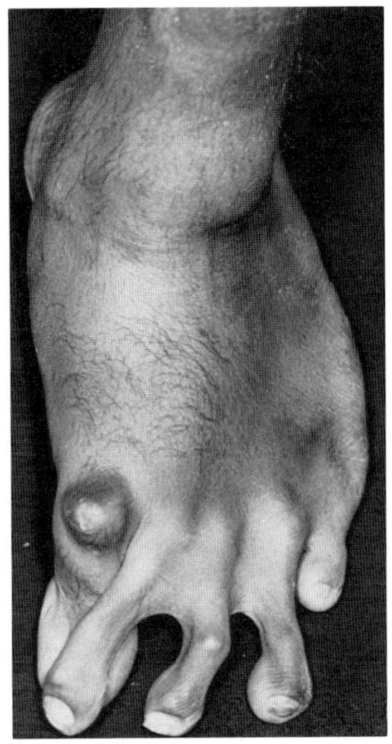

Fig. 13-36. *Ehlers–Danlos syndrome, Type VI.* Splayed toes with pseudotumor. (From *B Steinmann et al,* Helv Paediatr Acta **30**:255, 1975.)


5. Bertelsen TI: Dysgenesis mesodermalis corneae et sclerae. Rupture of both corneae in a patient with blue sclerae. *Acta Ophthalmol Scand* **46**:486–491, 1968.

6. Biering A, Iverson T: Osteogenesis imperfecta associated with Ehlers–Danlos syndrome. *Acta Paediatr* **44**:279–283, 1955.

7. Biglan AW et al: Keratoglobus and blue sclerae. *Am J Ophthalmol* **83**:225–233, 1979.

8. Cadle RG et al: Phenotypic Ehlers–Danlos, types VI with normal lysyl hydroxylase activity and macrocephaly. March of Dimes Birth Defects Conf, 10–13 July 1988, Baltimore, Maryland.

8a. Chamson A et al: Collagen biosynthesis and isomorphism in a case of Ehlers–Danlos syndrome type VI. *Arch Dermatol Res* **279**:303–307, 1987.

9. Dembure PP et al: Genotyping and prenatal assessment of collagen lysyl hydroxylase deficiency in a family with Ehlers–Danlos syndrome type VI. *Am J Hum Genet* **36**:783–790, 1984.

10. Durham DG: Cutis hyperelastica (Ehlers–Danlos syndrome) with blue scleras, microcornea and glaucoma. *Arch Ophthalmol* **49**:220–221, 1953.

11. Elsas LJ II et al: Inherited human collagen lysyl hydroxylase deficiency: Ascorbic acid response. *J Pediatr* **92**:378–384, 1978.

11a. Farag TI, Schimke RN: Ehlers–Danlos syndrome: A new oculo-scoliotic type with associated polyneuropathy? *Clin Genet* **35**:121–124, 1989.


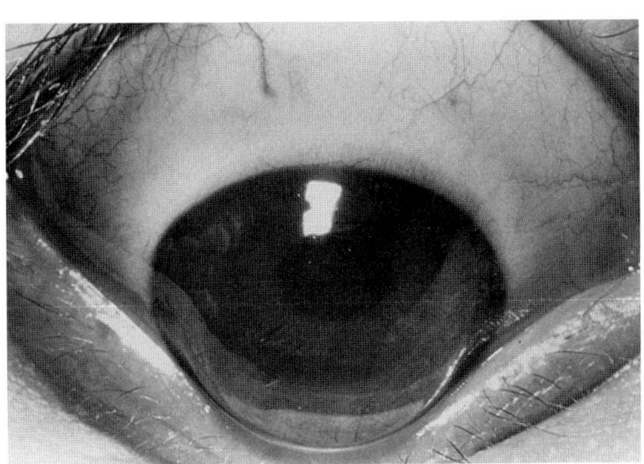
A

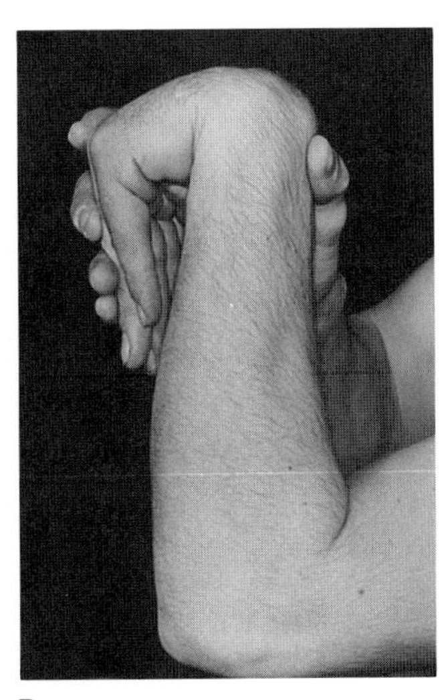
B

Fig. 13-37. *Ehlers–Danlos syndrome, Type VI.* (A) Keratoglobus. (B) Hypermobility at wrist. (A,B from *AW Biglan,* Am J Ophthalmol **83**:225, 1979.)

12. Greenfield G et al: Blue sclerae and keratoconus: Key features of a distinct heritable disorder of connective tissue. *Clin Genet* **4**:8–16, 1973.

13. Gregoratos N et al: Blue sclerae with keratoglobus and brittle cornea. *Br J Ophthalmol* **55**:424–426, 1971.

14. Hyams SW et al: Blue sclerae and keratoglobus. Ocular signs of a systemic connective tissue disorder. *Br J Ophthalmol* **53**:53–58, 1969.

15. Ihme A et al: Biochemical characterization of variants of Ehlers–Danlos syndrome type VI. *Eur J Clin Invest* **13**:357–362, 1983.

16. Judisch GF et al: Ocular Ehlers–Danlos syndrome with normal lysyl hydroxylase activity. *Arch Ophthalmol* **94**:1489–1491, 1976.

17. Krane SM et al: Lysyl-protocollagen hydroxylase deficiency in fibroblasts from siblings with hydroxylysine deficient collagen. *Proc Natl Acad Sci USA* **69**:2899–2903, 1972.

18. Krieg T et al: Biochemical characteristics of Ehlers–Danlos syndrome type VI in a family with one affected infant. *Hum Genet* **46**:41–49, 1979.

19. Kuming BS, Joffe L: Ehlers–Danlos syndrome associated with keratoconus. *S Afr Med J* **52**:403–405, 1977.

20. Neuhäuser G et al. Autosomal recessive syndrome of pseudogliomatous blindness, osteoporosis and mild mental retardation. *Clin Genet* **9**:324–332, 1976.

21. May MA, Beauchamp GR: Collagen maturation defects in Ehlers–Danlos keratopathy. *J Pediat Ophthalmol Strab* **24**:78–81, 1987.

22. Moestrup B: Tenuity of cornea with Ehlers–Danlos syndrome. *Acta Ophthalmol* **47**:704–708, 1969.

23. Pemberton JW et al: Familial retinal detachment and the Ehlers–Danlos syndrome. *Arch Ophthalmol* **76**:814–824, 1966.

24. Pinnell SR et al: A heritable disorder of connective tissue, hydroxylysine-deficient collagen disease. *N Engl J Med* **286**:1013–1020, 1972.

25. Quinn RS, Krane SM: Abnormal property of collagen lysyl hydroxylase from skin fibroblasts of siblings with hydroxylysine-deficient collagen. *J Clin Invest* **57**:83–89, 1976. (Same cases as Ref. 16.)

26. Robertson I: Keratoconus and the Ehlers–Danlos syndrome. A new aspect of keratoconus. *Med J Aust* **1**:571–573, 1975.

27. Sigurdson E et al: The Ehlers–Danlos syndrome and colonic perforation. *Dis Colon Rectum* **28**:962–966, 1985.

28. Stein R: Brittle cornea. A familial trait associated with blue sclera. *Am J Ophthalmol* **66**:67–69, 1968.

29. Steinmann B et al: Ehlers–Danlos syndrome in two siblings with deficient lysyl hydroxylase activity in cultured skin fibroblasts but only mild hydroxylysine deficit in skin. *Helv Paediatr Acta* **30**:255–274, 1975.

30. Sussman M et al: Hydroxylysine deficient skin collagen in a patient with a form of the Ehlers–Danlos syndrome. *J Bone Jt Surg* **56A**:1228–1234, 1974.

31. Thomas C et al: Les alterations oculaires de la maladie d'Ehlers–Danlos. *Arch Ophthalmol* **14**:691–697, 1954.

32. Tucker DP: Blue sclerotics syndrome simulating buphthalmos. *Am J Ophthalmol* **47**:345–348, 1959.

33. Wenstrup RJ et al: Ehlers–Danlos syndrome type VI. Clinical manifestations of collagen lysyl hydroxylase deficiency. *J Pediatr* **115**:405–409, 1989.

**Type VII (arthrochalasis multiplex congenita).** In 1973, Lichtenstein et al (7) identified a distinct type of EDS in three isolated patients. They presented with severe hypermobility of joints with multiple dislocations, especially of the hips, and scoliosis, which in large part accounted for the short stature (Fig. 13–38). The skin was soft, but only mild hyperelasticity and fragility of the skin were observed. The facies was characterized by epicanthal folds, depressed nasal bridge, and micrognathia. Similarly affected isolated examples had been described earlier by Hass and Hass (4). Several affected sibs in a large consanguineous kindred have also been documented (1).

The autosomal dominant forms are based on abnormal Type I procollagen maturation, that is, failure to remove the amino-terminal propeptide. Thus, Type VII has been subclassified into Types VIIA and VIIB depending on whether the pro$\alpha$1(I) or 2(I) chain is affected, respectively (2,3,6,10). Hood et al (5) demonstrated markedly variable clinical expression in individuals with the same structural mutation. Recent work (9) has demonstrated that the types are homogeneous, being caused by abnormal multiexon splicing.

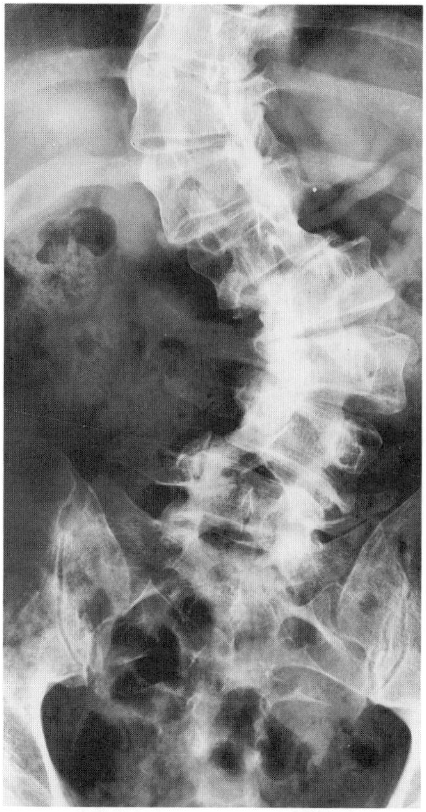

Fig. 13–38. *Ehlers–Danlos syndrome, Type VII.* Severe scoliosis. (Courtesy of *P Beighton,* Cape Town, South Africa.)

### References [Type VII (arthrochalasis multiplex congenita)]

1. Capotorti L, Antonelli M: Sindrome di Ehlers–Danlos. Quatro casi accertati e due probabli in una famiglia con piu matrimoni fra consanguinei. *Acta Genet Med Gemellol* **15**:273–295, 1966.

2. Cole WG et al: The clinical features of Ehlers–Danlos type VII due to a deletion of 24 amino acids from the pro$\alpha$(I) chain of type I procollagen. *J Med Genet* **24**:698–701, 1987.

3. Eyre DR: A heterozygous collagen defect in a variant of the Ehlers–Danlos syndrome type VII. *J Biol Chem* **260**:11322–11329, 1985.

4. Hass J, Hass R: Arthrochalasis multiplex congenita. *J Bone Jt Surg* **40A**:663–674, 1958.

5. Hood OJ et al: Clinical and etiologic heterogeneity in autosomal dominant Ehlers–Danlos syndrome, type VII. *Proc Greenwood Genet Ctr* **7**:170, 1988.

6. Hood OJ et al: Ehlers–Danlos syndrome VII caused by an apparent structural mutation in the carboxy portion of pro$\alpha$2 (I). *Am J Hum Genet* **41**(3):A100, 1987.

7. Lichtenstein JR et al: Defect in conversion of procollagen to collagen in a form of Ehlers–Danlos syndrome. *Science* **12**:298–300, 1973.

8. Prokop DJ et al: The biosynthesis of collagen and its disorders. *N Engl J Med* **301**:13–23, 77–85, 1979.

9. Ramirez F et al: Molecular characterization of the Ehlers–Danlos type VII phenotype. March of Dimes Birth Defects Conf, July, 1988, Baltimore, Maryland.

10. Steinmann B et al: Evidence for a structural mutation of procollagen type I in a patient with the Ehlers–Danlos syndrome type VII. *Eur J Pediatr* **130**:203, 1979, and *J Biol Chem* **255**:8887–8893, 1980.

**Type VIII (periodontal).** In 1972, McKusick (6) first identified an autosomal dominant form of EDS characterized by periodontal disease, scarring of pretibial skin unaccompanied by joint hypermobility, and hyperextensibility of skin. Stewart et al (9) and Hollister et al (3) reported additional kindreds. In one family, marfanoid habitus, unusual facies, severe skin fragility and bruisability, and evidence of visceral involvement (duodenal rupture) were noted. In the other family, normal body habitus and skin fragility but moderate to marked skin stretchability were evident. The family reported by Nelson and King (7) had normal facies and body habitus with only slight joint hypermobility. Other examples are less certain (5,8).

The onset of skin fragility is noted in childhood. Ecchymoses following slight trauma resolve except for the pretibial areas. These heal with tender yellow-brown atrophic wrinkled scars which somewhat

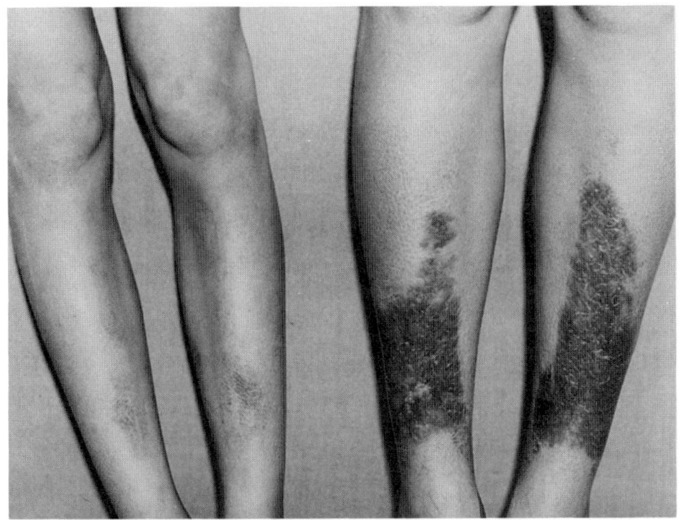

A

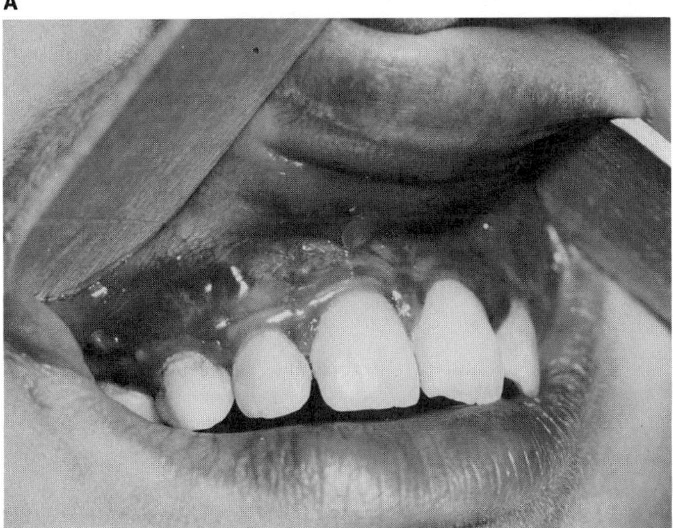

B

Fig. 13–39. *Ehlers–Danlos syndrome, Type VIII.* (A) Ecchymoses of pretibial areas showing variable involvement in sibs. (B) Premature periodontal destruction.

resemble necrobiosis lipoidica diabeticorum or venous stasis (Fig. 13–39A). Scarring becomes worse with age.

The periodontal disease appears after puberty and the permanent teeth are usually lost by the third decade (Fig. 13–39B).

Hartsfield et al (1,2) have questioned the existence of this type, since they found periodontitis in Type IV and Lapière and Nusgens (4) found reduced Type III collagen. *We believe that this type should be eliminated from the classification since all cases can be accepted as examples of Type IV.*

### References [Type VIII (periodontal)]


1. Hartsfield JK Jr et al: Periodontitis in Ehlers–Danlos syndrome type IV. *Am J Hum Genet* **41**(3):A67, 1987.
2. Hartsfield JK Jr et al: Ehlers–Danlos (EDS) types IV and VIII; two separate entities? March of Dimes Birth Defects Conf, July, 1988, Baltimore, Maryland and *Am J Med Genet* (in press).
3. Hollister DW et al: Ehlers–Danlos type VIII. *Clin Res* **28**:99A, 1980.
4. Lapière CM, Nusgens BV: Ehlers–Danlos (ED) type VIII skin has a reduced proportion of collagen type III. *J Invest Dermatol* **76**:422, 1981.
5. Linch DC, Acton CHC: Ehlers–Danlos syndrome presenting with juvenile destructive periodontitis. *Br Dent J* **147**:95–96, 1979.
6. McKusick VA: *Heritable Disorders of Connective Tissue,* C.V. Mosby, St. Louis, 1972, pp 292–371.
7. Nelson DL, King RA: Ehlers–Danlos syndrome type VIII. *J Am Acad Dermatol* **5**:297–303, 1981.
8. Piette E, Douniau R: Paradontolyse infantile symptomatique d'un syndrome d'Ehlers–Danlos, un cas sporadique? *Acta Stomatol Belg* **77**:217–229, 1980.
9. Stewart RE et al: A new variant of Ehlers–Danlos syndrome: An autosomal disorder of fragile skin, abnormal scarring, and generalized periodontitis. *Birth Defects* **13**(3B):85–93, 1977.


**Type X (fibronectin defect).** Arneson et al (1) in 1980 described four sibs affected with joint hyperextensibility, easy bruising, thin hyperextensible skin, and "fish-mouth scarring," but with normal texture. A platelet aggregation deficit resulted in petechiae. Although there was resemblance to EDS Types II and III, the characteristic velvety skin was absent and striae distensae were common. Petechiae were noted with mild upper respiratory tract infections and often with towelling. The authors suggested that the common link to joint hypermobility and platelet malfunction might be defective fibronectin. However, bleeding was not excessive and menses were normal. A narrow pelvis (3) and mitral valve prolapse (1,3) have been noted. It should be pointed out that platelet size and function have been demonstrated to be defective in various types of Ehlers–Danlos syndrome (2,4,5).

### References [Type X (fibronectin defect)]


1. Arneson MA et al: A new form of Ehlers–Danlos syndrome: Fibronectin corrects defective platelet function. *JAMA* **244**:144–147, 1980.
2. Estes JW: Platelet size and function in the heritable disorders of connective tissue. *Ann Intern Med* **68**:1237–1249, 1968.
3. Hammerschmidt DE et al: Maternal Ehlers–Danlos syndrome type X. *JAMA* **248**:2487–2488, 1982.
4. Kashiwagi H et al: Functional and ultrastructural abnormalities of platelets in Ehlers–Danlos syndrome. *Ann Intern Med* **63**:249–254, 1965.
5. Onel D et al: Platelet defect in a case of Ehlers–Danlos syndrome. *Acta Haematol* **50**:238–244, 1973.


**Type Hernández (progeroid form).** Hernández et al (1–3) described another variant of EDS. In addition to joint and skin hyperextensibility, increased bruisability, hernia, and papyraceous scars, the propositi had mental retardation, wrinkled facies, short stature, fine curly hair, scant eyebrows and lashes, telecanthus, prominent pinnae, periodontitis, winged scapulae, cryptorchidism, multiple nevi, and varicose veins. Another case appears to be that of Kresse et al (4).

All were isolated cases but advanced paternal age suggested autosomal dominant inheritance (2). The defect appears to result from a defect in proteodermatan sulfate biosynthesis (4).

### References [Type Hernández (progeroid form)]


1. Hernández A et al: A distinct variant of the Ehlers–Danlos syndrome. *Clin Genet* **16**:335–339, 1979.
2. Hernández A et al: Third case of a distinct variant of the Ehlers–Danlos syndrome (EDS). *Clin Genet* **20**:222–224, 1981.
3. Hernández A et al: Ehlers–Danlos syndrome featured with progeroid facies and mild mental retardation: Further delineation of the syndrome. *Clin Genet* **30**:456–461, 1986.
4. Kresse H et al: Glycosaminoglycan-free small proteoglycan core protein is secreted by fibroblasts from a patient with a syndrome resembling progeroid. *Am J Hum Genet* **41**:436–453, 1987.


**Type Friedman–Harrod.** Friedman and Harrod (1) reported a mother and son with severe scoliosis, severe hernias, mild joint hypermobility, periodontitis, aortic rupture, and allergic abnormalities (eczema, asthma).

### Reference (Type Friedman–Harrod)


1. Friedman JM, Harrod MJE: An unusual connective tissue disease in mother and son: A "new" type of Ehlers–Danlos syndrome? *Clin Genet* **21**:168–173, 1982.

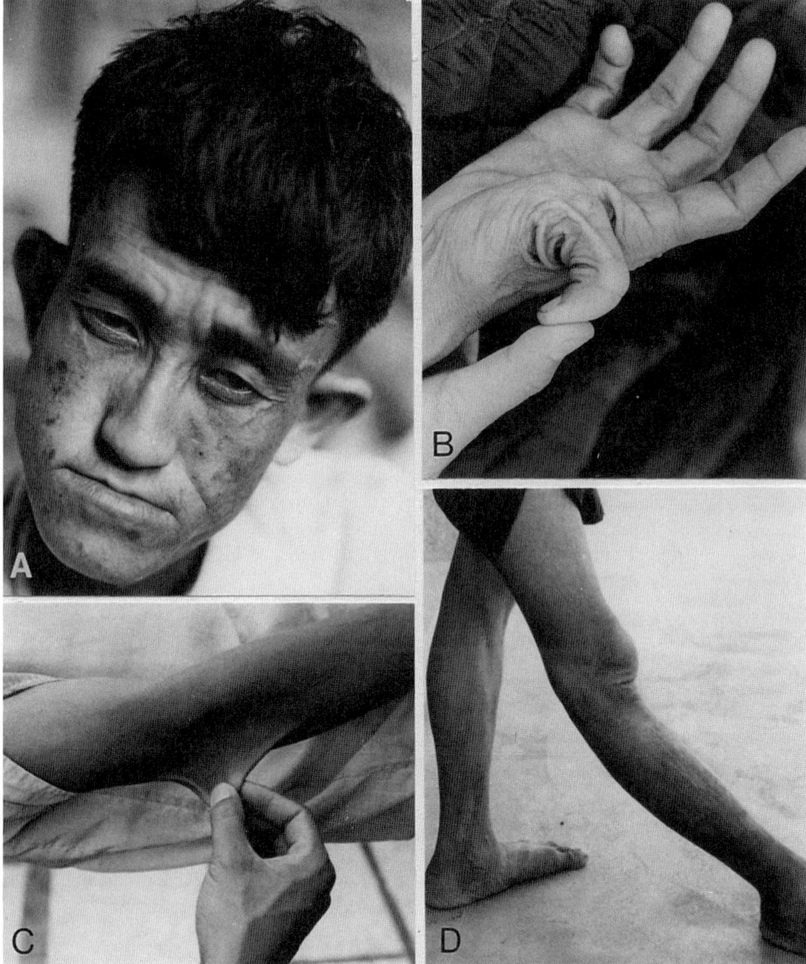


Fig. 13–40. *Ehlers–Danlos syndrome, Type Beasley–Cohen.* (A–D) Narrow face with midface deficiency, small eyes, hypermobility of joints, and hyperelasticity of skin.

**Type Beasley–Cohen.** In 1979, Beasley and Cohen noted two affected sibs with consanguineous parents. There were hyperelasticity of the skin of the hands, very mild scarring, generalized hyperextensibility of the joints with dislocation of hips, and decreased muscle mass. The face was narrow with midface deficiency, small eyes, and possible hearing deficit (Fig. 13–40A–D). One of the sibs had inguinal hernia, cataracts, laterally protruding pinnae, and diarrhea. Both were mentally retarded.

### Reference (Type Beasley–Cohen)

1. Beasley RP, Cohen MM Jr.: A new presumably autosomal recessive form of the Ehlers–Danlos syndrome. *Clin Genet* **16**:19–24, 1979.

**Type Viljoen.** Viljoen et al (1) reported a mother and four children with gross generalized joint laxity, multiple joint dislocations and subluxations, moderate skin hyperextensibility, mild connective tissue fragility, and wormian bones in the skull. The severity of the articular complications and the wormian bones differentiate this form from Type III.

### Reference (Type Viljoen)

1. Viljoen D et al: Ehlers–Danlos syndrome: Yet another type? *Clin Genet* **32**:196–201, 1987.

**Differential diagnosis.** The term ''cutis laxa'' has been applied by some authors to the Ehlers–Danlos syndromes. In true *cutis laxa,* the skin hangs in loose, inelastic folds, the voice is deep, and emphysema and hernia occur. Joint hypermobility may exist as an isolated finding or in association with various disorders such as familial generalized articular hypermobility (Fig. 13–41), *Marfan syndrome, osteogenesis imperfecta, Down syndrome, Larsen syndrome,* and a host of other syndromes. Familial joint laxity has been suggested as still a separate type of Ehlers–Danlos Type XI (18). The *occipital horn syndrome* is clearly distinctive. Probably many cases of acrogeria of Gottron represent EDS Type IV.

**Laboratory aids.** The basic defect can be demonstrated in several types as indicated. Joint mobility may be tested by the Ellis–Bundick method (12) and skin elasticity by the ''pinch meter'' of Olmstead et al (23) and by the method of Grahame and Beighton (14).

### References (Ehlers–Danlos syndromes)

General references (include introduction, differential diagnosis, and laboratory aids)

1. Beighton P: *The Ehlers–Danlos syndrome,* Heinemann, London, 1970.
2. Beighton P et al: Variants of the Ehlers–Danlos syndrome: Clinical, biochemical, haematological, and chromosomal features of 100 patients. *Ann Rheum Dis* **28**:228–245, 1969.
3. Beighton P et al: *Hypermobility of joints,* Springer-Verlag, Berlin–New York, 1983.
4. Black CM et al: The Ehlers–Danlos syndrome: An analysis of the structure of the collagen fibres of the skin. *Br J Dermatol* **102**:85–96, 1980.
5. Byers PH: Inherited disorders of collagen biosynthesis, in *Clinical Medicine,* Spittell JB (ed), Harper & Row, New York, 1983.
6. Byers PH, Holbrook KA: Molecular basis of clinical heterogeneity in the Ehlers–Danlos syndrome. *Ann NY Acad Sci* **460**:198–310, 1985.
7. Byers PH et al: Molecular mechanisms of connective tissue abnormalities in the Ehlers–Danlos syndrome. *Coll Res* **5**:475–489, 1981.

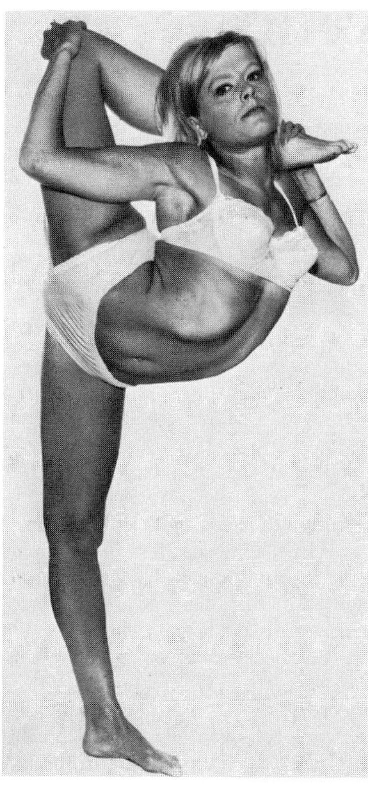

Fig. 13–41. *Familial generalized articular hypermobility*. Should not be confused with Ehlers–Danlos syndrome. Proposita is an acrobatic dancer. (From *P Beighton et al,* J Bone Jt Surg **52**:145, 1970.)


8. Cabeen WR et al: Mitral valve prolapse and conduction defects in Ehlers–Danlos syndrome. *Arch Intern Med* **137**:1227–1231, 1977.

9. Danlos H: Un cas de cutis laxa avec tumeurs par contusion chronique des coudes et des genoux. *Bull Soc Fr Dermatol Syph* **19**:70–72, 1908.

10. Denko CW: Chernogobuv's syndrome: A translation of the first modern case report of the Ehlers–Danlos syndrome. *J Rheumatol* **5**:347–352, 1978.

11. Ehlers E: Cutis laxa, Neigung zu Hemorrhagien in der Haut, Lockerung mehrerer Artikulationen. *Dermatol Z* **8**:173–174, 1901.

12. Ellis FE, Bundick WR: Cutaneous elasticity and hyperelasticity. *Arch Dermatol* **74**:22–32, 1956.

13. Fleischmajer R et al: Collagen fibrillogenesis in human skin. *Ann NY Acad Sci* **460**:246–257, 1985.

14. Grahame R, Beighton P: Physical properties of the skin in the Ehlers–Danlos syndrome. *Ann Rheum Dis* **28**:246–251, 1969.

15. Holbrook KA, Byers PH: Ultrastructural characteristics of the skin in a form of Ehlers–Danlos syndrome type IV: Storage in the rough endoplasmic reticulum. *Lab Invest* **44**:336–341, 1981.

16. Hollister DW: Heritable disorders of connective tissue: Ehlers–Danlos syndrome. *Ped Clin N Am* **25**:575–591, 1978.

17. Hollister DW et al: Genetic disorders of collagen metabolism. *Adv Hum Genet* **12**:1–87, 1982.

18. Horton WA et al: Familial joint instability syndrome. *Am J Med Genet* **6**:221–228, 1980.

19. Kobayasi T et al: Dermal changes in Ehlers–Danlos syndrome. *Clin Genet* **25**:477–484, 1984.

20. Krieg T et al: Molecular defects of collagen metabolism in the Ehlers–Danlos syndrome. *Int J Dermatol* **20**:415–425, 1981.

21. Leier CV et al: The spectrum of cardiac defects in the Ehlers–Danlos syndrome, types I and III. *Ann Intern Med* **92**:171–178, 1980.

22. McKusick VA: *Heritable Disorders of Connective Tissue,* 4th ed., C.V. Mosby, St. Louis, 1972.

23. Olmstead F et al: A device for objective clinical measurement of cutaneous elasticity. *Am J Med Sci* **222**:73–75, 1951.

24. Pinnell SR, McKusick VA: Heritable disorders of connective tissue with skin changes, in *Dermatology in General Medicine,* 3rd ed., Fitzpatrick TB et al (eds), McGraw-Hill, New York, 1978, chapter 149, pp 1775–1990.

25. Pope FM, Nicholls AC: Molecular abnormalities of collagen in human disease. *Arch Dis Childh* **62**:523–528, 1987.

26. Prockop DJ, Kivirikko KI: Heritable diseases of collagen. *N Engl J Med* **311**:376–386, 1984.

27. Sevenich M et al: Ehlers–Danlos syndrome: A disease of fibroblasts and collagen fibrils. *Arch Dermatol Res* **267**:237–251, 1980.

28. Shinkai H et al: Connective tissue metabolism in cultured fibroblasts of a patient with Ehlers–Danlos syndrome type I. *Arch Dermatol Res* **257**:113–122, 1976.

29. Sulica VI et al: Cutaneous histologic features in Ehlers–Danlos syndrome. *Arch Dermatol* **115**:40–42, 1979.

30. Taylor DJ et al: Ehlers–Danlos syndrome during pregnancy: A case report and review of the literature. *Obstet Gynecol Surv* **36**:277–281, 1981.

31. Vogel A et al: Abnormal collagen fibril structure in the gravis form (type I) of the Ehlers–Danlos syndrome. *Lab Invest* **40**:201–206, 1979.

32. Weber FP, Aitken JK: Nature of subcutaneous spherules in some cases of Ehlers–Danlos syndrome. *Lancet* **1**:198–199, 1938.


## Dyskeratosis congenita (Zinsser–Engman–Cole syndrome)

Dyskeratosis congenita is characterized by reticular skin hyperpigmentation, nail dystrophy, lacrimal duct obstruction, leukoplakia of mucous membranes, bone marrow hypofunction, and a predisposition to malignancy. It was first described by Zinsser (39) in 1906, with a second report by Engman (14) appearing in 1926, the author being unaware of Zinsser's earlier publication. In 1930, Cole et al (7) further delineated the condition and brought it to the attention of dermatologists. The syndrome has been seen in blacks, whites, and orientals (32). Excellent reviews are those of Sirinavin and Trowbridge (32) and Womer et al (37). Over 100 cases have been reported (32).

Dyskeratosis congenita is a misnomer since the syndrome is neither dyskeratotic nor congenital. Wilgram and Weinstock (36) pointed out that oral lesions are characterized by a decreased number of keratinosomes associated with decreased epithelial turnover. Furthermore, cutaneous manifestations of the disorder are not present during infancy, as a rule.

The syndrome is etiologically heterogeneous, most cases being compatible with X-linked recessive inheritance (4,9,10,19,32,38). Among 104 cases there were 92 males and 12 females (12a). The gene has been assigned to Xq28 (10). Several families have likely had autosomal dominant inheritance (10,34a) and in one of these (31a), male-to-male transmission was evident. In a few instances, a brother and sister have been affected and, on occasion, sporadic cases have involved females (1,17a,21a,22a-24,33). It is not known whether they represent lyonization of female heterozygotes or incomplete penetrance of the dominant form (or fresh mutations in the case of sporadic instances of males or females). The use of linked restriction fragment length polymorphisms may help resolve the question.

Although chromosomal breakage has been reported by several authors (1,4,19,25,29), these findings have been questioned, and most karyotypes have been normal with no evidence of chromosomal instability (9,37). Burgdorf et al (5) reported a modest increase in the rate of sister chromatid exchange, but this was not confirmed in two patients studied by Womer et al (37).

Clinical features in hemizygous males are usually manifest at 5–10 years. Skin, nail, and mucosal changes appear during the first decade. Patients are usually frail and have generalized growth retardation. Hematopoietic manifestations appear in about 50% and develop during the second and third decades. Neoplasms may evolve over the third, fourth or fifth decades. Mean age at death is about 25 years, usually from infections (50%), gastrointestinal bleeding (20%), or malignancy (30%). Findings in dyskeratosis congenita are summarized in Table 13–3.

**Skin and skin appendages.** The most prominent skin changes closely resemble those found in poikiloderma vasculare atrophicans, involving especially the face, neck, upper arms, and chest and appearing around 8–9 years of age. A prominent reticulated hyperpigmentation of the skin, usually described a gunmetal in color, involves the same areas (Fig. 13–42A,B). Microscopically, there is atrophy of

Table 13–3. Findings in dyskeratosis congenita[a]

| Findings | Percentage |
|---|---|
| Growth | |
| Hypasthenic build | 48 |
| Performance | |
| Subnormal intelligence | 42 |
| Skin | |
| Hyperpigmentation | 100 |
| Atrophy | 93 |
| Hyperhidrosis (palms and soles) | 89 |
| Hyperkeratosis (palms and soles) | 72 |
| Bullous eruptions | 78 |
| Acrocyanosis | 55 |
| Hair | |
| Alopecia | 51 |
| Nails | |
| Dystrophy | 98 |
| Mucosa | |
| Leukoplakia | 87 |
| Hematopoietic | |
| Anemia | 52 |
| Leukopenia and/or thrombocytopenia | 49 |
| Other | |
| Epiphora | 78 |
| Dysphagia | 59 |
| Small testes, small penis[b] | 40 |

[a]Based on Sirinavin and Trowbridge, *J Med Genet* **12**:339, 1975.
[b]Percentage adjusted for male sex.

the epidermis and subcutaneous tissues, accompanied by capillary hyperplasia. Melanin pigment is heavily deposited, especially near blood vessels. Characteristically, no inflammatory exudate is evident (8,11,15–18,21,30–32,37). Cutaneous macular amyloidosis has been noted in one instance (23).

Hyperhidrosis of the palms and soles but with generalized hypohidrosis elsewhere has been noted in about 90%, and palmar and plantar keratoses have occurred in approximately 70%. Cutaneous bullae may appear. Acrocyanosis of the hands and feet occurs in about 55% (8,11,15–18,21,30–32,37).

Nail dystrophy, initially longitudinal ridging and splitting, (Fig. 13–43A,B), found in 98%, is not of the same degree in all digits, and fingernails tend to be more severely affected than toenails (32). Nail changes are usually seen by 10 years of age. They become progressively dystrophic and are often lost.

The hair is thin and lusterless in 50%. Sparse body hair in some cases is related to hypogonadism. Eyelashes and eyebrows may be absent (20,30,32).

**Mucosal manifestations.** Leukoplakia, a feature in 87%, can occur on any mucosa, but most commonly affects the oral mucosa prior to puberty. Uncommonly, the urethral mucosa, rectoanal region, glans penis, and vagina are involved (32).

Crops of vesicles and bullae appear on the oral mucosa. They are recurrent and essentially painless. Because of moisture and maceration, they rupture early, leaving ulcerated areas, with epithelial tags along the margin. After several attacks, the mucosa becomes atrophic and the tongue loses its papillae and appears smooth. Under ultraviolet light, the normal orange fluorescence of the tongue is absent. Eventually, the mucosa becomes thickened, fissured, and white (Fig. 13–44). Mucosal atrophy or stenosis has been described in the mouth, esophagus, anus, urethra, and vagina (6,16,17,28,32).

**Eyes and ears.** Chronic blepharitis, ectropion (7,11,14,17,28,30), and profuse tearing due to keratinization with obstruction of the lacrimal points (17,21,28,30) around puberty have been described. Rare ophthalmologic findings, for example, congenital cataracts, are reviewed by Womer et al (37). Thinning of the eardrum, malformation of the middle ear, and sensorineural hearing deficit have been reported (9,11,25).

**Gastrointestinal features.** Although not often emphasized, many patients have had gastrointestinal problems. Dysphagia, occurring in about 60%, may result from esophageal strictures or diverticula. Se-

Fig. 13–42. *Dyskeratosis congenita.* (A) Whitish, irregular atrophic areas, increased pigmentation, and telangiectasia, producing appearance that resembles poikiloderma vasculare atrophicans. (B) Sparse scalp hair, blepharitis due to absence of lacrimal puncta.

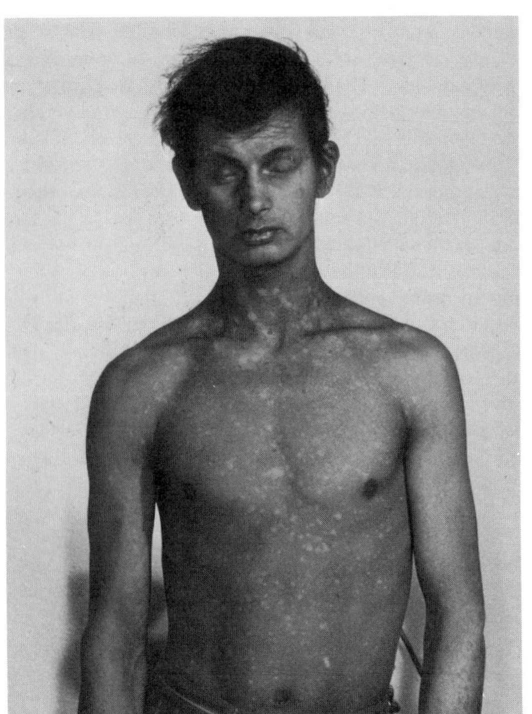

A

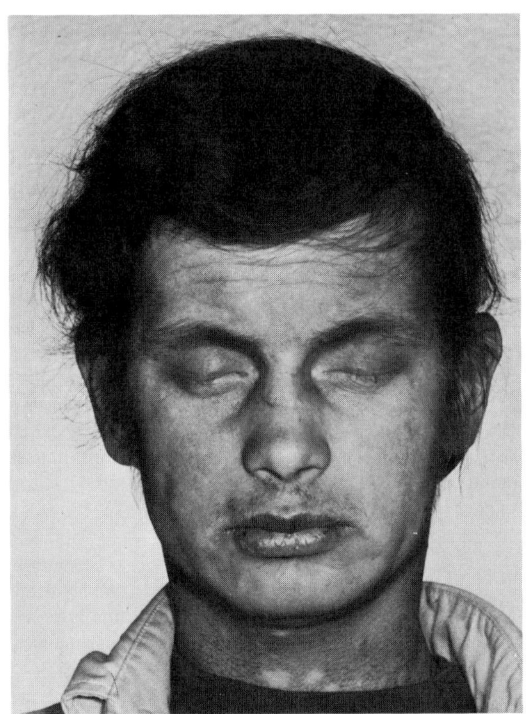

B

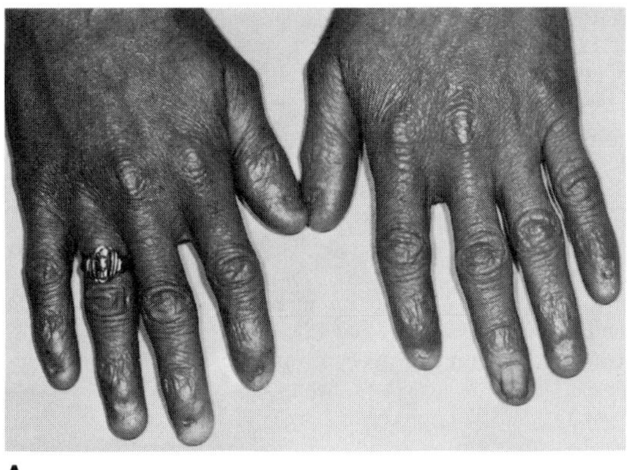

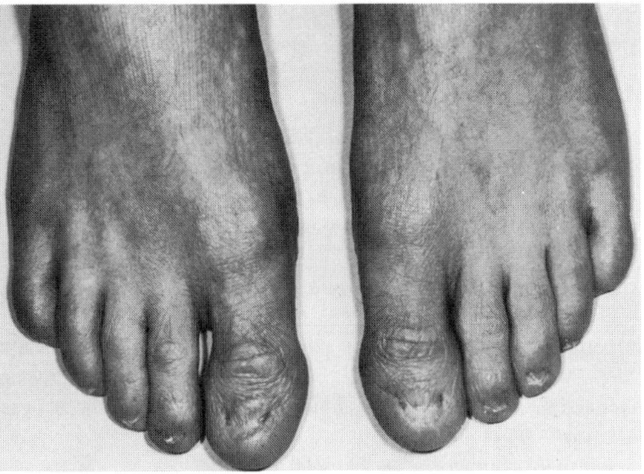

A B

Fig. 13–43. *Dyskeratosis congenita.* (A,B) Note shriveled, shrunken appearance of nails.

vere diarrhea, sometimes bloody, has been noted. Liver disease is relatively common, and cases of hepatic cirrhosis have been recorded (1,4,11,17,22,25,33,34,37).

**Hematopoietic manifestations.** Progressive pancytopenia is found in about half the cases. Age of onset and rapidity of progression are variable, but mean age of onset is about 15 years. A decline in the platelet count usually occurs first, followed by progressive anemia, and, last, granulocytopenia. Splenomegaly occurs in 45% and hepatomegaly is seen in about 15%, although they are not sites of extramedullary hematopoiesis (1–4,7,12,13,17,21,21a,22,32,34,35,37). Friedland et al (15a) consider the disorder a stem cell defect.

**Immunologic manifestations.** A wide range of immunoglobulin abnormalities may occur, including decreased IgG and IgM with normal IgA, normal IgG and IgA with slightly decreased IgM, and increased IgG. Abnormalities of cell-mediated immunity have also been noted. Autopsies have revealed a number of findings including lymphoid depletion with absence of the primary germinal follicles and occasionally fibrosis of lymph nodes. The high frequency of opportunistic infections is further evidence of abnormal immune function (20,29,32,37).

**Neoplasia.** Malignancy occurs in about 17%, the mean age being about 30 years. Carcinoma of the oral mucosa has been recorded at least six times. Other tumors have included carcinomas of the nasopharynx, esophagus, rectum, cervix, and vagina. Squamous cell carcinoma of the hand, adenocarcinoma of the pancreas, and Hodgkin's disease have each been reported once (1,6,8,9,11,16,25,30,32,37). Tschou and Kohn (34a) suggested an increased rate of cancer in unaffected sibs.

Fig. 13–44. *Dyskeratosis congenita.* Thick white plaques of labial and lingual mucosa.

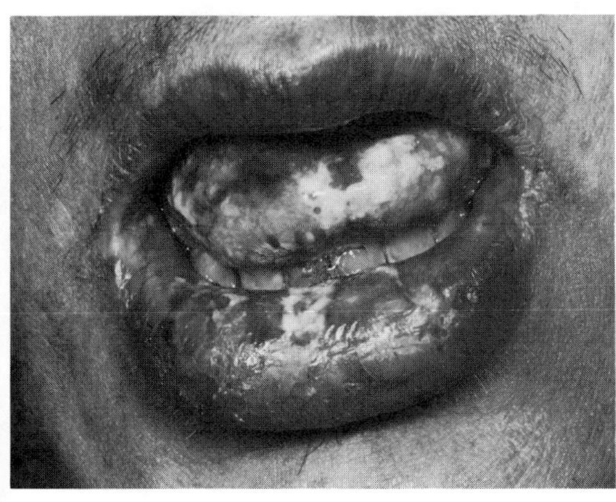

**Central nervous system.** Mental deficiency occurs in about 40% and is mild to moderate in degree. Schizophrenia has also been noted (11,17,32).

**Other findings.** Small testes and small penis have been observed in about 40% of males. Rarely, hypopituitarism, enlarged thyroid, and secondary hypogonadism have been recorded (32).

A variety of skeletal anomalies have been noted including joint deformities, incomplete closure of the vertebral arches, and other minor skeletal anomalies (33). Radiolucencies in the shafts of long bones with coarse trabeculation in the metaphyses have been reported (32,37). Osteoporosis, bone fragility, and aseptic necrosis of the hip have been encountered, usually attributable to steroid therapy (32). However, several patients did not experience steroid therapy (37). Intracranial calcifications were reported by several authors (21b,21c,26).

The teeth are said to be subject to early decay (30,32), periodontal disease (11,35,39), malformation, and malposition (11,30). These changes with rare exception have never been well documented and merit further investigation.

**Differential diagnosis.** Differential diagnosis includes Fanconi syndrome, *Rothmund–Thomson syndrome, Bloom syndrome, xeroderma pigmentosum,* and *focal dermal hypoplasia.* Similarities and differences between dyskeratosis congenita and Fanconi syndrome have been discussed by a number of authors (32,34). The condition described by Moon-Adams and Slatkin (27) is probably not dyskeratosis congenita.

**Laboratory aids.** The skin shows atrophy and pigment-laden macrophages.

Studies of bone marrow have shown a defect of stem cells (15a).

### References [Dyskeratosis congenita (Zinsser–Engman–Cole syndrome)]


1. Addison M, Rice MS: The association of dyskeratosis congenita and Fanconi's anemia. *Med J Aust* **1**:797–799, 1965.
2. Bazex A, Dupre A: Dyskeratose congénitale (type Zinsser–Cole–Engman) associée a une myelopathie constitutionelle (purpura thrombopenique et neutropenie). *Ann Dermatol Syph (Paris)* **84**:497–513, 1957.
3. Bodalski J et al: Fanconi's anemia and dyskeratosis congenita as a syndrome. *Dermatologica* **127**:330–342, 1963.
4. Bryan HG, Nixon RK: Dyskeratosis congenita and familial pancytopenia. *JAMA* **192**:203–208, 1965.
5. Burgdorf W et al: Sister chromatid exchange in dyskeratosis congenita lymphocytes. *J Med Genet* **14**:256–257, 1977.

6. Cannell H: Dyskeratosis congenita. *Br J Oral Surg* **9**:8–20, 1971.

7. Cole HN et al: Dyskeratosis congenita with pigmentation, dystrophia unguis and leukokeratosis oris. *Arch Dermatol Syph (Chic)* **21**:71–95, 1930.

8. Cole HN et al: Dyskeratosis congenita. *Arch Dermatol* **76**:712–719, 1957.

9. Connor JM, Teague RH: Dyskeratosis congenita. Report of a large kindred. *Br J Dermatol* **105**:321–325, 1981.

10. Connor JM et al: Assignment of the gene for dyskeratosis congenita to Xq28. *Hum Genet* **72**:348–351, 1986.

11. Costello MJ, Buncke CM: Dyskeratosis congenita. *Arch Dermatol* **73**:123–132, 1956.

12. De Boeck K et al: Thrombocytopenia: First symptom in a patient with dyskeratosis congenita. *Pediatrics* **67**:898–903, 1981.

12a. Davidson HR, Connor JM: Dyskeratosis congenita. *J Med Genet* **25**:843–846, 1988.

13. Dodd HJ et al: Dyskeratosis congenita with pancytopenia: Clinical meeting of the St. John's Hospital Dermatological Society: 3 November 1983. *Clin Experiment Dermatol* **10**:73–78, 1985.

14. Engman MF Jr: A unique case of reticular pigmentation of the skin with atrophy. *Arch Dermatol Syph (Chic)* **13**:685–687, 1926.

15. Engman MF Jr: Congenital atrophy of the skin, with reticular pigmentation. *JAMA* **105**:1252–1256, 1935.

15a. Friedland M et al: Dyskeratosis congenita with hypoplastic anemia: A stem cell defect. *Am J Hematol* **20**:85–87, 1985.

16. Garb J: Dyskeratosis congenita with pigmentation, dystrophia unguium and leukoplakia oris: A follow-up report of two brothers. *Arch Dermatol* **77**:704–712, 1958.

17. Garb J, Rubin D: Dyskeratosis congenita with pigmentation, dystrophia unguium and leukoplakia oris (Cole and others). *Arch Dermatol Syph (Chic)* **50**:191–198, 1944.

17a. Georgouras K: Dyskeratosis congenita. *Aust J Dermatol* **8**:36–43, 1965.

18. Grekin JN, Schwartz OD: Dyskeratosis congenita with pigmentation dystrophia unguium and leukokeratosis oris. *Arch Dermatol* **85**:124–125.

19. Gutman A et al: X-linked dyskeratosis congenita with pancytopenia. *Arch Dermatol* **114**:1667–1671, 1978.

20. Inoue S et al: Dyskeratosis congenita with pancytopenia. *Am J Dis Child* **126**:389–396, 1973.

21. Jansen LH: The so-called "dyskeratosis congenita". *Dermatologica* **103**:167–177, 1951.

21a. Juneja HS et al: Abnormality of platelet size and T-lymphocyte proliferation in an autosomal recessive form of dyskeratosis congenita. *Eur J Haematol* **39**:306–310, 1987.

21b. Kalb RE: Avascular necrosis of bone in dyskeratosis congenita. *Am J Med* **80**:511–513, 1986.

21c. Kelly TE, Stelling CB: Dyskeratosis congenita: Radiologic features. *Pediatr Radiol* **12**:31–36, 1982.

22. Koszewski BJ, Hubbard TF: Congenital anemia in hereditary ectodermal dysplasia. *Arch Dermatol* **74**:159–166, 1956.

22a. Ling NS et al: Dyskeratosis congenita in a girl simulating chronic graft-vs-host disease. *Arch Dermatol* **121**:1424–1428, 1985.

23. Llistosella E et al: Dyskeratosis congenita with macular cutaneous amyloid deposits. *Arch Dermatol* **120**:1381–1382, 1984.

24. Marshall J, van der Meulen H: Dyskeratosis congenita—its occurrence in the female. *Br J Dermatol* **77**:162, 1965.

25. Milgrom H et al: Dyskeratosis congenita. *Arch Dermatol* **89**:345–349, 1964.

26. Mills SE et al: Intracranial calcifications and dyskeratosis congenita. *Arch Dermatol* **115**:1437–1439, 1979.

27. Moon-Adams D, Slatkin MH: Familial pigmentation with dystrophy of the nails. *Arch Dermatol* **71**:591–598, 1955.

28. Orfanos C, Gartmann H: Leukoplakien, Pigmentverschiebungen und Nageldystrophie. *Med Welt* **48**:2589–2594, 1966.

29. Ortega JA et al: Congenital dyskeratosis. *Am J Dis Child* **124**:701–705, 1972.

29a. Pai GS et al: Etiologic heterogeneity of dyskeratosis congenita. *Am J Med Genet* **32**:63–66, 1989.

30. Pastinszky I et al: Ein Beitrag zur Pathologie der "Dyskeratosis congenita" Cole–Rauschkolb–Toomey. *Dermatol Wochenschr* **135**:587–593, 1957.

31. Ramos e Silva J: Syndrome de Zinsser-Fanconi. *Ann Dermatol Syph* **93**:497–502, 1966.

31a. Scoggins RB et al: Dyskeratosis congenita with Fanconi-type anemia: Investigations of immunologic and other defects. *Clin Res* **19**:409, 1971.

32. Sirinavin C, Trowbridge AA: Dyskeratosis congenita: Clinical features and genetic aspects. *J Med Genet* **12**:339–354, 1975.

33. Sorrow JM, Hitch JM: Dyskeratosis congenita. *Arch Dermatol* **88**:340–347, 1963.

34. Steier W et al: Dyskeratosis congenita: Relationship to Fanconi's anemia. *Blood* **39**:510–521, 1972.

34a. Tschou PK et al: Dyskeratosis congenita: An autosomal dominant disorder. *J Am Acad Dermatol* **6**:1034–1039, 1982.

35. Wald C, Diner H: Dyskeratosis congenita with associated periodontal disease. *Oral Surg* **37**:736–744, 1974.

36. Wilgram GF, Weinstock A: Advances in genetic dermatology. *Arch Dermatol* **94**:456–479, 1966.

37. Womer R et al: Dyskeratosis congenita: Two examples of this multisystem disorder. *Pediatrics* **71**:603–609, 1983.

38. Woog JJ et al: The role of aminocaproic acid in lacrimal surgery in dyskeratosis congenita. *Am J Ophthalmol* **100**:728–732, 1985.

39. Zinsser F: Atrophica cutis reticularis cum pigmentatione, dystrophia unguium et leukoplakia oris. *Ikonograph Derm Kioto* 219–223, 1906.

## Dyskeratosis benigna intraepithelialis mucosae et cutis hereditaria

In 1977, From et al (1) described a father and son with a hereditary syndrome that affected the skin, oral mucosa, and bulbar conjunctiva. The disorder was characterized by recurrent nonseasonal conjunctivitis, epiphora, and photophobia. Suppurating, red or brown, papular eruptions with central keratotic plugs (not greater than 10 mm in diameter) were found on the scrotum, buttocks, and body, the majority being on the lower limbs (Figs. 13–45 and 13–46). Trauma-provoked papular lesions and Koebner phenomenon were also noted. Verruca-like lesions were noted on palmar and plantar surfaces of the hands. Leukoplakic lesions occurred on the buccal mucosa. Total tooth loss (both primary and permanent) at an early age due to edematous hypertrophic gingivitis was noted (Fig. 13–47).

Pathologic findings in mucous membrane epithelium and epidermis included moderate to severe simple hyperplasia, acantholysis (with no

Fig. 13–45. *Dyskeratosis benigna intraepithelialis mucosae et cutis hereditaria.* Skins showing brownish papules with keratotic plugs. (From *E From et al,* J Cutan Pathol **5**:105, 1978.)

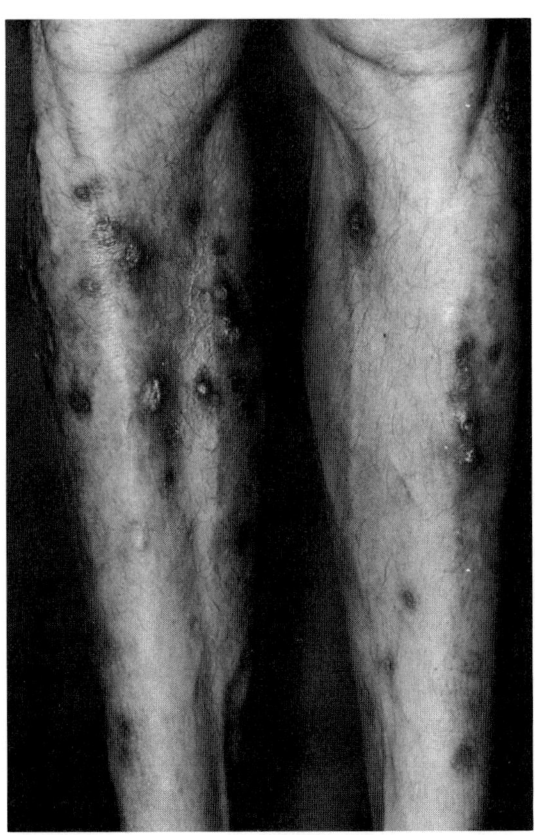

Fig. 13–46. *Dyskeratosis benigna intraepithelialis mucosae et cutis hereditaria.* Hyperkeratotic papules of scrotum. (From *E From et al,* J Cutan Pathol **5**:105, 1978.)

''corps ronds''), moderate hyperortho- and/or hyperparakeratosis, and benign dyskeratosis characterized by pronounced single cell keratinization. No true dysplasia was found (Fig. 13–48). Gingival biopsy showed focal intraepithelial microabscesses and spongiform edema.

The syndrome is distinguished from *Papillon–Lefèvre syndrome* and *dyskeratosis congenita.*

#### Reference (Dyskeratosis benigna intraepithelialis mucosae et cutis hereditaria)

1. From E et al: Dyskeratosis benigna intraepithelialis et cutis hereditaria. A report of this disorder in father and son. *J Cutan Pathol* **5**:105–115, 1978.

Fig. 13–47. *Dyskeratosis benigna intraepithelialis mucosae et cutis hereditaria.* Marked hypertrophic gingivitis at 2.5 years. (From *E From et al,* J Cutan Pathol **5**:105, 1978.)

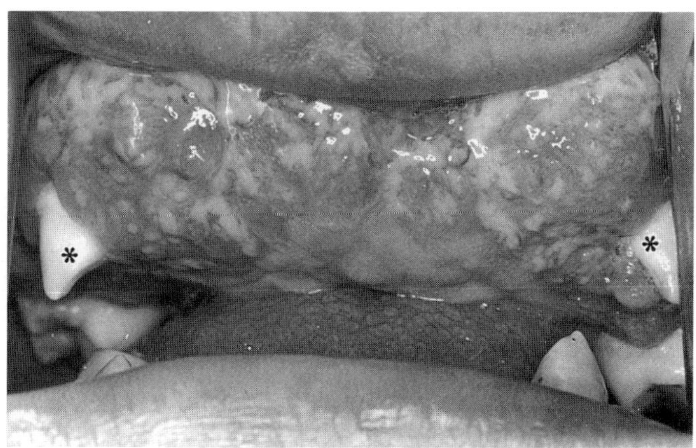


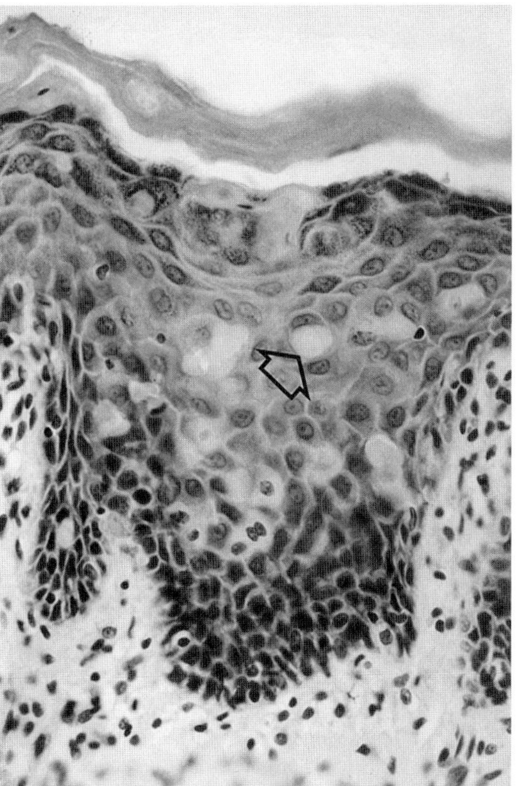

Fig. 13–48. *Dyskeratosis benigna intraepithelialis mucosae et cutis hereditaria.* Photomicrograph of buccal white lesion showing hyperparakeratosis and numerous dyskeratotic cells (arrow). (From *E From et al,* J Cutan Pathol **5**:105, 1978.)

### Pachyonychia congenita (Jadassohn–Lewandowsky syndrome, Jackson–Lawler syndrome)

In 1910, Jadassohn and Lewandowsky (24) described a patient with a syndrome of dystrophic fingernails and toenails, palmoplantar keratosis and hyperhidrosis, follicular keratosis, and oral leukokeratosis.

In 1951–1952, Jackson and Lawler (23) reported a family with pachyonychia congenita, palmoplantar hyperkeratosis, hyperhidrosis, and follicular keratosis. However, oral leukokeratosis was not observed. In addition, their patients had cutaneous cysts and natal teeth. However, Stieglitz and Centerwall (38) and Anneroth et al (4) reported kindreds with characteristics shared by both syndromes, suggesting that the differences originally observed represented variability of expression. Since we have seen similar overlap, the two disorders will be presented as a single disorder. The opposite point of view, that of separating the disorder into four subtypes, is that of Feinstein et al (15a).

The syndrome follows an autosomal dominant mode of transmission, with a 9:5 male predilection (31). The frequency is about 0.7/100,000 population (17). At least 150 cases have been reported.

**Skin and skin appendages.** At birth (12,16,18,19,39) or soon after (40), fingernails and toenails become thickened, tubular, and hard (Fig. 13–49A). Their undersurfaces fill with a horny, yellow-brown material that causes the nail to project upward from the nailbed at the free edge.

Hyperhidrosis of the palms and soles nearly always occurs but the rest of the skin is dry and may be described as ichthyotic (28).

Palmar and plantar hyperkeratoses are noted in 40–65% during the first few years of life. During warm weather, bullae appear on the feet, especially on the plantar surfaces of the toes and heels and along the sides (2). They burst, may become infected, are very painful, and

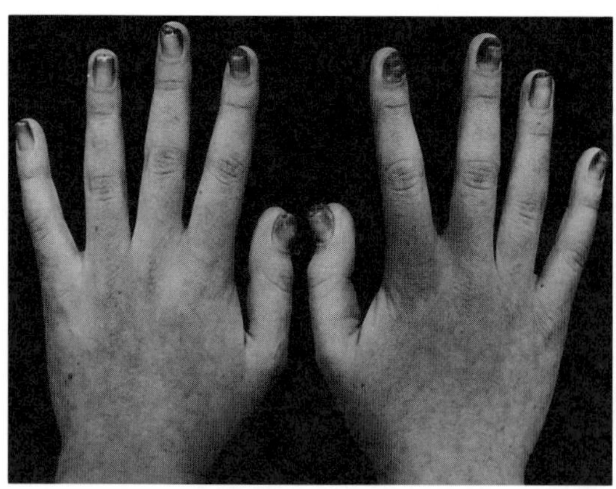

A

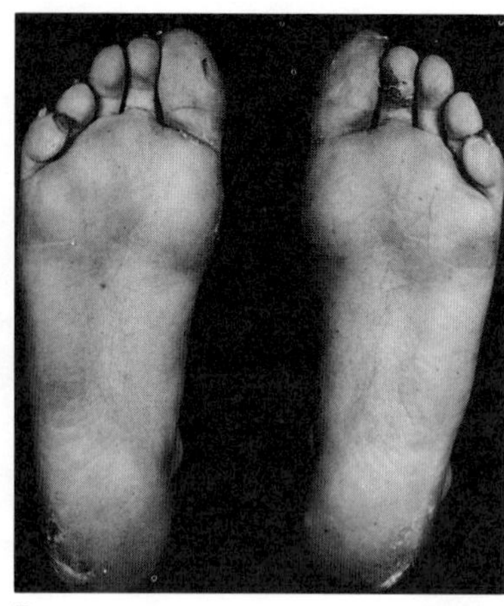

B

Fig. 13–49. *Pachyonychia congenita.* (A) Note thickening and elevation of fingernails at free edge in 13-year-old female. (B) Note ruptured blisters of toes and heels. Often there is severe hyperkeratosis of soles. (From *ADM Jackson* and *SD Lawler,* Ann Eugen **16**:141, 1951.)

often make walking difficult (3,18,28,35,39) (Fig. 13–49B). Diffuse hyperkeratosis of the perineum has also been described (7,12,18).

During the first few years of life, pinhead-sized follicular papules appear over the elbows, knees, popliteal areas, and buttocks (2,39). In the center of each papule, a horny plug is seen. Verrucous lesions may also occur in the same areas (18). The skin is thickened because of acanthosis and parakeratosis, especially about the pilosebaceous apparatus. Follicles and sweat pores are dilated and plugged with imperfectly cornified and partly degenerated horny material (2). The hair may be dry (9) and twisted (45), and alopecia has been reported (27,31). Cornoid lamellae have been described in a patient in whom oral leukokeratosis was not well documented (44).

Electron microscopic studies of involved skin from the knee (39) show increased number and thickness of tonofibrils, an increased number of desmosomes throughout the epidermis, intracellular and extracellular edema, and large, abnormal keratohyaline granules. These findings are consistent with a defect in keratinization. Similar findings were reported by Thormann and Kobayasi (40). Ultrastructural findings have also been described by Perrot et al (34) and Thomas et al (39). Cysts, especially of the head, neck, and upper chest, appear around puberty (5,8,13,15,23,31,32,36,37,42,43,46) (Fig. 13–50). Most reports do not adequately describe the histology of the skin cysts. However, Hodes and Norins (22) and Velasquez and Bustamante (42) determined that their patients had steatocystomas, whereas Clementi et al (11), Chapman (9), and Soderquist and Reed (37) determined that the cysts in their patients were epidermal inclusion cysts.

Fig. 13–50. *Pachyonychia congenita.* Multiple cysts of skin. (Courtesy of *RM Goodman,* Tel Aviv, Israel.)

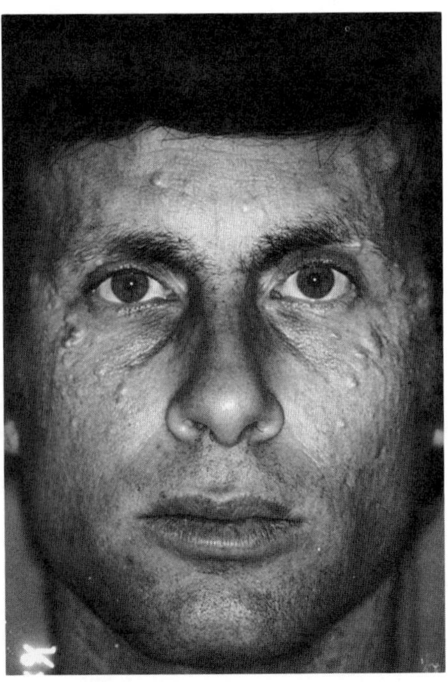

**Laryngologic abnormalities.** Hoarse voice and thickening of the posterior commissure of the larynx have been noted (1,11,12, 21,23,28,31). Patients may have multiple, white exophytic lesions involving the ventricles, true cords, subglottis, and interarytenoid area (11,19).

**Oral manifestations.** White patches are observed on the dorsum and lateral borders of the tongue (3,14,18,19,26,27,38,41,47). The buccal mucosa at the interdental line, the gingiva, and palate can also be involved (9,14,18,19,29,47,48) (Fig. 13–51). These lesions may be present as early as the first decade.

On microscopic examination, the oral mucosa is characterized by hyperparakeratosis, acanthosis, and generalized intracellular vacuolization of the epithelial cells (4,48).

Natal teeth are frequently present (4,5,7,22,23,33,35,37,42).

**Differential diagnosis.** Nail changes are distinctive. The oral leukokeratosis is nonspecific and is also seen in *dyskeratosis congenita, hereditary benign intraepithelial dyskeratosis,* white sponge nevus (25), and *leukoplakia, tylosis, and esophageal carcinoma.* Other clinical findings serve to distinguish these disorders from pachyonychia congenita. Patients with *endocrine-candidosis syndrome* also have diffuse white lesions of the oral mucous membranes.

The frequency of natal teeth unassociated with other abnormalities has been estimated to range from 1:700 to 1:30,000 white newborns (6,28). However, natal teeth occur in 1 in 28 Native Americans of Athabascan origin (34a). Natal teeth may also be found in *Ellis–van Creveld syndrome,* and in *Hallermann–Streiff syndrome.* McDonald and Reed (30) reported a kindred in which natal teeth and steatocystoma multiplex were inherited as a dominant condition; nails were normal.

Haber and Rose (20) reported two male sibs with bullae on the hands and feet, oral leukokeratosis, leukonychia of the fingernails, onycholysis of the fingernails and toenails with slight elevation by subungual debris, keratoderma of the palms and soles, and hyperkeratotic papules on the dorsa of the fingers and toes. There were no skin cysts, and natal teeth were not mentioned. Parents were normal but consanguineous. Although this disorder might be autosomal recessive, X-linked inheritance cannot be excluded. A 4-year-old patient, born to a consanguineous couple, was reported by Chong-Hai and Rajagopolan (10) as having the nail abnormalities of pachyonychia congenita, palmar and plantar hyperhidrosis, blistering of the feet, follicular

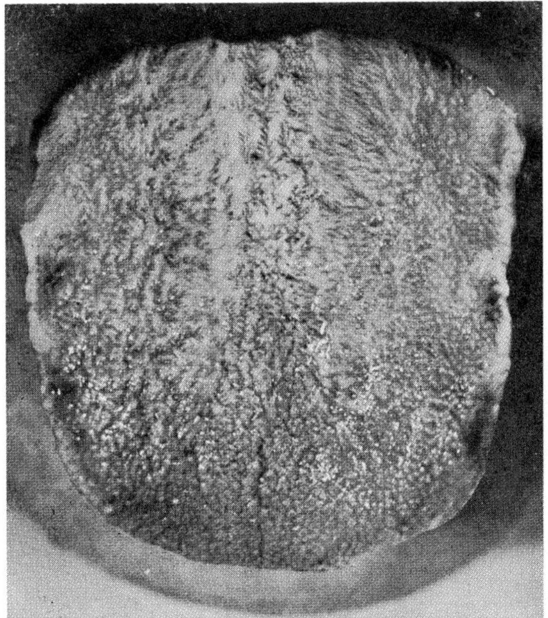

Fig. 13–51. *Pachyonychia congenita.* Thickening of oral mucosa, especially that of tongue and buccal mucosa.

keratoses, epidermal cysts, and oral leukokeratosis; that the patient represented a new dominant mutation cannot be excluded.

A dominant form associated with diffuse rippled pigmentation of the neck, axillae, waist, buttocks, and thighs that fades with age has been reported (8a,40a). The material in the skin appears to be amyloid. Oral involvement has not been documented.

### References [Pachyonychia congenita (Jadassohn–Lewandowsky syndrome, Jackson–Lawler syndrome)]

1. Åkesson HO: Pachyonychia congenita in six generations. *Hereditas* **58**:103–110, 1967.

2. Andrews GC: Pachyonychia congenita. *Arch Dermatol Syph* **33**:183–184, 1936.

3. Andrews GC, Strumwasser S: Pachyonychia congenita, *NYJ Med* **29**:747–749, 1929.

4. Anneroth G et al: Pachyonychia congenita. A clinical, histological and microradiographic study with special reference to oral manifestations. *Acta Derm Venereol* **55**:387–394, 1975.

5. Besser FS, Moynahan EJ: Pachyonychia congenita with epidermal cysts and teeth at birth: 4th generation. *Br J Dermatol* **84**:95–96, 1971. (Same family as in Refs. 23 and 36.)

6. Bjuggren G: Premature eruption in the primary dentition—a clinical and radiological study. *Swed Dent J* **66**:343–355, 1973.

7. Boxley JD, Wilkinson DS: Pachyonychia congenita and multiple epidermal hamartoma. *Br J Dermatol* **85**:298–299, 1971.

8. Brain RT: Pachyonychia congenita with ectodermal defect. *Proc 10th Int Congr Dermatol,* London, 1952, pp 507–508.

8a. Buckley WR, Cassuto J: Pachyonychia congenita. *Arch Dermatol* **85**:397–402, 1962.

9. Chapman RS: Pachyonychia congenita type 3 (Samman). *Br J Dermatol* **107** (Suppl) 22:91, 1982.

10. Chong-Hai T, Rajagopolan K: Pachyonychia congenita with recessive inheritance. *Arch Dermatol* **113**:685–686, 1977.

11. Clementi M et al: Pachyonychia congenita Jackson–Lawler type: A distinct malformation syndrome. *Br J Dermatol* **114**:367–370, 1986.

12. Cohn AM, McFarlane JR: Pachyonychia congenita with involvement of the larynx. *Arch Otolaryngol* **102**:233–235, 1976.

13. de Groot WP, van Ketel WG: Pachyonychia congenita with sebocystomatosis (Sertoli). *Dermatologica* **133**:344, 1966.

14. Dupré A et al: Pachyonychie congénitale. Description de 3 cas familiaux. Traitement par le rétinoïde aromatique (RO 10.9359). *Ann Dermatol Venereol* **108**:145–149, 1981.

15. Engel S, Pinzer B: Über die Kombinationsmöglichkeiten von Sebozystomatosis Günther mit anderen Erkrankungen. *Dermatol Mschr* **155**:687–699, 1969.

15a. Feinstein A et al: Pachyonychia congenita. *J Am Acad Dermatol* **19**:705–711, 1988.

16. Forslind B et al: Pachyonychia congenita. A histologic and microradiographic study. *Acta Derm Venereol (Stockholm)* **53**:211–216, 1973.

17. Franzot J et al: Pachyonychia congenita (Jadassohn–Lewandowsky syndrome. A review of 14 cases in Slovenia. *Dermatologica* **162**:462–472, 1981.

18. Goodman H: Pachyonychia congenita. *Urol Cutan Rev* **50**:465–467, 1946.

19. Gorlin RJ, Chaudhry AP: Oral lesions accompanying pachyonychia congenita. *Oral Surg Oral Med Oral Pathol* **11**:541–544, 1958.

20. Haber RM, Rose TH: Autosomal recessive pachyonychia congenita. *Arch Dermatol* **122**:919–923, 1986.

21. Hadida E, Maril FG: Pachyonychie congénitale avec kératodermie et kératoses disseminés de la peau et des muqueuses (syndrome de Jadassohn et Lewandowski). *Bull Soç Fr Dermatol Syph* **59**:236–237, 1952.

22. Hodes ME, Norins AL: Pachyonychia congenita and steatocystoma multiplex. *Clin Genet* **11**:359–364, 1977.

23. Jackson ADM, Lawler SK: Pachyonychia congenita: A report of six cases in one family. *Ann Eugen* **16**:142–146, 1951–1952. (Same family as in Refs. 5 and 36.)

24. Jadassohn J, Lewandowsky F: Pachyonychia congenita. Keratosis disseminata circumscripta (follicularis). Tylomata. Leukokeratosis linguae, in *Ikonographia Dermatologica,* Vol. 1, Neisser A, Jacobi E (eds), Urban and Schwarzenberg, Berlin, 1910, pp 29–31.

25. Jorgenson RJ, Levin LS: White sponge nevus. *Arch Dermatol* **117**:73–76, 1981.

26. Kelly EW Jr, Pinkus H: Report of case of pachyonychia congenita. *Arch Dermatol* **77**:724–728, 1958.

27. Kumer L, Loos HO: Über Pachyonychia congenita (typus Riehl). *Wien Klin Wochenschr* **48**:174–178, 1935.

28. Laing CR et al: Pachyonychia congenita. *Am J Dis Child* **111**:649–652, 1966.

29. Maser ED: Oral manifestations of pachyonychia congenita. Report of a case. *Oral Surg Oral Med Oral Pathol* **43**:373–378, 1977.

30. McDonald RM, Reed WB: Natal teeth and steatocystoma multiplex complicated by hidradenitis suppurativa. *Arch Dermatol* **112**:1132–1134, 1976.

31. Moldenhauer E, Ernst K: Das Jadassohn–Lewandowsky-Syndrom. *Hautarzt* **19**:441–447, 1968.

32. Moldennauer E, Seidel R: Ist die Sebozystomatose ein Symptom des Jadassohn–Lewandowski-Syndroms? *Dermatol Mschr* **159**:540–543, 1973.

33. Murray FA: Four cases of hereditary hypertrophy of the nail bed associated with a history of erupted teeth at birth. *Br J Dermatol* **33**:409–411, 1921.

34. Perrot H et al: Étude ultrastructurale des lesions cutanées du syndrome de Jadassohn–Lewandowsky. *Arch Dermatol Forsch* **246**:114–124, 1973.

34a. Ribnik LR, Hoyme HE: Natal teeth in Native Americans. David W. Smith Workshop on Malformations and Morphogenesis. Madrid, Spain, 23–29 May 6, 1989.

35. Schönfeld PH: The pachyonychia congenita syndrome. *Acta Dermatol Venereol* **60**:45–49, 1980.

36. Shrank AB: Pachyonychia congenita. *Proc R Soc Med* **59**:975–976, 1966. (Same family as in Refs. 5 and 23.)

37. Soderquist NA, Reed WB: Pachyonychia congenita with epidermal cysts and other congenital dyskeratoses. *Arch Dermatol* **97**:31–33, 1968.

38. Stieglitz JB, Centerwall WR: Pachyonychia congenita (Jadassohn–Lewandowsky syndrome): A seventeen-member, four-generation pedigree with unusual respiratory and dental involvement. *Am J Med Genet* **14**:21–28, 1983.

39. Thomas DR et al: Pachyonychia congenita. Electron microscopic and epidermal glycoprotein assessment before and during isotretinoin treatment. *Arch Dermatol* **120**:1475–1479, 1984.

40. Thormann J, Kobayasi T: Pachyonychia congenita Jadassohn–Lewandowsky: A disorder of keratinization. *Acta Dermatol Venereol* **57**:63–67, 1977.

40a. Tidman MJ et al: Pachyonychia congenita with cutaneous amyloidosis and hyperpigmentation—a distinct variant. *J Am Acad Dermatol* **16**:935–940, 1987.

41. Touraine A: Pachyonychie congénitale. *Presse Méd* **45**:1569–1572, 1937.

42. Velasquez JP, Bustamante J: Sebocystomatosis with congenital pachyonychia. *Int J Dermatol* **11**:77–81, 1972.

43. Vineyard WR, Scott, RA: Steatocystoma multiplex with pachyonychia congenita. Eight cases in four generations. *Arch Dermatol* **84**:824–827, 1961.

44. Wilkin JK: Cornoid lamella in pachyonychia congenita. *Arch Dermatol* **114**:1795–1796, 1978.

45. Wilkinson DS: Pachyonychia congenita with epidermal cysts. *Br J Dermatol* **103** (Suppl **18**):40–41, 1980.

46. Wolfshaut A, Cernaianu R: Steatocistom si keratodermie familiale. *Dermatol Venereol (Bucharest)* **13**:447–454, 1968.

47. Wright CS, Guequierre JP: Pachyonychia congenita. Report of two cases, with studies on therapy. *Arch Dermatol Syph* **55**:819–827, 1947.

48. Young LW, Lenox JA: Pachyonychia congenita, a long-term evaluation of associated oral and dermal lesions. *Oral Surg Oral Med Oral Pathol* **36**:663–666, 1973.

## Hereditary benign intraepithelial dyskeratosis (Witkop–Von Sallmann syndrome)

Hereditary benign intraepithelial dyskeratosis (HBID) was described in a North Carolina triracial isolate (white–black–Indian) in 1960 by Witkop et al (7) and by Von Sallmann and Paton (5). The patients reported by Yanoff (9) had already been included in the study by Witkop et al (7). The chief components are plaques of the bulbar conjunctiva and oral mucosal thickenings clinically similar to white folded hypertrophy.

The syndrome has autosomal dominant inheritance with complete penetrance (7). Attempts at linkage with several blood groups have met with negative results (1).

**Eyes.** About the limbus, both nasally and temporally, there are foamy gelatinous plaques, more superficial than pterygia, on a hyperemic bulbar conjunctiva (Fig. 13–52). The eye lesion is usually noted within the first year of life (5). There is vernal exacerbation with shedding in the summer or fall.

The dyskeratotic process may involve the cornea, producing blindness from shedding and resultant vascularization of this structure (2). Photophobia and itching of the eyes, especially in children, is common.

**Oral manifestations.** The oral mucosal thickenings are asymptomatic. They appear as soft white folds and plaques, resembling white sponge nevus (10,11). Though the thickenings appear at birth, they are mild, increasing in severity to about 15 years of age. There is no tendency for the plaques to undergo malignant degeneration (Fig. 13–53).

**Differential diagnosis.** White sponge nevus and oral lesions of *pachyonychia congenita* bear a distinct clinical resemblance to those of HBID.

**Laboratory aids.** Tissue sections of buccal mucosal or conjunctival scrapings treated with Giemsa stain are characteristic (Fig. 13–54). Acanthosis, vacuolization of the stratum spinosum, and intraepithelial dyskeratosis characterized by waxy eosinophilic cells called "tobacco cells" and a "cell-within-a-cell" pattern are noted. These latter changes are especially evident in Papanicolaou stained smears (5) (Fig. 13–55).

Witkop and Gorlin (8) found similarities in oral smears from HBID and keratosis follicularis (Darier–White disease). The grains of the

Fig. 13–52. *Hereditary benign intraepithelial dyskeratosis.* Superficial gelatinous plaques on hyperemic bulbar conjunctiva involving limbus and cornea. (Courtesy of *CJ Witkop Jr,* Minneapolis, Minnesota.)

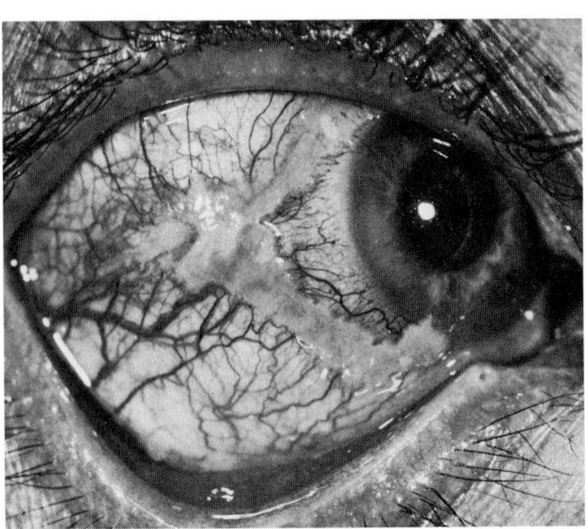

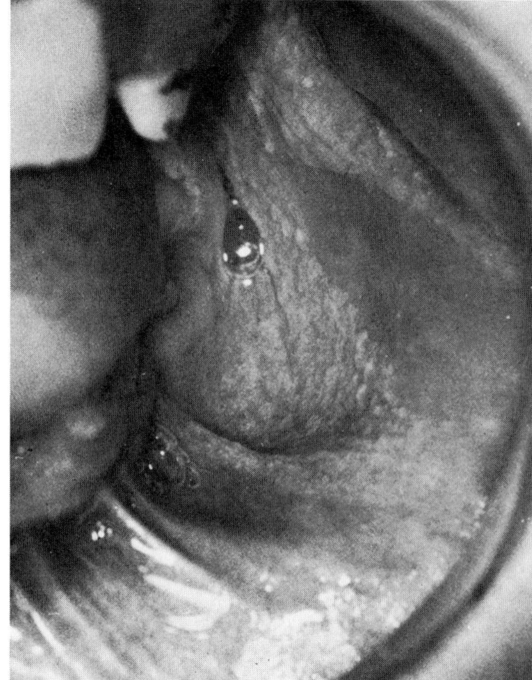

Fig. 13–53. *Hereditary benign intraepithelial dyskeratosis.* Leukokeratosis of buccal mucosa. (Courtesy of *CJ Witkop Jr,* Minneapolis, Minnesota.)

Fig. 13–54. *Hereditary benign intraepithelial dyskeratosis.* Section of buccal mucosa demonstrating dyskeratotic eosinophilic cells in acanthotic epithelium (Giemsa stain). (Courtesy of *CJ Witkop Jr,* Minneapolis, Minnesota.)

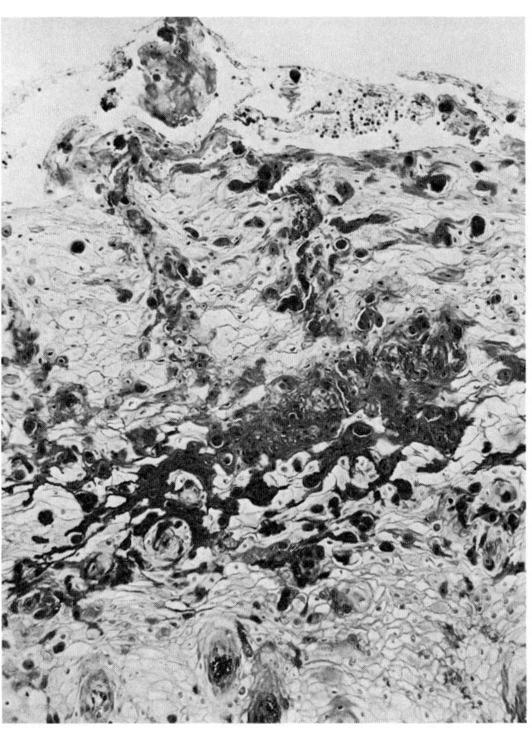

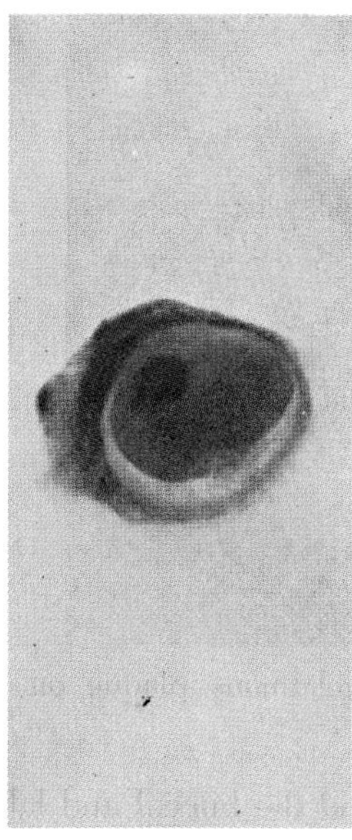

Fig. 13–55. *Hereditary benign intraepithelial dyskeratosis.* "Cell-within-a-cell" phenomenon, Papanicolaou smear. (Courtesy of *CJ Witkop Jr*, Minneapolis, Minnesota.)

latter resemble the so-called "tobacco cells" of the former, and the corps ronds of the latter resemble the "cell-within-a-cell" body seen in the syndrome under discussion. However, the cell-within-a-cell is far more common in oral smears of HBID, and, in addition, one rarely sees the small blue parabasilar cells so often noted in keratosis follicularis. Witkop (6) pointed out that patients receiving methotrexate and 5-fluorouracil also exhibit the "cell-within-a-cell" phenomenon in exfoliated buccal cells.

Ultrastructural studies have revealed numerous vesicular bodies in immature dyskeratotic cells, and the disappearance of cellular interdigitations and desmosomes in mature dyskeratotic cells (3). *Candida albicans* has been found in oral biopsies but its significance is moot (4).

#### References [Hereditary benign intraepithelial dyskeratosis (Witkop–Von Sallmann syndrome)]


1. Pollitzer WS et al: Hereditary benign intraepithelial dyskeratosis—a linkage study. *Am J Hum Genet* **17**:104–108, 1965.
2. Reed JW et al: Corneal manifestations of hereditary benign intraepithelial dyskeratosis. *Arch Ophthalmol* **97**:297–300, 1979.
3. Sadeghi EM, Witkop CJ: Ultrastructural study of hereditary benign intraepithelial dyskeratosis. *Oral Surg* **44**:567–577, 1977.
4. Sadeghi EM, Witkop CJ: The presence of *Candida albicans* in hereditary benign epithelial dyskeratosis. *Oral Surg* **48**:342–346, 1979.
5. Von Sallmann L, Paton D: Hereditary benign intraepithelial dyskeratosis: I. Ocular manifestations. *Arch Ophthalmol* **63**:421–429, 1960.
6. Witkop CJ Jr: Epithelial intracellular bodies associated with hereditary dyskeratoses and cancer therapy, in *Proc First Int Cong Exfoliative Cytology*, Wied GL (ed), Vienna, Austria, 1961. JB Lippincott, Philadelphia, 1962.
7. Witkop CJ et al: Hereditary benign intraepithelial dyskeratosis: II. Oral manifestations and hereditary transmission. *Arch Pathol* **70**:696–711, 1960.
8. Witkop CJ, Gorlin RJ: Four hereditary mucosal syndromes. *Arch Dermatol* **84**:762–771, 1960.
9. Yanoff M: Hereditary benign intraepithelial dyskeratosis. *Arch Ophthalmol* 79:291–293, 1968.
10. Zegarelli EV, Kutscher AH: Familial white folded hypertrophy of the mucous membranes. *Oral Surg* **10**:262–270, 1957.
11. Zegarelli EV et al: Familial white folded hypertrophy of the mucous membranes. *Arch Dermatol* **80**:97–103, 1959.


## Leukoplakia, tylosis, and esophageal carcinoma

In 1957 and 1959, Clarke et al (1,2) and Howel-Evans and others (4) reported the association of hyperkeratosis of the palms and soles (tylosis) and carcinoma of the esophagus in two families. Subsequently, Tyldesley and Hughes (11) and Tyldesley (10) reexamined members of these two kindreds and found, in addition, oral leukoplakia. At least four other kindreds with this constellation of abnormalities have been reported (3,6,7,9). Inheritance is autosomal dominant.

**Skin.** The age of onset of palmar and plantar hyperkeratosis is usually during the second decade of life (4,9) although onset between 1 and 5 years of age has been suggested (3) (Fig. 13–56A,B). The tylosis is diffuse, may be yellow and sometimes associated with hyperhidrosis. In some women, the feet alone have been reported affected (3,4). Ritter and Petersen (6) also described hyperkeratotic nodules on the distal and proximal interphalangeal joints and hyperpigmentation and hyperkeratosis on the upper parts of the arms.

**Gastrointestinal tract.** Squamous cell carcinoma of the esophagus develops in 95% of individuals by age 65. The age of onset may be as early as the fourth decade (4). Although the lower third of the esophagus is the most common site, the lesion may originate in the middle or upper third (4) (Fig. 13–57).

**Oral manifestations.** Clinical leukoplakia develops on the buccal mucosa (6,10) (Fig. 13–58) as early as 4 years of age as a diffuse gray white lesion. On microscopic examination, parakeratosis, spongiosis of the superficial layer of the epithelium, and acanthosis are noted; no atypia is seen. In older patients, lesions are smaller and more discrete; however, more diffuse regions may also be found. Parakeratosis or hyperorthokeratosis with epithelial atrophy or acanthosis are found, but microscopic features similar to the more diffuse lesions are also noted. No cytologic atypia is demonstrable. Ultrastructural studies demonstrate intranuclear electron-dense particles, similar to those found in nuclei of epithelium from the esophageal carcinomas (12).

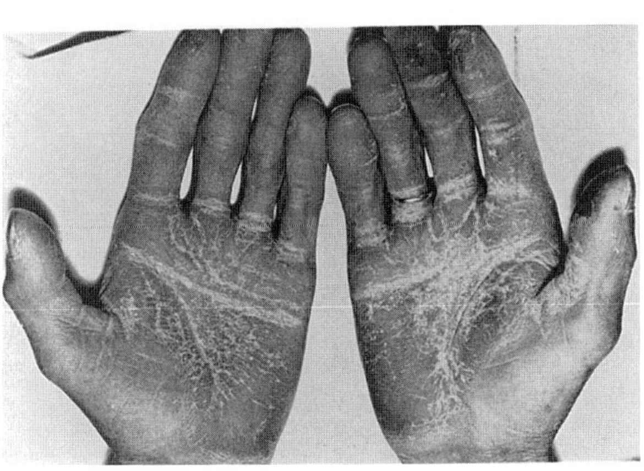

A

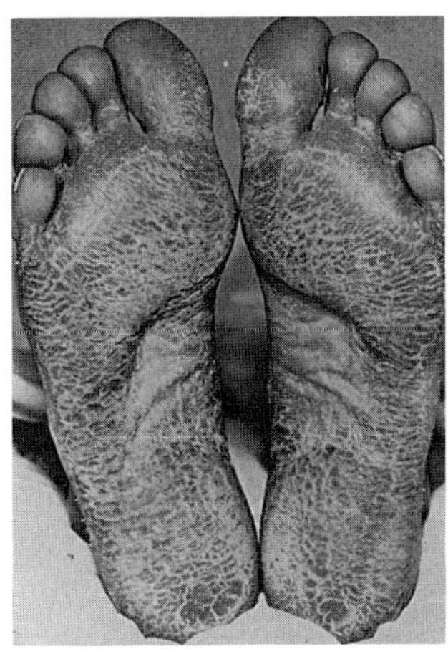

B

Fig. 13–56. *Leukoplakia, tylosis, and esophageal carcinoma.* (A) Palmar hyperkeratosis. (B) Plantar hyperkeratosis. (A,B from *W Howel-Evans*, Q J Med **27**:413, 1958.)

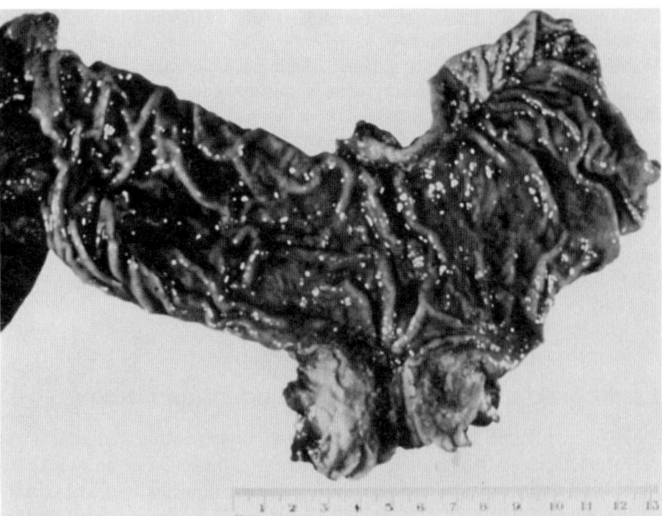

Fig. 13–57. *Leukoplakia, tylosis, and esophageal carcinoma.* Esophageal carcinoma. (Courtesy of *PS Harper,* Cardiff, Wales.)

**Differential diagnosis.** Many syndromes have associated hyperkeratosis of palms and soles. These include *hyperkeratosis palmoplantaris and periodontoclasia in childhood (Papillon–Lefèvre syndrome)* and *hyperkeratosis palmoplantaris and attached gingival hyperkeratosis.* Yesudian et al (13) reported a large kindred in which tylosis was present at birth. Inheritance was autosomal dominant. The tylosis was not associated with hyperhidrosis. Three members developed squamous cell carcinoma of the tylotic skin in the third decade of life; one of these three also developed squamous carcinoma of the lower end of the esophagus at age 40. The oral cavity was not described. Schöpf et al (8) reported two sisters, who, in addition to palmoplantar keratosis, had small cysts of the marginal eyelids, hypotrichosis, and onychodystrophy; hypodontia was reported but not well documented. Shine and Allison (9) reported a family with onset of hyperkeratosis of the palms and soles at adolescence; there was no associated hyperhidrosis. Dysphagia had been noted since infancy. The proband developed esophageal carcinoma at the site of an esophageal stricture. Inheritance was autosomal dominant.

White lesions of the oral mucous membranes, similar to those seen in this syndrome, are also found in white sponge nevus (5), *hereditary benign intraepithelial dyskeratosis, pachyonychia congenita,* and *dyskeratosis congenita.*

Fig. 13–58. *Leukoplakia, tylosis, and esophageal carcinoma.* Focal lesions of leukoplakia on the buccal mucosa.

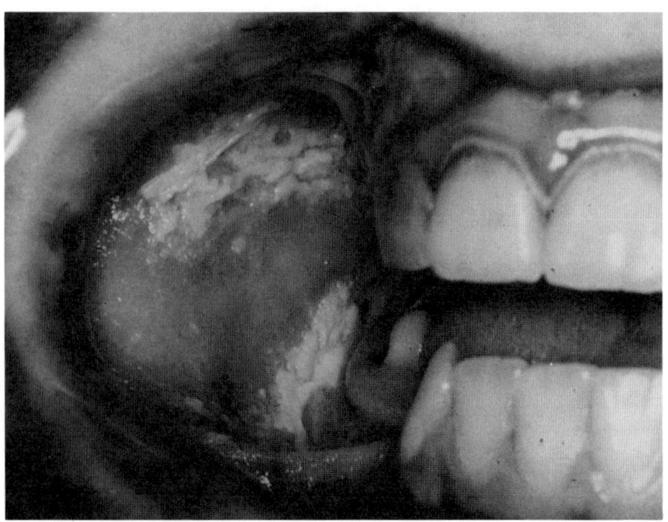

### References (Leukoplakia, tylosis, and esophageal carcinoma)


1. Clarke CA et al: Carcinoma of oesophagus associated with tylosis. *Br Med J* **1**:945, 1957 (letter).
2. Clarke CA et al: Carcinoma of oesophagus in association with tylosis. *Br Med J* **2**:1100, 1959.
3. Harper PS et al: Carcinoma of the oesophagus with tylosis. *Q J Med* **39**:317–333, 1970.
4. Howel-Evans W et al: Carcinoma of the oesophagus with keratosis palmaris et plantaris (tylosis). A study of two families. *Q J Med* **27**:413–429, 1958.
5. Jorgenson RJ, Levin LS: White sponge nevus. *Arch Dermatol* **117**:73–76, 1981.
6. Ritter SB, Petersen G: Esophageal cancer, hyperkeratosis, and oral leukoplakia. *JAMA* **235**:1723, 1976.
7. Ritter SB, Petersen G: Esophageal cancer, hyperkeratosis and oral leukoplakia: Follow-up family study. *JAMA* **236:**1844–1845, 1976 (letter).
8. Schöpf E: Syndrome of cystic eyelids, palmo-plantar keratosis, hypodontia and hypotrichosis as a possible autosomal recessive trait. *Birth Defects* **7**(8):219–221, 1971.
9. Shine I, Allison PR: Carcinoma of the oesophagus with tylosis (keratosis palmaris et plantaris). *Lancet* **1**:951–953, 1966.
10. Tyldesley WR: Oral leukoplakia associated with tylosis and esophageal carcinoma. *J Oral Pathol* **3**:62–70, 1974.
11. Tyldesley WR, Hughes RO: Tylosis, leukoplakia and esophageal carcinoma. *Br Med J* **4**:427, 1973.
12. Tyldesley WR, Kempson SA: Ultrastructure of the oral epithelium in leukoplakia associated with tylosis and esophageal carcinoma. *J Oral Pathol* **4**:49–58, 1975.
13. Yesudian P et al: Genetic tylosis with malignancy: A study of a South Indian pedigree. *Br J Dermatol* **102**:597–600, 1980.


## Mucoepithelial dysplasia syndrome (Witkop syndrome)

Witkop et al (2a–6), in 1978–1982, described an autosomal dominantly inherited syndrome that involved the skin and various mucous membranes, eyes, and lungs.

Other families were reported by Okamoto (1) and Robinow et al (2), and RJ Gorlin has seen another kindred.

The disorder is heralded in infancy by severe tearing and photophobia. Corneal vascularization, development of a pannus, nystagmus, and cataracts first appear at 4–6 years of age leading to blindness before puberty (Fig. 13–59A,B). The oral, nasal, vaginal, urethral, anal, and bladder mucosae are fiery red (Fig. 13–60A,B). Red, erosive or granular periorificial mucosal lesions, noted by the end of the first year of life, tend to persist. The skin of the trunk and extremities (but not the hands and feet) is rough (follicular keratosis) and the scalp

Fig. 13–59. *Mucoepithelial dysplasia syndrome.* (A,B) Sparse scalp hair, strabismus, photophobia, and cataracts. (Courtesy of *CJ Witkop Jr,* Minneapolis, Minnesota.)

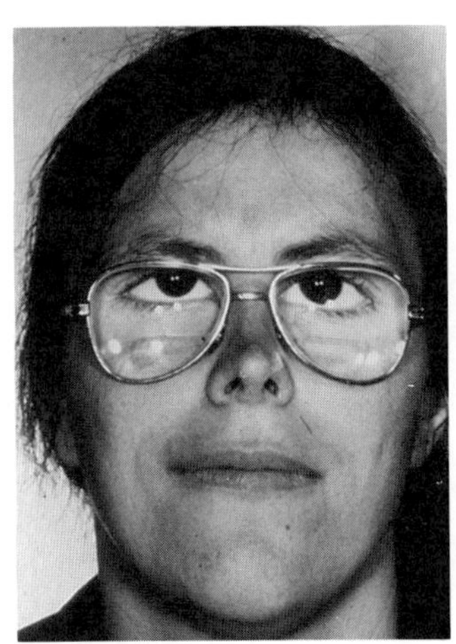

A

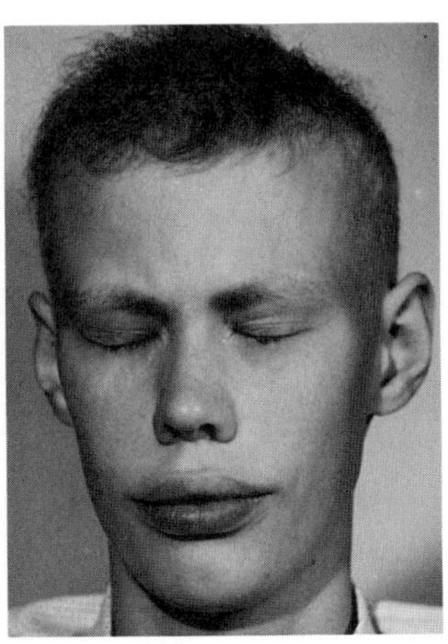

B

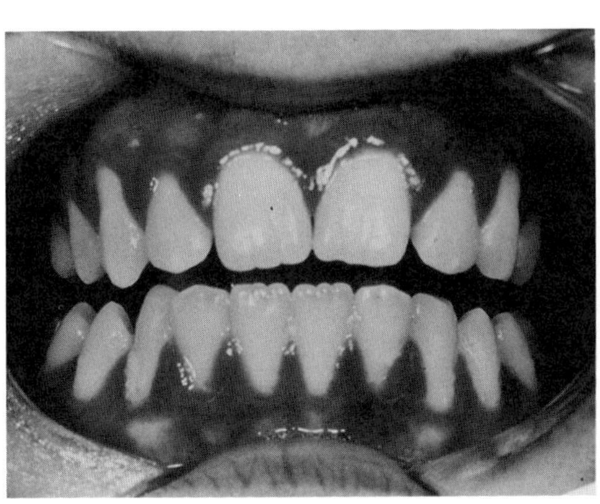
A

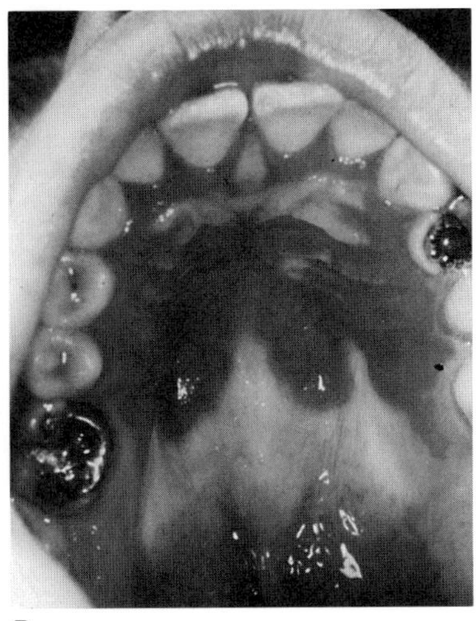
B

Fig. 13–60. *Mucoepithelial dysplasia syndrome.* (A) Fiery red gingival mucosa involving both free and attached gingiva. (B) Anterior palate involvement.

hair is scant (nonscarring alopecia). Patients often complain of easy burning on sun exposure.

Chronic rhinorrhea, upper respiratory infections, pneumonia, diarrhea, bladder infections with pyuria, and/or hematuria are common findings. Spontaneous pneumothorax is frequent, eventuating in terminal fibrocystic-type lung disease, cor pulmonale, and death in the third and fourth decades.

Mucosal biopsies show dyshesion, lack of keratinization, and dyskeratosis. Mucosal PAP smears principally exhibit numerous basal and parabasal cells (lack of epithelial maturation), nuclear atypia, and cytoplasmic vacuolization (Fig. 13–61). Ultrastructural changes suggest desmosomal and gap junction defects.

#### References [Mucoepithelial dysplasia syndrome (Witkop syndrome)]


1. Okamoto GA et al: New syndrome of chronic mucocutaneous candidosis. *Birth Defects* **13**(3B):117–125, 1977.

2. Robinow M et al: Hereditary mucoepithelial dysplasia—A new variant? David W. Smith Workshop on Malformations and Morphogenesis, Burlington, Vermont, August, 1986.

2a. Scheman AJ et al: Hereditary mucoepithelial dysplasia. *J Am Acad Dermatol* **21**:351–357, 1989.

3. Witkop CJ Jr et al: Clinical, histologic, cytologic and ultrastructural characteristics of the oral lesions from hereditary mucoepithelial dysplasia: Disease of gap junction and desmosomal formation. *Oral Surg* **46**:645–657, 1978.

4. Witkop CJ Jr et al: Hereditary mucoepithelial dysplasia: A disease apparently of desmosome and gap junction formation. *Am J Hum Genet* **31**:414–427, 1979.

5. Witkop CJ Jr: Disorders affecting cellular communications in oral tissues: Gap junctions. *Birth Defects* **16**(2):197–209, 1980.

6. Witkop CJ Jr: Hereditary mucoepithelial dysplasia, a disease of gap junctions and desmosome formation. *Birth Defects* **18**(6):493–511, 1982.


Fig. 13–61. *Mucoepithelial dysplasia syndrome.* Exfoliated cell smears show numerous immature epithelial cells, perinuclear vacuoles, and strand-shaped inclusions. Small dark cells are dyskeratotic cells that are orangeophilic. (Courtesy of *CJ Witkop Jr,* Minneapolis, Minnesota.)

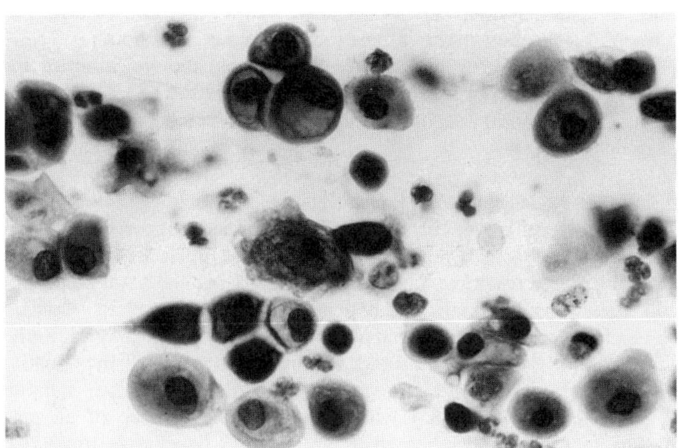

## Hypohidrotic ectodermal dysplasia

The syndrome is characterized chiefly by hypohidrosis, hypotrichosis, and hypodontia. Charles Darwin (25) cited Wedderburn as having found the disorder in an Indian from the subcontinent. Thurnam (109), in 1848, also described the condition. According to Perabo et al (81), it may have been recorded as early as 1792 by Danz. Christ (16), in 1913, further defined it as a ''congenital ectodermal defect,'' and Weech (117), in 1929, impressed by the depression of sweat gland function, coined the term ''anhidrotic ectodermal dysplasia.'' Felsher (35), in 1944, pointed out that the skin is rarely, if ever, completely anhidrotic, and suggested the adjective ''hypohidrotic,'' a far better choice. The reader is referred to the study of Perabo et al (81) for an analysis of sign and symptom frequency. The frequency of occurrence is about 1/100,000 births (17). A most comprehensive survey is that of Airenne (1).

The syndrome usually (vide infra) has X-linked recessive inheritance (47,112). As indicated under Laboratory Aids, the gene has been located at Xq13→q21 (51). Minimal expression of the gene in the form of hypodontia and/or teeth with conical crown form, and reduction in sweating and body hair in a patchy distribution may be observed in 60–75% of carrier females (15,19,61,76,85,87,88). About 80% of carriers have difficulty in nursing (19). Abnormalities in heterozygotes are consistent with lyonization (61). Sofaer (103,104) suggested that 1 in 500 females with hypodontia in the permanent dentition and 1 in 50 with hypodontia in the deciduous dentition may be carriers for hypohidrotic ectodermal dysplasia. A remarkable report describes the disorder in monozygous quadruplets (28). There may be a second mild X-linked form (98).

At least 35 females have the complete syndrome; it is likely that many of these patients have autosomal recessive hypohidrotic ectodermal dysplasia, a genocopy clinically indistinguishable from the X-linked disorder in males (3,7,23,30,47,50,62,64,80,108). Other cases of the autosomal recessive form have been cited by Jorgenson et al (58). Elejalde and de Elejalde (30) suggested that the autosomal recessive form exhibits less marked skin pigmentation than the X-linked

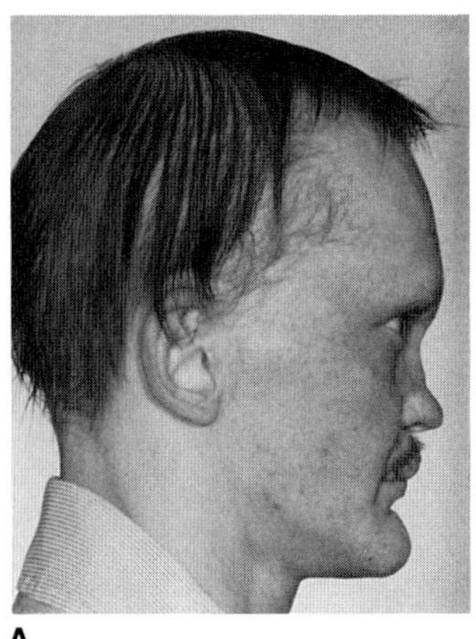
A

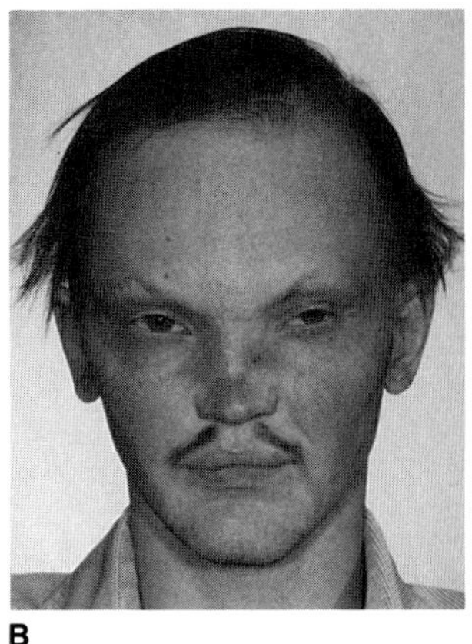
B

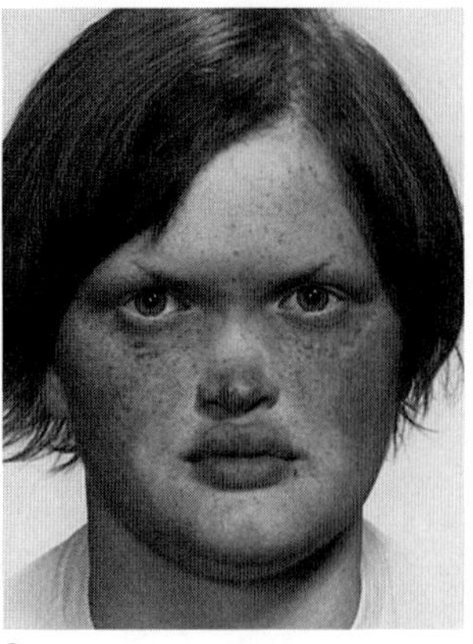
C

Fig. 13–62. *Hypohidrotic ectodermal dysplasia.* (A,B) X-linked recessive form. Face is characterized by frontal bossing, depressed nasal bridge, thin hair, and periorbital increased pigmentation. (C) Female with autosomal recessive form. Note sparse hair and fine linear wrinkling around eyes. (From *EC Passarge et al,* Humangenetik **3**:181, 1966.)

type. Sybert (106a) has suggested that the autosomal recessive form can be explained by marked lyonization of affected females.

**Facies.** Frontal bossing, usually marked, and a depressed nasal bridge give added emphasis to the small size of the face (99). The nasal bridge resembles the saddle nose of congenital syphilis. There are often fine linear wrinkles and pigmentation about the eyes (34,102). Due to absence of teeth, and resulting reduced vertical height, the lips are protuberant. The pinnae are often outstanding (81) (Fig. 13–62A–C).

Cephalometric and anthropometric studies of hemizygotes and heterozygotes have shown a small lower facial height and depth, small palatal and cranial base widths, small malar bones, small calvarial height, prominent forehead, depressed nasal bridge, and high-set orbits (9,93).

**Skin and skin appendages.** A most remarkable characteristic of this disorder is hypohidrosis. Because other physical features of this syndrome are not so apparent in the first year, the child may present with a "fever of unknown origin." Inability to sweat, because eccrine sweat glands are severely reduced in number (101), results in intolerance to heat, with severe incapacitation and hyperpyrexia after only mild exertion. Bizarre heat loss has been documented (91). In several newborns that we have seen, the skin has been somewhat scaly.

The skin is soft and thin. Dryness is often severe, because sebaceous glands are absent (60). Eczema is seen in 65%, especially during the early years of life (19,45). The body is usually devoid of lanugo hair. After puberty, the moustache and beard are described as normal (35), although axillary and pubic hair are frequently scant (54,73). Scalp hair is often blond, fine, stiff, and short. The eyelashes and, especially, the eyebrows are scanty or often missing. Female heterozygotes often exhibit sparse scalp hair, spotty sweating, random reduction in sweat pores and a mosaic, patchy distribution of body hair (85). Happle and Frosch (52) noted that the hypohidrosis in heterozygotes follows Blaschko lines.

Nails are usually normal, but may be spoon-shaped (117). Mammary glands may be aplastic or hypoplastic (14,34,73,77,101,112) and nipples may be rudimentary (12,34,73,77). Congenital absence of nipples in a putative heterozygote has been reported (13). Dermatoglyphic alterations have been described (89,114).

**Eyes.** Lacrimal gland function has been reported diminished in a few cases (49,54,68,75) and increased in others (86). There are no meibomian glands and, frequently, a punctate epithelial keratopathy has been found (29). Congenital glaucoma has been noted in some patients (57,96); this association may have been coincidental or the patients may have had a syndrome different from hypohidrotic ectodermal dysplasia.

**Otolaryngologic manifestations.** Mucous glands in portions of the respiratory tract have been described as absent by some investigators (14,27). Pharyngeal and laryngeal mucosa may be atrophic, resulting in dysphonia (29,82,102). On laryngologic examination, breathy voice quality has been noted (82). Vocal folds appeared to close completely, although they were dry (82). Abnormal voice quality on spectrographic analysis has also been documented (82). Several reports have described atrophy of the nasal mucosa associated with severe crusting and a fetid secretion (ozena) (29,56,68,81). These crusts may obstruct the nasal passages and make feeding difficult. Scintigraphic studies (71) have suggested an inflammatory process in the parotid glands.

**Other findings.** Several reports have noted that about 65% exhibit allergic disorders, especially eczema and asthma (8,19,102,113). Although rare, death may occur in infancy due to respiratory infection or hyperpyrexia (6,22,27). The latter may be accompanied by seizures. Pinheiro and Freire-Maia (86) noted deviation of some of the terminal phalanges of the hands and feet.

**Oral manifestations.** In hemizygotes, the most striking oral abnormality is absence of most deciduous and permanent teeth. Maxillary central incisors and canines usually have a conical crown form (Fig. 13–63A,B). Frequently one or more molars may be present. More rarely, one or both jaws may be edentulous (97). Female heterozygotes exhibit reduction in numbers of teeth and smaller crown size than hemizygous males (71,76). Because so many teeth are congenitally missing in hemizygous males, vertical dimension is reduced, and the lips are protuberant. The vermilion border is indistinct and pseudorhagades may be present (81,112).

The alveolar process does not develop in the absence of teeth and, hence, is missing (33,97). Cephalometric studies have been limited in number (67,97) but have indicated that apart from defective alveolar growth, jaw and facial development is essentially normal. However, Ward and Bixler (115) employing cephalometric analysis in 16 affected individuals found pronounced size reduction, compared to controls, in anteroposterior dimensions of the lower two-thirds of the face, in facial height, and in the size of the ears, nose, and mouth.

It has been stated by a few investigators that the oral mucosa appears dry (82), or that salivary secretion is diminished (33,42,101,102) Aplasia of intraoral mucous glands has been noted on microscopic examination (14,33).

**Differential diagnosis.** Freire-Maia and Pinheiro (37) have provided a comprehensive discussion of differential diagnosis. Although the physiognomy in hypohidrotic ectodermal dysplasia is distinctive, several features resemble those of other disorders. The nasal deformity and linear perioral scarring may suggest congenital syphilis. Conical or tapered teeth, as well as congenitally missing teeth, may be found in individuals who have no other abnormalities (isolated hypodontia, oligodontia); in many cases, this condition may be familial (5,31,36,41,102,107, LS Levin, unpublished). Absence of all permanent teeth may be inherited as an autosomal recessive trait (48).

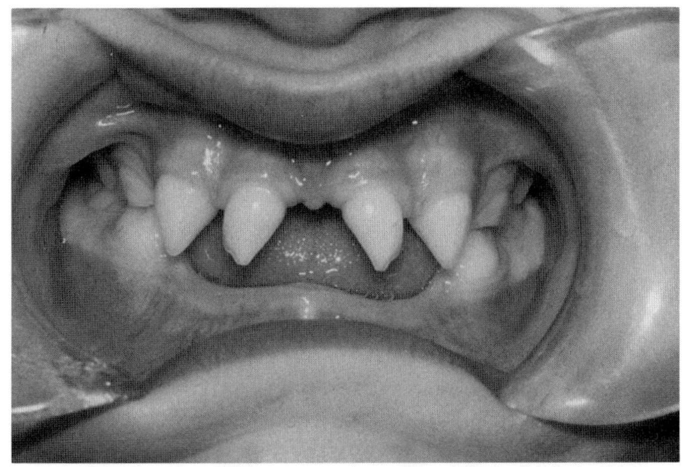

A

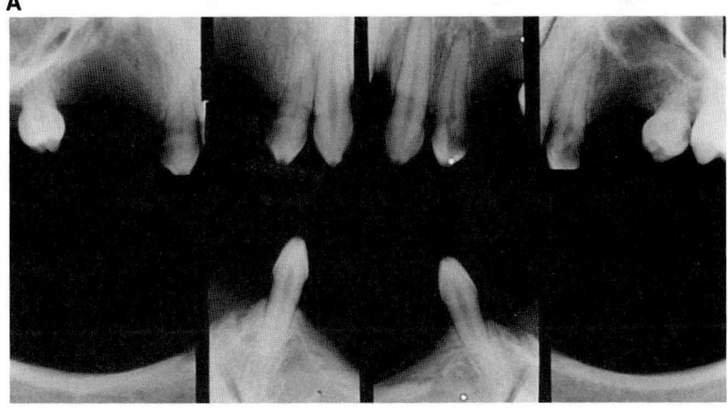

B

Fig. 13–63. *Hypohidrotic ectodermal dysplasia.* (A) Hypodontia and conical crown tooth form in 14-year-old female. No other stigmata present; possible carrier of trait. (B) Dental radiographs showing hypodontia and conical form of tooth crowns.

Conical teeth are also found in *acrodental dysostosis* (24,99). Congenital absence of teeth as well as conical teeth may be seen in *Ellis–van Creveld syndrome, Rieger syndrome,* and *incontinentia pigmenti;* however, other abnormalities should serve to distinguish these disorders from hypohidrotic ectodermal dysplasia.

In the trichodental syndrome (95), an autosomal dominant disorder, patients have fine scalp hair that grows slowly, and sparse eyebrows distally; males have a high anterior hairline. Teeth are congenitally absent, although not as many as in hypohidrotic ectodermal dysplasia. The nails are presumably normal and no abnormalities in sweating have been described. Pinheiro et al (84) reported a mother and two children with sparse, thin, and brittle scalp hair from birth, scanty axillary and pubic hair, sparse eyebrows and eyelashes, mild palmoplantar keratosis, multiple congenitally absent deciduous and permanent teeth, conically shaped maxillary central incisors, café-au-lait spots, and mildly dystrophic toenails. LS Levin has seen a 15-month-old white male with two erupted pegged deciduous incisors in the mandible and two in the maxilla. Hair was blond and slow growing; nails were also slow growing but not dysplastic. He tolerated heat well, and sweating was noted. The parents were normal. A dominant hypohidrotic ectodermal dysplasia has also been described (58). The reader is also referred to a paper by Solomon and Keuer (105) for a discussion of the ectodermal dysplasias.

Teeth are congenitally absent in patients with *Witkop tooth-nail syndrome* inherited as an autosomal dominant disorder. Fried's tooth and nail syndrome should also be excluded. LS Levin has seen two sibs with hypodontia, taurodontia, sparse hair, and spoon-shaped fingernails and toenails. Both parents were normal. Oligodontia or anodontia has been seen in association with facial clefts and nail dysplasia (116).

Cole et al (20) reported sisters with alopecia, small conical teeth, hypoplastic nails, hypohidrosis, mottled skin, lamellar cataracts, and mental retardation. Freire-Maia and Pinheiro (37) called the disorder tricho-odonto-onycho-hypohidrotic ectodermal dysplasia and suggested autosomal inheritance. Pabst et al (78), Morris et al (75a), and Pike et al (83) described hypohidrotic ectodermal dysplasia, freckling, enteropathy, onychodystrophy, atopy, and hypothyroidism in sibs. Pulmonary and upper respiratory infection due to a ciliary defect was found. They had normal facies and dentition. This combination has been called ANOTHER syndrome. Zadik et al (119) reported the combination of hypothyroidism, sparse hair, unusual facies, hypohidrosis, oligodontia, and dermoid cysts. One should exclude the recessive syndrome of cleft palate, congenital hypothyroidism, and spiked hair. Tuffli and Laxova (111) reported a mother and son with hypohidrosis, scalp defect at the vertex, hypoplasia of nipples, onychodysplasia, and possible abnormalities of dentition; in addition, the mother was unable to lactate.

Patients with *cleidocranial dysplasia* lack teeth on clinical examination. However, radiographically, a full complement of impacted teeth from the normal dentition as well as supernumerary teeth are found. Shokeir (100) described a family in which failure of most permanent teeth to erupt was inherited as an autosomal dominant disorder; Tipton and Gorlin (110) described an autosomal recessive syndrome of *growth retardation, alopecia, pseudoanodontia, and optic atrophy (GAPO syndrome)*.

An autosomal dominant hidrotic ectodermal dysplasia syndrome exists in a family of French extraction that migrated to Canada, Scotland, and northern United States (32,40,118). This disorder has also been described in a Chinese family (90). There are normal sweat and sebaceous gland function, sparse hair, severe nail dystrophy, palmar and plantar hyperkeratosis, skin hyperpigmentation, especially over the joints, and normal teeth. Oral leukoplakia has also been noted (40).

Another dominantly inherited hidrotic ectodermal dysplasia is characterized by sensorineural hearing impairment, polydactyly, syndactyly, nail dystrophy, and teeth that have conical crowns (69,92). The patients described by Mannkopf and Hanney (74) as having ectodermal dysplasia were of several types. Their case 7 had *focal dermal hypoplasia.* Patients possibly similarly involved are those described by Salamon and Miličevič (94) and by Friedrich and Seitz (39).

Anhidrosis and middle life onset neurolabyrinthitis has been described as an autosomal dominant disorder (55). Isolated anhidrosis may have autosomal recessive inheritance (72). Dominantly inherited hypohidrosis and hyperpigmentation with hyperkeratosis is a recognized disorder (106). Isolated amastia may be inherited as a dominant trait (46). A number of patients with so-called autosomal recessive HED undoubtedly represent still other disorders (20,64,65).

**Laboratory aids.** The dental radiograph is invaluable in determining whether teeth are congenitally absent, or whether they are present but unerupted.

Decreased sweating may be demonstrated by the starch-iodine method (Minor test) (33,35,59), by pilocarpine iontophoresis (85), by the use of agar plates containing silver nitrate and potassium chromate (44), by altered electrical resistance of the skin (64), or by use of *o*-phthaldialdehyde in xylene (114) (Fig. 13–64). Sweat pore counting may be carried out by direct observation or by use of silicone rubber (Fig. 13–65) or cellulose acetate (6,21,23,53,66,85,114). Lambert and Bilinski (65) advocated palmar skin biopsy. Happle and Frosch (52) suggested that the entire back be tested using the starch-iodine method, the areas of hypohidrosis following Blaschko lines.

Frias and Smith (38) advocated counting sweat pores per linear centimeter, the normal number decreasing from about 40 in infancy to 20 in old age. They suggested that carrier mothers had fewer sweat pores, a finding confirmed by Verbov (114). Although no sweat pores or intradermal eccrine glands or ducts are found in male hemizygotes with the X-linked form (65), a decreased number are noted in affected males and females with the autosomal recessive type (23,79). Hetero-

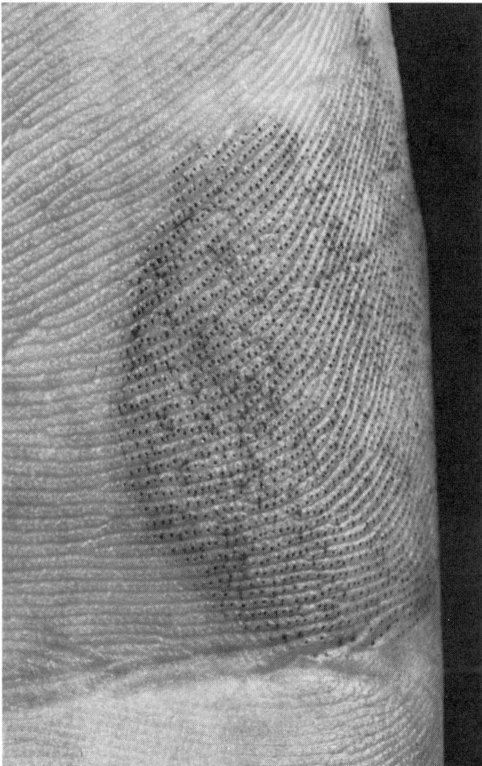

Fig. 13–64. *Hypohidrotic ectodermal dysplasia.* Sweating demonstrated by use of *o*-phthalaldialdehyde. (From *J Verbov,* Br J Dermatol **83**:341, 1970.)

zygotes with the autosomal recessive form have somewhat reduced pore counts (62,79).

Davis and Solomon (26) described some degree of cellular immunodeficiency in hypohidrotic ectodermal dysplasia. IgA has been normal in serum and saliva (102).

Hypohidrotic ectodermal dysplasia has been diagnosed prenatally by biopsy on fetoscopy (2,4,10,43).

A number of investigators, employing various DNA probes, have located the locus within the region Xq13-21.1 by linkage analysis (18,51,63,70,120). This should enable accurate carrier detection and prenatal diagnosis. It should also help determine whether there is an autosomal recessive form.

A

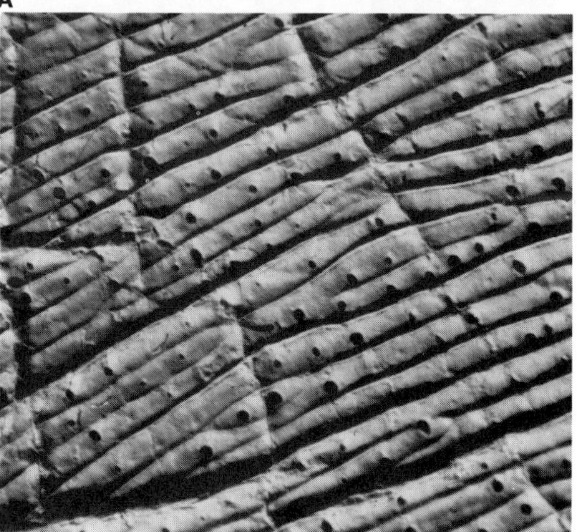

B

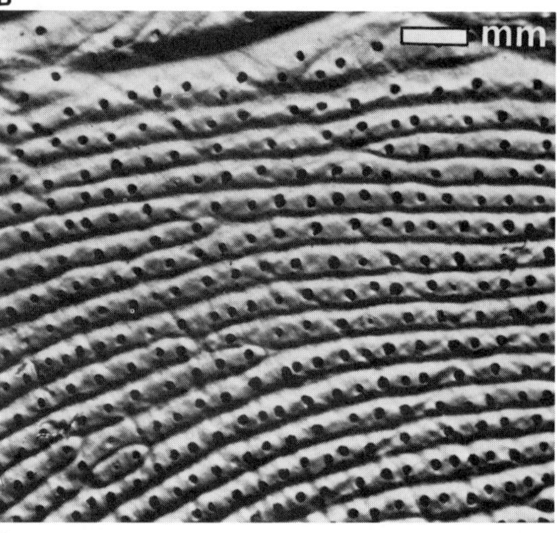


C

Fig. 13–65. *Hypohidrotic ectodermal dysplasia.* Sweating using silicone rubber technique. (A) Affected male. (B) Female heterozygote. (C) Normal individual. (From *J Verbov,* Br J Dermatol **83**:341, 1970.)

### References (Hyphidrotic ectodermal dysplasia)


1. Airenne P: X-linked hypohidrotic ectodermal dysplasia in Finland. *Proc Finn Dent Soc* **77**(Suppl 1):1–107, 1981.
2. Anton-Lamprecht I et al: Letter to the editor. *Hum Genet* **62**:180, 1982.
3. Anton-Lamprecht I et al: Autosomal recessive anhidrotic ectodermal dysplasia: Report of a case and description of diagnostic features. *Birth Defects* **24**(2):183–195, 1988.
4. Arnold M-L et al: Prenatal diagnosis of anhidrotic ectodermal dysplasia, *Prenatal Diagn* **4**:85–94, 1984.
5. Arya BS, Savara BH: Familial partial anodontia. *J Dent Child* **41**:47–54, 1974.
6. Awwaad S, El Essawy M: Hereditary anhidrotic ectodermal dysplasia. Review of literature and report of two cases. *Arch Pediatr* **77**:496–502, 1960.
7. Bartlett RC: Autosomal recessive hypohidrotic ectodermal dysplasia: Dental manifestations. *Oral Surg* **33**:736–742, 1972.
8. Beahrs JO: Anhidrotic ectodermal dysplasia: Predisposition to bronchial disease. *Ann Intern Med* **74**:92–96, 1971.
9. Bixler D et al: Characterization of the face in hypohidrotic ectodermal dysplasia by cephalometric and anthropometric analysis. *Birth Defects* **24**(2):197–203, 1988.
10. Blanchet-Bardon C, Nazzaro V: Use of morphological markers in carriers as an aid in genetic counseling and prenatal diagnosis. *Curr Probl Dermatol* **16**:109–119, 1987.
11. Bonora G et al: Displasia ectodermica ipoidrotica (DEA) nel sesso femminile. *Min Pediatr* **33**:911–916, 1981.
12. Borjian J: The effect of early dental treatment of anhidrotic ectodermal dysplasia. *J Am Dent Assoc* **61**:555–559, 1960.
13. Burck U, Held KR: Athelia in a female infant heterozygous for anhidrotic ectodermal dysplasia. *Clin Genet* **19**:117–121, 1981.
14. Capitanio MA et al: Congenital anhidrotic ectodermal dysplasia. *Am J Roentgenol* **103**:168–172, 1968.

15. Carter JW, Bordy MD: Ectodermal dysplasia and the Lyon hypothesis. *J Dent Child* **34**:265–268, 1967.

16. Christ J: Über die Korrelationen der kongenitalen Defekte des Ektoderms untereinander, mit besonderer Berücksichtingen ihrer Beziehungen zum Auge. *Zentralb Haut Geschlechtskr* **40**:1–21, 1932.

17. Clarke A: Hypohidrotic ectodermal dysplasia. *J Med Genet* **24**:659–663, 1987.

18. Clarke A et al: X-linked hypohidrotic ectodermal dysplasia: DNA probe linkage analysis and gene localization. *Hum Genet* **75**:378–380, 1987.

19. Clarke A et al: Clinical aspects of X-linked hypohidrotic ectodermal dysplasia. *Arch Dis Childh* **62**:989–996, 1987.

20. Cole HM et al: Congenital cataracts in sisters with congenital ectodermal dysplasia. *J Am Med Assoc* **129**:723–728, 1945.

21. Collard P, Dodinval P: Une nouvelle méthode pour prendre des empreintes digitales et palmaires. *J Génét Hum* **15**:21–26, 1966.

22. Cook WA, Kane FG: A family history of anhidrotic mesodermal-ectodermal dysplasia. *J Am Dent Assoc* **76**:1032–1037, 1968.

23. Crump JA, Danks DM: Hypohidrotic ectodermal dysplasia. *J Pediatr* **78**:466–473, 1971.

24. Curry CJR, Hall BD: Polydactyly, conical teeth, nail dysplasia, and short limbs: A new autosomal dominant malformation syndrome. *Birth Defects* **15**(5B):253–263, 1979.

25. Darwin C: *The Variations of Plants and Animals under Domestication*, Vol II, D. Appleton, New York, 1897, p 319.

26. Davis JR, Solomon LM: Cellular immunodeficiency in anhidrotic ectodermal dysplasia. *Acta Dermatovenereol* **56**:115–120, 1976.

27. de Jager H: Congenital anhidrotic ectodermal dysplasia. *J Pathol Bacteriol* **90**:321–322, 1965.

28. Delaire J et al: La dysplasie ectodermique anhidrotique. A propos d'une observation de quadruplés. *Rev Stomatol Chir Maxillofac* **85**:34–37, 1984.

29. Ekins MB, Waring GO: Absent meibomian glands and reduced corneal sensation in hypohidrotic ectodermal dysplasia. *J Pediatr Ophthalmol Strab* **18**(4):44–47, 1981.

30. Elejalde BR, de Elejalde MM: Pigmentary characteristics of the ectodermal dysplasia. *J Clin Dysmorphol* **1**(1):2–8, 1983.

31. Emery BJ: Partial anodontia of the deciduous dentition. *Br Dent J* **117**:487–488, 1964.

32. Escobar V et al: Clouston syndrome: An ultrastructure study. *Clin Genet* **24:**140–146, 1983.

33. Everett FG et al: Anhidrotic ectodermal dysplasia with anodontia: A study of two families. *J Am Dent Assoc* **44**:173–186, 1952.

34. Familusi JB et al: Hereditary anhidrotic ectodermal dysplasia in a Nigerian family. *Arch Dis Childh* **50**:642–647, 1975.

35. Felsher Z: Hereditary ectodermal dysplasia. Report of a case with experimental study. *Arch Dermatol Syph (Chic)* **49**:410–414, 1944.

36. Foster TD, Van Roey ORC: The form of the dentition in partial anodontia. *Dent Pract Dent Rec* **20**:163–160, 1970.

37. Freire-Maia N, Pinheiro M: *Ectodermal Dysplasias: A Clinical and Genetic Study*. Alan R. Liss, New York, 1984.

38. Frias JL, Smith DW: Diminished sweat pores in hypohidrotic ectodermal dysplasia: A new method of assessment. *J Pediatr* **72**:606–610, 1968.

39. Friederich HC, Seitz R: Über eine Form der ektodermalen Dysplasie unter dem Bilde der Pili torti mit Augenbeteiligung und Störungen der Schweisssekretion. *Dermatol Wochenschr* **131**:277–283, 1955.

40. George DI Jr, Escobar VH: Oral findings of Clouston's syndrome (hidrotic ectodermal dysplasia). *Oral Surg Oral Med Oral Pathol* **57**:258–262, 1984.

41. Gertzman GB: Genetics—A tool for the dentist. Report of a case of inherited oligodontia. *Clin Prev Dent* **4**:19–21, 1982.

42. Gibbs JH: Total absence of teeth in two brothers. *Dent Cosmos* **58**:352–353, 1916.

43. Gilgenkrantz S et al: Hypohidrotic ectodermal dysplasia. Clinical study of a family of 30 over three generations. *Hum Genet* **81**:120–122, 1989.

44. Glicklich LB, Rosenthal IM: Anhidrotic ectodermal dysplasia: Use of silver nitrate plate to detect anhidrosis. *J Pediatr* **54**:19–26, 1959.

45. Goepferd SJ, Carroll CE: Hypohidrotic ectodermal dysplasia: A unique approach to esthetic and prosthetic management. *J Am Dent Assoc* **102**:867–869, 1981.

46. Goldenring H, Crelin ES: Mother and daughter with bilateral congenital amastia. *Yale J Biol Med* **33**:466–467, 1961.

47. Gorlin RJ et al: Hypohidrotic ectodermal dysplasia in females: A critical analysis and argument for genetic heterogeneity. *Z Kinderheilkd* **108**:1–11, 1970.

48. Gorlin RJ et al: Complete absence of the permanent dentition: An autosomal recessive disorder. *Am J Med Genet* **5**:207–209, 1980.

49. Grant R, Falls HF: Anodontia: Report of a case associated with ectodermal dysplasia of the anhidrotic type. *Am J Orthodont* **30**:661–672, 1944.

50. Hall BD: Twenty-two year follow-up on original autosomal recessive hypohidrotic ectodermal dysplasia family. Southern Genetics Group, Navarre Beach, Florida, July, 1988.

51. Hanauer A et al: Genetic mapping of anhidrotic ectodermal dysplasia: DXS 159, a closely linked proximal marker. *Hum Genet* **80**:177–180, 1988.

52. Happle R, Frosch PJ: Manifestation of the lines of Blaschko in women heterozygous for X-linked hypohidrotic ectodermal dysplasia. *Clin Genet* **27**:468–471, 1985.

53. Harris DR et al: Evaluating sweat gland activity with imprint techniques. *J Invest Dermatol* **58**:78–84, 1972.

54. Hartwell SWJ, Pickrell K: Congenital anhidrotic ectodermal dysplasia. *Clin Pediatr* **4**:383–386, 1965.

55. Helweg-Larsen JH, Ludvigsen K: Congenital familial anhidrosis and neurolabyrinthitis. *Acta Dermatol Venereol* **26**:489–505, 1946.

56. Isaa H: Total anodontia with ectodermal dysplasia. *Br Dent J* **118**:537–544, 1965.

57. Jerndal T: Ectodermal dysplasia with infantile congenital glaucoma. *J Pediatr Ophthalmol* **7**:29–32, 1970.

58. Jorgenson RJ et al: Autosomal dominant ectodermal dysplasia. *J Craniofac Genet Dev Biol* **7**:403–412, 1987.

59. Juhlin L, Shelley WB: A stain for sweat pores. *Nature (London)* **312**:408, 1967.

60. Katz SI, Penneys NS: Sebaceous gland papules in anhidrotic ectodermal dysplasia. *Arch Dermatol* **103**:507–509, 1971.

61. Kerr CB et al: Genetic effect in carriers of anhidrotic ectodermal dysplasia. *J Med Genet* **3**: 169–176, 1966.

62. Kleinebrecht J et al: Sweat pore counts in ectodermal dysplasias. *Hum Genet* **57**:437–439, 1981.

63. Kølvraa TA et al: Close linage between X-linked ectodermal dysplasia and a cloned DNA sequence detecting a two allele restriction fragment length polymorphism in the region Xp11-q12. *Hum Genet* **74**:284–287, 1986.

64. Kratzsch R: Ectodermale Dysplasie von anhidrotischen Typ bei zwei Schwestern. *Klin Pädiatr* **184**:328–332, 1972.

65. Lambert WC, Bilinski DL: Diagnostic pitfalls in anhidrotic ectodermal dysplasia: Indications for palmar skin biopsy. *Cutis* **31**:182–187, 1983.

66. Laurent JM, Fontaine G: Le comptage des pores cutanés sur prise d'empréinte au silicum organique. *J Génét Hum* **29**:141–149, 1981.

67. Lipschutz A: Anhidrotic ectodermal dysplasia. *J Albert Einstein Med Ctr* **11:**33–37, 1963.

68. Lowenburg H, Grimes EL: Ectodermal dysplasia of the anhidrotic type. *Am J Dis Child* **63**:357–365, 1942.

69. Lowry RB et al: Hereditary ectodermal dysplasia. *Clin Pediatr* **5**:395–402, 1966.

70. MacDermot KD et al: Gene localisation of X-linked hypohidrotic ectodermal dysplasia (C-S-T syndrome). *Hum Genet* **74**:172–173, 1986.

71. Machtens E et al: Klinische Aspekte der ektodermalen Dysplasie. *Z Kinderheilkd* **112**:265–280, 1972.

72. Mahloudji J, Livingston KE: Familial and congenital simple anhidrosis. *Am J Dis Child* **113**:477–479, 1967.

73. Malagon V, Taveras JE: Congenital anhidrotic ectodermal and mesodermal dysplasia. Report on two cases with atrichia and amastia. *Arch Dermatol Syph* **74**:253–258, 1956.

74. Mannkopf H, Hanney F: Zum Erscheinungsbild der kongenitalen ektodermalen Dysplasien. *Albrecht Graefe's Arch Klin Exp Ophthalmol* **159**:643–661, 1957.

75. Martin-Pascual A et al: Anhidrotic ectodermal dysplasia. *Dermatologica* **154**:235–243, 1977.

75a. Morris CA et al: Another case of ANOTHER syndrome (hypohidrotic ectodermal dysplasia with hypothyroidism). *Proc Greenwood Genet Ctr* **6**:145–146, 1987.

76. Nakata M et al: A genetic study of anodontia in X-linked hypohidrotic ectodermal dysplasia. *Am J Hum Genet* **32**:908–919, 1980.

77. Osbourn RA: Congenital ectodermal dysplasia with amastia. *J Am Med Assoc* **148**:644–645, 1952.

78. Pabst HF et al: Hypohidrotic ectodermal dysplasia with hypothyroidism. *J Pediatr* **98**:233–237, 1981.

79. Passarge E, Fries E: Autosomal recessive hypohidrotic ectodermal dysplasia with subclinical manifestation in the heterozygote. *Birth Defects* **13**(3C):95–100, 1977.

80. Passarge EC et al: Anhidrotic ectodermal dysplasia as autosomal recessive trait in an inbred kindred. *Humangenetik* **3**:181–185, 1966.

81. Perabo F et al: Ektodermale Dysplasie von anhidrotischen Typus. *Helv Paediatr Acta* **11**:604–639, 1956.

82. Peterson-Falzone SJ et al: Abnormal laryngeal vocal quality in ectodermal dysplasia. *Arch Otolaryngol* **107**:300–304, 1981.

83. Pike MG et al: A distinctive type of hypohidrotic ectodermal dysplasia featuring hypothyroidism. *J Pediatr* **108**:109–111, 1986.

84. Pinheiro M et al: Trichodermodysplasia with dental alterations: An apparently new genetic ectodermal dysplasia of the tricho-odonto-onychial subgroup. *Clin Genet* **29**:332–336, 1986.

85. Pinheiro M, Freire Maia N: Christ-Siemens-Touraine syndrome—a clinical and genetic analysis of a large Brazilian kindred: I. Affected females. *Am J Med Genet* **4**:113–122, 1979.

86. Pinheiro M, Freire-Maia N: Christ-Siemens-Touraine syndrome—A clinical and genetic analysis of a large Brazilian kindred: II. Affected males. *Am J Med Genet* **4**:123–129, 1979.

87. Pinheiro M, Freire-Maia N: Christ-Siemens-Touraine syndrome—A clinical and genetic analysis of a large Brazilian kindred: III. Carrier detection. *Am J Med Genet* **4**:129–134, 1979.

88. Pinheiro M et al: Christ–Siemens–Touraine syndrome. Investigation on two large Brazilian kindreds with a new estimate of the manifestation rate among carriers. *Hum Genet* **57**:428–431, 1981.

89. Priest J: Dermatoglyphics in ectodermal dysplasia. *Lancet* **2**:1093, 1967.

90. Rajagopalan K, Tay CH: Hidrotic ectodermal dysplasia. *Arch Dermatol* **113**:481–485, 1977.

91. Rietschel RL, Wilmore DW: Heat loss in anhidrotic ectodermal dysplasia. *J Invest Dermatol* **71**:145–147, 1978.

92. Robinson GC et al: Familial ectodermal dysplasia with sensory neural deafness and other anomalies. *Pediatrics* **30**:797–802, 1962.

93. Saksena SS, Bixler D: Facial morphometrics in the identification of gene carriers for hypohidrotic ectodermal dysplasia. March of Dimes Birth Defects Conference, Baltimore, Maryland, July, 1988.

94. Šalamon T, Miličevič M: Über eine besondere Form der ektodermalen Dysplasie mit Hypohidrosis, Hypotrichosis, Hornhautveränderungen, Nagel- und anderen Anomalien bei einem Geschwisterpaar. *Arch Klin Exp Dermatol* **220**:564–575, 1964.

95. Salinas CF, Spector M: Tricho-dental syndrome, in *Hair, Trace Elements and Human Illness,* Brown AC, Crounse RG (eds), Praeger, New York, 1980, chapter 18, pp 240–256.

96. Samuelson G: Hypohidrotic ectodermal dysplasia. *Acta Paediatr Scand* **59**:94–99, 1970.

97. Sarnat BG et al: Fourteen year report of facial growth in case of complete anodontia with ectodermal dysplasia. *Am J Dis Child* **86**:162–169, 1953.

98. Settineri WM et al: X-linked anhidrotic ectodermal dysplasia with some unusual features. *J Med Genet* **13**:212–216, 1975.

99. Shapiro SD et al: Brief clinical report: Curry–Hall syndrome. *Am J Med Genet* **17**:579–583, 1984.

100. Shokeir MHK: Complete failure of eruption of all permanent teeth: An autosomal dominant disorder. *Clin Genet* **5**:322–326, 1974.

101. Smith J: Hereditary ectodermal dysplasia. *Arch Dis Childh* **4**:215–226, 1929.

102. Söderholm AL, Kaitila I: Expression of X-linked hypohidrotic ectodermal dysplasia in six males and in their mothers. *Clin Genet* **28**:136–144, 1985.

103. Sofaer JA: A dental approach to carrier screening in X-linked hypohidrotic ectodermal dysplasia. *J Med Genet* **18**:459–460, 1981.

104. Sofaer JA: Hypodontia and sweat pore counts in detecting carriers of X-linked hypohidrotic ectodermal dysplasia. *Br Dent J* **151**:327–330, 1981.

105. Solomon LM, Keuer EJ: The ectodermal dysplasias: Problems of classification and some newer syndromes. *Arch Dermatol* **116**:1295–1299, 1980.

106. Sparrow GP et al: Hyperpigmentation and hypohidrosis (the Naegeli–Franceschetti–Jadassohn syndrome). *Clin Exp Dermatol* **1**:127–140, 1976.

106a. Sybert VP: Hypohidrotic ectodermal dysplasia: Argument against an autosomal recessive form clinically indistinguishable from X-linked hypohidrotic ectodermal dysplasia (Christ-Siemens-Touraine syndrome). *Pediatr Dermatol* **6**:76–81, 1989.

107. Tal H: Familial hypodontia in the permanent dentition: A case report. *J Dent* **9**:260–264, 1981.

108. Thivolet JJ et al: Dysplasie ectodermique anhidrotique chez un nourisson de sexe féminin. *Ann Dermatovenereol* **104**:417–418, 1977.

109. Thurnam J: Two cases in which the skin, hair and teeth were very imperfectly developed. *Med Chir Trans* **31**:71–82, 1848.

110. Tipton RE, Gorlin RJ: Growth retardation, alopecia, pseudoanodontia, and optic atrophy—the GAPO syndrome. *Am J Med Genet* **19**:209–216, 1984.

111. Tuffli GA, Laxova R: New, autosomal dominant form of ectodermal dysplasia. *Am J Med Genet* **14**:381–384, 1983.

112. Upshaw BY, Montgomery H: Hereditary anhidrotic ectodermal dysplasia: A clinical and pathologic study. *Arch Dermatol Syph (Chic)* **60**:1170–1183, 1949.

113. Vanselow NA et al: The increased prevalence of allergic disease in anhidrotic congenital ectodermal dysplasia. *J Allergy* **45**:302–309, 1970.

114. Verbov J: Hypohidrotic (or anhidrotic) ectodermal dysplasia: An appraisal of diagnostic methods. *Br J Dermatol* **83**:341–348, 1970.

115. Ward R, Bixler D: The distinctive facial features in hypohidrotic ectodermal dysplasia. *Am J Hum Genet* **39**:86A, 1986 (Abstract).

116. Watson RM, Hardwick CE: Hypodontia associated with cleft palate. *Br Dent J* **130**:77–80, 1971.

117. Weech AA: Hereditary ectodermal dysplasia. *Am J Dis Child* **37**:766–790, 1929.

118. Williams M, Fraser FC: Hidrotic ectodermal dysplasia in Clouston's family revisited. *Can Med Assoc J* **96**:36–38, 1967.

119. Zadik Z et al: Case report 112. Dermoid cysts, hypothyroidism, cleft palate and hypodontia. *J Clin Dysmorphol* **1**(4):24–27, 1983.

120. Zonana J et al: X-linked hypohidrotic ectodermal dysplasia: Localization within the region Xq11-21.1 by linkage analysis and implications for carrier detection and prenatal diagnosis. *Am J Hum Genet* **43**:75–85, 1988.

## Hair-nail-skin-teeth dysplasias (dermo-odonto-dysplasia, pilo-dento-ungular dysplasia, odonto-onycho-dermal dysplasia, odonto-onychial dysplasia, tricho-dermo-dysplasia with dental alterations)

There are an extremely large number of rare disorders involving dysplasia of hair, nails, skin appendages, and teeth in binary, ternary, or quaternary combination (1,2,4–8). These have been dealt with exhaustively by Freire-Maia and Pinheiro (3). Only a few can be briefly described here (1,2,5–8).

Pinheiro et al (5) reported sibs with hair loss ranging from almost total alopecia involving scalp, eyebrows, lashes, and axillary and pubic hair to less severe involvement: finger and toenail dystrophy, palmoplantar hyperkeratosis, supernumerary nipples, and enamel hypoplasia (Fig. 13–66A,B). All had short stature and two sibs had a frontoparietal skull defect. Autosomal recessive inheritance was suggested. The disorder was termed *tricho-dermo-dysplasia with dental alterations.*

We are not entirely convinced that the disorder is different from *odonto-onycho-dysplasia* with alopecia also reported by Pinheiro et al (6), except for greater severity in the latter. They described sisters with probably consanguineous parents. At birth, both sibs had sparse scalp hair which soon fell out. The sparse body hair distribution was similar to that noted in tricho-odonto-onychial dysplasia. Hypodontia, microdontia, and enamel hypoplasia were found (Fig. 13–67A–C).

Pinheiro et al (7) described an autosomal dominant disorder with trichodysplasia, hypodontia, onychodysplasia, and bilateral inward deflection of the fourth toes. The disorder was termed *tricho-dermo-dysplasia,* and a four component one was entitled tricho-odonto-onycho-dermal syndrome (4a).

Fig. 13–66. *Tricho-dermo-dysplasia with dental alterations.* (A) Hypotrichosis of scalp, sparse lateral eyebrows. (B) Severe oligodontia in daughter of woman seen in A. (From *M Pinheiro et al,* Clin Genet **29**:332, 1986.)

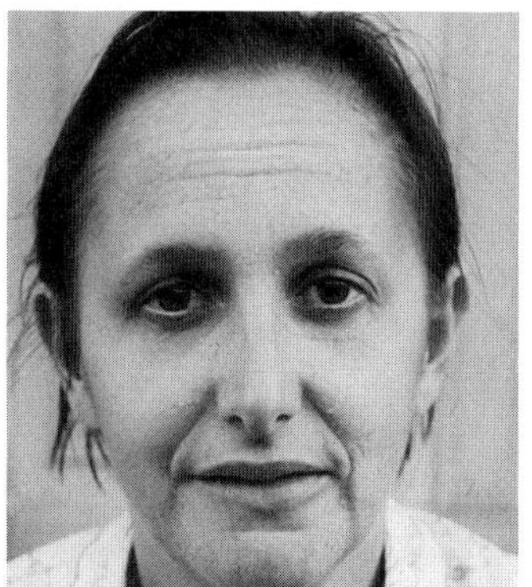
A

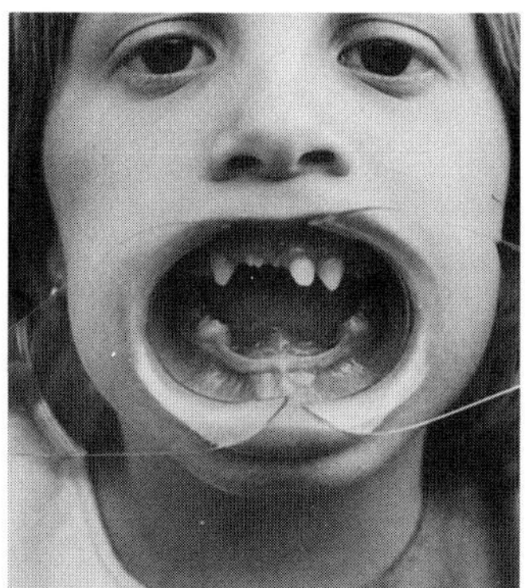
B

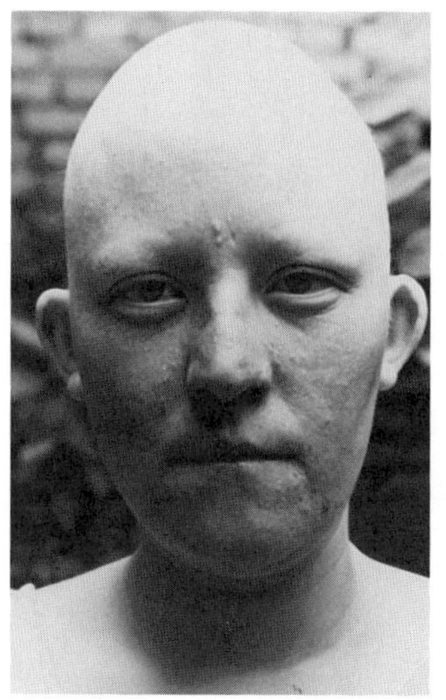
A

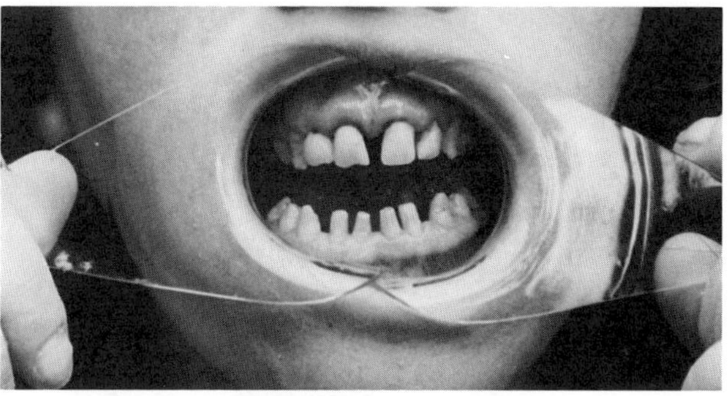
B

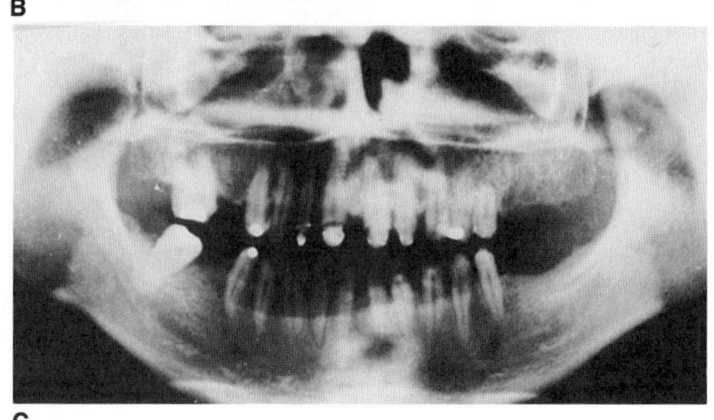
C

Fig. 13–67. *Odonto-onycho-dysplasia with alopecia.* (A) Total alopecia. (B,C) Oligodontia. (Courtesy of *M Pinheiro et al,* Curtitiba, Brazil.)

Pinheiro and Freire-Maia (4) reported a four generation autosomal dominantly inherited syndrome termed *dermo-odonto-dysplasia.* Thin, fragile, or brittle finger and toe nails, palmoplantar xeroderma, oligodontia, and/or microdontia with persistence of deciduous teeth were found (Fig. 13–68A,B). The hair was dry and the beard and axillary and pubic hair were slow growing. There was variable expressivity. Differentiation must be made from the *Witkop tooth–nail syndrome.*

Fahdil et al (2) described two families with a syndrome of hyperkeratosis of palms and soles with hyperhidrosis in these areas; somewhat scaly dry skin, erythema, and telangiectasia over the nose and malar areas; and peg-shaped maxillary central incisors. The term chosen was *odonto-onycho-dermal dysplasia.* Inheritance appears to be autosomal recessive (Fig. 13–68C–F).

### References [Hair-nail-skin-teeth dysplasias (dermo-odonto-dysplasia, pilo-dento-ungular dysplasia, odonto-onycho-dermal dysplasia, odonto-onychial dysplasia, tricho-dermo-dysplasia with dental alterations)]

1. Cecatto L et al: Trichodysplasia-dental anomalies-onychodysplasia. *Am J Med Genet* (in press).

2. Fahdil M et al: Odontoonychodermal dysplasia: A previously apparently undescribed ectodermal dysplasia. *Am J Med Genet* **14**:335–346, 1983.

3. Freire-Maia N, Pinheiro M: *Ectodermal Dysplasias—A Clinical and Genetic Study,* Alan R. Liss, New York, 1984.

4. Pinheiro M, Freire-Maia N: Dermoodontodysplasia: An eleven-member, four generation pedigree with an apparently hitherto undescribed pure ectodermal dysplasia. *Clin Genet* **24**:58–68, 1983.

4a. Pinheiro M et al: A previously undescribed condition: Tricho-odonto-onycho-dermal syndrome. A review of the tricho-odonto-onychal subgroup of ectodermal dysplasias. *Br J Dermatol* **105**:371–382, 1981.

5. Pinheiro M et al: Trichoodontoonychial dysplasia—a new meso-ectodermal dysplasia. *Am J Med Genet* **15**:67–70, 1983.

6. Pinheiro M et al: Odontoonychodysplasia with alopecia: A new pure ectodermal dysplasia with probable autosomal recessive inheritance. *Am J Med Genet* **20**:197–202, 1985.

7. Pinheiro M et al: Trichodermodysplasia with dental alterations: An apparently new genetic ectodermal dysplasia of the tricho-odonto-onychial subgroup. *Clin Genet* **29**:332–336, 1986.

8. Taraja EH et al: Pilodentoungular dysplasia with microcephaly: A new ectodermal dysplasia/malformation syndrome. *Am J Med Genet* **26**:153–156, 1987.

## Incontinentia pigmenti (Bloch–Sulzberger syndrome)

Although incontinentia pigmenti (IP) may have been first noticed by Garrod (16) in 1906 and Adamson (1) in 1908, credit is usually given to Bardach (3), Bloch (6), Sulzberger (67), and Siemens (62) for clearly defining it in the 1920s. The major features include vesicular, verrucous, and pigmented macular lesions of the skin. Over 50% of the patients have involvement of systems other than the integument including the eyes, central nervous system, and teeth (49). Person (48) raised the intriguing question of whether the disorder represents a failure of immune tolerance.

Familial occurrence has been found in 15%. Comprehensive reviews (37,41,49) conclude that the syndrome is an X-linked dominant disorder, essentially lethal in males. An affected four generation family has been reported (73). Autosomal dominant inheritance, however, has been suggested (35,64). About 2–3% are males (10,39,47,69), with phenotypes identical to those of affected females (4). Wieacker et al (72) demonstrated that preferential inactivation of the X-chromosome carrying the IP gene has a proliferative advantage in the cell population. Lenz (38) suggested that affected males may be the result of half-chromatid mutations, although Hecht and co-workers (26,27) disagreed. Langenbeck (36) considered a half-chromatid back mutation as an alternative hypothesis to explain affected males. Several investigators (11,17,28,32) have mapped the gene for IP to band Xp11.21, although this has recently been challenged (25,63). Selfiani et al (63) and Ciolla et al (11a) indicated that the gene for IP could not be located in the major part of the Xp arm but Gorski et al (18a) suggested two IP genes, one at Xp11.21. Kunze et al (34) and Ormerod et al (47) and Prendiville at al (50a) reported males with the disorder and XXY Klinefelter syndrome, also consistent with X-linked dominant inheritance, lethal in the male.

**Skin.** Cutaneous abnormalities serve as a basis for the three clinical stages of the condition. The first stage is present at birth or begins within the first few weeks of life. It is characterized by linear or grouped vesicles on the extremities; in some patients, other skin sites may be affected. By the end of the first month, the vesicles may disappear, recur, or be replaced by irregularly distributed violaceous papules and inflammatory lesions. The second stage is characterized by hyperkera-

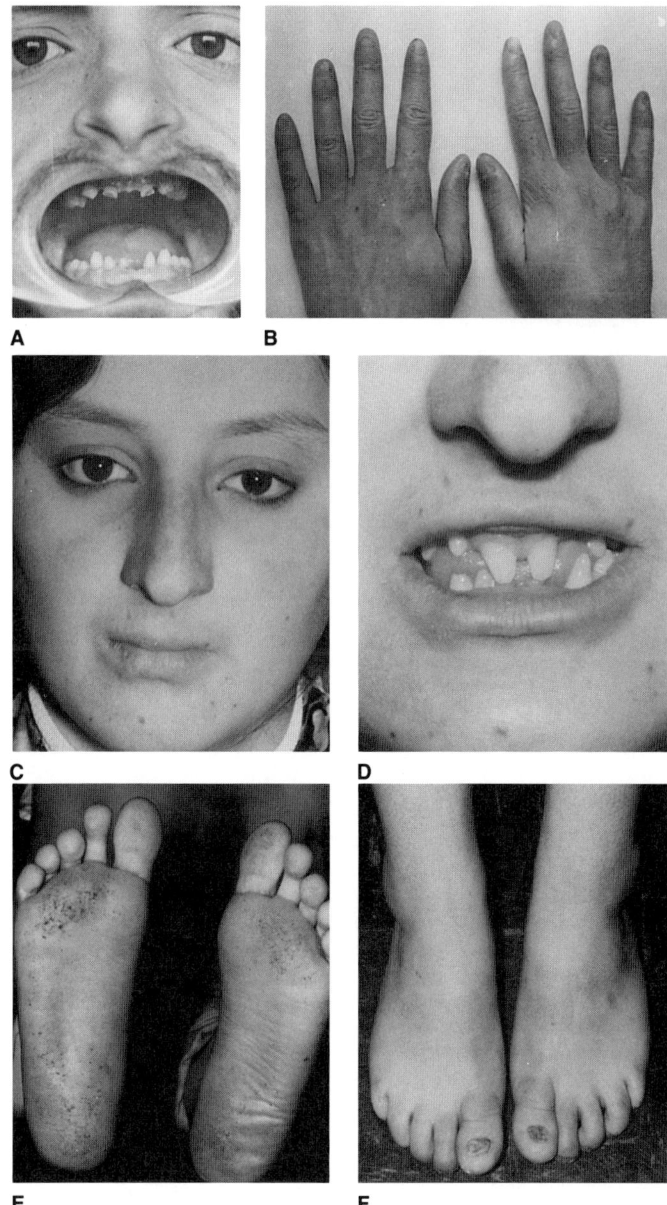


Fig. 13–68. *Dermo-odonto-dysplasia.* (A) Oligodontia and/or microdontia with persistence of deciduous teeth. (B) Hypoplastic thin, fragile fingernails. *Odonto-onycho-dermal dysplasia.* (C) Patient showing mild telangiectasia of nose and malar areas. (D) Peg-shaped anterior teeth. (E,F) Dystrophic nails and hyperkeratosis of soles. (A,B from *M Pinheiro* and *N Freire-Maia,* Clin Genet **24**:58, 1983; C–F from *M Fahdil et al,* Am J Med Genet **14**:335, 1983.)

totic, warty lesions on the dorsal surface of the digits, knuckles, joints, and limbs; these lesions are usually manifest at about 1 month of age, although they may be present at birth. Resolution is usually spontaneous but rarely may be recurrent throughout childhood (46).

The third stage usually has its onset between the third and sixth months of life, although it may be present at birth. It is characterized by brownish-gray macules arranged in a reticulated pattern or in streaks, whorls, or patches; although these lesions may occur in sites previously involved by vesicular or verrucous lesions, areas not significantly involved in either of the first two stages may be affected. The distribution of the pigmentation tends to follow Blaschko's lines. Lenz et al (39) reported unilateral involvement. Pigmentation may begin to fade at about 2 years, and may resolve so completely as to be unnoticeable, although some residuum is often present for life (Fig. 13–69). Depigmented lesions of the calves in adulthood may be the only residual sign (74).

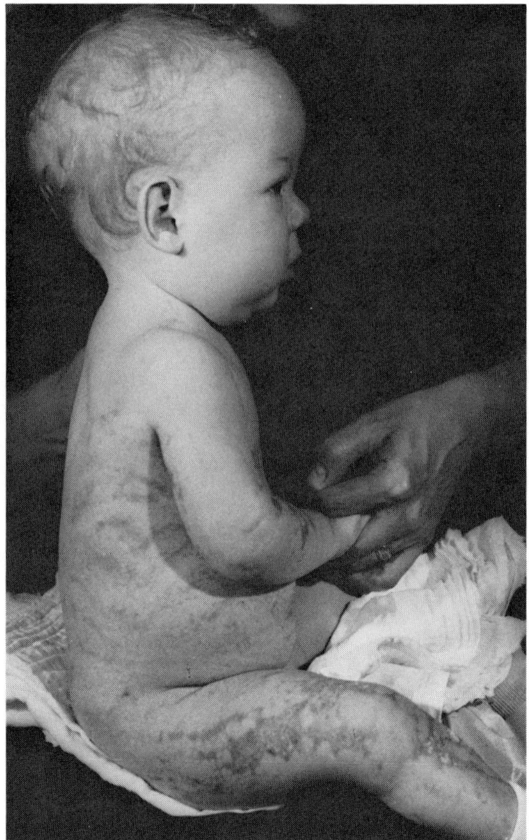

Fig. 13–69. *Incontinentia pigmenti.* Ten-month-old female manifesting bullae, verrucae, and whorled and linear distribution of pigment. Also note area of crusting and scarring on thigh.

Alopecia of the atrophic, scarring (pseudopelade) type is seen near the apex of the crown in most affected persons (Fig. 13–70). Rarely, fingernails are dystrophic and breasts asymmetric. Painful subungual tumors have been reported (42,63). Rott (54) reported partial or complete lack of sweat pores on the palms and fingers of 5 female probands, consistent with lyonization of the X-chromosome.

On light microscopic examination of the vesicular stage, intraepithelial vesicles containing eosinophils are found; spongiosis and individual dyskeratotic epithelial cells are also noted (40). Eosinophils are also found in the connective tissue. The second stage is characterized by hyperkeratosis, acanthosis, papillomatosis and epithelial dyskeratosis. Basal cells are vacuolated and their pigment granules are decreased in number. A mild chronic inflammatory cell infiltrate is seen in the connective tissue as well as in the epithelium. In the third stage, extensive melanin deposits are found within melanophages in the upper dermis, usually associated with a decrease in basal cell pigmentation. Ultrastructural changes in the skin have been described by several investigators (9,22,57,58,73).

**Eyes.** Ophthalmologic abnormalities are present in 25–35% (15,31,44,49,51,53,59). These abnormalities have been reviewed (15). The most common alterations include strabismus, cataract, optic atrophy, a retrolental mass (described as persistent hyperplastic primary vitreous, pseudoglioma, or retrolental fibroplasia), retinal detachment, microphthalmos, retinal telangiectasia and ectasia, and irregular hyperpigmentation of the conjunctiva, iris, and retina (15).

**Central nervous system.** Central nervous system involvement is seen in 35–40% (49). Mental retardation, microcephaly, hydrocephalus, spastic and lax paralysis, paresis of eye muscles, and convulsive

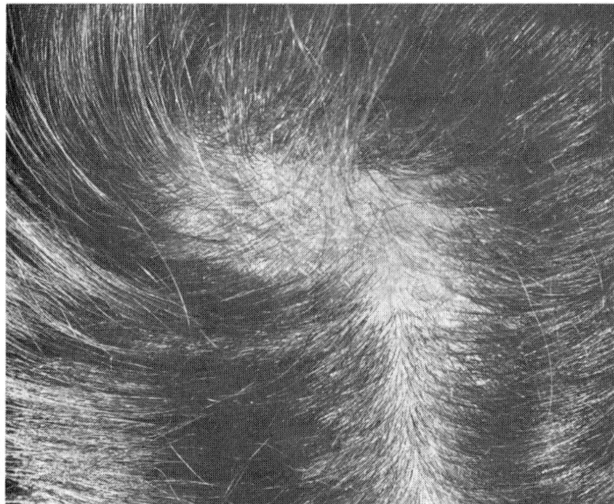

Fig. 13–70. *Incontinentia pigmenti.* Alopecia of pseudopelade type located at crown of head.

episodes have been reported (14,31). CT scans have demonstrated brain atrophy (2). Those with neonatal seizures appear to have a higher rate of mental retardation (46).

**Other abnormalities.** Several reports have described IP patients who develop many unusual and recurrent infections, suggesting that immunodeficiency may be characteristic. Abnormalities of one or more serum immunoglobulins have been found in some patients (12,13), as have defects in neutrophil chemotaxis (12,30) and lymphocyte transformation (30). No immunologic or leukocytic function abnormalities were found in the patient studied by Diamantopoulous et al (13). An increased number of chromosomal breaks has been noted by some investigators (8,20,29,33,70) but not by others (27).

**Oral manifestations.** Oral changes are limited to the teeth and have been noted in 90% (10,23). Pegged or conically crowned-teeth (30%) and congenitally missing teeth (40%) are the characteristic dental abnormalities (18,23,50,56,61,62,75) (Fig. 13–71A,B). Primary and permanent dentitions are affected. Congenitally missing teeth and malformed teeth create diastemas.

**Differential diagnosis.** Skin changes present in early infancy must be distinguished from those of congenital syphilis, *epidermolysis bullosa,* bullous impetigo, contact dermatitis, dermatitis herpetiformis, and verrucous nevus.

Incontinentia pigmenti should be differentiated from hypomelanosis of Ito (incontinentia pigmenti achromians) and Naegli syndrome.

Hypomelanosis of Ito (52,66,68) is a neurocutaneous syndrome characterized by development of linear areas of cutaneous hypopigmentation within the first year of life, as well as neurologic, ophthalmologic, and musculoskeletal anomalies. No preceding bullous, verrucous, or hyperpigmented lesions are noted. A consistent pattern of dental abnormalities has not been reported to be associated with this disorder. Talon cusps have been described in one patient (24). Browne and Byrne (7) described a patient whose anterior deciduous teeth were conical, and had pitted crowns that were yellow-brown; the other teeth were normal. On light microscopy of several of these teeth, a localized mass of irregularly formed coronal dentin was found. More than twice as many affected females as affected males have been reported. Male patients are no more severely affected than females, and multiple cases within a family, with male-to-male transmission, have been reported (55). Thus, autosomal inheritance has been suggested. There is mounting evidence that hypomelanosis of Ito represents a nonspecific sign of somatic mosaicism (68a).

Naegli syndrome (19,45), a rare disorder, is characterized by reticular pigmentation of the skin that develops at about 2 years of life and is not preceded by an inflammatory stage. Heat intolerance and moderate hyperkeratosis of the palms and soles are also noted. Dominant inheritance has been reported. Too few patients with the disorder have been reported to determine whether dental or other oral abnormalities are characteristic of the condition; yellow spots on the enamel have been noted. Sparrow et al (65) reported a family with what might also be Naegli syndrome. Affected individuals had diffuse or patchy, mottled hyperpigmentation of the skin that developed in childhood. There was also thickening of the fingernails, onycholysis and subungual keratosis, decreased sweating, punctate keratoses on the palms and soles, and hypoplastic or absent dermatoglyphic patterns. A few patients had blistering of the heels during the first week of life. Male-to-male transmission was described.

Dental anomalies in IP resemble those of other ectodermal dysplasias, such as *Ellis–van Creveld syndrome* and *hypohidrotic ectodermal dysplasia.* In congenital syphilis, the primary dentition is rarely involved (except possibly a deciduous molar) because of the inability of the spirochetes to pass the placental barrier until at least the eighteenth week of pregnancy. The incisors in congenital syphilis are never conical; rather, the incisal edge is narrower than the cervical portion of the crown. In hypohidrotic ectodermal dysplasia, many more teeth are congenitally missing and those present are more severely malformed than in IP. In Ellis–van Creveld syndrome other oral anomalies not present in IP are found: fusion of the lip with the adjacent alveolar ridges and notching of the mandibular alveolar process.

**Laboratory aids.** Blood eosinophilia during the vesicular stage or even later may be marked, in some cases reaching over 55% (12,50).

### References [Incontinentia pigmenti (Bloch–Sulzberger syndrome)]

1. Adamson HG: Congenital pigmentation with atrophic scarring associated with other congenital abnormalities. *Proc R Soc Med* **1**:9–10, 1908.

Fig. 13–71. *Incontinentia pigmenti.* (A) Missing teeth and teeth with conical crown form. (B) Dental radiographs of patient demonstrating missing teeth, impacted teeth, and conical crown form of several teeth. (From *RJ Gorlin* and *JA Anderson,* J Pediatr, **57**:78, 1960.)

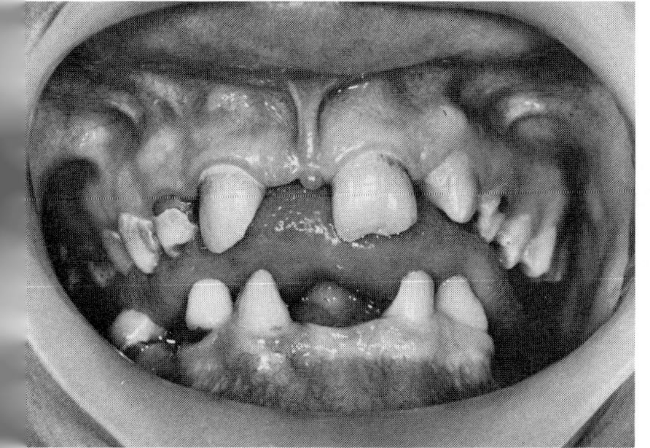

A

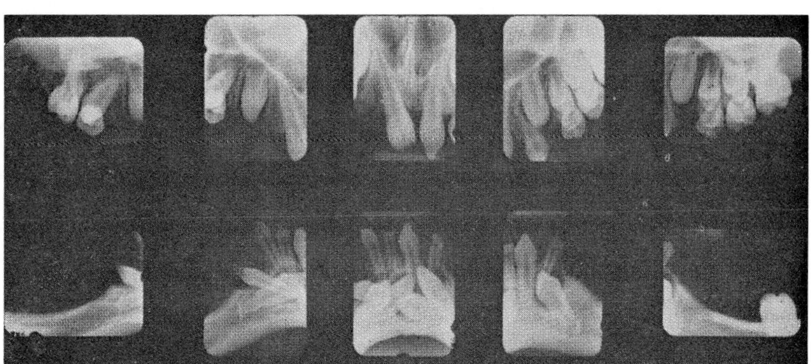

B

2. Avrahami E et al: Computer tomography demonstration of brain changes in incontinentia pigmenti. *Am J Dis Child* **139**:372–374, 1985.

3. Bardach M: Systematisierte Naevusbildungen bei einem eineiigen Zwillingspaar. *Z Kinderheilkd* **39**:542–550, 1925.

4. Bargman HB, Wyse C: Incontinentia pigmenti in a 21-year-old man. *Arch Dermatol* **111**:1606–1608, 1975.

5. Bjellerup M: Incontinentia pigmenti with dental anomalies: A three generation study. *Acta Dermatovenereol* **62**:262–264, 1982.

6. Bloch B: Eigentümliche, bisher nicht beschriebene Pigmentaffektion (Incontinentia pigmenti). *Schweiz Med Wochenschr* **7**:404–405, 1926.

7. Browne RM, Byrne JPH: Dental dysplasia in incontinentia pigmenti achromians (Ito). An unusual form. *Br Dent J* **140**:211–214, 1976.

8. Cantú JM et al: Chromosomal instability in incontinentia pigmenti. *Ann Génét* **16**:117–119, 1973.

9. Caputo R et al: Ultrastructural findings in incontinentia pigmenti. *Int J Dermatol* **14**:46–55, 1975.

10. Carney RG Jr: Incontinentia pigmenti. A world statistical analysis. *Arch Dermatol* **112**:535–542, 1976.

11. Cannizzaro LA, Hecht F: Gene for incontinentia pigmenti maps to band Xp11 with an (X;10)(p11q22) translocation. *Clin Genet* **32**:66–69, 1987.

11a. Ciolla JA et al: Incontinentia pigmenti and X-autosomal translocations. *Hum Genet* **81**:269–272, 1989.

12. Dahl MV et al: Incontinentia pigmenti and defective neutrophil chemotaxis. *Arch Dermatol* **111**:1603–1605, 1975.

13. Diamantopoulos N et al: Actinomycosis meningitis in a girl with incontinentia pigmenti. *Clin Pediatr* **24**:651–654, 1985.

14. Findlay GH: On the pathogenesis of incontinentia pigmenti: With observations on an associated eye disturbance resembling retrolental fibroplasia. *Br J Dermatol* **64**:141–146, 1952.

15. François J: Incontinentia pigment (Bloch–Sulzberger syndrome) and retinal changes. *Br J Ophthalmol* **68**:19–25, 1984.

16. Garrod AE: Peculiar pigmentations of the skin of an infant. *Trans Clin Soc London* **39:**216, 1906.

17. Gilgenkrantz S et al: Translocation (X;9)(p11;q35) in a girl with incontinentia pigmenti (IP): Implications for the regional assignment of the IP locus to Xp11? *Ann Génét* **28**:90–92, 1985.

18. Gorlin RJ, Anderson JA: The characteristic dentition of incontinenti pigmenti. *J Pediatr* 57:78–85, 1960.

18a. Gorski JL et al: Molecular analysis of incontinentia pigmenti (IP1) translocation chromosomes map two DNA sequences to within the IP1 locus. March of Dimes Clinical Genetics Conference, Boston, 9–12 July, 1989.

19. Greither A, Haensch R: Anhidrotische retikuläre Pigment-dermatose mit blasig erythematösem Anfangsstadium. *Schweiz Med Wochenschr* **100**:228–233, 1970.

20. de Grouchy J et al: Cassures chromosomiques dans l'incontinentia pigmenti. *Ann Génét* **16**:61–66, 1972.

21. de Grouchy J et al: Incontinentia pigmenti (Ip) and r(X). Tentative mapping of the Ip locus to the X juxtacentric region. *Ann Génét* **28**:86–89, 1985.

22. Guerrier CJW, Wong CK: Ultrastructural evolution of the skin in incontinentia pigmenti (Bloch–Sulzberger). Study of 6 cases. *Dermatologica* **149**:10–22, 1974.

23. Hagemann E: Zahnbefund bei der Incontinentia pigmenti. *Dtsch Zahnärztl Z* **18**:1198–1208, 1262–1268, 1963.

24. Happle R, Vakilzadeh F: Hamartomatous dental cusps in hypomelanosis of Ito. *Clin Genet* **21**:65–68, 1982.

25. Harris A et al: The gene for incontinentia pigmenti: Failure of linkage studies using DNA probes to confirm cytogenetic localization. *Clin Genet* **34**:1–6, 1988.

26. Hecht F et al: Incontinentia pigmenti in Arizona Indians including transmission from mother to son inconsistent with the half chromatid mutation model. *Clin Genet* **21**:293–296, 1982.

27. Hecht F, Hecht BK: The half chromatid mutation model and bidirectional mutation in incontinentia pigmenti. *Clin Genet* **24**:177–179, 1983.

28. Hodgson SV et al: Two cases of X/autosome translocation in females with incontinentia pigmenti. *Hum Genet* **71**:231–234, 1985.

29. Iáncu T et al: Incontinentia pigmenti. *Clin Genet* **7**:103–110, 1975.

30. Jesson RT et al: Incontinentia pigmenti. Evidence for both neutrophil and lymphocyte dysfunction. *Arch Dermatol* **114**:1182–1186, 1978.

31. Jones ST: Retrolental membrane associated with Bloch–Sulzberger syndrome (incontinentia pigmenti). *Am J Ophthalmol* **62**:330–334, 1966.

32. Kajii T et al: Translocation (X;13)(p11.21;q12.3) in a girl with incontinentia pigmenti and bilateral retinoblastoma. *Ann Génét* **28**:219–223, 1985.

33. Kelley TE et al: Incontinentia pigmenti: A chromosomal breakage syndrome. *J Hered* **67**:171–172, 1976.

34. Kunze J et al: Klinefelter's syndrome and incontinentia pigmenti Bloch–Sulzberger. *Hum Genet* **35**:237–240, 1977.

35. Kurczynski TW et al: Studies of a family with incontinentia pigmenti variably expressed in both sexes. *J Med Genet* **19**:447–451, 1982.

36. Langenbeck U: Transmission of incontinentia pigmenti from mother to son is consistent with a half chromatid back-mutation (reversion) model. *Clin Genet* **22**:290–291, 1982.

37. Lenz W: Zur Genetik der Incontinentia pigmenti. *Ann Paediatr (Basel)* **196:**149–165, 1961.

38. Lenz W: Half chromatid mutations may explain incontinentia pigmenti in males. *Am J Hum Genet* **27**:690–691, 1975 (letter).

39. Lenz W et al: Halbseitige Incontinentia pigmenti bei einem Mann. *Pädiatr Pädol* **17**:187–199, 1952.

40. Lever WF, Schaumburg-Lever G: *Histopathology of the Skin,* 6th ed., JB Lippincott, Philadelphia, 1983, pp 83–84.

41. Lucas D: Beitrag zur Genetik der Incontinentia pigmenti. *Klin Pädiatr* **186**:142–147, 1974.

42. Mascaro JM et al: Painful subungual keratotic tumors in incontinentia pigmenti. *J Am Acad Dermatol* **13**:913–918, 1985.

43. Miller CA, Parker WD Jr: Hypomelanosis of Ito: Association with a chromosomal abnormality. *Neurology* **35**:607–610, 1985.

44. Menshaha-Manhart O et al: Retinal pigment epithelium in incontinentia pigmenti. *Am J Ophthalmol* **79**:571–577, 1975.

45. Naegli: Familiärer Chromatophorennävus. *Schweiz Med Wochenschr* **57**:48, 1927.

46. O'Brien JE, Feingold M: Incontinentia pigmenti. *Am J Dis Child* **139**:711–712, 1985.

47. Ormerod AD et al: Incontinentia pigmenti in a boy with Klinefelter's syndrome. *J Med Genet* **24**:439–441, 1987.

48. Person JR: Incontinentia pigmenti: A failure of immune tolerance? *J Am Acad Dermatol* **13**:120–123, 1985.

49. Pfeiffer RA: Zur Frage der Vererbung der Incontinentia pigmenti (Bloch–Siemens). *Z Menschl Vererb Konstit Lehre* **35**:469–493, 1960.

50. Pollack JJ: Oral anomalies in incontinentia pigmenti: A report of a case. *J Md State Dent Assoc* **22**:12–16, 1979.

50a. Prendiville JS et al: Incontinentia pigmenti in a male infant with Klinefelter syndrome. *J Am Acad Dermatol* **20**:937–940, 1989.

51. Raab EL: Ocular lesions in incontinentia pigmenti. *J Pediatr Ophthalmol Strab* **20**:42–48, 1983.

52. Rosemberg S et al Hypomelanosis of Ito. Case report with involvement of the central nervous system and review of the literature. *Neuropediatrics* **15**:52–55, 1984.

53. Rosenfeld SI, Smith ME: Ocular findings in incontinentia pigmenti. *Ophthalmology* **92**:543–546, 1985.

54. Rott H-D: Partial sweat gland aplasia in incontinentia pigmenti Bloch–Sulzberger. *Clin Genet* **26**:36–38, 1984.

55. Rubin MB: Incontinentia pigmenti achromians. Multiple cases within a family. *Arch Dermatol* **105**:424–425, 1972.

56. Russell DL, Finn SB: Incontinentia pigmenti (Bloch–Sulzberger syndrome): A case report with emphasis on dental manifestations. *J Dent Child* **34**:494–500, 1967.

57. Schamburg-Lever G, Lever WF: Electron microscopy of incontinentia pigmenti. *J Invest Dermatol* **61**:151–158, 1973.

58. Schmidt H et al: Incontinentia pigmenti with associated lesions in two generations. Clinical, light microscopic, and electronmicroscopic examinations. *Acta Dermatovenereol* **52**:281–287, 1972.

59. Scott J et al: Ocular changes in the Bloch–Sulzberger syndrome (incontinentia pigmenti). *Br J Ophthalmol* **39**:276–282, 1955.

60. Sefiani A et al: Linkage studies do not confirm the cytogenetic location of incontinentia pigmenti in Xp11. *Hum Genet* **80**:282–286, 1988.

61. Shotts N, Emery AEH: Bloch–Sulzberger syndrome. *J Med Genet* **3**:148–152, 1966.

62. Siemens HW: Die Melanosis corii degenerativa, eine neue Pigmentdermatose. *Arch Dermatol Syph (Berlin)* **157**:382–391, 1929.

63. Simmons DA et al: Subungual tumors in incontinentia pigmenti. *Arch Dermatol* **122**:1431–1434, 1986.

64. Sommer A, Liu PH: Incontinentia pigmenti in a father and his daughter. *Am J Med Genet* **17**:655–659, 1984.

65. Sparrow GP et al: Hyperpigmentation and hypohydrosis. (The Naegeli–Franceschetti–Jadassohn syndrome): Report of a family and review of the literature. *Clin Exp Dermatol* **1**:127–140, 1976.

66. Stricker M et al: Congenital craniofacial dysmorphosis associated with Ito's syndrome (incontinentia pigmenti achromians). *Br J Plast Surg* **37**:472–476, 1984.

67. Sulzberger MB: Über eine bisher nicht beschriebene congenitale Pigmentanomalie (Incontinentia pigmenti). *Arch Dermatol Syph (Berlin)* **154**:19–32, 1928.

68. Takematsu H et al: Incontinentia pigmenti achromians (Ito). *Arch Dermatol* **119**:391–395, 1983.

68a. Thomas IT et al: Association of pigmentary anomalies with chromosomal and genetic mosaicism and chimerism. *Am J Hum Genet* **45**:193–205, 1989.

69. Trabelsi M et al: L'incontinentia pigmenti ou syndrome de Bloch–Sulzberger. *J Pédiatr* **35**:53–56, 1988.

70. Vissian L et al: Incontinentia pigmenti. Étude chromosomique d'une famille. *Ann Dermatol Venereol* **105**:119–121, 1978.

71. Wettke-Schaffer R, Kanter G: X-linked dominant inherited disease with lethality in hemizygous males. *Hum Genet* **64**:1–23, 1983.

72. Wieacker P et al: X inactivation patterns in two syndromes with probable X-linked dominant, male lethal inheritance. *Clin Genet* **28**:238–242, 1985.

73. Wiklund DA, Weston WL: Incontinentia pigmenti: A four generation study. *Arch Dermatol* **116**:701–703, 1980.

74. Wiley HE, Frias JL: Depigmented lesions in incontinentia pigmenti: A useful diagnostic sign. *Am J Dis Child* **128**:546–547, 1974.

75. Wong CK et al: An electron microscopical study of Bloch–Sulzberger syndrome (incontinentia pigmenti). *Acta Dermatol Venereol* **51**:161–168, 1971.

## Leopard syndrome (multiple lentigines syndrome, progressive cardiomyopathic lentiginosis)

The word *LEOPARD* is an acronym coined by Gorlin et al (10), in 1969, to serve as a mnemonic. The syndrome as originally described consisted of multiple *l*entigines, *e*lectrocardiographic conduction abnormalities, *o*cular hypertelorism, *p*ulmonic stenosis, *a*bnormal genitalia, *r*etardation of growth, and sensorineural *d*eafness (10). Several reviews have been published (5a,10–12,32,32a,47). Incompletely documented cases are known (1,8,14,15,26,44). A neural crest defect has been suggested to explain the clinical abnormalities (27,32). Over 100 cases have been tabulated by us.

Autosomal dominant inheritance with high penetrance and marked variation in expression has been demonstrated (6,10,12,21,24,32,37).

**Facies.** The face is usually triangular with biparietal bossing, hypertelorism, ptosis, epicanthal folds, and low-set ears (Figs. 13–72 to 13–74). Strabismus is seen in 20% and mild mandibular prognathism in 10%. In our experience, mild pterygium colli is frequent.

Fig. 13–72. *Leopard syndrome.* Ocular hypertelorism, numerous lentigines. (From *PE Polani* and *EF Moynahan,* Q J Med **41**:205, 1972.)

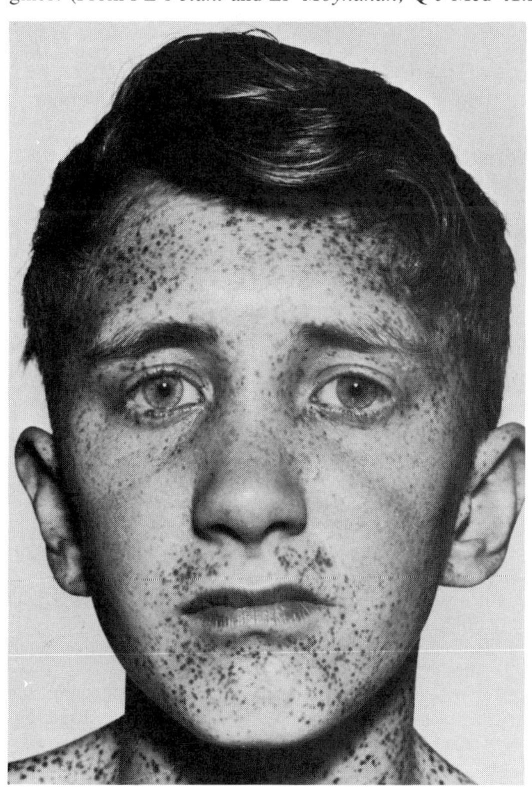

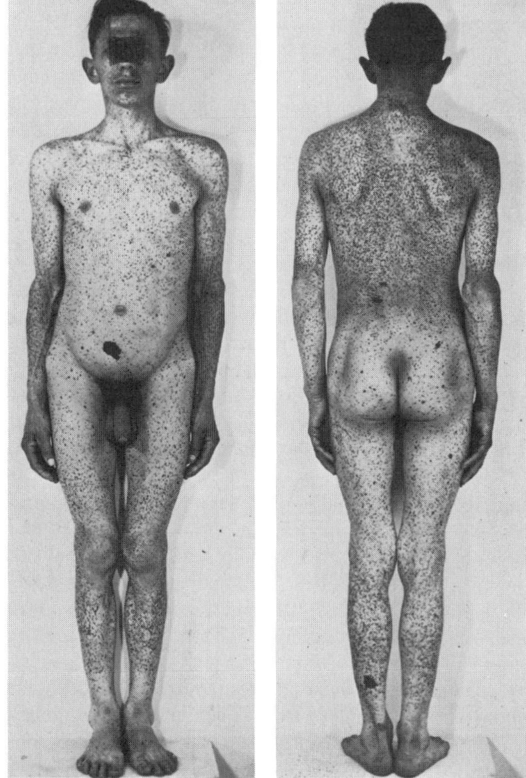

A B

Fig. 13–73. *Leopard syndrome.* (A,B) Twenty-seven-year-old with thousands of lentigines widely scattered over body. Lentigines first appeared at 4 years of age. Also note larger "café-noir" spots. (From *JJ Herzberg,* Z Kinderchir **2**:187, 1965.)

**Skin.** Many pinpoint black-brown macules may be found anywhere on the skin including the axillae, palms, soles, and genitalia (27,50). They may number in the thousands (21,32) and occasionally may be as large as 1–5 cm (café-noir spots) (10,11,21,27,35). Although strik-

Fig. 13–74. *Leopard syndrome.* Numerous lentigines of hands. (From *PE Polani* and *EF Moynahan,* Q J Med **41**:205, 1972.)

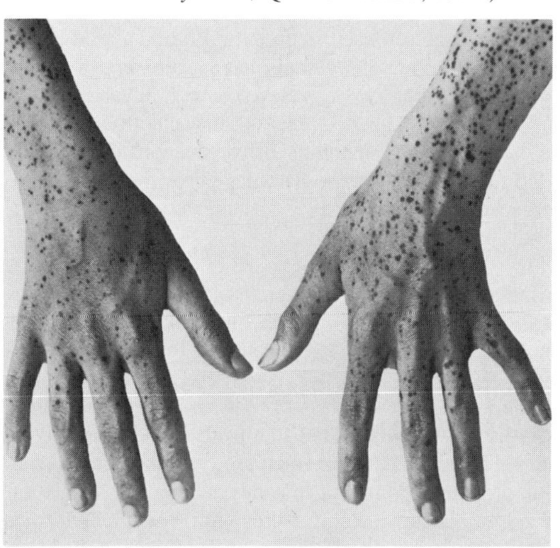

ing when present, some patients do not develop pigmented macules (10). They may be congenital (7,48) or appear soon after birth (22,27) and increase in number with age (23,24,27,32,35,40). On microscopic examination, the pigmented skin lesions are lentigines (10,18,23,32,36,38,47,50). Intracellular giant pigment granules similar to those in neurofibromatosis have been noted (2,30,33,50). Features of compound nevi have also been found (34). Granular cell schwannomas of the skin of the limbs and breast have been described (1a,19,28,35,41a,45). Café-au-lait spots are noted in about 20% (47).

**Cardiovascular system.** Cardiac defects have been reviewed by several authors (10,11,32,32b,39,40,47). A superiorly oriented mean QRS axis in the frontal plane, generally located between −60 and −120° ($S_1$, $S_2$, $S_3$ pattern), tends to characterize the syndrome, regardless of the type of cardiac malformation. This electrocardiographic finding has been demonstrable in about 45%, but is sometimes present even in patients with no structural abnormality of the heart. Complete heart block, complete bundle branch block, and ASD have been reported (37,39).

Valvular pulmonic stenosis, usually mild, the most common cardiac defect, occurs in 40% (16,47). Hypertrophic cardiomyopathy, primarily involving the interventricular septum and resulting in both subaortic and subpulmonic stenosis, has been reported in 20% (17,32,47). Right ventricular hypertrophy (22,25,40) and left ventricular hypertrophy (30,32) have been found. Other abnormalities have included infundibular or supravalvular pulmonic stenosis and muscular subaortic stenosis. Electrocardiogram and vectorcardiogram may show left axis deviation, a conductive defect interpreted as left anterior hemiblock, first degree AV block, left axis deviation, and low voltage and posterior rotation (27,38,50). Clinically detectable cardiac anomalies have been absent in some patients (18,32,50).

**Genitourinary system.** Hypospadias is present in 50% of males (10,11,26,32,48,49). Unilateral or bilateral cryptorchidism is frequent (26,30,32) and the penis may be small (26). Absence or hypoplasia of an ovary has been reported (10,32).

**Skeletal system.** Growth retardation is frequent; 85% are below the 25th percentile for height and weight (11), and 20% are below the third percentile in height (18,22,30,47). Pectus carinatum (18) or excavatum (20,24,27,38,50) has been noted in about 10% (47), as has winging of the scapulas (7,18,40). Scoliosis is present in about 10% (22). In old age, there is a tendency to develop thoracic kyphosis (17a). Spina bifida occulta, absent ribs, cubitus valgus, limitation of motion at the elbows, and deficiency in the outer table of the temporal bone have been described (11,32,40). Bone age may be retarded (6).

**Central nervous system.** Hearing loss has been documented as early as birth (18), 1 year (4,27) and early childhood (22,40). Although hearing loss has been described in about 25% (47) as sensorineural (4,22,27,50), the age of onset, degree, and type have not been well documented. Some young individuals have been reported with normal hearing (36,37,47), but loss may develop with advancing age.

Mild mental retardation (11,27,32,38,41) and electroencephalographic abnormalities (2,27,32) including diffuse encephalopathy (27) have been reported in about 30% (47). The degree of mental retardation has not been specified in one case (30) whereas in others, intelligence was normal (7,18,22,47).

**Oral manifestations.** Oral mucous membranes have not been involved (5,10,11,23,27,32,35,37,39,46,47). Cherubism has been described in one family (51). The reader is referred for further discussion to page 899.

**Differential diagnosis.** Lentigines differ from freckles in having darker color and on biopsy having an increased number of melanocytes per unit skin area and prominent rete ridges.

The leopard syndrome shares some features with *Noonan syndrome* and a Noonan-like syndrome with polyarticular pigmented villonodular synovitis and cherubism (see page 899). Characteristics in common include hypertelorism, ptosis, short stature, pulmonic stenosis, cryptorchidism, delayed development of secondary sexual characteristics, and pectus. Pulmonic stenosis, abnormal QRS axis, short stature, delayed puberty, and hearing loss may be observed in rubella embryopathy. Hypertelorism, ptosis, short stature, and cryptorchidism are present in *Aarskog syndrome*. Koroxenidis et al (16) and Lewis et al (20) reported a two generation black family in which a mother and four of her eight children had hypertelorism, broad nasal root, hypoplastic helices, early onset hearing impairment, congenital pulmonary valvular and/or infundibular stenosis, pectus excavatum or carinatum, and scoliosis; no unusual pigmented skin lesions were mentioned. Patients with *Turner syndrome* have multiple pigmented nevi. Swanson et al (43) reported an 18-year-old male with unilateral sensorineural hearing loss, nevi and freckles on his back, chest and face, mild hypertelorism, normal electrocardiogram, lack of sexual hair, small penis, and renal abnormalities; he likely had a form of Kallman syndrome. Pipkin and Pipkin (31) reported a family with generalized lentigines and nystagmus; inheritance was autosomal dominant.

Sutton et al (42) described 11 patients with multiple lentigines and hypertrophic obstructive cardiomyopathy; symptoms included chest pain and dyspnea on exertion and syncope. Hearing and genitalia were normal and there was no growth retardation. Family histories were negative for relatives with similar abnormalities.

Forney et al (9) described a dominantly inherited syndrome of short stature, bilateral conductive hearing impairment (due at least in part to stapes footplate fixation) with onset during the first decade of life, multiple freckles on the face and shoulders, and mitral insufficiency. In addition, there were skeletal anomalies including fusions of the cervical vertebrae and carpal and tarsal bones.

Halal et al (13) described a dominantly inherited disorder of multiple lentigines, café-au-lait spots, hypertelorism, myopia, and hiatal hernia/peptic ulcer.

Paver and Coleman (29) reported a male with generalized multiple lentigines, facial asymmetry and other craniofacial abnormalities, patchy sparse scalp hair, cicatricial alopecia, epicanthus, ophthalmologic abnormalities, and undescended testes.

Multiple lentigines have also been reported in the *macular cutaneous and mucosal pigmentation, myxoma, and endocrine neoplasia syndrome*. *Neurofibromatosis* should also be excluded.

**Laboratory aids.** Electrocardiogram and cardiac catheterization may be helpful. Biopsy may be used to confirm the diagnosis of lentigo.

### References [Leopard syndrome (multiple lentigines syndrome, progressive cardiomyopathic lentiginosis)]


1. Almkvist: Zwei Fälle mit reichlicher Ausbreitung von Lentigoflecken, "Lentiginose profuse." *Zentralbl Haut Geschlechtskr* **22**:320, 1927.

1a. Apted JH: Multiple granular-cell myoblastoma (schwannoma) in a child. *Br J Dermatol* **80**:257–280, 1968.

2. Bhawan J et al: Giant and "granular melanosomes" in leopard syndrome: An ultrastructural study. *J Cutan Pathol* **3**:207–216, 1976.

3. Butenschön H: Das LEOPARD-Syndrom. *Z Hautkr* **54**:613–618, 1979.

4. Capute AJ et al: Congenital deafness and multiple lentigines. A report of cases in a mother and daughter. *Arch Dermatol* **100**:207–213, 1969. (Same cases as in Ref. 5.)

5. Char F: Leopard syndrome in mother and daughter. *Birth Defects* **7**(8):234–235, 1971. (Same cases as in Ref. 4.)

6. Colomb D, Morel JP: Le syndrome des lentigines multiples. *Ann Dermatol Venereol* **111**:371–381, 1984.

7. Cronje RE, Feinberg A: Lentiginosis, deafness and cardiac abnormalities. *S Afr Med J* **47**:15–17, 1973.

8. Fabry J: Über einen seltener Fall von Naevus spilus. *Arch Dermatol Syph (Berlin)* **59**:217–228, 1902.

9. Forney WR et al: Congenital heart disease, deafness, and skeletal malformations: A new syndrome? *J Pediatr* **68**:14–26, 1966.

10. Gorlin RJ et al: Multiple lentigines syndrome. Complex comprising multiple lentigines, electrocardiographic conduction abnormalities, ocular hypertelorism, pulmonary stenosis, abnormalities of genitalia, retardation of growth, sensorineural deafness, and autosomal dominant hereditary pattern. *Am J Dis Child* **117**:652–662, 1969.

11. Gorlin RJ et al: The leopard (multiple lentigines) syndrome revisited. *Birth Defects* **7**(4):110–115, 1971.
12. Gorlin RJ et al: The leopard (multiple lentigines) syndrome revisited. *Laryngoscope* **81**:1674–1681, 1971.
13. Halal F et al: Gastro-cutaneous syndrome: Peptic ulcer/hiatal hernia, multiple lentigines/café-au-lait spots, hypertelorism, and myopia. *Am J Med Genet* **11**:161–176, 1982.
14. Herzberg JJ: Naevi, Tierfellnaevi, neurocutane Melanosen. *Z Kinderchir* **2**:187–201, 1965.
15. Jordan: Lentigo profuse Darier. *Dermatol Wochenschr* **73**:883, 1921.
16. Koroxenidis GT et al: Congenital heart disease, deaf-mutism, and associated somatic malformations occurring in several members of one family. *Am J Med* **40**:149–155, 1966. (Same case as Ref. 20.)
17. Kraunz RF, Blackmon JR: Cardiocutaneous syndrome continued. *N Engl J Med* **279**:325, 1968 (letter).
18. Lassonde M et al: Generalized lentigines associated with multiple congenital defects (leopard syndrome). *Can Med Assoc J* **103**:293–294, 1970.
19. Laude TA et al: Congenital lentiginosis. *Cutis* **19**:615–617, 1977.
20. Lewis SM et al: Familial pulmonary stenosis and deaf-mutism: Clinical and genetic considerations. *Am Heart J* **55**:458–462, 1958. (Same case as in Ref. 16.)
21. Loyd DW et al: A study of a family with leopard syndrome. *J Clin Psychiat* **43**:113–116, 1982.
22. MacEwen GD, Zaharko W: Multiple lentigines syndrome. A case report of a rare familial syndrome with orthopaedic considerations. *Clin Orthoped* **97**:34–37, 1973.
23. MacMillan DC: Profuse lentiginosis, minor cardiac abnormality, and small stature. *Proc R Soc Med* **62**:1011–1012, 1969.
24. Matthews NL: Lentigo and electrocardiographic changes. *N Engl J Med* **278**:780–781, 1968.
25. Moynahan EJ: Progressive cardiomyopathic lentiginosis. First report of autopsy findings in a recently recognized inheritable disorder (autosomal dominant). *Proc R Soc Med* **63**:445–451, 1970.
26. Moynahan EJ: Multiple symmetrical moles, with psychic and somatic infantilism and genital hypoplasia: First male case of a new syndrome. *Proc R Soc Med* **55**:959–960, 1962.
27. Nordlund JJ et al: The multiple lentigines syndrome. *Arch Dermatol* **107**:259–261, 1973.
28. Nurse DS: Granular cell myoblastoma. Lentiginosis profusa. *Aust J Dermatol* **15**:33–34, 1974.
29. Paver K, Coleman M: A unique combination of developmental defects. *Med J Aust* **2**:203–205, 1971.
30. Pickering D et al: Little Leopard syndrome. Description of three cases and review of 24. *Arch Dis Childh* **46**:85–90, 1971.
31. Pipkin AC, Pipkin SB: A pedigree of generalized lentigo. *J Hered* **41**:79–82, 1950.
32. Polani PE, Moynahan EJ: Progressive cardiomyopathic lentiginosis. *Q J Med* **41**:205–225, 1972.
32a. Ruiz-Maldonado R et al: Progressive cardiomyopathic lentiginosis: Report of six cases and one autopsy. *Pediatr Dermatol* **1**:146–153, 1983.
32b. Schmidt H: Hereditäre Syndrom mit Herzfehlern und Anomalien der Haut unter Berücksichtigung des LEOPARD-Syndroms. *Dermatol Mschr* **166**:622–629, 1980.
33. Selmanowitz VJ: Lentiginosis profusa syndrome. IV. Giant pigment granules (light microscopy). *Acta Dermatovenereol (Stockholm)* **55**:481–484, 1975. (Same cases as Refs. 34–36 and 45.)
34. Selmanowitz VJ: Lentiginosis profusa syndrome (multiple lentigines syndrome). II. Histological findings, modified Crowe's sign, and possible relationship to von Recklinghausen's disease. *Acta Dermatovenereol (Stockholm)* **52**:387–393, 1971. (Same cases as Refs. 33,35,36,45.)
35. Selmanowitz VJ, Ortenreich N: Lentiginosis profusa in daughter and mother: Multiple granular cell "myoblastomas" in the former. *Arch Dermatol* **101**:615–616, 1970. (Same cases as in Refs. 33,34,36,45.)
36. Selmanowitz VJ et al: Lentiginosis profusa syndrome (multiple lentigines syndrome). *Arch Dermatol* **104**:393–401, 1971. (Same cases as Refs. 33–35, and 45.)
37. Seuanez H et al: Cardio-cutaneous syndrome (the "LEOPARD" syndrome). Review of the literature and a new family. *Clin Genet* **9**:266–276, 1976.
38. Shelly WB: Factitial dermatitis as the presenting sign of multiple lentigines syndrome. *Arch Dermatol* **118**:260–262, 1982.
39. Smith RF et al: Generalized lentigo, electrocardiographic abnormalities, conduction disorders and arrhythmias in three cases. *Am J Cardiol* **25**:501–506, 1970.
40. Somerville J, Bonham-Carter RE: The heart in lentiginosis. *Br Heart J* **34**:58–66, 1972.
41. Sommer A et al: A family study of the leopard syndrome. *Am J Dis Child* **121**:520–523, 1971.
41a. Stavrianeas N et al: Association de multiples tumeurs d'Abrikossoff disséminées sur le corps et les membres avec des anomalies oculo-cardiaques chez une fille de 8 ans. *Ann Dermatol Venereol* **106**:609–610, 1979.
42. Sutton MGSt J et al: Hypertrophic obstructive cardiomyopathy and lentiginosis: A little known neural ectodermal syndrome. *Am J Cardiol* **47**:214–217, 1981.
43. Swanson SL et al: Multiple lentigines syndrome: New findings of hypogonadotrophism, hyposmia, and unilateral renal agenesis. *J Pediatr* **78**:1037–1039, 1971.
44. Touraine A: La melanoblastose neurocutanée. *Bull Soç Fr Dermatol Syph* **51**:421–431, 1941.
45. Van Voolen GA, Selmanowitz VJ: Lentiginosis profusa. III. Aesthetic and sanguineous aspects. *Cutis* **17**:325–329, 1976. (Same cases as Refs. 33–36.)
46. Vickers HR, Macmillan D: Profuse lentiginosis, minor cardiac abnormality and small stature. *Proc R Coll Med* **62**:1011–1012, 1969.
47. Voron DA et al: Multiple lentigines syndrome. Case report and review of the literature. *Am J Med* **60**:447–456, 1976.
48. Walther RJ et al: Electrocardiographic abnormalities in a family with generalized lentigo. *N Engl J Med* **275**:1220–1225, 1966.
49. Watson GH: Pulmonary stenosis, café-au-lait spots, and dull intelligence. *Arch Dis Childh* **42**:303–307, 1967.
50. Weiss LW, Zelickson AS: Giant melanosomes in multiple lentigines syndrome. *Arch Dermatol* **113**:491–494, 1977.
51. Wendt RG et al: Polyarticular pigmented villonodular synovitis in childhood: Evidence for a genetic condition. *J Rheumatol* **13**:921–926, 1986.

## Macular cutaneous and mucosal pigmentation, myxoma, and endocrine neoplasia syndrome (Carney syndrome, NAME syndrome, LAMB syndrome)

Frankenfeld et al (10a), in 1960, described the association of biatrial cardiac myxomas and freckles.

In 1973 Rees et al (16) reported the binary combination of cutaneous lentiginosis and left atrial myxoma. Subsequently, several other reports describing the condition in greater detail were published. Most were sporadic cases (1,10,12,13a,14,17) although several pedigrees were compatible with autosomal dominant inheritance (3,6–9,15,16, 19,21–23). Carney et al (4–7) and Vidaillet et al (22) provided comprehensive reviews of the literature. Because of the wide variety of associated abnormalities and because large kindreds have not been described in detail, the disorder may be heterogeneous.

**Skin.** Multiple dark-brown macules are found on the face, lips, hands (including the palms and ventral aspects of the fingers), neck,

Fig. 13–75. *Macular cutaneous and mucosal pigmentation, myxoma, and endocrine neoplasia syndrome.* (A) Macular pigmentation of face and lips. (B) Note similar facial and mucosal pigmentation. There is a small myxoma of the right lower eyelid. [A from *DL Atherton et al,* Br J Dermatol **103**:421, 1980. B from *JA Carney et al,* Medicine (Baltimore) **64**:270, 1985.]

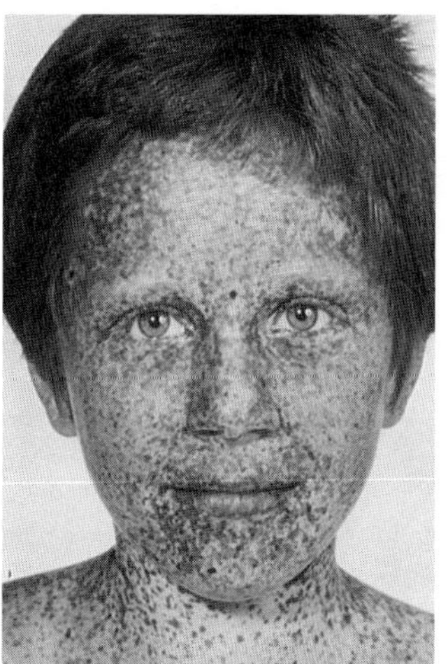

A

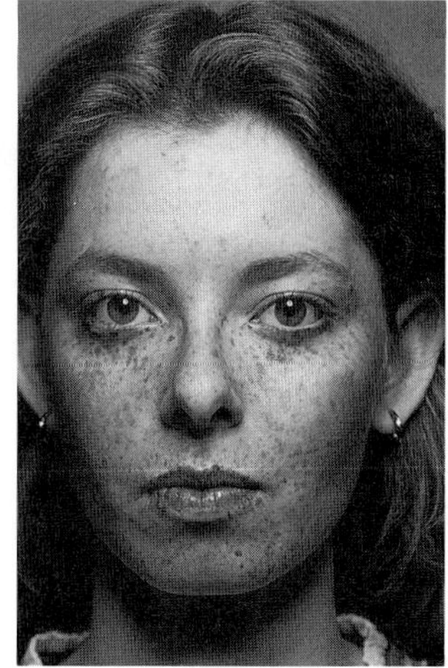

B

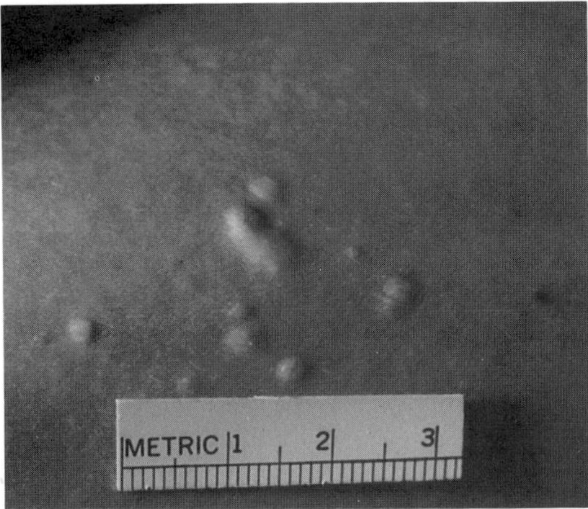


Fig. 13–76. *Macular cutaneous and mucosal pigmentation, myxoma, and endocrine neoplasia syndrome.* Myxomas on the skin of the arm. [From *JA Carney et al,* Medicine (Baltimore) **64**:270, 1985.]

shoulders, arms, trunk, and thighs in 70% (1,6,7,14,16) (Fig. 13–75A,B). They may be present at birth (1), with additional lesions developing thereafter (1,17). The lesions measure from a few millimeters to 1 cm in diameter (1). On microscopic examination, they are usually diagnosed as lentigines (6,14,16), although Atherton (1) reinterpreted the biopsy from Rees' patient (16) as an ephelis. Blue nevi, present from birth, have also been described (1,10,17).

About 60% may also develop multiple subcutaneous myxoid papules or nodules (1,5,11,17) with predilection for the eyelids, ears, and nipples. The lesions are usually 1 cm or less in diameter. Described as neurofibromas in some reports (1,20), in most publications the lesions are more consistent with myxoma (5) (Fig. 13–76).

**Cardiovascular system.** Multiple myxomas [biatrial (40%), ventricular (15%), and recurrent (20%)] are found in 50% (1,13, 16,17,19,20) (Fig. 13–77). Systemic embolization (13) from these tumors, as well as congestive heart failure (14,17), may occur. In 40% the cardiac myxomas are familial (22).

Fig. 13–77. *Macular cutaneous and mucosal pigmentation, myxoma, and endocrine neoplasia syndrome.* Myxoma of heart. [From *M Schweizer-Cagianut et al,* Virchows Arch (A) **397**:183, 1982.]

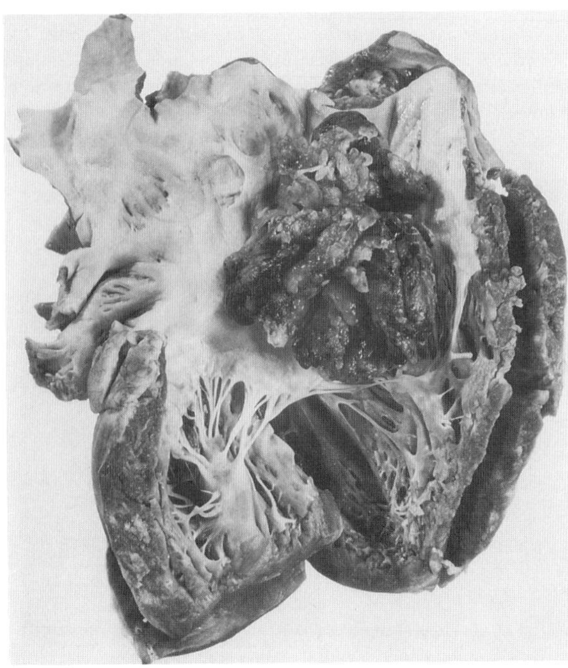

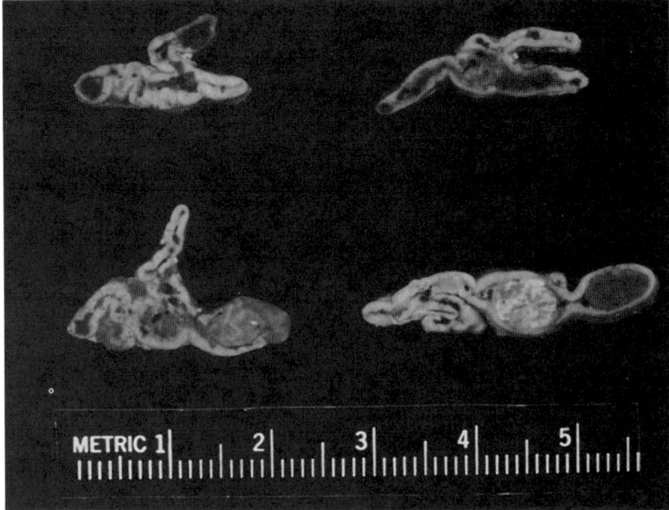


A

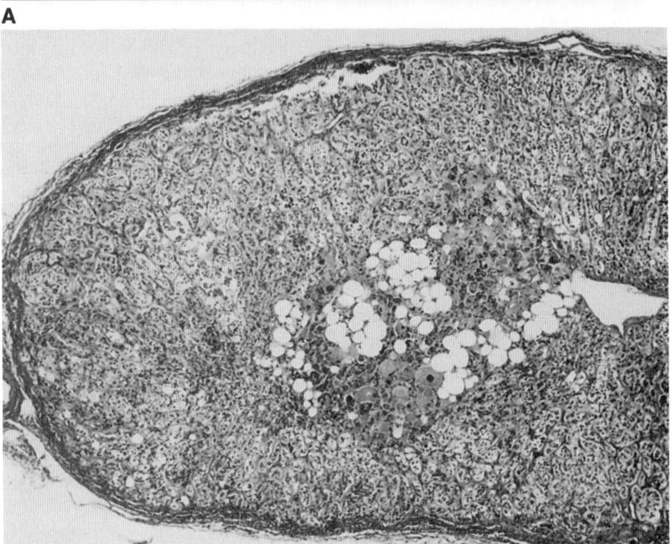

B

Fig. 13–78. *Macular cutaneous and mucosal pigmentation, myxoma, and endocrine neoplasia syndrome.* (A) Primary pigmented nodular adrenocortical disease. Note small pigmented nodules. (B) Primary pigmented nodular adrenocortical disease. Typical pigmented nodule composed of epithelial cells, fat cells, and a few lymphocytes, situated deep in cortex. The extra nodular cortex is disorganized and lacks normal zonation. (H&E, ×64). [A from *JA Carney et al,* Medicine (Baltimore) **64**:270, 1985. B courtesy of *JA Carney,* Rochester, Minnesota, unpublished, 1987.]

**Ophthalmologic abnormalities.** Macular pigmentation has been found on the conjunctiva (1,6) and sclerae (17). Myxomas frequently occur on the eyelids (5,6,11,19).

**Genitourinary tract.** Multiple black macules may be found on the vulva (6,14) and labia majora (6). Opalescent papules and dermal nodules, probably myxomas, have also been reported on the vulva (17) and on the perineum (6). Large cell calcifying Sertoli cell tumor of the testis has been described (6,15, see 19). Myxoma of the uterus has been reported (3).

**Other abnormalities.** Cushing syndrome with pigmented nodular adrenocortical disease (1,3,6,19,20), (Fig. 13–78A,B), pheochromocytoma (3), myxoid mammary fibroadenomas (6,8) (Fig. 13–79), be-

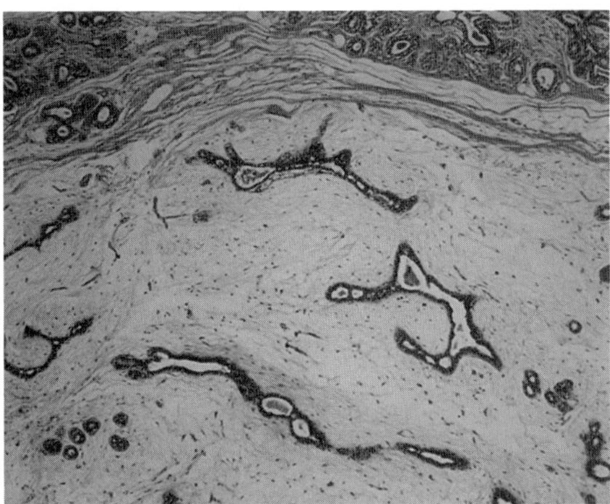

Fig. 13–79. *Macular cutaneous and mucosal pigmentation, myxoma, and endocrine neoplasia syndrome.* Unencapsulated myxoid fibroadenoma of breast. Note elongated proliferated ducts surrounded by hypocellular stroma.[From *JA Carney et al,* Medicine (Baltimore) **64**:270, 1985.]

nign papillary and follicular hyperplasia of the thyroid (17), fibrolaminar hepatoma (10), acromegaly, and thyroid adenoma have been reported in 15–30% (6,19).

**Oral manifestations.** Pigmented macules may be found on the perioral skin, lip mucosa, and buccal mucosa (1,6,8,19). The patient originally reported by Rees et al (16) later developed myxoid neurofibroma on the palate (1). The patient noted by Tway et al (21) had a neurofibroma on the hard palate; Cook et al (8) described a myxoma on the hard and soft palates of the same patient (Fig. 13–80). Myxomas have also been described on the tongue (17) (Fig. 13–81).

**Differential diagnosis.** The syndrome should be considered in any patient under 40 years old who has cardiac myxoma in other than the left atrium or who has more than one cardiac myxoma. The skin pigmentation should not be confused with that found in *Peutz–Jeghers syndrome.* The condition should also be differentiated from the *LEOPARD syndrome.* Isolated familial cardiac myxomas have been reported (19). Carney et al (4) reviewed the differences between nonfamilial and familial cardiac myxomas. Many of the following reports are likely examples: Roper et al (18) reported two unrelated patients each with intraatrial myxoma and diffuse fibroadenomas of the breast with myxoid stroma. According to Carney et al (6), one of these patients subsequently had skin myxomas, although no pigmented skin macules were found. Barlow et al (3) described a woman with myxomas of the right atrium and uterus, adrenal hyperplasia, and Cushing syndrome, an incidental finding of pheochromocytoma of one adrenal gland, and myxoid change in a leiomyoma; her sister had Cushing's disease, and on follow-up (6) was found to have cardiac myxomas, pigmented skin lesions, dermal myxomas, pigmented nodular adrenocortical disease, and myxoid mammary fibroadenomas. The patient reported by Crawford et al (9) with atrial myxoma was later (6) found to have pigmented nodular adrenocortical disease and mammary myxoid fibroadenomas but no pigmented skin lesions; her mother died from complications of a left atrial myxoma. Balk et al (2) described a 29-year-old male with myxomas of a shoulder and thigh, Hürthle cell adenoma of the thyroid, and cardiac myxomas; no abnormal skin pigmentation was described.

Fig. 13–80. *Macular cutaneous and mucosal pigmentation, myxoma, and endocrine neoplasia syndrome.* Palatal myxoma. Circumscribed by nonencapsulated myxoma of minor salivary glands of palate (Movat pentachrome stain, ×2). (From *CA Cook et al,* Oral Surg **63**:175, 1987.)

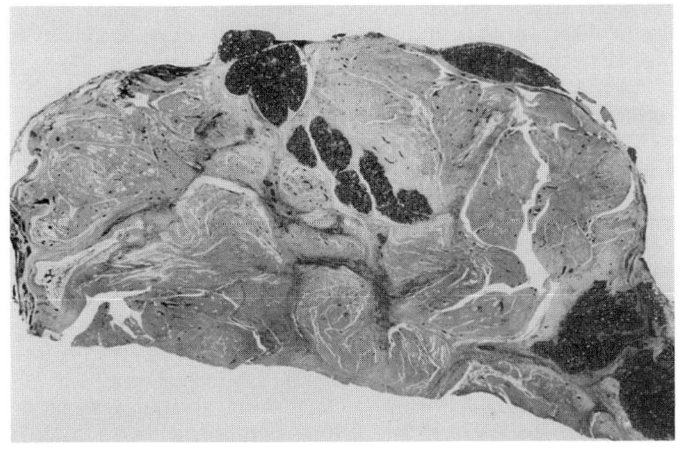

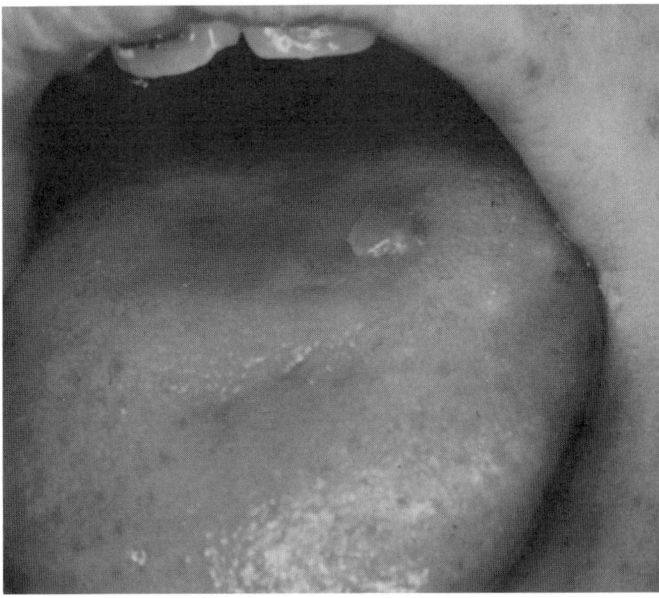

Fig. 13–81. *Macular cutaneous and mucosal pigmentation, myxoma, and endocrine neoplasia syndrome.* Myxoma on the tongue. (From *AR Rhodes et al,* J Am Acad Dermatol **10**:72, 1984.)

#### References [Macular cutaneous and mucosal pigmentation, myxoma, and endocrine neoplasia syndrome (Carney syndrome, NAME syndrome, LAMB syndrome)]


1. Atherton DL et al: A syndrome of various cutaneous pigmented lesions, myxoid neurofibromata and atrial myxoma: The NAME syndrome. *Br J Dermatol* **103**:421–429, 1980.
2. Balk AH et al: Bilateral cardiac myxomas and peripheral myxomas in a patient with recent myocardial infarction. *Am J Cardiol* **44**:767–770, 1979.
3. Barlow JF et al: Myxoid tumor of the uterus and right atrial myxomas. *S Dakota J Med* **36**:9–13, July, 1983.
4. Carney JA: Differences between nonfamilial and familial cardiac myxoma. *Am J Surg Pathol* **9**:53–55, 1985.
5. Carney JA et al: Cutaneous myxomas: A major component of the complex of myxomas, spotty pigmentation, and endocrine overactivity. *Arch Dermatol* **122**:790–798, 1986.
6. Carney JA et al: The complex of myxomas, spotty pigmentation and endocrine overactivity. *Medicine* **64**:270–283, 1985.
7. Carney JA et al: Dominant inheritance of the complex of myxomas, spotty pigmentation, and endocrine overactivity. *Mayo Clin Proc* **61**:165–172, 1986.
8. Cook CA et al: Mucocutaneous pigmented spots and oral myxomas: The oral manifestations of the complex of myxomas, spotty pigmentation, and endocrine overactivity. *Oral Surg Oral Med Oral Pathol* **63**:175–183, 1987.
9. Crawford FA et al: Unusual aspects of atrial myxoma. *Ann Surg* **188**:240–244, 1978.
10. Danoff A et al: Adrenal micronodular dysplasia, cardiac myxomas, lentigines, and spindle cell tumors. *Arch Intern Med* **147**:443–448, 1987.

10a. Frankenfeld RH et al: Bilateral myxomas of the heart. *Ann Intern Med* **53**:827–838, 1960.
11. Kennedy RH et al: Ocular pigmented spots and eyelid myxomas. *Am J Ophthalmol* **104**:533–538, 1987.
12. Lortscher RH et al: Left atrial myxoma presenting as rheumatic fever. *Chest* **66**:302–303, 1974.
13. McCarthy PM et al: The significance of multiple, recurrent and "complex" cardiac myxomas. *J Thorac Cardiovasc Surg* **91**:389–396, 1986.
13a. Okada N et al: Metastasizing atrial myxoma: A case with multiple subcutaneous tumours. *Br J Dermatol* **115**:239–242, 1986.
14. Peterson LL, Serrill WS: Lentiginosis associated with a left atrial myxoma. *J Am Acad Dermatol* **10**:337–340, 1984.
15. Proppe KH, Scully RE: Large-cell calcifying Sertoli cell tumor of the testis. *Am J Clin Pathol* **74**:607–619, 1980.
16. Rees JR et al: Lentiginosis and left atrial myxoma. *Br Heart J* **35**:874–876, 1973.
16a.Reinecke P et al: Kombination von Vorhofmyxom und multiplen Hautmyxomen: ein eigener Symptomenomplex? *Dtsch Med Wochenschr* **112**:1902–1905, 1987.
17. Rhodes AR et al: Mucocutaneous lentigines, cardiomucocutaneous myxomas, and multiple blue nevi: The "LAMB" syndrome. *J Am Acad Dermatol* **10**:72–82, 1984.
18. Roper CL et al: Atrial myxoma associated with fibroadenomas of the breast. *Missouri Med* **62**:113–116, 1965.
19. Schweizer-Cagianut M et al: Primary adrenocortical nodular dysplasia with Cushing's syndrome and cardiac myxomas. A peculiar familial disease. *Virchows Arch [A]* **397**:182–183, 1982.
20. Shenoy BV et al: Bilateral primary pigmented nodular adrenocortical disease. *Am J Surg Pathol* **8**:335–344, 1984.
21. Tway KP et al: Multiple biatrial myxomas demonstrated by two-dimensional echocardiography. *Am J Med* **71**:896–899, 1981.
22. Vidaillet HJ Jr et al: "Syndrome myxoma": A subset of patients with cardiac myxoma associated with pigmented skin lesions and peripheral and endocrine neoplasms. *Br Heart J* **57**:247–255, 1987.
23. Wilsher ML et al: A familial syndrome of cardiac myxomas, myxoid neurofibromata, cutaneous pigmented lesions and endocrine abnormalities. *Aust NZ J Med* **16**:393–396, 1986.

## Waardenburg syndrome

The syndrome so precisely described by Waardenburg in 1948 (59) was reported earlier by Mende (35), van der Hoeve (58), Klein (28), and others (41). The main features include lateral displacement of the medial canthi and lacrimal punctae with a broad nasal root, poliosis, heterochromia irides, hyperplasia of the medial portions of the eyebrows, and congenital sensorineural hearing loss. The reader is referred to a comprehensive review (41).

At least 1400 cases have been reported to date. During the last 15 years, a number of studies (3,23) have clearly separated Waardenburg syndrome into two forms, based on whether dystopia canthorum is present (Type I) or absent (Type II).* It has been estimated that Type II is 20 times as common as Type I (6). Several series have shown that hearing loss is much more common in Type II (about 50%) than in Type I (about 25%). Perhaps Type II is more common among blacks but more documentation is needed. Pigmentary disorders are more frequent in Type II. However, it has also been shown that patients with Type I and hearing loss more commonly have pigmentary disorders of the eyes, hair, and skin than those with Type I who have normal hearing. Arias (3) suggested a Type III that he designated "pseudo-Waardenburg syndrome," characterized by absence of dystopia canthorum but with one-sided ptosis of the upper lid. Whether this form is a distinct type needs further documentation. Bard (8) also suggested further heterogeneity, but we are not convinced that his patient is not an extreme example of Type II.

The syndromes exhibit autosomal dominant inheritance with essentially complete penetrance and variable expressivity. Estimates of prevalence suggest that the syndrome is found in 2–5% of all congenitally deaf persons (11,12,). Between 2 and 3/100,000 in the general population are affected in the Netherlands (59). However, since most probands have been recognized because of impaired hearing, there is ascertainment bias (41). There is possible loose linkage between the ABO locus on chromosome 9 and Waardenburg syndrome Type I (6,7,51). However, Ishikiriyama et al (26a) suggested the locus is located at 2q35 or 2q37.3. New mutation for Type I has been shown to be associated with advanced paternal age (27).

*In fairness, there had been numerous hints of genetic heterogeneity for several decades. Mounier-Kuhn et al (37) and Partsch (43), for example, pointed out that some kindreds did not exhibit dystopia canthorum. Unfortunately, even currently, this separation is not observed in reports (4).

**Facies.** Poliosis, synophrys, heterochromia irides, broad nasal root, hypoplastic alar cartilages, and mild mandibular prognathism produce a remarkably striking appearance in those with Type I syndrome (Fig. 13–82A–C).

**Eyes.** Although the interpupillary and outer canthal distances are normal in most affected individuals, an increased distance between the inner canthi (dystopia canthorum) is evident, resulting in blepharophimosis. This is the sine qua non in Type I patients. The sclerae may be covered medially, giving a false impression of esotropia.

Arias and Mota (5) pointed out that clinical judgment regarding the presence of dystopia canthorum is not always accurate and that a W-index value should be calculated to separate Type I from Type II. Preus et al (45) suggested modification of the formula and De Saxe et al (11) used a palpebral fissure index.

True ocular hypertelorism is present in about 10% (41,57). Normal inner canthal, outer canthal, and interpupillary values have been discussed by Pryor (46) and others (16,30,36).

The inferior lacrimal points are displaced laterally, sometimes as far as the cornea (Fig. 13–83A). There is an increased susceptibility to dacryocystitis (41). Hyperplasia of the medial portions of the eyebrows, or even confluence (synophrys), is present in 85% of Type I and 25% of Type II patients (23). This characteristic may be less evident in women, who often pluck these hairs (12,41). Heterochromia irides, or, more accurately, hypoplasia of the irides, is seen in about one-third of the cases in both types. It may be partial or total, both irides ranging from a mottled to blue appearance. Hypoplastic blue irides are seen in 10–15% of both types (23) although De Saxe et al (11) suggested that blacks with Type II more often have clear blue eyes. DiGeorge et al (12) and others (19) noted pigmentary mottling at the periphery of the fundi. The fundi are albinoid (10) (Fig. 13–83B). Other eye anomalies such as cataracts, microphthalmia, and ptosis have been reported in association with Type I and Type II patients (3,19,34a,42,57,59,61). Convergent strabismus has been found in 20% (10).

**Hair and skin.** White forelock (poliosis) is present in 30–40% of patients with either Type I or Type II (23). In some cases, only a few white hairs are evident. This feature may be present at birth but tends to disappear with age (11,14). In females, poliosis may not be evident because of the tendency to mask it by dyeing the hair (12). Premature (beginning in the 20s) graying of eyebrows, lashes, and hair occurs in 20–35% of Type I cases and in 5% of Type II cases (23). Pigmentary anomalies, such as vitiligo, may be found in 15–20% of affected persons of either type (41). Black forelock has also been reported in both Type I syndrome (3) and in Type II syndrome (11).

**Ears and nose.** Bilateral congenital sensorineural hearing loss has been observed in 20% of Type I patients (23,59) and in 55% of Type II patients (23). Little residual hearing is present for the lower frequencies. About 15% of Type I and 5% of Type II have unilateral hearing loss, rarely of severe degree, in which moderate uniform loss for the lower and middle ranges occurs with improvement in the higher range—often with normal hearing for 6000–8000 cycles per second (15,23).

Hageman (22) demonstrated that where there is unilateral hearing loss, there is no correlation between the laterality of the hearing loss and the side having the blue eye in case of heterochromia irides.

The hearing loss has been shown to be due to absence of the organ of Corti and stria vascularis and a sparcity of ganglion cells (15,47).

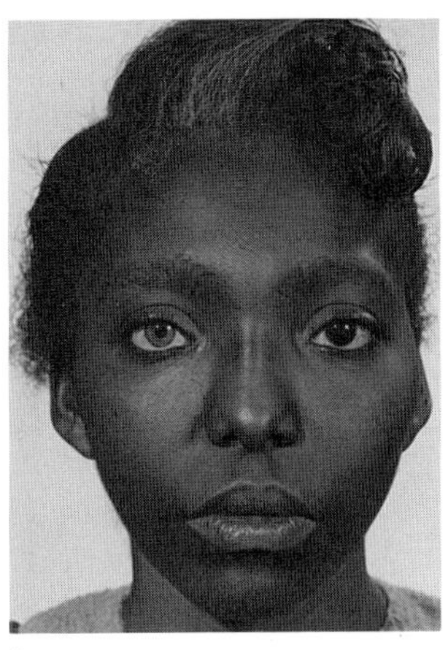
A

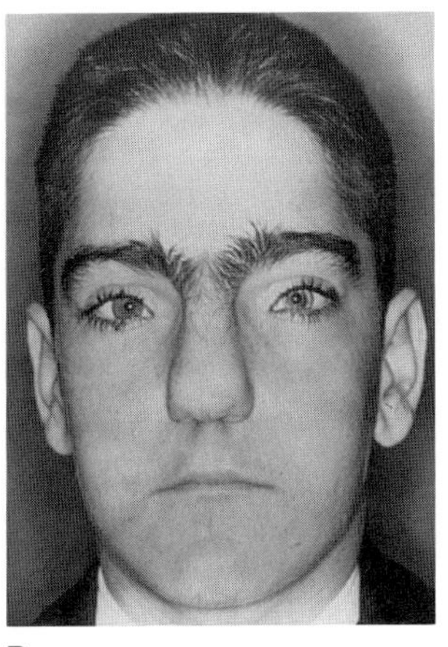
B

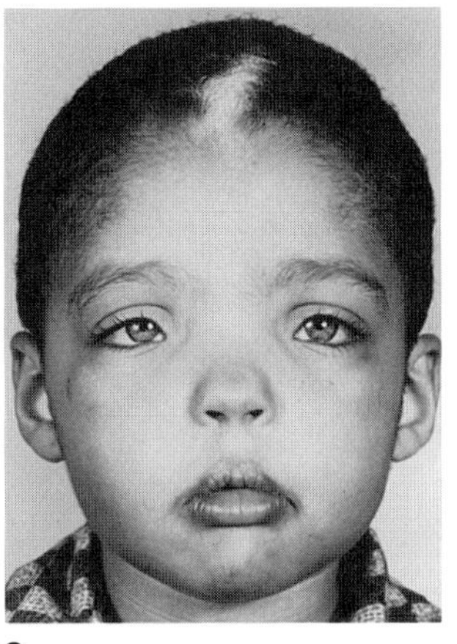
C

Fig. 13–82. *Waardenburg syndrome.* (A) Note white forelock, heterochromia irides, dystopia canthorum. (B) Note similar phenotype. (C) White forelock, synophrys, dystopia canthorum, hypoplastic nasal alae. (A from *AM DiGeorge et al,* J Pediatr **57**:649, 1960. B from *MW Partington,* Can Med Assoc J **90**:1008, 1964. C courtesy of *M Goldberg,* Chicago, Illinois.)

Vestibular function was shown to be normal in 25 patients with Type I syndrome (24). However, Marcus (33) noted abnormal vestibular function in Type II syndrome in blacks. Tomographic findings of the inner ears of 24 Type I patients were found to be normal by Nemansky and Hageman (38), but they pointed out that other investigators have discovered malformations of the semicircular canals.

The nasal root is broad, often with loss of the frontonasal angle. The alar cartilages are usually hypoplastic, resulting in narrow nostrils (Fig. 13–82B). The tip of the nose tends to be rounded and slightly upturned, revealing the columella (28).

Fig. 13–83. *Waardenburg syndrome.* (A) Lateral displacement of lacrimal points. (B) Albinoid fundus. (From *JW Delleman* and *MJ Hageman,* J Pediatr Ophthalmol **15**:341, 1978.)

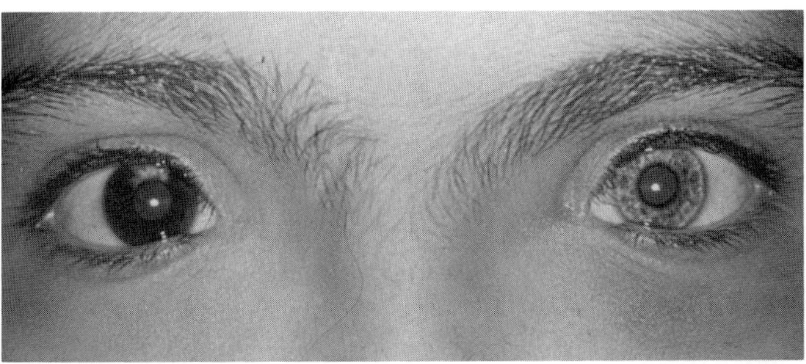
A

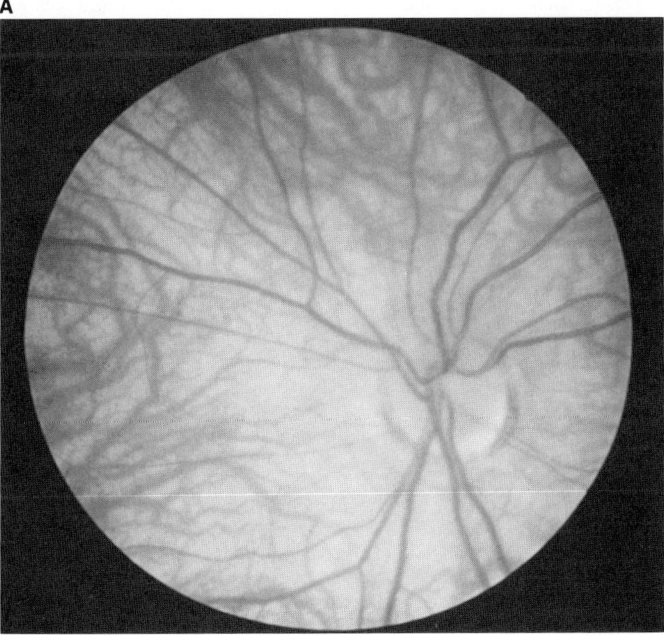
B

**Gastrointestinal system.** There have been numerous reports (1,9,13,34,40,47,50) of the association between Type I Waardenburg syndrome and Hirschsprung's disease, a disorder of intestinal motility characterized by absence of parasympathetic ganglion cells from the submucosal and myenteric plexuses of the gut (1,50,68). The association has also been documented in Type II disease (17,32a,34a).

Waardenburg syndrome, Type I, has been also reported in association with anal atresia (39) and with esophageal atresia (15, 39a), but these may well be chance findings.

**Skeletal system.** Few cases have been reported in which full radiographic surveys were carried out. Among the cases reported, a variety of skeletal anomalies such as the Sprengel deformity, skull anomalies, bony defects of the thorax, abnormal upper limb length, abnormal carpal bones, syndactyly, sacralization of the fifth lumbar vertebra, sacral dimple, and spina bifida have been noted (11,41,59).

Klein (28) reported a woman with Waardenburg syndrome and myoosteoarticular defects of the upper limb and pectoral region with absence of muscles, rigidity of joints, axillary webs, and camptodactyly of all fingers (Fig. 13–84A,B). Waardenburg, on seeing this patient, investigated several Dutch institutions for the deaf and subsequently defined Waardenburg syndrome, Type I. Waardenburg considered Klein's patient to have another disorder. Klein's view that his patient had Waardenburg syndrome has been subsequently vindicated by his finding of a father with essentially the same type of anomalies who has a child with full expression of Waardenburg syndrome, Type I. Senrui (49) reported flexion contractures and limited spination of forearms. We believe that the family reported by Sommer et al (53) with flexion contractures of the hands actually had Waardenburg syndrome and does not represent a new entity. Other cases have been analyzed by Goodman et al (20). Although Goodman et al (20) and Klein (29) have suggested that this represents still another type of Waardenburg syndrome (Type III), we believe that the muscular and skeletal defects in the upper limb represent an associated component of the disorder, much as Hirschsprung's disease. A patient described by Wilbrandt (61) who also had arthromyodysplasia was noted in a family having otherwise typical Waardenburg syndrome.

**Other findings.** Goodman et al (20a) described vaginal agenesis.

**Oral manifestations.** The lower lip is especially full and protrudes. The mandible seems to be prognathic in most patients, although this feature has been studied cephalometrically in few instances (12,41,57,59,61). Radiographic examination of the skull in a small series of patients has demonstrated greater lateral winging of the angle of the mandible, increasing the bigonial distance.

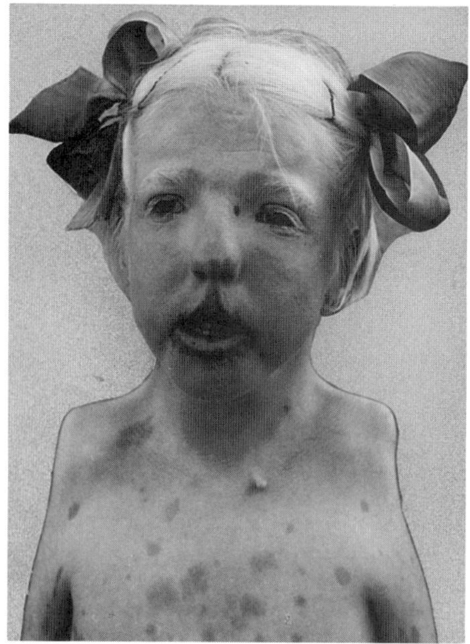

A

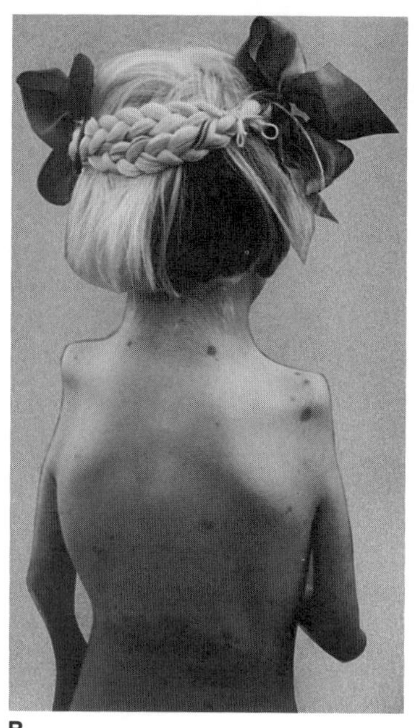

B

Fig. 13–84. *Waardenburg syndrome.* (A,B) Patient exhibiting hypoplasia of shoulder girdle, axillary webbing, and midline defect of upper lip. (From *D Klein,* Arch Klaus Stift Vererb Forsch **22**:336, 1947.)

Facial clefting as an associated finding was pointed out by Giacoia and Klein (18) and by Gorlin et al (21). Cleft lip and/or cleft palate is about eight times as frequent in Waardenburg syndrome as in the normal population, being seen in about 1% of Waardenburg, Type I population (2,18,19,21,59).

**Differential diagnosis.** Dystopia canthorum and hypertelorism may be seen in a variety of disorders (44). Synophrys is a feature of *de Lange syndrome* and may also occur with pilonidal cyst (48). Hypoplastic alar cartilages may be seen in the *OFD I syndrome.* Poliosis, with or without piebaldness, may be inherited as an autosomal dominant trait and may be seen with *pseudocleft of the upper lip, cleft lip-palate, and hemangiomatous branchial cleft.* Poliosis, vitiligo, and dysacousia may be seen in combination with alopecia and uveitis in Vogt–Koyanagi syndrome (25). Heterochromia irides may be acquired, inherited as an autosomal dominant trait (15), or associated with *Romberg syndrome.* Partial albinism with deafness may occur as an X-linked trait (62). Vitiligo and congenital sensorineural hearing loss have been reported as an autosomal recessive combination (56). An autosomal dominant syndrome of white forelock, leukoderma, cerebellar ataxia, poor motor coordination, and mild mental retardation was described by Telfer et al (54).

Hirschsprung's disease has been reported by several authors to be associated with congenital sensorineural hearing loss (31,52,60) and with facial clefting. It should be emphasized that there is no evidence to indicate that any of these patients had Waardenburg syndrome.

**Laboratory aids.** See the Appendix for normal inner canthal, outer canthal, and interpupillary distances and canthal indices. A rough estimate of whether a person has dystopia canthorum may be made either on the person or a photograph employing an inexpensive caliper. The inner canthal distance divided by the interpupillary is greater than 0.6 with dystopia canthorum (42).

## References (Waardenburg syndrome)


1. Ambani LM: Waardenburg and Hirschsprung syndromes. *J Pediatr* **102**:802, 1983.

2. Amoni SS, Abdurrahman MB: Waardenburg's syndrome: Case reports in two Nigerians. *J Pediatr Ophthalmol Strab* **16**:172–175, 1979.

3. Arias S: Genetic heterogenity in the Waardenburg syndrome. *Birth Defects* **7**(4):87–101, 1971.

4. Arias S: Waardenburg syndrome—two distinct types. *Am J Med Genet* **6**:99–100, 1980.

5. Arias S, Mota M: Apparent nonpenetrance for dystopia in Waardenburg syndrome Type I, with some hints on the diagnosis of dystopia canthorum. *J Génét Hum* **26**:103–131, 1978.

6. Arias S, Mota M: Current status of the ABO-Waardenburg syndrome Type I linkage. *Cytogenet Cell Genet* **22**:291–294, 1978.

7. Arias S et al: Probable loose linkage between the ABO locus and Waardenburg syndrome Type I. *Humangenetik* **27**:145–149, 1975.

8. Bard LA: Heterogeneity and Waardenburg's syndrome. *Arch Ophthalmol* **96**:1193–1198, 1978.

9. Branski D et al: Hirschsprung's disease and Waardenburg's syndrome. *Pediatrics* **63**:803–805, 1979.

10. Delleman JW, Hageman MJ: Ophthalmological findings in 34 patients with Waardenburg syndrome. *J Pediatr Ophthalmol* **15**:341–345, 1978.

11. De Saxe M et al: Waardenburg syndrome in South Africa. *S Afr Med J* **66**:256–261, 1984.

12. DiGeorge AM et al: Waardenburg's syndrome. *J Pediatr* **57**:649–669, 1960.

13. Farndon PA, Bianchi A: Waardenburg's syndrome associated with total aganglionosis. *Arch Dis Childh* **58**:932–933, 1983.

14. Feingold M et al: Waardenburg's syndrome during the first year of life. *J Pediatr* **71**:874–876, 1967.

15. Fisch L: Deafness as part of an hereditary syndrome. *J Laryngol Otolaryngol* **73**:355–383, 1959.

15a. François J: Waardenburg's syndrome. *Int Ophthalmol* **5**:3–13, 1982.

16. Freihofer HPM: Inner intercanthal and interorbital distances. *J Maxillofac Surg* **8**:324–326, 1980.

17. Fried K, Beer S: Waardenburg's syndrome and Hirschsprung's disease in the same patient. *Clin Genet* **18**:91–92, 1980.

18. Giacoia JP, Klein SW: Waardenburg's syndrome with bilateral cleft lip. *Am J Dis Child* **117**:344–348, 1969.

19. Goldberg MF: Waardenburg's syndrome with fundus and other anomalies. *Arch Ophthalmol* **76**:797–810, 1966.

20. Goodman RM et al: Upper limb involvement in the Klein–Waardenburg syndrome: Evidence for further genetic heterogeneity. *Am J Med Genet* **11**:425–433, 1982.

20a. Goodman RM et al: Absence of vagina and right-sided adnexa uteri in the Waardenburg syndrome: A possible clue to the embryological defect. *J Med Genet* **25**:355–357, 1988.

21. Gorlin RJ et al: Facial clefting and its syndromes. *Birth Defects* **7**(7):3–49, 1971.

22. Hageman MJ: Audiometric findings in 34 patients with Waardenburg syndrome. *J Laryngol Otolaryngol* **91**:575–584, 1977.

23. Hageman MJ, Delleman JW: Heterogeneity in Waardenburg syndrome. *Am J Hum Genet* **29**:468–485, 1977.

24. Hageman MJ, Oosterveld WJ: Vestibular findings in 25 patients with Waardenburg's syndrome. *Arch Otolaryngol* **103**:648–652, 1977.

25. Hague EB: Uveitis, dysacousia, alopecia, poliosis and vitiligo. *Arch Ophthalmol* **31**:520–538, 1944.

26. Houghton NI: Waardenburg syndrome with deafness as a presenting syndrome. Report of two cases. *N Z Med J* **63**:83–89, 1964 (Case 1).

26a. Ishikiriyama S et al: Waardenburg syndrome type I in a child with de novo inversion 2(q35q37.3). *Am J Med Genet* 33:505–507, 1989.

27. Jones KL et al: Older paternal age and fresh gene mutation: Data on additional disorders. *J Pediatr* **86**:84–88, 1975.

28. Klein D: Albinisme partiel (leucisme) accompagne de surdimutité, d'osteomyodysplasie, de raideurs articulaire congénitales multiples et d'autres malformations congénitales. *Arch Klaus Stift Vererb Forsch* **22**:336–342, 1947.

29. Klein D: Historical background and evidence for dominant inheritance of the Klein–Waardenburg syndrome (Type III). *Am J Med Genet* **14**:231–239, 1983.

30. Laestadius ND et al: Normal inner canthal and outer orbital dimensions. *J Pediatr* **74**:465–468, 1969.

31. Lowry RB: Hirschsprung's disease and congenital deafness. *J Med Genet* **12**:114–115, 1975.

32. Mahakrishnan A, Srinivasan MS: Piebaldness with Hirschsprung's disease. *Arch Dermatol* **116**:1102, 1980.

32a. Mallory SB et al: Waardenburg's syndrome with Hirschsprung's disease: A neural crest defect. *Pediatr Dermatol* **3**:119–124, 1986.

33. Marcus RE: Vestibular function and additional findings in Waardenburg's syndrome. *Acta Otolaryngol Suppl* **229**:1–30, 1968.

34. McKusick VA: Congenital deafness and Hirschsprung's disease. *N Engl J Med* **288**:691, 1968.

34a. Meire F et al: Waardenburg syndrome. Hirschsprung megacolon and Marcus Gunn ptosis. *Am J Med Genet* **27**:683–686, 1987.

35. Mende I: Über eine Familie hereditär-degenerativer Taubstummer mit mongoloidem Einschlag und teilweisen Leukismus der Haut und Haare. *Arch Kinderheilkd* **79**:214–222, 1926.

36. Metres K, Kutzvegli F: Inner canthal and intermammillary indices in the newborn infant. *J Pediatr* **85**:90–92, 1975.

37. Mounier-Kuhn P et al: Albinisme partiel et surdi-mutité. *J Fr Otorhinolaryngol* **7**:915–919, 1958.

38. Nemansky J, Hageman MJ: Tomographic findings of the inner ears of 24 patients with Waardenburg's syndrome. *Am J Roentgenol* **124**:250–255, 1975.

39. Nutman J et al: Anal atresia and the Klein–Waardenburg syndrome. *J Med Genet* **18**:239–241, 1981.

39a. Nutman J et al: Possible Waardenburg syndrome with gastrointestinal anomalies. *J Med Genet* **23**:175–178, 1986.

40. Omenn GS, McKusick VA: The association of Waardenburg syndrome and Hirschsprung megacolon. *Am J Med Genet* **3**:217–223, 1979.

41. Pantke OA, Cohen MM Jr.: The Waardenburg syndrome. *Birth Defects* **7**(7):147–152, 1971.

42. Partington MW: Waardenburg's syndrome and heterochromia iridum in a deaf school population. *Can Med Assoc J* **90**:1008–1017, 1964.

43. Partsch CJ: Hereditäre Taubheit beim Syndrom nach Waardenburg-Klein. *Z Laryngol Rhinol Otol* **41**:752–762, 1962.

44. Peterson MA et al: Comments on frontonasal dysplasia, ocular hypertelorism and dystopia canthorum. *Birth Defects* **7**(7):120–124, 1971.

45. Preus M et al: Waardenburg syndrome—penetrance of major signs. *Am J Med Genet* **15**:383–388, 1983.

46. Pryor HB: Objective measurement of interpupillary distance. *Pediatrics* **44**:973–977, 1969.

47. Rarey KE, Davis LE: Inner ear anomalies in Waardenburg's syndrome with Hirschsprung's disease. *Int J Pediatr Otorhinolaryngol* **8**:181–189, 1984.

48. Sebrechts PH: A significant sign of pilonidal cyst. *Dis Colon Rectum* **4**:56–59, 1961.

49. Senrui H: Congenital clasped thumbs with Waardenburg syndrome in three generations of one family: An undescribed congenital anomalies complex. *J Pediatr Orthoped* **4**:472–476, 1984.

50. Shah KN et al: White forelock, pigmentary disorder of irides, and long segment Hirschsprung disease: Possible variant of Waardenburg syndrome. *J Pediatr* **99**:432–434, 1981.

51. Simpson JL et al: Analysis for possible linkage between the loci for the Waardenburg syndrome and various blood groups and serological traits. *Humangenetik* **23**:45–50, 1974.

52. Skinner R, Irvine D: Hirschsprung's disease and congenital deafness. *J Med Genet* **10**:337–339, 1973.

53. Sommer A et al: Previously undescribed syndrome of craniofacial, hand anomalies and sensorineural deafness. *Am J Med Genet* **15**:71–77, 1983.

54. Telfer MA et al: Dominant piebald trait (white forelock and leukoderma) with neurological impairment. *Am J Hum Genet* **23**:383–389, 1971.

55. Thorkilgaard O: Waardenburg's syndrome in father and daughter. *Acta Ophthalmol* **40**:590–599, 1962.

56. Thurmon TF et al: Deafness and vitiligo. *Birth Defects* **12**(5): 315–320, 1976.

57. Ulivelli A, Silenzi M: Hypertelorism and Waardenburg's syndrome. *Helv Paediatr Acta* **24**:123–126, 1969.

58. van der Hoeve J: Abnorme Länge der Tränenröhrchen mit Ankyloblepharon. *Klin Monatsbl Augenheilkd* **56**:232–238, 1916.

59. Waardenburg PJ: A new syndrome combining developmental anomalies of the eyelids, eyebrows and nose root with congenital deafness. *Am J Hum Genet* **3**:195–253, 1951.

60. Weinberg AG et al: Hirschsprung's disease and congenital deafness. *Hum Genet* **38**:157–161, 1977.

61. Wilbrandt HR, Ammann F: Nouvelle observation de la forme grave du syndrome de Klein-Waardenburg. *Arch Julius Klaus Stift Vererbungsforsch* **39**:80–92, 1964.

62. Ziprkowski L et al: Partial albinism and deafmutism, due to a recessive sex-linked gene. *Arch Dermatol* **86**:530–536, 1962.

## Ataxia-telangiectasia

Delineation of the syndrome of ionizing radiation sensitivity characterized by progressive cerebellar ataxia, oculocutaneous telangiectasia, recurrent sinopulmonary infections, high incidence of neoplasia, variable immunodeficiency and chromosomal instability is usually credited to Mme. Louis-Bar (25) in 1941, although earlier examples were reported by Syllaba and Henner (50). Boder and Sedgwick (6) coined the term ''ataxia-telangiectasia'' and clearly defined the syndrome in 1958. Boder (4,5) published exhaustive reviews in 1975 and 1985. Recent excellent summaries are those of McKinnon (28) in 1987 and Cohen and Levy (9a) in 1989. Over 300 cases have been described. There have also been recent texts on the subject (7,13).

The syndrome has autosomal recessive inheritance. Its frequency in the United States, based on case reports, has been estimated as ranging between 3 and 11/million births (49), but its true frequency is probably greater. Heterozygote frequency is possibly as high as 3% (49). The most common ancestral countries of origin are England and Wales, Ireland, Eastern Europe, and Germany. Five different complementation groups have been demonstrated (22,32,49) and clinical heterogeneity has been shown by some (52) but denied by other investigators (20).

Death usually occurs during the second decade from either overwhelming pulmonary infection or cancer, the disorder possibly being more lethal in blacks (31). There is heightened sensitivity to radiation and to radiomimetic drugs (3,51). Although there is an altered DNA repair defect, it is not known whether this is primary or secondary (28).

**Facies and appearance.** The face is described as thin, and the expression as relaxed, dull, or sad (Fig. 13–85). The patient often stoops, with the shoulders drooped and the head held to one side. Mild athetoid movements may be noted around the shoulders. Growth is markedly diminished in over 65% of patients (41).

**Nervous system.** The ataxia is of the cerebellar type and usually becomes apparent between 3 and 6 years of age. The truncal ataxia is slowly progressive. Eventually there may be dyssynergia and intention tremor of the upper extremities. Hypotonia and diminished tendon reflexes are noted. Speech is slow, slurred, and often scanning.

Mental deficiency is usually not apparent until the child reaches the age of 9 years, when mental development begins to taper off. This has been noted in about 30%.

Pneumoencephalographic study shows there is cerebellar atrophy, and, on necropsy, thinning of the granular layer of the cerebellum, with diminished numbers of Purkinje cells without reactive gliosis (54).

**Eyes.** Usually at 4–6 years, fine, symmetric, bright-red streaks are noted in the temporal and nasal areas of the conjunctiva. The telan-

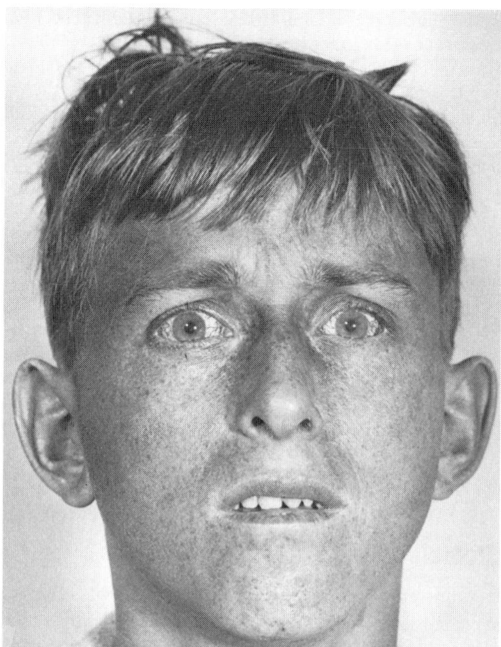

Fig. 13–85. *Ataxia-telangiectasia.* Note rigid facies marked by staring expression and prominent bulbar conjunctival vessels. (From *SJ Miller* and *W Gooddy,* Brain **87**:581, 1964.)

giectasia has been shown to be venous and nonhemorrhagic (19) (Fig. 13–86).

Strabismus, poor convergence, photophobia, and nystagmus are commonly found. Movement of the eyes is slow and interrupted, halting midway on lateral and upward gaze (apraxia). This movement is often accompanied by rapid blinking. Fixation nystagmus is present in over 80% (41).

**Skin.** The cutaneous telangiectasia is first noted on the ears, butterfly area of face, bridge of nose, and periorbitally (Fig. 13–87). The distribution is not always correlated with sun exposed areas (9). With time it extends to the neck, antecubital and popliteal areas, and the dorsum of hands and feet. Heterozygotes have increased numbers of telangiectases (9).

With continued sun exposure and/or aging, the skin tends to become sclerodermatous in appearance, with a mottled pattern of hyperpigmentation and hypopigmentation (poikiloderma). Café-au-lait spots have been noted in 35% (9). Diffuse graying of the scalp hair is noted in about 40%, even in young patients (9). Seborrheic dermatitis, follicular keratosis, and hirsutism of the arms and legs are constant features (4,6,41). A few patients have acanthosis nigricans (9), possibly related to insulin resistance (40). About 15% are generally hirsute (9).

Fig. 13–86. *Ataxia-telangiectasia.* Telangiectasia of bulbar conjunctiva. (From *G Karapati et al,* Am J Dis Child **110**:51, 1965.)

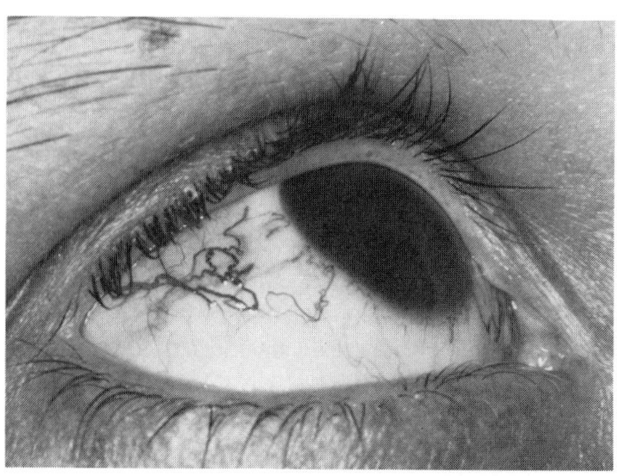

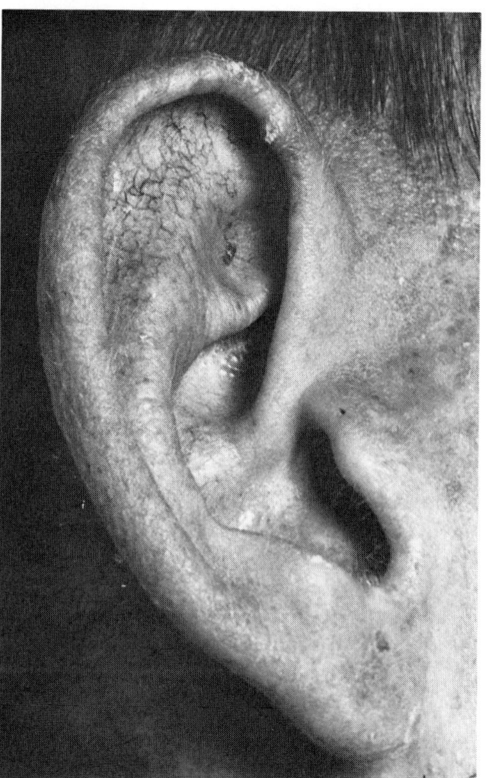

Fig. 13–87. *Ataxia-telangiectasia.* Telangiectasia of pinna. (From *M Ruiter,* Hautarzt **15**:667, 1964.)

**Respiratory system.** Recurrent sinopulmonary infections have occurred in 75–80% (6,37). These may vary in severity from acute rhinitis with otitis media to chronic bronchitis, recurrent pneumonia, and bronchiectasis.

**Neoplasia.** There is a high frequency (about 20%) of malignancy, usually lymphoreticular of T-cell type (9,11,19,29,36,39,47,48): non-Hodgkin's lymphoma (45%), leukemia (25%), carcinoma (20%), Hodgkin's disease (10%). Various types of cancer have included adenocarcinoma of the stomach, medulloblastoma, glioma, and carcinoma of the skin, gallbladder, liver, larynx, ovary, breast, and parotid gland (9,15,18,24,31,45,48,57). There is some evidence that young parents of affected children have a 5-fold higher rate of cancer and autoimmune disease than do control parents (8,9,48).

**Oral manifestations.** Telangiectasia of the hard and soft palate has also been reported (4,6,37,55). Nasal mucosal involvement with telangiectasia may be inferred from recurrent episodes of epistaxis (37). Drooling has been noted in the vast majority of patients (6,41), but its cause has not been established (4). Speech, as noted, is often of the bulbar type. Oropharyngeal lymphosarcoma of the palate has been noted in several cases (16,44).

**Differential diagnosis.** Ordinarily considered in differential diagnosis are cerebral palsy, structural anomaly or neoplasm of the posterior fossa or foramen magnum, or any of several degenerative or metabolic disorders such as Friedreich's ataxia, hepatolenticular degeneration, Pelizaeus–Merzbacher disease, and Hallevorden–Spatz disease. Jaspers et al (23) and Seemanová et al (42) have reviewed a possibly related autosomal recessive syndrome of microcephaly, immunodeficiency, and chromosome instability seen in several kindred (Nijmegen syndrome). This is still different from a disorder of immunodeficiency, chromosome instability, microcephaly, defects of skin pigmentation, and anal stenosis/atresia described by Conley et al (9b) and Wegner et al (58).

Various other genetic cutaneous disorders are reviewed in Table 13–1.

**Laboratory aids.** Synthesis of antibodies and certain immunoglobulin subclasses is disrupted due to disorders of B-cell and helper T-cell function. This has been characterized as a decreased production of $\gamma$-globulin, particularly IgA and IgE in over 60% (4). $IgG_2$ is usually decreased (33). Probably all patients have $IgA_2$ deficiency (59).

In the blood, relative and absolute numbers of T cells are usually decreased whereas the number of B cells is usually elevated. T-helper cells are diminished and defective (14,53). Functionally, there is a defect in cellular immunity. The defect in T cells is manifested by poor stimulation to mitogens, antigens, and allogenic cells, that is, there is failure to respond to phytohemagglutinin (PHA), delayed homograft rejection, an inability to sensitize to dinitrochlorobenzene, and decrease in *in vitro* lymphocyte transformation (4,16,29,36). Serum $\alpha$-fetoprotein and carcinoembryonic antigen levels are elevated (14,46,56,57). There is an altered cell cycle (28) and an elevated frequency of in vivo somatic cell mutations (2a).

Primary gonadal failure is indicated by hypoplastic uterus and absence of ovarian follicles (60). Bizarre nuclear change or cytomegaly has been noted in myocardium, endocrine organs, liver, and central nervous system.

The thymus usually appears vestigial, lacking corticomedullary differentiation and Hassall's corpuscles. Lymph nodes exhibit depletion of lymphocytes in the deep cortex (T-dependent areas) whereas follicles remain intact. The tonsils are hypoplastic, often exhibiting poor germinal follicles and secondary crypt formation.

Over 50% with ataxia-telangiectasia have abnormal glucose tolerance, but rarely with glycouria and never with ketosis, implying insulin resistance (2,29,40).

Fibroblastoid and lymphoblastoid cells are hypersensitive to the cytotoxic and clastogenic action of ionizing, and radiomimetic chemical agents such as bleomycin, adriamycin, and hydrogen peroxide, reflecting a defect in response to DNA damage or a relative resistance of DNA synthesis (7,10,11,13,21,34,35). There is radiosensitivity to the chromosomes of heterozygotes (34a).

Initially random chromosome breakage followed by a translocation involving the long arm of chromosome 14 and either the short or long arm of chromosome 7 has been found. In the final stage, a (14;14) translocation may be identified in some patients (1,20,26,27,29). Increased chromosome breakage has been used in prenatal diagnosis of this disorder (43).

An exfoliated epithelial cell micronucleus test obtained from the mouth or urine yields 4- to 14-fold values in homozygotes and elevated levels in 70% of heterozygotes (38).

CAT scans and magnetic resonance imaging (MRI) should be of help in diagnosing cerebellar atrophy.

### References (Ataxia-telangiectasia)


1. Aurias A: Analyse cytogénétique de 21 cas d'ataxie—telangiectasie. *J Génét Hum* **29**:235–247, 1981.

2. Bar RS et al: Extreme insulin resistance in ataxia-telangiectasia. *N Engl J Med* **298**:1164–1171, 1978.

2a. Bigbee WL et al: Evidence for an elevated frequency of in vivo somatic cell mutations in ataxia telangiectasia. *Am J Hum Genet* **44**:402–408, 1989.

3. Bridges BA, Harden DG: Untangling ataxia-telangiectasia. *Nature (London)* **289**:223–224, 1981.

4. Boder E: Ataxia-telangiectasia: Some historic, clinical and pathologic observations. *Birth Defects* **11**(1):255–270, 1975.

5. Boder E: Ataxia-telangiectasia: An overview, in *Ataxia-Telangiectasia: Genetics, Neuropathology and Immunology of a Degenerative Disease of Childhood,* Gatti RA et al (eds), Alan R. Liss, New York, 1985, pp 1–63.

6. Boder E, Sedgwick RP: Ataxia-telangiectasia: A familial syndrome of progressive cerebellar ataxia, oculocutaneous telangiectasia and frequent pulmonary infection. *Pediatrics* **21**:526–554, 1958.

7. Bridges BA, Harnden DG (eds). *Ataxia-telangiectasia,* Alan R. Liss, New York, 1985.

8. Chen PC et al: Identification of ataxia-telangiectasia heterozygotes, a cancer-prone population. *Nature (London)* **274**:484–486, 1978.

9. Cohen LE et al: Common and uncommon cutaneous findings in patients with ataxia-telangiectasia. *J Am Acad Dermatol* **10**:431–438, 1978.

9a. Cohen MM Jr, Levy HP: Chromosome instability syndrome. *Adv Hum Genet* **18**:43–149, 1989.

9b. Conley ME et al: A chromosome breakage syndrome with profound immunodeficiency. *Blood* **67**:1251–1256, 1986.

10. Cox R et al: Ataxia-telangiectasia. *Am J Dis Child* **53**:386–390, 1978.

11. Gatti RA, Good RA: Occurrence of malignancy in immunodeficiency disease. *Cancer* **28**:89–98, 1971.

12. Gatti RA, Hall K: Ataxia-telangiectasia-search for a control hypothesis, in *Chromosome Mutation and Neoplasia,* German J (ed), Alan R. Liss, New York, 1983, pp 23–41.

13. Gatti RA, Swift M (eds): *Ataxia-Telangiectasia,* Alan R. Liss, New York, 1985.

14. Gatti RA et al: Ataxia-telangiectasia: A multiparameter analysis of eight families. *Clin Immunol Immunopathol* **23**:501–516, 1982.

15. German J: Patterns of neoplasia associated with the chromosome breakage syndromes, in *Chromosome Mutation and Neoplasia,* German J (ed), Alan R. Liss, New York, 1983, pp 97–134.

16. Gimeno A et al: Ataxia-telangiectasia with absence of IgG. *J Neurol Sci* **8**:545–554, 1969.

17. Gschnait F et al: Ataxia-telangiectatica (Louis-Bar-Syndrom). *Hautarzt* **30**:527–531, 1979.

18. Haerer A et al: Ataxia-telangiectasia with gastric adenocarcinoma. *JAMA* **210**:1884–1887, 1970.

19. Harris VJ, Seeler RA: Ataxia-telangiectasia and Hodgkin's disease. *Cancer* **32**:1415–1420, 1973.

20. Hecht F, Kaiser-McCaw B: Chromosome instability syndromes, in *Genetics of Human Cancer,* Mulvihill JJ et al (eds), Raven Press, New York, 1977, pp 105–123.

21. Huang PC, Sheridan RB: Genetics and biochemical studies with ataxia-telangiectasia. *Hum Genet* **59**:1–9, 1981.

22. Jaspers NG, Bootsma D: Genetic heterogeneity in ataxia-telangiectasia studied by cell fusion. *Proc Natl Acad Sci USA* **779**:2641–2644, 1982.

23. Jaspers NG et al: Patients with an inherited syndrome characterized by immunodeficiency, microcephaly, and chromosomal instability: Genetic relationship to ataxia-telangiectasia. *Am J Hum Genet* **42**:66–73, 1988.

24. Kumar GK et al: Ataxia-telangiectasia and hepatocellular carcinoma. *Am J Med Sci* **278**:157–160, 1979.

25. Louis-Bar (Mme.): Sur un syndrome progressif comprenant des telangiectasies capillaires cutanées et conjonctivales symetriques, a disposition naevoide et des troubles cerebelleux. *Confin Neurol* **4**:32–42, 1941.

26. McCaw B, Hecht F: The interrelationships in ataxia-telangiectasia of immune deficiency, chromosome instability and cancer, in *Chromosome Mutation and Malignancy,* German J (ed), Alan R. Liss, New York, 1983, pp 193–202.

27. McCaw B et al: Somatic rearrangement of chromosome 14 in human lymphocytes. *Proc Natl Acad Sci USA* **72**:2071–2075, 1975.

28. McKinnon PJ: Ataxia-telangiectasia: An inherited disorder of ionizing-radiation sensitivity in man. Progress in the elucidation of the underlying biochemical defect. *Hum Genet* **75**:197–208, 1987.

29. McFarlin DE et al: Ataxia-telangiectasia. *Medicine* **51**:281–305, 1972.

30. Miller ME, Chatten J: Ovarian changes in ataxia-telangiectasia. *Acta Paediatr Scand* **56**:559–561, 1967.

31. Morrell D et al: Mortality and cancer incidence in 263 patients with ataxia-telangiectasia. *J Natl Cancer Inst* **77**:89–92, 1986.

32. Murnane JP, Painter RB: Complementation of the defects in DNA synthesis in irradiated and unirradiated ataxia-telangiectasia cells. *Proc Natl Acad Sci USA* **79**:1960–1963, 1982.

33. Oxelius VA et al: IgG2 deficiency in ataxia-telangiectasia. *N Engl J Med* **306**:515–517, 1982.

34. Painter RB, Young BR: Radiosensitivity in ataxia-telangiectasia: A new explanation. *Proc Natl Acad Sci USA* **77**:7315–7317, 1980.

34a. Parshad R et al: Chromosomal radiosensitivity of ataxia-telangiectasia heterozygotes. *Cancer Genet Cytogenet* **14**:163–168, 1985.

35. Patterson MC et al: Defective excision repair of gamma ray-damaged DNA in human (ataxia-telangiectasia) fibroblasts. *Nature (London)* **260**:444–446, 1976.

36. Peterson RDA et al: Lymphoid tissue abnormalities associated with ataxia-telangiectasia. *Am J Med* **41**:341–359, 1966.

37. Reye C: Ataxia-telangiectasia. *Am J Dis Child* **99**:238–241, 1960.

38. Rosin MP, Ochs HD: In vivo chromosomal instability in ataxia-telangectasia homozygotes and heterozygotes. *Hum Genet* **74**:335–340, 1986.

39. Saxon A et al: Ataxia-telangiectasia and chronic lymphocytic leukemia: Evolution of a T lymphocyte clone with differentiation to both helper and suppressor T lymphocytes. *N Engl J Med* **300**:700–704, 1979.

40. Schalch DS et al: An unusual form of diabetes mellitus in ataxia-telangiectasia. *N Engl J Med* **282**:1396–1402, 1970.

41. Sedgwick RP, Boder E: Progressive ataxia in childhood with particular reference to ataxia-telangiectasia. *Neurology* **10**:705–715, 1960.

42. Seemanová et al: Familial microcephaly with normal intelligence, immunodeficiency and risk for lymphoreticular malignancies. *Am J Med Genet* **20**:639–648, 1985.

43. Shaham M et al: Prenatal diagnosis of ataxia-telangiectasia. *J Pediatr* **100:**134–137, 1982.

44. Smeby B: Ataxia-telangiectasia. *Acta Paediatr Scand* **55**:239–243, 1966.

45. Spector BD et al: Epidemiology of cancer in ataxia-telangiectasia, in *Ataxia-Telangiectasia: A Cellular and Molecular Link between Cancer, Neuropathy and Immune Deficiency,* Bridges BA, Harnden DG (eds), Wiley, New York, 1982, pp 103–138.

46. Sugimoto T et al: Plasma levels of carcinoembryonic antigen in patients with ataxia-telangiectasia. *J Pediatr* **92**:436–439, 1978.

47. Sugimoto T et al: Ataxia-telangiectasia associated with non-T, non-B cell acute leukemia. *Acta Paediatr Scand* **71**:509–510, 1982.

48. Swift M et al: Malignant neoplasms in the families of patients with ataxia-telangiectasia. *Cancer Res* **36:**209–215, 1976.

49. Swift M et al: The incidence and gene frequency of ataxia-telangiectasia in the United States. *Am J Hum Genet* **39**:573–586, 1986.

50. Syllaba L, Henner K: Contribution a l'independence de l'athétose double idiopathique et congénital atteinte familiale, syndrome dystrophique, signe du reseau vasculaire conjunctival intégrité psychique. *Rev Neurol* **15**:541–562, 1926.

51. Taylor AMR et al: Ataxia-telangiectasia: A human mutation with abnormal radiation sensitivity. *Nature (London)* **258**:427–429, 1975.

52. Taylor AMR et al: Variant forms of ataxia-telangiectasia. *J Med Genet* **24**:669–677, 1987.

53. Trompeter RS et al: Primary and secondary abnormalities of T cell subpopulations. *Clin Exp Immunol* **34**:388–392, 1978.

54. Terplan KL, Krauss RF: Histopathologic brain changes in association with ataxia-telangiectasia. *Neurology* **19**:446–454, 1969.

55. Utian HL, Plit J: Ataxia-telangiectasia. *J Neurol Neurosurg Psych* **27**:38–40, 1964.

56. Waldmann TA, McIntire KR: Serum alpha-fetoprotein levels in patients with ataxia-telangiectasia. *Lancet* **2**:1112–1115, 1972.

57. Waldmann TA et al: Ataxia-telangiectasia: A multisystem hereditary disease with immunodeficiency, impaired organ maturation, x-ray hypersensitivity, and a high incidence of neoplasia. *Ann Intern Med* **99**:367–379, 1983.

58. Wegner RD et al: A new chromosomal instability disorder confirmed by complementation studies. *Clin Genet* **33**:20–32, 1988.

59. Yount WJ: IgG2 deficiency and ataxia-telangiectasia. *N Engl J Med* **306**:541–543, 1982.

60. Zadik Z et al: Gonadal dysfunction in patients with ataxia-telangiectasia. *Acta Paediatr Scand* **67**:477–479, 1978.

## Focal dermal hypoplasia (Goltz–Gorlin syndrome)

Goltz et al (15) in 1962 and Gorlin et al (17) in 1963 defined a syndrome consisting of asymmetry of face, trunk, and extremities; atrophy, telangiectasia, and linear hyperpigmentation of the skin; localized cutaneous deposits of superficial fat; multiple papillomas of mucous membranes and periofacial skin; and abnormalities of the skeleton, especially the extremities. The earliest case report appears to be that of Jessner (25) in 1921. An excellent survey of early cases was carried out by Braun-Falco and Hofmann (2). The reader is referred to several comprehensive reviews (2,14,16,19,33a,51–54).

Over 200 cases have been reported under a variety of names. About 95% have been sporadic examples and only about 10% have occurred in males (3,10–13,16,19,30,35a,38,43,45,48,52).

There have been a few familial examples, mostly mother–daughter (15,36,38,46,54,55), but father–daughter transmission has been described (4,30). The transmitting parent is usually only mildly affected. The evidence suggests that the disorder is X-linked dominant, lethal in male hemizygotes, and with markedly reduced fertility in the female (8,54). Happle and Lenz (20) suggested that affected males might be explained by gametic half-chromatid mutations and that the asymmetry and osteopathia striata may be explained on the basis of lyonization. However, we have observed these same findings in affected males.

**Facies.** Mild microcephaly is common. The face and/or skull are asymmetric and the scalp hair sparse and brittle in about 30% (19). The nasal alae may be notched or underdeveloped (4). The pinnae are usually outstanding and poorly modeled. The helices have hypoplastic cartilages. The auditory canals may be narrow. Due to midface hypoplasia there is often mild relative mandibular prognathism. (Fig. 13–88).

**Skin and skin appendages.** Asymmetric skin lesions, present at birth in 50–75%, include thin skin with linear or reticular hyperpigmentation or hypopigmentation, and telangiectasia (19,44). There is focal absence of skin at birth in 15% (19). The iliac crest area, groin, and posterior aspects of the thigh frequently show subcutaneous fatty collections, varying in color from tan to yellow or pinkish-yellow (Fig. 13–89A,B). Supernumerary nipples are common (53). The breasts are often significantly asymmetric (26). In childhood, arborescent papillomas appear in the axillae, in the periumbilical area, or around the anus and vulva in 50–60% (19,33,44). Those in the genital area may be mistaken for condyloma acuminatum (5).

Spotty hypotrichosis of the scalp or pubic hair has been found in 25%. The fingers and toenails may be absent, poorly developed, or spooned or grooved in 50% (Fig. 13–90A,B). The epidermal ridges of the fingertips may be hypoplastic (57). Hidrocystomas (56) and apocrine nevi (49) have also been reported.

**Eyes.** Approximately 40% have anomalies of the eyes. These have been extremely well reviewed in a number of articles (3,47,53). Among the more common findings are coloboma of the iris, choroid, retina and optic nerve (40%), microphthalmia (15%) and strabismus (15%). Unilateral anophthalmia (18,24), obstructed tear ducts, with associated photophobia (27,37,40), ectopia lentis (6%), optic atrophy (65), aniridia (3%), nystagmus, and conjunctival papillomas have also been reported.

Fig. 13–88. *Focal dermal hypoplasia.* Note multiple frambesiform lesions on lips, coloboma of iris, strabismus, thin hair.

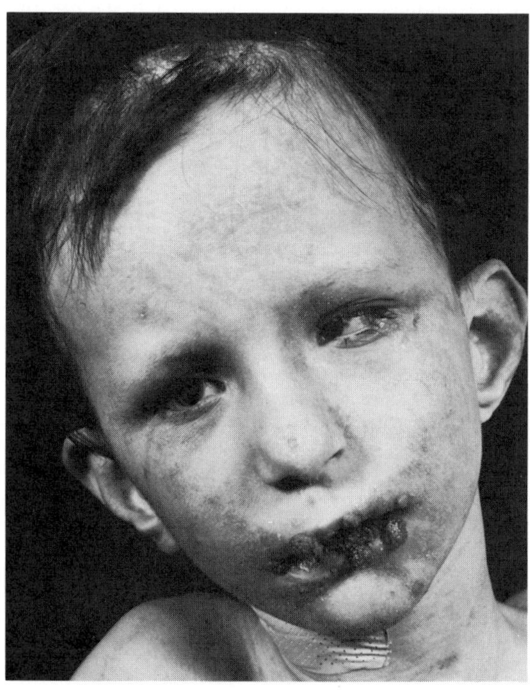

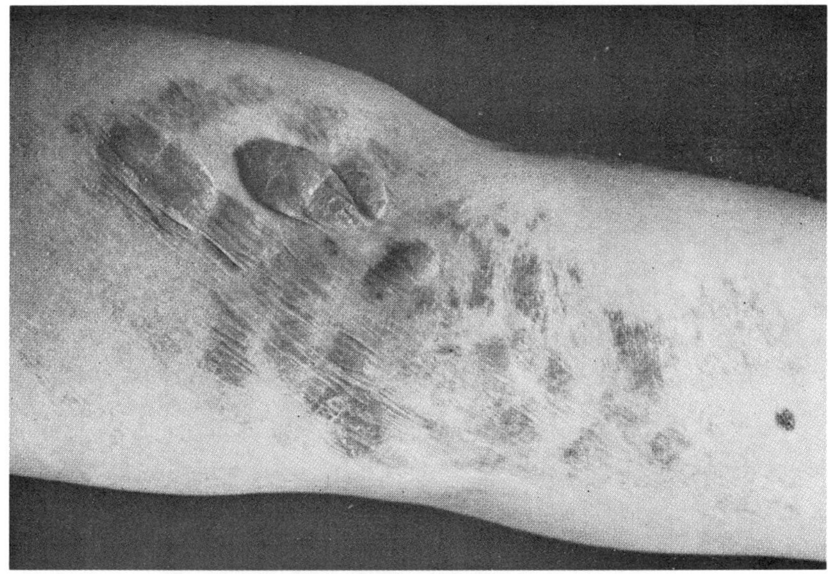

A

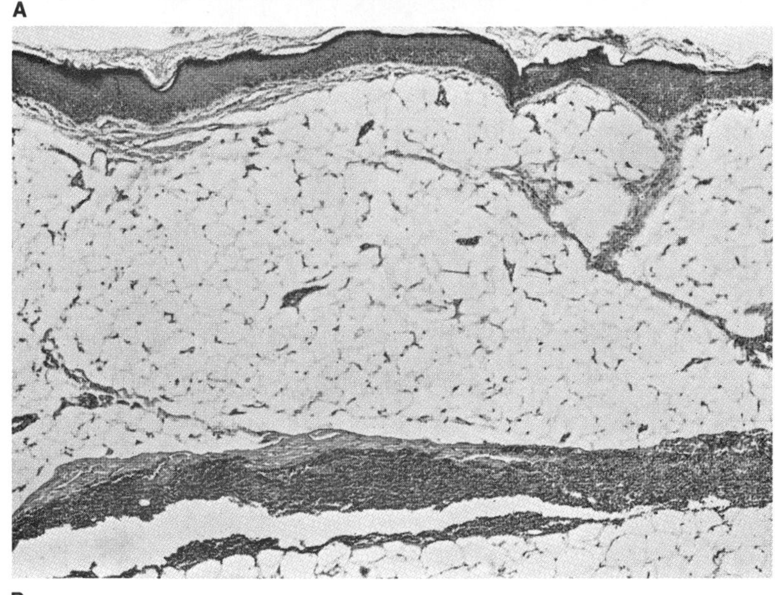

B

Fig. 13–89. *Focal dermal hypoplasia.* (A) Multiple sacular growths presenting in antecubital area. (B) Photomicrograph of focal dermal defect showing subcutaneous fat covered only by epithelium.

**Central nervous system.** Mental retardation, usually mild, has been documented in about 15% (5). Mixed hearing loss has been manifested in a number of cases (11,14,17,19,28,33,38,48). Myelomeningocele, hydrocephalus, and Arnold–Chiari malformation have been found (1).

**Musculoskeletal system.** Short and asymmetric stature is found in at least 25%. Syndactyly (75%), brachydactyly (60%), and oligodactyly (45%) are extremely common (Fig. 13–90A,B). Postaxial polydactyly and clinodactyly have been found in a number of cases. The extremities, or parts thereof, are often asymmetric. Complete hemimelia of a limb has been reported (1,24). Brachydactyly can result from shortened phalanges, metacarpals, or metatarsals. Rib anomalies have been reported (11). Midclavicular aplasia or hypoplasia (pseudarthrosis), apparently limited to the right side, is common (16).

Scoliosis (25%), spina bifida (7%), and congenital hip dislocation (9,16,18,21) are found. Osteopathia striata has been found in at least 80% (1b,4,10,12,23,28,30,35,51) (Fig. 13–91). Several authors have noted what appears to be multiple giant cell lesions of long bones (6,26,32,41). Umbilical hernia and/or omphalocele has been reported in 10–15% (2,5,8,12,14,16,21,27,38,39,50). A rudimentary tail (14) and a split sternum (7) have been documented.

**Genitourinary system.** Hydronephrosis (3,19) bifid ureter and renal pelvis (28), horseshoe kidney (30), hypoplasia of clitoris and labia (31,53), and cryptorchidism (11,43) have been sporadically reported.

**Miscellaneous findings.** Diaphragmatic hernia (29), diastasis recti (2), and anterior displacement of anus (53) have been found. The reader is referred to a detailed comprehensive article for other anomalies (19).

**Oral manifestations.** The most frequent dental findings is hypodontia or oligodontia although supernumerary teeth have been reported (19). The teeth may be small with dysplastic enamel, irregularly spaced, maloccluded, and with delayed eruption (3a,48a) (Fig. 13–92A,B).

Arborescent papillomas of the lips, gingiva, base of tongue, and perioral skin have been noted in 40–50% (27,53). They have even been reported in the esophagus (56).

Cleft lip and/or cleft palate have also been described (1,6a,11,16,37,44,50), as well as lateral oral cleft (1). The reader is referred to the detailed article of Hall and Terezhalmy (19) for other minor oral anomalies.

**Differential diagnosis.** Nevus lipomatosis superficialis (Hoffmann–Zurhelle) is characterized by neutral fat-bearing cells adjacent

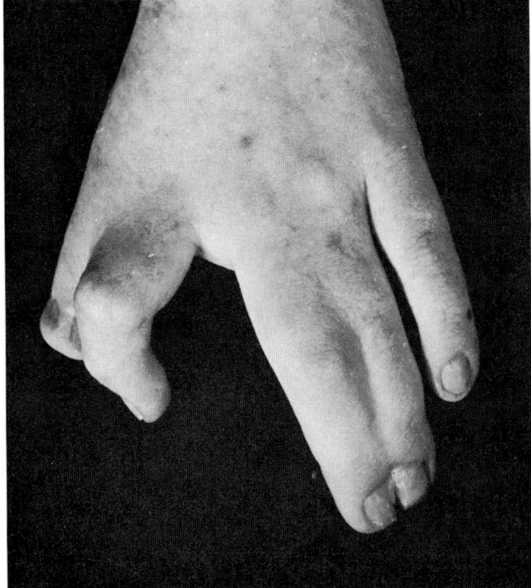

A

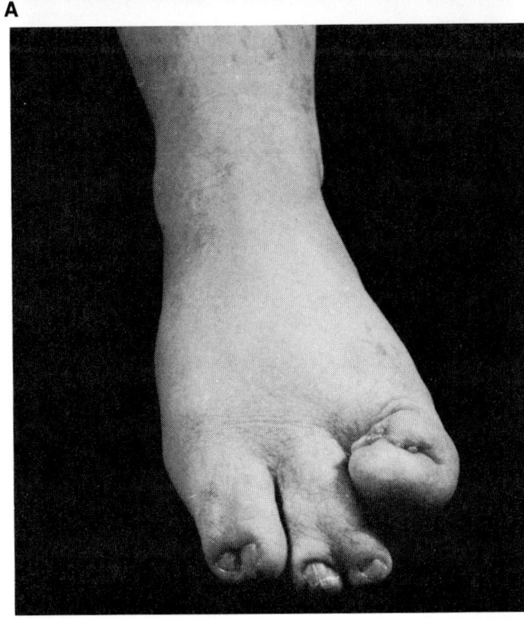

B

Fig. 13–90. *Focal dermal hypoplasia.* (A,B) Syndactyly of digits, bizarre nail formation.

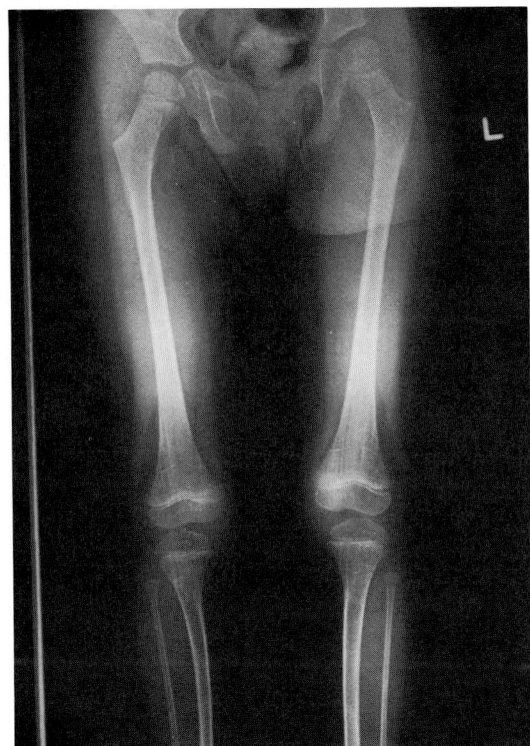



Fig. 13–91. *Focal dermal hypoplasia.* Osteopathia striata is constant finding.

to blood vessels in the middle and upper layers of the skin (22). In differential diagnosis, *incontinentia pigmenti* and *Rothmund–Thomson syndrome* should be excluded (2).

Clavicular pseudarthrosis as an isolated finding is a right-sided anomaly in 90% of the cases and may be inherited as an autosomal dominant trait. Rarely it is bilateral (34). It is also found in the *Floating–Harbor syndrome.*

**Laboratory aids.** Skin biopsy of a fatty lesion demonstrates normal adipose cells in the corium. Less specific changes are found in the scar-like areas (14,21). Radiographic studies show osteopathia striata and numerous skeletal changes previously described.

### References [Focal dermal hypoplasia (Goltz–Gorlin syndrome)]


1. Alemeda L et al: Myelomeningocele, Arnold-Chiari anomaly and hydrocephalus in focal dermal hypoplasia. *Am J Med Genet* **30**:917–923, 1988.

1a. Beganović N, Lommer EJP: A case of focal dermal hypoplasia (Goltz's syndrome) with some new aspects. *Acta Paediatr Scand* **66**:255–256, 1977.

1b. Boothroyd AE, Hall CM: The radiological features of Goltz syndrome: Focal dermal hypoplasia. A report of two cases. *Skel Radiol* **17**:505–508, 1988.

2. Braun-Falco O, Hofmann C: Das Goltz-Gorlin-Syndrom: Ubersicht und Kasuistik. *Hautarzt* **26**:393–400, 1975.

3. Broughton WL et al: Focal dermal hypoplasia: Ocular manifestations in a male. *J Pediatr Ophthalmol Strab* **19**:315–317, 1982.

3a. Bucci E et al: Oral and dental anomalies in Goltz syndrome. *J Pedodont* **13**:161–168, 1989.

4. Burgdorf WHC et al: Focal dermal hypoplasia in a father and daughter. *J Am Acad Dermatol* **4**:273–277, 1981.

5. Contarini O et al: Focal dermal hypoplasia (Goltz's syndrome) manifesting as condyloma acuminatum. Report of a case and review of the literature. *Dis Col Rect* **20**:43–48, 1977.

6. Cottenot F et al: Hypoplasie dermique en aires associée à une dysplasie fibreuse des os. *Ann Dermatol Venereol* **106**:167–169, 1979.

6a. Delaire J et al: Goltz syndrome with facial clefts. *Ann Pédiatr* **32**:847–854, 1985.

7. Derks B et al: Focal dermal hypoplasia (Goltz syndrome). *S Afr Med J* **54**:27–29, 1978.

7a. Dick GF et al: Focal dermal hypoplasia in a father and daughter. *J Am Acad Dermatol* **4**:273–277, 1981.

8. Fattah AA: Focal dermal hypoplasia. *Dermatologica* **138**:105–115, 1969.

9. Fazekas A et al: Goltzsches Syndrom—Osteo-, Okulo-, Dermale Dysplasie. *Z Haut Geschlechtskr* **48**:307–316, 1973.

10. Feinberg A, Menter MA: Focal dermal hypoplasia (Goltz syndrome) in a male. *S Afr Med J* **50**:554–555, 1971.

11. Ferrara A: Goltz's syndrome. *Am J Dis Child* **123**:263, 1972.

12. Fjellner B: Focal dermal hypoplasia in a 46,XY male. *Int J Dermatol* **18**:812–815, 1985.

13. Fryns JP et al: Focal dermal hypoplasia (Goltz syndrome) in a male. *Acta Paediatr Belg* **31**:37–39, 1978.

14. Ginsburg LD et al: Focal dermal hypoplasia syndrome. *Am J Roentgenol* **110**:561–571, 1970.

15. Goltz RW et al: Focal dermal hypoplasia. *Arch Dermatol* **86**:708–717, 1962.

16. Goltz RW et al: Focal dermal hypoplasia syndrome. *Arch Dermatol* **101**:1–11, 1970.

17. Gorlin RJ et al: Focal dermal hypoplasia syndrome. *Acta Derm Venereol* **43**:421–440, 1963.

18. Gottlieb SK et al: Focal dermal hypoplasia. *Arch Dermatol* **108**:551–553, 1973.

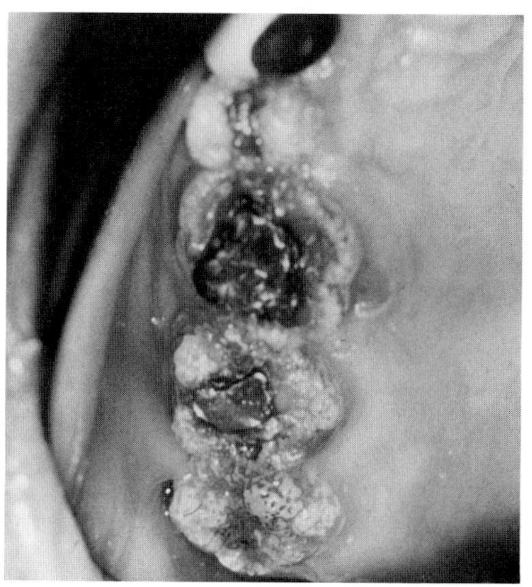

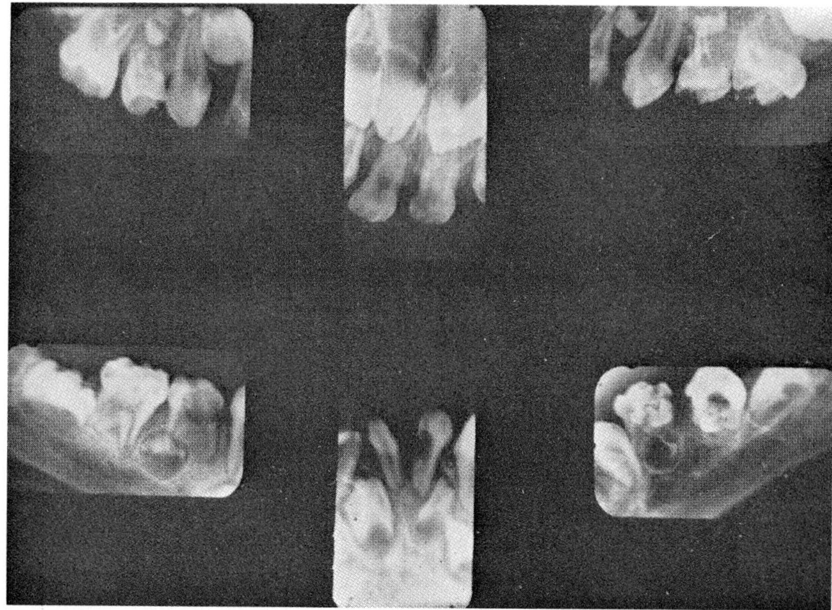

A B

Fig. 13–92. *Focal dermal hypoplasia.* (A) Severely hypoplastic enamel. (B) Roentgenogram exhibiting severe hypoplasia of teeth.


19. Hall EH, Terezhalmy GT: Focal dermal hypoplasia syndrome. *J Am Acad Dermatol* **9**:443–451, 1983.

20. Happle R, Lenz W: Striations of bones in focal dermal hypoplasia: Manifestations of functional mosaicism? *Br J Dermatol* **96**:133–138, 1977.

21. Holden JD, Akers WA: Goltz's syndrome: Focal dermal hypoplasia: Combined mesoectodermal dysplasia. *Am J Dis Child* **114**:292–300, 1967.

22. Howell JB: Nevus angiolipomatosus vs. focal dermal hypoplasia. *Arch Dermatol* **92**:238–248, 1965.

23. Howell JB, Reynolds J: Osteopathia striata. *Trans St John's Hosp Dermatol Soc* **60**:178–182, 1974.

24. Ishibashi A, Kurihawa Y: Goltz's syndrome: Focal dermal hypoplasia syndrome. *Dermatologica* **144**:156–167, 1972.

25. Jessner M: Falldemonstration Breslauer dermatologische Vereinigung. *Arch Dermatol Syph (Berlin)* **133**:48, 1921.

26. Joannides T et al: Giant cell tumour of bone in focal dermal hypoplasia. *Br J Radiol* **56**:684–685, 1983.

27. Kleinhans D: Incontinentia pigmenti mit multipler Fehlbildungssymptomatik bei einer Erwachsen. *Hautarzt* **21**:133–136, 1970.

28. Knockaert D, Dequeker J: Osteopathia striata and focal dermal hypoplasia. *Skelet Radiol* **4**:223–227, 1979.

29. Kunze J et al: Diaphragmatic hernia in a female newborn with focal dermal hypoplasia and marked asymmetric malformations (Goltz–Gorlin syndrome). *Eur J Pediatr* **131**:213–218, 1979.

30. Larrègue M et al: L'osteopathie striée, symptome radiologique de l'hypoplasie dermique en aires. *Ann Radiol (Paris)* **15**:287–295, 1972.

31. Liebermann SL: Atrophodermia linearis maculosa et papillomatosis congenitalis. *Acta Dermatol Venereol* **16**:476–484, 1935.

32. Lynch RD et al: Focal dermal hypoplasia syndrome (Goltz's syndrome) with an expansile iliac lesion. *J Bone Jt Surg* **63A**:470–473, 1981.

33. Mair IWS, Høy C: Focal dermal hypoplasia. *J Laryngol Otolaryngol* **85**:853–856, 1971.

33a. Mallory SB, Krafchik BR: Goltz syndrome. *Pediatr Dermatol* **6**:251–253, 1989.

34. March HC: Congenital pseudarthrosis of the clavicle. *J Can Assoc Radiol* **33**:35–36, 1982.

35. Nickel WR, Lockwood JH: Congenital ectodermal dysplasia with associated mesodermal defects. *Arch Dermatol* **74**:327, 1956.

35a. Reber T et al: Goltz-Gorlin-Syndrom bei einem Mann. *Hautarzt* **38**:218–223, 1987.

36. Rochiccioli P et al: Hypoplasie dermique en aire, osteopathie striée et nanisme. *Pédiatrie* **30**:271–280, 1975.

37. Rodermund OE, Heiusmann D: Kasuistischer Beitrag zum Goltz–Gorlin Syndrom. *Hautarzt* **28**:37–38, 1977.

38. Ruiz-Maldonado R et al: Focal dermal hypoplasia. *Clin Genet* **6**:36–45, 1974.

39. Samejima N et al: Omphalocele and focal dermal hypoplasia. *Z Kinderchir* **34**:284–289, 1981.

40. Scott CI, Moyer FG: Focal dermal hypoplasia (Goltz syndrome): A follow-up. *Birth Defects* **7**(8):240–241, 1971.

41. Selzer G et al: Goltz syndrome with multiple giant-cell tumor-like lesions in bones. *Ann Intern Med* **80**:714–717, 1974.

42. Solomon LM, Bolet I: Focal dermal hypoplasia (Goltz's syndrome). *Arch Dermatol* **111**:272, 1975.

43. Stalder JF et al: Syndrome de Goltz unilatéral chez un garçon. *Ann Dermatol Venereol* **111**:829–830, 1984.

44. Stalder JF et al: Syndrome de Goltz: deux cas feminins. *Ann Dermatol Venereol* **111**:833–834, 1984.

45. Staughton RCD: Focal dermal hypoplasia (Goltz's syndrome) in a male. *Proc R Soc Med* **69**:232–233, 1976.

46. Stewart WM et al: Hypoplasie dermique en aires. *Bull Soç Fr Dermatol Syph* **76**:812–814, 1969.

47. Thomas JV et al: Ocular manifestations of focal dermal hypoplasia syndrome. *Arch Ophthalmol* **95**:1997–2001, 1977.

48. Toro-Sola MA et al: Focal dermal hypoplasia syndrome in a male. *Clin Genet* **7**:325–327, 1975.

48a. Ureles SD, Needleman HL: Focal dermal hypoplasia (Goltz syndrome). *Pediatr Dent* **8**:239–244, 1986.

49. Vakilzadeh F, Happle R: Fokale dermale Hypoplasie mit apokrinen Naevi und streifenförmiger Anomalie der Knochen. *Arch Dermatol Res* **256**:189–195, 1976.

50. Valerius NH: A case of focal dermal hypoplasia syndrome (Goltz) with bilateral cheilo-gnatho-palatoschisis. *Acta Paediat Scand* **63**:287–288, 1974.

51. Verger P et al: Hypoplasie dermique en aires et ostéopathie striée. *Ann Pédiatr* **16**:349–354, 1975.

52. Walbaum R et al: Syndrome de Goltz chez un garçon. *Pédiatrie* **25**:911–920, 1970.

53. Warburg M: Focal dermal hypoplasia. Ocular and general manifestations with a survey of the literature. *Acta Ophthalmol* **48**:525–536, 1970.

53a. Wechsler MA et al: Variable expression in focal dermal hypoplasia. *Am J Dis Child* **142**:297–300, 1988.

53b. Welsh O: Focal dermal hypoplasia. *Mod Probl Paediatr* **17**:25–43, 1975.

54. Wettke-Schäfer R, Kantner G: X-linked dominant inherited diseases with lethality in hemizygous males. *Hum Genet* **64**:1–23, 1983.

54a. Willetts GS: Focal dermal hypoplasia. *Br J Ophthalmol* **58**:620–624, 1974.

55. Wodniansky P: Über die Formen der congenitalen Poikilodermie. *Arch Klin Exp Dermatol* **205**:331–342, 1957.

56. Zala L et al: Fokale dermale Hypoplasie mit Keratokonus, Ösophaguspapillomen und Hidrokystomen. *Dermatologica* **150**:176–185, 1975.

## Hereditary hemorrhagic telangiectasia (Osler–Rendu–Parkes Weber syndrome)

The syndrome characterized by familial occurrence of multiple capillary and venous dilatations of skin and mucous membranes with repeated hemorrhage was mentioned in 1864 by Sutton (53) and in 1865 by Babington (4). Rendu (44), in 1896, reported a patient with recurrent epistaxis and numerous dilated vessels on the face and oral mucosa. Osler (36), in 1901, gave a full account of the syndrome, and Parkes Weber (37), in 1907, pointed out that the condition becomes worse with age. The descriptive term "hereditary hemorrhagic telangiectasia" was suggested by Hanes (21), in 1909, and is now widely used, although various eponyms are still employed.

In 1950, Garland and Anning (17), in an extensive review of the literature, found 264 affected families, with approximately 1500 persons involved. The frequency in Germany has been estimated at about 1 or 2/100,000 population (55). The syndrome is recognized more commonly among whites, but has been reported among all ethnic groups (40). It is rarely recognized in infants (32).

Inheritance is autosomal dominant (8,17,55) with almost complete penetrance (39a). In our opinion the lethal homozygotic state (52) is less than certain.

Hereditary hemorrhagic telangiectasia should be considered a generalized pleomorphic angiodysplasia of almost any organ. The lesions are bright red, violaceous, or purple, and pinpoint, spider-like, or nodular. When a glass slab is pressed on them, they blanch. Often the patients are pale, and they may have a history of fatigue and weakness caused by bleeding from the telangiectases, with resultant anemia. The hemorrhage, often nontraumatic, is the most severe complication and becomes more frequent with advancing age. It has been postulated that the hemorrhages are aggravated by anemia, thus producing a vicious cycle.

**Skin.** Telangiectases are observed on the facial skin (35%), especially on the cheeks, ears, and nasal orifices (Fig. 13–93). They may also occur on fingers, toes, and nail beds in about 40%. Usually appearing in the second to third decades of life and increasing in number and size with age, rare cases without external signs of the syndrome occur. The lesions may be purpuric. In elderly persons, spider-like configurations may be seen.

Fig. 13–93. *Hereditary hemorrhagic telangiectasia.* Note numerous telangiectases on face and lips.

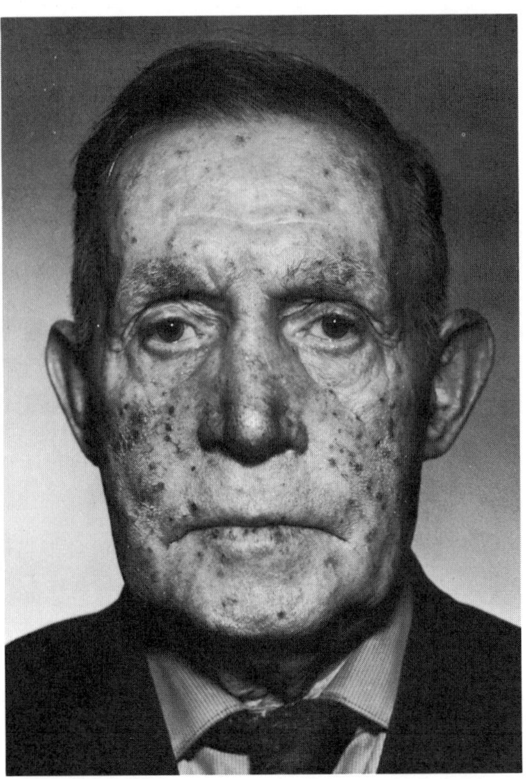

**Nasal mucosa.** Telangiectasis of the nasal mucosa is common. About 95% experience recurrent epistaxis which tends to become more severe with time (39a,43,50). Epistaxis usually precedes the appearance of telangiectasia on the skin, appearing in childhood or young adulthood in 50% (39a). Bleeding from the nose may be oozing, sometimes persisting continuously for several days. Profuse hemorrhage may be initiated by sneezing or coughing. Death due to hemorrhage has occurred in 2–4% (30).

**Other mucous membranes.** The gastric mucosa is often involved, resulting in melena and hematemesis which increase with age in 20–45% (8,43). Other areas of the gastrointestinal tract (pharynx, esophagus, jejunum, sigmoid colon), conjunctiva (35%), or retina (10%) (9,10a,13,50) may be sites of telangiectasia. When located in the bladder, vagina, or uterus, telangiectases may lead to genitourinary bleeding (22,50).

**Other findings.** Almost any organ may be affected by angiodysplasias: bone, liver, brain, spinal cord, heart, and lungs (14,29,31, 43,48). Pulmonary arteriovenous fistulas occur in 15–25% or more (38a,43). Román et al (48) found more than 200 reported cases with involvement of the nervous system. About 60% of these were secondary to pulmonary arteriovenous fistulas. Adams et al (1) stated that about 5% of patients with arteriovenous fistulas develop cerebral abscesses. Conversely, virtually all patients reported with hereditary hemorrhagic telangiectasia and cerebral abscess had pulmonary arteriovenous fistula. Cerebral abscess must be considered a serious complication of hereditary hemorrhagic telangiectasia (39,48). Intrahepatic arteriovenous fistulas, also, have been reported to cause encephalopathy (23). Duodenal ulcer has been found in about 5% but in almost 20% of those with gastrointestinal bleeding.

Hereditary hemorrhagic telangiectasia has been reported concomitantly with thrombocytomegaly, primary thrombocytemia (12), hemophilia A (28), but most often with Von Willebrand's disease (2,11).

**Oral manifestations.** The lips and tongue (especially the dorsum and tip) are sites of telangiectasia in about 60% (Fig. 13–94). The palate, gingiva, buccal mucosa, and mucocutaneous junctions may be similarly affected in about 20%. Recognition of oral lesions has often led to the correct diagnosis. Bleeding from the mouth is second in frequency to epistaxis. It has been noted in about 20% (8). Hemorrhage from the gingiva and buccal mucosa occurs less frequently than from the lips and tongue. A dramatic report of bleeding from telangiectasia of the tongue was given by Bird et al (8). Philips (38) mentioned a fatal outcome after gingival bleeding. Gingival hemorrhages may become manifest at a rather late age. Labial lesions are more common (about 85%) in those with gastrointestinal bleeding. Dental prophylaxis has caused bleeding from lip telangiectases (38). Even toothbrushing should be carried out with great care in patients with gingival telangiectasia.

**Pathology.** Angiographic studies have demonstrated three different vascular abnormalities: aneurysms, arteriovenous communications, and angiomas (19). Histologically, the telangiectases are found just beneath the epidermis. Their walls are extremely thin, consisting of a single layer of endothelial cells on a continuous basement membrane. The telangiectases are not associated with new formation of vessels but consist solely in marked dilatation. A detailed analysis of the course of the dilated vessels has been given by Nödl (33). Jahnke (25) and Hashimoto and Pritzker (22) studied these vessels ultrastructurally, suggesting that they were primarily weak, rather than thin, because of defective overlapping of terminal villi of extended cytoplasm of endothelial cells, with resultant gaps.

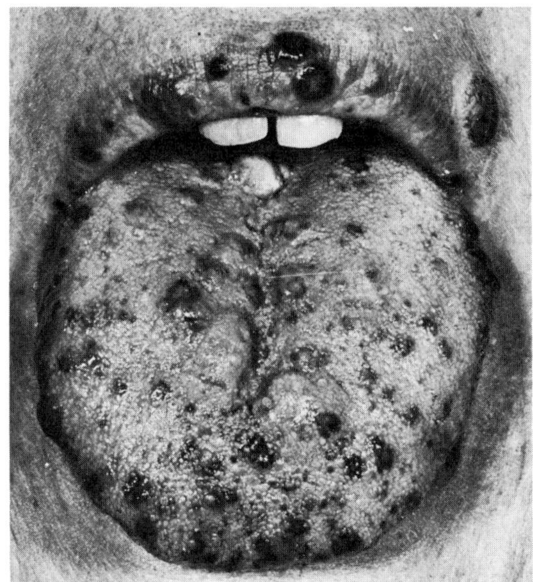

Fig. 13–94. *Hereditary hemorrhagic telangiectasia.* Close-up showing lesions of lips and tongue.

**Differential diagnosis.** Scrotal angiokeratomas are present in 10–20% of normal males over 50 years of age (6,40). Sublingual phlebectases (caviar spots) are present in about 15% of healthy males under 40 years (41) and in 50% of males and 20% of females between 50 and 60 years of age. Caviar spots increase with age, so that 60% of males and 75% of females between 70 and 79 years have them (6). Spider nevi have been noted in over 45% of healthy school children (3) and in about 15% of normal adults (6). Cherry angiomas (de Morgan spots) are seen in 90% of patients over 40 years of age (51). Phlebectasia of the lips or oral mucosa was noted by Bean (6) in about 50% of normal individuals over 40 years of age. Autosomal dominant phlebectasia of the lips was described by Reed (42). There is a plethora of types of angiomatosis that may affect oral structures. The interested reader should carefully peruse the works of Bean (6,7) and Mulliken and Young (32a). Labial involvement is common in *Fabry syndrome,* and oral angiomas occasionally occur in *Maffucci syndrome.* Multiple phlebectasia of the oral cavity (lower lip, oral floor, buccal mucosa, palate), jejunum, and scrotum is a nongenetic, age-dependent association first described by Rappaport and Shiffman (41). Most patients have been males over 40 years of age. The scrotal lesions (Fordyce angiokeratomas) are round, raised, red to black, well-defined lesions less than 4 mm in diameter. Phlebectatic lesions are present in the wall of the jejunum, less often of the ileum or cecum. A number of patients have also had gastric or duodenal ulcer (41,51). Conversely, Gius et al (18) found that 25% of patients with peptic ulcers exhibit phlebectatic lesions of the lower lip.

Bean (6,7) defined the blue rubber bleb nevus syndrome as multiple, vascular, nipple, or bladder-like lesions of the skin and mucous membranes of the gastrointestinal tract that may be accompanied by gastrointestinal bleeding and hepatic and pulmonary angiomas. The lungs, liver, spleen, joints, and subcutaneous tissues are occasionally involved. The disorder has autosomal dominant inheritance. Oral angiomas have been noted in several cases (5,15,24a,25a,34,46,47).

Autosomal dominant multiple glomus tumors exhibit only cutaneous involvement (49). Multilocular hemangiomatosis, separable from the above by the tendency of these lesions to disappear before the second year of life, may also involve the mouth (18a,24). Katz and Askin (26) described oral lesions in a patient with multiple hemangiomas with thrombocytopenia, a variant of the Kasabach–Merritt syndrome. Also present were intraosseous angiomas and hepatosplenomegaly. LaDow et al (27) noted a maxillary angioma in a patient with von Hippel syndrome, and Tasanen (54) described oral, retinal, and encephalic angiomatosis in a patient with thrombocytopenia. Arteriovenous aneurysms of the mandible and retina with hematemesis and epistaxis were described by Bower et al (10). Vascular malformations (cavernous and capillary angiomas, AV malformations), especially involving the oral mucous membranes and the skin and soft tissues of the extremities, were reported in an extremely large kindred (37a).

Telangiectasia is also seen in the CREST syndrome (*c*alcinosis, *R*aynaud's phenomenon, *e*sophageal dysfunction, *s*clerodactyly, and *t*elangiectasia), a form of scleroderma. This nonhereditary syndrome has marked female predilection. The various components of the syndrome characteristically appear at various times in a patient's life. Only rarely does the disorder present prior to puberty; in these cases Raynaud's phenomenon first becomes evident (56). Calcinosis is usually limited to the areas of sclerodactyly. Telangiectasias of the face (especially the cheeks, nose, and lips), oral mucosa, hands, and upper trunk are the most common sites. The nasal mucosa is characteristically spared. Primary biliary cirrhosis has been described in several individuals (55). CREST syndrome patients usually exhibit antinuclear and anticentromeric antibodies (16).

**Laboratory aids.** The hemoglobin and erythrocyte count may be lowered because of hemorrhage. When telangiectases are suspected in the gastrointestinal tract, gastroscopy and sigmoidoscopy may be done. To diagnose arteriovenous fistulas, angiography, computed tomography, and nuclear magnetic resonance (NMR) imaging are extremely helpful (19,34).

### References [Hereditary hemorrhagic telangiectasia (Osler–Rendu–Parkes Weber syndrome)]


1. Adams HP et al: Neurologic aspects of hereditary hemorrhagic telangiectasia. *Arch Neurol* **34**:101–104, 1977.
2. Ahr DJ et al: Von Willebrand's disease and hemorrhagic telangiectasia. *Am J Med* **62**:452–458, 1977.
3. Alderson MR: Spider naevi: Their incidence in healthy school-children. *Arch Dis Childh* **38**:286–288, 1963.
4. Babington BG: Hereditary epistaxis. *Lancet* **2**:362–363, 1865.
5. Baiocco FA et al: Blue rubber bleb nevus syndrome: A case with predominantly ENT localization. *J Laryngol Otolaryngol* **98**:317–319, 1984.
6. Bean WB: *Vascular Spiders and Related Lesions of the Skin,* Charles C Thomas, Springfield, Illinois, 1958.
7. Bean WB: *Rare Diseases and Lesions,* Charles C Thomas, Springfield, Illinois, 1967.
8. Bird RM et al: A family reunion: A study of hereditary hemorrhagic telangiectasia. *N Engl J Med* **257**:105–109, 1957.
9. Blodi FC et al: Netzhautveränderungen beim Osler-Syndrom. *Klin Mbl Augenheilkd* **179**:251–253, 1981.
10. Bower LE et al: Arteriovenous angioma of mandible and retina with pronounced hematemesis and epistaxis. *Am J Dis Child* **64**:1023–1029, 1942.

10a. Brant AM et al: Ocular manifestations in hereditary hemorrhagic telangiectasia. *Am J Opthalmol* **107**:642–646, 1989.

11. Conlon CL et al: Telangiectasia and von Willebrand's disease in two families. *Am Intern Med* **89**:921–924, 1978.
12. Cronstedt J et al: Coexistent hereditary hemorrhagic telangiectasia and primary thrombocythaemia—coincidence or syndrome? *Acta Med Scand* **212**:261–265, 1982.
13. Ecker JA et al: Gastrointestinal bleeding in hereditary hemorrhagic telangiectasia. *Am J Gastroenterol* **33**:411–421, 1960.
14. Fisher M, Zito JL: Focal cerebral ischemia distal to a cerebral aneurysm in hereditary hemorrhagic telangiectasia. *Stroke* **14**:419–421, 1983.
15. Fretzin DF, Potter B: Blue rubber bleb nevus. *Arch Intern Med* **116**:924–949, 1965.
16. Fritzler MJ et al: Hereditary hemorrhagic telangiectasia versus CREST syndrome: Can serology aid diagnosis? *J Am Acad Dermatol* **10**:192–196, 1984.
17. Garland HG, Anning ST: Hereditary hemorrhagic telangiectasia: A genetic and bibliographical study. *Br J Dermatol* **62**:289–310, 1950.
18. Gius JA et al: Vascular formations of the lip and peptic ulcer. *JAMA* **183**:725–729, 1963.

18a. Golitz LE et al: Diffuse neonatal hemangiomatosis. *Pediatr Dermatol* **3**:145–152, 1986.

19. Gutierrez FR et al: NMR imaging of pulmonary arteriovenous fistulae. *J Comput Assist Tomogr* **8**:750–752, 1984.

20. Halpern M et al: Hereditary hemorrhagic telangiectasia: An angiographic study of abdominal visceral angiodysplasia associated with gastrointestinal hemorrhage. *Radiology* **90**:1143–1149, 1968.

21. Hanes FM: Multiple hereditary telangiectases causing hemorrhage (hereditary hemorrhagic telangiectasia). *Bull Johns Hopkins Hosp* **20**:63–73, 1909.

22. Hashimoto K, Pritzker MS: Hereditary hemorrhagic telangiectasia. *Oral Surg* **34**:751–768, 1972.

23. Henderson JM et al: Liver involvement in hereditary hemorrhagic telangiectasia. *J Comput Assist Tomogr* **5**:773–776, 1981.

24. Holden KL, Alexander F: Diffuse neonatal hemangiomatosis. *Pediatrics* **46**:411–421, 1970.

24a. Ishii T et al: Blue rubber bleb naevus syndrome. *Virchows Arch [A] Pathol Anat* **413**:485–490, 1988.

25. Jahnke V: Ultrastructure of hereditary telangiectasia. *Arch Otolaryngol* **91**:262–265, 1970.

25a. Jennings NM et al: Blue rubber bleb naevus disease: An unusual cause of gastrointestinal tract bleeding. *Gut* **29**:1408–1412, 1988.

26. Katz HP, Askin J: Multiple hemangiomata with thrombopenia. *Am J Dis Child* **115**:351–357, 1968.

27. LaDow CS et al: Central hemangioma of the maxilla with von Hippel's disease. *J Oral Surg* **22**:252–259, 1964.

28. Maggi CA et al: Hemorrhagic telangiectasia and hemophilia A: An occasional association. *Haematologica (Pavia)* **68**:399–406, 1983.

29. Martini GA: The liver in hereditary hemorrhagic telangiectasia: An inborn error of vascular structure with multiple manifestations: A reappraisal. *Gut* **19**:531–537, 1978.

30. McCue CM et al: Pulmonary arteriovenous malformations related to Rendu–Osler–Weber syndrome. *Am J Med Genet* **19**:19–27, 1984.

31. Merry GS, Appleton DB: Spinal arterial malformation in a child with hereditary hemorrhagic telangiectasia. *J Neurosurg* **44**:613–616, 1976.

32. Mestre JR, Andres JM: Hereditary hemorrhagic telangiectasia causing hematemesis in an infant. *J Pediatr* **101**:577–579, 1982.

32a. Mulliken JB, Young AE: *Vascular Birthmarks, Hemangiomas and Malformations.* W. B. Saunders, Phila, 1988.

33. Nödl F: Zur Histopathogenese der Telangiectasia hereditaria hämorrhagica Rendu-Osler. *Arch Klin Exp Dermatol* **204**:213–235, 1957.

34. Nyman U: Angiography in hereditary hemorrhagic telangiectasia. *Acta Radiol (Diagn) (Stockholm)* **18**:581–592, 1977.

35. Olsen TG et al: Laser surgery for blue rubber bleb nevus. *Arch Dermatol* **115**:81–82, 1979.

36. Osler W: On a family form of recurring epistaxis, associated with multiple telangiectases of the skin and mucous membranes. *Bull Johns Hopkins Hosp* **12**:333–337, 1901.

37. Parkes Weber F: Multiple hereditary developmental angiomata (telangiectases) of the skin and mucous membranes associated with recurring hemorrhages. *Lancet* **2**:160–162, 1907.

37a. Pasyk KA et al: Familial vascular malformations. Report of 25 members of one family. *Clin Genet* **26**:221–227, 1984.

38. Philips S: Multiple telangiectases. *Proc R Soc Med 1 (Clin Sec)*:**64**, 1908.

38a. Piepras U, Sielecki S: Die Lungenbeteiligung bei Morbus Osler und ihre zerebralen Komplikationen. *Radiologe* **26**:27–30, 1986.

39. Pitlik SD et al: Cerebral abscess in hereditary hemorrhagic telangiectasia. *NY State J Med* **84**:197–201, 1984.

39a. Plauchu H et al: Age-related clinical profile of hereditary hemorrhagic telangiectasia in an epidemiologically recruited population. *Am J Med Genet* **32**:291–297, 1989.

40. Posner DB, Sampiliner RE: Hereditary hemorrhagic telangiectasia in three black men. *Am J Gastroenterol* **70**:389–392, 1978.

41. Rappaport I, Shiffman MA: Multiple phlebectasia involving jejunum, oral cavity and scrotum. *JAMA* **185**:437–440, 1963.

42. Reed WB: Hereditary phlebectasia of the lips. *Arch Dermatol* **112**:712–714, 1976.

43. Reilly PJ, Nostraut TT: Clinical manifestations of hereditary hemorrhagic telangiectasia. *Am J Gastroenterol* **79**:363–367, 1984.

44. Rendu M: Epistaxis répétées chez un sujet porteur de petits angiomes cutanés et muqueux. *Bull Soç Méd (Paris)* **13**:731–733, 1896.

45. Reynolds TB et al: Primary biliary cirrhosis with scleroderma, Raynaud's phenomenon, and telangiectasia. *Am J Med* **50**:302–312, 1971.

46. Rice JS, Fischer DS: Blue rubber bleb nevus syndrome. *Arch Dermatol* **86**:503–511, 1962.

47. Richter G: "Blue rubber-bleb nevus Syndrom" als Anämieursache. *Z Haut Geschlechtskr* **39**:256–261, 1965.

48. Román G et al: Neurological manifestations of hereditary hemorrhagic telangiectasia. *Ann Neurol* **4**:130–144, 1978.

49. Rycroft RJG et al: Hereditary multiple glomus tumors. *Trans St John's Hosp Dermatol Soc* **61**:70–81, 1975.

50. Saunders WH: Hereditary hemorrhagic telangiectasia. *Arch Otolaryngol* **76**:245–260, 1962.

51. Shiffman MA, Rappaport I: Multiple phlebectasia. *Arch Surg* **94**:771–775, 1967.

52. Snyder LH, Doan CA: Studies in human inheritance: Is homozygous form of multiple telangiectasia lethal? *J Lab Clin Med* **29**:1211–1216, 1944.

53. Sutton HG: Epistaxis as an indication of impaired nutrition and of degeneration of the vascular system. *Med Mirror* **1**:769–781, 1864.

54. Tasanen A: Bonnet–Dechaume–Blanc syndrome accompanied by mandibular angiomatosis and thrombocytopenia. *Br J Oral Surg* **4**:213–217, 1967.

55. Tünte W: Klinik und Genetik der Oslerschen Krankheit. *Z Menschl Vererb Konstit Lehre* **37**:221–250, 1964.

56. Velayos EE et al: The "CREST" syndrome. Comparison with systemic sclerosis (scleroderma). *Arch Intern Med* **139**:1240–1249, 1979.

## Acrolabial telangiectasia

Millns and Dickens (1) described blue lips, blue nails, and blue nipples in a mother and two daughters. There was also discrete telangiectasia of the chest, elbows, and dorsa of hands, varicosities of the lower legs, and migraine headaches. The syndrome is probably different from the hereditary benign telangiectasia reported by Ryan and Wells (2).

### References (Acrolabial telangiectasia)

1. Millns JL, Dickens CH: Hereditary acrolabial telangiectasia: A report of familial blue lips, nails, and nipples. *Arch Dermatol* **115**:474–478, 1979.

2. Ryan TJ, Wells RS: Hereditary benign telangiectasia. *Trans St. John's Hosp Dermatol Soc* **57**:148–156, 1971.

## Pseudoxanthoma elasticum

Originally described by Rigal (45), in 1881, and Balzer (3), in 1884, and grouped with the xanthomatoses, the syndrome of alterations of the skin, recurrent severe gastrointestinal hemorrhages, weak peripheral pulses, and failing vision was identified as a separate nonxanthomatous entity and named by Darier (14) in 1896. Angioid streaks were added to the syndrome by Grönblad (20) and Strandberg (51) in 1929, although they had been described as early as 1889 by Doyne (15). Thorough reviews are those of McKusick (31) and Neldner (35a). An excellent historic review was written by Pope (38a).

Pseudoxanthoma elasticum is a genetic heterogeneity but most cases have autosomal recessive inheritance (31). Pope (39–41) suggested that there are two recessive and two dominant forms. The alleged clinical differences are delineated in Table 13–4. However, Stutz et al (52), because of marked variability in expression, could not support two autosomal recessive types. Neldner (35a) could not separate distinct phenotypes among 100 patients.

Viljoen (56a) surveyed 81 patients with pseudoxanthoma elasticum (PXE) in South Africa and Zimbabwe. Three had autosomal dominant PXE, Type I, 3 AD, Type II, 19 autosomal recessive Type I, and 6 AR Type II disease. However, 50 individuals of Dutch–French Huguenot extraction had a distinct clinical subgroup characterized by autosomal recessive inheritance, mild to moderate cutaneous and cardiovascular changes, but severe visual impairment by age 50 due to macular neovascularization and hemorrhage.

The data presented represent the average of the various types. If ascertainment is made from skin changes, then 65% of patients are female, although this may be due to cosmetic concerns. However, it has also been suggested that there may be an estrogen effect, based on the aggravation of cardiac symptoms during pregnancy (47). If angioid streaks are used for discernment, there is no sex predilection. Consanguinity has been found in at least 20% of recessive cases. The combined prevalence of all types has been estimated at 1/40,000 (1).

The primary alteration in PXE is calcification of elastic fibers due to the accumulation of polyanions within the fibers that may play a

Table 13–4. Comparison of clinical findings of autosomal dominant and recessive forms of pseudoxanthoma elasticum[a]

| Characteristics | Dominant I (%) | Dominant II (%) | Recessive I (%) | Recessive II (%) |
|---|---|---|---|---|
| Cutaneous changes | | | | |
| Classical peau d'orange and flexural rash | 100 | 25 | 75 | |
| Macular rash | | 70 | 15 | |
| General increase of extensibility | 10 | 65 | 10 | |
| General cutaneous PXE | | | | 100 |
| Vascular disease | | | | |
| Angina | 55 | | | |
| Claudication | 55 | | | |
| Hypertension | 75 | 10 | 20 | |
| Hematemesis | 10 | 5 | 15 | |
| Ophthalmic abnormalities | | | | |
| Severe choroiditis | 75 | 10 | 35 | |
| Angioid streaks | 35 | 50 | 50 | |
| Washed-out pattern | | 15 | 2 | |
| Prominent choroidal vessels | | 20 | | |
| Myopia | 25 | 50 | 5 | |
| Blue sclerae | 10 | 40 | 10 | |
| Other findings | | | | |
| High-arched palate | | 55 | 15 | |
| Joint hypermobility | | 35 | 5 | |

[a]Based on data presented by FM Pope, *Arch Dermatol* **110**:209, 1974.

role in increased calcium binding (30). Gordon et al (18) found evidence that increased, abnormal catabolism of proteoglycans occurred because of increased secretion of serine protease.

Elevated lysine and hydroxylysine levels in collagen (5) and raised glutamine, asparagine, glutamic acid, and aspartic acids in the elastin (6) have been discovered. There is some evidence that a relationship may exist between *hyperphosphatasemia* and PXE (33). Also see *Differential diagnosis.*

**Skin.** The skin becomes thickened and the markings are accentuated by raised, yellowish, flat papules, especially about the mouth, neck, axillas, elbows, groin, and periumbilical area. A few yellowish papules on the neck or axilla to virtually general cutaneous involvement have been described in association with calcification of the subcutis (24). These changes, usually recognized after the second decade, may appear as early as the third or fourth years of life. The yellowish skin becomes inelastic, redundant, thickened, and grooved, the normal skin marking becoming accentuated so that the skin resembles coarse-grained leather, especially on the neck and in the axilla (Fig. 13–95). Miescher elastoma has also been noted (31,48), and epidermal perforation was found in one woman with hypertension and chronic renal failure who was subsequently diagnosed as having PXE (36). Acne was present in 42% of patients in one study (57).

Ultrastructural changes consist of granular changes in elastin fibers. They exhibit an increased affinity for cationic stains (46). Thready material found in lesions has been found to consist of fibrinogen, collagenous protein, and glycoprotein (61).

**Eyes.** Visual disturbances have been noted in almost 30% of older patients (16). Funduscopic examination demonstrates atypical drusen and white, brown, or gray streaking (angioid streaks) in over 85% (9). The streaks probably appear in the second decade or later. They present as a ring around the optic disc, extending outward beneath the retinal vessels. Although resembling vessels, they actually represent discontinuity in Bruch's membrane (Fig. 13–96). Retinal hemorrhage, followed by organization, eventuates in loss of visual acuity. Less often there is senile macular degeneration or central chorioretinitis (7,31,56). Other eye findings have included myopia in 26% and bilateral cataracts in 5% of 19 patients with PXE (57).

**Cardiovascular system.** By the third decade, there may be weakness or absence of pulses in the arms and legs, accompanied by variable degrees of intermittent claudication in about 20% (9). Radiographic examination has revealed calcification of peripheral arteries, especially those of the lower extremities, in about 15% (16). Renovascular hypertension has been noted in about 20% (16). Angina pectoris occurs in at least 50% of the patients (9), and abdominal angina due to stenosis of the celiac artery is common (31). Congestive cardiac failure is found in about 70% (43).

Lebwohl et al (27) reported that 71% had echocardiographic evidence of mitral valve prolapse. Pyeritz et al (44) found mitral valve prolapse in only 18%; however, they suggested that mitral valve prolapse may be more common in the dominant form of PXE.

Hemorrhage, especially gastrointestinal, occurs in 15% (12) because of involvement of small blood vessels. A number of patients have had peptic ulcer (1). Bleeding may also occur in the retina, kidney, uterus, or bladder (1). Epistaxis (9) and hemarthrosis have been recorded. Aneurysms of the internal carotid artery have also been reported, and there is one report of ruptured aneurysm of the anterior spinal artery (26).

**Other findings.** Neurologic deficits may result from complications of hypertension and cardiovascular disease: aneurysmal dilation of vessels, subarachnoid and intracerebral hemorrhage, bilateral carotid occlusion, psychiatric disorders, and seizures (9,16,22).

Mamtora and Cope (29) described two cases of pulmonary opacities, and cited two other cases. Jackson and Loh (23) reported a case of pulmonary involvement; lung biopsy showed deposition of calcium in arteries, arterioles, and venules, as well as in alveolar lumina and septa.

Prick and Thyssen (42) described several skeletal anomalies, including dysplastic vertebrae and fibroosseous dysplasia and transverse sclerotic bands of the metaphyseal and diaphyseal regions of long bones. Joint hyperextensibility has also been noted in a few cases (44).

Hashimoto's thyroiditis, hypothyroidism, hyperthyroidism, and diabetes mellitus have also been occasionally reported (35).

Berde et al (4) summarized the pregnancy histories in seven women with PXE and noted that five suffered severe gastrointestinal hemorrhage during one or more of their pregnancies. In four, the bleeding

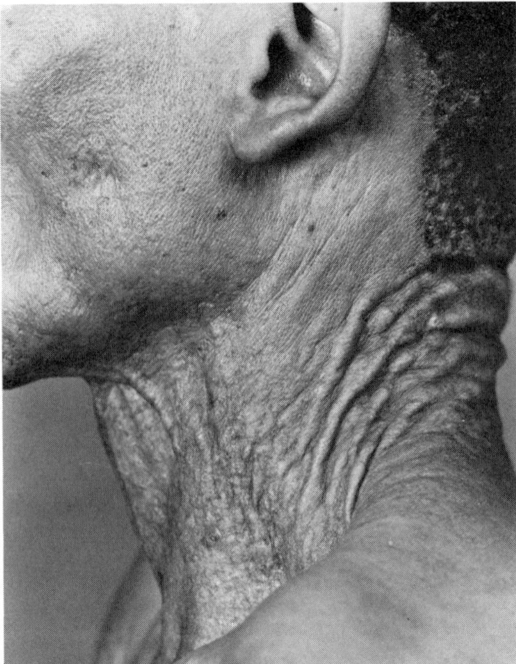

Fig. 13–95. *Pseudoxanthoma elasticum.* Skin becomes thickened and markings are accentuated by yellowish papules, especially around neck, axilla, elbows, and groin. (From *T Heyl,* Arch Dermatol **96**:528, 1967.)

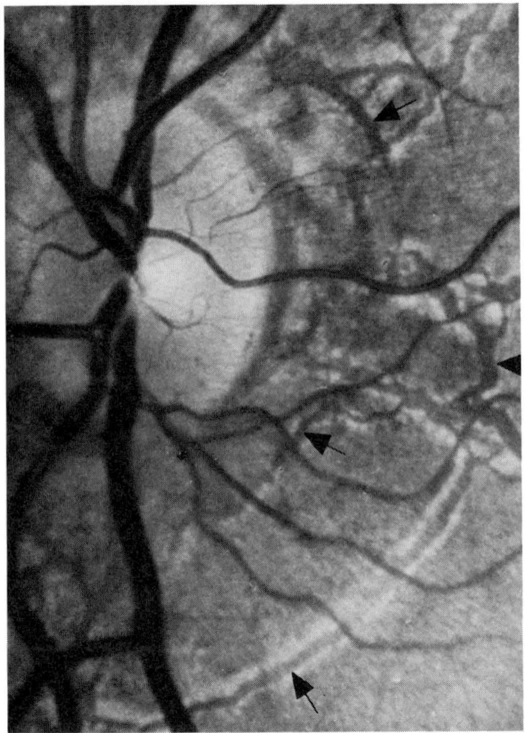

Fig. 13–96. *Pseudoxanthoma elasticum.* Arrows point to angioid streaks. Egg shell "fractures" in Bruch's membrane, usually grouped around optic disks, mimic appearance of blood vessels. (Courtesy of *WF Hoyt,* San Francisco, California.)

that occurred during pregnancy was the first such episode in their lives. In one instance of intrauterine growth retardation, it was suggested that maternal vascular compromise led to chronic uteroplacental insufficiency (8). Miscarriage during the first trimester of women with PXE is increased (57a).

**Oral manifestations.** The skin about the mouth may become redundant, the nasolabial folds and skin creases becoming accentuated and producing a saggy appearance. The mucosal surface of the lips, especially the lower one, may exhibit yellowish intramucosal nodules (14,16a,31) in about 10%. The buccal mucosa, soft or hard palate, and tonsillar areas may be similarly affected (9,34,39,40) (Fig. 13–97). The vaginal, gastric, and rectal mucosas have also been involved (9,17). Ultrastructural study of oral lesions showed large numbers of thickened and twisted collagen fibers (13). Highly arched palate has been occasionally noted, particularly in the autosomal dominant form (44).

**Differential diagnosis.** Angioid streaks may also be seen in Paget's disease of bone (21). A combination of Paget's disease and PXE has been described by Woodcock (60), Shaffer et al (49), and others (31). Angioid streaks have also been seen in sickle cell disease, possibly because of iron deposits in Bruch's membrane (20), and in osteoectasia with macrocranium (hyperphosphatasia). Angioid streaks in combination with tumoral calcinosis and hyperphosphatemia have been noted in several families (2). The clinical skin changes in senile elas-

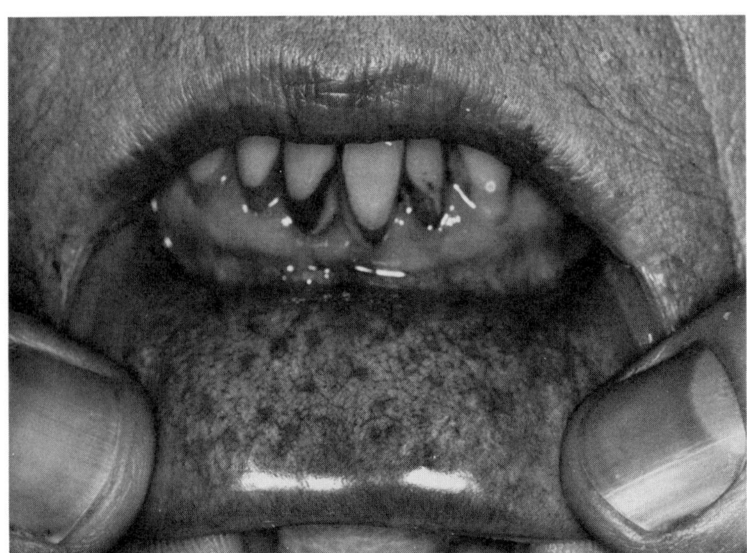

Fig. 13–97. *Pseudoxanthoma elasticum.* Mucosal surface of lower lip exhibits yellowish intramucosal nodules. (Courtesy of *RM Goodman,* Tel Aviv, Israel.)

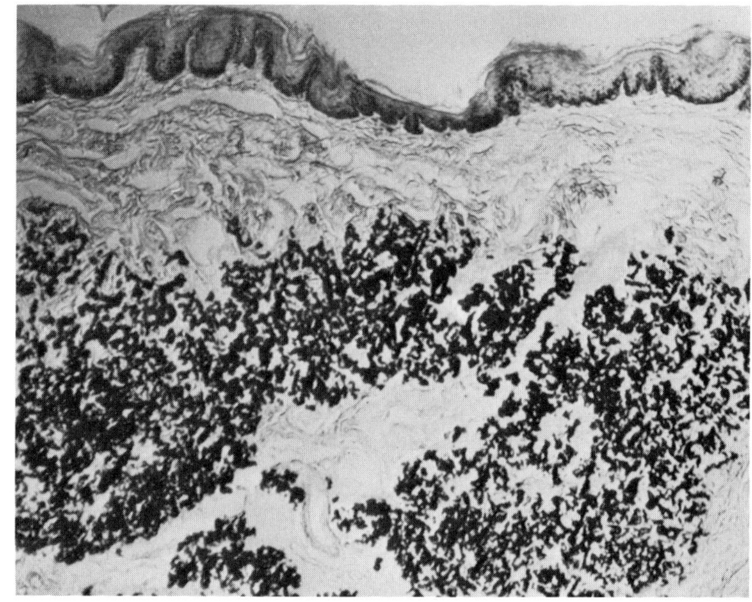

Fig. 13–98. *Pseudoxanthoma elasticum.* Elastic tissue appears fragmented and granular because of presence of calcium salts (von Kossa stain). (From *T Heyl,* Arch Dermatol **96**:528, 1967.)

tosis somewhat resemble those in PXE. They may be seen in the *Ehlers–Danlos syndrome,* following trauma, and in other disorders (21). Pseudoxanthoma elasticum-like changes have also been found in patients treated with high doses of D-penicillamine (55).

**Laboratory aids.** The involved skin or mucous membrane presents a characteristic microscopic picture. The changes are observed in the middle and lower part of the corium. The elastic tissue appears as fragmented masses, presenting a granular structure that is due at least in part to the presence of calcium salts demonstrated by von Kossa staining (Fig. 13–98). The amount of normal collagen fibers is reduced, whereas reticulum fibers are present in large amounts. The finding of calcium salts in ostensibly normal elastic fibers indicates that calcification is the primary event in PXE (13). Increased amounts of calcium have also been found in the skin on microincineration (28). Photographs of the fundus following injection of fluorescent dye clearly demonstrate the angioid streaks (50).

Using light and electron microscopy, Vogel et al (59) noted differences in appearance between elastic fibers in patients with the dominant and recessive forms. In addition, patients with the dominant condition had only an unusual aggregation of small and large collagen fibrils. However, Pierard (38) did not find any significant differences between dominant Type I and recessive Type I skin biopsies, stating that variations among individuals were greater than variation between the two types.

### References (Pseudoxanthoma elasticum)


1. Altman LK et al: Pseudoxanthoma elasticum. *Arch Intern Med* **134**:1048–1054, 1974.
2. Baldursson H et al: Tumoral calcinosis with hyperphosphatemia. *J Bone Jt Surg* **51A**:913–925, 1969.
3. Balzer F: Recherches sur les caracteres anatomiques du xanthelasma. *Arch Physiol* **4**:65–80, 1884.
4. Berde C et al: Pregnancy in women with pseudoxanthoma elasticum. *Obstet Gynecol Survey* **38**:339–344, 1983.
5. Blumenkrantz N et al: Biosynthesis of collagen in pseudoxanthoma elasticum. *Acta Dermatol Venereol* **53**:429–434, 1973.
6. Blumenkrantz N et al: Biosynthesis of elastin in pseudoxanthoma elasticum. *Acta Dermatol Venereol* **53**:435–438, 1973.
7. Britten MJ: Unusual traumatic retinal hemorrhage associated with angioid streaks. *Br J Ophthalmol* **50**:540–542, 1966.
8. Broekhuizen FF, Hamilton PR: Pseudoxanthoma elasticum and intrauterine growth retardation. *Am J Obstet Gynecol* **148**:112–114, 1984.
9. Carlborg V et al: Vascular studies in pseudoxanthoma elasticum with a series of color photographs of the eyeground lesions. *Acta Med Scand* (Suppl) **350**:1–84, 1959.
10. Connor PO et al: Pseudoxanthoma elasticum and angioid streaks: A review of 106 cases. *Am J Med* **30**:537–543, 1961.
11. Cunningham JR et al: Pseudoxanthoma elasticum: Treatment of gastrointestinal hemorrhage by arterial embolization and observation on autosomal dominant inheritance. *Johns Hopkins Med J* **147**:168–173, 1980.
12. Danielsen L, Kobayasi T: Pseudoxanthoma elasticum. *Acta Dermatol Venereol* **54**:121–128, 173–176, 1974.
13. Danielsen L et al: Pseudoxanthoma elasticum, a clinico-pathological study. *Acta Dermatol Venereol* **50**:355–373, 1970.
14. Darier J: Pseudoxanthoma elasticum. *Monatsh Prakt Dermatol* **23**:609–616, 1896.
15. Doyne RW: Choroidal and retinal changes: The result of blows on the eyes. *Trans Ophthalmol Soc UK* **9**:128, 1889.
16. Eddy DD, Farber EM: Pseudoxanthoma elasticum: Internal manifestations. A report of cases and a statistical review of the literature. *Arch Dermatol* **86**:729–740, 1962.
16a. Goette DK, Carpenter WM: The mucocutaneous marker of pseudoxanthoma elasticum. *Oral Surg* **51**:68–72, 1981.
17. Goodman RM et al: Pseudoxanthoma elasticum: A clinical and histopathological study. *Medicine (Baltimore)* **42**:297–334, 1963.
18. Gordon SG et al: Evidence for increased protease activity secreted from cultured fibroblasts from patients with pseudoxanthoma elasticum. *Connect Tissue Res* **6**:61–68, 1978.
19. Goto K: Involvement of central nervous system in pseudoxanthoma elasticum. *Fol Psychiatr Neurol* **29**:263–277, 1975.
20. Grönblad E: Angioid streaks—pseudoxanthoma elasticum: Vorläufige Mitteilung. *Acta Ophthalmol (Kbh)* **7**:329, 1929.
21. Hogan JF, Heaton CL: Angioid streaks and systemic disease. *Br J Dermatol* **89**:411–416, 1973.
22. Iqbal A et al: Pseudoxanthoma elasticum: A review of neurological complications. *Ann Neurol* **4**:18–20, 1978.
23. Jackson A, Loh CL: Pulmonary calcification and elastic tissue damage in pseudoxanthoma elasticum. *Histopathology* **4**:607–611, 1980.
24. James AE et al: Roentgen findings in pseudoxanthoma elasticum (PXE). *Am J Roentgenol* **106**:642–647, 1969.
25. Kaplan EN, Henjyoji EY: Pseudoxanthoma elasticum: A dermal elastosis with surgical implication. *Plast Reconstr Surg* **58**:595–599, 1976.
26. Kito K et al: Ruptured aneurysm of the anterior spinal artery associated with pseudoxanthoma elasticum. *J Neurosurg* **58**:126–128, 1983.
27. Lebwohl MG et al: Pseudoxanthoma elasticum and mitral-valve prolapse. *N Engl J Med* **307**:228–231, 1982.
28. Lobitz WC Jr, Osterberg AE: Pseudoxanthoma elasticum: Microincineration. *J Invest Dermatol* **15**:297–298, 1950.
29. Mamtora H, Cope V: Pulmonary opacities in pseudoxanthoma elasticum: Report of two cases. *Br J Radiol* **54**:65–67, 1981.

30. Martinez-Hernandez A, Huffer WE: Pseudoxanthoma elasticum: Dermal polyanions and mineralization of elastic fibers. *Lab Invest* **31**:181–186, 1974.

31. McKusick VA: *Heritable Disorders of Connective Tissue*, 4th ed., C.V. Mosby, St. Louis, 1972.

32. Mendelsohn G et al: Cardiovascular manifestations of pseudoxanthoma elasticum. *Arch Pathol Lab Med* **102**:298–302, 1978.

33. Mitsudo SM: Chronic idiopathic hyperphosphatasia associated with pseudoxanthoma elasticum. *J Bone Jt Surg* **53A**:303–314, 1971.

34. Miyaka H et al: A new finding in oral cavity in pseudoxanthoma elasticum. *Nagoya J Med Sci* **29**:251–259, 1967.

35. Muratani H et al: Pseudoxanthoma elasticum associated with Hashimoto's thyroiditis. *Jpn J Med* **21**:223–226, 1982.

35a. Neldner KH: Pseudoxanthoma elasticum. *Clin Dermatol* **6**:1–159, 1988.

36. Nickoloff BJ et al: Perforating pseudoxanthoma elasticum associated with chronic renal failure and hemodialysis. *Arch Dermatol* **121**:1321–1322, 1985.

37. Pasquali-Ronchetti I et al: Pseudoxanthoma elasticum: Biochemical and ultrastructural studies. *Dermatologica* **163**:307–325, 1981.

38. Pierard GE: Le pseudoxanthome elastique. Etude morphologique et biomecanique des formes dominante type I et recessive type I. *Ann Dermatol Venereol* **111**:111–116, 1984.

38a. Pope FM: Pseudoxanthoma elasticum: An historical survey. *Trans St John's Hosp Dermatol Soc* **58**:235–250, 1972.

39. Pope FM: Autosomal dominant pseudoxanthoma elasticum. *J Med Genet* **11**:152–157, 1974.

40. Pope FM: Two types of autosomal recessive pseudoxanthoma elasticum. *Arch Dermatol* **110**:209–212, 1974.

41. Pope FM: Historical evidence for the genetic heterogeneity of pseudoxanthoma elasticum. *Br J Dermatol* **92**:498–510, 1975.

42. Prick JJG, Thijssen HOM: Radiodiagnostic signs in pseudoxanthoma elasticum generalisatum (dysgenesis elastofibrillaris mineralisans). *Clin Radiol* **28**:549–554, 1977.

43. Przybojewski JZ et al: Pseudoxanthoma elasticum with cardiac involvement. *S Afr Med J* **59**:268–275, 1981.

44. Pyeritz RE et al: Pseudoxanthoma elasticum and mitral-valve prolapse. *N Engl J Med* **307**:1451–1452, 1982.

45. Rigal D.: Observation pour servir à l'histoire de la cheloide diffuse xanthelasmique. *Ann Dermatol Syph (Paris)* **2**:491–501, 1881.

46. Ross R et al: Fine structure alterations of elastic fibers in pseudoxanthoma elasticum. *Clin Genet* **13**:213–223, 1978.

47. Schaber DC et al: Grönblad–Strandberg syndrome: Report of a case and review of the literature. *J Am Osteopath Assoc* **79**:326–331, 1980.

48. Schutt D: Pseudoxanthoma elasticum and elastosis perforans serpingosa. *Arch Dermatol* **91**:151–152, 1965.

49. Shaffer B et al: Pseudoxanthoma elasticum: A cutaneous manifestation of a systemic disease: Report of a case of Paget's disease and a case of calcinosis with arteriosclerosis as manifestations of this syndrome. *Arch Dermatol Syph* **76**:622–633, 1957.

50. Smith JL et al: Fluorescein fundus photography of angioid streaks. *Br J Ophthalmol* **48**:517–521, 1964.

51. Strandberg J: Pseudoxanthoma elasticum. *Z Haut Geschlechtskr* **31**:689, 1929.

52. Stutz SB: Pseudoxanthoma elasticum—Groenblad–Strandberg syndrome—Elastorrhexia systemisata. Eine Literaturanalyse. *Zbl Haut Geschlechtskr* **148**:115–126, 1982.

53. Stutz SB et al: Zur Klinik und Genetik des Pseudoxanthoma elasticum (PXE). *Hautarzt* **36**:265–268, 1985.

54. Tay CH: Pseudoxanthoma elasticum. *Postgrad Med J* **46**:97–108, 1970.

55. Thomas RHM, Kirby JDT: Elastosis perforans serpingosa and pseudoxanthoma elasticum-like change due to penicillamine. *Clin Exp Dermatol* **10**:386–391, 1985.

56. Van Balen AT, Houtsmuller AJ: Syndrome of Gronblad–Strandberg–Touraine. *Ophthalmologica* **149**:246–247, 1965.

56a. Viljoen DL: Heterogeneity of pseudoxanthoma elasticum in 80 southern African patients. March of Dimes Birth Defects Conf, Baltimore, Maryland, July, 1988.

57. Viljoen DL et al: Pseudoxanthoma elasticum in South Africa—genetic and clinical implications. *S Afr Med J* **66**:813–816, 1984.

57a. Viljoen DL et al: The obstetric and gynaecological implications of pseudoxanthoma elasticum. *Br J Obstet Gynaecol* **94**:1–5, 1987.

58. Vitto J: Biochemistry of the elastic fibers in normal connective tissues and its alteration in diseases. *J Invest Dermatol* **72**:1–8, 1979.

59. Vogel A et al: Zur Licht und Elektronenmikroskopie des Pseudoxanthoma Elasticum (PXE). *Hautarzt* **36**:269–273, 1985.

60. Woodcock CW: Pseudoxanthoma elasticum, angioid streaks of retina and osteitis deformans. *Arch Dermatol Syph (Chic)* **65**:623, 1952.

61. Yamamura T, Sano S: Ultrastructural and histochemical analysis of thready material in pseudoxanthoma elasticum. *J Cutan Pathol* **11**:282–291, 1984.

## Progeria (Hutchinson–Gilford syndrome)

Progeria, consisting of dwarfism, early aging, and characteristic craniofacial appearance, was first described by Hutchinson (28) in 1886 and Gilford (21,22) in 1897 and 1904. Approximately 75 cases have been recorded to date (3,39). The frequency of progeria is about 1 per 4 to 8 million live births with a 2:1 male/female ratio (4). Autopsies have been performed in fewer than a dozen cases (2,21,31,35,48,55). Affected sibs were reported by Franklyn (16), Lejman (34), Mostfa and Gabr (40) and Rava (46), and affected cousins by Khalija (30). Consanguinity has been noted in a few instances (16,30,40). Monozygotic twins concordant for progeria were reported by Viegas et al (57). Autosomal recessive inheritance is debatable, as most cases have been sporadic and the consanguinity rate is so low. It may be dominant, the few familial cases representing gonadal mosaicism. Fathers of affected children tend to be older (28a). The seriousness of the condition with its early demise during teenage years may have an effect on many parents as far as having further offspring (7). A recent review is that of Badame (2a).

A number of cases are not included in our tabulation because we feel that they either represent other conditions or are not well enough documented over time to be sure. Paterson (44) reported two sibs resulting from a consanguineous union as examples of progeria. However, both patients were mentally retarded. The case reported by Green (25) in a 22-year-old woman does not look like progeria. Furthermore, the patient had pulmonic stenosis, carotid artery aneurysms, lacked secondary sexual characteristics, and was not bald. A case reported as progeria in two sisters (45) is now known to be an example of the *Wiedemann–Rautenstrauch syndrome* (12,52,59). The case of Grossman et al (26) may represent an example of *Seckel syndrome*. The patient reported by Feingold and Kidd (14), followed only until 22 months of age, had prominent eyes and prognathism, although early scleroderma, anterior hair loss, and a horse-riding stance certainly suggest progeria.

Since the aging process is accelerated in progeria, cultured fibroblasts from affected individuals possibly might show an earlier appearance of altered gene products or a more severe disturbance of such products. Enzyme thermolability studies from cultured fibroblasts (26) were consistent with both possibilities, showing a significantly higher percentage of heat-labile glucose-6-phosphate dehydrogenase, 6-phosphogluconate dehydrogenase, and hypoxanthine-guanine phosphoribosyltransferase. No significant association between HLA type and progeria has been detected (6). Sister chromatid exchange frequencies in progeria have not differed significantly from normal (9).

Aminoaciduria has been documented (53). Hypermetabolism has been suggested by some (53,55), but denied by others (29,48). Many patients have exhibited elevated cholesterol or phospholipid levels (47,58). Rosenthal et al (49) suggested that progeria was a disease of intermediate lipoprotein metabolism. Keay et al (29) found the pattern of circulatory lipids and lipoproteins similar to that observed in adult coronary sclerosis. Villee et al (58) reported unresponsiveness to growth hormone, relative insulin resistance, and highly cross-linked collagen. Hyperglycemia and insulin resistance were documented by Rosenbloom et al (49).

**Growth.** Average birthweight is 2.7 kg. Growth proceeds almost normally until the first year, when it essentially plateaus until about 10 years of age. By the end of the first decade, height is approximately that of a normal 3-year-old; only rarely do patients exceed 110 cm in height or 15 kg in weight (10,36). There is no sexual maturation.

**Performance.** Intelligence is normal. The voice is high-pitched and squeaky (20,36,48,50). Hearing may be deficient (41).

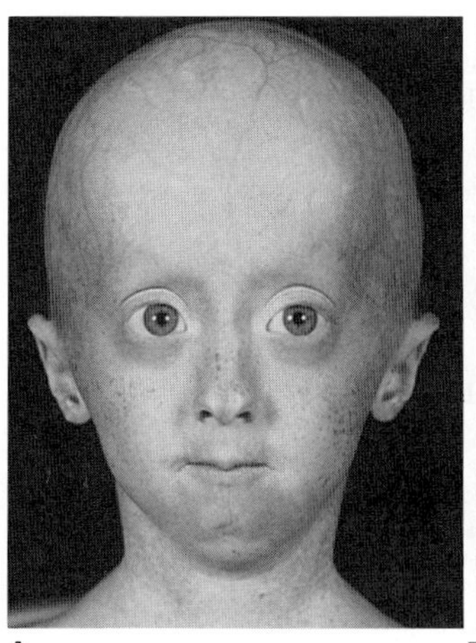

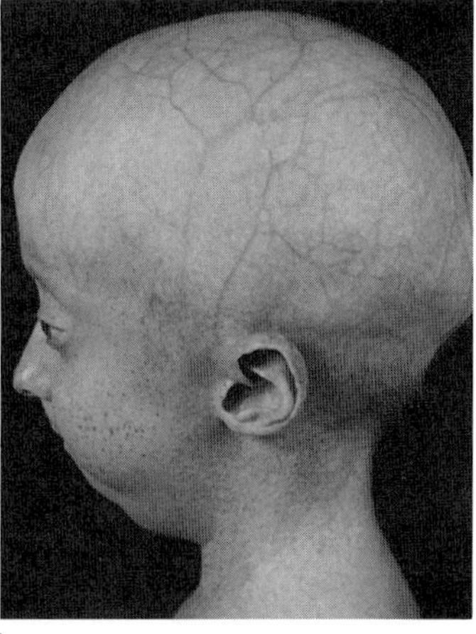

A B

Fig. 13–99. *Progeria.* (A,B) Face is small in comparison with cranial vault. Lashes are absent, brows are sparse. Note beaklike nose, mandibular hypoplasia, absent earlobes, prominent scalp veins. (From *L Atkins,* N Engl J Med **250**:1065, 1954.)

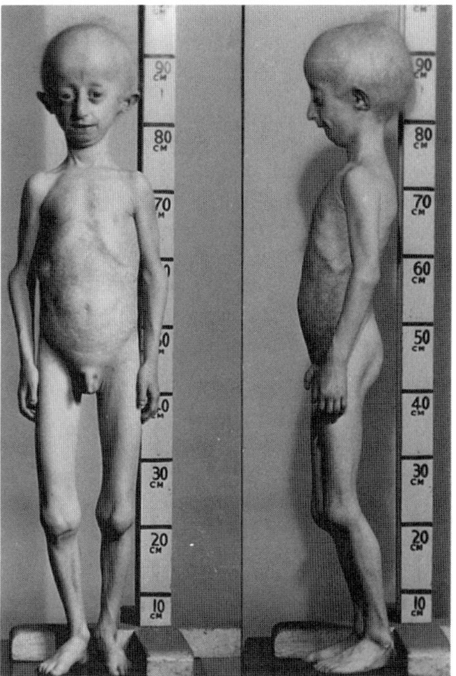

Fig. 13–100. *Progeria.* Clavicles are short, hands, elbows, and knees contracted, giving "horse-riding" stance. Note senile appearance. (From *L Atkins,* N Engl J Med **250**:1065, 1954.)

**Craniofacial appearance.** During the second year of life, scalp hair is lost and replaced by downy fuzz. Eyebrows and, occasionally, eyelashes are lost. The face is disproportionately small in relation to the cranium although, in fact, head circumference is usually 2–4 cm smaller than average. Frontal and parietal bossing occur together with prominent scalp veins. The ears are small without lobules and the nose is thin and rather beaked (Fig. 13–99 A,B). Micrognathia and delayed dental eruption of the permanent teeth are common features, the deciduous dentition often being retained. The mandibular angle may be increased. The plane angle is steep and the anterior fontanel is open. There are no frontal sinuses and mastoid development is poor. Teeth may be crowded because of mandibular hypoplasia with normal size teeth. However, teeth have been irregular in form, small, discolored, and deficient in number in most cases (27). Microscopic evidence of senile changes in the dental pulp has been reported (1,8,10,17,36,48).

**Musculoskeletal system.** The chest is narrow or pyriform and the abdomen protuberant. Mild flexion of the knees results in a horse-riding stance (Fig. 13–100). The bones are delicate and osteoporotic, although bone maturation is normal. The joints of the extremities become prominent and limited in extension from periarticular fibrosis, which appears in some patients as early as the sixth year. Less frequently affected are the spine, elbows, and knees. Hip pain, subluxation and eventual hip dislocation result from degenerative changes in the acetabulum (19,39).

Roentgenographically, terminal phalanges undergo progressive osteolysis and become abnormally short and taper abruptly to pointed ends (Fig. 13–101A). The clavicles also undergo progressive osteolysis. Coxa valga and acetabular degeneration are constant findings. There may be some predilection for fracture of the humeral shaft. There is also loss of muscular and subcutaneous fat. The calvaria is remarkably thin, the anterior fontanel is open, and frontal sinuses are often absent. The neurocranium is normal in size and configuration. Mastoid development is poor (1,2,16,18,32,36–39,43). Osteosarcoma has been reported by King et al (31).

**Cardiovascular system.** Cardiac murmurs appear after 5 years of age followed by diastolic systemic hypertension and cardiomegaly. Atherosclerosis is early and severe, and anginal attacks and cerebro-

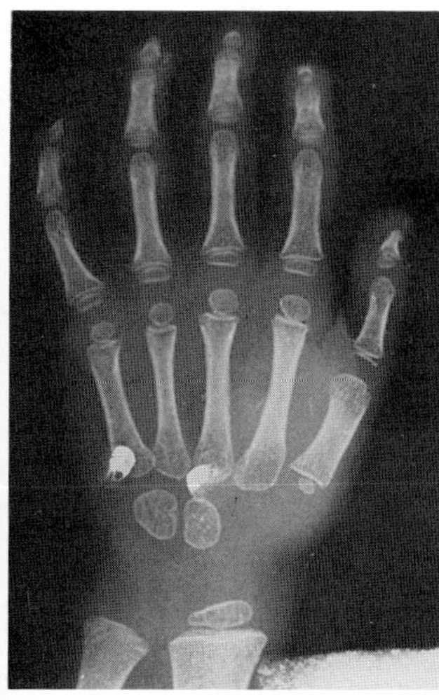

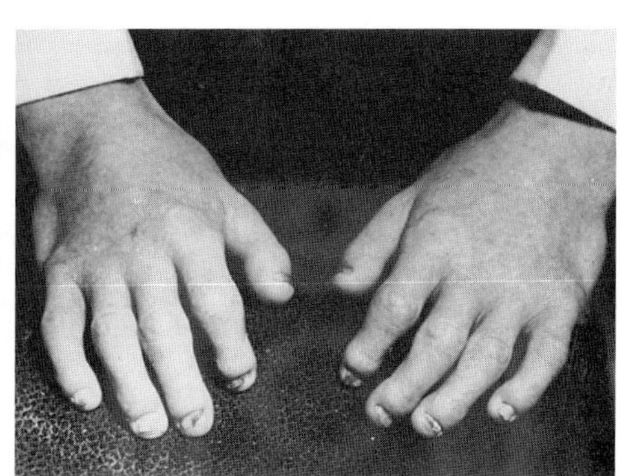

A B

Fig. 13–101. *Progeria.* (A) Terminal phalanges are short and taper abruptly. (B) Skin of hands is dry, taut, and mottled; fingers are short with enlarged joints. Nails are dry, brittle, and hypoplastic. (From *MM Album* and *JW Hope,* Oral Surg **11**:985, 1958.)

vascular accidents have been experienced as early as 7 years of age, but, in most instances, death occurs by approximately 14 years of age (3,13,35,55) from myocardial infarction or congestive heart failure. However, a 45-year-old man has been documented (42). Baker et al (3) reviewed the cardiovascular abnormalities.

**Skin and skin appendages.** The skin is thin, atrophic, and often pigmented, with sparse subcutaneous fat. Scleroderma-like changes have been recorded in some patients. Veins are especially prominent over the scalp and thighs. Electron microscopic studies of scalp hairs reveal abnormal longitudinal depressions with minor cuticular defects. The nails are thin, yellow, atrophic (Fig. 13–101B), or absent (2, 8,15,36,48,54,56,61).

**Differential diagnosis.** Several patients with *Hallermann–Streiff syndrome* have been erroneously labeled as having progeria. Occasionally patients with *Bloom syndrome* or *Cockayne syndrome* have also mistakenly been thought to have progeria. *Werner syndrome* is also a condition of premature aging. Other progeroid syndromes have been discussed elsewhere including the *Ehlers–Danlos syndromes, Wiedemann–Rautenstrauch* (neonatal progeroid) *syndrome* (12,45, 52,59), *De Barsy syndrome* (33), acrogeria (11), metageria (23), and a unique disorder described by Ruvalcaba et al (51).

**Laboratory aids.** Marquart et al (38) reported increased secretion of fibronectin and collagen by fibroblasts of an affected patient. There is reported increase in urinary hyaluronic acid (60). Other inconsistent findings have been reported (24).

### References [Progeria (Hutchinson–Gilford syndrome)]


1. Album MM, Hope JW: Progeria. *Oral Surg* **11**:985–998, 1958.

2. Atkins L: Progeria: Report of a case with post-mortem findings. *N Engl J Med* **250**:1065–1069, 1954.

2a. Badame AJ: Progeria. *Arch Dermatol* **125**:540–544, 1989.

3. Baker PB et al: Cardiovascular abnormalities in progeria. *Arch Pathol Lab Med* **105**:384–386, 1981.

4. Beauregard S, Gilchrest B: Syndromes of premature aging. *Dermatol Clin* **5**:109–121, 1987.

5. Bhahoo ON et al: Progeria with unusual ocular manifestation. *Indian Pediatr* **2**:164–169, 1965.

6. Brown T et al: Detection of HLA antigens on progeria syndrome fibroblasts. *Clin Genet* **17**:213–219, 1980.

7. Brown WT: Genetics of human aging. *Rev Biol Res Aging* **2**:105–114, 1985.

8. Cooke JV: The rate of growth in progeria, with a report of two cases. *J Pediatr* **42**:26–37, 1953.

9. Darlington GJ et al: Sister chromatid exchange frequencies in progeria and Werner syndrome patients. *Am J Hum Genet* **33**:762–766, 1981.

10. DeBusk FL: The Hutchinson–Gilford progeria syndrome. *J Pediatr* **80**:697–724, 1972.

11. De Groot WP et al: Familial acrogeria (Gottron). *Br J Dermatol* **103**:213–222, 1980.

12. Devos EA et al: The Wiedemann–Rautenstrauch or neonatal progeroid syndrome: Report of a patient with consanguineous parents. *Eur J Pediatr* **136**:245–248, 1981.

13. Dyck JD et al: Management of coronary artery disease in Hutchinson-Gilford syndrome. *J Pediatr* **111**:407–409, 1987.

14. Feingold M, Kidd R: Progeria and scleroderma in infancy. *Am J Dis Child* **122**:61–62, 1971.

15. Fleischmajer R, Nedwich A: Progeria (Hutchinson–Gilford). *Arch Dermatol* **107**:253–258, 1973.

16. Franklyn PP: Progeria in siblings. *Clin Radiol* **27**:327–333, 1976.

17. Gabr M: Progeria: Review of the literature with report of a case. *Arch Pediatr* **71**:35–46, 1954.

18. Gabr M et al: Progeria, a pathologic study. *J Pediatr* **57**:70–77, 1960.

19. Gamble JG: Hip disease in Hutchinson–Gilford progeria syndrome. *J Pediatr Orthoped* **4**:585–589, 1984.

20. Ghosh S, Varma K: Progeria: Report of a case with review of the literature. *Indian Pediatr* **1**:146–155, 1964.

21. Gilford H: On a condition of a mixed premature and immature development. *Trans Med-Chir Soc Edinburgh* **80**:17–45, 1987.

22. Gilford H: Progeria, a form of senility. *Practitioner* **73**:188–217, 1904.

23. Gilkes JJH et al: The premature aging syndromes: Report of eight cases and description of a new entity named metageria. *Br J Dermatol* **91**:243–262, 1974.

24. Goldstein S, Moerman E: Heat-labile enzymes in skin fibroblasts from subjects with progeria. *N Engl J Med* **292**:1305–1309, 1975.

25. Green LN: Progeria with carotid artery aneurysms: Report of a case. *Arch Neurol* **38**:659–661, 1981.

26. Grossman HJ et al: Progeroid syndrome, report of a case of pseudosenilism. *Pediatrics* **15**:413–423, 1955.

27. Hasty MF, Yann WF Jr: Progeria in a pediatric dental patient. Literature review and a case report. *Pediatr Dent* **10**:314–319, 1988.

28. Hutchinson J: Congenital absence of hair with atrophic condition of skin. *Trans Med-Chir Soc Edinburgh* **69**:473–477, 1886.

28a. Jones KL et al: Older paternal age and fresh gene mutation: Data on additional disorders. *J Pediatr* **86**:84–88, 1975.

29. Keay AJ et al: Progeria and atherosclerosis. *Arch Dis Childh* **30**:410–414, 1955.

30. Khalifa MM: Hutchinson–Gilford progeria syndrome: Report of a Libyan family and evidence of autosomal recessive inheritance. *Clin Genet* **35**:125–132, 1989.

31. King CR et al: Osteosarcoma in a patient with Hutchinson–Gilford progeria. *J Med Genet* **15**:481–484, 1978.

32. Kozlowski K: Étude radiologique d'un cas de nanisme sénile (progéria). *Ann Radiol* **8**:92–96, 1965.

33. Kunze J et al: De Barsy syndrome—an autosomal recessive, progeroid syndrome. *Eur J Pediatr* **144**:348–354, 1985.

34. Lejman K: Progeria in sibs. *Hautarzt* **13**:187–188, 1962.

35. Makous N et al: Cardiovascular manifestations in progeria. Report of clinical and pathologic finding in a patient with arteriosclerotic heart disease and aortic stenosis. *Am Heart J* **64**:334–336, 1962.

36. Manschot WA: A case of progeronanism (progeria of Gilford). *Acta Paediatr Scand* **39**:158–164, 1950.

37. Margolin FR, Steinbach HL: Progeria: Hutchinson–Gilford syndrome. *Am J Roentgenol* **103**:173–178, 1968.

38. Marquart FX et al: Increased secretion of fibronectin and collagen by progeria (Hutchinson–Gilford) fibroblasts. *Eur J Pediatr* **147**:442, 1988.

39. Moen C: Orthopaedic aspects of progeria. *J Bone Jt Surg* **64A**:542–546, 1982.

40. Mostafa AH, Gabr M: Heredity in progeria with follow-up of two affected sisters. *Arch Pediatr* **71**:163–172, 1954.

41. Nelson M: Progeria, audiologic aspects. *Ann Otol Rhinol Laryngol* **74**:376–385, 1965.

42. Ogihara T: Hutchinson–Gilford progeria syndrome in a 45-year-old man. *Am J Med* **81**:135–138, 1986.

43. Ozonoff MB, Clemett AR: Progressive osteolysis in progeria. *Am J Roentgenol* **100**:75–79, 1967.

44. Paterson D: Case of progeria. *Proc R Soc Med* **16**:42, 1923.

45. Rautenstrauch T, Snigula F: Progeria: A cell culture study and clinical report of familial incidence. *Eur J Pediatr* **124**:101–111, 1977.

46. Rava G: Su un nucleo familiare di progeria. *Minerva Med* **58**:1502–1509, 1967.

47. Reichel W, Garcia-Bunuel R: Pathologic findings in progeria. *Am J Clin Pathol* **53**:243–253, 1970.

48. Rosenbloom AL et al: Progeria: Report of a case with cephalometric roentgenograms and abnormally high concentrations of lipoproteins in the serum. *Pediatrics* **18**:565–577, 1956.

49. Rosenthal IM et al: Progeria: Insulin resistance and hyperglycemia. *J Pediatr* **102**:400–401, 1983.

50. Runge P et al: Hutchinson–Gilford progeria syndrome. *South Med J* **71**:877–879, 1978.

51. Ruvalcaba RHA et al: Children who age rapidly—progeroid syndromes. *Clin Pediatr* **16**:248–252, 1977.

52. Snigula F, Rautenstrauch T: Letter to the editor: A new neonatal progeroid syndrome. *Eur J Pediatr* **136**:325–326, 1981.

53. Steinberg AH et al: Amino-aciduria and hypermetabolism in progeria. *Arch Dis Childh* **32**:401–403, 1957.

54. Strunz F: Ein Fall von Progeria, beginnend mit ausgedehnter Sklerodermie. *Z Kinderheilkd* **47**:401–416, 1929.

55. Talbot NB et al: Progeria, clinical, metabolic and pathologic studies on a patient. *Am J Dis Child* **69**:267–279, 1945.

56. Thomson J, Forfar JO: Progeria (Hutchinson–Gilford syndrome): Report of a case and review of the literature. *Arch Dis Childh* **25**:224–234, 1950.

57. Viegas J et al: Progeria in twins. *J Med Genet* **11**:384–385, 1974.

58. Villee DB et al: Metabolic studies in two boys with classical progeria. *Pediatrics* **43**:207–216, 1969.

59. Wiedemann H-R: An unidentified neonatal progeroid syndrome: Follow-up report. *Eur J Pediatr* **130**:65–70, 1979.

60. Zebrower M et al: Urinary hyaluronic acid elevation in Hutchinson–Gilford progeria syndrome. *Mech Aging Dev* **35**:39–46, 1986.

61. Zeder E: Über Progerie, eine seltene Form des hypophysären Zwergwuchses mit diffuser Sklerodermie. *Monatsschr Kinderheilkd* **81**:167–205, 1940.

## Werner syndrome

The syndrome, first delineated by Werner (37), in 1904, consists of shortness of stature, premature graying of hair (canities), baldness, scleropoikiloderma, trophic leg ulcers, juvenile cataracts, hypogonadism, diabetes mellitus, calcification of blood vessels, and osteoporosis. Confusion with Rothmund–Thomson syndrome occurred until Oppenheimer and Kugel (23), Thannhauser (34), and Greither (15) clearly separated the syndromes.

Autosomal recessive inheritance has been established (7,14). The frequency of the disorder was estimated to be between 1 and 20 cases per million population in the United States (7). It appears to have a high frequency in Sardinia (2a). Over 150 cases have been reported (7,10a,18a,38).

The disorder becomes apparent in the third and fourth decades of life. Graying of the hair occurs at about 20 years, skin changes and loss of hair by 25, cataracts by 30, diabetes mellitus by 35, and death at about 45 years.

Short stature, due to arrest of growth at puberty, is a constant feature. Mean height is 157 cm for males and 146 cm for females (7,39). Weight is low, relative even to the short stature. Thinness of arms and legs and diminution in size of hands and feet are striking. In contrast, the trunk is often stocky and the abdomen protuberant. Patients characteristically develop abnormally high-pitched, sometimes squeaky, or, less often, hoarse voices due to vocal cord atrophy.

**Facies.** The combination of atrophic skin changes, baldness, and graying hair gives these patients, even while young, an appearance of being 20–30 years older than their age. Although the cheeks are full, the nose assumes a pinched or beaked appearance due to hypoplasia of nasal cartilages (Fig. 13–102).

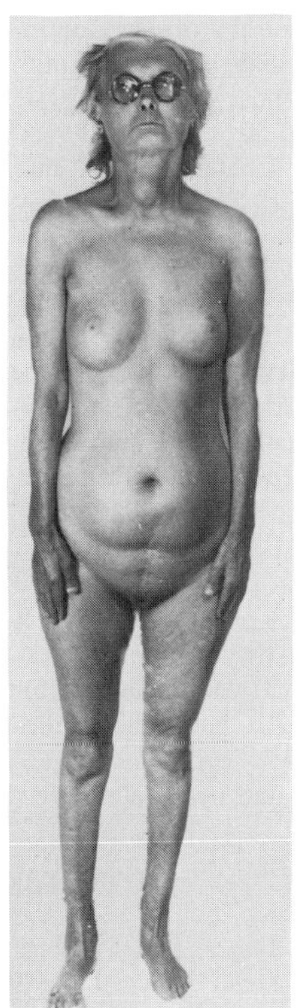

Fig. 13–102. *Werner syndrome*. Aged appearance together with marked thinness of arms and legs. (Courtesy of *S Jablonska*, Warsaw, Poland.)

**Skin.** The principal areas of skin involvement are the face and distal extremities, especially the feet. There is atrophy of the skin over areas depleted of adipose tissue, connective tissue, and musculature. This results in shiny smooth skin that adheres to the underlying tissues, giving a sclerodermoid appearance. Commonly there is ischemic ulceration. The eyes appear protuberant because of loss of circumorbital tissue. The ears may become stiff and inelastic. Circumscribed hyperkeratoses develop over bony prominences and on the soles, and may become ulcerated, with slow or no healing (27).

Localized or, less commonly, generalized hyperpigmentation, depigmentation, and telangiectasia occur on the arms and legs in some patients, the term ''poikilodermatous'' then being applied.

**Eyes.** Senile cataracts, another cardinal feature, are invariably bilateral, usually posterior, cortical, or subcapsular, and homogeneous or striate. They have abrupt onset during the third or fourth decade. Retinitis pigmentosa, senile macular degeneration, chorioretinitis, and corneal calcification have also been described (16a,24,27).

**Hair.** Premature loss and graying of scalp hair with onset usually before 20 years is characteristic (Fig. 13–103). Generalized hair loss involving the scalp, eyebrows, eyelashes, and body (axillary and pubic) hair may be secondary to hypogonadism.

**Musculoskeletal.** Osteoporosis and profound wasting of the musculature of the legs, feet, and hands are characteristic (Fig. 13–104A,B). The feet are usually flat. Soft-tissue calcification has been observed in one-third of the cases, especially in the tendons, ligaments and synovia of the knees, elbows, and ankles (7). The tissues adjacent to these areas may also exhibit soft-tissue calcification similar to that observed in scleroderma (26).

Fig. 13–103. *Werner syndrome*. Premature graying of hair, aging of skin. (Courtesy of *S Jablonska*, Warsaw, Poland.)

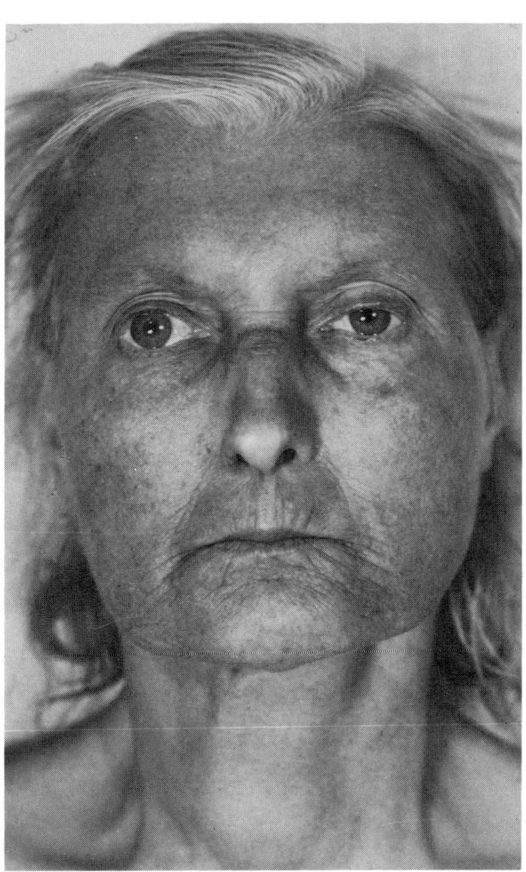

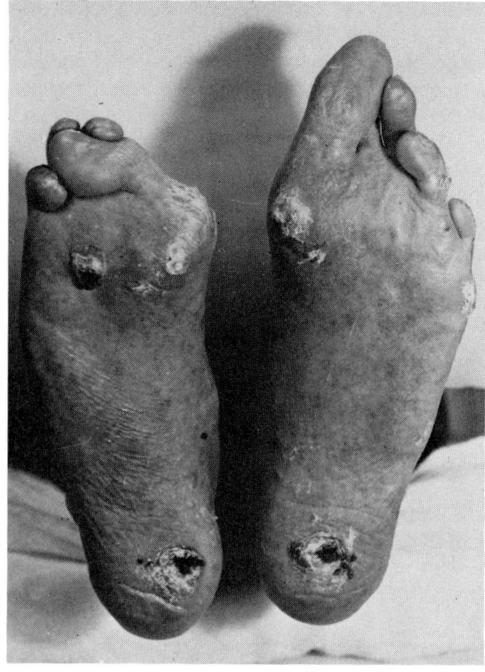

A

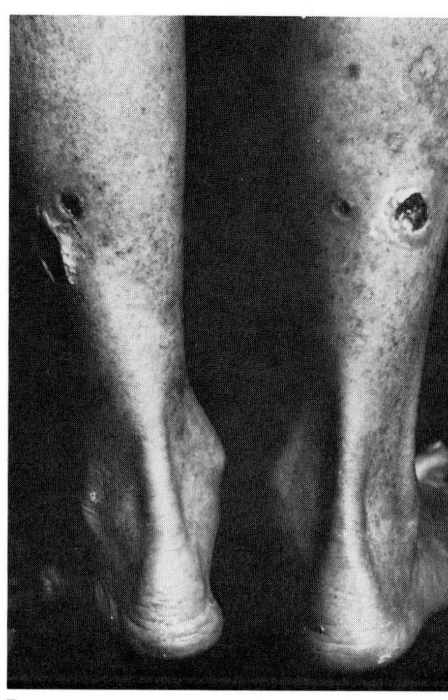

B

Fig. 13–104. *Werner syndrome.* (A) Hyperkeratosis and ulceration of soles. (B) Ulceration and pigmentation of ankles. (A courtesy of *S Jablonska,* Warsaw, Poland. B from *W Knoth et al,* Hautarzt **14**:145, 1963.)

**Cardiovascular system.** Patients develop severe, often generalized atherosclerosis. Calcification occurs in peripheral (Mönckeberg) vessels of the legs and in the mitral and aortic valves and coronary vessels (7,26). Peripheral vascular disease further complicates or promotes atrophy and ulceration of the skin.

**Endocrine system.** Diabetes mellitus has been recognized in 45% (7). However, the usual complications of diabetes (nephropathy, retinopathy, neuropathy) have not been observed. Neither the vascular disease nor the cataracts are well correlated with the diabetes mellitus.

Hypogonadism occurs in both sexes. Most males have small testes and penis, with diminished pubic hair. Women have poorly developed genitalia and breasts, and some never develop secondary sex characteristics. Menses are sparse and irregular. However, there are valid reports of men who have sired children and women who have borne children.

**Central nervous system.** Some patients may be mentally retarded (26), and about one-third may have mild neurologic deficits, such as loss of deep-tendon reflexes, paresthesias, and dizziness (7).

**Oral manifestations.** The skin around the mouth is often radially ridged (38). Frenkel (9) described atrophy of the oral mucosa.

**Neoplasia.** Over 10% have an unusual array of both benign and malignant neoplasms: meningioma, paraganglioma, adenomas of the pituitary, thyroid, and adrenal glands, various malignant epithelial neoplasms (basal and squamous carcinomas of the skin, melanoma, adenocarcinoma of the thyroid, stomach, ovary, and liver), and malignant connective tissue tumors (fibrosarcoma, neurofibrosarcoma, leiomyosarcoma, osteosarcoma) (1a,7,16,19,23,25-27,33,36,38). German (11) presented a critical analysis of reported tumors.

**Differential diagnosis.** In *Rothmund–Thomson syndrome,* the onset of skin changes and cataracts occurs during the first 10 years of life. Photosensitivity is present in about 35%. The skin also shows telangiectasia, scaling, and pigmentary changes. Skin cancer is a prominent feature.

*Progeria,* in contrast to Werner syndrome, has very early onset. There is severe dwarfing, nearly complete alopecia, absence of subcutaneous fat, and characteristic facies. Generalized atherosclerosis leads to early death.

*Acrogeria* refers to thin skin largely limited to the distal extremities (4). The subcutaneous vascular pattern is usually evident over the trunk. Several so-diagnosed patients really have *Ehlers–Danlos syndrome.* A definite inheritance pattern has not been established. Small stature and micrognathia have been present in several cases. Mulvihill and Smith (18) reported a syndrome of *premature aging, microcephaly, unusual facies, multiple nevi, and mental retardation.*

*Mandibuloacral dysplasia* is characterized by early onset, small mandible, and brachytelephalangy. Inheritance is autosomal recessive.

Metageria, an autosomal recessive disorder apparent at birth, is characterized by normal height and diabetes mellitus of early onset. There is no loss of hair and no cataracts. The skin is dry, atrophic, and mottled. The face is pinched with a beaked nose (12).

*Myotonic dystrophy* frequently has its onset in the second and third decades. Cataracts, diabetes mellitus, premature frontal baldness, and atrophy of the skin are features of the disorder. Testicular atrophy occurs, but the external genitalia are otherwise normal. Inheritance is autosomal dominant.

**Laboratory aids.** Karyotypes of long-term cultured skin fibroblasts show multiple, variable, predominantly stable, essentially random translocations that are clonal in nature (variegated translocation mosaicism) (28–30,32). In contrast, lymphocyte cultures are only rarely reported to be abnormal (31). However, an increased number of chromosome breaks has been found (22).

Cultured fibroblasts have a greatly reduced potential for *in vitro* growth and a reduced life-span (29). Bauer et al (1) found enhanced collagen synthesis. There is a slower rate of DNA replication (10,21), but repair function appears to be normal (19). Sister chromatid exchange frequency has been reported as normal (3) and enhanced (6). An altered immune mechanism has been noted (5,20), but further studies need to be carried out.

Increased urinary hyaluronic acid has been found by several investigators (2,8,13,17,35).

### References (Werner syndrome)

1. Bauer EA et al: Werner's syndrome. Evidence for preferential regional expression of a generalized mesenchymal cell defect. *Arch Dermatol* **124**:90–101, 1988.

1a. Björnberg A: Werner's syndrome and malignancy. *Acta Dermatovenereol* **56**:149–150, 1976.
2. Brown WT: Werner's syndrome, in *Chromosome Mutation and Neoplasia,* German J (ed), Alan R. Liss, New York, 1983, pp 85–93.
2a. Ceremele D et al: High prevalence of Werner's syndrome in Sardinia. Description of six patients and estimate of the gene frequency. *Hum Genet* **62**:25–30, 1982.
3. Darlington GJ et al: Sister chromatid exchange frequencies in progeria and Werner syndrome patients. *Am J Hum Genet* **33**:762–766, 1981.
4. DeGroot WP et al: Familial acrogeria (Gottron). *Br J Dermatol* **103**:213–223, 1980.
5. Djawari D et al: Altered cellular immunity in Werner's syndrome. *Dermatologica* **161**:233–237, 1980.
6. Elli R et al: Cytogenetic investigations on Werner's syndrome, acrogeria, and keratosis palmo-plantaris. *J Génét Hum* **31**:211–221, 1983.
7. Epstein CJ et al: Werner's syndrome. *Medicine* **45**:177–221, 1966.
8. Fleischmajer R, Nedwich A: Werner's syndrome. *Am J Med* **54**:111–118, 1973.
9. Frenkel G: Schleimhautatrophie unter besonderer Berücksichtigung des Werner-Syndroms. *Dtsch Zahnärztl Z* **25**:1026–1039, 1970.
10. Fujiwara Y et al: A retarded rate of DNA replication and normal level of DNA repair in Werner's syndrome: Fibroblasts in culture. *J Cell Physiol* **92**:365–374, 1977.
10a. Gaetani SA et al: Case report 485. *Skel Radiol* **17**:298–301, 1988.
11. German J: Patterns of neoplasia associated with the chromosome breakage syndromes, in *Chromosome Mutation and Neoplasia,* German J (ed), Alan R. Liss, New York, 1983, pp 97–134.
12. Gilkes JJH et al: The premature aging syndromes. Report of eight cases and description of a new entity named metageria. *Br J Dermatol* **91**:243–262, 1974.
13. Goto M, Murata K: Urinary excretion of macromolecular acidic glycosamino-glycans in Werner's syndrome. *Clin Chim Acta* **85**:101–106, 1978.
14. Goto M et al: Family analysis of Werner's syndrome: A survey of 42 Japanese families with a review of the literature. *Clin Genet* **19**:8–15, 1981.
15. Greither A: Über das Rothmund-und das Werner-Syndrom. *Arch Klin Exp Dermatol* **201**:431–445, 1955.
16. Hrabko RP et al: Werner's syndrome with associated malignant neoplasms. *Arch Dermatol* **118**:106–108, 1982.
16a. Kremer I et al: Corneal metastatic calcification in Werner's syndrome. *Am J Ophthalmol* **106**:221–226, 1988.
17. Mackawa Y, Hayashibara T: Determination of hyaluronic acid in the urine of a patient with Werner's syndrome. *J Dermatol* **8**:467–472, 1981.
18. Mulvihill JJ, Smith DW: Another disorder with premature shortness of stature and premature aging. *Birth Defects* **11**(2):368–371, 1975.
18a. Murata K, Nakashima H: Werner's syndrome: 24 cases with a review of the Japanese medical literature. *J Am Geriatr Soc* **30**:303–309, 1982.
19. Nakao Y et al: Werner's syndrome. In vivo and in vitro characteristics as a model of aging. *Am J Med* **65**:919–932, 1978.
20. Nakao Y et al: Immunological studies in Werner's syndrome. *Clin Exp Immunol* **43**:10–19, 1980.
21. Nienhaus AJ et al: Fibroblast culture in Werner's syndrome. *Humangenetik* **13**:244–246, 1971.
22. Norwood TH et al: Cellular aging in Werner's syndrome: A unique phenotype. *J Invest Dermatol* **73**:92–96, 1979.
23. Oppenheimer BS, Kugel VH: Werner's syndrome: Report of the first necropsy and of findings in a new case. *Am J Med Sci* **202**:629–642, 1941.
24. Petrohelos MA: Werner's syndrome. *Am J Ophthalmol* **56**:941–953, 1963.
25. Rabbiosi G, Borroni G: Werner's syndrome: Seven cases in one family. *Dermatologica* **158**:355–360, 1979.
26. Riley TR et al: Werner's syndrome. *Ann Intern Med* **63**:285–294, 1965.
27. Rosen RS et al: Werner's syndrome. *Br J Radiol* **43**:193–198, 1970.
28. Salk D: Werner's syndrome: A review of recent research with an analysis of connective tissue metabolism, growth control of cultured cells and chromosomal aberrations. *Hum Genet* **62**:1–16, 1982.
29. Salk D et al: Systemic growth studies, cocultivation and cell hybridization of Werner syndrome cultured skin fibroblasts. *Hum Genet* **58**:310–316, 1981.
30. Salk D et al: Cytogenetics of Werner's syndrome cultured skin fibroblasts; variegated translocation mosaicism. *Cytogenet Cell Genet* **30**:92–107, 1981.
31. Scappaticci S et al: Clonal structural chromosomal rearrangements in primary fibroblast cultures and in lymphocytes of patients with Werner's syndrome. *Hum Genet* **62**:16–24, 1982.
32. Schonberg S et al: Werner's syndrome: Proliferation in vitro of clones of cells bearing chromosome translocations. *Am J Hum Genet* **36**:387–397, 1984.
33. Tao LC et al: Werner's syndrome and acute myeloid leukemia. *Can Med Assoc J* **105**:951–954, 1971.
34. Thannhauser SJ: Werner's syndrome (progeria of the adult) and Rothmund's syndrome: Two types of closely related heredo-familial atrophic dermatosis with juvenile cataracts and endocrine features. *Ann Intern Med* **23**:559–626, 1945.
35. Tokunaga M et al: Werner's syndrome as "hyaluronuria." *Clin Chim Acta* **62**:89–96, 1975.
36. Usui M et al: The occurrence of soft tissue sarcomas in three siblings with Werner's syndrome. *Cancer* **54**:2580–2586, 1984.
37. Werner O: *Über Katarakt in Verbindung mit Sklerodermie.* Thesis, 1904. Kiel, Schmidt & Klaunig, Kiel, West Germany.
38. Zalla JA: Werner's syndrome. *Cutis* **25**:275–278, 1980.
39. Zucker-Franklin D et al: Werner's syndrome. An analysis of ten cases. *Geriatrics* **23**(8):123–135, 1968.

## Premature aging, microcephaly, unusual facies, multiple nevi, and mental retardation

Mulvihill and Smith (3) described a male with low birthweight, somatic and mild to moderate mental retardation, multiple pigmented nevi of the face, neck, and trunk, micropenis, and hypospadias (Fig. 13–105A–C). A similar example in a female was reported by Wong et al (5). However, intelligence was normal. Other examples have been described (1,2,4).

The craniofacies was characterized by microcephaly, broad forehead, fine scalp hair, sparse facial hair, disproportionately small face with reduced lower facial height, small pointed chin, and marked micrognathia. Facial subcutaneous fat was deficient. There was moderate progressive sensorineural hearing loss. The voice was high pitched and hoarse. The ears were lobeless. Oligodontia was a constant feature.

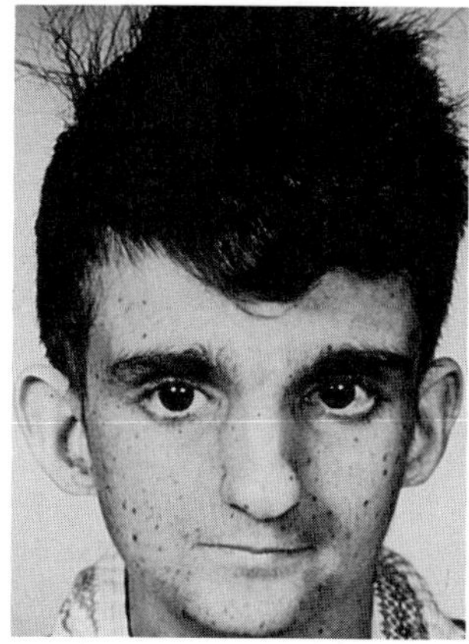
A

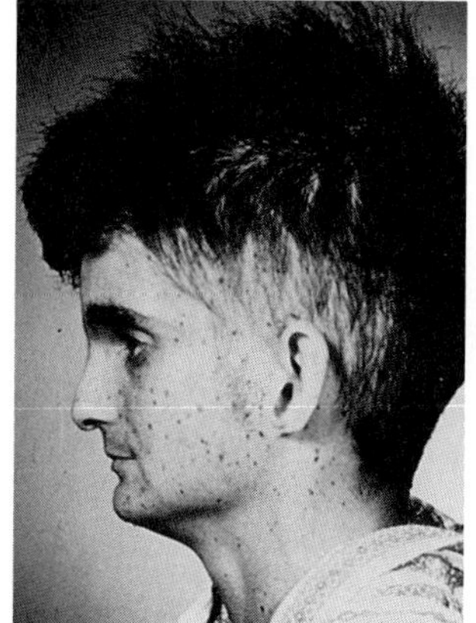
B

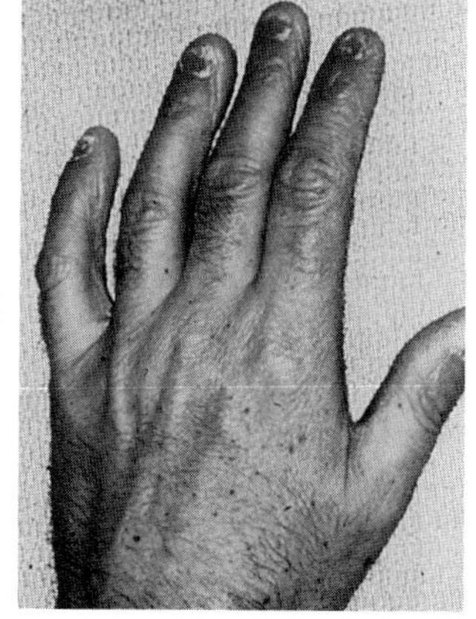
C

Fig. 13–105. *Premature aging, microcephaly, unusual facies, multiple nevi, and mental retardation.* (A,B) Seventeen-year-old male with microcephaly, bird-like facies, and multiple pigmented nevi. (C) Hand showing pigmented nevi, dysplastic fingernails. (Courtesy of *JT Mulvihill,* Bethesda, Maryland.)

### References (Premature aging, microcephaly, unusual facies, multiple nevi, and mental retardation)

1. Baraitser M et al: A recognizable short stature syndrome with premature aging and pigmented nevi. *J Med Genet* **25**:53–56, 1988.
2. Elliott DE: Undiagnosed syndrome of psycho-motor retardation, low birthweight, dwarfism, skeletal, dental, dermal and genital anomalies. *Birth Defects* **11**(2):364–367, 1975.
3. Mulvihill JJ, Smith DW: Another disorder with prenatal shortness of stature and premature aging. *Birth Defects* **11**(2):368–371, 1975. (Same patient as in Ref. 4.)
4. Shepard MK: An unidentified syndrome with abnormality of skin and hair. *Birth Defects* **7**(8):353–354, 1971.
5. Wong W et al: Case report for syndrome identification. *Cleft Palate J* **16**:286–290, 1979.


## Wiedemann–Rautenstrauch syndrome

In 1977, Rautenstrauch et al (3), believing that they were dealing with progeria in sibs, described a new progeroid syndrome that was recognized as such by Wiedemann (6). An aged facies is present from birth. Children are born small for dates and subsequent growth is slow. Frontal and biparietal bossing, together with small facial bones, gives the child a hydrocephalic appearance. The anterior fontanel is large and closes late. The sutures tend to persist. Scalp hair tends to be sparse but is of normal consistency. The scalp veins are prominent. The nose is small and beaked and the ears are low set and posteriorly angulated. The facial bones are hypoplastic. The mouth is small and the chin is somewhat prominent (4,5). Natal incisors are consistently present (4) (Fig. 13–106A,B).

Subcutaneous fat is generally produced allowing the veins and muscles to become prominent, but caudal fat accumulates during infancy. The abdomen is prominent. The hands and feet are large with long fingers and toes.

Mental and, especially, motor development are usually deficient. Horizontal nystagmus is noted with limited visual acuity (4). There is progressive neurologic degeneration by the age of 5 years. Lack of head control, hypotonia, severe truncal ataxia, dysmetria, and intension tremor are common. Martin et al (2) found extensive demyelinization of the central nervous system with an occasional tigroid pattern and large amounts of neutral fat and myelin breakdown in macrophages, a picture characteristic of pure sudanophilic leukodystrophy. Recurrent respiratory tract infections are common. Most of the children die prior to the age of 5 years.

Inheritance is clearly autosomal recessive (1,3). To date, study of collagen has not been fruitful (4).


### References (Wiedemann–Rautenstrauch syndrome)

1. Devos EA et al: Wiedemann–Rautenstrauch or neonatal progeroid syndrome: Report of a patient with consanguineous parents. *Eur J Pediatr* **136**:245–248, 1981.
2. Martin JJ et al: The Wiedemann–Rautenstrauch or neonatal progeroid syndrome—neuropathological study of a case. *Neuropediatrics* **15**:43–48, 1984.
3. Rautenstrauch T et al: Progeria: A cell culture study and clinical report of familial incidence. *Eur J Pediatr* **124**:101–111, 1977.
4. Rudin C et al: The neonatal pseudo-hydrocephalic progeroid syndrome (Wiedemann–Rautenstrauch). *Eur J Pediatr* **147**:433–438, 1988.
5. Snigula F, Rautenstrauch T: A new neonatal progeroid syndrome. *Eur J Pediatr* **136**:325, 1981.
6. Wiedemann H-R: An unidentified neonatal progeroid syndrome: Follow-up report. *Eur J Pediatr* **130**:65–70, 1979.


## Rothmund–Thomson syndrome

The syndrome of infantile poikiloderma, cataracts, and hypogonadism was described independently by Rothmund (35), in 1868, and Thomson (41), in 1923. Over 130 cases have been reported to date (28a). Especially good reviews of the syndrome are those of Taylor (40), Greither and Dyckerhoff (13), Rodermund and Hausmann (30), and Vanscheidt et al (43a).

The syndrome has autosomal recessive inheritance. A significant female excess (26,39,40) has been reported but was not supported by our survey. Consanguinity has been present in about 16% (2, 14,25,26,33,35,37). The syndrome has been found in identical twins (11). Report of the condition in mother and son (16) raises the question of genetic heterogeneity. However, this may represent pseudodominance, that is, a female homozygote marrying a carrier male and producing an affected child. There also appears to be a verrucous type with warty hyperkeratoses that has a tendency to malignant degeneration (10,12,31,33,45) (vide infra).

**Craniofacial features.** Several authors have noted a large head with frontal bossing and broad low nasal bridge (1a,30,38,41,42). However, microcephaly has also been mentioned (10,18,32,41,42). The eyebrows and lashes are sparse in at least 50% (Figs. 13–107, 13–109, and 13–110).

**Skin and skin appendages.** Dermatologic features have been extensively discussed by Rook et al (33). The skin of the cheeks and ears, uninvolved at birth, becomes red and swollen at about 3–6 months of age. The buttocks and the extensor surfaces of the hands, forearms, legs, and thighs then become involved but usually to a lesser degree, the exposed surfaces being more severely affected. The trunk is often spared. The inflammatory phase soon subsides and leaves areas that manifest varying combinations of pigmentation, depigmentation, atrophy, and telangiectasia. The dull-brown, irregular macular or reticular pigmentation usually follows the appearance of the atrophy and telangiectasia.

Sensitivity to sunlight in the form of blister production is seen in at least 33% (10,11,25,33,38,40). It is usually more severe in early life. Scalp, pubic, and axillary hair is often somewhat sparse, and may be almost absent in some patients (4.32). Eyebrows and lashes are fre-

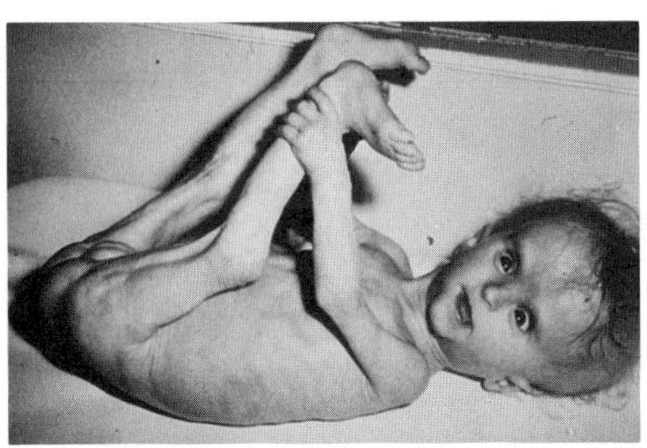
A

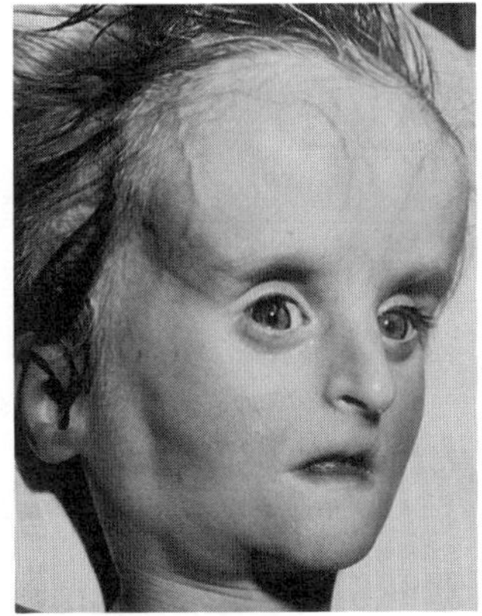
B

Fig. 13–106. *Wiedemann–Rautenstrauch syndrome.* (A) Frontal and biparietal bossing together with small facial bones causes 8-month-old child to appear hydrocephalic. Hair is sparse and scalp veins are prominent. The anterior fontanel is large. Nose is small and beaked, mouth is small. (B) Same child at 6 years. Note frontal bossing, pronounced vascular pattern on forehead, hydrocephalic appearance, small facial bones. (A from *T Rautenstraunch et al,* Eur J Pediatr **124**:101, 1977; B courtesy of *T Rautenstrauch,* München, West Germany.)

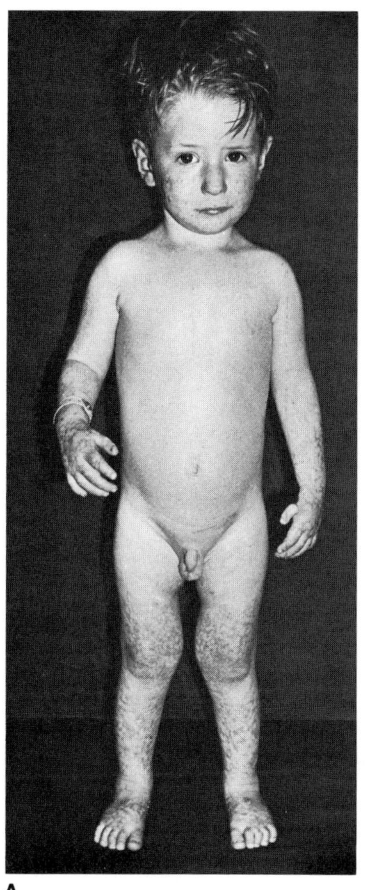

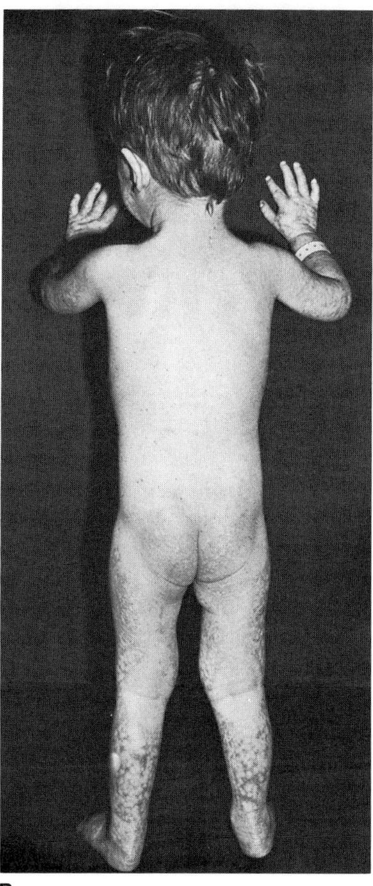

A B

Fig. 13–107. *Rothmund–Thomson syndrome.* (A,B) Skin of cheeks and ears, uninvolved at birth, assumes a poikilodermatous appearance about the third to sixth month of life. There is similar involvement of legs, thighs, and buttocks.

quently missing or severely diminished (2,11,19,25,26,32,38,40–42) (Figs. 13–107, 13–109, and 13–110).

Warty hyperkeratoses in 25%, especially over the joints (17,26, 32,37a,41,42,45), appear in 25% between 6 and 10 years. Late developing squamous cell carcinoma and palmoplantar keratoses have been found (6,10,16,24a,26,32,37,44). This may represent a separate entity, which we will call verrucous Rothmund–Thomson syndrome. Nail dystrophy has been noted in at least 25% (2,19,25,26,38,40).

**Endocrine system.** An endocrine disorder, most frequently hypogonadism, occurs in about 25% (2,19,26,32,36,38). Scanty menstruation is common, and few affected women have borne children. Micropenis has been noted (38).

**Musculoskeletal system.** At least 65% have been markedly short. The growth retardation has been proportionate (2,5,10,15,16,19, 32,35,36,38). The limbs are often slender or delicate, the terminal phalanges abbreviated, and acrocyanosis may be severe. Soft tissue contractures may be an important cause of disability in the older patient (15).

Absence of thumbs and occasionally rudimentary ulnae and radii have been reported (1a,23,28,33,37a,41,42,43a) (Fig. 13–108). Jäckli

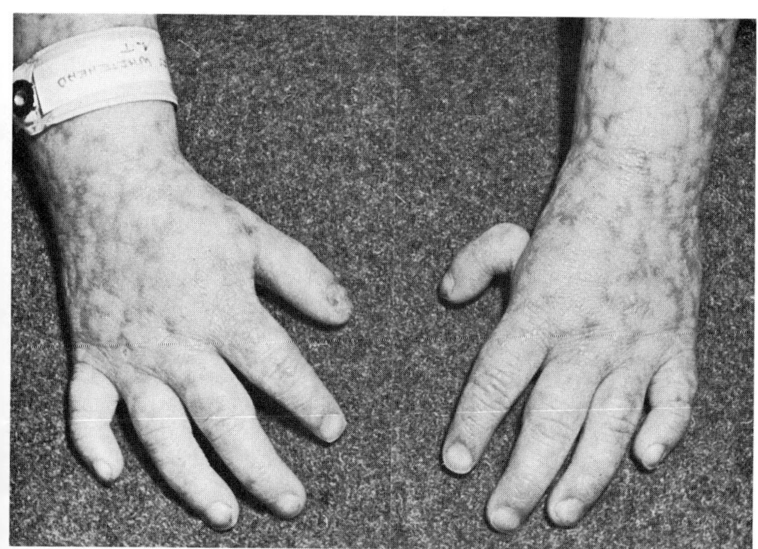

Fig. 13–108. *Rothmund–Thomson syndrome.* Similar involvement of hands with hypoplasia of thumbs. (From *RK Oates et al,* Aust Paediatr J **7**:103, 1971.)

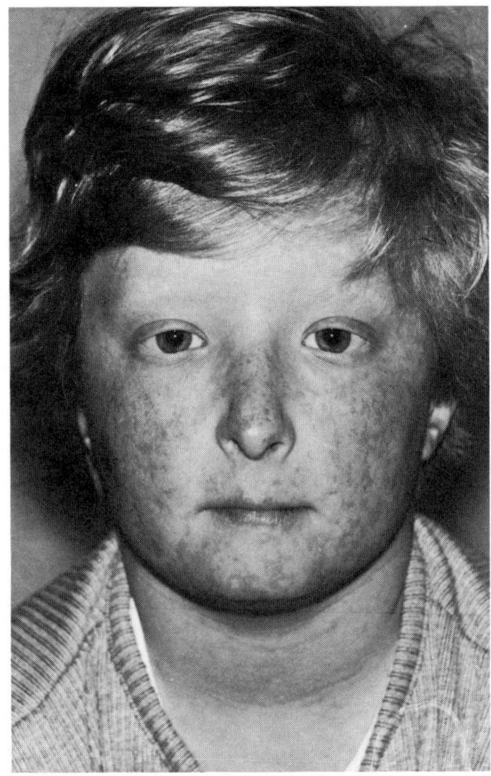

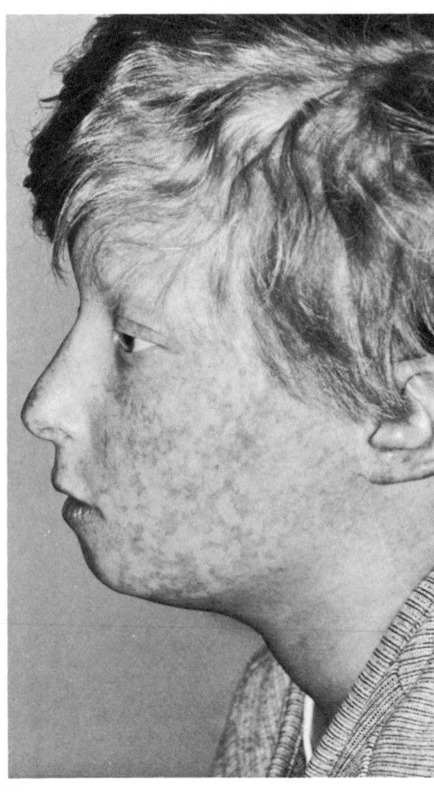

A B

Fig. 13–109. *Rothmund–Thomson syndrome.* (A,B) Similar facial involvement in another patient. (Courtesy of *DW Smith,* Seattle, Washington.)

(19) described bipartite patella and bone sclerosis. Several authors (5,14,23,35,40,42) noted cystic areas similar to fibrous dysplasia; others have reported osteogenesis imperfecta (29,34). Phalangeal tufts may be resorbed (27).

**Eyes.** Anterior subcapsular, perinuclear, and posterior stellate cataracts have been described in 50–75% (2,22,25,26,35,38,40). The cataracts are bilateral and usually appear between the fourth and seventh years (26), although they may appear earlier (38) (Fig. 13–111). They are complete and semisolid and produce loss of vision within weeks.

Various eye anomalies less frequently encountered include band keratopathy, microcornea, and strabismus (24).

**Neoplasia.** Osteosarcoma (1,8,23,28a,34,39,43), fibrosarcoma (6), and late developing squamous cell carcinoma (vide supra) have been described.

**Other findings.** Gastric carcinoma (8) and hypertension have been noted (7). Low vitamin A levels have been reported by Sexton (37) and Whittle (45). Taylor (40) found an abnormal peak in the $\alpha_1$-globulin fraction of the blood.

**Oral manifestations.** Microdontia, multiple crown malformations, delayed and ectopic eruption, supernumerary and congentially missing teeth, and short conical roots have been mentioned (3,19,26). Bifid uvula has also been noted (18,22).

**Differential diagnosis.** A somewhat similar syndrome in adults is *Werner syndrome* consisting of short stature, premature graying of hair, scleropoikiloderma, trophic ulcers of legs, arteriosclerosis, juvenile cataracts, hoarse and high-pitched voice, hypogonadism, osteoporosis, and diabetic tendency. It also has autosomal recessive inheritance. In contrast to Rothmund–Thomson syndrome (Table 13–5), patients with Werner syndrome are essentially normal until 20–30 years of age, when the hair becomes gray. Skin changes and cataracts develop after the hair changes color. Short stature and atrophy of muscle and subcutaneous fat of the distal extremities are more pronounced in Werner syndrome. Arteriosclerosis, diabetic tendency, and osteoporosis are not seen in Rothmund–Thomson syndrome. Skin involvement in Werner syndrome is different from that in Rothmund–Thomson syndrome. In the former, the forearms, hands, and face are chiefly involved. Hyperkeratotic areas are present on the soles and ulcers are present over the heels, toes, and ankles.

In *gerodermia osteodysplastica* there is premature senescence of the skin, nanism, and various anomalies of the eye (including microcornea and congenital corneal opacities), bony anomalies, and yellowish teeth, presumably due to an enamel defect. *Cockayne syndrome* consists of primordial short stature, flexion deformities, kyphosis, hyperostosis of skull bones, sensitivity to sunlight, and various eye anomalies such as optic atrophy, retinal degeneration, and cataract. The child appears prematurely senile, with sunken eyes, prognathic mandible, and carious teeth. *Bloom syndrome* is characterized by dwarfism, sunlight sensitivity, chromosomal breakage, a tendency to develop leukemia, and autosomal recessive inheritance. There is some superficial resemblance of Rothmund–Thomson syndrome to *focal*

Table 13–5. Clinical features of Rothmund–Thomson syndrome[a]

| Feature | Percentage |
|---|---|
| Poikiloderma | 100 |
| Short stature | 100 |
| Musculoskeletal abnormalities | 95 |
| Hypogonadism | 94 |
| Thin hair | 65 |
| Abnormal teeth | 59 |
| Cataracts | 45 |
| Dysplastic nails | 42 |

[a] Adapted from DG Starr et al, *Clin Genet* **27**:102–104, 1985.

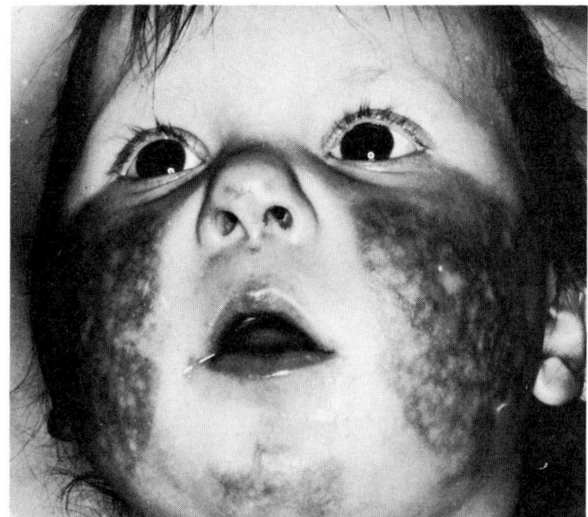

A

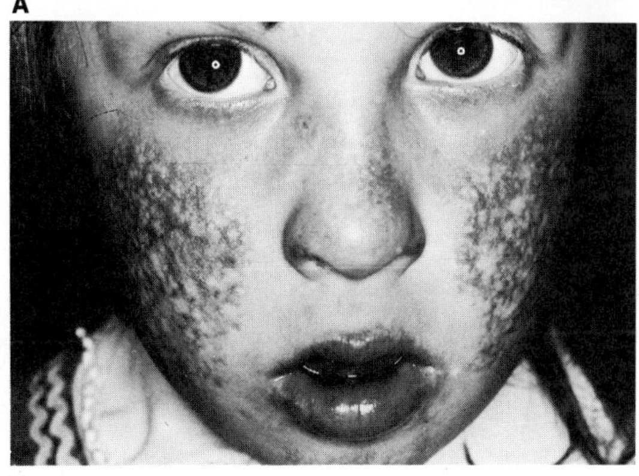

B

Fig. 13–110. *Rothmund–Thomson syndrome.* (A,B) Affected 2- and 4-year-old sibs showing characteristic facial poikiloderma. (Courtesy of *S Alexander,* Barking, Essex, England.)

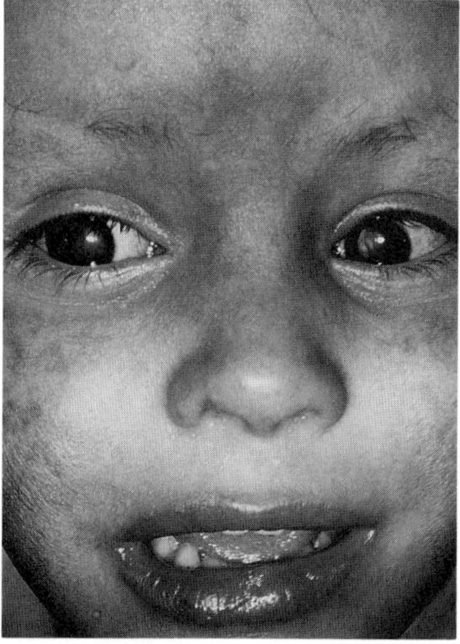

Fig. 13–111. *Rothmund–Thomson syndrome.* Bilateral cataracts. (Courtesy of *GB Sexton,* London, Ontario, Canada.)

*dermal hyperplasia,* a disorder that also has poikilodermatous changes and for which it has been mistaken (20).

The reader is referred to the comprehensive analysis by Kaufmann et al (21) of disorders associated with growth hormone deficiency.

**Laboratory aids.** Growth hormone deficiency has been reported (21).

### References (Rothmund–Thomson syndrome)


1. Baró PR et al: Case report 529. *Skel Radiol* **18**:136–139, 1989.

1a. Blinstrub RS et al: Poikiloderma congenitale: Report of two cases. *Arch Dermatol* **89**:659–664, 1964.

2. Bloch B, Stauffer H: Skin diseases of endocrine origin (dyshormonal dermatosis). *Arch Dermatol Syph* **19**:22–34, 1929.

3. Bottomley WK, Box JM: Dental anomalies in the Rothmund–Thomson syndrome. *Oral Surg* **41**:321–326, 1976.

4. Braun W, Unger C: Zur Frage des Rothmund–Thomson-Syndroms. *Dermatol Wochenschr* **151**:1189–1198, 1965.

5. Castel Y et al: Une observation de poikilodermie congénitale (syndrome de Rothmund–Thomson) avec dystrophies majeures, nanisme tres severe et intolerance digestive. *Ann Pédiatr* **24**:647–651, 1977.

6. Davies MG: Rothmund–Thomson syndrome and malignant disease. *Clin Exp Dermatol* **7**:455–457, 1982.

7. Dechenne CH et al: A Rothmund–Thomson case with hypertension. *Clin Genet* **24**:266–272, 1983.

8. Dick DC et al: Rothmund–Thomson case and osteogenic sarcoma. *Clin Exper Dermatol* **7**:119–123, 1982.

9. Diem E: Rothmund–Thomson-Syndrom. Kasuistischer Beitrag. *Hautarzt* **26**:425–429, 1975.

10. Dowling CB: Congenital developmental malformations. *Br J Dermatol* **48**:645–647, 1936.

11. Feldreich H: Poikiloderma congenitale in twins. *Acta Derm Venereol* **35**:86–87, 1955.

12. Greither A: Über eine mit Keratosen und Pigmentstörung einhergehende erbliche Dysplasie der Haut. *Hautarzt* **9**:364–369, 1958.

13. Greither A, Dyckerhoff D; Über das Rothmund- und das Werner-Syndrom. *Arch Klin Exp Dermatol* **201**:411–445, 1955.

14. Haberman P, Fleck M: Über das Rothmund-Syndrom. *Z Kinderheilkd* **77**:306–321, 1955.

15. Hall JG et al: Rothmund–Thomson syndrome with severe dwarfism. *Am J Dis Child* **134**:165–169, 1980.

16. Hallman N, Pätiälä R: Congenital poikiloderma atrophicans vasculare in a mother and her son. *Acta Derm Venereol* **31**:401–406, 1951.

17. Haneke E, Gutschmidt E: Warty hyperkeratoses, cellular immune defect, and tapetoretinal degeneration in poikilodermia congenitale. *Dermatologica* **52**:331–336, 1976.

18. Heuyer G et al: Nanisme avec infantilisme, microcephalie, malformations osseuse et cutanées du type de nanisme senile ou progeria chez deux freres. *Bull Soc Pédiatr* **34**:159–170, 1936.

19. Jäckli W: Ein Fall von infantiler Poikilodermie (Atrophodermia reticularis cum Incontinentia pigmenti kombiniert mit Alopecie, Mikrodontie und frühzeitiger Cataracta complicata). *Monatsschr Kinderheilkd* **78**:73–81, 1939.

20. Kassner EG et al: Rothmund–Thomson syndrome (poikiloderma congenitale) associated with mental retardation, growth disturbance and skeletal features. *Skel Radiol* **2**:99–103, 1977.

21. Kaufmann S et al: Growth hormone deficiency in the Rothmund–Thomson syndrome. *Am J Med Genet* **23**:861–868, 1986.

22. Kirkham TH, Werner EB: The ophthalmic manifestations of Rothmund's syndrome. *Can J Ophthalmol* **10**:1–14, 1975.

23. Kozlowski K et al: Osteosarcoma in a boy with Rothmund–Thomson syndrome. *Pediatr Radiol* **10**:42–45, 1980.

24. Kraus BS et al: The dentition in Rothmund's syndrome. *J Am Dent Assoc* **81**:894–915, 1970.

24a. Lindlbauer R et al: Syndrome de Rothmund–Thomson. *Ann Dermatol Venereol* **107**:1045–1049, 1980.

25. Lutz W: Poikiloderma atrophicans. *Schweiz Med Wochenschr* **45**:1118–1119, 1928.

26. Maeder G: Le syndrome de Rothmund et le syndrome de Werner. *Ann Ocul (Paris)* **182**:809–854, 1949.

27. Maurer RM, Langford OL: Rothmund's syndrome: A cause of resorption of phalangeal tufts and dystrophic calcification. *Radiology* **89**:706–708, 1967.

28. Oates RK et al: The Rothmund–Thomson syndrome. *Aust Paediatr J* **7**:103–107, 1971. (Same case as Ref. 23.)

28a. Rebaud P et al: Observation familiale de syndrome de Rothmund–Thomson. Observations en faveur de l'unicite du syndrome. Association à un ostéosarcome. *Pédiatrie* **40**:487–492, 1985.

29. Reid J: Congenital poikiloderma with osteogenesis imperfecta. *Br J Dermatol* **79**:243–244, 1967.

30. Rodermund OH, Hausmann D: Das Rothmund-Syndrom-Ein Beitrag zu den kongenitalen Poikilodermien. *Z Hautkr* **52**:129–144, 1977.

31. Rodermund OE, Hausmann D: Typus verrucosus des Thomson-Syndroms. Ein Beitrag zu den kongenitalem Poikilodermien. *Hautarzt* **28**:308–313, 1977.

32. Rook A, Whimster I: Congenital cutaneous dystrophy (Thomson's type). *Br J Dermatol* **61**:197–205, 1949.

33. Rook A et al: Poikiloderma congenitale: Rothmund–Thomson syndrome. *Acta Dermatol Venereol* **39**:392–420, 1959.

34. Roschlau G: Rothmund-Syndrom, kombiniert mit Osteogenesis imperfecta tarda und Sarcom des Oberschenkels. *Z Kinderheilkd* **86**:289–298, 1962.

35. Rothmund A: Über Cataracten in Verbindung mit einer eigentümlichen Hautdegeneration. *Albrecht von Graefes Arch Klin Ophthalmol* **14**:159–182, 1868.

36. Schneider WF: Über Katarakt im Kindesalter bei gleichzeitigem Vorkommen von Poikilodermia atrophicans. *Schweiz Med Wochenstr* **65**:719–721, 1935.

37. Sexton G: Thomson's syndrome. *Can Med Assoc J* **70**:662–665, 1954.

37a. Shuttleworth D, Marks R: Epidermal dysplasia and skin deformity in congenital poikiloderma (Rothmund–Thomson syndrome). *Br J Dermatol* **117**:377–384, 1987.

38. Silver HK: Rothmund–Thomson syndrome: An oculocutaneous disorder. *Am J Dis Child* **111**:182–190, 1966.

39. Starr DG et al: Non-dermatological complications and genetic aspects of the Rothmund–Thomson syndrome. *Clin Genet* **27**:102–104, 1985.

40. Taylor WB: Rothmund's syndrome-Thomson's syndrome. *Arch Dermatol* **75**:236–244, 1957.

41. Thomson MS: An hitherto undescribed familial disease. *Br J Dermatol* **35**:455–461, 1923.

42. Thomson MS: Poikiloderma congenitale. *Br J Dermatol* **48**:221–234, 1936.

43. Tokunaga M et al: Rothmund–Thomson syndrome associated with osteosarcoma. *J Jpn Orthop Assoc* **50**:287–293, 1976.

43a. Vanscheidt E et al: Rothmund-Syndrom oder Thomson-Syndrom. *Mschr Kinderheilkd* **136**:264–269, 1988.

44. Werder EA et al: Hypogonadism and parathyroid adenoma in congenital poikiloderma (Rothmund–Thomson syndrome). *Clin Endocrinol* **4**:75–82, 1975.

45. Whittle CH: Poikiloderma congenitale (Thomson). *Proc R Soc Med* **40**:499–500, 1947.


## Cockayne syndrome

Cockayne syndrome (CS) is characterized by cachectic dwarfism, premature aging, mental deficiency, microcephaly, intracranial calcifications, neurologic deficits, retinal pigmentary abnormalities, sensorineural hearing loss, and photosensitivity. It manifests defective repair of damage induced in DNA by ultraviolet light.

Cockayne (5), in 1936, described proportionate dwarfism with retinal atrophy and deafness in two sibs, and 10 years later published a brief review of their progress (6). The syndrome becomes apparent at around the second year of life, but pre- and perinatal onset has been observed (15,23,28,37,42,43). Based on this difference of onset, it has been proposed that classical examples of CS, with onset at age 2 years or later, be classified as CS Type I and the prenatal or neonatal cases be CS Type II (22). The latter has been called CAMFAK syndrome (43). Patton et al (31) dispute this belief. The disorder has been reported in a female patient with normal intelligence and onset at age 19 years, with worsening at age 23 during a second pregnancy (16).

The disorder has autosomal recessive inheritance (5,6,11,15,23,30–32). Familial cases as well as consanguinity among parents of affected children have been observed. Over 70 cases have been reported (30). Complementation studies indicate genetic heterogeneity (19).

**Facies.** Lack of subcutaneous facial fat, particularly of the cheeks, gives prominence to the facial bones. This feature, combined with microcephaly, sunken eyes, thin, often beaklike nose, and large ears, gives the patient a birdlike appearance (Fig. 13–112).

**Musculoskeletal alterations.** Proportionate dwarfism is the most prominent feature of the disorder. Growth retardation becomes evident during the second year of life after a normal gestation, birthweight, and infancy. Kyphosis and osteoporosis are frequent (11,24,31,33,45).

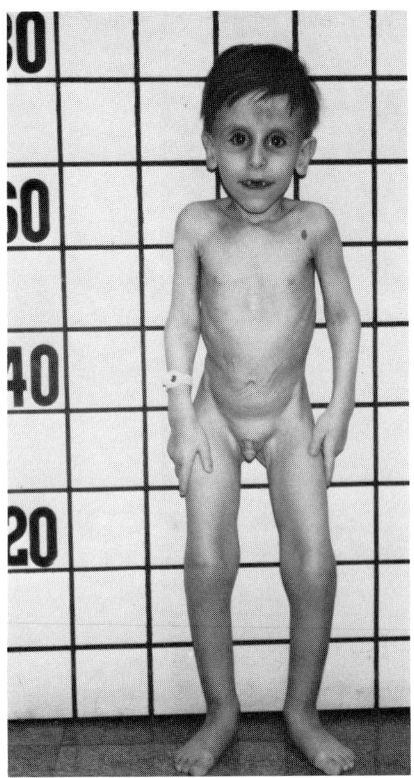

Fig. 13–112. *Cockayne syndrome*. Note marked enophthalmos and horse-riding stance. (Courtesy of *RL Summitt*, Nashville, Tennessee.)

The limbs are disproportionately long, and the hands and feet disproportionately large (21,37). Flexion contractures may involve the ankles, knees, and elbows. The interphalangeal joints of the hands and feet may show periarticular thickening (Fig. 13–113).

Roentgenographic studies show abnormalities in the skull, extremities, and spine. There is increased thickening of bones, particularly the skull base and calvaria, most noticeable in the frontal and parietooccipital regions (15,24,33,37). Often there is associated osteoporosis (33). Platyspondyly with tongue-like protrusion of the anterior aspects of the vertebral bodies, hypoplasia of the iliac wings and acetabular roofs, brachydactyly and marble epiphyses in the terminal phalanges of the hands have been described as characteristic (1,39).

**Skin.** Photosensitivity is a prominent feature (15). Dermatitis appears on the sun-exposed parts of the body by the second year of life, with a butterfly arrangement on the face (21,45). The forehead is spared, but the pinnae and chin are involved. The photodermatitis may result in scarring and pigmentary changes in older patients (24). Seborrheic dermatitis may also be present. The scalp hair and, sometimes, the eyebrows are diminished (11). Subcutaneous fat appears to be decreased throughout the body except for the suprapubic areas (11). Microdissection studies of the eccrine sweat glands have demonstrated these glands to be abnormally small for the patient's age (17).

**Eyes.** Enophthalmos is a virtually constant feature. The retina is studded with fine, speckled pigment of salt-and-pepper type, with the greatest concentration in the macular area. Optic atrophy and arteriolar narrowing are common (4,6,15,24,25,45). Cataracts develop by adolescence (6,7,12,15). A poor response to mydriasis with homatropine or Neosynephrine has been noted (7). Corneal dystrophy with recurrent epithelial erosions, nystagmus, and photophobia are less frequently observed (3,25). Histopathologic examination of the eyes demonstrated optic nerve atrophy, retinal pigmentation, loss of nerve fibers and myelin sheaths, and atrophy of retinal nerve fibers and ganglion cell layers, findings consistent with those observed in demyelinating disease (20).

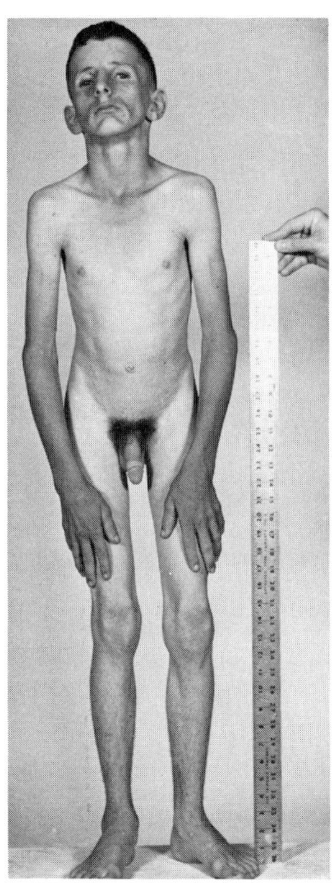

Fig. 13–113. *Cockayne syndrome.* Sixteen-year-old male who experienced marked mental and somatic retardation, eye difficulties, deafness. Note wizened appearance, photodermatitis of sun-exposed areas, horse-riding stance. (Courtesy of *RM Paddison,* New Orleans, Louisiana.)

**Nervous system.** Progressive neurologic signs include cerebellar ataxia, choreoathetosis, moderate to severe mental deficiency, sensorineural deafness, and blindness. (38). Neuropathologic studies of the brain have shown microcephaly, wide-spread mineralization in the cortex, basal ganglia, and cerebellum, and patchy demyelinization, often severe, in the subcortical white matter (10,15,26,35). Biopsy of the sural nerve showed the main pathologic change to be segmental demyelination and remyelination with onion-bulb formation and moderate decrease in the number of myelinated fibers (29). Normal intelligence has been reported in one patient (18). CAT scans show strio-pallido-dentate calcinosis (Fig. 13–114).

**Other findings.** Hypertension and renal disease are frequent complications. Elevated peripheral vein renin and deposits of immunoglobulins and complement in the kidney vessels and glomeruli have been reported (13). Undescended testes (12), small breasts, and oligomenorrhea (16) have also been noted.

**Oral manifestations.** An increase in dental caries has been reported by most authors (6,12,21,25,35). In some patients, numerous permanent teeth were congenitally absent (11,35–37). Atrophy of alveolar processes (11,37), condylar hypoplasia (9), and short conical roots (36) have also been observed.

**Differential diagnosis.** CS is one of a group of disorders that has certain common characteristics. The photosensitivity may suggest *Bloom syndrome,* Hartnup disease, *xeroderma pigmentosum,* or erythropoietic porphyria. Bloom syndrome has many of the features of CS except for the neurologic and ophthalmologic findings. Both have reduced IgA levels (5). No patient with CS has been reported with leukemia or lymphoma, and chromosomal breakage is absent (31,45). Patton et al (31) have suggested that Cockayne syndrome overlaps strongly with *cerebro-oculo-facial-skeletal syndrome. Rothmund–Thomson syndrome* is not a neurologic disease; although cataracts are seen in about half the patients, there are no retinal changes. *Xeroderma pigmentosum* is characterized by multiple cutaneous malignancies in light-exposed areas. Skin cancers have not been reported thus far in the photosensitive areas in patients with CS. Cases of CS with xeroderma pigmentosum have been described (27,34). *Progeria* does not feature photosensitivity, neurologic disease, or retinal changes, and the phenotype is quite different from that of CS.

**Laboratory aids.** Cells from patients with CS demonstrate defective repair of damage due to ultraviolet (UV) radiation and mitomycin C, but they have normal unscheduled DNA synthesis and repair normally in the presence of caffeine (caffeine interferes with normal postreplication repair). This suggests a defect in either replication or excision repair and not in postreplication (14,32,32a,35a). A most complete discussion of UV-induced damage is that of Seguin et al (37a). The degree of hypersensitivity to UV radiation varies from patient to patient, possibly reflecting clinical and cellular heterogeneity for CS (31,42). Studies of complementation groups in patients with CS have shown patients to belong to groups A, B, and C (19) (see the following section on xeroderma pigmentosum). UV treatment of fibroblasts obtained from obligate heterozygotes has yielded conflicting results (32,44). Thus, no reliable method to identify heterozygotes presently exists (30).

Thymic hormone serum level has been found to be undetectable or

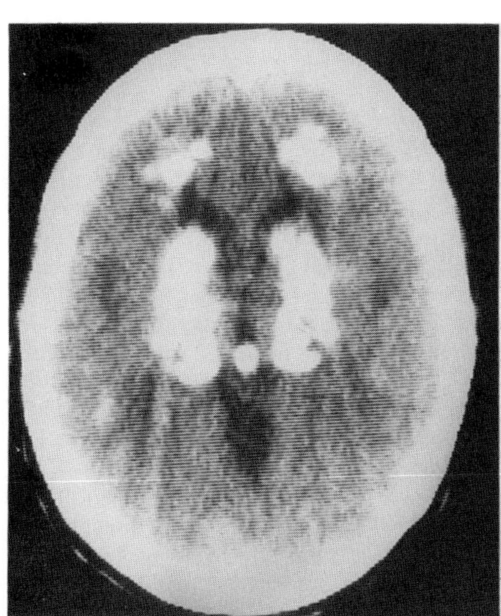

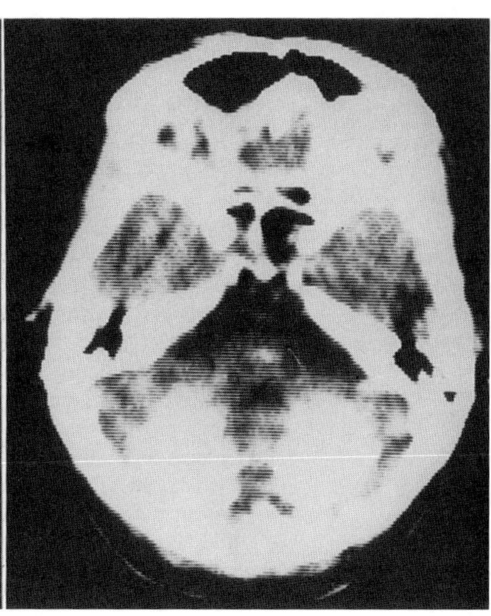

Fig. 13–114. *Cockayne syndrome.* CAT scan showing calcification of basal ganglia and dentate nuclei. (From *MG Smits,* Clin Neurol Neurosurg **85**:145, 1983.)

significantly reduced in patients with CS, whereas T lymphocyte and mixed lymphocyte functions are normal (2). These findings indicate premature immunological aging.

Prenatal diagnosis is possible using an assay of the colony-forming ability of UV irradiated cells, obtained from the amniotic fluid (41). CS cells are significantly more sensitive to UV radiation than normal control cells.

Hyperinsulinemia and hyperlipoproteinemia but normal growth hormone levels have been noted in several patients (8,9,11).

Autopsy findings have included pancreatic islet hyperplasia, atrophy and fibrosis of ovaries, and atrophic epidermis with small hair follicles and sweat glands (40).

Prenatal diagnosis has been effected by Lehmann et al (19a). Cultured amniocytes exhibit significantly reduced RNA synthesis after irradiation with UV lights.

### References (Cockayne syndrome)


1. Bensman A et al: The spectrum of x-ray manifestations in Cockayne's syndrome. *Skel Radiol* **7**:173–177, 1981.
2. Bensman A et al: Decrease of thymic hormone serum level in Cockayne syndrome. *Pediatr Res* **16**:92–94, 1982.
3. Brodrick JD, Dark AJ: Corneal dystrophy in Cockayne's syndrome. *Br J Ophthalmol* **57**:391–399, 1973.
4. Civantos F: Human chromosomal abnormalities. *Bull Tulane Med Fac* **20**:241–253, 1961.
5. Cockayne EA: Dwarfism with retinal atrophy and deafness. *Arch Dis Childh* **11**:1–8, 1936.
6. Cockayne EA: Dwarfism with retinal atrophy and deafness. *Arch Dis Childh* **21**:52–54, 1946.
7. Coles WH: Ocular manifestations of Cockayne's syndrome. *Am J Ophthalmol* **67**:762–764, 1969.
8. Conly PW et al: Hyperinsulinemia in Cockayne's syndrome. *South Med J* **63**:1491–1495, 1970.
9. Cotton RB et al: Abnormal blood glucose regulation in Cockayne's syndrome. *Pediatrics* **46**:54–60, 1970.
10. Crome L, Kanjilal GC: Cockayne's syndrome. *J Neurol Neurosurg Psychiatr* **34**:171–178, 1971.
11. Fujimoto WY et al: Cockayne's syndrome, report of a case with hyperlipoproteinemia, hyperinsulinemia, renal disease and normal growth hormone. *J Pediatr* **75**:881–884, 1969.
12. Guzetta F: La sindrome di Cockayne. *Minerva Pediatr* **19**:891–895, 1967.
13. Higginbottom MC et al: The Cockayne syndrome: An evaluation of hypertension and studies of renal pathology. *Pediatrics* **64**:929–934, 1979.
14. Hoar DI, Waghorne C: DNA repair in Cockayne syndrome. *Am J Hum Genet* **30**:590–601, 1978.
15. Houston CS et al: Identical male twins and brother with Cockayne syndrome. *Am J Med Genet* **13**:211–223, 1982.
16. Kennedy RM et al: Cockayne syndrome, an atypical case. *Neurology* **30**:1268–1272, 1980.
17. Landing BH et al: Eccrine sweat gland anatomy in Cockayne syndrome. *Pediatr Pathol* **1**:349–353, 1983.
18. Lanning M, Simila S: Cockayne's syndrome: Report of a case with normal intelligence. *Z Kinderheilkd* **109**:70–75, 1970.
19. Lehmann AR: Three complementation groups in Cockayne syndrome. *Mutat Res* **106**:347–356, 1982.

19a. Lehmann AR et al: Prenatal diagnosis of Cockayne's syndrome. *Lancet* **1**:486–488, 1985.

20. Levin PS et al: Histopathology of the eye in Cockayne's syndrome. *Arch Ophthalmol* **101**:1093–1097, 1983.
21. Lieberman WJ et al: Cockayne's disease. *Am J Ophthalmol* **52**:116–118, 1961.
22. Lowry RB: Early onset of Cockayne syndrome. *Am J Med Genet* **13**:209–210, 1982.
23. Lowry RB et al: Cataracts, microcephaly, kyphosis and limited joint movements in two siblings: A new syndrome. *J Pediatr* **79**:282–284, 1971.
24. Macdonald WB et al: Cockayne's syndrome. *Pediatrics* **25**:997–1007, 1960.
25. Marie J et al: Nanisme avec surdi-mutité et rétinité pigmentaire: Syndrome de Cockayne. *Arch Fr Pédiatr* **15**:1101–1103, 1958.
26. Moosa A, Dubowitz V: Peripheral neuropathy in Cockayne's syndrome. *Arch Dis Childh* **45**:674–677, 1970.
27. Moshel AN et al: A new patient with both xeroderma pigmentosum and Cockayne syndrome establishes the new xeroderma pigmentosum complementation group H, in *Cellular Responses to DNA Damage,* Friedeberg E, Bridges B (eds), Alan R. Liss, New York, 1983, pp 209–213.
28. Moyer DB et al: Cockayne syndrome with early onset of manifestations. *Am J Med Genet* **13**:225–230, 1982.
29. Ohnishi A et al: Primary segmental demyelination in the sural nerve in Cockayne's syndrome. *Muscle Nerve* **10**:163–167, 1987.
30. Otsuka F, Robbins JG: The Cockayne syndrome. An inherited multisystem disorder with cutaneous photosensitivity and defective repair of DNA. *Am J Dermatopathol* **7**:387–392, 1985.
31. Patton MA et al: Early onset Cockayne's syndrome: Case reports with neuropathological and fibroblast studies. *J Med Genet* **26**:154–159, 1989.
32. Proops R et al: A clinical study of a family with Cockayne's syndrome. *J Med Genet* **18**:288–293, 1981.

32a. Rainbow AJ, Howes MA: A deficiency in the repair of UV and gamma-ray damaged DNA in fibroblasts from Cockayne's syndrome. *Mutat Res* **93**:235–247, 1982.

33. Riggs W Jr, Seibert J: Cockayne's syndrome. Roentgen findings. *Am J Roentgenol* **116**:623–633, 1972.
34. Robbins JH et al: Xeroderma pigmentosum: An inherited disease with sun sensitivity, multiple cutaneous neoplasms, and abnormal DNA repair. *Ann Intern Med* **80**:221–248, 1974.
35. Rowlatt U: Cockayne's syndrome: Report of a case with necropsy findings. *Acta Neuropathol* **14**:52–61, 1969.

35a. Schmickel RD et al: Cockayne syndrome: A cellular sensitivity to ultraviolet light. *Pediatrics* **60**:135–139, 1977.

36. Schneider PE: Dental findings in a child with Cockayne's syndrome. *J Dent Child* **50**:58–64, 1983.
37. Scott-Emuakpor AB et al: A syndrome of microcephaly and cataract in four siblings: A new genetic syndrome. *Am J Dis Child* **131**:167–169, 1977.

37a. Seguin LR et al: Ultraviolet light-induced chromosomal aberrations in culture cells from Cockayne syndrome and complementation group C xeroderma pigmentosum patients: Lack of correlation with cancer susceptibility. *Am J Hum Genet:* **42**:468–475, 1988.

38. Shemen LJ et al: Cockayne syndrome, an audiologic and temporal bone analysis. *Am J Otolaryngol* **5**:300–307, 1984.
39. Silengo MC et al: Distinctive skeletal dysplasia in Cockayne syndrome. *Pediatr Radiol* **16**:264–266, 1986.
40. Sugarman GI et al: Cockayne syndrome: Clinical study of two patients and neuropathologic findings in one. *Clin Pediatr* **16**:225–232, 1977.
41. Sugita T et al: Prenatal diagnosis of Cockayne syndrome using assay of colony-forming ability in ultraviolet light irradiated cells. *Clin Genet* **22**:137–142, 1982.
42. Sugita K et al: Cockayne syndrome with delayed recovery of RNA synthesis after ultraviolet irradiation but normal ultraviolet survival. *Pediatr Res* **21**:34–37, 1987.
43. Talwar D, Smith SA: CAMFAK syndrome: A demyelinating inherited disease similar to Cockayne syndrome. *Am J Med Genet* **34**:194–198, 1989.
44. Wade MH, Chu EHY: Effects of DNA damaging agents on cultured fibroblasts derived from patients with Cockayne syndrome. *Mut Res* **59**:49–60, 1979.
45. Windmiller J et al: Cockayne syndrome with chromosomal analysis. *Am J Dis Child* **105**:204–208, 1963.


## Xeroderma pigmentosum

Xeroderma pigmentosum (XP), originally described by Hebra and Kaposi (25) in 1874 and named by Kaposi (34) in 1882, is characterized by defective excision repair of DNA damaged by ultraviolet light (UV). When exposed to sunlight, the UV radiation induces thymine dimers and other damage in the DNA of human cells (48–57). Normal individuals have several mechanisms for repairing damaged DNA. These mechanisms are absent to various degrees in patients affected with XP. Not only UV radiation but certain chemicals, carcinogens, and ionizing radiation can induce severe damage to patients with XP (52). Based upon the cellular ability to repair DNA after UV irradiation of cultured fibroblasts (see Laboratory Aids), XP has been divided into nine subgroups, all having autosomal recessive inheritance. However, *vide infra.* These subgroups have been classified as complementation groups A through I and the XP "variant." Groups A through I are defective in excision repair of DNA following UV damage. The XP "variant" has normal excision repair but has abnormal replication of UV-damaged DNA (7,43). The XP "variant" in Japan has been subdivided into caffeine-resistant and caffeine-susceptible subgroups based on the capability of XP cells to achieve caffeine potentiation of UV lethality. Patients with caffeine-resistant cells did not develop neoplasms (28). An excellent recent survey is that of Cohen and Levy (8a).

Division into the various complementation groups is done based on (a) the rate of UV-induced unscheduled DNA synthesis (UDS) in fibroblast colonies, measured in percentage of the value obtained from normal fibroblasts, and (b) the UV dose required to reduce fibroblast colony-forming ability *in vitro*. The degree of sun sensitivity, neurological abnormalities, and other clinical manifestations associated with XP varies from group to group. Experimental studies based on protein synthesis tend to suggest that complementation groups A, C, D, and H carry mutations at different loci (15a,18,29a,61a).

De Sanctis and Cacchione (10) described a form of XP associated with severe neurologic abnormalities. This phenotype and variants can be associated with any of the complementation groups, most often group A and, occasionally, group D. The complementation groups tend to segregate differently in various populations. There is only one patient reported in group B (52); group E consists of two related European patients (2); group F comprises four Japanese kindreds (40); group G consists of two unrelated European patients (35); and in groups H and I there is only one patient each (40,44) as of this writing.

Jung (30,32), using the term "pigmented xerodermoid," described a group of patients presenting solar cutaneous degeneration with eventual development of skin neoplasms after the age of 30. Cleaver et al (6) suggested that cells from patients with pigmented xerodermoid have the same post-UV synthesis and repair abilities as cells from patients with XP "variant." Pigmented xerodermoid may well represent XP "variant" with late onset (2). Follow-up studies of one patient, originally reported by Jung (30), showed that in that kindred the condition had autosomal recessive inheritance (26). Anderson and Berg (1) reported a family with a variety of XP possibly having autosomal dominant inheritance. Other authors have reported similar families (30,40). This autosomal dominant variety represents a mild form of XP, possibly within the range of complementation group E. Imray et al (29) suggest 100% penetrance. Affected individuals present marked xeroderma, dark pigmentation, and freckling with severe erythematous reaction when exposed to sunlight. Tumors seem to be rare in these patients and non neurological abnormalities have been reported (29,40). Laboratory studies demonstrate decreased cell survival, reduced DNA repair synthesis, increased chromosome breaks, but normal sister chromatid exchange (29).

In Japan, most patients with XP belong to group A and the XP "variant." In Europe and in the United States, group C is the largest (36), followed by groups D and A (2). Combining all varieties, XP occurs in 1 in 250,000 in the general population of the United States and Europe (2,52), with the prevalence in Japan estimated as 1 in 40,000 (2). It has been described in all races (2). The degree of parental consanguinity is quite high (53).

**Facies and skin.** Marked sensitivity to sunlight, especially wavelengths from 2800 to 3100 Å, is noted early in life, following minimal sun exposure. The majority of children with the various types of XP experience severe sunburn reactions, and this tapers off with adulthood. The skin not exposed to the sun is generally within normal limits. The areas exposed to the sun first develop freckles, generally before the second year of life (13,14,16,53,63,64). With progressive exposure to the sun, the freckles become larger and vary from light to dark-brown in color (Figs. 13–115 and 13–116). In addition, macular areas of hypopigmentation occur, resulting in a "salt-and-pepper" appearance of the skin. The second stage of skin involvement is characterized by atrophy and telangiectasia with the development of scaly patches and small spider-like hemangiomas. Skin atrophy on the face can be so severe as to interfere with normal opening of the mouth or eyes. Loss of scalp hair is frequent (64). Some patients, especially those with severe neurological manifestations, present loss of subcutaneous fat. This seems to be a generalized process not related to sun

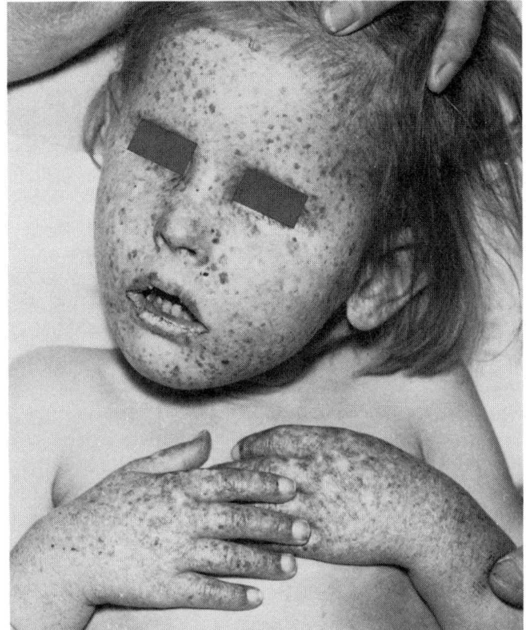

Fig. 13–115. *Xeroderma pigmentosum.* Note skin lesions. (Courtesy of *OP Hornstein*, Erlangen, Germany.)

exposure (2). Reaction to sunlight, ranging from mild to severe, varies from patient to patient. Those presenting acute sensitivity are more prone to develop neurological abnormalities (2). The third stage is the development of actinic keratoses, papillomatous growths, and neoplasms as early as 2 years of age (36). The neoplasms encompass both ectodermal and mesodermal tumors: keratoacanthomas (53), squamous cell carcinoma, basal cell carcinoma, melanoma (42,43), fibrosarcoma and angiosarcoma, as well as benign lesions such as angiomas and fibromas (2). The most frequent neoplasms are basal cell carcinoma followed by squamous cell carcinoma and melanoma. The tumors tend to arise in areas of solar keratoses. This specificity indicates a direct mutagenic effect of sunlight and possibly other carcinogens (5).

Basal and squamous cell carcinomas have been reported in 45%,

Fig. 13–116. *Xeroderma pigmentosum.* Another individual with xeroderma pigmentosum.

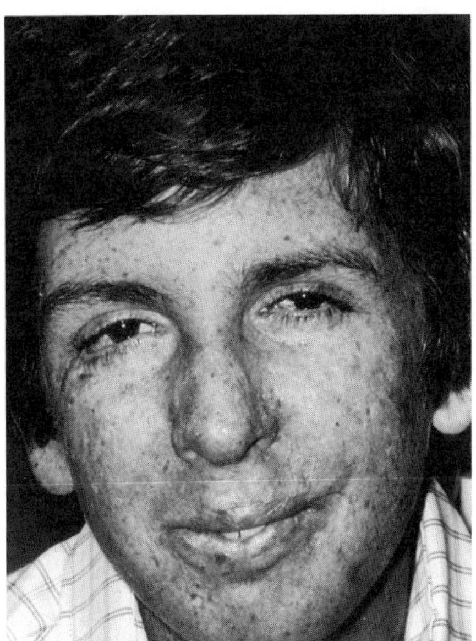

and melanoma in 5% (40). In about half of these cases, the melanomas are multiple (58). However, the frequency of melanoma has varied from 0% in a Japanese series (55) to around 50% in the United States (51). The latter probably represents a highly skewed sample. Sixty-five percent of the melanomas and 97% of the basal and squamous cell carcinomas arise in the face, head, and neck (40). Melanoma and its metastases have been reported to undergo spontaneous regression in some patients (43). The propensity to develop skin neoplasms varies with the complementation group. Patients in group A tend to present basal cell carcinomas very early in life, whereas patients with the XP ''variant'' develop the same tumors after the second decade (36). In general, more than 50% of the patients develop cutaneous neoplasms before age 10, in contrast to the general population in which over 50% of patients with skin neoplasms are over 60 years old (37,39). There is marked affinity for patients in group A to develop multiple squamous cell carcinoma and, with less frequency, basal cell carcinoma and rarely malignant melanoma. Similarly, patients in complementation group E tend to develop basal cell carcinoma in large proportions. The D group is more prone to have multiple melanomas. This will explain the fact that melanomas in Japanese patients are extremely rare, because group D in Japan is very rare (15,31). Heterozygotes are clinically normal and, in general, do not present sun sensitivity, but skin tumors seem to be four times higher than in the general population (54).

**Eyes.** Ocular involvement occurs in 40% (40). Most patients have marked photophobia and conjunctivitis with profuse lacrimation (53,63,64). The bulbar conjunctiva may exhibit pigmentation and telangiectasia. With severe involvement, the eyelids often lose their lashes, and ectropion, entropion, or symblepharon may complicate the clinical picture (2,12,13,37,38,40). Squamous cell carcinoma of the conjunctiva, ulceration of the cornea, and iritis with synechia or iris atrophy have been described (16,24) (Fig. 13–117).

**Central nervous system.** About 20% have neurologic abnormalities, the onset generally being after age 5 (40). Neurological abnormalities are observed largely in patients from complementation groups A, B, D, and G. Patients in group A tend to develop neurologic problems before age 7 whereas those in group D exhibit neurologic damage after this age (38). Only one patient from group C has presented severe CNS complications (22). This patient also had systemic lupus erythematosus. Rarely patients from the other groups have CNS manifestations (5). The majority are in group A, with the more severe cases classified as De Sanctis–Cacchione syndrome (31). Clinically, there may be spastic-ataxic gait and athetoid movements of the head and arms of variable degree (27,48). Areflexia, apparently due to lower motor neuron involvement, has been described (52). Pyramidal and extrapyramidal signs are common, and speech is often disturbed (10,27). Sensorineural hearing loss has been noted in several cases (40). Patients with De Sanctis–Cacchione syndrome present marked mental retardation and retarded physical development. The mental retardation is progressive and becomes evident during the first year of life (27,49). Abnormal electroencephalographic changes (diffuse dysrhythmia with poorly developed alpha rhythm and with occasional paroxysmal burst-like slow-wave discharges) have been noted (8,27). Microcephaly, small brain, and diffuse mild cerebral and olivopontocerebellar neuronal loss have been found at autopsy (8,27,48,50,56). Nonaffected relatives exhibit a higher incidence of mental retardation and microcephaly than the general population (60). Brain atrophy, ventricular dilation, marked cortical atrophy, and thickening of the calvaria with excessive pneumatization of the frontal sinuses have been found by computed tomography in patients with De Sanctis–Cacchione syndrome (20,23). Peripheral neuropathy has been described and sural nerve biopsies have suggested severe damage in both myelinated and unmyelinated fibers (20).

**Other findings.** Nuclear atypia of hepatic and pancreatic cells (14,56) and gonadal hypoplasia (10,27) have been reported.

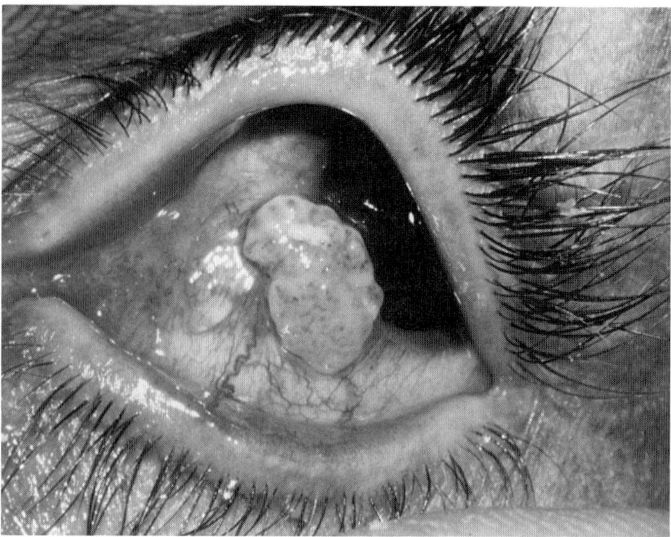

Fig. 13–117. *Xeroderma pigmentosum.* Squamous cell carcinoma arising on palpebral conjunctiva.

**Oral manifestations.** Areas of hyperpigmentation occur on the lips and anterior tongue (2). Squamous cell carcinoma of the lips, developing as early as age 14 (64), is frequent. Neoplasms of the tip of the tongue have been described (24,59,63). A 1987 survey of the literature (40) showed a total of 13 XP patients with squamous cell carcinoma of the tip of the tongue, two tumors arising in the gingiva and one in the palate (Fig. 13–118). Nearly 50% were below 14 years of age. Patients as young as 3 years of age have been reported with malignant oral neoplasms (39). These findings contrast with those in the general population of the United States, in which less than 1% of patients with oral or pharyngeal cancer are younger than 20 years (9).

A so-called ''cheilo-nasal prominence'' form of XP, characterized by profuse involvement of lips and nose starting in early childhood, has been described by Kraemer and Slor (38). This represents complementation group C, with marked actinic cheilitis, which eventually develops into carcinoma.

**Histologic features.** Skin changes consist of hyperkeratosis, acanthosis, and areas of epidermal atrophy (53). A chronic inflammatory

Fig. 13–118. *Xeroderma pigmentosum.* Squamous cell carcinoma arising on tongue tip. *From H Plotnick* and *A Lupulescu,* J Am Acad Dermatol **9**:876, 1983.)

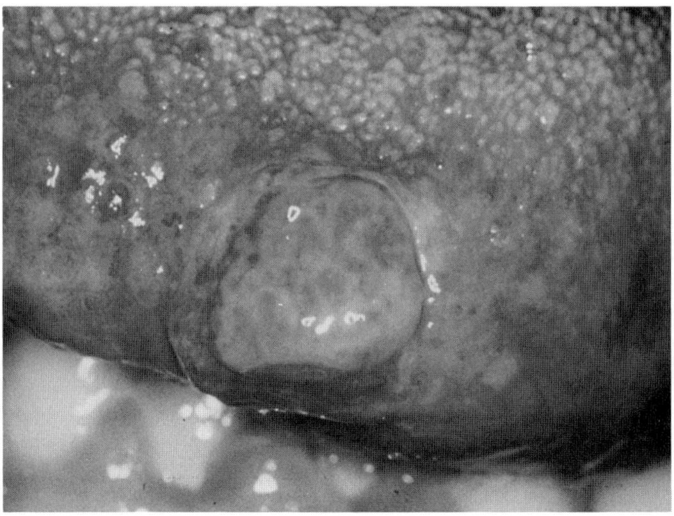

infiltrate of the epidermis can be observed as well as abnormal accumulation of melanin in the basal cell layer. With increasing age, these changes become accentuated and eventually neoplasia occurs. The histopathology of the neoplasms is identical to that of the same neoplasms arising in the general population.

Electron microscopic studies of sun-exposed and nonexposed skin reveal various abnormalities in melanin synthetic activity as well as phagocytosis of melanosomes by fibroblast-like cells (19,46).

**Differential diagnosis.** To be excluded are *Cockayne syndrome* (65), *Bloom syndrome,* and *Rothmund–Thomson syndrome,* drug-induced and allergen-induced photosensitivity (41), as well as erythropoietic protoporphyria and polymorphic light eruption (Hutchinson summer prurigo). Cockayne syndrome and xeroderma pigmentosum have been reported to occur simultaneously in two patients. (44,52). One patient belonged to complementation group B and the other to group H. Thrust et al (56) reported sibs with an unclassified form of XP characterized by progressive dementia, chorea, sensorineural hearing loss, corticospinal tract degeneration, peripheral neuropathy, and skeletal abnormalities.

**Laboratory aids.** Autoradiographic determination of UDS is used to demonstrate DNA-excision repair of UV damage to DNA. Autoradiographs show a positive reaction in the nuclei of UV-irradiated cells after incorporation of tritium-labeled thymidine during the non-S phase. If fused fibroblasts from two different XP patients exhibit the reaction, they belong to two different complementation groups (2).

Fibroblasts from patients with XP have 10–20% of the normal ability to repair DNA probably due to an abnormal endonuclease activity (52). Increased chromosal breakage has been found in fibroblasts, but not in lymphocyte cultures (46). Abnormalities in C and D chromosome groups (17,27) have been found in lymphocytes and cultured nontreated fibroblasts from patients with XP. Sister chromatid exchanges (SCE) are increased in cultured lymphoblastoid cells and fibroblasts following UV radiation (33). The number of SCEs has been found to correlate with the onset, severity, and prevalence of clinical symptoms and neoplasia. This reflects the mutation rate and the mutagenicity induced by UV radiation (33). However, it should be pointed out that similar changes occur in Cockayne syndrome, a disorder not prone to cancer production (53a). Immunologic studies have demonstrated that some patients exhibit delayed rejection of skin grafts and reduced induction of lymphocytes by PHA. A decrease in T3-positive lymphocytes has also been noted (62). Elevation of IgE level has been noted (47).

Prenatal diagnosis is possible (3,21,45,47,51). UDS studies after UV radiation in cultured fibroblasts obtained from obligate carriers have generally shown normal DNA repair levels (40,45,61). Reduced repair activity in heterozygotes has also been reported (4,11). There is a higher than expected (about 80%) association with blood group O (14).

Sural nerve biopsy of two sibs with type A showed marked reduction in the number of large and small myelinated fibers (44a).

### References (Xeroderma pigmentosum)


1. Anderson TE, Berg M: Xeroderma pigmentosum of mild type. *Br J Dermatol* **62**:402–407, 1950.

2. Andrews AD: Xeroderma pigmentosum, in *Chromosome Mutation and Neoplasia,* German J (ed), Alan R. Liss, New York, 1983, pp 63–83.

3. Barthelemy H et al: Diagnostic anténatal du xeroderma pigmentosum. *Arch Fr Pédiatr* **40**:198, 1983.

4. Cleaver JE: DNA repair and radiation sensitivity in human (xeroderma pigmentosum) cells. *Int J Radiat Biol* **18**:557–565, 1971.

5. Cleaver JE: DNA repair and replication in xeroderma pigmentosum. *Basic Life Sci* **39**:425–438, 1986.

6. Cleaver JE et al: Similar defects in DNA repair and replication in the pigmented xerodermoid and the xeroderma pigmentosum variants. *Carcinogenesis* **1**:647–655, 1980.

7. Cleaver JE et al: Xeroderma pigmentosum variants. *Cytogenet Cell Genet* **31**:188–192, 1981.

8. Clodi PH et al: Xeroderma pigmentosum mit körperlichem sowie geistigem Entwicklungsrückstand und intermittierender Aminoacidurie. Ein neues Syndrom? *Z Kingerkeilk* **93**:223–235, 1965.

8a. Cohen MM, Levy HP: Chromosome instability syndromes. *Adv Hum Genet* **18**:43–149, 1989.

9. Cutler SJ, Young JL: Third national cancer survey: Incidence data. *Natl Cancer Inst Monogr* #41, table 19a, p 100.

10. De Sanctis C, Cacchione A: L'idiozia xerodermica. *Riv Sper Freniatr* **56**:269–292, 1932.

11. Der Kaloustian VM et al: The genetic defect in the De Sanctis–Cacchione syndrome. *J Invest Dermatol* **63**:392–396, 1974.

12. El-Hefnawi H, Mortada A: Ocular manifestations of xeroderma pigmentosum. *Br J Dermatol* **77**:261–276, 1965.

13. El-Hefnawi H et al: Xeroderma pigmentosum. *Br J Dermatol* **74**:201–221, 1962.

14. El-Hefnawi H et al: Xeroderma pigmentosum—its inheritance and relationship to the ABO blood group system. *Ann Hum Genet* **28**:273–290, 1965.

15. Fischer E et al: Xeroderma pigmentosum patients from Germany: Clinical symptoms and DNA repair characteristics. *Arch Dermatol Res* **274**:229–247, 1982.

15a. Friedberg EC et al: Molecular aspects of DNA repair. *Mutat Res* **184**:67–86, 1987.

16. Gaasterland DE et al: Ocular involvement in xeroderma pigmentosum. *Ophthalmology* **89**:980–986, 1982.

17. German J et al: Mutant karyotypes in cultured cells from a man with xeroderma pigmentosum. *Ann Génét* **16**:23–27, 1973.

18. Giannelli F et al: Differences in patterns of complementation of the more common groups of xeroderma pigmentosum: Possible implications. *Cell* **29**:451–458, 1982.

19. Guerrier CJ et al: An electron microscopical study of the skin in 18 cases of xeroderma pigmentosum. *Dermatologica* **146**:211–221, 1973.

20. Hakamada S et al: Xeroderma pigmentosum: Neurological, neurophysiological and morphological studies. *Eur Neurol* **21**:69–76, 1982.

21. Halley DJJ et al: Prenatal diagnosis of xeroderma pigmentosum (group C) using assays of unscheduled DNA synthesis and postreplication repair. *Clin Genet* **16**:137–146, 1979.

22. Hananian J, Cleaver JE: Xeroderma pigmentosum exhibiting neurological disorders and systemic lupus erythematosus. *Clin Genet* **17**:39–45, 1980.

23. Handa J et al: Cranial computed tomography findings in xeroderma pigmentosum with neurologic manifestations (De Sanctis-Cacchione syndrome). *J Comp Asst Tomogr* **2**:456–459, 1978.

24. Harper JI, Copemen PWM: Carcinoma of the tongue in a boy with xeroderma pigmentosum. *Clin Exp Dermatol* **6**:601–604, 1981.

25. Hebra F, Kaposi M: *On Disease of the Skin Including the Exanthemata,* Vol 3, Tay W (trans), New Sydenham Society, London, 1874, pp 252–258.

26. Hofmann H et al: Pigmented xerodermoid: First report of a family. *Bull Cancer* **65**:347–350, 1978.

27. Hokkanen E et al: Zu den neurologischen Manifestationen des Xeroderma pigmentosum. *Dtsch Z Nervenheik* **196**:206–216, 1969.

28. Ichihashi A, Fujiwara Y: Clinical and pathobiological characteristics of Japanese xeroderma pigmentosum variant. *Br J Dermatol* **105**:1–12, 1981.

29. Imray FP et al: Sensitivity to ultraviolet radiation in a dominantly inherited form of xeroderma pigmentosum. *J Med Genet* **23**:72–78, 1986.

29a. Johnson RT et al: Lack of complementation between xeroderma pigmentatosum complementation groups D and H. *Hum Genet* **81**: 203–210, 1989.

30. Jung EG: New form of molecular defect in xeroderma pigmentosum. *Nature (London)* **228**:361–362, 1970.

31. Jung EG: Xeroderma pigmentosum. *Int J Dermatol* **25**:629–633, 1986.

32. Jung EG, Bantle K: Xeroderma pigmentosum and pigmented xerodermoid. *Birth Defects* **7**(8):125–128, 1971.

33. Jung EG et al: Heterogeneity of xeroderma pigmentosum (XP); variability and stability within and between the complementation groups C,D,E,I, and variants. *Photodermatol* **3**:125–132, 1986.

34. Kaposi M: Xeroderma pigmentosum. *Med Jahrb (Wien)* **13**:619–633, 1882.

35. Keijzer W et al: A seventh complementation group in excision-deficient xeroderma pigmentosum. *Mut Res* **62**:183–190, 1979.

36. Kobayashi M et al: Skin tumors of xeroderma pigmentosum. *J Dermatol (Tokyo)* **9**:319–322, 1982.

37. Kraemer KH: *Xeroderma Pigmentosum in Clinical Dermatology,* Vol. 4, Demis DJ et al (eds), Harper & Row, Hagerstown, 1980, pp 1–33.

38. Kraemer KH, Slor H: Xeroderma pigmentosum. *Clin Dermatol 3*:33–69, 1985.

39. Kraemer KH et al: Early onset of skin and oral cavity neoplasms in xeroderma pigmentosum. *Lancet* **2**:56–57, 1982.

40. Kraemer KH et al: Xeroderma pigmentosum. *Arch Dermatol* **123**:241–259, 1987.

41. Lorette G et al: Xeroderma pigmentosum-like changes. *Arch Dermatol* **119**:873, 874, 1983.

42. Lynch HT et al: Cancer control in xeroderma pigmentosum. *Arch Dermatol* **113**:193–195, 1977.

43. Lynch HT et al: Xeroderma pigmentosum. Complementation group C and malignant melanoma. *Arch Dermatol* **120**:175–179, 1984.

44. Moshel AN et al: A new patient with both xeroderma pigmentosum and Cockayne syndrome establishes the new xeroderma pigmentosum complementation group H, in *Cellular Responses to DNA Damage,* Friedberg E, Bridges B (eds), Alan R. Liss, New York, 1983, pp 209–213.

44a. Origuchi Y et al: Quantitative histologic study of sural nerves in xeroderma pigmentosum. *Pediatr Neurol* **3**:356–359, 1987.

45. Pawsey SA et al: Clinical, genetic and DNA repair studies on a consecutive series of patients with xeroderma pigmentosum. *Q J Med* **48**:179–210, 1979.

46. Plotnick H, Lupulescu A: Ultrastructural studies of xeroderma pigmentosum. *J Am Acad Dermatol* **9**:876–882, 1983.

47. Ramsay CA et al: Prenatal diagnosis of xeroderma pigmentosum. *Lancet* **2**:1109–1112, 1974.

48. Reed WB: Xeroderma pigmentosum, clinical and laboratory investigation of its basic defect. *JAMA* **207**:2073–2079, 1969.

49. Reed WB et al: Xeroderma pigmentosum with neurological complications: The De Sanctis–Cacchione syndrome. *Arch Dermatol* **91**:224–226, 1965.

50. Reed WB et al: De Sanctis–Cacchione syndrome. *Arch Dermatol* **113**:1561–1563, 1977.

51. Regan JD et al: Xeroderma pigmentosum: A rapid sensitive method for prenatal diagnosis. *Science* **174**:147–150, 1971.

52. Robbins JH et al: Xeroderma pigmentosum: An inherited disease with sun sensitivity, multiple cutaneous neoplasms, and abnormal DNA repair. *Ann Intern Med* **80**:221–248, 1974.

53. Romagnolo J et al: Le xeroderma pigmentosum: revue de la clinique, de l'histologie, de l'etiopatogenie et de la therapeutique (a propos d'un cas clinique). *Acta Stomatol Belg* **80**:101–120, 1983.

53a. Seguin LR et al: Ultraviolet light-induced chromosomal aberrations in culture cells from Cockayne syndrome and complementation group C xeroderma pigmentosum patients: Lack of correlation with cancer susceptibility. *Am J Hum Genet* **42**:468–475, 1988.

54. Swift M, Chase C: Cancer in families with xeroderma pigmentosum. *J Natl Cancer Inst* **62**:1415–1421, 1979.

55. Takebe H et al: DNA repair characteristics and skin cancers of xeroderma pigmentosum patients in Japan. *Cancer Res* **37**:490–495, 1977.

56. Thrush DC et al: Neurological manifestations of xeroderma pigmentosum in two siblings. *J Neurol Sci* **22**:91–104, 1974.

57. Trosko JE et al: Sunlight induced pyrimidine dimers in human cells in vitro. *Nature (London)* **228**:358–359, 1970.

58. Tulles GD et al: Multiple melanomas occurring in a patient with xeroderma pigmentosum. *J Am Acad Dermatol* **11**:364–367, 1984.

59. Wade MH, Plotnick H: Xeroderma pigmentosum and squamous cell carcinoma of the tongue. *J Am Acad Dermatol* **12**:515–521, 1985.

60. Welshimer K, Swift M: Congenital malformations and developmental disabilities in ataxia-telangiectasia, Fanconi anemia and xeroderma pigmentosum families. *Am J Hum Genet* **34**:781–793, 1982.

61. Went M et al: Study of DNA repair on a xeroderma pigmentosum patient and his heterozygotic parents. *Arch Dermatol Res* **270**:291–297, 1981.

61a. Wood RD et al: Complementation of xeroderma pigmentosum DNA repair defect in cell-free extracts. *Cell* **53**:97–106, 1988.

62. Wysenbeek AJ et al: Immunologic alterations in xeroderma pigmentosum patients. *Cancer* **58**:219–221, 1986.

63. Yagi K et al: Carcinoma of the tongue in a patient with xeroderma pigmentosum. *Int J Oral Surg* **10**:73–76, 1981.

64. Yagi K, Prabhu SR: Carcinoma of the lip in xeroderma pigmentosum. *J Oral Med* **38**:97–98, 1983.

65. Yoder FW et al: A comparison of some typical manifestations of Cockayne syndrome and xeroderma pigmentosum. *Clin Res* **24**:624A, 1977.

## Endocrine–candidosis syndrome (candidosis, idiopathic hypoparathyroidism, and hypoadrenalcorticism)

The association of mucocutaneous candidosis with polyglandular autoimmune disease, especially idiopathic hypoparathyroidism and idiopathic hypoadrenalcorticism, has been known for many years, the first reference probably being that of Thorpe and Handley (31) in 1929. However, it has required almost five decades for analysis of enough cases to see recognizable patterns of occurrence.

The syndrome is manifested as variable combination of (in approximate order of decreasing frequency) superficial mucocutaneous candidosis, hypoparathyroidism, hypoadrenocorticism, ovarian atrophy, keratoconjunctivitis, alopecia, intestinal malabsorption, pernicious anemia, diabetes mellitus, hepatitis, and thyroiditis (3,11,13,30).

Neufeld et al (21) divided autoimmune polyglandular syndromes into three distinct types, plus a miscellaneous type. Type I includes at least two of the following: chronic mucocutaneous candidosis, acquired hypoparathyroidism, and idiopathic hypoadrenalcorticism with or without other autoimmune disease. The complete triad is present in only one-third of the cases. Type II includes idiopathic hypoadrenalcorticism plus autoimmune thyroid disease and/or insulin-requiring diabetes, but not hypoparathyroidism or chronic mucocutaneous candidosis. It subsumes the so-called Schmidt syndrome (9). In contrast to Type I, which has early onset (12 years), Type II typically becomes evident in mid-life (30 years). Type III patients manifest autoimmune thyroid disease *without* Addison's disease but with at least one associated organ-specific autoimmune disease, such as insulin-dependent diabetes, pernicious anemia, vitiligo, and/or alopecia. We will not be concerned with Type III disease here.

The cell-mediated immune defects may well be unique. In nearly all cases, there is limited ability of lymphocytes to proliferate in response to *Candida* antigen. Response to mitogens and other antigens is essentially normal. It is possible that the defect is mediated by a humoral factor in Type I patients (2,4,15,17,18,33). At least one patient has died of fulminating chicken pox (13). There are several excellent reviews of these autoimmune syndromes (6,8,11,17,21).

Whereas Type I has autosomal recessive inheritance (1,2,7,18a, 23,26,30,35–37), Type II appears to be HLA linked (B8,Dw3,DR3) and exhibits familial aggregation, but does not appear to have single gene inheritance (6,21). A high frequency of Type I has been found in Finland (19).

Candidosis is usually the first component to appear, most observers reporting its presence during the first 6 years of life. It is followed from 3 months to 13 years later by the other components. Idiopathic hypoparathyroidism and hypoadrenalcorticism become manifest most frequently during the prepubescent period, the former at a mean of about 5 years, the latter at about 8.5 years.

**Endocrine systems.** Idiopathic hypoparathyroidism, eventually present in about 80%, is commonly asymptomatic initially. Serum calcium is reduced and serum phosphorus is elevated in the absence of significant disease of the genitourinary and gastrointestinal systems. Tetany is not evident unless calcium levels are significantly reduced. Furthermore, parathormone injection results in increased serum calcium and in elevated urinary phosphorus excretion. Candidosis does not complicate the hypocalcemia seen in chronic tetany owing to inadvertent parathyroidectomy in the course of thyroidectomy (11). At necropsy, absence of the parathyroid glands or their replacement with fat has been noted.

Hypoadrenalcorticism is found in about 65% of Type I patients. Often initially silent, it appears soon after the signs of hypoparathyroidism (21). Lassitude, anorexia, progressive weakness, hypotension and progressive pigmentation of the skin and mucous membranes signal its onset. Laboratory findings confirm the low serum sodium and high potassium levels. Death from adrenal crisis may result, and necropsy shows adrenocortical atrophy of the "cytotoxic type."

Autoimmune thyroid disease, also initially subtle, has been diagnosed in only 10% of Type I cases but in 70% of Type II examples (21). It has been shown to be primary, since thyroid-stimulating hormone (TSH) does not produce a rise in thyroid $^{131}I$ uptake or alter the protein-bound iodine levels.

Insulin-requiring diabetes mellitus is found in only 2–4% of Type I cases, but in over 50% of Type II cases (21). Diabetes insipidus has been reported in sibs with the disorder (13). The relation of *Candida* infection to the endocrine glands is not known. It has been suggested that candidosis is a superficial expression of undetermined abnormalities already present in the host that favor the development of the infection as well as the endocrinopathies.

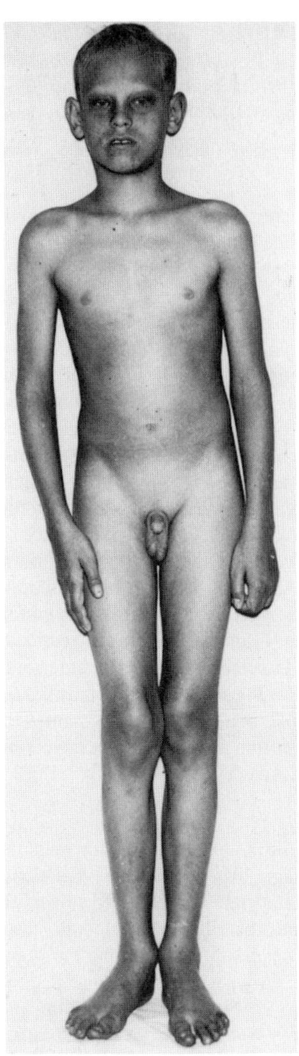

Fig. 13–119. *Endocrine–candidosis syndrome.* Note generalized skin pigmentation, especially marked over face, midtrunk, and knees. (From *JA Whitaker,* J Clin Endocrinol **16**:1374, 1956.)

Primary gonadal failure (almost exclusively ovarian) has been reported in 10–15% of Type I cases and in 3–4% of Type II patients (3,5,21,26).

**Skin and skin appendages.** The skin is dry, and the hair of both body and scalp usually brittle and diminished (Fig. 13–119). Eyebrows and axillary and pubic hair are remarkably sparse. Total alopecia may develop spontaneously in 25–30% of Type I cases (13,21). Fingernails and toenails are frequently thin, ridged, and brittle, and often are the site of *Candida* infections (Fig. 13–120). If infected, the nails become brown, irregular, and thickened with a crumbled outer edge. Vitiligo is present in 10% of Type I patients and in 5% of Type II patients (7,21). Premature graying of the hair has also been noted.

**Central and peripheral nervous system.** Increased cerebrospinal fluid pressure, associated with papilledema, intracranial calcification, epileptiform seizures, and mental retardation have been noted (8,24,32) (Fig. 13–121). Muscle twitchings and cramps, tetany, abdominal pain, paresthesias, and rigidity are seen. The neurologic findings have been especially well reviewed by Jordan and Kelsall (14).

**Ocular findings.** Several patients have manifested photophobia, blepharospasm, recurrent blepharitis, corneal pannus, keratoconjunctivitis, corneal scarring, loss of eyebrows and/or eyelashes, and cataracts (8,13,21,34) (Fig. 13–122).

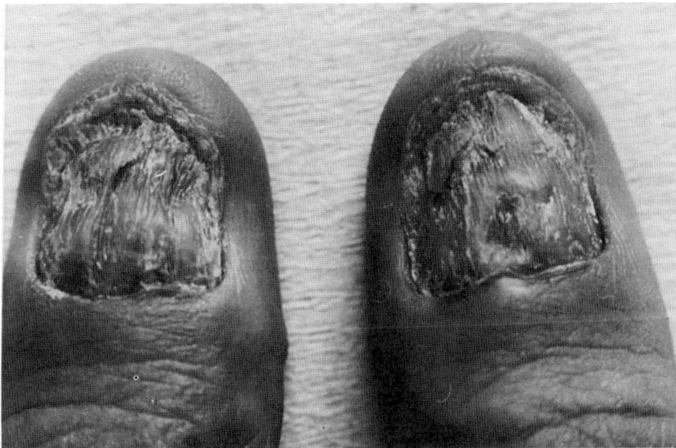

Fig. 13–120. *Endocrine–candidosis syndrome.* Candidosis of fingernails. (From *KH Sjöberg,* Acta Med Scand **179**:157, 1966.)

**Gastrointestinal findings.** Malabsorption has been noted in 25% of Type I cases (21). Autoimmune chronic active hepatitis occurs in 10–15% of Type I cases, but not in Type II cases (5,12,21,25,36).

**Blood findings.** Pernicious anemia is found in about 15% of Type I cases but very rarely in Type II patients (21).

**Oral manifestations.** Candidosis, present in about 75% of Type I patients (19,21,24a), may be superficial or deep seated and may involve the lips, tongue, buccal mucosa, palate, and larynx with thick, creamy-white plaques (Fig. 13–123A,B). The mucosa between the plaques is often hyperemic, and the tongue may be smooth and devoid of papillae. Perleche, or angular stomatitis, may extend over a considerable portion of the perioral skin. Similar involvement of the anal and vaginal mucosae has been described.

With the advent of Addison's disease, areas of splotchy melanin pigment are seen in the mouth, especially on the buccal mucosa and palate.

The teeth are chalky, with pitted crowns or transverse hypoplastic bands of enamel in 80% (8,19,20,28,29) (Fig. 13–123B). Involvement, however, is usually of mild degree (10,11,14,36) and appears to be independent of the hypoparathyroidism (19), probably being the

Fig. 13–121. *Endocrine–candidosis syndrome.* Skull roentgenogram exhibiting extensive basal ganglion calcification. (From *JDM Gass,* Am J Ophthalmol **54**:660, 1962.)

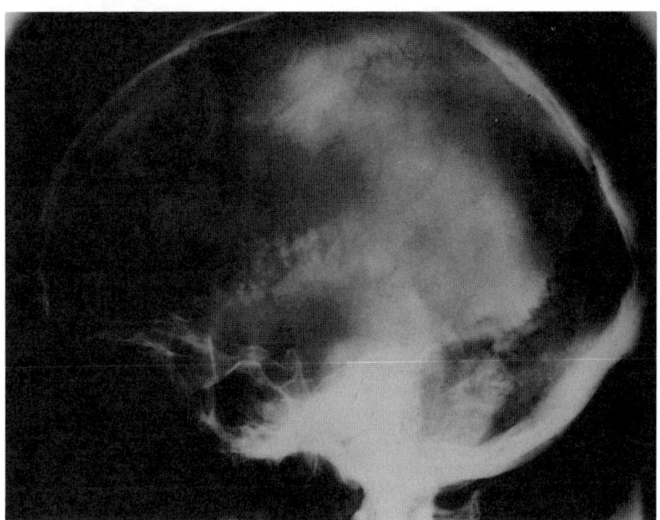

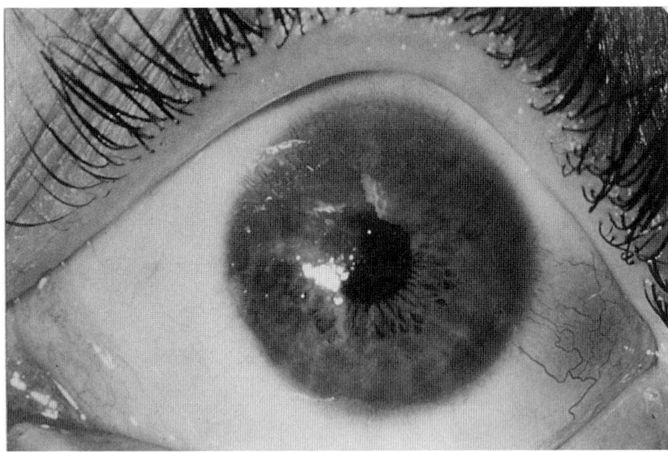

Fig. 13–122. *Endocrine–candidosis syndrome*. Keratoconjunctivitis. Note peripheral corneal vascularization and irregular, slightly raised, confluent nodular opacities in midperiphery of cornea. Lower paracentral cornea is not involved. (From *JDM Gass*, Am J Ophthalmol **54**:660, 1962.)

Fig. 13–123. *Endocrine–candidosis syndrome*. (A) Marked angular cheilosis due to monilia. (B) Note enamel deficiency related to hypoparathyroidism and monilial granulomas of tongue. (A courtesy of *W Reither*, München, West Germany.)

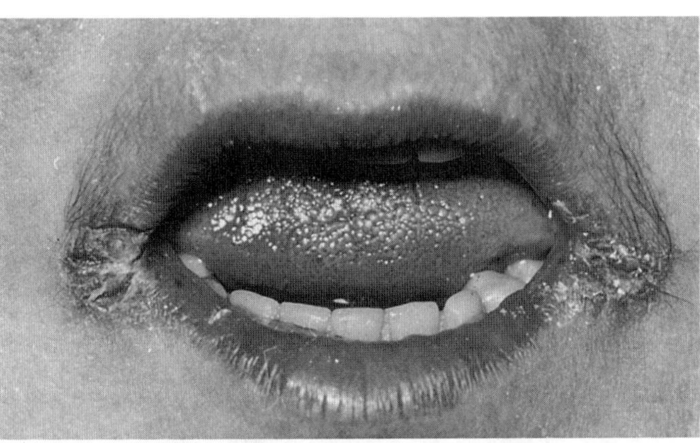

A

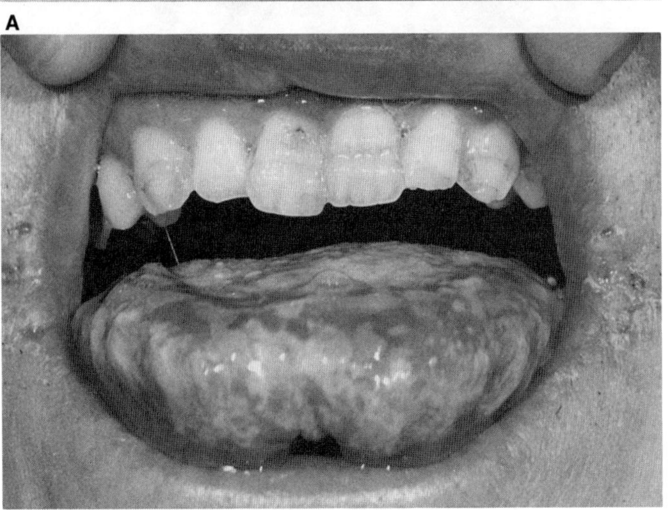

B

result of autoimmune abnormalities. Dental age is probably normal (19). Late dental eruption has been found in some patients (14,26) but not in others (19). Impacted teeth (19,28) have also been found. The occurrence of squamous cell carcinoma of the oral cavity has been reported (18b,26).

**Differential diagnosis.** Because of the wide spectrum of this syndrome, differential diagnosis must include the differential diagnosis of each of its components. This runs an exceedingly wide gamut.

Oral and perioral candidosis, possibly as a secondary factor, may also be seen in association with an underlying malignant process (leukemia, lymphoma, iatrogenically prolonged antibiotic or corticosteroid therapy). It is also seen in acrodermatitis enteropathica and in zinc deficiency. Chronic mucocutaneous candidosis is also seen in association with primary immunodeficiencies, both those with predominantly T lymphocyte defects (AIDS, severe combined immunodeficiency, *DiGeorge sequence*), and occasionally with other disorders (myeloperoxidase deficiency, chronic granulomatous disease). Chronic mucocutaneous candidosis can also occur by itself in a late onset or familial form or it can be found in association with thymoma with or without myasthenia gravis (16,27). An autosomal dominant endocrine–candidosis syndrome was reported by Okamoto et al (22) which is the same as *Witkop syndrome*. Both T and B cells were reduced. Autoantibody abnormalities were not detected, but adrenal deterioration was noted. Pisanty and Garfunkel (24) reported five families in which hypoparathyroidism and candidosis were found with mental retardation. It was not determined whether this autosomal recessive syndrome is the same as endocrine candidosis, Type I.

**Laboratory aids.** There are no specific diagnostic aids for the syndrome other than those used in the diagnosis of the individual components. Laboratory examinations to determine calcium and phosphorus levels, tests for kidney function, and the Ellsworth–Howard tests and levels of serum parathormone are necessary to establish the diagnosis of idiopathic hypoparathyroidism. Chvostek and Trousseau signs are positive when the calcium level is sufficiently low. Calcium and phosphate levels should be checked every 6 months after the diagnosis of mucocutaneous candidosis is made. Hypoadrenalcorticism is established on the basis of the clinical findings, low serum sodium and high potassium levels, low plasma cortisol and/or aldosterone levels, and adrenal antibodies. Sodium and potassium levels should be periodically checked after the diagnosis of hypoparathyroidism has been established. Hypothyroidism is established on finding low serum thyroxine and high serum thyrotropin. It, like the diabetes, is often overlooked.

Blood should be examined for evidence of pernicious anemia. Ophthalmologic study should be carried out for corneal surface abnormalities by use of rose bengal, fluorescein, or Schirmer testing.

Immunologic abnormalities should be investigated: autoantibodies (both organ and nonorgan specific), cell mediated immunity, and HLA linkage (33,37).

### References [Endocrine–candidosis syndrome (candidosis, idiopathic hypoparathyroidism, and hypoadrenalcorticism)]

1. Ahonen P: Autoimmune polyendocrinopathy-candidosis-ectodermal dystrophy (APECED): Autosomal recessive inheritance. *Clin Genet* **27**:535–542, 1985.

2. Arulanantham K et al: Evidence for defective immunoregulation in the syndrome of familial candidiasis endocrinopathy. *N Engl J Med* **300**:164–168, 1979.

3. Blizzard RM, Gibbs JH: Candidosis: Studies pertaining to its association with endocrinopathies and pernicious anemia. *Pediatrics* **42**:231–237, 1968.

4. Castells S et al: Familial moniliasis, defective delayed hypersensitivity and adrenocorticotropic hormone deficiency. *J Pediatr* **79**:72–79, 1971.

5. Craig JM et al: Chronic moniliasis associated with Addison's disease. *Am J Dis Child* **89**:669–684, 1955.

6. Eisenbarth G et al: The polyglandular failure syndrome: Disease inheritance, HLA type, and immune function. *Ann Intern Med* **91**:528–533, 1979.

7. Fisher M, Fitzpatrick TB: Candidiasis, vitiligo, thyroid disease and autoimmunity. *Arch Dermatol* **102**:110–112, 1970.

8. Gass JDM: The syndrome of keratoconjunctivitis, superficial moniliasis, idiopathic-hypoparathyroidism and Addison's disease. *Am J Ophthalmol* **54**:660–674, 1962.

9. Genant HK et al: Addison's disease and hypothyroidism (Schmidt's syndrome). *Metabolism* **16**:189–194, 1967.

10. Greenberg MS et al: Idiopathic hypoparathyroidism, chronic candidosis and dental hypoplasia. *Oral Surg* **28**:42–53, 1969.

11. Hermans PE, Ritts RE: Chronic mucocutaneous candidiasis: Its association with immunologic and endocrine abnormalities. *Minn Med* **53**:75–80, 1970.

12. Hetzel BS, Robson HN: The syndrome of hypoparathyroidism, Addison's disease and moniliasis. *Aust Ann Med* **7**:27–33, 1958.

13. Hung SO, Patterson A: Ectodermal dysplasia associated with autoimmune disease. *Br J Ophthalmol* **68**:367–369, 1984.

14. Jordan A. Kelsall AR: Observations on a case of idiopathic hypoparathyroidism. *Arch Intern Med* **87**:242–258, 1951.

15. Kaffe S et al: Variable cell-mediated immune defects in a family with "candida endocrinopathy syndrome." *Clin Exp Immunol* **20**:397–408, 1975.

16. Kaneko F et al: Clinical observations on a case of immunodeficiency and thymoma (Good's syndrome) associated with chronic mucocutaneous candidiasis. *J Dermatol* **9**:355–365, 1982.

17. Kirkpatrick CH et al: Chronic mucocutaneous moniliasis with impaired delayed hypersensitivity. *Clin Exp Immunol* **6**:375–385, 1970.

18. Kirkpatrick CH et al: Chronic mucocutaneous candidiasis: Model-building in cellular immunity. *Ann Intern Med* **74**:955–978, 1971.

18a. MacLeod RI, Bird AG: Chronic mucocutaneous candidosis in monozygotic twins. *J Oral Maxillofac Surg* **45**:616–617, 1987.

18b. Messenger AG, Harrington CI: Candida-endocrinopathy syndrome. Squamous cell carcinoma of the tongue. *Br J Dermatol* **115**(Suppl **30**):100–101, 1986.

19. Myllarniemi S, Perheentupa J: Oral findings in the autoimmune polyendocrinopathy-candidosis (APECS) and other forms of hypoparathyroidism. *Oral Surg* **45**:721–729, 1978.

20. Nally FF: Idiopathic juvenile hypoparathyroidism with superficial moniliasis. *Oral Surg* **30**:356–365, 1970.

21. Neufeld M et al: Autoimmune polyglandular syndrome. *Pediatr Ann* **9**:154–162, 1980.

22. Okamoto GA et al: New syndrome of chronic mucocutaneous candidiasis. *Birth Defects* **13**(3B):117–125, 1977.

23. Perheentupa J, Hiekkala H: Twenty cases of the syndrome of autoimmune endocrinopathy and candidiasis. *Acta Paediatr Scand* **62**:110–111, 1973.

24. Pisanty S, Garfunkel A: Familial hypoparathyroidism with candidiasis and mental retardation. *Oral Surg* **44**:374–383, 1977.

24a. Porter SR, Scully C: Candida endocrinopathy syndrome. *Oral Surg* **61**:573–578, 1986.

25. Price ML, McDonald DM: Candida endocrinopathy syndrome. *Clin Exp Dermatol* **9**:105–109, 1984.

26. Richman RA et al: Candidiasis and multiple endocrinopathies. *Arch Dermatol* **111**:625–627, 1975.

27. Rycroft RJG et al: Chronic muco-cutaneous candidiasis of late onset, thymoma and myopathy. *Clin Exp Dermatol* **1**:59–74, 1976.

28. Schmid-Meier E: Zahnstatus beim Syndrom Hypoparathyreoidismus. Morbus Addison, Moniliasis. *Schweiz Mschr Zahnheilk* **91**:409–417, 1981.

29. Sjöberg KH: Moniliasis—an internal disease. *Acta Med Scand* **179**:157–166, 1966.

30. Spinner MW et al: Clinical and genetic heterogeneity in idiopathic Addison's disease and hypoparathyroidism. *J Clin Endocrine Metab* **28**:795–804, 1968.

31. Thorpe ES, Handley HE: Chronic tetany and chronic mycelial stomatitis in a child aged four and one-half years. *Am J Dis Child* **38**:328–338, 1929.

32. Traboulsi EI et al: Ocular findings in the candidiasis-endocrinopathy syndrome. *Am J Ophthalmol* **99**:486–487, 1985.

33. Trence DL et al: Polyglandular autoimmune syndrome. *Am J Med* **77**:107–115, 1985.

34. Wagman RD et al: Keratitis associated with the multiple endocrine deficiency autoimmune disease and candidosis syndrome. *Am J Ophthalmol* **103**:569–575, 1987.

35. Wells RS et al: Familial chronic muco-cutaneous candidiasis. *J Med Genet* **9**:302–310, 1972.

36. Whitaker J et al: The syndrome of familial juvenile hypoadrenalcorticism, hypoparathyroidism and superficial moniliasis. *J Clin Endocrinol* **16**:1374–1387, 1956.

37. Wirfält A: Genetic heterogeneity in autoimmune polyglandular failure. *Acta Med Scand* **210**:7–13, 1981.

## Epidermolysis bullosa

Epidermolysis bullosa (EB) includes several hereditary vesicular disorders that involve skin and often oral and other mucosas. The vesicles usually arise at points of trauma. For some types, however, heat may be the precipitating factor, or blisters may arise spontaneously.

The name "epidermolysis bullosa" is usually improper because cytolysis of the epidermis is observed in only three or four of the subtypes. Nevertheless it is preserved because of its many years of usage.

EB is presently classified according to various combination of inheritance, scarring, and location of the bullae. Because many of the clinical manifestations in the various subtypes of EB can be similar if not identical, it is imperative that proper typing of patients with EB be accomplished with electron microscopic, immunofluorescent, and immunohistochemical studies (26,30). The interested reader is referred to the works and reviews of Bergenholtz and Olsson (8), Haber et al (30), Gedde-Dahl and Anton-Lamprecht (26), Gedde-Dahl (25), Cooper and Bauer (16), Fine (23), and Tabas et al (55).

The present classification of EB is based primarily on the level at which the cleavage that forms the bulla takes place. This is combined with the resultant clinical signs (e.g. scarring) and the type of inheritance (35). Thus, the basic classification is as follows (also see Table 13–6):

*Type I:* Intraepidermal forms, characterized as nonscarring varieties having autosomal dominant, recessive (vide infra) and X-linked forms.

*Type II:* Junctional forms, characterized primarily by skin atrophy, with autosomal recessive inheritance.

*Type III:* Dermal forms, characterized by atrophy and scarring, with autosomal dominant and recessive varieties.

*Type IV:* Acquired (nonhereditary) variety known as EB acquisita.

Here only those types presenting oral mucosal and/or dental manifestations will be described. It is possible that Cockayne–Touraine, Pasini, and Bart types represent different expressions of the same mutant gene (10).

### EB simplex Koebner type

Systemic manifestations. Sites or friction or trauma are most frequently involved. Nails are affected in about 20%. Scarring and/or pigmentation do not result after healing. The disorder, which usually appears neonatally or during infancy when the child begins to crawl, involves principally the feet, hands, and neck, but rarely the ankles, knees, trunk, and elbows (Fig. 13–124A). After the third year of life, usually only the hands and feet are affected. The nails are normal. The condition generally improves markedly at puberty. Heat also seems to be an important precipitating factor in blister production (14). This type of EB must be differentiated from the Weber–Cockayne type in which only the feet are affected. This latter form of EB more commonly (70 vs. 40%) appears at less than 1 year of age, and does not present oral involvement.

Oral manifestations. Occasional intraoral blister formation is seen. The lesions are not as pronounced as those in the more severe types of EB. They can occur at any intraoral location. Tooth alterations are not present (25,29,53). (Fig. 13–124B).

Histologic study reveals cleavage through the basal layer, with nuclei on the blister floor, that is, above the PAS-positive basement membrane. Adjacent to areas of separation, the basal cells are vacuolated, with displacement of the nucleus to the epidermal end of the cell. These alterations have been confirmed ultrastructurally (46,47). No histochemical abnormality has been observed (39).

**EB simplex with muscular dystrophy.** Several authors (24b,37a,43a) have described an autosomal recessively inherited EB simplex with muscular dystrophy or congenital myasthenia gravis. Appearing at birth, the disorder becomes generalized but more severely involve the extremities. Milia are common. Oral mucosal involvement is frequent. A mild atropic scarring has been seen in most patients.

Table 13–6. Epidermolysis bullosa

| Type | Onset | Feet | Hand | Face | Other | Corneal erosions | Oral mucosa | Teeth | Nails | Seasonal variation |
|---|---|---|---|---|---|---|---|---|---|---|
| I: Intraepidermal blister formation | | | | | | | | | | |
| A. Nonscarring EB with AD inheritance | | | | | | | | | | |
| 1. EB simplex generalized Koebner type (D-EBS-K) | Birth to ½ year | Yes | Yes | ± | Yes | No | Some | No | No | Warm weather |
| 2. EB simplex localized Weber–Cockayne type (D-EBS-WC) | First to third decade | Yes | Yes | No | No | No | No | No | No | Summer |
| 3. EB simplex with mottled pigmentation (D-EBS-M) | Birth to infancy | Yes | Yes | No | Yes | No | No | No | Yes | No |
| 4. EB simplex Ogna (D-EBS-0) | ? | Yes | Yes | No | Rarely | No | No | No | Yes | Summer |
| 5. EB herpetiformis Dowling–Meara (D-EBH-DM) | First month | Yes | Yes | Yes | Yes | Some | No | Some | Yes | No |
| 6. EB Bart (congenital, localized absence of skin (D-EB-B) | At birth | Yes | Yes | No | Yes | No | No | No | Yes | No |
| B. Nonscarring EB with X-linked inheritance | | | | | | | | | | |
| 1. Dystrophia bullosa hereditaria macular type Mendes da Costa (X-DBM-MC) | First to third year | Yes | Yes | No | No | No | No | No | ? | |
| C. Atrophic scarring EB with AR inheritance | | | | | | | | | | |
| 1. EB with muscular dystrophy or myasthenia gravis | At birth | Yes | Yes | Yes | Yes | No | Yes | No | Yes | No |
| II: Junctional blister formation with hemidesmosome defect atrophicans group | | | | | | | | | | |
| A. Nonscarring EB with AR inheritance | | | | | | | | | | |
| 1. EB atrophicans generalisata gravis Herlitz (R-EBA-GH) | First days | Yes at infancy | Yes at infancy | Yes | Yes | Rarely | Yes | Yes | Yes | No |
| 2. EB atrophicans generalisata mitis (R-EBA-MITIS) | At birth | Yes | Yes | No | Yes | No | Yes | Yes | Yes | Summer |
| 3. EB atrophicans localisata (R-EBA-L) | At birth | Yes | No | No | Yes | No | No | Yes | Yes | No |
| 4. EB atrophicans inversa (R-EBA-I) | First days | No | No | No | Yes | Yes | Rarely | Yes | Yes | No |
| 5. EB progressiva (R-EBP) juvenile onset deposit type | 5–8 years | Yes | Yes | No | Yes | No | Yes | No | No | No |
| III. Dermolytic blister formation dystrophic group | | | | | | | | | | |
| A. Scarring EB with AD inheritance | | | | | | | | | | |
| 1. EB dystrophica Cockayne–Touraine type (D-EBD-CT) | Birth to fifth year | Yes | Yes | No | Yes | No | Rarely | No | Yes | No |
| 2. EB dystrophica albopapuloidea Pasini (D-EBD-P) | First weeks | Yes | Yes | Yes | Yes | No | Yes | No | Yes | No |
| 3. EB pretibial (D-EBO-KP) (Kuske–Portugal disease) | 11–24 years | No | No | No | Yes | No | No | No | Yes | No |
| B. Scarring EB with AR inheritance | | | | | | | | | | |
| 1. EB dystrophica Hallopeau–Siemens (R-EBD-HS) | | | | | | | | | | |
| a. Local | First weeks | Yes | Yes | Yes | Yes | No | Yes | Slight | Slight | No |
| b. Generalized | First weeks | Yes | Yes | Yes | Yes[a] | No | Yes[a] | Yes[a] | Yes | No |
| c. Mutilans | First weeks | Yes[a] | Yes[b] | Yes | Yes[b] | Rarely | Yes[b] | Yes[b] | Yes[b] | No |
| 2. EB dystrophica inversa (R-EBD-I) | Neonatal | No | No | No | Yes | No | Yes | No | No | No |
| IV. Acquired phenocopy of dystrophic epidermolysis bullosa | | | | | | | | | | |
| EB acquisita (A-EB) | Adult | Yes | Yes | No | Yes | No | Slight | No | Yes | No |

[a] Mild.
[b] Moderate.

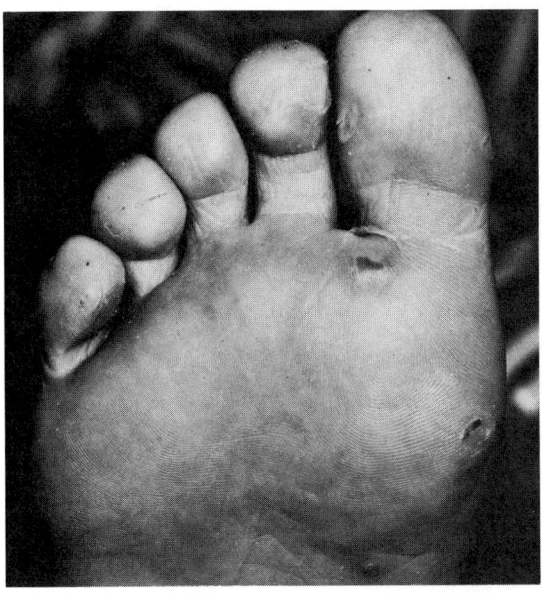

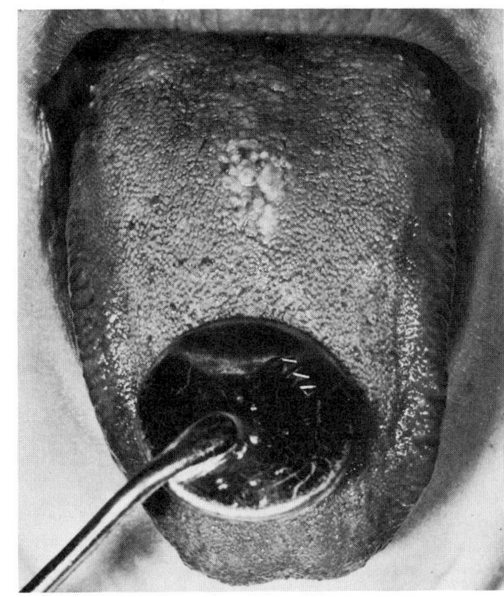

A B

Fig. 13–124. *Epidermolysis bullosa, dominant simplex type.* (A) Non-scarring blisters of feet. (B) Oral vesicles seen in about 2% of patients.

**EB atrophicans generalisata gravis (Herlitz type)**

Systemic manifestations. This type of EB, inherited as autosomal recessive, was originally described by Herlitz under the name of EB letalis. As a number of infants with this form survive into adulthood, the term letalis should be discarded. Because affected infants have survived the first months of life, it may be assumed that in the past several cases have been misdiagnosed (18).

Usually the vesicles, often hemorrhagic, are noted at the base of the fingernails within the first few hours of life. The nails soon become loose and are shed (28). This is followed by involvement of the trunk, umbilicus, face, scalp, and extremities. The palms and soles are never affected. There seems to be an absence of reaction to traumatic provocation. A number of affected infants die within the first few months of life, usually from secondary septic infections (Fig. 13–125). The patients that survive do not present syndactyly. Their hands and feet are affected only to a minor degree (30). Anemia, growth retardation, and nail dystrophy are frequently present (26,30). With age the healing process is retarded or arrested, leaving large excoriated areas.

Fig. 13–125. *Epidermolysis bullosa, recessive letalis type.* Extensive skin involvement. (From *EB Brain* and *JS Wigglesworth,* Br Dent J **124**:255, 1968.)

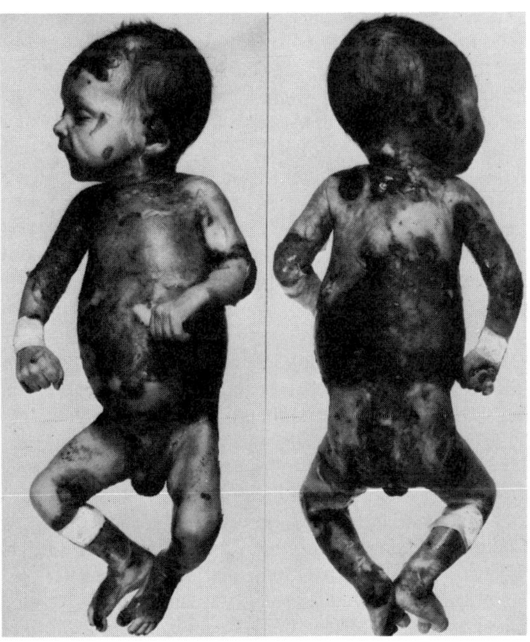

Light microscopic examination demonstrates epidermal–dermal cleavage that follows the rete ridge contour. Inflammatory changes are not originally present (40,48). Vesicles at the dermoepidermal junction in the stratum germinativum of intact skin have been demonstrated (8,39).

Electron microscopic appearance, originally reported by Pearson (46), demonstrated separation between the plasma membrane of the basal cell and the basement membrane. The basic defect is hypoplasia of hemidesmosomes and lack of the subbasal dense plate (31,32).

Oral manifestations. Bullae, which are remarkably fragile and hemorrhagic, are found in nearly all patients, especially at the junction between the hard and soft palates (6). Oral ulcerations as well as blistering can be observed at birth or shortly thereafter (32). Hypoplastic and pitted enamel affecting principally the molars is also seen. Enamel formation is abnormal leading to extensive caries (15). Histologic study reveals cleavage between the epithelium and connective tissue. The cells of the basal layer seem quite palisaded as a result of extracellular vacuolization. Perioral and perinasal crusted and granular hemorrhagic lesions tend to develop between the sixth and the twelfth month of life (26,30,60). These lesions are believed to be pathognomonic for the Herlitz type of EB in older patients (46).

Arwill and co-workers (6) ultrastructurally investigated skin as well as oral mucosa in four cases of the Herlitz form. In contrast to findings in the normal infant, the gingiva and skin exhibited fewer hemidesmosomes and tonofilaments. The initial changes consisted of edema of the subepithelial connective tissue, degeneration of mitochondria, and electron-opaque perinuclear zone that, upon dissolution, frees remnants of tonofibrils and mitochondria, and widened intercellular spaces between epithelial basal cells. Although the desmosomes remained intact, minute vesicles were found between the basement membrane and the basal cell membrane in the intermediate zone. The lamellated pattern of the hemidesmosomes disappeared, replaced by a granular substance.

Arwill and colleagues (5) appear to have been the first to have performed microscopic and microradiographic studies on the teeth in the Herlitz form of the disease. The enamel, lamellar in appearance, varied in thickness from 50 to 400 μm, the outer and inner zones being more markedly mineralized than the intermediate zone. Outside and at varied distances from the surface were free globular structures exhibiting variable mineralization. The cervical enamel was more highly mineralized than the incisal edge. In the permanent teeth there was folding at the dentinoenamel junction, which was normal in the deciduous teeth.

Decalcified sections showed intense proliferation of the dental lamina and the inner and outer enamel epithelium. The latter also manifested metaplasia to stratified squamous epithelium. Numerous vacuoles containing epithelium were observed. The enamel stroma was hyalinized, lamellar, and, at times, globular. Vascular proliferation was marked about the tooth germs, and hemorrhage was noted in the dental sac and stellate reticulum.

Brain and Wigglesworth (11) also investigated the deciduous dentition of an infant with the Herlitz form of EB, noting that the enamel was only 40 μm thick. There was no organized prism structure, and some parts appeared laminated. Globules resembling enamel were distributed between the hypoplastic layer of enamel and the outer enamel epithelium. The reduced enamel epithelium was extensively disorganized. Vesicle-like structures were present adjacent to tooth crowns.

### Dominant dystrophic (hypertrophic) form (Cockayne–Touraine type)

Systemic manifestations. This form of epidermolysis bullosa is characterized by flat, pink, scar-producing bullae of the ankles, knees, hands, elbows, and feet, in decreasing order of frequency (19). Milia are common but less numerous than in the recessive dysplastic type. The nails are usually (in 80%) thick and dystrophic. In contrast to recessive dystrophic cases, the conjunctiva and cornea are never involved. About 20% show changes before the age of 1 year. Improvement seems to occur with age. Hyperhidrosis of palms and soles may also occur.

Oral manifestations. The consensus favors the view that teeth are not affected. Touraine's often quoted statement (57) that 20% manifest oral bullae would appear to be essentially substantiated by other studies (19), but series have been small. Oral milia were noted by Andreasen and associates (3) (Fig. 13–126). These milia are not retention cysts, but epidermoid cysts that originate from detached islands of epithelium in areas of earlier bulla formation.

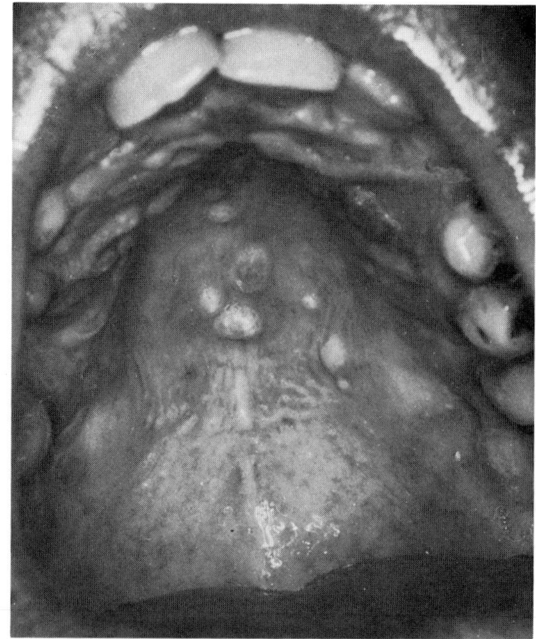

Fig. 13–126. *Epidermolysis bullosa, dominant dystrophic type.* Milia of palate. (From *JO Andreasen et al,* Acta Pathol Microbiol Scand **63**:37, 1965.)

### Scarring EB with dermolytic blisters (Hallopeau–Siemens type)

Systemic manifestations. This type of EB presents various subtypes that are difficult to differentiate (see Table 13–6). All four subtypes have autosomal recessive inheritance. The extensive studies of Gedde-Dahl (25) and Gedde-Dahl and Anton-Lamprecht (26) suggest that the disorder has genetic heterogeneity.

Bullae are usually manifested at or shortly after birth, arising at sites of pressure or trauma or appearing spontaneously. In infants, the most commonly affected areas are the feet, buttocks, scapulas, elbows, fingers, and occiput. In older children, the hands, feet, knees, and elbows are most often involved. When a bulla ruptures or its roof peels off, a raw painful surface is evident. The fluid contained in the bulla is at first sterile but may become secondarily infected and bloody. Upon healing, the bullae often are followed by keloidal scars, causing contraction, and by various degrees of pigmentation or depigmentation. They are frequently associated with milium-like cysts. The scars may lead to loss of bony structures or to interference with growth and resultant dwarfism. Formation of clawhand and the enclosure of the hand in a glove-like epidermal sac have been noted frequently (12) (Fig. 13–127A). The Nikolsky sign is often present. The nails may be extremely involved, often being dystrophic or absent (25) (Fig. 13–127B). Hyperhidrosis of palms and soles is an inconsistent finding but may be marked. Hair may be deficient.

The eyes have been subject to a number of changes: essential shrinkage of the conjunctiva, nonspecific blepharitis, symblepharon, conjunctivitis, and keratitis with corneal opacity and vesicle formation (25).

Hoarseness and, rarely, aphonia and dysphagia may occur as a result of bullae of the larynx or pharynx. Laryngeal stenosis may eventuate from the scarring. Involvement of the esophagus may result in complete obstruction (7,33).

Changes are essentially limited to the hands, feet, and esophagus. The esophagus (most often the upper half) may become segmentally stenotic in childhood, with consequent dysphagia (7,58). The metacarpals become slender and overconstricted and distal phalanges become pointed and clawlike (12).

Oral manifestations. Although a considerable number of authors have remarked on the teeth having hypoplastic enamel with great susceptibility to dental caries, delayed eruption, and frequent retention, there is little documentation concerning frequency of these manifestations (22,36,41,59). We share with Rodermund (50) the belief that there is no correlation between the degree of cutaneous involvement and the degree of dental involvement. Pockmarked alteration of the enamel was handsomely illustrated by Rodermund (50) (Fig. 13–127G). Excellent histologic studies of unerupted teeth have been published by Delaire and colleagues (21) and by Arwill and co-workers (5). Both groups noted hypoplasia of enamel with absence of prismatic structure. Extracted teeth were examined by Crawford et al (18) who found accentuation of enamel tufts formation extending from the dentinoenamel junction to the enamel surface. Additionally irregular dentinal foldings and indentations were found at the dentinoenamel junction.

Oral mucosal involvement occurs soon after birth, vesicles apparently forming from the negative pressure involved in the sucking reflex (Fig. 13–127D). Although oral bullae are said to occur in at least 16% (53), our impression is that the percentage is much higher. The lingual mucosa appears thick, gray, and smooth and may become bound down (Fig. 13–127E,F). Repeated blistering with scar formation may lead to ankyloglossia, tongue atrophy, elimination of buccal and vestibular sulci, and perioral stricture (9,18) (Fig. 13–127F). Severe periodontal disease with alveolar bone resorption had also been noted (22). Even routine dental management may cause the eruption of bullae on the lips and oral mucosa (9). Diminished oral opening following scarring after blistering as well as ankyloglossia due to the same cause have been reported (9). The slightest abrasion from normal

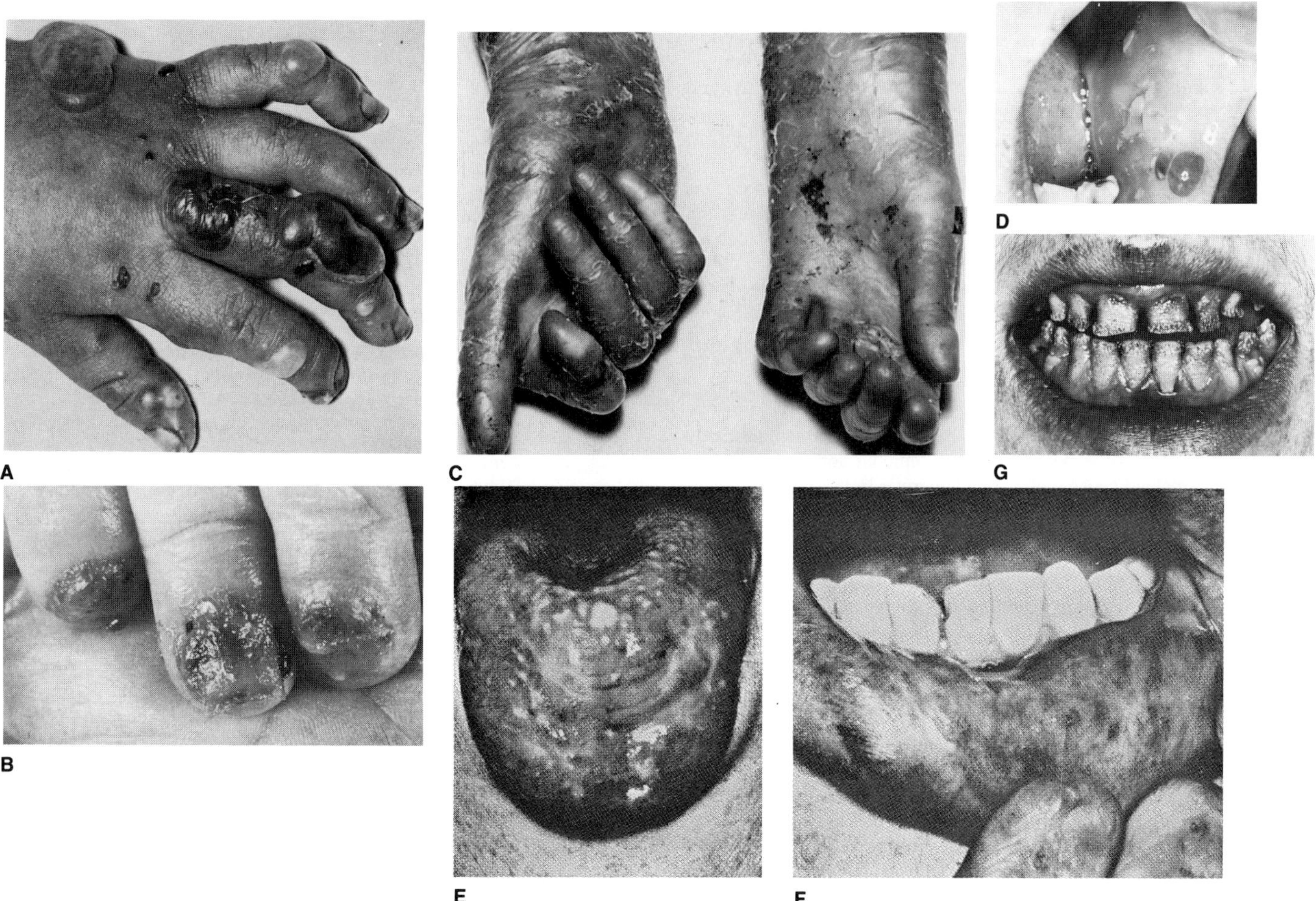

Fig. 13–127. *Epidermolysis bullosa, recessive dystrophic type.* (A) Extensive blister formation of hand. (B) Destructive lesions of fingernails in newborn. (C) Severe involvement of hands. Note dystrophy of nails. (D) Bullae of buccal mucosa. (E) Dorsum of tongue exhibiting numerous scars and complete loss of papillae. Often mobility is limited. (F) Labial mucosa showing extensive scarring. Also note nail dystrophy. (G) Pitted enamel. (A courtesy of *A Lodin*, Stockholm Sweden. C,E and F from *D Winstock*, Br J Dermatol **74**:431, 1962. G from *OE Rodermund*, Dermatol Wochenschr **153**:350, 1967.)

toothbrushing may cause serious sequelae. Other oral changes, the frequencies of which have not been established, are atrophy of the maxilla with resultant relative mandibular prognathism, increased mandibular angle (1,12), and oral carcinoma (37,49,52). Histopathologic changes in the oral mucosa were well demonstrated by Arwill and associates (4). The bullae occur below the PAS-positive basement membrane (39). Hemidesmosomes and tonofibrils are absent or decreased in numbers. Lowe (39) also noted an increase in elastic and preelastic fibers in the corium. Hitchin (34) described defective cementum.

**Oral manifestations of other forms of EB.** *Epidermolysis bullosa acquisita* rarely presents oral manifestations, but they have been described (43). This form of EB is nonhereditary and it manifests in adulthood with blister formation, especially in areas of trauma. It may be associated with other systemic diseases such as amyloidosis, multiple myeloma, diabetes mellitus, and inflammatory bowel disease. The last association may represent a gastrointestinal cutaneous syndrome with an immune pathogenesis (56).

*Scarring or cicatricial junctional type of EB* has autosomal recessive inheritance. Patients present manifestations from birth with skin blisters and lack of nails. There is marked disfigurement, especially of the hands and feet. The skin blisters heal with scar formation producing marked atrophy, especially around the nares. Occasionally a diminished oral opening is noted. Patients present intraoral blisters that are found at any site, but are not as profuse and severe as those in some other forms of EB. All patients with this condition have been reported as having either extensive dental caries, or as being edentulous. Further studies are needed to assess whether there is some form of enamel defect present. Since this type of EB has definite ectodermal alterations (anonychia), an enamel defect is within the realm of possibility (28).

*EB nonscarring atrophicans generalisata mitis (R-EBA-MITIS)* inherited as an autosomal dominant disorder presents oral and dental alterations. Oral mucosa can be the site of blisters, especially trauma related; nodular excrescences on the palate and gingiva have also been reported (27). Teeth can present pitting of enamel, conducive to extensive caries (45).

A possible new variety of EB has been reported by Nielsen and Sjölund (42) in two sisters, offspring of nonconsanguinous parents. These patients presented hemorrhagic blisters localized to hand and feet associated with partial alopecia, onychogryphosis, anodontia, and amelogenesis imperfecta-like condition. Light microscopy showed intraepithelial bullae formation. No EM studies were reported. Further reports will be needed to add this condition to the EB classification. Epidermolysis bullosa has been described in association with pyloric stenosis in two, possibly three sibs. Autosomal recessive inheritance is suggested (14,17,20,38).

**Differential diagnosis.** Recognition of the disease is seldom difficult, even without assessing its entire clinical, genetic, and histologic character. In infants and children, the syndrome may be occasionally confused with conditions such as bullous impetigo (so-called "pemphigus neonatorum"), Ritter disease, porphyria congenita, congenital syphilis, or juvenile bullous dermatitis herpetiformis. In older patients, differential diagnosis includes pemphigus, drug eruptions, dermatitis herpetiformis, and bullous erythema multiforme.

It has been noted that in EB, pulling up of a strip of cellophane adhesive tape will cause the top layers of the skin to adhere to the tape. This phenomenon is not seen in Ritter disease or in generalized impetigo. The lesions of porphyria are generally confined to chronically sun-exposed areas. Only the exposed skin is fragile; epidermolysis bullosa shows no localization of skin fragility.

The differential diagnosis should include Kindler syndrome characterized by bullae formation, since birth, on pressure areas that on healing leave atrophic scars. In addition, affected individuals present severe photosensitivity and poikiloderma. The oral mucosas are atrophic and present multiple white macules (54).

**Laboratory aids.** Microscopically, the bullae are intraepidermal in the nonscarring forms, at the level of the epithelial dermal junction in the trophic forms and subepithelial or intradermal in the scarring forms. Unfortunately light microscopy is not useful in differentiating between the severe types of EB. Periodic acid–Schiff (PAS) stain is useful in subdividing the various type of EB according to the location of the basement membrane, but is unreliable since during the process of blister formation there will be retractions in the actual position of the basal membrane. The basal membrane remains attached to the dermis in the EB simplex and junctional types whereas it remains adherent to the epidermis in the dystrophic types of EB.

Ultrastructural examinations are essential for an accurate diagnosis of EB types; neither regular light microscopy nor clinical presentation is sufficient or adequate for a proper diagnosis.

Tonofilaments, which are the intercellular bindings of the basal cells, have been shown to be defective in the intraepidermal types of EB. The junctional types of EB are characterized by defects in anchoring filaments of hemidesmosomes whereas primary or secondary destruction of the anchoring fibrils that extend from the basal lamina into the dermis are observed in the dermal types of EB (31).

Other pathogenetic mechanisms for certain types of EB include degradation of collagen through collagenase and gelatinase (51). Excessive secretion of collagenase by dystrophic fibroblasts from patients with the simplex and the dystrophic forms of EB has been reported (44).

Immunofluorescent mapping studies are used to differentiate among various types of EB by detecting the presence of four structural components within skin basement membrane: bullous pemphigoid antigen, laminin, Type IV collagen, and lamina densa antigen-1 (LDA-1). The diagnosis is made by identifying the location of these antigens and their relation to the level of cleavage in a freshly induced blister (24). Prenatal diagnosis of the various types of EB is possible through fetoscopy with fetal skin sampling, electron microscopic studies (2), and immunofluorescent mapping (24a).

The severe autosomal recessive dystrophic form has been found to be associated with Type VII collagen deficiency (13), but this needs confirmation.

It has been reported (SK Tyring, personal communication, 1989) that children with the autosomal recessive form of EB exhibit homozygosity for HLA-DR4. Obviously, this needs to be confirmed.

### References (Epidermolysis bullosa)


1. Alpert M: Roentgen manifestations of epidermolysis bullosa. *Am J Roentgenol* **78**:66–72, 1957.

2. Anton-Lamprecht I et al: Prenatal diagnosis of epidermolysis bullosa dystrophica Hallopeau-Siemens with electron microscopy of fetal skin. *Lancet* **2**:1077–1079, 1981.

3. Andreasen JO et al: Milia formation in oral lesions in epidermolysis bullosa. *Acta Pathol Microbiol Scand* **63**:37–41, 1965.

4. Arwill T et al: Epidermolysis bullosa hereditaria. IV. Histologic changes of the oral mucosa in the polydysplastic, dystrophica, and the letalis forms. *Odontol Rev* **16**:101–111, 1965.

5. Arwill T et al: Epidermolysis bullosa hereditaria, III. Histologic study of changes in teeth in the polydysplastic, dystrophic, and lethal forms. *Oral Surg* **19**:723–744, 1965.

6. Arwill T et al: Epidermolysis bullosa hereditaria. V. The ultrastructure of oral mucosa and skin in four cases of the letalis form. *Acta Pathol Microbiol Scand* **74**:311–324, 1968.

7. Becker MH, Swinyard CA: Epidermolysis bullosa dystrophica in children; radiologic manifestations. *Radiology* **90**:124–128, 1968.

8. Bergenholtz A, Olsson O: Epidermolysis bullosa hereditaria, I Epidermolysis bullosa hereditaria letalis; a survey of the literature and report of 11 cases. *Acta Derm Venereol* **48**:220–241, 1968.

9. Block MS, Gross BD: Epidermolysis bullosa dystrophica recessiva. *J Oral Maxillofac Surg* **40**:753–758, 1982.

10. Bowwes Bavinck JN et al: Autosomal dominant epidermolysis bullosa dystrophica: Are the Cockayne-Touraine, the Pasini and the Bart-types different expressions of the same mutant gene? *Clin Genet* **31**:416–424, 1987.

11. Brain EB, Wigglesworth JS: Developing teeth in epidermolysis bullosa hereditaria letalis; a histologic study. *Br Dent J* **124**:255–260, 1968.

12. Brinn LB, Khilnam MT: Epidermolysis bullosa with characteristic hand deformities. *Radiology* **89**:272–274, 1967.

13. Bruckner-Tuderman L et al: Lack of type VII collagen in unaffected skin of patients with severe recessive dystrophic epidermolysis bullosa. *Dermatologica* **176**:57–64, 1988.

13a. Bruckner-Tuderman L et al: Epidermolysis bullosa simplex with mottled pigmentation. JADA **21**:425–432, 1989.

14. Bull MJ et al: Epidermolysis bullosa–pyloric stenosis. *Am J Dis Child* **137**:449–451, 1983.

15. Carroll DL et al: Epidermolysis bullosa—review and report of case. *J Am Dent Assoc* **107**:749–751, 1983.

16. Cooper TW, Bauer EA: Epidermolysis bullosa—a review. *Ped Dermatol* **1**:181–188, 1984.

17. Cowton JAL et al: Epidermolysis bullosa in association with aplasia cutis congenita and pyloric atresia. *Acta Paediatr Scand* **71**:155–160, 1982.

18. Crawford EG et al: Hereditary epidermolysis bullosa: Oral manifestations and dental therapy. *Oral Surg* **42**:490–500, 1976.

19. Davidson BCC: Epidermolysis bullosa. *J Med Genet* **2**:233–242, 1965.

20. De Groot WG et al: Familial pyloric atresia associated with epidermolysis bullosa. *J Pediatr* **92**:429–431, 1978.

21. Delaire J et al: Manifestation buccodentaires des epidermolyses bulleuses. *Rev Stomatol (Paris)* **61**:189–200, 1960.

22. D'Angelo M: Le lesioni orali nell'epidermolisi bollosa. *Min Stomatol* **30**:169–174, 1981.

23. Fine JD: Epidermolysis bullosa. *Int J Dermatol* **25**:143–157, 1986.

24. Fine JD, Gay S: LDA-1 monoclonal antibody. *Arch Dermatol* **122**:48–51, 1986.

24a. Fine JD et al: Prenatal diagnosis of dominant and recessive dystrophic epidermolysis bullosa. Application and limitations in the use of KF-1 and LH 7:2 monoclonal antibodies and immunofluorescent mapping technics. *J Invest Dermatol* **91**:465–471, 1988.

24b. Fine JD et al: Autosomal recessive epidermolysis bullosa simplex. *Arch Dermatol* **125**:931–938, 1989.

24c. Fischer T, Gedde-Dahl TJ: Epidermolysis bullosa simplex and mottled pigmentation: A new dominant syndrome. *Clin Genet* **15**:228–238, 1979.

25. Gedde-Dahl T Jr: *Epidermolysis Bullosa: A Clinical, Genetic and Epidemiologic Study,* Johns Hopkins Press, Baltimore, 1971.

26. Gedde-Dahl T Jr, Anton-Lamprecht I: Epidermolysis bullosa, in *Principles and Practice in Medical Genetics,* Emery AEH, Rimoin DL (eds), Churchill Livingstone, Edinburgh, 1983.

27. Giansanti JS: Oral nodular excrescences in epidermolysis bullosa. *Oral Surg* **40**:385–390, 1975.

28. Haber RM et al: Cicatricial junctional epidermolysis bullosa. *J Am Acad Dermatol* **12**:836–844, 1985.

29. Haber RM et al: Epidermolysis bullosa simplex with keratoderma of the palms and soles. *J Am Acad Dermatol* **12**:1040–1044, 1985.

30. Haber RM et al: Hereditary epidermolysis bullosa. *J Am Acad Dermatol* **13**:252–278, 1985.

31. Hashimoto et al: Ultrastructural studies in epidermolysis bullosa hereditaria, IV. Recessive dystrophic type with junctional blistering. *Arch Dermatol Res* **257**:17–32, 1976.

32. Hashimoto I et al: Epidermolysis bullosa hereditaria letalis: Report of a case and probable ultrastructural defects. *Helv Paediatr Acta* **30**:543–552, 1976.

33. Hillemeier C et al: Esophageal web: A previously unrecognized complication of epidermolysis bullosa. *Pediatrics* **67**:678–682, 1981.

34. Hitchin AD: The defects of cementum in epidermolysis bullosa dystrophica. *Br Dent J* **135**:437–441, 1973.

35. Kero M, Niemi KM: Epidermolysis bullosa. *Int J Dermatol* **25**:75–82, 1986.

36. Kinast H, Schuh E: Die Epidermolysis bullosa und ihre orale Erscheinungsform. *Oest Z Stomatol* **70**:166–175, 1973.

37. Klausner E: Zungenkrebs als Folgestand bei einem Falle von Epidermolysis bullosa (dystrophische Form). *Arch Dermatol Syph (Berlin)* **116**:71–78, 1913.

37a. Kletter G et al: Congenital muscular dystrophy and epidermolysis bullosa simplex. *J Pediatr* **114**:104–107, 1989.

38. Korber JS, Glasson MJ: Pyloric atresia associated with epidermolysis bullosa. *J Pediatr* **90**:600–601, 1977.

39. Lowe LB: Hereditary epidermolysis bullosa. *Arch Dermatol* **95**:587–595, 1967.

40. Maddison TG, Barter RA: Epidermolysis bullosa hereditaria letalis. *Arch Dis Childh* **36**:337–339, 1961.

41. Matras H: Ein Beitrag zur Epidermolysis bullosa dystrophica mit Zahn– und Mundschleimhautveränderungen. *Oest Z Stomatol* **60**:138–145, 1963.

42. Neilsen PG, Sjölund E: Epidermolysis bullosa simplex localisata associated with anodontia, hair and nail disorders: A new syndrome. *Acta Dermatol Venereol* **65**:526–530, 1985.

43. Nielsen R et al: Oral lesions of epidermolysis bullosa acquista. *Oral Surg* **45**:749–754, 1978.

43a. Niemi KM et al: Epidermolysis bullosa simplex with muscular dystrophy with recessive inheritance. *Arch Dermatol* **124**:551–554, 1988.

44. Oakley CA, Priestley GC: Collagen synthesis and degradation by epidermolysis bullosa fibroblasts. *Acta Dermatol Venereol* **65**:277–281, 1985.

45. Paller AS et al: The generalized atrophic benign form of junctional epidermolysis bullosa. *Arch Dermatol* **122**:704–710, 1986.

46. Pearson RW: Epidermolysis bullosa hereditaria letalis. *Arch Dermatol* **109**:349–355, 1974.

47. Ritzenfeld P: Zur Histogenese und Differentialdiagnose hereditären Epidermolysen. *Arch Klin Exp Dermatol* **224**:128–137, 1966.

48. Robert MH et al: Epidermolysis bullosa letalis, report of three cases with particular reference to the histopathology of the skin. *Pediatrics* **25**:283–290, 1960.

49. Röckl H: Carcinom bei Epidermolysis bullosa dystrophica. *Hautarzt* **65**:477–483, 1963.

50. Rodermund OE: Zahnveränderungen bei Epidermolysis bullosa. *Dermatol Wochenschr* **153**:350–357, 1967.

51. Sanchez G et al: Generalized dominant epidermolysis bullosa simplex. Decreased activity of a gelati-olytic protease in cultured fibroblasts as a phenotypic marker. *J Invest Dermatol* **81**:576–579, 1981.

52. Schiller F: Zungencarcinom bei Epidermolysis bullosa dystrophica. *Arch Klin Exp Dermatol* **209**:643–651, 1960.

53. Schuermann H: *Krankheiten der Mundschleimhaut und der Lippen,* 2nd ed, Urban & Schwarzenberg, Berlin, 1958.

54. Shoshana Hachman Z, Garfunkel AA: Kindler syndrome in two related Kurdish families. *Am J Med Genet* **20**:43–48, 1985.

55. Tabas M et al: The mechanobullous disease. *Dermatol Clinics* **5**:123–136, 1987.

56. Thomas LR et al: Epidermolysis bullosa acquisita and inflammatory bowel disease. *J Am Acad Dermatol* **6**:242–252, 1982.

57. Touraine MA: Classification des epidermolyses bulleuses. *Ann Dermatol Syph (Paris)* **2**:309–312, 1942.

58. Wey W, Schnyder UW: Über Ösphagustenosen bei Epidermolysis bullosa hereditaria und ihre Busehandlung. *Dermatologica* **128**:173–183, 1964.

59. Winstock D: Oral aspects of epidermolysis bullosa. *Br J Dermatol* **74**:431–438, 1962.

60. Wright JR: Epidermolysis bullosa: Dental and anesthetic management of two cases. *Oral Surg* **57**:155–157, 1984.

## Hyalinosis cutis et mucosae (lipoid proteinosis, Urbach–Wiethe syndrome)

The syndrome, consisting of yellow nodular infiltration of skin and mucous membranes and intracranial calcifications, was first described by Siebenmann (47) in 1908. Other cases were reported in the 1920s, but credit is usually given to Urbach and Wiethe (52,53,56) who defined the condition and applied the terms "lipoidosis cutis et mucosae" and "lipoid proteinosis" (33). The history of the syndrome has been reviewed by Laymon and Hill (33) and Gordon et al (16). About 200 patients have been described.

The syndrome is transmitted as an autosomal recessive trait. It has appeared in sibs (12,21,28,45,50,53); in about 20% of cases there has been parental consanguinity (4,5,31,34). Many patients have been of Dutch or German ancestry.

**Skin and skin appendages.** Discrete or confluent yellow-ivory nodules, from pinhead to matchhead in size, are found on the face, neck, axillas, and hands early in life. The perineum and scrotum have also been involved (16). Some patients develop bullous lesions of the face and upper extremities that have sometimes been described as impetigenous (9,16,22,30,45); on healing, these lesions leave hyperpigmented, pitted, atrophic patches (10,16,27) (Fig. 13-128). Brown-yellow, wartlike hyperkeratotic lesions occur on the knees, elbows, and interarticular surfaces of the fingers (1,10,16,27,38,39,40,46) (Fig. 13-129B,C). With time, there is generalized skin thickening.

Diffuse or patchy alopecia of the scalp hair is common (4–6, 17,19,30,36,37,40,49,50).

**Ophthalmologic abnormalities.** Bead-like excrescences (moniliform blepharosis) are found on the upper and lower eyelid margins in almost all patients (26,27), followed in some cases by loss of cilia (19,33,40) (Fig. 13-129A). Fundus changes, consisting of sharp or poorly demarcated small, round, grey-white or yellow-white lesions were seen in 9 of 20 patients described by Heyl (22); in addition, vascular tortuosity was found. Finlay (10) described significant vascular abnormalities in a 16-year-old legally blind patient who had visual problems since age 7. Other investigators noted drusen-like changes in patients as young as 23 and 34 years of age (42,44). On histologic examination, Marquardt (35) felt these lesions differed from classic drusen because they extended into the choriocapillaris. Histochemically they were identical to the skin lesions. Glaucoma was seen in several cases (14,35).

**Central nervous system.** Bilateral intracranial calcifications, found in at least 70%, have been located in the hypothalamus (37) and above the pituitary fossa in the hippocampus, falx cerebri, or temporal lobes (5,10,15–17,28,33,37,38,45,53a,55,59); these abnormalities are more often seen in individuals over 10 years of age (16). Calcification has been localized to the head of the caudate nucleus, globus pallidus, and amygdaloid nucleus (38). Seizures have been reported (5,22,33,34,38) (Fig. 13-130A,B). Emsley and Paster (7) reported two patients with long-standing memory impairment, paranoia, and bilateral temporal lobe calcification.

**Larynx and other mucosal involvement.** The voice in almost all patients is hoarse from birth (10,17,26,54), infancy (11,46), or from

Fig. 13–128. *Hyalinosis cutis et mucosae.* Note numerous raised yellowish nodules. (From *KH Holtz,* Arch Klin Exp Dermatol **214**:289, 1962.)

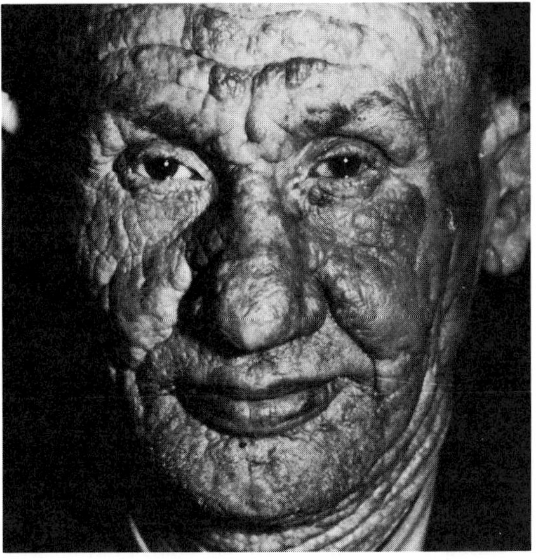

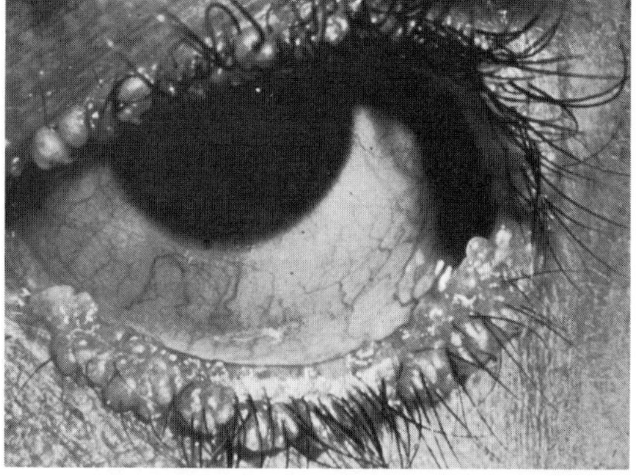

A

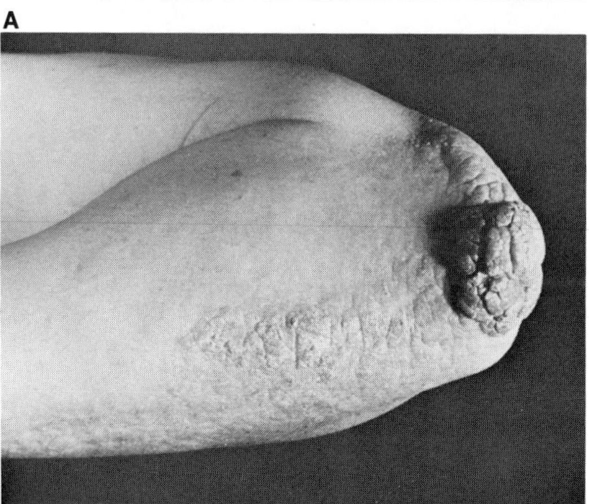

B

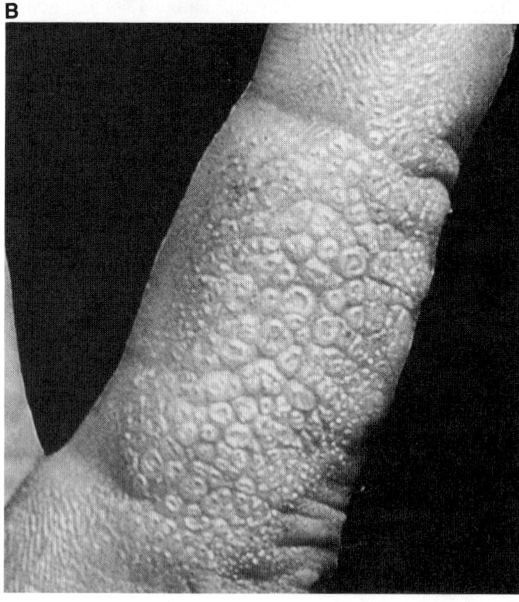

C

Fig. 13–129. *Hyalinosis cutis et mucosae.* (A) Beadlike excrescences among eyelid margins. (B) Verrucous lesions of elbow. (C) Verrucous lesions of fingers. (A from *AR Rosenthal* and *JR Duke,* Am J Ophthalmol **64**:1120, 1967; B from *RC Juberg et al,* J Med Genet **12**:110, 1975.)

the first few years of life (9,24,46) and a low-pitched cry has been described (6,38); these findings testify to early laryngeal involvement. Laryngoscopic examination reveals yellow-white plaques on the epiglottis, aryepiglottic folds, and interarytenoid regions (6,9,16,26,40,41). The cords are thickened (6,46,59) and nodular (13,18,59). Extension of the infiltration to the tonsils and pharynx may result in severe dysphagia (10,38) and dyspnea necessitating tracheostomy (3,16).

Other mucosal surfaces, such as the vulva, esophagus, stomach, and rectum, are also affected (5,16,17,23,53). Wart-like lesions have been found in the main bronchi (39).

**Oral manifestations.** Oral manifestations have been summarized (1,25). Oral mucous membranes are extensively infiltrated with yellow-white, elevated granules and pea-size plaques, which usually appear before puberty and gradually increase in severity (25). The lower lip is usually more severely affected than the upper, and has a cobblestone appearance (Fig 13-131A). Radiating fissures may appear at the angles of the mouth (9,30,33), and on the lips (58) and tongue (17,26).

The tongue becomes firm or woody, thick, large, and restricted in mobility (1,3,5,34,40). The dorsum loses its papillae (9,26,31,38,44). There is usually marked infiltration of the lingual frenum, which becomes inelastic (3). Ulcers may develop (4,10,17,22,54,57) (Fig. 13-131B–D).

With infiltration of the buccal mucosa, the opening of the parotid duct may become stenotic, with ensuing retrograde parotitis (1, 4,22,23,30,33,34,38,39,49). The soft palate and uvula may also be involved (48) (Fig. 13-132). Affected individuals may not be able to open their mouths widely (16).

Specific dental abnormalities have not been documented. However, patients have been described as having lost their teeth at an early age (10); perhaps this loss was due to dental caries, resulting from obstruction of the orifices of major salivary gland ducts, as well as hyalinization of minor salivary glands.

**Differential diagnosis.** Conditions to be considered include amyloidosis, colloid milium, adenoma sebaceum, *pseudoxanthoma elasticum,* and porphyria cutanea tarda.

**Laboratory aids.** Slight elevation in phosopholipid level has been reported in some patients (17). Several investigators noted an increase in $\alpha_2$- and $\gamma$-globulin levels and a diminution in the concentration of albumins (16,17,28), but these changes have been denied by others (33).

On light microscopic examination of the skin, there is hyalinization of the superficial connective tissue, beginning about the small and medium sized blood vessels and sweat glands. Concentric laminar thickening of the stroma around these structures is noted (8). Hyalinization later extends more deeply (2) (Fig. 13-133A,B). Biopsy of oral mucous membrane has also shown hyalinization, which may replace mucous gland acini (18,48). Histochemical study (17) has revealed the presence of carbohydrate but little or no lipids. The hyaline material stains intensely with the PAS procedure (46,51), thus appearing to be a glycoprotein either free or loosely bound to collagen (12,23,46). Ultrastructural studies have been carried out on the hyaline material (6a,21,32,37a,43). Several authors (21,37a,38) have suggested that it represents an altered collagen. Ishibashi (29), by electron microscopy, determined that the hyalinization was an accumulation of Type IV collagen and laminin. Concentric layers of basal laminae are seen surrounding blood vessels in an onionskin arrangement (20,37a).

On biochemical analysis, Harper et al (20) found reduced amounts of collagen in skin; however, the proportion of Type III to Type I collagen was elevated. In addition, the amount of Type V collagen was five times normal. Immunologic studies showed a patchy, diffuse, widely distributed Type III collagen, with increases in Types IV and V collagen. An abnormal pattern of collagen cross-linkage was also found. Fleischmajer et al (13) concluded that the hyaline material originated from overproduction of noncollagenous proteins.

Results of studies of Bauer et al (2) suggested that the disorder was a lysosomal storage disease; cultured fibroblasts displayed a 3- to 4-fold increase in total hexuronic acid. Studies of cultured fibroblasts by Shore et al (46) found a lipid content similar in quantity and distribution to that of control cells.

## References [Hyalinosis cutis et mucosae (lipoid proteinosis, Urbach–Wiethe syndrome)]

1. Bergenholtz H et al: Oral, pharyngeal and laryngeal manifestations in Urbach–Wiethe disease. *Ann Clin Res* **9**:1–7, 1977.

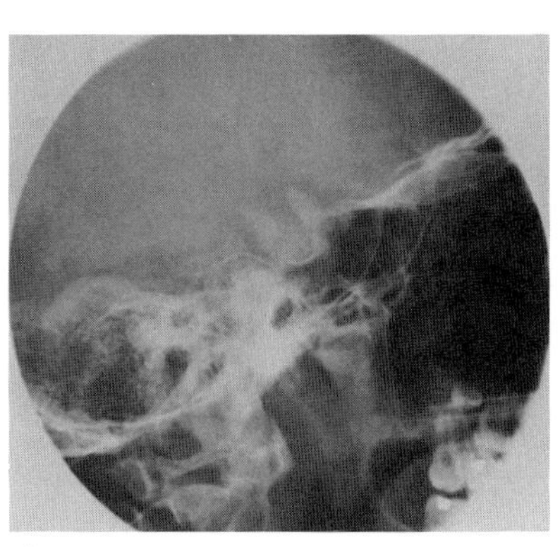

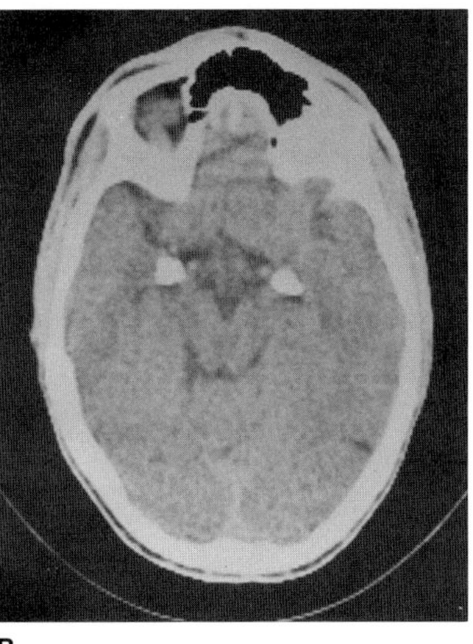

A B

Fig. 13–130. *Hyalinosis cutis et mucosae.* (A,B) Radiograph and CAT scan showing calcification of hippocampal gyri. (From *K Fochem et al,* Roefo **138**:265, 1983.)


2. Bauer EA et al: Lipoid proteinosis: In vivo and in vitro evidence for a lysosomal storage disease. *J Invest Dermatol* **76**:119–125, 1981.

3. Blodi FC et al: Lipid-proteinosis (Urbach–Wiethe) involving the lids. *Trans Am Ophthalmol Soc* **63**:155–166, 1960.

4. Burnett JW, Marcy SM: Lipoid proteinosis. *Am J Dis Child* **105**:81–84, 1963.

5. Caplan RM: Visceral involvement in lipoid proteinosis. *Arch Dermatol* **95**:149–155, 1967.

6. Cowan MA et al: Case of lipoid proteinosis. *Br Med J.* **2**:557–560, 1961.

6a. Dupre A et al: Etude ultrastructurale d'un cas de hyalinosis cutanéo-muqueuse. *Ann Dermatol Venereol* **108**:1003–1010, 1981.

7. Emsley RA, Paster L: Lipoid proteinosis presenting with neuropsychiatric manifestations. *J Neurol Neurosurg Psychiat* **48**:1290–1292, 1985.

8. Fabrizi G et al: Urbach–Wiethe disease: Light and electron microscopic study. *J Cutan Pathol* **7**:8–20, 1980.

9. Feiler-Ofry V et al: Lipoid proteinosis (Urbach–Wiethe syndrome). *Br J Ophthalmol* **63**:694–698, 1979.

10. Findlay GH: Hyalinosis cutis et mucosae (lipoid proteinosis). *S Afr Med J.* **34**:189–195, 1960.


Fig. 13–131. *Hyalinosis cutis et mucosae.* (A) Extensive infiltration of lower lip. (B,C) Tongue is thickened and indurated with plaques on dorsum, sides, and undersurface. (D) Advance infiltration of tongue in adult. (A from *JA Keipert,* Aust Paediatr J **6**:135, 1970; B,C from *RF Dickey* and *S Davis,* Ann Otol Rhinol Laryngol **73**:287, 1964; D from *RC Juberg et al,* J Med Genet **12**:110, 1975.)

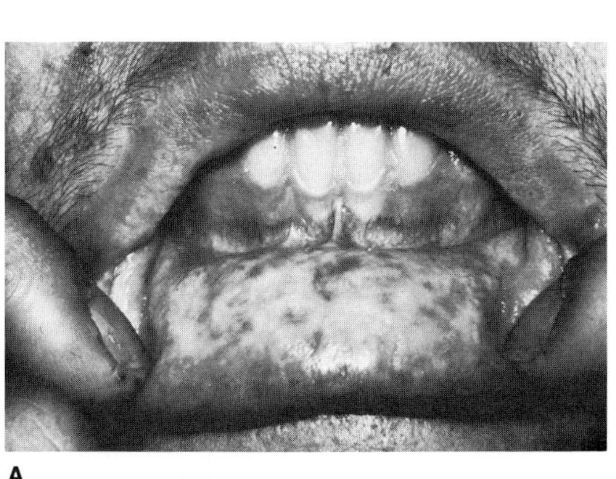

A

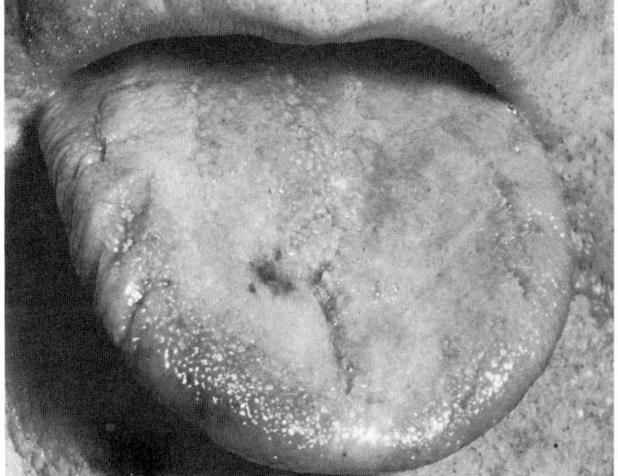

B

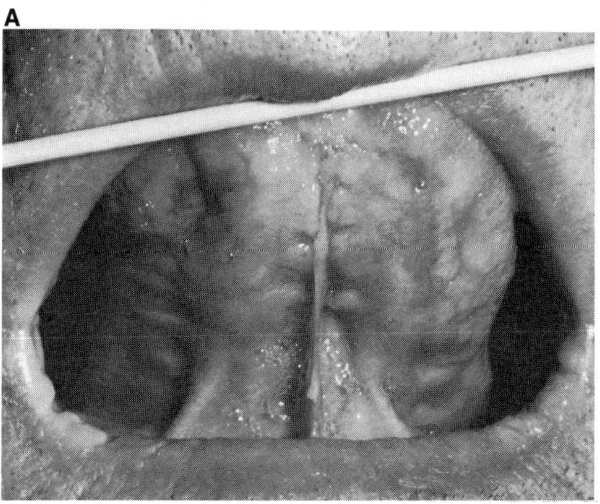

C

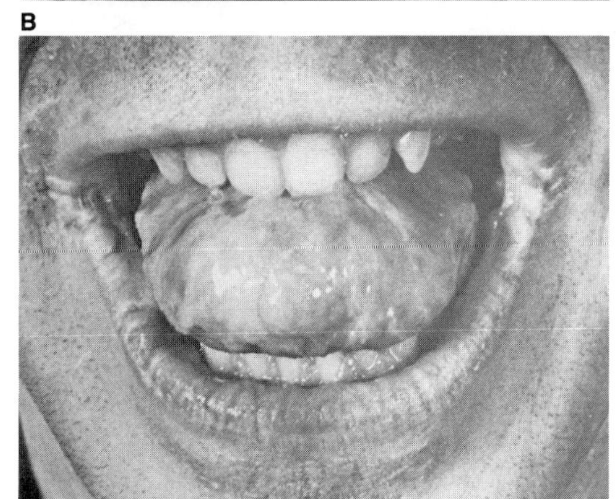

D

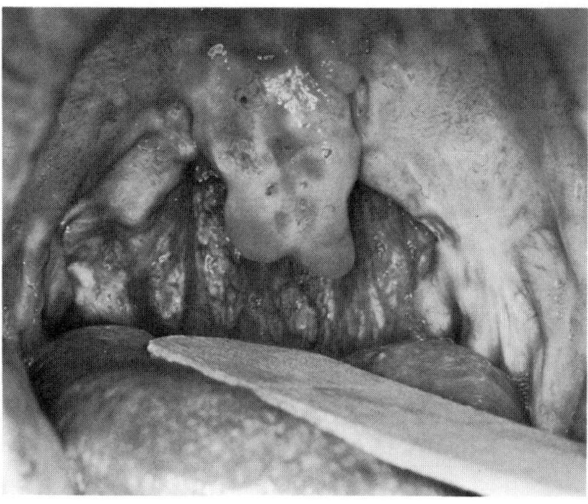

Fig. 13–132. *Hyalinosis cutis et mucosae.* Infiltrated mucosa of soft palate and uvula. (Courtesy of *W Mootz,* Homburg, Saar, Germany.)

11. Finkelstein MW et al: Hyalinosis cutis et mucosae. *Oral Surg Oral Med Oral Pathol* **54**:49–58, 1982.

12. Fleischmajer R et al: Hyalinosis cutis et mucosae: A histochemical staining and analytical biochemical study. *J Invest Dermatol* **52**:495–503, 1969.

13. Fleischmajer R et al: Ultrastructure and composition of connective tissue in hyalinosis cutis et mucosae skin. *J Invest Dermatol* **82**:252–258, 1984.

14. François J et al: Manifestations oculaires du syndrome d'Urbach–Wiethe. Hyalinosis cutis et mucosae. *Ophthalmologica* **155**:433–448, 1968.

Fig. 13–133. *Hyalinosis cutis et mucosae.* (A) Deposition of hyaline material around sweat glands in lower dermis. (B) Amorphous deposits lying directly beneath epithelium. (A from *R Shore,* Arch Dermatol **110**:591, 1974; B from *AR Rosenthal* and *JR Duke,* Am J Ophthalmol **64**:1120, 1967.)

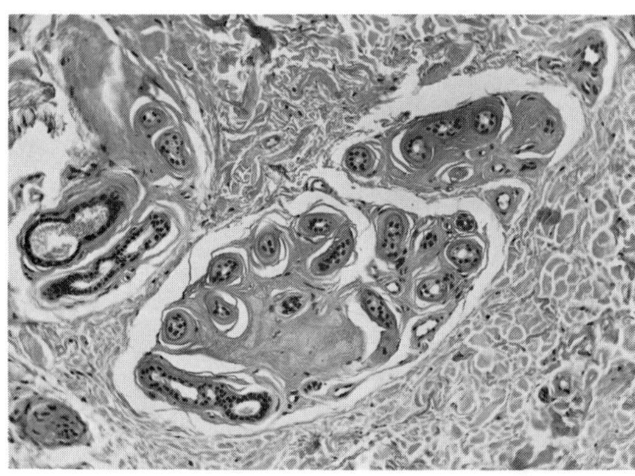

A

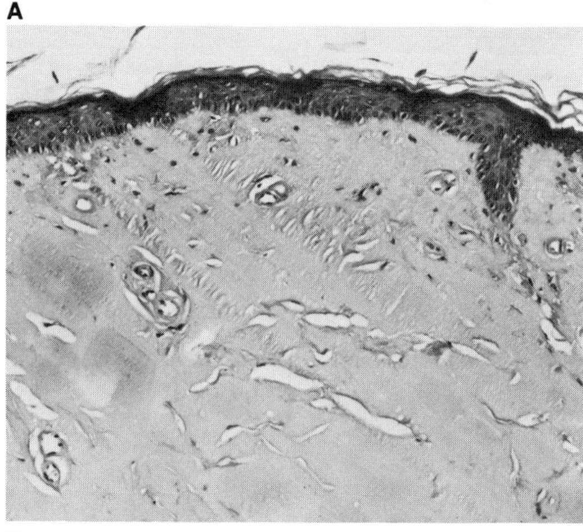

B

15. Friedman L et al: Radiographic and computed tomographic findings in lipid proteinosis. A case report. *S Afr Med J* **65**:734–735, 1984.

16. Gordon H et al: Lipoid proteinosis. *Birth Defects* **7**(8):164–177, 1972.

17. Grosfeld JCM et al: Hyalinosis cutis et mucosae (Lipoid proteinosis Urbach–Wiethe). *Dermatologica* **130**:239–266, 1965.

18. Haneke E et al: Hyalinosis cutis et mucosae in siblings. *Hum Genet* **68**:342–345, 1984.

19. Hardcastle SW, Rosenstrauch WJ: Lipoid proteinosis. A case report. *S Afr Med J* **66**:273–274, 1984.

20. Harper JI et al: Lipoid proteinosis: An inherited disorder of collagen metabolism? *Br J Dermatol* **113**:145–151, 1985.

21. Hashimoto K et al: Hyalinosis cutis et mucosae: An electron microscopic study. *Acta Dermatol Venereol* **52**:179–195, 1972.

22. Heyl T: Lipoid proteinosis. Clinical picture. *Br J Dermatol* **75**:465–472, 1963.

23. Hofer P-Å: Urbach–Wiethe disease (lipoglycoproteinosis; lipoid proteinosis; hyalinosis cutis et mucosae). A review. *Acta Dermatol Venereol* **53** (Suppl **71**):1–52, 1973.

24. Hofer P-Å: Urbach–Wiethe disease (lipoglycoproteinosis; lipoid proteinosis; hyalinosis cutis et mucosae). A clinico-genetic study of 14 families from northern Sweden. *Hereditas* **77**:209–218, 1974.

25. Hofer P-Å, Bergenholtz A: Oral manifestations in Urbach–Wiethe disease (lipoglycoproteinosis; lipoid proteinosis; hyalinosis cutis et mucosae). *Odontol Rev* **26**:39–57, 1975.

26. Hofer P-Å, Öhman J: Laryngeal lesions in Urbach–Wiethe disease. *Acta Pathol Microbiol Scand* **82A**:547–558, 1974.

27. Hofer P-Å et al: A clinical and histopathological study of twenty-seven cases of Urbach–Wiethe disease. *Acta Pathol Microbiol Scand* A (Suppl **245**): 7–87, 1974.

28. Holtz KH: Über Gehirn- und Augenveränderungen bei Hyalinosis cutis et mucosae mit Autopsiebefund. *Arch Klin Exp Dermatol* **214**:289–306, 1962.

29. Ishibashi A: Hyalinosis cutis et mucosae. Defective digestion and storage of basal lamina glycoprotein synthesized by smooth muscle cells. *Dermatologica* **165**:7–15, 1982.

30. Jensen A: Lipoidosis cutis et mucosae (lipoid proteinosis) in 2 brothers. *Acta Dermatol Venereol* **42**:164–166, 1962.

31. Juberg RC et al: A case of hyalinosis cutis et mucosae (lipoid proteinosisof Urbach and Wiethe) with common ancestors in four remote generations. *J Med Genet* **12**:110–112, 1975.

32. König WF: Elektronenmikroskopische Untersuchungen bei Hyalinosis cutis et mucosae. *Z Laryngol Rhinol* **45**:672–675, 1966.

33. Laymon CW, Hill EM: An appraisal of hyalinosis cutis et mucosae. *Arch Dermatol* **75**:55–65, 1957.

34. Macleod W: Hyalinosis cutis et mucosae. *Ann Radiol* **9**:305–310, 1966.

35. Marquardt R: Manifestation der Hyalinosis cutis et mucosae (Lipoidproteinose Urbach–Wiethe) am Auge. *Klin Monatsbl Augenheilkunde* **140**:684–692, 1962.

36. McCusker J, Caplan RM: Lipoid proteinosis (lipoglycoproteinosis): A histochemical study of two cases. *Am J Pathol* **40**:599, 1962.

37. Meenan FOC et al: Lipoid proteinosis: A clinical, pathological and genetic study, *Q J Med* **47**:549–561, 1978.

37a. Moy LS et al: Lipoid proteinosis: Ultrastructural and biochemical studies. *J Am Acad Dermatol* **16**:1193–1201, 1987.

38. Newton FH et al: Neurologic involvement in Urbach–Wiethe's disease (lipoid proteinosis). *Neurology (Minneap)* **21**:1205–1213, 1971.

39. Ramsey ML et al: Lipoid proteinosis. *Int J Dermatol* **24**:230–232, 1985.

40. Rasiewicz W et al: Hyalinosis cutis et mucosae Urbach–Wiethe. Report of a case and results of histochemical studies. *Dermatologica* **130**:145–157, 1965.

41. Richards SH, Bull PD: Lipoid proteinosis. *J Laryngol Otolaryngol* **87**:187–190, 1973.

42. Rodermund OE: Zur Hyalinosis cutis et mucosae (Urbach–Wiethe). *Z Haut Geschl Krkh* **43**:493–503, 1968.

43. Rodermund OE, Klingmüller G: Elektronenmikroskopische Befund des Hyalins bei Hyalinosis cutis et mucosae. *Arch Klin Exp Dermatol* **236**:238–249, 1970.

44. Rosenthal AR, Duke JR: Lipoid proteinosis: Case report of direct lineal transmission. *Am J Ophthalmol* **64**:1120–1125, 1967.

45. Scott FP, Findlay GH: Hyalinosis cutis et mucosae (lipoid proteinosis). *S Afr Med J* **34**:189–195, 1960.

46. Shore RN et al: Lipoid proteinosis. *Arch Dermatol* **110**:591–594, 1974.

47. Siebenmann F: Über Mitbeteilingung der Schleimhaut bei allgemeiner Hyperkeratose der Haut. *Arch Laryngol* **20**:101–109, 1908.

48. Simpson HE: Oral manifestations in lipoid proteinosis. *Oral Surg Oral Med Oral Pathol* **33**:528–532, 1972.

49. Tompkins J, Weinstein IM: Lipoid proteinosis: Two case reports includ-

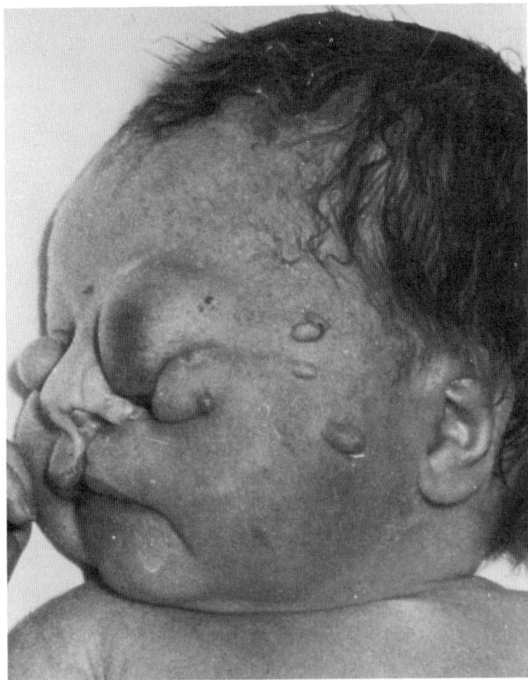

Fig. 13–134. *Oculocerebrocutaneous syndrome.* Note orbital cyst in addition to cleft lip and numerous skin tags. (From *JW Delleman et al,* Clin Genet **25**:470, 1984.)

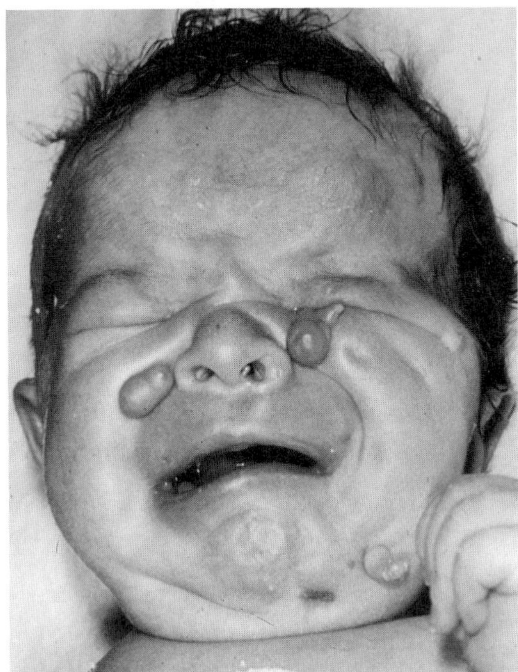

Fig. 13–135. *Oculocerebrocutaneous syndrome.* Note multiple skin tags, depressions in skin of chin and left cheek. (Courtesy of *I Schafer,* Cleveland, Ohio.)

ing liver biopsies, special blood lipid analysis and treatment with a lipotropic agent. *Ann Intern Med* **41**:163–171, 1954.

50. Tripp RN: Lipoidosis cutis et mucosae. *NY State J Med* **36**:619–626, 1936.

51. Ungar H, Katzenellenbogen I: Hyalinosis of skin and mucous membrane (Urbach–Wiethe's lipoid proteinosis). *Arch Pathol* **63**:65–74, 1957.

52. Urbach E: Über eine familiäre lokale Lipoidose der Haut und der Schleimhäute auf Grundlage einer diabetischen Stoffwechselstörung. *Arch Dermatol Syph (Berlin)* **159**:451–466, 1929.

53. Urbach E, Wiethe C: Lipoidosis cutis et mucosae. *Virchows Arch Pathol Anat* **273**:285–319, 1929.

53a. Walter E, Petersen D: Röntgenbefunde bei der Lipoid proteinose. *Roefo* **135**: 556–560, 1981.

54. Ward WQ et al: Lipoid proteinosis. *Birth Defects* **7**(8): 288–291, 1971.

55. Weidner WA et al: Roentgenographic findings in lipoid proteinosis. *Am J Roentgenol* **110**:457–461, 1970.

56. Wiethe C: Kongenitale diffuse Hyalinablagerungen in den obernen Luftwegen, familiärer auftretend. *Z Hals Nas Ohrenheilkd* **10**:359–362, 1924.

57. Wile WJ, Snow JS: Lipoid proteinosis. *Arch Dermatol Syph (Chic)* **43**:134–144, 1941.

58. Wood MG et al: Histochemical study of a case of lipoid proteinosis. *J Invest Dermatol* **26**:263–274, 1956.

59. Yakout YM et al: Radiological findings in lipoid proteinosis. *J Laryngol Otolaryngol* **99**:259–265, 1985.

## Oculocerebrocutaneous syndrome (Delleman syndrome)

Delleman et al (3,4) first delineated the syndrome of orbital cysts, cerebral malformations, accessory skin tags, and focal cutaneous hypoplasia. A few other cases have been published (1,2,5,6,7–10,13) and several unreported (6a,11,12). All have been isolated examples. We believe that this disorder has been underreported since many examples of cystic eye represent incomplete forms.

The face is often asymmetric. Unilateral or bilateral orbital cysts (most with microphthalmia), upper or lower eyelid coloboma, persistent hyaloid artery, and hamartomas have been reported (Fig. 13-134).

Skeletal defects include underdeveloped orbits, zygomas, and the sphenoid part of the lateral orbital wall on the side of the involved eye as well as scoliosis and rib malformation (1,3,5,7). Some patients have exhibited body asymmetry.

Cutaneous appendages have largely been periorbital but postauricular, malar, and even labial examples have been noted (1–4,6,8,11,12) (Fig. 13-135). Focal skin lesions have been of three types: aplasia, hypoplasia, and/or punch-like defects (Fig. 13-136A,B). Multiple trichofolliculomas have been reported in one case (6).

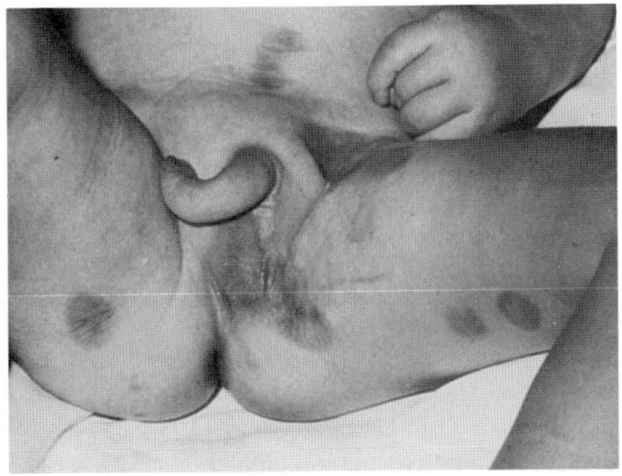

A

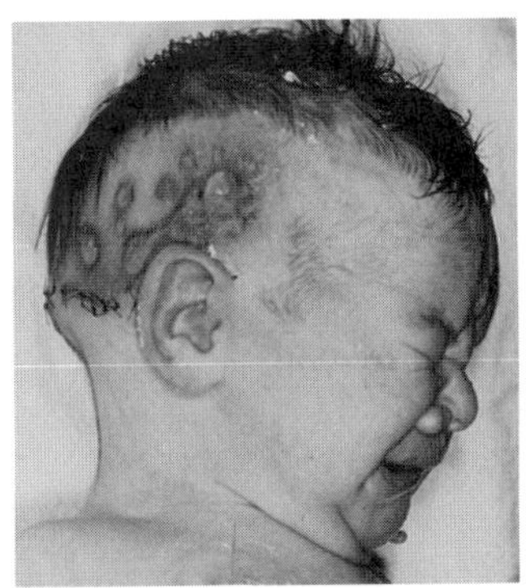

B

Fig. 13–136. *Oculocerebrocutaneous syndrome.* (A) Multiple focal cutaneous defects and underdeveloped scrotum. (B) Numerous cutaneous defects of scalp and facial skin tags. (A from *JW Delleman* and *JWE Oorthuys,* Clin Genet **19**:191, 1981.)

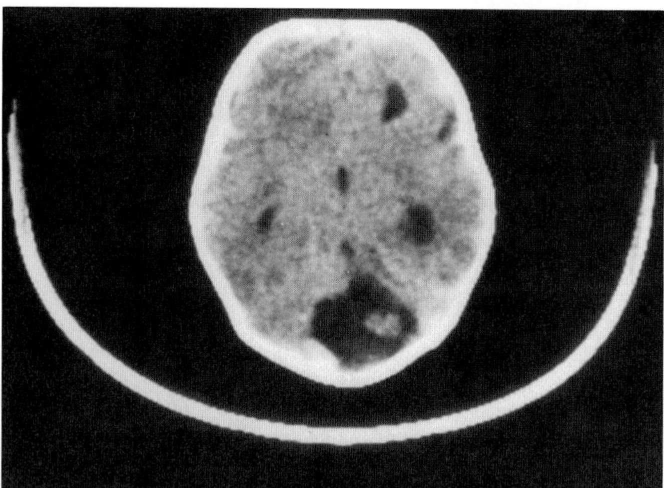

Fig. 13–137. *Oculocerebrocutaneous syndrome.* Fluid filled spaces in cortex.

Most patients have exhibited severe psychomotor retardation and seizures, but some have been only mildly retarded (6a). CAT scans have revealed agenesis of the corpus callosum and multiple fluid-filled spaces in the cortex (Fig. 13-137). Occipital meningoencephalocele, obstructive hydrocephaly, and spastic tetraplegia have also been reported.

### References [Oculocerebrocutaneous syndrome (Delleman syndrome)]


1. Al-Gazali LI et al: The oculocerebrocutaneous (Delleman) syndrome. *J Med Genet* **25**:773–778, 1988.

2. Braun-Vallon S et al: Un cas de teratome de l'orbite. *Bull Soç Ophtal Fr* **58**:805–809, 1958.

3. Delleman JW, Oorthuys JWE: Orbital cyst in addition to congenital cerebral and focal dermal malformations: A new entity? *Clin Genet* **19**:191–198, 1981.

4. Delleman JW et al: Orbital cyst in addition to congenital cerebral and focal dermal malformations: A new entity. *Clin Genet* **25**:470–472, 1984.

5. Dollfus MA et al: Congential cystic eyeball. *Am J Ophthalmol* **66**:504–509, 1968.

6. Ferguson JW et al: Ocular, cerebral and cutaneous malformations: Confirmation of an association. *Clin Genet* **25**:464–469, 1984.

6a. Hennekam R: Personal communication, 1989. Utrecht, The Netherlands.

7. Giorgi PL et al: Oculo-cerebro-cutaneous syndrome. Description of a new case. *Eur J Pediatr* **148**:325–326, 1989.

8. Ladenheim J, Metrick S: Congenital microphthalmos with cyst formation. *Am J Ophthalmol* **41**:1059–1062, 1956.

9. Renard G. et al: Microphthalmie bilatérale avec kystes orbitaires associée à des appendices faciaux surnuméraires. *Bull Mém Soç Fr Ophtalmol* **77**:297–320, 1964.

10. Saraux H, Ofret H: Les choristomes corneo-conjonctivaux. *J Génét Hum* **24**(Suppl):1–22, 1976 (Fig. 1).

11. Say B: Personal communication, 1987, Tulsa, Oklahoma.

12. Shafer I: Personal communication, 1986, Cleveland, Ohio.

13. Wilson RD et al: Oculocerebrocutaneous syndrome. *Am J Ophthalmol* **99**:142–148, 1985.


## Restrictive dermopathy (late fetal epidermal dysplasia, type II)

Witt et al (11) and Toriello (9), in 1986, defined a lethal syndrome characterized by rigid, tightly adherent skin with generalized joint contractures, unusual facies, pulmonary hypoplasia, abnormally large placenta, short umbilical cord, and skeletal alterations. Toriello et al (10) reported two earlier examples in 1983 and Lowry et al (8) and Schnur et al (8a) probably described the same condition. RJ Gorlin has recently diagnosed an unpublished example.

Inheritance is autosomal recessive. There may be a Hutterite type and a non-Hutterite type (2,4).

Prematurity (about 32 weeks), polyhydramnios and decreased fetal movement late in pregnancy, and short umbilical cord are common features.

The fixed facies manifests hypertelorism, with or without ectropion, small nose, apparently low-set anomalous pinnae, small tight open mouth, and micrognathia. Submucous cleft palate was described in one case (10) (Fig. 13-138A,B).

In some examples, the skin is a rigid, thick, smooth, and shiny shell-like structure whereas in others there are extensive erosions (Fig. 13-139). Histologic skin changes include orthokeratotic hyperkeratosis, hypoplastic pilosebaceous structures and sweat glands, absence of dermal rete ridges, and thin dermis (11). Fetal hyperkeratinization of the skin is probably the primary defect (3,6).

The chest is small and barrel shaped with increased AP diameter. There are dorsal kyphosis and rocker-bottom feet (Fig. 13-140).

Radiologic changes include wide fontanels and sutures, and overtubulation of all long bones with modeling defects of the proximal humeri and distal radii and ulnae (Fig. 13-141A,B). There is deficient ossification of the midportion of the clavicles and generalized fixed flexion contractures at all joints, including fingers and mandible.

The lungs are hypoplastic. The internal organs are otherwise normal.

Differential diagnosis includes sclerema neonatorum, scleredema, scleroderma, subcutaneous fat necrosis, stiff skin syndrome, Parana hard skin syndrome, *epidermolysis bullosa* with pyloric atresia, *fetal akinesia sequence,* and *Neu-Laxova syndrome.*

### References [Restrictive dermopathy (late fetal epidermal dysplasia, type II)]


1. Carey JC et al: Aplasia cutis congenita—the Carmi syndrome: Confirmation of a neonatal generalized skin disorder. *Proc Greenwood Genet Ctr* **2**:116–117, 1983.


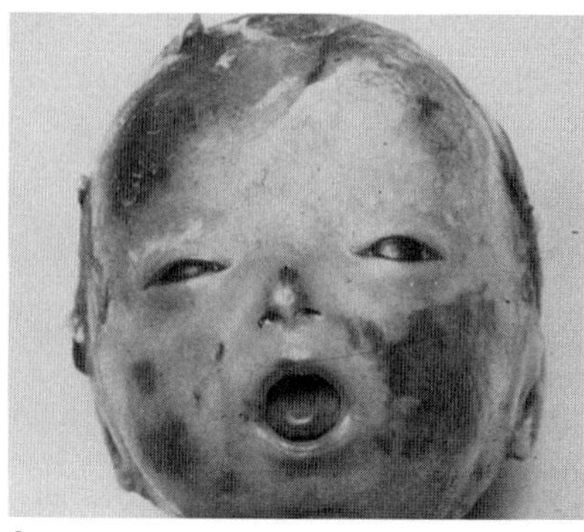

A

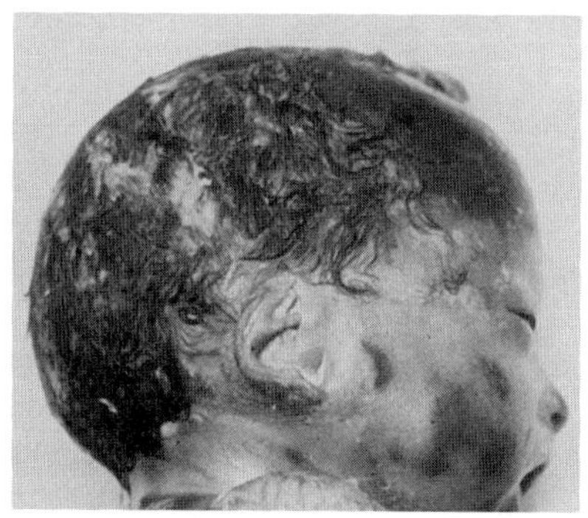

B

Fig. 13–138. *Restrictive dermopathy.* (A,B) Fixed facies with hypertelorism, small nose, low-set dysmorphic pinna, small tight open mouth, and micrognathia.

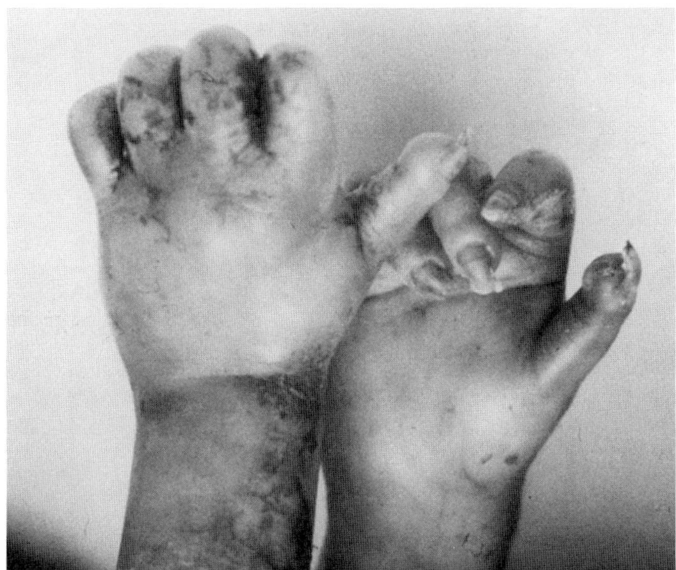

Fig. 13–139. *Restrictive dermopathy*. Edematous rigid skin of hands, flexion contraction of fingers.

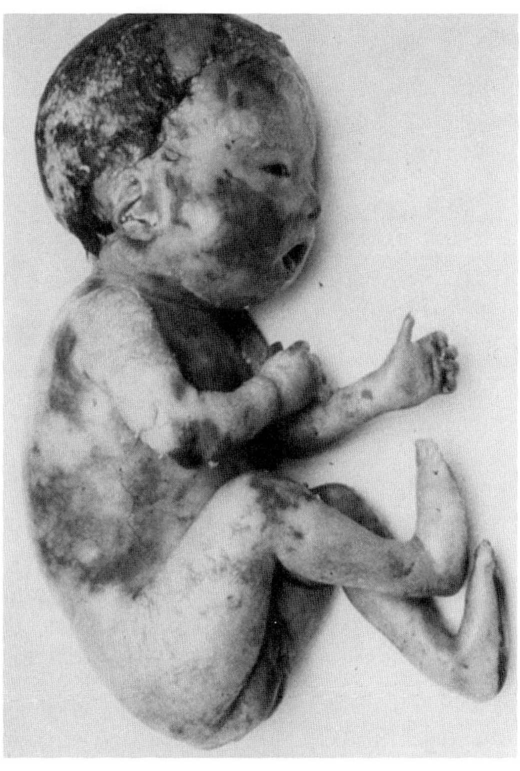

Fig. 13–140. *Restrictive dermopathy*. Note kyphosis and rocker-bottom feet.

2. Carmi R et al: Aplasia cutis congenita in two sibs discordant for pyloric stenosis. *Am J Med Genet* **11**:319–328, 1982.

3. Dale BA et al: Abnormal keratinization in restrictive dermopathy. *Curr Probl Dermatol* **17**:45–51, 1987.

4. De Groot WG et al: Familial pyloric atresia associated with epidermolysis bullosa. *J Pediatr* **92**:429–431, 1978.

5. Gillerot Y, Koulischer L: Restrictive dermopathy. *Am J Med Genet* **27**:239–240, 1987.

6. Holbrook KA et al: Arrested epidermal morphogenesis in three newborn infants with a fatal genetic disorder (restrictive dermopathy). *J Invest Dermatol* **88**:330–339, 1987.

7. Leschot NJ et al: Severe congenital skin defects in a newborn. *Eur J Obstet Gynecol Reprod Biol* **10**:381–388, 1980.

8. Lowry RB et al: Congenital contractures, edema, hyperkeratosis and intrauterine growth retardation: A fatal syndrome in Hutterite and Mennonite kindreds. *Am J Med Genet* **22**:531–543, 1985.

8a. Schnur RE et al: A lethal ichthyotic variant with arthrogryposis. *Am J Hum Genet* **37**:76A, 1985.

9. Toriello HV: Restrictive dermopathy and report of another case. *Am J Med Genet* **24**:625–629, 1986.

10. Toriello HV et al: Autosomal recessive aplasia cutis congenita—report of two affected sibs. *Am J Med Genet* **15**:153–156, 1983.

11. Witt DR et al: Restrictive dermopathy: A newly recognized autosomal recessive skin dysplasia. *Am J Med Genet* **24**:631–648, 1986.

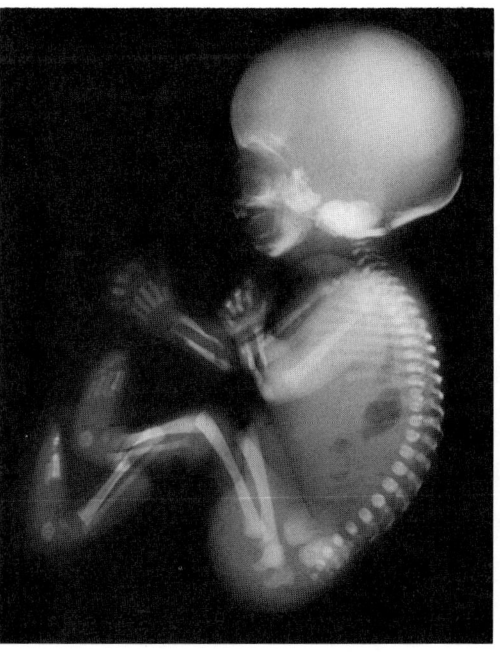

A

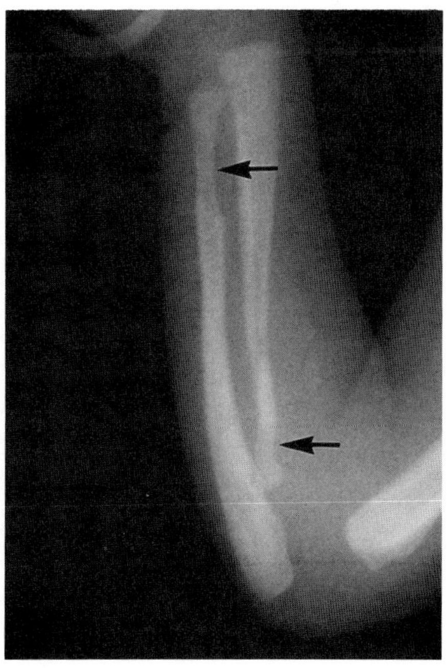

B

Fig. 13–141. *Restrictive dermopathy*. (A) Note wide fontanels and sutures, modeling defects of radii and ulnae. (B) Note bony defects in distal ulna and proximal radius. (B courtesy of *HV Toriello*, Grand Rapids, Michigan.)

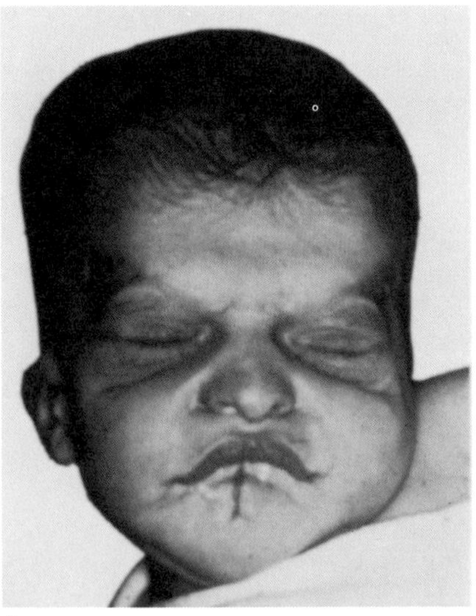
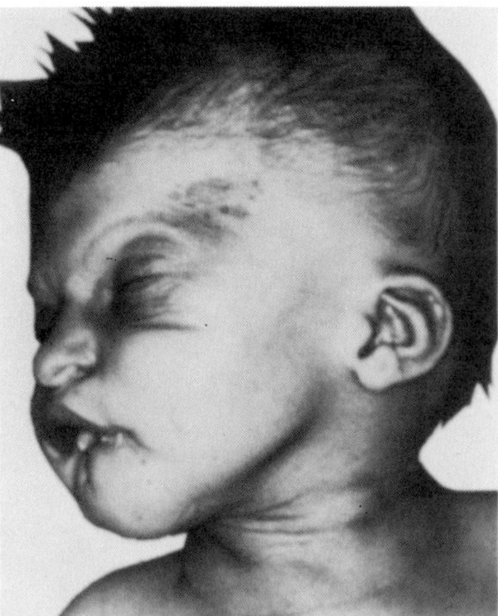
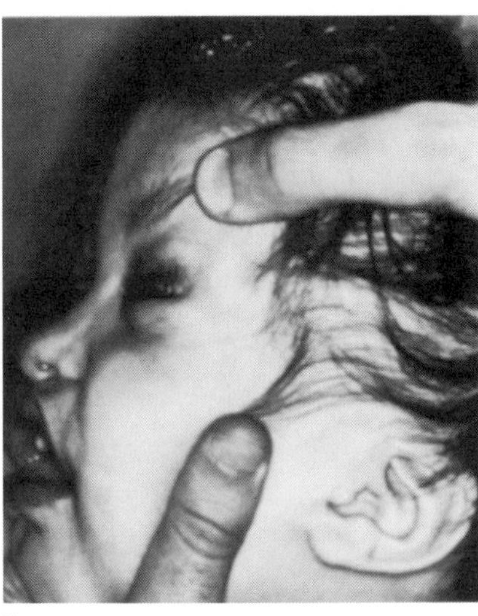

A B C

Fig. 13–142. *Setleis syndrome.* (A,B) Temporal defects, eyelash abnormalities, central defect of chin. (C) Multiple rows of lashes on upper lids, absence on lower lids. (From *H Setleis et al,* Pediatrics **32**:540, 1963.)

## Setleis syndrome

Setleis syndrome is characterized by bitemporal scarring that resembles forceps marks, periorbital puffiness with wrinkling of facial skin, abnormalities of eyebrows and lashes, flat nasal bridge with bulbous nasal tip, and increased mobility of facial skin associated with severe redundancy of soft tissues. Eight patients have been reported (3,5–7) and, to date, nearly all have been from isolated towns in Puerto Rico. A few others have been of Basque and Italian origin (1a). A ninth patient has been seen by D Daentl (personal communication, 1975). Two sets of affected sibs and consanguinity indicate autosomal recessive inheritance (3).

**Craniofacial features.** Features include "forceps marks" scarring bitemporally, periorbital puffiness, eyebrows that angle sharply upward and outward but are deficient laterally, distichiasis of upper lids and astichiasis of lower lids (Fig. 13–142A–C), flat nasal bridge, bulbous nasal tip, nasal septum extending below alae nasi, redundant skin and soft tissues, alopecia, thin scalp hair, low frontal hairline, and, in some instances, downslanting palpebral fissures (50%), epicanthic folds, chronic blepharitis, thick lips and cleft chin (3,5,7).

**Other findings.** Hypo- and hyperpigmentation of the skin have been recorded in 65%. Abnormal palmar creases have also been noted in 35% (3).

**Differential diagnosis.** A disorder known as focal facial dermal dysplasia has been documented with autosomal dominant inheritance (1,2,4). Real forceps marks, of course, are not accompanied by other features of Setleis syndrome.

### References (Setleis syndrome)

1. Brauer A: Hereditärer symmetrischer systematisierter Naevus aplasticus bei 38 Personen. *Derm Wschr* **89**:1163–1168, 1929.

1a. Clark RD et al: Expanded phenotype and ethnicity in Setleis syndrome. *Am J Med Genet* **34**:355–357, 1989.

2. Jensen NE: Congenital ectodermal dysplasia of the face. *Br J Dermatol* **84**:410–416, 1971.

3. Marion RW et al: Autosomal recessive inheritance in the Setleis bitemporal "forceps marks" syndrome. *Am J Dis Child* **41**:895–897, 1987.

4. McGeoch AH, Reed WB: Familial focal facial dermal dysplasia. *Arch Dermatol* **107**:591–595, 1973.

5. Rudolph RI et al: Bitemporal aplasia cutis congenita: Occurrence with other cutaneous abnormalities. *Arch Dermatol* **110**:615–618, 1974.

6. Rudolph RI et al: Emendation to bitemporal aplasia cutis congenita (letter to the editor). *Arch Dermatol* **110**:636, 1974.

7. Setleis H et al: Congenital ectogermal dysplasia of the face. *Pediatrics* **32**:540–547, 1963.

## Goeminne syndrome

Goeminne (1) described a family in which males had congenitally progressive muscular torticollis with facial asymmetry, multiple spontaneous keloids that appeared at puberty, cryptorchidism, chronic progressive pyelonephritis with hypertension, varicose veins of the legs, and multiple pigmented cutaneous nevi (Fig. 13-143). Female carriers were less severely affected. The literature on the familial occurrence of each component was extremely well reviewed by Goeminne (1). The gene has been mapped at Xq28, distal to the G6PD locus. It probably represents the most distal gene on the X chromosome (3).

Fig. 13–143. *Goeminne syndrome.* Facial asymmetry, torticollis, numerous keloids of arms and thorax. (From *L Goeminne,* Acta Genet Med (Roma) **17**:439, 1968.)

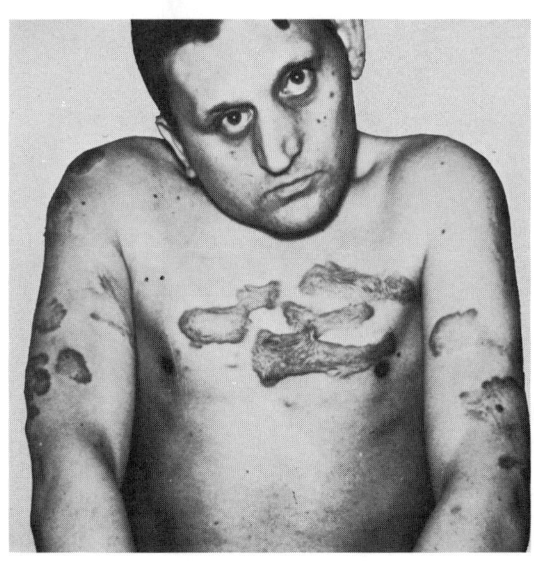

Sands (2) reported minor stigmata in a female with a translocation involving this area of the X chromosome.

### References (Goeminne syndrome)

1. Goeminne LA: New probably X-linked inherited syndrome: Congenital muscular torticollis, multiple keloids, cryptorchidism and renal dysplasia. *Acta Genet Med (Roma)* **17**:439–467, 1968.

2. Sands ME: Mental retardation in association with a balanced X-autosomal translocation and random inactivation of the X-chromosome. *Clin Genet* **17**:309–316, 1980.

3. Zuffardi O, Fraccaro M: Gene mapping and serendipity: The locus for torticollis, keloids, cryptorchidism, and renal dysplasia (31430, McKusick) is at Xq28, distal to the G6PD locus. *Hum Genet* **62**:280–281, 1982.

Fig. 13–144. *Tongue atrophy, linear skin atrophy, scarring alopecia, and anonychia.* (A) Note sparcity of hair and skin lesions over scalp and forehead. (B) Linear/reticular array of shiny, wrinkled and pitted skin of forearm and dorsa of hands. (C) Similar lesions of lower legs. (A–C from *J Sequeiros* and *GH Sack Jr,* Am J Med Genet **21**:669, 1985.)

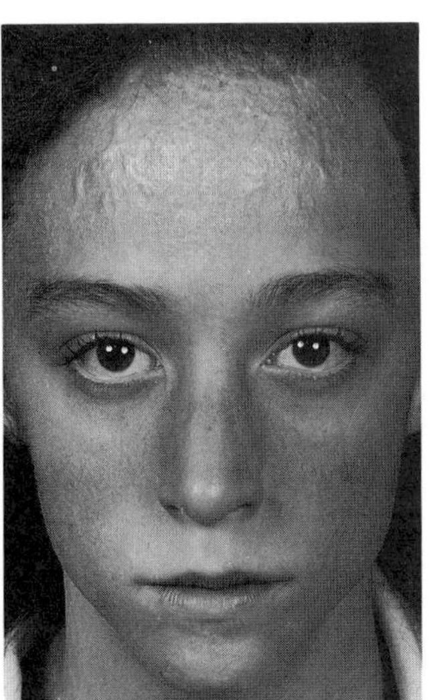

A

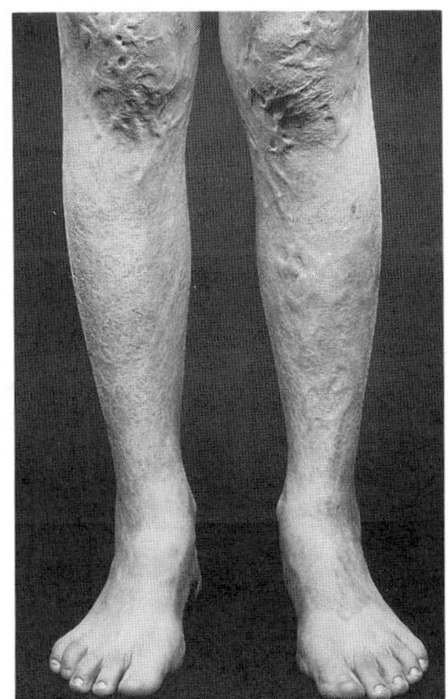

C

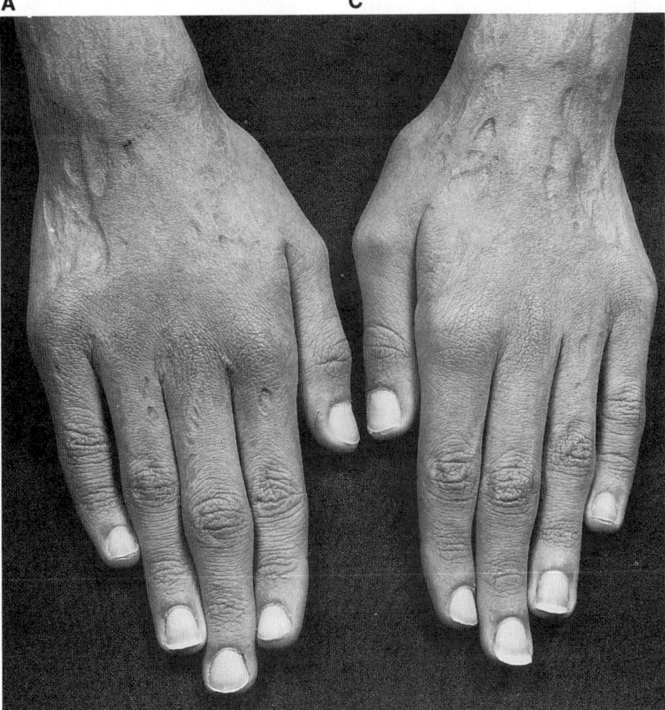

B

## Tongue atrophy, linear skin atrophy, scarring alopecia, and anonychia

The syndrome was independently described by Sequeiros and Sack (2) and Cohen et al (1) in 1985. The patients were isolated examples, the first being one of monozygous twins. All three were premature. Hence, it can be assumed that the condition is not a single gene disorder but may result from some form of intrauterine injury.

The condition is characterized by skin atrophy with linear alternation of depressed scar-like areas and intervening ridges of essentially normal skin. At birth the involved skin areas are friable with a vesiculobullous eruption followed by gradual scabbing. Symmetrical reticulated scarring involved at least 75% of the cutaneous surface. Scars on the trunk and head were cobblestone-like and, in some areas, were oriented along the skin cleavage lines. On the limbs, the reticulated pattern followed the long axes of the extremities. The palms, soles and face were not involved. Scarring alopecia and congenital absence of one or more toenails were evident. Abnormal sweating, heat intolerance, and febrile convulsions were seen in all three patients. A scar-like lesion of the dorsum of the tongue was found in two of the three (Fig. 13-144A–C).

Histologically, all showed thickening of the dermis, increase in collagen density, or a decrease in elastic fibers.

### References (Tongue atrophy, linear skin atrophy, scarring alopecia, and anonychia)

1. Cohen BA et al: Congenital erosive and vesicular dermatosis healing with reticulated supple scarring. *Arch Dermatol* **121**:361–367, 1985.

2. Sequeiros J, Sack GH Jr: Linear skin atrophy, scarring alopecia, anonychia, and tongue lesion: A ''new'' syndrome? *Am J Med Genet* **21**:669–680, 1985.

## Congenital erythrokeratoderma, keratitis, and sensorineural hearing loss (KID syndrome)

Burns (6), in 1915, reported a syndrome characterized by congenital atypical ichthyosiform erythrokeratoderma, palmoplantar keratosis, and sensorineural hearing loss. At least 30 cases have been reported. Some have been well documented (1–3,5,7,8,14,15a,15b,17,17a,18,22, 23,27,28,30–33,35,37) and others less well (6,10,15,16,19,25, 34). All have been isolated examples save one report in a father and daughter (12). There is no sex predilection. Skinner et al (33) suggested the term, KID syndrome, as an acronym to refer to *k*eratitis, *i*chthyosis and *d*eafness.

Postnatal growth deficiency has been evident in about 50% of the cases.

**Skin.** Congenital erythrokeratoderma principally involves the cheeks, nose, chin, ears, and limbs, especially the extensor and flexor surfaces of the elbows and joint areas of the knees, hands, and feet, and on the buttocks. Eyebrows, lashes, and scalp hair are scant. In some cases, scalp involvement results in partial or complete alopecia (Fig. 13-145A–C). The nails are frequently thickened and dystrophic. There is a palmar and plantar hyperkeratosis characterized by a stippled or dotted pattern (18,29) (Fig. 13-145D). Microscopically, the skin changes most clearly resemble lamellar ichthyosis, the keratin layer having a basket weave pattern. A prominent granular layer is present. Recurrent bacterial and mycotic (especially candidal) infections of the skin are found in 40%. Cutaneous squamous cell carcinoma has been noted (12,17b,30).

**Ears.** Hearing loss, found in nearly all patients, is sensorineural, usually bilateral, congenital, and often severe. The patient of Cram et al (7) had only mild hearing loss while one patient (19) had conductive deafness.

**Eyes.** Severe bilateral vascularized keratitis, preceded by photophobia, appears in early childhood and not uncommonly is compli-

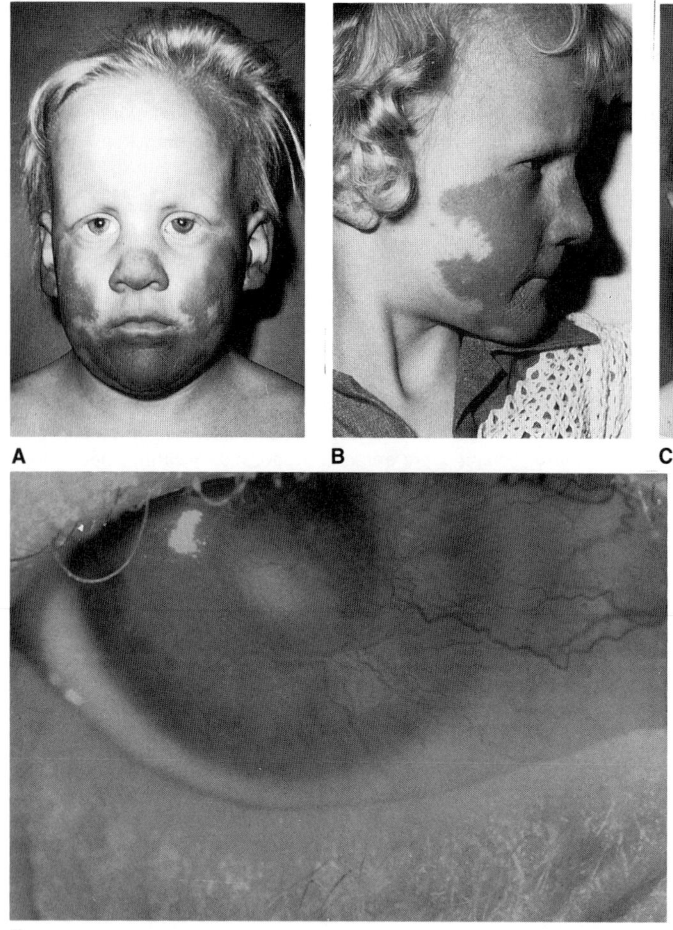

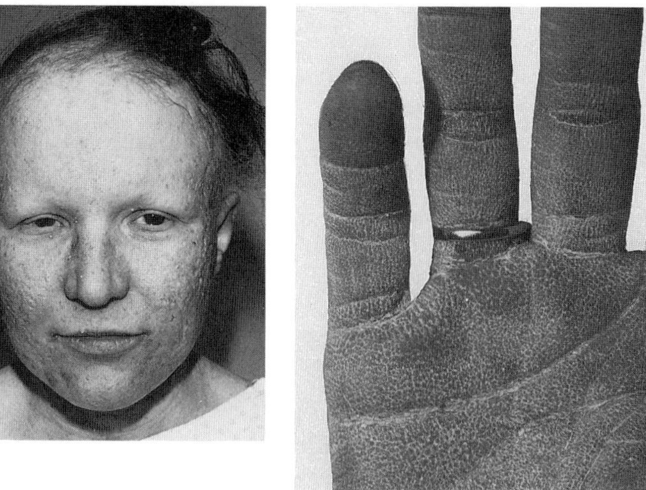

Fig. 13–145. *Congenital erythrokeratoderma, keratitis, and sensorineural hearing loss.* (A–C) Congenital erythrokeratoderma involves cheeks, nose, and chin. Some patients may have partial or even complete alopecia of scalp hair. (D) Palmar hyperkeratosis characterized by stippled or dotted pattern. (E) Vascularized keratitis. (A,B,D from *DL Cram et al,* Arch Dermatol **115**:465, 1979. C from *BA Skinner et al,* Arch Dermatol **117**:285, 1981. E courtesy of *J Milot,* Montreal, Canada.)

cated by corneal ulceration, which leads eventually to at least partial blindness (Fig. 13-145E).

**Nervous system.** Intelligence in normal. Reflexes appear to be decreased, and often there are motor disturbances, especially fine motor. Tight heel cords, pes cavus, and contractures of the knees and elbows are seen in about half the patients (7,22,24,31,32).

**Oral manifestations.** Occasionally the oral mucosa is the site of generalized or linear leukokeratosis or, occasionally, furrowing. Squamous cell carcinoma of the tongue (1,17) and oral leukoplakia (12,18) have been reported. Perhaps proclivity to oral cancer reflects a complication of mucocutaneous candidosis (18a).

**Other findings.** Cerebellar hypoplasia has been noted (15a,15b).

**Differential diagnosis.** Cases with overlap, but far less certain diagnosis, are those of Ruzicka et al (26) in which the patient had brachydactyly, clinodactyly, accessory cervical ribs, and carcinoma of the thyroid.

The patient with ichthyosis vulgaris, sensorineural hearing loss, pili torti, and tooth anomalies reported by Braun-Falco and Landthaler (41) must represent another entity. It may well be that the deafness is fortuitous. A syndrome of severe ichthyosis hystrix and sensorineural hearing loss reported by Gülzow and Anton-Lamprecht (13) may be related to the disorder reported by Morris et al (20) and Myers et al (21). Possibly, the patient reported by Freire-Maia et al (11) also belongs in this category. There is some overlap with *endocrine candidosis.*

The kindred described by Desmons et al (9) must have a different condition. They reported three sibs with congenital ichthyosiform erythrokeratoderma and profound congenital sensorineural hearing loss, but all had hepatomegaly. Autosomal recessive inheritance is likely in the kindred of Desmons et al (9).

**Laboratory findings.** Hyperimmunoglobulin E has been found by a few investigators (14,24). Absence of lymphocyte stimulation by *Candida* has been described (5,14).

### References [Congenital erythrokeratoderma, keratitis, and sensorineural hearing loss (KID syndrome)]


1. Baden HP, Alper JC: Ichthyosiform dermatosis, keratitis and deafness. *Arch Dermatol* **113**:1701–1704, 1977.
2. Badillet G et al: Etude mycologique de trois cas d'erythrodermie avec keratite et surdité. *Bull Soç Mycol Méd* **11**:191–198, 1982.
3. Beare JM et al: Atypical erythrokeratodermia with deafness, physical retardation and peripheral neuropathy. *Br J Dermatol* **87**:308–314, 1972.
4. Braun-Falco A, Landthaler M: Ichthyosis vulgaris, Taubheit, Pili torti und Zahnanomalien. *Haurtarzt* **29**:276–280, 1978.
5. Britton H et al: Keratosis follicularis spinulosa decalvans. An infant with failure to thrive, deafness and recurrent infections. *Arch Dermatol* **114**:761–764, 1978.
6. Burns FS: A case of generalized congenital keratoderma with unusual involvement of the eyes, ears and nasal and buccal mucous membrane. *J Cutan Dis* **33**:255–260, 1915.
7. Cram DL et al: A congenital ichthyosiform syndrome with deafness and keratitis. *Arch Dermatol* **115**:467–471, 1979.
8. Cremers CWRJ et al: Deafness, ichthyosiform erythroderma, corneal involvement, photophobia and dental dysplasia. *J Laryngol Otol* **91**:585–590, 1977.
9. Desmons F et al: Erythrodermie ichthyosiforme congénitale seche, sordi-

mutité, hepatomégalie de transmission recessive autosomique. *Bull Soç Fr Dermatol Syphil* **78**:585–591, 1971.

10. Fishman JE, Crystal N: Sensorineural deafness: Familial evidence and additional defects. Study of a school for deaf children. *Am J Med Sci* **266**:111–117, 1973.

11. Freire-Maia N et al: An ectodermal dysplasia syndrome of alopecia, onychodysplasia, hypohidrosis, hyperkeratosis, deafness and other manifestations. *Hum Hered* **27**:127–133, 1977.

12. Grob JJ et al: Keratitis, ichthyosis, and deafness (KID) syndrome: Vertical transmission and death from multiple squamous cell carcinomas. *Arch Dermatol* **123**:777–782, 1987.

13. Gülzow J, Anton-Lamprecht I: Ichthyosis hystrix gravior Typus Rheydt: ein otologisch-dermatologisches Syndrom. *Laryngol Rhinol Otol* **56**:949–955, 1977.

14. Harms M et al: KID syndrome (Keratitis, Ichthyosis, and Deafness) and chronic mucocutaneous candidiasis: Case report and review of the literature. *Pediatr Dermatol* **2**:1–7, 1984.

15. Hauxthausen H: Hyperkeratosis ichthyosiformis? Acanthosis nigricans in a 4-year-old girl with congenital deafness. *Acta Dermatovenereol* **35**:191–192, 1955.

15a. Hsu H-C et al: Keratitis, ichthyosis and deafness (KID) syndrome with cerebellar hypoplasia. *Int J Dermatol* **27**:695–697, 1988.

15b. Jurecka W et al: Keratitis, ichthyosis and deafness syndrome with glycogen storage. *Arch Dermatol* **121**:799–801, 1985.

16. Koblenzer PJ: Ichthyosis congenita: Deafness. *Arch Dermatol* **92**:743, 1965.

17. Lancaster L Jr, Fournet LF: Carcinoma of the tongue in a child: Report of case. *J Oral Surg* **27**:269–270, 1969.

17a. Legrand J et al: Un syndrome rare oculo-auriculo-cutanée (syndrome, de Burns). *J Fr Ophtalmol* **5**:441–445, 1982.

17b. Madariaga J et al: Squamous cell carcinoma in congenital ichthyosis with deafness and keratitis. *Cancer* **57**:2026–2029, 1986.

18. Marghescu S et al: Kongenitale Erythrokeratodermie mit Taubheit Schnyder. *Hautarzt* **33**:416–419, 1982.

18a. McGurk M, Holmes M: Chronic muco-cutaneous candidiasis and oral neoplasia. *J Laryngol Otol* **102**:643–645, 1988.

19. Milot J, Sheridan S: Corneal leukomas in ichthyosis. *J Pediatr Ophthalmol* **11**:209–212, 1974.

20. Morris J et al: Generalized spiny hyperkeratosis and universal alopecia. *Arch Dermatol* **100**:692–698, 1969.

21. Myers EN et al: Congenital deafness, spiny hyperkeratosis, and unusual alopecia. *Arch Otolaryngol* **93**:68–74, 1971.

22. Ochs HD et al: Ichthyosiform erythroderma and congenital sensorineural deafness. *J Pediatr* **93**:330, 1978 (Same case as Ref. 24).

23. Oikarinen A et al: A congenital ichthyosiform syndrome with deafness and elevated serum steroid disulphate levels. *Acta Dermatol Venereol (Stockholm)* **60**:503–507, 1980.

24. Pincus et al: Defective neutrophil chemotaxis with variant ichthyosis, hyperimmunoglobulinemia E, and recurrent infections. *J Pediatr* **87**:908–911, 1975.

25. Pindborg JH, Gorlin RJ: Oral changes in acanthosis nigricans (juvenile type). *Acta Derm Venereol (Stockholm)* **42**:63–71, 1962.

25a. Ruiz-Maldonado R: Neuroichthyosis. *In Neurocutaneous Diseases—A Practical Approach,* MR Gomez (ed), Butterworths, Boston, 1987, pp. 214–218.

26. Ruzicka T et al: Syndrome of ichthyosis congenita, neurosensory deafness, oligophrenia, dental aplasia, brachydactyly, clinodactyly, accessory cervical ribs and carcinoma of the thyroid. *Dermatologica* **162**:124–136, 1981.

27. Rycroft RJC et al: Atypical ichthyosiform erythroderma, deafness and keratitis. *Br J Dermatol* **94**:211–217, 1976.

28. Salamon T et al: Erythrodermia ichthyosiformis congenita mit Hypotrichosis, Anhidrose, Taubstummheit and verminderter Ausscheidung einiger Aminosäuren im Harn. *Hautarzt* **25**:448–453, 1974.

29. Schnyder UW et al: Eine weitere Form von atypischer Erythrokeratodermie mit Schwerhörigkeit and cerebraler Schädigung. *Helv Paediatr Acta* **23**:220–230, 1968.

30. Senter TP: Carcinoma in atypical ichthyosiform dermatosis. *Arch Dermatol* **115**:1034–1035, 1979.

31. Senter TP et al: Atypical ichthyosiform erythroderma and congenital neurosensory deafness—a distinct syndrome. *J Pediatr* **92**:68–71, 1978.

32. Silvestri DL: Ichthyosiform dermatosis and deafness. *Arch Dermatol* **114**:1243–1244, 1987 (same case as Ref. 24).

33. Skinner BA et al: The keratitis, ichthyosis, and deafness (KID) syndrome. *Arch Dermatol* **117**:285–289, 1981.

34. Taussig LR: Ichthyosis with involvement of the cornea. *Arch Dermatol Syphil* **40**:504–505, 1939.

35. Van Everdingen JJE et al: Normal sweating and tear production in congenital ichthyosiform erythroderma with deafness and keratitis. *Acta Derm Venereol (Stockholm)* **62**:76–78, 1982 (same case as Ref. 8).

36. Voigtländer V: Hereditäre Verhornungsstörungen und Taubheit. *Z Hautkr* **52**:1017–1025, 1977 (same case as Ref. 29).

33. Wilson FM et al: Corneal changes in ectodermal dysplasia. *Am J Ophthalmol* **75**:17–27, 1973.

# Chapter 14
# Syndromes with Craniosynostosis: General Aspects and Well-known Syndromes

## Introduction

The craniosynostoses are etiologically and pathogenetically heterogeneous. Premature sutural fusion may occur alone or together with other anomalies, making up various syndromes. About 70 syndromes are known, not including secondary or occasional synostosis. The most exhaustive reviews are those of Cohen (8,10,11).

Hunter and Rudd (17) calculated a population frequency of 0.4/1000. Chung and Myrianthopoulos (7) gave a similar estimate. In a prospective study of isolated nonsyndromic craniostenosis, Shupe et al (34) reported a frequency of 0.6/1000. Craniosynostosis has been observed in white, black, and oriental populations (8). Sagittal synostosis is the most common isolated type, with percentages ranging from 56 to 58%. Coronal synostosis occurs less frequently, with estimates varying from 18 to 29%. With sagittal involvement, the male-to-female preponderance is 3:1. With coronal involvement, there is usually but not always a slight predilection for females. Metopic and lambdoidal synostoses occur less commonly (8,10).

Most cases of isolated craniostenosis are sporadic, but familial instances are known, autosomal dominant patterns occurring much more frequently than autosomal recessive patterns. Associated anomalies are more frequent in coronal series than in sagittal series. With syndromic craniostenosis, the types of anomalies most commonly associated are limb defects, ear anomalies, and cardiovascular malformations (8,9).

Table 14–1. Syndromes with craniosynostosis

| Syndromes | Pages |
|---|---|
| In this chapter | |
| Apert syndrome | 520 |
| Crouzon syndrome | 524 |
| Pfeiffer syndrome | 526 |
| Saethre–Chotzen syndrome | 527 |
| Jackson–Weiss syndrome | 529 |
| Carpenter syndrome | 531 |
| Antley–Bixler syndrome | 533 |
| Baller–Gerold syndrome | 535 |
| Cloverleaf skull syndrome | 536 |
| In chapters other than chapter 14 | |
| Baraitser syndrome | 745 |
| Cole–Carpenter syndrome type osteogenesis imperfecta | 161 |
| Craniofrontonasal dysplasia | 789 |
| C syndrome (Opitz trigonocephaly syndrome) | 889 |
| Elejalde syndrome | 342 |
| FG syndrome | 882 |
| Greig cephalopolysyndactyly syndrome | 799 |
| Michels syndrome | 766 |
| Osteoglophonic dysplasia | 214 |
| Seckel syndrome | 313 |
| Stanescu type osteosclerosis | 287 |
| Tricho-dento-osseous syndrome | 861 |
| Miscellaneous syndromes | 541 (see Table 15–1) |

Table 14-1 lists the main craniosynostosis syndromes discussed in this chapter and gives page numbers for syndromes that appear in other chapters. Chromosomal syndromes with appropriate references appear in Table 14-2. Environmentally induced craniostenosis and pertinent references are provided in Table 14-3. Table 14-4 lists conditions with secondary or occasional craniosynostosis. Finally, the miscellaneous syndromes discussed in the next chapter are given in Table 15-1.

### References (Introduction)


1. Alfi OS et al: The 9p− syndrome. *Ann Génét (Paris)* **19**:11–16, 1976.
2. Ardinger HH et al: Verification of the fetal valproate syndrome phenotype. *Am J Med Genet* **29**:171–185, 1988.
3. Berry R et al: Cytogenetic abnormalities in 6 patients with craniosynostosis. Paper presented at the March of Dimes Clinical Genetics Conference, Minneapolis, July 19–22, 1987.
4. Buchsinger G et al: Familial chromosome translocation t(3;18) (p21;p11). *J Med Genet* **18**:119–123, 1981.
5. Cassidy SB et al: Trigonocephaly and the 11q− syndrome. *Ann Génét* **20**:67–69, 1977.


Table 14–2. Chromosomal craniosynostosis

| Chromosomal anomaly | References |
|---|---|
| del(1q) | 12,26 |
| del(2q) | 18a |
| del(3q) | 30 |
| dup(3p) | 4 |
| dup(3q) | 39,40 |
| dup(5p) | 23 |
| dup(6p) | 26b |
| del(6q) | 3,26a |
| dup(6q) | 28,31,38 |
| del (7p) | 20,22 |
| dup(7p) | J Willner, personal communication, 1979 |
| dup(8q) | 37 |
| del(9p) | 1 |
| del(9q) | 3 |
| del(11q) | 5,18 |
| del(12p) | 25 |
| del(13q) | 24 |
| dup(13q) | 16 |
| del(15q) | 3 |
| dup(15q) | 27 |
| del(17p) | 35,36 |
| 45,X | 3,11a |
| 45,X/46,X, + frag | 3 |
| 46,fraX(q27.3)Y | 3 |
| 47,XX, + cen frag/46,XX | 41 |
| Triploidy | 29 |
| Tetrasomy 14q | 22a |

Table 14–3. Environmentally induced craniosynostosis

| Etiology | Reference |
|---|---|
| Mechanical | |
| Deformational | 11,13 |
| Drug related | |
| Aminopterin, methotrexate | 21 |
| Fetal hydantoin effects | 6 |
| Retinoic acid embryopathy | 14 |
| Valproic acid embryopathy | 2 |
| Suggested or suspected teratogens | |
| Oxymetazoline | 15 |

6. Char F et al: Patterns of malformation in infants exposed to gestational anticonvulsants. March of Dimes Birth Defects Meeting, San Francisco, June 1978.

7. Chung CS, Myrianthopoulos NC: Factors affecting risk of congenital malformations. I. Epidemiological analysis. *Birth Defects* **11**(10):1–22, 1975.

8. Cohen MM Jr: Craniosynostosis and syndromes with craniosynostosis: Incidence, genetics, penetrance, variability, and new syndrome updating. *Birth Defects* **15**(5B):13–63, 1979.

9. Cohen MM Jr: Perspectives on craniosynostosis. *West J Med* **132**:507–513, 1980.

10. Cohen MM Jr: *Craniosynostosis: Diagnosis, Evaluation, and Management.* Raven Press, New York, 1986.

11. Cohen MM Jr. Craniosynostosis update 1987. *Am J Med Genet Suppl* **4**:99–148, 1988.

11a. Cohen MM Jr: Craniosynostosis in the Turner syndrome. *Am J Med Genet* **34**, 1989.

12. Falk RE et al: Profound growth failure and minor congenital anomalies in a child with a small chromosomal deletion: 46,XX,del(1)(pter→qter). *Clin Res* **25**(2):176A, 1977.

13. Graham JM et al: Sagittal craniostenosis: Fetal head constraint as one possible cause. *J Pediatr* **95**:747–750, 1979.

14. Happle R et al: Teratogene Wirkung von Etretinat beim Menschen. *Dtsch Med Wschr* **109**:1476–1480, 1984.

15. Holm IA, Clarren SK: An unusual pattern of malformation associated with gestational exposure to nasal spray. *J Pediatr* **106**:860–861, 1985.

16. Hornstein L, Soukup S: A recognizable phenotype in a child with partial duplication 13q in a family with t(10q:13q). *Clin Genet* **19**:81–86, 1981.

17. Hunter AGW, Rudd NL: Craniosynostosis. I. Sagittal synostosis: Its genetics and associated clinical findings in 214 patients who lacked involvement of the coronal suture(s). *Teratology* **14**:185–193, 1976.

18. Lippe BM et al: Craniosynostosis and syndactyly: Expanding the 11q− chromosomal deletion phenotype. *J Med Genet* **18**:480–483, 1981.

Table 14–4. Conditions with secondary or occasional craniosynostosis[a]

| | |
|---|---|
| Hematologic disorders | Oligosaccharidoses |
| Thalassemia | Pseudo-Hurler polydystrophy |
| Sickle cell anemia | Iatrogenic disorders |
| Polycythemia vera | Hydrocephaly with shunt |
| Congenital hemolytic icterus | Occasional craniosynostosis |
| Malformations | Ataxia-telangiectasia |
| Microcephaly | Jeune syndrome |
| Encephalocele | Job's syndrome |
| Holoprosencephaly | Epidermal nevus syndrome |
| Metabolic disorders | Lowe syndrome |
| Hyperthyroidism | Pseudohypoparathyroidism |
| Idiopathic hypercalcemia | Rubella embryopathy |
| Vitamin D deficiency rickets | Williams syndrome |
| Vitamin D-resistant rickets | |
| Hypophosphatasia | |
| Mucopolysaccharidoses | |
| Hurler syndrome | |
| Morquio syndrome | |
| $\beta$-Glucuronidase deficiency | |

[a] From MM Cohen, Jr. (8). See for specific references.

18a. Lucas J et al: Délétion interstitielle, de novo, du bras long d'un chromosome 2: 46,XX,del(2)(q14q12), associée à une craniosynostose prématurée. *Ann Génét* **30**:33–38, 1987.

19. McCorquodale M et al: Kleeblattschädel anomaly and partial trisomy for chromosome 13(47,XY,+der(13),t(3,13)(q24:q14). *Clin Genet* **17**:409–414, 1980.

20. McPherson E et al: Chromosome 7 short arm deletion and craniosynostosis. A 7p− syndrome. *Hum Genet* **35**:117–123, 1976.

21. Milunsky A et al: Methotrexate-induced congenital malformations. *J Pediatr* **72**:790–795, 1968.

22. Motegi T et al: A craniosynostosis in a boy with a del(7)(9p15.3p21.3): Assignment by deletion mapping of the critical segment for craniosynostosis to the mid-portion of 7p21. *Hum Genet* **7**:160–162, 1985.

22a. Nuessle NS, Miles JH: Proximal 14q tetrasomy syndrome. 39th Annual Meeting, American Society of Human Genetics, New Orleans, October 12–15, 1988 (Abstract 249).

23. Opitz JM, Patau K: A partial trisomy 5p syndrome. *Birth Defects* **11**(5):191–200, 1975.

24. Orbeli DJ et al: The syndrome associated with the partial D-monosomy. Case report and review. *Hum Genet* **13**:296–308.

25. Orye E, Craen M: Short arm deletion of chromosome 12. Report of two new cases. *Hum Genet* **28**:335–342, 1975.

26. Pablo CE de et al: Interstitial deletion in the long arms of chromosome 1:46,XY,del(1)(pter→q22::q25→qter). *J Med Genet* **17**:483–486, 1980.

26a. Park JP et al: A *de novo* interstitial deletion of chromosome 6(q22. 2q23.1). *Clin Genet* **33**:65–68, 1988.

26b. Pearson G et al: Inversion duplication of chromosome 6 with trisomic codominant expression of HLA antigens. *Am J Hum Genet* **31**:29–34, 1979.

27. Pedersen C: Partial trisomy 15 as a result of an unbalanced 12/15 translocation in a patient with cloverleaf skull anomaly. *Clin Genet* **9**:378–380, 1976.

28. Robertson KP et al: Acrocephalosyndactyly and partial trisomy 6. *Birth Defects* **11**(5):267–271, 1975.

29. Saath AA et al: Triploidy syndrome. A report of two live-born (69,XXY) and one still-born (69,XXX) infants. *Clin Genet* **9**:43–50, 1976.

30. Sargent C et al: Trigonocephaly and the Opitz C syndrome. *J Med Genet* **22**:39–45, 1985.

31. Schmid W et al: Trisomy 6q25→6qter in a severely retarded 7-year-old boy with turricephaly, bow-shaped mouth, hypogenitalism and club feet. *Hum Genet* **46**:279–284, 1979.

32. Schömig-Spingler M et al: Chromosome 7 short arm deletion, 7p21→pter. *Hum Genet* **74**:323–325, 1986.

33. Schutten W et al: Partial trisomy of chromosome 13: Case report and review of literature. *Ann Génét* **21**:95–99, 1978.

34. Shupe A et al: The incidence of craniosynostosis in the newborn infant. *Am J Dis Child* **139**:85–86, 1985.

35. Smith ACM et al: Interstitial deletion of (17)(p11.2p11.2) in nine patients. *Am J Med Genet* **24**:393–414, 1986.

36. Stratton RF et al: Interstitial deletion of (17)(p11.2p11.2): Report of six additional patients with a new chromosome deletion syndrome. *Am J Med Genet* **24**:421–432, 1986.

37. Sujansky E et al: Familial pericentric inversion of chromosome 8. *Am J Med Genet* **10**:229–235, 1981.

38. Turleau C, de Grouchy J: Trisomy 6qter. *Clin Genet* **19**:202–266, 1981.

39. Wilson GN et al: The association of chromosome 3 duplication and the Cornelia De Lange syndrome. *J Pediatr* **93**:783–788, 1978.

40. Wilson GN et al: Further delineation of the dup(3q) syndrome. *Am J Med Genet* **22**:117–123, 1985.

41. Wilson GN et al: Craniosynostosis, choanal atresia, cerebral anomalies, digital anomalies and chromosome fragment mosaicism. Paper presented at the March of Dimes Clinical Genetics Conference, Minneapolis, July 19–22, 1987.

## Apert syndrome (acrocephalosyndactyly)

Apert syndrome is characterized by craniosynostosis, midfacial malformations, and symmetric syndactyly of the hands and feet, minimally involving digits 2, 3, and 4 (4,6,7). Apert (2) is credited with the discovery of the syndrome, although the condition was reported earlier by Wheaton (36) and others. Park and Powers (21) reported an extensive study of the disorder in 1920. The most extensive reviews are those of Cohen (8) and Kreiborg and Cohen (16b). In 1960, Blank (4), in examining 54 cases, divided acrocephalosyndactyly into typical and atypical clinical categories. Typical, or Apert type, included only those patients who had a middigital hand mass consisting of osseous and soft tissue syndactyly of digits 2, 3, and 4. More than 300 cases have been reported to date.

Blank (4) noted that Apert syndrome was found once in 160,000 births and, because of the high mortality rate in the neonatal period, occurred once in 2 million in the general population. The frequency of Apert syndrome was established to be 1 in 100,000 births by Tünte and Lenz (32). Although most cases are sporadic, representing fresh mutations, autosomal dominant transmission with complete penetrance has been reported eight times (3,8c,17,25,33,34). Male transmission was recorded by Rollnick (25b). Germinal mosaicism has been noted in one family (1). Increased paternal age has been associated with sporadic cases (11). An extensive study of genetics and prevalence is underway (8c,16b).

There is no compelling evidence for genetic heterogeneity in Apert syndrome, that is, there is no autosomal recessive type, despite suggestions to the contrary (19a,27a).

**Craniofacial features.** During infancy, there is a complete, widely gaping midline calvarial defect that extends almost from the root of the nose to the posterior fontanel via the metopic and sagittal ''suture'' areas and the anterior fontanel. The defect is widely patent during infancy and only gradually fills in completely during the third year of life. Bony islands form within the calvarial defect, grow, and coalesce until the gap is completely covered by bone. ''Sutural synostosis'' is probably an inaccurate term since proper sutures do not appear in the calvarial midline. Rather, ossification centers seem to form and gradually bridge the gap. The coronal area is fused at birth. Only the lambdoidal suture appears to be a true suture that forms interdigitations visible on radiographs and on dry skulls. The head circumference at birth is normal.

Hyperacrobrachycephaly is commonly observed, and the occiput is flattened. The forehead is steep, and during infancy a horizontal groove that disappears with age may be present above the supraorbital ridges (Fig. 14-1A). Bulging at the bregma or slightly anterior to the bregma may be noted in some cases. The cranial base is malformed and often asymmetric (20a). The anterior cranial fossa is very short. Shallow orbits and, frequently, expanded ethmoidal labyrinth, orbital hypertelorism, and ethmoidal prolapse are associated. The sella is enlarged, and the clivus and anterior cranial fossa are very short. The lesser sphenoidal wings slope upward and laterally. The great wings of the sphenoid are protruded. The cranial base angle is variable, but platybasia occurs most commonly. Cloverleaf skull may be observed on occasion. An encephalocele has rarely been noted (8).

The middle third of the face is retruded and commonly hypoplastic, resulting in relative mandibular prognathism. The nasal bridge is depressed, and the nose is beaked and humped (Fig. 14-1B). The nasal septum is often deviated (6,21).

Hypertelorism, proptosis, downslanting palpebral fissures, and, frequently, strabismus are observed. Absence of the superior rectus muscle has been reported (9,35). Structural alterations of the extraocular muscles have also been noted (18), indicating that ocular motility disturbances in the Apert syndrome may not be caused solely by mechanical factors. Albinoid findings have also been observed. Iris transillumination and depigmentation of the fundus are associated with absent or diffuse foveal reflexes. Unlike classical oculocutaneous albinism, however, visual acuity is not severely impaired, and pendular nystagmus is not observed. In some Apert syndrome patients, light hair color and pale skin are also noted (19). Optic atrophy occurs in some instances and, rarely, luxation of the eye globes, congenital glaucoma, keratoconus, and ectopia lentis have been reported (6,20, 21).

The ears may appear low set in some instances. Minor anomalies of the ear may be noted on occasion. Otitis media is common in Apert syndrome, although the exact frequency is unknown. Middle ear disease is related to the high frequency of cleft palate (6) and may also be related to eustachian tube dysfunction compounded by obstruction of the epipharyngeal space (24). Congenital fixation of the stapedial foot plate is also believed to be a frequent finding. The possibility of significant hearing loss should always be kept in mind and not overlooked because of preoccupation with other more obvious handicaps of the Apert syndrome. The otologic manifestations have been reviewed by Bergstrom et al (3).

In the relaxed state, the lips frequently assume a trapezoidal configuration. The palate is highly arched, constricted, and usually has a median furrow (Fig. 14-2) (6,27). Lateral palatal swellings are present, which increase in size with age. These swellings have been shown to have excess mucopolysaccharide content, predominantly hyaluronic acid and, to a lesser extent, sulfated mucopolysaccharides (30). Cleft soft palate, or bifid uvula, is observed in 30% ($N=55$) of cases (6,7,22). The hard palate is shorter than normal, but the soft palate is both longer and thicker than normal (22,23). Alterations in the nasopharyngeal architecture consist of reduction in pharyngeal height, width, and depth. The combination of reduced nasopharyngeal dimensions and decreased patency of the posterior nasal choanae poses the possible threat of respiratory embarrassment and cor pulmonale, especially in the young child (23).

The maxillary dental arch is V-shaped with severely crowded teeth and bulging alveolar ridges. The maxillary hypoplasia so frequently present in Apert syndrome leads to compression of the dental arch, which becomes V-shaped. This results in irregular positioning of teeth and marked thickening of the alveolar process. Class III malocclusion is commonly present, with anterior open bite as well as anterior and posterior crossbite (6,22). Delayed dental eruption is a common finding, and supernumerary teeth have been noted on occasion (6).

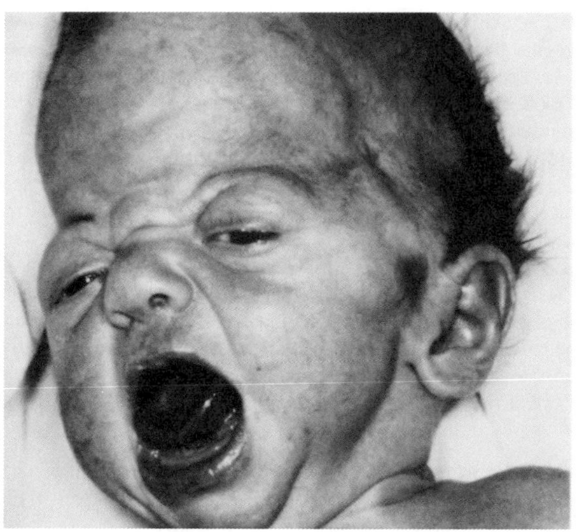
A

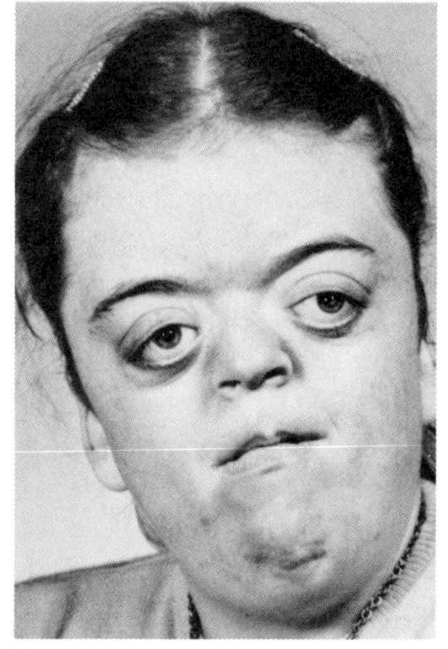
B

Fig. 14–1. *Apert syndrome.* (A) Note turribrachycephaly, proptosis, and horizontal grooves above supraorbital ridges. (B) Note turribrachycephaly, hypertelorism, downslanting palpebral fissures, beaked nose, midface hypoplasia, and relative mandibular prognathism. (B from *A Kahn Jr* and *J Fulmer,* N Engl J Med **252:**379, 1955.)

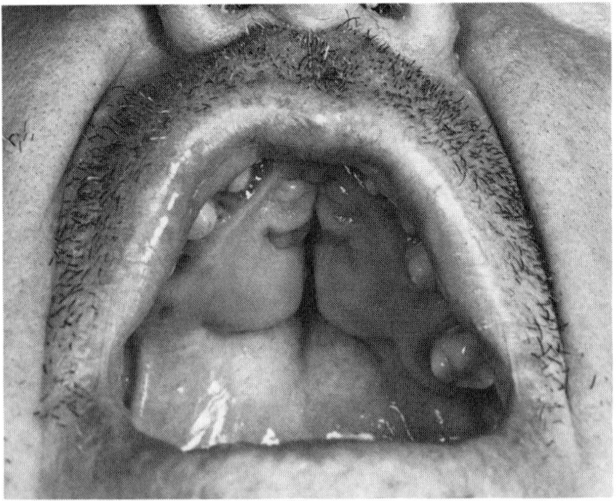

Fig. 14–2. *Apert syndrome.* Note Byzantine-arch-shaped palate.

**Growth.** The growth pattern in Apert syndrome is unique. Length and weight at birth tend to be increased and head circumference is approximately normal. Birth measurements are explained by megalencephaly, dramatically shortened cranial base, fusion of the coronal area, and the wide calvarial midline defect which result in a tall, wide, and large head (8b,16a).

The growth pattern in infancy and childhood consists of a gradual decrease in height so that most values fall between the 5th and 50th centiles. From adolescence to adulthood, the decrease in centiles becomes more pronounced. This two-step deceleration in height results from rhizomelic shortening of the lower limbs, and is more exaggerated in females than in males.

**Central nervous system.** Only in recent years has the central nervous system in the Apert syndrome come into focus. The subject has been extensively reviewed by Cohen and Kreiborg (8a).

A significant proportion of patients are mentally retarded. IQ has been studied at two large craniofacial centers. Lefebvre et al (16c) assessed 20 children with a full battery of psychometric tests. Mean IQ was 73.6 with a range of 52–89. Patton et al (21a) found that approximately half of their patients had an IQ greater than 70, although none had an IQ above 100 ($n=29$). Of 14 patients with IQs above 70, seven found suitable training or employment. Only 7% had IQs lower than 35. In patients who had craniectomies performed during the first year of life, no significant differences were found between retarded vs. nonretarded outcome. On the other hand, patients with normal intelligence and above average intelligence have been observed (8).

Cohen and Kreiborg (8a) indicated that central nervous system malformations may be responsible for many instances of mental deficiency. Combining their data with what was available in the literature, they found a large number of cases associated with absent or defective corpus callosum, defects of the limbic structures, or both. Other frequent findings in their neuropathology series included megalencephaly and gyral abnormalities. Hypoplasia of cerebral white matter and heterotopic gray matter were found in some cases. Encephalocele and pyramidal tract abnormalities were noted occasionally (8a). At present, no denominator figures are available since different patients had different kinds of workups and many were not complete by today's standards.

The observation of an "anomalous fifth ventricle" (10, cases 4,5) appears to be erroneous; the findings are most compatible with an interpretation of septum pellucidum cyst.

Cohen and Kreiborg (8a) further observed that progressive hydrocephaly appears to be uncommon and has frequently been confused with nonprogressive ventriculomegaly (12,15), which is very common and which may possibly occur as an embryonic malformation or possibly as a distortion caused by abnormal brain shaping by the malformed skull.

**Hands and feet.** A middigital hand mass minimally involving the second, third, and fourth fingers is always observed (Fig. 14-3A,B). Associated synonychia is variable in degree. The first and fifth fingers may be joined to the middigital hand mass or may be separate. When the thumb is free, it is broad and deviates radially. Some degree of brachydactyly involving all five fingers is usually present. The interphalangeal joints become stiff by 4 years of age (13,16).

Roentgenographically, the first metacarpal is normal. The proximal phalanx of the thumb is short, frequently narrow, and sometimes delta-shaped. The distal phalanx of the thumb is enlarged and trapezoidal in shape. In approximately half the cases, only the distal phalanx is present in the thumb. The second to fifth metacarpals are shorter than normal and tend to be uniform in length. The proximal ends of the fourth and fifth metacarpals are frequently fused (Fig. 14-3C). Symphalangism of the proximal interphalangeal joints occurs by 4–6 years of age. The distal interphalangeal joints are less frequently fused. Synostosis of adjacent distal phalanges occurs with age, most frequently between the third and fourth distal phalanges, but also between the second to fourth or between the second to fifth distal phalanges. Fusion of carpal bones, especially the hamate and capitate, may be observed. Other bony abnormalities of the hands have also been noted (16,26).

In the feet, syndactyly involves the second, third, and fourth toes (Fig. 14-3D). The first and fifth toes are sometimes free and sometimes joined by soft tissue union to the second and fourth toes, respectively. Toenails may be separate or partially continuous. The great toes are broad, and hallux varus is commonly observed (6).

The distal phalanx of the great toe is enlarged and trapezoidal. The proximal phalanx of the great toe is malformed. The second phalanges of the second to fifth toes are often absent. The first metatarsal is broad, shortened in some instances, and may exhibit partial or complete duplication. Symphalangism, fusion of the tarsal bones, six metatarsals, and other bony abnormalities may be observed in the feet (26).

Progressive calcification and fusion of the bones of the hands, feet, and cervical spine occur with age in all Apert syndrome patients (26). Fusion of the proximal interphalangeal joints is evident by 4–6 years, the fingers becoming gradually stiff (16).

**Other findings.** Final height attainment in the Apert syndrome population falls somewhat below the average for the general population. This may be due in part to mild rhizomelia of the lower limbs. Rhizomelia of the upper limbs may be more pronounced. In Cohen's series, 4% ($N=55$) exhibited severe rhizomelia (7). This occurred only in Apert syndrome patients of Italian genetic background. Progressive generalized bony dysplasia with ankylosis and progressive limitation of motion at the elbows, shoulders, hips, and knees have also been documented (14).

A large number of low-frequency visceral anomalies have been reported (4,5,27). Although individually uncommon in Apert syndrome, in the aggregate they were present in 5 of 11 autopsied cases recorded by Blank (4). Frequently, several anomalies were present in the same patient. In Cohen's series of 55 patients with Apert syndrome (5), 9% had cardiovascular defects of various kinds. Details are presented elsewhere (8).

Acne vulgaris with unusual extension to the forearms may be seen in more than 70% ($N=19$) of patients at adolescence and thereafter. Frank comedones and pustules occurring on the face, chest, back, and upper arms may be very severe (28,29). Hyperseborrhea is especially common (31). Because acneiform lesions of the forearms are unusual in the general population but common in the Apert syndrome population and because sebaceous glands are known to be highly sensitive to androgens, the end-organ responsiveness of the skin to steroid hormones may be a fruitful area for investigation in Apert syndrome.

**Differential diagnosis.** Apert syndrome should be distinguished from *Pfeiffer syndrome, Saethre–Chotzen syndrome, Jackson–Weiss syndrome, Crouzon syndrome,* and *Carpenter syndrome.*

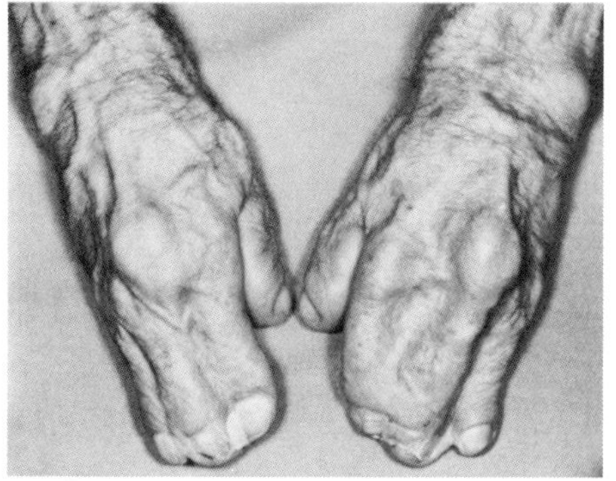

A

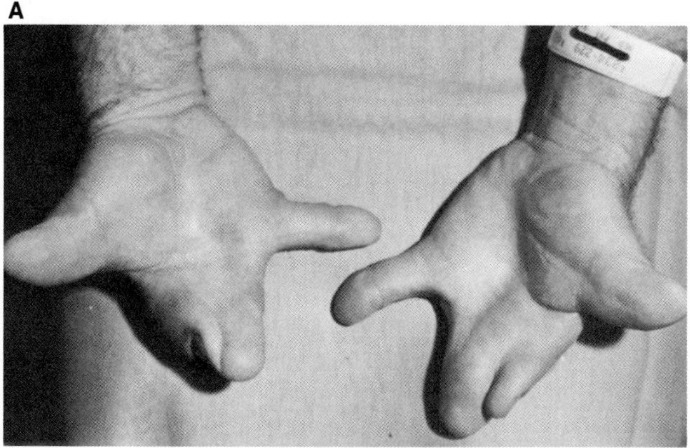

B

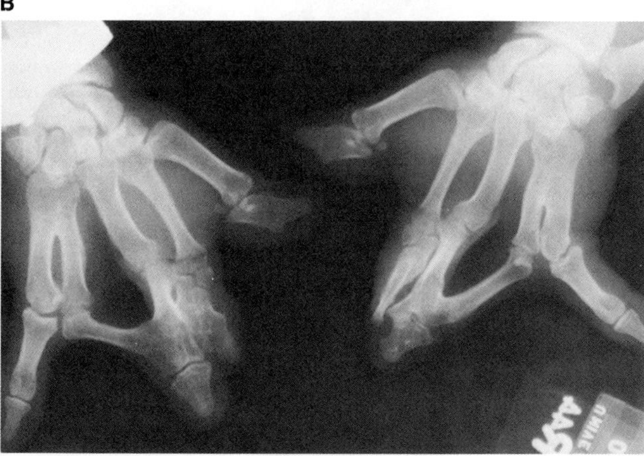

C

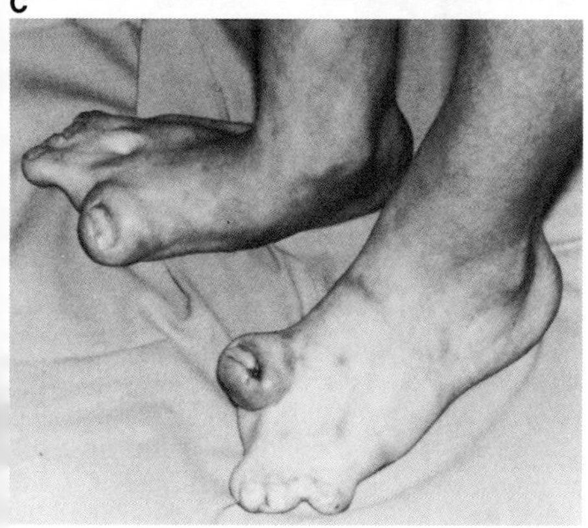

D

Fig. 14–3. *Apert syndrome.* (A) Syndactyly of second to fourth fingers. Note broad, radially deviated thumbs. (B) Mid-digital hand mass with first and fifth digits free. (C) Radiograph showing bony synostoses. (D) Syndactyly of second to fourth toes. Note broad, proximally placed halluces. (B and C from *JM Opitz,* Helena, Montana. D courtesy of *R Bauer,* Innsbruck, Austria.)

**Prenatal diagnosis.** Prenatal diagnosis of Apert syndrome has been reported (14a,17,25a).

### References [Apert syndrome (acrocephalosyndactyly)]

1. Allanson JE: Germinal mosaicism in Apert syndrome. *Clin Genet* **29**:429–433, 1986.

2. Apert E: De l'acrocéphalosyndactylie. *Bull Soc Méd Paris* **23**:1310–1330, 1906.

3. Bergstrom L et al: Otologic manifestations of acrocephalosyndactyly. *Arch Otolaryngol* **96**:117–123, 1972.

4. Blank CE: Apert's syndrome (a type of acrocephalosyndactyly): Observations on a British series of thirty-nine cases. *Ann Hum Genet* **24**:151–164, 1960.

5. Cohen MM Jr: Cardiovascular anomalies in Apert type acrocephalosyndactyly. *Birth Defects* **8**(5):132–133, 1972.

6. Cohen MM Jr: An etiologic and nosologic overview of craniosynostosis syndromes. *Birth Defects* **11**(2):137–189, 1975.

7. Cohen MM Jr: Craniosynsostosis and syndromes with craniosynostosis: Incidence, genetics, penetrance, variability, and new syndrome updating. *Birth Defects* **15**(5B):13–63, 1979.

8. Cohen MM Jr. *Craniosynostosis: Diagnosis, Evaluation, and Management.* Raven Press, New York, 1986.

8a. Cohen MM Jr, Kreiborg S: The central nervous system in the Apert syndrome. *Am J Med Genet,* in press.

8b. Cohen MM Jr, Kreiborg S: The growth pattern in the Apert syndrome. *Am J Med Genet,* in press.

8c. Cohen MM Jr, Kreiborg S: Studies of genetics and prevalence in the Apert syndrome, *Am J Med Genet* **35**:36–45, 1990.

9. Cuttone JM et al: Absence of the superior rectus muscle in Apert's syndrome. *J Pediatr Ophthalmol* **16**:349–354, 1980.

10. Dodge HW Jr et al: Craniofacial dysostosis: Crouzon's disease. *Pediatrics* **23**:98–106, 1959.

11. Erickson JD, Cohen MM Jr: A study of parental age effects on the occurrence of fresh mutations for the Apert syndrome. *Ann Hum Genet (London)* **38**:89–96, 1974.

12. Fishman MA et al: The concurrence of hydrocephalus and craniosynostosis. *J Neurosurg* **34**:621–629, 1971.

13. Green SM: Pathological anatomy of the hands in Apert's syndrome. *J Hand Surg* **7**:450–453, 1982.

14. Harris V et al: Progressive generalized bony dysplasia in Apert syndrome. *Birth Defects* **14**(6B):175, 1977.

14a. Hill LM et al: The ultrasound detection of Apert syndrome. *J Ultrasound Med* **6**:601–604, 1987.

15. Hogan GR, Bauman ML: Hydrocephalus in Apert's syndrome. *J Pediatr* **79**:782–787.

16. Hoover GH, et al: The hand and Apert's syndrome. *J Bone Jt Surg* **52A**:878–895, 1970.

16a. Kreiborg S, Cohen MM Jr: The midline calvarial defect during infancy in the Apert syndrome. *J Craniofac Genet Dev Biol,* in press.

16b. Kreiborg S, Cohen MM Jr: The Apert syndrome: An analysis of 113 cases, in progress.

16c. Lefebvre A et al: A psychiatric profile before and after reconstructive surgery in children with Apert's syndrome. *Br J Plast Surg* 39:510–513, 1986.

17. Leonard CO et al: Prenatal fetoscopic diagnosis of the Apert syndrome. *Am J Med Genet* **11**:5–9, 1982.

18. Margolis S et al: Structural alterations of extraocular muscle associated with Apert syndrome. *Br J Ophthalmol* **61**:683–689, 1977.

19. Margolis S et al: Depigmentation of hair, skin, and eyes associated with the Apert syndrome. *Birth Defects* **14**(6B):341–360, 1978.

19a. Maroteaux P, Fonfria MC: Apparent Apert syndrome with polydactyly: Rare pleiotropic manifestation or new syndrome? *Am J Med Genet* **28**:153–158, 1987.

20. Mathur JS et al: Apert's anomaly—A transition. *Acta Ophthalmol (Copenhagen)* **46**:1041–1045, 1968.

20a. Ousterhout D, Melsen B: Cranial base deformity in Apert's syndrome. *Plast Reconstr Surg* **69**:254–263, 1982.

21. Park EA, Powers GF: Acrocephaly and scaphocephaly with symmetrically distributed malformations of the extremities. *Am J Dis Child* **20**:235–315, 1920.

21a. Patton MA et al: Intellectual development in Apert's syndrome: A long term follow up of 29 patients. *J Med Genet* **25**:164–167, 1988.

22. Peterson SJ, Pruzansky S: Palatal anomalies in the syndromes of Apert and Crouzon. *Cleft Palate J* **11**:394–403, 1974.

23. Peterson-Falzone SJ et al: Nasopharyngeal dysmorphology in the syndromes of Apert and Crouzon. *Cleft Palate J* **18**:237–250, 1981.

24. Pruzansky S: Clinical investigation of the experiments of nature. Orofacial anomalies: Clinical and research implications. *ASHA Rept* **8**:62–94, 1975.

25. Roberts KB, Hall JG: Apert's acrocephalosyndactyly in mother and daughter: Cleft palate in the mother. *Birth Defects* **7**(7):262–263, 1971.

25a. Robinson LK et al: Prenatal diagnosis of the Apert syndrome. Ninth Annual David W. Smith Workshop on Malformations and Morphogenesis, Oakland, August 7–10, 1988.

25b. Rollnick BR: Male transmission of Apert syndrome. *Clin Genet* **33**:87–90, 1988.

26. Schauerte EW, St-Aubin PM: Progressive synosteosis in Apert's syndrome (acrocephalosyndactyly) with a description of roentgenographic changes in the feet. *Am J Roentgenol* **97**:67–73, 1966.

27. Scott CI: The Apert syndrome with coarctation of the aorta. *Birth Defects* **8**(5):239–241, 1972.

27a. Sidhu SS, Deshmukh R: Recessive inheritance of apparent Apert syndrome with polydactyly? *Am J Med Genet* **31**:179–180, 1988.

28. Solomon LM et al: Pilosebaceous abnormalities in Apert type acrocephalosyndactyly. *Birth Defects* **7**(8):193–195, 1971.

29. Solomon LM et al: Pilosebaceous abnormalities in Apert's syndrome. *Arch Dermatol* **102**:381–385, 1970.

30. Solomon LM et al: Apert syndrome and palatal mucopolysaccharides. *Teratology* **8**:287–292, 1973.

31. Tessier P: The definitive plastic surgical treatment of the severe facial deformities of craniofacial dysostosis. Crouzon's and Apert's diseases. *Plast Reconstr Surg* **48**:419–442, 1971.

32. Tünte W, Lenz W: Zur Häufigkeit und Mutationsrates des Apert-Syndroms. *Hum Genet* **4**:104–111, 1967.

33. Van den Bosch J, cited in Blank CE: Apert's syndrome (a type of acrocephalosyndactyly): Observations on a British series of thirty-nine cases. *Ann Hum Genet* **24**:151–164, 1960.

34. Weech AA: Combined acrocephaly and syndactylism occurring in mother and daughter. *Bull Johns Hopkins Hosp* **40**:73–76, 1927.

35. Weinstock FJ, Hardesty HH: Absence of superior recti in craniofacial dysostosis. *Arch Ophthalmol* **74**:152–153, 1965.

36. Wheaton SW: Two specimens of congenital cranial deformity in infants associated with fusion of fingers and toes. *Trans Pathol Soc London* **45**:238–241, 1894.

## Crouzon syndrome (craniofacial dysostosis)

Crouzon syndrome is characterized by craniosynostosis, maxillary hypoplasia, shallow orbits, and ocular proptosis (Fig. 14-4A–F). First described by Crouzon (5) in 1912, 86 published cases were reviewed by Atkinson (1) by 1937. Numerous other publications on Crouzon syndrome have appeared (2–22,25,27–31), with the most complete study being Kreiborg's 1981 monograph (17), which analyzes 61 cases. Cohen (4) provided an exhaustive review in 1986.

Crouzon syndrome follows an autosomal dominant mode of transmission. In Atkinson's review (1), 67% of the cases were familial and 33% were sporadic, representing fresh mutations. In Kreiborg's monograph (17), 44% were familial and 56% occurred sporadically. Increased paternal age at the time of conception was a statistically significant factor in fresh mutations in Kreiborg's study (17). Rollnick (26a) reported an example of germinal mosaicism. Crouzon syndrome has been said to occur with a frequency of 1/25,000 in the general population (4). Variability of expression characterizes Crouzon syndrome. Nowhere is this more apparent than in the pedigree reported by Shiller (27; see also 15). The proband, the most severely affected member of the family, presented with cloverleaf skull. Several sibs manifested classic Crouzon syndrome. The affected mother and various other members of the family exhibited ocular proptosis and midface deficiency without craniosynostosis.

The clinical findings of Crouzon syndrome are summarized in Table 14-5.

**Craniofacial features.** Cranial malformation in Crouzon syndrome depends on the order and rate or progression of sutural synostosis. Brachycephaly is most commonly observed, but scaphocephaly, trigonocephaly, and, as already indicated, the cloverleaf skull may be observed. Craniosynostosis commonly begins during the first year of life and is usually complete by 2–3 years of age. In some cases, craniosynostosis may be evident at birth (3). Occasionally, no sutural

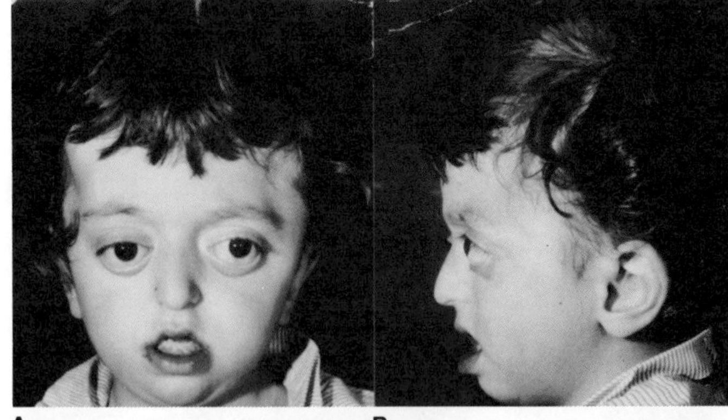
A B

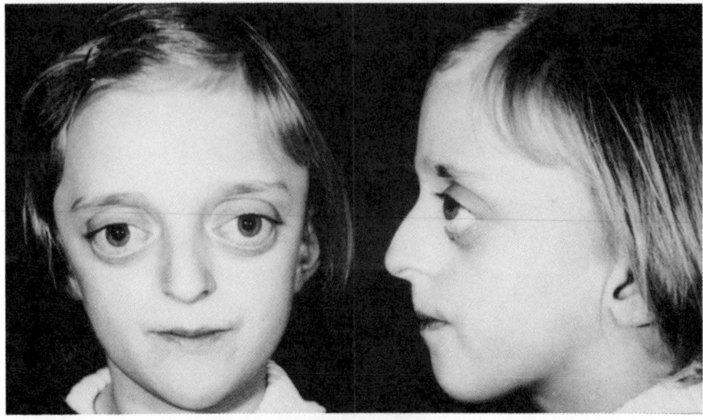
C D

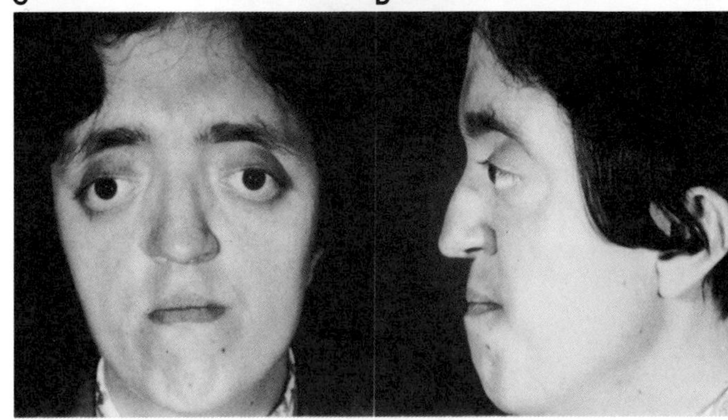
E F

Fig. 14–4. *Crouzon syndrome.* (A–F) Note various craniofacial shapes, shallow orbits, proptosis, divergent strabismus, hypertelorism, short upper lip, and midface hypoplasia. (Courtesy of *P Tessier,* Paris, France.)

involvement may be noted. Shallow orbits with ocular proptosis are an essential diagnostic feature for Crouzon syndrome. Diagnosis may be evident at birth or during the first year of life. On occasion, the phenotypic features of Crouzon syndrome may be absent and evolve gradually during the first few years of life. Some affected individuals with minor stigmata of Crouzon syndrome are detected only because they are related to a severely affected proband (3,4,17). Various sutures may be prematurely synostosed, and multiple sutural involvement is found eventually in most cases. Increased digital markings are common on skull radiographs (3,17). An extensive anthropometric study of the skull and face of 61 patients was carried out by Kolar et al (16a).

Ocular proptosis, a constant feature occurring in 100% of the cases ($N=61$), is secondary to shallow orbits and results in a high frequency of exposure conjunctivitis or keratitis. Luxation of the eyeglobes has been observed in some cases. Exotropia is an extremely common finding (77%, $N=60$) (2,3,17). Poor vision occurs in approximately 46%

Table 14–5. Clinical findings in Crouzon syndrome[a]

| Features | Total number of patients ascertained for specific abnormality | Cases with abnormality (%)[b] |
|---|---|---|
| Craniosynostosis | 47 | 100[c] |
| Coronal | 47 | 2 |
| Coronal and sagittal | 47 | 19 |
| Coronal, sagittal, and lambdoidal | 47 | 75 |
| Sagittal and lambdoidal | 47 | 4 |
| Neurologic | | |
| Frequent headaches | 52 | 29 |
| Seizures | 52 | 12 |
| Marked mental deficiency | 61 | 3 |
| Agenesis of corpus callosum | — | 1 Case |
| Ophthalmologic | | |
| Ocular proptosis | 61 | 100 |
| Exotropia | 60 | 77 |
| Poor vision | 54 | 46 |
| Optic atrophy | 45 | 22 |
| Blindness | 61 | 7 |
| Nystagmus | 52 | 12 |
| Iris coloboma | 60 | 2 |
| Exposure conjunctivitis | 52 | 52 |
| Exposure keratitis | 52 | 12 |
| Aural | | |
| Mild to moderate hearing deficit | 49 | 55 |
| Atresia of external auditory canal | 53 | 13 |
| Oral | | |
| Cleft lip | 61 | 2 |
| Cleft palate | 61 | 3 |
| Bifid uvula | 53 | 9 |
| Lateral palatal swellings | 54 | 50 |
| Obligatory mouthbreathing | 53 | 32 |
| Other | | |
| Calcification of stylohyoid ligament | 50 | 88 |
| Cervical spine anomalies[d] | 60 | 30 |

[a]From S Kreiborg, *Scand J Plast Reconstr Surg Suppl* 18, 1981.
[b]Percentages have been rounded to the nearest percent.
[c]Although 100% of Kreiborg's patients had craniosynostosis, a few instances of Crouzon syndrome without craniosynostosis have been reported elsewhere in the literature.
[d]Fused vertebrae account for 83% of all cervical spine anomalies of which the following are seen: ($C_2$–$C_3$ fusion—73%, $C_2$–$C_3$ fusion combined with $C_5$–$C_6$ fusion—13%, $C_2$–$C_3$–$C_4$ fusion combined with $C_5$–$C_6$ fusion—7%, and $C_5$–$C_6$ fusion—7%; occipitalization of $C_2$—11%; and atlas assimilation—6%).

($N=54$), with optic atrophy found in 22% ($N=45$) and blindness in 7% ($N=61$) (17). Low-frequency findings include nystagmus, iris coloboma, aniridia, anisocoria, corectopia, microcornea, megalocornea, keratoconus, cataract, ectopia lentis, blue sclera, and glaucoma (1,3,4,17).

Approximately 50% ($N=54$) have lateral palatal swellings but only in a few instances are they large enough to produce the median pseudocleft appearance found so frequently in Apert syndrome (17,23). Cleft lip and cleft palate are anomalies of low frequency (17). Because of maxillary hypoplasia in Crouzon syndrome, the anteroposterior dimension of the maxillary dental arch is shortened. Dental arch width is also reduced, and the constricted arch gives the appearance of highly arched palate, although palatal height is normal by measurement. Unilateral or bilateral posterior crossbite is evident in two-thirds of Crouzon syndrome patients. Crowding of maxillary teeth is common, and ectopic eruption of maxillary first molars occurs in approximately 47% ($N=17$). Anterior open bite, mandibular overjet, and crowding of mandibular anterior teeth are also commonly observed. Aplasia of single teeth (rarely more), shovel-shaped maxillary incisors, and abnormal premolar morphology have been observed in Crouzon syndrome with the same low frequency found in the general population (17,23,24).

**Central nervous system.** Frequent headaches were found in 29% ($N=52$) in Kreiborg's series (17). Seizures occurred in 12% ($N=52$), and marked mental deficiency was found in only 3% ($N=61$). Agenesis of the corpus callosum was documented in one case (17). The association of increasing hydrocephaly is rare, but has been recorded in some instances (4,14).

**Other findings.** Conductive hearing deficit is found in 55% ($N=49$) and atresia of the external auditory canals occurs in 13% ($N=53$) (17). Deviation of the nasal septum was observed in 33% ($N=60$) of Kreiborg's series (17). Calcification of the stylohyoid ligament was especially common, being found in 88% ($N=60$). Cervical spine anomalies were also common (30%, $N=60$). Fusion of $C_2$ with $C_3$ occurred with the highest frequency, but in some instances $C_5$ and $C_6$ were fused, and in still other instances, $C_2$–$C_3$ and $C_5$–$C_6$ fusions occurred in the same patient (17). In Kreiborg's series (17), stiffness of the elbows was noted in 16% ($N=61$), and in two of these cases, subluxation of the radial heads was evident. Cubitus valgus was observed in one patient by VA McKusick (personal communication, 1982). Ectrodactyly was recorded in one instance (13). Sacral, vertebral, and rib anomalies were also noted by Golabi et al (14).

Acanthosis nigricans has been observed in association with Crouzon syndrome by Suslak (28), Reddy (26), Breitbart et al (2a), and by other craniofacial surgeons (I Munro, Dallas, personal communication, 1986). Cementomas were observed in the patient reported by Suslak (28), but as the patient was a black female, the finding is not significant.

**Differential diagnosis.** Crouzon syndrome should be distinguished from simple craniosynostosis, *Apert syndrome, Pfeiffer syndrome, Saethre–Chotzen syndrome,* and *Jackson–Weiss syndrome.*

## References [Crouzon syndrome (craniofacial dysostosis)]

1. Atkinson FRB: Hereditary craniofacial dysostosis, or Crouzon's disease. *Med Press Circular* **195**:118–124, 1937.
2. Bertelsen TI: The premature synostosis of the cranial sutures. *Acta Ophthalmol* **51**(Suppl): 1–176, 1958.
2a. Breitbart AS et al: Crouzon's syndrome associated with acanthosis nigricans. Ramifications for the craniofacial surgeon. *Ann Plast Surg* **22**:310–315, 1989.
3. Cohen MM Jr: An etiologic and nosologic overview of craniosynostosis syndromes. *Birth Defects* **11**(2):137–189, 1975.
4. Cohen MM Jr: *Craniosynostosis: Diagnosis, Evaluation, and Management.* Raven Press, New York, 1986.
5. Crouzon O: Dysostose cranio-faciale héréditaire. *Bull Soc Méd Hôp Paris* **33**:545–555, 1912.
6. Crouzon O: Dysostose cranio-faciale héréditaire. *Arch Méd Enf* **18**:545–555, 1915.
7. Crouzon O: Une nouvelle famille atteinte de dysostose cranio-faciale héréditaire. *Arch Méd Enf* **18**:540, 1915.
8. Devine P et al: Completely cartilaginous trachea in a child with Crouzon syndrome. *Am J Dis Child* **138**:40–43, 1984.
9. Dodge HW Jr et al: Craniofacial dysostosis: Crouzon's disease. *Pediatrics* **23**:98–106, 1959.
10. Don N, Sigger DC: Cor pulmonale in Crouzon's disease. *Arch Dis Childh* **46**:394–397, 1971.
11. Flippen JH Jr: Cranio-facial dysostosis of Crouzon: Report of a case in which the malformation occurred in four generations. *Pediatrics* **50**:90–96, 1950.
12. Fogh-Andersen P: Dysostosis craniofacialis (Crouzon) som dominant arvelig lidelse. *Nord Med* **18**:993–996, 1942.
13. Garcin M et al: Sur un cas isolé de dysostose craniofaciale (maladie de Crouzon) avec ectrodactylie. *Bull Soc Méd Hôp Paris* **56**:1458–1466, 1932.
14. Golabi M et al: Radiographic abnormalities of Crouzon syndrome. A survey of 23 cases. *Proc Greenwood Genet Ctr* **3**:102, 1984.
15. Hall BD et al: Kleeblattschädel (cloverleaf) syndrome: Severe form of Crouzon's disease? *J Pediatr* **80**:526–528, 1972.

16. Heuyer G: À propos de la dysostose cranio-faciale (ou maladie de Crouzon). *Actualités Odonto-stomatol* **16**:413, 1951.

16a. Kolar JC et al: Patterns of dysmorphology in Crouzon syndrome: An anthropometric study. *Cleft Palate J* **25**:235–244, 1988.

17. Kreiborg S: Crouzon syndrome. *Scand J Plast Reconstr Surg Suppl* 18, 1–198, 1981.

18. Kreiborg S: Craniofacial growth in plagiocephaly and Crouzon syndrome. *Scand J Plast Reconstr Surg* **15**:187–197, 1981.

19. Kreiborg S, Bjork A: Description of a dry skull with Crouzon syndrome. *Scand J Plast Reconstr Surg* **16**:245–253, 1982.

20. Kreiborg S, Jensen BL: Variable expressivity of Crouzon's syndrome within a family. *Scand J Dent Res* **85**:175–184, 1977.

21. Kreiborg S, Pruzansky S: Craniofacial growth in premature craniofacial synostosis. *Scand J Plast Reconstr Surg* **15**:171–186, 1981.

22. Lake MS, Kuppinger JC: Craniofacial dysostosis (Crouzon's disease): Report of three cases. *Arch Ophthalmol* **44**:37–46, 1950.

23. Peterson SJ, Pruzansky S: Palatal anomalies in the syndromes of Apert and Crouzon. *Cleft Palate J* **11**:394–403, 1974.

24. Peterson-Falzone SJ et al: Nasopharyngeal dysmorphology in the syndromes of Apert and Crouzon. *Cleft Palate J* **18**:237–250, 1981.

25. Pinkerton OD, Pinkerton FJ: Hereditary craniofacial dysplasia. *Am J Ophthalmol* **35**:500–506, 1952.

26. Reddy BSN: An unusual association of acanthosis nigricans and Crouzon's disease. *J Dermatol* **12**:85–90, 1985.

26a. Rollnick BR et al: Germinal mosaicism in Crouzon syndrome. *Clin Genet* **33**:145–150, 1988.

27. Shiller JG: Craniofacial dysostosis of Crouzon: A case report and pedigree with emphasis on heredity. *Pediatrics* **23**:107–112, 1959.

28. Suslak E: Crouzon's craniofacial dysostosis, periapical cemental dysplasia, and acanthosis nigricans: The pleiotropic effect of a single gene? *Soc Craniofacial Genet*, Denver, June 17, 1984.

29. Vuilliamy DG, Normandale PA: Craniofacial dysostosis in a Dorset family. *Arch Dis Childh* **41**:275–382, 1966.

30. Weigand R: Dysostosis craniofacialis (Morbus Crouzon, 1912) mit beidseitiger (häutiger) Gehörgangsatresie. *Arch Ohr Nas Kehlk Heilkd* **166**:128–139, 1954.

31. Wood-Smith D et al: Transcranial decompression of the optic nerve in the osseous canal in Crouzon's disease. *Clin Plast Surg* **3**:621–623, 1976.

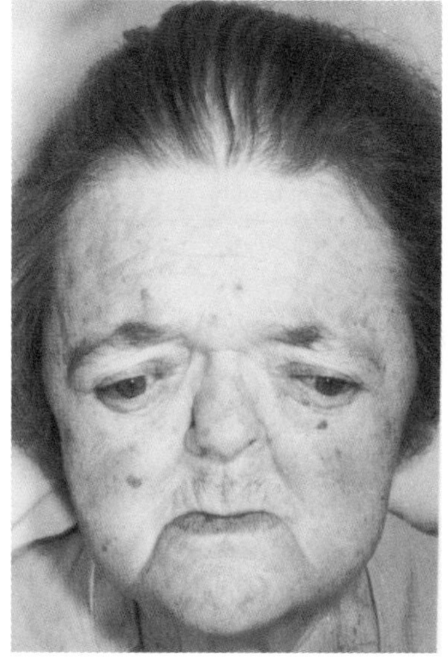

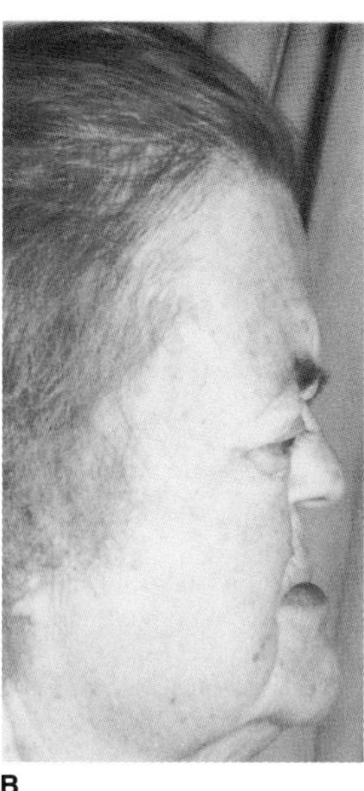

A B

Fig. 14–5. *Pfeiffer syndrome.* (A,B) Hypertelorism, downslanting palpebral fissures, and midface deficiency.

## Pfeiffer syndrome

In 1964, Pfeiffer (11) described a syndrome consisting of craniosynostosis, broad thumbs, broad great toes, and a variable feature, partial soft tissue syndactyly of the hands. Eight affected individuals in three generations were noted in Pfeiffer's report. Other pedigrees consistent with autosomal dominant transmission were noted by several authors (10,12,14,17). Penetrance has been complete, and expressivity has been very variable. Sporadic cases have also been noted (1,5,6,9,16). The most exhaustive review is that of Cohen (4).

The affected mother and daughter reported by Sanchez and De Negrotti (15) exhibited variable expressivity. However, in our opinion, the family reported by Baraitser et al (2) as an example of Pfeiffer syndrome represents Jackson–Weiss syndrome. And we are not convinced of the mild expression of Pfeiffer syndrome reported in a mother by Rasmussen and Frias (13). The only feature the mother had was a unilateral stub thumb which had the same width as the normal thumb.

**Craniofacial features.** The skull is usually turribrachycephalic. Craniofacial asymmetry may be present in some instances. Maxillary hypoplasia and relative mandibular prognathism are observed. The nasal bridge is depressed. Hypertelorism, downslanting palpebral fissures, ocular proptosis, and strabismus are common (3,10,14) (Fig. 14-5A,B). The nose may be beaked. The palate is highly arched, the alveolar ridges are broad, and teeth are crowded (3). Some affected individuals are fair and have prominent veins. Results of craniofacial surgery have been surprisingly disappointing in some instances (JG McCarthy, personal communication, 1986); to date, the reasons for this are not known.

**Performance and central nervous system.** In the common form of Pfeiffer syndrome, intelligence is usually normal (1,10,11), but mental deficiency has been observed in a number of cases (3). Hydrocephaly, Arnold–Chiari malformation, and seizures have been noted (14). The syndrome has been frequently reported in association with the cloverleaf skull anomaly (7,8,8a). All such cases to date have been sporadic, and none has occurred within an affected family. Various other anomalies, not usually found with the usual form of Pfeiffer syndrome, may be observed in Pfeiffer cloverleaf patients. The prognosis for mental development is poor, and early demise during infancy is common, even when extensive craniofacial surgery is performed (4).

Fig. 14–6. *Pfeiffer syndrome.* (A) Broad radially deviated thumbs, brachydactyly, and clinodactyly of terminal phalanges. (B) Broad halluces and crooked toes.

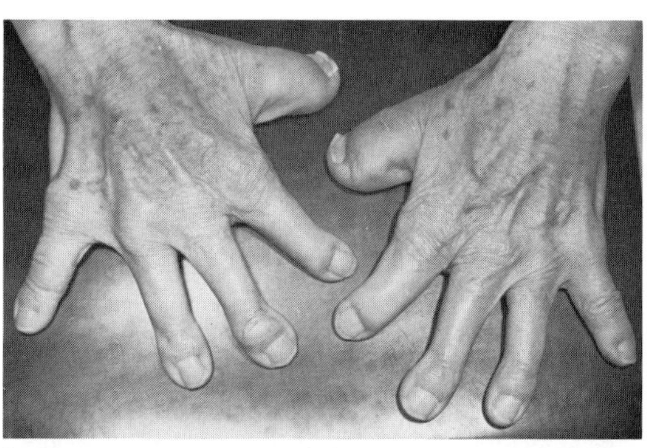

A

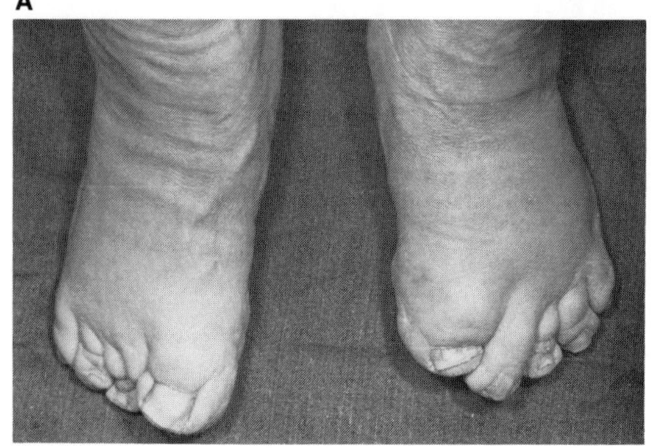

B

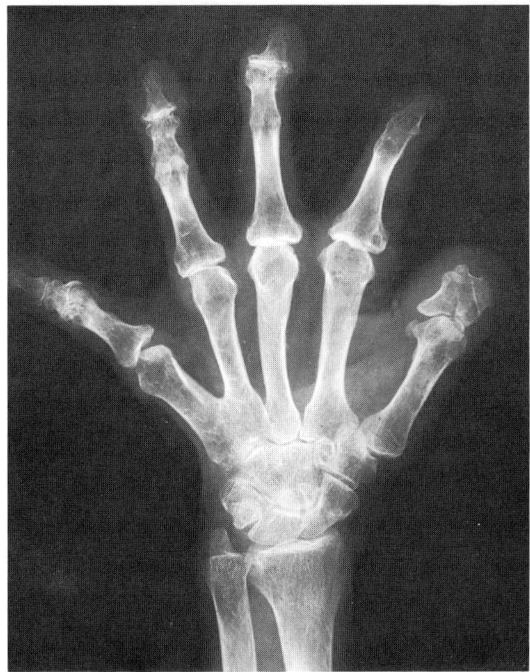

Fig. 14–7. *Pfeiffer syndrome.* Roentgenogram showing malformed fused phalanges of thumbs, brachymesophalangy, symphalangism, fusion of proximal ends of fourth and fifth metacarpals.

**Hands and feet.** The thumbs and great toes are broad (Fig. 14-6A,B), usually with varus deformity (11). Mild soft tissue syndactyly may especially involve digits 2 and 3 and sometimes digits 3 and 4 of both hands and feet (3,11). Partial soft tissue syndactyly between toes 1 and 2 has also been reported (1,3,17). Brachydactyly may be observed and, in some cases, syndactyly is absent (3). Clinodactyly has also been noted (3,9a,14).

Brachymesophalangy of both hands and feet is frequently present. Middle phalanges may be absent in some cases. The distal phalanx of the great toe is broad, and the proximal phalanx malformed. The first metatarsal is broad, may be shortened, and may be duplicated in some instances (3,9a,10,11).

Accessory epiphyses in the first and second metatarsals and double ossification centers in the proximal phalanx of the great toe have been reported. Partial duplication of the great toe may be observed occasionally. Symphalangism of both hands and feet has been reported. Fusion of carpals and tarsals, in some instances involving the proximal ends of the metacarpals and metatarsals, respectively, has also been noted (3,10,12,17) (Fig. 14-7).

**Other findings.** Fused cervical vertebras and lumbar vertebras have been described. Shortened humerus, cubitus valgus, radiohumeral and radioulnar synostosis, abnormalities of the pelvis, coxa valga, and talipes calcaneovarus have been reported occasionally (3,10,14). Low-frequency abnormalities have included pyloric stenosis, umbilical hernia, malpositioned anus at the scrotal base, bifid scrotum, widely spaced nipples, ptosis of eyelids, corectopia, scleralization of the cornea, optic nerve hypoplasia, choanal atresia, preauricular tag, absent external auditory canals, hearing deficit, bifid uvula, supernumerary teeth, and gingival hypertrophy (1,3,5,10,14).

**Differential diagnosis.** Pfeiffer syndrome should be distinguished from *Apert syndrome, Crouzon syndrome, Saethre–Chotzen syndrome,* and *Jackson–Weiss syndrome.*

### References (Pfeiffer syndrome)

1. Asnes RS, Morehead CD: Pfeiffer syndrome. *Birth Defects* **5**(3):199–203, 1969.

2. Baraitser M et al: Pitfalls of genetic counseling in Pfeiffer syndrome. *J Med Genet* **17**:250–256, 1980.

3. Cohen MM Jr: An etiologic and nosologic overview of craniosynostosis syndromes. *Birth Defects* **11**(2):137–189, 1975.

4. Cohen MM Jr: *Craniosynostosis: Diagnosis, Evaluation, and Management.* Raven Press, New York, 1986.

5. Cremers CWRJ: Hearing loss in Pfeiffer's syndrome. *Int J Pediatr Otorhinolaryngol* **3**:343–353, 1981.

6. Degenhardt KH: Zum entwicklungsmechanischen Problem der Akrocephalosyndaktylie. *Z Menschl Vererbung Konstit Lehre* **29**:719–819, 1950.

7. Eaton AP et al: The Kleeblattschädel anomaly. *Birth Defects* **11**(2):238–246, 1975.

8. Hodach RJ et al: Studies of malformation syndromes in man XXXVI: The Pfeiffer syndrome, association with Kleeblattschädel and multiple visceral anomalies. Case report and review. *Z Kinderheilkd* **119**:87–103, 1975.

8a. Kroczek RA et al: Cloverleaf skull associated with Pfeiffer syndrome: Pathology and management. *Eur J Pediatr* **145**:442–445, 1986.

9. Lenz W: Zur Diagnose und Ätiologie der Akrocephalosyndaktylie. *Z Kinderheilkd* **79**:546–558, 1957.

9a. Manouvrier-Hanu S et al: L'acrocephalosyndactylie de type V (syndrome de Pfeiffer). *Arch Fr Pédiatr* **46**:433–437, 1989.

10. Martsolf JT et al: Pfeiffer syndrome. *Am J Dis Child* **121**:257–262, 1971.

11. Pfeiffer RA: Dominant erbliche Akrocephalosyndaktylie. *Z Kinderheilkd* **90**:301–320, 1964.

12. Pfeiffer RA: Associated deformities of the head and hands. *Birth Defects* **5**(3):18–34, 1969.

13. Rasmussen SA, Frias JL: Mild expression of the Pfeiffer syndrome. *Clin Genet* **33**:5–10, 1988.

14. Saldino RM et al: Familial acrocephalosyndactyly (Pfeiffer syndrome). *Am J Roentgenol* **116**:609–622, 1972.

15. Sanchez JM, De Negrotti TC: Variable expression in Pfeiffer syndrome. *J Med Genet* **18**:73–75, 1981.

16. Van Dyke DG et al: Clinical observations: Ocular abnormalities in a patient with Pfeiffer syndrome (acrocephalosyndactyly, type V). *J Clin Dysmorphol* **1**(4):2–3, 1983.

17. Zippel H, Schuler KH: Dominant vererbte Akrozephalosyndaktylie (ACS). *Fortschr Röntgenstr* **110**:2340–2345, 1969.

## Saethre–Chotzen syndrome

Saethre–Chotzen syndrome is characterized by a broad and variable pattern of malformations, including craniosynostosis, low-set frontal hairline, facial asymmetry, ptosis of eyelids, deviated nasal septum, brachydactyly, partial cutaneous syndactyly, especially of the second and third fingers, and various skeletal anomalies. Autosomal dominant inheritance is evident with a high degree of penetrance and variable expressivity (5,6). Incomplete penetrance was documented by Carter et al (3).

First recognized as an entity by Saethre (22) in 1931 and by Chotzen (4) in 1932, many authors have since reported affected families (1–6,8–25). The most extensive discussions have been published by Pantke et al (19) and Cohen (7).

**Craniofacial features.** Craniosynostosis is a facultative rather than an obligatory abnormality. When present, the time of onset and degree of craniosynostosis are quite variable. Brachycephaly or acrocephaly with coronal sutural synostosis is frequently observed, and involvement is often asymmetric, producing plagiocephaly and facial asymmetry (Fig. 14-8A-D). Trigonocephaly has also been observed. Frontal bossing, parietal bossing, and flattened occiput have been reported in various cases. Large late-closing fontanels, large parietal foramina, ossification defects of the calvaria, and enlargement of the sella turcica have also been recorded. Calvarial hyperostosis was noted in one family (1,4,5,11,14,19).

Low-set frontal hairline is commonly observed. Ptosis of eyelids, hypertelorism, and strabismus are common. Occasionally, the forehead may be protuberant in association with metopic synostosis and hypotelorism. Blepharophimosis is evident in some cases. Dacryostenosis may be a feature. Low-frequency findings have included epicanthic folds, optic atrophy, downslanting palpebral fissures, irregular eyelid margins, and sparse eyebrows medially with heavy eyebrows laterally (2,4,5,10,11,19).

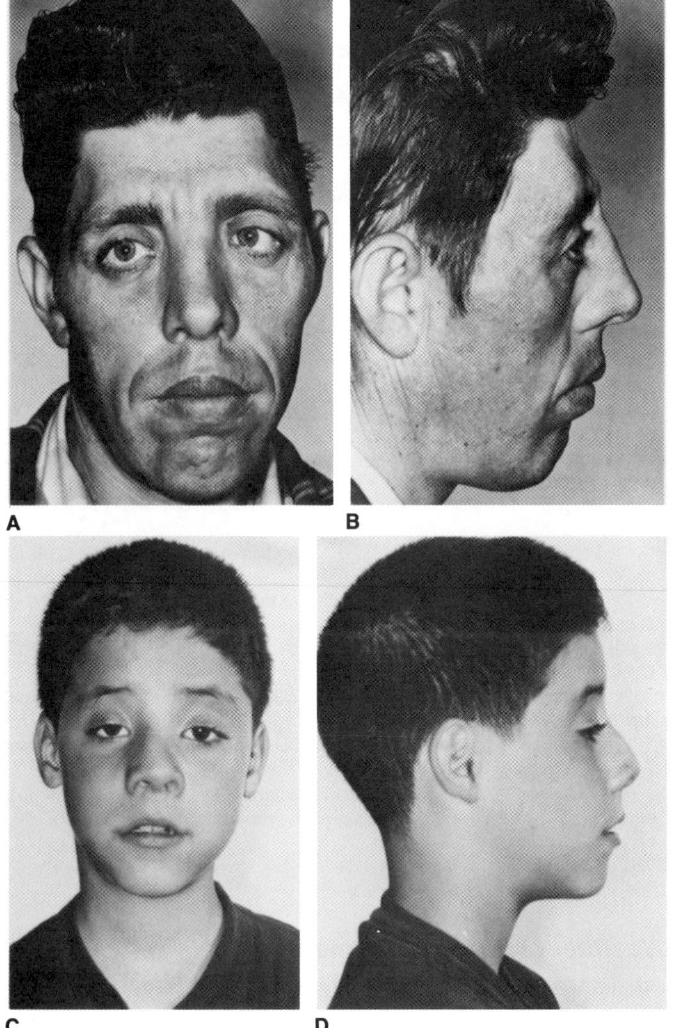

Fig. 14–8. *Saethre–Chotzen syndrome.* (A,B) Acrocephaly, facial asymmetry, low hairline. (C,D) Son of man seen in A and B. Note eyelid ptosis.

The ears may be low set, small, posteriorly angulated, or may have folded helices or prominent antihelical crura. Mild hearing deficit is common (5,19).

The nasofrontal angle may be flattened in some instances. Maxillary hypoplasia with relative mandibular prognathism may be evident. The midface may be broad and flat in some cases. The nose is often beaked, and deviation of the nasal septum is common (5,11,19).

Oral anomalies include narrow or highly arched palate, cleft palate on occasion, malocclusion, supernumerary teeth, enamel hypoplasia, and other dental defects (2,5,11,19).

**Performance and central nervous system.** Intelligence is usually normal (19). However, mild to moderate mental retardation has been observed in a number of cases (2,4). Moderate dilatation of the lateral ventricles was demonstrated in one case (2). Epilepsy and schizophrenia have also been noted (2,4,22).

**Hands and feet.** Some degree of brachydactyly may be observed (5,19). Partial cutaneous syndactyly is present in some instances, most frequently between the second and third fingers (Fig. 14-9), but sometimes extending from the second to fourth fingers (2,4,18,19,22). Clinodactyly, especially of the fifth finger, has been found in some cases (1,11). The distal phalanges may be hypoplastic, and, on occasion, finger-like thumbs have been noted (11). Dermatoglyphic findings have included single palmar creases, distally placed axial triradii, an increased frequency of thenar and hypothenar patterns, an increased frequency of fingertip arch patterns, and low total ridge count (1,2,11,19).

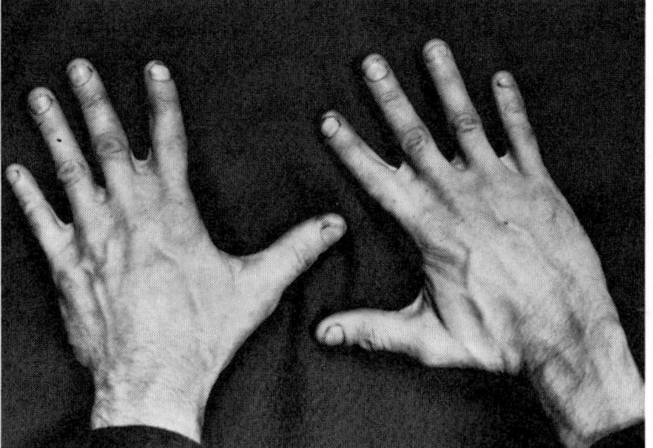

Fig. 14–9. *Saethre–Chotzen syndrome.* Soft tissue syndactyly between second and third fingers. (From *CS Bartsocas et al,* J Pediatr **77:**267, 1967).

Partial cutaneous syndactyly between the second and third toes, but occasionally involving other toes, has been reported (2,4,22). Broad great toes and hallux valgus have been noted in some instances (11, 19, 22) and dorsiflexion of the fourth toes has been observed (2).

**Other findings.** Short stature has been documented in some instances (11,19). Defects of the cervical and lumbar spine have been reported by several authors (4,19,22). Radioulnar synostosis was noted by Bartsocas et al (2) and shortened fourth metacarpals were reported by Aase and Smith (1). Friedman et al (11) documented short clavicles, limitation of motion at the elbows, small ilia, large ischia, and coxa valga. Other findings have included cryptorchidism (2,4), renal anomalies (2), and congenital cardiac defects (1). Escobar et al (8) reported a case with congenital adrenal hyperplasia. McKeen et al (17) observed an affected family in which multiple malignancies occurred, including nasopharyngeal carcinoma, Hodgkin's disease, and embryonal cell carcinoma of the testis; however, the Saethre–Chotzen syndrome and the familial malignancies were probably inherited independently.

**Differential diagnosis.** Saethre–Chotzen syndrome is frequently confused with simple craniosynostosis. It should also be distinguished from *Crouzon syndrome, Pfeiffer syndrome, Apert syndrome,* and *Jackson–Weiss syndrome.* Because syndactyly is not an obligatory anomaly of the Saethre–Chotzen syndrome, a sporadic case without this finding can present difficulties in diagnosis.

### References (Saethre–Chotzen syndrome)


1. Aase JM, Smith DW: Facial asymmetry and abnormalities of palms and ears: A dominantly inherited developmental syndrome. *J Pediatr* **76**:928–930, 1970.
2. Bartsocas CS et al: Chotzen's syndrome. *J Pediatr* **77**:267–272, 1970.
3. Carter CO et al: A family study of craniosynostosis, with probable recognition of a distinct syndrome. *J Med Genet* **19**:280–285, 1982.
4. Chotzen F: Eine eigenartige familiäre Entwicklungsstörung. (Akrocephalosyndaktylie, Dysostosis craniofacialis und Hypertelorismus). *Monatschr Kinderheilkd* **55**:97–122, 1932.
5. Cohen MM Jr: An etiologic and nosologic overview of craniosynostosis syndromes. *Birth Defects* **11**(2):137–189, 1975.
6. Cohen MM Jr: Craniosynostosis and syndromes with craniosynostosis: Incidence, genetics, penetrance, variability, and new syndrome updating. *Birth Defects* **15**(5B):85–89, 1979.
7. Cohen MM Jr. *Craniosynostosis: Diagnosis, Evaluation, and Management.* Raven Press, New York, 1986.
8. Escobar V et al: Unusual association of Saethre–Chotzen syndrome and congenital adrenal hyperplasia. *Clin Genet* **11**:365–381, 1977.
9. Evans CA, Christiansen RL: Cephalic malformations in Saethre–Chotzen syndrome. *Radiology* **121**:399–403, 1976.

10. Francke U et al: Saethre–Chotzen syndrome (SCS) with additional abnormalities in a Mexican family. *Birth Defects* **13**(3B):241, 1977.
11. Friedman JM et al: Saethre–Chotzen syndrome: A broad and variable pattern of skeletal malformations. *J Pediatr* **91**:929–933, 1977.
12. Gallazzi F et al: La sindrome di Saethre–Chotzen. *Minerva Pediatr* **32**:325–328, 1980.
13. Gellis SS, Feingold M: In: *Atlas of Mental Retardation Syndromes,* Figs. 3 and 4, Publ 25, US Dept of Health, Education, and Welfare, Washington, DC.
14. Hunter AGW et al: Case report: Trigonocephaly and associated minor anomalies in mother and son. *J Med Genet* **13**:77–79, 1976.
15. Kopyse Z et al: The Saethre–Chotzen syndrome with partial bifidity of the distal phalanges of the great toes. *Hum Genet* **56**:195–204, 1980.
16. Kreiborg S et al: The Saethre–Chotzen syndrome. *Teratology* **6**:287–294, 1972.
17. McKeen EA et al: The concurrence of Saethre–Chotzen syndrome and malignancy in a family with in vitro immune dysfunction. *Cancer* **54**:2946–2951, 1984.
18. Moffie D: Une famille avec oxycéphalie et acrocéphalosyndactylie. *Rev Neurol* **83**:306–312, 1950.
19. Pantke OA et al: The Saethre-Chotzen syndrome. *Birth Defects* **11**(2):190–225, 1975.
20. Pfeiffer RA: Associated deformities of the head and hands. *Birth Defects* **5**(3):18–34, 1969.
21. Pruzansky S et al: Roentgencephalometric studies of the premature craniofacial synostoses: Report of a family with the Saethre–Chotzen syndrome. *Birth Defects* **11**(2):226–237, 1975.
22. Saethre H: Ein Beitrag zum Turmschädelproblem. (Pathogenese, Erblichkeit und Symptomologie). *Dtsch Z Nervenheilkd* **117**:533–555, 1931.
23. Waardenburg PJ: Verschillende torenschedelvorman en daarbij voorkomende ogkas-en oogverschijnselen. *Ned Tijdschr Geneeskd* **78**:1700, 1934.
24. Waardenburg PJ et al: *Genetics and Ophthalmology,* Vol 1, Charles C Thomas, Springfield, 1961.
25. Young I, Harper PS: An unusual form of familial acrocephalosyndactyly. *J Med Genet* **19**:286–288, 1982.

## Jackson–Weiss syndrome

Jackson and Weiss reported an autosomal dominant syndrome consisting of craniosynostosis, midfacial hypoplasia, and abnormalities of the feet (Figs. 14-10 and 14-11) (14,18,27). Eighty-eight affected individuals were examined by Jackson and Weiss, and another 50 individuals were reliably reported to be affected (14), thus making a total of 138. Penetrance was extremely high, and great variability in expression was observed. Several other families with Jackson–Weiss syndrome have been reported, and a number of other papers attempt to deal with the genetic nosology of the acrocephalosyndactylies (1–3,5,6,8–12,14,16–19,22,23,25,27,28). The most exhaustive analysis is that of Cohen (6).

Fig. 14–10. *Jackson–Weiss syndrome.* (A) Hypertelorism, proptosis, and midface deficiency. (B) Severe acrocephaly. (A,B from *CE Jackson et al,* J Pediatr **88:**963, 1976).

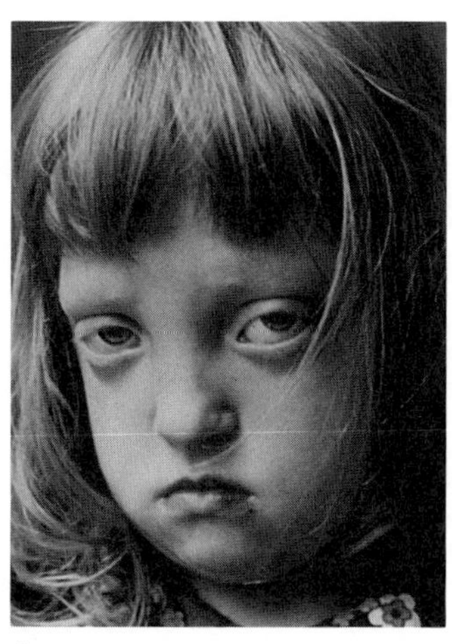
A

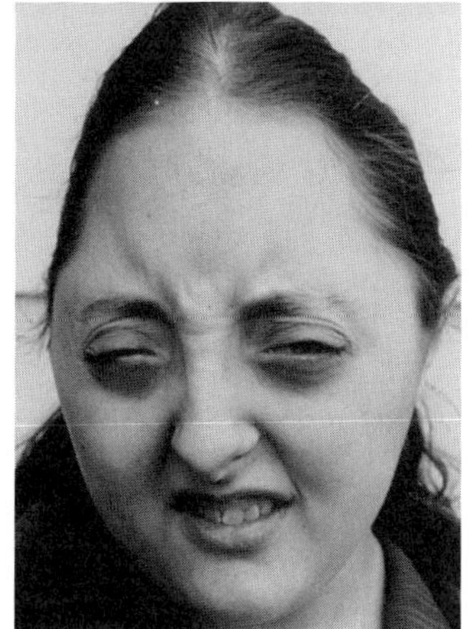
B

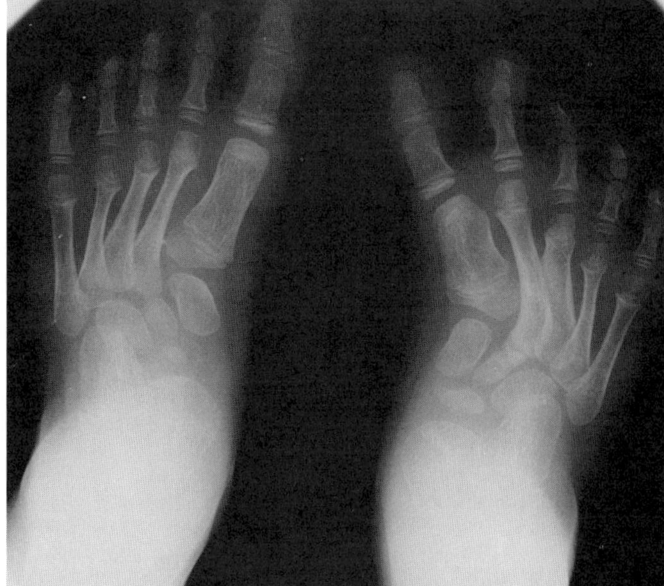
A

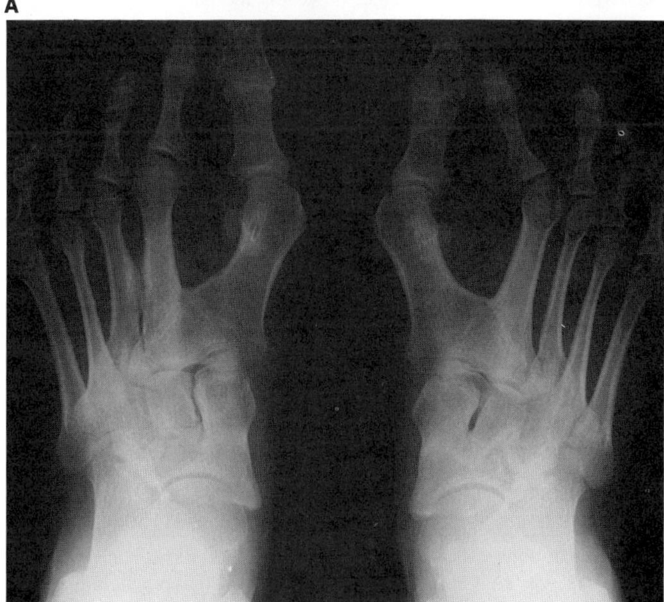
B

Fig. 14–11. *Jackson–Weiss syndrome.* (A) Short, broad first metatarsals and abnormally shaped tarsal bones. Oblique views of patient's feet (not illustrated) showed calcaneocuboid fusion. (B) Short, broad first metatarsals, fused first and second metatarsals, partially fused second and third left metatarsals, abnormally shaped tarsal bones, and fused navicular and first cuneiform bones. Calcaneocuboid coalition is also present, although it cannot be seen on radiograph. In this patient, the face and skull were normal. (A,B from *CE Jackson et al,* J Pediatr **88:**963, 1976).

**Clinical and pedigree findings.** In the Jackson–Weiss family, some affected individuals had mild cutaneous syndactyly of the second and third toes and broad great toes that deviated medially. No affected individual had broad thumbs. The only consistent manifestation was some abnormality in the clinical or radiographic appearance of the feet. Minimal manifestations included short, broad first metatarsal, abnormally shaped tarsal bone, and calcaneocuboid fusion. In some instances, the first and second metatarsals were fused, the second and third metatarsals were partially fused, the tarsal bones were abnormally shaped, and fusion of the navicular and first cuneiform bone occurred. Some affected individuals showed no clinical or radiographic abnormalities of the face or skull. Only one instance each of webbing between the third and fourth fingers, preaxial polydactyly of the feet, and fusion of the phalanges in the thumb was observed. No mentally retarded individuals were detected in this family.

The variability of expression in this kindred was so marked that Jackson and Weiss felt that the entire spectrum of craniofacial dysostosis-acrocephalosyndactylies, except for Apert syndrome, had been observed. Jackson–Weiss syndrome is definitely a distinct nosologic and genetic entity. However, for many reasons, we think the condition is caused by a *different,* possibly allelic, gene. Several large families with many affected individuals through three or four generations are known to "breed true" for Crouzon syndrome (15), Saethre–Chotzen syndrome, and Pfeiffer syndrome (4–6). Also, *the variation observed in 138 affected members of the family studied by Jackson and Weiss did not once allow for the expression of a broad thumb or for the expression of mental deficiency.* Finally, we are unimpressed with the clinical photographs shown as being Crouzon-like. Pronounced ocular proptosis is not evident.

Escobar and Bixler (9) reported an autosomal dominant pedigree in which "classic forms of the Pfeiffer and Apert syndromes as well as transitional types . . . occurred . . . Several members of the family . . . [had] . . . a facial resemblance to Crouzon disease." Penetrance was very high. We regard this dominant pedigree and the phenotypic variability as genuine; however, the pedigree does not do away with various acrocephalosyndactylies as separate nosologic entities. The condition described by Escobar and Bixler is caused by a different, possibly allelic, gene. At the present time, there is even reason to believe that the condition may possibly differ from that described by Jackson and Weiss. *In no instance in the family reported by Escobar and Bixler was there any evidence of the radiographic abnormalities of the feet* (D Bixler, personal communication, 1978) *found so consistently in the family reported by Jackson and Weiss. Furthermore, in the latter family, the variation observed in 138 affected members did not once allow for the expression of an Apert-like phenotype.* Several of the same comments about the Jackson–Weiss family apply to the Escobar–Bixler family as well.

**Differential diagnosis.** Jackson–Weiss syndrome is a separate entity from *Apert syndrome, Crouzon syndrome, Pfeiffer syndrome,* and *Saethre–Chotzen syndrome* and there has been much confusion about presumed overlap between these separate entities. *Crouzon syndrome* is too frequently diagnosed in the absence of really *significant ocular proptosis.* This is a fundamental criterion in just about all cases of Crouzon syndrome. The condition is known to "breed true" through several generations without overlap with the other acrocephalosyndactylies. This was true in Kreiborg's study of 61 cases, including 12 familial instances (15). This was apparently the case also with 23 Crouzon syndrome patients studied by Golabi et al (13). In their abstract, they noted that all patients had a complete genetic and dysmorphology evaluation.

Classic *Pfeiffer syndrome* has *both broad thumbs and broad great toes.* The broad great toes have a squared configuration and are usually in varus position. On occasion, straight great toes or mild valgus positioning (7) can be observed, but most commonly this is associated with significant recession of the great toes usually resulting from shortened first metatarsals. Certainly, broad thumbs and/or broad great toes are not obligatory anomalies and either, or occasionally both, might be absent because of variability in expression as, for example, in the pedigree by Sanchez and Negrotti (25) in which the proband had classic Pfeiffer syndrome and the proband's mother had craniofacial involvement but normal hands and feet. The large kindred of Jackson et al (14)—a kindred that to some suggests overlap between Pfeiffer and Saethre–Chotzen phenotypes by demonstrating variable expression of one and the same disorder—contains *only instances of broad great toes and no instances of broad thumbs.* The fact that broad thumbs were not expressed even *once* in 138 affected individuals in the pedigree of Jackson and Weiss is especially significant; some affected individuals manifest some Pfeiffer-like characteristics, but the *full Pfeiffer phenotype is never expressed.* On the other hand, several Pfeiffer syndrome pedigrees breed true for the full phenotype through several generations (4,6).

Several kindreds (12,19,23) have what initially appeared to be *Saethre–Chotzen syndrome,* but because broad great toes were encountered in some but not all affected individuals [the exception being the small kindred reported by Naveh and Friedman (19)], some clinicians become concerned since they regard broad great toes as part of Pfeiffer syndrome, not Saethre–Chotzen syndrome. However, such broad great toes are different from Pfeiffer great toes. They have a more triangular and/or bulbous shape and are in valgus position, and hallux valgus is known to be a feature of Saethre–Chotzen syndrome (2). Furthermore, in two previously reported cases of classic Saethre–Chotzen syndrome (21,24), the great toes appear slightly broad in the photographs. Broad great toes of triangular and/or bulbous configuration in valgus position are low-frequency findings in Saethre–Chotzen syndrome. In the pedigree of Robinow and Sorauf (23), radiographic study of the feet showed broad first metatarsals with duplication or partial duplication of the terminal phalanges which should be regarded as low-frequency features that accompany some instances of triangular and/or bulbous broad great toes.

As Cohen indicated (5), and as Schinzel et al (26) have also cautioned, assignment to one of the acrocephalosyndactyly categories on the basis of craniofacial shape and degree of asymmetry can be very misleading because of the variation that occurs both within and between different diagnostic categories. However, it is important to *distinguish between individual cases and syndrome populations.* Ocular proptosis tends to be more severe in Crouzon syndrome, for example, whereas craniofacial asymmetry tends to be more severe in Apert syndrome. On the other hand, an individual Apert syndrome patient may, on occasion, appear Crouzonoid.

**Laboratory aids.** Radiographic abnormalities of the feet were the most consistent findings, even in seemingly normal affected individuals in the large pedigree of Jackson and Weiss (14).

### References (Jackson–Weiss syndrome)

1. Baraitser M: Pitfalls of genetic counselling in Pfeiffer's syndrome. *J Med Genet* **17**:250–256, 1980.

2. Bull M et al: Phenotype definition and recurrence risk in the acrocephalosyndactyly syndrome. *Birth Defects* **15**(5B):65–74, 1979.

3. Carter CO et al: A family study of craniosynostosis, with probable recognition of a distinct syndrome. *J Med Genet* **19**:280–285, 1982.

4. Cohen MM Jr: An etiologic and nosologic overview of craniosynostosis syndromes. *Birth Defects* **11**(2):137–189, 1975.

5. Cohen MM Jr: Craniosynostosis and syndromes with craniosynostosis: Incidence, genetics, penetrance, variability, and new syndrome updating. *Birth Defects* **15**(5B):13–63, 1979.

6. Cohen MM Jr: *Craniosynostosis: Diagnosis, Evaluation, and Management.* Raven Press, New York, 1986.

7. Cremers CWRJ: Hearing loss in Pfeiffer's syndrome. *Int J Pediatr Otorhinolaryngol* **3**:343–353, 1981.

8. Eastman JR et al: Linkage analysis in dominant acrocephalosyndactyly. *J Med Genet* **15**:292–293, 1978.

9. Escobar V, Bixler D: Are the acrocephalosyndactyly syndromes variable expressions of a single gene defect? *Birth Defects* **13**(3C):139–154, 1977.

10. Escobar V, Bixler D: The acrocephalosyndactyly syndromes: A metacarpophalangeal pattern profile analysis. *Clin Genet* **11**:295–305, 1977.

11. Escobar V, Bixler D: Metacarpophalangeal pattern profile analysis as a tool to identify gene carriers in the acrocephalosyndactyly syndromes. *Birth Defects* **13**(3C): 299, 1977.

12. Friedman JM et al: Saethre–Chotzen syndrome: A broad and variable pattern of skeletal malformations. *J Pediatr* **91**:929–933, 1977.

13. Golabi M et al: Radiographic abnormalities of Crouzon syndrome: A survey of 23 cases. *Proc Greenwood Genet Ctr* **3**:102–103, 1984.

14. Jackson CE et al: Craniosynostosis, mid-facial hypoplasia, and foot abnormalities: An autosomal dominant phenotype in a large Amish kindred. *J Pediatr* **88**:963–968, 1976.

15. Kreiborg S: Crouzon syndrome: A clinical and roentgencephalometric study. *Scand J Plast Reconstr Surg Suppl* **18**:1–98, 1981.

16. Manns KJ, Bopp KP: Dysostosis craniofacialis Crouzon mit digitaler Anomalie. *Med Klin* **60**:1899–1910, 1965.

17. Mathur JS et al: Apert's anomaly—A transition. *Acta Ophthalmol (Kbh)* **46**:1051–1056, 1968.

18. McKusick VA (ed): *Medical Genetic Studies of the Amish.* Johns Hopkins University Press, Baltimore, 1978.

19. Naveh Y, Friedman A: Pfeiffer syndrome: Report of a family and review of the literature. *J Med Genet* **13**:277–280, 1976.

20. Pantke OA et al: The Saethre–Chotzen syndrome. *Birth Defects* **11**(2):190–225, 1975.

21. Pfeiffer RA: Associated deformities of the head and hands. *Birth Defects* **5**(3):18–34, 1969.

22. Robinow M: Comments on Pfeiffer syndrome and case report 113. *J Clin Dysmorphol* **2**(1):7, 1984.

23. Robinow M, Sorauf TJ: Acrocephalopolysndactyly, type Noack, in a large kindred. *Birth Defects* **11**(5):99–106, 1975.

24. Saethre H: Ein Beitrag zum Turmschädelproblem. (Pathogenese, Erblichkeit und Symptomologie). *Dtsch Z Nervenheilkd* **117**:533–555, 1931.

25. Sanchez JM, Negrotti TC: Variable expression in Pfeiffer syndrome. *J Med Genet* **18**:73–75, 1981.

26. Schinzel A et al: Acrocephalosyndactyly: How many different syndromes? *Proc Greenwood Genet Ctr* **3**:102, 1984.

27. Smith DW: Syndromes II. In: *Birth Defects: Proceedings of the Fourth International Conference,* AG Motulsky, W Lenz (eds), Excerpta Medica, Amsterdam, 1974, p 309.

28. Young I, Harper PS: An unusual form of familial acrocephalosyndactyly. *J Med Genet* **19**:286–288, 1982.

## Carpenter syndrome (acrocephalopolysyndactyly)

Carpenter syndrome is characterized by craniosynostosis, commonly but not always preaxial polysyndactyly of the feet, short fingers with clinodactyly, and variable soft tissue syndactyly, sometimes postaxial polydactyly, and other abnormalities, such as congenital heart defects, short stature, obesity, and mental deficiency (Figs. 14-12 to 14-14).

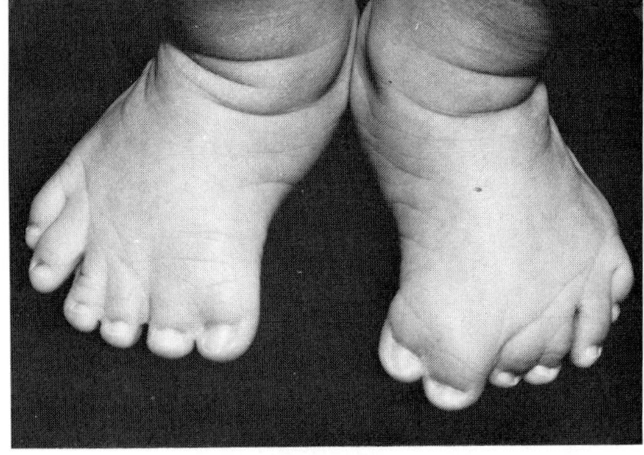

Fig. 14–13. *Carpenter syndrome.* General webbing as well as bilateral polysyndactyly of halluces. (From *V Der Kaloustian et al,* Am J Dis Child **124:**716, 1972).

First described by Carpenter in 1901 (3) and 1909 (4), the syndrome was not recognized as a distinct nosologic and genetic entity until Temtamy's report (27) in 1966. The most extensive reviews are by Cohen (8) and Cohen et al (5). Approximately 40 cases have been recorded to date (1–6,9–11,13,14a,15–23,26–29). Affected sibs have been reported by Carpenter (3,4), Frias et al (11), McLoughlin et al (15), Nørvig (16), Pfeiffer et al (19), Rudert (21), Schönenberg and Scheidhauer (23), and Cohen et al (5); and sporadic occurrences have been described by several authors (6,9–11,17,19,26–29). Consanguinity was noted in the case recorded by Der Kaloustian et al (9) and may also have been present in the sibs described by Rudert (21). Carpenter syndrome clearly has autosomal recessive inheritance.

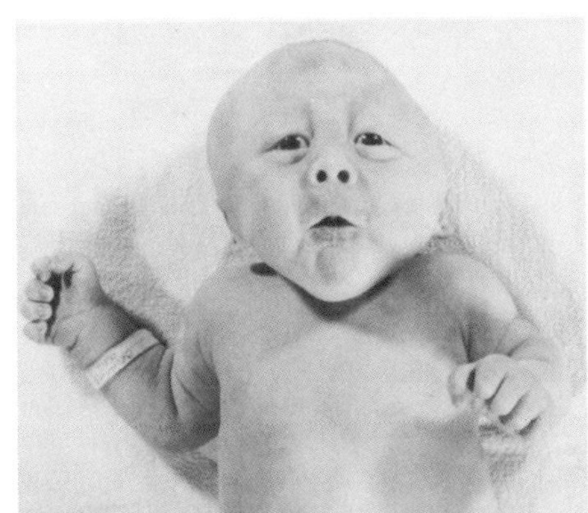

A

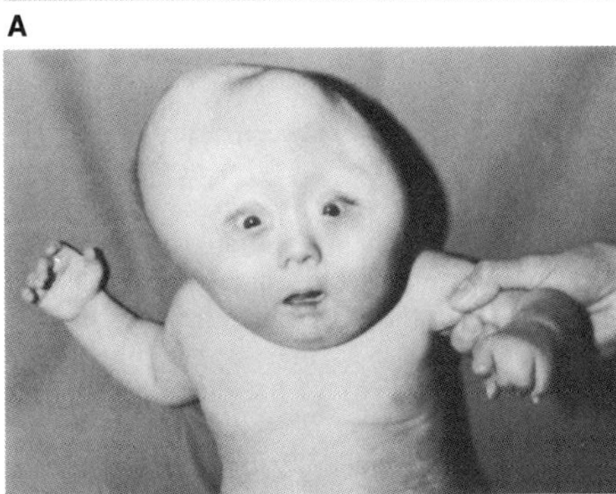

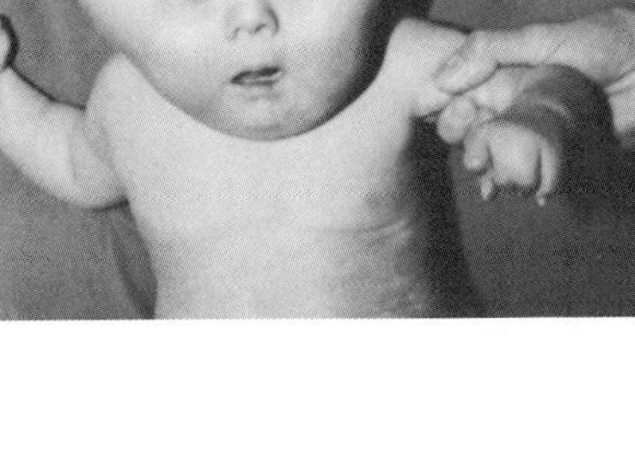

B

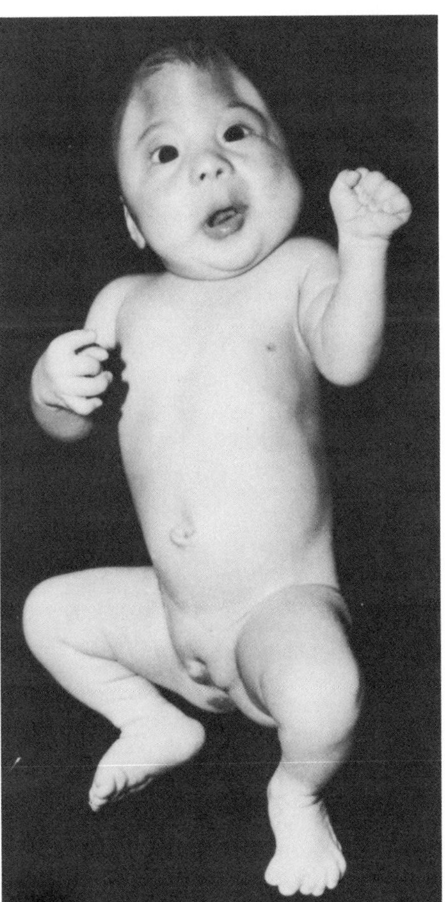

C

Fig. 14–12. *Carpenter syndrome.* (A) Asymmetric tower-shaped skull, low-set ears, and short neck. (B) Asymmetric malformation of skull and soft tissue syndactyly of third and fourth digits. (C) Compare with A and B. (A from *E Sunderhaus* and *JR Wolter,* J Pediatr Ophthalmol **5:**118, 1968. C from *V Der Kaloustian et al,* Am J Dis Child **124:**716, 1972).

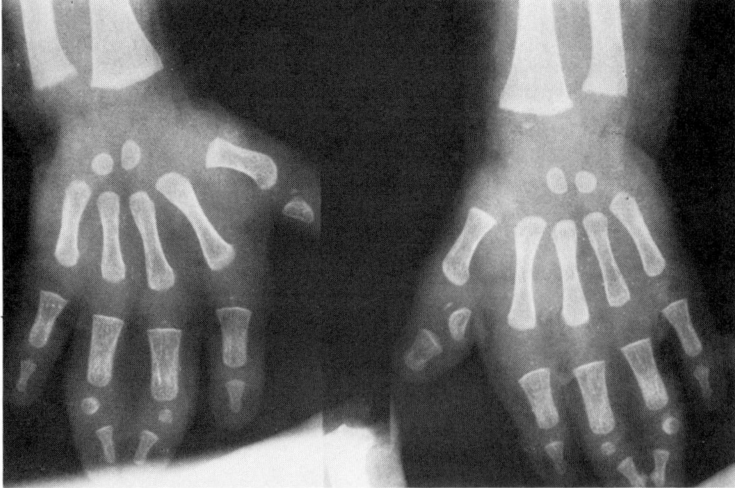

A

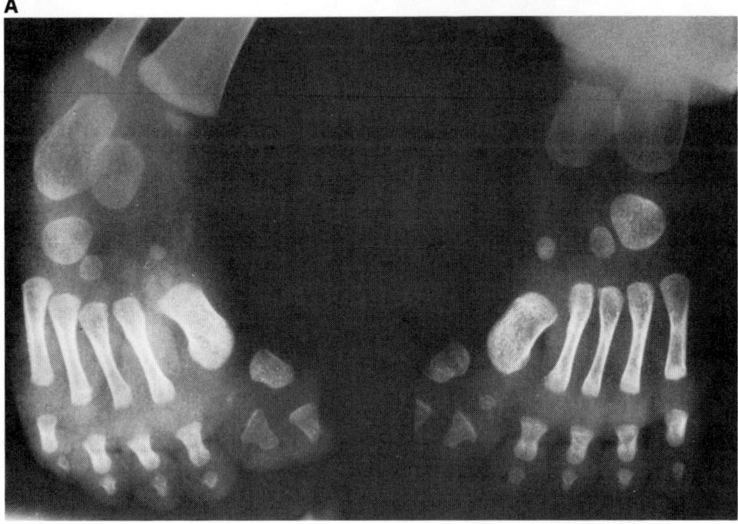

B

Fig. 14–14. *Carpenter syndrome.* (A,B) Note especially brachymesophalangy and agenesis of middle phalanges together with soft tissue syndactyly and clinodactyly of the third and fourth digits; also polysyndactyly of halluces, varus deformity, broadening of the first metatarsal, and absent middle phalanges. (From *H Schönenberg* and *E Scheidhauser,* Monatsschr Kinderheilkd **114:**322, 1966).

Height is usually below the 25th percentile, but in some instances may be above the 60th percentile. Weight is often above average. Obesity of the trunk, proximal limbs, face, and neck is common (27).

**Craniofacial features.** Craniosynostosis usually involves the sagittal and lambdoidal sutures first, the coronal being last to close. The calvaria may be grossly malformed in some instances, but variable in shape (Fig. 14-12). In many cases, unilateral involvement of the coronal or lambdoidal suture produces marked cranial asymmetry (6,8). Wormian bones were noted in the anterior fontanel in one case (17). The cloverleaf skull anomaly has been observed in association with Carpenter syndrome (6). One instance of otherwise classic Carpenter syndrome did not exhibit craniosynostosis (8). Dystopia canthorum and mildly downslanting palpebral fissures are observed. Epicanthic folds, microcornea, corneal opacity, slight optic atrophy, and blurring of the disc margins have been reported in some instances (3,4,15,27). The ears appear low set (26,27) and the neck is short. Preauricular fistulas have been noted (29). The mandible may be somewhat small, and the palate may be narrow or highly arched (8).

**Performance and central nervous system.** Mental deficiency has been a feature in many patients with Carpenter syndrome, but normal intelligence is also encountered (5,11,12,24,25,30), and two others are known to have had normal intelligence (AGW Hunter, personal communication, 1978). Warkany's patient was estimated to have an IQ of 70 (29). Early craniectomy may possibly prevent mental deficiency in some cases. However, normal intelligence has been observed without neurosurgical intervention. Furthermore, mental deficiency has been noted despite early craniectomy, suggesting a primary brain abnormality in at least some cases of Carpenter syndrome.

**Hands and feet.** The hands are short and the fingers stubby. Marked soft tissue syndactyly may be present, especially between the third and fourth fingers with less marked involvement between other fingers (Fig. 14-13). In some cases, minimal or no syndactyly may be evident. Clinodactyly of the fingers, single flexion crease, and sometimes postaxial polydactyly may be observed. Roentgenographically, brachymesophalangy, or agenesis of the middle phalanges is evident (Fig. 14-14). A tongue-shaped projection may extend from the radial side of the epiphysis of the proximal phalanx of the second finger. Two ossification centers in the proximal phalanx of the thumbs have been reported (27,29).

Abnormalities of the feet include bilateral varus deformities and commonly but not always preaxial polydactyly with duplication of the second toes or halluces. Soft tissue syndactyly may be observed between the toes, and only two phalanges are present in each toe (27).

**Cardiovascular anomalies.** Congenital heart defects have been reported in approximately 33%. Anomalies have included VSD, ASD,

PDA, pulmonic stenosis, and tetralogy of Fallot. Transposition of the great vessels and anomalous duplication of the superior vena cava have also been noted (7,8).

**Other abnormalities.** Genua valga and lateral displacement of the patellae are common. Other skeletal findings have included decreased hip-joint mobility, flaring of the ilia with poor development of the acetabula, coxa valga, absent coccyx, spina bifida occulta, and scoliosis (16,23,26,27). Umbilical hernia or omphalocele has been a feature in a number of cases. Other findings have included cryptorchidism, inguinal hernia, hydronephrosis, hydroureter, and accessory spleens (3,4,6,7,10,15,17,23,26,27,29).

**Differential diagnosis.** Carpenter syndrome is sometimes confused with *Apert syndrome, Biedl–Bardet syndrome,* and *Sakati syndrome.* The conditions known as Summitt syndrome (24,25) and Goodman syndrome or acrocephalopolysyndactyly type IV (12) represent examples of Carpenter syndrome (8,14). This problem is thoroughly analyzed elsewhere (8).

#### References [Carpenter syndrome (acrocephalopolysyndactyly)]


1. Aschner B, Engelmann G: Konstitutionspathologie in der Orthopädie. Erbbiologie des peripheren Bewegungsapparates. Springer-Verlag, Vienna, 1928.
2. Bartkowiak D et al: Zespol Carpentera u 11-letniej dziewczynki. *Pediatr Pol* **47**:349–351, 1972.
3. Carpenter G: Two sisters showing malformations of the skull and other congenital abnormalities. *Rep Soc Study Dis Child (London)* **1**:110–118, 1901.
4. Carpenter G: Case of acrocephaly with other congenital malformations. *Proc Roy Soc Med II (Part 1)*:45–53, 199–201, 1909.
5. Cohen DM et al: Acrocephalopolysyndactyly type II—Carpenter syndrome: Clinical spectrum, natural history, and an attempt at unification with Goodman and Summitt syndrome. *Am J Med Genet* **28**:311–324, 1987.
6. Cohen MM Jr: An etiologic and nosologic overview of craniosynostosis syndromes. *Birth Defects* **11**(2):137–189, 1975.
7. Cohen MM Jr: Craniosynostosis and syndromes with craniosynostosis: Incidence, genetics, penetrance, variability, and new syndrome updating. *Birth Defects* **15**(5B):13–63, 1979.
8. Cohen MM Jr: *Craniosynostosis: Diagnosis, Evaluation, and Management.* Raven Press, New York, 1986.
9. Der Kaloustian V et al: Acrocephalopolysyndactyly type II (Carpenter's syndrome). *Am J Dis Child* **124**:716–718, 1972.
10. Eaton AP et al: Carpenter syndrome—acrocephalopolysyndactyly, type II. *Birth Defects* **10**(9):249–260, 1974.
11. Frias JL et al: Normal intelligence in two children with Carpenter syndrome. *Am J Med Genet* **2**:191–199, 1978.
12. Goodman RM et al: Acrocephalopolysyndactyly type IV: A new genetic syndrome in 3 sibs. *Clin Genet* **15**:209–214, 1979.
13. Gorlin RJ et al: *Syndromes of the Head and Neck,* 2nd ed. McGraw-Hill, New York, 1976.
14. Hall JG et al: Autosomal recessive acrocephalosyndactyly revisited. *Am J Med Genet* **5**:423–424, 1980.
14a. Kapras J et al: Acrocephalosyndactylia. *Čs Pediatr* **29**:279–282, 1974.
15. McLoughlin TG et al: Heart disease in the Laurence–Moon–Biedl–Bardet syndrome. A review and report of three brothers. *J Pediatr* **65**:388–399, 1966.
16. Nørvig J: To tilfaelde af akrocephalosyndaktylie hos søskende. *Hospitalstidende* **72**:165–178, 1929.
17. Owen RH: Acrocephalosyndactyly. A case with congenital cardiac abnormalities. *Br J Radiol* **25**:103–106, 1952.
18. Palacios E, Schimke RN: Craniosynostosis-syndactylism. *Am J Roentgenol* **106**:144–155, 1969.
19. Pfeiffer RA et al: Akrozephalopolysyndaktylie (Akrozephalosyndaktylie, Typ II McKusick) (Carpenter-Syndrom). Bericht über 4 Fälle und eine Beobachtung des Typs von Marshall-Smith. *Klin Pädiatr* **189**:120–130, 1977.
20. Robinson LK et al: Carpenter syndrome: Natural history and clinical spectrum. *Am J Med Genet* **20**:461–469, 1985.
21. Rudert I: Über die Vererblichkeit der präaxialen Polydaktylie. Dissertation, Göttingen, 1938.
22. Saxena S et al: Carpenter's syndrome. Report of a case. *Indian J Pediatr* **37**:627–628, 1981.
23. Schönenberg H, Scheidhauer E: Über zwei ungewöhnlich Dyscranio-Dysphalangien bei Geschwistern (Atypische Akrocephalosyndaktylie) und fragliche Dysencephalia splanchnocystica. *Monatsschr Kinderheilkd* **114**:322–327, 1966.
24. Sells CJ et al: The Summitt syndrome: Observation on a third case. *Am J Med Genet* **3**:27–33, 1979.
25. Summitt RL: Recessive acrocephalosyndactyly with normal intelligence. *Birth Defects* **5**(3):35–38, 1969.
26. Sunderhaus E, Wolter JR: Acrocephalosyndactylism. *J Pediatr Ophthalmol* **5**:118–120, 1968.
27. Temtamy SA: Carpenter's syndrome: Acrocephalopolysyndactyly. An autosomal recessive syndrome. *J Pediatr* **69**:111–120, 1966.
28. Turner H: Endocrine clinic; diabetes insipidus, Laurence–Moon–Biedl syndrome. *J Okla State Med Assoc* **24**:148–155, 1931.
29. Warkany J et al: The Laurence–Moon–Biedl syndrome. *Am J Dis Child* **53**:455–470, 1937.
30. White J et al: Carpenter syndrome with normal intelligence and precocious growth. *Acta Neurochir* **57**:43–49, 1987.


## Antley–Bixler syndrome

The Antley–Bixler syndrome is a disorder consisting of craniosynostosis, dysplastic ears, arachnodactyly, radiohumeral synostosis, femoral bowing, and joint contractures (Figs. 14–15 and 14–16). Antley and Bixler (2) reported a distinctive patient in 1975. Lacheretz et al (9) reported an earlier case under the heading of acrocephalosynankie. However, this term has been used to describe another patient (1) who does not have the disorder in question. At least a dozen cases have been observed to date (2–3,6–7a,9–14), although not all have yet been published (J Willner, personal communication, 1985). The syndrome has also been known as trapezoidocephaly-multiple synostosis syndrome (3), multisynostotic osteodysgenesis (6), and acrocephalosynankie (9,14). Cohen introduced the term Antley–Bixler syndrome in 1977 (4) and again in 1979 (5), and most authors (10–13) have used it. The syndrome has been observed in affected sibs by Schinzel et al (13) and by J Willner (personal communication, 1985). Consanguinity was noted by Lacheretz et al (9) and Yasui et al (15). Thus, autosomal recessive inheritance is supported. Escobar et al (6a) reported a case and summarized the findings of nine previous cases from the literature.

Prenatal diagnosis has made by ultrasound, with immobility and flexion at the elbows being essential findings (12). Polyhydramnios has been noted as a feature of pregnancy. Most patients have died early in life following apneic episodes.

**Craniofacial features.** Features of the Antley–Bixler syndrome are summarized in Table 14–6. Craniosynostosis usually involves the coronal and lambdoid sutures, resulting in a brachycephalic head shape. When viewed from the top, head shape may be trapezoidal. Craniofacial features include large anterior fontanel, frontal bossing, ocular proptosis, low nasal bridge, large nose, midface hypoplasia, choanal atresia or stenosis, and low-set protruding ears (Fig. 14–15) (5a).

Fig. 14–15. *Antley–Bixler syndrome.* (A,B) Brachycephaly, biparietal bossing, malformed ears, and low nasal bridge.

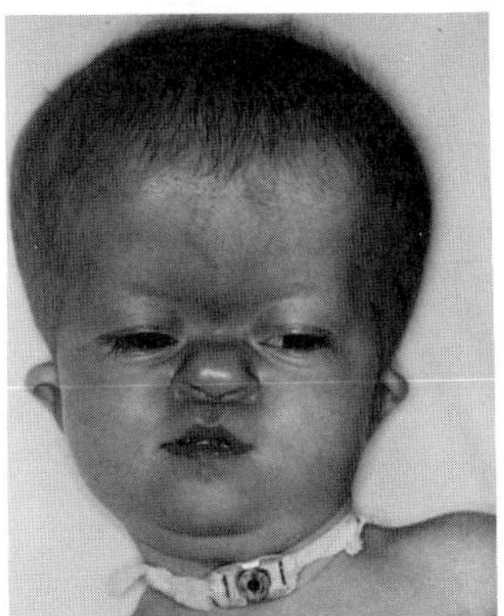
A

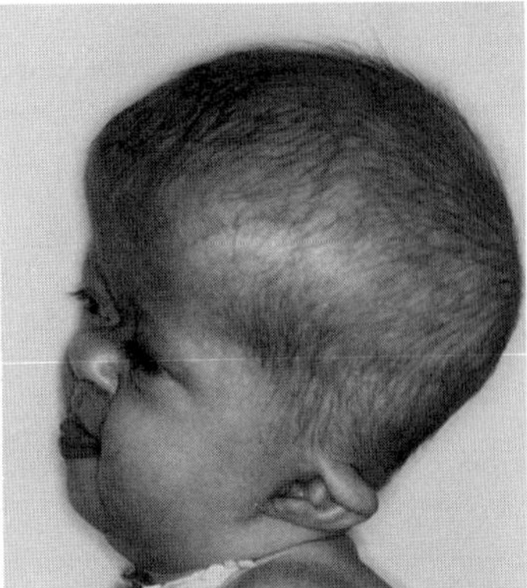
B

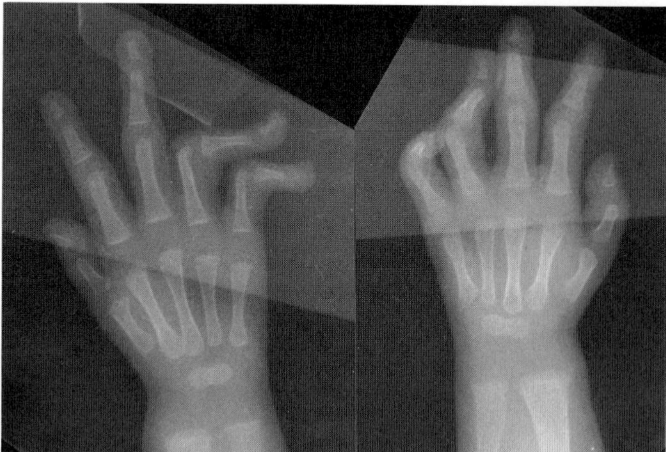

A

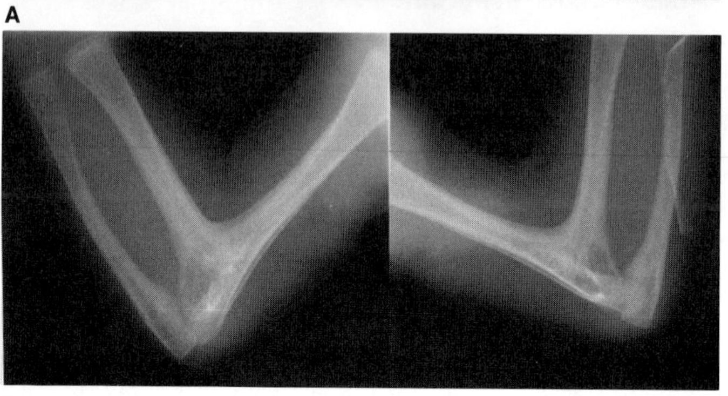

B

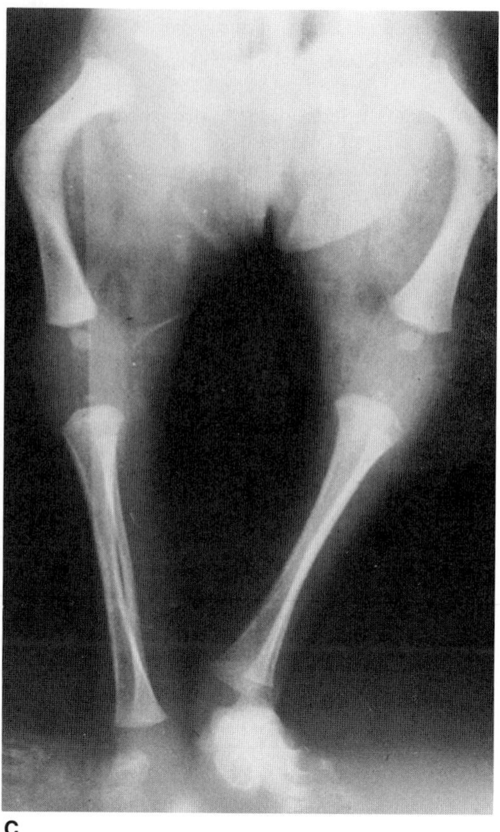

C

Fig. 14–16. *Antley–Bixler syndrome.* Roentgenograms. (A) Contractures, hypoplasia of terminal phalanges, and hamate-capitate fusion. (B) Bilateral radiohumeral synostosis. (C) Pronounced femoral bowing in a 6-week-old infant. (A,B from *R Antley* and *D Bixler,* Birth Defects **11**(2):397, 1975. C from *LK Robinson et al,* J Pediatr **101:**201, 1982.)

Table 14–6. Features of Antley–Bixler syndrome[a]

| Features | Frequency |
|---|---|
| Craniofacial | |
| Brachycephaly | 10/10 |
| Frontal bossing | 10/10 |
| Ocular proptosis | 10/10 |
| Midface hypoplasia | 10/10 |
| Dysplastic ears | 10/10 |
| Low nasal bridge | 10/10 |
| Large anterior fontanel | 9/10 |
| Choanal atresia/stenosis | 8/10 |
| Craniosynostosis | 7/10 |
| Stenotic external auditory canals | 5/10 |
| Skeletal | |
| Radiohumeral synostosis | 10/10 |
| Femoral bowing | 10/10 |
| Femoral fractures | 5/10 |
| Multiple joint contractures | 10/10 |
| Narrow chest and pelvis | 2/10 |
| Limb | |
| Camptodactyly | 8/10 |
| Long palms and fingers | 7/10 |
| Rocker-bottom feet | 6/10 |
| Partial cutaneous syndactyly | 2/10 |
| Urogenital | |
| Hypoplastic labia majora | 3/10 |
| Fused labia minora | 2/10 |
| Large clitoris | 2/10 |
| Duplication of kidney | 2/10 |

[a]Adapted from LF Escobar et al, *Am J Med Genet* **29**:829, 1988.

Respiratory obstruction often leads to an early demise within the first 8 months of life (2,6,10,11).

**Skeletal system.** Trunk abnormalities include tower-shaped rib cage, gracile ribs, narrow pelvis, vertically slanted ilia, limited hip abduction, and/or hip dislocation. Limb anomalies include radiohumeral synostosis, joint contractures, medially bowed ulnae, synostosis of carpal bones, arachnodactyly with camptodactyly, femoral bowing, femoral fractures, synostosis of tarsal bones, rocker bottom feet, and camptodactyly of the toes (Fig. 14–16) (2,3,6,10,11,13).

**Other abnormalities.** Congenital heart defects and urogenital anomalies (absent and/or ectopic kidney, duplication of kidney and ureter, dilated ureter, hypoplasia of labia majora, fused labia minora, lower vaginal atresia and/or urogenital sinus and clitoromegaly) have been observed in some instances. Low-frequency abnormalities include atresia of the external auditory canals, absent nipples, imperforate anus, vertebral anomalies, and lumbar meningomyelocele (7,10,13; J Willner, personal communication, 1985). An increased frequency of fingertip whorl patterns has been noted (9a).

**Differential diagnosis.** Bowing of long bones is also found with *campomelic syndrome* and *short limb campomelia with cloverleaf skull.* Keutel et al (8) reported two male sibs with radiohumeral synostosis, mesomelia, microcephaly, patent sutures, downslanting palpebral fissures, broad nasal bridge, micrognathia, dysplastic ears, micropenis, and cryptorchidism. Consanguinity was established.

### References (Antley–Bixler syndrome)

1. Allain D et al: Acrocephalosynankie, pseudohermaphrodisme feminin et nephropathie hypertensive. *Ann Pédiatr (Paris)* **23**:277–284, 1976.

2. Antley R, Bixler: Trapezoidocephaly, midfacial hypoplasia and cartilage abnormalities with multiple synostoses and skeletal fractures. *Birth Defects* **11**(2):397–401, 1975.

3. Antley RM, Bixler D: Invited editorial comment: Developments in the trapezoidocephaly-multiple synostosis syndrome. *Am J Med Genet* **14**:149–150, 1983.

4. Cohen MM Jr: Genetic perspectives on craniosynostosis and syndromes with craniosynostosis. *J Neurosurg* **47**:886–898, 1977.

5. Cohen MM Jr: Craniosynostosis and syndromes with craniosynostosis: Incidence, genetics, penetrance, variability, and new syndrome updating. *Birth Defects* **15**(5B):13–63, 1979.

5a. DeLozier-Blanchet CD: Antley–Bixler syndrome from a prognostic perspective. *Am J Med Genet* **32**:262–263, 1989.

6. DeLozier CD et al: The syndrome of multisynostotic osteodysgenesis with long-bone fractures. *Am J Med Genet* **7**:391–403, 1980.

6a. Escobar LF et al: Antley–Bixler syndrome from a prognostic perspective: Report of a case and review of the literature. *Am J Med Genet* **29**:829–836, 1988.

7. Herva R, Seppänen U: Multisynostotic osteodysgenesis. *Pediatr Radiol* **15**:63–64, 1985.

7a. Hurst JA et al: The Antley–Bixler syndrome. Third Manchester Birth Defects Conference, Manchester, October 25–28, 1988.

8. Keutel J et al: Eine wahrscheinlich autosomal recessive vererbte Skelettmissbildung mit Humeroradialsynostose. *Humangenetik* **9**:43–53, 1970.

9. Lacheretz M et al: L'acrocephalo-synankie. A propos d'une observation avec synostoses multiples. *Pédiatrie* **29**:169–177, 1974.

9a. Reed T, Escobar LF: Dermatoglyphics in the Antley–Bixler syndrome. *Dysmorphol Clin Genet* **2**(1):6–8, 1988.

10. Robert E et al: Le syndrome d'Antley–Bixler. *J Génét Hum* **32**:291–298, 1984.

11. Robinson LK et al: The Antley–Bixler syndrome. *J Pediatr* **101**:201–205, 1982.

12. Savoldelli G, Schinzel LA: Prenatal ultrasonic detection of humero-radial synostosis in a case of Antley–Bixler syndrome. *Prenat Diagn* **2**:219–223, 1982.

13. Schinzel A et al: Antley–Bixler syndrome in sisters: A term newborn and a prenatally diagnosed fetus. *Am J Med Genet* **14**:139–147, 1983.

14. Walbaum R: Antley–Bixler syndrome. *J Pediatr* **104**:799, 1983.

15. Yasui Y et al: The first case of the Antley–Bixler syndrome with consanguinity in Japan. *Jpn J Hum Genet* **28**:215–220, 1983.

## Baller–Gerold syndrome

Baller in 1950 (2) and Gerold in 1959 (3) reported a syndrome of craniosynostosis and radial reduction defects (Fig. 14–17). To date, 13 cases have been recorded (1–7) and the syndrome has been further delineated to include growth deficiency, abnormal ears, palatal defects, imperforate or anteriorly placed anus, various skeletal defects, and other anomalies. The Baller–Gerold syndrome has autosomal recessive inheritance. Affected sibs were reported by Gerold (3) and consanguinity was noted in the cases of Baller (2) and Pelias et al (5). Like-sexed twins were reported by Woon et al (7). Huson et al (personal communication, 1989) have told us that their patient (4a) showed chromosome changes typical of Roberts–pseudothalidomide syndrome.

**Craniofacial features.** Clinical features are summarized in Table 14–7. Craniosynostosis is a variable feature and may involve a single suture or multiple sutures. The coronal is most frequently synostosed, but lambdoid, sagittal, and metopic synostosis have also been noted. Facial features commonly include malformed or low set ears and frequently cleft palate, bifid uvula, or highly arched palate. Other facial features may include epicanthic folds, prominent nasal bridge, midline capillary hemangioma, and either micrognathia or a prominent mandible (1,5,6).

**Performance and central nervous system.** Sudden infant death has been recorded in three instances (4,7). In one of the cases, polymicrogyria was evident at autopsy. In the other two cases, brain architecture was distorted by synostosis involving the cranial base and vault, but was otherwise normal. Psychomotor and/or mental retardation has been a feature in a few instances, but mental status was not mentioned in 3 of 10 cases. Intelligence has been normal occasionally (1a).

**Skeletal system.** Radial reduction defects are variable in degree and may include absent thumbs, radial hypoplasia or aplasia, and associated upper limb defects, including postaxial abnormalities such as

A

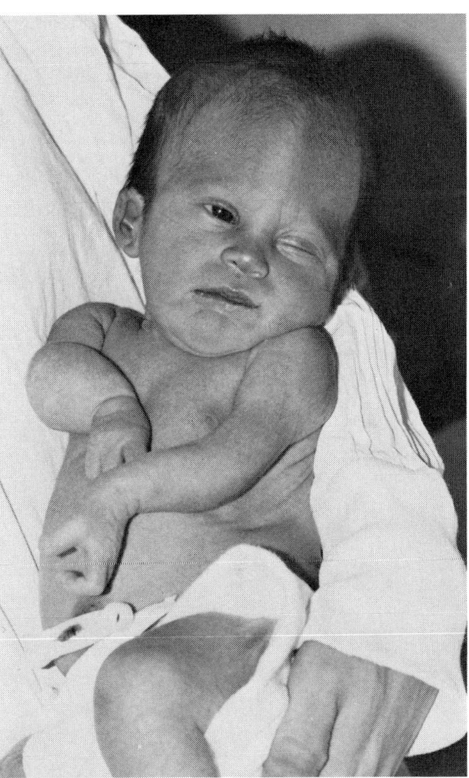

B

Fig. 14–17. *Baller–Gerold syndrome.* (A) Craniosynostosis and radial aplasia. (B) Note metopic ridging, short forearms (absent radii), and three fingers on each hand. (A from *F Baller,* Z Menschl Vererb Konstit-Lehre **29**:782, 1950.)

Table 14–7. Features of the Baller–Gerold syndrome[a]

| Features | Frequency |
|---|---|
| Growth | |
| Deficiency | 8/9 |
| Performance | |
| Sudden infant death | 3/10 |
| Mental and/or motor retardation | 2/5 |
| Cranial | |
| Craniosynostosis | 10/10 |
| Polymicrogyria | 1/10 |
| Facial | |
| Epicanthic folds | 3/8 |
| Prominent nasal bridge | 3/9 |
| Midline capillary hemangioma | 2/7 |
| Malformed ears | 5/9 |
| Low-set ears | 2/9 |
| Cleft palate | 2/10 |
| Bifid uvula | 2/10 |
| Highly arched palate | 2/10 |
| Micrognathia | 2/10 |
| Prominent mandible | 2/10 |
| Upper limbs | |
| Absent thumbs, radial hypoplasia or aplasia, associated limb defects | 10/10 |
| Other anomalies | |
| Cardiac[b] | 2/10 |
| Renal[c] | 2/6 |
| Anal[d] | 4/7 |
| Skeletal[e] | 7/9 |

[a] From MM Cohen Jr., *Craniosynostosis: Diagnosis, Evaluation, and Management.* Raven Press, New York, 1986.
[b] Ventricular septal defect; subaortic valvular hypertrophy.
[c] Pelvic kidney; crossed renal ectopia.
[d] Anteriorly placed anus; imperforate anus; imperforate anus with perineal fistula.
[e] Lordosis at junction of skull and cervical spine; limited range of motion at shoulders, elbows, and knees; rib fusions and flat vertebrae in the dorsal area; absent vertebral bodies; sacral anomaly; anomalies of pelvic girdle and lower limbs; absent middle phalanges in fingers 2 and 5; absent middle phalanges in toes 2–5; metatarsus adductus.

curved ulna, clinodactyly of the fifth finger, or absent middle phalanx of the fifth finger. A variety of low frequency skeletal defects, such as vertebral anomalies, are listed in footnote *e* of Table 14–7. Growth deficiency resulting in short stature has been commonly observed (1,5,6).

**Other findings.** Anal anomalies such as imperforate anus, sometimes in association with a perineal fistula and anteriorly placed anus, are frequently noted. Cardiac defects, such as VSD, PDA, and subaortic valvular hypertrophy, and renal anomalies, such as crossed renal ectopia, have been documented (1,4,4a,5).

**Differential diagnosis.** The Baller–Gerold syndrome should be distinguished from radial aplasia-thrombocytopenia (TAR) syndrome, Holt–Oram syndrome, and Fanconi syndrome.

### References (Baller–Gerold syndrome)


1. Anyane-Yeboa K et al: Baller–Gerold syndrome: Craniosynostosis-radial aplasia syndrome. *Clin Genet* **17**:161–166, 1980.

1a. Arena JFP, Carlin ME: A further delineation of Baller–Gerold syndrome. March of Dimes Clinical Genetics Conference, Baltimore, July 10–13, 1988.

2. Baller F: Radiusaplasie und Inzucht. *Z Menschl Vererb Konstit-Lehre* **29**:782–790, 1950.

3. Gerold M: Frakturheilung bei einen seltenen Fall kongenitaler Anomalie der oberen Gliedmässen. *Zentralbl Chir* **84**:831–834, 1959.

4. Greitzer LJ et al: Craniosynostosis-radial aplasia syndrome. *J Pediatr* **84**:723–724, 1974.

4a. Huson SM et al: The Baller–Gerold syndrome: Report of two cases and further delineation of the syndrome. Third Manchester Birth Defects Conference, Manchester, October 25–28, 1988.

5. Pelias M et al: Brief clinical report: A sixth report (eighth case) of craniosynostosis-radial aplasia (Baller–Gerold) syndrome. *Am J Med Genet* **16**:133–139, 1981.

6. Sklower SL et al: Craniosynostosis-radial aplasia: Baller–Gerold syndrome. *Am J Dis Child* **133**:1279–1280, 1979.

7. Woon KC et al: Craniosynostosis with associated cranial base anomalies: A morphologic and histologic study of affected like-sexed twins. *Teratology* **22**:23–35, 1980.


## Cloverleaf skull (Kleeblattschädel)

Classically, the cloverleaf skull anomaly or Kleeblattschädel consists of a trilobular skull with craniosynostosis. However, the degree of severity varies and different sutures may be involved in different patients (Figs. 14–18 to 14–21). Holtermüller and Wiedemann (24) first identified the condition in 1960 and noted several earlier cases. Approximately 90 publications on the subject have appeared since then (1–11,14–49). The most extensive reviews are those of Cohen (12) and Cohen et al (13). An early case was described by Vrolik (45) in 1849. Another possible early example is that of Nettleship (35a).

Synostosis may involve the coronal, lambdoidal, and metopic sutures, with bulging of the cerebrum through an open sagittal suture or, in some cases, through open squamosal sutures. Synostosis of the sagittal and squamosal sutures with cerebral eventration through a widely patent anterior fontanel may also be observed. Finally, a trilobular skull may occur with complete synostosis of all cranial sutures in some instances or with widely patent sutures and no evidence of craniosynostosis at birth in other instances.

When the cloverleaf skull malformation is severe, the ears are displaced downward, facing the shoulders. Midface hypoplasia and relative mandibular prognathism are frequently encountered. The nasal bridge is low and, in some cases, the nose may be beak-like. Severe proptosis, hypertelorism, and downslanting palpebral fissures are common (Fig. 14–18). The eyelids may fail to close, leading to corneal ulceration and clouding. Venous distention of the scleras, eyelids, periorbital areas, and scalp has been noted (1,6,14,18,20,24,34).

Roentgenographically, the skull has a trilobular contour with marked

Fig. 14–18. *Cloverleaf skull.* Classical trefoiling of skull. Note subluxation of eyeballs. (Courtesy of *RT Guzman,* Mexico City, Mexico.)

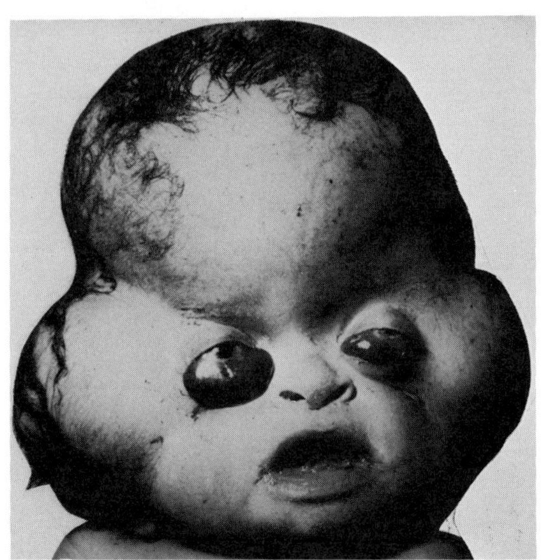

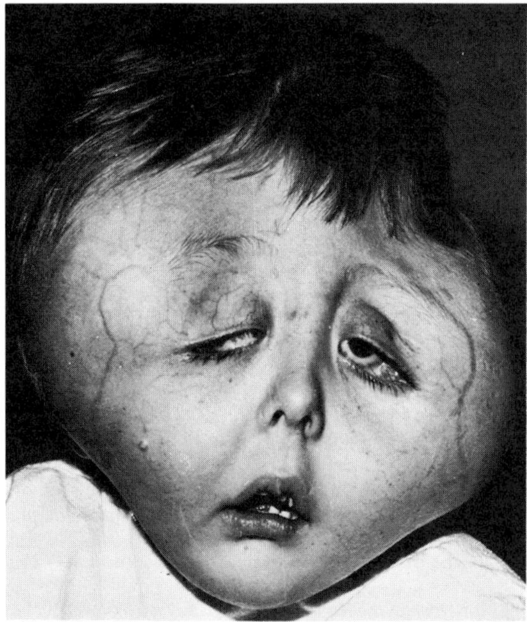

Fig. 14–19. *Cloverleaf skull.* Asymmetric involvement of skull. Note severe downslanting of palpebral fissures. (From *DG Wollin et al,* J Can Assoc Radiol **19:**148, 1968.)

convolutional impressions and a thin, distorted vault, giving a honeycomb appearance (Fig. 14–21). Synostosis and shortening of the cranial base have been reported. The posterior cranial fossa is small (3,40,47,48). In some cases, shortening of the cranial base with concomitant distortion of the hindbrain may be responsible for cerebrospinal fluid obstruction at the level of the fourth ventricle (1,24), leading to hydrocephaly, which is commonly found with cloverleaf skulls. Obstruction of the cerebral aqueduct has been demonstrated in one instance (18). Hypoplasia and malformation of the orbits, sphenoid, ethmoid, nasal bones, and maxilla are common. In one case (14), the orbits and zygomatic arches were absent.

Fig. 14–20. *Cloverleaf skull.* Cloverleaf skull in infant with bilateral oblique facial clefts. Child also had intrauterine amputation of digits. (From *A Schuch* and *HJ Pesch,* Z Kinderheilkd **109:**187, 1971).

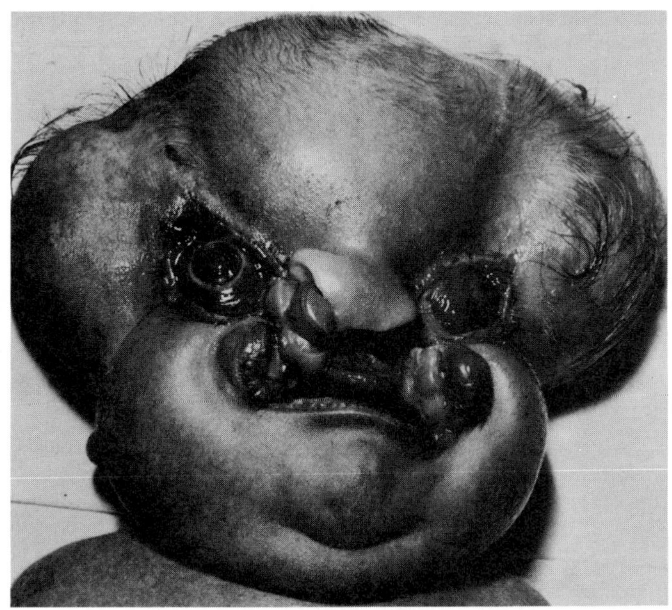

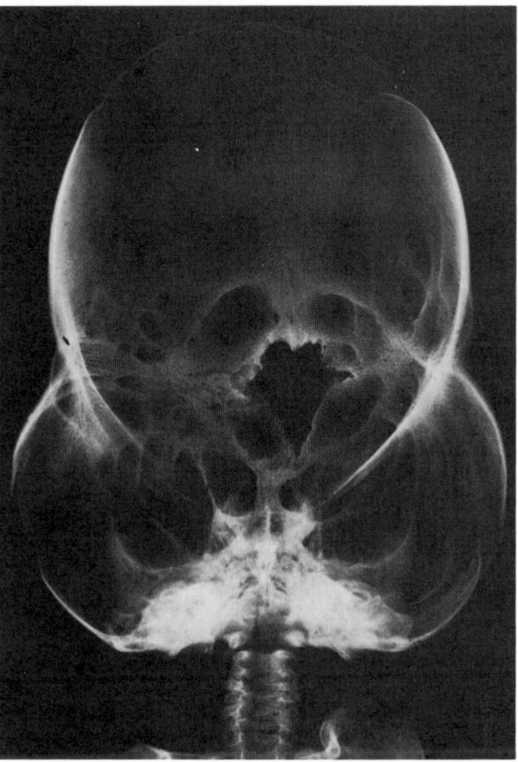

Fig. 14–21. *Cloverleaf skull.* Roentgenogram showing classic trilobular skull. (From *MW Partington et al,* Arch Dis Child **46:**656, 1971.)

Early demise has been noted in most cases, although there have been exceptions (18,47). One patient is known to have been 14 years old (18). Changes in skull shape were noted with age. At birth the skull was trilobular, but by 14 years of age, skull shape had changed dramatically.

As mentioned above, hydrocephaly is common (1,14,24). Psychomotor retardation (1,46), small cerebellum (1), cerebellar herniation, polymicrogyria (23), and various other brain anomalies (25) have been reported. Other findings have included iris colobomas, blindness, obstructed nasolacrimal ducts, absent external auditory canals, macrostomia, macroglossia, oblique facial clefting, cleft lip–palate, bifid uvula, highly arched palate, natal teeth, malocclusion, PDA, ASD, bicuspid aortic valve, common mesentery, absent lesser omentum, hypoplastic gallbladder, single umbilical artery, and omphalocele (1,8,14,15,17,23,46). Still other malformations associated with cloverleaf skull anomaly make up various syndromes, which are considered in the following.

The cloverleaf skull anomaly is known to be both etiologically and pathogenetically heterogeneous (4,7–10,16,23,31,38). Kokich and coworkers (31) demonstrated heterogeneity at the anatomic and histologic levels in two different cloverleaf specimens. Although both exhibited premature fusion of the coronal, sagittal, and lambdoidal sutures, the interrelationships and spacial orientation of their respective articulations and skeletal components differed markedly.

Etiologic heterogeneity has been demonstrated repeatedly (7–10,23). The cloverleaf skull malformation is nonspecific—that is, the condition may be observed as an isolated anomaly or together with other anomalies making up various syndromes known to have different causes. Diagnosis is made on the basis of the overall pattern of anomalies. Syndromes associated with the cloverleaf skull anomaly include amniotic rupture sequence (Fig. 14–20) (41–43), Apert syndrome [12 (p. 351),17,39], Beare–Stevenson cutis gyratum syndrome (22), campomelic dysplasia of the craniosynostotic type (30), Carpenter syndrome (8), Crouzon syndrome (21,44), COH syndrome (12), osteoglophonic dysplasia (29), dup(13q) syndrome (35), dup(15q) syndrome (37), Pfeiffer syndrome (15,23), Saethre–Chotzen syndrome (6a), Say–

Table 14–8. Conditions with cloverleaf skull anomaly[a]

| Etiology | Text pages | References |
|---|---|---|
| Known | | |
| Monogenic | | |
| Apert syndrome | 520 | 12 (p. 351), 17, 39 |
| Carpenter syndrome | 531 | 8 |
| Crouzon syndrome | 524 | 21,44 |
| Osteoglophonic dysplasia | 214 | 29 |
| Chromosomal | | |
| dup(13q) | 40 | 35 |
| dup(15q) | 90 | 37 |
| Disruptive | | |
| Amniotic bands | 11 | 41–43 |
| Iatrogenic | | |
| Bilateral subtemporal decompression procedures for hydrocephaly | | 8 |
| Uncertain or unknown | | |
| Beare–Stevenson cutis gyratum syndrome | 542 | 22 |
| Campomelic dysplasia (craniosynostotic type)[b] | 181 | 30 |
| COH syndrome | 546 | 12 |
| Isolated cloverleaf skull anomaly | 536 | 8,9 |
| Pfeiffer syndrome[c] | 526 | 15,23 |
| Say–Poznanski syndrome | 560 | 42a |
| Thanatophoric dysplasia, Type 2[d] | 223 | 33,36 |

[a]Modified from MM Cohen Jr., *Proc Greenwood Genet Ctr* **6**:186–187, 1987.
[b]Some examples may represent Antley–Bixler syndrome.
[c]All Pfeiffer cloverleafs sporadic to date.
[d]Dominant mutation?

Poznanski syndrome (42a), Type 2 thanatophoric dysplasia (33,36), and bilateral subtemporal decompression procedures for hydrocephaly (8).

All conditions with the cloverleaf skull anomaly known to date are summarized in Table 14–8 (12a), with appropriate pages for detailed information in the text together with pertinent references. It has been estimated that approximately 40% ($N = 51$) of all cloverleaf skull malformation cases represent thanatophoric dysplasia and about 20% represent Pfeiffer syndrome (23). The great variety of other malformations reported in association with the cloverleaf skull anomaly suggests that several unknown genesis syndromes in this group may become further delineated in the future (8,10). Ultrasonographic evaluation of affected fetuses may permit correct antenatal diagnosis of the cloverleaf skull anomaly (2,5).

### References [Cloverleaf skull (Kleeblattschädel)]


1. Angle CR et al: Cloverleaf skull: Kleeblattschädel-deformity syndrome. *Am J Dis Child* **114**:198–202, 1967.

2. Banna M et al: The cloverleaf skull. *Br J Radiol* **53**:730–732, 1980.

3. Bloomfield JA: Cloverleaf skull and thanatophoric dwarfism. *Australas Radiol* **14**:429–434, 1970.

4. Bonucci E, Nardi F: The cloverleaf skull syndrome. Histological, histochemical and ultrastructural findings. *Virchows Arch (Pathol Anat)* **357**:119–212, 1972.

5. Brahman S et al: Sonographic in-utero appearance of Kleeblattschädel syndrome. *Clin Ultrasound* **7**:481–484, 1979.

5. Büttner HH: Über Ankylosen beim "Kleeblatt"-Schädelsyndrom. *Zentralbl Allg Pathol* **113**:383–390, 1970.

6a. Clark RD, Eteson D: Kleeblattschädel association with Saethre-Chotzen syndrome. Third Manchester Birth Defects Conference, Manchester, October 25–28, 1988.

7. Cohen MM Jr: The Kleeblattschädel phenomenon—sign or syndrome? *Am J Dis Child* **124**:944, 1972.

8. Cohen MM Jr: An etiologic and nosologic overview of craniosynostosis syndromes. *Birth Defects* **11**(2):137–189, 1975.

9. Cohen MM Jr: The cloverleaf skull malformation. In: *Development of the Basicranium*, Bosma JF (ed), US Dept Health, Education, and Welfare, Publication No. (NIH)76-989, Bethesda, MD, 1976, pp 373–382.

10. Cohen MM Jr: Craniosynostosis and syndromes with craniosynostosis: Incidence, genetics, penetrance, variability, and new syndrome updating. *Birth Defects* **15**(5B):13–63, 1979.

11. Cohen MM Jr: *The Child with Multiple Birth Defects*. Raven Press, New York, 1982.

12. Cohen MM Jr: *Craniosynostosis: Diagnosis, Evaluation and Management*. Raven Press, New York, 1983.

12a. Cohen MM Jr: Cloverleaf syndrome update. *Proc Greenwood Genet Ctr* **6**:186–187, 1987.

13. Cohen MM Jr et al: *Perspectives on Cloverleaf Skulls,* in progress.

14. Comings DE: The Kleeblattschädel syndrome—a grotesque form of hydrocephalus. *J Pediatr* **67**:126–129, 1965.

15. Eaton A et al: The Kleeblattschädel anomaly. *Birth Defects* **11**(2):238–246, 1975.

16. Esbaugh D: Relation of the changes in the brain to those in the skull of Crouzon's and similar diseases, with report of a case. *J Neuropathol Exp Neurol* **7**:328–343, 1948.

17. Feingold M et al: The demise of a syndrome? *Syndrome Ident* **1**(2):21, 1973.

18. Feingold M et al: Kleeblattschädel syndrome. *Am J Dis Child* **118**:589–594, 1969.

19. Gathmann HA, Meyer RD: *Der Kleeblattschädel. Ein Beitrag zur Morphogenese.* Springer, Berlin, 1977.

20. Gruber GB: Über einen akrocephalen Reliefschädel. Ein Beitrag zur Frage der partiellen Chondrodystrophie. *Beitr Path Anat Allg Path* **97**:9, 1936.

21. Hall BD et al: Kleeblattschädel (cloverleaf) syndrome: Severe form of Crouzon's disease? *J Pediatr* **80**:526–527, 1972.

22. Hall BD et al: The Beare–Stevenson cutis gyratum syndrome. *Am J Med Genet,* in progress.

23. Hodach RJ: Studies of malformation syndromes in man. XXXVI. The Pfeiffer syndrome: Association with Kleeblattschädel and multiple visceral anomalies. *Z Kinderheilkd* **119**:87–103, 1975.

24. Holtermüller K, Wiedemann H-R: Kleeblattschädel-Syndrom. *Med Monatsschr* **14**:439–446, 1960.

25. Hori A et al: Ventricular diverticles with localized dysgenesis of the temporal lobe in cloverleaf skull anomaly. *Acta Neuropathol (Berlin)* **60**:132–136, 1983.

26. Horton WA: Discordance for the Kleeblattschädel anomaly in monozygotic twins with thanatophoric dysplasia. *Am J Med Genet* **15**:97–101, 1983.

27. Iannacone G, Gerlini G: The so-called "cloverleaf skull syndrome." *Pediatr Radiol* **2**:174–184, 1974.

28. Isaacson G et al: Thanatophoric dysplasia with cloverleaf skull. *Am J Dis Child* **137**:896–898, 1983.

29. Kelley RI: Osteoglophonic dwarfism in two generations. *J Med Genet* **20**:436–440, 1983.

30. Khajavi A et al: Heterogeneity in the campomelic syndromes: Long and short bone varieties. *Birth Defects* **12**(6):93–100, 1976.

31. Kokich VG et al: The cloverleaf skull anomaly: An anatomic and histologic study of two specimens. *Cleft Palate J* **19**:89–99, 1982.

32. Kremens B et al: Thanatophoric dysplasia with cloverleaf-skull. *Eur J Pediatr* **139**:298–303, 1982.

33. Langer LO Jr et al: Thanatophoric dysplasia and cloverleaf skull. A report of nine new cases and review of the literature. *Am J Med Genet Suppl* **3**:167–179, 1987.

34. Liebaldt G: "Kleeblatt" Schädel-Syndrom, als Beitrag zur formalen Genese der Entwicklungsstörung des Schädeldaches. *Ergeb Allg Pathol* **45**:23–38, 1964.

35. McCorquodale M et al: Kleeblattschädel anomaly and partial trisomy for chromosome 13(47,XY, + der(13),t(3,13)(q24:q14). *Clin Genet* **17**:409–414, 1980.

35a. Nettleship E: A case of multiple symmetrical hyperostoses of skull with post-papillary atrophy of optic nerve. *Trans Ophthalmol Soc United Kingdom* **7**:222–224, 1887.

36. Partington MW et al: Cloverleaf skull and thanatophoric dwarfism. Report of four cases, two in the same sibship. *Arch Dis Childh* **46**:656–664, 1971.

37. Pedersen C: Partial trisomy 15 as a result of an unbalanced 12/15 translocation in patient with cloverleaf skull anomaly. *Clin Genet* **9**:378–380, 1976.

38. Pena SDJ: Kleeblattschädel anomaly—which one? *Syndrome Ident* **2**(1):27, 1974.

39. Rosenbaum KN, Weisskopf B: Kleeblattschädel syndrome. *J Kentucky Med Assoc* **69**:594–597, 1971.

40. Ryan J, Kozlowski K: Radiography of stillborn infants. *Australas Radiol* **15**:213–226, 1971.

41. Saul RA: Severe facial clefting and Kleeblattschädel anomaly. *Proc Greenwood Genet Ctr* **1**:3–5, 1982.

42. Saul RA: Facial clefting/Kleeblattschädel anomaly—secondary to amniotic bands? *Proc Greenwood Genet* **4**:77, 1985.

42a. Say B, Poznanski AK: Cloverleaf skull associated with unusual skeletal anomalies. *Pediatr Radiol* **17**:93–96, 1987.

43. Schuch A, Pesch HJ: Beitrag zum Kleeblattschädel-Syndrom. *Z Kinderheilkd* **109**:187–198, 1971.

44. Shiller JG: Craniofacial dysostosis of Crouzon. A case report and pedigree with emphasis on heredity. *Pediatrics* **23**:107–112, 1959.

45. Vrolik W: *Tabulae ad illustrandam embryogenesin Hominis et Mammalium.* GMP, London, 1849.

46. Welter H: Zur Frage des Hydrocephalus chondrodystrophicus congenitus. *Beitr Path Anat Allg Path* **97**:1–8, 1936.

47. Wollin DG et al: Cloverleaf skull. *J Can Assoc Radiol* **19**:148–154, 1968.

48. Young RW et al: Thanatophoric dwarfism and cloverleaf skull (''Kleeblattschädel''). *Radiology* **106**:401–490, 1973.

49. Zuleta A, Basauri L: Cloverleaf skull syndrome. *Child's Brain* **11**:418–427, 1984.

# Chapter 15
# Syndromes with Craniosynostosis: Miscellaneous Syndromes

All miscellaneous syndromes listed in Table 15–1 are covered in this chapter except *Baraitser syndrome* (6), *Cole–Carpenter type osteogenesis imperfecta* (19), *familial osteodysplasia* (3,10,18,71), *Michels syndrome* (58), *osteoglophonic dysplasia* (8,22,44,45), *Stanescu type osteosclerosis* (21,32,57,77), and *tricho-dento-osseous dysplasia* (53,73), which are discussed elsewhere in this text (see Table 14–1). The subject has been addressed exhaustively by Cohen (18,18a).

Table 15–1. Miscellaneous craniosynostosis syndromes

| |
|---|
| Acrocephalosynankie |
| Acrocraniofacial dysostosis |
| Armendares syndrome |
| Auralcephalosyndactyly |
| Auro-digital-anal syndrome |
| Beare–Stevenson cutis gyratum syndrome |
| Berant syndrome |
| Calabro syndrome |
| Calvarial hyperostosis |
| Christian syndrome |
| COH syndrome |
| Cranioectodermal dysplasia |
| Craniofacial dyssynostosis |
| Craniotelencephalic dysplasia |
| Curry–Jones syndrome |
| Cutis aplasia and cranial stenosis (Spear–Mickle syndrome) |
| Fontaine–Farriaux syndrome |
| Frydman trigonocephaly syndrome |
| Gomez–López-Hernández syndrome |
| Gorlin–Chaudhry–Moss syndrome |
| Hall syndrome |
| Herrmann syndrome |
| Hersh syndrome |
| Hunter syndrome |
| Hurst syndrome |
| Hyper-IgE syndrome (Job's syndrome) |
| Hypomandibular faciocranial dysostosis |
| Idaho syndrome |
| Ives–Houston syndrome |
| Jones syndrome |
| Lampert syndrome |
| Lin–Gettig syndrome |
| Lowry syndrome |
| Lowry–MacLean syndrome |
| Marfanoid features and craniostenosis (includes Shprintzen–Goldberg syndrome) |
| Pfeiffer type cardiocranial syndrome |
| Pfeiffer type dolichocephalosyndactyly |
| Sakati syndrome |
| Salinas syndrome |
| San Francisco syndrome |
| Say–Barber syndrome |
| Say–Meyer trigonocephaly syndrome |
| Say–Poznanski syndrome |
| Thanos syndrome |
| Ventruto syndrome |
| Wisconsin syndrome |

## Acrocephalosynankie

Lacheretz et al (49) and Allain and co-workers (2) described a syndrome of craniosynostosis, radiohumeral synostosis, and fusion of the cuboid and third cuneiform bones. In the patient described by Lacheretz et al, an unusual facial appearance and consanguinity were noted. In the female reported by Allain et al, female pseudohermaphroditism was documented. Both sets of authors referred to the condition as acrocephalosynankie.

The condition may represent a recurrent-pattern syndrome. Conversely, the patient reported by Lacheretz et al resembles the *Antley–Bixler syndrome*. Keutel et al (46) reported two male sibs with radiohumeral synostosis, mesomelia, microcephaly, patent sutures, downslanting palpebral fissures, broad nasal bridge, micrognathia, dysplastic ears, micropenis, and cryptorchidism. Consanguinity was established.

## Acrocraniofacial dysostosis

Recently, Kaplan et al (42) reported a newly recognized oculo-oto-palato-digital syndrome in two sisters. Features consisted of short stature, craniosynostosis involving the coronal suture, ocular hypertelorism, ocular proptosis, ptosis of the eyelids, downslanting palpebral fissures, high nasal bridge, anteverted nostrils, short philtrum, cleft palate, micrognathia, mixed hearing loss, proximally placed thumbs and great toes, bulbous digits, metatarsus adductus, and other abnormalities (Fig. 15–1A–C). Consanguinity was noted, strongly suggesting autosomal recessive inheritance.

## Armendares syndrome

Armendares and co-workers (4,5) described a heritable syndrome consisting of growth deficiency, craniosynostosis, retinitis pigmentosa, and other anomalies. Because birth lengths were unavailable, it is not known if the postnatally observed growth deficiency had prenatal onset, although birth weights were known to be normal. Bone age was delayed. Intelligence was noted to be normal.

Craniosynostosis involved both the coronal and sagittal sutures. Craniofacial features included microcephaly, cranial asymmetry, small face, scant eyebrows, short obtuse nose, micrognathia, highly arched palate, and prominent or malformed auricles. Ocular findings included

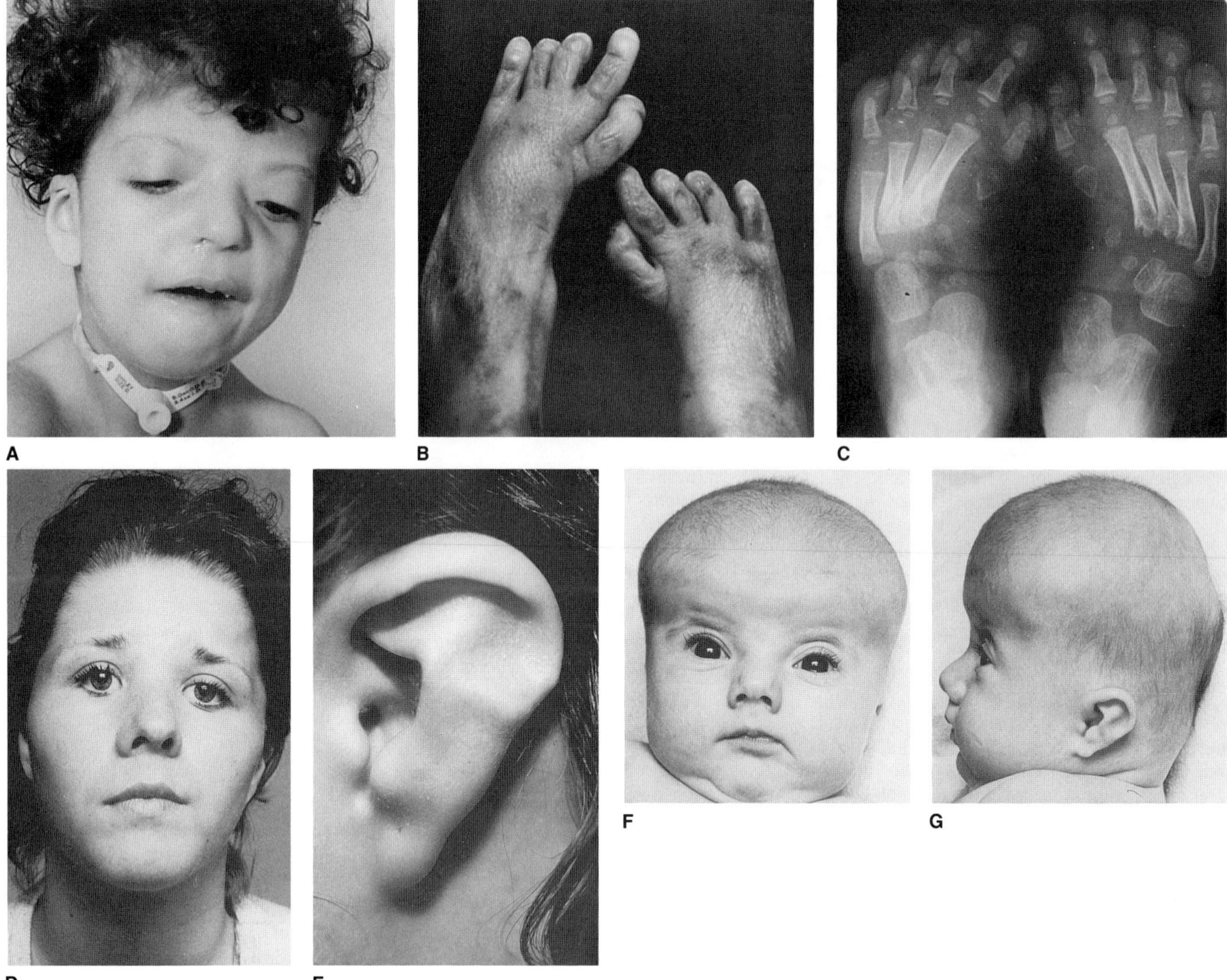

Fig. 15–1. (A–C) *Acrocraniofacial dysostosis.* (A) One of two sisters with acrocephaly, hypertelorism, shallow orbits, ptosis, downslanting palpebral fissures, dysmorphic pinnae. Both had cleft palate and pectus excavatum. (B) Feet of one sib showing metatarsus adductus and fingerlike halluces. (C) Radiograph of feet of older sib showing metatarsus adductus and severe hypoplasia of first metatarsals and digital phalanges. (D–G) *Auralcephalosyndactyly.* (D) Affected mother with craniosynostosis and facial asymmetry. (E) Mother's ear shaped like question mark. (F,G) Affected daughter with coronal synostosis and characteristic ear anomaly. (A–C from *P Kaplan,* Am J Med Genet **29**:95, 1988; D–G from *TW Kurczynski* and *SM Casperson,* J Med Genet **25**:491, 1988.

ptosis of eyelids, epicanthic folds, primary telecanthus, downslanting palpebral fissures, diminution of visual acuity, irregular dispersion of retinal pigment resulting in atypical retinitis pigmentosa, constriction of the visual fields, and narrowing of retinal arteries. Electroretinograms disclosed diminution of the *b* wave. Limb anomalies consisted of short fifth fingers with clinodactyly, wide fourth interdigital spaces, and single palmar creases.

The syndrome was reported in three male sibs from a Mexican mestizo family, and consanguinity was evident. Thus, autosomal recessive inheritance seems likely. However, X-linked recessive inheritance cannot be ruled out at present. The reporting of other affected families should resolve the mode of inheritance.

## Auralcephalosyndactyly

Kurczynski and Casperson (48) reported an autosomal dominantly inherited syndrome consisting of craniosynostosis involving the coronal suture, unusual ears shaped like question marks, cutaneous syndactyly of the fourth and fifth toes, and hearing deficit (Fig 15–1D–G).

## Auro-digital-anal syndrome

The auro-digital-anal syndrome consists of sensorineural hearing loss, bifid and/or triphalangeal thumbs, and either imperforate or anteriorly placed anus. Coronal synostosis appears to be a variable feature. The syndrome has autosomal dominant inheritance (74,81).

## Beare–Stevenson cutis gyratum syndrome

First reported by Beare et al (7) and Stevenson et al (78), the syndrome is characterized by cutis gyratum, acanthosis nigricans, hypertelorism, cleft palate, bifid scrotum, large umbilical stump, and other anomalies (Fig. 5–2). Hall and Cadle (30) reported a third case with craniosynostosis. MM Cohen (personal observation) and M Golabi (personal communication, 1987) have also observed cases with cra-

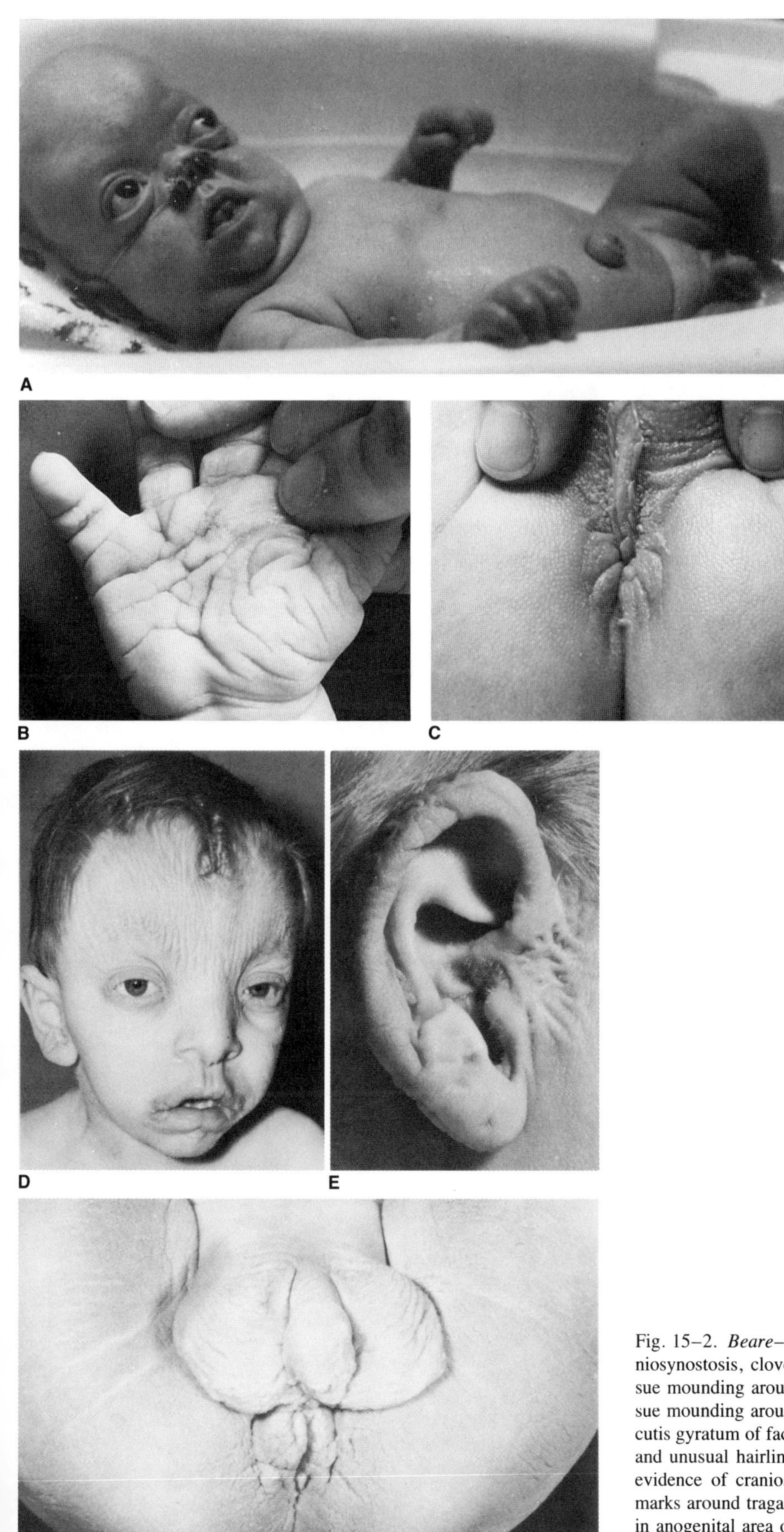

Fig. 15–2. *Beare–Stevenson cutis gyratum syndrome*. (A) Infant with craniosynostosis, cloverleaf skull, cutis gyratum of palms and soles, and tissue mounding around anus. (B) Note palmar cutis gyratum. (C) Note tissue mounding around anus and thickened raphe. (D) Another patient with cutis gyratum of facial skin, acanthosis nigricans. Note broad nasal bridge, and unusual hairline. Patient also had cleft palate and natal teeth, but no evidence of craniosynostosis. (E) Malformed pinna showing indentation marks around tragal area in same patient. (F) Marked acanthosis nigricans in anogenital area of same patient. (A–C from MM Cohen Jr, Am J Med Genet Suppl **4**:99, 1988. D–F from *JM Beare et al,* Br J Dermatol **81**:241, 1969.)

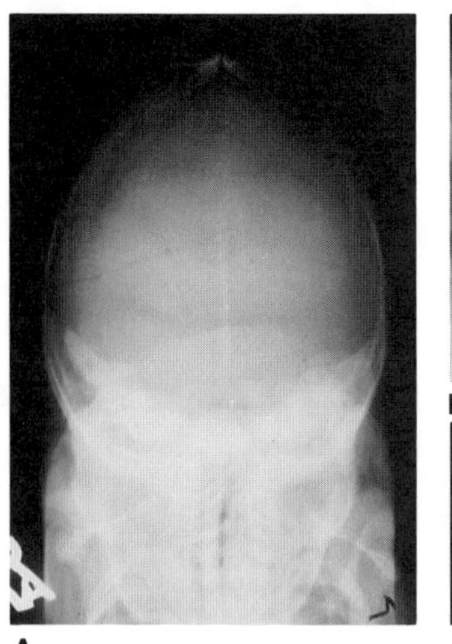


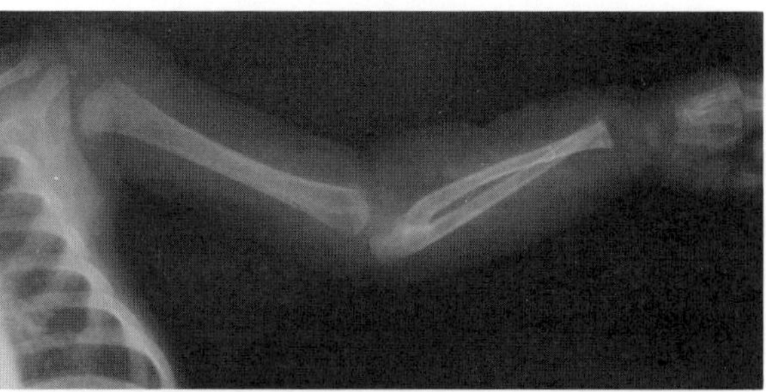


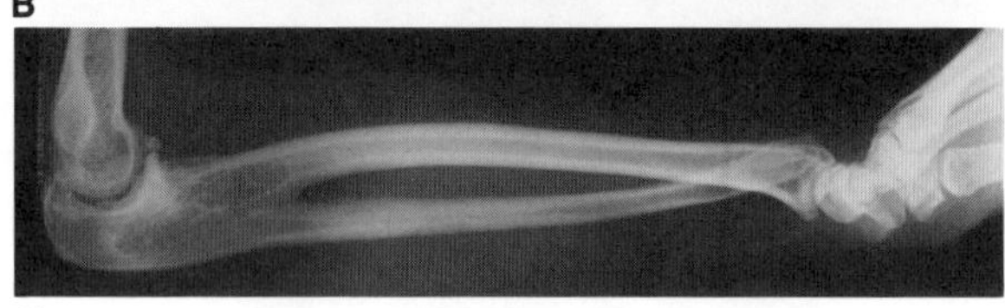


Fig. 15–3. *Berant syndrome.* (A) Note midline sagittal ridge. (B,C) Radioulnar synostosis in two affected members of the same family. (A–C from *M Berant* and *N Berant,* J Pediatr **83**:88, 1973.)

niosynostosis. Of the five known cases to date, skin furrows have occurred in variable combinations on the face, hands, feet, axilla, and/or perineum. Two patients had cloverleaf skulls, one had craniosynostosis without cloverleafing, and two had patent sutures. All five cases were sporadic. Hall et al (31) have extensively reviewed all known cases.

## Berant syndrome

In 1973, Berant and Berant (9) described a syndrome consisting of craniosynostosis involving the sagittal suture and radioulnar synostosis (Fig. 15–3). Autosomal dominant inheritance with variable expressivity seems likely. Radioulnar synostosis was unilateral in some affected individuals and bilateral in others. By way of differential diagnosis, isolated sagittal synostosis, the most common type of isolated synostosis, may occasionally follow an autosomal dominant mode of inheritance. Proximal radioulnar synostosis may occur as an isolated anomaly, may be inherited as an autosomal dominant trait, or may occur as part of a broader pattern of anomalies, as in XXXXY or XXXXX syndrome.

## Calabro syndrome

Calabro et al (11) reported a sporadic instance of a patient affected with synostosis of the coronal and metopic sutures, unilateral ulnar aplasia and oligodactyly, downslanting palpebral fissures, micrognathia, short neck, micropenis, cryptorchidism, and pulmonic stenosis (Fig. 15–4). This is the first reported patient with the combination of craniosynostosis and ulnar aplasia. Radial aplasia occurs with secondary ulnar involvement in *Baller–Gerold syndrome.* Fibular aplasia without ulnar involvement occurs in *Lowry syndrome.* Radial aplasia and shortened ulnas are observed in *Herrmann syndrome* (18).

## Calvarial hyperostosis

Pagon et al (62) reported an X-linked recessive form of calvarial thickening (Fig. 15–5). The cranial base, maxilla, and mandible were not affected. Although the coronal and sagittal sutures closed early, producing a Lückenschädel appearance, there was no evidence of increased intracranial pressure. Calvarial bone biopsy showed vacuolated histiocytes, suggesting a storage disease; however, at 18 years of age, one patient showed no evidence of neurologic deterioration or hepatosplenomegaly.

## Christian syndrome

In 1971, Christian et al (13) reported an autosomal recessive disorder consisting of microcephaly, craniosynostosis, arthrogryposis, and cleft palate (Fig. 15–6). To date, the disorder has usually been referred to as the "adducted thumbs syndrome" (13,23,47), a term that we find too nonspecific. Many features of the syndrome are secondary to abnormal development of the central nervous system. Ten cases have been recorded to date (13,23,47). Findings are summarized in Table 15–2.

Table 15–2. Christian syndrome[a]

| Findings | Frequency |
|---|---|
| Performance | |
| Hypotonia | 4/4 |
| Seizures | 6/10 |
| Mental deficiency | 5/5 |
| Early demise | 7/10 |
| Ororespiratory | |
| Dysphagia | 10/10 |
| Respiratory problems | 7/7 |
| Pneumonia | 5/5 |
| Laryngomalacia | 2/2 |
| Craniofacial | |
| Microcephaly | 8/10 |
| Craniosynostosis | 7/10 |
| Prominent occiput | 5/7 |
| Myopathic face | 3/3 |
| Downslanting palpebral fissures | 3/6 |
| Ophthalmoplegia | 6/6 |
| Cleft palate, bifid uvula, or highly arched palate | 9/9 |
| Micrognathia | 4/4 |
| Abnormal ear placement | 9/9 |
| Torticollis | 2/4 |
| Limbs | |
| Adducted thumbs | 10/10 |
| Camptodactyly | 3/3 |
| Short first metacarpal | 1/3 |
| Limited extension at elbows, wrists, and knees | 4/6 |
| Talipes equinovarus or calcaneovalgus | 9/10 |
| Other | |
| Pectus excavatum | 3/5 |
| Hirsutism | 6/6 |

[a]From MM Cohen Jr, *Craniosynostosis: Diagnosis, Evaluation, and Management.* Raven Press, New York, 1986.

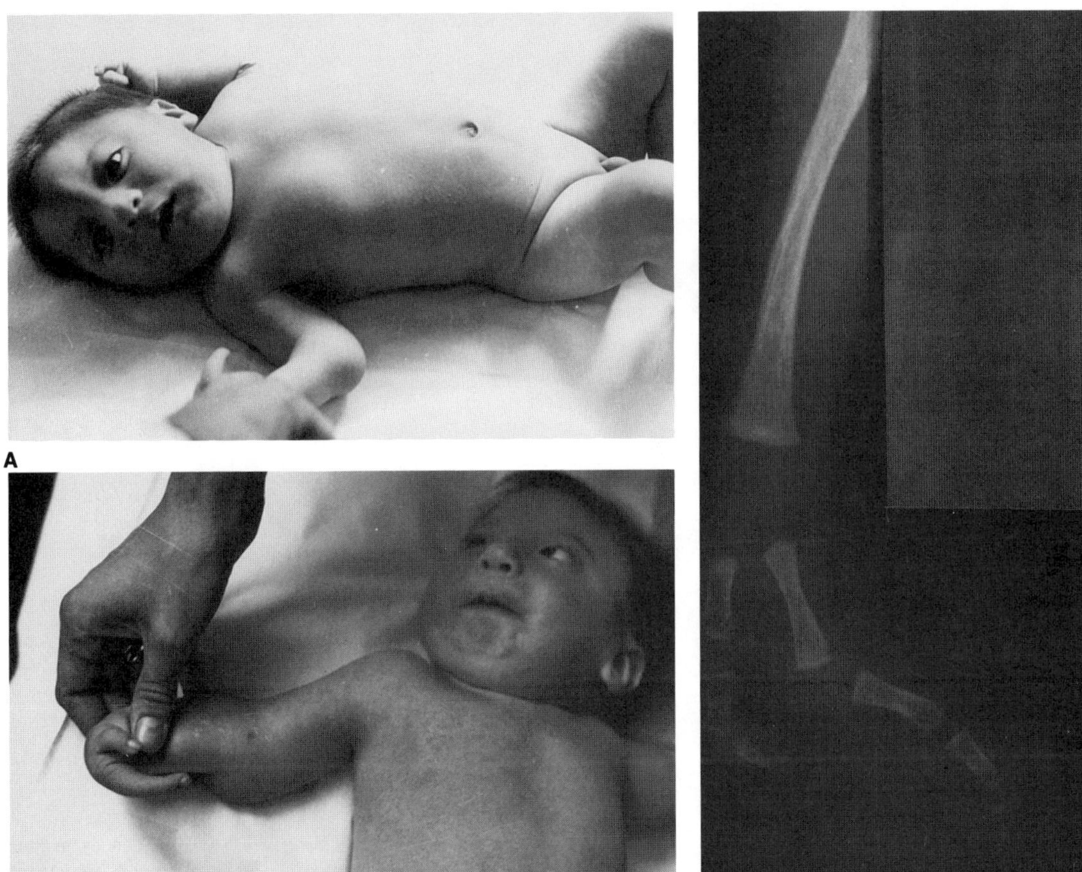

Fig. 15–4. *Calabro syndrome.* (A) Infant at 5 months. Note prominence of metopic suture. (B) Note right ulnar aplasia and oligodactyly. (C) Radiograph of right hand and forearm. Note ulnar aplasia and oligodactyly. (A–C from *A Calabro et al,* Am J Med Genet **20**:203, 1985).

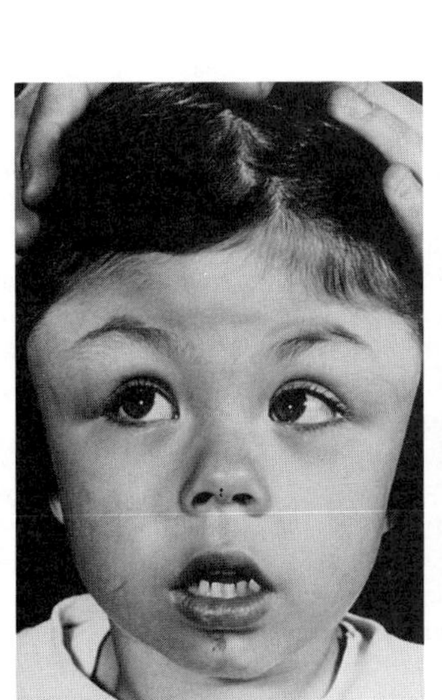

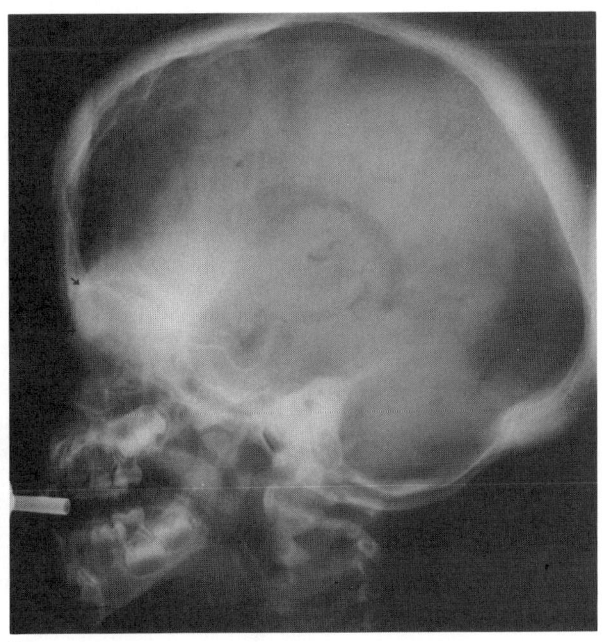

Fig. 15–5. *Calvarial hyperostosis.* (A) Age 3½ years. Note lateral frontal prominences. (B) Radiograph showing thickening in the basifrontal area (arrow) and supraoccipital bone. (From *RA Pagon et al,* Clin Genet **29**:73, 1986.)

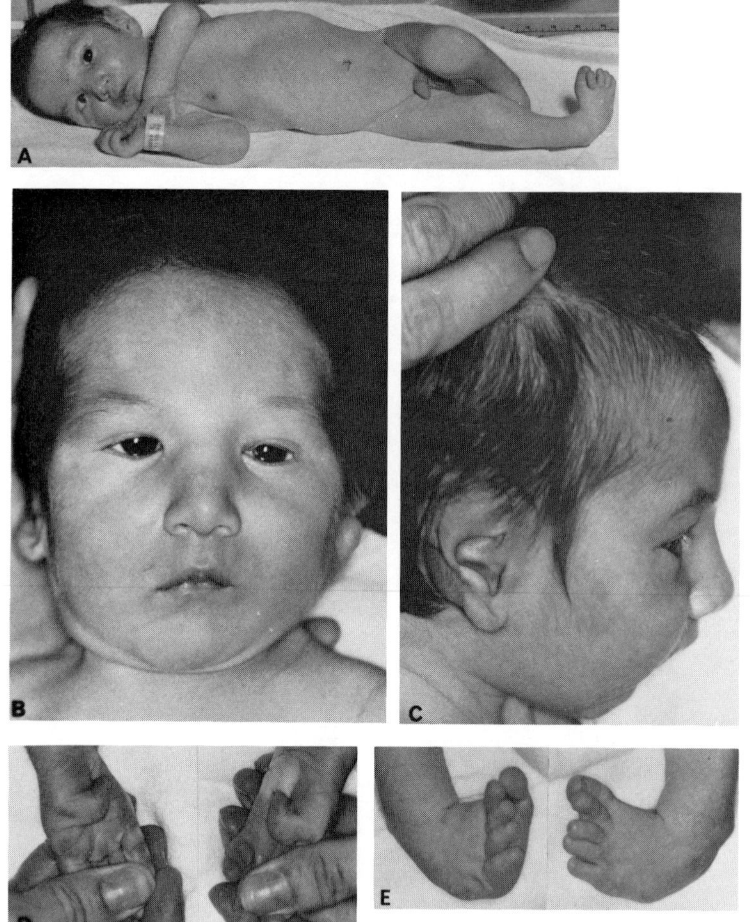

Fig. 15–6. *Christian syndrome*. (A–E) Pectus excavatum, facial asymmetry with telecanthus and downslanting palpebral fissures, hirsutism, micrognathia, posteriorly angulated ears, adducted thumbs, and talipes equinovarus. (From *JC Christian et al,* Clin Genet **2**:95, 1971.)

Microcephaly, secondary craniosynostosis, and prominent occiput are accompanied by hypotonia, seizures, and mental deficiency. Ororespiratory problems include dysphagia, laryngomalacia, and pneumonia. An early demise is common. Facial features include a myopathic face, downslanting palpebral fissures in some instances, cleft palate or bifid uvula, micrognathia, low-set and posteriorly angulated ears, and torticollis. Limb abnormalities include adducted thumbs, camptodactyly, limited extension at the elbows, wrist, and knees, and talipes equinovarus or calcaneovalgus. Hirsutism is common.

Neuropathologic studies of one affected infant (13) revealed dysmyelination with widespread glial proliferation in both the spinal cord and the brain. However, muscle biopsy in another patient (47) showed changes consistent with a myopathy rather than a neurogenic atrophy. Further study is necessary.

Differential diagnosis of the Christian syndrome from other syndromes with adducted thumbs or with more generalized arthrogrypotic changes are discussed by Fitch and Levy (23) and by Kunze et al (47).

## COH syndrome

COH syndrome represents a sporadically occurring case of cloverleaf skull, polymicrogyria, absent olfactory tracts and bulbs, proptosis, low nasal bridge, short upturned nose, downturned mouth, narrow palate, thumb duplication, small fifth fingers, micropenis, bifid scrotum, agenesis of the cervical thymic lobes, and bilobed lungs (Fig. 15–7). Chromosomes were normal. The mother, an unmarried primigravida, was normal and knew of no relatives affected with a similar set of anomalies. COH stands for Children's Orthopedic Hospital, where the patient was seen as a newborn (17).

## Cranioectodermal dysplasia

Reported earlier in synoptic form by Sensenbrenner et al (72) in 1975, cranioectodermal dysplasia was reported in detail by Levin and coworkers (52) in 1979. Striking features include dolichocephaly, sagittal synostosis, frontal bossing, sparse, slow-growing hair, microdontia and hypodontia, short, narrow thorax, short limbs, brachydactyly, and soft tissue syndactyly of the fingers and toes (Fig. 15–8). The syndrome has autosomal recessive inheritance and should be distinguished from *achondroplasia, Ellis–van Creveld syndrome, cartilage–hair hypoplasia,* peripheral dysostosis, and certain of the spondyloepiphyseal dysplasias, especially pseudoachondroplastic dysplasia.

Of the five reported patients with cranioectodermal dysplasia, three weighed more than 4000 g at birth. Rhizomelic shortening of the limbs is much more pronounced in the upper extremities than in the lower extremities. Stature tends to be in the low–normal range (third to tenth percentiles). To date, dolichocephaly and frontal bossing have been constant features. Sagittal synostosis was evident in only four of seven affected individuals. Other findings include sparse, slow growing, thin hair; epicanthic folds; hypotelorism; variably hyperopia, nystagmus, and myopia; anteverted nostrils; everted lower lip; dental anomalies, such as hypodontia, fused deciduous teeth, microdontia, taurodontism, enamel dysplasia, and widely spaced teeth; short, narrow thorax; pectus excavatum; joint laxity; brachydactyly; soft tissue syndactyly of the fingers and/or toes; clinodactyly; and single palmar crease.

On radiographs, mild osteoporosis is found. The metaphyseal cutback zones of the long bones are less abrupt than normal. The medullary cavities are slightly wide and the cortices, thin. Many epiphyseal ossification centers are flat. The femoral capital epiphyses are small and round, and ossify late. Fibulas are short, and the semilunar

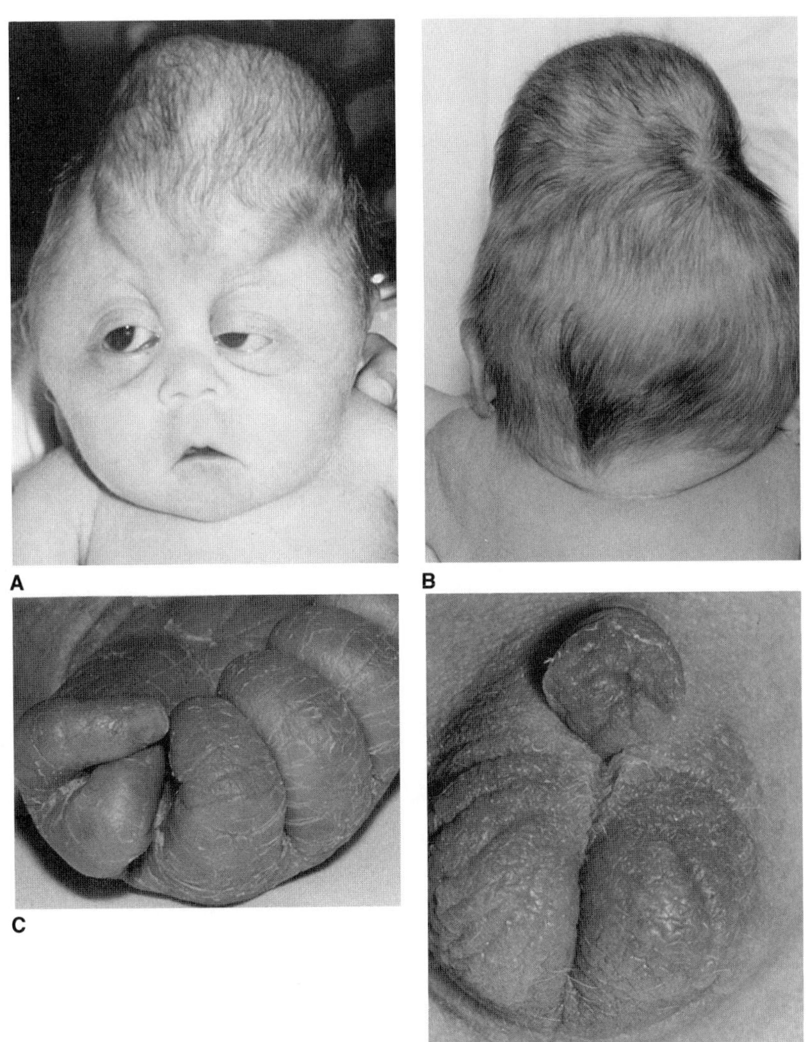


Fig. 15–7. *COH syndrome.* (A,B) Craniosynostosis, cloverleaf skull, and facial anomalies. (C) Preaxial polydactyly. (D) Micropenis and bifid scrotum. (From *MM Cohen Jr,* The Child With Multiple Birth Defects, Raven Press, New York, 1982.)

notches of the ulnas are small. Abnormalities of the vertebral bodies include short pedicles and convex superior and inferior surfaces. The epiphyses of all long bones are flattened. There may be premature fusion of the sutures of the body of the sternum. Sacral laminas are irregular and frequently cleft in the midline.

Of the five originally reported patients, three died by age 7 years; one died of heart failure, one of interstitial pneumonitis, and the third of unknown cause. An additional patient was reported by Gellis and Feingold (26a); this patient has been lost to followup (M Feingold, personal communication, 1985). ID Young (83) and LS Levin have seen other patients with this condition.

## Craniofacial dyssynostosis

Neuhäuser et al (60) described two affected sisters and five sporadic instances of craniosynostosis involving the lambdoidal and posterior sagittal sutures together with growth disturbance of the cranial base and minor facial anomalies. (Fig. 15–9). They termed the condition craniofacial dyssynostosis. One of the seven patients had synostosis of the coronal suture as well. Four exhibited mental deficiency. Agenesis of the corpus callosum and interventricular lipoma were noted in one patient. Still another had PDA with possible VSD. All seven patients were of short stature. Spanish ancestry in four of seven patients and the occurrence of the disorder in two sibs suggest autosomal recessive inheritance.

## Craniotelencephalic dysplasia

Jabbour and Taybi (40) and Daum et al (20) reported sporadic instances of a condition known as craniotelencephalic dysplasia. In both cases, premature closure of the sagittal, coronal, and metopic sutures occurred together with protuberance of the frontal bone and retarded neurologic development (Fig. 15–10). A paramedian frontal encephalocele was present in one case (20).

Two affected sisters were reported by Hughes et al (37). One had septooptic dysplasia, absent olfactory nerves, agenesis of the corpus callosum, and lissencephaly. The occurrence of craniotelencephalic dysplasia in association with such central nervous system anomalies suggests that this combination probably represents a distinct autosomal recessive syndrome.

Because detailed studies of the brain were not possible in the two sporadic cases of craniotelencephalic dysplasia reported by Jabbour and Taybi (40) and Daum et al (20), it is not known whether they had similar central nervous system anomalies and therefore represent the same overall condition reported by Hughes et al (37). Final judgment as to the specificity or nonspecificity of craniotelencephalic dysplasia will have to await further studies.

## Curry–Jones syndrome

Curry et al (19a) reported a patient with unilateral coronal synostosis and plagiocephaly, unilateral microphthalmia, iris coloboma, broad

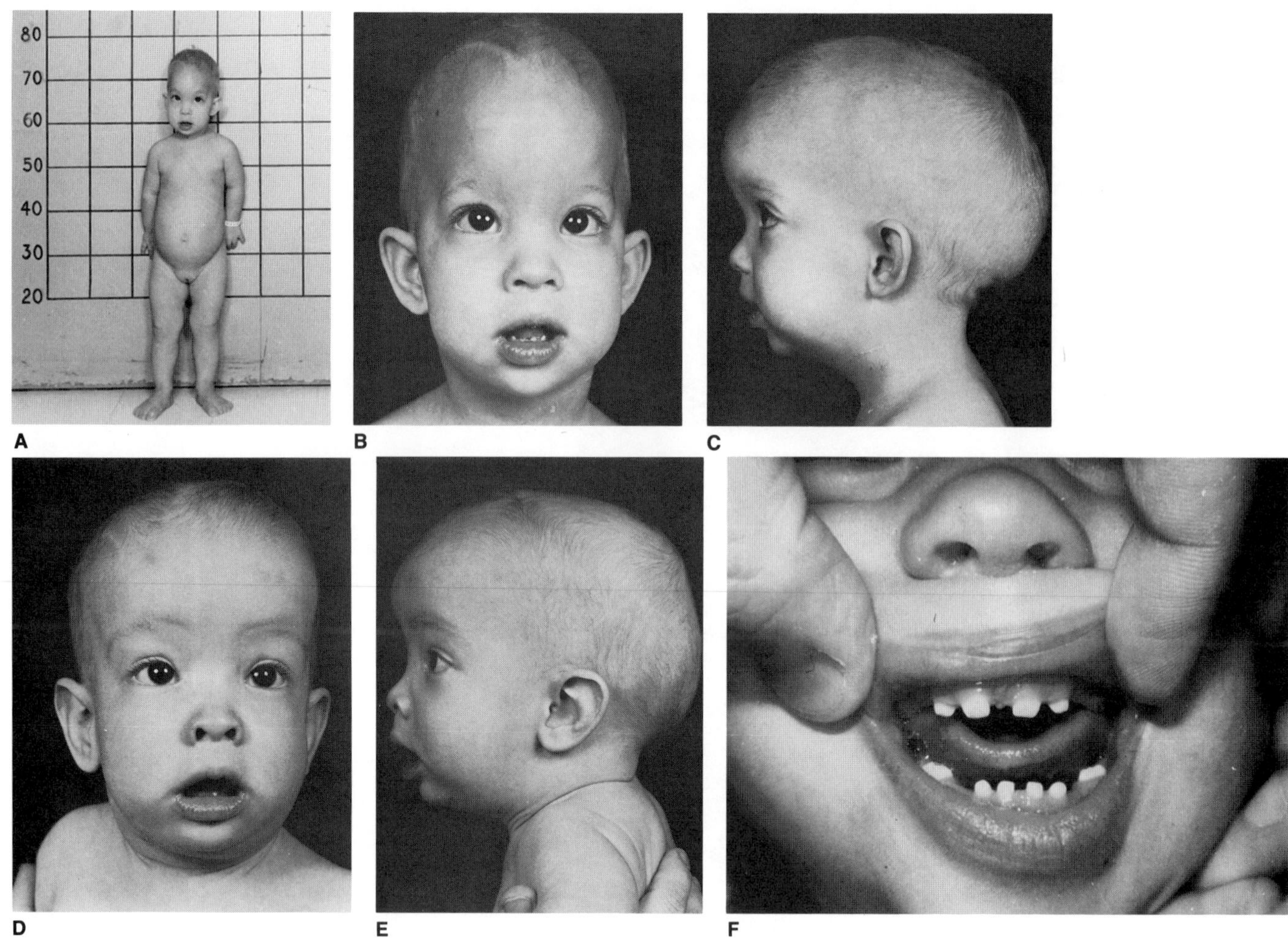


A B C D E F

thumbs, complete 1–4 syndactyly of one hand, with partial 1–3 syndactyly of the other, bilateral preaxial polydactyly of the feet, and unusual skin lesions (Fig. 15–11). Scalp defects were present at birth. Other skin abnormalities included erosive skin lesions over the toes, scarring of the dorsa of both feet, and scarlike patterning of the axilla. With time, diffuse, silver, streaklike scars appeared over the trunk, with hyperpigmented, freckle-like spotting of both feet. Marked hypotonia and moderate developmental delay were noted. Chromosomes were normal. Consanguinity was denied, and the family history was negative for birth defects. A similar case of this distinctive syndrome was observed by MC Jones (personal communication).

## Cutis aplasia and cranial stenosis (Spear–Mickle syndrome)

Spear and Mickle (76) reported a patient with several soft tissue tumors in the left frontal area of the scalp together with unilateral coronal synostosis and plagiocephaly, bifid nose, highly arched palate, lumbar meningomyelocele, cryptorchidism, thoracic kyphoscoliosis, hip dislocation, equinovarus deformity of one foot with calcaneovalgus deformity of the other, and elbow contracture. Chromosomes were normal and there were no other affected individuals in the family. The mother had been a frequent user of alcohol and heroin and the pregnancy had been ectopic.

## Fontaine–Farriaux syndrome

In 1977, Fontaine et al (24) described a sporadic occurrence of a newly recognized syndrome consisting of craniosynostosis, phalangeal hypoplasia with anonychia of the third, fourth, and fifth fingers, patches of lipodystrophy at the antecubital and popliteal fossae, aplasia of the abdominal muscles, genital hypoplasia, hypospadias, and cryptorchidism. Growth deficiency of prenatal onset and an early demise at 3.5 months were reported. Karyotype was normal. Autopsy findings included VSD, absence of the common mesentery, lissencephaly, absent interhemispheric fissure, and disorganization of the basal ganglia. Fontaine–Farriaux syndrome is named after the first two authors of the report (24) to clearly distinguish it from another condition already known as *Fontaine syndrome,* an autosomal dominant disorder characterized by micrognathia, dysplastic ears, ectrodactyly and syndactyly of the feet, and, in some instances, submucous cleft palate and mental deficiency.

## Frydman trigonocephaly syndrome

In 1984, Frydman et al (25) reported a newly recognized autosomal dominantly-inherited trigonocephaly syndrome. The proband had trigonocephaly, premature metopic synostosis, ridging of the metopic suture, mild synophrys, hypotelorism, S-curved lower eyelids, preauricular tag, omphalocele, large phallus, and hemivertebra at L-5. Mental development was normal. The father of the proband had trigonocephaly with marked ridging of the metopic suture, mild microcephaly, mild synophrys, S-curved lower eyelids, and hypotelorism. Intelligence was normal. Four other affected members of the family had trigonocephaly and normal intelligence; they were interviewed, but they refused permission for physical examination. Although Frydman trigonocephaly syndrome was recognized in six affected family mem-

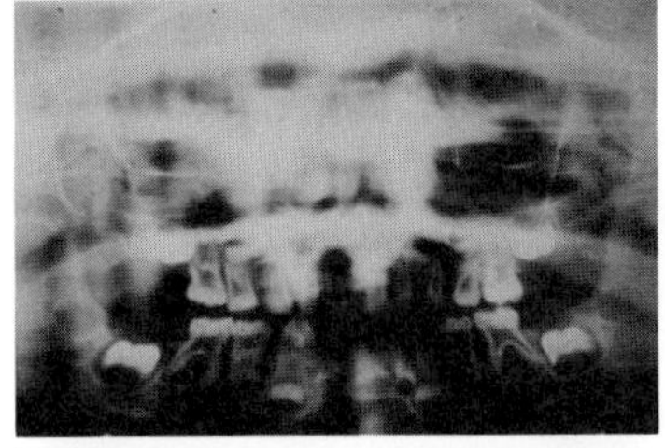

G

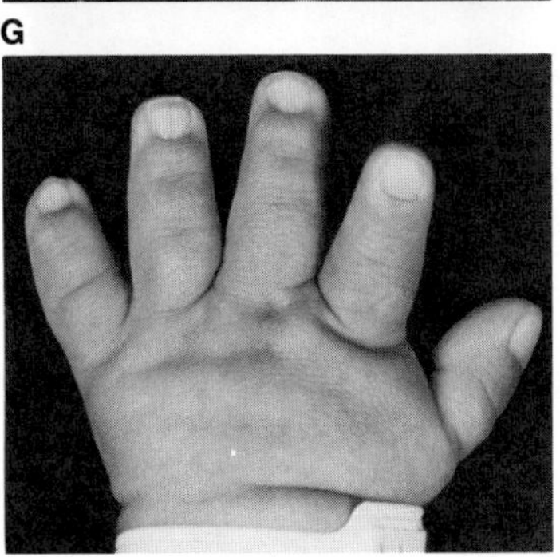

H

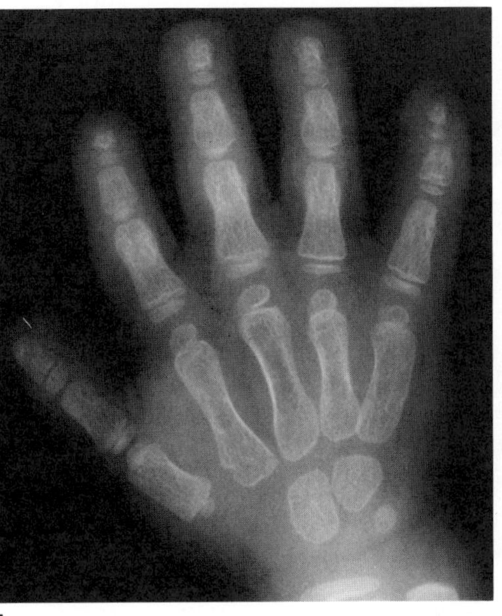

I

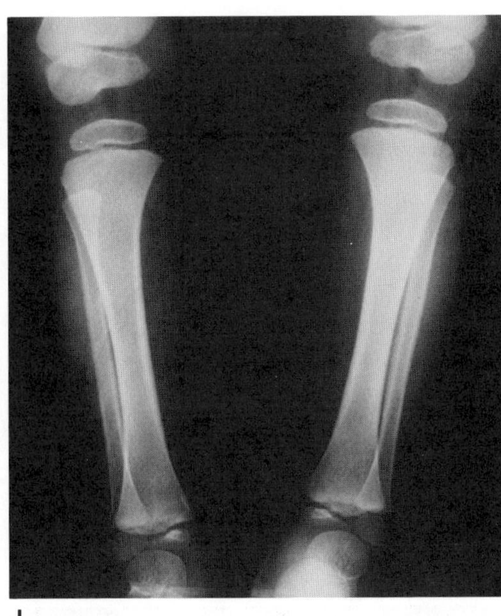

J

Fig. 15–8. *Cranioectodermal dysplasia.* (A) Age 23 months. Note rhizomelic shortening of limbs, especially upper limbs; small, narrow, thorax with mild pectus excavatum; and dolichocephaly with short, thin hair. (B–E) Sibs exhibiting dolichocephaly, short thin hair, frontal bossing, epicanthic folds, hypertelorism, and everted lower lip. (F) Microdontia. (G) Taurodontism. (H) Brachydactyly. (I) Hand-wrist roentgenogram at age 4 years. Brachydactyly, osteoporosis with thin cortices and coarse trabeculae and mild clinodactyly of the fifth finger. (J) Radiograph showing flat irregular epiphyses and shortening of fibulae, both proximally and distally. (A–J from *LS Levin et al,* J Pediatr **90**:55, 1977.)

bers, the pattern of anomalies was based on the findings in the two examined patients. Thus, the condition needs further delineation.

The syndrome should be distinguished from *Opitz trigonocephaly syndrome (C syndrome)* and *Say–Meyer trigonocephaly syndrome.* Trigonocephaly may occur with mental deficiency in various chromosomal syndromes, most commonly with del(9p) [1] and del(11q) [12] and less commonly with del(13q) [61] and dup(13q) [36]. Trigonocephaly may be a feature of alobar *holoprosencephaly,* either chromosomal or nonchromosomal, or may occur as an isolated form of craniosynostosis.

## Gomez–López-Hernández syndrome

Gomez–López-Hernández syndrome is a distinctive condition consisting of craniosynostosis, ataxia, trigeminal anesthesia, pons-vermis fusion, mental deficiency, short stature, parietal alopecia, ocular hypertelorism, corneal opacities, midface hypoplasia, low-set posteriorly angulated ears, clinodactyly, and hypoplastic labia majora (Fig. 15–12). Two unrelated Mexican girls were reported by López-Hernández (54). Chromosomes were normal in each girl. In both families, the parents were normal, consanguinity was denied, and there were no affected sibs. Another earlier case was reported by Gomez (28) who recognized the disorder as a distinct entity and named it cerebellotrigeminodermal dysplasia. Two patients are compared in Table 15–3. Gomez (27) recently reviewed the subject extensively. To date, six sporadic cases have been described, and males as well as females have been affected (28,54,63,43,66). Only two of the six affected individuals had craniostenosis, which appears to be secondary to microcephaly, a variable feature of the syndrome.

Fig. 15–9. *Craniofacial dyssynostosis.* Posterior sagittal and lambdoidal synostosis. (Courtesy of *JM Opitz,* Helena, Montana.)

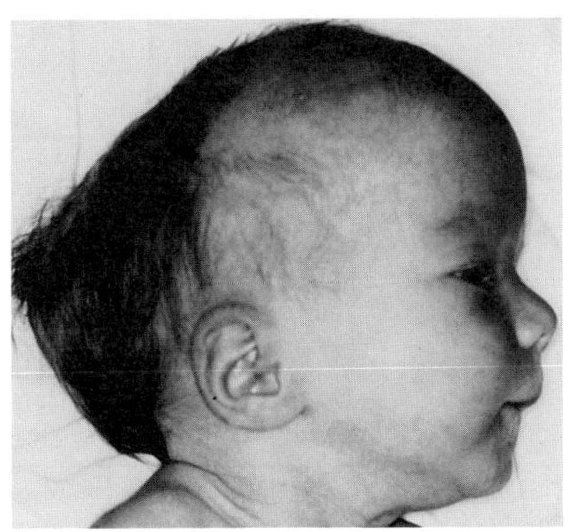

## Gorlin–Chaudhry–Moss syndrome

In 1960, Gorlin et al (29) described a syndrome in two female sibs consisting of craniosynostosis, midface hypoplasia, hypertrichosis, and anomalies of the eyes, teeth, heart, and external genitalia (Fig. 15–13). There was no parental consanguinity. Autosomal recessive inheritance seemed likely. A sporadic instance was observed in a young girl by W Lenz (personal communication, 1983) and another female patient was observed by PL Ippel (personal communication, 1985). The condition should be distinguished from *Crouzon syndrome.*

The original sibs were short and of stocky build. Both held their heads in mild anteflexion when walking. Pronounced midface hypoplasia and depressed supraorbital ridges were observed but were more pronounced in the older sib. Hypertrichosis of the scalp, arms, legs, and back was noted. The scalp hairline was lower than normal. Downslanting palpebral fissures, inability to fully open or close the eyes, upper eyelid colobomas, microphthalmia, and hyperopia were reported. The younger sib had unilateral persistence of the iridopupillary membrane. Bilateral conductive hearing deficit was noted in both sibs.

Oral anomalies consisted of class III malocclusion, highly arched narrow palate, hypodontia, microdontia, and abnormally shaped teeth.

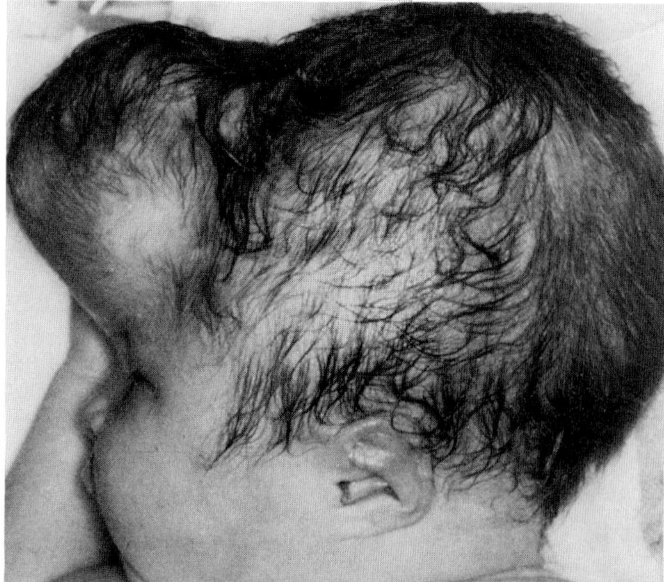

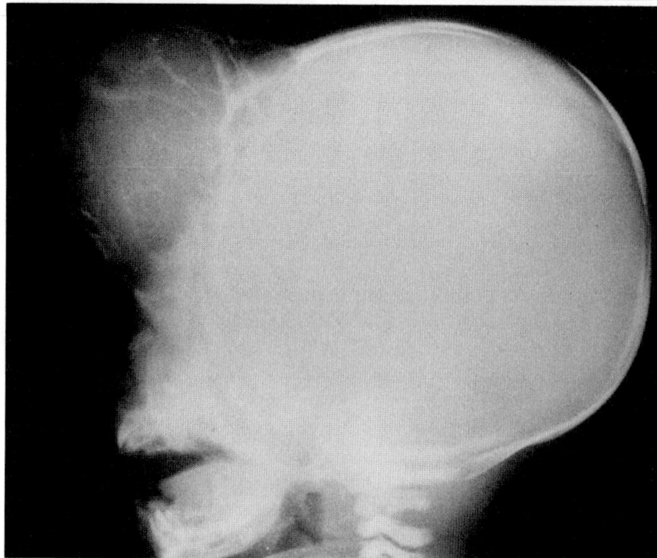

Fig. 15–10. *Craniotelencephalic dysplasia.* Premature closure of the sagittal, coronal, and metopic sutures with protuberance of frontal bone. (From *HE Hughes et al,* Am J Med Genet **14**:557, 1983.)

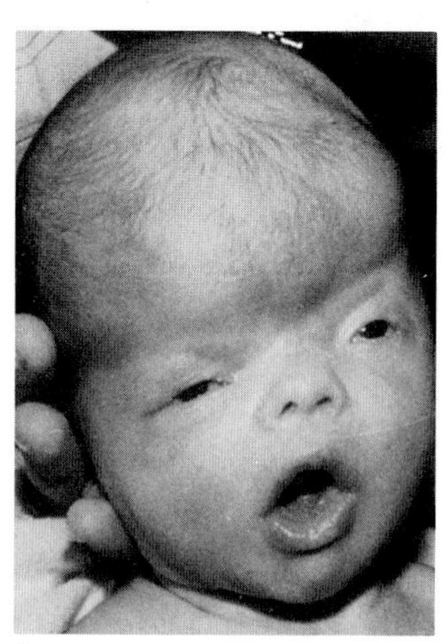

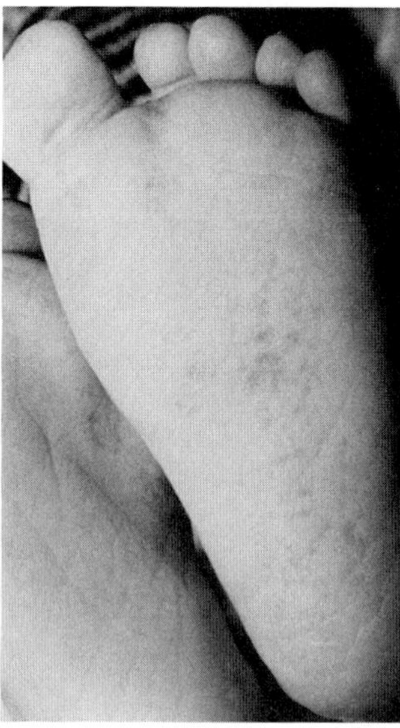

A B

Fig. 15–11. *Curry–Jones syndrome.* (A) Brachyplagiocephaly, unilateral microphthalmia. (B) Hyperpigmented freckle-like spotting of foot. (Courtesy of *CJR Curry,* Fresno, California.)

No similar condition was known in any member of the family. Both parents were normal, and consanguinity was denied. Chromosomes were normal. However, the proband was born following six second-trimester spontaneous abortions.

## Herrmann syndrome

Herrmann et al (33) described a provisionally unique pattern syndrome consisting of craniosynostosis, severe symmetrically malformed limbs, and cleft lip–palate (Fig. 15–15). A chromosome study was normal.

The skull was microbrachycephalic. Synostosis involved mainly the coronal suture, the sagittal and lambdoidal sutures remaining partially open. The metopic suture was patent. Hypertelorism, occipital capillary hemangioma, protruding dysplastic ears, deviation of the nasal

Table 15–3. Gomez–López–Hernández syndrome[a]

| Features | Patient 1 | Patient 2 |
|---|---|---|
| Growth | | |
| Short stature | + | + |
| Performance | | |
| Mental deficiency | + | + |
| Central nervous system | | |
| Pons-vermis fusion | + | + |
| Atresia of fourth ventricle | + | + |
| Cerebellar ataxia | + | + |
| Trigeminal anesthesia | + | + |
| Craniofacial | | |
| Craniosynostosis | + | + |
| Parietal alopecia | + | + |
| Ocular hypertelorism | + | − |
| Corneal opacities | + | + |
| Midface deficiency | + | + |
| Low-set, posteriorly angulated ears | + | + |
| Limbs | | |
| Clinodactyly | + | + |
| Genitalia | | |
| Hypoplastic labia majora | + | + |

From MM Cohen Jr, *Craniosynostosis: Diagnosis, Evaluation, and Management.* Raven Press, New York, 1986.

Other findings included patent ductus arteriosus, pronounced hypoplasia of the labia majora, and umbilical hernia.

Roentgenographic examination of the skull revealed premature synostosis of the coronal suture, brachycephaly, hypoplastic maxillary and nasal bones, ocular hypertelorism, lordosis of the petrous ridges, clival hypoplasia, and elevation of the lesser sphenoidal wings.

## Hall syndrome

In 1974, JG Hall (personal communication) observed a patient with craniosynostosis and a Turner-like phenotype (Fig. 15–14). Growth and development, however, were rapid, both height and weight being at the 97th percentile. Mild mental deficiency was evident.

Significant dysmorphic features included unilateral synostosis involving the coronal sutures, craniofacial asymmetry, downslanting and narrow palpebral fissures, midface deficiency, small ears, mild webbing of the neck with redundant skin, broad chest, low-placed and widely set nipples, and pudgy hands with mild hypoplasia of the nails.

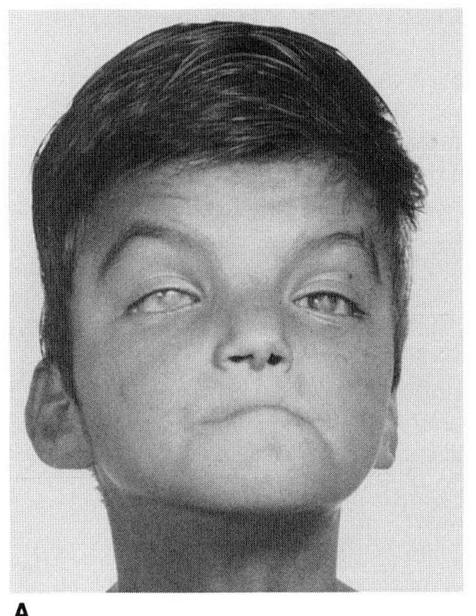

A

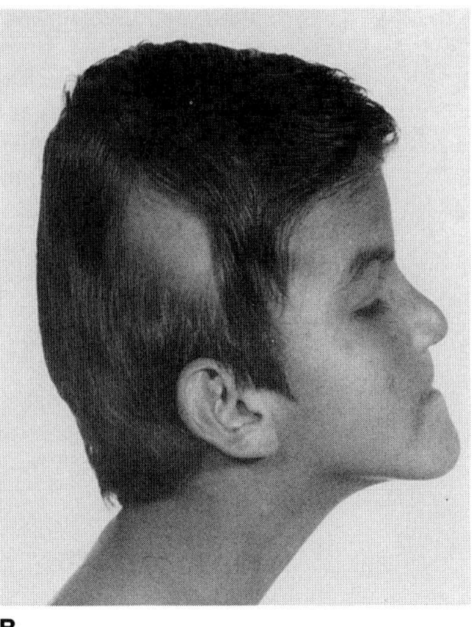

B

Fig. 15–12. *Gomez–López-Hernández syndrome.* (A,B) Craniosynostosis, focal parietal alopecia, low-set posteriorly angulated ears, and midface deficiency with relative mandibular prognathism. Note corneal opacities. (From *A López-Hernández,* Neuropediatrics **13**:99, 1982.)

Fig. 15–13. *Gorlin–Chaudhry–Moss syndrome.* (A–C) Note hypertrichosis, severe midface hypoplasia, and eyelid ptosis. Patient in C is reminiscent of de Lange syndrome. (D) Hypoplasia of labia majora. (A,B,D from *RJ Gorlin et al,* J Pediatr **56**:778, 1960. C courtesy of *W Lenz,* Münster, Germany.)

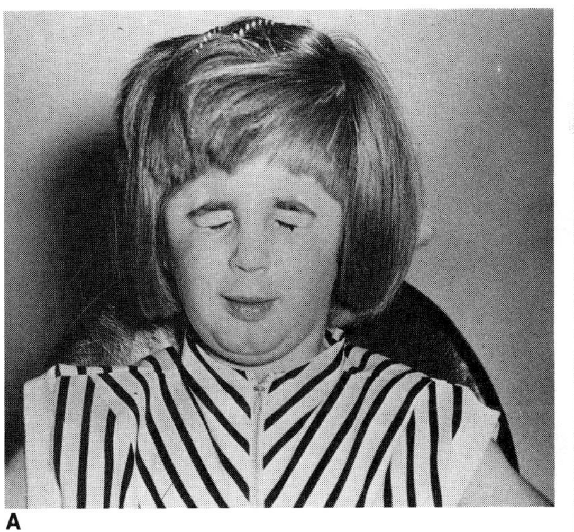

A

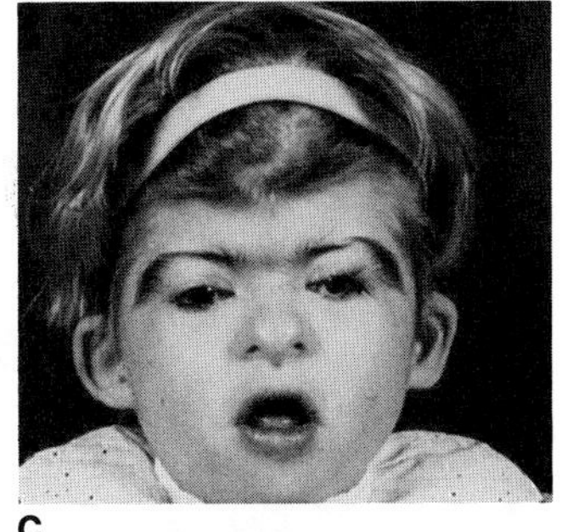

C

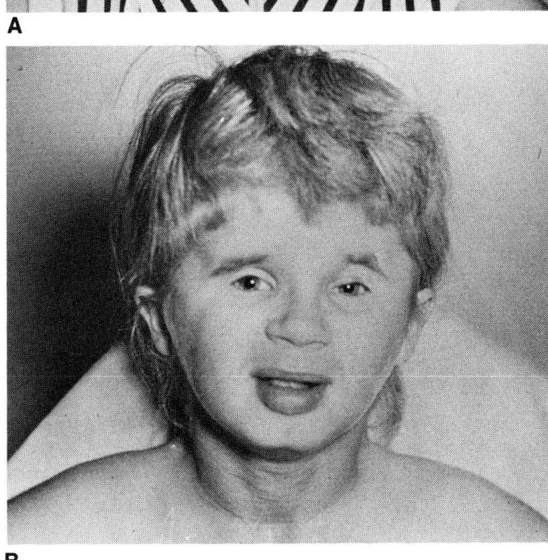

B

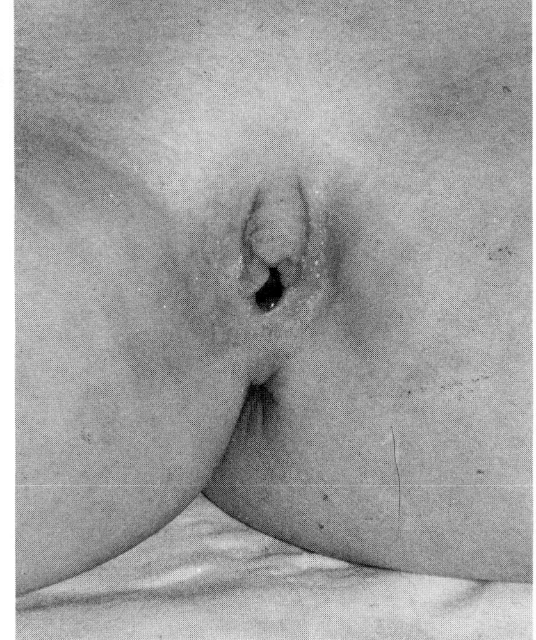

D

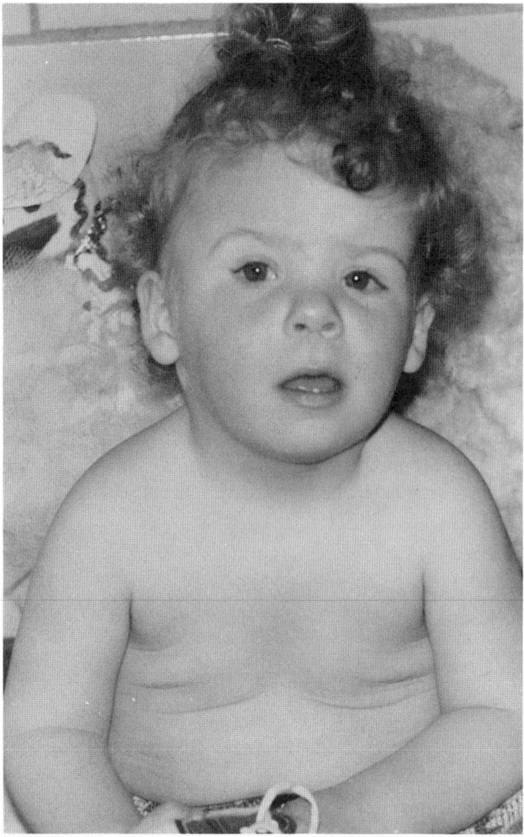

Fig. 15–14. *Hall syndrome.* Unilateral synostosis involving coronal suture, craniofacial asymmetry, downslanting and narrow palpebral fissures, midface deficiency, mild neck webbing, and broad chest. Note Turner-like phenotype. (Courtesy of *JG Hall,* Vancouver, British Columbia.)

septum, and cleft lip–palate were reported. Severe mental retardation was noted.

Radial aplasia and shortened ulna were observed bilaterally. The hands were in valgus position. Both fused and missing carpal bones were evident. Three metacarpals and three fingers were present on each hand, the fourth and fifth metacarpals and fingers being absent. Two epiphyses were present in each metacarpal. The third finger was split with two separate fingernails.

Congenital hip dislocation, dysplasia of the femoral heads and necks, ankylosis of the knees, and absent fibulae were reported. The feet were held in varus position, and the third and fourth toes on each foot were hypoplastic.

Other findings included narrow shoulders and thorax, slight depression of the lower sternum, and cryptorchidism. A remarkably similar patient was reported by Ladda et al (50).

## Hersh syndrome

Hersh et al (34) described a syndrome in a brother and sister consisting of coronal synostosis in the brother, dolichocephaly in the sister, and shared features including sensorineural hearing deficit, hypertelorism, flat nasal bridge, broad nasal tip, micrognathia, and sparse curly hair. There was no evidence of syndactyly. Intellectual abilities and language skills were normal in the sister. The brother had low–average nonverbal intelligence with significant language delay and autistic-like behavior. Consanguinity was evident; the parents were first cousins. Thus, autosomal recessive inheritance of the Hersh syndrome seems likely. Recently, it has become evident that there is no sagittal synostosis in the sister, although the skull shape is dolichocephalic (J Hersh, personal communication, 1987).

## Hunter syndrome

Hunter et al (38) described a syndrome in 1977 consisting of a characteristic facial appearance together with microcephaly, mental retar-

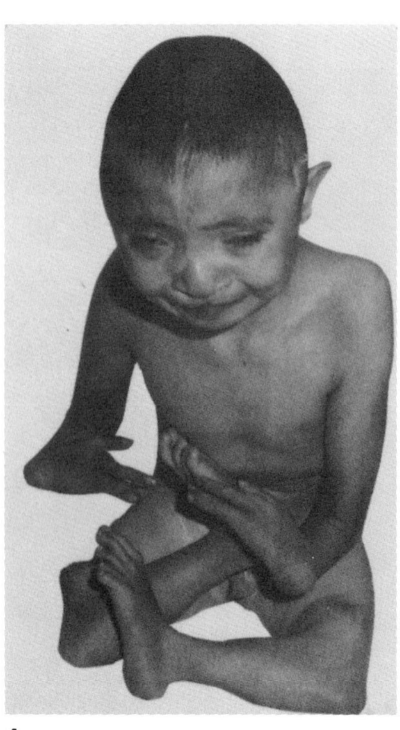

A

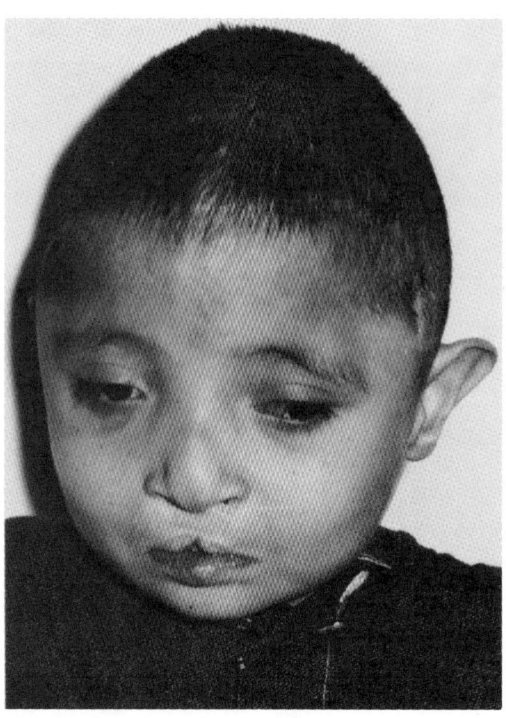

B

Fig. 15–15. *Herrmann syndrome.* (A,B) Note symmetric limb malformations, microbrachycephaly, hypertelorism, repaired cleft lip. (From *J Herrmann et al,* Rocky Mt Med J **66**:45, 1969.)

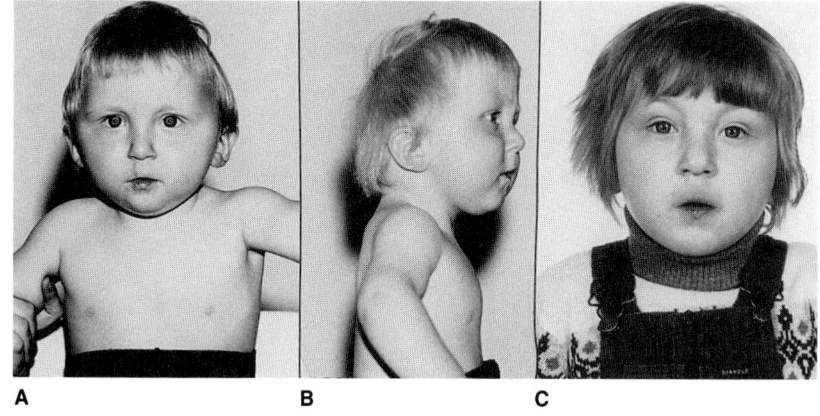


Fig. 15–16. *Hunter syndrome*. (A–C) Characteristic facial appearance in same patient at different ages. (Courtesy of *W Lenz*, Münster, Germany.)

dation, short stature, minor acral skeletal anomalies, and, less commonly, craniosynostosis and congenital heart defects (Fig. 15–16). Autosomal dominant inheritance with reduced penetrance was evident.

Facial features included almond-shaped eyes, small nose, and small downturned mouth. When craniosynostosis occurred, the coronal and lambdoidal sutures were involved. Strabismus and mildly low-set ears were encountered in some instances. Skeletal anomalies included brachydactyly, cone-shaped epiphyses of the middle phalanges, clinodactyly, incomplete extension at the elbows, and spina bifida occulta in the lumbosacral region. Low-frequency findings included ASD and inguinal hernia.

The degree of mental deficiency ranged from mild with minor motor and speech delay to severe retardation requiring institutional care. Affected individuals were frequently below the third percentile for head circumference and below the third percentile for height as well.

## Hurst syndrome

Hurst et al (38a) reported two unrelated patients with a distinctive syndrome of short stature, delayed bone age, microcephaly, craniosynostosis, small abnormally modeled ears with atretic external auditory meati, microstomia, small mandible, slender bones, hooked clavicles, dislocated radial heads, and camptodactyly. Although similarities to Type II osteodysplastic primordial dwarfism were noted, Hurst et al (38a) regarded their condition as a separate entity.

## Hyper-IgE syndrome (Job's syndrome)

Hyper-IgE syndrome *(Job's syndrome)* is characterized by recurrent pyogenic infections of the skin and lower respiratory tract together with increased serum IgE levels. The condition has been associated with a variety of skeletal findings including growth retardation, severe osteoporosis, and osteogenesis imperfecta (35). Four instances of craniosynostosis have been reported (35,75). To date, nearly all cases have been male.

## Hypomandibular faciocranial dysostosis

Neidich et al (59a) described a patient with almost complete absence of the mandible; hypoplasia of the maxillae and zygomas, with deficiency of the orbits and midface; and coronal synostosis. The mouth aperture was minute, with pursing of the lips, the gingivae were fused, the buccal cavity was small, and the buccopharyngeal membrane persisted. Palatal structures and tongue were not identified (Fig. 15–17). The only extracephalic anomaly was an atrial septal defect. Chromosomes were normal, and family history was noncontributory.

## Idaho syndrome

Idaho syndrome is a provisionally unique pattern syndrome consisting of premature fusion of the sagittal suture, micrognathia, umbilical hernia, anomalous pulmonary venous return, complete anterior dislocation of the tibias, contractures at the proximal interphalangeal joints, and umbilical hernia (Fig. 15–18) (15,17). A blood chromosome study was normal. Both parents were normal and consanguinity was denied. The father and mother were 33 years of age at the time of conception. There were no full sibs. Each parent had two normal children by previous marriages.

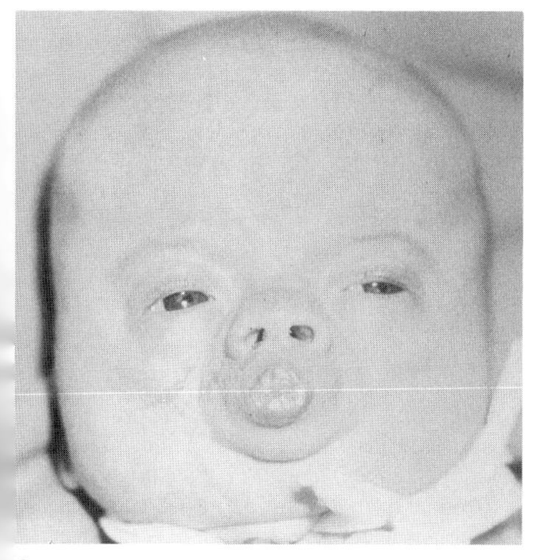


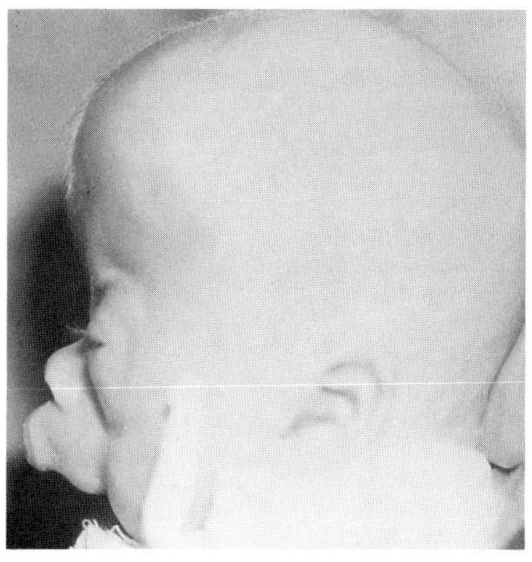


Fig. 15–17. *Hypomandibular faciocranial dysostosis*. (A,B) Absence of mandible, minute mouth aperture with pursing of lips, hypoplasia of maxilla and zygomatic arches, deficiency of orbits and midface, and coronal synostosis. (From *JA Neidich et al,* Am J Med Genet Suppl **4**:161, 1988.)

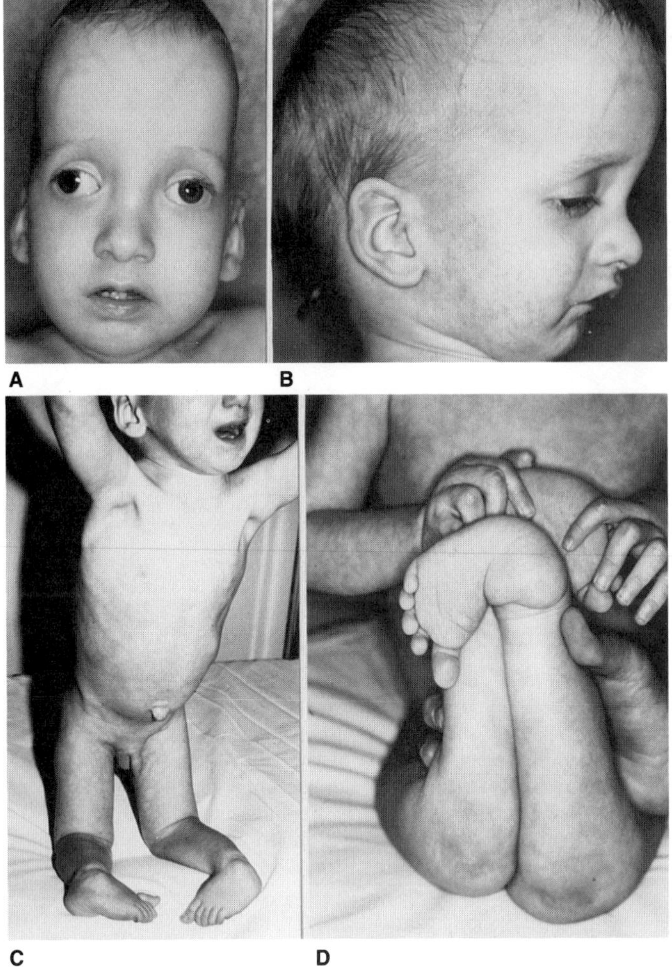

Fig. 15–18. *Idaho syndrome*. (A–D) Craniosynostosis involving sagittal suture, scaphocephaly, prominent veins, strabismus, overfolded helices, micrognathia, umbilical hernia, complete anterior dislocation of tibiae and fibulae with absent patellae allowing the knees to be flexed with the feet against the ventral surface of the body, and foot deformities. (From *MM Cohen Jr,* The Child With Multiple Defects, Raven Press, New York, 1982.)

## Ives–Houston syndrome

Ives and Houston (39), in 1980, reported a lethal, autosomal recessive syndrome consisting of intrauterine growth retardation, marked microcephaly with craniosynostosis, and severe malformations of the limbs, such as greatly shortened forearms, fused elbows, and the presence of two to four abnormal fingers (Fig. 15–19). The syndrome was discovered in a highly inbred, predominantly Cree Indian community in Northern Saskatchewan with 14 known instances.

Growth deficiency of prenatal onset was marked; birth weights varied from 450 to 1390 g. At least six infants were in breech position at delivery. Four infants were stillborn, seven expired within 20 minutes following severe gasping, and three died of recurrent apneic episodes.

Severe microcephaly with complete craniosynostosis was evident in every case. Each had a small brain having only primary sulci and gyri. Absence of the corpus callosum was noted in three instances. Facial features included short palpebral fissures, small eyes, prominent broad nose, micrognathia, and microstomia. Cleft soft palate was reported in one instance.

Upper limb anomalies were severe and consisted of fixed elbows

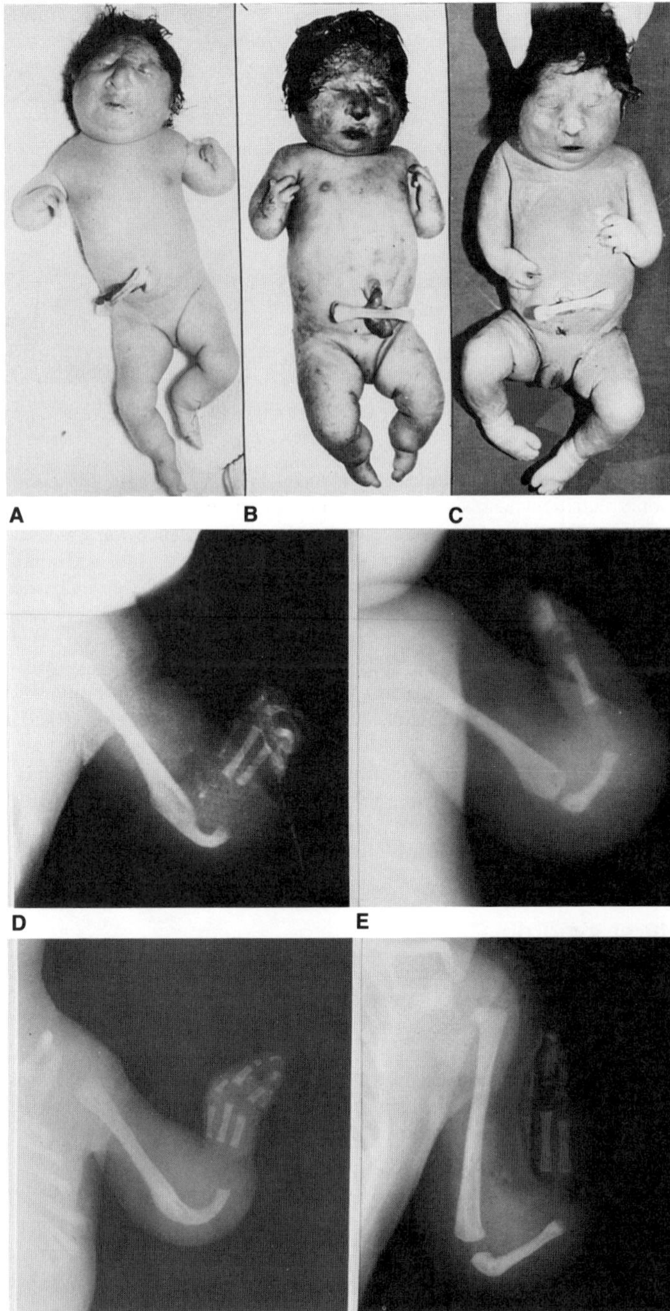

Fig. 15–19. *Ives–Houston syndrome*. (A–C) Microcephaly, short palpebral fissures, small eyes, broad nose, micrognathia, microstomia, fixed elbows, markedly short forearms, and malformed hands with 2–4 digits. (D) Radiohumeral synostosis with absence of first and fifth digits. (E) Single short forearm bone with only two metacarpal rays. (F) Radiohumeral synostosis with absence of first and fifth digits. (G) Single short forearm bone with absence of first and fifth digits. (From *EJ Ives* and *CS Houston,* Am J Med Genet **7**:351, 1980.)

and markedly shortened forearms. In four cases, only a single forearm bone was present, and it was not possible to determine whether it was a radius or an ulna. An abnormal and variable pattern characterized the hands. Two to four malformed digits were present. Most commonly, three digits were observed; two were well-developed central digits with a less well-developed index finger. A fifth finger, when present, had a tendency to arise from a bifid fourth metacarpal. A thumb was noted in only one instance. The legs were less severely affected. First metacarpals and great toes were shortened. The fourth and fifth toes were either hypoplastic or absent. Short legs and absence of the fibulae were noted in one case. Other findings included two or more narrow ribs, sacral segmentation defects, and, in some instances, hip dislocation.

The Ives–Houston syndrome should be distinguished from other syndromes with small head circumference and limb anomalies, such as *Baller–Gerold syndrome, Roberts syndrome,* and *Neu–Laxova syndrome.* However, the Ives–Houston syndrome is quite distinctive because of its attendant intrauterine growth retardation, severe microcephaly, and incompatibility with life.

## Jones syndrome

In 1977, KL Jones of San Diego observed a mother and son with craniosynostosis and Dandy–Walker malformation. The mother had involvement of both the coronal and sagittal sutures and her son had only sagittal synostosis. The condition may possibly have autosomal dominant inheritance. The mother's father was 42 years of age at the time of conception, suggesting the possibility of the mother representing a fresh mutation for the disorder. A number of monogenic, chromosomal, and environmentally-induced syndromes are known to be associated with Dandy–Walker malformation, and these are discussed by Murray et al (59) and Golden et al (26b).

## Lampert syndrome

Lampert et al (51) reported a sporadic instance of a patient with craniosynostosis, long thin face, midface hypoplasia, mild micrognathia, long tapering fingers, postaxial polydactyly, congenital hip dislocation, hallux valgus, and absent uterus.

## Lin–Gettig syndrome

Lin and Gettig (53a) reported a syndrome of craniosynostosis, agenesis of the corpus callosum, severe mental deficiency, hypogonadism, and distinctive facial appearance. Two affected brothers had midline sutural involvement—sagittal in one, metopic in the other, ptosis of eyelids, blepharophimosis, slanted palpebral fissures, epicanthic folds, thin lips, and long hypoplastic philtrum. Hypertonia, long slender fingers, and contractures were also evident. Third degree hypospadias was striking. Karyotypes were normal. The syndrome may be autosomal recessive, although X-linked inheritance cannot be ruled out. One of the sibs resembled *Opitz trigonocephaly (C) syndrome,* although microcephaly, abnormal palate with midline furrow, oral frenula, loose skin, and digital and limb malformations were absent.

## Lowry syndrome

Lowry (55) described a syndrome in two male sibs consisting of craniosynostosis and bilateral fibular aplasia (Fig. 15–20). Chromosome studies of one sib were normal. The other sib died shortly after birth of respiratory distress, and no chromosome studies were carried out. Consanguinity was also noted, suggesting an autosomal recessive mode of transmission.

In one patient, both coronal and sagittal sutures were synostosed. In the other patient, only the coronal suture was involved. The eyes were prominent in both cases, and strabismus was noted in one case. A partial cleft of the hard palate was observed in one patient, a highly arched palate in the other.

The fingers were normal, and bilateral single palmar creases were found in both cases. Bilaterally absent fibulae, talipes equinovarus, and normal toes were reported in both cases.

Other findings included cryptorchidism in both sibs; short sternum,

Fig. 15–20. *Lowry syndrome.* (A) Craniosynostosis with tower skull, ptosis of eyelids, and strabismus. (B) Radiograph showing fibular aplasia and talipes equinovarus. (From *RB Lowry,* J Med Genet **9**:227, 1972.)

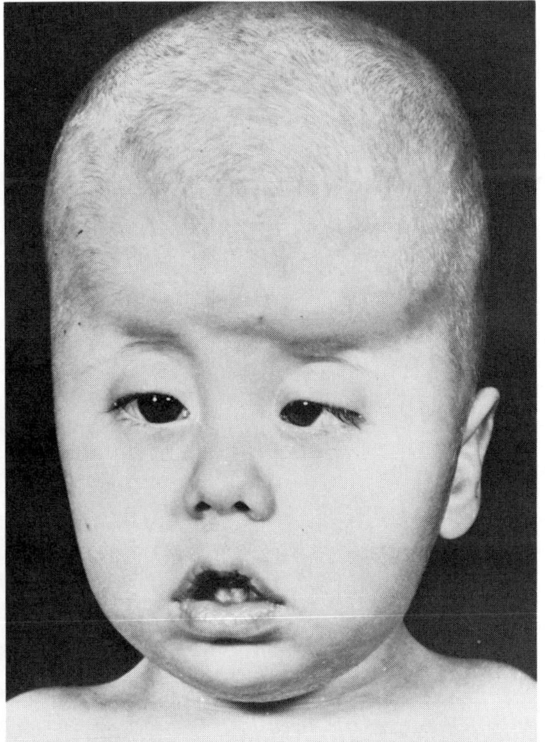

A

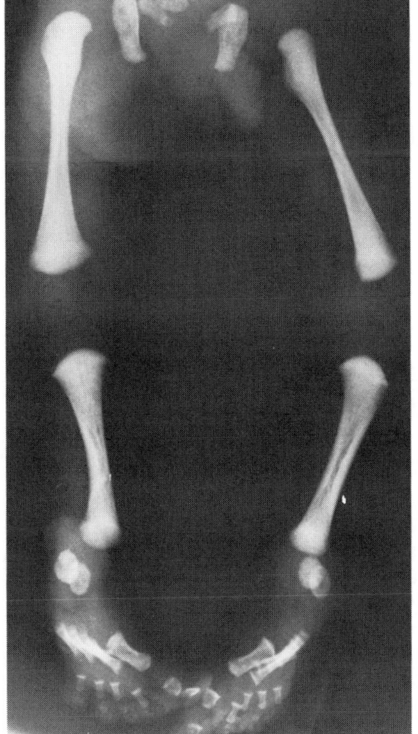

B

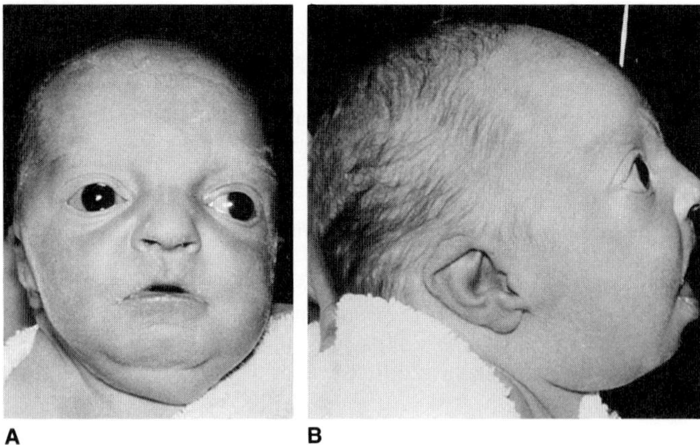


Fig. 15–21. *Lowry–MacLean syndrome.* (A,B) Ocular proptosis, divergent squint, mild downslanting of palpebral fissures, prominent nose, and low-set posteriorly angulated ears with abnormal helix and preauricular fistula. [From *RB Lowry* and *JR MacLean,* Birth Defects **13**(3B):203, 1977.]

pilonidal dimple, and normal intelligence in one sib; and large posterior fontanel, wormian bones, low-set ears with pointed helices, short webbed neck, and mild chordee in the other sib.

## Lowry–MacLean syndrome

In 1977, Lowry and MacLean (56) reported a sporadic occurrence of a syndrome characterized by mental retardation, ocular proptosis, glaucoma, cleft palate, eventration of diaphragm, congenital heart defect, growth failure, and craniosynostosis. (Fig. 15–21). Chromosomes were normal, consanguinity was denied, and there was no family history of similar anomalies. The Lowry–MacLean syndrome should be distinguished from *Crouzon syndrome* and *Christian syndrome.*

Head circumference was below the 10th percentile, weight was just above the 10th percentile, and length was just below the 50th percentile at birth. The postnatal course was characterized by seizures, mental retardation, and growth deficiency. Pneumonia coupled with further seizures and hyponatremia, suggestive of inappropriate antidiuretic hormone production, led to the patient's demise at 29 months of age. At that time height, weight, and head circumference were all below the third percentile.

Craniosynostosis developed during the second year of life. Craniofacial features included ocular proptosis, glaucoma developing during the second year, downslanting palpebral fissures, strabismus, Brushfield spots, prominent beaked nose, abnormal ears, incomplete clefting of the palate, and delayed dental eruption. Left hemiparesis, including left facial palsy, became evident during the second year of life. The fingernails were narrow and hyperconvex. Autopsy findings included hypoplasia of the corpus callosum, eventration of the left diaphragm, ASD, and bicuspid pulmonary valve.

## Marfanoid features and craniostenosis (including Shprintzen–Goldberg syndrome)

Several patients with marfanoid features and craniostenosis have been reported. Further delineation is necessary to determine how many distinct entities may be involved.

Shprintzen and Goldberg (74a) described a syndrome consisting of craniosynostosis, ocular proptosis, downslanting palpebral fissures, Byzantine arch palate, micrognathia, arachnodactyly, camptodactyly, inguinal and umbilical hernias, hypotonia, and mental deficiency (Fig. 15–22). A third patient, with the same overall pattern of anomalies but without craniosynostosis, was reported by Sugarman and Vogel (79). The three patients represented sporadic occurrences in their respective families and had normal chromosomes. The etiology is unknown. Elsewhere, Cohen (18) has referred to this condition as Shprintzen–Goldberg syndrome; earlier, he used the term Montefiore syndrome (16) provisionally naming the condition after the hospital where Shprintzen and Goldberg had acquainted him with their two original patients long before their paper was published. Clinical features are summarized in Table 15–4.

Arachnodactyly occurs together with craniosynostosis, dysplastic ears, radiohumeral synostosis, femoral bowing, and joint contractures in the autosomal recessively inherited *Antley–Bixler syndrome* (18).

Furlong et al (26) reported a patient with typical features of Marfan syndrome including dolichocephaly, dolichostenomelia, scoliosis, pectus carinatum, dilated aortic root, myopia, recurrent inguinal hernias, and sudden death from aortic dissection. However, craniosynostosis was present, involving the sagittal and lambdoidal sutures and part of the coronal suture. Intelligence was normal. The further findings of ptosis of the eyelids, hypospadias, spondylolisthesis, and absence of ectopia lentis suggest a distinctive connective tissue disorder separate from the Marfan syndrome. The family history was noncontributory.

MM Cohen reported a sporadic instance of sagittal synostosis, marfanoid features, and other anomalies (15). Cohen (16) later referred to the condition as Idaho syndrome II. The patient had pronounced dolichocephaly, severe mental deficiency, seizures, small low-set posteriorly angulated ears, preauricular tags, tortuous ear canals, downslanting palpebral fissures, dystopia canthorum, stellate iris pattern, highly arched palate, and micrognathia. He had long neck, sloping shoulders, narrow thorax, pectus carinatum, inguinal hernia, mild winging of scapulas, spina bifida at L-5, cubitus valgus, mild genua valga, single palmar crease on one hand, arachnodactyly, and bifid renal pelvis bilaterally. Pregnancy had been complicated by hydramnios.

Table 15–4. Shprintzen–Goldberg syndrome[a]

| Findings | Relative frequency |
|---|---|
| Performance | |
| Hypotonia | 3/3 |
| Mental deficiency | 3/3 |
| Craniofacial | |
| Craniosynostosis | 2/3 |
| Ocular hypertelorism | 3/3 |
| Ocular proptosis | 3/3 |
| Strabismus | 3/3 |
| Downslanting palpebral fissures | 2/3 |
| Maxillary hypoplasia | 3/3 |
| Obstructive apnea | 2/3 |
| Byzantine arch palate | 3/3 |
| Micrognathia | 3/3 |
| Low-set anomalous ears | 3/3 |
| Limbs | |
| Arachnodactyly | 3/3 |
| Camptodactyly | 3/3 |
| Talipes equinovarus | 3/3 |
| Genua valga | 1/3 |
| Hypermobile joints | 1/3 |
| Other | |
| Mitral valve prolapse | 1/3 |
| Pectus excavatum | 1/3 |
| Pectus carinatum | 1/3 |
| Scoliosis | 1/3 |
| Hyperelastic skin | 1/3 |
| Umbilical hernia | 3/3 |
| Inguinal hernia | 3/3 |
| Cryptorchidism | 1/3 |
| Small external genitalia | 1/3 |

[a]From MM Cohen Jr., *Craniosynostosis: Diagnosis, Evaluation, and Management.* Raven Press, New York, 1986.

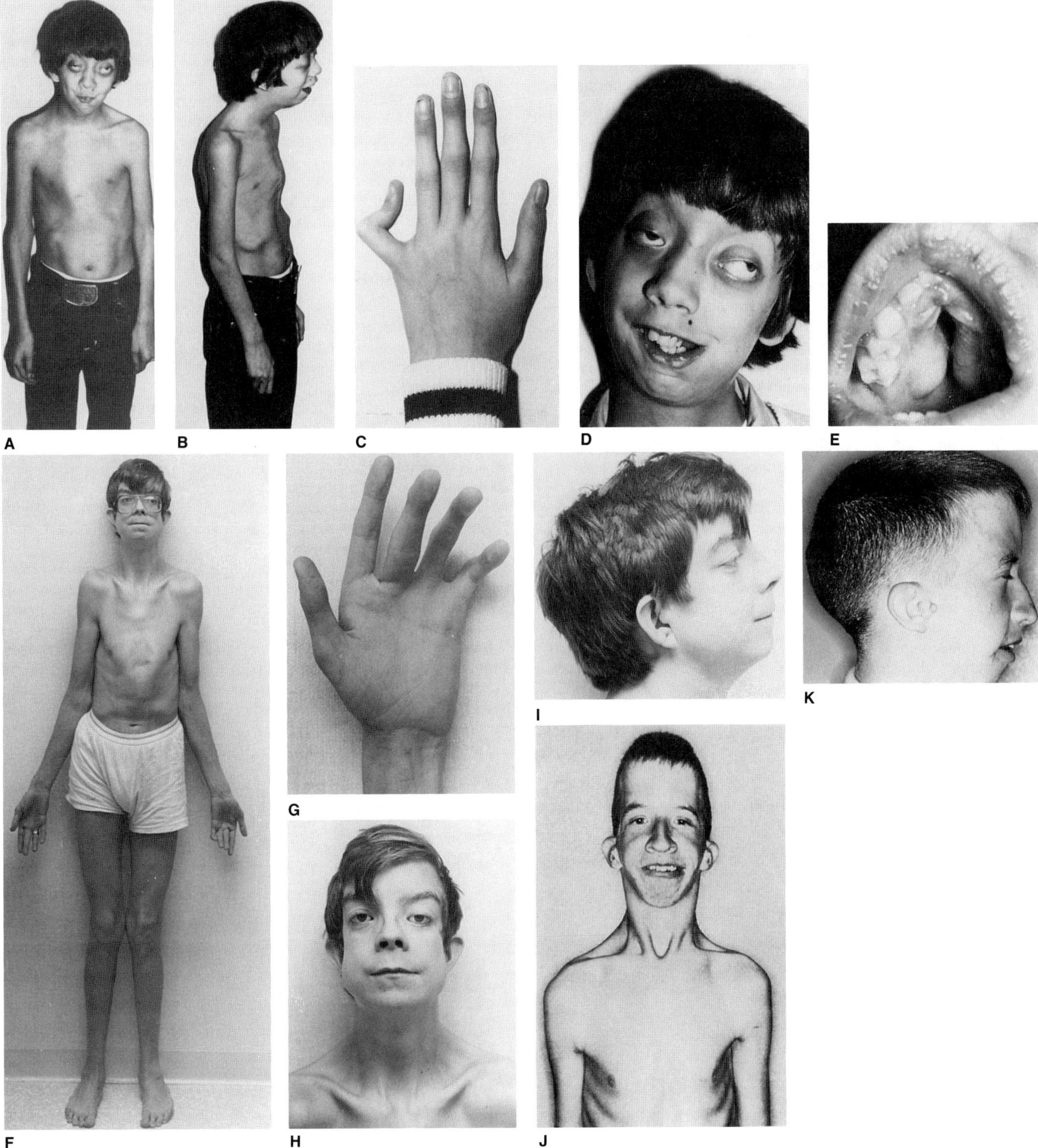

Fig. 15–22. *Marfanoid features and craniostenosis.* (A–E) *Shprintzen–Goldberg syndrome.* Scaphocephaly, ocular proptosis, strabismus, micrognathia, pectus excavatum, arachnodactyly, camptodactyly of fifth fingers, and enlargement of palatal shelves. (F–I) *Marfanoid syndrome with craniosynostosis.* Dolichostenomelia, pectus carinatum, single palmar crease, flexion contractures of proximal interphalangeal joints of fourth and fifth fingers, craniosynostosis, narrow forehead, ptosis of eyelids, mild micrognathia, and prominent ears. (J,K) Idaho syndrome II. Craniosynostosis with dolichocephaly, downslanting palpebral fissures, malformed ears, long neck, sloping shoulders, narrow thorax, pectus carinatum, and long arms. Note low-set, posteriorly angulated ears with preauricular tags, and micrognathia. (A–E from *RJ Shprintzen* and *RB Goldberg,* J Cranifac Genet Dev Biol **2**:65, 1982. F–I from *J Furlong et al,* Am J Med Genet **26**:599, 1987. J,K from *MM Cohen Jr,* J Neurosurg **47**:886, 1977).

Two reports of patients with Marfanoid features without craniosynostosis deserve mention. Clunie and Mason (14) described three sibs of a consanguineous union in which a Marfanoid habitus was combined with multiple inguinal and femoral hernias. Features of the Marfan syndrome such as ectopia lentis and cardiovascular abnormalities were absent. No specific mention was made of craniosynostosis. Jaffer and Beighton (41) reported a patient with arachnodactyly, pectus carinatum, spondylolisthesis, joint laxity, and mental deficiency, but craniosynostosis was not evident.

## Pfeiffer type cardiocranial syndrome

Pfeiffer et al (64) reported two sibs with sagittal synostosis, facial dysmorphism, and complex cardiovascular malformations. Both sibs failed to thrive and had severe developmental delay, mild contractures of the large joints, and mandibular ankylosis. Craniofacial features included dolichocephaly, hypertelorism, strabismus, micrognathia, submucous cleft palate with absent uvula in one sib, and low-set dysplastic ears. Cardiovascular anomalies consisted of VSD, ASD, and PDA in one sib, and ASD, tetralogy of Fallot, peripheral pulmonic stenosis, and an azygous connection instead of a normal inferior vena cava in the other sib. Hypogenitalism and zygodactyly were present in one sib; hypoplastic kidneys and anomalous fourth and fifth ribs were present in the other.

Chromosome studies were normal. The family history was noncontributory except for a first cousin on the paternal side who had metopic synostosis with trigonocephaly but no other morphological abnormalities. Mental development was normal. Pfeiffer et al (64) thought that trigonocephaly in the cousin probably resulted from intrauterine constraint and bore no causal relationship to the condition in the two propositi. Stratton and Parsons (78a) described a male with sagittal synostosis and pointed occiput, ASD, midthoracic kyphosis, microphallus, cryptorchidism, bilateral palmar creases, and bilateral talipes. Autosomal recessive inheritance is possible, but the syndrome needs to be further delineated.

## Pfeiffer type dolichocephalosyndactyly

Pfeiffer et al (65) reported a distinctive, sporadically occurring syndrome consisting of sagittal synostosis, facial dysmorphism, genitourinary anomalies, and limb defects (Fig. 15–23). Craniofacial features included dolichocephaly, upslanting palpebral fissures, inverse epicanthic folds, bilateral iris colobomas, atypical fundus vessels, possibly mild optic atrophy, highly arched palate, and micrognathia. Severe mental deficiency was evident. Perineoscrotal hypospadias and bilateral vesicoureteral reflux were also noted. Ectrodactyly of one hand and bilateral 2–5 syndactyly of the toes were observed. The karyotype was normal and the family history noncontributory.

## Sakati syndrome

Sakati et al (67) described a provisionally unique-pattern syndrome consisting of acrocephalopolysyndactyly, short limbs, congenital heart defect, ear anomalies, and skin defects (Fig. 15–24). A chromosome study was normal. The condition should be distinguished from *Carpenter syndrome*.

The calvaria was large and the face was disproportionately small. The orbits were shallow and the eyes prominent. All cranial sutures were synostosed. The ears were dysplastic and low set. A unilateral ear tag was noted. Patches of alopecia with atrophic skin were present above the ears. Linear scar-like lesions were observed in the submental area. The palate was narrow and highly arched. Maxillary hypoplasia and crowding of the upper teeth were reported. The neck was short and the hairline was low.

Short arms, cubitus valgus, and short broad hands were evident. Fusion of the proximal interphalangeal joints of the fourth fingers and proximally placed thumbs were observed. A sixth digit had been surgically removed from one hand.

Extreme shortening of the legs and adducted feet were seen together with polysyndactyly (seven toes on one foot and six toes on the other). Roentgenographic examination revealed bilateral coxa valga, lateral bowing of both femora, hypoplastic tibiae, deformity and displace-

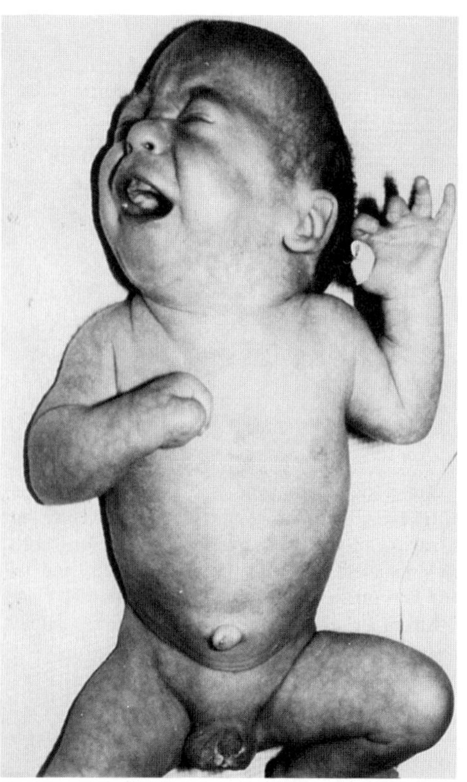
A

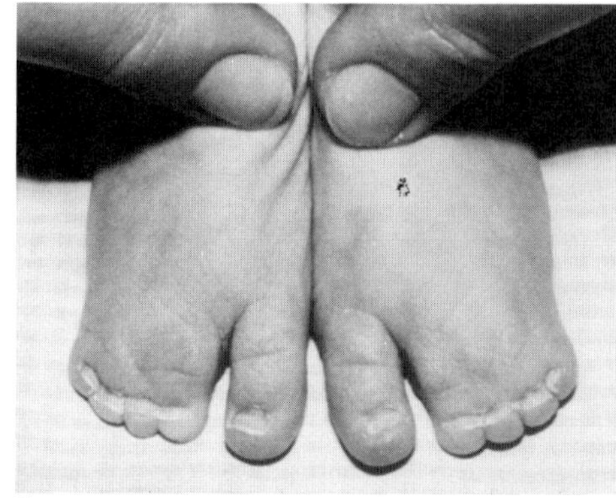
B

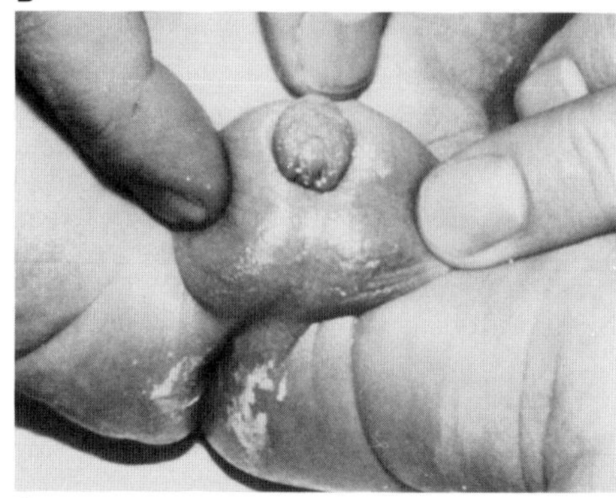
C

Fig. 15–23. *Pfeiffer type dolichocephalosyndactyly.* (A) Note dolichocephaly, micrognathia, and unilateral ectrodactyly. (B) Note bilateral syndactyly of second, third, fourth, and fifth toes. (C) Note perineoscrotal hypospadias. (From *RA Pfeiffer et al,* Eur J Pediatr **146**:74, 1987.)

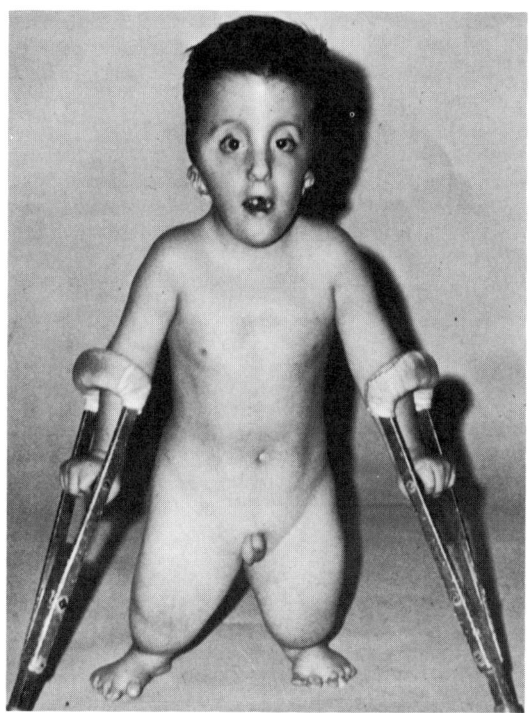

Fig. 15–24. *Sakati syndrome.* Acrocephalopolysyndactyly, short limbs, congenital heart defect, ear anomalies, and skin defects. (*From N Sakati et al,* J Pediatr **79**:104, 1971.)

ment of the fibulae, and abnormal malposed tarsals and metatarsals. There were six metatarsals in each foot.

Other findings included congenital heart defect, widely spaced nipples, cryptorchidism, small phallus, and inguinal hernia. Intelligence was normal.

## Salinas syndrome

Salinas et al (68) reported a sporadic instance of multiple anomalies confined to the head and face. Features included acrocephaly with prominent coronal suture (thought to be fused), coarse dry hair, malar hypoplasia with downslanting palpebral fissures, colobomas of lower eyelids, small upturned nose, choanal atresia, cleft palate, midline cleft of the lower vermilion, and clefting of the mandibular symphysis.

## San Francisco syndrome

In 1978, BD Hall (personal communication) observed a striking syndrome of craniosynostosis, midface hypoplasia, ptosis of eyelids, bulbous nose, and small ears. The disorder followed an autosomal dominant mode of transmission.

## Say–Barber syndrome

Say et al (70) reported two brothers with short stature, developmental delay, microcephaly, craniosynostosis in one of them, sloping forehead, beaked nose with high nasal bridge, highly arched palate, micrognathia, large protruding ears, flexion contractures, anal stenosis, hypoplastic patellas, scoliosis, small testes, and decreased subcutaneous fat (Fig. 15–25). Both brothers developed eczema during in-

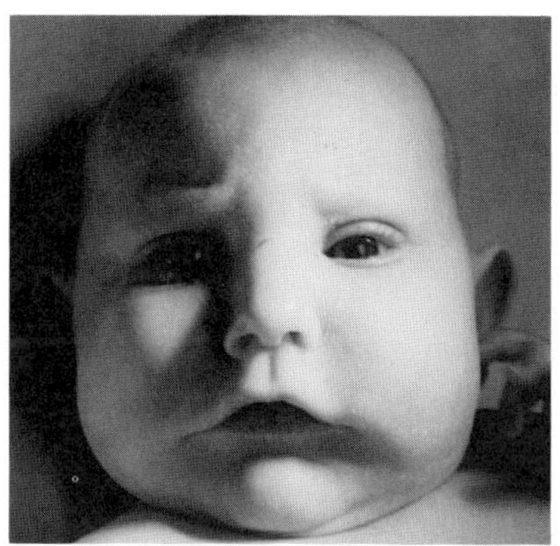

A

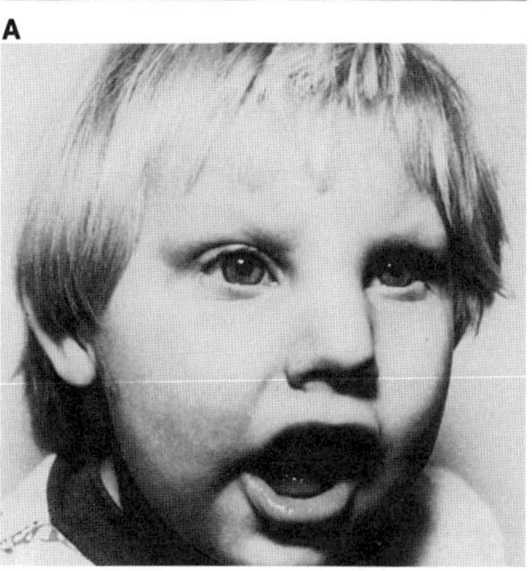

B

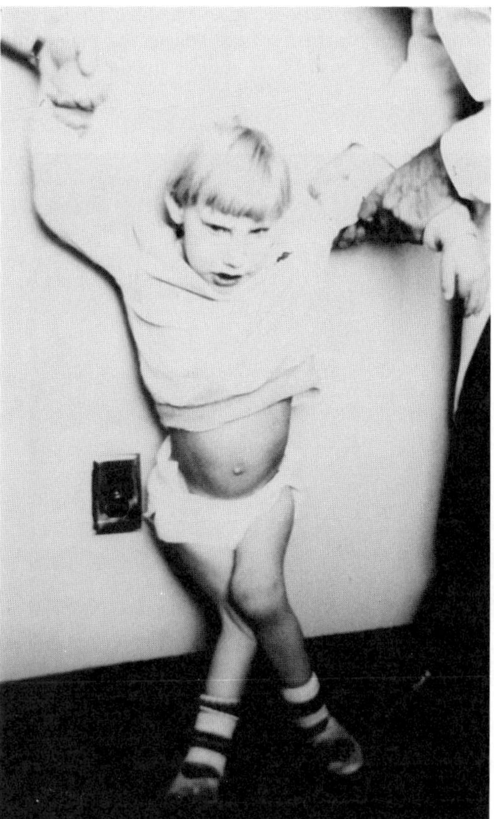

C

Fig. 15–25. *Say–Barber syndrome.* (A) Age 3 months. Note microcephaly, large protruding ears, carp-shaped mouth, and micrognathia. (B) Age 6.5 years. Compare with A. (C) Age 6.5 years. Note elbow contractures, dislocated hips, marked laxity of knees with absent patellae, and inward deviation of lower ends of femora. (From *B Say et al,* J Med Genet **23**:355, 1986.)

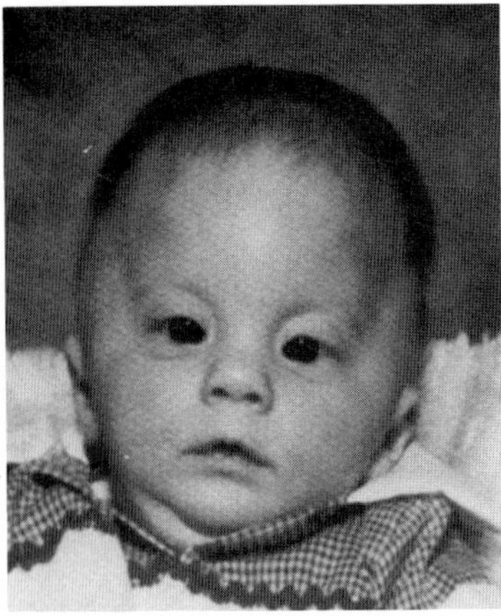

Fig. 15–26. *Say–Meyer trigonocephaly syndrome.* Prominent vertical ridge over frontal bone, narrow forehead, hypotelorism associated with trigonocephaly. Short stature and developmental delay were also present. (From *B Say* and *J Meyer,* Am J Dis Child **135**:711, 1981.)

fancy, had recurrent infections, and transient hypogammaglobulinemia.

Chromosomes were normal. Etiology is as yet unclear. From the available data, X-linked inheritance or autosomal recessive inheritance is equally possible. However, an undetectable chromosomal abnormality is a real possibility, especially in view of some alleged family history that was not possible to verify. The father's half-brother's child reportedly had craniostenosis, anal stenosis, and contractures, but normal growth and development.

## Say–Meyer trigonocephaly syndrome

In 1981, Say and Meyer (69) described an X-linked recessively inherited syndrome consisting of craniosynostosis especially involving the metopic suture, developmental delay, and short stature (Fig. 15–26). All three patients had vertical ridging of the metopic area together with hypotelorism. In two patients, the lambdoid sutures were also prematurely synostosed, and, in one patient, the sagittal suture was additionally involved. One of the three affected individuals had seizures. Low-frequency findings also included highly arched palate, secondary alveolar ridges, clinodactyly, and VSD. The syndrome needs to be further delineated.

Differential diagnosis includes *Frydman trigonocephaly syndrome* and *Opitz trigonocephaly syndrome (C syndrome).* Trigonocephaly may occur with mental deficiency in various chromosomal syndromes, most commonly with del(9p) [1], del(11q) [12], and less commonly with del(13q) [61] and dup(13q) [36].

## Say-Poznanski syndrome

Say and Poznanski (69a) reported a patient with cloverleaf skull with polydactyly of the hands and feet, grossly abnormal metacarpals and metatarsals, angular ulnae with probable fusion to the midportion of the radial bones, short, wide clavicles, winged scapulae, unusually shaped ribs with abnormal spacing between them and with prominent costovertebral junctions, and widely separated ischia (Fig. 15–27).

## Thanos syndrome

In 1977, Thanos and co-workers (80) described a syndrome of craniosynostosis, bony hyperostoses, epibulbar dermoids, linear verrucous epidermal nevus, and mental deficiency (Fig. 15–28). Two other cases are known (R Pauli and R Stewart, personal communications) but are unpublished to date. All three instances have occurred sporadically, and the syndrome needs to be further delineated. Similarity to *Proteus syndrome* is obvious. However, limbs have been normal to date. No full-length journal article on this distinctive syndrome has yet appeared in the literature. The term Thanos syndrome is proposed as a provisional name on the basis of the only known synoptic description of a single case provided by Thanos et al (80).

Craniosynostosis and craniofacial asymmetry develop between the fourth and seventh months. Progressive hyperostoses of the skull also occur and especially involve the frontal, parietal, and occipital bones as well as the mandibular symphysis. Bony deposition at the sites of sutural fusion and apparently at sites of minor or moderate trauma

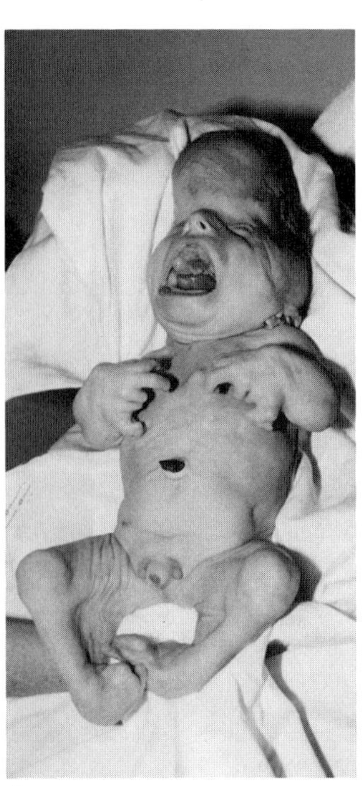

A

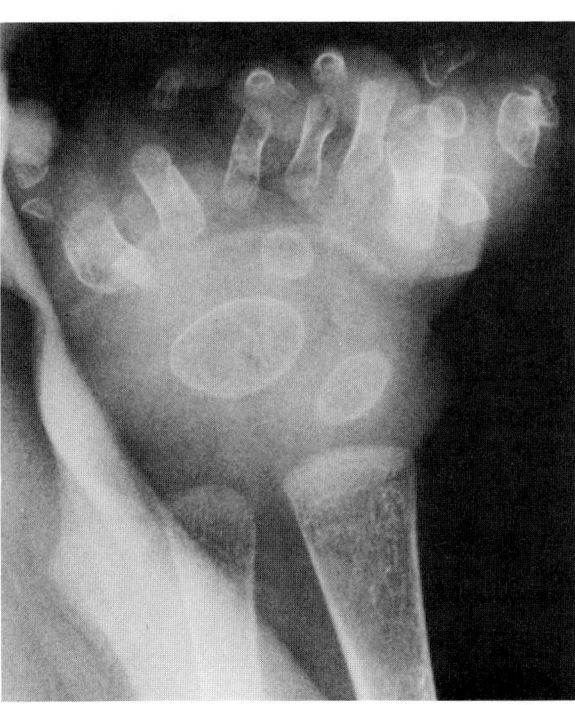

B

Fig. 15–27. *Say–Poznanski syndrome.* (A) Note cloverleaf skull and bizarre polydactyly of hands and feet. (B) Hand radiograph showing polydactyly with many ossicles. (From *B Say* and *AK Poznanski,* Pediatr Radiol **17**:93, 1987.)

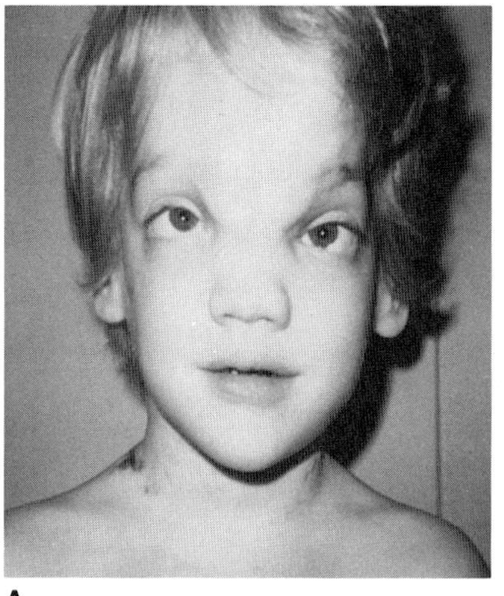
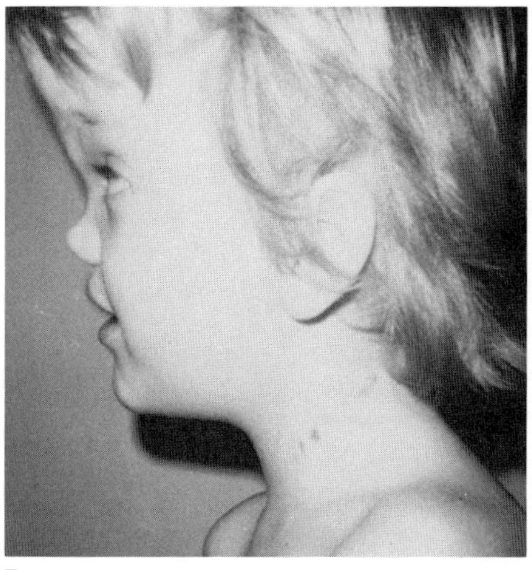
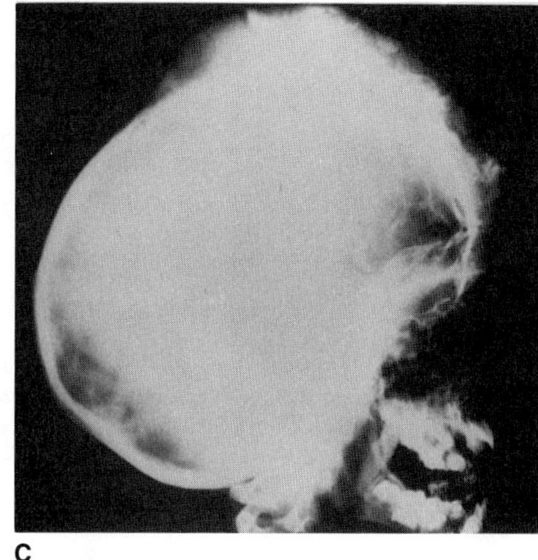

Fig. 15–28. *Thanos syndrome.* (A,B) Hypertelorism, broad nasal root, broad nose with prominent alar cartilages, and epidermal nevi on neck. (C) Radiograph showing hyperostosis of frontal and vertex regions. [From *C Thanos et al,* Syndrome Ident **5**(1):19, 1977.]

may be very marked and protrusive. Ocular findings include hypertelorism, epibulbar dermoids, strabismus, myopia, and visual deterioration. Pauli's patient was functionally blind at 21 years of age. Mental deficiency has been a feature in all three cases.

The nose is broad with prominent alar cartilages. The ears may be mildly dysmorphic. Linear verrucous epidermal nevi, a feature in two of the three cases, have involved the neck, axilla, and thorax. Kyphoscoliosis has also been noted.

## Ventruto syndrome

In 1976, Ventruto et al (82) reported a distinctive autosomal dominantly inherited syndrome consisting of brachydactyly, aplastic or hypoplastic nails, symphalangism, carpal and tarsal fusion, dysplastic hip joints, and craniosynostosis (Fig. 15–29). Five affected individuals were reported through three generations. The proband was first seen for acute leukemia.

Craniosynostosis involved the coronal suture. Strabismus was commonly observed. Anomalies of the hands and feet varied in severity and were not symmetric in all affected individuals. The thumbs and great toes were not involved. Absence or hypoplasia of the middle and distal phalanges of the fingers was variable in degree and was associated with absence or hypoplasia of the nails. In some instances, both the middle and the distal phalanges were involved. In other instances, only the distal phalanges were affected. Symphalangism, when present, involved the proximal interphalangeal joints. First metacarpals showed brachymetapody. Carpal fusions involved the hamate and capitate or there was a more complex type of fusion of the trapezium, trapezoid, capitate, hamate, and lunate.

Fusion in the tarsals involved the second and third cuneiform bones and/or other fusions such as between the first cuneiform and scaphoid, the calcaneus and cuboid, or the second cuneiform and second metatarsal. In some affected individuals, one or more cuneiform bones were absent. Brachymetapody involved the first metatarsal in several patients. The pattern of absence or hypoplasia of the middle and distal phalanges of the toes was similar to the pattern observed in the fingers. In the feet, symphalangism also involved the proximal interphalangeal joints. Pelvic radiographs showed osteochondritis or sclerosis involving the acetabular roofs, but sparing the femoral necks.

Differential diagnosis includes isolated type-C brachydactyly and type-C brachydactyly with associated abnormalities. Type-C brachydactyly involves the proximal and middle phalanges of the second and third fingers, hypersegmentation of the proximal phalanges, brachymetapody, and symphalangism. Type-C brachydactyly also occurs with Legg–Perthes disease of the hip as a familial syndrome.

## Wisconsin syndrome

Wisconsin syndrome consists of craniosynostosis, mental deficiency, upslanting palpebral fissures, microtia, and short fourth metatarsals with recessed fourth toes (Fig. 15–30). Chromosomes were normal and the patient was the only affected member of the family (JM Opitz, personal communication, 1976).

### References (Miscellaneous craniosynostosis syndromes)

1. Alfi OS et al: The 9p− syndromes. *Ann Génét (Paris)* **19**:11–16, 1976.
2. Allain D et al: Acrocephalosynankie, pseudohermaphrodisme feminin et nephropathie hypertensive. *Ann Pédiatr (Paris)* **23**:277–284, 1976.
3. Anderson LG et al: Familial osteodysplasia. *JAMA* **220**:1687–1693, 1972.
4. Armendares S et al: A newly recognized inherited syndrome of dwarfism, craniosynostosis, retinitis pigmentosa, and multiple congenital malformations. *J Pediatr* **85**:872–873, 1974.
5. Armendares S et al: A newly recognized inherited syndrome of dwarfism, craniosynostosis, retinitis pigmentosa, and multiple congenital malformations. *Birth Defects* **11**(5):49–53, 1975.
6. Baraitser M et al: A new craniosynostosis/mental retardation syndrome diagnosed by fetoscopy. *Clin Genet* **22**:12–15, 1982.
7. Beare JM et al: Cutis gyratum, acanthosis nigricans, and other congenital anomalies: A new syndrome. *Br J Dermatol* **81**:241–247, 1969.
8. Beighton P et al: Osteoglophonic dwarfism. *Pediatr Radiol* **10**:46–50, 1980.
9. Berant M, Berant N: Radioulnar synostosis and craniosynostosis in one family. *J Pediatr* **83**:88–90, 1973.
10. Buchignani JS et al: Roentgenographic findings in familial osteodysplasia. *Am J Roengenol* **116**:602–608, 1972.
11. Calabro A et al: Letter to the editor: Craniosynostosis and unilateral ulnar aplasia. *Am J Med Genet* **20**:203–204, 1985.
12. Cassidy SB et al: Trigonocephaly and the 11q− syndrome. *Ann Génét* **20**:67–69, 1977.
13. Christian JC et al: The adducted thumbs syndrome. An autosomal recessive disease with arthrogryposis, dysmyelination, craniostenosis, and cleft palate. *Clin Genet* **2**:95–103, 1971.
14. Clunie GJA, Masion JM: Visceral diverticula and the Marfan syndrome. *Br J Surg* **50**:51–52, 1962.
15. Cohen MM Jr: Genetic perspectives on craniosynostosis and syndromes with craniosynostosis. *J Neurosurg* **47**:886–898, 1977.
16. Cohen MM Jr: Craniosynostosis and syndromes with craniosynostosis: Incidence, genetics, penetrance, variability, and new syndrome updating. *Birth Defects* **15**(5B):13–63, 1979.
17. Cohen MM Jr: *The Child with Multiple Birth Defects.* Raven Press, New York, 1982.
18. Cohen MM Jr: *Craniosynostosis: Diagnosis, Evaluation, and Management.* Raven Press, New York, 1986.
18a. Cohen MM Jr: Craniosynostosis update 1987. *Am J Med Genet Suppl* **4**:99–148, 1988.

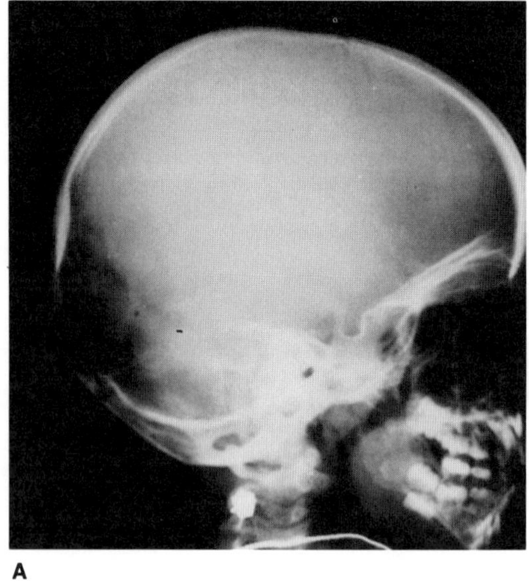

A

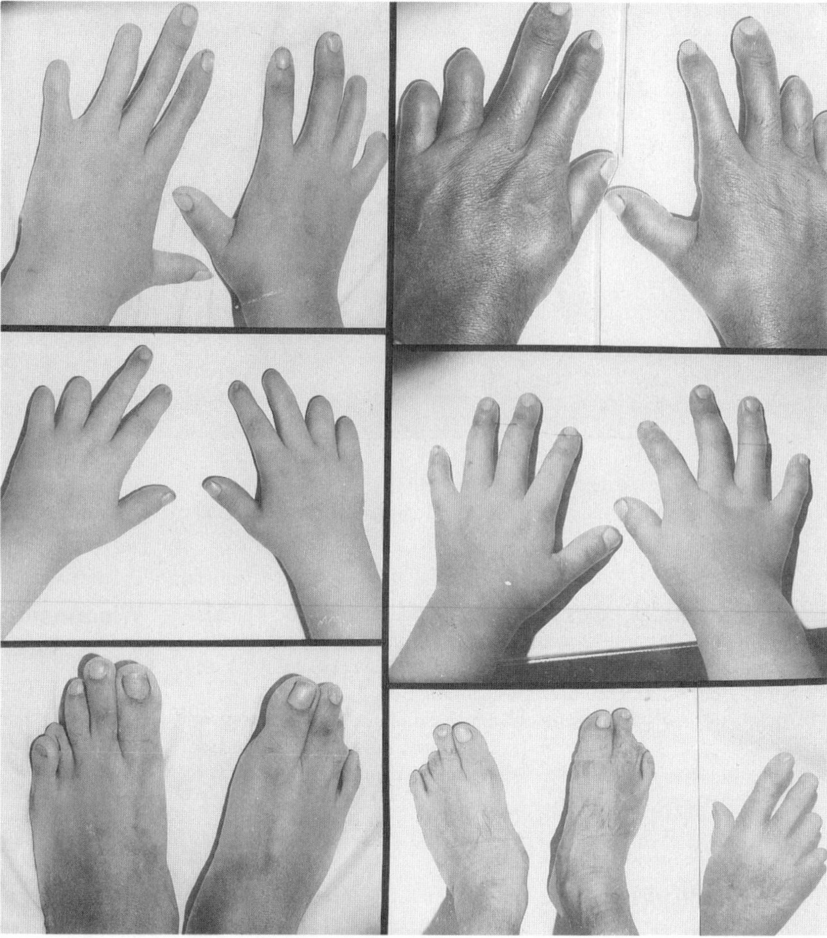

B

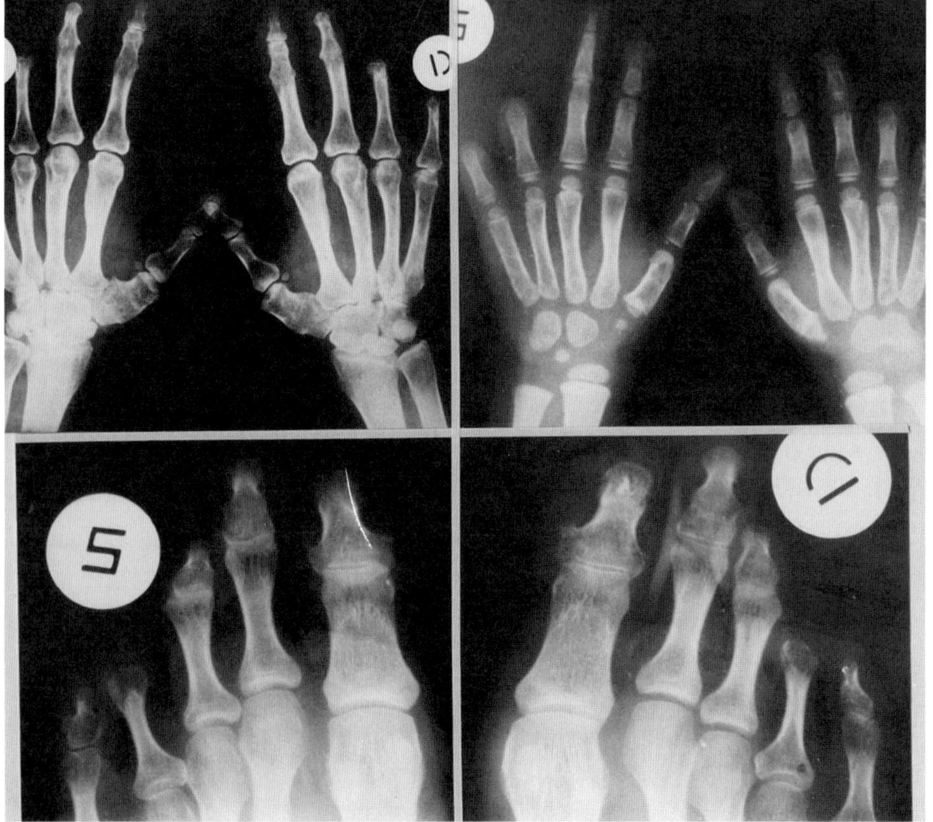

C

Fig. 15–29. *Ventruto syndrome.* (A) Craniosynostosis. (B) Hands and feet of affected individuals showing brachydactyly, absence or hypoplasia of middle and distal phalanges, and aplastic or hypoplastic nails. (C) Radiographic changes in hands and feet of affected individuals. Note absence or hypoplasia of middle and distal phalanges and symphalangism. (From *V Ventruto et al,* J Med Genet **13**:394, 1976.)

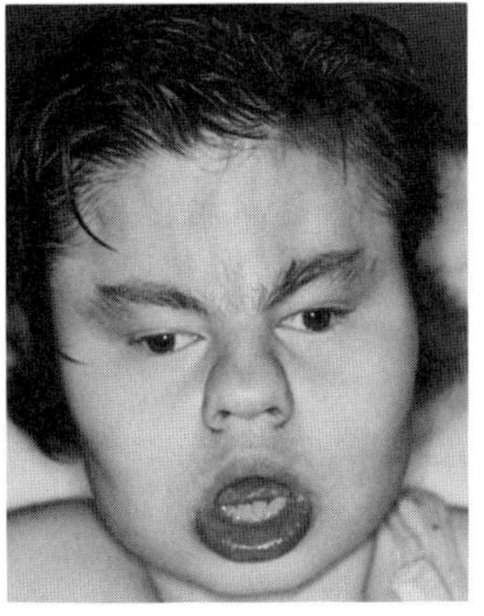

A

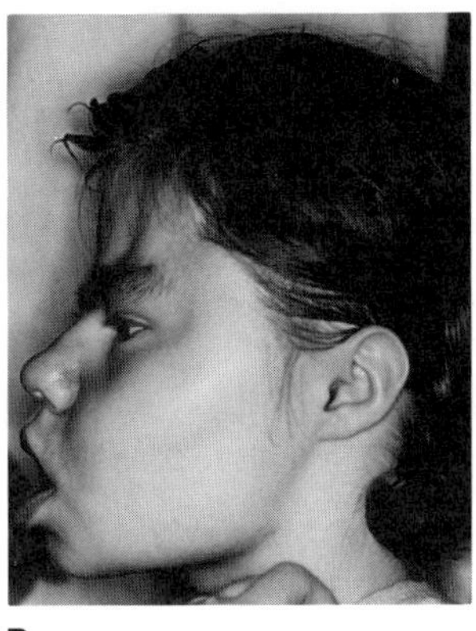

B

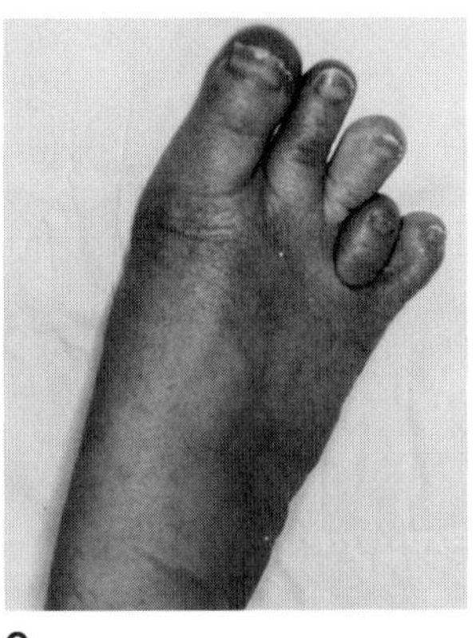

C

Fig. 15–30. *Wisconsin syndrome.* (A,B) Craniosynostosis, sloped forehead, upslanting palpebral fissures, microtia, and hypoplastic helix. (C) Short fourth metatarsal with recessed fourth toe. (A–C courtesy of *JM Opitz,* Helena, Montana.)


19. Cole DEC, Carpenter TO: Bone fragility, craniosynostosis, ocular proptosis, hydrocephalus and distinctive facial features: A newly recognized type of osteogenesis imperfecta. *J Pediatr* **110**:76–80, 1987.

19a. Curry CJR et al: Unknown case 1. Paper presented at the *David W. Smith 8th Annual Workshop on Malformations and Morphogenesis,* Greenville, SC, August 15–19, 1987.

20. Daum S et al: Dysplasie télencéphalique avec excrossance de l'os frontal. *Sem Hôp Paris* **34**:893–896, 1958.

21. Dipierri JE, Guzman JD: A second family with autosomal dominant osteosclerosis-type Stanescu. *Am J Med Genet* **18**:13–18, 1984.

22. Fairbank T: *An Atlas of General Affections of the Skeleton.* Case 84, ES Livingstone, Edinburgh/London, 1951.

23. Fitch N, Levy EP: Adducted thumb syndromes. *Clin Genet* **8**:190–198, 1975.

24. Fontaine G et al: Un nouveau syndrome polymalformatif complexe. *J Génét Hum* **25**:109–119, 1977.

25. Frydman M et al: Trigonocephaly: A new familial syndrome. *Am J Med Genet* **18**:55–59, 1984.

26. Furlong J et al: New Marfanoid syndrome with craniosynostosis. *Am J Med Genet* **26**:599–604, 1987.

26a. Gellis SS, Feingold M: Cranioectodermal dysplasia. *Am J Dis Child* **133**:1275–1276, 1979.

26b. Golden JA et al: Dandy–Walker syndrome and associated anomalies. *Pediatr Neurosurg* **13**:38–44, 1987.

27. Gomez MR: Cerebello-trigemino-dermal dysplasia, in *Neurocutaneous Diseases I,* Gomez MR (ed), Raven Press, New York, 1987, Ch. 38, pp 145–348.

28. Gomez MR: Cerebellotrigeminal and focal dermal dysplasia: A newly recognized neurocutaneous syndrome. *Brain Dev* **1**:253–256, 1979.

29. Gorlin RJ et al: Craniofacial dysostosis, patent ductus arteriosus, hypertrichosis, hypoplasia of labia majora, dental and eye anomalies. *J Pediatr* **56**:778–785, 1960.

30. Hall BD, Cadle RG: Cutis gyratum, acanthosis nigricans and a specific pattern of multiple anomalies: Report of a third case. *Proc Greenwood Genet Ctr* **6**:182, 1987.

31. Hall BD et al: The Beare–Stevenson cutis gyratum syndrome, in progress.

32. Hall JG: Craniofacial dysostosis—either Stanescu dysostosis or a new entity. *Birth Defects* **10**(12):521–523, 1974.

33. Herrmann J et al: Craniosynostosis syndromes. *Rocky Mt Med J* **66**:45–56, 1969.

34. Hersh JH et al: Craniosynostosis, sensorineural hearing loss and craniofacial abnormalities in siblings. *Proc Greenwood Genet Ctr* **5**:186, 1986.

35. Hoger PH et al: Craniosynostosis in hyper-IgE-syndrome. *Eur J Pediatr* **144**:414–417, 1985.

36. Hornstein L, Soukup S: A recognizable phenotype in a child with partial duplication 13q in a family with t(10q;13q). *Clin Genet* **19**:81–86, 1981.

37. Hughes HE et al: Craniotelencephalic dysplasia in sisters: Further delineation of a possible syndrome. *Am J Med Genet* **14**:557–565, 1983.

38. Hunter AGW et al: A "new" syndrome of mental retardation with characteristic facies and brachyphalangy. *J Med Genet* **14**:430–437, 1977.

38a. Hurst JA et al: Distinctive syndrome of short stature, craniosynostosis, skeletal changes, and malformed ears. *Am J Med Genet* **29**:107–115, 1988.

39. Ives, EJ, Houston CS: Autosomal recessive microcephaly and micromelia in Cree Indians. *Am J Med Genet* **7**:351–360, 1980.

40. Jabbour JT, Taybi H: Craniotelencephalic dysplasia. *Am J Dis Child* **108**:627–632, 1964.

41. Jaffer Z, Beighton P: Case report 98—Arachnodactyly, joint laxity, and spondylolisthesis. *J Clin Dysmorphol* **1**:14–18, 1983

42. Kaplan P et al: A new acro-cranio-facial dysostosis syndrome in sisters. *Am J Med Genet* **29**:95–106, 1988.

43. Kayser B: Ein Fall von angeborener Trigeminuslähmung und angeborenem totalem Tränenmangel. *Klin Mbl Augenheilkd* **66**:652–654, 1921.

44. Keats TE et al: Craniofacial dysostosis with fibrous metaphyseal defects. *Am J Roentgenol* **124**:271–275, 1975.

45. Kelley RI et al: Osteoglophonic dwarfism in two generations. *J Med Genet* **20**:436–440, 1983.

46. Keutel J et al: Eine wahrscheinlich autosomal recessiv vererbte Skelettmissbildung mit Humeroradialsynostose. *Humangenetik* **9**:43–53, 1970.

47. Kunze J et al: Adducted thumb syndrome: Report of a new case and a diagnostic approach. *Eur J Pediatr* **141**:122–126, 1983.

48. Kurczynski TW, Casperson SM: Auralcephalosyndactyly: A new hereditary craniosynostosis syndrome. *J Med Genet* **25**:491–493, 1988.

49. Lacheretz M et al: L'acrocéphalo-synankie. A propos d'une observation avec synostoses multiples. *Pédiatrie* **29**:169–177, 1974.

50. Ladda RL et al: Craniosynostosis associated with limb reduction malformations and cleft lip/palate: A distinct syndrome. *Pediatrics* **61**:169–177, 1974.

51. Lampert R et al: S.S. 11 8/12 year old white female. *Proc Greenwood Genet Ctr* **3**:70, 1984.

52. Levin LS et al: A variable syndrome of craniosynostosis, short thin hair, dental abnormalities, and short limbs: Cranioectodermal dysplasia. *J Pediatr* **90**:55–61, 1977.

53. Lichtenstein J et al: The tricho-dento-osseous (TDO) syndrome. *Am J Hum Genet* **24**:569–582, 1972.

53a. Lin AE, Gettig E: A possible newly recognized syndrome? Craniosynostosis, agenesis of the corpus callosum, severe mental retardation, distinctive facies, and hypogonadism. *Am J Med Genet* (in press).

54. López-Hernández A: Craniosynostosis, ataxia, trigeminal anesthesia and parietal alopecia with pons-vermis fusion anomaly (atresia of the fourth ventricle). Report of two cases. *Neuropediatrics* **13**:99–102, 1982.

55. Lowry RB: Congenital absence of the fibula and craniosynostosis in sibs. *J Med Genet* **9**:227–229, 1972.

56. Lowry RB, MacLean JR: Syndrome of mental retardation, cleft palate, eventration of diaphragm, congenital heart defect, glaucoma, growth failure and craniosynostosis. *Birth Defects* **13**(3B):203–228, 1977.

57. Maximilian C et al: Syndrome de dysostose cranio-faciale avec hyperplasie. *J Génét Hum* **29**:129–139, 1981.

58. Michels VV et al: A clefting syndrome with ocular anterior chamber defect and lid anomalies. *J Pediatr* **93**:444–446, 1978.

59. Murray JC et al: Dandy-Walker malformation: Etiologic heterogeneity and empiric recurrence risks. *Clin Genet* **28**:272–283, 1985.

59a. Neidich JA et al: Aglossia with congenital absence of the mandibular rami and other craniofacial abnormalities. *Am J Med Genet Suppl* **4**:161–166, 1988.

60. Neuhäuser G et al: A craniosynostosis-craniofacial dysostosis syndrome with mental retardation and other malformations. Craniofacial dyssynostosis. *Eur J Pediatr* **123**:15–28, 1976.

61. Orbeli DJ et al: The syndrome associated with the partial D-monosomy. Case report and review. *Hum Genet* **13**:296–308, 1971.

62. Pagon RA et al: Calvarial hyperostosis: A benign X-linked recessive disorder. *Clin Genet* **29**:73–78, 1986.

63. Pascuel-Castroviejo I: Displasia cerebelotrigeminal. *Neurologia Infantil,* Vol 1. Editorial cientifico-medica, Barcelona.

64. Pfeiffer RA et al: Sagittal craniostenosis, congenital heart disease, men-

tal deficiency and various dysmorphies in two sibs—a "new" syndrome? *Eur J Pediatr* **146**:75–78, 1987.

65. Pfeiffer RA et al: An unusual, possibly "new" MA/MR syndrome with sagittal craniosynostosis. *Eur J Pediatr* **146**:74–75, 1987.

66. Pillat A: Wiener Ophthalmologische Gesellschaft. Epithelschädigung der Hornhaut bei angeborener Trigeminushypoplasie. *Wien Klin Wochenschr* **61**:605, 1949.

67. Sakati N et al: A new syndrome with acrocephalopolysyndactyly, cardiac disease, and distinctive defects of the ear, skin and lower limbs. *J Pediatr* **79**:104–109, 1971.

68. Salinas CF et al: Case report 38: Colobomas of lower lids, malar hypoplasia, antimongoloid slant and clefting. *Syndrome Ident* **4**(1):5–6, 1976.

69. Say B, Meyer J: Familial trigonocephaly associated with short stature and developmental delay. *Am J Dis Child* **135**:711–712, 1981.

69a. Say B, Poznanski AK: Cloverleaf skull associated with unusual skeletal anomalies. *Pediatr Radiol* **17**:93–96, 1987.

70. Say B et al: Microcephaly, short stature, and developmental delay associated with a chemotactic defect and transient hypogammaglobulinaemia in two brothers. *J Med Genet* **23**:355–359, 1986.

71. Schendel SA, Delaire J: Familial osteodysplasia. *Head Neck Surg* **4**:335–343, 1982.

72. Sensenbrenner JA et al: New syndrome of skeletal, dental, and hair anomalies. *Birth Defects* **11**(2):372–379, 1975.

73. Shapiro SD et al: Tricho-dento-osseous syndromes: Heterogeneity or clinical variability. *Am J Med Genet* **16**:225–236, 1983.

74. Sheffield LJ et al: Auro-digital-anal syndrome. *Pathology* **13**:175–176, 1980.

74a. Shprintzen RJ, Goldberg RB: A recurrent pattern syndrome of craniosynostosis associated with arachnodactyly and abdominal hernias. *J Craniofac Genet Dev Biol* **2**:65–74, 1982.

75. Smithwick EM et al: Cranial synostosis in Job's syndrome. *Lancet* **1**:826, 1978.

76. Spear SI, Mickle JP: Simultaneous cutis aplasia congenita of the scalp and cranial stenosis. *Plast Reconstr Surg* **71**:413–417, 1983.

77. Stanesco V et al: Syndrome héréditaire dominant, reunissant une dysostose craniofaciale de type particulier, une insuffisance de croissance d'aspect chondrodystrophique et un epaississement massif de la corticale des os longs. *Rev Franc Endocrinol Clin* **4**:219–231, 1963.

78. Stevenson RE et al: Cutis gyratum and acanthosis nigricans associated with other anomalies: A distinctive syndrome. *J Pediatr* **99**:950–952, 1978.

78a. Stratton RF, Parsons DS: Confirmation of a new craniosynostosis syndrome. *Am J Med Genet* (in press).

79. Sugarman G, Vogel MW: Case report 76: Craniofacial and musculoskeletal abnormalities—a questionable connective tissue disease. *Syndrome Ident* **7**(1):16–17, 1981.

80. Thanos C et al: Craniosynostosis, bony exostoses, epibulbar dermoids, epidermal nevus and slow development. *Syndrome Ident* **5**(1):19–21, 1977.

81. Thong YH et al: Abnormal neutrophil chemotaxis in a syndrome of unusual facies, disproportionate small stature and sensorineural deafness-mutism. *Acta Paediatr Scand* **67**:383–387, 1979.

82. Ventruto V et al: Family study of inherited syndrome with multiple congenital deformities: Symphalangism, carpal and tarsal fusion, brachydactyly, craniosynostosis, strabismus, hip osteochondritis. *J Med Genet* **13**:394–398, 1976.

83. Young ID: Cranioectodermal dysplasia. *J Med Genet* **26**:393–396, 1989.

# Chapter 16
# Syndromes of Abnormal Craniofacial Contour

## Anencephaly

Anencephaly is characterized by an open neural tube in the cephalic region with an exposed mass of degenerate neural tissue at birth. The cranial vault is absent, producing characteristic bulging of the eyes and absence of the neck. Both the membranous neurocranium and the chondrocranium are grossly malformed (Figs. 16–1 and 16–2). Anencephaly is classified anatomically as meroacrania if the defect does not involve the foramen magnum, holoacrania if the defect extends through the foramen magnum, and holoacrania with rachischisis if spina bifida accompanies anencephaly (24).

The reader is referred to the following sources for extensive coverage: general aspects (22,24), classification (22,24,25), anatomy and morphology (5,7,14,25,27,28,29a), pathogenesis (24), epidemiology (1a,2,12,13,26,36,39,41–43,45,47,52), twin studies (11,35,38,46,53, 54), etiologic factors (4,8,10,19–21,34,37,44,48,50,54), associated anomalies (1b,3,6,29,51), prenatal diagnosis (15,16,31–33), and recurrence risk (17,18,20).

Most epidemiologic studies include both anencephaly and spina bifida (ASB) because both arise by nonclosure of the neural tube, the closing of the anterior and posterior neuropores being 24 and 26 days, respectively. Neural tube defects (NTD) in spontaneous abortions is about 10 times higher than NTDs found at term. Of nine specimens in the series of Byrne and Warburton (2), ASB accounted for four of the cases and chromosomal abnormalities were noted in two. Wide divergence in birth prevalence has been recorded. Average prevalence is approximately 1/1000 births in the eastern United States, Denmark, Kenya, Malaysia, India, Taiwan, and South Africa. High prevalence has been observed in Ireland and Wales (3.05–6.79/1000), the remaining British Isles (approximately 2–3/1000), Egypt, and Lebanon. Low prevalence (0.1–0.6/1000) has been noted in France, Norway, Hungary, Czechoslovakia, Japan, and Yugoslavia. Overall, an east–west gradient occurs with higher prevalence in the British Isles followed by eastern North America, then western North America, and finally the Far East (22). Some studies have shown decreasing rates with time (12,26,36,39,42,52), although the reasons for this are not known.

Wide ethnic differences also occur. U.S. whites have a rate of over 1/1000; native Americans, 0.32/1000; and blacks, 0.1–0.4/1000. Striking variation exists in Singapore with the Chinese having a rate of 0.62/1000, Eurasians 1.45/1000, Europeans 2.15/1000, and Sikhs 6.5/1000 (22).

Most studies show a higher frequency of anencephaly in females than in males, although the sex ratio has been close to 1:1 in some studies (22,24). The prevalence in twins is significantly higher than in singletons. The twin female-to-male sex ratio is lower than the singleton female-to-male sex ratio. Among twins, concordance for anencephaly is low (approximately 3.7%) (53,54). Anencephaly appears to occur with higher frequency among monozygotic than among dizygotic twins as same-sexed twins have a higher frequency than opposite-sexed twins. Furthermore, anencephaly seems to be relatively common in conjoined twins (11).

A consistently reported epidemiologic factor has been a higher prevalence of ASB among poorer social classes (22). An epidemiologic study that strongly suggests etiologic heterogeneity is that of Khoury et al (13). After excluding cases of known cause, they divided their sample into two groups—isolated and syndromic (multiple malformations)—and found marked differences in the epidemiology of the two groups. The isolated group showed marked predominance of females and whites, geographic variation with an east-to-west gradient, and decreasing rates over time. The group with multiple malformations showed no female predilection and occurred less often in whites, showed no geographical variation, and demonstrated little or no downward trend over time.

The etiology of most cases is unknown, although various factors are known to be responsible for some. The condition is clearly etiologically heterogeneous. Most instances occur sporadically, but a tendency toward familial aggregation of ASB is observed in some pedigrees. In other kindreds, instances of encephalocele or hydrocephaly may also be noted. Most cases appear to be multifactorially determined. Chromosomal, monogenic, and environmental causes have been reviewed by Holmes et al (8), Khoury et al (13), Toriello et al (50), and Farag et al (4). It has been suggested that nongenetic factors fall into two categories: immediate factors related to pregnancy and intergenerational factors related to a woman's growth and development (42).

Autosomal recessive and X-linked families have been recorded (4,50). The Meckel syndrome is most commonly observed with occipital encephalocele, but anencephaly has been noted on rare occasions (8). Holmes et al (8), Khoury et al (13), and Maldergem et al (26a) have reviewed known chromosomal conditions in which ASB have been reported. Known teratogens include aminopterin (anencephaly or encephalocele in some cases), thalidomide (rarely anencephaly and/or spina bifida), and diabetic pregnancies (8,37). Postulated environmental causes have included influenza A, hyperthermia, alcohol ingestion during pregnancy, water softness, maternal zinc deficiency, organic solvents, dietary factors, and vitamin deficiency (10,19,21,34,44,50). Intrauterine and maternal physiologic factors have also been considered as possible causes: previous abortion, subfertility, pituitary dysfunction, and twin–twin interaction (50).

Theories of pathogenesis have been reviewed by Lemire et al (24). The primary defect is either in the neuroepithelium itself or in the surrounding mesoderm. Some lesions may be caused by reopening of a closed neural tube secondary to either an increase in the intraluminal pressure or to some defect in the neuroepithelium. Most experimental evidence favors the concept that anencephaly arises from a neural tube that fails to close (22).

**Craniofacial features.** Variation in central nervous system pathology is related to the extent of the neural tube defect as well as the gestational age. Generally, the more complete the absence of the cranial vault, the less likely nervous tissue is to be found on the cranial base. The cerebral hemispheres are soft, red-to-purple in color, and formless, being composed of thin-walled vascular channels distended with blood and separated by irregular masses of brain tissue. A thin covering of squamous epithelium and choroid plexus-like tissue may be present in some cases. The cerebellum may be normal or malformed (24). In some instances of meroacrania, brain tissue may bulge out, resembling an encephalocele. As a result, meroacrania has been misdiagnosed on occasion in the literature. Absence of skin covering separates meroacrania from encephalocele (22). Anencephaly and holo-

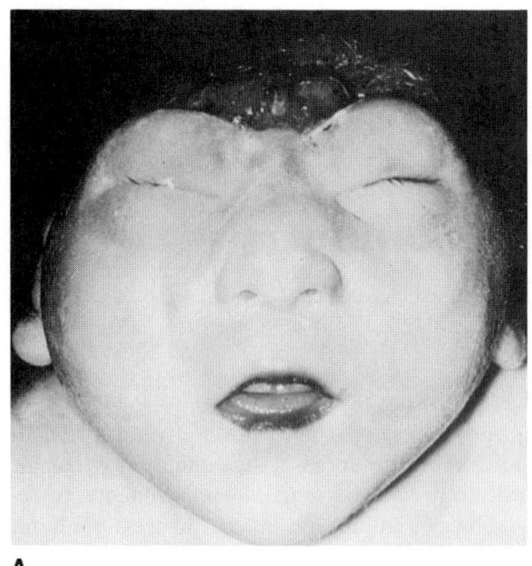

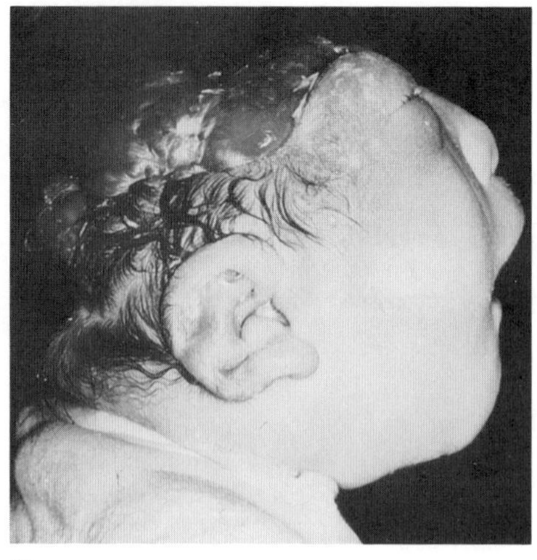

A B

Fig. 16–1. (A,B) *Anencephaly.*

prosencephaly may occur together (25a). Diprosopus and facial clefting may also be associated (7a).

The pituitary gland is almost always present, despite the assertion in the literature that it may be absent. The neurohypophysis is almost always hypoplastic, distorted, and frequently found in the craniopharyngeal canal, which usually remains open in anencephaly. The hypopothalamus is absent as are the neural connections between the adenohypophysis and the central nervous system (24).

Craniofacial malformations may include hypoplastic and/or low-set ears, anomalous ear folds, absent helices, exotropia, ocular proptosis, small palpebral fissures, microphthalmia, clinical anophthalmia, subcutaneous nasal cleft, absent or incompletely defined philtrum, anomalous palate with midline grooving, cleft lip, cleft palate, bifid uvula, microstomia, prominent maxillary alveolar ridges, and mandibular prognathism (24). Short neck is common. The thyroid gland may be normal, small, or increased in size (24).

The craniofacial skeleton has been studied by a number of authors (5,7,14,27,28,29a). In the facial skeleton, the frontal bone is severely affected with a marked increase in angulation. The superior orbital rims are directed posterolaterally. The orbital plate of the frontal bone usually cannot be identified. The zygomatic bones are severely affected and, in lateral view, have a rhomboid shape. Cephalometric study shows the mandible to be prognathic in every case (29a).

In large cranial defects, the interparietal portion of the occipital bone is relocated anteriorly to approximate the frontal bone. Occipital components are rotated anterolaterally and inferiorly with lack of fusion of the chondrocranium posterior to the foramen magnum. The squamae of the frontal bone are collapsed horizontally and reduced in size, lying peripheral to the anterior cranial fossa to form most of the orbital roofs (7).

Changes in the cranial floor are caused by alterations in the size, form, or duration of the neural functional matrix. Sphenoid bone alterations result in a reduced cranial floor angle and a more vertically oriented clivus. Schematic drawings of the cranial fossae in meroacrania and holoacrania are compared with normal in Figure 16–2. In both types, the anterior and middle cranial fossae are constricted laterally and the posterior cranial fossa is increased in lateral extension, but constricted anteroposteriorly. The lesser wings of the sphenoid have an anteroposterior orientation; the greater wings of the sphenoid have a reduced transverse dimension; and the lateral end of the petrous portion of the temporal bone is positioned more anteriorly (5).

Fig. 16–2. *Anencephaly.* Schematic drawing of the bones of the cranial floor in (a) a normal fetus, (b) a fetus with meroacrania, and (c) a fetus with holoacrania. Note the anteroposterior position of the lesser wings of the sphenoid and the reduced transverse dimension of the greater wings of the sphenoid (b and c) relative to the normal (a). The more anterior position of the lateral end of the petrous portion of the temporal bone is also ev1lent in anencephaly. In holoacrania, the supraoccipital components are nonunited and widely divergent (c). ACF, MCF, PCF, anterior, middle, and posterior cranial fossa. (From *HW Fields et al,* Teratology **17**:57, 1978.)

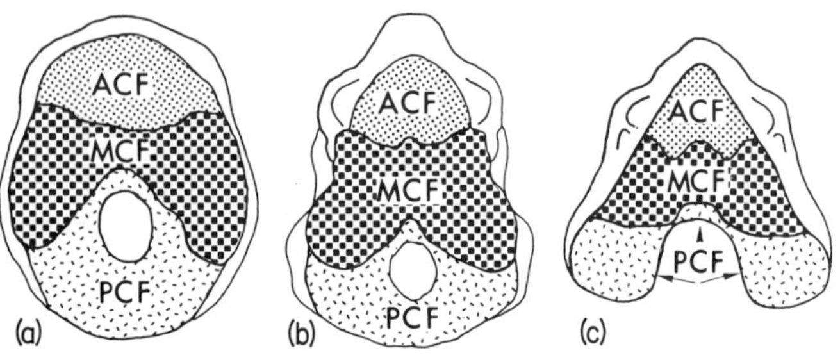


**Associated anomalies.** David and Nixon (3) in a study of 294 cases of anencephaly found that the most frequent single malformations were hydronephrosis (8%), cleft palate (8%), diaphragmatic hernia (5%), omphalocele (5%), cleft lip (4%), and horseshoe kidney (4%). Lemire et al (24) studied associated malformations in 68 cases and Melnick and Myrianthopoulos (29) reported a series of 36 cases. Especially significant are spinal retroflexion, scoliosis, talipes equinovarus or calcaneovalgus, increased anteroposterior chest diameter, high diaphragm, eventration of the diaphragm, absent diaphragm, patent foramen ovale, hypoplastic heart, various other cardiac defects, anomalous lobation of lungs, hypoplastic lungs, hydroureter, hydronephrosis, hypertrophy of the bladder wall, hypoplastic ureters and/or bladder, hyperlobulated kidneys, absent ureters or bladder, undescended cecum, omphalocele, short intestines, and Meckel diverticulum (1,3,24,26a,29). The adrenals are characteristically small. The frequency of the sternalis muscle may be as high as 50% in some series. In contrast, the frequency varies between 1 and 4% in the general population. The upper limbs are excessively long in relation to the lower limbs. On the average, arm length is about 12% greater than normal and a proximal-to-distal gradient exists, the upper arm being increased by 24%, the forearm by 16%, and the hand by 2% (24).

**Differential diagnosis.** Some cases of *amnion rupture sequence* may resemble anencephaly. Since the former is associated with limb amputations, ring constrictions, distal syndactyly, and tissue bands, in practice little difficulty should be experienced in differentiating the two. Some authors discuss iniencephaly and anencephaly together (3,25). Both share a female preponderance; iniencephaly appears to be more common in areas where anencephaly is common, and both may be accompanied by spina bifida. However, the time of onset for each of these malformations is different and families have not been observed with affected members of both types. Anencephaly may occur together with *holoprosencephaly* (25a) and all such cases to date have been sporadic. Townes et al (50a) reported a sibship with anencephalic male twins and a female infant with alobar holoprosencephaly, radial aplasia, and oligodactyly.

Wynne-Davies (55) carried out a family survey of 337 patients with multiple vertebral anomalies without apparent spina bifida. She found that vertebral anomalies were etiologically related to anencephaly and spina bifida, carrying a 5–10% risk to subsequent sibs. On the other hand, some families with spina bifida and/or other vertebral anomalies are known to occur without anencephaly (9,40,49), suggesting further heterogeneity.

**Diagnosis and laboratory aids.** Diagnosis is based on the overall pattern of anomalies and family history. An attempt should be made to rule out known syndromes and environmental causes. Certainly chromosome study is indicated. Radiographic studies of the vertebral column in relatives of the proband are also indicated. The presence of an NTD in a first degree relative, second degree relative, or third degree relative of a fetus is an indication for amniotic fluid $\alpha$-fetoprotein determination, although there is little evidence of an increased risk unless the affected relative is a parent or sib. Approximately 60% of newborns with NTDs have anencephaly. The risks for a neural tube defect in relatives of affected persons are as follows: father's siblings, 7%; mother's sister's children 13%; mother's brother's children 3%; father's sister's children 6%; and father's brother's children 4%. Maternal serum $\alpha$-fetoprotein determination can be used as a screening device since 95% of affected infants occur from pregnancies not recognized to be at increased risk. The success of maternal serum $\alpha$-fetoprotein screening requires a well-organized program with strict adherence to protocol. In pregnancies with elevated amnionic $\alpha$-fetoprotein level and a nondiagnostic ultrasound, acetylcholinesterase assay of amnionic fluid assumes major importance. In a pregnant woman with a previously affected child or an elevated maternal serum or amnionic fluid $\alpha$-fetoprotein level, ultrasonography can be made longitudinally and transversely to specifically identify the NTD (15,16,17,30–33).

### References (Anencephaly)


1. Ardinger HH et al: Association of neural tube defects with omphalocele in chromosomally normal fetuses. *Am J Med Genet* **27**:135–142, 1987.

1a. Baird PA: Neural tube defects in the Sikhs. *Am J Med Genet* **16**:49–56, 1983.

1b. Baird PA: Neural tube defects and tracheo-oesophageal dysraphism. *Lancet* **2**:615, 1982.

2. Byrne J, Warburton D: Neural tube defects in spontaneous abortions. *Am J Med Genet* **25**:327–333, 1986.

3. David TJ, Nixon A: Congenital malformations associated with anencephaly and iniencephaly. *J Med Genet* **13**:263–265, 1976.

4. Farag TI et al: Brief clinical report: Nonsyndromal anencephaly: Possible autosomal recessive variant. *Am J Med Genet* **24**:461–464, 1986.

5. Fields HW et al: The craniofacial skeleton in anencephalic human fetuses. I. Cranial floor. *Teratology* **17**:57–65, 1978.

6. Fraser FC et al: Increased frequency of neural tube defects in sibs of children with other malformations. *Lancet* **2**:144–145, 1982.

7. Garol JD et al: The craniofacial skeleton in anencephalic human fetuses. II. Calvarium. *Teratology* **17**:67–82, 1978.

7a. Gorlin RJ: Disprosopus, anencephaly, and facial clefting. *Am J Med Genet* **30**:845, 1988.

8. Holmes LB et al: Etiologic heterogeneity of neural-tube defects. *N Engl J Med* **294**:365–369, 1976.

9. James HE, Walsh JW: Spinal dysraphism, in *Current Problems in Pediatrics*, L Gluck (ed), Vol 11, Year Book Medical Publishers, Chicago, 1981, pp 1–25.

10. James N et al: Diet as a factor in aetiology of neural tube defects. *Z Kinderchir* **31**:302–308, 1980.

11. James WH: Differences between the events preceding spina bifida and anencephaly. *J Med Genet* **18**:17–21, 1981.

12. Janerich DT, Piper J: Shifting genetic patterns in anencephaly and spina bifida. *J Med Genet* **15**:101–105, 1978.

13. Khoury MJ et al: Etiologic heterogeneity of neural tube defects: Clues from epidemiology. *Am J Epidemiol* **115**:538–548, 1982.

14. Kokich VG et al: Craniofacial development in anencephaly: A morphological and histological evaluation, in *Advances in the Study of Birth Defects*, Persaud, TUN (ed), MTP Press Ltd., England, *CNS and Craniofacial Malformations*, Ch. 3, 1981, pp 47–74.

15. Laurence KM: Genetics and prevention of neural tube defects, in *Principles and Practice of Medical Genetics*, AEH Emery and DL Rimoin, eds, Churchill Livingstone, Edinburgh, 1983, Vol 1, pp 231–249.

16. Laurence KM: Prevention of neural tube malformation by genetic counselling, and prenatal diagnostic surveillance. *J Génét Hum* **27**:289–299, 1979.

17. Laurence KM: Recurrence risk in spina bifida cystica and anencephaly, in *Developmental Medicine and Child Neurology*, Vol II, Suppl No 20, Studies in Hydrocephalus and Spina Bifida, 1968, pp 23–30.

18. Laurence KM, Morris J: The effect of the introduction of prenatal diagnosis on the reproductive history of women at increased risk from neural tube defects. *Prenatal Diag* **1**:51–60, 1981.

19. Laurence KM et al: Double-blind randomised controlled trial of folate treatment before conception to prevent recurrence of neural-tube defects. *Br Med J* **282**:1509–1511, 1981.

20. Laurence KM et al: Increased risk of recurrence of pregnancies complicated by fetal neural tube defects in mothers receiving poor diets, and possible benefit of dietary counselling. *Br Med J* **281**:1–8, 1980.

21. Layde PM et al: Maternal fever and neural tube defects. *Teratology* **21**:105–108, 1980.

22. Lemire RJ: Anencephaly, in *Handbook of Clinical Neurology: Malformations*, Vinken PJ, Bruyn GW, Klawans HL (eds), Vol 6, Elsevier Science Publishers, New York, Ch. 4, 1987, pp 71–95.

23. Lemire RJ: Neural tube defects. *JAMA* **259**:558–562, 1988.

24. Lemire RJ et al: *Anencephaly*. Raven Press, New York, 1978.

25. Lemire RJ et al: Iniencephaly and anencephaly with spina retroflexion. A comparative study of eight human specimens. *Teratology* **6**:27–36, 1971.

25a. Lemire RJ et al: The facial features of holoprosencephaly in anencephalic human specimens. I. Historical review and associated malformations. *Teratology* **23**:297–303, 1981.

26. Lorber J, Ward AM: Spina bifida—A vanishing nightmare? *Arch Dis Childh* **60**:1086–1091, 1985.

26a. Maldergem, L van et al: Neural tube defects and omphalocele in trisomy 18. *Clin Genet* **35**:77–79, 1989.

27. Marin-Padilla M: Morphogenesis of anencephaly and related malformations. *Curr Top Pathol* **51**:145–174, 1970.

28. Marin-Padilla M: Study of the skull in human cranioschisis. *Acta Anat* **62**:1–20, 1965.

29. Melnick M, Myrianthopoulos NC: Studies in neural tube defects. II. Pathologic findings in a prospectively collected series of anencephalics. *Am J Med Genet* **26**:797–810, 1987.

29a. Metzner L et al: The craniofacial skeleton in anencephalic human fetuses. III. Facial skeleton. *Teratology* **17**:75–82, 1978.

30. Milunsky A: Prenatal detection of neural tube defects. VI. Experience with 20,000 pregnancies. *JAMA* **244**:2731–2735, 1980.

31. Milunsky A, Neff RK: Prenatal diagnosis of neural tube defects. IV. Maternal serum alpha-fetoprotein screening. *Obstet Gynecol* **55**:60–66, 1980.

32. Milunsky A, Sapirstein VS: Prenatal diagnosis of open neural tube defects using the amniotic fluid acetylcholinesterase assay. *Obstet Gynecol* **59**:1–5, 1982.

33. Milunsky A et al: Prenatal diagnosis of neural tube defects. VIII. The importance of serum alpha-fetoprotein screening in diabetic pregnant women. *Am J Obstet Gynecol* **142**:1030–1032, 1982.

34. Oakley GP et al: Vitamins and neural tube defects. *Lancet* **2**:798–799, 1983.

35. Putz B, Rehder H: Anencephaly in one monoamniotic-monochorionic twin and encephalocele in the other. *Am J Med Genet* **21**:631–635, 1985.

36. Romijn JA, Treffers PE: Anencephaly in the Netherlands: A remarkable decline. *Lancet* **1**:64–65, 1983.

37. Rosengarten AM, Martin TM: Fetal anencephaly in a pregnant diabetic. *BC Med J* **24**:406–407, 1982.

38. Rowlatt U: Anencephaly and unilateral cleft lip and palate in conjoined twins. *Cleft Palate J* **15**:73–75, 1978.

39. Seller MJ, Hancock PC: Is recurrence rate of neural tube defects declining? *Lancet* **1**:175, 1985.

40. Sever LE: Case report: A case of meningomyelocele in a kindred with multiple cases of spondylolisthesis and spina bifida occulta. *J Med Genet* **11**:94–96, 1974.

41. Sever LE: An epidemiologic study of neural tube defects in Los Angeles

County. II. Etiologic factors in an area with low prevalence at birth. *Teratology* **25**:323–334, 1982.

42. Sever LE, Strassburg MA: Epidemiologic aspects of neural tube defects in the United States: Changing concepts and their importance for screening and prenatal diagnostic programs. *New York State Department of Health Birth Defects Institute Birth Defects Symposium XIV* "Alpha Fetoprotein and Congenital Disorders," Albany, New York, October 3–4, 1983.

43. Sever LE et al: An epidemiologic study of neural tube defects in Los Angeles County I. Prevalence at birth based on multiple sources of case ascertainment. *Teratology* **25**:315–321, 1982.

44. Smithells RW et al: Apparent prevention of neural tube defects by periconceptional vitamin supplementation. *Arch Dis Childh* **56**:911–918, 1981.

45. Strassburg MA et al: A population-based case-control study of anencephalus and spina bifida in a low-risk area. *Dev Med Child Neurol* **25**:632–641, 1983.

46. Symonds EM: Anencephaly and hydrocephaly in a twin pregnancy: Report of a case. *Obstet Gynecol* **26**:414–416, 1965.

47. Timson J: A study of the first degree relatives of the parents of spina bifida children. *Clin Genet* **3**:99–102, 1972.

48. Toriello HV, Higgins JV: Occurrence of neural tube defects among first-, second-, and third degree relatives of probands: Results of a United States study. *Am J Med Genet* **15**:601–606, 1983.

49. Toriello HV, Higgins JV: Possible causal heterogeneity in spina bifida cystica. *Am J Med Genet* **21**:13–20, 1985.

50. Toriello HV et al: Possible X-linked anencephaly and spina bifida—Report of a kindred. *Am J Med Genet* **6**:119–121, 1980.

50a. Townes PL et al: XK aprosencephaly and anencephaly in sibs. *Am J Med Genet* **29**:523–528, 1988.

51. Windham GC, Bjerkedal T: Malformation frequencies in sibs of atresia cases. *Lancet* **2**:816–817, 1982.

52. Windham GC, Edmonds LD: Current trends in the incidence of neural tube defects. *Pediatrics* **70**:333–337, 1982.

53. Windham GC, Sever LE: Neural tube defects among twin births. *Am J Hum Genet* **34**:988–998, 1982.

54. Windham GC et al: The association of twinning and neural tube defects: Studies in Los Angeles, California, and Norway. *Acta Genet Med Gemellol* **31**:165–172, 1982.

55. Wynne-Davies R: Congenital vertebral anomalies: Aetiology and relationship to spina bifida cystica. *J Med Genet* **12**:280–288, 1975.

## Encephaloceles

Encephaloceles may occur as isolated malformations or together with other anomalies making up various syndromes or associations. The literature has been reviewed extensively elsewhere (21,99). Although encephalocele is the most commonly used term for all such lesions, cranium bifidum and cephalocele more properly describe the spectrum of malformations that includes both encephalocele and cranial meningocele (95).

Encephaloceles (Fig. 16–3 to 16–5) can be classified as occipital, parietal, or anterior (Table 16–1). Further subdivision of anterior encephaloceles can be made: visible (sincipital) and not directly visible (basal) (2,88). About 75% of all encephaloceles occur in the occipital region (6,55). Prevalence is said to vary from 1 in 2000 to 1 in 5000 live births. They are more common in African and Oriental populations (25,29). Parietal encephaloceles account for approximately 10–14% of all encephaloceles (6,55,73). Frontal encephaloceles are much less common than occipital encephaloceles in the United States, the ratio being 1:6 (29). African studies, on the other hand, show a 1:1 ratio. They are very common in Thailand, Burma, and India (15), prevalence being about 1/3500 live births (37). The frequency of basal encephalocele appears to be comparatively low, an undocumented prevalence being 1/35,000 live births (93).

At present, it is not possible to know how frequently other anomalies occur with encephaloceles because surveys in the literature have focused primarily on signs and symptoms, natural history, treatment, and treatment outcome (6,14,23,55,68,73,74). Although associated anomalies are mentioned and even tabulated, such data cannot be used for several reasons. First, associated anomalies are not always properly broken down. In some series, anomalies that occur with encephalocele and with spina bifida are lumped together. In other surveys that deal only with encephaloceles, associated anomalies cannot al-

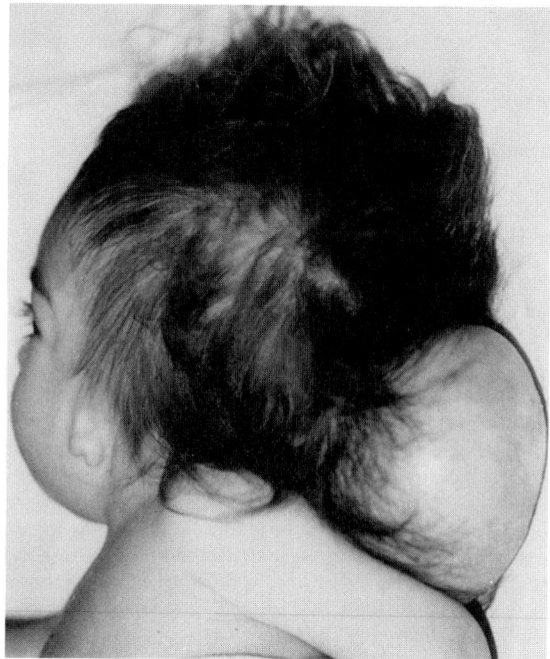

Fig. 16–3. *Encephalocele.* Occipital encephalocele with hemifacial microsomia. [From *MM Cohen Jr,* Birth Defects **7**(7):103, 1971.]

Table 16–1. Classification of sincipital and basal encephaloceles[a]

| Location | Boundaries (bones) | Presentation |
|---|---|---|
| Sincipital | | |
| Frontoethmoidal | | |
| Nasofrontal | Cribriform plate of ethmoid<br>Nasal process of frontal<br>Frontal process of maxillary | Midline at root of nose |
| Nasoethmoidal | Ethmoid<br>Frontal<br>Nasal | On one or both sides of nose |
| Nasoorbital | Lacrimal<br>Frontal | Superior medial angle of orbital cavity; displaces eye laterally; may be bilateral |
| Interfrontal | | |
| Craniofacial cleft | | |
| Basal | | |
| Sphenoorbital | Superior orbital fissure | Enters orbit, causing exophthalmos |
| Sphenomaxillary | Superior orbital, then inferior orbital fissure | Pterygopalatine fossa |
| Transethmoidal | Defect in cribriform plate | Anterior nasal fossae |
| Sphenoethmoidal | Sphenoethmoidal suture | Posterior nasal fossae |
| Sphenopharyngeal | Defect in sphenoid | Rhinopharynx, sphenoid sinus |

[a]Adapted from G Avanzini and G Crivelli, *Acta Neurochir (Wien)* **22**:205, 1970; C Suwanwela and N Suwanwela, *J Neurosurg* **36**:201, 1972; J Warkany et al, *Mental Retardation and Congenital Malformations of the Central Nervous System,* Year Book Publishers, Chicago, 1981.

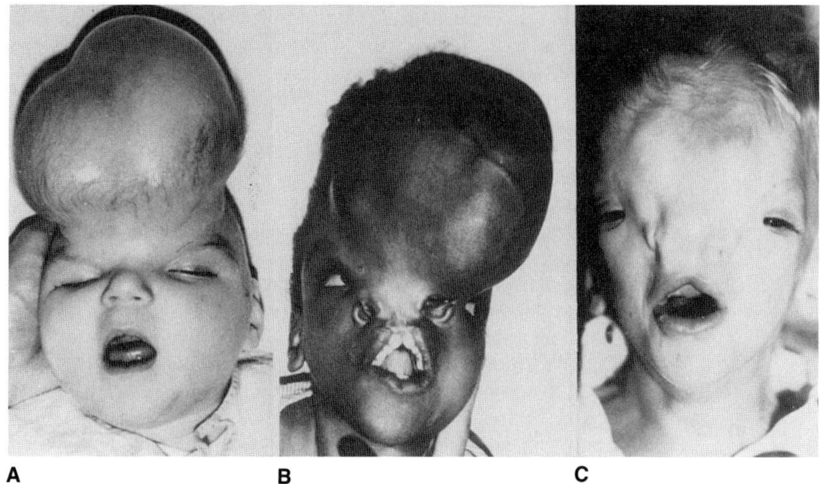


Fig. 16–4. *Encephalocele.* (A–C) From left to right: frontal encephalocele, frontonasal encephalocele, and frontonasal malformation with flattened encephalocele. (From *MM Cohen Jr* and *RJ Lemire,* Teratology **25**:161, 1982.)

ways be sorted on the basis of the anatomic location of the encephalocele. Second, associated anomaly categories may be too inclusive. Primary and secondary anomalies are not always separated. Thus, the total number of associated anomalies is confounded in patients with only one anomaly in the total number, and there is no way to break the anomalies down on a per patient basis. Finally, there are a number of problems with ascertainment bias in some surveys and problems with sample size in others (21).

Conditions with encephaloceles together with appropriate references and, when relevant, pages in the text are listed in Table 16–2. They are discussed under the following headings: chromosomal syndromes, monogenic syndromes, teratogenic causes, disruptive causes, conditions of unknown genesis, various associations, and miscellaneous conditions occasionally associated with encephaloceles.

**Chromosomal syndromes.** Encephalocele has been reported with several chromosomal syndromes, including trisomy 18 on occasion (4,95). It has also been noted with Turner syndrome (11), but this finding can be questioned. With some Turner syndrome abortuses, severe lymphedema of the posterior neck area can mimic cephaloceles and care should be taken not to confuse the two. An encephalocele was said to be present in one patient with dup(lq) karyotype (40), but a photograph of the affected infant shows apparent anencephaly of the meroacrania type. Moeschler et al (78) reported a patient with del(13q) syndrome with a large occipital encephalocele, severe microcephaly, and Type 1 Arnold–Chiari malformation. Schinzel (84) cited a case with meningocele.

Fig. 16–5. *Encephalocele.* Transsphenoidal encephalocele with herniation into the posterior nasopharynx. (From *JM Lieblich et al,* Ann Intern Med **89**:910, 1978.)

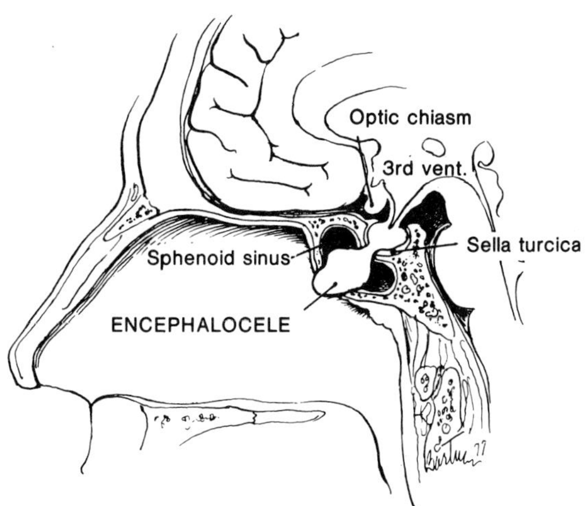


Hsia et al (54) reported a sporadic instance of a syndrome with many clinical similiarities to the Meckel syndrome but with some differences. The condition was thought to be caused by a chromosomal anomaly. Features of the pseudo-Meckel syndrome included occipital encephalocele, absent olfactory tracts and bulbs, absent corpus callosum, Arnold–Chiari malformation, cleft palate, wide aorta originating from the right ventricle, atretic pulmonary valve, narrow pulmonary artery arising from a blind lumen to the left of the aortic origin, patent ductus arteriosus, ventricular septal defect, single coronary artery, accessory spleen, talipes equinovarus, and hallucal hammertoes. No polydactyly was observed and retinal dysplasia, a consistent feature of the Meckel syndrome (70), was conspicuously absent. Lymphocyte study of 15 metaphase spreads showed a karyotype of 46,XX,t(3p+;?). The proximal part of the elongated short arm of chromosome 3 showed the usual banding pattern with quinacrine fluorescence. The origin of the translocated distal portion could not be determined with certainty, although it was thought to have arisen from the long arm of either chromosome 10 or chromosome 6. Karyotype was normal in both parents and in a normal brother of the proband.

**Monogenic syndromes.** The Meckel syndrome is the best known malformation syndrome with occipital encephalocele and has been reviewed extensively elsewhere (45,53,70,75,76,80). Features include polydactyly, polycystic kidneys, holoprosencephaly, microphthalmia, retinal dysplasia, various cardiac anomalies, orofacial clefting, ambiguous external genitalia, and various other abnormalities. Occipital encephalocele occurred in 39 of 49 cases tabulated from the literature (75). The syndrome has autosomal recessive inheritance.

The cryptophthalmos syndrome consists of extension of the skin of the forehead covering one or both eyes, unusual hairline, ear anomalies, total or partial soft tissue syndactyly of the fingers and toes, notching of the wings of the nose, and genital anomalies such as micropenis, hypospadias, clitoromegaly, vaginal atresia, and incomplete development of the labia. The syndrome has autosomal recessive inheritance. Approximately 10% of reported cases have been associated with encephalocele. In some instances, osseous defects of the calvaria occur without encephalocele (90).

Dyssegmental dysplasia is etiologically heterogeneous with an autosomal recessive severe form (type Silverman–Handmaker) and an autosomal recessive milder form (type Rolland–Desbuquois). Both types are usually lethal. Facial features in both consist of mild blepharophimosis, flat nasal bridge, hypoplastic supraorbital ridges, micrognathia, and cleft palate. Occipital encephalocele, occipital bone defect, and hydrocephaly have been noted in some instances. The neck is short. Radiographic findings include short, broad tubular bones with metaphyseal widening, accelerated carpal bone maturation, bowing of the

Table 16–2. Conditions with encephaloceles

| Condition | Reference (page in text) |
|---|---|
| Chromosomal | |
| Trisomy 18 syndrome | 4,95 (43) |
| del(13q) syndrome | 78,84 (51) |
| Pseudo-Meckel syndrome | 54 (577) |
| Monogenic | |
| Meckel syndrome | 45,53,70,75,76,80 (724) |
| Cryptophthalmos syndrome | 90 (816) |
| Dyssegmental dysplasia | |
| Type Silverman–Handmaker | 1,46,48 (198) |
| Type Rolland–Desbuquois | 1,28 (200) |
| Knobloch syndrome | 63 |
| Roberts syndrome | 39 (735) |
| Walker–Warburg syndrome | 3,10,12,31,34,82,103 (592) |
| Teratogenic | |
| Fetal hyperthermia | 35 (31) |
| Warfarin embryopathy | 83,89,96,97 (23) |
| Disruptive | |
| Amnion rupture sequence | 47,50,56,61,69,91 (11) |
| Unknown genesis | |
| Frontonasal malformation | 20,22,27,85,98 (785) |
| von Voss syndrome | 13,58,94 |
| Associations | |
| Encephalocele/absent corpus callosum | 2,33,66,73,79 |
| Encephalocele/clefting | 23,44 (571) |
| Encephalocele/craniostenosis | 6,24,43,68,80,101, JL Loeser, personal communication (571) |
| Encephalocele/Dandy–Walker; Arnold–Chiari | 6,73,81 |
| Encephalocele/hypothalamic–pituitary dysfunction | 66 |
| Encephalocele/ectrodactyly | JM Opitz, personal communication |
| Encephalocele/iniencephaly; Klippel–Feil | 7,23,52,65,68,74,86 (571, 886) |
| Encephalocele/meningomyelocele | 68,102 |
| Encephalocele/oculo-auriculo-vertebral spectrum | 17,18,49 (571, 641) |
| Miscellaneous and/or occasional | |
| Congenital heart defects | 23,36,55,68,74, BD Hall, personal communication |
| Ehlers–Danlos syndrome | 30 (429) |
| Infants of diabetic mothers | 32 |
| Marfan syndrome | 72 |
| Neurofibromatosis | 51, JW Hanson, personal communication (392) |
| Poland anomaly | M Bull, personal communication |
| Polysplenia syndrome | KL Jones, personal communication |
| Rubella syndrome | 8,60 |
| TAR syndrome | RJ Gorlin, personal observation |
| Tuberous sclerosis | 9,57 (410) |
| With eyelid ptosis and single, maxillary central incisor | MJ Stephan, personal communication |

legs as well as the thighs and forearms, short broad pelvis with widely flared iliac wings, vertebral anomalies, and small thorax (1,28,46,48).

In 1971, Knobloch and Layer (63) reported a syndrome in 5 of 10 siblings. Features included high myopia, vitreoretinal degeneration, retinal detachment, and occipital encephalocele. It is known that all affected family members had normal intelligence (see 82). Therefore, the diagnosis of encephalocele can be questioned. Occipital meningocele seems more likely. Autosomal recessive inheritance is presumed. The father and five of his siblings were normal. However, the finding of encephalocele in a sixth sibling of the father may be more than coincidence.

Roberts syndrome is characterized by growth deficiency, severe hypomelia, cleft lip–palate, prominent eyes with hypertelorism, large phallus, and other anomalies. Inheritance is autosomal recessive, and heterochromatic puffing and centromere separation have been observed cytogenetically. Encephalocele was a feature of one case reported by Freeman et al (39). However, care should be taken not to confuse nuchal cystic hygroma in the fetus, also observed in some cases, with encephalocele.

Walker–Warburg syndrome, also known as Warburg syndrome, HARD±E syndrome, and Chemke syndrome, consists of hydrocephaly, agyria, occipital encephalocele in about 50%, microphthalmia, and/or anterior and posterior chamber abnormalities including corneal opacity, cataract, persistent primary vitreous, and retinal dysplasia. Genitourinary abnormalities are present in some cases. Inheritance is autosomal recessive (3,10,12,21,31,34,82,103). An early demise is characteristic, and recognition of the syndrome is essential for proper genetic counseling of families at risk. Although agyria is a feature of the syndrome, the condition should not be confused with Miller–Dieker lissencephaly syndrome (87,92). In the latter disorder, the cerebral cortex is thickened. In Walker–Warburg syndrome, the cortex is atrophic and disorganized. Walker–Warburg syndrome and Meckel syndrome share in common both occipital encephalocele and retinal dysplasia. However, the neuropathologic malformations are entirely different, and polydactyly and polycystic kidneys, which are common with Meckel syndrome, are not features of Walker–Warburg syndrome.

**Teratogenic causes.** Maternal hyperthermia may be implicated as one cause of occipital encephalocele. Of 17 patients with isolated posterior encephalocele, Fisher and Smith (35) found that four of the mothers gave a history of at least 1.5°C above normal from prolonged fever early during pregnancy. Since Fraser and Skelton (38) implicated maternal hyperthermia as a teratogen that can produce abnormalities from 3 to 14 weeks of gestation, it seems possible that at least some instances of multiple anomalies with encephalocele arising around the same embryonic time may occur on this basis.

Warfarin treatment during pregnancy has been associated with a spectrum of malformations, most commonly nasal hypoplasia and bone stippling. The condition has been reviewed elsewhere (83,97). Various other abnormalities have been observed including limb shortening, low birthweight, optic atrophy, mental deficiency, hypotonia, seizures, and various other defects. Occipital encephalocele has been observed in two instances. Tejani (89) described a patient with nasal hypoplasia, occipital encephalocele, hydrocephaly, microphthalmia, and persistent truncus arteriosus. Warkany and Bofinger (97) reported two affected children with hydrocephaly. One had a huge occipital encephalocele and renal malformations.

**Disruptive causes.** Features of the amnionic rupture sequence have been reviewed elsewhere (47,50,56,61,69,91). Most instances involve digits or limbs with ring constrictions or amputations. In some cases, an affected newborn will have unusual orofacial clefts, encephaloceles, and other facial anomalies. Amnionic bands become entangled in the oral and nasal orifices and sometimes become attached directly to the face or cranium, producing bizarre disruptions. These encephaloceles are unusual because of their tendency to have (1) irregular surfaces, (2) asymmetric placement with respect to the midsagittal plane, (3) predominantly anterior placement, and (4) multiple sites of involvement. Other anomalies may also occur. Besides disruptions, deformations and malformations have been noted in some instances. Some reported malformations are not secondary to amnionic bands or oligohydramnios. Although the overwhelming majority of cases are sporadic, occasional familial instances have been recorded.

**Unknown genesis.** Frontonasal malformation (Fig. 16–4C) or median cleft face syndrome has been reviewed elsewhere (20,22,27,85).

The condition is characterized by a flattened nasofrontal encephalocele, hypertelorism, widow's peak, and wide-set nostrils with lack of elevation of the nasal tip. Occasionally encountered are notching or clefting of the nostrils or an unusual form of median cleft lip. A variety of extracephalic anomalies may be present in some cases. The etiology of frontonasal dysplasia is unknown and virtually all reported cases are sporadic except for the two severely affected half-sisters reported by Warkany et al (98).

von Voss et al (94), Cherstvoy et al (13), and Kahler (58) reported a recurrent-pattern syndrome in three sporadically occurring patients. All had occipital encephalocele, aplasia or hypoplasia of the corpus callosum, phocomelia, and urogenital anomalies. Thrombocytopenia was present in two. Some features were variable, including various central nervous system anomalies. Hypoplastic right lung and dextroposition of the heart were present in Kahler's patient.

**Encephalocele/absent corpus callosum association.** The association of encephalocele and absent corpus callosum has been observed repeatedly both as a binary combination and together with various other anomalies. McLaurin (73) noted absence of the corpus callosum in four of 13 parietal encephaloceles. Lieblich et al (66) and Avanzini and Crivelli (2) reported agenesis of the corpus callosum with transsphenoidal encephalocele. A posterior orbital encephalocele was observed with agenesis of the corpus callosum and cerebellar atrophy in a patient reported by Fargueta et al (33). Other cases of encephalocele and absent corpus callosum have been described by Mood (79). The combination also has been reported in frontonasal malformation (98) and in von Voss syndrome (13).

Absent corpus callosum is part of the spectrum of holoprosencephalic disorders in some instances but not in others. The latter is assumed to be the case in the association discussed. The relationship of absent corpus callosum to holoprosencephaly has been discussed extensively by Marburg (71) and Loeser and Alvord (67). Encephalocele and agenesis of the corpus callosum as part of holoprosencephaly may occur in the Meckel syndrome (53,80) and has been observed with cyclopia in a patient with r(18) karyotype (16). Encephalocele and holoprosencephaly can be produced by the same teratogen in animals. DeMyer (26) was able to induce both malformations in rat fetuses with vincristine injected intramuscularly into pregnant rats 8.5 days following mating. The sporadic occurrence of cyclopia and encephalocele has been reported in the goat (59).

**Encephalocele/clefting association.** The binary combination of encephalocele and orofacial clefting has been reported on a number of occasions and the literature has been reviewed elsewhere (44). In some cases, a sphenopharyngeal encephalocele is associated with cleft palate. In these instances, it is possible to speculate that early encephalocele formation mechanically interfered with closure of the secondary palate. More commonly, sphenopharyngeal encephalocele is associated wtih cleft lip with or without cleft palate. The association of occipital encephalocele and cleft lip–palate has been noted (23) and the combination of encephalocele and oblique facial clefting has also been observed. Unusual orofacial clefts and encephalocele can be part of the amnionic rupture sequence. Unusual median clefts are observed occasionally with frontonasal dysplasia.

**Encephalocele/craniostenosis association.** Craniostenosis is found occasionally as a low-frequency anomaly with encephalocele. Lorber (68) noted two instances of premature sutural fusion in 20 cases of encephalocele, and Barrow and Simpson (6) found one instance among 14 cases. Craniostenosis has been observed occasionally with the Meckel syndrome (80) in which occipital encephalocele is a common feature. Encephalocele has also been noted in acrocephaly (43), craniotelencephalic dysplasia (24), and Apert syndrome (101, J Loeser, personal communication).

Three possible theories can be invoked to explain the association of encephalocele with craniostenosis, and each of the three probably has some validity, depending upon the particular case. In some instances, "blow-out" lesions such as encephaloceles may result in lack of growth stretch across the sutures producing craniosynostosis on this basis. When hydrocephaly is associated with encephalocele, surgically placed shunts with low-pressure systems may be implicated in which growth stretch at the sutural areas suddenly becomes totally deficient. Finally, in some patients with multiple anomaly syndromes, both encephalocele and craniostenosis may occur on a primary basis (19,21,62).

**Encephalocele with Dandy–Walker and Arnold–Chiari malformations.** Dandy–Walker defect has been reported in association with parietal encephalocele (73) but also may be observed with occipital encephalocele (81). Arnold–Chiari defect has been noted with occipital encephalocele by several authors (6,81), and several instances of combined Dandy–Walker and Arnold–Chiari defects with occipital encephalocele in the same patient have been noted (81).

**Encephalocele/ectrodactyly association.** The combination of occipital encephalocele and ectrodactyly has been observed in two sporadic instances (JM Opitz, personal communication). Mental deficiency was present in both cases. No other major anomalies were present and the condition is distinct from von Voss syndrome in which encephalocele and ectrodactyly also may occur.

**Encephalocele/hypothalamic–pituitary dysfunction association.** Transsphenoidal encephaloceles (Fig. 16–5) herniate through a defect in the floor of the sella turcica and present as epipharyngeal masses. In some instances, they may be small and not always readily appreciated. The encephalocele may occur by itself or together with other craniofacial anomalies, including additional encephaloceles, optic nerve abnormalities, hypertelorism, and cleft lip–palate. Several patients have been reported with hypothalamic–pituitary dysfunction but with different patterns of involvement. Growth hormone deficiency, hypothyroidism, central hypogonadism, central adrenocortical insufficiency, vasopressin deficiency, and diabetes insipidus have been documented (66).

**Encephalocele/Klippel–Feil/iniencephaly association.** Occipital encephalocele has been reported in a number of instances of Klippel–Feil anomaly (23,68,74), and in iniencephaly apertus it is constant (7,52,65,86). Iniencephaly consists of a defect in the occipital bone around the foramen magnum, spina bifida of considerable degree, and retroflexion of the head on the spine. Lewis (65) divided various cases by the presence or absence of occipital encephalocele, calling the former iniencephaly apertus, the latter iniencephaly clausus. Gilmour (41) suggested that iniencephaly and Klippel–Feil anomaly are part of a spectrum. Both conditions seem to have several features in common (64). Further evidence of the association of these defects is found in the experimental work of Warkany and Takacs (100) who produced the lesion in rats with streptonigran.

**Encephalocele/myelomeningocele association.** Lorber (68) and Mealey and his co-workers (74) have observed several instances of occipital encephalocele associated with myelomeningocele. DeMyer (26) was able to produce encephalocele in some rat fetuses and spinal dysraphism in others by administering vincristine to pregnant rats at 8.5 days. The same combination of defects was produced by the administration of salicylates to pregnant rats (42). Wilner (102) described a mother with meningocele who gave birth to a girl with myelomeningocele. Lorber (68) noted that eight siblings of seven infants with occipital encephalocele had myelomeningocele.

**Encephalocele/oculo-auriculo-vertebral spectrum.** Oculo-auriculo-vertebral spectrum is a well-known condition affecting aural, oral, and mandibular growth together with anomalies of the eyes and vertebral column. It has been observed with occipital encephalocele occasionally (Fig. 16–3), and such cases have been reviewed by Cohen (17) and Herrmann and Opitz (49). Minne and Gernez (77) reported occipital encephalocele with absent lung, which is intriguing because hemifacial microsomia has also been observed with absent lung. Frontonasal dysplasia, which occurs with cranium bifidum and flattened encephalocele, has been reported on a number of occasions with dysplastic ears, ear tags, and epibulbar dermoids.

**Miscellaneous and/or occasional involvement.** A number of miscellaneous conditions have occasionally been associated with encephalocele. Several authors have documented the binary combination of encephalocele with various congenital heart defects (23,36,55,68,74, BD Hall, personal communication). In a case associated with Ehlers–Danlos syndrome (30), the encephalocele was not congenital, but acquired. Rarely, occipital encephalocele has been recorded in infants of diabetic mothers (32). Other conditions occasionally reported with encephalocele have included Marfan syndrome (72), neurofibromatosis (51, JW Hanson, personal communication), Poland anomaly (M Bull, personal communication), polysplenia syndrome (KL Jones, personal communication), rubella syndrome (8,60), TAR syndrome (RJ Gorlin, personal observation), lateral nasal proboscis (1a), and tuberous sclerosis (9,57). Patients with the combination of encephalocele, ptosis of the eyelids, epicanthic folds, and single maxillary central incisor have been observed (MJ Stephan, personal communication).

### References (Encephaloceles)


1. Aleck KA et al: Dyssegmented dysplasias: Clinical, radiographic, and morphologic evidence of heterogeneity. *Am J Med Genet* **27**: 295–312, 1987.

1a. Antoniades K, Baraitser M: Proboscis lateralis: A case report. *Teratology* **40**:193–197, 1989.

2. Avanzini G, Crivelli G: A case of sphenopharyngeal encephalocele. *Acta Neurochir (Wien)* **22**:205–215, 1970.

3. Ayme S, Mattei J-F: Brief clinical report: HARD (± E) syndrome: Report of a sixth family with support for autosomal-recessive inheritance. *Am J Med Genet* **14**:759–766, 183.

4. Babini B et al: Su di un caso di trisomia del gruppo cromosomici 16–18. *Boll Soc Ital Biol Sper* **39**:332–335, 1963.

5. Bach C: Puberté précose avec tumeur du diencéphale. Association à un syndrome de Dandy-Walker et à ses anomalies dysraphiques. *Ann Pédiatr* **16**:762–767, 1969.

6. Barrow N, Simpson DA: Cranium bifidum. Investigation, prognosis and management. *Aust Pediatr J* **2**:20–26, 1966.

7. Brodsky I: Four examples of iniencephalus, with statistical review of the literature. *Med J Aust* **2**:795–800, 1939.

8. Browder JA: Occipital encephalocele after probable prenatal rubella. *Lancet* **2**:383–384, 1971.

9. Browder J, De Veer JA: Rhino-encephalocele. *Arch Pathol* **18**:647–659, 1934.

10. Burton BK et al: Brief clinical report: Walker–Warburg syndrome with cleft lip and cleft palate in two sibs. *Am J Med Genet* **27**:537–541, 1987.

11. Carr DH: Chromsome abnormalities and their influence on the development of the nervous system, in *Int Copenhagen Congress for the Scientific Study of Mental Retardation,* 1964, pp 132–138.

12. Chemke JB et al: A familial syndrome of central nervous system and ocular malformation. *Clin Genet* **7**:1–7, 1975.

13. Cherstvoy EG et al: Syndrome of multiple congenital malformations including phocomelia, thrombocytopenia, encephalocele, and urogenital abnormalities. *Lancet* **2**:485, 1980.

14. Chervenak FA et al: The diagnosis and management of fetal cephalocele. *Obstet Gynecol* **64**:86–91, 1984.

15. Chohan BS et al: Anterior orbital meningoencephalocele. *Am J Ophthalmol* **68**:144, 1969.

16. Cohen MM et al: A ring chromosome (No. 18) in a cyclops. *Clin Genet* **3**:249–252, 1972.

17. Cohen MM Jr: Variability versus "incidental findings" in the first and second branchial arch syndrome: Unilateral variants with anophthalmia. *Birth Defects* **7**(7):103–108, 1971.

18. Cohen MM Jr: *The Child with Multiple Birth Defects.* Raven Press, New York, 1982.

19. Cohen MM Jr: Perspectives on craniosynostosis. *West J Med* **132**:507–513, 1980.

20. Cohen MM Jr: Ocular hypertelorism and syndromes with ocular hypertelorism. In progress.

21. Cohen MM Jr, Lemire RJ: Syndromes with cephaloceles. *Teratology* **25**:161–172, 1982.

22. Cohen MM Jr et al: Frontonasal dysplasia (median cleft face syndrome): Comments on etiology and pathogenesis. *Birth Defects* **7**(7):117–119, 1971.

23. Cohn JA, Hamby WB: Surgery of cranium bifidum and spina bifida: A follow-up report of 64 cases. *J Neurosurg* **10**:297–300, 1953.

24. Daum S et al: Dysplasie télécephalique avec excroissance de l'os frontal. *Sem Hôp Paris* **34**:1893–1896, 1958.

25. Deklerk DJJ, DeVilliers JC: Frontal encephaloceles. *S Afr Med J* **47**:1350, 1973.

26. DeMyer W: Cleft lip and jaw induced in fetal rats by vincristine. *Arch Anat Histol Embryol* **48**:181–186, 1965.

27. DeMyer W: The median cleft face syndrome: Differential diagnosis of cranium bifidum occultum, hypertelorism, and medium cleft nose, lip and palate. *Neurology (Minn)* **17**:961–967, 1967.

28. Dinno ND et al: Chondrodysplastic dwarfism, cleft palate and micrognathia in a neonate, a new syndrome? *Eur J Pediatr* **123**:39–42, 1976.

29. Dodge HW Jr, Love JG: Intranasal encephalomeningoceles associated with cranium bifidum. *Arch Surg* **79**:87, 1959.

30. Dodge JA, Shillito J Jr: Ehlers–Danlos syndrome associated with acquired encephalocele. *J Pediatr* **66**:1061–1067, 1965.

31. Donnai D, Farndon PA: Walker–Warburg syndrome (Warburg syndrome, HARD ± E syndrome). *J Med Genet* **23**:200–203, 1986.

32. Evrard P, Caviness VS Jr: Extensive developmental defect of the cerebellum associated with posterior fossa ventriculocele. *J Neuropathol Exp Neurol* **33**:385–399, 1974.

33. Fargueta JS et al: Posterior orbital encephalocele with anophthalmos and other brain malformations. Case report. *J Neurosurg* **38**:215–217, 1973.

34. Farrell SA et al: Prenatal diagnosis of retinal detachment in Walker–Warburg syndrome. *Am J Med Genet* **28**:619–624, 1987.

35. Fisher NL, Smith DW: Occipital encephalocele and early gestational hyperthermia. *Pediatrics* **68**:480–483, 1981.

36. Fisher RG et al: Spina bifida and cranium bifidum: Study of 530 cases. *Mayo Clin Proc* **27**:33–38, 1952.

37. Flatz G, Sukthomya C: Fronto-ethmoidal encephalomeningoceles in the population of Northern Thailand. *Humangenetik* **11**:1–8, 1970.

38. Fraser FC, Skelton J: Possible teratogenicity of maternal fever. *Lancet* **2**:634, 1978.

39. Freeman MVR et al: The Roberts syndrome. *Clin Genet* **5**:1–16, 1974.

40. Gardner RJM et al: Are lq + chromosomes harmless? *Clin Genet* **6**:383–393, 1974.

41. Gilmour JR: The essential identity of Klippel–Feil syndrome and iniencephaly. *J Pathol Bacteriol* **53**:117–131, 1941.

42. Goldman AS, Yakovac WC: The enhancement of salicylate teratogenicity by maternal immobilization in the rat. *J Pharmacol Exp Ther* **142**:351–357, 1963.

43. Gördüren S: Orbital encephalocele associated with acrocephaly. *Br J Ophthalmol* **36**:151–154, 1952.

44. Gorlin RJ et al: Facial clefting and its syndromes. *Birth Defects* **7**(7):3–49, 1971.

45. Gruber GB: Beiträge zur Frage "gekoppelter" Missbildungen (Akrocephalo-Syndactylie und Dysencephalia splanchnocystica). *Beitr Pathol Anat* **93**:459, 1934.

46. Gruhn JG et al: Dyssegmental dwarfism. *Am J Dis Child* **132**:382–386, 1978.

47. Hall BD: Syndromes and situations simulated by amniotic bands. *March of Dimes Birth Defects Conference,* Chicago, June 24–27, 1979.

48. Handmaker SD et al: Dyssegmental dwarfism: A new syndrome of lethal dwarfism. *Birth Defects* **13**(3D):79–90, 1978.

49. Herrmann J, Opitz JM: A dominantly inherited first arch syndrome. *Birth Defects* **5**(2):110–112, 1969.

50. Higginbottom MC et al: The amniotic band disruption complex: Timing of amnion rupture and variable spectra of consequent defects. *J Pediatr* **95**:544–549, 1979.

51. Holt J: Neurofibromatosis in children. *Am J Roentgenol* **130**:615–639, 1978.

52. Howkins J, Lawrie RS: Iniencephaly. *J Obstet Gynaecol Br Commonw* **46**:25–31, 1939.

53. Hsia YE et al: Genetics of the Meckel syndrome (dysencephalia splanchnocystica). *Pediatrics* **48**:237–247, 1971.

54. Hsia YE et al: Chromosomal abnormality (46,XX,3p + ) in a case of the Meckel syndrome. *Birth Defects* **10**(8):19–25, 1974.

55. Ingraham FD, Swan H: Spina bifida and cranium bifidum. I. Survey of five hundred and forty-six cases. *N Engl J Med* **228**:559–563, 1943.

56. Jones KL et al: A pattern of craniofacial and limb defects secondary to aberrant tissue bands. *J Pediatr* **84**:90–95, 1974.

57. Kak VK, Gulati DR: The familial occurence of frontoethmoidal encephalomeningocele. *Neuro India* **21**:41–43, 1973.

58. Kahler SG: Phocomelia, encephalocele, and urogenital abnormalities. *Proc Greenwood Genet Ctr* **2**:134, 1983.

59. Kalter H: *Teratology of the Central Nervous System.* University of Chicago Press, Chicago, 1968, pp 257, 336.

60. Kappers JA: Developmental disturbance of the brain induced by German measles in an embryo of the 7th week. *Acta Anat* **31**:1, 1957.

61. Keller H et al: "ADAM complex" (amniotic deformity, adhesions, mutilations)—A pattern of craniofacial and limb defects. *Am J Med Genet* **2**:81–98, 1978.

62. Kloss JL: Craniosynostosis secondary to ventriculoatrial shunt. *Am J Dis Child* **116**:315–317, 1968.

63. Knobloch WH, Layer JM: Retinal detachment and encephalocele. *J Pediatr Ophthalmol* **8**:181–184, 1971.

64. Lemire RJ et al: *Normal and Abnormal Development of the Human Nervous System.* Harper & Row, Hagerstown, 1975, pp 306–307.

65. Lewis HF: Iniencephalus. *Am J Obstet* **35**:11–53, 1897.

66. Lieblich JM et al: The syndrome of basal encephalocele and hypothalamic-pituitary dysfunction. *Ann Intern Med* **89**:910–916, 1978.

67. Loeser JD, Alvord EC Jr: Agenesis of the corpus callosum. *Brain* **91**:553–570, 1968.

68. Lorber J: The prognosis of occipital encephalocele. *Dev Med Child Neurol Suppl* **13**:75–86, 1967.

69. Lubinsky M et al: Familial amniotic bands. *Am J Med Genet* **14**:81–87, 1983.

70. MacRae DW et al: Ocular manifestations of the Meckel syndrome. *Arch Ophthalmol* **88**:106–113, 1972.

71. Marburg O: So-called agenesia of the corpus callosum (callosal defect). *Arch Neurol Psychiatr* **61**:297–312, 1949.

72. McKusick VA: *Heritable Disorders of Connective Tissue,* 4th ed. C.V. Mosby, St. Louis, 1972, p 140.

73. McLaurin RL: Parietal cephaloceles. *Neurology* **14**:764–772, 1964.

74. Mealey J Jr et al: The prognosis of encephaloceles. *J Neurosurg* **32**:209–218, 1970.

75. Mecke S, Passarge E: Encephalocele, polycystic kidneys, and polydactyly as an autosomal recessive trait simulating certain other disorders: The Meckel syndrome. *Ann Génét* **14**:97–103, 1971.

76. Meckel JF: Beschreibung zweier durch sehr ähnliche Bildungsabweichung entstellter Geschwister. *Dtsch Arch Physiol* **7**:99–172, 1822.

77. Minne J, Gernez L: Agénésie complète du poumon gauche chez un nouveau-né exencéphale. *Ann Anat Pathol* **10**:503, 1933.

78. Moeschler JB et al: Neural tube defects in 13q− syndrome. *Proc Greenwood Genet Ctr* **6**:140, 1987.

79. Mood GF: Congenital anterior herniations of the brain. *Ann Otol Rhinol Laryngol* **47**:391–401, 1938.

80. Opitz JM, Howe JJ: The Meckel syndrome (dysencephalia splanchnocystica, the Gruber syndrome). *Birth Defects* **5**(2):167–179, 1969.

81. Padget DH, Lindenberg R: Inverse cerebellum morphogenetically related to Dandy–Walker and Arnold–Chiari syndromes: Bizarre malformed brain with occipital encephalocele. *Johns Hopkins Med J* **131**:228–246, 1972.

82. Pagon RA et al: Hydrocephalus, agyria, retinal dysplasia, encephalocele—(HARD ± E) syndrome—an autosomal recessive condition. *Birth Defects* **14**(6B):233–241, 1978.

83. Pauli RM et al: Warfarin therapy initiated during pregnancy and phenotypic chondrodysplasia punctata. *J Pediatr* **88**:506–508, 1976.

84. Schinzel A: *Catalogue of Unbalanced Chromosome Aberrations in Man.* Walter de Gruyter, Berlin and New York, 1984, pp 482–493.

85. Sedano HO et al: Frontonasal dysplasia. *J Pediatr* **76**:906–913, 1970.

86. Stark AM: A report of two cases of iniencephalus. *J Obstet Gynaecol Br Commonw* **58**:462–464, 1951.

87. Stratton DF et al: New chromosome syndrome: Miller–Dieker syndrome and monosomy 17p13. *Hum Genet* **67**:193–200, 1984.

88. Suwanwela C, Suwanwela N: A morphological classification of sincipital encephalomeningoceles. *J Neurosurg* **36**:201, 1972.

89. Tejani N: Anticoagulant therapy with cardiac valve prosthesis during pregnancy. *Obstet Gynecol* **42**:785–793, 1973.

90. Thomas IT et al: Isolated and syndromic cryptophthalmos: A review of the Fraser syndrome. *Am J Med Genet* **25**:85–98, 1986.

91. Torpin R: *Fetal Malformations.* C.C. Thomas, Springfield, Illinois, 1968.

92. Van Allen M, Clarren SK: A spectrum of gyral anomalies in Miller–Dieker (lissencephaly) syndrome. *J Pediatr* **102**:559–564, 1983.

93. Van Nouhuys JM, Bruyn GW: Nasopharyngeal transsphenoidal encephalocele, craterlike hole in the optic disc and agenesis of the corpus callosum: Pneumoencephalographic visualization in a case. *Psychiatr Neurol Neurochir* **67**:243, 1964.

94. von Voss et al: *Klinische Genetik in der Pädiatrie,* Kiel, Thieme, Stuttgart, 1979, p 70.

95. Warkany J: *Congenital Malformations. Notes and Comments.* Year Book Medical Publishers, Chicago, 1971.

96. Warkany J: Warfarin embryopathy. *Teratology* **14**:205–210, 1976.

97. Warkany J, Bofinger M: Le rôle de la coumadine dans les malformations congénitales. *Méd Hyg* **33**:1454–1457, 1975.

98. Warkany J et al: Median facial cleft syndrome in half-sisters. Dilemmas in genetic counseling. *Teratology* **8**:273–286, 1973.

99. Warkany J et al: *Mental Retardation and Congenital Malformations of the Central Nervous System.* Year Book Medical Publishers, Chicago, 1981.

100. Warkany J, Takacs E: Congenital malformations in rats from streptonigran. *Arch Pathol* **79**:65–79.

101. Waterson JR et al: Brief clinical report: Apert syndrome with frontonasal encephalocele. *Am J Med Genet* **21**:777–783, 1985.

102. Wilner IA: Familial repetition of myelomeningocele. *J Michigan Med Soc* **48**:727, 1949.

103. Whitley CB et al: Warburg syndrome: Lethal neurodysplasia with autosomal recessive inheritance. *J Pediatr* **102**:547–551, 1983.

## The holoprosencephalic disorders

Holoprosencephaly is a malformation sequence in which impaired midline cleavage of the embryonic forebrain is the basic feature. The prosencephalon fails to cleave sagittally into cerebral hemispheres, transversely into telencephalon and diencephalon, and horizontally into olfactory and optic bulbs (Fig. 16–6). The condition can be graded according to the degree of severity as alobar, semilobar, or lobar holoprosencephaly. Various gradations of facial dysmorphism are commonly associated with holoprosencephaly including cyclopia, ethmocephaly, cebocephaly, median cleft lip, and less severe facial dysmorphism (Fig. 16–7) (Table 16–3). Holoprosencephaly and facial dysmorphism have extremely heterogeneous causes.

Exhaustive reviews and bibliographies are those of Cohen (29,31,32), Cohen and Sulik (35), DeMyer (42), and Siebert et al (141). The reader is referred to St. Hilaire (144), Förster (52), Ahlfeld (2), Vrolik (159), Otto (113), Dareste (41), Taruffi (151), Ballantyne (7), Hannover (64), and Warkany (160) for historical coverage; Esser (48) for mythology; Cohen (29–31,36) and Münke (106a) for etiology, associated anomalies, and syndromes; Adelmann (1), Binns and Keeler (16,78,79), Kalter (77), Sulik and Johnston (149), and Wright and Wagner (166) for animal studies; Yakovlev (167), DeMyer (42,43), and others (87,124) for central nervous system pathology; and De-

Table 16–3. Holoprosencephalic facies[a]

| Facial type[b] | Main facial features | Brain |
|---|---|---|
| Cyclopia | Median monophthalmia, synophthalmia, or anophthalmia; proboscis may be single or absent; hypognathism in some cases | Alobar holoprosencephaly |
| Ethmocephaly | Ocular hypotelorism with proboscis | Alobar holoprosencephaly |
| Cebocephaly | Ocular hypotelorism and blind-ended, single nostril nose | Usually alobar holoprosencephaly |
| Median cleft lip | Ocular hypotelorism, flat nose, and median cleft lip | Usually alobar holoprosencephaly |
| Less severe facial dysmorphism | Variable features including ocular hypotelorism or hypertelorism, flat nose, unilateral or bilateral cleft lip, iris coloboma, or other anomalies; minimal facial dysmorphism in some cases | Semilobar or lobar holoprosencephaly |

[a]Modified after WE DeMyer et al, *Pediatrics* **34**:256, 1964.
[b]Transitional facial forms are known to occur.

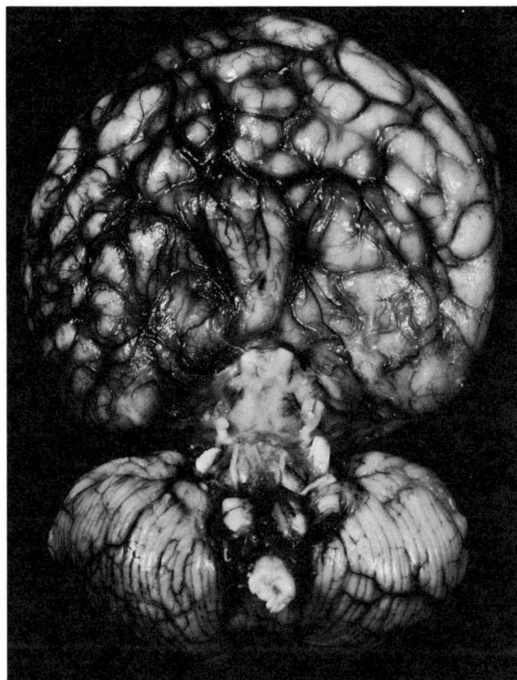

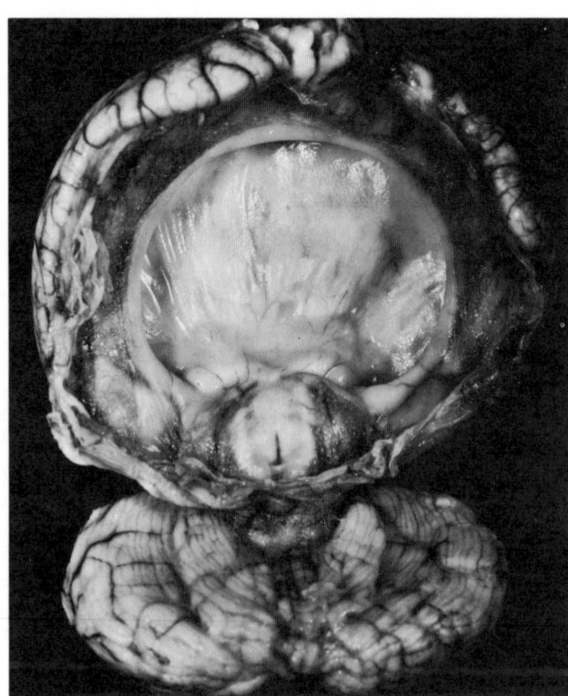

A B

Fig. 16–6. *The holoprosencephalic disorders.* Holoprosencephalic brain, alobar type. (A) Dorsal view showing pancake-like structure without corpus callosum or septum pellucidum. Diencephalon and telencephalon are uncleft. (B) Ventral view showing absent olfactory bulbs and tracts and no interhemispheric fissure. (From *W DeMyer* and *PT White,* Arch Neurol **11**:507, 1964.)

Myer et al (44), Kokich et al (81), Torczynski et al (153), van Duyse (156), Pettersen (119), and Mieden (104) for holoprosencephalic facial anatomy. Various authors have addressed epidemiologic aspects (102,125), extended family study (125), and twin study (22,31,97,125).

Holoprosencephaly is found universally in white, black, and oriental populations (31,33,47,125). Prevalence estimates have been critically examined by Cohen (31). A general prevalence estimated for newborns may be approximately 1.26/10,000 (31). Some estimates have been high with fluctuations in time of holoprosencephalic births (86,130); some have been very low (108). The frequency of holoprosencephalic abortuses is high, being approximately 40/10,000 (102). Some authors have suggested that approximately half of all liveborns and stillborns with holoprosencephaly are associated with chromosomal anomalies and half are associated with normal chromosome complements (86,105), although this has been questioned (31). Birthweights tend to be low (125) except in holoprosencephalic infants of diabetic mothers (8). The sex ratio in alobar holoprosencephaly has a 3:1 female predilection. In contrast, the sex ratio is 1:1 in lobar holoprosencephaly (125). Major facial dysmorphism occurs more frequently in females than in males, especially with cyclopia (31,42,139). Among the major facial types associated with holoprosencephaly (Fig. 16–7) (Table 16–3), the frequency of occurrence of each type is inversely related to its severity, ethmocephaly, the rarest type, being the one exception (31).

The overwhelming majority of cases are sporadic. To date, 66 distinctive conditions with holoprosencephaly or arhinencephaly are known including 34 chromosomal syndromes (31), 18 monogenic conditions (31), 1 environmentally induced condition (8), and 12 distinctive conditions of unknown genesis (31,35). In addition, several environmental teratogens have been suggested or suspected (12,24,84,91,92,98, 127,145, M Shokeir, personal communication, 1980). All known conditions with holoprosencephaly are listed in Table 16–4 together with characteristic clinical features, etiology, and pertinent references. Diagnosis is based on the overall pattern of abnormalities, chromosome analysis, and family history. Careful examination of the proband's family for microforms of holoprosencephaly is essential, especially for anosmia or hyposmia, mild hypotelorism, microcephaly, or single maxillary central incisor (4,14,15,25,28,33,74,93). Cohen and Gorlin (33) noted that in the presence of a holoprosencephalic proband, sibs with cleft lip–palate but without holoprosencephaly per se should probably be regarded as microforms of the disorder. A family history of short stature, endocrinopathy, or various types of central nervous system anomalies other than holoprosencephaly may be significant (21,31,125,126).

Even when confronted with nonsyndromic holoprosencephaly, it is important to rule out a chromosomal etiology because chromosomal causes are sometimes observed with holoprosencephaly with no, few, or trivial extracephalic anomalies (31). For nonsyndromic cases, if the findings are consistent with autosomal dominant holoprosencephaly with incomplete penetrance and variable expressivity, the risk for an obligate carrier is on the order of 16–21%; the risk for an incomplete form or microform is on the order of 13–14%; and the overall effect for some risk, either mild or severe, is on the order of 29–35% (31). Affected sibs, in the absence of familial microforms in more than one generation, are consistent with autosomal recessive inheritance (20,33,36). A sporadic case with negative family history except for consanguinity certainly suggests that the recurrence risk may be as high as 25% (31). For sporadic, nonchromosomal, nonsyndromic holoprosencephaly, a recurrence risk of approximately 6% may be given. This risk is based on the genetic study of Roach et al (125) of 30 families with holoprosencephalic probands; only two of their pedigrees represented examples of familial holoprosencephaly.

**Central nervous system.** In alobar holoprosencephaly (Fig. 16–6), a small monoventricular cerebrum lacking interhemispheric division is present. The thalami and the corpora striata are undivided across the midline. Olfactory tracts and bulbs are always absent as is the corpus callosum, although a few commissural fibers may cross the midline (43). DeMyer (42) further subdivided alobar holoprosencephaly depending upon the degree to which the dorsal lip of the holotelecephalon rolls over to cover the membranous ventricular roof. The three common external configurations of alobar holoprosencephaly are the pancake type, the cup type, and the ball type.

In semilobar holoprosencephaly, rudimentary cerebral lobes are

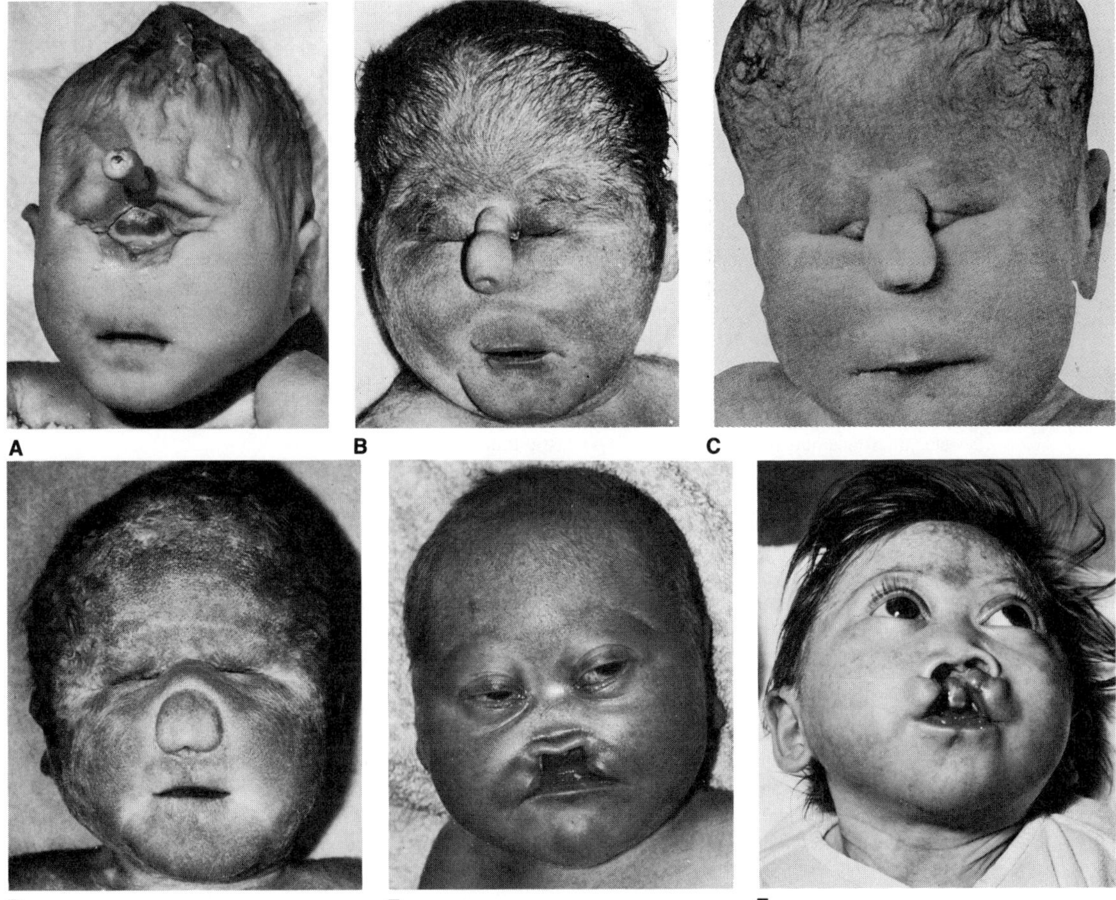


Fig. 16–7. *The holoprosencephalic disorders.* (A) Cyclopia. Note central eye where roof of nose is normally located. Proboscis with single opening is located above eye. (B) Cyclopia-ethmocephaly. Intermediate form having separate palpebral fissures but single bony orbit; infant had trisomy 13. (C) Ethmocephaly. Separate eye sockets, proboscis located above eye. (D) Cebocephaly. Microphthalmia, hypotelorism, single-nostril nose. (E) Median cleft lip. Hypotelorism, upslanting palpebral fissures, absence of nasal bones and cartilages, and agenesis of philtral area. (F) Partial median cleft. Note hypotelorism, hypoplasia of nasal cartilages and philtral area. (B from *K Taysi* and *K Tinaztepe,* Am J Dis Child **124**:170, 1972. C from *WB Davis,* Surg Gynecol Obstet **61**:209, 1935. D from *PE Conen,* Can Med Assoc J **87**:709, 1962. F from *W DeMyer et al* Pediatrics **34**:256, 1964.)

present and although the interhemispheric fissure is never complete, it may be present posteriorly. Commonly, the olfactory tracts and bulbs are absent, but in some instances, they may be hypoplastic. The corpora striata are continuous across the midline. The corpus callosum is not a distinct bundle, although some commissural fibers may cross the midline (42).

In lobar holoprosencephaly, the brain has well formed lobes which may be of normal size. Although a distinct interhemispheric fissure is present, there may be some midline continuity of the cingulate gyrus. The olfactory tracts and bulbs may be absent or hypoplastic and the corpus callosum may be absent, hypoplastic, or normal. Midline cleavage of the thalami and the corpora striata may be incomplete (42).

At the mild end of the holoprosencephalic spectrum are malformations such as absence of the corpus callosum and arhinencephaly (absence of the olfactory tracts and bulbs) (42,43). Defects of the corpus callosum result from a variety of different mechanisms (99). In some instances, absence of the corpus callosum clearly represents the holoprosencephalic spectrum. In other instances, isolated absence of the corpus callosum is not related to the holoprosencephalic spectrum (42,99).

Other central nervous system anomalies may sometimes be present. Although the hindbrain usually develops normally, it may be hypoplastic since the descending fiber tracts fail to develop (124). Absence of the inferior vermis has also been observed in a number of cases (61).

Microcephaly, hydrocephaly, and trigonocephaly are all related to holoprosencephaly. A small brain is characteristic of alobar and semilobar holoprosencephaly. With lobar holoprosencephaly and with less severe malformations at the mild end of the holoprosencephalic spectrum, the head may be microcephalic or normocephalic. Holoprosencephaly and microcephaly are obviously not synonymous since microcephaly can occur without any of the anomalies found in holoprosencephaly. Trigonocephaly may be found with some cases of holoprosencephaly, but obviously can occur without it. Hydrocephaly may accompany holoprosencephaly occasionally, and it should be obvious that most instances of hydrocephaly have nothing to do with holoprosencephaly (27,40). Anencephaly and encephalocele have also been reported with holoprosencephaly on occasion (90). Some families have been reported in which various central nervous system anomalies have occurred in relatives of holoprosencephalic probands (21,125).

With alobar holoprosencephaly, patients have apneic episodes, seizures, poikilothermia, failure to thrive, and failure to make developmental progress. Such infants do not seem conscious of their environment nor do they make purposeful, goal-directed movements (42). There is no evidence of useful vision or hearing. The neurological findings in semilobar and lobar holoprosencephaly are not as well

Table 16–4. Conditions with holoprosencephaly[a]

| Condition | Striking features | References |
|---|---|---|
| Chromosomal holoprosencephaly | | |
| Well-established types | | |
| Trisomy 13 | Holoprosencephaly common, full spectrum of facial dysmorphism; most common of chromosomal syndromes with holoprosencephaly. Growth deficiency, failure to thrive, microcephaly, microphthalmia, iris coloboma, malformed ears, glabellar hemangioma, parietooccipital scalp defects, micrognathia, postaxial polydactyly, hyperconvex fingernails, prominent calcaneus, rocker-bottom feet, congenital heart defects (ASD, PDA, VSD, dextroposition), accessory spleen, cryptorchidism, bicornuate uterus, hypoplastic ovaries | 37,69,152 |
| del(13q) | Holoprosencephaly common, semilobar or lobar holoprosencephaly without extremely severe facial dysmorphism. Microcephaly, mental deficiency, microphthalmia, iris coloboma, retinoblastoma, cleft palate, micrognathia, malformed ears, hypoplastic thumbs, imperforate anus, hypospadias, cryptorchidism, congenital heart defects | 111,112,136,165 |
| del(18p) | Holoprosencephaly in 10%, full spectrum of facial dysmorphism. Variable degrees of mental deficiency, microcephaly, hypotonia, growth deficiency; variable phenotype sometimes with Turner-like features including webbed neck, shield chest with widely spaced nipples, and lymphedema of hands and feet; variable facial dysmorphism, which may include hypertelorism, epicanthic folds, strabismus, large dysmorphic ears, small mandible, and occasionally holoprosencephalic facies | 49,60,94,136 |
| Trisomy 18 | Holoprosencephaly uncommon, full spectrum of facial dysmorphism. Severe mental deficiency, hypertonia, growth retardation, failure to thrive, feeding difficulties, limited hip abduction, ulnar deviation of fingers, short sternum, congenital heart defects (VSD, PDA, ASD), short dorsiflexed halluces, rocker-bottom feet, calcaneovalgus deformity, cryptorchidism, prominent occiput, low-set malformed ears, micrognathia | 69,73,85,135,152 |
| Triploidy | A number of cases reported. Mental retardation, hypotonia, growth deficiency, asymmetry, microphthalmia, iris and choroid colobomas, mild hypertelorism, anomalous low-set ears, micrognathia, syndactyly of third and fourth fingers, single palmar creases, clubfoot, congenital heart defects, other anomalies | 83,169 |
| del(2p) | Cyclopia, median cleft lip | 32a,60a,107,164a |
| dup(3p) | Holoprosencephaly in 10% of cases (cyclopia, cebocephaly). Frontal bossing, square face, prominent cheeks, ocular hypertelorism, short nose with large tip, downturned mouth, cleft lip–palate, prominent philtrum, micrognathia, congenital heart defects, hypoplastic male genitalia, other anomalies | 55,57,82,101 |
| del(7)(pter→q32) | Holoprosencephaly (cyclopia, cebocephaly, median cleft lip) in about 20%. Growth retardation, microcephaly, microphthalmia, upslanting palpebral fissures, ocular hypertelorism in some cases, flat nasal bridge, short bulbous nose, small mandible, ear anomalies, distal limb anomalies, male genital hypoplasia | 16a,53,132,134,138, 142 |
| Uncommon and less consistent types | | |
| del(1q) | Ocular hypertelorism, cleft palate, micrognathia, low-set triangular ears, micropenis, complex congenital heart defects | 67 |
| dup(5q),del(5p) | Holoprosencephaly with median cleft lip and hypotelorism in a single case. No dysmorphic features of trunk and extremities. Esophageal atresia, tetralogy of Fallot | 137 |
| t(6;18)(p24.1;p11.21) | Two sibs with agnathia, holoprosencephaly, facial dysmorphism. Karyotype not reported in original article, but by personal communication | 81a,115, RM Pauli, personal communication, 1987. |
| del(6p),r(6) | Hypoplastic irides, severe glaucoma, complex congenital heart defects, variable phenotypic expression, semilobar or lobar holoprosencephaly, hypertelorism, cleft palate | 66b,117 |
| dup(9p),dup(11q) | Wide anterior fontanel, hypertelorism, downslanting palpebral fissures, cleft lip–palate, micrognathia, low-set dysplastic ears, gastroschisis, small lungs, patent foramen ovale, dilation of coronary sinus, hydronephrosis and hydroureter, cryptorchidism | 45 |
| del(11q) | Cyclopia, ear anomalies, widely-spaced nipples, congenital heart defect. One case of holoprosencephaly among 27 reported. Mosaicism present | 66a |

| Condition | Striking features | References |
|---|---|---|
| dup(11q) | Short nose, long philtrum, micrognathia, retracted lower lip, micropenis, bicornuate uterus, accessory lung lobes, congenital heart defects | 123 |
| dup(14)(pter→q24) | Small flat nose, downturned mouth, cleft palate, low-set crumpled ears, narrow chest, hypoplastic larynx and lungs, unilateral renal agenesis, unilateral cryptorchidism | 39 |
| Trisomy 20 | Holoprosencephaly | 10 |
| Trisomy 21 | Holoprosencephaly rare. Mental deficiency, hypotonia, microcephaly, atrophic CNS changes of Alzheimer's disease, hyperextensibility of joints, loose skin on posterior part of neck, flat facial profile, upslanting palpebral fissures, Brushfield spots, epicanthic folds, short ears with overhanging helices, short neck, congenital heart defects (especially atrioventricularis communis, VSD, PDA, ASD, and aberrant subclavian artery), gastrointestinal malformations, dysplastic pelvis, brachydactyly, clinodactyly, single palmar creases, increased ulnar loops, short stature | 122,155 |
| r(21) | Semilobar holoprosencephaly, anteverted nares, mild retrognathia, hemivertebra. Holoprosencephaly rare | 5,70 |
| Trisomy 22 | Hypertelorism, broad flat nose, large mouth, prominent upper lip, micrognathia, low-set dysplastic ears, short webbed neck, hypoplastic nails, genital anomalies, anal atresia, tetralogy of Fallot, hypoplastic bilobed lungs, hypoplastic diaphragm, intestinal malrotation, hypoplastic kidneys | 158 |
| dup(22q) | Macrocephaly, prominent occiput, hypertelorism, upslanting palpebral fissures, short nose, prominent upper lip, cleft palate, micrognathia, short neck, micropenis, hypospadias, bifid scrotum, hypoplastic nails, tetralogy of Fallot | 133 |
| del(22)(pter→q11) | Trisomy 13 syndrome-like phenotype | 6 |
| 46,XX/46,X,del(X) (p11) | Holoprosencephaly, median cleft lip, short limbs. Autopsy refused so ovaries not examined. Chromosomal deletions of this type usually associated with Turner syndrome phenotype | 130 |
| 47,XXX | Holoprosencephaly | 120 |
| Prebanding era karyotypes | | |
| 47,XY, + B | Cyclopia | 18 |
| 46,XY/45,XY, − G | Holoprosencephaly with cyclopia in one case. Trigger thumb, hydroureters | 27 |
| 46,XX,t(3p;?) | Meckel-like phenotype | 72 |
| 47,XX, + C | Holoprosencephaly with cebocephaly in one case. Malformed ears, micrognathia, elongated thorax, enlarged liver | 89 |
| 46,XY/47,XY, + frag | Holoprosencephaly with cyclopia in four members of a family | 121 |
| Chromosomal arhinencephaly | | |
| dup(1q) | Variable phenotype depending on duplication. Arhinencephaly uncommon, anencephaly (meroacrania) uncommon. Relative macrocephaly, small or downslanting palpebral fissures, hypertelorism, cleft lip, cleft or highly arched palate, micrognathia, ear anomalies, long fingers and toes, involuted or absent thymus, cardiac anomalies, absent gallbladder, dysplastic kidneys | 54,103 |
| dup(6p) | Frontal bossing, hypotelorism, hypoplastic midface, cardiovascular defects, small kidneys, common mesentery, absent uterus | 116,143 |
| dup(6q) | Hydrocephaly, microphthalmia, broad flat nasal bridge, cleft lip–palate, bow-shaped mouth, micrognathia, low-set dysplastic ears, short webbed neck, prominent clitoris, congenital heart defects, unlobed lungs, malrotation of bowel, hypoplastic kidneys, hypoplastic ovaries | 109 |
| i(12p) | Hypertelorism, short nose with anteverted nostrils, macrostomia, low-set dysplastic ears, edematous webbed neck, brachymelia, hypoplastic thumbs, camptodactyly, short broad palms, anal atresia, malrotation of gut, stenotic ureters, hydronephrosis, duplication of kidney | 146 |
| dup(16q) | Brachycephaly, downslanting palpebral fissures, cleft palate, micrognathia, malformed low-set ears, low hairline, short neck, contractures of fingers, toes, hips, and knees, congenital heart defects | 19 |
| 49,XXXXY | Large head circumference, seizures, downslanting palpebral fissures, Brushfield spots, wide flat nose, low-set ears, flexed hip, knee and finger joints, genua valga, low total ridge count, small penis, small testes, unilateral cryptorchidism, renal hypoplasia | 114 |

Table 16–4. Conditions with holoprosencephaly[a] (*cont'd.*)

| Condition | Striking features | References |
|---|---|---|
| **Monogenic holoprosencephaly** | | |
| Meckel syndrome | May have holoprosencephaly with median or lateral cleft lip. Occipital encephalocele, polydactyly, polycystic kidneys, congenital heart defects, genital anomalies, other anomalies. Autosomal recessive inheritance | 71 |
| Autosomal recessive holoprosencephaly[b] | Holoprosencephaly with full spectrum of facial dysmorphism; may vary within families | 33,36 |
| Autosomal dominant holoprosencephaly | Holoprosencephaly with incomplete penetrance and remarkably variable expressivity; widely affected individuals may have only slight hypotelorism, anosmia, hyposmia, single maxillary central incisor, or lateral cleft lip. Others have the severe facial dysmorphism of alobar holoprosencephaly | 15,25,29,74 |
| Holoprosencephaly–fetal hypokinesia syndrome | Microcephaly, holoprosencephaly, sloping forehead, flattened nasal tip, micrognathia, multiple contractures, talipes equinovarus, X-linked recessive inheritance | 68,106 |
| Martin syndrome | Holoprosencephaly with median cleft lip as a variable feature, microcephaly, mental deficiency, hypotelorism, downslanting palpebral fissures, cleft lip, large ears, clubfoot, spinal anomalies, chronic constipation. Autosomal dominant inheritance with incomplete penetrance and remarkably variable expressivity | 100 |
| Velocardiofacial syndrome | Holoprosencephaly with median cleft lip in 1 of 61 cases. Long face, narrow palpebral fissures, prominent nose with square nasal root and narrow alar base, long philtrum, cleft or submucous cleft palate, small ears, cardiac anomalies, learning disability. Autosomal dominant inheritance with variable expression | 165 |
| Grote syndrome | Holoprosencephaly, hydrocephaly, octodactyly, absent tibial bones, cardiac malformations. Autosomal recessive inheritance likely | 59 |
| Steinfeld syndrome | Holoprosencephaly with median cleft lip. Short forearms, absent thumbs, cardiac defects, renal anomalies, absent gallbladder. Most likely autosomal dominant inheritance minimally involving the limb defects | 147 |
| **Monogenic arhinencephaly** | | |
| Kallmann syndrome, Type I | Anosmia, hypogonadotropic hypogonadism, mental deficiency, ocular hypotelorism, unilateral renal agenesis, other anomalies. X-linked or sex-limited autosomal dominant inheritance | 161,162 |
| Kallmann syndrome, Type II | Anosmia, hypogonadotropic hypogonadism, other anomalies. Diabetes mellitus or hearing deficit in some patients. Autosomal recessive inheritance | 162 |
| Isolated anosmia, Type I | Anosmia. X-linked inheritance | 56 |
| Isolated anosmia, Type II | Anosmia. Autosomal dominant inheritance | 96 |
| Perrin syndrome[c] | Anosmia, hypogonadotropic hypogonadism, mental deficiency, congenital ichthyosis. X-linked recessive inheritance | 118 |
| Johnson syndrome | Anosmia or hyposmia, hypogonadotropic hypogonadism, alopecia, conductive deafness, protruding dysplastic ears, other anomalies | 75 |
| Váradi syndrome | Arhinencephaly, cleft lip–palate, lingual nodule, growth deficiency, psychomotor retardation, duplicated halluces, supernumerary fingers, congenital heart defect, cryptorchidism. Autosomal recessive inheritance | 157 |
| Campomelic dysplasia | Macrodolichocephaly, arhinencephaly, slit-like wide set eyes, flat nasal bridge, long philtrum, micrognathia, small mouth, cleft palate, abnormal ears, bell-shaped chest, short bowed femora and tibiae, various roentgenographic abnormalities, pretibial skin dimples, talipes equinovarus, hallux deviation, mild brachydactyly, hypotonia. Autosomal recessive inheritance | 62 |
| Fitch syndrome | Arhinencephaly, absent corpus callosum, hydrocephaly, absent left hemidiaphragm, ventricular septal defect, absent fifth fingernails. Autosomal recessive inheritance likely | 50 |
| Anosmia/radiohumeral synostosis syndrome | Anosmia, Robin sequence, radiohumeral synostosis. Autosomal dominant inheritance most likely | JW Hanson and MM Cohen, Jr, personal observation |
| **Environmentally induced[d]** | | |
| Infants of diabetic mothers | Holoprosencephaly with full spectrum of facial dysmorphism uncommon. Other anomalies | 8 |

| Condition | Striking features | References |
|---|---|---|
| Unknown genesis | | |
| Aicardi syndrome | Flexion spasms, chorioretinal lacunae, vertebral anomalies, mental deficiency, arhinencephaly, agenesis of the corpus callosum, other central nervous system anomalies, rarely holoprosencephaly and/or cleft lip | 129 |
| Aprosencephaly syndrome | Aprosencephaly with facial features of holoprosencephaly, radial aplasia, genital anomalies. One sibship of anencephalic twins and infant with aprosencephaly | 32,95,154 |
| Hall–Pallister syndrome | Congenital hypothalamic hamartoblastoma, hypopituitarism, imperforate anus, postaxial polydactyly, other anomalies including holoprosencephaly, which occurs rarely on a disruptive basis | 63 |
| Hartsfield syndrome | Lobar holoprosencephaly, premature synostosis of coronal and metopic sutures, ocular hypertelorism, cleft lip–palate, ectrodactyly | 65 |
| Holoprosencephaly, ectopia cordis, and embryonal neoplasms | Alobar holoprosencephaly, cleft lip, ectopia cordis, complex cardiac anomalies, neuroblastoma, germ cell tumor of the ovary | KA Aleck, personal communication, 1988 |
| Majewski-like chondrodysplasia | Majewski-like chondrodysplasia with cebocephaly | 110 |
| Pseudotrisomy 13 syndrome | Holoprosencephaly, severe facial dysmorphism, congenital heart defect, postaxial polydactyly, normal karyotype | 32a,140,168 |
| COH syndrome | Arhinencephaly, polymicrogyria, hydrocephaly, cloverleaf skull, duplication of thumb, micropenis, bifid scrotum | 30 |
| Associations | | |
| Holoprosencephaly/neural tube defect association | Anencephaly with holoprosencephalic facies; all cases sporadic to date. Sibship of anencephalic twins and female infant with aprosencephaly. Encephalocele associated with holoprosencephaly but usually as part of specific diagnostic category such as Meckel syndrome or del(18p) | 34,90,154 |
| Holoprosencephaly/frontonasal association malformation | Holoprosencephaly, frontonasal dysplasia, agenesis of the corpus callosum, hydrocephaly, choanal atresia | 17,128 |
| Arhinencephaly/DiGeorge association | Arhinencephaly, absent or hypoplastic parathyroids, absent or hypoplastic thymus, cardiac defects, other anomalies | 38 |
| Goldenhar spectrum/arhinencephaly association | Ocular, auricular, mandibular, and vertebral anomalies. Various central nervous system anomalies, congenital heart defects, and other abnormalities | 3,163 |

[a]From MM Cohen, Jr, *Teratology* **40**:211, 1989.
[b]Sibs with holoprosencephaly and normal facies have been reported (80).
[c]Sunohara et al. (150) reported a condition in which many features are identical to those observed in Perrin syndrome, although differences included nystagmus, decreased visual acuity, strabismus, hypopigmented iris, and mirror movements of the hands and feet in the former. Steroid sulfatase and arylsulfatase C activities in leukocytes and fibroblasts were markedly diminished in affected individuals.
[d]Other suspected or suggested teratogens not included in the table are alcohol (98,127), diphenylhydantoin (M Shokeir, personal communication, 1980), salicylate (12), retinoic acid (84), cytomegalovirus (24), Toxoplasma (92), viral etiology (91), and estroprogestins (145).

documented as the findings in the alobar type (42,43). Spasticity and athetoid movements may occur. Although seizures are common to all patients with alobar holoprosencephaly, they may not be a feature in some patients at the milder end of the phenotypic spectrum. Patients with well-developed brains and absence or hypoplasia of the olfactory tracts and bulbs have anosmia or hyposmia (42).

In alobar holoprosencephaly, electroencephalographic study shows various abnormal waves, ranging from asynchrony between the right and left sides of the head to synchronous and asynchronous spikes. In lobar holoprosencephaly, multifocal spikes are common (42,43).

Patients with alobar holoprosencephaly are completely amented and do not survive infancy. With semilobar holoprosencephaly, patients are more or less severely amented and some survive into childhood. With lobar holoprosencephaly, most patients are severely mentally retarded and may survive infancy and childhood. At the less severe end of the holoprosencephalic spectrum, patients may be mildly or moderately retarded and some have sufficient intelligence to live freely in society (42,43).

**Face.** In a classic article entitled "The Face Predicts the Brain," DeMyer et al (44) discussed the graded series of facial anomalies that occur with holoprosencephaly (Fig. 16–7A–F). In cyclopia, the most extreme variant, a single median eye with varying degrees of doubling of the intrinsic ocular structures is associated with arhinia and usually with proboscis formation. In ethmocephaly, two separate hypoteloric eyes are associated with arhinia and proboscis formation. In cebocephaly, hypotelorism is associated with a blind-ended, single-nostril nose. With median cleft lip, hypotelorism is associated with a flat nose and a median cleft due to agenesis of the primary palate. Less severe facial dysmorphism may include hypotelorism or hypertelorism, lateral cleft lip, iris coloboma, or single maxillary central incisor (14,28,30,36,44,66,85) (Table 16–3).

Although the face–brain correlation within this spectrum is strongly positive, DeMyer et al (44) noted that the correlation weakened at the more normal end of the spectrum. It is now apparent that even at the severe end of the phenotypic spectrum, alobar holoprosencephaly may be observed with a normal face. DeMyer (42) indicated that 17% of his personally studied cases had a nondiagnostic face. Furthermore, severe facial dysmorphism, particularly median cleft lip, may rarely occur with perfectly normal brain development (9). Thus, there are exceptions to the strong face–brain correlation in some severe cases as well as in some mild cases (31,42,44).

Facial variability can be observed with syndromic or with nonsyndromic holoprosencephaly. In familial nonsyndromic holoprosencephaly, major facial dysmorphism may be the same among different family members or may differ among various family members (36,170). Microforms of holoprosencephaly may occur in some. Affected family members with holoprosencephaly and normal faces have also been reported (80). For *agnathia-holoprosencephaly,* the reader is referred to the section on *otocephaly,* Table 16–4, and specific references (88,115). For *aprosencephaly,* the reader should consult Table 16–4 or specific references (95,154).

**Differential diagnosis.** Hypotelorism may be observed with craniostenotic trigonocephaly (29,35). Isolated median cleft lip with normal brain development has been reported (13). Some instances of lateral nasal proboscis represent an accessory proboscis in the presence of both a normal nose and a normal brain (29,35). In *Binder syndrome,* hypoplasia of the anterior nasal spine occurs with normal intelligence; the facial appearance may appear "arhinencephaloid" (29,35). Agenesis of the corpus callosum may occur as an isolated anomaly independent of holoprosencephaly or may occur as a component in a variety of nonholoprosencephalic syndromes (29,35,42, 46,99). The relationship between holoprosencephaly and septooptic dysplasia or with *Rieger syndrome* remains speculative (29,35). Unilateral arhinencephaly has been noted in a patient with *oculo-auriculo-vertebral spectrum* (3). Single maxillary central incisor may occur as an isolated defect, with hypothalamic hamartoma and sexual precocity, and also with coloboma and hypomelanosis of Ito (9,35,140).

**Laboratory aids.** Any patient suspected of having holoprosencephaly by physical examination and neurologic observation and testing is a candidate for CT scanning (23,29,51). All patients with holoprosencephaly should have a banded chromosome study. Antenatal diagnosis with ultrasound during the third trimester has been reported. The diagnosis of alobar holoprosencephaly is virtually certain when hypotelorism is found in association with absence of the midline echo. Sonographic demonstration of orofacial clefting may be helpful. In semilobar and lobar holoprosencephaly, sonographic diagnosis is difficult or impossible (26).

### References (The holoprosencephaly disorders)


1. Adelmann HB: The problem of cyclopia. *Q Rev Biol* **11**:161–182, 284–304, 1936.

2. Ahlfeld F: *Die Missbildungen des Menschen.* FW Grunow, Leipzig, 1880–1882.

3. Aleksic S et al: Unilateral arhinencephaly in Goldenhar–Gorlin syndrome. *Dev Med Child Neurol* **17**:498–504, 1975.

4. Ardinger HH, Bartley JA: Microcephaly in familial holoprosencephaly. *J Craniofac Genet Dev Biol* **8**:53–61, 1988.

5. Aronson DC et al: A male infant with holoprosencephaly, associated with a ring chromosome 21. *Clin Genet* **31**:48–52, 1987.

6. Back E et al: Partial monosomy 22 pter→q11 in a newborn with the clinical features of trisomy 13 syndrome. *Ann Génét* **23**:244, 1980.

7. Ballantyne JW: *Antenatal Pathology and Hygiene,* Green W (ed). Edinburgh, 1904.

8. Barr M et al: Holoprosencephaly in infants of diabetic mothers. *J Pediatr* **102**:565–568, 1983.

9. Bartholomew DW et al: Single maxillary central incisor and coloboma in hypomelanosis of Ito. *Clin Genet* **32**:370–373, 1987.

10. Becker, LE, Takada K: Structural malformations of the cerebral hemispheres. In *Disorders of the Developing Central Nervous System: Diagnosis and Treatment.* Hoffman HJ, Epstein F (eds), Blackwell Scientific Publications, Boston, 1986, pp 191–223.

11. Begleiter ML, Harris DJ: Brief clinical report: Holoprosencephaly and endocrine dysgenesis in brothers. *Am J Med Genet* **7**:315–318, 1980.

12. Benawra R et al: Cyclopia and other anomalies following maternal ingestion of salicylates. *J Pediatr* **96**:1069, 1980.

13. Ben-Hur N et al: An unusual case of median cleft lip with orbital hypotelorism—a missing link in the classification. *Cleft Palate J* **15**:365–368, 1978.

14. Benke PJ, Cohen MM Jr: Recurrence of holoprosencephaly in families with a positive history. *Clin Genet* **24**:324–328, 1983.

15. Berry SA et al: Single central incisor in familial holoprosencephaly. *J Pediatr* **104**:877–880, 1984.

16. Binns W et al: A congenital cyclopian-type malformation in lambs induced by maternal ingestion of a range plant, *Veratrum californicum. Am J Vet Res* **24**:1164–1175, 1963.

16a. Bogart MH: Terminal deletions of the long arm of chromosome 7: Five new cases. *Am J Med Genet,* in press.

17. Bomelburg T et al: Median cleft face syndrome in association with hydrocephalus, agenesis of the corpus callosum, holoprosencephaly and choanal atresia. *Eur J Pediatr* **146**:301–302, 1987.

18. Boué J, Albert DM: The eyes of embryos with chromosomal abnormalities. *Am J Ophthalmol* **78**:167, 1974.

19. Buckton KE, Barr DGD: Partial trisomy for long arm of chromosome 16. *J Med Genet* **18**:483, 1981.

20. Burck U: Genetic counseling in holoprosencephaly. *Helv Paediat Acta* **37**:231–237, 1982.

21. Burck U et al: Occurrence of cyclopia, myelomeningocele, deafness, and abducens paralysis in siblings. *Am J Med Genet* **11**:443–448, 1982.

22. Burck U et al: Brief clinical report: Holoprosencephaly in monozygotic twins—clinical and computer tomographic findings. *Am J Med Genet* **9**:13–17, 1981.

23. Byrd SE et al: Computed tomography evaluation of holoprosencephaly in infants and children. *J Comp Assist Tomog* **1**:445–463, 1977.

24. Byrne PJ et al: Cyclopia and congenital cytomegalovirus infection. *Am J Med Genet* **28**:61–65, 1987.

25. Cantu JM: Dominant inheritance of holoprosencephaly. *Birth Defects* **14**(6B):215–220, 1978.

26. Chervenak FA et al: The obstetric significance of holoprosencephaly. *Obstet Gynecol* **63**:115–121, 1984.

27. Cohen MM Jr: Chromosomal mosaicism associated with a case of cyclopia. *J Pediatr* **69**:793–798, 1966.

28. Cohen MM Jr: Holoprosencephaly revisited. *Am J Dis Child* **127**:597, 1974.

29. Cohen MM Jr: Holoprosencephaly: Cyclopia series, in *Mental Retardation and Congenital Malformations of the Central Nervous System.* Warkany J, Lemire JT, Cohen MM Jr (eds), Year Book Medical Publishers, Chicago, 1981, pp 1176–1190.

30. Cohen MM Jr: An update on the holoprosencephalic disorders. *J Pediatr* **101**:865–869, 1982.

31. Cohen MM Jr: Perspectives on holoprosencephaly. Part I. Epidemiology, genetics, and syndromology. *Teratology* **40**:211–236, 1989.

32. Cohen MM Jr: Perspectives on holoprosencephaly. Part III. Spectra, distinctions, continuities, and discontinuities. *Am J Med Genet* **34**:271–288, 1989.

32a. Cohen MM Jr: Holoprosencephaly and cytogenetic findings: Further information. *Am J Med Genet* **34**:265, 1989.

33. Cohen MM Jr, Gorlin RJ: Genetic considerations in a sibship of cylopia and clefts. *Birth Defects* **5**(2):113–118, 1969.

34. Cohen MM Jr, Lemire RJ: Syndromes with cephaloceles. *Teratology* **25**:161–172, 1982.

35. Cohen MM Jr, Sulik KK: Perspectives on holoprosencephaly. Part II. Central nervous system, craniofacial anatomy, syndrome commentary, diagnostic approach, and experimental studies. *J Craniofacial Genet Dev Biol,* in press.

36. Cohen MM Jr et al: Holoprosencephaly and facial dysmorphia: Nosology, etiology and pathogenesis. *Birth Defects* **7**(7):125–135, 1971.

37. Conen PE et al: The "D" syndrome. *Am J Dis Child* **111**:236–247, 1966.

38. Conley ME et al: The spectrum of DiGeorge syndrome. *J Pediatr* **94**:883–890, 1979.

39. Cottrall K et al: A case of proximal 14 trisomy with pathological findings. *J Ment Def Res* **25**:1–6, 1981.

40. Dallaire L et al: Familial holoprosencephaly. *Birth Defects* **7**(7):136–142, 1971.

41. Dareste C: Mode de la formation de la cyclopie. *Ann d'Ocul* **106**:171–182, 1891.

42. DeMyer W: Holoprosencephaly (cyclopia-arhinencephaly), in *Handbook of Clinical Neurology,* Vinken PJ, Bruyn GW (eds), North-Holland Publishing Co, Amsterdam, 1977, Ch. 18, pp 431–478.

43. DeMyer W et al: Familial alobar holoprosencephaly (arhinencephaly) with median cleft lip and palate. *Neurology (Minneap)* **13**:913–918, 1963.

44. DeMyer WE et al: The face predicts the brain: Diagnostic significance of median facial anomalies for holoprosencephaly (arhinencephaly). *Pediatrics* **34**:256–263, 1964.

45. Dinno ND et al: 47,XY,t(9p+;11q+) in a male infant with multiple malformations. *Clin Genet* **6**:125–131, 1974.

46. Dionisi Vici C et al: Agenesis of the corpus callosum, combined immunodeficiency, bilateral cataract, and hypopigmentation in two brothers. *Am J Med Genet* **29**:1–8, 1988.

47. Emanuel I et al: The incidence of congenital malformations in a Chinese population. *Teratology* **5**:159–169, 1974.
48. Esser AAM: Kyklopenauge, Kyklopen und Arimasper. *Klin Monatsbl Augenheilkd* **79**:398–400, 1927.
49. Faust J et al: The 18p− syndrome. *Eur J Pediatr* **123**:59–66, 1976.
50. Fitch N et al: Absent left hemidiaphragm, arhinencephaly, and cardiac malformations. *J Med Genet* **15**:399–401, 1978.
51. Fitz CR: Holoprosencephaly and related entities. *Neuroradiology* **25**:225–238, 1983.
52. Förster A: *Die Missbildungen des Menschen: Systematisch Dargestellt,* Druck und Verlag von Friedrich Mavke, Jena, 1865.
53. Friedrich U et al: A girl with karyotype 46,XX,del(7)(pter→q32:). *Hum Genet* **51**:231–235, 1979.
54. Gardner R et al: Are 1q+ chromosomes harmless? *Clin Genet* **6**:383–393, 1974.
55. Gillerot Y et al: Brief clinical report: Prenatal diagnosis of a dup(3p) with holoprosencephaly. *Am J Med Genet* **26**:225–227, 1987.
56. Glaser O: Hereditary deficiencies in the sense of smell. *Science* **48**:647–648, 1918.
57. Gimelli G et al: dup(3)(p2→pter) in two families, including one infant with cyclopia. *Am J Med Genet* **20**:341–348, 1985.
58. Gorlin RJ et al: Short arm deletion of chromosome 18 in cebocephaly. *Am J Dis Child* **115**:473–476, 1968.
59. Grote W et al: Prenatal diagnosis of a probable hereditary syndrome with holoprosencephaly, hydrocephaly, octodactyly, and cardiac malformations. *Eur J Pediatr* **143**:155–157, 1984.
60. Grouchy J de: The 18p−, 18q−, and 18r syndromes. *Birth Defects* **5**(5):74–87, 1969.
60a. Grundy HO et al: Prenatal detection of cyclopia associated with interstitial deletion of 2p. *Am J Med Genet* **34**:268–270, 1989.
61. Habedank M, Ekkehard T: Clinical and neuropathological investigations of four cases of holoprosencephaly with arhinencephaly. *Neuropädiatrie* **2**:144–163, 1970.
62. Hall BD, Spranger JW: Campomelic dysplasia: Further elucidation of a distinct entity. *Am J Dis Child* **134**:285, 1980.
63. Hall JG et al: Congenital hypothalmic hamartoblastoma, hypopituitarism, imperforate anus, and postaxial polydactyly: A new syndrome? I. Clinical, causal, and pathogenetic considerations. *Am J Med Genet* **7**:47–74, 1980.
64. Hannover A: *Der Menneskelige Hjerneskals Bygning ved Zyklopie,* Copenhagen, 1882.
65. Hartsfield JK et al: Syndrome identification case report 119: Hypertelorism associated with holoprosencephaly and ectrodactyly. *J Clin Dysmorphol* **2**(2):27–31, 1984.
66. Hattori H et al: Brief clinical report: Single central maxillary incisor and holoprosencephaly. *Am J Med Genet* **28**:483–487, 1987.
66a. Helmuth RA et al: Holoprosencephaly, ear abnormalities, congenital heart defect, and microphallus in a patient with 11q− mosaicism. *Am J Med Genet* **32**:178–181, 1989.
66b. Hess LE et al: Semilobar holoprosencephaly associated with 6p− aneuploidy (in preparation).
67. Hill LM et al: Ultrasonic findings with holoprosencephaly. *J Reprod Med* **27**:172–175, 1982.
68. Hockey A et al: Microcephaly, holoprosencephaly, hypokinesia—second report of a new syndrome. *Prenatal Diagnosis* **8**:683–686, 1988.
69. Hodes ME et al: Trisomy 18 (29 cases) and trisomy 13 (19 cases): A summary. *Birth Defects* **14**(6C):377–382, 1978.
70. Hoovers JMN, Jansweijer MCE: Holoprosencephaly associated with ring chromosome 21. *Clin Genet* **32**:207, 1987.
71. Hsia YE et al: Genetics of the Meckel syndrome (dysencephalia splanchnocystica). *Pediatrics* **48**:237–247, 1971.
72. Hsia YE et al: Chromsomal abnormality (46,XX,3p+) in a case of the Meckel syndrome. *Birth Defects* **10**(8):19–25, 1974.
73. Hunter AGW et al: Cebocephaly in an infant with trisomy 18. *J Med Genet* **14**:291–292, 1977.
74. Jaramillo C et al: Autosomal dominant inheritance of the DeMyer sequence. *J Craniofac Genet Dev Biol* **8**:199–204, 1988.
75. Johnson VP et al: A newly recognized neuroectodermal syndrome of familial alopecia, anosmia, deafness, and hypogonadism. *Am J Med Genet* **15**:497–506, 1983.
76. Kallmann FJ et al: The genetic aspects of primary eunuchoidism. *Am J Ment Defic* **48**:203–236, 1944.
77. Kalter H: *Teratology of the Central Nervous System.* University of Chicago Press, Chicago, 1968.
78. Keeler RF: Teratogenic effects of cyclopamine and jervine in rats, mice and hamsters. *Proc Soc Exp Biol Med* **149**:302–306, 1975.
79. Keeler RF: Teratogenic compounds of *Veratrum californicum* (Durand). X. Cyclopia in rabbits produced by cyclopamine. *Teratology* **3**:175–180, 1970.
80. Khan M et al: Familial holoprosencephaly. *Dev Med Child Neurol* **12**:71–76, 1970.
81. Kokich VG et al: Cyclopia: An anatomic and histologic study of two specimens. *Teratology* **26**:105–113, 1982.
81a. Krassikoff N, Sekhon GS: Familial agnathia–holoprosencephaly due to an inherited unbalanced translocation and not autosomal recessive inheritance. *Am J Med Genet* **34**:255–257, 1989.
82. Kurtzman DN et al: Brief clinical report: Duplication 3p21→3pter and cyclopia. *Am J Med Genet* **27**:33–37, 1987.
83. Lambert JC et al: Triploidy with cyclopia and identical HLA alleles in the parents. *J Med Genet* **21**:63–66, 1984.
84. Lammer EJ et al: Retinoic acid embryopathy. *N Engl J Med* **313**:837–841, 1985.
85. Lang AP et al: Trisomy 18 and cyclopia. *Teratology* **14**:195–204, 1976.
86. Laurence KM, Ishmael J: Arhinencephaly and 13-15(D) trisomy. Oxford Chromosome Conf. Chromosomes Today, Proc. 2, 1969, pp 86–89.
87. Leech RW, Shuman RM: Holoprosencephaly and related midline cerebral anomalies: A review. *J Child Neurol* **1**:3–18, 1986.
88. Leech RW et al: Agnathia, holoprosencephaly and situs inversus: Report of a case. *Am J Med Genet* **29**:483–490, 1988.
89. Lejeune J et al: Sur trois cas de trisomie C. *Ann Génét* **12**:28–35, 1969.
90. Lemire RJ et al: The facial features of holoprosencephaly in anencephalic human specimens. I. Historical review and associated malformations. *Teratology* **23**:297–303, 1981.
91. Lillard RL: The cyclops deformity. Case report. *Rocky Mtn Med J* **61**:32–34, 1964.
92. Lison MP et al: Arhinencephalia: Considerations à propos a case diagnosed during life. *Acta Neurol Belg* **67**:25–36, 1967.
93. Lowry RB: Holoprosencephaly. *Am J Dis Child* **128**:887, 1974.
94. Lurie IW, Lazjuk GJ: Partial monosomies 18: Review of cytogenetical and phenotypical variants. *Humangenetik* **15**:203–222, 1972.
95. Lurie I et al: Brief clinical reports: Aprosencephaly-atelencephaly and the aprosencephaly (XK) syndrome. *Am J Med Genet* **3**:303–309, 1979.
96. Lygonis CS: Familial absence of olfaction. *Hereditas* **61**:415–416, 1969.
97. Machin GA et al: Monozygotic twin aborted fetuses discordant for holoprosencephaly/synotia. *Teratology* **31**:203–215, 1985.
98. Majewski F: Alcohol embryopathy: Some facts and speculations about pathogenesis. *Neurobehav Toxicol Teratol* **3**:129–144, 1981.
99. Marburg O: So-called agenesia of the corpus callosum (callosal defect). *Arch Neurol Psychiat* **61**:297–312, 1949.
100. Martin AO et al: An autosomal dominant midline cleft syndrome resembling familial holoprosencephaly. *Clin Genet* **12**:65–72, 1977.
101. Martin NJ, Steinberg BG: The dup(3)(p25→pter) syndrome: A case with holoprosencephaly. *Am J Med Genet* **14**:767–772, 1983.
102. Matsunaga E, Shiota K: Holoprosencephaly in human embryos: Epidemiologic studies of 150 cases. *Teratology* **16**:261–272, 1977.
103. Michels VV et al: Duplication of part of chromosome 1q: Clinical report and review of literature. *Am J Med Genet* **18**:125–134, 1984.
104. Mieden GD: An anatomical study of three cases of alobar holoprosencephaly. *Teratology* **26**:123–133, 1982.
105. Ming PM et al: Cytogenetic variants in holoprosencephaly. *Am J Dis Child* **130**:864–867, 1976.
106. Morse RP et al: Prenatal diagnosis of a new syndrome: Holoprosencephaly with hypokinesia. *Prenatal Diagnosis* **7**:631–638, 1987.
106a. Münke M: Clinical cytogenetic and molecular approaches to the genetic heterogeneity of holoprosencephaly. *Am J Med Genet* **34**:237–245, 1989.
107. Münke M et al: Holoprosencephaly: Association with interstitial deletion of 2p and review of the cytogenetic literature. *Am J Med Genet* **30**:929–938, 1988.
108. Myrianthopoulos NC, Chung CS: Congenital malformations in singletons: Epidemiologic survey. *Birth Defects* **10**(11):1–58, 1974.
109. Neu RL et al: An infant with trisomy 6q21→qter. *Ann Génét* **24**:167–169, 1981.
110. Nivelon-Chevallier A et al: Chondrodysplasie letale à côtes, courtes, type Majewski diagnostic in utero. *Pédiatrie* **37**:453–460, 1982.
111. Opitz JM et al: Report of a patient with a presumed Dq− syndrome. *Birth Defects* **5**(5):93–99, 1969.
112. Orbeli DJ et al: The syndrome associated with the partial D-monosomy. Case report and review. *Humangenetik* **13**:296–308, 1971.
113. Otto AW: *Lehrbuch der pathologischen Anatomie.* Rücker, Berlin, 1830.
114. Pallister PD: 49,XXXXY syndrome. *Am J Med Genet* **13**:337–339, 1982.
115. Pauli RM et al: Familial agnathia—holoprosencephaly. *Am J Med Genet* **14**:677–698, 1983.

116. Pearson G et al: Inversion duplication of chromosome 6 with trisomic codominant expression of HLA antigens. *Am J Hum Genet* **31**:29–34, 1979.

117. Peeden JN et al: Ring chromosome six: Presentation of three cases and literature review. *Am J Hum Genet* **34** (Suppl):104, Abstract No. 286, 1982.

118. Perrin JCS et al: X-linked syndrome of congenital ichthyosis, hypogonadism, mental retardation and anosmia. *Birth Defects* **7**(5):267–274, 1976.

119. Pettersen JC: An anatomical study of two cases of cebocephaly, in *Development of the Basicranium*. Bosma JF (ed), US Dept Health Ed Welfare No. (NIH) 760989, Bethesda, 1976, pp 240–265.

120. Pfeiffer RA: *Karyotyp und Phänotyp der autosomalen Chromosomenaberrationen beim Menschen*. Stuttgart: Gustav Fischer, 1968.

121. Pfitzer P, Müntefering H: Cyclopism as a hereditary malformation. *Nature (London)* **217**:1071–1072, 1968.

122. Pi SY et al: Brief clinical reports: Holoprosencephaly in a Down syndrome child. *Am J Med Genet* **5**:201–206, 1980.

123. Pihko E et al: Partial 11q trisomy syndrome. *Hum Genet* **58**:129–134, 1981.

124. Probst EP: *The Prosencephalies: Morphology, Neuroradiological Appearance and Differential Diagnosis*. Springer-Verlag, Berlin, 1979.

125. Roach E et al: Holoprosencephaly: Birth data, genetic and demographic analysis of 30 families. *Birth Defects* **11**(2):294–313, 1975.

126. Romshe C, Sotos JF: Hypothalamic-pituitary deformation in siblings of patients with holoprosencephaly. *J Pediatr* **83**:1088–1089, 1973.

127. Ronen GM, Andrews WL: Holoprosencephaly as a possible embryonic alcohol effect. *Child Neurology Society Meeting*. San Diego, October 20, 1987.

128. Roubicek M et al: Frontonasal dysplasia as an expression of holoprosencephaly. *Eur J Pediatr* **137**:229–231, 1981.

129. Sato N et al: Aicardi syndrome with holoprosencephaly and cleft lip and palate. *Pediatr Neurol* **3**:114–116, 1987.

130. Saunders ES et al: What is the incidence of holoprosencephaly? *J Med Genet* **21**:21–26, 1984.

131. Schinzel A: *Catalogue of Unbalanced Chromosome Aberrations in Man*. Walter de Gruyter, Berlin and New York, 1984.

132. Schinzel A: Cyclopia and cebocephaly in two newborn infants with unbalanced segregation of a familial translocation rcp(1;7)(q32;q34) familial translocation. *Am J Med Genet* **18**:153–161, 1984.

133. Schinzel A: Incomplete trisomy 22. II. Familial trisomy of the distal segment of chromosome 22q in two brothers from a mother with a translocation, t(6;22)(q27:q13). *Hum Genet* **56**:263–268, 1981.

134. Schinzel A: Letter to the editor: A further case of cyclopia due to unbalanced segregation of a previously reported rcp(1;7)(q32;q34) familial translocation. *Am J Med Genet* **24**:205–206, 1986.

135. Schinzel A, Schmidt W: Trisomie 18. *Helv Paediatr Acta* **26**:673, 1971.

136. Schinzel A et al: Prenatal ultrasonographic diagnosis of holoprosencephaly, two cases of cebocephaly, and two of cyclopia. *Arch Gynecol* **236**:47–53, 1984.

137. Schroeder HW et al: Recombination aneusomy of chromosome 5 associated with multiple severe congenital malformations. *Clin Genet* **30**:285–292, 1986.

138. Schwartz S et al: Brief clinical report: Cebocephaly-holoprosencephaly in a newborn girl with a terminal 7q deletion [46,XX,del(7)(pter→q31:]. *Am J Med Genet* **15**:141–144, 1983.

139. Sedano HO, Gorlin RJ: The oral manifestations of cyclopia. *Oral Surg* **16**:823–838, 1963.

140. Shiota K, Tanimura T: Holoprosencephaly, ventricular septal defect, and postaxial polydactyly in a human embryo. *J Med Genet,* **25**:502–503, 1988.

141. Siebert JR et al: *Holoprosencephaly*. Alan R. Liss, New York, in press.

142. Smart RD et al: Brief clinical report: Cyclopia as a result of an unbalanced familial translocation, rcp(7;18)(q34;q21). *Am J Med Genet* **24**:269–272, 1986.

143. Smith BS, Pettersen JC: An anatomical study of a duplicaton 6p based on two sibs. *Am J Med Genet* **20**:649–663, 1985.

144. St. Hilaire IG: *Historie Générale et Particulière des Anomalies de l'Organisation chez l'Homme et les Animaux*. JB Baillere, Paris, 1832.

145. Stabile M et al: A case of suspected teratogenic holoprosencephaly. *J Med Genet* **21**:147–149, 1984.

146. Steinbach P, Rehder H: Tetrasomy for the short arm of chromosome 12 with accessory isochromosome [+(12p)] and a marked LDH-B gene dosage effect. *Clin Genet* **32**:1–4, 1987.

147. Steinfeld HJ: Case report 81: Holoprosencephaly and visceral defects with familial limb abnormalities. *Syndrome Ident* **8**(1):1–2, 1982.

148. Steward VW, Jevtic MM: Derangement of neuronal migration in a child with multiple congenital anomalies, two congenital neoplasms, without apparent chromosomal abnormalities. *J Neuropathol Exp Neurol* **38**:259–285, 1979.

149. Sulik KK, Johnston MC: Embryonic origin of holoprosencephaly: Interrelationship of the developing face and brain. *Scanning Electron Microsc* **1**:309–322, 1982.

150. Sunohara N et al: A new syndrome of anosmia, ichthyosis, hypogonadism, and various neurological manifestations with deficiency of steroid sulfatase and arylsulfatase C. *Ann Neurol* **19**:174–181, 1986.

151. Taruffi C: *Storia della Teratologia*. Regia Typographia, Bologna, 1894.

152. Taylor AI: Autosomal trisomy syndromes: A detailed study of 27 cases of Edwards' syndrome and 27 cases of Patau's syndrome. *J Med Genet* **5**:227–252, 1968.

153. Torczynski E et al: Synophthalmia and cyclopia: A histopathologic, radiographic and organogenetic analysis. *J Doc Ophthalmol* **44**:311–378, 1977.

154. Townes PL et al: XK aprosencephaly and anencephaly in sibs. *Am J Med Genet* **29**:523–528, 1988.

155. Urioste M et al: Holoprosencephaly and trisomy 21 in a child born to a nondiabetic mother. *Am J Med Genet* **30**:925–928, 1988.

156. van Duyse D: Pathogénie de la cyclopie. *Arch Ophtalmol* **18**:481–508, 581–601, 623–637, 1898.

157. Váradi V et al: Syndrome of polydactyly, cleft lip/palate or lingual lump, and psychomotor retardation in endogamic gypsies. *J Med Genet* **17**:119–122, 1980.

158. Voiculescu I et al: Trisomy 22 in a newborn with multiple malformations. *Hum Genet* **76**:298–301, 1987.

159. Vrolik W: *Tabulae ad Illustrandam Embryogenesin Hominis et Mammalium*. G.M.P. Londonck, Amstelodami, 1849.

160. Warkany J: *Congenital Malformations: Notes and Comments*. Year Book Medical Publishers, Chicago, 1971.

161. Wegenke JD et al: Familial Kallmann syndrome with junilateral renal aplasia. *Int J Clin Genet* **7**:368–381, 1975.

162. White BJ et al: The syndrome of anosmia with hypogonadotropic hypogonadism: A genetic study of 18 new families and a review. *Am J Med Genet* **15**:417–435, 1983.

163. Wilson GN: Cranial defects in the Goldenhar syndrome. *Am J Med Genet* **14**:435–443, 1983.

164. Wilson GN et al: Occurrence of holoprosencephaly in chromosome 13 disorders cannot be explained by duplication/deficiency of a single locus. *Am J Med Genet Suppl* **2**:65–72, 1986.

164a. Wilson GN et al: Brief clinical report: Holoprosencephaly and interstitial deletion of 2(p2101p2109). *Am J Med Genet* **34**:252–254, 1989.

165. Wraith JE et al: Velo-cardio-facial syndrome presenting as holoprosencephaly. *Clin Genet* **27**:408–410, 1985.

166. Wright S, Wagner K: Types of subnormal development of the heads from inbred strains of guinea pigs and their bearing on the classification and interpretation of vertebrate monsters. *Am J Anat* **54**:383–447, 1934.

167. Yakovlev PI: Pathoarchitectonic studies of cerebral malformations. III. Arrhinencephalies (holotelencephalies). *J Neuropathol Exp Neurol* **18**:22–55, 1959.

168. Young ID, Madders DJ: Unknown syndrome: Holoprosencephaly, congenital heart defects, and polydactyly. *J Med Genet* **24**:714–716, 1987.

169. Zergollern L et al: A liveborn infant with triploidy (69,XXX). *Z Kinderheilk* **112**:293–300, 1972.

170. Zwetsloot CP et al: Holoprosencephaly: Variation of expression in face and brain in three sibs. *J Med Genet* **26**:274–276, 1989.

## Octocephaly (agnathia)

The primary malformation in otocephaly is agnathia, with malplacement of the external ears with or without fusion, microstomia, and persistence of the buccopharyngeal membrane likely being secondary effects of absence or hypoplasia of the mandibular arch (Fig. 16–8A–E). The condition is lethal since oropharyngeal abnormalities result in an impatent or poorly functioning airway. Agnathia may occur alone (2–6,11,13–15,17–21,23,25,26,28,30–32) or together with holoprosencephaly (1,7,7a,9,12,16,22,24,27,29). By the end of the last century, 22 cases of isolated agnathia and 29 cases of agnathia with cyclopia had been recorded. Possibly the first case was described by Kerckring in 1717 (see 23). Within the last 25 years, 24 cases have been reported, including 15 cases of isolated agnathia and 9 cases of agnathia with holoprosencephaly (5). An excellent anatomic study was carried out by Lawrence and Bersu (18). The maximal prevalence is almost certainly less than 1/70,000 births (23).

The terms agnathia, synotia, and otocephaly, for all practical purposes. have approximately equivalent meanings, even though the first term specifically refers to absent mandible, the second term to fused

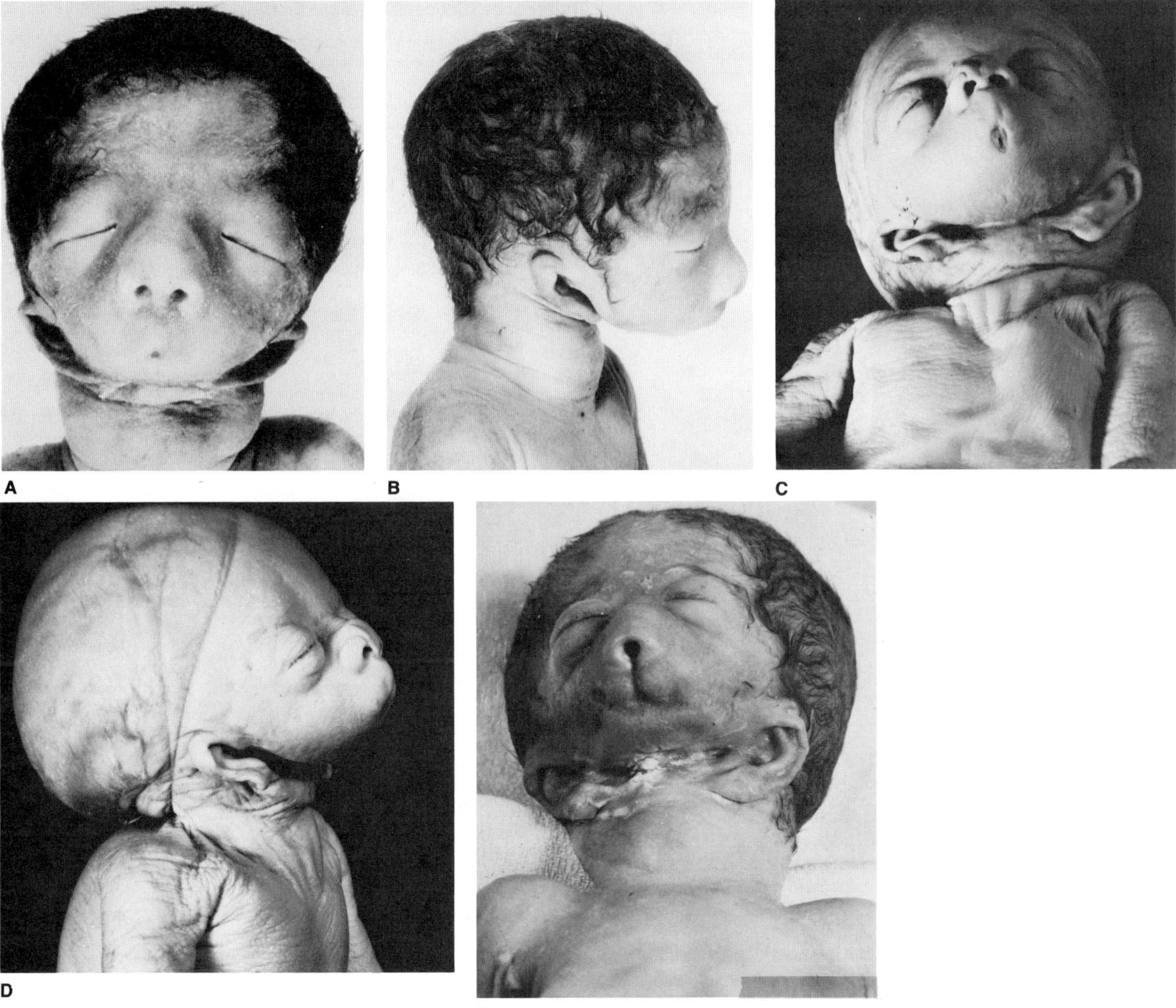


Fig. 16–8. *Otocephaly.* (A–E) Note tendency of ears to fuse in midline, absence of mandible. Note separate but tiny mouth in A and C. Note unified nose and mouth in E. (C–E courtesy of *HW Edmonds,* Washington, DC.)

ears. Agnathia with cyclopia has been called cyclocephaly, cyclopia hypognathus, and agnathia-holoprosencephaly (23,24).

To date, all cases of isolated agnathia have been sporadic. All cases of hypognathic cyclopia have also been sporadic. One instance of familial agnathia-holoprosencephaly has been reported in which one sib had cebocephaly with agnathia and the other had a nondiagnostic face for holoprosencephaly but with agnathia and absent olfactory tracts and bulbs (24). In this family, the condition was caused by a chromosomal anomaly—46,XX,t(6;18)(p24.1;p1.21)—not originally reported (RM Pauli, personal communication, 1987;16a). Sewall Wright (33) showed that with inbred guinea pigs, there was a holoprosencephalic continuum of face–brain defects that included agnathia. Since genetic forms have been reported in other species, a similar spectrum has been postulated for humans (5). However, in our opinion, the autosomal recessive form of isolated holoprosencephaly should be distinguished from agnathia-holoprosencephaly because, to date, none of the many reported sibships of the former has had agnathia. Thus, the continuum has never been recorded within the same sibship (8). Pauli et al (23) reported a recurrent-pattern syndrome in two nonrelated infants who had agnathia, situs inversus, unilateral renal agenesis, renal ectopia, absent ribs, and vertebral anomalies. Some of these defects have been reported individually in association with agnathia in a few other instances (23). Thus, the validity of this early recurrent pattern syndrome depends on future reports. Leech et al (19a) reported a sporadic case of agnathia-holoprosencephaly with situs inversus.

With isolated agnathia, the average gestation is 32.7 weeks and birthweights are commonly below the tenth centile. Polyhydramnios has been a feature of most pregnancies (5). Affected infants die within the first few hours or days of life (23). The findings in 13 cases of isolated agnathia, based on observations of Bixler et al (5), are summarized in Table 16–5.

**Craniofacial features.** The mandible is either absent or hypoplastic. Although some authors have claimed that the mandible is never completely absent (14,31), radiography and dissection have failed to reveal any evidence of a mandible in a number of cases (7,23,26,28). The oral aperture may be absent (astomia), but more often is a minute opening 2–3 mm in diameter with its long axis usually rotated 90°. The buccopharyngeal membrane may persist, and microglossia of extreme degree may be present low in the pharynx. The external ears

Table 16–5. Published cases of isolated agnathia (1961–1986)[a]

| Feature | Frequency |
|---|---|
| Craniofacial | |
| Mandible | |
| Agnathia | 12/13 |
| Micrognathia | 1/13 |
| Microstomia | 13/13 |
| Aglossia or extreme microglossia | 9/10 |
| Blind mouth | 10/11 |
| Ears | |
| Low-set (not fused) | 2/13 |
| Approach midline (not fused) | 7/13 |
| Fused | 4/13 |
| Middle ear anomalies | 4/5 |
| Clefts | (1CL + 4CP)/9[b] |
| Downslanting palpebral fissures | 12/12 |
| General | |
| Sex | (6M + 6F)/12[c] |
| Race | (5W + 4B)/9[d] |
| Gestational age | $\bar{x}$ = 32.7 weeks, $n$ = 13 |
| Low birthweight (<10th percentile) | 7/11 |
| Polyhydramnios | 10/11 |

[a]Thirteen total cases based on Bixler et al (5), Black et al (6), Johnson and Cook (11), Lawrence and Bersu (18), Leckie (19), Le Marec et al (20), Libersa and Heritier (21), Scholl (26), Ursell (28), Van de Sande (30), Woon and Tan (32).
[b]CL, cleft lip; CP, cleft palate.
[c]M, male; F, female.
[d]W, white; B, black.

are malformed, ventrally placed, and approach each other or fuse in the midline (Fig. 16–8A–E). Ear canals may be atretic and middle ear anomalies may occur in some cases. Rarely, the external ears may be absent. The palate may be cleft. The palpebral fissures slant downward (5,23).

Other craniofacial features have been reported including hypertelorism, proptosis, blepharophimosis, epibulbar dermoid, synophrys, cleft lip, choanal atresia, abnormal inner ear and vestibular apparatus, bilobed epiglottis, and rudimentary vocal cords (23).

When agnathia is associated with holoprosencephaly, most instances represent cyclopia. Hypognathic cyclopia differs from classic cyclopia in that the proboscis located above the single median eye in the latter is usually absent in the former (8). Uncommonly, other facial forms of holoprosencephaly are associated with agnathia. For example, Pauli et al (24) observed two agnathic-holoprosencephalic sibs—one cebocephalic infant and one with a nondiagnostic face. Leech et al (19a) recorded agnathic cebocephaly.

**Central nervous system.** In hypognathic cyclopia, alobar holoprosencephaly is present. In the agnathic-holoprosencephalic sibs reported by Pauli et al (24), one had alobar holoprosencephaly and the other had absent olfactory tracts and bulbs. Central nervous system abnormalities have been recorded, on occasion, with isolated agnathia: cerebellar hypoplasia, septum pellucidum cavum, and agenesis of the left calcar avis (23).

**Other abnormalities.** Other reported anomalies have included situs inversus totalis, unilateral renal agenesis, renal ectopia, cystic kidney, Müllerian duct agenesis, cryptorchidism, vertebral anomalies, absent ribs, Sprengel deformity, talipes equinovarus, VSD, pulmonic stenosis, PDA, patent foramen ovale, and unlobated lungs (5,10,11,23).

**Differential diagnosis.** Other forms of *holoprosencephaly* should be distinguished from agnathia-holoprosencephaly (8).

### References [Otocephaly (agnathia)]


1. Allan R: Dissection of human astomatous cyclops. *Lancet* **1**:227–228, 1848.
2. Altman F: The ear in severe malformations of the head. *Arch Otolaryngol* **66**:7–25, 1957.
3. Arey LB et al: Correlated defects in human agnathus. *Anat Rec* **94**:414, 1946.
4. Arnold J: Beschreibung einer Missbildung und Hydropsie der gemeinsamen Schlundtrommelhöhle. *Virchows Arch Pathol Anat* **38**:145–172, 1867.
5. Bixler D et al: Agnathia-holoprosencephaly: A developmental field complex including face and brain. Report of 3 cases. *J Craniofacial Dev Biol (Suppl)* **1**:241–249, 1985.
6. Black FO et al: Aplasia of the first and second branchial arches. *Arch Otolaryngol* **98**:124–128, 1973.
7. Blanc L: Sur l'otocéphalie et la cyclopia. *J Anat (Paris)* **31**:187–218, 288–309, 1895.
7a. Carles D et al: Cyclopia-otocephaly association: A new case of the most severe variant of agnathia-holoprosencephaly complex. *J Craniofacial Genet Dev Biol* **7**:107–113, 1987.
8. Cohen MM Jr: Perspectives on holoprosencephaly. Part III. Spectra, distinctions, continuities, and discontinuities. *Am J Med Genet,* in press.
9. Gaba AR et al: Alobar holoprosencephaly and otocephaly in a female infant with a normal karyotype and placental villitis. *J Med Genet* **19**:78, 1982.
10. Hersh JH et al: Otocephaly–midline malformation association. *Am J Med Genet* **34**:246–249, 1989.
11. Johnson WW, Cook, JB: Agnathia associated with pharyngeal isthmus atresia and hydramnios. *Arch Pediatr* **78**:211–217, 1961.
12. Jones KL et al: Determining role of the optic vesicle in orbital and periocular development and placement. *Pediatr Res* **14**:703–708, 1980.
13. Josephy H: Missbildungen des Halses: Otocephaly, in E Schwalbe, *Morphologie der Missbildungen der Menschen und der Tiere.* Vol 3, part 1, G Fischer, Jena, 1909, p 24.
14. Keen JA: A case of agnathia with a note on the development of the maxillary process. *S Afr J Lab Clin Med* **1**:197–202, 1955.
15. Keith A: Congenital malformation of palate, face and neck. *Br Med J* **2**:363–367, 1909.
16. Koogler MA: A report of three human monstrosities (Case 1). *Am J Med Sci* **84**:129–132, 1882.
16a. Krasnikoff N, Sekhon GS: Familial agnathia–holoprosencephaly due to an inherited unbalanced translocation and not autosomal recessive inheritance. *Am J Med Genet* **34**:255–257, 1989.
17. Kuse: Über Agnathie und die dabei zu erhabenden Zungenbefunde. *Münch Med Wochenschr* **48**:890–893, 1901.
18. Lawrence DK, Bersu ET: An anatomical study of otocephalic malformation. *Teratology* **25**:58A, 1982.
19. Leckie GB: Aplasia of the first and second branchial arches. *J Laryngol Otol* **89**:1263–1269, 1975.
19a. Leech RW et al: Agnathia, holoprosencephaly, and situs inversus: Report of a case. *Am J Med Genet* **29**:483–490, 1988.
20. Le Marec B et al: A case of otocephaly. *J Génét Hum* (Suppl) **24**:253–260, 1976.
21. Libersa JC, Heritier M: Orofacial anomalies in an otocephalic fetus. *Rev Stomatodontol Nord Fr* **29**:101–105, 1974.
22. Mollica F et al: A case of cyclopia: Role of environmental factors. *Clin Genet* **16**:69–71, 1979.
23. Pauli RM et al: Agnathia, situs inversus and associated malformations. *Teratology* **23**:85–93, 1981.
24. Pauli RM et al: Familial agnathia-holoprosencephaly. *Am J Med Genet* **14**:677–698, 1983.
25. Rogers RMW: A case of agnathia or congenital absence of the lower jaw. *J Pathol Bacteriol* **5**:137–142, 1898.
26. Scholl HW: In utero diagnosis of agnathia, microstomia and synotia. *Obstet Gynecol* **49** *(Suppl* **1**):815–835, 1977.
27. Smith RM, Parker AJ: Dissection of a human otocephalic cyclops monstrosity. *Am J Med Sci 84:*132–140, 1882.
28. Ursell W: Hydramnios associated with congenital microstomia, agnathia and synotia. *J Obstet Gynaec Br Commonw* **79**:185–186, 1972.
29. Usandizasa JA: An unusual case of otocephalic cyclopic fetus of 7 months. *Acta Gynaecol Obstet Hisp Lusit* **15**:18–28, 1966.
30. Van de Sande PL: A case of otocephalus. *Ned Tijdschr Verloskel Gynaecol* **66**:471–474, 1966.
31. Winckel F: Aetiologische Untersuchungen über einige sehr seltene tötale Missbildungen. *Münch Med Wochenschr* **43**:423–429, 1896.
32. Woon KY, Tan KL: Aplasia of the first and second branchial arches. *Aust Paediatr J* **15**:275–277, 1979.
33. Wright S, Wagner K: Types of subnormal development of the head from inbred strains of guinea pigs and their bearing on the classification and interpretation of vertebrate monsters. *Am J Anat* **54**:383–448, 1934.

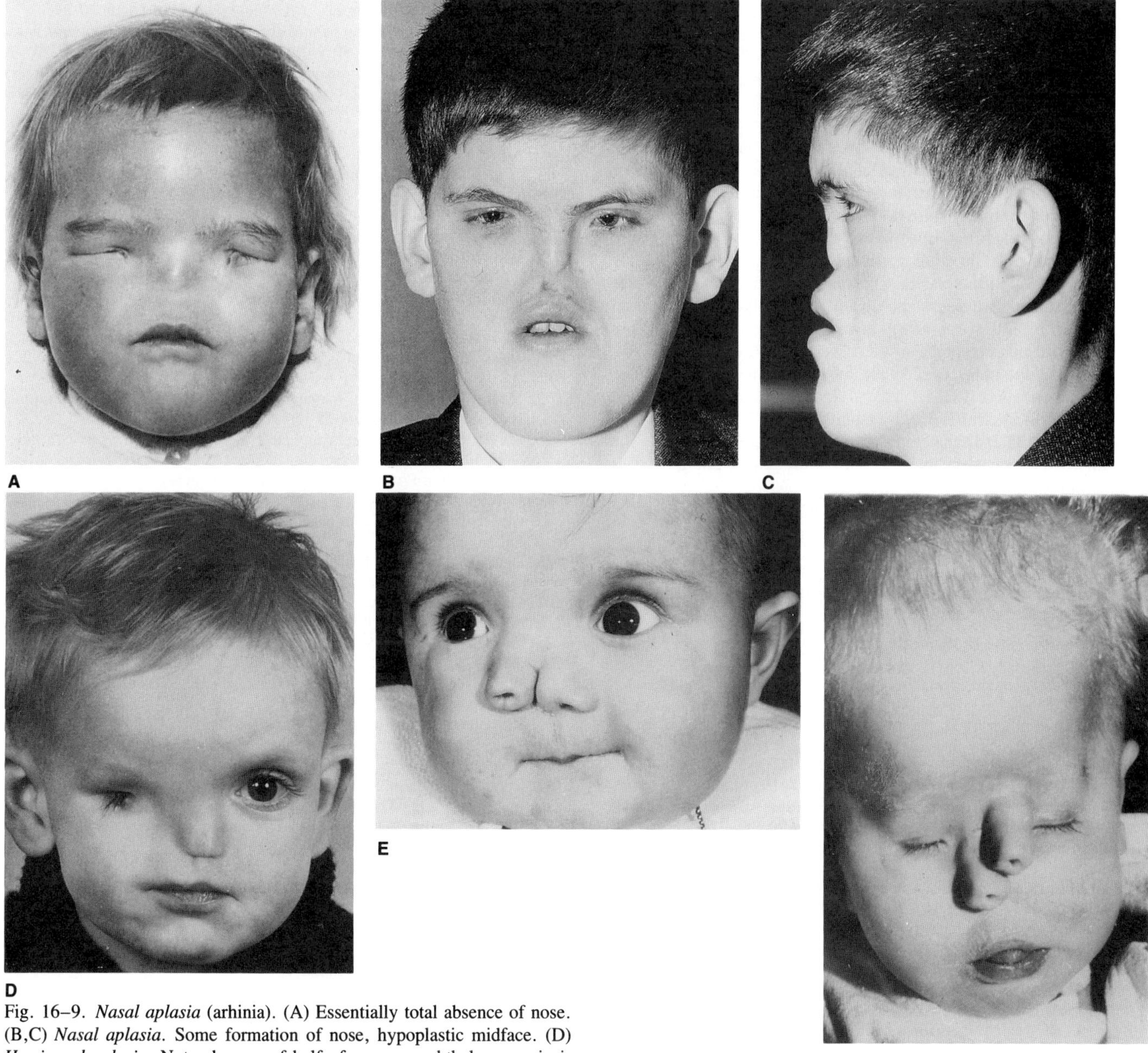


Fig. 16–9. *Nasal aplasia* (arhinia). (A) Essentially total absence of nose. (B,C) *Nasal aplasia.* Some formation of nose, hypoplastic midface. (D) *Heminasal aplasia.* Note absence of half of nose, anophthalmus on ipsilateral side. (E) *Unilateral proboscis.* Observe ipsilateral location of proboscis. (F) *Bilateral nasal proboscides.* (B,C from *V Lütolf,* J Maxillofac Surg **4**:245, 1976. E courtesy of P Tessier, Paris, France. F courtesy of *AJ Cassini,* Cincinnati, Ohio.)

## Nasal aplasia, heminasal aplasia with or without proboscis, and their associations

The nose begins its development by the appearance bilaterally of the nasal or olfactory placodes on the lower outer portion of the frontonasal prominence around the 28th day (37a). Proliferation of ectomesenchyme around the placodes and programmed cell death result in the formation of primitive nostrils (nasal pits) that ultimately extend in a funnellike (choanal) tube to the stomodeum. The horseshoe-shaped ridge on either side of the nostrils becomes the medial and lateral nasal prominences. Due to selective growth of the facial prominences, the medial nasal prominences move toward one another, finally forming the nasal bridge, philtrum, primary palate, and columellar area. The lateral nasal prominences form the lateral sides of the nose. If there is agenesis of both nasal placodes, there is nasal aplasia; if one placode is missing, heminasal aplasia results. The reader is referred to Rontal and Duritz (32) and Leperchey et al (17) for discussion of the pathoembryology.

Complete aplasia of the nose has been recorded on several occasions (1b,3,4,8,9,11a,11b,14,18,21,28,34,36,37) (Fig. 16–9A–C). Severe hypoplasia of the nose has been reported in sisters who, in addition, had anosomia, submucous cleft palate, Peters anomaly, iris coloboma, and microphthalmus (34). All other examples have been isolated cases. Arhinia has been seen in association with *mandibulofacial dysostosis* (4).

Aplasia of half the nose (heminasal aplasia) may be found without a nasal proboscis (Fig. 16–9D) (5,11,14a,16,35,36,38,41) or, more often, found in association with other anomalies of the facial region. All have been isolated cases. Frequently there is a blind ended lateral nasal proboscis (Fig. 16–9E) (1,2,5–13,15,16,19–27,29–32,38–42).

For examples not cited here, the reader is referred to two publications (10,21). The proboscis, about 2–4 cm in length and 0.5–1 cm in width, depends from an area just lateral to the nasal root. Rarely, a proboscis may be located in the midline with a normal nose (13), attached to an eyelid (2), or lateral to the eye (1a). There are examples of bilateral proboscides (33) (Fig. 16–9F).

Bony abnormalities in the involved area include lack of cribriform plate, nasal septum deviated to unaffected side, loss of structures of the lateral wall of the nasal chamber on the affected side, absence or hypoplasia of the paranasal sinuses on the affected side, absent nasal bone, and disrupted lacrimal bone (32).

In addition to cleft lip and/or palate or submucous cleft palate (2,4,25,35,37), also discussed in Chapter 23, anomalies of the eye region are common with nasal aplasia but especially with heminasal aplasia. The ipsilateral eye is nearly always laterally displaced. Other eye anomalies include coloboma of the lower lid (1,6,12,20,24,27,30–32), coloboma of the upper eyelid (30,38), coloboma of the iris, choroid, and optic nerve (2,13,14,19,23,34,35,38), microphthalmia and anophthalmia (2,14,24,25,33–35), nystagmus (2,18,35), Peters anomaly (34), cataract (13,35,36), and doubling of lens or globe (19,22). The nasolacrimal canal is disturbed in its formation and epiphora or dacryocystitis is common.

Intelligence is normal with rare exceptions (22).

### References (Nasal aplasia, heminasal aplasia with or without proboscis, and their associations)


1. Alves D'Assumpção E: Proboscis lateralis. *Plast Reconstr Surg* **55**:494–497, 1975.

1a. Antoniades K, Baraitser M: Proboscis lateralis: A case report. *Teratology* **40**:193–197, 1989.

1b. Banks PA, Robinson MJ: Nasal agenesis—a modified oral appliance to aid neonatal airway patency and to support oro-enteric intubation. *Eur J Orthodont* **10**:137–142, 1988.

2. Beg MHA: Half nose. *J Laryngol Otol* **98**:915–916, 1984.

3. Berger M, Martin C: L'arhinogénésie totale (absence congénitale chez du nez et des fosses nasales). *Rev Laryngol Otol Rhinol* **90**:300–319, 1969.

4. Berndorfer A: Über die seitliche Nasenspalte. *Acta Otolaryngol* **55**:163–174, 1962 (Cases 4 and 5).

5. Biber JJ: Proboscis lateralis: A rare malformation of the nose: Its genesis and treatment. *J Laryngol Otol* **63**:734–741, 1949.

6. Binns JH: Congenital tubular nostril (proboscis lateralis). *Br J Plast Surg* **22**:265–268, 1969.

7. Bishop BW: Lateral nasal proboscis. *Br J Plast Surg* **17**:18–21, 1964.

8. Blair VP: Congenital atresia or obstruction of the nasal air passages. *Ann Otol Rhinol Laryngol* **40**:1021–1035, 1931.

9. Blair VP, Brown JB: Nasal abnormalities, fanciful and real. *Surg Gynecol Obstet* **53**:797–819, 1931.

10. Boo-Chai K: The proboscis lateralis—a 14-year follow up. *Plast Reconstr Surg* **75**:569–577, 1985.

11. Coetzee T: Proboscis lateralis: A rare facial anomaly. *S Afr J Lab Clin Med* **10**:81–83, 1964.

11a. Cohen D, Goitein K: Arhinia revisited. *Rhinology* **24**:287–292, 1986, **25**:237–244, 1987.

11b. Cole RR et al: Congenital absence of the nose: A case report. *Int J Ped Otorhinolaryngol* **17**:171–177, 1989.

12. Cotin G et al: Le proboscis lateralis. Malformation faciale rare. *Ann Oto-laryngol (Paris)* **100**:353–356, 1983.

13. Dasgupta G et al: Proboscis. *J Laryngol Otol* **85**:401–406, 1971.

14. Gifford GH Jr et al: Congenital absence of the nose and anterior nasopharynx. *Plast Reconstr Surg* **50**:5–12, 1972.

14a. Kernahan DA: Developmental defects of the nose, in *Plastic Surgery of the Head and Neck,* Stark RB, (ed) Churchill Livingstone, New York, 1987, pp 558–559.

15. Kirkpatrick TJ: Lateral nasal proboscis. *J Laryngol Otol* **84**:83–89, 1970.

16. Kukreja HK: Half nose. *J Laryngol Otol* **87**:599–602, 1973.

17. Leperchey F et al: Problemes embryologiques soulevés par le proboscis lateralis. *Ann Oto-laryngol (Paris)* **100**:357–364, 1983.

18. Lütolf V: Bilateral aplasia of the nose. *J Maxillofac Surg* **4**:245–249, 1976.

19. Lyford JH, Roy FH: Arhinencephaly unilateralis, uveal coloboma, and lens reduplication. *Am J Ophthalmol* **77**:315–318, 1977.

20. Mahindra S et al: Lateral nasal proboscis. *J Laryngol Otol* **87**:177–181, 1973.

21. Mazzola RF: Congenital malformations in the frontonasal area: Their pathogenesis and classification. *Clin Plast Surg* **3**:573–609, 1976 (Figs. 22–25).

22. Meeker LH, Aebli R: Cyclopean eye and lateral proboscis with normal one-half face. *Arch Ophthalmol* **38**:159–173, 1947.

23. Meyer R: Über angeborene äussere Nasendeformitäten. *Pract Otorhinolaryngol* **18**:399–416, 1956 (Case 1).

24. Mugaddu EG: Proboscis lateralis—a rare congenital anomaly. *S Afr Med J* **68**:45, 1985.

25. Nessel E: Proboscis lateralis als seltene Nasenmissbildungen. HNO **7**:25–26, 1958.

26. Ogura Y, Onoda M: Half nose: A very rare congenital anomaly. *J Otolaryngol Soc Jpn* **70**:2037–2043, 1967.

27. Ohura T: Reconstructive surgery of the nose in non-Caucasians. *Clin Plast Surg* **1**:93–120, 1974 (Fig. 25).

28. Palmer CR, Thompson HG: Congenital absence of the nose. *Can J Surg* **10**:83–86, 1976.

29. Ployet MJ et al: Deux malformations exceptionelles du nez. *Ann Otolaryngol (Paris)* **100**:365–369, 1983.

30. Rao PB: Proboscis lateralis. *J Laryngol Otol* **77**:1028–1031, 1963.

31. Recamier J, Florentin M: Un type exceptionelle de malformation congénitale du nez. *Ann Chir Plast* **2**:13–16, 1957.

32. Rontal M, Duritz G: Proboscis lateralis: A case report and embryologic analysis. *Laryngoscope* **87**:996–1006, 1977.

33. Rosen Z, Gitlin G: Bilateral nasal proboscis. *Arch Otolaryngol* **70**:545–550, 1959.

34. Ruprecht KW, Majewski F: Familiäre Arhinie mit Petersscher Anomalie und Kiefermissbildungen, ein neues Fehlbildungssyndrom? *Klin Mbl Augenheilkd* **172**:708–715, 1978.

35. Schweckendiek W, Hein H: Seltene Nasenfehlbildungen. Einseitige Aplasie in Kombination mit Gaumenspalten. *HNO* **21**:73–76, 1973.

36. Šercer A: Das Syndrom der Dysplasie der medianen Linie des Kopfes. *Acta Otolaryngol* **36**:289–295, 1948 (cases 1,2).

37. Shubich I, Sanchez C: Nasal aplasia associated with meningocele and submucous cleft palate. *ENT J* **64**:259–260, 1985.

37a. Sperber GH: *Craniofacial Embryology* (2nd ed), J Wright and Sons Ltd, Bristol, 1981.

38. Tiefenthal G: Total Aplasie einer Nasenhälfte. *Mschr Ohrenheilkd* **44**:1071–1075, 1910.

39. Walker DG: *Malformations of the Face,* ES Livingstone, London, 1961.

40. Wang S et al: Proboscis lateralis, microphthalmos, and cystic degeneration of the optic nerve. *Ann Ophthalmol* **15**:756–758, 1983.

41. Wicke W: Aplasie einer Nasenhälfte. *Monatschr Ohrenheilk Laryngol* **106**:438–441, 1972.

42. Young F: Surgical repair of nasal deformities. *Plast Reconstr Surg* **4**:59–91, 1949 (Fig. 22).

# Chapter 17
# Syndromes Affecting the Central Nervous System

## Myotonic dystrophy (Steinert syndrome)

Early reports of myotonic dystrophy date from Hoffman (21), in 1896, and from Rossolimo (37), in 1902, but delineation of myotonic dystrophy as a condition clearly distinct from myotonia congenita (Thomsen's disease) was first made by Steinert (41) and Batten and Gibb (2) in 1909. The adult form of the disease is characterized by myotonia, progressive muscle wasting, and variable mental deficiency. The diverse clinical manifestations relate to multiple system involvement that includes the eye, heart, endocrine system, gastrointestinal tract, skeleton, and skin. An excellent comprehensive and historically interesting monograph on myotonic dystrophy has been written by Harper (20).

Myotonic dystrophy has autosomal dominant inheritance with complete penetrance and extremely variable clinical presentation and course. The prevalence is estimated at 4–15/100,000 population (25,40,46). This may be an underestimate because of the high frequency of stillbirth and neonatal death in families with this disorder.

Bundey (7) suggested that myotonic dystrophy occurs in two genetically distinct clinical types: Type I, with more severe limb weakness and symmetric facial involvement, and Type II, with milder limb weakness but severe facial weakness, preceding limb weakness by several years. Mental retardatioin, infertility, and neonatally affected offspring are more common in Type II patients.

Congenital myotonic dystrophy was first described by Vanier (44) in 1960 and has subsequently been delineated in several clinical reports (3,14,19,31,41,43). Infants with the neonatal form vary widely in clinical presentation. Problems vary from mild hypotonia, feeding problems, and talipes to severe respiratory insufficiency, leading to death. Other findings may include hydramnios, decreased fetal activity, facial diplegia, cleft palate, cryptorchidism, hip dislocations, and thin ribs on chest radiograph. Impaired response of fetal breathing to maternal intravenous infusion of glucose has been demonstrated by Vilos et al (45). Several infants have been reported presenting with pleural effusions and hydrops fetalis (12,38,41). In those who survive, mental retardation is the usual sequela. Myotonia is rarely elicited in the newborn period, and EMG and muscle biopsy changes are inconsistent in this age group.

Infants with the neonatal form are almost exclusively born to mothers with myotonic dystrophy who may manifest few or no symptoms of their disorder at the time of their infant's birth. Thus, diagnosis in an infant with a neonatal presentation may be quite difficult. The empiric risk for a mother with myotonic dystrophy who has not had a neonatally affected child is estimated at between 6 and 9% (15,20). Mothers who have had a neonatally affected child have a significantly higher risk for recurrence of the neonatal presentation (29%). The risk for a clinically normal child born in a family with one affected sib to develop myotonic dystrophy has been estimated at approximately 21% before age 16 and at between 0 and 7% after age 16 (29).

The basic defect in myotonic dystrophy remains unknown, although a generalized disorder of the plasma membrane has been postulated. There are reports of increased fragility of the erythrocyte membrane, decreased red blood cell survival, increased accumulation of intracellular calcium, decreased solidity and decreased polarity of the red cell membrane, and abnormal configuration of plasma proteins (9,33,42). The significance of these findings is unclear and they may be nonspecific indicators of a general membrane abnormality.

The disorder has multiple and variable manifestations affecting every organ system. The usual age of onset is about 20–25 years, but isolated symptoms may precede definitive diagnosis by several years. These estimates do not reflect those with neonatal presentation and those whose symptoms are so mild as to escape detection. Average age of death has been estimated at between 45 and 50. Again, neonatal deaths were not considered and mildly affected persons were not ascertained (4,24). Death generally results from respiratory illness or cardiac failure.

**Facies.** Facial weakness is an early and consistent feature of myotonic dystrophy. The face changes with time, becoming narrow, expressionless, and masklike with progressive muscle wasting. There is diminished ability to close the eyes, inability to wrinkle the forehead, whistle, or puff out the cheeks with air. The face in congenital myotonic dystrophy is particularly notable with a triangular open mouth due to jaw muscle weakness (Fig. 17–1A,B). Gazit et al (14a) discussed the cephalometric aspects of myotonic dystrophy.

**Extremities.** The distal limb muscles of arms and legs are involved first in myotonic dystrophy in contrast to initial proximal muscle involvement in Duchenne muscular dystrophy and most other muscular dystrophies. Atrophy of intrinsic hand muscles occurs late. Myotonia is the most common neurologic finding in both symptomatic and asymptomatic patients, although some patients often are not greatly bothered by the symptom and may deny its presence. Cold aggravates the myotonia and warmth diminishes it. Myotonia is rarely present in infants with neonatal presentation but appears later in childhood (Fig. 17–2).

**Eyes.** Cataracts are observed in over 85% of older individuals with myotonic dystrophy. Slit lamp examinations in younger individuals may reveal iridescent dust-like subcapsular lens opacities useful in presymptomatic detection. Cataracts are rare in congenital myotonic dystrophy. Retinal abnormalities consisting of peripheral pigmentary changes, macular degeneration, and abnormal dark adaptation are quite common (8). Ptosis is extremely common, but not invariable, even in severely affected patients. Other less common eye abnormalities include blepharospasm, anomalies of extraocular muscles, keratoconjunctivitis, and decreased intraocular pressure.

**Hair and skin.** Early frontal balding is very common in males, the extent appearing to parallel the severity of the muscle disease (20). A very unusual benign skin tumor, calcifying epithelioma of Malherbe or pilomatrixoma, has been observed to segregate with myotonic dystrophy in several reported families (13). The nature of this association is unclear but unlikely to be coincidental.

**Central nervous system.** The exact frequency of mental deficiency in myotonic dystrophy has been difficult to estimate and may, in the past, have been overestimated because of the patients' lack of facial expression and general weakness (34). There seems little doubt, however, that survivors of the congenital presentation are generally retarded, although usually this is mild in degree and does not progress. Many authors comment on the cheerfulness and lack of initiative in myotonic dystrophy patients. This does not appear to be ex-

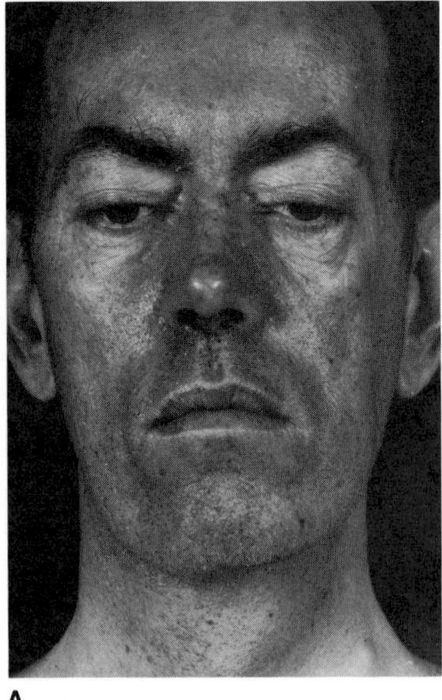

A

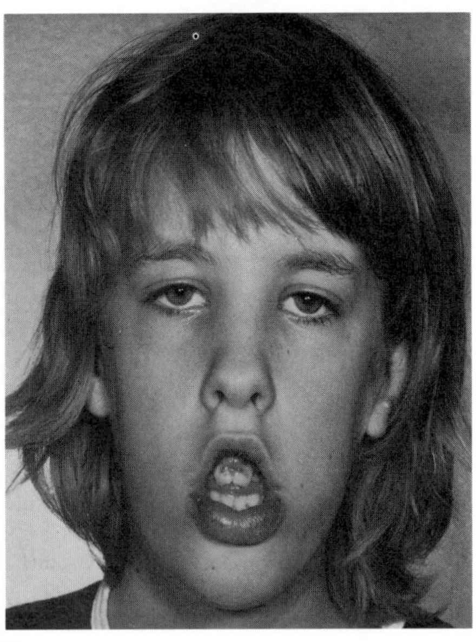

B

Fig. 17–1. *Myotonic dystrophy.* (A) Characteristic expressionless face, ptosis of eyelids, temporal and buccal hollows, and slack mandible. (From *HH Thayer* and *J Crenshaw,* J Am Dent Assoc **72**:1405, 1966.) (B) Compare facies in 12-year-old.

plained by the degree of intellectual impairment. Some degree of progression of mental handicap appears to accompany general disease progression in the adult disorder.

**Cardiac and respiratory.** Cardiac involvement in myotonic dystrophy has been recognized since 1911 (17). This usually includes conduction defects and/or atrial arrythmias. Ventricular arrythmias have been rarely reported (32). Cardiac dysfunction is an integral part of myotonic dystrophy, and sudden death unaccompanied by other symptoms of myotonic dystrophy may occur (18,23a). Progressive respiratory distress and aspiration pneumonia are problems in the later stages of the disease due to weakened chest wall and diaphragmatic musculature as well as to possible central nervous system factors. Moderate restriction and lung volume with mild hypoxemia can be observed in myotonic dystrophy patients without clinical dyspnea (23).

Hypersomnia has been a frequently noted clinical symptom, occasionally attributed to alveolar hypoventilation. Sleep apnea can be a serious complication and may be worsened by procedures such as a pharyngeal flap designed to reduce velopharyngeal insufficiency, which is also common (10). Anesthesia, which may pose special risks in the infant, pregnant patient, and adult with myotonic dystrophy, has been the topic of several reviews (1,6,30).

Fig. 17–2. *Myotonic dystrophy.* (A,B) Myotonic response to percussion of thenar eminence and biceps muscle.

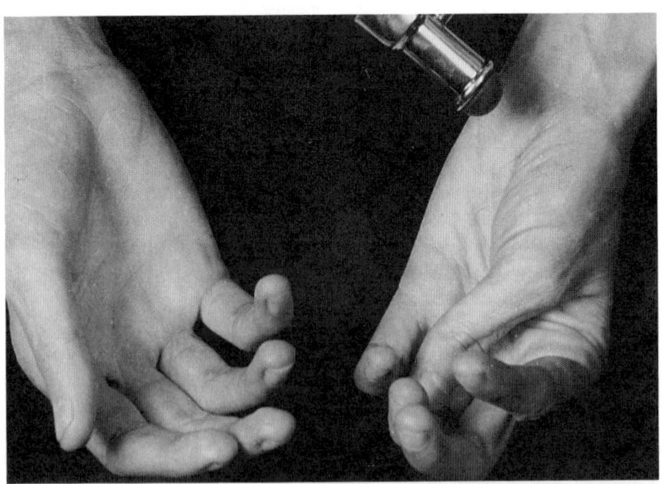

A

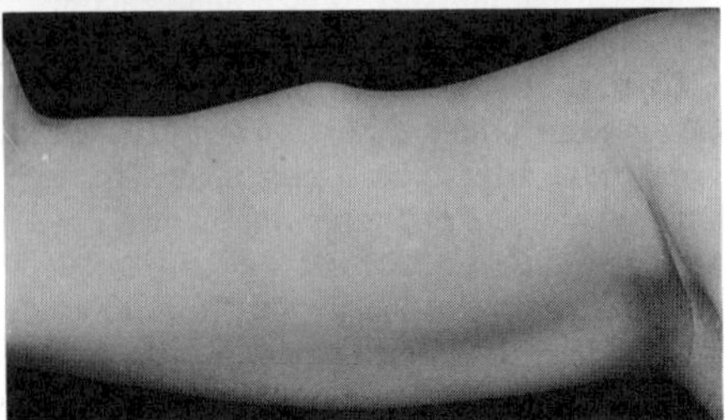

B

**Gastrointestinal system.** Multiple dysfunctional aspects of gastrointestinal motility have been repeatedly observed with a resultant wide array of symptoms including esophageal reflux, gastric retention, megacolon, cholelithiasis, malabsorption, and constipation in 80% (23a). Manometric studies of intestinal motility have confirmed abnormal motility patterns (16,28). Infants with neonatal presentation may exhibit poor sucking and swallowing, delayed gastric emptying, and recurrent vomiting (5).

**Endocrine system.** Multiple endocrine abnormalities have been noted with testicular atrophy, infertility, and hypogonadism very commonly observed in affected males. Females may have dysmenorrhea and symptoms of ovarian dysfunction. Pregnancy and its complications were reviewed by Jaffe et al (22). Postprandial hyperinsulinemia is a common finding, probably being secondary to total body insulin resistance (27). The risk for diabetes mellitus is estimated to be increased 4-fold, but this association is not nearly as strong as once supposed (20). Colloid goiter and hypothyroidism are occasional associated findings.

**Skeletal abnormalities.** Radiographic changes include calvarial hyperostosis and large paranasal sinuses, small sella turcica, prognathism, or micrognathia. Other changes such as thin ribs in infancy, kyphoscoliosis, and talipes equinovarus are probably secondary to inactivity and muscle weakness (Fig. 17–3).

**Oral manifestations.** Lingual myotonia and/or velopharyngeal incompetence resulting in slow nasal and indistinct speech may be the initial sign (37a). Retention of saliva in the oral cavity and a high rate of dental caries are common (35). Gazit et al (14a) studied a series of

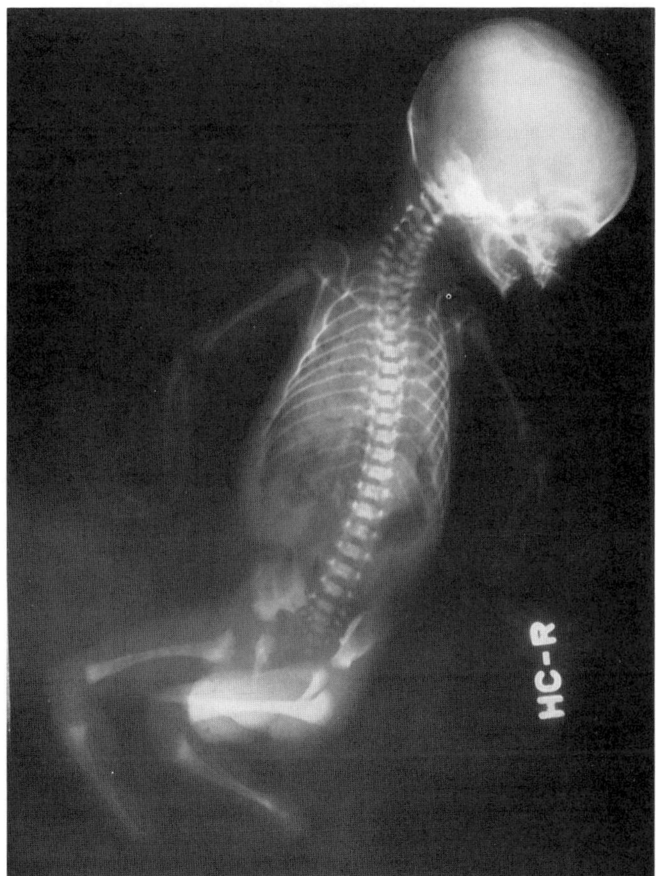


Fig. 17–3. *Myotonic dystrophy.* Radiograph of newborn of affected mother. Note remarkably thin or gracile ribs.

patients cephalometrically and found increased interocclusal distance, increased face height, increased height of the palatal vault, narrow maxillary arch, increased crossbite, open bite, tongue thrust, and mouth breathing (Fig. 17–4). Recurrent jaw dislocation is common (23a).

**Differential diagnosis.** Congenital myotonia is rare in myotonic dystrophy as in other myotonic syndromes but can be occasionally seen in myotonia congenita, a heterogeneous phenotype with both dominant and recessive modes of inheritance. The differential diagnosis of the neonatal presentation of myotonic dystrophy may include congenital myasthenia gravis, *Moebius syndrome,* and *Prader–Willi syndrome.* In the older patients differential diagnosis includes various myotonia congenita syndromes, paramyotonia congenita, *Schwartz–Jampel syndrome,* different forms of periodic paralysis, and other types of muscular dystrophy (4).

Fig. 17–4. *Myotonic dystrophy.* Pronounced anterior open bite. (From *HH Thayer* and *J Crenshaw,* J Am Dent Assoc **72**:1405, 1966.)

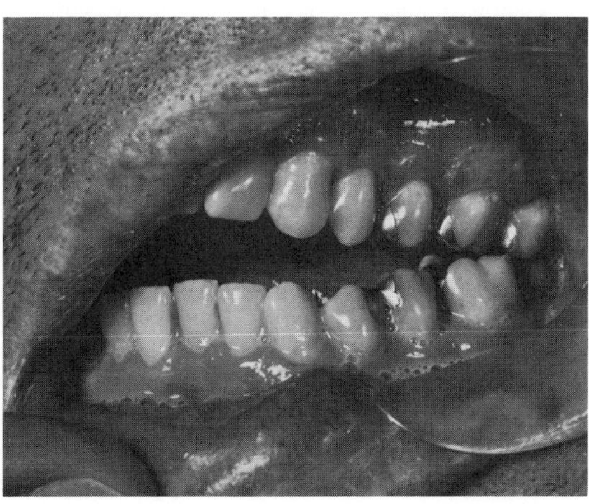

**Laboratory aids.** Diagnosis is aided by electromyographic studies, particularly of forearm and hand musculature. Occasionally muscle biopsy may be helpful. Slit-lamp evaluation of the eyes is valuable in the diagnosis of at-risk family members, and electroretinogram abnormalities may be even more sensitive in the detection of asymptomatic affecteds. Linked DNA markers will undoubtedly become increasingly clinically helpful in presymptomatic diagnosis and in prenatal diagnosis. In view of these advances, DNA banking of blood on affected family members should be considered.

The gene of myotonic dystrophy is located just below the centromeric region of chromosome 19. A gene probe for apolipoprotein C2 is apparently a close marker for the myotonic dystrophy gene with a recombination fraction of 4% (26,36,39). Four common polymorphisms are known at the apo C2 locus with at least 80% of individuals polymorphic for at least one polymorphism. Prenatal diagnosis can be performed on infants in informative families using chorionic villus biopsy. DNA techniques will replace the utilization of secretor status that was useful for prenatal diagnosis in a low percentage of families. In families who are not informative for linked DNA markers or secretor status, monitoring by ultrasound of the infant's biophysical profile (i.e., fetal breathing and fetal activity) may be of use in predicting fetal status in at-risk families.

Although no specific microscopic feature is pathognomonic for myotonic dystrophy, combined histologic changes on muscle biopsies are characteristic in myotonic dystrophy. Biopsy is rarely necessary for the diagnosis of this disorder, but may be helpful in patients with minimal myotonia and progressive muscle wasting, in those with minimal weakness and a negative family history, and in children in whom the differential diagnosis of congenital hypotonia may be extensive. Microscopic features include increased central nuclei, nuclear chaining, ringed fibers, Type I fiber atrophy, terminal innervation, sarcoplasmic masses, and muscle spindle abnormalities (20).

An abnormal electroretinogram has proven useful in distinguishing asymptomatic gene carriers (11).

#### References [Myotonic dystrophy (Steinert syndrome)]


1. Aldridge LM: Anaesthetic problems in myotonic dystrophy. *Br J Anestheseol* **57**:1119–1130, 1985.
2. Batten FE, Gibb HP: Myotonia atrophica. *Brain* **32**:187–205, 1909.
3. Bell DB, Smith DW: Myotonic dystrophy in the neonate. *J Pediatr* **81**:83–86, 1972.
4. Bodensteiner JB: Congenital myopathies. *Neurol Clin* **6**:499–518, 1988.
5. Bodensteiner JB, Grunow JE: Gastroparesis in neonatal myotonic dystrophy. *Muscle Neurol* **7**:486–487, 1984.
6. Bray RJ, Inkster JS: Anaesthesia in babies with congenital dystrophia myotonica. *Anaesthesia* **39**:1007–1011, 1984.
7. Bundey S: Clinical evidence for heterogeneity in myotonic dystrophy. *J Med Genet* **19**:341–348, 1982.
8. Burian HM, Burns CA: Ocular changes in myotonic dystrophy. *Am J Ophthalmol* **63**:22–24, 1967.
9. Butterfield A: Myotonic muscular dystrophy: Time dependent alterations in erythrocyte membrane solidity. *J Neurol Sci* **52**:61–67, 1981.
10. Cadieux RJ et al: Sleep apnea precipitated by pharyngeal surgery in a patient with myotonic dystrophy. *Arch Otolaryngol* **110**:611–613, 1984.
11. Creel DJ et al: Identification of minimal expression in myotonic dystrophy using electroretinogram. *Electroencephalogr Clin Neurophysiol* **61**:229–235, 1985.
12. Curry CJR, Finer NA: Hydrops and pleural effusions in congenital myotonic dystrophy. *J Pediatr* **113**:555–557, 1988.
13. Delfino M et al: Multiple familial pilomatricomas: A cutaneous marker for myotonic dystrophy. *Dermatologica* **170**:128–132, 1985.
14. Farnath HB et al: Clinical effects of myotonic dystrophy on pregnancy and the neonate. *Arch Neurol* **33**:459–465, 1976.

14a. Gazit E et al: The stomatognathic system in myotonic dystrophy. *Eur J Orthodont* **9**:160–164, 1987.

15. Glanz A, Fraser FC: Risk estimates for neonatal myotonic dystrophy. *J Med Genet* **21**:186–188, 1984.

16. Goldberg HI, Sheft DJ: Esophageal and colon changes in myotonia dystrophica. *Gastroenterology* **63**:134–139, 1972.

17. Griffith TW: On myotonia. *Q J Med* **5**:229–249, 1911.

18. Grigg LE et al: Ventricular tachycardia and sudden death in myotonic dystrophy: Clinical, electrophysiologic and pathologic features. *J Am Coll Cardiol* **6**:254–256, 1985.

19. Hanson PA: Myotonic dystrophy in infancy and childhood. *Pediatr Ann* **13**:123–126, 1984.

20. Harper PS: *Myotonic Dystrophy—Major Problems in Neurology*. W.B. Saunders Company, Philadelphia, 1979.

21. Hoffmann J: Ein Fall von Thomsen'scher Krankheit, complicirt durch Neuritis multiplex. *Dtsch Z Nervenheilkd* **9**:272–278, 1896.

22. Jaffe R et al: Myotonic dystrophy and pregnancy: A review. *Obstet Gynecol Surv* **41**:272–278, 1978.

23. Jammes Y et al: Pulmonary function and electromyographic study of respiratory muscles in myotonic dystrophy. *Muscle Neurol* **8**:586–594, 1985.

23a. Jozefowicz RF, Griggs RC: Mytonic dystrophy. *Neurol Clin* **6**:455–472, 1988.

24. Klein D: La dystrophie myotonique (Steinert) et la myotonie congénitale (Thomsen) en Suisse. *J Génét Hum* (*Suppl* **7**):1–328, 1958.

25. Lotz BP, Van Der Meyden CH: Myotonic dystrophy. Part I: A genealogical study in the northern Transvaal. *S Afr Med J* **67**:812–817, 1985.

26. Meredith AL et al: Application of a closely linked restriction fragment linked polymorphism to genetic counseling and prenatal prediction in families with myotonic dystrophy. *Am J Hum Genet* **39**:No 3 Suppl (Abstract #533), 1986.

27. Moxley RT et al: Whole body insulin resistance in myotonic dystrophy. *Ann Neurol* **15**:157–162, 1984.

28. Nowak TV et al: Small intestinal motility in myotonic dystrophy patients. *Gastroenterology* **86**:808–813, 1984.

29. O'Brien T et al: Outlook for a clinically normal child in sibship with congenital myotonic dystrophy. *J Pediatr* **103**:762–763, 1983.

30. Patterson RA et al: Caesarean section for twins in a patient with myotonic dystrophy. *Can Anaesth Soc J* **32**:418–421, 1985.

31. Pearse RG, Howeler CJ: Neonatal form of dystrophia myotonica. *Arch Dis Childh* **54**:331–338, 1979.

32. Perloff JK et al: Cardiac involvement in myotonic muscular dystrophy (Steinert's disease): A prospective study of 25 patients. *Am J Cardiol* **54**:1074–1081, 1984.

33. Phishker G et al: Myotonic muscular dystrophy—altered calcium transport in erythrocytes. *Science* **200**:323–325, 1978.

34. Portwood MM et al: Psychometric evaluation in myotonic muscular dystrophy. *Arch Phys Med Rehabil* **65**:533–536, 1984.

35. Pruzanski W: Variants of myotonic dystrophy in preadolescent life. *Brain* **89**:563–568, 1956.

36. Roses AD et al: A new tightly linked DNA probe for myotonic dystrophy. *Neurology* **36**:1146, 1986.

37. Rossolimo G: Atrophische Form der Thomsen'schen Krankheit. *Neurol Zentralbl* **21**:135–136, 1902.

37a. Salomonson J et al: Velopharyngeal incompetence as the presenting symptom of myotonic dystrophy. *Cleft Palate J* **25**:296–300, 1988.

38. Sarnat HB et al: Clinical effects of myotonic dystrophy on pregnancy and the neonate. *Arch Neurol* **33**:459–465, 1976.

39. Shaw DJ et al: The apolipoprotein CII gene: Subchromosomal localisation and linkage to the myotonic dystrophy locus. *Hum Genet* **70**:271–273, 1985.

40. Slatt B: Myotonia dystrophia: A review of 17 cases. *Can Med Assoc J* **85**:250–261, 1961.

41. Steinert H: Myopathologische Beiträge. 1. Über das klinische und anatomische Bild des Muskelschwunds der Myotoniker. *Dtsch Z Nervenheilkd* **37**:58–104, 1909.

42. Sydow O et al: Abnormal erythrocyte survival in patients with myotonic dystrophy. *Acta Neurol Scand* **72**:522–524, 1985.

43. Thurson H et al: Neonatal form of dystrophia myotonica. *Arch Dis Childh* **54**:331–338, 1979.

44. Vanier TM: Dystrophia myotonica in childhood. *Br Med J* **2**:1284–1288, 1960.

45. Vilos UA et al: Absence or impaired response of fetal breathing to intravenous glucose as associated with pulmonary hypoplasia in congenital myotonic dystrophy. *Am J Obstet Gynecol* **148**:558–562, 1984.

46. Wielander L: Genetic research in muscular disease in Sweden. *Int Cong Ser #32, 2nd Int Conf Hum Genet*, Vermont, 1961, p E88.

## Miller–Dieker syndrome (type I lissencephaly)

Originally buried in the literature under lissencephaly, an etiologically nonspecific condition, a syndrome was defined by Miller (13) in 1963 and Dieker et al (4a,6) in 1969. Miller described autopsy findings of lissencephaly and multiple anomalies in two sibs. In 1966, Daube and Chou (3) observed similar findings in two unrelated infants. Dieker expanded the latter report to include a male sib and female first cousin of one of the patients. A characteristic phenotype was defined that included microcephaly, bitemporal hollowing, long philtrum with thin prominent upper lip, mild micrognathia, abnormal ears, anteverted nares, and growth retardation. These investigators thought that antemortem diagnosis was possible and that inheritance was probably autosomal recessive. Since then, a number of other reports have appeared (1a,2, 5–7,9–12,14,16,17). The syndrome has been especially well reviewed by Dobyns et al (6) and Stratton et al (17). At least two dozen cases have been documented to date. We cannot accept the cases of Van Allen and Clarren (18) as examples of the syndrome. Dobyns et al (6) and Stratton et al (17) found no evidence to support inclusion of those patients as having any form of lissencephaly syndrome.

The Miller–Dieker syndrome has been referred to as a microdeletion syndrome, a small deletion syndrome (8), and as a contiguous gene syndrome (15). Affected sibs (4,9,10,13,14) have been reported, but del(17p13) has been seen (1,3a,5,6,16–17) in over half of the cases. Other examples have had normal karyotypes (6,9,14). Evidence is mounting that the syndrome is caused by a microdeletion within the 17p13 region *even in cases without visible chromosomal defects*. Using DNA probes for genes known to be located in the 17p13 region, van Tuinen et al (18a) and Schwartz et al (16a) were able to demonstrate absence of these genes in Miller–Dieker patients with normal appearing, high resolution banded chromosomes. However, this should not be taken as evidence that either isolated or syndromic lissencephaly other than Miller–Dieker syndrome is caused by a defect in the 17p13 region.

Polyhydramnios occurs in over 30% and fetal movement is decreased. The infant is small for gestational age with severe postnatal growth deficiency.

**Central nervous system.** Lissencephaly has been discussed by many authors (7,19,20). It has been especially well discussed by Dobyns (7). Lissencephaly may occur alone as a malformation sequence, as part of a complex malformation syndrome, or may be associated with other structural defects of the brain and craniofacial region not explicable by a migration defect. Dobyns (7) defined lissencephaly as complete (grade 1), nearly complete (grade 2), or as mixed agyria/pachygyria (grade 3). Manifestations of the lissencephaly sequence are found in Table 17–1. Features of the Miller–Dieker syndrome are listed in Table 17–2.

Neonatal apneic spells and cyanosis, poor feeding, and profound mental deficiency are characteristic. Perinatal hypotonia occurs, but by 6 weeks, hypertonia associated with sustained opisthotonus becomes evident. Most infants develop seizures before the ninth week. Death has occurred before 6 months of age in over half the patients.

Lissencephaly, resulting from deficiency in neuronal migration (neuroglial heterotopia), is a constant feature. The ventricles are dilated and the cerebral cortex thickened with decreased white matter. The corpus callosum is thinned or absent. A frequent finding is midline calcification. Less often there are abnormalities in development of the medulla, especially the inferior olivary nucleus, and the cerebellum.

**Craniofacial features.** Facial findings are subtle. Features include microcephaly, prominent occiput, and large anterior fontanel. The forehead is bitemporally hollowed. The suprametopic skin is vertically furrowed in over 50%. There are roving eye movements. The upper lids are mildly ptotic. The nose is small with a wide bridge and anteverted nostrils. The ears are low set and/or posteriorly angulated with abnormal helices. The upper lip is long and thin and the chin is small

Table 17–1. Lissencephaly sequence[a]

| Features | Manifestations |
|---|---|
| Frequent | |
| Brain | Type I or cerebrocerebellar lissencephaly, hypoplastic opercula, subcortical heterotopias, colpocephaly, hypoplasia of the corticospinal tracts, pontine nuclei, inferior olives, olivary heterotopias |
| Brain function | Severe or profound mental retardation, decreased spontaneous activity, early hypotonia, subsequent hypertonia, seizures, poor feeding |
| Craniofacial | Microcephaly, bitemporal hollowing, prominent occiput, micrognathia |
| Occasional | |
| Brain | Small cerebellum with cortical dysplasia, hypoplasia of corpus callosum |
| Brain function | Infantile spasms |
| Prenatal | Polyhydramnios, decreased fetal movement |
| Perinatal | May require resuscitation, contractures |

[a]From WB Dobyns, *Birth Defects* **23**(1):225–241, 1987.

Table 17–2. Features of Miller–Dieker syndrome[a]

| Features | Manifestations |
|---|---|
| Frequent | |
| Brain | Type I (grade 1–2) lissencephaly, lissencephaly sequence, midline calcification |
| Brain function | Lissencephaly sequence |
| Craniofacial | Lissencephaly sequence, relatively high forehead. Abnormal irides, tortuous fundal vessels. Low nasal bridge, telecanthus, epicanthal folds, upturned nares. Long, thin upper lip, late eruption of primary teeth, prominent alveolar ridges |
| Body | Abnormal palmar creases, clinodactyly, camptodactyly. Cryptorchidism, sacral dimples |
| Growth | Small at birth, severe postnatal growth deficiency |
| Occasional | |
| Brain and brain function | Lissencephaly sequence |
| Prenatal, perinatal | Lissencephaly sequence, prolonged neonatal jaundice |
| Craniofacial | Vertical wrinkling of forehead |
| Body | Congenital heart defect, polydactyly, visceral anomalies, sacral tail |
| Cause | |
| Chromosomes | del(17)(p13.3) in >50% |
| Other | Unknown |

[a]From WB Dobyns, *Birth Defects* **23**(1):225–241, 1987.

(Figs. 17–5 and 17–6). Late eruption of teeth, prominent palatine ridges (6), and cleft palate (12) have been reported.

**Other findings.** Congenital heart defects of various types have occurred, most frequently PDA and VSD, in about 65%. Urogenital anomalies have included cryptorchidism, pilonidal sinus, unilateral renal agenesis, bilateral collecting system, and hydronephrosis (6). Other findings have been noted such as circumscribed corneal clouding (4), unilateral polydactyly, and camptodactyly (4, 13).

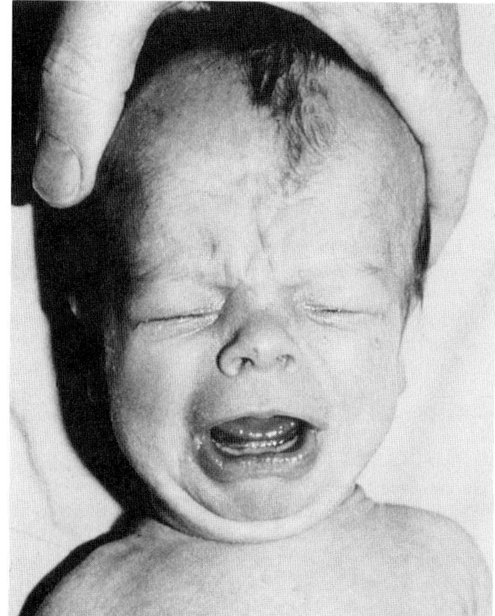

Fig. 17–5. *Miller–Dieker syndrome.* Microcephaly, high forehead with glabellar wrinkles, mild downward slant, palpebral fissures, small nose with anteverted nostrils and micrognathia. [From *H Dieker et al,* Birth Defects **5**(2):53, 1969.]

**Differential diagnosis.** Lissencephaly may occur as an isolated finding or as part of Norman–Roberts syndrome, *Walker–Warburg syndrome, Neu–Laxova syndrome, Zellweger syndrome, fetal akinesia sequence,* cerebrocerebellar lissencephaly, *isotretinoin embryopathy, fetal alcohol syndrome,* cerebro-oculo-muscular syndrome, muscle–eye–brain syndrome, *aprosencephaly syndrome,* cerebral perfusion failure, and anoxia (7).

Fig. 17–6. *Miller–Dieker syndrome.* Note bitemporal hollowing, anteverted nares. (From *WB Dobyns et al,* J Pediatr **102**:552, 1983.)

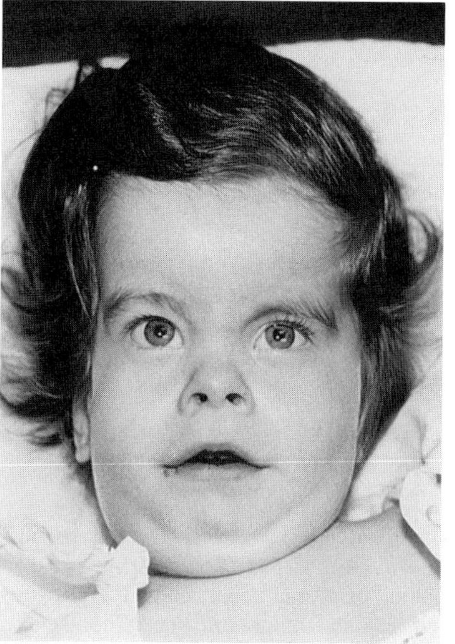

**Laboratory aids.** Brain abnormalities may be detected by computerized tomographic scan (20) or by magnetic resonance imaging. Chromosome analysis has shown del(17p13) in over 50% of the cases (16a,16b,18a) and can be helpful to separate the syndrome from isolated lissencephaly (1a). For those with an inherited chromosomal abnormality, prenatal diagnosis is possible (17).

### References [Miller–Dieker syndrome (type I lissencephaly)]


1. Batanian JR et al: Rapid diagnosis of Miller–Dieker and isolated lissencephaly by the polymerase chain reaction. March of Dimes Clinical Genetics Conference, Boston, 9–12 July 1989.

1a. Carpenter NJ et al: An infant with ring 17 chromosome and unusual dermatoglyphics: A new syndrome? *J Med Genet* **18**:234–236, 1981.

2. Dambska M et al: Lissencephaly: Two distinct clinico-pathological types. *Brain Dev* **5**:302–310, 1983.

3. Daube JR, Chou SM: Lissencephaly: Two cases. *Neurology (Minneap)* **16**:179–181, 1966.

3a. Dhellemmes C et al: Agyria-polygyria and Miller-Dieker syndrome: Clinical, genetic and chromosome studies. *Hum Genet* **79**:163–167, 1988.

4. Dieker H et al: The lissencephaly syndrome. *Birth Defects* **5**(2):53–64, 1969.

4a. Dobyns WB: Developmental aspects of lissencephaly and the lissencephaly syndromes. *Birth Defects* **23**(1):225–241, 1987.

5. Dobyns WB et al: Miller-Dieker syndrome: Lissencephaly and monosomy 17p. *J Pediatr* **102**:552–558, 1983.

6. Dobyns WB et al: Syndromes with lissencephaly. I. Miller–Dieker and Norman–Roberts syndrome and isolated lissencephaly. *Am J Med Genet* **18**:509–526, 1984.

7. Dobyns WB: Developmental aspects of lissencephaly and the lissencephaly syndromes. *Birth Defects* **23**(1):225–241, 1987.

8. Epstein CJ: *The Consequences of Chromosome Imbalance: Principles, Mechanisms, and Models,* Cambridge University Press, Cambridge, 1986.

9. Garcia CA et al: The lissencephaly (agyria) syndrome in siblings. *Arch Neurol* **35**:608–611, 1978.

10. Greenberg F et al: Familial Miller–Dieker syndrome with pericentric inversion of chromosome 17. *Am J Med Genet* **23**:853–860, 1986.

11. Jellinger K, Rett A: Agyria-pachygyria (lissencephaly syndrome). *Neuropädiatrie* **7**:66–91, 1976.

12. Jones KL et al: The Miller–Dieker syndrome. *Pediatrics* **66**:277–281, 1980.

13. Miller JQ: Lissencephaly—two siblings. *Neurology (Minneap)* **13**:841–850, 1963.

14. Norman MG et al: Lissencephaly. *Can J Neurol Sci* **3**:39–46, 1976.

15. Schmickel RD: Contiguous gene syndromes: A component of recognizable syndromes. *J Pediatr* **109**:231–241, 1986.

16. Schroer RJ, Phelan MC: Miller–Dieker syndrome with ring 17, deletion 17p and monosomy 17. *Proc Greenwood Genet Ctr* **6**:41–43, 1987.

16a. Schwartz CE et al: Detection of sub-microscopic deletions in band 17p13 in patients with the Miller–Dieker syndrome. *Am J Hum Genet* **43**:597–604, 1988.

16b. Selypes A, Laszlo A: Miller-Dieker syndrome and monosomy 17p13: A new case. *Hum Genet* **80**:103–104, 1988.

17. Stratton RF et al: Miller–Dieker syndrome and monosomy 17p13. *Hum Genet* **67**:193–200, 1984.

18. Van Allen M, Clarren SK: A spectrum of gyral anomalies in Miller–Dieker (lissencephaly) syndrome. *J Pediatr* **102**:559–564, 1983.

18a. van Tuinen P et al: Molecular detection of microscopic and submicroscopic deletions associated with Miller–Dieker syndrome. *Am J Hum Genet* **43**:587–596, 1988.

19. Warkany J et al: *Mental Retardation and Congenital Malformations of the Central Nervous System.* Year Book Medical Publishers, Chicago, 1971.

20. Williams JP, Joslyn JN: Lissencephaly: Computed tomographic diagnosis. *J Comput Assist Tomogr* **7**:141–144, 1983.


## Walker–Warburg syndrome (HARD ± E syndrome, Warburg syndrome, type II lissencephaly)

Walker (18), in 1942, first reported a syndrome of lissencephaly (agyria), hydrocephaly, microphthalmia, and retinal dysplasia. Krause (13) appears to have reported the same disorder in 1946. Warburg (19), in 1971, reviewing the combination of retinal nonattachment and hydrocephaly, suggested autosomal recessive inheritance, and Chemke et al (5) reported affected sibs in a consanguineous family. Pagon et al (16) introduced the term HARD ± E for **h**ydrocephaly, **a**gyria, **r**etinal **d**ysplasia with or without **e**ncephalocele. For a complete discussion of various forms of and diagnostic criteria for Type II lissencephaly, see Dobyns et al (8,9,9a) and Dambska et al (7). Inheritance is clearly autosomal recessive, several affected sibs (1,5,8,16,17) having been described. The disorder is lethal, death occurring within the first year of life, with about 65% dying within the first 3 months.

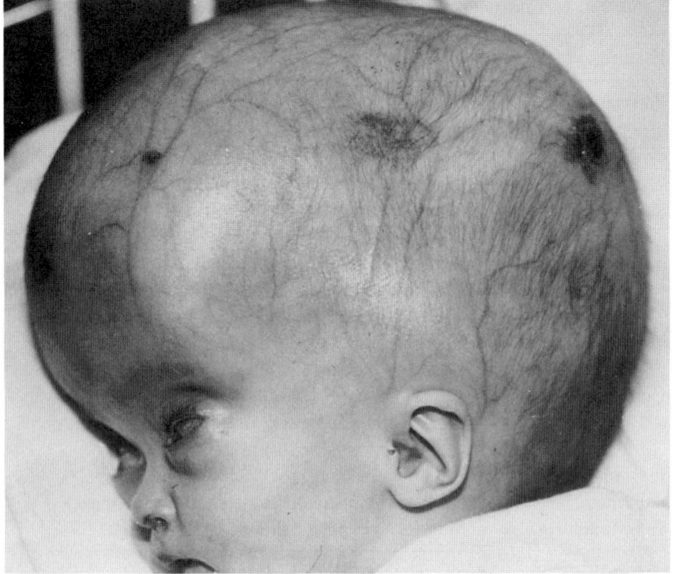

Fig. 17–7. *Walker–Warburg syndrome.* Hydrocephalus. (From *S Aymé* and *J–F Mattei,* Am J Med Genet **14**:759, 1983.)

Fig. 17–8. *Walker–Warburg syndrome.* Agyria-smooth lateral surfaces of cerebral cortex with absence of normal convolutions. Only narrow Sylvian fissure is identified.

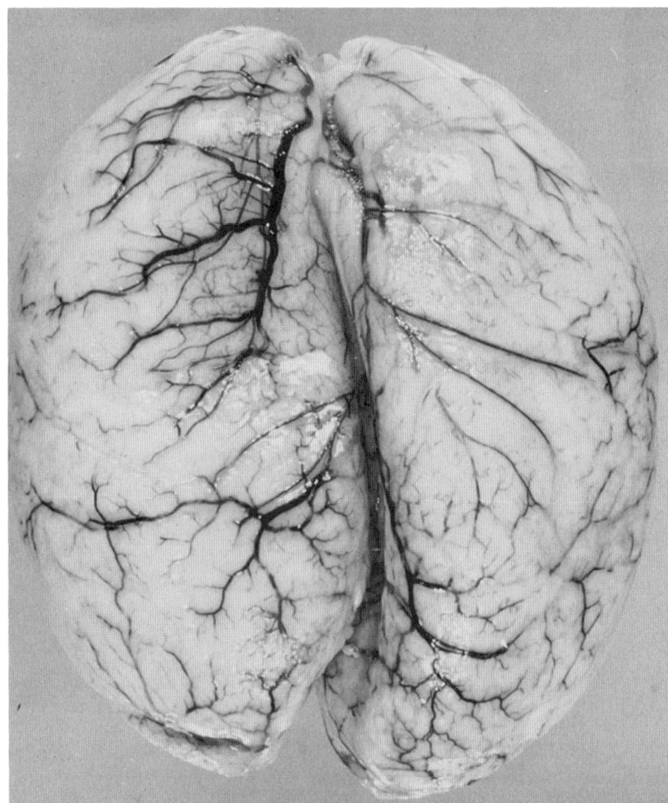

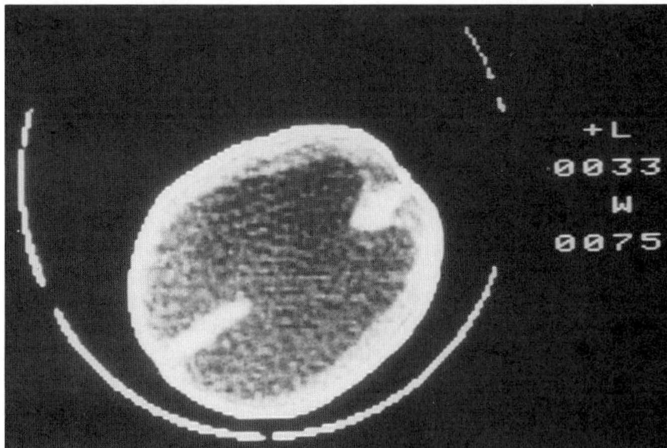


Fig. 17–9. *Walker–Warburg syndrome*. CT scan showing ventriculomegaly and thin cortical mantle. (From *D Donnai* and *PA Farndon,* J Med Genet **23**:200, 1986.)

An unusual but not striking facies results from the eye anomalies, hydrocephaly, hypertelorism, and micrognathia (Figs. 17–7 to 17–9).

Essentially constant features include agyria, thickened cortex but absent cortical layers, hypoplasia or agenesis of olfactory structures and optic pathways, absent or small corpus callosum and septum pellucidum, gliotic fused hemispheres, hydrocephaly, hypoplasia of the cerebellar vermis, and abnormal cerebellar folia. Dandy–Walker malformation (4,5,8,9,12) and occipital meningocele or encephalocele (1,2,5,12,24) have been found in about 40–50%. Microscopic changes include heterotopic neurons, poor myelinization, and glial and vascular proliferation.

Eye anomalies have included unilateral or bilateral microphthalmia, angle anomalies, iris hypoplasia, corneal opacity, cataracts, persistent papillary membrane, persistent hyperplastic primary vitreous, persistent posterior hyaloid artery, retinal detachment and/or dysplasia, and optic nerve hypoplasia (2,4,5,8,14,17,18,24) (Fig. 17–10A,B).

Sibs with cleft lip–palate have been described (3).

Genitourinary abnormalities are less frequently found: micropenis, cryptorchidism, and secondary hydronephrosis.

Differential diagnosis includes other syndromes with lissencephaly (8) and various intrauterine infections (toxoplasmosis, cytomegalovirus, etc.) that cause hydrocephaly and eye problems (retinal nonattachment, retinal dysplasia). Dandy–Walker malformation occurs in an inordinately large number of combinations reviewed by Murry et al (14a).

Prenatal diagnosis has been accomplished ultrasonographically (6,11).

### References [Walker–Warburg syndrome (HARD ± E syndrome, Warburg syndrome, type II lissencephaly)]

1. Aymé S, Mattei J-F: HARD (E) syndrome: Report of a sixth family with support for autosomal-recessive inheritance. *Am J Med Genet* **14**:759–766, 1983.

2. Bordarier C et al: Congenital hydrocephalus and eye abnormalities with severe developmental brain defects: Warburg's syndrome. *Ann Neurol* **16**:60–65, 1984.

3. Burton BK et al: Walker–Warburg syndrome with cleft lip and cleft palate in two sibs. *Am J Med Genet* **27**:537–542, 1987.

4. Chan CC et al: Oculocerebral malformations: A reappraisal of Walker's lissencephaly. *Arch Neurol* **37**:104–108, 1980.

5. Chemke J et al: A familial syndrome of central nervous system and ocular malformations. *Clin Genet* **7**:1–7, 1975.

6. Crowe C et al: The prenatal diagnosis of the Walker–Warburg syndrome. *Prenat Diag* **6**:177–186, 1986.

7. Dambska M et al: Lissencephaly: Two distinct clinico-pathological types. *Brain Dev* **5**:302–310, 1983.

8. Dobyns WB et al: Syndromes with lissencephaly. II. Walker–Warburg and cerebro-oculo-muscular syndromes and a new syndrome with type II lissencephaly. *Am J Med Genet* **22**:157–195, 1985.

9. Dobyns WB et al: Further comments on the lissencephaly syndromes. *Am J Med Genet* **22**:197–211, 1985.

9a. Dobyns WB et al: Diagnostic criteria for Walker–Warburg syndrome. *Am J Med Genet* **32**:195–210, 1989.

10. Donnai D, Farndon PA: Walker–Warburg syndrome (Warburg syndrome, HARD ± E syndrome). *J Med Genet* **23**:200–203, 1986.

11. Farrell SA et al: Prenatal diagnosis of retinal detachment in Walker–Warburg syndrome. *Am J Med Genet* **28**:619–624, 1987.

12. Halsey JH et al: The morphogenesis of hydranencephaly. *J Neurol Sci* **12**:187–217, 1971.

13. Krause AC: Congenital encephalo-ophthalmic dysplasia. *Arch Ophthalmol* **36**:387–444, 1946.

14. Levine RA: Warburg syndrome. *Ophthalmology* **90**:1600–1603, 1983.

14a. Murry JC et al: Dandy–Walker malformation: Etiologic heterogeneity and empiric recurrence risks. *Clin Genet* **28**:272–283, 1985.

15. Pagon RA, Clarren SK: HARD ± E: Warburg's syndrome. *Arch Neurol* **38**:66, 1981.

16. Pagon RA et al: Hydrocephalus, agyria, retinal dysplasia, encephalocele (HARD ± E) syndrome: An autosomal recessive condition. *Birth Defects* **14**(6B):232–241, 1978.

17. Pagon RA et al: Autosomal recessive eye and brain anomalies: Warburg syndrome. *J Pediatr* **102**:542–546, 1983.

18. Walker AE: Lissencephaly. *Arch Neurol Psychol* **48**:13–29, 1942.

19. Warburg M: The heterogeneity of microphthalmia in the mentally retarded. *Birth Defects* **7**(3):136–154, 1971.

20. Warburg M: Hydrocephaly, congenital retinal nonattachment and congenital falciform fold. *Am J Ophthalmol* **85**:88–94, 1978.

21. Warburg M: Ocular malformations and lissencephaly. *Eur J Pediatr* **146**:450–452, 1987.

22. Whitley CB et al: Warburg syndrome: Lethal neurodysplasia with autosomal recessive inheritance. *J Pediatr* **102**:547–551, 1983.

23. Williams R et al: Cerebro-ocular dysgenesis (Walker–Warburg syndrome): Neuropathic and etiologic analysis. *Neurology* **34**:1531–1541, 1984.

24. Winter RM, Garner A: Hydrocephalus, agyria, pseudoencephalocele, retinal dysplasia, and anterior chamber anomalies. *J Med Genet* **18**:314–317, 1981.

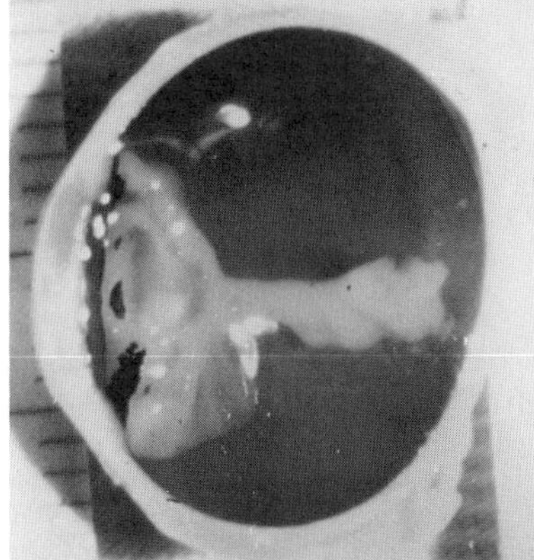

A

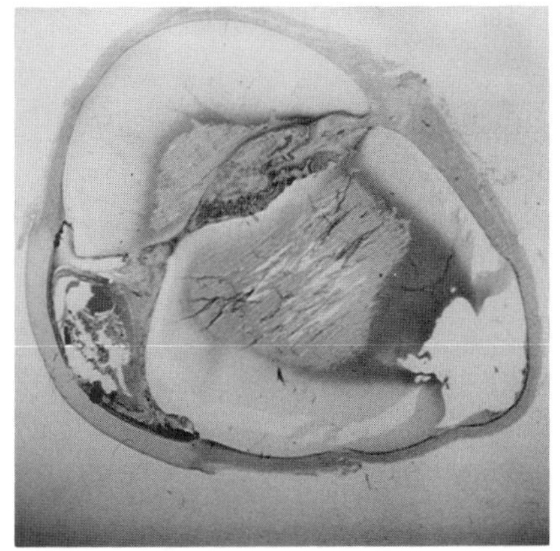

B

Fig. 17–10. (A) Section of eye showing retinal detachment. (B) Sagittal section of eye showing anterior and posterior chamber anomalies (corneal opacity, cataract, persistent primary vitreous) and retinal dysplasia. (A from *J Chemke et al,* Clin Genet **7**:1, 1975; B from *D Donnai* and *PA Farndon,* J Med Genet **23:**200, 1986.)

## Familial dysautonomia (Riley–Day syndrome)

Riley et al (29), in 1949, described a syndrome involving autonomic, sensory, and motor dysfunctions: absence of overflow tears, vasomotor instability, hypoactive deep-tendon reflexes, relative indifference to pain, feeding difficulties, and absence of lingual fungiform papillae. Subsequently, Riley et al (27,28) and others (3,7,8,13,16,17,20) presented thorough reviews of its clinical picture. An earlier report is that of Aring and Engel (2). The syndrome has autosomal recessive inheritance. Nearly all patients have Ashkenazic Jewish ancestry, the great majority stemming from eastern Europe (Galicia, Bukovina, Ukraine, Romania, and parts of Austro-Hungary) (7). Both in Israel and in the United States the frequency of the disorder has been about 1 in 10,000 to 20,000 Ashkenazic Jews (17), but a recent report suggests 1/3700 in Israel (14a). Cases in non-Jews are rare, but well documented (14,18,19,21). Parental consanguinity has been noted in at least 5%. Carriers have no clinical signs of the disorder.

About 30% have breech presentation in contrast to 3–8% of normal infants (5). Premature rupture of membranes occurs in about 30% (5). Meconium staining (35%) is about 3-fold normal (5). Stature is small even at birth (3). About 25% die of pulmonary infection by the age of 20 years. Recurrent bronchopneumonia is common, probably because food is frequently aspirated. Respiratory arrest during sleep is more common during adolescence (2a). Diagnosis is extremely difficult in the neonate, especially in the premature infant (25).

**Facies.** A fixed, sad, empty or frightened expression with a slit-like mouth (grimace) and a peculiar lingual rolling movement are typical (3,16,28) (Figs. 17–11 and 17–13). The face is thin, frequently asymmetric, with a pale to grayish color except during excitement. External strabismus is common.

**Nervous system.** In the newborn, about 60% have poor or no suck with 25–30% exhibiting hypotonia and/or hypothermia. Difficulty in swallowing and regurgitation appear soon after birth, and the infant fails to produce overflow tears with the usual stimuli (15). Especially marked are lethargy, somnolence, and unresponsiveness (25). Other virtually constant signs are absent to diminished deep-tendon reflexes, motor retardation and/or motor incoordination, hypotonia, breath-holding spells, hypersalivation, dizziness or ataxia on change in position, postural hypotension, paroxysmal hypertension, emotional lability, and relative indifference to pain (28). On the other hand, many of these children do not like their feet or scalp to be touched (dysesthesia). The indifference to pain may result in Charcot joints as the child ages. Older patients tend to have increased dysfunction in pain sensation, proprioception, and vibratory sense due to neuron depletion (5,24). There is relative insensitivity to hypercapnia and hypoxia (8). Many manifest alternate bouts of diarrhea and constipation. Skin blotching, abnormal sweating, erratic temperature control, and cyclic vomiting are frequently seen (15). About 80% exhibit growth retardation. Severe progressive scoliosis, found in 55%, appears around the eighth or ninth year of life (20,28). Intelligence is normal (3,36). Speech is often monotonous, slurred, and dysarthric with an unusual nasal quality (7).

Defective nerve condition (6) and a reduction in the number of unmyelinated nerve fibers, no myelinated nerve fibers greater than 12 $\mu$m in diameter, and abnormally short internodal length of small myelinated nerves have been demonstrated in the sural nerve (1,5,23). Sensory neurons in the spinal ganglia are decreased (24). There is a paucity of cells in the geniculate and scarpa ganglia (34).

**Skin.** A macular erythema, varying in color from pink to bright orange and located principally on the trunk and limbs, especially during periods of emotional excitement or eating, is common. Cold hands and feet, verging on acrocyanosis, and excessive sweating after the first month of life are constant features (5,20). Severe burns have resulted from indifference to pain in some patients.

**Eyes.** Constant features are decreased tearing, absent corneal reflex, and immediate papillary constriction in response to subconjunctival instillation of methacholine (3,28). Neuroparalytic keratitis sicca or corneal ulceration has been noted in at least 30%. The eye changes can be so severe that blindness may result (9). Myopia and retinal vascular tortuosity are also frequently observed (7,13). Optic atrophy has been reported (30).

**Oral manifestations.** Quite characteristically, the mouth is transversely elongated into a horizontal slit (Fig. 17–13). Smith et al (32)

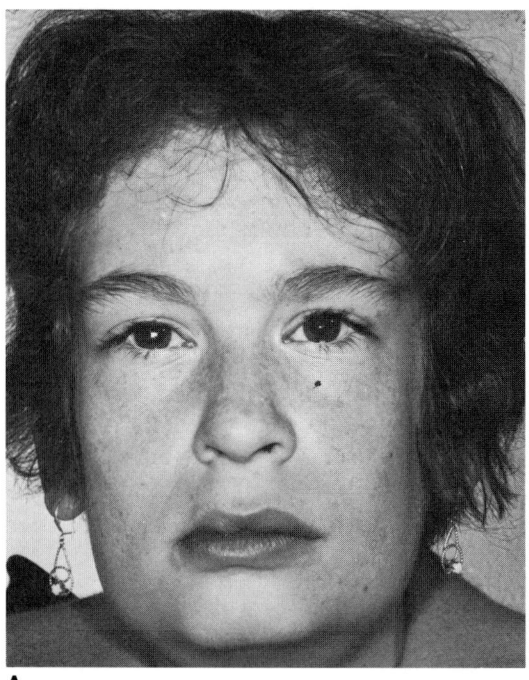

A

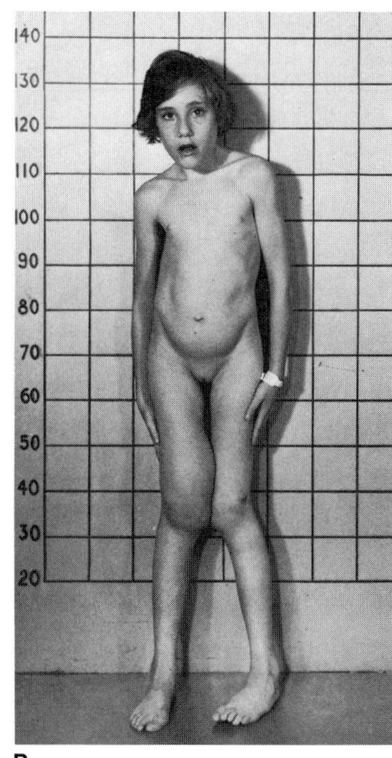


B

Fig. 17–11. *Familial dysautonomia.* (A, B) Fixed expression, asymmetric thorax, scoliosis, genu valgum due to Charcot joint-like changes. (B courtesy of *VA McKusick*, Baltimore, Maryland.)

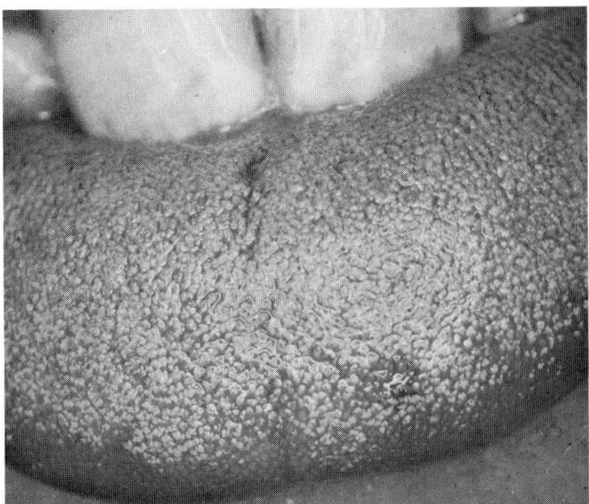

A

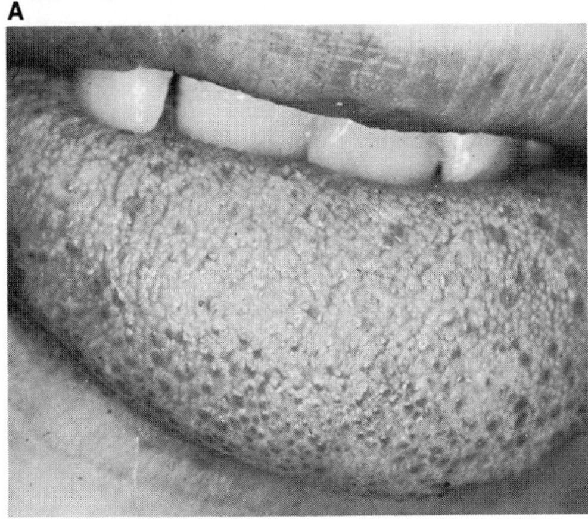

B

Fig. 17–12. *Familial dysautonomia.* (A) Tongue in familial dysautonomia. Note absence of fungiform papillae. (B) Normal tongue for comparison. Observe presence of fungiform papillae. (Courtesy of *AA Smith,* New York.)

Fig. 17–13. *Familial dysautonomia.* Characteristic slitlike mouth. (Courtesy of *AA Reitman,* Garden City, New York.)

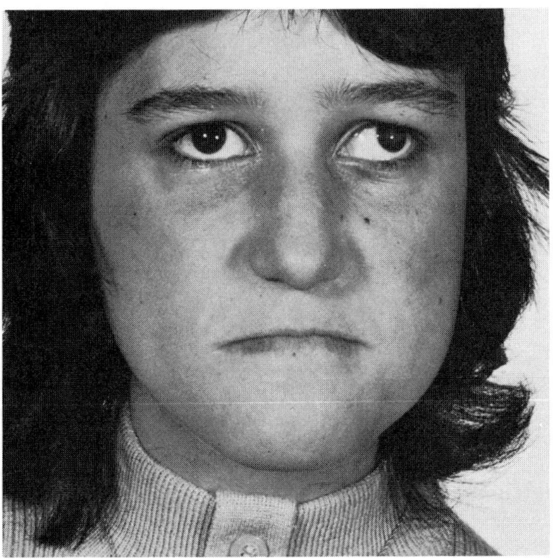

reported absence of fungiform papillae on the tongue (Fig. 17–12A,B). Rarely, some are present but are atrophic (10). Circumvallate papillae are also absent or greatly reduced in number (12,20). Sensitivity to sweet and bitter taste is diminished. Pearson et al (22) were unable to find nerve fibers traveling to the rudimentary papillae. The parents of some patients may have decreased numbers of fungiform papillae (22,28).

Excessive drooling and diminished gag reflex or swallowing disturbance occur in about 80% (20).

Dental caries is infrequent, perhaps because of reduced taste sensibility and consequent lowered desire for sweets (26). However, periodontal disease, malocclusion, and dental arch crowding are common (7).

**Differential diagnosis.** No disorder exactly mimics familial dysautonomia, but many progressive sensory neuropathies have similar features (7) (Table 17–3). Absence of fungiform and circumvallate papillae is an important specific clinical finding. Fungiform papillae may also be absent in Behçet syndrome.

There is some resemblance to *congenital sensory neuropathy with anhidrosis,* a disorder characterized by sensory deficit (diminished knee jerks, deficient taste discrimination, failure of axon flare after histamine injection) but differing from familial dysautonomia in that there are normal tear secretion and failure to produce miosis or to alter taste thresholds after methacholine administration. Suzuki et al (33) summarized cases of dysautonomia in non-Jewish infants. Most were atypical examples, probably representing other disorders.

**Laboratory aids.** Patients show a wheal but little to no pain or axon flare on intracutaneous injection of 0.01 ml of 1:10,000 histamine phosphate and a strong reaction to norepinephrine and related drugs (3). Some children fail to respond to a pinprick. There is immediate pupillary constriction with 0.0625% pilocarpine eyedrops. This test is helpful but not pathognomonic and there are 5–15% false positive responses. Taste acuity for salty and sweet foods, absent in familial dysautonomia, has become normal with parentally administered methacholine (methacholine is no longer commercially available) (12). An increased homovanillic acid to vanilmandelic acid (HVA/VMA) ratio has been reported in the urine of affected patients (11,17,28). Heterozygotes have normal ratios (11). There may be inappropriate or prolonged hypoglycemic response to insulin (3). Weinshilboum and Axelrod (35) noted decreased plasma dopamine–$\beta$-hydroxylase activity in about 25% of children with this disorder. The mothers of these children had decreased activity of the enzymes. Ziegler et al (37) demonstrated that there is no norepinephrine response on exercise and Siggers et al (31) described a several-fold increase in serum antigen to nerve growth factor.

### References [Familial dysautonomia (Riley–Day syndrome)]


1. Aguayo AJ et al: Peripheral nerve abnormalities in the Riley–Day syndrome. *Arch Neurol* **24**:106–116, 1971.

2. Aring CD, Engel GL: Hypothalmic attacks with thalamic lesions. *Arch Neurol Psychiatr* **54**:37–50, 1945.

2a. Axelrod FB, Abularrage JJ: Familial dysautonomia: A prospective study of survival. *J Pediatr* **101**:234–236, 1982.

3. Axelrod FB et al: Familial dysautonomia: Diagnosis, pathogenesis and management. *Adv Pediatr* **21**:75–96, 1974.

4. Axelrod FB et al: Progressive sensory loss in familial dysautonomia. *Pediatrics* **67**:517–522, 1981.

5. Axelrod FB et al: Neonatal recognition of familial dysautonomia. *J Pediatr* **110**:946–948, 1987.

6. Brown JC: Nerve condition in familial dysautonomia (Riley–Day syndrome). *JAMA* **201**:200–203, 1967.

7. Brunt PW, McKusick VA: Familial dysautonomia: A report of genetic and clinical studies with a review of the literature. *Medicine* **49**:343–374, 1970.

8. Dancis, J, Smith AA: Familial dysautonomia. *N Engl J Med* **274**:207–209, 1966.

9. Edwards TS: Familial dysautonomia, a rare cause of corneal ulcers. *J Pediatr Ophthalmol* **1**:51–56, 1964.

Table 17–3. Comparison of familial dysautonomia with other conditions in which pain sensation is impaired

| Characteristic | Familial dysautonomia | Congenital indifference to pain | Asymbolia for pain | Radicular sensory neuropathy | Sensory neuropathy with anhidrosis | Congenital sensory neuropathy |
|---|---|---|---|---|---|---|
| Inheritance | Autosomal recessive | Autosomal recessive | None | Autosomal dominant | Autosomal recessive | Mostly sporadic |
| Features present at birth | Yes | Yes | Probably not | No | Yes | Yes |
| Sensory loss other than pain | | | | | | |
| Temperature | Abnormal | Normal | Normal | Abnormal | Abnormal | Abnormal |
| Touch | Abnormal | Normal | Normal | Abnormal | Abnormal | Abnormal |
| Physiologic reactions to pain | Reduced | Normal | Normal | Reduced | Reduced | Reduced |
| Axon reflex | Absent | Normal | Normal | Absent | Absent | Absent |
| Distribution of sensory loss | Variable | Universal | Universal | Predominantly distal limbs | Incomplete | Incomplete |
| Intelligence | Dull to normal | Dull to normal | Normal | Normal | Defective | Dull to normal |
| Demonstrable pathologic changes | None consistent | None | Parietal lobe | Abnormal nerve roots | Possibly changes in Lissauer's tract | Absence of dermal nerve networks |

10. Gadoth N et al: Presence of fungiform papillae in classic dysautonomia. *Johns Hopkins Med J* **151**:298–301, 1982.

11. Gitlow SE et al: Excretion of catecholamine metabolites by children with familial dysautonomia. *Pediatrics* **46**:513–522, 1970.

12. Henkin R, Kopin I: Abnormalities of taste and smell thresholds in familial dysautonomia: Improvement with methacholine. *Life Sci* **3**:1319–1325, 1964.

13. Keith CG: Riley–Day syndrome. *Br J Ophthalmol* **49**:667–672, 1965.

14. Levine SL et al: Familial dysautonomia—unusual presentation in an infant of non-Jewish ancestry. *J Pediatr* **90**:79–81, 1977.

14a. Maayan C et al: Incidence of familial dysautonomia in Israel 1977–1981. *Clin Genet* **32**:106–108, 1987.

15. Mahloudji M et al: Clinical neurological aspects of familial dysautonomia. *J Neurol Sci* **11**:383–395, 1970.

16. Metha K: Familial dysautonomia in a Hindu boy. *Am J Dis Child* **132**:719, 1978.

17. McKendrick T: Familial dysautonomia. *Arch Dis Childh* **33**:465–468, 1958.

18. McKusick VA et al: The Riley–Day syndrome: Observations on genetics and survivorship. *Isr J Med Sci* **3**:372–379, 1967.

19. Mensher JH: Familial dysautonomia. Report of a case in a non-Jewish child. *J Pediatr Ophthalmol* **12**:40–48, 1975.

20. Moses SW et al: A clinical, genetic and biochemical study of familial dysautonomia in Israel. *Isr J Med Sci* **3**:358–369, 1967.

21. Orbeck H, Oftedal G: Familial dysautonomia in a non-Jewish child. *Acta Paediatr Scand* **66**:777–781, 1977.

22. Pearson J et al: The tongue and taste in familial dysautonomia. *Pediatrics* **45**:739–745, 1970.

23. Pearson J et al: The sural nerve in familial dysautonomia. *J Neuropathol Exp Neurol* **34**:413–424, 1975.

24. Pearson J et al: Quantitative studies of dorsal root ganglia and neuropathologic observations on spinal cords in familial dysautonomia. *J Neurol Sci* **35**:77–92, 1978.

25. Perlman M et al: Neonatal diagnosis of familial dysautonomia. *Pediatrics* **63**:238–241, 1979.

26. Reitman AA et al: Clinical evaluation of the dental aspects of familial dysautonomia. *J Am Dent Assoc* **71**:1436–1446, 1965.

27. Riley CM: Familial dysautonomia. *Adv Pediatr* **9**:157–190, 1957.

28. Riley CM, Moore RH: Familial dysautonomia differentiated from related disorders. *Pediatrics* **37**:435–446, 1966.

29. Riley CM et al: Central autonomic dysfunction with defective lacrimation. *Pediatrics* **3**:468–481, 1949.

30. Rizzo JF et al: Optic atrophy in familial dysautonomia. *Am J Ophthalmol* **102**:463–467, 1986.

31. Siggers DC et al: Increased nerve growth factor beta-chain cross-reacting material in familial dysautonomia. *N Engl J Med* **295**:629–634, 1976.

32. Smith AA et al: Tongue in familial dysautonomia. *Am J Dis Child* **110**:152–154, 1965.

33. Suzuki T et al: A case of new dysautonomia-like disorder found in Japan. *Eur Neurol* **14**:146–160, 1976.

34. Tokita N et al: Familial dysautonomia (Riley–Day syndrome): Temporal bone findings and otolaryngological manifestations. *Ann Otol Rhinol Laryngol (Suppl)* **46**:1–12, 1978.

35. Weinshilboum RM, Axelrod J: Reduced dopamine-beta-hydroxylase activity in familial dysautonomia. *N Engl J Med* **285**:938–942, 1971.

36. Welton W et al: Intellectual development and familial dysautonomia. *Pediatrics* **63**:708–712, 1979.

37. Ziegler M et al: Deficient sympathetic nervous response in familial dysautonomia. *N Engl J Med* **294**:630–633, 1976.

## Lesch–Nyhan syndrome (hypoxanthine–guanine phosphoribosyltransferase deficiency)

Lesch-Nyhan syndrome is an X-linked disorder resulting from deficiency of the enzyme hypoxanthine–guanine phosphoribosyltransferase (HPRT). It is characterized by hyperuricemia, spasticity, choreoathetosis, mental retardation, and compulsive, aggressive self-mutilating behavior. Although the disorder had been noted earlier (2,27), Lesch and Nyhan (18) first described the complete syndrome in 1964 and Hoefnagel et al (11) clearly demonstrated X-linkage soon after.

More than 150 patients have been reported (14,22,24) and several multiple case series have been published (3,36,38).

**Genetics and biochemistry.** The locus for the human HPRT gene has been mapped to the distal end of the long arm of the X chromosome (Xq26–Xq27.3). Using cloned DNA probes, molecular analysis has shown that the gene is present as a single copy of approximately 40 kilobases (kb) in length (32,38). Processing of the mRNA transcript results in excision of eight introns varying in length from 0.2 to 10.8 kb pairs to produce a mature RNA message 1.3 kb in length. The message is translated into a single peptide of 217 amino acids, which then undergoes glycosylation and several other posttranslational modifications. The processed peptide aggregates to form a catalytically active tetramer (molecular weight 98,000) with four identical subunits (38,39). Gibbs and Caskey (8) described HPRT mutation detection by ribonuclease A cleavage (8).

The HPRT enzyme catalyzes the transfer of the phosphoribosyl moiety of 5-phosphoribosyl-1-pyrophosphate (PRPP) to either hypoxanthine or guanine to form inosine monophosphate (IMP—a purine nucleotide precursor) or guanosine monophosphate (GMP). These reactions are considered to be "salvage pathways" as the alternate metabolic routes for hypoxanthine and guanine lead to formation of uric acid—the end-product of purine catabolism—which is subsequently excreted unchanged in the urine (14). Substrates and products of the reactions are regulators of other purine metabolizing enzymes; hence, HPRT influences multiple steps in purine metabolism (32).

Complete deficiency of HPRT activity is the usual finding in those expressing the Lesch–Nyhan phenotype, although some neurological deficits are present in about 20% of patients with significant partial HPRT deficiency (3,5,9,12,14,20,26). In other individuals with partial deficiency, hyperuricemia and gout may be the only manifestations. Mutations responsible for HPRT deficiency are quite heterogeneous and may affect enzyme expression, catalytic activity, or protein stability *in vivo* (15). As predicted for genetically lethal X-linked disorders, most cases have been shown to represent new mutations by molecular analysis (38).

**Pathogenesis.** In spite of extensive studies of the biochemical consequences of HPRT deficiency, the pathogenesis of the neurological deficit remains unclear, but there is a partial defect in the adrenocortical 11$\beta$-hydroxylation of steroids (37). Particularly impressive are the lack of discernible lesions in the basal ganglia and the absence of other histopathologic changes, apart from generalized cortical atrophy (3,14,36). Recent neurochemical studies suggest that abnormalities in the dopaminergic pathways of the basal ganglia may be instrumental in the development of behavioral changes. This area of the brain is highly dependent on HPRT activity to maintain purine pools, and synaptic function is regulated by purine nucleotides (14,16,19,32).

Patients with HPRT deficiency also develop gout. This is directly related to increased *de novo* purine synthesis that arises from lack of feedback inhibition from the products of the HPRT reaction. The excess purines are converted to uric acid which, being of limited solubility, crystallize *in situ*. Tissues with high urate concentrations, particularly the kidney and joints, are prone to crystal-induced damage.

**Growth.** Patients are normal at birth but growth failure is evident at the time neurological changes appear—after the second year of life (37). The height and weight of older patients are usually below the third percentile. Delayed osseous maturation may occur, but is usually less severe than delay in linear growth. The onset of puberty may also be delayed. Christie et al (3) found microcephaly in only 6 of 16 patients (39%) for whom head circumferences were reported.

**Neurological features.** Patients are usually not identified clinically until neurological symptoms are noted. Some delay in motor development and hypertonicity becomes significant at about 3–5 months of age. Choreoathetosis develops in all patients, but other nonspecific neurological features are also found, including opisthotonus (89%), scissoring (63%), upgoing toes (58%), and, less frequently, pure dystonia (21%) (3,14). Although self-mutilation is said to be present in all cases, exceptions have been reported (18,36). The average age of onset for self-mutilating behaviors is about 2 years but varies widely (4 months to 4 years) and has been delayed until the second decade in some instances (3,17,31,33). Self-mutilation most often involves biting with lips the most common target, followed by fingers and shoulders (Figs. 17–14 and 17–15). Other destructive behaviors include head-banging or chin-banging, which may lead to mutilation of the ears or nose, and hitting out at other people. Mental deficit varies widely and normal intelligence has been recorded. Most children will talk but speech is always dysarthric, presumably on the basis of choreoathetosis (3).

**Hyperuricemia and gout.** Urinary uric acid concentrations are always increased. Retrospectively, most parents (95%) have reported that "orange sand" or "orange crystals" were present in their infants' diapers even before the onset of neurological problems (3). If untreated, hyperuricosuria may progress to symptomatic nephrolithiasis, obstructive uropathy, and permanent renal damage. This has been described as the commonest cause of death in untreated patients in the first decade of life (14). Gouty arthritis is uncommon (14), but other features of gout including auricular tophi (24,29) and subcutaneous nodules (30) have been described.

**Other features.** Megaloblastic anemia observed in some patients has been associated with low serum folate (14). An increased rate of infection has also occurred, but this may be more a consequence of

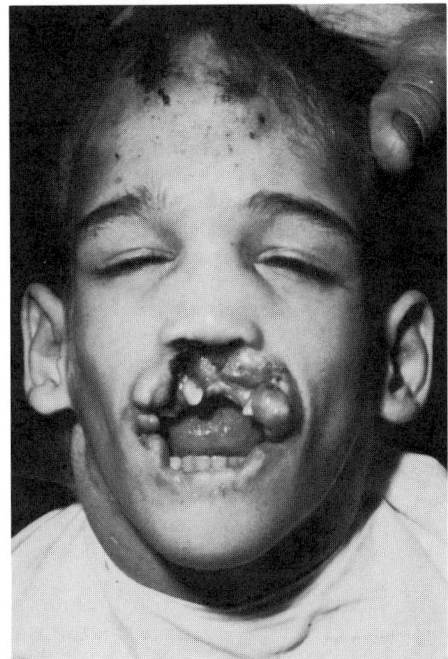

Fig. 17–14. *Lesch–Nyhan syndrome.* Self-mutilation of lips. (Courtesy of *D Hoefnagel,* Hanover, New Hampshire.)

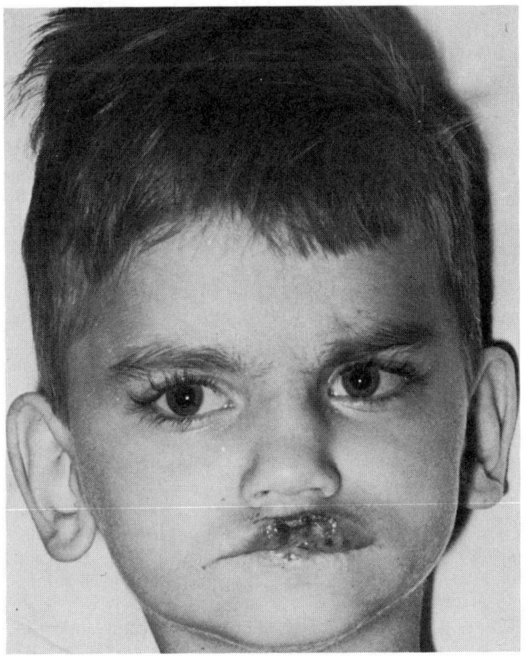

Fig. 17–15. *Lesch–Nyhan syndrome.* This boy and his brother have Lesch–Nyhan syndrome. Brother is more severely affected neurologically. (Courtesy of *B ter Haar,* Nijmegen, The Netherlands.)

the neurological handicap than of the biochemical defect (3,14). Testicular atrophy is common (3,7).

**Differential diagnosis.** If self-multilation occurs, there should be little difficulty recognizing *Lesch–Nyhan syndrome*. Among the mentally handicapped, however, some degree of self-mutilation is not that uncommon, particularly if autism is also present (36). Self-mutilation has been associated with *de Lange syndrome* and is seen in those with *congenital indifference to pain syndromes*. Individuals with conditions related to Lesch–Nyhan syndrome with similar clinical findings but without an enzymatic defect have been described (14,21,28). Conversely, in ostensibly normal patients with hematuria, episodes of oliguria, and azotemia, a variant of HPRT has been found (35). An unusual variant with mild mental retardation, spastic gait, pyramidal tract signs, short stature, proximally placed thumbs, and clinodactyly of the fifth fingers has been reported (25).

**Laboratory aids.** Assay of HPRT enzyme in skin fibroblasts, hair root lysates, or lymphocyte cloning may be used to identify heterozygous or hemizygous states. Similarly, assays of amniocytes or chorionic villus tissue provide a reliable basis for prenatal diagnosis (1,4,6,7,10,14,34,40).

In affected children, serum urate (normally 2–5 to 6.0 mg/dl), is usually markedly elevated. The urinary uric acid/creatinine ratio is a useful screening test to detect individuals with undifferentiated neurological signs. Although the normal ratio varies with age and individual results must be compared with those of age-matched controls, the urinary ratio reliably identifies Lesch–Nyhan patients and other hyperuricemic conditions (12,23).

### References [Lesch–Nyhan syndrome (hypoxanthine–guanine phosphoribosyltransferase deficiency)]


1. Bakay B et al: Detection of Lesch–Nyhan syndrome carriers: Analysis of hair roots for HPRT by agarose gel electrophoresis and autoradiography. *Clin Genet* **17**:369–374, 1980.
2. Catel W, Schmidt J: Über familiäre gichtische Diathese in Verbindung mit zerebralen and renalen Symptomen bei einem Kleinkind. *Dtsch Med Wochenschr* **84**:2145–2148, 1959.
3. Christie R et al: Lesch–Nyhan disease: Clinical experience with nineteen patients. *Dev Med Child Neurol* **24**:293–306, 1982.
4. Dempsey JL et al: Detection of the carrier state for an X-linked disorder, the Lesch–Nyhan syndrome, by the use of lymphocyte cloning. *Hum Genet* **64**:288–290, 1983.
5. Emmerson BT, Thompson L: The spectrum of hypoxanthine guanine phosphoribosyltransferase deficiency. *Q J Med* **166**:423–440, 1973.
6. Francke U et al: Detection of heterozygous carriers of Lesch–Nyhan syndrome by electrophoresis of hair root lysates. *J Pediatr* **82**:472–478, 1973.
7. Gibbs DA et al: First-trimester diagnosis of Lesch–Nyhan syndrome. *Lancet* **2**:1180–1183, 1984.
8. Gibbs RA, Caskey CT: Identification and localization of mutations of the Lesch–Nyhan locus by ribonuclease A cleavage. *Science* **236**:303–305, 1987.
9. Greene ML: Clinical features of patients with "partial" deficiency of the X-linked uricaciduria enzyme. *Arch Intern Med* **130**:193–198, 1972.
10. Halley V, Heukels-Dully MJ: Rapid prenatal diagnosis of the Lesch–Nyhan syndrome. *J Med Genet* **14**:100–102, 1977.
11. Hoefnagel D et al: Hereditary choreoathetosis, self-mutilation and hyperuricemia in young males. *N Engl J Med* **273**:130–135, 1965.
12. Kaufman JM et al: Urine uric acid to creatinine ratio—a screening test for inherited disorders of purine metabolism. *J Pediatr* **73**:583–592, 1968.
13. Kelley WN: Biochemistry of the X-linked uricaciduria enzyme defect and its genetic variants. *Arch Intern Med* **130**:199–206, 1972.
14. Kelley WN, Wyngaarden JB: Clinical syndromes associated with hypoxanthine-guanine phosphoribosyltransferase deficiency, in *The Metabolic Basis of Inherited Disease,* 5th ed. Stanbury JB et al (eds), McGraw-Hill, New York, 1982, pp. 1115–1143.
15. Keram E et al: Impaired kinetic properties of hypoxanthine-guanine phosphoribosyl transferase as a cause of uric and nephropathy in early infancy. *Eur J Pediatr* **146**:595–597, 1987.
16. Kopin IJ: Neurotransmitters and the Lesch–Nyhan syndrome. *N Engl J Med* **305**:1148–1150, 1981.
17. LaBanc J, Epker BN: Lesch–Nyhan syndrome: Surgical treatment in a case with lip chewing. *J Maxillofac Surg* **9**:64–67, 1981.
18. Lesch M, Nyhan WL: A familial disorder of uric acid metabolism and central nervous system function. *Am J Med* **36**:561–570, 1964.
19. Lloyd KG et al: Biochemical evidence of dysfunction of brain neurotransmitters in the Lesch–Nyhan syndrome. *N Engl J Med* **305**:1106–1111, 1981.
20. Lorentz WB Jr et al: Failure to thrive, hyperuricemia and renal insufficiency in early infancy secondary to partial hypoxanthine-guanine phosphoribosyl transferase deficiency. *J Pediatr* **104**:94–97, 1984.
21. Meigel W, Braun-Falco O: Automutilation der Unterlippe, verbunden mit Athetose und Oligophrenie ohne Purinstoffwechelstörung (pseudo-Lesch–Nyhan-Syndrom). *Hautarzt* **24**:158–160, 1973.
22. Michener WM: Hyperuricemia and mental retardation with athetosis and self-mutilation. *Am J Dis Child* **113**:195–206, 1967.
23. Mitchell G, McInnes RR: Differential diagnosis of cerebral palsy: Lesch–Nyhan syndrome without self-mutilation. *Can Med Assoc J* **130**:1323–1324, 1984.
24. Nyhan WL: Clinical features of the Lesch–Nyhan syndrome. *Arch Intern Med* **130**:186–192, 1972.
25. Page T et al: Syndrome of mild mental retardation, spastic gait, and skeletal malformations in family with partial deficiency of hypoxanthine-guanine phosphoribosyltransferase. *Pediatrics* **79**:713–717, 1987.
26. Rijksen G et al: Partial hypoxanthine-guanine phosphoribosyl transferase deficiency with full expression of the Lesch–Nyhan syndrome. *Hum Genet* **57**:39–47, 1981.
27. Riley LD: Gout and cerebral palsy in a three-year-old boy. *Am J Dis Child* **35**:293–295, 1960.
28. Rosenberg AL et al: Hyperuricemia and neurological deficits. *N Engl J Med* **282**:992–996, 1970.
29. Rosenbloom FM et al: Inherited disorders or purine metabolism. *JAMA* **202**:175–177, 1967.
30. Sass JK et al: Juvenile gout with brain involvement. *Arch Neurol* **13**:639–655, 1965.
31. Scully C: The orofacial manifestations of the Lesch–Nyhan syndrome. *Int J Oral Surg* **10**:380–383, 1981.
32. Singh S et al: A case of Lesch–Nyhan syndrome with delayed onset of self-mutilation: Search for abnormal biochemical, immunological and cell growth characteristic in fibroblasts and neurotransmitters in urine. *Adv Exp Biol Med* **195A**:205–210, 1985.
33. Stout JT, Caskey CT: HPRT: Gene structure, expression and mutation. *Annu Rev Genet* **19**:127–148, 1985.
34. Stout JT et al: First trimester diagnosis of Lesch–Nyhan syndrome: Applications to other disorders of purine metabolism. *Prenat Diag* **5**:183–189, 1985.
35. Sweetman L et al: A distinct human variant of hypoxanthine-guanine phosphoribosyl transferase. *J Pediatr* **92**:385–389, 1978.
36. Watts RWE et al: Clinical, post-mortem, biochemical and therapeutic observations on the Lesch–Nyhan syndrome with particular reference to the neurological manifestations. *Q J Med* **51**:43–78, 1982.
37. Watts RWE et al: Lesch–Nyhan syndrome: Growth delay, testicular atrophy and a partial failure of the 11$\beta$ hydroxylation of steroids. *J Inherit Metabol Dis* **10**:210–223, 1987.
38. Wilton JM, Kelley, WN: Molecular genetics of hypoxanthine-guanine phosphoribosyltransferase deficiency in man. *Arch Intern Med* **145**:1895–1900, 1985.
39. Wilson JM et al: A molecular survey of hypoxanthine-guanine phosphoribosyltransferase deficiency in man. *J Clin Invest* **77**:188–195, 1986.
40. Zoref-Shani E et al: Prenatal diagnosis of Lesch–Nyhan syndrome: Experience with three fetuses at risk. *Prenat Diag* **9**:657–661, 1989.


## Hereditary sensory syndromes with oral mutilation

In 1932, Dearborn (6) described a syndrome in which there was congenital indifference to pain, resulting in injury to both hard and soft tissues of the extremities, lips, and tongue with subsequent severe scarring. Trauma to bones caused aseptic necrosis, osteomyelitis, or fractures. His patient worked in a sideshow as a "human pincushion." This disorder, originally termed congenital analgia, has come to be known as congenital indifference or insensitivity to pain. Early case reports were tabulated by Fanconi and Ferrazzini (10). Heterogeneity became evident and Swanson (36) and Brown and Podosin (3) described another childhood sensory syndrome with oral mutilation called congenital sensory neuropathy with anhidrosis. Further studies of sensory and autonomic nerve functions as well as biopsies of peripheral nerves, mainly sural nerve, have led to identification of various conditions associated with lack of pain sensation.

Dyck (7) and Dyck et al (8) proposed a classification of conditions having analgesia based on mode of inheritance, natural history, and population of neurons or axons affected, under the name "hereditary sensory and autonomic neuropathy" (HSAN). They divided the disorders into five forms. HSAN Type I, autosomal dominantly inherited, was previously reported under a large variety of names, among them hereditary perforating ulcers of the feet and hereditary sensory radicular neuropathy. Symptoms start between the second and fourth decades. The main manifestations are in feet and legs and particularly with variable degree of peroneal muscular atrophy. Biopsy of peripheral nerves demonstrates diminished numbers of both myelinated (MF) and unmyelinated fibers (UF), especially of cutaneous nerves. Types II, III, IV, and V are congenital and have autosomal recessive inheritance. Type II is characterized by pansensory neuropathy mostly manifested in the limbs, with less tendency to affect the face and trunk. It is an abnormality of cutaneous rather than deep nerves, which are small and formed mostly of unmyelinated fibers that are atrophic and decreased in number. In these patients, MF tend to be absent from cutaneous nerves. Consanguinity among the parents of these patients (30) as well as affected sibs (30,40) has been reported. Type III is *familial dysautonomia* associated with various degrees of pansensory disturbance. In familial dysautonomia, biopsy demonstrates altered afferent neurons and gamma motor neurons with great reduction in UF and an absence of MF greater than 12 $\mu$m in diameter. Type IV has been described as congenital insensitivity to pain with anhidrosis. In this type, biopsy of peripheral nerves demonstrates complete absence of UF but normal or near normal large MF. Autopsy, performed in one patient, demonstrated the brain to be normal, but small neurons in the dorsal root ganglia were absent and the dorsolateral fascicular tract in the spinal cord was also absent as well as the spinothalamic tract (37). Consanguinity (2a,18a,20a,22,28,41,42) as well as affected sibs [2a,28,33(cases 2 and 3),36,41,43] have been found. Type V is similar to Type IV, but without mental retardation (Table 17–4). Histology demonstrates marked decrease in A delta fibers and minimal decrease of C fibers. Consanguinity was suspected in Dyck's (8) case but not proven. Congenital indifference to pain was not included in Dyck's classification because it does not present neuropathic abnormalities. Good reviews are those of Axelrod and Pearson (1) and Donaghy et al (6a).

Most patients reported earlier as having indifference or insensitivity to pain [2,6,10(case 3),14,17,21,29,31,35,38] have not been evaluated by present day methods. Therefore, it is nearly impossible to assign those cases according to Dyck's classification. Earlier cases of indifference to pain may have had Type IV or Type V HSAN (15,19,20) due to the fact that the neuropathy was not recognized or because of faulty diagnosis. Other patients may have had true universal lack of pain sensation [1,10 (cases 1 and 2),12,20,23,25 (case 1),32].

The usage of the terms indifference and insensitivity to pain should be discouraged because neither reflects the true nature of the conditions being discussed. For the purpose of clarification, we would like to introduce the term "hereditary nonneuropathic analgia" (HNNA), to refer to the condition characterized by universal lack of pain sensation but with perception of nociceptive stimulation, and with no alterations in sweating or other autonomic functions. Additionally, nerve biopsy should demonstrate normal peripheral nerve histology. Consanguinity [9,10(case 1),23,25(case 1),27] and affected sibs [10(case 1),12,20,23] have been reported in HNNA. Among those cases not classified, consanguinity (14,15,38) and affected siblings (14) have also been observed.

The following division may aid in understanding these conditions:

### Hereditary sensory syndromes

1. Hereditary sensory and autonomic neuropathy (HSAN) Types I, II, III, IV, and V
2. Hereditary nonneuropathic analgia (HNNA) (formerly congenital indifference to pain)

Type I does not present oral manifestations and Type III is discussed in this chapter. Types II, IV, V, and HNNA will be discussed here because they present variable degrees of oral and/or paraoral mutilation. Table 17–4 shows a schematic differentiation among the entities discussed.

**Facies and general appearance.** Frequent scarring of the face, with mutilation of the lips, arms, and legs, as well as phalangeal am-

Table 17–4. Hereditary sensory syndromes

| Condition | Inheritance | Mental retardation | Oral lesions | Deep tendon reflexes | Touch sensation | Temperature sensation | Axon reflex | Tears | Sweat | Abnormal histology | Childhood manifestations | Other manifestations |
|---|---|---|---|---|---|---|---|---|---|---|---|---|
| HSAN Type I, hereditary radicular neuropathy | AD | No | No | ± | Reduced+ | Reduced++ | No | Normal | Distal loss | Yes | No | Feet and legs peroneal atrophy |
| HSAN Type II, congenital sensory neuropathy | AR | No | Yes | Absent or reduced | Reduced++ | Reduced+ | No | Diminished | Distal loss | Yes | Yes | Mostly arms and legs |
| HSAN Type III, familial dysautonomia (Riley–Day) | AR | Could be retarded | Yes | Decreased | Normal | Reduced± | Yes | Absent | Erratic | Yes | Birth | Yes |
| HSAN Type IV, congenital sensory neuropathy + anhidrosis | AR | Could be retarded | Yes | Decreased | Reduced±, some normal | Reduced± | Yes | Normal | Absent or reduced | Yes | Birth | Yes |
| HSAN Type V | AR | No | Yes | Normal | Reduced++ | Reduced± | Yes | Normal | Absent peripheral | Yes | Birth | Yes |
| HNNA | AR | Normal or reduced | Yes | Normal | Normal | Normal | Yes | Normal | Normal | No | Birth | Yes |

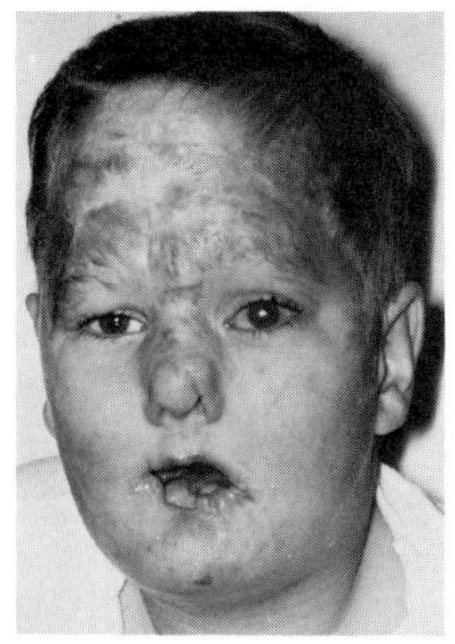

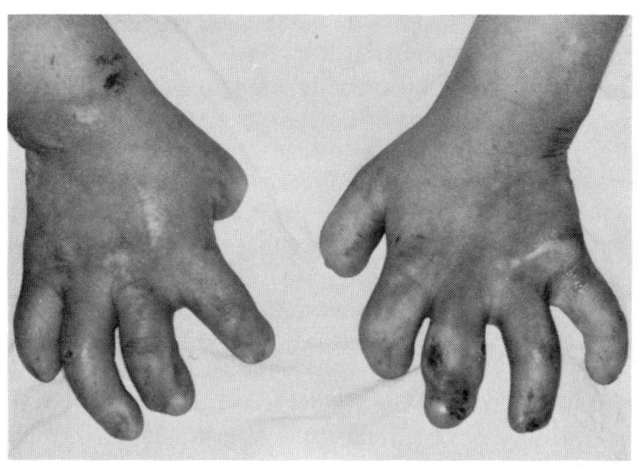

A B

Fig. 17–16. *Hereditary sensory syndromes with oral mutilation.* (A) Five-year-old male chewed off anterior tongue and lower lip. Repeatedly traumatized nose, eye, forehead. Parents were first cousins. (B) Hands of same child.

putation due to self-mutilation have been noted in Type II HSAN (8,30,40,43), Type IV HSAN (3,22,25,26,33,36), Type V HSAN (7), and HNNA (1,10,12,20,23) (Fig. 17–16A,B).

**Musculoskeletal system.** Osteomyelitis, aseptic necrosis, fracture, Charcot joint, and distal necrosis with spontaneous resorption of toes and fingers are virtually constant features in all these conditions (1–3,6–8,10–12,15–34,36,38–43). The complications have been ascribed to lack of pain sensation, the patient being unable to recognize trauma. Several patients have exhibited loose joints.

**Nervous system.** In Type II HSAN there is absence of reaction to painful stimuli most marked in the extremities. In some patients the lack of sensation may affect almost the entire body but present islands of normal sensation (7,8,43). Patients may have occasional seizures. Temperature sensation is reduced and touch sensation is markedly decreased. There is also lack of sweating over arms and legs (43). Anosmia has also been reported (40).

In Type IV HSAN patients present with various degrees of mental retardation. Lack of pain sensation is almost universal but islands of normal sensation can be found. Temperature and touch sensation are slightly reduced. Hypohidrosis, but with variable degree, is typical (2a,3,5a,8,13,25,26,36,43). Horner syndrome has been occasionally documented (3,33), as well as absence of Lissauer's tract and of small dorsal root ganglion cells and axons (18a,37).

Type V HSAN has been described in only a few patients (8,24,32a). These patients have universal lack of pain sensation, markedly reduced temperature sensation, moderate reduction in touch sensation, and peripheral hypohidrosis. The main difference with the other types of HSAN and HNNA is found in the abnormal histology of peripheral nerves (8).

In HNNA there is total absence of reaction to painful stimuli over the entire body. Deep-tendon reflexes are intact. The corneal and gag reflexes are often absent or diminished (12,21,23,25,27); taste, touch, sweat, temperature sensation, joint position sense, tickling, itching, and vibration are generally normal (1,8,12,21,25,27).

Intelligence has ranged from dull to normal. Patients with mental retardation generally represent examples of Type IV HSAN (see Table 17–4). Recent studies (9) tend to support the theory that HNNA is not due to neuropathic alterations but possibly to an endogenous analgesic peptide disorder in which the patient produces abnormally high levels of opiate and nonopiate analgesic substances.

**Oral manifestations.** Lack of pain sensation results in severe oral mutilation (Fig. 17–16A). The tongue and lips are especially subject to injury, with resultant scarring. This has been seen in Type II HSAN (8,30,40,43), Type IV HSAN [13 (case 1),22,25 (cases 2 and 4),28 (case 1),33,34,36 (case 1),39,41 (case 1, same as 28),42,43 (case 2, same as 39)], Type V HSAN [8 (case 1)], and in HNNA [1,10 (cases 1 and 2),12 (cases 1 and 2),23,32 (same as 1)]. These types of oral lesions have also been found in unclassified cases [10 (case 3),14, 15,17–19,30,38].

Oral involvement becomes apparent as soon as the teeth appear and may lead to early diagnosis (12,39). Extensive decay is not accompanied by toothache, and teeth may be lost early on this account (14). In many cases, patients have painlessly extracted their own teeth (39). Thrush (40) described a fibrous cord buccal to the teeth in each of his Type II HSAN patients.

**Differential diagnosis.** Lack of pain sensation also occurs in leprosy, oligophrenia, hysteria, multiple sclerosis, postleukotomy state, and asymbolia (40). Asymbolia for pain is a form of aplasia associated with lesions of the supramarginal gyrus of the parietal lobe, usually due to trauma, tumor, or infection (1). Analgia may also be seen in syringobulbia and syringomyelia, but in these conditions temperature sense is also lost. Proper classification of patients presenting with congenital absence of pain sensation can be achieved only by careful clinical and neurological study, including testing of sensory and autonomic functions, as well as by histologic studies of peripheral nerves (sural nerve biopsy). The reader interested in a detailed differentiation of these entities is referred to the excellent work of Dyck et al (7,8).

Patients with *Lesch–Nyhan syndrome* also show signs of self-mutilation, especially of lips and hands, but they usually have severe mental retardation, choreoathetosis, and hyperuricemia (4,35a). Critchley et al (5) described an autosomal recessively inherited syndrome characterized by acanthocytosis and neurologic disorder without $\beta$-lipoproteinemia. Tongue, lip, and buccal mucosal biting leading to severe mutilation resulted from uncontrollable, jerky movements which generally occurred at night. Patients responded normally to pain stimulation, but deep-tendon reflexes were absent.

**Laboratory aids.** Histaminic flare is normal in HNNA, but is diminished or absent in HSAN Types II, IV, and V (8,41). Fabbri et al (9) have recently found high levels of analgesic peptides such as opioid and calcitonin in the cerebrospinal fluid of a patient with HNNA. Injection of this lyophilized fluid in the lateral cerebral ventricles of rats induced deep analgesia. Haaxma et al (17) have shown insufficient calcium absorption in the digestive tract with an excess of osteoid tissue formation in their two patients with possible HNNA.

#### References (Hereditary sensory syndromes with oral mutilation)

1. Axelrod FB, Pearson J: Congenital sensory neuropathies. *Am J Dis Child* **138**:947–949, 1984.

2. Boyd DH Jr, Nie LW: Congenital universal indifference to pain. *Arch Neurol Psychiatr* **61**:402–412, 1949.

2a. Brahim JS et al: Oral and maxillofacial complications associated with congenital sensory neuropathy with anhidrosis. *J Oral Maxillofac Surg* **45**:331–334, 1987.

3. Brown JW, Podosin R: A syndrome of the neural crest. *Arch Neurol* **15**:294–301, 1966.

4. Christie R et al: Lesch–Nyhan disease: Clinical experience with 19 patients. *Dev Med Child Neurol* **24**:293–306, 1982.

5. Critchley EMR et al: Acanthocytosis and neurological disorder without betalipoproteinemia. *Arch Neurol* **18**:134–140, 1968.

5a. Daniel A et al: Congenital sensory neuropathy with anhidrosis. *Dev Behav Pediatr* **1**:49–53, 1980.

6. Dearborn G: A case of congenital general pure analgesia. *J Nerv Ment Dis* **75**:612–615, 1932.

6a. Donaghy M et al: Hereditary sensory neuropathy with neurotrophic keratitis. *Brain* **110**:563–583, 1987.

7. Dyck PJ: Neuronal atrophy and degeneration predominantly affecting peripheral sensory and autonomic neurons, in *Peripheral Neuropathy,* Dyck PJ et al (ed), W.B. Saunders, Philadelphia, 1984, pp. 1557–1599.

8. Dyck PJ et al: Not "indifference to pain" but varieties of hereditary sensory and autonomic neuropathy. *Brain* **106**:373–390, 1983.

9. Fabbri A et al: Intracerebroventricular injection of cerebrospinal fluid (CSF) from a patient with congenital indifference to pain induces analgesia in rats. *Experientia* **40**:1365–1366, 1985. (Same cases as Ref. 27.)

10. Fanconi G, Ferrazzini F: Kongenitale Analgie (Kongenitale generalisierte Schmerzindifferenz). *Helv Paediatr Acta* **12**:79–115, 1957.

11. Fath ME et al: Congenital absence of pain. *J Bone Jt Surg* **65B**:186–188, 1983.

12. Gaudier B et al: L'indifférence congénitale a la douleur. A propos de deux nouvelles observations. *Arch Fr Pédiatr* **26**:1027–1040, 1969.

13. Goebel HH et al: Confirmation of virtual unmyelinated fiber absences in hereditary sensory neuropathy type IV. *J Neuropathol Exp Neurol* **39**:670–675, 1980.

14. Guillermo A, Grinspan GA: Indifférénce congénitale a la douleur, a propos d'un cas avec antécédents de consanguinité. *Rev Neurol (Paris)* **123**:434–435, 1970.

15. Gutman D et al: Congenital analgesia. *Oral Surg* **39**:867–869, 1975.

16. Gwathmey FW, House JH: Clinical manifestations of congenital insensitivity of the hand and classification of syndromes. *J Hand Surg* **9**:863–869, 1984.

17. Haaxma R et al: Congenital indifference to pain associated with a defect in calcium metabolism. *Acta Neurol Scand* **47**:194–208, 1971.

18. Ingwersen OS: Congenital indifference to pain. *J Bone Jt Surg* **49B**:704–709, 1967.

18a. Ishii N et al: Congenital sensory neuropathy with anhidrosis. *Arch Dermatol* **124**:544–546, 1988.

19. Jewsbury ECO: Insensitivity to pain. *Brain* **74**:336–353, 1951.

20. Khan SA, Peterkin GAG: Congenital indifference to pain. *St Johns Hosp Dermatol Soc Trans* **56**:122–130, 1970.

20a. Kouvelas N, Terzoglou C: Congenital insensitivity to pain with anhidrosis. *Pediatr Dent* **11**:47–51, 1989.

21. Lamy M et al: L'analgésie generalisée congénitale. *Arch Fr Pédiatr* **15**:433–448, 1958.

22. Lee EL et al: Congenital sensory neuropathy with anhidrosis. *Pediatrics* **57**:259–262, 1976.

23. Lievre JA et al: Analgésie generalisée congénitale (indifference congénitale a la douleur) chez deux freres issu de cousins germains. *Soç Méd Hôp Paris* **119**:447–456, 1968.

24. Low PA et al: Congenital sensory neuropathy with selective loss of small myelinated fibers. *Ann Neurol* **3**:179–182, 1978.

25. MacEwen GD, Floyd GC: Congenital insensitivity to pain and its orthopedic implications. *Clin Orthop Rel Res* **68**:100–107, 1970.

26. Makoto M et al: Congenital insensitivity to pain with anhidrosis in a 2-month-old boy. *Neurology* **31**:1190–1192, 1981.

27. Manfredi M et al: Congenital absence of pain. *Arch Neurol* **38**:507–511, 1981. (Same cases as in Ref. 9.)

28. Mazar A et al: Congenital sensory neuropathy with anhidrosis. *Clin Orthop Rel Res* **118**:184–187, 1976. (Same cases as in Ref. 41.)

29. Murray RO: Congenital indifference to pain with special reference to skeletal changes. *Br J Radiol* **30**:2–6, 1957.

30. Murray TJ: Congenital sensory neuropathy. *Brain* **97**:387–394, 1973.

31. Nissler K, Parnitzke KH: Fehlen der Schmerzempfindung bei einem Kinde. *Dtsch Med Wochenschr* **76**:861–863, 1951.

32. Petrie JG: A case of progressive joint disorders caused by insensitivity to pain. *J Bone Jt Surg* **35B**:399–401, 1953.

32a. Pinckers A et al: Congenital sensory neuropathy. *Acta Ophthalmol* **66**:293–298, 1988.

33. Pinsky L, Di George AM: Congenital familial sensory neuropathy with anhidrosis. *J Pediatr* **68**:1–13, 1966.

34. Rafel E et al: Congenital insensitivity to pain with anhidrosis. *Muscle Nerve* **3**:216–220, 1980.

35. Saldanha PH et al: A genetical investigation of congenital analgesia. *Acta Genet (Basel)* **14**:143–158, 1964.

35a. Scully C: The orofacial manifestations of the Lesch–Nyhan syndrome. *Int J Oral Surg* **10**:380–383, 1981.

36. Swanson AG: Congenital insensitivity to pain with anhidrosis. *Arch Neurol* **8**:299–306, 1963.

37. Swanson AG et al: Anatomic changes in congenital insensitivity to pain. *Arch Neurol* **12**:12–18, 1965.

38. Thiemann HH: Analgesia congenita (angeborene universelle Schmerzindifferenz). *Arch Kinderheilkd* **164**:255–262, 1961.

39. Thompson CC et al: Oral manifestations of the congenital insensitivity-to-pain syndrome. *Oral Surg* **50**:220–225, 1980. (Same cases as in Ref. 43.)

40. Thrush DC: Congenital insensitivity to pain. A clinical genetic and neurophysiological study of four children from the same family. *Brain* **96**:369–386, 1973.

41. Vardy PA et al: Congenital insensitivity to pain with anhidrosis. *Am J Dis Child* **133**:1153–1155, 1979. (Same cases as in Ref. 28.)

42. Vassella F et al: Congenital sensory neuropathy with anhidrosis. *Arch Dis Childh* **43**:124–130, 1968.

43. Verity CM et al: Children with reduced sensitivity to pain: Assessment of hereditary sensory neuropathy types II and IV. *Dev Med Child Neurol* **24**:785–797, 1982. (Same cases as in Ref. 39.)

## Cranial nerve syndromes involving III, IV, V, VI

These arise from areas in which cranial nerves are geographically located in close approximation. Lesions in the following areas cause recognizable syndromes:

| | |
|---|---|
| Orbital apex syndrome | (Rollet syndrome) |
| Superior orbital fissure syndrome | (Rochon-Duvigneau syndrome) |
| Sphenoidal fissure syndrome | (no eponym, same as above) |
| Petrosphenoidal space syndrome | (Jacod syndrome) |
| Cavernous sinus syndrome | |
| a. Lateral wall of cavernous sinus | (Foix syndrome) |
| b. Cavernous sinus cavity | (Jefferson syndrome) |
| Apex of petrous bone syndrome | (Gradenigo syndrome) |

The signs and symptoms result from an intracranial lesion arising from the floor of the middle cranial fossa. It could involve the nerves that pass through the foramen ovale, foramen rotundum, and superior orbital fissure. Tumors extend along any of these paths to produce ophthalmoplegia by involvement of the third, fourth, and sixth nerves and blindness or optic atrophy by spread to the optic nerve itself.

**Anatomical guidelines.** The foramen ovale transmits the mandibular division of the fifth cranial nerve ($V_3$); the foramen rotundum, the maxillary division of the trigeminal ($V_2$); and the superior orbital fissure, the III, IV, VI as well as the first or ophthalmic division of the trigeminal nerve ($V_1$).

The apex of the orbit encompasses the area of the superior orbital fissure plus the optic foramen. There is considerable overlap between the syndromes as the same nerves traverse different anatomical locations.

It should be noted that there is not always a clear difference between the orbital apex syndrome and the superior orbital fissure syndrome.

The superior orbital fissure syndrome was first described by Rochon-Duvigneau (25) in 1896. If the lesion is more anteriorly situated and also includes the optic foramen and hence the optic nerve, the condition is called the orbital apex syndrome of Rollet (also called Jacod syndrome, but see below).

### Clinical features

**Orbital apex (Rollet) syndrome.** The syndrome is characterized by pain in the region of $V_1$ plus involvement of III, IV, and VI due to juxtaposition. Paralysis occurs either simultaneously or in succession. Resultant ptosis, proptosis, ophthalmoplegia, fixation and dilation of the pupil, blindness, and anesthesia of the upper eyelid and forehead occur. It differs from Jacod (petrosphenoidal) syndrome in which pain is largely over the $V_2$ area. Blindness arises from optic nerve involvement (15,33).

The syndrome can be difficult to differentiate from the superior orbital fissure syndrome that shares manifestations but in which there is no blindness.

Superior orbital fissure (Rochon-Duvigneau) syndrome. There is total ophthalmoplegia due to involvement of III, IV, and VI as well as sensory changes due to involvement of $V_1$. Hence, ptosis, proptosis of eyeball, fixation and dilatation of pupil, and anesthesia of upper eyelid and forehead are present, as noted above. Rarely, $V_1$ and $V_2$ are both involved.

The syndrome may be secondary to fracture of the zygomatic complex with edema and bleeding or facial fractures of the Le Fort II and III types, infections of the retrobulbar space, meninges, cavernous sinus and CNS, a neoplasm arising in the nose and extending through the orbital floor after traversing the pterygopalatine fossa (Behr pterygopalatine fossa syndrome or Dejean orbital floor syndrome), or a retrobulbar hematoma in the cavernous sinus and in the orbital muscle cone space. Lesions causing the superior orbital fissure syndrome can easily extend to involve the cavernous sinus (11,18,23,26,31,32).

Petrosphenoidal space (Jacod) syndrome. The syndrome may be confused with the superior orbital fissure syndrome. Further confusion occurs because the orbital apex syndrome is often used synonymously. The main clinical difference from the orbital apex syndrome is that pain is in the $V_2$ distribution.

Cavernous sinus (Foix) syndrome. This disorder is due to tumors (rarely pituitary), intracranial aneurysms, or thrombosis involving the cavernous and/or lateral sinuses. Ophthalmoplegia results from involvement of III, IV, and VI. Although pain is experienced in both the $V_1$ and $V_2$ areas, it is most severe over $V_1$ (4a,7,9,16).

When thrombosis occurs, the patient is acutely ill with fever, headaches, and pain around the eye that is proptosed and swollen because of conjunctival edema. There is usually complete ophthalmoplegia with a fixed pupil.

Apex of the petrous bone (Gradenigo) syndrome. This consists of sixth nerve palsy, diplopia, strabismus, lacrimation, and pain behind the eye, that is, in the $V_1$ region (8a,10). The syndrome is uncommon since the antibiotic era (mastoid infection), but has been reported following gunshot wound (2) or surgery (12). Horner syndrome may be present, but is rare.

#### Other syndromes involving eye movements

Duane syndrome. Named after Duane who fully described it in 1905, the syndrome is characterized by limitation of eye movement. On lateral gaze, the abducted eye becomes retracted and the eye opening narrows. In 15–20%, the condition is bilateral.

Etiology is unknown although aberrant peripheral innervation has been postulated. The clinical features include profound loss of abduction, partial deficiency of adduction, ocular retraction on attempted adduction, oblique movement of the eye (both up and down) on adduction, narrowing of palpebral fissures on adduction, and defect of convergence.

In the study of 62 cases by O'Malley et al (22), 62% were female, and 65% involved the left eye. Esotropia occurred in 53% and amblyopia in 20%. Nearly all cases were isolated. In one of these, transmission was from mother to daughter. The other involved concordance in monozygotic twins. About 15% had anisometropia.

The coexistence of Duane syndrome and *Marcus Gunn phenomenon* has been noted (14).

Fisher variant of Landry–Guillain–Barré syndrome. In 1956, Fisher (5) described three patients with acute external ophthalmoplegia, ataxia, and hyporeflexia. Ptosis and some degree of internal ophthalmoplegia (facial and bulbar weakness) can occur, but characteristically there is impairment of conjugate upward and lateral gaze, often leading to complete ophthalmoplegia. Usually it is benign with recovery complete. There is often a preceding upper respiratory tract infection but etiology is unknown. Hypersensitive response to a virus is probable.

Tolosa–Hunt syndrome. The main criteria for diagnosis (13,29) have been summarized by Dornan et al (4): steady nonthrobbing pain, often preceding ophthalmoplegia by several days, and paresis of any of the nerves passing through the cavernous sinus, that is, III, IV, VI, and $V_1$ (sympathetic nerves and II occasionally involved). Symptoms may last for days or weeks. There is no obvious lesion and attacks may be repeated.

Remission is either total or subtotal. Etiology is unknown (28).

**Syndromes of V.** It must be emphasized that a single cranial nerve is rarely involved in a pathological process. Because of the intimate association of V with III, IV, and VI in the cavernous sinus and elsewhere, a host of eponymic syndromes exist. Diffuse processes, such as brain tumors at the base of the skull or tumors that invade the cranium from the nasopharynx, may involve the last four cranial nerves in combination with the fifth, but this is less common.

See the previous discussion for nerve V involvement in the orbital apex syndrome, superior orbital fissure syndrome, and syndromes of the cavernous sinus. Trigeminal neuralgia is discussed in "Some Disorders of Facial Pain" later in this chapter.

#### Other syndromes involving the Vth cranial nerve during its peripheral course

Raeder (paratrigeminal) syndrome. This syndrome is characterized by unilateral facial pain/headache with a trigeminal distribution. The pain is mostly in and around the eye. In addition, there is ptosis, miosis, and hyperemia because of an oculosympathetic defect (3,8,17,19,24,27).

Multiple parasellar cranial nerve involvement of III, IV, V, and VI can occur. Four of five patients originally described by Raeder had other cranial nerve involvement. Decreased sensation and muscular weakness in all three divisions of V have been described.

The syndrome is a combination of a Horner-like syndrome with ipsilateral motor and sensory V dysfunction. Neoplasms of the middle cranial fossa or in the space adjacent to V—hence paratrigeminal—are often responsible. Facial anhidrosis is absent. It is often seen in males in the fourth to sixth decades, but trauma has also been implicated.

There must be involvement of V. Evidence for this may be decreased sensation in the distribution of V, weakness, pain of the tic douloureux type and distribution, and pain over the ipsilateral part of the face. Facial pain in the distribution of V alone, although it may be present, is not sufficient (20).

Pterygopalatine fossa (Behr) syndrome. This syndrome, resembling that of Trotter (see Table 17–5), is due to a tumor or metastases in the pterygopalatine fossa, located behind the maxillary sinus and below the inferior surface of the sphenoid. It contains $V_2$, the sphenopalatine ganglion, and the internal maxillary artery. The main symptoms are pain in the upper teeth, loss of sensation over the inferior orbital foramen, paralysis of the pterygoids, and blindness.

Trotter syndrome. First described by Trotter (30) in 1911, the syndrome consists of unilateral deafness, pain over $V_3$, ipsilateral defective mobility of the soft palate, and subsequent trismus. A sensation of ear pressure, distention of the tympanic membrane, or repeated middle ear infections are common. The syndrome results from invasion of the lateral wall of the nasopharynx by tumor, usually an anaplastic carcinoma. Immediately below the sinus is the attachment of the levator veli palatini muscle, just laterally is the foramen ovale, and posteriorly is the opening to the eustachian tube. As the tumor extends, it compresses the tube (causing deafness) and invades the palatal musculature (decreasing pain in the temporal area, ear, lower

Table 17–5. Cranial nerve syndromes involving midbrain or pons

| Syndrome | Location | Signs |
|---|---|---|
| Midbrain | | |
| Weber | Base of midbrain anterior to red nucleus | Ipsilateral III<br>Contralateral hemiparesis |
| Claude | Tegmentum, red nucleus, and brachium conjunctiva | Ipsilateral III (sometimes IV), sometimes reduction of conjugate gaze to opposite side.<br>Contralateral ataxia dysmetria, dysdiadochokinesia |
| Benedict | Tegmentum; destruction of red nucleus where third nerve crosses it | Ipsilateral third plus contralateral movement disorder—Parkinsonian tremor, athetosis, hemibalismus |
| Nothnagel | Tectum | Paralysis of lateral gaze plus ataxia—unilateral or bilateral |
| Parinaud | Pretectum (pinealomas, craniopharyngiomas) | Paralysis of upward gaze and of convergence<br>Pupils are responsive |
| Pons | | |
| Millard–Gubler | Ventral pons<br>VI, VII<br>corticofacial and corticospinal fibers before decussation | Facial paralysis, paralysis of adduction of eye<br>Contralateral hemiparesis |
| Inferior syndrome of Foville | Caudate one-third of pons<br>Supranuclear VI and VII medial lemniscus spinal root of inferior cerebellar peduncle | Lateral conjugate gaze toward side of lesion<br>Crossed hemiparesis with diminished pain and temperature sensation<br>Homolateral cerebellar signs |
| Superior syndrome of Raymond-Cestan or superior type of Foville | Rostral pons<br>Includes medial lemniscus, spinothalamic tract<br>Superior cerebellar peduncle | Paralysis of conjugate gaze to size of lesion<br>Diminished pain, touch, temperature, vibration and position sense<br>Cerebellar signs, (ataxia) athetotic or ballistic movement, all on the opposite side |
| Brissaud and Sicard | Anterolateral and inferior pons, pyramidal tract before it crosses VII nucleus | Hemifacial spasms, crossed hemiparesis |

jaw, teeth, and tongue, and resulting in anesthesia over the mental area). Trismus results from extension of the tumor into the pterygoids. The syndrome is seen most often in males during the third and fourth decades. There are usually no symptoms due to nasopharyngeal obstruction (1,21,30). The use of computer tomography has vastly aided diagnosis (6).

Godtfredsen (cavernous sinus-nasopharyngeal tumor) syndrome. Tumors arising in the nasopharynx, in addition to extending into the skull and producing ophthalmoplegia and trigeminal neuralgia, may also be associated with paralysis of the tongue. This is produced by compression of the hypoglossal nerve by enlarged, involved, retropharyngeal lymph nodes (9).

### References (Cranial nerve syndromes involving III, IV, V, VI)

1. Asherton N: Trotter's syndrome and associated lesions. *J Laryngol Otol* **65**:349–366, 1951.

2. Cappina AH: Traumatic intracranial aneurysm and Gradenigo's syndrome secondary to gunshot wound. *Surg Neurol* **22**:263–266, 1984.

3. Cohen DN et al: Paratrigeminal syndrome. *Am J Ophthalmol* **79**:1044–1049, 1975.

4. Dornan TL et al: Remittent painful ophthalmoplegia: The Tolosa–Hunt syndrome? *J Neurol Neurosurg Psychiatr* **42**:270–275, 1979.

4a. Ferrari G et al: Foix–Chavany syndrome: CT study and clinical report of three cases. *Neuroradiology* **18**:41–42, 1979.

5. Fisher M: An unusual variant of acute idiopathic polyneuritis (syndrome of ophthalmoplegia, ataxia, and areflexia). *N Eng J Med* **255**:57–65, 1956.

6. Flynn TR, Eisenberg E: Computerized tomography and Trotter's syndrome in the diagnosis of maxillofacial pain. *Oral Surg* **61**:440–447, 1986.

7. Foix C: Syndrome de la paroi externe du sinus caverneux. *Rev Neurol (Paris)* **38**:827–832, 1922.

8. Ford FR, Walsh FB: Raeder's paratrigeminal syndrome: A benign disorder, possibly a complication of migraine. *Bull Johns Hopkins Hosp* **103**:296–298, 1958.

8a. Gillanders DA: Gradenigo's syndrome revisited. *J Otolaryngol* **12**:169–174, 1983.

9. Godtfredsen E: Studies on the cavernous sinus syndrome. *Acta Neurol Scand* **40**:69–75, 1964.

10. Gradenigo G: A special syndrome of endocranial otitic complications (paralysis of the motor oculi externus of otitic origin). *Ann Otol* **13**:637, 1904.

11. Hedstrom J et al: Superior orbital fissure syndrome. *J Oral Surg* **32**:198–201, 1979.

12. Herceg SJ, Harding RL: Gradenigo's syndrome following correction of a posterior choanal atresia. *Plast Reconstr Surg* **48**:181–183, 1971.

13. Hunt WE et al: Painful ophthalmoplegia: Its relation to indolent inflammation of the cavernous sinus. *Neurology* **11**:56–62, 1961.

14. Isenberg S, Blechman B: Marcus Gunn jaw winking and Duane's retraction syndrome. *J Pediat Ophthalmol Strab* **20**:235–237, 1983.

15. Jacod M: Sur la propagation intracranienne des sarcomes de la trompe d'eustache syndrome du carrefour petro-sphenoidal paralysie des 2e3e4e5e et 6e paires craniennes, Thesis, Univ Paris, 1921.

16. Jefferson G: Cavernous sinus syndrome. *Trans Ophthalmol Soc UK* **73**:117, 1953.

17. Klingon GH, Smith WM: Raeder's paratrigeminal syndrome. *Neurology* **6**:750–753, 1956.

18. Kurzer A, Patel MP: Superior orbital fissure syndrome associated with fractures of the zygoma and orbit. *Plast Reconstr Surg* **64**:715–719, 1979.

19. Minton LR, Bounds GW: Raeder's paratrigeminal syndrome. *Am J Ophthalmol* **58**:271–275, 1964.

20. Mokrii B: Raeder's paratrigeminal syndrome. Original concept and subsequent deviations. *Arch Neurol* **39**:395–399, 1982.

21. Olivier RM: Trotter's syndrome. *Oral Surg* **15**:527–530, 1962.

22. O'Malley ER et al: Duane's retraction syndrome—plus. *J Ped Ophthalmol Strab* **19**:161–165, 1982.

23. Pogrel MA: The superior orbital fissure syndrome. *J Oral Surg* **38**:215–218, 1980.

24. Raeder JG: "Paratrigeminal" paralysis of the oculopupillary sympathetic. *Brain* **47**:149–158, 1924.

25. Rochon-Duvigneau: Quelques cas de paralysie de tous les nerfs orbitaires (ophtalmoplégie totale avec amaurose et anesthesie dans le domaine de l'ophtalmique) d'origine syphilitique. *Arch Ophtalmol* **16**:746–760, 1896.

26. Sieverink NP, van der Wal KG: Superior orbital fissure syndrome in a 7-year-old boy. *Int J Oral Surg* **9**:216–220, 1980.

27. Smith JL: Raeder's paratrigeminal syndrome. *Am J Ophthalmol* **46**:194–199, 1958.

28. Takeoka T et al: Tolosa–Hunt syndrome. *Arch Neurol* **35**:219–223, 1978.

29. Tolosa E: Periasteritic lesions of the carotid siphon with the clinical features of a carotid infraclinoidal aneurysm. *J Neurol Neurosurg Psychiat* **17**:300–302, 1954.

30. Trotter W: On clinically obscure malignant tumors of the nasopharyngeal wall. *Br Med J* **2**:1057–1059, 1911.

31. Zachariades N: The superior orbital fissure syndrome: Review of the literature and report of a case. *Oral Surg* **53**:237–240, 1982.

32. Zachariades N et al: The superior orbital fissure syndrome. *J Maxillofac Surg* **13**:125–128, 1985.

33. Zachariades N et al: Orbital apex syndrome. *Int J Oral Maxillofac Surg* **16**:352–354, 1987.

## Marcus Gunn phenomenon [jaw-winking and winking-jaw syndromes (Marcus Gunn and inverse Marcus Gunn syndromes), Marin Amat syndrome]

The short communication by Marcus Gunn (21) (often erroneously hyphenated), in 1883, describing the syndrome of unilateral congenital ptosis and rapid exaggerated elevation of the ptotic lid on moving the lower jaw to the contralateral side, stimulated immediate interest in this problem. By 1895, 33 cases had been reviewed by Sinclair (28) and since then at least another 200 cases have been published (2,3,6,15,25,27,29–31). The syndrome is seen in about 5% of cases of congenital ptosis of the eyelid (25).

The name "jaw-wink" syndrome is not well chosen for the symptom is not a wink but an exaggerated opening of the eye. However, the term has been used so long and extensively that until the cause is clarified, its use will probably be continued. Although usually sporadic, there have been a few familial cases involving two generations (8,14,17,18,25). There is neither sex nor side predilection.

The cause is unknown, but it was originally assumed that the syndrome was based on aberrant innervation of the levator palpebrae superioris from the motor branch of V because of the close approximation of the nuclei of III and V. However, a supranuclear, or at least a combined supranuclear–nuclear, involvement has been suggested, and the view has gained support (5,9,25,31). A good review of theories of etiology has been compiled by Simpson (27).

The ptosis is congenital in over 90%. It may, however, arise spontaneously in older persons (14,20,27). Ordinarily the lid cannot be raised to any significant degree. Seven percent of patients have experienced the syndrome without having ptosis (25), and 3% have bilateral ptosis (3,6).

Superior rectus palsy has been found in 25% and double elevator palsy in another 25%. Between 35 and 60% exhibit amblyopia or strabismus (6,25). Correction of the strabismus was accompanied by opening the mouth in one patient (23). The pupillary and corneal reflexes are normal. The pupil in the ptotic eye may be smaller, but true Horner syndrome is not present.

About 40% manifest the syndrome both on depressing the mandible and on moving it to the side opposite the ptotic eye (Fig. 17–17A,B). Another 40% need only to have the jaw depressed. However, in some individuals the syndrome may be produced by movement of the lips, whistling, smiling, clenching the teeth, chewing, puffing out the cheeks, protrusion of the tongue, or sucking (4,25). It may even be precipitated by swallowing or, rarely, by moving the jaw to the ipsilateral side (11), or to both sides (12). The syndrome has also been triggered by closing the other eye. The syndrome may even temporarily disappear. Several muscles and cranial nerves come into play during these acts; the most important appears to be the external pterygoid. That the masseter and temporal muscles play no significant role was demonstrated by Dupuy-Dutemps (7). A comprehensive review of critical jaw movements was made by Simpson (27). Cleft lip has been found but is probably a chance association (2,25).

The so-called *winking-jaw syndrome* (inverse Marcus Gunn syndrome, corneomandibular reflex, pterygocorneal reflex) is manifested by a brisk movement of the mandible to the contralateral side when the cornea is touched. The jaw may also be thrust forward (13,32). The term "inverse Marcus Gunn syndrome" has also been used to refer to brisk ptosis of the upper eyelid on moving the mandible to the contralateral side (19) (Fig. 17–18A,B). Both jaw-winking and winking-jaw syndromes are considered to be release phenomena due to a supranuclear lesion, caused either by malformation or by severe cerebral or brainstem lesions, that reunites the orbicularis oculi and external pterygoid muscles, which phylogenetically belong together (31,32).

The so-called *Marin Amat (19) syndrome* appears to be an intrafacial reaction on the part of regenerating facial nerve following facial palsy (1,10,16,26,31,32). A partially ptosed eyelid droops further when the mandible moves from side to side.

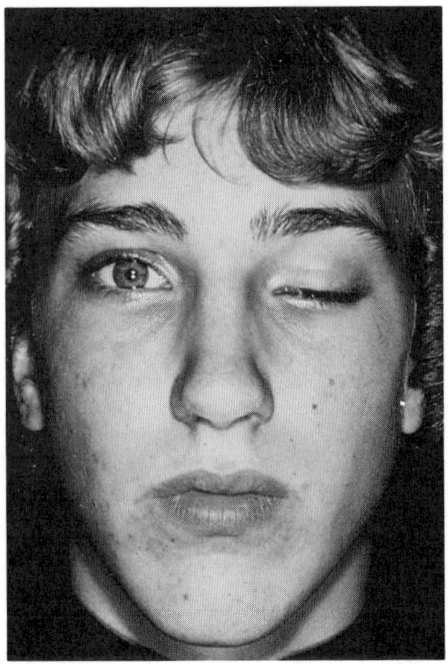

A

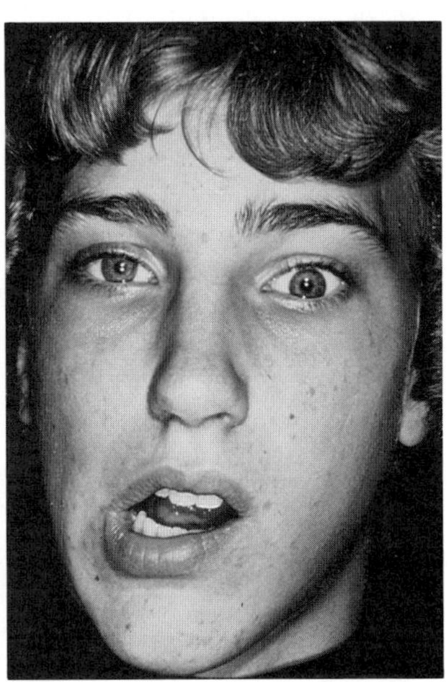

B

Fig. 17–17. *Marcus Gunn phenomenon.* (A,B) Ptotic left eye exhibits exaggerated opening upon deviation of mandible. (From *JD Bullock,* J Pediatr Ophthalmol Strab **17**:375, 1980.)

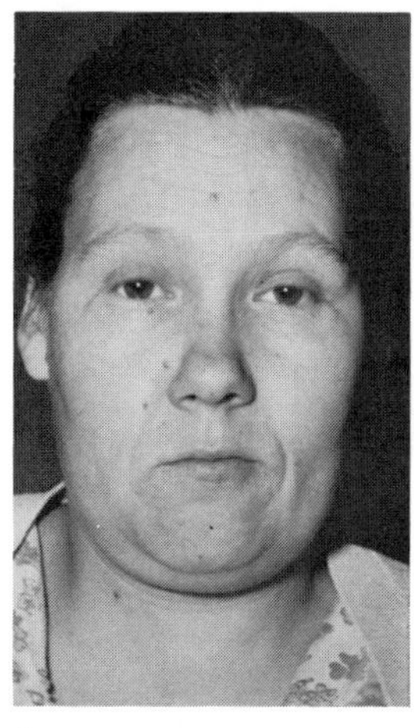
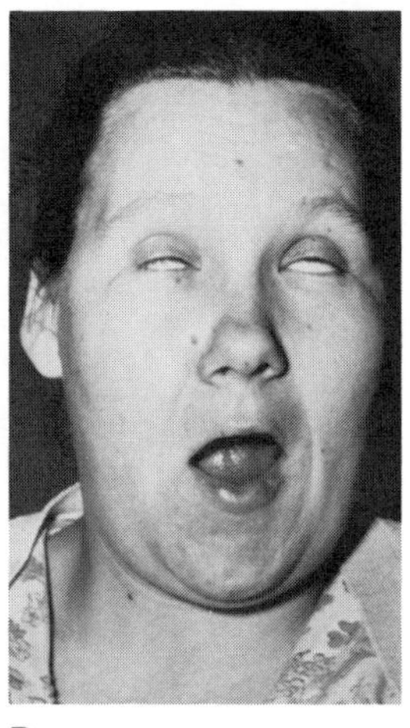

A B

Fig. 17–18. *Inverse Marcus Gunn phenomenon.* (Inverse jaw winking). (A) Note slight ptosis of single eyelid. (B) On opening jaw, there is drooping of eyelids with conjugate upward deviation of eyes. (From *J Jancar,* Ophthalmologica **151**:548, 1966.)

Eye bobbing (elevation and depression of the globe) may also occur with mandibular movement. Synkinetic stimulation of the muscles of mastication and the superior rectus muscles has been suggested (24).

#### References [Marcus Gunn phenomenon (jaw-winking and winking-jaw syndromes (Marcus Gunn and inverse Marcus Gunn syndromes), Marin Amat syndrome)]


1. Abraham JE, Selvam ET: Inverse Marcus Gunn phenomenon. *Br J Ophthalmol* **46**:186–187, 1962.
2. Awan KJ: Marcus Gunn (jaw-winking syndrome). *Am J Ophthalmol* **82**:503–504, 1976.
3. Bullock JD: Marcus Gunn jaw-winking ptosis: Classification and surgical management. *J Pediatr Ophthalmol Strab* **17**:375–379, 1980.
4. Creel D: Problems of ocular miswiring in albinism, Duane's syndrome and Marcus Gunn phenomenon. *Int Ophthalmol Clin* **24**:165–176, 1984.
5. Domke H, Habild D: Zur Kiefer-Lidsynkinese im Kindesalter (Marcus Gunn Phänomen). *Mschr Kinderheilkd* **116**:517–522, 1968.
6. Doucet TW, Crawford JS: The quantification, natural course, and surgical results in 57 eyes with Marcus Gunn (jaw-winking) syndrome. *Am J Ophthalmol* **92**:702–707, 1981.
7. Dupuy-Dutemps L: Ptosis congénital et phénomène de Marcus Gunn. *Bull Soç Ophtalmol (Paris)* **41**:136–140, 1929.
8. Falls HF et al: Three cases of Marcus Gunn phenomenon in two generations. *Am J Ophthalmol* **32**:53–59, 1949.
9. Feric-Seiwerth F et al: Marcus Gunn-Syndrom. *Klin Mbl Augenheilkd* **154**:519–524, 1969.
10. Freuh BR: Associated facial contractions after seventh nerve palsy mimicking jaw winking. *Ophthalmology* **90**:1105–1109, 1983.
11. Friedenwald H: On movements of the eyelids associated with movements of the jaws with lateral movements of the eyeballs. *Bull J Hopkins Hosp* **7**:134–136, 1896.
12. Fromaget C, Brun C: Phénomène de Marcus Gunn, compliqúe. *Bull Soç Ophtalmol (Paris)* **38**:153, 158, 1926.
13. Gordon RM, Binder MB: The corneomandibular reflex. *J Neurol Neurosurg Psychiatr* **34**:236–242, 1971.
14. Grant FC: The Marcus Gunn phenomenon: Report of a case with suggestions as to relief. *Arch Neurol Psychiatr* **35**:487–500, 1936.
15. Hepler RS et al: Paradoxical synkinetic levator inhibition and excitation. *Arch Neurol* **18**:416–424, 1968.
16. Jancar J: Inverse jaw-winking with exaggerated Bell's phenomenon. *Ophthalmologica* **151**:548–554, 1966.
17. Kirkham TH: Familial Marcus Gunn phenomenon. *Br J Ophthalmol* **53**:282–283, 1969.
18. Kuder GG, Laws HW: Hereditary Marcus Gunn phenomenon. *Can J Ophthalmol* **3**:97–98, 1968.
19. Lubkin V: The inverse Marcus Gunn phenomenon. *Arch Neurol* **35**:249, 1978.
20. Lutz A: Jaw-winking phenomenon and its explanation. *Arch Ophthalmol* **48**:144–158, 1919.
21. Marcus Gunn: Congenital ptosis with peculiar associated movements of the affected lid. *Trans Ophthalmol Soc UK* **3**:283–287, 1883.
22. Marin Amat M: Sur le syndrome ou phénomène de Marcus Gunn. *Ann Ocul (Paris)* **156**:513–528, 1919.
23. Mosavy SH, Hornat M: Marcus Gunn phenomenon associated with synkinetic oculopalpebral movements. *Br Med J* **2**:675–676, 1976.
24. Oesterle CS et al: Eye bobbing associated with jaw movements. *Ophthalmology* **89**:63–67, 1982.
25. Pratt SG et al: The Marcus Gunn phenomenon: A review of 71 cases. *Ophthalmology* **91**:27–30, 1984.
26. Rana PVS, Wadia RS: The Marin Amat syndrome: An unusual facial synkinesia. *J Neurol Neurosurg Psychiat* **48**:939–941, 1985.
27. Simpson DG: Marcus Gunn phenomenon following squint and ptosis surgery. Definition and review. *Arch Ophthalmol* **56**:743–748, 1956.
28. Sinclair WW: Abnormal associated movements of the eyelids. *Ophthalmol Rev* **14**:307–319, 1895.
29. Speath EB: The Marcus Gunn phenomenon: Discussion, presentation of four instances and consideration of its surgical correction. *Am J Ophthalmol* **30**:143–158, 1947.
30. Villard H: Le phénomène de Marcus Gunn. *Bull Soç Fr Ophtalmol* **38**:725–753, 1925.
31. Wartenberg R: Winking-jaw phenomenon. *Arch Neurol Psychiatr* **59**:734–753, 1948.
32. Wartenberg R: Inverted Marcus Gunn phenomenon (so-called Marin Amat syndrome). *Arch Neurol Psychiatr* **60**:584–596, 1948.


## Some disorders of facial pain

Because of the plethora of possible combinations of involvement of cranial nerves, both within and outside the cranial cavity, a considerable number of eponymic syndromes have been created to describe these conditions. The interested reader is referred to some extremely comprehensive reviews (1–8). Burton (3), for example, lists 36 causes for facial pain! Those considered here are those we have deemed to have greater importance.

**Trigeminal neuralgia (tic douloureux).** Trigeminal neuralgia is characterized by extreme severe recurrent episodic attacks of unilateral pain distributed over a branch or, after many years, more than one branch of V (Fig. 17–19A). There may be over 7000 new cases each year in the United States. The pain may occur spontaneously or may be initiated by stimulation of a trigger zone, most often extending from the lateral border of the nose to the angle of the mouth (Fig. 17–19B). The pain may vary in intensity but, along with pain resulting from kidney stones, may be one of the most dreadful and dreaded pains experienced by humans. It occurs suddenly, without warning, a single spasm often lasting from 30 to 60 seconds. Because these brief paroxysms are repeated at short intervals, the pain may be stated to have lasted for a longer duration. The severe paroxysms not uncommonly are preceded by electric shock sensations lasting for but a few seconds. The severe pain rapidly disappears, but occasionally a patient may have a mild residual ache or burning that persists for several minutes. With time, a chronic burning facial pain may gradually develop (14).

Trigger zones are not necessarily present, but when they occur, they will often aid in diagnosis. The trigger may be tripped by mastication, talking, shaving, a draft of air on the face, or even the vibration from a loud noise. Within the mouth, a trigger zone may be located on the gingiva or at the base of a single tooth, giving the patient the erroneous notion that it is dental in origin causing, at times, unnecessary extraction of teeth. We have seen several patients who became edentulous. The patient may avoid brushing the teeth or applying cosmetics so as to avoid these areas. Talking may so often trigger the pain that the individual often presents with an immobilized face, talking through the teeth, with eyes expressing fear, and pointing to but not touching the affected area. In extreme cases, the patient may even be afraid to eat or drink.

The pain usually begins in the second or third branch of the trigeminal nerve, but rarely may progress to involve all three branches. Tri-

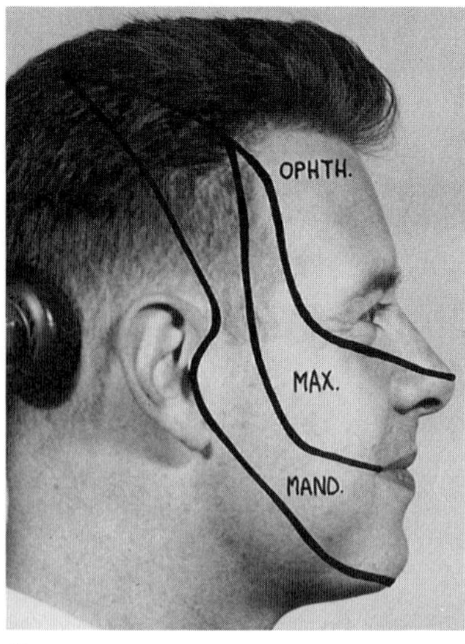

A

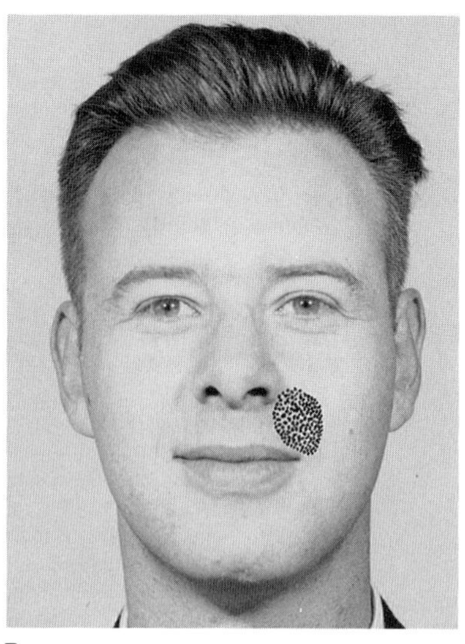
B

Fig. 17–19. (A) *Trigeminal nerve.* Areas supplied by the three sensory branches of the trigeminal nerve. (B) *Trigeminal neuralgia.* Most common "trigger zone" area.

geminal neuralgia occurs more often on the right side. Approximately 5% of patients have bilateral involvement. In these cases the disorder usually starts with unilateral involvement, only rarely being followed by involvement of the other side by months or even years. It occurs twice as commonly in women as in men, and over 70% of the cases are seen in persons over 50 years of age. However, about 15% experience the onset prior to the age of 40. More than half of the patients have remission of 6 months or more, but the time between remissions gradually decreases (9–11,17,18).

Diagnosis depends on the history and description of the pain. There is no sensory loss. No physical signs or laboratory tests are diagnostic. Ordinarily, there is no difficulty in making the correct diagnosis.

In most cases, there is compression or distortion of the nerve root by an aberrant arterial loop and/or venous channel (14,16). Perhaps in 5–10% neoplasms or arterial aneurysms that impinge on the trigeminal nerve may result in trigeminal neuralgia (9,14,16,20). It has also been estimated that approximately 2% of cases of trigeminal neuralgia have been associated with multiple sclerosis, but usually there are other signs of neurologic deficit present in addition to the pain and the patients are usually in their third or fourth decade (19). If suspected, serum IgG should be measured. Rarely, dental disorders such as split tooth, interradicular periodontal abscess, or pulpitis may simulate trigeminal neuralgia. Facial pains of nondental origin that may occasionally resemble trigeminal neuralgia include periodic migrainous neuralgia, antral carcinoma, temporomandibular joint syndrome (myofascial pain dysfunction), temporal (giant cell) arteritis, post-zoster neuralgia, benign brain tumors, and atypical facial pain with personality changes.

There are several published cases of familial aggregation (13).

**Glossopharyngeal neuralgia.** Much rarer than trigeminal neuralgia (1:100 cases), glossopharyngeal neuralgia is characterized by unilateral paroxysmal stabbing pain lasting from a few seconds to a minute, followed by a burning sensation that may last from 2 to 5 minutes in the posterior one-third of the tongue, pharynx, larynx, and soft palate (25,28,29). The pain usually extends to the ear. As in the case of trigeminal neuralgia, the pain is associated with a trigger zone that is usually located on the lateral wall of the pharynx, at the base of the tongue, or in the external ear area posterior to the mandibular ramus. Swallowing and talking usually trip the trigger, although it may also be precipitated by yawning, coughing, or consuming cold drinks. Trigger zones can often be demonstrated in the tonsillar or pharyngeal area by touching these regions with a probe.

The syndrome seems to occur with equal frequency in both sexes, but involvement is slightly more common on the left side than on the right. Bilateral involvement is extremely rare, being seen in less than 1% of a series of 116 patients reported by Rushton et al (29).

The pain is usually of an intense, severe quality and occurs most often at night. It is followed by homolateral tearing and salivation caused by neuronal bombardment of the lacrimal and parotid glands. Between attacks, tinnitus, partial deafness, and facial tics may be experienced. Cocainization (10% cocaine or similar surface anesthetic agent) of the lateral pharyngeal wall and/or tonsil is ordinarily performed to ascertain whether one is dealing with true glossopharyngeal neuralgia, for this brings about immediate but temporary relief. Associated with the neuralgia may be bradycardia, hypotension, syncope, and cardiac arrest (23,26). These symptoms apparently result from activation of the glossopharyngeal–vagus complex that eventuates in cardiac slowing. This in turn will result in syncopy or rarely cardiac arrest. Coughing is common during an attack of pain (24,27,30).

About 55% occur after the fifth decade. When found in a younger age group (about 25% are younger than 40 years), it may be due to anomalous arteries at the cerebellopontine angle, inflammation of the arachnoid membrane, tonsillitis, perineural fibrosis, viral infection, or elongation of the styloid process (Eagle syndrome). Like trigeminal neuralgia, glossopharyngeal neuralgia may be associated with unpredictable remissions and recurrences.

**Post-zoster neuralgia (geniculate ganglion syndrome, Ramsay Hunt syndrome, cephalic zoster syndrome).** Herpes zoster (shingles) is an acute self-limited disease due to the varicella-zoster virus. Although herpes zoster may occur even in the newborn (assuming an intrauterine chicken pox infection), most cases are seen in older patients. Men and women are affected with equal frequency (32,33,38).

Injury to the nervous sytem is most severe in one or two dorsal root ganglia or in the corresponding sensory ganglia of the cranial nerves. There is inflammation of the white matter of the dorsal horn and, to a lesser extent, of the anterior horn. Similar involvement of the gray matter of the brainstem often accompanies involvement of the sensory ganglia of the cranial nerves.

Infection of the geniculate ganglion results in vesicular eruption of the external ear and oral mucosa, facial palsy of the lower motor neuron type, severe pain in the external auditory canal and pinna, and ipsilateral loss of taste and lacrimation. The syndrome comprises about 3–9% of cases of facial palsy (Fig. 17–20A,B). By measuring tear production, submandibular salivary flow, and evoked electromyography, one can estimate prognosis (36). There has been a consensus to limit use of the term Ramsay Hunt syndrome to those with involvement of the conchal zone and to exclude cases with involvement of other cranial or cervical nerves or ganglia.

Aleksic et al (31) have presented good evidence to support the contention of Ramsay Hunt that the syndrome arises from geniculate ganglionitis. The facial paralysis apparently results from interstitial neu-

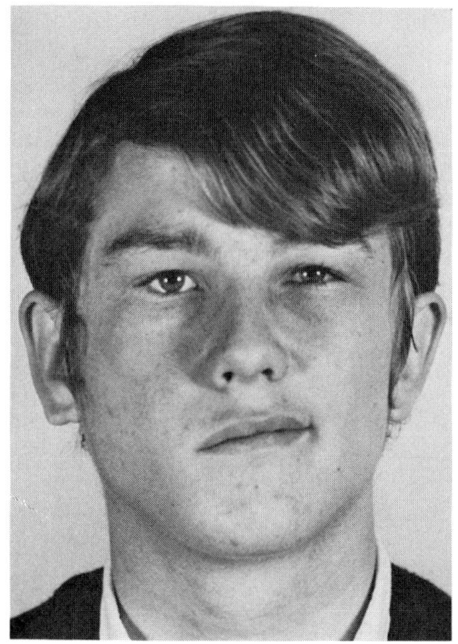

A

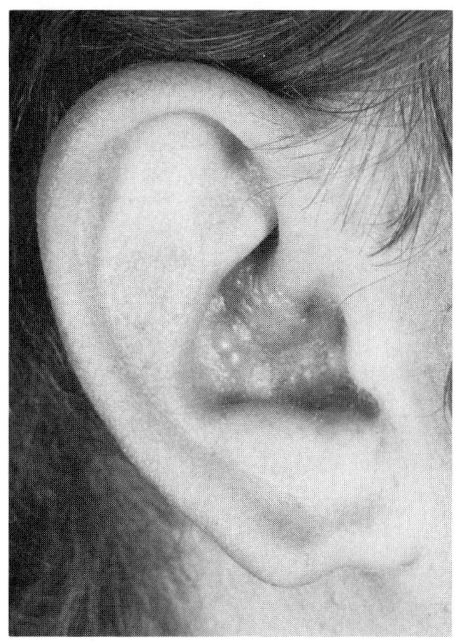

B

Fig. 17–20. *Cephalic zoster.* (A,B) Note peripheral right facial weakness and vesicular lesions of conchal area. (From *SG Harner,* Arch Otolaryngol **92**:632, 1970.)

ritis with attendant edema within the bony canal. The combined loss of taste of the anterior two-thirds of the tongue and loss of lacrimation can be explained by a lesion at or proximal to the geniculate ganglion. Round cell infiltration of the facial nerve has been repeatedly demonstrated.

Pain in the ear is severe and paroxysmal and, as with zoster infection of other areas, it often precedes the cutaneous and mucosal lesions by a week. The ear becomes red and swollen followed by vesicular eruption of the auditory meatus and tympanic membrane. The concha, tragus, antitragus, anthelix and, rarely, the lobule may be involved. Rarely, there is no auricular eruption. Associated signs and symptoms may include diminished lacrimation, tinnitus, hearing loss, vertigo, nystagmus, hoarseness, and loss of superficial and deep sensation of the face. The regional lymph nodes are enlarged and tender. Pain persists for weeks after the eruption has disappeared. Among those 40 years or older, about one-third experience pain for months or even years.

Oral signs and symptoms include vesicles or eroded areas in the peritonsillar region, oral pharynx, and posterolateral third of the tongue, and pain in the same areas. Occasional loss of taste, diminished salivation, and palatal paralysis may be present. These changes may rarely precede the auricular signs (35).

Prior to the appearance of the cutaneous or mucosal lesions there is often a stinging sensation experienced in the area subsequently involved by the vesicles. The pustules form crusts that are shed after 2–3 weeks and generally the skin resumes its normal appearance. The trigeminal nerve is involved in 15–20% of patients with herpes zoster, with the ophthalmic branch affected about 10 times as often as the maxillary and mandibular branches combined. The seventh, ninth, and tenth nerves may rarely (less than 1%) be involved (34,37).

Approximately 65% who manifest post-zoster neuralgia are over 70 years. It is rare in individuals less than 30 years of age. The neuralgia may last for many months or, in fact, even for years. Patients describe it as burning, aching, or boring. Not uncommonly it is associated with paresthesia in the involved area. The pain and paresthesia are often present throughout the day and may seriously interfere with sleep. In some cases, the patient may experience lancinating pain reminiscent of trigeminal neuralgia, but there is no trigger zone.

**Atypical facial neuralgias.** Horton's headache (histaminic cephalgia, cluster headache, petrosal neuralgia). Horton's headache occurs at least eight times as commonly in men as in women. It usually appears between the ages of 30 and 45 years. It may be precipitated by alcohol which is a vasodilator. The pain is unilateral, occurs with great regularity, and involves the region of the orbit and temple with occasional spread to the neck, upper jaws, or occiput. Its onset is rapid and may reach excruciating intensity and subside after a period varying from a quarter of an hour to 2 hours. The pain becomes so dreaded that in spite of being free of pain, the patient becomes emotionally exhausted. The attacks may occur as often as 5–10 times during a day, and are so severe that they may awaken the patient from sleep in the early morning hours (''alarm clock headache''). The patient then arises because attacks are more painful when the patient is recumbent. The pain is boring, burning, and is associated with hyperemia of the eye, profuse lacrimation, and stuffiness or obstruction of the nostril on the ipsilateral side. Often there is edema of the eyelid. Usually the attacks subside spontaneously after several weeks, although remissions are common. Narrowing of the pupil and palpebral fissure has been noted in approximately 15% of patients (39,40,42, 43,47).

Atypical facial pain. Atypical facial pain is usually described by exclusion (44). It is not trigeminal neuralgia, glossopharyngeal neuralgia, post-zoster neuralgia, or pain secondary to disorders of the teeth, pharynx, ears, or eyes. It is usually poorly localized and often described as ''deep.'' Frequently, the pain does not follow normal anatomic distribution of the trigeminal nerve, commonly crossing the midline or involving portions of the face not limited to the sensory distribution of a single nerve. The pain often is described as lasting for weeks, months, or years and being constant in quality. Rushton et al (51), in a study of 100 patients, classified atypical facial pain under three headings: (1) psychogenic, (2) organic, and (3) indeterminate. Patients with psychogenic facial pain constituted over half of their 100 patients. Although the patient complained primarily of facial pain, there were associated depressive reactions, schizophrenia, or conversion hysteria. Most were in the fourth and fifth decades of life and there was a marked preponderance of female patients. About half attributed the onset of the facial pain to a definite incident such as a dental procedure or being struck in the face. The diagnosis is justified when the pain does not follow normal anatomic distribution (46). There are no signs of organic disease and often there is a history of emotional illness (49,52,53).

Organic facial pain. In one-third of their series of 100 patients, Rushton et al (51) further classified organic facial pain into five subgroups: (1) vasodilating pain, (2) dental disease, (3) neuritis, (4) neoplasms, and (5) miscellaneous organic conditions. Vasodilating facial pain resembles that of Horton's headache. The onset, character, and duration of pain are similar. However, the location is atypical, the pain being in the lower portion of the face in contrast to the head and eye pain of typical histaminic headache. But otherwise the duration of the pain and the numerous recurrent episodes within a 24-hour period closely simulate histaminic cephalgia. Furthermore, ergotamine tartrate stops the pain of a typical episode.

Dental disease. Eight patients had atypical facial pain caused by previously unrecognized dental disease. The pain was poorly localized in the face and in no patient was it limited to the involved tooth until late in the course of the disorder. Clinical and radiographic studies were normal. Only late in the disorder was the final diagnosis of pulpitis made. Extraction of the responsible tooth stopped the pain. Several other authors have attributed a series of atypical facial neuralgia to incompletely healed tooth sockets (48,50).

Neuritis. Eight of the 100 patients described the facial pain as burning, itching, sticking, or jabbing. A slight but definite sensory loss limited to the division of the trigeminal nerve was exhibited in each instance.

Indeterminate cause. Fifteen of the 100 patients described by Rushton et al (51) were classified as having atypical facial pain of indeterminate cause. The duration of the pain was short compared with that in those with psychogenic facial pain. In at least half the patients, the pain started after the age of 50 years. Remissions were common and most of the patients derived substantial, although temporary, relief with the use of ordinary analgesics. Chewing, touching, or rubbing the face seemed to cause the pain to become worse, but other characteristics of trigeminal neuralgia were not present.

Angina pectoris. We have seen several patients who described pain in the mandible and neck upon exertion. This preceded pain in the chest by several weeks (45).

Trigeminal neuropathy. Trigeminal neuropathy is a protracted disorder of sensation confined to the distribution of the fifth cranial nerve either unilaterally or bilaterally. The term trigeminal neuritis is avoided here because that implies an inflammatory reaction for which it is extremely difficult to present histologic proof. It is to be contrasted with trigeminal neuralgia that is characterized by brief attacks of very sharp pain without sensory impairment. In trigeminal neuropathy, the pain is described as an ache or as burning, boring, pulling, or drawing (54,60). Some patients complain of paresthesia, others of pain involving the area supplied by any of the three divisions of the trigeminal nerve or combinations thereof. The discomfort may persist for hours, days, or weeks. In a study of 61 patients with trigeminal neuropathy, Goldstein et al (57) considered six possible etiologies: dental surgical procedures, mandibular denture pressure, nondental surgical procedures, mechanical trauma, stilbamide intoxication, and a miscellaneous group of causes. Approximately one-third of the patients noted onset of symptoms secondary to a dental surgical procedure, in most cases following tooth removal. Presumably this is due to injury to the sensory fibers of the trigeminal nerve resulting from needle puncture during anesthesia. Goldstein et al (57) suggested that it is advisable to caution patients with deeply impacted teeth that a sensory disturbance such as trigeminal neuropathy may follow removal of impacted third molars. Almost 5% experience dysesthesia after mandibular third molar extraction (58). They suggested that the patient should be seen frequently during the course of the neuropathy, especially if the patient becomes alarmed during nerve regeneration when a sensation of "pins and needles" or tingling may be manifest. Trigeminal neuritis may also occur following alveolar osteitis ("dry socket"). Pressure from a mandibular denture on the mental nerve at its exit from the mental foramen may produce trigeminal neuropathy. This usually results from extensive alveolar resorption due to periodontal destruction, causing the nerve at the mental foramen to lie on top of the alveolar ridge. Pressure from movement of a mandibular denture may produce pain or paresthesia over the distribution of the mental nerve. It is highly recommended that the reader peruse in detail the paper of Goldstein et al (57) for a variety of other causes for trigeminal neuropathy including such rare findings as intracranial aneurysms. The series of patients reported by Spillane and Wells (60) are extremely valuable. The combination of paresthesia and facial pain should cause the clinician to consider trigeminal neuropathy, tumors of the Gasserian ganglion or its sensory root, pontine tumors, and vascular abnormalities in the region of the ganglion. One should also consider multiple sclerosis. This has already been discussed in association with trigeminal neuralgia, but occasionally the only complaint is paresthesia of the face.

Nasopharyngeal and cranial lesions producing facial pain. Excellent reviews of the neurologic manifestations of malignant nasopharyngeal tumors are those of Thomas and Walz (61) and Schnetler and Hopper (51a). Prior to the onset of neurologic signs or symptoms, patients often present with pains in the head, face, or ear, swelling of cervical lymph nodes, and nasal or pharyngeal symptoms. These may antedate such obvious neurologic complications as diplopia, visual loss, or facial paresthesia by several years. Frequently several cranial nerves are involved that give rise to a host of eponymic syndromes. If only a single nerve is involved, Thomas and Walz (61) found the sixth nerve to be involved in over two-thirds of their patients, the fifth nerve in approximately half, and the ninth nerve in over a third. The most common binary combination is cranial nerves V and VI (Gradenigo syndrome). Somewhat less common are conditions discussed earlier such as the syndrome of the parapharyngeal space involving cranial nerves IX, X, and XII; the superior orbital fissure syndrome (III, IV, V, and VI); the petrosphenoidal syndrome of Jacod (II, III, IV, V, and VI); the Collet–Sicard syndrome (IX–XII); and unilateral involvement of all cranial nerves, the so-called Garcin syndrome. Most patients have either squamous cell carcinoma or lymphoepithelioma. The 5-year survival rate is about 25%. Encroachment of the tumor on the nasopharynx may be manifest as trismus. One of the most important symptoms is increased pain during the night. If the pain awakens the patient, it is almost certain that a nasopharyngeal tumor is present (55).

Dental pain. Hyperemia of the pulp and associated early inflammation are extremely difficult to localize. With progression, however, the clinical signs become more obvious. Extension of the inflammatory process to the periodontal ligament makes the diagnosis easier. An early sign of pulpitis is increased response to cold stimulation. With increased degeneration of the pulp, the response to cold appears to decrease and eventually disappears. This is followed by increased pain with application of heat that may be extremely severe. Patients will frequently put cold water in their mouths to assuage the pain. Subsequently, pulpal degeneration eliminates this symptom. The patient commonly will no longer complain about thermal stimuli producing toothache. In the absence of clinical or radiologic evidence of dental caries, there is always the possibility of a hairline fracture of a tooth. This is usually not evident on the radiograph or on casual visual examination. At times, the use of a disclosing solution may reveal the crack. There may be painful response to percussion. Ordinarily, percussion is of little diagnostic aid in the early course of pulpitis, but as the inflammatory process extends through the apex of the tooth, painful response to percussion is nearly always noted. Referred dental pain often taxes the clinician's skills. Such pain rarely crosses the midline except in such cases in which the affected tooth is in the midline. Pain from either the maxillary or mandibular teeth may be referred to the ear region, whereas referred pain from infected teeth is nearly always limited to the quadrant of the jaw in which the tooth is located.

**Symptomatic elongated styloid process (Eagle syndrome).** The styloid process may exceed its mean length of 25 mm or the stylohyoid ligament may be mineralized in 4–28% of the general population. Involvement is nearly always bilateral. Rarely does an elongated styloid process produce pain, while some of those of normal length may be symptomatic. Although it appears that a true clinical entity exists, there is no satisfactory pathophysiologic explanation. Eagle (65–67) focused attention on the elongated styloid process as a source of pain. There are many comprehensive reviews of the literature (61a–63,69–71,73,74,76).

The syndrome presents clinically with pain in the throat which is frequently referred to the ear. Usually there is a sensation of a foreign body in the tonsillar area producing dysphagia. It is believed to be due to physical irritation of nerves or vessels passing near the elongated styloid tip (68); in the absence of an elongated process, it has been ascribed to insertion tendinosis at the junction of the lesser horn and stylohyoid ligament.

The diagnosis is established by palpation of the styloid process through the tonsillar fossa, which in the case of a symptomatic styloid process produces pain. Symptoms similar to those of the classic elongated styloid process syndrome may also be observed in fractures of the styloid process (64,75).

**Glossodynia (glossopyrosis, burning tongue, oral-lingual paresthesia).** Alteration in sensation of the oral mucous membrane in the absence of observable alterations is of considerable problem to the clinician because it is so frequent and because the etiology is so rarely found. Although the lateral anterior border of the tongue is most frequently the site of the paresthetic alteration, other portions of the oral cavity may be involved, so technically the terms "stomatopyrosis," "stomatodynia," or "burning mouth" may be used.

As the tissues involved almost invariably look normal, there is only the statement of the patient to describe the sensation experienced. This may be reported as burning, itching, or pain of the mucous membranes. Our experience is consonant with that of other clinicians with regard to its marked predilection for women, middle-aged or older. Although there have been an inordinate number of local and systemic disorders suggested (various deficiency states such as pernicious anemia and sundry vitamin deficiencies), it seems apparent that these are rarely etiologic. Fear of oral cancer or other psychological factors appears to play a major role. The patient commonly complains that spicy foods are causative, but they seem rather to make the condition more apparent to the patient, and are not etiologic. Several have experienced glossodynia following antibiotic therapy, the prolonged use of antihistaminics, and/or diuretics, but the precise relationship is not evident (77–83).

## References (Some disorders of facial pain)


1. Alling CC (ed): *Facial Pain,* 2nd ed., Lea & Febiger, Philadelphia, 1977.
2. Bell WE: *Orofacial Pains: Differential Diagnosis,* 3rd ed., Year Book Medical Publishers, Chicago, 1985.
3. Foley KM, Payne RM: *Current Therapy of Pain,* B. C. Decker, Inc., Toronto, 1989.
4. Hurwitz LJ: Facial pain of non-dental origin. *Br Dent J* **124**:167–171, 1968.
5. Miller H: Pain in the face. *Br Med J* **2**:577–580, 1969.
6. Mumford JM: *Orofacial Pain: Aetiology, Diagnosis and Treatment,* 3rd ed., Churchill Livingstone, Edinburgh, 1982.
7. Needham CW: Major cranial neuralgias and surgical treatment of headache. *Med Clin N Am* **62**:545–558, 1978.
8. Poser CM: Facial pain. *Geriatrics* **30**:110–115, 1975.

Trigeminal neuralgia (tic douloureux)

9. Bayer DB, Stenger TG: Trigeminal neuralgia: An overview. *Oral Surg* **48**:393–399, 1979.
10. Gardner WJ: Trigeminal neuralgia. *Clin Neurosurg* **15**:1–56, 1968.
11. Graham JG: Trigeminal neuralgia. *Practitioner* **198**:497–504, 1967.
12. Henderson WR: Trigeminal neuralgia. The pain and its treatment. *Br Med J* **1**:7–15, 1967.
13. Herzberg L: Familial trigeminal neuralgia. *Arch Neurol* **37**:285–286, 1980.
14. Janetta PJ: Trigeminal neuralgia, in *Oral and Maxillofacial Surgery,* Archer WH (ed.), 5th ed., W.B. Saunders, Philadelphia, 1975.
15. Kerr FW, Miller RH: The pathology of trigeminal neuralgia. *Arch Neurol* **15**:308–319, 1966.
16. Lazar ML: Current treatment of tic douloureux. *Oral Surg* **50**:504–508, 1980.
17. Peet MM, Schneider RC: Trigeminal neuralgia: A review of six hundred and eighty-nine cases with a follow-up study on sixty five percent of the group. *J Neurosurg* **9**:367–377, 1952.
18. Rothman KJ, Monsor RR: Epidemiology of trigeminal neuralgia. *J Chron Dis* **26**:3–12, 1973.
19. Rushton JG, Olafson RA: Trigeminal neuralgia associated with multiple sclerosis. *Arch Neurol* **13**:383–386, 1965.
20. Sweet WH: Controversies in the treatment of trigeminal neuralgia (tic douloureux). *N Engl J Med* **315**:174–177, 1986.
21. Tew JM et al: Treatment of trigeminal neuralgia in the aged by a simplified surgical approach (percutaneous electrocoagulation). *J Am Geriat Soc* **23**:426–430, 1975.
22. Voorhies R, Patterson RH: Management of trigeminal neuralgia (tic douloureux). *JAMA* **245**:2521–2523, 1981.

Glossopharyngeal neuralgia

23. Alpert JN et al: Glossopharyngeal neuralgia, asystole and seizures. *Arch Neurol* **34**:233–235, 1977.
24. Bohm E, Strang RR: Glossopharyngeal neuralgia. *Brain* **85**:371–388, 1962.
25. Holt GW: Clinical syndromes of the glossopharyngeus: Practical neurologic concepts. *Am J Med Sci* **238**:85–96, 1959.
26. Jacobson RR, Russell RWW: Glossopharyngeal neuralgia with cardiac arrhythmia: A rare but treatable cause of syncope. *Br Med J* **1**:379–380, 1979.
27. Orton CJ: Glossopharyngeal neuralgia. *Br J Oral Surg* **9**:228–232, 1972.
28. Peet MM: Glossopharyngeal neuralgia. *Ann Surg* **101**:256–268, 1935.
29. Rushton JG et al: Glossopharyngeal (vagoglossopharyngeal) neuralgia: A study of 217 cases. *Arch Neurol* **38**:201–205, 1981.
30. Uhlein A et al: Intracranial section of the glossopharyngeal nerve. *Arch Neurol Psychiatr* **74**:320–324, 1955.

Post-zoster neuralgia and Ramsay Hunt syndrome

31. Aleksic SN et al: Herpes zoster oticus and facial paralysis (Ramsay Hunt syndrome). *J Neurol Sci* **20**:149–159, 1973.
32. Blackley B et al: Herpes zoster auris associated with facial nerve palsy and auditory nerve syndrome. *Acta Oto-laryngol* **63**:533–550, 1967.
33. Crabtree JA: Herpes zoster oticus. *Laryngoscope* **78**:1853–1878, 1968.
34. Devriese PP: Facial paralysis in cephalic herpes zoster. *Ann Otol Rhinol Laryngol* **77**:1101–1119, 1968.
35. Dworkin H, Imburgia R: Ramsay Hunt syndrome presenting symptoms of acute pharyngitis. *NY State J Med* **62**:3976–3977, 1962.
36. May M, Blumenthal F: Herpes zoster oticus: Surgery based upon prognostic indicators and results. *Laryngoscope* **92**:65–67, 1982.
37. Myall RWT, Smith MJA: The Ramsay Hunt syndrome: A dynamic demonstration of applied anatomy. *J Oral Surg* **35**:663–666, 1977.
38. Payten RJ, Dawes JDK: Herpes zoster of the head and neck. *J Laryngol Otol* **86**:1031–1055, 1972.

38a. Portenoy RK et al: Acute herpetic and postherpetic neuralgia: Clinical review and current management. *Ann Neurol* **20**:651–664, 1986.

Atypical facial neuralgias (Horton's headache, atypical facial pain, organic facial pain, trigeminal neuropathy, nasopharyngeal and cranial lesions, etc.)

39. Brook RI: Periodic migrainous neuralgia. *Oral Surg* **46**:511–516, 1978.
40. Eversole LR, Stone CE: Vagogenic facial pain (cluster headache). *Int J Oral Maxillofac Surg* **16**:25–35, 1987.
41. Harris M: Psychiatric aspects of facial pain. *Br Dent J* **136**:199–202, 1974.
42. Kudrow L. *Cluster headache: Mechanisms and Management.* Oxford University Press, London, 1980.
43. Medina JL et al: The nature of cluster headache. *Headache* **19**:309–332, 1979.
44. Moore DS, Nally FF: Atypical facial pain: An analysis of 100 cases with discussion. *J Can Dent Assoc* **7**:396–401, 1975.
45. Natkin E et al: Anginal pain referred to the teeth. *Oral Surg* **40**:678–680, 1975.
46. Paulson GW: Atypical facial pain. *Oral Surg* **43**:338–341, 1977.
47. Rapidis AD et al: Horton's syndrome. *Int J Oral Surg* **6**:321–327, 1977.
48. Ratner EJ et al: Jaw bone cavities and trigeminal and atypical facial neuralgias. *Oral Surg* **48**:3–20, 1979.
49. Remick RA et al: Ineffective dental and surgical treatment associated with atypical facial pain. *Oral Surg* **55**:355–358, 1983.
50. Roberts AM, Person P: Etiology and treatment of idiopathic trigeminal and atypical facial neuralgias. *Oral Surg* **48**:298–308, 1979.
51. Rushton JG et al: Atypical face pain. *JAMA* **171**:545–548, 1959.

51a. Schnetler J, Hopper C: Intracranial tumours presenting with facial pain. *Br Dent J* **166**:80–83, 1989.

52. Smith DP et al: A psychiatric study of atypical facial pain. *Can Med Assoc J* **100**:286–291, 1969.

53. Spearland B et al: Intractable facial pain. *Br J Oral Surg* **17**:166–178, 1979.

54. Blau JN: Trigeminal sensory neuropathy. *N Engl J Med* **281**:873–876, 1969.

55. Devine KD et al: Diagnostic problems in surgery of the head and neck. *Surg Clin N Amer* **41**:1049–1059, 1961.

56. Foster JB: Facial pain. *Br Med J* **4**:667–669, 1969.

57. Goldstein NP et al: Trigeminal neuropathy and neuritis. *JAMA* **184**:458–462, 1963.

58. Kipp DP: Dysesthesia after third molar surgery. *JADA* **100**:185–192, 1980.

59. Rushton JG et al: Atypical face pain. *JAMA* **171**:545–548, 1959.

60. Spillane JD, Wells CEC: Isolated trigeminal neuropathy: A report of 16 cases. *Brain* **82**:391–416, 1959.

61. Thomas JE, Waltz AG: Neurological manifestations of nasopharyngeal malignant tumors. *JAMA* **192**:95–98, 1965.

Symptomatic elongated styloid process (Eagle syndrome)

61a. Camarda AJ et al: Styloid chain ossification: A discussion of etiology. *Oral Surg* **67**:508–520, 1989.

62. Christiansen TA et al: Styloid process neuralgia: Myth or fact. *Arch Otolaryngol* **101**:120–122, 1975.

63. Correll RW et al: Mineralization of the stylohyoid-stylomandibular ligament complex. *Oral Surg* **49**:288–291, 1979.

64. Douglas BL et al: Atypical facial neuralgia resulting from fractured styloid process of the temporal bone. *Oral Surg* **6**:1199–1201, 1957.

65. Eagle WW: Elongated styloid process: Report of two cases. *Arch Otolaryngol* **25**:584–587, 1937.

66. Eagle WW: Elongated styloid process: Further observations and a new syndrome. *Arch Otolaryngol* **47**:630–640, 1948.

67. Eagle WW: Symptomatic elongated styloid process. *Arch Otolaryngol* **49**:490–503, 1949.

68. Frommer J: Anatomic variations in the stylohyoid chain and their possible clinical significance. *Oral Surg* **38**:659–664, 1974.

69. Gossman JR Jr et al: The styloid-stylohyoid syndrome. *J Oral Surg* **35**:555–560, 1977.

70. Marano PD et al: Eagle's syndrome necessitating bilateral styloid amputation. *Oral Surg* **33**:874–878, 1972.

71. Messer EJ et al: The stylohyoid syndrome. *J Oral Surg* **33**:664–667, 1975.

72. Moffat DA et al: The styloid process syndrome. *J Laryngol Otol* **91**:279–294, 1977.

73. Phillips JD et al: Prosthetic implications of Eagle's syndrome. *J Prosthet Dent* **34**:614–616, 1975.

74. Pirruccello FM et al: Ossified stylohyoid ligament (epihyal bone): An unusual case. *Dent Dig* **78**:126–129, 1972.

75. Reichart PA et al: Fracture of the styloid process of the temporal bone: An unusual complication of dental treatment. Report of a case. *Oral Surg* **42**:150–154, 1976.

76. Winklmair M. Styloid-Syndrom durch arthrotische Deformierung. *Stoma (Heidelb)* **24**:205–209, 1971.

Glossodynia (glossopyrosis, burning tongue, oral-lingual paresthesia)

77. Glick E et al: Relation between idiopathic glossodynia and saliva flow rate and content. *Int J Oral Surg* **5**:161–165, 1976.

78. Grushka M: Clinical features of burning mouth syndrome. *Oral Surg* **63**:30–36, 1987.

79. Haneke E: Psychische Aspekte der Glossodynie. *Dtsch Med Wochenschr* **103**:1302–1305, 1978.

80. Haneke E: Glossodynie (Glossodynia). *HNO* **30**:208–212, 1982.

81. Main DMG, Basker RM: Patients complaining of a burning mouth. *Br Dent J* **154**:206–211, 1983.

82. Van der Ploeg HM et al: Psychological aspects of patients with burning mouth syndrome. *Oral Surg* **63**:664–668, 1987.

83. Zegarelli DJ: Burning mouth: An analysis of 57 patients. *Oral Surg* **58**:34–38, 1984.

## Oral mandibular dystonia with blepharospasm

Blepharospasm, oral mandibular dystonia syndrome appears to be a disorder of basal ganglia. It presents in adult life, is usually nonprogressive, and is not genetic, although it has been reported once in sisters (3). It has been referred to both as Meige syndrome and Brueghel syndrome, since it has been alleged that Peter Brueghel the Elder illustrated the syndrome in the sixteenth century in one of his paintings. It most commonly occurs in the middle aged or elderly. There is a 2:1 female predilection (1). About 125 cases have been reported (1–7).

It is characterized by spontaneous, repetitive, nonrhythmic symmetric dystonic spasms, first of the orbicularis oculi muscles (blepharospasm) that begin unilaterally but soon become bilateral and progress to the muscles of the lower face, jaw, and tongue (Figs. 17–21A,B). The blepharospasm is often preceded by or associated with photophobia or tearing. The dyskinesia may fluctuate considerably from day to day, being aggravated by emotional stress and fatigue and disappearing with sleep. With time, the lower facial musculature becomes involved, being manifested as yawning, jaw opening, and abnormal tongue movements. It resembles the oral–facial dyskinesia that is phenothiazine induced. Rarely do the muscles other than those of the head and neck become dystonic.

### References (Oral mandibular dystonia with blepharospasm)

1. Jankovic J, Ford J: Blepharospasm and orofacial—cervical dystonia: Clinical and pharmacological findings in 100 patients. *Ann Neurol* **13**:402–411, 1983.

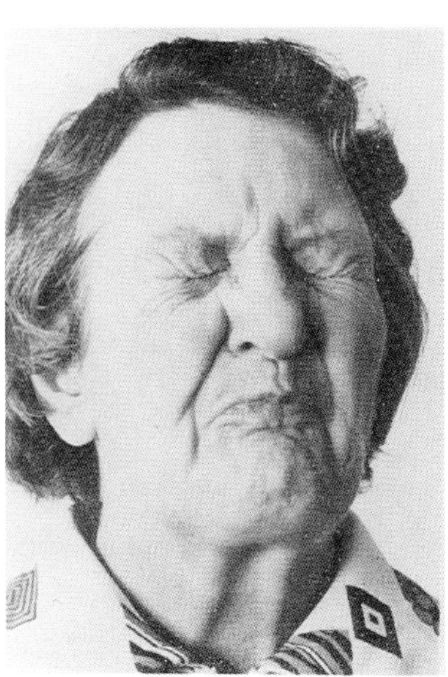

A

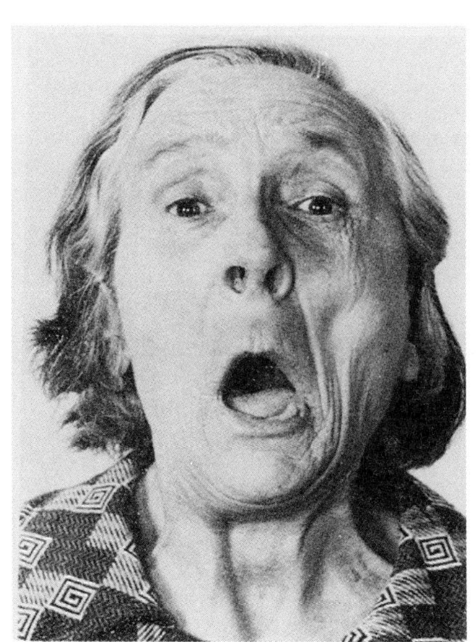

B

Fig. 17–21. *Oral mandibular dystonia with blepharospasm.* (A) Blepharospasm and spasm of jaw closure and mouth-pursing. (B) Spasm of jaw opening and mouth retraction. (From *CD Marsden,* J Neurol Neurosurg Psychol **39**:1204, 1976.)

2. Marsden CD: Blepharospasm—oral mandibular dystonia syndrome (Brueghel's syndrome). *J Neurol Neurosurg Psychol* **39**:1204–1209, 1976.
3. Nutt JG, Hammerstad JP: Blepharospasm and oromandibular dystonia. (Meige's syndrome) in sisters. *Ann Neurol* **9**:189–191, 1981.
4. Paulson GW: Meige's syndrome. *Geriatrics* **27**(8):69–73, 1972.
5. Stevens MR, Wong ME: Meige syndrome: An unusual cause of involuntary facial movements. *Oral Surg Oral Med Oral Pathol* **66**:427–429, 1988.
6. Sutcher HD et al: Orofacial dyskinesia—a dental dimension. *JAMA* **216**:1459–1463, 1971.
7. Tolosa ES: Clinical features of Meige's disease (idiopathic orofacial dystonia): A report of 17 cases. *Arch Neurol* **38**:147–151, 1981.

## Syndromes involving the VIIth nerve

We deal here with only the Melkersson–Rosenthal syndrome and asymmetric crying facies.

**Melkersson–Rosenthal syndrome.** The syndrome consists of unilateral or bilateral facial paralysis, chronic swelling of the face (especially the lips) and lingua plicata (scrotal tongue). Familial occurrence has been described, but most cases have been sporadic (1,4). The syndrome is remarkably variable in its expression (3a). Mair et al (5), studying 23 patients, found the triad in only 9, 2 signs in 13 patients, and a single component in the remaining patient.

The swelling begins suddenly, in most cases prior to facial paralysis but sometimes after or simultaneously with it. It generally begins during childhood. The upper lip is most affected (Figs. 17–22 and 17–23). The edema assumes a reddish-brown appearance. The swollen lip may assume large proportions—at times being three to four times thicker than normal. The edema recurs in most patients, resulting in permanent enlargement of the lips. The eyelids, nose, and chin may also be affected.

The facial palsy begins suddenly in childhood or adolescence. It is present in about 20%. It is peripheral and clinically indistinguishable from Bell's palsy. It may be partial, complete, or occasionally bilateral. Relapses often occur, but most patients ultimately recover. The paralysis may precede attacks of swelling by years. The paralyzed side does not always correspond to that of the swelling. Occasionally, defects in taste over the anterior two-thirds of the tongue are noted (2,3, 6–9).

Simultaneously with the swelling of the skin, the oral mucosa is often affected in a similar way. The swollen buccal mucosa is cushion-like, divided by furrows of varying depth. The buccal, palatal, and sublingual mucosa may be involved (Figs. 17–24 and 17–26). Approximately 40% manifest lingua plicata, a condition found in 1–5% of the general population (Fig. 17–25). The gingiva are involved in 15% (10).

Histopathologic changes include edematous alterations and/or sarcoid-like granulomatous picture (Figs. 17–27 and 17–28).

The *Hughes syndrome* should be excluded.

Fig. 17–22. *Melkersson–Rosenthal syndrome.* Swelling of upper lip of 20-year-old man with the syndrome.

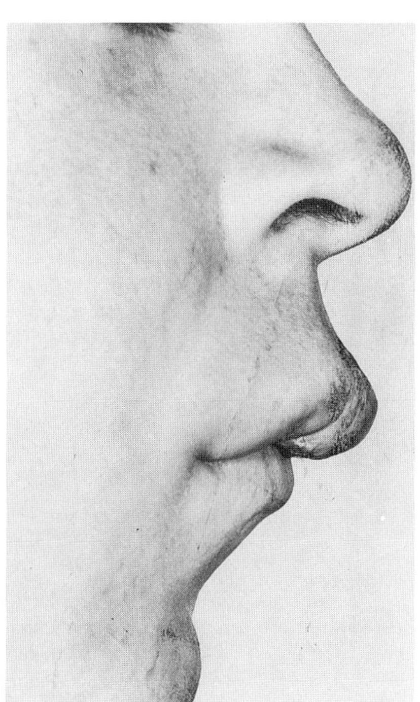

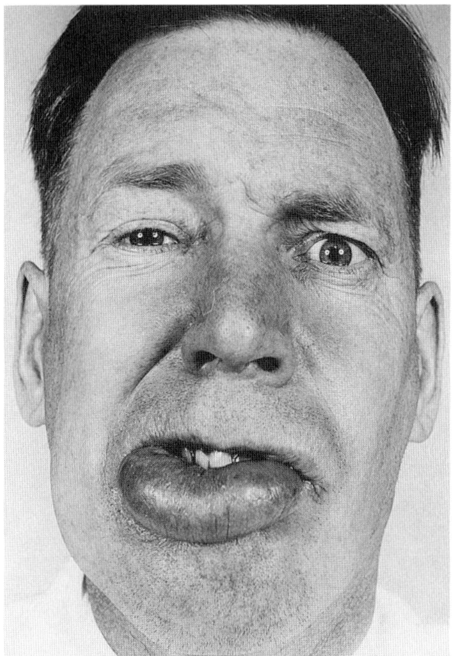

Fig. 17–23. *Melkersson–Rosenthal syndrome.* Markedly swollen lower lip; left-sided facial palsy.

Fig. 17–24. *Melkersson–Rosenthal syndrome.* Plicated swelling of the buccal mucosa in a patient with Melkersson–Rosenthal syndrome. (Courtesy of *H Schuermann,* Bonn, Germany.)

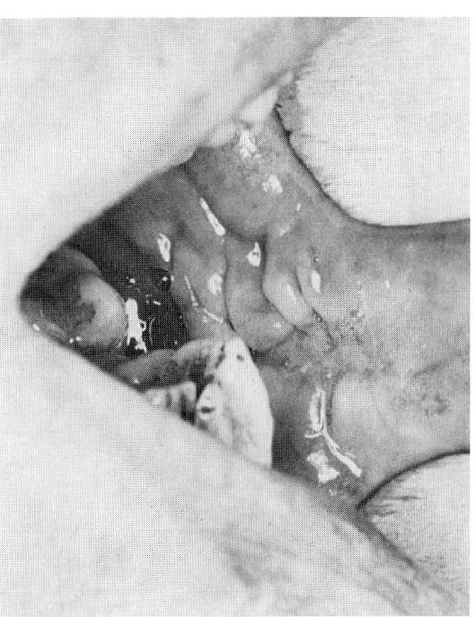

Fig. 17–25. *Melkersson–Rosenthal syndrome.* Plicated tongue of the patient shown in Fig. 17–22.

### References (Melkersson–Rosenthal syndrome)


1. Gorlin RJ et al: Melkersson–Rosenthal syndrome, in *Syndromes of the Head and Neck,* ed. 2, McGraw-Hill Company, New York, 1976.

2. Graaf–Radford SB: Melkersson–Rosenthal: A review of the literature and a case report. *S Afr Med J* **60**:71–74, 1981.

3. Hornstein O, Schuermann H: Das sogenannte Melkersson-Rosenthal–Syndrom (ein schliesslich "Cheilitis granulomatosa" Miescher). *Ergeb Inn Med Kinderheilkd* **17**:190–263, 1962.

3a. Langervitz P et al: Melkersson–Rosenthal syndrome: An oligosymptomatic form. *South Med J* **79**:1159–1160, 1986.

4. Lygidakis C et al: Melkersson–Rosenthal's syndrome in four generations. *Clin Genet* **15**:189–192, 1979.

5. Mair IWS et al: Clinical manifestations of the Melkersson–Rosenthal syndrome. *Can J Otolaryngol* **3**:123–131, 1974.

6. Nally FF: Melkersson–Rosenthal syndrome. *Oral Surg* **29**:694–703, 1970.

7. Rintala A et al: Cheilitis granulomatosa. The Melkersson–Rosenthal syndrome. *Scand J Plast Surg* **7**:130–136, 1973.


Fig. 17–26. *Melkersson–Rosenthal syndrome.* Plicated palatal mucosa in another patient with the syndrome. (Courtesy of *A Lodin,* Stockholm, Sweden.)

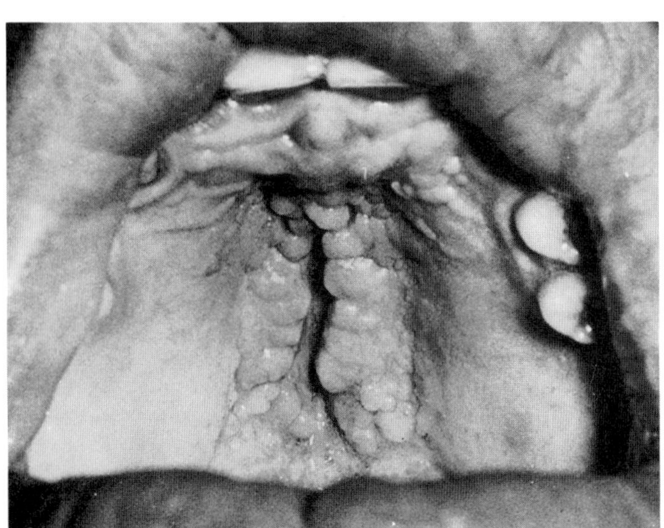

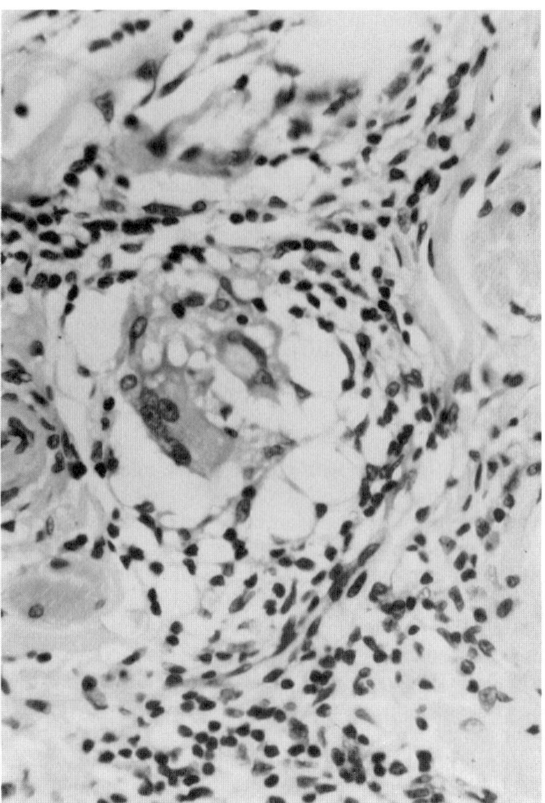

Fig. 17–27. *Melkersson–Rosenthal syndrome.* Giant cell in a sarcoid-like granuloma in lip.

Fig. 17–28. *Melkersson–Rosenthal syndrome.* Replacement of muscle fibers with edematous fibrous tissue in lip of patient seen in Fig. 17–22.

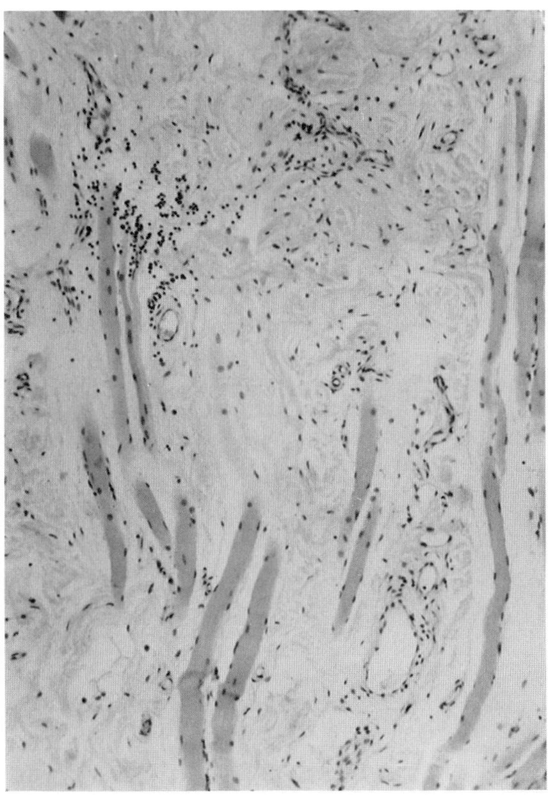

8. Wadlington WB et al: The Melkersson–Rosenthal syndrome. *Pediatrics* **73**:502–506, 1984.

9. Worsaae N, Pindborg JJ: Granulomatous gingival manifestations in Melkersson–Rosenthal syndrome. *Oral Surg* **49**:131–138, 1980.

10. Worsaae N et al: Melkersson–Rosenthal syndrome and cheilitis granulomatosa. *Oral Surg* **54**:404–413, 1982.

**Asymmetric crying facies (cardiofacial syndrome)** In 1967 Cayler (2) was the first to point out the association of asymmetric crying facies with other congenital anomalies, especially those of the cardiovascular, genitourinary, and respiratory systems. Employing the term ''cardiofacial syndrome,'' Cayler (3) and others (1,9) suggested that between 5 and 10% of infants with asymmetric crying facies had congenital heart disease.

The frequency of the disorder in various series has ranged from 1 in 120 to 1 in 160 newborns (1,9,11,12). Some authors have found a left-sided predilection, whereas others have found none. Several authors (3,5,6,7,9) reported familial occurrence, and monogenic inheritance has been suggested by some (6). The spectrum of anomalies remarkably overlaps that seen in the *oculoauriculovertebral spectrum* and may in some instances be the minimal expression of it (4,12).

Most authors attribute the defect to unilateral agenesis or hypoplasia of the anguli oris depressor muscle (8). However, Monreal (7) suggested that the side of the lip that pulls lowest is the abnormal one. Because the muscles involved act only to depress the lower lip margin, the defect does not interfere with sucking or smiling and does not foster drooling. The lips are symmetric at rest, the asymmetric facies becoming apparent only during crying. In most cases, lower facial weakness persists, but occasionally it may diminish to some degree (Fig. 17–29A,B). Pape and Pickering (10) noted that associated defects tended to occur on the same side as the partial lower facial palsy, but exceptions are many. The cardiovascular anomalies most frequently found include VSD, PDA, and tetralogy of Fallot. Also to be found may be right aortic arch, double aortic arch, pulmonic stenosis, coarctation of aorta, ASD, atrioventricularis communis, tricuspid atresia, single ventricle, hypoplastic right ventricle, hypoplastic pulmonary arteries, and bicuspid aortic valve (5,11,11a,13).

Skeletal anomalies may include hemivertebra, fused vertebrae, sternal and rib anomalies, and hypoplastic or absent radius and/or thumb (10).

Genitourinary anomalies may include absent or hypoplastic kidney, ectopic kidney, polycystic kidneys, bifid scrotum, hypogonadism, cryptorchidism, and hypospadias (10). Neuroblastoma has also been reported on one occasion (4a).

Tracheoesophageal atresia or fistula, laryngomalacia, bronchial stenosis, absent bronchus, absent lung lobe, and a variety of other anomalies such as anal stenosis, imperforate anus, and absent thymus have also been described.

In one study approximately 10% had either cleft lip/palate or isolated cleft palate (10).

#### References [Asymmetric crying facies (cardiofacial syndrome)]

1. Alexiou D et al: Frequency of other malformations in congenital hypoplasia of depressor anguli oris muscle syndrome. *Arch Dis Childh* **51**:891–893, 1976.

2. Cayler GG: An ''epidemic'' of congenital facial paresis and heart disease. *Pediatrics* **40**:666–668, 1967.

3. Cayler GG: Cardiofacial syndrome: Congenital heart disease and facial weakness, a hitherto unrecognized association. *Arch Dis Childh* **44**:69–75, 1969.

4. Dorchy H et al: Association of asymmetric crying facies: Malformation of the ear and pulmonary agenesis. *Acta Paediatr Belg* **29**:255–256, 1976.

4a. Lenarsky C et al: Occurrence of neuroblastoma and asymmetric crying facies. Case report and review of the literature. *J Pediatr* **107**:268–270, 1986.

5. Levin SE et al: Hypoplasia or absence of the depressor anguli oris muscle and congenital abnormalities, with special reference to the cardiofacial syndrome. *S Afr Med J* **61**:227–231, 1982.

6. Miller M, Hall JG: Familial asymmetric crying facies. *Am J Dis Child* **133**:743–746, 1979.

7. Monreal FJ: Asymmetric crying facies: An alternative interpretation. *Pediatrics* **65**:146–149, 1980.

8. Nelson KB, Eng PD: Congenital hypoplasia of the depressor anguli oris muscle: Differentiation from congenital facial palsy. *J Pediatr* **81**:16–20, 1972.

9. Papadatos C et al: Congenital hypoplasia of depressor anguli oris muscle. A genetically determined condition? *Arch Dis Childh* **49**:927–931, 1974.

10. Pape KE, Pickering D: Asymmetric crying facies: An index of other congenital anomalies. *J Pediatr* **81**:21–30, 1972.

11. Perlman M, Reisner SH: Asymmetric crying facies and congenital anomalies. *Am J Dis Child* **48**:627–629, 1973.

11a. Perrin P et al: Le syndrome cardio-faciale de Cayler. A propos de 19 observations. *Arch Fr Pediatr* **46**:257–261, 1989.

12. Singhi S et al: Congenital asymmetrical crying facies. *Clin Pediatr* **19**:673–678, 1980.

13. Strong WB, Silbert DR: Paralysis of the facial nerve and congenital heart defects. *Am Heart J* **78**:279–280, 1969.

### Gustatory lacrimation (paroxysmal lacrimation, crocodile tears syndrome, Bogorad syndrome, and gustolacrimal reflex)

The syndrome of lacrimation that accompanies eating was first described by Oppenheim (see 5) in 1913, and in far more detail, by Bogorad (3), in 1928, who called it the ''syndrome of crocodile tears,''

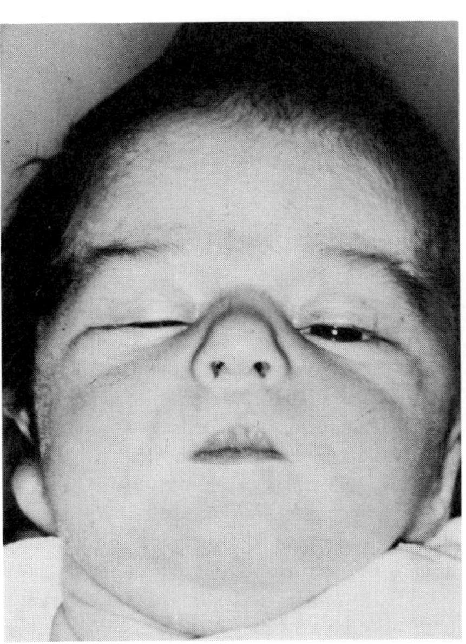
A

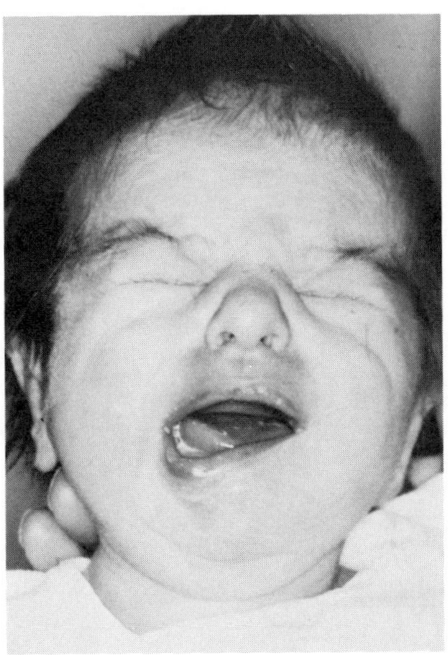
B

Fig. 17–29. *Asymmetric crying facies.* (A) Face at rest. (B) Asymmetry during crying. (From *G Cayler*, Sacramento, California.)

referring to the tale that the crocodile weeps hypocritical tears while devouring its victims (5). Since the original report, over 125 cases have been recorded. Ford (6,7) wrote the first report in English and linked the syndrome with gustatory sweating and flushing (auriculotemporal or Frey syndrome), both conditions caused by misdirected regrowth of nerve fibers. Ford used the term "paroxysmal lacrimation" to describe the disorder. To parallel the use of the term "gustatory sweating and flushing," the present authors suggest the use of "gustatory lacrimation." Axelsson and Laage-Hellman (2) have employed "gustolacrimal reflex."

In so-called acquired cases, the syndrome has followed traumatic or inflammatory conditions of the facial or greater superficial petrosal nerves (1,8,12). By the time motor function is becoming restored, the patient usually notices that when eating, tears flow from the eye on the affected side. The syndrome is not associated with the common form of Bell's palsy, in which the lesion is distal to the geniculate ganglion, but occurs only when the lesion is proximal to the ganglion (6). Rarely the syndrome is bilateral and/or congenital (9–11,16).

Congenital cases have usually been associated with abducens palsy or Duane syndrome (2,14,15). Ramsay and Taylor (13) postulated that the association of two conditions would be a lesion causing nuclear degeneration or dysgenesis in the vicinity of the abducens nucleus. The paradoxical aspects result from substitute innervation of the lateral rectus by fibers from the oculomotor nerve and lacrimal gland by fibers subserving salivation.

The facial nerve has several components: a motor component supplying the muscles of facial expression; a sensory component, which carries gustatory impulses from the anterior two-thirds of the tongue and proprioceptive impulses from the face; and an autonomic–parasympathetic component. These latter fibers arise in the superior salivary nucleus and provide secretory and vasodilatory function to the submandibular and sublingual glands via the nerve of Wrisberg, the chorda tympani, and the lingual nerve, as well as to the lacrimal gland via the nerve of Wrisberg, the greater superficial petrosal nerve, the vidian nerve, the sphenopalatine ganglion, and the zygomaticotemporal nerve. Proximal to the geniculate ganglion, all these components are in contiguity.

If a lesion occurs proximal to the geniculate ganglion, then during regeneration, fibers destined for the submandibular and sublingual glands may become partially interchanged with those destined for the lacrimal gland. Thus, when the gustatory stimulus is evoked, lacrimation is produced (Fig. 17–30).

In the exceptional cases cited by Boyer and Gardner (4), the syndrome appeared after surgery on the greater superficial petrosal nerve for relief of headache. During the course of surgery, the lesser superficial petrosal nerve was injured, as it runs only 2 mm lateral to the greater superficial petrosal nerve in the middle cranial fossa. In regeneration, the fibers of the two nerves became interchanged, producing the syndrome. This was proved by resection of the glossopharyngeal nerve, which abolished the disorder. The syndrome has also been seen in association with neurosyphilis, acoustic neuroma, vascular disease, and facial palsy in association with herpes zoster of the ear.

Facial contracture or diffuse facial muscle response is often associated with tearing. If, for example, the teeth are shown, the forehead may wrinkle and the eyelid close. Conversely, wrinkling the forehead or closing the eye may cause the corner of the mouth to be retracted and the nasolabial fold to deepen. This indicates that impulses formerly directed toward isolated muscle groups are diffusely distributed over the face, thus bolstering the concept of misdirected fiber regeneration. This has been called "the Marin Amat phenomenon."

This syndrome should not be confused with lacrimation seen in association with facial paralysis, where the tearing is due to ectropion, which permits tears to flow out of the conjunctival sac. The latter is not associated with hypersecretion of the lacrimal gland and is unaffected by eating.

The condition has been found in association with the *branchio-oto-renal syndrome* (12a).

#### References [Gustatory lacrimation (paroxysmal lacrimation, crocodile tears syndrome, Bogorad syndrome, and gustolacrimal reflex)]


1. Axelsson A, Laage-Hellman JE: The gusto-lachrymal reflex: The syndrome of crocodile tears. *Acta Otolaryngol (Stockholm)* **54**:239–254, 1962.
2. Biedner B et al: Bilateral Duane's syndrome associated with crocodile tears. *J Pediatr Ophthalmol* **16**:113–114, 1979.
3. Bogorad FA: Symptom of crocodile tears. *Vračh Dielo* **11**:1328–1330, 1928.
4. Boyer FC, Gardner WJ: Paroxysmal lacrimation (syndrome of crocodile


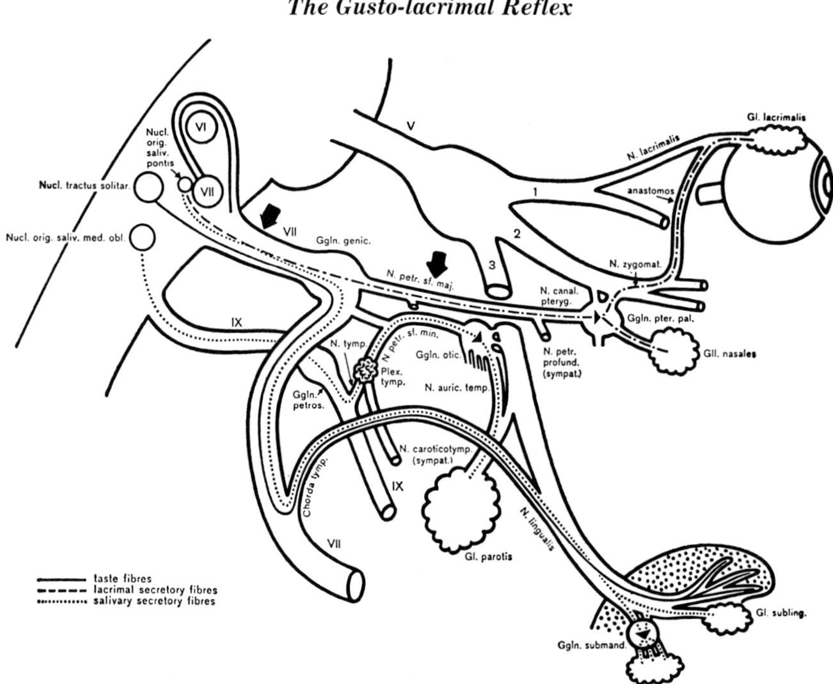


Fig. 17–30. *Gustatory lacrimation.* Diagram shows two possible mechanisms for production of the syndrome. Most often, lesion is proximal to geniculate ganglion and in regeneration, fibers destined for the submandibular and sublingual salivary glands may become interchanged with those for the lacrimal gland. The lesion may also be distal to the geniculate ganglion and involve interchange of fibers of the greater and lesser superficial petrosal nerves. [From *A Axelsson* and *JE Laage-Hellman,* Acta Otolaryngol (Stockholm) **54**:239, 1962.]

tears) and its surgical treatment: Relation to auriculotemporal syndrome. *Arch Neurol Psychiatr* **61**:56–64, 1949.

5. Chorobski J: The syndrome of crocodile tears. *Arch Neurol Psychiatr* **65**:299–318, 1951.
6. Ford FR: Paroxysmal lacrimation during eating as sequel of facial palsy: Syndrome of crocodile tears. *Arch Neurol Psychiatr* **29**:1279–1288, 1933.
7. Ford FR, Woodhall B: Phenomena due to misdirection of regenerating fibers of cranial, spinal and autonomic nerves: Clinical observations. *Arch Surg* **36**:480–496, 1938.
8. Golding-Wood PH: Crocodile tears. *Br Med J* **1**:1518–1521, 1963.
9. Jacklin HN: The gusto-lacrimal reflex (syndrome of crocodile tears). *Arch Ophthalmol* **61**:1521–1526, 1966.
10. Jampel RS, Titone C: Congenital paradoxical gustatory lacrimal reflex and lateral rectus paralyses. *Arch Ophthalmol* **67**:123–126, 1962.
11. Lutman FC: Paroxysmal lacrimation when eating. *Am J Ophthalmol* **30**:1583–1585, 1947.
12. McCoy FJ, Goodman RC: The crocodile tear syndrome. *Plast Reconstr Surg* **63**:58–62, 1979.

12a. Priesch JW et al: Gustatory lacrimation in association with the branchio-oto-renal syndrome. *Clin Genet* **27**:506–509, 1985.

13. Ramsay J, Taylor D: Congenital crocodile tears: A key to the aetiology of Duane's syndrome. *Br J Ophthalmol* **64**:518–522, 1980.
14. Regenbogen L, Stein R: Crocodile tears associated with homolateral Duane's syndrome. *Ophthalmologica* **156**:353-360, 1968.
15. Rosen G, Konak S: Le neurectomie tympanique dans le syndrome de "larmes de crocodile." *Ann Oto-laryngol (Paris)* **90**:723–726, 1973.
16. Spiers ASD: Syndrome of "crocodile tears": Pharmacological study of a bilateral case. *Br J Ophthalmol* **54**:330–334, 1970.

## Gustatory sweating and flushing–auriculotemporal and chord tympani syndromes (Frey syndrome)

The earliest references to the syndrome of gustatory sweating and flushing appear to be those of Duphenix (4,21), in 1757, and Dupuy (5), in 1816. In 1897, Weber (31) recognized that the syndrome was related to the auriculotemporal nerve. However, the syndrome was brought to the attention of the medical public in 1923 by Frey (7). The syndrome is often referred to under her name. Several hundred cases have been published subsequently. Laage-Hellman (14–17) showed that the syndrome, far from being rare, is a virtually constant complication of conservative parotidectomy. Other investigators have found a 60% complication rate (12a). Hogeman (12) presented an excellent historical survey.

It should be pointed out that the term "auriculotemporal syndrome" is misleading, as the skin innervated by the greater auricular nerve, the lesser occipital nerve, the long buccal nerve, or any cutaneous branch of the cervical plexus may be involved.

The syndrome is clinically manifested by sweating, flushing, a sense of warmth, and, sometimes, mild pain in the preauricular and temporal areas during eating of foods that produce a strong salivary stimulus (29). The syndrome not only accompanies conservative parotidectomy (14) but may follow suppurative parotitis (5–7,18,23) or direct trauma to the parotid region or mandibular condylar head (20,22,26,27). Some patients exhibit only sweating or flushing (12,14). The syndrome occasionally occurs in infants (2,3,24).

The etiology in most cases is damage to the auriculotemporal nerve. This nerve, in addition to supplying sensory fibers to the preauricular and temporal areas, carries parasympathetic fibers to the parotid gland and sympathetic vasomotor and sudomotor fibers to the skin of the same area. Injury to the auriculotemporal nerve denervates the sweat glands and vessels of the skin over its distribution, in addition to producing sensory disturbances.

Both the parasympathetic and sympathetic nerves of the face are cholinergic, hence compatible, and in the process of regeneration, parasympathetic fibers become misdirected and grow along sympathetic pathways (8,14–17). Thus, a gustatory stimulus produces sweating and flushing.

That the syndrome is caused by misdirection of parasympathetic fibers is shown by the ability of procaine, injected over the auriculotemporal nerve, to abolish the syndrome. Severing of the auriculotemporal, glossopharyngeal (8), or Jacobson (11) nerve, local atropine injection, blockage of the otic ganglion (9,12), and local application of 3% scopolamine hydrobromide cream (15) or 1% glycopyrrolate cream (10) have all been shown to inhibit or abolish the reaction. Conversely, acetylcholine injection increases the phenomenon (6,9).

The time of the first appearance of the fully developed syndrome varies considerably. Most observers have noted its occurrence within 2 months to 2 years (average, 9 months) postoperatively (9,14–17,26,29). There are cases in which the syndrome has appeared in a few days or as much as 17 years after surgery (26). Misdirected regrowth of fibers cannot explain the few cases in which the syndrome has appeared a few days postoperatively (25); the alternate possibility of transaxonal excitation exists.

As a rule, once the syndrome appears, the area of skin involved increases (16) and remains increased for life (12). However, about 5% of patients (14) experience regression and disappearance of the symptoms. The area affected varies both in degree and in extent, in some persons being seen at the corner of the mouth or extending down to the angle of the mandible, an area supplied by the greater auricular nerve (14) (Fig. 17–31A). RJ Gorlin saw a patient whose sole area of gustatory sweating was a half-dollar-sized area near the corner of his mouth where he had driven a scissor blade many years prior to development of the syndrome. Presumably parasympathetic fibers came from those supplying the minor salivary glands of the cheek (Fig. 17–31B). A similar example was reported by Storrs (27). RJ Gorlin has seen two children with gustatory sweating involving the maxillary division of the trigeminal. In neither case was there a history of prior trauma.

Related to the auriculotemporal syndrome is the *chorda tympani syndrome* (1,30,32) in which the sweating and flushing are limited to the skin of the chin and submental region. This syndrome may accompany surgery or injury to the submandibular gland (14). It has been shown to be abolished by blockage of the lingual nerve proximal to the chorda tympani.

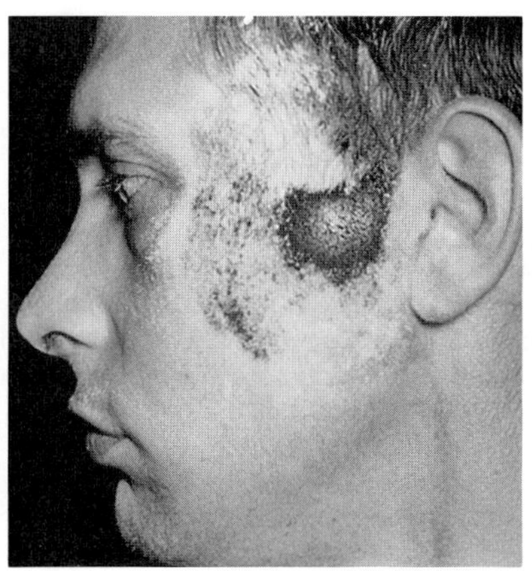
A

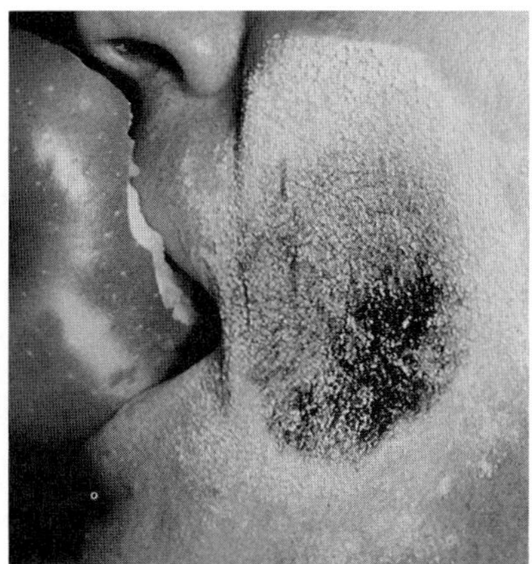
B

Fig. 17–31. *Gustatory sweating and flushing–auriculotemporal syndrome.* (A) Following meals there is profuse sweating immediately in front of ear. This is demonstrated by starch–iodine test. (B) Patient drove scissors into cheek as a child. Severe sweating and flushing are noted at mealtime.

The auriculotemporal syndrome should not be confused with excessive sweating sometimes seen in persons who have just eaten spicy or sour foods (19). In this case, the sweating appears limited to the forehead, tip of nose, and upper lip. Neither should it be confused with signs or symptoms of hysteria or postsympathectomy sweating.

To demonstrate sweating, the simplest method is application of the starch–iodine test (15 ml of 1% iodine, 5 ml castor oil, 80 ml of 95% alcohol) (14). A more quantitative estimate can be made by using plastic tape (9,13,28).

#### References [Gustatory sweating and flushing–auriculotemporal and chord tympani syndromes (Frey syndrome)]

1. Bailey BMW, Pearce DE: Gustatory sweating following submandibular gland removal. *Br Dent J* **158**:17–18, 1985.

2. Balfour HH Jr, Bloom JE: The auriculotemporal syndrome beginning in infancy. *J Pediatr* **77**:872–874, 1970.

3. Davis RS, Strunk RC: Auriculotemporal syndrome in childhood. *Am J Dis Child* **135**:832–833, 1981.

4. Duphenix: Sur une playe compliquée à la joue, ou le canal salivaire fut déchiré. *Mém Acad R Chir* **3**:431–437, 1757.

5. Dupuy: Sur l'enlèvement des ganglions gutturaux des nerfs tresplanchinques sur des chevaux. *J Méd Chir Pharmacol* **37**:340–350, 1816.

6. Freedberg AS et al: The auriculotemporal syndrome—a clinical and pharmacologic study. *J Clin Invest* **27**:669–676, 1948.

7. Frey L: Le syndrome du nerf auriculo-temporal. *Rev Neurol (Paris)* **2**:97–104, 1923.

8. Gardner WJ, McCubbin JW: Auriculotemporal syndrome: Gustatory sweating due to misdirection of regenerating nerve fibers. *JAMA* **160**:272–277, 1956.

9. Glaister DH et al: The mechanism of post-parotidectomy gustatory sweating (the auriculo-temporal syndrome). *Br Med J* **2**:942–946, 1958.

9a. Harper KE, Spielvogel RL: Frey's syndrome. *Int J Dermatol* **25**:524–526, 1986.

10. Hays LL: The Frey syndrome: A review and double blind evaluation of the topical use of a new anticholinergic agent. *Laryngoscope* **88**:1796–1824, 1978.

11. Hemenway WG: Gustatory sweating and flushing: The auriculo-temporal syndrome: Frey's syndrome. *Laryngoscope* **70**:84–90, 1960.

12. Hogeman KE: Surgical-orthopedic correction of mandibular protrusion. *Acta Chir Scand (Suppl)* **159**:94–111, 1951.

12a. Hüttenbrink KB, Hüttenbrink B: Das gustatorische Schwitzen nach Parotidektomie. *Laryngol Rhinol Otol* **65**:130–134, 1986.

13. Karat ABA: Quinarizin compound applied on cellulose tape as a test of sudomotor function in leprosy. *Lancet* **1**:651, 1969.

14. Laage-Hellman JE: Gustatory sweating and flushing after conservative parotidectomy. *Acta Otolaryngol* **48**:234–252, 1957.

15. Laage-Hellman JE: Treatment of gustatory sweating and flushing. *Acta Otolaryngol* **49**:132–143, 1958.

16. Laage-Hellman JE: Aetiologic implications of latent period and mode of development after parotidectomy. *Acta Otolaryngol* **49**:306–314, 1958.

17. Laage-Hellman JE: Aetiologic implications of response of separate sweat glands to various stimuli. *Acta Otolaryngol* **49**:363–374, 1958.

18. Langenskiöld A: Gustatory local hyperhidrosis following injuries in the parotid region. *Acta Chir Scand* **93**:294–306, 1946.

19. Mailander JC: Hereditary gustatory sweating. *JAMA* **201**:203–204, 1967.

20. Martis C, Athanassiades S: Auriculotemporal syndrome (Frey's syndrome) secondary to fracture of the mandibular condyle. *Plast Reconstr Surg* **44**:603–604, 1969.

21. Nicolai JPA: An early account of gustatory sweating (Frey's syndrome): A chance observation 250 years ago. *Br J Plast Surg* **38**:122–123, 1985.

22. Olson RE et al: Gustatory sweating caused by blunt trauma. *J Oral Surg* **35**:306–308, 1977.

23. Payne RT: Pneumococcal parotitis and antecedent auriculotemporal syndrome. *Lancet* **1**:634–636, 1940.

24. Pfeffer W Jr, Gellis SS: Auriculotemporal syndrome: Report of a case developing in early childhood with a review of the literature. *Pediatrics* **7**:670–678, 1951.

25. Smith RO et al: Jacobson's neurectomy for Frey's syndrome. *Am J Surg* **120**:478–480, 1970.

26. Spiro RH, Martin H: Gustatory sweating following parotid surgery and radical neck dissection. *Ann Surg* **165**:118–127, 1967.

27. Storrs TJ: A variation of the auriculotemporal syndrome. *Br J Oral Surg* **11**:236–242, 1974.

28. Thomson ML, Sutarman: The identification and enumeration of active sweat glands in man from plastic impressions of the skin. *Trans R Soc Trop Med* **47**:412–417, 1953.

29. Turner JC Jr et al: The auriculotemporal (Frey) syndrome occurring after parotid surgery. *Surg Gynecol Obstet* **111**:564–568, 1960.

30. Urprus V et al: Localized abnormal flushing and sweating on eating. *Brain* **57**:443–453, 1934.

31. Weber F: A case of localized sweating and blushing on eating, possibly due to temporary compression of vasomotor fibers. *Trans Chir Soc London* **31**:277–280, 1897–1898.

32. Young AG: Unilateral sweating of the submental region after eating (chorda tympani syndrome). *Br Med J* **2**:976–979, 1956.

### Gustatory rhinorrhea

Rhinorrhea following parotid surgery was reported by Boddie et al (1) and Stevens and Doyle (2). Vollrath (3) reported unilateral tearing and homolateral rhinorrhea accompanied by sweating 2 years following midface fracture. The etiology appears to be regeneration of damaged nerve fibers of the lesser superficial petrosal nerve. These fibers contain parasympathetic components that lead to the parotid gland. The regeneration occurs through abnormal connections via the greater superficial petrosal nerve to the vidian nerve, the sphenopalatine ganglion, and the long sphenopalatine nerve to reach the nasal mucous glands.

#### References (Gustatory rhinorrhea)

1. Boddie AW Jr et al: Gustatory rhinorrhea developing after radical parotidectomy—a new syndrome. *Arch Otolaryngol* **102**:248–250, 1976.

2. Stevens HE, Doyle PJ: Bilateral gustatory rhinorrhea following bilateral parotidectomy: A case report. *J Otorhinolaryngol* **17**:191–193, 1988.

3. Vollrath M: Das Schwitz-Tränen-ein neues Syndrom? *HNO* **32**:476–478, 1984.

### Angelman syndrome

The syndrome was first described by Angelman (1) in 1965. Most authors have used the name "happy puppet syndrome." We prefer the eponym Angelman syndrome due to the pejorative nature of the other term. Over 50 examples have been described to date (2–26). Nearly all have been sporadic. The syndrome has been reported in sibs (2,3,16,20,23,24), and autosomal recessive inheritance has been questioned. Monozygotic twins have been described (13). A recurrence risk of about 5% has been proposed (10). There is no sex predilection. A deletion in the 15q11–q13 chromosome region has been found in 40% (11,22,26). The same deletion is seen in Prader–Willi syndrome, but in Angelman syndrome, the deleted chromosome is maternal in origin (7,15) in most but not all (27).

Although patients resemble one another markedly, early diagnosis is often difficult (7). There is microbrachycephaly due to occipital flattening. The hair tends to be fair. Prognathism, midfacial retrusion, frequent extrusion of the tongue with drooling, and macrostomia are characteristic (Fig. 17–32A–C). The teeth are widely spaced, probably due to tongue pressure.

Ophthalmologic examination shows decreased pigmentation of the choroid in 70% and optic pallor or atrophy in over 40%. About 30% have Brushfield spots. Blue irises are more common than expected.

Patients have severe mental retardation with intelligence quotients below 40. There is no speech. They exhibit unprovoked and prolonged bursts of laughter. Motor development is retarded and muscular tone decreased, producing an awkward gait. Deep tendon reflexes are often brisk. One is impressed by their constant useless activity and ataxic uncoordinated movements. They begin to exhibit epilepsy in the form of hypsarrhythmia or grand mal seizures during the first few years of life.

Pathologic alterations are unknown, but the symptoms suggest brain damage, probably involving the cerebellum. Electroencephalographic studies have shown slow wave or spike activity. Pneumonencephal-

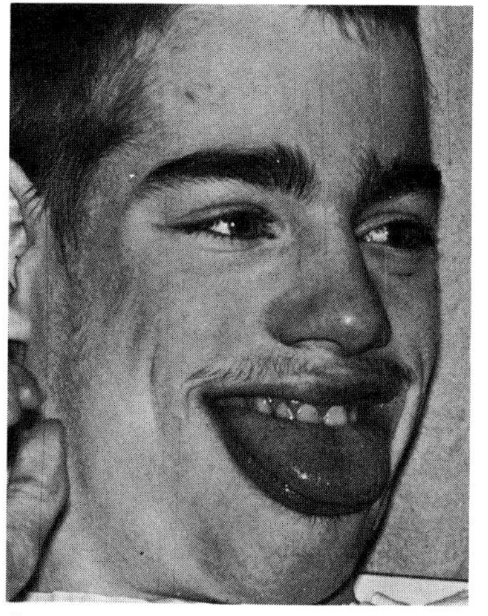
A
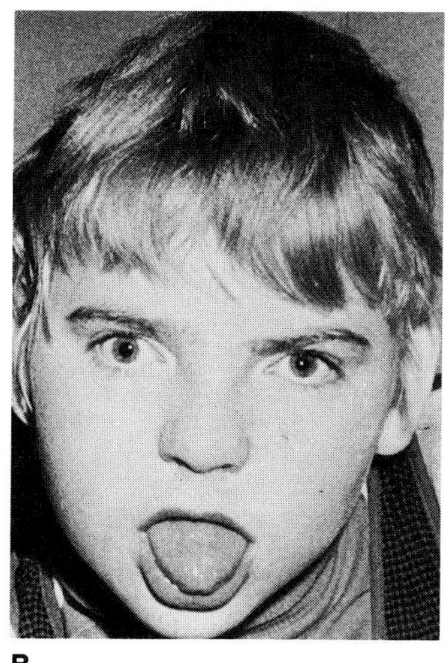
B
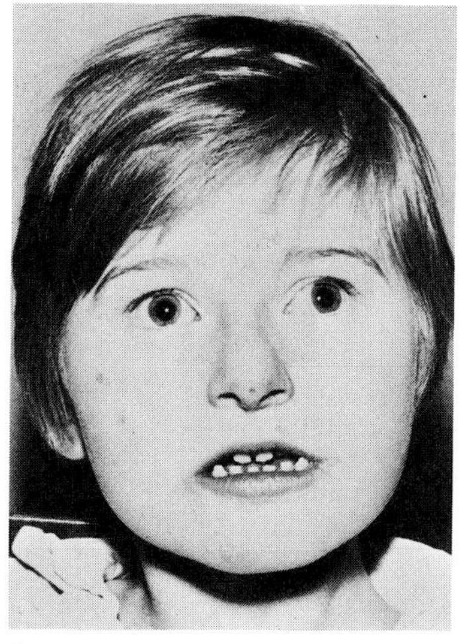
C

Fig. 17–32. *Angelman syndrome.* (A–C) Note protruding tongue, mandibular prognathism. Patients exhibit pointless activity, ataxic uncoordinated puppetlike movements, and unprovoked prolonged paroxysms of laughter. (A from *C Howry,* Yakima, Washington; B from *M Warburg,* Gentofte, Denmark; C from *O Mayo et al,* Dev Med Child Neurol **15:**63, 1973.)

ography has demonstrated cortical atrophy or dilation of the ventricular system.

Roentgenographically, there is flattening in the occipital region and tilting of the cranial base so that the sella turcica is obliquely positioned. CT scans show diffuse brain strophy (7).

#### References (Angelman syndrome)


1. Angelman H: "Puppet" children. A report on 3 cases. *Dev Med Child Neurol* **7**:681–685, 1965.
2. Baraitser M et al: The Angelman (happy puppet) syndrome: Is it autosomal recessive? *Clin Genet* **3**:323–330, 1987.
3. Berg JM, Pakula Z: Angelman's "happy puppet" syndrome. *Am J Dis Child* **123**:72–74, 1972.
4. Bjerre I et al: The Angelman or "happy puppet" syndrome. *Acta Paediatr Scand* **73**:398–402, 1984.
5. Dooley JM et al: The puppet-like syndrome of Angelman. *Am J Dis Child* **135**:621–624, 1981.
6. Donlon TA: Similar molecular deletions on chromosome 15q11.2 are encountered in both the Prader–Willi and Angelman syndromes. *Hum Genet* **80**:322–328, 1988.
7. Dörries A et al: Angelman ("happy puppet") syndrome—seven new cases documented by cerebral computed tomography. Review of the literature. *Eur J Pediatr* **148**:270–273, 1988.
8. Eber SW et al: Das Angelman-Syndrom. *Mschr Kinderheilk* **134**:158–160, 1986.
9. Elian M: Fourteen happy puppets. *Clin Pediatr* **14**:902–908, 1975.
10. Fisher JA et al: Angelman (happy puppet) syndrome in a girl and her brothers. *J Med Genet* **24**:294–298, 1987.
11. Fitchett M et al: The association of Angelman's syndrome with deletions within 15q11-13. *J Med Genet* **26**:73–77, 1989.
12. Halal F, Chagnon J: Le syndrome de la "marionette joyeuse." *Union Méd Can* **105**:1077–1083, 1976.
13. Hersh J et al: Behavioural correlates in the happy puppet syndrome: Characteristic profile? *Dev Med Child Neurol* **23**:792–800, 1973.
14. Kibel MA, Burness FR: The "happy puppet" syndrome. *Cent Afr J Med* **19**:91–93, 1973.
15. Knoll JH et al: Angelman and Prader–Willi syndromes share a common chromosome 15 deletion but differ in parental origin of the deletions. *Am J Med Genet* **32**:285–290, 1989.
16. Kuroki Y et al: The happy puppet syndrome in two siblings. *Hum Genet* **56**:227–229, 1980.
17. Massey JY, Roy FA: Ocular manifestations of the happy puppet syndrome. *J Pediatr Ophthalmol* **10**:282–284, 1973.
18. Mayo O et al: Three more happy puppets. *Dev Med Child Neurol* **15**:63–69, 1973.
19. Moore JR, Jeavons PM: The "happy puppet" syndrome: Two new cases and a review of five previous cases. *Neuropädiatrie* **4**:172–179, 1973.
20. Pashayan H et al: The Angelman syndrome in two brothers. *Am J Med Genet* **13**:295–298, 1982.
21. Pele S et al: Happy puppet syndrome on syndrome du "patin hilaire." *Helv Paediatr Acta* **31**:183–188, 1976.
22. Pembry M et al: The association of Angelman syndrome and deletions within 15q11-13. *J Med Genet* **25**:274, 1988.
23. Robb SA et al: The "happy puppet" syndrome of Angelman. Review of the clinical features. *Arch Dis Childh* **64**:83–86, 1989.
24. Willems PJ et al: Recurrence risk in the Angelman ("happy puppet") syndrome. *Am J Med Genet* **27**:773–780, 1987.
25. Williams CA, Frias JL: The Angelman (happy puppet) syndrome. *Am J Med Genet* **11**:453–460, 1982.
26. Williams CA et al: Incidence of 15q deletions in the Angelman syndrome: A survey of 12 affected persons. *Am J Med Genet* **32**:339–345, 1989.
27. Williams CA et al: Angelman syndrome in a daughter with del(15)(q11q13) associated with brachycephaly, hearing loss, enlarged foramen magnum, and ataxia in the mother. *Am J Med Genet* **32**:333–338, 1989.


## Syndromes of lower cranial nerves and medulla

In both nuclear and infranuclear lesions, simultaneous involvement of IX, X, and XI is common. The association of these three cranial nerves in various combinations with XII, the cervical sympathetic chain, and rarely the spinothalamic tracts has led to a plethora of eponymic syndromes, only a few of which are discussed here (Fig. 17–33). They may be roughly divided into central medullary lesions (about 10%) and peripheral lesions (about 90%). The central lesions may be vascular in origin (Avellis syndrome, Cestan–Chenais syndrome, Wallenberg syndrome, and Dejerine syndrome), or they may be due to other causes, such as syringobulbia, multiple sclerosis, poliomyelitis, neoplasms, or trauma (1,3,14,18,26).

Wallenberg syndrome was originally considered different from the other medullary syndromes, that is, those of Babinski–Nageotte and Cestan–Chenais, by the absence of motor signs in trunk and limbs. However, a unilateral infarction can extend contralaterally to catch the already crossed corticospinal fibers, and there may be little justification for designating these as separate disorders.

In spite of the long list of eponyms, only two syndromes are of practical importance: (1) medial medullary syndrome—rare, and (2) lateral medullary syndrome—common. The other syndromes do not help with localization nor do they aid in identifying the vessels involved. Because they still appear in the literature, some of these will be mentioned.

The lateral syndrome is that of Wallenberg, the medial is Dejerine

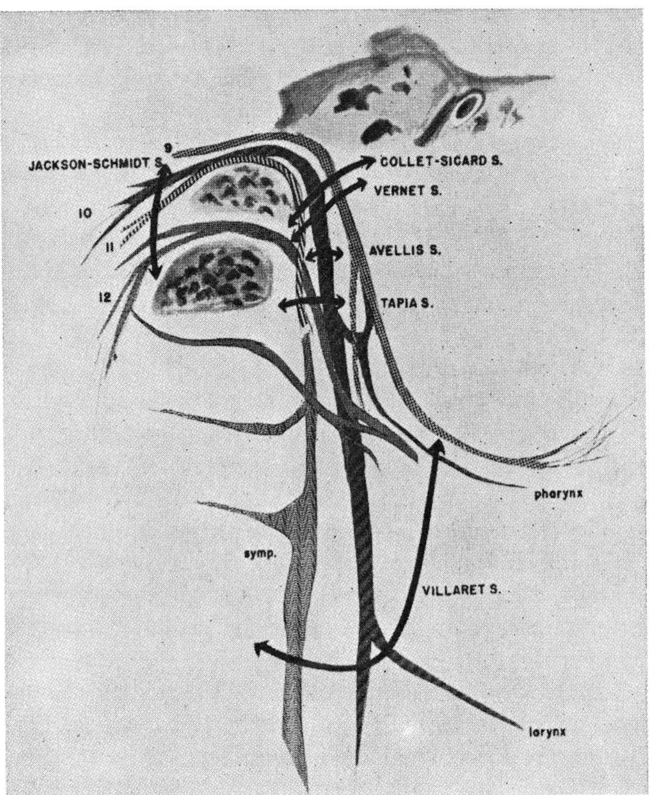


Fig. 17–33. *Syndromes of lower cranial nerves and medulla.* Diagram showing involvement of last four cranial nerves in several syndromes. (From *SG Harner,* Arch Otolaryngol **92**:632, 1970.)

Table 17–6. Syndromes involving the nuclei in the medulla

| Syndrome | Other name | Comment |
|---|---|---|
| Avellis | | Corticospinal tract, and X nucleus |
| Schmidt | Vago-accessory syndrome | Location uncertain; more than one lesion probable |
| Jackson–Mackenzie | Vago-accessory-hypoglossal syndrome | Corticospinal tract, and X, XII nuclei |
| Wallenberg | Lateral medullary syndrome | Spinal V, IX, X, XI lateral spino-thalamic spino-cerebello pupil-lodilatory fibers |
| Cestan–Chenais | Lateral medulla syndrome plus medial medulla syndrome | |
| Babinski–Nageotte | | Location uncertain; more than one lesion present |
| Dejerine | Anterior bulbar syndrome or medial medul-lary syndrome | Pyramids, medial lemniscus, hypoglossal fibers |

syndrome, and where both medial and lateral are involved, Cestan–Chenais syndrome is the appropriate designation. Syndromes involving the medulla are listed in Table 17–6.

**Avellis syndrome.** This syndrome is usually due to occlusion of the vertebral artery. It is characterized by combined involvement of the nucleus ambiguus and spinothalamic tracts with ipsilateral paralysis and anesthesia of the soft palate, pharynx, and larynx, contralateral loss of pain and temperature sensation of the trunk and extremities, frequent contralateral loss of proprioception and tactile senses, and, rarely, ipsilateral Horner syndrome (4,5,7,10,12).

**Schmidt syndrome.** This is now thought not to be a simple syndrome. The original report (one paragraph) did not localize the pathology with any clarity (28).

**Jackson–Mackenzie syndrome.** The original patient had multiple infarctions—so that the localizing value of the syndrome is limited.

**Wallenberg syndrome.** The main clinical features are ipsilateral paralysis and anesthesia of the soft palate, pharynx, and larynx, ipsilateral loss of the corneal reflex and anesthesia of the face for pain and temperature, ipsilateral Horner syndrome, ipsilateral ataxia, contralateral loss of temperature and pain sensation of the trunk and extremities, and, rarely, involvement of VI–VIII (17,37).

**Dejerine syndrome.** Hemorrhage or thrombosis of the anterior spinal artery (which supplies the pyramids, medial lemniscus, and emerging hypoglossal fibers) produces ipsilateral paralysis of the tongue, contralateral pyramidal paralysis of the arm and leg, and, occasionally, contralateral loss of proprioceptive and tactile sensation.

**Cestan–Chenais syndrome.** This syndrome indicates involvement of the lateral and medial medulla. It is caused by thrombosis of the vertebral artery below the point of origin of the posterior inferior cerebellar artery. It differs from the Wallenberg syndrome by the presence of pyramidal tract signs (hemiplegia and contralateral diminution of touch and proprioceptive senses). Pain and temperature sensations are normal (6).

**Vernet (jugular foramen) syndrome.** The syndrome is characterized by ipsilateral paralysis of IX, X, and XI with resultant paralysis of the soft palate, larynx, pharynx, sternocleidomastoid muscle, and part of the trapezius muscle, anesthesia of the posterior soft palate, and loss of sensation and taste over the posterior third of the tongue (33,34). About one-half of the cases have been attributed to brain tumors. However, more recently, among 17 cases of Vernet syndrome, only two patients had neoplasms (31). In most there was no obvious cause. Full or partial recovery occurred within a month.

**Villaret (retroparotid space) syndrome.** The syndrome combines ipsilateral paralysis of the last four cranial nerves and Horner syndrome (enophthalmos, ptosis, miosis) (23,25). Occasionally, VII is also involved. The syndrome is due to a lesion in the retroparotid space, which is bounded posteriorly by the cervical vertebrae, medially by the pharynx, anteriorly by the parotid gland, laterally by the sternocleidomastoid muscle, and superiorly by the skull near the jugular foramen.

**Collet–Sicard syndrome.** The disorder is manifest by unilateral involvement of the last four cranial nerves. It was described in the injured of World War I by Collet and later by Sicard. The syndrome may also occur as a result of tumors, inflammatory lesions, aneurysm of the internal or external carotid (30), or fibromuscular dysplasia of the carotid (15). The lesion is typically extracranial and involves the jugular foramen and the anterior condylar or hypoglossal foramen through which XII passes. Currie (8) expressed doubt whether this syndrome should be preserved. The original patient might have had a lateral medullary syndrome with contralateral weakness and sensory loss due to an earlier infarction higher in the brain stem or in the medial medulla.

**Tapia (vagohypoglossal) syndrome.** The main features include unilateral paralysis of the muscles of the tongue, and unilateral paralysis of the vocal cord and soft palate (33). The pharynx, sternocleidomastoid, and trapezius muscles are usually spared. There is dysphonia but no dysphagia. The lesion is localized to the crossing of the X–XI nerves below the nodose ganglia (29). A finding of trauma is more common than tumors (2).

**Garcin (half-base) syndrome.** The syndrome is usually due to a malignant tumor invading the base of the skull, but meningiomas, basal meningitis, and trauma have all been implicated. Typical progression may be to complete ophthalmoplegia (13,20,22,23,25,27,36). It may be somewhat more common in Chinese due to the high incidence of nasopharyngeal carcinoma. Tumors of the nasopharynx can cause all or nearly all cranial nerves on one side to become involved without signs of motor or sensory disturbances of the extremities and without evidence of increased intracranial pressure (papilledema or spinal fluid changes). Radiographic alterations are often apparent in the skull base.

**Progressive bulbar palsy.** In adults, we shall briefly consider Duchenne syndrome, while in children, Fasio-Londe syndrome will be described.

**Duchenne syndrome (progressive bulbar palsy, glossopharyngolabial paralysis).** The condition is principally seen in individuals in their sixth and seventh decades. The syndrome can occur in amyotrophic lateral sclerosis in which it is characterized by a slowly ascending process that involves the motor nuclei of the medulla and often the pons and midbrain (9,21). XII is usually involved first. Early signs are bilateral fasciculations and atrophy of the tongue. X is then affected with resultant dysphagia for both liquids and solids, dysarthria, and thickened nasal speech. Atrophy and fasciculations of the palatal and pharyngeal musculature may be seen. Involvement of the nuclei of V and VII causes paralysis and wasting of the muscles of facial expression and mastication. The nucleus of the accessory nerve is less often affected (13a,32).

**Fasio–Londe syndrome.** Childhood onset of progressive bulbar palsy may be part of infantile and childhood spinal muscular atrophies. Age of onset varies between 2 and 10 years. Unilateral facial weakness, dysphagia, and chewing difficulties are early signs. There is often spread to involve most lower cranial nerves. It is doubtful whether this condition is a pure progressive bulbar palsy distinct from the more frequent group of spinal muscular atrophies.

**Horner syndrome.** Horner syndrome is characterized by enophthalmos, miosis, partial ptosis, ipsilateral vascular dilatation, and anhidrosis of the face, neck, and arms. The syndrome may also be seen in association with involvement of V and X in Wallenberg syndrome. *Alternating Horner syndrome* has been described following cervical spinal cord injury (24). The cause is not known but movement of fluid within an intramedullary cystic cavity (posttraumatic) has been postulated (11). This stimulates the sympathetic nerve, causing the signs to alternate. Dural adhesions following a cervical spine injury were postulated by Ottomo and Heimburger (24).

### References (Syndromes of lower cranial nerves and medulla)


1. Adams RD, Victor M: *Principles of Neurology,* ed 2, McGraw-Hill, New York, 1981, pp 929–939.
2. Andrioli G et al: Tapia's syndrome caused by a neurofibroma of the hypoglossal and vagus nerves. *J Neurosurg* **52**:730–732, 1980.
3. Astor FC et al: Incidence of cranial nerve dysfunction following carotid endarterectomy. *Head and Neck Surg* **6**:660–663, 1983.
4. Avellis G: Klinische Beiträge zur halbseitigen Kehlkopflähmung. *Berl Klin* **41**:1–26, 1891.
5. Burger H: Classification and nomenclature of associated paralysis of vocal cords. *Acta Otolaryngol* **19**:389–403, 1934.
6. Cestan R, Chenais L: Du myosis dans certaines lésions bulbaires en foyer (hemiplégie du type Avellis associée au syndrome oculaire sympathique). *Gaz Hôp (Paris)* **76**:1229–1233, 1903.
7. Cody CC Jr: Associated paralyses of the larynx. *Ann Otol Rhinol Laryngol* **55**:549–561, 1946.
8. Currie RD: Syndromes of the medulla oblongata. *Handbook of Clinical Neurology,* Vol 2, North-Holland Publishing Company, Amsterdam, 1969, pp 217–235.
9. Duchenne GBA: Glosso-labial-laryngeal paralysis: Selections from the clinical works of Dr. Duchenne (edited by GV Poore). *New Sydenham Soc* **105**:143–172, 1883.
10. Fox SL, West GB: Syndrome of Avellis: A review of the literature and report of one case. *Arch Otolaryngol* **46**:773–778, 1947.
11. Freeman LW, Russell JR: The significance of temporary and alternating ptosis, meiosis and anhidrosis. *J Neurosurg* **12**:584–590, 1955.
12. Furstenberg AC, Magielski JE: A motor function of the nucleus ambiguus: Its clinical significance. *Ann Otol Rhinol Laryngol* **64**:788–793, 1955.
13. Garcin R: Le syndrome paralytique unilatérale globale des nerfs crâniens. Thesis, Paris, 1927.
13a. Gelmers, HJ: Tapia's syndrome after thoracotomy. *Arch Otolaryngol* **109**:622–623, 1983.
14. Grundy DJ et al: Cranial nerve palsies in cervical injuries. *Spine* **9**:339–343, 1984.
15. Havelius U et al: Carotid fibromuscular dysplasia and paresis of lower cranial nerves (Collet–Sicard syndrome). *J Neurosurg* **56**:850–853, 1982.
16. Haymaker W, Kuhlenbeck H: Disorders of the brainstem and its cranial nerves, in *Clinical Neurology,* Vol 3, Baker AB, Baker LH (eds), Harper & Row, New York, 1977, p 55.
17. Hörnsten G: Wallenberg's syndrome. *Acta Neurol Scand* **50**:434–468, 1974.
18. Jannetta PJ: Cranial nerve vascular compression syndromes (other than tic douloureux and hemifacial spasm). *Clin Neurosurg* **28**:445–456, 1981.
19. Kay MC et al: Multiple cranial nerve palsies in late metastasis of midline malignant reticulosis. *Am J Ophthalmol* **88**:1087–1090, 1979.
20. Kono C: A case of Garcin's syndrome associated with carcinoma of the middle ear. *Nagoya J Med Sci* **33**:47–54, 1970.
21. Madsen E: Dysphagia in bulbar and pseudobulbar lesions and its similarity to esophageal carcinoma. *Acta Radiol (Stockholm)* **41**:517–524, 1954.
22. Massey EW et al: Cylindroma causing Garcin's syndrome. *Arch Neurol* **37**:786, 1980.
23. New GB: Laryngeal paralysis associated with the jugular foramen syndrome and other syndromes. *Am J Med Sci* **165**:727–737, 1923.
24. Ottomo M, Heimburger RF: Alternating Horner's syndrome and hyperhidrosis due to dural adhesions following cervical spinal cord injury. *J Neurosurg* **53**:97–100, 1980.
25. Pisi E: Contributo anatomo-clinico alla conoscenza della sindrome di Garcin. *Arch Patol Clin Med* **30**:1–39, 1952.
26. Rosa L et al: Multiple cranial nerve palsies due to a hyperextension injury to the cervical spine. *J Neurosurg* **61**:172–173, 1984.
27. Schiffer KH: Das Halbbasissyndrom (Garcin), seine Diagnostik und sein Genese. *Arch Psychiatr* **186**:298–326, 1951.
28. Schmidt A: Casuistische Beiträge zur Nervenpathologie: II: Doppelseitige Accessoriuslähmung bei Syringomyelie. *Dtsch Med Wochenschr* **18**:606–608, 1892.
29. Schoenberg BS. Tapia's syndrome: The erratic evolution of an eponym. *Arch Neurol* **36**:257–260, 1979.
30. Svien HJ et al: Jugular foramen syndrome and allied syndromes. *Neurology* **13**:797–809, 1963.
31. Tanaka M et al: Jugular foramen syndrome. *Neurology* **33**:119–120, 1983.
32. Tapia AG: Un nouveau syndrome: Quelques cas d'hemiplégia du larynx et de la langue avec ou sans paralysie du sterno-cléïdo-mastoïden et du trapèze. *Arch Int Laryngol* **22**:780–785, 1906.
33. Vernet M: Sur le syndrome des quatre dernières paires crâniennes d'après une observation personelle chez un blessé de guerre. *Bull Soç Méd Paris* **40**: Ser 3:210–223, 1916.
34. Vernet M: The classification of the syndromes of associated laryngeal paralysis. *J Laryngol* **33**:354–365, 1918.
35. Villaret M: Sur le syndrome du trou déchiré postérieur. *Paris Méd* **23**:78–81, 1917.
36. Weiss H: Halbbasissyndrom (Syndrom Garcin) bei Glomustumoren. *Nervenarzt* **26**:289–291, 1955.
37. Wilkins RH, Brody LA: Wallenberg's syndrome. *Arch Neurol* **22**:379–382, 1970.
38. Wilson H et al: Jugular foramen syndrome as a complication of metastatic cancer of the prostate. *South Med J* **77**:92–93, 1984.

# Chapter 18
# Syndromes with Contractures

## Fetal akinesia sequence (Pena–Shokeir I syndrome)

Pena and Shokeir (20) and Punnett et al (22) in 1974, described children with intrauterine growth retardation, multiple congenital joint contractures, facial abnormalities, and pulmonary hypoplasia (Fig. 18–1). About 60 cases have subsequently been reported. The frequency has been estimated as 1/10,000 births. Shokeir (26) proposed that the syndrome be referred to as Pena–Shokeir syndrome, Type I and that the cerebro-oculo-facial syndrome (COFS) be called Pena–Shokeir syndrome, Type II. Moessinger (18) and Toriello et al (27) proposed abandonment of the term and replacement with *fetal akinesia sequence*. The disorder has been classified as a form of arthrogryposis. It surely is heterogeneous, the phenotype probably reflecting decreased fetal movement from many different causes (12,27): neurogenic atrophy (2,17), congenital myopathy (16), maternal myasthenia gravis (23), and fetal edema (6). The most critical survey is that of Hall (7).

The phenotypic features have been found in sibs (20,27) and parental consanguinity (22,26) has been reported. Inheritance for at least some examples appears to be autosomal recessive (21,25), although more than half of the reported cases are sporadic.

In most cases, the infant is either stillborn or dies within the neonatal period. Only a few have survived (11a,16,24). Polyhydramnios, short gut syndrome, small placenta or short umbilical cord have been noted in 50% (3,11a,25). Hydrops fetalis is a common finding. Most have intrauterine growth retardation that may have resulted from reduced muscle and bone mass secondary to prolonged immobilization and disuse.

**Facies.** Hypertelorism, high nasal bridge, depressed nasal tip, low-set malformed pinnae, and microretrognathia are present in nearly all cases. Epicanthal folds are present in about 20%. The neck is short in 60% (Fig. 18–2A,B).

**Musculoskeletal system.** There are moderate contractures at the hips, elbows, wrists, knees, and ankles due to reduced intrauterine movement. The chest is small with an increased AP diameter. The lungs are severely hypoplastic. Camptodactyly and clubfeet or rocker-bottom feet are essentially constant features. Mild webbing of the neck or elbows has been seen in a few cases (1,17,28) (Fig. 18–2A).

Radiographic changes include micrognathia, long and narrow tubular bones, multiple joint dislocations, and malposition of fingers, especially angular subluxations at the interphalangeal joints (11,14). A few infants have exhibited variable polydactyly (11,22,27) (Fig. 18–3A,B).

**Genitourinary abnormalities.** Probably all affected males have cryptorchidism and 20% have hypospadias. Clitoral enlargement and/or labial abnormalities are present in 15%. Various urinary tract anomalies occur in 30% (13).

**Cardiac anomalies.** Various congenital heart anomalies have been described in 25% (13).

**Oral manifestations.** Small mandible and cleft palate are frequent (27).

**Pathology.** Apart from measurable lung hypoplasia, present in nearly all cases, internal organs have been essentially normal. Muscle biopsy may show focal atrophy and replacement with fibrous connective tissue or fat or may be normal. Neuropathologic findings consist of a marked paucity of anterior horn cells in the spinal cord and widespread neuronal loss in the central nervous system (1,4,17,28), or may be normal. Pathogenesis appears to be related to a number of different mechanisms in different families, resulting in muscle weakness and intrauterine fetal immobility. This is implied by the congenital joint contractures and webbing, and the facial features (17). If any one of the phenotypic features is seen in a patient, the others should be looked for.

**Differential diagnosis.** Arthrogryposis, a group of conditions of diverse etiology and characterized by congenital joint contractions and web formation, must be excluded (11). The anomalies of the extremities somewhat resemble those in *trisomy 18*. The syndrome of multiple ankyloses and facial anomalies must also be excluded (29).

Fig. 18–1. *Fetal akinesia sequence.* Main dysmorphic features. (Adapted from *SDJ Pena* and *MHK Shokeir,* J Pediatr **85:**373, 1974.)

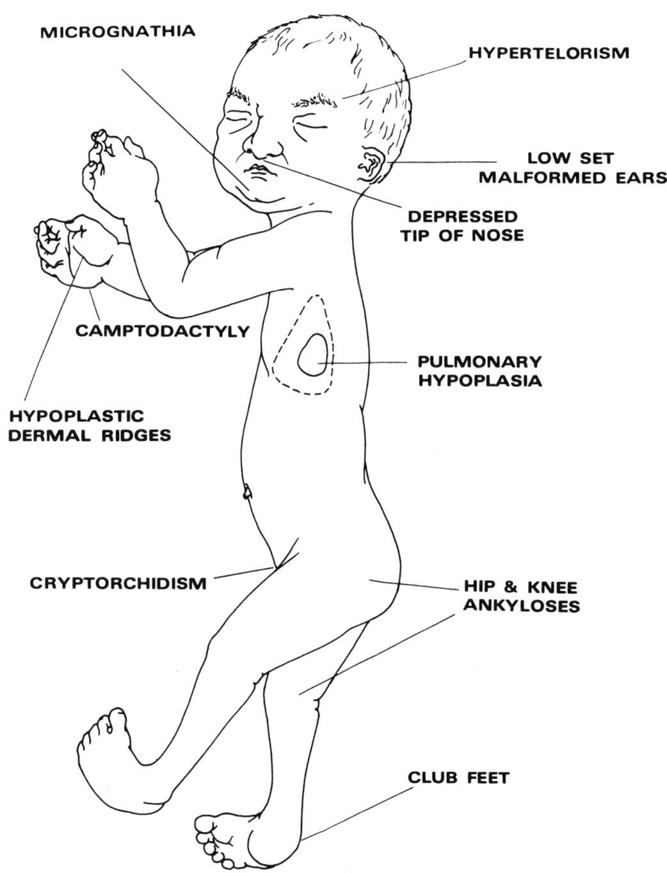

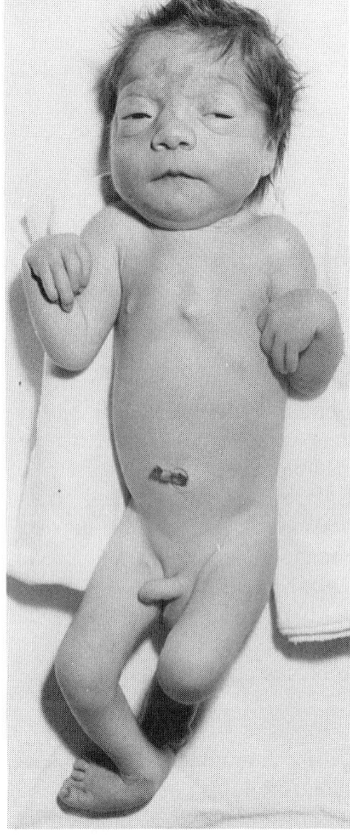

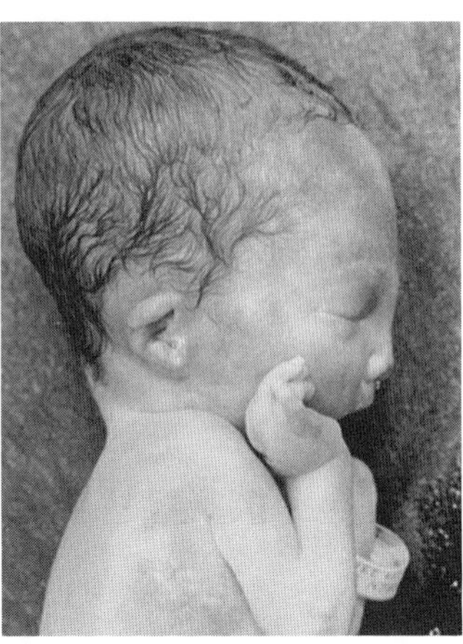

A B

Fig. 18–2. *Fetal akinesia sequence.* (A) Note suborbital creases, hypertelorism, nevus flammeus of forehead, micrognathia, arthrogryposis (camptodactyly, club feet), and pectus carinatum. (From *HH Punnett et al,* J Pediatr **85:**375, 1974.) (B) Note downturned pinna, camptodactyly, and micrognathia. (Courtesy of *MHK Shokeir,* Winnipeg, Manitoba, Canada.)

A lethal, recessively inherited *multiple pterygium syndrome* has been reported. In addition to the ankyloses, camptodactyly, lung hypoplasia, and facial abnormalities, multiple joint webs and/or cardiac hypoplasia were found. This condition shares many features of fetal akinesia sequence (1,5,6,9) because both have decreased movement. There is some facial resemblance to *Potter sequence.*

Hall et al (8) reported an autosomal dominantly inherited disorder characterized by small mouth and jaw, limited jaw movement in infancy, mild short stature, mild microcephaly, large ears without an anthelix, and severe flexion contractures of the hands and feet that led to subluxation of fingers and club feet. A dominantly inherited syndrome of similar anomalies of the hands and rocker-bottom feet due to vertical talus (digitotalar syndrome) but with no facial anomalies was described by Sallis and Beighton (23).

**Laboratory aids.** Sonography has been employed for prenatal diagnosis based on associated hydrops (15,19,25).

### References [Fetal akinesia sequence (Pena–Shokeir I syndrome)]


1. Bisceglia M et al: Pathologic features in two siblings with the Pena–Shokeir I syndrome. *Eur J Pediatr* **146**:283–287, 1987.
2. Chen H et al: The Pena–Shokeir syndrome: Report of five cases and further delineation of the syndrome. *Am J Med Genet* **16**:213–224, 1983.
3. Chen H et al: Syndrome of multiple pterygia, camptodactyly, facial anomalies, hypoplastic lungs and heart, cystic hygroma, and skeletal anomalies: Delineation of a new entity and review of lethal forms of multiple pterygium syndrome. *Am J Med Genet* **17**:809–826, 1984.
4. Dimmick JE et al: Syndrome of ankylosis, facial anomalies, and pulmonary hypoplasia: A pathologic analysis of one infant. *Birth Defects* **13**(3D):133–137, 1977.
5. Gillin ME, Pryse-David J: Pterygium syndrome. *J Med Genet* **13**:249–251, 1976.
6. Hall J: The lethal multiple pterygium syndromes. *Am J Med Genet* **17**:803–807, 1984.
7. Hall J: Analysis of Pena–Shokeir phenotype. *Am J Med Genet* **25**:99–117, 1986.
8. Hall JD et al: A new arthrogryposis syndrome with facial and limb anomalies. *Am J Dis Child* **129**:120–122, 1975.
9. Hall JD et al: Limb pterygium syndromes: A review and report of eleven patients. *Am J Med Genet* **12**:377–409, 1982.
10. Holmes LB et al: Contractures in an infant of a mother with myasthenia gravis. *J Pediatr* **96**:1067–1069, 1980.


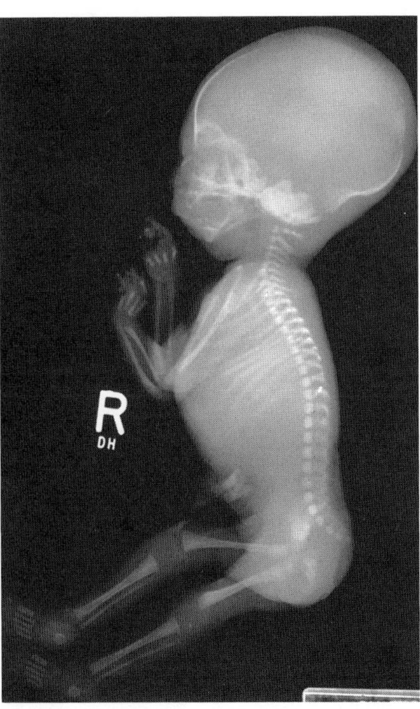


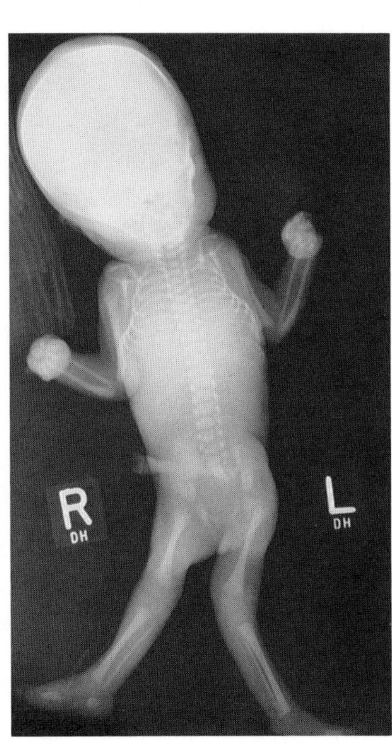


A B

Fig. 18–3. *Fetal akinesia sequence.* (A,B) Anteroposterior and lateral views of stillborn infant showing gracile bones, fixed deformity of many joints, hip dislocation, and micrognathia. (From *CS Houston* and *MHK Shokeir,* J Can Assoc Radio **32**:215, 1981.)

11. Houston CS, Shokeir MHK: Separating Pena–Shokeir I syndrome from the "arthrogryposis basket." *J Can Assoc Radiol* **32**:215–219, 1981.

11a. Katzenstein M, Goodman RM: Pre- and postnatal findings in Pena–Shokeir I syndrome: Case report and a review of the literature. *J Craniofac Genet Develop Biol* **8**:111–126, 1988.

12. Lazjuk GI et al: Pulmonary hypoplasia, multiple ankyloses and camptodactyly: One syndrome or some related forms? *Helv Paediatr Acta* **33**:73–79, 1978.

13. Lindhour D et al: The Pena–Shokeir syndrome: Report of nine Dutch cases. *Am J Med Genet* **21**:655–668, 1985.

14. Mailhes JB et al: "Pena–Shokeir syndrome" in a newborn male infant. *Am J Dis Child* **131**:1419–1420, 1977.

15. MacMillan RH et al: Prenatal diagnosis of Pena–Shokeir syndrome type I. *Am J Med Genet* **21**:279–284, 1985.

16. Mease AD et al: A syndrome of ankylosis, facial anomalies and pulmonary hypoplasia secondary to fetal neuromuscular dysfunction. *Birth Defects* **12**(5):193–200, 1976.

17. Moerman P et al: Multiple ankyloses, facial anomalies and pulmonary hypoplasia associated with severe antenatal spinal muscular atrophy. *J Pediatr* **103**:238–241, 1983.

18. Moessinger AC: Fetal akinesia deformation sequence: An animal model. *Pediatrics* **72**:857–863, 1983.

19. Ohlsson A et al: Prenatal sonographic diagnosis of Pena–Shokeir syndrome type I, or fetal akinesia deformation sequence. *Am J Med Genet* **29**:59–66, 1988.

20. Pena SDJ, Shokeir MHK: Syndrome of camptodactyly, multiple ankyloses, facial anomalies, and pulmonary hypoplasia: A lethal condition. *J Pediatr* **85**:373–375, 1974.

21. Pena SDJ, Shokeir MHK: Syndrome of camptodactyly, multiple ankyloses, facial anomalies and pulmonary hypoplasia: Further delineation and evidence for autosomal recessive inheritance. *Birth Defects* **12**(5):201–208, 1976.

22. Punnett HH et al: Syndrome of ankylosis, facial anomalies, and pulmonary hypoplasia. *J Pediatr* **85**:375–377, 1974.

23. Sallis JG, Beighton P: Dominantly inherited digito-talar dysmorphism. *J Bone Jt Surg* **54B**:509–515, 1972.

24. Say B et al: Ankyloses, facial anomalies, and pulmonary hypoplasia syndrome. *Am J Dis Child* **133**:1196–1197, 1979.

25a. Shenker L et al: Syndrome of camptodactyly, ankyloses, facial anomalies and pulmonary hypoplasia (Pena–Shokeir syndrome): Obstetric and ultrasound aspects. *Am J Obstet Gynecol* **152**:303–307, 1985.

26. Shokeir MHK: Multiple ankyloses, camptodactyly, facial anomalies, and hypoplasia (Pena–Shokeir syndrome). *Handbook Clin Neurol* **45**(11):341–343, 1982.

27. Toriello HV et al: Sibs with the fetal akinesia sequence, fetal edema, and malformations: A new syndrome? *Am J Med Genet* **21**:271–277, 1985.

28. Williams J et al: Syndrome of camptodactyly, facial anomalies, and pulmonary hypoplasia. *J Pediatr* **93**:151–152, 1978.

29. Williams RS, Holmes LB: The syndrome of multiple ankyloses and facial anomalies. *Acta Neuropathol (Berlin)* **50**:175–179, 1980.

## Cerebro-oculo-facial-skeletal syndrome (COFS syndrome, Pena–Shokeir syndrome, type II)

In 1974, Pena and Shokeir (9) reported a disorder consisting of microcephaly, cataracts and other eye anomalies, characteristic facies, flexion contractures, and generalized osteoporosis. It probably was described earlier in sibs by Lowry et al (7) although we cannot completely exclude the possibility that the sibs had Cockayne syndrome (2).

Affected sibs and parental consanguinity indicate autosomal recessive inheritance (9,13). About 25 cases have been reported to date.

About half of the infants are small for dates. There is failure to thrive. All exhibit progressive postnatal growth deficiency and progressive demyelinization of the central nervous system. Average life span is about 4 years.

**Facies.** Head circumference is small. The nasal root is prominent and the pinnae are large. Almost all have microphthalmia and blepharophimosis. About 75% exhibit cataracts (6).

There is a prominent nasal root. The pinnae are large and the upper lip overhangs the lower one. The chin is small. The neck is not short (Fig. 18–4).

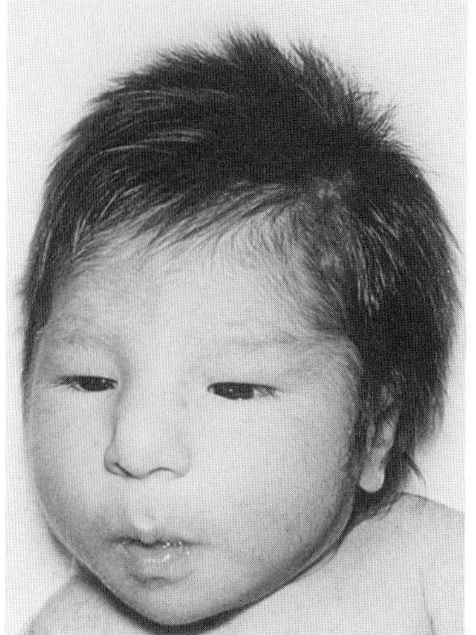

Fig. 18–4. *Cerebral-oculo-facial-skeletal syndrome.* Facies showing microcephaly, blepharophimosis, prominent nasal root, small mouth, micrognathia. (Courtesy of *SDJ Pena*, Winnipeg, Manitoba, Canada.)

**Musculoskeletal.** Kyphosis and/or scoliosis and camptodactyly with overlapping fingers are virtually constant features. Flexion contractures, especially of the elbows and knees, are common. About 60% manifest rocker-bottom feet due to vertical talus. Dislocated hips, acetabular dysplasia, coxa valga, and osteoporosis are also frequent features. The second metatarsal may be proximally placed due to hypoplasia of the second cuneiform bones (Fig. 18–5A,B).

**Central nervous system.** All exhibit microcephaly and mental retardation. Perhaps 40% of those autopsied have had hypoplasia or agenesis of the corpus callosum, cerebellar hypoplasia, and intracranial calcification (6,13).

**Other findings.** The nipples are widely set in about 50%. Single palmar creases have been seen in about 30% and longitudinal foot groove folding in 70%.

**Differential diagnosis.** There appears to be a form characterized by early death. The infants exhibit the facial and musculoskeletal anomalies of "classic" COFS syndrome. Low birthweight is common. Visceral anomalies include agenesis of the corpus callosum, cerebellar hypoplasia, renal hypoplasia or agenesis, horseshoe kidney, absence of hemidiaphragm, malrotation of colon, hypoplasia of spleen, and multiple foci of calcification of heart and kidneys (8,11). Whether this represents a different disorder or an allelic form is not known.

We are not entirely convinced that the COFS syndrome differs from the Bowen–Conradi syndrome found in Hutterites (1,4). The latter disorder is stated not to be associated with microphthalmia, blepharophimosis, cataracts, or large pinnae while COFS syndrome patients do not have dolichocephaly or cryptorchidism. There is otherwise a remarkable overlap in phenotypes.

Several authors have suggested that the COFS syndrome represents a severe allelic form of *Cockayne syndrome.* The usual pattern in Cockayne syndrome is normal growth until 2 years of age, then gradual delay in overall development, small head, and unsteady gait. Joint problems and kyphosis are mild. Arthrogryposis and *trisomy 18* must be excluded. Winter et al (15) noted an overlap with the *Neu–Laxova syndrome.*

**Laboratory aids.** Hwang et al (5) found extensive necrosis of iliac crest cartilage cells. However, necrotic chondrocytes can also be seen in various chondrodysplasias: *achondroplasia, diastrophic dysplasia,* and *Dyggve–Melchior–Clausen syndrome.*

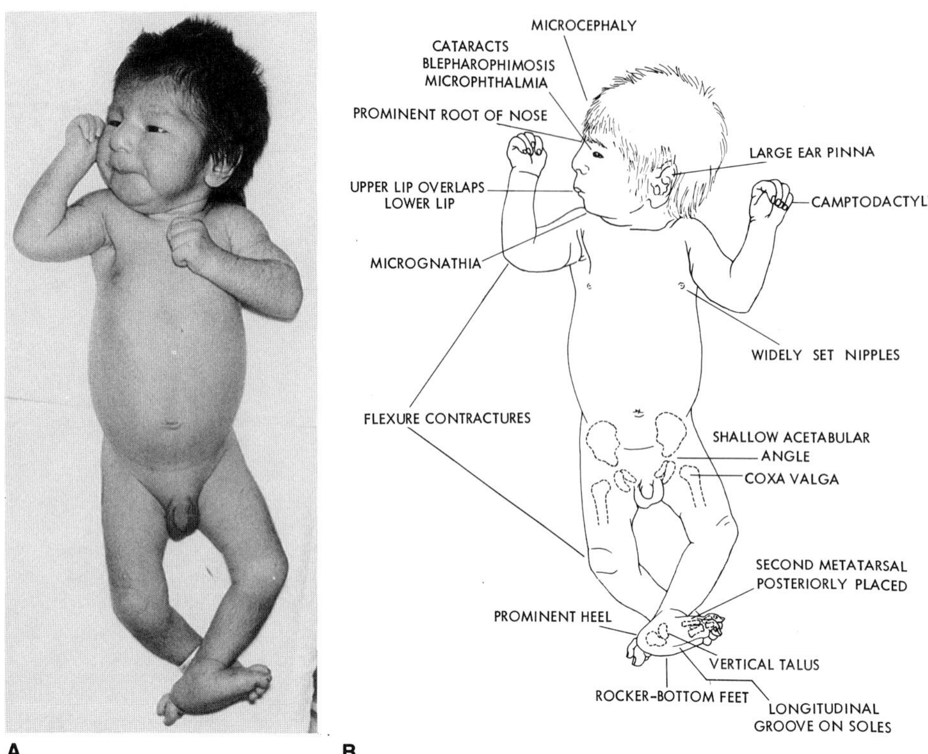


Fig. 18–5. *Cerebral-oculo-facial-skeletal syndrome.* (A) Note camptodactyly, widely set nipples, rocker-bottom feet. (B) Diagram depicting most important characteristics of COFS syndrome. [From *SDJ Pena et al,* Birth Defects **14**(6B):205, 1978.]

### References [Cerebro-oculo-facial-skeletal syndrome (COFS syndrome, Pena–Shokeir syndrome, type II)]

1. Bowen P, Conradi GJ: Syndrome of skeletal and genitourinary anomalies with unusual facies and failure to thrive in Hutterite sibs. *Birth Defects* **12**(6):101–108, 1976.
2. Dolman CL, Wright VJ: Necropsy of original cases of Lowry's syndrome. *J Med Genet* **15**:227–229, 1978.
3. Grizzard WS et al: The cerebro-oculo-facial syndrome. *Am J Ophthalmol* **89**:293–298, 1980.
4. Hunter AGW et al: The Bowen–Conradi syndrome in a highly lethal autosomal recessive syndrome of microcephaly, micrognathia, low birthweight and joint deformity. *Am J Med Genet* **3**:269–279, 1979.
5. Hwang WS et al: Chondro-osseous changes in the cerebro-oculo-facial-skeletal (COFS) syndrome. *J Pathol* **138**:33–40, 1982.
6. Linna SL et al: Intracranial calcifications in cerebro-oculofacio-skeletal (COFS) syndrome. *Pediat Radiol* **12**:28–30, 1982.
7. Lowry RB et al: Cataracts, microcephaly, kyphosis and limited joint movement in two siblings. *J Pediatr* **79**:282–284, 1982.
8. Lurie IW et al: Further evidence for the autosomal recessive inheritance of the COFS syndrome. *Clin Genet* **10**:343–346, 1976.
9. Pena SDJ, Shokeir MHK: Autosomal recessive cerebro-oculo-facio-skeletal (COFS) syndrome. *Clin Genet* **5**:285–293, 1974.
10. Pena SDJ et al: COFS syndrome revisited. *Birth Defects* **14**(6B):205–213, 1978.
11. Preus M, Fraser FC: The cerebro-oculo-facio-skeletal syndrome.*Clin Genet* **5**:194–297, 1974.
12. Preus M et al: Renal anomalies and oligohydramnios in the cerebro-oculo-facio-skeletal syndrome. *Am J Dis Child* **131**:62–64, 1977.
13. Scott-Emuakpor AB et al: A syndrome of microcephaly and cataracts in four siblings. A new genetic syndrome? *Am J Dis Child* **131**:167–169, 1977.
14. Surana RB et al: The cerebro-oculo-facio-skeletal syndrome. *Clin Genet* **13**:486–488, 1978.
15. Winter RM: Syndromes of microcephaly, microphthalmia, cataracts, and joint contractures. *J Med Genet* **18**:129–133, 1981.

## Neu–Laxova syndrome

Neu et al (10), in 1971, and Laxova et al (6), in 1972, independently reported sibs with a lethal syndrome of marked intrauterine growth retardation, marked microcephaly, characteristic facies and flexion deformities. At least 25 examples have been published. Cases resembling the COFS syndrome have been described (7,16). Both consanguinity (2,3,5,6,10,14,18) and affected sibs (5,6,10,12,18) have been reported. Inheritance clearly is autosomal recessive. It may have a higher frequency in Pakistanis (8).

A history of multiple spontaneous miscarriages and polyhydramnios is frequent. The placenta is remarkably small and there is a short umbilical cord (13). In some cases, the umbilical cord has only two vessels. The condition is characterized by severe intrauterine growth retardation. At full term the infant usually weighs between 850 and 1500 g. Crown–heel length ranges from 25 to 35 cm. Nearly all are stillborn or die within hours after birth. The longest survivor lived approximately 6 weeks.

Many of the findings such as polyhydramnios, intrauterine growth retardation, generalized edema, deformed swollen limbs, osteopenia, and short survival can be explained by protein loss through skin fissures. Some structural deformities may be secondary to intrauterine mobility, similar to those seen in *restrictive dermopathy.*

**Facies.** Facial appearance is striking. There is microcephaly (OFC at birth 23–26 cm) and a markedly sloping forehead. There appears to be premature closure of sutures and fontanels, probably secondary to small brain size (5–7,9–11,13). The nose is flat and simple with a broad nasal bridge. The nares are usually small or, rarely, almost absent. The cheeks are full. The upper and lower eyelids exhibit extreme ectropion and are very hypoplastic, giving the impression that they are completely missing. Apparent hypertelorism and microphthalmia have been noted in a few cases (10,12,20). The pinnae are low set and very often large and deformed. The lips are everted. Bilateral cleft lip–palate has been seen (9,16). Micrognathia has been present in about one half the cases. The head appears to sit directly upon the shoulders (Fig. 18–6A,B).

**Musculoskeletal alterations.** Most patients have short contracted limbs. The hands and feet are swollen and the digits hypoplastic, with or without nails. Frequently the fingers overlap. All have rocker-bottom feet. Syndactyly of fingers and toes is extremely common. A

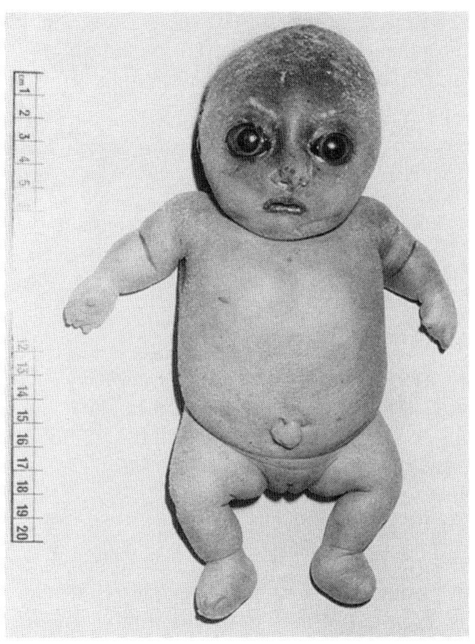


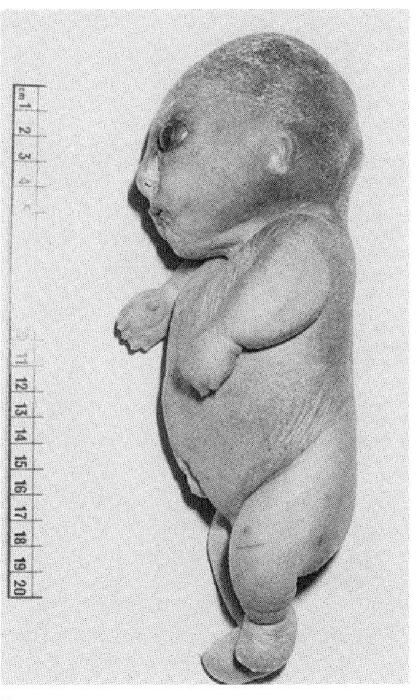


Fig. 18–6. *Neu–Laxova Syndrome.* (A,B) Characteristic facies with globular head, widely opened palpebral fissures, protuberant globes, squashed nasal tip, rudimentary pinnae, small mouth, and short neck. Limbs are short; hands and feet are swollen with absent or hypoplastic digits without nails. (From *CE Scott et al,* Am J Med Genet **9**:165, 1981.)

detailed anatomic study of the upper limbs was carried out by Shved et al (15).

The bones have a stick-like appearance due to undertubulation. There is underossification of the bones of the hands and feet. The ilia are dysplastic and there is poor ossification of pubic bones with lack of ossification of sacrum and coccyx. Possibly the condition termed by Spranger et al (16) as "cerebroarthrodigital syndrome" is not specific for a single disorder since their patient 3 appears to have Neu–Laxova syndrome (13).

**Central nervous system.** Marked microcephaly with variable anomalies such as brain atrophy, agenesis of the corpus callosum, hydrocephaly, lissencephaly, and holoprosencephaly, agenesis of vermis, hypoplasia of corpora quadrigemina, and hypoplasia of the cerebellum have been found (7,8,10,10a). There are reduced numbers of nuclei in the pons, cerebellum, and brainstem. The cerebellar cortex has a two layer lamination. The anterior and lateral columns of the spinal cord are absent (7) (Fig. 18–7).

Fig. 18–7. *Neu–Laxova syndrome.* Forebrain grossly maldeveloped with bilaterally symmetrical unconvoluted and unilobed cerebral masses. (From *CE Scott et al,* Am J Med Genet **9**:165, 1981.)

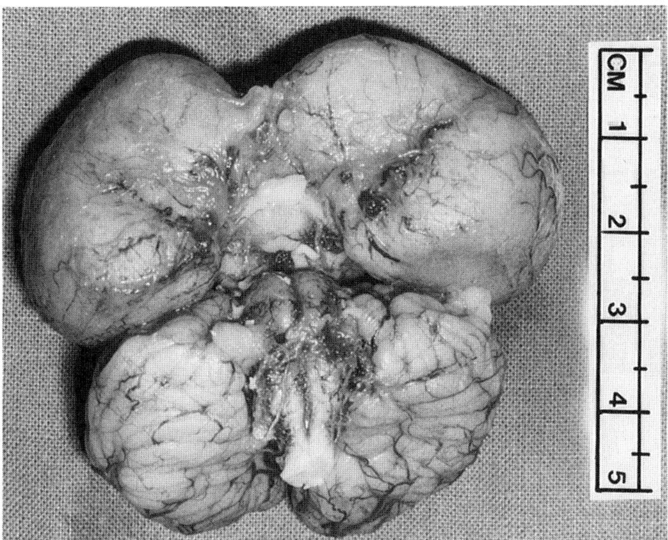


**Urogenital.** Both unilateral renal agenesis and hypoplasia of external genitalia have been reported (6,7,12,13).

**Skin.** Scaly, yellowish ichthyotic skin has been described in most affected infants. There is massive accumulation of adipose tissue beneath the epidermis, surrounding atrophic muscle bundles (5).

**Cardiac anomalies.** Various anomalies have included ASD, VSD, PDA, and transposition of great vessels.

**Oral manifestations.** Cleft lip–palate has been reported in a few cases (5,11,12,19).

**Differential diagnosis.** There is some similarity (intrauterine growth retardation, lissencephaly, agenesis of corpus callosum, cataracts, joint contractures, overlapping fingers) to patients with *cerebro-oculo-facial* (COFS) *syndrome* that die at an early age (19). Silengo et al (16) suggested that they are the same disorder. Collodion baby and severe congenital ichthyosis must be excluded (4,14).

The skeletal alterations as noted above are seen in several clinically different disorders (13). Perhaps the common denominator is a primary neural tube–neural crest dysplasia.

*Restrictive dermopathy* and harlequin fetus should be excluded.

**Laboratory aids.** The disorder has been diagnosed prenatally by ultrasonography at 6–8 weeks for accurate dating, at 12–16 weeks for reduced fetal limb movement, and at 16–24 weeks for facial and skeletal abnormalities, the detection of intrauterine growth retardation, and polyhydramnios (9,12a,18).

## References (Neu–Laxova syndrome)


1. Curry CJR: Further comments on the Neu–Laxova syndrome. *Am J Med Genet* **13**:441–444, 1982.
2. Ejeckam GG et al: Neu–Laxova syndrome: Report of two cases. *Pediatr Pathol* **5**:295–306, 1986.
3. Fitch N et al: The Neu–Laxova syndrome: Comments on syndrome identification. *Am J Med Genet* **13**:445–452, 1982.
4. Goldsmith LA: The ichthyoses. *Prog Med Genet* **1**:185–210, 1976.
5. Karimi-Nejad MH et al: Neu–Laxova syndrome: Report of a case and comments. *Am J Med Genet* **28**:17–24, 1987.
6. Laxova R et al: Further example of a lethal autosomal recessive condition in sibs. *J Ment Defic Res* **16**:139–143, 1972.
7. Lazjuk GI et al: Brief clinical observations: The Neu–Laxova syndrome—a distinct entity. *Am J Med Genet* **3**:261–267, 1979.

8. Mueller RF et al: Neu–Laxova syndrome: Two further case reports and comments on proposed subclassification. *Am J Med Genet* **16**:645–649, 1983.

9. Muller LM et al: A case of the Neu–Laxova syndrome: Prenatal ultrasonographic monitoring in the third trimester and the histopathological findings. *Am J Med Genet* **26**:421–429, 1987.

10. Neu RL et al: A lethal syndrome of microcephaly with multiple congenital anomalies in three siblings. *Pediatrics* **47**:611–612, 1971.

10a. Ostrovskaya TI, Lazjuk GI: Cerebral abnormalities in the Neu–Laxova syndrome. *Am J Med Genet* **30**:747–756, 1988.

11. Paes B et al: Craniofacial and visceral defects with congenital contractures: A variant of the Neu–Laxova syndrome. *J Clin Dysmorphol* **3**(1):21–25, 1985.

12. Póvyšilová V et al: Letálni syndrom mnohočetných malformaci u tři sourozencú. sourozencú. *Čs Pediatr* **31**:190–194, 1976.

12a. Russo R et al: Neu–Laxova syndrome: Pathological, radiological, and prenatal findings in a stillborn female. *Am J Med Genet* **32**:136–139, 1989.

13. Scott CI et al: Comments on the Neu–Laxova syndrome and CAD complex. *Am J Med Genet* **9**:165–175, 1981.

14. Seemanová E, Rudolph R: The Neu–Laxova syndrome. *Am J Med Genet* **20**:13–16, 1985.

15. Shved IA et al: Elaboration of the phenotypic changes of the upper limbs in the Neu–Laxova syndrome. *Am J Med Genet* **20**:1–11, 1985.

16. Silengo MC et al: The Neu–COFS (cerebro-oculo-facio-skeletal) syndrome. *Clin Genet* **25**:201–204, 1984.

17. Spranger JW et al: Cerebroarthrodigital syndrome: A newly recognized formal genesis syndrome in three patients with apparent arthromyodysplasia and sacral agenesis and digital dysplasia. *Am J Med Genet* **5**:13–24, 1980.

18. Tolmie JL et al: The Neu–Laxova in female sibs: Clinical and pathological features with prenatal diagnosis in the second sib. *Am J Med Genet* **27**:175–182, 1987.

19. Turkel SB et al: Additional manifestations of the Neu–Laxova syndrome. *J Med Genet* **20**:227–229, 1983.

20. Winter RM et al: Syndromes of microcephaly, micrognathia, cataracts, and joint contractures. *J Med Genet* **18**:129–133, 1981.

## Multiple pterygium syndrome

Multiple pterygium syndrome was buried in the early literature, most often under the nonspecific diagnosis of arthrogryposis (22,28,40) or *Noonan syndrome* (14,37). The syndrome consists of growth retardation, multiple pterygia involving the neck, fingers, and antecubital, popliteal, and intercrural areas, and cleft palate. Credit should go to Matolcsy (30) for clearly defining the entity. Approximately 60 cases have been reported to date. Thompson et al (42) have elegantly reviewed the evolution of the phenotype.

Affected sibs (3,6,13,24,25,30–32,40–42) as well as parental consanguinity (14,17,42) have been reported. The syndrome has autosomal recessive inheritance. Recurrent chest infections, probably the results of congenital kyphoscoliosis, may result in early death (6,15,28,42).

**Facies.** The palpebral fissures have mild downslanting in 85%. There is mild ptosis of the lids in 30% and epicanthal folds in 75% (Fig. 18–8A,B). About one-half of the patients have puffiness about the eyes. The mandible is frequently small. In some patients there is a central neck web (vide infra).

**Musculoskeletal alteration.** Growth is usually retarded below the third percentile, the patient's adult height rarely exceeding 135 cm.

The popliteal pterygia, noted in 60%, may markedly inhibit walking and, even if only mild, produce bizarre stance and gait. Pterygia occur in the cervical area in 85%. They may resemble those seen in Turner and Noonan syndromes. Rarely, they completely surround the neck (Fig. 18–9). In about 15%, a pterygium extends from the chin to the sternum (16,33,45). In 50%, there is mild soft-tissue syndactyly between the fingers.

Flexion deformity of the digits, the thumbs being flexed and apposed, occurs in 70% (1,26,32,38) (Fig. 18–10). Talipes calcaneovalgus, either unilateral or bilateral, and rocker-bottom feet have been noted in 65% (1,5). Kyphoscoliosis, often congenital, and other vertebral anomalies (failure of posterior fusion of vertebrae, fusion of cervical vertebrae) are found in 60%. The axillae and antecubital fossae are the sites of webs in about 50%. Intercrural webs are found in 40% of both males and females. In the presence of an intercrural web, the penis and scrotum are retropositioned. This may be associated

Fig. 18–8. *Multiple pterygium syndrome.* (A,B) Note antimongoloid obliquity of palpebral fissures, pterygium colli, low-set posterior hairline. [From *C Scott,* Birth Defects **5**(2):231, 1969.]

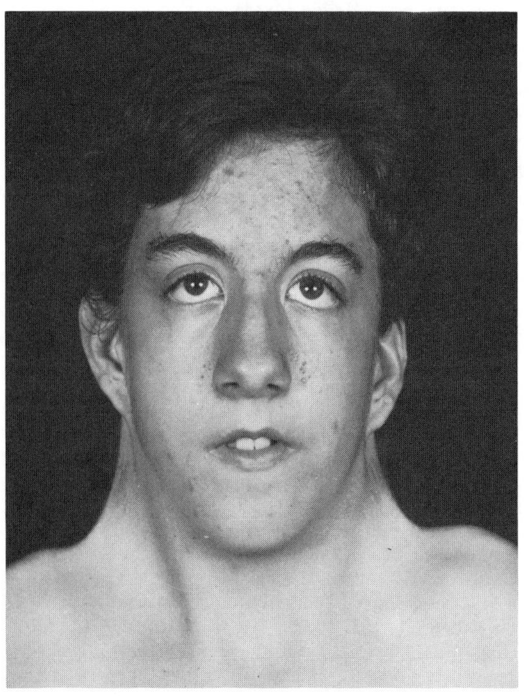

A

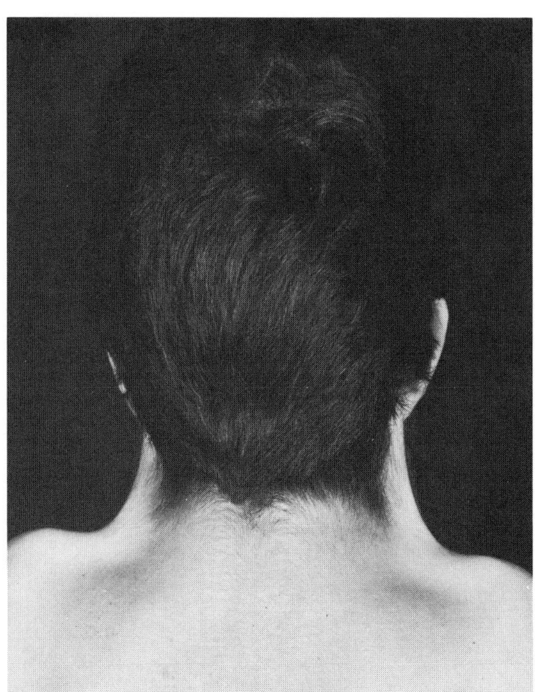

B

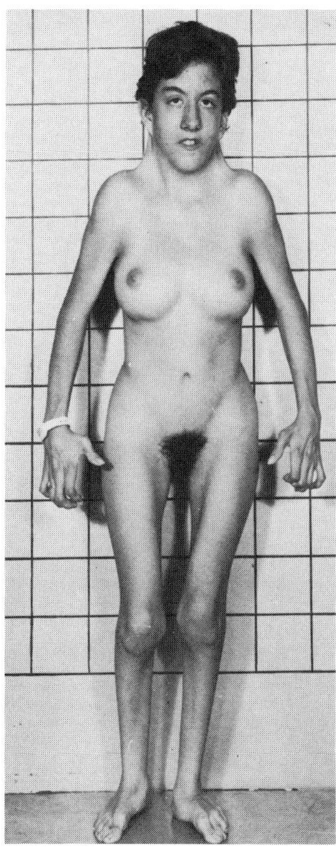

Fig. 18–9. *Multiple pterygium syndrome.* Pterygia of neck, axillas, and popliteal areas. [From *C Scott,* Birth Defects **5**(2):231, 1969.]

with cryptorchidism and/or inguinal hernia (30,32). In females, the labia majora may be absent (1,44).

Radiographic changes include vertebral segmentation anomalies such as fusion of cervical vertebrae and rib anomalies (25%), tall narrow vertebral bodies (30%), vertical talus with talipes calcaneovalgus or equinovarus (65%), and camptodactyly of fingers and toes (65%) (42). The patellae may be absent.

**Oral manifestations.** Cleft palate has been found in about 35% (2,5,6,9,24,27,35), and a hearing deficit has been noted in several patients (35a). R Hennekam (personal communication, 1989) told us of three affected sibs whose tongues had flattened spoonlike tips.

**Differential diagnosis.** The syndrome should clearly be distinguished from autosomal dominantly inherited *popliteal pterygium syndrome.* Although both share popliteal pterygium, the cervical, axillary, and elbow webbing as well as the finger contractures and vertebral anomalies and difference in inheritance pattern should furnish a clear distinction. There is also overlap with the *femoral hypoplasia-unusual facies syndrome.*

Fig. 18–10. *Multiple pterygium syndrome.* Marked flexion deformity of digits. [From *C Scott,* Birth Defects **5**(2): 231, 1969.]

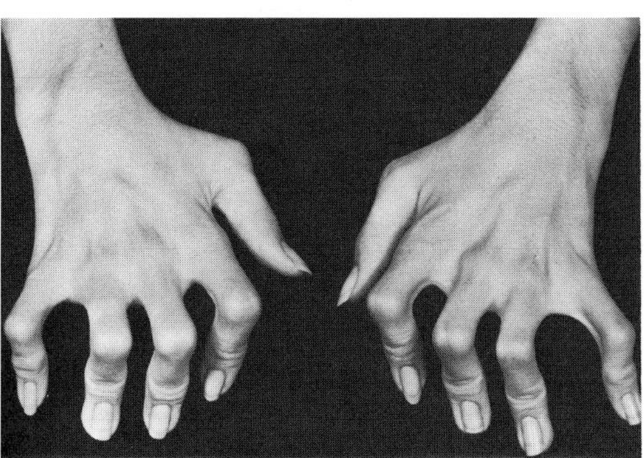

Pterygia about the neck may be seen in *Turner syndrome, LEOPARD syndrome, Noonan syndrome,* and in some cases of *craniocarpotarsal dysplasia.* Pashayan et al (34) described multiple pterygium syndrome in a patient with mosaic *Klinefelter syndrome.* The patient also exhibited cataract and glaucoma. Some patients with *fetal akinesia sequence* have pterygium formation.

A dominantly inherited syndrome of multiple pterygia, ptosis, and skeletal abnormalities has been reported by Frias et al (10). Short stature was not present. The skeletal anomalies included malsegmentation of vertebrae, pelvic dysplasia, and some carpal and tarsal bone fusion.

Kawira and Bender (23) described a dominantly inherited disorder of short stature, cervical vertebral anomalies, scoliosis, nuchal and axillary pterygia, and distal arthrogryposis.

A new syndrome of severe mental retardation, postnatal growth retardation, multiple pterygia with distal muscular wasting, hypogonadism, and characteristic facies (trigonocephaly, frontal bossing, epicanthal folds, low-set angulated pinnae, microretrognathia) has been described (17a, 37a). Cleft palate was noted in two of four children.

Four or more lethal autosomal recessively inherited multiple pterygium syndromes have been reported. The pterygia may result from embryonic onset fetal akinesia (8). All are characterized by death or miscarriage during the second trimester and by the presence of multiple congenital contractures with skin webs across joints, apparent hypertelorism, cystic hygromatous masses in the neck with residual edema and loose skin, thin crowded ribs, and hypoplastic lungs and hearts. The pregnancies frequently are complicated by polyhydramnios. Cleft palate may be present. The conditions may not always be diagnosable prenatally (20).

Distinction among the types is based on age of onset of intrauterine growth retardation, the extent of neck swellings, and whether and where bony fusions and modeling errors of the bones occur. Hall (15) presented a very lucid discussion of the problem. Martin et al (29) have denied that such division can be made.

The first of these disorders was described by Gillin and Pryse-David (12), Isaacson et al (21), and Fryns et al (11) (Fig. 18–11A). It is characterized by intrauterine growth retardation that is recognized at birth. There are hypertelorism, epicanthal folds, short flat nose, tented upper lip, long philtrum, cleft palate, micrognathia, mild neck edema, and small chest. The abdominal musculature is decreased and there may be genital anomalies and atresia of the gut. The pterygia are tightly flexed. There is no bony fusion of the extremities.

The second form, illustrated by cases 2 and 6 of Chen et al (7) and by Serville et al (39), has short broad limbs and hypertelorism. There is marked cystic hygroma formation that extends from the neck to the mid-back (Fig. 18–11B,C). Generalized scalp edema is also evident. Intrauterine growth retardation is recognized at 20–30 weeks of gestation. Often there is hydramnios and swollen placenta. Fusion of bones (humerus–ulna, radius–ulna, spinous processes of vertebrae, epiphyseal cartilages of long bones) is prominent.

The third form [7(cases 4,5),44] is characterized by hypertelorism, upward slanting palpebral fissures, open eyes, open mouth, hypoplastic nose, absent mandibular angle, micrognathia, and redundant skin at the nape. Some have cleft lip and/or cleft palate. The legs are thin with decreased muscle mass. There are fused long bone cartilages and failure of normal modeling of the femur and tibia. Intrauterine growth retardation is recognized at 19–25 weeks.

The fourth form (18,19), so far limited to fetuses from Finland, has intrauterine growth retardation, fetal hydrops, multiple contractures, and abnormal facies, especially micrognathia (Fig. 18–12A). Pulmonary hypoplasia and muscular atrophy, thin tubular bones, and paucity of anterior horn cells were evident at autopsy (Fig. 18–12B). A possible fifth entity was reported by Robinson et al (36). An X-linked recessive form was characterized by defectively modelled long bones and broad ribs, clavicles, scapulae, and cystic hygroma (43).

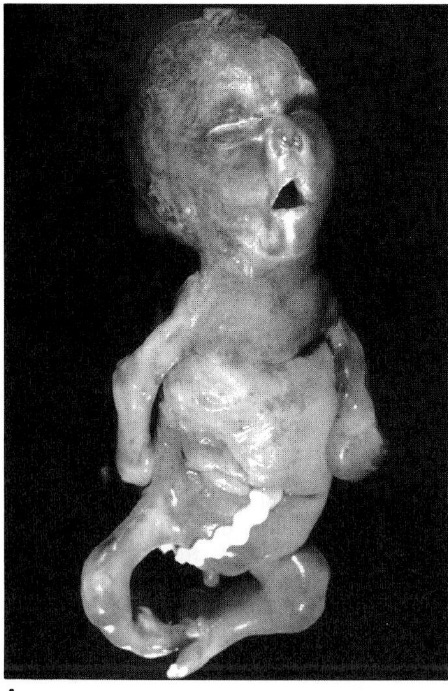
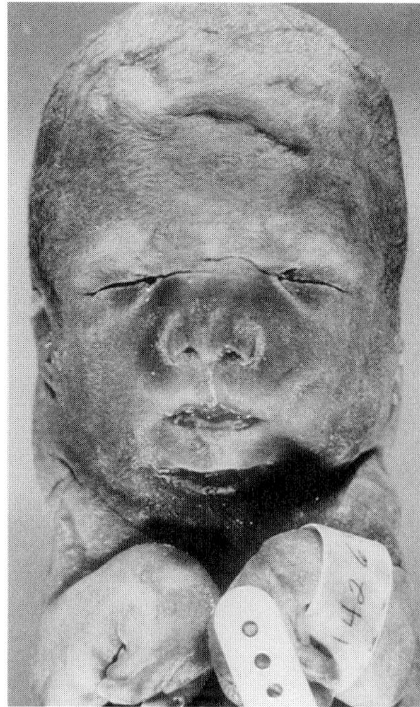
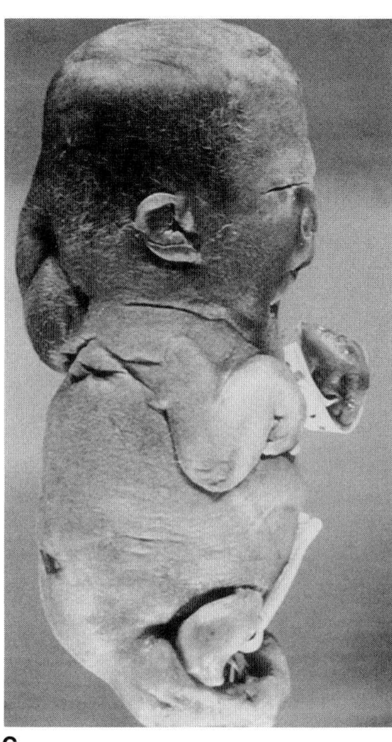

A B C

Fig. 18–11. *Multiple pterygium syndrome.* (A) Eighteen-week fetus showing hydrops, nuchal edema, flat upturned nose, long philtrum, gaping mouth, receding chin and multiple pterygia. (B,C) Note cystic hygromas, small mouth, flat nose, talipes, and hand contractures. (A from *G Isaacson et al,* Am J Med Genet **17:**835, 1984; B and C from *H Chen et al,* Am J Med Genet **17:**809, 1984.)

Prenatal diagnosis is possible after the sixteenth week of pregnancy with ultrasound because of cystic hygroma and hydrops, which will exclude *fetal akinesia sequence* that shares many clinical features (4,28a). Infants with the latter often survive the third trimester or are born alive. The placenta is small and the infants exhibit ulnar deviation of the hands and talipes equinovarus.

### References (Multiple pterygium syndrome)

1. Aarskog D: Pterygium syndrome. *Birth Defects* **7**(6):232–234, 1971.

2. Addison A, Webb PJ: Flexion contractures of the knee associated with popliteal webbing. *J Pediatr Orthop* **3**:376–379, 1983.

3. Beckerman RC, Buchino JJ: Arthrogryposis multiplex congenita as part of an inherited symptom complex. *Pediatrics* **61**:417–422, 1978.

4. Bendon R et al: Prenatal diagnosis of arthrogryposis multiplex congenita. *J Pediatr* **111**:942–947, 1987.

5. Bettman AG: Congenital bands about the shoulder girdle. *Plast Reconstr Surg* **1**:205–215, 1946.

6. Chen H et al: Multiple pterygium syndrome. *Am J Med Genet* **7**:91–102, 1980.

7. Chen H et al: Syndrome of multiple pterygia, camptodactyly, facial anomalies, hypoplastic lungs and heart, cystic hygroma, and skeletal anomalies. *Am J Med Genet* **17**:809–826, 1984.

8. Davis JE, Kalousek DK: Fetal akinesia deformation sequence in previable fetuses. *Am J Med Genet* **29**:77–88, 1988.

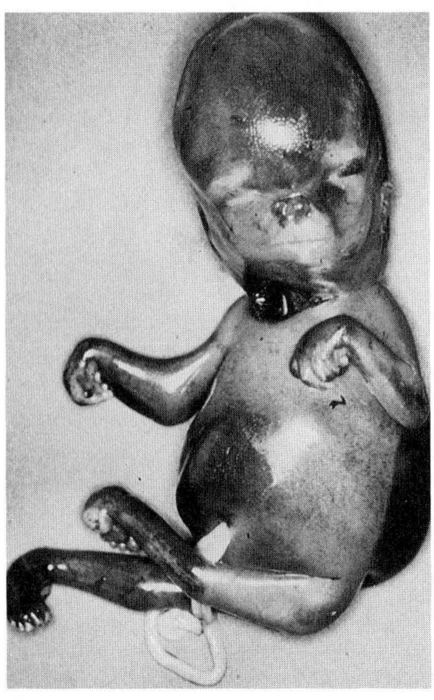
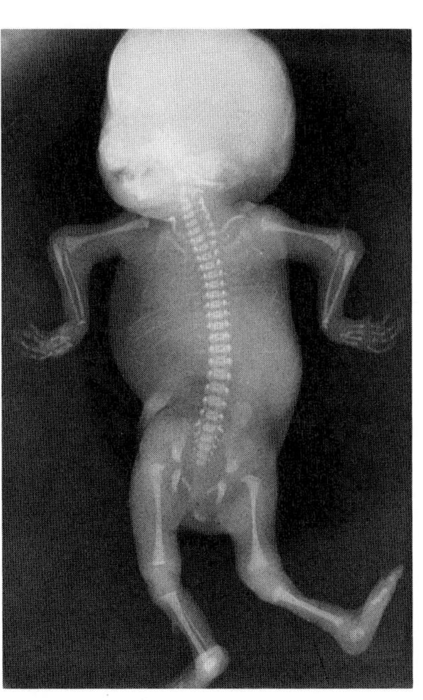

A B

Fig. 18–12. *Multiple pterygium syndrome.* (A) Fetal hydrops, multiple pterygia, and talipes. (B) Extremely thin fishbone-like ribs, narrow clavicles, overtubulation of long bones. (From *R Herva et al,* Am J Med Genet **20:**431, 1985.)

9. Escobar V et al: Multiple pterygium syndrome. *Am J Dis Child* **132**:609–611, 1978.

10. Frias JL et al: An autosomal dominant syndrome of multiple pterygium, ptosis and skeletal abnormalities (#41). *Int Conf Birth Defects (Vienna),* 1973, Excerpta Medica p 19.

11. Fryns JP et al: Cystic hygroma and multiple pterygium syndrome. *Ann Génét* **27**:252–253, 1984.

12. Gillin ME, Pryse-Davis J: Pterygium syndrome. *J Med Genet* **13**:249–251, 1976.

13. Golden RL, Lakin H: The forme fruste in Marfan's syndrome. *N Engl J Med* **260**:797–801, 1959.

14. Guinand-Doniol J: Observations nouvelles sur le status Bonnevie-Ullrich. *Ann Paediat* **169**:317–324, 1947 (Case 1).

15. Hall JG: The lethal multiple pterygium syndrome. *Am J Med Genet* **17**:803–807, 1984.

16. Hall JG et al: The distal arthrogryposes: Delineation of new entities—review and nosologic discussion. *Am J Med Genet* **11**:185–239, 1982.

17. Hall JG et al: Limb pterygium syndromes: A review and report of eleven patients. *Am J Med Genet* **12**:377–409, 1982.

17a. Haspelagh M et al: Mental retardation with pterygia, shortness and distinct facial appearance. A new MCA/MR syndrome. *Clin Genet* **28**:550–555, 1985.

18. Herva R et al: A lethal autosomal recessive syndrome of multiple congenital contractures. *Am J Med Genet* **20**:431–440, 1985.

19. Herva R et al: A syndrome of multiple congenital contractures: Neuropathological analysis on five fetal cases. *Am J Med Genet* **29**:67–76, 1988.

20. Hogge WA et al: The lethal multiple pterygium syndromes: Is prenatal detection possible? *Am J Med Genet* **20**:441–442, 1985.

21. Isaacson G et al: Lethal multiple pterygium syndrome in an 18-week fetus with hydrops. *Am J Med Genet* **17**:835–839, 1984.

22. Jahn M et al: Arthrogryposis multiplex congenita in a one-month-old child. *Čs Pediatr* **20**:150–153, 1965.

23. Kawira EL, Bender HA: An unusual distal arthrogryposis. *Am J Med Genet* **20**:425–429, 1985.

23a. Kok JG et al: Multiple pterygium syndrome. *Clin Neurol Neurosurg* **86**:101–105, 1984.

24. Kopits E: Die als "Flughaut" bezeichneten Missbildungen und deren operative Behandlung. *Arch Orthop Unfall-Chir* **37**:538–549, 1937.

25. Krieger I, Espiritu CE: Arthrogryposis multiplex congenita and the Turner phenotype. *Am J Dis Child* **123**:263–268, 1972 (Cases 3,4).

26. Kroll W et al: Beitrag zur Pterygoathromyodysplasia congenita (Rossi-Syndrom). *Z Kinderheilk* **116**:263–268, 1974.

27. Lamy M et al: Le pterygium colli congenitum et le syndrome de Bonnevie-Ullrich. *Sem Hôp Paris* **50**:2356–2360, 1974.

28. Lang K et al: Beitrag zum Bilde der Pterygomyodysplasia arthrogrypotica generalis. *Mschr Kinderheilk* **108**:248–252, 1959.

28a. Lockwood C et al: The prenatal sonographic diagnosis of lethal multiple pterygium syndrome. A heritable cause of recurrent abortion. *Am J Obstet Gynecol* **159**:474–476, 1988.

29. Martin NJ et al: Lethal multiple pterygium syndrome: Three consecutive cases in one family. *Am J Med Genet* **24**:295–304, 1986.

30. Matolcsy T: Über die chirurgische Behandlung der angeborene Flughaut. *Arch Klin Chir* **185**:675–681, 1936.

31. Naguib KK: Multiple pterygium syndrome in five Arab sibs. *Ann Génét* **30**:122–125, 1978.

32. Norum RA: Pterygium syndrome in three children in a recessive pedigree pattern. *Birth Defects* **5**(2):233–235, 1969.

33. O'Brien B et al: Multiple congenital skin webbing with cutis laxa. *Br J Plast Surg* **23**:329–336, 1970.

34. Pashayan H et al: Bilateral aniridia, multiple webs and severe mental retardation in a 47,XXY/48,XXXY mosaic. *Clin Genet* **4**:125–129, 1973.

35. Penchaszadeh VB, Salzberg B: Multiple pterygium syndrome. *J Med Genet* **18**:451–455, 1981.

35a. Ramer JC et al: Multiple pterygium syndrome. *Am J Dis Child* **142**:794–798, 1988.

36. Robinson LK et al: Multiple pterygium syndrome: A case complicated by malignant hyperthermia. *Clin Genet* **32**:5–9, 1987.

37. Rossi E, Caflisch A: Le syndrome du pterygium: status Bonnvie-Ullrich, dystrophia brevicolli congenita, syndrome de Turner et arthromyodysplasia congenita. *Helv Paediatr Acta* **6**:119–148, 1951.

37a. Schrander-Stumpel C et al: Mental retardation with pterygia, shortness and distinct facial appearance. Confirmation of a new MCA/MR syndrome. *Clin Genet* **34**:279–281, 1988.

38. Scott CI: Pterygium syndrome. *Birth Defects* **5**(2):231–232, 1969.

39. Serville F et al: Syndrome léthal des pterygium multiples et hygroma de la nuque. *J Génét Hum* **35**:409–414, 1987.

40. Srivastava R: Arthrogryposis multiplex congenita—case report of two siblings. *Clin Pediatr* **7**:691–694, 1968.

41. Stoll C et al: Familial pterygium syndrome. *Clin Genet* **18**:317–320, 1980.

42. Thompson EM et al: Multiple pterygium syndrome: Evolution of the phenotype. *J Med Genet* **24**:733–749, 1987.

43. Tolmie JL et al: A lethal multiple pterygium syndrome with apparent X-linked inheritance. *Am J Med Genet* **27**:913–920, 1987.

44. Van Regemorter N et al: Lethal multiple pterygium syndrome. *Am J Med Genet* **17**:827–834, 1984.

45. Zuccati CG et al: La sindrome degli pterigi multiple. *Pathologica* **73**:23–29, 1981.

## Popliteal pterygium syndrome (facio-genito-popliteal syndrome)

The first recorded case would appear to be that of Trélat (35) in 1869. Fewer than 80 cases have been described to date. There have been many extensive reviews of the syndrome (7,10,14,16,28).Prevalence is about 1 per 300,000 births. Rintala and Lahti (30) suggested the term "facio-genito-popliteal syndrome."

Although most cases have been isolated examples, probably because of hypoplastic genitalia. The condition has been transmitted from an affected parent to one or more children (10,13,18–22,28,29). There have been affected sibs with normal parents (7). Nevertheless, massive evidence suggests autosomal dominant inheritance with variable expressivity and incomplete penetrance (8b,10,14,18,28,32a).

**Facies.** The most obvious facial alterations are cleft lip and/or cleft palate and filiform adhesions between the eyelids (20%) (2a,10,16,19,21a,26,34a).

**Cutaneous and musculoskeletal anomalies.** Certainly the most striking and most common (70%) component, the popliteal web, extends from the heel to the ischial tuberosity, limiting extension and abduction as well as rotation of the leg (Fig. 18–13A,B). In all but a few cases (12,21,28,30,34,37), these webs have been bilateral. Along the free edge of the pterygium runs a hard, inelastic, subcutaneous cord or fibrous band or calcaneoischiadicus muscle (2). The sciatic nerve lies free within the web, deep to the fibrous band about halfway between the free edge and the apex, being covered by a fibromuscular septum. Special caution must be taken in repair of the skin fold. The popliteal vessels are normally situated deep in the popliteal space. In many cases, muscle groups are absent or muscle insertions are abnormal.

About 20% exhibit hypoplasia or agenesis of digits (8,15,17,18,37), 20% have varus or valgus deformities of the foot (6,12,15,18,34,35), and over 50% manifest variable syndactyly of the second to fifth toes, less often the fingers (4,6–8,10,11,15,17,27–29,32,34,37). Ectrodactyly may rarely occur (2a). Spina bifida occulta has been documented in 20% and scoliosis or lordosis in 10% (15,21,37). Bipartite or absent patella has been documented rarely (28,37).

The skin over the hallux may have a somewhat pyramidal form, one vertex extending over the nail in about 40%. If present in a child with cleft lip and/or palate, *even in the absence of a distinct popliteal pterygium,* we believe that this anomaly is sufficiently distinct to make the diagnosis. The toenails, most often the second, are hypoplastic and triangular in about 30% (13,14,17,19,25,27,32,34a) (Fig. 18–13C).

**Genitourinary system.** Anomalies in the male have included cryptorchidism in 40% (1,10,13,32), absent, cleft, or ectopic scrotum in 35% (4,6,15,27,30,32), small penis in 40%, and inguinal hernia in 10% (14,23,32,34) (Fig. 18–14B). In the female, absence or displacement of the labia majora in 60% (1,12,15,18,28,32) and enlarged clitoris in 20% (15,18,20,33) have been noted (Fig. 18–14A). Hypoplastic uterus has also been described (15).

An intercrural pterygium extending between the thighs, ventral to the anus, is found in about 20% (8,11,14) (Fig. 18–14C).

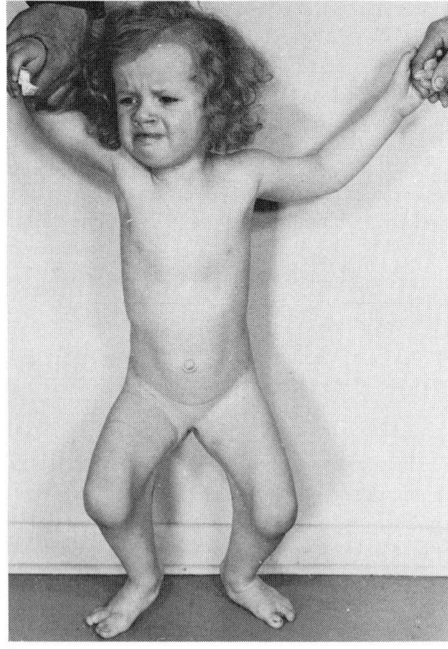

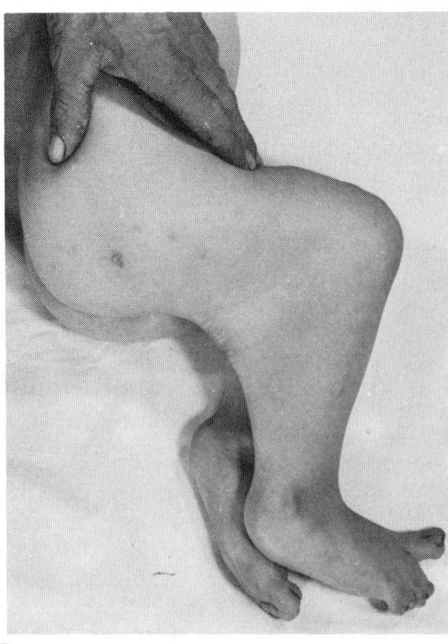

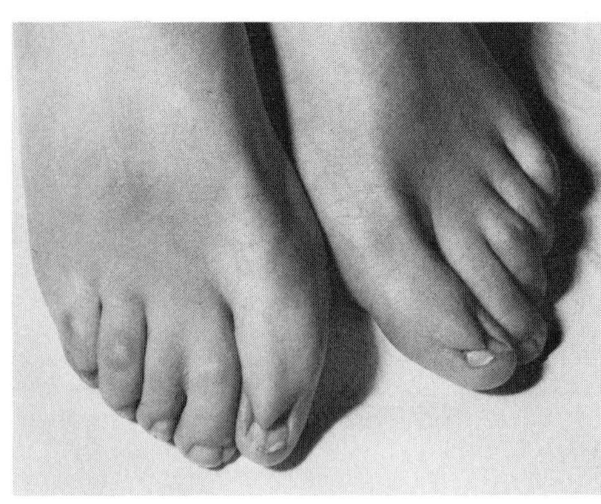

A B C

Fig. 18–13. *Popliteal pterygium syndrome.* (A) Child with repaired bilateral cleft lip–palate, bilateral popliteal pterygia, and hypoplasia of external genitalia. (B) Side view showing popliteal pterygium. Sciatic nerve is located in web under free margin. (C) Pyramidal skin folds extending to free edge of hallucal nails. (A,B from *G Dahman,* Z Orthopäd **95**:112, 1962; C from *D Klein* J Génét Hum **11**:65, 1962.)

**Oral manifestations.** Cleft lip and/or cleft palate have been reported in about 85%. Pits or sinuses of the lower lip (10,11,14–21a, 28,29,35) have been noted in about 60% (Fig. 18–15A). Congenital bands or threads of mucous membrane extending between the jaws (syngnathia) have been reported in about 35% (2–4,6,8b,17,18,21a, 27,28,30,34a,38) (Fig. 18–15B).

**Differential diagnosis.** The popliteal pterygia make this syndrome distinctive. It is possible that the patients described by Neuman and Schulman (24) with cleft palate, lip pits, oral mucosal adhesions, and ankyloblepharon represent incomplete expression of the syndrome. Lip pits and ankyloblepharon may each occur as an isolated phenomenon or in combination with cleft lip and/or cleft palate. Various extreme types of bifid scrotum may occur as isolated anomalies. The *Hay–Wells syndrome (AEC)* includes both ankyloblepharon and cleft lip or cleft palate. Pterygia of the axilla, neck and antecubital, and popliteal areas constitute the *multiple pterygium syndrome.* Pterygia-like alterations of the lower extremities occur as a part of the caudal regression complex.

A possibly separate disorder, *popliteal pterygia with ectodermal defects,* has been reported (23,31). In addition to the stigmata found in the classic autosomal dominant form of the disorder, these patients have hypoplastic or aplastic thumbs, vertebral anomalies, and various ectodermal abnormalities (subungual keratoses, striated nails, supernumerary nipples, eczematous skin changes and short woolly hair). Inheritance may be autosomal recessive (31).

Bartsocas and Papas (5) and other authors (8a,9,16,25) described a lethal autosomal recessively inherited popliteal pterygium syndrome. Findings included low birthweight, mental retardation, mild growth retardation, microcephaly, filiform adhesions of eyelids, corneal ulceration, hypoplastic nasal tip, microstomia, cleft lip–palate, soft-tissue syngnathia, micrognathia, supernumerary nipples, aplastic labia majora, bicornuate uterus, and lanugo hair. Anomalies of the extremities comprised popliteal pterygia, pes equinovarus, hypoplastic or absent phalanges, syndactyly of fingers and toes, and synostosis of hand bones. There were no vertebral anomalies.

### References [Popliteal pterygium syndrome (facio-genito-popliteal syndrome)]


1. Aberle-Horstenegg A: Flughautbildung zwischen Ober-und Unterschenkel mit abnorme Muskelbildung (M. ischiosuralis). *Z Orthopäd* **67**:21–29, 1937.
2. Addison A, Webb PJ: Flexion contractures of the knee associated with popliteal webbing. *J Pediatr Orthoped* **3**:376–379, 1983 (Cases 3–5).


Fig. 18–14. *Popliteal pterygium syndrome.* (A) Agenesis of labia major and enlarged clitoris. (B) Testicle displaced to upper right thigh. (C) Intercrural pterygium. (B from *A Rintala* and *A Lahti,* Scand J Plast Reconstr Surg **4**:67, 1970.)

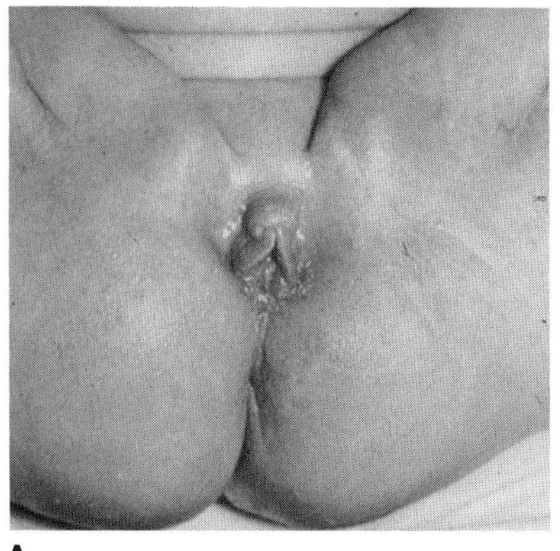

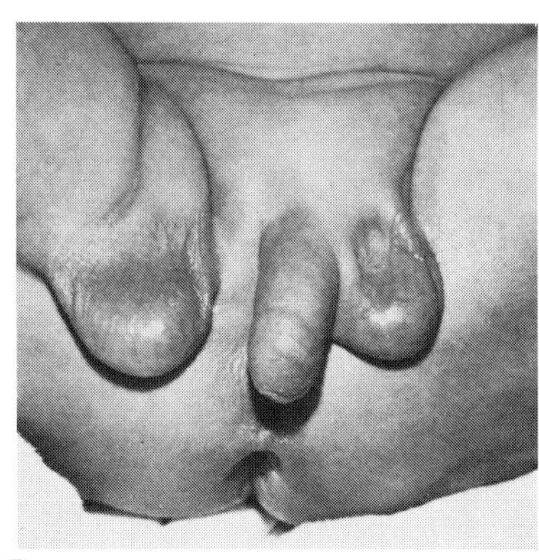

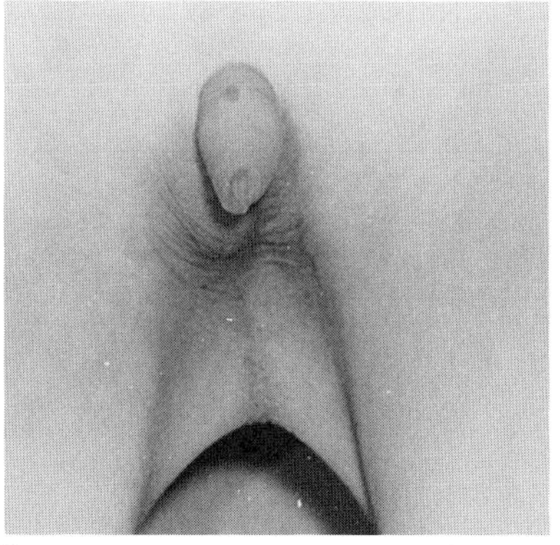

A B C

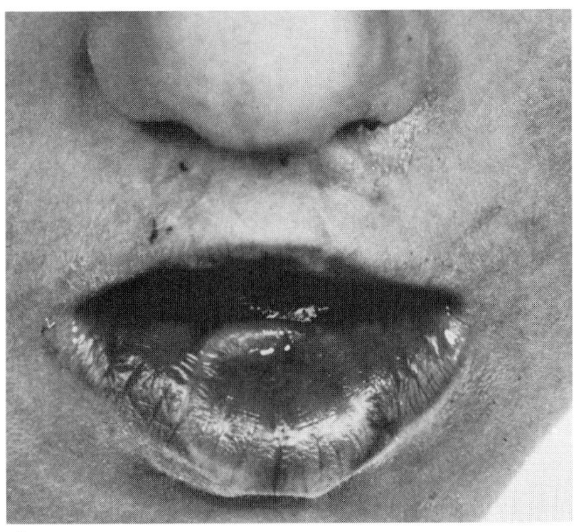
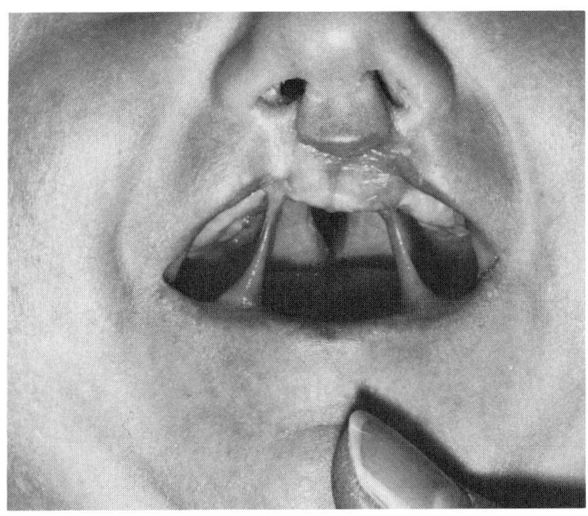

A B

Fig. 18–15. *Popliteal pterygium syndrome.* (A) Repaired bilateral cleft lip, lower lip pits. (B) Bilateral synechiae connecting upper and lower jaws. (From *A Rintala* and *A Lahti,* Scand J Plast Reconstr Surg **4**:67, 1970.)


2a. Aron NK et al: The popliteal web syndrome. *Eur J Plast Surg* **11**:93–94, 1988.

3. Audino G et al: Popliteal pterygium syndrome present with orofacial abnormalities. *J Maxillofac Surg* **12**:174–177, 1984.

4. Baja PS, Bailey BM: Ectopic scrotum: A case report. *Br J Plast Surg* **22**:87–89, 1969.

5. Bartsocas CS, Papas CV: Popliteal pterygium syndrome: Evidence for a severe autosomal recessive form. *J Med Genet* **9**:222–226, 1972.

6. Basch K: Ein weiterer Fall von sog. Flughautbildung. *Prag Med Wschr* **16**:572–573, 1981 and **17**:287–291, 1892.

7. Bixler D et al: Phenotypic variation in the popliteal pterygium syndrome. *Clin Genet* **4**:200–228, 1973.

8. Champion R, Cregan JCF: Congenital popliteal webbing in siblings: Report of two cases. *J Bone Surg* **41B**:355–359, 1969.

8a. Cheirif S et al: Bartsocas–Papas syndrome. *NY State Dent J* **51**:101, 1985.

8b. Deskin RW, Sawyer DG: Popliteal pterygium syndrome. *Int J Pediatr Otorhinol* **15**:11–22, 1988.

9. Distefano G, Romeo MG: La sindrome delle pterigo popliteo. *Riv Ped Sci* **29**:54–75, 1974.

10. Escobar V et al: Popliteal pterygium syndrome: A phenotypic and genetic analysis. *J Med Genet* **15**:35–42, 1978.

11. Fèvre M, Languepin A: Les brides cruro-jambières contenant le nerf sciatique: le syndrome bride poplitée et malformations multiples (division palatine, fistules de la lèvre inférieure, syndactylie des orteils). *Presse Méd* **70**:615-618, 1962.

12. Fisher H: Ein Fall von sogenannter Flughautbildung zwischen Ober- und Unter-schenkel. *Prag Med Wochenschr* **18**:579, 1893. (See Ref. 6.)

13. Frohlich GS et al: Popliteal pterygium syndrome: Report of a family. *J Pediatr* **90**:91–93, 1977.

14. Gorlin RJ et al: Popliteal pterygium syndrome: A syndrome comprising cleft lip/palate, popliteal and intercrural pterygia, digital and genital anomalies. *Pediatrics* **41**:503–509, 1968.

15. Hackenbroch M: Über einen Fall von kongenitaler Kontraktur der Kniegelenke mit Flughautbildung. *Z Orthopäd* **43**:508–524, 1923.

16. Hall JG et al: Limb pterygium syndromes: A review and report of eleven patients. *Am J Med Genet* **12**:377–409, 1982.

16a. Hamamoto J, Matsumoto T: A case of facio-genito-popliteal syndrome. *Ann Plast Surg* **13**:224–229, 1984.

16b. Hammer J et al: The popliteal pterygium syndrome: Distinct phenotype variation in two families. *Helv Paediatr Acta* **43**:507–514, 1988.

17. Hansson LI et al: Popliteal pterygium syndrome in a 74-year-old woman. *Acta Orthop Scand* **47**:525–533, 1976.

18. Hecht F, Jarvinen JM: Heritable dysmorphic syndrome with normal intelligence. *J Pediatr* **70**:927–935, 1967.

19. Khan SN et al: Intrafamilial variability of popliteal pterygium syndrome: A family description. *Cleft Palate J* **23**:233–236, 1986.

20. Kind HP: Popliteales Pterygium-Syndrom. *Helv Paediatr Acta* **25**:508–516, 1970.

21. Klein D: Un curieux syndrome héréditaire: cheilo-palatoschizis avec fistules de la lèvre inférieure associé a une syndactylie, une onychodysplasie particuliere, un ptérygion poplité unilatérale et des pieds varus équins. *J Génét Hum* **11**:65–71, 1962.

21a. Leck GD, Aird JC: An incomplete form of the popliteal pterygium syndrome. *Br Dent J* **157**:318–319, 1984.

22. Lewis E: Congenital webbing of the lower limbs. *Proc R Soc Med* **41**:864, 1948.

23. Marquardt W: Die angeborene Flughautbildung und ihre konservative Behandlung. *Z Orthopäd* **67**:379–386, 1937.

24. Neuman Z, Shulman J: Congenital sinuses of the lower lip. *Oral Surg* **14**:1415–1420, 1961.

25. Papadia F et al: The Bartsocas–Papas syndrome: Autosomal recessive form of popliteal pterygium in a male infant. *Am J Med Genet* **17**:841–848, 1984.

26. Pandya A et al: A multiple malformations syndrome with cleft lip and palate and ankyloblepharon filiforme adnatum. *Ind J Pediatr* **52**:667–670, 1985.

27. Pashayan HM, Lewis MB: A family with the popliteal pterygium syndrome. *Cleft Palate J* **17**:48–51, 1980.

28. Pfeiffer RA et al: Das Kniepterygium-Syndrom. Ein autosomal dominant vererbtes Missbildungssyndrom. *Z Kinderheilkd* **108**:103–116, 1970.

29. Raithel H et al: Popliteal pterygium syndrome in 3 generations. *Z Kinderchir* **26**:56–62, 1979.

30. Rintala A, Lahti A: The facio-genito-popliteal syndrome. *Scand J Plast Reconstr Surg* **4**:67–71, 1970.

31. Rosselli D, Guilienetti R: Ectodermal dysplasia. *Br J Plast Surg* **14**:190–204, 1961.

32. Rydgier L: Demonstration von Abbildungen seltener Fälle von Missbildungen. *Arch Klin Chir* **42**:769, 1891.

32a. Salis-Soglio G et al: Das Fèvre–Languepin-Syndrom. *Z Orthopäd* **124**:144–147, 1986.

33. Schönenberg H: Über die Kombination von Lippen–Kiefer–Gaumen–Spalten mit Extremitätenmissbildung. *Z Kinderheilkd* **79**:90, 1955.

34. Scramm G: Über die angeborene Flughautbildung. *Z Orthopäd* **70**:189–195, 1940.

34a. Steinberg B, Saunders V: Popliteal pterygium syndrome. *Oral Surg* **63**:17–20, 1987.

35. Trélat U: Sur un vice conformation tres rare de la lèvre inférieure. *J Méd Chir Prat* **40**:442–445, 1869.

36. Williams DW: Anomaly of scrotum and testes: Simple plastic repair. *J Urol* **89**:860–863, 1963.

37. Wolff J: Über einen Fall von angeborener Flughautbildung. *Arch Klin Chir* **38**:66–73, 1889.

38. Wynne JM et al: Massive oral membrane in the popliteal web syndrome. *J Pediat Surg* **17**:59–60, 1982.


## Chondrodystrophic myotonia (Schwartz–Jampel syndrome, osteochondromuscular dystrophy)

In 1962, Schwartz and Jampel (30) described a syndrome characterized by growth retardation, osteochondrodysplasia, continuous muscle

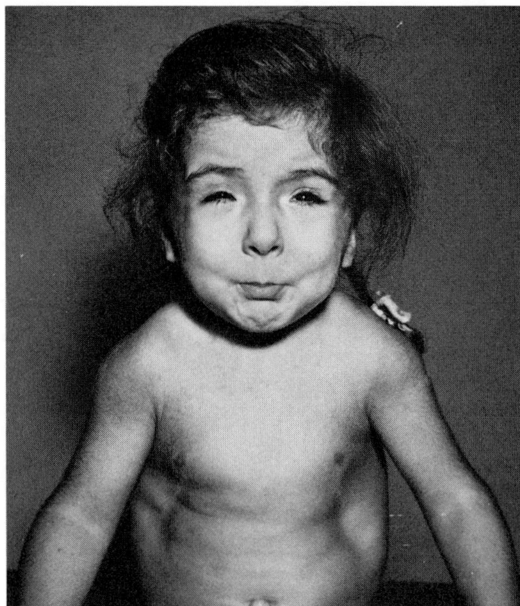

Fig. 18–16. *Chondrodystrophic myotonia.* Three-year-old female. Note blepharophimosis, rigid facial expression, pigeon-breast deformity with lateral chest constriction. (From *O Schwartz* and *RS Jampel,* Arch Ophthalmol **68**:52, 1962.)

Fig. 18–17. *Chondrodystrophic myotonia.* Note puckered mouth, mildly narrowed palpebral fissures, flexion at hips, knees, and elbows. (From *WM Fowler Jr et al,* J Neurol Sci **22**:127, 1974.)

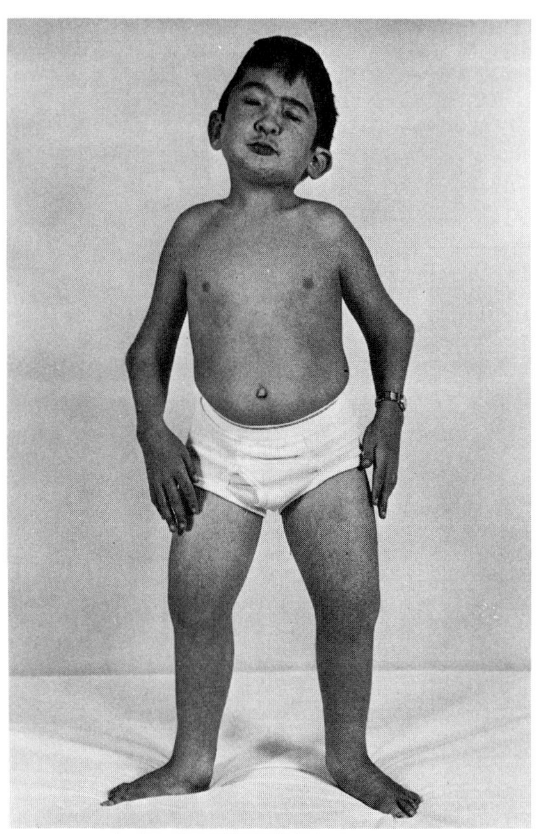

contracture at rest, and unusual facies. The same children were reported by Aberfeld et al (1,2). An earlier case is that of Catel (6). Additional cases of the disorder were described by a number of authors (3,4,7–9,11–16,18,21–23,25–29,31–33). We cannot classify the cases of Cao et al (5).

The syndrome has autosomal recessive inheritance. Affected sibs (3,6,7,12,13,15,21,24,25,33,34) and parental consanguinity (3,13, 23,28) have been noted. About 50 cases have been reported.

**Facies.** The facies, usually normal at birth, is usually clinically recognizable at 1–3 years of age (Figs. 18–16 and 18–17). It has been documented during the neonatal period (8a). There is then progressive tonic contraction of facial muscles into a pinched or frozen smile with puckered lips and narrow palpebral fissures due to medial displacement of the outer canthi. Ptosis of eyelids, micrognathia, and short neck are evident.

**Eyes.** In addition to ptosis, blepharospasm, and blepharophimosis, myopia has been noted in about 50%. In a few cases, congenital cataracts have been observed.

**Musculoskeletal alterations.** Height is reduced below the 10th percentile. Limitation of motion at the hips usually is evident within the first 6 months of life but rarely is there prenatal onset (8a). Gait becomes waddling and progressively difficult because of stiff hips and knees and the child fatigues easily (Figs. 18–16 and 18–17). There is hyporeflexia. Pectus carinatum, kyphoscoliosis, lumbar lordosis, iliac base shortening with acetabular hypoplasia, and osteopenia are common features (Fig. 18–18). The long bones may be bowed with metaphyseal widening. Occasionally cervical kyphosis may be marked.

Choking on cold liquids and mild muscular hypertrophy have been described during early childhood (2,21). Weakness is not a prominent feature. Repetitive contracture decreases the myotonia. There is continuous muscle fiber activity at rest. It has been suggested that the abnormal discharges originate in the muscle component of the neuromuscular junction rather than the nerve, perhaps in the sarcolemmal membrane (11,31). There is no true myotonia since the continuous repetitive discharges are abolished with curare (11).

**Other findings.** High-pitched voice and moderate generalized hirsutism have been noted (1,2,21,24,25,33,34). Mental retardation has been observed in 25%.

**Oral manifestations.** The lips are pursed, that is, the perioral muscles are contracted. The palate is highly arched.

Fig. 18–18. *Chondrodystrophic myotonia.* Severe coxa vara. (From *PR Huttenlocher et al,* Pediatrics **44**:945, 1969.)

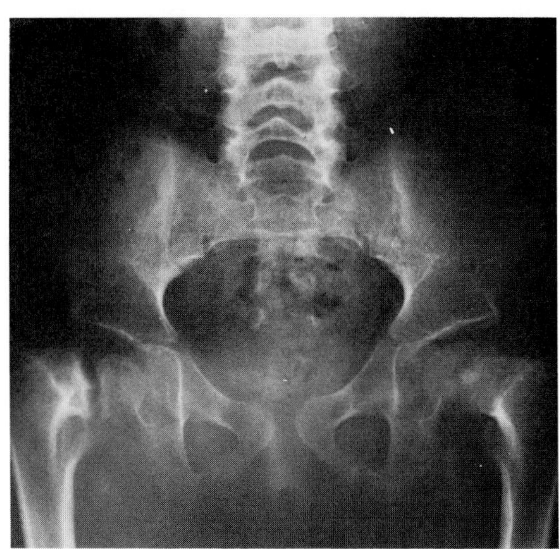

**Differential diagnosis.** The facies resembles to some degree that of *craniocarpotarsal dysplasia.* There is some overlap in phenotype with *Marden–Walker syndrome* (20).

Continuous muscle contraction at rest can also be seen in Isaacs–Mertens syndrome, stiff-man syndrome, and other disorders (4). Familial myotonia may be seen in several conditions (2).

**Laboratory aids.** Ultrastructural studies of muscle have shown inconsistent findings.

Standards for normal palpebral fissure size according to age have been established (10,19). Mollica et al (22a) found reduced numbers of B and T cells.

Prenatal diagnosis has been accomplished (14a).

### References [Chondrodystrophic myotonia (Schwartz–Jampel syndrome, osteochondromuscular dystrophy)]


1. Aberfeld DC et al: Myotonia, dwarfism, diffuse bone disease and unusual ocular and facial abnormalities (a new syndrome). *Brain* **88**:313–322, 1965.
2. Aberfeld DC et al: Chondrodystrophic myotonia: Report of two cases. Myotonia, dwarfism, diffuse bone disease and unusual ocular and facial abnormalities. *Arch Neurol* **22**:455–462, 1972.
3. Beighton P: The Schwartz syndrome in Southern Africa. *Clin Genet* **4**:548–555, 1973.
4. Blank NK et al: Persistent motor neuron discharges of central origin in the resting state. *Neurology* **24**:277–281, 1974.
5. Cao A et al: Schwartz–Jampel syndrome: Clinical electrophysiological and histopathological study of a severe variant. *J Neurol Sci* **35**:175–187, 1978.
6. Catel W: *Differentialdiagnostische Symptomatologie von Krankheiten des Kindesalters,* 2d ed, Thieme, Stuttgart, 1951, pp. 48–51.
7. Desbois JC et al: Chondrodystrophie myotoniques (ou syndrome de Schwartz–Jampel). Etude d'une fratrie et revue de la littérature. *Ann Pédiatr* **24**:563–574, 1977.
8. Edwards WC, Root AW: Chondrodystrophic myotonia (Schwartz–Jampel syndrome). *Am J Med Genet* **13**:51–56, 1982.
8a. Farrell SA et al: Neonatal manifestations of Schwartz–Jampel syndrome. *Am J Med Genet* **27**:799–805, 1987.
9. Ferrannini E et al: Schwartz–Jampel syndrome with autosomal dominant inheritance. *Eur Neurol* **21**:137–146, 1982 (Case 1).
10. Fox SA: The palpebral fissure. *Am J Ophthalmol* **62**:73–78, 1966.
11. Fowler WM JR et al: The Schwartz–Jampel syndrome. Its clinical, physiological and histological expressions. *J Neurol Sci* **22**:127–146, 1974.
12. Gellis S, Finegold M: Schwartz–Jampel syndrome. *Am J Dis Child* **126**:339–340, 1973.
13. Greze J et al: Dystrophie osteo-chondromusculaire de Schwartz–Jampel: deux cas famileaux. *Arch Fr Pédiatr* **32**:59–75, 1975.
14. Horan F, Beighton P: Orthopaedic aspects of the Schwartz syndrome. *J Bone Jt Surg* **57A**:542–544, 1975.
14a. Hunziker UA et al: Prenatal diagnosis of Schwartz–Jampel syndrome with early manifestation. *Prenatal Diag* **9**:127–131, 1989.
15. Huttenlocher PR et al: Osteo-chondro-muscular dystrophy: A disorder manifested by multiple skeletal deformities, myotonia, and dystrophic changes in muscle. *Pediatrics* **44**:945–958, 1969.
16. Keating PD, Hepler RS: Blepharophimosis in acquired somato-facial dysmorphism. *Arch Ophthalmol* **82**:1–3, 1969.
17. Kirschner BS, Pachman LM: IgA deficiency and recurrent pneumonia in the Schwartz–Jampel syndrome. *J Pediatr* **88**:1060–1061, 1976.
18. Kozlowski K, Wise G: Spondylo-epi-metaphyseal dysplasia with myotonia. A radiographic study. *Radiol Diagn (Berlin)* **15**:817–824, 1974.
19. Lopez-Terradas JM et al: Sindrome de Schwartz–Jampel. Aportacion de un caso y revision de la literatura. *An Esp Pediatr* **12**:345–348, 1979.
20. Marden PM, Walker WA: A new generalized connective tissue syndrome: Association with multiple congenital anomalies. *Am J Dis Child* **112**:225–228, 1966.
21. Mereu TR et al: Myotonia, shortness of stature, and hip dysplasia. Schwartz–Jampel syndrome. *Am J Dis Child* **117**:470–478, 1969.
22. Milachowski K et al: Das Schwartz-Jampel-Syndrom. *Z Orthopäd* **120**:657–661, 1982.
22a. Mollica F et al: Immuno-deficiency in Schwartz–Jampel syndrome. *Acta Paediatr Scand* **68**:133–135, 1979.
23. Pavone L et al: Schwartz–Jampel syndrome in two daughters of first cousins. *J Neurol Neurosurg Psychiat* **41**:161–169, 1978.
24. Pavone L et al: Immunologic abnormalities in Schwartz–Jampel syndrome. *J Pediatr* **98**:512, 1981.
25. Pfeiffer RA et al: Das Syndrom von Schwartz–Jampel (myotonia chondrodystrophica). *Helv Paediatr Acta* **32**:251–261, 1977.
26. Plasencia A et al: Sindrome de Schwartz–Jampel. Revision bibliographica y aportacion de dos neuvos cases. *An Esp Pediatr* **13**:1031–1042, 1980.
27. Rosignoli R, Zanini F: Sindrome di Schwartz–Jampel. *Minerva Pediatr* **35**:509–513, 1983.
28. Saadat M et al: Schwartz syndrome: Myotonia with blepharophimosis and limitation of joints. *J Pediatr* **81**:348–350, 1972.
29. Scaff M et al: Chondrodystrophic myotonia: Electromyographic and cardiac features of a case. *Acta Neurol Scand* **60**:243–249, 1979.
30. Schwartz O, Jampel RS: Congenital bleopharophimosis associated with a unique generalized myopathy. *Arch Ophthalmol* **68**:52–57, 1962.
31. Scribanu N, Ionanescu V: Schwartz–Jampel syndrome. Stimulating effect of calcium and A23187 calcium ionophore for protein synthesis in muscle cell culture. *Eur Neurol* **20**:46–51, 1981.
32. Seay AR, Ziter FA: Malignant hyperpyrexia in a patient with Schwartz–Jampel syndrome. *J Pediatr* **93**:83–84, 1983.
33. Taylor RG et al: Continuous muscle fiber activity in the Schwartz–syndrome. *Electroencephal Clin Neurophys* **33**:497–509, 1972.
34. Van Huffelin AC et al: Chondrodystrophia myotonia: A report of two unrelated Dutch patients. *Neuropädiatrie* **5**:71–90, 1974 (Case 1).


## Marden–Walker syndrome

Marden and Walker (10) first reported a syndrome of mental retardation, failure to thrive, microcephaly, immobility of facial muscles, absent Moro and decreased deep-tendon reflexes, blepharophimosis, strabismus, micrognathia, anteverted nostrils, pectus carinatum or excavatum, muscle weakness and reduced muscle mass, multiple congenital joint contractures, arachnodactyly, kyphoscoliosis, and transverse palmar creases. Other investigators have reported similarly affected children (1,3–7,10,12–14). Some examples are less certain (2,12,15). Inheritance is autosomal recessive as affected sibs and parental consanguinity are documented (7,8,14).

Failure to thrive and psychomotor delay have been evident in nearly all examples (8). Stature is short.

**Facies.** The face appears immobile with sagging cheeks. Blepharophimosis is a constant feature. Ptosis and exotropia/esotropia are present in about 35%. The ears are generally lowset and malformed (Figs. 18–19A,B and 18–20).

**Musculoskeletal.** Arachnodactyly (1,2,8,10,11,13), flexion contractures of the digits (2,8,10), talipes equinovarus (2,4,8,10,11), reduced muscle mass (4,8,10,11,13–15), contractures of the elbow, hip, and knee joints (1–15), pectus excavatum and/or carinatum (8,11,14),

Fig. 18–19. *Marden–Walker syndrome.* (A,B) Facies of sibs. In addition to mental retardation, both had tall forehead, blepharophimosis, hypertelorism, and immobile facies. (From *F Howard* and *P Rowlandson,* J Med Genet **18**:50, 1981.)

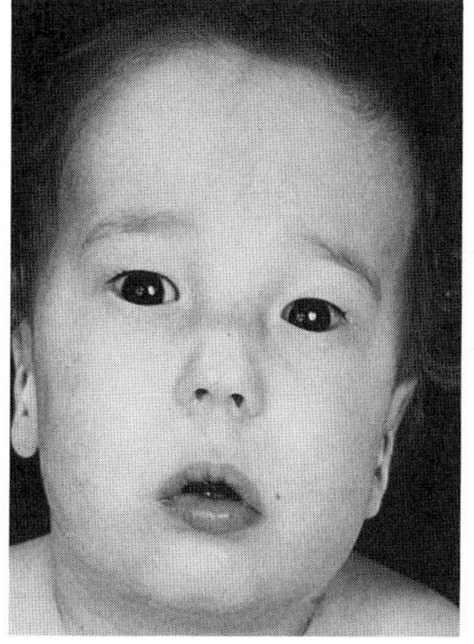

A

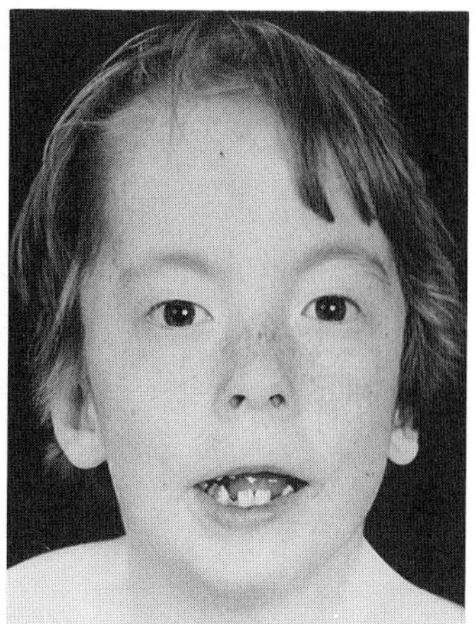

B

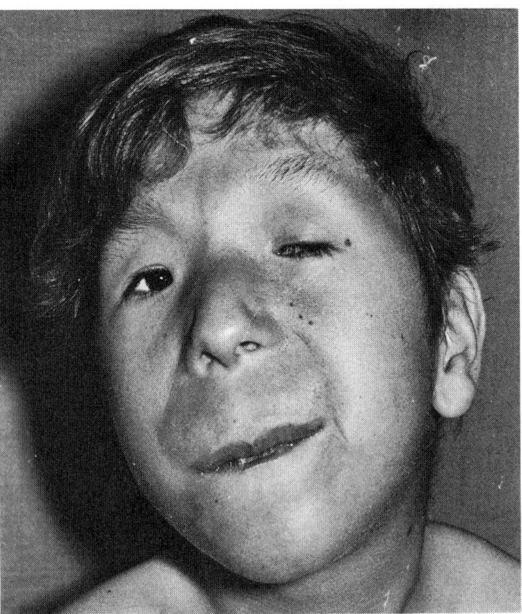

Fig. 18–20. *Marden–Walker syndrome.* Blepharophimosis, unusual facies, ptosis of lids. [From *E Passarge,* Birth Defects **11**(2):470, 1975.]

and kyphoscoliosis (1,4,8,11,12,14,15) have been documented. The joint contractures tend to resolve during the first year of life.

**Heart anomalies.** Various inconsistent congenital cardiac anomalies have been found (1,6,10,14).

**Central nervous system.** Microcephaly has been noted in about half the cases (8,10,13–15). Decreased deep tendon reflexes are common.

**Oral manifestations.** Cleft palate has been found in about 30% (1,11,12,15).

**Differential diagnosis.** There is some similarity to *Schwartz–Jampel syndrome* but, in the Marden–Walker syndrome, failure to thrive, mental retardation, myopathy, and contractures are congenital rather than presenting at 1.5–2 years and myotonia and generalized bone diseases are absent.

**Laboratory aids.** Fitch et al (4) found reduction in the size of scattered muscle fibers of both histochemical types. No percussion myotonia or myotonic discharges on the electromyogram were elicitable. Pneumoencephalographic study indicated partial or complete agenesis of the cerebellum and brainstem.

### References (Marden–Walker syndrome)


1. Abe K et al: Zollinger–Ellison syndrome with Marden–Walker syndrome. *Am J Dis Child* **133**:735–738, 1979.
2. Ealing MI: Amyoplasia congenita causing malpresentation of the foetus. *J Obstet Gynaecol Br Emp* **51**:144–146, 1944.
3. Ferguson SD et al: Congenital myopathy with oculofacial and skeletal abnormalities. *Dev Med Child Neurol* **23**:232–242, 1981.
4. Fitch N et al: Congenital blepharophimosis, joint contractures and muscular hypotonia. *Neurology* **21**:1214–1220, 1971.

4a. Gellis SS: *The Year Book of Pediatrics.* Year Book Medical Publishers, Chicago, 1963–4, p 193.

5. Gossage D et al: A 26-month-old child with Marden–Walker syndrome and pyloric stenosis. *Am J Med Genet* **26**:915–920, 1987.
6. Howard FM, Rowlandson P: Two brothers with the Marden–Walker syndrome: Case report and review. *J Med Genet* **18**:50–53, 1981.
7. Jaatoul NY et al: The Marden–Walker syndrome. *Am J Med Genet* **11**:259–271, 1982.
8. Jancar J, Mlele TJ: The Marden–Walker syndrome: A case report and review of the literature. *J Ment Defic Res* **29**:63–70, 1985.
9. King CR, Magenis E: The Marden–Walker syndrome. *J Med Genet* **15**:366–369, 1978.
10. Marden PM, Walker WA: A new generalized connective tissue syndrome: Association with multiple congenital anomalies. *Am J Dis Child* **112**:225–228, 1966.
11. Passarge E: Marden–Walker syndrome. *Birth Defects* **11**(2):470–471, 1975.
12. Simpson JL, Degnon M: A child with facial and skeletal dysmorphism reminiscent of Schwartz syndrome. *Birth Defects* **11**(2):456–458, 1975.
13. Temtamy SA et al: Probable Marden–Walker syndrome: Evidence for autosomal recessive inheritance. *Birth Defects* **11**(2):104–108, 1975.
14. Younessian S, Amman F: Deux cas de malformation cranio-faciales. *Ophthalmologica* **147**:108–117, 1964 (Case 1).


## Craniocarpotarsal dysplasia (whistling face syndrome, Freeman–Sheldon syndrome)

In 1938, Freeman and Sheldon (13) described a syndrome characterized by microstomia, flat midface, talipes equinovarus, and ulnar deviation of the fingers. Unaware of this report, Burian (5), in 1963, described the same complex, utilizing the term "whistling face syndrome." A number of authors have tabulated and analyzed cases (2,4,10,25,33,44). Over 65 patients have been reported (2,25,27).

Although most cases have been sporadic, there are several examples of the disorder in two or more generations (2,4,10,12,15,16,20,24,25,

Fig. 18–21. *Craniocarpotarsal dysplasia.* Sunken eyes, chin grooves, overlapping fingers, talipes equinovarus, and inguinal hernia. (From *AE Rintala,* Acta Paediatr Scand **57**:553, 1968.)

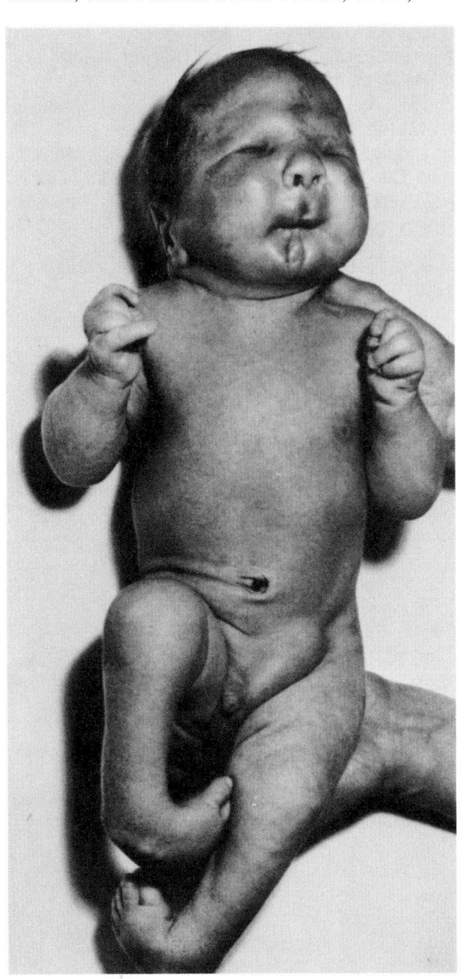

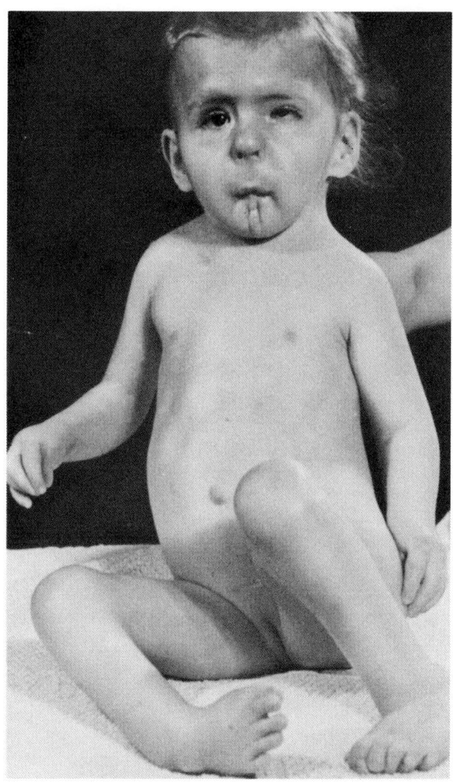

Fig. 18–22. *Craniocarpotarsal dysplasia.* Deep-set eyes, lid ptosis, convergent strabismus, H-shaped chin grooves, overlapping fingers, and talipes. (Courtesy of *J Külz,* Rostock, East Germany.)

27,31,32,34,38,42,45). The condition has also been described in sibs with normal, in some cases, consanguineous parents (3,10a,11,18,21, 26,35,43). Hashemi (17) found parental consanguinity in an isolated example. This seems to suggest genetic heterogeneity but separation of phenotypes has not been possible. Sauk et al (36) pointed out the possible depiction of the condition in a pre-Columbian vase.

The disorder has been suggested to be one of congenital myopathy (40).

**Facies.** The stiff immobile, flat midface, full cheeks, small nose, long philtrum, puckered mouth, and H-shaped defect on the chin are striking but extremely variable features. Microstomia and long philtrum are constant, but such features as small nose with broad bridge, pronounced supraorbital ridges, deep-set eyes, micrognathia, and short broad neck are less frequent. The typical facies is present in only about 50% (4) (Fig. 18–21 to 18–23B).

**Eyes.** Strabismus, epicanthus, and hypertelorism have been noted in over 60% (4,8,25–27,34). About half of the patients have exhibited downslanting palpebral fissures and ptosis (9,38,44) (Fig. 18–23B).

**Nose.** The nose is usually small. The nostrils are narrow, the alae being bent in about half the cases, simulating nostril colobomas. Near the tip, the alae are of normal thickness, but they thin dorsally to be inserted close to the columella. The nasolabial folds are evident only near the sides of the nose. The philtrum is long (Fig. 18–23B).

**Musculoskeletal system.** Growth has been retarded below the third percentile in about 30% of the cases (16). There is ulnar deviation of the fingers without bony abnormalities. Flexion contractures of the fingers are a common feature, the thumbs being adducted at the metacarpophalangeal joints. It should be emphasized, however, that the hands may be normal (30,41).

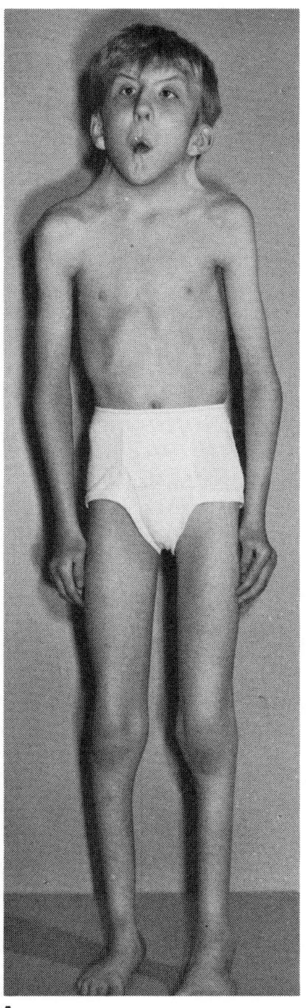

A

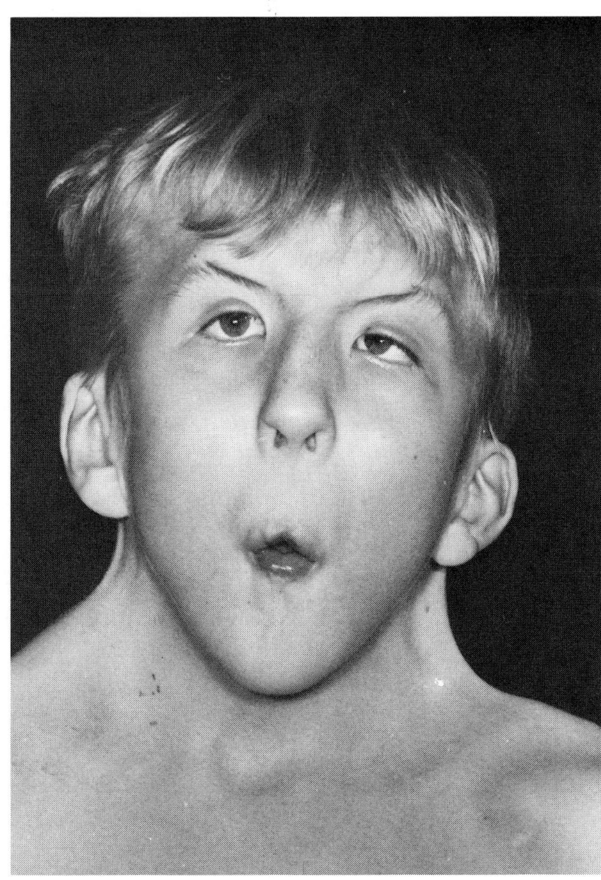

B

Fig. 18–23. *Craniocarpotarsal dysplasia.* Whistling facies is produced by small puckered mouth. Note pterygium colli and notched nostrils.

Talipes equinovarus, occasionally unilateral, has been found in approximately 80% (Figs. 18–21 and 18–22). Moderate to severe kyphoscoliosis has been documented in about 60%. Frequently there is atrophy of forearm and leg muscles, especially below the knees (5,12,23,31,32,37,42,44) (Figs. 18–21 and 18–23A). Other less frequent anomalies include spina bifida occulta (7,28,37,39,44,45) and inguinal hernia (5,11,13,26,34).

The facial skeleton is small. The anteroposterior lengths of both base and cranium are short while facial height is relatively great. Most patients have exhibited a steep cranial base (34,44). In a few, the ribs have been broad and tapering near the costovertebral angle (12,23,30).

Other joints have occasionally shown limitation. Hall et al (16) found approximately 15% had limited movement at the hips and 5% at the elbow. Dislocation of the radial heads has been noted (11). Pterygia of the neck (24,44) or axillae (11) have also been described.

**Other findings.** A few children have suffered aspiration pneumonia or respiratory death (12,23,34), and a few were mentally retarded (4,11,34).

**Oral manifestations.** Microstomia may be marked, the intercommissural distance being about 65% of that for a child of the same age (9). The lips are pursed or held as in whistling in about half the cases. The hard palate is often highly arched and the mandible and tongue tend to be small. Frequently there is nasal speech. Extending from the middle of the lower lip to the chin is a fibrous band or elevation that is demarcated by two paramedian grooves forming an H- or V-shaped scarlike structure in about 30%.

**Differential diagnosis.** Differential diagnosis includes arthrogryposis, *Schwartz–Jampel syndrome,* and Burton syndrome. The Burton syndrome is an autosomal recessive condition characterized by microstomia, pursed lips, ectopia lentis, and skeletal changes similar to those of *Kniest syndrome* and *dyssegmental dysplasia, Rolland–Desbuquois type*. However, no coronal clefts were noted. Scattered dense patches of extremely broad collagen fibers were found scattered throughout the cartilage matrix (6). We believe the constellation of anomalies seen here is sufficiently characteristic to cause little confusion with other entities. It should be pointed out that ulnar deviation of the fingers (windmill fingers) may have autosomal dominant inheritance (1,19,22). However, it is possible that windmill fingers is the same as craniocarpotarsal dysplasia in which craniofacial expression has been minimal.

**Laboratory aids.** Electromyography has shown hypoplasia of the musculature of the buccinator and some of the hand and foot muscles (5,16,36,37). Biopsy of buccinator muscle has shown fibrous connective tissue replacement of muscle bundles (36,39).

### References [Craniocarpotarsal dysplasia (whistling face syndrome, Freeman–Sheldon syndrome)]


1. Aalam M, Kühhirt M: Angeborene Windmühlenflügeldeformität der Finger. *Z Orthopäd* **110**:395–398, 1972.
2. Aldinger G, Eulert J: Das Freeman-Sheldon-Syndrom. *Z Orthopäd* **121**:630–633, 1983.
3. Alves AF, Azevedo ES: Recessive form of Freeman–Sheldon's syndrome or "whistling face." *J Med Genet* **14**:139–141, 1977.
4. Antley RM et al: Diagnostic criteria for the whistling face syndrome. *Birth Defects* **11**(5):161–168, 1975.
5. Burian F: The "whistling face" characteristic in a compound craniofacio-corporal syndrome. *Br J Plast Surg* **16**:140–163, 1963.
6. Burton BK et al: A new skeletal dysplasia: Clinical, radiologic and pathologic findings. *J Pediatr* **109**:642–648, 1986.
7. Burzynski NJ et al: Craniocarpotarsal dysplasia syndrome (whistling face syndrome). *Oral Surg* **39**:893–900, 1975.
8. Call WH, Strickland JW: Functional hand reconstruction in the whistling-face syndrome. *J Hand Surg* **6**:148–151, 1981.
9. Cervenka J et al: Cranio-carpo-tarsal dysplasia or the whistling face syndrome. II. Oral intercommissural distance in children. *Am J Dis Child* **117**:434–435, 1969.
10. Cervenka J et al: Craniocarpotarsal dysplasia or whistling face syndrome. *Arch Otolaryngol* **91**:183–187, 1974.
10a. Dallapiccola B et al: Autosomal recessive form of whistling face syndrome. *Am J Med Genet* **33**:542–544, 1989.
11. Fitzsimmons JS et al: Genetic heterogeneity in the Freeman–Sheldon syndrome: Two adults with probable autosomal recessive inheritance. *J Med Genet* **21**:364–368, 1984.
12. Fraser FC et al: Cranio-carpo-tarsal dysplasia. *JAMA* **211**:1374–1376, 1970.
13. Freeman EA, Sheldon JH: Cranio-carpo-tarsal dystrophy—an undescribed congenital malformation. *Arch Dis Childh* **13**:277–283, 1938.
14. Grasshoff H et al: Beitrag zum Freeman-Sheldon-Syndrom. *Beitr Orthop Traumatol* **25**:241–247, 1978.
15. Gross-Kieselstein E et al: Familial occurrence of the Freeman–Sheldon syndrome; cranio-carpo-tarsal dysplasia. *Pediatrics* **47**:1064–1067, 1971.
16. Hall JG et al: The distal arthrogryposes: Delineation of new entities, review and nosologic discussion. *Am J Med Genet* **11**:185–239, 1982.
17. Hashemi G: The whistling face syndrome. *Indian J Pediatr* **40**:23–24, 1973.
18. Kousseff BG et al: Autosomal recessive type of whistling face syndrome in twins. *Pediatrics* **69**:328–330, 1982.
19. Jacquemain B: Die angeborene Windmühlenflügelsstellung als erbliche Kombinationsmissbildung. *Z Orthopäd* **102**:146–154, 1966.
20. Jorgenson RJ: Craniocarpotarsal dystrophy (whistling face syndrome) in two families. *Birth Defects* **10**(5):237–242, 1974.
21. Kousseff B et al: Autosomal recessive type of whistling face in twins. *Pediatrics* **69**:328–331, 1982.
22. Lundblom A: On congenital ulnar deviation of the fingers of familial occurrence. *Acta Orthopaed Scand* **3**:393–404, 1932.
23. MacLeod P, Patriquin H: The whistling face syndrome—cranio-carpotarsal dysplasia. *Clin Pediatr* **13**:184–189, 1974.
24. Malkawi H, Tarawneh M: The whistling face syndrome or craniocarpotarsal dysplasia: Report of two cases in a father and son and review of the literature. *J Pediatr Orthoped* **3**:364–369, 1983.
25. Martini K et al: Die Handdeformitäten beim Freeman–Sheldon-Syndrom und ihre operative Behandlung. *Z Orthopäd* **121**:623–629, 1983.
26. O'Connell DJ, Hall CM: Cranio-carpo-tarsal dysplasia. A report of seven cases. *Radiology* **123**:719–722, 1977.
27. O'Keefe M et al: Ocular abnormalities in the Freeman–Sheldon syndrome. *Am J Ophthalmol* **102**:346–348, 1986.
28. Otto FMG: Die "Cranio-carpo-tarsal Dystrophie" (Freeman and Sheldon): Ein kasuistischer Beitrag. *Z Kinderheilkd* **73**:240–250, 1953.
29. Pahor AL: Whistling face syndrome. *Ear Nose Throat J* **59**:232–234, 1980.
30. Patel M, Mahmud F: A case of Freeman–Sheldon syndrome (craniocarpotarsal dysplasia with spatulate ("canoe-paddle") ribs. *Br J Radiol* **56**:50–51, 1983.
31. Pfeiffer RA et al: Das Syndrom von Freeman und Sheldon. *Z Kinderheilkd* **112**:43–53, 1972.
32. Pitanguy I, Bisaggio S: A chiro-cheilo-podalic syndrome. *Br J Plast Surg* **22**:79–85, 1969.
33. Rinsky LA, Bleck EE: Freeman–Sheldon (whistling-face) syndrome. *J Bone Jt Surg* **58A**:148–150, 1976.
34. Rintala AE: Freeman–Sheldon's syndrome: Cranio-carpo-tarsal dystrophy. *Acta Paediatr Scand* **57**:553–556, 1968.
35. Sánchez JM, Kaminker CP: New evidence for genetic heterogeneity of the Freeman–Sheldon syndrome. *Am J Med Genet* **25**:507–512, 1986.
36. Sauk JJ et al: Electromyography of oral-facial musculature in craniocarpaltarsal dysplasia (Freeman–Sheldon syndrome). *Clin Genet* **6**:132–137, 1974.
37. Sharma RN, Tandon SN: "Whistling Face" deformity in cranio-faciocorporal syndrome. *Br Med J* **4**:33, 1970.
38. Temtamy SA: *Genetic factors in hand malformations.* Thesis. Johns Hopkins Univ. Press, Baltimore, 1966.
39. Vaitiekaitis AS et al: A new surgical procedure for correction of eye deformity in cranio-carpo-tarsal dysplasia (whistling face syndrome). *J Oral Surg* **37**:669–672, 1979.
40. Vanek J et al: Freeman–Sheldon syndrome: A disorder of congenital myopathic origin? *J Med Genet* **23**:231–236, 1986.
41. Walbaum R et al: Le syndrome de Freeman–Sheldon (syndrome du siffleur). *Ann Pédiatr* **20**:357–364, 1973.
42. Walker BA: Whistling face-windmill vane hand syndrome. *Birth Defects* **5**(2):228–230, 1969.
43. Wang TR, Lin SJ: Further evidence for genetic heterogeneity of whis-

Fig. 18–24. *Trismus-pseudocampylodactyly syndrome.* Kindred of affected father and several affected and normal children showing attempt to open mouth fully. [From *F Hecht* and *RK Beals,* Birth Defects **5**(3): 96, 1969.)

tling face or Freeman–Sheldon syndrome in a Chinese family. *Am J Med* **28**:471–476, 1987.

44. Weinstein S, Gorlin RJ: Cranio-carpo-tarsal dysplasia or the whistling face syndrome. I. Clinical considerations. *Am J Dis Child* **117**:427–433, 1969.
45. Wettstein A et al: A family with whistling face syndrome. *Hum Genet* **55**:177–189, 1980.
46. Zizka J et al: Syndrome of cranio-carpo-tarsal dysplasia in a 6-year-old girl (whistling-face syndrome). *Čs Pediatr* **34**:96–97, 1979.

## Trismus-pseudocampylodactyly syndrome

The syndrome of limited oral opening and curvature of the fingers at all interphalangeal joints on dorsiflexion of the wrists (pseudocampylodactyly) due to shortened flexor muscle–tendon units was described independently in 1969 by Hecht and Beals (3) and by Wilson et al (17). The family originated in the Netherlands (1,6,13,16). The disorder has autosomal dominant inheritance with variable expressivity (1,3,4,10,13,17,18). Hall et al (2) classified the disorder as a form of distal arthrogryposis. Linkage studies (10) done on the family reported by Yamashita and Arnet (18) were not fruitful.

The patients are normally proportioned, but stature is often reduced to between the 3rd and 25th percentile.

**Facies.** De Jonge (1) and Mercuri (8) described blepharochalasis in older individuals, and a quilted appearance of the cheeks (Fig. 18–24).

**Musculoskeletal alterations.** In addition to mildly reduced stature and limited oral opening, the distinguishing features of the syndrome are limited to abnormalities of the extremities that result from shortness of various skeletal muscles or flexor tendons.

The hands may be clenched at birth but loosen during infancy and childhood (16). On dorsiflexion of the wrist, due to shortening of the flexor tendon–muscle unit, curvature of all fingers at each interphalangeal joint is noted. Upon volar flexion, all fingers are completely extended (Fig. 18–25A,B). There is no associated muscular weakness, but there is mild limitation of dorsiflexion and supination at the wrist.

Leg or foot problems are found in at least 10%. Short leg muscles result in a variety of mild foot deformities: talipes equinovarus or calcaneovalgus, metatarsus varus, etc. Shortened hamstring muscles produce pelvic tilt with straight leg raising (12). Pes planus, hammer toes, and torticollis have also been noted (4,17).

**Oral manifestations.** The maximum aperture between the incisal edges of the upper and lower incisors ranges from 3 to 18 mm. This inhibits mastication, an affected individual requiring about twice as much time to consume a meal. Some infants cannot use a conventional nipple to feed. There may be difficulty with endotracheal intubation during general anesthesia.

The temporomandibular joints have been shown to be normal (4). The coronoid processes may be enlarged or distorted, unilaterally or bilaterally, and are often responsible for the inability to open the mouth widely due to their impingement on the zygomatic bones (Fig. 18–26). Fibrous bands extend from the maxilla to the mandible in the retromolar areas. The decreased opening may be secondary to shortening of the muscles of mastication, which in turn causes elongation of the coronoid processes (13). A hardened nodular cord has been noted in the region of the nasolabial furrow. Some patients have micrognathia (13).

**Differential diagnosis.** Inability to open the mouth must be differentiated from microstomia seen in a variety of disorders *[craniocarpotarsal dysplasia, chondrodystrophic myotonia (Schwartz–Jampel syndrome),* etc.]. Abnormalities, either congenital or acquired, of the

Fig. 18–25. *Trismus-pseudocampylodactyly syndrome.* (A) On volar flexion, all fingers can be completely extended. (B) On dorsiflexion of wrist, there is flexion of fingers at interphalangeal joints. [From *F Hecht* and *RK Beals,* Birth Defects **5**(3):96, 1969.]

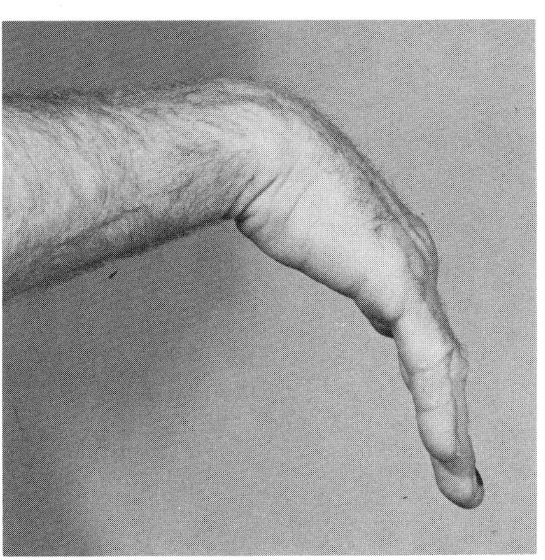

A

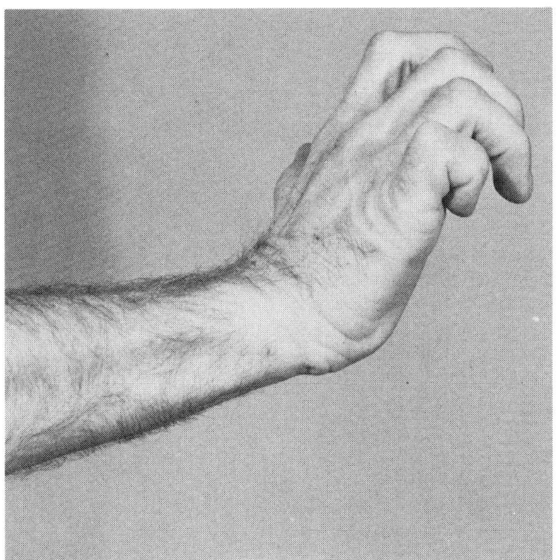

B

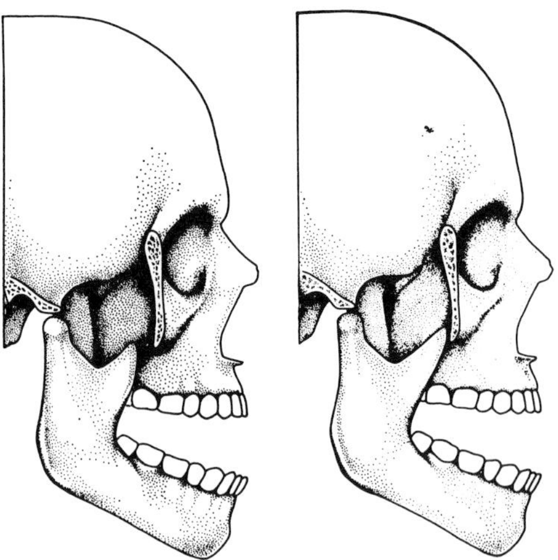

Fig. 18–26. *Trismus-pseudocampylodactyly syndrome.* Left, normal relationship. Right, enlarged coronoid process that prevents normal opening of jaws. (Courtesy of *RF Van Hoof,* Leiden, The Netherlands.)

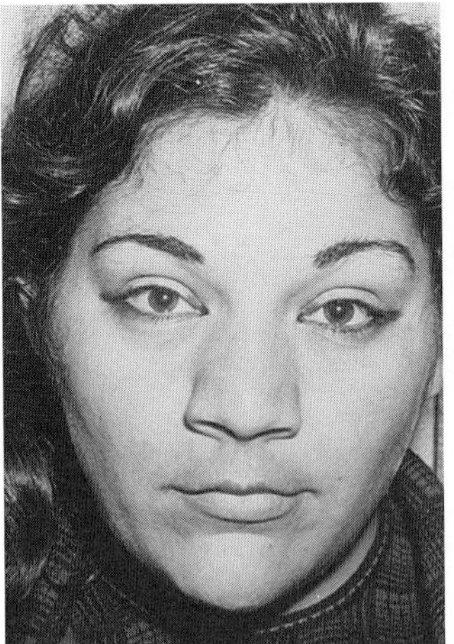

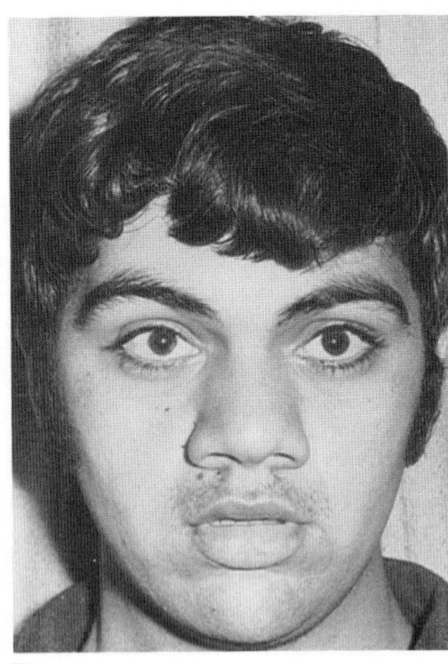

A B

Fig. 18–27. *Unusual facies, camptodactyly with fibrous tissue hyperplasia, knuckle pads, and skeletal dysplasia.* Note similar broad nose with round nostrils in sibs.

temporomandibular joint may result in reduced jaw opening. Patients with the temporomandibular joint dysfunction syndrome cannot open their mouths widely, but this disorder can easily be differentiated from the syndrome under discussion.

**Laboratory aids.** Average opening between edges of incisor teeth is 49 mm (range 30–70 mm) in normal adults. From 8–12 years of age, the normal vertical distance is about 40 mm (range 25–55 mm) 3,5–7). This has also been our experience, but not that of Sheppard and Sheppard (11).

Fig. 18–28. *Unusual facies, camptodactyly with fibrous tissue hyperplasia, knuckle pads, and skeletal dysplasia.* Note camptodactyly and knuckle pads. (From *R Goodman et al,* J Med Genet **9:**203, 1972.)

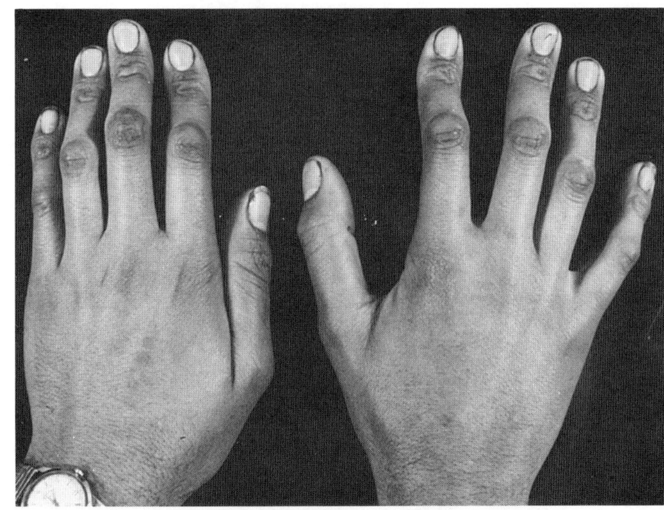

A

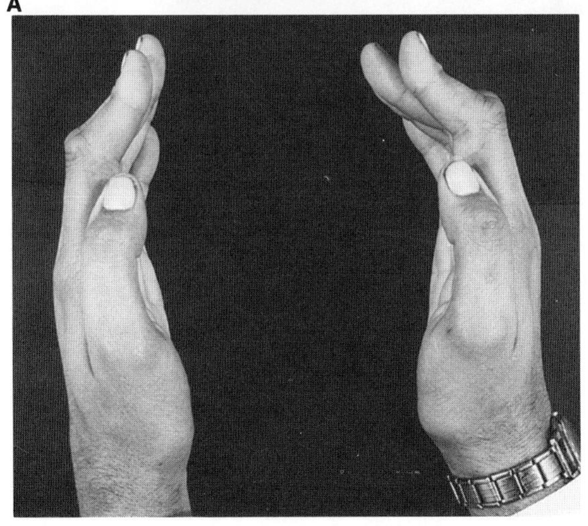

B

### References (Trismus-pseudocampylodactyly syndrome)


1. De Jonge JGY: A family showing strongly reduced ability to open the mouth and limitation of some movements of the extremities. *Humangenetik* **12**:210–217, 1971. (Same family as described in Ref. 13.)
2. Hall JG et al: The distal arthrogryposes. *Am J Med Genet* **11**:185–239, 1982.
3. Hecht F, Beals RK: Inability to open the mouth fully: An autosomal dominant phenotype with facultative campylodactyly and short stature. *Birth Defects* **5**(3):96–98, 1969.
4. Horowitz SL et al: Limited intermaxillary opening—an inherited trait. *Oral Surg* **36**:490–492, 1973.
5. Lignell L, Ransjö K: Maximal gapformaga. *Sverig Tandläk Forb Tidn* **59**:859–862, 1967.
6. Mabry CC et al: Trismus-pseudocamptodactyly syndrome. *J Pediatr* **85**:503–508, 1974.
7. Markus AF: Limited mouth opening and shortened flexor muscle-tendon units. *Br J Oral Maxillofac Surg* **24**:137–142, 1986.
8. Mercuri LG: The Hecht, Beale and Wilson syndrome. *J Oral Surg* **39**:53–56, 1981.
9. O'Brien PJ et al: Orthopaedic aspects of the trismus-pseudocamptodactyly syndrome. *J Pediatr Orthoped* **4**:469–471, 1984.
10. Robertson RD et al: Linkage analysis with the trismus-pseudocamptodactyly syndrome. *Am J Med Genet* **12**:115–120, 1982.
11. Sheppard IM, Sheppard SM: Maximal incisal opening—diagnostic index? *J Dent Med* **20**:13–15, 1965.
12. Surana RB, Ives EJ: Shortened flexor tendons in hands and feet associated with limitation of mouth opening. *4th Int Congr Hum Genet.* Int Congress Ser No 235. 173, Excerpta Medica, Paris, 1971.
13. Ter Haar BG, Van Hoof RF: The trismus-pseudocampylodactyly syndrome. *J Med Genet* **11**:41–49, 1974.
14. Tsukahara M et al: Trismus-pseudocamptodactyly syndrome in a Japanese family. *Clin Genet* **28**:247–250, 1985.
15. Uhlemann HJ, Doring G: Über angeborene symmetrische Gelenkskontrakturen. *Dtsch Z Nervenheilkd* **163**:329–336, 1950.
16. Van Hoof RF: *Enlargement of the Coronoid Process with Special Reference to the Trismus-Pseudocampylodactyly Syndrome.* Stafleu and Tholen, Leiden, 1973.

17. Wilson RV et al: Autosomal dominant inheritance of shortening of the flexor profundus muscle-tendon unit with limitation of jaw excursion. *Birth Defects* **5**(3):99–102, 1969.

18. Yamashita DKR, Arnet GF: Trismus-pseudocamptodactyly syndrome. *J Oral Surg* **38**:625–630, 1980.

## Unusual facies, camptodactyly with fibrous tissue hyperplasia, knuckle pads, and skeletal dysplasia

Goodman et al (1) described a consanguineous kindred of Jewish Iranian ancestry. Three sibs, two female and one male, manifested a broad nose with flared nostrils (Fig. 18–27A,B). The hands and feet were especially large for stature. All digits of the hand except the thumbs were camptodactylous, the onset of which began around 10 years. Knucklepads were noted on the second, third, and fourth fingers bilaterally (Fig. 18–28A,B). Hammertoes were bilateral. All had mild thoracic scoliosis. Each presented dull expression with low normal intelligence. Inheritance appears to be autosomal recessive.

### Reference (Unusual facies, camptodactyly with fibrous tissue hyperplasia, knuckle pads, and skeletal dysplasia)

1. Goodman RM et al: Camptodactyly: Occurrence in two new genetic syndromes and its relationship to other syndromes. *J Med Genet* **9**:203–212, 1972.

# Chapter 19
# Branchial Arch and Oro-Acral Disorders

## Oculo-auriculo-vertebral spectrum (hemifacial microsomia, Goldenhar syndrome)

In the 1960s, hemifacial microsomia was defined as a condition affecting primarily aural, oral, and mandibular development. The disorder varied from mild to severe, and involvement was limited to one side in many cases, but bilateral involvement was also known to occur, with more severe expression on one side. Goldenhar syndrome was considered a variant of this complex, characterized additionally by vertebral anomalies and epibulbar dermoids (37). The condition is now known to be extremely complex and heterogeneous. Thus, in the present edition we have employed the term oculo-auriculo-vertebral (OAV) spectrum.

The first recorded cases may have been those of Canton (16), in 1861, and von Arlt, in 1881 (5). An excellent early review is that of Geissmar (33). There are numerous important reviews in modern times (4,6,8,15,18,18a,20,28,31,39,48,59,62,69,90–92,95,101,106,113).

The many terms used for this complex indicate the wide spectrum of anomalies described and emphasized by various authors. The complex has been known as hemifacial microsomia, oculo-auriculo-vertebral dysplasia, Goldenhar syndrome, Goldenhar–Gorlin syndrome, first arch syndrome, first and second branchial arch syndrome, lateral facial dysplasia, unilateral craniofacial microsomia, otomandibular dysostosis, unilateral mandibulofacial dysostosis, unilateral intrauterine facial necrosis, auriculo-branchiogenic dysplasia, and facio-auriculo-vertebral malformation complex (49b). These terms have been reviewed elsewhere (18a).

While there are no agreed upon minimal diagnostic criteria, the facial phenotype is characteristic when enough manifestations are present. In some instances, isolated microtia or auricular or preauricular abnormality may represent the mildest manifestation (7,38,39, 61,92,113). Unilateral microtia or ear abnormality, including preauricular tags, has been suggested as a mandatory feature by some authors (92). Involvement is not limited to facial structures; cardiac, renal, skeletal, and other anomalies may also occur. In particular, the so-called "expanded Goldenhar complex" has demonstrated a wide spectrum of central nervous system malformations not previously appreciated (1–4,77a,118). It has been suggested by several authors that invalid nosologic splitting, on the one hand, and inappropriate lumping on the other, complicate our understanding of this complex (18,18a, 38,46,73,77,113). Those with the Goldenhar end of the spectrum constitute only about 10% of the cases (92).

Poswillo (84) suggested a frequency of 1/3500 births, although he presented no data to support his conjecture. Grabb (39) estimated 1 case/5600 births. Stoll (107) noted a prevalence of 1/19,500 consecutive births. Melnick (62) recorded a frequency of 1/26,550 live births in a prospective newborn study. Our impression is that the true frequency is near that offered by Grabb (39). The male/female ratio is at least 3:2 (38,91,105,118); there is also a 3:2 predilection for right-sided ear involvement (17,91).

Poswillo (83–85), using an animal model, showed that early vascular disruption with expanding hematoma formation *in utero* resulted in destruction of differentiating tissues in the region of the ear and jaw. The severity appeared to be related to the degree of local destruction. Another possible mechanism is that disturbances in the branchial arches or various populations of neural crest cells may impede development of adjacent medial or frontonasal processes. The constellation of anomalies suggests their origin about 30–45 days of gestation. This has been confirmed in humans by demonstration of disruption of vascular supply (37,89). Overripeness ovopathy as a possible cause has been raised (49a,122).

First and second branchial arch anomalies often combined with facial palsy have been observed in infants born to pregnant women exposed to thalidomide (56,65,94), primidone (43a), and retinoic acid (55). The OAV phenotype has also been noted in infants born to diabetic mothers (41,49).

Several chromosomal anomalies have been associated including del(5p) (23,54,72), del(6q) (39a), trisomy 7 mosaicism (47), del(8q) (116), trisomy 9 mosaicism (119), trisomy 18 (9a,39a), recombinant chromosome 18 (109), del(18q) (22), ring 21 chromosome (39a), del(22q) (45a), 49,XXXXY (53), and 47,XXY (82).

Discordance in monozygotic twins has been reported frequently (11,11a,15,19,21,24,36,39,74,78,92,100,107). Rarely, concordance with variable expression has been documented in monozygotic twins (90,96a,112). Discordance has been reported in dizygotic twins (39,44,86,102), in twins of undetermined zygosity (10,35,75), and in triplets (111). The rarity of reports of concordance of the defect in twins supports the suggestion that the condition is sporadic in most families. The interested reader is referred to the excellent analysis of Burck (15).

Most cases are sporadic, but familial instances may also be observed. Within families, expression varies. For example, there are reports of ear and mandibular involvement in two first-degree relatives, and reports of isolated microtia or preauricular tags in one first-degree relative of a patient with ear and mandibular involvement (92). These reports support the suggestion that isolated microtia or preauricular tags may represent the mildest expression in some families (38,39,61, 91,92,113). Affected individuals in successive generations have been observed (18,39,46,68,76,87,88,93,100,110,111,114). Affected sibs with normal parents have also been reported (39,51,52,99). Consanguinity was noted in a single sporadic instance (77). Although autosomal dominant and autosomal recessive inheritance have been suggested to explain the rare familial occurrence, etiologic heterogeneity is a more likely explanation for most cases. Overall, it would be compatible with a low empiric recurrence risk of 2–3% (39,92), and the usual discordance and rare concordance between monozygotic twins. However, some families (46,87,110), probably representing about 1–2% of the cases, suggest that the disorder has an autosomal dominant form. RJ Gorlin has documented a large family.

A branchial arch anomaly should be regarded as a nonspecific symptom complex that is etiologically and pathogenetically heterogeneous. Evaluation of first-degree relatives is important to exclude mild facial manifestations of the defect and various extracranial anomalies. Recurrence risk counseling should be provided on an individual family basis.

Extreme variability of expression is characteristic (Figs. 19-1 to 19-3). About 50% have been noted to have other anomalies in addition to the principal features of the defect (91). Not uncommonly the infants are small for dates and may have feeding difficulties that may be related to cleft lip and/or cleft palate or an anatomically narrow pharyngeal airway. Obstructive sleep apnea diagnosed by all-night polysomnography and nasopharyngoscopy has been described (102).

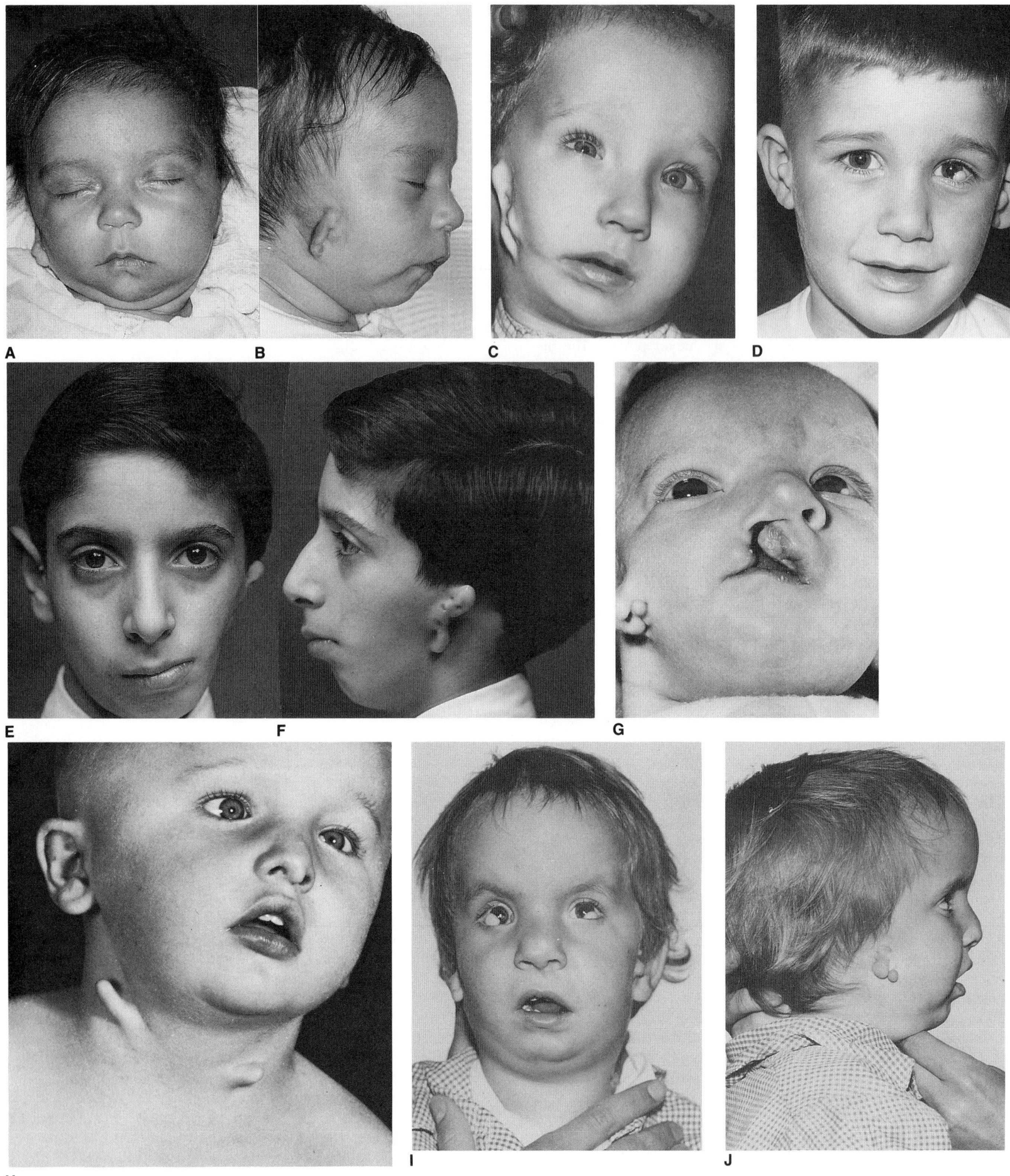

Scrupulous medical and nursing care in the first few weeks can often prevent the need for surgical gastrostomy (101).

**Facies.** Marked facial asymmetry is present in 20% but some degree of asymmetry is evident in 65% (104). This is due to both displacement and/or abnormality of the pinna and also to underlying abnormalities of skeletal support. The asymmetry may not be apparent in the infant or young child but can surface with growth, usually becoming apparent by age four. Adequate overlying soft tissues may mask the skeletal asymmetry (Fig. 19-1A–N). Conversely, a deficient soft tissue mass overlying adequate bony structures may result in facial asymmetry (28).

The maxillary, temporal, and malar bones on the more severely involved side are somewhat reduced in size and flattened (96). Asym-

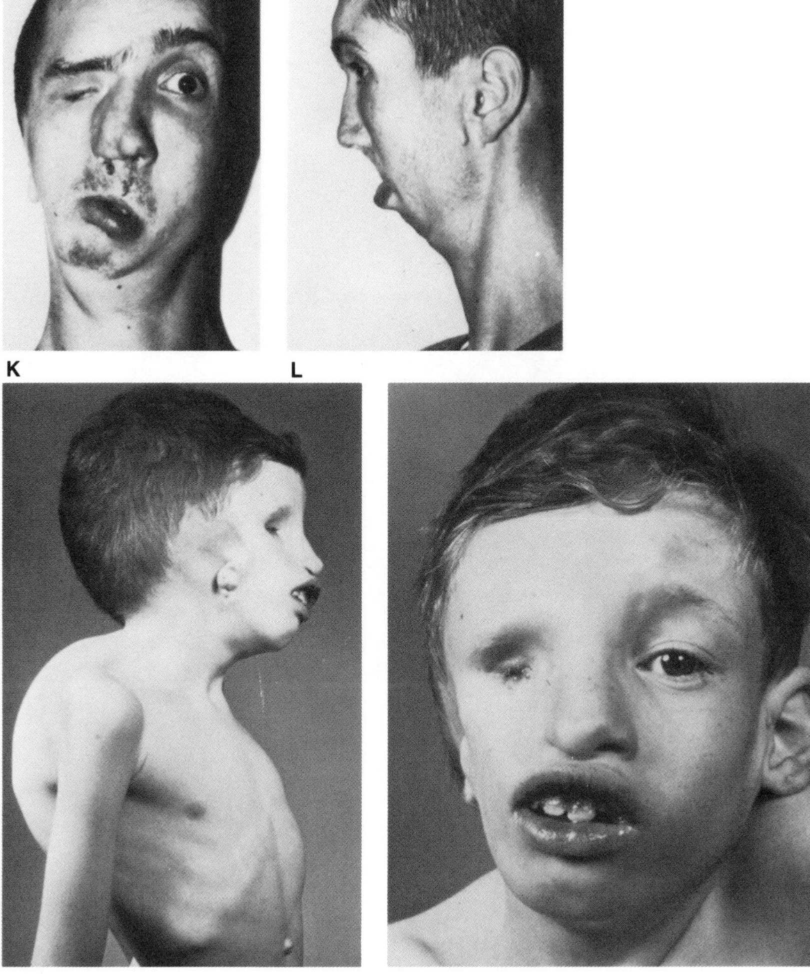


Fig. 19–1. *Oculo-auriculo-vertebral spectrum.* (A–N) Variable degrees of severity. Compare facies. Note cleft lip–palate, ear anomalies, and epibulbar dermoid in G. Note unusual cervical structures containing cartilage, epibulbar dermoid, and strabismus in H. Note bilateral epibulbar dermoids in I and J. Note anophthalmia in K–N. [A,B from *MM Cohen Jr,* The Child with Multiple Birth Defects, Raven Press, New York, 1982, p 62. D courtesy of *M Lejour,* Brussels, Belgium. E,F courtesy of *WC Grabb,* Ann Arbor, Michigan. G courtesy of *BGA ter Haar,* Nijmegen, The Netherlands. H courtesy of *S Budden,* Vancouver, British Columbia. K,L from *MM Cohen Jr,* Birth Defects **7**(7):103, 1971. M,N from *CH Ide et al,* Arch Ophthalmol **84**:427, 1970.]

metry may result from aplasia or hypoplasia of the mandibular ramus and condyle. Some patients manifest mild pneumatization of the mastoid region. About 10–33% of patients have bilateral involvement (15,39,91). The disorder is nearly always more severe on one side. Among 193 unilaterally affected patients studied by Rollnick et al (91), the right side was involved in over 60%. Frontal bossing can be noticeable at birth but becomes less apparent with time (37). The reader is referred to various anthropometric studies (25,26,50,104).

**Eye.** Blepharoptosis or narrowing of the palpebral fissure occurs on the affected side in about 10%. Unilateral palpebral shortening was noted in 6 of 57 patients studied by Mansour et al (59). Clinical anophthalmia or microphthalmia has been described in several patients who also manifested severe mental retardation (1,6,17,60,101,118). Various retinal abnormalities have been reported (60). Epibulbar tumors (Fig. 19–2A) are found in about 35% (6,35,59). They appear as solid yellowish or pinkish white ovoid masses varying in size from that of a pinhead to 8–10 mm in diameter. They occur most often at the inferotemporal quadrant at the limbus. The surface is usually smooth and frequently has fine hairs. They can occur at any location on the globe or in the orbit and can be dermoid (white solid masses), lipodermoid (yellow, movable, conjunctival), or dermis-like or complex (mesoectodermal). Unilateral epibulbar dermoids are seen in 50% (6). Bilateral lesions occur in 25% and have a tendency to be symmetrically placed in both eyes. However, multiple lesions can occur in the same eye. Lipodermoids are less common (25%), more often bilateral (20%), and more often are superotemporal or inferotemporal (6). Vision may be impaired by encroachment on the pupillary axis, by lipid infiltration of the cornea, or as a result of astigmatism (6). Removal of the masses can lead to scar formation with resultant leukoma.

Patients with epibulbar dermoids have a higher frequency of eyelid, extraocular, and lacrimal drainage abnormalities (6,43,59), microcornea, blepharoptosis (6), microphthalmia, and clinical anophthalmia (59,91). Unilateral colobomas of the upper lid are noted in about 20%, bilateral in possibly 3% (Fig. 19–2B) (6). Other ocular motility disorders, found in up to 25%, include esotropia, exotropia, and Duane syndrome (2,6,66,80). Choroidal or retinal colobomas, congenital cystic eye, and various other anomalies may occasionally be found (6).

Skin tags are the only characteristic of the syndrome that correlates with epibulbar dermoids for laterality (59). This association/localizing significance might be related to the fact that both are histologically choristomas.

**Ear.** Abnormalities ranging from anotia to an ill-defined mass of tissue that is displaced anteriorly and inferiorly, to a mildly dysmorphic ear are found in over 65% (Fig. 19–2C–E) (17). Farkas et al (25) showed no direct relationship between the degree of microtia and the extent of facial defect, whereas other authors have shown moderate correlation (28). Occasionally, bilateral anomalous pinnae are noted. Preauricular tags of skin and cartilage are extremely common, and may be unilateral or bilateral. Supernumerary ear tags may occur anywhere from the tragus to the angle of the mouth. They are more com-

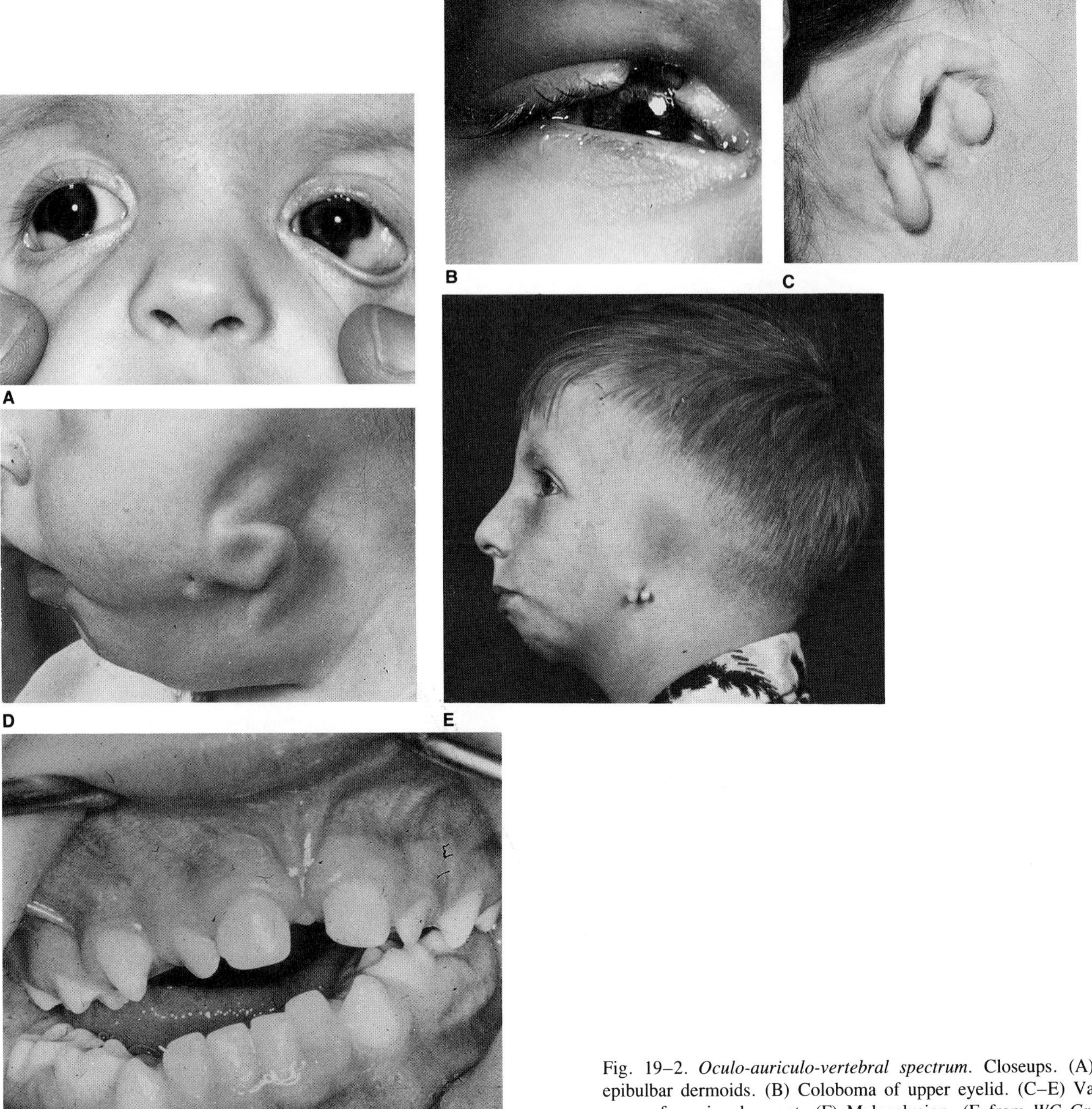

Fig. 19–2. *Oculo-auriculo-vertebral spectrum.* Closeups. (A) Bilateral epibulbar dermoids. (B) Coloboma of upper eyelid. (C–E) Variable degrees of ear involvement. (F) Malocclusion. (E from *WC Grabb,* Plast Reconstr Surg **36**:485, 1965.)

monly seen in patients with macrostomia and/or aplasia of the parotid gland and epibulbar dermoids. Preauricular sinuses and/or tags are noted in over 40%. Narrow external auditory canals are found in more mild cases; atretic canals are seen in more severe cases. At times small auricles with normal architecture are seen. Isolated microtia is considered to be a microform of OAV spectrum (7). Rollnick et al (91) require at least some ear involvement to confirm the diagnosis. Many previous studies included patients without ear involvement (38,39,95,97,113). Both conductive and, less frequent (15%), sensorineural hearing loss due to lesions of the middle and external ears, hypoplasia or agenesis of ossicles, aberrant facial nerves, patulous eustachian tube, and skull base anomaly have been reported in over 50% (5b,6,14,20,67,117,123).

**Central nervous system.** A wide range of central nervous system defects may be associated, and reported frequencies of mental deficiency appear to have ascertainment bias. For example, series preselected for severe central nervous system malformations have an inordinately high frequency of mental retardation, whereas those reported as treated patients from various craniofacial centers tend to be low. Most estimates range from 5 to 15% (48,101). Higher frequencies are those reported by Tenconi and Hall (37%) (113) and Wilson (82%) (118). Cohen (18) suggested an association between microphthalmia or clinical anophthalmia and mental retardation, as the eye is an outpouching of the primitive brain. However, patients with normal intelligence have been observed also.

Nearly all cranial nerves have occasionally been involved (4). Lower

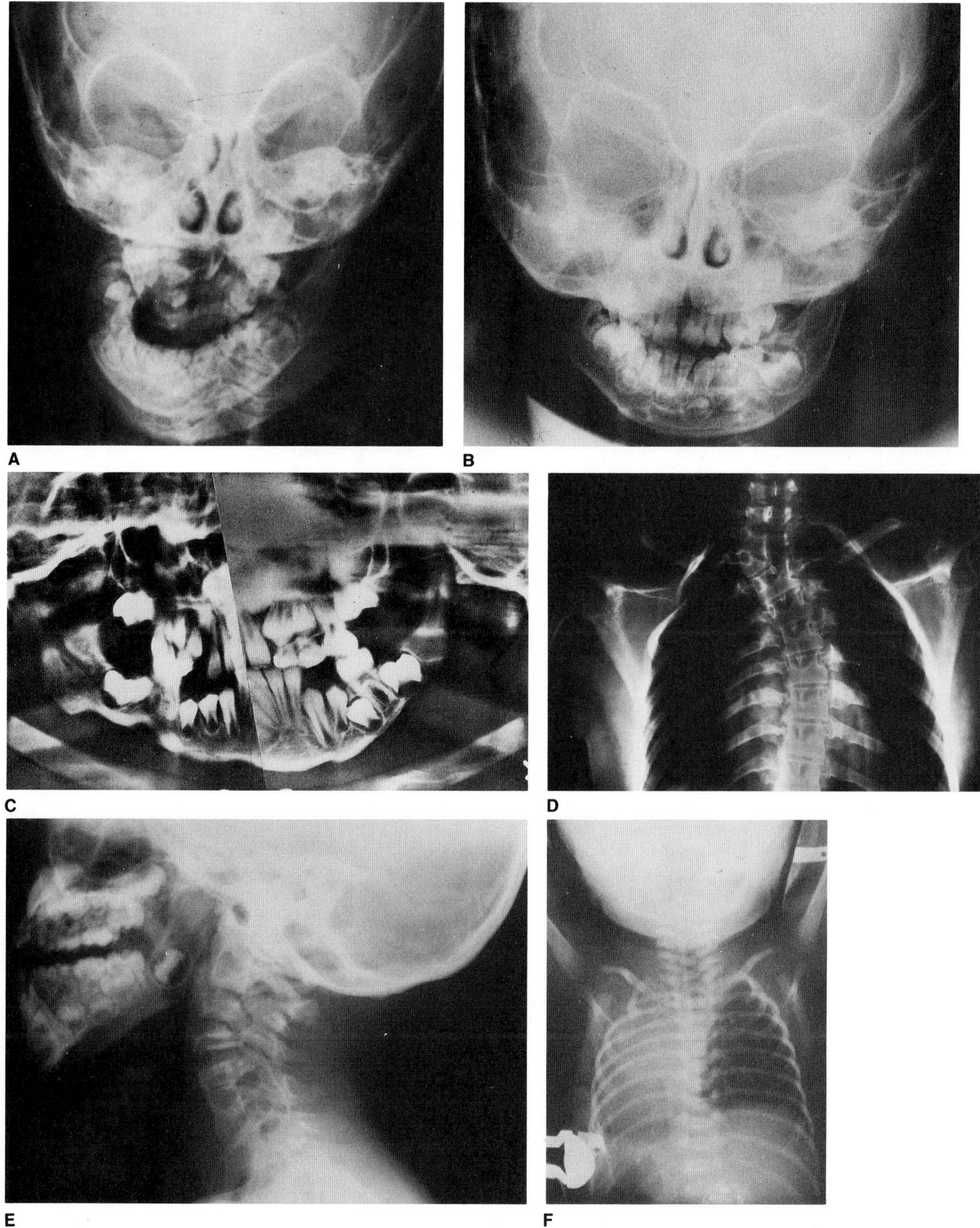


Fig. 19–3. *Oculo-auriculo-vertebral spectrum.* Roentgenograms. (A) Asymmetry and unilateral hypoplasia of mandibular ramus. (B) Absent ramus and condyle on affected side. (C) Panorex X ray showing more severe involvement on one side. Compare mandibular angle and ramus on both sides. (D) Hemivertebra in thoracic spine. (E) Note fusion in cervical area. (F) Unilateral absence of lung.

facial weakness occurs in 10–20%, probably being related to bony involvement in the region of the facial canal (5b,39). Abnormal course of the seventh cranial nerve (122) and unilateral aplasia of the trigeminal nuclei and the facial nerve (4,24) have been described, as well as trigeminal anesthesia (13,14,24,120). Other cranial nerves involved have included I (1), II (60), III, IV, and VI (2), and VIII, IX, and X (4).

The range of skull defects includes cranium bifidum, microcephaly, dolichocephaly, and plagiocephaly (4,14,38).

For the expanded oculo-auriculo-vertebral spectrum, brain malformations have been especially well discussed by Aleksic et al (4). Intracranial anomalies may include occipital and frontal encephaloceles, hydrocephaly, lipoma, dermoid cyst, teratoma, Arnold–Chiari malformation, lissencephaly, arachnoid cyst, holoprosencephaly, unilateral arhinencephaly, and hypoplasia of the corpus callosum (1,3,4,6a,18, 32a,38,43,43a,46,64,70,77a,101,114a,118).

When flattened frontal encephalocele occurs, the patient appears to have frontonasal malformation together with ear tags, other ear anomalies, and even epibulbar dermoids. Since encephalocele is more common in the occipital region than in the frontal region, we suggest that cases of so-called frontonasal malformation with epibulbar dermoids and ear tags may represent oculo-auriculo-vertebral spectrum with the encephalocele expressed anteriorly (30,43).

**Lung.** Pulmonary anomalies range from incomplete lobulation to hypoplasia to agenesis, unilateral or bilateral, the absent lung usually being ipsilateral to the facial anomalies (Fig. 19–3F) (12,38,60,73,81). Tracheoesophageal fistula has also been documented (12).

**Kidney.** A variety of renal abnormalities have been reported: absent kidney, double ureter, crossed renal ectopia, anomalous blood supply to the kidney, hydronephrosis, hydroureter, and other defects (13,91,101).

**Gastrointestinal anomalies.** Imperforate anus with or without rectovaginal fistula has been described (13).

**Heart.** A range of from 5 to 58% of patients with various forms of heart anomalies have been reported (32,38,40,81,91,101,113,118). VSD and tetralogy of Fallot with or without right aortic arch account for 50% of the anomalies, although no single cardiac lesion is characteristic. A wide variety of associated cardiac findings including transposition of the great vessels, tubular hypoplasia of the aortic arch associated with mild coarctation of the aorta, cardiomegaly, rare isolation of the left innominate artery with bilateral PDA (isolated or in combination with other anomalies), pulmonary stenosis, dextrocardia, and other aortic arch abnormalities have been described (81,91). Hypoplasia of the external carotid artery has also been noted (37).

**Skeletal alterations.** Facial anteroposterior and vertical dimensions are reduced on the affected side, especially in the lower face toward the otocephalic center (Fig. 19–3A–C). The temporomandibular joint is anteroinferiorly displaced The orbit is reduced in size and often vertically dislocated (104). Cervical spine and cranial base anomalies occur with increased frequency. Skull defects have also been noted (18,43,64,70). This group seem to have a worse prognosis (118). Cervical vertebral fusions (Fig. 19–3E) occur in 20–35%, whereas platybasia and occipitalization of the atlas are found in about 30% (5a,27). Spina bifida, hemivertebrae, butterfly, fused and hypoplastic vertebrae, Klippel–Feil anomaly, scoliosis, and anomalous ribs (agenesis, bifidity, fusion, supernumerary) occur in at least 30% (Fig. 19–3D) (5a,45). Talipes equinovarus has been reported in about 20% (38). Radial limb anomalies have been noted in about 10% (95). These may take the form of hypoplasia or aplasia of radius and/or thumb and bifid or digitalized thumb (13,37,58,63,68,108,118).

**Oral manifestations.** At least 35% with agenesis of the mandibular ramus have associated macrostomia or pseudomacrostomia, that is, lateral facial cleft, usually of mild degree (38). In the presence of epibulbar dermoids, the frequency of macrostomia may be somewhat higher (38). It is nearly always unilateral and on the side of the more affected ear. Occasionally, there may be agenesis of the ipsilateral parotid gland, displaced salivary gland tissue, or salivary fistulas (38). Intraorally, there is decreased palatal width from the midline palatal raphe to the lingual surface of the teeth on the affected side. The palatal and tongue muscles may be unilaterally hypoplastic and/or paralyzed. Unilateral or bilateral cleft lip and/or cleft palate occurs in 7–15% of patients (29,91). Cleft palate is twice as common as cleft lip with or without cleft palate (91). Tooth development tends to be somewhat delayed and third molars and other teeth tend to be missing on the affected side (24a). Canting of the occlusal plane and malocclusion are common (Fig. 19–2F). The occurrence of a tonguelike structure projecting from the tonsillar pillar in place of a tonsil has been reported by several authors (63a,99a,99b). In some cases there was partial aplasia of the soft palate, ear atresia, and facial palsy.

About 35% have velopharyngeal insufficiency (57a). Shprintzen (103) reported 12 patients with velopharyngeal insufficiency, and 2 of the 12 had cleft palate. Among the 10, the insufficiency resulted from asymmetry of movement of the lateral pharyngeal wall and dyssynchrony of palatal motion. Tongue restriction and hypernasality have been studied (97).

**Differential diagnosis.** The OAV spectrum is phenotypically variable and causally heterogeneous (90). It is essential to exclude various chromosome disorders and several syndromes with overlapping features. *Townes–Brocks syndrome* (115) consists of dysplastic ears, ear tags, and hearing loss in addition to thumb anomalies, anal defects, and renal anomalies. It is an autosomal dominant condition. The *branchio-oto-renal (BOR) syndrome* is associated with mixed hearing loss, preauricular pits, branchial fistulas or cysts, anomalous pinnae, malformed middle or inner ear, lacrimal duct stenosis/aplasia, and/or renal dysplasia. There is autosomal dominant inheritance with high penetrance and variable expression. Families have been described in which first-degree relatives have varying features of hemifacial microsomia and/or BOR, suggesting that in some instances hemifacial microsomia may constitute a component toward the severe end of the spectrum of the BOR syndrome (93). Epibulbar dermoid may be an isolated finding. Corneal dermoids are usually isolated (42). Preauricular skin tags or nodules occur in about 1% of the normal population (91).

The characteristic features of OAV spectrum are distinguishable from *mandibulofacial dysostosis, maxillofacial dysostosis, Nager acrofacial dysostosis,* and *postaxial acrofacial dysostosis.* Facial involvement in the OAV spectrum is usually asymmetric, with one side of the face more severely involved (91). Furthermore, there is far less hypoplasia of the malar bones present. Partial to total absence of the lower eyelashes has not been reported in the OAV spectrum. Preauricular tags rarely occur in mandibulofacial dysostosis.

Colobomas in OAV spectrum affect the upper eyelid; in mandibulofacial dysostosis and Nager acrofacial dysostosis the notch is in the outer one-third of the lower lid.

Characteristics of the VATER association (vertebral anomalies, ventricular septal defect, anal atresia, T-E fistula with esophageal atresia, and radial and renal dysplasia), the *CHARGE association* (**c**oloboma, **h**eart disease, **a**tresia choanae, **r**etarded growth and development, **g**enital anomalies, and **e**ar anomalies and/or hearing loss), and the MURCS association (Müllerian duct aplasia, renal aplasia, and cervicothoracic vertebral dysplasia) overlap with the OAV spectrum. It has already been pointed out that some patients share stigmata of both OAV spectrum and *frontonasal malformation* (18,30,43,43a,71, 113).

**Laboratory aids.** Prenatal diagnosis by ultrasonography in severe hypoplasia of mandible, severe abnormality of the auricle, and cleft lip and/or cleft palate may be helpful in some cases (6b,110a). Polyhydramnios may be present. CT is essential to see the middle ear bones when evaluating hearing. Radiographic analysis for skeletal findings is also indicated. The descent of the tegmen may, in association with microtia, be the most reliable clinical marker of OAV spectrum. The level of the tegmen below the base of the middle cranial fossa appears to bear a direct relationship to the severity of the disorder (79).

### References [Oculo-auriculo-vertebral spectrum (hemifacial microsomia, Goldenhar syndrome)]


1. Aleksic S et al: Unilateral arhinencephaly in Goldenhar–Gorlin syndrome. *Dev Med Child Neurol* **17**:498–504, 1975.
2. Aleksic S et al: Congenital ophthalmoplegia in oculoauriculovertebral dysplasia in hemifacial microsomia (Goldenhar–Gorlin syndrome). *Neurology* **26**:638–644, 1976.
3. Aleksic S et al: Encephalocele (cerebellocele) in the Goldenhar–Gorlin syndrome. *Eur J Pediatr* **140**:137–138, 1983.
4. Aleksic S et al: Intracranial lipomas, hydrocephalus and other CNS

anomalies in oculoauriculo-vertebral dysplasia (Goldenhar–Gorlin syndrome). *Child's Brain* **11**:285–297, 1984.

5. Arlt F von: *Klinische Darstellung der Krankheiten des Auges*. W. Braunmüller, Wien, 1881.

5a. Avon SW, Shively JL: Orthopaedic manifestations of Goldenhar syndrome. *J Pediatr Orthop* **8**:683–686, 1988.

5b. Bassila MK, Goldberg R: The association of facial palsy and/or sensorineural hearing loss in patients with hemifacial microsomia. *Am J Med Genet* **26**:287–291, 1989.

6. Baum JL, Feingold M: Ocular aspects of Goldenhar's syndrome. *Am J Ophthalmol* **75**:250–257, 1973.

6a. Beltinger C, Saule H: Imaging of lipoma of the corpus callosum and intracranial dermoids in the Goldenhar syndrome. *Pediatr Radiol* **18**:72–73, 1988.

6b. Benacerraf BR, Frigoletto FD Jr: Prenatal ultrasound recognition of Goldenhar's syndrome. *Am J Obstet Gynecol* **159**:950–952, 1988.

7. Bennum RD et al: Microtia: A microform of hemifacial microsomia. *Plast Reconstr Surg* **76**:859–863, 1985.

8. Berkman MD, Feingold M: Oculoauriculovertebral dysplasia (Goldenhar's syndrome). *Oral Surg* **25**:408–417, 1968.

9. Bersu ET, Ramirez-Castro JL: Anatomical analysis of the developmental effects of aneuploidy in man: The 18 trisomy syndrome: I. Anomalies of the head and neck. *Am J Med Genet* **1**:173–193, 1977.

10. Berthelon S, Cremer N: Syndrome de Goldenhar avec malformations associeés. *Bull Soç Belg Ophtalmol* **149**:509–513, 1969.

11. Bock RH: Ein Fall von epibulbärem Dermolipome mit Missbildungen einer Gesichtshälfte. Diskordantes Vorkommen bei einem eineiigen Zwillingspaar. *Ophthalmologica* **155**:86–90, 1961.

11a. Boles DJ et al: Goldenhar complex in discordant monozygotic twins: A case report and review of the literature. *J Med Genet* **28**:103–109, 1987.

12. Bowen AD, Parry WH: Bronchopulmonary foregut malformation in the Goldenhar anomalad. *Am J Roentgenol* **134**:186–188, 1980.

13. Bowen DI et al: Clinical aspects of oculo-auriculo-vertebral dysplasia. *Br J Ophthalmol* **55**:145–154, 1971.

14. Budden SS, Robinson GC: Oculoauricular vertebral dysplasia. *Am J Dis Child* **125**:431–433, 1973.

15. Burck U: Genetic aspects of hemifacial microsomia. *Hum Genet* **64**:291–296, 1983.

16. Canton E: Arrest of development of the left perpendicular ramus of the lower jaw, combined with malformation of the external ear. *Tr Path Soc London* **12**:237–238, 1861.

17. Coccaro PJ et al: Clinical and radiographic variations in hemifacial microsomia. *Birth Defects* **11**(2):314–324, 1975.

18. Cohen MM Jr: Variability versus "incidental findings" in the first and second branchial arch syndrome: Unilateral variants with anophthalmia. *Birth Defects* **7**(7):103–108, 1971.

18a. Cohen MM Jr et al: Oculoauriculovertebral spectrum: An updated critique. *Cleft Palate J* **26**:276–286, 1989.

19. Connor JM, Fernandez C: Genetic aspects of hemifacial microsomia. *Hum Genet* **68**:349, 1984.

20. Converse JM et al: On hemifacial microsomia. The first and second branchial arch syndrome. *Plastr Reconstr Surg* **51**:268–279, 1973.

21. Cordier J et al: Syndrome de Franceschetti–Goldenhar discordant chez deux jumelles monozygotes. *Arch Ophtalmol (Paris)* **30**:321–328, 1970.

22. Curran JP et al: Partial deletion of the long arm of chromosome E-18. *Pediatrics* **46**:721–729, 1970.

23. Dyggve HV, Mikkelsen M: Partial deletion of the short arms of a chromosome of the 4–5 group (Denver). *Arch Dis Childh* **40**:82–85, 1965.

24. Ebbesen F, Petersen W: Goldenhar's syndrome: Discordance in monozygotic twins and unusual anomalies. *Acta Paediatr Scand* **71**:685–687, 1982.

24a. Farias M, Vargervik K: Dental development in hemifacial microsomia. I. Eruption and agenesis. *Pediatr Dent* **10**:140–143, 1988.

25. Farkas LG et al: Anthropometry of the face in lateral facial dysplasia. The bilateral form. *Cleft Palate J* **14**:41–51, 1977.

26. Farkas LG, James JS: Anthropometry of the face in later facial dysplasia. The unilateral form. *Cleft Palate J* **14**:193–199, 1977.

27. Figueroa AA, Friede H: Craniovertebral malformations in hemifacial microsomia. *J Craniofac Genet Dev Biol, Suppl* **1**:167–178, 1985.

28. Figueroa AA, Pruzansky S: The external ear, mandible and other components of hemifacial microsomia. *J Maxillofac Surg* **10**:200–211, 1982.

29. Feingold M, Baum J: Goldenhar's syndrome. *Am J Dis Child* **132**:136–138, 1978.

30. Fleischer-Peters A: Goldenhar-Syndrom und Kiefermissbildungen. *Dtsch Zahnärztl Z* **24**:545–551, 1969.

31. François J, Haustrate L: Anomalies colobomateuses du globe oculaire et syndrome du premier arc. *Ann Ocul* **187**:340–368, 1954.

32. Friedman S, Saraclar M: The high frequency of congenital heart disease in oculo-auriculo-vertebral dysplasia (Goldenhar's syndrome). *J Pediatr* **85**:873–874, 974.

32a. Gaupp R, Janz H: Zur Kasuistik der Balkenlipome. *Nervenarzt* **45**:321–325, 1941.

33. Geissmar F: Zur Kasuistik der kongenitalen Liddefekte. *Beitr Augenheilkd* **4**:467–479, 1899.

34. Godel V et al: Autosomal dominant Goldenhar syndrome. *Birth Defects* **18**(6):621–628, 1982.

35. Goldenhar M: Associations malformatives de l'oeil et de l'oreille, en particulier le syndrome dermöide epibulbaire-appendices auriculaires-fistula auris congenita et ses relations avec la dysostose mandibulo-faciale. *J Génét Hum* **1**:243–282, 1952.

36. Gomez Garcia A et al: Sindrome de Goldenhar. Discordancia en gemelos monocigotos. *An Esp Pediatr* **20**:400–402, 1984.

37. Gorlin RJ et al: Oculoauriculovertebral dysplasia. *J Pediatr* **63**:991–999, 1963.

38. Gorlin RJ et al: Oculoauriculovertebral dysplasia, in *Syndromes of the Head and Neck,* ed. 2, McGraw-Hill, New York, 1976, pp 546–552.

39. Grabb WC: The first and second branchial arch syndrome. *Plast Reconst Surg* **36**:485–508, 1965.

39a. Greenberg F et al: Chromosome abnormalities associated with facio-auriculo-vertebral dysplasia. *Clinical Genetics Conference: Neural Crest and Craniofacial Disorders*. March of Dimes Birth Defects Meeting. Minneapolis, July 19–22, 1987.

40. Greenwood RD et al: Cardiovascular malformations in oculoauriculovertebral dysplasia. *J Pediatr* **85**:816–818, 1974.

41. Grix A Jr: Malformations in infants of diabetic mothers. *Am J Med Genet* **13**:131–137, 1982.

42. Guizar Vazquez J et al: Corneal dermoids and short stature in brother and sister—a new syndrome? *Am J Med Genet* **8**:229–234, 1981.

43. Gupta JS et al: Oculoauricular-cranial dysplasia. *Br J Ophthalmol* **52**:346–347, 1968.

43a. Gustavson EE, Chen H: Goldenhar syndrome, anterior encephalocele and aqueductal stenosis following fetal primidone exposure. *Teratology* **32**:13–17, 1985.

44. Heimann K von: Beitrag zur Klinik des Goldenhar-Syndroms. *Klin Monatsbl Augenheilkd* **152**:686–692, 1968.

45. Helmi C, Pruzansky S: Craniofacial and extracranial malformations in the Klippel–Feil syndrome. *Cleft Palate J* **17**:65–88, 1980.

45a. Herman GE et al: Brief clinical observation: Multiple congenital anomaly/mental retardation (MCA/MR) syndrome with Goldenhar complex due to a terminal del(22q). *Am J Med Genet* **29**:909–915, 1988.

46. Herrmann J, Opitz JM: A dominantly inherited first arch syndrome. *Birth Defects* **5**(2):110–112, 1969.

47. Hodes ME et al: Trisomy 7 mosaicism and manifestations of Goldenhar syndrome with unilateral radial hypoplasia. *J Craniofac Genet Dev Biol* **1**:49–55, 1981.

48. Hollwich F, Verbeck B: Zur Dysplasia oculoauricularis (Franceschetti–Goldenhar). *Klin Monatsbl Augenheilkd* **154**:430–443, 1969.

49. Ide CH et al: Familial facial dysplasia. *Arch Ophthalmol* **84**:427–433, 1970.

49a. Israel J et al: Goldenhar syndrome (oculoauricular vertebral dysplasia) in an infant of a diabetic: Case report. *Clinical Genetics Conference: Neural Crest and Craniofacial Disorders*. March of Dimes Birth Defects Meeting. Minneapolis, July 19–22, 1987.

49b. Jongbloet PH: Goldenhar syndrome and overlapping dysplasias. *J Med Genet* **24**:616–620, 1987.

50. Kaban LB et al: Three dimensional approach to analysis and treatment of hemifacial microsomia. *Cleft Palate J* **18**:90–99, 1981.

50a. Kay ED, Kay CN: Dysmorphogenesis of the mandible, zygoma, and middle ear ossicles in hemifacial microsomia and mandibulofacial dysostosis. *Am J Med Genet* **32**:27–31, 1989.

51. Kirke DK: Goldenhar's syndrome: Two cases of oculo-auriculo-vertebral dysplasia occurring in full-blood Australian aboriginal sisters. *Aust Paediatr J* **6**:213–214, 1970.

52. Krause VH: The syndrome of Goldenhar affecting two siblings. *Acta Ophthalmol (Kbh)* **48**:494–499, 1970.

53. Kushnick T, Colondrillo M: 49,XXXXY patient with hemifacial microsomia. *Clin Genet* **7**:442–448, 1975.

54. Ladekarl S: Combination of Goldenhar's syndrome with the cri-du-chat syndrome. *Acta Ophthalmol (Kbh)* **46**:605–610, 1968.

55. Lammer ES et al: Retinoic acid embryopathy. *N Engl J Med* **313**(14):837–841, 1985.

56. Livingston G: Congenital ear abnormalities due to thalidomide. *Proc R Soc Med* **58**:493–497, 1965.

57. Loevy HT, Shore SW: Dental maturation in hemifacial microsomia. *J Craniofac Genet Dev Biol, Suppl* **1**:267–272, 1985.

57a. Luce EA et al: Velopharyngeal insufficiency in hemifacial microsomia. *Plast Reconstr Surg* **60**:602–606, 1977.

58. Mandelcorn MS et al: Goldenhar's syndrome and phocomelia. *Am J Ophthalmol* **72**:618–621, 1971.

59. Mansour AM et al: Ocular findings in the facio-auriculovertebral sequence (Goldenhar–Gorlin syndrome). *Am J Ophthalmol* **100**:555–559, 1985.

60. Margolis S et al: Retinal and optic nerve findings in Goldenhar–Gorlin syndrome. *Ophthalmology* **91**:1327–1333, 1984.

61. Melnick M, Myrianthopoulos NC: *External Ear Malformations: Epidemiology, Genetics and Natural History*. Alan R. Liss, New York, 1979, p 21.

62. Melnick M: The etiology of external ear malformations and its relation to abnormalities of the middle ear, inner ear and other organ systems. *Birth Defects* **16**(4):303–331, 1980.

63. Mendelberg A et al: Tracheo-oesophageal anomalies in the Goldenhar anomalad. *J Med Genet* **22**:149–150, 1985.

63a. Menzel KM: Über noch nicht beschriebenes Vorkommen einer Lingua accessoria. *Virchow Arch Path Anat* **288**:253–268, 1933.

64. Michaud C, Sheridan S: Goldenhar's syndrome associated with cranial and neurological malformations. *Can J Ophthalmol* **9**:347–350, 1974.

65. Miehlke A, Partsch CJ: Ohrmissbildung, Facialis- und Abducenslähmung als Syndrom der Thalidomidschädigung. *Arch Ohrenheilkd* **181**:154–174, 1963.

66. Miller MT: Association of Duane retraction syndrome with craniofacial malformations. *J Craniofac Genet Dev Biol Suppl* **1**:273–282, 1985.

67. Miyamoto RT et al: Goldenhar syndrome associated with submandibular gland hyperplasia and hemihypoplasia of the mobile tongue. *Arch Otolaryngol* **102**:313–314, 1976.

68. Moeschler J, Clarren SK: Familial occurrence of hemifacial microsomia with radial limb defects. *Am J Med Genet* **12**:371–375, 1982.

69. Mounoud RL et al: A propos d'un cas de syndrome de Goldenhar. *J Génét Hum* **23**:135–154, 1975.

70. Murphy MJ et al: Intracranial dermoid cyst in Goldenhar's syndrome. *J Neurosurg* **53**:408–411, 1980.

71. Musarella MA, Young ID: A patient with median cleft face anomaly and bilateral Goldenhar anomaly. *Am J Med Genet Suppl* **2**:135–141, 1986.

72. Neu KW et al: Hemifacial microsomia in cri du chat (5p−) syndrome. *J Craniofac Genet Dev Biol* **2**:295–298, 1982.

73. Opitz JM, Faith GC: Visceral anomalies in an infant with the Goldenhar syndrome. *Birth Defects* **5**(2):104–105, 1969.

74. Papp Z et al: Probably monozygotic twins with discordance for Goldenhar syndrome. *Clin Genet* **5**:86–90, 1974.

75. Par MM et al: Syndrome de Goldenhar et gemellite. *Bull Soç Ophtalmol Fr* **12**:1134–1145, 1967.

76. Par MM et al: A propos d'une observation familiale de syndrome de Franceschetti–Goldenhar. *Bull Soç Ophtalmol Fr* **63**:705–707, 1963.

77. Pashayan H et al: Hemifacial microsomia: Oculo-auriculo-vertebral dysplasia. A patient with overlapping features. *J Med Genet* **7**:185–188, 1970.

77a. Pauli R et al: Goldenhar association and cranial defects. *Am J Med Genet* **15**:177–179, 1983.

78. Perez Alvarez F et al: Sindrome otocraneofacial asimetrico (microsomia hemifacial) en gemelos monocigoticos discordantes. Aspectos otologicos. *An Esp Pediatr* **21**(8):769–773, 1984.

79. Phelps PD et al: The ear deformity in craniofacial microsomia and oculo-auriculo-vertebral dysplasia. *J Laryngol Otol* **97**:995–1005, 1983.

80. Pieroni D: Goldenhar's syndrome associated with bilateral Duane's retraction syndrome. *J Pediatr Ophthalmol* **6**:16–18, 1969.

81. Pierpont MEM et al: Congenital cardiac, pulmonary and vascular malformations in oculoauriculovertebral dysplasia. *Ped Cardiol* **2**:297–302, 1982.

82. Poonawalla HH et al: Hemifacial microsomia in a patient with Klinefelter syndrome. *Cleft Palate J* **17**:194–196, 1980.

83. Poswillo D: The pathogenesis of the first and second branchial arch syndrome. *Oral Surg* **35**:302–329, 1973.

84. Poswillo D: Otomandibular deformity: Pathogenesis as a guide to reconstruction. *J Maxillofac Surg* **2**:64–72, 1974.

85. Poswillo D: Hemorrhage in development of the face. *Birth Defects* **11**(7):61–81, 1975.

86. Rapin I, Ruben RJ: Patterns of anomalies in children with malformed ears. *Laryngoscope* **86**:1469–1502, 1976.

87. Regenbogen L et al: Further evidence for an autosomal dominant form of oculoauriculovertebral dysplasia. *Clin Genet* **21**:161–167, 1982.

88. Robinow M et al: Hemifacial microsomia, ipsilateral facial palsy, and malformed auricle in two families: An autosomal dominant malformation. *Am J Med Genet Suppl* **2**:129–133, 1986.

89. Robinson L et al: The vascular pathogenesis of unilateral craniofacial defects. *J Pediatr* **111**:236–239, 1987.

90. Rollnick BR: Oculoauriculovertebral anomaly: Variability and causal heterogeneity. *Am J Med Genet Suppl* **4**:41–53, 1988.

91. Rollnick BR et al: Oculoauriculovertebral dysplasia and variants: Phenotypic characteristics of 294 patients. *Am J Med Genet* **26**:361–375, 1987.

92. Rollnick BR, Kaye CI: Hemifacial microsomia and variants: Pedigree data. *Am J Med Genet* **15**:233–253, 1983.

93. Rollnick BR, Kaye CI: Hemifacial microsomia and the branchio-oto-renal syndrome. *Am J Med Genet Suppl* **1**:287–295, 1985.

94. Rosenal TH: Aplasia-hypoplasia of the otic labyrinth after thalidomide. *Acta Radiol* **3**:225–236, 1965.

95. Ross RB: Lateral facial dysplasia (first and second branchial arch syndrome, hemifacial microsomia). *Birth Defects* **11**(7):51–59, 1975.

96. Rune B et al: Growth in hemifacial microsomia studied with the aid of roentgen stereophotogrammetry and metallic implants. *Cleft Palate J* **17**:128–146, 1981.

96a. Ryan CA et al: Discordance of signs in monozygotic twins concordant for the Goldenhar anomaly. *Am J Med Genet* **29**:755–761, 1988.

97. Samuels et al: Characteristics of patients with hemifacial microsomia at the Craniofacial Center, in *Treatment of Hemifacial Microsomia,* Harvold EP (ed), Alan R. Liss, New York, 1983, pp 51–55.

98. Sando I et al: Temporal bone histopathology findings in oculo-auriculo-vertebral dysplasia. Goldenhar's syndrome. *Ann Otol Rhinol Laryngol* **95**:396–400, 1986.

99. Saraux MH, Besnainou L: Les syndrome maxillooculaires. *Ann Ocul* **198**:953–964, 1965.

99a. Satake T: Ein Fall von Missbildung der Zunge. *Zbl Hals Nas Ohrenheilk* **28**:288, 1937.

99b. Sercer A: Komplette Anotie kombiniert mit Aplasie der Tonsille. *Z Hals Nas Ohrenheilk* **22**:75–78, 1928.

100. Setzer ES et al: Etiologic heterogeneity in the oculoauriculovertebral syndrome. *J Pediatr* **98**:88–91, 1981.

101. Shokeir MHK: The Goldenhar syndrome: A natural history. *Birth Defects* **13**(3C):67–83, 1977.

102. Shprintzen RJ: Palatal and pharyngeal anomalies in craniofacial syndromes. *Birth Defects* **18**(1):53–78, 1982.

103. Shprintzen RJ et al: Velopharyngeal insufficiency in the facio-auriculo-vertebral malformation complex. *Cleft Palate J* **17**:132–137, 1980.

104. Smakel Z: Craniofacial changes in hemifacial microsomia. *J Craniofac Genet Develop Biol* **6**:151–170, 1986.

105. Smith DW: Facio-auriculo-vertebral spectrum, in *Recognizable Patterns of Human Malformation,* 3rd ed., W. B. Saunders, Philadelphia, 1982, pp 497–500.

106. Stark RB, Saunders DE: The first branchial syndrome: The oral-mandibular-auricular syndrome. *Plast Reconstr Surg* **29**:229–239, 1967.

107. Stoll C: Discordance for skeletal and cardiac defect in monozygotic twins. *Acta Genet Med Gemellol* **33**:501–504, 1984.

108. Sugiura Y: Congenital absence of the radius with hemifacial microsomia, VSD, and crossed renal ectopia. *Birth Defects* **7**(7):109–115, 1971.

109. Sujansky E, Smith ACM: Recombinant chromosome 18 in two male sibs with first and second branchial arch syndrome. *Amer Soc Hum Genet* (Abstr) 92A, 1981.

110. Summitt R: Familial Goldenhar syndrome. *Birth Defects* **5**(2):106–109, 1969.

110a. Tamas DE et al: Prenatal sonographic diagnosis of hemifacial microsomia (Goldenhar–Gorlin syndrome). *J Ultrasound Med* **5**:461–463, 1986.

111. Taysi K et al: Familial hemifacial microsomia. *Cleft Palate J* **2**:47–53, 1983.

112. ter Haar B: Oculo-auriculo-vertebral dysplasia (Goldenhar's syndrome). Concordant in identical twins. *Acta Med Genet (Roma)* **21**:116–124, 1972.

113. Tenconi R, Hall BD: Hemifacial microsomia: Phenotypic classification, clinical implications and genetic aspects, in *Treatment of Hemifacial Microsomia,* Harvold EP (ed), Alan R. Liss, New York, 1983, pp 39–49.

114. Thomas P: Goldenhar syndrome and hemifacial microsomia: Observations on three patients. *Eur J Pediatr* **133**:287–292, 1980.

114a. Thommen L et al: Balkenlipom als Bestandteil des Goldenhar-Syndroms. *Mschr Kinderheilkd* **134**:541–543, 1986.

115. Townes PL, Brocks ER: Hereditary syndrome of imperforate anus with hand, foot and ear anomalies. *J Pediatr* **81**:321–326, 1972.

116. Townes PL, White MR: Inherited partial trisomy 8q(22 qter). *Am J Dis Child* **132**:498–501, 1978.

117. Wells MD et al: Oculo-auriculo-vertebral dysplasia. *J Laryngol Otol* **97**:689–696, 1983.

118. Wilson GN: Cranial defects in Goldenhar's syndrome. *Am J Med Genet* **14**:435–443, 1983.

119. Wilson GN, Barr M Jr: Trisomy 9 mosaicism: Another etiology for the manifestations of Goldenhar syndrome. *J Craniofac Genet Dev Biol* **3**:313–316, 1983.

120. Yanagihara N et al: Goldenhar's syndrome associated with anomalous internal auditory meatus. *J Laryngol Otol* **93**:1217–1222, 1979.

121. Yovich SL et al: Fetal abnormality (Goldenhar syndrome) occurring in one of triplet infants derived from in vitro fertilization with possible monozygotic twinning. *In Vitro Fertil Embryo Transfer* **2**(1):27–32, 1985.

122. Yovich J et al: Goldenhar syndrome and overlapping dysplasias, in vitro fertilisation and ovopathy. *J Med Genet* **24**:616–620, 1987.

123. Zeitzer LD, Lindeman RC: Multiple branchial arch anomalies. *Arch Otolaryngol* **93**:562–567, 1971.

## Mandibulofacial dysostosis (Treacher Collins syndrome, Franceschetti–Zwahlen–Klein syndrome)

Mandibulofacial dysostosis (Figs. 19–4 to 19–6) involves structures derived from the first and second pharyngeal arch, groove, and pouch. Although the syndrome was probably first described by Thomson (34) and Toynbee (34a) in 1846–1847, credit for its discovery is usually given to Berry (4) or, especially, to Treacher Collins (35) (name often erroneously hyphenated), who described the essential components of the syndrome. Franceschetti and co-workers (8,9), during the 1940s, published extensive reviews of the disorder and coined the term *mandibulofacial dysostosis*.

The syndrome has autosomal dominant inheritance with variable expressivity. A number of families have expressed the syndrome in several generations (7,29,32,36). Franceschetti et al (9), in a review of 63 cases, found 27 with a family history of the syndrome. Among the roughly 60% that represent new mutations, it has been shown that the fathers tend to be older (15). Perhaps 400 or more cases have been published. Lowry et al (17a) described a recessive syndrome which we believe is a different disorder.

Poswillo (26), Sulik et al (33), and Wiley et al (37) employing mouse, rat, and hamster models gave teratogenic doses of vitamin A and isotretinoin and produced malformations of the craniofacial skeleton that closely resembled the syndromic features of human mandibulofacial dysostosis. Histological studies indicated that abnormalities of neural crest development occurred in these phenocopies but whether the principal defect is in migration or differentiation of the ectomesenchymal cells or their death in the condensing trigeminal ganglia or results from the effects on the first and second arch placodal cells following their release from the neural folds of neural crest cells into the developing cranial regions remains unclear (2,13,33).

**Facies.** The facial appearance is characteristic. Abnormalities are bilateral and usually symmetric (Fig. 19–4A–E). The nose appears large but really is not (15a); the appearance is secondary to hypoplastic supraorbital rims and hypoplastic zygomas. The face is narrow. Downward-sloping palpebral fissures, depressed cheekbones, malformed pinnae, receding chin, and large downturned mouth are characteristic. About 25% manifeset a tongue-shaped process of hair that extends toward the cheek (28).

**Skull.** The calvaria is essentially normal, but radiographic studies reveal that the supraorbital ridges are poorly developed (20,23,32). The body of the malar bones may be totally absent but more often is grossly and symmetrically underdeveloped, with nonfusion of the zygomatic arches. The zygomatic process of the frontal bone is hypoplastic as are the lateral pterygoid plates and muscles. The mastoids are not pneumatized and are frequently sclerotic. The paranasal sinuses are often small and may be completely absent. The orbits are hyperteloric (15a). The lower margin of the orbit may be defective and the infraorbital foramen is usually absent. The cranial base is progressively kyphotic (24a). The mandibular condyle is severely hypoplastic (Fig. 19–6A,B) and is covered with hyaline cartilage rather than fibrocartilage. There have been several excellent anatomic studies (2,6,13,17,21).

**Eyes.** The palpebral fissures are short and slope laterally downward and, often (75%), there is a coloboma in the outer third of the lower lid (Fig. 19–5A,B). We have seen several patients in whom the downward slope is asymmetric. About half of the patients have deficiency of cilia medial to the coloboma. Iridial coloboma may also occur. The lower lacrimal points may be absent as well as the Meibomian glands and intermarginal strip (19). The eye anomalies have been especially well reviewed by Franceschetti et al (10).

**Ears.** The pinnae are often malformed, crumpled forward, or misplaced toward the angle of the mandible. In the survey of Stovin et al (32), 51 of 63 patients had anomalous pinnae with meatal atresia and over one-third had absence of the external auditory canal or ossicle defect accompanied by bilateral conductive hearing loss. Kolar et al (15a) found microtia in 60%. The auditory ossicles and cochlear and vestibular apparatus have been observed to be absent or severely malformed (14,16,19,21,25a). Radiographic and surgical studies have shown agenesis or hypoplasia of the mastoid and mastoid antrum, absence of the external auditory canal, narrowing or agenesis of the middle ear cleft, agenesis or malformation of the malleus and/or incus, monopodal stapes, absence of the stapes and oval window, ankylosis of the stapes in the oval window, deformed suprastructure of the stapes, complete absence of the middle ear and epitympanic space. The space may be filled with connective tissue. The inner ears are normal (14,16,19,21).

Extra ear tags and blind fistulas may occur anywhere between the tragus and the angle of the mouth. In one case, blind fistulas were found behind the ear lobes (12).

**Nose.** The nasofrontal angle is usually obliterated, and the bridge of the nose raised. The nose appears large (15a) because of the lack of malar development and hypoplastic supraorbital ridges. The nares are often narrow, and the alar cartilages hypoplastic. Choanal atresia has been reported (18,22,30).

**Mental status.** Intelligence is usually normal. However, Stovin et al (32) reviewed 63 patients and found four who were mentally deficient. Other investigators have also noted mild mental retardation in their patients (12). We suspect that retardation may be secondary to hearing loss.

**Oral findings.** The palate is cleft in about 35% (8,23,25,32). Congenital palatopharyngeal incompetence (agenesis of soft palate, foreshortened soft palate, submucous palatal cleft, immobile soft palate) has been found in an additional 30–40% (25). Rarely cleft lip–palate has been noted. Macrostomia, observed in about 15%, may be unilateral or bilateral. The elevator muscles of the upper lip are deficient (13). The parotid salivary glands are absent or hypoplastic (13,19a,21). Pharyngeal hypoplasia, a constant finding, may explain cases of neonatal death (31).

Radiographic studies have shown that the neck is short and the condyle is malformed. The undersurface of the body of the mandible is often quite concave. The angle is more obtuse than normal and the ramus is often deficient. The coronoid and condyloid processes are flat or even aplastic. There is no articular eminence and the articular area is atypically medial. Grönvall and Olsson (12) reconstructed a model of the mandible. Detailed cephalometric studies have been carried out by Garner (11) and Roberts et al (27).

**Differential diagnosis.** *Oculoauriculovertebral spectrum* is easily excluded. The facies is so characteristic that little difficulty in diagnosis should be experienced. However, affected relatives may have minimal signs of the syndrome and must be closely examined. A

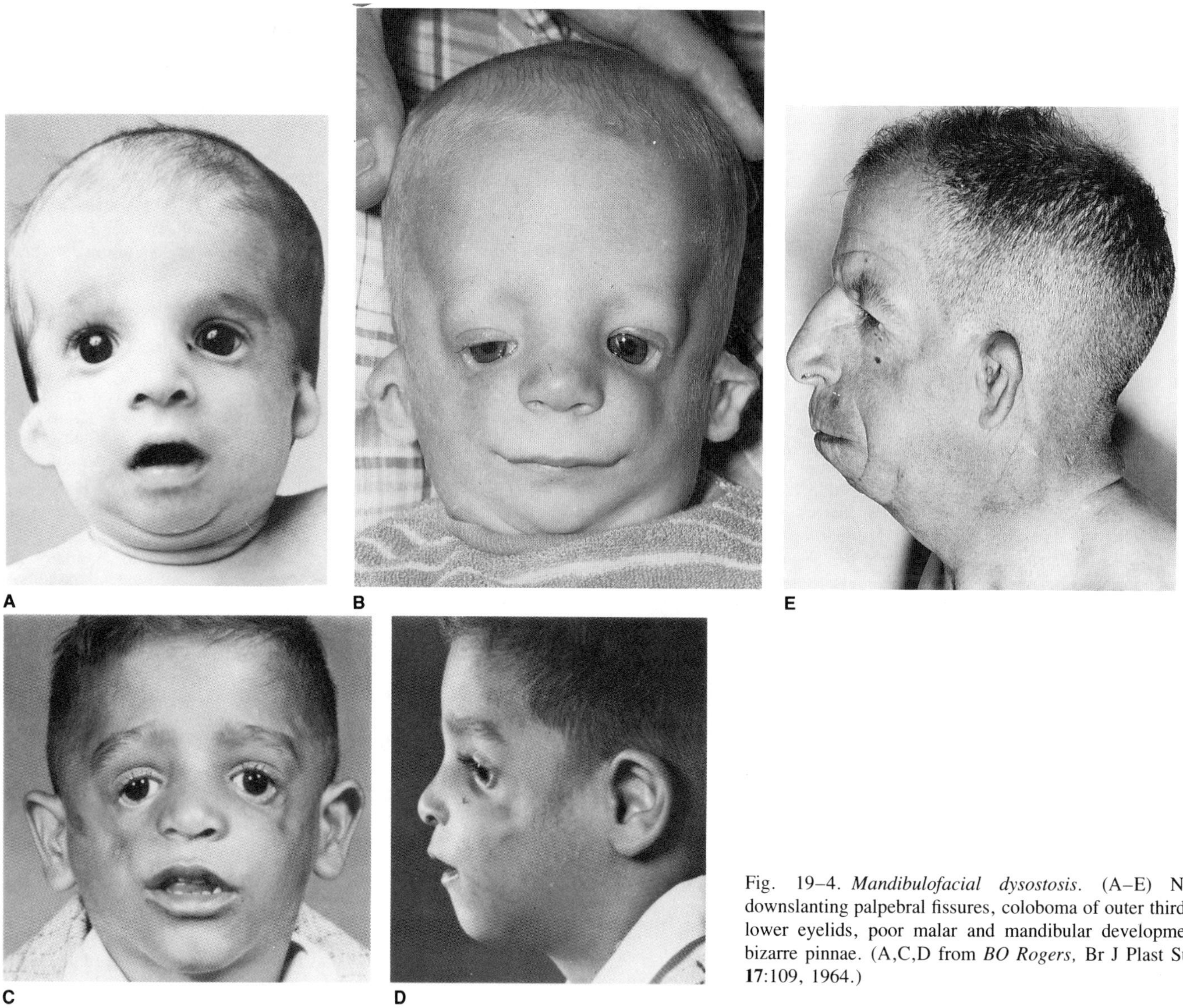

A B E

C D

Fig. 19–4. *Mandibulofacial dysostosis.* (A–E) Note downslanting palpebral fissures, coloboma of outer third of lower eyelids, poor malar and mandibular development, bizarre pinnae. (A,C,D from *BO Rogers,* Br J Plast Surg **17**:109, 1964.)

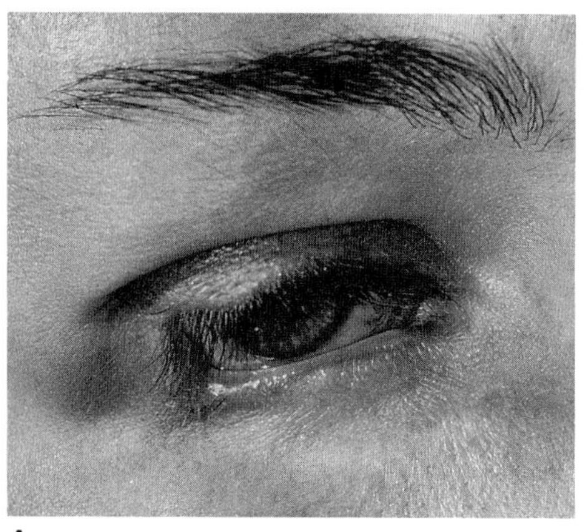

A

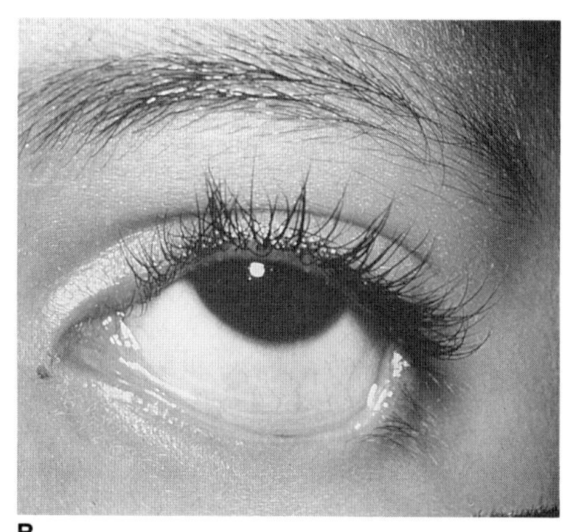

B

Fig. 19–5. *Mandibulofacial dysostosis.* (A,B) Close-up of eye in different patients showing absence of cilia on lower eyelid. Note coloboma of lower lid in A.

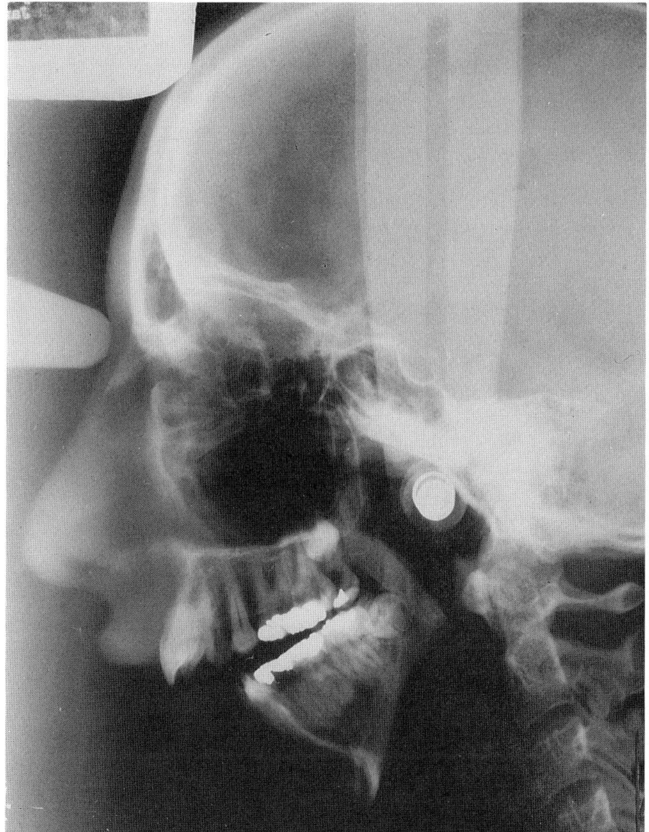

A

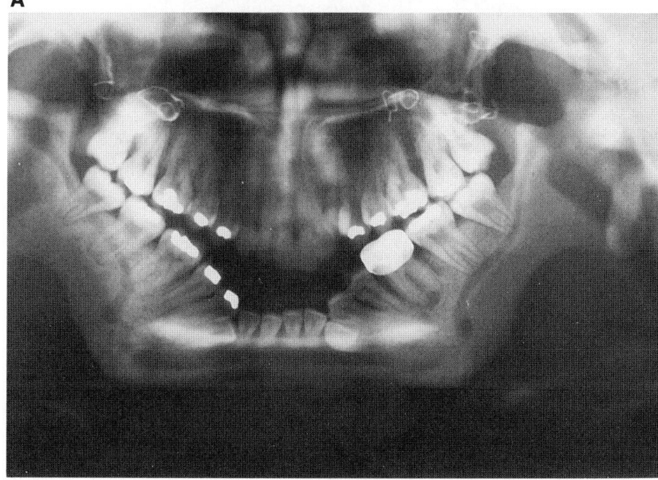

B

Fig. 19–6. *Mandibulofacial dysostosis.* Radiographs. (A) Lateral cephalogram showing extremely steep mandibular plane angle, antegonial notching, and mandibular deficiency. Soft tissue demonstrates prominent nasal contour, acute nasolabial angle, and lip incompetence. (B) Panorex X-ray showing prominent antegonial notching and open bite.

sporadic case with minimal involvement can present difficulties in diagnosis.

*Nager acrofacial dysostosis* closely resembles mandibulofacial dysostosis. The thumbs are hypoplastic or absent, the radius and ulna may be fused or there may be absence or hypoplasia of the radius and/or one or more metacarpals. Lower lid colobomas are rarer, cleft palate more frequent, and the mandible more severely retarded in growth than in mandibulofacial dysostosis.

Dominantly inherited and X-linked *maxillofacial dysostosis* consists of bilateral hypoplasia of malar bones, downward-slanting palpebral fissures without colobomas, maxillary hypoplasia, open bite, and relative mandibular prognathism.

A similar facies has been seen in an autosomal recessive disorder found in Hutterites (17a), in an autosomal dominant osteosclerotic disorder (15b), and in a father and son with ectrodactyly (see EEC syndrome).

**Laboratory aids.** Midtrimester sonographic diagnosis has been accomplished (2,3,5). Fetoscopy has also been employed (24).

### References [Mandibulofacial dysostosis (Treacher Collins syndrome, Franceschetti–Zwahlen–Klein syndrome)]


1. Axelsson A et al: Dysostosis mandibulofacialis. *J Laryngol Otol* **77**:575–592, 1963.
2. Behrents RG et al: Prenatal mandibulofacial dysostosis (Treacher Collins in man). *Arch Oral Biol* **20**:265–282, 1975.
3. Behrents RG et al: Prenatal mandibulofacial dysostosis (Treacher Collins syndrome). *Cleft Palate J* **14**:13–34, 1977.
4. Berry GA: Note on a congenital defect (coloboma?) of the lower lid. *R Lond Ophthalmol Hosp Rep* **12**:255–257, 1889.
5. Crane JP, Beaver HA: Midtrimester sonographic diagnosis of mandibular dysostosis. *Am J Med Genet* **25**:251–255, 1986.
6. Dahl E et al: A morphologic description of a dry skull with mandibulofacial dysostosis. *Scand J Dent Res* **83**:257–266, 1975.
7. Fazen LE et al: Mandibulofacial dysostosis. *Am J Dis Child* **113**:405–410, 1967.
8. Franceschetti A, Klein D: Mandibulo-facial dysostosis: New hereditary syndrome. *Acta Ophthalmol (Kbh)* **27**:143–224, 1949.
9. Franceschetti A et al: Dysostose mandibulo-facial unilatérale avec déformations multiples du squelette (Processus paramastöide, synostose des vertèbres, sacralisation, etc.) et torticolis clonique. *Ophthalmologica* **118**:796–814, 1949.
10. Franceschetti A et al: La dysostose mandibulo-faciale dans le cadre des syndrome du premier arc branchial. *Schweiz Med Wochenschr* **89**:478–483, 1959.
11. Garner LD: Cephalometric analysis of Berry–Treacher Collins syndrome. *Oral Surg* **23**:320–327, 1967.
12. Grönvall H, Olsson Y: Dysostosis mandibulofacialis. *Acta Ophthalmol (Kbh)* **31**:245–252, 1953.
13. Herring SE et al: Anatomical abnormalities in mandibulofacial dysostosis. *Am J Med Genet* **3**:225–259, 1979.
14. Hutchinson JC Jr et al: The otologic manifestations of mandibulofacial dysostosis. *Tr Am Acad Ophthalmol* **84**:520–528, 1977.
15. Jones KL et al: Older paternal age and fresh gene mutation: Data on additional disorders. *J Pediatr* **86**:84–88, 1976.

15a. Kolar JC et al: Surface morphology in Treacher Collins syndrome: An anthropometric study. *Cleft Palate J* **22**:266–274, 1985.

15b. Lehman RAW: Familial osteosclerosis with abnormalities of the nervous system and meninges. *J Pediatr* **90**:49–54, 1977.

16. Lloyd GAS, Phelps PD: Radiology of the ear in mandibulo-facial dysostosis-Treacher Collins syndrome. *Acta Radiol Diag* **20**:233–240, 1979.
17. Lockhart RD: Variants coincident with congenital absence of zygoma (zygomatic process of temporal bone). *J Anat* **63**:233–236, 1928–1929.

17a. Lowry RB et al: Mandibulofacial dysostosis in Hutterite sibs: A possible recessive trait. *Am J Med Genet* **22**:501–512, 1985.

18. Lübke F von: Über die Beobachtung einer Dysostosis mandibulofacialis. *Z Geburtsh Gynäkol* **156**:235–246, 1961.
19. Mann I, Kilner TP: Deficiency of the malar bones with defect of the lower lids. *Br J Ophthalmol* **27**:13–20, 1943.

19a. Markitzin A et al: Major salivary glands in branchial arch syndromes. *Oral Surg* **58**:672–677, 1984.

20. Marsh JL et al: The skeletal anatomy of mandibulofacial dysostosis. (Treacher Collins syndrome). *Plast Reconstr Surg* **78**:460–468, 1986.
21. McKenzie J, Craig J: Mandibulo-facial dysostosis (Treacher Collins syndrome). *Arch Dis Childh* **30**:391–395, 1955.
22. McNeill KA, Wynter-Wedderbum L: Choanal atresia: A manifestation of the Treacher Collins syndrome. *J Laryngol Otol* **67**:365–369, 1953.
23. Nager FR, deReynier JP: Das Gehörorgan bei den angeborenen Kopfmissbildungen. *Pract Otorhinolaryngol (Basel) Suppl 2,* **10**:1–128, 1948.
24. Nicolaides KH et al: Prenatal diagnosis of mandibulofacial dysostosis. *Prenat Diag* **4**:201–205, 1984.

24a. Peterson-Falzone S, Figueroa AA: Longitudinal changes in cranial base angulation in mandibulofacial dyostosis. *Cleft Palate J* **26**:31–35, 1989.

25. Peterson-Falzone S, Pruzansky S: Cleft palate and congenital palato-

pharyngeal incompetency in mandibulofacial dysostosis: Frequency and problems in treatment. *Cleft Palate J* **13**:354–360, 1976.

25a. Phelps PD et al: The ear deformities in mandibulofacial dysostosis. *Clin Otolaryngol* **6**:15–28, 1981.

26. Poswillo D: The pathogenesis of the Treacher Collins syndrome (mandibulo-facial dysostosis). *Br J Oral Surg* **13**:1–26, 1975.

27. Roberts FG et al: An X-radiocephalometric study of mandibulofacial dysostosis in man. *Arch Oral Biol* **20**:265–282, 1975.

28. Rogers BO: Berry–Treacher Collins syndrome: A review of 200 cases. *Br J Plast Surg* **17**:109–137, 1964.

29. Rovin S et al: Mandibulofacial dysostosis: A familial study of five generations. *J Pediatr* **65**:215–221, 1964.

30. Sahawi E: Beitrag zur Dysostosis mandibulofacialis. *Z Kinderheilkd* **94**:195–201, 1965.

31. Shprintzen RJ et al: Pharyngeal hypoplasia in Treacher Collins syndrome. *Arch Otolaryngol* **105**:127–131, 1979.

32. Stovin JJ et al: Mandibulofacial dysostosis. *Radiology* **74**:225–231, 1960.

33. Sulik K et al: Mandibulofacial dysostosis (Treacher Collins syndrome): A new proposal for its pathogenesis. *Am J Med Genet* **27**:359–372, 1987.

34. Thomson A: Notice of several cases of malformation of the external ear, together with experiments on the state of hearing in such persons. *Monthly J Med Sci* **7**:420, 1846–1847, cited by Klein D, Dysostose mandibulofaciale. *Prat Odontol-Stomatol* No. 487–490, pp 1–8, 1953.

34a. Toynbee J: Description of a congenital malformation in the ears of a child. *Monthly J Med Sci* **1**:738–739, 1847.

35. Treacher Collins E: Cases with symmetrical congenital notches in the outer part of each lid and defective development of the malar bones. *Trans Ophthalmol Soc UK* **20**:190–192, 1960.

36. Wildervanck LS: Dysostosis mandibulo-facialis (Franceschetti–Zwahlin) in four generations. *Acta Genet Med (Roma)* **9**:447–451, 1960.

37. Wiley MJ et al: Effects of retinoic acid on the development of the facial skeleton in hamsters: Early changes involving cranial neural crest cells. *Acta Anat* **116**:180–192, 1983.

## Preaxial acrofacial dysostosis (Nager syndrome)

Nager syndrome was reported by Slingenberg (21) in 1908 but was first recognized by Nager and de Reynier (15) in 1948 who employed the term *acrofacial dysostosis* to separate the disorder from mandibulofacial dysostosis. The same patient was independently described by Franceschetti and Klein (5) in 1949 as an atypical example of mandibulofacial dysostosis. We have used the term *preaxial acrofacial dysostosis* to distinguish it from the Wildervanck–Smith syndrome. The infant reported by van Goethem et al (23) does not fit into either group.

Earlier reports have been summarized by Halal et al (8) and Pfeiffer and Stoess (19) in 1983. To date there have been approximately 35 documented cases (1–9,11–16,18–22,24,27). Others have been poorly documented (17).

Most cases have been sporadic. Affected sibs (2a,8a,16,25) have been recorded and consanguineous parents (2,19) have been noted, suggesting the possibility of autosomal recessive inheritance. The syndrome has occurred in one of dizygotic twins on two occasions (2,13). The argument for autosomal dominant inheritance is not convincing in some papers (13,27), although vertical transmission is convincing in others (1). Although genetic aspects have been discussed in a number of papers (1,2,2a,8a,8b,13,16,25,27), the inheritance pattern needs to be established. Nosology has been discussed by Opitz (17a). Sulik et al (21a) have suggested that the pathogenesis of the syndrome results from involvement of the apical ectodermal ridge of limb buds.

The facies is similar to that of *mandibulofacial dysostosis* (Fig. 19–7A,B). The zygomatic hypoplasia results in downslanting palpebral fissures. The lower eyelids exhibit colobomas in 20% and reduced numbers of eyelashes in 60%. Velopharyngeal insufficiency has been noted (1,13). Micrognathia is usually more marked than in mandibulofacial dysostosis. External ear defects and cleft palate are more common while lower eyelid colobomas are less common than in mandibulofacial dysostosis. Laryngeal and epiglottal hypoplasia has been reported (12). Speech and hearing characteristics have been well documented by Meyerson and Nisbet (13a).

Stature is often reduced. Hip dislocation has been described on several occasions (11,12,14). Rib anomalies are rare (8,12).

The thumb, by definition until recently, is hypoplastic or aplastic,

Fig. 19–7. *Preaxial acrofacial dysostosis (Nager).* (A,B) Note malar hypoplasia, slight downslanting of palpebral fissures, almost total absence of eyelashes, low-set, cup-shaped ears, and micrognathia. [A from *FA Walker,* Birth Defects **10**(8):135, 1974. B from *P Bowen* and *F Harley,* Birth Defects **10**(5):109, 1974.]

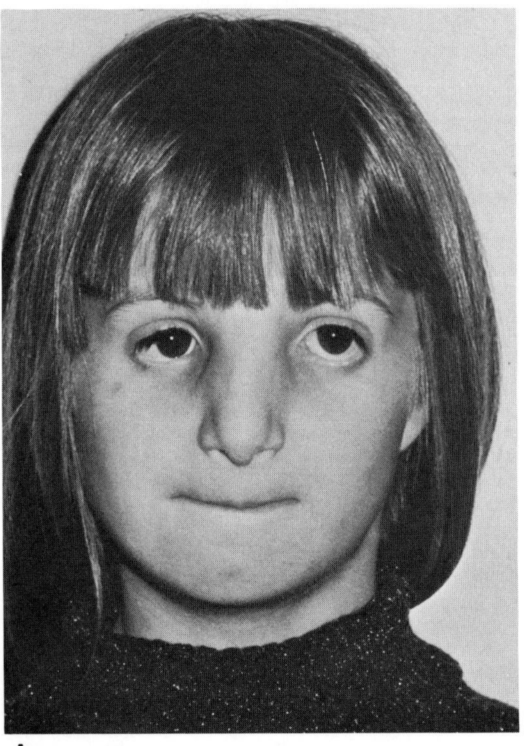

A

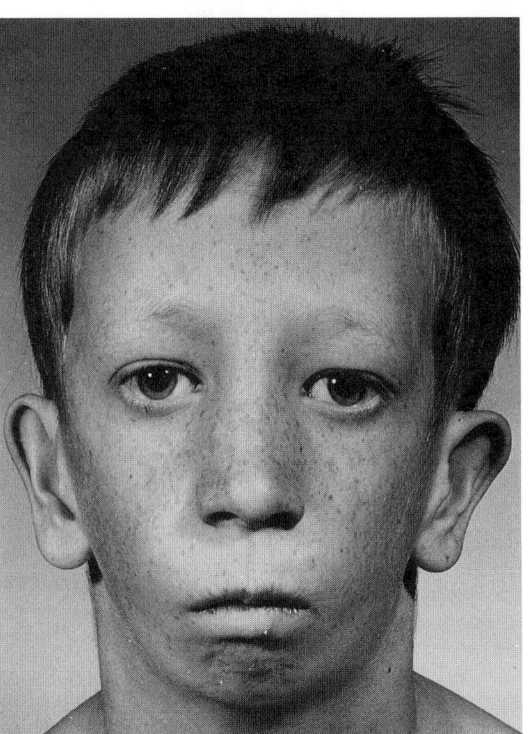

B

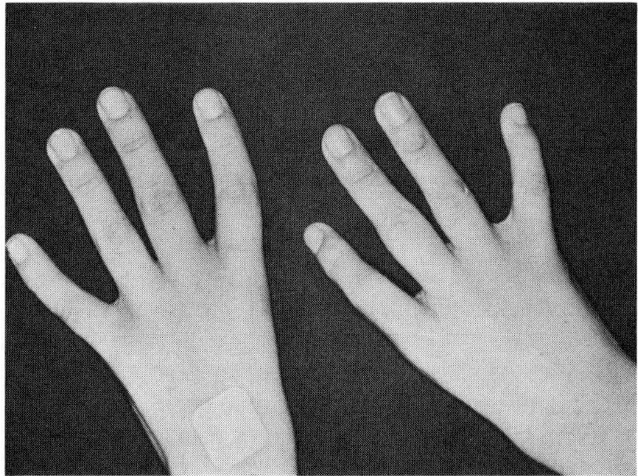

A

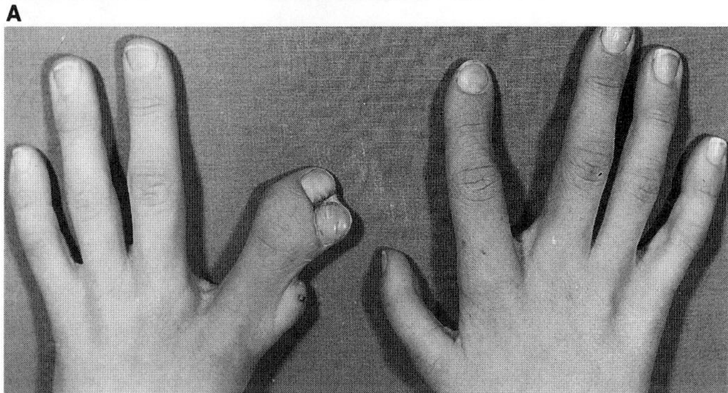

B

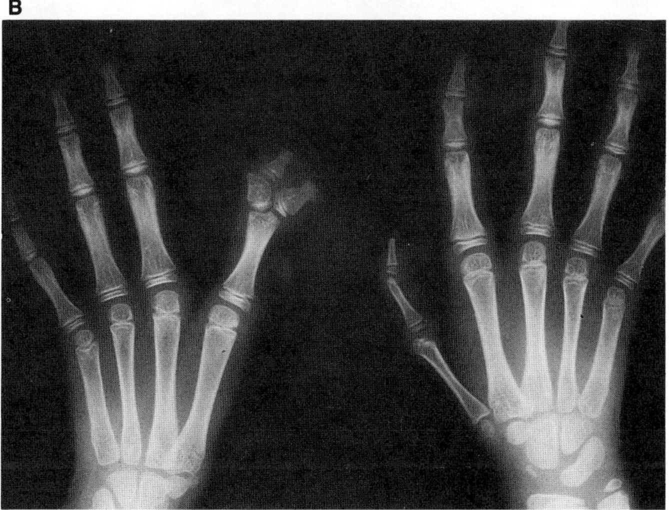

C

Fig. 19–8. *Preaxial acrofacial dysostosis (Nager).* (A) Absence of thumbs. (B,C) Preaxial involvement with missing phalanges and syndactyly. [A from *FA Walker,* Birth Defects **10**(8):135, 1974. B,C from *P Bowen* and *F Harley,* Birth Defects **10**(5):109, 1974.]

but an occasional patient will have duplication or triphalangy of the thumbs (1,7,16). The thumb anomalies are usually asymmetric. About 50% have associated radial hypoplasia or aplasia or proximal radioulnar synostosis that is often unilateral. The radial defect never occurs without agenesis of the thumb. Some patients have limited extension at the elbow. Rarely, the index finger is missing or has soft tissue syndactyly with the third finger (Fig. 19–8A–C).

A few patients have exhibited mild mental retardation (8,11,26).

Various genitourinary abnormalities have been encountered: unilateral agenesis of kidney, malposed kidney, duplicated calyx, and bicornuate uterus (8,18–20). Tetralogy of Fallot has been noted (17a).

A particularly severe form has been described by Kawira et al (10) and by Goldstein and Mirkin (7a). Coloboma of the lids, lateral and oblique facial clefts, bilateral cleft lip, and absent forearms and thumbs were recorded. The lower limbs were reduced with the feet attached to the femora or directly to the trunk. Parental consanguinity was noted in one case. RJ Gorlin has seen a similar example in Minneapolis and another in consultation for Dr. U. Froster-Iskenius from Lübeck, West Germany. It is possible that this represents a new disorder rather than severe Nager syndrome.

Patients with dup(2q) have some overlap in phenotype with Nager syndrome such as downward-slanting palpebral fissures, dysplastic pinnae, and micrognathia (25,28). However, malar hypoplasia, atresia of the external auditory canals, cleft palate, and hypoplastic thumbs are not characteristic of dup(2q).

Prenatal ultrasonography has been employed successfully on one occasion (8a).

### References [Preaxial acrofacial dysostosis (Nager syndrome)]


1. Aylesworth AS, Friedman PA: Father to son transmission of the Nager syndrome. *8th Annual Workshop on Malformations and Morphogenesis.* Greenville, South Carolina, August 15–19, 1987.

1a. Bowen P, Harley F: Mandibulofacial dysostosis with limb malformations (Nager's acrofacial dysostosis). *Birth Defects* **10**(5):109–115, 1974.

2. Burton BK, Nadler HL: Nager acrofacial dysostosis. *J Pediatr* **91**:84–86, 1977.

2a. Chemke J et al: Autosomal recessive inheritance of Nager acrofacial dysostosis. *J Med Genet* **25**:230–232, 1988.

3. Chou YC: Mandibulofacial dysostosis. *Chinese Med J* **80**:373–375, 1960.

4. Fernandez AO, Ronis ML: The Treacher Collins syndrome. *Arch Otolaryngol* **80**:502–520, 1964 (Case 5).

5. Franceschetti A, Klein D: The mandibulo-facial dysostosis. A new hereditary syndrome. *Acta Ophthalmol* **27**:143–224, 1949 (Case 4).

6. Gellis SS et al: Nager's syndrome. *Am J Dis Child* **132**:519–520, 1978.

7. Giugliani R, Pereira CH: Nager's acrofacial dysostosis with thumb duplication: Report of a case. *Clin Genet* **26**:228–230, 1984.

7a. Goldstein DJ, Mirkin LD: Nager acrofacial dysostosis: Evidence for apparent heterogeneity. *Am J Med Genet* **30**:741–746, 1988.

8. Halal F et al: Differential diagnosis of Nager acrofacial dysostosis syndrome: Report of four patients with Nager syndrome and discussion of other related syndromes. *Am J Med Genet* **14**:209–224, 1983.

8a. Hall BD: Nager acrofacial dysostosis: Autosomal dominant inheritance in mild to moderately affected mother and lethally affected phocomelic son. *Am J Med Genet* **33**:394–397, 1989.

8b. Hecht JT et al: Brief clinical report: The Nager syndrome. *Am J Med Genet* **27**:965–969, 1987.

8c. Jackson IT et al: A significant feature of Nager's syndrome: Palatal agenesis. *Plast Reconstr Surg* **84**:219–226, 1989 (Cases 1,2,3,5).

9. Jones RG: Mandibulofacial dysostosis. *Ctr Afr J Med* **14**:193–200, 1968.

10. Kawira EI et al: Acrofacial dysostosis with severe facial clefting and limb reduction. *Am J Med Genet* **17**:641–647, 1984.

11. Klein D et al: Sur une forme extensive de dysostose mandibulofaciale (Franceschetti) accompagnée de malformations des extremities. *Rev Oto Neuro Ophtal* **42**:432–440, 1970.

12. Krauss CM et al: Anomalies in an infant with Nager acrofacial dysostosis. *Am J Med Genet* **21**:761–764, 1985.

12a. La Merrer M et al: Acrofacial dysostoses. *Am J Med Genet* **33**:318–322, 1989.

13. Lowry B: The Nager syndrome (acrofacial dysostosis): Evidence for autosomal dominant inheritance. *Birth Defects* **13**(3C):195–202, 1977.

13a. Meyerson MD, Nisbet JB: Nager syndrome: An update of speech and hearing characteristics. *Cleft Palate J* **24**:142–151, 1987.

14. Meyerson MD et al: Nager acrofacial dysostosis: Early intervention and longterm planning. *Cleft Palate J* **14**:35–40, 1977.

15. Nager FR, de Reynier JP: Das Gehörorgan bei den angeborenen Kopfmissbildungen. *Pract Otorhinolarngol (Suppl 2)* **10**:1–128, 1948.

16. Neidhart E: *Die Dysostosis mandibulo-facialis in Kombination mit Missbildungen der obern Extremitäten,* Thesis, Zurich, 1968.

17. O'Connor GB, Conway ME: Treacher Collins syndrome (dysostosis mandibulofacialis). *Plast Reconstr Surg* **5**:419–425, 1950 (Case 1).

17a. Opitz JM: Editorial comment: Nager "syndrome" versus "anomaly" and its nosology with the postaxial acrofacial dysostosis syndrome of Genée and Wiedemann. *Am J Med Genet* **27**:959–963, 1987.

18. Pfeiffer RA: Associated deformities of the head and hand. *Birth Defects* **5**(3):18–34, 1969 (Case 17).

19. Pfeiffer RA, Stoess H: Acrofacial dysostosis (Nager syndrome): Synopsis and report of a case. *Am J Med Genet* **15**:255–260, 1983.

20. Schönenberg H: Die Differentialdiagnose der radialen Defektbildungen. *Pädiatr Prax* **7**:455–467, 1968.

21. Slingenberg B: Missbildungen von Extremitäten. *Virchows Arch (Pathol Anat)* **193**:1–91, 1908 (Case 10).

21a. Sulik KK et al: Pathogenesis of cleft palate in Treacher Collins, Nager, and Miller syndromes. *Cleft Palate J* **26**:209–216, 1989.

22. Temtamy S, McKusick VA: Genetics of hand malformations. *Birth Defects* **14**(3):92–95, 1978 (same as Ref. 1a).

22a. Thompson E et al: The Nager acrofacial syndrome and the tetralogy of Fallot. *J Med Genet* **22**:408–410, 1985.

23. van Goethem H et al: Nager's acrofacial dysostosis. *Acta Paediatr Belg* **34**:253–256, 1981.

24. Vatré JL: Etude génétique et classification clinique de 154 cas de dysostose mandibulo-faciale (syndrome de Franceschetti) avec description de leurs associations malformatives. *J Génét Hum* **19**:17–100, 1971 (same as Ref. 11).

25. Wagner SF, Cole J: Nager syndrome with partial duplication of the long arm of chromosome 2. *Am J Hum Genet* **31**:116A, 1979.

26. Walker FA: Apparent autosomal recessive inheritance of the Treacher Collins syndrome. *Birth Defects* **10**(8):135–139, 1974.

27. Weinbaum M et al: Autosomal dominant transmission of Nager acrofacial dysostosis. *Am J Hum Genet* **33**:93A, 1981.

28. Zankl M et al: Distal 2q duplication. Report of two familial cases and an attempt to define a syndrome. *Am J Med Genet* **4**:5–16, 1979.

## Postaxial acrofacial dysostosis (Miller syndrome, Wildervanck–Smith syndrome, Genée–Wiedemann syndrome)

The syndrome was defined by Miller et al (9), in 1979, although several authors had earlier described patients with the syndrome (5,11,14,16). Since then, a number of reports have been published (1–4,6–8,10,11,13,15). Affected sibs have been observed (3,7) and Allanson and McGillivray (1) have reported a large kindred with autosomal dominant inheritance and variable expression. Miller et al (9) did not include the case of Wiedemann (15) because the patient was retarded. The patient of Piper (12) and another seen by RJ Gorlin, through the courtesy of Dr. Mette Warburg, Copenhagen, Denmark, had no digital anomalies. Neither did those in the affected kindred described by Allanson and McGillivray (1). The syndrome needs to be further delineated. Sulik et al (14a) postulated involvement of apical ectodermal ridge of limb buds.

The facies has some resemblance to that of *mandibulofacial dysostosis* (Fig. 19–9A,B), but there is postaxial limb deficiency. The malar bones are often hypoplastic with resultant downslanting palpebral fissures. The eyelids may exhibit coloboma and usually there is ectropion of the lower lids. Cleft lip and/or cleft palate are common. The pinnae tend to be cup-shaped. The external auditory canals and middle ears are often malformed. Conical teeth have been reported (1).

Postaxial agenesis of a digit of the hands and feet is common (Fig. 19–10A,B). Preaxial anomalies are less frequently present. Abnormal thumbs occur in about 50%. The radius and ulna tend to be short and, in some cases, there is radioulnar synostosis. One patient had absent fibulae and short tibiae.

Various congenital heart defects have been documented (PDA, VSD, ostium primum with endocardial cushion defect). Genitourinary anomalies (cryptorchidism, micropenis) as well as accessory nipples have been reported (2,7,8,10).

It is possible that the autosomal dominant disorder reported by Falace and Hall (2a) and by Allanson and McGillivray (1) involving ectropion, clefting, and conical teeth represents a separate disorder.

### References [Postaxial acrofacial dysostosis (Miller syndrome, Wildervanck–Smith syndrome, Genée–Wiedemann syndrome)]

1. Allanson JE, McGillivray BC: Familial clefting syndrome with ectropion and dental anomaly without limb anomalies. *Clin Genet* **27**:426–429, 1985.

1a. Barbuti D et al: Postaxial acrofacial dysostosis or Miller syndrome. *Eur J Pediatr* **148**:445–446, 1989.

1b. Chrzanowska KH et al: Phenotypic variability in the Miller acrofacial dysostosis syndrome. Report of two further patients. *Clin Genet* **35**:157–160, 1989.

2. Donnai D et al: Postaxial acrofacial dysostosis (Miller) syndrome. *J Med Genet* **24**:422–425, 1987.

2a. Falace PB, Hall BD: Congenital euryblepharon with ectropion and dental anomaly: An autosomal dominant clefting disorder with marked variability of expression. Annual David W. Smith Workshop on Malformations and Morphogenesis, Oakland, August, 1988.

3. Fineman RM: Recurrence of the postaxial acrofacial dysostosis syndrome in a sibship: Implications for genetic counseling. *J Pediatr* **98**:87–88, 1981.

4. Fryns JP, Van den Berghe H: Brief clinical report: Acrofacial dysostosis with postaxial limb deficiency. *Am J Med Genet* **29**:205–208, 1988.

5. Genée E: Une forme extensive de dysostose mandibulofaciale. *J Génét Hum* **17**:45–52, 1969.

5a. Jackson IT et al: A significant feature of Nager's syndrome: Palatal agenesis. *Plast Reconstr Surg* **84**:219–226, 1989 (Case 4).

6. Lenz W: Genetische Syndrome mit Aplasie ulnare und/oder fibularer Randstrahlen. *Klinische Genetik in der Pädiatrie,* 2nd Symposium, Mainz, 1979.

7. Meinecke P, Rauschkolb R: Autosomal rezessive mandibulo-faziale Dysostose mit Aplasie des 5. Strahls an Händen und Füssen. *Klinische Genetik in der Pädiatrie,* 3rd Symposium, Kiel, 1982.

8. Meinecke P, Wiedemann H-R: Letter to the editor: Robin sequence and oligodactyly in mother and son—probably a further example of the postaxial acrofacial dysostosis syndrome. *Am J Med Genet* **27**:953–956, 1987.

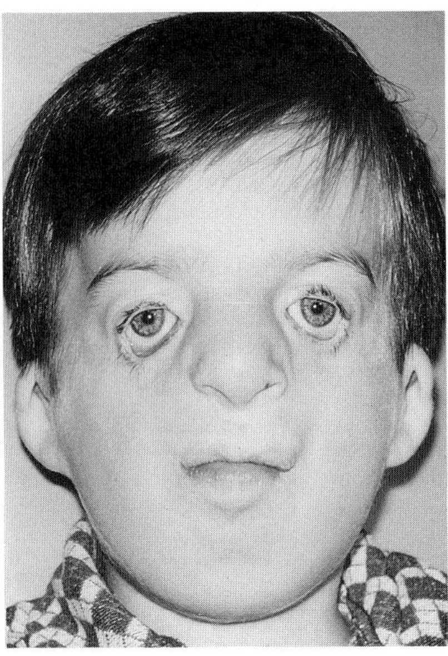

A

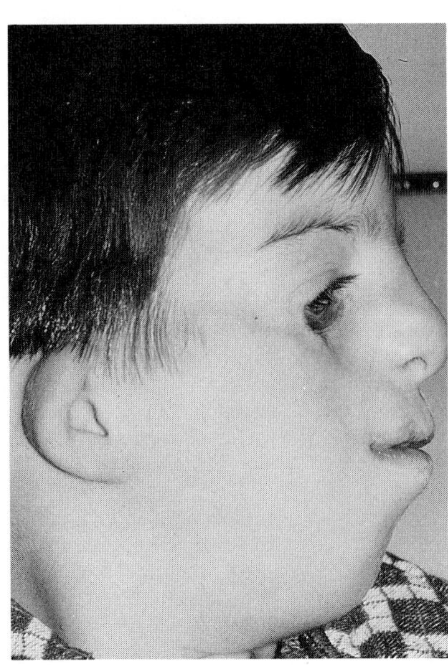

B

Fig. 19–9. *Postaxial acrofacial dysostosis (Miller or Wildervanck–Smith).* (A,B) Characteristic facial appearance with downslanting palpebral fissures, ectropion of lower lids, repaired cleft lip, and malformed ears. (From *MM Cohen Jr,* Dysmorphic syndromes with craniofacial manifestations, in Oral Facial Genetics, *RE Stewart* and *GH Prescott,* eds, C.V. Mosby, St. Louis, 1976.)

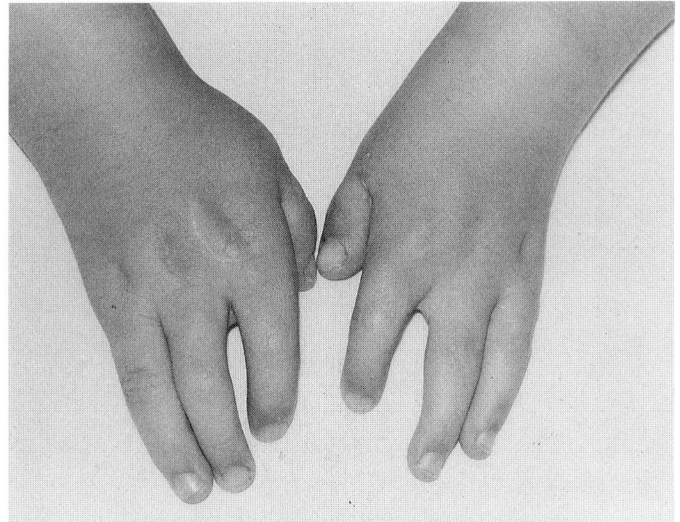

A

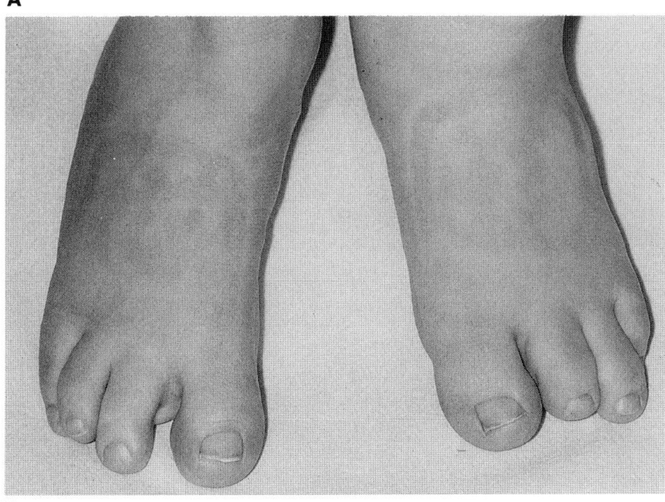

B

Fig. 9–10. *Postaxial acrofacial dysostosis (Miller* or *Wildervanck–Smith).* (A,B) Postaxial agenesis of digits. (From *MM Cohen Jr,* Dysmorphic syndromes with craniofacial manifestations, in Oral Facial Genetics, *RE Stewart* and *GH Prescott,* eds, C.V. Mosby, St. Louis, 1976.)


9. Miller M et al: Postaxial acrofacial dysostosis syndrome. *J Pediatr* **95**:970–975, 1979.

10. Opitz JM, Stickler GB: Letter to the editor: The Genée-Wiedemann syndrome, an acrofacial dysostosis—further observation. Am J Med Genet **27**:971–975, 1987.

11. Pashayan H, Feingold M: Case report 28. *Syndrome Ident* **3**(1):9–10, 1975.

12. Piper HF; Augenärztliche Befunde bei frühkindlicher Entwicklungsstörungen. *Mschr Kinderheilk* **105**:170–176, 1957.

13. Robinow M et al: Reply to Drs. Meinecke and Wiedemann. *Am J Med Genet* **27**:957, 1987.

14. Smith DW, Jones KL: Case report 28. *Syndrome Ident* **3**(1):7–8, 1975.

14a. Sulik KK et al: Pathogenesis of cleft palate in Treacher Collins, Nager, and Miller syndromes. *Cleft Palate J* **26**:209–216, 1989.

15. Wiedemann H-R: Missbildungs-Retardierungs-Syndrom mit Fehlen des 5. Strahl an Händen und Füssen, Gaumenspalte, dysplastischen Ohren und Augenlidern und radioulnaren Synostose. *Klin Pädiatr* **185**:181–186, 1973.

16. Wildervanck LS: Case report 28. *Syndrome Ident* **3**(1):11–13, 1975.


## Maxillofacial dysostosis

Peters and Hövels (3), in 1960, described autosomal dominant inheritance of maxillary hypoplasia, downward-slanting palpebral fissures, minor changes in the pinnae, and abnormal and retarded speech (Fig. 19–11). They employed the term maxillofacial dysostosis to differentiate the condition from mandibulofacial dysostosis. An earlier report by Villaret and Desoille (4) probably represents the same disorder. Other more recent reports (1,2) have confirmed the existence of the syndrome. Our combined experience suggests that the disorder is more frequent than the sparse number of reports indicate.

### References (Maxillofacial dysostosis)


1. Escobar V et al: Maxillofacial dysostosis. *J Med Genet* **14**:355–358, 1977.

2. Melnick M, Eastman JR: Autosomal dominant maxillofacial dysostosis. *Birth Defects* **13**(3B):39–44, 1977.

3. Peters A, Hövels O: Die Dysostosis maxillo-facialis, eine erbliche, typische Fehlbildung des 1. Visceralbogens. *Z Menschl Vererb Konstit Lehre* **35**:434–444, 1960.

4. Villaret M, Desoille H: L'hypoplasie primitive familiale du maxillaire supérieur. *Ann Méd (Paris)* **32**:378–381, 1932.


## Maxillofacial dysostosis, X-linked

Toriello et al (1) reported two male sibs and their male first cousin with somewhat short stature, microcephaly, mild mental retardation, downslanting palpebral fissures due to malar hypoplasia, sparse lateral

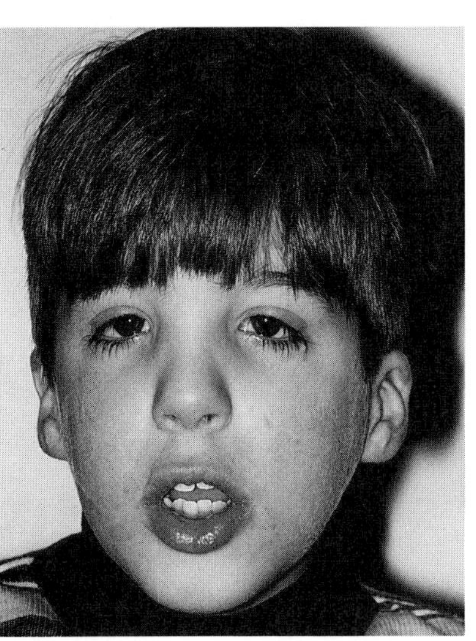

A

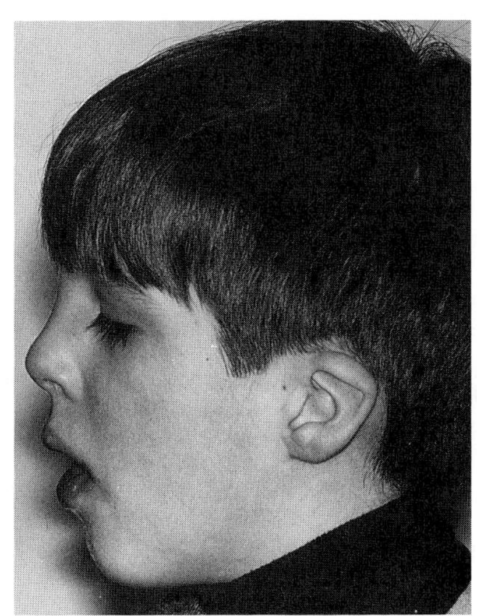

B

Fig. 19–11. *Maxillofacial dysostosis.* (A,B) Maxillary hypoplasia, downslanting palpebral fissures, minor ear anomalies. [From M Melnick and *JR Eastman,* Birth Defects **13**(3B):39, 1977.]

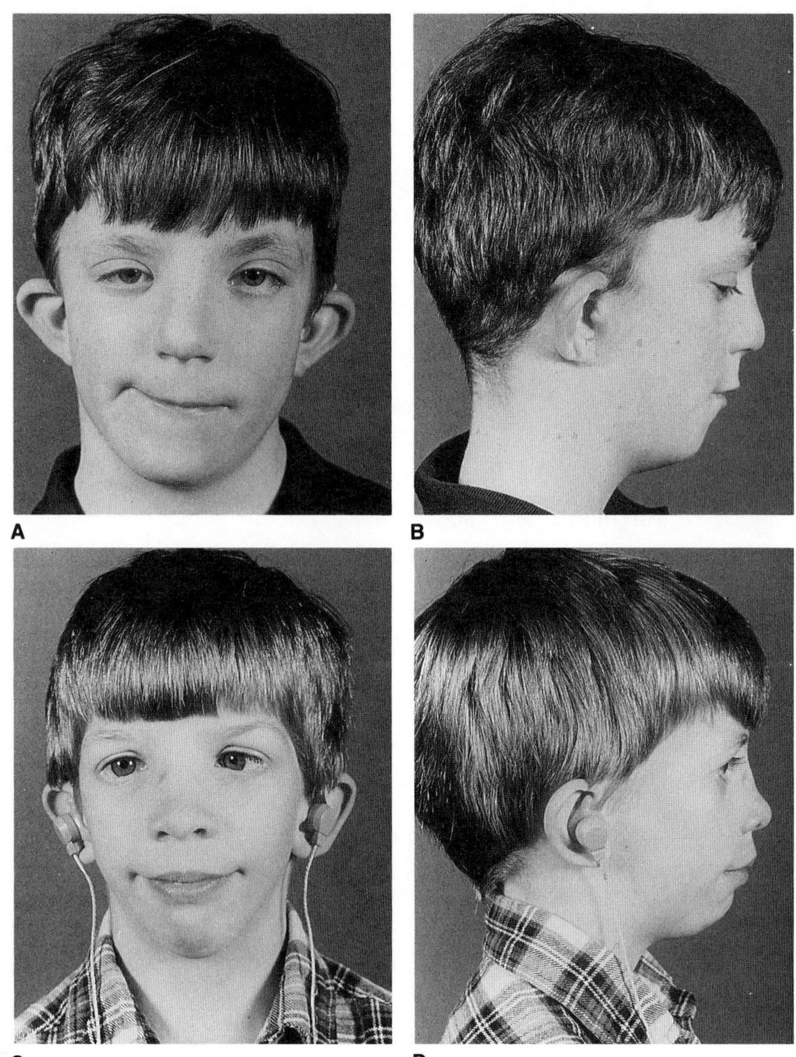


Fig. 19–12. *Maxillofacial dysostosis, X-linked.* (A–D) Two male sibs with microcephaly, downslanting palpebral fissures, sparse eyebrows laterally, outstanding pinnae, mixed hearing loss, mild micrognathia, and slightly webbed neck. (From *HV Toriello et al,* Am J Med Genet **21**:137, 1985.)

eyebrows, outstanding pinnae, mixed hearing loss, mild micrognathia, slightly webbed neck (Fig. 19–12A–D), and cryptorchidism. One child has subvalvular pulmonic stenosis. Inheritance may be X-linked. There is some resemblance to *maxillofacial dysostosis.*

#### Reference (Maxillofacial dysostosis, X-linked)

1. Toriello HV et al: X-linked syndrome of branchial arch and other defects. *Am J Med Genet* **21**:137–142, 1985.

## Acrofacial dysostosis (Reynolds syndrome)

Reynolds et al (1) described a family manifesting autosomal dominant inheritance of a previously undescribed acrofacial dysostosis syndrome. The craniofacial manifestations were those of mild mandibulofacial dysostosis and included prominent forehead, ptosis, downslanting palpebral fissures, malar hypoplasia, highly arched palate with dental malocclusion, and micrognathia (Fig. 19–13A,B). The pinnae were normal. However, mild congenital mixed type hearing loss was a feature.

The variable acral abnormalities predominantly affected the radial ray, manifesting as mild hypoplasia of the first metacarpal and first proximal phalanx. This was more evident on metacarpal–phalangeal pattern profile than on clinical examination in some affected individuals.

There is some resemblance of the facies to that of *maxillofacial dysostosis.*

#### Reference [Acrofacial dysostosis (Reynolds syndrome)]

1. Reynolds JF et al: A new autosomal dominant acrofacial dysostosis syndrome. *Am J Med Genet Suppl* **2**:143–150, 1986.

## Acrofacial dysostosis, cleft lip–palate, and triphalangeal thumbs

Richieri-Costa et al (1) described two daughters of a normal nonconsanguineous couple. The facies of the elder daughter was typical of mandibulofacial dysostosis. One ear was small. Both sisters had bilateral cleft lip and bifid uvula. The thumbs were hypoplastic and triphalangeal. The first metacarpals were also hypoplastic. The younger daughter had only mild malar hypoplasia. She had bilateral cleft lip–palate and a hypoplastic thumb.

#### Reference (Acrofacial dysostosis, cleft lip–palate, and triphalangeal thumbs)

1. Richieri-Costa A et al: Syndrome of acrofacial dysostosis, cleft lip-palate and triphalangeal thumb in a Brazilian family. *Am J Med Genet* **14**:225–229, 1983.

A

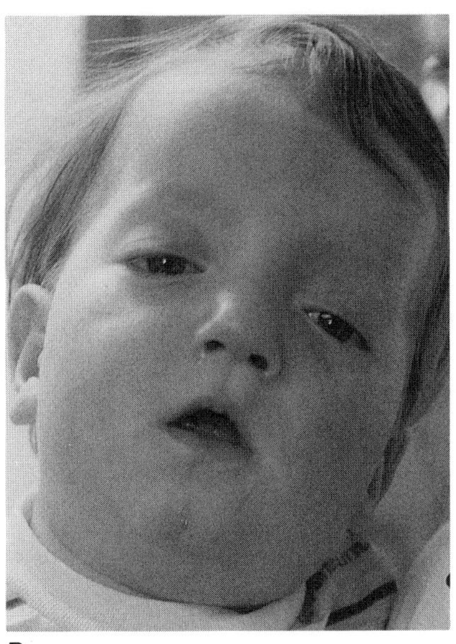
B

Fig. 19–13. *Acrofacial dysostosis (Reynolds syndrome).* Affected grandfather and grandson. (A,B) Craniofacial features are those of mild mandibulofacial dysostosis and include prominent forehead, eyelid ptosis, downslanting palpebral fissures. Note normal ears in both and micrognathia in B. (From *JF Reynolds et al,* Am J Med Genet Suppl **2**:143, 1986.)

## Branchio-oto-renal syndrome (BOR syndrome)

The term branchio-oto-renal syndrome (BOR) was first used by Melnick et al (31) to refer to patients with **b**ranchial cleft fistulas or cysts, **o**tologic anomalies including hearing loss, anomalies of the pinnae and prehelical pits, and **r**enal anomalies. Other associated findings have been noted (7). An early report was that of Heusinger (20) in 1864.

Initially, the branchio-oto (BO) syndrome was considered to be distinct from the BOR syndrome based on the absence of reported renal anomalies and the type of hearing loss in the former (25,30,32). The ear pit-deafness syndrome has also been considered a distinct entity (28). This distinction was based largely on early reports of patients in whom complete renal and/or branchial evaluations were not undertaken (12,14,29,42,50). The BO and BOR syndromes began to be considered as one condition with variable expression as families were reported where individual members had either branchio-oto involvement or branchio-oto-renal involvement, and had either conductive, sensorineural or mixed hearing loss.

The BOU syndrome (**b**ranchio-**o**to-**u**reteral = double collecting system) (17) may also represent variable expression of the BOR syndrome based on the pedigree of Heimler and Lieber (19) in which several affected individuals in a family with the BOR syndrome had double collecting systems, and other affected family members had other renal anomalies. There are many case reports (1,3,4,9,18,21,22,27, 29,33–37,42,46–48,51) and several extensive reviews (8,15).

The syndrome follows an autosomal dominant mode of inheritance (15,22,24,31,33). Expression is variable. Penetrance is very high (7,15,16), but probably not complete (19). Prevalence is estimated at 1/40,000 (16). Conversely, a newborn with a pretragal pit has 1/400 chance of having the syndrome (16). Attempts have been made to localize the gene, but no linkage has been found (7). No association is reported between the BOR syndrome and a particular HLA haplotype (10).

Pathogenesis is thought to involve an ectomesenchymal deficiency in the area of the first and second pharyngeal arches that results in aberrant inductive interaction between the ureteric bud and the metanephric blastema (32). The relationship between the inner ear pathology and the renal pathology is thought to be explained partly by an antigen shared between the kidney and the cochlea, especially between the stria vascularis of the inner ear and the renal glomeruli (32). On necropsy, Saito (43) observed renal agenesis and absence of the stria vascularis in the basal turn of the cochlea in a neonate with Potter sequence. Fitch et al (12) observed atrophy and dysplasia of the stria vascularis in histopathologic examination of the temporal bones.

Variable expression of the phenotype should be expected. The presence of one primary manifestation in a first-degree relative of a known gene carrier is generally sufficient to make the diagnosis.

**Branchial cysts or fistulas.** About 60% have branchial fistulas (7,15). These fistulas (pits, sinuses) are located on the external lower third of the neck, usually on the median border of the sternomastoid muscle. The fistulas may open internally into the tonsillar fossa. The fistulas may drain fluid or become infected. They are usually bilateral, but may be unilateral.

**Otologic findings.** External ear anomalies include malformed pinnae, other pinnal ear anomalies such as lop, cup, flap, flat, or hypoplastic pinna, narrow or atretic ear canals, cartilaginous preauricular appendages, and helical or preauricular pits (Fig. 19–14A–C). These pits are shallow, pinhead-sized, blind depressions in the helix of the ear near the upper attachment, or in the skin anterior to this site (6,7,19,31). About 70–80% exhibit preauricular pits, and about 60% have anomalies of the pinna (7,15).

Anomalies of the middle ear have been described. Malformations of the ossicles include unconnected or fused stapes and incus, and abnormal stapedial reflexes (7,8,19). Temporal bone anomalies (13) and a small mastoid process with a reduced number of aerated air cells (15) have been reported. A mother and daughter had congenital cholesteatoma (26).

Inner ear malformations include unilateral or bilateral malformed cochlea, unilateral or bilateral reduced or absent irritability of the labyrinth, and abnormal internal auditory canals (7,8). Mondini dysplasia of the inner ear has been reported (18,31).

Hearing loss has been reported in about 75% (7,15). Of these, about 30% have conductive hearing loss, 20% have sensorineural loss, and 50% have mixed loss. All three types of hearing loss have been observed in different members of the same family affected with variable manifestations of the BOR syndrome, including individuals with only branchial and otologic manifestations. These observations led to the conclusion that there is no distinction between the BO and the BOR syndrome based on type of hearing loss (7,15,19). The hearing loss is usually not progressive (7), but exceptions have been described (18).

**Renal anomalies.** Between 9 and 12% of affected individuals have diagnosed structural anomalies of the renal system (7,15). This is probably underreporting as the renal anomalies may be so subtle that they can be missed on routine intravenous pyelography (8,19). For accurate readings it is advisable to inform the radiologist of potential

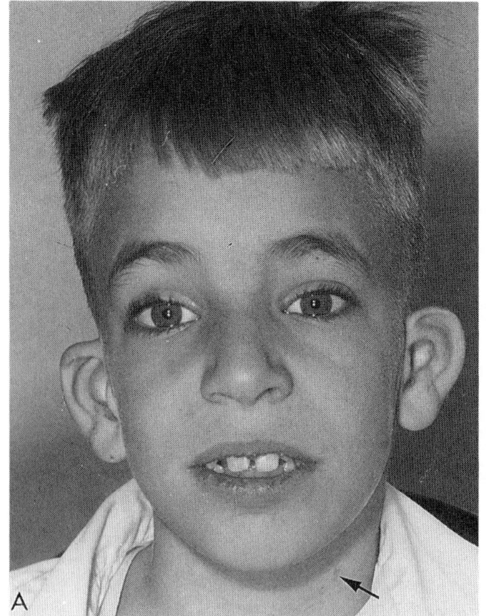


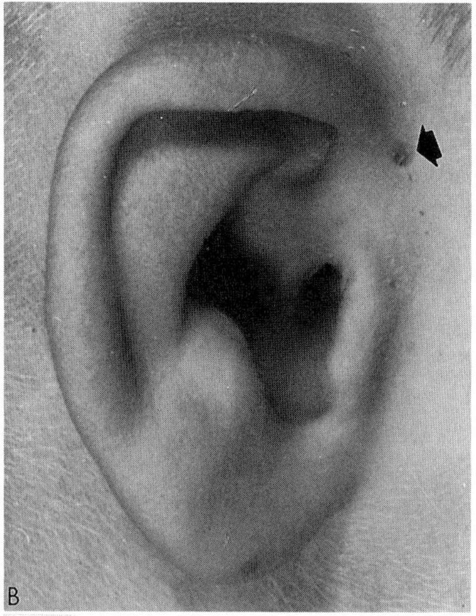


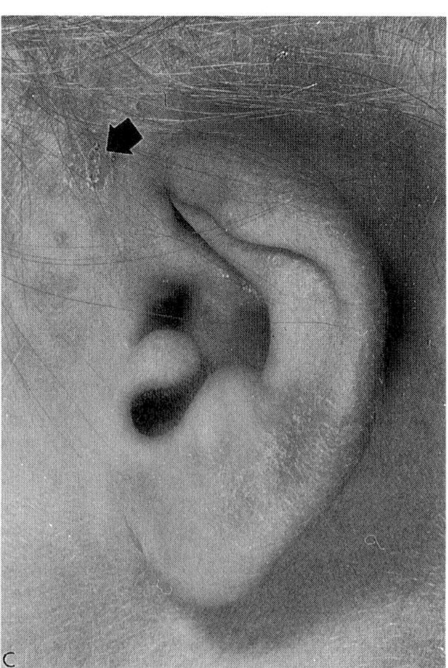


Fig. 19–14. *Branchio-oto-renal syndrome.* (A) Outstanding malformed pinnae. (B) Preauricular pit, marked by arrow. (C) Auricular appendage, grossly malformed ear, and preauricular pit.

manifestations of the syndrome (19). One systematic study of 19 patients by intravenous pyelography showed 75% to have a structural anomaly of the renal system, and 33% to have functional anomalies of the renal system (7). Another systematic study of 16 patients demonstrated 100% to have structural or functional renal manifestations (49). Renal anomalies often remain asymptomatic as most are minor (7). Only about 6% have severe renal involvement (16). If renal agenesis or severe hypoplasia is not present in infancy, the anomalies are not progressive (49).

Severe renal anomalies include bilateral renal agenesis (5,13,32), polycystic kidneys (7,31), and enlarged blunted kidneys (49). Other structural anomalies of the renal system that can range from mild to severe include hypoplastic kidneys (13,32,49), unilateral renal agenesis (11), vesicoureteral reflux (6,19), crossed renal ectopia (5), bilateral bifid renal pelvis (17,19), uretero-pelvic junction obstruction (19), duplication of ureter and collecting system (15,17,19), extrarenal pelvis (19), fetal lobulation (19), abnormal rotation of the kidney (49), calyceal diverticuli or distorted calyceal system (15,31,49), abnormal renal parenchymal thickness (49), and tapered superior pole of the kidney (30,31).

Mild structural anomalies of the renal system include slight blunting of the calices, blunted calyceal fornices without pyelonephritis or papillary necrosis, segmented hypoplasia of the superior pole, reduced renal parenchymal volume (7), and out-pouching of the renal pelvis on the medial border of the kidney (7,31).

Renal function studies have shown normal urinary sediment in all patients (49). A small number of patients have disturbed concentration capacity and proteinuria, or reduced clearance of creatinine and diminished glomerular filtration rates (7,49). Prominent glomerular lesions have been noted (10). Segmental and focal hyalinization with dense immunoglobulin deposits of IgG, IgM, IgA, and C3 along the basement membrane and in the mesangium has been observed (10). Affected individuals can have hydronephrosis, pyelonephritis, and urinary tract infection (19). Anaphylaxis during hemodialysis has been reported (23).

**Other findings.** Aplasia or stenosis of the lacrimal ducts has been found in 9% (7,15,32). Preisch et al (38) reported that some patients with the clinical features of lacrimal duct stenosis really have gustatory lacrimation secondary to misdirected VII cranial nerve innervation. Strabismus has been noted (6,15,19). High arched palate, cleft palate, or bifid uvula can be present (6,7). Overriding bite (6,31) and retrognathia (7,31) have been observed. Intelligence is normal (10).

Facial or mandibular asymmetry (19,33,40,41) and facial nerve deficiency (7,19,40) have been reported. Rollnick and Kaye (40,41) described a family and two additional probands with both the BOR syndrome and hemifacial microsomia (HFM). They hypothesized that the HFM phenotype may constitute a severe form of BOR in some families. Lending support to this hypothesis, Heimler and Lieber (19) reported an individual in a large BOR pedigree affected with both BOR and HFM.

**Differential diagnosis.** Some manifestations of the *cat eye syndrome* overlap with the BOR syndrome. These include ocular coloboma, imperforate anus, cardiac malformations, mild to moderate mental retardation, urinary tract anomalies, and preauricular tags or fistulas. A chromosomal anomaly, tetrasomy 22pter→q11 or other acrocentric chromosome, has been demonstrated (44). Cooper and Jabs (5a) reported two sisters with posteriorly angulated ears with overlapping helices and aural atresia, midface deficiency, VSD, anteriorly placed anus, long fifth fingers, proximally placed thumbs, and camptodactyly of the interphalangeal joints.

Branchial cysts, sinuses, fistulas, and/or branchial skin tags containing cartilage have been reported as isolated defects that occur in some families with autosomal dominant transmission (45).

Autosomal dominant preauricular pits or sinuses can occur as isolated findings. They occur in about 1% of the population and are significantly more frequent in blacks than in whites (30).

Autosomal dominant preauricular pits and sensorineural hearing loss (14), autosomal dominant malformed auricles, preauricular tags, or preauricular pits and moderate conductive hearing loss (50), and branchial fistulas, malformed auricles, and hearing loss (29) have been reported. These may represent variable expression of the BOR syndrome, not distinct entities.

Several syndromes have been described with ear and renal anomalies including the autosomal dominant dysmorphic pinna–polycystic kidney syndrome (21), the autosomal dominant dysmorphic pinna, hypospadias, renal dysplasia syndrome (21), and the autosomal recessive oto-renal-genital syndrome (51).

**Laboratory aids.** Audiologic evaluation is indicated to assess hearing status. Tomography of the middle and inner ear is indicated to identify structural alterations.

Intravenous pyelogram, ultrasonography of the renal system, and renal function studies should be performed to assess renal status.

Prenatal diagnosis of at-risk pregnancies by direct real-time ultrasonography may identify more severe cases with renal involvement (18a). Diagnostic errors in non-real-time ultrasonography have been noted (5). Measurement of maternal serum $\alpha$-fetoprotein as an indicator of fetal renal agenesis has been reported (2), but its reliability as a prenatal diagnostic technique for this defect is doubtful.

### References [Branchio-oto-renal syndrome (BOR syndrome)]


1. Bailleul JP et al: Surdité et fistules auriculaires congénitales familiales. *Pédiatrie* **27**:937–747, 1972.
2. Balfour RP, Laurence KM: Raised serum AFP levels and fetal renal agenesis. *Lancet* **1**:317, 1980.
3. Bourguet J et al: De L'atteinte des deux premières fentes et des deux premiers arcs branchiaux. *Rev Oto-Neuro-Ophtal* **38**:161–175, 1966.
4. Brusis T: Gleichzeitiges Vorkommen von degenerativer Innenohrschwerhörigkeit, Vestibularisstörung, beiderseitigen Ohr-und lateralen Hals-fisteln bei mehreren Mitgliedern einer Familie. *Laryngol Rhinol Otol (Stuttg)* **53**:131–139, 1974.
5. Carmi R et al: The branchio-oto-renal (BOR) syndrome: Report of bilateral renal agenesis in three sibs. *Am J Med Genet* **14**:625–627, 1983.

5a. Cooper LF, Jabs EW: Aural atresia associated with multiple congenital anomalies and mental retardation: A new syndrome. *J Pediatr* **110**:747–750, 1987.

6. Côté A, O'Regan S: The branchio-oto-renal syndrome. *Am J Nephrol* **2**:144–146, 1982.
7. Cremers CWRJ, Fikkers-von Noord M: The earpits-deafness syndrome. Clinical and genetic aspects. *Int J Ped Otorhinolaryngol* **2**:309–322, 1980.
8. Cremers CWRJ et al: Otological aspects of the earpit-deafness syndrome. *ORL* **43**:223–239, 1981.
9. Desnos J et al: Le syndrome malformatif branchio-oto-rénal. *Ann Oto-Laryngol (Paris)* **96**:849–861, 1979.
10. Dumas R et al: Glomerular lesions in the branchio-oto-renal (BOR) syndrome. *Int J Ped Nephrol* **3**:67–70, 1982.
11. Fara M et al: Dismorphia otofaciocervicalis familiaris. *Acta Chir Plast (Praha)* **9**:255–268, 1967.
12. Fitch N et al: The temporal bone in the preauricular pit, cervical fistula, hearing loss syndrome. *Ann Otol (St. Louis)* **85**:268–275, 1976.
13. Fitch N, Srolovitz H: Severe renal dysgenesis produced by a dominant gene. *Am J Dis Child* **130**:1356–1357, 1976.
14. Fourman P, Fourman J: Hereditary deafness in family with ear-pits (fistula auris congenita). *Br Med J* **2**:1354–1356, 1955.
15. Fraser FC et al: Genetic aspects of the BOR syndrome-branchial fistulas, ear pits, hearing loss, and renal anomalies. *Am J Med Genet* **2**:241–252, 1978.
16. Fraser FC et al: Frequency of the of the branchio-oto-renal (BOR) syndrome in children with profound hearing loss. *Am J Med Genet* **7**:341–349, 1980.
17. Fraser FC: Autosomal dominant duplication of the renal collecting system, hearing loss, and external ear anomalies: A new syndrome? *Am J Med Genet* **14**:473–478, 1983.
18. Gimsing S, Dyrmose J: Branchio-oto-renal dysplasia in three families. *Ann Otol Rhinol Laryngol* **95**:421–426, 1986.

18a. Greenberg CR et al: The BOR syndrome and renal agenesis—Prenatal diagnosis and further clinical delineation. *Prenatal Diag* **8**:103–108, 1988.

19. Heimler A, Lieber E: Branchio-oto-renal syndrome: Reduced penetrance and variable expressivity in four generations of a large kindred. *Am J Med Genet* **25**:15–27, 1986.
20. Heusinger CF: Hals-Kiemen-Fisteln von noch nicht beobachteter Form. *Virchows Arch Pathol Anat Physiol* **29**:358–380, 1864.
21. Hilson D: Malformation of ear as sign of malformation of genito-urinary tract. *Br Med J* **2**:785–789, 1957.
22. Hunter AGW: Inheritance of branchial sinuses and preauricular fistulae. *Teratology* **9**:225–228, 1974.
23. Hurley RM: Anaphylaxis during hemodialysis. *Int J Ped Nephrol* **5**:53–54, 1984.
24. Karmody CS, Feingold M: Autosomal dominant first and second branchial arch syndrome. A new inherited syndrome? *Birth Defects* **10**(7):31–40, 1974.
25. Konigsmark BW, Gorlin RJ: *Genetic and Metabolic Deafness.* W. B. Saunders, Philadelphia, 1976.
26. Lipkin DF et al: Hereditary congenital cholesteatoma. *Arch Otolaryngol Head Neck Surg* **112**:1097–1100, 1986.
27. Martins AG: Lateral cervical and preauricular sinuses: Their transmission as dominant characters. *Br Med J* **1**:255–256, 1961.
28. McKusick VA: *Mendelian Inheritance in Man: Catalogs of Autosomal Dominant, Autosomal Recessive and X-linked Phenotypes,* 8th ed. The Johns Hopkins University Press, Baltimore, 1988.
29. McLaurin JW et al: Hereditary branchial anomalies and associated hearing impairment. *Laryngoscope* **76**:1277–1288, 1966.
30. Melnick M, Myrianthopoulos NC: *External ear malformations: Epidemiology, genetics, and natural history. Birth Defects* **15**(9):22–23, 1979.
31. Melnick M et al: Familial branchio-oto-renal dysplasia: A new addition to the branchial arch syndromes. *Clin Genet* **9**:25–34, 1976.
32. Melnick M et al: Branchio-oto-renal dysplasia and branchio-oto dysplasia: Two distinct autosomal dominant disorders. *Clin Genet* **13**:425–442, 1978.
33. Muckle TJ: Hereditary branchial defects in a Hampshire family. *Br Med J* **1**:1297–1299, 1961.
34. Nance WE, Sweeney A: Genetic factors in deafness of early life. *Otolaryngol Clin N Am* **1**:19–48, 1975.
35. Nevin NC: Hereditary deafness associated with branchial fistulae and external ear malformations. *J Laryngol Otol* **91**:709–716, 1977.
36. Ombredanne M: Fistules congénitales cervico-facialies et aplasies de l'oreille. *Ann Chir Plast* **7**:29–33, 1962.
37. Přecechtěl A: Pedigree of anomalies in the first and the second branchial cleft, inherited according to the laws of Mendel, and a contribution to the technique of the extirpation of congenital lateral fistulae colli. *Acta Oto-Laryngol* **11**:23–30, 1927.
38. Preisch JW et al: Gustatory lacrimation in association with the branchio-oto-renal syndrome. *Clin Genet* **27**:506–509, 1985.
39. Rockman-Greenberg C, Evans JA: Renal agenesis and the BOR syndrome. Abstract, March of Dimes Birth Defects Meeting, 1980, New York.
40. Rollnick BR, Kaye CI: Hemifacial microsomia and the branchio-oto-renal syndrome. *J Craniofac Genet Develop Biol, Suppl* **1**:287–295, 1985.
41. Rollnick BR, Kaye CI: Letter: Branchio-oto-renal syndrome: Reduced penetrance and variable expressivity in four generations of a large kindred. *Am J Med Genet* **27**:233, 1987.
42. Rowley PT: Familial hearing loss associated with branchial fistulas. *Pediatrics* **44**:978–985, 1969.
43. Saito R: Anomalies of the auditory organ in Potter's syndrome. Histopathological findings in the temporal bone. *Arch Otolaryngol* **108**:484–488, 1982.
44. Schinzel A et al: The "cat eye syndrome": Dicentric small marker chromosome probably derived from a No. 22 (tetrasomy 22pter→q11) associated with a characteristic phenotype. *Hum Genet* **57**:148–158, 1981.
45. Sedwick CE, Walsh JF: Branchial cysts and fistulas—study of seventy-five cases relative to clinical aspects and treatment. *Am J Surg* **83**:3–8, 1952.
46. Shenoi PM: Wildervanck's syndrome. *J Laryngol Otol* **86**:1121–1135, 1972.
47. Smith PG et al: Clinical aspects of the branchio-oto-renal syndrome. *Otolaryngol Head Neck Surg* **92**:468–475, 1984.
48. Wheeler CE et al: Branchial anomalies in three generations of one family. *Arch Dermatol* **77**:715–719, 1958.
49. Wildervanck LS: Hereditary malformations of the ear in three generations. *Acta Oto-Laryngol* **54**:533–560, 1962.
50. Wirth G: Aurikularanhänge auf dominanten Erbgrundlage. *Z Laryngol Rhinol Otol* **41**:656–659, 1962.
51. Won KH et al: Genetic hearing loss with preauricular sinuses and branchiogenic fistula. *Arch Otolaryngol* **103**:676–680, 1977.


## Cervico-oculo-acoustic syndrome (Wildervanck syndrome)

The major clinical features of this condition are fused cervical vertebrae (Klippel–Feil anomaly), sensorineural hearing loss, and abducens palsy with retracted globe (Duane syndrome). The condition was first described as a specific disorder by Wildervanck (29), but an earlier report was by Baumann (1). In 1978, Wildervanck (32) published an extensive review. The inheritance pattern is poorly understood. All cases with this triad of abnormalities are sporadic. The overwhelming majority of affected individuals reported are female; a few males with the condition have been described (9,10). X-linked dominant inheritance with lethality in males (27), autosomal dominant inheritance with incomplete penetrance and variable expression (15), multifactorial inheritance (17,31), and polygenic inheritance with limitation to females (32) have been proposed. Nongenetic causes cannot be excluded (4).

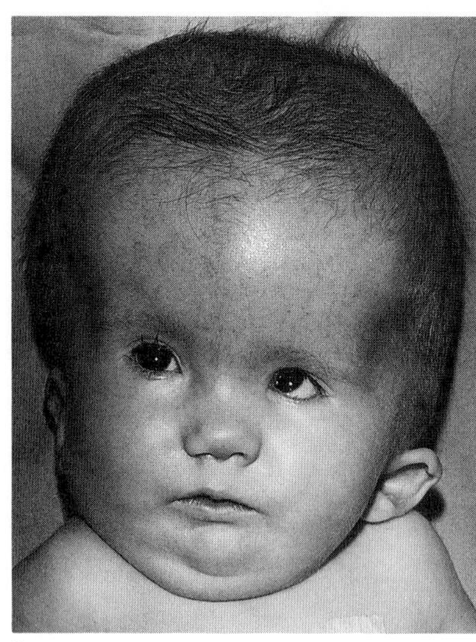

A

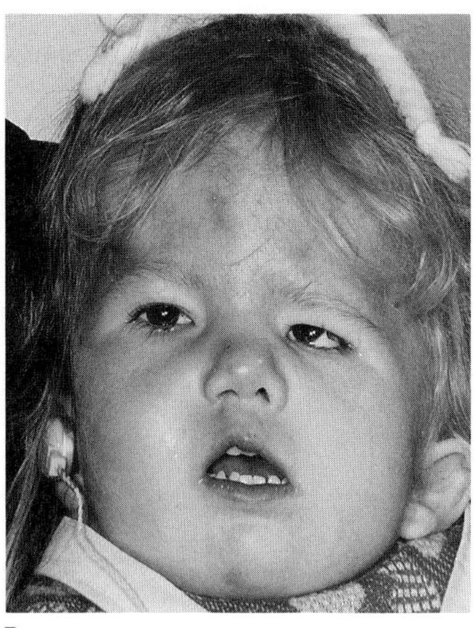

B

Fig. 19–15. *Cervico-oculo-acoustic syndrome (Wildervanck syndrome).* (A,B) Same patient at different ages showing asymmetry of head, severe cervical vertebral involvement, strabismus, arrested hydrocephaly, and sensorineural deafness.

Some investigators suggest that patients with Klippel–Feil anomaly and either retracted globe or hearing loss nonetheless have the Wildervanck syndrome (32). Other reports describe pedigrees in which the index case has Wildervanck syndrome, while relatives have sensorineural hearing loss or Duane syndrome only (14,16,30). Further studies will be necessary to determine whether there is indeed a difference among these phenotypes.

**Facies.** There may be facial asymmetry (Fig. 19–15A,B) (2,9, 10,11,26). Nonprogressive hemifacial weakness, present since birth, has been described (6). Skin tags on the cheek have also been found (13).

**Ophthalmologic manifestations.** Unilateral or bilateral Duane syndrome is a major feature (5,8,11,13,15,20,21,26,29). Duane syndrome consists of abducens paralysis that prevents external rotation of the affected eye. On adduction, the lid fissure of the affected eye narrows, and the globe retracts. Abducens paralysis without retraction has also been described (1,2,6). Pseudopapilledema (13), unilateral epibulbar dermoids (5,9,13), and bilateral temporal subluxation of the lens (26) have been reported.

**Otolaryngologic manifestations.** Hearing loss may be sensorineural (9–11,13,15,21,26), conductive (2,6), or mixed (4,9). Although the age of onset is usually in the first decade (4,13,28,29) and may be profound (13,15), the severity and age of onset have not been well documented. The loss may be unilateral (5,8,26a). In addition, preauricular tags (5,9,13), malformation of the pinna (2,13), atresia or absence of the external auditory canal (5,6), abnormal and absent ossicles (5,18), ossified stapedial tendon (5), stenotic or short internal auditory meatus (8), stapes fixation (4), abnormal semicircular canals (8,18,31), and underdevelopment of the bony labyrinth (Mondini deformity) (23,24,26a,31,33) have also been described. Marked positional vertigo and an abnormal vestibular response were found in the one patient formally tested (6).

An anterior glottic web has been noted (6).

**Skeletal abnormalities.** Klippel–Feil anomaly, consisting of fusion of one or more cervical and sometimes thoracic vertebrae, is characteristic (1,2,4–6,9,13,15,29). The neck is short, thick, and webbed, and the head appears to sit directly on the trunk (1, 2,5,8,10,11,13,26). Flexion, extension, and lateral mobility of the neck are severely restricted (1,2,6,8,10,11,13,15) and there may be torticollis (1,9,10,20). Spina bifida occulta (1,6,9–11,20), Sprengel deformity (13), hemivertebrae (5), fusion of ribs (5), absent ribs (5), kyphosis (5,7,10,11,20), scoliosis (5,8,9,20), and basilar impression (2,6,8–10) have been reported.

**Central nervous system.** A few reports describe mild (2) or severe (11) mental retardation. Facial paralysis has been noted (26). Mirror movements have been described (1,6).

**Other findings.** The neck is webbed (2,6) and the posterior hairline is low (1,2,5,10).

**Oral manifestations.** Cleft palate may be present (2,11,15).

**Differential diagnosis.** Differential diagnosis is complicated because many patients with Klippel–Feil anomaly have been incompletely described; some may indeed have the Wildervanck syndrome. Adequate family studies have not been performed. *Klippel–Feil anomaly* and associated abnormalities have been reviewed by Helmi and Pruzansky (12).

Everberg (7) reported a boy with Klippel–Feil anomaly and thoracic scoliosis; the patient lacked the left thumb, radius, first metacarpal, major multangular and navicular. There was left-sided blindness and squint without abducens paralysis or eyeball retraction. The left condyle and mandibular fossa were absent. Abnormalities of the left external, middle, and inner ear were also found; there was a 60–80 dB mixed hearing loss in the left ear. Hearing in the other ear was normal. Intelligence was normal. Sherk and Nicholson (25) reported a girl with Duane syndrome, epicanthal folds, auricular atresia, normal hearing, Klippel–Feil anomaly, low posterior hair line, webbed neck, mild mental retardation, and bilateral radial hemimelia.

Livingstone and Delahunty (19) noted two unrelated patients with unilateral external ear abnormalities and bilateral abducens nerve palsy; one had unilateral conductive hearing loss, while the other had mixed loss. Neither investigator mentioned Klippel–Feil anomaly. Kirkham (15) described two girls with cleft palate, profound hearing loss, and Duane syndrome as well as a girl with cleft palate, bilateral Duane syndrome, but normal hearing; Klippel–Feil anomaly was not mentioned. Brik and Athayde (3) described a girl with Duane syndrome, paroxysmal lacrimation, and Klippel–Feil anomaly. Pintucci and di Tizio (22) described a man with Klippel–Feil anomaly and Duane syndrome; his nephew had Duane syndrome.

Witzel (34) reported a male with Klippel–Feil anomaly, cleft palate, hemivertebrae, spina bifida occulta, profound sensorineural hearing loss in the left ear, total paralysis of conjugate gaze movements to the right and left, and absent vestibular movements.

There is overlap with *oculo-auriculo-vertebral spectrum* (5,9,13).

**References [Cervico-oculo-acoustic syndrome (Wildervanck syndrome)]**

1. Bauman GI: Absence of the cervical spine. Klippel–Feil syndrome. *JAMA* **98**:129–132, 1932.
2. Bintliff SJ: Klippel-Feil syndrome. *J Am Med Women's Assoc* **20**:547–550, 1965.
3. Brik M, Athayde A: Bilateral Duane's syndrome, paroxysmal lacrimation and Klippel–Feil anomaly. *Ophthalmologica* **167**:1–8, 1973.
4. Cremers CWRJ et al: Hearing loss in the cervico-oculo-acoustic (Wildervanck) syndrome. *Arch Otolaryngol* **110**:54–57, 1984.
5. Cross HE, Pfaffenbach DD: Duane's retraction syndrome and associated congenital malformations. *Am J Ophthalmol* **73**:442–449, 1972.
6. Eismann ML, Sharma GK: The Wildervanck syndrome: Cervico-oculo-acoustic dysplasia. *Otolaryngol Head Neck Surg* **87**:892–897, 1979.
7. Everberg G: Congenital absence of the oval window. *Acta Otolaryngol (Stockholm)* **66**:320–332, 1968.
8. Everberg G et al: Wildervanck's syndrome. Klippel–Feil's syndrome associated with deafness and retraction of the eyeball. *Br J Radiol* **36**:562–567, 1963.
9. Franceschetti A, Klein D: Dysmorphie cervico-oculo-faciale avec surdité familiale. (Klippel–Feil, retractio bulbi, asymétrie cranio-faciale et autres anomalies congénitales). *J Génét Hum* **3**:176–13, 1954.
10. Franceschetti A et al: An extensive form of cervico-oculo-facial dysmorphia (Wildervanck–Franceschetti–Klein). *Acta Fac Med Univ Brun* **25**:53–61, 1965.
11. Fraser WI, MacGilivray RC: Cervico-oculo-acoustic dysplasia (''the syndrome of Wildervanck''). *J Ment Defic Res* **12**:322–329, 1968.
12. Helmi C, Pruzansky S: Craniofacial and extracranial malformations in the Klippel–Feil syndrome. *Cleft Pal J* **17**:65–88, 1980.
13. Kirkham TH: Cervico-oculo-acousticus syndrome with pseudopapilloedema. *Arch Dis Childh* **44**:504–508, 1969.
14. Kirkham TH: Duane's syndrome and familial perceptive deafness. *Br J Ophthalmol* **53**:335–339, 1969.
15. Kirkham TH: Duane's retraction syndrome and cleft palate. *Am J Ophthalmol* **70**:209–212, 1970.
16. Kirkham TH: Inheritance of Duane's syndrome. *Br J Ophthalmol* **54**:323–329, 1970.
17. Konigsmark BW, Gorlin RJ: *Genetic and Metabolic Deafness*. W.B. Saunders, Philadelphia, 1976, pp 188–191.
18. Lindsay JR: Inner ear histopathology in genetically determined congenital deafness. *Birth Defects* **7**(4):21–32, 1971.
19. Livingstone G, Delahunty JE: Malformation of the ear associated with congenital ophthalmic and other conditions. *J Laryngol Otol* **82**:495–504, 1968.
20. Magnus JA: Congenital paralysis of both external recti treated by transplantation of eye muscles. *Br J Ophthalmol* **28**:241–245, 1949.
21. McLay K, Maran AGD: Deafness and the Klippel–Feil syndrome. *J Laryngol* **83**:175–184, 1969.
22. Pintucci F, di Tizio A: Dysmorphie cervico-oculo-faciale, *J Génét Hum* **10**:156–171, 1961. (Cited by Wildervanck LS: The cervico-oculo-acusticus syndrome, *Handbook of Clinical Neurology*, Vol 32, Vinken PJ, Bruyn GW, Myrianthopoulos NC (eds), North-Holland Publishing, Amsterdam, 1978, pp 123–130.)
23. Regenbogen L, Godel V: Cervico-oculo-acoustic syndrome. *Ophthal Paediatr Genet* **6**:183–187, 1985.
24. Schild JA et al: Wildervanck syndrome—the external appearance and radiographic findings. *Int J Pediatr Otorhinolaryngol* **7**:305–310, 1984.
25. Sherk HH, Nicholson JT: Cervico-oculo-acusticus syndrome. *J Bone Jt Surg* **54A**:1776–1778, 1972.
26. Strisciuglio P et al: Wildervanck's syndrome with bilateral subluxation of lens and facial paralysis. *J Med Genet* **20**:72–73, 1983.
26a. West PDB et al: Wildervanck's syndrome—unilateral Mondini dysplasia identified by computed tomography. *J Laryngol Otol* **103**:408–411, 1989.
27. Wettke-Schafer R, Kantner G: X-linked dominant inherited diseases with lethality in hemizygous males. *Hum Genet* **64**:1–23, 1983.
28. Wildervanck LS: Een cervico-oculo-acusticussyndroom. *Ned Tijdschr Geneeskd* **104**:2600–2605, 1960.
29. Wildervanck LS: Een geval van aandoening van Klippel-Feil gecombineerd met abducensparalyse, retractio bulbi en doofst omheid. *Ned Tijdschr Geneesk* **96**:2752–2756, 1952. (Cited by Helmi C, Pruzansky S: Craniofacial and extracranial malformations in the Klippel-Feil syndrome. *Cleft Palate J* **17**:65–88, 1980.)
30. Wildervanck LS: A cervico-oculo-acousticus syndrome belonging to the status dysraphicus, *Proc 2nd Int Cong Hum Genet (Rome)* 1961, **3**:E186–E187, 1963 (Abstr).
31. Wildervanck LS et al: Radiological examination of the inner ear of deaf-mutes presenting the cervico-oculo-acousticus syndrome. With a summary of roentgenologic and pathologico-anatomical findings in other endogenous forms of deafness. *Acta Otolaryngol (Stockholm)* **61**:445–453, 1966.
32. Wildervanck LS: The cervico-oculo-acusticus syndrome, in *Handbook of Clinical Neurology*, Vol 32, Vinken PJ, Bruyn GW, Myrianthopoulos NC (eds), North-Holland Publishing, Amsterdam, 1978, pp 123–130.
33. Windle-Taylor PC et al: Ear deformities associated with the Klippel–Feil syndrome. *Ann Otol Rhinol Laryngol* **90**:210–246, 1981.
34. Witzel SH: Congenital paralysis of lateral conjungate gaze. Occurrence in a case of Klippel–Feil syndrome. *Arch Ophthalmol* **59**:463–464, 1959.


## Townes–Brocks syndrome

In 1972, Townes et al (8) reported a syndrome of sensorineural hearing loss, imperforate anus, ''lop'' ears, abnormal thumbs (especially triphalangeal), pes planus, and anomalies of the metacarpals, metatarsals, and digits (Fig. 19–16A–F). Since then, about 20 cases of the syndrome have been reported (1–3,5,7–10). Inheritance is autosomal dominant with complete penetrance and variable expressivity (3,8,9).

The patient reported by Reid and Turner (7) expressed only three of the signs found in the syndrome—preauricular tags, bifid thumbs, and imperforate anus. Less certain is the sporadic case with ASD reported by Monteiro de Pina-Neto (5).

**Clinical manifestations.** Ear anomalies are variable. Observed may be the satyr form of lop ear, hypoplastic ears, and preauricular tags. By history, large ears were reported by Kurnit et al (3). Other types of ear dysgenesis may also be observed such as thickened superior helix, microtia, external auditory atresia, pretragal or intraauricular pit, and cheek skin tags (3). Hearing loss is sensorineural (3,8–10). Various skeletal defects can be observed, especially in the preaxial part of the distal upper limb. Features may include triphalangeal or hypoplastic thumbs, bifid thumbs, supernumerary thumbs, medial deviation of the distal phalanges of the thumbs, elongated and deformed first metacarpal, pseudoepiphysis of the second metacarpal, pes planus, clinodactyly of the fifth toes, fusion of metatarsals, absent or hypoplastic third toes, absent triquetral or navicular bone, fusion of triquetral and hamate, and cone-shaped epiphyses (3,8,10).

Imperforate anus, anal stenosis, and anteriorly placed anus have

Fig. 19–16. *Townes–Brocks syndrome.* (A,B) Radial dysplasia. Note fingerlike thumb in A. Note vestigial supernumerary thumb in B. (C,D) Right and left ears of same patient. Note differences. (E,F) Right and left ears of a second patient. Note differences. (A–F from *DM Kurnit et al,* J Pediatr **93**:270, 1978.)

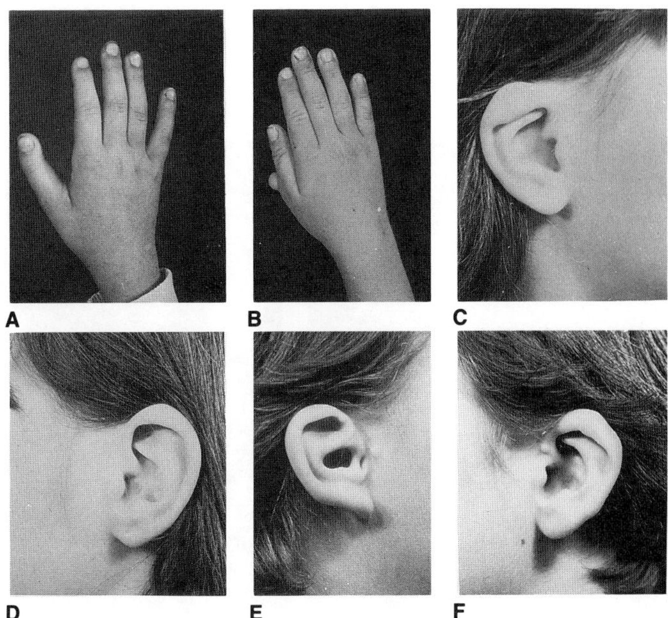

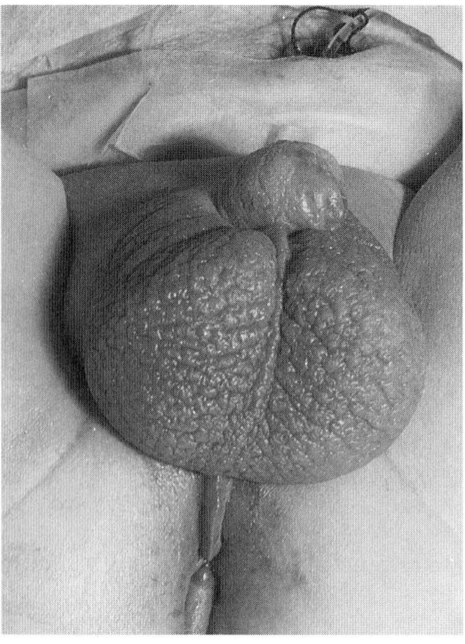

Fig. 19–17. *Townes–Brocks syndrome.* Hypospadias, bifid scrotum. (From *Van der Weerd MA et al,* Clin Genet **34**:195, 1988.)

### References (Townes-Brocks syndrome)


1. Barakat AY et al: Townes–Brocks syndrome: Report of three additional patients with previously undescribed renal and cardiac abnormalities. *Dysmorphol Clin Genet* **2**:104–108, 1988.

1a. Hersh JH et al: Townes syndrome: A distinct multiple malformation syndrome resembling VACTERL association. *Clin Pediatr* **25**:100–102, 1986.

2. Johnson JP, Sherman S: Townes–Brocks syndrome: Three generations with variable expression. Ninth Annual David W. Smith Workshop on Malformations and Morphogenesis, Oakland, August 7–10, 1988.

3. Kurnit DM et al: Autosomal dominant transmission of a syndrome of anal, ear, renal, and radial congenital malformations. *J Pediatr* **93**:270–273, 1978.

4. Lowe J et al: Dominant ano-rectal malformation, nephritis and nerve-deafness: A possible new entity? *Clin Genet* **24**:191–193, 1983.

5. Monteiro de Pina-Neto J: Phenotypic variability in Townes–Brocks syndrome. *Am J Med Genet* **18**:147–152, 1984.

6. Pinsky L: The syndromology of anorectal malformation (atresia, stenosis, ectopia). *Am J Med Genet* **1**:461–474, 1978.

7. Reid JS, Turner G: Familial anal abnormality. *J Pediatr* **88**:992–994, 1976.

8. Townes PL et al: Hereditary syndrome of imperforate anus with hand, foot, and ear anomalies. *J Pediatr* **81**:321–326, 1972.

9. Van der Weerd MA et al: A new family with the Townes–Brocks syndrome. *Clin Genet* **34**:195–200, 1988.

10. Walpole IR and Hockey A: Syndrome of imperforate anus, abnormalities of hands and feet, satyr ears, and sensorineural deafness. *J Pediatr* **100**:250–252, 1982.


been recorded. Also observed have been rectovaginal fistula and rectoperineal fistula (3,8). Congenital heart anomalies have been found (1,5).

Kurnit et al (3) reported hypoplastic kidneys, ureterovesical reflux, posterior urethral valves, and meatal stenosis. Hypospadias and bifid scrotum have also been noted (Fig. 19–17) (1,8). A prominent raphe has been documented (3,7,9).

**Differential diagnosis.** To be excluded are *branchio-oto-renal syndrome* and *oculo-auriculo-vertebral spectrum.* Lowe et al (4) reported dominantly inherited anorectal malformation, nephritis, and sensorineural hearing loss. No preaxial limb anomalies were observed. The syndromology of anorectal malformations and the VATER/VACTERL association has been discussed by Pinsky (6).

**Laboratory aids.** Audiologic evaluation is indicated to assess hearing status. Radiographs of the hands and feet may be helpful. Intravenous pyelogram or ultrasonography of the renal system should be performed to assess renal status.

## Mandibulofacial dysostosis, microtia, talipes, and agenesis of the patellae

Meier et al (2), in 1959, reported a male child with micrognathia, microtia, midface hypoplasia, ankylosis of the knees, cryptorchidism, and agenesis of the patellae. The parents were consanguineous. Gorlin et al (1) described a male with microtia, absent patellae, and micrognathia (Fig. 19–18A,B). As an infant, the patient had talipes equinovarus and "his knees bent backwards." He had unilateral cryptorchidism. Radiographic examination showed Blount osteochondritis dessicans and bilateral aseptic necrosis of the lateral femoral condyles. There was no parental consanguinity.

### References (Mandibulofacial dysostosis, microtia, talipes, and agenesis of the patellae)


1. Gorlin RJ et al: A selected miscellany. *Birth Defects* **11**(2):39–50, 1975.

2. Meier Z et al: Ein Fall von Arthrogryposis multiplex congenita kombiniert mit Dysostosis mandibulofacialis (Franceschetti-Syndrom). *Helv Paediatr Acta* **14**:213–216, 1959.


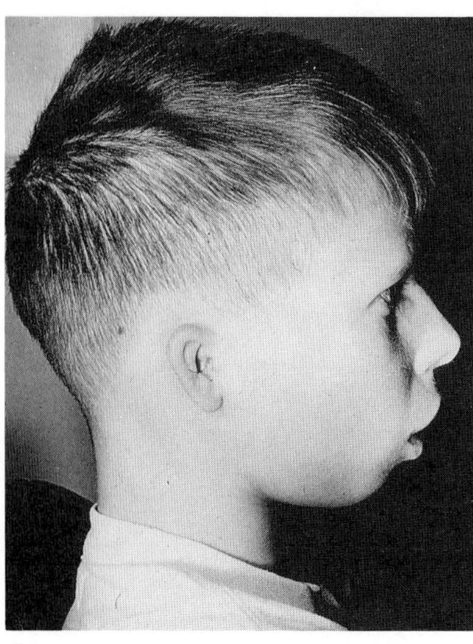

A

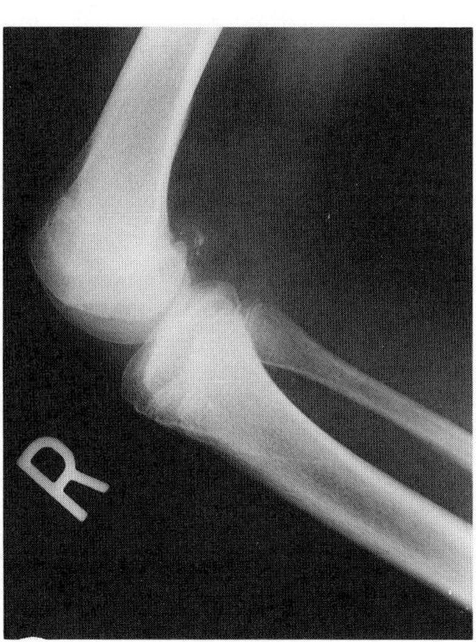


B

Fig. 19–18. *Mandibulofacial dysostosis, microtia, talipes, agenesis of patellae.* (A) Microtia, micrognathia. (B) Agenesis of patellae. [From *RJ Gorlin et al,* Birth Defects **11**(2):39, 1975.]

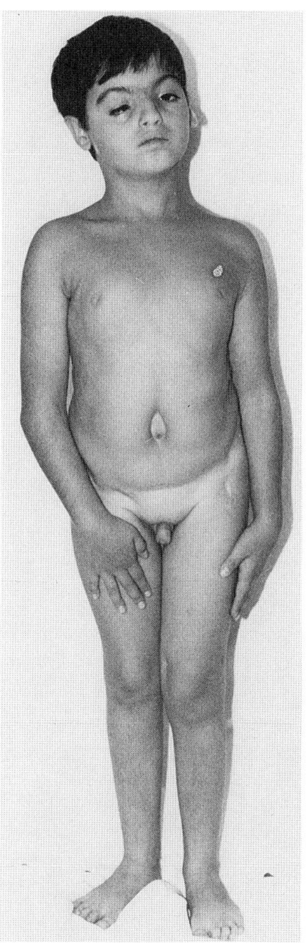

Fig. 19–19. *Carnevale syndrome.* Note ptosis, strabismus, limited extension at elbow, partial agenesis of abdominal musculature with diastasis, and cryptorchidism. (Courtesy of *F Carnevale,* Bari, Italy.)

Fig. 19–20. *Carnevale syndrome.* Two male sibs, the offspring of a consanguineous marriage. Note downslanting palpebral fissures, upper eyelid ptosis, and convergent strabismus. (Courtesy of *F Carnevale,* Bari, Italy.)

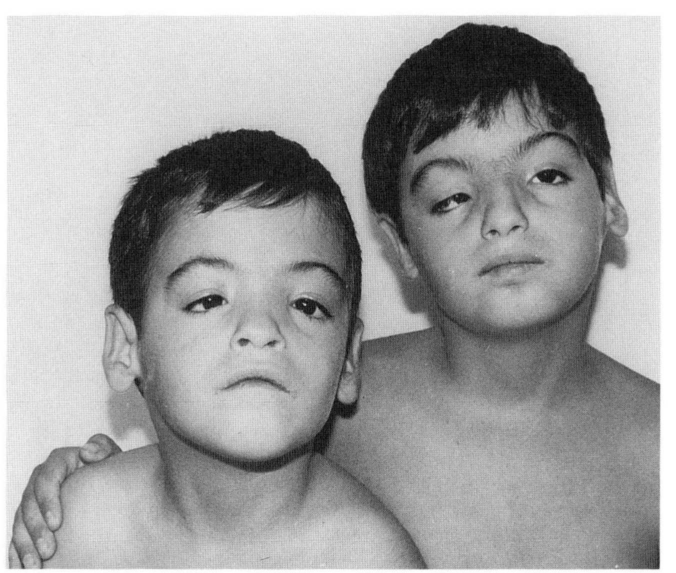

## Carnevale syndrome

Carnevale (1) described two male sibs, the offspring of a consanguineous marriage, who manifested downslanting palpebral fissures, ptosis of the upper eyelids, and convergent strabismus. They had limited extension of the forearms and hip dysplasia. Both exhibited mild mental deficiency, cryptorchidism, and partial agenesis of the abdominal musculature with diastasis (Figs. 19–19 and 19–20).

### Reference (Carnevale syndrome)


1. Carnevale F et al: New syndrome: Ptosis of eyelids, strabismus, diastasis recti, hip defect, cryptorchidism, and developmental delay in two sibs. *Am J Med Genet* **33**:186–189, 1989.


## Hypertelorism–microtia–clefting syndrome (Bixler syndrome)

Bixler et al (2), in 1969, described two sisters with hypertelorism, markedly hypoplastic pinnae, short stature, mild psychomotor retardation, and cleft lip–palate (Fig. 19–21A–D). Tomographic examination revealed hypoplasia of the auditory ossicles and atresia of the canals. Mild microcephaly was present, as well as ectopic kidneys and congenital heart defects (ASD, endocardial cushion defect). Growth was below the third percentile. There was mild thenar hypoplasia and short fifth fingers. The parents were normal, suggesting autosomal recessive inheritance.

The disorder has also been reported by Schweckendiek et al (5,6), Baraitser (1), and Mourad-Baars and Maaswinkel-Mooij (4). Cardiac anomalies were not present (1,5,6). The so-called variant of Bixler syndrome (3) is clearly *oculo-auriculo-vertebral spectrum.*

### References [Hypertelorism–microtia–clefting syndrome (Bixler syndrome)]


1. Baraitser M: The hypertelorism-microtia-clefting syndrome. *J Med Genet* **19**:387–388, 1982.
2. Bixler D et al: Hypertelorism, microtia, and facial clefting: A newly described inherited syndrome. *Am J Dis Child* **118**:495–498, 1969; and *Birth Defects* **5**(2):77–81, 1969.
3. Ionasescu V, Roberts RJ: Variant of Bixler syndrome. *J Génét Hum* **22**:133–138, 1974.
4. Mourad-Baars PEC, Maaswinkel-Mooij PD: A new patient with HMC syndrome. *Am J Med Genet,* in press.
5. Schweckendiek W, Hillig U: Neue Beobachtungen von Spaltbildungen bei eineiigen Zwillingen. *Acta Stomat Belg* **73**:57–62, 1962.
6. Schweckendiek W et al: HMC syndrome in identical twins. *Hum Genet* **33**:315–318, 1976.


## DiGeorge sequence

Depending on etiology, severity, and time of embryonic insult during the fourth to seventh week of gestation, DiGeorge sequence can be minimally expressed as the III–IV pharyngeal pouch complex in which there is absence or hypoplasia of the thymus and/or parathyroid glands (Fig. 19–22) (22). On the other hand, it may be maximally expressed with cardiovascular anomalies, especially interrupted aortic arch and truncus arteriosus, and various craniofacial anomalies (6,25). The neural crest plays a critical role in the development of the thymus and parathyroid glands, the aortic arches, and the conotruncal part of the heart (2a,3a,21,42,45). DiGeorge (9) first called attention to the association of absence of the thymus with absence of the parathyroid glands and described human thymic deficiency. Good (see 40) coined the term DiGeorge syndrome. Although we and others (14) use the term DiGeorge sequence, the condition has also been known as III–IV pharyngeal pouch syndrome (32–34,40), DiGeorge syndrome (1–7, 15,16,19,23,24,26,26a,36,37,46), and DiGeorge anomaly (18,21,28a, 42).

Clinical and genetic aspects of DiGeorge sequence are extensively

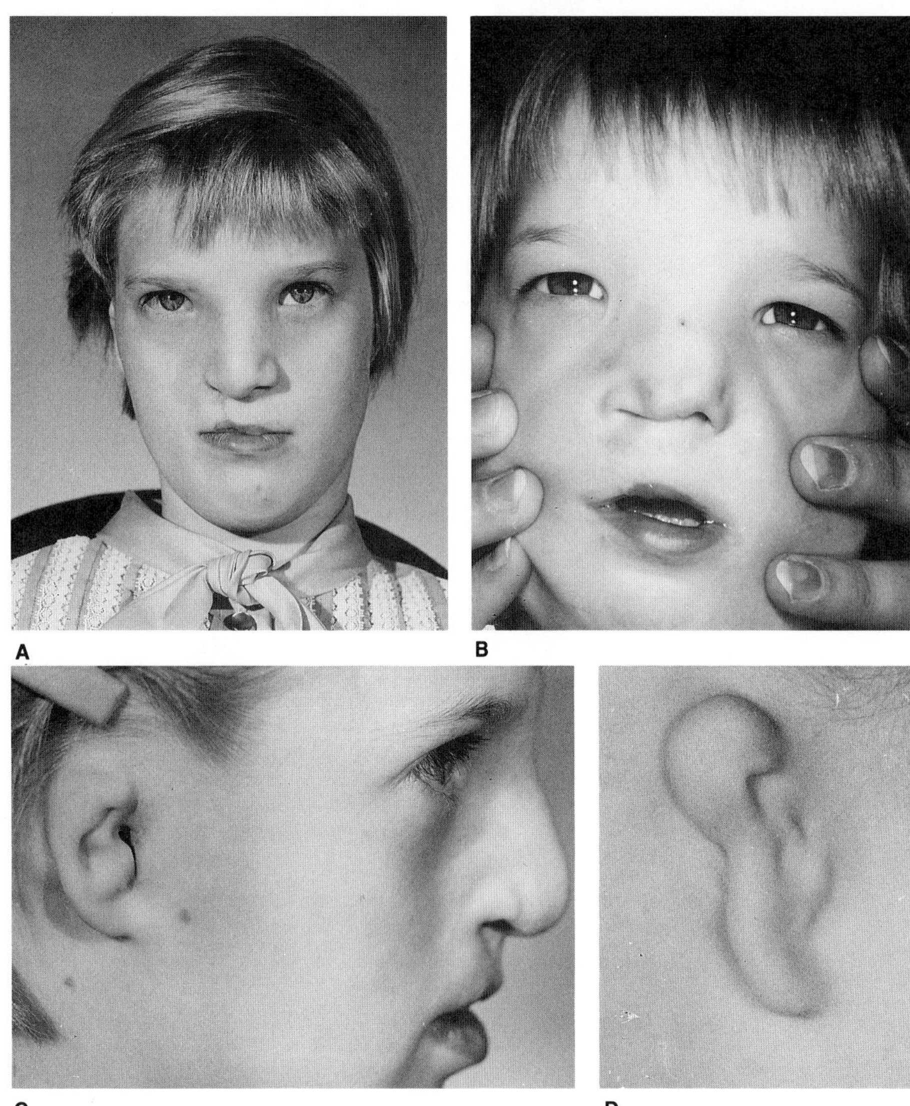


Fig. 19–21. *Hypertelorism-microtia-clefting syndrome*. (A,B) Note hypertelorism, repaired cleft lip in sisters. (C,D) Dysplastic pinna. [A–D from *D Bixler et al,* Birth Defects **5**(2):77, 1969.]

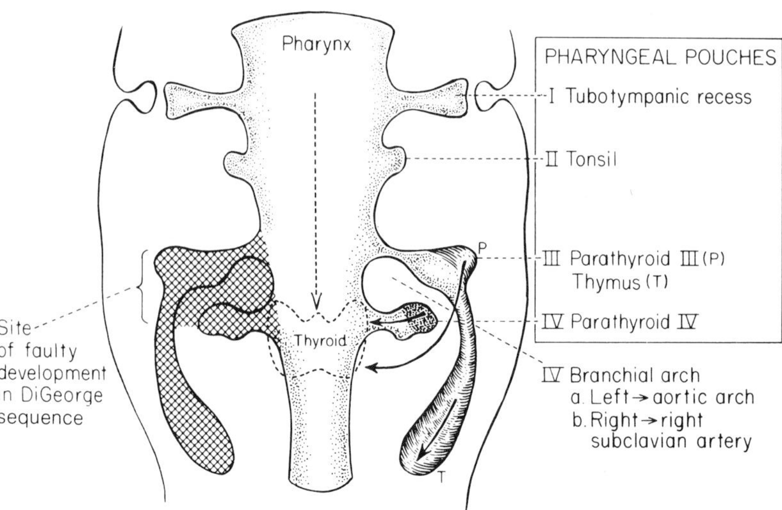


Fig. 19–22. *DiGeorge sequence*. Schematic appearance of anterior foregut and its derivatives around fifth week of development showing site of DiGeorge defect. III–IV pharyngeal pouch complex. Thymus and inferior parathyroids develop from third pouch. Superior parathyroids develop from fourth pouch. Aortic arch from IV may also be involved. Left involvement results in type B interrupted aortic arch. Involvement of the right aortic arch results in aberrant right subclavian artery. (From *DW Smith,* Recognizable Patterns of Human Malformation, 3rd ed, W.B. Saunders, Philadelphia, 1982, p 470.)

reviewed by Lammer and Opitz (21), Müller et al (26a), and Belohradsky (2a). Conley et al (6) reported necropsy data on 25 patients. Miller et al (25) reviewed 156 patients with idiopathic hypoparathyroidism and found 40 different kinds of branchial arch and pharyngeal pouch anomalies. Thomas et al (42) have elegantly reviewed these variations.

Most instances of DiGeorge sequence occur sporadically, but occasionally autosomal recessive (32,27,46) and autosomal dominant (39) inheritance have been documented. The possibility of X-linked inheritance still remains a theoretical possibility. DiGeorge sequence has also been observed in monozygotic twins (24).

Clinical and pathologic evidence indicates that DiGeorge sequence is etiologically heterogeneous (42). The sequence may occur as part of a broader pattern of abnormalities, and various syndromes and associations have been reported (5,21) including del(22q) and various other chromosomal disorders (7,8,16,16a,18b,23a,41–43), four monogenic conditions (6,13,14,17,18a,31,32,37,39), three teratogenic conditions (1,3,15,20), and three associations (6,27,30,38). These are listed together with appropriate references in Table 19–1.

**Clinical history.** Infants with complete DiGeorge sequence have a poor prognosis. Patients may have hypocalcemia with seizures, murmurs, or cardiac failure, or, after the first few weeks of life, may develop infections with purulent rhinorrhea, maculopapular rashes, failure to thrive, and development delay. Those with partial DiGeorge sequence who survive infancy are moderately to severely mentally retarded (6,26a).

**Parathyroids.** Hypocalcemia is a variable feature and patients with severe hypocalcemia are likely to have complete absence of the parathyroids. In most cases, one or two parathyroids or ectopic parathyroids are present (6).

**Thymus.** Defective cell-mediated immunity is attributable to deficiency of thymic tissue. The thymus gland may be absent (complete DiGeorge sequence) or hypoplastic (partial DiGeorge sequence). In the necropsy series of Conley et al (6), partial absence occurred in 58%. Three patterns were observed: cervical lobes only, median thymic tissue situated above the left innominate vein, or small lateralized accessory thymic lobules located in the neck (6).

Table 19–1. Conditions with DiGeorge sequence*

| Condition | References |
|---|---|
| Chromosomal | |
| dup(1q) | 44 |
| del(5p) | 41 |
| dup(8q) | 43 |
| del(10p) | 16a,18b |
| del(17p) | 16b |
| del(22q) | 7,8,16,18,35 |
| Monogenic | |
| Isolated autosomal dominant | 18a,39 |
| Isolated autosomal recessive[a] | 32,27 |
| Velocardiofacial syndrome | 14 |
| Zellweger syndrome | 6,13,17,31 |
| Teratogenic[b] | |
| Diabetic embryopathy | 3,15 |
| Fetal alcohol syndrome | 1 |
| Retinoid embryopathy | 20 |
| Associational | |
| Arhinencephaly/DiGeorge[c] | 6 |
| CHARGE/DiGeorge | 30,38 |
| Cardiofacial/DiGeorge | 27 |

*From MM Cohen, Jr. and DEC Cole, *J Pediatr* **115**:161, 1989.
[a]Affected male sibs have been reported on three occasions (2,11,46), suggesting that X-linked inheritance is still a theoretical possibility.
[b]Bisdiamine, a drug initially studied as a male contraceptive and not approved for human use in the United States, is known to produce a DiGeorge-like pattern of defects in experimental animals (28,29).
[c]Including Kallman syndrome (36).

**Cardiovascular anomalies.** Only 5% have a normal heart. Symptomatic heart defects are common. Among 161 cases, type B interrupted aortic arch was found in 30% of clinically symptomatic cases and in 50% of necropsy cases (6,45). This may occur with or without aberrant right subclavian artery. Persistent truncus arteriosus was found in 24% (45). Pulmonary arteries in these patients are of variable size. Approximately 12% of cardiovascular anomalies are of other types, most commonly tetralogy of Fallot (6%) (5,25,45).

**Craniofacial anomalies.** Craniofacial anomalies occur in approximately 60%. The most frequent features include micrognathia, deep-set, low-set, small, posteriorly angulated ears that are sometimes pointed, anteverted nostrils, blunted nose, clefting or indentation of the nose, and hypertelorism (6). Short philtrum, cleft palate or bifid uvula, highly arched palate, choanal atresia, abnormalities of the middle ear, various eye anomalies, and central nervous system malformations have also been noted (6,22,25,26a).

**Other abnormalities.** A variety of other abnormalities have been reported including absent thyroid lobe, accessory thyroid tissue, esophageal atresia, anomalous speech musculature, diaphragmatic defects, Meckel diverticulum, abnormal bowel rotation, imperforate anus, and hydronephrosis (6,25).

**Diagnosis and differential diagnosis.** DiGeorge sequence includes patients with two of the three following findings: (1) cellular immune deficiency or demonstrated absence of all or part of the thymus gland or both, (2) symptomatic hypocalcemia or anatomic deficiency of the parathyroids or both, and (3) congenital heart defects. DiGeorge sequence should be suspected in any neonate with interrupted aortic arch or truncus arteriosus, an infant with hypocalcemia, failure to thrive, chronic purulent rhinitis, or a mildly retarded child with manifestations of immunodeficiency. Table 19–1 should be consulted when DiGeorge sequence occurs as part of a broader pattern of abnormalities.

Several conditions have features in common with DiGeorge sequence. *Oculo-auriculo-vertebral spectrum* has overlapping features with DiGeorge sequence. However, to our knowledge, the finding of T-cell deficiency and/or hypoparathyroidism has not been reported to date. Characteristics of *trisomy 18 syndrome* include low birthweight, triangular facies, low-set pointed ears, micrognathia, congenital heart defects, renal anomalies, and central nervous system malformations. The thymus is frequently noted to be small on necropsy, and absence of the parathyroid glands has also been reported (6).

Studies of calcium metabolism, cell-mediated immunocompetence, and cardiovascular function are indicated. Immunoregulatory disturbances (hypergammaglobulinemia, high titers of specific antibody production) prevail in partial DiGeorge sequence (26a).

### References (DiGeorge sequence)


1. Ammann AJ et al: The DiGeorge syndrome and the fetal alcohol syndrome. *Am J Dis Child* **136**:906–908, 1982.

2. Atkin JF et al: Familial DiGeorge syndrome in 7 children. *Am J Hum Genet* **34**:80A, 1982.

2a. Belohradsky BH: Thymusaplasie und - hypoplasie mit Hypoparathyreoidismus. Herz- und Gefässmissbildungen-DiGeorge-Syndrom. *Ergeb Inn Med Kinderheilkd* **54**:36–105, 1985.

3. Black FO et al: Aural abnormalities in partial DiGeorge syndrome. *Arch Otolaryngol* **101**:129–134, 1975.

3a. Bockman DE, Kirby ML: Dependence of thymus development on derivatives of the neural crest. *Science* **223**:498–500, 1984.

4. Burgio GR, Ugazio AG: Annotation: Errors of morphogenesis and inborn errors of immunity 20 years after the discovery of the DiGeorge anomaly. *Eur J Pediatr* **144**:9–12, 1985.

5. Carey JC: Spectrum of the DiGeorge "syndrome." *J Pediatr* **96**:955–956, 1980.

6. Conley ME et al: The spectrum of the DiGeorge syndrome. *J Pediatr* **94**:883–890, 1979.

7. De la Chapelle A et al: A deletion in chromosome 22 can cause DiGeorge syndrome. *Hum Genet* **57**:253–256, 1981.

8. DeCicco F et al: Monosomy of chromosome no. 22: A case report. *J Pediatr* **83**:836–838, 1973.

9. DiGeorge AM: Discussion on a new concept of the cellular basis of immunology. *J Pediatr* **67**:907, 1965.

10. DiGeorge AM: Congenital absence of the thymus and its immunologic consequences: Concurrence with congenital hypoparathyroidism. *Birth Defects* **4**(1):116–121, 1968.

11. Forsythe R: Thymic aplasia in siblings. *Pediatr Res* **15**:561 (Abstr), 1981.

12. Freedom RM et al: Congenital cardiovascular disease and anomalies of the third and fourth pharyngeal pouch. *Circulation* **46**:165–172, 1972.

13. Gilchrist DW et al: Immunodeficiency in the cerebro-hepato-renal syndrome of Zellweger. *Lancet* **1**:164–165, 1974.

14. Goldberg R et al: Phenotypic overlap between velo-cardio-facial syndrome and the DiGeorge sequence. *Am J Hum Genet* **37**:54A, 1985.

15. Gosseye S et al: Association of bilateral renal agenesis and DiGeorge syndrome in an infant of a diabetic mother. *Helv Paediatr Acta* **37**:471–474, 1982.

16. Greenberg F et al: Familial DiGeorge syndrome and associated partial monosomy of chromosome 22. *Hum Genet* **65**:317–319, 1984.

16a. Greenberg F et al: Hypoparathyroidism and T cell immune defect in a patient with 10p deletion syndrome. *J Pediatr* **109**:489–492, 1987.

16b. Greenberg F et al: Prenatal diagnosis of deletion 17p13 associated with DiGeorge anomaly. *Am J Med Genet* **31**:1–4, 1988.

17. Hong R et al: The cerebro-hepato-renal syndrome of Zellweger: Similarity to and differentiation from the DiGeorge syndrome. *Thymus* **3**:97–104, 1981.

18. Kelley RI et al: The association of the DiGeorge anomaly with partial monosomy of chromosome 22. *J Pediatr* **101**:197–200, 1982.

18a. Keppen LD et al: Confirmation of autosomal dominant transmission of the DiGeorge malformation complex. *J Pediatr* **113**:506–508, 1988.

18b. Koenig R et al: Hypoparathyroidism with partial monosomy 10p. *J Pediatr* **111**:310, 1987.

19. Kretschmer R et al: Congenital aplasia of the thymus gland (DiGeorge's syndrome). *N Engl J Med* **279**:1295–1301.

20. Lammer EJ et al: Retinoic acid embryopathy. *N Engl J Med* **313**:837–841, 1985.

21. Lammer EJ, Opitz JM: The DiGeorge anomaly as a developmental field defect. *Am J Med Genet (Suppl)* **2**:113–127, 1986.

22. Lischner HW: DiGeorge syndrome(s). *J Pediatr* **81**:1042–1044, 1972.

23. Marmon LM et al: Congenital cardiac anomalies associated with the DiGeorge syndrome: A neonatal experience. *Ann Thoracic Surg* **38**:146–150, 1984.

23a. Mascarello JT et al: Interstitial deletion of chromosome 22 in a patient with the DiGeorge malformation sequence. *Am J Med Genet* **32**:112–114, 1989.

24. Miller JD et al: DiGeorge's syndrome in monozygotic twins. *Am J Dis Child* **137**:438–440, 1983.

25. Miller MJ et al: Branchial anomalies in idiopathic hypoparathyroidism: Branchial dysembryogenesis. *Henry Ford Hosp Med J* **20**:3–14, 1972.

26. Moerman P et al: Cardiovascular malformations in DiGeorge syndrome, (congenital absence or hypoplasia of the thymus). *Br Heart J* **44**:452–459, 1980.

26a. Müller W et al: The DiGeorge syndrome. I. Clinical evaluation and course of partial and complete forms of the syndrome. *Eur J Pediatr* **147**:496–502, 1988.

27. Ohaki Y et al: Thymic abnormalities in newborns with cardiofacial syndrome. *Teratology* **26**:35A, 1982.

28. Okishima T et al: Immunological defects in a bis (dichloroacetyl) diamine induced malformation complex in rats closely resembling the DiGeorge syndrome. *Cong Anom* **25**:29–44, 1984.

28a. Opitz JM, Lewin SO: The developmental field concept in pediatric pathology—especially with respect to fibular A/hypoplasia and the DiGeorge anomaly. *Birth Defects* **23**:(1)277–292, 1987.

29. Oster G et al: Chemically induced congenital thymic dysgenesis in the rat: A model of the DiGeorge syndrome. *Clin Immunol Immunopathol* **28**:128–134, 1983.

30. Pagon R et al: Coloboma, congenital heart disease, and choanal atresia with multiple anomalies: CHARGE association. *J Pediatr* **99**:223–227, 1981.

31. Patton RG et al: Cerebro-hepato-renal syndrome of Zellweger. *Am J Dis Child* **124**:840–844, 1972.

32. Raatika M et al: Familial third and fourth pharyngeal pouch syndrome with truncus arteriosus: DiGeorge syndrome. *Pediatrics* **67**:173–175, 1981.

33. Robinson HB: DiGeorge's or the III-IV pharyngeal syndrome: Pathology and a theory of pathogenesis. *Perspect Ped Path* **2**:173–206, 1975.

34. Rohn RD et al: Familial third-fourth pharyngeal pouch syndrome with apparent autosomal dominant transmission. *J Pediatr* **105**:47–51, 1984.

35. Rosenthal IM et al: Multiple anomalies including thymic aplasia associated with monosomy 22. *Pediatr Res* **6**:358A, 1972.

36. Shen SC et al: Coexistence of Kallman syndrome and DiGeorge syndrome. *Clin Res* **27**:119A (Abstr), 1979.

37. Shepard MK et al: Familial thymic aplasia with intrauterine growth retardation and fetal death: A new syndrome or a variant of DiGeorge syndrome, in *Growth Problems and Clinical Advances*. Bergsma D, Schimke RN (eds), *Birth Defects* **12**(6):123–125, 1976.

38. Siebert J et al: Pathologic features of the "CHARGE" association: Support for involvement of the neural crest. *Teratology* **31**:331–336, 1984.

39. Steele RW et al: Familial thymic aplasia: Attempted reconstitution with fetal thymus in a millipore diffusion chamber. *N Engl J Med* **287**:787–791, 1972.

40. Taitz LS et al: Congenital absence of the parathyroid and thymus glands in an infant (III and IV pharyngeal pouch syndrome). *Pediatrics* **38**:412, 1966.

41. Taylor MJ, Josifek K: Multiple congenital anomalies, thymic dysplasia, severe congenital heart disease, and oligosyndactyly, with a deletion of the short arm of chromosome 5. *Am J Med Genet* **9**:5–11, 1981.

42. Thomas RA et al: Embryologic and other developmental considerations of thirty-eight possible variants of the DiGeorge anomaly. *Am J Med Genet Suppl* **3**:43–66, 1987.

43. Townes PL, White MR: Inherited partial trisomy 8q(22→qter). *Am J Dis Child* **132**:498–501, 1978.

44. van den Berghe et al: Partial trisomy 1, karyotype 46,XY,12−t(1q, 1wp)+. *Humangenetik* **18**:225–230, 1973.

45. Van Mierop LHD, Kutsche LM: Cardiovascular anomalies in DiGeorge syndrome and importance of neural crest as a possible pathogenetic factor. *Am J Cardiol* **58**:133–137, 1986.

46. Winter WE et al: Familial DiGeorge syndrome with tetraology of Fallot and prolonged survival. *Eur J Pediatr* **141**:171–172, 1984.

## Oromandibular-limb hypogenesis syndromes

### General aspects

**Nosology.** Syndromes of oromandibular-limb hypogenesis are confusing. Hall (21) called attention to the number of ambiguous and overlapping entities that exist in the literature on the subject. The problem is twofold. First, because almost all cases reported to date are seemingly sporadic, it has been extremely difficult to define syndrome boundaries among this group. Entities have been delimited on a somewhat arbitrary basis, and cases can be graded with various degrees of overlap between defined syndrome entities. Second, Hall (21) noted that the terminology used to describe these disorders shows many discrepancies, thus compounding the problem. He proposed a classification of syndromes with oromandibular-limb hypogenesis based on the presence of hypoglossia (Table 19–2). It should be noted that mild degrees of hypoglossia are difficult to detect and may, in fact, go unnoticed.

This chapter considers six defined syndrome entities, including Moebius syndrome, hypoglossia–hypodactylia syndrome, Hanhart syndrome, glossopalatine ankylosis syndrome, limb deficiency–splenogonadal fusion syndrome, and Charlie M. syndrome. All are very uncommon except the Moebius syndrome. After an extensive introductory section, each condition is discussed separately. However, this is done for convenience in documenting the vast literature on the subject. We do not recognize all the conditions as separate entities because of significant overlap among them. Furthermore, to speak of them as syndromes, as the literature does, may be misleading because some may represent sequences.

The etiology of this group of conditions is unknown (11a). Because sporadicity is seemingly apparent in almost all cases and because the clinical features, especially limb anomalies, are remarkably variable, and, furthermore, because clinically graded overlap occurs between defined entities, it is not known how many entities there are or how etiologically heterogeneous or homogeneous this group is. Kaplan et

Table 19–2. Syndromes of oromandibular-limb hypogenesis[a]

| |
|---|
| Type I |
| Hypoglossia |
| Aglossia |
| Type II |
| Hypoglossia–hypodactylia |
| Hypoglossia–hypomelia |
| Type III |
| Glossopalatine ankylosis |
| With hypoglossia |
| With hypoglossia–hypodactylia |
| With hypoglossia–hypomelia |
| With hypoglossia–hypodactylomelia |
| Type IV |
| Intraoral bands and fusion |
| With hypoglossia |
| With hypoglossia–hypodactylia |
| With hypoglossia–hypomelia |
| With hypoglossia–hypodactylomelia |
| Type V |
| Hanhart syndrome |
| Charles M. syndrome |
| Robin sequence |
| Moebius syndrome |
| Amnionic rupture sequence |

[a]Modified after BD Hall, *Birth Defects* **7**(7):233, 1971.

al (29) considered them to be a syndrome community in which the overlapping phenotypes of the defined syndromes reflect underlying developmental relationships.

Documentation of phenotypic overlap is extensive. Spivak and Bennett (45) suggested that the glossopalatine ankylosis syndrome and the hypoglossia–hypodactylia syndrome were closely related. Cohen et al (12) reported a patient with a minute tongue, limb deficiency, and Moebius involvement; they suggested that hypoglossia–hypodactylia syndrome, Hanhart syndrome, and glossopalatine ankylosis syndrome might represent the same recurrent pattern syndrome. Bökesoy et al (9) reported a patient with features of hypoglossia–hypodactylia and glossopalatine ankylosis and called the condition Hanhart syndrome. Pauli and Greenlaw (38) and Barr (4) questioned whether Hanhart syndrome and limb deficiency–splenogonadal fusion might not be variants of the same condition. Tsingoglou and Wilkinson (53) and Watson (59) reported micrognathia and splenogonadal fusion; in the latter case, microstomia and cleft palate also occurred. Sieber (44) observed Moebius syndrome with splenogonadal fusion. The association of Moebius syndrome with Poland anomaly has been seen by a number of authors (25,47).

Herrmann et al (25) analyzed 7 personally studied and 62 previously reported cases of oromandibular-limb hypogenesis. They found that severity of upper limb involvement and especially malformation of the feet, but not the presence of cranial nerve palsies, was significant in differentiating cases, and that the group of patients with cranial nerve palsies included some with limb defects similar to those observed in Hanhart syndrome and others with Poland anomaly. Finally, cases with cranial nerve palsies without limb involvement were documented. Herrmann et al (25) concluded that most reported cases of "aglossia–adactylia," "ankyloglossia superior," and Moebius syndrome represented instances of Hanhart syndrome. Second, they concluded that cases of Moebius syndrome combining a chest defect and/or symbrachydactyly represented Poland–Moebius syndrome. Finally, they noted that cranial nerve palsies obviously occurred in several etiologically distinct conditions.

Temtamy and McKusick (49a) recognized three clinical entities within the Moebius syndrome category: (1) sixth and seventh cranial nerve palsy alone, (2) sixth and seventh nerve palsy with limb anomalies, and (3) sixth and seventh nerve palsy with arthrogryposis.

Etiology. Teratogenic etiology has been suspected by some authors, but reports to date have generally been anecdotal in nature. Gorlin and Pindborg (19) first suggested that cases of the hypoglossia–hypodactylia syndrome should be examined carefully for intrauterine environmental factors. Torpin (52) noted that rupture of the amnion during early pregnancy may produce membranous strands that constrict or amputate limbs and that may also lead to oral anomalies by ingestion of amnionic strands which interfere with orofacial development. Hall (21) noted in his case that Tigan was given to the mother during the critical period of facial and limb development and cited another unpublished case in which the mother received Tigan. In the patient reported by Bökesoy et al (9), meclizine hydrochloride usage during pregnancy was noted. Putschar and Manion (40) postulated that limb deficiency–splenogonadal fusion syndrome is secondary to some environmental teratogen. Promethazine was prescribed during the sixth week of pregnancy in the case reported by Pauli and Greenlaw (38). In a case of glossopalatine ankylosis syndrome, Nevin et al (37a) found that the mother had taken imipramine, diazepam, chlorpromazine, and meclizine throughout the pregnancy. Drug usage during gestation has also been recorded in other oromandibular-limb hypogenesis cases (25,45,54).

Graham et al (20a) reported three cases of Moebius syndrome in which gestational hyperthermia toward the end of the first trimester or early second trimester may have been the etiology. Superneau and Wertelecki (48) reported a patient whose mother had a febrile illness during early pregnancy. They suggested similarities to the experimental effects of hyperthermia as a teratogen. In an animal model of the Moebius syndrome (32a), when pregnant rats were subjected to hyperthermia (or uterine trauma), hemorrhage occurred in the distal limbs and brain stem. Human neuropathologic studies have demonstrated brain stem tegmental necrosis affecting multiple cranial nuclei (50).

Bavinck and Weaver (5) hypothesized subclavian artery disruption sequence as a cause for Moebius syndrome. Possible causes for interruption or reduction of blood flow in the subclavian arteries and their branches include mechanical factors such as blood vessel edema, thrombi, or emboli, hemorrhage, early intrauterine compression, delayed or abnormal formation of blood vessels, disruption of newly formed vessels, or premature disappearance of transient vessels (55).

Genetic causes have been postulated by a number of authors. However, almost all cases of oromandibular-limb hypogenesis have occurred sporadically. Genetic interpretations seem to depend on the presence of consanguinity or the rare presence of a relative of the proband with partial (never complete) manifestation of the condition or the presence of mendelian inheritance in a disorder that mimics oromandibular-limb hypogenesis syndromes but clearly differs from them.

In a case of hypoglossia–hypodactylia, Temtamy and McKusick (49) noted that the father and paternal aunt of the proband had hypodontia. They further observed in their study of several patients with the hypoglossia–hypodactylia syndrome or glossopalatine ankylosis syndrome that orofacial anomalies were observed in various relatives, but in no instance were limb abnormalities observed in a relative. They suggested either multifactorial inheritance or autosomal dominant transmission with reduced penetrance and extreme variability in expression. Since most cases are seemingly sporadic, the possibility that some examples represent autosomal dominant mutations with reduced genetic fitness could not be ruled out. In Turkey, one case of hypoglossia–hypodactylia and two instances of isolated hypoglossia were each associated with consanguinity (54). However, the consanguinity rate is known to be high in the Turkish population and all three cases occurred sporadically within their respective families. Dellagrammaticas (14a) reported two sporadic cases of hypoglossia–hypodactylia, referring to them as examples of Hanhart syndrome. Both sets of parents were Gypsies in whom the consanguinity rate is known to be high. Dellagrammaticas (14a) established definite consanguinity in one of the families and suggested the possibility of autosomal recessive inheritance, a suggestion with which we do not agree. In two published papers on Hanhart syndrome (22,34), distant consanguinity was noted, also suggesting autosomal recessive inheritance. However, affected sibs have never been reported.

Although autosomal dominant inheritance (17,35,57), autosomal recessive inheritance (6,30,51), affected first cousins (23), and instances of consanguinity [11,23,28 (case 3)] have been reported for some cases of the Moebius syndrome, critical review of such cases strongly suggests that they represent entities distinct from the Moebius syndrome. In every instance, only the seventh cranial nerve was affected, and none of these patients had other manifestations of the Moebius syndrome (minimally sixth nerve palsy in addition). Furthermore, patients in this genetically determined group also had a high frequency of hearing deficit and ear anomalies and were usually of normal intelligence. In the family reported by Becker-Christensen and Lund (7), only the proband had sixth and seventh nerve palsy. It is also clear that the dominantly inherited conditions described by Legum et al (33), Rubinstein et al (42), and Ziter et al (61) as Moebius syndrome represent other entities. Wishnick et al (60) reported bilateral sixth and seventh nerve palsies in six family members of two generations. However, affected individuals also had arthrogryposis and except for mild soft tissue syndactyly and talipes equinovarus, no limb defects were noted, especially reductive.

Using the original criteria of Moebius syndrome as sixth and seventh cranial nerve involvement with or without limb defects, almost all reported cases have occurred sporadically except for the dominantly inherited binary combination of abducens and facial paralysis reported by Krueger and Friedrich (31) and the mother–son involvement described by Hicks (27). Collins and Schimke (13) observed a Moebius syndrome proband whose father had a transverse defect of one arm with absent radius and ulna just below the elbow but no cranial nerve involvement. Mitter and Chudley (37) reported a Moebius syndrome proband whose sister had isolated oligosyndactyly; the mother was noted to have bilateral facial weakness.

Pathogenesis. In the Moebius syndrome, combined palsy of the sixth and seventh cranial nerves occurs (21,46). The basic cause remains to be clarified and is probably heterogeneous. Postmortem studies have shown nuclear agenesis (26), prenatal encephalomalacic lesions with mineralized necrotic foci in brain stem nuclei (50), neuron hypoplasia (41), absence or hypoplasia of various cranial nerve trunks and nerves (1a), and hypoplasia or absence of the facial muscles (39). Electromyograms and muscle biopsies have been of little value in determining etiology, although results have usually indicated a neural deficit (56). The most complete discussion of pathologic findings and theories of pathogenesis has been presented by Pitner and associates (39).

In a detailed anatomic study of two infants with Hanhart syndrome, Bersu et al (8) indicated that phenotypic features probably resulted from incomplete rather than abnormal morphogenesis. They postulated that oral and limb anomalies resulted from deficient mesodermal proliferation caused by disturbances in ectoderm–mesoderm interactions beginning about the fourth week of intrauterine development. In glossopalatine ankylosis, the mechanism of fusion between intraoral structures such as the tongue and the palate is not yet clear. It is known that their epithelial coverings must come into contact, fuse, and disintegrate (43a).

In subclavian artery disruption as a postulated cause for Moebius syndrome, vascular compromise at specific locations around the sixth week of embryonic development could result in predictable patterns. Involvement of the subclavian artery distal to the origin of the internal thoracic artery could cause terminal transverse limb defects. Interruption of the subclavian artery proximal to the internal thoracic artery but distal to the origin of the vertebral artery could lead to Poland anomaly. Moebius involvement could result from premature regression of the primitive trigeminal arteries and/or delayed formation of, or obstruction in, the basilar and/or vertebral arteries (5).

In the limb deficiency–splenogonadal fusion syndrome, splenic and gonadal anlage may combine by adherence of the left surface of the dorsal mesogastrium (which becomes the dorsal border of the spleen following rotation) to the dorsal coelomic wall or ventral border of the developing mesonephric–gonadal structures. The presence of intraovarian splenic nodules found in some cases cannot be explained by adhesion, but by migrating splenic cells (20). Splenogonadal fusion may be of two types: the continuous variant type, in which a cord connects the spleen with the mesonephric–gonadal structures, and the discontinuous type, in which the splenogonadal fusion has no cord connected to the spleen itself (40). Fetal activity and mobility may be important in development of the two types (20).

Differential diagnosis. As we have already pointed out, there may be variable degrees of overlap between defined syndrome entities of oromandibular-limb hypogenesis. It is not essential to distinguish between the variants, except for Moebius syndrome, but it is essential to distinguish conditions that do not belong to this group. Limb defects have the same range of variability in all six syndromes, except for limb deficiency–splenogonadal fusion syndrome in which extensive lower limb defects predominate. Limb defects may be observed with isolated Poland anomaly, amnion rupture sequence, autosomal recessive acheiropody, and other syndromes with reductive limb anomalies.

Isolated hypoglossia is known to occur (14,15). Hypoglossia has also been observed in association with transposed viscera and dextrocardia (58). Hypoglossia has further been noted to occur with intraoral bands (16,32). Cleft palate has been reported in association with intraoral bands through three generations (24). Such bands have also been documented in some instances of the *popliteal pterygium syndrome*. Sato et al (43) reported a family with ankyloglossia (in which the lingual frenulum is adhered to the tongue tip) associated with heterochromia irides and congenital clasped thumbs.

Superficially, syndromes of oromandibular-limb hypogenesis may be confused with *Robin sequence, oculo-auriculo-vertebral spectrum,* and *mandibulofacial dysostosis.*

Isolated seventh nerve palsy may result from birth trauma in some instances and has also been observed to follow autosomal dominant and autosomal recessive modes of inheritance. Familial juvenile onset of Bell's palsy (1) and familial palsy in Kawasaki disease (23a) have been reported. Families with hereditary seventh nerve palsy without sixth nerve involvement with or without other cranial nerve palsies and with or without other types of anomalies (6,17,30,35,51,57) should not be confused with Moebius syndrome. Some Moebius-like cases eventuate as facioscapulohumeral dystrophy or as infantile myotonic dystrophy, both autosomal dominant disorders (2). Families with total external ophthalmoplegia and not abducens palsies (33) should not be confused with Moebius syndrome (3). Ziter et al (60) reported a three-generation pedigree of what they called a Moebius syndrome variant; patients had seventh nerve palsy, normal sixth nerve, finger contractures, mild mental deficiency, and a reciprocal translocation between chromosomes 1 and 13. Rubenstein et al (42) reported a patient with congenital paresis of the third, fourth, and seventh nerves, progressive peripheral neuropathy, anosmia, and hypogonadism. The *cardiofacial "syndrome"* (really an association) combines seventh nerve involvement as asymmetric crying facies with congenital heart defects, skeletal anomalies, and renal defects. Some cases of *thalidomide embryopathy* (28,36) have been noted to mimic the Moebius syndrome.

Recurrence risk. Since syndromes of oromandibular-limb hypogenesis occur sporadically in almost all instances reported to date, recurrence risk is essentially negligible. Moebius syndrome should be separated from other syndromes of oromandibular-limb hypogenesis syndromes and should especially be separated from high-risk congenital myopathies, from seventh cranial nerve palsy without sixth nerve involvement, and from various other pseudo-Moebius syndromes (for differential diagnosis, vide supra). Several papers have dealt with recurrence risk of the Moebius syndrome (2,3,18). Baraitser (2) studied sibs and parents of 15 cases and concluded that when Moebius syndrome was accompanied by skeletal anomalies, recurrence risk was only about 2%, whereas bilateral facial palsy with or without abducens palsy without skeletal anomalies had a greater risk of recurrence. Defining skeletal anomalies as a feature of Moebius syndrome helped exclude high risk monogenic disorders of muscle and anterior horn cell, which facially appear Moebius-like in infancy.

Careful family history and examination are essential in Moebius syndrome patients. Although recurrence risk is negligible, rare familial full (27,31) or partial (13,37) instances have been recorded. Establishing a diagnosis of Moebius syndrome either with or without limb anomalies is important because assessment of psychomotor development tends to be erroneous. Early relative delay tends to disappear with age. Poor feeding during the first year of life, often resulting in poor growth, tends to improve. Some degree of mental deficiency occurs in only 10–15%.

### References (Oromandibular-limb hypogenesis syndrome, general aspects)


1. Amit R: Familial juvenile onset of Bell's palsy. *Eur J Pediatr* **146**:608–609, 1987.

1a. Balint A: Angeborenen Fazialis- und Abducenslähmung. *Arch Kinderheilkd* **147**:256–259, 1936.

2. Baraitser M: Genetics of Möbius syndrome. *J Med Genet* **14**:415–417, 1977.

3. Baraitser M: Letters to the editors: Heterogeneity and pleiotropism in the Moebius syndrome. *Clin Genet* **21**:290–291, 1982.

4. Barr M: Amelia-splenogonadal fusion and Hanhart syndrome: The same or different conditions? *Proc Greenwood Genet Ctr* **6**:180–181, 1987.

5. Bavinck JNB, Weaver DD: Subclavian artery supply disruption sequence: Hypothesis of a vascular etiology for Poland, Klippel–Feil, and Möbius anomalies. *Am J Med Genet* **23**:903–918, 1986.

6. Beetz P: Beitrag zur Lehre von den angeborenen Beweglichkeitsdefekten im Bereich der Augen-, Gesichts-, und Schultermuskulatur (infantiler Kernschwund Moebius). *J Psychol Neurol (Lpz)* **20**:137–171, 1913.

7. Becker-Christensen F, Lund HT: A family with Möbius syndrome. *Pediatrics* **84**:115–117. 1974.

8. Bersu ET et al: Studies of malformation syndromes of man XXXXIA: Anatomical studies in the Hanhart syndrome—a pathogenetic hypothesis. *Eur J Pediatr* **122**:1–17, 1976.

9. Bökesoy I et al: Oromandibular limb hypogenesis/Hanhart's syndrome: Possible drug influence on the malformation. *Clin Genet* **24**:47–49, 1983.

10. Bosch-Banyeras JM et al: Poland–Möbius syndrome associated with dextrocardia. *J Med Genet* **21**:70–71, 1984.

11. Cadwalader WB: A clinical report of two cases of agenesis (congenital paralysis) of the cranial nerves. *Am J Med Sci* **163**:744–748, 1922.

11a. Chicarelli ZN, Palayes IM: Oromandibular limb hypogenesis syndromes. *Plast Reconstr Surg* **76**:13–24, 1985.

12. Cohen MM Jr et al: Nosologic and genetic considerations in the aglossy-adactyly syndrome. *Birth Defects* **7**(7):237–240, 1971.

13. Collins DL, Schimke RN: Moebius syndrome in a child and extremity defect in her father. *Clin Genet* **22**:312–314, 1982.

14. de Jussieu M: Observation sur la maniere dont une fille sans langue s'acquitte des fonctions qui dependent de cet organe. *Hist Acad R Sci Paris Mem* **6–14**, 1718–1719.

14a. Dellagrammaticas HM: Hanhart syndrome: Possibility of autosomal recessive inheritance, in *Skeletal Dysplasias,* CJ Papadatos, CS Bartsocas (eds), Alan R. Liss, New York, 1982, pp 299–304.

15. Eskew HA, Shepard EE: Congenital aglossia. *Am J Orthodont* **35**:116–119, 1949.

16. Farrington RK: Aglossia congenita: Report of a case with other congenital manifestations. *North Carolina Med J* **8**:24–26, 1947.

17. Fortanier AH, Speijer N: Eine Erblichkeitsforschung bei einer Familie mit angeborenen Beweglichkeits-störungen der Hirnnerven (infantiler Kernschwund von Moebius). *Genetica* **17**:471–486, 1935.

18. Gorlin RJ: Risk of recurrence in usually nongenetic malformation syndromes. *Birth Defects* **15**(5C):181–188, 1979.

19. Gorlin RJ, Pindborg JJ: *Syndromes of the Head and Neck,* First Edition. McGraw-Hill, New York, 1964.

20. Gouw ASH et al: The spectrum of splenogonadal fusion: Case report and review of 84 reported cases. *Eur J Pediatr* **144**:316–323, 1985.

20a. Graham JM Jr et al: Gestational hyperthermia as a cause for Moebius syndrome. *Teratology* **37**:461A, 1988.

21. Hall BD: Aglossia-adactylia. *Birth Defects* **7**(7):233–236, 1971.

22. Hanhart E: Über die Kombination von Peromelie mit Mikrognathie: Ein neues Syndrom beim Menschen, entsprechend der Akroteriasis congenita von Wreidt und Mohr beim Rinde. *Arch Julius Klaus Stift Vererbungsforsch* **25**:531–540, 1950.

23. Harrison M, Parker N: Congenital facial diplegia. *Med J Aust* **1**:650–653, 1960.

23a. Hattori T et al: Facial palsy in Kawasaki disease. *Eur J Pediatr* **146**:601–602, 1987.

24. Hayward JR, Avery JK: A variation in cleft palate. *J Oral Surg* **15**:320–324, 1957.

25. Herrmann J et al: Studies of malformation syndromes of man XXXXIB: Nosologic studies in the Hanhart and the Möbius syndrome. *Eur J Pediatr* **122**:19–55, 1976.

26. Heubner O: Über angeborenen Kernmangel (infantiler Kernschwund, Moebius). *Charité-Ann* **25**:211–243, 1900.

27. Hicks AM: Congenital paralysis of lateral rotations of eyes with paralysis of muscles of face. *Arch Ophthalmol* **30**:38–42, 1943.

28. Horstmann W: Hinweise auf zentralnervöse Schäden im Rahmen der Thalidomid-Embryopathie. Pathologisch-anatomische, elektrencephalographische und neurologische Befunde. *Z Kinderheilkd* **96**:291–307, 1966.

28a. Johnson GF, Robinow M: Aglossia-adactylia syndrome, *Radiology* **128**:127–132, 1978.

29. Kaplan P et al: A "community" of face-limb malformation syndromes. *J Pediatr* **89**:241–247, 1976.

30. Koster (1902), cited by Henderson JL: The congenital facial diplegia syndrome: Clinical features, pathology, and aetiology. *Brain* **62**:381–403, 1939.

31. Krueger KE, Friedrich D: Familiäre kongenitale Motilitätsstörungen der Augen. *Klin Monatsbl Augenheilkd* **142**:101–117, 1963.

32. de Lamothe D: Un cas d'absence congénitale de la langue avec persistance de la membrane orale. *Ann Mal Oreil Larynx* **49**:717–719, 1930.

33. Legum C et al: Heterogeneity and pleiotropism in the Moebius syndrome. *Clin Genet* **20**:254–259, 1981.

33a. Lipson T et al: Animal model—human correlations for the Moebius syndrome. 10 cases. *Teratology* **37**:474A, 1988.

34. Martius G, Walter S: Peromelie und Mikrognathie als Missbildungskombination (Hanhartsches Syndrom). *Geburtsh Frauenheilkd* **14**:558–563, 1954.

35. Masaki S: Congenital bilateral facial paralysis. *Arch Otolaryngol* **94**:260–263, 1971.

36. Miehlke A: Anatomy and clinical aspects of the facial nerve. *Arch Otolaryngol* **81**:444–445, 1965.

37. Mitter NS, Chudley AE: Facial weakness and oligosyndactyly: ? independent variable features of familial type of the Möbius syndrome. *Clin Genet* **24**:350–354, 1983.

37a. Nevin NC et al: Ankyloglossum superior syndrome. *Oral Surg* **50**:254–256, 1980.

38. Pauli RM, Greenlaw A: Limb deficiency and splenogonadal fusion. *Am J Med Genet* **13**:81–90, 1982.

39. Pitner SE et al: Observations on the pathology of the Moebius syndrome. *J Neurol Neurosurg Psychiatr* **28**:362–374, 1965.

40. Putschar WGJ, Manion WC: Splenic-gonadal fusion. *Am J Pathol* **32**:15–35, 1956.

41. Richter RB: Congenital hypoplasia of the facial nucleus. *J Neuropathol Exp Neurol* **17**:520A, 1958.

42. Rubinstein AE et al: Moebius syndrome in association with peripheral neuropathy and Kallmann syndrome. *Arch Neurol* **32**:480–482, 1975.

43. Sato T et al: A family with hereditary ankyloglossia complicated by heterochromia irides and a congenital clasped thumb. *Int J Oral Surg* **12**:359–362, 1983.

43a. Shah RM: Palatomandibular and maxillo-mandibular fusion, partial aglossia and cleft palate in a human embryo. Report of a case. *Teratology* **15**:261–272, 1977.

44. Sieber WK: Splenotesticular cord (splenogonadal fusion) associated with inguinal hernia. *J Pediatr Surg* **4**:208–210, 1969.

45. Spivak J, Bennett E: Glossopalatine ankylosis. *Plast Reconstr Surg* **42**:129–136, 1968.

46. Steigner M et al: Combined limb deficiencies and cranial nerve dysfunction: Report of six cases. *Birth Defects* **11**(5):133–141, 1975.

47. Sugarman GI, Stark HH: Möbius syndrome with Poland's anomaly. *J Med Genet* **10**:192–196, 1973.

48. Superneau DW, Wertelecki W: Brief clinical report: Similarity of effects—experimental hyperthermia as a teratogen and maternal febrile illness associated with oromandibular and limb defects. *Am J Med Genet* **21**:575–580, 1985.

49. Temtamy S, McKusick VA: Synopsis of hand malformations with particular emphasis on genetic factors. *Birth Defects* **5**(3):125–184, 1969.

49a. Temtamy S, McKusick VA: The genetics of hand malformations. *Birth Defects* **14**(3):81–85, 1978.

50. Thakkar N et al: Möbius syndrome due to brain stem tegmental necrosis. *Arch Neurol* **34**:124–126, 1977.

51. Thomas HM: Congenital facial paralysis. *J Nerv Ment Dis* **25**:571–593, 1898.
52. Torpin R: *Fetal Malformations: Caused by Amnion Rupture during Gestation.* Charles C. Thomas, Springfield, Illinois, 1968.
53. Tsingoglou S, Wilkinson AW: Splenogonadal fusion. *Br J Surg* **63**:297–298, 1976.
54. Tuncbilek E et al: Aglossia-adactylia syndrome (special emphasis on the inheritance pattern). *Clin Genet* **11**:421–423, 1977.
55. Van Allen MI: Fetal vascular disruptions: Mechanisms and some resulting birth defects. *Pediatr Ann* **10**:219–233, 1981.
56. Van Allen MW, Blodi FC: Neurological aspects of the Möbius syndrome: A case study with electromyography of the extraocular and facial muscles. *Neurology (Minneap)* **10**:249–259, 1960.
57. Van der Wiel HJ: Hereditary congenital facial paralysis. *Acta Genet (Basel)* **7**:348, 1957.
58. Watkin HG: Congenital absence of the tongue. *Int J Orthodont* **11**:941–943, 1925, and *Dent Rec* **44**:486–488, 1924.
59. Watson RJ: Splenogonadal fusion. *Surgery* **63**:853–858, 1968.
60. Wishnick MM et al: Mobius syndrome and limb abnormalities with dominant inheritance. *Ophthalmic Paediatr Genet* **2**:77–81, 1983.
61. Ziter FA et al: Three-generation pedigree of a Möbius syndrome variant with chromosome translocation. *Arch Neurol* **34**:437–442, 1977.

**Moebius syndrome.** Moebius (22,23), in attempting to classify multiple congenital cranial nerve palsies, created a division in which palsies of the sixth and seventh cranial nerves were combined. Subsequent authors have broadened this classification to include involvement of other cranial nerves (15,31a). Additional abnormalities may include reductive limb anomalies, defects of the chest wall, and mental retardation (Fig. 19–23A–C).

If the minimal criteria for Moebius syndrome include palsies of the

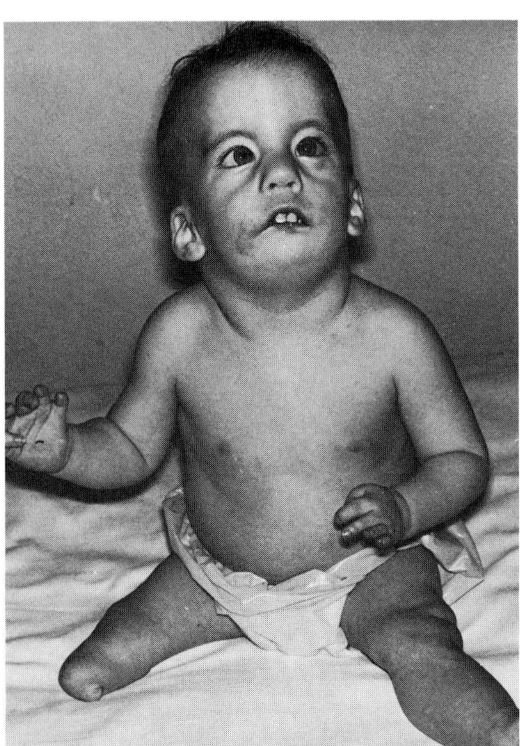
A

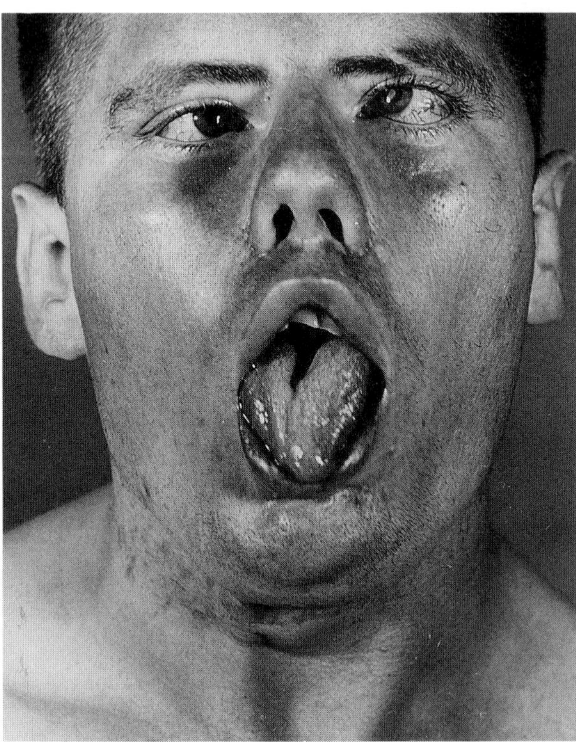
C

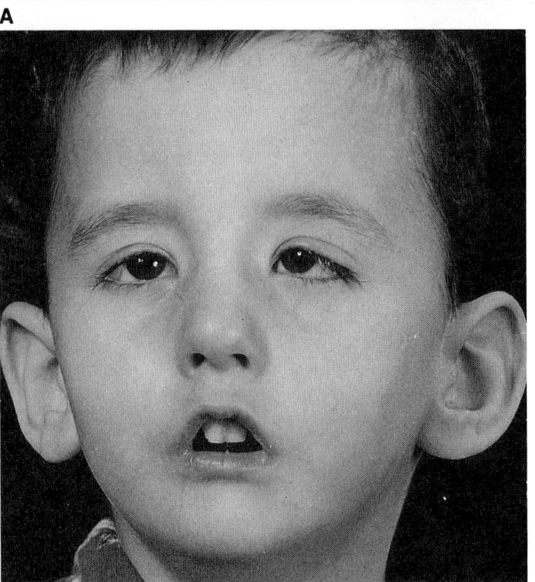
B

Fig. 19–23. *Moebius syndrome.* (A) Variable degrees of limb hypogenesis. Strabismus, micrognathia, mask-like facies. Child had small tongue. (From *SM Harwin* and *LC Lorinsky,* New Britain, Connecticut.) (B) Patient had clubfoot, inability to execute horizontal eye movements, and expressionless face. (From *R Sogg,* Arch Ophthalmol **65:**16, 1961.) (C) Similar anomalies in a 23-year-old male with defective ocular rotation, bilateral facial paralysis, and atrophy of tongue evident from birth. (From *MW Van Allen,* Fig. 64, Grinker's Neurology, Charles C. Thomas, Springfield, Illinois, 1960.)

sixth and seventh cranial nerves, then almost all examples are apparently sporadic with few exceptions (8,17,18,21). Approximately 160 cases had been published by 1970 (13) and over 200 cases have been recorded to date. The ratio of affected males to females is 1:1. Extensive reviews include those of Henderson (15) and Danis (9). Various topics have already been addressed in the first section of this chapter: pathogenesis (1,16,25,28,33,36); pseudo-Moebius genetic entities with autosomal dominant inheritance (11,19,20,29,35,37), with autosomal recessive inheritance (3,4,34), with affected first cousins (14), and with consanguinity [6,14,28a (case 3)]; rare truly inherited cases (17,18); cases with partial manifestation in a first degree relative (8,21); and recurrence risk (2).

Masklike facies may be obvious in the newborn, but may often go unrecognized at this time (10,12,15). Bilateral facial paralysis usually imparts a symmetric appearance to the face, but variance in the degree of involvement on each side of the face or upper and lower portions of the face can cause significant asymmetry. Occasionally, only unilateral facial nerve palsy is present (10). Sixth nerve palsy is the most common ocular finding (31). Nearly every remaining cranial nerve can be affected in addition, but of these, the third, fifth, ninth, and twelfth nerves predominate (10,30,31a).

Most patients cannot abduct either eye beyond the midline. However, unilateral palsy does occur (17). Ptosis, nystagmus, or strabismus may accompany the above findings (5,13,15). Epicanthal folds are frequent. Some patients may be unable to close their eyes either during sleep or while awake, resulting in conjunctivitis or corneal ulceration (17,26).

The nasal bridge is often high and broad, particularly during infancy and early childhood (26). The broadness of the nasal bridge extends downward in a parallel fashion to include the nasal tip (15). Thus, an upper midfacial prominence usually results.

The mouth aperture is often small. The angles of the mouth droop and may allow saliva to escape (5,26). Attempts at opening the mouth further result in little change. As the child grows older, a definite but slow improvement can be observed in the ability both to open the mouth and to feed adequately (10,15). Poor feeding during the first year of life frequently results in poor growth.

Unilateral tongue hypoplasia is frequent (10), but bilateral hypoplasia can occur (10,13,26). The degree of hypoplasia may, on occasion, be extreme (7). Muscular fasciculations of the tongue may be present (12). Poor palatal mobility, inefficient sucking and swallowing, coarse voice, and speech impairment are often present (26).

Frequently, the mandible is mildly to moderately hypoplastic (10,26), and when combined with the small mouth aperture, imparts the appearance of a lower midfacial hypoplasia.

The pinnas may be normal. However, they may be large, may be deficient in cartilage, and may protrude laterally (10). Occasionally, they are hypoplastic, but most instances of severe hypoplasia probably do not represent true examples of the Moebius syndrome. Likewise, the hearing deficits that are reported occasionally with these hypoplastic ears are very infrequent findings in the Moebius syndrome.

Bilateral, unilateral, or asymmetric hypoplasia or aplasia of the pectoralis major muscle or complete Poland anomaly (32) occurs in 15%. Polythelia or athelia may be present (15). Spinal curvatures are occasionally noted but, as a rule, only in those few instances associated with arthrogryposis (13).

Limb defects occur in approximately 50%. Thirty percent constitute talipes deformities (15). The remaining 20% primarily include hypoplasia of digits, syndactyly, or more severe reduction deformities of the limbs. Clinodactyly, polydactyly, joint contractures, and congenital hip dislocation occur less frequently (15,27).

Other findings may rarely include congenital heart defects, urinary tract abnormalities, hypogenitalism, and hypogonadism (13,24,32). The authors have seen two examples with congenital heart defects.

Ten to 15% are mentally retarded (9,15). The degree of mental retardation is usually mild (15). Estimates of early psychomotor development tend to be erroneously low (5,10). The severe physical deformities often cause a relative delay in apparent psychomotor development that tends to disappear as the child grows older.

### References (Moebius syndrome)


1. Balint A: Angeborenen Fazialis- und Abducenslähmung. *Arch Kinderheilkd* **147**:256–259, 1936.

2. Baraitser M: Genetics of Möbius syndrome. *J Med Genet* **14**:415–417, 1977.

3. Becker-Christensen F, Lund HT: A family with Möbius syndrome. *Pediatrics* **84**:115–117, 1974.

4. Beetz P: Beitrag zur Lehre von den angeborenen Beweglichkeitsdefekten im Bereich der Augen-, Gesichts-, und Schultermuskulatur (infantiler Kernschwund Moebius). *J Psychol Neurol (Lpz)* **20**:137–171, 1913.

5. Bonar BE, Owens RW: Bilateral congenital facial paralysis. *Am J Dis Child* **38**:1256–1272, 1929.

6. Cadwalader WB: A clinical report of two cases of agenesis (congenital paralysis) of the cranial nerves. *Am J Med Sci* **163**:744–748, 1922.

7. Cohen MM Jr et al: Nosologic and genetic considerations in the aglossy-adactyly syndrome. *Birth Defects* **7**(7):237–240, 1971.

8. Collins DL, Schimke RN: Moebius syndrome in a child and extremity defect in her father. *Clin Genet* **22**:312–314, 1982.

9. Danis P: Les paralysies oculo-facialis congénitales (à propos de trois observations nouvelles). *Ophthalmologica* **110**:113–137, 1945.

10. Evans PR: Nuclear agenesis. Moebius syndrome: The congenital facial diplegia syndrome. *Arch Dis Childh* **30**:237–243, 1955.

11. Fortanier AH, Speijer N: Eine Erblichkeitsforschung bei einer Familie mit angeborenen Beweglichkeits störungen der Hirnnerven (infantiler Kernschwund von Moebius). *Genetica* **17**:471–486, 1935.

12. Gutman D et al: Moebius syndrome. *Br J Oral Surg* **11**:20–24, 1973.

13. Hanissian AS et al: Moebius syndrome in twins. *Am J Dis Child* **120**:472–475, 1970.

14. Harrison M, Parker N: Congenital facial diplegia. *Med J Aust* **1**:650–653, 1960.

15. Henderson JL: The congenital facial diplegia syndrome: Clinical features, pathology, and aetiology. A review of sixty-one cases. *Brain* **62**:381–403, 1939.

16. Heubner O: Über angeborenen Kernmangel (infantiler Kernschwund, Moebius). *Charité-Ann* **25**:211–243, 1900.

17. Hicks AM: Congenital paralysis of lateral rotations of eyes with paralysis of muscles of face. *Arch Ophthalmol* **30**:38–42, 1943.

18. Krueger KE, Friedrich D: Familiäre kongenitale Motilitätsstörungen der Augen. *Klin Monatsbl Augenheilkd* **142**:101–117, 1963.

19. Legum C et al: Heterogeneity and pleiotropism in the Moebius syndrome. *Clin Genet* **20**:254–259, 1981.

20. Masaki S: Congenital bilateral facial paralysis. *Arch Otolaryngol* **94**:260–263, 1971.

21. Mitter NS, Chudley AE: Facial weakness and oligosyndactyly: ? independent variable features of familial type of the Möbius syndrome. *Clin Genet* **24**:350–354, 1983.

22. Moebius PJ: Über angeborene doppelseitige Abducens-Facialis Lähmung. *Münch Med Wochenschr* **35**:91–94, 108–111, 1888.

23. Moebius PJ: Über infantilen Kernschwund. *Münch Med Wochenschr* **39**:17–21, 41–43, 55–58, 1892.

24. Olson WH et al: Moebius syndrome: Lower motor neuron involvement and hypogonadotrophic hypogonadism. *Neurology (Minneap)* **20**:1002–1008, 1970.

25. Pitner SE et al: Observations on the pathology of the Moebius syndrome. *J Neurol Neurosurg Psychiatr* **28**:362–374, 1965.

26. Reed H, Grant W: Möbius syndrome. *Br J Ophthalmol* **41**:731–739, 1957.

27. Richards RN: The Moebius syndrome. *J Bone Jt Surg* **35A**:437–444, 1953.

28. Richter RB: Congenital hypoplasia of the facial nucleus. *J Neuropathol Exp Neurol* **17**:520A, 1958.

29. Rubinstein AE et al: Moebius syndrome in association with peripheral neuropathy and Kallmann syndrome. *Arch Neurol* **32**:480–482, 1975.

30. Sprofkin BE, Hillman JW: Moebius' syndrome: Congenital oculofacial paralysis. *Neurology (Minneap)* **6**:50–54, 1956.

31. Steigner M et al: Combined limb deficiencies and cranial nerve dysfunction: Report of six cases. *Birth Defects* **11**(5):133–141, 1975.

31a. Sudarshan A, Goldie WD: The spectrum of congenital facial diplegia (Moebius syndrome). *Pediatr Neurol* **1**:180–184, 1985.

32. Sugarman GI, Stark HH: Möbius syndrome with Poland's anomaly. *J Med Genet* **10**:192–196, 1973.

33. Thakkar N et al: Möbius syndrome due to brain stem tegmental necrosis. *Arch Neurol* **34**:124–126, 1977.

34. Thomas HM: Congenital facial paralysis. *J Nerv Ment Dis* **25**:571–593, 1898.

35. Van der Wiel HJ: Hereditary congenital facial paralysis. *Acta Genet (Basel)* **7**:348, 1957.

36. Van Allen MW, Blodi FC: Neurological aspects of the Möbius syndrome: A case study with electromyography of the extraocular and facial muscles. *Neurology (Minneap)* **10**:249–259, 1960.

37. Ziter FA et al: Three-generation pedigree of a Möbius syndrome variant with chromosome translocation. *Arch Neurol* **34**:437–442, 1977.

**Hypoglossia–hypodactylia syndrome.** In a critical review of the literature in 1971, Hall (10) was able to confirm only nine cases of the hypoglossia–hypodactylia syndrome (Fig. 19–24A–C) including one of his own (4,7,8,10,13,14,18,22,24). Since then, many other cases have been recorded (1,2,5,9,11,12,15,17,23,25). Of six patients reported by Johnson and Robinow (12), four had hypoglossia–hypodactylia and two had Moebius syndrome. The cases reported by Temtamy and McKusick (23) had either hypoglossia–hypodactylia syndrome or glossopalatine ankylosis syndrome. The case described by Schuhl (20) is an example of Moebius syndrome. To date, about 30 cases have been recorded, although a few of these, as already mentioned (23), had glossopalatine ankylosis. The condition is frequently called aglossia–adactylia, a misleading term, in our opinion, because the tongue is never completely absent, and adactylia does not convey the variation in limb defects of affected individuals.

All cases of the hypoglossia–hypodactylia syndrome have been sporadic to date. Etiologic, pathogenetic, and nosologic considerations have already been addressed in the first section of this chapter.

The mandible is small and the chin recedes. Hypoglossia is variable in degree, being extreme in the cases of Rosenthal (18), Fulford et al (7), and Thornton (24), but the tongue was reduced only 25% in Hall's case (10). Mandibular incisors have been noted to be absent, with concomitant atrophy of the associated alveolar ridge (10,18). Marked enlargement of the sublingual muscular ridges and hypertrophy of the sublingual and submandibular glands have been described (7,18). A mild defect of the lower lip was noted by Rosenthal (18). Fusion of the anterior alveolar processes and cleft palate have been noted in several cases (2,11,17). In some patients, the size of the oral opening is markedly reduced.

Limb anomalies are extremely variable. Any limb may be affected, and variability may be observed in the same patient as well as in different patients. For example, hemimelia of all four limbs was noted by Temtamy and McKusick (23). Involvement of both left limbs with normal right limbs was reported by Nevin et al (15). Oligodactyly and syndactyly of both hands and complete adactyly of one foot with absence of the hallux on the other foot were observed by Kelln et al (13).

Speech is surprisingly good, and patients have been of normal intelligence. Other findings have included fused labia majora (13), unilateral renal agenesis [12 (case 5)], and congenital occlusion of the superior mesenteric artery [12 (case 2)]. Because of associated abducens and oculomotor palsies, the patient reported by Ernst and Meinhold (6) represents a nosologic bridge. The patient reported by Cohen et al (3) had Moebius facies with sixth and seventh nerve palsies, limb anomalies, and severe microglossia.

Isolated hypoglossia was first reported by de Jussieu (4a) in 1718. Other such cases have been described by several authors (19,25,27,28). Shah (21) tabulated 30 cases of hypoglossia and found cleft palate in 11 of these and other malformations in 24. Watkin (26) reported hypoglossia in association with transposed viscera and dextrocardia. Oulis and Thornton (16) observed a patient with severe hypoglossia, micrognathia, oligodontia, enamel hypoplasia, situs inversus, dextrocardia, and asplenia.

### References (Hypoglossia–hypodactylia syndrome)

1. Alvarez GE: The aglossia-adactylia syndrome. *Br J Plast Surg* **29**:175–178, 1976.

2. Bernard R et al: Syndrome aglossie adactylie avec synostose bimaxillaire antêrieure. *Pédiatrie* **26**:877–883, 1971.

3. Cohen MM Jr et al: Nosologic and genetic considerations in the aglossy-adactyly syndrome. *Birth Defects* **7**(7):237–240, 1971.

4. Cosman B, Crikelair GF: Midline branchiogenic syndromes. *Plast Reconstr Surg* **44**:41–48, 1969.

4a. de Jussieu M: Observation sur la maniere dont une fille sans langue s'acquitte des fonctions qui dependent de cet organe. *Hist Acad R Sci Paris Mem* **6–14**, 1718–1719.

5. Elzay RP et al: Oromandibular-limb hypogenesis syndrome: Type II C, hypoglossia-hypodactylomelia. *Oral Surg* **48**:146–149, 1979.

6. Ernst T, Meinhold G: Ein Beitrag zur angeborenen Aglossie. *Dtsch Zahn Mund Kieferheilkd* **43**:375–384, 1964.

7. Fulford GE et al: Aglossia congenita. Cineradiographic findings. *Arch Dis Childh* **31**:400–407, 1956.

8. Gorlin RJ, Pindborg JJ: *Syndromes of the Head and Neck* (1st ed). McGraw-Hill, New York, 1964.

9. Grislain J et al: Aglossie-adactylie et syndrome d'Hanhart. *Pédiatrie* **26**:353–364, 1971.

10. Hall BD: Aglossia-adactylia. *Birth Defects* **7**(7):233–236, 1971.

11. Hoggins GS: Aglossia congenita with bony fusion of the jaws. *Br J Oral Surg* **7**:63–65, 1969.

12. Johnson GF, Robinow M: Aglossia-adactylia. *Radiology* **128**:127–132, 1978.

13. Kelln EE et al: Aglossia-adactylia syndrome. *Am J Dis Child* **116**:549, 1968.

Fig. 19–24. *Hypoglossia–hypodactylia syndrome.* (A,B) Small mandible and variable limb deformities. (C) Attachment of anterior oral floor to lower lip. Small tongue was located in posterior part of mouth. (A from *NC Nevin et al,* Oral Surg **29**:443, 1970. B courtesy of *N Freire-Maia*, Curitiba, Brazil, and *JM Opitz*, Helena, Montana. C from *WH Boon* and *CT Seng,* J Singapore Paediatr Soc **8**:75, 1966.)

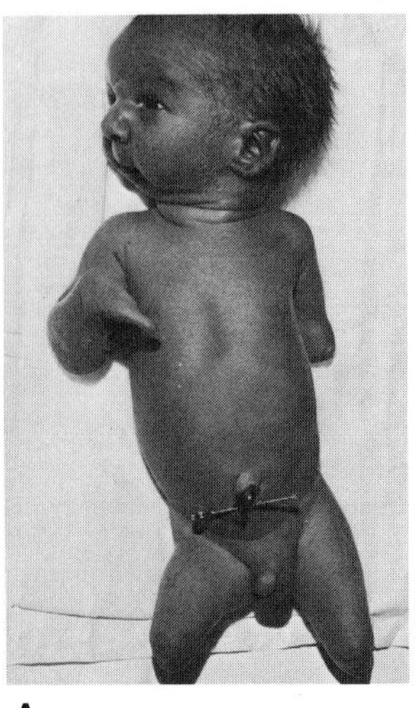

A

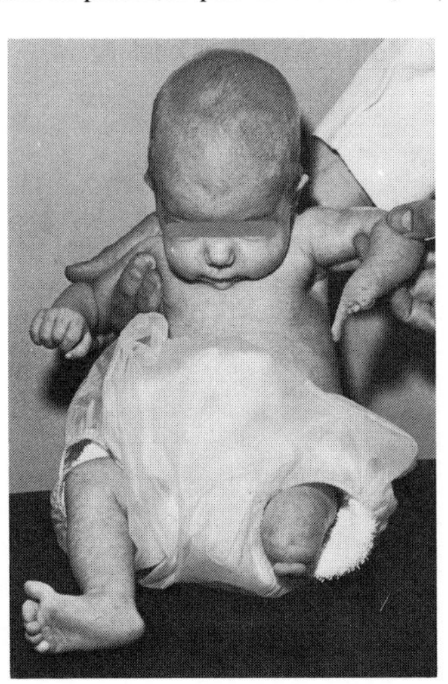

B

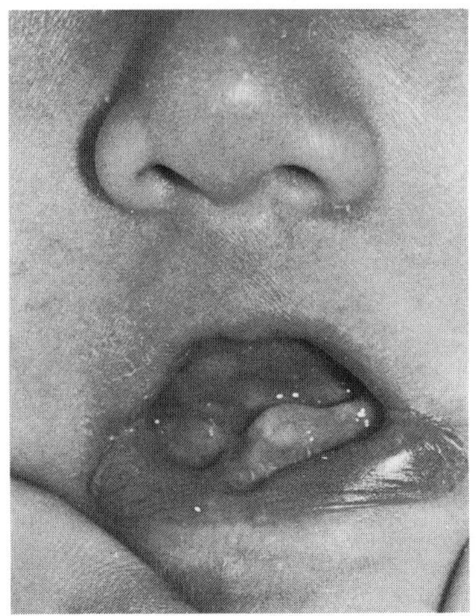

C

14. Lorinsky LC: Aglossia-adactylia syndrome. *Am J Dis Child* **119**:255, 1970.
15. Nevin NC et al: Aglossia-adactylia syndrome. *J Med Genet* **12**:89–93, 1975.
16. Oulis CJ, Thornton JB: Severe congenital hypoglossia and micrognathia with other multiple birth defects. *J Oral Pathol* **11**:276–282, 1982.
17. Pettersson G: Aglossia congenita with bony fusion of the jaws. *Acta Chir Scand* **122**:93–95, 1961.
18. Rosenthal R: Aglossia congenita: A report of a case of the condition combined with other congenital malformations. *Am J Dis Child* **44**:383–389, 1932.
19. Roth JB et al: Microglossia-micrognathia. *Clin Pediatr* **11**:357–359, 1972.
20. Schuhl JF: L'aglossie-adactylie. *Ann Pédiatr (Paris)* **33**:137–140, 1986.
21. Shah RM: Palatomandibular and maxillo-mandibular fusion, partial aglossia and cleft palate in a human embryo. Report of a case. *Teratology* **15**:261–272, 1977.
22. Shear M: Congenital underdevelopment of the maxilla associated with partial adactylia, partial anodontia, and microglossia. *J Dent Assoc S Afr* **11**:78–83, 1956.
23. Temtamy S, McKusick VA: Synopsis of hand malformations with particular emphasis on genetic factors. *Birth Defects* **5**(3):125–184, 1969.
24. Thornton M: Pierre Robin syndrome: A single clinical history. *J Dent Child* **33**:27–32, 1966.
25. Tuncbilek E et al: Aglossia-adactylia syndrome (special emphasis on the inheritance pattern). *Clin Genet* **11**:421–423, 1977.
26. Watkin HG: Congenital absence of the tongue. *Int J Orthodont* **11**:941–943, 1925, and *Dent Rec* **44**:486–488, 1924.
27. Weinberg B, Paras N: Speech intelligibility of a seven-year-old girl with severe congenital hypoplasia of the tongue. *Cleft Palate J* **7**:436–442, 1969.
28. Weinberg B et al: Clinical reports: Severe hypoplasia of the tongue. *J Speech Hearing Disorders* **34**:157–168, 1969.

**Hanhart syndrome.** Hanhart syndrome as originally defined consisted of micrognathia and peromelia (Fig. 19–25A–C). Using this definition, about 12 cases have been described (1,2,5–8,10–12). The patient reported by Bökesoy et al (3) had microglossia and glossopalatine ankylosis. The two patients reported by Dellagrammaticas et al (4) had severe microglossia and fit the original definition of the hypoglossia–hypodactylia syndrome. Using a broader definition of Hanhart syndrome that encompasses several conditions with oromandibular-limb hypogenesis, Herrmann et al (8) tabulated 31 cases from the literature and presented four of their own. To date, all cases have been sporadic. Nosologic, etiologic, and pathogenetic considerations have previously been addressed. American clinicians should carefully note that in the European literature, although four conditions are called Hanhart syndrome (9), they are nosologically unrelated and are as distinct as von Recklinghausen's neurofibromatosis and von Recklinghausen's disease of bone. None of the other three conditions named after Hanhart has oligodactyly or partial syndactyly, despite a report to the contrary [5 (case 1)].

Micrognathia has been a feature in all cases. In Case 2 of Garner and Bixler (5), a fibrous band was noted connecting the maxilla and mandible on both sides but not anteriorly. The tongue was noted to be relatively normal in size in their patient. Most authors have not commented upon tongue size except Herrmann et al (8) who specifically mentioned microglossia. In one case, the tongue was fused to the lower lip, alveolus, and the floor of the mouth. Because of the degree of micrognathia expressed in some cases of Hanhart syndrome, we feel that hypoglossia should certainly be looked for in any suspected case. Hypodontia has been reported (7,8,11). Assemany et al (1) reported Robin sequence in their case. Absence of major salivary glands has been noted (2).

Limb anomalies range from stunted digits and oligodactyly to more severe peromelia. Any limb may be affected, and variability of involvement in the same patient and in different patients is well illustrated by Hanhart (7). Coxa valga was noted in one instance (7). Imperforate anus has also been recorded (2).

### References (Hanhart syndrome)

1. Assemany SR et al: Syndrome of phocomelia with mandibular hypoplasia. *Helv Paediat Acta* **26**:403–409, 1971.
2. Bersu ET et al: Studies of malformation syndromes of man XXXXIA: Anatomical studies in the Hanhart syndrome—a pathogenetic hypothesis. *Eur J Pediatr* **122**:1–17, 1976.
3. Bökesoy I et al: Oromandibular-limb hypogenesis/Hanhart's syndrome: Possible drug influence on the malformation. *Clin Genet* **24**:47–49, 1983.
4. Dellagrammaticas HM et al: Hanhart syndrome: Possibility of autosomal recessive inheritance, in *Skeletal Dysplasias,* Papadatos CJ, Bartsocas CS (eds), Alan R. Liss, New York, 1982, pp 299–304.
5. Garner LD, Bixler D: Micrognathia, an associated defect of Hanhart's syndrome, Types II and III. *Oral Surg* **27**:601–606, 1969.
6. Grislain J et al: Aglossie-adactylie et syndrome d'Hanhart. *Pédiatrie* **26**:353–364, 1971.
7. Hanhart E: Über die Kombination von Peromelie mit Mikrognathie: Ein neues Syndrom beim Menschen, entsprechend der Akroteriasis congenita von Wreidt und Mohr beim Rinde. *Arch Julius Klaus Stift Vererbungsforsch* **25**:531–540, 1950.
8. Herrmann J et al: Studies of malformation syndromes of man XXXXIB: Nosologic studies in the Hanhart and the Möbius syndrome. *Eur J Pediatr* **122**:19–55, 1976.
9. Leiber B, Olbrich G: *Die klinischen Syndrome,* (6th ed). Urban & Schwarzenberg, Munich-Baltimore. 1981.
10. Martius G, Walter S: Peromelie und Mikrognathie als Missbildungskombination (Hanhartsches Syndrom). *Geburtsh Frauenheilkd* **14**:558–563, 1954.
11. Test D III et al: Congenital micrognathia and partial adactylia. *J Oral Surg* **34**:559–565, 1976.
12. Wexler MR, Novark BW: Hanhart's syndrome. *Plast Reconstr Surg* **54**:99–101, 1974.

**Glossopalatine ankylosis syndrome.** This syndrome, first described in 1887 by Illera (6) and early in the twentieth century by Kettner (8) and Kramer (9), is rare. About 15 cases have been reported (1–18). To date, all cases have been sporadic. Nosologic, etiologic, and pathogenetic considerations have been previously discussed. Shah (14) published a summary of cases of intraoral attachments (Table 19–3). In 14 cases with glossopalatine ankylosis, cleft palate

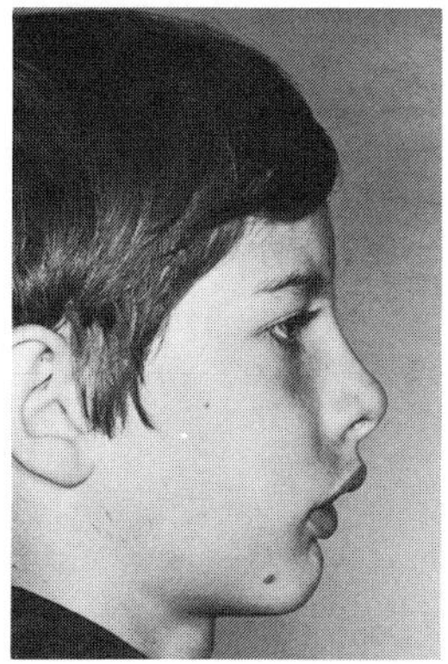
A

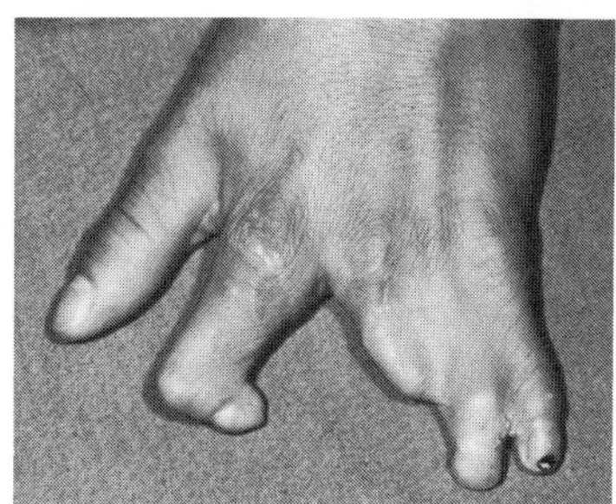
B

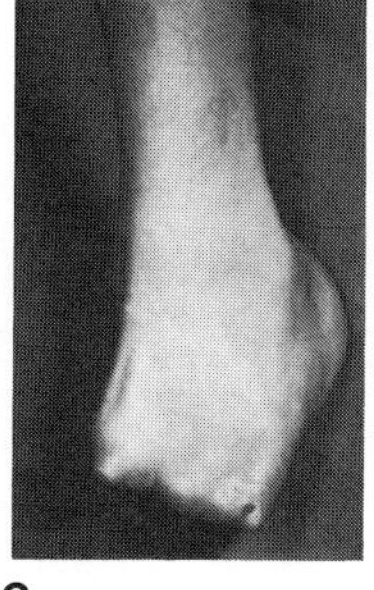
C

Fig. 19–25. *Hanhart syndrome.* (A) Micrognathia. (B,C) Oligodactyly of hands and feet.

Table 19–3. Summary of published cases of intraoral attachments[a]

| Type of intraoral attachment | Number of cases | Presence of cleft palate | Presence of other malformations |
|---|---|---|---|
| Tongue–palate | 14 | 6 | 4 |
| Tongue–other structure[b] | 5 | 2 | 2 |
| Maxilla–mandible | 9 | 4 | 6 |
| Other oral structural attachments[c] | 10 | 6 | 6 |

[a] Adapted from RM Shah, *Teratology* **15**:261, 1977.
[b] Nasal septum, gingiva, anterior pillars, maxilla, mandible, lower lip.
[c] Uvula–tonsil, gingiva–buccal mucosa–soft palate–pharynx, palate–nasopharynx, palate–floor of mouth, palate–lingual frenum, buccal mucosa–alveolar process, maxilla–lower lip.

was present in six and other types of malformations (e.g., limb anomalies) in four.

The tongue is usually attached to the hard palate (Fig. 19–26) (1,4), but may also be adherent to the maxillary alveolar ridge (4). Highly arched palate has been documented in some instances (2). Cleft palate has also been observed (7,17) and, in these cases, the tongue may be attached to the lower edge of the nasal septum. In glossopalatine ankylosis, palatal attachment usually occurs to the anterior portion of the tongue. The tongue tip has been noted to be mildly cleft (1). Its mobility is reduced, and its extension beyond the teeth is limited (1,2). The mandible may be hypoplastic, and the central portion of the upper lip has also been described as hypoplastic (1,2). Hypodontia principally affects the incisor teeth (1,2). Ankylosis of the temporomandibular joint has also been recorded (1). Facial paralysis has been noted in several instances (1,9,15). Limb anomalies are extremely variable and may affect both hands and feet. Oligodactyly, syndactyly, polydactyly, and more severe peromelia have been observed (1,1a,2,7,13,15,17,18).

### References (Glossopalatine ankylosis syndrome)

1. Bünnige M: Angeborene Zungen-Munddach-Verwachsung. *Ann Paediatr* **195**:173–180, 1960.

1a. Chicarelli ZN, Palayes IM: Oromandibular limb hypogenesis syndromes. *Plast Recontr Surg* **76**:13–24, 1985.

2. Cosack G: Die angeborene Zungen-Munddach-Verwachsung als Leitmotiv eines Komplexes von multiplen Abartungen (zur Genese des Ankyloglossum superius). *Z Kinderheilkd* **72**:240–257, 1952–1953.

Fig. 19–26. *Glossopalatine ankylosis syndrome.* White scar (see arrow) shows point of congenital linguopalatal adhesion. (From *M Bünnige,* Ann Paediatr **195**:175, 1960.)

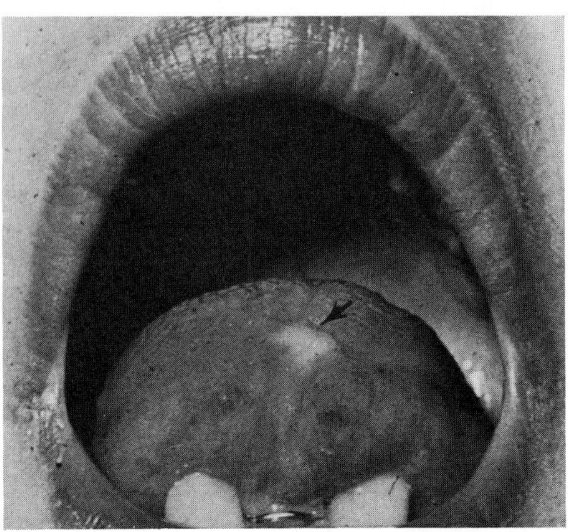

3. De Conick A et al: Congenital palatolingual ankylosis with abnormalities of the extremities. *Acta Stomatol Belg* **72**:89–96, 1975.

4. Esau P: Seltene angeborene Missbildungen: Verwachsungen der Zungenspitze mit dem harten Gaumen. *Arch Klin Chir* **118**:817–820, 1921.

5. Gima H et al: Ankyloglossum superius syndrome. *J Oral Maxillofac Surg* **45**:158–160, 1987.

6. Illera MD: Congenital occlusion of the pharynx. *Lancet* **1**:742, 1887.

7. Keith A: Concerning the origin and nature of certain malformations of the face, head and foot. *Br J Surg* **28**:173–192, 1940.

8. Kettner: Kongenitaler Zungendefekt. *Dtsch Med Wochenschr* **33**:532, 1907.

9. Kramer W: Zur Entstehung der angeborenen Gaumenspalte. *Zbl Chir* **38**:385–387, 1911.

10. Lekkas C, Bruaset I: Ankyloglossia superior. *Oral Surg* **55**:556–557, 1983.

11. Marden P: The syndrome of ankyloglossia superior. *Minn Med* **49**:1223–1225, 1966.

12. Mokhtar Farag M et al: Rare midline congenital anomalies of the face. *J Laryngol Otol* **102**:1040–1043, 1988 (case 2).

13. Nevin NC et al: Ankyloglossum superius syndrome. *Oral Surg* **50**:254–256, 1980.

14. Shah RM: Palatomandibular and maxillo-mandibular fusion, partial aglossia and cleft palate in a human embryo. Report of a case. *Teratology* **15**:261–272, 1977.

15. Spivak J, Bennett E: Glossopalatine ankylosis. *Plast Reconstr Surg* **42**:129–136, 1968.

16. Superneau DW, Wertelecki W: Similarity of effects-experimental hyperthermia as a teratogen and maternal febrile illness associated with oromandibular and limb defects. *Am J Med Genet* **21**:575–580, 1985.

17. Wehinger H: Kiefermissbildung und Peromelie. *Z Kinderheilkd* **108**:46–53, 1970.

18. Wilson RA et al: Ankyloglossia superior (palatoglossal adhesion in the newborn infant). *Pediatrics* **31**:1051–1054, 1963.

**Limb deficiency–splenogonadal fusion syndrome.** Splenogonadal fusion was mentioned by Bostroem (4) as early as 1883 and described first in detail by Pommer (16) in 1889. By 1985, 84 cases were on record (8). Putschar and Manion (17) divided splenogonadal fusion into two types: a continuous type (Fig. 19–27), in which a cord-like structure connects the spleen with the gonadomesonephric structures, and a discontinuous type, in which the splenogonadal fusion has no structural connection with the spleen itself. Approximately 48% of continuous type fusions are associated with malformations; in contrast, only 9% of the discontinuous type show this association. Limb defects occur in about 34% of patients with continuous type splenogonadal fusion (8). Although there appears to be a strong male predilection [13:1 (8)], strong ascertainment bias (23) is evident (vide infra).

Fifteen cases of splenogonadal fusion have been reported in association with limb deficiency (1–3,6,7,9–15,18,21). Two patients have been reported twice each (1,2,12,14). A third patient has been cited as a personal communication (14). The combination of micrognathia, limb reduction defects, and splenogonadal fusion has been observed in seven of the 15 recorded cases (3,6,14–17,21). To date, all cases have been sporadic. Nosologic (3,14,20,22,23), etiologic (17), and pathogenetic (8) concepts have been dealt with previously. An excellent review of the limb deficiency–splenogonadal fusion syndrome is that of Pauli and Greenlaw (14). Gouw et al (8) provided an extensive analysis of splenogonadal fusion and its associated anomalies.

Craniofacial anomalies have been reported in some patients, but less frequently than micrognathia. Findings have included plagiocephaly (14), malformed and/or low-set ears (3,17), deviated nose (15), V-shaped palate (14), microglossia (3), and unerupted teeth (14). Splenogonadal fusion has also been associated with oral anomalies in patients with normal limbs. Tsingoglou and Wilkinson (22) reported micrognathia and Watson (23) described micrognathia, microstomia, and cleft soft palate. Sieber (20) noted a patient with splenogonadal fusion and Moebius syndrome.

Limb reduction defects have great variability. Involvement appears to be more severe more frequently than in other syndromes of oromandibular-limb hypogenesis. On the average, lower limbs are more severely affected than upper limbs and there appears to be a predilec-

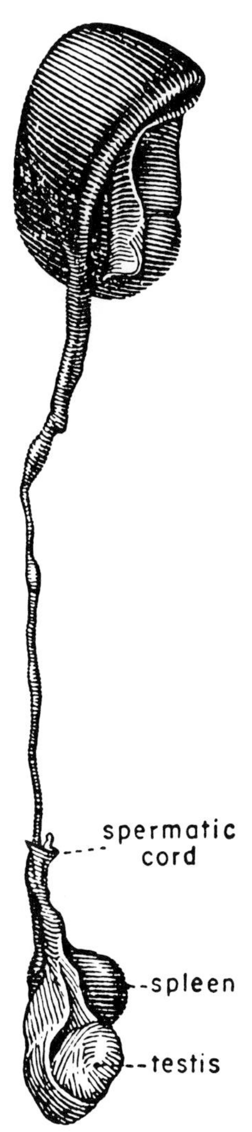


Fig. 19–27. *Limb deficiency-splenogonadal fusion syndrome* (continuous type). (From *WA Sheath,* J Anat Physiol **47**:340, 1912–13. Redrawn by *WGJ Putschar* and *WC Manion,* Am J Pathol **32**:15, 1956.)

tion for the right side (14). It remains unknown whether these trends are significant until a larger series of patients are recorded.

The male:female ratio of 13:1 is corrected to a ratio of 4:1 by leaving out those reported males who were discovered because of genital complications such as scrotal enlargement, inguinal hernia, or cryptorchidism (8). In males, fusion almost always involves the left gonad. In females, the left ovary or left adnexal structure is usually involved as well (8). In the patient reported by Barr (3), four accessory spleens were found along the descending colon and a fifth accessory spleen was fused to the left ovary.

In males, correct preoperative diagnosis is rarely made, differential diagnosis usually including supernumerary testis, epididymitis, testicular tumor, hydrocele, and angioma or cyst of the lower spermatic cord (5,19). Splenogonadal fusion should be suspected in any patient with limb reduction defects and either a scrotal mass or left cryptorchidism. A bloody tap from a punctured scrotal mass should arouse suspicion of an aberrant spleen even more. Paratesticular spleen occasionally may produce clinical symptoms such as pain and swelling of the scrotum (17,21). In rare instances, a continuous splenic-gonadal band may cross ventral to the intestinal loops, resulting in bowel obstruction (11). In Sommer's case (21), a unique alternate drainage of considerable magnitude shunted blood directly into the inferior vena cava via the spermatic and left renal vein, bypassing the liver.

Various other anomalies have also been reported. Significantly, imperforate anus and other anorectal abnormalities have been noted (13,15,16), a finding also reported in one instance of Hanhart syndrome. Low-frequency findings have included hypoplastic lungs (3,15), abnormal lung fissures (14), agenesis of the left diaphragm (15), congenital heart defects (3,15), spina bifida (21), and a coccygeal defect (16). Central nervous system abnormalities have been variable and nonspecific (1,2,6,12,14,21). Polymicrogyria was described by Fritzsche (6). Pauli and Greenlaw (14) reported mild mental deficiency.

### References (Limb deficiency–splenogonadal fusion syndrome)

1. Arnett JC: Ectromelia: Case presentation. *W Va Med J* **47**:328–330, 1951.
2. Arnett JC: Ectromelia: Case presentation (final report) *W Va Med J* **49**:316–317, 1953.
3. Barr M: Amelia-splenogonadal fusion and Hanhart syndrome: The same or different conditions? *Proc Greenwood Genet Ctr* **6**:180–181, 1987.
4. Bostroem: *Gesellschaft deutscher Naturforscher und Aerzte,* Verhandlungen der 56 Versammlung, Freiburg, 1883, p 149.
5. Daniel DS: An unusual case of ectopic splenic tissue resembling a third testicle. *Ann Surg* **145**:960–962, 1957.
6. Fritzsche F: Lien caudatus mit eigenartiger Implantation des oberen Milzpols in die Leber. *Virchows Arch (Pathol Anat)* **329**:35–45, 1956.
7. Given HF, Guiney EJ: Splenic-gonadal fusion. *J Pediatr Surg* **13**:341, 1978.
8. Gouw ASH et al: The spectrum of splenogonadal fusion: Case report and review of 84 reported cases. *Eur J Pediatr* **144**:316–323, 1985.
9. Guarin U et al: Splenogonadal fusion—A rare congenital anomaly demonstrated by 99mTc-sulfur colloid imaging: Case report. *J Nucl Med* **16**:922–924, 1975.
10. Halvorsen JF, Stray O: Splenogonadal fusion. *Acta Paediatr Scand* **67**:379–381, 1978.
11. Hines JR, Eggum PR: Splenic-gonadal fusion causing bowel obstruction. *Arch Surg* **83**:887–889, 1961.
12. May JE, Bourne CW: Ectopic spleen in the scrotum: Report of two cases. *J Urol* **111**:120–123, 1974.
13. McNamara JH: Splenic-gonadal fusion: Case report. *Pathology* **1**:331–332, 1969.
14. Pauli RM, Greenlaw A: Limb deficiency and splenogonadal fusion. *Am J Med Genet* **13**:81–90, 1982.
15. Poidevin LOS: Amelia: Review of the literature and report of a case. *J Obstet Gynecol* **60**:922–926, 1953.
16. Pommer G: Verwachsung des linken kryptorchischen Hodens und Nebenhodens mit der Milz in einer Missgeburt mit zahlreichen Bildungsdefecten. *Ber Naturwissen Med Ver Innsbruck* **17–19**:141–148, 1887/9.
17. Putschar WGJ, Manion WC: Splenic-gonadal fusion. *Am J Pathol* **32**:15–35, 1956.
18. Schworzoff M: Zur Frage über die den Descensus testiculorum bewirkenden Kräfte. *Virchows Arch (Pathol Anat)* **250**:636–640, 1924.
19. Settle EB: The surgical significance of accessory spleens with report of two cases. *Am J Surg* **50**:22–26, 1940.
20. Sieber WK: Splenotesticular cord (splenogonadal fusion) associated with inguinal hernia. *J Pediatr Surg* **4**:208–210, 1969.
21. Sommer JR: Continuous splenic-gonadal fusion with ectromelia: Teratogenesis. *Pediatrics* **22**:1183–1189, 1958.
22. Tsingoglou S, Wilkinson AW: Splenogonadal fusion. *Br J Surg* **63**:297–298, 1976.
23. Watson RJ: Splenogonadal fusion. *Surgery* **63**:853–858, 1968.

**Charlie M. syndrome.** In 1969 Gorlin (1) observed that the Moebius syndrome category had been overworked as a repository for several different disorders and cited the Charlie M. syndrome as an example (Fig. 19–28A–E). The condition is characterized by hypertelorism, facial paralysis in some instance, absent or conically crowned incisors, cleft palate, and variable degrees of hypodactyly of the hands and feet. We have observed several cases and, in each instance, the affected individual has represented a sporadic occurrence. Another patient was reported by Kaplan et al (2).

### References (Charlie M. syndrome)

1. Gorlin RJ: Some facial syndromes. *Birth Defects* **5**(2):65–76, 1969.
2. Kaplan P et al: A "community" of face-limb malformation syndromes. *J Pediatr* **89**:241–247, 1976.

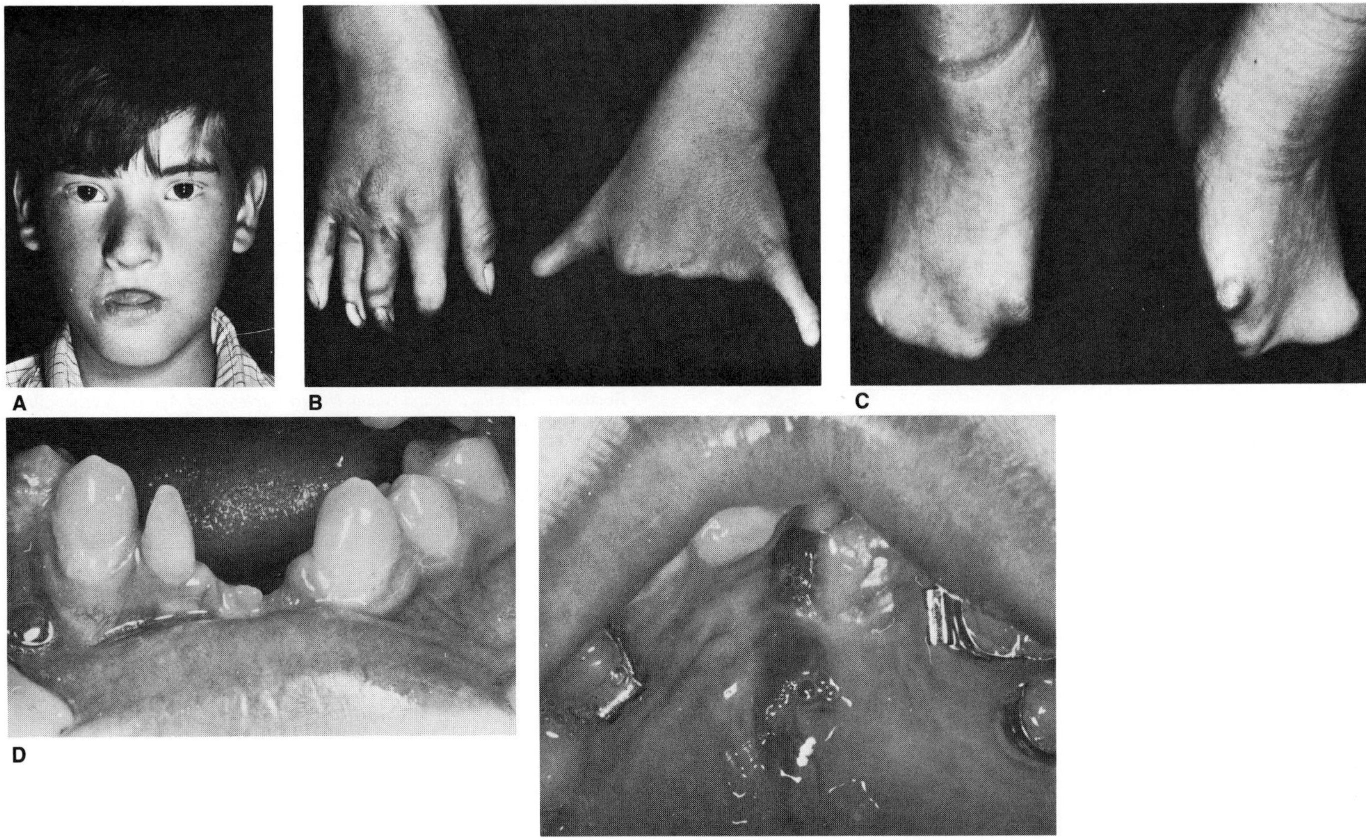

Fig. 19–28. *Charlie M. syndrome.* (A) Downslanting palpebral fissures, ocular hypertelorism, lower facial palsy. (B,C) Bizarre digital agenesis and malformations. (D) Absence of three lower incisors, conical crown form of remaining incisor. (E) Repaired cleft of anterior palate. [From *RJ Gorlin,* Birth Defects **5**(2):65, 1969.]

## Oral–facial–digital syndromes

It is becoming increasingly apparent that striking heterogeneity exists among the oral–facial–digital (OFD) syndromes. In common to all are minor facial anomalies, oral findings (e.g. cleft or lobulated tongue, oral frenulae, and/or cleft palate), and digital anomalies, including brachydactyly, syndactyly, clinodactyly, and polydactyly. Other anomalies are often present, which make it difficult to ascertain whether cases with unusual other anomalies indicate variable expressivity of OFD I or II, or represent other conditions, thus indicating heterogeneity. Often these cases are sporadic occurrences in otherwise normal families; in the few instances in which affected sibs are reported, information is often incomplete on all but the propositus. Classification of many of the new OFD syndromes is therefore tentative, and is usually based on the consistent occurrence of an unusual anomaly in all affected cases (1).

### Reference (Oral–facial–digital syndromes)

1. Toriello HV: Heterogeneity and variability in the oral–facial–digital syndromes. *Am J Med Genet (Suppl* **4**):149–159, 1988.

**Type I (Papillon-Léage–Psaume syndrome).** In 1954, Papillon-Léage and Psaume (16) defined a syndrome of hyperplastic frenula, multilobulated tongue, hypoplasia of nasal alar cartilages, median pseudocleft of upper lip, asymmetric cleft palate, various malformations of digits and mild mental retardation.

Similar cases had been reported under a variety of names as early as 1860. To date, more than 200 cases have appeared in the literature. Comprehensive reviews are by Gorlin et al (8,9), Fuhrmann et al (7), Stahl and Fuhrmann (23), and Melnick and Shields (15). Wahrman et al (30) suggested that the incidence of the syndrome is probably about 1/50,000 live births.

OFD I syndrome has dominant X-linked inheritance. It is limited to females and lethal in males (5–7,9,14,26). It has been described in a male with XXY Klinefelter syndrome (30). One of the most perplexing pedigrees is that of Vaillaud et al (28,29). In this family the disorder appears to have been transmitted through unaffected males. Several other reported ''affected'' males probably had Mohr syndrome (OFD II syndrome).

Facies. The facies is remarkably distinctive. Frontal bossing has been documented in 30%. Usually there is euryopia, but in about 35% dystopia canthorum (lateral displacement of the inner canthi) is evident (14). Some aquiline thinning of the nose, due at least in part to hypoplasia of alar cartilages, and a pseudocleft in the midline of the upper lip are present in about 35% (Fig. 19–29A,B). The upper lip is usually short, and the nasal root is broad. One nostril may be smaller than the other, and there may be flattening of the nasal tip. Because of zygomatic hypoplasia, the midfacial region is flattened in about 25% (15).

Skin and skin appendages. Commonly there are evanescent milia of the face and ears (Fig. 19–30). These usually disappear before the third year of life (22). About 65% have dryness, brittleness, and/or alopecia of the scalp hair (17).

Dermatoglyphics have been studied by Doege et al (6), Reinwein et al (17), and Co-Te et al (3). A preponderance of whorls has been noted. However, Kernohan and Dodge (12) described 10 arches in their patient.

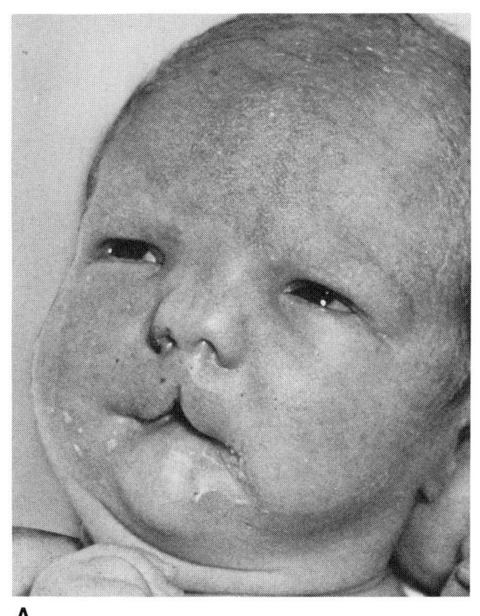
A

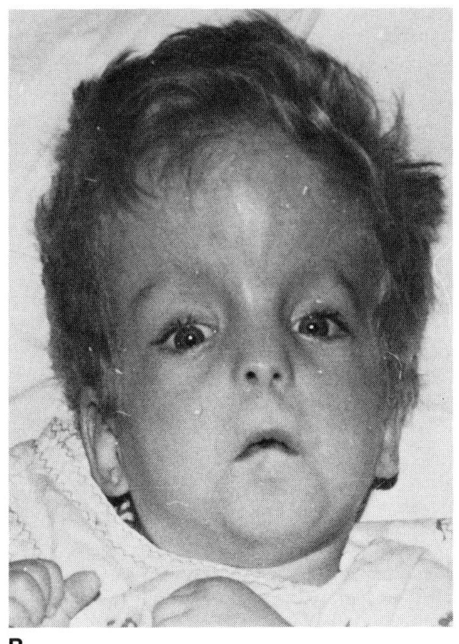
B

Fig. 19–29. *Oral–facial–digital syndrome, Type I.* (A) Pseudocleft in midline of upper lip, lack of nasal alar flare. (B) Note frontal bossing, widely spaced eyes, hypoplastic alar cartilages, and downturned mouth. (A from *H Reinwein et al,* Humangenetik **2**:165, 1966.)

Skeletal manifestations. Malformations of the fingers, seen in 50–70% include clinodactyly, syndactyly, and brachydactyly of digits 2–5 (Fig. 19–31A,B). Toe malformations, noted in 25%, include unilateral hallucal polysyndactyly, syndactyly, and brachydactyly (Fig. 19–31C). Bilateral polydactyly of halluces has been documented on one occasion (25). The hallux is often bent in a fibular direction, with brachydactyly and hypoplasia of the second to fifth toes. Occasionally, there is a postminimus finger or toe (5,6,14,17,22,23,28).

The nasion–sella–basion (cranial base) angle is increased, being about 144° and exceeding the normal value of 131° (SD=4.5°) by almost 3 SD (1,4,9,23) in about one-half the patients.

Radiographic examination shows the short tubular bones of the hands and feet to be irregularly short and thick. Irregular reticular pattern of radiolucency and/or spiculelike formation are observed in metacarpals and, especially, phalanges (31). Some patients have cone-shaped epiphyses in the fingers. Irregularities of the long bones were also reported by Stapleton et al (25).

Fig. 19–30. *Oral–facial–digital syndrome, Type I.* Milia of pinna, which ordinarily disappear by the third year of life. (From *F Majewski et al,* Z Kinderheilkd **112**:89, 1972.)

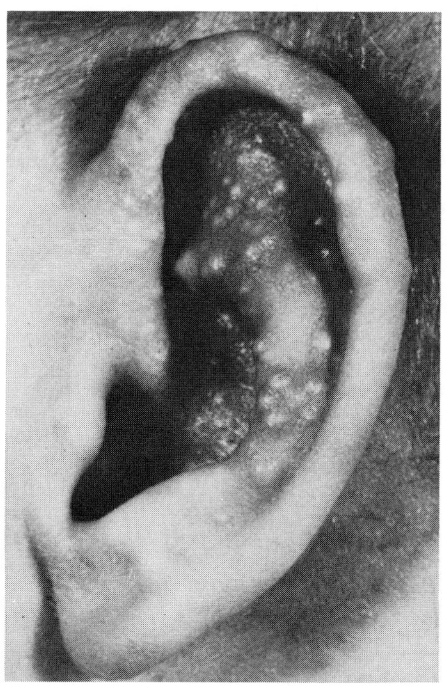

Central nervous system. Mild mental retardation is seen in about 40% (14,15). The intelligence quotient usually ranges from 70 to 90. Various central nervous system alterations have been described: hydrocephaly, hydranencephaly, porencephaly, and partial agenesis of corpus callosum (Fig. 19–32) (2a,3,8,9,14,17,22,29). Precocious puberty has been noted in association with hypothalamic hamartoma (22a). These and other anomalies have been reviewed by Wood et al (31) and Baraton (2). Decreased hearing has also been noted (5,6,11).

Urinary system. Adult onset bilateral polycystic kidneys, usually asymptomatic, have been reported in several cases (2a,5, 6a,11,14,27,30). In the case reported by Stapleton et al (25), the kidneys were normal at 1 year; reevaluation at 11 years demonstrated bilateral polycystic kidneys.

Oral manifestations. The most striking oral manifestations are the "clefts" associated with hyperplasia of frenula (Fig. 19–33A–J). There is often (in about 45%) a small midline "cleft" in the upper lip extending through the vermilion border. Upon retraction of the short upper lip, a wide, thickened, or hyperplastic reduplicated frenum is seen to be associated with the pseudocleft. This, in part, eradicates the mucobuccal fold in the area. Because of these bands, complete retraction of the lip is often not possible. Thick frenula are seen in virtually all patients (7,8,14).

The palate is cleft laterally, deep bilateral grooves extending medially from the maxillary buccal frenula, dividing the palate into an anterior segment (containing the incisors and canines) and two lateral palatal processes. The soft palate is completely and asymmetrically cleft in at least 80% (7,15). In some persons, a large bony ridge extends from the alveolar crest medially to the midline in the canine–premolar area, somewhat resembling a misplaced torus.

Numerous thick fibrous bands are evident in the lower mucobuccal fold in 75%. Cleft tongue with two lobes is seen in 30% with three or more lobes in 45%. On the ventral surface of the tongue, between the tongue halves or lobules, a small whitish hamartomatous mass is seen in about 70% (14,15). This consists of fibrous connective tissue, salivary gland tissue, fat, a few striated or smooth muscle fibers, and,

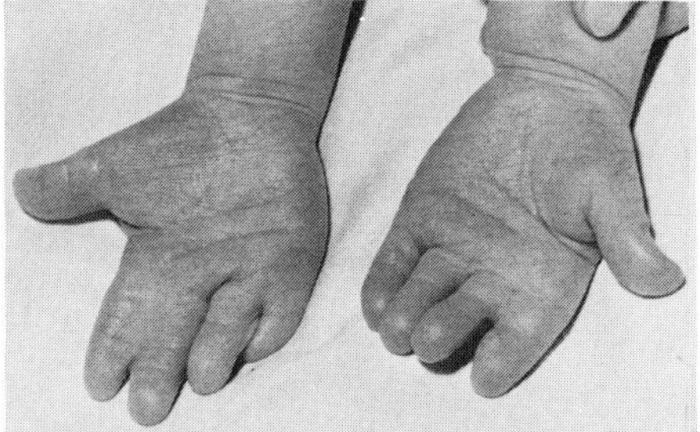

A

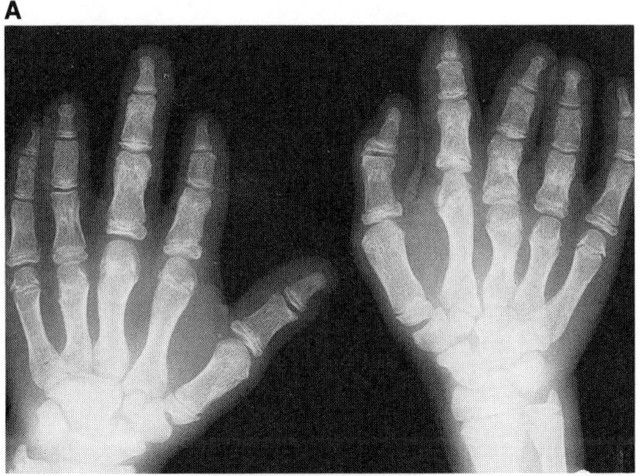

B

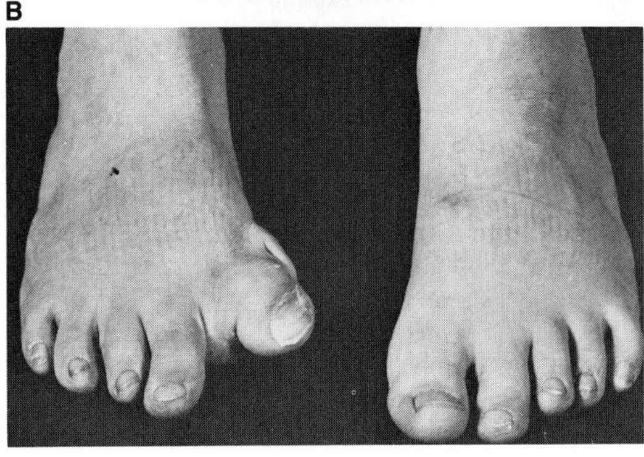

C

Fig. 19–31. *Oral–facial–digital syndrome, Type I.* (A) Asymmetric soft-tissue syndactyly, abbreviation of fingers. (B) Shortened metacarpals, malformed, abbreviated phalanges. (C) Unilateral preaxial polysyndactyly of hallux. (A from *H Reinwein et al,* Humangenetik **2**:165, 1966. B from *JA Dodge* and *DC Kernohan,* Arch Dis Childh **42**:214, 1967. C from *JW Curtin,* Plast Reconstr Surg **34**:579, 1964.)

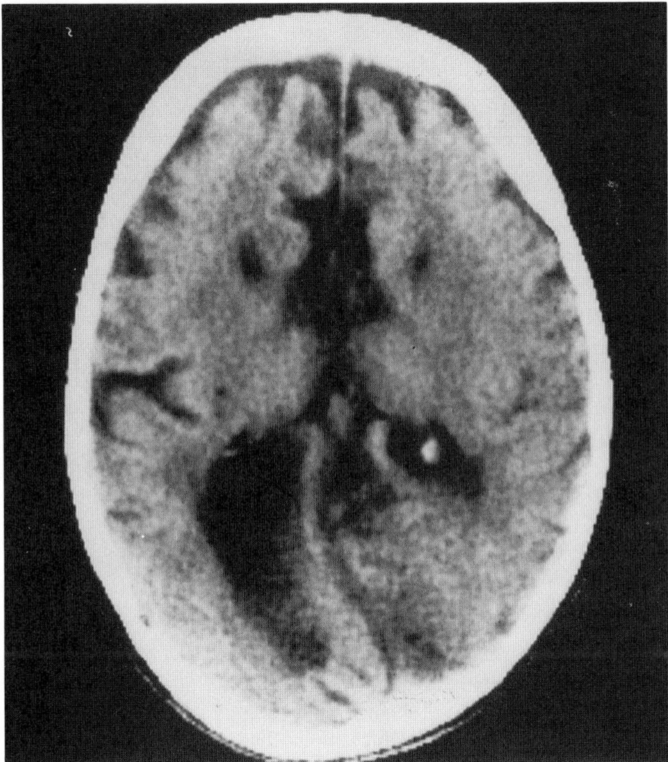

Fig. 19–32. *Oral–facial–digital syndrome, Type I.* Agenesis of corpus callosum and cerebral atrophy. (From *AA Connacher et al,* J Med Genet **24**:116, 1987.)

rarely, cartilage (15). Ankyloglossia or tongue-tie of a diffuse nature is present in at least 30% (14,15).

Malposition of the maxillary canine teeth, supernumerary maxillary deciduous canines and premolars, and infraocclusion are common. Supernumerary canines, often separated by the clefts, are noted in about 20%. The canine crown form is often T-shaped. Aplasia of mandibular lateral incisors occurs in about 50% and appears to be predicated on the effect of the fibrous bands on developing tooth germs. The mandible is small or hypoplastic with a short ramus (13).

Differential diagnosis. Several of these components may be seen in other syndromes. For example, in *Ellis–van Creveld syndrome,* the upper or, occasionally, the lower lip may be attached to the gingiva, there being no mucobuccal fold anteriorly. Associated with this may be a mild mid-upper lip defect. Alar cartilage hypoplasia and dystopia canthorum occur in *Waardenburg syndrome,* but to a greater degree. Postaxial polydactyly of the hands and feet, unusual facial appearance (metopic prominence, upward-slanting palpebral fissures, epicanthal folds, anteverted nostrils), labiogingival frenula, highly arched palate, short stature, skin laxity, and failure to thrive occur in the *Opitz trigonocephaly (C) syndrome.*

Most closely related to this syndrome is OFD II syndrome. It has autosomal recessive inheritance. These patients also have midline cleft of the upper lip, lobate tongue with small excrescences, bilateral manual hexadactyly, and bilateral polysyndactyly of the halluces. In contrast to patients with OFD I syndrome, these patients may be male. They do not have hyperplastic frenula, the lower central incisors may be absent, and occasionally there is a hearing defect. Other oral–facial–digital syndromes can also usually be distinguished on clinical grounds. Hallucal polydactyly can be found in many syndromes: *Greig syndrome, acrocallosal syndrome, Apert syndrome, Pfeiffer syndrome, Jackson–Weiss syndrome, Saethre–Chotzen syndrome, craniofrontonasal dysplasia, Rubinstein–Taybi syndrome, femoral hypoplasia–unusual facies syndrome, del(4p) syndrome,* and *del(7p) syndrome.*

Laboratory Aids. CAT scan and/or NMI can be employed for brain and kidney abnormalities.

### References [Type I (Papillon-Léage–Psaume syndrome)]

1. Aduss H, Pruzansky S: Postnatal craniofacial development in children with the OFD syndrome. *Arch Oral Biol* **9**:193–203, 1954.

2. Baraton J: Les malformations cérébrales dans le syndrome oro-facio-digital de Papillon-Leage et Psaume. *J Radiol Électr* **58**:103–110, 1977.

2a. Connacher AA et al: Orofaciodigital syndrome type I associated with polycystic kidneys and agenesis of the corpus callosum. *J Med Genet* **24**:116–122, 1987.

3. Co-Te P et al: Oral-facial-digital syndrome. *Am J Dis Child* **119**:280–283, 1970.

4. Dodge JA, Kernohan DC: Oral-facial-digital syndrome. *Arch Dis Childh* **42**:214–219, 1967.

5. Doege TC et al: Studies of a family with the oral-facial-digital syndrome. *N Engl J Med* **271**:1073–1080, 1964.

6. Doege TC et al: Mental retardation and dermatoglyphics in a family with the oral-facial-digital syndrome. *Am J Dis Child* **116**:615–622, 1968.

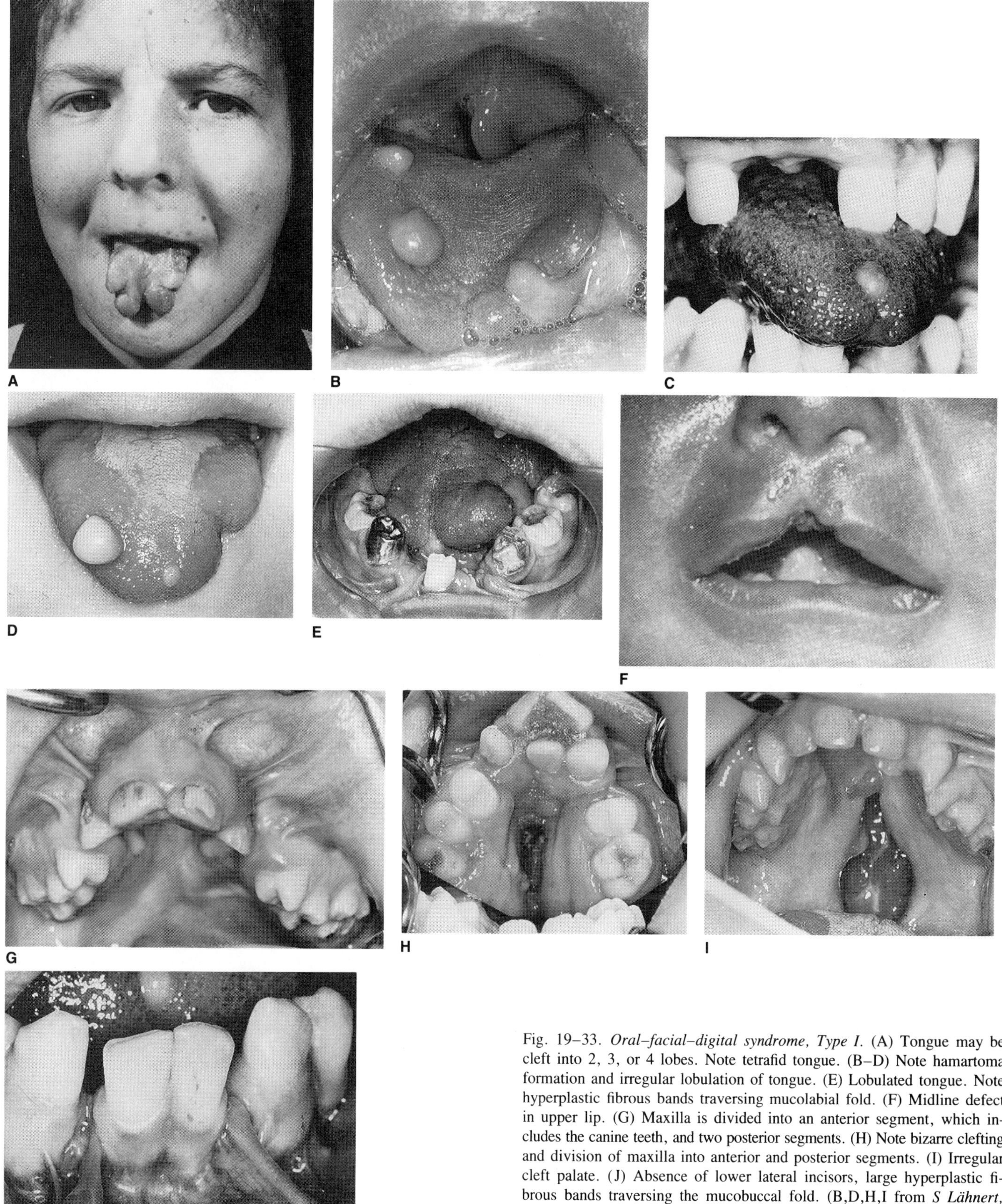


Fig. 19–33. *Oral–facial–digital syndrome, Type I.* (A) Tongue may be cleft into 2, 3, or 4 lobes. Note tetrafid tongue. (B–D) Note hamartoma formation and irregular lobulation of tongue. (E) Lobulated tongue. Note hyperplastic fibrous bands traversing mucolabial fold. (F) Midline defect in upper lip. (G) Maxilla is divided into an anterior segment, which includes the canine teeth, and two posterior segments. (H) Note bizarre clefting and division of maxilla into anterior and posterior segments. (I) Irregular cleft palate. (J) Absence of lower lateral incisors, large hyperplastic fibrous bands traversing the mucobuccal fold. (B,D,H,I from *S Lähnert,* Dtsch Zahnärztl Z **24**:993, 1969. F,G from *W Fuhrmann et al,* Humangenetik **2**:133, 1966.)

6a. Donnai D et al: Familial orofaciodigital syndrome I presenting as adult polycystic kidney disease. *J Med Genet* **24**:84–87, 1987.

7. Fuhrmann W et al: Das oro-facio-digitale Syndrom. *Humangenetik* **2**:133–164, 1966.

8. Gorlin RJ: The oral-facial-digital (OFD) syndrome. *Cutis* **4**:1345–1349, 1968.

9. Gorlin RJ, Psaume J: Orodigitofacial dysostosis: A new syndrome. A study of 22 cases. *J Pediatr* **61**:520–530, 1962.

10. Harrod MJE et al: Polycystic kidney disease in a patient with the oral-facial-digital syndrome—Type I. *Clin Genet* **9**:183–186, 1976.

11. Hauenstein P: Beziehungen zwischen Anomalien der bukkalen Frenuli der Kiefer und Gesichtsdeformitäten. *Fortschr Kieferorthop* **22**:100–107, 1961.

12. Kernohan DC, Dodge JA: Further observations on a pedigree of the oral-facial-digital syndrome. *Arch Dis Childh* **44**:729–731, 1969.

13. Lauterstein A, Pruzansky S: Tooth anomalies in the oral-facial-digital syndrome. *Teratology* **2**:137–146, 1969.

14. Majewski F et al: Das oro-facio-digitale Syndrom. *Z Kinderheilkd* **112**:89–112, 1972.

15. Melnick M, Shields ED: Orofaciodigital syndrome, type I: A phenotypic and genetic analysis. *Oral Surg* **40**:599–610, 1975.

16. Papillon-Léage (Mme), Psaume J: Une malformation héréditaire de la muqueuse buccale et freins anormaux. *Rev Stomatol (Paris)* **55**:209–227, 1954.

17. Reinwein H et al: Untersuchungen an einer Familie mit Oral-facial-digital Syndrom. *Humangenetik* **2**:165–177, 1966.

18. Ruess AL et al: The oral-facial-digital syndrome. *Pediatrics* **29**:985–995, 1962.

19. Ruess AL: Intellectual development and the OFD syndrome—a review. *Cleft Palate J* **2**:350–356, 1965.

20. Schwarz E, Fish A: Roentgenographic features of a new congenital dysplasia. *Am J Roentgenol* **84**:511–517, 1960.

21. Simon P et al: Association de la polykystose renale et du syndrome oral-facial-digital. *Nouv Presse Méd* **7**:122, 1978.

22. Solomon L et al: Pilosebaceous dysplasia in the OFD syndrome. *Arch Dermatol* **102**:598–602, 1970.

22a. Somer M et al: Precocious puberty associated with oral-facial-digital syndrome type I. *Acta Paediatr Scand* **75**:672–675, 1986.

23. Stahl A, Fuhrmann W: Oro-facio-digitales Syndrom. *Dtsch Med Wochenschr* **93**:1224–1228, 1968.

24. Stahl A et al: Beitrag zur Genetik des orofazialendigitalen Syndroms. *Fortschr Kieferorthop* **26**:455–464, 1965.

25. Stapleton FB et al: Cystic kidneys in a patient with oral-facial-digital syndrome, type I. *Am J Kidney Dis* **1**:288–293, 1982.

26. Thuline HC: Current status of a family previously reported with the oral-facial-digital syndrome. *Birth Defects* **5**(2):102–104, 1969.

27. Townes PL et al: Further heterogeneity of the oral-facial-digital syndromes. *Am J Dis Child* **130**:548–554, 1976.

28. Vaillaud JC et al: Le syndrome oro-facio-digital. Etude clinique et génétique à propos de 10 cas dans une même famille. *Rev Pédiatr* **4**:383–392, 1968.

29. Vissian L, Vaillaud JC: Le syndrome oro-facio-digital. *Ann Dermatol Syphiligr (Paris)* **99**:5–20, 1972.

30. Wahrman J et al: The oral-facial-digital syndrome: A male lethal condition in a boy with 47/XXY chromosome. *Pediatrics* **37**:812–821, 1966.

31. Wood BP et al: Cerebral abnormalities in the oral-facial-digital syndrome. *Pediatr Radiol* **3**:130–136, 1975.

**Type II (Mohr syndrome)** In 1941, Mohr (19) described a family later expanded upon by Claussen (7). Several sibs and a cousin exhibited lobed tongue, manual polydactyly, and bilateral polysyndactyly of the halluces (Fig. 19–34A–G). Earlier cases may have been described by Otto (20), in 1841, and Lyons (16), in 1939. Subsequently, a number of well-defined examples have been published (5,8,11,14,17). Less certain are several other cases (2,3). Although some examples previously described as Mohr syndrome have been subsequently recognized as being distinct conditions (vide infra), further heterogeneity within this group is very likely.

The syndrome clearly has autosomal recessive inheritance. Parental consanguinity has been demonstrated (7) and there have been several examples in sibs (7,20). Its frequency may be about 1/300,000 live births (1).

Facies. Frequently there is a midline cleft of the upper lip. Hypertelorism and micrognathia are also common (1,6,15). The ears may be lowset and/or posteriorly angulated (1).

Skeletal alterations. Bony changes appear to be limited to hands and feet. Bilateral manual hexadactyly and bilateral polysyndactyly of the halluces are characteristics (1,5,8,11,14,17). The hallucal polysyndactyly in the patients described by Beaudry (3) and Mayer (18) was, however, unilateral. Christophorou and Nicolaidou (6) reported duplicated thumbs in their patient. Patients described by several authors (3,7,16,17) also had one or more postminimus digits.

Bimanual hexadactyly apparently is not requisite for the diagnosis of the syndrome, since in some cases there have been five fingers with ulnar deviation of the fifth finger, 3–4 syndactyly with extra bones in the web, or hexadactyly of only one hand (7,15,16).

Central nervous system. Mental retardation has been reported in several cases (14,20). Various other neurologic anomalies have been described, including microcephaly (14), porencephaly (14), internal hydrocephalus (8), conductive hearing loss (15), choroid coloboma (8,18), and muscular hypotonia with poor coordination (7,14,17).

Other findings. There appears to be increased susceptibility to respiratory infection, which in several patients has resulted in death during infancy (2,7,8). Tachypnea has also been reported (7). Cryptorchidism (3,7) and inguinal hernia (14) have been noted.

Oral manifestations. Cleft tongue is probably a constant feature of the syndrome, and several authors have spoken of general ankyloglossia (2,3). In a few cases, cleft or highly arched palate (1,2,15–17) has been mentioned. However, in most patients, the palate has been intact (14,17). A small median cleft of the upper lip is a relatively common feature, but was missing in the case described by Mohr (19) and Claussen (7). Multiple frenula are occasionally present (1,14,17), but far less frequent than in OFD I syndrome. Fatty hamartomas on the dorsum of the tongue have been found in several cases (2,3,7,17,18,20).

Differential diagnosis. Many cases of "median cleft" of the upper lip are not true median clefts but examples of the median cleft lip form of *holoprosencephaly,* characterized by agenesis of the septum pellucidum and corpus callosum, agenesis of the prolabium, premaxilla, and nasal bones, microcephaly, and hypotelorism (23). True median cleft may be seen in association with bifid nose and hypertelorism in frontonasal malformation (12,22). An infant that exhibited certain features of both *frontonasal malformation* and median clefting holoprosencephaly was described by Bell and Van Allen (4). Median cleft of the upper lip is also seen in other *OFD syndromes* (vide infra).

### References [Type II (Mohr syndrome)]

1. Ajacques JC: Le syndrome de Mohr. *Rev Stomatol Chir Maxillofacial* **82**:234–240, 1981.

2. Barling G: Cleft tongue with median lobe and cleft palate. *Br Med J* **2**:1061, 1885.

3. Beaudry GO: Polydactylisme et malconformation de langue. *Union Méd Can* **4**:342–343, 1875.

4. Bell WE, Van Allen MW: Agenesis of corpus callosum with associated facial anomalies. *Neurology (Minneap)* **9**:694–698, 1959.

5. Burian F: *Chirurgie der Lippen und Gaumenspalten.* VEB-Verlag Volk und Gesundheit, Berlin, 1963.

6. Christophorou MN, Nicolaidou P: Median cleft lip, polydactyly, syndactyly, and toe anomalies in a non-Indian infant. *Br J Plast Surg* **36**:447–448, 1983.

7. Claussen O: Et arveligt syndrom omfattende tungemissdannelse og polydaktyli. *Nord Med* **30**:1147–1151, 1946.

8. Dittmer J: Über 2 Fälles des Papillon-Léage–Psaume Syndroms (case 2). *Pädiatr Grenzgeb* **6**:35–42, 1967.

9. Dreibholz E: *Beschreibung einer sogenannter Phokomele.* Thesis, Berlin, 1873.

10. Froriep LF, Froriep R: Missbildung (Monstrum per excessum). *Neue Notizen Geb Natur Heilkd* **4**:8, 1837.

11. Huguet C: *Contribution à l'étude du syndrome de Papillon-Léage et Psaume: A propos d'un cas chez un enfant de sexe masculin.* Thesis, Nantes, 1967–1968.

12. Kazanjian VH, Holmes EM: Treatment of median cleft lip associated with bifid nose and hypertelorism. *Plast Reconstr Surg* **24**:582–587, 1959.

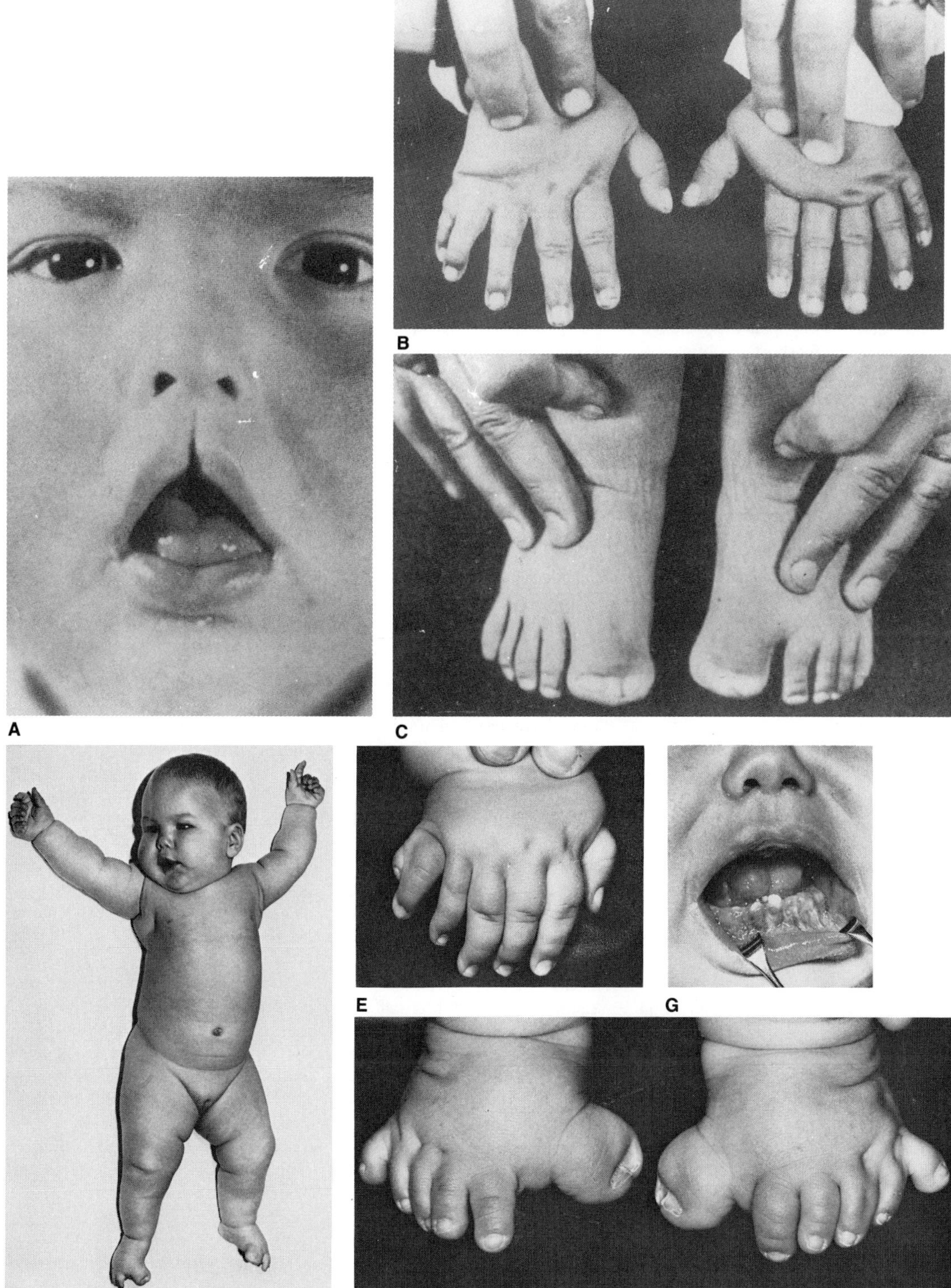


Fig. 19–34. *Oral–facial–digital syndrome, Type II.* (A–C) Note median cleft of upper lip and bimanual hexadactyly. Child also had bilateral polysyndactyly of halluces. (D–G) Compare with patient in A–C. Note bilateral manual hexadactyly, polysyndactyly of halluces, defect in middle of upper lip, hamartoma of tongue. (A–C courtesy of *F Burian,* Prague, Czechoslovakia. B from *RA Pfeiffer et al,* Klin Pädiatr **185**:224, 1973.)

13. Kölliker T: Über das Os intermaxillare des Menschen und die Anatomie der Hasenscharte und des Wolfsrachens. *Nova Acta Academiae Caesareae Leopoldino-Carolinae Germanicae Nature Curiosorum* **43**:325–396, 1882.

14. Kushnick T et al: Orofaciodigital syndrome in a male. *J Pediatr* **63**:1130–1134, 1963.

15. Levy EP, Fletcher BD, Fraser FC: Mohr syndrome with subclinical expression of the bifid great toe. *Am J Dis Child* **128**:531–533, 1974.

16. Lyons DC: Skeletal anomalies associated with cleft palate and harelip. *Am J Orthodont* **25**:895–897, 1939.

17. Mandell F et al: Oral-facial-digital syndrome in a chromosomally normal male. *Pediatrics* **40**:63–68, 1967.

18. Mayer UM, Schnidder AC: Les symptomes oculaires dans le syndrome oro-facio-digital type II de Mohr-Claussen. *Bull et Mem SFO* **95**:520–524, 1984.

19. Mohr OL: A hereditary sublethal syndrome in man. *Nor Vidensk Akad Oslo I Mat Naturv Klasse* **14**:3–18, 1941.

20. Otto G: *Monstrorum sexcentorum descriptio anatomica*. SF Hirt, Vratislaviae, 1841.

21. Pfeiffer RA et al: Das Syndrom von Mohr und Claussen. *Klin Pädiatr* **185**:224–229, 1973.

22. Sedano HO et al: Frontonasal dysplasia. *J Pediatr* **76**:906–913, 1970.

23. Weaver DF, Bellinger DH: Bifid nose associated with midline cleft of the upper lip. *Arch Otolaryngol* **44**:480–482, 1946.

**Type III [see-saw winking with oral–facial–digital anomalies and postaxial polydactyly of the hands and feet (Sugarman syndrome)].** Sugarman et al (1) reported female sibs with a syndrome similar to, but distinct from OFD I and II syndromes. Both sisters had lobulated, hamartomatous tongue, dental anomalies, including malocclusion and extra teeth, bifid uvula and normal palate, postaxial hexadactyly of hands and feet, pectus excavatum, short sternum, kyphosis, and profound mental retardation. In addition, the younger sister had frontal bossing, hypertelorism, downslanting palpebral fissures, esotropia, apparently low-set ears, brachydactyly with hyperconvex nails, and continuous alternating winking of the eyelids which the authors termed "see-saw" winking (Fig. 19–35). The older sister did not exhibit this phenomenon, but did have blepharospasm. McKusick (personal communication, 1986) has seen a similar case. Inheritance is likely autosomal recessive.

### Reference [Type III (see-saw winking with oral–facial–digital anomalies and postaxial polydactyly of the hands and feet [Sugarman syndrome])]

1. Sugarman GI et al: See-saw winking in a familial oral-facial-digital syndrome. *Clin Genet* **2**:248–254, 1971.

**Type IV (Oral–facial–digital syndrome with tibial dysplasia).** There have been 15 reported cases (1–7,9–14) of patients with oral, facial, and digital anomalies who have also had tibial dysplasia and/or mesomelia (Fig. 19–36A–G). Three reports have involved affected sibs (3,7,9); the remaining reports have been sporadic cases. Sensenbrenner and Jorgenson (10) reported a case they described as having Mohr syndrome and skeletal dysplasia; Temtamy and McKusick (13) later described the same case and suggested this entity represented an overlap of Majewski and Mohr syndromes. Baraitser et al (1) also reported a case with features of both Mohr and Majewski syndromes; Burn et al (3) later described an affected sib of the proband in Baraitser's report and suggested this entity be named OFD IV.

Fig. 19–35. *Oral–facial–digital syndrome, Type III.* See-saw winking, strabismus, pectus excavatum, short sternum, polydactyly, profound mental deficiency. (Courtesy of *GI Sugarman,* Los Angeles.)

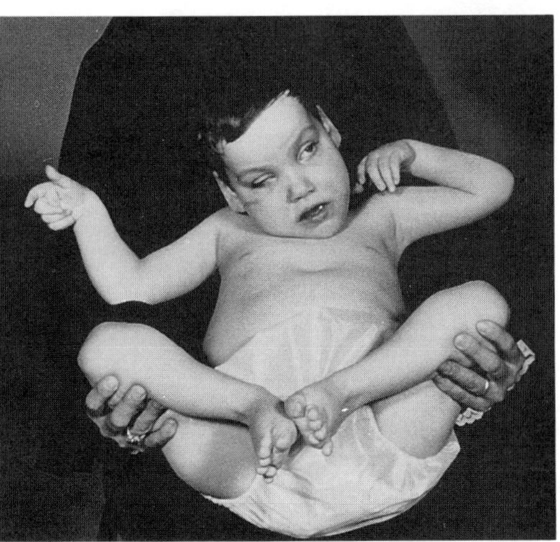

Facial anomalies have included broad nasal root (2,9,11,13), broad nasal tip (5,7,9,10,12), hypertelorism or telecanthus (2,3,9,11,13), micrognathia (3,9,10,12,13), hypoplastic mandible (2,3,9), and low-set ears (3,7). Numerous oral anomalies are present, including cleft lip (5,7,9,13), cleft or highly arched palate (3,5,7,8a,9,10,12), bifid uvula (12), cleft or hypoplastic maxillary and/or mandibular alveolar ridge (3,7,9,12,13), oral frenulae (6,8), and lingual hamartoma (2,3,5,8a,10,12,13). Dental anomalies are also common, and generally include absent and supernumerary teeth (5,7,9–13). Goldstein and Medina (7) further described the teeth as being small, and exhibiting abnormal crown and root morphology (hypertaurodontism and talonism). Absent or hypoplastic epiglottis has been found (8a,13).

Digital anomalies were varied, but generally present. Both pre- and postaxial polydactyly of the hands have been described (6–9); syndactyly (5,6,12), clinodactyly (2,6,7,9,11,13), and brachydactyly (6,9,11–13) can also occur. Preaxial polydactyly of the feet was described in three cases (2,9,13) and both pre- and postaxial polydactyly in five cases (2,8). One patient had six toes on each foot, but whether pre- or postaxial was not specified (6). Neither of the sibs reported by Goldstein and Medina (7) had polydactyly of either hands or feet, so this finding is apparently not consistent.

Talipes equinovarus has also been described (3,8a,12). Mesomelia of varying degrees is present in most cases, and often the tibia is described as dysplastic. Several authors (3,6,12) described the tibia as being short with midshaft bowing. Büttner and Eysholdt (4) and Fenton and Watts-Smith (5) reported the tibia as pseudoarthrotic. Other tibial anomalies include proximal metaphyseal or epiphyseal flattening (3,7,9,12) and/or metaphyseal flaring (2,7,9). Forearms can also be short (10,11) and stature, when noted, is below the third percentile (7,9,10–12).

Conductive hearing loss occurs in some patients (3,7,8a,9,12). Although intellect has been reported as normal in several cases (9–11), most have been mentally retarded (2,3,5). We have also seen a child with OFD IV with retardation secondary to porencephaly.

The chest is sometimes small (12–14) and pectus carinatum or excavatum has also been reported (9,10).

Two additional cases of OFD IV may be the sibs reported by Michels et al (8). These sibs had oral, facial, and digital anomalies, and were thought to have OFD II; however, one or both sibs also had clubfeet, pectus excavatum, and short stature, which is more consistent with OFD IV. Since it was not noted whether X rays of the long bones were done, classification of these cases as examples of OFD IV is tentative.

In conclusion, all cases of OFD IV have tibial defects in common, ranging from metaphyseal and epiphyseal abnormalities detected by X ray (2,9) to dysplasia severe enough to necessitate amputation (11). Tibial defects may be unilateral (5) or bilateral. The presence of postaxial polydactyly and pectus excavatum in some cases also helps distinguish this condition from OFD II. However, it is possible that this group is still heterogeneous, and resolution of this question awaits description of further cases of affected siblings, and delineation of degree of intrafamilial variability. Inheritance at this point seems autosomal recessive, as evidenced by the presence of affected sibs (3,7,9) and consanguinity in parents (2).

### References [Type IV (oral–facial–digital syndrome with tibial dysplasia)]

1. Baraitser M et al: A female infant with features of Mohr and Majewski syndromes: Variable expression, a genetic compound, or a distinct entity? *J Med Genet* **20**:65–67, 1983.

2. Bonioli E et al: The heterogeneity of oral-facial-digital anomalies. *Panminerva Med* **21**:127–130, 1979.

3. Burn J et al: Orofacial digital syndrome with mesomelic limb shortening. *J Med Genet* **21**:189–192, 1984.

4. Büttner A, Eysholdt KG: Die angeborenen Verbiegungen und Pseudoarthrosen des Unterschenkels (case 14). *Ergeb Chir Orthoped* **36**:165–222, 1950.

5. Fenton OM, Watt-Smith SR: The spectrum of the oro-facial-digital syndrome. *Br J Plast Surg* **38**:532–539, 1985.

6. Fontaine G et al: Incurvation femoro-tibiale unilaterale avec duplication du gros orteil: Syndrome de Mohr? *Ann Pédiatr* **26**:564–568, 1979.

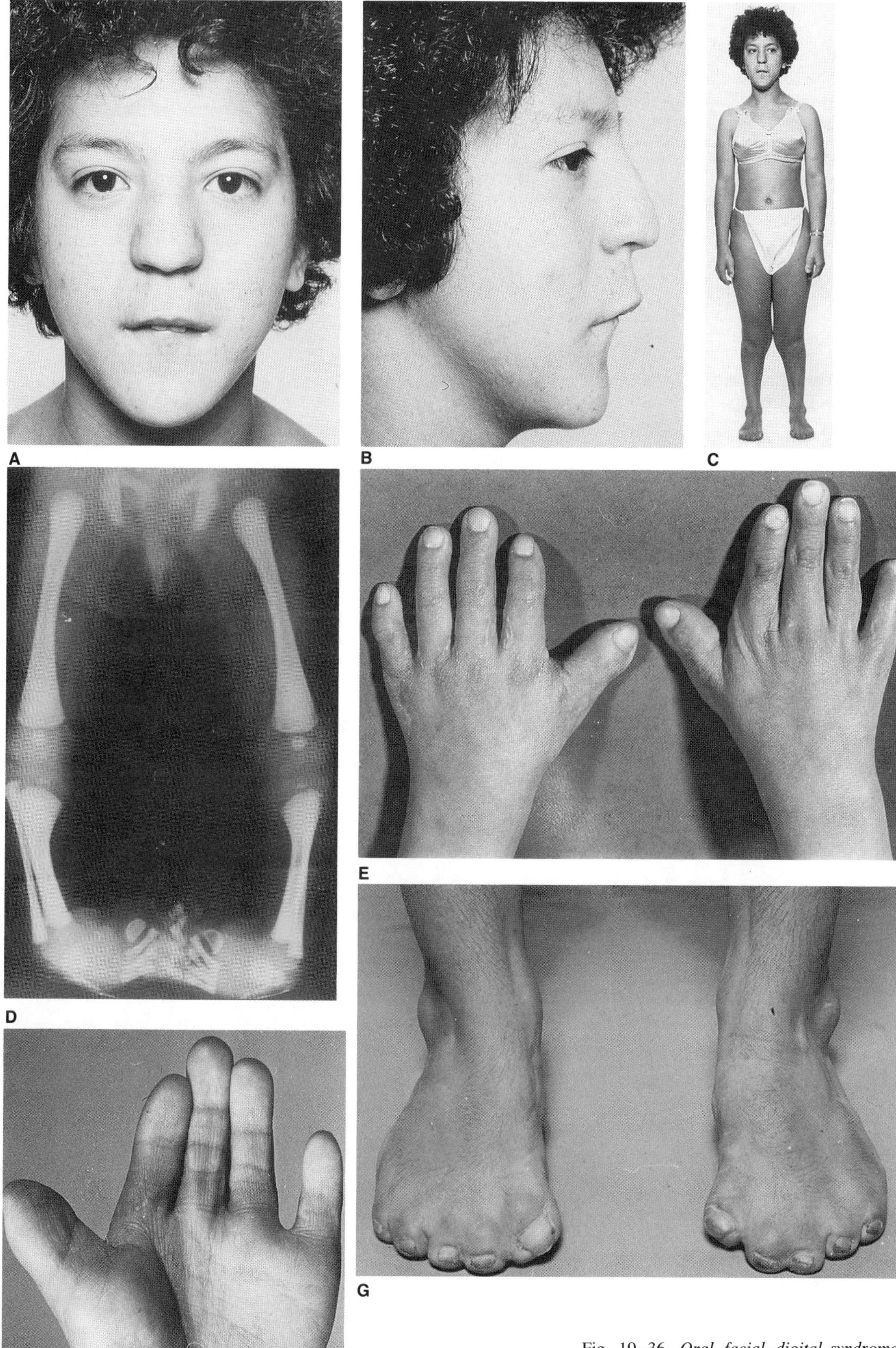


Fig. 19–36. *Oral–facial–digital syndrome, Type IV.* (A,B) Broad nasal root, broad nasal tip, micrognathia. (C) Short stature with short tibiae. (D) Note short tibiae with midshaft bowing, talipes equinovarus. (E,F) Brachydactyly. Extra postaxial digit surgically removed from each hand. (G) Syndactyly. Excision of postaxial digit from each foot. (A–C, E–G from *GI Sugarman,* J Clin Dysmorph **1**:16, 1983. D from *J Burn et al,* J Med Genet **21**:189, 1984.)

7. Goldstein E, Medina JL: Mohr syndrome or OFD II: Report of two cases. *J Am Dent Assoc* **89**:377–382, 1974.

8. Michels VV et al: Polysyndactyly in the oro-facial digital syndrome, type II. *J Clin Dysmorphol* **3**:2–9, 1985.

8a. Nevin NC, Thomas PS: Orofaciodigital syndrome type IV: Report of a patient. *Am J Med Genet* **32**:151–154, 1989.

9. Rimoin DL, Edgerton MT: Genetic and clinical heterogeneity in the oral-facial-digital syndrome. *J Pediatr* **71**:94–102, 1975.

10. Sensenbrenner JA, Jorgenson RJ: Mohr syndrome and skeletal dysplasia, new syndrome? *Birth Defects* **11**(2):384–390, 1975.

11. Shaw M et al: Oral facial digital syndrome—case report and review of the literature. *Br J Oral Surg* **19**:142–147, 1981.

12. Sugarman GI: Orofacial defects and polysyndactyly. Syndrome identification case report 91. *J Clin Dysmorphol* **1**:16–19, 1983.

13. Temtamy S, McKusick VA: The genetics of hand malformations. *Birth Defects* **14**(3):434, 1978.

14. Wolf H: Mediane Oberlippenspalte mit Persistenz des Frenulum tectolabiale. *Dtsch Zahnärztl Z* **7**:373–379, 1952.

**Type V [median cleft of upper lip and postaxial polydactyly of hands and feet (Thurston syndrome)].** Thurston (6), in 1909, appears to be the first to describe a syndrome of median cleft of the vermilion of the upper lip and postaxial polydactyly of the hands and feet (Fig. 19–37A,B). All reported patients have been from India (1–6). Inheritance is autosomal recessive.

Although described as a separate entity in the 1964 edition of this text, it was considered to be incomplete expression of the OFD II syndrome in our 1976 edition. Reexamination of the evidence suggests that it can stand on its own.

Thurston's patients were brothers. In one brother, the extra digit was of normal morphology, but of somewhat smaller size. The extra toe was of normal size and appearance. The other brother had seven fingers on each hand. There were six toes on one side and five on the other. No other anomalies were present.

Chowdhury (2) described nine cases in one family. Two sets of second cousins sibs were affected. Consanguinity was noted. Another large kindred was reported by Chaurasia and Goswami (1). Two sibs had heptadactyly.

Khoo and Saad (4) and Sidhu and Grewal (5) reported isolated cases in Indian males. Another Indian male was noted by Gopalakrishna and Thatte (3). There was hexadactyly of one hand and heptadactyly of the other. Both feet exhibited hexadactyly. A sister had similar hand abnormalities. The upper median frenum was duplicated.

### References [Type V (median cleft of upper lip and postaxial polydactyly of hands and feet [Thurston syndrome])]

1. Chaurasia BD, Goswami HK: A contribution to the genetics of oral-facial-digital (OFD) syndrome. *Jpn J Hum Genet* **18**:294–299, 1973.

2. Chowdhury J: A study of five siblings with median cleft lips and polydactyly. *Trans 6th Int Cong Plast Reconstr Surg (Paris)* 1975, pp 208–211.

3. Gopalakrishna A, Thatte RL: Median cleft lip associated with bimanual hexadactyly and bilateral accessory toes: Another case. *Br J Plast Surg* **35**:354–355, 1982.

4. Khoo CT, Saad MN: Median cleft of the upper lip in association with bilateral hexadactyly and accessory toes. *Plast Reconstr Surg* **33**:407–409, 1980.

5. Sidhu SS, Grewal MS: Cleft lip, notched alveolus, postaxial polydactyly. Case 64. *Syndrome Ident* **6**(1):12–13, 1980.

6. Thurston EO: A case of median hare-lip associated with other malformations. *Lancet* **2**:996–997, 1909.

**Type VI [polydactyly, cleft lip/palate, lingual lump, and cerebellar anomalies (Váradi syndrome)].** Váradi et al (12) reported a syndrome resembling Mohr and trisomy 13 syndromes in a family of endogamic Gypsies. Mattei and Aymé (8) reported a similar case, and suggested this condition should be distinguished from other OFD syndromes. There are likely at least 10 other cases in the literature (1,3–6,9), some of which are reported as examples of Mohr syndrome or Joubert–Boltshauser syndrome with polydactyly. Other possible examples were reviewed by Münke et al (9).

Facial features are often described as consisting of hypertelorism (1,4–6,9), epicanthal folds (1,5,9), broad nasal tip (4,6,12), and cleft

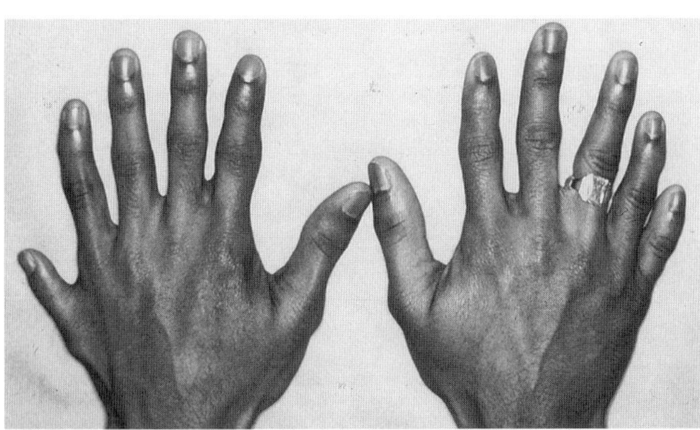

A

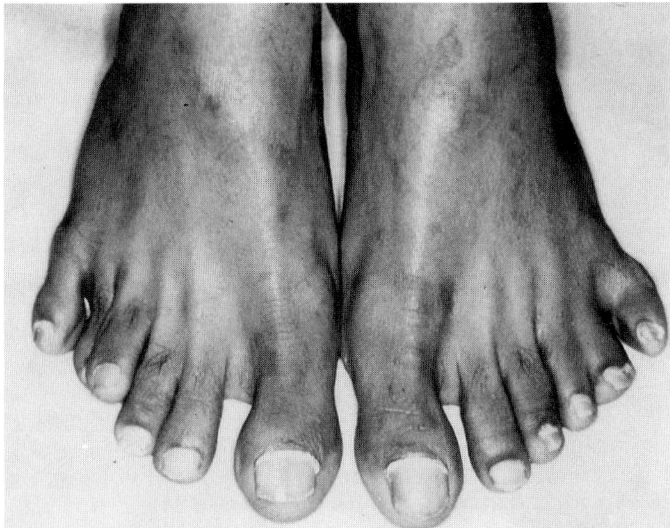

B

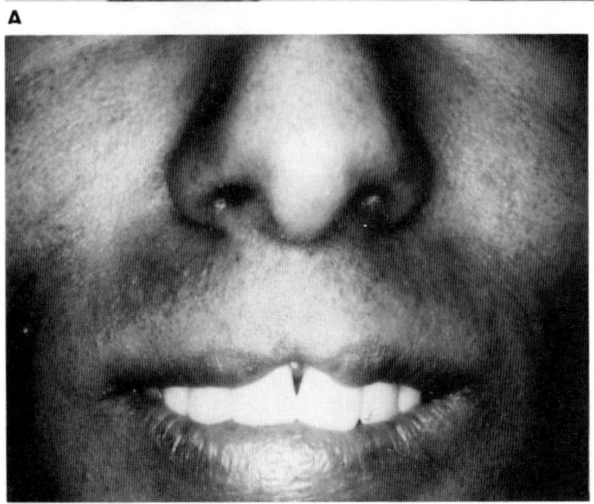

C

Fig. 19–37. *Oral–facial–digital syndrome, Type V.* (A,B) Postaxial polydactyly of hands and feet. (C) Typical midline cleft of upper lip. (From *CT Khoo* and *MN Saad,* Plast Reconstr Surg **33**:407–409, 1980.)

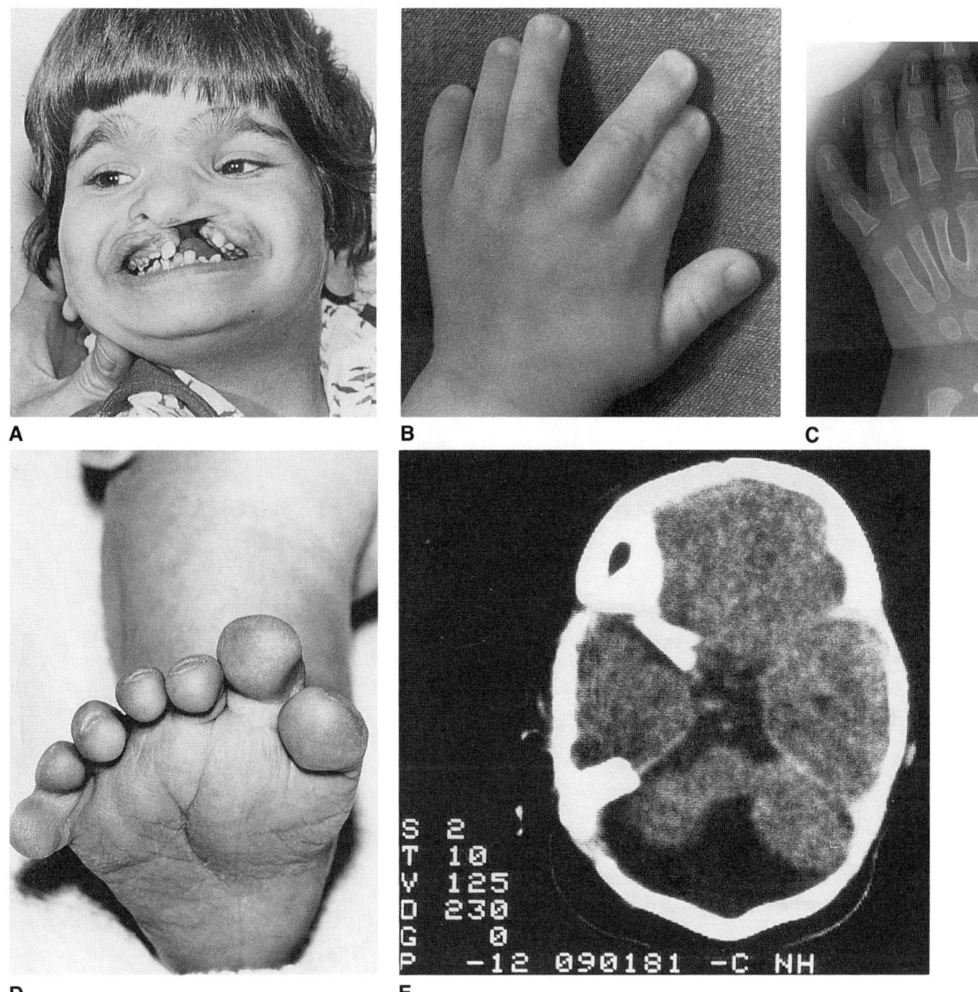


Fig. 19–38. *Oral–facial–digital syndrome, Type VI.* (A) Note hypertelorism, epicanthic folds, broad nasal tip, cleft lip. (B,C) Manual polysyndactyly with forked or bifid metacarpals. (D) Hallucal polysyndactyly. (E) Extensive cystic fourth ventricle occupies enlarged posterior fossa. Anterior vermis and both cerebellar hemispheres appear reduced in size. (A from *V Váradi et al,* J Med Genet **117**:119, 1980. B,C from *G Annerén et al,* Clin Genet **26**:178, 1984. E from *D Haumont* and *S Pelc,* Clin Genet ***24***:41, 1983.)

lip (Fig. 19–38A) (1,8,9,12). In some cases there is also a mild midline cleft of the upper lip. Microphthalmia was present in one case (4). Rotary nystagmus, oculomotor apraxia, and esotropia are common (5,9,12). Posteriorly angulated and/or apparently low-set ears were reported in several examples (4,6–8,12).

Both intraoral frenulae (4,6,9) and lingual or sublingual lumps (4,5,8,9,12) have been noted. The palate can be highly arched or cleft (1,4–6,12). Dental anomalies are generally not present, although one patient had enamel hypoplasia (12).

Postaxial polydactyly of the hands is common (Fig. 19–38 B,C) (1,3–6,8–10,12). Few cases had preaxial polydactyly of the hands (9). Clinodactyly and syndactyly can also occur (4–6,8,9,12). Most striking, however, and present in all cases thus far examined, is the presence of a central Y-shaped or forked metacarpal noted on X-ray.

The feet usually have bilateral preaxial polydactyly (Fig. 19–38D) (1,4–6,8,9,11,12), although four patients had an accessory fifth toe as well (3,5,9). Cutaneous syndactyly has also been reported (5).

Several internal anomalies have also been described. Cardiac defects included aortic stenosis in one case (12) and common atrioventricular canal and coarctation of the aorta in the other (5). Unilateral renal agenesis was reported in one affected sib by Mattei and Aymé (8). One patient manifested hypogonadotrophic hypogonadism with micropenis (9).

However, the most distinguishing feature of this condition seems to be various degrees of cerebellar defects, including absent or hypoplastic cerebellar vermis (Fig. 19–38E) or features of the Dandy–Walker malformation (3,6), or evidence of cerebellar defects such as motor incoordination and delayed or absent speech (1,4,5). Recurrent episodes of tachypnea and hyperpnea are common (9). One child reported by Váradi et al (12) also had arhinencephaly. Although Baraitser (2) considers some of these cases (3,5,6) to be examples of Joubert–Boltshauser syndrome with polydactyly, we feel it is more appropriate to classify these cases as examples of Váradi syndrome because of the overall phenotype.

Nearly all affected children have severe mental retardation with rare exception (9); postnatal growth retardation (5,8,9,12), hypotonia and/or gait disturbance (5,9,12), and deafness (5,8,9) can also occur. The presence of consanguinity in one family (12) and affected sibs in other families (3,4,6,8) suggest autosomal recessive inheritance.

Another possible example of Váradi syndrome may be the sibs reported by Hooft and Jongbloet (7). One or both had oral frenulae, ocular coloboma, microphthalmia, palatal malformations, and postaxial polydactyly. One sib had a tumor at the base of the brain, although no additional data are given regarding the nature of the tumor.

Dandy–Walker malformation (hydrocephalus, incomplete cerebellar vermis, and posterior fossa cyst) may occur as an isolated finding or in association with myriad conditions, including Joubert–Boltshauser syndrome, *Coffin–Siris syndrome, Meckel syndrome, Ellis–van Creveld syndrome,* and *cryptophthalmos syndrome* (10).

Fetal ultrasonography may possibly be used to detect the anomalies of the extremities, the cleft lip and palate, and the partial agenesis of the cerebellar vermis.

### References [Type VI (polydactyly, cleft lip/palate, lingual lump, and cerebellar anomalies [Váradi syndrome])]

1. Annerén G et al: Oro-facio-digital syndromes I and II: Radiologic methods for diagnosis and the clinical variations. *Clin Genet* **26**:178–186, 1984.

2. Baraitser M: The orofaciodigital (OFD) syndromes. *J Med Genet* **23**:116–119, 1986.

3. Egger J et al: Joubert-Boltshauser syndrome with polydactyly in siblings. *J Neurol Neurosurg Psychiatr* **45**:737–739, 1982.

4. Gencík A, Gencíkova A: Mohr syndrome in two sibs. *J Génét Hum* **31**:307–315, 1983.

5. Gustavson KH et al: Syndrome characterized by lingual malformation, polydactyly, tachypnea, and psychomotor retardation (Mohr syndrome). *Clin Genet* **2**:261–266, 1971. (Same as patient 3 in Ref. 1.)

6. Haumont, D, Pele SC: The Mohr syndrome: Are there two variants? *Clin Genet* **24**:41–46, 1983.

7. Hooft C, Jongbloet P: Syndrome oro-digito-facial chez deux freres. *Arch Fr Pédiatr* **21**:729–740, 1964.

8. Mattei JF, Aymé S: Syndrome of polydactyly, cleft lip, lingual hamartomas, renal hypoplasia, hearing loss, and psychomotor retardation: Variant of the Mohr syndrome or a new syndrome? *J Med Genet* **20**:433–435, 1983.

9. Münke M et al: Oral-facial-digital syndrome Type VI (Váradi syndrome): Further clinical delineation. *Am J Med Genet* (in press).

10. Murray JC et al: Dandy–Walker malformation: Etiologic heterogeneity and empiric recurrence risks. *Clin Genet* **28**:272–283, 1985.

11. Silengo MC et al: Oro-facial-digital syndrome II. Transitional type between the Mohr and the Majewski syndromes: Report of the two new cases. *Clin Genet* **31**:331–336, 1987.

12. Váradi V et al: Syndrome of polydactyly, cleft lip/palate or lingual lump, and psychomotor retardation in endogamic Gypsies. *J Med Genet* **17**:119–122, 1980.

**Type VII [facial asymmetry, pseudocleft lip, lobulated tongue, and hydronephrosis (Whelan syndrome)].** Whelan et al (1) reported a mother and daughter with similar findings, which the authors suggested represented OFD I syndrome. However, some of the findings are inconsistent with that diagnosis, and this likely represents a distinct entity.

Both mother and daughter had apparent hypertelorism, with pseudocleft of the upper lip, highly arched palate, bifid tongue, and facial asymmetry. The daughter also had a low-set, posteriorly angulated ear, preauricular tags, and mucosal frenulae. Polydactyly was not present; the only digital anomaly was clinodactyly in both mother and daughter.

In addition, both had bilateral hydronephrosis. The daughter was developmentally delayed, whereas the mother had low–normal intelligence. Inheritance is either X-linked or autosomal dominant.

#### Reference [Type VII (facial asymmetry, pseudocleft lip, lobulated tongue, and hydronephrosis [Whelan syndrome])]

1. Whelan DT et al: The oro-facial-digital syndrome. *Clin Genet* **8**:205–212, 1975.

**Type VIII.** Edwards et al (1), in 1988, described a family with an X-linked form of oral–facial–digital syndrome with tongue lobulations and hamartomas, median cleft upper lip, dystopia canthorum, broad or bifid nasal tip, bilateral preaxial and postaxial polydactyly of the hands and feet, duplication of the halluces, shortened tibiae and/or radii, short stature, recurrent aspiration pneumonia, and developmental delay. There was forking of metatarsals 6–7 in one boy. RJ Gorlin knows of another affected family in Brazil.

There were four affected males in three generations. Female heterozygotes had mild expression (tongue hamartoma, ankyloglossia, mild tongue clefting).

Death occurred in three of the affected males during infancy, probably from aspiration pneumonia.

#### Reference (Type VIII)

1. Edwards M et al: X-linked recessive inheritance of an orofaciodigital syndrome with partial expression in females and survival of affected males. *Clin Genet* **34**:325–332, 1988.

**Miscellaneous or unclassifiable syndromes.** Tucker et al (6) reported a patient with oral, facial, and digital anomalies. In addition to hypertelorism, micrognathia, oral frenulae, bifid tongue, and an extra finger, the deceased newborn female had posterior encephalocele, omphalocele, and talipes valgus. Autopsy revealed cerebral microgyria, partial hypoplasia of parietal lobes, and multicystic liver and kidneys. Perhaps this child really had *Meckel syndrome*.

There have also been three cases of an oral–facial–digital syndrome with cardiac, but not brain, abnormalities. The first case (3), in addition to midline cleft upper lip, had lobulated tongue, pseudocleft palate, and pre- and postaxial polydactyly of hands and feet, a complex cardiac defect, including common atrium, single atrioventricular valve, single ventricle, and truncus arteriosus, small epiglottis, anteriorly placed anus, and absent lobation of the left lung. Iaccarino et al (3) reported sibs each with cardiac, as well as oral, facial, and digital anomalies. The first sib had cleft lip, polydactyly of hands and feet, VSD, and mitral atresia; the second sib had preaxial polysyndactyly of the hands and feet, and complete atrioventricular canal. Whether oral, facial, and digital anomalies with cardiac defects represent yet another type of OFD syndrome remains to be seen; however, the presence of cardiac defects in both sibs reported by Iaccarino et al (4) suggests that this is a distinct possibility. A similar condition has been noted by Bonneau et al (2).

Toriello et al (5) reported an infant with features of *frontonasal malformation,* as well as polydactyly of the hands and feet, agenesis of corpus callosum, Dandy–Walker malformation, persistent fetal circulation, short limbs, and cryptorchidism. X-ray examination demonstrated narrow ribcage, elongated trunk, and long bones which were shorter than average (65–90% for age). The tibiae were most severely affected. This case has features of both OFD IV and Váradi syndrome, and may be a unique condition. This disorder is discussed in chapter 24.

Berberich et al (1) reported a patient with centrally notched upper lip, frenulae extending from the alveolar ridge to the buccal mucosa, lobulated tongue, partial cleft palate, thoracic dystrophy, preaxial polysyndactyly involving one thumb and both halluces, mild shortening of the humeri, bowed tibiae, short penis, and imperforate anus. The condition was compatible with survival during the first year of life.

Differential diagnosis. There are a significant number of disorders that have midline defect of the upper lip ranging from isolated midline cleft to *encephalocele* to *Ellis–van Creveld syndrome* to *short rib-polydactyly syndrome, Majewski type.* None of these disorders should present problems in differential diagnosis for the clinician.

#### References (Miscellaneous or unclassifiable syndromes)

1. Berberich MS: A syndrome of thoracic dystrophy, bone dysplasia, and microcephaly, with preaxial polydactyly, intra-oral abnormalities, micropenis and membranous imperforate anus, compatible with survival in the first year of life. Ninth Annual David W. Smith Workshop on Malformations and Morphogenesis, Oakland, August 3–7, 1988.

2. Bonneau JC et al: Polysyndactylie avec cardiopathie complexe à propos de trois cas dans une même fratrie. *J Génét Hum* **31**:93–105, 1983.

3. Cordero JF, Holmes LB: Heart malformation as a feature of the Mohr syndrome. *J Pediatr* **91**:683–684, 1977.

4. Iaccarino M et al: Prenatal diagnosis of Mohr syndrome by ultrasonography. *Prenatal Diag* **5**:415–418, 1985.

5. Toriello HV et al: Frontonasal "dysplasia," cerebral anomalies, and polydactyly: Report of a new syndrome and discussion from a developmental field perspective. *Am J Med Genet (Suppl* **2**):89–96, 1986.

6. Tucker CC et al: Oral-facial-digital syndrome with polycystic kidneys and liver: Pathological and cytogenetic findings. *J Med Genet* **3**:145–147, 1966.

## Otopalatodigital syndrome, Type I

In 1962, Taybi (22) first described the otopalatodigital (OPD) syndrome type I. In 1967, Dudding et al (3) expanded the phenotypic spectrum to include characteristic facies, conductive hearing loss, short stature, cleft palate, and generalized bone dysplasia.

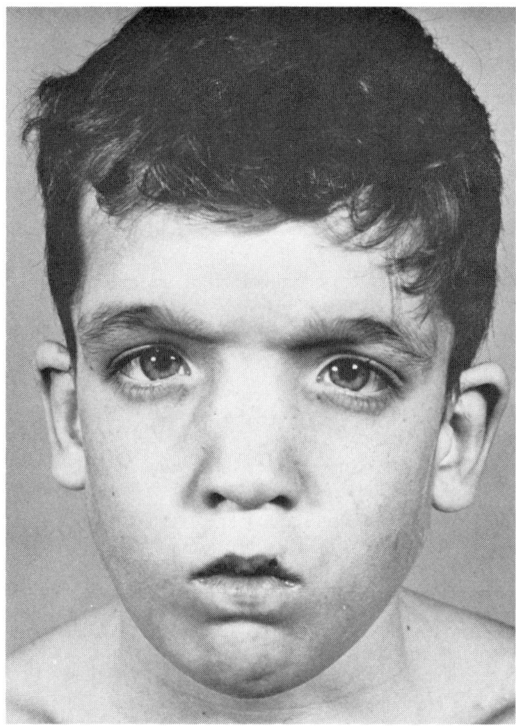

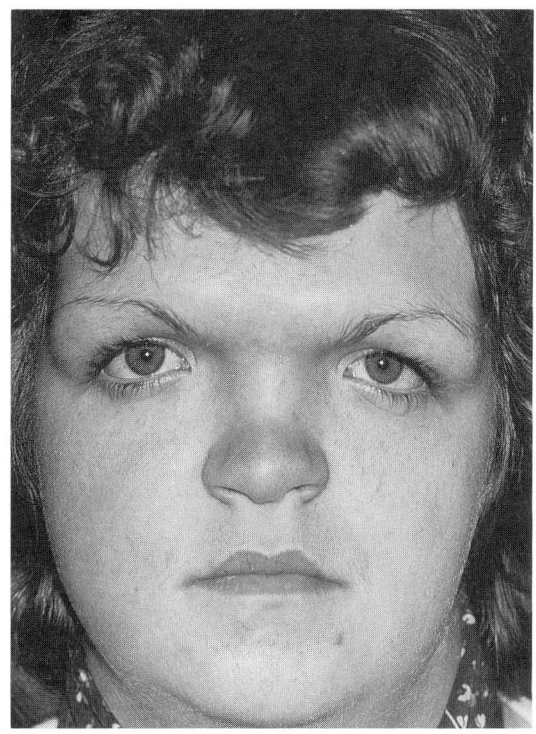

A B

Fig. 19–39. *Otopalatodigital syndrome, Type I.* (A) Overhanging brow with prominent supraorbital ridge and wide nasal bridge resulting in pugilistic appearance. (B) An affected female. (A from *BA Dudding et al,* Am J Dis Child **113**:214, 1967. B courtesy of *HD Rott,* Nürnberg, West Germany.)

Although most patients are males, females with almost complete expression of the syndrome have been described (8,10,14,20). Most pedigrees are compatible with X-linked inheritance (8,23). However, Salinas et al (18) reported male-to-male transmission and concluded that the syndrome is inherited as an autosomal dominant disorder. There are some differences however. The patients reported by Salinas et al (18) had distal displacement of thumbs, fusion or hypoplasia of carpal bones, normal distal phalanges, curved tibias, and delayed bone age.

**Facies.** The facies in males is distinctive (3,13,14). An overhanging brow, prominent supraorbital ridges, and downslanting palpebral fissures are noted. Hypertelorism and a broad nasal root give the patient a pugilistic appearance (Fig. 19–39A,B). Slight notching may be noted at the junction between the medial third and lateral two-thirds of the upper eyelid margin in some affected males (5). The corners of the mouth are often downturned. Facial features in affected females are more variable and usually more mild than in affected males. The most constant features in affected females are overhanging brow, hyperteloric appearance, prominent lateral supraorbital ridges, depressed nasal bridge, and flat midface (5).

**Central nervous system.** Most male patients have mild mental retardation, intelligence quotients ranging between 75 and 90. Speech development is slow. Mental retardation and slow speech development may be related to the conductive hearing loss that, in the few patients tested, ranged from 30 to 90 dB (1,2,12,15,22). Abnormally shaped middle ear ossicles and small external auditory canals have been found (3,8,12,15).

**Skeleton.** Skeletal growth is retarded; all male patients are below the tenth percentile and may be below the third percentile for height (5) (Fig. 19–40). The trunk is small, and there is pectus excavatum (3,20). Limited elbow extension and wrist supination have been noted in several patients, some of whom have subluxation of the radial heads (3,11,19,22). Hands and feet are striking (3). Thumbs and halluces are spatulate and especially abbreviated. Clefting between the hallux and the other toes is exaggerated. The toes and fingers are irregular in form and direction of curvature. The second and third fingers may deviate to the ulnar side, while the fifth finger often bends to the radial side (Fig. 19–41A,B). Affected females are not unusually short; although only mild abnormalities may be observed in the hands, the feet usually have more obvious abnormalities (5).

Roentgenographic alterations are marked. Frontal and occipital bossing

Fig. 19–40. *Otopalatodigital syndrome, Type I.* Skeletal growth is retarded. Three affected sibs, ages 8, 7, and 6 years, are flanked by their two normal brothers, ages 10 and 4 years. Note subluxation of elbows of affected. (From *BA Dudding et al,* Am J Dis Child **113**:214, 1967.)

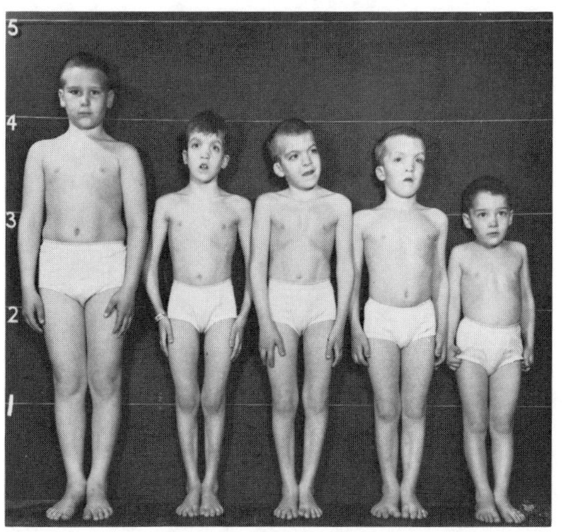

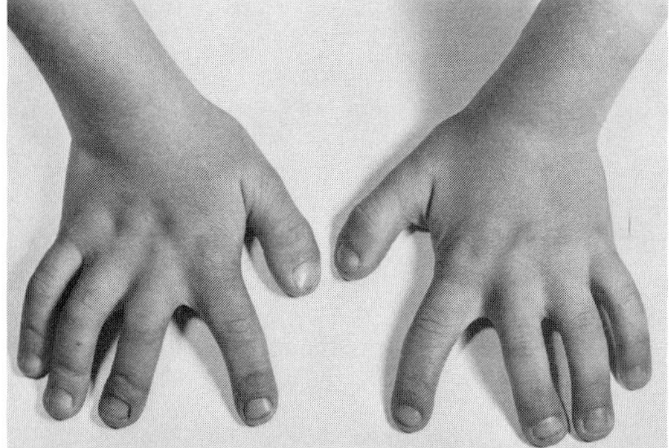

A

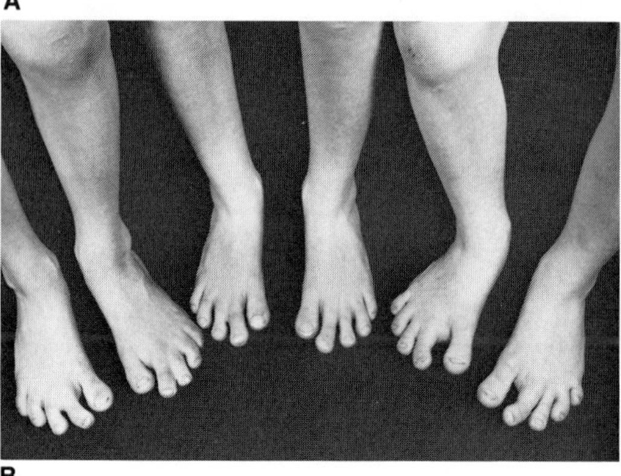

B

Fig. 19–41. *Otopalatodigital syndrome, Type I.* (A) Mild syndactyly and clinodactyly of digits. Note abbreviated and flattened terminal phalanges. (B) Short big toes, variable soft-tissue syndactyly and clinodactyly, gap between hallux and second toes. (A from *A Prader,* Zürich, Switzerland. B from *BA Dudding et al,* Am J Dis Child **113**:214, 1967.)

and thickening give the skull a mushroom-like appearance. The skull base is thick, the facial bones are hypoplastic, and the paranasal sinuses and mastoids are poorly pneumatized. The nasion–sella–basion angle is about 116° (normal mean = 132°), and the mandibular plane angle is increased (5,11,18). The clivus, or basisphenoid, lies further posterior than normal in relation to the cervical spine. These changes are essentially limited to affected males (Fig. 19–42A).

Iliac bones are small, with decreased flare. Coxa valga is a common finding. The lower tibia is laterally bowed. Failure of fusion of several vertebral arches is common.

Distinctive changes in the hands of males include shortening of the radial side of the middle phalanx of the fifth finger, clinodactyly, short distal phalanx of thumb (which during development has a cone-shaped epiphysis), accessory ossification center in second metacarpal, tear-drop-shaped trapezium, a transverse capitate, and trapezium-scaphoid fusion (5,11,17,21). Females may have greater multangular-scaphoid fusion (Fig. 19–42B).

In affected males, radiologic abnormalities in the feet include short phalanges and metatarsals of the great toes. The second and third metatarsals are long and abnormally shaped because of their fusion with the cuneiform bones. The fifth metatarsal may be prominent, with an extra ossification center. Tarsal fusions are common, and males usually have two ossification centers in the navicular bone (Fig. 19–42C). Langer (11), Gall et al (5), and Poznanski et al (16,17) have described the radiologic findings in detail.

**Oral manifestations.** All but one male patient reported have had cleft palate; none was seen in the boy reported by Podoshin et al (15). Cleft palate has not been found in affected females. The mandible is small and the mandibular angle, more obtuse than normal (5,11,18).

**Differential diagnosis.** *Larsen syndrome* shares a number of features, such as cleft palate and joint dislocations, with the OPD I syndrome. However, patients with Larsen syndrome have a different facies, multiple carpal bones, juxtacalcaneal bone, and false flexion creases of the fingers. It is difficult to classify the female patient described by Jäger and Refior (10), since she has features of both OPD syndrome and Larsen syndrome. The patients described by Weinstein and Cohen (24) as having X-linked cleft palate are examples of OPD I syndrome.

FIg. 19–42. *Otopalatodigital syndrome, Type I.* Roentgenograms. (A) Vertical basisphenoid. (B) Abnormal middle phalanx of fifth finger, shortened terminal phalanges, capitate–hamate complex, accessory ossification center of second metacarpal. (C) Short hallux, fusion between middle and lateral cuneiforms and corresponding metatarsals, forming paddle-shaped structures. (B,C from *BA Dudding et al,* Am J Dis Child **113**:214, 1967.)

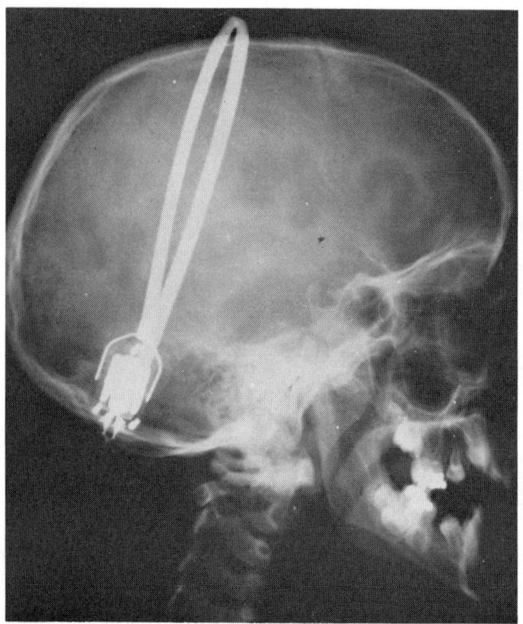

A

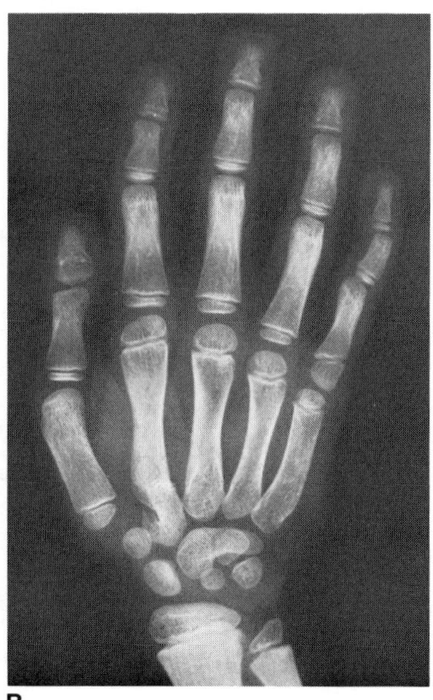

B

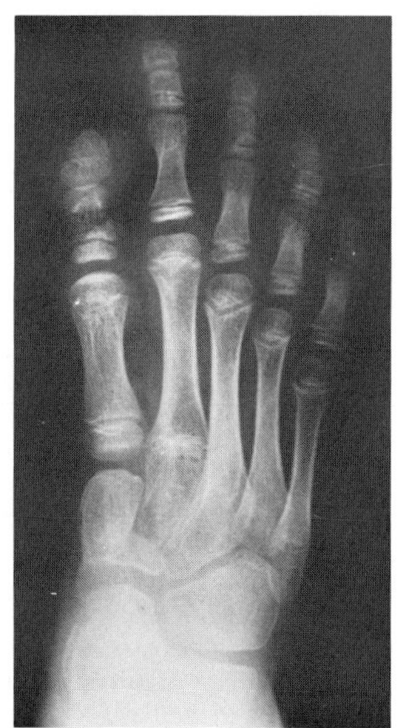

C

*Otopalatodigital syndrome, type II* is also X-linked and far more severe in its skeletal manifestations. *Acro-cranio-facial dysostosis syndrome* is also more severe in its expression.

### References (Otopalatodigital syndrome, Type I)


1. Aase JM: Oto-palato-digital syndrome. *Birth Defects* **5**(3):43–44, 1969.
2. Buran DJ, Duvall AJ III: The oto-palato-digital (OPD) syndrome. *Arch Otolaryngol* **85**:394–399, 1967. (Same cases reported in Refs. 3,6,8, and 11.)
3. Dudding BA et al: The oto-palato-digital syndrome. A new symptom-complex consisting of deafness, dwarfism, cleft palate, characteristic facies, and a generalized bone dysplasia. *Am J Dis Child* **113**:214–221, 1967. (Same cases reported in Refs. 2,6,8, and 11.)
4. Fryns JP et al: The otopalatodigital syndrome. *Acta Paediatr Belg* **31**:159–163, 1978.
5. Gall JC Jr et al: Oto-palato-digital syndrome: Comparison of clinical and radiographic manifestations in males and females. *Am J Hum Genet* **24**:24–36, 1972. (Same cases reported in Ref. 24.)
6. Gorlin RJ: Discussion on oto-palato-digital syndrome. *Birth Defects* **5**(3):45–47, 1969. (Same cases reported in Refs. 2,3,8, and 11.)
7. Gorlin RJ: Comment by Dr. Gorlin. *Am J Dis Child* **119**:377, 1970 (letter).
8. Gorlin RJ et al: The oto-palato-digital (OPD) syndrome in females. *Oral Surg Oral Med Oral Pathol* **35**:218–224, 1973. (Same cases reported in Refs. 2,3,6 and 11.)
9. Ichimura K, Hoshino T: Otological findings in oto-palato-digital syndrome. *Jibiinkoka* **53**:287–293, 1981.
10. Jäger M, Refior HJ: Ein Knochendysplasie-Syndrom. *Z Orthop* **105**:196–208, 1968.
11. Langer LO Jr: The roentgenologic features of the oto-palato-digital (OPD) syndrome. *Am J Roentgenol* **100**:63–70, 1967. (Same cases reported in Refs. 2,3,6, and 8.)
12. Nager GT, Char F: The otopalatodigital (OPD) syndrome: (Conductive deafness, cleft palate and anomaly of digits). *Birth Defects* **7**(7):273–274, 1971.
13. Pazzaglia VE, Giampiero B: Oto-palato-digital syndrome in four generations of a large family. *Clin Genet* **30**:338–344, 1986.
14. Plenier V et al: Le syndrome oto-palato-digital. A propos de trois cas feminins. *Rev Stomatol Chir Maxillofac* **84**:322–329, 1983.
15. Podoshin L et al: The oto-palato-digital syndrome. *J Laryngol Otol* **90**:407–411, 1976.


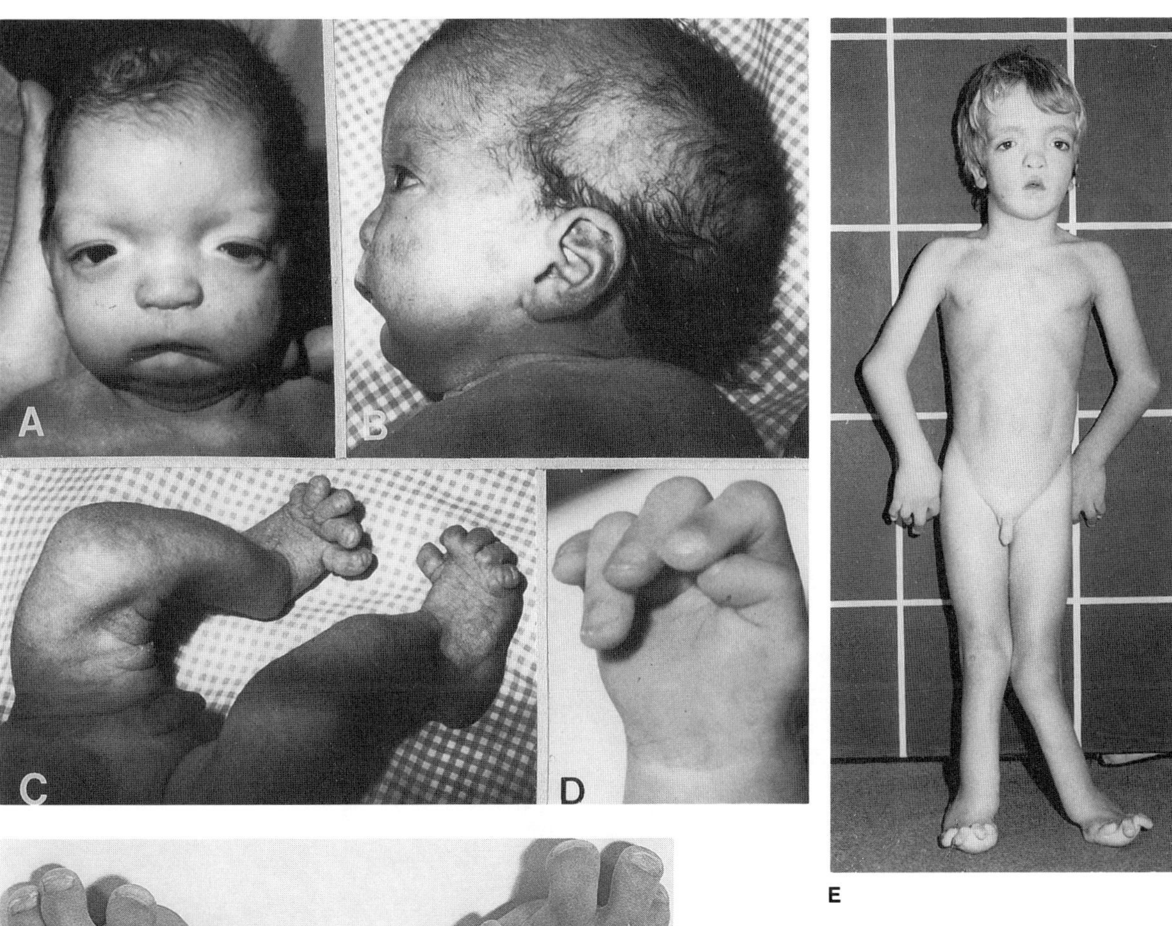


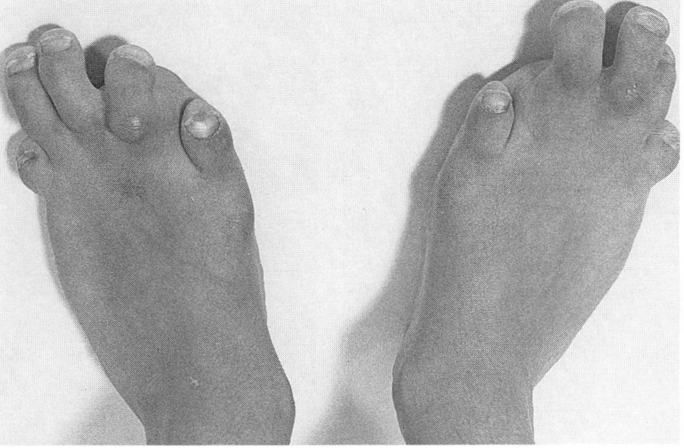


Fig. 19–43. *Otopalatodigital syndrome, Type II.* (A) Hypertelorism, downslanting palpebral fissures, small mouth. (B) Posteriorly angulated ears, micrognathia. (C) Curvature of lower limbs, fanned-out toes. (D) Abnormal position of fingers. (E) Flattened face, dislocated elbows and hips, fingers held in flexed overlapping position. (F) Severely abbreviated halluces and fifth toes, and camptodactyly. (A–D from *M André et al,* J Pediatr **98**:747, 1981. E,F from *K Kozlowski,* Pediatr Radiol **6**:97, 1977.)

16. Poznanski AK et al: Otopalatodigital syndrome: Radiologic findings in the hand and foot. *Birth Defects* **10**(5):125–149, 1974.

17. Poznanski AK et al: The hand in the oto-palato-digital syndrome. *Ann Radiol* **16**:203–209, 1973.

18. Salinas C et al: Variable expression in otopalatodigital syndrome; cleft palate in females. *Birth Defects* **15**(5B):329–345, 1979.

19. Singh SD et al: Oto-palato-digital syndrome. *Indian J Pediat* **37**:112–114, 1970.

20. Sirvent Gomez J et al: Sindrome oto-palato-digital. Aportacion de un caso en el sexo femenino. *An Esp Pediatr* **20**:905–910, 1984.

21. Takato T et al: Otopalatodigital syndrome. *Ann Plast Surg* **14**:371–374, 1985.

22. Taybi H: Generalized skeletal dysplasia with multiple anomalies. A note on Pyle's disease. *Am J Roentgenol* **88**:450–457, 1962.

23. Turner G: Inheritance of the oto-palato-digital syndrome. *Am J Dis Child* **119**:377, 1970 (letter).

24. Weinstein ED, Cohen MM: Sex-linked cleft palate. Report of a family and review of 77 kindreds. *J Med Genet* **3**:17–22, 1966. (Same cases reported in Ref. 5.)

## Otopalatodigital syndrome, Type II

This skeletal dysplasia is characterized by short stature, unusual facies, cleft palate, and multiple skeletal anomalies (Fig. 19–43 to 19–45). Inheritance is X-linked. Skeletal manifestations are far more severe in this syndrome than they are in otopalatodigital syndrome, type I.

Death has occurred, usually due to respiratory infection, within the first 5 months of life in seven of 12 males with the disorder (1–3,8); another died of similar causes at age 2.5 years (8).

**Facies.** Hypertelorism, frontal bossing, broad nasal bridge, downslanting palpebral fissures, midface hypoplasia, low-set ears, and marked mandibular micrognathia are noted (1,2,8) (Fig. 19–43A,B). Carrier females have mild facial abnormalities (1,2,8).

**Skeletal changes.** There is apparent hypertelorism. The anterior fontanel may be large (3). The base of the skull is sclerotic (3). Clavicles are thin and wavy. The thorax is small, and there may be pectus excavatum. Ribs are wide posteriorly and anteriorly, while narrow in their middle portions. Vertebrae are flattened. The humeri, radii, ulnae, femora, and tibiae are bowed; curving of long bones may disappear early in life (4). Radial heads may be dislocated. Fibulae are small or absent. Ilia are hypoplastic. Thumbs are broad and short, as are the great toes. Fingers are held in a flexed, overlapping position. Tubular bones of the hands and feet are deformed and bones of the wrists and ankles are hypoplastic and malformed (Figs. 19–43C to

Fig. 19–44. *Otopalatodigital syndrome, Type II*. Roentgenograms. (A) Curvature of humerus and radius, anomalies of metacarpals and phalanges. (B) Small first metacarpal, large proximal phalanges. (C) Hip dislocation, curvature of femora, hypoplastic fibulae. Advanced bone age. (D) Curvature of tibia. (E) Absence of ossification of first metatarsal and its phalanges, short fifth metatarsal. (F) Anomalies of metacarpals and phalanges. Note extra bone between proximal phalanx and metacarpal on fifth finger, enlarged epiphyses, capitate–hamate complex. (G) Compare this patient with patient in B and patient in F. Note unusual proximal phalangeal epiphyses, transverse capitate, extra bone in carpal complex. (A–E from *M André et al,* J Pediatr **98**:747, 1981. F from *K Kozlowski,* Pediatr Radiol **6**:97, 1977. G from *N Fitch et al,* Am J Med Genet 15:655, 1983.)

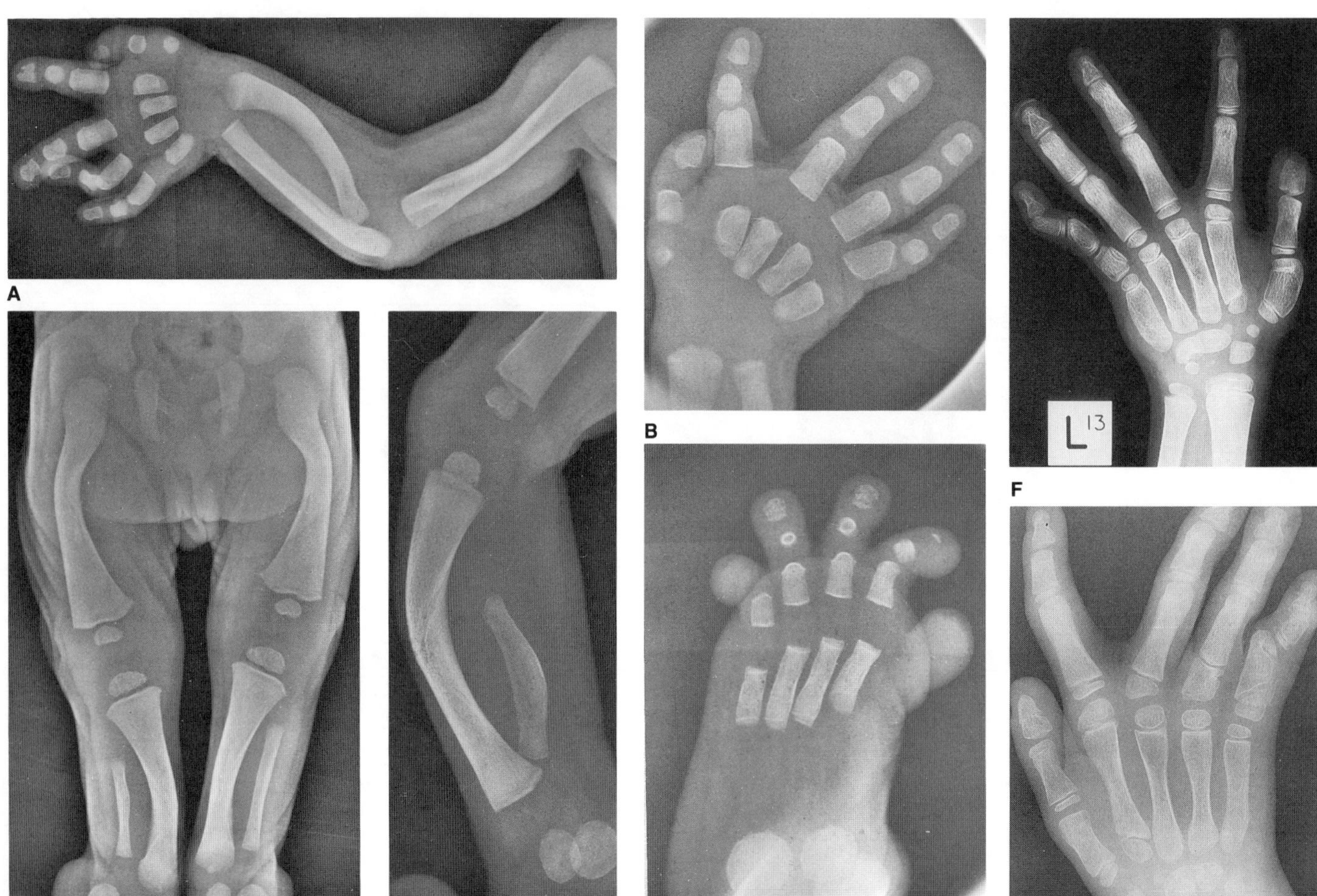

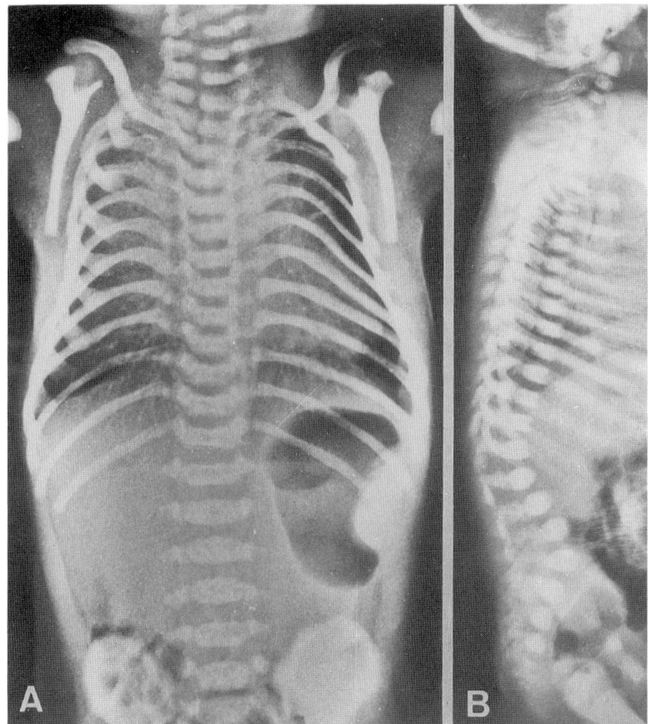


Fig. 19–45. *Otopalatodigital syndrome, Type II.* (A) Steeply sloping sinuous clavicles, sinuous ribs. (B) Dorsolumbar kyphosis, increased lumbar lordosis. (From *M André et al,* J Pediatr **98**:747, 1981.)

19–45B). Rocker-bottom feet may be present (1,2,6). Carrier females may have abnormalities of the hands and feet (3,8).

Growth plates of the vertebrae and costochondral junction were normal in the one patient examined (2).

**Ears.** Hearing loss has been described in several patients (4–6); one of them (4) had aseptic meningitis twice during the first year of life. Bilateral conductive hearing loss was also described in a carrier female (1); malformed ossicles were noted on surgery. The 2.5-year-old male reported by Shi (8) likely had OPD II (H Schucknecht, personal communication, 1985). Histologic studies of a temporal bone from this child revealed malformed ossicles and abnormalities of the bony labyrinth.

**Other findings.** Psychomotor development and intelligence appear normal in some patients (4), whereas others are mentally retarded (6).

**Oral manifestations.** Cleft palate is found in affected males, and bifid uvula in a carrier female has been noted (3). In some instances, frank Robin sequence may be present (1).

**Differential diagnosis.** This syndrome should be distinguished from *otopalatodigital syndrome, type I, campomelic dysplasia,* and *trisomy 18.* There is similarity to acro-coxo-melic dysplasia, an autosomal recessive disorder described by Plauchu et al (7).

**Laboratory aids.** Prenatal diagnosis has been done by ultrasound (9).

#### References (Otopalatodigital syndrome, Type II)

1. André M et al: Abnormal facies, cleft palate, and generalized dysostosis: A lethal X-linked syndrome. *J Pediatr* **98**:747–752, 1981.
2. Brewster TG et al: Oto-palato-digital syndrome, type II—An X-linked skeletal dysplasia. *Am J Med Genet* **20**:249–254, 1985.
3. Fitch N et al: A familial syndrome of cranial, facial, oral, and limb abnormalities. *Clin Genet* **10**:226–231, 1976.
4. Fitch N et al: The oto-palato-digital syndrome, proposed type II. *Am J Med Genet* **15**:655–664, 1983.
5. Kaplan J, Maroteaux P: Syndrome oto-palato-digital de type II. *Ann Génét* **27**:79–82, 1984.
6. Kozlowski K et al: Oto-palato-digital syndrome with severe x-ray changes in two half-brothers. *Pediatr Radiol* **6**:97–102, 1977.
7. Plauchu H et al: Le nanisme acro-coxo-mésomélique: Variété nouvelle de nanisme récessif autosomique. *Ann Génét* **27**:83–87, 1984.
8. Shi S-R: Temporal bone findings in a case of otopalatodigital syndrome. *Arch Otolaryngol* **111**:119–121, 1985.
9. Vigneron J et al: Le syndrome oto-palato-digital de type II. Diagnostique prénatal par echographie. *J Génét Hum* **35**:69, 1987.

# Chapter 20
# Orofacial Clefting Syndromes: General Aspects

## Facial clefts and associated anomalies

The first evidence of clefting is in an Egyptian mummy dating from 2400–1300 B.C. (48). A 2000-year-old ceramic statue of a king with a cleft has been reported from Colombia (89), and an African mask showing cleft lip and palate has been described (20). It is curious that in spite of the frequent occurrence of facial clefts, Greek and Roman physicians made no mention of them. Treatment of cleft lip was apparently first carried out by a Chinese physician in 390 A.D. (15).

Before cleft lip and cleft palate syndromes are considered, the embryologic, epidemiologic, and genetic aspects of isolated cleft lip and cleft palate will be reviewed. Readers wishing to examine the subject in greater detail should consult Grabb et al (48), Ross and Johnston (100), Fraser (44), Millard (81), Melnick et al (76), Dronemaraju (33), and Bear (7).

Clefts of the primary and/or secondary palates are included among the more common congenital anomalies. Clinically there is great variability in degree of cleft formation. Minimal involvement includes bifid uvula, linear lip indentations with or without nostril deformity (intrauterine-healed clefts), lateral upper lip fistula, and submucous palatal cleft (16,39a,53,54,65,71,118b,125).

A cleft may involve only the upper lip or may extend to involve the nostril and the hard and soft palates. An isolated palatal cleft may be limited to the uvula (bifid uvula) or may be more extensive, cleaving the soft palate or both the hard and soft palates. A combination of cleft lip and cleft palate is most common. Breakdown according to type has differed somewhat in several large surveys. Roughly, however, cleft lip with cleft palate comprises about 45%, isolated cleft palate about 30%, and cleft lip about 25% (123). The rare instance of separate clefts of the lip and palate represents a variant of cleft lip–palate (95).

**Embryology.** Neural crest cells play an integral part in facial morphogenesis (62a,118a,129). Just prior to the time when the neural folds fuse to form the neural tube, ectodermal cells adjacent to the neural plate migrate into the underlying regions where they act as mesenchyme (ectomesenchyme). Crest cells in the head and face form essentially all the skeletal and connective tissues of the face: bone, cartilage, fibrous connective tissue, and all dental tissues except enamel.

By the end of the fourth week, the anterior neuropore has closed. The face, if one can call it that at this time, consists of a large frontal prominence that constitutes the upper boundary of the stomodeum (primitive mouth) and results from ectodermal proliferation ventral to the developing brain (85). The primary mouth is separated from the foregut by the buccopharyngeal membrane that undergoes selective cell death and ruptures at this time in development. On each side of the lower aspect of the frontonasal prominence, the nasal placodes are forming. These bilateral structures located just above the primary mouth are represented by local thickenings of the surface ectoderm. Rapid proliferation of ectomesenchyme results in nasal swellings both lateral to and medial to the nasal placodes. By means of selective cell death and proliferation of tissues, the nasal or olfactory pits are formed and extend into the primitive mouth. These are the primitive nostrils (Fig. 20–1). Sulik and Schoenwolf (118a) have elegantly illustrated these changes by SEM.

Extremely active growth occurs during the fifth and sixth weeks.

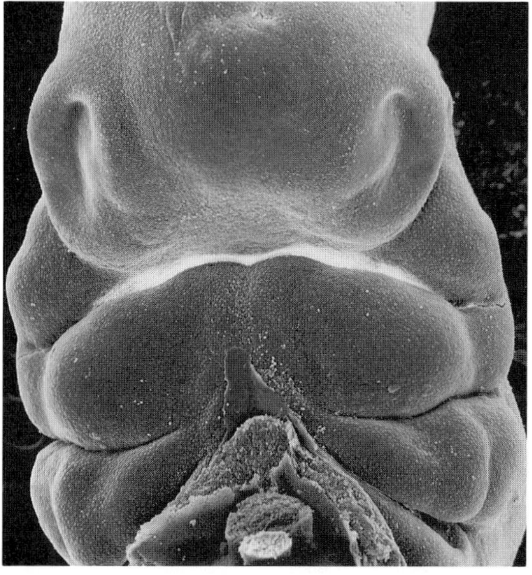

Fig. 20–1. *Facial clefts and associated anomalies.* SEM of face of 5-week-old embryo. Note olfactory pits developing on lateral portions of frontonasal prominence. (Courtesy of *K Sulik,* Chapel Hill, North Carolina.)

The maxillary swellings, which represent the upper anterior portion of the first pharyngeal arch, enlarge considerably, again through ectomesenchymal proliferation. During the sixth and seventh weeks, the medial nasal prominences merge with each other and the bilateral maxillary processes (85). Thus, the upper lip is formed laterally by the maxillary prominences and medially by the fused medial nasal prominences (Fig. 20–2). It should be emphasized that the lateral nasal prominences play no role in formation of the upper lip but form the alae or wings of the nose. The philtral ridges have nothing to do with facial prominences. They do not appear until the third to fourth embryonal month (83).

The primary palate consists of the two merged medial nasal prominences that form the intermaxillary segment. This, in turn, consists of two portions: (1) a labial component that forms the philtrum of the upper lip, that is, the indented area flanked by roughly parallel ridges that run from the columella of the nose to the middle of the upper lip, and (2) the triangular palatal component of bone that includes the four maxillary incisor teeth. The primary palate extends posteriorly to the incisive foramen or, clinically, to the incisive papilla (116). The rare paramedian pit or fistula of the upper lip occurs at the line of fusion of the median nasal prominence and the maxillary process (16,39a,54,71,118b).

The so-called secondary palate comprises at least 90% of the hard and soft palates, that is, all except the anterior portion that holds the incisor teeth. Its development appears to be somewhat more complicated than originally thought. The palatal shelves originate as swellings or shelf-like burgeonings of the medial surfaces of the maxillary prominences. They appear in the sixth week and grow downward, lateral to the tongue. Elevation of the palatal processes to a horizontal

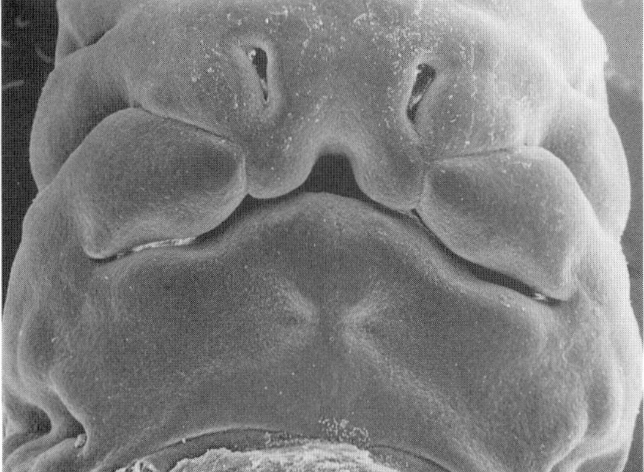

Fig. 20–2. *Facial clefts and associated anomalies.* SEM of 6-week-old embryo. Upper lip formed from fusion between notched median nasal prominence has not yet fused with maxillary processes. (Courtesy of *K Sulik,* Chapel Hill, North Carolina.)

plane is more vigorous anteriorly, that is, in the part nearer the primary palate. Elevation begins in the seventh week (40). Final closure, by fusion, occurs somewhat later in females than in males (93) (Figs. 20–3A,B). The soft palate and uvula are formed by merging, not fusion, of two growth centers at the caudal end of the hard palate. An excellent review of secondary palate formation is that of Greene and Pratt (50).

What promotes the elevation has been called "intrinsic shelf force" and has a complex biochemical and physicochemical basis. The tongue appears to play little role in shelf elevation. Elevation of the shelves is associated with widening and forward growth of the mandible. Once the shelves are elevated to the horizontal plane, there is programmed cell death of the medial edge. When the shelf edges touch, the epithelium has thinned and has undergone advanced stages of degeneration allowing mesenchyme from each side to join in the midline (69a). Complete fusion is effected by the tenth week (93). The soft palate and uvula are formed from secondary growth centers by successive merging. In some infants, there is cystic degeneration of the epithelial remnants that produces evanescent midline palatal microcysts (Epstein's pearls).

Knowledge concerning mechanisms involved in regulation of embryonic growth is, at best, sparse. The blueprints for the rate at which facial prominences burgeon and for the direction in which they grow are in part determined by the genetic template, the DNA code located in the cell nucleus. But hereditary factors do not act alone. Growth patterns can also be influenced by environmental factors. For example, there are several mouse models for cleft lip and for cleft palate. There are both susceptible and nonsusceptible strains and a large list of teratogenic substances that produce clefting. However, there is little evidence that these agents play a significant role in cleft production in humans (61), the exceptions being phenytoin, folic acid antagonists, and retinoic acid (see Chapter 2).

Various genetic and/or environmental factors may inhibit the flow of or decrease the number of neural crest cells or may affect their mass so that contact between prominences is impossible or inadequate (61,62). The epithelium covering the mesenchyme may not undergo programmed cell death so that fusion cannot take place. Exact timing and positioning play critical roles. For example, Johnston and Sulik (62) showed that if the bilaterally formed nasal placodes contact one another, there is severe to complete suppression of development of medial nasal prominences.

Altered positioning of the nasal placodes or abnormal directional growth of the facial prominences may be reasonable for cleft of the primary palate (cleft lip). This has been supported not only by animal models but by the clinical observation that parents of children with cleft lip exhibit some degree of reduction in overall midface size.

There may be many mechanisms by which a cleft of the secondary palate (cleft palate) occurs. Failure of growth of the mandible (micrognathia) may inhibit elevation of the shelves. For example, the *Robin sequence* is associated with U-shaped cleft palate.

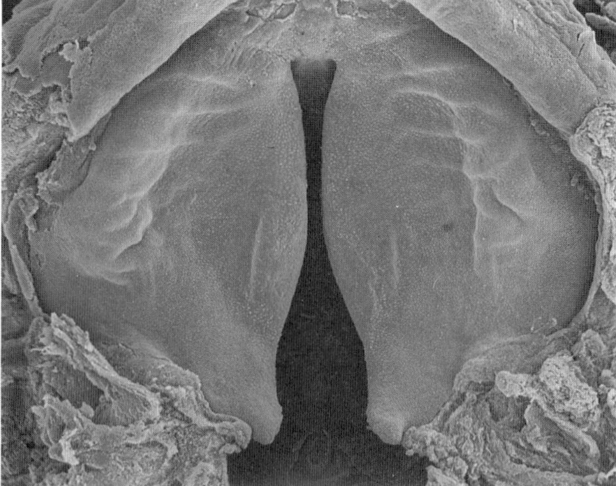

A

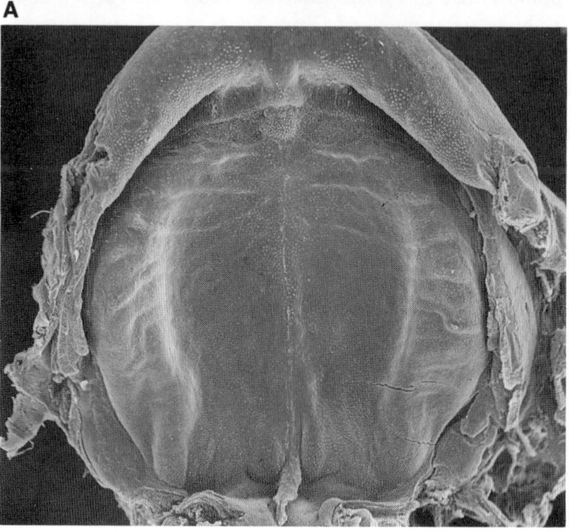

B

Fig. 20–3. *Facial clefts and associated anomalies.* (A) SEM of secondary palate in 53-day-old embryo. Fusion with primary palate has occurred. (B) At 59 days, complete fusion of secondary palate has occurred. (Courtesy of *L Russell,* Chapel Hill, North Carolina.)

Clefts of the primary *and* secondary palates occur together in about one-half the cases. A common mechanism of production has been sought. It has been suggested that the tongue tip becomes wedged in the labial cleft, produced by failure of fusion of the medial nasal and maxillary prominences. This, it has been argued, would not allow the tongue to drop and, consequently, would interfere with midline contact and inhibit fusion. This explanation appears to be simplistic. It would seem that reduction in size of both the labial maxillary prominence *and* the palatine process of the maxillary prominences is a more reasonable explanation.

Clefts of the secondary palate likely result from either hypoplasia of the shelves or delay in timing of shelf elevation. Experiments carried out on susceptible strains of mice have suggested that both mechanisms are operative, but at different times in gestation. For example, large doses of vitamin A given early in gestation inhibit palatal shelf growth whereas cortisone given later in gestation inhibits palatal shelf elevation (87a).

Recent evidence for reduction of neural tube defects (anencephaly, spina bifida) by administration of dietary supplements of folic acid has caused several teams of investigators to see whether a similar reduction in cleft lip and/or cleft palate might be effected. Preliminary results suggest that there is indeed some reduction in the rate of clefting (123).

### Epidemiology

**Cleft lip with or without cleft palate.** Cleft lip with or without cleft palate [CL(P)] occurs in about 1/1000 white births (range 0.7–1.3), although data differ somewhat from study to study depending on source of information (birth records, birth certificates, pregnancy follow-up data, surgical records, hospital records, multiple sources), whether the data were obtained retrospectively or prospectively, and whether stillbirths or syndromal disorders were included (1,6,13,23,27,31,38,64b, 67,69,75,91,110,113,123,127,131). The problems of data gathering were elegantly discussed by Vanderas (127) and Sayetta et al (104a). Greene (49) has shown that birth certificate data are extremely inaccurate.

There is marked racial predilection. The frequency of CL(P) in American Indians appears to be the highest of any group in the world (over 3.6/1000 live births) (38,58,69,88). It is also high in Japanese (about 2.1/1000 births) (23,38,63,69,86,87,119) and Chinese (about 1.7/1000 births) (37,55,69) and, in contrast, it is lower among blacks (approximately 0.3/1000 births) (2,26,56,127) and Maori (22). Data from African blacks are cited by Cervenka (20).

In general, the more severe the defect, the greater the proportion of males affected [i.e. CLP (2M:1F) is greater than CL (1.5M:1F) and bilateral CL is greater than unilateral CL (45,91,123,124)]. However, not all races have similar sex predilections; for example, most Japanese studies have shown that females with CL outnumber males. However, males with CLP outnumbered females, as in whites.

Isolated CL may be unilateral (80%) or bilateral (20%) (131) (Figs. 20–4A,B, 20–5A–C, and 20–6). However, when only cleft lip is involved, bilateral cases amount to about 10%; when cleft lip is combined with cleft palate, about 25% are bilateral (31,75,76,124). When unilateral, the cleft is more common on the left side (about 70%), although no more extensive (124). About 85% of cases of bilateral CL and 70% of unilateral CL are associated with cleft palate. Cleft lip is not always complete (i.e., extending into the nostril). In about 10%, the cleft is associated with skin bridges. Minimal involvement is manifested by linear lip indentation(s) (53,65) or, very rarely, lateral fistulas of the upper lip (16,54,118b) (Fig. 20–7).

Fig. 20–4. *Facial clefts and associated anomalies.* (A,B) Unilateral cleft upper lip showing various degrees of completeness. [From *W Hoppe,* Arch Kinderheilkd **171** (Suppl **52**):1, 1965.)

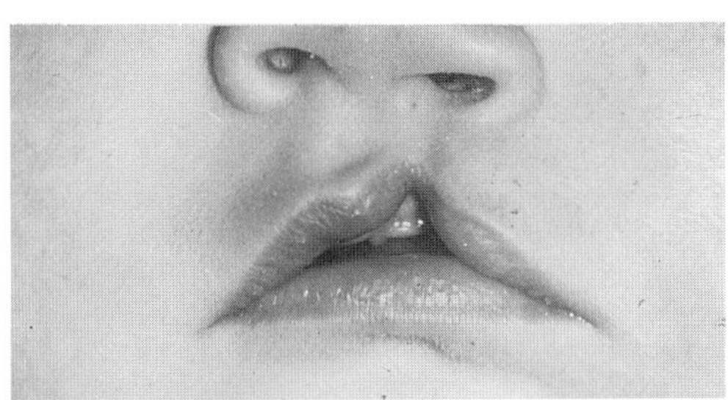

A

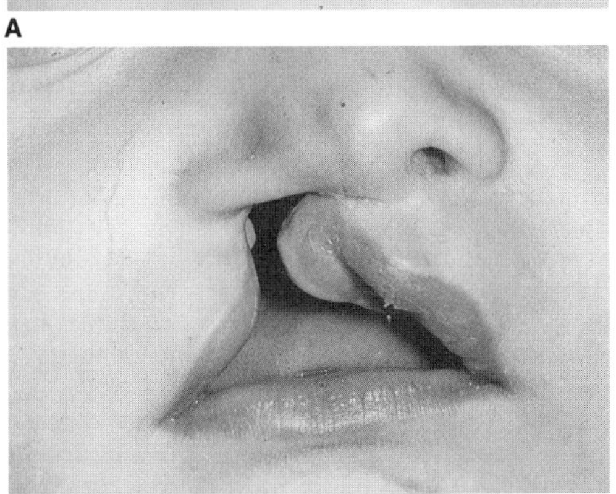

B

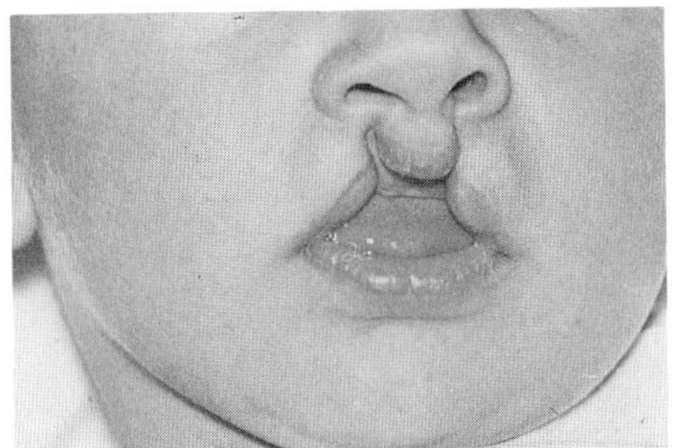

A

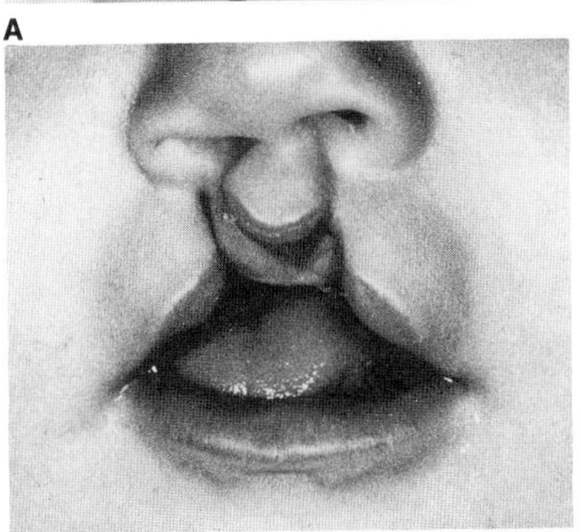

B

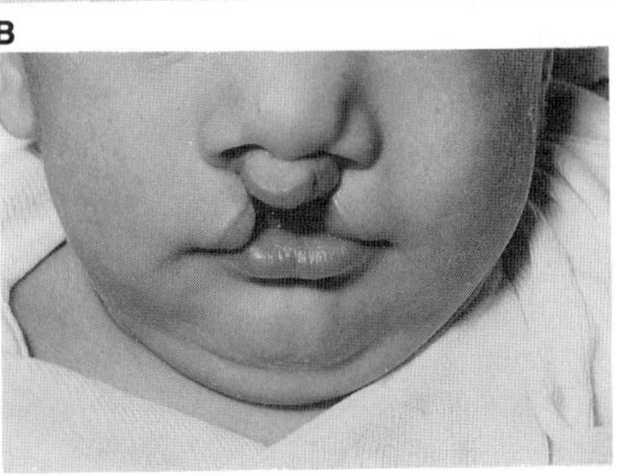

C

Fig. 20–5. *Facial clefts and associated anomalies.* (A,C). Bilateral cleft lip showing varying degrees of completeness. (From *T Skoog,* Plast Reconstr Surg **35**:34, 1965.)

**Isolated cleft palate.** Isolated CP appears to be an entity separate from CL(P) (Fig. 20–8). Sibs of patients with CL(P) have an increased frequency of the same anomaly but not of isolated CP and vice versa (45).

The frequency of CP among whites and blacks would appear to be approximately 0.4/1000 births and is more common in females (123).

Fig. 20–6. *Facial clefts and associated anomalies.* Bilateral cleft with "floating" primary palate. (Courtesy of *W Hoppe,* Lübeck, Germany.)

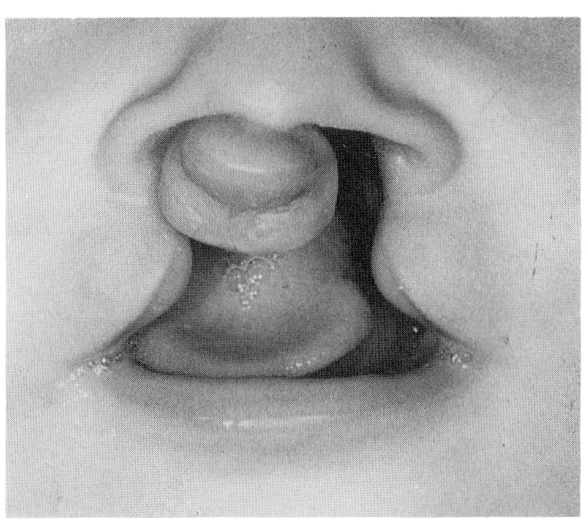

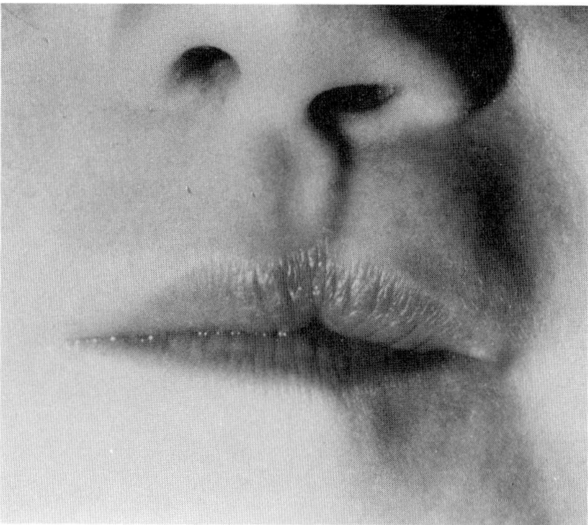

Fig. 20–7. *Facial clefts and associated anomalies.* Minimal cleft lip showing groove extending from vermilion to nostrils representing defect in orbicularis oris. (From *FR Heckler et al,* Cleft Palate J **16**:240, 1979.)

CP among the Maori of New Zealand is approximately 1.9/1000 births (22). When cases of CP are broken down according to extent, a 2:1 female predilection for complete clefts of the hard and soft palate is clearly indicated, but the ratio approaches 1:1 for clefts of the soft palate only (80,86a). This is consistent with the observation that the soft palate forms by merging, after closure of the hard palate.

The incidence of cleft uvula (1/80 white individuals) is much higher than that for cleft palate (1/2500 births) (27,77,108,125). By endoscopy, the frequency may be as high as 2–4% (112) or even higher. Like cleft of the soft palate, cleft uvula approaches a 1:1 sex ratio (24,67,80). The frequency of cleft uvula among various American Indian and Inuit groups is high, ranging from 1/9 to 1/14 individuals (21,52,57,67,107) and in orientals in 1/10 to 1/25 (67). In blacks, it is comparatively rare (1/250) (67,97,105,107).

Submucous palatal cleft refers to the condition in which there is imperfect muscle union across the velum but an intact mucosal surface (93) (Fig. 20–9A). An incidence of approximately 1/1200 to 1/2000 births has been reported (5,117,130). Over 50% occur in males, an unusual finding since cleft palate has a distinct female predilection. The palate is short in about 60% with poor mobility demonstrated in 20%. This makes velopharyngeal closure incompetent, resulting in hypernasal speech. There is frequently a notch in the bone at the posterior edge of the hard palate and bifid uvula is found in most but not all patients (66). Conversely, bifid uvula is usually accompanied by submucous cleft palate (112) and the former has been considered a "marker" for that condition. Rarely, the hard palate is entirely perforated (39,64,82) (Fig. 20–9B). Kono et al (64) estimated that about 13% of patients with clefts of the primary palate have submucous cleft palate.

Fig. 20–8. *Facial clefts and associated anomalies.* Complete isolated cleft palate. [From *W Hoppe,* Arch Kinderheilkd **171** (Suppl **52**):1, 1965.]

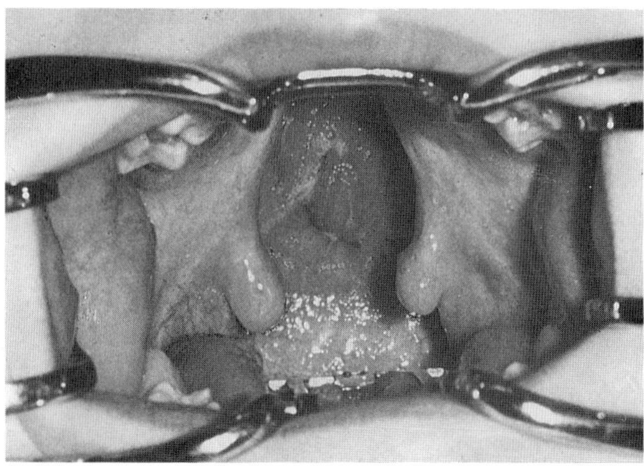

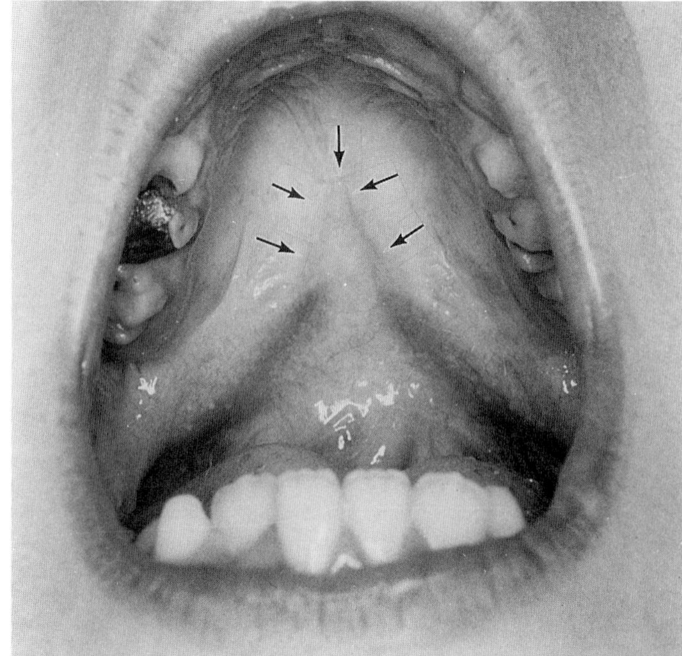

A

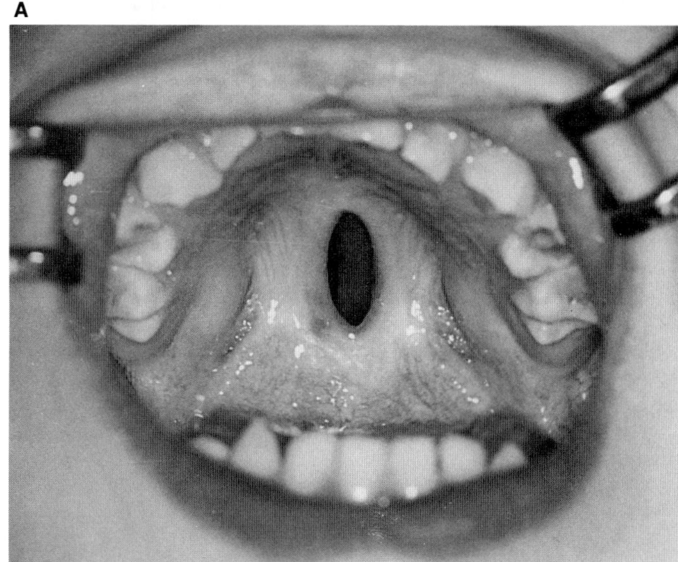

B

Fig. 20–9 *Facial clefts and associated anomalies.* (A) Submucous palatal cleft with arrows showing V-shaped groove at posterior edge of hard palate. (Courtesy of *HW Smith,* New Haven, Connecticut.) (B) Perforated soft palate representing extreme form of submucous cleft palate. (From *M Fára,* Plast Reconstr Surg **48**:44, 1971.)

**Skull, facial bones, and dentition.** Postnatal studies of patients with facial clefts have shown that growth of the face proceeds in harmony with the deviation initially induced by the cleft. However, overall head size is slightly smaller and the length of the cranial base slightly shorter than normal in individuals with CP. A complete review of the subject is that of Ross and Johnston (100).

The anomalous growth associated with CL(P) is generally considered to be limited to the middle part of the face. Bony interorbital distance is slightly increased. About one-third of those with CP exhibit an anteroposterior deficiency of the middle third of the face, and over 50% with bilateral clefts manifest this alteration (15a,32,53a).

Nasopharyngeal widths are greater than normal in patients with clefts. The mandible is shorter in length with a shorter ramus, thus reducing posterior facial height. The mandibular plane angle is increased and the gonial angle more obtuse.

The teeth in the cleft area, most commonly the maxillary lateral incisors, are commonly missing or supernumerary. In general, the more severe the cleft, the more teeth are missing or eruption delayed. In patients with clefts of the lip and alveolus, a median fissural tooth is found in almost 50% in the primary dentition and in about 25% in the secondary dentition (94). Corresponding figures for a distal fissural tooth are 75% and 45%, in the primary and secondary dentitions, respectively. Agenesis of teeth in the cleft area is found in 15% in the primary and 45% in the secondary dentition, and both fissural teeth are present in about 35% in the primary and 15% in the secondary dentition. Hypodontia is not uncommon outside the area of clefting in children with CL(P). Over 40% have missing teeth, more often in the upper jaw. Even in isolated CP, which does not involve a tooth-bearing area, hypodontia is present in about 35%. The teeth most commonly missing are premolars and maxillary lateral incisors. The size of the permanent teeth tends to be smaller and there is often metric asymmetry (94,95a,131a).

**Genetics.** During the early decades of this century, it became evident that recurrence risks for clefting did not correspond to any simple mendelian pattern of inheritance. This was bolstered by twin studies that indicated the relative roles of genetic and nongenetic influences in cleft production. We have tabulated twin studies in 1988 (Table 20–1). In twins with CL(P), concordance is far greater for monozygotic twins (36%) than for dizygotic twins (4.7%). In twins with isolated CP, the difference in concordance is not quite as great between the groups (MZ—22%; DZ—4.6%), suggesting a stronger genetic basis for CL(P) than for isolated CP. Twin studies are, at best, of limited value; the concordance of 4.6% for dizygotic twins (greater than that for sibs) implies that maternal factors are operative. Microforms such as submucous palatal cleft could alter these data markedly. An unusual report is that of clefts in three of quintuplets (133).

In the 1970–1980s, enthusiasm waxed regarding multifactorial inheritance of CL(P) and CP. This view was supported by Wolff et al (132), Carter (17), Mendell et al (78), and Tolarová (121,123,126). Fraser (41–46), however, while generally supporting the multifactorial-threshold inheritance of CL(P), expressed skepticism that CP obeyed the rules and suggested that environmental factors played a major role. Bixler and co-workers (10), in a reexamination of the Danish data of Fogh-Andersen, noted that there was little support for the sex-modified threshold polygenic model for CL(P), but also found little evidence for major gene effects. They agreed that CP did not fit the polygenic threshold concept and suggested genetic heterogeneity (110). For CP, they proposed three categories: syndromal, autosomal dominant, and environmental etiologies. Czeizel (31), in his analysis of Hungarian patients, found CL(P) in agreement with polygenic multifactorial inheritance but did not find similar support for CP. Melnick et al (75), Shields et al (109), and Marazita et al (72), reassessing the Danish data, indicated that CL(P) and isolated CP data were not compatible with the multifactorial threshold model, and posed an alternate hypothesis that CL(P) and CP result from the effects of a single major gene (most likely recessive) with reduced penetrance. The same was true for studies of CL(P) families in England (73) and in Shanghai, China (77). Chung et al (27), employing complex segregation analysis, could not distinguish between major gene versus multifactorial inheritance on Hawaiian data, but later Chung et al (29) suggested that a major gene with low penetrance fits the Danish material but that a multifactorial model best fit the Japanese data. Eiberg et al (36) proposed major gene action combined with multifactorial inheritance and that linkage studies indicated that the major gene locus was on the distal part of the long arm of chromosome 6. An earlier linkage study on van der Woude syndrome (115) also suggested chromosome 6. Until corroborative data are presented, however, we remain skeptical. Ardinger et al (3a) reported association of cleft lip and palate with transforming growth factor alpha which is at chromosome 2p13.

Recently, Kurnit et al (64a) revolutionized our thinking about the role that chance may play in the occurrence of malformations. Their model may be applicable to malformations that are presently thought to have "low risk multifactorial recurrence." They suggested that variability may be inherent in the role that chance itself plays in the process of development. Based on computer simulations, a stochastic single gene model generated a continuous liability curve resembling that obtained from a multifactorial threshold model. Outcomes were quite variable despite using identical genotypes and environments. Thus, segregation of a given malformation may be explained by a single defective gene that predisposes to, but does not necessarily result in, the malformation. Their model was successfully applied to VSD. Whether the model can be used for other malformations, such as clefting, remains to be determined.

It certainly appears that clefting is heterogeneous, the variation in liability probably being determined by a number of major genes, minor genes, environmental insults, and a developmental threshold. Within recent years, recurrence risk calculations have purposely excluded syndromal clefting, and lethal disorders such as trisomy 13, from the data. Recurrence risk is thereby related to both the genome and environmental exposure. We would agree that both CL(P) and CP do not fit the original multifactorial model very well. Further, CP is probably more heterogeneous than CL(P).

Since one cannot at this time distinguish clearly among the mixed model, a major gene with reduced penetrance, and an admixture of sporadic cases in most instances, it seems foolish to argue heatedly about one *or* the other (44). Until more evidence has been presented, we can see little error being made using empiric risk data or, in fact, even mathematical model data such as that of Spence et al (114) or Bonaiti-Pellié (14). While one may fault the models, the figures derived are very close to the empiric risks found.

It should be pointed out that a few families appear to have X-linked inheritance of isolated cleft palate (9,11,51,68,84,98,102). The gene has been localized on the long arm at Xq13–21 (56a,84). One family apparently had autosomal dominant inheritance of cleft palate (59) and a few cleft lip-palate (119a). Autosomal dominant inheritance of velopharyngeal incompetence has been observed in a large kindred by Andres et al (3), but we wonder whether that family had *velocardiofacial syndrome*.

Table 20–1. Concordance in facial clefts—twin studies (1922–1988)[a]

| | Monozygotic twins | | Dizygotic twins | |
|---|---|---|---|---|
| | CL(P), $n=84$ | CP, $n=27$ | CL(P), $n=191$ | CP, $n=44$ |
| Concordance ($n$) | 34 | 6 | 9 | 2 |
| % | 36.2 | 22.2 | 4.7 | 4.6 |

[a]Based on Gorlin et al (47) together with cases of Blake and Wreakes (12), Schweckendiek (106), Shields et al (109), Czeizel (31), Davies and Thompson (32a), Tolarová (personal communication, 1988), and other authors.

**Recurrence risks.** Data regarding recurrence risks of CL(P) and CP for an affected parent or in case of normal parents who have an affected child have been ascertained in numerous studies. In general, the data do not differ drastically from the Falconer estimation, that is, the recurrence risk being the square root of the frequency of the disorder in the population. In the case of CL(P), $\sqrt{1/900} = 1/30 = 3.3\%$ and for isolated CP, $\sqrt{1/2500} = 1/50 = 2\%$. Of course, many factors can modify the data: similarly affected sib(s) or parent, racial group, sex of proband, severity of cleft, etc. These have been dealt with by an inordinate number of investigators (17,18,31,41–46,91,96, 107,110,114,121,124,126,132). Tenconi et al (120) have also devised a theoretical recurrence risk.

For CL(P) and CP, if the proband has no other affected first or second degree relatives, the empiric risk is 3–5%. However, for CL(P)

Table 20–2. Comparative empiric risks for cleft lip–palate and cleft palate[a]

| | CL±P | | CP | |
|---|---|---|---|---|
| | % Proband | | % Proband | |
| Relative | Male | Female | Male | Female |
| Brother | 6.7 | 6.8 | 1.8 | 2.8 |
| Sister | 2.8 | 4.4 | 3.7 | 1.7 |
| Son | 6.7 | 2.4 | 11.5 | 6.0 |
| Daughter | 4.0 | 8.7 | 5.6 | 17.2 |

[a]Modified from M Melnick et al, *Clinical Dysmorphology of Oral-Facial Structures*, John Wright, Boston, 1982.

if the proband has other affected first degree relatives, the risk to sibs or offspring is 10–20% (121,132, Melnick unpublished data, 1987). In the rare case of both parents being affected, empiric risk data range from 25 to 50% (121). For these and other situations, see Tolarová (unpublished data, 1987), Melnick et al (75), and Wolff (132) (Tables 20–2 and 20–3). Regarding sex and severity, Tolarová (124) calculated the following risk figures (%) for sibs of those with CL±P: unilateral male—2.9±0.5; unilateral female—4.6±0.9; bilateral male—6.4±1.7; bilateral female—6.8±2.4. Similar data may be found in Woolf (132). Naturally, the usual caveat applies; *one must exclude cleft syndromes*. The number of cleft syndromes has grown from about 75 in our second edition to over 250 in this edition of our text. They have been reviewed by a number of authors (30,47,70,92).

**Associated anomalies.** Many investigators have tabulated the frequency and types of anomalies that accompany facial clefts (31,111). For a thorough critical review of the earlier data, see Shprintzen et al (111). Data secured at birth differ from those on patients who have undergone surgery, since those with more serious defects often succumb prior to surgery. Recent studies have suggested that as high as 44–64% of cleft patients have associated anomalies (99,111). When the data are broken down according to subtype, there is general agreement that isolated CP (13–50%) is far more often associated with other congenital defects than either isolated CL (7–13%) or CLP (2–11%) (41).

More malformations have been found in infants with bilateral CL(P) than with unilateral CL. The more malformations a child has, the lighter the birth weight. Clefts are more frequently found in spontaneously aborted and in stillborn infants, reflecting, in part, chromosomal aneuploidy. Associated anomalies are more frequently noted in patients without a family history of clefts than in patients with affected relatives. Conversely, the recurrence risk is lower in sibs of patients with, than without, associated malformations. Congenital velopharyngeal incompetence has been noted to be frequently associated with cervical spine anomalies (90). Congenital heart defects have been found in 3–7% of cleft cases (111,131). Strunz et al (118) noted an increased rate of urinary tract anomalies, especially in association with CP. Biewald et al (8) found about 40% of patients with facial clefts to have minor anomalies of the urinary tract. Azmy et al (4), examining 114 patients with esophageal atresia, found 5% with CL(P). Sandham (103), studying cervical vertebral anomalies in cleft lip and palate patients, found 13% to have either posterior arch deficiency or fusion anomalies versus 0.8% in the normal population. Richman et al (97a) described olfaction defects in about 50% of males with cleft palate.

Table 20–3. Recurrence risk for clefts in sibs of propositi

| Affected parent | CL(P) % | CP % |
|---|---|---|
| Mother affected | | |
| Affected sibs | | |
| 0 | 2.7 | 2.3 |
| 1 | 9.9 | 11.2 |
| 2 | 18.3 | 21.1 |
| Father affected | | |
| Affected sibs | | |
| 0 | 2.3 | 5.0 |
| 1 | 9.3 | 14.4 |
| 2 | 17.6 | 23.9 |
| Both parents affected | | |
| Affected sibs | | |
| 0 | 24.0 | 45.0 |
| 1 | 31.7 | 51.6 |
| 2 | 37.6 | 54.5 |

*Source:* M Tolarová, Ph.D. Thesis, Charles University, Prague, Czechoslovakia, 1984.

Dronemaraju and Bixler (34,35) noted a greater fetal mortality in sibs of probands with CL(P) than in those with isolated CL but this has been denied by Bear (7) and Tolarová (122).

Rudman et al (101) found that children with isolated cleft lip–palate have short stature four times and growth hormone deficiency 40 times more often than those without cleft lip–palate. Short stature was also found by Shprintzen et al (111).

**Laboratory aids.** Ultrasonography has been used to detect clefting prenatally (25,104).

### References (Facial clefts and associated anomalies)


1. Åbyholm FE: Cleft lip and palate in Norway. *Scand J Plast Reconstr Surg* **12**:29–34, 1978.

2. Altemus LA: The incidence of cleft lip and cleft palate among North American negroes. *Cleft Palate J* **3**:357–361, 1966.

3. Andres R et al: Dominant inheritance of velopharyngeal incompetence. *Clin Genet* **19**:443–447, 1981.

3a. Ardinger HH et al: Association of genetic variation of the transforming growth factor alpha gene with cleft lip and palate. *Am J Hum Genet* **45**:348–354, 1989.

4. Azmy AF et al: Orofacial clefts and oesophageal atresia. *Arch Dis Childh* **58**:639–641, 1983.

5. Bagatin M: Submucous cleft palate. *J Maxillofac Surg* **13**:37–38, 1985.

6. Bear JC: A genetic study of facial clefting in Northern England. *Clin Genet* **9**:277–284, 1976.

7. Bear JC: Liability to cleft lip and palate. Interpreting the human data. *Issues Rev Teratology* **4**:163–203, 1988.

8. Biewald W et al: Die Häufigkeit von Harnwegsanomalien bei Lippen-Kiefer-Gaumenspalten. *Zbl Chir* **108**:790–792, 1983.

9. Bixler D: X-linked cleft palate. *Am J Med Genet* **28**:503–506, 1987.

10. Bixler D et al: Incidence of cleft lip and palate in the offspring of cleft parents. *Clin Genet* **2**:155–159, 1971.

11. Björnsson Å et al: X-linked cleft palate and ankyloglossia in an Icelandic family. *Cleft Palate J* **26**:3–8, 1989.

12. Blake GB, Wreakes G: Clefts of the lip and palate in twins. *Br J Plast Surg* **25**:155–164, 1972.

13. Bonaiti C et al: An epidemiological and genetic study of facial clefting in France. I. Epidemiology and frequency in relatives. *J Med Genet* **19**:8–15, 1982.

14. Bonaiti-Pellié C, Smith C: Risk tables for genetic counselling in some common congenital malformations. *J Med Genet* **11**:374–377, 1974.

15. Boo-Chai K: An ancient Chinese text on cleft lip. *Plast Reconstr Surg* **38**:89–91, 1966.

15a. Burdi AR et al: Prenatal pattern emergency in early human facial development. *Cleft Palate J* **25**:8–15, 1988.

16. Calderon S, Garlick JA: Surgical excision of a congenital lateral fistula of the upper lip. An intraoral approach. *J Cranio-Maxillofac Surg* **16**:46–48, 1988.

17. Carter CO: Genetics of common single malformations. *Br Med Bull* **32**:21–26, 1976.

18. Carter CO et al: A three generation family study of cleft lip with or without cleft palate. *J Med Genet* **19**:246–261, 1982.

19. Carter CO et al: A family study of isolated cleft palate. *J Med Genet* **19**:329–331, 1982.

20. Cervenka J: African mask with cleft lip and palate. *Cleft Palate J* **21**:38–40, 1984.

21. Cervenka J, Shapiro B: Cleft uvula in Chippewa Indians: Prevalence and genetics. *Hum Biol* **42**:47–52, 1970.

22. Chapman CJ: Ethnic differences in the incidence of cleft lip and/or cleft palate in Auckland, 1960–1976. *N Z Med J* **96**:327–329, 1983.

23. Ching GH, Chung CS: A genetic study of cleft lip and palate in Hawaii. I: Interracial crosses. *Am J Hum Genet* **26**:162–176, 1974.

24. Chosack A, Eidelman E: Cleft uvula: Prevalence and genetics, *Cleft Palate J* **15**:63–67, 1978.

25. Christ JE, Meininger MG: Ultrasound diagnosis of cleft lip and cleft palate before birth. *Plast Reconstr Surg* **68**:854–859, 1981.

26. Chung CS, Myrianthopoulos NC: Factors affecting risks of congenital malformations. I. Epidemiologic analysis. *Birth Defects* **11**(10):1–38, 1975.

27. Chung CS et al: A genetic study of cleft lip and palate in Hawaii. II. Complex segregation analysis and genetic risks. *Am J Hum Genet* **26**:177–188, 1974.

28. Chung CS et al: Population and family studies of cleft lip and palate, in *Etiology of Cleft Lip and Cleft Palate,* Melnick M, Bixler D, Shields E (eds), Alan R. Liss, New York, 1980, pp 325–352.

29. Chung CS et al: Segregation analysis of cleft lip with or without cleft palate: A comparison of Danish and Japanese data. *Am J Hum Genet* **39**:603–611, 1986.

30. Cohen MM Jr: Syndromes with cleft lip and cleft palate. *Cleft Palate J* **15**:306–328, 1978.

31. Czeizel A: Studies of cleft lip and cleft palate in east European populations, in *Etiology of Cleft Lip and Cleft Palate,* Melnick M, Bixler D, Shields E (eds), Alan R. Liss, New York, 1980, pp 249–291.

32. Dahl E: Craniofacial morphology in congenital clefts of the lip and palate. *Acta Odontol Scand* **22**: Suppl **57**:1–167, 1970.

32a. Davies TM, Thompson RPJ: Cleft of the lip and/or palate in monozygotic twins. *Eur J Orthodont* **100**:52–61, 1988.

33. Dronemaraju KR: *Cleft Lip and Palate: Aspects of Reproductive Biology.* Charles C. Thomas, Springfield, Illinois, 1986.

34. Dronemaraju KR, Bixler D: Fetal mortality and cleft lip with or without cleft palate. *Clin Genet* **23**:38–40, 1983.

35. Dronemaraju KR, Bixler D: Fetal mortality in oral cleft families. Data from Indiana and Montreal. *Clin Genet* **23**:350–353, 1983.

36. Eiberg H et al: Suggestion of linkage of a major locus for nonsyndromic orofacial cleft with F13A and tentative assignment to chromosome 6. *Clin Genet* **32**:129–132, 1987.

37. Emanuel I et al: The incidence of congenital malformations in a Chinese population: The Taipei collaborative study. *Teratology* **5**:159–170, 1972.

38. Emanuel I et al: The further epidemiological differentiation of cleft lip and palate: A population study of clefts in King County, Washington, 1956–1965. *Teratology* **7**:271–282, 1973.

39. Fára M: Congenital defects in the hard palate. *Plast Reconstr Surg* **46**:44–47, 1971.

39a. Fenner W, van der Leyen UE: Über die kongenitale Oberlippenfistel. *Dtsch Zahnärztl Z* **24**:963–966, 1969.

40. Ferguson MWJ: The mechanism of palatal shelf elevation and the pathogenesis of cleft palate. *Virchows Arch (Pathol Anat)* **375**:97–113, 1977.

41. Fraser FC: The genetics of cleft lip and cleft palate. *Am J Hum Genet* **22**:336–352, 1970.

42. Fraser FC: Updating the genetics of cleft lip and palate. *Birth Defects* **10**(8):107–111, 1974.

43. Fraser FC: The multifactorial/threshold concept—uses and misuses. *Teratology* **14**:267–280, 1976.

44. Fraser FC: Evolution of a palatable multifactorial threshold model. *Am J Hum Genet* **32**:796–813, 1980.

45. Fraser FC: The genetics of cleft lip and palate: Yet another look, in *Current Research Trends in Prenatal Craniofacial Development,* Pratt RM, Christiansen RL (eds), Elsevier North Holland, Amsterdam, 1980.

46. Fraser FC: The genetics of common familial disorders—major genes or multifactorial? *Can J Genet Cytol* **23**:1–8, 1981.

47. Gorlin RJ et al: Facial clefting and its syndromes. *Birth Defects* **7**(7):3–49, 1971.

48. Grabb WC et al (eds: *Cleft Lip and Palate,* Little, Brown & Co., Boston, 1971.

49. Greene HG et al: Accuracy of birth certificate data for detecting focal cleft defects in Arkansas children. *Cleft Palate J* **16**:167–170, 1979.

50. Greene RM, Pratt RM: Developmental aspects of secondary palate formation. *J Embryol Exp Morphol* **36**:225–245, 1976.

51. Hall BD: A further X-linked isolated nonsyndromic cleft palate family with a nonexpressing obligate affected male. *Am J Med Genet* **26**:239–240, 1987.

52. Heathcote GM: The prevalence of cleft uvula in an Inuit population. *Am J Phys Anthropol* **41**:433–437, 1973.

53. Heckler FR et al: The minimal cleft lip revisited. *Cleft Palate J.* **16**:240–247, 1979.

53a. Horswell BB, Levant BA: Craniofacial growth in unilateral cleft lip and palate: Skeletal growth from eight to eighteen years. *Cleft Palate J* **25**:114–121, 1988.

54. Hosokawa K et al: A congenital lateral sinus of the upper lip. *Ann Plast Surg* **11**:69–70, 1982.

55. Hu DN et al: Genetics of cleft lip and cleft palate in China. *Am J Hum Genet* **34**:999–1002, 1982.

56. Iregbulem LM: The incidence of cleft lip and palate in Nigeria. *Cleft Palate J* **19**:201–205, 1982.

56a. Ivens A et al: X-linked cleft palate: The gene is localized between polymorphic DNA markers DXYS12 and DXS17. *Hum Genet* **78**:356–358, 1988.

57. Jaffe BF, DeBlanc GB: Cleft palate, cleft lip and cleft uvula in Navajo Indians. *Cleft Palate J* **7**:301–305, 1972.

58. Jarvis A, Gorlin RJ: Minor orofacial abnormalities in an Eskimo population. *Oral Surg* **33**:417–427, 1972.

59. Jenkins M, Stady C: Dominant inheritance of cleft of the soft palate. *Hum Genet* **53**:341–342, 1980.

60. Johnston M: The neural crest in abnormalities of the face and brain. *Birth Defects* **11**:(7):1–18, 1975.

61. Johnston MC et al: The embryology of cleft lip and palate. *Clin Plast Surg* **2**:195–203, 1975.

62. Johnston MC, Sulik KK: Some abnormal patterns of development in the craniofacial region. *Birth Defects* **15**(8):23–42, 1979.

62a. Johnston MC, Sulik KK: Embryology of the head and neck, in: Serafin D, Georiade NG (eds): *Pediatric Plastic Surgery,* C.V. Mosby Co., St. Louis, 1984, pp 184–215.

63. Koguchi H: Population data on cleft lip and cleft palate in the Japanese, in *Etiology of Cleft Lip and Cleft Palate,* Melnick M, Bixler D, Shields E (eds), Alan R. Liss, New York, 1980, pp 297–323.

64. Kono D et al: The association of submucous cleft palate. *Cleft Palate J* **18**:207–209, 1981.

64a. Kurnit DM et al: Genetics, chance and morphogenesis. *Am J Hum Genet* **41**:979–995, 1987.

64b. Leck I: The geographical distribution of neural tube defects and oral clefts. *Br Med Bull* **40**:390–395, 1984.

65. Lehman JA, Artz JS: The minimal cleft lip. *Plast Reconstr Surg* **58**:306–309, 1976.

66. Lewin ML et al: Velopharyngeal insufficiency due to hypoplasia of the musculus uvulae and occult submucous cleft palate. *Plast Reconstr Surg* **65**:585–591, 1980.

67. Lindemann G et al: Prevalence of cleft uvula among 2,732 Danes. *Cleft Palate J* **14**:266–269, 1977.

68. Lowry RB: Sex-linked cleft palate in a British Columbia Indian family. *Pediatrics* **46**:123–128, 1970.

69. Lowry RB, Trimble BK: Incidence rates for cleft lip and palate in British Columbia 1952–1971 for North American Indian, Japanese, Chinese and total populations: Secular trends over twenty years. *Teratology* **16**:277–283, 1977.

69a. Luke DA: Development of the secondary palate in man. *Acta Anat* **94**:596–608, 1976.

70. Lynch HT, Kimberling WJ: Genetic counseling in cleft lip and cleft palate. *Plast Reconstr Surg* **68**:800–815, 1981.

71. Mahler D, Karev A: Lateral congenital sinus of the upper lip. *Br J Plast Surg* **28**:203–204, 1975.

72. Marazita ML et al: Genetic analysis of cleft lip with or without cleft palate in Danish kindreds. *Am J Med Genet* **19**:9–18, 1984.

73. Marazita ML et al: Cleft lip with or without cleft palate: Reanalysis of a three generation family study from England. *Genet Epidem* **3**:335–342, 1986.

74. Melnick M et al: Facial clefting: An alternative biologic explanation for its complex etiology. *Birth Defects* **13**(3A):93–112, 1977.

75. Melnick M et al: Cleft lip ± cleft palate: An overview of the literature and an analysis of Danish cases born between 1941 and 1968. *Am J Med Genet* **6**:83–97, 1980.

76. Melnick M, Bixler D, Shields E (eds): *Etiology of Cleft Lip and Cleft Palate.* Alan R. Liss, New York, 1980.

77. Melnick M et al: Genetic analysis of cleft lip with or without cleft palate in Chinese kindreds. *Am J Med Genet Suppl* **2**:193–190, 1986.

78. Mendell NR et al: Multifactorial/threshold models and their applications to cleft lip and cleft palate, in *Etiology of Cleft Lip and Cleft Palate,* Melnick M, Bixler D, Shields E (eds), Alan R. Liss, New York, 1980, pp 387–406.

79. Meskin LH et al: Abnormal morphology of the soft palate. I. The prevalence of cleft uvula. *Cleft Palate J* **1**:324–346, 1964.

80. Meskin LH et al: An epidemiologic investigation of factors related to the extent of facial clefts. *Cleft Palate J* **5**:23–29, 1968.

81. Millard DR Jr: *Cleft Craft: The Evolution of Its Surgery*, 3 vol. Little, Brown, & Co., Boston, 1977.

82. Mitts TF et al: Cleft of the hard palate with soft palate integrity. *Cleft Palate J* **18**:204–206, 1981.

83. Monie IW, Cacciatore A: The development of the philtrum. *Plast Reconstr Surg* **30**:313–321, 1962.

84. Moore GE et al: Linkage of an X-chromosome cleft palate gene. *Nature* (London) **326**:91–92, 1987.

85. Moore KL: *The Developing Human*, 4th ed. W.B. Saunders, Philadelphia, 1988.

86. Natsume N, Kawai T: Incidence of cleft lip and cleft palate in 39,696 Japanese babies born during 1983. *Int J Oral Surg* **15**:565–568, 1986.

86a. Natsume N et al: Effect of sexual differences on the development of cleft palate in the human. *Plast Reconstr Surg* **84**:854–855, 1989.

87. Natsume N et al: The prevalence of cleft lip and palate in the Japanese: Their birth prevalence in 40,304 infants born during 1982. *Oral Surg Oral Med Oral Pathol* **63**:421–423, 1987 and *Br J Oral Maxillofac Surg* **26**:232–236, 1988.

87a. Newell DR, Edwards JRG: The effect of vitamin A on fusion of mouse palates. I. Retinyl palmitate and retinoic acid in vivo. *Teratology* **23**:115–124, 1985.

88. Niswander JD et al: Congenital malformations in the American Indians. *Soc Biol* **22**:203–209, 1975.

89. Orticochea M: The harelipped king: A pre-Columbian ceramic statue over 2000 years old. *Br J Plast Surg* **36**:392–394, 1983.

90. Osborne GS et al: Upper cervical spine anomalies and osseous nasopharyngeal depth. *J Speech Disord* **14**:14–22, 1971.

91. Owens JR et al: Epidemiology of facial clefting. *Arch Dis Childh* **60**:521–524, 1985.

92. Pashayan HM: What else to look for in a child born with a cleft of the lip and/or palate. *Cleft Palate J* **20**:54–82, 1983.

93. Poswillo D: The pathogenesis of submucous cleft palate. *Scand J Plast Reconstr Surg* **8**:34–41, 1979.

94. Ranta R: A review of tooth formation in children with cleft lip/palate. *Am J Orthod Dentofac Orthop* **90**:11–18, 1986.

95. Ranta R, Rintala A: Separate clefts of the lip and the palate. *Scand J Plast Reconstr Surg* **18**:233–235, 1984.

95a. Ranta R, Tulensalo T: Symmetry and combinations of hypodontia in non-cleft and cleft palate children. *Scand J Dent Res* **96**:1–8, 1988.

96. Reich T et al: The use of multiple thresholds and segregation analysis in analyzing the phenotypic heterogeneity of multifactorial traits. *Ann Hum Genet* **42**:371–390, 1979.

97. Richardson ER: Cleft uvula: Incidence in Negroes. *Cleft Palate J* **7**:669–672, 1970.

97a. Richman RA et al: Olfaction defects in boys with cleft palate. *Pediatrics* **82**:840–844, 1988.

97b. Rivron RP: Bifid uvula: Prevalence and association in children admitted for routine audiologic operations. *J Laryngol Otol* **103**:249–252, 1989.

98. Rollnick BR: Kaye CI: Mendelian inheritance of isolated nonsyndromic cleft palate. *Am J Med Genet* **24**:465–473, 1986.

99. Rollnick BR, Pruzansky S: Genetic services at a center for craniofacial anomalies. *Cleft Palate J* **20**:54–82, 1983.

100. Ross RB, Johnston MC: *Cleft Lip and Palate*. Williams & Wilkins, Baltimore, 1972.

101. Rudman D et al: Prevalence of growth hormone deficiency in children with cleft lip or palate. *J Pediatr* **93**:378–382, 1978.

102. Rushton AR: Sex-linked inheritance of cleft palate. *Hum Genet* **48**:179–181, 1979.

103. Sandham A: Cervical vertebral anomalies in cleft lip and palate. *Cleft Palate J* **23**:206–214, 1986.

104. Savoldelli G et al: Prenatal diagnosis of cleft lip and palate by ultrasound. *Prenat Diagn* **2**:313–317, 1982.

104a. Sayetta RB et al: Incidence and prevalence of cleft lip and palate: What we think we know. *Cleft Palate J* **26**:242–248, 1989.

105. Schaumann BF et al: Minor craniofacial anomalies among a Negro population. *Oral Surg* **29**:566–575, 1970.

106. Schweckendiek W: Further findings concerning the occurrence of clefts in twins. *J Maxillofac Surg* **1**:109–112, 1973.

107. Shapiro BL: The genetics of cleft lip and palate, in *Oral Facial Genetics*, Stewart RE, Prescott GH (eds), C.V. Mosby, St. Louis, 1976, pp 473–499.

108. Shapiro BL et al: Cleft uvula, a microform of facial clefts and its genetic basis. *Birth Defects* **7**(7):80–82, 1971.

109. Shields ED et al: Facial clefts in Danish twins. *Cleft Palate J* **16**:1–6, 1979.

110. Shields ED et al: Cleft palate: A genetic and epidemiologic investigation. *Clin Genet* **20**:13–24, 1981.

111. Shprintzen RJ et al: Anomalies associated with cleft lip, cleft palate or both. *Am J Med Genet* **20**:585–595, 1985.

112. Shprintzen RJ et al: Morphologic significance of bifid uvula. *Pediatrics* **75**:553–561, 1985.

113. Siegel B: A racial comparison of cleft patients in a clinic population. Associated anomalies and recurrence rates. *Cleft Palate J* **16**:193–197, 1979.

114. Spence MA et al: Estimation of polygenic recurrence risk for cleft lip and palate. *Hum Hered* **26**:327–336, 1976.

115. Spence MA et al: Genetic linkage studies with cleft lip and palate: Report of two family studies. *J Craniofac Genet Dev Biol* **3**:207–212, 1983.

116. Sperber GH: *Craniofacial Embryology* (ed. 2), John Wright, Bristol, England, 1981.

117. Stewart J et al: Submucous cleft palate. *Birth Defects* **7**(7):64–66, 1971.

118. Strunz V et al: Incidence of malformations of the urinary tract in association with clefts of lip, alveolus and palate. *J Maxillofac Surg* **12**:14–16, 1984.

118a. Sulik KK, Schoenwolf GC: Highlights of craniofacial morphogenesis in mammalian embryos as revealed by scanning electron microscopy. *Scanning Electron Microsc* **4**:1735–1752, 1985.

118b. Takenoshita Y: Congenital lateral fistulas of the upper lip. *J Craniomaxillofac Surg* **17**:186–188, 1989.

119. Tanaka T: A clinical and epidemiologic study of cleft lip and/or cleft palate. *Jpn J Hum Genet* **16**:278–308, 1972.

119a. Temple K et al: Dominantly inherited cleft lip and palate in two families. *J Med Genet* **26**:386–389, 1989.

120. Tenconi R et al: Theoretical recurrence risks for cleft lip derived from a population of consecutive newborns. *M Med Genet* **25**:243–246, 1988.

121. Tolarová M: Empirical recurrence risks figures for genetic counseling of clefts. *Acta Chir Plast (Praha)* **14**:234–235, 1972.

122. Tolarová M: Spontaneous abortions and facial clefts. *Clin Genet* **26**:77–79, 1984.

123. Tolarová M: Orofacial clefts in Czechoslovakia. Incidence, genetics and prevention of cleft lip and palate over a 19-year period. *Scand J Plast Reconstr Surg* **21**:19–25, 1987.

124. Tolarová M: A study of the incidence, sex ratio, laterality and clinical severity in 3,660 probands with facial clefts in Czechoslovakia. *Acta Chir Plast (Praha)* **29**:77–87, 1987.

125. Tolarová M et al: Distribution of signs considered as microforms of lip and/or palatal clefts in normal population. *Acta Chir Plast* **9**:1–14, 189–194, 1967.

126. Trasler DG: Pathogenesis of cleft lip and its relation to face shape in A/J and C57BL mice. *Teratology* **1**:33–49, 1968.

127. Vanderas AP: Incidence of cleft lip, cleft palate, and cleft lip and palate among races: A review. *Cleft Palate J* **24**:216–225, 1987.

128. Van Limborgh J et al: Cleft lip and palate due to deficiency of mesencephalic neural crest cells. *Cleft Palate J* **20**:251–259, 1983.

129. Verwoerd DCA, Van Oostrom CG: Cephalic neural crest and placodes. *Adv Anat Embryol Exper Morphol* **58**:1–75, 1978.

130. Weatherly-White RCA et al: Submucous cleft palate. *Plast Reconstr Surg* **49**:297–304, 1972.

131. Welch J, Hunter AGW: An epidemiological study of facial clefting in Manitoba. *J Med Genet* **17**:127–132, 1980.

131a. Werner SP, Harris EF: Odontometrics of the permanent teeth in cleft lip and palate: Systemic size reduction and amplified asymmetry. *Cleft Palate J* **26**:34–41, 1989.

132. Woolf CM: Congenital cleft lip: A genetic study of 496 propositi. *J Med Genet* **8**:65–83, 1971.

133. Zilberman Y et al: Clefts in quintuplets: A case report. *Cleft Palate J* **17**:58–61, 1980.

## Robin sequence

The well-recognized combination of micrognathia, cleft palate, and glossoptosis, known as Robin sequence, was named after Pierre Robin (33) whose first report appeared in 1923. However, the condition was described earlier by St. Hilaire (see 38) in 1822, Fairbairn (15) in 1846, and Shukowsky (41) in 1910. Historic development is thoroughly discussed by Dennison (11), Grimm et al (18), and Randall et al (31).

Birth prevalence estimates have varied from 1/2000 (28) to 1/30,000; Bush and Williams (4) suggested 1/8500. Definitions of Robin sequence are variable. Our own definition is based on the triad of micrognathia, cleft palate, and glossoptosis. However, others have been advocated. For example, one definition permits the presence of submucous cleft palate (32). Another consists of micrognathia ± cleft palate ± glossoptosis (27). Such a view allows isolated cleft palate or isolated micrognathia to qualify as Robin sequence as well as the binary combination of cleft palate and micrognathia. Williams et al (47) have cautioned against describing every infant with cleft palate and micrognathia as an example of Robin sequence, indicating that respiratory difficulty is an essential component of the condition. It is obvious that differences in definition result in differences in estimated birth prevalence and differences in the list of syndromes said to be associated with Robin sequence. An excellent discussion of this problem is that of Shprintzen (40a).

An animal model based on intrauterine mandibular constraint resulting in failure of the tongue to descend and resultant cleft palate has been described by a number of investigators (7,28). Explaining Robin sequence on a deformational basis has been supported by several authors (8,24,27,28). Others have considered absent or delayed lowering of the embryonic tongue as the primary pathogenetic event (2). Still others have suggested that growth disturbance affecting both the maxilla and mandible can result directly in cleft palate and micrognathia (14). Cohen in 1979 (9) and in 1982 (10) indicated that a unitary etiologic and pathogenetic hypothesis was unlikely, noting that clinical evidence was most consistent with etiologic and pathogenetic heterogeneity.

Table 20–4 lists some representative syndromes in which Robin sequence may be a feature with variable frequency. In Stickler syndrome (37,46), Robin sequence may occur in some patients. Since abnormalities of bones and joints are characteristic, the major pleiotropic effect appears to be on connective tissue. Thus, in this condition, Robin sequence may result from intrinsic mandibular hypoplasia and failure of connective tissue penetration across the palate.

Another condition that may have Robin sequence is dup(11q) syndrome. With the hypoplastic growth that accompanies most chromosomal syndromes, there may not be significant mandibular catch-up growth in infants with dup(11q) syndrome who survive. Therefore, to include such patients in a mandibular growth study of Robin sequence would be to study "fruit" since "oranges" are being confused with "apples."

Some instances of Robin sequence have been associated with oligohydramnios. It is thought that reduced amnionic fluid results in compression of the chin against the sternum, restricting mandibular growth and impacting the tongue between the palatal shelves. Because micrognathia is based on intrauterine molding, mandibular catch-up growth is expected after birth when intrauterine deforming forces are no longer acting. Poswillo (28) produced a phenocopy of Robin sequence in rats by puncturing the amniotic sac prior to palatal closure. Some experimental animals also had anomalies of the limbs, ranging from clubfoot to ring constrictions and intrauterine amputations. Such limb abnormalities have also been observed with Robin sequence in humans.

The association of amputations and/or limb reduction defects in some human cases suggests that Robin sequence may occur on a disruptive basis. For example, an amnionic tear may cause oligohydramnios that can result in severe compressive disruption, causing limb reduction defects, and bands, causing amputations. Such primary disruption with oligohydramnios can also cause secondary deformation such as mandibular constraint, leading to Robin sequence.

Finally, Robin sequence has been associated with congenital hypotonia. If neurogenic hypotonia occurred prior to complete closure of the palate, it is conceivable that Robin sequence might result from lack of mandibular exercise. Different etiologic and pathogenetic possibilities are summarized diagramatically in Fig. 20–10. Further evidence for etiologic and pathogenetic heterogeneity was provided by Carey et al (5).

Hanson and Smith (19) found Robin sequence in specific syndromes in 25% of their patients. Another 35% had multiple anomalies, but no specific syndrome was recognized. The remaining 40% were instances

Table 20–4. Some representative conditions associated with Robin sequence

| Condition | References |
|---|---|
| Monogenic | |
| Beckwith–Wiedemann syndrome | 8,9,10 |
| Campomelic syndrome | 8,9,10 |
| Carey neuromuscular syndrome | 5 |
| Catel–Manzke syndrome | 16,22,44,45 |
| Cerebrocostomandibular syndrome | 8,9,10 |
| Congenital myotonic dystrophy | 8,9,10 |
| Diastrophic dysplasia | 8,9,10 |
| Donlan syndrome | 10 |
| Mandibulofacial dysostosis | 40a |
| Miller–Dieker syndrome | 40a |
| Nager acrofacial dysostosis | 5 |
| Otopalatodigital syndrome II | 1 |
| Persistent left superior vena cava syndrome | 17 |
| Popliteal pterygium syndrome | 40a |
| Postaxial acrofacial dysostosis | 5 |
| Radiohumeral synostosis syndrome | 19 |
| Robin–oligodactyly syndrome | 34 |
| Spondyloepiphyseal dysplasia congenita | 8,9,10,22,40a |
| Stickler syndrome | 37,40a,46 |
| Velocardiofacial syndrome | 40,40a |
| Chromosomal | |
| del(4q) syndrome | 5 |
| del(6q) syndrome | 40a |
| dup(11q) syndrome | 8,9,10 |
| Teratogenically induced | |
| Fetal alcohol syndrome | 19,40a |
| Fetal hydantoin syndrome | 19 |
| Fetal trimethadione syndrome | 19 |
| Disruption | |
| Amniotic band sequence | 21,40a |
| Unknown genesis | |
| CHARGE association | 26 |
| Femoral dysgenesis–unusual facies syndrome | 8,9,10,40a |
| Martsolf syndrome | 10 |
| Moebius sequence | 5,42 |
| Robin/amelia association | 8,9,10,22 |
| Sickle-shaped scapulae and club feet | 3a |

of isolated Robin sequence. In the study of Williams et al (47), only 26% of children with Robin sequence had other anomalies including both recognized and unrecognized syndromes. Among 64 patients analyzed by Sheffield et al (39a), 16 died and 12 of these were thought to have a syndrome. Of those that lived, 26% had an underlying syndrome, one-half of these having Stickler syndrome. There was no recurrence in sibs of those with nonsyndromal Robin sequence. Conditions associated with Robin sequence listed in Table 20–4 are derived primarily from Cohen (8–10), Hanson and Smith (19), Carey et al (5), and Shprintzen (40a). The reader is referred to several articles on some of these conditions (1,16,17,21,26,34,37,40,44–46). With hindsight, it is fascinating to note that of the six patients with Robin sequence and multiple anomalies reported by Holthusen (22) in 1972, two had spondyloepiphyseal dysplasia congenita, two had Catel–Manzke syndrome, one had unusual facies/femoral dysgenesis syndrome, and one had Robin/amelia association. It should be recognized that the list of conditions associated with Robin sequence found in Table 20–4, although representative, is not complete; Robin sequence may be found with other conditions. Since there is further ill-defined heterogeneity, the reader may expect to find Robin sequence with various other anomalies, especially involving the eye, ear, heart, and limbs.

Genetic factors are clearly implicated in many examples of Robin sequence (2,6,16,17,25,34,36,37,46). The most common genetic syndrome associated with Robin sequence is Stickler syndrome (37,46).

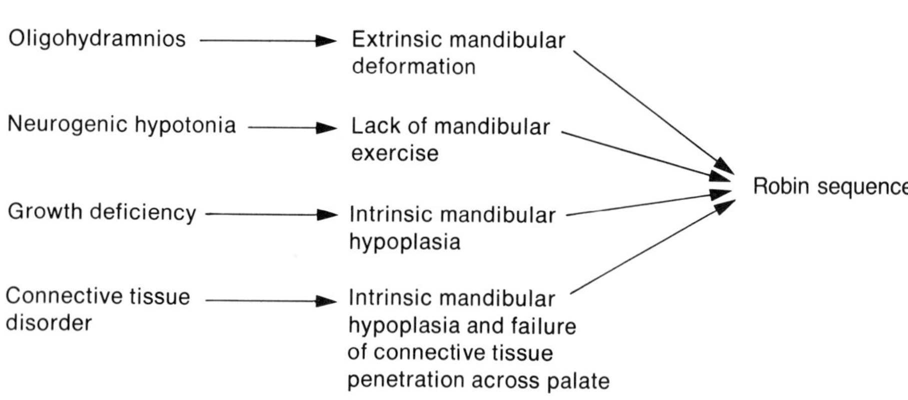


Fig. 20–10. *Robin sequence*. Etiologic heterogeneity suggests pathogenetic heterogeneity in Robin sequence. The following pathogenetic possibilities should be considered. (I) Oligohydramnios results in decreased amnionic fluid, compressing the chin against the sternum and thus restricting mandibular growth. (II) If hypotonia restricts mouth opening during early fetal life prior to complete palatal closure, Robin sequence might result from lack of mandibular exercise. (III) Growth deficiency, as observed in chromosomal syndromes such as dup(11q) syndrome, may produce Robin sequence by intrinsic mandibular hypoplasia. (IV) In a connective tissue disorder such as Stickler syndrome, Robin sequence may result from intrinsic hypoplasia and failure of connective tissue penetration across the palate. (From *MM Cohen Jr*, The Child with Multiple Birth Defects, Raven Press, New York, 1982.)

Active detection of myopia should be sought because blindness, resulting from retinal detachment, can probably be prevented in this syndrome.

**Clinical manifestations.** The facies is striking at birth. The mandible is small and symmetrically receded (Figs. 20–11 and 20–12). Commonly the base of the nose is flattened. The palatal cleft may be U-shaped (Fig. 20–13) (19). Rintala et al (32) found that both U- and V-shaped clefts (Fig. 20–14) were observed in Robin sequence and in isolated cleft palate with equal frequency. However, on the average, the width of the cleft was greater in Robin sequence than in isolated cleft palate.

Difficulty in the inspiratory phase of respiration is apparent, with periodic cyanotic attacks, labored breathing, and recession of the sternum and ribs. This becomes especially apparent when the child is in the supine position. Respiratory difficulty is usually evident at birth, although it may not be severe for the first week. Only rarely is its initiation delayed until the first month (35,38).

Although there is no complete agreement concerning the exact mechanism by which respiratory and feeding difficulties are produced, the classic explanation suggests that the micrognathia makes for little support of the tongue musculature. This allows the tongue to fall downward and backward (glossoptosis) into the lower postpharyngeal space, obstructing the epiglottis. In this position, the tongue permits egress of air but prevents inhalation, acting much as a ball valve, causing periodic cyanosis and sternal retraction. Feeding problems are thought to be due to inadequate control of the tongue. Nursing, even when performed in a favorable position, is an ordeal (12).

Routledge (35), on the other hand, suggested that the respiratory difficulty might be due, in large part, to impaction of the tongue tip

Fig. 20–11. *Robin sequence*. Characteristic facial features. (From *M Gewitz et al*, J Med Genet **15**:162, 1978.)

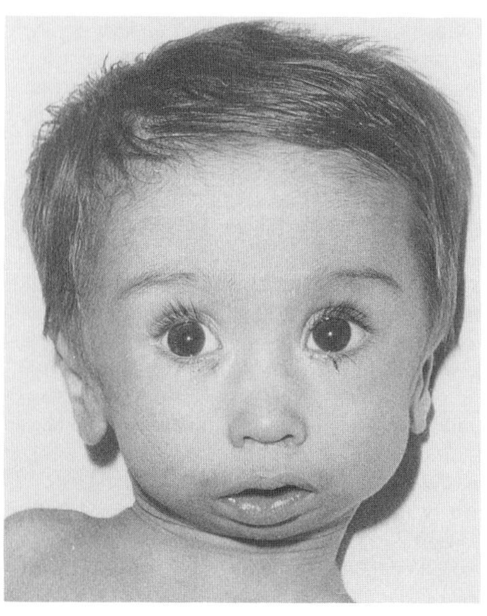

Fig. 20–12. *Robin sequence*. Severe micrognathia. (From *M Gewitz et al*, J Med Genet **15**:162, 1978.)

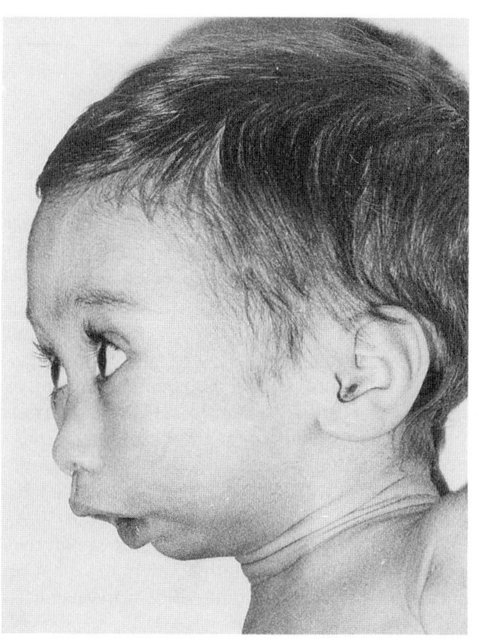

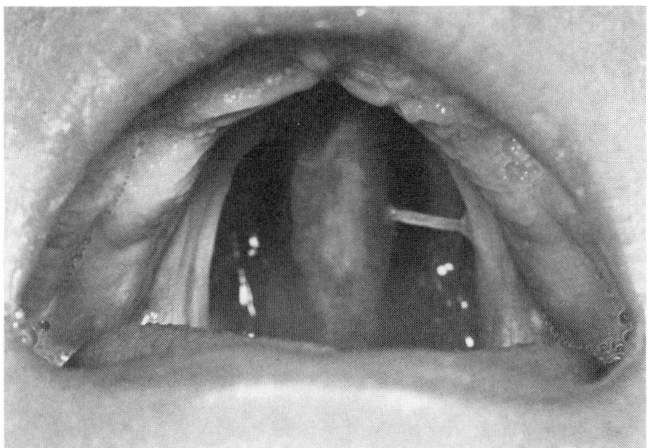

Fig. 20–13. *Robin sequence.* U-shaped cleft palate.

in the palatal cleft, from which it is not easily disengaged. Violent muscular contractions resulting from efforts to free it cause the tongue to bulge into the nasopharynx resulting in asphyxia. Routledge realized that his theory did not explain those cases in which surgical repair of the palate failed to correct the respiratory difficulty or those cases associated with ankyloglossia, which does not permit upward movement of the tongue tip.

It has also been shown, most notably by Pruzansky and his coworkers (3,29,30) and Stellmach and Schettler (43), that mandibular catch-up growth results in a normal profile by 4–6 years of age (Figs. 20–15 and 20–16). Pruzansky (29) described the mandible as having a foreshortened body with a characteristic ratio of the ramus to mandibular body length.

The tongue has been stated by some investigators to be small, by others, normal, and by still others, large. Depending on particular syndromic diagnoses, all may be correct. For example, the tongue may be small in Moebius sequence but large in Beckwith–Wiedemann syndrome. It has also been suggested that there may be disproportionate growth of the tongue (35). Pruzansky and Richmond (30) concluded from cephalometric studies that micrognathia, *sui generis,* is not sufficient to produce respiratory embarrassment unless the tongue is normal in size or enlarged. Ankyloglossia is a commonly associated complication. For additional discussion of cephalometric findings, see Ranta (31a).

Growth deficiency, which sometimes accompanies Robin sequence, may be related to severity of airflow obstruction or to the overall syndromic condition if it is caused by a chromosomal abnormality or teratogenic agent. Heaf et al (20) found that failure to thrive was significantly correlated with severity of airflow obstruction. Failure to thrive was reversed in infants managed with a nasopharyngeal airway. In a large series of patients reported by Williams et al (47), 26% died within the first 3 months. Of those who survived, 13% showed delayed language development. Other authors have indicated that as many as 20% exhibit major mental deficiency, and various anomalies may be associated such as microcephaly and hydrocephaly (42).

Congenital murmurs and/or heart defects have been observed in 15–

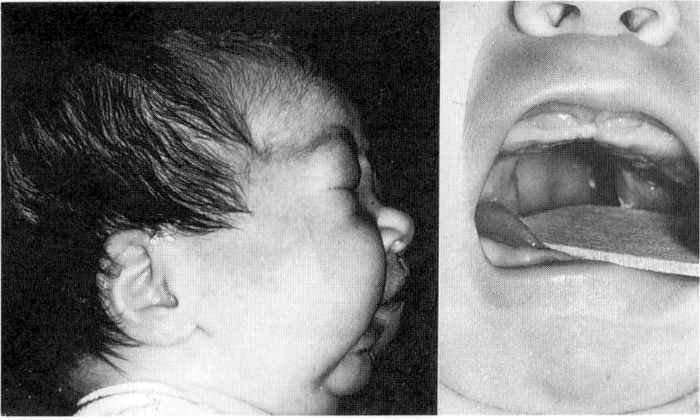

A

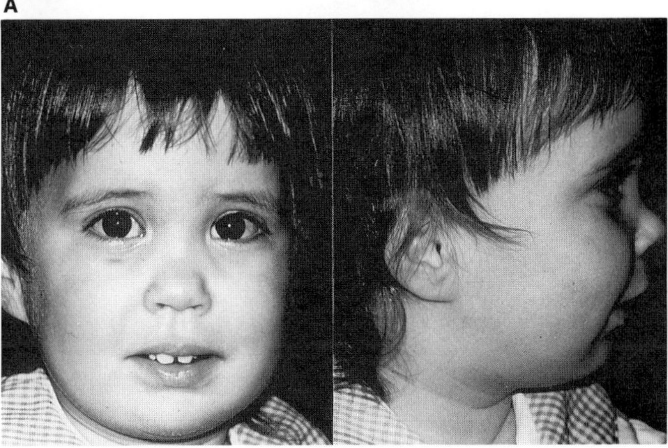

B

Fig. 20–15. *Robin sequence.* (A) Patient with Robin sequence showing micrognathia and U-shaped cleft palate. (B) Same patient showing mandibular catch-up growth. (From *MM Cohen Jr,* The Child with Multiple Birth Defects, Raven Press, New York, 1982.)

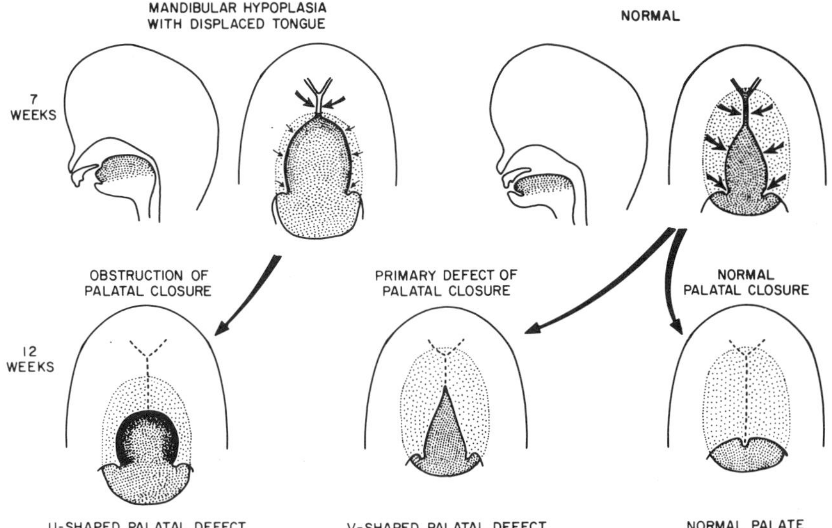


Fig. 20–14. *Robin sequence.* At left, small mandible results in posteriorly placed tongue partially interposed between palatal shelves. This prevents closure and posterior growth of soft palate, producing U-shaped cleft palate. V-shaped defect at right is frequently seen in primary defects of palatal closure not secondary to mandibular involvement. (From *JW Hanson* and *DW Smith,* J Pediatr **87**:30, 1975.)

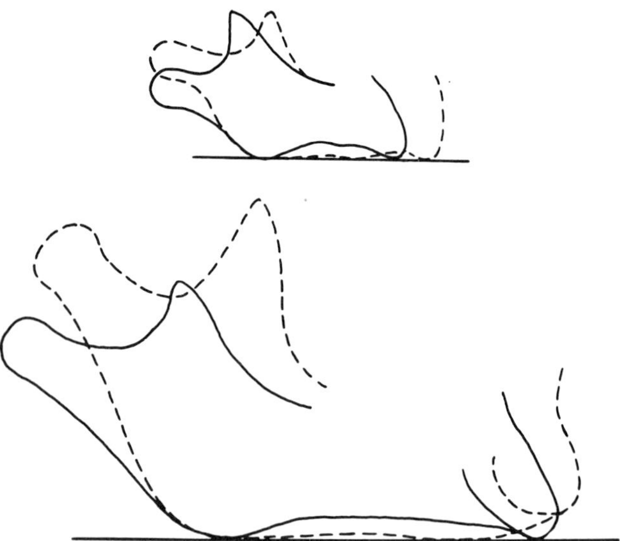

Fig. 20–16. *Robin sequence.* Superimposed tracings from normal (broken line) and Robin (solid line) mandibles. Note differences in height of ramus, length of body, gonial angle, and inclination of condyle to ramus. Note also catch-up growth. [From *S Pruzansky,* Birth Defects **5**(2):120, 1969.]

25% of patients who die in early infancy (39). Necropsy has revealed PDA, patent foramen ovale, ASD, VSD, ventricular hypertrophy, cor triloculare, coarctation of the aorta, biventricular aorta, and dextrocardia (11,12,35,38,39,42). Congestive cardiac failure with signs of cor pulmonale has been recorded by a number of authors (13,23). Nasopharyngeal intubation or tracheostomy may reverse the signs of cor pulmonale.

Many striking abnormalities are features of the syndromes listed in Table 20–4. Other anomalies occur in unrecognized patterns. Ocular findings are particularly common. Smith and Stowe (42) found 13 eye anomalies in nine patients including esotropia, glaucoma, and microphthalmia. Ear anomalies occur less frequently, but especially include low-set ears and malformed ears (18,42).

Limb anomalies may be particularly striking and occur with some frequency. Wood and Sandlin (48), in a review of eight patients with Robin sequence, found three cases with limb anomalies including syndactyly, hypoplastic digits, and Poland anomaly. Williams et al (47) noted that peripheral limb defects were particularly common. Congenital amputations, limb reduction defects, talipes equinovarus, and congenital hip dislocation have been reported. Rib and sternal anomalies have also been noted (18,19,34,35,42).

### References (Robin sequence)


1. André M et al: Abnormal facies, cleft palate, and generalized dysostosis: A lethal X-linked syndrome. *J Pediatr* **98**:747–752, 1981.

2. Becker VR, Palm D: Zur kausalen und formal Genese des Pierre Robin-Syndroms. *Dtsch Zahnärztl Z* **21**:1321–1338, 1966.

3. Beers MD, Pruzansky S: The growth of the head of an infant with mandibular micrognathia: Glossoptosis and cleft palate following the Beverly Douglas operation. *Plast Reconstr Surg* **16**:189–193, 1955.

3a. Bezirdjian DR, Szucs R: Sickle-shaped scapulae in a patient with the Pierre Robin syndrome. *Br J Radiol* **62**:171–173, 1989.

4. Bush PG, Williams AJ: Incidence of the Robin anomalad (Pierre Robin syndrome). *Br J Plast Surg* **36**:434–437, 1983.

5. Carey JC et al: The Robin sequence as a consequence of malformation, dysplasia, and neuromuscular syndromes. *J Pediatr* **101**:858–864, 1982.

6. Carroll DB et al: Hereditary factors in the Pierre Robin syndrome. *Br J Plast Surg* **24**:43–47, 1971.

7. Cocke WJ: Experimental production of micrognathia and glossoptosis associated with cleft palate (Pierre Robin syndrome). *Plast Reconstr Surg* **38**:395–403, 1966.

8. Cohen MM Jr: The Robin anomalad—its nonspecificity and associated syndromes. *J Oral Surg* **34**:587–593, 1976.

9. Cohen MM Jr: Syndromology's message for craniofacial biology. *J Maxillofac Surg* **7**:89–109, 1979.

10. Cohen MM Jr: *The Child with Multiple Birth Defects.* Raven Press, New York, 1982.

11. Dennison WM: The Pierre Robin syndrome. *Pediatrics* **36**:336–341, 1965.

12. Douglas B: The treatment of micrognathia with obstruction by a plastic operation. *Lyon Chir* **52**:420–431, 1956.

13. Dykes EH et al: Pierre Robin syndrome and pulmonary hypertension. *J Pediatr Surg* **20**:49–52, 1985.

14. Edwards JRG, Newall DR: The Pierre Robin syndrome reassessed in the light of recent research. *Br J Plast Surg* **38**:339–342, 1985.

15. Fairbairn P: Suffocation in an infant from retraction of the base of the tongue, connected with defect of the frenum. *Month J Med Sci* **6**:280–281, 1846.

16. Gewitz M et al: Cleft palate and accessory metacarpal of index finger syndrome: Possible familial occurrence. *J Med Genet* **15**:162–164, 1978.

17. Gorlin RJ et al: Robin's syndrome: A probably X-linked recessive subvariety exhibiting persistence of left superior vena cava and atrial septal defect. *Am J Dis Child* **119**:176–178, 1970.

18. Grimm G et al: Die klinische Bedeutung des Pierre Robin-Syndroms und seine Behandlung. *Dtsch Zahn Mund Kieferheilkd* **43**:385–416, 1964,

19. Hanson J, Smith DW: U-shaped palatal defect in the Robin anomaly: Developmental and clinical relevance. *J Pediatr* **87**:30–33, 1975.

20. Heaf DP et al: Nasopharyngeal airways in Pierre Robin syndrome. *J Pediatr* **100**:698–703, 1982.

21. Higginbottom MC et al: The amniotic band disruption complex: Timing of amniotic rupture and variable spectra of consequent defects. *J Pediatr* **95**:544–549, 1979.

22. Holthusen W: The Pierre Robin syndrome: Unusual associated developmental defects. *Ann Radiol* **15**:253–262, 1972.

23. Jersty RM et al: Pierre Robin syndrome. *Am J Dis Child* **117**:710–716, 1969.

24. Latham RA: The pathogenesis of cleft palate associated with the Pierre Robin syndrome. *Br J Plast Surg* **19**:205–214, 1966.

25. Opitz JM: Familial anomalies in the Pierre Robin syndrome. *Birth Defects* **5**(2):119, 1969.

26. Pagon RA et al: Coloboma, congenital heart disease, and choanal atresia with multiple anomalies: CHARGE association. *J Pediatr* **99**:223–227, 1981.

27. Pashayan HM, Battle CU: Isolated Robin anomalad: Height, weight and head circumference as a measure of growth and development. Unpublished data, 1986.

28. Poswillo D: The aetiology and surgery of cleft palate with micrognathia. *Ann R Coll Surg Eng* **43**:61–88, 1968.

29. Pruzansky S: Not all dwarfed mandibles are alike. *Birth Defects* **5**(2):120–129, 1969.

30. Pruzansky S, Richmond JB: Growth of the mandible in infants with micrognathia. *Am J Dis Child* **88**:29–42, 1954.

31. Randall P et al: Pierre Robin and the syndrome that bears his name. *Cleft Palate J* **2**:237–246, 1965.

31a. Ranta R et al: Cephalometric comparison of the cranial base and face in children with the Pierre Robin anomalad and isolated cleft palate. *Proc Finn Dent Soc* **81**:82–90, 1985.

32. Rintala A et al: On the pathogenesis of cleft palate in the Pierre Robin syndrome. *Scand J Plast Reconstr Surg* **18**:237–240, 1984.

33. Robin P: La chute de la base de la langue considerée comme une nouvelle cause de gene dans la respiration naso-pharyngienne. *Bull Acad Méd (Paris)* **89**:37–41, 1923.

34. Robinow M et al: Robin sequence and oligodactyly in mother and son. *Am J Med Genet* **25**:293–297, 1986.

35. Routledge RT: The Pierre Robin syndrome: A surgical emergency in the neo-natal period. *Br J Plast Surg* **13**:204–218, 1960.

36. Russo G et al: Robin's syndrome in three children of consanguineous parents. *Acta Genet Med Gemellol* **21**:349–353, 1972.

37. Schreiner RL et al: Stickler syndrome in a pedigree of the Pierre Robin syndrome. *Am J Dis Child* **126**:86–91, 1973.

38. Schönenberg H, Lautermann R: Das Robin-Syndrom. *Z Kinderheilkd* **97**:326–346, 1966.

39. Shah CV et al: Cardiac malformations with facial clefts. *Am J Dis Child* **119**:238–244, 1970.

39a. Sheffield LJ et al: A genetic follow-up study of 64 patients with the Pierre Robin complex. *Am J Med Genet* **28**:25–36, 1987.

40. Shprintzen RJ: The velo-cardio-facial syndrome: A clinical and genetic analysis. *Pediatrics* **67**:167–172, 1981.

40a. Shprintzen RJ: Pierre Robin, micrognathia and airway obstruction: The dependency of treatment on accurate diagnosis. *Int Anesthesiology Clin* **26**:64–71, 1988.

41. Shukowksy WP: Zur Ätiologie des Stridor inspiratorius congenitus. *Z Kinderheilkd* **73**:459, 1911.

42. Smith JL, Stowe FR: The Pierre Robin syndrome (glossoptosis, micrognathia, cleft palate): A review of 39 cases with emphasis on associated ocular lesions. *Pediatrics* **27**:128–133, 1961.

43. Stellmach R, Schettler D: Beobachtungen zum Robin-Syndrom und kephalometrische Untersuchungen bei 12 Behandlungsfällen. *Dtsch Zahn Mund Kieferheilkd* **49**:137–149, 1967.

44. Sundaram V et al: Hyperphalangy and clinodactyly of the index finger with Pierre Robin anomaly: Catel–Manzke syndrome. A case report and review of the literature. *Clin Genet* **21**:407–410, 1982.

45. Thompson EM et al: A male infant with the Catel–Manzke syndrome and dislocatable knees. *J Med Genet* **23**:271–274, 1986.

46. Turner G: The Stickler syndrome in a family with the Pierre Robin syndrome and severe myopia. *Aust Paediatr J* **10**:103–108, 1974.

47. Williams AJ et al: The Robin anomalad (Pierre Robin syndrome)—a follow-up study. *Arch Dis Childh* **56**:663–668, 1981.

48. Wood VE, Sandlin C: The hand in the Pierre Robin syndrome. *J Hand Surg* **8**:273–276, 1983.

## Noncleft palatal anomalies

The palate may or may not be highly arched, but this can only be known by measurement of palatal height. The term "highly arched palate" is most commonly used as a subjective description. Objective measurements often belie subjective clinical impression. For example, patients with trisomy 21 syndrome are commonly described as having highly arched palates. However, in the metric study of Shapiro et al (9), palatal height tended to be in the normal range and was not markedly high. On the other hand, palatal width tended to be more narrow than in normal controls, resulting in the illusion of highly arched palate (Fig. 20–17). Surprisingly, palatal length was found to be dramatically shorter than normal, and in the overwhelming majority of cases, trisomy 21 syndrome patients and normal controls could be distinguished on the basis of this measurement (Fig. 20–18).

Prominent lateral palatine ridges (Fig. 20–19) are a nonspecific feature of various conditions (Table 20–5). Two types of disorders predominate: those with neuromotor dysfunction and those with primary palatal malformation. In both types, proper tongue thrust into the palatal vault is prevented. This may occur in patients with neuromuscular dysfunction of long standing or with malformations such as Byzantine arch palate. Hanson et al (4) suggested that long-standing deficit of tongue thrust is the common pathogenetic mechanism. In the Byzantine arch palate of Apert syndrome, lateral palatal swellings are present which increase in size with age (Fig. 20–20). These swellings have been shown to contain excess mucopolysaccharide content, predominantly hyaluronic acid, and, to a lesser extent, sulfated mucopolysaccharides (2,10).

Acquired palatal groove (Fig. 20–21) has been observed in infants with prolonged orotracheal intubation. The palatal groove appears to

Fig. 20–17. *Noncleft palatal anomalies.* Cross-sections through palate. (A) Normal. (B) Narrow palate of trisomy 21 syndrome giving the illusion of being high. (From *BL Shapiro et al,* N Engl J Med **276**:1460, 1967.)

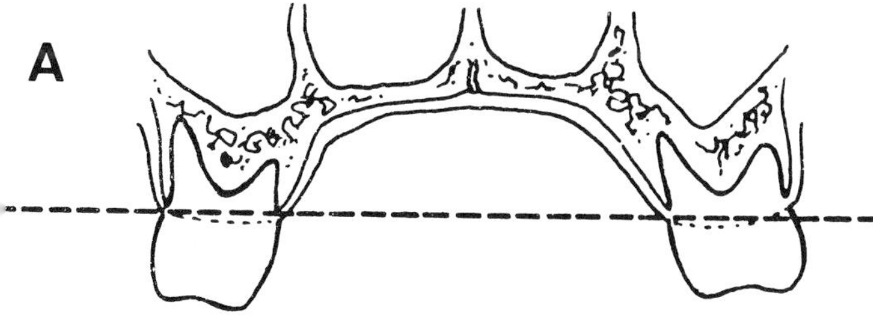


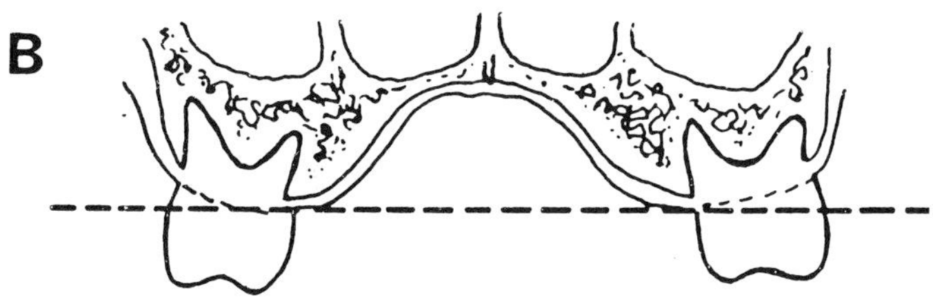


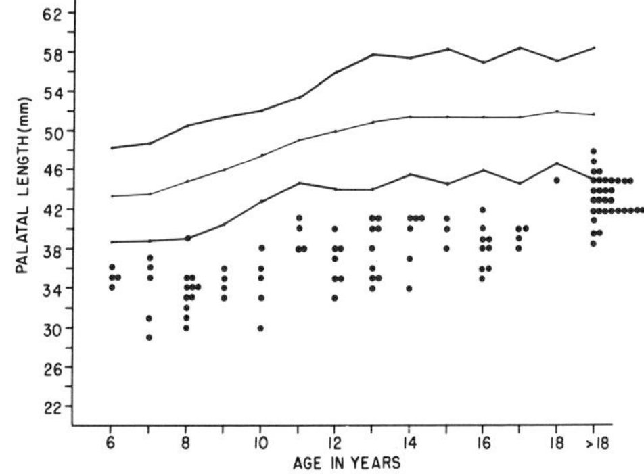


Fig. 20–18. *Noncleft palatal anomalies.* Palatal length of male trisomy 21 syndrome patients plotted on a graph with normal mean ± 2 standard deviations for reference. Palatal length is dramatically short. (From *BL Shapiro et al,* N Engl J Med 276:1460, 1967).

Fig. 20–19. *Noncleft palatal anomalies.* Examples of unusually prominent lateral palatine ridges. (A) Fetal alcohol syndrome (16 months). (B) Trisomy 21 syndrome (3 years, 9 months). (C) Sotos syndrome (10 months). (D) Smith–Lemli–Opitz syndrome (12 months). (From *JW Hanson et al,* J Pediatr **89**:54, 1976.)

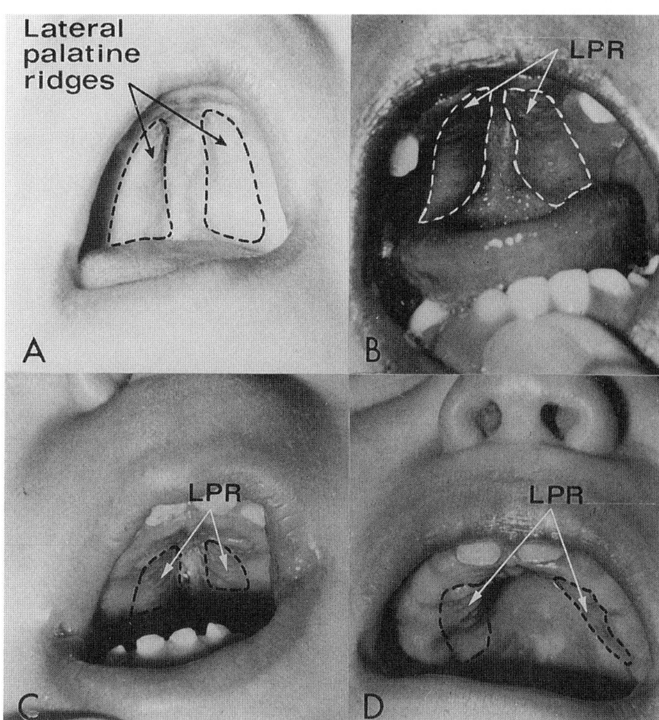

Table 20–5. Conditions with altered neuromuscular function or malformations associated with prominent lateral palatal ridges[a]

| Clinical sign of neuromuscular dysfunction | Condition |
|---|---|
| Hypotonia | Trisomy 21 syndrome |
| | Prader–Willi syndrome |
| | Zellweger syndrome |
| Hypertonia | Smith–Lemli–Opitz syndrome |
| | Menkes syndrome |
| Other | Sotos syndrome |
| | Fetal alcohol syndrome |
| Syndrome | Abnormality proposed to interfere with palatal molding by tongue |
| Apert syndrome | Extremely narrow midpalate |
| Glossopalatine ankylosis syndrome | Mechanically immobilized tongue |
| Hypoglossia–hypodactylia syndrome | Small tongue |

[a]Modified from JW Hanson et al, *J Pediatr* **89**:990–991, 1976.

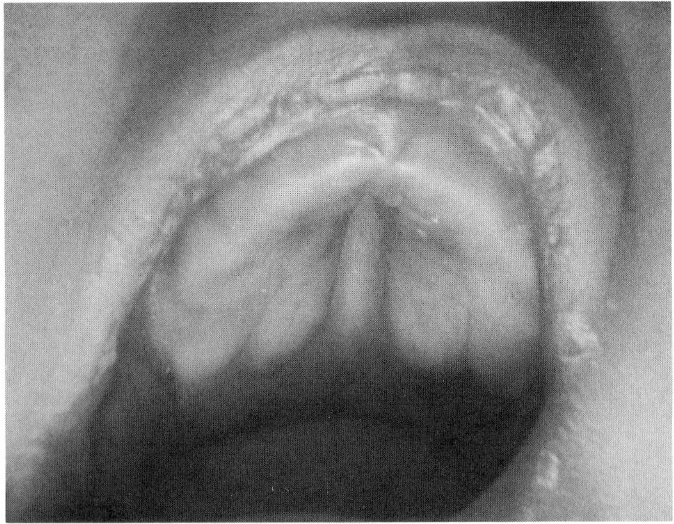

Fig. 20–21. *Noncleft palatal anomalies.* Acquired palatal groove from orotracheal intubation. Infant intubated for 70 days. (From *BS Saunders et al,* J Pediatr **89**:988, 1976.)

Fig. 20–20. *Noncleft palatal anomalies.* Palatal swellings in Apert syndrome. (A) Younger patient. (B) Older patient. Note change in size of palatal swellings with age. (From *MM Cohen Jr,* Craniosynostosis: Diagnosis, Evaluation, and Management, Raven Press, New York, 1986.)

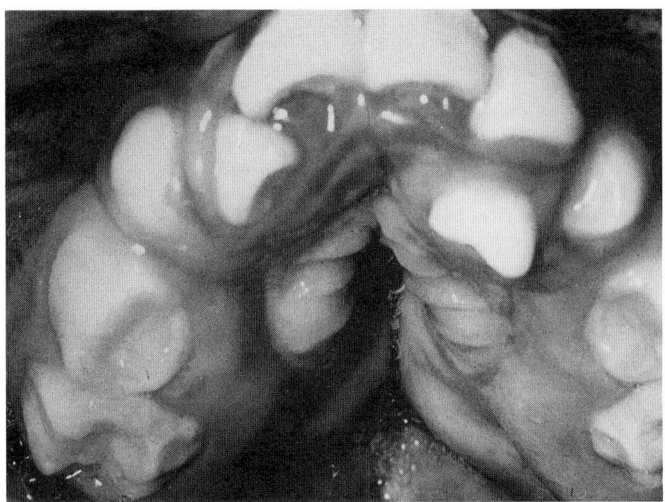

A

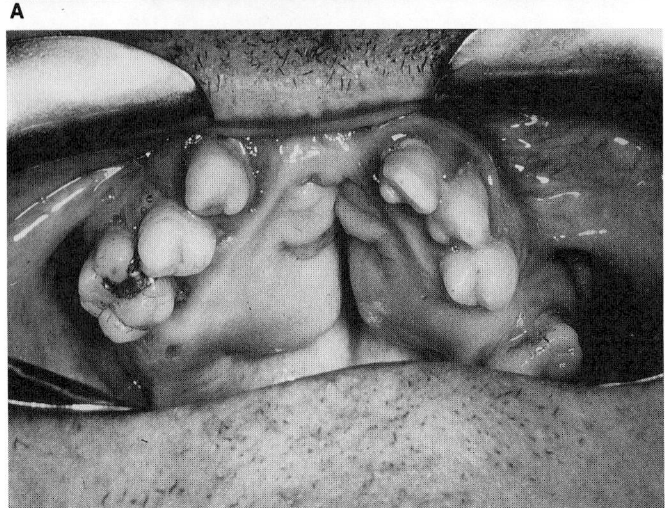

B

be related to the duration of intubation and to vigorous sucking during that period of time (3,8).

An unusual palatal anomaly is found in anencephaly (5–7) and holoprosencephaly (1). Although the palate is closed, deep depressions are found extending the length of the hard palate. The abnormality appears to be caused by incomplete downgrowth of the central nasal process. The appearance of such a palate in a case of cyclopia is shown in Fig. 20–22.

### References (Noncleft palatal anomalies)

1. Cohen MM Jr, Gorlin RJ: Genetic implications of a sibship of cyclopia and clefts. *Birth Defects* **5**(2):113–118, 1969.

Fig. 20–22. *Noncleft palatal anomalies.* Unusual palatal form in cyclopia. Arrow demonstrates bifid uvula. [From *MM Cohen Jr* and *RJ Gorlin,* Birth Defects **5**(2):113, 1969.]

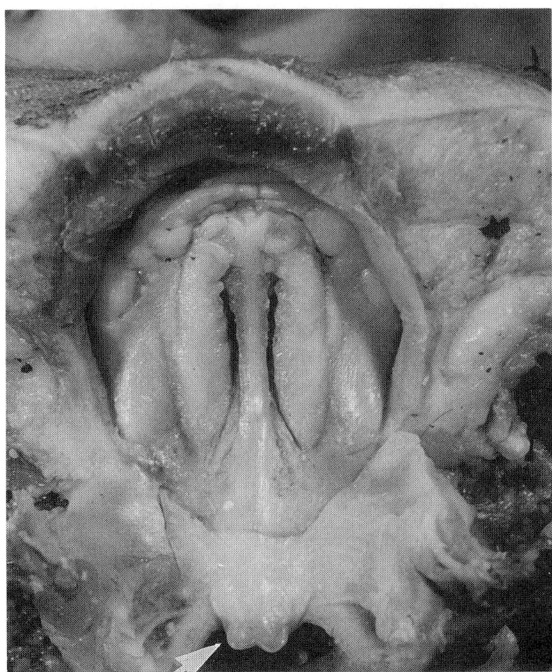

2. Cohen MM Jr: *Craniosynostosis: Diagnosis, Evaluation, and Management*. Raven Press, New York, 1986.

3. Duke PM et al: Cleft palate associated with prolonged orotracheal intubation in infancy. *J Pediatr* **89**:990–991, 1976.

4. Hanson JW et al: Prominent lateral palatine ridges: Developmental and clinical relevance. *J Pediatr* **89**:54–58, 1976.

5. Lemire RJ et al: *Anencephaly*. Raven Press, New York, 1977.

6. Marin-Padilla M: Study of the skull in human cranioschisis. *Acta Anat* **62**:1–20, 1965.

7. Potter EL, Craig JM: *Pathology of the Fetus and the Infant*, Third Edition. Year Book Medical Publishers, Chicago, 1975, pp 524–525.

8. Saunders BS et al: Acquired palatal groove in neonates. *J Pediatr* **89**:988–989, 1976.

9. Shapiro BL et al: The palate and Down's syndrome. *N Engl J Med* **276**:1460–1463, 1967.

10. Solomon LM et al: Apert syndrome and palatal mucopolysaccharides. *Teratology* **8**:287–292, 1973.

## Tessier clefting system

Many classifications for orofacial clefting have been used (2,4) and more elaborate classifications have been proposed to deal with extensive craniofacial clefting (3,5–7,9). Tessier (8) described a classificatory system in which clefts are situated along definite axes. His system assigns numbers to various sites of clefting, depending on their relationships to the sagittal midline (Fig. 20–23). Bone and soft tissues are both involved, but rarely to the same extent. From the sagittal midline to the infraorbital foramen, abnormalities of soft tissue predominate. From the infraorbital foramen to the temporal bone, however, osseous defects are more severe than those of soft tissue, a notable exception being the ear. Clefts through the orbit use the lower eyelid as an equator. Cleft numbered lines may be either northbound (cranial) or southbound (facial). Cranial numbered lines have facial numbered counterparts, yet these numbers are different to avoid the implication that they necessarily have the same etiopathogenesis.

In some instances, overlying soft tissue defects predict the possibility of underlying bony clefts. Features such as colobomatous notching of the upper or lower eyelid and nostril or interruption of eyelashes and eyebrows have been referred to as Tessier signs (1). Such patients should be examined radiographically to rule out possible underlying bony clefts. Several types of osseous clefting are illustrated in Fig. 20–24 and clinical examples of the Tessier clefting system are shown in Fig. 20–25 to 20–27.

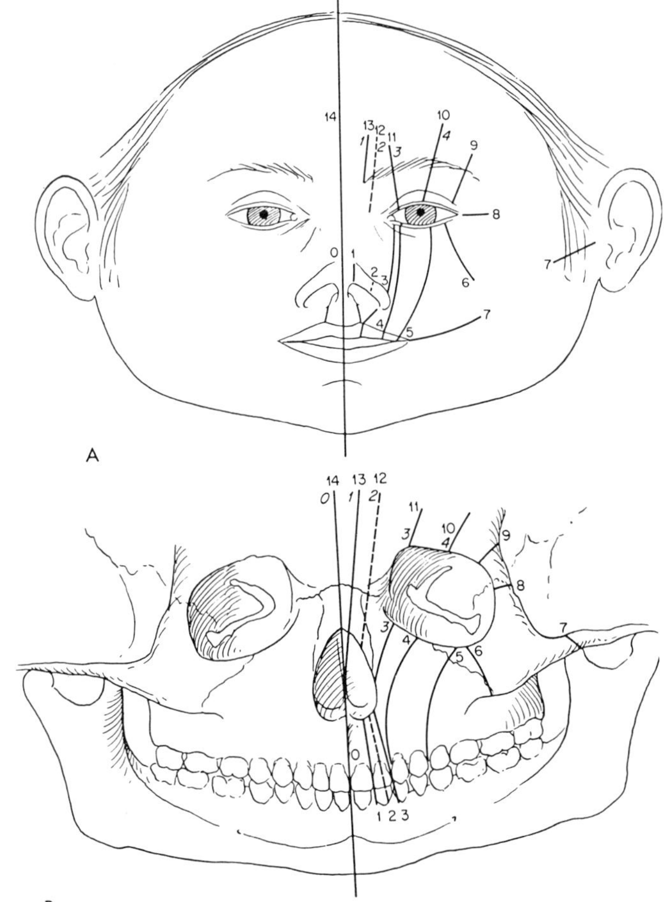


Fig. 20–23. *Tessier clefting system.* (A) Soft tissue clefts. (B) Bony clefts. Dotted lines represent uncertain localization or uncertain clefting. Note that northbound cranial line has different number than its counterpart southbound facial line. Thus, system is descriptive and anatomic, and avoids etiologic and/or pathogenetic speculation. For example, the cause of a No. 10 cleft may possibly be different than the cause of a No. 4 cleft. (From *P Tessier*, J Maxillofac Surg **4**:69, 1976.)

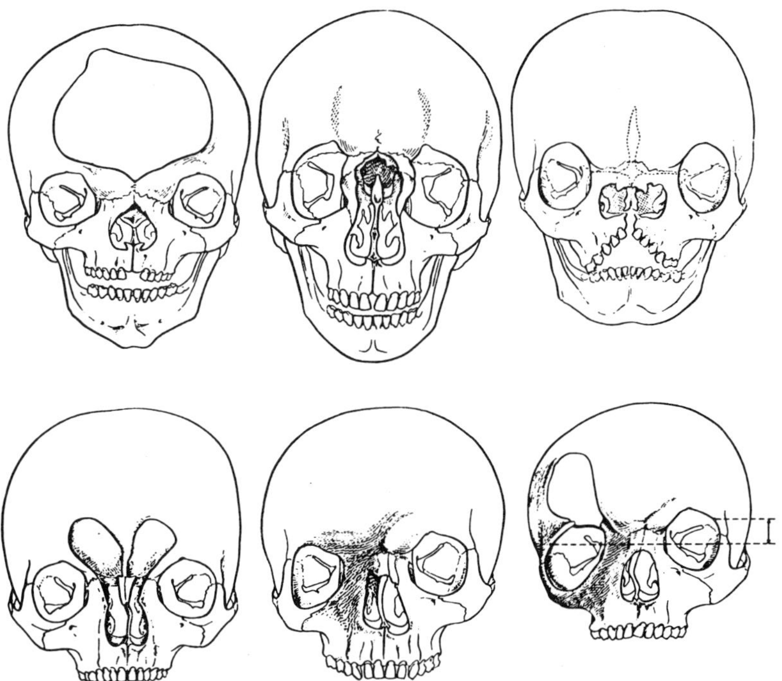

Fig. 20–24. *Tessier clefting system.* Upper left: Cleft Nos. 0, 14. Median craniofacial dysraphism involving encephalocele, hypertelorism, bifid nose, and diastema between maxillary central incisors. Upper middle: Cleft Nos. 0, 14. Median frontonasal encephalocele and dystopia canthorum. Note sparing of upper cranial region and tooth bearing part of maxilla. Upper right: Cleft Nos. 0, 14. Median craniofacial dysraphism with encephalocele or calcification of the falx cerebri or both, duplication of the crista galli, hypertelorism, bifid nose, absence of vomer, and keel-shaped maxillary dental arch. Lower left: Cleft Nos. 1, 13. Bilateral paramedian encephaloceles and hypertelorism. This case was associated with colobomatous notching of both nostrils. Lower middle: Cleft Nos. 2, 12. Unilateral involvement affecting facial component more severely than cranial component. Cleft is either through frontal process of maxilla or between maxilla and nasal bone. Note unilateral hypertelorism. Patient had cleft through medial part of the right nostril. Lower right: Cleft No. 10. Unilateral large defect of frontal bone involving supraorbital rim and orbital roof with encephalocele, unilateral hypertelorism, and dystopia with lateral rotation of the affected orbit. (From *P Tessier*, J Maxillofac Surg **4**:69, 1976.)

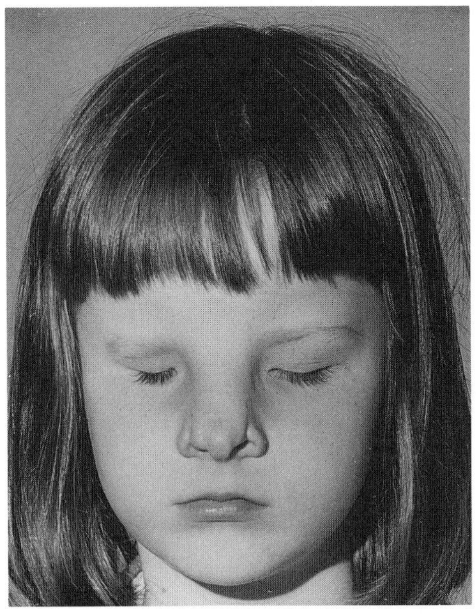

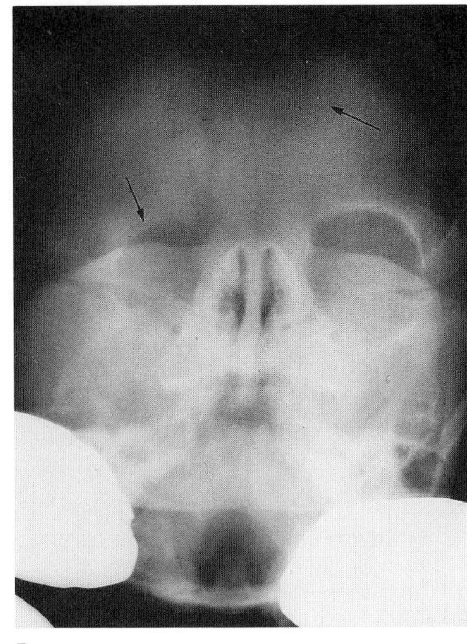

Fig. 20–25. *Tessier clefting system.* (A,B) Cleft Nos. 0, 14. Note midline cranium bifidum, hypertelorism, unilateral microphthalmia with small orbit, and colobomatous notching of nostrils. (From *HO Sedano et al,* J Pediatr **76**:906, 1970.)

The Tessier classification is anatomic and descriptive, thereby avoiding terms that imply, often erroneously, etiopathogenetic mechanisms. The causes of most such clefts are unknown. The overwhelming majority occur sporadically. Many cannot be explained embryologically, suggesting the possibility of disruptive factors. David et al (1a) have elegantly illustrated each type and have added three dimensional reconstruction.

### References (Tessier clefting system)

1. Cohen MM Jr: Syndromology: An updated conceptual survey. II. Syndrome classifications. *Int J Oral Maxillofac Surg* **18**:223–228, 1989.

1a. David DJ et al: Tessier clefts with a third dimension. *Cleft Palate J* **26**:163–183, 1989.

2. Harkins CS: Proposed morphological classification of congenital cleft lip and cleft palate. *Cleft Palate J* **10**:11–12, 1960.

3. Karfik V: Oblique facial clefts. *Transact 4th Int Congress Plastic Surg, Rome.* Excerpta Medica Found, 1967.

4. Kernahan DA et al: A new classification for cleft lip and cleft palate. *Plast Reconstr Surg* **22**:435–441, 1958.

5. Mazzola R: Congenital malformations in the frontonasal area: Their pathogenesis and classification. *Clin Plast Surg* **3**:573–609, 1976.

6. Morian, R: Über die schräge Gesichtsspalte. *Arch Klin Chir* **35**:245–288, 1887.

7. Sanvenero-Roselli G: Developmental pathology of the face and the dysraphic syndrome. *Plast Reconstr Surg* **11**:36–38, 1953.

Fig. 20–26. *Tessier clefting system.* (A–C) Cleft No. 10. Note unilateral hypertelorism with downward displacement of the orbit, interruption of eyebrow on affected side, and repaired cleft lip (CCFA #1240♀, age 14 yrs, 7 mos). [From *S Pruzansky,* in *JF Bosma,* Symposium on Development of the Basicranium, HEW Publication No. (NIH) 76-989, Bethesda, 1976, pp 278–300.]

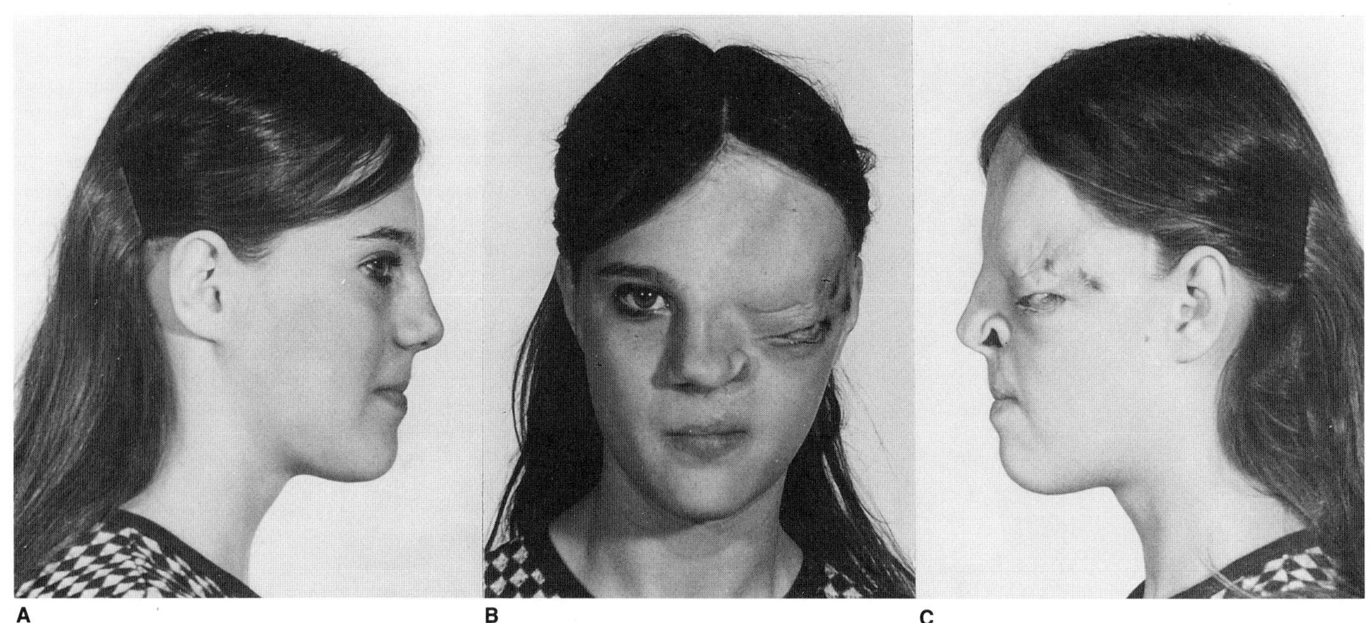

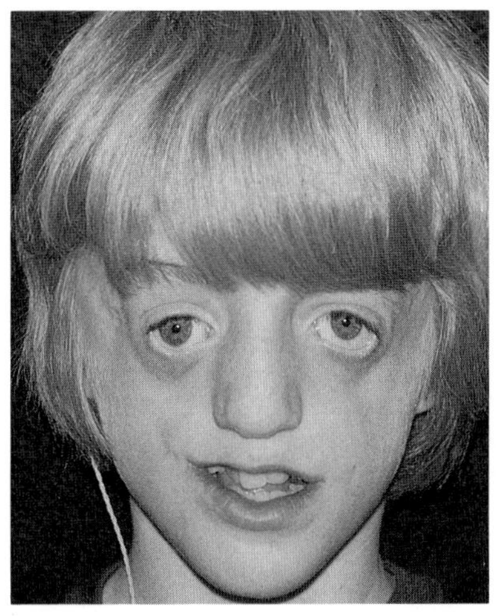


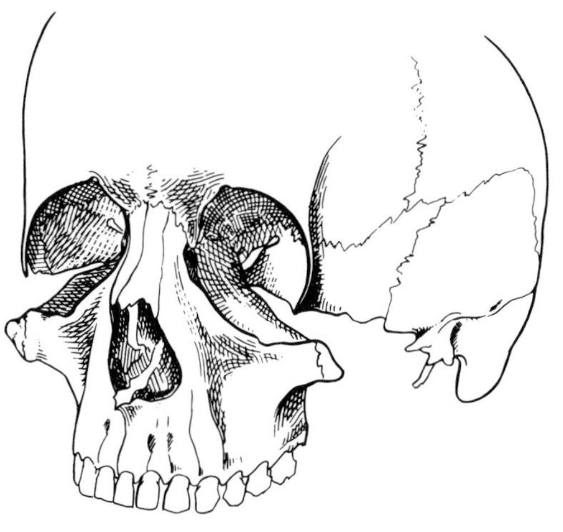


Fig. 20–27. *Tessier clefting system.* Cleft Nos. 6, 7, 8. Mandibulofacial dysostosis. (A) Clinical appearance showing downslanting palpebral fissures and malar deficiency. Note abnormal shape of lower eyelids which often have missing lashes medially. (From *MM Cohen Jr,* The Child with Multiple Birth Defects, Raven Press, New York, 1982, p. 106.) (B) Bony clefting of ovoid-shaped orbits with absence of zygomatic arches. (From *P Tessier,* J Maxillofac Surg **4**:69, 1976.)

8. Tessier P: Anatomical classification of facial, cranio-facial and latero-facial clefts. *J Maxillofac Surg* **4**:69–92, 1976.

9. Van der Meulen JC et al: A morphogenetic classification of craniofacial malformations. *Plast Reconstr Surg* **71**:560–572, 1983.

## Lateral facial clefts

Lateral facial cleft (macrostomia) may be an isolated phenomenon. More often, however, it occurs in association with other disorders: *mandibulofacial dysostosis, oculo-auriculo-vertebral spectrum, Nager acrofacial dysostosis, amniotic rupture sequence,* etc.

Various authors have estimated the frequency of isolated lateral facial cleft as 1/100 to 350 cases of cleft lip and/or palate, that is, about 1/50,000 to 175,000 live births (4–6,11). Like oblique facial cleft, the isolated lateral cleft does not appear to have a genetic basis. It has been reported in one of monozygous twins (7a). Especially bizarre are the severe cases in sibs whose father had cleft lip and cleft palate (14) and in three half sibs, each having a different father (7).

The cleft may represent failure of penetration of ectomesenchyme between the developing maxillary and mandibular prominences. Multiple growth centers are involved in each facial prominence. Failures of these centers to merge could eventuate in macrostomia. On the other hand, the variable placement of the lateral cleft in relation to the ear in different cases (Fig. 20–28 and 20–29) suggests the possibility of disruptive factors in many cases.

The lateral facial cleft may be unilateral or bilateral, partial or rarely complete, extending from the angle of the mouth to the ear. It appears to be somewhat more common in males, and, when unilateral, is more frequent on the left side. Minimally, it may involve only a groove-like thinning of the cheek or mild lateral displacement of the commissure. The external defect is always accompanied by an underlying muscle defect (Fig. 20–29A–E). The reader is referred to *Tessier clefting system.*

Associated anomalies have been reported including syndactyly or absence of digits (4), micrognathia (3), nasal dermoid (4), epignathus (1), bifid uvula and redundant tooth, and alveolar bone formation (13).

### References (Lateral facial clefts)

1. Ahlfeld F: Beiträge zur Lehre von den Zwillingen. *Arch Gynäkol* **7**:210–286, 1875.

2. Bauer BS et al: Incorporation of the W-plasty in repair of macrostomia. *Plast Reconstr Surg* **70**:752–756, 1982.

3. Benavent WJ, Ramos-Oller A: Micrognathia: Report of twelve cases. *Plast Reconstr Surg* **22**:486–490, 1958.

4. Blackfield HM, Wilde NJ: Lateral facial clefts. *Plast Reconstr Surg* **6**:62–78, 1950.

5. Boo-Chai K: The transverse facial cleft: Its repair. *Br J Plast Surg* **22**:119–124, 1969.

6. Fogh-Andersen P: Rare clefts of the face. *Acta Chir Scand* **129**:275–281, 1965.

7. Grünberg KL: Missbildungen des Kopfes, in Schwalbe, *Die Morphologie der Missbildungen des Menschen und der Tiere. III Die Einzelmissbildungen,* Fischer, Jena, 1908.

7a. Hartsfield JK, Bixler D: Bilateral macrostomia in one of monozygotic twins. *Oral Surg* **57**:648–651, 1984.

8. Hawkins DB et al: Bilateral macrostomia as an isolated deformity. *J Laryngol Otol* **87**:309–313, 1973.

9. Jaworski S: Macrostomia: A modified technic of surgical repair. *Acta Chir Plast* **18**:117–121, 1976.

Fig. 20–28. *Lateral and oblique facial clefts.* The oblique facial cleft does not follow a uniform pattern. It may involve the nostril (A), or it may skirt the ala (B). Transverse facial cleft (macrostomia) is similarly variable (C). (From *M Grob,* Lehrbuch der Kinderchirurgie, Georg A Thieme, Stuttgart, 1957, p 101.)

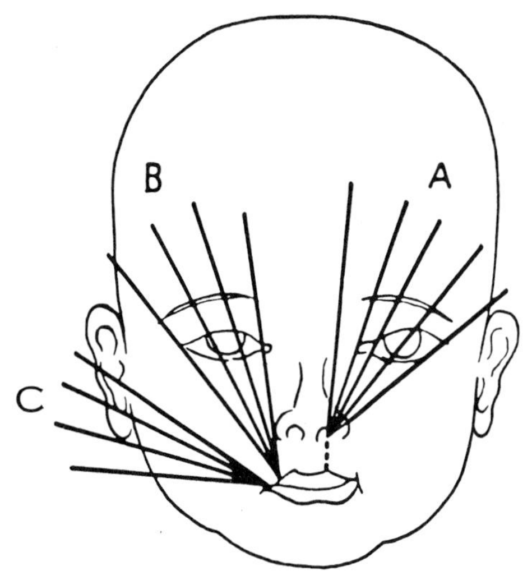

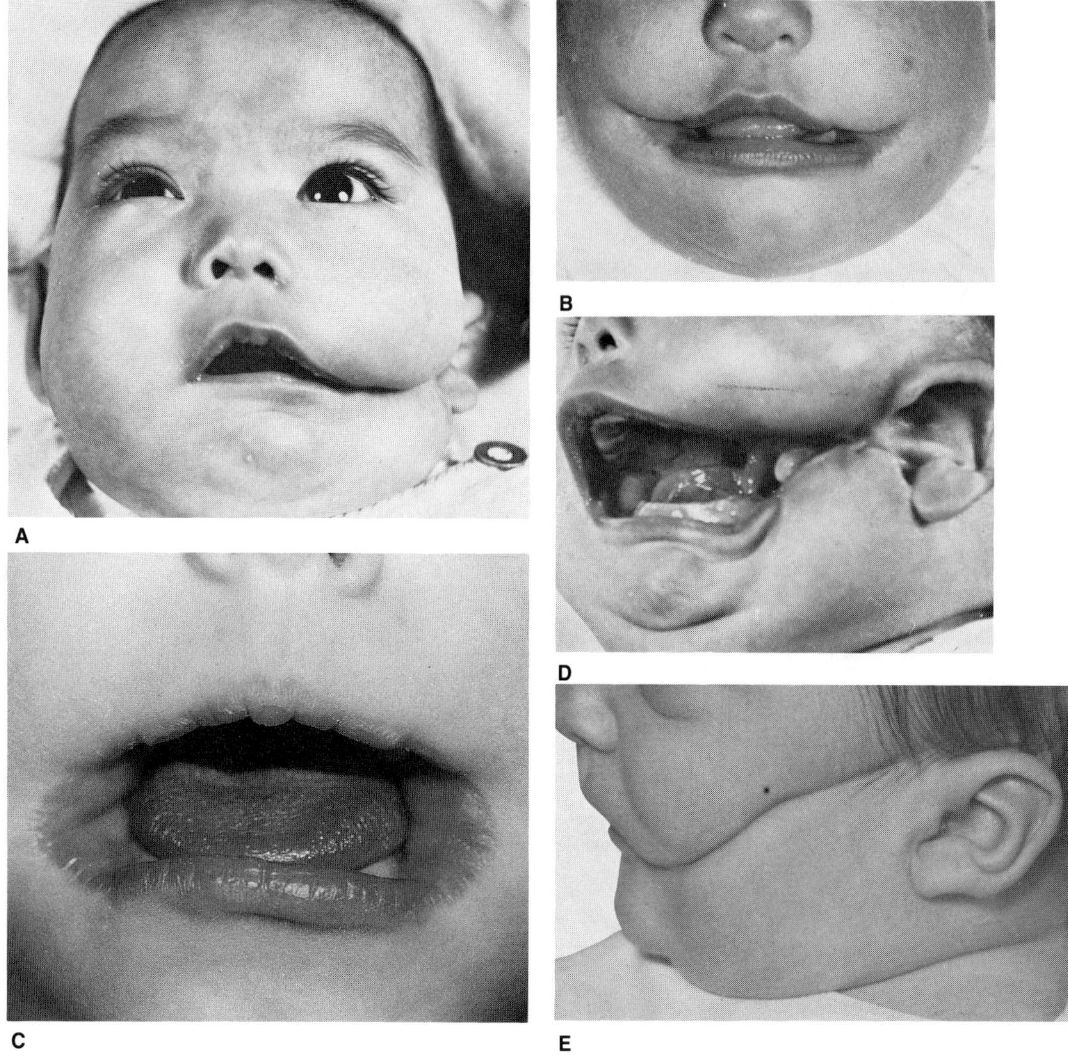


Fig. 20–29. *Lateral facial clefts.* (A) Unilateral facial cleft. (B) Bilateral lateral facial cleft. (C) Bilateral macrostomia. (D) Unilateral lateral facial cleft extending to tragus. (E) Lateral facial cleft extending above helix of low-set, posteriorly angulated ear. (A,D from *P Fogh-Andersen,* Acta Chir Scand **129**:275, 1964. B courtesy of *K Schuchardt,* Hamburg, Germany. C from *BS Bauer et al,* Plast Reconstr Surg **70**:752, 1982. E from *HM Blackfield* and *NJ Wilde,* Plast Reconstr Surg **6**:62, 1950.)

10. Kaplan EN: Commissuroplasty and myoplasty for macrostomia. *Ann Plast Surg* **7**:136–144, 1981.

11. Ogo K et al: Ten year survey of macrostomia. *Jpn J Plast Reconstr Surg* **16**:431–438, 1973.

12. Powell WJ, Jenkins HP: Transverse facial clefts. *Plast Reconstr Surg* **42**:454–459, 1968 (Case 2).

13. Rushton MA, Walker FA: Unilateral secondary facial cleft with excess tooth and bone formation. *Proc R Soc Med* **30**:79–82, 1936.

14. Scholz W: Quere Gesichtsspalte (Makrostoma) bei einem Geschwisterpaar, in *Dysostosen,* Wiedemann HR (ed), G. Fisher Verlag, Stuttgart, 1966, pp 82–84.

15. Shetty DK et al: Klinik und Therapie des Makrostoma. *Zahn Mund Kieferheilkd* **62**:745–749, 1974.

## Oblique facial clefts

Oblique facial cleft (meloschisis) is exceedingly rare. Fogh-Andersen (7) reported 1 case/1300 cases of facial cleft. Gunter (8) found 4 among 900 cases of facial cleft. Wilson et al (23) estimated that they represented 0.25% of clefts.

From the approximately 100 published examples that we have tabulated, the cleft appears to be bilateral in about 20% and more often on the right side when unilateral (2). The cleft is nearly always associated with cleft lip and palate. Occasionally lateral facial cleft is also present (2,4,21). There is no sex predilection. All known cases are sporadic. For historic discussion, see Morian (12) and Boo-Chai (2).

Oblique facial cleft is extremely variable in degree. It may extend through the upper lip to the nose (as a typical cleft lip) to involve the eye, at times even reaching the brow or temple (2,6,17), or it may arise lateral to the philtrum and extend to the eye by skirting the nose (Figs. 20–28, 20–30 and 20–31). The more severe types are, not uncommonly, incompatible with life.

An early attempt was made by Morian (12) to classify the oblique facial cleft into two types: (1) *naso-ocular cleft,* extending from the nostril to the lower eyelid border with possible extension to the temporal region, that is, along the line of closure of the nasolacrimal groove, and (b) *oro-ocular cleft,* extending from the eye to the lip. This latter form, the more common (60%) type, was subdivided into ($b_1$) oromedial canthal type and ($b_2$) orolateral canthal type (1–3,10,22). Tessier (20) proposed a more elaborate and surgically oriented numerical classification of rare facial clefts. Those that involve the nostril are called Type 3; the ones that skirt the nostril are Type 4, and those that extend from the oral commissure to the eye are Type 5. These classifications are important descriptive observations which do not

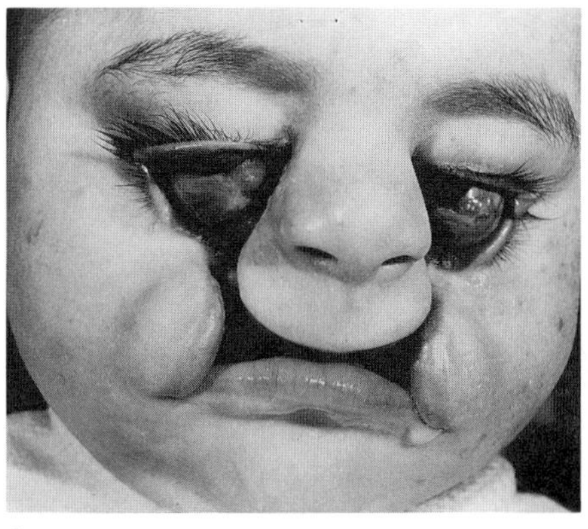
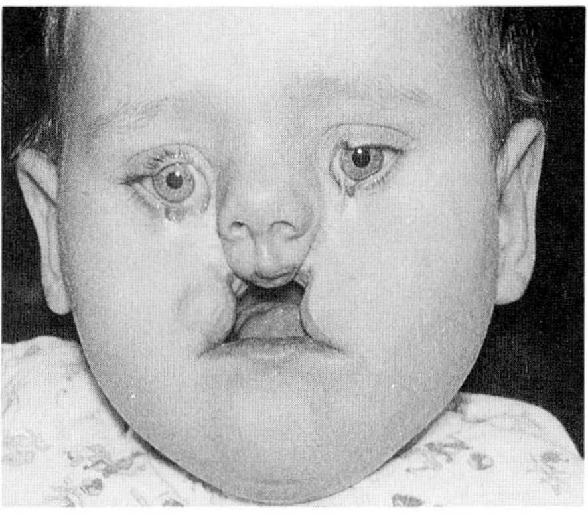
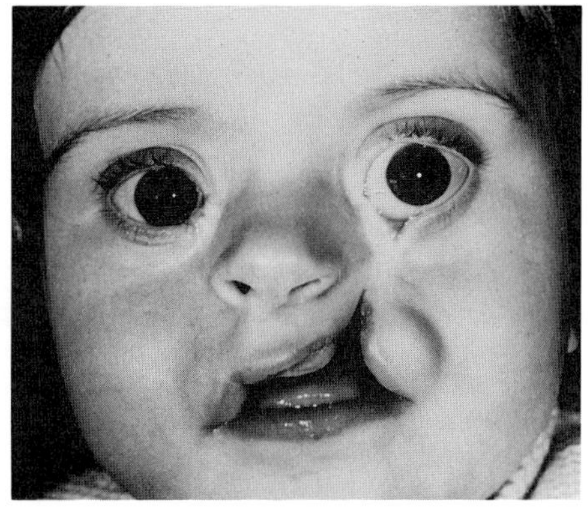

A B C

Fig. 20–30. *Oblique facial clefts.* Bilateral clefts (A,B) and unilateral cleft (C) bypass nostrils and extend roughly along line of closure of nasolacrimal canals. (A,C courtesy of *P Tessier,* Paris, France.)

consider etiopathogenesis. The reader is referred to *Tessier clefting system* for further discussion.

The cleft has been stated to represent failure of penetration of ectomesenchyme between the lateral nasal and maxillary prominences with resultant failure of coverage of the nasolacrimal groove. We believe that most examples are due to tears in the ectomesenchyme related to the *amniotic rupture sequence.* This appears to be clearly evident in those cases associated with constriction rings or bands, intrauterine amputation of digits, or aplasia cutis congenita (2,3,7,10, 11,17,21). In some cases, an amniotic strand is presumably swallowed and becomes attached to the pharynx, slicing the ectomesenchyme of the facial processes.

Other so-called associated anomalies may be interpreted as being secondary to the cleft, such as abnormal eyebrow and/or hairline, facial asymmetry, choanal atresia, absence of nasal ala, agenesis of lacrimal puncta with or without formation of nasolacrimal duct, eyelid colobomas as well as sundry eye anomalies (choroid coloboma, microphthalmia), supernumerary teeth, and dental malocclusion (1a,3–5,14–19).

Various other associated anomalies may occur including hernia (15), genitourinary abnormalities (8), talipes (8,11), spinal and costal defects (4), encephalocele (9,10), and hydrocephalus (12).

Fig. 20–31. *Oblique facial clefts.* Clefts extending from angles of mouth to outer canthi of eyes, following no line of embryonic fusion of facial processes. (From *J Pintanguy* and *T Franco,* Plast Reconstr Surg **39**:569, 1967.)

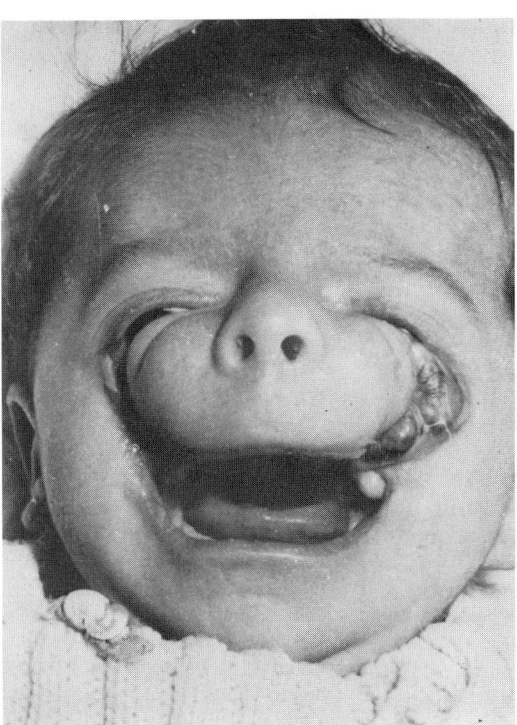

### References (Oblique facial clefts)


1. Bartels RJ et al: Naso-ocular cleft. *Plast Reconstr Surg* **47**:351–353, 1971.

1a. Bhattacharaya S et al: Rare craniofacial clefts with duplication of alveolar arch. *Eur J Plast Surg* **11**:93–94, 1988.

2. Boo-Chai K: The oblique facial cleft: A report of 2 cases and a review of 41 cases. *Br J Plast Surg* **23**:352–359, 1970.

3. Dey DL: Oblique facial clefts. *Plast Reconstr Surg* **52**:258–263, 1973.

4. Dodds GE: A case showing partial deficient fusion of a maxillary process with lateral nasal process on one side. *Br J Ophthalmol* **27**:414–415, 1943.

5. Eckert HA: An unusual bilateral oblique facial cleft. *J Oral Surg* **8**:618–619, 1970.

6. Ergin NO: Naso-ocular cleft. *Plast Reconstr Surg* **38**:573–575, 1966.

7. Fogh-Andersen P: Rare clefts of the face. *Acta Chir Scand* **129**:275–281, 1965.

8. Gunter GS: Nasomaxillary cleft. *Plast Reconstr Surg* **32**:637–645, 1963.

9. Harkins C et al: A classification of cleft lip and cleft palate. *Plast Reconstr Surg* **29**:31–39, 1962.

10. Kubácek V, Pénkava J: Oblique clefts of the face. *Acta Chir Plast* **16**:152–163, 1974.

11. Mayou BJ, Fenton OM: Oblique facial clefts caused by amniotic bands. *Plast Reconstr Surg* **68**:675–681, 1981.

12. Morian R: Über die schräge Gesichtsspalte. *Arch Klin Chir* **35**:245–288, 1887.

13. Onizuka T et al: Naso-ocular clefts. *Plast Reconstr Surg* **61**:118–122, 1978.

14. Ortega J, Flor E: Incomplete naso-ocular cleft. *Plast Reconstr Surg* **43**:630–632, 1969.

15. Potter J: A case of bilateral cleft of the face. *Br J Plast Surg* **3**:209–213, 1951.

16. Rintala A et al: Oblique facial clefts. *Scand J Plast Surg* **14**:291–297, 1980.

17. Sakurai EH et al: Bilateral oblique facial clefts and amniotic bands. *Cleft Palate J* **3**:181–185, 1966.

18. Schlenker JD et al: Classification of oblique facial clefts with microphthalmia. *Plast Reconstr Surg* **68**:675–681, 1981.

19. Tange I, Murofushi H: Six cases with oblique facial cleft. *Jpn J Plast Surg* **9**:114–119, 1966.

20. Tessier P: Anatomical classification of facial, craniofacial and laterofacial clefts. *J Maxillofac Surg* **4**:69–92, 1976.

21. Tower P: Coloboma of lower lid and choroid, with facial defects and deformity of hand and forearm. *Arch Ophthalmol* **50**:333–343, 1953.

22. Van der Meulen JCH: Oblique facial clefts: Pathology, etiology and reconstruction. *Plast Reconstr Surg* **76**:212–224, 1985.

23. Wilson LF et al: Reconstruction of oblique facial clefts. *Cleft Palate J* **9**:109–114, 1972.

## Median cleft of the upper lip

Most median clefts of the upper lip are not true median clefts. They represent agenesis of the primary palate associated with holoprosencephaly with or without trisomy 13.

True median cleft of the upper lip is due to failure of the lowest part of the median nasal prominences to approximate in the midline (10). It is among the rarest of facial clefts, constituting only about 0.2% (4). The median cleft usually involves only the vermilion of the upper lip but occasionally extends between the central incisors into the alveolar process (Fig. 20–32). Often there is duplication of the frenum of the upper lip with a small pit between the mucosal folds. The philtral columns are somewhat widened (1–3,5–7,9,10).

Median cleft of the upper lip may be associated with a number of syndromes such as the *oral-facial-digital syndromes, Ellis–van Creveld syndrome, Majewski syndrome,* and *median cleft of upper lip, double frenum, and hamartoma of columella and/or anterior alveolar ridge.*

Fogh-Andersen (4) very briefly reported male sibs with midline cleft of the upper lip and dysplastic terminal phalanges of the hands (digital amputation and syndactyly). A singular case is one of median cleft of the upper lip, lipomas of the central nervous system, and skin tags (8).

### References (Median cleft of the upper lip)

1. Bishara SE et al: Dentofacial findings in a child with unrepaired median cleft of the lip at 4 years of age. *Am J Orthodont* **88**:157–162, 1985.

1a. Braitwaite F, Watson J: A report on three unusual cleft lips. *Br J Plast Surg* **2**:38–49 (Case 3).

2. Dun RC: Two cases of median hare-lip. *Br Med J* **2**:761, 1909.

3. Fára M: Medial cleft lips. *Acta Chir Plast (Praha)* **19**:1–9, 1977 (Cases S and M).

4. Fogh-Andersen P: Rare clefts of the face. *Acta Chir Scand* **129**:275–291, 1965.

5. Iregbulem LM: Midline clefts of the upper lip. *Br J Plast Surg* **31**:63–65, 1978 (Case 2).

6. Lehman JA Jr, Cuddapah S: The true hare lip—a case report. *Cleft Palate J* **11**:497–498, 1974.

7. Millard R, Williams S: Median clefts of the upper lip. *Plast Reconstr Surg* **42**:4–14, 1961 (Case 11).

8. Pai GS et al: Median cleft of the upper lip associated with lipomas of the central nervous system and cutaneous polyps. *Am J Med Genet* **26**:921–924, 1987.

9. Pěnkava J: Median cleft of the upper lip and jaw. *Acta Chir Plast (Praha)* **16**:201–208, 1974 (Cases P.A. and V.J.).

10. Wiemer DR et al: Anatomic findings in median cleft of upper lip. *Plast Reconstr Surg* **62**:866–869, 1978 (Cases 1,2).

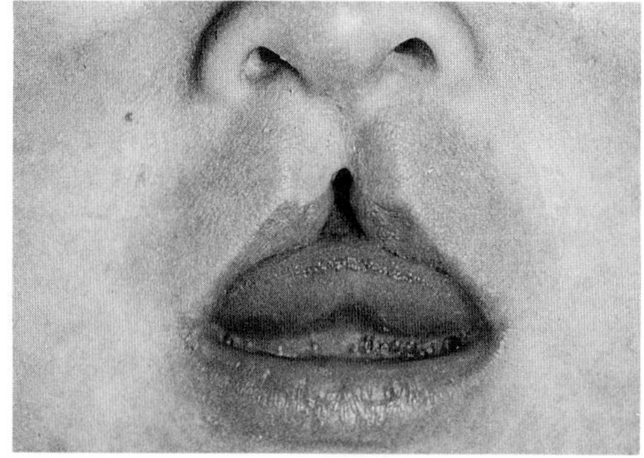

Fig. 20–32. *Median cleft of the upper lip.* Median cleft upper lip. [From *W Hoppe,* Arch Kinderheilkd **171** (Suppl **52**):1, 1965.]

## Median cleft of upper lip, double frenum, and hamartoma of columella and/or anterior alveolar ridge

Several authors (7,8,10–12) have described patients with a median cleft of the upper lip extending through the vermilion. The cleft is associated with duplication of the maxillary median frenum. A pedunculated club-shaped mass has extruded through a nostril, its base attached to the nasal septum while a similar mass was attached to the anterior alveolar process (Figs. 20–33 and 20–34). This represents a malformation of the primary palate. Probable cases include those of Fára (2), Iregbulem (5) and others (1,4a). No other congenital abnormalities are present.

Most cases of isolated median cleft upper lip are associated only with doubling of the frenum (1,4,6,9,14). Midline clefts of the upper lip can be seen in other disorders such as *holoprosencephaly,* palate duplication (3), various *oral–facial–digital syndromes, Ellis–van Creveld syndrome, Majewski syndrome,* and bifid nose (13).

### References (Median cleft of upper lip, double frenum, and hamartoma of columella and/or anterior alveolar ridge)

1. Das BC, Majumder NK: Rare case of congenital malformation of the nose (appendix septi congenita). *Antiseptic* **67**:303–304, 1970.

1a. Dick BM: A case of median cleft of the upper lip. *Edinb Med J* **34**:45–47, 1927.

2. Fára M: Medial cleft lips. *Acta Chir Plast (Praha)* **19**:1–9, 1977 (Case B).

3. Fish J: Median cleft lip with dental arch and palatal duplication. *Br Dent J* **140**:143–145, 1976.

4. Grenman R et al: Midline sinuses of the upper lip. *Scand J Plast Surg* **19**:215–219, 1985 (Case 3).

4a. Krajena Z et al: A rare case of congenital malformation of the nose (appendix septi congenita). *J Laryngol Otol* **79**:1005–1007, 1965.

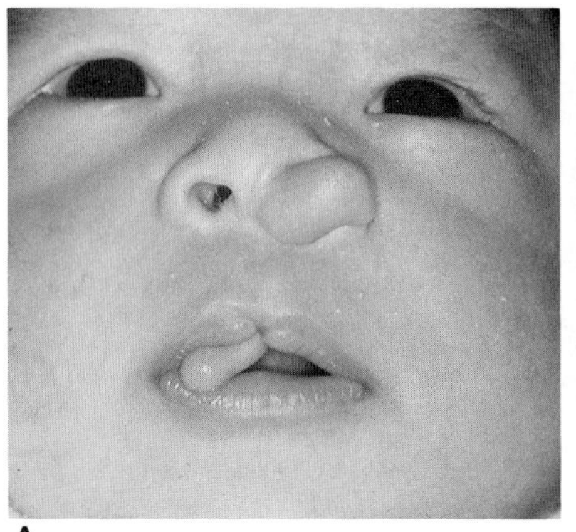

A

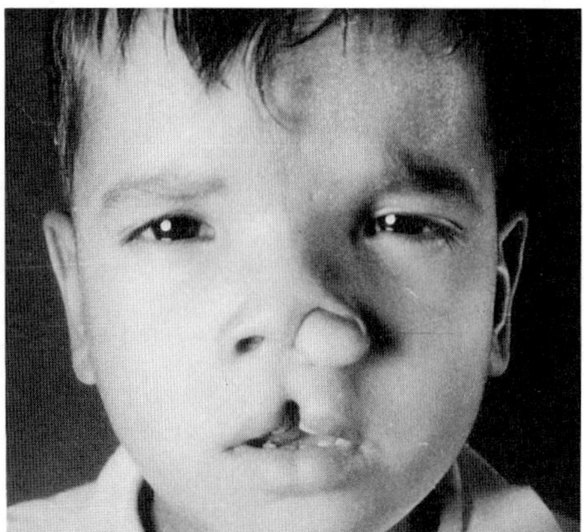

B

Fig. 20–33. *Median cleft of upper lip, double frenum, and hamartoma of columella and/or anterior alveolar ridge.* (A) Median cleft of upper lip, median mass extending from alveolar process, nasal mass extending from columella. (B) Compare anomalies seen in A with those seen in second patient. (A from *J Nakamura et al,* Plast Reconstr Surg **75**:727, 1985. B courtesy of *LK Sharma,* Nagpur, India.)

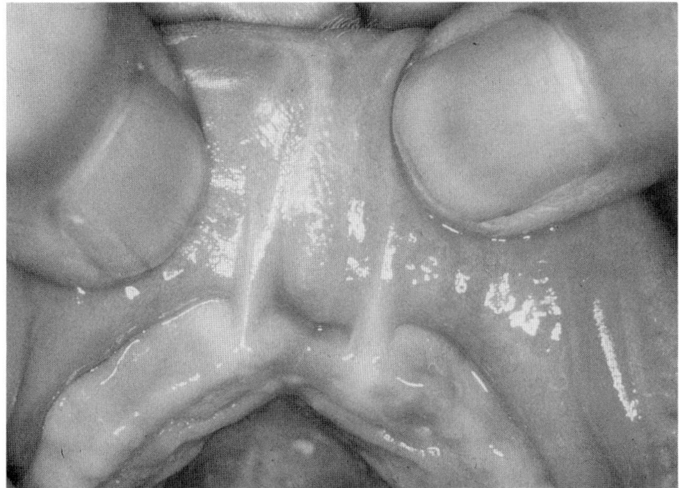

A

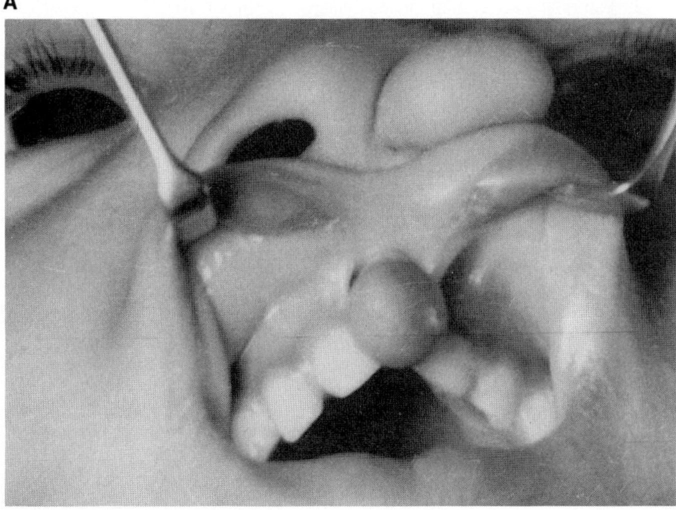

B

Fig. 20–34. *Median cleft of upper lip, double frenum, and hamartoma of columella and/or anterior alveolar ridge.* (A) Duplication of maxillary median frenum. (B) Pedunculated mass attached to frenum area. (A from *J Nakamura et al,* Plast Reconstr Surg **75**:727, 1985. B courtesy of *LK Sharma,* Nagpur, India.)


5. Iregbulem LM: Midline clefts of the upper lip. *Br J Plast Surg* **31**:63–65, 1978 (Case 3).

6. Lehman JA Jr, Cuddapah S: The true hare lip. *Cleft Palate J* **11**:497–498, 1974.

7. Lewin ML: Median cleft lip with pedunculated skin masses. *Plast Reconstr Surg* **79**:843–844, 1987.

8. Nakamura J et al: True median cleft of the upper lip associated with three pedunculated club-shaped skin masses. *Plast Reconstr Surg* **75**:727–731, 1985.

9. Pěnkava J: Median cleft of the upper lip and jaw. *Acta Plast Chir* **16**:201–208, 1974.

10. Ponniah RD: Midline cleft lip with associated abnormalities. *J Laryngol Otol* **91**:177–181, 1977.

11. Scrimshaw GC: Previous reports on median cleft lip. *Plast Reconstr Surg* **77**:159–160, 1986.

12. Sharma LK: Median cleft of the upper lip. *Plast Reconstr Surg* **53**:155–157, 1979.

13. Weaver DF, Bellinger DH: Bifid nose associated with midline cleft of the upper lip. *Arch Otolaryngol* **44**:480–482, 1946.

14. Wiemer DR et al: Anatomic findings in median cleft of upper lip. *Plast Reconstr Surg* **62**:866–869, 1978.


## Median mandibular clefts

This rare anomaly is usually an isolated finding. Redard and Michel (16) reviewed early examples. The earliest may be that of Couronne (3) in 1819. Although Rey et al (17), in 1982, listed 64 examples, they included a number we cannot accept. Perhaps 50 true cases have been reported. The disorder is not hereditary and there is no sex predilection. Its frequency is about 4 to 5 cases/million births (13,14).

Some examples are mild, involving only the lower lip and sparing the bone; some are a bit more extensive (1,20) (Fig. 20–35). In approximately 35%, the tongue is significantly cleft. In nearly all cases there is ankyloglossia, the tongue tip being attached to the median cleft by a broad frenum (1a). In others, there is more extensive ankyloglossia. Only rarely is the tongue spared. The cleft may be very severe, bifurcating the mandible, tongue, and structures of the midneck down to the region of the hyoid bone (Figs. 20–36 and 20–37). Often the hyoid bone and occasionally the manubrium of the sternum are absent. Some patients have a midline cervical cord (4,12). Ranta (15) described 12 cases of minimal cleft of the lower lip with cleft palate in 7, Robin sequence in 3, and/or agenesis of lower central incisor teeth in 2 patients.

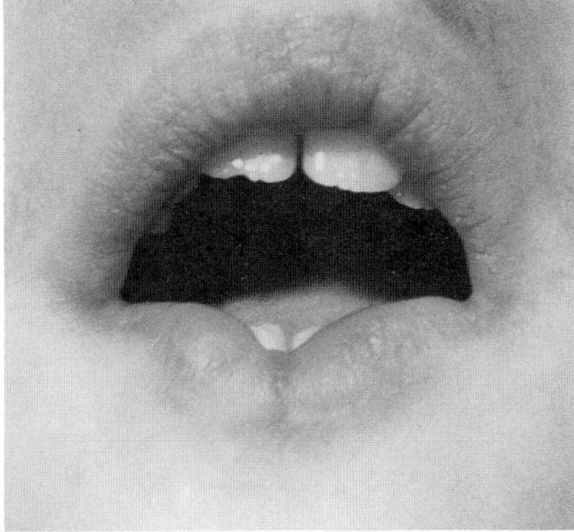

Fig. 20–35. *Median mandibular clefts.* Median cleft of lower lip. Mild cleft involving only soft tissues. (From *R Ranta,* Int J Oral Surg **13**:555, 1984.)

Fig. 20–36. *Median mandibular clefts.* Incomplete cleft of lower lip and chin with complete cleft of mandible. (From *CC Knowles et al,* Br Dent J **127**:337, 1969.)

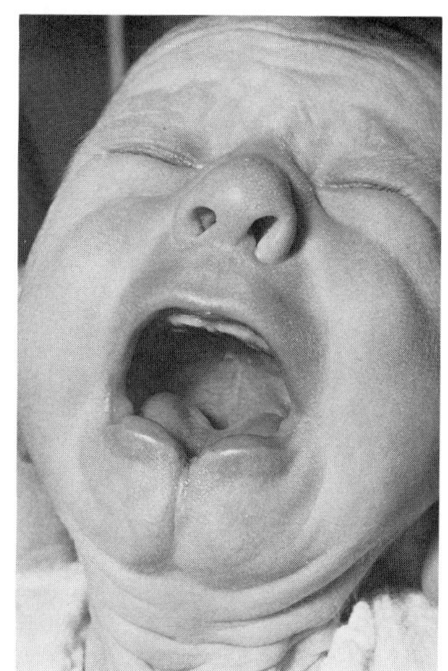

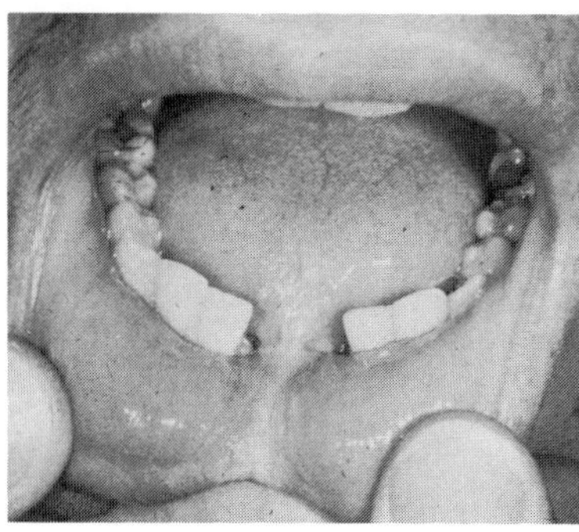

Fig. 20–37. *Median mandibular clefts.* Median cleft of lower lip with tongue tip attached to mandible. (From *HA Ecker,* Am J Surg **96**:815, 1958.)

The anomaly probably results from failure of merging of the paired first and second arches (10). Giroud and Martinet (5) reported an animal model.

Various associated congenital abnormalities appear to be adventitious (4,8,17,19).

Cleft tongue can be seen in association with the *oral–facial–digital syndromes.* It may be an isolated phenomenon (7). It should not be mistaken for true doubling or replication of the tongue (6,18).

#### References (Median mandibular clefts)


1. Ajacques JC et al: Un cas de fente médiane labio-mandibulaire totale. *Actual Odontostomatol* **116**:657–667, 1976.

1a. Chidzonga MM, Shija JK: Congenital median cleft of the lower lip, bifid tongue with ankyloglossia, cleft palate and submental epidermal cyst. *J Oral Maxillofac Surg* **46**:809–812, 1988.

2. Constantinides CG, Cywes S: Complete median cleft of the mandible and aplasia of the epiglottis. *S Afr Med J* **64**:293–294, 1983.

3. Couronne A: Observations des bec de lievre. *Ann Clin Soc Prat Montpellier* **6**(Ser 2):107–111, 1819.

4. Fujino H et al: Median cleft of the lower lip, mandible and tongue with midline cervical cord. *Cleft Palate J* **7**:679–684, 1970.

5. Giroud A, Martinet M: Fissures médianes de la machoire inférieure. *Rev Stomatol (Paris)* **65**:796–804, 1964.

6. Greig DM: Schistoglossus and double tongue. *Edinburgh Med J* **32**:1–16, 1925.

7. Hubinger HL: Bifid tongue. *J Oral Surg* **10**:64–66, 1952.

8. Iregbulem LM: Median cleft of the lower lip. *Plast Reconstr Surg* **61**:787–789, 1978.

9. Marcadé E et al: Fente médiane de la lèvre inférieure. *Ann Chir Plast* **26**:148–150, 1981.

10. Millard DR et al: Median cleft of the lower lip and mandible. *Br J Plast Surg* **24**:391–395, 1971.

11. Millard DR et al: Median cleft of the lower lip and mandible: Correction of the mandibular defect. *Br J Plast Surg* **32**:345–347, 1979.

12. Monroe CW: Midline cleft of the lower lip, mandible and tongue with flexion contracture of the neck. *Plast Reconstr Surg* **38**:312–319, 1966.

13. Ploner J: Betrachtung zur medianen Unterlippen-Kieferspalte. *Fortschr Kiefer Gesichts Chir* **2**:263–265, 1956.

14. Petit P, Psaume J: Fente médiane de la lèvre inférieure. *Ann Chir Plast* **10**:91–96, 1965.

15. Ranta R: Incomplete median cleft of the lower lip with cleft palate, the Pierre Robin anomaly or hypodontia. *Int J Oral Surg* **13**:555–558, 1984.

16. Redard P, Michel F: Sillon médiane profond de la lèvre inférieure et du menton. *Presse Méd* **7**:191–193, 1899.

17. Rey A et al: Fentes labio-mandibulaires. *Rev Stom Chir Maxillofac* **83**:39–44, 1982.

18. Seitz A: A case of double tongue. *Med Rec* **61**:474–475, 1902.

19. Sherman JE, Goulian D: The successful one stage surgical management of a midline cleft of the lower lip, mandible and tongue. *Plast Reconstr Surg* **66**:756–759, 1980.

20. Weinberg S et al: Midline cleft of the mandible. *J Oral Surg* **30**:143–148, 1972.

# Chapter 21
# Orofacial Clefting Syndromes: Common and/or Well-Known Syndromes

## Cerebro-costo-mandibular syndrome (rib-gap syndrome)

The syndrome originally described by Smith et al (18), in 1966, consists of both intrauterine and postnatal growth disturbance, cerebral maldevelopment or malfunction, abnormal rib development, cleft palate, and micrognathia. Although about 40 cases have been reported, it is possible that the correct diagnosis of other examples may not have been made due to failure to perform autopsy or to take a radiograph of the chest.

The inheritance pattern is unknown (17). There is no sex predilection. McNichol et al (10) and Trautman et al (21) reported affected sibs. Leroy et al (9) described the disorder in a mother and in her two children by different fathers. Merlob et al (12) reported the disorder in a father and daughter. Hennekam et al (4) noted affected sibs, one of them having only mild expression. In three cases, relatives had some stigmata (3,13,16). All other examples have been isolated. Consanguinity was noted in only one case (1a). Chromosome studies have been normal (15). We are not certain how to classify one example (13). The child, the product of an incestuous union, exhibited cleft palate, hypoplasia of the sternum, clavicles and pubis, semirudimentary ribs, and porencephalic cyst.

Polyhydramnios has been reported in a few cases (3,10,12). About 35% have died prior to the first year of life.

Microcrania has been reported in about 35% (11,17), and mental retardation, ranging from moderate to severe, has been documented in about 35% of those that have survived. Specifically, normal mentation has been described (8,11,15). A consistent feature, micrognathia is usually severe, several patients requiring tracheotomy (Fig. 21–1A).

The constant feature is bilateral rib gaps between the costovertebral junction area and the lateral arc. They are most frequent between the third and seventh thoracic segments, although all 12 levels have been affected. However, the lower ribs tend to be spared. Rudimentary ribs may occur at the second level with absent twelfth ribs in about one-half the cases (11) (Fig. 21–1B). Follow-up reveals that the gaps are ultimately joined by bony bridges (2,23).

Respiratory difficulties are evident in about 75% in early infancy.

Miscellaneous findings have included pterygium colli (2,6,17,18), and an assortment of orthopedic anomalies: clubfoot, scoliosis (14), hip dislocation, stippled epiphyses (1), elbow dysplasia (3,6,8,10, 19,23), and conductive hearing loss (2,9,17).

The mandible has been small in all cases. Cleft palate (65%) and glossoptosis (50%) have been noted and several infants were classified as examples of Robin sequence (11,15,22).

### References [Cerebro-costo-mandibular syndrome (rib-gap syndrome)]

1. Burton EM, Oestreich AE: Cerebro-costo-mandibular syndrome with stippled epiphyses and cystic fibrosis. *Pediatr Radiol* **18**:365–367, 1988.

1a. Clarke EA, Nguyen VD: Cerebro-costo-mandibular syndrome with consanguinity. *Pediatr Radiol* **15**:264–266, 1985.

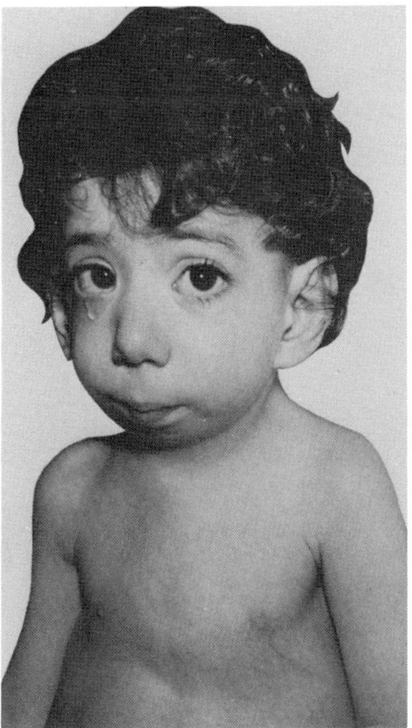

A

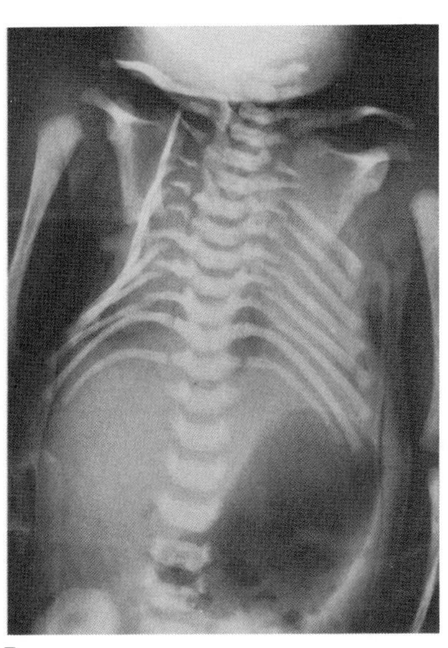

B

Fig. 21–1. *Cerebro-costo-mandibular syndrome.* (A) One of three sibs with microcephaly, thoracic deformity with rib gaps, and vertebral anomalies. (B) Roentgenogram showing severely narrowed upper part of the thorax and bizarre vertebral and rib anomalies. (From *B McNicholl et al,* Arch Dis Childh **5**:421, 1970.)

2. Fauré C et al: Le syndrome cérébro-costo-mandibulaire. Trois nouvelles observations. *Nouv Presse Méd* **7**:445–448, 1978.

3. Harris DJ, Fellows RA: The course of the cerebrocostomandibular syndrome. *Birth Defects* **13**(3C):117–130, 1977.

4. Hennekam RC et al: The cerebro-costo-mandibular syndrome: Third report of familial occurrence. *Clin Genet* **28**:118–121, 1985.

5. Kemperdick H, Lemburg P: Rippenserienfrakturen oder Rippenfehlbildung? *Klin Pädiatr* **188**:278–280, 1976.

6. Kringelbach J, Henriksen K: Anomalies of the ribs combined with other mesodermal developmental defects—a new syndrome? *Dan Med Bull* **15**:139–142, 1968.

7. Kuhn JP et al: Cerebro-costo-mandibular syndrome. A case with cardiac anomaly. *J Pediatr* **86**:243–244, 1975.

8. Langer LO Jr, Herrmann J: The cerebro-costo-mandibular syndrome. *Birth Defects* **10**(7):167–170, 1974.

9. Leroy JG et al: Cerebro-costo-mandibular syndrome with autosomal dominant inheritance. *J Pediatr* **99**:441–443, 1981.

10. McNicholl B et al: Cerebro-costo-mandibular syndrome: New familial developmental disorder. *Arch Dis Childh* **45**:421–424, 1970.

11. Meinecke P et al: Cerebro-costo-mandibuläres Syndrom ohne cerebrale Beteiligung bei einem 4 jährigen Jungen. *Mschr Kinderheilkd* **135**:54–58, 1987.

12. Merlob P et al: Autosomal dominant cerebro-costo-mandibular syndrome: Ultrasonographic and clinical findings. *Am J Med Genet* **26**:195–202, 1987.

13. Miller KE et al: Rib gap defects with micrognathia. The cerebro-costomandibular syndrome—a Pierre Robin-like syndrome with rib dysplasia. *Am J Roentgenol* **114**:253–256, 1972.

14. Mohan M, Mandalam KR: Cerebro-costo-mandibular syndrome. *Indian Pediatr* **19**:97–98, 1982.

15. Nicholls S, Fletcher E: Congenital rib defect with the Pierre Robin syndrome. *Pediat Radiol* **1**:246–247, 1973.

16. Schroer RJ, Meyer LC: Cerebro-costo-mandibular syndrome. *Proc Greenwood Genet Ctr* **4**:55–59, 1985.

17. Silverman FN et al: Cerebro-costo-mandibular syndrome. *J Pediatr* **97**:406–415, 1980.

18. Smith DW et al: Rib gap defect with micrognathia, malformed tracheal cartilages and redundant skin: New pattern of development. *J Pediatr* **69**:799–803, 1966.

19. Smith KG, Sekar KC: Cerebrocostomandibular syndrome. *Clin Pediatr* **24**:223–225, 1985.

20. Tachibana K et al: Cerebro-costo-mandibular syndrome. *Hum Genet* **54**:283–286, 1980.

21. Trautman MS et al: Cerebro-costo-mandibular syndrome: A familial case consistent with autosomal recessive inheritance. *J Pediatr* **107**:990–991, 1985.

22. Walizadeh GR: Pierre Robin–Syndrom mit Rippenanomalien. *Röfo* **129**:275–276, 1978.

23. Williams HJ, Sane SM: Cerebro-costo-mandibular syndrome: Long term follow-up of a patient and review of the literature. *Am J Roentgenol* **126**:1223–1228, 1976.

## EEC syndrome (ectrodactyly–ectodermal dysplasia–clefting syndrome)

The first report of the syndrome of lobster-claw anomaly of the hands and feet, nasolacrimal duct obstruction, and cleft lip–palate was likely that of Eckoldt and Martens (16) in 1804. Another early description was by Cruveilhier (14). In excess of 100 cases have been reported (30). Most reports describe sporadic examples of the disorder (4–6,10,19,25–27,37,41–43,49,51,56–58,61,62). However, several investigators have noted transmission of the disorder from an affected parent to one or more children (8,12,13,25,30,35,45,47,53,58,61,64,70); others have reported affected sibs born to presumably normal parents (1,49,65). Thus, genetic heterogeneity may exist. There is also considerable variation in expression among affected members of the same kindred (35,58). Pedigrees with skipped generations have been noted (13,17,25,65).

More families with vertical transmission have been reported than with horizontal transmission. Thus, the condition may have autosomal dominant inheritance with low penetrance and variable expressivity. Alternately, genetic heterogeneity may exist.

**Facies.** The facies is characterized by dacryocystitis, keratoconjunctivitis, tearing, photophobia, and cleft lip. Scalp hair, lashes, and eyebrows are nearly always sparse (6,19,41) (Figs. 21-2 and 21–3).

**Extremities.** Lobster-claw anomaly (ectrodactyly) involves all four extremities in 90% (51); however, one hand may be normal (13). Other patients have been described without ectrodactyly (30) (Fig. 21–4). Occasionally, soft tissue syndactyly, especially of the toes, occurs (1,10,43).

**Eyes.** Absent lacrimal punctas have been noted in about 90% (Fig. 21–5). This abnormality is associated with tearing, blepharitis, dacryocystitis, keratoconjunctivitis, and photophobia (3,10,13,19,26,28a,31,34,37,43,53,55,57,61–63,65–70). Corneal ulcers and scarring (10,19,43,49) and primary telecanthus (6) have been observed. The number of meibomian orifices is reduced (36,43,49,61,64).

**Ears.** Conductive hearing loss has been documented in about 30% (49,51,53).

**Skin, hair, and nails.** Hypopigmentation of the skin and hair has been noted in most white patients (1,6,10,19,43,56), but black patients have normal pigmentation. Scalp hair, eyebrows, and lashes are

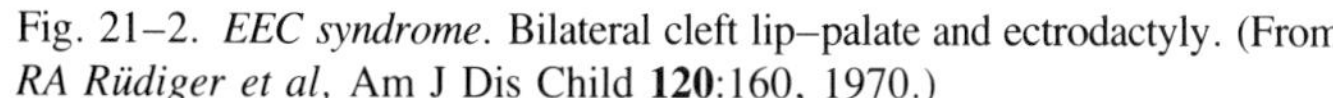
Fig. 21–2. *EEC syndrome.* Bilateral cleft lip–palate and ectrodactyly. (From *RA Rüdiger et al,* Am J Dis Child **120**:160, 1970.)

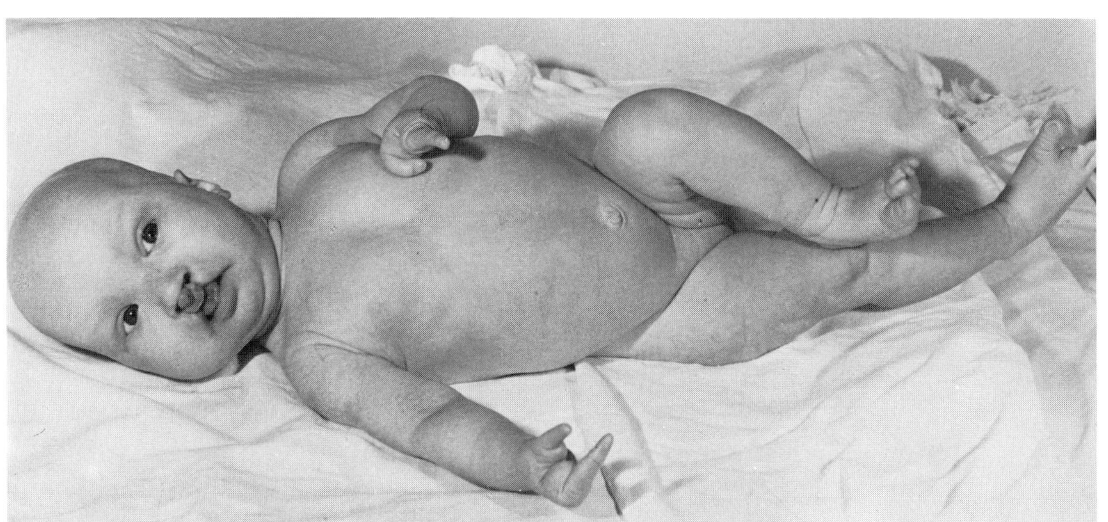

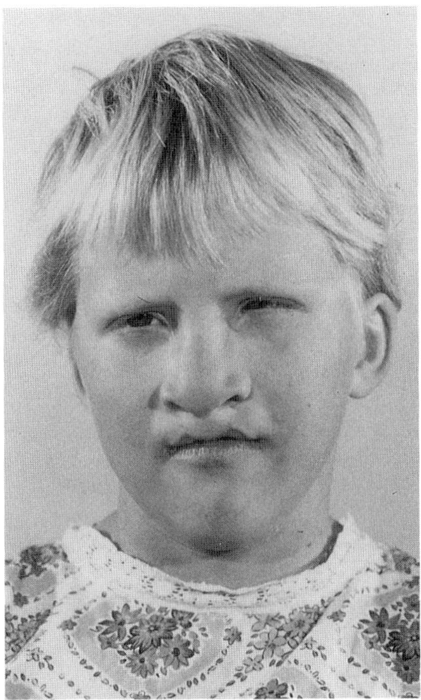

Fig. 21–3. *EEC syndrome.* Patient has photophobia, clefting, and sparse hair.

sparse (8,10,19,43,56,57,62); on light microscopic examination in two patients, the hair was found to be normal (43,61). Nails are nearly always dystrophic (6,26,43,51,55,56,62). Absence or sparse sebaceous glands have been observed on skin biopsy (43,49,61,64). About 12% have many pigmented nevi (10,19,43,49), and widespread comedone nevi have been described (31).

**Central nervous system.** Microcephaly and mental retardation have been reported in about 10% (1,4,6,19,20,27,43,57) but there may have been an ascertainment bias. Growth hormone deficiency has been described (28).

**Genitourinary system.** Kidney and ureter malformations (duplication of the kidney, collecting system and ureter, absent kidney, small dysplastic kidney, hydronephrosis, and hydroureter) have been described in at least 20% (2,6–8,19,20,24,26,29,33,34,48,49,59,64–66) as has hypospadias (15,51,53,61). Cryptorchidism (2,15,21, RJ Gorlin, unpublished) and prune belly (24) have been noted.

**Otolaryngologic manifestations.** Conductive hearing loss has been reported (7,10,12,26,43,49,53,62). Absence of the stapes and part of the incus was found in one patient (12). Breathy voice has been observed in some patients (46, RJ Gorlin, personal observation). Laryngoscopic examination showed no visible form of incomplete closure along the vocal folds, although the folds were dry, suggesting that reduction in lubrication resulted in an incomplete seal between the folds during phonation. Spectrographic analysis showed abnormal voice quality (46).

**Oral manifestations.** Cleft lip–palate, often bilateral, has been described in about 75% (6,10,12,13,19,26,31,34,42,43,49,51,53, 57,61,62,64). In possibly 10%, cleft palate without cleft lip was noted (8,46,48). Clefting is absent in other patients (8,12,15,31, 34,43,48,72). Congenitally missing permanent teeth and coniform teeth (31) are common. Occasionally, the maxillary deciduous first molars are missing (50). Enamel dysplasia (6,9,18,49,55,62, 63,66,67) has been described. Xerostomia (43,49), requiring large amounts of water while eating, and enamel dysplasia may contribute to high dental caries rate. Parotid duct atresia has been documented (43,49). A deep anteroposterior furrow in the midline of the dorsum of the tongue has also been described (43,49). Candidal cheilitis and candidal perleche have been reported (43,49).

**Differential diagnosis.** Wildervanck (71) described two sibs with severe sensorineural hearing impairment and ectrodactyly of the hands and feet, born to presumably normal parents; no other abnormalities were recorded. Patterson and Stevenson (44) reported a patient with ectrodactyly and craniofacial findings reminiscent of *mandibulofacial dysostosis.* Reed et al (52) described a mother and daughter with ectrodactyly, lacrimal duct obstruction, early graying of the hair with pili torti, subtotal alopecia, and atrophic pigmented macules on the extensor surfaces of the body. The syndrome of *cleft lip–palate, hypohidrosis, thin, wiry hair, and dystrophic nails (Rapp–Hodgkin syndrome)* is inherited as an autosomal dominant disorder. The *ECP syndrome* shares similar features with the EEC syndrome, but in one large family there was only cleft palate and ectrodactyly (40). The *odontotrichomelic* syndrome *(cleft lip–palate, tetraperomelia, de-*

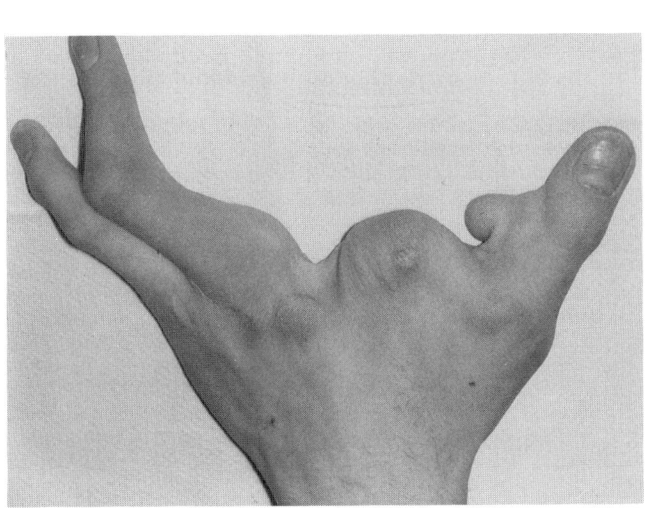

A

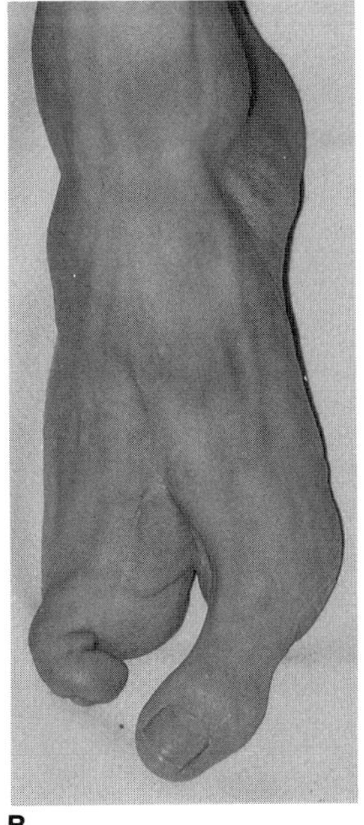

B

Fig. 21–4. *EEC syndrome.* (A,B) Ectrodactyly of hands and feet.

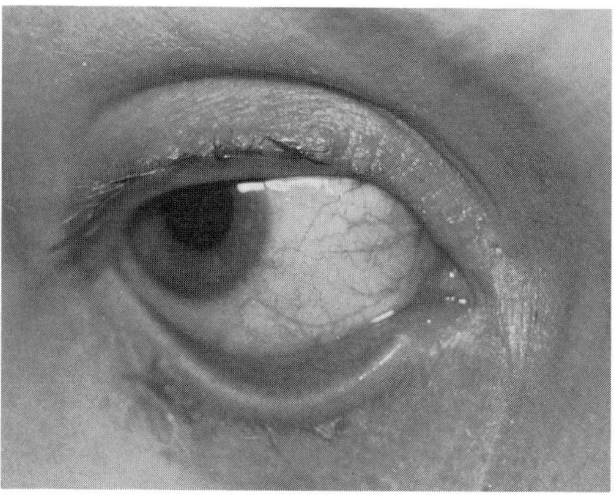

Fig. 21–5. *EEC syndrome.* Absence of lacrimal point.

*formed pinnas, and ectodermal dysplasia)* and *Bowen–Armstrong syndrome* appear to follow an autosomal recessive mode of inheritance (7). Patients with the syndrome of *craniosynostosis, severe symmetrically malformed extremities, and cleft lip–palate (Herrmann syndrome)* also resemble those with the EEC syndrome. Stewart et al (60) reported cleft lip–palate with ectrodactyly and hypomelanosis of Ito. Findings that overlap the EEC syndrome may be seen in *fetal alcohol syndrome* (22) and *odontotrichomelic syndrome.* Ohdo et al (39) reported a kindred with ectrodactyly, syndactyly, sparse, thin hair, and macular dystrophy; inheritance was autosomal recessive. Cat et al (11) described an autosomal recessive disorder characterized primarily by hypohidrosis, hypotrichosis, hypodontia, cleft lip, dysplastic nails, reduction deformities of all four limbs, and thin skin. Although lacrimal duct obstruction occurs in 1–2% of the general childhood population, it has been found in about 10% of patients with isolated cleft lip–palate (68). Hoyme et al (23) described autosomal dominant ectrodactyly and absence of long bones of upper or lower extremities. Ogur and Yüksel (38) described two male sibs, the product of a consanguineous union, with tetramelic syndactyly, ectodermal dysplasia, cleft lip and palate, renal anomalies, and mental retardation. Wallis (66a) reported four generations of ectrodactyly with ectodermal dysplasia but without clefting. Atkin and Patil (2a) reported a male infant and his maternal uncle with microphthalmia, cloudy corneae, and a CNS anomaly. The infant also had ectrodactyly, cleft lip, brain cyst, renal anomaly, hypospadias, and cryptorchidism. A host of additional ectrodactyly syndromes has been reviewed by Schroer (59a). MM Cohen, Jr has seen an EEC-like disorder with hypertelorism and some band-like disruptions of the limbs.

### References [EEC syndrome (ectrodactyly–ectodermal dysplasia–clefting syndrome)]


1. Ahrens K: Chromosomale Untersuchungen bei craniofacialen Missbildungen (Cases 3A, B). *HNO* **15**:106–109, 1967.

2. Aldenhoff P et al: Das EEC-Syndrom. *Mschr Kinderheilk* **126**:575–578, 1978.

2a. Atkin JF, Patil S: Apparently new oculo-cerebro-acral syndrome. *Am J Med Genet* **19**:585–587, 1984.

3. Baum JL, Bull MJ: Ocular manifestations of the ectrodactyly, ectodermal dysplasia, cleft lip-palate syndrome. *Am J Ophthalmol* **78**:211–216, 1974.

4. Berendt H: Case report of partial anodontia connected with missing and stunted phalanges of hands and feet. *Oral Surg* **1**:283–290, 1948.

5. Berndorfer A: Gesichtsspalten gemeinsam mit Hand-und Fussspalten. *Z Orthopäd* **107**:344–354, 1970.

6. Bixler D et al: The ectrodactyly–ectodermal dysplasia–clefting (EEC) syndrome. *Clin Genet* **3**:43–51, 1971.

7. Bowen P, Armstrong HB: Ectodermal dysplasia, mental retardation, cleft lip/palate and other anomalies in three sibs. *Clin Genet* **9**:35–42, 1976.

8. Brill CB et al: The syndrome of ectrodactyly, ectodermal dysplasia and cleft lip and palate. *Clin Genet* **3**:293–302, 1972.

9. Brunn C: Et tilfaelde af dysplasia acro-dentalis. *Tandlaegebladet* **72**:162–167, 1968.

10. Bystrom EB et al: The syndrome of ectrodactyly, ectodermal dysplasia and clefting (EEC). *J Oral Surg* **33**:192–198, 1975.

11. Cat I et al: Odontotrichomelic hypohidrotic dysplasia: A clinical reappraisal. *Hum Hered* **22**:91–95, 1972.

12. Chiang TP, Robinson GC: Ectrodactyly, ectodermal dysplasia, and cleft lip/palate syndrome: The importance of dental anomalies. *J Dent Child* **41**:38–42, 1974. (Same family as reported by Robinson el al, Ref. 53.)

13. Cockayne EA: Cleft palate, hare lip, dacryocystitis and cleft hand and feet. *Biometrika* **28**:60–63, 1936.

14. Cruveilhier J: *Anatomie Pathologique du Corps Humaine. Maladies des Extremities,* Vol. II, part 38, S. Bailliere, Paris, 1829–1842..

15. Curran AS, Curran JP: Associated acral and renal malformations: A new syndrome? *Pediatrics* **49**:716–725, 1972.

16. Eckoldt JG, Martens FH: *Über eine sehr komplicierte Hasenscharte.* Steinacker, Leipzig, 1804.

17. Fraser FC: Genetic counseling. *Hosp Pract* **6**:49–56, 1971.

18. Freuenthaller P: Kephalometrische Untersuchungen und Berichte über Fälle von Anodontie und Oligodontie. *Schweiz Mschr Zahnheilkd* **76**:484–489, 1966.

19. Fried K: Ectrodactyly–ectodermal dysplasia–clefting (EEC) syndrome. *Clin Genet* **3**:396–400, 1972.

20. Gehler J, Grosse R: Fehlbildungs-Retardierungs-Syndrom mit Spalthänden-Spaltfüssen, Iriskolobom, Nierenagenesie und Ventrikelseptumdefekt. *Klin Pädiatr* **184**:389–392, 1972.

21. Hecht F: Updating a diagnosis: The EEC/EECUT syndrome. *Am J Dis Child* **139**:1185, 1985.

22. Herrmann J et al: Tetraectrodactyly and other skeletal manifestations in the fetal alcohol syndrome. *Eur J Pediatr* **133**:221–226, 1980.

23. Hoyme HE et al: Autosomal dominant ectrodactyly and absence of long bones of upper or lower limbs: Further clinical delineation. *J Pediatr* **111**:538–543, 1987.

24. Ivarrson S et al: Coexisting ectrodactyly–ectodermal dysplasia–clefting (EEC) and prune belly syndromes. *Acta Radiol Diag* **23**:287–292, 1982.

25. Jaworska M, Popiolek J: Genetic counseling in lobster claw anomaly: Discussion of variability of genetic influence in different families. *Clin Pediatr* **7**:396–399, 1968.

26. Kaiser-Kupfer M: Ectrodactyly, ectodermal dysplasia and clefting syndrome. *Am J Ophthalmol* **76**:992–998, 1973.

27. Kellner AW: Über Spalthand und-Fuss mit Oligodaktylie. *Klin Wochenschr* **13**:1507–1509, 1934.

28. Knudtzon J, Aarskog D: Growth hormone deficiency associated with ectodactyly–ectodermal dysplasia–clefting syndrome and isolated absent septum pellucidum. *Pediatrics* **79**:410–412, 1987.

28a. Koniszewski G et al: Augenbeteiligung bei ektodermaler Dysplasie. *Klin Monatsbl Augenheilkd* **190**:519–523, 1987.

29. Küster W: Further reports of urinary tract involvement in EEC syndrome. *Am J Dis Child* **140**:411, 1986.

30. Küster W et al: EEC syndrome without ectrodactyly? Report of 8 cases. *Clin Genet* **28**:130–135, 1985.

31. Leibowitz MR, Jenkins T: A newly recognized feature of ectrodactyly, ectodermal dysplasia, clefting (EEC) syndrome: Comedone naevus. *Dermatologica* **169**:80–85, 1984.

32. Lewis MB, Pashayan HM: Ectrodactyly, cleft lip and palate in two half sibs. *J Med Genet* **18**:394–396, 1981.

33. London R et al: Urinary tract involvement in EEC syndrome. *Am J Dis Child* **139**:1191–1193, 1985.

34. Maisels DO: Lobster-claw deformities of the hand and feet. *Br J Plast Surg* **23**:269–282, 1970.

35. Majewski F, Küster W: EEC syndrome sine sine? *Clin Genet* **33**:69–72, 1988.

36. Mondino BT et al: Absent meibomian glands in the ectrodactyly-ectodermal dysplasia-clefting syndrome. *Am J Ophthalmol* **97**:496–501, 1984.

37. Nazif M: Hypodontia, anomalies of the extremities, and associated stenosis of lacrimal ducts. *J Dent Child* **40**:55–57, 1973.

38. Ogur G, Yüksel M: Association of syndactyly, ectodermal dysplasia, and cleft lip and palate: Report of two sibs from Turkey. *J Med Genet* **25**:37–40, 1988.

39. Ohdo S et al: Association of ectodermal dysplasia, ectrodactyly, and macular dystrophy: The EEM syndrome. *J Med Genet* **20**:52–57, 1983.

40. Opitz JM et al: The ECP syndrome, another autosomal dominant cause of monodactylous ectrodactyly. *Eur J Pediatr* **133**:217–220, 1980.

41. Parent P et al: Le syndrome EEC. *Ann Pédiatr* **34**:293–300, 1987.

42. Parkash H et al: Ectrodactyly, ectodermal dysplasia, cleft lip and palate (EEC)—a rare syndrome. *Ind J Pediatr* **50**:337–340, 1983.

43. Pashayan HM et al: The EEC syndrome. *Birth Defects* **10**(7):105–127, 1974.

44. Patterson TJS, Stevenson AC: Craniofacial dysostosis and malformations of feet. *J Med Genet* **1**:112–114, 1964.

45. Penchaszadeh VB, deNegrotti TC: Ectrodactyly–ectodermal dysplasia–clefting (EEC) syndrome, dominant inheritance and variable expression. *J Med Genet* **13**:281–284, 1976.

46. Peterson-Falzone SJ et al: Abnormal laryngeal vocal quality in ectodermal dysplasia. *Arch Otolaryngol* **107**:300–304, 1981.

47. Pfeiffer RA, Verbeck C: Spalthand und Spaltfuss, ectodermale Dysplasie und Lippen–Kiefer–Gaumen–Spalte: ein autosomal dominant vererbtes Syndrom, *Z Kinderheilkd* **115**:235–244, 1973.

48. Preus M, Fraser FC: The lobster claw defect with ectodermal defects, cleft lip-palate, tear duct anomaly and renal anomalies. *Clin Genet* **4**:369–375, 1973.

49. Pries C et al: The EEC syndrome. *Am J Dis Child* **127**:840–844, 1974.

50. Psaume J et al: Douze observations du syndrome Ectrodactylie Dysplasie Ectodermique. Fente faciale, syndrome E.E.C. *Rev Stomatol* **82**:226–229, 1981.

51. Rasmussen SA et al: Further delineation of the EEC syndrome. *Proc Greenwood Genetics Ctr,* in press.

52. Reed WB et al: The REEDS syndrome. *Birth Defects* **10**(8):61–73, 1974.

53. Robinson GC et al: Ectrodactyly, ectodermal dysplasia and cleft lip–palate syndrome. *J Pediatr* **82**:107–109, 1973. (Same cases as reported by Chiang and Robinson, Ref. 12.)

54. Rollnick BR, Hoo JJ: Genitourinary anomalies are a component manifestation in the ectodermal dysplasia, ectrodactyly, cleft lip/palate (EEC) syndrome. *Am J Med Genet* **29**:131–136, 1988.

55. Rosenmann A et al: Ectrodactyly, ectodermal dysplasia and cleft palate (EEC syndrome). *Clin Genet* **9**:347–353, 1976.

56. Rosselli D, Gulienetti R: Ectodermal dysplasia. *Br J Plast Surg* **14**:190–204, 1961 (Case 1).

57. Rüdiger RA et al: Association of ectrodactyly, ectodermal dysplasia and cleft lip-palate: The EEC syndrome. *Am J Dis Child* **120**:160–163, 1970.

58. Schmidt R, Nitowsky HM: Split hand and foot deformity and the syndrome of ectrodactyly, ectodermal dysplasia, and clefting (EEC). A report of five patients. *Hum Genet* **39**:15–25, 1977.

59. Schnitzler L et al: Le syndrome de Rüdiger syndrome (EEC). *Ann Dermatol Venereol (Paris)* **105**:201–206, 1978.

59a. Schroer RJ: Split-hand/split-foot. *Proc Greenwood Genet Ctr* **5**:65–75, 1986.

60. Stewart RE et al: A malformation complex of ectrodactyly, clefting and hypomelanosis of Ito (incontinentia pigmenti achromians). *Cleft Palate J* **16**:358–362, 1979.

61. Summitt RL, Hiatt RL: Hypohidrotic ectodermal dysplasia with multiple associated anomalies. *Birth Defects* **7**(8):121–124, 1971.

62. Swallow JN et al: Ectrodactyly, ectodermal dysplasia and cleft lip and cleft palate (EEC syndrome). *Br J Dermatol* **89***(Suppl* **9***)*:54–56, 1973.

63. Temtamy S, McKusick VA: Synopsis of hand malformations with particular emphasis on genetic factors. *Birth Defects* **5**(3):125–184, 1969.

64. Temtamy SA, McKusick VA: The genetics of hand malformations. *Birth Defects* **14**(3):157–172, 1978.

65. Walker JC, Clodius L: The syndromes of cleft lip, cleft palate and lobster-claw deformities of hands and feet. *Plast Reconstr Surg* **32**:627–636, 1963.

66. Walker JC et al: Ectodermal dysplasia, *Trans 4th Int Cong Plast Reconstr Surg,* Rome, 1967, pp 122–128.

66a. Wallis CE: Ectrodactyly (split-hand/split-foot) and ectodermal dysplasia with normal lip and palate in a four generation kindred. *Clin Genet* **34**:252–257, 1988.

67. Wegner J: Über Hypdontia vera der Milch- und Ersatz-zähne bei vererbter Ekto-und Mesoderm Dysplasie. *Dtsch Zahnärztl Z* **13**:1019–1026, 1958.

68. Whitaker L et al: The lacrimal apparatus in patients with cleft lip and palate. *Proc 2nd Int Cong Cleft Palate,* Copenhagen, 1973, p 167.

69. Wiedemann H-R, Dibbern H: EEC-Syndrom. *Med Welt* **31**:1862–1863, 1980.

70. Wiegmann OA, Walker FA: The syndrome of lobster claw deformity and nasolacrimal obstruction. *J Pediatr Ophthalmol* **7**:79–85, 1970.

71. Wildervanck LS: Perceptive deafness associated with split-hand and -foot, a new syndrome? *Acta Genet (Basel)* **13**:161–169, 1963.

## ECP syndrome (ectrodactyly–cleft palate syndrome)

In 1980, Opitz et al (2) reported an autosomal dominantly inherited syndrome of cleft palate and ectrodactyly involving the hands and feet in a large five generation pedigree (Figs. 21–6A,B and 21–7). Features of the *EEC syndrome* such as cleft lip, hypotrichosis, nail dysplasia, and defects of the lacrimal canal were not observed. Other less common manifestations of the EEC syndrome such as mental retardation, microcephaly, conductive hearing loss, renal anomalies, and cryptorchidism were also not observed. The ECP syndrome is, in addition, distinct from *Fontaine syndrome* (1) consisting of ectrodactyly of the feet, micrognathia, submucous cleft palate, dysplastic ears and, in some instances, mental deficiency. Inheritance is autosomal dominant.

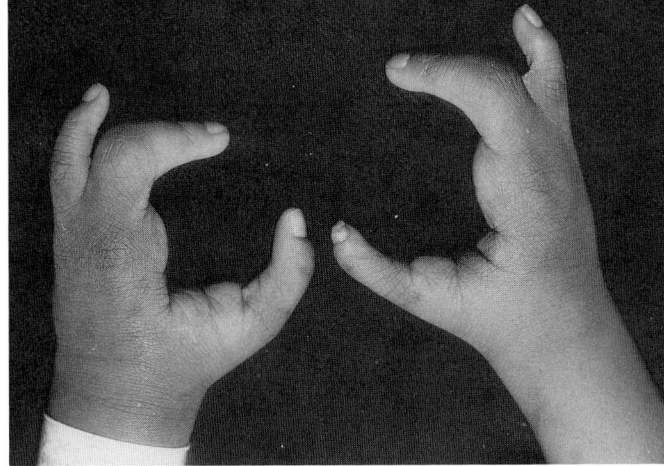

A

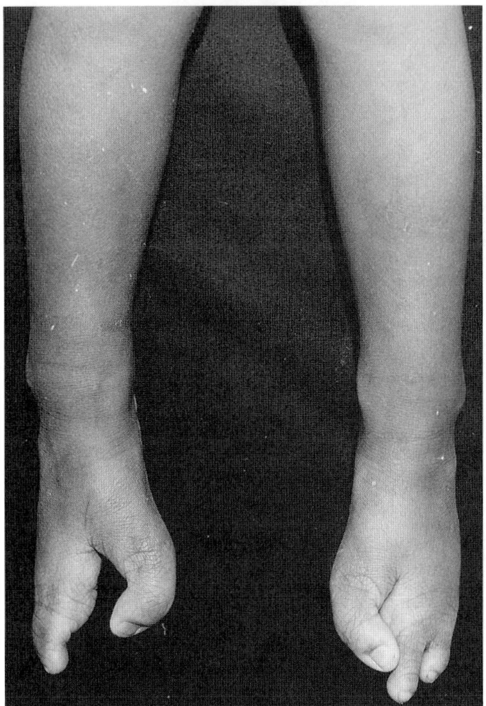

B

Fig. 21–6. *ECP syndrome.* (A) Ectrodactyly of the hands. (B) Ectrodactyly of the feet. (Courtesy of *JM Opitz,* Helena, Montana.)

### References [ECP syndrome (ectrodactyly–cleft palate syndrome)]

1. Fontaine G et al: Une observation familiale du syndrome ectrodactylie et dysostose mandibulofaciale. *J Génét Hum* **22**:289–307, 1974.

2. Opitz JM et al: The ECP syndrome, another autosomal dominant cause of monodactylyous ectrodactyly. *Eur J Pediatr* **133**:217–220, 1980.

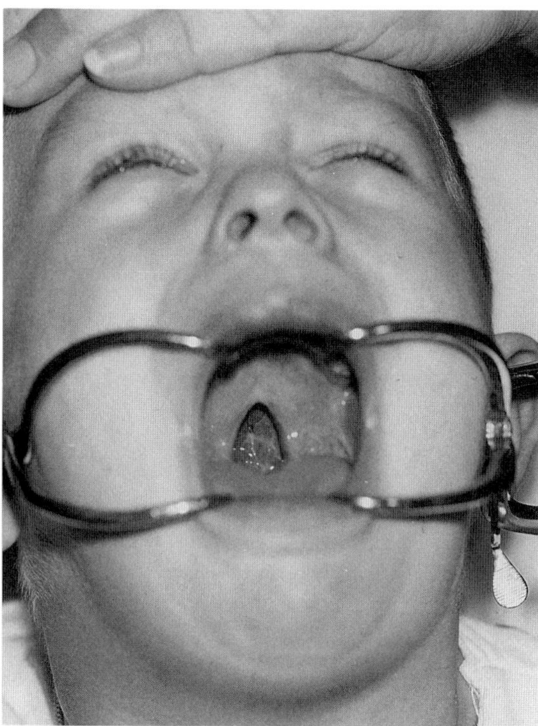

Fig. 21–7. *ECP syndrome*. Cleft palate. (From *JM Opitz* et al, Eur J Pediatr **133**:217, 1980.)

## Ectrodactyly of the feet and cleft palate (Fontaine syndrome)

Fontaine et al (1) in 1974 reported a kindred of four individuals having autosomal dominant inheritance of ectrodactyly and syndactyly of the feet, microretrognathia, dysmorphic pinnae, and submucous cleft palate and cleft uvula. One of four affected patients had moderate mental retardation and seizures while two exhibited mild retardation. This does not appear to be the *EEC syndrome*.

### Reference [Ectrodactyly of the feet and cleft palate (Fontaine syndrome)]

1. Fontaine G et al: Une observation familiale du syndrome ectrodactylie et dysostose mandibulo-faciale. *J Génét Hum* **22**:289–307, 1974.

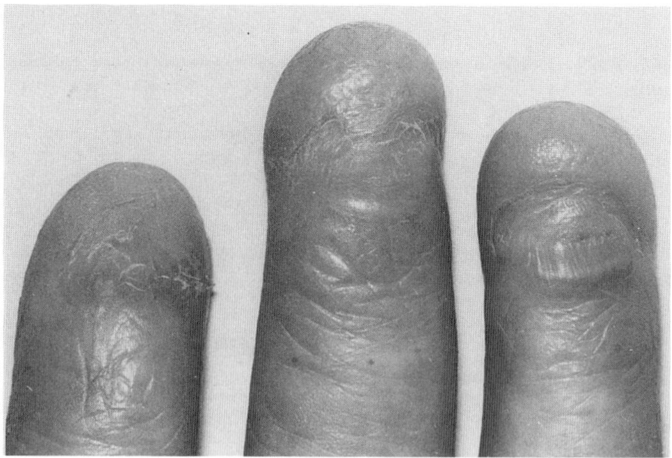

Fig. 21–9. *Hay–Wells syndrome*. Congenitally absent fingernails. (From *RJ Hay* and *RS Wells*, Br J Dermatol **94**:277, 1976.)

## Hay–Wells syndrome (AEC syndrome)

In 1976, Hay and Wells (3) described seven patients from four families with autosomal dominant inheritance of congenital ankyloblepharon, which principally involved the lateral lid margins, absent or dystrophic nails, coarse wiry sparse hair, scalp infections, sparse to absent eyelashes, deficient teeth and teeth with conical crown form, maxillary hypoplasia, and mild hypohidrosis (Figs. 21–8A,B and 21–9). Several patients had lacrimal duct atresia and/or absence of lacrimal punctae with or without photophobia. Cleft palate, rarely cleft lip, was present in most of their patients. Spiegel and Colton (6) added another two patients. In addition, one of their patients had cupped pinnae and stenotic external auditory canals, and her son had hypospadias. Schwayder et al (5) and Greene et al (2) added several cases.

Ankyloblepharon may be a dominantly inherited isolated finding, or found in binary combination with cleft lip and/or cleft palate in the *popliteal pterygium syndrome,* in association with clefts, lip pits, and syngnathia (4) and with curly hair and nail dysplasia (1).

### References [Hay–Wells syndrome (AEC syndrome)]

1. Baughman FA: CHANDS: The curly hair–anklyoblepharon–nail dysplasia syndrome. *Birth Defects* **7**(8):100–102, 1971.
2. Greene SL et al: Variable expression in ankyloblepharon–ectodermal defects-cleft lip and palate syndrome. *Am J Med Genet* **27**:207–212, 1987.

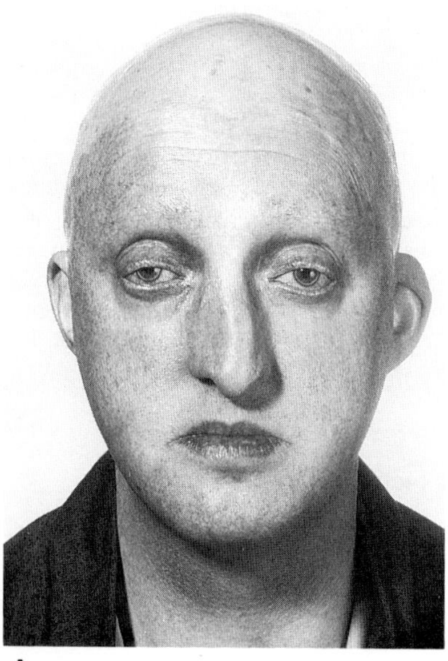

A

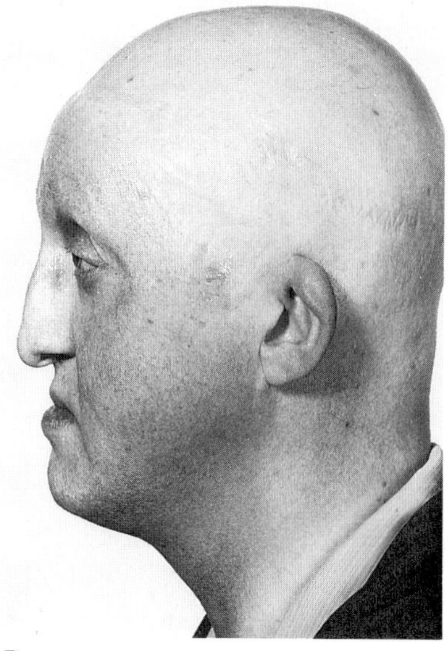

B

Fig. 21–8. *Hay–Wells syndrome*. (A,B) Frontal and lateral facial views of 38-year-old male patient showing lack of scalp hair, eyebrows, and lashes, midfacial hypoplasia, dysmorphic left pinna. Patient has cleft palate. (From *RJ Hay* and *RS Wells*, Br J Dermatol **94**:277, 1976.)

3. Hay RJ, Wells RS: The syndrome of ankyloblepharon, ectodermal defects and cleft lip and palate: An autosomal dominant condition. *Br J Dermatol* **94**:277–289, 1976.

4. Neuman Z, Shulman J: Congenital sinuses of the lower lip. *Oral Surg* **14**:1415–1420, 1961.

5. Shwayder TA et al: Hay–Wells syndrome. *Pediatr Dermatol* **3**:399–402, 1986.

6. Spiegel J, Colton A: AEC syndrome: Ankyloblepharon, ectodermal defects, and cleft lip and palate. *J Am Acad Dermatol* **12**:810–815, 1985.

## Bowen–Armstrong syndrome (cleft lip–palate, ectodermal dysplasia, and mental retardation)

Bowen and Armstrong (1) described three female sibs with mild to moderate mental retardation, ectodermal dysplasia, and cleft lip and/or palate. It is uncertain whether these children had incomplete expression of the *EEC syndrome,* had the *AEC (Hay–Wells syndrome),* or represented a unique disorder.

The hair was sparse and fair, with the brows and lashes absent (Fig. 21–10A). One child had congenital skin defects on the buttocks (Fig. 21–10B). Mottled brown pigmentation subsequently appeared on the genital area, axillae, and buttocks (Fig. 21–10C). Blepharitis and photophobia were present in one child and ankyloblepharon filiforme adnatum in two of the sibs. The extremities exhibited variable soft tissue syndactyly of the feet. There was some degree of bony syndactyly. The nails were somewhat hypoplastic. Genital anomalies included hypoplasia of the labia minora with absence of vaginal opening. Dental abnormalities were absence of primary and secondary maxillary incisors and canines and enamel hypoplasia (Fig. 21–10D).

### Reference [Bowen–Armstrong syndrome (cleft lip–palate, ectodermal dysplasia, and mental retardation)]

1. Bowen P, Armstrong HB: Ectodermal dysplasia, mental retardation, cleft lip/palate and other anomalies in three sibs. *Clin Genet* **9**:35–42, 1976.

## Filiform adhesions of the eyelids and cleft lip–palate

The association of cleft lip with or without cleft palate and congenital filiform fusion of the eyelids (ankyloblepharon filiforme adnatum) appears to be inherited as an autosomal dominant trait with extremely variable expressivity (1–14). In several families, there were relatives with facial clefting but no eyelid anomalies (2,6,8,13,14). There has been one association with isolated cleft palate (4). The disorder may be more common in India.

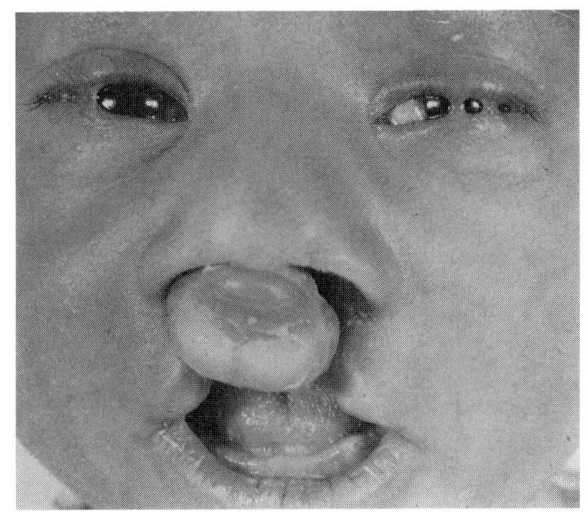

Fig. 21–11. *Filiform adhesions of the eyelids and cleft lip–palate.* Note bilateral involvement of eyes and lip. (From *JC Long* and *SE Blandford,* Am J Ophthalmol **53**:126, 1962.)

Multiple connective tissue bands, 0.3 to 5.0 mm in width, extend from the white line of one lid to that of the other lid, posterior to the cilia and anterior to the Meibomian orifices (Fig. 21–11). No associated anomalies are found in the eyeballs. The filiform adhesions may be an isolated finding or in combination with the *cleft lip–palate and paramedian lower lip pits,* in the *popliteal pterygium syndrome* and in the *Hay–Wells (AEC) syndrome.* Associated anomalies that appear to be aleatory are spina bifida, imperforate anus, superior adherence of pinna, and syndactyly of fourth and fifth toes.

### References (Filiform adhesions of the eyelids and cleft lip–palate)

1. Akkermans CH, Stern LM: Ankyloblepharon filiforme adnatum. *Br J Ophthalmol* **63**:129, 1969.

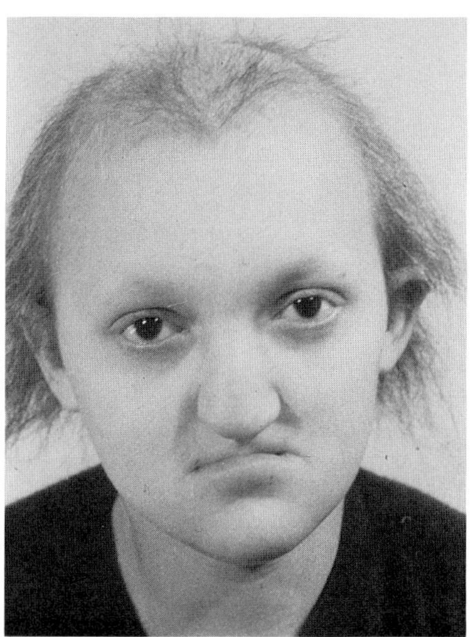

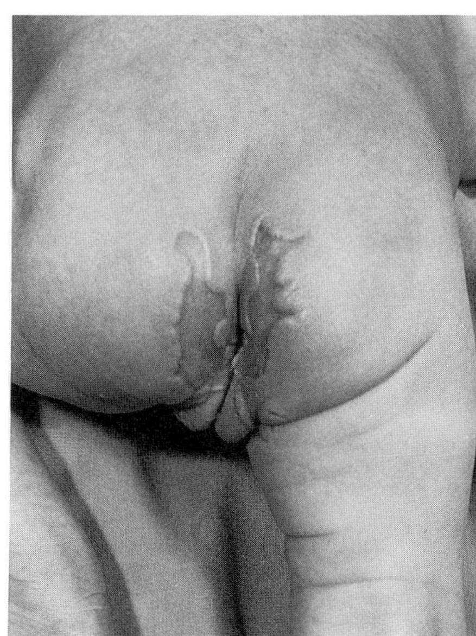

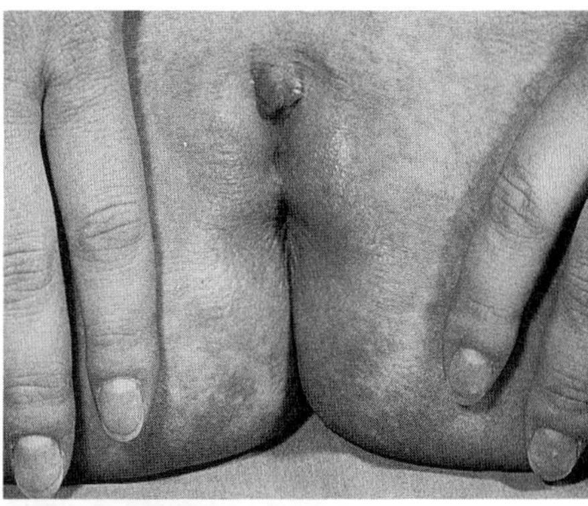

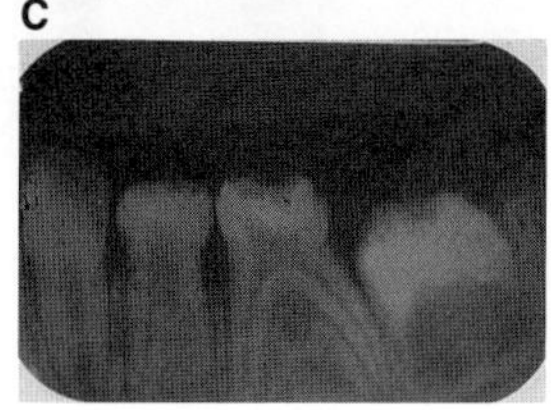

A B C D

Fig. 21–10. *Bowen–Armstrong syndrome.* (A) Note unusual facies, wiry, scaly hair, sparse eyebrows and lashes. (B) Congenital skin defects on buttocks. (C) Absence of labia and vaginal introitus and mottled pigmentation of area. (D) Thin dysplastic enamel. (From *P Bowen* and *HB Armstrong,* Clin Genet **9**:35, 1976.)

2. Ehlers N, Jensen IK: Ankyloblepharon filiforme congenitum associated with hare-lip and cleft palate. *Acta Ophthalmol* **48**:465–467, 1970.

3. Gupta SP, Saxena R: Ankyloblepharon filiforme adnatum. *J All-India Ophthalmol Soc* **10**:19–21, 1962.

4. Kapoor MS et al: Unilateral ankyloblepharon filiforme adnatum. *Indian J Ophthalmol* **25**(1):43–44, 1977.

5. Kazarian EL, Goldstein P: Ankyloblepharon filiforme adnatum with hydrocephalus, meningomyelocele, and imperforate anus. *Am J Ophthalmol* **84**:355–357, 1977.

6. Khanna VD: Ankyloblepharon filiforme adnatum. *Am J Ophthalmol* **43**:774–777, 1957.

7. Lemtis H, Neubauer H: Ankyloblepharon filiforme et membraniforme adnatum. *Klin Monatsbl Augenheilkd* **135**:510–516, 1959.

8. Long JC, Blandford SE: Ankyloblepharon filiforme adnatum with cleft lip and palate. *Am J Ophthalmol* **53**:126–129, 1962.

9. Modi AJ: A multiple malformation syndrome with ankyloblepharon filiforme adnatum with cleft lip and palate. *Indian J Ophthalmol* **33**:129–131, 1985.

10. Pahwa JM, Sud SD: Ankyloblepharon filiforme adnatum. *Orient Arch Ophthalmol* **4**:170–172, 1966.

11. Pandya A et al: The multiple malformation syndrome with cleft lip and palate and ankyloblepharon filiforme adnatum. *Indian J Pediatr* **52**:667–670, 1985.

12. Rogers JW: Ankyloblepharon filiforme adnatum. *Arch Ophthalmol* **65**:114–117, 1961.

13. Rosenman Y et al: Ankyloblepharon filiforme adnatum. *Am J Dis Child* **134**:751–753, 1980.

14. Sood NM et al: Ankyloblepharon filiforme adnatum with cleft lip and palate. *J Pediatr Ophthalmol* **5**:30–32, 1968.

## Larsen syndrome

The syndrome, first recognized as a distinct entity by Larsen et al (20) in 1950, includes flattened facies, multiple congenital joint dislocations, clubfoot deformities, and, frequently, cleft palate. However, the syndrome was surely described earlier, having been buried under a number of diagnoses (17,26,27). Silverman (39) discussed several other possible early examples. At least 75 well documented cases have been reported to date. We are not able to precisely categorize the case of Anderson et al (1).

The syndrome has been described in sibs whose parents were apparently normal (3,4,6,42). Many cases are consistent with autosomal dominant transmission (11,13,21,27,31,35,37,41,48,50). Numerous sporadic examples have also been published (8,10,12,16,17,19,20,28, 30,34,40,43,47). Careful examination of so-called recessive, dominant, and sporadic cases has not permitted differentiation. It is our opinion that there is little evidence for recessive inheritance. Because of the extremely variable expression of the disorder, it is not unlikely that patients with so-called recessive cases had a mildly affected parent. This appears to be bolstered by the apparent lack of parental consanguinity in "recessive" cases as well as in sporadic cases (12).

**Facies.** The face is typically flattened and there is a depressed nasal bridge. The eyes appear widely spaced. Frontal bossing may be marked (Fig. 21–12A,B).

**Skeletal system.** Bony alterations include bilateral anterior dislocation of the radial head (70%), bilateral anterior dislocation of the tibia on the femur (80%), hip dislocation (80%), and talipes equinovarus or equinovalgus (85%). The patella is often dislocated laterally. The joint dislocations become less frequent with age. Adult height is reduced, the patient usually being less than 152 cm tall (Fig. 21–13A,B). The fingers are long and cylindric (pseudoclubbing) with extra creases in about 75% (Fig. 21–14). The thumb is spatulate. A number of authors have reported mild thoracolumbar scoliosis (8,10, 11,16,17,19,21,22,27,34,42,48), which is absent at birth but becomes evident during infancy.

Radiographic changes include dislocations, as previously noted. The distal phalanges are abbreviated. The metacarpals and metatarsals are relatively shortened. Bone age is retarded. When the carpal bones appear, they are irregular. Not uncommonly there are from one to four supernumerary carpal bones (Fig. 21–15A). Abnormal segmentation and/or hypoplasia of cervical and thoracic vertebrae are common (8,11,17,19–21,24,28,31,42,45,47,48). This may be associated with marked cervical kyphosis (16,47). In some cases, cervical vertebral instability has led to quadriplegia or death (3a,20a). The vertebral bodies may be flattened (9,19,21,24,45). A juxtacalcaneal accessory bone is frequently present (20,38,45). It coalesces with the calcaneus by 5–8 years of age forming a bifid calcaneus which is evident in lateral view (Fig. 21–15B). Additional centers of ossification may be seen at the elbow (12,16,29,39,43). The proximal tibial epiphysis is often cone-shaped during development.

**Cardiovascular anomalies.** ASD and VSD as well as aortic dilatation and valvular anomalies are seen (18,26,34,44,46).

**Central nervous system.** Mental retardation has been noted in about 15% (37). Sensorineural or mixed hearing loss has been described (15,33,40,50).

**Oral manifestations.** Oral changes are essentially limited to cleft palate, which occurs in about 30% (43). This may be limited to the soft palate or uvula. Some patients have laryngotracheomalacia

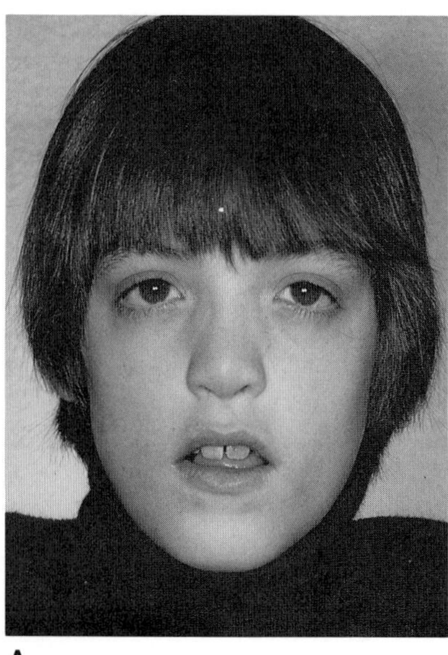

A

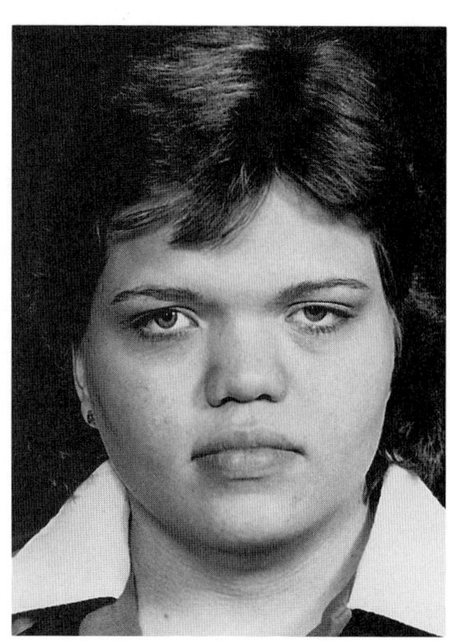

B

Fig. 21–12. *Larsen syndrome*. (A,B) Flattened midface with broad nasal bridge in male and female patients.

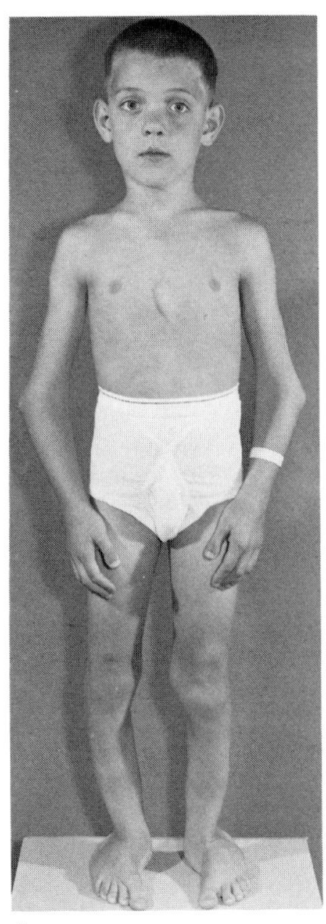

A

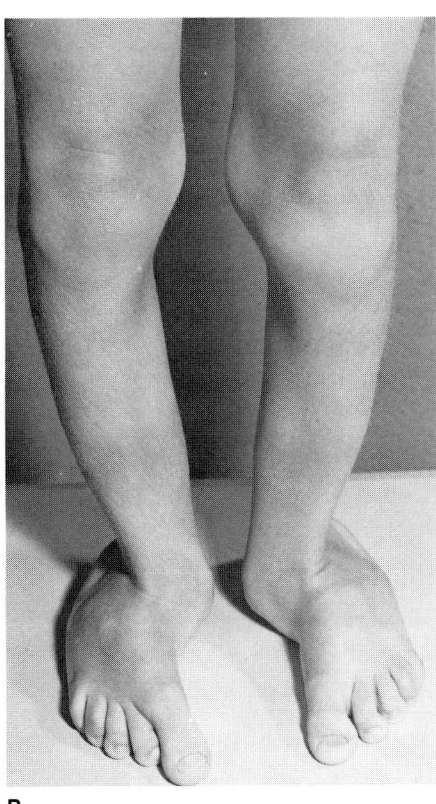

B

Fig. 21–13. *Larsen syndrome.* (A,B) Note similar facies, subluxation of radial heads, sternal anomaly, dislocated knees, "windmill" feet, spatulate thumbs.

(7,10,21,35). Collapse of the flabby laryngeal cartilages may lead to death. Others may have laryngeal stenosis (7a,47). Cephalometric study has been reported (49).

**Differential diagnosis.** *Otopalatodigital syndrome* is most often mistaken for Larsen syndrome. In contrast, these patients exhibit a pugilistic facies, hearing loss, paddle-shaped metatarsal bones, no juxtacalcaneal bone, and no supernumerary carpal bones. Differential diagnosis also includes arthrogryposis, *fetal akinesia sequence, monosomy 21, Ehlers–Danlos syndromes, COFS syndrome, cleft palate,*

Fig. 21–14. *Larsen syndrome.* Note extra flexion creases of fingers, spatulate thumb.

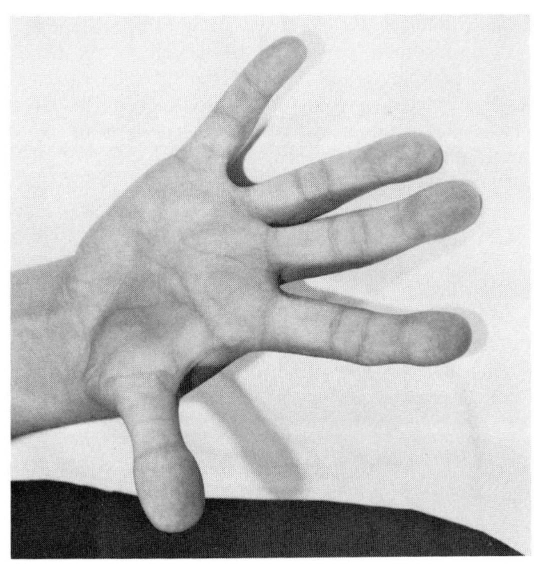

Fig. 21–15. *Larsen syndrome.* (A) The carpal bones, late to appear, are supernumerary. (B) Juxtacalcaneal bone fuses with calcaneus just prior to puberty. Also note unusual relationship of metatarsals to tarsal bones.

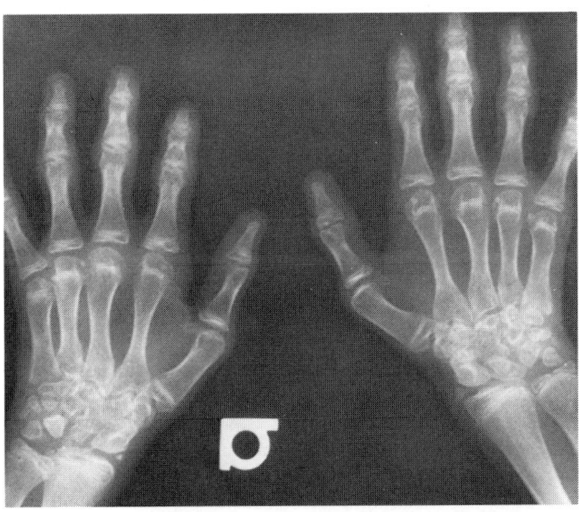

A

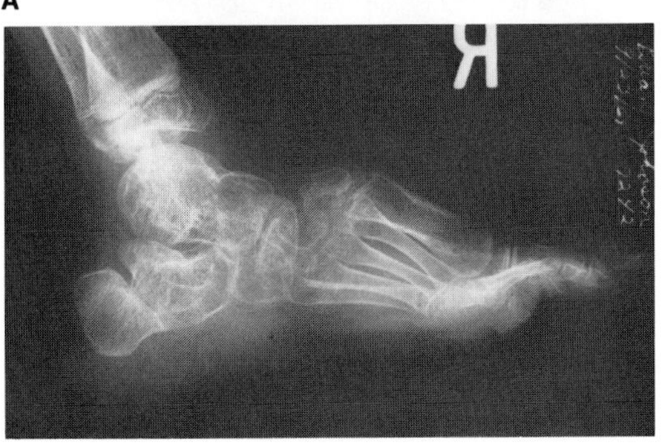

B

*short stature, depressed nasal bridge, and sensorineural hearing loss* (16), and *SPONASTRIME dysplasia*.

Schröder (38) reported a syndrome in five sibs. There were bizarre pinnae, bilateral luxation of the radius, radiohumeral ankylosis, dislocation of hips, shoulders, and knees, genua valga, and camptodactyly.

Matsoukas et al (25) described an autosomal dominantly inherited syndrome of joint dislocations (hips, elbows, wrists), short stature, mental retardation (IQ—65), and various ocular anomalies (myopia, microphthalmia, corneal sclerosis).

Although Chen et al (7) have suggested that their patients had a different disorder, we believe that they have Larsen syndrome.

Multiple congenital dislocations may occur as an isolated finding (14,32).

### References (Larsen syndrome)


1. Anderson CE et al: A syndrome of short stature, joint laxity and developmental delay. *Clin Genet* **22**:40–46, 1982.
2. Bartsocas CS, Dimitriou JK: Multiple joint dislocation in mother and child. *J Pediatr* **80**:299–301, 1972.
3. Beluffi G, Segré A: Syndrome di Larsen in due fratelli. *Minerva Pediatr* **30**:293–294, 1978.
4. Block C, Peck HM: Bilateral congenital dislocation of the knees. *J Mt Sinai Hosp* **32**:607–608, 1965.
5. Bowen J et al: Spinal deformities in Larsen's syndrome. *Clin Orthop* **197**:159–163, 1985.
6. Buxton SJD: Multiple deformity in two sisters. *Proc R Soc Med* **32**:284–286, 1939.
7. Chen H et al: A lethal Larsen-like multiple joint dislocation syndrome. *Am J Med Genet* **13**:149–161, 1982.
7a. Crowe AV et al: Tracheal stenosis in Larsen's syndrome. *Arch Otolaryngol* **115**:626, 1989.
8. Curtis BH, Fisher RL: Heritable congenital tibio-femoral subluxation (Case 5). *J Bone Jt Surg* **52A**:1104–1114, 1970.
9. Galanski M, Statz A: Radiologische Befunde beim Larsen-Syndrom. *Roefo* **128**:536–537, 1978.
10. Haarmeyer A: Larsen-Syndrom. Symptomatik und Therapie. *Z Orthop* **116**:802–809, 1978.
11. Habermann ET et al: Larsen's syndrome: A heritable disorder. *J Bone Jt Surg* **58A**:558–561, 1976.
12. Hackenbroch M: Multiple kongenitale Gelenkmissbildungen. *Z Orthop Chir* **45**:467–476, 1924.
13. Harris R, Cullen CH: Autosomal dominant inheritance in Larsen's syndrome. *Clin Genet* **2**:87–90, 1971.
14. Hayashi K, Matsuoko M: Angeborene Missbildungen kombiniert mit der kongenitalen Huftverrenkung. *Z Orthop Chir* **31**:369–399, 1913.
15. Herrmann HC et al: The association of a hearing deficit with Larsen's syndrome. *J Otolaryngol* **10**:45–48, 1981.
16. Houston CS et al: Separating Larsen syndrome from the "arthrogryposis basket". *J Can Assoc Radiol* **32**:206–223, 1981.
17. Kaijser R: Über kongenitale Kniegelenksluxationen (Case 6). *Acta Orthop Scand* **6**:1–20, 1935.
18. Kiel EA et al: Cardiovascular malformations in the Larsen syndrome. *Pediatrics* **71**:942–946, 1983.
19. Kozlowski K et al: Radiographic findings in Larsen's syndrome. *Australas Radiol* **18**:336–344, 1974.
20. Larsen JH et al: Multiple congenital dislocations associated with characteristic facial abnormality. *J Pediatr* **37**:574–581, 1950.
21. Latta RJ et al: Larsen's syndrome: A skeletal dysplasia with multiple joint dislocation and unusual facies. *J Pediatr* **78**:291–298, 1971.
22. Lee PA: Multiple joint dislocations with peculiar facies. *Am J Dis Child* **126**:828–830, 1973.
23. Lefort G et al: Dislocation du rachis cervical supérieur dans le syndrome de Larsen. *Chir Pédiatr* **24**:211–212, 1983.
24. Masson A et al: Le syndrome de Larsen. *Pédiatrie* **33**:775–785, 1978.
25. Matsoukas J et al: A newly recognized dominantly inherited syndrome: Short stature, ocular and articular anomalies, mental retardation. *Helv Paediatr Acta* **28**:383–386, 1973.
26. McFarland BL: Congenital dislocation of the knee (Case 3). *J Bone Jt Surg* **11**:281–285, 1929.
27. McFarlane AL: A report on four cases of congenital genu recurvatum occurring in one family. *Br J Surg* **34**:388–391, 1947.
28. Micheli LJ et al: Spinal instability in Larsen's syndrome. Report of three cases. *J Bone Jt Surg* **58A**:562–565, 1976.
29. Muzumdar AS et al: Quadriplegia in Larsen syndrome. *Birth Defects* **13**(3C):202–211, 1977.
30. Niebauer JJ, King DE: Congenital dislocation of the knee (Case 2). *J Bone Jt Surg* **42A**:207–225, 1960.
31. Oki T et al: Clinical features and treatment of joint dislocations in Larsen's syndrome. Report of three cases in one family. *Clin Orthop Rel Res* **119**:206–210, 1976.
32. Provenzano RW: Congenital dislocation of the knee. *N Engl J Med* **236**:360–362, 1947.
33. Renault F et al: Le syndrome de Larsen. *Arch Fr Pédiatr* **39**:35–38, 1982.
34. Robertson FW et al: Larsen's syndrome. *Clin Pediatr* **14**:53–60, 1975.
35. Rønningen H, Bjerkrein I: Larsen's syndrome. *Acta Orthop Scand* **49**:138–142, 1978.
36. Rotter W, Erb W: Über eine Systemerkrankung des Mesenchyms mit multiplen Luxationen aus angeborener Gelenkschlaffheit und über Wirbelbogenspalten. *Virchows Arch Pathol Anat* **315**:233–263, 1949.
37. Samuel AW, Davies DRA: The Larsen syndrome with multiple congenital dislocations and a normal facies. *Int Orthop (SICOT)* **5**:229–231, 1981.
38. Schröder CH: Familiäre kongenitale Luxationen. *Z Orthop Chir* **57**:580–596, 1932.
39. Silverman FN: Larsen's syndrome: Congenital dislocation of the knees and other joints, distinctive facies and, frequently, cleft palate. *Ann Radiol* **15**:297–328, 1972.
40. Stanley CS et al: Mixed hearing loss in Larsen syndrome. *Clin Genet* **33**:395–398, 1988.
41. Stanley D, Seymour N: The Larsen syndrome occurring in four generations of one family. *Int Orthop* **8**:267–272, 1985.
42. Steel HH, Kohl EJ: Multiple congenital dislocations associated with other skeletal anomalies (Larsen's syndrome) in three siblings. *J Bone Jt Surg* **54A**:75–82, 1972.
43. Stevenson RE el al: Larsen syndrome with mental retardation. *Proc Greenwood Genet Ctr* **2**:39–43, 1983.
44. Strisciuglio P et al: Severe cardiac anomalies in sibs with Larsen syndrome. *J Med Genet* **20**:422–424, 1983.
45. Sugarman GI: The Larsen syndrome, autosomal dominant form. *Birth Defects* **11**(2):121–129, 1975.
46. Swenson RE et al: Striking aortic root dilatation in a patient with the Larsen syndrome. *J Pediatr* **86**:914–915, 1975.
47. Szabo L, Perjés K: Über die Differenzierung der Arthrogryposis multiplex congenita und des Larsen-Syndroms. *Z Orthop* **112**:1275–1281, 1974.
48. Trigueros AP et al: Larsen's syndrome: Report of three cases in the one family: Mother and two offspring. *Acta Orthop Scand* **49**:582–588, 1978.
49. Tsang MCK et al: Oral and craniofacial morphology of a patient with Larsen syndrome. *J Craniofac Genet Dev Biol* **6**:357–362, 1986.
50. Ventruto V et al: Larsen syndrome in two generations of an Italian family. *J Med Genet* **13**:538–539, 1976.


## Meckel syndrome (Gruber syndrome, dysencephalia splanchnocystica)

Meckel (15), in 1822, delineated a lethal syndrome of microcephaly, encephalocele, microphthalmia, congenital heart defects, postaxial polydactyly, polycystic kidneys, and clefts of the lip and/or palate. The condition has been forgotten, and periodically rediscovered. We refer the reader to several comprehensive historical reviews of cases published prior to 1970 (5,8,14,17,19).

Inheritance is autosomal recessive (8,12). Parental consanguinity has been noted in approximately 30% of cases. Estimates of frequency in various parts of the world range between 1 in 3000 to 1 in 50,000 births (12,24,25). Perhaps 200 cases have been reported to date.

Breech presentation occurs in about half the cases. Oligohydramnios has been noted in over 60%. The thorax is short and bell-shaped and the abdomen bulges. The syndrome is lethal, death usually occurring at birth or in the first few hours of life. The placenta is grossly enlarged in approximately 25%.

**Facies.** Microcephaly, sloping forehead, sunken eyes, occasional Potter facies, large lowset dysmorphic pinnae, macrostomia, clefting, and micrognathia present a striking appearance (Figs. 21–16 and 21–17A,B).

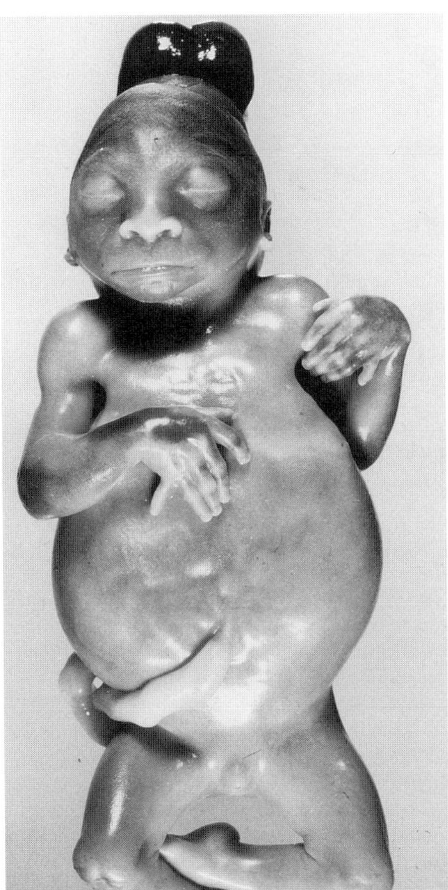

Fig. 21–16. *Meckel syndrome.* Male fetus at 23 weeks gestation. Note occipital encephalocele, microcephaly, polydactyly, enlarged abdomen due to polycystic kidneys. (From *NC Nevin et al,* Clin Genet **15**:1, 1979.)

**Eyes.** Various eye anomalies, seen in about 30%, include anophthalmia, microphthalmia, iris coloboma, retinal dysplasia, and hypoplasia of optic nerves (13).

**Central nervous system.** Microcephaly/anencephaly (50–65%), absent olfactory bulbs (80%), and double occipital encephalocele (65–90%) are striking (Figs. 21–16 and 21–17A,B). The larger encephalocele occurs rather high in the occipital bone. The smaller is low in the basal occipital bone resulting in a cleft in the occipital foramen magnum and extending onto the first vertebral arches (cerebellar encephalocele) (19a). It is covered by a focal depression of skin. It has been estimated that Meckel syndrome may account for 5% of neural tube defects. Microgyria or polygyria, heterotopias, fusion of frontal lobes, agenesis or hypoplasia of the optic tracts, olfactory lobes, corpus callosum and septum pellucidum, cerebellum and brain stem, neurorosettes, and internal hydrocephalus have been found in 15–20% (19b). Cervical rachischisis occurs in about 5% (16). Craniosynostosis is a rare finding.

**Kidneys, liver, pancreas, and lungs.** Huge cystic and dysplastic kidneys (10–20 times larger than normal) are a constant feature (21a). The spherical cysts, ranging in size from 1 to 10 mm, are smallest in the subcapsular area and largest near the renal pelvis. Glomeruli are grossly deficient in numbers, interstitial fibrosis is marked, and the finding of cartilage is frequent (2a). The renal pelves, calices, ureters, and urinary bladder are often hypoplastic (1,2,4,7,11,16,21–23) (Fig. 21–18A,B).

The liver exhibits macroscopic cysts in 15% with fibrotic changes in the portal areas due to hamartomatous proliferation and dilatation of bile ducts in nearly all patients (2a). The pancreas is fibrotic and

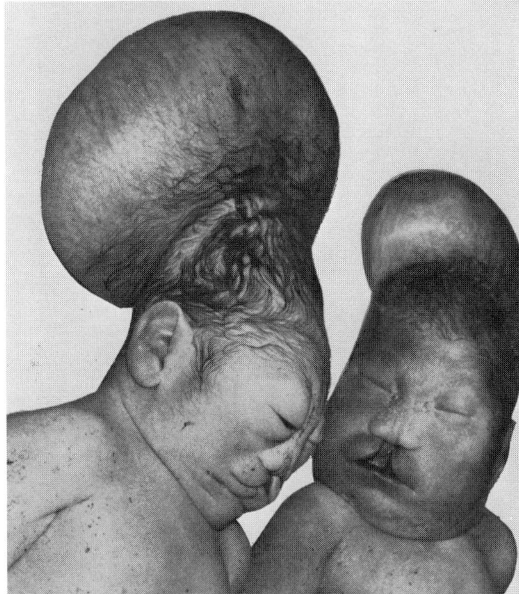

A

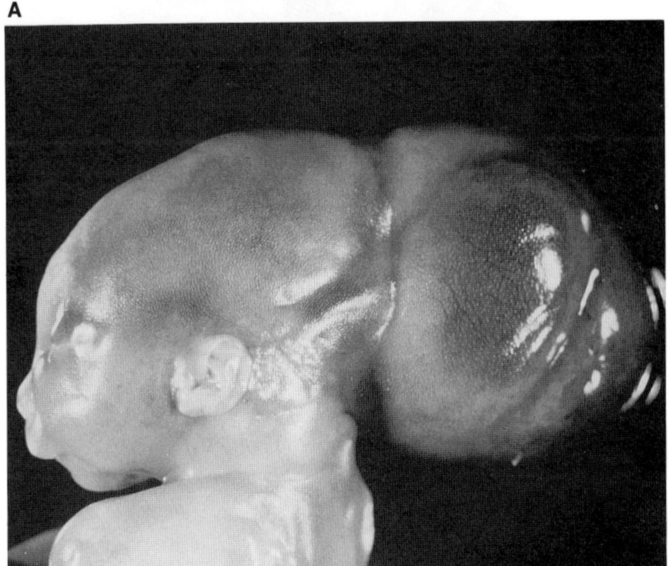

B

Fig. 21–17. *Meckel syndrome.* (A) Microcephaly, occipital exencephalocele, microphthalmia, ocular hypotelorism, premaxillary agenesis in twins. (B) Note microcephaly, occipital encephalocele, dysmorphic pinna. (A from *YE Hsia et al,* Pediatrics **48**:237, 1971.)

the ducts dilated in 30% (2,4,11,21a,22). The lungs are uniformly hypoplastic.

**Skeletal changes.** The most common skeletal alteration is bilateral postaxial polydactyly (55–75%). The hands are more often involved than the feet. There may be six or more digits (Figs. 21–16 and 21–19). There are usually six metacarpal bones or, if five, the middle one is often bifurcated. Talipes equinovarus has been noted in 30–50%. The tibiae may be short and angulated and the upper extremities are shortened (4,20,23,28).

**Genital anomalies.** Hypoplastic penis and cryptorchidism in males are almost constant features (21a) (Fig. 21–20). Males often have Müllerian duct remnants (20%) and epididymal cysts (40%) (21a). Septate vagina and hypoplastic or bicornuate uterus in females have been observed in 20–25% (2,4,11,16,23,27). On occasion, ambiguous external genitalia have been noted (19).

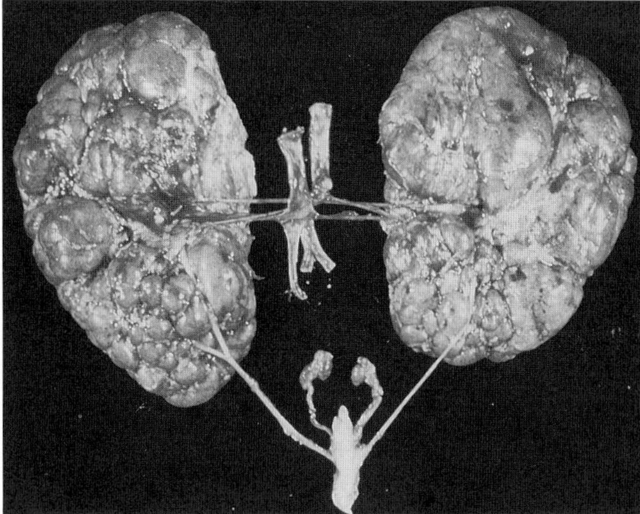

A

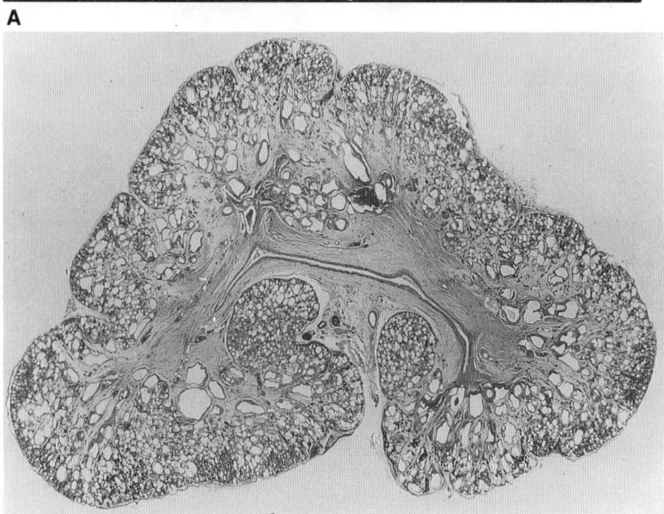

B

Fig. 21–18. *Meckel syndrome.* (A) Huge multicystic dysplastic kidneys. Note tiny ureters and hypoplastic bladder. (B) Multicystic dysplastic kidney. (A from *P Moerman et al,* Hum Genet **62**:240, 1982. B from *U Friedrich et al,* Clin Genet **15**:278, 1979.)

**Other findings.** Malrotation of the intestines, situs inversus, and a wide array of congenital/cardiac anomalies (VSD, ASD, common atrium, absent mitral valve, preductal coarctation, persistence of left superior vena cava) have been found in 30% (4,11,16,23).

**Oral manifestations.** Cleft lip (20%) and/or, more often, cleft palate (45%) and various malformations of the tongue (ankyloglossia, bifid tongue, anterior marginal hamartomas) have been noted. The larynx is frequently hypoplastic (23).

**Differential diagnosis.** Marked variability makes for difficulty in diagnosis. Meckel syndrome may be confused with *trisomy 13* and occasionally severe *Smith–Lemli–Opitz syndrome, C syndrome,* and *hydrolethalus syndrome* (10). Although not obligatory, the triad of occipital encephalocele, postaxial polydactyly, and polycystic kidneys is persuasive in making a diagnosis of Meckel syndrome. Seller (27) indicated that the triad is present in 60–70% of the cases. In approximately 15%, there is binary combination of either encephalocele or polydactyly with polycystic kidneys. Among the remainder, there is only a single component present. In studies of affected sibs, approximately 50% have identical anomalies while the other half do not. The *short-rib–polydactyly syndromes* (Majewski type, Saldino–Noonan type, Naumoff type, etc.) can easily be separated from Meckel syndrome.

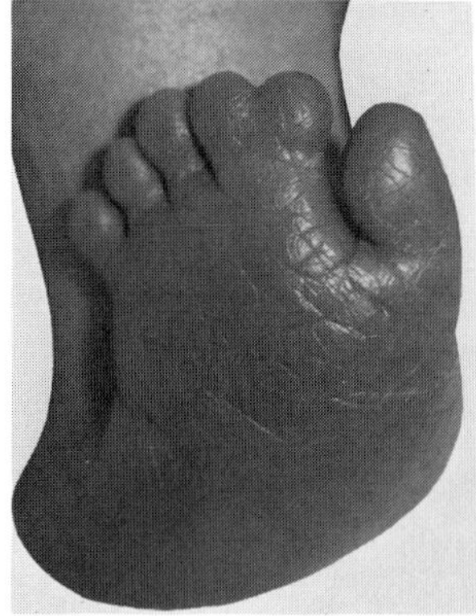

Fig. 21–19. *Meckel syndrome.* Polydactyly, syndactyly, abbreviation of hallux, talipes. [From *JM Opitz* and *JJ Howe,* Birth Defects **5**(2):167, 1969.]

**Laboratory aids.** Chromosome studies will exclude *trisomy 13.* Prenatal diagnosis may be made by ultrasonography (oligohydramnios, microcephaly, anencephaly, occipital encephalocele, internal hydrocephalus, large kidneys, empty bladder, reduced femur length). Ultrasonography should be employed early and serially (19a). Amniotic fluid $\alpha$-fetoprotein levels are elevated in approximately 70% (open CNS lesions, cystic kidneys) at around week 17–18 of pregnancy. Various other backup tests may be employed: study of amniotic fluid for acetylcholinesterase levels, elevated pregnancy specific $\beta_1$-glycoprotein, positive $\beta$-trace protein, human chorionic gonadotropin levels, and amniotic cell culture (3,6,9,18,22,26).

## References [Meckel syndrome (Gruber syndrome, dysencephalia splanchnocystica)]


1. Altmann P et al: A casuistic report on the Gruber or Meckel syndrome. *Hum Genet* **38**:357–362, 1977.
2. Anderson VM: Meckel syndrome: Morphologic considerations. *Birth Defects* **18**(3B):145–160, 1982.
2a. Blankenberg TA et al: Pathology of renal and hepatic anomalies in Meckel syndrome. *Am J Med Genet Suppl* **3**:395–410, 1987.


Fig. 21–20. *Meckel symdrome.* Hypoplastic penis, hypospadias. [From *JM Opitz* and *JJ Howe,* Birth Defects **5**(2):167, 1969.]

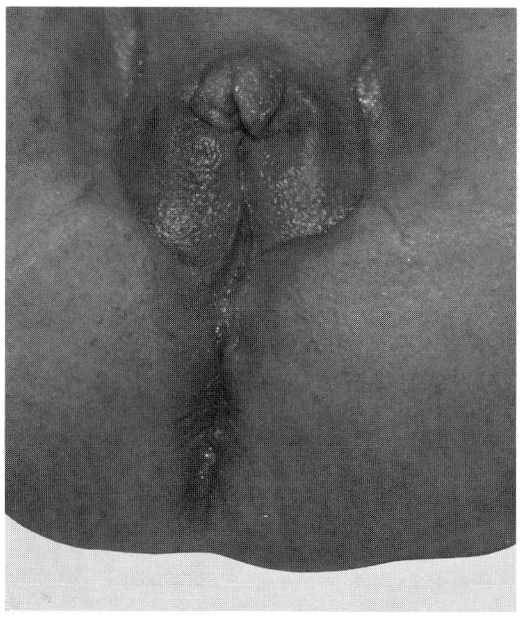

3. Chemke J et al: Prenatal diagnosis of Meckel syndrome: Alpha-fetoprotein and beta-trace protein in amniotic fluid. *Clin Genet* **11**:285–289, 1977.

4. Fraser FC, Lytwyn A: Spectrum of anomalies in the Meckel syndrome, or: "Maybe there is a malformation syndrome with at least one constant anomaly." *Am J Med Genet* **9**:67–73, 1981.

5. Fried K et al: Polycystic kidneys associated with malformations of brain, polydactyly and other birth defects in newborn sibs. *J Med Genet* **8**:285–290, 1971.

6. Friedrich U et al: Prenatal diagnosis of polycystic kidneys and encephalocele (Meckel syndrome). *Clin Genet* **15**:278–286, 1979.

7. Göcke H et al: Das Meckel-Syndrom bei einem Neugeborenen. *Geburtsch Frauenheilk* **42**:602–604, 1982.

8. Hsia YE et al: Genetics of the Meckel syndrome (dysencephalia splanchnocystica). *Pediatrics* **48**:237–247, 1971.

9. Johnson VP, Holzwarth DR: Prenatal diagnosis of Meckel syndrome. *Am J Med Genet* **18**:699–711, 1984.

10. Lowry RB: Variability of the Smith–Lemli–Opitz syndrome; overlap with the Meckel syndrome. *Am J Med Genet* **14**:429–434, 1983.

11. Lowry RB et al: Survival and spectrum of anomalies in the Meckel syndrome. *Am J Med Genet* **14**:417–421, 1983.

12. Lurie IW: Meckel syndrome in different populations. *Am J Med Genet* **18**:661–669, 1984.

13. MacRae DW et al: Ocular manifestations of the Meckel syndrome. *Arch Ophthalmol* **88**:106–113, 1972.

14. Mecke S, Passarge E: Encephalocele, polycystic kidneys and polydactyly as an autosomal recessive trait simulating certain other disorders: The Meckel syndrome. *Ann Génét* **14**:97–103, 1971.

15. Meckel JF: Beschreibung zweier durch sehr ähnliche Bildungsabweichung entstellter Geschwister. *Dtsch Arch Physiol* **7**:99–172, 1822.

16. Moerman P et al: The Meckel syndrome: Pathological and cytogenetic observations in eight cases. *Hum Genet* **62**:240–245, 1982.

17. Naffah J et al: A propos de trois nouveaux cas dans une même fratrie du syndroms de Meckel ou dysencéphalie splanchocystique de Gruber. *Arch Fr Pédiatr* **29**:1069–1081, 1972.

18. Nevin NC et al: Prenatal diagnosis of the Meckel syndrome. *Clin Genet* **15**:1–4, 1979.

19. Opitz JM, Howe JJ: Meckel's syndrome: Dysencephalia splanchnocystica, Gruber's syndrome. *Birth Defects* **5**(2):167–179, 1969.

19a. Pachi A et al: Meckel–Gruber syndrome: Ultrasonographic diagnosis at 13 weeks' gestation in an at-risk case. *Prenat Diagn* **9**:187–190, 1989.

19b. Paetau A et al: Brain pathology in the Meckel syndrome: A study of 59 cases. *Clin Neuropathol* **4**:56–62, 1985.

20. Pettersen JC: Gross anatomical studies of a newborn infant with the Meckel syndrome. *Am J Med Genet* **18**:649–659, 1984.

21. Plauchu H et al: Le syndrome de Meckel. Sa variabilité d'expression peut faire obstacle au diagnostic prénatal. *J Génét Hum* **29**:431–440, 1981.

21a. Rapola J, Salonen R: Visceral anomalies in the Meckel syndrome. *Teratology* **31**:193–202, 1985.

22. Rehder H, Labbé F: Prenatal morphology in Meckel's syndrome. *Prenat Diagn* **1**:161–172, 1981.

23. Salonen R: The Meckel syndrome: Clinicopathologic findings in 67 patients. *Am J Med Genet* **18**:671–689, 1984.

24. Salonen R, Norio R: The Meckel syndrome in Finland: Epidemiologic and genetic aspects. *Am J Med Genet* **18**:691–698, 1984.

25. Schurig V et al: The Meckel syndrome in the Hutterites. *Am J Med Genet* **5**:373–381, 1980.

26. Seller MJ: Meckel syndrome and the prenatal diagnosis of neural tube defects. *J Med Genet* **15**:462–465, 1978.

27. Seller MJ: Phenotypic variation in Meckel syndrome. *Clin Genet* **20**:74–77, 1981.

28. Seppänen U, Herva R: Roentgenologic features of the Meckel syndrome. *Pediatr Radiol* **13**:329–331, 1983.

29. Young ID: High incidence of Meckel's syndrome in Gujarati Indians. *J Med Genet* **22**:301–304, 1985.

## Hydrolethalus syndrome

Salonen et al (9), in 1981, reported 28 patients from 18 families with a lethal syndrome characterized by polyhydramnios, variable degrees of hydrocephalus, crossed polydactyly, talipes, congenital heart anomalies (most frequently VSD and common AV canal), and severe micrognathia. About 50 patients have been described to date (2,6,11). A mild example was reported (3). Nearly all have been aborted or stillborn.

Inheritance is clearly autosomal recessive (2,6). To date, almost all certain reports emanate from Finland (8). It is possible that the patient described by Fitch and Pinsky (4) may have the disorder, but one cannot exclude *Meckel syndrome* or, more likely, *Smith–Lemli–Opitz syndrome*. The sibs reported by Adetoro et al (1) appear to have some new form of short rib–polydactyly syndrome.

The facies is rather striking. The eyes are often deeply set and occasionally may be microphthalmic. The tongue is usually malformed, small, or absent. About 40% have cleft palate. An equal number have smaller clefts of the mid-upper and lower lip. Cleft lip–palate has been noted (2). Micrognathia is marked (Fig. 21–21).

In addition to the hydrocephalus, various other brain abnormalities found in about 25% include widely separated hemispheres and agenesis of the corpus callosum and septum pellucidum and, occasionally, the pituitary gland (Fig. 21–22A,B).

The larynx, trachea, and/or bronchi are stenotic or, less often, dilated. There is reduced lobation of the lungs. Various other abnormalities include incomplete rotation of the gut (25%), accessory spleens (15%), cryptorchidism (75%), and bicornuate or duplicated uterus (30%).

Radiographically, there is obvious hydrocephalus with a midline defect in the occipital bone posterior to the foramen magnum. Nonfusion of one of the dorsal arches of the first few cervical vertebrae has been noted. This "keyhole" defect in the base of the skull has been termed occipitoschisis (7,10). Among those with polydactyly (60%), the extra digits are postaxial in the hands and preaxial in the feet. Rarely, there are postaxial toes. The upper limbs are short in about 15%. The tibiae exhibit proximal hypoplasia. Prenatal diagnosis is possible (5).

Fig. 21–21. *Hydrolethalus syndrome*. Small wide set eyes, malformed nose, midline cleft of upper lip, micrognathia. (From *R Salonen*, The Meckel and hydrolethalus syndromes. Diagnostic, genetic, and epidemiologic studies. Academic dissertation, Med Fac of Helsinki, 1986.)

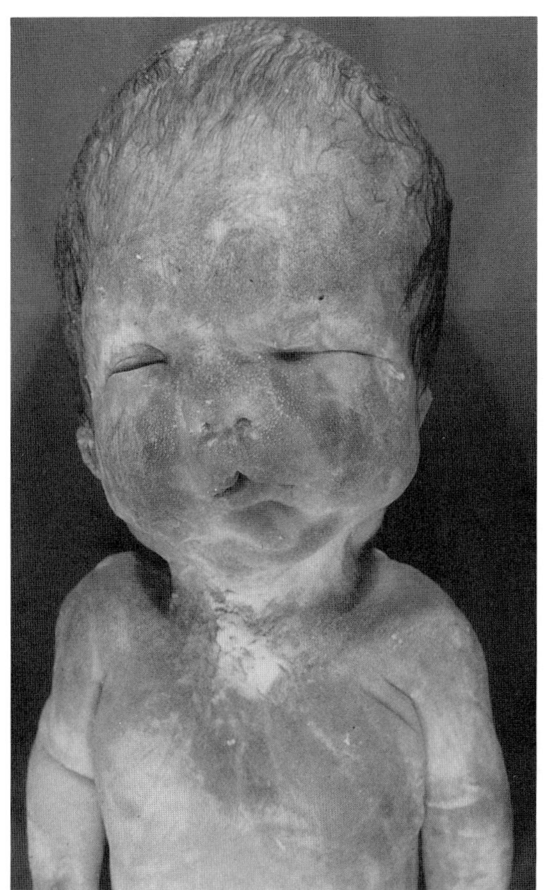

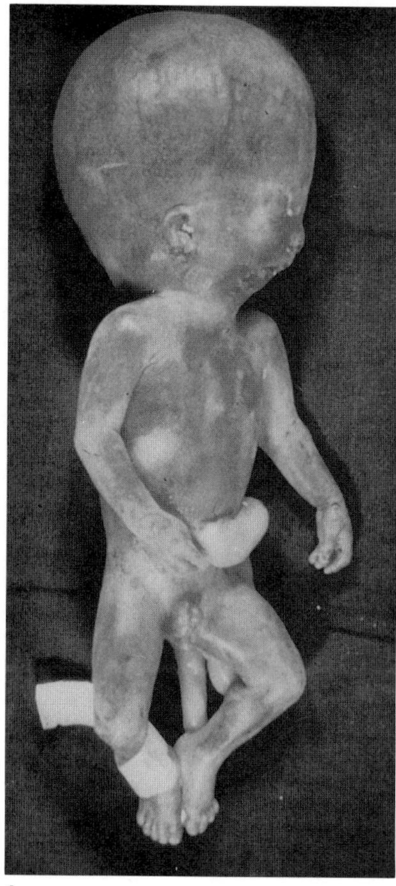

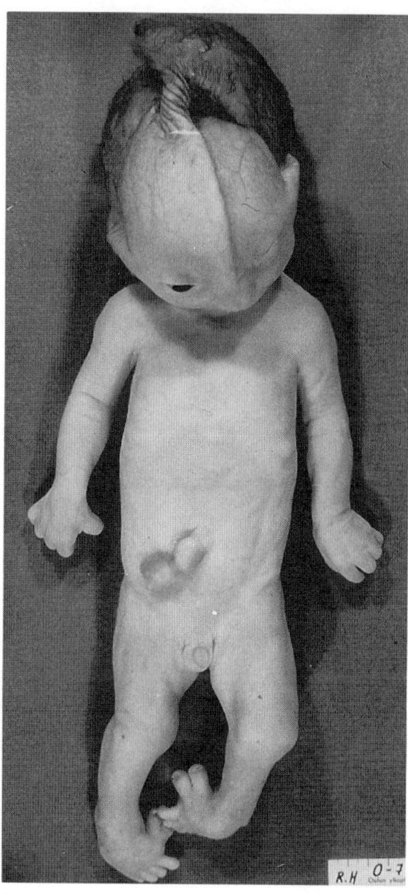

A B

Fig. 21–22. *Hydrolethalus syndrome.* (A,B). Hydrocephalus, small mandible, short lower limbs, and polydactyly. (From *R Salonen,* The Meckel and hydrolethalus syndromes. Diagnostic, genetic, and epidemiologic studies. Academic dissertation, Med Fac of Helsinki, 1986.)

In hydrolethalus syndrome, in contrast to the *Meckel syndrome,* one finds hydrocephaly rather than microcephaly and/or occipital encephalocele. The polydactyly is crossed. The kidneys are normal, the tibiae exhibit proximal hypoplasia, and there is polyhydramnios rather than oligohydramnios. Macrocephaly, hallucal duplication and postaxial polydactyly of the hands are seen in the *acrocallosal syndrome* but hydrocephalus does not occur. Hydrocephalus, cerebellar hypoplasia, cardiac and genital anomalies, cleft palate, and polydactyly are seen in *Smith–Lemli–Opitz syndrome, Type II* and hydrocephalus and agenesis of midline structures are found in *Walker–Warburg syndrome.*

#### References (Hydrolethalus syndrome)


1. Adetoro OO et al: Hydrolethalus syndrome in consecutive African siblings. *Pediatr Radiol* **14**:422–424, 1984.
2. Anyane-Yeboa K et al: Hydrolethalus (Salonen–Herva–Norio) syndrome: Further clinicopathological delineation. *Am J Med Genet* **26**:899–908, 1987.
3. Aughton DJ, Cassidy SB: Hydrolethalus syndrome: Report of an apparent mild case, literature review and differential diagnosis. *Am J Med Genet* **27**:935–942, 1987.
4. Fitch N, Pinsky L: The Meckel syndrome with limited expression in relatives. *Clin Genet* **4**:33–37, 1973.
5. Hartikainen-Sorri AL et al: Prenatal detection of hydrolethalus syndrome. *Prenat Diag* **3**:219–224, 1983.
6. Herva R, Seppänen U: Roentgenologic findings of the hydrolethalus syndrome. *Pediatr Radiol* **14**:41–43, 1984.
7. Kirkinen P et al: Intra-uterine growth and fatal fetal abnormality. *Acta Obstet Gynecol Scand* **62**:43–47, 1983.

7a. Krassikoff N et al: The hydrolethalus syndrome. *Birth Defects* **23**(1):411–419, 1987.

8. Salonen R: *The Meckel and Hydrolethalus Syndromes. Diagnostic, Genetic, and Epidemiologic Studies.* Academic Dissertation, Med Fac of Helsinki, 1986.
9. Salonen R et al: The hydrolethalus syndrome: Delineation of a "new" malformation syndrome based on 28 patients. *Clin Genet* **19**:321–330, 1981.
10. Sepänen U: The value of perinatal postmortem radiography—experience of 514 cases. *Ann Clin Res* **17** (*Suppl* **44**):1–59, 1985.
11. Toriello HV, Bauserman SC: Bilateral pulmonary agenesis: Association with the hydrolethalus syndrome and review of the literature from a developmental field perspective. *Am J Med Genet* **21**:93–103, 1985.


## Pseudocleft of the upper lip, cleft lip–palate, and hemangiomatous branchial cleft (branchio-oculo-facial syndrome)

Hall et al (4) reported a syndrome of branchial clefts, pseudocleft of the upper lip, and congenital nasolacrimal duct obstruction. Lee et al (6) and Fujimoto et al (3) demonstrated autosomal dominant inheritance. Earlier sporadic cases have been noted (2,5). Perhaps there is reduced fertility (3).

The forehead tends to be small. The scalp hair begins to gray at puberty (3,6). Hypertelorism (3), unilateral microphthalmus (3,4), anophthalmus (2,8), myopia (3), cataracts (3,6), strabismus (6), colobomas of the iris, choroid retina, and optic nerve (3,4), and upslanting palpebral fissures (3) have been noted. The pinnae are posteriorly angulated with a thin helix, prominent anthelix, and upturned lobules. Nasolacrimal duct obstruction results in dacryocystitis which, in turn, eventuates in thickened lower eyelids (1,3,4). The nose is broad and misshapen with a wide bridge and indented nasal tip. Pre- and postauricular pits have been documented (3). Conductive hearing loss has been reported (3–6). The philtrum is narrow and prominent with thick fibrous vertical ridges producing a pseudocleft (Fig. 21–23A). The palate tends to be highly arched (3). Cleft lip–palate was found (1,8), and RJ Gorlin has seen several patients with bilateral cleft lip–palate and hemangiomatous blebs of the cervical region. Another likely case

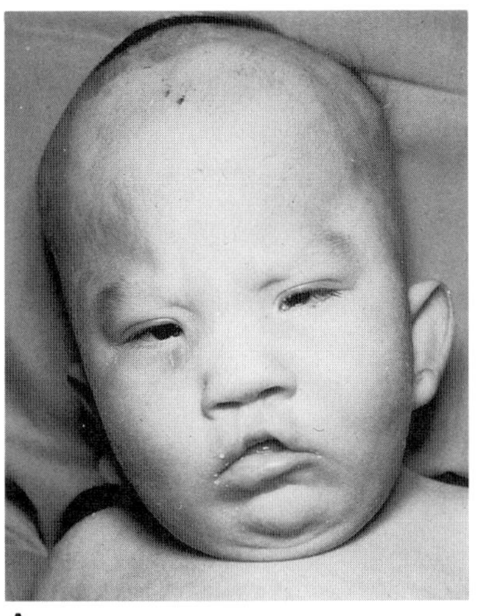


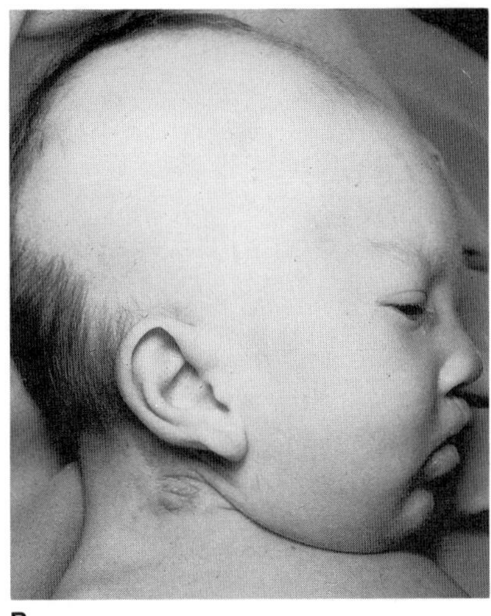


Fig. 21–23. *Pseudocleft of the upper lip, cleft lip–palate, and hemagiomatous branchial cleft.* (A,B) Apparent hypertelorism, pseudocleft in upper lip, branchial cleft covered by thinned skin, posteriorly angulated pinna. (From *A Fujimoto et al,* Am J Med Genet **27**:943, 1987.)

is that of an infant with bilateral cleft lip–palate and aplasia cutis congenita with neck fistulas reported by Schweckendiek et al (7). The child's mother also had cleft lip–palate. Paramedian upper lip pits, seen in one patient, correspond to the fusion sites of the median nasal and maxillary prominences (3). In one case (1), the lateral neck defects were interpreted as cervical thymus. The pathologic findings in the case of Schweckendiek et al (7) appear similar to those of Çiv̌i et al (1).

The hemangiomatous postauricular clefts or thinned epidermis extend along the sternocleidomastoid muscle (Figs. 21–23B and 21–24). Other findings have included prenatal growth deficiency, hypoplastic or polydactylous thumbs, clinodactyly of fifth fingers, and transverse palmar creases (3,6). Cystic kidney was found by Schweckendiek et al (7) and mental retardation and thoracic kyphosis by Tom et al (8).

### References [Pseudocleft of the upper lip, cleft lip–palate, and hemangiomatous branchial cleft (branchio-oculo-facial syndrome)]


1. Çiv̌i I et al: Bilateral thymus found in association with unilateral cleft lip and palate. *Plast Reconstr Surg* **83**:143–147, 1989.
2. Farmer AW, Maxmen MD: Congenital absence of skin. *Plast Reconstr Surg* **25**:291–297, 1960.
3. Fujimoto A et al: New autosomal dominant branchio-oculo-facial syndrome. *Am J Med Genet* **27**:943–951, 1987.
4. Hall BD et al: A new syndrome of hemangiomatous branchial clefts, lip pseudoclefts, and abnormal facial appearance. *Am J Med Genet* **14**:135–138, 1983.
5. Harrison MS: The Treacher Collins–Franceschetti syndrome. *J Laryngol Otol* **71**:597–603, 1957.
6. Lee WK et al: Bilateral branchial cleft sinuses associated with intrauterine and postnatal growth retardation, premature aging and unusual facial appearance: A new syndrome with dominant transmission. *Am J Med Genet* **11**:345–352, 1982.
7. Schweckendiek W et al: Doppelseitige retro-und subaurikuläre Fisteln mit Ektropium des Fistelgangepithels. *Laryngol Rhinol Otol* **56**:795–800, 1977.
8. Tom DWK et al: Inflammatory linear epidermal nevus caused by branchial cleft sinuses in a woman with numerous congenital anomalies. *Pediatr Dermatol* **2**:318–321, 1985.


Fig. 21–24. *Pseudocleft of the upper lip, cleft lip–palate, and hemangiomatous branchial cleft.* Note pouting upper lip due to pseudoclefting and aplasia cutis congenita of lateral neck.

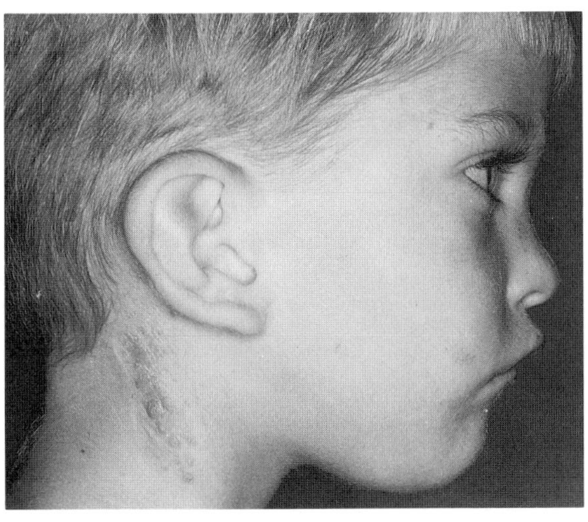

## Rapp–Hodgkin syndrome (cleft lip–palate, hypohidrosis, thin wiry hair, dystrophic nails, oligodontia, and hypospadias)

The syndrome described by Rapp and Hodgkin (7) in 1968 and by Wannarachue et al (15) in 1972 consists of cleft lip and/or cleft palate, sparse hair, small mouth, small stature, and hypospadias in males. Several other families (1,9,10,12–14), isolated examples (4,5), and possible cases (9) have been described. The condition reported by Bonafé et al (2) probably represents this syndrome. Autosomal dominant inheritance is evident.

The hair is slow growing, blond, wiry or uncombable, and sparse. Silengo et al (12) reported pili torti. Sparse lateral eyebrows and eyelashes, photophobia, blepharitis, and agenesis or atresia of lacrimal puncta, and oligodontia have been described. In some patients the ear canals have been noted to be slit-like (9,14). The nose tends to be narrow. The teeth are deficient in number or have conical crown form. Sweat glands are decreased (10).

Probably there is etiologic heterogeneity. Watson and Hardwick (16) described a girl with bilateral cleft lip–palate, hypoplastic pinnae, anodontia, and dystrophic nails (Figs. 21–25A,B and 21–26A,B). A sib who died soon after birth had cleft lip–palate and congenital heart anomalies. The mother and a sister exhibited oligodontia. The parents were consanguineous. Shelley and Shelley (11) described sibs with uncombable sparse blond hair, scarring alopecia of the scalp at the occiput, absence of tears, sparse eyelashes, cup-shaped pinnae, dysplastic ridged nails, and short palate with bifid uvula or cleft palate. The nose was narrow. The incisors were anomalous. One sib had clubfoot and another mesodermal dysgenesis of the anterior segment of both eyes. Their father had syndactyly of the right index and mid-

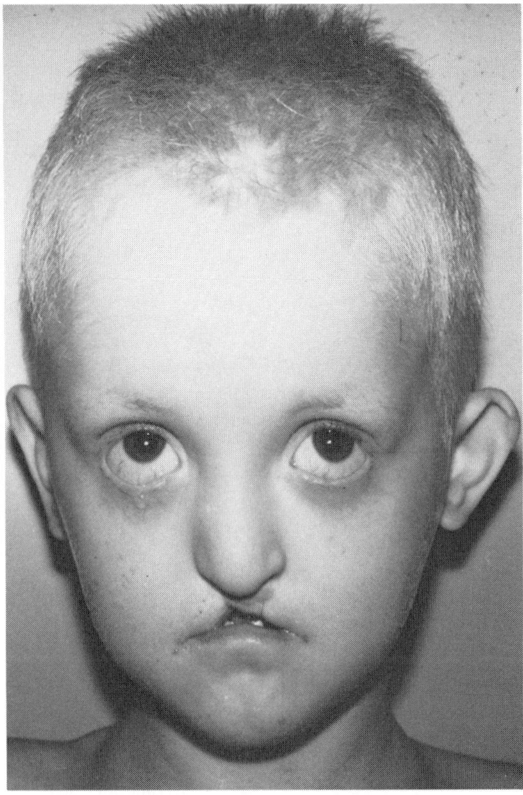

A

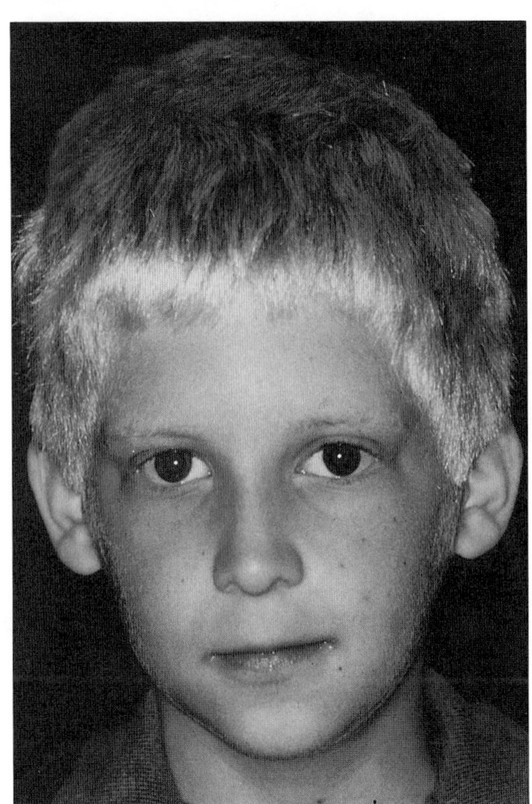

B

Fig. 21–25. *Rapp–Hodgkin syndrome.* (A) Sparse wiry hair, absent lashes and lacrimal puncta, ectropion of lower lids, hypoplasia of nasal alae, abnormal pinnae, bilateral cleft lip–palate. (B) Boy has cleft palate and blond, wiry hair. [A from *RL Summitt* and *RL Hiatt,* Birth Defects **7**(8):121, 1971. B courtesy of *C Salinas,* Charleston, South Carolina.]

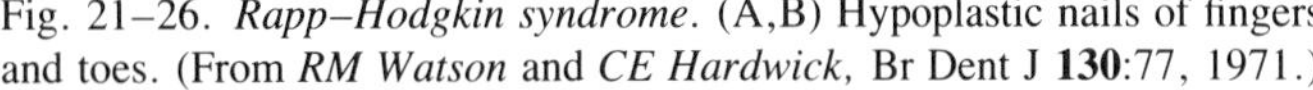

Fig. 21–26. *Rapp–Hodgkin syndrome.* (A,B) Hypoplastic nails of fingers and toes. (From *RM Watson* and *CE Hardwick,* Br Dent J **130**:77, 1971.)

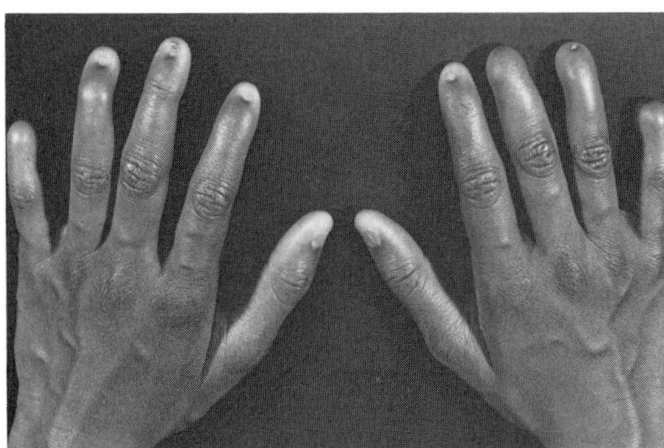

A

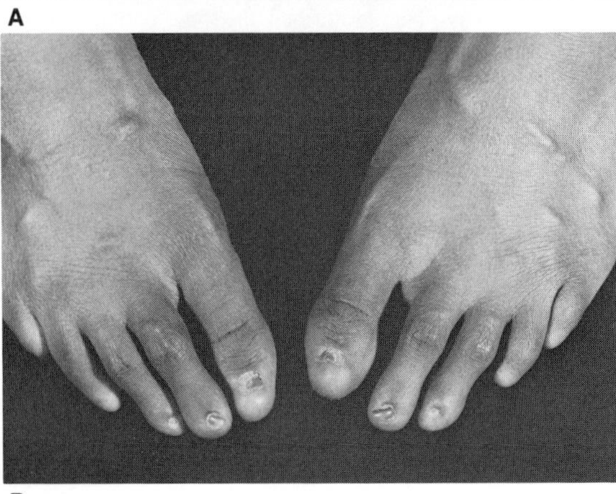

B

dle fingers without a nail on the middle finger. The child described by Crawford et al (3) had hypoplastic nasal alae, absent lacrimal puncta, epiphora, photophobia, cleft soft palate, sparse eyebrows, coarse kinky scalp hair, dysplastic nails, reduced sweat pores, hypoplastic dermal ridges, and oligodontia. Moynahan (6) reported sibs with what he called XTE syndrome. The parents were consanguineous. The syndrome consisted of cleft palate, hypohidrosis, dry skin, dry coarse hair, nail deformities, evanescent cutaneous bullae, absence of eyelashes, small lacrimal punctae, talipes, and defective enamel.

Rosselli and Gulienetti (8) described what is probably a unique syndrome. Their cases 2 and 3 had wooly hair, cleft lip and palate, dystrophic nails, subungual keratoses, dystrophic facial skin, a papulofollicular dermatosis of the trunk, and hypoplasia or aplasia of the thumbs and popliteal pterygia.

Perhaps some patients have incomplete expression of the *EEC syndrome* or the *AEC syndrome.* The hair is triangular or ellipsoidal in cross-section. SEM has demonstrated canal-like grooves along the axis (9,11) (Fig. 21–27). Sweat pore outlets appear to be decreased in size and number (3,9).

### References [Rapp–Hodgkin syndrome (cleft lip–palate, hypohidrosis, thin wiry hair, dystrophic nails, oligodontia, and hypospadias)]

1. Beckerman BL: Lacrimal anomalies in anhidrotic ectodermal dysplasia. *Am J Ophthalmol* **75**:728–730, 1973.

2. Bonafé JL et al: Association dysplasia ectodermique-division palatine-cheveu-chiendent. *Ann Dermatol Venereol* **106**:989–993, 1979.

3. Crawford PJM et al: Rapp-Hodgkin syndrome: An ectodermal dysplasia involving the teeth, hair, nails and palate. *Oral Surg Oral Med Oral Pathol* **67**:50–62, 1989.

4. Fára M: Regional ectodermal dysplasia with total bilateral cleft. *Acta Chir Plast* **13**:100–105, 1971.

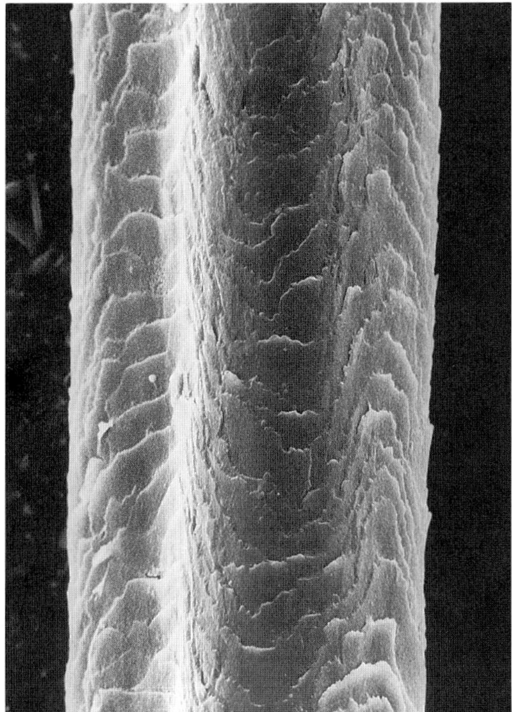

Fig. 21–27. *Rapp–Hodgkin syndrome.* SEM of hair showing characteristic grooving.


5. Massengill R et al: An abnormal speech pattern associated with an orofacial anomaly. *Acta Otolaryngol* **68**:537–542, 1969.

6. Moynahan EJ: XTE (Xeroderma, Talipes and Enamel defect): A new heredofamilial syndrome. Two cases. Homozygous inheritance of a dominant gene. *Proc R Soc Med* **63**:447–448, 1970.

7. Rapp RS, Hodgkin WE: Anhidrotic ectodermal dysplasia: Autosomal dominant inheritance with palate and lip anomalies. *J Med Genet* **5**:269–272, 1968.

8. Rosselli D, Gulienetti R: Ectodermal dysplasia. *Br J Plast Surg* **14**:190–204, 1961.

9. Salinas C, Montes GM: Rapp–Hodgkin syndrome: Observations on ten cases and characteristic hair changes (pili canaliculi). *Birth Defects* **24**(2):149–168, 1988.

10. Schroeder HW Jr, Sybert VP: Rapp–Hodgkin ectodermal dysplasia. *J Pediatr* **110**:72–75, 1987.

11. Shelley WB, Shelley ED: Uncombable hair syndrome: Observations on response to biotin and occurrence in siblings with ectodermal dysplasia. *J Am Acad Dermatol* 13:97–102, 1985.

12. Silengo MC et al: Distinctive hair change (pili torti) in Rapp–Hodgkin ectodermal dysplasia syndrome. *Clin Genet* **21**:297–300, 1982.

13. Stasiowska B: Rapp–Hodgkin ectodermal dysplasia syndrome. *Arch Dis Childh* **56**:793–795, 1981.

14. Summitt RL, Hiatt RL: Hypohidrotic ectodermal dysplasia with multiple associated anomalies. *Birth Defects* **7**(8):121–124, 1971.

15. Wannarachue N et al: Ectodermal dysplasia and multiple defects (Rapp–Hodgkin type). *J Pediatr* **81**:1217–1218, 1972.

16. Watson RM, Hardwick CE: Hypodontia associated with cleft palate. *Br Dent J* **130**:77–80, 1971.


## Femoral hypoplasia–unusual facies syndrome

Daentl et al (6), in 1975, defined the femoral hypoplasia–unusual facies syndrome. However, a number of other authors reported earlier examples (1,2,8,16,21) or, at least, possible examples (22). About 35% of those with the syndrome are infants of diabetic mothers (15,22,23), thus making the syndrome analogous to caudal regression (sacral agenesis) with which one can also find cleft lip and/or cleft palate (3,8). However, 65% of the cases result from other causes. Although the condition has been reported in two generations (17), its inheritance pattern, if any, would more likely be multifactorial than autosomal dominant (18).

The facies is characterized by upward slanting palpebral fissures (35%), short nose with hypoplastic alae (65%), mildly dysmorphic pinnae (45%), flat philtrum, thin vermilion of upper lip (65%), and micrognathia (80%). The ear canals may be narrowed. Cleft palate is common (80%); cleft lip–palate is rare (14,24) (Fig. 21–28).

Stature is markedly reduced due to abbreviation of the lower extremities (Figs. 21–29A,B and 21–30A). The upper extremities and shoulder girdle are affected in 75%. The humerus may be mildly shortened and there is limited motion of the shoulders and elbows. Sprengel deformity is rather frequent (1,9). Syndactyly of toes and broad halluces have been noted. Inguinal hernia is common. Cardiac anomalies, documented in 20%, include VSD, pulmonary stenosis, and persistent truncus arteriosus.

Genitourinary anomalies have been noted in 60% (absent or hypoplastic labia, cryptorchidism, hypoplastic penis, polycystic kidneys, absent kidney, abnormal collecting system) (4,6,10,12,14) (Fig. 21–30A).

Radiographic changes are often asymmetric. The femora, fibulae, and acetabulae range from being hypoplastic to aplastic (Fig. 21–30B). Vertebral anomalies, seen in 35%, include scoliosis, hemivertebrae, anterior synostosis of lumbar vertebrae, spina bifida occulta, and malsegmentation of the sacrum. The ribs may taper, and some may be fused or missing. The pelvis may be hypoplastic. Derangement of elbow joints (radiohumeral synostosis, radioulnar synostosis) and talipes are frequent. The great toes may be bifid and, rarely, there is polydactyly or agenesis of a toe (9,10,22).

#### References (Femoral hypoplasia–unusual facies syndrome)


1. Assemany SR et al: Syndrome of phocomelic diabetic embryopathy (caudal dysplasia). *Am J Dis Child* **123**:489–491, 1972.


Fig. 21–28. *Femoral hypoplasia–unusual facies syndrome.* Note shortened nose, long philtrum, and micrognathia. (From *RB Goldberg et al,* Cleft Palate J **15**:386, 1978.)

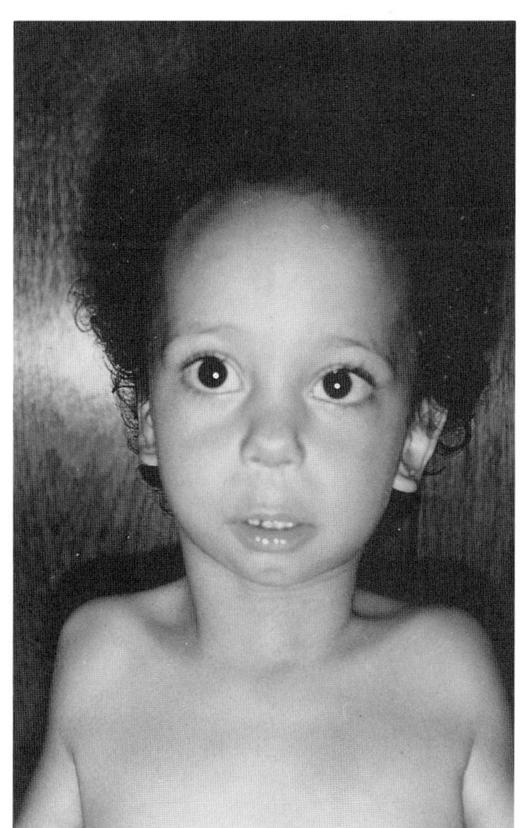

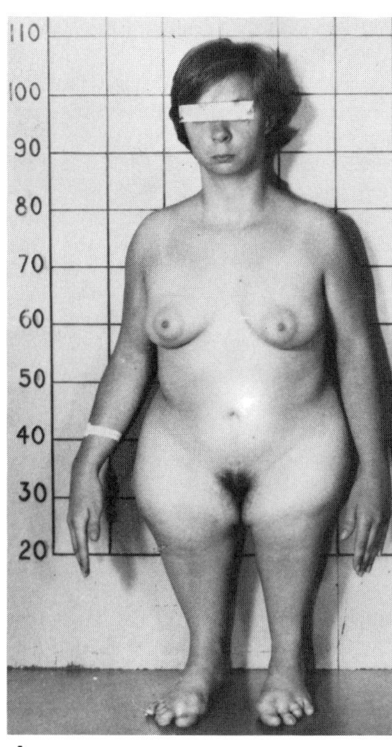

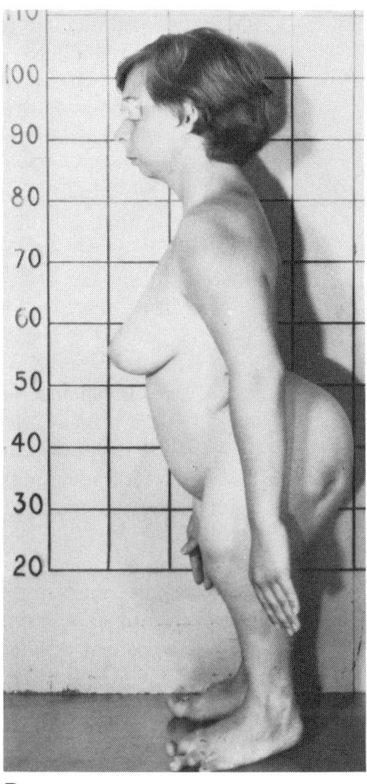

A B

Fig. 21–29. *Femoral hypoplasia–unusual facies syndrome.* (A,B) Severe abbreviation of lower limbs. Patients had dislocated and hypoplastic patellas, abnormal toes. (From *JA Bailey* and *P Beighton,* Clin Pediatr **9**:668, 1970.)

2. Bailey JA, Beighton P: Bilateral femoral dysgenesis. *Clin Pediatr* **9**:668–674, 1970.

3. Blumel J et al: Partial and complete agenesis or malformation of the sacrum with associated anomalies. *J Bone Jt Surg* **41A**:497–518, 1959.

4. Burck U et al: Bilateral femoral dysgenesis with micrognathia, cleft palate, abnormalities of the spine and pelvis and foot deformities: Clinical and radiological findings. *Helv Paediatr Acta* **36**:473–482, 1981.

5. Burn J et al: The femoral hypoplasia-unusual facies syndrome. *J Med Genet* **21**:331–340, 1984.

6. Daentl DL et al: Femoral hypoplasia–unusual facies syndrome. *J Pediatr* **86**:107–111, 1975.

6a. De Palma L et al: Femoral hypoplasia–unusual facies syndrome: Autopsy findings in an unusual case. *Pediatr Pathol* **5**:1–8, 1986.

7. Eastman JR, Escobar V: Femoral hypoplasia–unusual facies syndrome—a genetic syndrome? *Clin Genet* **13**:72–76, 1978.

Fig. 21–30. *Femoral hypoplasia–unusual facies syndrome.* (A) Lower torso showing hypoplastic external genitalia, shortened legs, residual varus deformity, and polysyndactyly. (B) Roentgenogram showing minuscule femurs, absent acetabula, hypoplastic pelvis. (A from *RB Goldberg et al,* Cleft Palate J **15**:386, 1978. B from *JA Bailey* and *P Beighton,* Clin Pediatr **9**:668, 1970.)

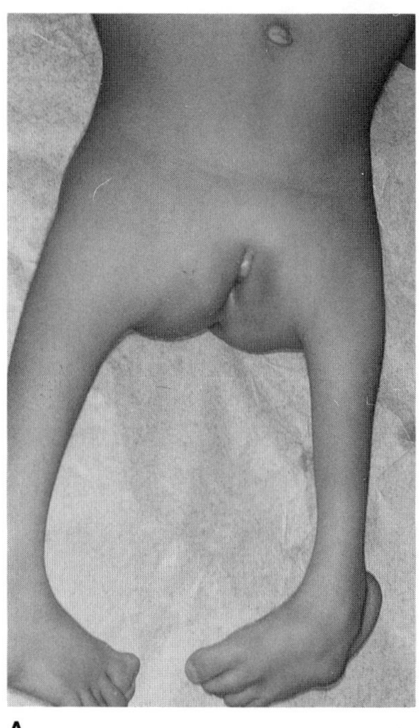

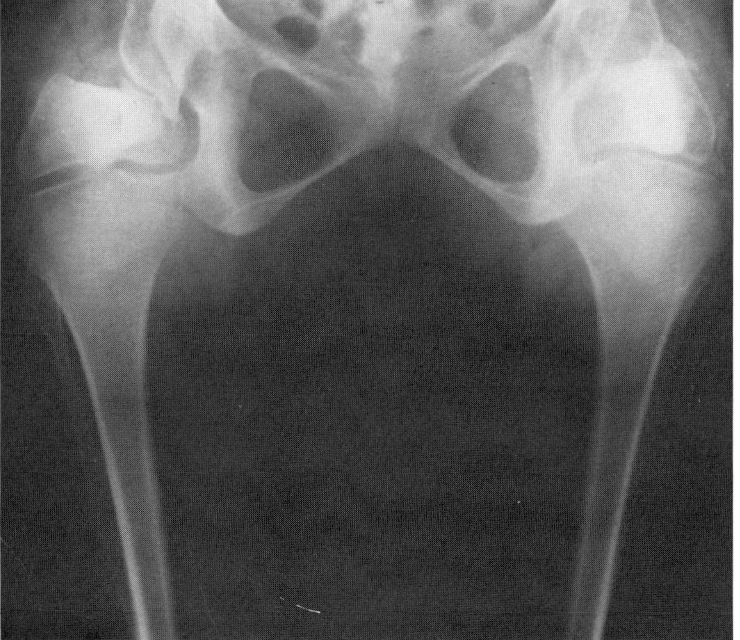

A B

8. Frantz CH, O'Rahilly R: Congenital skeletal limb deficiencies. *J Bone Jt Surg* **43A**:1202–1284, 1961.

9. Gleiser S et al: Femoral hypoplasia–unusual facies syndrome from another viewpoint. *Eur J Pediatr* **128**:1–5, 1978.

10. Goldberg RB et al: Bilateral femoral dysgenesis syndrome. *Cleft Palate J* **15**:386–391, 1978.

11. Graviss ER et al: Proximal femoral focal deficiency associated with the Robin anomalad. *J Med Genet* **17**:390–392, 1980.

12. Holmes LB: Femoral hypoplasia–unusual facies syndrome. *J Pediatr* **87**:668–669, 1975.

13. Holthusen W: The Pierre Robin syndrome: Unusual associated developmental defects. *Ann Radiol* **15**:252–262, 1972 (Case 2).

14. Hurst D, Johnson DF: Femoral hypoplasia–unusual facies syndrome. *Am J Med Genet* **5**:255–258, 1980.

15. Johnson JP et al: Femoral hypoplasia–unusual facies syndrome in infants of diabetic mothers. *J Pediatr* **102**:866–872, 1982.

16. Kučera J et al: Missbildungen der Beine und der kaudalen Wirbelsäule bei Kindern diabetischer Müttern. *Dtsch Med Wschr* **90**:901–905, 1965.

17. Lampert RP: Dominant inheritance of femoral hypoplasia–unusual facies syndrome. *Clin Genet* **17**:255–258, 1980.

18. Lord J, Beighton P: The femoral hypoplasia–unusual facies syndrome: A genetic entity? *Clin Genet* **20**:267–275, 1981.

19. Maisel DO, Stillwell JH: The Pierre Robin syndrome associated with femoral dysgenesis. *Br J Plast Surg* **33**:237–241, 1980.

20. Mastroiacovo P et al: Femoral hypoplasia–unusual facies: A rare malformation syndrome without mental retardation. *Riv Ital Pediatr* **5**:347–353, 1979.

21. Passarge E: Congenital malformation and maternal diabetes. *Lancet* **1**:324–325, 1965.

22. Pitt DB et al: Femoral hypoplasia–unusual facies syndrome. *Aust Pediatr J* **18**:63–66, 1982.

23. Riedel F, Froster-Iskenius U: Caudal dysplasia and femoral hypoplasia–unusual facies syndrome: Different manifestations of the same disorder? *Eur J Pediatr* **144**:80–82, 1985.

24. Walden RH et al: Pierre Robin syndrome in association with combined congenital lengthening and shortening of the long bones. *Plast Reconstr Surg* **48**:80–82, 1971.

25. Williamson DAJ: A syndrome of congenital malformations possibly due to maternal diabetes. *Dev Med Child Neurol* **12**:145–152, 1970.

## Fryns syndrome (cleft palate, diaphragmatic hernia, coarse facies, and acral hypoplasia)

Fryns (6), in 1979, described an essentially lethal syndrome of diaphragmatic hernia, coarse facies, acral hypoplasia, and cleft palate. About 45 cases have been described (1–3a,6,7,9–11,13–15). The infant reported by Fitch et al (5) possibly had the disorder. Inheritance is clearly autosomal recessive (1,2,9,13,14). Aymé et al (1) estimated the prevalence as 0.7/10,000 births.

Polyhydramnios developing in the second trimester has been present in about 60% of the cases (1,2). Death often occurs at about 30 weeks gestation. Birth length, weight, and OFC tend to be normal.

The face is coarse and hirsute with narrow palpebral fissures, broad flat nasal bridge, large upturned nose, large dysplastic pinnae with hypoplastic attached lobes, short upper lip, macrostomia, microretrognathia, and short broad neck (Fig. 21–31A–D). Dandy–Walker malformation (1) and agenesis of the corpus callosum have been noted in a few cases (9,11). Arhinencephaly and agenesis of olfactory bulbs have been described (1). The corneae are cloudy in about 15% (1). Retinal dysplasia has been reported (1).

Diaphragmatic defects are a virtually constant feature. They are often large, allowing herniation of the abdominal contents into the thoracic cavity. They are usually unilateral (85%) and left-sided (70%). Fryns et al (6) suggested that the diaphragmatic hernia was secondary to the primary defect in lung development. Digestive tract anomalies, secondary to the diaphragmatic defect, are found in 60%. They include intestinal malrotation, anomalous attachment of the gut, omphalocele, duodenal atresia, and esophageal atresia. The lungs are usually hypoplastic and abnormally lobulated. The anus may be imperforate, atretic, or especially anteriorly displaced. Agangliosis of the bowel and bladder have been found (2).

Various renal abnormalities (renal agenesis, cystic kidney, hydronephrosis, ureteral cysts, misplaced megaureters, duplicate and atretic ureters) have been reported in 55%. Genital anomalies, noted in 60%, consist of bicornuate uterus, hypoplastic upper vagina, hypoplastic ovaries, cryptorchidism, and saddle scrotum (1,2,6,7,9,10,14).

Assorted cardiac defects (VSD, ASD, overriding aorta) were noted in 55% (1,2).

Distal limb hypoplasia (brachytelephalangy) with nail deficiency is found in 75%. The terminal digits may be broad. Camptodactyly of the fifth finger is frequent. Other fingers may be axially deviated. Other skeletal changes are minimal, consisting of broad medial ends of the clavicles (2).

Isolated cleft palate has been present in 30% (1,2,11,13) and cleft lip with cleft palate has been found in 25% (1,2,9,10). Macrostomia is a constant feature.

Prenatal diagnosis has been made at 19–20 weeks by ultrasonographic demonstration of diaphragmatic hernia and elevated alphafetoprotein levels (1,2,11).

Diaphragmatic hernia is ordinarily considered to be a multifactorial trait with a 3:1 male predilection occurring in 1 in 2000 births. In a study of associations among 143 cases of diaphragmatic hernia, David and Illingworth (4) noted two patients with cleft palate and micrognathia (Robin sequence) and three with cleft lip. Crane (3) reported a different, probably X-linked lethal syndrome of cleft lip–palate, polyhydramnios, diaphragmatic hernia, and various urogenital anomalies. Lowry and MacLean (8) described a female child with cleft palate, proptosis of eyeballs, divergent squint, glaucoma, multiple congenital heart defects, mental retardation, diaphragmatic eventration, and craniosynostosis *(Lowry–MacLean syndrome).* Saal et al (12) reported monozygotic twins with cleft palate, agenesis of hemidiaphragm, and ambiguous genitalia but no distal limb anomalies.

Also to be excluded are *trisomy 18* and *Zellweger syndrome.* There appears to be considerable overlap with *Rüdiger syndrome.* Diaphragmatic defects are also seen in *de Lange syndrome.*

#### References [Fryns syndrome (cleft palate, diaphragmatic hernia, coarse facies, and acral hypoplasia)]

1. Aymé S et al: Fryns syndrome. Report on 8 new cases. *Clin Genet* **35**:191–201, 1989.

2. Bamforth JS et al: Congenital diaphragmatic hernia, coarse facies, and acral hypoplasia: Fryns syndrome. *Am J Med Genet* **32**:93–99, 1989.

3. Crane JP: Familial congenital diaphragmatic hernia: Prenatal diagnostic approach and analysis of twelve families. *Clin Genet* **16**:244–252, 1978.

3a. Cunniff C et al: Fryns syndrome: An autosomal recessive disorder associated with craniofacial dysmorphism, diaphragmatic hernia, and distal digital hypoplasia. *Pediatrics* (in press).

4. David RJ, Illingworth CA: Diaphragmatic hernia in the southwest of England. *J Med Genet* **13**:253–262, 1976.

5. Fitch N et al: Absent left hemidiaphragm, arhinencephaly, and cardiac formations. *J Med Genet* **15**:399–401, 1978.

6. Fryns JP: A new lethal syndrome with cloudy cornea, diaphragmatic defects, and distal limb deformities. *Hum Genet* **50**:65–70, 1979.

7. Goddeeris P et al: Diaphragmatic defects, craniofacial dysmorphism, cleft palate and distal limb deformities: A new lethal syndrome. *J Génét Hum* **28**:57–60, 1980.

8. Lowry RB, MacLean JR: Syndrome of mental retardation, cleft palate, eventration of diaphragm, congenital heart disease, glaucoma, growth failure, and craniosynostosis. *Birth Defects* **13**(3B):203–210, 1977.

9. Lubinsky M et al: Fryns syndrome: A new variable multiple congenital anomalies (MCA) syndrome. *Am J Med Genet* **14**:461–466, 1983.

10. Meinecke P, Fryns JP: The Fryns syndrome: Diaphragmatic defects, craniofacial dysmorphism, and distal digital hypoplasia. Further evidence for autosomal recessive inheritance. *Clin Genet* **28**:516–520, 1985.

11. Moerman P et al: The syndrome of diaphragmatic hernia, abnormal face, and distal limb anomalies (Fryns syndrome): Report of two sibs with further delineation of this multiple congenital anomaly (MCA) syndrome. *Am J Med Genet* **31**:805–814, 1988.

12. Saal HM et al: Ambiguous genitalia, agenesis of the left hemidiaphragm and other malformations in monozygotic twins. *Proc Greenwood Genet Ctr* **7**:142, 1988.

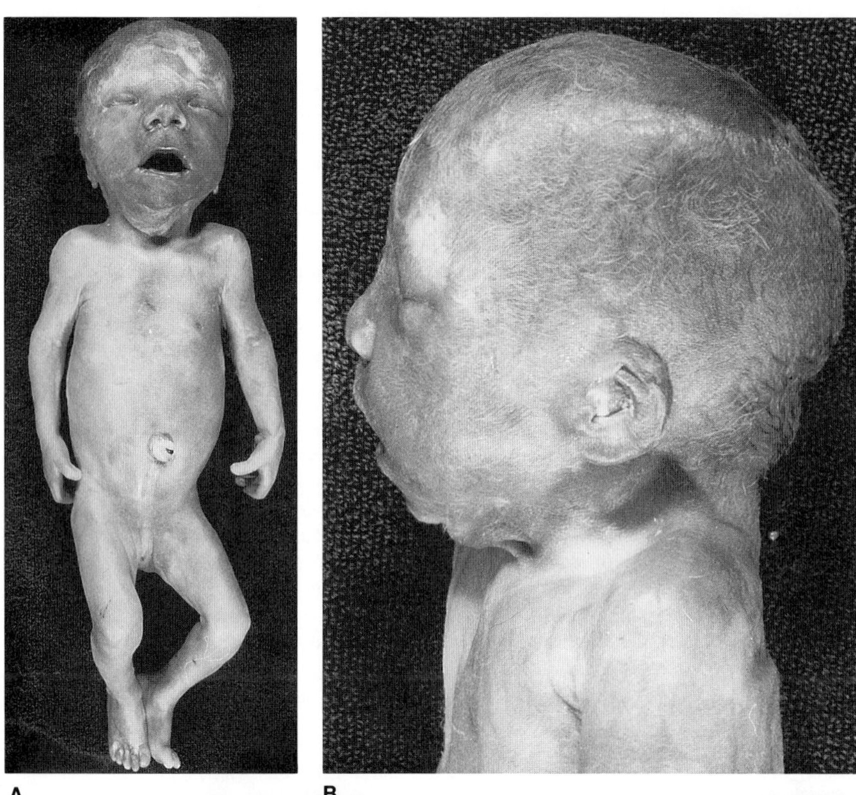

A B

Fig. 21–31. *Fryns syndrome.* (A) Infant showing coarse facies, small eyes, flat nasal bridge, large upturned nose, short upper lip, and macrostomia. The thorax is small, the nipples hypoplastic. The thumbs are digitalized and proximally implanted. (B) The pinnae are low set, poorly lobed, and posteriorly rotated. (C) The terminal digits are hypoplastic. (D) Radiograph showing agenesis of terminal phalanges. (From *JP Fryns et al,* Hum Genet **50**:65, 1979.)

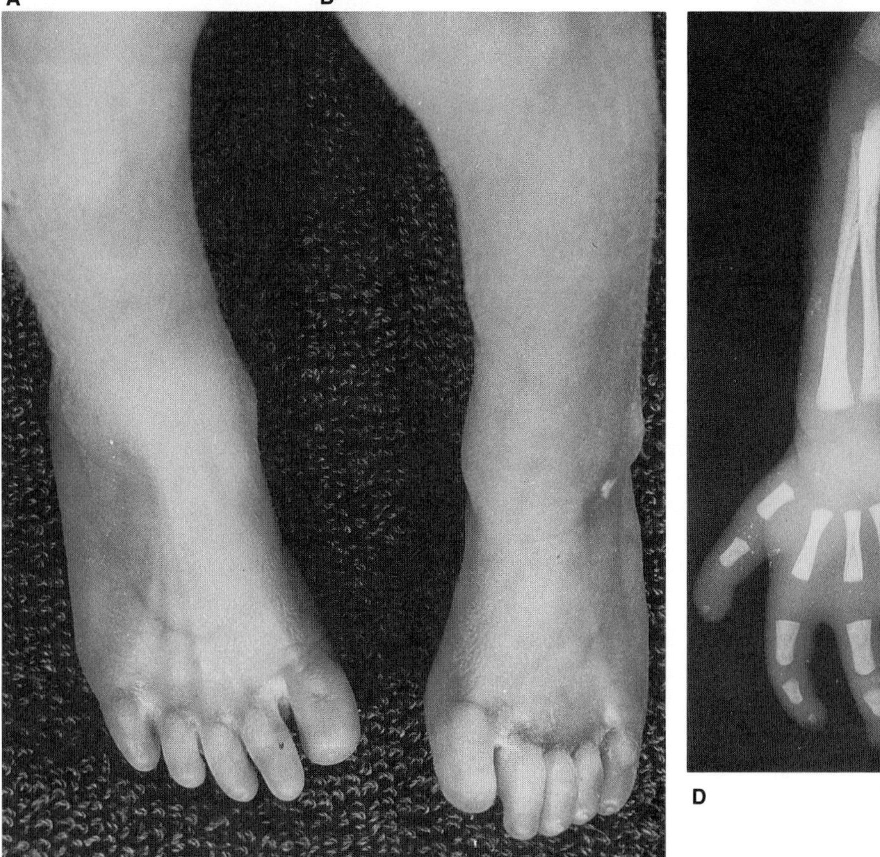

C D

13. Samueloff A et al: Fryns syndrome: A predictable lethal pattern of multiple congenital anomalies. *Am J Obstet Gynecol* **156**:86–88, 1987.
14. Schwyzer U et al: Fryns syndrome in a girl born to consanguineous parents. *Acta Paediatr Scand* **76**:167–171, 1987.
15. Young ID et al: A case of Fryns syndrome. *J Med Genet* **23**:82–88, 1986.

## Roberts–pseudothalidomide syndrome (Roberts syndrome, SC phocomelia syndrome)

The syndrome of cleft lip–palate and tetraphocomelia has a fascinating history, having been discovered independently by a number of authors. The first description was that by Roberts (32) in 1919. There were earlier and later scattered reports (4,5,9,11,18,20,28,33,35,39,41) but the condition was largely forgotten until Appelt et al (1), in 1966, rediscovered it and Freeman et al (6) extensively summarized case reports. Meanwhile, Herrmann et al (14), in 1969, reported the pseudothalidomide or SC syndrome, believing it to be a separate entity. Analysis of reported cases by several authors (6,10,13,32b,43) has provided convincing evidence that the two disorders represent extreme ends of a phenotypic spectrum of a single syndrome. A very mildly affected patient was documented by Petrinelli et al (29). We cannot identify with certainty the disorder described by Keutel et al (19).

Inheritance is autosomal recessive. Parental consanguinity and occurrence in sibs have been reported in over 40% of the cases (3, 6,7,14,16,21,25,27,28,31,32,35,36,42,43). It has even been described in identical twins (7).

Birthweight is often less than 2250 g with continued growth failure postnatally. Life expectancy is often short. At least 20% are stillborn. Among those that survive infancy, most have exhibited failure to thrive.

**Facies.** There is a tendency to microcephaly. Among those that die within the first month of life, there is hypertelorism and proptosis due to shallow orbits. The nose is often flattened and the nostrils are small. The pinnae are lowset, posteriorly angulated, dysmorphic, and lobeless (Fig. 21–32A,B).

Bilateral cleft lip–palate is extremely common while isolated cleft palate occurs rarely (Figs. 21–33A,B and 21–34A,B). (7,9,10,14,30).

In those that survive early infancy, the scalp hair is often sparse and silvery blond (10,12,24,30), but it may be initially dark (2). The nasal bridge is narrow and alar cartilages are hypoplastic, producing thin nares and a pointed tip.

In both groups, one often notes superficial capillary hemangiomas on the midface, forehead, and ears (2,3,6,12,14,15,21,25,34,42,43). Cataracts and corneal opacities resulting in blindness have been observed more in those that survive infancy, but can be found in both groups (6,21,43).

**Musculoskeletal.** The skull tends to be microbrachycephalic. There are reduction anomalies of variable degree in the tubular bones of all extremities but the arms are especially shortened. The hands may be radially clubbed and single palmar creases are common. The number of digits is reduced, more frequently in the hands (thumb, fifth finger) than in the feet. The index fingers may be clinodactylous. Thumbs are often missing, but rarely polydactyly of toes (6,34) has been reported. Various joints exhibit flexion contractures, especially the elbows and knees. Talipes and soft tissue syndactyly are common.

Radiographically, Wormian bones may be seen in the lambdoidal sutures. There are more severe skeletal anomalies among those that die in early infancy. Absence or severe hypoplasia of the radius, ulna, tibia, fibula, or femur is seen. The thumb phalanges and first and fifth metacarpals are hypoplastic or missing. The carpal bones are irregular and reduced in number. Humeroradial or humeroulnar synostosis, talipes calcaneovalgus, and absence or posterior displacement and/or bilateral deformity of the fibulae are commonly reported (3,7,15) (Fig. 21–35A,B). Less frequently, shortened or curved humerus and tibia, fusion of the fourth and fifth metacarpals, and femorotibial synostosis have been described (2,14,25,31). The ribs are occasionally reduced in number (7).

**Central nervous system.** Among those that survive, about 50% are mentally retarded. Miscellaneous findings include hydrocephalus (6), calcification of basal ganglia (16), seizures (14), encephalocele (6,27), absent olfactory lobes (27), and cranial nerve paralysis (28a).

**Urogenital.** Various findings include enlarged penis (3,6,9, 27,32,34,39) or clitoris (1,27,32,33,42), ambiguous genitalia (27), hypospadias (40), enlarged or cleft labia minora (1), septate vagina (6), and bicornuate uterus (6). Cryptorchidism is a frequent feature in males.

Polycystic kidneys, hydronephrosis, ureteral stenosis, and horseshoe kidney have been described (6,27,40).

**Cardiac.** Miscellaneous anomalies include VSD (17,24,34,40), ASD (6), and PDA (6,40,42).

**Miscellaneous.** Thrombocytopenia has been reported in a few cases (3,40). Melanoma (28a) has also been noted.

**Differential diagnosis.** Reduction anomalies occur in a number of disorders: *thalidomide embryopathy,* Holt–Oram syndrome, thrombocyte aplasia radius (TAR) syndrome, VACTERL association, Fanconi anemia, *trisomy 18, Nager syndrome,* etc. (30). A case having some similarities was reported by Ladda et al (22). However, one of us suspects that this patient has Roberts–pseudothalidomide syndrome.

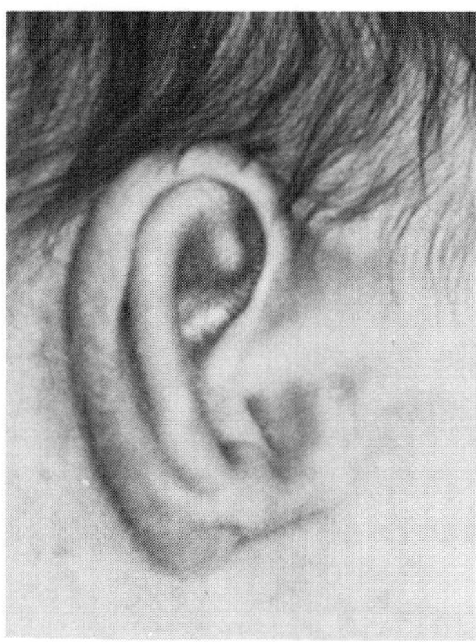
A

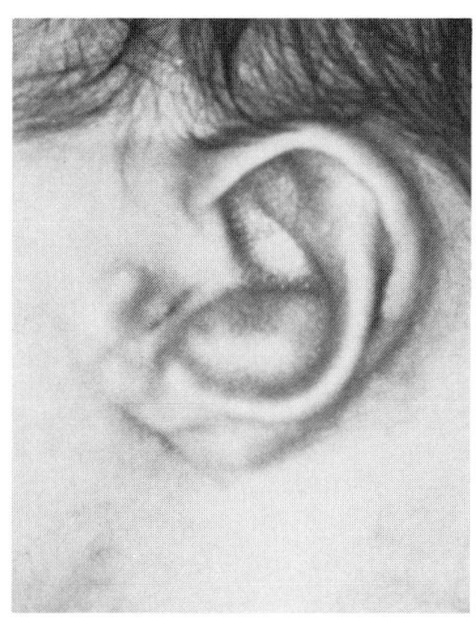
B

Fig. 21–32. *Roberts–pseudothalidomide syndrome.* (A,B) Close-up of ears in affected child. (Courtesy of *S Pruzansky,* Chicago, Illinois.)

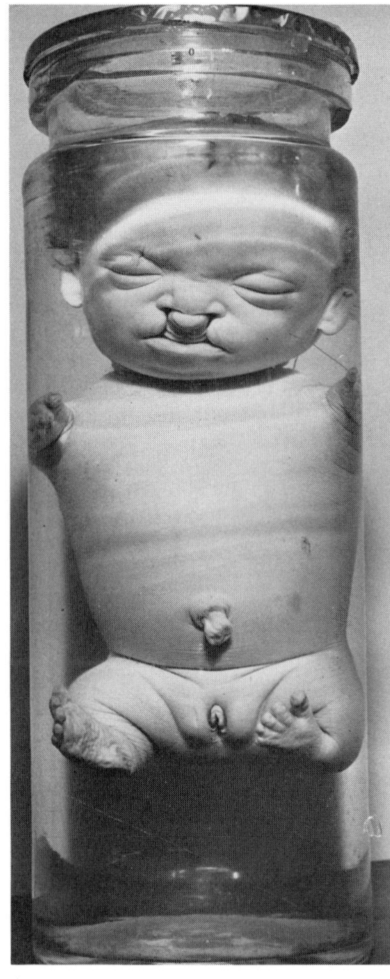

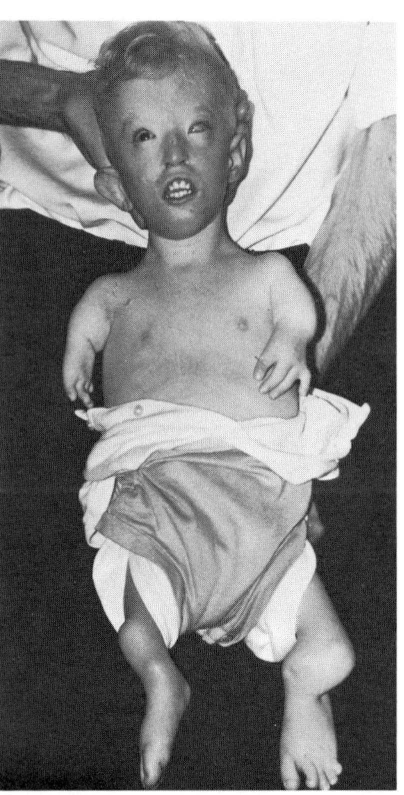

A B

Fig. 21–33. *Roberts–pseudothalidomide syndrome.* (A) Ocular hypertelorism, reduction in digits, clitoral enlargement, bizarre pinnae. (B) Thin, pinched nose, abnormal pinnae, convergent strabismus, symmetric reduction of arms and legs with decreased number of fingers. [A from *J Appelt,* Pädiatr Pädol **2**:119, 1966. B from *J Hermann et al,* Birth Defects **5**(3):81, 1969.]

Fig. 21–34. *Roberts–pseudothalidomide syndrome.* (A,B) Compare with other cases. (From *BD Hall* and *MH Greenberg,* Am J Dis Child **123**:602, 1972.)

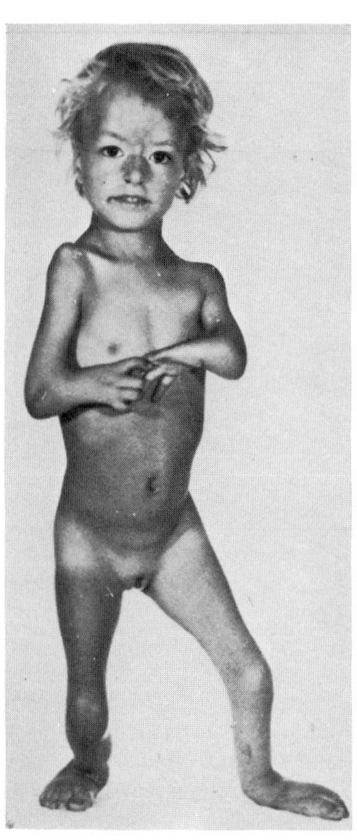

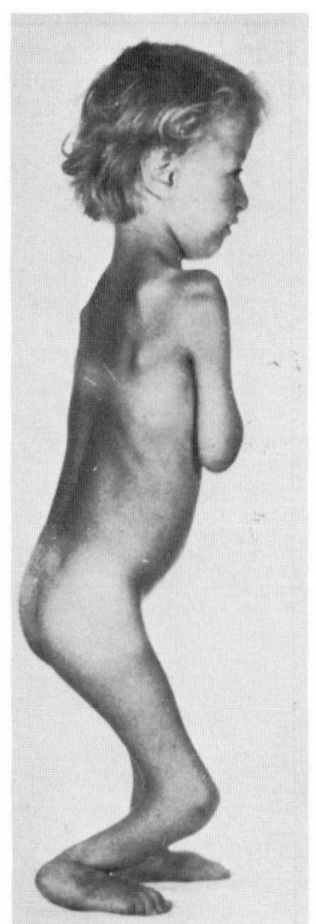

A B

This is discussed further under *Herrmann syndrome.* An X-linked lethal disorder (tetra-amelia, hydrocephalus, facial clefts, eye anomalies) has some overlap (44).

**Laboratory aids.** Approximately half of the patients exhibit a striking chromosomal change involving the heterochromatic C-banding regions of most chromosomes (2,6–8,16,24,26,27,31,38,42,43). The heterochromatic regions around the centromeres and nucleolar organizers have a puffed appearance. The heterochromatin of the long arms of the Y chromosome is often widely separated in metaphase spreads (26). It has been suggested that these changes manifest repulsion or lack of attraction between the chromatids in these regions, which in turn leads to premature separation during prophase and metaphase (Fig. 21–36). Fibroblasts exhibit unusual growth characteristics. Their mitotic processes are also altered (37). Prenatal diagnosis has been accomplished (32a).

### References [Roberts–pseudothalidomide syndrome (Roberts syndrome, SC phocomelia syndrome)]

1. Appelt H et al: Tetraphokomelie mit Lippen–Kiefer–Gaumenspalte und Clitorishypertrophie-ein Syndrom. *Pädiatr Pädol* **2**:119–124, 1966.

2. Bökesoy I et al: A case of SC–phocomelia syndrome with nonrandom centromere separation. *Prog Clin Biol Res* **104**:351–356, 1982.

3. Da Silva EO, Bezerra LHGE: The Roberts syndrome. *Hum Genet* **61**:372–374, 1982.

4. Doerffer C: Ein Fall von Phokomelie. *Mschr Geburtsh Gynäk* **72**:195–198, 1926.

5. Engelhart E, Pischinger A: Über eine durch Röntgenstrahlen verursachte menschliche Missbildung. *Münch Med Wochenschr* **86**:1315–1316, 1939.

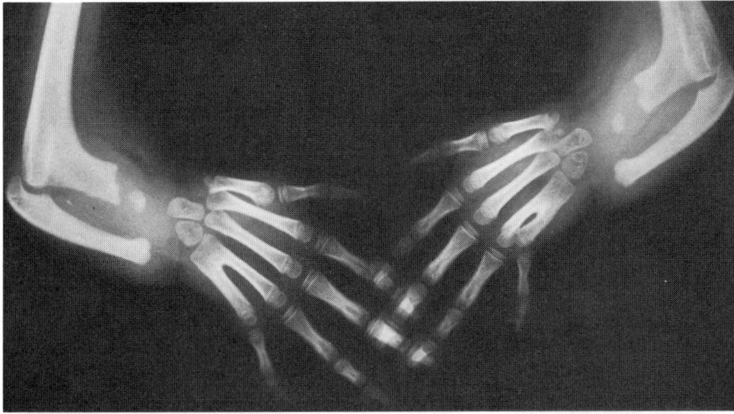

A

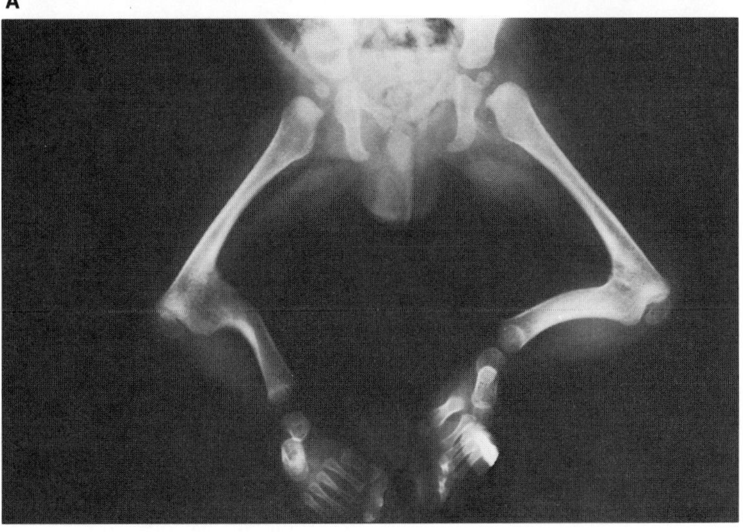

B

Fig. 21–35. *Roberts–pseudothalidomide syndrome.* (A,B) Radiohumeral synostosis with marked shortening of humerus, synmetacarpalism, femorotibial fusion. (From *M Levy et al,* Ann Pédiatr **19**:313, 1972.)


6. Freeman MVR et al: The Roberts syndrome. *Clin Genet* **5**:1–16, 1974.

7. Fryns H et al: The tetraphocomelia–cleft palate syndrome in identical twins. *Hum Genet* **53**:278–281, 1980.

8. German J: Roberts syndrome: Cytological evidence for a disturbance in chromatid pairing. *Clin Genet* **16**:441–447, 1979.

9. Grillo RA: Über einen Fall von Phokomelie. *Dtsch Med Wschr* **62**:1332–1333, 1936.

10. Grosse FR et al: The tetraphocomelia–cleft palate syndrome. *Humangenetik* **28**:353–356, 1975.

11. Gruber G: Die Entwicklungsstörungen der menschlichen Gliedmassen. In: E Schwalbe and G Gruber (eds): *Die Morphologie der Missbildungen der Menschen und der Tiere* III, G Fischer, Jena, 1937, p 310.

12. Hall BD, Greenberg MH: Hypomelia–hypotrichosis–facial hemangioma syndrome. *Am J Dis Child* **123**:602–604, 1972.

13. Herrmann J, Opitz JM: The SC phocomelia and the Roberts syndrome: Nosologic aspects. *Eur J Pediatr* **125**:117–134, 1977.

14. Herrmann J et al: A familial dysmorphogenetic syndrome of limb deformities, characteristic facial appearance and associated anomalies: The "pseudothalidomide" or SC-syndrome. *Birth Defects* **5**(3):81–89, 1969.

15. Hunter AGW et al: The genetics of and associated clinical findings in humeroradial synostosis. *Clin Genet* **9**:470–478, 1976 (Cases 3, 4).

16. Judge C: A sibship with the pseudothalidomide syndrome and an association with Rh incompatibility. *Med J Austral* **2**:280–281, 1973.

17. Jurenka SB: Variations in IQ of two patients with pseudothalidomide syndrome. *Dev Med Child Neurol* **18**:525–527, 1976.

18. Kaul A: *Über eine besondere Form der Phokomelie verbunden mit Hasenscharte und Wolfsrachen nebst Bemerkungen über die Ätiologie dieser Missbildung.* Thesis, Würzburg, 1899.

19. Keutel J et al: Eine wahrscheinlich autosomal recessive vererbte Skelettmissbildung mit Humeroradialsynostose. *Humangenetik* **9**:43–53, 1970.

20. Krueger R: *Die Phocomelie und ihre Übergange* (Case 87), August Hirschwald, Berlin, 1906, p 92.

21. Kucheria K et al: A familial tetraphocomelia syndrome involving limb deformities, cleft lip, cleft palate, and unassociated anomalies—a new syndrome. *Hum Genet* **33**:323–326, 1976.

22. Ladda RL et al: Craniosynostosis associated with limb reduction malformations and cleft lip/palate: A distinct syndrome. *Pediatrics* **61**:12–15, 1978.

23. Lenz WD et al: Pseudothalidomide syndrome. *Birth Defects* **10**(5):97–107, 1974.

24. Leonard P et al: Roberts'–SC syndrome with cytogenetic findings. *Hum Genet* **60**:379–380, 1982.

25. Levy M et al: Malformations graves des membres et oligophrenie dans une famille (avec études chromosomiques). *Ann Pédiatr* **19**:313–320, 1972.

26. Louie E., German J: Roberts' syndrome: Aberrant Y-chromosome behavior. *Clin Genet* **19**:71–74, 1981.

27. Mann NP et al: Roberts syndrome: Clinical and cytogenetic aspects. *J Med Genet* **19**:116–119, 1982.

28. O'Brien HR, Mustard HS: An adult living case of total phocomelia. *JAMA* **77**:1964–1967, 1921.

28a. Parry DM et al: SC phocomelia syndrome premature centromere separation, and congenital cranial nerve paralysis in two sisters, one with malignant melanoma. *Am J Med Genet* **24**:653–672, 1986.

29. Petrinelli P et al: Premature centromeric splitting in a presumptive mild form of Roberts syndrome. *Hum Genet* **66**:96–99, 1984.

30. Pfeiffer RA, Zwerner H: Das Roberts-Syndrom. *Mschr Kinderheilk* **130**:296–298, 1982.

31. Qazi QH et al: The SC–phocomelia syndrome: Report of two cases with cytogenetic abnormalities. *Am J Med Genet* **4**:231–238, 1979.

32. Roberts JB: A child with double cleft of lip and palate, protrusion of the intermaxillary portion of the upper jaw and imperfect development of the bones of the four extremities. *Ann Surg* **70**:252, 1919.

32a. Robins DB et al: Prenatal detection of Roberts–SC phocomelia syndrome: Report of 2 sibs with characteristic manifestations. *Am J Med Genet* **32**:390–394, 1989.

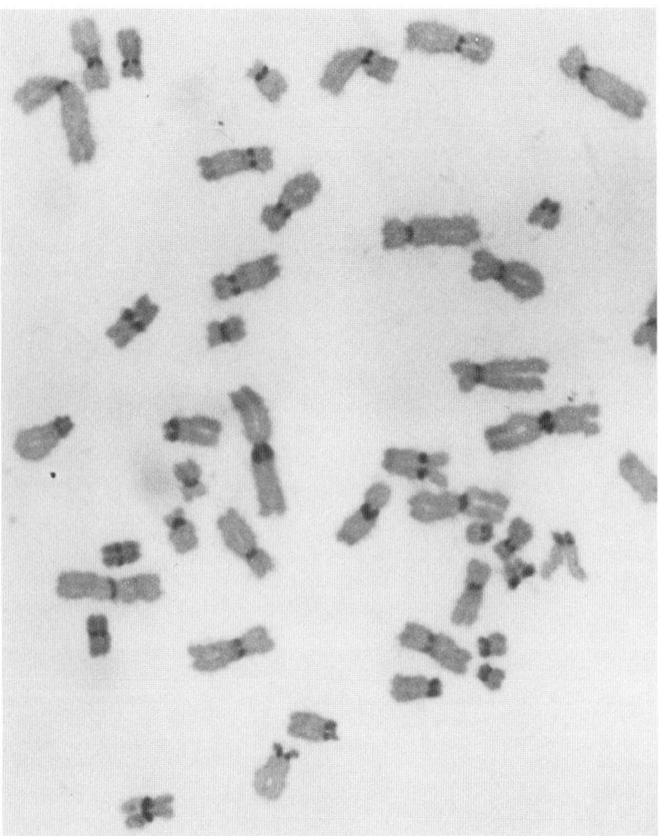

Fig. 21–36. *Roberts–pseudothalidomide syndrome*. Note centromeric separation with C-banding. (Courtesy of *W Wertelecki*, Mobile, Alabama.)


32b. Römke C et al: Roberts syndrome and SC phocomelia: A single genetic entity. *Clin Genet* **31**:170–177, 1987.

33. Slingenberg B: Missbildungen von Extremitäten. *Virchows Arch Pathol Anat* **193**:1–92, 1908 (Case 11).

34. Stoll C et al: Roberts' syndrome and clonidine. *J Med Genet* **16**:486–488, 1979.

35. Ströer WFH: Über das Zusammentreffen von Hasenscharte mit ernsten Extremitätenmissbildungen. *Erbarzt* **7**:101–107, 1939.

36. Temtamy SA, Loutfi AH: Some genetic and surgical aspects of cleft lip/cleft palate problem in Egypt. *Cleft Palate J* **7**:578–594, 1970.

37. Tomkins DJ, Sisken JE: Abnormalities in the cell-division cycle in Roberts syndrome fibroblasts: A cellular basis for the phenotypic characteristics. *Am J Med Genet* **36**:1332–1340, 1984.

38. Tomkins D et al: Cytogenetic findings in Roberts'–SC phocomelia syndrome(s). *Am J Med Genet* **4**:17–26, 1979.

39. Virchow R: Die Phokomelien und das Bärenweib. *Z Ethnol* **30**:55–61, 1898.

40. Waldenmaier C et al: The Roberts' syndrome. *Hum Genet* **40**:345–349, 1978.

41. Werthemann A: Die Entwicklungsstörungen der Extremitäten, in *Handbuch der speziellen pathologischen Anatomie und Histologie,* Vol IX, Lubarsch O et al (eds), Springer-Verlag, Berlin, 1952.

42. Zergollern L, Hitrec V: Three siblings with Roberts' syndrome. *Clin Genet* **9**:433–436, 1976.

43. Zergollern L, Hitrec V: Four siblings with Roberts syndrome. *Clin Genet* **21**:1–6, 1982.

44. Zimmer EZ et al: Tetra-amelia with multiple malformations in six male fetuses in one kindred. *Eur J Pediatr* **144**:412–414, 1985.


## Van der Woude syndrome (cleft lip–palate and paramedian sinuses of the lower lip)

The first report of congenital sinuses of the lower lip in combination with cleft lip and/or cleft palate appears to be that of Demarquay (5) in 1845. It was independently reported by Murray (15) in 1860. Over 850 cases have been published (2). The syndrome occurs in roughly 1–2% of patients with facial clefts (2,22) or roughly 1:35,000 to 1:100,000 in the white population (1a,4,22). Numerous authors (4,9,10,12,14,35) have carried out extensive reviews and analyses of published cases. The most complete survey is that of Burdick et al (2).

The syndrome has autosomal dominant inheritance with variable expressivity. Penetrance ranges between 0.89 and 0.99 (1a,2,11,29). The estimate of 80% penetrance (4) is too low, probably due to failure to include submucous cleft palate in the calculations. Between 30 and 50% represent new mutations (19,26). The mutation rate has been calculated to be $1.8 \times 10^{-5}$ (2). Linkage studies have not been fruitful (6), but a deletion of 1q32→41 has been reported (1) as well as that of 2q34→q36 (10a).

Manifestations of the syndrome in other than the oral or facial areas are unusual.

**Oral manifestations.** Usually bilateral, often symmetrically placed depressions are observed on the vermilion portion of the lower lip, one on each side of the midline (Fig. 21–37A.B). These dimples are usually circular but they may be transverse slits or sulci (22). On occasion, they may be located at the apex of nipple-like elevation. Rarely, the elevations may fuse in the midline, producing a snoutlike structure (17). The depressions represent blind sinuses that descend through the orbicularis oris muscle to a depth of 1 mm to 2.5 cm and communicate with the underlying minor salivary glands through their excretory ducts. The sinuses often transport a viscid saliva to the surface, either spontaneously or on pressure.

They may be so small as barely to permit a hair probe, or they may be as wide as 3 cm or more. Nipple-like processes occasionally occur without demonstrable fistulas (12,33). Although usually bilateral and symmetrically placed, an asymmetric single pit (12,14,24,31,33,34), a central single pit (23,24,33), or bilateral asymmetric pits (23) may occur.

Microforms exist such as transverse mucosal ridges of the lower lip and bilateral somewhat conical elevations of the lower lip mucosa (22). The latter are associated with cleft palate but not with cleft lip (21).

Sicher and Pohl (30) and Warbrick et al (34) suggested that the congenital fistulas arise from arrested development, that is, persistence of a median and/or lateral sulcus or sulci that normally are evanescent structures. These median and lateral grooves appear in the 5–6 mm embryo and disappear at the 10–16 mm stage (Fig. 21–37C). It should be pointed out that the grooves disappear at about the same time that fusion occurs between several facial prominences. This probably accounts for the simultaneous occurrence of sinuses and facial cleft. Although surveys have differed considerably due to various biases (4,11,17,22,29,32,35), roughly 33% have pits without clefts, 33% have pits with cleft lip–palate, and 33% have pits with cleft palate or submucous cleft palate. The combination of pits with cleft lip alone is very rare. Conversely, about 10% of those with the syndrome do not exhibit lip pits. Burdick (1a) calculated that the risk of inheriting a cleft from an affected parent is about 20%; the risk of inheriting lip pits only or being nonpenetrant is about 25%.

Adhesions between maxilla and mandible (syngnathia) have been noted (13,16,28). Absence of maxillary and mandibular second premolars has been described as part of the syndrome in 10–20% of cases (20,25).

**Extremities.** Talipes equinovarus (8,10,19,31) and 3–4 syndactyly of the fingers (3,15,19) have been reported.

**Other findings.** Ankyloblepharon has been described in association with the condition (13,16,17). These patients, however, may have had mild expression of the *popliteal pterygium syndrome*. Accessory nipples (19), congenital heart defects (18), and Hirschsprung disease (7,27) have also been reported.

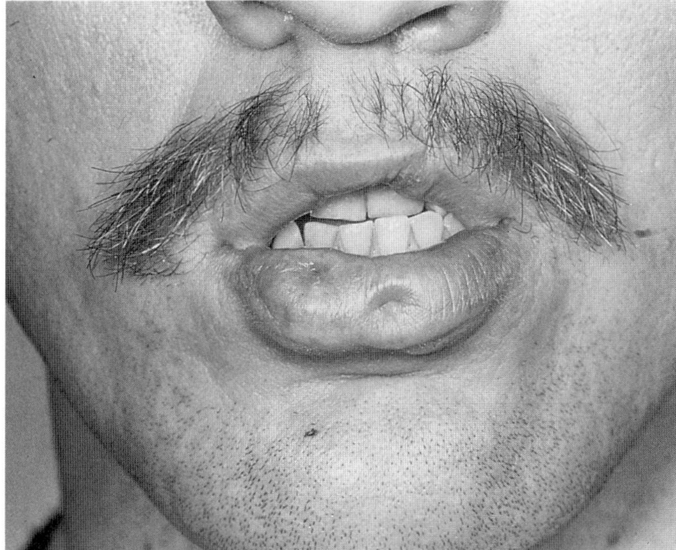

A

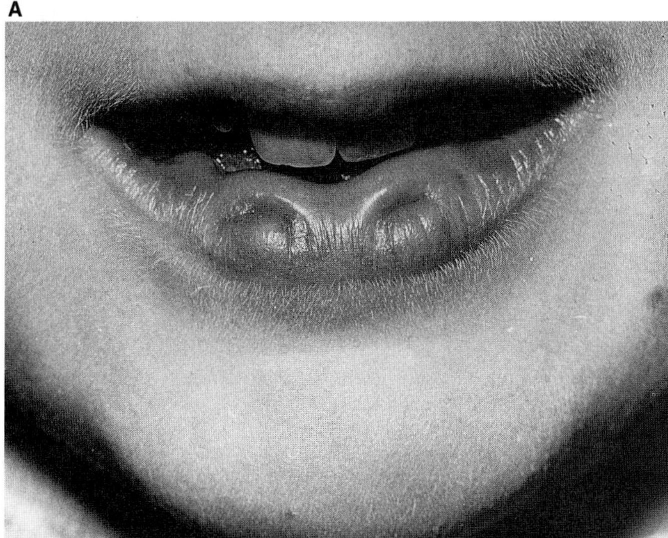

B

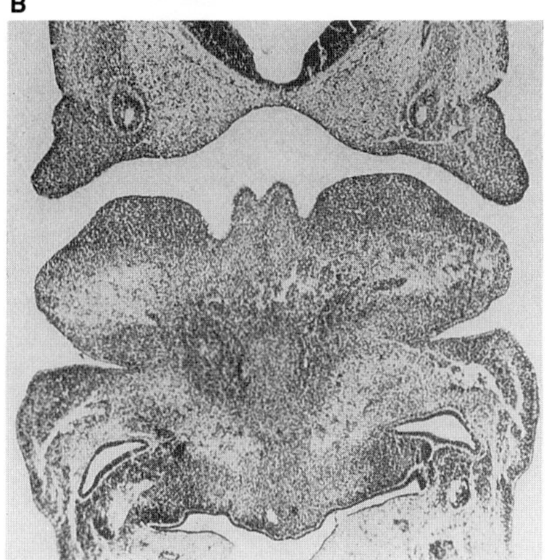

C

Fig. 21–37. *Van der Woude syndrome.* (A) Note repaired unilateral cleft lip, asymmetric sinuses of lower lip. (B) Symmetrical paramedian sinuses of lower lip. (C) Photomicrograph of 7.5-mm embryo demonstrating three invaginations in mandibular process. These disappear by the 14-mm stage. (From *JG Warbrick et al,* Br J Plast Surg **4**:254, 1952.)

**Differential diagnosis.** Because of variable expressivity, affected family members may exhibit clefts without pits. This may greatly complicate genetic counseling as the chances for transmitting cleft lip or palate vary depending on whether the anomaly is associated with lip pits (50%) or is an isolated phenomenon (2–5%). Lip pits may also occur in the *popliteal pterygium syndrome* and with *aganglionic megacolon and cleft lip and/or palate,* if this is a separate entity. The type of pits under discussion should not be confused with commissural lip pits at the corners of the mouth which are seen in 10–20% of the population.

### References [Van der Woude syndrome (cleft lip–palate and paramedian sinuses of the lower lip)]


1. Bocian M, Walker AP: Lip pits and deletion 1q32→41. *Am J Med Genet* **26**:437–443, 1987.

1a. Burdick AB: Genetic epidemiology and control of genetic expression in van der Woude syndrome. *J Craniofac Genet Dev Biol Suppl* **2**:99–105, 1986.

2. Burdick AB et al: Genetic analysis in families with Van der Woude syndrome. *J Craniofac Genet Dev Biol* **5**:181–208, 1985.

3. Calnan J: Congenital double lip: Record of a case with a note on the embryology. *Br J Plast Surg* **5**:197–202, 1952.

4. Cervenka J et al: The syndrome of pits of the lower lip and cleft lip and/or palate: Genetic considerations. *Am J Hum Genet* **19**:416–432, 1967.

4a. Cheney ML et al: Familial incidence of labial pits. *Am J Otolaryngol* **7**:311–313, 1986.

5. Demarquay JN: Quelques considerations sur le bec-de-lièvre. *Gaz Méd (Paris)* **13**:52–54, 1845.

6. Eastman JR et al: Linkage studies in Van der Woude syndrome. *J Med Genet* **15**:217–218,1978.

7. Goldberg RB, Shprintzen RJ: Hirschsprung megacolon and cleft palate in two sibs. *J Craniofac Genet Dev Biol* **1**:185–189, 1981.

8. Hamilton E: Congenital deformity of the lower lip. *Dublin J Med Sci* **72**:1–2, 1881.

9. Heiner H, Leutert G: Zur Klinik and Genese der kongenitalen Unterlippenfisteln. *Arch Klin Chir* **299**:775–799, 1962.

10. Hilgenreiner H: Die angeborenen Fisteln bzw. Schleimhauttaschen der Unterlippe. *Dtsch Z Chir* **188**:273–309, 1924.

10a. Hughes M et al: Does van der Woude syndrome map to chromosome 2q34–36? *Proc Greenwood Genet Ctr* **7**:176, 1988.

11. Janku P et al: The Van der Woude syndrome in a large kindred: Variability, penetrance, genetic risks. *Am J Med Genet* **5**:117–123, 1980.

12. Koberg W: Zur Kenntnis der kongenitalen Unterlippenfisteln. *Oest Z Stomatol* **63**:60–74, 1966.

13. Leck GD, Aird JC: An incomplete form of the popliteal pterygium syndrome? *Br Dent J* **157**:318–319, 1984.

14. Ludy JB, Shirazy E: Concerning congenital fistulae of the lips; their mooted significance; review of the literature; and report of a family with congenital fistulae of the lower lip. *Int Clin* **3**:75–88, 1938.

15. Murray JJ: Undescribed malformation of the lower lip occurring in four members of one family. *Br Foreign Med Chir Rev* **26**:502–509, 1860.

16. Neuman Z, Shulman J: Congenital sinuses of the lower lip. *Oral Surg* **14**:1415–1420, 1961.

17. Oberst: Über die angeborenen Unterlippenfisteln. *Beitr Klin Chir* **68**:795–801, 1910.

18. Pauli RM, Hall JG: Lip pits, cleft lip and/or palate and congenital heart disease. *Am J Dis Child* **134**:293–295, 1980.

19. Pohl L: Letter to the editor. *Dent Pract Dent Rec* **7**:112, 1956.

20. Ranta R, Rintala AE: Tooth anomalies associated with congenital sinuses of the lower lip and cleft lip/palate. *Angle Orthod* **52**:212–221, 1982.

21. Ranta R, Rintala AE: Correlations between microforms of the Van der Woude syndrome and cleft palate. *Cleft Palate J* **20**:158–162, 1983.

22. Rintala AE, Ranta R: Lower lip sinuses. Epidemiology, microforms and transverse sulci. *Br J Plast Surg* **34**:25–30, 1980.

23. Rintala AE, Ranta R: Congenital sinuses of the lower lip in connection with cleft lip and palate. *Cleft Palate J* **7**:336–346, 1970.

24. Ruppe C, Magdeleine J: Fistules muqueuses congénitales des levres. *Rev Stomatol (Paris)* **29**:1–8, 1927.

25. Schneider EL: Lip pits and congenital absence of second premolars. Varied expression of the lip pits syndrome. *J Med Genet* **10**:346–349, 1973.

26. Schinzel A, Kläusler M: The van der Woude syndrome (dominantly inherited lip pits and clefts). *J Med Genet* **23**:291–294, 1986.

27. Schwarz KB et al: Congenital lip pits and Hirschsprung disease. *J Pediatr Surg* **14**:162–164, 1979.

28. Shaw WC, Simpson JP: Oral adhesions associated with cleft lip and palate and lip fistulae. *Cleft Palate J* **17**:127–131, 1980.

29. Shprintzen RJ et al: The penetrance and variable expression of the Van der Woude syndrome: Implications for genetic counseling. *Cleft Palate J* **17**:52–57, 1980.

30. Sicher H, Pohl L: Zur Entstehung des menschlichen Unterkiefers (ein Beitrag zur Entstehung des Unterlippenfisteln). *Oest Z Stomatol* **32**:552–560, 1934.

31. Taylor WB, Lane DK: Congenital fistulas of the lower lip: Associations with cleft lip–palate and anomalies of the extremities. *Arch Dermatol* **94**:421–424, 1966.

32. Van der Woude A: Fistula labii inferioris congenita and its association with cleft lip and palate. *Am J Hum Genet* **6**:244–256, 1954.

33. Wang MKH, Macomber WB: Congenital lip sinuses. *Plast Reconstr Surg* **18**:319–328, 1956.

34. Warbrick JG et al: Remarks on etiology of congenital bilateral fistulae of the lower lip. *Br J Plast Surg* **4**:254–262, 1952.

35. Watanabe Y et al: Congenital fistulas of the lower lip. *Oral Surg* **4**:709–722, 1951.

## Velocardiofacial syndrome (Shprintzen syndrome, Sedlačková syndrome)

In 1978, Shprintzen et al (12) reported a syndrome of typical facies, prominent nose, retruded mandible, cardiovascular anomalies, cleft palate, and learning disabilities. Stern et al (14), Kinouchi et al (7), and Kimura et al (6) probably described the same entity. Sedlačková and co-workers (3,10,16) included several examples of the syndrome under the title of *velofacial hypoplasia* as early as 1955. In 1968, Strong (15) described an autosomal dominant syndrome of right-sided aorta, facial dysmorphism, and mental retardation in four members of a family in two generations. Cleft palate was not cited, but Strong (personal communication, 1984) related that patients had "cleft palate speech."

Inheritance is autosomal dominant with variable expressivity (13,15,17). Over 60 cases have been described (17). We believe that many examples are overlooked.

Hypotonia and upper respiratory illnesses in infancy and childhood are frequent. Failure to thrive in infancy is exhibited by about 25%, and about 35% manifest short stature (17). Obstructive sleep apnea has been found in about one half of neonates with the disorder.

**Facies.** Approximately 40% are microcephalic. Scalp hair is abundant in over 50%. The face is long with vertical maxillary excess (85%), malar flatness (70%), and mandibular retrusion. About 15% exhibit *Robin sequence*. Conversely, about 15% of infants with Robin sequence have the velocardiofacial syndrome. The nose is prominent with squared nasal root, hypoplastic alae nasi, and narrow nasal passages in 75% (1). The philtrum is long and the upper lip is thin. The mouth is often held open (Figs. 21–38 and 21–39A,B). The adenoids are hypoplastic in almost 85% (18). Kimura (6), examining 73 patients with velopharyngeal insufficiency without cleft palate, found birth asphyxia in 33, congenital heart anomalies in 16, seizures in 13, and speech retardation in 66. Forty-two of these 73 patients resembled one another. They divided the facial groups into two types: humorous (round eyes, pursed lips) and nervous (plump upper eyelids, upslanting palpebral fissures) forms.

**Eyes.** Narrow palpebral fissures with blue suborbital coloring or "allergic shiners" occur in 35%. Fitch (4) noted small optic discs and tortuous retinal vessels. Beemer et al (2) found blue sclerae, retinal coloboma, cataracts, and tortuous vessels. Tortuous retinal vessels were observed by Williams et al (17) in 50%. Mansour et al (8a) estimated that retinal vascular tortuosity was present in about 35%. Posterior embryotoxon has also been noted (8a).

**Ears.** Small auricles and minor thickening of the helical rims have been seen in about 60% (Fig. 21–39B). Intermittent conductive hearing loss, noted in 75%, is probably secondary to frequent bouts of serous otitis media and cleft palate. Marked reduction in size of the nasopharyngeal lumen of the Eustachian tube has been noted. One child had sensorineural hearing loss (13).

**Cardiovascular aspects.** Multiple cardiac anomalies are present in over 80%, especially VSD (65%), right-sided aortic arch (35%), tetralogy of Fallot (20%), and aberrant left subclavian artery (20%) (11,17,20). Aortic valve disease is not uncommonly found. Of interest and diagnostic significance, right-sided aortic arch was seen in 30% of patients with VSD, in 80% of those with tetralogy of Fallot and in nearly 50% of patients with otherwise normal hearts. MacKenzie et al (8), Shprintzen (personal communication, 1987), and D'Antonio and Marsh (2a) noted enlargement, medial displacement, and tortuosity of the internal carotid arteries in the posterior pharyngeal wall (Fig. 21–40A,B).

A number of patients have exhibited Raynaud's phenomenon.

**Mental status.** Learning disability is experienced by 100%. Developmental milestones are mildly delayed. Language development is

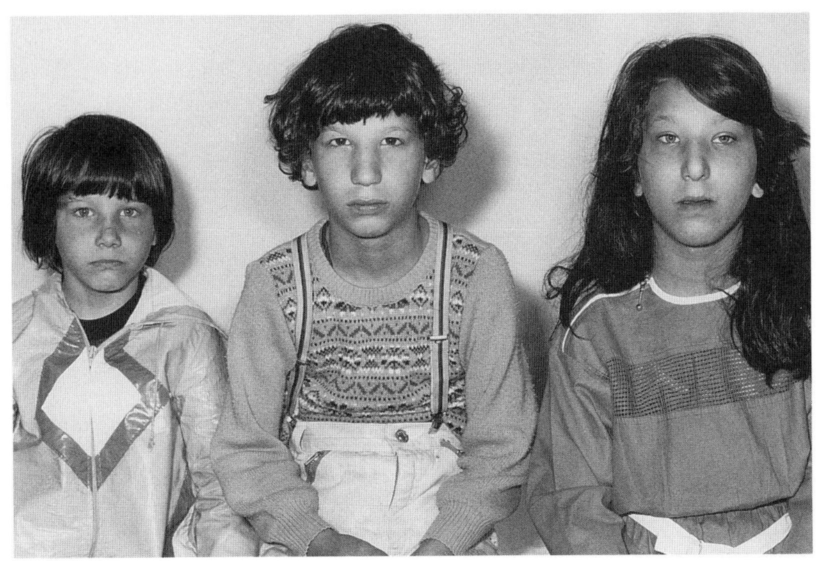

Fig. 21–38. *Velocardiofacial syndrome.* Two affected patients to the right contrasting sharply with their 6-year-old normal sister. Note narrow receding forehead, broad and prominent nasal bridge, upslanting short palpebral fissures, and epicanthal folds. (From *P Meinecke et al,* Eur J Pediatr **145**:539, 1986.)

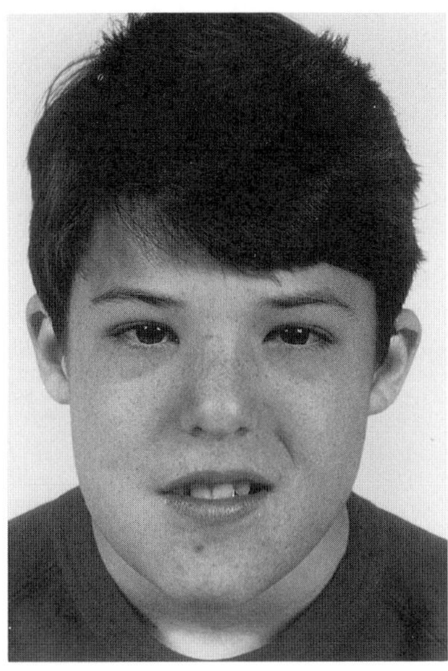
A

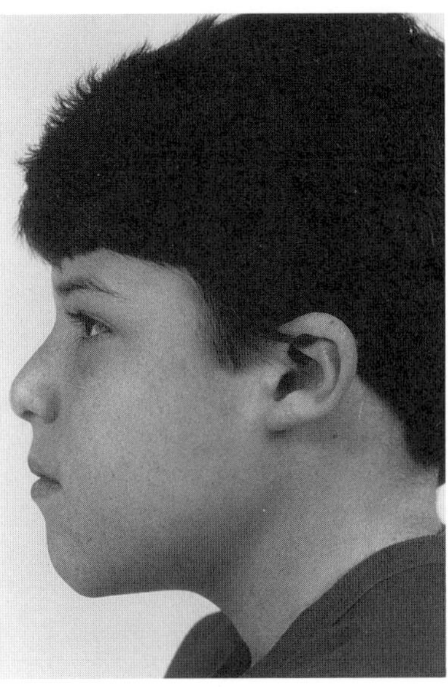
B

Fig. 21–39. *Velocardiofacial syndrome*. (A,B) Note similar facies as well as small pinnae. (Courtesy of *K MacKenzie*, Toronto, Ontario, Canada.)

often slow. Speech is usually characterized by hypernasality with or without overt clefting of the palate, but this may be related to adenoidal hypoplasia (18). The voice is high-pitched. However, only 40% are considered to have mild to moderate mental retardation (17). In secondary school age children, IQ scores generally have ranged from 69 to 87 on verbal scale and 55 to 78 on performance scale (5). An association with holoprosencephaly has been noted in one case (19).

A characteristic personality was found among children studied by Golding-Kushner et al (5). At all ages, the children tended to have bland affect. Social interaction was poor with respect to quality and

Fig. 21–40. *Velocardiofacial syndrome*. (A,B) Carotid arteriograms showing tortuous right internal carotid artery and smaller than usual left internal carotid artery. (Courtesy of *K MacKenzie*, Toronto, Ontario, Canada.)

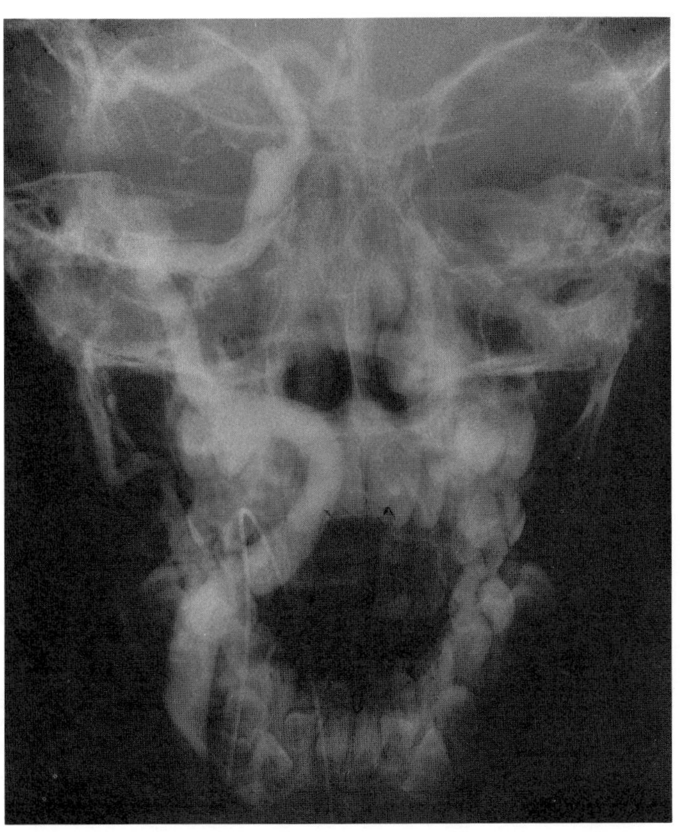
A

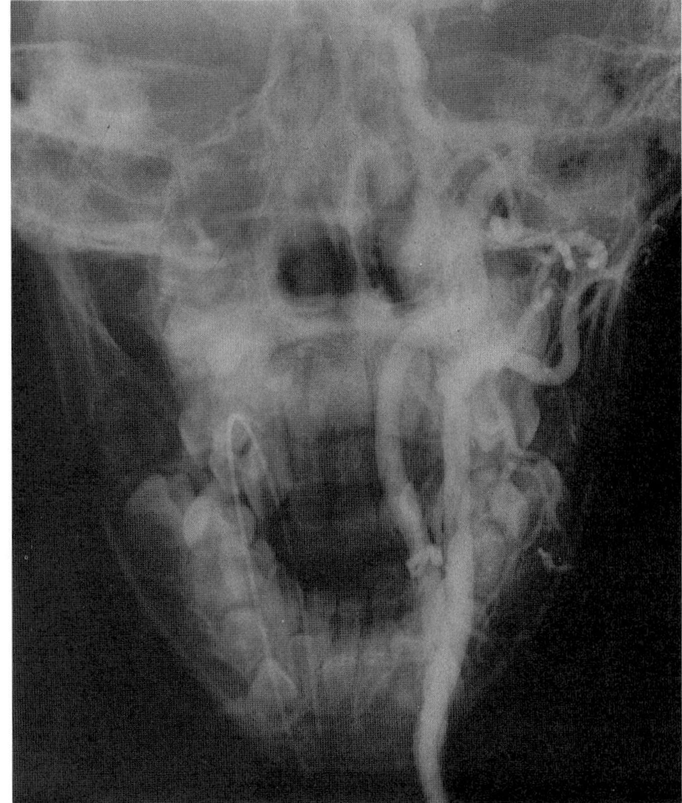
B

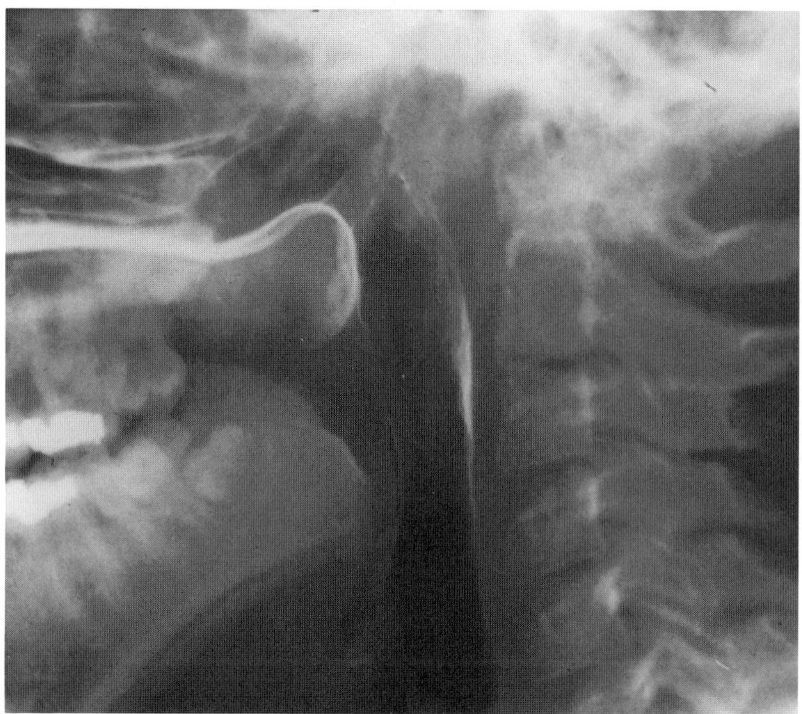

Fig. 21–41. *Velocardiofacial syndrome.* Videofluoroscopy showing inadequate velopharyngeal closure during speech. (Courtesy of *K MacKenzie,* Toronto, Ontario, Canada.)

quantity. The children were described as demonstrating extremes of behavior, being disinhibited, impulsive, and very affectionate or serious and shy. RJ Shprintzen (personal communication, 1989) has seen a number of examples of psychosis, especially phobic responses, in adolescents and adults.

**Musculoskeletal findings.** About 25% have umbilical or inguinal hernia and 15% have scoliosis (17). The hands and fingers appear slender and tapered in about 60%. The digits may be hyperextensible. Radiographically, there is platybasia in 75%.

**Immunologic findings.** Absent or small thymus (10%), tonsils (50%), and adenoids (85%) and hypocalcemia in infancy noted in 10% indicate an association with the *DiGeorge sequence.*

**Oral manifestations.** Cleft palate (35%), submucous cleft palate (33%), and occult submucous cleft palate or velar paresis (33%) resulting in hypernasal speech have been found in nearly all patients, but this may reflect ascertainment bias. Class I malocclusion is a common finding. The pharynx is hypotonic (Fig. 21–41).

#### References [Velocardiofacial syndrome (Shprintzen syndrome, Sedlačková syndrome)]

1. Arvystas M, Shprintzen RJ: Craniofacial morphology in the velocardiofacial syndrome. *J Craniofac Genet Dev Biol* **4**:39–46, 1984.

2. Beemer F et al: Additional findings in a girl with the velo-cardio-facial syndrome. *Am J Med Genet* **24**:541–542, 1986.

2a. D'Antonio LL, Marsh JL: Abnormal carotid arteries in the velocardiofacial syndrome. *Plast Reconstr Surg* **80**:471–472, 1987.

3. Fára N et al: Submucous cleft palates. *Acta Chir Plast (Praha)* **13**:221–234, 1971.

4. Fitch N: Letter to the editor: Velo-cardio-facial and eye abnormality. *Am J Med Genet* **15**:669, 1983.

5. Golding-Kushner K et al: Velo-cardio-facial syndrome: Language and psychological profiles. *J Craniofac Genet Dev Biol* **5**:259–264, 1985.

6. Kimura A: Congenital velopharyngeal insufficiency without cleft palate, involving typical facies. *Proc Congr Int Assoc Logoped Phoniatr* 3–7 August 1986, Tokyo, Japan.

7. Kinouchi A et al: Facial appearance of patients with conotruncal anomalies. *Pediatr Jpn* **17**:84–87, 1976.

8. MacKenzie K et al: Abnormal carotid arteries in the velocardiofacial syndrome: A report of three cases. *Plast Reconstr Surg* **80**:347–351, 1987.

8a. Mansour AM et al: Ocular findings in the velocardiofacial syndrome. *J Pediatr Ophthalmol Strab* **24**:263–266, 1987.

9. Meinecke P et al: The velo-cardio-facial (Shprintzen) syndrome. Clinical variability in eight patients. *Eur J Pediatr* **145**:539–544, 1986.

10. Sedlačková E: The syndrome of the congenitally shortened velum. The dual innervation of the soft palate. *Folia Phoniat* **19**:441–443, 1967.

11. Shaw CV et al: Cardiac malformation in facial clefts. *Am J Dis Child* **119**:238–244, 1977.

12. Shprintzen RJ et al: A new syndrome involving cleft palate, cardiac anomalies, typical facies, and learning disabilities: Velo-cardio-facial syndrome. *Cleft Palate J* **15**:56–62, 1978.

13. Shprintzen RJ et al: The velo-cardio-facial syndrome: A clinical and genetic analysis. *Pediatrics* **67**:167–172, 1981.

14. Stern AM et al: An association of aorticotruncoconal abnormalities, velopalative incompetence and unusual spine fusion. *Birth Defects* **13**(3B):259, 1977.

15. Strong WB: Familial syndrome of right-sided aortic arch, mental deficiency, and facial dysmorphism. *J Pediatr* **13**:882–888, 1967.

16. Vrtička K et al: Klinische, histologische und histomorphometrische Befunde bei velofazialer Hypoplasie (Sedlačková-Syndrom) *Sprache Stimme Gehör* **10**:109–112, 1986.

17. Williams MA et al: Male-to-male transmission of the velo-cardio-facial syndrome: A case report and review of 60 cases. *J Craniofac Genet Dev Biol* **5**:175–180, 1985.

18. Williams MA et al: Adenoid hypoplasia in the velo-cardio-facial syndrome. *J Craniofac Genet Dev Biol* **7**:23–26, 1987.

19. Wraith JE et al: Velo-cardio-facial syndrome presenting as holoprosencephaly. *Clin Genet* **27**:408–410, 1985.

20. Young D et al: Cardiac malformations in the velo-cardio-facial syndrome. *Am J Cardiol* **46**:643–648, 1980.

# Chapter 22
# Orofacial Clefting Syndromes: Other Syndromes

## Aase–Smith syndrome I

Aase and Smith (1) in 1968 and Patton et al (2) in 1985 reported two generation inheritance of hydrocephalus, severe contractures of joints, dysmorphic pinnae, and cleft palate (Fig. 22–1). In each case, the parent was less severely affected (Fig. 22–2).

All extremities were contracted, including the fingers (Fig. 22-3). There were thoracic scoliosis and dislocated hips (1). Talipes was seen in the other family (2).

The facies was normal but the pinnae were dysmorphic in the patients of Aase and Smith (1). However, in those of Patton et al (2) there were broad forehead, short palpebral fissures, and broad nasal bridge. Cleft palate was found in most cases (1,2).

Dermal ridge patterning of palms and fingertips was hypoplastic.

Autopsy showed Dandy–Walker formation in the infants but not in the affected parent. A neuroblastoma was found in the adrenal in one case (1).

Dandy–Walker malformation may be an isolated finding or seen in combination with many disorders such as Joubert syndrome, *Walker–Warburg syndrome,* the *oral-facial-digital syndromes,* and, rarely, *Ellis–van Creveld syndrome.*

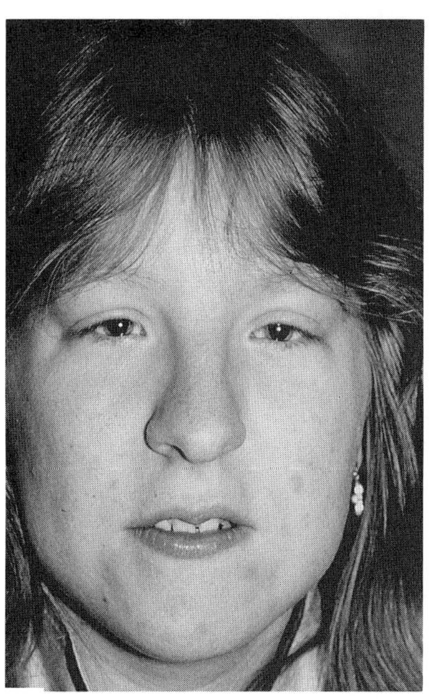

Fig. 22–2. *Aase–Smith syndrome I.* Mother of similarly affected child. Note broad forehead, broad nasal bridge, and short palpebral fissures. (From *MA Patton et al,* Clin Genet **28**:521, 1985.)

Fig. 22–1. *Aase–Smith syndrome I.* Hydrocephalus, cleft palate, malformed pinnae, joint contractures, hip dislocation. (From *JM Aase* and *DW Smith,* J Pediatr **73**:606, 1968.)

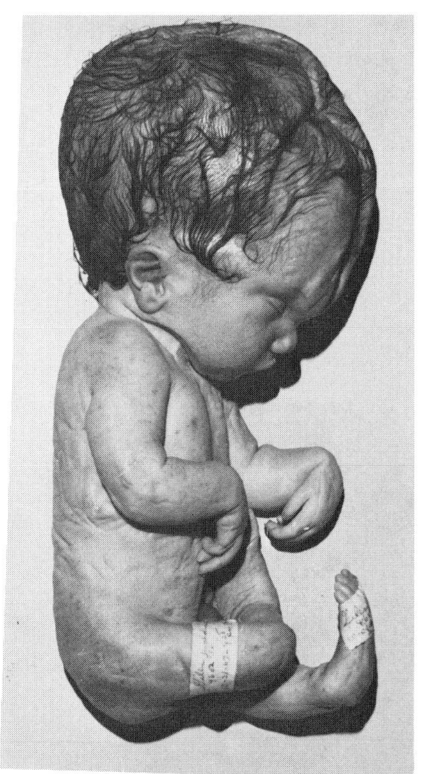

### References (Aase–Smith syndrome I)


1. Aase JM, Smith DW: Dysmorphogenesis of joints, brain and palate: A new dominantly inherited syndrome. *J Pediatr* **73**:606–609, 1968.
2. Patton MA et al: The Aase–Smith syndrome. *Clin Genet* **28**:521–525, 1985.


Fig. 22–3. *Aase–Smith syndrome I.* Long thin fingers. Note absence of finger creases. Patient was unable to make a fist. (From *MA Patton et al,* Clin Genet **28**:521, 1985.)

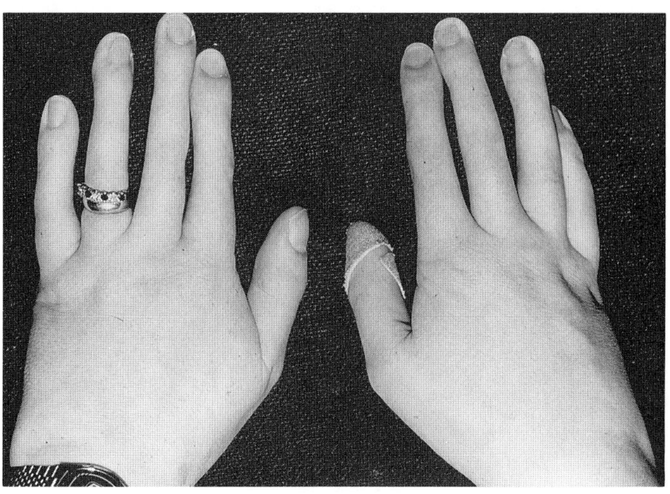

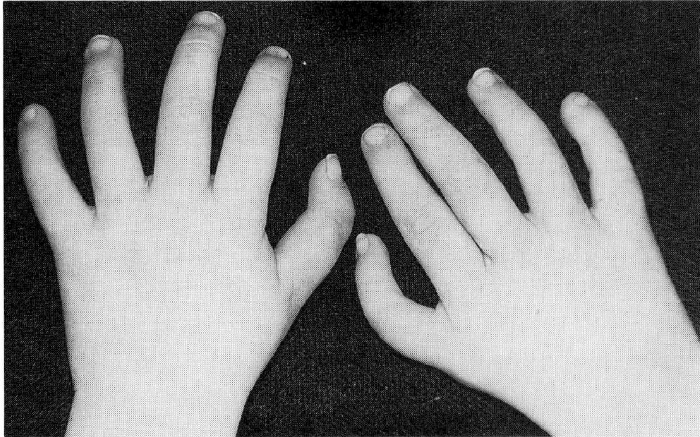

Fig. 22–4. *Aase–Smith syndrome II.* Digitalization of the thumbs. (Courtesy of *JM Aase,* Albuquerque, New Mexico.)

## Aase–Smith syndrome II (Blackfan–Diamond syndrome)

Aase and Smith (1) reported male sibs with congenital hypoplastic anemia, triphalangeal thumbs, hypoplastic radii, and cleft lip (Fig. 22–4). One sib had narrow shoulders. Murphy and Lubin (9) described a male child with cleft lip–palate, congenital erythroid hypoplasia, and triphalangeal thumbs. Also noted were somatic and mental retardation, narrow shoulders, abnormal pigmentation of the skin, and short webbed neck. There was a family history of a similar thumb anomaly. Muis et al (8) reported a male infant with cleft lip–palate, congenital hypoplastic anemia, and triphalangeal thumbs. Others (5–7,10–13) have reported similar examples but without cleft lip and/or palate. RJ Gorlin has seen a child with cleft palate, strabismus, and Blackfan–Diamond syndrome.

We consider the combination of anomalies described by Aase and Smith to be Blackfan–Diamond syndrome (3,4). All of the anomalies reported by Aase and Smith have been described as part of the syndrome. These have been well tabulated by Alter (2). Included are congenital hypoplastic anemia, triphalangeal thumbs, webbed neck, narrow shoulders, and cleft lip and/or palate.

### References [Aase–Smith syndrome II (Blackfan–Diamond syndrome)]


1. Aase JM, Smith DW: Congenital anemia and triphalangeal thumbs: A new syndrome. *J Pediatr* **74**:471–474, 1969.
2. Alter BP: Thumbs and anemia. *Pediatrics* **62**:613–614, 1978.
3. Diamond LK et al: Congenital (erythroid) hypoplastic anemia. *Am J Dis Child* **102**:403–415, 1961.
4. Diamond LK et al: Congenital hypoplastic anemia. *Adv Pediatr* **22**:349–378, 1976.
5. Harvey DR: Congenital hypoplastic anemia. *Proc R Soc Med* **59**:490–491, 1966.
6. Higginbottom MC et al: The Aase syndrome in a female infant. *J Med Genet* **15**:484–486, 1978.
7. Jones B et al: Triphalangeal thumbs associated with hypoplastic anemia. *Pediatrics* **52**:609–612, 1973.
8. Muis N et al: The Aase syndrome: Case report and review of the literature. *Eur J Pediatr* **145**:153–157, 1986.
9. Murphy SH, Lubin B: Triphalangeal thumbs and congenital erythroid hypoplasia: Report of a case with unusual features: *J Pediatr* **81**:987–989, 1972.
10. Paradis D, Simard H: Erythroblastopenie congénitale, phalange surnumeraire an pouce et luxation de la hanche. *Union Méd Can* **103**:1599–1601, 1974.
11. Pfeiffer RA, Ambs E: Das Aase-Syndrom: Autosomal-rezessiv vererbte, konnatale insuffiziente Erythropoese und Triphalangie der Daumen. *Mschr Kinderheilkd* **131**:235–237, 1983.
12. Terheggen HG: Hypoplastische Anämie mit dreigliedrigem Daumen. *Z Kinderheilkd* **118**:71–80, 1974.
13. Van Weel-Sipmann M et al: A female patient with "Aase syndrome." *J Pediatr* **91**:753–755, 1977.


## Ablepharon–macrostomia syndrome

In 1977, McCarthy and West (5) reported a syndrome in two unrelated male children. Hornblass and Reifler (3) described the same condition. Common features were triangular facies, hypertelorism, late development of sparse thin hair, absence of upper and lower eyelids, eyebrows, and eyelashes, and internal strabismus. The pinnae were low set and rudimentary in form with collapsed canals. The nose was small with some degree of coloboma formation of the mid-alae producing triangular nostrils. The malar eminences were flattened and the labial commissures were not adequately fused, producing macrostomia and thin vermilion (Fig. 22–5A).

The skin was dry and coarse with excess skin folds, especially over

Fig. 22–5. *Ablepharon-macrostomia syndrome.* (A) Shortening of eyelids and perforation of left cornea with vitreous loss and central corneal opacity of right eye. Also note abnormally shaped cartilages of alae, nose, and fishlike mouth. (B) Eyelids have been constructed. Note pronounced vascularity of skin of abdomen, dysplastic ears. (C) Ambiguous genitalia with posteriorly displaced micropenis, cryptorchidism, and absent scrotum. (A courtesy of *A Hornblass,* New York, New York. B,C courtesy of *S Sklower,* Staten Island, New York.)

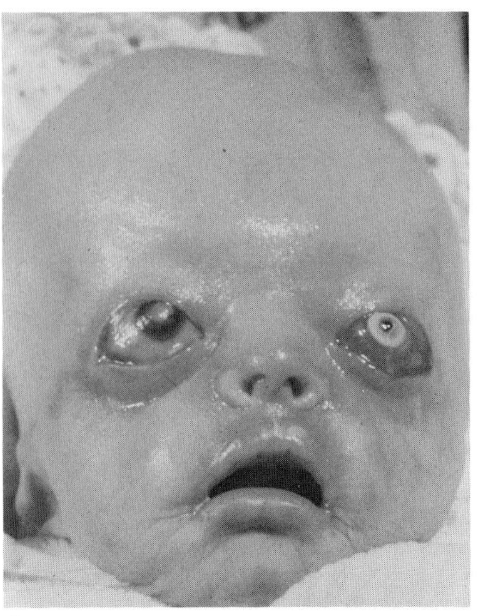

A

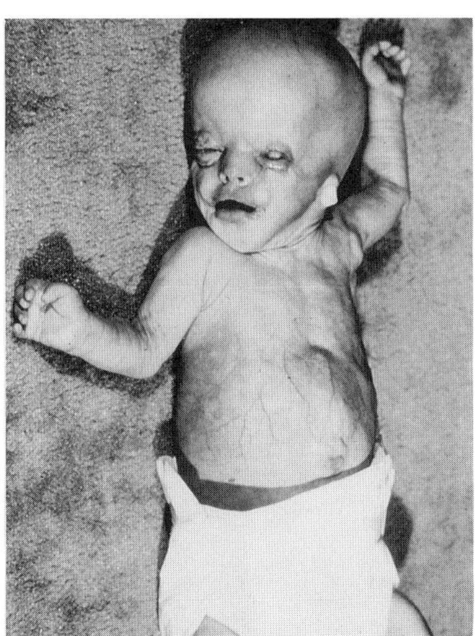

B

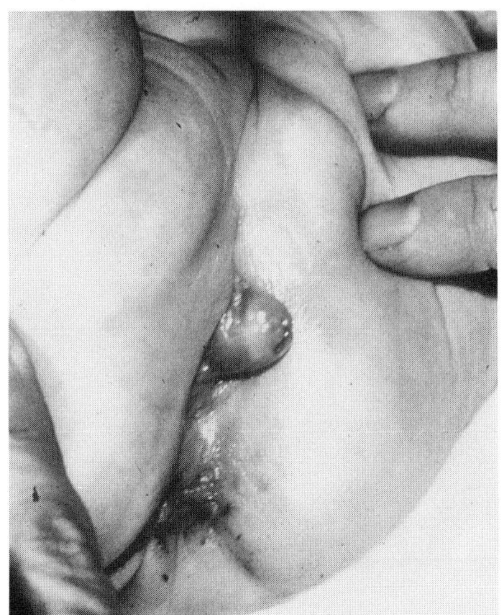

C

the hands and feet, buttocks, popliteal fossae, and neck. There was webbing between the proximal phalanges of the fingers and limitation of finger extension due to tight skin over the interphalangeal joints. The nipples were absent or hypoplastic (Fig. 22–5B). The genitalia were ambiguous with posteriorly displaced micropenis, cryptorchidism, and absent scrotum (Fig. 22–5C). There was umbilical hernia or omphalocele in two of three cases.

Intelligence was normal in one patient but moderately delayed in the other two. To date, all cases have been isolated examples.

Children having several of the features seen in the aforementioned cases were reported by Barber et al (1), Cesarino et al (2), Matalon (4), and Price et al (6). In the child reported in Barber et al (1) the pinnae were less dysmorphic, the eyelids more developed, and the child was hirsute. She was mildly retarded and there were no limb anomalies. RJ Gorlin has seen a similarly affected patient.

#### References (Ablepharon–macrostomia syndrome)

1. Barber N et al: Macrostomia, ectropion, atrophic skin, hypertrichosis and growth retardation. *Syndrome Ident* **8**(1):6–9, 1982.

2. Cesarino EJ et al: Lid agenesis-psychomotor retardation-forehead hypertrichosis—a new syndrome? *Am J Med Genet* **31**:299–304, 1988.

3. Hornblass A, Reifler DM: Ablepharon macrostomia syndrome. *Am J Ophthalmol* **99**:552–556, 1985.

4. Matalon R: (Personal communication, 1988).

5. McCarthy GT, West CM: Ablepharon macrostomia syndrome. *Dev Med Child Neurol* **19**:659–663, 1977.

6. Price NJ et al: The ablepharon–macrostomia syndrome—a neonatal oculoplastic emergency. *Third Manchester Birth Defects Conf,* Manchester, UK, October, 1988.

### Aganglionic megacolon and cleft lip and/or palate

Goldberg and Shprintzen (3) described sibs with unusual facies, neonatal hypotonia, short stature, aganglionic megacolon of the short type, and submucous cleft palate. Sibs were also reported by Kumasaka and Clarren (6). One had cleft lip–palate, the other cleft palate. Hurst et al (4) described three examples. A less likely example is that of Brunoni et al (2).

The facies was characterized by microcephaly, fine scalp hair, hypertelorism, narrow palpebral fissures, synophrys, thick curly eyelashes, prominent nose, and maxillary hypoplasia (Fig. 22–6A,B).

Aganglionic megacolon has also been seen with submucous *cleft palate and paramedian pits of the lower lip* (9), with bilateral cleft lip–palate (8), with cleft palate and hearing loss (1), *Aarskog syndrome,* and *Waardenburg syndrome.*

Several authors have reviewed the many associations of aganglionic megacolon (5–9).

#### References (Aganglionic megacolon and cleft lip and/or palate)

1. Bodian M, Carter CO: A family study of Hirschsprung's anomaly. *Ann Hum Genet* **26**:261–277, 1963.

2. Brunoni D et al: Hirschsprung megacolon, cleft lip and palate, mental retardation and minor congenital malformations. *J Clin Dysmorphol* **1**(1):20–22, 1983.

3. Goldberg RB, Shprintzen RJ: Hirschsprung megacolon and cleft palate in two sibs. *J Craniofac Genet Dev Biol* **1**:185–189, 1981.

4. Hurst JA et al: Unknown syndrome: Hirschsprung's disease, microcephaly and iris coloboma: A new syndrome of defective neuronal migration. *J Med Genet* **25**:494–500, 1988.

5. Kilcoyne RF, Taybi H: Conditions associated with congenital megacolon *Am J Roentgenol* **108**:615–620, 1970.

6. Kumasaka K, Clarren SK: Familial patterns of central nervous system dysfunction, growth deficiency, facial clefts and congenital megacolon. A specific disorder? *Am J Med Genet* **31**:465–466, 1985.

7. Prevot J et al: Hirschsprung's disease with total colonic involvement. *Prog Pediatr Surg* **4**:63–89, 1970.

8. Sane SM, Girdany BE: Total aganglionosis coli. *Radiology* **107**:397–404, 1973.

9. Schwarz KB et al: Congenital lip pits and Hirschsprung's disease. *J Pediatr Surg* **14**:162–164, 1979.

### Baraitser syndrome and cleft lip–palate

Baraitser syndrome is a presumed autosomal recessive pattern of anomalies defined on the basis of a 5½-year-old female of normal karyotype and an 18-week-old male fetus (1). There was one normal sib. Both parents were normal and consanguinity was denied.

The 5½-year-old proband manifested craniosynostosis involving the coronal suture, mental retardation, seizures, choroidal colobomas, mild hypertelorism, beaked nose, cleft lip–palate, protuberant ears, mild mesomelic shortening of the upper and lower extremities, short broad fingers, and cystic dysplastic kidneys.

A second case was diagnosed prenatally by fetoscopy at 16 weeks. The presence of cleft lip was indicative with high probability of the full syndrome. The pregnancy was terminated and postmortem examination at 18 weeks showed cleft lip–palate, small posterior fontanel, and findings consistent with choroidal coloboma.

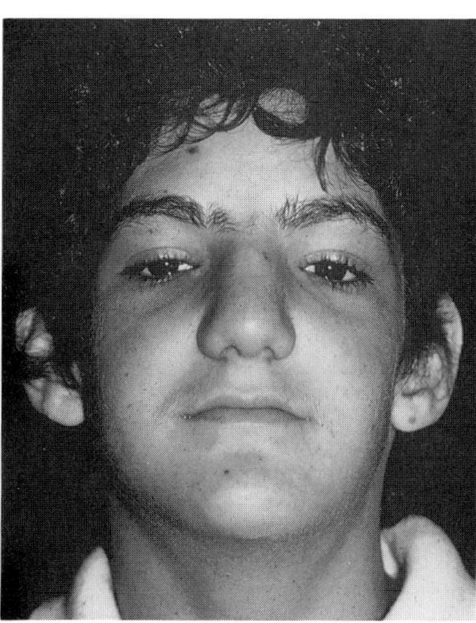

A

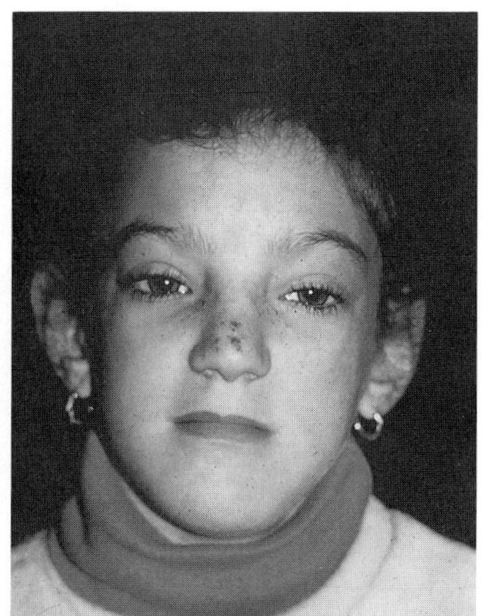

B

Fig. 22–6. *Aganglionic megacolon and cleft lip and/or palate.* (A,B) Sibs manifesting microcephaly, fine scalp hair, apparent hypertelorism, narrow palpebral fissures, synophrys, thick curly eyelashes, prominent nose, and maxillary hypoplasia. (From *RB Goldberg* and *RJ Shprintzen.* J Craniofac Genet Dev Biol **1**:185, 1981.)

### Reference (Baraitser syndrome and cleft lip–palate)

1. Baraitser M et al: A new craniosynostosis/mental retardation syndrome diagnosed by fetoscopy. *Clin Genet* **22**:12–15, 1982.

## Bencze syndrome

The Bencze syndrome (1,2) consists of mild facial asymmetry, esotropia, amblyopia, and submucous cleft palate. The syndrome has autosomal dominant inheritance with variable expression. On the hypoplastic side the palpebral fissure is usually upslanting (Figs. 22–7 and 22–8). Other body proportions are symmetric.

### References (Bencze syndrome)

1. Bencze J et al: Dominant inheritance of hemifacial hypoplasia associated with strabismus. *Oral Surg* **35**:489–500, 1973.

2. Kurnit D et al: An autosomal dominantly inherited syndrome of facial asymmetry, esotropia, and amblyopia, and submucous cleft palate (Bencze syndrome). *Clin Genet* **16**:301–304, 1979.

## Catel–Manzke syndrome (palatodigital syndrome)

Catel (2), in 1961, and Manzke (8), in 1966, reported the same patient with cleft palate, micrognathia, and bilateral clinodactyly (radial angulation) of the index finger (Fig. 22–9). Approximately 25 cases have been documented to date, almost all in males. RJ Gorlin has seen a case of the syndrome in a female patient. Although there has been some familial aggregation (1,5,10), in our view X-linked recessive inheritance is not tenable and recurrence risk is low.

Radiographically, the clinodactyly is seen to result from an accessory bone, irregular in form, located between the shortened second metacarpal and the corresponding shortened proximal phalanx of the index finger. With time, the supernumerary bone fuses with both the radially deviated proximal phalanx and the metacarpal (6) (Fig. 22–10A,B). Miscellaneous bony anomalies have included bifurcated first metacarpal (6) and multiple vertebral anomalies (10).

The cranium may be rather globular. Over 90% have cleft palate and micrognathia. Talipes equinovarus has been noted in 30% (4,6). Dislocatable knees have also been found (12). Congenital heart defects, most often VSD, have occurred in about one-half of the cases (4,5,9–11).

Fig. 22–7. *Bencze syndrome.* Primary telecanthus, strabismus, malocclusion. (From *D Kurnit et al,* Clin Genet **16**:301, 1979.)

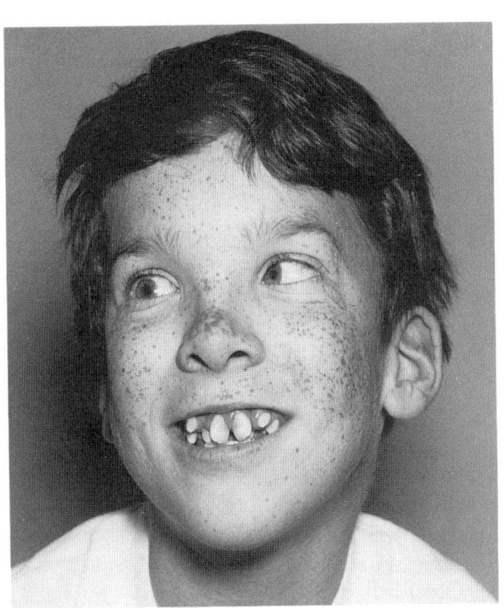

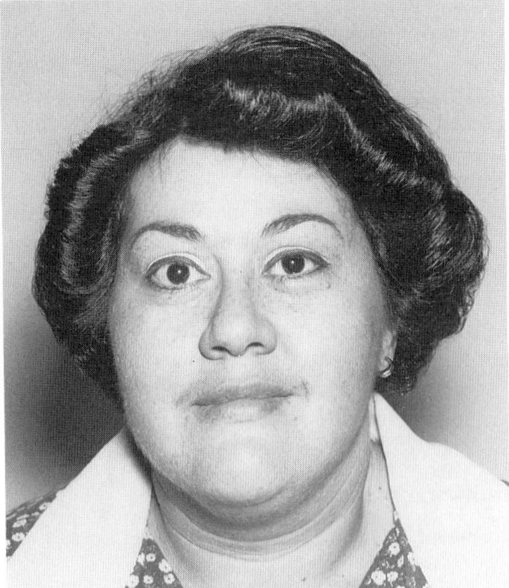

Fig. 22–8. *Bencze syndrome.* Aunt of proband showing facial asymmetry and upslanting palpebral fissure on hypoplastic side. (*From D Kurnit et al,* Clin Genet **16**:301, 1979.)

### References [Catel–Manzke syndrome (palatodigital syndrome)]

1. Brude E: Pierre Robin sequence and hyperphalangy—a genetic entity (Catel–Manzke syndrome). *Eur J Pediatr* **142**:222–223, 1984.

2. Catel W: *Differentialdiagnose von Krankheitssymptomen bei Kindern und Jugendlichen,* ed 3, Vol 1, G Thieme, 1961, pp 218–220.

3. Dignan PS et al: Pierre Robin anomaly with an accessory metacarpal of the index fingers: The Catel–Manzke syndrome. *Clin Genet* **29**:168–173, 1986.

4. Farnsworth PB, Pacik PT: Glossoptotic hypoxia and micrognathia—the Pierre Robin syndrome reviewed. *Clin Pediatr* **10**:600–606, 1971.

5. Gewitz M et al: Cleft palate and accessory metacarpal of index finger syndrome: A possible familial occurrence. *J Med Genet* **15**:162–164, 1978.

5a. Gorlin RJ: Type $A_2$ brachydactyly syndrome. *Birth Defects* **2**(2):41–42, 1975.

Fig. 22–9. *Catel–Manzke syndrome.* Note marked clinodactyly of index finger. (Courtesy of *E Brude,* Eur J Pediatr **142**:1984.)

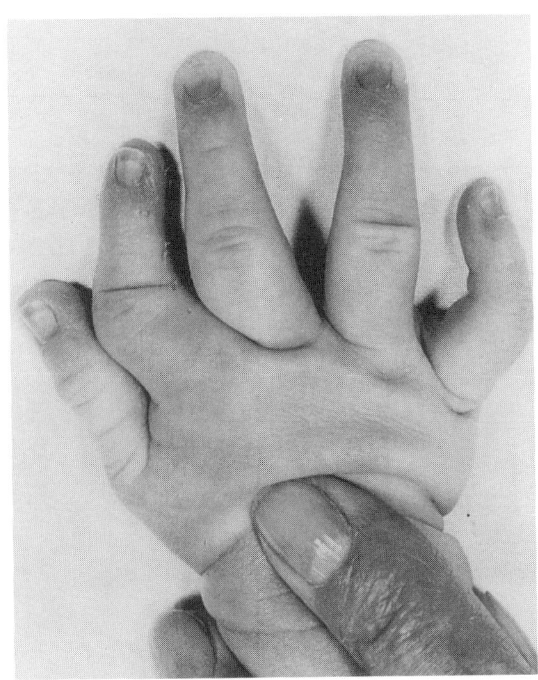

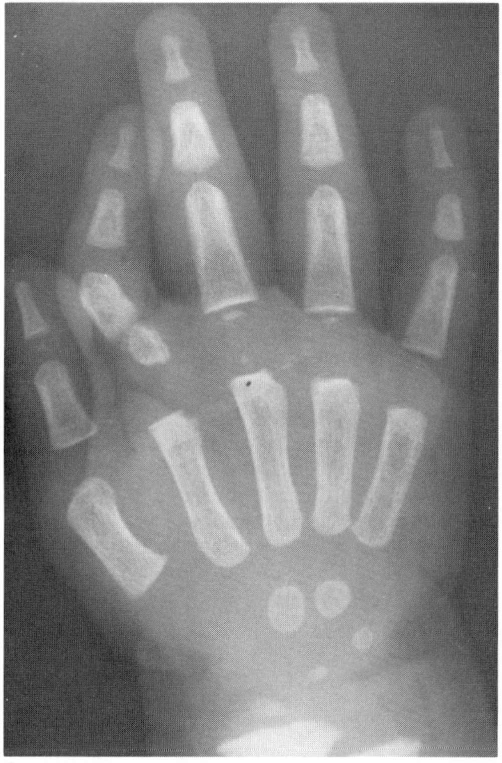
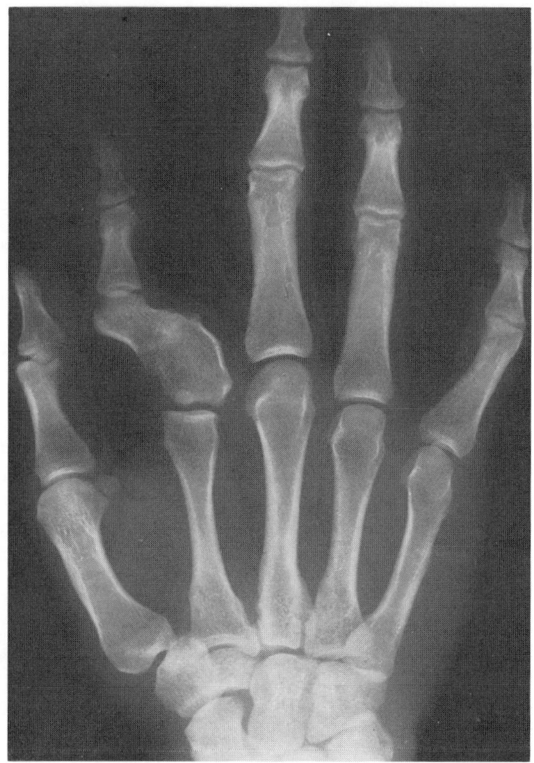

A B
Fig. 22–10. *Catel–Manzke syndrome.* (A) Note separate bone at base of proximal phalanx of index finger. (B) Same anomaly in older patient, but fusion has taken place. (From *W Holthusen,* Ann Radiol **15**:253, 1972.)

6. Holthusen W: The Pierre Robin syndrome: Unusual associated developmental defects. *Ann Radiol* **15**:253–262, 1972 (Cases 5,6).

7. Klug MS et al: Symmetric hyperphalangism of the index finger in the palatodigital syndrome. *J Hand Surg* **8**:599–603, 1983.

8. Manzke H: Symmetrische Hyperphalangie des zweiten Fingers durch ein akzessorisches Metacarpale. *Roefo* **105**:425–427, 1966.

9. Silengo MC et al: Pierre Robin syndrome with hyperphalangism-clinodactylism of the index fingers: A possible new palato-digital syndrome. *Pediatr Radiol* **6**:178–180, 1977.

9a. Skinner SA et al: Catel–Manzke syndrome. *Proc Greenwood Genet Ctr* **8**:60–63, 1989.

10. Stevenson RE et al: A digitopalatal syndrome with associated anomalies of the heart, face, skeleton. *J Med Genet* **17**:238–242, 1980.

11. Sundaram V et al: Hyperphalangy and clinodactyly of the index finger with Pierre Robin anomaly: Catel–Manzke syndrome: A case report and review of the literature. *Clin Genet* **21**:407–410, 1982.

12. Thompson EM et al: A male infant with the Catel–Manzke syndrome and dislocatable knees. *J Med Genet* **23**:271–274, 1986.

## Cleft lip and palate, contractures, and agenesis of fibula and ulna

Pfeiffer et al (2) described two sibs of Turkish origin. The parents were consanguineous. They cited the report of Fuhrmann et al (1) of three somewhat similarly affected sibs. Inheritance is autosomal recessive.

Radiographically, the fibula was aplastic in all cases and the ulna aplastic or hypoplastic. The radius was present. There was oligodactyly in one set of sibs (2), polydactyly in another set (1), and aplasia of nails. The humerus and tibia were normal but the femur was markedly angulated and the pelvis abnormally shaped. All joints were somewhat contracted.

Cleft lip and palate were noted only in the sibs reported by Pfeiffer et al (2).

### References (Cleft lip and palate, contractures, and agenesis of fibula and ulna)

1. Fuhrmann W et al: Poly-, syn- and oligodactyly, aplasia or hypoplasia of fibula, hypoplasia of pelvis and bowing of femora in three sibs—a new autosomal recessive syndrome. *Eur J Pediatr* **133**:123–129, 1980.

2. Pfeiffer RA et al: Absence of fibula and ulna with oligodactyly, contractures, right-angle bowing of femora, abnormal facial morphology, cleft lip/palate and brain malformation in two sibs: A possibly new lethal syndrome. *Am J Med Genet* **29**:901–908, 1988.

## Cleft lip and palate, pili torti, malformed ears, partial syndactyly of fingers and toes, and mental retardation

Zlotogora et al (3), in 1987, reported two sibs in a consanguineous family. Both had downslanting palpebral fissures, sparse outer eyebrows, cleft lip and palate, micrognathia, malformed protruding pinnae, sparse scalp hair with pili torti, and widely spaced teeth. There was partial syndactyly of the third and fourth fingers and second and third toes. Transverse palmar creases or Sydney lines were present (Fig. 22–11A,B). Motor development was slow. The surviving child at 3½ years performed at approximately a 24 month level and spoke few words. Additional findings included wide-spaced nipples, hypoplastic scrotum, and cryptorchidism.

Falace and Hall (1) described a child with cleft lip–palate, sparse hair, high nasal bridge, small simple pinnae, and 3–4 finger and 2–3, 4–5 toe soft tissue syndactyly. There was significant postnatal growth retardation. Intelligence was normal. The parents were first cousins. Another pair of male sibs was described by Oğur and Yüksel (2).

There is considerable overlap with the *AEC, EEC, Rapp–Hodgkin,* and *Bowen–Armstrong syndromes.*

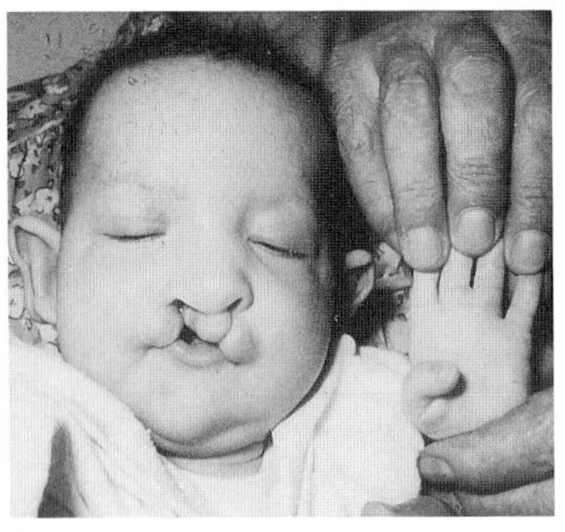
A

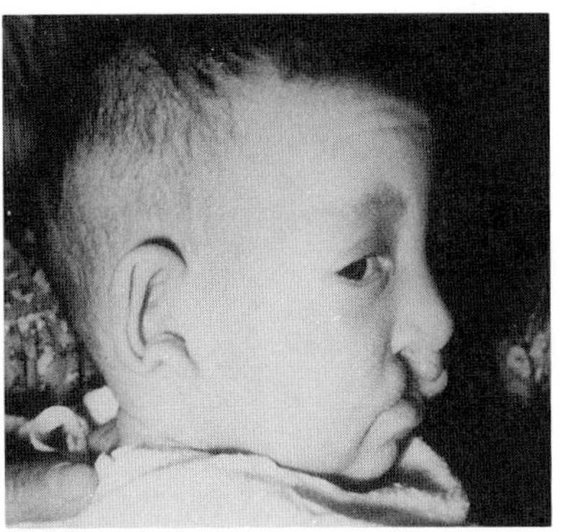
B

Fig. 22–11. *Cleft lip and palate, pili torti, malformed ears, partial syndactyly of fingers and toes, and mental retardation.* (A,B) Note high forehead, sparse hair, dysmorphic pinnae, bilateral cleft lip, partial soft tissue syndactyly of fingers 3–4, and transverse palmar crease. (From *J Zlotogora et al,* J Med Genet **24**:291, 1987.)

### References (Cleft lip and palate, pili torti, malformed ears, partial syndactyly of fingers and toes, and mental retardation)

1. Falace PB, Hall BD: A new ectodermal dysplasia syndrome with syndactyly cleft lip/palate and short stature. *Proc Greenwood Genet Ctr* **7**:214–215, 1988.
2. Oğur G, Yüksel M: Association of syndactyly, ectodermal dysplasia, and cleft lip and palate: Report of two sibs from Turkey. *J Med Genet* **25**:37–40, 1988.
3. Zlotogora J et al: Cleft lip and palate, pili torti, malformed ears, partial syndactyly of fingers and toes and mental retardation: A new syndrome? *J Med Genet* **24**:291–293, 1987.

## Cleft lip and palate, prominent eyes, and congenital heart disease

Dinno (1), in 1987, described three generations with cleft lip–palate, large ears, prominent eyes, high forehead, hypertelorism, and a beak-shaped nose with a short philtrum (Fig. 22–12). One of the sibs had ASD.

### Reference (Cleft lip and palate, prominent eyes, and congenital heart disease)

1. Dinno ND: Cleft lip and cleft palate, high forehead, abnormal ears, prominent eyes and congenital heart disease in three generations; A new syndrome. *March of Dimes Clinical Genetics Conference,* Minneapolis, July, 1987.

Fig. 22–12. *Cleft lip and palate, prominent eyes, and congenital heart disease.* Note cleft lip, very prominent eyes, hypertelorism, short philtrum. (Courtesy of *ND Dinno,* Seattle, Washington.)

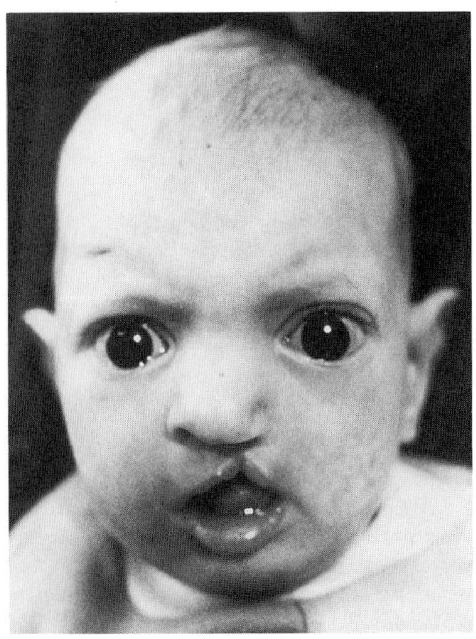

Fig. 22–13. *Cleft lip and palate, abnormal ears, congenital heart defect, and skeletal abnormalities.* (A,B) Radiographs show short femur and humerus, absence of one metacarpal and one metatarsal, tapering of proximal femur with absence of femoral head and absence of calcaneus. (From *SP Verloove-Vanhorick et al,* Acta Paediatr Scand **70**:767, 1981.)

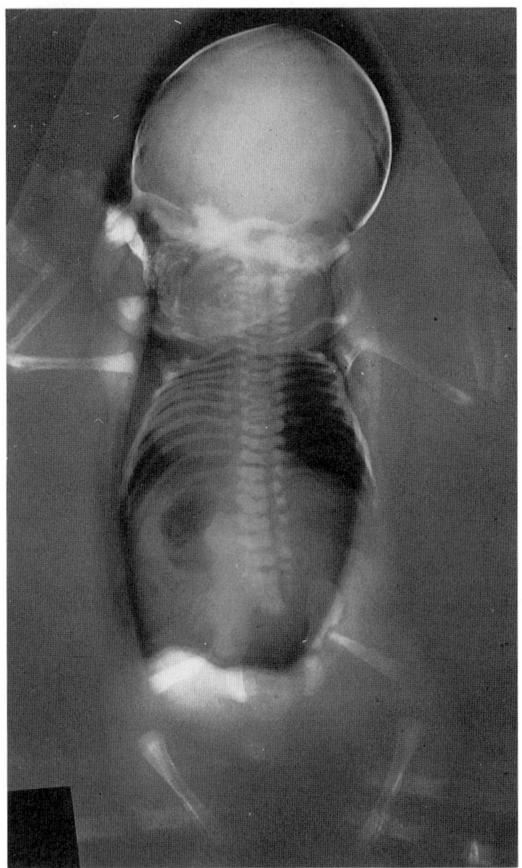
A

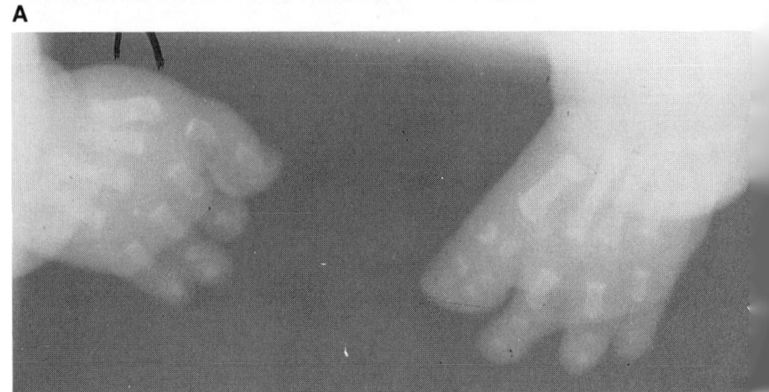
B

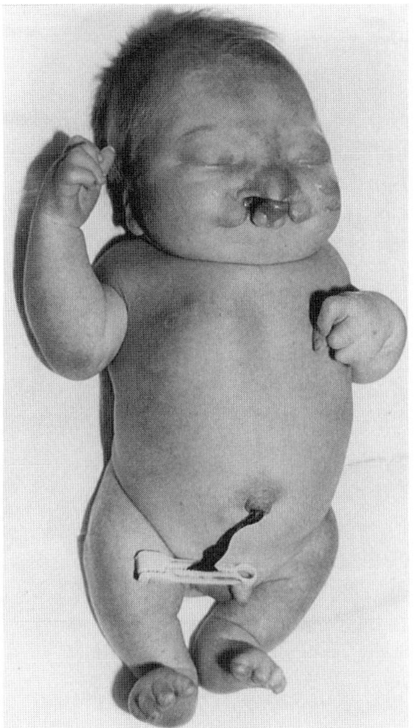

Fig. 22–14. *Cleft lip and palate, abnormal ears, congenital heart defect, and skeletal abnormalities.* One of two sibs with bilateral cleft lip–palate and short extremities. (From *SP Verloove-Vanhorick et al,* Acta Paediatr Scand **70**:767, 1981.)

## Cleft lip and palate, abnormal ears, congenital heart defect, and skeletal abnormalities

The sibs described by Verloove-Vanhorick et al (1) had bilateral short femur and humerus, absence of one metacarpal and one metatarsal with partial soft tissue syndactyly of fingers and toes, four lumbar vertebrae, tapering of proximal femur with absence of femoral head, and absence of calcaneus (Fig. 22–13A,B).

There was bilateral cleft lip–palate, micrognathia, and malformed pinnae with absent external meatus (Fig. 22–14). Internal anomalies included cryptorchidism, persistent truncus arteriosus, bilobed lungs, and parathyroid aplasia.

This lethal condition possibly has autosomal recessive inheritance.

### Reference (Cleft lip and palate, abnormal ears, congenital heart defect, and skeletal abnormalities)

1. Verloove-Vanhorick SP et al: Extensive congenital malformations in two siblings. *Acta Paediatr Scand* **70**:767–769, 1981.

## Cleft lip–palate, hypospadias, and inguinal hernia

RJ Gorlin has seen two male siblings with the following findings: large anterior fontanel, patent posterior fontanel, superficial capillary angioma of forehead, mild hypertelorism, absent eyebrows, small palpebral fissures, cleft lip and palate in one sib, cleft palate in the second, broad nasal base, hypospadias, large inguinal hernia, and atrial septal defect in both. One sib died at the age of 4 months and the other at 2 weeks from infection. Only one of the sibs had the following: two-vessel cord, metatarsus adductus, digitalized thumbs, proximally placed thumbs, bilateral coloboma of iris, unilateral ear tags, and double urinary collecting system. We were unable to find any additional examples of this syndrome in the literature.

## Cleft lip–palate, microcephaly, and hypoplasia of radius and thumb (Juberg–Hayward syndrome, orocraniodigital syndrome)

Juberg and Hayward (1), in 1969, reported three of six sibs with cleft lip–palate, mild microcephaly, hypoplastic and distally positioned thumbs, limited extensions of the elbows due to anterior displacement of the radial head, and short stature (Fig. 22–15A,B and 22–16). Birthweight was low. Nevin et al (4) and Kingston et al (3) described isolated examples. The child reported by Nevin et al (4) was found to have absence of the pituitary fossa and flattened vertebral bodies. There were no hormone deficiencies. Kingston et al (3) noted isolated growth hormone deficiency, micropenis, and bilateral thumb aplasia in their patient. Kato (2) tabulated several case reports, some of which may represent the condition.

### References [(Cleft lip–palate, microcephaly, and hypoplasia of radius and thumb (Juberg–Hayward syndrome, orocraniodigital syndrome)]

1. Juberg RC, Hayward JR: A new familial syndrome of oral, cranial and digital anomalies. *J Pediatr* **74**:755–762, 1969.
2. Kato K: Congenital absence of the radius with review of literature and report of three cases. *J Bone Jt Surg* **6**:589–626, 1924.
3. Kingston HM et al: Orocraniodigital (Juberg–Hayward) syndrome with growth hormone deficiency. *Arch Dis Childh* **57**:790–792, 1981.
4. Nevin NC et al: A case of the orocraniodigital (Juberg–Hayward) syndrome. *J Med Genet* **18**:478–480, 1981.

## Cleft lip–palate, microcephaly, mental retardation, and holoprosencephaly

Martin et al (3) reported three generations of a large kindred in which there was variable expression of cleft lip and/or cleft palate, micro-

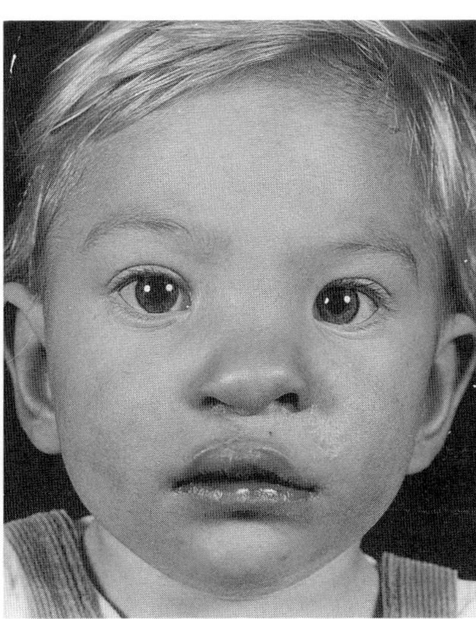

A

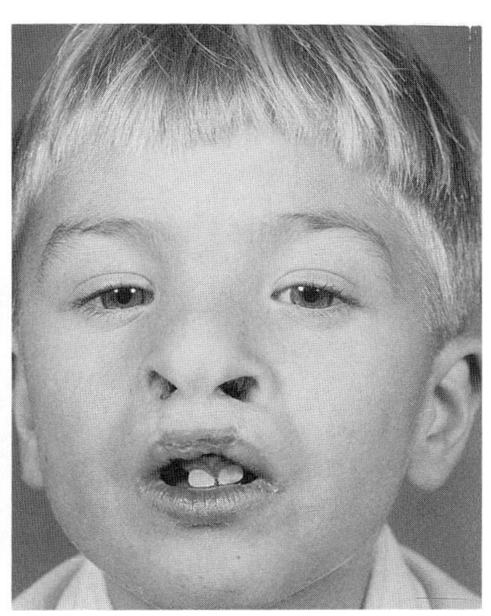

B

Fig. 22–15. *Cleft lip-palate, microcephaly, and hypoplasia of radius and thumb.* (A,B) Sibs with cleft lip, hypertelorism, epicanthal folds, broad nasal bridge, and hypoplastic columella. (From *RC Juberg* and *JR Hayward,* J Pediatr **74**:755, 1969.)

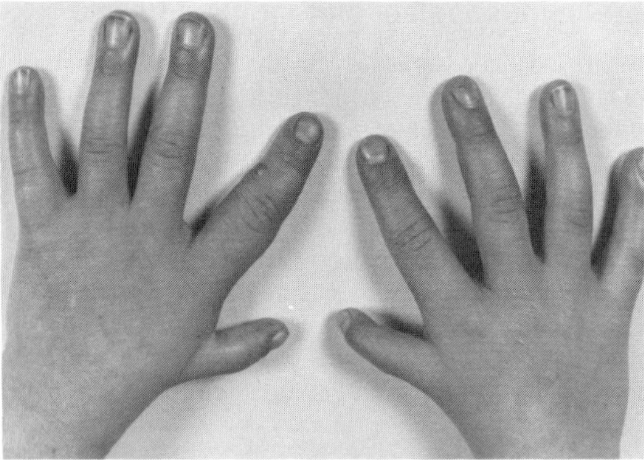

Fig. 22–16. *Cleft lip–palate, microcephaly, and hypoplasia of radius and thumb.* Severe hypoplasia of thumbs. (From *RC Juberg* and *JR Hayward*, J Pediatr **74**:755, 1969.)

cephaly, mental retardation, hypotelorism, downslanting palpebral fissures, large pinnae, skeletal anomalies including scoliosis and talipes, and chronic constipation. One of the affected clearly had the premaxillary agenesis form of holoprosencephaly. Two others had agenesis of maxillary central and lateral incisors. Three affected males lived past 20 years of age while three affected females died early in infancy.

Martin et al (3) separated this disorder from familial isolated holoprosencephaly (1,2) because of the skeletal anomalies found in this family. See *holoprosencephaly* for more extensive discussion.

#### References (Cleft lip–palate, microcephaly, mental retardation, and holoprosencephaly)

1. Benke PJ, Cohen MM Jr: Recurrence of holoprosencephaly in families with a positive history. *Clin Genet* **24**:324–328, 1983.

2. Cantú JM et al: Dominant inheritance of holoprosencephaly. *Birth Defects* **14**(6B):215–220, 1978.

3. Martin AO et al: An autosomal dominant midline cleft syndrome resembling familial holoprosencephaly. *Clin Genet* **12**:65–72, 1977.

### Cleft lip–palate, posterior keratoconus, short stature, mental retardation, and genitourinary anomalies

Young et al (8) described two sibs with keratoconus posticus circumscriptus, retinal colobomas, short stature, moderate mental retardation, vertebral segmentation and fusion anomalies, tight heel cords, and cleft lip–palate. The nose was prominent. The neck was mildly webbed with low posterior hairline (Fig. 22–17A,B and 22–18). Extension at the elbows was mildly limited. Haney and Falls (3) and Streeten et al (6) described somewhat similarly affected sibs.

It may be argued that posterior keratoconus is a mild form of the Peter anomaly (developmental defect of Descemet membrane and deep stromal layers of the cornea with or without iris–cornea adhesions) that has also been reported with cleft lip–palate (1,2,4,5,7). (See section on *eye coloboma/cleft lip–palate association* for overlap).

#### References (Cleft lip–palate, posterior keratoconus, short stature, mental retardation, and genitourinary anomalies)

1. Anyane-Yeboa K et al: Cleft lip and palate, corneal opacities and profound psychomotor retardation: A newly recognized genetic syndrome. *Cleft Palate J* **20**:246–250, 1983.

2. Bettini F et al: La sindrome di Peters. *Pathologica* **72**:699–705, 1980.

3. Haney WP, Falls HF: The occurrence of congenital keratoconus posticus circumscriptus in two siblings presenting a previously unrecognized syndrome. *Am J Ophthalmol* **7**:841–842, 1975.

4. Harcourt B: Anterior chamber cleavage syndrome associated with Weill-Marchesani syndrome and craniofacial dysplasias. *J Pediatr Ophthalmol* **7**:24–29, 1970.

5. Ide CH et al: Dysgenesis mesodermalis of the cornea (Peter's anomaly) associated with cleft lip and palate. *Ann Ophthalmol* **7**:841–842, 1975.

6. Streeten BW et al: Posterior keratoconus associated with systemic abnormalities. *Arch Ophthalmol* **101**:616–622, 1983.

7. Van Schoonveld MJ et al: Peters'-plus: A new syndrome. *Ophthalmol Paediatr Genet* **4**:141–146, 1984.

8. Young ID et al: Keratoconus posticus circumscriptus, cleft lip and palate, genitourinary abnormalities, short stature, and mental retardation in sibs. *J Med Genet* **19**:332–336, 1982.

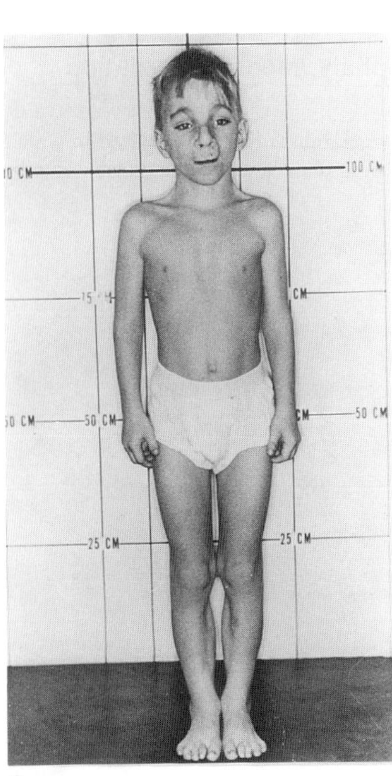

A

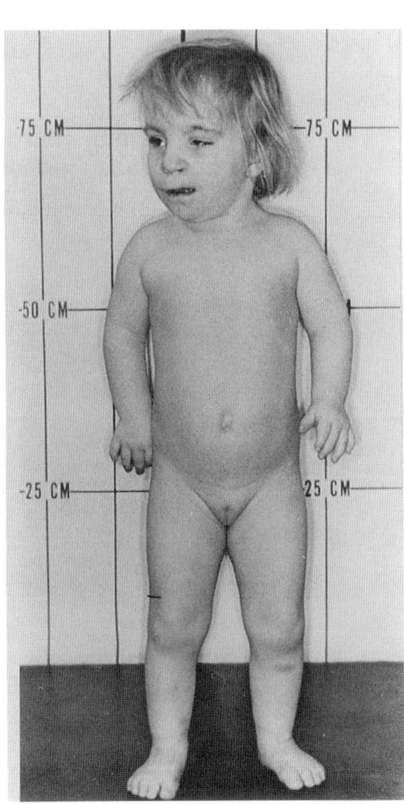


B

Fig. 22–17. *Cleft lip–palate, posterior keratoconus, short stature, mental retardation, and genitourinary anomalies.* (A,B) Mentally retarded brother and sister with cleft lip–palate, short stature, mild neck webbing, small hands and feet. (Courtesy of *ID Young*, Leicester, England.)

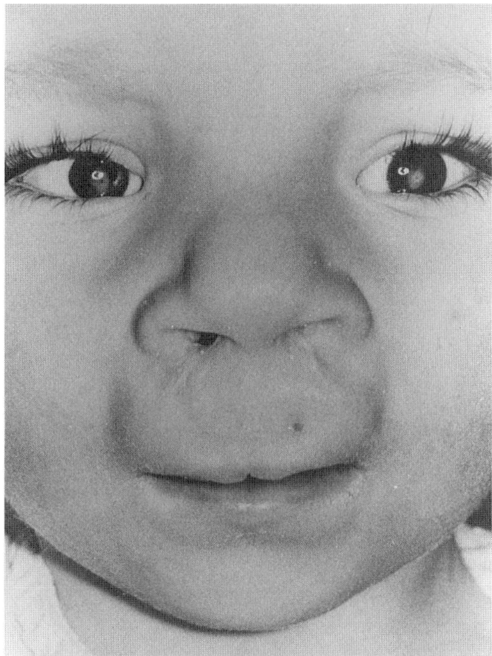

Fig. 22–18. *Cleft lip–palate, posterior keratoconus, short stature, mental retardation, and genitourinary anomalies.* Boy with central corneal opacities (Peter-like anomaly). (Courtesy of *ID Young,* Leicester, England.)

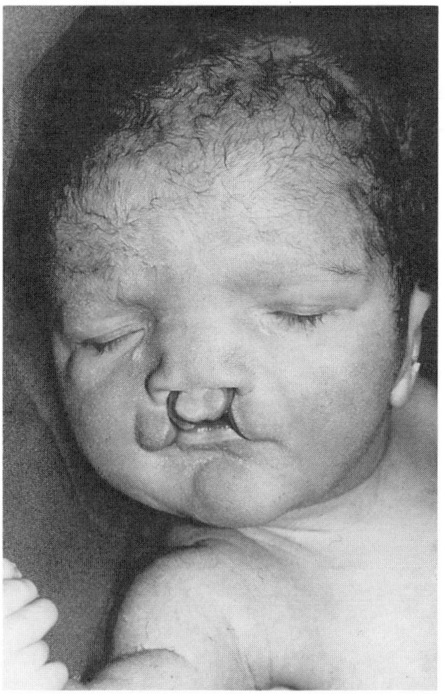

Fig. 22–19. *Cleft lip–palate, preaxial and postaxial polydactyly of hands and feet, congenital heart defect, and genitourinary anomalies.* Cleft lip–palate and hypertelorism. (From *U Töllner et al,* Eur J Pediatr **136**:207, 1981.)

## Cleft palate, agenesis of the corpus callosum, and unusual facies

Toriello and Carey (1) reported four children, three of whom were sibs, with telecanthus, short palpebral fissures, small nose with anteverted nares, cleft palate, micrognathia, mildly dysmorphic pinnae, excess neck skin, congenital heart anomalies (ASD, PDA), cryptorchidism, brachydactyly, hypotonia, and laryngeal anomaly. Agenesis of the corpus callosum was found in three affected from both kindreds.

The disorder appears to have autosomal recessive inheritance.

### Reference (Cleft palate, agenesis of the corpus callosum, and unusual facies)

1. Toriello HV, Carey JC: Corpus callosum agenesis, facial anomalies, Robin sequence, and other anomalies: A new autosomal recessive syndrome? *Am J Med Genet* **31**:17–23, 1988.

## Cleft lip–palate, preaxial and postaxial polydactyly of hands and feet, congenital heart defect, and genitourinary anomalies

Töllner et al (1), in 1981, reported a child with bilateral cleft lip–palate, preaxial and postaxial polydactyly of the hands and feet, internal hydrocephalus, ependymal cysts of the lateral ventricle, complex malformations of the heart and great vessels, horseshoe kidney, and micropenis (Figs. 22–19 to 22–22). The karyotype was normal.

### Reference (Cleft lip–palate, preaxial and postaxial polydactyly of hands and feet, congenital heart defect, and genitourinary anomalies)

1. Töllner U et al: Heptocarpo-octatarso-dactyly combined with multiple malformations. *Eur J Pediatr* **136**:207–210, 1981.

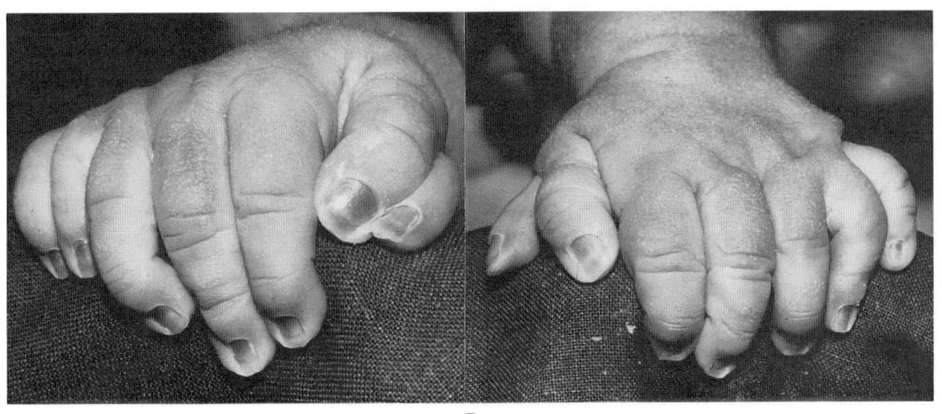


Fig. 22–20. *Cleft lip–palate, preaxial and postaxial polydactyly of hands and feet, congenital heart defect, and genitourinary anomalies* (A,B) Heptadactyly of hands. (From *U Töllner et al,* Eur J Pediatr **136**:207, 1981.)

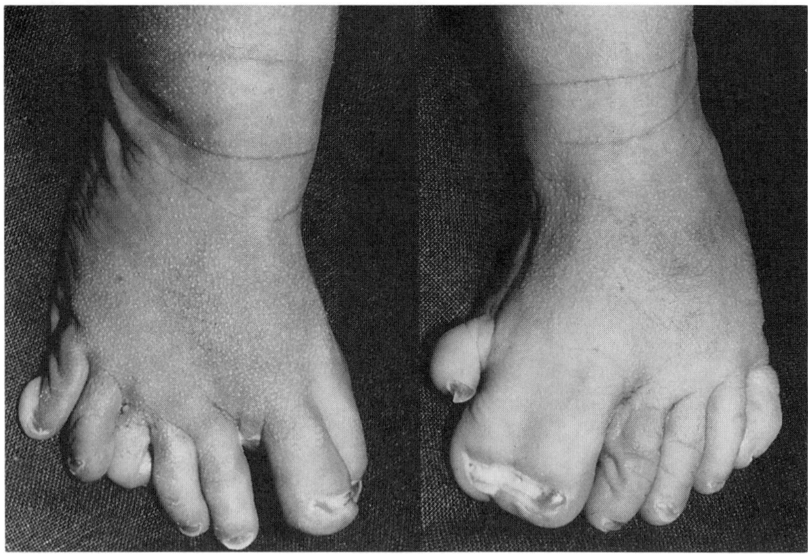

Fig. 22–21. *Cleft lip–palate, preaxial and postaxial polydactyly of hands and feet, congenital heart defect, and genitourinary anomalies.* (A,B) Octodactyly of feet. (From *U Töllner et al,* Eur J Pediatr **136**:207, 1981.)

## Cleft lip–palate, urogenital anomalies, somatic and mental retardation

Malpuech et al (1) reported six sibs in a highly inbred Gypsy family with cleft lip–palate, somatic retardation (−3 to −5 SD), moderate mental retardation, and various urogenital anomalies. There was marked hypertelorism (Fig. 22–23A–D). The urogenital abnormalities consisted of unilateral aplasia or bilateral hypoplasia of the kidney, ectopia or malrotation of the kidney, vesicoureteral reflex, bladder diverticulosis, micropenis, penoscrotal hypospadias, and ectopic testes. Three of the sibs were stillborn. Their cousin and seven other second or third degree relatives were similarly affected.

### Reference (Cleft lip–palate, urogenital anomalies, somatic and mental retardation)

1. Malpuech G et al: A previously undescribed autosomal recessive multiple congenital anomalies/mental retardation (MCA/MR) syndrome with growth failure, lip/palate cleft(s) and urogenital anomalies. *Am J Med Genet* **16**:475–480, 1983.

Fig. 22–22. *Cleft lip–palate, preaxial and postaxial polydactyly of hands and feet, congenital heart defect, and genitourinary anomalies.* Micropenis and penis palmatus. (From *U Töllner et al,* Eur J Pediatr **136**:207, 1981.)

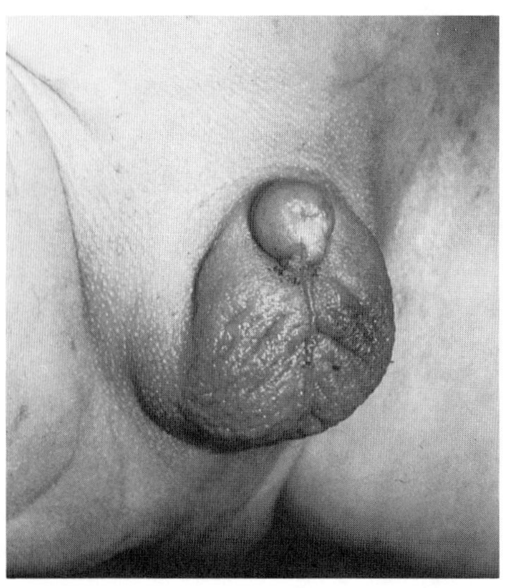

Fig. 22–23. *Cleft lip–palate, urogenital anomalies, somatic and mental retardation.* (A–D) Four of six sibs exhibiting cleft lip–palate, somatic retardation, moderate mental retardation, and various urogenital anomalies (unilateral aplasia or bilateral hypoplasia of kidney, reflux, bladder diverticulosis, micropenis, hypospadias). (From *G Malpuech et al,* Am J Med Genet **16**:475, 1983.)

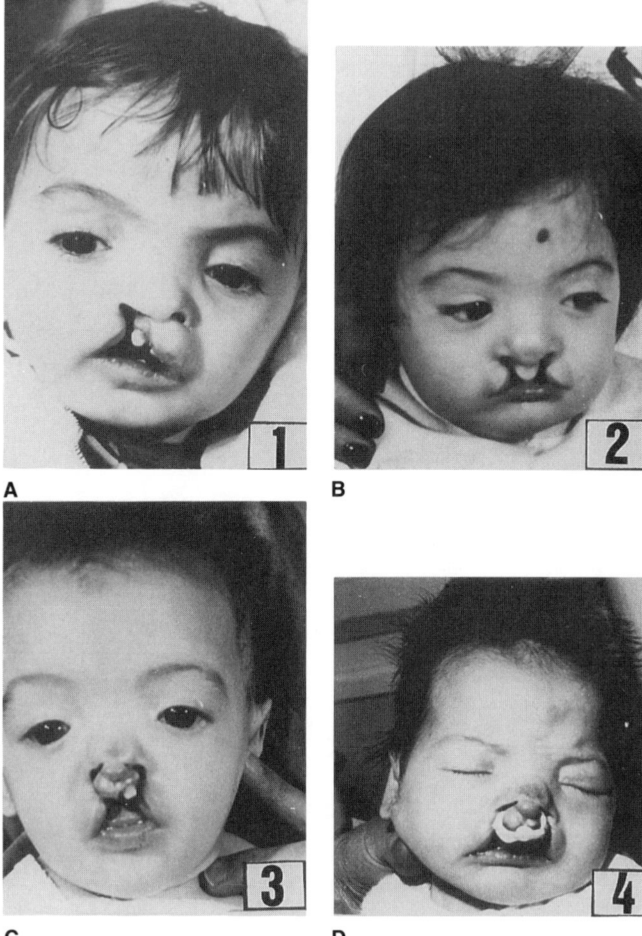

## Cleft palate, absent tibiae, preaxial polydactyly of the feet, and congenital heart defect

Ho et al (1) reported a child with lower legs symmetrically curved so that the feet were oriented cranially with the plantar surfaces facing dorsally. There was preaxial polydactyly of the halluces. The fingers were long with ulnar deviation. The child also had cleft palate, micrognathia, VSD, and bilateral single palmar creases (Fig. 22–24). Radiographically, wormian bones, absent tibias, and fibulas curved at midshaft were evident (Fig. 22–25). The authors cited several reports of agenesis of the tibia with preaxial polydactyly but indicated that none was truly similar to their case.

### Reference (Cleft palate, absent tibiae, preaxial polydactyly of the feet, and congenital heart defect)

1. Ho CK et al: Cleft palate, congenital heart disease, absent tibiae and polydactyly. *Am J Dis Child* **129**:714–716, 1975.

## Cleft palate, craniofacial sclerosis, and musculoskeletal abnormalities

Currarino and Friedman (1) reported a family having variable expression of a syndrome of abnormal facies, cleft palate, and various musculoskeletal abnormalities.

The facies in the boy was characterized by large calvaria with biparietal bossing, thick arched eyebrows, synophrys, long upper lip with Cupid's bow form, cleft palate, and thick dental arches. There was delayed shedding of deciduous teeth. A speech defect and conductive hearing loss were noted.

Musculoskeletal changes include short stature, inguinal hernia, digital anomalies, progressive lumbar lordosis, dislocation of radial heads, and hallux valgus (Fig. 22–26A,B).

Radiographically, enlarged calvaria, vertical clivus, cervical kyphosis, severe lumbar lordosis, and delta phalanx of middle phalanges of the index fingers and proximal phalanges of halluces were found.

Fig. 22–24. *Cleft palate, absent tibiae, preaxial polydactyly of the feet, and congenital heart defect.* Small mandible, ulnar deviation of fingers, bowing of lower extremities, polydactyly. (From *CK Ho* et al, **129**:714, 1975.)

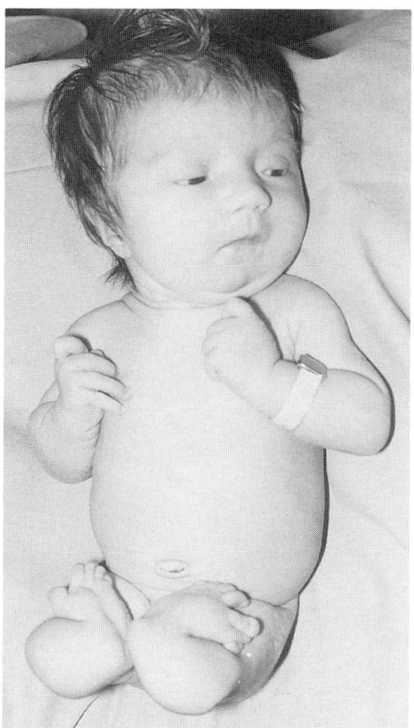

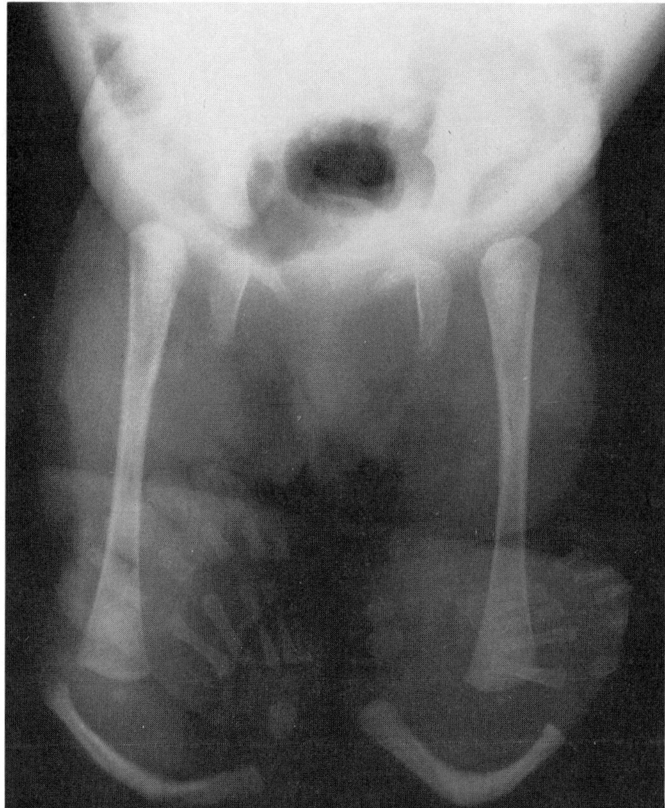

Fig. 22–25. *Cleft palate, absent tibiae, preaxial polydactyly of the feet, and congenital heart defect.* Absent tibiae, dislocated and bowed fibulae, dislocated hips, polydactyly. (From *CK Ho* et al, **129**:714, 1975.)

The mother exhibited osteopathia striata, enlarged head, highly arched eyebrows, and Cupid's bow mouth. Another child who died during infancy had enlarged head, cleft palate, large soft pinnae, omphalocele, and hypospadias.

Although the mother has findings compatible with the *cleft palate, osteopathia striata, cranial sclerosis, and hearing loss,* the findings in her children are too severe for that diagnosis to be considered.

### Reference (Cleft palate, craniofacial sclerosis, and musculoskeletal abnormalities)

1. Currarino G, Friedman JM: Severe craniofacial sclerosis with multiple anomalies in a boy and his mother. *Pediatr Radiol* **16**:441–447, 1986.

## Cleft palate, dwarfism, trichodysplasia, and unusual facies

Jorgenson and Bick (1) recently described two brothers with an apparently new malformation syndrome. At birth, both were below the third percentile for length, weight, and head circumference. The elder was still dwarfed when examined at age three. Both had sparse, coarse, and brittle hair. Light and scanning electron microscopy of the hair showed fractures, thin shafts, and a decreased number of cuticles. The palpebral fissures were narrow (below the third percentile by measurement), and the elder had bilateral entropion of the lashes. Both had prominent pinnae, micrognathia, and cleft palate. The older had telecanthus, abnormally shaped incisors, bilateral inguinal herniae, and bilateral clinodactyly of the fifth fingers. The younger had bilateral single palmar flexion creases. Sensory, motor, and cerebellar functions appeared normal in each, although the older used only 10 words at age three. Skeletal radiographs were unremarkable.

Parental consanguinity was denied, although both parents were from

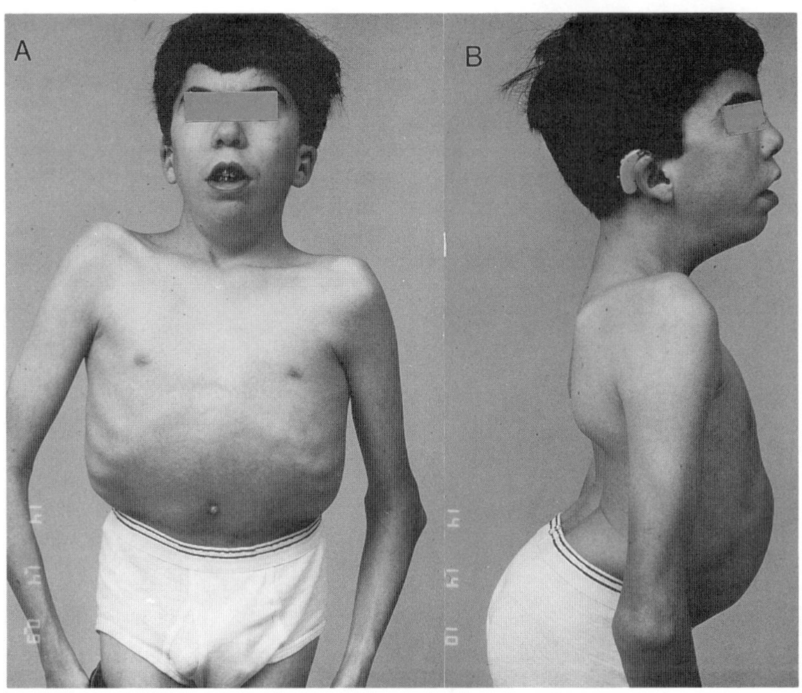


Fig. 22–26. *Cleft palate, craniofacial sclerosis, and musculoskeletal abnormalities.* (A,B) Note large calvaria, biparietal bossing, thick arched eyebrows, long upper lip with cupid's bow form, conductive hearing loss, short stature, progressive lumbar lordosis, dislocation of radial heads. (From *G Currarino* and *JM Friedman,* Pediatr Radiol **16**:441, 1986.)

a small village near Sabinas, Mexico. Autosomal recessive inheritance is likely.

#### Reference (Cleft palate, dwarfism, trichodysplasia, and unusual facies)

1. Jorgenson RJ, Bick D: A new syndrome of dwarfism, trichodysplasia, cleft palate and unusual facies. *March of Dimes Clinical Genetics Conference,* Baltimore, Maryland, 1988.

Fig. 22–27. *Cleft palate, dysmorphic facies, and digital defects.* Child has cleft palate, micrognathia, and rhizomelic dwarfism. [From *J Martsolf,* Syndrome Ident **5**(1):14, 1977.]

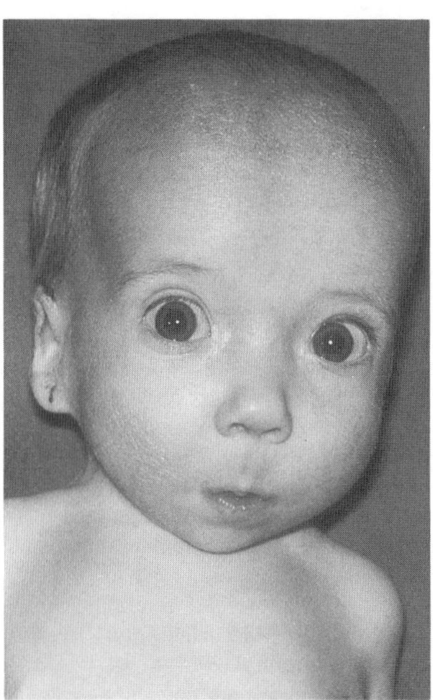

## Cleft palate, dysmorphic facies, and digital defects

Martsolf et al (1), in 1977, reported a male infant with a rhizomelic dwarfism, polydactyly, and Robin sequence. Length at birth was below the third percentile.

At 18 months, he had a prominent, square forehead. The anterior fontanel was still widely patent with apparent hypertelorism and marked micrognathia (Fig. 22–27).

The thumbs were broad with varus deflexion, and there was hypoplasia of the third and fifth middle phalanges and absence of the middle phalanx of the index fingers (Fig. 22–28). The halluces were broad with valgus deflexion and there was bilateral postaxial hexadactyly of

Fig. 22–28. *Cleft palate, dysmorphic facies, and digital defects.* Broad thumb with varus deflection. Index finger is short due to agenesis of middle phalanx. Note single palmar flexion crease. [From *J Martsolf,* Syndrome Ident **5**(1):14, 1977.]

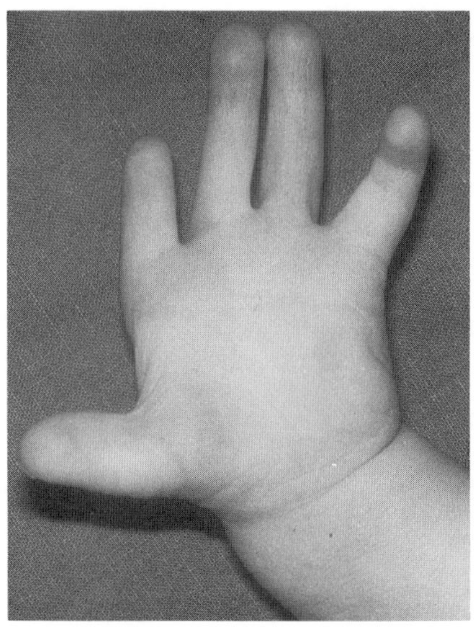

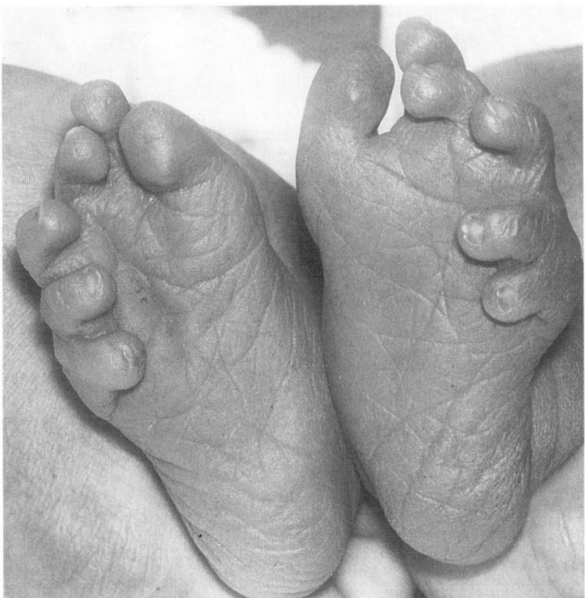

Fig. 22–29. *Cleft palate, dysmorphic facies, and digital defects.* Postaxial hexadactyly. [From *J Martsolf,* Syndrome Ident **5**(1):14, 1977.]

the feet (Fig. 22–29). Radiographic studies showed reversal of normal curvature of the spine, increased sacral angle, abnormal acetabula, short tubular bones, and tibial bowing.

We were not able to place this in any of the *oral–facial–digital syndromes.*

### Reference (Cleft palate, dysmorphic facies, and digital defects)

1. Martsolf JT et al: Case report 56: Skeletal dysplasia, Robin anomalad, and polydactyly. *Syndrome Ident* **5**(1):14–18, 1977.

## Cleft palate, ectodermal dysplasia, growth failure, and apparent pancreatic insufficiency

Donlan (1) described a brother and sister with V-shaped cleft palate, relative micrognathia, growth failure, eczema, and apparent pancreatic insufficiency (Fig. 22–30A,B). There was diarrhea with absence of stool trypsin. The skin appeared thin and there was a deficiency of teeth and enamel hypoplasia. Inheritance is probably autosomal recessive. In spite of overlap of signs and symptoms, there was no facial resemblance to *Johanson–Blizzard syndrome.*

### Reference (Cleft palate, ectodermal dysplasia, growth failure, and apparent pancreatic insufficiency)

1. Donlan MA: Growth failure, cleft palate, ectodermal dysplasia and apparent pancreatic insufficiency—a new syndrome. *Birth Defects* **13**(3B):230–231, 1977.

## Cleft palate, eye coloboma, short stature, and hypospadias (Abruzzo–Erickson syndrome)

Cleft palate in association with short stature, hypospadias, hearing loss, eye coloboma, and radial synostosis was described in three males by Abruzzo and Erickson (1). Inheritance is X-linked (2).

The midface was somewhat flat. The pinnae were soft and outstanding. Hearing loss was sensorineural in some and mixed in others. Hypospadias ranged from mild to moderately severe with chordee. The coloboma involved the iris, choroid, and retina. There was wide spacing between the second and third digits of the hands. Radial synostosis prevented external rotation of the forearms. Stature was below the third percentile. Intelligence was normal.

### References [(Cleft palate, eye coloboma, short stature, and hypospadias (Abruzzo–Erickson syndrome)]

1. Abruzzo MA, Erickson RP: A new syndrome of cleft palate associated with coloboma, hypospadias, deafness, short stature and radial synostosis. *J Med Genet* **14**:76–80, 1977.

2. Abruzzo MA, Erickson RP: Re-evaluation of new X-linked syndrome for evidence of CHARGE syndrome or association. *Am J Med Genet* **34**:397–400, 1989.

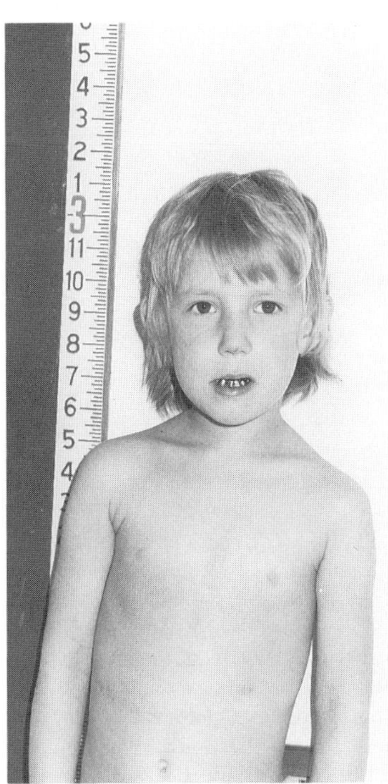

A

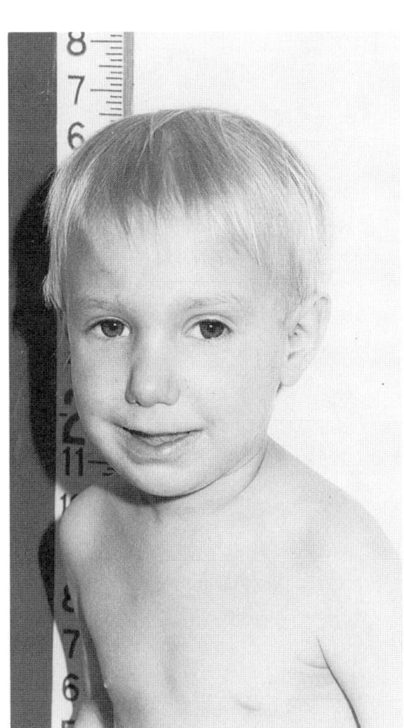

B

Fig. 22–30. *Cleft palate, ectodermal dysplasia, growth failure, and apparent pancreatic insufficiency.* (A,B) Sibs affected with cleft palate, short stature, mild micrognathia, diarrhea with absence of stool trypsin. Teeth were deficient in number and enamel. [From *MA Donlan,* Birth Defects **13**(3B):230, 1977.]

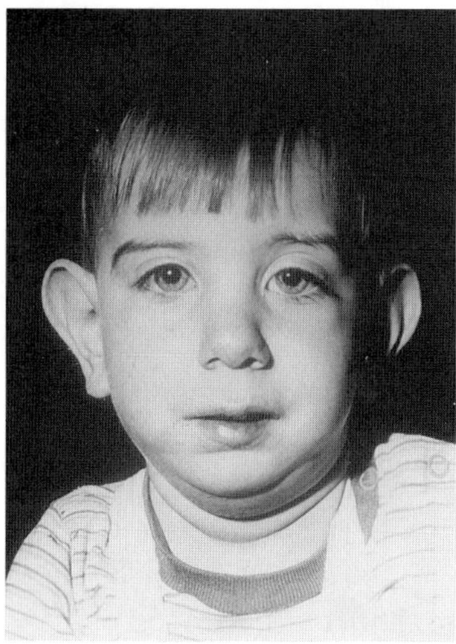

Fig. 22–31. *Cleft palate, microcephaly, large ears, and short stature.* Child with microcephaly, outstanding pinnae, and cleft palate. Mother, mother's sister, and maternal grandfather had similar findings but also had hypoplastic second and fourth digits. (From *B Say et al,* Humangenetik **26**:267, 1975.)

## Cleft palate, hypotonia, and mental retardation

Davis and Lafer (1) reported a brother and sister with cleft palate, mental retardation, hypotonia, failure to thrive, frontal bossing, and epicanthal folds. The male sib had hypospadias, cryptorchidism, and inguinal hernias.

### Reference (Cleft palate, hypotonia, and mental retardation)

1. Davis JG, Lafer C: A possible new mental retardation syndrome. *Birth Defects* **12**(5):235–238, 1976.

## Cleft palate, macular coloboma, short stature, and skeletal abnormalities (Phillips–Griffiths syndrome)

Phillips and Lloyd Griffiths (1) reported male and female sibs with bilateral macular colobomas, cleft palate, hallux valgus, somatic retardation, and flexion deformities of distal interphalangeal joints of the fifth fingers. The male sib also had mental retardation, dislocation of patellae, coxa valga, and genua valga.

### Reference [Cleft palate, macular coloboma, short stature, and skeletal abnormalities (Phillips–Griffiths syndromes)]

1. Phillips CI, Lloyd Griffiths D: Macular coloboma and skeletal abnormalities. *Br J Ophthalmol* **53**:346–349, 1969.

## Cleft palate, microcephaly, large ears, and short stature

Say et al (1) reported a dominantly inherited syndrome of cleft palate associated with microcephaly, large ears, and short stature (Fig. 22–31).

The mother and her sister had distally tapering fingers with hypoplastic distal phalanges involving the second and fourth digits bilaterally, ulnar deviation of the middle fingers, low-set thumbs, and bilateral acromial dimples.

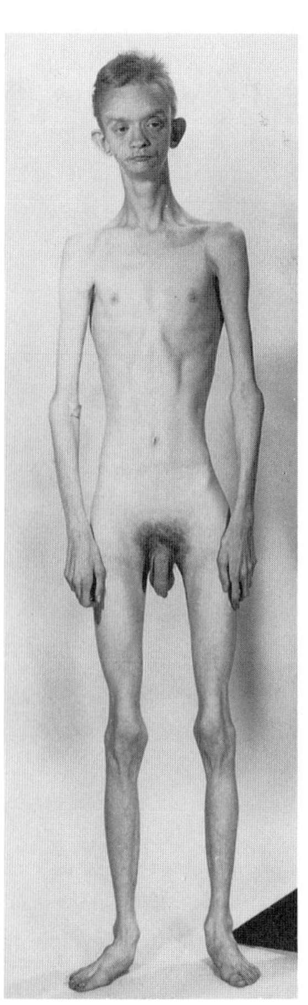

A

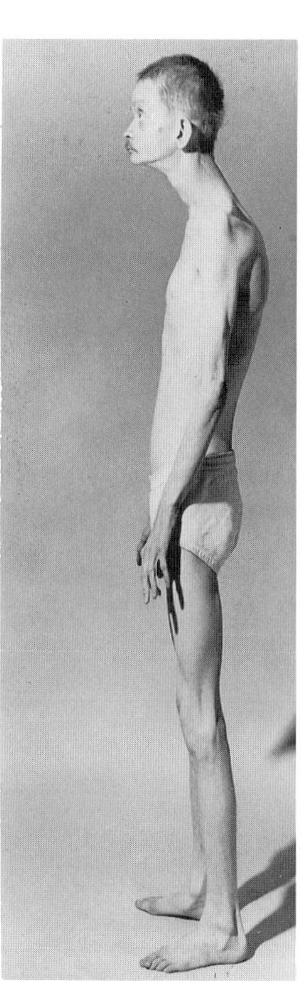

B

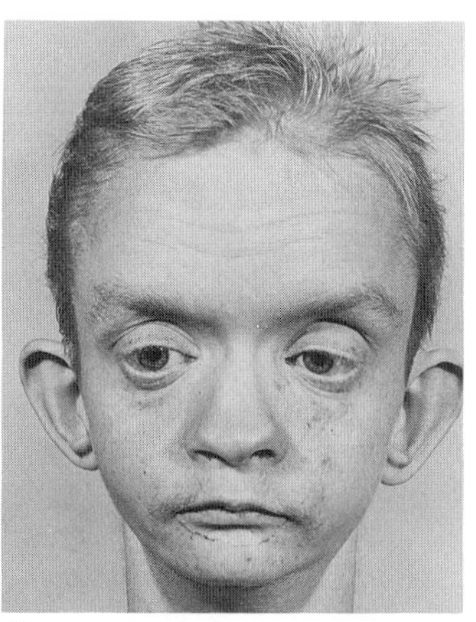

C

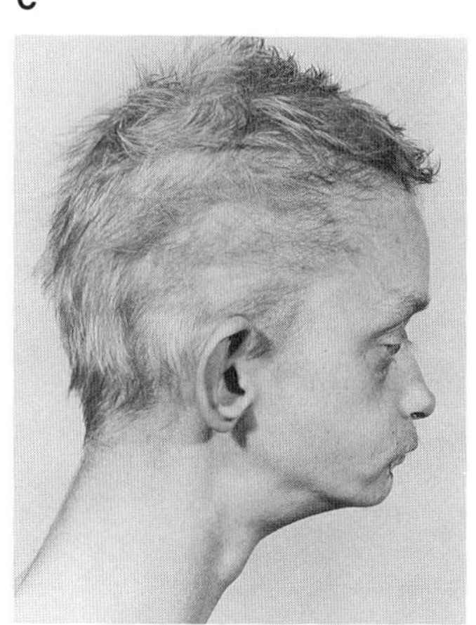

D

Fig. 22–32. *Cleft palate, microcephaly, mental retardation, and musculoskeletal mass deficiency.* (A,B) Very thin limbs with normal leg and trunk proportions. Long neck. (C,D) Microcephaly, cupped pinnae, broad nose, midfacial hypoplasia, small mouth with downturned corners, micrognathia, and long thin neck. [From *DD Weaver* and *CPS Williams,* Birth Defects **13**(3B):69, 1977.]

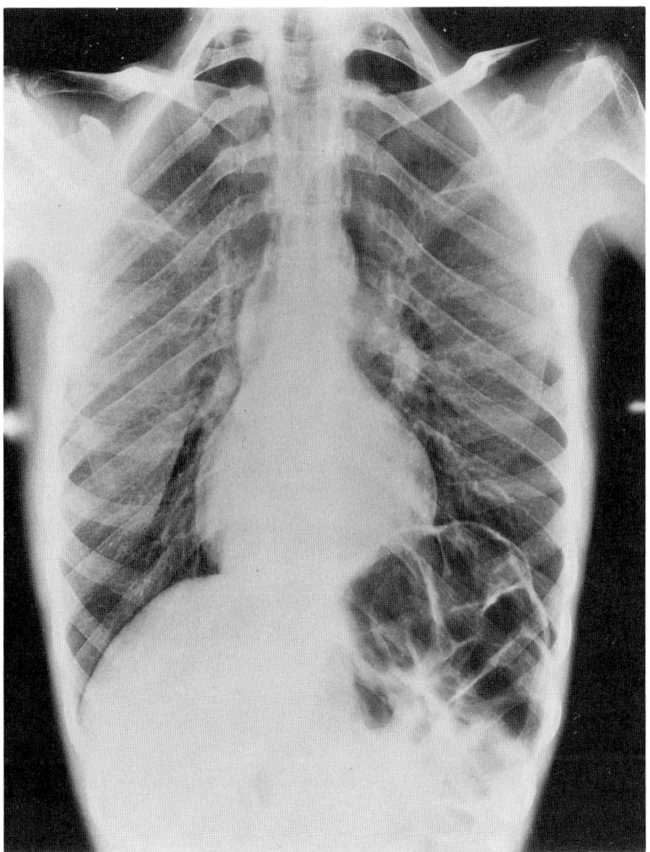

Fig. 22–33. *Cleft palate, microcephaly, mental retardation, and musculoskeletal mass deficiency.* Markedly down-sloping ribs with diminished muscles of chest and shoulders. Note elevated left hemidiaphragm. [From *DD Weaver* and *CPS Williams*, Birth Defects **13**(3B):69, 1977.]

#### Reference (Cleft palate, microcephaly, large ears, and short stature)

1. Say B et al: A new dominantly inherited syndrome of cleft palate. *Humangenetik* **26**:267–269, 1975.

## Cleft palate, microcephaly, mental retardation, and musculoskeletal mass deficiency (Weaver–Williams syndrome)

Weaver and Williams (1) described male and female sibs with mental retardation, microcephaly, unusual facies, and cleft palate. Autosomal recessive inheritance is possible.

There was marked deficiency in muscle mass (Fig. 22–32A,B). Weight was −4 SD. Height was less significantly reduced. Head circumference was −5 SD, but because the facial bones were small, it did not appear severe. There was midface hypoplasia and the mouth was small and downturned. The neck was long and was held in an extended position (Fig. 22–32C,D). The incisors had a somewhat conical crown form.

Both sibs manifested moderate to severe mental retardation with incomprehensible speech.

The limbs were extremely thin. The fingers exhibited clinodactyly.

Radiographically, there was generalized bone hypoplasia with increased tubulation, tall narrow vertebral bones, downward sloping ribs, short ulnas and fibulas, and delayed bone age (Fig. 22–33).

#### Reference [Cleft palate, microcephaly, mental retardation, and musculoskeletal mass deficiency (Weaver–Williams syndrome)]

1. Weaver DD, Williams CPS: A syndrome of microcephaly, mental retardation, unusual facies, cleft palate and weight deficiency. *Birth Defects* **13**(3B):69–84, 1977.

## Cleft palate, micrognathia, talipes equinovarus, atrial septal defect, and persistence of left superior vena cava

Gorlin et al (1) presented a kindred exhibiting cleft palate, micrognathia, talipes equinovarus, atrial septal defect, persistence of the left superior vena cava, a cardiac conduction anomaly, and death in infancy. The occurrence of the syndrome in several male children of sisters suggests X-linked recessive inheritance.

It is possible that other examples have been reported. Smith and Stowe (3), in their survey of 39 patients with Robin sequence, noted that five died of congenital heart anomaly and that three had clubfeet. No mention was made of the sex of the affected. Sachtleben (2) described a male infant with ASD of the primum type, cleft palate, inguinal hernia, and talipes equinovarus. He further noted two brothers with this combination.

#### References (Cleft palate, micrognathia, talipes equinovarus, atrial septal defect, and persistence of left superior vena cava)

1. Gorlin RJ et al: Robin's syndrome: A probably X-linked subvariety exhibiting persistence of left superior vena cava and atrial septal defect. *Am J Dis Child* **119**:176–178, 1970.
2. Sachtleben P: Zur Pathogenese und Therapie des Pierre Robin–Syndroms. *Arch Kinderheilkd* **171**:55–63, 1964.
3. Smith JL, Stowe FR: The Pierre Robin syndrome: A review of 39 cases with emphasis on ocular lesions. *Pediatrics* **27**:128–133, 1961.

## Cleft palate, stapes fixation, and oligodontia

Gorlin et al (1) described two sisters, the offspring of a consanguineous marriage, who had stapes fixation, cleft palate, and oligodontia (1). Neither girl had ever had more than three or four deciduous teeth, and those had conical crown form. No permanent teeth were ever present, and alveolar ridges were absent.

There was coalition of all cuneiform bones, as well as coalition of the navicular and talus, the talus and calcaneus, and the first cuneiform with the first metatarsal. The talus was malformed, having a superior and medial hump. The carpal navicular was sharply wedge-shaped, lacking its normal convexity (Fig. 22–34A–C).

#### Reference (Cleft palate, stapes fixation, and oligodontia)

1. Gorlin RJ et al: Stapes fixation and oligodontia—a new autosomal recessively inherited syndrome. *Birth Defects* **7**(7):87–88, 1971.

## Cleft palate, unusual facies, conduction hearing loss, and male pseudohermaphroditism

Ieshima et al (1) described male and female sibs with cleft of the soft palate, intrauterine growth retardation and failure to thrive, severe mental retardation, microcephaly, hypotonia, repeated respiratory infections, conduction deafness, and pulmonary hypertension with PDA.

The facies was characterized by asymmetry, arched eyebrows, hypertelorism, broad nasal bridge, small nose, anteverted nostrils, microtia, and micrognathia (Fig. 22–35A,B).

The male had hypospadias, micropenis, cryptorchidism and shawl scrotum (Fig. 22–36).

Inheritance is probably autosomal recessive.

#### Reference (Cleft palate, unusual facies, conduction hearing loss, and male pseudohermaphroditism)

1. Ieshima A et al: Peculiar facies, deafness, cleft palate, male pseudohermaphroditism, and growth and psychomotor retardation; a new autosomal recessive syndrome? *Clin Genet* **30**:136–141, 1986.

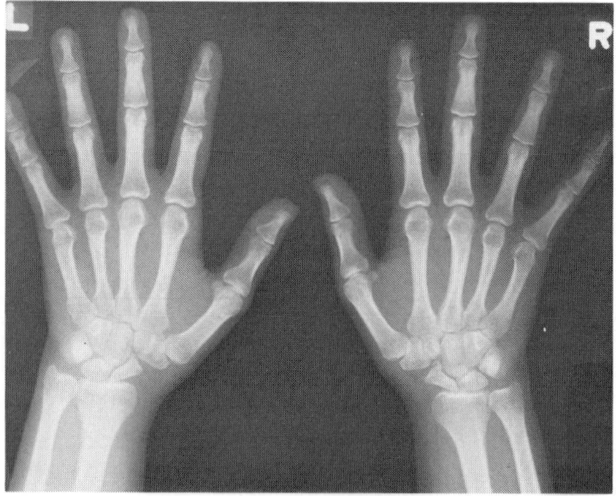


A

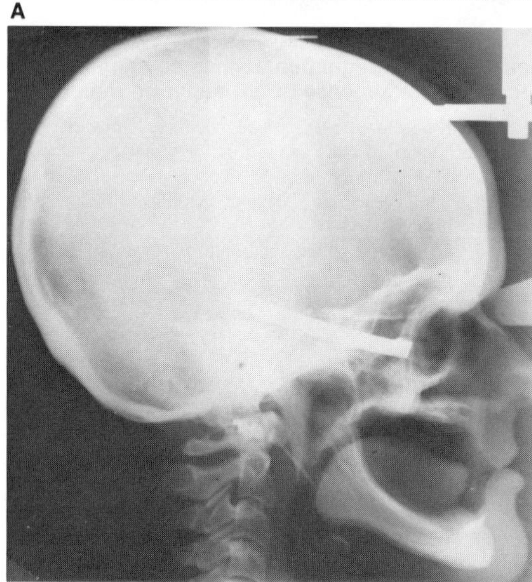

B

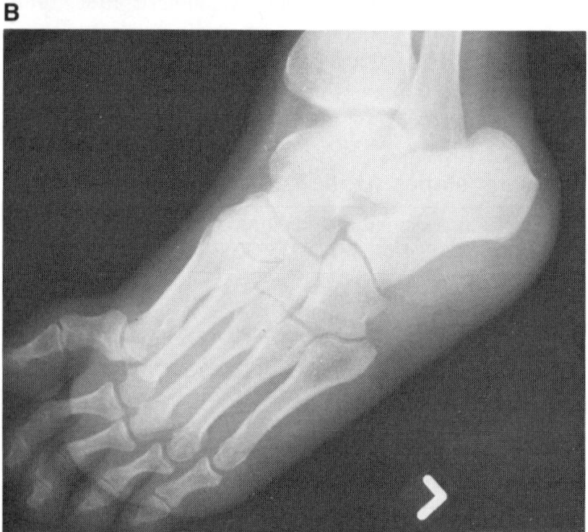

C

Fig. 22–34. *Cleft palate, stapes fixation, and oligodontia.* (A) Hypoplasia of navicular bones. (B) Absence of alveolar processes due to severe oligodontia. (C) Talocalcaneal fusion with unusual tibiotalar articulation and talar hump. [From *RJ Gorlin et al,* Birth Defects **7**(7):87, 1971.]

## Cleft palate, unusual facies, mental retardation, and limb abnormalities

Palant et al (1) described sisters with mild microcephaly, short stature, mental retardation, almond-shaped deep-set eyes, bulbous nasal tip, cleft palate, clinodactyly of toes, and firm nonbony prominence of the anteromedial aspect of the wrists. The syndrome probably has autosomal recessive inheritance (Fig. 22–37A–C).

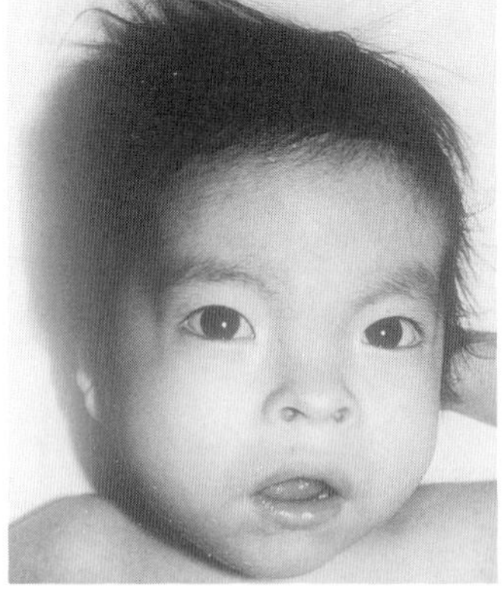

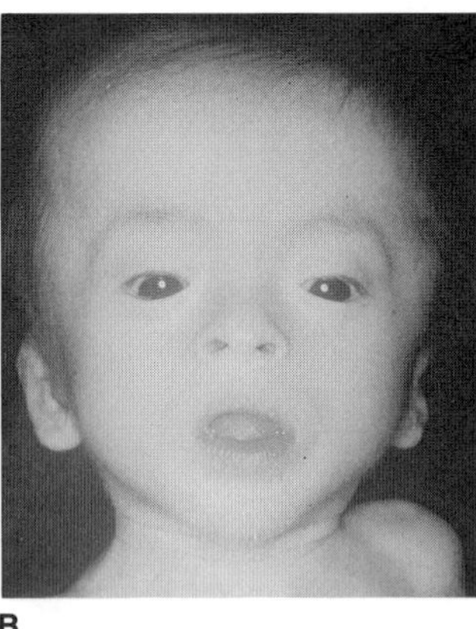

A B

Fig. 22–35. *Cleft palate, unusual facies, conduction hearing loss, and male pseudohermaphroditism.* (A,B). One of two sibs having hypertelorism, arched eyebrows, broad flat nasal bridge, profound hearing loss, severe psychomotor retardation, cleft palate, and short neck. (From *A Ieshima et al,* Clin Genet **30**:136, 1986.)

### Reference (Cleft palate, unusual facies, mental retardation, and limb abnormalities)

1. Palant DI et al: Unusual facies, cleft palate, mental retardation and limb abnormalities in siblings—a new syndrome. *J Pediatr* **78**:686–689, 1971.

## Cleft palate, ventricular septal defect, persistent truncus arteriosus, and intrauterine death

Lowry and Miller (1) noted sibs with cleft palate, persistent truncus arteriosus, and abnormal right pulmonary artery. The left pulmonary artery came off the truncus. Presumably the syndrome has autosomal recessive inheritance.

### Reference (Cleft palate, ventricular septal defect, persistent truncus arteriosus, and intrauterine death)

1. Lowry RB, Miller JR: Cleft palate and congenital heart disease. *Lancet* **1**:1302–1303, 1971.

Fig. 22–36. *Cleft palate, unusual facies, conduction hearing loss, and male pseudohermaphroditism.* Male pseudohermaphroditism (hypospadias, micropenis, cryptorchidism, small scrotum). (Courtesy of *A Ieshima,* Yonago, Japan.)

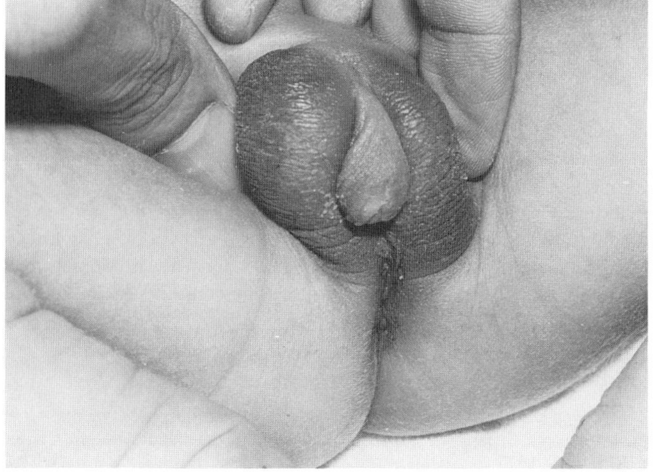

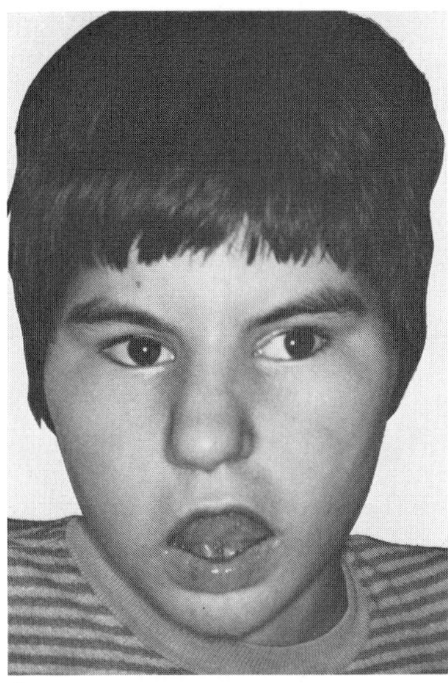

A

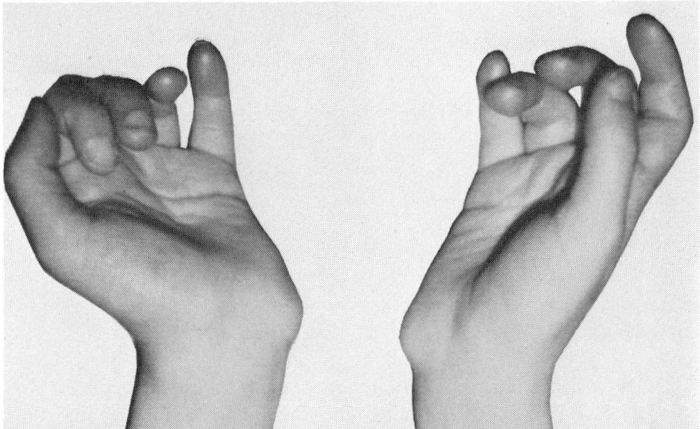

B

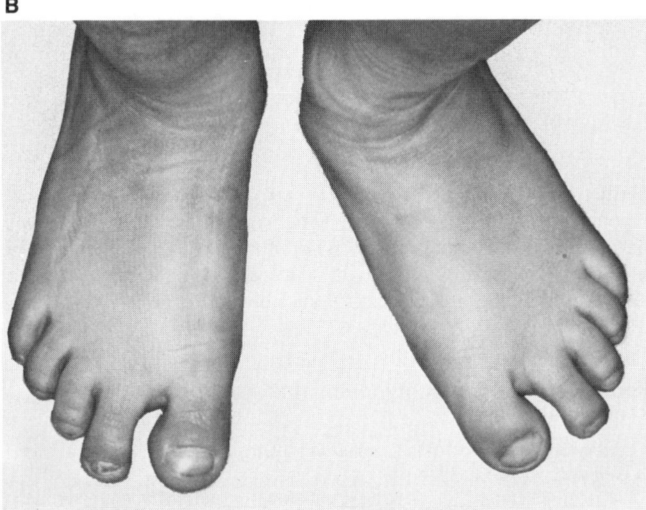

C

Fig. 22–37. *Cleft palate, unusual facies, mental retardation, and limb abnormalities.* (A) One of the two female sibs with cleft palate, short stature, bulbous nasal tip. (B) Firm nonbony prominences of anteromedial aspects of wrists. (C) Short halluces with space between hallux and second toe. (From *DI Pallant et al,* J Pediatr **78**:686, 1971.)

## Cleft uvula, deafness, nephrosis, congenital urinary tract, and digital anomalies

The syndrome was found by Braun and Bayer (2) in five of seven boys and in none of five girls in a sibship. Inheritance appears to be X-linked or, less likely, autosomal recessive.

The hearing defect proved to be a conduction deficit in all but one child who had mixed hearing loss. The digital anomaly consisted of shortening and broadening of the distal portion of the thumbs and great toes. Radiographic examination revealed that the distal phalanges were rudimentary with bifid ends. The urinary tract anomalies present in two children were bandlike constrictions of the ureter and duplication of the renal pelvis and ureter. The nephrosis appeared in infancy.

The uvula was bifid in two of five affected males. This appears to be statistically significant, since the prevalence of bifid uvula in whites is about 1 in 70. The frequency of duplication of ureter and renal pelvis is about 1 in 100.

Four male sibs were reported by Bakarat et al (1) with nephrosis, sensorineural hearing loss, and hypoparathyroidism. However, this appears to be a different disorder.

### References (Cleft uvula, deafness, nephrosis, congenital urinary tract, and digital anomalies)

1. Bakarat AY et al: Familial nephrosis, nerve deafness, and hypoparathyroidism. *J Pediatr* **91**:61–64, 1977.

2. Braun FC Jr, Bayer JF: Familial nephrosis associated with deafness and congenital urinary tract anomalies in siblings. *J Pediatr* **60**:33–41, 1962.

## Cleft uvula, preaxial and postaxial polysyndactyly, and somatic and motor retardation

Engelhard and Yatziv (1) reported a male with short stature, microcephaly, psychomotor retardation, eyelid ptosis, stenosis of lacrimal points, asymmetric manual preaxial and postaxial polysyndactyly, brachyphalangy, and kyphosis (Fig. 22–38A–C). The uvula was bifid. The mandibular incisors were fused.

The parents were cousins. The authors considered acropectorovertebral (F) syndrome (2) in differential diagnosis.

### References (Cleft uvula, preaxial and postaxial polysyndactyly, and somatic and motor retardation)

1. Engelhard D, Yatziv S: Pre- and postaxial polysyndactyly, microcephaly and ptosis. *Eur J Pediatr* **130**:47–51, 1979.

2. Grosse FR et al: The F-form of acro-pectorovertebral dysplasia: The F syndrome. *Birth Defects* **5**(3):48–63, 1969.

## Congenital myopathy, malignant hyperthermia, and cleft palate

Stewart et al (8) reported six children of similar ethnic origin (Lumbee Indians of south central North Carolina) with congenital myopathy, malignant hyperthermia, skeletal anomalies, and cleft palate (Fig. 22–39A,B). Affected sibs suggest autosomal recessive inheritance.

The facies was characterized by myotonia and eyelid ptosis. The mouth appeared small.

Musculoskeletal anomalies consisted of short stature, generalized weakness, myotonia, congenital kyphoscoliosis, and talipes. Tendon reflexes were hypoactive to absent. Plantar responses were flexor. The hyperthermia makes anesthesia hazardous.

This disorder has marked overlap with King syndrome (short stature, motor delay, slowly progressive myopathy, downslanting palpebral fissures, posteriorly rotated pinnae, mandibular hypoplasia, scoliosis, pectus carinatum, cryptorchidism, and susceptibility to malignant

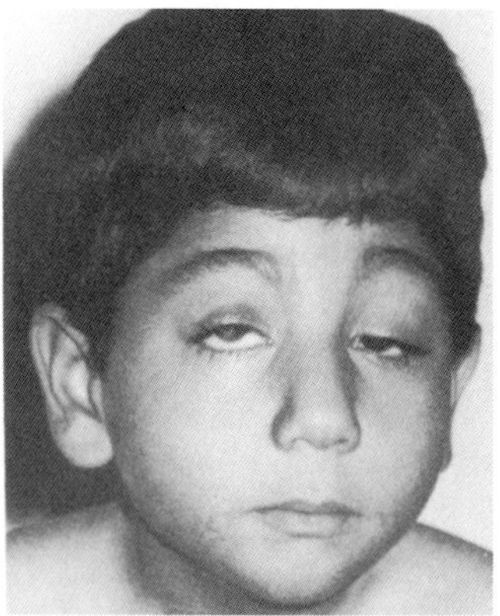

A

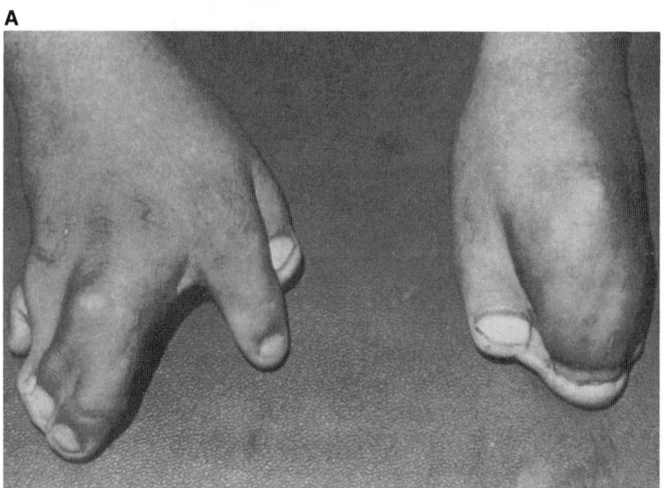

B

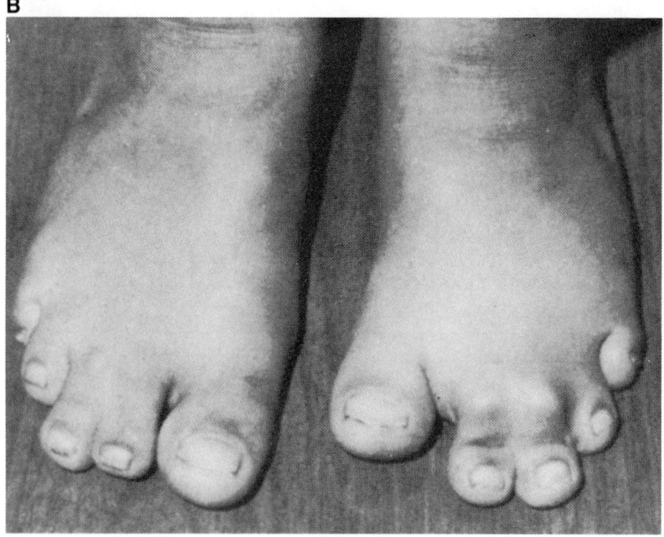

C

Fig. 22–38. *Cleft uvula, preaxial and postaxial polysyndactyly, and somatic and motor retardation.* (A) Five-year-old male with bilateral ptosis, congenital bilateral dacryostenosis, and bilateral leukoma. (B) Asymmetric cutaneous syndactyly of fingers. (C) Bilateral syndactyly of toes 2–3 with widened space between hallux and rest of toes. (From *D Engelhard* and *S Yatziv, Eur J Pediatr* **130**:47, 1979.)

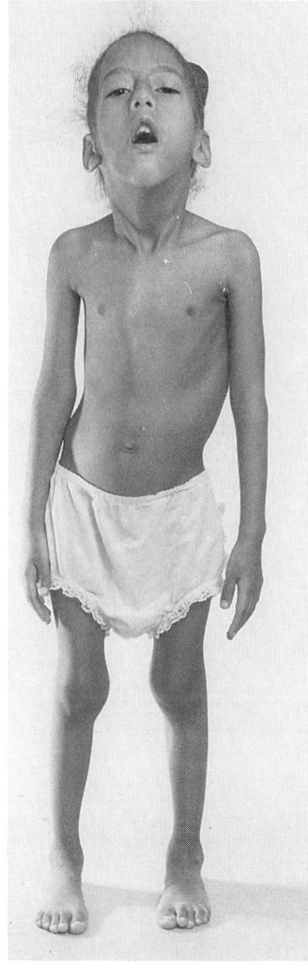

A

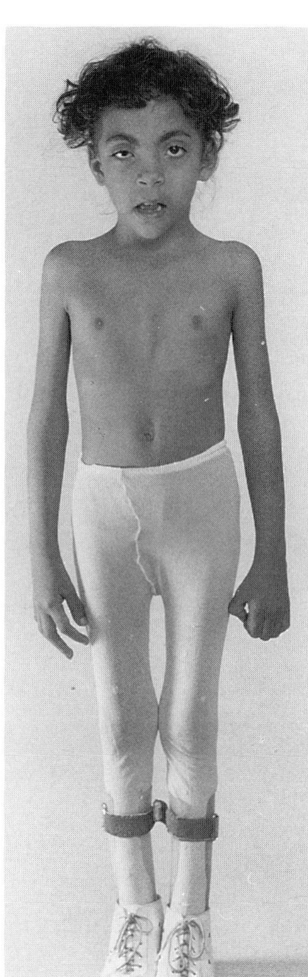

B

Fig. 22–39. *Congenital myopathy, malignant hyperthermia, and cleft palate.* (A,B) Two children of Lumbee Indian extraction exhibiting congenital myopathy, malignant hyperthermia, scoliosis, and cleft palate. Note eyelid ptosis and facial resemblance of child shown in B to that of Noonan syndrome.

hyperthermia), a sporadic disorder (2–7). Cleft palate is not a frequent component of King syndrome (3). King syndrome, in turn, has been alleged to be a subset of *Noonan syndrome* (1), but we remain skeptical of creating separate nosologic status although we recognize the hazard that hyperthermia creates for anesthesia.

### References (Congenital myopathy, malignant hyperthermia, and cleft palate)


1. Allanson JE et al: Noonan syndrome: The changing phenotype. *Am J Med Genet* **21**:507–514, 1985.
2. Heiman-Paterson T et al: King–Denborough syndrome: Contracture testing and literature review. *Pediatr Neurol* **2**:175–177, 1986.
3. Kaplan A et al: Malignant hyperthermia associated with myopathy and normal muscle enzymes. *J Pediatr* **91**:431–434, 1977.
4. King JO, Denborough MA: Anesthetic-induced malignant hyperthermia in children. *J Pediatr* **83**:37–40, 1973.
5. McPherson EW, Taylor CA: The King syndrome: Malignant hyperthermia, myopathy, and multiple anomalies. *Am J Med Genet* **8**:159–165, 1981.
6. Saul RA et al: A female with the King syndrome in a family with elevated CPK levels. *Proc Greenwood Genet Ctr* **3**:7–10, 1984.
7. Steenson AJ, Torkelson RD: King syndrome with malignant hyperthermia. *Am J Dis Child* **141**:271–273, 1987.
8. Stewart CR et al: Congenital myopathy with malignant hyperthermia and cleft palate. King syndrome? *Neurology* **37**(Suppl 1):201, 1987.

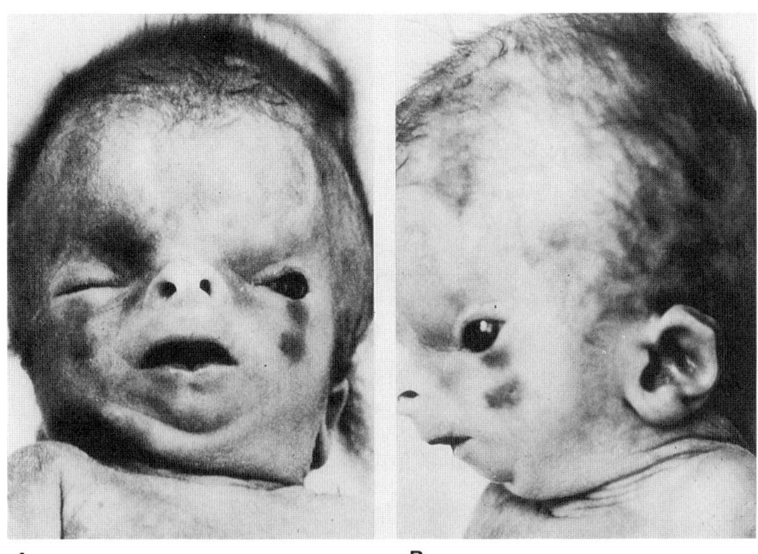


Fig. 22–40. *Crane–Heise syndrome.* (A,B) Postmortem view of one of three sibs showing large head, depressed nasal bridge with anteverted nares, apparent hypertelorism, hypoplastic helices, and short neck. (From *JP Crane* and *RL Heise,* Pediatrics **68**:235, 1981.)

## Crane–Heise syndrome (cleft lip–palate, agenesis of clavicles and cervical vertebrae, and talipes equinovarus)

Crane and Heise (1) described three sibs having disproportionately large head, small face, hypertelorism, low-set dysmorphic pinnae, depressed nasal bridge with upturned nares, micrognathia, cleft lip and/or cleft palate, and short neck (Fig. 22–40A,B).

In addition, they had intrauterine growth retardation, partial soft tissue syndactyly of the second to fourth fingers and toes, talipes equinovarus, short penis, and cryptorchidism. Radiographic studies revealed poorly mineralized calvaria, apparently absent cervical vertebrae and clavicles, small scapulae, and dislocated radial heads (Fig. 22–41A,B).

Agenesis of the corpus callosum was noted in one sib and single palmar creases in two sibs. This apparently lethal syndrome probably has autosomal recessive inheritance.

### Reference [Crane–Heise syndrome (cleft lip–palate, agenesis of clavicles and cervical vertebrae, and talipes equinovarus)]

1. Crane JP, Heise RL: New syndrome in three affected siblings. *Pediatrics* **68**:235–237, 1981.

## Genito-palato-cardiac syndrome (Gardner–Silengo–Wachtel syndrome)

Greenberg et al (4) coined the term genito-palato-cardiac syndrome to describe a disorder of cleft palate, male pseudohermaphroditism (46,XY gonadal dygenesis), and conotruncal anomalies. They cited a number of earlier cases (1–3,5–8) as examples of the disorder. Inheritance is probably autosomal recessive (1).

There has been debate as to whether this is a disorder separate from *Smith–Lemli–Opitz syndrome, Type II.* Like Opitz and Lowry (5), we reserve judgment.

Most of the infants succumbed during the neonatal period.

Hydrocephalus was found in two cases (4,8). A few manifest downslanting palpebral fissures. The pinnae were low set in nearly all cases. Cleft palate (1–4,7,8) or cleft lip (4) and micrognathia were virtually constant findings.

Musculoskeletal defects included flexion deformities of thumb and great toes (4), camptodactyly (6), dysplastic ribs (6), postaxial polydactyly (2), prominent heels (2), and clubfeet (6,8). Congenital heart defects such as double outlet right ventricle and/or VSD, right aortic arch, transposition of great vessels, and tetralogy of Fallot were documented (1,2,4).

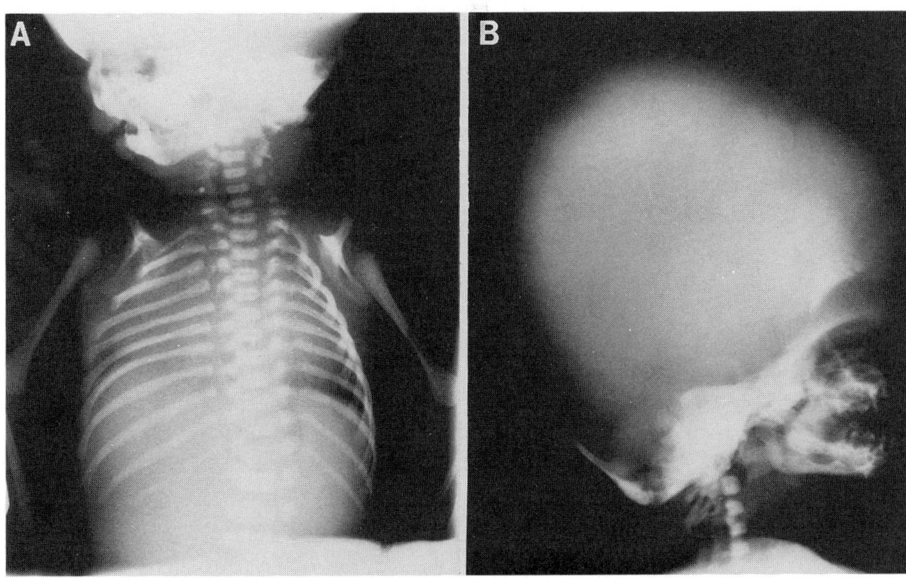


Fig. 22–41. *Crane–Heise syndrome.* (A,B) Note hypoplasia of cervical vertebrae, absent clavicles, small scapulae, markedly deficient mineralization of calvaria. (From *JP Crane* and *RL Heise,* Pediatrics **68**:235, 1981.)

Although the phenotypic sex is nearly always female, the chromosomal sex is 46,XY. Thus, there is sex reversal. Hypospadias was found in one infant with male phenotype (4). The genitalia have usually been normal female with rare exception, when gonadal dysgenesis has been found (2,7). Miscellaneous abnormalities have included cystic kidneys (4), dysgenesis of urinary bladder (4), agenesis of gallbladder (4), and intestinal malrotation (4).

#### References [Genito-palato-cardiac syndrome (Gardner–Silengo–Wachtel syndrome)]

1. Beemer FA, Ertbruggen IV: Peculiar facial appearance, hydrocephalus, double-outlet right ventricle, genital anomalies, and dense bones with lethal outcome. *Am J Med Genet* **19**:391–394, 1984.

2. Bernstein R et al: Female phenotype and multiple abnormalities in sibs with a Y chromosome and partial X chromosome duplication: H-Y antigen and Xg blood group findings. *J Med Genet* **17**:291–300, 1980.

3. Gardner LI et al: 46,XY female: Anti-androgenic effects of oral contraceptive? *Lancet* **2**:667–668, 1970.

4. Greenberg F et al: The Gardner–Silengo–Wachtel or genito-palato-cardiac syndrome: Male pseudohermaphroditism with micrognathia, cleft palate, and conotruncal cardiac defect. *Am J Med Genet* **26**:59–64, 1987.

5. Opitz JM, Lowry RB: Lincoln vs. Douglas again; comments on the papers by Curry et al, Greenberg et al, and Belmont et al. *Am J Med Genet* **26**:69–72, 1987.

6. Robinow M et al: Case report 118: Costal dysgenesis in male pseudohermaphroditism: A new syndrome? *J Clin Dysmorphol* **2**(2):23–26, 1984.

7. Silengo M et al: A 46,XY infant with uterus, dysgenetic gonads and multiple anomalies. *Hum Genet* **25**:65–68, 1974.

8. Wolman SR et al: Aberrant testicular differentiation in 46,XY gonadal dysgenesis: Morphology, endocrinology, serology. *Hum Genet* **55**:321–325, 1980.

## Gordon syndrome (cleft palate, camptodactyly, and clubfoot)

Gordon et al (2) first recognized the syndrome of camptodactyly, talipes, and cleft palate. Several additional families have been described (3,5,7,8,10). Hall et al (4) designated the Gordon syndrome as dominant distal arthrogryposis, Type 2A. The patient reported by Bijlsma (1) also had spasticity and mental retardation and cannot be accepted as a valid example of this disorder. The patients reported by McCormack et al (6) and Sack (9) probably have Gordon syndrome.

There is autosomal dominant inheritance with variable expressivity and incomplete penetrance, possibly being more reduced in females.

The camptodactyly affects only the PIP joints, sparing the thumbs. Cleft palate occurs in approximately 20%. Other patients have exhibited submucous cleft palate or bifid uvula.

Additional components of the syndrome appear to be short stature, hip dislocation, various vertebral abnormalities (lumbar lordosis, kyphoscoliosis), and short neck with mild pterygia. Some patients have ptosis of eyelids, epicanthal folds, and nevus flammeus of the face (4,7,8,10). Cryptorchidism may be part of the syndrome (2,8).

#### References [Gordon syndrome (cleft palate, camptodactyly, and clubfoot)]

1. Bijlsma JB: Cleft palate, flexion contracture, aqueductal stenosis and mental retardation. *Syndrome Ident* **6**(1):4–5, 1980.

2. Gordon H et al: Camptodactyly, cleft palate and club foot: Syndrome showing the autosomal dominant pattern of inheritance. *J Med Genet* **6**:266–274, 1969.

3. Halal F, Fraser FC: Camptodactyly, cleft palate and club foot (the Gordon syndrome): A report of a large pedigree. *J Med Genet* **16**:149–150, 1979.

4. Hall JG et al: The distal arthrogryposes: Delineation of new entities—review and nosologic discussion. *Am J Med Genet* **11**:185–239, 1982.

5. Higgins JV et al: A second family with cleft palate, club foot and camptodactyly. *Am J Hum Genet* **24**:58A, 1972.

6. McCormack MK et al: Autosomal-dominant inheritance of distal arthrogryposis. *Am J Med Genet* **6**:163–169, 1980.

7. Moldenhauer E: Zur Klinik des Nielson-Syndroms. *Derm Wochenschr* **150**:599–601, 1964.

8. Robinow M, Johnson GF: The Gordon syndrome: Autosomal dominant cleft palate, camptodactyly and club feet. *Am J Med Genet* **9**:139–146, 1981.

9. Sack GF Jr: A dominantly inherited form of arthrogryposis multiplex congenita with unusual dermatoglyphics. *Clin Genet* **14**:317–323, 1978.

10. Say B et al: The Gordon syndrome. *J Med Genet* **17**:405, 1980.

## Kabuki make-up syndrome

In 1981, Niikawa et al (4) and Kuroki et al (3) independently reported a syndrome characterized by mild to moderate mental retardation, postnatal progressive growth retardation, and strikingly unusual facies reminiscent of the make-up used in Kabuki theatre. All of more than 60 cases reported to date have been isolated examples (6). Of 62 cases, 58 were Japanese. The disorder has no sex predilection (1–10). The condition is more difficult to diagnose in non-Japanese (1b,3a,8a).

The palpebral fissures are long with eversion of the lateral third of the lower eyelids. The eyebrows are arched and tend to be diminished laterally. The lashes are heavy and long. The sclerae are blue in 25%. Strabismus and epicanthal folds are present in 50–60%. The ears are outstanding with large prominent lobes in 85%. The anthelix tends to be hypoplastic. About 40% have a pretragal pit. The nose is broad with a depressed tip (80%) and/or short septum (90%). The teeth are widely spaced in 65%. Cleft lip and/or cleft palate or submucous cleft palate have been noted in 40%. The occipital hairline is often low. One-third have retrognathia (Fig. 22–42A,B).

Other frequent anomalies include short stature (80%), short fifth fingers (90%), scoliosis (50%—with about one-third with sagittally cleft vertebrae), dislocation of the hip (35%). Various rib anomalies (20%) and spina bifida occulta (15%) have been noted (6). Otitis media is extremely frequent during childhood (60%) and there is hearing loss in 25%. Early breast development is found in 25%. Congenital heart anomalies have been reported in about 30%: ASD, VSD, single ventricle with common atrium, PDA, and coarctation of the aorta (3,7).

Abnormal dermatoglyphic patterns include an excess of ulnar loops and hypothenar loops (70%). There are absence of the *c* and/or *d* digital triradius (75%) and persistence of fetal fingertip pads (75%) (6).

Halal et al (1a) have reported an autosomal dominantly inherited syndrome that has somewhat similar facies.

#### References (Kabuki make-up syndrome)

1. Braun OH, Schmid E: Kabuki make-up syndrome (Niikawa–Kuroki syndrome) in Europe. *J Pediatr* **105**:849–850, 1984.

1a. Halal F et al: Autosomal dominant inheritance of the Kabuki make-up (Niikawa–Kuroki) syndrome. *Am J Med Genet* **33**:376–381, 1989.

1b. Kaiser-Kupfer MI et al: The Niikawa–Kuroki (Kabuki make-up) syndrome in an American black. *Am J Ophthalmol* **102**:667–668, 1986.

2. Koutras A, Fisher S: Niikawa–Kuroki syndrome: A new malformation

Fig. 22–42. *Kabuki make-up syndrome.* (A,B) Note long palpebral fissures, highly arched abnormal eyebrows, large pinnae, depressed nasal tip, and repaired cleft lip. (Courtesy of *Y Kuroki*, Yokahama, Japan.)

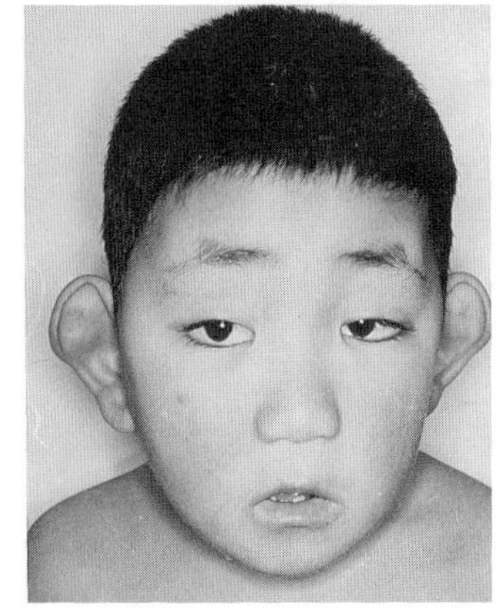
A

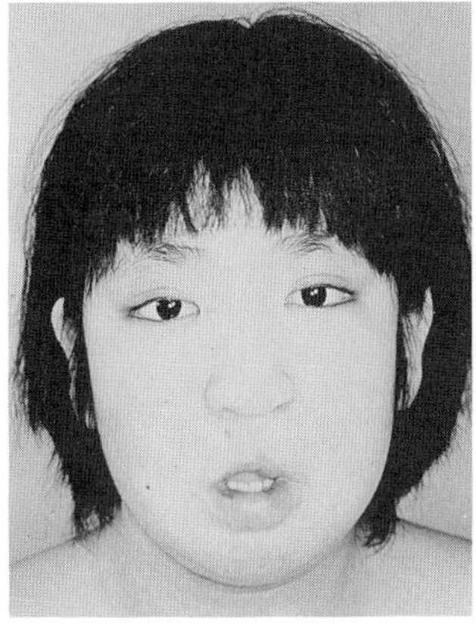
B

syndrome of post-natal growth dwarfism, mental retardation, unusual face and protruding ears. *J Pediatr* **101**:417–419, 1982.

3. Kuroki Y et al: A new malformation syndrome of long palpebral fissures, large ears, depressed nasal tip and skeletal anomalies associated with post-natal dwarfism and mental retardation. *J Pediatr* **99**:570–573, 1981.

3a. Mulvihill J, Kaiser-Kupfer MI: Niikawa–Kuroki (Kabuki Make-up) syndrome. *Am J Med Genet* **33**:425–426, 1989. (Same case as in Ref. 1a).

4. Niikawa N et al: Kabuki make-up syndrome: A syndrome of mental retardation, unusual facies, large and protruding ears and post-natal growth deficiency. *J Pediatr* **99**:565–569, 1981.

5. Niikawa N et al: The dermatoglyphic pattern of the Kabuki make-up syndrome. *Clin Genet* **21**:315–320, 1982.

6. Niikawa N et al: Kabuki make-up (Niikawa–Kuroki) syndrome: A study of 62 patients. *Am J Med Genet* **31**:565–589, 1988.

7. Ohdo S et al: Kabuki make-up syndrome (Niikawa–Kuroki syndrome) associated with congenital heart disease. *J Med Genet* **22**:126–127, 1985.

8. Pagon RA et al: Kabuki make-up syndrome in a caucasian. *Ophthalmol Paediatr Genet* **7**:97–100, 1980.

8a. Pe Benito R, Ferreti C: Kabuki make-up syndrome (Niikawa–Kuroki syndrome) in a black child. *Ann Ophthalmol* **21**:312–315, 1989.

9. Santos H et al: Kabuki make-up syndrome: A Portuguese case. *Third Manchester Birth Defects Conf,* Manchester, UK, October, 1988.

10. Turnbow JM, Pai B: Niikawa–Kuroki syndrome: Report of an American black child with a review of the literature. *Pediatrics,* in press.

## Lateral palatal synechiae and cleft palate

Unilateral or bilateral bands of mucosa may extend from the edges of a palatal cleft to the lateral margins of the tongue or oral floor (Fig. 22–43A–D). Several authors have reported this autosomal dominantly inherited syndrome with variable expressivity (1–5). RJ Gorlin has seen the disorder in two generations. Also see *syngnathia and cleft palate*.

Fig. 22–43. *Lateral palatal synechiae and cleft palate.* (A) Syngnathia; arrows point to fibrous bands joining upper and lower gingiva. (B) Autosomal dominantly inherited synechiae. (C,D) Cleft palate and persistence of buccopharyngeal membrane. (A courtesy of *H Weyers,* Cuxhaven, Germany. B courtesy of *M Mazaheri,* Lancaster, Pennsylvania. C,D from *M Hub* and *J Jirásek,* Čas Lék Čes **99**:1297, 1960.)

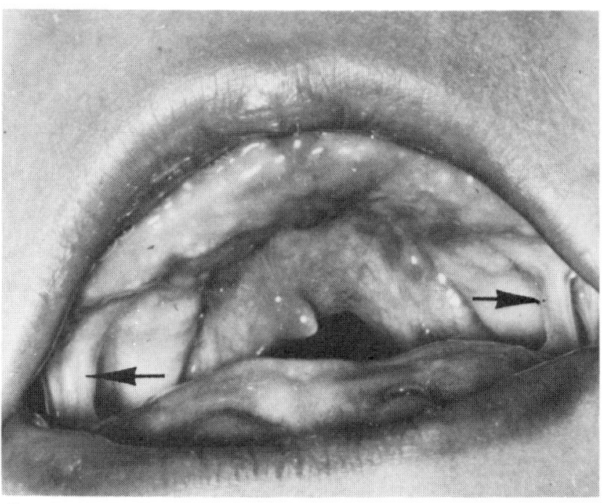

A

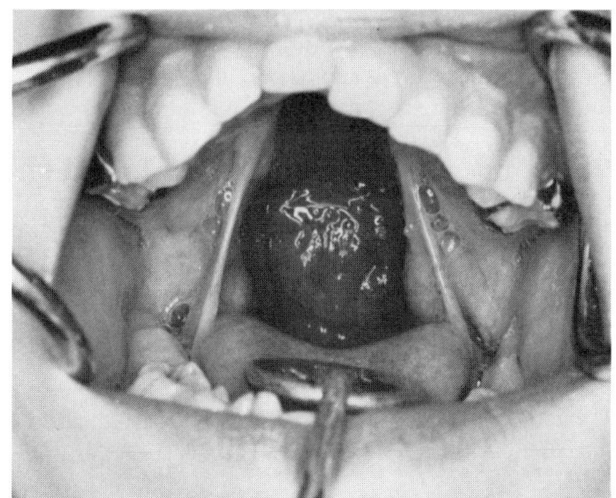

B

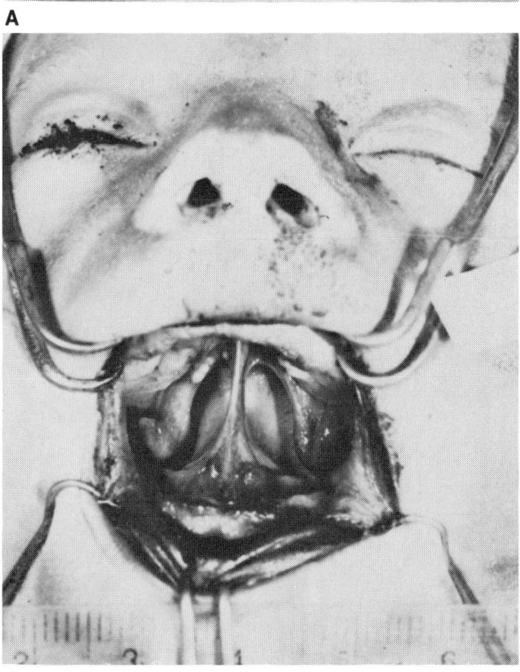

C

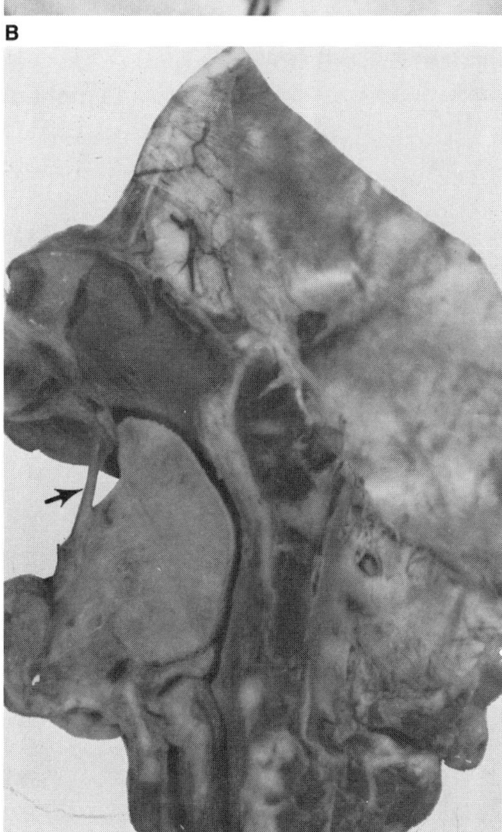

D

### References (Lateral palatal synechiae and cleft palate)

1. Berendes J: Angeborene Synechie zwischen der Mundbodenschleimhaut und den Oberkieferfortsätzen am Rande einer Gaumenspalte. *HNO* **9**:180–182, 1961.
2. Fuhrmann W et al: Autosomal dominante Vererbung von Gaumenspalte und Synechien zwischen Gaumen und Mundboden oder Zunge. *Humangenetik* **14**:196–203, 1972.
3. Gassner I et al: Familial occurrence of syngnathia congenita syndrome. *Clin Genet* **15**:241–244, 1979.
4. Hayward JR, Avery JK: A variation in cleft palate. *J Oral Surg* **15**:320–324, 1957.
5. Mazaheri M: Personal communication, 1974.


## Posterior palatal synechiae and cleft palate

The buccopharyngeal membrane is an evanescent structure that undergoes dissolution in the fourth embryonal week. It separates the primitive mouth from the oral pharynx. In a few cases, there has been membranous extension of the cleft palate to the base of the tongue (1,4–6,8,10,12) (Fig. 22–44A–C). All examples have been isolated. It has been suggested that this represents persistence of the buccopharyngeal membrane. We are skeptical, since the tongue develops behind the buccopharyngeal membrane. Cases of subglossopalatal membrane and cleft palate (18,14) appear to be more legitimate examples of persistence of the membrane. Associated anomalies included ankylosis of the DIP joints of the fifth fingers, absence of the distal phalanx of the fifth toes (11), and toenail hypoplasia, umbilical hernia, and hypospadias (14). We do not know how to classify the patient described by Wassermann (13). Seghers (12), Flannery (5), and RJ Gorlin have seen rib and vertebral changes similar to those in *cerebro-costo-mandibular syndrome*. Complete failure of development of the mandible has been associated with cleft palate and persistence of the buccopharyngeal membrane (3).

It is difficult to know at this writing whether the so-called Holzgreve–Wagner–Rehder syndrome (bilateral renal agenesis, buccopharyngeal membrane, cleft palate, congenital heart anomalies, mesoaxial hexadactyly, and bifid metacarpal) is a single entity. However, several examples have been reported (2,7,9).


### References (Posterior palatal synechiae and cleft palate)

1. Arcand P, Haikal J: Persistent buccopharyngeal membrane. *J Otolaryngol* **17**:125–127, 1988.
2. Bonnet J et al: Hypoplasie rénale, polydactylie, cardiopathie: Un nouveau syndrome. *J Génét Hum* **35**:279–289, 1987.
3. Brecht K, Johnson CM III: Complete mandibular agenesis. *Arch Otolaryngol* **111**:132–134, 1985.
4. Chandra R et al: Persistent buccopharyngeal membrane. *Plast Reconstr Surg* **54**:678–679, 1974.
5. Flannery DB: Syndrome of imperforate oropharynx with costovertebral and auricular anomalies. *Am J Med Genet* **32**:189–191, 1989.
6. Hoffman RA: Persistent pharyngeal membrane. *Arch Otolaryngol* **105**:286–287, 1979.
7. Holgreve W et al: Bilateral renal agenesis with Potter phenotype, cleft palate, anomalies of the cardiovascular system, skeletal anomalies including hexadactyly and bifid metacarpals. A new syndrome. *Am J Med Genet* **18**:177–182, 1984.
8. Hub M, Jirasek JE: Persistence of the middle portion of the buccopharyngeal syndrome. *Cas Lek Česk* **99**:1297–1300, 1960.
9. Legius E et al: Holgreve–Wagner–Rehder syndrome: Potter sequence associated with persistent buccopharyngeal membrane. A second observation: *Am J Med Genet* **31**:269–272, 1988.
10. Longacre JJ: Congenital atresia of the oropharynx. *Plast Reconstr Surg* **8**:341–348, 1951.
11. Nakajima T et al: Subglossopalatal membrane. *Plast Reconstr Surg* **63**:574–576, 1979.
12. Seghers MJ: Une malformation rare: L'imperforation oropharyngiene. *Acta Paediatr Belg* **20**:130–137, 1966.
13. Wassermann M: Ein kongentiales Diaphragma pharyngopalatinum. *Arch Laryngol Rhinol* **15**:611–612, 1904.
14. Zalzal GH et al: Subglossopalatal membrane. *Arch Otolaryngol Head Neck Surg* **112**:1101–1103, 1986.


## Lethal omphalocele and cleft palate

Czeisel (1) reported three female sibs with cleft palate and omphalocele. One sib had bicornuate uterus, another internal hydrocephalus, and the third a hypoplastic uterus. The parents were nonconsanguineous. All three children died in infancy, presumably from pneumonia.


### Reference (Lethal omphalocele and cleft palate)

1. Czeisel A: New lethal omphalocele—cleft palate syndrome? *Hum Genet* **64**:99, 1983.


## Lymphedema of pubertal onset and cleft palate

Figueroa et al (2) reported the association of lymphedema of pubertal onset (Meige type) with cleft palate. Inheritance is probably autosomal

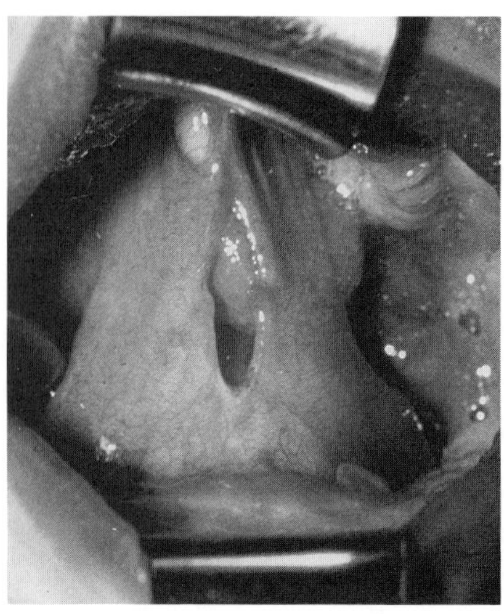

A

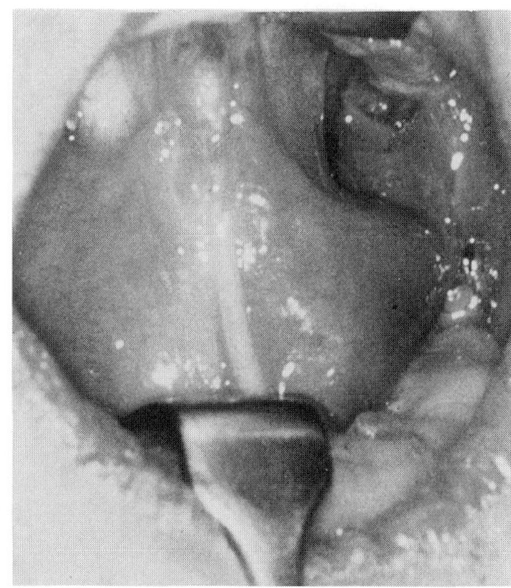

B

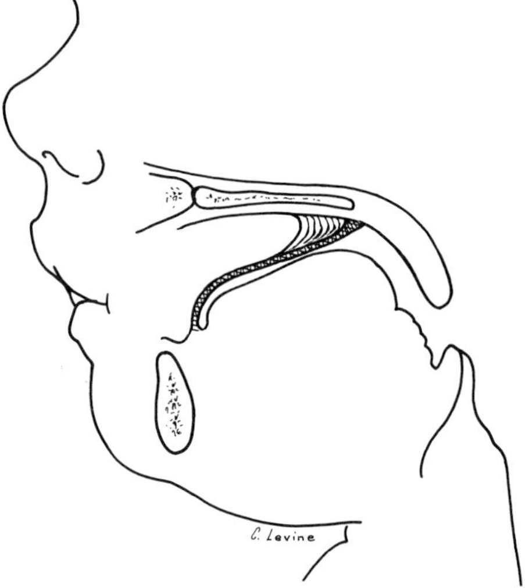


C

Fig. 22–44. *Posterior palatal synechiae and cleft palate.* (A) Oral web extending from subglossal region to palate. Surgical slit is present in midline. (B) Compare to similar case. (C) Schematic representation of subglossopalatal membrane. (A,C from *GH Zalzal et al,* Arch Otolaryngol Head Neck Surg **112**:1101, 1986. B from *T Nakajima et al,* Plast Reconstr Surg **8**:341, 1951.)

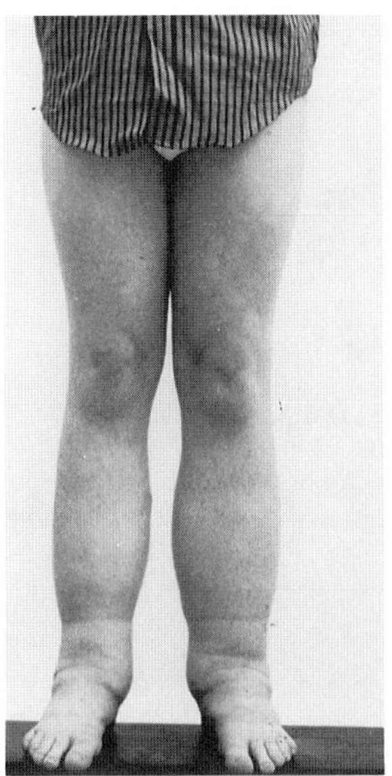
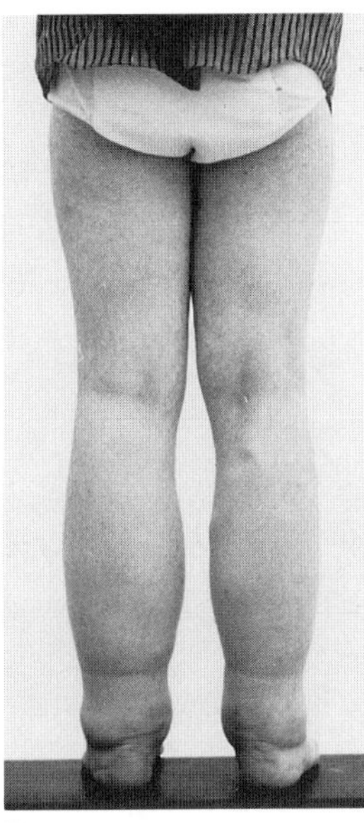

A B

Fig. 22–45. *Lymphedema of pubertal onset and cleft palate.* (A,B) Severe lymphedema affecting lower extremities. (From *AA Figueroa et al,* Cleft Palate J **20**:151, 1983.)

dominant with variable expressivity (1,3–5,8). The lymphedema is largely limited to the legs (Fig. 22–45A,B). The arms, hands, and feet may also be involved. Among the latter group, the genitalia and intestines are more commonly affected. The etiology is thought to be hypoplasia or late development of the superficial lymphatic channels that inhibits effective drainage of a body part. There is a host of syndromes involving lymphedema (distichiasis, yellow nails, ptosis, partial lower eyelid ectropion, spinal extradural cysts, vertebral anomalies, *Noonan syndrome, Turner syndrome,* etc.) (6,7). The patient reported by Jester (5) and Bartley and Jackson (1) also had distichiasis of the upper eyelids. *Hennekam syndrome,* which should be excluded, has autosomal recessive inheritance.

#### References (Lymphedema of pubertal onset and cleft palate)

1. Bartley GB, Jackson IT: Distichiasis and cleft palate. *Plast Reconstr Surg* **84**:129–132, 1989.
2. Figueroa AA et al: Meige disease (familial lymphedema praecox) and cleft palate: Report of a family and review of the literature. *Cleft Palate J* **20**:151–157, 1983.
3. Higgins HL: Two patients in the same family with hereditary edema of the legs (Milroy's disease). *J Med* **8**:199–200, 1927.
4. Jennett JH: Persistent hereditary edema of the legs—Milroy's disease. *J Mo Med Assoc* **28**:601–605, 1931.
5. Jester HG: Lymphedema-distichiasis: A rare hereditary syndrome. *Hum Genet* **39**:113–116, 1977.
6. Pap Z et al: Syndrome of lymphedema and distichiasis. *Hum Genet* **53**:309–310, 1980.
7. Schwartz JF et al: Hereditary spinal arachnoidal cysts, distichiasis and lymphedema. *Ann Neurol* **7**:340–344, 1980.
8. Strauss AE: Hereditary edema (Milroy's disease). *J Mo Med Assoc* **26**:242–247, 1929.

### Median cleft upper lip, hypertelorism, and holoprosencephaly

Bell and Van Allen (1) described an infant with marked hypertelorism and unusual median cleft lip with cleft palate. There were microcephaly, agenesis of the corpus callosum, and ear tags (Fig. 22–46).

In what may be a second example, Ohtsuka (4) reported an 11-year-old female with hypertelorism and similar facial abnormalities. The child had seizures since birth and unilateral microphthalmos. Hydranencephaly and holoprosencephaly were demonstrated. Cleft palate was also present. Arthrogryposis was probably secondary to central nervous system alterations. Bony abnormalities included lumbar scoliosis, bilateral dislocation of hips, underdeveloped pelvic bones, and slender femora.

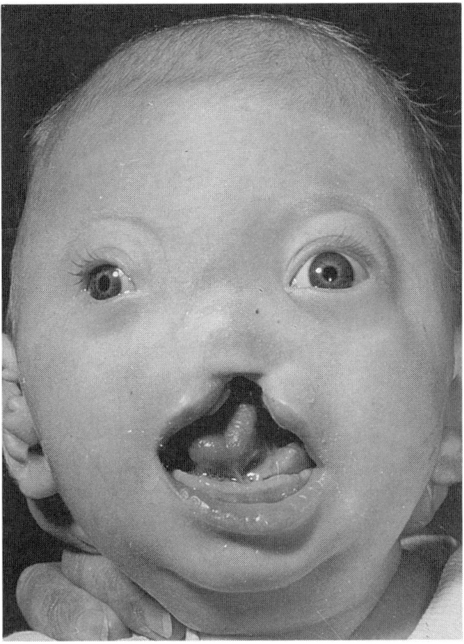

Fig. 22–46. *Median cleft upper lip, hypertelorism, and holoprosencephaly.* Median cleft upper lip and palate. (From *WE Bell* and *WM Van Allen,* Neurology **9**:694, 1959.)

Both were isolated cases but consanguinity was documented by Ohtsuka (4). Hypertelorism is also known to be associated with semilobar holoprosencephaly with *del(18q)* and with *frontonasal malformation.* See *holoprosencephaly* for more detail. The combination of holoprosencephaly and hypertelorism has been described by Hartsfield et al (2) and Moore et al (3).

#### References (Median cleft upper lip, hypertelorism, and holoprosencephaly)

1. Bell WE, Van Allen WM: Agenesis of the corpus callosum with associated facial anomalies. *Neurology* **9**:694–698, 1959.
2. Hartsfield JK et al: Hypertelorism associated with holoprosencephaly and ectrodactyly. *J Clin Dysmorphol* **2**(2):27–31, 1984.
3. Moore CA et al: An unusual craniofacial condition: Holoprosencephaly, ocular hypertelorism, and anterior encephalocoele associated with tissue bands. *Dysmorphol Clin Genet* **2**(3):68–71, 1988.
4. Ohtsuka H: A rare patient with a false median cleft lip associated with multiple congenital anomalies. *Ann Plastic Surg* **17**:155–160, 1986.

### Median cleft upper lip, mental retardation, and pugilistic facies (W syndrome)

Pallister et al (2) described a mother and three children. The mother and daughter were more mildly affected. The disorder may have X-linked inheritance.

The facies was characterized by frontal prominence, anterior cowlick, hypertelorism, downward slanting palpebral fissures, and alternating internal strabismus (Figs. 22–47A,B and 22–48A,B). The nasal bridge was flattened. The upper lip had a midline notch due to a bifid frenum (Fig. 22–49). The hard palate exhibited anterior palatal cleft. The upper central incisors were missing.

Mental retardation, seizures, and mild spasticity were noted in a few individuals.

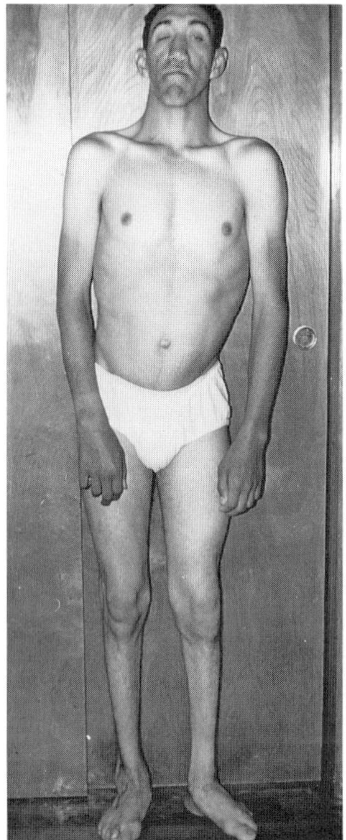

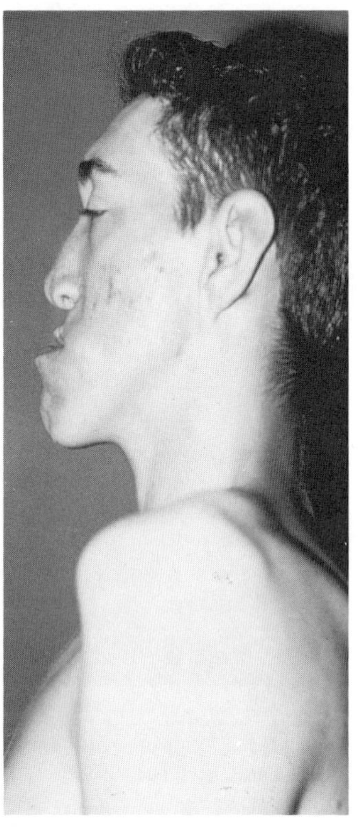

A B

Fig. 22–47. *Median cleft upper lip, mental retardation, and pugilistic facies.* (A,B) Note frontal prominence, internal strabismus, flattened nasal bridge, midline notch of upper lip, Compare with Fig. 22–48A,B. Although facies are similar, this probably is not the same disorder. The latter appears to have *cleft lip–palate, mental retardation, and holoprosencephaly syndrome.* [From *PD Pallister et al,* Birth Defects **10**(7):51, 1974.]

Musculoskeletal anomalies consisted of hernias, cubitus valgus with subluxation of the proximal radioulnar joints, lateral bowing of the radii, mild clinodactyly and camptodactyly of the fourth and fifth fingers, and pes cavus.

The syndrome should be differentiated from *oto-palatal-digital syndrome, Type I,* autosomal dominant *holoprosencephaly,* and *cleft lip-palate, microcephaly, mental retardation, and holoprosencephaly* (1).

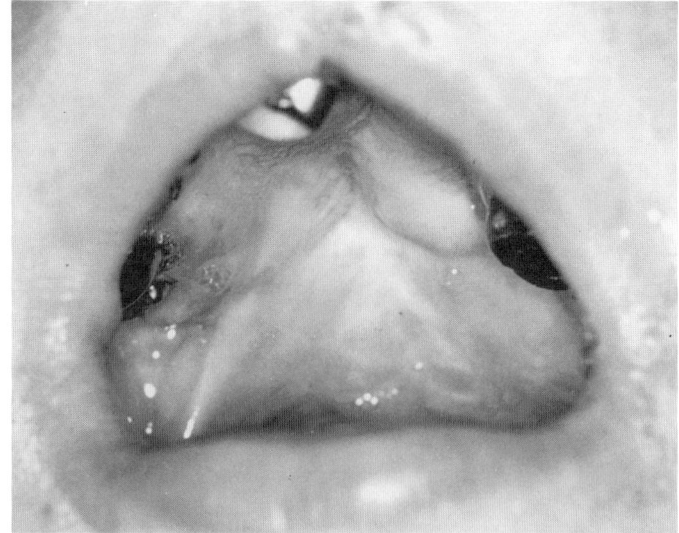

Fig. 22–49. *Median cleft upper lip, mental retardation, and pugilistic facies.* Midline cleft of upper lip, deep anterior palatal groove. [From *PD Pallister et al,* Birth Defects **10**(7):51, 1974.]

### References [Median cleft upper lip, mental retardation, and pugilistic facies (W syndrome)]

1. Martin AO et al: An autosomal dominant midline cleft syndrome resembling familial holoprosencephaly. *Clin Genet* **12**:65–72, 1977.
2. Pallister PD et al: The W syndrome. *Birth Defects* **10**(7):51–60, 1974.

## Michels syndrome

Michels et al (1) reported four patients with an autosomal recessive syndrome consisting of anterior chamber eye abnormalities, other craniofacial anomalies, and skeletal defects.

Short stature with height below the third percentile was evident in three of the four affected individuals. In one instance, growth deficiency of prenatal onset was noted. Head circumference below the third percentile and mild mental deficiency were reported in two instances.

Eye anomalies included blepharoptosis, blepharophimosis, epicanthus inversus, mild ocular hypertelorism, disturbances of eye motility, and corneal stromal opacities (Fig. 22–50). Craniosynostosis involv-

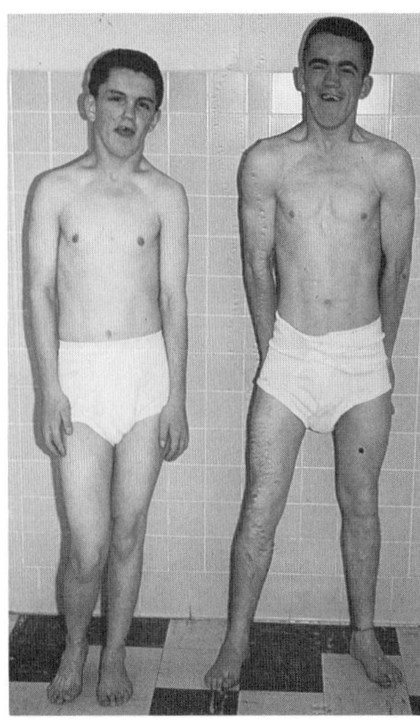

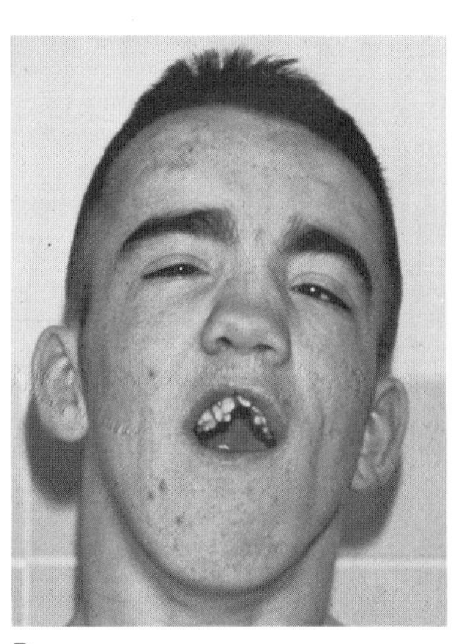

A B

Fig. 22–48. *Median cleft upper lip, mental retardation, and pugilistic facies.* (A) Note downslanting palpebral fissures, pugilistic appearance, median cleft lip, hypoplasia of lower leg muscles. (B) Hypoplastic midface. (Courtesy of *J Perrin,* Birmingham, Michigan.)

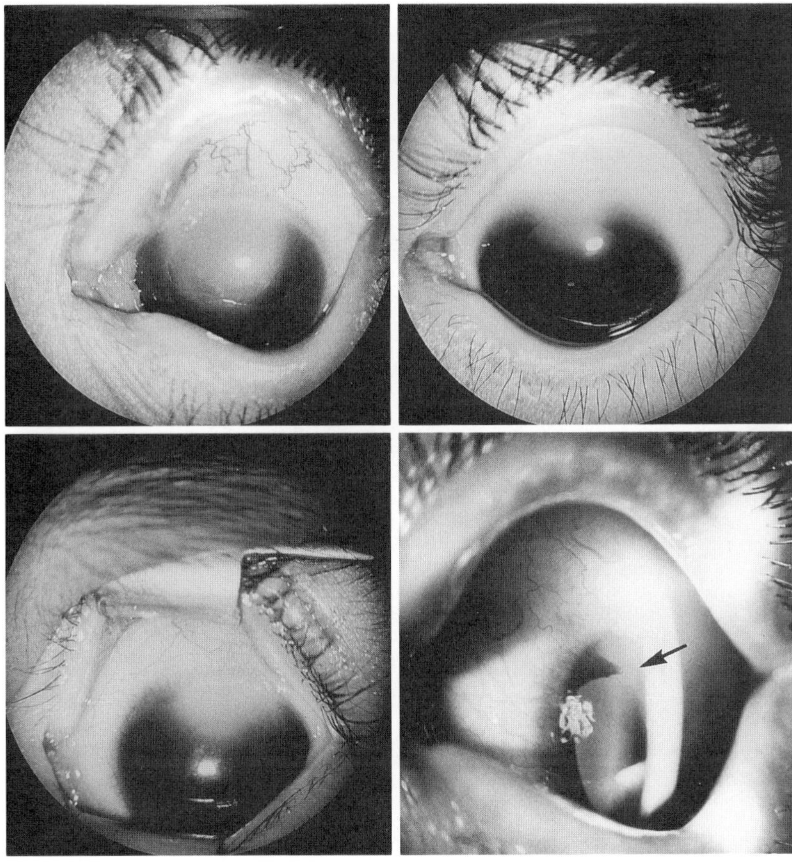

Fig. 22–50. *Michels syndrome.* Corneal opacities. Arrows point to iris adhesion in lower right. (From *VV Michels et al,* J Pediatr **93**:444, 1978.)

ing the lambdoidal sutures was observed in two of four affected individuals. Cleft lip and palate were reported in two, and hearing deficit was noted in three (Fig. 22–51).

Skeletal defects consisted of short fifth fingers, clinodactyly, and low-frequency anomalies, such as radioulnar synostosis and spina bifida occulta.

Differential diagnosis includes the triad of blepharoptosis, blepharophimosis, and epicanthus inversus, which is a recognized entity with autosomal dominant inheritance and high penetrance. Michels syndrome should also be distinguished from the anterior chamber cleavage syndrome and finally, from the *Rieger syndrome,* which consists of eye anomalies, including anterior chamber defects, hypoplastic teeth, missing teeth, and, in some instances, short stature.

### Reference (Michels syndrome)

1. Michels VV et al: A clefting syndrome with ocular anterior chamber defect and lid anomalies. *J Pediatr* **93**:444–446, 1978.

## Microcephaly and cleft palate

Halal (1) reported cleft palate and microcephaly in two sisters and their mother. The facies was somewhat unusual. There was mild hypotelorism and maxillary hypoplasia.

Abnormal retinal pigmentation with normal ERG findings was demonstrated. The microcephaly was about at the third percentile. Mental retardation was mild. The mother had euthyroid goiter and bilateral camptodactyly at the DIP joints of fingers 2 to 5.

A similar combination of autosomal dominantly inherited cleft palate and microcephaly was reported by Say et al, but the patients also had short stature and large pinnae (see *cleft palate, microcephaly, large ears, and short stature*). Microcephaly, mental retardation, persistent primary vitreous, and short stature have been reported as a recessive syndrome by Frydman (see *oculo-palato-cerebral dwarfism).*

### Reference (Microcephaly and cleft palate)

1. Halal F: Dominantly inherited syndrome of microcephaly and cleft palate. *Am J Med Genet* **15**:135–140, 1983.

## Oculo-palato-cerebral dwarfism

Frydman et al (1985) described a syndrome in three children of consanguineous parents. The syndrome consisted of cleft palate, microcephaly, mental retardation, spasticity, short stature, asthma, and persistent hypertrophic primary vitreous of the eye. Inheritance appears to be autosomal recessive.

Gestational hypertension was noted with all three sibs. Birthweight was approximately −2 SD. Stature fell to −2.5 to −4 SD. Hypotonia was noted in the neonatal period.

The facies was characterized by coarse hair, full cheeks, and deeply set eyes (Figs. 22–52 and 22–53). Hands and feet were small. The joints were mildly hypermobile. The skin was soft with a doughlike texture. The veins were visible beneath the skin.

Leukocoria became evident during the first month of life. Primary persistent hyperplastic primary vitreous and microphthalmia were evident.

Mental retardation ranged from mild to severe with microcephaly −3 to −5.5 SD. Spasticity ranged from a moderate degree to complete quadriplegia. Hearing loss was mild. All patients had severe asthma, pneumonitis, otitis media, pectus excavatum, and cleft palate.

A combination of microcephaly and cleft palate is seen in the syndromes described by Halal (see *microcephaly and cleft palate*) and by

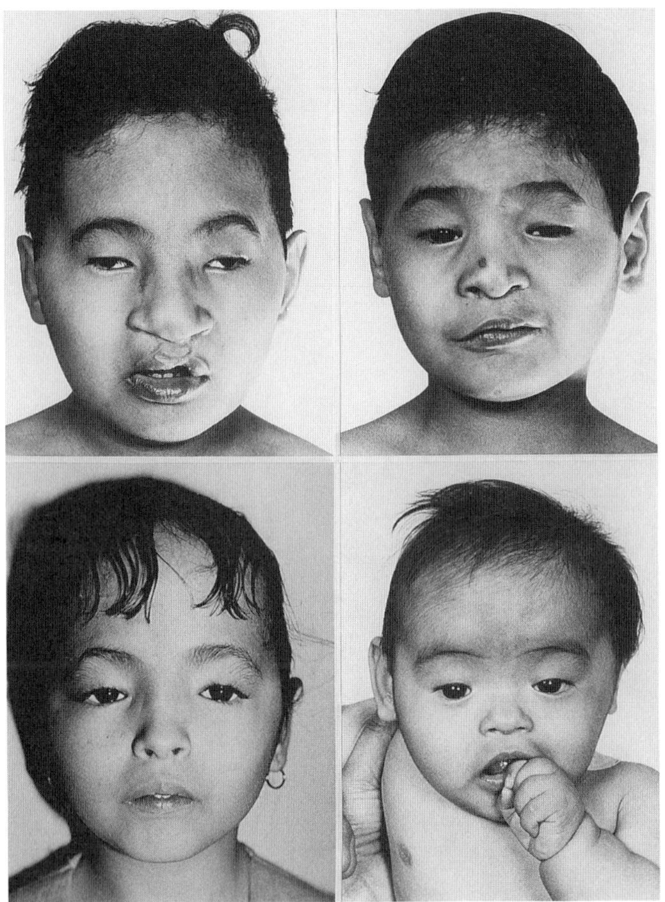

Fig. 22–51. *Michels syndrome.* Blepharoptosis, blepharophimosis, and epicanthus inversus. Note repaired cleft lip in upper two patients. (From *VV Michels et al,* J Pediatr **93:**444, 1978.)

Fig. 22–52. *Oculo-palato-cerebral dwarfism.* Fifteen-year-old with microcephaly, mild mental retardation, hypotonia, cleft palate, short stature, long thin neck, and sloping shoulders. Patient had persistent hypertrophic primary vitreous, microphthalmus, mild hearing loss, small hands and feet, ankle spasticity, and asthma. (From *M Frydman et al,* Clin Genet **27**:414, 1985.)

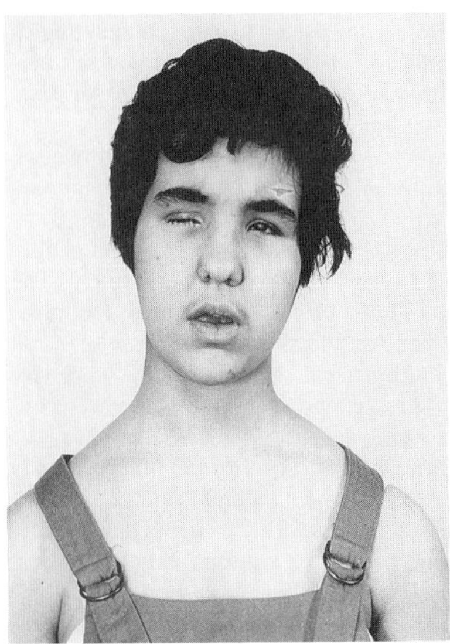

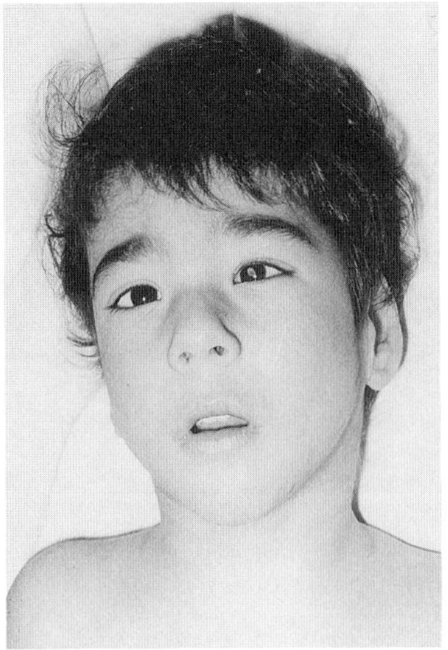

Fig. 22–53. *Oculo-palato-cerebral dwarfism.* Sib having similar features: short stature, severe mental retardation, cleft palate, microphthalmus, persistent hypertrophic primary vitreous of left eye, quadriplegia, and asthma. (From *M Frydman et al,* Clin Genet **27**:414, 1985.)

Say et al (see *cleft palate, microcephaly, large ears, and short stature*).

#### Reference (Oculo-palato-cerebral dwarfism)

1. Frydman M et al: Oculo-palato-cerebral dwarfism: A new syndrome. *Clin Genet* **27**:414–419, 1985.

### Odontotrichomelic syndrome

Freire-Maia and co-workers (1–4) described a form of ectodermal dysplasia with tetramelic deficiency in four of eight Brazilian sibs. The parents were normal and apparently not consanguineous but were from a highly inbred region. Rather extensive bone deficiencies involved all extremities. Various skeletal abnormalities were reported. There was marked reduction in the amount of scalp and body hair, eyebrows, and lashes. The ears were large, thin, outstanding, and deformed. Oligodontia and conical crown form were noted. The nipples and areolas were hypoplastic, and hypogonadism was found, as well as mild mental retardation. Two of four affected sibs died in infancy. Cleft lip occurred in only one boy (Fig. 22–54). Some of these patients resemble those with *EEC syndrome* (6).

Herrmann et al (5) described a mentally retarded Amerindian boy with some of the stigmata seen in the syndrome noted above. However, their case probably represents another syndrome which is discussed under *Herrmann syndrome.*

#### References (Odontotrichomelic syndrome)

1. Cat I et al: Odontotrichomelic hypohidrotic dysplasia: A clinical reappraisal. *Hum Hered* **22**:91–95, 1972.

2. Chautard EA, Freire-Maia N: Dermatoglyphic analysis in a highly mutilating syndrome. *Acta Genet Med Gemell* **19**:421–424, 1970.

3. Freire-Maia N: A newly recognized genetic syndrome of tetramelic deficiencies, ectodermal dysplasias, deformed ears and other anomalies. *Am J Hum Genet* **22**:370–377, 1970.

4. Freire-Maia N, Pinheiro M: *Ectodermal Dysplasias: A Clinical and Genetic Study.* Alan R. Liss, New York, 1984, pp 106–108.

5. Herrmann J et al: Craniosynostosis and craniosynostosis syndromes. *Rocky Mt Med J* **66**:145–56, 1969.

6. Pavone L et al: A case of the Freire–Maia odontotrichomelic syndrome: Nosology with the EEC syndrome. *Am J Med Genet* **33**:190–193, 1989.

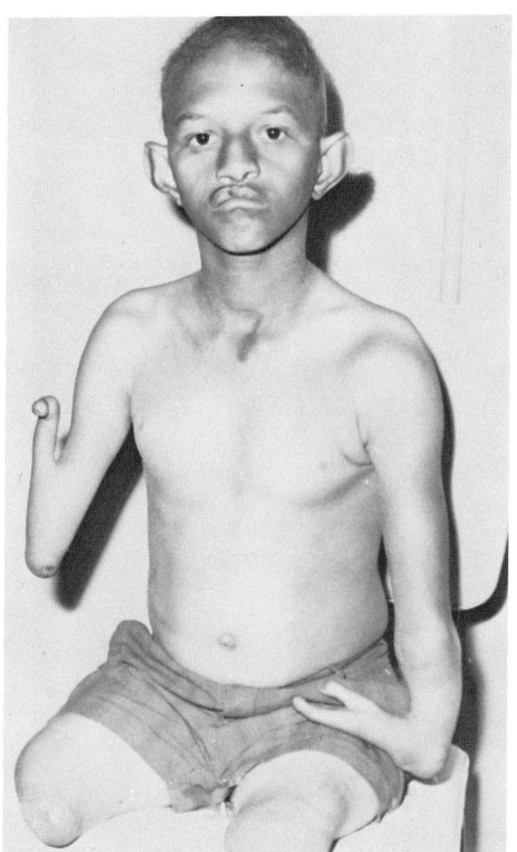

Fig. 22–54. *Odontotrichomelic syndrome.* Sparsity of hair, deformed ears, hypoplastic nipples, deficient teeth, and cleft lip–palate. A sister was similarly affected. (From *N Freire-Maia,* Am J Hum Genet **22**:370, 1970.)

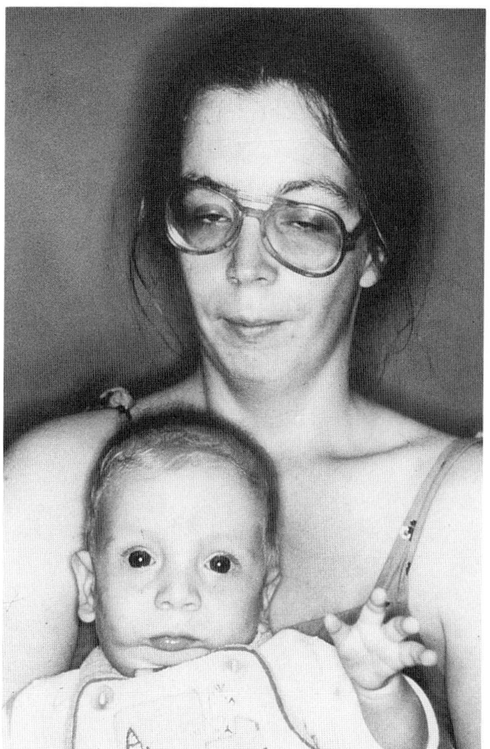

Fig. 22–55. *Oligodactyly and cleft palate.* Mother and son with cleft palate and micrognathia. (From *M Robinow et al,* Am J Med Genet **25**:293, 1986.)

## Oligodactyly and cleft palate

Robinow et al (2), in 1986, reported a mother and son with cleft palate, microretrognathia, and anomalies of the hands and feet. The first ray in the hands was hypoplastic and the fifth ray was absent. The fourth metacarpal was somewhat broadened. In the feet of the boy, the fifth rays were hypoplastic (Figs. 22–55A–C and 22–56). The mother had no abnormality of the toes. We are not aware of similar cases. Meinecke and Wiedemann (1) suggested that the syndrome represents incomplete expression of *postaxial acrofacial dysostosis.* Both Robinow et al (3) and RJ Gorlin express skepticism.

### References (Oligodactyly and cleft palate)

1. Meinecke P, Wiedemann H-R: Robin sequence and oligodactyly in mother and son—probably a further example of the postaxial acrofacial dysostosis syndrome. *Am J Med Genet* **27**:953–956, 1987.

2. Robinow M et al: Robin sequence and oligodactyly in mother and son. *Am J Med Genet* **25**:293–298, 1986.

3. Robinow M et al: Reply to Drs. Meinecke and Wiedemann. *Am J Med Genet* **27**:957, 1987.

## Recurrent brachial plexus neuropathy, characteristic facies, and cleft palate

Jacob et al (9) were probably the first to recognize the syndrome of recurrent brachial neuropathy, characteristic facies, and cleft palate. There have been a considerable number of subsequent reports (1–8,10–15). Inheritance is autosomal dominant with variable expressivity and nearly complete penetrance.

The facies has been likened to that of a Modigliani portrait (2). The face is somewhat narrow and often there is mild facial asymmetry. The eyes appear deeply set and hypoteloric. There are epicanthal folds and mild upslanting of the palpebral fissures (Fig. 22–57A–D). A number of patients have cleft palate or bifid uvula (1–5,7,9,11). Some patients experience hoarseness, implying involvement of the recurrent laryngeal nerve (7,9,10).

Usually beginning in the fourth to eighth year of life, the patient experiences sudden attacks of pain in the shoulder that radiate to the arms and hands. A few days after the pain subsides, paresthesia, weakness, and, later, atrophy of the upper extremities occur. Some patients exhibit wrist drop, while others cannot lift the arm above the horizontal level. Reflexes are reduced and there may be some transient sensory loss over the forearms and hands. Attacks may be more common during pregnancy, or following parturition, infectious disorders, or strenuous exercise. Recurrences occur at intervals of months or years. Electromyographic studies have shown partial denervation of several muscles of the upper extremities corresponding to a lesion of the brachial plexus. Stature is often small. Limitation of extension at the elbow and winging of the scapulae have also been noted. Rarely, other peripheral nerves or lower cranial nerves are involved. In most cases, there is complete resolution, but long-standing residual weakness and atrophy may occur.

The facies is remarkably similar to that described by Schilbach and Rott (12) in a dominantly inherited syndrome of hypotelorism, submucous cleft palate, and hypospadias. However, in that family, no brachial neuropathy was mentioned.

### References (Recurrent brachial plexus neuropathy, characteristic facies, and cleft palate)

1. Airaksinen EM et al: Hereditary recurrent brachial plexus, neuropathy with dysmorphic features. *Acta Neurol Scand* **71**:309–316, 1985.

2. Arts WFM et al: Hereditary neurologic amyotrophy. Clinical, genetic, electrophysiological and histopathological studies. *J Neurol Sci* **62**:261–279, 1983.

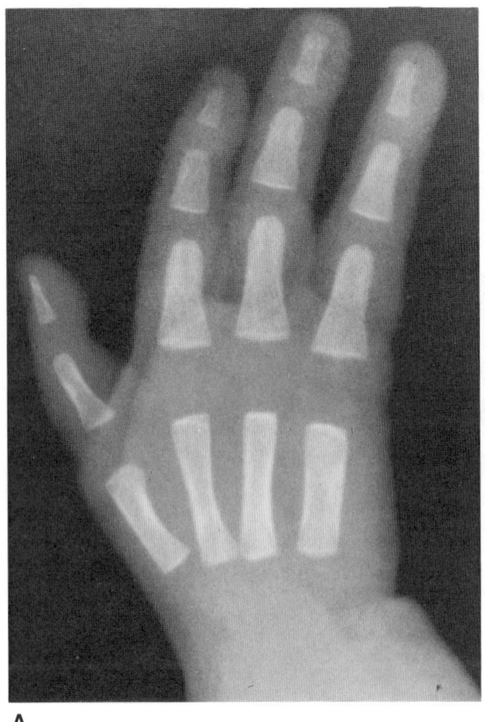
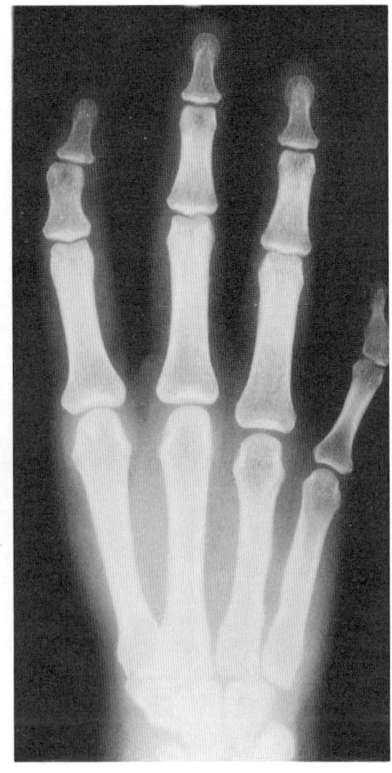
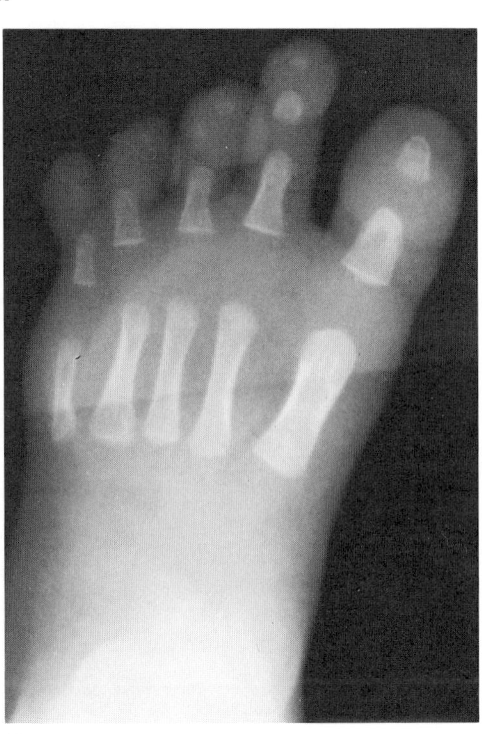

A B C

Fig. 22–56. *Oligodactyly and cleft palate.* (A,B) Both have hypoplastic or aplastic first and fifth rays of hands. (C) Child has hypoplastic fifth rays of feet. (From *M Robinow et al,* Am J Med Genet **25**:293, 1986.)

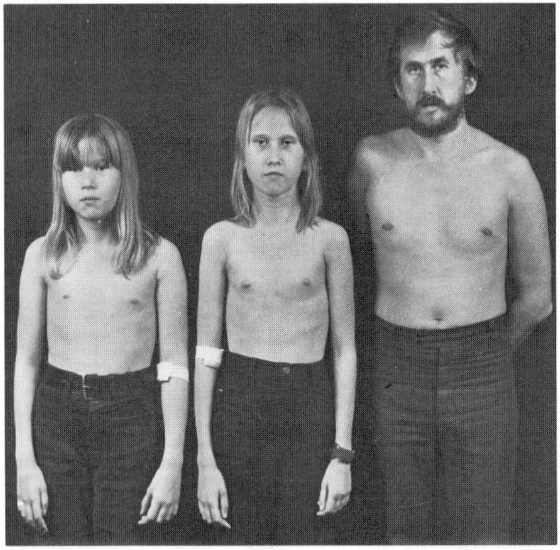
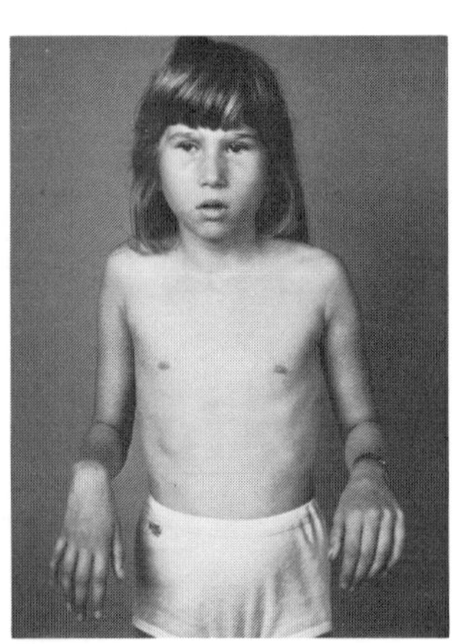

A B

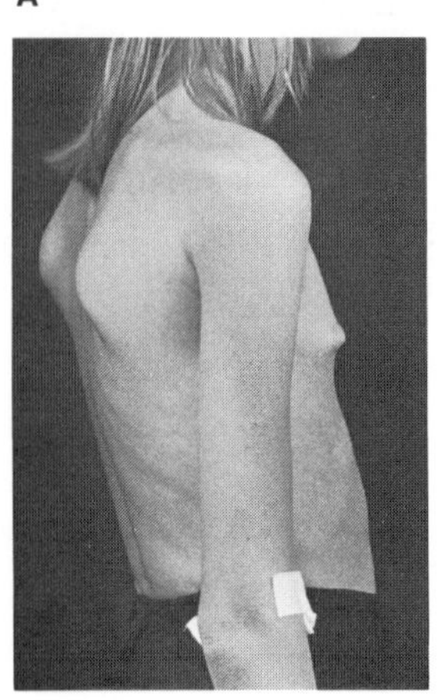
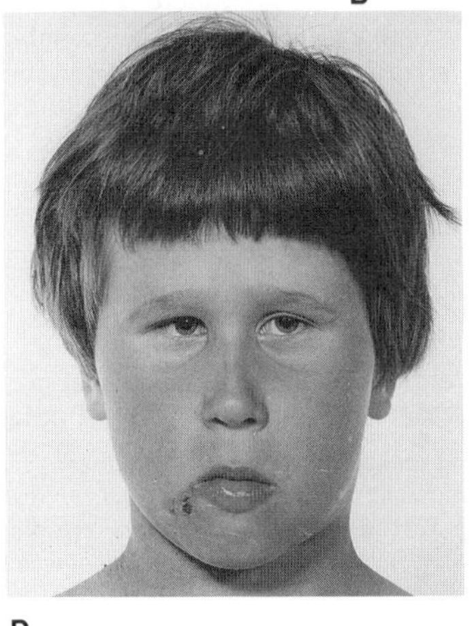

C D

Fig. 22–57. *Recurrent brachial plexus neuropathy, characteristic facies, and cleft palate.* (A) Father and two daughters. Note deeply set eyes, ocular hypotelorism, and mongoloid obliquity of palpebral fissures. (B) Patient exhibiting wrist drop during attack of recurrent brachial neuritis. (C) Winged scapulae are clearly evident. (D) Typical facies. (A–C from *A Erikson,* Acta Paediatr Scand **63**:885, 1974. D from *EM Airaksinen et al,* Acta Neurol Scand 71:309, 1985.)

3. Bradley WG et al: Recurrent brachial plexus neuropathy. *Brain* **98**:381–398, 1975.

4. Dunn HG et al: Heredofamilial brachial plexus neuropathy (hereditary neuralgic amyotrophy with brachial predilection) in childhood. *Dev Med Child Neurol* **20**:28–46, 1978.

5. Erikson A: Hereditary syndrome consisting in recurrent attacks resembling brachial plexus neuritis, special facial features and cleft palate. *Acta Paediatr Scand* **63**:885–888, 1974.

6. Gardner JH, Maloney W: Hereditary brachial and cranial neuritis genetically linked with ocular hypotelorism and syndactyly. *Neurology* **18**:278, 1968.

7. Geiger LR et al: Familial neurologic amyotrophy: Report of three families with review of the literature. *Brain* **97**:87–102, 1974.

8. Guillozet N, Mercer RD: Hereditary brachial neuropathy. *Am J Dis Child* **125**:884–887, 1973.

9. Jacob JC et al: Heredofamilial neuritis with brachial predilection. *Neurology* **11**:1025–1033, 1961.

10. Poffenbarger AL: Heredofamilial neuritis with brachial predilection. *W Virginia Med J* **64**:425–429, 1968.

11. Rogers RC, Hamilton TB: Recurrent brachial neuritis. *Proc Greenwood Genet Ctr* **4**:5–7, 1985.

12. Schilbach U, Rott HD: Ocular hypotelorism, submucous cleft palate, and hypospadias: A new autosomal dominant syndrome. *Am J Med Genet* **31**:863–870, 1988.

13. Smith BH et al: Familial brachial neuropathy. *Neurology* **21**:941–945, 1971.

14. Taylor RA: Heredofamilial mononeuritis multiplex. *Brain* **83**:113–137, 1960.

15. Ungley CC: Recurrent polyneuritis in pregnancy and puerperium affecting 3 members of a family. *J Neurol Psychopathol* **14**:15–26, 1933.

## Ring-shaped skin creases and cleft palate

Kunze and Riehm (1) reported two families each with autosomal dominant transmission of ring-shaped skin creases (Fig. 22–58). In one family, a son had cleft palate and in the other family a daughter had cleft palate. Other patients with ring-shaped skin creases have not had cleft palate (2,3). We view the term "Michelin tire baby" as pejorative.

Fig. 22–58. *Ring-shaped skin creases and cleft palate.* Multiple benign ring-shaped skin creases. (From *J Kunze* and *H Riehm,* Eur J Pediatr **138**:301, 1982.)

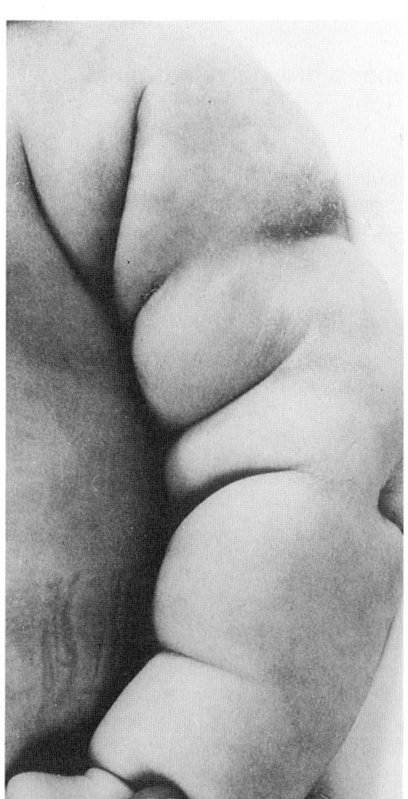

### References (Ring-shaped skin creases and cleft palate)

1. Kunze J, Riehm H: A new genetic disorder: Autosomal-dominant multiple benign ring-shaped skin creases. *Eur J Pediatr* **138**:301–303, 1982.

2. Niikawa N et al: The "Michelin tire baby" syndrome—an autosomal dominant trait. *Am J Med Genet* **22**:637–638, 1985.

3. Ross CM: Generalized folded skin with an underlying lipomatous nevus: The Michelin tire baby. *Arch Dermatol* **100**:320–323, 1969.

## Rüdiger syndrome

Rüdiger et al (1) described a lethal, probably autosomal recessive, syndrome in sibs, the infants succumbing within the first year of life. The disorder is characterized by somatic retardation, flexion contractures of hands with thick, single palmar creases, small fingers and nails, and ureterovesical stenosis. The male sib had micropenis and inguinal hernias; the female sib had bicornuate uterus. Arches were noted on all fingers. The facies was coarse with prominent forehead, flat nasal bridge, stubby nose, and prominent upper lip. The soft palate was cleft (Fig. 22–59A,B).

### Reference (Rüdiger syndrome)

1. Rüdiger RA et al: Severe developmental failure with coarse facial features, distal limb hypoplasia, thickened palmar creases, bifid uvula and ureteral stenosis: A previously undescribed familial disorder with lethal outcome. *J Pediatr* **79**:977–981, 1971.

## Short stature and cleft palate

Gareis and Smith (1) reported a kindred in which several affected members had stature below the third percentile because of relative shortness of the extremities. All affected males and about half the affected females had cleft palate or submucous cleft palate and bifid uvula. Micrognathia was present in about half those affected. The condition is dominantly inherited, possibly X-linked.

### Reference (Short stature and cleft palate)

1. Gareis FJ, Smith DW: Diminished stature-defective palate syndrome: A dominantly inherited disorder. *J Pediatr* **79**:470–472, 1971.

## Submucosal cleft palate, hypotelorism, and hypospadias

Schilbach and Rott (1) described 10 affected individuals in five generations of a family. All exhibited hypotelorism and submucosal cleft palate. In addition, the facies was characterized by epicanthic folds, blepharophimosis, and upslanting palpebral fissures (Fig. 22–60A,B). Males exhibited hypospadias of variable degrees.

There was very mild cutaneous syndactyly of the third and fourth fingers as well as the second and third toes. In the hands, the syndactyly correlated with an abnormally low ridge count of digital triradii *b* and *c*.

The face is quite similar to the facies in *recurrent brachial plexus neuropathy.*

Inheritance is clearly autosomal dominant.

### Reference (Submucosal cleft palate, hypotelorism, and hypospadias)

1. Schilbach U, Rott H-D: Ocular hypotelorism, submucosal cleft palate, and hypospadias: A new autosomal dominant syndrome. *Am J Med Genet* **31**:863–870, 1988.

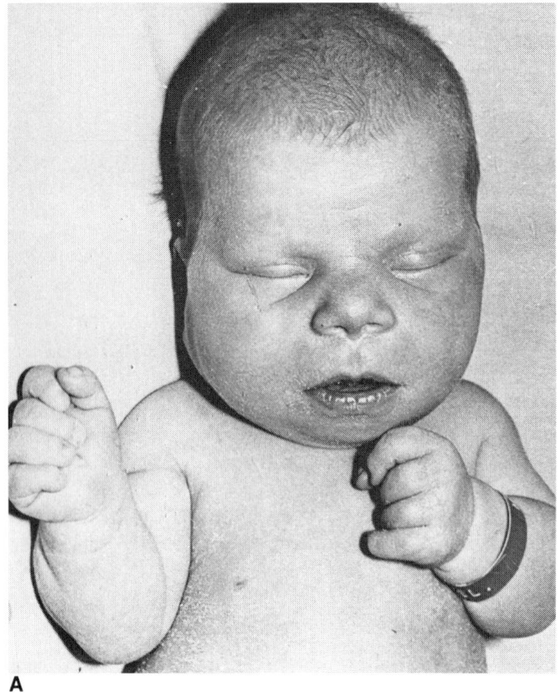

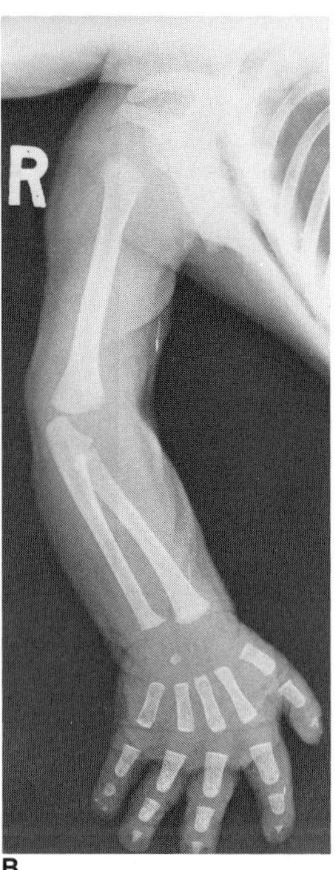


A B

Fig. 22–59. *Rüdiger syndrome.* (A) Coarse facies, finger contractures. (B) Small middle phalanges and small triangular terminal phalanges. (From *RA Rüdiger et al,* J Pediatar **79**:977, 1971.)

## Cleft palate, congenital hypothyroidism, and spiky hair

Bamforth et al (1), in 1989, described two male sibs with congenital athyroidal hypothyroidism, spiky hair, choanal atresia, and bifid epiglottis. Both had cleft palate. Polyhydramnios became evident during the third trimester. At 3 years the elder child was found to be performing at a 2-year level, delay being most marked in the cognitive and language fields. The second child was too young for evaluation. Their mother had unilateral choanal atresia.

A probably different condition was described by Zadik et al (2). They reported a female child with congenital hypothyroidism, dacryostenosis, and dermoid cysts in the subpalpebral region of both upper eyelids. Oligodontia was marked, and cleft palate was present. Sweating was normal.

Hypothyroidism has been seen with various forms of ectodermal dysplasia. The reader is referred to the *differential diagnosis* section of *hypohidrotic ectodermal dysplasia.*

#### References (Cleft palate, congenital hypothyroidism, and spiky hair)

1. Bamforth JS et al: Congenital hypothyroidism, spiky hair, and cleft palate. *J Med Genet* **26**:49–51, 1989.

2. Zadik Z et al: Syndrome identification case report 112. Dermoid cysts, hypothyroidism, cleft palate, and hypodontia. *J Clin Dysmorphol* **1**(4):24–25, 1983.

## Cleft lip–palate, mental and growth retardation, sensorineural hearing loss, and postaxial polydactyly

RJ Gorlin and J Cervenka have seen a family with autosomal dominant inheritance of cleft lip–palate, micrognathia, sensorineural hearing loss, mental and somatic retardation, and postaxial polydactyly.

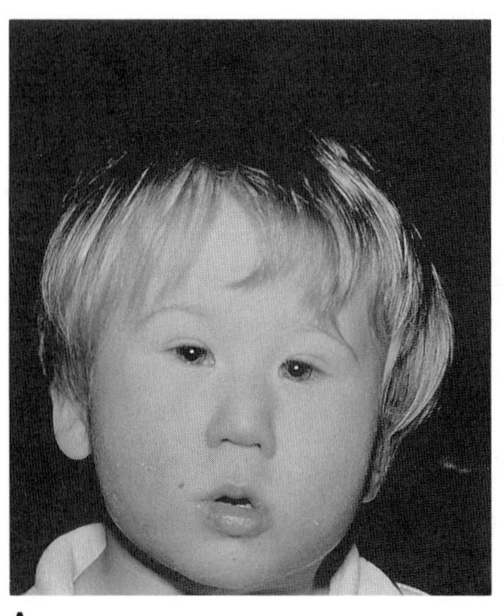

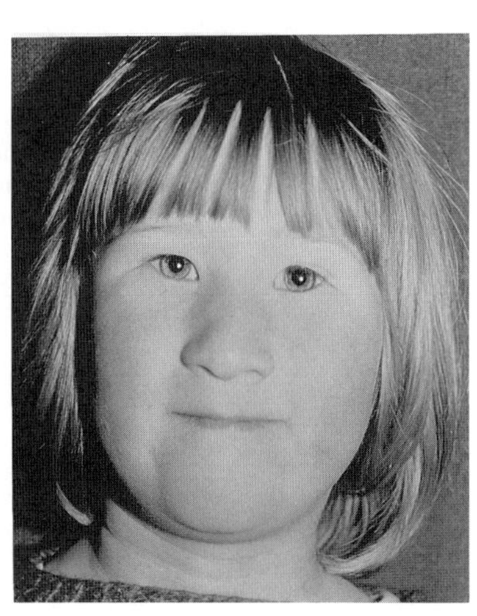

A B

Fig. 22–60. *Submucosal cleft palate, hypotelorism, and hypospadias.* (A,B) Affected sibs with blepharophimosis, upslanting palpebral fissures, and hypotrichosis. (From *U Schilbach* and *H-D Rott,* Am J Med Genet **31**:863–870, 1988.)

## Cleft lip–palate, sensorineural hearing loss, sacral lipomas, and aberrant fingerlike appendages

RB Lowry (personal communication, 1989) observed two male Chinese siblings with cleft lip and palate, profound congenital sensorineural hearing loss, and sacral lipomas. In one child there was connection of the lipoma with the spinal cord. The other child had two separate aberrant fingerlike appendages of the heel and thigh. Intelligence is probably normal.

Inheritance appears to be either X-linked or autosomal recessive. Parents were nonconsanguineous.

Perhaps there is some relationship to the mouse mutant "disorganization" described by Winter and Donnai (1).

### Reference (Cleft lip-palate, sensorineural hearing loss, sacral lipomas, and aberrant fingerlike appendages)

1. Winter RM, Donnai D: A possible human homologue for the mouse mutant disorganization. *J Med Genet* **26**:417–420, 1989.

## Cleft lip–palate, uveal colobomata, and mental retardation (Kingston syndrome)

Kingston et al (1) reported autosomal dominant inheritance of various eye anomalies (uveal coloboma, microphthalmia, cataracts, retinal detachment, posterior embryotoxon, glaucoma, ptosis) with cleft lip and palate.

### Reference [Cleft lip–palate, uveal colobomata, and mental retardation (Kingston syndrome)]

1. Kingston HM et al: An autosomal dominant syndrome of uveal coloboma, cleft lip and palate and mental retardation. *J Med Genet* **19**:444–446, 1982.

## Cleft palate, macrosomia, and microphthalmia

Teebi et al (1), in 1989, reported a new syndrome in five of ten siblings whose parents were consanguineous.

All exhibited macrosomia, newborn weights ranging from 4050–5100 g. Severe bilateral microphthalmia, wide anterior fontanel, protuberant abdomen, hepatomegaly, repeated respiratory infection with resultant death within the first months of life were characteristic. Three of the sibs manifested cyanotic apneic spells.

Three had cleft palate.

Inheritance is autosomal recessive.

### Reference (Cleft palate, macrosomia, and microphthalmia)

1. Teebi AS et al: Macrosomia, microphthalmia, ± cleft palate and early infant death: A new autosomal rcessive syndrome. *Clin Genet* **36**:174–177, 1989.

## Cleft palate, radial and patellar aplasia/hypoplasia, and short stature

Kääriänen et al (1) reported a syndrome of radial aplasia or hypoplasia, absence of thumbs, absent or hypoplastic patellae, dislocation of joints, diarrhea in infancy, small stature, and normal intelligence. The face was elongated with a long slender nose and small chin. The voice was strikingly high-pitched. Two of the four patients had cleft palate.

They suggested the acronym RAPADILINO syndrome: *RA* (*ra*dial anomalies), *PA* (absent/hypoplastic *pa*tellae, and cleft *pa*late), *DI* (*di*arrhea, *di*slocated joints), *LI* (*li*ttle size, *li*mb formations), *NO* (long thin *no*se and *no*rmal intelligence).

Inheritance appears to be autosomal recessive.

### Reference (Cleft palate, radial and patellar aplasia/hypoplasia, and short stature)

1. Kääriänen H et al: RAPADILINO syndrome with radial and patellar aplasia/hypoplasia as main manifestations. *Am J Med Genet* **33**:346–351, 1989.

# Chapter 23
# Orofacial Clefting Syndromes: Associations

## Aicardi syndrome and cleft lip–palate

Aicardi syndrome consists of flexion spasms, callosal agenesis, and ocular abnormalities (choreoretinopathy, microphthalmia) occurring in retarded females. The syndrome has been reported in association with porencephaly, corticoheterotopias, Arnold–Chiari malformation, papilloma of the choroid plexus, lissencephaly, polygyria, microgyria, arhinencephaly, and Dandy–Walker malformation. It has been infrequently reported in association with cleft lip–palate (1,2) and *holoprosencephaly* (2).

### References (Aicardi syndrome and cleft lip–palate)


1. Robinow M et al: Aicardi syndrome, papilloma of the choroid plexus, cleft lip and cleft of the posterior palate. *J Pediatr* **104**:404–405, 1984.

2. Sato N et al: Aicardi syndrome with holoprosencephaly and cleft lip and palate. *Pediatr Neurol* **3**:114–116, 1987.


## Anencephaly and/or cleft lip–palate

Anencephaly, a disorder due to failure of closure of the anterior neuropore and resultant lack of brain development, is characterized by absence of cranial vault and brain, hypoplastic and/or deformed pituitary gland, and absent hypothalamus. Associated are protruding eyeballs and tongue, aquiline nose, malformations of the cranial base, short neck, and changes secondary to lack of neural connections between the pituitary and the generally absent hypothalamus, such as hypoplasia of the adrenal glands and gonads (4). The protruding tongue may be due to an abnormally small oral cavity. The mandible is relatively large. A general account of *anencephaly* appears in chapter 16. The reader is referred to the comprehensive work of Lemire et al (14) for detailed information.

The association of clefting and anencephaly is certainly higher than chance would dictate. Khoury et al (13a) found a relationship with neural tube defects and CL(P) especially significant. Heintel (11) stated that every anencephalic had submucous cleft palate. Although this may be an exaggeration, palatal form is certainly bizarre with a pronounced midline palatal raphe, suggesting to us altered temporal fusion of the palatal shelves with the nasal septum. Similar palatal changes are seen in holoprosencephaly (see Fig. 20–22). Among 114 infants with isolated cleft palate, MacMahon and McKeown (15) found one anencephalic, but three with this condition were noted among 105 infants with cleft lip and palate among 34 Japanese anencephalics. A number of authors (7,8,18) have also noted the association. Conversely, several investigators examining infants with anencephaly have found cleft lip–palate (2,3,5,6,12,19) or cleft palate or uvula (1,9,12,13,16,19). However, since no systematic large survey has been done, clefting has ranged from 2 to 20% (12,19). Gleeson and Stovin (10) described an anencephalic child with cleft palate, Klippel–Feil-like anomaly, and neuroenteric cyst.

### References (Anencephaly and/or cleft lip–palate)


1. Allen JP et al: Endocrine function in an anencephalic infant. *J Clin Endocrinol Metab* **38**:94–98, 1974.

2. Benirschke K: Adrenals in anencephaly and hydrocephaly. *Obstet Gynecol* **8**:412–425, 1956.

3. Broder SB: An anencephalic monster with "rhinodymie" and other anomalies. *Am J Pathol* **11**:761–774, 1935.

4. Brown FJ: The anencephalic syndrome in its relation to apituitarism. *Edinb Med J* **25**:296–307, 1920.

5. Ch'in KY: The endocrine glands of anencephalic foetuses. A quantitative and morphological study of 15 cases. *Chin Med J* **2**:63–90, 1938.

6. Christakos AC, Simpson JL: Anencephaly in three siblings. *Obstet Gynecol* **33**:267–270, 1969.

7. Deppe B: *Beitrag zum anatomischen Wesen der Anencephalie, Amyelie und Craniorhachischisis.* Thesis, Göttingen, 1935.

8. Fogh-Andersen P: *Inheritance of Harelip and Cleft Palate.* Nyt Nordisk Forlag, Copenhagen, 1942.

9. Frutiger P: Zur Frage der Arhinencephalie. *Acta Anat* **73**:410–430, 1969.

10. Gleason JA, Stovin PGI: Mediastinal enterogenous cysts associated with vertebral anomalies. *Clin Radiol* **12**:41–48, 1961.

11. Heintel H: Die Palatoschisis und ihre Beziehung zum Anencephalie-Syndrom. *Stoma (Heidelberg)* **9**:204–211, 1956.

12. Jones WR: Anencephalus, a 23-year survey in a Sydney hospital. *Med J Aust* **1**:104–105, 1967.

13. Kappers JA: Developmental disturbance of the brain induced by German measles in embryo of the 7th week. *Acta Anat* **31**:1–20, 1957.

13a. Khoury MJ et al: Selected midline defect association: A population study. *Pediatrics* **84**:266–272, 1989.

14. Lemire RJ et al: *Anencephaly.* Raven Press, New York, 1978.

15. MacMahon B, McKeown T: Incidence of harelip and cleft palate. Related to birth rank and maternal age. *Am J Hum Genet* **5**:176–183, 1953.

16. Mufarri IK, Kilejian VO: An analysis of anencephalic births and report of a case of repeated anencephaly. *Obstet Gynecol* **22**:657–661, 1963.

17. Neel JV: A study of major congenital defects in Japanese infants. *Am J Hum Genet* **10**:398–445, 1958.

18. Roches P: *Zur Frage der Anenkephalie.* Thesis, Basel, 1951.

19. Smithells RW et al: Anencephaly in Liverpool. *Dev Med Child Neurol* **6**:231–240, 1964.


## Lateral proboscis and cleft lip–palate

Lateral proboscis usually occurs in association with unilateral absence of the nose on the ipsilateral side. The tube represents the missing half of the nose. The proboscis is a blind-ended tubular structure usually arising above the inner canthus. The eye is characteristically laterally displaced. The proboscis also may be laterally displaced, above and to the side of the eye. In the rare example, the proboscis is located in the midline with a normal nose (2). The nasolacrimal canal on the involved side is often missing. Colobomas of the iris and especially of the lower eyelid are the more commonly associated findings (1a). Unilateral absence of the olfactory bulb and cribriform plate on the side of the proboscis, that is, unilateral arhinencephaly has been demonstrated. In some cases, the nose appears almost normal, especially in those with cleft lip (Fig. 23–1). For detailed discussion on the embryology and associated findings, the reader is referred to several articles (1a,6,14).

We have surveyed over 50 published cases. Cleft lip–palate or cleft palate has been found in approximately 8%. Cleft of the lip has been unilateral and ipsilateral (3,8,15,16) and may be incomplete (7,15). In a few cases, there has been cleft palate or submucous cleft palate (3,4,8,9). Bilateral cleft lip and cleft palate have been seen in association with anencephaly and lateral proboscis (5,11,12). Lateral proboscis has also been seen in association with cleft mandible (13). A proboscis lateral to the eye has been seen with oblique facial cleft and

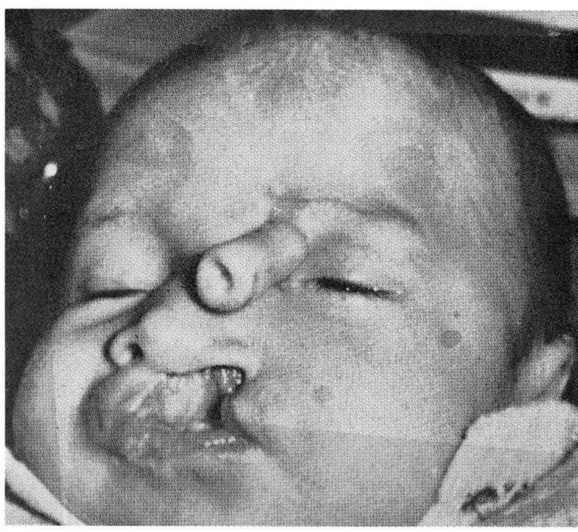

Fig. 23–1. *Lateral proboscis and cleft lip–palate.* (Courtesy of *LR McLaren,* Br J Plast Surg **8**:57, 1955.)

arhinencaphaly. See *holoprosencephaly* and *nasal aplasia, heminasal aplasia with or without nasal proboscis* for related information.

#### References (Lateral proboscis and cleft lip–palate)

1. Antoniades K, Baraitser M: Proboscis lateralis. *Teratology* **40:**193–198, 1989.

1a. Boo-Chai K: The proboscis lateralis—a 14 year follow-up. *Plast Reconstr Surg* **75**:569–577, 1985.

2. Dasgupta G et al: Proboscis. *J Laryngol Otol* 85:401–406, 1971.

3. Kirchmayr L: Ein Beitrag zu den Gesichtmissbildungen. *Dtsch Z Chir* **81**:70–81, 1906.

4. Kirkpatrick TJ: Lateral nasal proboscis. *J Laryngol Otol* **84**:83–89, 1970.

5. Lang B: Über seltene Nasenmissbildungen. *Virchows Arch Pathol Anat* **285**:93–111, 1932.

6. Lieuw Kie Song SH, Been W: Median facio-cerebral anomalies in chick embryos resulting from local destruction of the anteriormost parts of the early neural plate and neural crest. *Acta Morphol Neerl-Scand* **18**:231–252, 1980.

7. Mahindra S et al: Lateral nasal proboscis. *J Laryngol Otol* **87**:177–181, 1973.

8. McLaren LR: A case of cleft lip and palate with polypoid nasal tubercle. *Br J Plast Surg* **8**:57–59, 1955–56.

9. Meeker LH, Aebli R: Cyclopian eye and lateral proboscis with normal one-half face. *Arch Ophthalmol* **38**:159–173, 1947.

10. Nessel E: Proboscis lateralis als seltene Nasenmissbildung. *HNO* **7**:25–26, 1958.

11. Peters A: Über eine ungewöhnlich Gesichtsmissbildung bei Anencephalie. *Virchows Arch Pathol Anat* **288**:223–242, 1933.

12. Rating B: Über die bei Missbildungen des Gesichtes vorkommende Rüsselbildung. *Ber Versamml Ophthalmol Gesellsch* **36**:163–174, 1911.

13. Récamier J, Florentin M: Un type exceptionnel de malformation congénitale du nez. *Ann Chir Plast* **2**:13–16, 1957.

14. Rontal M, Duritz G: Proboscis lateralis: Case report and embryologic analysis. *Laryngoscope* **87**:996–1006, 1977.

15. Tendlau A: Ein Fall von Proboszis lateralis. *Albrecht von Graefes Arch Klin Ophthalmol* **95**:135–144, 1918.

16. Van Duyse GM: Proboscide latérale et colobome oculaire atypique avec lenticone postérieur. *Arch d'Ophtal* **36**:463–501, 555–575, 1919.

## Duane anomaly and cleft palate

The Duane retraction syndrome is a congenital eye movement disorder characterized by marked limitation of abduction, limitation of adduction, palpebral fissure narrowing, and globe retraction on attempted adduction. The anomaly may be unilateral or bilateral. The sixth nerve nucleus or nuclei have been found to be hypoplastic, the lateral rectus fibrotic, and the medial rectus hypertrophic.

Occasionally, Duane anomaly is associated with cleft palate. Although some patients clearly have Klippel–Feil anomaly or Wildervanck syndrome (Klippel–Feil anomaly, Duane anomaly, and sensorineural hearing loss), others had only the binary combination (1,2). Kirkham (3), describing patients with Duane anomaly, cleft palate, and profound hearing loss, suggested that the patients had incomplete expression of the wider spectrum. Konigsmark and Gorlin (4) have reviewed the general problem.

#### References (Duane anomaly and cleft palate)

1. Bintliff SJ: Klippel–Feil syndrome. *J Am Med Women's Assoc* **20**:547–550, 1965.

2. Fraser WI, MacGillivray RC: Cervico-oculo-acoustic dysplasia (the syndrome of Wildervanck). *J Ment Defic Res* **12**:322–329, 1968.

3. Kirkham TH: Duane's retraction syndrome and cleft palate. *Am J Ophthalmol* **70**:209–212, 1970.

4. Konigsmark BW, Gorlin RJ: *Genetic and Metabolic Deafness.* W.B. Saunders, Philadelphia, 1976, pp 188–191.

## Retinoblastoma and cleft palate

Bonaiti-Pellie et al (1) reported two sibs with cleft palate and retinoblastoma. Their mother had retinoblastoma but no cleft palate. The children also had severe micrognathia. Jensen and Miller (2) noted a child with Robin sequence and retinoblastoma; and Walker (3) described identical twins, one having cleft palate and the other retinoblastoma.

#### References (Retinoblastoma and cleft palate)

1. Bonaiti-Pellie C et al: Associated congenital malformations in retinoblastoma. *Clin Genet* **7**:37–39, 1975.

2. Jensen RD, Miller RW: Retinoblastoma: Epidemiologic characteristics. *N Engl J Med* **285**:307–311, 1971.

3. Walker NF: Discordant monozygotic twins with retinoblastoma and cleft palate. *Am J Hum Genet* **2**:375–384, 1950.

## Aplasia of the trochlea and cleft palate

Mead and Martin (1) described cleft palate in a patient who also had aplasia of the trochlea. Conceivably the association is one of chance. However, because of the rarity of aplasia of the trochlea, the finding may be significant.

#### Reference (Aplasia of the trochlea and cleft palate)

1. Mead CA, Martin M: Aplasia of the trochlea in an original mutation. *J Bone Jt Surg* **45A**:379–383, 1963.

## Cleft larynx, laryngeal web, and cleft palate

Cleft lip–palate is rarely associated with cleft larynx with or without extension into the trachea and/or esophagus (3,6–8,10). It presents with stridor, respiratory distress, or choking during feeding. Several patients have had multiple malformations (malformed pinnae, hypertelorism, tracheoesophageal fistula, congenital heart anomalies, urogenital anomalies). Many have succumbed in early infancy. Some surely represent examples of the *G syndrome* (10). Laryngeal cleft is also seen in the *Pallister–Hall syndrome.* Lim et al (8) and Tyler (11) listed other associated anomalies.

Laryngeal webs, most often (75%) glottic, may rarely be found with cleft lip–palate or submucous palatal cleft (1,2,4,5). The clinical manifestations are hoarseness, dyspnea, and inspiratory stridor. Excellent reviews of the early and recent literature have been published (9,10).

Both combinations are usually sporadic. Cleft larynx is probably a developmental field defect, occurring with separation of the larynx and esophagus and closure of the larynx.

#### References (Cleft larynx, laryngeal web, and cleft palate)

1. Benjamin B: Congenital laryngeal webs. *Ann Otol Rhinol Laryngol* **92**:317–326, 1983.

2. Bhatia ML et al: Congenital laryngeal web (case 3). *Indian J Otolaryngol* **20**:107–111, 1962.

3. Cameron AH, Williams TC: Cleft larynx: A case of laryngeal obstruction and incompetence. *J Laryngol Otol* **76**:381–387, 1962.

4. Cavanagh F: Congenital laryngeal web. *Proc R Soc Med* **58**:272–277, 1965.

5. Cohen SR: Congenital glottic webs in children. A retrospective review of 51 patients. *Ann Otol Rhinol Laryngol* **94**(Suppl 121):2–6, 1985.

6. Franchi L: Contributo alla casuistica dei restringimenti della laringe vizio conformazione. *Gior Med Mil* **56**:666–668, 1908.

7. Kingston HG et al: Laryngotracheoesophageal cleft—a problem of airway management. *Anesth Analg* **62**:1041–1043, 1983.

8. Lim TA et al: Laryngeal clefts. *Ann Otol Rhinol Laryngol* **88**:837–845, 1979.

9. McHugh HE, Loch WE: Congenital webs of the larynx. *Laryngoscope* **52**:43–65, 1942.

10. Roth B et al: Laryngo-tracheo-esophageal cleft: Clinical features, diagnosis and therapy. *Eur J Pediatr* **140**:41–46, 1983.

11. Tyler DC: Laryngeal cleft: Report of eight patients and a review of the literature. *Am J Med Genet* **21**:61–75, 1985.

## Eye coloboma/cleft lip–palate association

The association of cleft lip–palate with ocular colobomas is extremely heterogeneous. Kingston et al (6) reported autosomal dominant inheritance of various eye anomalies (uveal coloboma, microphthalmia, cataracts, retinal detachment, posterior embryotoxon, glaucoma, ptosis) with cleft lip and palate. Mental retardation was found in 4 of 11 affected members. Atkin and Patil (3) reported a child and his maternal uncle with cleft lip–palate, microphthalmia, cloudy cornea, congenital heart defect, hemivertebrae, ureteral anomalies, hypospadias, cryptorchidism, and ectrodactyly. Pinsky et al (9) reported three sisters with microphthalmia, cloudy cornea, cleft lip, variable anomalies of the pinnae, mental retardation, and spasticity. Their mother had microphthalmia. Ho et al (4) noted the association of various forms of ocular coloboma with congenital heart anomalies, mental retardation, skeletal defects, and cleft palate. Pilotto et al (8) described a child with brachycephaly, cleft lip–palate, high nasal bridge, mildly dysmorphic pinnae, congenital heart defect, vertebral anomalies, short stature, and mental retardation. A similarly affected child was noted by Pena (7). The latter patient had microphthalmia, coloboma of the optic discs and choroid, and scoliosis. Van Schooneveld et al (10) reported 11 patients with Peters anomaly (congenital central corneal leukoma, posterior corneal defect including corneal stroma and Descemet membrane and tiny iris filaments crossing central iris to cornea) in combination with short stature, short fifth fingers with clinodactyly, mental retardation, low positioned and/or dysplastic pinnae, and cleft lip/palate; they suggested autosomal recessive inheritance. S Kapur (personal communication, 1989) described male and female sibs with cleft lip–palate, bulbous nasal tip, microphthalmia, iris coloboma, cataract, ASD, VSD, tetralogy of Fallot, displaced kidneys, hypoplastic thumbs, finger contractures, anal stenosis, and severe mental retardation.

The association of clefting with eye colobomas is also found in *Abruzzo–Erickson syndrome, Baraitser syndrome, Phillips–Griffiths syndrome,* and *Michels syndrome.* Female sibs with cleft lip–palate, corneal opacities, optic nerve coloboma, profound psychomotor retardation, and congenital heart defects were reported by Anyane-Yeboa et al (1). James et al (5) noted the association of cleft lip and/or palate with uveal colobomas. The eye coloboma/cleft lip–palate association overlaps with *CHARGE association* (2).

### References (Eye coloboma/cleft lip–palate association)

1. Anyane-Yeboa K et al: Cleft lip and palate, corneal opacities and profound psychomotor retardation. *Cleft Palate J* **20**:246–250, 1983.

2. Awrich PD et al: CHARGE association anomalies in siblings. *Am J Hum Genet* **34**:80A, 1982.

3. Atkin JF, Patil S: Apparently new oculo-vertebro-acral syndrome. *Am J Med Genet* **19**:585–587, 1984.

4. Ho CK et al: Ocular colobomas, cardiac defect and other anomalies. A study of seven cases including two sibs. *J Med Genet* **12**:289–292, 1975.

5. James PLM et al: Systemic association of uveal colobomata. *Br J Ophthalmol* **58**:917–921, 1974.

6. Kingston HM et al: An autosomal dominant syndrome of uveal colobomata, cleft lip and palate and mental retardation. *J Med Genet* **19**:444–446, 1982.

7. Pena SDJ: Cleft lip and palate, congenital heart disease, scoliosis, short stature, and mental retardation: The Pilotto syndrome. *Birth Defects* **18**(3B):183–186, 1982.

8. Pilotto RF et al: Study of a child with an idiopathic malformation/mental retardation syndrome. *Birth Defects* **11**(2):51–53, 1975.

9. Pinsky L et al: Microphthalmia, corneal opacity, mental retardation and spastic cerebral palsy: An oculocerebral syndrome. *J Pediatr* **67**:387–398, 1965.

10. Van Schooneveld MJ et al: Peters'-plus: A new syndrome. *Ophthalmol Paediatr Genet* **4**:141–146, 1984.

## Oral duplication and cleft palate

The occurrence of supernumerary mouth and/or jaws is rare indeed. Although these cases have differed enough from one another to make each one relatively unique, for purposes of general discussion, we will briefly consider four forms: (a) single mouth with duplication of maxillary arch, (b) supernumerary mouth laterally placed with rudimentary mandible, (c) single mouth with replication of mandibular segments, and (d) true facial duplication (diprosopus) with or without anencephaly. Duplication, in most cases, probably arises from forking of the notochord, partial doubling of the prosencephalon or olfactory placodes, or duplication of growth centers around the stomodeal plate (24).

Our interest here will largely be limited to the first and last forms. The child reported by Avery and Hayward (2) had hypertelorism and a single mouth with two palates, each of which was cleft. There were two arcades of maxillary teeth and a bilaterally grooved or bifid or trifid tongue. We believe that the third lobe represents the tuberculum impar (Fig. 23–2A,B). She was also moderately mentally retarded and had tetralogy of Fallot. Quite similarly affected children have been been described (5,7,10,11a,16,17,20,22,24). Probably two pituitary glands are found in these children (4,20,23). Two of the children had hamartomatous tissue hanging from the palate. An unusual variant of the condition was reported by Fish (13). This child also had a nasal dermoid, a median cleft of the upper lip, and a fleshy hamartomatous mass (fat, cartilage, nevus cells) on the chin (see p. 712). Similar cases but without cleft palate have been described (9,14,23).

A child with two mouths, one of which had cleft palate, was documented by Bacsich et al (4). We are also aware of several examples of children with an extra tongue. The children have other stigmata which would classify them in the *oculo–auriculo–vertebral spectrum.* Some of these infants had cleft palate.

Diprosopus is the most extreme form of facial doubling. There may be two lateral eyes, a median eye, and doubling of the hypophysis, mouth, and nose. Clefting of the lip or palate is common. There have been several examples, some of which have been anencephalic (1,3,6,11,15,18,19,21).

### References (Oral duplication and cleft palate)

1. Andres R, Bixler D: Anencephaly with facial duplication. *J Craniofac Genet Dev Biol* **1**:245–252, 1981.

2. Avery JK, Hayward JR: Duplication of oral structures with cleft palate. *Cleft Palate J* **6**:505–516, 1969.

3. Ayer AA, Mariappa D: Hare-lip and cleft lip in double monsters. *Indian Acad Sci Proc* **38**:153–156, 1953 (Case 2).

4. Bacsich P et al: A rare case of duplicitas anterior: A female infant with two mouths and two pituitaries. *J Anat* **98**:292–293, 1964.

5. Bell RC: A child with two tongues. *Br J Plast Surg* **24**:193–196, 1971.

6. Broder SB: An anencephalic monster with "rhinodymie" and other anomalies. *Am J Pathol* **11**:761–774, 1935.

7. Bumm E: Ein Fall von Verdoppelung des Oberkiefers. *Arch Klin Chir* **135**:506–519, 1925.

8. Burckhard G: Diprosopus distomus. *Z Geburtsch Gynäk* **40**:20–33, 1899.

9. Chandra R: Congenital duplication of lip, maxilla and palate. *Br J Plast Surg* **31**:46–47, 1978.

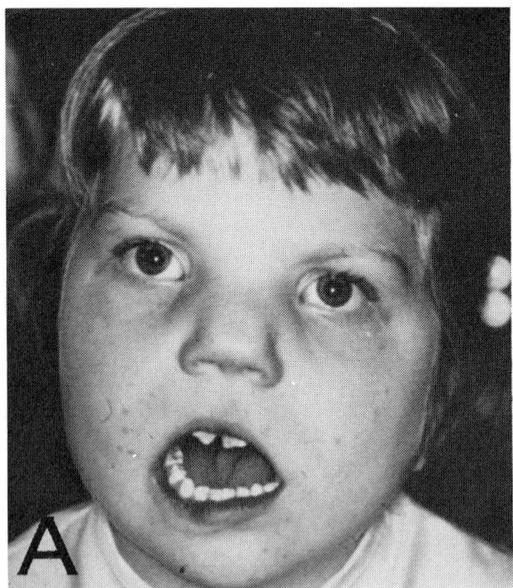


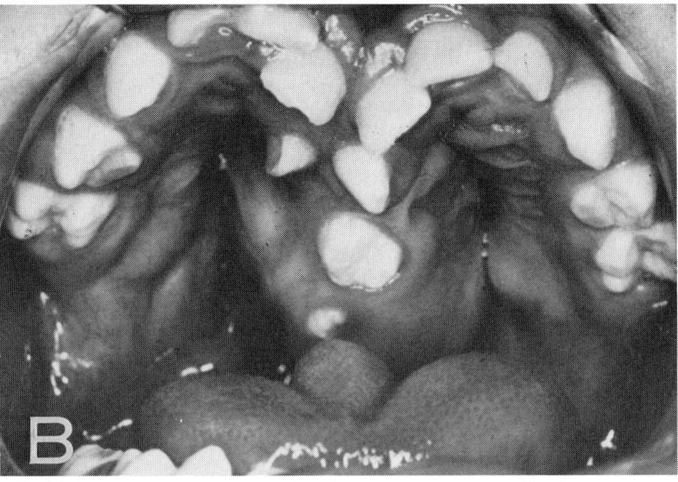


Fig. 23–2. *Oral duplication and cleft palate.* (A,B) Ocular hypertelorism, two hard palates, and two dental arches. Note two cleft palates and partial duplication of the tongue. (From *JK Avery* and *JR Hayward,* Cleft Palate J **6**:505, 1969.)


10. Clausnitzer KH: Über eine Diprosopie bei geschlossenem Neurocranium. *Zbl Allg Pathol* **95**:329–336, 1956.

11. Dodinval P, Humblet J: Etude anatomique d'un cas de prosopie. *J Génét Hum* **22**:225–230, 1974.

11a. Fearon JA, Mulliken JB: Midfacial duplication: A rare malformation sequence. *Plast Reconstr Surg* **79**:260–264, 1987.

12. Feller A: Über geringe Grade von Prosopie. *Z Anat Entwickl Gesch* **94**:180–205, 1931.

13. Fish J: Median cleft lip with dental arch and palate duplication. *Br Dent J* **140**:143–145, 1976.

14. Gupta DS: Double palate. *Oral Surg* **40**:53–55, 1975.

15. Hayk H van: Über die Beziehungen des Epignathus zum Diprosopus. *Zbl Allg Pathol* **85**:171–175, 1949.

16. Koblin I: Zwischenkieferverdoppelung mit normotoper und heterotoper Zahnüberzahl sowie Gaumenspalte. *Dtsch Zahn Mund Kieferheilkd* **54**:204–210, 1970.

17. Kohnz R: Erstbeobachtung einer Zwischenkieferverdoppelung beim Menschen mit Epignathus, Zungenkörper- und Gaumenspalte. *Dtsch Zahn Mund Kieferheilkd* **35**:18–32, 1961.

18. Latteier KK, Anderson RT: Diprosopus. *Med Radiogr Photog* **28**:22–24, 1952.

19. Marzels G: Parallel duplication of the face in an anencephalic foetus. *J Obstet Gynecol Br Emp* 45:679–682, 1938.

20. Morton WRM: Duplication of the pituitary and stomatodaeal structures in a 38-week male infant. *Arch Dis Childh* **32**:135–141, 1957.

20a. Obwegeser HL et al: Facial duplication. The unique case of Antonio. *J Maxillofac Surg* **6**:179–188, 1978.

21. Pavone L et al: Diprosopus with association malformations: Report of two cases. *Am J Med Genet* **28**:85–88, 1987.

22. Robertson NRE: Apparent duplication of the upper alveolar process and dentition. *Br Dent J* **12**:333–334, 1970.

23. Steinhilber W: Partielle Doppelanlage des Unterkiefers mit Epignathus, Zwischenkieferbürzel und Zungenspalten. *Z Kinderheilkd* **112**:171–176, 1972.

24. Wittkampf ARM, van Limborgh J: Duplication of structures around the stomatodeum. *J Maxillofac Surg* **12**:17–20, 1984.

25. Wolf H, Dubrauszky V: Verdoppelung der Kiefer in einem Falle von Anencephalie mit Epignathus. *Dtsch Zahnärztl Z* **4**:65–75, 1949.


## Thoracopagous twins and cleft lip–palate

Knowledge of conjoined twins dates back to the Neolithic period (29). There are early reports from the sixteenth century (5,7). The four basic types of conjoined twins: (1) thoracopagus (73%), exhibiting various degrees of thoracic fusion, (2) pygopagus (19%), generally fused at the level of the sacral spine, (3) ischiopagus (6%), in which each twin has a single leg and a common fused leg, and (4) craniopagus (2%), exhibiting fusion of the skulls (24).

Investigators agree that conjoined twins result from incomplete separation of a single fertilized ovum, prior to the end of the third week (19,20,31). The degree and site of incomplete splitting determine the type of conjoined twins. It has been suggested that a defect in the primitive streak, aging of the ovum, and environmental factors may play a causative role. Experimental manipulation of the primitive streak in amphibian and chick embryos has induced complete and incomplete twinning (31).

Epidemiologic studies, based on data from the Birth Defects Monitoring Program of the CDC, indicated that 81 sets of conjoined twins were found among 7,903,000 births from 1970 to 1977, representing an incidence of roughly 10 per million births. Thoracoomphalopagus represented 28%, followed by thoracopagus 18%, omphalopagus 10%, parasitic twins 10%, and craniopagus 6%. There was an approximately 2F:1M breakdown. The birth prevalence among nonwhites (16/million) was greater than among whites (9/million). No maternal age effect was found. Among the 81 cases, six presented orofacial clefts, five being cleft lip–palate and one isolated cleft palate (9). The six cases of clefts all occurred in thoracopagous twins, representing 7% of all cases or 13% of thoracopagous cases. Other studies have shown similar data (2,19,20). Thoracopagous twins thus have a higher rate of facial clefting than that expected by chance (9,11). Mirror image cleft lip has been reported in several instances (10,15,18–20,25) (Fig. 23–3). Discordance for cleft lip in thoracopagous twins has also been found (19,20,22,24). Cleft palate (4,8) and cleft lip and palate (12) have been reported in cephalothoracopagus. Ayer and Mariappa (3) observed unilateral cleft lip and palate in both thoracopagous twins and bilateral cleft lip and palate in both faces of a diprosopic child. Similar cases were documented by Latteier and Anderson (16), Ueo et al (26), and Pavone et al (21). Ivy (14) described a two-headed fetus, one with cleft lip–palate and the other normal. Buchta (6) described anencephaly in thoracopagous twins with facial clefting. Cleft palate has been reported in craniopagus (30). Rowlatt (23) noted a double tongue and a single pituitary fossa containing two pituitary glands in thoracocephalopagous twins.

### References (Thoracopagous twins and cleft lip–palate)


1. Aird I: Conjoined twins—further observations. *Br Med J* **1**:1313–1315, 1959.

2. Apuzzio JJ et al: Prenatal diagnosis of conjoined twins. *Am J Obstet Gynecol* **148**:343–344, 1984.

3. Ayer AA, Mariappa D: Hare-lip and cleft palate in double monsters. *Indian Acad Sci* **38**:153–156, 1953.

4. Bartlett RC: Cephalothoracopagus. *Arch Pathol* **68**:292–298, 1959.

5. Bertsch M et al: Les monstres a deux têtes. A propos d'une nouvelle observation. *Pédiatrie* **31**:473–486, 1976.

6. Buchta RM: Anencephaly in female thoracopagous conjoined twins. *Clin Pediatr* **12**:598–599, 1973.

7. Calagan JL: The conjoined twins born near Worms, 1495. *J Hist Med Allied Sci* **38**:450–451, 1983.

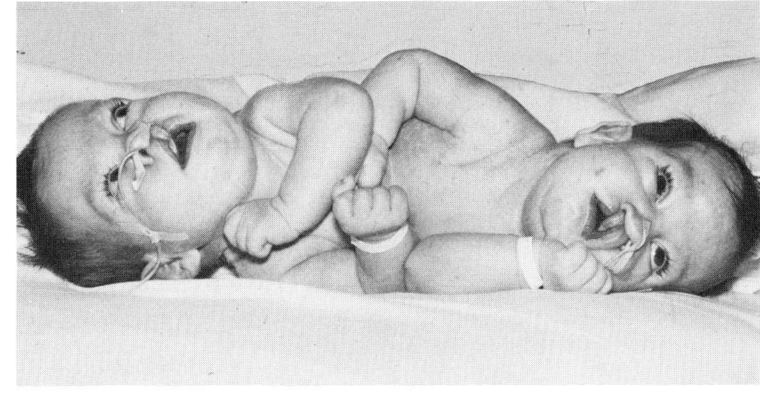

Fig. 23–3. *Thoracopagous twins and cleft lip–palate.* Conjoined twins with mirror image clefts. (From *R Stellmach* and *G Frenkel,* Dtsch Zahnärztl Z **25**:28, 1970.)


8. Delpram WJ, Baird PJ: Cephalothoracopagus syncephalus: A case report with previously unreported anatomical abnormalities and chromosomal analysis. *Teratology* **29**:1–9, 1984.

9. Edmonds LD, Layde PM: Conjoined twins in the United States, 1970–1977. *Teratology* **25**:301–308, 1982.

10. Edwards WD et al: Conjoined thoracopagus twins. *Circulation* **56**:491–497, 1977.

11. Guttmacher AF, Nichols BL: Teratology of conjoined twins. *Birth Defects* **3**(1):2–9, 1967.

12. Herbst E, Apffelstaedt M: *Atlas und Grundriss der Missbildungen der Kiefer und Zähne.* J.F. Lehmann, München, 1928.

13. Herring SW, Rowlatt UF: Anatomy and embryology in cephalothoracopagus twins. *Teratology* **23**:159–173, 1981.

14. Ivy RJ: A curiosity in the area of cleft lip–cleft palate. *Plast Reconstr Surg* **42**:160, 1968.

15. Lange VB et al: Zur Geburt siamesischer Zwillinge. *Zbl Gynäkol* **105**:105–109, 1983.

16. Latteier KK, Anderson RT: Diprosopus. *Med Radiogr Photog* **28**:22–24, 1952.

17. Markovec MD: Conjoined twins with mirror-image clefts of lip and palate. *Cleft Palate J* **7**:690–695, 1970.

18. Nichols BL et al: General clinical management of thoracopagous twins. *Birth Defects* **3**(1):38–51, 1967.

19. Ornoy A et al: Conjoined twins. A report of a case of thoracoomphalopagus. *Isr J Med Sci* **7**:679–684, 1971.

20. Ornoy A et al: Asymmetry and discordance for congenital anomalies in conjoined twins: A report of six cases. *Teratology* **22**:145–154, 1980.

21. Pavone L et al: Diprosopus with associated malformations: Report of two cases. *Am J Med Genet* **28**:85–88, 1987.

22. Robertson GSA, McKenzie J: Thoracopagous twins with different first arch defects. *Br J Surg* **51**:362–363, 1964.

23. Rowlatt U: Anencephaly and unilateral cleft lip and palate in conjoined twins. *Cleft Palate J* **15**:73–75, 1978.

24. Sangvichien S: A thoracopagus, one with harelip and cleft palate. *Anat Rec* **67**:157–158, 1936.

25. Stellmach R, Frenkel G: Über das Vorkommen von spiegelbildlich konkordanten vollständigen einseitigen Lippen-Kiefer-Gaumenspalten bei einem siamesichen Zwillingspaar. *Dtsch Zahnärztl Z* **25**:28–31, 1970.

26. Ueo K et al: A case of thoracopagus with clefts of lip and palate. *J Jpn-Stomatol Soc* **19**:252–258, 1970.

27. Vestergaard P: Triplets pregnancy with a normal foetus and a dicephalus dibrachius sirenomelus. *Acta Obstet Gynecol Scand* **51**:93–94, 1972.

28. Votteler TP: Surgical separation of conjoined twins. *AORN J* **35**:35–46, 1982.

29. Warkany J: *Congenital Malformations: Notes and Comments.* Year Book Medical Publishers, Chicago, 1971.

30. Wieczorkiewicz B, Komraus-Gatniejewski M: Craniopagus. *Z Kinderchir* **14**:207–209, 1974.

31. Zimmerman AA: Embryologic and anatomic considerations of conjoined twins. *Birth Defects* **3**(1):18–27, 1967.


## Oral teratoma, encephalomeningocele, and cleft palate

Epignathus in a term used to describe a congenital teratoma protruding from the mouth of an infant (1–9,13,15,16,20,21,26,27). It has been shown that these growths usually develop in the area of Rathke's pouch and project and fill the pharynx or oral cavity between the palatal shelves (7,26) (Fig. 23–4). Among 81 cases of teratomas seen in infancy and childhood, cleft palate was found in 1.2% of the 81 cases (12). Affected infants rarely survive the neonatal period.

Erich (7), reviewing 22 cases of oral teratomas, noted that 11 were attached to the sphenoid bone, six to the lateral wall of the epipharynx near the opening of the eustachian tube, two to the soft palate, and one to the hard palate. They may even be attached to the lower jaw (10b,22). The size of the teratoma varies from several centimeters to huge parasitic masses. The smaller and simpler cases are compatible with life and, as a rule, arise further back in the oral cavity (3,15). The larger examples often cause deformity of the mandible due to mechanical interference in growth (15). Congenital oral teratomas are far more common in females.

The cleft, in most cases, has involved both hard and soft palates, but in others only the soft palate is cleft (3,15,20). The cleft may be submucosal (26). Hirshowitz and Mahler (11) reported a case of congenital intraoral teratoma with *frontonasal malformation.* The teratoma was attached to the area normally occupied by the maxillary central incisors.

Histologically the teratomas can be solid and/or cystic. They consist of tissues derived from the three germ layers, including skin and adnexa, mucous and nonmucous secreting glands, muscle, cartilage, lymphoid tissue, nervous tissue, and bone. Some are covered with hair (1,3,20,27). Wynn et al (27) found a perfectly formed finger with

Fig. 23–4. *Oral teratoma and cleft palate.* (Courtesy of *CO Dummett,* Los Angeles, California.)

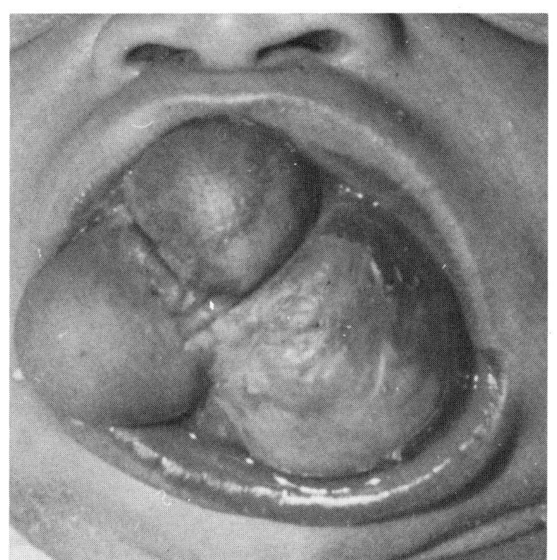

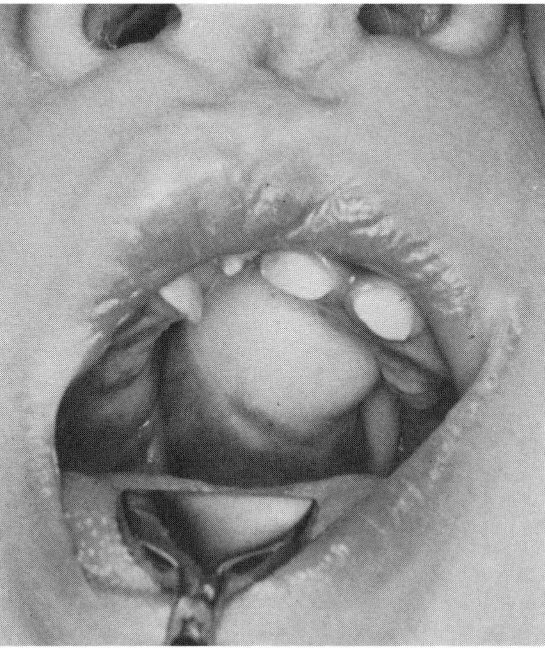

Fig. 23–5. *Encephalomeningocele and cleft palate.* Repaired cleft lip, unrepaired cleft palate.

bones, joints and nail inside the mass. Malignant changes have not been observed in these tumors (15).

Cleft lip and/or cleft palate has also been described in association with glioma or encephalomeningocele which herniates through the sphenoid bone to present in the mouth (4,10,14,17,18,21,22,24,25,28) (Fig. 23–5). The encephalocele may also present as a bulge on the frontal bone (17) and may be associated with ocular hypertelorism.

The cleft can involve both lip and palate (9,17,21,22) or only the palate (10,18,23,28). Exner (8) illustrated a most unusual midline cleft of the upper lip and a separate cleft of the hard and soft palates.

Naidich et al (19a), in a survey of 30 cases of basal (sphenoidal, sphenoethmoidal, ethmoidal) encephaloceles, noted that in addition to optic nerve dysplasia (rarely microphthalmia) (40%) and callosal agenesis (rarely lipoma) (40%), about 50% were associated with median cleft of the upper lip. Conversely, Grogono's (10a) analysis of 50 patients with callosal agenesis showed one (2%) with median cleft upper lip. Carpenter and Druckemiller (3a), surveying 45 patients with callosal agenesis, also found midline cleft upper lip, cleft lip, and cleft palate. It would, thus, appear that midline cleft of the upper lip in the presence of hypertelorism should suggest further examination of the child for midline dysraphic anomalies such as basal encephalocele, callosal agenesis, or optic nerve dysplasia.

Histologically, these masses are composed of glial tissue or choroid plexus (10,28) or even melanocytes (10).

### References (Oral teratoma, encephalomeningocele, and cleft palate)


1. Agris J et al: Teratomatous choristoma and its associated syndromes. *Plast Reconstr Surg* **58**:232–238, 1976.

2. Arnold J: Über behaarte Polyposen der Rachen-Mundhöhle und deren Stellung zur den Teratomen. *Virchows Arch Pathol Anat* **111**:176–210, 1888.

3. Bennett JP: A case of epignathus with long term survival. *Br J Plast Surg* **23**:360–364, 1970.

3a. Carpenter MB, Druckemiller WB: Agenesis of the corpus callosum diagnosed during life; review of the literature and presentation of two cases. *Arch Neurol Psychiatr* **69**:305–322, 1953.

4. Crockford DA et al: Some observations on congenital teratoma and ectopic cerebromata. *Plast Reconstr Surg* **24**:31–42, 1971.

5. Dohlman G, Sjövall A: Large epignathous teratoma successfully operated upon immediately after birth. *Glasgow Med J* **34**:122–126, 1953.

6. Dummett CO et al: Epignathoid teratoma. *J Can Dent Assoc* **29**:788–791, 1963.

7. Erich WE: Teratoid parasites of the mouth (episphenoids, epipalati, epignathia). *Am J Orthodont* **31**:650–659, 1945.

8. Exner A: Über basale Cephalocele. *Dtsch Z Chir* **90**:23–41, 1907.

9. Fogh-Andersen P: *Inheritance of Harelip and Cleft Palate.* Nyt Nordisk Forlag, Copenhagen, 1942, pp 26–38.

10. Gold AH et al: Central nervous system heterotopia in association with cleft palate. *Plast Reconstr Surg* **66**:434–441, 1980.

10a. Grogono JL: Children with agenesis of the corpus callosum. *Dev Med Child Neurol* **10**:613–616, 1968.

10b. Hirabayashi S, Ueda K: Nasopharyngeal teratoma attached to the lower jaw. *Plast Reconstr Surg* **76**:939–941, 1985.

11. Hirshowitz B, Mahler D: Epignathus and bifid nose: A report of a rare congenital anomaly. *Br J Plast Surg* **26**:271–273, 1973.

12. Hossein Mahour G et al: Teratoma in infancy and childhood. Experience with 81 cases. *Surgery* **76**:309–318, 1974.

13. Hoyek H von: Über die Beziehungen des Epignathus zum Diprosopus. *Zentralbl Allg Pathol* **85**:171–175, 1949.

14. Ingraham FD, Matson DD: Spina bifida and cranium bifidum. IV. An unusual nasopharyngeal encephalocele. *N Engl J Med* **228**:815–820, 1943.

15. Keswani RK et al: Epignathus: A case report. *Br J Plast Surg* **21**:355–359, 1968.

16. Kraus EJ: Über ein epignathisches Teratom der Hypophysengegend. *Virchows Arch Pathol Anat* **271**:546–555, 1929.

17. Lewin ML, Shuster MM: Transpalatal correction of basilar meningocoele with cleft palate. *Arch Surg* **90**:687–693, 1965.

18. Low NL et al: Brain tissue in the nose and throat (Case 3). *Pediatrics* **18**:254–259, 1956.

19. Modesti LM et al: Spheno-ethmoidal encephalocele: A case report and review of the literature. *Child's Brain* **3**:140–153, 1977.

19a. Naidich TP et al: Midline craniofacial dysraphism: Midline cleft upper lip, basal encephalocele, callosal agenesis and optic nerve dysplasia. *Concepts Pediatr Neurosurg* **4**:186–207, 1983.

20. Nase H: Behaarter Rachenpolyp als Ursache einer Gaumenspaltbildung. *Zentralbl Chir* **66**:29–32, 1939.

21. Neumann H, Peters UH: Frontale Cephalocele kombiniert mit Kiefergaumenlippenspalte und psychomotorischen Anfällen. *Arch Psychiatr Nervenkr* **200**:531–540, 1960.

22. Rintala A, Ranta R: Separate epignathi of the mandible and the nasopharynx with cleft palate. *Br J Plast Surg* **27**:103–106, 1974.

23. Shapiro MJ, Mix BS: Heterotopic brain tissue of the palate. A report of two cases. *Arch Otolaryngol* **87**:522–526, 1968.

24. Stuart EA: An otolaryngologic aspect of frontal meningocoele. *Arch Otolaryngol* **40**:171–174, 1944.

25. Vistnes LM et al: Nasal glioma: Case report. *Plast Reconstr Surg* **43**:195–197, 1969.

26. Wilson JW, Gehweiler JA: Teratoma of the face associated with a patent canal extending into the cranial cavity (Rathke's pouch) in a three-week-old child. *J Pediatr Surg* **5**:349–359, 1970.

27. Wynn SK et al: Epignathus. *Am J Dis Child* **91**:495–497, 1956.

28. Zarem HA et al: Heterotopic brain in the nasopharynx and soft palate: Report of two cases. *Surgery* **61**:483–486, 1967.


## Hypogonadotropic hypogonadism and cleft lip and/or palate

Cleft lip and/or cleft palate have been reported in individuals, both male and female, with low urinary gonadotropin levels. They have normal sexual differentiation but manifest oligomenorrhea or infantile testes with aspermia and absent Leydig cells. Most had anosmia (Kallmann syndrome) (1–6,9); others did not (7,8). Conversely, it has been estimated that about 15% of patients with Kallmann syndrome have clefts. See *holoprosencephaly* for related material.

### References (Hypogonadotropic hypogonadism and cleft lip and/or palate)


1. Christian JC et al: Hypogonadotropic hypogonadism with anosmia. The Kallmann syndrome (Case 4). *Birth Defects* **7**(6):166–171, 1971.

2. Henkin RI et al: Improvement of recognition of familial hyposmia in patients with facial hypoplasia and growth retardation. *Clin Res* **14**:236, 1966.

3. Lieblich JM et al: Syndrome of anosmia with hypogonadotropic hypo-

gonadism (Kallmann syndrome): Clinical and laboratory studies in 23 cases. *Am J Med* **73**:506–519, 1982.
4. Rosen SW: The syndrome of hypogonadism, anosmia, and midline cranial anomalies. *Proc 47th Meet Endocrinol Soc,* New York, 1965.
5. Santen RJ, Paulsen CA: Hypogonadotropic eunuchoidism. I. Clinical study of the mode of inheritance. *J Clin Endocrinol* **36**:47–54, 1973.
6. Sparkes RS et al: Familial hypogonadotropic hypogonadism and anosmia. *Arch Intern Med* **121**:534–538, 1968.
7. Tuohy PG, Franklin RC: Cleft palate and gonadotropin deficiency. *Arch Dis Childh* **59**:580–582, 1984.
8. Turner RC et al: Cryptorchidism in a family with Kallmann's syndrome. *Proc R Soc Med* **67**:33–35, 1974.
9. White BJ et al: The syndrome of anosmia with hypogonadotropic hypogonadism: A genetic study of 18 new families and a review. *Am J Med Genet* **15**:417–435, 1983.

## Hypophyseal short stature and cleft lip–palate

Francés et al (2) and others (3,5–9) described cleft lip and/or palate associated with either panhypopituitarism or isolated growth hormone deficiency. A converse study was done in which 200 children with isolated clefts of the lip, palate, or both were examined. Among those that were short, about 30% showed partial or complete growth hormone deficiency. However, normal values were found by Köster et al (4) in an unselected group of 200 patients with clefts. The second patient reported by Francés et al (2) appeared to have some degree of the premaxillary agenesis form of holoprosencephaly.

The patient described by Zuppinger et al (10) also exhibited colobomata of the choroid and optic nerve with resultant retinal detachment. Brewer (1) described bilateral cleft lip and agenesis of the pituitary gland. Also see *hypogonadotropic hypogonadism and cleft lip and/or cleft palate* and *holoprosencephaly.*

Growth hormone deficiency may be found in association with a number of congenital malformations: *anencephaly, holoprosencephaly, frontonasal malformation, Rieger syndrome,* microcephaly, *single central incisor, EEC syndrome,* and septo-optic dysplasia.

#### References (Hypophyseal short stature and cleft lip–palate)

1. Brewer DB: Congenital absence of the pituitary gland and its consequences. *J Pathol Bact* **73**:59–67, 1957.
2. Francés JM et al: Hypophysärer Zwergwuchs bei Lippen-Kiefer-Spalte. *Helv Paediatr Acta* **21**:315–322, 1966.
3. Gendrel D et al: Les hypopituitarismes congénitaux par anomalies de la ligne médiane. *Arch Fr Pédiatr* **38**:227–232, 1981.
4. Köster K et al: Wachstum und Wachstumshormon bei Kindern mit angeborenen Lippen-Kiefer-Gaumenspalten. *Klin Pädiatr* **196**:304–309, 1984.
5. Laron Z et al: Pituitary growth hormone insufficiency associated with cleft lip and palate. *Helv Paediatr Acta* **24**:576–581, 1969.
6. Prüsener L: Hypophysäre Wachstumshemmung mit Kachexie beim Kinde. *Z Mensch Vererb Konstitl* **17**:215–224, 1932–3.
7. Roitman A, Laron Z: Hypothalmo-pituitary hormone insufficiency associated with cleft lip and palate. *Arch Dis Childh* **53**:952–955, 1978.
8. Rudman D et al: Prevalence of growth hormone deficiency in children with cleft lip or palate. *J Pediatr* **93**:378–382, 1978.
9. Zimmerman TS et al: Hypopituitarism with normal or increased height. *Am J Med Genet* **42**:146–150, 1967.
10. Zuppinger KA et al: Cleft lip and choroidal coloboma associated with multiple hypothalamo-pituitary dysfunctions. *J Clin Endocrinol Metab* **33**:934–939, 1971.

## Amelia and cleft palate

In a comprehensive literature survey of amelia, Ilberg (4) did not note an association with facial clefting. Centa and Rovei (1) noted amelia, micrognathia, and cleft palate in a child described as having the Robin sequence. Holthusen (3) reported a less severely affected child. The arms, femurs, and fibulas were absent. Perhaps this case should be classified with absent femurs and cleft palate. Frantz and O'Rahilly (2) noted the association between amelia and cleft lip and cleft palate.

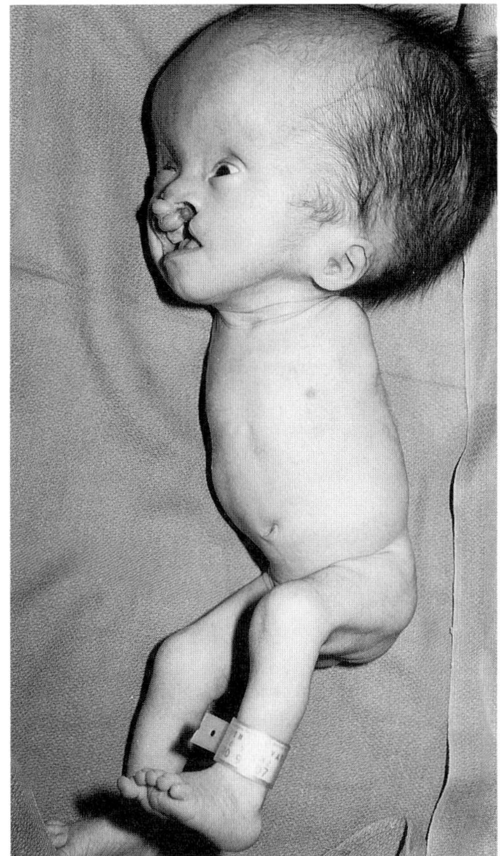

Fig. 23–6. *Amelia and cleft palate.* Complete brachial amelia with hydrocephaly and bilateral cleft lip–palate. [From *DKC Yim* and *AJ Ebbin,* Syndrome Ident **8**(1):3, 1982.]

Yim and Ebbin (5) reported brachial amelia with cleft lip–palate and hydrocephalus (Fig. 23–6).

#### References (Amelia and cleft palate)

1. Centa A, Rovei S: Dismelia grave non talidomidica associata a sindrome di Pierre Robin. *Pathologica* **57**:245–250, 1965.
2. Frantz CH, O'Rahilly R: Congenital skeletal limb deficiencies. *J Bone Jt Surg* **43A**:1202–1224, 1961.
3. Holthusen W: The Pierre Robin syndrome: Unusual associated developmental defects. *Ann Radiol (Paris)* **15**:253–262, 1972.
4. Ilberg G: Foetus ohne Arme und Beine (Amelos): Zusammenstellung von Beobachtungen über Amelie. *Z Geburtsch Gynäk* **114**:174–184, 1936.
5. Yim DKC, Ebbin AJ: Case Report 82. Bilateral brachial amelia with cleft lip and palate and hydrocephaly. *Syndrome Ident* **8**(1):3–5, 1982.

## Cleft lip–palate and congenital heart defects

Clefting can be associated with a wide spectrum of congenital heart defects. Boeson et al (2) noted 11 cases of cleft lip and/or cleft palate among 516 patients with congenital heart anomalies. The cardiac defects were of various types, the most common being ASD and VSD. Rabl and Schulz (7) noted the association of cleft lip–palate with pulmonary valvular atresia and septal defect in twins. They also described two sibs with clefts. Geis et al (4) found that 6% of their cleft lip–palate population had congenital heart defects of various forms. Baetz-Greenwalt et al (1) found 14% of infants with hypoplastic right-sided heart to have cleft palate.

Shah et al (8), in an extensive review of congenital cardiac malformations in children with facial clefts, found no consistent pattern of heart abnormalities. Among 32 children with cleft lip or cleft palate who died, cardiac anomalies were present in 66%. Seventeen had cleft palate, six had Robin sequence, four had tetralogy of Fallot, four had tricuspid stenosis, three had coarctation of the aorta, and three had ASD of the secundum type. There were no cases of transposition of the great vessels or examples of a single ventricle among the group. Similar observations were made by Okada et al (6) who also noted that the Robin sequence was associated with coarctation of the aorta and tetralogy of Fallot.

An equally diverse group with congenital heart anomalies and hypoplastic, aplastic, or triploid thumbs is described in Holt–Oram syndrome (3,5). The reader is also referred to *cleft lip–palate and extrathoracic ectopia of the heart*.

#### References (Cleft lip–palate and congenital heart defects)

1. Baetz-Greenwalt B et al: Hypoplastic right sided heart complex: A cluster of cases with associated congenital birth defects—a new syndrome? *J Pediatr* **103**:399–401, 1983.

2. Boeson I et al: Extracardiac congenital malformations in children with congenital heart diseases. *Acta Paediatr Scand (Suppl)* **146**:28–33, 1963.

3. Brans YW et al: The upper limb-cardiovascular syndrome. *Am J Dis Child* **124**:779–783, 1972.

4. Geis N et al: The prevalence of congenital heart disease among the population of a metropolitan cleft lip and palate clinic. *Cleft Palate J* **18**:19–23, 1981.

5. Holt M, Oram S: Familial heart disease with skeletal malformations. *Br Heart J* **22**:236–242, 1960.

6. Okada R et al: Extracardiac malformations associated with congenital heart disease. *Arch Pathol* **85**:649–657, 1968.

7. Rabl R, Schulz F: Angeborene Herzfehler und Lippen-Kiefer-Gaumenspalten bei Zwillingen. *Virchows Arch Pathol Anat* **305**:505–520, 1939.

8. Shah CV et al: Cardiac malformations with facial clefts: With observation on the Pierre Robin syndrome. *Am J Dis Child* **119**:238–244, 1970.

## Enlarged parietal bone fenestrae and cleft lip and/or palate

Enlarged parietal bone defects (Catlin marks), usually bilateral, frequently represent an isolated autosomal dominant trait. They may be associated with clefting of the lip or palate with a possibly greater-than-chance frequency.

The nature of these cranial defects has been discussed by numerous authors (5–8). The fenestrae are usually symptom free, but occasionally there is associated headache (7). The frequency of the parietal bone defects would appear to be less than 1 in 25,000 individuals. Over 300 cases have been described. Miscellaneous other malformations may be associated with enlarged parietal fenestrae. Cases in which clefting has been noted are those of Irvine and Taylor (3), James (4), and Hollender (2). In the case of Hässler (1), a maternal uncle had cleft lip.

#### References (Enlarged parietal bone fenestrae and cleft lip and/or palate)

1. Hässler E: Fenestrae parietales symmetricae. *Mschr Kinderheilkd* **64**:337–340, 1936.

2. Hollender L: Enlarged parietal foramina. *Oral Surg* **23**:447–453, 1967.

3. Irvine ED, Taylor FW: Hereditary and congenital large parietal foramina. *Br J Radiol* **9**:456–462, 1936.

4. James FE: Hypertelorism associated with poor frontal development of skull and bilateral Sprengel's shoulders. *Br Med J* **1**:1019–1020, 1959.

5. Mükke GC, Poppe W: Fenestrae parietales symmetricae unter phylogenetischem Aspekt. *Fortschr Roentgenstr* **111**:300–303, 1969.

6. O'Rahilly R, Twohig MJ: Foramina parietalia permagna. *Am J Roentgenol* **67**:551–561, 1952.

7. Pang D, Lin A: Symptomatic large parietal foramina. *Neurosurgery* **11**:33–37, 1982.

8. Pepper OHP, Pendergrass EP: Hereditary occurrences of enlarged parietal foramina. *Am J Roentgenol* **35**:1–8, 1936.

## Extrathoracic ectopia of the heart and cleft lip–palate

Millhouse and Joos (15) tabulated reported cases of extrathoracic ectopia of the heart and found cleft lip and/or cleft palate in 21 of 27 cases (Fig. 23–7). Our independent tally does not confirm their data; but among the 100 cases that we reviewed, there is surely a high correlation between ectopia of the heart and cleft lip–palate (5,9,12,14,15,17,21,22) or cleft palate (2,6,13). Many examples clearly represent the amniotic band sequence following early rupture of the chorion or yolk sac (1,3,6,7,10,11,18,20,20a). Numerous additional examples have been cited by Kaplan et al (8). See *amnion rupture sequence* for detailed discussion. In one case there was cleft mandible

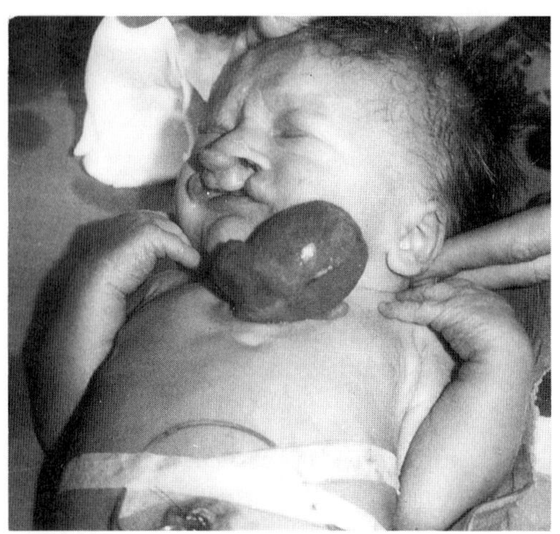

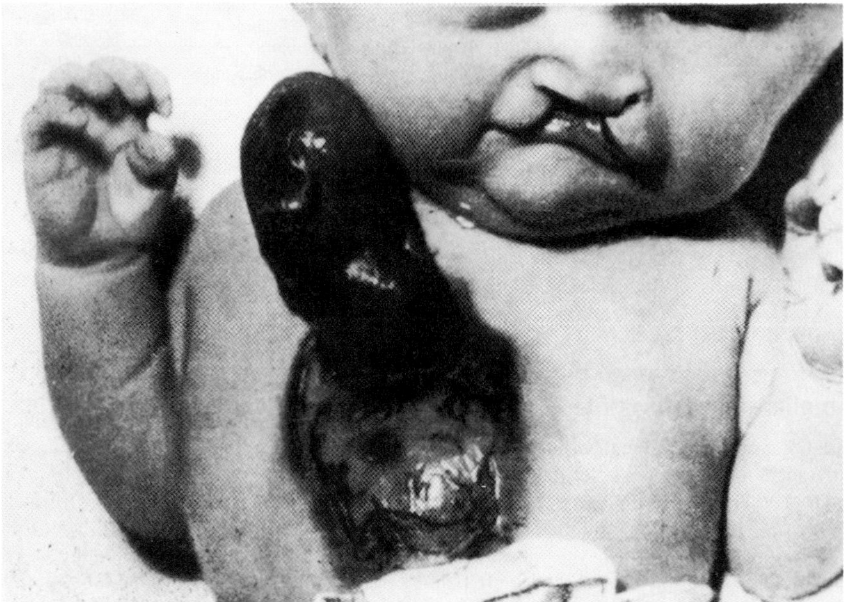

A B

Fig. 23–7. *Extrathoracic ectopia of the heart and cleft lip–palate.* (A,B) Note bilateral cleft lip–palate and extrathoracic position of the heart. (A from *F Didier et al,* Z Kinderchir **14:**252, 1974. B from *ER Meitner,* Anat Anz **107**:222, 1959.)

(9). Cleft sternum, which may represent the same anomaly, has been described with cleft lip (9).

#### References (Extrathoracic ectopia of the heart and cleft lip–palate)

1. Bieber FR et al: Amniotic band sequence associated with ectopia cordis in one twin. *J Pediatr* **105**:817–819, 1984.

2. Cosgrove SA, St. George AV: Report of a case of ectopia cordis. *Am J Dis Child* **27**:594–597, 1924.

3. Crelin ES: An unusual ectopia cordis in a human stillborn infant. *Yale J Biol Med* **30**:38–42, 1957.

4. Didier F et al: Un cas d'ectopie cardiaque extrathoracique. *Z Kinderchir* **14**:252–259, 1974.

5. George JP: Ectopia cordis. *Can Med Assoc J* **53**:167–168, 1945.

6. Hurwitt ES, Lebendiger A: Ectopia cordis in a twin. *Arch Surg* **78**:197–202, 1959.

7. Kanagasuntheram R, Verzin JA: Ectopia cordis in man. *Thorax* **17**:150–167, 1962.

8. Kaplan LC et al: Ectopia cordis and cleft sternum: Evidence for mechanical teratogenesis following rupture of the chorion or yolk sac. *Am J Med Genet* **21**:187–199, 1985.

9. Kellett CE: Ectopia cordis cum sterni fissura. *Arch Dis Childh* **11**:205–214, 1936.

10. Lachmann D et al: Spaltbildung der vorderen Leibeswand mit kompletter Ektopie des Herzens. *Pädiatr Pädol* **6**:181–187, 1971.

11. Ledényi J: Ectopia cordis toracica und Theorien ihrer Entstehung. *Anat Anz* **79**:277–287, 1935.

12. Lilwall-Cormac J: Fissure of the sternum: Ectopia of the heart. *Br Med J* **1**:1320, 1903.

13. Lintgen C: A case of ectopia cordis. *Am J Obstet Gynecol* **25**:449–453, 1933.

14. Mattenucci: Un caso di cordis congenita in feta vivo. *Zbl Inn Med* **25**:25–26, 1904.

15. Meitner ER: Über die Ectopia cordis. *Anat Anz* **107**:222–224, 1959.

16. Millhouse RF, Joos HA: Extrathoracic ectopia cordis: Report of cases and review of literature. *Am Heart J* **57**:470–476, 1959.

16a. Moore KL: *The Developing Human,* 3rd ed. W.B. Saunders, Philadelphia, 1982, p. 320.

17. Nguyen-Huu, Cazes C: L'ectocardie thoracique avec fissure sternale. *Arch Fr Pédiatr* **9**:915–933, 1952.

18. Puddu V, Cammarella C: Un cas d'ectopia cordis. *Arch Mal Coeur* **31**:862–872, 1938.

19. Roth F: Morphologie und Pathogenese der Ektopia cordis congenita. *Frankf Pathol* **53**:60–100, 1939.

20. Stiemens MJ: Een geval van ectopia cordis. *Nederl Tijd Geneesk* **73**:425–435, 1929.

20a. Van Allen MI, Myhre S: Ectopia cordis thoracalis with craniofacial defects resulting from early amnion rupture. *Teratology* **32**:19–24, 1985.

21. Von Braun-Fernwald R: Fall von Ectopia Cordis completa congenita. *Wein Med Blatt* **17**:183–184, 1894.

22. Wilson EJ: A case of ectopia cordis surviving nine hours. *Br Med J* **1**:1175–1176, 1952.

## Sprengel shoulder and cleft palate

Hodgson and Chin (1) reported apparent autosomal dominant inheritance of elevated scapula (Sprengel deformity) with cleft palate. One member had Klippel–Feil syndrome. Probably this family represents variable expression of the latter syndrome rather than being a separate entity.

#### Reference (Sprengel shoulder and cleft palate)

1. Hodgson SV, Chiu DC: Dominant transmission of Sprengel's shoulder and cleft palate. *J Med Genet* **18**:263–265, 1981.

## Syngnathia and cleft palate

A combination of cleft palate and soft tissues or bony adhesions between the upper and lower jaws has been reported (Fig. 23–8). Presumably the disorder results from primary fusion between gingiva or alveolar processes that in turn leads to cleft palate due to failure of the tongue to drop. It may also be related to the small size of the mandible, which would lead to fibrous ankylosis of the temporomandibular joint because of restricted movement during the developmental period (9).

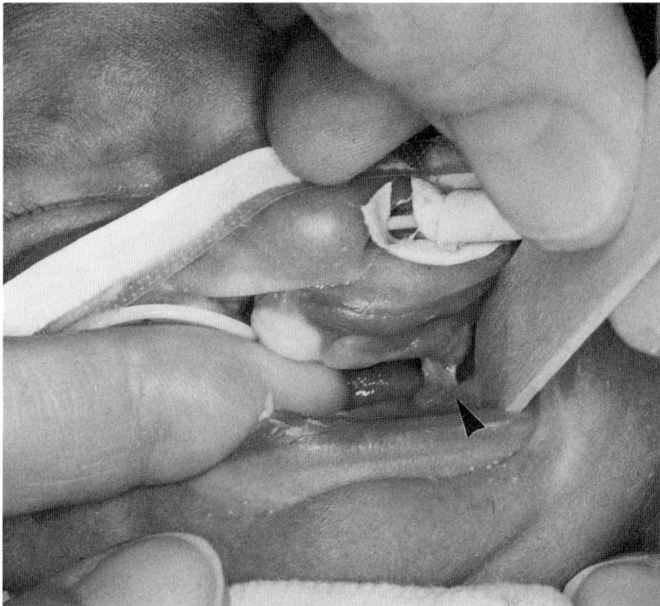

Fig. 23–8. *Syngnathia and cleft palate.* Arrow points to attachment between alveolar ridges. (From *P Randall,* Plast Reconstr Surg **74**:686, 1984.)

Since not all cases have been well documented, it is difficult to know how many are legitimate examples (1a,2–7,10). Gassner (3) described the occurrence in mother and child, while Sternberg et al (9) reported the combination in sibs. The other cases have been isolated examples. In a few cases there has been congenital fusion of the jaws but no associated anomalies (1,8). Shah (6) reviewed cases of interjaw fusion.

Syngnathia also occurs with *popliteal pterygium syndrome* and *Van der Woude syndrome.*

#### References (Syngnathia and cleft palate)

1. Alfery DD et al: Airway management for neonate with congenital fusion of the jaws. *Anesthesiology* **51**:340–342, 1979.

1a. Dinardo NM et al: Cleft palate lateral synechia syndrome. *Oral Surg Oral Med Oral Pathol* **68**:565–566, 1989.

2. Dobrow B: Syngnathia and multiple defects. *J Clin Dysmorphol* **1**(2):5–7, 1983.

3. Gassner I: Familial occurrence of syngnathia congenita syndrome. *Clin Genet* **15**:241–244, 1979.

4. Mathis H: Über einen Fall von Ernährungsschwierigkeit bei connataler Syngnathie. *Dtsch Zahnärztl Z* **17**:1167–1171, 1962.

5. Randall P: Cleft palate and congenital alveolar synechia syndrome. *Plast Reconstr Surg* **74**:686, 1984.

6. Shah RM: Palatomandibular and maxillo-mandibular fusion, partial aglossia and cleft palate in a human embryo. *Teratology* **15**:261–272, 1977.

7. Simpson JR, Maves MD: Congenital syngnathia or fusion of the gums and jaws. *Otolaryngol Head Neck Surg* **93**:96–99, 1985.

8. Snijman PC, Prinsloo JG: Congenital fusion of the gums. *Am J Dis Child* **112**:593–595, 1966.

9. Sternberg N et al: Congenital fusion of the gums with bilateral fusion of the temporomandibular joints. *Plast Reconstr Surg* **72**:385–387, 1983. (Same cases reported in *Ann Ophthalmol* **15**:822–823, 1983.)

10. Verdi GB, O'Neal B: Cleft palate and congenital alveolar synechia syndrome. *Plast Reconstr Surg* **74**:684–685, 1984.

# Chapter 24
# Syndromes with Unusual Facies: Well-Known Syndromes

## Frontonasal malformation (frontonasal dysplasia, median cleft face syndrome)

Frontonasal malformation, probably first described by Hoppe (22) in 1859, consists of hypertelorism, broad nasal root, lack of nasal tip, widow's peak, and anterior cranium bifidum occultum. Associated defects may include midline clefting of nose and/or upper lip, rarely, the palate, and unilaterally or bilaterally, the nasal alae.

The term *median cleft face syndrome* was introduced by DeMyer (10) to describe the disorder in 1967; Rohasco and Massa (38), in 1968, proposed *frontonasal syndrome,* and Sedano et al (40), in 1970, recognizing that this condition did not fulfill the definition of a syndrome (a concurrence of signs and symptoms that constitutes a disease), that it was a primary malformation, and that it likely resulted from an "abnormal tissue development" (definition of dysplasia), coined the name *frontonasal dysplasia.*

The primary set of anomalies may be explained as a single malformation. If the nasal capsule fails to develop properly, the primitive brain vesicle fills the space normally occupied by the capsule, thus producing anterior cranium bifidum occultum, a morphokinetic arrest in the positioning of the eyes, and lack of formation of the nasal tip. The widow's peak scalp-hair anomaly results from ocular hypertelorism, since the two fields of hair-growth suppression are also further apart than usual. Thus, the fields fail to overlap sufficiently high on the forehead, resulting in widow's peak formation (7,35,41).

Experimental models have been produced by various teratogenic agents in animals. Burck and Sadler (2a) concluded that cell death in the frontonasal process midline mesenchyme and in the neural epithelium and/or increased facial width were underlying factors in diazo-oxo-norleucine-induced median cleft face in mice. Darab et al (8a), studying methotrexate-induced median cleft face in mice, implicated damage to the blood vessels in the frontonasal process (dilated and congested blood vessels) and to distention of the developing brain.

We consider frontonasal malformation to be a nonspecific developmental field defect in which there is midfacial malformation and a host of low-frequency anomalies, resulting in broader patterns of abnormalities. Based on this concept, Sedano and Gorlin (39) proposed the term frontonasal malformation.

Most cases of frontonasal malformation are sporadic (10,40). Reports citing familial aggregation represent other entities that have frontonasal malformation as a component. Several case reports do not represent examples of frontonasal malformation (1,15–17,30). Warkany et al (45) reported what appears to be the only familial cases of frontonasal malformation in two half sisters. However, these children had bilateral polydactyly of halluces, making them atypical (vide infra). The same facies may be observed with large anterior encephalocele, hamartoma, frontal lipoma, frontal teratoma, and intracranial cysts (21a,28). Chen et al (4) described *dup(2q) syndrome* with frontonasal malformation.

Toriello (43) described a distinct subgroup associated with the severe (D) form of frontonasal malformation. This syndrome includes epibulbar dermoids, agenesis of the corpus callosum, Dandy–Walker malformation, tibial aplasia, and bilateral polydactyly of the halluces. This combination in part or in full has been reported by a large number of authors (3,12,13,15,18a,31,35a,36,40,43,45). Warkany et al (45) reported two half-sisters with this combination, and parental consanguinity (13) has been reported. It has been found in identical twins (A Perez Aytes et al, personal communication, 1989).

In addition to this distinct syndrome that has epibulbar dermoids as a feature, the combination of oculo-auriculo-vertebral spectrum may also occur with ocular hypertelorism and epibulbar dermoids (oculo-auriculo-frontonasal syndrome) [14,15(case 2),19,21,30a,32]. When flattened encephalocele occurs, the patient appears to have frontonasal malformation together with ear tags, other ear anomalies, and epibulbar dermoids. Because encephalocele occurs more commonly in the occipital region than in the frontal region, we suggest that such cases represent oculo-auriculo-vertebral spectrum with the encephalocele expressed anteriorly.

The number of instances of twinning is greater in families with frontonasal malformation than in the general population (40). We have no explanation for this phenomenon. Some clinicians have maintained that frontonasal malformation represents an incomplete form of twinning. Twinning of the head results from anterior duplication of the notochord. Doubling of the hypophysis constitutes the mildest form of anterior duplication. In diprosopia, a more extensive duplication, there may be doubling of the hypophysis, mouth, and nose. Doubling may lead to formation of two lateral eyes and a median eye. To our knowledge, there is no solid evidence of duplication of any structure in frontonasal malformation. Although Hori (23) described two hypophyses, this case may have been an example of facial duplication.

**Facies.** The clinical appearance of the face has been classified on a somewhat different basis by DeMyer (10) and Sedano et al (40). Facies can be graded from mild (Type A) to severe (Type D) (Figs. 24–1 to 24–8).

**Eyes.** Hypertelorism is a constant feature. Secondary telecanthus may be observed in severe examples. Epibulbar dermoids have been noted (7,12). Rare findings include congenital cataracts (40), upper eyelid colobomas (15,18a,27,41), and symblepharon (27). Iris colobomas have been recorded in a number of instances (27,42a). Temple et al (42a) reported four children with iris colobomas and mental retardation, suggesting that those with frontonasal malformation and iris colobomas might be at increased risk for mental deficiency.

**Nose.** In severe cases, the nose may be flattened with widely spaced nostrils and broad nasal root. In other cases, clefting of the nose or notching or clefting of the nasal alae may be seen. When notching occurs bilaterally, the nose appears square (Fig. 24–6). Nose tags have also been noted (38).

**Ears.** Preauricular tags (2), low-set ears, absent tragus, and conductive hearing loss (40) have been reported.

**Central nervous system.** Mental deficiency has been found occasionally (12,18a,46,47). Of clinical importance is DeMyer's (10) observation that when extracephalic anomalies occur or when hyper-

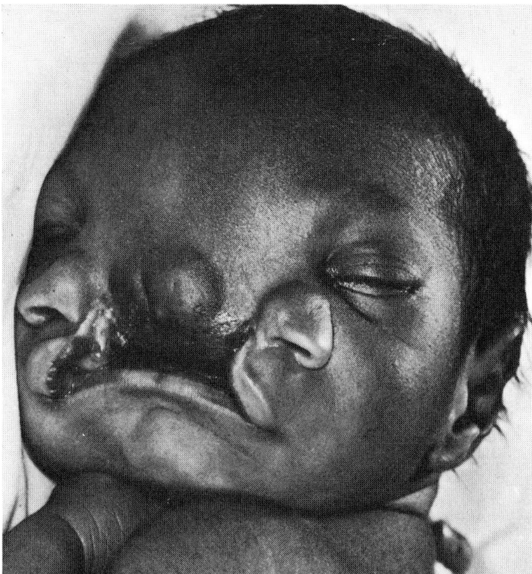

Fig. 24–1. *Frontonasal malformation.* Most extreme example showing marked hypertelorism, separated nasal halves, rectangular oronasal opening, absence of premaxilla. (From *W DeMyer,* Neurology **17**:961, 1967.)

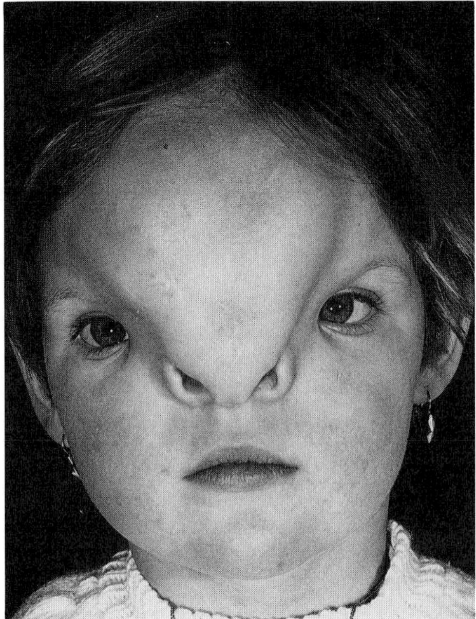

Fig. 24–3. *Frontonasal malformation.* Note similarity to embryonal face. (Courtesy of *P Tessier,* Paris, France.)

telorism is severe, mental deficiency appears to be increased. Conversely, when extracephalic anomalies are absent and hypertelorism is mild, the probability of mental deficiency is low.

Radiographically, there are anterior cranium bifidum (24) and hypoplastic frontal sinuses (25,44). Other findings include absence of the corpus callosum (10,15,18a,24,45), but these cases may represent examples of the newly described syndrome previously discussed (15,43). About 50% of those with more severe facies B-D, on MR and CT examination, have dense calcification of the falx and interhemispheric lipoma (30a). The latter is sometimes mistaken for callosal agenesis (32). Hydrocephaly (40), occipital encephalocele (31), early occlusive anterior and middle cerebral artery disease (RJ Gorlin and MM Cohen Jr, personal observations), as well as mild (pseudohemispheric) holoprosencephaly (37) have been documented.

Fig. 24–2. *Frontonasal malformation.* Hypertelorism, frontal encephalomeningocele, separated nostrils, cleft lip-palate. (From *W DeMyer,* Neurology **17**:961, 1967.)

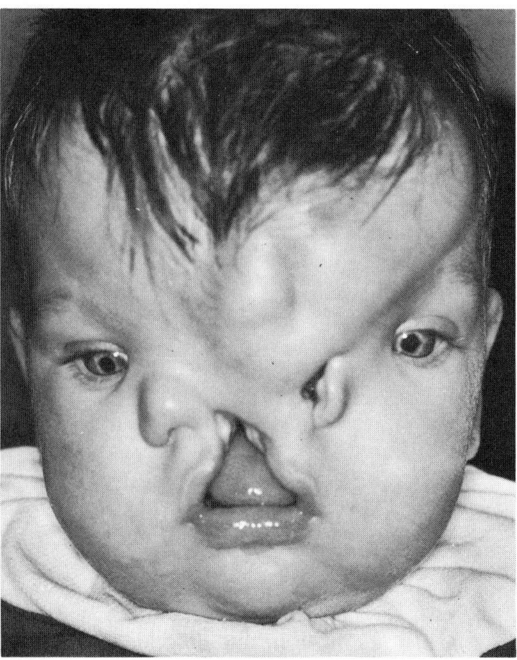

**Other findings.** As noted above, preaxial polydactyly with or without tibial hypoplasia is associated only with the severe (Type D) form of frontonasal malformation. Several examples of the combination have been reported (3,12,13,40,45). Consanguinity has been noted (13). Preaxial polydactyly alone has also been found with the severe (Type D) form (31,36,45).

Clinodactyly (10,46,47), brachydactyly (RJ Gorlin and MM Cohen Jr, personal observations), parietal foramina (31), micropenis (18a), cryptorchidism (10,40, RJ Gorlin, personal observation), Poland anomaly (K Temple, personal communication, 1989), congenital heart anomalies (9a,30a), and other anomalies (10,40) have been documented.

**Oral manifestations.** Median cleft of the upper lip was present, especially in the more severe Type D, as noted above (40). Rarely, cleft lip and/or palate is observed (23,40).

**Differential diagnosis.** Bifid nose may occur without hypertelorism, and familial instances are known (18,46). Since epibulbar dermoids, and sometimes upper eyelid colobomas and preauricular tags, may occur in association with frontonasal malformation, differentiation from *oculo-auriculo-vertebral spectrum* should be kept in mind (vide infra). *Frontofacionasal dysostosis* should be considered in differential diagnosis because of clinical similarities in facial appearance (20). This condition apparently has autosomal recessive inheritance. Keith and Macomber (26) described a case of "hypertelorism" occurring in two sisters that we cannot classify among known conditions. Fragoso et al (17) reported a patient who, in addition to frontonasal malformation, exhibited fusion of the second and third cervical vertebrae and pedal postaxial polydactyly. It was labeled frontonasal malformation in combination with Klippel–Feil anomaly. The patient was female and may represent an example of craniofrontonasal dysplasia, although craniosynostosis was not mentioned.

Various conditions recently isolated from the frontonasal malformation pool should be excluded such as autosomal dominant *Greig cephalopolysyndactyly* (6,11). *Craniofrontonasal dysplasia* has a similar facies. Patients also present craniosynostosis. This disorder has dominant inheritance and seems to occur mostly in females (8,9,

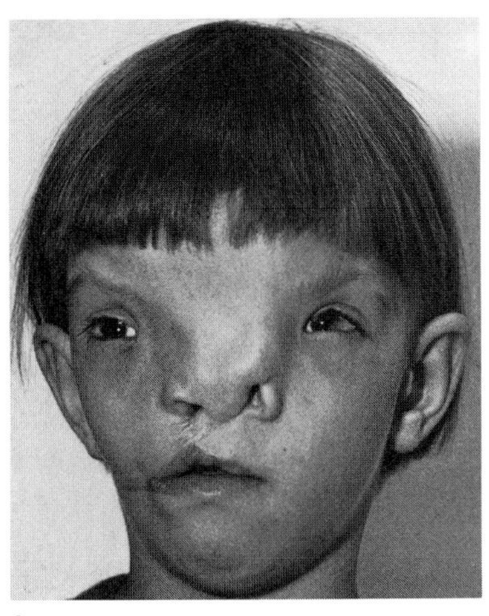
A

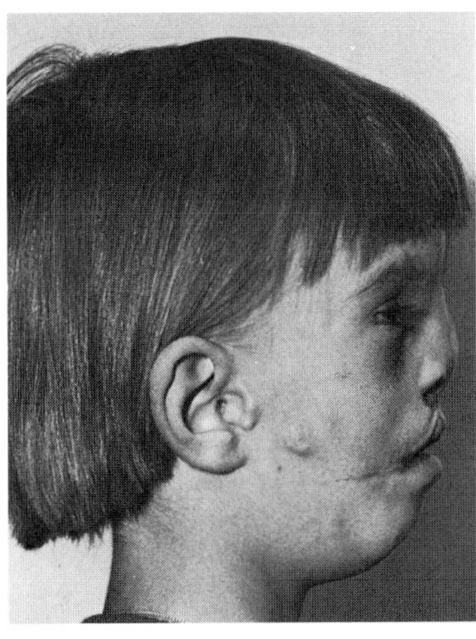
B

Fig. 24–4. *Frontonasal malformation.* (A,B) Facies similar to that in Fig. 24–3 but with coloboma of upper left eyelid and epibulbar dermoid, repaired cleft lip, repaired macrostomia, remnants of ear tags (oculo-auriculo-vertebral spectrum). (From *A Fleischer-Peters,* Dtsch Zahnärztl Z **24**:545, 1969.)

12,33,34). *Ophthalmofrontonasal dysplasia* should also be considered. In addition to the classic facial findings of frontonasal malformation, there is severe eye disease ranging from micro- to anophthalmia and clefts of the alae nasi (9,15,40). *Oculo-auriculo-frontonasal syndrome* consists of frontonasal malformation, midline defects, and features typical of *oculo-auriculo-vertebral spectrum* (19). Other examples of this variant have been published under other names [14,15(case 2),21,30b,32].

Widow's peak may be an isolated finding or a component of several syndromes: *craniofrontonasal dysplasia, Aarskog syndrome, acro-fronto-facio-nasal dysostosis, Opitz oculo-genital-laryngeal syndrome,* and *Waardenburg syndrome.*

**Laboratory aids.** Frontonasal malformation has been diagnosed prenatally (5).

### References [Frontonasal malformation (frontonasal dysplasia, median cleft face syndrome)]


1. Bakken A, Aabyholm G: Frontonasal dysplasia. Possible hereditary connection with other congenital defects. *Clin Genet* **10**:214–217, 1976.

2. Burian F: Median clefts of the nose. *Acta Chir Plast (Praha)* **2**:180–189, 1960.

2a. Burk D, Sadler TW: Morphogenesis of median facial clefts in mice treated with diazo-oxo-norleucine (DON). *Teratology* **27**:385–394, 1983.

3. Calli LJ Jr: Ocular hypertelorism and nasal agenesis (midface syndrome) with limb anomalies. *Birth Defects* **7**(7):268, 1971.

4. Chen H et al: Frontonasal dysplasia and partial trisomy 2q syndrome. *March of Dimes Clin Genet Conf,* Minneapolis, July 19–22, 1987.

5. Cherrenak FA et al: Antenatal diagnosis of median cleft face syndrome, sonographic demonstration of cleft lip and hypertelorism. *Am J Obstet Gynecol* **149**:94–97, 1984.

6. Cocozza G, Ferola R: Considerazioni sulla sindrome ipertelorica di Greig nei suoi rapporti con altre craniosinostosis patologiche. *Pediatria (Napoli)* **66**:592–613, 1958.

7. Cohen MM Jr et al: Frontonasal dysplasia (median cleft face syndrome): Comments on etiology and pathogenesis. *Birth Defects* **7**(7):117–119, 1971.

8. Cohen MM Jr: Craniofrontonasal dysplasia. *Birth Defects* **15**(5B):85–89, 1979.

8a. Darab DJ et al: Pathogenesis of median facial clefts in mice treated with methotrexate. *Teratology* **36**:77–86, 1987.

9. Day D: Unique subpopulations in the diagnostic category of frontonasal dysplasia. *Birth Defects* **19**(5):182–183, 1983.

9a. De Moor MMA et al: Frontonasal dysplasia associated with tetralogy of Fallot. *J Med Genet* **24**:607–609, 1987.


Fig. 24–5. *Frontonasal malformation.* Hypertelorism, bifid nose. (From *W DeMyer,* Neurology **17**:961, 1967.)

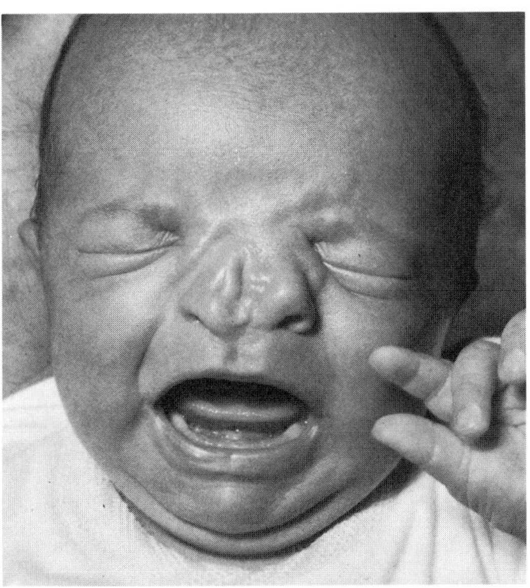

Fig. 24–6. *Frontonasal malformation.* Colobomas of nostrils, wide nasal bridge, anophthalmia on right, microphthalmia on left.

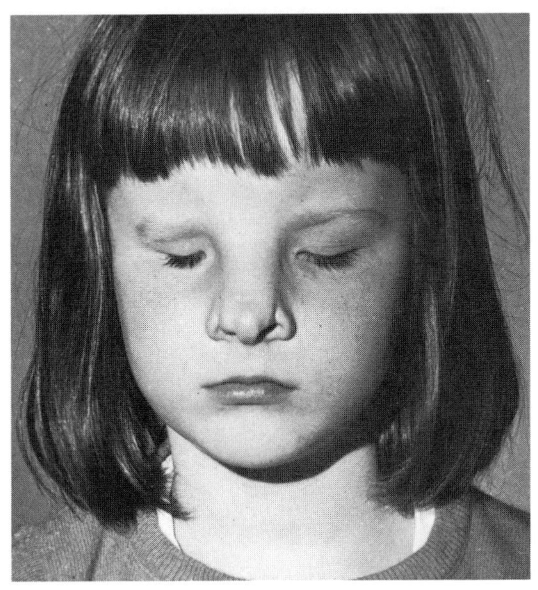

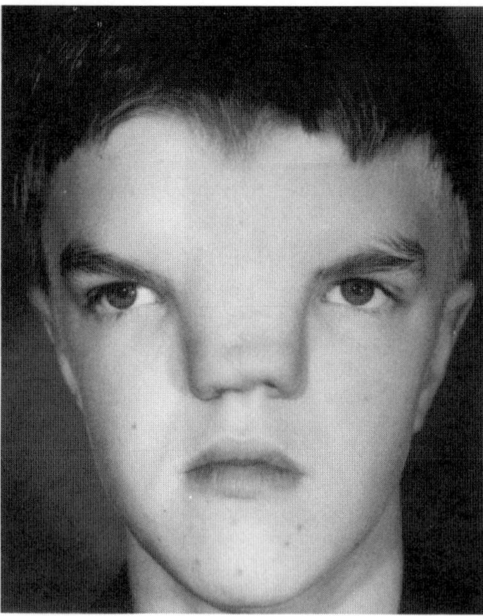

Fig. 24–7. *Frontonasal malformation.* Hypertelorism, broad nose. (Courtesy of *P Tessier,* Paris, France.)


10. DeMyer W: The median cleft face syndrome: Differential diagnosis of cranium bifidum occultum, hypertelorism, and median cleft nose, lip and palate. *Neurology (Minneap)* **17**:961–971, 1967.

11. Duncan PA et al: Greig cephalopolysyndactyly syndrome. *Am J Dis Child* **133**:818–821, 1979.

12. Edwards WC et al: Median cleft face syndrome. *Am J Ophthalmol* **72**:202–205, 1971.

13. Eilers BLC et al: Tibial deficiency, hallucal polydactyly, and uplifted slit-like nares in frontonasal dysplasia, a specific clinical phenotype with significant genetic implications. *March of Dimes Birth Defects Conf,* Birmingham, June 13–16, 1982, p. 104.

14. Fleischer-Peters A: Goldenhar Syndrom und Kiefermissbildungen. *Dtsch Zahnärztl Z* **24**:545–551, 1969.


Fig. 24–8. *Frontonasal malformation.* Least severely affected example. (Courtesy of *S Cocuzza* and *E Zoratto,* Torino, Italy.)

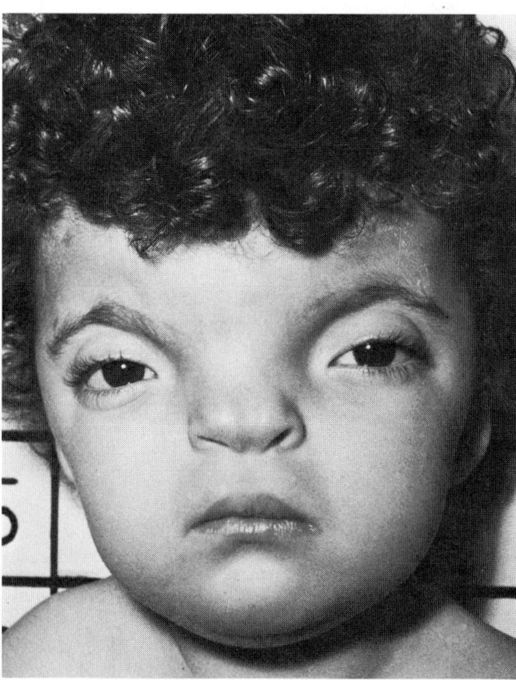


15. Fontaine G et al: La dysplasie fronto-nasale (A propos de quatre observations). *J Génét Hum* **31**:351–365, 1983.

16. Fox FW et al: Frontonasal dysplasia with alar clefts in two sisters. *Plast Reconstr Surg* **57**:553–561, 1976.

17. Fragoso R et al: Frontonasal dysplasia in the Klippel–Feil syndrome: A new associated malformation. *Clin Genet* **22**:270–273, 1982.

18. Francesconi G, Fortunato G: Median dysraphia of the face. *Plast Reconstr Surg* **43**:481–491, 1969.

18a. François J et al: Agenesis of the corpus callosum in the median facial cleft syndrome with associated ocular malformations. *Am J Ophthalmol* **76**:241–245, 1973.

19. Golabi M et el: A new syndrome of oculoauriculovertebral dysplasia and midline craniofacial defect: The oculoauriculofrontonasal syndrome. Two new cases in sibs. *Birth Defects* **19**(5):183–184, 1983.

20. Gollop TR: Fronto-facio-nasal dysostosis. A new autosomal recessive syndrome. *Am J Med Genet* **10**:409–412, 1981.

21. Gupta JS et al: Oculo-auricular-cranial dysplasia. *Br J Ophthalmol* **52**:346–347, 1968.

21a. Hennekam R et al: Congenital hypothalamic hamartoma associated with severe midline defect: A developmental field defect. Report of a case. *Am J Med Genet Suppl* **2**:45–52, 1986.

22. Hoppe I: Eine angeborene Spaltung der Nase. *Preuss Med-Ztg Berl* **2**:164–165, 1859.

23. Hori A: A brain with two hypophyses in median cleft syndrome. *Acta Neuropathol (Berlin)* **59**:150–154, 1983.

24. Jaouen H et al: Malformation cranio-faciale médiane: La schizoprosopie. A propos de quatre cas. *Ann Otolaryngol* (Paris) **101**:267–275, 1984.

25. Kazanjian V, Holmes E: Treatment of median cleft lip associated with bifid nose and hypertelorism. *Plast Reconstr Surg* **24**:582–587, 1959.

26. Keith MJ, Macomber WB: Hypertelorism. A report of two cases. *N Carolina Med J* **16**:259–261, 1955.

27. Kinsey JA, Streeten BW: Ocular abnormalities in the median cleft face syndrome. *Am J Ophthalmol* **83**:261–266, 1977.

28. Kitlowski EA: Congenital anomaly of the face: Inclusion cyst arising from mucous membrane of pharynx with deformity of nose and cleft lip and palate. *Plast Reconstr Surg* **23**:64–68, 1959.

29. Kurlander GF et al: Roentgenology of the median cleft face syndrome. *Radiology* **88**:473–478, 1967.

30. Kwee ML, Lindhout D: Frontonasal dysplasia, coronal craniosynostosis, pre- and postaxial polydactyly and split nails: A new autosomal dominant mutant with reduced penetrance and variable expression? *Clin Genet* **24**:200–205, 1983.

30a. Meinecke P, Blunck W: Frontonasal dysplasia, congenital defect and short stature: A further observation. *J Med Genet* **26**:408–409, 1989.

30b. Naidich TP et al: Median cleft face syndrome: MR and CT data from 11 children. *J Comp Assist Tomog* **12**:57–64, 1988.

31. Neidich JA et al: Frontonasal dysplasia: Two new syndromes. *March of Dimes Clin Genet Conf,* Minneapolis, July 19–22, 1987.

32. Pascual-Castroviejo I et al: Fronto-nasal dysplasia and lipoma of the corpus callosum. *Eur J Pediatr* **144**:66–71, 1985.

33. Pellerin P et al: Cranio-facio-sténose asymétrique avec syndactylie: Syndrome de Saethre-Chotzen? *Rev Stomatol* **82**:247–250, 1981.

34. Pendl G, Zimprich H: Ein Beitrag zum Syndrom des Hypertelorismus Greig. *Helv Paediatr Acta* **26**:319–325, 1971.

35. Peterson MO et al: Comments on frontonasal dysplasia, ocular hypertelorism and dystopia canthorum. *Birth Defects* **7**(7):120–124, 1971.

35a. Prescott K. Personal communication, 1989.

36. Reich E. Personal communication, 1987.

37. Roubicek M et al: Frontonasal dysplasia as an expression of holoprosencephaly. *Eur J Pediatr* **137**:229–231, 1981.

38. Rohasco SA, Massa JL: Frontonasal syndrome. *Br J Plast Surg* **21**:244–249, 1968.

39. Sedano HO, Gorlin RJ: Frontonasal malformation as a field defect and in syndromal associations. *Oral Surg* **65**:704–710, 1988.

40. Sedano HO et al: Frontonasal dysplasia. *J Pediatr* **76**:906–913, 1970.

41. Smith DW, Cohen MM Jr: Widow's peak, scalp-hair anomaly and its relation to ocular hypertelorism. *Lancet* **2**:1127–1128, 1973.

42. Stewart, FE: Craniofacial malformations: Clinical and genetic consideration. *Pediatr Clin N Am* **25**:485–515, 1978.

42a. Temple IK et al: Midline facial defects with ocular colobomata. *Am J Med Genet* (in press).

43. Toriello HV: Frontonasal "dysplasia," cerebral anomalies, and polydactyly: Report of a new syndrome and discussion from a developmental field perspective. *Am J Med Genet Suppl* **2**:89–96, 1986.

44. Viezens A, Willenberg W: Zwei seltene Gesichtsmissbildungen im Hals-

Nasen- Ohrenbereich: Doppelnase und Dysostosis mandibulofacialis. *Dtsch Med J* 7:457–461, 1956.

45. Warkany J et al: Median facial cleft syndrome in half-sisters. Dilemmas in genetic counseling. *Teratology* **8**:273–286, 1973.

46. Webster J, Deming E: The surgical treatment of the bifid nose. *Plast Reconstr Surg* **6**:1–37, 1950.

47. Zunin C: Hipertelorismo di Greig e suoi rapporti con le sindromi di disostosicraniche e facciali. *Minerva Pediatr* **7**:71–79, 1955.

## Craniofrontonasal dysplasia

A subpopulation of frontonasal malformation patients was identified by Cohen in 1979 (1) as representing a heritable condition which he called craniofrontonasal dysplasia. An earlier series of patients was reported in 1977 by Reich et al (18). The syndrome consists of frontonasal malformation, craniosynostosis (especially including the coronal suture), and various skeletal and soft tissue abnormalities. To date over 90 cases have been identified (1–9,13–25,27,28). The mothers and daughters reported by McCowatt Montford (11), Ogilvie and Posel (14), and Friede (5) probably had the disorder, although craniosynostosis was not specifically mentioned. The patients with pre- and postaxial polydactyly, hypertelorism, and coronal synostosis reported by Kwee and Lindhout (10) probably have *Greig cephalopolysyndactyly syndrome* in which craniosynostosis can occur, albeit with low frequency. The family with three affected males and three affected females reported by Morris et al (12) may represent some other condition. In this family, short stature, delayed bone age, shawl scrotum, and hypospadias were prominent features in affected males, and only one of three affected females had craniosynostosis. The female patient reported by Fragoso et al (4a) with frontonasal dysplasia and Klippel–Feil anomaly probably had craniofrontonasal dysplasia; however, craniostenosis was not mentioned.

Vertical transmission of craniofrontonasal dysplasia has been observed through three generations (6,9,17,20,24,25) and females are affected much more frequently and severely than males (6,22). In an unpublished study of 21 patients, Reich et al (19), in an abstract, noted two cases of male-to-male transmission, and these cases need to be documented. Although X-linked dominant inheritance, metabolic interference, and dominant mutation analogous to the T-semilethal allele of the mouse T locus have all been proposed (17,20–25), the condition probably represents a unique X-linked disorder having marked female expression but with mild male expression (hypertelorism, exotropia) in most cases. However, several severely affected males with craniostenosis have been noted (2, H Bruner, personal communication, 1989). An extensive review is that of Grutzner and Gorlin (6).

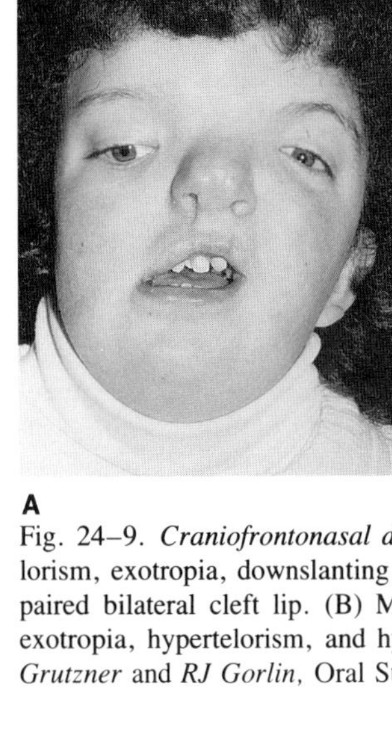
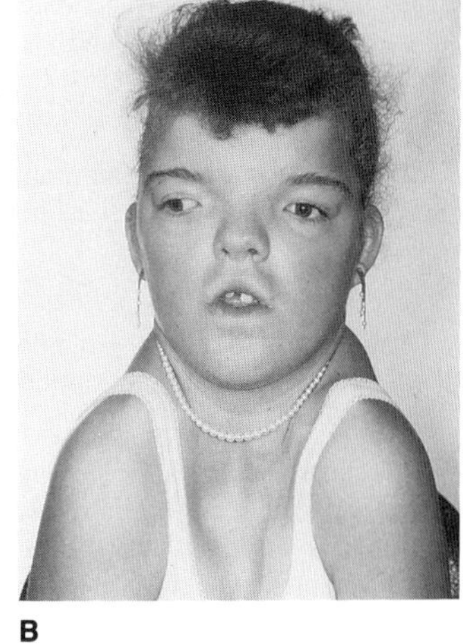

A B

Fig. 24–9. *Craniofrontonasal dysplasia.* (A) Facial asymmetry, hypertelorism, exotropia, downslanting palpebral fissures, bifid nasal tip, and repaired bilateral cleft lip. (B) More mildly affected sister of A showing exotropia, hypertelorism, and hypermobility of shoulder girdle. (From *E Grutzner* and *RJ Gorlin,* Oral Surg Oral Med Oral Pathol **65:**436, 1988.)

**Craniofacial features.** Craniofacial features include brachycephaly, coronal synostosis, frontal bossing, hypertelorism, and broad nasal bridge with broad bifid nasal tip. An increased bony interorbital distance noted radiographically is essential to diagnosis the "carrier male." Strabismus, downslanting palpebral fissures, and facial asymmetry have been frequently reported. Unilateral or bilateral cleft lip has occasionally been observed. Other findings have included highly arched palate, malocclusion, and pterygium colli (1,4–6,9,11a–17,20,25) (Figs. 24–9A,B and 24–10A,B).

**Limbs.** Longitudinally split nails have been observed in many cases (Fig. 24–11). Other findings have included hyperextensible joints, broad

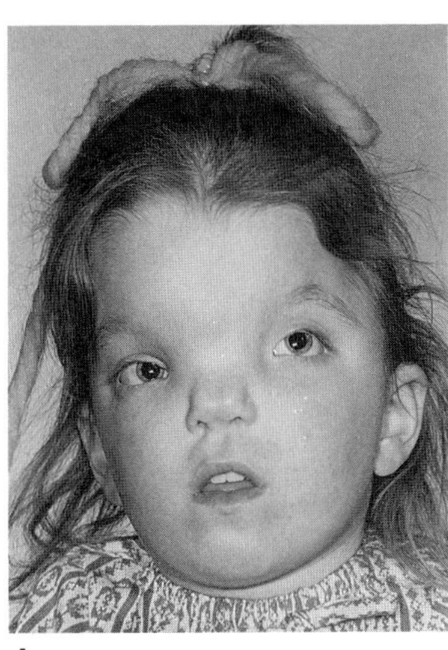
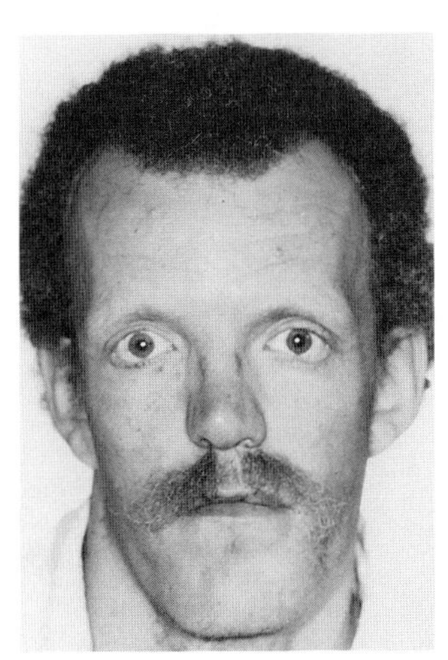
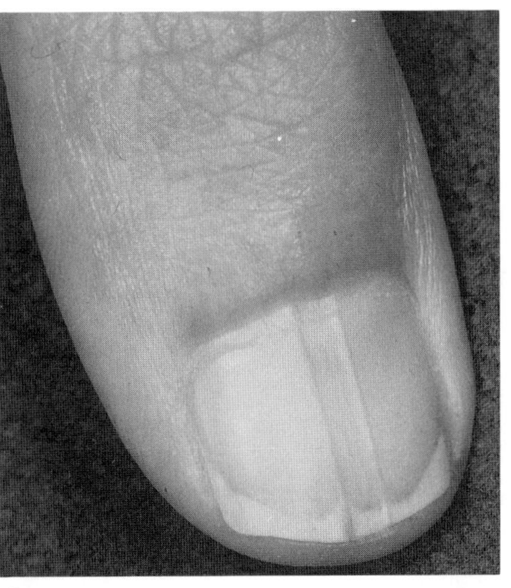

A B C

Fig. 24–10. *Craniofrontonasal dysplasia.* (A) Similar facies in unrelated patient. (B) Male hemizygote. Note only hypertelorism, mild downslanting palpebral fissures, and repaired cleft lip. (C) Vertical striping of nail. (From *E Grutzner* and *RJ Gorlin,* Oral Surg Oral Med Oral Pathol **65:**436, 1988.)

great toes, partial soft tissue syndactyly, short fifth fingers, clinodactyly, gap between the first and second toes, long fingers and toes, and transverse palmar creases (1,4–6,9,13–17, 20,23,25).

**Intelligence.** Although normal intelligence has been usually been found, mental deficiency has been noted in a few females (6,9, 17,24,25).

**Other abnormalities.** Rounded sloping shoulders, scoliosis, vertebral anomalies, and other minor skeletal defects have been recorded.

**Differential diagnosis.** *Frontonasal dysplasia, fronto-facio-nasal dysostosis, Greig cephalopolysyndactyly syndrome,* and other craniosynostosis syndromes must be excluded. Teebi (26) described a syndrome having some resemblance to craniofrontonasal dysplasia but without bifid nose, craniosynostosis, pterygium colli, rounded sloping shoulders, or nail abnormalities. Some of his patients had shawl scrotum. This may be the same disorder as that described by Morris et al (12) as their two male patients also had shawl scrotum. We have called the disorder *brachycephalofrontonasal dysplasia*.

### References (Craniofrontonasal dysplasia)


1. Cohen MM Jr. Craniofrontonasal dysplasia. *Birth Defects* **15**:(5B):85–89, 1979.

2. David JD et al: *The Craniosynostoses: Causes, Natural History, and Management*. Springer-Verlag, New York, 1982 (see p. 73: male sex established by personal communication, 1989).

3. Day D: Unique subpopulation in the diagnostic category of frontonasal dysplasia. *Birth Defects* **19**(5):182–183, 1982.

4. Edwards WC et al: Median cleft face syndrome. *Am J Ophthalmol* **72**:202–205, 1971 (case 2).

4a. Fragoso R et al: Frontonasal dysplasia in the Klippel–Feil syndrome: A new assorted malformation. *Clin Genet* **22**:270–273, 1982.

5. Friede R: Uber physiologische Euryopie und pathologischen Hypertelorismus ocularis. *Graefes Arch Ophthalmol* **155**:359–385, 1954.

6. Grutzner E, Gorlin RJ. Craniofrontonasal dysplasia phenotypic expression in females and males and genetic considerations. *Oral Surg Oral Med Oral Pathol* **65**:436–444, 1988.

7. Hunter AGW, Rudd NL: Craniosynostosis. II. Coronal synostosis: Its familial characteristics and associated clinical findings in 109 patients lacking bilateral polysyndactyly or syndactyly. *Teratology* **15**:301–310, 1977 (pedigree 1).

7a. Hurst J, Baraitser M: Craniofrontonasal dysplasia. *J Med Genet* **25**:133–134, 1988.

8. Keith MJ, Macomber WB: Hypertelorism without mental deficiency. *Arch Dis Childh* **4**:381–384, 1929.

9. Kumar D et al: A family with craniofrontonasal dysplasia, and fragile site 12q13 segregating independently. *Clin Genet* **29**:530–537, 1986.

10. Kwee ML, Lindhout D: Frontonasal dysplasia, coronal craniosynostosis, pre- and postaxial polydactyly and split nails: A new autosomal dominant mutant with reduced penetrance and variable expression. *Clin Genet* **24**:200–225, 1983.

11. McCowatt Montford T: Hereditary hypertelorism without mental deficiency. *Arch Dis Childh* **4**:381–384, 1929.

11a. Michels VV et al: Craniofrontonasal dysostosis with deafness and axillary pterygia. *Am J Med Genet* **34:**445–450, 1989.

12. Morris CA et al: Delineation of the male phenotype in craniofrontonasal dysplasia. *Am J Med Genet* **27**:623–632, 1987.

13. Pellerin P et al: Cranio-facio-stenose asymetrique avec syndactylie: syndrome de Saethre-Chotzen? *Rev Stomatol* **81**:247–250, 1981.

14. Ogilvie AG, Posel MM: Scaphocephaly, oxycephaly and hypertelorism. *Arch Dis Childh* **2**:146–154, 1927 (Cases 2 and 3).

15. Pendl G, Zimprich H: Ein Beitrag zum Syndrom des Hypertelorimus Greig. *Helv Paediatr Acta* **26**:319–325, 1971.

16. Pruzansky S: Time: The fourth dimension in syndrome analysis applied to craniofacial malformations. *Birth Defects* **13**(3C):3–28, 1977.

17. Pruzansky S et al: Radiocephalometric findings in a family with craniofrontonasal dysplasia. *Birth Defects* **18**(1):120–138, 1982.

18. Reich EW et al: A new heritable syndrome with frontonasal dysplasia and associated extracranial anomalies, in *Fifth International Conference on Birth Defects,* 1977, Littlefield JW (ed.), Abstr Excerpta Medica, Amsterdam, p 243.

19. Reich EW et al: Craniofrontonasal dysplasia: Clinical delineation. *Am J Hum Genet Suppl* **37**:A72, 1985.

19a. Reilly WA: Hypertelorism: Report of four cases. *JAMA* **96:**1929–1933, 1931.

20. Reynolds JF et al: Craniofrontonasal dysplasia in a three-generation kindred. *J Craniofac Genet Dev Biol* **2**:233–238, 1982.

21. Rollnick BR et al: A pedigree: Possible evidence for the metabolic interference hypothesis. *Am J Hum Genet* **33**:823–826, 1981.

22. Sax CM et al: Craniofrontonasal dysplasia: A semilethal mutation with similarities to the mouse T-locus. *Am J Hum Genet Suppl* **36**:73S, 1984.

23. Sax CM, Flannery DD: Craniofrontonasal dysplasia: Clinical and genetic analysis. *Clin Genet* **29**:77–84, 1986.

24. Slover R, Sujansky E: Frontonasal dysplasia with coronal craniosynostosis in three sibs. *Birth Defects* **15**(5B):75–83, 1979.

25. Sršen S. Prispevok k frontonasálnej dysplázii syndrómu stredneho obličajového rázštepu. *Čs Pediatr* **41**:523–525, 1986.

26. Teebi AS: New autosomal dominant syndrome resembling craniofrontonasal dysplasia. *Am J Med Genet* **28**:581–592, 1987.

27. Webster JP, Deming EG: The surgical treatment of the bifid nose. *Plast Reconstr Surg* **6**:1–37, 1950.

28. Young ID, Moore JR: Craniofrontonasal dysplasia—a distinct entity with lethality in the male? *Clin Genet* **25**:473–475, 1984.


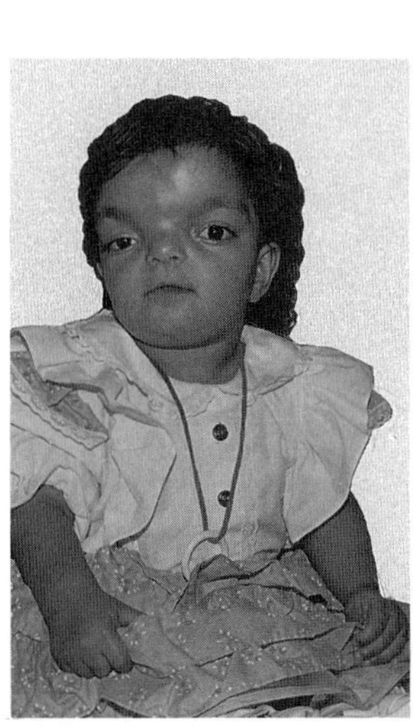
A

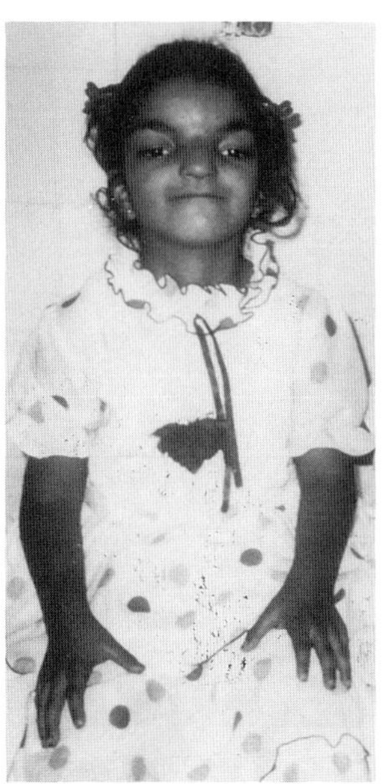
B

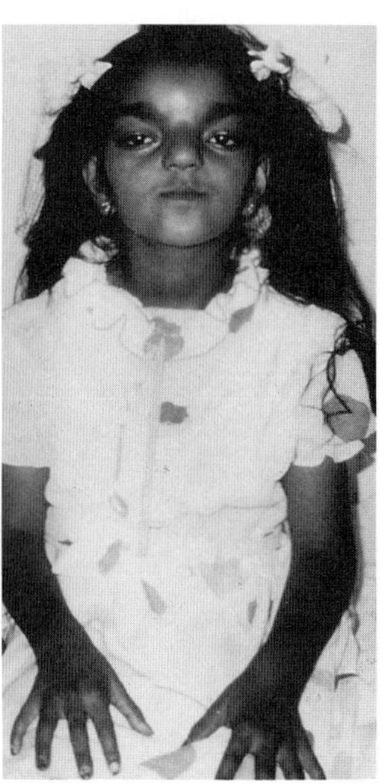
C

Fig. 24–11. *Brachycephalofrontonasal dysplasia.* (A) Note round face with mild asymmetric, prominent forehead with widow's peak, thick eyebrows, hypertelorism, long philtrum. (B,C) Sibs of child seen in A showing similar phenotype. (From *AS Teebi,* Am J Med Genet **28**:581, 1987.)

## Brachycephalofrontonasal dysplasia

Teebi (2) described an apparently new autosomal dominantly inherited syndrome with features of craniofrontal dysplasia but with normal or slightly broad nasal tip and without evidence of craniosynostosis or nail abnormalities. The face was round with prominent forehead, widow's peak, pronounced hypertelorism, mild downslanting palpebral fissures, very heavy and broad eyebrows, ptosis of lids, broad and/or depressed nasal bridge, short nose, fleshy ear lobules and/or prominent anthelix, hypoplastic maxilla, long deep philtrum, dental malocclusion, horizontal thin upper lip and/or pouty lower lip, and prominent lower jaw (Fig. 24–11A–C). It may be the same as the disorder reported by Morris et al (1).

The hands were small and broad with mild interdigital webbing and clinodactyly of the fifth fingers. All of the males had a shawl scrotum.

### References (Brachycephalofrontonasal dysplasia)

1. Morris CA et al: Delineation of the male phenotype in craniofrontonasal dysplasia. *Am J Med Genet* **27**:623–632, 1987.

2. Teebi AS: New autosomal dominant syndrome resembling craniofrontonasal dysplasia. *Am J Med Genet* **28**:581–591, 1987.

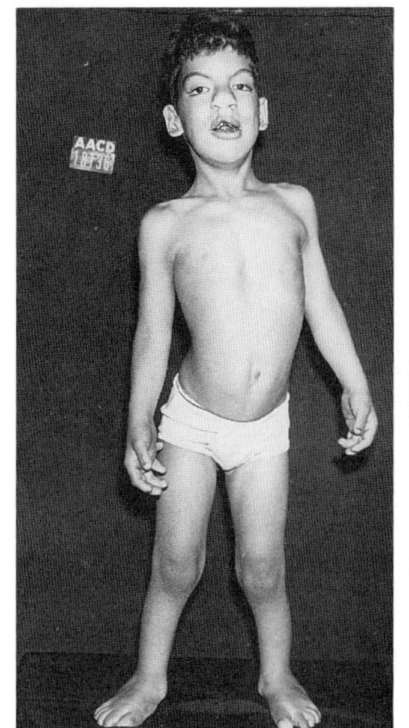


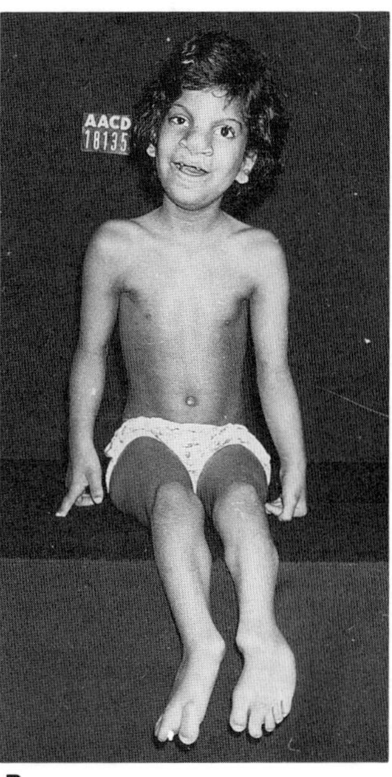


A B

Fig. 24–12. *Acro-fronto-facio-nasal dysostosis.* (A,B) Sibs with bilateral cleft lip–palate, severe mental retardation, and limb anomalies. (From *A Richieri-Costa et al,* Am J Med Genet, **20**:631, 1985.)

## Acro-fronto-facio-nasal dysostosis

Richieri-Costa et al (1), in 1985, described two sibs born to first cousins. In addition to bilateral cleft lip–palate, they shared severe mental retardation, hypertelorism, widow's peak, S-shaped palpebral fissures, ptosis, long eyelashes and eyebrows, broad notched nasal tip, macrostomia, anteverted pinnae, postaxial polysyndactyly of hands, hypoplasia of distal phalanges of digits 1–5, PIP camptodactyly of digits 2–5, short legs, and talipes equinovarus (Fig. 24–12A,B). Radiographic changes included brachycephaly, vertical clivus, capitate–hamate fusion, abnormal ossification of patellae, curved or hypoplastic fibulae, brachymetacarpalia, and iliac hypoplasia (Figs. 24–13 to 24–15).

In 1989, Richieri–Costa et al (2) reported still another type of acrofronto-facio-nasal dysostosis associated with microbrachycephaly, wide forehead, marked hypertelorism, broad nose with midline groove, soft tissue syndactyly of fingers 3-4, and hypospadias. Affected sibs and parental consanguinity suggest autosomal recessive inheritance.

### References (Acro-fronto-facio-nasal dysostosis)

1. Richieri-Costa A et al: A previously undiscussed autosomal recessive multiple congenital anomalies/mental retardation (MCA/MR) syndrome with fronto-nasal dysostosis, cleft lip–palate, limb hypoplasia, and postaxial polysyndactyly: Acro-fronto-facio-nasal dysostosis syndrome. *Am J Med Genet* **20**:631–638, 1985.

2. Richieri-Costa A et al: Autosomal recessive acro-fronto-facio-nasal dysostosis associated with genitourinary anomalies. *Am J Med Genet* **33:**121–124, 1989.

## Fronto-facio-nasal dysplasia (cleft lip–palate, blepharophimosis, lagophthalmos, and hypertelorism)

In 1981, Gollop (1) described two sibs with consanguineous parents. The sibs had cleft lip–palate, blepharophimosis, lagophthalmos, primary telecanthus, S-shaped palpebral fissures, and eyelid coloboma.

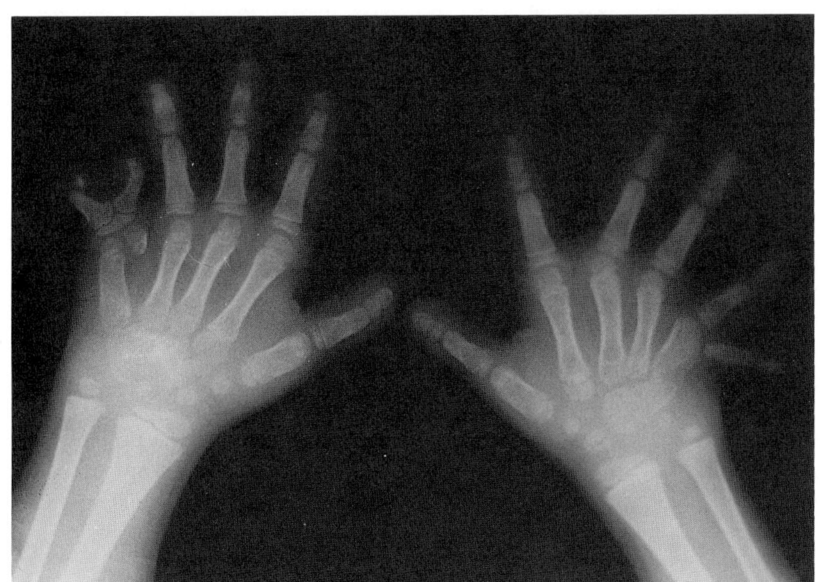

Fig. 24–13. *Acro-fronto-facio-nasal dysostosis.* Radiograph of hands showing campobrachypolysyndactyly, hamate–capitate fusion. (From *A Richieri-Costa et al,* Am J Med Genet, **210**:631, 1985.)

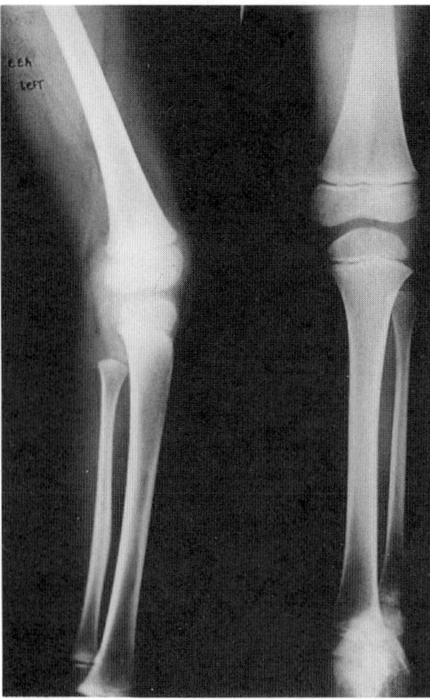

Fig. 24–14. *Acro-fronto-facio-nasal dysostosis.* Hypoplastic fibula. (From *A Richieri-Costa et al,* Am J Med Genet **20**:631, 1985.)

Gollop et al (2) and V Felix (personal communication, 1989) have seen additional patients. The patients had a variety of abnormalities: encephalocele, cranium bifidum occultum, widow's peak, frontal lipoma, limbic dermoid of eye, microphthalmia, absent eyelashes, and iris coloboma (Fig. 24–16).

Inheritance appears to be autosomal recessive. We see no relationship to *craniofrontonasal dysplasia.*

Fig. 24–15. *Acro-fronto-facio-nasal dysostosis.* Tibiotalar dislocation, abnormally modeled tarsal bones, metatarsus adductus, fibular deviation of toes 4–5, hypoplasia of distal phalanges. (From *A Richieri-Costa et al,* Am J Med Genet, **20**:631, 1985.)

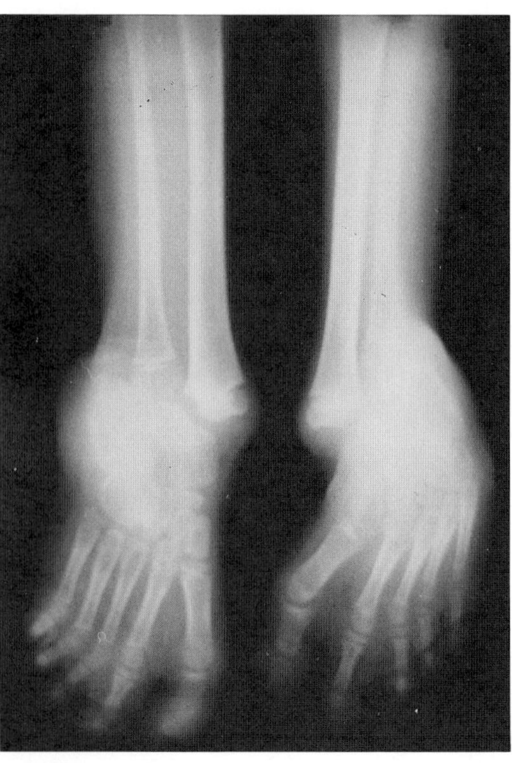

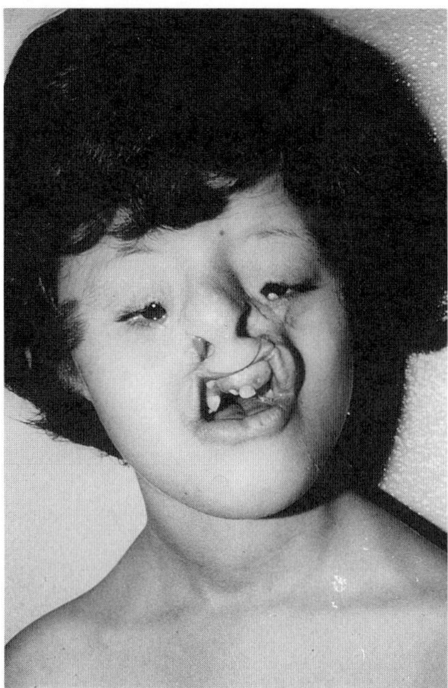

Fig. 24–16. *Fronto-facio-nasal dysplasia.* Facies of one of two sibs with bilateral facial clefts, blepharophimosis, lagophthalmos, telecanthus, S-shaped palpebral fissures, and eyelid coloboma. (From *TR Gollop,* Am J Med Genet **10**:409, 1981.)

#### References [Fronto-facio-nasal dysplasia (cleft lip–palate, blepharophimosis, lagophthalmos, and hypertelorism)]

1. Gollop TR: Fronto-facio-nasal dysostosis—a new autosomal recessive syndrome. *Am J Med Genet* **10**:409–412, 1981.

2. Gollop TR et al: Frontofacionasal dysplasia: Evidence for autosomal recessive inheritance. *Am J Med Genet* **19**:301–305, 1984.

## Blepharonasofacial syndrome

Pashayan et al (1) and Putterman et al (2) described autosomal dominant inheritance of a syndrome of telecanthus and lateral displacement and stenosis of lacrimal puncta, bulky nose with broad nasal bridge, masklike face, midfacial hypoplasia, longitudinal cheek furrows, and trapezoidal upper lip (Fig. 24–17A–C). Joints were hyperextensible. All patients exhibited a positive Babinski reflex, poor coordination, torsion dystonia, mild soft-tissue syndactyly of the fingers, and mental retardation.

#### References (Blepharonasofacial syndrome)

1. Pashayan H et al: A family with blepharo-naso-facial malformation *Am J Dis Child* **125**:389–393, 1973.

2. Putterman AM et al: Eye findings in the blepharo-naso-facial malformation syndrome. *Am J Ophthalmol* **76**:825–831, 1973.

## Opitz oculo-genito-laryngeal syndrome (Opitz BBB/G compound syndrome)

In 1965, Opitz et al (22) reported a syndrome consisting of hypertelorism, hypospadias, and other anomalies. Subsequently, Opitz et al (24), employing the name BBB syndrome, presented their findings in detail. The acronym BBB represented the initials of the last name of each of the three originally reported families. Opitz et al (23), in the same issue, described the G syndrome, again using the initial letter of the family name to designate the disorder. The G syndrome was described as being composed of apparent hypertelorism, mild downslanting palpebral fissures, epicanthal folds, hypospadias, and laryn-

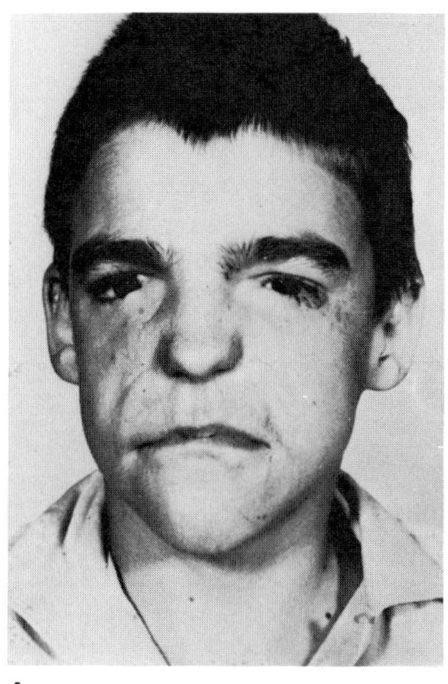
A

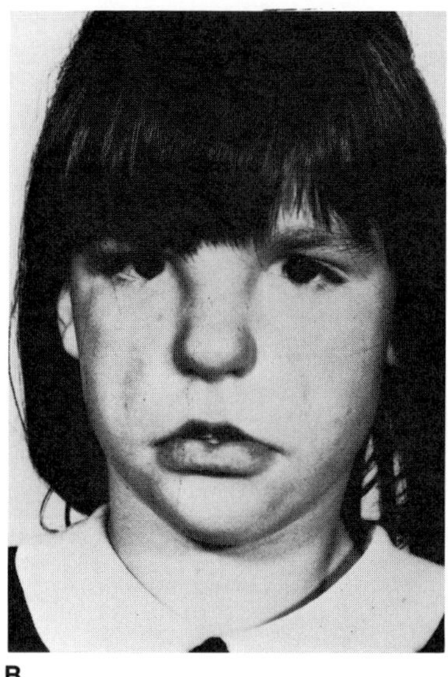
B

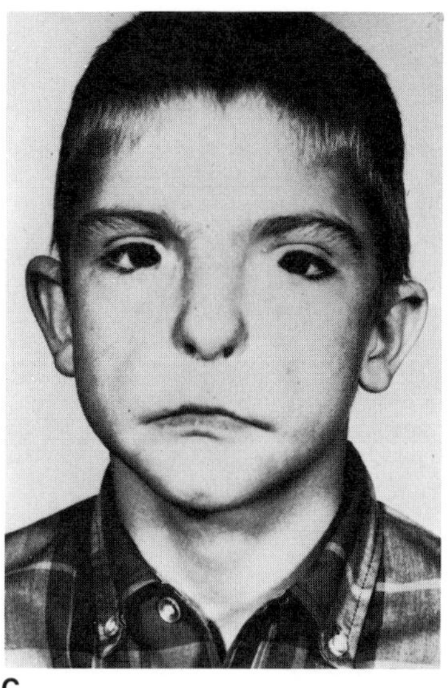
C

Fig. 24–17. *Blepharonasofacial syndrome.* (A–C) Facies of three mentally retarded sibs with torsion dystonia, showing telecanthus with temporal displacement of puncta, downslanting palpebral fissures. Nose is fleshy, midface hypoplastic, upper lip trapezoidal, lower lip pouty. (From *H Pashayan et al,* Am J Dis Child **125**:389, 1973.)

gotracheoesophageal (LTE) defects. Both of these conditions have autosomal dominant inheritance with the most severe manifestations of both of them being observed in males. Hypertelorism may be the only expression in affected females, and some female transmitters of the gene(s) appear normal (29b). Affected females, as a rule, have normal genitalia. Male-to-male transmission has been reported (2a,7,24,30). Subsequent publications (1,5,9,12,14,16,20,25), noting similarities and differences between both disorders, proposed various names: hypertelorism–hypospadias, telecanthus–hypospadias, and Opitz syndrome for the BBB syndrome and hypospadias–dysphagia, Opitz–Frias syndrome, Opitz G syndrome, and hypertelorism with esophageal abnormality and hypospadias for the G syndrome. Cappa et al (2) suggested the eponym Opitz syndrome to unify both entities, but this eponym has been previously employed by Gorlin et al (12) and Smith (29a) to refer to the BBB syndrome.

Opitz (21) reported that a more recently born nephew of the original propositus in a BBB family was presumably affected with G syndrome. Opitz et al (24) had noted that an affected member of the same family had died suddenly at age of eight weeks of "suffocation," and that an affected member of another BBB family, at age 3 years, had "laryngeal stridor." These statements further suggested laryngeal complications in patients originally reported as examples of the so-called BBB syndrome. Phenotypic similarities of the two entities have been well demonstrated (5) as well as some differences in facial features (9). We believe that the marked overlap of clinical manifestations, the same inheritance pattern, and male preponderance in these assumedly different syndromes suggest unification. However, until RFLP linkage studies on G and BBB families are undertaken, it cannot be known whether the G and BBB syndromes are the same condition or are allelic (21). We propose the term *Opitz oculo-genito-laryngeal syndrome (Opitz BBB/G compound syndrome)* to unify the two syndromes, subdividing them into Type I for the BBB variety and Type II for the G variety.

**Facies.** Various degrees of hypertelorism/telecanthus are observed in 90% of both types as well as slight upslanting or downslanting palpebral fissures, epicanthic folds, and strabismus (3,11,20,22,24,29), but to a markedly lesser degree in Type II (4,5,9,15,34) Figs. 24–18 and 24–19A,B). Some patients have exhibited relative entropion of the lower eyelid. The nasal bridge is elevated in Type I while in Type II it is flattened with anteverted nostrils (Fig. 24–19) and the philtrum is rather flat and inapparent (11). Usually there is mild micrognathia (6). The pinnae may exhibit mild posterior rotation and occasionally abnormal modeling, primarily affecting the helix in both types (4,6,8,17,24). Type I presents cranial asymmetry in 20% (3,27) and brachycephaly, prominent forehead, open fontanels (6,8,18,26), prominent metopic suture, widow's peak, and low-set anterior and posterior scalp line have been observed occasionally (5,7,9,11,15). Prominence of occipital and parietal eminences is present in Type II (5,9,24). Patients of either type can present normal intelligence or mild to moderate mental deficiency (2,17,20,23,32).

**Genitourinary system.** Hypospadias is a constant feature in affected males with either phenotype (Fig. 24–20). Renal anomalies, ureteral stenosis, and inguinal hernia have been noted in about 40% of those with the Type II phenotype (2,3,5,9,11,17,20,23,24).

The degree of hypospadias varies from a mild coronal to a scrotal

Fig. 24–18. *Opitz oculo-genito-laryngeal syndrome.* Widely-spaced eyes. [From *JM Opitz* et al, Birth Defects **5**(2):86, 1969.]

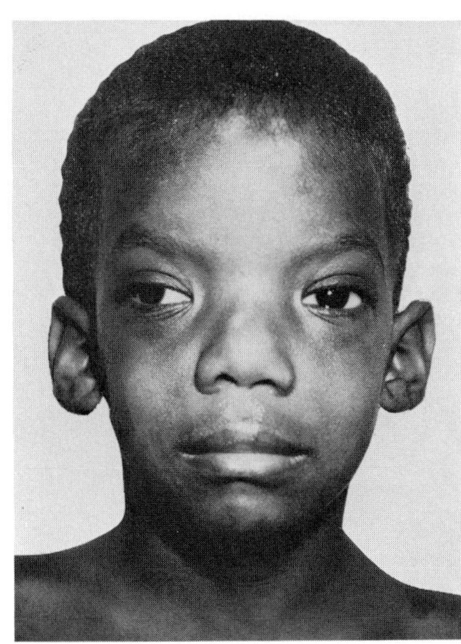

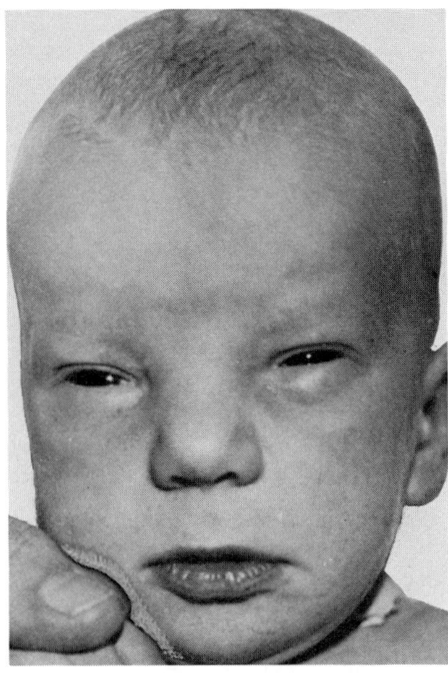

A

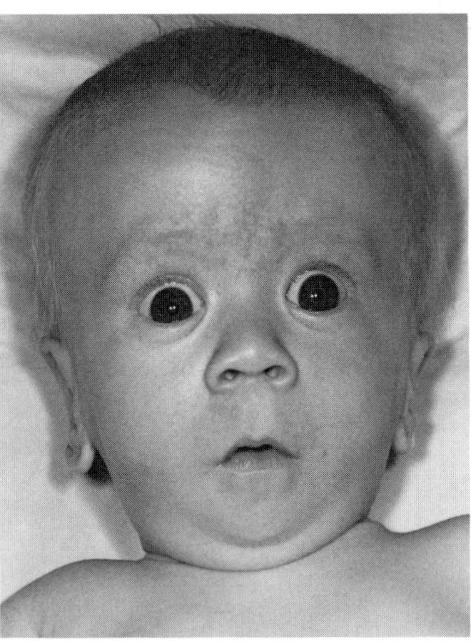

B

Fig. 24–19. *Opitz oculo-genito-laryngeal syndrome.* (A) Hypertelorism, narrow palpebral fissures, anteverted nostrils. (B) Patient of one of the authors. [A from *JM Opitz et al,* Birth Defects **5**(2):95, 1969.]

type with an associated ventral urethral groove (11). In mild cases, the scrotum appears normal; in severe examples, the scrotum is cleft and chordee may be so severe as to draw the tip of the glans to the anterior edge of the anus (Fig. 24–21A,B). Cryptorchidism was found in about 30% of those with Type I form (29a).

Imperforate or ectopic anus with rectourethral fistula has been reported (5,7,23,26,31) as well as bilateral ureteral reflex. Affected females have normal genitalia, but splayed posterior labia majora, anteriorly placed anus, and imperforate anus have also been rarely noted (21).

**Respiratory system.** Type I seems not to be associated with LTE defects, but indications that minor respiratory anomalies might be present in patients with this type have been published (21). In Type II, respiration usually begins spontaneously at birth, and the infants generally have normal Apgar scores (5). In some infants, stridorous respiration and a hoarse cry may be noted even before the first feeding. In severe cases, the infant sucks eagerly but seems to have difficulty in swallowing, manifest by choking, coughing, and resultant cyanosis (12b, 18). This may be followed by respiratory distress, increase in stridor, apparent aspiration pneumonia (5), patchy atelectasis and emphysema, and, in chronic cases, bronchiectasis and anesthetic risk. Cinefluoroscopic studies have shown apparent neuromuscular dysfunction of the swallowing mechanism, with up to half of each mouthful entering the tracheobronchial tree with gastroesophageal reflex (4) (Fig. 24–22A,B). About 40% of patients with Type II present with apparent clefting or malformation of the upper gastrointestinal tract or respiratory tract (33,34). Hypoplasia of vocal cords has been found at autopsy (6). Esophageal dysmotility is present in about 70% (34). In one family, this was associated with complex malformations: short trachea with high carina and supraclavicular tracheal bifurcation, severe hypoplasia of a lung, epiglottis, vocal cords, and larynx, with absence of the dorsal portions of the larynx and first tracheal rings and a single cavity extending from the epiglottis and hypopharynx superiorly to the trachea and inferiorly to the esophagus (15).

Fig. 24–20. *Opitz oculo-genito-laryngeal syndrome.* Hypospadias. [From *JM Opitz et al,* Birth Defects **5**(2):86, 1969.]

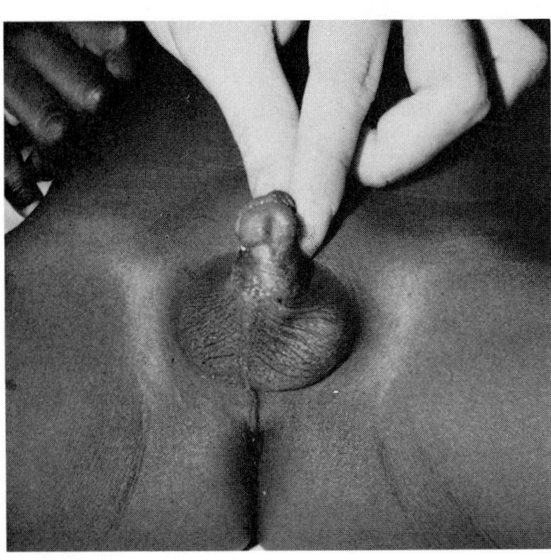

**Cardiovascular findings.** Congenital heart defects (CA, ASD) has been estimated to occur in 20–25% of patients with both Types I and II (20,27,29a,34). In Type II, anomalous venous return to the heart has been noted, as well as midline position of the heart. PDA has been reported in both types (5,10,30).

**Other findings.** Diastasis recti, multiple lipomas, and slight hyperextensibility of joints have been observed (3,7,24,29). Cutaneous pedal syndactyly of toes 2, 3, and 4 has been noted in Type II (9). Distally placed axial triradii, increased number of arches, and bridged palmar creases also have been reported in both types (9,22,30). Unlobed lungs as well as bifid renal pelvis with double ureter, Meckel diverticulum, and absence of the gallbladder and duodenal stricture have been seen in Type II (13). Fetal hydrops has been reported prenatally in a patient subsequently diagnosed as Type II (26). Agenesis of the corpus callosum has been observed in a female patient with Type II (19).

**Oral manifestations.** Cleft lip–palate and micrognathia have been observed in 25–35% of both types (2,3,5,9,11,17,20,24,27,29,29b,31, 32,34). Fused and supernumerary teeth and malocclusion have been noted in Type I (3,17,20,24,29,29b). Patients with Type II have shown broad or bifid uvula and ankyloglossia, bifid tongue, or shortened lingual frenulum (1a,6,7,8,10).

**Differential diagnosis.** Hypertelorism and hypospadias may each be observed as isolated anomalies or in association with various other

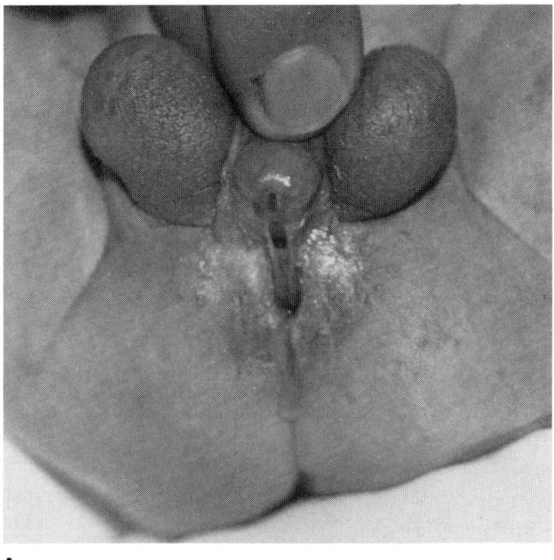

A

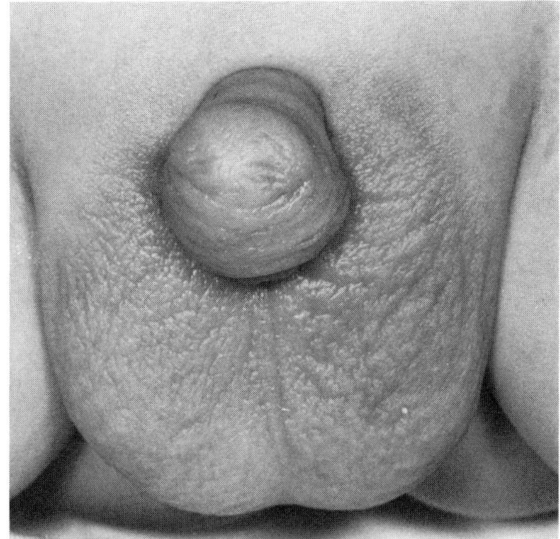

B

Fig. 24–21. *Opitz oculo-genito-laryngeal syndrome.* (A) Cleft scrotum with anterior displacement, severe chordee deformity of hypoplastic phallus. (B) Hypospadias, chordee. [A from *JM Opitz et al,* Birth Defects 5(2):95, 1969.]

malformation syndromes. Reed et al (28) erroneously reported three brothers with the *branchio-skeleto-genital syndrome* as examples of the Type I syndrome.

**Laboratory aids.** Cinefluoroscopic study of swallowing should be carried out to determine LTE defects. The disorder has been diagnosed prenatally by ultrasound (12a).

### References [Opitz oculo-genito-laryngeal syndrome (Opitz BBB/G compound syndrome)]

1. Allanson JE: G syndrome: An unusual family. *Am J Med Genet* **31**:637–642, 1988.

1a. Arya S et al: The G syndrome—additional observations. *Am J Med Genet* **5**:321–324, 1960.

1b. Bolsin SN, Gillbe C: Opitz–Frias syndrome. *Anesthesia* **40**:1189–1193, 1985.

2. Cappa M et al: The Opitz syndrome: A new designation for the clinically indistinguishable BBB and G syndromes. *Am J Med Genet* **28**:303–309, 1987.

2a. Chemke J et al: Male to male transmission of the G syndrome. *Clin Genet* **26**:164, 1984.

3. Christian JC et al: Familial telecanthus with associated congenital anomalies. *Birth Defects* **5**(2):82–85, 1969.

4. Coburn TP: G syndrome. *Am J Dis Child* **120**:466, 1970.

5. Cordero JF, Holmes LB: Phenotypic overlap of the BBB and G syndromes. *Am J Med Genet* **2**:145–152, 1978.

6. Côté GB et al: The G syndrome of dysphagia, ocular hypertelorism and hypospadias. *Clin Genet* **19**:473–478, 1981.

6a. DaSilva EO: The hypertelorism-hypospadias syndrome. *Clin Genet* **23**:30–34, 1983.

7. Farndon PA, Donnai D: Male to male transmission of the G syndrome. *Clin Genet* **24**:446–448, 1983.

8. Frias JL, Rosenbloom AL: Two familial cases of the G syndrome. *Birth Defects* **11**(2):54–57, 1975.

9. Funderburk SJ, Stewart R: The G and BBB syndromes: Case presentations, genetics, and nosology. *Am J Med Genet* **2**:131–144, 1978.

Fig. 24–22. *Opitz oculo-genito-laryngeal syndrome.* (A,B) Artist's conception of defects of soft palate, larynx, and hypopharynx. Note absence of posterior wall of larynx and presence of single, large laryngopharyngeal cavity. Arrows show that solid substances are as likely to enter trachea as to enter esophagus. (From *EF Gilbert et al,* Z Kinderheilkd **111**:290, 1972.)

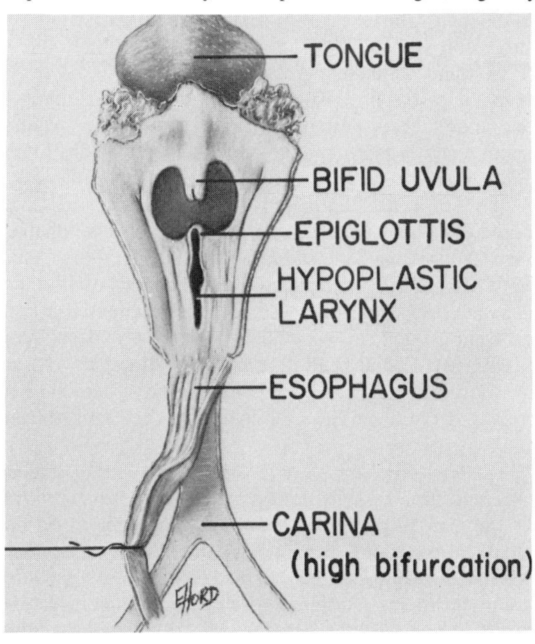


A

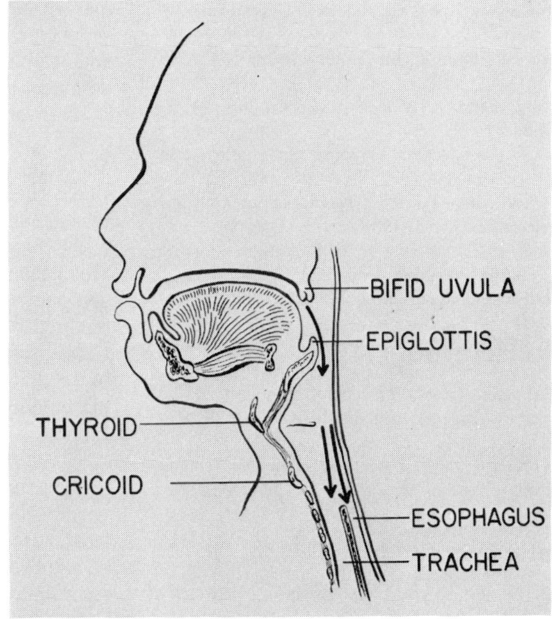


B

10. Gilbert EF et al: The pathologic anatomy of the G syndrome. *Z Kinderheilkd* **111**:290–298, 1972.

11. Gonzalez CH et al: Studies of malformation syndromes of man VB: The hypertelorism-hypospadias (BBB) syndrome. *Eur J Pediatr* **125**:1–13, 1977.

12. Gorlin RJ, Pindborg JJ, Cohen MM Jr. (eds): *Syndromes of the Head and Neck,* 2nd ed, McGraw-Hill, New York, 1976.

12a. Hogdahl C et al: Prenatal diagnosis of hypertelorism-hypospadias syndrome by ultrasound. *March of Dimes Clin Genet Conf,* Minneapolis, July 19–22, 1987.

12b. Howell L, Smith JD: G syndrome and its otolaryngologic manifestations. *Ann Otol Rhinol Laryngol* **98:**185–190, 1989.

13. Kasner J et al: The G syndrome—further observations. *Z Kinderheilkd* **118**:81–86, 1974.

14. Kimmelman CP, Denneny JC: Opitz (G) syndrome. *Int J Pediatr Otorhinolaryngol* **4**:343–347, 1982.

15. Little JR, Opitz JM: The G syndrome. *Am J Dis Child* **121**:505–507, 1971.

16. McKusick VA: *Mendelian Inheritance in Man: Catalogs of Autosomal Dominant, Autosomal Recessive, and X-Linked Phenotypes,* 7th ed, Johns Hopkins University Press, Baltimore [Entry #30710: Hypertelorism with esophageal abnormality and hypospadias (G syndrome; hypospadias-dysphagia syndrome)], 1986, pp 1391–1392.

17. Michaelis E, Mortimer W: Association of hypertelorism and hypospadias in the BBB syndrome. *Helv Paediatr Acta* **27**:575–581, 1972.

18. Miller PR et al: Hypertelorism-hypospadias syndrome with a laryngotracheoesophageal cleft. *J Pediatr* **90**:157–158, 1977.

19. Neri G et al: A girl with G syndrome and agenesis of the corpus callosum. *Am J Med Genet* **28**:287–291, 1987.

20. Noe HN et al: Hypertelorism-hypospadias syndrome. *J Urol* **132**:951–952, 1984.

21. Opitz JM: Editorial comment: G syndrome (hypertelorism with esophageal abnormality and hypospadias or hypospadias-dysphagia, or "Opitz-Frias" or "Opitz-G" syndrome)—perspective in 1987 and bibliography. *Am J Med Genet* **28**:275–285, 1987.

22. Opitz JM et al: Hypertelorism and hypospadias. A newly recognized hereditary malformation syndrome. *J Pediatr* **67**:968, 1965.

23. Opitz JM et al: The G syndrome of multiple congenital anomalies. *Birth Defects* **5**(2):95–101, 1969.

24. Opitz JM et al: The BBB syndrome. Familial telecanthus with associated congenital anomalies. *Birth Defects* **5**(2):86–94, 1969.

25. Parisian S, Toomey KE: Features of the G (Opitz–Frias) and BBB (hypospadias-hypertelorism) syndrome in one family—are they a single disorder? *Am J Hum Genet* **390**:72A, 1978.

26. Patton MA et al: Prenatal treatment of fetal hydrops associated with the hypertelorism-dysphagia syndrome (Opitz-G syndrome). *Prenat Diag* **6**:109–115, 1986.

27. Peeden JN et al: The telecanthus-hypospadias syndrome revisited. *Proc Greenwood Genet Ctr* **2**:131–132, 1983.

28. Reed MH et al: The hypertelorism-hypospadias syndrome. *J Assoc Can Radiol* **26**:240–248, 1975.

28a. Saer H et al: Laryngo-tracheo-oesophageal cleft and G-syndrome. *Z Kinderchir* **32:**29–37, 1981.

29. Sanchez Cascos A: Genetics of atrial septal defect. *Arch Dis Childh* **47**:58j1–588, 1972.

29a. Smith DW: *Recognizable Patterns of Human Malformations,* W.B. Saunders, 2nd ed, 1976, Philadelphia.

29b. Stevens CA, Wilroy RS Jr.: The telecanthus-hypospadias syndrome. *J Med Genet* **25**:536–542, 1988.

30. Stoll C et al: Male-to-male transmission of the hypertelorism-hypospadias (BBB) syndrome. *Am J Med Genet* **20**:221–225, 1985.

31. Tolmie JL et al: Congenital anal anomalies in two families with the Opitz G syndrome. *J Med Genet* **24**:688–691, 1987.

32. Van Biervliet JP, van Hemmel JO: Familial occurrence of the G syndrome. *Clin Genet* **7**:238–244, 1975.

32a. Verloes A et al: BBBC syndrome or Opitz syndrome. *Am J Med Genet* **34:**313–316, 1989.

33. Williams CA, Frias JL: Apparent G syndrome presenting as neck and upper limb dystonia and severe gastroesophageal reflux. *Am J Med Genet* **28**:297–302, 1987.

34. Wilson GN, Oliver WJ: Further delineation of the G syndrome: A manageable genetic cause of infantile dysphagia. *J Med Genet* **25**:157–163, 1988.

## Robinow syndrome (fetal face syndrome)

Robinow et al (18), in 1969, reported a syndrome consisting primarily of characteristic facies, mesomelic brachymelia of arms, short fingers, and hypoplastic genitalia. Since then, over 40 other patients have been described (2–5,7–12,14–17,19–31). Robinow (personal communication, 1987) indicated that he had seen over 30 unpublished examples. We cannot accept some cited cases (6,13).

In some families (18,24,25), the syndrome has autosomal dominant inheritance, and there is increased paternal age (17a). However, affected sibs with normal parents (20,23,28–30, JM Opitz, personal communication, 1975) indicate an autosomal recessive type. Although the dominant and recessive forms have phenotypic overlap, the recessive type is more severe. Both share the typical facies, short stature, hypoplastic genitalia, and normal intelligence, but in the recessive form short stature and brachymelia appear to be more marked and there are numerous rib and vertebral anomalies and radial head dislocation (35%) (1,17). However, the ability to separate the forms has been denied by Butler and Wadlington (2) who suggested that in an isolated case, a recurrence risk of 25% should be given. Robinow and Markert (17a) aver that in about 20%, separation into dominant and recessive forms cannot be made.

Birthweight and birth length are somewhat greater in the dominant type. Death occurred before 3 years in about 10% with the recessive form (14,23,26,30); two died of pneumonia while the cause of death in the third case was not given.

**Facies.** The facies resembles that of a fetus at 8 weeks: disproportionately large neurocranium, dolichocephaly, bulging forehead, hypertelorism, wide palpebral fissures, S-shaped lower eyelids, and short broad and upturned nose with anteverted nostrils. There is a long philtrum with a broad horizontal upper lip in the dominant form (Fig. 24–23A–C and 24–24A–C). A short philtrum is seen in the recessive form with an inverted V-shaped or tented upper lip (Fig. 24–23D,E). In both forms, there are frequent overfolding of the helices and a small chin (Fig. 24–25). Cephalometric study has been carried out by Israel and Johnson (7a).

**Central nervous system.** Intelligence is usually normal but developmental delay and mental retardation have been found in about 20%. Severe mental retardation was noted in the sibs reported by Seemanová et al (23) and Baxová et al (1a). The macrocephaly is not associated with hydrocephaly nor is there any dramatic change in the slope of head circumference with age.

**Musculoskeletal alterations.** Stature is −1.5 SD in the dominant form and −4.9 SD in the recessive form. A few authors (2,24) described adults with the dominant type having normal height. Forearm brachymelia is found to some degree in nearly all patients (8,10,24,26) making the height/span ratio excessive. Limitation of extension and supination or pronation at the elbows has been noted (8,28). Robinow and Markert (17a) found radius/humerus length to be −3.6 SD in the dominant form in contrast with −10.6 SD in the recessive form.

Hemivertebrae and/or fused vertebrae (65%), progressive scoliosis (50%), and displaced and/or fused and abnormal numbers of ribs (35%) are found in the recessive form (Fig. 24–26A–C). In 10%, there is a single butterfly vertebra in the dominant form. Short radius and short ulna are also found. The phalanges are short, and the terminal phalanx of the thumbs and halluces may be bifid (5–8,15,21) (Fig. 24–27A,B). Small hands with clinodactyly and brachymesophalangy of fifth fingers are noted in over 85%. An increase in bone sclerosis has been reported (8,24). Delayed bone age is evident in 35%. Dysplastic nails occur in about 50% (2). Radial head dislocation is almost never seen in the dominant form but is usual in the recessive form. Inguinal hernia has been reported (24,30). The umbilicus may be highly placed (3a).

**Genitourinary system.** In spite of the genital malformations frequently present, sex can be properly assigned in the newborn period. Unnecessary delay in evaluating ambiguous external genitalia should be avoided since sex determination can be made by examination alone regardless of the malformation. Micropenis, a very common feature, is usually apparent at birth (18,23,30) (Fig. 24–28). Cryptorchidism

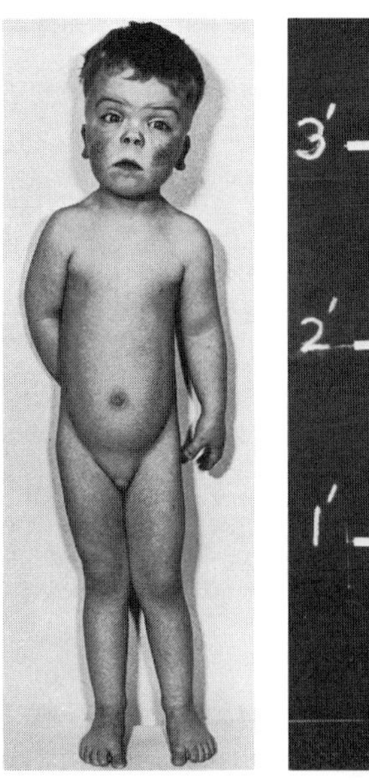
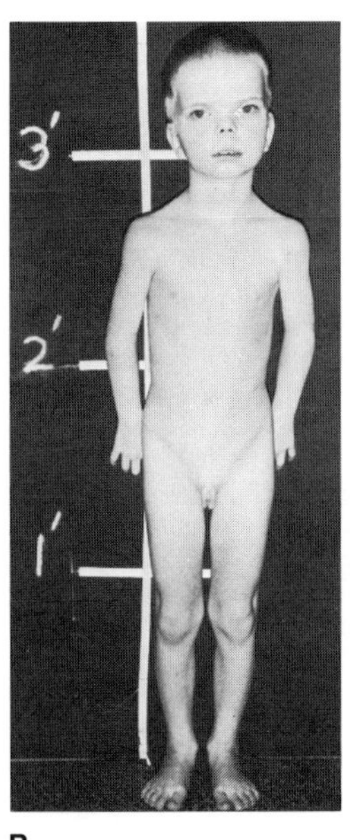

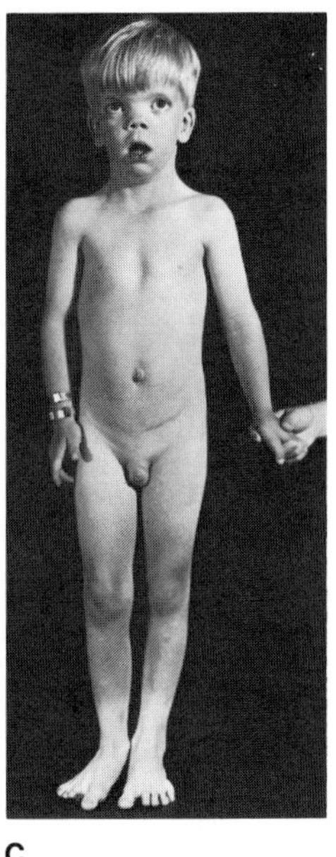
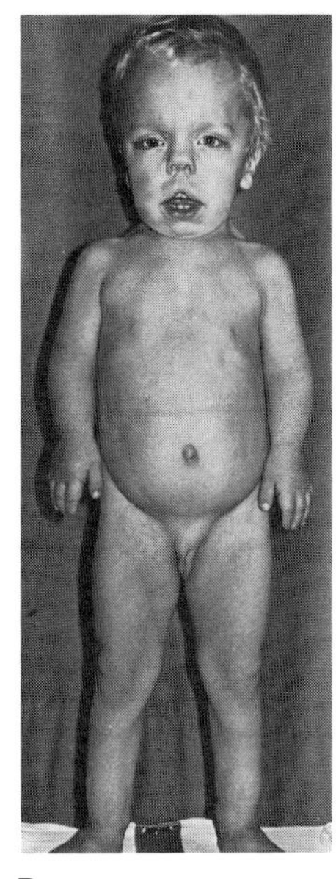
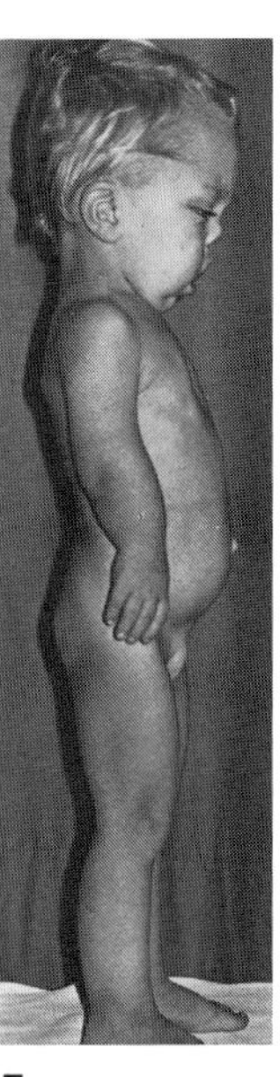

A B C D E

Fig. 24–23. *Robinow syndrome.* (A) Flattened facies, micropenis. (B) *Dominant form.* Short stature, mesomelia, and micropenis. (C) Compare with boy seen in A. (D,E) *Recessive form.* Note short philtrum and tented upper lip. (A from *RA Pfeiffer* and *H Müller,* Pädiatr Pädol **6**:262, 1971. B courtesy of *M Robinow,* Yellow Springs, Ohio. C courtesy of *HN Needleman,* Boston, Massachusetts. D,E from *RE Seel et al,* Monatschr Kinderheilkd **122**:663, 1974.)

has been found in 65% (2,7,18,24) and absent penis (26,30) has been observed. Lee et al (11) evaluated testicular function and pubertal development in four males with the syndrome and found partial primary hypogonadism, normal pubertal virilization with persistence of micropenis, and normal $5\alpha$-reductase and androgen receptor activity in genital skin fibroblasts. The clitoris and labia minor are hypoplastic in affected females (16,18,30). Bilateral renal duplication and bilateral hydronephrosis have been described (26,30).

**Oral manifestations.** The maxillary arch is trapezoidal, and the teeth are almost always crowded. An upper lip with inverted V-shape has been described in the recessive form (1a,17a,18,30). There is gingival enlargement in 65% (2,8,10,15,18,21,23). Cleft lip and cleft palate (23,24) are found in 10% and bifid or hypoplastic uvula (8,24) in 20%. The palate is short (7a). Minor clefting of the lower lip has been noted (30) as has ankyloglossia (7a).

**Differential diagnosis.** The combination is so striking and characteristic that no other condition with the possible exception of *Aarskog syndrome, achondroplasia,* and *pseudohypoparathyroidism* with severe facial and acral involvement need be considered. Differential diagnosis has been discussed extensively by Giedion et al (5).

### References [Robinow syndrome (fetal face syndrome)]

1. Bain MD et al: Robinow syndrome without mesomellic 'brachymelia': A report of five cases. *J Med Genet* **23**:350–354, 1986.

Fig. 24–24. *Robinow syndrome.* (A) Close-up of face showing fetal proportions. (B,C) Sister of boy seen in A. Note S-shaped lower eyelids, ocular hypertelorism, and midface hypoplasia. (Courtesy of *M Robinow,* Yellow Springs, Ohio.)

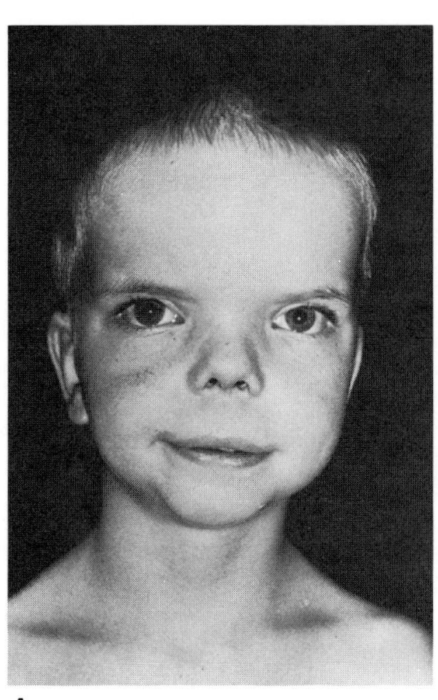
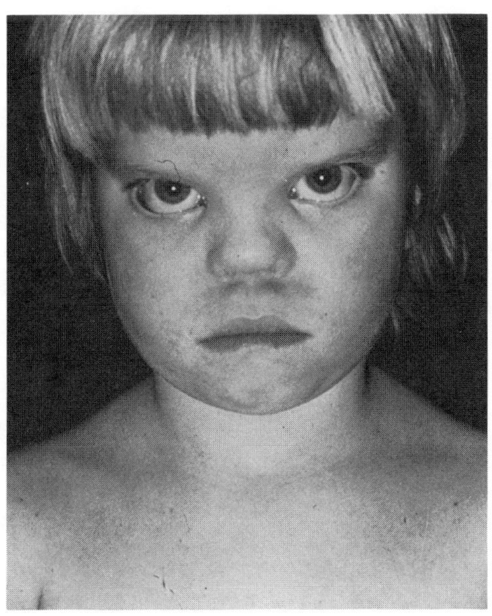
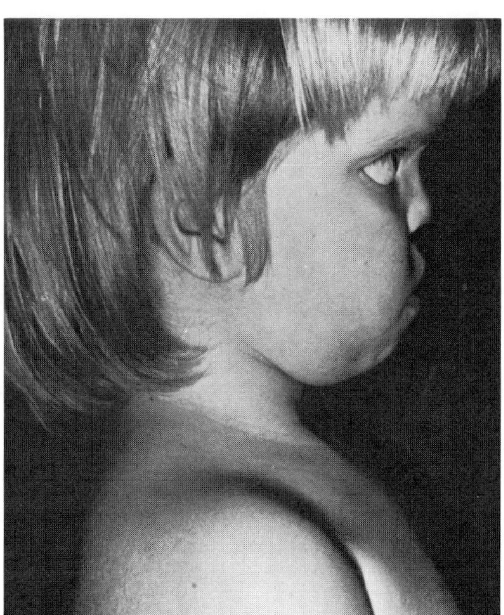

A B C

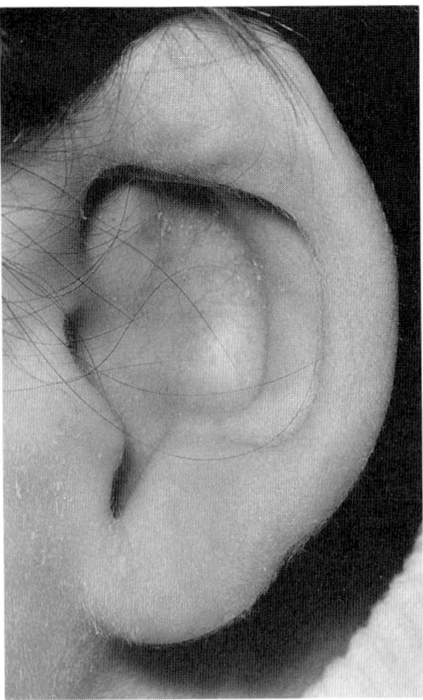

Fig. 24–25. *Robinow syndrome*. Ear with overfolded helix.

1a. Baxová A et al: Two cases of Robinow's syndrome with mental retardation. *Čs Pediatr* **44:**543–546, 1989.

2. Butler MG, Wadlington WB: Robinow syndrome: Report of two patients and review of literature. *Clin Genet* **31**:77–85, 1987.

3. Feingold M, Bull M: Case Report 4. *Syndrome Ident* **1**(1):14–16, 1973.

3a. Friedman JM: Umbilicus dysmorphology. The importance of contemplating the belly button. *Clin Genet* **28**:343–347, 1985.

4. Gellis SS et al: Fetal face syndrome (Robinow syndrome). *Am J Dis Child* **129**:351–352, 1975.

5. Giedion A et al: The radiological diagnosis of the fetal-face (Robinow) syndrome (mesomelic dwarfism and small genitalia). Report of three cases. *Helv Paediatr Acta* **30**:409–423, 1975.

6. Hanssler H, Schwanitz G: Bericht über eine Sonderform des Fetalgesicht-Minderwuchs-Syndroms. *Klin Pädiatr* **187**:274–277, 1975.

7. Harms S. Kracht H: Seltene Fehlbildung des Genitale bei einem Jungen mit multiplen Anomalien im Bereich des Skelett-und Muskelsystems. *Z Kinderchir* **12**:128–134, 1973. (Same case reported in Ref. 15.)

7a. Israel H, Johnson GF: Craniofacial pattern similarities and additional orofacial findings in siblings with the Robinow syndrome. *J Craniofac Genet Dev Biol* **8**:63–73, 1988. (Same family as in Ref. 18.)

8. Kelly TE et al: The Robinow syndrome. *Am J Dis Child* **129**:383–386, 1975.

9. Khayat D et al: Syndrome de Robinow. *Rev Stomatol* **84**:222–224, 1983.

10. Khayat D et al: Syndrome de Robinow. A propos d'un cas avec thrombopenie. *Arch Fr Pédiatr* **40**:327–330, 1983.

11. Lee PA et al: Robinow's syndrome: Partial primary hypogonadism in pubertal boys, with persistence of micropenis. *Am J Dis Child* **136**:327–330, 1982.

12. Marni E et al: Sindrome della faccia fetale a sindrome de Robinow. *Minerva Pediatr* **32**:47–52, 1980.

13. Menon PS et al: The Robinow syndrome. *Indian Pediatr* **20**:783–787, 1983.

14. Petit P et al: The Robinow syndrome. *Ann Génét* **23**:221–223, 1980.

15. Pfeiffer RA, Müller H: Ein Komplex multipler Missbildungen bei zwei nicht verwandten Kindern. *Pädiatr Pädol* **6:**262–267, 1971.

16. Portnoy Y: Robinow syndrome. *Clin Pediatr* **18**:707–708, 1979.

17. Robinow M: A syndrome's progress. *Am J Dis Child* **126**:150, 1973.

17a. Robinow M, Markert RJ: The fetal face (Robinow) syndrome: Delineation of the dominant and recessive phenotypes. *Proc Greenwood Genet Ctr* **7**:144, 1988.

18. Robinow M et al: A newly recognized dwarfing syndrome. *Am J Dis Child* **117**:645–651, 1969.

19. Rodriquez-Costa T et al: Robinow syndrome: Presentation of a case and review of literature. *An Esp Pediatr* **20**:55–61, 1984.

20. Saal HM et al: Autosomal recessive Robinow syndrome. *Am J Hum Genet* **37**:74A, 1985.

21. Schinzel A et al: Fetal face syndrome with acral dysostosis. *Helv Paediatr Acta* **29**:55–60, 1974.

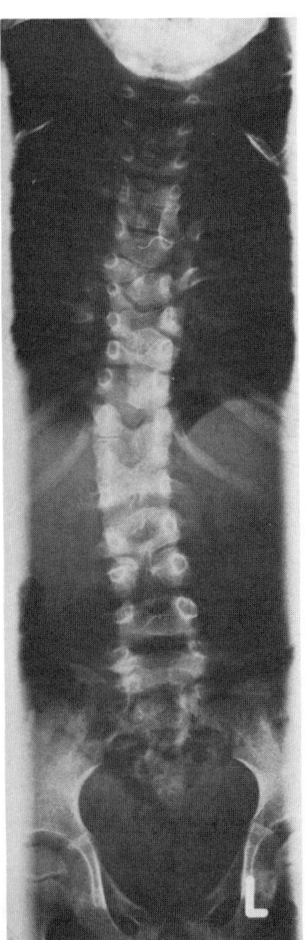


A

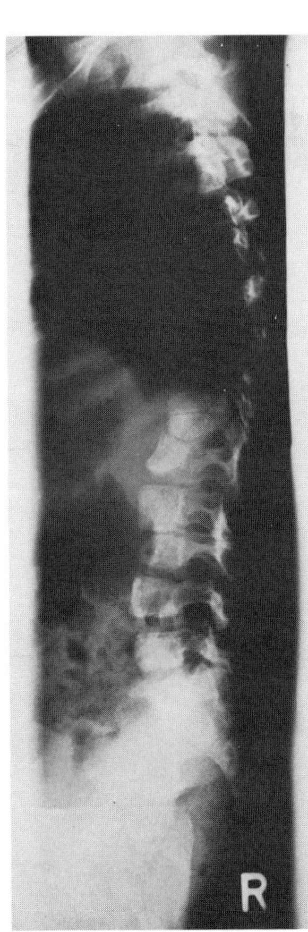


B

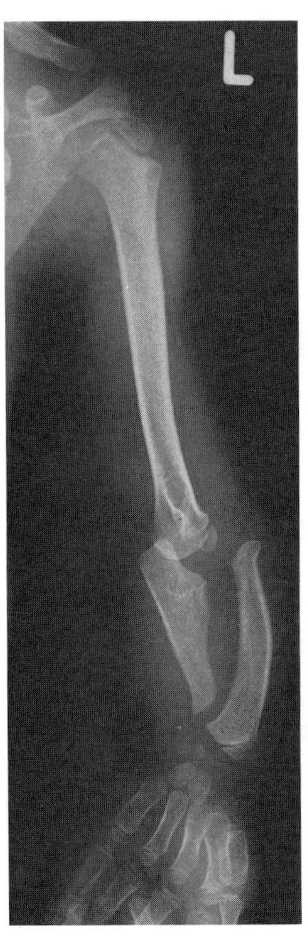


C

Fig. 24–26. *Robinow syndrome*. Radiographs showing keel-shaped vertebrae, hemivertebrae, and fused vertebrae. Note short, plump, and bent radius and ulna but normal humerus. (From *RE Seel et al,* Monatschr Kinderheilkd **122**:663, 1974.)

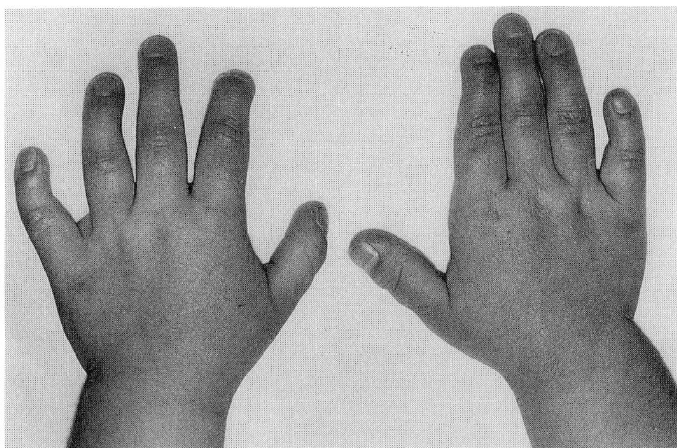
A

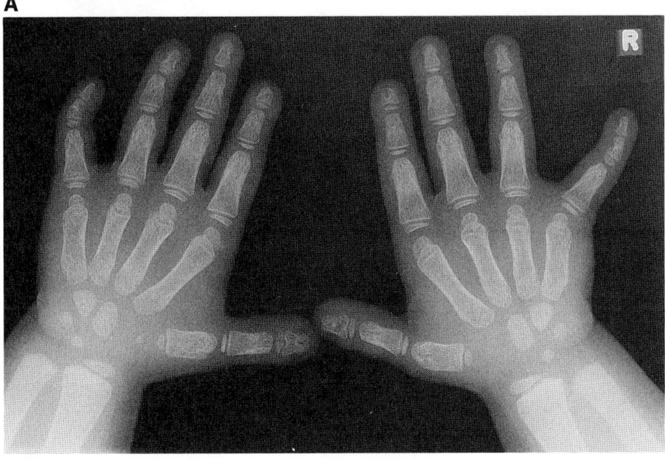

B

Fig. 24–27. *Robinow syndrome.* (A,B) Note shortened terminal phalanx of second digit and middle phalanx of fifth digit. Also note bifid terminal phalanges of thumb and right second digit.

22. Seel RE et al: Das Fetalgesicht-Minderwuchs-Syndrom nach Robinow. *Monatschr Kinderheilk* **122**:663–664, 1974.

23. Seemanová E et al: Fetal face syndrome with mental retardation. *Humangenetik* **23**:79–81, 1974.

24. Shprintzen RJ et al: Male–male transmission of Robinow's syndrome. Its occurrence in association with cleft lip and cleft palate. *Am J Dis Child* **136**:594–597, 1982.

25. Vallee L et al: Robinow's syndrome with dominant transmission. *Arch Fr Pédiatr* **39**:447–448, 1982.

26. Vera-Roman JM: Robinow dwarfism accompanied by penile agenesis and hemivertebrae. *Am J Dis Child* **126**:206–208, 1973.

27. Vogt J et al: Das Robinow-Syndrom. *Pädiatr Prax* **21**:103–105, 1979.

28. Wadia RS: Recessively inherited costavertebral segmentation defect with mesomelia and peculiar facies (Covesdem syndrome). A new genetic entity? *J Med Genet* **15**:123–127, 1978.

29. Wadia RS: Covesdem syndrome (letter). *J Med Genet* **16**:162, 1979.

30. Wadlington WB et al: Mesomelic dwarfism with hemivertebrae and small genitalia (the Robinow syndrome). *Am J Dis Child* **126**:202–205, 1973.

31. Ziška J et al: Fetal face syndrome. *Čs Pediatr* **36**:328–330, 1981.

Fig. 24–28. *Robinow syndrome.* Micropenis. (Courtesy of *M Robinow*, Yellow Springs, Ohio.)

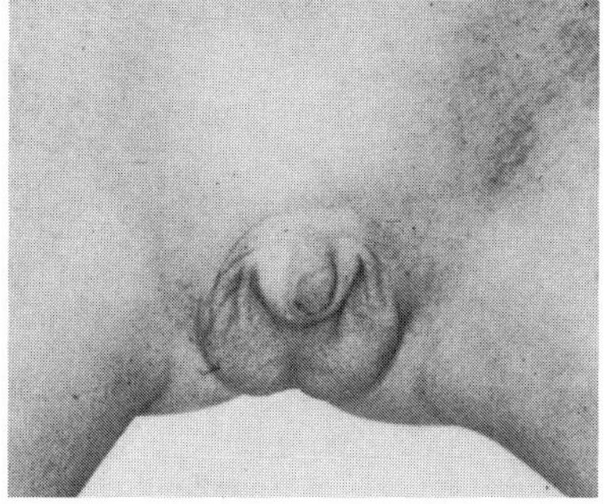

## Greig cephalopolysyndactyly syndrome

The combination of frontal bossing, scaphocephaly, hypertelorism, broad thumbs and halluces, pre- and postaxial polydactyly of hands and feet, and variable syndactyly of fingers and toes was probably first described by Greig (10) in 1926. Over 40 patients have been described subsequently (3–9,11–15,17–19,25, 26, 28) under a variety of terms, including frontodigital syndrome (17), Hootnick–Holmes syndrome (11), Noack-type acrocephalopolysyndactyly (6), frontonasal dysplasia (15), median cleft face syndrome (12), and acrofacial dysostosis (13). Greig cephalopolysyndactyly syndrome is the most commonly used designation and appears to be most appropriate. The designation of cephalopolysyndactyly is too nonspecific. Furthermore the term Greig syndrome, used in the past to refer to hypertelorism, describes a symptom complex now known to be etiologically and pathogenetically heterogeneous. Hence, the term Greig syndrome bears no relationship to Greig cephalopolysyndactyly syndrome and should be abandoned.

Autosomal dominant inheritance with markedly variable expressivity has been demonstrated in several families. The gene has been mapped to 7p13 (1a,28). The patient of Töllner et al (27), in spite of having cleft lip–palate, may have the syndrome.

**Facies.** About 50% manifest frontal bossing, increased head circumference, and broad forehead. Broad nasal bridge is seen in 85% (Fig. 24–29A). In children, the sutures may be broad and close late. The facies is not always striking. Hypertelorism is less common than the literature would suggest (3,4,18). Craniosynostosis has been reported on occasion (11,15).

**Extremities.** Broad thumbs (60%) and broad halluces (40%) are rather frequent. Soft tissue syndactyly of fingers (70%) and toes (95%) is variable in degree. Although postaxial polydactyly of the hands is common (65%), postaxial polydactyly of the feet is relatively infrequent (10%). Preaxial polysyndactyly of the thumbs varies from broad terminal phalanx (common) to bifid distal phalanx to complete duplication (rare) (26). Preaxial duplication of the thumbs always seems associated with bifid hallux (Fig. 24–29B). In the feet, the most common anomaly is preaxial polydactyly (75%). Postaxial polydactyly is rare (10%). Syndactyly, nearly a constant feature in the feet, may be present in the absence of polydactyly (Fig. 24–29C). It is unilateral in 40%. The halluces and thumbs may be medially deviated. Radiographic changes in the halluces are bifid terminal phalanges (30%) and duplication of both phalanges (55%). The first metacarpal is duplicated in about 15%. Bony fusion of the third to fifth distal phalanges of the hands has been reported (5,6,28). Bone age has been somewhat advanced (18,27,28). Characteristic dermatoglyphic changes have been reported. Several digital triradii are missing. Whorl patterns are present on the thumbs with the loops extending beyond the end of the thumb (12). A t′ axial triradius is often seen.

**Muscular system.** Inguinal and, rarely, umbilical hernia have been noted.

**Central nervous system.** Intelligence is normal with rare exception (6,11). Agenesis of the corpus callosum (11) and hydrocephaly (1,11) have been reported.

**Differential diagnosis.** The *acrocallosal syndrome* markedly overlaps with Greig cephalopolysyndactyly syndrome but differs in that agenesis of the corpus callosum and severe mental retardation are rare in Greig cephalosyndactyly syndrome. Furthermore, the latter

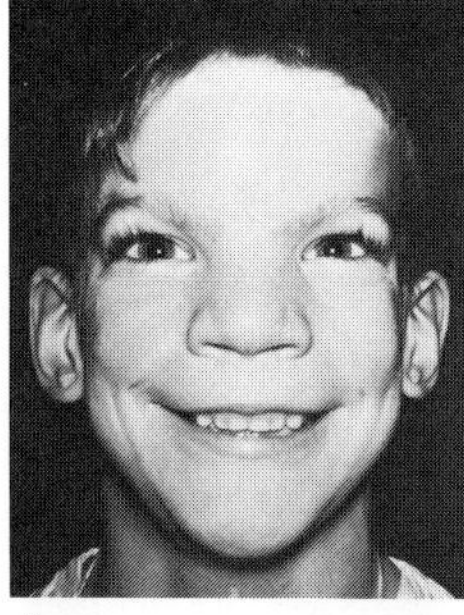

A

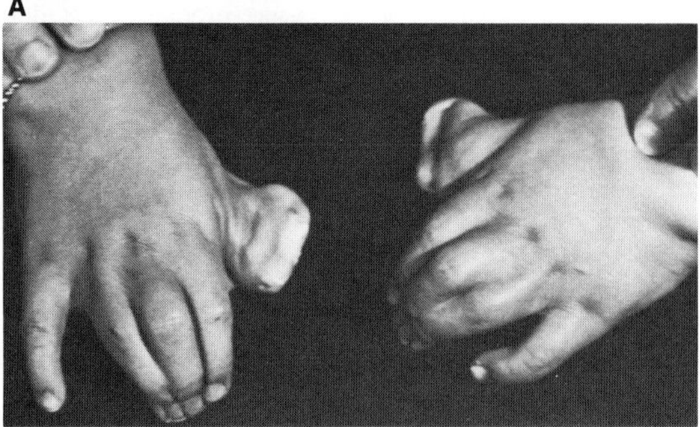

B

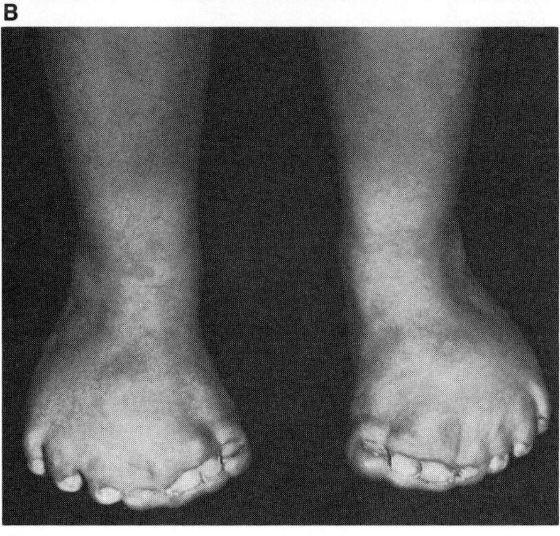

C

Fig. 24–29. *Greig cephalopolysyndactyly syndrome.* (A) Note broad forehead, broad nasal bridge, and apparent hypertelorism. (B) Deviated and duplicated thumbs with 2–4 soft tissue syndactyly. Postaxial digit removed earlier. (C) Preaxial duplication or triplication and syndactyly. (From *D Hootnick* and *LB Holmes,* Clin Genet **3**:128, 1972.)

clearly has autosomal dominant inheritance. Acrocallosal syndrome possibly had autosomal recessive inheritance (16,20–24).

*Craniofrontonasal dysplasia* has a somewhat similar facies. The *OFD syndromes* (cleft palate, tongue hamartoma, polysyndactyly) have similar hands and feet.

### References (Greig cephalopolysyndactyly syndrome)

1. Baraitser M et al: Greig cephalosyndactyly: Report of 13 affected individuals in three families. *Clin Genet* **24**:257–265, 1983.

1a. Brueton L et al: Direct DNA analysis demonstrates that Greig cephalosyndactyly maps to 7p13. *Am J Med Genet* **31**:799–804, 1988.

2. Chudley AE, Houston CS: The Greig cephalopolysyndactyly syndrome in a Canadian family. *Am J Med Genet* **13**:269–276, 1982.

3. Duncan PA et al: Greig cephalopolysyndactyly syndrome. *Am J Dis Child* **133**:818–821, 1979.

4. Fryns JP et al: The Greig polysyndactyly-craniofacial dysmorphism syndrome. *Eur J Pediatr* **126**:283–287, 1977.

5. Fryns JP et al: The Greig craniofacial dysmorphism syndrome: Variable expression in a family. *Eur J Pediatr* **136**:217–220, 1981.

6. Gnamey D, Farriaux JP: Syndromes dominant associant polysyndactylie, pouces en spatule, anomalies faciales et retard mental une forme particulière de l'acrocéphalo-polysyndactylie de type Noack. *J Génét Hum* **19**:299–316, 1971.

7. Goldberg MJ, Pashayan HM: Hallux syndactyly–ulnar polydactyly–abnormal ear lobes: A new syndrome. *Birth Defects* **12**(5):255–266, 1976.

8. Gollop TR, Fontes LF: The Greig cephalopolysyndactyly syndrome: Report of a family and review of the literature. *Am J Med Genet* **22**:59–68, 1985.

9. Goodman RM: Family with polysyndactyly and other anomalies. *J Hered* **56**:37–38, 1965.

10. Greig DM: Oxycephaly. *Edinb Med J* **33**:189–218, 1926.

11. Hootnick D, Holmes LB: Familial polysyndactyly and craniofacial anomalies. *Clin Genet* **3**:128–134, 1972.

12. Ide C, Holt JE: Median cleft face syndrome associated with orbital hypertelorism and polysyndactyly. *EENT Month* **54**:150–151, 1975.

13. Korting GW, Ruther H: Ichthyosis vulgaris und akrofaciale Dysostosis. *Arch Dermatol Syph (Berlin)* **197**:91–104, 1954.

14. Kunze J, Kaufmann HJ: Greig cephalopolysyndactyly syndrome. *Helv Paediatr Acta* **40**:489–495, 1985.

15. Kwee ML, Lindhout D: Frontonasal dysplasia, coronal craniosynostosis, pre- and postaxial polysyndactyly and split nails: A new autosomal dominant mutant with reduced penetrance and variable expression. *Clin Genet* **24**:200–205, 1983.

16. Legius E et al: Schinzel acrocallosal syndrome: A variant example of the Greig syndrome? *Ann Génét* **28**:239–240, 1985.

17. Marshall RE, Smith DW: Frontodigital syndrome: A dominantly inherited disorder with normal intelligence. *J Pediatr* **77**:129–133, 1970.

18. Merlob P et al: A newborn infant with craniofacial dysmorphism and polysyndactyly (Greig's syndrome). *Acta Paediatr Scand* **70**:275–277, 1981.

19. Merlob P et al: Le syndrome de Greig. *J Radiol Electrol* **65**:187–189, 1984.

20. Nelson MM, Thompson AJ: The acrocallosal syndrome. *Am J Med Genet* **12**:195–199, 1982.

20a. Philip N et al: The acrocallosal syndrome. *Eur J Pediatr* **147**:206–208, 1988.

21. Schinzel A: Postaxial polydactyly, hallux duplication, absence of the corpus callosum, macroencephaly and severe mental retardation: A new syndrome? *Helv Paediatr Acta* **34**:141–146, 1979.

22. Schinzel A: Acrocallosal syndrome. *Am J Med Genet* **12**:201–203, 1982.

22a. Schinzel A: The acrocallosal syndrome in first cousins: Widening of the spectrum of clinical features and further support for autosomal recessive inheritance. *J Med Genet* **25**:332–336, 1988.

23. Schinzel A, Kaufmann U: The acrocallosal syndrome in sisters. *Clin Genet* **30**:399–405, 1986.

24. Schinzel A, Schmidt N: Hallux duplication, postaxial polydactyly, absence of the corpus callosum, severe mental retardation, and additional anomalies in two unrelated patients: A new syndrome? *Am J Med Genet* **6**:241–249, 1980.

25. Temtamy S, McKusick VA: Synopsis of hand malformations with particular emphasis on genetic factors. *Birth Defects* **5**(3):125–184, 1969.

26. Temtamy SA, Loutfy AH: Polysyndactyly in an Egyptian family. *Birth Defects* **10**(5):207–215, 1974.

27. Töllner U et al: Heptacarpo-octatarso-dactyly combined with multiple malformations. *Eur J Pediatr* **136**:207–210, 1981.

28. Tommerup N, Nielsen F: A familial reciprocal translocation t(3:7) (p21.1;p13) associated with the Greig polysyndactyly—craniofacial anomalies syndrome. *Am J Med Genet* **16**:313–321, 1983.

## Acrocallosal syndrome

The acrocallosal syndrome, first reported by Schinzel (9) in 1979, is characterized by post- and/or preaxial polydactyly and syndactyly of fingers or toes, severe mental retardation, agenesis or hypoplasia of the corpus callosum, and growth retardation. Schinzel et al (10–13), in a series of papers, developed and expanded the phenotype. Less than 20 cases have been reported (1,3–14). A good survey is that of Casamassima et al (1).

There has been no sex predilection. The disorder has been described in sisters and in cousins (11,12). All other cases have been sporadic. Parental consanguinity has been demonstrated (6,7,14). While autosomal recessive inheritance has been suggested, we remain skeptical.

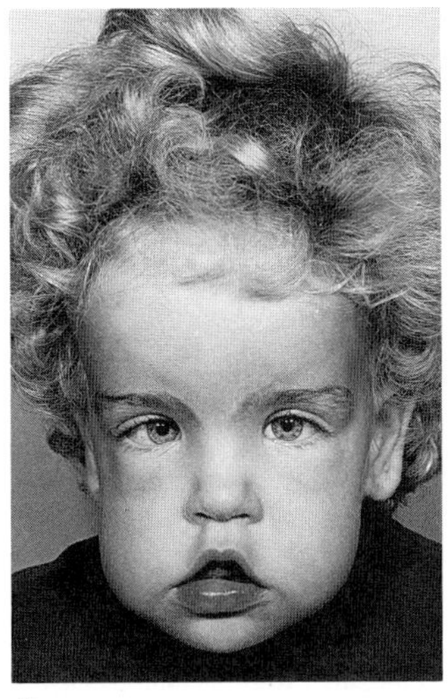
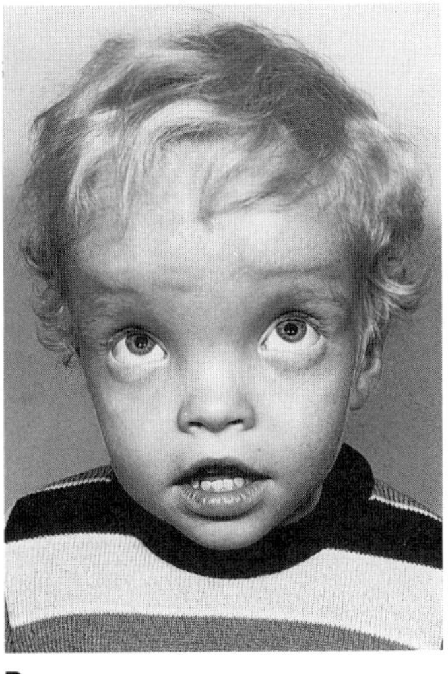

A B

Fig. 24–30. *Acrocallosal syndrome.* (A,B) Unrelated patients exhibiting large head, prominent forehead, broad nasal bridge. (From *A Schinzel* and *N Schmidt,* Am J Med Genet **6**:241, 1980.)

**Craniofacial findings.** Macrocephaly with bulging and high broad forehead, prominent occiput, large fontanel, hypertelorism, downslanting palpebral fissures, ptosis, epicanthal folds, convergent strabismus, hypoplastic midface, posteriorly rotated and somewhat dysmorphic pinnae, short nose, and protruding lips constitute the facies (Fig. 24–30A,B). Cleft lip and palate has been found in about 15%.

**Growth and development.** Severe mental retardation is a constant finding. About 25% exhibit seizures. Growth retardation has been noted in 55%.

**Musculoskeletal findings.** Duplicated halluces (65%), postaxial polydactyly of toes (40%), fully formed postaxial polydactyly of fingers (70%), bifid terminal phalanges of thumbs (20%), inguinal (55%) or umbilical (20%) hernia, and epigastric hernia (15%) have been found. Partial syndactyly of the toes occurred in about 65% (Fig. 24–31A).

**Central nervous system.** Severe mental retardation and total or partial agenesis of the corpus callosum are constant findings (Fig. 24–31B). About 75% exhibit seizures. Congenital cyst of the brain (12) and central diffuse cerebral cortical atrophy (7) have been demonstrated.

Fig. 24–31. *Acrocallosal syndrome.* (A) Duplication, syndactyly, and fusion of nails, and halluces. (B) Agenesis of corpus callosum. (A from *A Schinzel* and *N Schmidt,* Am J Med Genet **6**:241, 1980. B from *A Schinzel* and *U Kaufmann,* Clin Genet **30:**399, 1986.)

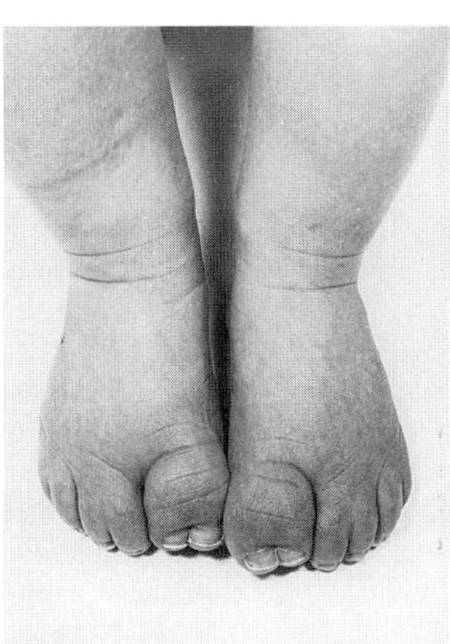
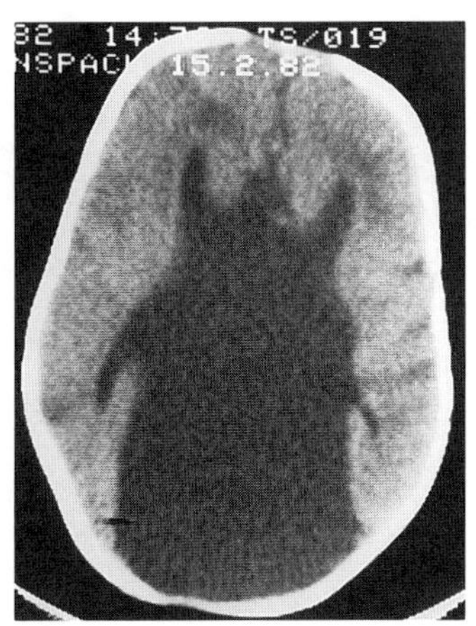

A B

**Other findings.** Miscellaneous findings include: cryptorchidism (20%) (14), micropenis and hypospadias (7), broad nasal bridge, and prominent occiput.

**Differential diagnosis.** *Greig cephalopolysyndactyly syndrome* overlaps considerably, but in that disorder there are few reports of agenesis of the corpus callosum or mental retardation (2). Moreover, postaxial manual digits are only minimally found. In contrast to acrocallosal syndrome, inheritance is clearly autosomal dominant.

Agenesis of the corpus callosum is a field defect and is found in a host of chromosomal and single gene disorders listed extensively by Temtamy and Meguid (14). Among these are the *FG syndrome,* Aicardi syndrome, *Neu–Laxova syndrome, cerebro-oculo-facio-skeletal syndrome, Opitz G syndrome, hydrolethalus syndrome* and several of the *oral-facial-digital syndromes.*

It should be noted that there is a severe form of *frontonasal malformation* in which duplication of halluces and agenesis of the corpus callosum are found. This has been called Toriello syndrome.

### References (Acrocallosal syndrome)


1. Casamassima AC et al: Acrocallosal syndrome: Additional manifestations. *Am J Med Genet* **32:**311–317, 1989.
2. Hootnick D, Holmes LB: Familial polysyndactyly and craniofacial anomalies. *Clin Genet* **3:**128–134, 1972.
3. Legius E et al: Schinzel acrocallosal syndrome: A variant example of the Greig syndrome? *Ann Génét* **28:**239–240, 1985.
4. Moeschler JB et al: Acrocallosal syndrome: New findings. *Am J Med Genet* **32:**306–310, 1988.
5. Nelson MM, Thompson AJ: The acrocallosal syndrome. *Am J Med Genet* **12:**195–199, 1982.
6. Philip N et al: The acrocallosal syndrome. *Eur J Pediatr* **147:**206–208, 1988.
7. Salgado LJ et al: Acrocallosal syndrome in a girl born to consanguineous parents. *Am J Med Genet* **32:**298–300, 1989.
8. Sanchis A et al: Duplication of hands and feet, multiple joint dislocations, absence of corpus callosum and hypsarrhythmia: Acrocallosal syndrome? *Am J Med Genet* **20:**123–130, 1985.
9. Schinzel A: Postaxial polydactyly, hallux duplication, absence of the corpus callosum, macrencephaly and severe mental retardation: A new syndrome? *Helv Paediatr Acta* **34:**141–146, 1979.
10. Schinzel A: Acrocallosal syndrome. *Am J Med Genet* **12:**201–203, 1982.
11. Schinzel A: The acrocallosal syndrome in first cousins: Widening of the spectrum of clinical features and further support for autosomal recessive inheritance. *J Med Genet* **25:**332–336, 1988.
12. Schinzel A, Kaufmann U: The acrocallosal syndrome in sisters. *Clin Genet* **30:**399–405, 1986.
13. Schinzel A, Schmidt N: Hallux duplication, postaxial polydactyly, absence of the corpus callosum, severe mental retardation, and additional anomalies in two unrelated patients: A new syndrome? *Am J Med Genet* **6:**241–249, 1980.
14. Temtamy SA, Meguid NA: Hypogenitalism in the acrocallosal syndrome. *Am J Med Genet* **32:**301–305, 1989.


## Branchio-skeleto-genital syndrome (Elsahy–Waters syndrome)

Elsahy and Waters (1) described three male sibs who exhibited seizures, moderate mental retardation (IQ 45–60), pectus excavatum, and penoscrotal hypospadias. Family 1, in the article by Reed et al (2), also had the syndrome. The detailed account of the oral and dental findings presented by Wedgwood et al (4) represent the same family reported by Elsahy and Waters (1). A very doubtful example was reported by Shafai et al (3).

The facies was characterized by brachycephaly, frontal bossing, marked midfacial hypoplasia, hypertelorism, divergent strabismus, nystagmus, and mild ptosis. The nasal bridge was broad with a wide nasal tip and flared alae. There was relative mandibular prognathism and facial acne (Fig. 24–32A,B).

Radiographic study showed brachycephaly, mild microcephaly, extensive development of the mastoids, hypertelorism, short cranial base,

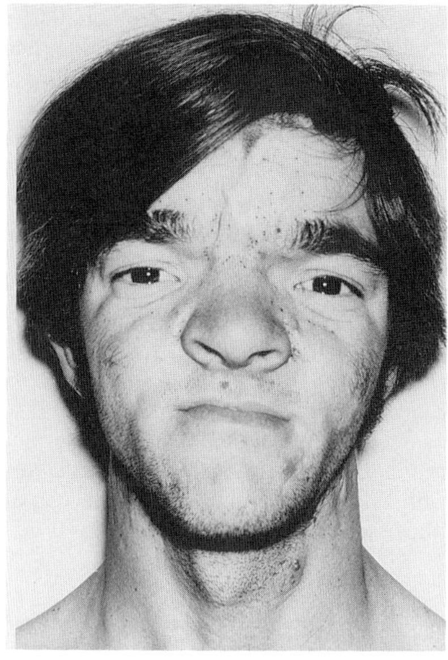
A

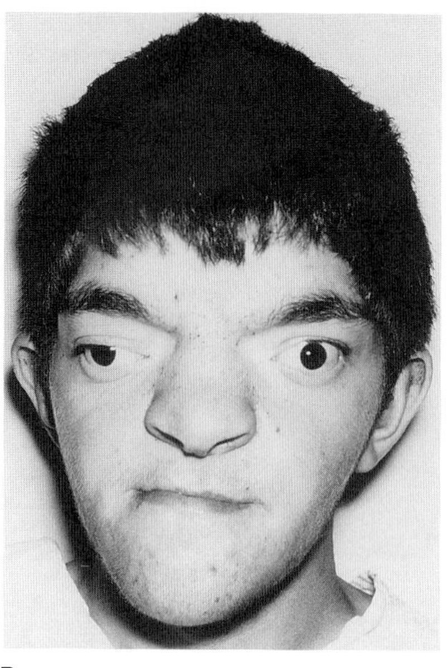
B

Fig. 24–32. *Branchio-skeleto-genital syndrome.* (A,B) Two affected sibs with hypertelorism, exotropia, midfacial hypoplasia, extensive acne, and relative mandibular prognathism. (From *MH Reed et al,* J Can Assoc Radiol **26**:240, 1975.)

bony projections above the external occipital protuberance (2), and fusion of the second and third cervical vertebrae in two brothers. Schmorl nodules were evident in the vertebral bodies (Fig. 24–33).

Hypospadias involving the glans, penile shaft, and proximal scrotum were present in all three sibs but to various degrees (Fig. 24–34).

Oral manifestations included multiple dentigerous cysts, cleft uvula or cleft soft palate, and teeth having dysplastic dentin with partial obliteration of pulp chambers resembling that seen in radicular dentin dysplasia (5) (Fig. 24–35A,B). The dental anomalies were also described by Wedgwood et al (4).

Fig. 24–33. *Branchio-skeleto-genital syndrome.* Note maxillary hypoplasia, mastoid hyperpneumatization, and bony excrescences at occipital protuberance. (From *MH Reed et al,* J Can Assoc Radiol **26**:240, 1975.)

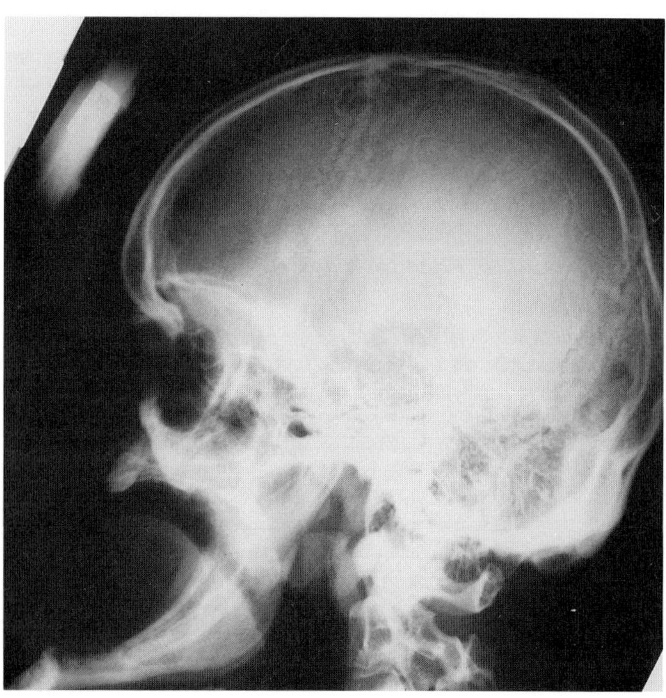

#### References [Branchio-skeleto-genital syndrome (Elsahy–Waters syndrome)]

1. Elsahy NI, Waters WR: The branchio-skeleto-genital syndrome. *Plast Reconstr Surg* **48**:542–550, 1971.
2. Reed MH et al: The hypertelorism-hypospadias syndrome. *J Can Assoc Radiol* **26**:240–248, 1975 (Family 1).
3. Shafai J et al: The branchioskeletogenital syndrome. *Birth Defects* **18**(3B):193–196, 1982.
4. Wedgwood DL et al: Cranio-facial and dental anomalies in the branchio-skeleto-genital (BSG) syndrome with suggestions for more appropriate nomenclature. *Br J Oral Surg* **21**:94–102, 1983.
5. Witkop CJ Jr: Hereditary defects of dentin. *Dent Clin N Am* **19**:25–45, 1975.

Fig. 24–34. *Branchio-skeleto-genital syndrome.* Penoscrotal hypospadias. (From *NI Elsahy* and *WR Waters,* Plast Reconstr Surg **48**:542, 1971.)

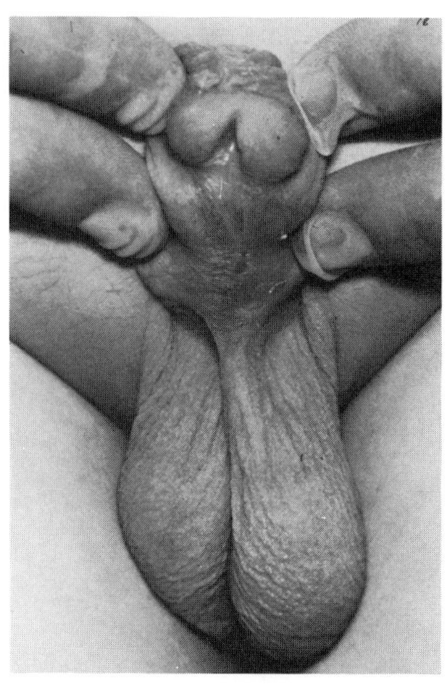

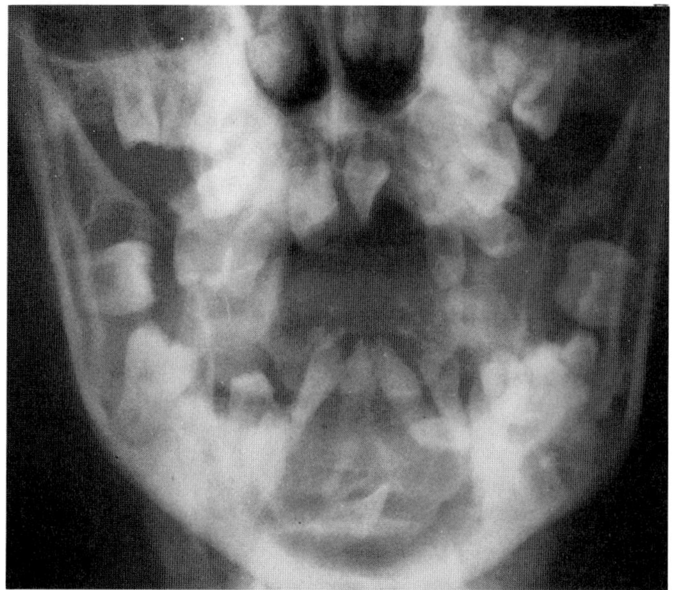

A

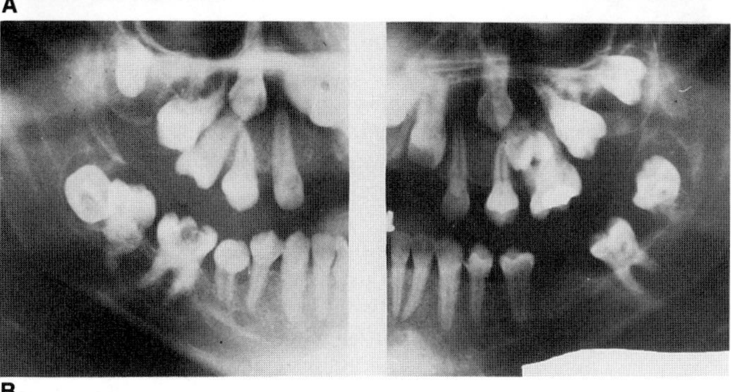

B

Fig. 24–35. *Branchio-skeleto-genital syndrome.* (A,B) Dentigerous cysts. (From *MH Reed et al,* J Can Assoc Radiol **26**:240, 1975.)

## Nasopalpebral lipoma–coloboma syndrome (Penchaszadeh syndrome)

Penchaszadeh et al (1), in 1982, reported an apparently unique autosomal dominantly inherited syndrome in eight members of a Venezuelan kindred.

The facies is striking. It is characterized by congenital symmetrical round to ovoid lipomas of the upper eyelids. More extensive, less circumscribed lipomas of the nasopalpebral regions may cover the root of the nose and extend symmetrically toward the inner canthi that are laterally displaced, covering the medial part of the irides. There are bilateral symmetrical colobomas of both upper and lower eyelids at the junction of the inner and middle third of the lids. The eyelashes are long and curly lateral to the lid defect but absent medially. The upper and, less often, lower lacrimal puncta are malpositioned or occasionally absent with resultant epiphora. Divergent strabismus is marked in all patients. The conjunctivae are hyperemic and several patients have corneal opacities, probably secondary to exposure (Fig. 24–36A,B).

Broad forehead, widow's peak, medial outflaring of the eyebrows, short philtrum, and maxillary hypoplasia add to the unusual appearance. Radiographic studies reveal maxillary hypoplasia and normal bony interorbital distances.

### Reference [Nasopalpebral lipoma–coloboma syndrome (Penchaszadeh syndrome)]

1. Penchaszadeh VB et al: The nasopalpebral lipoma–coloboma syndrome: A new autosomal dominant dysplasia-malformation syndrome with congenital nasopalpebral lipomas, eyelid colobomas, telecanthus and maxillary hypoplasia. *Am J Med Genet* **11**:397–410, 1982.

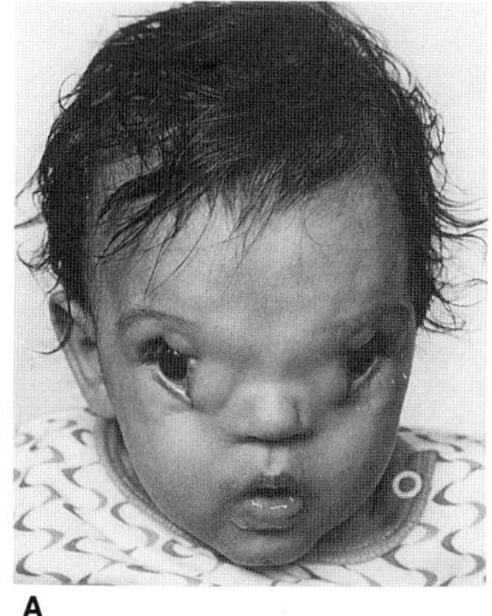

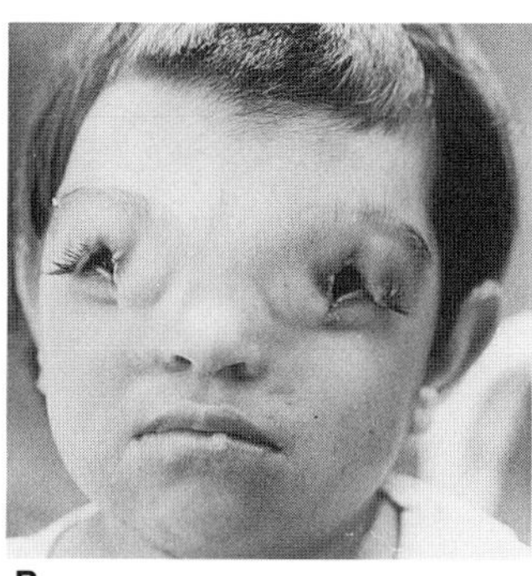

A B

Fig. 24–36. *Nasopalpebral lipoma–coloboma syndrome.* (A,B) Facies showing symmetric round to ovoid lipomas of upper eyelids and nasopalpebral regions, lateral displacement of inner canthi, colobomas of upper and lower eyelids. Also note widow's peak. (Courtesy of *VB Penchaszadeh,* New York, New York.)

## Noonan syndrome

Noonan syndrome is characterized by short stature, various congenital heart defects, broad or webbed neck, chest deformity, hypertelorism with characteristic facial appearance and, in some cases, mild mental deficiency. Described by Noonan and Ehmke in 1963 (42), the syndrome was first reported by Kobylinski in 1883 (33). Over 300 cases have been recorded and a number of excellent reviews are available (1,11,38,44,46). Birth prevalence is estimated to be between 1/1000 and 1/2500 (43).

**Etiology and pathogenesis.** Although most of the early cases appear to be sporadic, recent surveys indicate transmission from parent to child in 30–75% (1,11,38,43,44,46). Autosomal dominant inheritance with highly variable expression is evident. Improved recognition of the adult phenotype may further reduce the frequency of sporadic cases. Maternal transmission of the gene is three times more common than paternal transmission, most likely because of associated cryptorchidism and male infertility (1).

Several pathogenetic hypotheses have been proposed. Sanchez-Cascos (52) noted that the abnormalities of the head, neck, and heart that occur in Noonan syndrome could be produced by altered embryonic branchial arch structures. Another hypothesis implicates lymphedema. Intrauterine cystic hygroma resulting in pterygium colli and disruption of tissue migration and organ placement could possibly explain hypertelorism, downslanting palpebral fissures, low-set posteriorly angulated ears, prominent trapezius, widely spaced nipples, cryptorchidism, and abnormal dermatoglyphics. Clark (13,14) proposed that lymphatic obstruction might reduce right-sided cardiac blood flow and cause pulmonic stenosis.

**Craniofacial features.** Facial characteristics change with age (2). In the newborn, features include hypertelorism, downslanting palpebral fissures (95%), deeply grooved philtrum with high, wide peaks of the vermilion border (95%), highly arched palate (45%), micrognathia (25%), low-set and posteriorly angulated ears with thick helices (90%), and excessive nuchal skin with low posterior hairline (55%). During infancy, the head is relatively large with a turricephalic configuration. Hypertelorism, prominent eyes, and thick hooded eyelids are characteristic. The nasal bridge is low and the nose has a wide base with bulbous tip. During childhood, the face may appear coarse

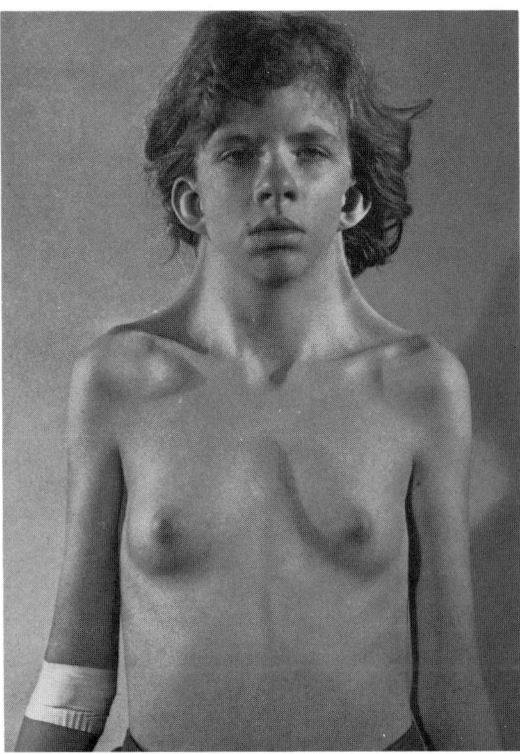

Fig. 24–37. *Noonan syndrome.* Pterygium colli and prominent ears. Note normal breast development.

or myopathic. Facial contour becomes more triangular with age. During adolescence and young adulthood, the eyes become less prominent, and the nose has a thin, high bridge and a wide base. The neck appears longer with accentuated webbing or prominent trapezius (90%). In older adults, the nasolabial folds are prominent, the anterior hairline is high, and the skin appears wrinkled and transparent (1,2) (Figs. 24–37 and 24–38A–C).

Features present regardless of age include blue-green irides, halo iris, arched eyebrows, and low-set posteriorly angulated ears with thick helices (2,10). The hair may be wispy during infancy and curly or woolly in older childhood and adolescence. Malocclusion occurs in 35% (1,2). Allanson (1a) discussed the changing facies of Noonan syndrome with age.

**Cardiovascular anomalies.** Congenital heart defects occur in two-thirds of the patients. Common anomalies include pulmonic valvular stenosis (50%), ASD (10%), asymmetric septal hypertrophy (10%), VSD (5%), and PDA (3%). Low-frequency abnormalities include pulmonary artery branch stenosis, mitral valve prolapse, Ebstein anomaly, and single ventricle. ECG commonly demonstrates a wide QRS complex, left axis deviation, giant Q waves, and the left precordial leads have a negative pattern (52).

**Growth and skeletal findings.** Average birth length is 47 cm and birthweight is usually normal (40%) but may be increased by subcutaneous edema. Childhood growth tends to parallel the third percentile (60%), usually with normal growth velocity. The growth spurt during puberty is frequently reduced or absent (55). Delay in osseous maturation has been recorded in approximately 20%. Growth hormone levels have been normal with a slight rise in somatomedin in some patients (19). Growth curves specifically for Noonan syndrome are available (49a,58). Average adult height in males is 162.5 and in females is 152.7 cm (49a).

Characteristic are pectus carinatum superiorly and pectus excavatum inferiorly (70%). The thorax is broad. During childhood, the upper chest appears to lengthen; the nipples appear low set and axillary webbing, which persists into adulthood, is present. Common features include rounded shoulders, cubitus valgus (50%), clinobrachydactyly and blunt fingertips (30%), and vertebral and sternal anomalies (25%).

**Central nervous system.** Failure to thrive in infancy occurs in 40%. Other abnormalities include psychomotor delay (25%), learning disability with specific visual–constructional problems and verbal performance discrepancy (15%), and language delay (20%), which may be secondary to perceptual motor disabilities, mild hearing deficit (10%), or articulation abnormalities (70%). Mild mental retardation, found more often in males (49a), may occur in up to 35%. However, IQ has ranged from 64 to 127 (1,38,39,44,46).

**Genitourinary system.** In males, pubertal development varies from normal virilization with subsequent fertility to delay but normal pub-

Fig. 24–38. *Noonan syndrome.* (A–C) Similar facies showing ptosis, widespread eyes, pterygium colli, and dysmorphic pinnae. [C from *F Char et al,* Birth Defects **8**(5):110, 1972.]

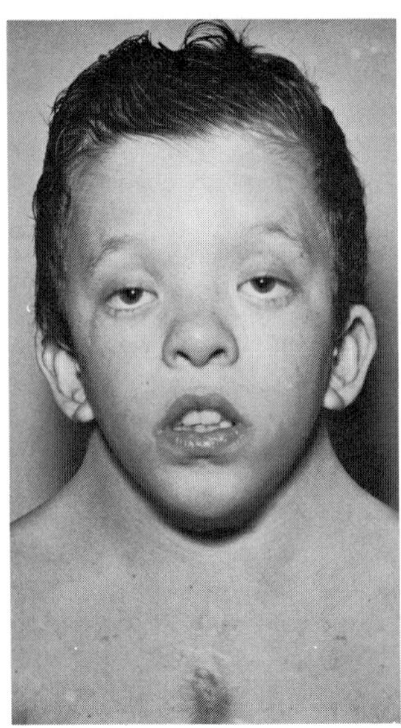

A

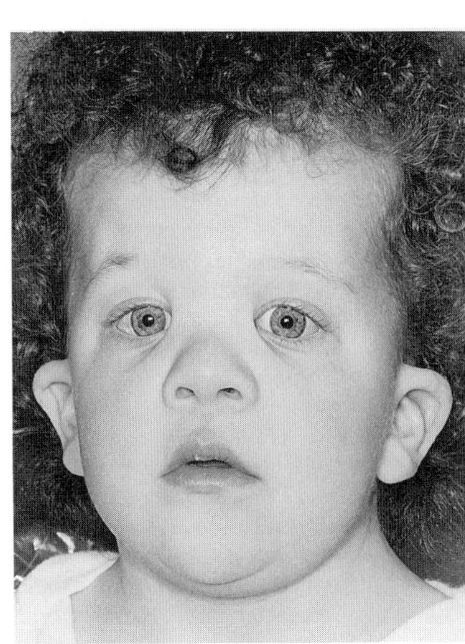

B

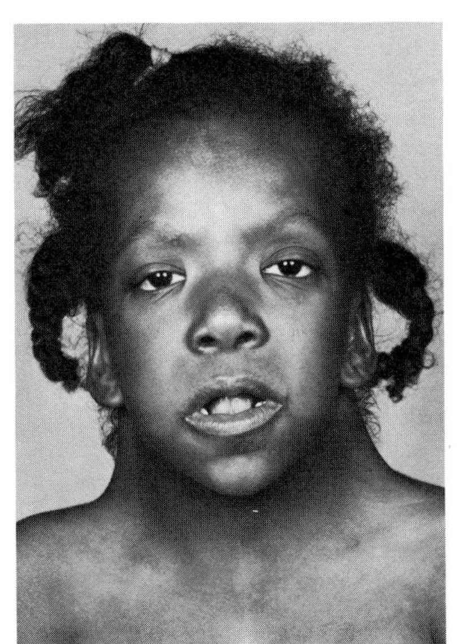

C

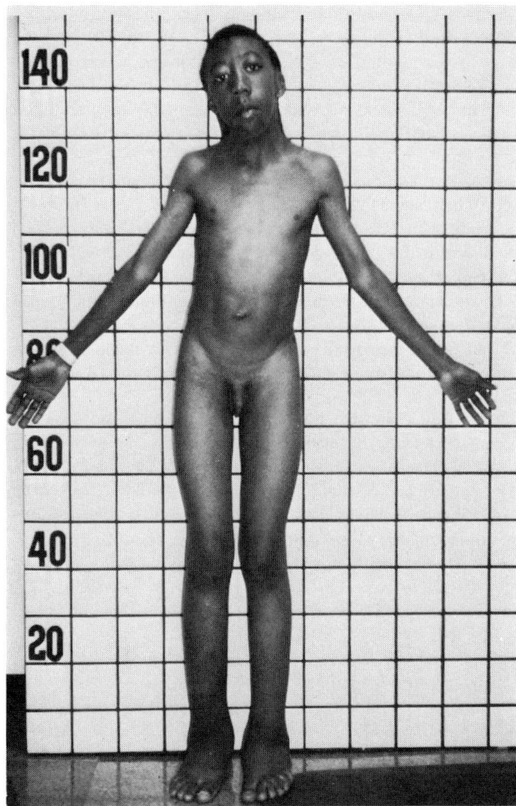


Fig. 24–39. *Noonan syndrome.* Note lymphangiectatic edema of feet and ankles. (Courtesy of *RL Summitt,* Memphis, Tennessee.)

ertal development to inadequate sexual development associated with early cryptorchidism (60%) and later deficiency in spermatogenesis. Most females are fertile, and puberty may be either normal or delayed (1).

**Skin.** A number of skin manifestations have been reported including café-au-lait spots (10%), pigmented nevi (25%), and lentigines (2%) (see *Differential diagnosis*). Keratosis pilaris atrophicans faciei has been noted in several instances (47).

**Lymphatics.** Hypoplasia or aplasia of the lymphatic channels (20%) may result in generalized lymphedema, peripheral lymphedema, pulmonary lymphangiectasia, intestinal lymphangiectasia, hydrops fetalis, and cystic hygroma (7,59) (Fig. 24–39).

**Hematologic abnormalities.** Bleeding abnormalities (60) are found in 20% and include factor XI deficiency (32), von Willebrand disease, and platelet dysfunction, which may be associated with trimethylaminuria (25,57). Low-frequency abnormalities such as congenital bone marrow hypoplasia (21) and congenital hypoplastic anemia (35) have been reported.

**Malignant hyperthermia.** Malignant hyperthermia has been reported in a number of cases (26,27,31,51,54a). Some authors have designated this as King syndrome, but we feel separate nosologic status is unnecessary as long as the possibility of malignant hyperthermia is recognized.

**Other findings.** Various low-frequency abnormalities have included autoimmune thyroiditis (12,56), pheochromocytoma (8), ganglioneuroma (30), malignant schwannoma (28), arthrogryposis (34), Arnold–Chiari malformation (5), lateral meningocele (24), xanthomas of the skin and oral mucous membranes (41), odontogenic keratocysts (16), polydactyly (22), and vasculitis (9). L Kaplan (personal communication, 1987), C Dunlap (18a), Chuong et al (12a), Hoyer and Neukam (23a), Weldon and Cozzi (56a), and C Hall (personal communication, 1988) found central giant cell lesions of the jaws (cherubism?).

**Differential diagnosis.** Differential diagnosis includes *Turner syndrome, Aarskog syndrome* (20), *Williams syndrome* (17), *fetal alcohol syndrome* (23,54), and *primidone embryopathy* (15,40). There is also overlap with several cardiocutaneous syndromes including *LEOPARD syndrome,* Watson syndrome (4) and, especially, *neurofibromatosis* (3,24,29,36,37,45,48,49,53). Opitz and Weaver (45) enumerated four possible interpretations: (1) chance concurrence of Noonan syndrome and neurofibromatosis, (2) neurofibromatosis/Noonan syndrome as an unusual variant of Noonan syndrome, (3) neurofibromatosis/Noonan syndrome as an unusual type of neurofibromatosis, and (4) neurofibromatosis/Noonan syndrome as a newly recognized entity. Watson syndrome has been demonstrated to probably be allelic with the NF1 gene on chromosome 17 (1a). A similar relationship exists for Noonan syndrome–LEOPARD syndrome–cherubism–polyarticular pigmented villonodular synovitis. We suspect that this represents a contiguous gene syndrome. Cherubism may occur as an autosomal dominant disorder, in combination with *neurofibromatosis* or in the *Ramon syndrome* with juvenile rheumatoid arthritis (polyarticular pigmented villonodular synovitis?). Also see p. 899.

Reynolds et al (50) reported six patients with *cardiofaciocutaneous syndrome.* Features include mental deficiency, growth retardation, cardiac anomalies, a Noonanoid facial appearance, and abnormalities of the hair and skin. Facial features were coarse. Sparse hair broke easily, leaving patches of alopecia, and skin abnormalities varied from follicular hyperkeratosis to an ichthyosis hystrix-like picture. Possibly the children reported by Baraitser and Patton (6) have the same syndrome.

**Laboratory aids.** If cystic hygromas are present on ultrasound and polyhydramnios and normal karyotype are present, a tentative diagnosis of Noonan syndrome is indicated (8a).

### References (Noonan syndrome)

1. Allanson JE: Noonan syndrome. *J Med Genet* **24**:9–13, 1987.
1a. Allanson JE: Time and natural history: The changing face. *J Craniofac Genet Develop Biol* **9**:21–28, 1989.
2. Allanson JE et al: Noonan syndrome: The changing phenotype. *Am J Med Genet* **21**:507–514, 1985.
3. Allanson JE et al: Noonan phenotype associated with neurofibromatosis. *Am J Med Genet* **21**:457–462, 1985.
4. Allanson JE, Watson GH: Watson syndrome: Nineteen years on. *Proc Greenwood Genet Ctr* **6**:173, 1987.
5. Ball MJ, Peiris A. Chiari (type I) malformation and syringomyelia in a patient with Noonan's syndrome. *J Neurol Neurosurg Psychiat* **45**:753–754, 1982.
6. Baraitser M, Patton MA: A Noonan-like short stature syndrome with sparse hair. *J Med Genet* **23**:161–164, 1986.
7. Bawle EV, Black V: Nonimmune hydrops fetalis in Noonan's syndrome. *Am J Dis Child* **140**:758–760, 1986.
8. Becker CE et al: Pheochromocytoma and hyporesponsiveness to thyrotropin in a 46,XY male with features of Turner phenotype. *Ann Intern Med* **70**:325–333, 1969.
8a. Benacerraf BR et al: The prenatal sonographic features of Noonan's syndrome. *J Ultrasound Med* **8**:59–64, 1989.
9. Berberich MS, Hall JG: Noonan syndrome—an unusual family with above average intelligence, a high incidence of cancer and rare type of vasculitis. *Birth Defects* **12**(1):181–186, 1976.
10. Char FC: The halo iris of the Noonan syndrome. *Proc Greenwood Genet Ctr* **6**:159, 1987.
11. Char FC et al: The Noonan-syndrome—a clinical study of forty-five cases. *Birth Defects* **8**(5):110–118, 1972.
12. Chaves-Carballo E, Hayles AB: Ullrich–Turner syndrome in the male: Review of the literature and report of a case with lymphocytic (Hashimoto's) thyroiditis. *Mayo Clin Proc* **41**:843–854, 1966.

12a. Chuong R et al: Central giant cell lesions of the jaws: A clinicopathologic study. *J Oral Maxillofac Surg* **44**:708–713, 1986.

13. Clark EB: Mechanisms in the pathogenesis of congenital heart defects. *Proc Greenwood Genet Ctr* **4**:80, 1985.

14. Clark EB: Neck web and congenital heart defects: A pathogenic association in 45,XO Turner syndrome? *Teratology* **29**:355–361, 1984.

15. Clericuzio C: Fetal primidone effects. *Proc Greenwood Genet Ctr* **4**:93, 1985.

16. Connor JM et al: Multiple odontogenic keratocysts in a case of the Noonan syndrome. *Br J Oral Surg* **20**:213–216, 1982.

17. Cortada X et al: Familial Williams syndrome. *Clin Genet* **18**:173–176, 1980.

18. Duncan WJ et al: A comprehensive scoring system for evaluating Noonan syndrome. *Am J Med Genet* **10**:37–50, 1981.

18a. Dunlap C et al: The Noonan syndrome/cherubism association. *Oral Surg Oral Med Oral Pathol* **67**:698–705, 1989.

19. Elders MJ, Char F: Possible etiologic mechanisms of the short stature in Noonan syndrome. *Birth Defects* **12**(6):127–133, 1976.

20. Escobar V, Weaver DD: Aarskog syndrome. New findings and genetic analysis. *JAMA* **240**:2638–2641, 1978.

21. Feldman KW et al: Congenital stem cell dysfunction associated with Turner-like phenotype. *J Pediatr* **88**:979–998, 1976.

22. Grix A, Hall BD: Noonan phenotype with polydactyly. *Birth Defects* **15**(5B):313–319, 1979.

23. Hall BD: Noonan's phenotype in offspring of alcoholic mothers. *Lancet* **1**:680, 1974.

23a. Hoyer PF, Neukam FW: Cherubism-eine osteofibröse Kiefererkrankung im Kindesalter. *Klin Pädiatr* **194:**128–131, 1982.

24. Hughes HE et al: Noonan syndrome and lateral meningoceles: Another link with neurofibromatosis. *Proc Greenwood Genet Ctr* **6**:159, 1987.

25. Humbert JR et al: Trimethylaminuria: The fish-odour syndrome. *Lancet* **2**:770–771, 1970.

26. Hunter A, Pinsky L: An evaluation of the possible association of malignant hyperpyrexia with Noonan syndrome using serum creatine phosphokinase levels. *J Pediatr* **96**:412–415, 1975.

27. Kaplan AM et al: Malignant hyperthemia associated with myopathy and normal muscle enzymes. *J Pediatr* **91**:431–433, 1977.

28. Kaplan MS et al: Noonan's syndrome: A case with elevated serum alkaline phosphatase levels and malignant schwannoma of the left forearm. *Am J Dis Child* **116**:359–366, 1968.

29. Kaplan P, Rosenblatt B: A distinctive facial appearance in neurofibromatosis von Recklinghausen. *Am J Med Genet* **21**:463–470, 1985.

30. Khodadoust A, Paton D: Turner's syndrome in a male. Report of a case with myopia, retinal detachment, cataract and glaucoma. *Arch Ophthalmol* **77**:630–634, 1967.

31. King JO, Denborough MA: Anesthetic-induced malignant hyperpyrexia in children. *J Pediatr* **83**:37–40, 1973.

32. Kitchens CS, Alexander JA: Partial deficiency of coagulation factor XI as a newly recognized feature of Noonan syndrome. *J Pediatr* **102**:224–227, 1983.

33. Kobylinski O: Über eine flughautähnliche Ausbreitung am Halse. *Arch Anthropol* **14**:342–348, 1883.

34. Krieger I, Espiritu CE: Arthrogryposis multiplex congenita and the Turner phenoype. *Am J Dis Child* **123**:141–144, 1972.

35. Krishan EU et al: Congenital hypoplastic anemia terminating in acute promyelocytic leukemia. *Pediatrics* **61**:898–901, 1978.

36. Meinecke P: Evidence that the "Neurofibromatosis-Noonan syndrome" is a variant of von Recklinghausen neurofibromatosis. *Am J Med Genet* **26**:741–745, 1987.

37. Mendez HMM: The neurofibromatosis–Noonan syndrome. *Am J Med Gen* **21**:471–476, 1985.

38. Mendez HMM, Opitz JM: Noonan syndrome: A review. *Am J Med Genet* **21**:493–506, 1985.

39. Money J, Kalus ME: Noonan's syndrome: IQ and specific disabilities. *Am J Dis Child* **133**:846–850, 1979.

40. Myhre SA, Williams R: Teratogenic effects associated with maternal primidone therapy. *J Pediatr* **99**:160–162, 1981.

41. Nelson JF et al: Noonan's syndrome: Report of a case with oral findings. *J Oral Med* **33**:94–96, 1978.

42. Noonan JA, Ehmke DA: Associated noncardiac malformations in children with congenital heart disease. *J Pediatr* **63**:468–470, 1963.

43. Nora JJ, Fraser FC. *Medical Genetics: Principles and Practice,* 2nd ed. Lea & Febiger, Philadelphia, 1981.

44. Nora JJ et al: The Ullrich–Noonan syndrome (Turner phenotype). *Am J Dis Child* **127**:48–55, 1974.

45. Opitz JM, Weaver DD: Editorial comment: The neurofibromatosis–Noonan syndrome. *Am J Med Genet* **21**:477–490, 1985.

46. Pearl W: Cardiovascular anomalies in Noonan's syndrome. *Chest* **71**:677–679, 1977.

47. Pierini DO, Pierini AM: Keratosis pilaris atrophicans faciei (ulerythema ophryogenes): A cutaneous marker in the Noonan's syndrome. *Br J Dermatol* **100**:409–416, 1979.

48. Proud VK et al: Complications of Noonan syndrome–neurofibromatosis in a family with neurofibromatosis. *Proc Greenwood Genet Ctr* **6**:160–161, 1987.

49. Quattrin T et al: Vertical transmission of the neurofibromatosis/Noonan syndrome. *Am J Med Genet* **26**:645–649, 1987.

49a. Ranke MB et al: Noonan syndrome: Growth and clinical manifestations in 144 cases. *Eur J Pediatr* **148**:220–227, 1988.

50. Reynolds JF et al: Cutaneous manifestations of the cardio-facio-cutaneous (CFC) syndrome: Differentiation from the Noonan syndrome. *Proc Greenwood Genet Ctr* **6**:171, 1987.

51. Rissam HS et al: Post-operative hyperpyrexia in a case of Noonan's syndrome. *Indian Heart J* **34**:180–182, 1982.

52. Sanchez-Cascos A: The Noonan syndrome. *Eur Heart J* **4**:223–229, 1983.

53. Saul RS: Noonan syndrome in a patient with hyperplasia of the myenteric plexuses and neurofibromatosis. *Am J Med Genet* **21**:491–492, 1985.

54. Spiegel PG et al: The orthopedic aspects of the fetal alcohol syndrome. *Clin Orthoped* **139**:58–63, 1979.

54a. Steenson AJ, Torkelson RD: King's syndrome with malignant hyperthermia. *Am J Dis Child* **141**:271–273, 1987.

55. Thientz G, Savage MO: Growth and pubertal development in five boys with Noonan's syndrome. *Arch Dis Childh* **57**:13–17, 1982.

56. Vesterhus P, Aarskog D: Noonan's syndrome and autoimmune thyroiditis. *J Pediatr* **83**:237–240, 1973.

56a. Weldon L, Cozzi G: Multiple giant cell lesions of the jaws. *J Oral Maxillofac Surg* **40:**520–522, 1982.

57. Witt DR et al: Bleeding disorders in 7 cases of Noonan syndrome: Further evidence of heterogeneity. *Am J Hum Genet* **37**:83A, 1985.

58. Witt DR et al: Growth curves for height in Noonan's syndrome. *Clin Genet* **30**:150–153, 1986.

59. Witt DR et al: Lymphedema in Noonan syndrome: Clues to pathogenesis and premature diagnosis and review of the literature. *Am J Med Genet* **27**:841–856, 1987.

60. Witt DR et al: Bleeding diathesis in Noonan syndrome: A common association. *Am J Med Genet* **31**:305–317, 1988.

## Trichorhinophalangeal syndrome, Type I

The disorder was described initially by Klingmüller (37) in 1956, although it was mentioned even earlier (65). The name trichorhinophalangeal syndrome was first applied by Giedion (19,20). The condition is characterized by cone-shaped epiphyses, sparse fine hair, bulbous nose with lack of alar flare, and variable growth retardation. Although the majority of reported kindreds are consistent with autosomal dominant inheritance (1,3,10–12,14,15,17,18,22,23,25,30,41–43,47,48, 50,51,53,56,67–69), in some cases autosomal recessive inheritance could not be excluded (32,37,65). Finally, it has been conclusively demonstrated that TRP syndrome, Type I, is due to a deletion of band 8q24.12. The more severe Type II syndrome is due to a larger deletion (6).

**Facies.** The nose is pear-shaped (40–50) and the alae are retropositioned. The philtrum is elongated (31,46,53) and the upper lip is thin (25,46). There may be a small protuberance below the lower lip (25,32,51,53). Cephalometric analysis (32,55a) has documented a shortened posterior facial height associated with short mandibular ramus and reduced and superiorly deflected posterior cranial base. The lower border of the mandible is more steeply inclined, and the gonial angle is more obtuse than normal. Ears are often large and outstanding (7,10,12,25,39,56,66,68) (Fig. 24–40A,B).

**Hair and nails.** Scalp hair may be sparse from birth (29,52,68) and is especially scant in the frontotemporal areas, simulating male baldness pattern. Its texture is fine and brittle, and growth is slow

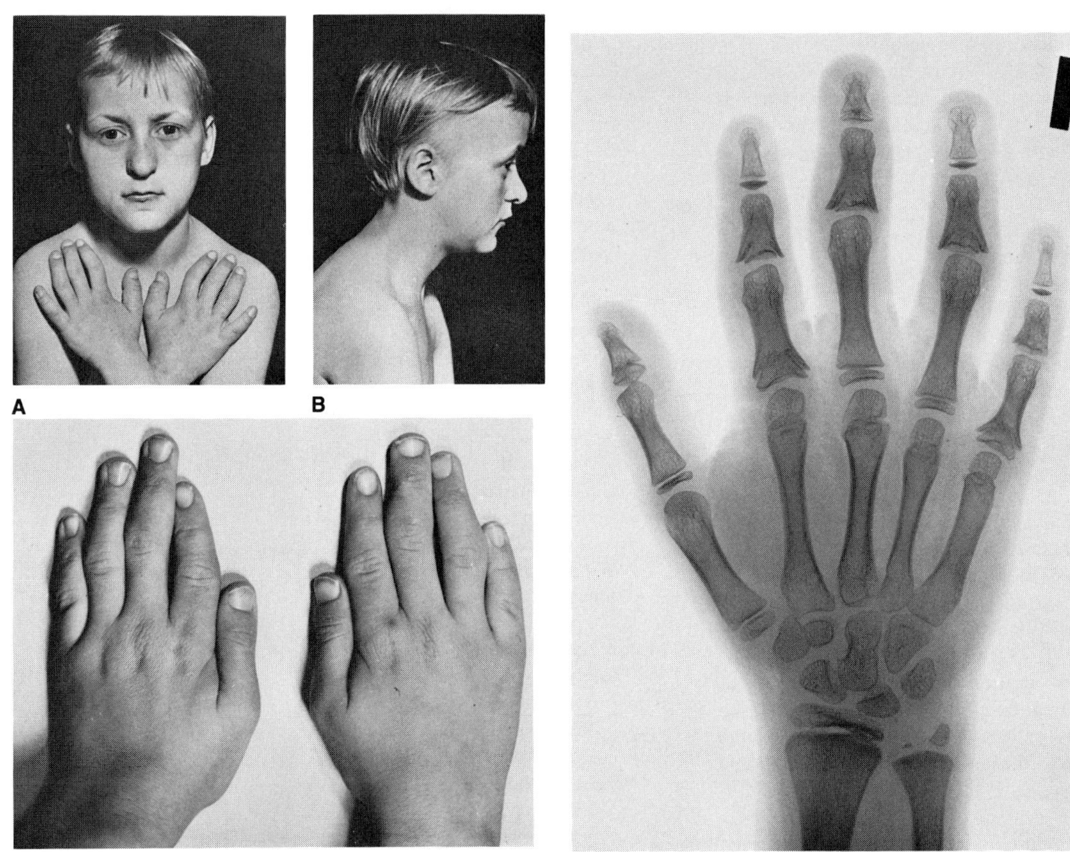

Fig. 24–40. *Trichorhinophalangeal syndrome, Type I.* (A,B) Pear-shaped nose, frontal bossing with high lateral hairline, fingers deviated at proximal interphalangeal joints. (C) Hands showing deviation of fingers at proximal interphalangeal joints. (D) Cone-shaped epiphyses (Type 12) and eburnated epiphyses. (A, B, and D from *A Giedion,* Fortschr Roentgenstr **110**:507, 1969. C from *A Giedion,* Helv Paediatr Acta **21**:475, 1966.)

(31,51,54,64). Scalp hair is more like vellus than terminal hair (51). In some patients, hair falls out during the mid-second decade of life (2,31,35,46,49,64); male (23) and female (63) patients may become almost totally bald prematurely. Eyebrows are thick medially and sparse laterally (11,23,25,35,37–39,50–53). Cilia and pubic and axillary hair may be scant (18,37,51). Nails may be thin (19,29,68); koilonychia (25,52,53) and leukonychia (25,52,53,64) have been described. The diameter of the hair shaft is markedly decreased (51,55) and there are abnormalities of the cuticle cells (55). Van Neste and Dumortier (66) noted that hair diameter was thinner than normal, but that the cuticular pattern was normal; they also reported that hair follicles had a relatively hypoplastic papilla when inserted into a well-developed hair matrix.

**Skeletal system.** Height is below the third percentile in 40% and seldom above the twenty-fifth percentile (19,23,32,54,65). Bone age is often several years behind chronologic age. Trigonocephaly has been noted in one report (25).

Characteristically, there is swelling of the proximal interphalangeal joints, resulting in clinobrachydactyly (53,68) (Fig. 24–40C). The distal phalanges in both thumbs and halluces are usually short (29,52,53). Toes as well as fingers may be abbreviated. Pes planus has been noted in several cases (19,38). Limitation of movement and bone pain can occur in the hips (19,23,31,35,38,39,50) and hands (39,50,62). Joint hypermobility at the elbows (7) and fingers (29,40,50) has been noted. On occasion, severe arthritic involvement has been found (BD Hall, personal communication, 1984).

Roentgenographically, the mandibular condyles may be flattened (44). Cone-shaped epiphyses (Type 12) are most frequently present in the middle and less often the proximal phalanges of the second, third, and fourth fingers (21,53) and there are eburnated epiphyses of the distal phalanges in the fingers 2–4–5 (Fig. 24–40D). The fifth fingers as well as both phalanges of the thumbs may be similarly affected. Metacarpals, especially the fourth and fifth, are shortened (12) in about 50% (23). Multiple cone-shaped epiphyses may also be noted in the toes (10,51). Scoliosis and lordosis have been described by several authors (10,19,25,29,32). The scapulas are often winged (48). Legg–Perthes-like changes have been noted in the hips (3,7,10,18, 31,48,50,53,62), and degenerative arthritis in the spine, hips, knees, elbows, and sacroiliac joints (7). The ulnas may be short. The capital femoral epiphyses may be abnormal, either unilaterally or bilaterally. Irregular and delayed mineralization of the ossification center has been described (10). Schlessinger and Poznanski (61) documented flattened distal femoral epiphyses. Pectus carinatum has been found in 40% (7,10). Pattern profile analysis of the hands without typical cone deformity in clinically normal individuals at risk for the syndrome has been reported (10,46). Findings suggested that these individuals indeed had the disorder, and that the condition has wide variation in expression; however, none had affected children. Frias et al (13) described a woman who had no clinical signs of the disorder, yet her mother was affected; her son had marked changes in the capital femoral epiphyses but normal radiographs of the hands. Metacarpophalangeal pattern profiles resemble those of Type II (9).

**Other findings.** Frequent upper respiratory tract infections have been described (19,29,37,50,59) but are not well documented. Hussels (32) cited repeated cyanotic attacks caused by glossoptosis during infancy.

Giedion et al (23) suggested possible increased frequency of congenital heart defects and renal disorders. Floppy mitral valve has been described by several authors (14,23,34). Intelligence is normal (11,24,27,31,32,40,45,46,54), although mild mental retardation has been described (25,48). Severe mental retardation was found by Yamamoto et al (70) in a patient with a significant interstitial deletion around 8q24.12.

**Oral manifestations.** Malocclusion (12,51) and supernumerary teeth (19,29,40,50,57,64) have been documented.

**Differential diagnosis.** Cruz and Francés (8) described a mother and daughter with abnormalities similar to those observed in the trichorhinophalangeal syndrome but without characteristic nasal alterations. The disorder described by Bellini and Bardara (2) represents a different entity, with cone-shaped epiphyses (not Type 12) and involvement of the epiphyses of both knees.

Cone-shaped epiphyses in the hand may occur as an isolated finding in approximately 1% of children in southwest Ohio and in almost 9% in Guatemala (28). They may be associated with a plethora of disorders, among which are *de Lange syndrome, cleidocranial dysplasia,* asphyxiating thoracic dystrophy, *Ellis–van Creveld syndrome, cartilage–hair hypoplasia,* and *pseudohypoparathyroidism.*

Absence of exostoses should distinguish this disorder from *trichorhinophalangeal syndrome, Type II.* (Table 24–1). However, Jorgenson et al (33) discussed the great variation that may exist in both of these conditions and emphasized that there may be phenotypic overlap between the two. This presumably reflects the extent of chromosome deletion. The autosomal dominant cartilaginous exostoses syndrome also exists without unusual facies, hair abnormalities, or cone epiphyses.

Short stature, sparse hair, and short fingers may also be observed in *cartilage-hair hypoplasia.*

**Laboratory aids.** Using high resolution G-banding, a number of investigators (5,6,16,24,26) have found a microdeletion of 8q23q24 while others (4,18,50) did not. Sanchez et al (58) described a boy with presumed trichorhinophalangeal syndrome, Type I, who was severely mentally retarded and who, by age 9, had no exostoses; his karyotype showed a complex, apparently balanced translocation involving break points in bands 3q13, 8p22, 8q13, 11p12, and 11q21.

Table 24–1. Comparison of features in TRP I and TRP II and additional features of TRP II[a]

| Features | TRP I | TRP II |
|---|---|---|
| Features in common | | |
| Recurrent respiratory infections | + | + |
| Sparse scalp hair | + | + |
| Large, laterally protruding ears | + | + |
| Bulbous nose with tented alae | + | + |
| Prominent elongated philtrum | + | + |
| Thin upper lip | + | + |
| Apparent mandibular micrognathia | + | + |
| Clinobrachydactyly | + | ± |
| Cone-shaped epiphyses (Type 12) of the hands | + | + |
| Winged scapulae | + | + |
| Short stature | + | + |
| Additional features of the Langer–Giedion syndrome | | |
| Multiple exostoses | − | + |
| Redundant and/or loose skin in infancy and early childhood | − | + |
| Laxity or hypermobility at joints | − | + |
| Microcephaly | − | + |
| Mental retardation | − | + |
| Significantly delayed onset of speech | − | + |
| Skin nevi | − | + |

[a]Modified after BD Hall et al, *Birth Defects* **10**(12):147, 1974.

## References (Trichorhinophalangeal syndrome, Type I)


1. Beals RK: Tricho-rhino-phalangeal dysplasia: Report of a kindred. *J Bone Jt Surg* **55A**:821–826, 1973.
2. Bellini F, Bardara M: Su un caso di disostosi periferica. *Minerva Pediatr* **18**:106–110, 1966.
3. Besser F, Wells RS: The tricho-rhino-phalangeal syndrome (a family with three affected members). *Br J Dermatol* **95** *(Suppl* **14**):39–41, 1976.
4. Booth CW, Maurer WF: De novo 9:11 translocation in a sporadic case of tricho-rhino-phalangeal (I) syndrome. *Pediatr Res* **15**:559, 1981 (Abstr).
5. Bühler EM, Malik NJ: The tricho-rhino-phalangeal syndrome(s): Chromosome 8 long arm deletion: Is there a shortest region of overlap between reported cases? TRP I and TRP II syndromes: Are they separate entities? *Am J Med Genet* **19**:113–119, 1984.
6. Bühler EM et al: A final word on the tricho-rhino-phalangeal syndromes. *Clin Genet* **31**:273–275, 1987.
7. Cope R et al: The trichorhinophalangeal dysplasia syndrome: Report of eight kindreds, with emphasis on hip complications, late presentations, and premature osteoarthrosis. *J Ped Orthoped* **6**:133–138, 1986.
8. Cruz M, Francés JM: Le syndrome tricho-phalangien: Une forme nouvelle de dysostose peripherique. *Arch Fr Pédiatr* **27**:649–656, 1970.
9. Dijkstra PF: Analysis of metacarpophalangeal pattern profiles. *Roefo* **139**:158–159, 1983.
10. Felman AH, Frias JL: The trichorhinophalangeal syndrome: Study of 16 patients in one family. *Am J Roentgenol* **129**:631–638, 1977. (Same family reported in Ref. 13.)
11. Ferrández A et al: The trichorhinophalangeal syndrome. Report of 4 familial cases belonging to 4 generations. *Helv Paediatr Acta* **35**:559–567, 1980.
12. Fontaine G et al: Le syndrome tricho-rhino-phalangien. *Arch Fr Pédiatr* **27**:635–647, 1970.
13. Frias JL et al: Variable expressivity in the trichorhinophalangeal syndrome type I. *Birth Defects* **15**(5B): 361–372, 1979. (Same family reported in Ref. 10.)
14. Frisch H, Vormittag W: Tricho-rhino-phalangeales Syndrom mit autosomal dominanten Erbgang. *Z Kinderheilkd* **120**:141–150, 1975.
15. Fruchter Z et al: Radiodiagnostic et pathogenie de la dysostose phalangienne. *J Radiol Électrol* **47**:653–656, 1966.
16. Fryns JP, Van der Berghe H: 8q24.12 interstitial deletion in trichorhinophalangeal syndrome type 1. *Hum Genet* **74**:188–189, 1986.
17. Fukushima N et al: Tricho-rhino-phalangeal syndrome. The first case in Japan. *Hum Genet* **32**:207–210, 1976.
18. Gaardsted C et al: A Danish kindred with tricho-rhino-phalangeal syndrome type I. *Eur J Pediatr* **139**:84–87, 1982.
19. Giedion A: Das tricho-rhino-phalangeale Syndrom. *Helv Paediatr Acta* **21**:475–482, 1966.
20. Giedion A: Cone-shaped epiphyses of the hands and their diagnostic value. The tricho-rhino-phalangeal syndrome. *Ann Radiol* **10**:322–329, 1967.
21. Giedion A: Zapfenepiphysen. Naturgeschichte und diagnostische Bedeutung einer Störung des enchondralen Wachstums. *Ergeb Med Radiol* **1**:59–124, 1968.
22. Giedion A: Die periphere Dysostoses (PD)-ein Sammelbegriff. *Fortschr Roentgenstr* **110**:507–525, 1969.
23. Giedion A et al: Autosomai dominant transmission of the tricho-rhino-phalangeal syndrome. Report of 4 unrelated families, review of 60 cases. *Helv Paediatr Acta* **28**:249–259, 1973.
24. Goldblatt J, Smart RD: Tricho-rhino-phalangeal syndrome without exostoses, with an interstitial deletion of 8q23. *Clin Genet* **29**:434–438, 1986.

25. Goodman RM et al: New clinical observations in the trichorhinophalangeal syndrome. *J Craniofac Genet Dev Biol* **1**:15–29, 1981.

26. Hamers A et al: Micro-cytogenetics of chromosome 8q. Poster presented at the 8th International Chromosome Conference, Lübeck, September, 1983 (Abstr 2–8). (Cited by Goldblatt and Smart, Ref. 24.)

27. Hansen DD, Shewmake SW: Tricho-rhino-phalangeal syndrome. *Int J Dermatol* **18**:561–564, 1979.

28. Hertzog KP, et al: Cone shaped epiphyses in the hand. Population frequencies, anatomic distribution, and developmental stages. *Invest Radiol* **3**:433–441, 1968.

29. Hobolth N, Mune D: Dysostosis epiphysaria peripherica. *Acta Rheumatol Scand* **9**:269–276, 1963.

30. Hornsby VPL, Pratt AE: The tricho-rhino-phalangeal syndrome. *Clin Radiol* **35**:243–247, 1984.

31. Howell CJ, Wynne-Davies R: The tricho-rhino-phalangeal syndrome: A report of 14 cases in 7 kindreds. *J Bone Jt Surg* **68B**:311–314, 1986.

32. Hussels I: Trichorhinophalangeal syndrome in two sibs. *Birth Defects* **7**(7):301–302, 1971.

33. Jorgenson RJ et al: Heterogeneity in the trichorhinophalangeal syndromes. *Birth Defects* **19**(1):167–179, 1983.

34. Jüngst BK, Spranger J: Ballonierende Mitralklappe bei Tricho-rhinophalangealem Syndrome I. *Mschr Kinderheilkd* **124**:538–541, 1976.

35. Kennedy C: Tricho-rhino-phalangeal syndrome. *Br J Dermatol* **101** (*Suppl* **17**):36–38, 1979.

36. King GJ, Frias JL: A cephalometric study of the craniofacial skeleton in trichorhinophalangeal syndrome. *Am J Orthodont* **75**:70–77, 1979.

37. Klingmüller G: Über eigentümliche Konstitutions-anomalien der zwei Schwestern und ihre Beziehungen zu neueren entwicklungspathologischen Befunden. *Hautarzt* **7**:105–113, 1956.

38. Kozlowski K et al: Tricho-rhino-phalangeal syndrome. *Australas Radiol* **16**:411–416, 1972.

39. Kozlowski K et al: Tricho-rhino-phalangeal syndrome with Perthes disease-like changes and coxa vara (report of a case). *Australas Radiol 23*:170–172, 1979.

40. Kuna GB et al: Trichorhinophalangeal dysplasia (Giedion syndrome). *Clin Pediatr* **17**:96–98, 1978.

41. Langer LO Jr: The thoracic-pelvic phalangeal dystrophy. *Birth Defects* **5**(4):55–64, 1969.

42. Lemaitre G et al: Syndrome tricho-rhino-phalangien. Étude de deux fratries. *J Radiol Électrol* **51**:429–432, 1970.

43. Liess G: Aförmige Epiphysen an Händen und Füssen (periphere Dysostosen) (Case 2). *Fortschr Roentgenstr* **81**:173–191, 1954.

44. Madoule PH et al: Syndrome tricho-rhino-phalangien. A propos de quatre cas découverts chez l'adulte. *J Radiol* **65**:85–88, 1984.

45. Marchand C et al: Syndrome trichorhinophalangien. A propos d'une observations. *Ann Dermatol Venereol* **112**:973–980, 1985.

46. McCloud DJ, Solomon LM: The tricho-rhino-phalangeal syndrome. *Br J Dermatol* **96**:403–407, 1977.

47. McKusick VA: *Heritable Disorders of Connective Tissue,* ed 4, C.V. Mosby, St. Louis, 1972, pp 801,807,810–812. (Same case as Ref. 49.)

48. Morris L et al: Tricho-rhino-phalangeal syndrome I. (Report of 8 cases). *Australas Radiol* **29**:167–173, 1985.

49. Murdoch JL: Tricho-rhino-phalangeal dysplasia with possible autosomal dominant transmission. *Birth Defects* **5**(3):218–219, 1969. (Same case as Ref. 47.)

50. Noltorp S et al: Trichorhinophalangeal syndrome type I: Symptoms and signs, radiology and genetics. *Ann Rheum Dis* **45**:31–36, 1986.

51. Pashayan HM et al: The tricho-rhino-phalangeal syndrome. *Am J Dis Child* **127**:257–261, 1974.

52. Patzer H: Periphäre Dysostosen an den Fingern. *Kinderärztl Prax* **32**:231–235, 1964.

53. Peltola J, Kuokkanen K: Tricho-rhino-phalangeal syndrome in five successive generations: Report on a family in Finland. *Acta Derm Venereol* **58**:65–68, 1978.

54. Poznanski AK et al: The hand in the trichorhinophalangeal syndrome. *Birth Defects* **10**(9):209–217, 1974.

55. Prens EP et al: Clinical and scanning electron microscopic findings in a solitary case of trichorhinophalangeal syndrome type 1. *Acta Derm Venereol* **64**:249–253, 1984.

56. Ranke MB, Heitkamp H-Ch: Tricho-rhino-phalangeales Syndrom. Bericht über 4 Fälle in drei Generationen. *Monatschr Kinderheilkd* **128**:208–211, 1980.

57. Rosenbaum KN et al: Supernumerary teeth in the tricho-rhino-phalangeal phalangeal (TRP) syndrome. *Proceedings of the 1978 March of Dimes Birth Defects Conference,* San Francisco, California, p 178.

58. Sanchez JM et al: Complex translocation in a boy with trichorhinophalangeal syndrome. *J Med Genet* **22**:314–318, 1985.

59. Say B et al: Pattern profile analysis of the hand in trichorhinophalangeal syndrome. *Pediatrics* **59**:123–126, 1977.

60. Scheffer P et al: Syndrome tricho-rhino-phalangien. *Rev Stomatol Chir Maxillofac* **82**:230–233, 1981.

61. Schlessinger AE, Poznanski AK: Flattening of the distal femoral epiphyses in the trichorhinophalangeal syndrome. *Pediatr Radiol* **16**:498–500, 1986.

62. Sugiura Y: Tricho-rhino-phalangeal syndrome associated with Perthes-disease-like bone change and spondylolisthesis. *Jpn J Hum Genet* **23**:23–30, 1978.

63. Sugiura Y et al: Tricho-rhino-phalangeal syndrome: Report on three unrelated families. *Jpn J Hum Genet* **21**:13–22, 1976.

64. Thomas SE et al: Tricho-rhino-phalangeal syndrome. *Br J Dermatol* **115** (*Suppl* **32**):45–47, 1986.

65. Van der Werff ten Bosch JJ: The syndrome of brachymetacarpal dwarfism (''pseudo-pseudo-hypoparathyroidism'') with and without gonadal dysgenesis. *Lancet* **1**:69–71, 1959 (Cases 5 and 8).

66. Van Neste D, Dumortier M: Tricho-rhino-phalangeal syndrome. Disturbed geometric relationship between hair matrix and dermal papilla in scalp hair bulbs. *Dermatologica* **165**:16–23, 1982.

67. Verona R, Zoratto E: La sindrome di Giedion. Illustrazione di un caso con rassegna della litterature. *Minerva Pediatr* **26**:313–321, 1974.

68. Voigtländer V, Osswald F: Tricho-rhino-phalangeal syndrome. Report of a family with 3 affected members, *Dermatologica* **156**:34–39, 1978.

69. Weaver DD et al: The tricho-rhino-phalangeal syndrome. *J Med Genet* **11**:312–314, 1974.

70. Yamamoto Y et al: Tricho-rhino-phalangeal syndrome type I with severe mental retardation due to interstitial deletion of 8q23.3–24.13. *Am J Med Genet* **32**:133–135, 1989.

## Trichorhinophalangeal syndrome, Type II (Langer–Giedion syndrome)

Credit for having first described this disorder with multiple exostoses and many features of trichorhinophalangeal (TRP) syndrome, Type I is given to Langer (24) and Giedion (17). The condition was thus termed the ''Langer–Giedion syndrome'' by Hall et al (20). However, an earlier description was by Alè and Calò (1). Most reported cases have been sporadic. However, father-to-son transmission has been documented (27) and it has been seen in monozygotic twins (20,22a). Published cases have been reviewed in detail by Langer et al (25).

Cases with del(8)(q24.11→q24.13) have been reported by a number of authors (3,4,6–8,11–13,15,21,25,27a,27b,29,30,34,35,38,39) while others have found normal karyotypes (17,19,25,27,32,34,37). Bühler et al (6,8) have concluded that TRP Types I and II differ only in the size of the deletion, TRP Type I having the smaller.

**Facies.** The craniofacial appearance has some resemblance to that in trichorhinophalangeal syndrome, Type I, but the nose tends to be less bulbous and the philtral area more prominent. The upper lip is thin. The eyebrows are broad and the eyes deeply set. Mild microcephaly has been a feature in about 60% (16,25,26,29,36,38). Exotropia or exophoria has been documented in about 40% (20,25) (Fig. 24-41A–D).

**Central nervous system.** Mild to moderate mental retardation has been present in 75% (18,20,23,25,33,34,38). Severe mental retardation is rare, but has been reported (15,16,36). Some patients may not be mentally retarded (19a,25,28). Delayed onset of speech (18,20,38) and, less often, hearing deficit have been reported, although the frequencies affected, age of onset, and degree of severity have not been well documented (20,23,25).

**Skeletal system.** In addition to skeletal abnormalities similar to those found in trichorhinophalangeal syndrome, Type, I, multiple exostoses are found and are characteristic (5,15,33,35) (Figs. 24-41E,F and 24-42A,B). They are usually present by the third or fourth year

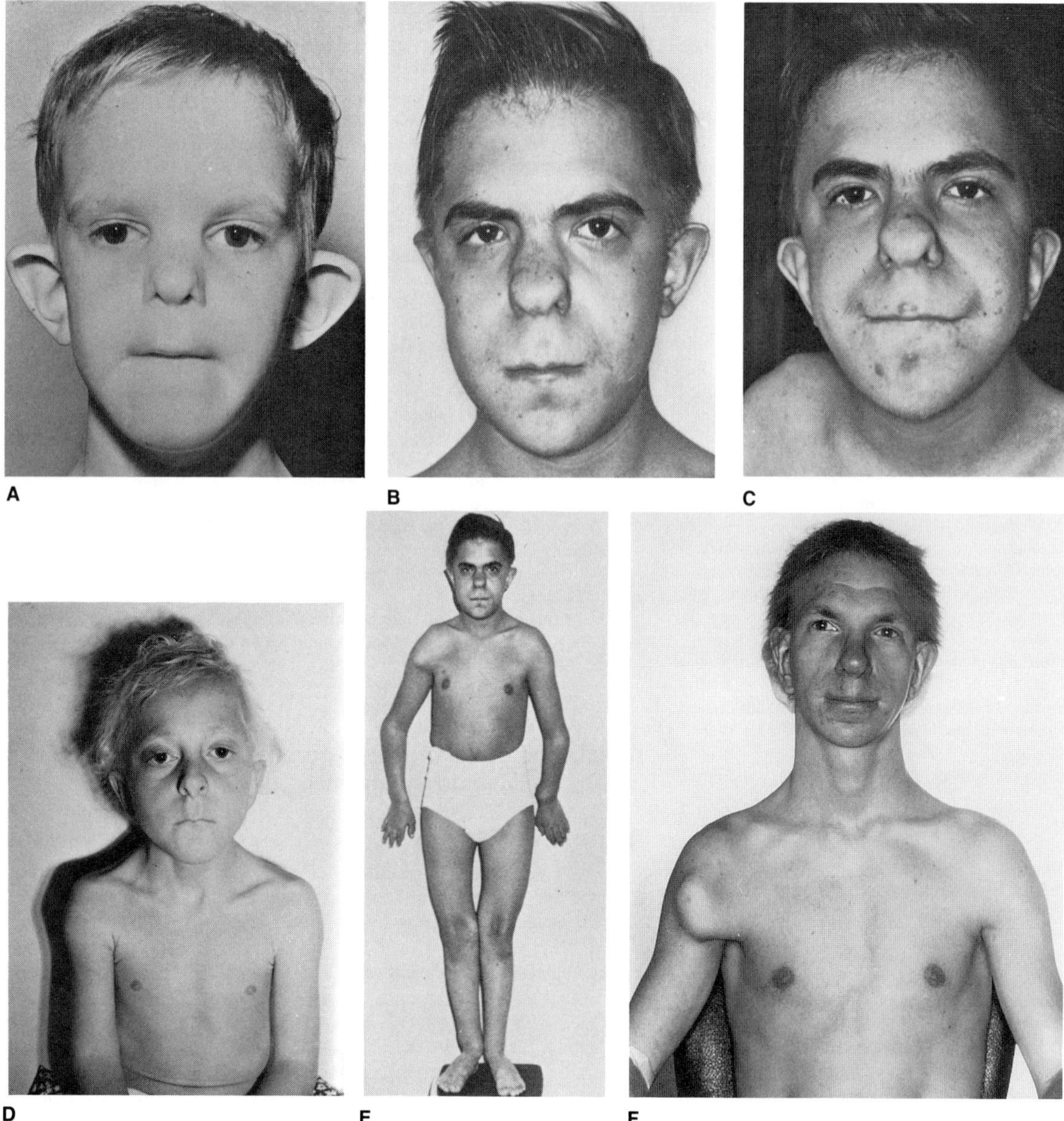


Fig. 24–41. *Trichorhinophalangeal syndrome, Type II.* (A–D) Compare facies of four children with the disorder. All have mild microcephaly, outstanding ears. (E,F) Multiple exostoses become evident by the fourth year of life. [A–C,E from *BD Hall et al,* Birth Defects **10**(12):147, 1974. D from *RJ Gorlin et al,* Am J Dis Child **118**:595, 1969.]

of life, although they may be found as early as the end of the first year (28). They occur in the long and short tubular bones of the limbs, primarily at the metaphyses (14,16,20,26,32). Cone-shaped epiphyses are common in both the proximal and middle phalanges of the hands. The ribs, which are narrow posteriorly (6,15,18,23,32), and scapulae, which are winged (15,33), may also be affected. Growth retardation, postnatal in onset (20,25), is essentially a constant finding, and seems to antedate multiple exostoses. Laxity or hypermobility of joints has been observed in 65% (20,23,25). The metacarpal phalangeal pattern profile differs mildly from that of Type I (9a).

**Skin and hair.** During infancy and early childhood, the skin is loose or redundant in 75% (14,15,20,25,28,32). Multiple nevi are found on the face and extremities in over 50% (20,25,32,35). Scalp hair is sparse, but the eyebrows are full (25). Brittle nails are found in 70% (25).

**Other findings.** Recurrent respiratory tract infections are found in 60% in infancy (18,20,25,35). Infantile feeding problems are seen in over 90% (25). Miscellaneous findings have been meticulously documented by Langer et al (25).

**Oral manifestations.** Micrognathia and retrognathia (18,25) have been reported and the mandibular angle may be more obtuse than normal (20); however, measurements have not been made to confirm these subjective observations. Supernumerary central incisors (15) as well as congenitally missing teeth (15) have been reported.

**Differential diagnosis.** The differential diagnosis includes *trichorhinophalangeal syndrome, Type I.* Cone-shaped epiphyses occur in the general population as an isolated finding. They may also be associated with *de Lange syndrome, cleidocranial dysplasia,* asphyxiating thoracic dystrophy, *Ellis–van Creveld syndrome, cartilage–hair hypoplasia,* and *pseudohypoparathyroidism.* Multiple cartilaginous exostoses in the absence of other abnormalities constitute a well-established autosomal dominant entity.

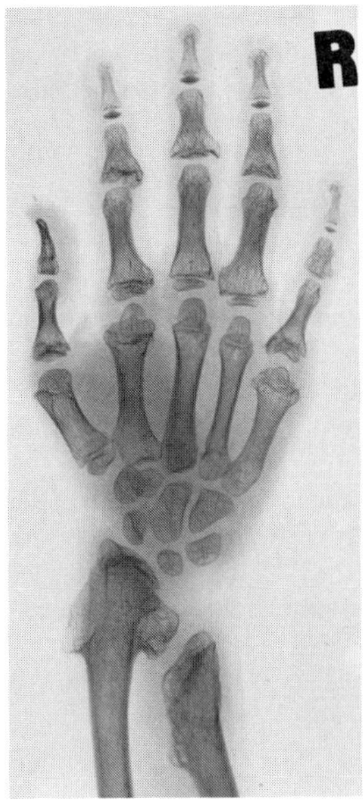


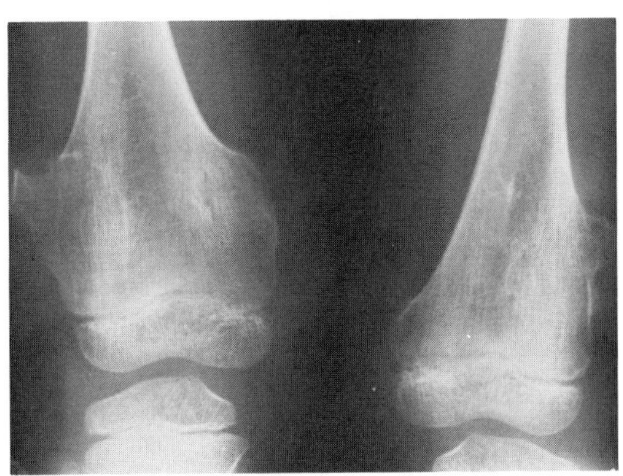

A B

Fig. 24–42. *Trichorhinophalangeal syndrome, Type II.* (A,B) Multiple exostoses, cone-shaped epiphyses. (A from *A Giedion,* Fortschr Roentgenstr **110**:507, 1969. B from *RJ Gorlin et al,* Am J Dis Child **118:**595, 1969.)

Shabtai et al (31) reported a mother and two sons with features of trichorhinophalangeal syndrome, Type II. They lacked mental retardation, microcephaly, and cone-shaped epiphyses but had a paracentric inversion of chromosome 8, inv8(q11.23; q21.1).

Deletions in 8q have been reported in patients without the trichorhinophalangeal syndrome, Type II phenotype (2,9,33). At age 5½ years the patient reported by Dallapiccola et al (9) still had no exostoses (10).

Lüdecke et al (25a) described an error in maternal gametogenesis in deletion of 8q23–24.1.

**Laboratory aids.** Karyotype analysis should be carried out to demonstrate the del(8)(q24.11→q24.13) abnormality.

#### References [Trichorhinophalangeal syndrome, Type II (Langer–Giedion syndrome)]


1. Alè G, Calò S: Su di un caso di disostosi periferica associata con esostosi osteogeniche multiple ed iposomia disuniforme e disarmonica. *Ann Radiol Diagn (Bologna)* **34**:376–385, 1961.

2. Beighle C et al: Small structural changes of chromosome 8. Two cases with evidence for deletion. *Hum Genet* **38**:113–121, 1977.

3. Bowen P, Biederman B: Tricho-rhino-phalangeal syndromes: Search for clinical/cytogenetic correlations in Type II (Langer–Giedion). *Proc Greenwood Genet Ctr* **3**:125, 1984 (Abstr).

4. Bowen P et al: The critical segment for the Langer–Giedion syndrome: 8q24.11→q24.12. *Ann Génét* **28**:224–227, 1985.

5. Bühler EM: Langer–Giedion syndrome and 8q – deletion. *Am J Med Genet* **11**:359, 1982 (editorial).

6. Bühler EM, Malik NJ: The tricho-rhino-phalangeal syndrome(s): Chromosome 8 long arm deletion: Is there a shortest region of overlap between reported cases? TRP I and TRP II syndromes: Are they separate entities? *Am J Med Genet* **19**:113–119, 1984.

7. Bühler EM et al: Terminal or interstitial deletion in chromosome 8 long arm in Langer–Giedion syndrome (TRP II syndrome)? *Hum Genet* **64**:163–166, 1983.

8. Bühler EM et al: A final word on the tricho-rhino-phalangeal syndromes. *Clin Genet* **31**:273–275, 1987.

9. Dallapiccola B et al: Deletion of the long arm of chromosome 8 resulting from a de novo translocation of t(4;8)(q13;q213). *Hum Genet* **38**:125–130, 1977.

9a. Dirkstra PF: Analysis of metacarpophalangeal pattern profiles. *Roefo* **139**:158–159, 1983.

10. Frontali M et al: 'Microgenetics' and Langer–Giedion syndrome. *J Med Genet* **19**:390–391, 1982.

11. Fryns JP et al: Interstitial deletion of the long arm of chromosome 8. Karyotype 46,XY,del(8)(q21). *Hum Genet* **48**:127–130, 1979.

12. Fryns JP et al: Langer–Giedion syndrome and deletion of the long arm of chromosome 8. *Hum Genet* **58**:231–232, 1981.

13. Fryns JP et al: Langer–Giedion syndrome and deletion of the long arm of chromosome 8: Confirmation of the critical segment to 8q23. *Hum Genet* **64**:194–195, 1983.

14. Fryns JP et al: Tricho-rhino-phalangeal syndrome type II: Langer–Giedion syndrome in a 2.5 year old boy. *J Génét Hum* **28**:53–56, 1980.

15. Fukushima Y et al: Two cases of the Langer–Giedion syndrome with the same interstitial deletion of the long arm of chromosome 8:46,XY or XX,del(8)(q23.3q24.13). *Hum Genet* **64**:90–93, 1983.

16. Gericke GS, Fialkov J: The Langer–Giedion phenotype associated with a unique skeletal finding in a mentally retarded adolescent male. A case report. *S Afr Med J* **57**:548–549, 1980.

17. Giedion A: Die periphere Dysostose (PD)—ein Sammelbegriff. *Fortschr Röntgenstr* **110**:507–534, 1969.

18. Gorlin RJ et al: Tricho-rhino-phalangeal syndrome. *Am J Dis Child* **118**:595–599, 1969.

19. Gorlin RJ et al: No chromosome deletion found on prometaphase banding in two cases of Langer–Giedion syndrome. *Am J Med Genet* **13**:345–347, 1982 (letter).

19a. Grover SC, Bamforth S: Intellectual competence in a girl with Langer–Giedion syndrome. *Am J Med Genet* **34:**456–457, 1989.

20. Hall BD et al: Langer–Giedion syndrome. *Birth Defects* **10**(12):147–164, 1974.

21. Harnsberger J et al: Interstitial deletion of the long arm of the number 8 chromosome and the Langer–Giedion syndrome (LG). *Birth Defects* **19**(5):175–176, 1982 (Abstract).

22. Howell CJ, Wynne-Davies R: The tricho-rhino-phalangeal syndrome. A report of 14 cases in 7 kindreds. *J Bone Jt Surg* **68**(B):311–314, 1986.

22a. Keret D et al: The Alè–Calò syndrome in monozygotic twins associated with bilateral cryptorchidism. *Z Kinderchir* **39**:145–146, 1984.

23. Kozlowski K et al: Multiple exostoses-mental retardation syndrome (Alè-Calò or M.E.M.R. syndrome). Description of two childhood cases. *Clin Pediatr* **16**:219–224, 1977.

24. Langer LO Jr: The thoracic-pelvic-phalangeal dystrophy. *Birth Defects* **5**(4):55–64, 1969.

25. Langer LO Jr et al: The tricho-rhino-phalangeal syndrome with exostoses (or Langer–Giedion syndrome): Four additional patients without mental retardation and review of the literature. *Am J Med Genet* **19**:81–111, 1984.

25a. Lüdecke H-J et al: Maternal origin of a de novo chromosome 8 deletion in a patient with Langer–Giedion syndrome. *Hum Genet* **82:**327–329, 1989.

26. Murachi S et al: Tricho-rhino-phalangeal syndrome type II. The Langer-Giedion syndrome. *Jpn J Hum Genet* **24**:27–36, 1979.

27. Murachi S et al: Familial tricho-rhino-phalangeal syndrome type II. *Clin Genet* **19**:149–155, 1981.

27a. Naselli A et al: La sindrome tricorinofalangea con esostosi. *Minerva Pediatr* **39**:25–31, 1987.

27b. Okuno T et al: Langer–Giedion syndrome with del 8(q24.13–q24.22). *Clin Genet* **32**:40–45, 1987

28. Oorthuys JWE, Beemer FA: The Langer–Giedion-syndrome (tricho-rhino-phalangeal syndrome, type II). *Eur J Pediatr* **132**:55–59, 1979.

29. Pfeiffer RA: Langer–Giedion syndrome and additional congenital malformations with interstitial deletion of the long arm of chromosome 8 46,XY, del8(q13–22). *Clin Genet* **18**:142–146, 1980.

30. Schwartz S et al: A complex rearrangement, including a deleted 8q, in a case of Langer–Giedion syndrome. *Clin Genet* **27**:175–182, 1985.

31. Shabtai F et al: Familial syndrome with some features of the Langer–Giedion syndrome, and paracentric inversion of chromosome 8, inv8 (q11.23; q21.1). *Clin Genet* **27**:600–605, 1985.

32. Stoltzfus E et al: Langer–Giedion syndrome: Type II tricho-rhino-phalangeal dysplasia. *J Pediatr* **91**:277–280, 1977.

33. Taysi K et al: Presumptive long arm deletion of chromosome 8: A new syndrome? *Hum Genet* **51**:49–53, 1979.

34. Turleau C et al: Langer–Giedion syndrome with and without del 8q. Assignment of critical segment to 8q23. *Hum Genet* **62**:183–187, 1982.

35. Wilson GW et al: Intersitial deletion of 8q. Occurrence in a patient with multiple exostoses and unusual facies. *Am J Dis Child* **137**:444–448, 1983.

36. Wilson WG et al: The Langer–Giedion syndrome: Report of a 22-year-old woman. *Pediatrics* **64**:542–545, 1979.

37. Wilson WG et al: Interstitial deletion of 8q in a patient with multiple exostoses and developmental delay. *Am J Hum Genet* **33**:96A, 1981 (Abstr).

38. Zabel BU, Bauman WA: Langer–Giedion syndrome with interstitial 8q– deletion. *Am J Med Genet* **11**:353–358, 1982.

39. Zaletajev DV, Marincheva GS: Langer–Giedion syndrome, in a child with complex structural aberration of chromosome 8. *Hum Genet* **63**:178–182, 1983.

## Johanson–Blizzard syndrome

Johanson and Blizzard (7), in 1971, reported three unrelated children with characteristic facies, severe mental and somatic retardation, sensorineural hearing loss, and malabsorption due to pancreatic insufficiency. The patients described in 1967 by Morris and Fischer (12) and Townes (18) in 1972 clearly had the same disorder. About 30 patients have been reported (6a,7a). We have classified a case as probable (8) and another as an unlikely (5) example. Parental consanguinity (2,9,16,17) and affected sibs (4,6,10,15) have been reported, indicating autosomal recessive inheritance.

Birthweight has been low in about one-third of the cases (11). There is failure to thrive; somatic retardation is usually severe. At least 70% remain well below the third percentile for height, weight, and head circumference with true microcephaly in 35%. Generalized edema (anasarca) due to protein loss is evident from birth.

**Facies.** Aplastic nasal alae producing a beaklike nose is the most striking and constant feature of the syndrome. Aplasia of lacrimal puncta or cutaneolacrimal fistula is often noted (4,12,16). The anterior fontanel is open (12). Midline skin dimples and/or defects are noted in 90% over the anterior and posterior fontanels (4a,6a,7,11,16). The scalp hair is often blond, sparse, dry, and coarse. The hair in the frontal area has a marked upsweep and there is extension of the lateral hairline onto the forehead (Fig. 24-43A–C).

Roentgenocephalometric study has been carried out on two patients (13). While microcephaly has been present in 35%, this was not borne out in patients studied by Motohashi et al (13) or by Sismanis et al (17). Facial, skeletal and cranial base measurements were reduced, especially the length of the maxilla.

**Musculoskeletal.** Severe hypotonia with hyperextensibility of joints is noted in 80% (7,11,12,16). Pitting edema of the hands and feet due to protein loss may be striking (12,14). Bone age is delayed in about 80% (13,17,18).

**Genitourinary and gastrointestinal.** Single urogenital orifice with infantile ovaries and double or septate vagina, clitoromegaly, micropenis, cryptorchidism, or urethrovaginal fistulae are noted in 30% and imperforate or anteriorly placed anus in about 50% (2,4,4a,6,7,8–10,17,19). Kristjansson et al (7a) have suggested that some examples of genital hypoplasia are secondary to hypopituitarism. Pancreatic insufficiency, failure to thrive, and malabsorption have been constant features (11).

**Central nervous system.** Bilateral sensorineural hearing loss has been noted in about 60% (4,4a,6a,7,10,11,16). Speech is rudimentary both from hearing loss and psychomotor retardation. About 60% of patients are mentally retarded (IQ 35–50), but intelligence may be normal (10,15) or only mildly delayed (4,6a,13,18).

**Thyroid.** Hypothyroidism has been found in about 30% (4a,6a,10, 11,20). The reader is also referred to several possible examples of hypothyroidism (3,4,7,13a).

Fig. 24–43. *Johanson–Blizzard syndrome.* (A–C) Children exhibiting similar facies characterized by hypoplasia of nasal alae. They also have hypotonia, genital anomalies, and trypsinogen deficiency. (*A* from *M Baraitser* and *SV Hodgson*, J Med Genet **19**:302, 1981. B and C from *A Johanson* and *R Blizzard*, J Pediatr **79**:982, 1971.)

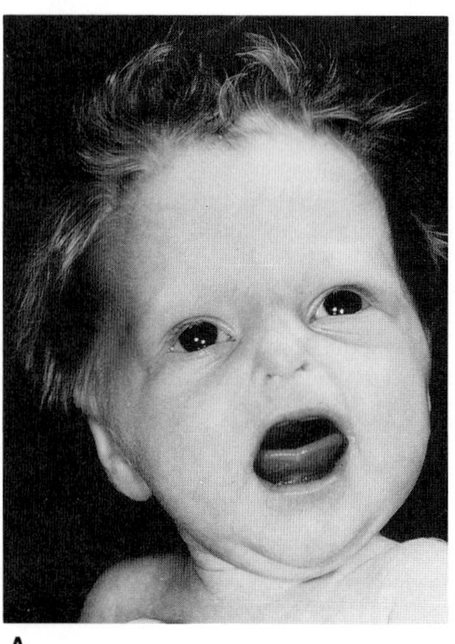
A

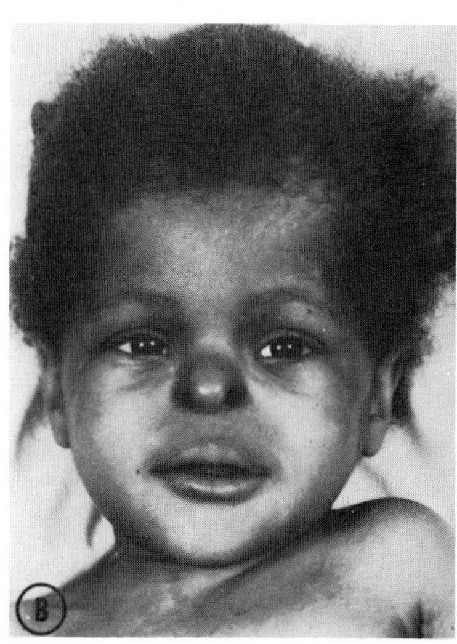

B

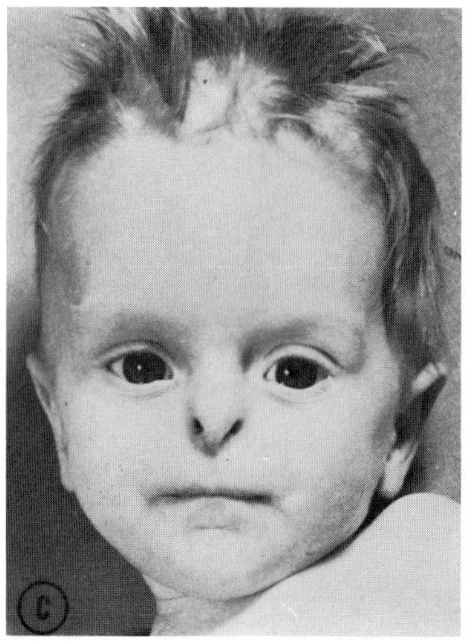

C

**Skin.** Nipples and areolae may be hypoplastic.

**Heart.** ASD, VSD, and situs inversus have been reported in 15% of the cases (6,6a).

**Oral manifestations.** There is severe microdontia in both dentitions and, not uncommonly, the permanent dentition may be entirely absent except for the first permanent molars (20). The roots of the deciduous teeth are short, irregular, and deformed. The crowns of the few secondary teeth are frequently reduced in form, incisors being conical, maxillary molars having only three cusps. The tooth pulps are large. Permanent first molars are somewhat taurodont (13a,15).

**Pathology.** Pancreatic changes include dramatic fatty replacement with a paucity of acini but without concomitant hypoglycemia (4a,11). The brain exhibits abnormal gyral formation and cortical neuronal disorganization (2,3).

**Differential diagnosis.** Johanson–Blizzard syndrome shares proteolytic deficiency features with other disorders: cystic fibrosis, Schwachman syndrome, *cartilage–hair hypoplasia,* trypsinogen deficiency disease, and intestinal enterokinase deficiency. Their differentiation and the use of electrophoretic studies of pancreatic enzymes have been discussed at length by Townes (18).

While the signs and symptoms manifested by the sisters described by Reichart et al (15) are similar to those of Johanson–Blizzard syndrome, they probably represent another disorder.

Several other syndromes are characterized by varying degrees of hypoplasia of nasal alae: *trichorhinophalangeal syndrome, oculodentoosseous dysplasia, craniocarpotarsal dysplasia,* and *Roberts pseudothalidomide syndrome. Aplasia cutis congenita* usually occurs as an isolated finding but may be seen with *trisomy 13, focal dermal hypoplasia, Sakati syndrome,* etc. *Aplasia cutis congenita* of the scalp is associated with a plethora of disorders.

**Laboratory aids.** Iron deficiency, low hemoglobin, fat, foul, bulky stools, hypocalcemia, and low total serum proteins are evident in early infancy (20).

There is a combined proteolytic, lipolytic, and amylolytic defect manifested by complete absence of trypsin, chymotrypsin, amylase, carboxypeptidase, and lipase activities. Activation studies have proven negative. Fat and nitrogen balance studies confirm the enzyme deficiencies.

### References (Johanson–Blizzard syndrome)


1. Baraitser M, Hodgson SV: The Johanson–Blizzard syndrome. *J Med Genet* **19**:302–303, 1981.

2. Bresson JL et al: Le syndrome de Johanson–Blizzard. Une autre cause de lipomatose pancreatique. *Arch Fr Pédiatr* **37**:21–24, 1980.

3. Daentl DL et al: The Johanson–Blizzard syndrome: Case report and autopsy findings. *Am J Med Genet* **3**:129–135, 1979.

4. Day DW, Israel JN: Johanson–Blizzard syndrome. *Birth Defects* **14**(6B):275–287, 1978.

4a. Gould NS et al: Johanson–Blizzard syndrome: Clinical and pathological findings in 2 sibs. *Am J Med Genet* **33**:194–199, 1989.

5. Grand RJ et al: Unusual case of XXY Klinefelter's syndrome with pancreatic insufficiency, hypothyroidism, deafness, chronic lung disease, dwarfism and microcephaly. *Am J Med* **41**:478–485, 1966.

6. Helin I, Jödal U: A syndrome of congenital hypoplasia of the alae nasi, situs inversus, and severe hypoproteinemia in two siblings. *J Pediatr* **99**:932–934, 1981.

6a. Hurst JA, Baraitser M: Johanson–Blizzard syndrome. *J Med Genet* **26**:45–48, 1989.

7. Johanson A, Blizzard R: A syndrome of congenital aplasia of the alae nasi, deafness, hypothyroidism, dwarfism, absent permanent teeth and malabsorption. *J Pediatr* **79**:982–987, 1971.

7a. Kristjansson K et al: Johanson–Blizzard syndrome and hypopituitarism. *J Pediatr* **113**:851–853, 1988.

8. Lumb G, Beautyman W: Hypoplasia of the exocrine tissue of the pancreas. *J Pathol Bact* **64**:679–686, 1952.

9. Mardini MK et al: Johanson–Blizzard syndrome in a large inbred kindred with three involved members. *Clin Genet* **14**:247–250, 1978.

10. Moeschler JB, Lubinsky MS: Johanson–Blizzard syndrome with normal intelligence. *Am J Med Genet* **22**:69–73, 1985.

11. Moeschler JB et al: The Johanson–Blizzard syndrome: A second report of full autopsy findings. *Am J Med Genet* **26**:133–138, 1987.

12. Morris MD, Fisher DA: Trypsinogen deficiency disease. *Am J Dis Child* **114**:203–208, 1967.

13. Motohashi N et al: Roentgenocephalometric analysis of craniofacial growth in the Johanson–Blizzard syndrome. *J Craniofac Genet Dev Biol* **1**:57–72, 1981.

13a. Ono K et al: Oral findings in Johanson–Blizzard syndrome. *J Oral Med* **42**:14–16, 1987.

14. Park IJ et al: Special female hermaphroditism associated with multiple disorders. *Obstet Gynecol* **39**:100–106, 1972.

15. Reichart P et al: Ektodermale Dysplasie und exokrine Pankreasinsuffizienzein erblich bedingtes Syndrom. *Dtsch Zahnärztl Z* **34**:263–265, 1979.

16. Schussheim A et al: Exocrine pancreatic insufficiency with congenital anomalies. *J Pediatr* **89**:782–784, 1976.

17. Sismanis A et al: Rare congenital syndrome associated with profound hearing loss. *Arch Otolaryngol* **105**:222–224, 1979.

18. Townes PL: Trypsinogen deficiency and other proteolytic deficiency disease. *Birth Defects* **8**(2):95–101, 1972.

19. Townes PL, White MR: Identity of two syndromes: Proteolytic, lipolytic and amylolytic deficiency of the exocrine pancreas with congenital anomalies. *Am J Dis Child* **135**:248–250, 1981.

20. Zerres K, Holtgrave EA: The Johanson–Blizzard syndrome: Report of a new case with special reference to the dentition and review of the literature. *Clin Genet* **30**:177–183, 1986.


## Binder syndrome (maxillonasal dysplasia)

In 1962, Binder (2) reported a condition which he called maxillonasal dysostosis. He thought the condition was a form of arhinencephaly, which it is not. Noyes (12) reported a case as early as 1939 and Zuckerkandl (21), in 1882, described a related malformation. The condition has also been called facies scaphoidea and congenitally flat nose syndrome (6).

Fifty patients were reported by Holström (6) and well over 100 examples have been described, including several large series of cases (1,3–11,13–21). We believe the condition is rather common, but we do not regard it as a true syndrome, agreeing with Toriello (20) that it is a nonspecific abnormality of the nasomaxillary complex. Its basic significance is as a surgical entity, not as a genetic, nosologic, or syndromic entity. It is neither a dysplasia nor a syndrome. Although most cases involve only the nasomaxillary complex, a variety of other anomalies have been recorded including especially cervical vertebral anomalies, but also various other skeletal defects, cardiac anomalies, orofacial clefting, strabismus, mental retardation, and other abnormalities (3,6,13,20).

Almost all cases are sporadic (14b). Affected parent and child have been noted twice (6,11) and affected sibs have been observed on four occasions (6). Consanguinity has been recorded in a few instances (3,19). Holström (6) suggested that 16% of his sample exhibited "hereditary factors." We interpret familial examples reported to date not as resulting from monogenic transmission, but as resulting from complex genetic factors similar to those involved in producing malocclusion.

**Facies.** The nasofrontal angle is absent and the nose is hypoplastic with flattened alae and nasal tip (Figs. 24-44A,B and 24-45A,B). There is a palpable depression in the anterior nasal floor and localized maxillary hypoplasia in the alar base region (6). The nostrils are half moon-shaped when viewed from below (Fig. 24-45A). The nasal mucosa has been described as atrophic, but the sense of smell is normal (2). The philtral crests may be bow-shaped, rising vertically to the columella without convergence.

Radiographically, aplasia or hypoplasia of the anterior nasal spine, thinness of the labial plate of alveolar bone over the upper incisors, and an increase in the nasomaxillary angle have been observed

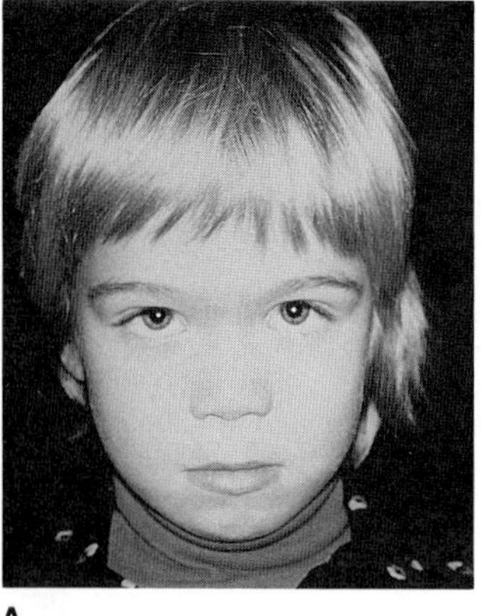
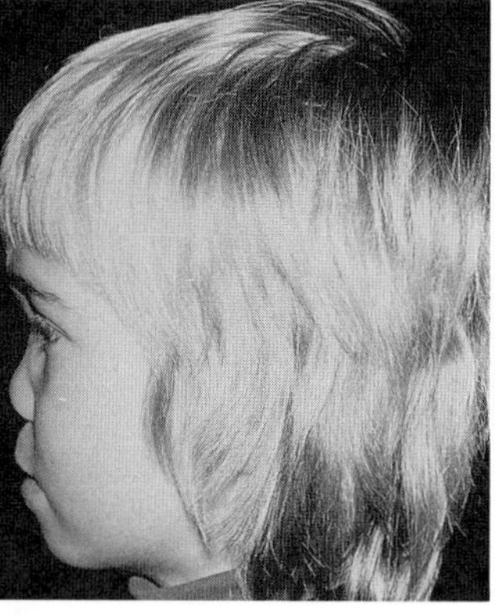

A B

Fig. 24–44. *Binder syndrome.* (A,B) Typical facies, frontal and lateral views. (Courtesy of *M Olow-Nordenram,* Linköping, Sweden.)

(3,4,7,9,10,14a) (Fig. 24–46). The frontal sinuses are often hypoplastic or absent (2,3). Cephalometric studies have shown increased gonial angle and proclination of the incisors (3,4,14). There are decreased anterior cranial base measurements and a smaller maxilla vertically and anteroposteriorly. The cervical spine is anomalous in 40–50%. The atlas and axis are most frequently involved. Various anomalies include short posterior arch, block vertebrae, separate odontoid, spina bifida occulta, and persistence of chorda dorsalis (15%) either isolated or in various combinations. Calcifications extending from the anterior arch of the atlas have been noted as well as mild scoliosis and/or kyphosis (3,13).

**Other findings.** Strabismus has been found occasionally as well as mild mental retardation (3,4,8), but they probably have no statistical significance. Nonspecific congenital heart defects have been found in 5% (6,8). Hearing loss occurs in about 5% (4,8).

**Oral manifestations.** The upper lip has a convex contour with poorly developed philtrum. The premaxillary area is hypoplastic, with flattening of the maxillary base and sagittal shortening of the dental arch. The mandible is of normal width, but the gonial angle is increased and the chin is flattened (3,4,14). Because of maxillary shortening, all patients have a relative mandibular prognathism with reverse overbite (class III malocclusion) (Fig. 24–47). Rarely there is cleft lip (3,6) or cleft palate (4,6,8).

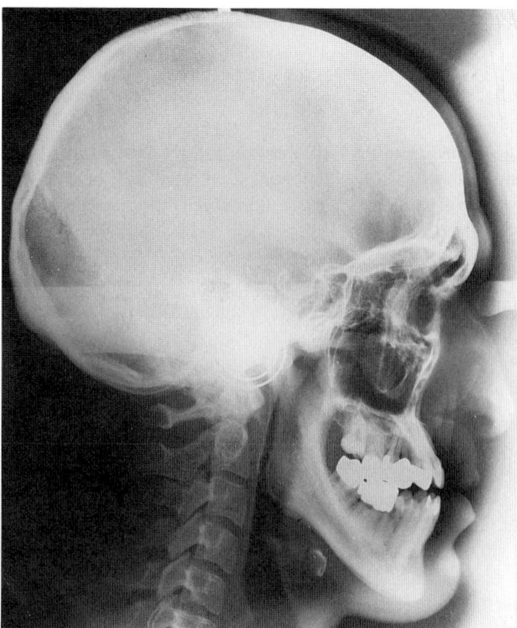

Fig. 24–46. *Binder syndrome.* Radiograph showing aplasia of anterior spine, thinness of labial plate of alveolar bone over upper incisors, and increase in nasomaxillary angle. (Courtesy of *M Olow-Nordenram,* Linköping, Sweden.)

**Differential diagnosis.** Although the facies may appear ''arhinencephaloid,'' no brain abnormalities have been observed in this condition. The sense of smell is completely normal. Absent frontal sinuses and relative mandibular prognathism can be associated with a host of conditions and may be seen in otherwise normal individuals. Nasal hypoplasia can be seen in *pseudohypoparathyroidism, Stickler syndrome,* and *Warfarin embryopathy.*

Fig. 24–45. *Binder syndrome.* (A,B) Flattened nose with depressed subnasal or alar base area and hypoplasia of premaxillary area of upper jaw. Upper lip has convex contour; nostrils are half-moon-shaped. (Courtesy of *H Holmstrom,* Göteborg, Sweden.)

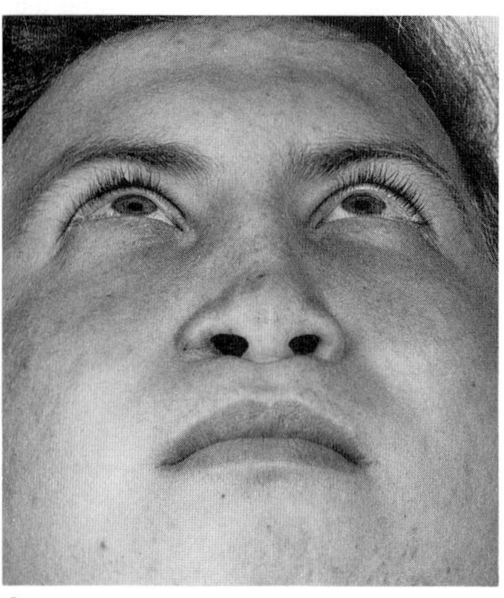
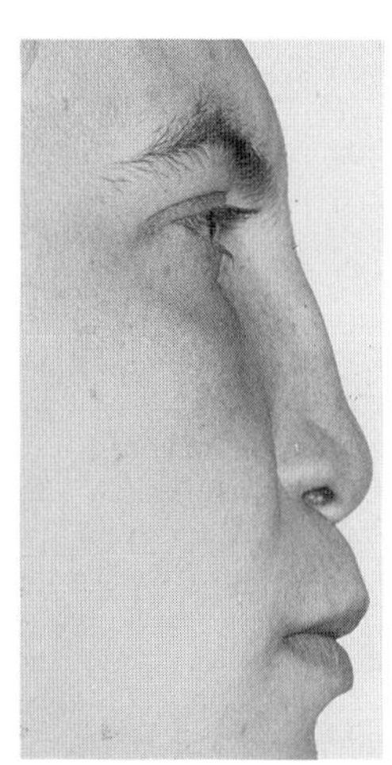

A B

Fig. 24–47. *Binder syndrome.* Relative mandibular prognathism with reverse overbite. (Courtesy of *I Jackson,* Rochester, Minnesota.)

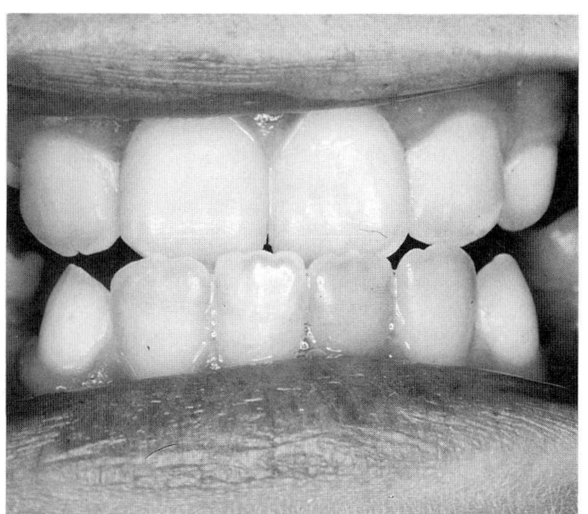

### References [Binder syndrome (maxillonasal dysplasia)]

1. Bimler HP: Über die microrhine Dysplasie. *Fortschr Kieferorthopäd* **26**:417–434, 1965.
2. Binder KH: Dysostosis maxillo-nasalis, ein arhinencephaler Missbildungskomplex. *Dtsch Zahnärztl Z* **17**:438–444, 1962.
3. Delaire J et al: Clinical and radiologic aspects of maxillonasal dysostosis (Binder's syndrome). *Head Neck Surg* **3**:105–122, 1980.
4. Ferguson JW, Thompson RPJ: Maxillonasal dysostosis (Binder syndrome): A review of the literature and case reports. *Eur J Orthodont* **7**:145–148, 1985.
5. Henderson D, Jackson IT: Naso-maxillary hypoplasia: The Le Forte II osteotomy. *Br J Oral Surg* **11**:77–93, 1973.
6. Holström H: Clinical and pathologic features of maxillonasal dysplasia (Binder's syndrome): Significance of the prenasal fossa on etiology. *Plast Reconstr Surg* **78**:559–567, 1986.
7. Hopkin GB: Hypoplasia of the middle third of the face associated with congenital absence of the anterior nasal spine, depression of the nasal bones, and Angle class III malocclusion. *Br J Plast Surg* **16**:146–153, 1963.
8. Horswell BB et al: Maxillonasal dysplasia (Binder's syndrome): A critical review and case study. *J Oral Maxillofac Surg* **45**:114–122, 1987.
9. Jackson IT et al: Total surgical management of Binder's syndrome. *Ann Plast Surg* **7**:25–34, 1981.
10. McWilliam J, Linder-Aronson S: Hypoplasia of the middle third of the face. *Angle Orthodont* **46**:260–267, 1976.
11. Munro IR et al: Maxillonasal dysplasia (Binder's syndrome). *Plast Reconstr Surg* **63**:657–663, 1979.
12. Noyes FB: Case report. *Angle Orthodont* **9**:160–165, 1939.
13. Olow-Nordenram MAK, Redberg CT: Maxillo-nasal dysplasia (Binder syndrome) and associated malformations of the cervical spine. *Acta Radiol (Diagn)* **25**:353–360, 1984.
14. Olow-Nordenram MAK, Thilander B: The cranio-facial morphology in individuals with maxillo-nasal dysplasia (Binder syndrome). A cephalometric study. *Eur J Orthodont* **8**:53–60, 1986 and **9**:224–236, 1987.

14a. Olow-Nordenram MAK, Thilander B: The craniofacial morphology in persons with maxillonasal dysplasia (Binder syndrome). *Am J Orthodont* **95**:148–158, 1989.

14b. Olow-Nordenram MAK, Valentin J: An etiologic study of maxillonasal dysplasia–Binder's syndrome. *Scand J Dent Res* **96**:69–74, 1988.

15. Pathenheimer F: Angeborene Fehlbildungen und Anomalien. 10. Mitteilung: Maxillo-nasales Syndrom. *Med Mschr* **20**:232–233, 1966.
16. Resche F et al: Craniospinal and craniocervical malformations associated with maxillonasal dysostosis (Binder syndrome). *Head Neck Surg* **3**:123–131, 1980.
17. Rintala A, Ranta R: Maxillonasal hypoplasia-Binder's syndrome. Morphology and treatment of two separate varieties. *Scand J Plast Surg* **19**:127–134, 1985.
18. Rival JM et al: Dysostose maxillo-nasale de Binder. A propos de 10 cas. *J Génét Hum* **22**:263–268, 1974.
19. Tessier P et al: Therapeutic aspects of maxillonasal dysostosis (Binder syndrome). *Head Neck Surg* **3**:207–215, 1980.
20. Toriello HV et al: Is Binder syndrome really a syndrome? *David W. Smith Workshop on Malformations and Morphogenesis,* Greenville, South Carolina, August 15–19, 1987.
21. Zuckerkandl E: Fossae praenasales. Normale und pathologische. *Anatomie der Nasenhöhle und ihrer pneumatischen Anhänge*. W. Braunmüller, Wien, 1882.


## Lenz microphthalmia syndrome

In 1955, Lenz (9) described a syndrome consisting of microphthalmia, skeletal anomalies of the hands and clavicles, renal anomalies, genital abnormalities, and defects of the dentition. Several other kindred have been reported (1–6,11,13). Four cases in a family described by Hoefnagel et al (7) possibly represent the same syndrome. We are less certain about the family reported by Cuendet (2) and that of James et al (8).

X-linked recessive transmission seems apparent. Minor malformations have been noted in heterozygous carriers in some instances (6,9). Female carriers have also been noted to have an increased abortion rate, suggesting that many cases may appear to be only sporadic (3,9). Linkage studies have not been helpful (12).

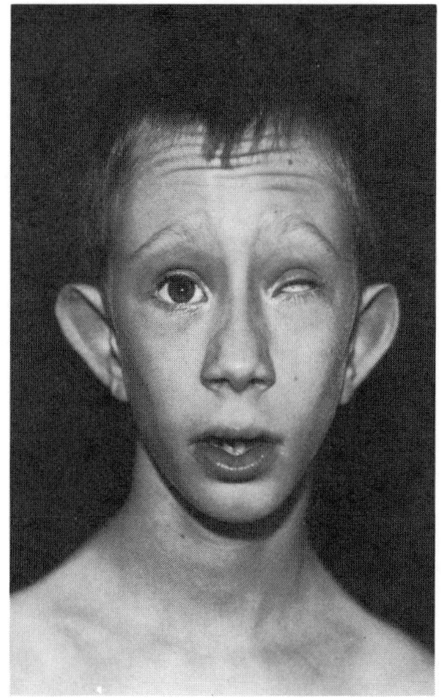

Fig. 24–48. *Lenz microphthalmia syndrome.* Eye defects range from microphthalmia to clinical anophthalmia and may be unilateral or bilateral. Note outstanding ears. [From *J Herrmann* and *JM Opitz,* Birth Defects **5**(2):138, 1969.]

**Facies.** The forehead is high. Unilateral or bilateral eye defects ranging from colobomatous (usually) microphthalmia to clinical anophthalmia have been observed (1–9,11). Upward slanting of palpebral fissures may be present (6). Microcornea (6), strabismus (9), nystagmus (5,6), epicanthal folds (5), and blepharoptosis (1,3–5,7,16) as well as other defects have been reported in over 80%. The ears are asymmetric, dysplastic, hypoplastic, or protuberant in 80% (1–9,11,16) (Fig. 24–48). Micrognathia has also been noted (6).

**Central nervous system.** Mental retardation and microcephaly have been reported in over 90% (1–6,11,13,16).

**Musculoskeletal system.** Camptodactyly of the fifth fingers (6, 9,13), clinodactyly of the second fingers (6), hypoplastic thumb (16), duplication of the thumb (9,16), cutaneous syndactyly of the third and fourth toes, clinodactyly of toes (16), wide gap between first and second toes, pseudoclubbing of the toes, flat foot, calcaneovalgus deformity (6,11), and varus deformity (7) have been observed in about 60%.

Short stature (1–6,9,10,12), cylindrical thorax with sloping shoulders and clavicular defects (3,6,9,11,13), kyphosis (3), low scapulas, notching of the vertebral bodies, mild cubitus valgus, limited extension in both hip joints, and mild genua valga with internally rotated knees and prominent fibulas (6) have been reported (Figs. 24–49 and 24–50).

**Genitourinary system.** Unilateral renal agenesis (9), bilateral renal agenesis (7), renal dysgenesis, hydroureters (6), cryptorchidism (6, 9,11,13), and hypospadias (3) were described in about 50%.

**Other findings.** Congenital heart defect (9), atresia of the ileum (7), umbilical hernia, unusual dermatoglyphics (6), defective speech (6,7), and hirsutism of lower back (3,6,13) have been noted.

**Oral manifestations.** Highly arched palate, cleft palate (1,2,4,6), crooked anterior teeth (6,9), and agenesis of the permanent maxillary lateral incisors (6,13,14) have been reported in about 65%.

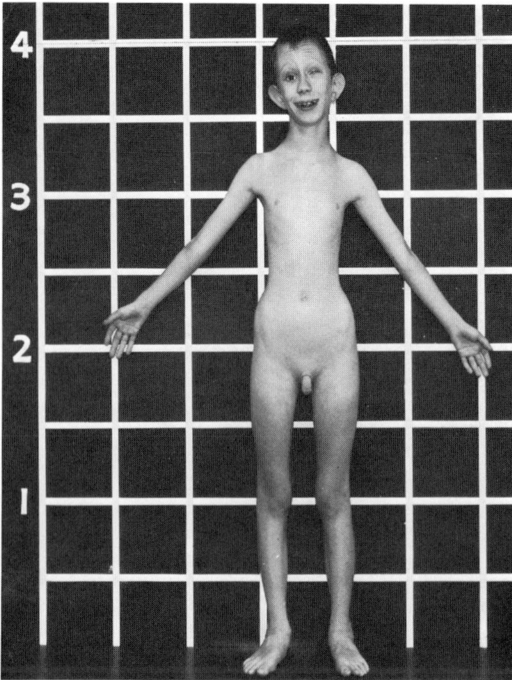


Fig. 24–49. *Lenz microphthalmia syndrome.* Patients are short with cylindric thorax and internally rotated knees. [From *J Herrmann* and *JM Opitz,* Birth Defects **5**(2):138, 169.]

Fig. 24–50. *Lenz microphthalmia syndrome.* Roentgenogram showing cylindric thorax, sloping shoulders, and clavicular defects. [From *J Herrmann* and *JM Opitz,* Birth Defects **5**(2):138, 1969.]

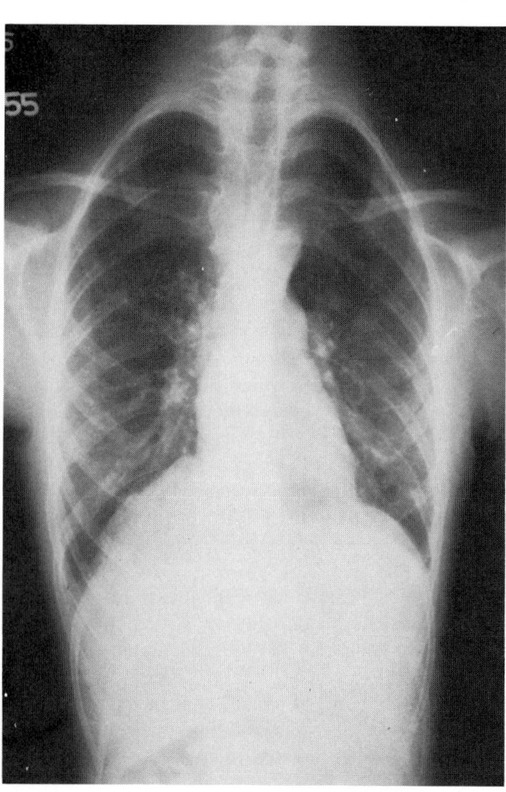


**Differential diagnosis.** Isolated bilateral anophthalmia is inherited as an autosomal recessive trait. Microphthalmia may be inherited as an autosomal dominant, autosomal recessive, or X-linked recessive trait. Microphthalmia may also occur as an integral part of several different disorders (10). There is some overlap with the *Nance–Horan syndrome* (17) and an X-linked syndrome consisting of microcephaly, microphthalmia, corneal opacities, hypospadias, and cryptorchidism (15). Similar changes are also seen in *dup(10q) syndrome.*

**Laboratory aids.** Ultrasound studies are indicated in affected individuals.

### References (Lenz microphthalmia syndrome)

1. Baraitser M et al: Lenz microphthalmia. *Clin Genet* **22**:99–101, 1982.

1a. Brunquell PJ et al: Sex-linked hereditary bilateral anophthalmos. *Arch Ophthalmol* **102:**108–113, 1984.

2. Cuendet JF: La microphtalmie compliquée. *Ophthalmologica* **141**:380–385, 1961.

3. Dinno N et al: Bilateral microcornea, coloboma, short stature and other skeletal anomalies—a new hereditary syndrome. *Birth Defects* **12**(6):109–114, 1976.

4. Glanz A et al: Lenz microphthalmia: A malformation syndrome with variable expression of multiple congenital anomalies. *Can J Ophthalmol* **18**:41–44, 1983.

5. Goldberg MF, McKusick VA: X-linked colobomatous microphthalmos and other congenital anomalies. *Am J Ophthalmol* **71**:1128–1138, 1971.

6. Herrmann J, Opitz JM: The Lenz microphthalmia syndrome. *Birth Defects* **5**(2):138–143, 1969.

7. Hoefnagel D et al: Heredofamilial bilateral anophthalmia. *Arch Ophthalmol* **69**:760–764, 1963.

8. James PML et al: Systemic associations of uveal coloboma. *Br J Ophthalmol* **58**:917–921, 1974 (Case 13).

9. Lenz W: Recessiv-geschlechtsgebundene Mikrophthalmie mit multiplen Missbildungen. *Z Kinderheilkd* **77**:384–390, 1955.

10. McKusick VA: *Mendelian Inheritance in Man,* 7th ed. Johns Hopkins Press, Baltimore, 1988.

11. Manzitti E, Alezzandrini A: Syndrome dyscephalique de François (Case 2). *Ann Ocul* **196**:456–465, 1963.

12. Ogunye OO et al: Linkage studies in Lenz microphthalmia. *Hum Hered* **25**:493–500, 1975. (Same family as in Ref. 7.)

13. Pallotta R: The Lenz microphthalmia syndrome. *Ophthalmol Paediatr Genet* **3**:103–108, 1983.

14. Pallotta R et al: An additional contribution to the syndrome with dental agenesis. *Acta Stomatol Belg* **80**:215–218, 1983.

15. Siber M: X-linked recessive microencephaly, microphthalmia with corneal opacities, spastic quadriplegia, hypospadias and cryptorchidism. *Clin Genet* **26**:453–456, 1984.

16. Traboulsi EI et al: The Lenz microphthalmia syndrome. *Am J Ophthalmol* **105**:40–45, 1988.

17. Van Dorp DB, Delleman JW: A family with X-chromosomal recessive congenital cataract, microphthalmia, a peculiar form of the ear and dental anomalies. *J Pediat Ophthalmol Strab* **16**:166–171, 1979.

## Cryptophthalmos syndrome (Fraser syndrome)

The cryptophthalmos syndrome is characterized by cryptophthalmos, syndactyly, abnormal genitalia, and other variable defects such as nasal colobomas, malformed ears, cleft lip–palate, renal agenesis, and mental deficiency. The syndrome was first described by Zehender (39), in 1872, although it has been suggested that Pliny the Elder may have reported it in the first century A.D. (14). Approximately 100 cases have been described with slightly over two dozen cases of isolated cryptophthalmos in addition (1–3,5–39). The most extensive review is that of Thomas et al (31). Several other reviews are also available (12,14,21,28–30). Histopathologic documentation of the ocular manifestations is especially well reviewed by François (12), and detailed postmortem studies have been reported (2a,25,31).

The syndrome has autosomal recessive inheritance. Among 86 cases, 48 were familial and 38 were sporadic. Consanguinity has been reported in 6 of 42 families in which published pedigree data are available (31). Because the syndrome is variably expressed—including some infants without cryptophthalmos (8,23,27) and some of isolated crypt-

Table 24–2. Diagnostic criteria for the cryptophthalmos syndrome[a]

Major
- Cryptophthalmos
- Syndactyly
- Abnormal genitalia
- Sibling with cryptophthalmos syndrome

Minor
- Congenital malformation of nose
- Congenital malformation of ears
- Congenital malformation of larynx
- Cleft lip and/or palate
- Skeletal defects
- Umbilical hernia
- Renal agenesis
- Mental retardation

[a]From IT Thomas et al, *Am J Med Genet* **25**:85, 1986.

ophthalmos occurring as a separate entity—Thomas et al (31) proposed diagnostic criteria of the syndrome (Table 24–2). Cases were diagnosed on the basis of at least two major criteria and one minor criterion or on the basis of one major criterion and at least four minor criteria. The frequencies of individual syndrome manifestations are listed in Table 24–3. Gattuso et al (14) surveyed 68 cases. Figures differ somewhat from those of Thomas et al (31).

About 50% survive for a year or more, 25% are stillborn, and 20% die within the first year of life with most deaths occurring within the first week of life. Death is usually associated with renal agenesis and laryngeal stenosis (2a,14).

**Craniofacial features.** The face is asymmetric in about 10% (14). The eyebrows may be completely or partially missing (40). The globes can be seen and felt beneath the skin covering which extends from the forehead over the eyes. Cryptophthalmos is found in 85%, and in approximately 72% of these there is bilateral involvement (31) (Fig. 24–51A–C). Exposure to strong light may induce reflex wrinkling of the skin because of contraction of the orbicularis muscles (14). With unilateral cryptophthalmos, the opposite eye may exhibit upper lid coloboma, microphthalmia, epibulbar dermoid, or supernumerary eyebrows (30,37). The conjunctival sac is partially or completely obliterated and lashes, meibomian glands, and lacrimal glands are absent. The cornea is differentiated from the sclera. The lens may be absent, hypoplastic, or calcified and displaced. Orbicularis and levator palpebrae muscles are, however, normal (40). Rarely there is Potter facies (14).

Table 24–3. Frequency of individual manifestations in cryptophthalmos syndrome[a]

| Features | Percentage | (sample size) |
|---|---|---|
| Craniofacial | | |
| Cryptophthalmos | 85 | (93) |
| Abnormal nose | 84 | (62) |
| Abnormal ears | 83 | (64) |
| Cleft lip–palate | 9 | (65) |
| Abnormal larynx | 81 | (27) |
| Limbs | | |
| Syndactyly | 78 | (78) |
| Urogenital | | |
| Abnormal genitalia | 80 | (71) |
| Renal agenesis | 84 | (37) |
| Performance | | |
| Mental deficiency | 81 | (21) |
| Other | | |
| Skeletal defects | 70 | (33) |
| Umbilical hernia | 28 | (60) |

[a]From IT Thomas et al, *Am J Med Genet* **25**:85, 1986.

The hairline is bizarre in at least 25%, extending over the entire temple area and tapering to a point on the skin of the forehead overlying the eyes.

Nasal abnormalities are seen in approximately 85%. Most frequently the nose is broad with a low nasal bridge. Colobomas of the alae nasi, either unilateral or bilateral, may be observed, and midline grooving of the nasal tip has been noted (1,17,18,30,31,35,39).

Cleft lip and/or cleft palate and ankyloglossia have been found in about 10% (14). Malocclusion and, on occasion, hypoplastic teeth have been reported (1,12,18,21,29,40). Laryngeal atresia or hypoplasia occurs in approximately 80% (1,31,35).

The ears are abnormal in about 85%. They may be malformed and/or low set (30%), small (10%), or with fusion of the superior margin of the helix to the scalp (6%) with stenotic external ear canals (6–15%) or with some combination of these (14,17,21,28,31,35) (Fig. 24–51D). Conductive hearing deficit is common and malformed ossicles have been noted (14,21).

**Urogenital anomalies.** Abnormal genitalia found in about 80% include cryptorchidism (7%), hypospadias, chordee, hypoplastic penis (20%), large clitoris (20%), large labia majora, vaginal atresia (15%), bicornuate or rudimentary uterus (20%), and ambiguous external genitalia (1,2,7,10,14,21,22,24,25,28,31,32,34,36,39) (Fig. 24–51E). Greenberg et al (16) reported gonadal dysgenesis and gonadoblastoma. Renal agenesis occurs in 85% and may be either unilateral (37%) or bilateral (47%) (2,8,10,12,21,24,25,27,28,31). Abnormalities of the ureter and/or bladder are found in 10–15% (14).

**Extremities.** Marked soft-tissue syndactyly has been observed in 60–75% (1,14,21,30,31,37,39,40). In most cases, both fingers and toes are involved, although in some instances, either fingers or the toes alone may be affected (Fig. 24–52A–C).

**Central nervous system.** Mental deficiency is observed in about 80% of those patients whose intelligence has been documented. Malformations of the central nervous system occur in approximately 20% and may include encephalocele (10%) or occasional anomalies such as Dandy–Walker cyst or absent left ventricle. Some patients have calcification of the falx cerebri (12,15,20,21,24,30,31,39,40).

**Other abnormalities.** Skeletal defects occurring in about 70% include small orbits, skull asymmetry, abnormal facial bones and sinuses, deformed optic foramen, absent sphenoid wing, parietal lucent defects, and diastasis of the symphysis pubis (20,21,25,28,31,34,35,39). Other low-frequency skeletal anomalies have been summarized by Thomas et al (31).

Umbilical hernia occurs in 15–30% (14,53). Omphalocele or umbilical hernia has been noted in 10% (14,39). Congenital heart defects have been recorded in about 5% and other less commonly occurring anomalies are listed elsewhere (14,31).

**Differential diagnosis.** Isolated cryptophthalmos may be either unilateral or bilateral and may, on occasion, have midline nasal grooving. Approximately two-thirds of these cases are sporadic, one-third occurring as an autosomal dominant trait (31). Brownstein et al (4) reported a patient with bilateral corneal and lenticular opacities, malformed nose, anomalous ears, bilateral renal agenesis, and cryptorchidism. Turner (33) and Winter et al (38) described a syndrome characterized by renal agenesis, genital abnormalities, anomalous ears with stenotic ear canals, ossicle defects and, in some instances, an abnormal nose with grooving of the nasal tip. The cryptophthalmos syndrome should easily be distinguished from *ablepharon-macrostomia* syndrome (26).

**Laboratory aids.** Prenatal diagnosis has been made using fetoscopy and ultrasound (2a,11,14).

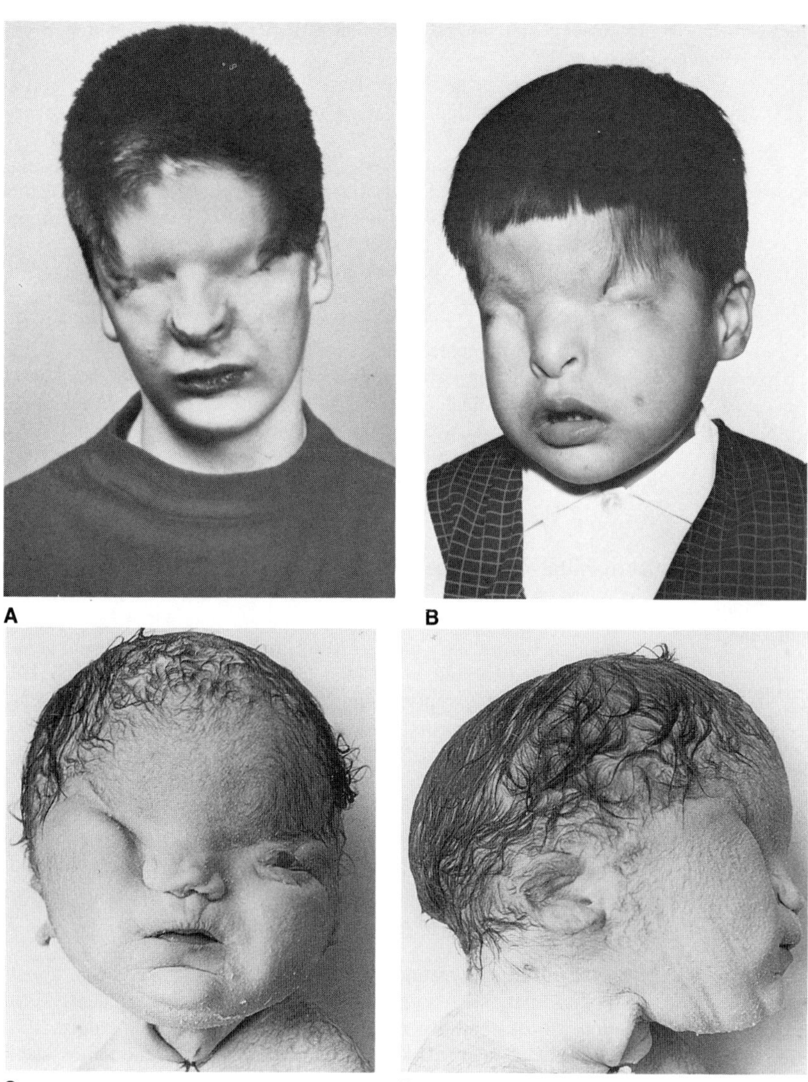


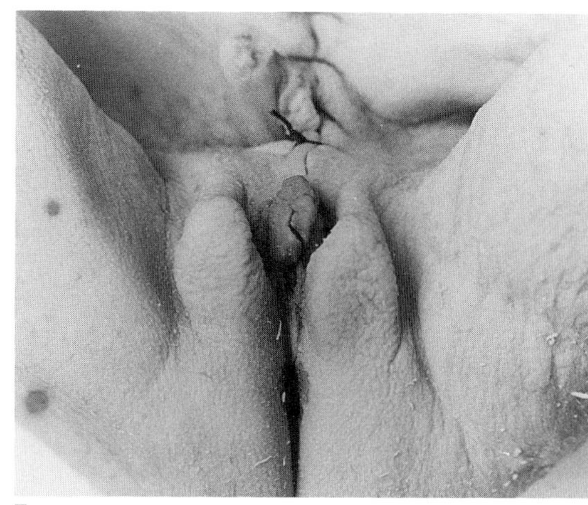


Fig. 24–51. *Cryptophthalmos syndrome.* (A) Skin extends from forehead covering eyes. Note tongue-shaped extensions. (B) Similar alterations in Greenland boy. (C,D) Asymmetric cryptophthalmos. Also note dysmorphic, low-set pinna. (E) Clitoral enlargement with pseudoscrotal enlargement of labia majora and vaginal atresia. [A from *CH Ide* and *PB Wollschlaeger,* Columbia, Missouri. B from *M Warburg,* Birth Defects **7**(3):136, 1971. C–E from *JM Emberger et al,* J Génét Hum (Suppl) **24**:23–29, 1976.]

### References [Cryptophthalmos syndrome (Fraser syndrome)]


1. Azevedo ES et al: Cryptophthalmos in two families from Bahia, Brazil. *J Med Genet* **10**:389–392, 1973.

2. Barry DR, Shortland-Webb WR: A case of cryptophthalmos syndrome. *Ophthalmologica (Basel)* **180**:234–240, 1980.

2a. Boyd PA et al: Fraser syndrome (cryptophthalmos-syndactyly syndrome): A review of eleven cases with postmortem findings. *Am J Med Genet* **31**:159–168, 1988.

3. Brazier DJ et al: Cryptophthalmos: Surgical treatment of the congenital symblepharon variant. *Br J Ophthalmol* **70**:391–395, 1986.

4. Brownstein S et al: Bilateral renal agenesis with multiple congenital ocular anomalies. *Am J Ophthalmol* **82**:770–774, 1976.

5. Burn J, Marwood RP: Fraser syndrome presenting as bilateral renal agenesis in three sibs. *J Med Genet* **19**:360–361, 1982.

6. Butler MG et al: Cryptophthalmos with an orbital cyst and profound mental and motor retardation. *J Pediatr Ophthalmol Strab* **15**:233–235, 1978.

7. Chiari H: Congenitales Ankylo- et Synblepharon und congenitale Atresia laryngis bei einem Kinde mit mehrfachen anderweitigen Bildungsanomalien. *Prag Z Heilk* **4**:143–154, 1883.

8. Codère F et al: Cryptophthalmos syndrome with bilateral renal agenesis. *Am J Ophthalmol* **91**:737–742, 1981.

9. Dinno ND et al: The cryptophthalmos-syndactyly syndrome. *Clin Pediatr* **13**:219–224, 1974.

10. Emberger JM et al: Étude anatomique d'une observation de cryptophtalmie familiale *J Génét Hum (Suppl)* **24**:23–29, 1976.

11. Feldman E et al: Microphthalmia-prenatal ultrasonic diagnosis. *Prenatal Diag* **5**:205–207, 1985.

12. François J: Syndrome malformatif avec cryptophtalmie. *Acta Genet Med (Roma)* **18**:18–50, 1969.

13. Fraser CR: Our genetic 'load.' A review of some aspects of genetical variation. *Ann Hum Genet* **25**:387–415, 1962.

14. Gattuso J et al: The clinical spectrum of the Fraser syndrome: Report of three new cases and review. *J Med Genet* **24**:549–555, 1987.

15. Goldhammer Y, Smith JL: Cryptophthalmos syndrome with basal encephaloceles. *Am J Ophthalmol* **80**:146–149, 1975.

16. Greenberg F et al: Gonadal dysgenesis and gonadoblastoma in situ in a female with Fraser (cryptophthalmos) syndrome. *J Pediatr* **108**:952–954, 1986.

17. Gupta SP, Saxena RC: Cryptophthalmos. *Br J Ophthalmol* **46**:629–632, 1962.

18. Guttmann A: Einseitiger Kryptophthalmos. *Zentralbl Prakt Augenheilkd* **233**:264, 1909.

19. Hancheng Z: Cryptophthalmos: A report on three sibling cases. *Br J Ophthalmol* **70**:72–74, 1986.

20. Howard RO et al: Unilateral cryptophthalmia. *Am J Ophthalmol* **87**:556–560, 1979.

21. Ide CH, Wollschlaeger PB: Multiple congenital abnormalities associated with cryptophthalmia. *Arch Ophthalmol* **81**:640–644, 1969.

22. Kapoor S: Otolaryngological features of 'malformation syndrome with cryptophthalmos.' *J Laryngol Otol* **93**:519–525, 1979.

23. Koenig R, Spranger J: Cryptophthalmos-syndactyly syndrome without cryptophthalmos. *Clin Genet* **29**:413–416, 1986.

23a. Konrad G et al: Bilateraler vollständiger Kryptophthalmus. *Klin Monatsbl Augenheilkd* **190**:121–124, 1987.

24. Levine RS et al: The cryptophthalmos syndrome. *Am J Roentgenol* **143**:375–376, 1984.

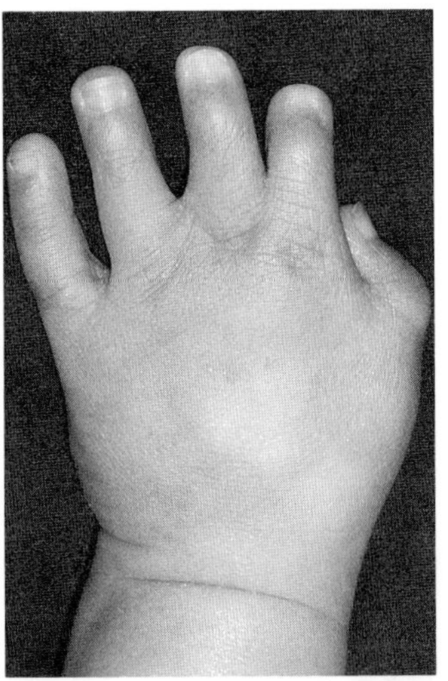
A

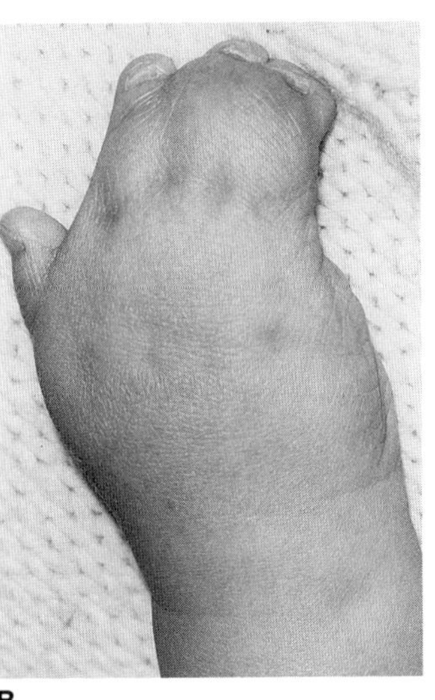
B

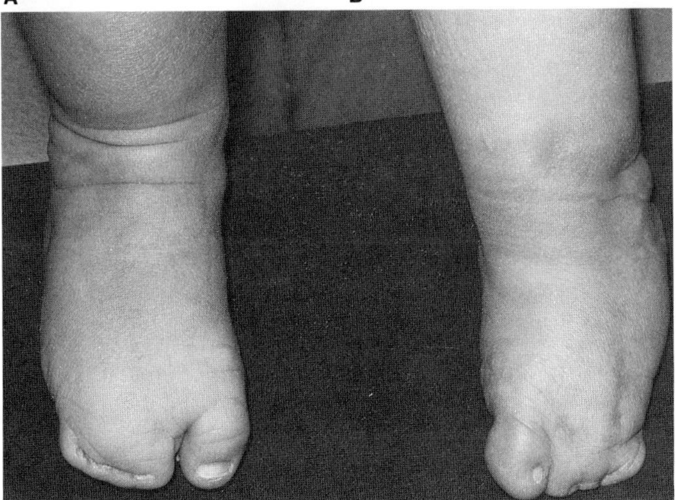
C

Fig. 24–52. *Cryptophthalmos syndrome.* (A,B) Varying degrees of syndactyly. Note similarity of changes in B to Apert syndrome. (C) Marked syndactyly of toes, somewhat resembling that seen in Apert syndrome. [From *N Dinno et al,* Birth Defects **12**(6):109, 1976.]


25. Lurie IW, Cherstvoy ED: Renal agenesis as a diagnostic feature of the cryptophthalmos-syndactyly syndrome. *Clin Genet* **25**:528–532, 1984.

26. McCarthy GT, West LM: Ablepharon macrostomia syndrome. *Dev Med Child Neurol* **19**:659–661, 1977.

27. Mortimer G et al: Fraser syndrome presenting as monozygotic twins with bilateral renal agensis. *J Med Genet* **22**:76–78, 1985.

28. Otradovec J, Janovsky M: Cryptophthalmos. *Čs Oftal* **18**:128–138, 1962.

29. Schönenberg H: Kryptophthalmus-Syndrom. *Klin Pädiatr* **185**:165–172, 1973.

30. Sugar HS: The cryptophthalmos-syndactyly syndrome. *Am J Ophthalmol* **66**:897–899, 1968.

31. Thomas IT et al: Isolated and syndromic cryptophthalmos. *Am J Med Genet* **25**:85–98, 1986.

32. Thorp AT: Bilateral cryptophthalmos and syndactylism. *NC Med J* **6**:484, 1945.

33. Turner G: A second family with renal, vaginal, and middle ear anomalies. *J Pediatr* **76**:641, 1970.

34. Varnek L: Cryptophthalmos, dyscephaly, syndactyly and renal aplasia. *Acta Ophthalmol* **56**:302–313, 1978.

35. Viallefont H et al: Un cas de cryptophtalmie. *Bull Soç Ophtal Fr* **65**:329–332, 1965.

36. Walbaum R et al: Un cas de syndrome cryptophtalmie-syndactylie. *J Génét Hum Suppl* **24**:31–34, 1976.

37. Waring GO, Shields JA: Partial unilateral cryptophthalmos with syndactyly, brachydactyly, and renal anomalies. *Am J Ophthalmol* **79**:437–440, 1975.

38. Winter JSD et al: A familial syndrome of renal, genital, and middle ear anomalies. *J Pediatr* **72**:88–93, 1968.

39. Zehender W: Eine Missgeburt mit hautüberwachsenen Augen oder Kryptophthalmus. *Klin Monatsbl Augenheilkd* **10**:225–234, 1872.

40. Zinn S: Cryptophthalmia. *Am J Ophthalmol* **40**:219–223, 1955.


## Nasal alar colobomas, mirror hands and feet, and talipes

Supernumerary digits of hands and feet with syndactyly and duplication of carpal and tarsal bones (mirror hands and feet) in combination with talipes equinovarus, dislocated knees, and hypoplasia and colobomas of the nasal alae and columella were documented in a father and daughter by Sandrow et al (1) (Fig. 24–53A–C). A similarly affected father and daughter were seen by KL Jones (personal communication, 1989).

### Reference (Nasal alar colobomas, mirror hands and feet, and talipes)


1. Sandrow RE et al: Hereditary ulnar and fibular dimelia with peculiar facies. *J Bone Jt Surg* **52A**:367–370, 1970.


## Romberg syndrome (progressive hemifacial atrophy)

Romberg syndrome consists of slowly progressive atrophy of the soft tissues of essentially half the face accompanied most frequently by contralateral Jacksonian epilepsy, trigeminal neuralgia, and changes in the eyes and hair. Romberg (27) is usually credited for delineation of the disorder, but Parry (23) discussed the disorders as early as 1825. The condition has been called progressive facial hemiatrophy (1,24,33,38–40), hemifacial atrophy (3,11,13,19), and Parry–Romberg syndrome (16). Comprehensive surveys have been carried out by a number of authors (1,3,6,8,16,17,20,25,36,37). Surgical aspects are particularly well covered by Rees (25,26).

Nearly all cases have been sporadic, but several familial instances have been recorded (12,15,38). Familial aspects have been particularly well reviewed by Lewkonia and Lowry (16). Theories of pathogenesis have been numerous (1,37), most emphasis being placed on alterations in the peripheral trophic sympathetic system (38). Some affected individuals have had a history of prior trauma (7,14). Moss and Crikelair (18) carried out unilateral cervical sympathectomy in rats and produced a condition that resembled hemifacial atrophy in man.

**Face, skin, and hair.** In advanced cases, the face is quite distinct (Fig. 24–54). The ear may become misshapen and smaller than normal or, because of lack of supporting tissues, may project from the head (13). Basilar kyphosis, an alteration in the basal skull angle, has been described (7). An extensive cephalometric study was carried out by Berkman (3). Early facial change, usually appearing during the first decade (Fig. 24–54), involves the paramedian area of the face and slowly spreads, resulting in atrophy of underlying muscle, bone, and cartilage. First to be involved is usually the area covered by the temporal or buccinator muscles. The process extends to involve the brow, angle of the mouth, neck, or even half the body (1,37) (Fig. 24–55). There is a marked predilection for left-sided involvement of the face. The overlying skin often becomes darkly pigmented. The condition slowly progresses for several years (about three) and then usually becomes stationary for life (7,13,30).

There has been a long-standing debate about the relationship between progressive hemifacial atrophy and scleroderma, several authors stoutly defending the position that the coup-de-sabre form of scleroderma is only a special type of progressive hemifacial atrophy (1,36,38). Others indicate that the two conditions may occur simultaneously (21). Lewkonia and Lowry (16) suggested that progressive hemifacial atrophy be considered an anatomically limited form of linear scleroderma

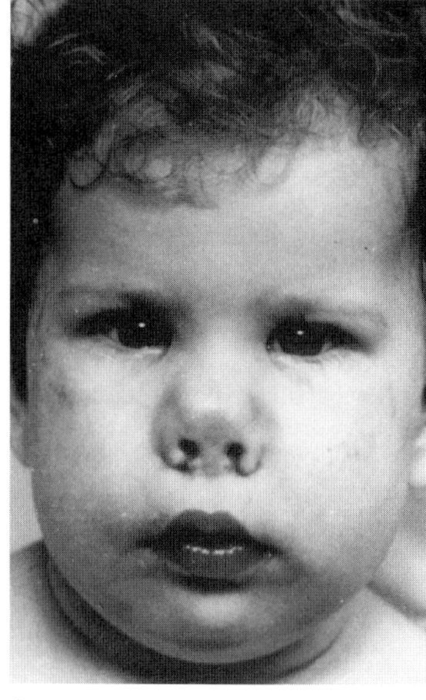

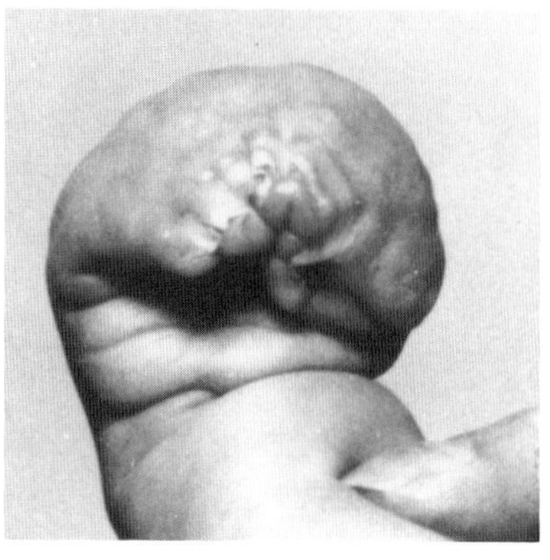

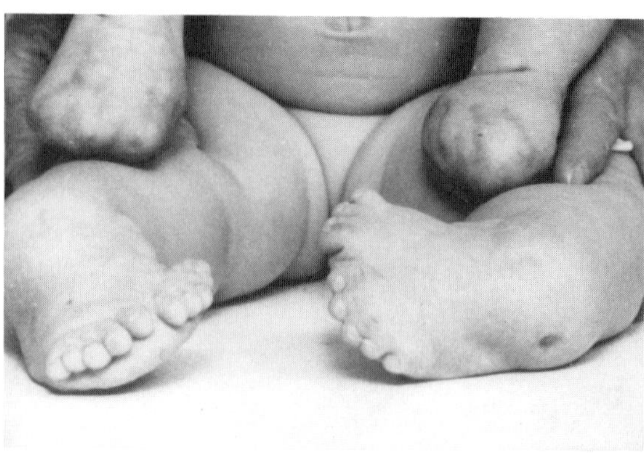

A B C

Fig. 24–53. *Nasal alar colobomas, mirror hands and feet, and talipes.* (A) Unusual facies marked by short nose, hypoplasia of nasal alae, and defective columella. (B) Mirror duplication of hands and extensive syndactyly leads to rosebud appearance. (C) Similar mirror duplication in feet. Note deep skin dimple, and talipes equinovarus. (Courtesy of *HH Steele,* Philadelphia, Pennsylvania.)

without extensive limb or trunk involvement. It should be pointed out that, in some instances, the skin is spared while only fat and subcutaneous tissue disappears. Occasionally, vitiligo accompanies the disorder (31).

Changes in the hair may precede those of the skin (1,38,39). The scalp on the affected side may exhibit circumscribed but complete alopecia limited to the paramedian area, eyelashes, and median portion of the eyebrows (38). Poliosis, or blanching of the hair, has also been noted (37).

**Ocular manifestations.** These are particularly well documented by Muchnick et al (19). Loss of periorbital fat results in enophthalmos (36). Underlying bone may disappear causing the outer canthus to be displaced downward. Muscular paralysis has been noted by several authors (10,12), as well as lagophthalmos and ptosis (1,10,38). Horner syndrome, heterochromia iridis, upper eyelid coloboma, and dilated and fixed pupil have also been reported. Common inflammatory processes involving the eyes include neuroparalytic keratitis, iritis, iridocyclitis, choroiditis, and cataract (12,37). In some instances, ocular findings may affect the eye on the contralateral side (12).

**Oral manifestations.** Atrophy of half of the upper lip and tongue are characteristic (Fig. 24–56). Maxillary teeth on the involved side are exposed. Spontaneous fracture on the affected side of the mandible has also been noted (4,10,35,39).

Roentgenographically, the body and ramus of the mandible are shorter on the involved side, and delayed development of the mandibular angle may be observed, resulting in malocclusion. Teeth on the affected

Fig. 24–54. *Romberg syndrome.* (A–C) Increased severity of disorder in patients illustrated from left to right. (A,B, from *D Glass,* Br J Oral Surg **1**:194, 1963. C courtesy of *P Cernea,* Paris, France.)

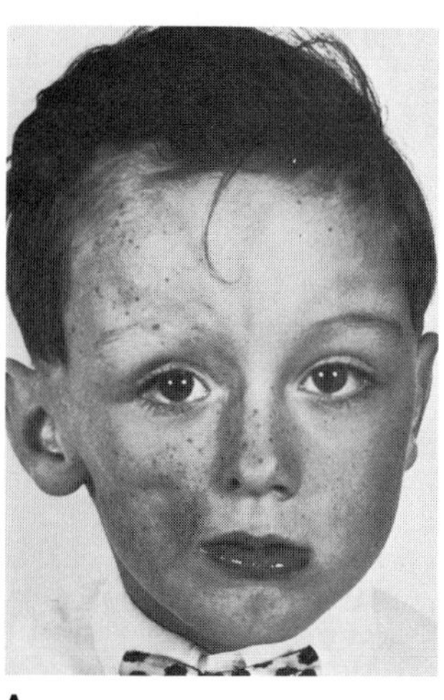

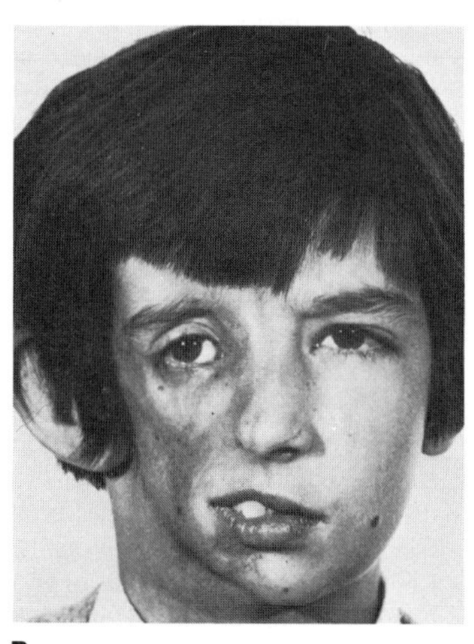

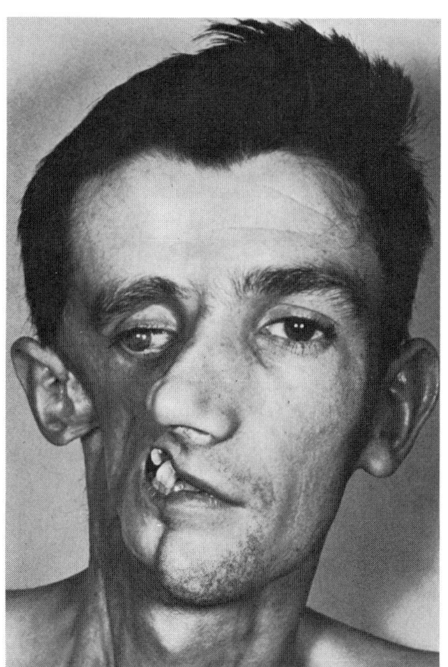

A B C

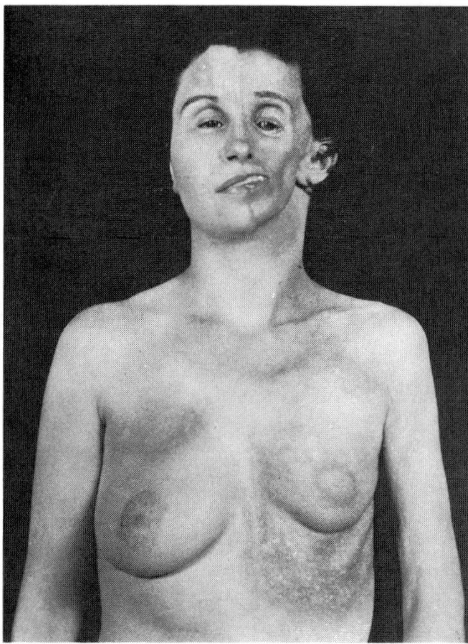

Fig. 24–55. *Romberg syndrome.* Involvement of half of body. (From *AJ Barsky et al,* Principles and Practice of Surgery, 2d ed, McGraw-Hill, New York, 1964.)

side occasionally are delayed in eruption or have atrophic roots (11,13,28,29,36).

**Central nervous system.** The most common neurologic finding is epilepsy, often of the sensory Jacksonian type, which frequently appears late (1,1a,9,24,35,37,39). However, trigeminal neuralgia and/or facial paresthesia may appear early, preceding other neurologic changes. Migraine has also been a common finding (9,37,39). Paradise et al (22) reported a child with progressive facial hemiatrophy, seizures, and Ewing's sarcoma.

**Differential diagnosis.** Asymmetric facial deformities have a diversity of manifestations and causes and these have been discussed by Souyris et al (34). Congenital facial hypoplasia can be diagnosed by its presence at birth and diminution of tooth size on the affected side following eruption (2,5,29). Scleroderma (32), fat necrosis, and *oculoauriculovertebral spectrum* should also be considered.

Fig. 24–56. *Romberg syndrome.* Atrophy of right side of tongue.

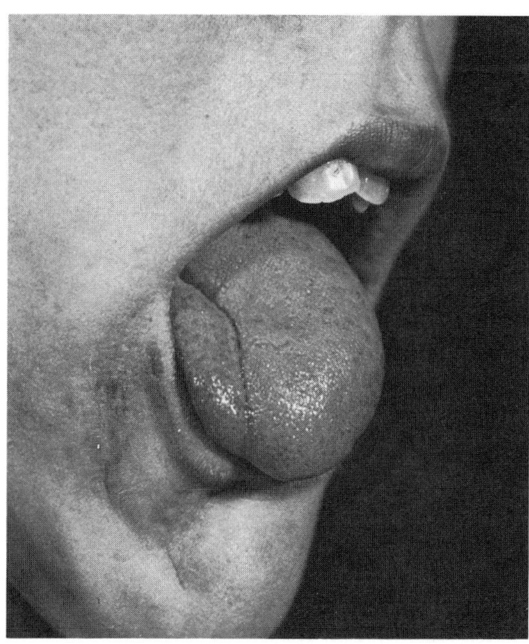

### References [Romberg syndrome (progressive hemifacial atrophy)]


1. Archambault L, Fromm NK: Progressive facial hemiatrophy. *Arch Neurol Psychiatr* **27**:529–584, 1932.

1a. Asher SW, Berg BO: Progressive hemifacial atrophy. *Arch Neurol* **39**:44–49, 1982.

2. Bates JF: Unilateral facial hypoplasia. *Br Dent J* **104**:453–454, 1958.

3. Berkman MD: Craniofacial changes in progressive hemifacial atrophy. Thesis, Columbia University, 1972.

4. Bromley P, Forbes A: A case of progressive hemiatrophy presenting with spontaneous fractures of the lower jaw. *Br Med J* **1**:1476–1478, 1960.

5. Burke PH: Unilateral facial hypoplasia affecting tooth size. *Br Dent J* **103**:41–44, 1957.

6. Calmettes L et al: L'hemiatrophie faciale (maladie de Romberg) et ses manifestations oculaires. *Rev Otoneuroophtalmol* **31**:215–241, 1959.

7. Crikelair GF et al: Facial hemiatrophy. *Plast Reconstr Surg* **29**:5–13, 1962.

8. Dechaume M et al: L'hemiatrophie faciale progressive. *Rev Stomatol (Paris)* **55**:12–44, 1954.

9. Dieler T: A case exhibiting symptoms of facial hemiatrophy and Jacksonian sensory epilepsy. *J Nerv Ment Dis* **20**:284–289, 1895.

10. Finesilver B, Rosow H: Total hemiatrophy. *JAMA* **110**:366–368, 1938.

11. Foster TD: The effects of hemifacial atrophy on dental growth. *Br Dent J* **146**:148–150, 1979.

12. Franceschetti A, Koenig H: L'importance du facteur heredodegeneratif dans l'hemiatrophie faciale progressive (Romberg). Étude des complications oculaires dans ce syndrome. *J Génét Hum* **1**:27–64, 1952.

13. Glass D: Hemifacial atrophy. *Br J Oral Surg* **1**:194–199, 1963–1964.

14. Hall BD, Lessing S: Progressive hemifacial atrophy (Romberg syndrome) with "coup de sabre" forehead and scalp lesions: Report of two cases. *Proc Greenwood Genet Ctr* **6**:100–101, 1987.

15. Klingman T: Facial hemiatrophy. *JAMA* **49**:1888–1891, 1907.

16. Lewkonia RM, Lowry RB: Progressive hemifacial atrophy (Parry–Romberg syndrome) report with review of genetics and nosology. *Am J Med Genet* **14**:385–390, 1983.

17. Meline F: Hemiatrophie faciale progressive. *Ann Chir Plast* **7**:49–60, 1962.

18. Moss ML, Crikelair GF: Progressive facial hemiatrophy following cervical sympathectomy in the rat. *Arch Oral Biol* **1**:254–258, 1960.

19. Muchnick RS et al: Ocular manifestations and treatment of hemifacial atrophy. *Am J Ophthalmol* **88**:889–897, 1979.

20. Mussinelli G et al: L'emiatrophia facciale progressiva. Malattia de Parry–Romberg. *Rev Otoneuroophtalmol* **39**:3–52, 283–303, 477–526, 583–638, 1964; **40**:105–197, 1965.

21. Osborne ED: Morphea associated with hemiatrophy of the face. *Arch Dermatol Syph (Chic)* **6**:27–34, 1922.

22. Paradise JE et al: Progressive facial hemiatrophy. *Am J Dis Child* **134**:1065–1067, 1980.

23. Parry CH: Collections from unpublished papers, Vol I, Underwood, London, 1825, p 478.

24. Pollock LJ: Progressive facial hemiatrophy. *Arch Neurol Psychiatr* **33**:888–889, 1935.

25. Rees TD: Facial atrophy. *Clin Plast Surg* **3**:637–646, 1976.

26. Rees TD et al: Silicone fluid injections for facial atrophy. A ten-year study. *Plast Reconstr Surg* **52**:118–127, 1973.

27. Romberg MH: Trophoneurosen. *Klinische Ergebnisse.* A. Forstner, Berlin, 1846, pp 75–81.

28. Rushton MA: An early case of facial hemiatrophy. *Oral Surg* **4**:1457–1460, 1951.

29. Rushton MA: Asymmetry of tooth size in congenital hypoplasia of one side of the body. *Br Dent J* **95**:309–311, 1953.

30. Schnall BS, Smith DW: Nonrandom laterality of malformations in paired structures. *J Pediatr* **85**:509–511, 1974.

31. Schneider G: Ein Beitrag zur Hemihypertrophia und hemiatrophia faciei. *Dtsch Zahn Mund Kieferheilkd* **12**:43–66, 1949.

32. Singh G, Bajpai HS: Progressive facial hemiatrophy. *Dermatologica* **138**:288–291, 1969.

33. Smith B, Guberina C: Coloboma in progressive hemifacial atrophy. *Am J Ophthalmol* **84**:85–89, 1977.

34. Souyris F et al: Facial asymmetry of developmental etiology. *Oral Surg* **56**:113–124, 1983.

35. Tauber EB, Goldman L: Hemiatrophia faciei progressiva. *Arch Dermatol Syph* **39**:696–704, 1939.

36. Walsh FB: Facial hemiatrophy. *Am J Ophthalmol* **22**:1–10, 1939.

37. Wartenberg R: Progressive facial hemiatrophy. *Arch Neurol Psychiatr* **54**:75–96, 1945.

38. Wolfe MO, Weber ML: Progressive facial hemiatrophy. *J Nerv Ment Dis* **91**:595–607, 1940.

39. Wolff HG: Progressive facial hemiatrophy. *Arch Otolaryngol* **7**:580–582, 1928.

## Leprechaunism (Donohue syndrome)

Described originally by Donohue (10) in 1948, the syndrome consists of failure to thrive, unusual facies, sexual precocity, retarded bone age, and insulin resistance with glucose intolerance and hyperinsulinemia. The term "leprechaunism" was introduced by Donohue and Uchida (11) in 1954.

The disorder has autosomal recessive inheritance, but there has been a 2:1 female predilection (4). Affected sibs (9,30,31) and increased (30%) parental consanguinity (4,5a,8,9,11,21,30,32) have been noted. About 50 examples have been reported (4,13).

The basic defect appears to be in a component of the cellular insulin receptor that is common to both the receptor and the insulin-like growth factor. The component apparently is involved in coupling both receptors to the glucose incorporation pathway (28), a mechanism that could explain both the laboratory and clinical findings. Maassen et al (24b) described a leucine to proline mutation in the insulin receptor. Elsas et al (13,13a) opined that two different recessive mutations impair high-affinity insulin receptor binding and that leprechaunism represents a compound heterozygote for these mutations. Target cell resistance to insulin varies according to age of onset. In leprechaunism, there is severe insulin resistance that results in small babies with almost an entire absence of subcutaneous fat and poor muscular development, poor early growth, and early death (28).

Gestation tends to be short. Nearly all infants have been marasmic, and over 60% have died at less than 2 years of age (4,5,12,18,31).

Fig. 24–57. *Leprechaunism.* Note hirsutism, large hands, feet, and penis.

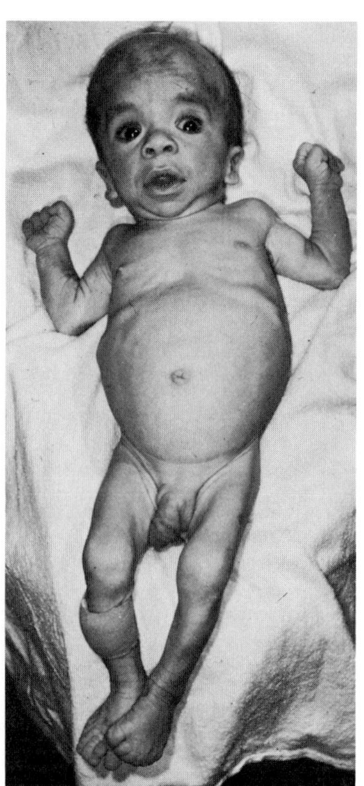

**Facies.** The face is gaunt, with large ears and lips. Facial hirsutism may be marked (11,21,24,30–33) (Fig. 24–57).

**Musculoskeletal changes.** Muscle mass is severely wasted. The abdomen is distended. Bone age is usually retarded (1). The hands and feet appear disproportionately large (3,8,11,14,30,32) (Fig. 24–57).

**Skin.** Acanthosis nigricans has been observed in most cases (3,12,14,24,30,31). Hirsutism is usually marked. Periorificial skin rugosities and dysplastic nails have been noted (31).

**Endocrine changes.** The breasts and clitoris or penis have been enlarged in over 50% and 75%, respectively (4). There have been cystic alterations in the ovaries or testes (1,5,11,14,18,21,31). Rogers (29) described basophilic hyperplasia of the pituitary gland. Pancreatic islet cell hyperplasia (nesidioblastosis) with low fasting blood sugar level is common (1,11,14,17,31,32,34).

**Oral manifestations.** Large mouth, thick lips, and gingival hypertrophy are present at birth (9,12,30,31). Roth's (31) patient at age 5 had a dental age of 14 years. Whether other children had advanced dental age is not known.

**Other findings.** Hepatic cholestasis and fibrosis with excessive deposits of iron in the liver are frequent findings. There is paucity of lymphatic tissue (tonsils, thymus, mesenteric nodes, Peyer patches). The gut is often dilated. Microscopic foci of nephrocalcinosis have been noted in several cases (25).

**Differential diagnosis.** It is possible that the condition has been overdiagnosed, since many marasmic infants or those with emotional deprivation may develop a similar phenotype (20). There is clinical resemblance between leprechaunism and *lipoatrophic diabetes.* These similarities may possibly be explained by shared mechanisms involving both insulin and insulin-like growth factor cell receptors (28).

The Patterson–David syndrome has been confused with leprechaunism. The patients of both Patterson (26,27) and David (6,7) presented large redundant loose folds of skin of the hands and feet since birth, marked generalized bronzed hyperpigmentation, hirsutism largely confined to the limbs, severe mental retardation, bony deformities, and skeletal dysplasia (Fig. 24–58A,B). The bony alterations are characterized by profound disorder of endochondral ossification with marked retardation of skeletal maturation (6,7). The ends of long bones are swollen, especially those of the knee, ankles, and wrists (Fig. 24–59A–C). Similar changes can be observed in hands and feet. The thickening and deformity affect other bones, the calvaria, maxilla, ethmoids, and mandibular condyles. The occipital bone may exhibit a defect. The vertebral bodies are deformed as well as ribs, clavicles, and scapulae. Kyphoscoliosis, genua valga, and valgus deformities of the ankles are common features. The skeletal changes increase in severity with age (6,7). David's (6,7) patient had premature adrenarche.

**Laboratory aids.** Biochemical evidence for insulin resistance includes markedly elevated plasma insulin, impaired glucose homeostasis, absence of antiinsulin receptor antibodies, and altered insulin–receptor binding and/or response to insulin by patients' cells (9,12,13,19,22,23,28,30,31,35,36). Chromosomal alterations have been reported in two infants with the syndrome (2,15). The patients reported by Roth (31) was further seen at age 8 by Frindik et al (16) and was found to have markedly elevated urinary levels of epidermal growth factor.

The insulin-binding defect has allowed for prenatal diagnosis (24a).

### References [Leprechaunism (Donohue syndrome)]

1. Adams JM et al: Leprechaunism (Donohue's syndrome) in a low birth weight infant. *South Med J* **70**:998–1001, 1977.

2. Ayraud N et al: Délétion interstitielle du bras long d'un chromosome 7 chez une enfant lepréchaune. *Ann Génét* **19**:265–268, 1976.

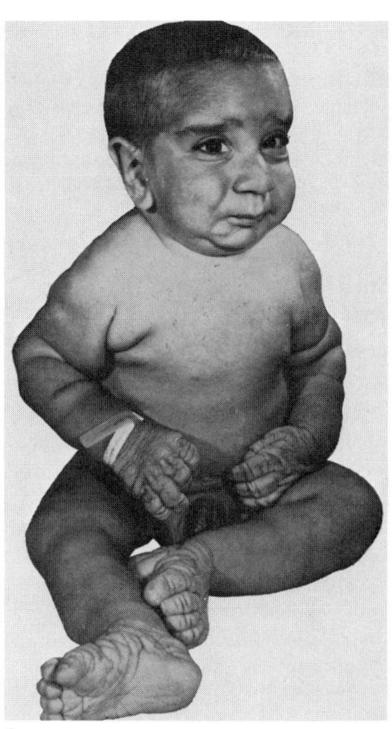
A

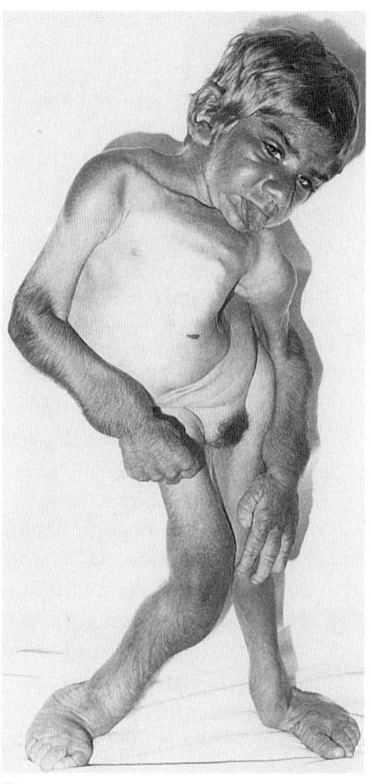
B

Fig. 24–58. *Patterson–David syndrome.* (A) Large hands and feet in 3-year-old male, with redundancy and wrinkling of overlying skin. (B) Mentally retarded 12-year-old with coarse facies, bronze pigmentation, hirsutism, large hands and feet with redundant skin, premature sexual maturity, seizures, and generalized skeletal dysplasia, especially affecting the metaphyses and epiphyses. [A from *JH Patterson,* Birth Defects **5**(4):117, 1969. B from *TJ David et al,* J Med Genet **18**:294, 1981.]

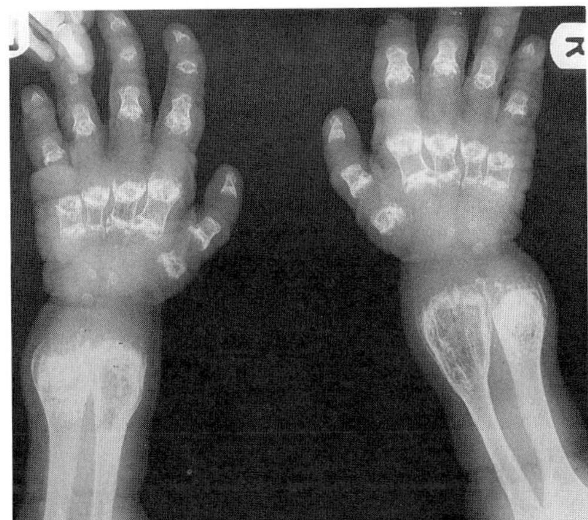
A

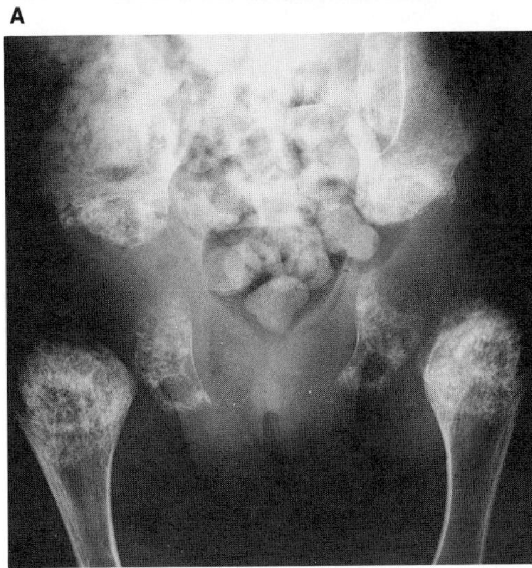
B

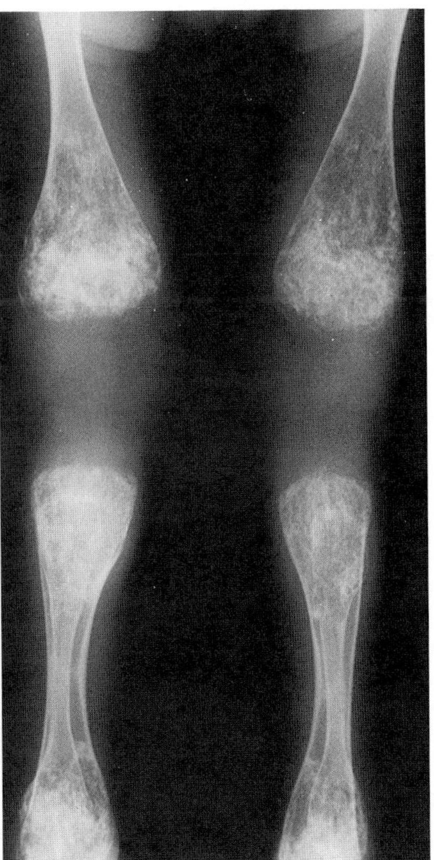
C

Fig. 24–59. *Patterson–David syndrome.* (A–C) Swollen metaphyses, late developing epiphyses, delayed bone age, especially involving the knees, ankles, and wrists. (From *TJ David et al,* J Med Genet **18**:294, 1981.)

3. Canlorbe P et al: Le léprechaunism: Nouvelle observation et review de la littérature. *Ann Pédiatr* **15**:282–291, 1968.

4. Cantani A et al: A rare polydysmorphic syndrome. Leprechaunism. Review of the 49 cases reported in the literature. *Ann Génét* **30**:221–227, 1987.

5. Dallaire L: Leprechaunism. *Birth Defects* **5**(4):121, 1969.

5a. David R, Goodman RM: Leprechaunism in four sibs: New pathologic and electron microscopic findings. *Hum Hered* **27**:172A, 1977.

6. David TJ: Skeletal dysplasia, hyperpigmentation, cutis laxa, endocrine abnormality, and mental retardation—the Patterson syndrome. *Prog Clin Biol Res* **104**:331–337, 1982.

7. David TJ et al: The Patterson syndrome, leprechaunism and pseudoleprechaunism. *J Med Genet* **18**:294–298, 1981.

8. Der Kaloustian VM et al: Leprechaunism: A report of two new cases. *Am J Dis Child* **122**:442–445, 1971.

9. D'Ercole AJ et al: Leprechaunism: Studies of the relationship among hyperinsulinemia, insulin resistance, and growth retardation. *J Clin Endocrinol* **48**:495–502, 1979.

10. Donohue WL: Dysendocrinism. *J Pediatr* **32**:739–748, 1948.

11. Donohue WL, Uchida I: Leprechaunism, a euphemism for a rare familial disorder. *J Pediatr* **45**:505–518, 1954.

12. Elders MJ et al: Endocrine-metabolic relationships in patients with leprechaunism. *Nat Med Assoc J* **74**:1195–1210, 1982.

13. Elsas LJ et al: Leprechaunism: An insulin defect in a high-affinity insulin receptor. *Am J Hum Genet* **37**:73–88, 1985.

13a. Endo F et al: Structural analysis of normal and mutant insulin receptors in fibroblasts cultured from families with leprechaunism. *Am J Med Genet* **41**:402–417, 1987.

14. Evans PR: Leprechaunism. *Arch Dis Childh* **30**:479–483, 1955.

15. Ferguson-Smith MA et al: An abnormal metacentric chromosome in an infant with leprechaunism. *Ann Génét* **11**:195–200, 1968.

16. Frindik JP et al: Phenotypic expression in Donahue syndrome (leprechaunism): A role for epidermal growth factor. *J Pediatr* **107**:428–430, 1985.

17. Fürste HO et al: Leprechaunism (Donohue-Syndrom). *Helv Paediatr Acta* **39**:95–104, 1984.

18. Gillerot Y et al: Leprechaunism (Donahue syndrome): A case with growth hormone deficiency. *Acta Paediatr Belg* **32**:279–282, 1979.

19. Grigorescu F et al: Defects in insulin binding and autophosphorylation of erythrocyte insulin receptors in patients with syndromes of severe insulin resistance and their parents. *J Clin Endocrinol Metab* **64**:549–556, 1987.

20. Hopwood NJ, Powell GF: Emotional deprivation: Report of a case with features of leprechaunism. *Am J Dis Child* **127**:892–894, 1974.

21. Kálló A et al: Leprechaunism. *J Pediatr* **66**:372–379, 1965.

22. Kashiwa H et al: Insulin resistance in an infant with leprechaunism. *Acta Paediatr Scand* **73**:701–704, 1984.

23. Kobayashi M et al: Insulin resistance due to a defect distal to the insulin receptor: Demonstration in a patient with leprechaunism. *Proc Natl Acad Sci USA* **75**:3469–3473, 1978.

24. Kuhlkamp F, Helwig H: Das Krankheitsbild des kongenitalen Dysendokrinismus oder Leprechaunismus. *Z Kinderheilkd* **109**:50–63, 1970.

24a. Lindhout D et al: Leprechaunism: From phenotype to genotype. *David W. Smith Conference on Malformations and Morphogenesis*. Madrid, Spain, May, 1989.

24b. Naassen JA et al: A leucine to proline mutation in the insulin receptor leading to insulin resistance. *March of Dimes Clinical Genetics Conference*, Boston, 9–12, 1989.

25. Ordway NK, Stout LC: Intrauterine growth retardation, jaundice and hypoglycemia in a neonate. *J Pediatr* **83**:867–874, 1973.

26. Patterson JH: Presentation of a patient with leprechaunism. *Birth Defects* **5**(4):117–121, 1969.

27. Patterson JH, Watkins WL: Leprechaunism in a male infant. *J Pediatr* **60**:730–739, 1962.

28. Rechler MM: Leprechaunism and related syndromes with primary insulin resistance: Heterogeneity of molecular defects. *Prog Clin Biol Res* **97**:245–281, 1982.

29. Rogers DR: Leprechaunism (Donohue's syndrome): A possible case with emphasis on changes in the adenohypophysis. *Am J Clin Pathol* **45**:614–619, 1966.

30. Rosenberg AM et al: A case of leprechaunism with severe hyperinsulinemia. *Am J Dis Child* **134**:170–175, 1980.

31. Roth SI et al: Cutaneous manifestations of leprechaunism. *Arch Dermatol* **117**:531–535, 1981.

32. Salmon MA, Webb JN: Dystrophic changes associated with leprechaunism in a male infant. *Arch Dis Childh* **38**:530–535, 1963.

33. Summitt RL, Favara B: Leprechaunism (Donohue's syndrome). *J Pediatr* **74**:601–610, 1969.

34. Tsujino G, Yoshinaga T: A case of leprechaunism as an analysis of some clinical manifestations of this syndrome. *Z Kinderheilkd* **118**:347–360, 1975.

35. Van Obberghen-Schilling EE et al: Receptors for insulin-like growth factor I are defective in fibroblasts cultured from a patient with leprechaunism. *J Clin Invest* **68**:1356–1365, 1981.

36. Vesely DL et al: Decreased cyclic guanosine 3′,5′ monophosphate and guanylate cyclase activity in leprechaunism: Evidence for a postreceptor defect. *Pediatr Res* **20**:329–331, 1986.

## Lipoatrophic diabetes (Berardinelli syndrome, Seip syndrome, generalized lipodystrophy)

The syndrome consists of generalized disappearance of body fat, increased rate of skeletal growth, acanthosis nigricans, enlarged external genitalia, hepatomegaly, and insulin-resistant diabetes. Although Berardinelli (2) and Seip (27) are usually credited with first delineating the disorder, Miescher (21), in 1921, clearly described the condition in sibs. For a historic review, see Seip (28).

The syndrome has autosomal recessive inheritance. Parental consanguinity has been noted in about 25% (2,3,28). The sublethal nature of the gene is suggested by the frequent history of miscarriage or death of affected children within the neonatal period.

The disorder may be due to disturbed insulin binding to membrane receptors, but possibly is of secondary origin, that is, disruption by a circulating factor (8,15,24,26). The best evidence suggests that the mechanism is via an autoantibody to the membrane receptor site for insulin.

A generalized loss of all subcutaneous fat causes the muscles to appear enlarged and the veins to stand out prominently. This is usually evident at birth (28). There is also absence of mesenteric and perinephric fat (28).

Females have a definitely masculinized body build, although the lipodystrophy does not affect breast development when the onset is after puberty. Skeletal growth is often accelerated during the first 10 years of life, during which time patients achieve 90% of their growth. However, growth soon tapers off, and they have average or even short stature after puberty. The phenotype is further altered by the enlarged joints of the hands and feet.

Long-term prognosis is rather poor, mainly due to diabetes with its complications (29).

**Facies.** The facies is quite distinctive. Loss of subcutaneous adipose tissue, especially the buccal fat pad, causes the cheeks to have a gaunt appearance (Fig. 24–60A,B). This, combined with large ears, hypertelorism, and hirsutism, is characteristic.

**Viscera.** Hepatomegaly with abdominal prominence may be pronounced in infancy; liver biopsies, taken later in life, have shown fatty infiltration with moderate early fibrosis and increased gylcogen deposits. The fatty infiltration is secondary to hypertriglyceridemia and to lack of functional peripheral fat storage depots.

Cardiac murmurs and cardiomegaly are not uncommon (27,28). In most patients, the kidneys are enlarged without apparent histologic cause.

**Nervous system.** Mental retardation of variable degree has been noted in about 50% (3,22,27,28,30).

**Musculoskeletal alterations.** Muscular prominence, due in part to lack of subcutaneous fat, may be absolute, as some patients exhibit increased urinary creatinine levels (Fig. 24–61A,B). Increased skeletal maturation is a common feature during the first 4 years of life, and several patients have exhibited increased bone density (9,28,32). These strongly suggest an androgen excess. An increase in subcutaneous fat may occasionally be noted after puberty. Cystic angiomatosis of bone has been found in a few families (4,11,31).

**Genital organs.** Enlarged penis or clitoris and polycystic ovaries are common features (2,12,28,30). Oligomenorrhea is frequent.

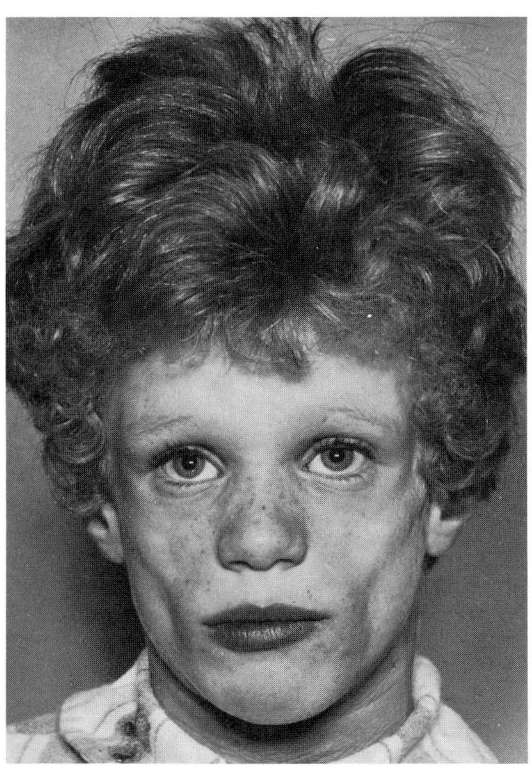

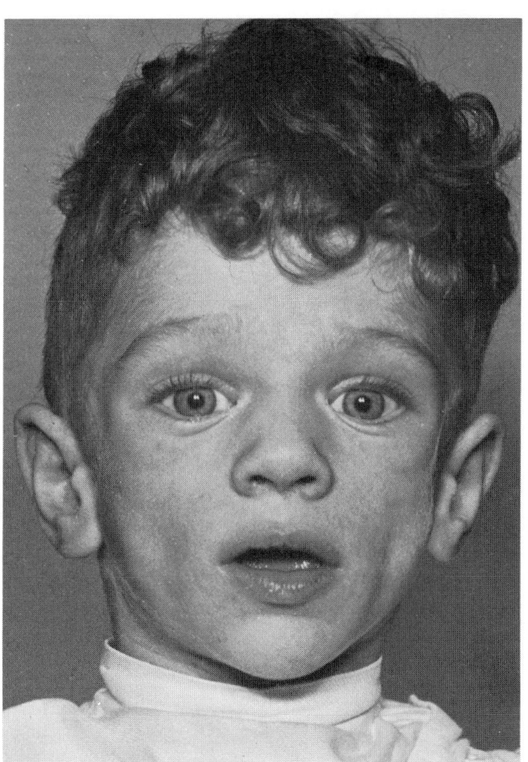

A B

Fig. 24–60. *Lipoatrophic diabetes.* (A,B) Absence of subcutaneous fat, especially buccal fat pad, produces gaunt appearance.

Fig. 24–61. *Lipoatrophic diabetes.* (A,B) Marked muscular prominence, due in part to lack of subcutaneous fat. Also note evidence of axillary acanthosis nigricans in female. Hirsutism may be marked in some patients. (From *M Seip,* Acta Paediatr Scand **48**:555, 1959.)

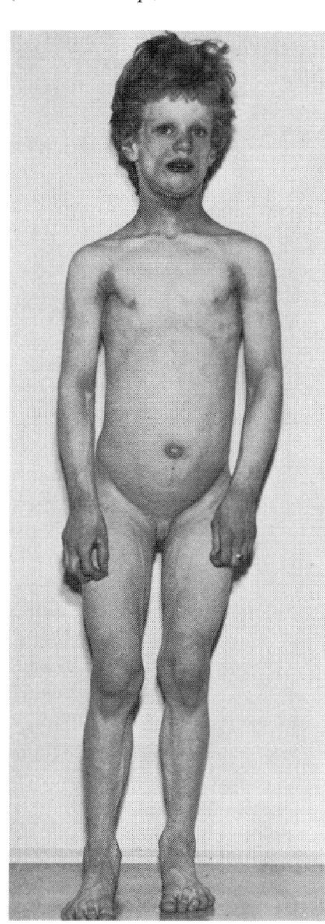

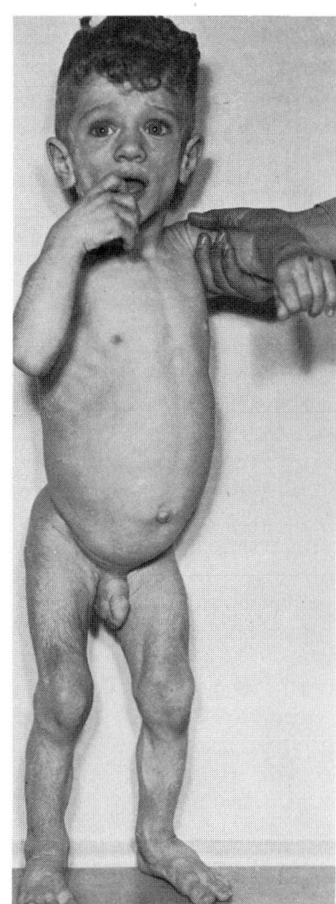

A B

**Skin.** Hirsutism of face, neck, arms, and legs may be present at birth and increases with age. With the onset of the lipodystrophy, the scalp hair may become excessively curly and thick, the hair growing nearly to the eyebrows.

Acanthosis nigricans is a prominent feature in almost all patients. It is especially common in the axillae and on the wrists and ankles. It tends to diminish with increasing age and may disappear after puberty (3) (Fig. 24–62).

Subcutaneous angiomatosis has been reported (4).

**Oral manifestations.** Enlarged tonsils and adenoids are common in the congenital type (3,28). The reported association with sicca syndrome is probably aleatory (13).

**Differential diagnosis.** The Dunnigan syndrome exhibits many overlapping features (*acanthosis nigricans,* decreased subcutaneous fat, enlarged clitoris, insulin-resistant diabetes, autosomal recessive inheritance), but with additional components such as thickened nails, pineal hyperplasia, premature eruption of teeth, open bite and macrodontia of the upper central incisors with palatal cusps arising from the cingulum, and enlargement of the filiform and fungiform papillae of the tongue (1,5,10,20,25,33,34).

In the so-called partial or acquired lipodystrophy, there is symmetric disappearance of facial fat, the patient appearing gaunt and cadaverous (Voltaire-like) (16). Fat may also disappear from the arms, chest, abdomen, and hips. The characteristic adiposity of pelvis and legs is seen only in postpubertal females (30). We are not convinced that there is other than relative adiposity of the buttocks. Onset is usually noted between 5 and 15 years of age. The wasting time ranges between 1.5 and 6 years. Renal problems (mesangiocapillary glomerulonephritis progressing to chronic sclerosing glomerulonephritis) have been noted in 25–50% about 5–20 years after the appearance of the lipodystrophy (30). The ratio of females to males is 4:1 (3). The disorder is sporadic. It is more common than lipoatrophic diabetes. A

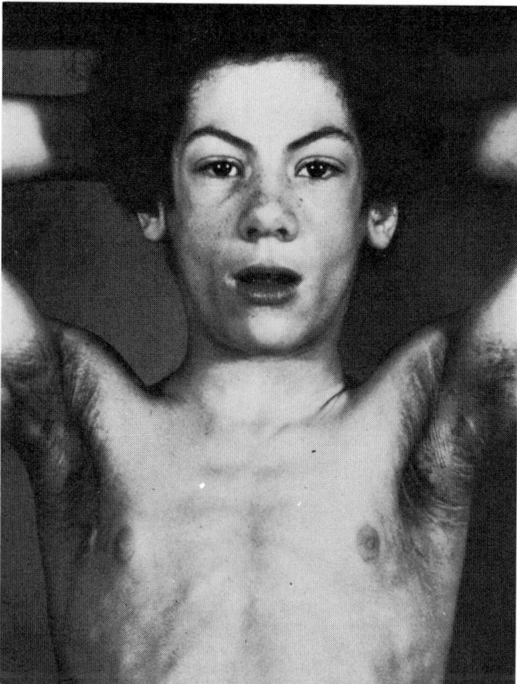

Fig. 24–62. *Lipoatrophic diabetes*. Acanthosis nigricans. (Courtesy of *W Reed*, Burbank, California.)

deficiency in the third component of complement (C3) and C3 activating factor is probably responsible for the disorder (16,19,35).

There are two X-linked dominant, lethal in the male, forms of partial lipodystrophy with onset at puberty. In one, the fat loss is confined to the limbs, sparing the face and trunk. In the other, the trunk is also affected but the vulva is spared, giving the appearance of labial hypotrophy. Diabetes mellitus, tubero-eruptive xanthomata, insulin-resistant diabetes mellitus, hyperlipoproteinemia Type V, and acanthosis nigricans frequently accompany the disorders (7,14,16,18).

Lawrence (17) reported a sporadically acquired lipoatrophic diabetes of adolescence or early adult life. There is female preponderance. Hepatosplenomegaly with cirrhosis, acanthosis nigricans, hyperlipidemia, and insulin-resistant diabetes mellitus were also found.

Many features of congenital lipodystrophy have been noted in *leprechaunism:* liver enlargement, lipodystrophy, enlarged clitoris or penis, neurologic damage, hirsutism, and insulin resistance. These are all evident at, or shortly following, birth. It is conceivable that leprechaunism represents a lethal form of lipoatrophic diabetes.

The so-called *SHORT syndrome* manifests partial lipodystrophy.

The diencephalic syndrome caused by a tumor in the region of the anterior hypothalamus is manifested by profound emaciation with accelerating early growth, increased motor activity, and euphoria (6). The affected infants are normal at birth but become symptomatic with age. Progressive emaciation leads to a loss of subcutaneous fat but without the muscularity of lipodystrophic patients. The hands and feet are often large. Hepatomegaly, genital enlargement, and hyperlipemia have not been reported.

*Progeria* and *Werner syndrome* patients also exhibit loss of body fat and an increased incidence of diabetes mellitus.

**Laboratory aids.** Lipoatrophic diabetes is clearly distinct from the common form of diabetes mellitus. The diabetes, which is insulin resistant and nonketotic, appears after the onset of the lipodystrophy, usually between 6 and 20 years of age (15,30). Normal glucose tolerance is observed during the first few years of life. Hyperglucagonemia has been found and may contribute to the insulin resistance (11).

Hyperlipemia, especially of the triglyceride fraction, is a constant feature. It usually precedes the onset of hyperglycemia. The serum of these patients is intermittently turbid or milky, with increased triglycerides, and contains very low-density lipoprotein. In addition, there is slow turnover of $^{14}$C-labeled triglycerides and cholesterol and slow clearance of ingested triglycerides with failure of adipocytes to incorporate triglyceride fatty acids from the blood.

#### References [Lipoatrophic diabetes (Berardinelli syndrome, Seip syndrome, generalized lipodystrophy)]


1. Barnes ND et al: Insulin resistance, skin changes and virilization: A recessively inherited syndrome possibly due to pineal gland dysfunction. *Diabetologia* **10**:285–289, 1974.
2. Berardinelli W: An undiagnosed endocrine-metabolic syndrome. *J Clin Endocrinol* **14**:193–204, 1954.
3. Brubaker MM et al: Acanthosis nigricans and congenital total lipodystrophy. *Arch Dermatol* **91**:320–325, 1965.
4. Brunzell JD et al: Congenital generalized lipodystrophy accompanied by cystic angiomatosis. *Ann Intern Med* **69**:501–516, 1968.
5. Burn J, Baraitser M: Partial lipoatrophy with insulin resistant diabetes and hyperlipidaemia (Dunnigan syndrome). *J Med Genet* **23**:128–130, 1986.
6. Burr IM et al: Diencephalic syndrome revisited. *J Pediatr* **88**:439–444, 1976.
7. Dunnigan MG et al: Familial lipoatrophic diabetes with dominant transmission. *Q J Med* **43**:33–48, 1974.
8. Flier JS et al: Receptors, antireceptor antibodies and mechanisms of insulin resistance. *N Engl J Med* **300**:413–419, 1979.
9. Gold RH, Steinbach HL: Lipoatrophic diabetes mellitus (generalized lipodystrophy): Roentgen findings in two brothers with congenital disease. *Am J Roentgenol* **101**:884–896, 1967.
10. Holmes J, Tanner MS: Premature eruption and macrodontia associated with insulin resistant diabetes and pineal hyperplasia. *Br Dent J* **141**:280–284, 1976.
11. Huseman C et al: Congenital total lipodystrophy: An endocrine study in 3 siblings. *J Pediatr* **93**:221–226, 1978.
12. Huseman C et al: Congenital lipodystrophy: Association with polycystic ovarian disease. *J Pediatr* **95**:72–74, 1979.
13. Ipp MM et al: Sicca syndrome and total lipodystrophy. *Ann Intern Med* **85**:443–446, 1976.
14. Köbberling J, Dunnigan MG: Familial partial lipodystrophy: Two types of an X-linked dominant syndrome; lethal in the hemizygous state. *J Med Genet* **23**:120–127, 1986.
15. Kodama S et al: Congenital generalized lipodystrophy with insulin-resistant diabetes. *Eur J Pediatr* **127**:111–119, 1978.
16. Kraai EJ, Van Son W: Partial lipodystrophy (facies Voltarien) and hypocomplementaemic nephritis. *Br J Dermatol* **108**:115–116, 1983.
17. Lawrence RD: Lipodystrophy and hepatomegaly with diabetes, lipaemia and other metabolic disturbances. *Lancet* **1**:724–731, 1946.
18. Lillystone D, West RJ: Lipodystrophy of limbs associated with insulin resistance. *Arch Dis Childh* **50**:737–739, 1975.
19. McLean RH, Hoefnagel D: Partial lipodystrophy and familial C3 deficiency. *Hum Hered* **30**:149–154, 1980.
20. Mendenhall EN: Tumor of the pineal body with high insulin resistance. *J Indiana State Med Assoc* **43**:32–36, 1950.
21. Miescher G: Zwei Fälle von congenitales familiärer Acanthosis nigricans, kombiniert mit Diabetes mellitus. *Dermatol Z* **31**:276–305, 1921.
22. Najjar SS et al: Congenital generalized lipodystrophy. *Acta Paediatr Scand* **64**:273–279, 1975.
23. Oseid S: Studies in congenital generalized lipodystrophy (Seip–Berardinelli syndrome): Development of diabetes. *Acta Endocrinol* **72**:475–494, 1973.
24. Oseid S et al: Decreased binding of insulin to its receptor in patients with congenital generalized lipodystrophy. *N Engl J Med* **296**:245–248, 1977.
25. Rabson SM, Mendenhall EN: Familial hypertrophy of pineal body, hyperplasia of adrenal cortex and diabetes mellitus. *Am J Clin Pathol* **26**:283–290, 1956.
26. Rosenbloom AL et al: Normal insulin binding to cultured fibroblasts from patients with lipoatrophic diabetes. *J Clin Endocrinol Metab* **44**:803–806, 1977.
27. Seip M: Lipodystrophy and gigantism with associated endocrine manifestations. *Acta Paediatr Scand* **48**:555–574, 1959.
28. Seip M: Generalized lipodystrophy. *Ergeb Inn Med Kinderheilkd* **31**:59–95, 1971.
29. Seip M, Oseid S: The pathogenesis of diabetes mellitus in generalized lipodystrophy. *Mod Prob Paediatr* **12**:180–188, 1975.
30. Senior B, Gellis SS: The syndromes of total lipodystrophy and of partial lipodystrophy. *Pediatrics* **33**:593–612, 1964.
31. Van Leeuwen HC: Über familiäres Vorkommen von Lipodystrophia

progressiva zusammen mit Otosklerose, Knochencysten und geistiger Debilität. *Z Klin Med* **123**:534–547, 1933.

32. Wesenberg RL et al: The roentgenographic findings in total lipodystrophy. *Am J Roentgenol* **103**:154–164, 1968.

33. West RJ et al: Familial insulin-resistant diabetes, multiple somatic anomalies and pineal hyperplasia. *Arch Dis Childh* **50**:703–708, 1975. (Same as Ref. 10).

34. West RJ, Leonard JV: Familial insulin resistance with pineal hyperplasia, metabolic studies and effect of hypophysectomy. *Arch Dis Childh* **55**:619–621, 1980.

35. Zafarulla MYM: Lipodystrophy: A case of partial lipodystrophy. *Br J Oral Maxillofac Surg* **23**:52–57, 1985.

## SHORT syndrome (unusual facies, lipoatrophy, short stature, and Rieger anomaly)

Gorlin et al (2) and Sensenbrenner et al (4) reported male sibs and a female singleton, respectively, with a striking phenotype. Gorlin et al (2) suggested the acronym SHORT (S—short stature, H—hyperextensibility of joints/hernia, O—ocular depression, R—Rieger anomaly, T—teething delay).

The lower face and upper limbs appeared thin due to lipoatrophy. The eyes were deep set and Rieger anomaly was evident. The pinnae were somewhat outstanding. The nasal alae were hypoplastic with wide nasal bridge (Fig. 24–63A,B). Stature was reduced. There were poor weight gain and frequent illness during infancy. Joints were hyperextensible and there was inguinal hernia in two cases. Speech was delayed in spite of normal intelligence. Dental eruption was delayed. Lipson et al (3) have seen a similar patient. Toriello et al (5) added another case, but her patient had profound sensorineural hearing loss and did not exhibit Rieger anomaly.

Inheritance appeared to be autosomal recessive. However, Aarskog et al (1) described what seems to be a similar but dominantly inherited disorder in four individuals in three generations. All had Rieger anomaly. Two members manifested diabetes mellitus at 14 and 39 years, respectively, and another, glucose intolerance at 55 years. The lipodystrophy affected not only the face but the buttocks. Additional findings included retarded bone age, hypospadias, and hypotrichosis.

### References [SHORT syndrome (unusual facies, lipoatrophy, short stature, and Rieger anomaly)]

1. Aarskog D et al: Autosomal dominant partial lipodystrophy associated with Rieger anomaly, short stature and insulinopenic diabetes. *Am J Med Genet* **15**:29–38, 1983.

2. Gorlin RJ et al: Rieger anomaly and growth retardation (the S-H-O-R-T syndrome). *Birth Defects* **11**(2):46–48, 1975.

3. Lipson A et al: The SHORT syndrome: Further delineation and natural history. *J Med Genet*, **26**:473–475, 1989.

4. Sensenbrenner JA et al: A low birthweight syndrome? Rieger syndrome. *Birth Defects* **11**(2):423–426, 1975.

5. Toriello HV et al: Report of a case and further delineation of the SHORT syndrome. *Am J Med Genet* **22**:311–314, 1985.

## Coffin–Lowry syndrome

Coffin et al (2), in 1966, and Lowry et al (11), in 1971, independently reported patients with mental and somatic retardation, characteristic facies, large soft hands with distally tapering fingers, and various skeletal anomalies. However, it was Temtamy et al (15), in 1975, who recognized the unity of the reports and coined the name Coffin–Lowry syndrome. An early report is that of Martinelli and Campailla (12).

The disorder has X-linked inheritance. The gene locus is on the short arm of the X chromosome at Xp22.2 (5,12a). Female heterozygotes commonly express the condition to a less severe degree. About 60 cases have been reported (4a,18).

**Facies.** Characteristic facial changes become more marked with age but are apparent by the second year of life. Hair is straight and coarse in males. The forehead is prominent and broad and the occiput is somewhat flat. There are prominent supraorbital ridges, hypertelorism, downslanting narrow palpebral fissures, heavy arched eyebrows,

Fig. 24–63. *SHORT syndrome*. (A,B) Brothers exhibiting severe growth retardation, disproportionately small face, sunken eyes, Rieger anomaly, and delayed tooth eruption. [From *RJ Gorlin et al*, Birth Defects **11**(2):46, 1975.]

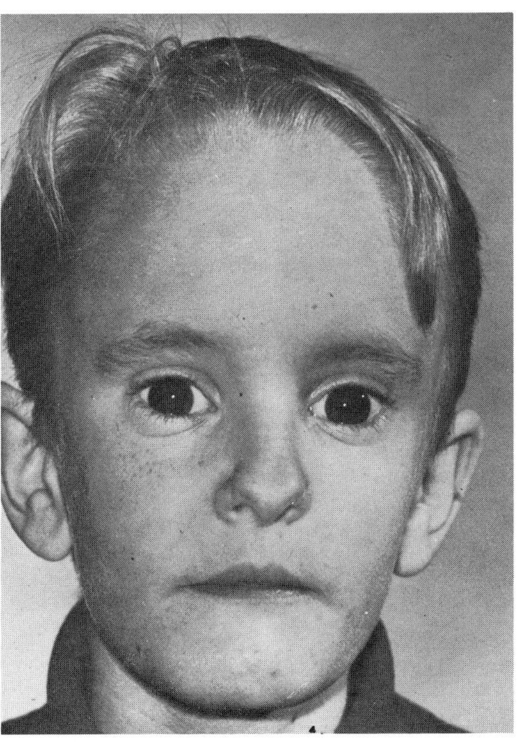

A

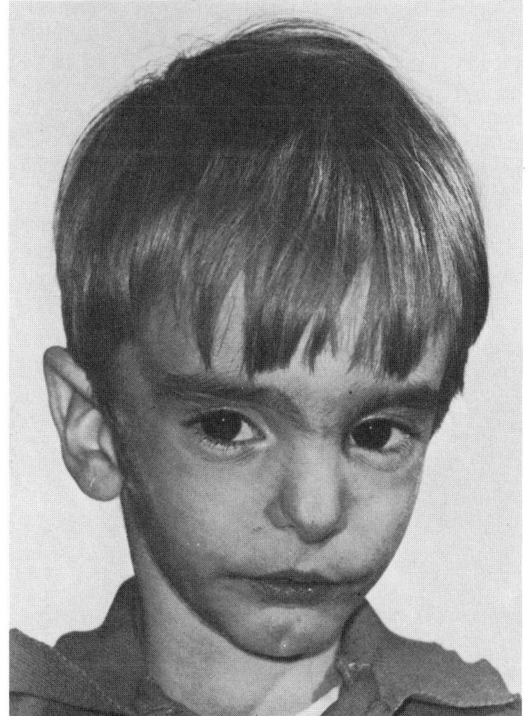

B

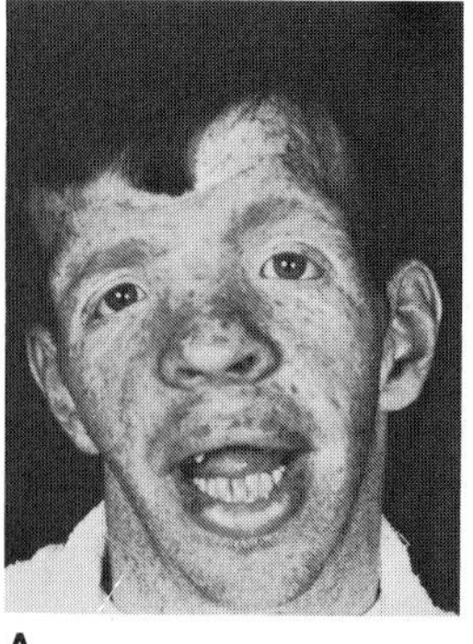

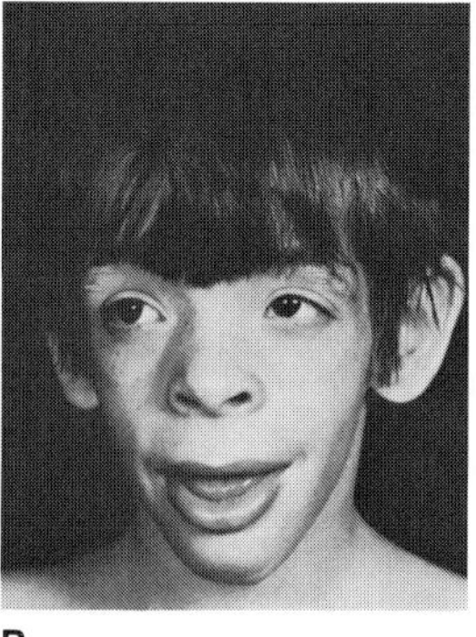

A B

Fig. 24–64. *Coffin–Lowry syndrome.* (A,B) Twenty-three- and ten-year old affected males. Note facial similarities: marked hypertelorism, prominent supraorbital ridges and outer margins, short upturned nostrils, open mouth with pouty lower lip, thick bulging chin, and enlarged protuberant ears. Facial features are more coarse in the older affected male. All males had straight coarse hair. (From *SA Temtamy et al,* J Pediatr **86**:724, 1975.)

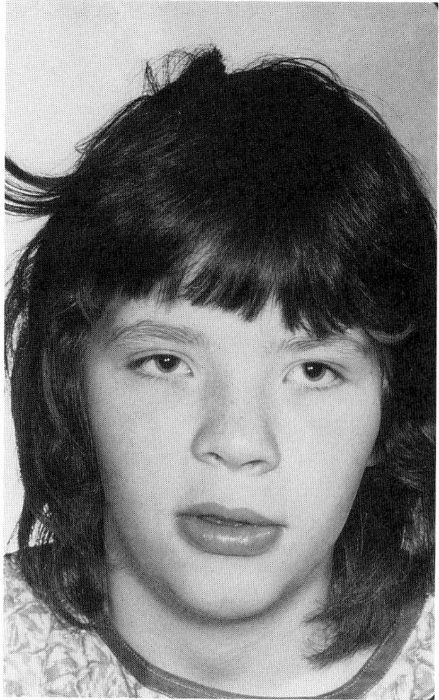

Fig. 24–65. *Coffin–Lowry syndrome.* Female heterozygote exhibiting many of same features.

ptotic upper eyelids, and somewhat hypoplastic midfacial development with relative mandibular prognathism. The nose is large with a broad base and flared alae. The nostrils are anteverted and the pinnae are prominent. The philtrum appears somewhat elongated and the nasal septum is wide. The lips are thick and pouting and the mouth is usually held open (Figs. 24–64A,B and 24–65).

**Musculoskeletal.** Although usually normal at birth, height and weight become reduced below the third percentile in hemizygotes and in 50% of heterozygotes. There is delayed ambulation and a clumsy broad-based gait. At birth, hypotonia and/or loose ligaments with pes planus and inguinal hernia may be noted. The hands are extremely large and soft with distal tapering fingers in both sexes (Fig. 24–66A). This is the most striking feature at birth (16,17). The nails are short and broad. The finger joints are hyperextensible. Radiographically, the calvaria is thickened, especially the frontal tables, in 60%. The anterior fontanel is large and suture closure is markedly delayed (16). Pectus carinatum or excavatum found in 80% of hemizygotes and in 30% of heterozygotes is associated with thoracolumbar kyphosis/scoliosis. Commonly there is dysplasia of the vertebral bodies at the thoracolumbar junction. The sternum is short and often bifid. Bone age is retarded in both sexes. Pseudoepiphyses at the base of each metacarpal may be seen in males during childhood. The distal phalanges are short or tufted. The middle phalanges are poorly modeled. The iliac wings may be narrowed. Female heterozygotes tend to be obese. Distal muscle atrophy is not uncommon.

**Central nervous system.** Intelligence quotients in males have ranged from 5 to 50 but speech is severely retarded. Female carriers have varied in intelligence from normal (20%) to 60 IQ. Another 20% of heterozygotes are severely retarded (18). Internal communicating hydrocephaly or ventricular dilatation has been noted (2,6,8,11,15) in hemizygotes and over 40% exhibit severe generalized seizures (4).

**Cardiovascular system.** Various heart anomalies have been described (2,8,10,13,15).

**Skin.** The skin appears loose and easily stretched. Cutis marmorata, dependent acrocyanosis, and varicose veins are frequent.

**Oral manifestations.** The lips are large, thick, and pouting. The tongue may have a deep midline furrow (8). Torus palatinus has been stated to be more common (15) but we have not found this to be true. The lower permanent incisors are frequently absent or have reduced crown form in 80% of affected males and in 20% of females (Fig. 24–67). Malocclusion with overjet and/or overbite appears to be a nearly constant feature. Early tooth loss due to periodontal disease has been noted.

**Pathology.** Ultrastructural studies of conjunctiva and skin have exhibited changes suggestive of a lysosomal storage disease but this needs confirmation (1).

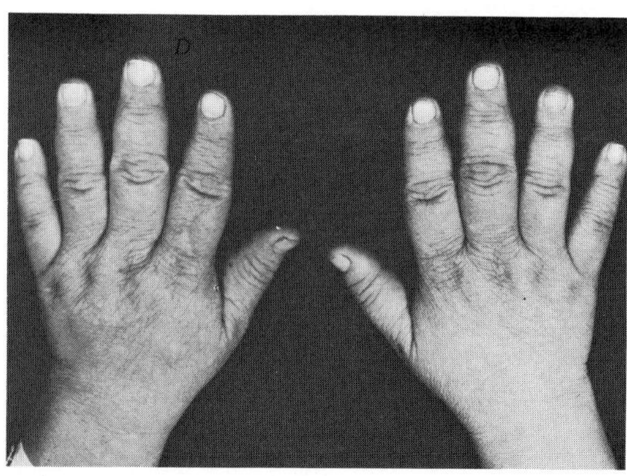

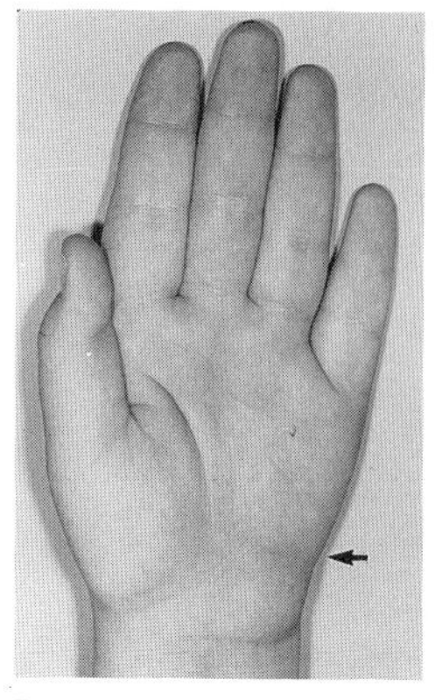

A B

Fig. 24–66. *Coffin–Lowry syndrome.* (A) Hands are spadelike; digits are thick at base and taper distally. (B) Flexion crease in hypothenar area at arrow. (A from *SA Temtamy et al,* J Pediatr **86**:724, 1975.)

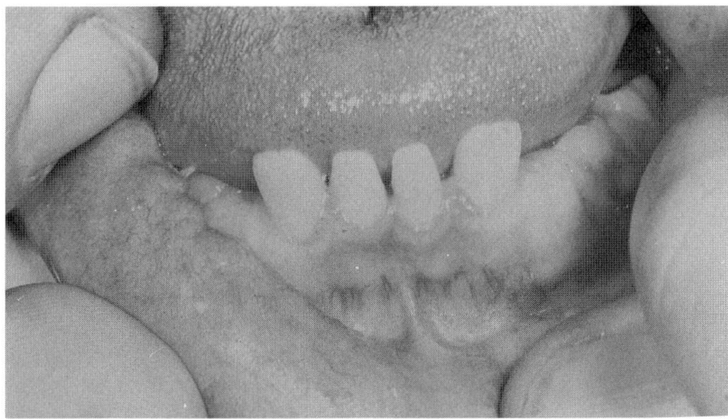

Fig. 24–67. *Coffin–Lowry syndrome.* Reduced crown form of lower anterior incisors.

**Differential diagnosis.** During infancy, several children have presented with a diagnosis of possible hypothyroidism. With age, diagnosis becomes easier but coarseness of features may suggest a *mucopolysaccharidosis* or *oligosaccharidosis*.

**Laboratory aids.** Dermatoglyphic changes include a characteristic horizontal hypothenar crease in both sexes (2,7,13,15) (Fig. 24–66B). The "atd" angle is increased and transverse palmar creases and Sydney lines are frequent. The finger tip ridge counts are decreased. Increased c–d and reduced b–c triradii distances appear significant (14). Abnormal proteodermatan sulfate storage has been found (1,15a).

### References (Coffin–Lowry syndrome)


1. Beck M et al: Abnormal proteodermatan sulfate in three patients with Coffin–Lowry syndrome. *Pediatr Res* **17**:926–929, 1983.

2. Coffin GS et al: Mental retardation with osteocartilaginous anomalies. *Am J Dis Child* **112**:205–213, 1966.

3. Collacott RA et al: Coffin–Lowry syndrome and schizophrenia: A family report. *J Ment Defic Res* **31**:199–207, 1987.

4. Fryns JP et al: The Coffin syndrome. *Hum Genet* **36**:271–276, 1977.

4a. Gilgenkrantz S et al: Coffin–Lowry syndrome: A multicenter study. *Clin Genet* **34**:230–245, 1988.

5. Hanauer A et al: Probable localization of the Coffin–Lowry locus in Xp22.p22.1 by multipoint linkage analysis. *Am J Med Genet* **30**:523–530, 1988.

6. Haspeslagh M et al: The Coffin–Lowry syndrome. *Eur J Pediatr* **143**:82–86, 1984.

7. Hersh J et al: Forearm fullness in Coffin–Lowry syndrome: A misleading yet possible early diagnostic clue. *Am J Med Genet* **18**:185–189, 1984.

8. Hunter AGW et al: The Coffin–Lowry syndrome: Experience from four centres. *Clin Genet* **21**:321–335, 1982.

9. Jammes J et al: Syndrome of facial abnormalities, kyphoscoliosis and severe mental retardation. *Clin Genet* **4**:203–209, 1973.

10. Krajewska-Walasek M et al: Cardiac involvement in Coffin–Lowry syndrome. *Eur J Pediatr* **147**:448, 1988.

11. Lowry B et al: A new dominant gene mental retardation syndrome. *Am J Dis Child* **121**:491–500, 1971.

12. Martinelli B, Campailla E: Contributo alla conoscenze delle syndrome di Coffin, Siris, and Wegienka. *G Psichiat Neuropatol* **97**:449–451, 1969.

12a. Partington MW et al: A family with the Coffin–Lowry syndrome revisited: Localization of CLS to Xp21-pter. *Am J Med Genet* **30**:509–521, 1988.

13. Procopis PG, Turner B: Mental retardation, abnormal fingers, and skeletal anomalies, Coffin's syndrome. *Am J Dis Child* **124**:258–265, 1974.

14. Temtamy SA et al: The Coffin–Lowry syndrome: A simple inherited trait comprising mental retardation, facio-digital anomalies and skeletal involvement. *Birth Defects* **9**(6):133–152, 1975.

15. Temtamy SA et al: The Coffin–Lowry syndrome: An inherited faciodigital mental retardation syndrome. *J Pediatr* **86**:724–731, 1975.

15a. Vine DT et al: Etiology of the weakness in Coffin–Lowry syndrome. *Am J Hum Genet* **38**:85A, 1986.

16. Vles JSH et al: Early clinical signs in Coffin–Lowry syndrome. *Clin Genet* **26**:448–452, 1984.

17. Wilson WG, Kelly T: Early recognition of the Coffin–Lowry syndrome. *Am J Med Genet* **8**:215–220, 1981.

18. Young ID: The Coffin–Lowry syndrome. *J Med Genet* **25**:344–348, 1988.

# Chapter 25
# Syndromes with Unusual Facies: Other Syndromes

## Coffin–Siris syndrome (absent fifth finger- and toenails, unusual facies, and mental deficiency)

In 1970, Coffin and Siris (4) reported three children with coarse facial features, low birthweight, retarded somatic and mental growth, and hypoplastic to absent fifth finger- and toenails. About 30 cases (1–3,6–11,13–15,17–19) have been documented. Less certain are the patients of Senior (16) and Mace and Gotlin (12), patients not as severely affected who may represent a separate entity. Although most have been isolated examples, affected sibs (1,3,7,10,14) have been described. About 85% are females, possibly implying lethality in males. Mild expression in a parent has also been noted (4,10,17,19). Perhaps some examples really have *fetal hydantoin syndrome* or *dup(9p) syndrome*.

Feeding problems and recurrent respiratory infections are common during infancy. Redundant gastric mucosa at the antrum has been reported (2a). Frequent findings include microcephaly (65%), coarse facial features (100%), sparse scalp hair (80%), bushy eyebrows, flat nasal root, bulbous upturned nose, thick lips, wide mouth, mental deficiency (100%), hypotonia (100%), growth deficiency (85%), short stature, hypoplastic to absent fifth finger- and toenails and terminal phalanges with less severe involvement of other digits (100%), lumbosacral hirsutism (100%), lax joints, dislocation of the radial heads, coxa valga, retarded bone age, and small to absent patellae (Fig. 25–1A–D). Various low-frequency anomalies have included eyelid ptosis, strabismus, cleft palate (4,17,18), and various congenital heart anomalies (30%). There appears to be a characteristic hand profile (1). Central nervous system anomalies were Dandy–Walker malformation (4,17), partial agenesis of the corpus callosum, and agenesis of the anterior commissure (5,17).

Fig. 25–1. *Coffin–Siris syndrome.* (A) Broad nose with anteverted nostrils. (B,C) Hypoplasia of terminal fifth phalanges. (D) Hypoplastic patella. (Courtesy of *E Siris*, Eldridge, California.)

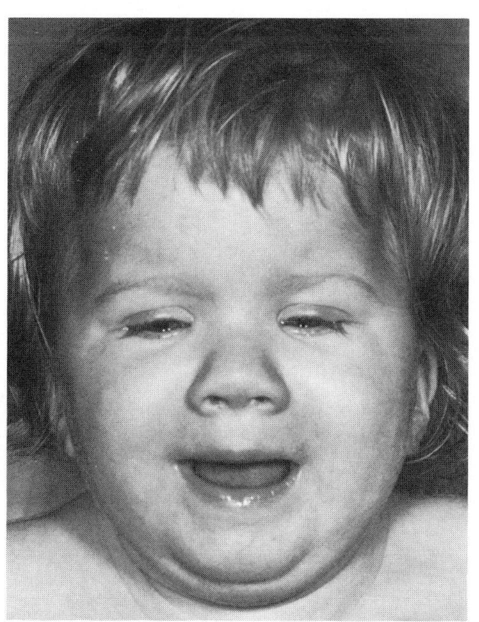

A

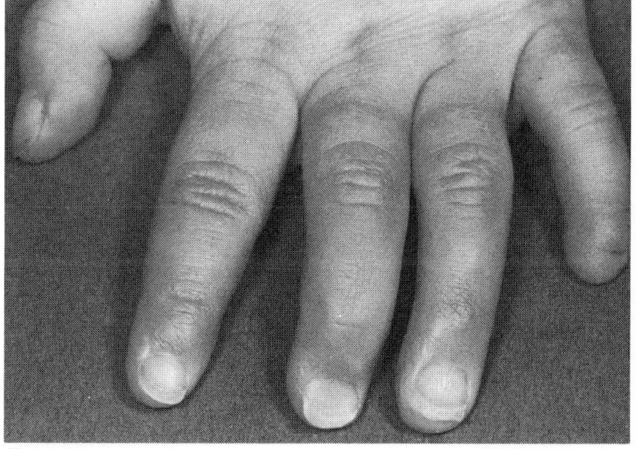

B

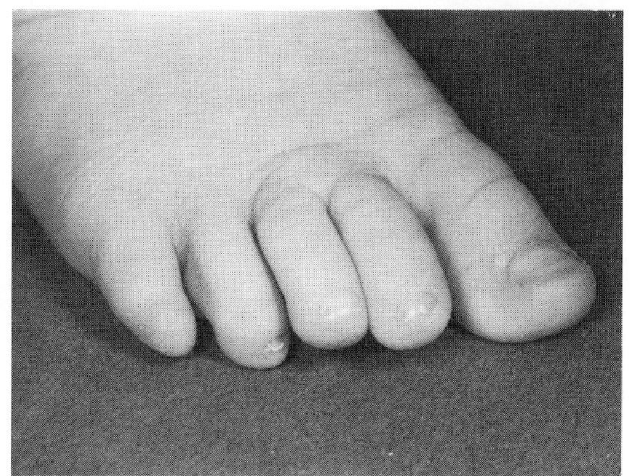

C

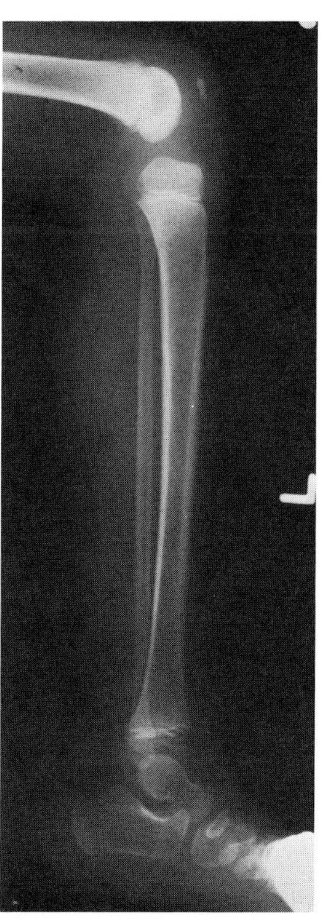

D

#### References [Coffin–Siris syndrome (absent fifth finger- and toenails, unusual facies, and mental deficiency)]

1. Barber N, Say B: Coffin–Siris syndrome. *Am J Dis Child* **132**:1044, 1978.

2. Bartsocas CS, Tsiantos AK: Mental retardation with absent fifth fingernail and terminal phalanx. *Am J Dis Child* **120**:493–494, 1970.

2a. Bodurtha J et al: Distinctive gastrointestinal anomaly associated with Coffin–Siris syndrome. *J Pediatr* **109**:1015–1017, 1986.

3. Carey JC, Hall BD: The Coffin–Siris syndrome: Five new cases including two siblings. *Am J Dis Child* **132**:667–671, 1978.

4. Coffin GS, Siris E: Mental retardation with absent fifth fingernail and terminal phalanx. *Am J Dis Child* **119**:433–439, 1970.

5. De Bassio WA et al: Coffin–Siris syndrome. Neuropathological findings. *Arch Neurol* **42**:350–353, 1985.

5a. D'Elia R et al: Considerazione cliniche intorne ad un probabile nuovo caso di sindrome di Coffin–Siris. *Minerva Pediatr* **33**:1021–1024, 1981.

6. Foasso MF et al: Le syndrome de Coffin–Siris. *Pédiatrie* **38**:111–117, 1983.

7. Franceschini P et al: The Coffin–Siris syndrome in two siblings. *Pediatr Radiol* **16**:330–333, 1986.

8. Gellis SS, Feingold M: Coffin–Siris syndrome. *Am J Dis Child* **132**:1213–1214, 1978.

9. Giovannucci Uzielli ML et al: La sindrome di Coffin–Siris. *Minerva Pediatr* **32**:245–254, 1980.

10. Haspeslagh M et al: The Coffin–Siris syndrome: Report of a family and further delineation. *Clin Genet* **26**:374–378, 1984.

11. Lucaya J et al: The Coffin–Siris syndrome: A report of four cases and a review of the literature. *Pediatr Radiol* **11**:35–38, 1981.

12. Mace JW, Gotlin RW: Short stature and onychodysplasia. *Am J Dis Child* **125**:114–116, 1973.

13. Mastroiacovo P et al: La sindrome di Coffin–Siris. *Minerva Pediatr* **29**:773–778, 1977.

14. Mattei JF et al: Coffin–Lowry syndrome in sibs. *Am J Med Genet* **8**:315–320, 1981 (error in title).

14a. Meinecke P et al: Coffin–Siris-Syndrome bei einem 5 jährigen Mädchen. *Mschr Kinderheilkd* **134**:692–695, 1986.

15. Schinzel A: The Coffin–Siris syndrome. *Acta Pediatr Scand* **68**:449–452, 1979.

16. Senior B: Impaired growth and onychodysplasia. *Am J Dis Child* **122**:7–9, 1971.

17. Tunnessen WW et al: The Coffin–Siris syndrome. *Am J Dis Child* **132**:393–395, 1978.

18. Ueda K et al: The Coffin–Siris syndrome. *Helv Paediatr Acta* **35**:385–390, 1980.

19. Weiswasser WH et al: Coffin–Siris syndrome. *Am J Dis Child* **125**:838–840, 1973 (Case 1).

## Ruvalcaba syndrome

In 1971, Ruvalcaba et al (4) described male sibs and first cousins with severe mental and somatic retardation, microcephaly, hypoplastic skin, hypoplastic genitalia, delayed adolescence, cryptorchidism, and unusual facies characterized by downslanted palpebral fissures, small narrow beaked nose with hypoplastic alae, microstomia, narrow maxilla, and crowded teeth (Fig. 25–2). Other cases have been reported (1–3,5). We suspect marked heterogeneity. In spite of normal karyotype, the patient of Bialer et al (1) facially resembles *del(2q) syndrome,* and we are reluctant to include the cases of Hunter et al (3).

Musculoskeletal anomalies include short stature, narrow trunk, pectus carinatum, kyphoscoliosis and/or scoliosis with osteochondritis of spine, limitation of elbow extension, prominent elbows, short limbs, short hands and feet, short metapodial bones and phalanges, broad hips, and inguinal hernia.

Present in at least 50% of the cases were pubertal delay and renal anomalies (1).

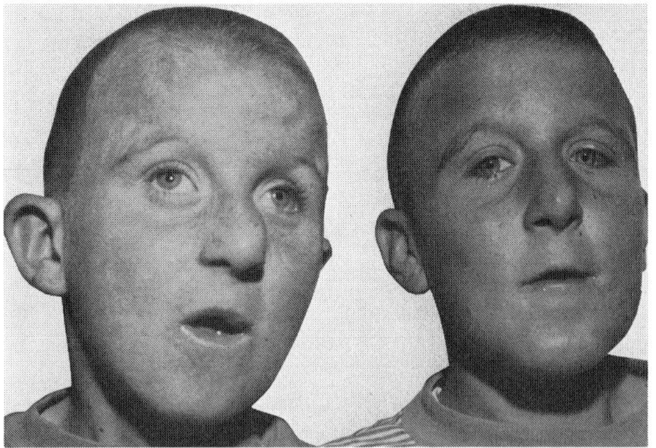

Fig. 25–2. *Ruvalcaba syndrome.* Downslanting palpebral fissures, small narrow nose. (From *R Ruvalcaba et al,* J Pediatr **79**:450, 1971.)

#### References (Ruvalcaba syndrome)

1. Bialer MG et al: Apparent Ruvalcaba syndrome with genitourinary abnormalities. *Am J Med Genet* **33**:314–317, 1989.

1a. Bianchi E et al: Ruvalcaba syndrome: A case report. *Eur J Pediatr* **142**:301–303, 1984.

2. Geormaneanu M et al: Über ein "neues Syndrom" in Verbindung mit familiärer Translokation 13/14. *Klin Pädiatr* **190**:500–506, 1978.

3. Hunter AGW et al: A "new" syndrome of mental retardation with characteristic facies and brachyphalangy. *J Med Genet* **14**:430–437, 1977.

4. Ruvalcaba RHA et al: A new familial syndrome with osseous dysplasia and mental deficiency. *J Pediatr* **79**:450–455, 1971.

5. Sugio Y, Kajii T: Ruvalcaba syndrome: Autosomal dominant inheritance. *Am J Med Genet* **19**:741–753, 1984.

## Schinzel–Giedion syndrome

In 1978, Schinzel and Giedion (7) described a syndrome of midface retraction, hirsutism, multiple skeletal anomalies, mental retardation, and seizures. Additional patients have been reported (1–6). The disorder has occurred in sibs (7). Autosomal recessive inheritance is probable.

Most exhibit severe somatic and mental retardation, spasticity, and seizures (1). Death has usually occurred before 18 months. Recurrent apneic spells are common.

The facies is coarse with frontal bossing and midface retraction. The skull sutures are widely patent and there is ocular hypertelorism. The nose is short, saddled, and upturned. Choanal stenosis has been noted in all cases. The ears are low set and the teeth delayed in eruption. The neck is short and the skin is redundant (Figs. 25–3 and 25–4A,B).

Dermatologic anomalies include hypoplastic dermal ridges, hypoplastic nipples, transverse palmar creases, hyperconvex nails, and marked hypertrichosis (Fig. 25–5). Genitourinary anomalies include congenital hydronephrosis, hypospadias, short penis, and deep interlabial sulcus. ASD has been reported in 50%.

Skeletal anomalies seen are orbital hypertelorism, short deep skull base, wide occipital synchondrosis, multiple Wormian bones, broad ribs, mild mesomelic brachymelia, hypoplasia of distal phalanges, hypoplasia of pubic bones, and talipes.

Congenital hydronephrosis is extremely rare and has been seen in few syndromes: *Johanson–Blizzard syndrome, trisomies 13 and 18, triploidy, Turner syndrome,* and *urofacial syndrome.* The facies is reminiscent of infants with a storage disease.

#### References (Schinzel–Giedion syndrome)

1. Al-Gazali LI et al: The Schinzel–Giedion syndrome: Clinical features and natural history. *J Med Genet* **25**:275, 1988.

2. Burck V: Mittelgesichtshypoplasie, Skelettanomalien, Apnoen, Retardierung-eine weitere Beobachtung. *Klin Genet Pädiatr* **3**:352–358, 1982.

3. Donnai D, Harris R: A further case of a new syndrome including midface retraction, hypertrichosis and skeletal anomaly. *J Med Genet* **16**:483–486, 1979.

4. Hall BD: Schinzel–Giedion syndrome: Report of the fifth case. *David W. Smith Workshop on Malformations and Morphogenesis.* Madrid, Spain, 23–29 May, 1989.

5. Kelley RI et al: Congenital hydronephrosis, skeletal dysplasia and severe

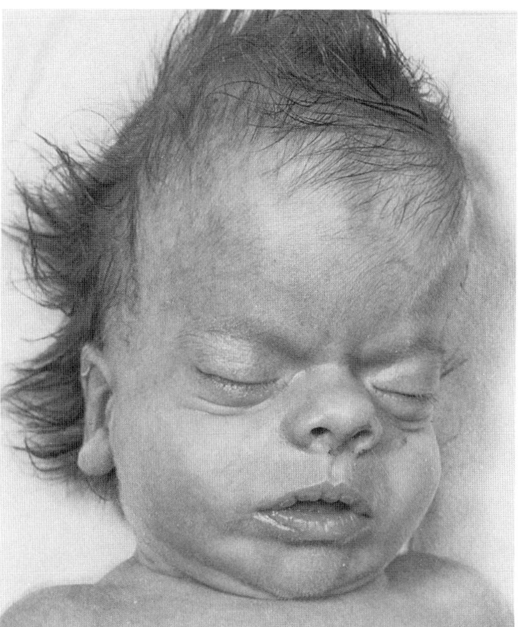

Fig. 25–3. *Schinzel–Giedion syndrome.* Note high bulging forehead, depressed nasal bridge with short root, protrusion of eyes, short neck, and hypertrichosis. (From *A Schinzel* and *A Giedion,* Am J Med Genet **1**:361, 1978.)

developmental retardation: The *Schinzel–Giedion* syndrome. *J Pediatr* **100**:943–945, 1982.

6. Leonard CO: Schinzel–Giedion or Rüdiger syndrome: Additional case discussion. *Proc Greenwood Genet Ctr* **2**:126–127, 1983.

7. Schinzel A, Giedion A: A syndrome of mid-face retraction, multiple skull anomalies, club feet, and cardiac and renal malformations in sibs. *Am J Med Genet* **1**:361–375, 1978.

## Growth retardation, alopecia, pseudoanodontia, and optic atrophy (GAPO syndrome)

Andersen and Pindborg (1), in 1947, were the first to publish a case of what Tipton and Gorlin (9) chose to term the GAPO (**G**rowth retardation, **A**lopecia, **P**seudoanodontia, **O**ptic atrophy) syndrome. A Brazilian patient was described in 1977 (2), as an example of Rothmund–Thomson syndrome, and republished in 1982 under a corrected title (10). To date, about a dozen patients have been described (1,2,4–10). An unpublished patient is being followed by S Kreiborg (personal communication, 1989). Inheritance is clearly autosomal recessive, there being affected sibs and parental consanguinity.

The phenotype is easily recognized. Scalp hair is present at birth, but, after the first years of life, the hair is completely lost and never regrows (Fig. 25–6A,B). The jaw bones are crowded with both dentitions, neither of which erupts (Fig. 25–7A,B). Birthweight is normal, but birth length is somewhat reduced. Growth retardation becomes quite evident at the sixth month check-up. Body build is proportional.

Frontal bossing, high forehead, dilated scalp veins, mild midfacial hypoplasia, and delayed closure of the anterior fontanel are common. Craniosynostosis was found in one case (1). Bone age is significantly retarded. Otherwise skeletal changes are not remarkable. Optic atrophy is variable in onset (1,4,5,7,8) and not a common feature. Other changes include mild pectus excavatum, umbilical hernia, hypogonadism, absent or occluded transverse and sigmoid sinuses (3,6–8,11),

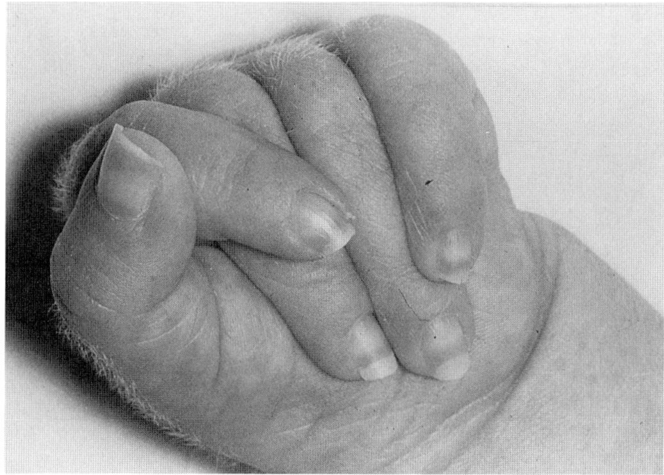

Fig. 25–5. *Schinzel–Giedion syndrome.* Narrow hyperconvex nails.

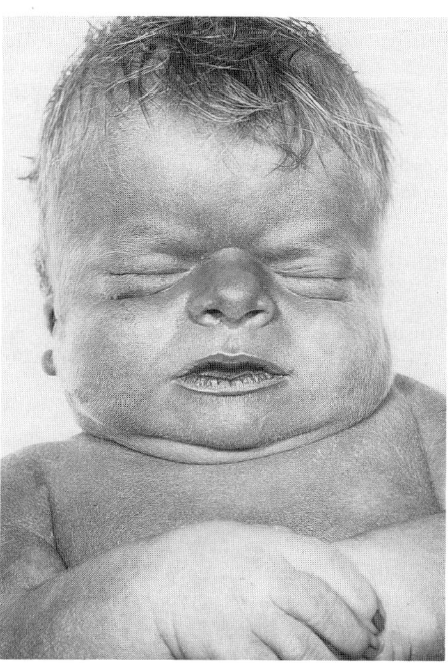

A

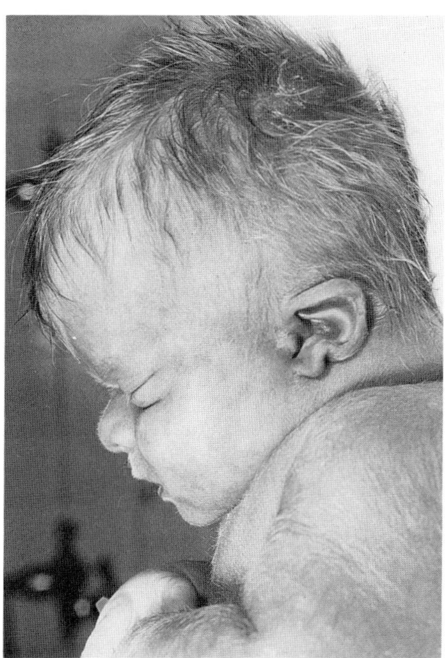

B

Fig. 25–4. *Schinzel–Giedion syndrome.* (A, B) Sib of child shown in Fig. 25–3. Note bulging forehead, short nose with low bridge, short neck with redundant skin fold, small broad pinna, and hypertrichosis. (From *A Schinzel* and *A Giedion,* Am J Med Genet **1**:361, 1978.)

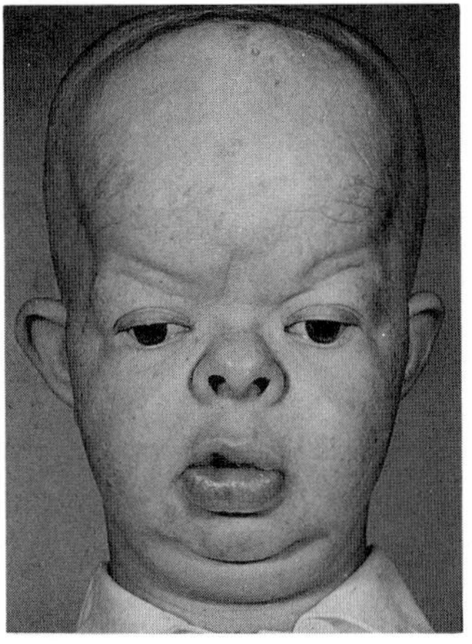

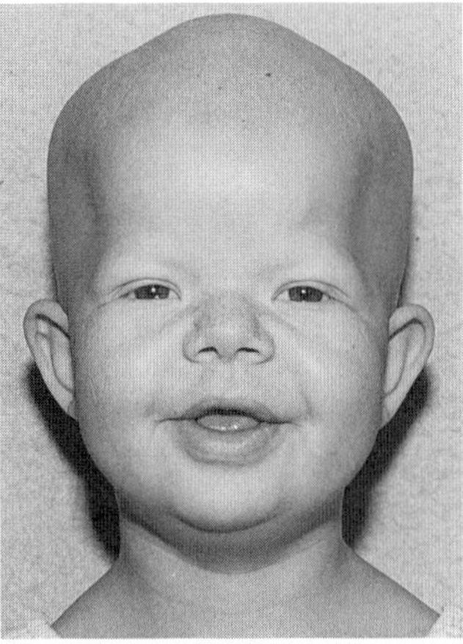

A B

Fig. 25–6. *Growth retardation, alopecia, pseudoanodontia, and optic atrophy (GAPO syndrome).* (A) Note 26-year-old female with alopecia. Patient had optic atrophy. (B) Compare facies with that in A. (A from *TH Andersen* and *JJ Pindborg,* Odont T **55**:484, 1947.)

nephrocalcinosis (1,7), polycystic kidneys (1), keratoconus, and glaucoma (11).

Differential diagnosis of disorders with severe alopecia is elegantly discussed by Feinstein et al (3).

#### References [Growth retardation, alopecia, pseudoanodontia, and optic atrophy (GAPO syndrome)]


1. Andersen TH, Pindborg JJ: Et tilfaelde af "pseudo-anodonti" i forbindelse med kraniedeformitet, dvaergvaekst og ektodermal displasi. *Odont T* **55**:484–493, 1947.

2. Epps DR et al: Poiquiloderma congenito familiar (S. de Rothmund–Thomson). *Ciencia Cultura* **29** *(Suppl):*740, 1977.

3. Feinstein A et al: Genetic disorders associated with severe alopecia in children: A report of two unusual cases and a review. *J Craniofac Genet Dev Biol* **7**:301–310, 1987.

4. Fuks A: Pseudoanodontia, cranial deformity, blindness, alopecia and dwarfism: A new syndrome. *J Dent Child* **45**:155–157, 1978.

5. Gagliardi ART et al: GAPO syndrome: Report of three affected brothers. *Am J Med Genet* **19**:217–224, 1984.

6. Gorlin RJ: Thoughts on some new and old bone dysplasias, in *Skeletal Dysplasias,* Papadatos CH, Bartsocas CS (eds), Alan R. Liss, New York, 1982, p 49.

7. Manouvrier-Hanu S et al: The GAPO syndrome: A new case. *Am J Med Genet* **26**:683–688, 1987 and *J Génét Hum* **36**:373–378, 1988.

8. Shapira Y et al: Case report 85: Growth retardation, alopecia, pseudoanodontia and optic atrophy. *Syndrome Ident* **8**(1):14–16, 1982.

8a. Silva EO: Dwarfism, alopecia, pseudoanodontia and other anomalies. *Rev Brasil Genet* **7**:743–747, 1984.

9. Tipton RE, Gorlin RJ: Growth retardation, alopecia, pseudo-anodontia, and optic atrophy—the GAPO syndrome. *Am J Med Genet* **19**:209–216, 1984.

10. Wajntal A et al: Nova sindrome de displasia ectodermica: Nanismo, alopecia, anodontia e cutis laxa. *Ciencia Culture* **34** *(Suppl):*705, 1982.

11. Wajntal A et al: GAPO syndrome (McKusick 23074)—a connective tissue disorder: Report on two affected sibs and the pathologic findings in the older. *Am J Med Genet* (in press).


## Brachymetapody, anodontia, hypotrichosis, and albinoid trait (Tuomaala–Haapanen syndrome)

Tuomaala and Haapanen (1) reported three sibs affected with short stature and shortening of all digits other than thumbs and halluces due to brachymetapody. Hypoplastic maxilla, anodontia, generalized hypotrichosis, hypoplastic breasts and genitalia, convergent strabismus, distichiasis, hypoplastic tarsus, cataracts, myopia, irregular nystagmus, and albinoid skin were found (Fig. 25–8A–D). The syndrome presumably has autosomal recessive inheritance.

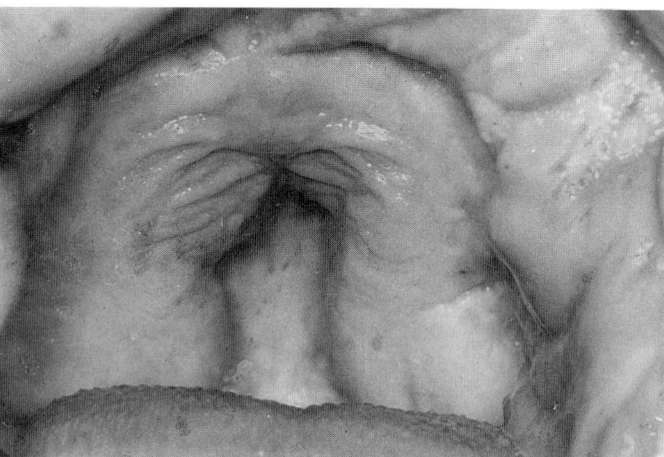

A

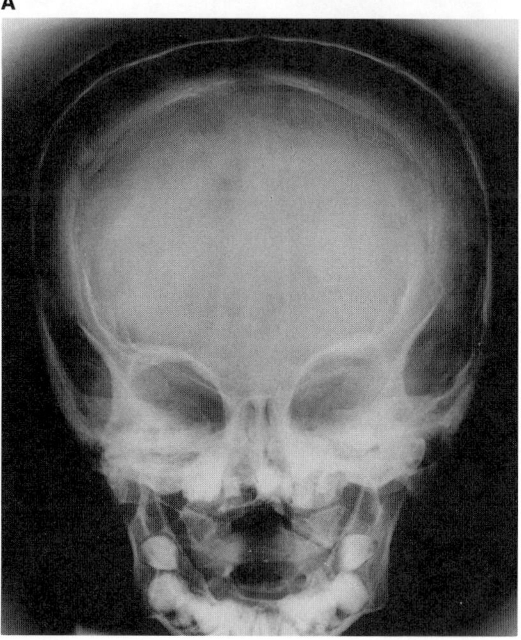

B

Fig. 25–7. *Growth retardation, alopecia, pseudoanodontia, and optic atrophy (GAPO syndrome).* (A) Edentulous jaws. (B) Note numerous unerupted teeth. (From *TH Andersen* and *JJ Pindborg,* Odont T **55**:484, 1947.)

#### Reference [Brachymetapody, anodontia, hypotrichosis, and albinoid trait (Tuomaala–Haapanen syndrome)]


1. Tuomaala P, Haapanen E: Three siblings with similar anomalies in the eyes, bones, and skin. *Acta Ophthalmol* **46**:365–371, 1968.


## Broad terminal phalanges and facial abnormalities (Keipert syndrome)

Male sibs with severe sensorineural hearing loss, unusual facies, and broad terminal phalanges were described by Keipert et al (1973). Autosomal or X-linked recessive inheritance is likely.

The facies was marked by a large nose with a high bridge, large rounded columella, and prominent nasal alae. The upper lip protruded, with a cupid's bow configuration overlapping the rather straight lower lip laterally (Fig. 25–9A,B). The distal phalanges of the thumbs, the first, second, and third fingers, and all the toes were remarkably broad. The fifth fingers were short and clinodactylous. The toes were rotated medially (Fig. 25–9C,D). Roentgenographically, one sib had bifid terminal phalanges in both index fingers. In the halluces of both patients, the proximal phalanges were short and the terminal phalanges were short with large, rounded epiphyses.

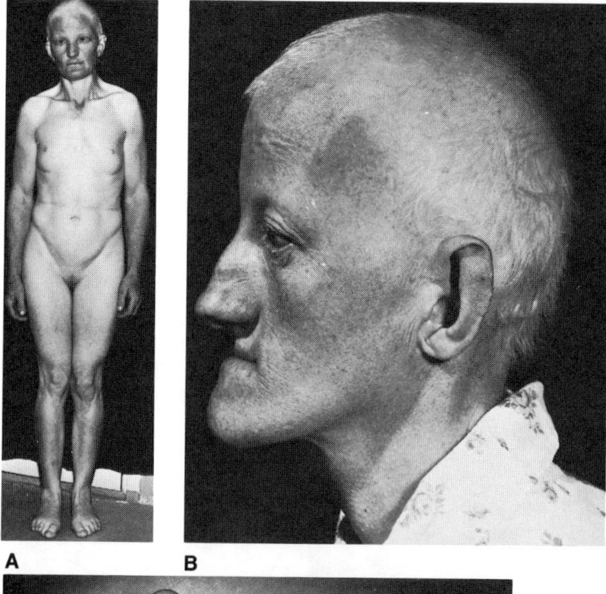

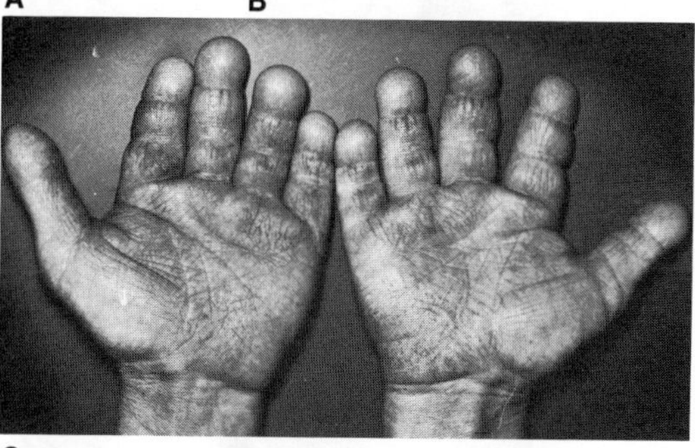

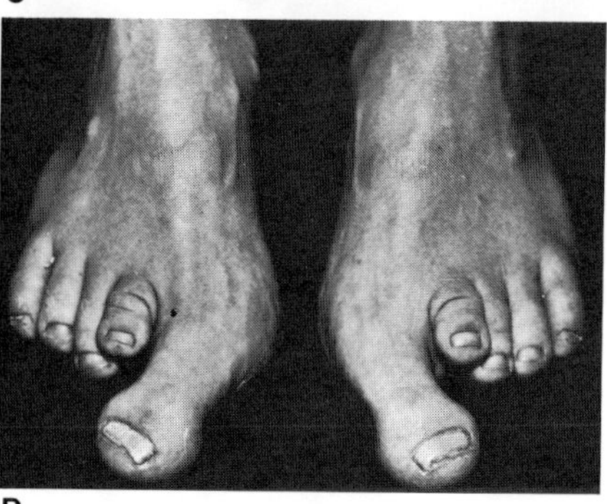

Fig. 25–8. *Brachymetapody, anodontia, hypotrichosis, and albinoid trait (Tuomaala-Haapanen syndrome).* (A) Albinoid skin, hypoplastic breasts, generalized hypotrichosis. (B) Sunken midface with loss of vertical height due to anodontia. (C,D) Shortening of all digits except thumbs and halluces. (From *P Tuomaala* and *E Haapanen,* Acta Ophthalmol **46**:365, 1968.)

#### Reference [Broad terminal phalanges and facial abnormalities (Keipert syndrome)]

1. Keipert JA et al: A new syndrome of broad terminal phalanges and facial abnormalities. *Aust Paediatr J* **9**:10–13, 1973.

## Cardio-facio-cutaneous syndrome (CFC syndrome)

In 1986, Reynolds et al (8) described eight patients having a characteristic facial appearance, ectodermal abnormalities, growth failure, and variable cardiac defects. Additional patients were reported (1–7,9,10). RJ Gorlin and CE Scott (personal communication, 1987) have seen similarly affected patients.

The disorder may have dominant inheritance (3,7).

The facies was characterized by high forehead with bitemporal constriction, hypoplasia of supraorbital ridges, ptosis, downslanting palpebral fissures, low nasal bridge, and posteriorly angulated pinnae with prominent helices. Hair was sparse and friable (Fig. 25–10A–C). Skin changes varied from patchy hyperkeratosis to a severe generalized ichthyosis-like condition. The facies is reminiscent of that seen in *Noonan syndrome*.

In three of eight cases there was polyhydramnios. Postnatal growth retardation was present in six of eight patients, but prenatal growth was usually normal. The head appeared disproportionately large. The cranial vault was high and somewhat boxy in appearance. The hair was sparse and rather curly with hypoplasia or absence of eyebrows. The dry hyperkeratotic skin was noted especially over the extensor surfaces of the limbs. Cutaneous hemangiomas were common. Esotropia or nystagmus has been present in half the patients. Submucous cleft palate has been noted (2).

Developmental disability or mental retardation was a constant feature and hypotonia was common. A mild degree of neck webbing was present in four of eight children. Splenomegaly was noted in half the patients. The most common cardiac anomalies were pulmonic stenosis and ASD.

CT scan may show ventricular enlargement, and cortical atrophy (2,6,8). Disturbed EEG patterns have been found.

#### References [Cardio-facio-cutaneous syndrome (CFC syndrome)]

1. Baraitser M, Patton MA: A Noonan-like short stature syndrome with sparse hair. *J Med Genet* **23**:161–164, 1986.

2. Chrzanowska K et al: Cardio-facio-cutaneous (CFC) syndrome: Report of a new patient. *Am J Med Genet* **33**:471–473, 1989.

3. Hughes HE et al: CFC and Noonan syndromes: Same or different? *David W. Smith Workshop on Malformations and Morphogenesis,* Madrid, Spain, 23–29 May, 1989.

4. Lupi M et al: A new case of CFC syndrome with autosomal dominant inheritance. *David W. Smith Workshop on Malformations and Morphogenesis,* Madrid, Spain, 23–29 May, 1989.

5. Mucklow ES: A case of cardio-facio-cutaneous syndrome. *Am J Med Genet* **33**:474–475, 1989.

6. Neri G et al: The CFC syndrome—report of the first two cases outside the United States. *Am J Med Genet* **27**:767–772, 1987.

7. Pierini DO, Pierini AM: Keratosis pilaris atrophicans faciei (ulerythema ophyrogenes): A cutaneous marker in the Noonan syndrome. *Br J Dermatol* **100**:409–416, 1979.

8. Reynolds JF et al: New multiple congenital anomalies—mental retardation syndrome with cardio-facio-cutaneous involvement—the CFC syndrome. *Am J Med Genet* **25**:413–428, 1986.

9. Sorge G et al: CFC syndrome: Report on three additional cases. *Am J Med Genet* **33**:476–478, 1989.

10. Verloes A et al: CFC syndrome: A syndrome distinct from Noonan syndrome. *Ann Génét* **31**:230–234, 1988.

## Coarse face, mental retardation, sensorineural hearing loss, and skeletal abnormalities (Fountain syndrome)

Fountain (1), in 1974, briefly described four sibs. Two exhibited generalized facial edema with massive swelling of the upper and lower lips. All were mentally retarded with profound congenital sensorineural hearing loss. One sib died in early infancy, but had profound hearing loss, mental retardation, and spina bifida. Three sibs had granulomatous swelling of the lips (Fig. 25–11A,B). Three of four showed gross thickening of the calvaria (Fig. 25–12). In 1986, Fryns et al (3) confirmed the syndrome by reporting sibs with hearing loss, mental retardation, seizures, and round and coarse face with mild swelling of subcutaneous tissues, particularly those of the cheeks and lips. The hands and feet were short and plump with broad and short terminal phalanges (2).

Radiographically, there was marked thickening of the calvaria and

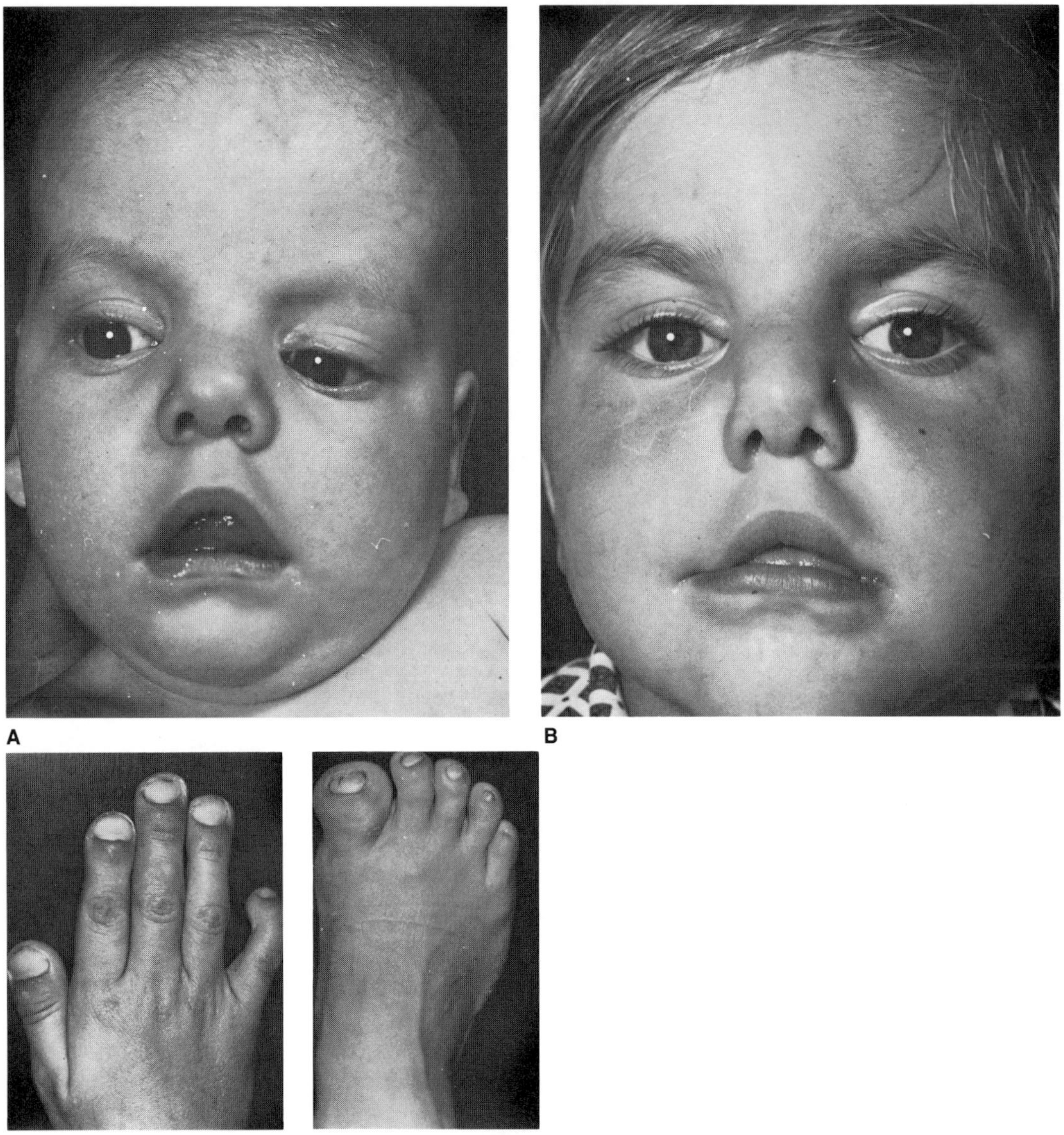

A B

C D

Fig. 25–9. *Broad terminal phalanges and facial abnormalities (Keipert syndrome).* (A,B) Nine-month- and three-year-old sibs exhibiting large nose with high bridge, large rounded columella, and prominent alae. Upper lip protrudes with cupid's bow configuration. (C) Terminal phalanges are short and broad. Fifth fingers are short and clinodactylous. (D) Toes short and broad, especially hallux. (From *JA Keipert et al,* Aust Paediatr J **9**:10, 1973.)

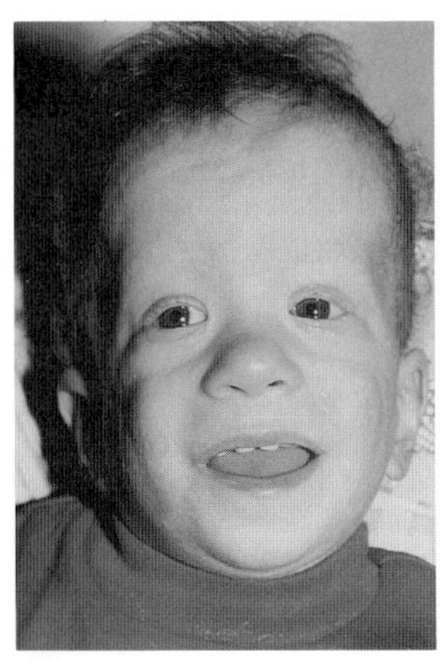

A

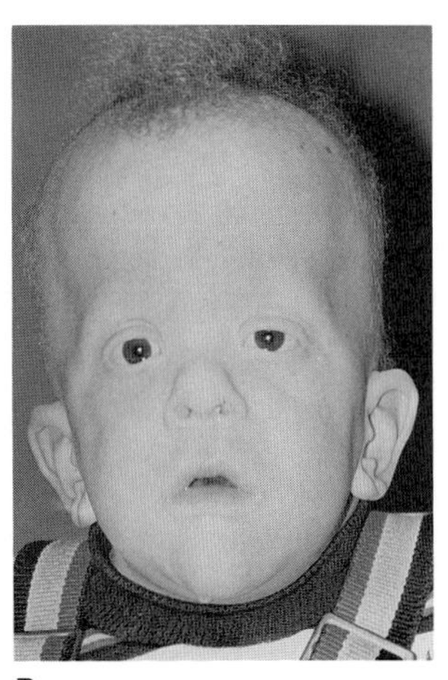

B

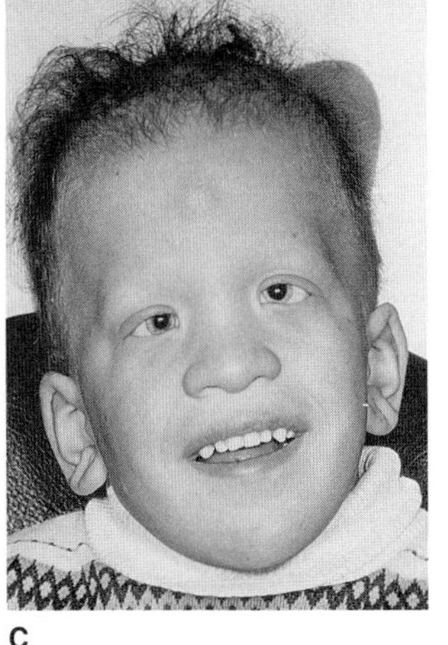

C

Fig. 25–10. *Cardio-facio-cutaneous syndrome.* (A,B) High forehead, bitemporal constriction, sparse friable curly hair, hypoplasia of supraorbital ridges, low nasal bridge, dysmorphic pinnae, short neck. Patients had dry hyperkeratotic skin over extensor surface of limbs, pulmonary stenosis, and atrial septal defect. (C) Compare facies of patient seen by RJ Gorlin. (A,B from *JF Reynolds et al,* Am J Med Genet **25**:413, 1986.)

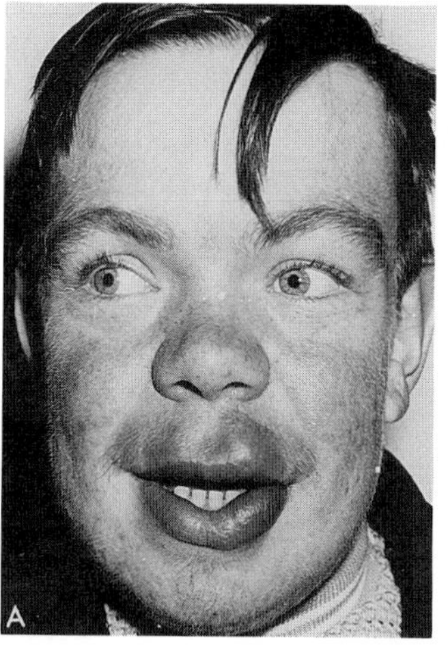


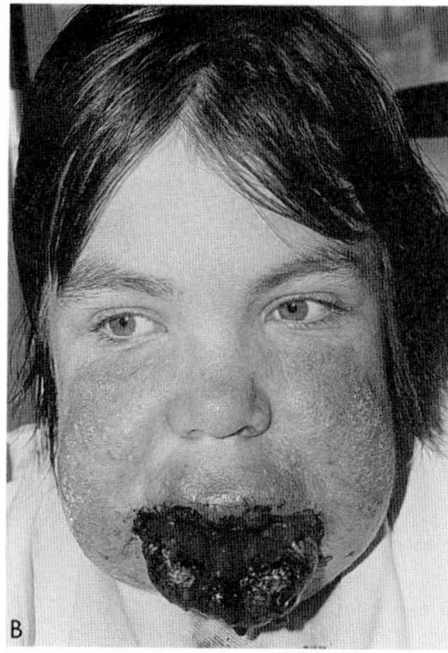


Fig. 25–11. *Coarse face, mental retardation, sensorineural hearing loss, and skeletal abnormalities (Fountain syndrome).* (A,B) Affected sibs exhibiting coarse facies and skin granulomata. (From *RB Fountain,* Proc R Soc Med **67**:878, 1974.)

thickened cortices of the bones of the hands and feet. Tomography of the temporal bones showed absence of cochlear windings.

Inheritance is presumably autosomal recessive.

There is some physical similarity to *mannosidosis* and *aspartylglycosaminuria,* but biochemical studies have been normal.

### References [Coarse face, mental retardation, sensorineural hearing loss, and skeletal abnormalities (Fountain syndrome)]


1. Fountain RB: Familial bone abnormalities, deaf mutism, mental retardation, and skin granulomas. *Proc R Soc Med* **67**:878–879, 1974.
2. Fryns JP: Fountain's syndrome: Mental retardation, sensorineural deafness, skeletal anomalies and coarse facies with full lips. *J Med Genet* **26**:722–724, 1989.
3. Fryns JP et al: Mental retardation, deafness, skeletal abnormalities and coarse face with full lips: Confirmation of the Fountain syndrome. *Am J Med Genet* **26**:551–556, 1987.


Fig. 25–12. *Coarse face, mental retardation, sensorineural hearing loss, and skeletal abnormalities (Fountain syndrome).* Radiograph showing marked thickening of calvaria. (From *RB Fountain,* Proc R Soc Med **67**:878, 1974.)

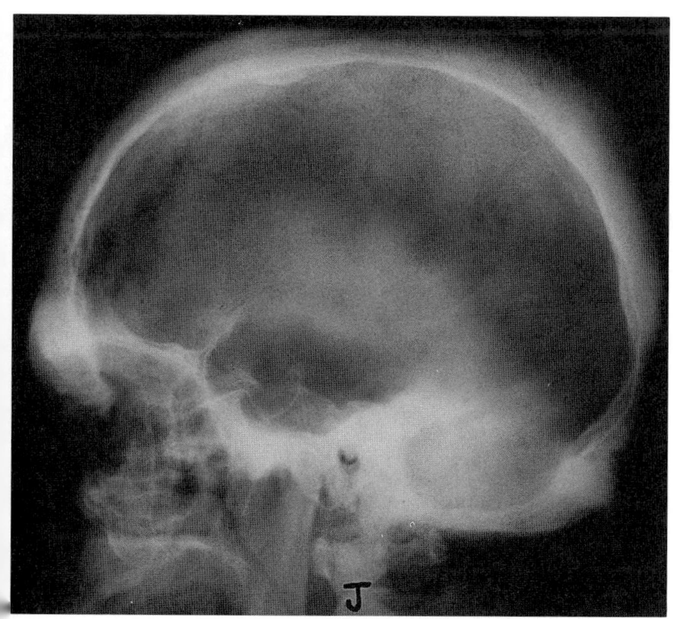

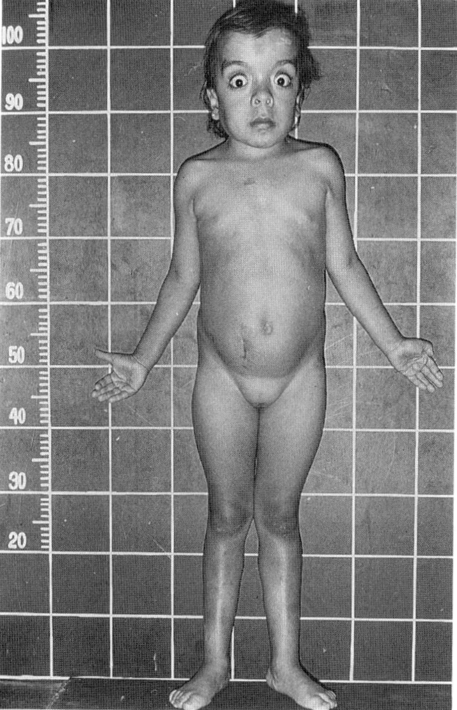

Fig. 25–13. *Cranio-facio-cardio-skeletal dysplasia.* Child with short stature and delayed psychomotor development. Note macrodolichocephaly with prominent forehead, scanty hair, hypertelorism, flat nasal bridge, exophthalmus, short nose with anteverted nostrils, long philtrum, and short neck. The abdomen was prominent and there was cubitus valgus. (From *JM Cantú,* Clin Genet **22**:172, 1982.)

## Cranio-facio-cardio-skeletal dysplasia

Cantú et al (1) reported four unrelated females with short stature (below the third percentile), delayed psychomotor development (60–70 IQ), macrodolichocephaly (above 90th percentile) with prominent forehead, scanty thin and hypopigmented hair, coarse facies, hypertelorism with flat nasal bridge, exophthalmos, short nose with anteverted nares, long philtrum, low-set ears, short neck, short wide thorax, prominent abdomen, cubitus valgus, cutis laxa with joint hyperelasticity, and wrinkled palms and soles (Fig. 25–13).

The ribs were slender but wider in the middle. The long bones were also slender and the vertebral bodies somewhat small.

### Reference (Cranio-facio-cardio-skeletal dysplasia)


1. Cantú et al: Individualization of a syndrome with mental deficiency, macrocranium, peculiar facies, and cardiac and skeletal anomalies. *Clin Genet* **22**:172–179, 1982.


## Familial osteodysplasia (Anderson syndrome)

Under this unfortunately nonspecific title, Anderson et al (1) reported four sibs with an apparently unique disorder. The parents were consanguineous. Inheritance is autosomal recessive.

The facies was characterized by prominent brows, flat nasal bridge, marked midface hypoplasia with relative mandibular prognathism, pointed chin, and large ear lobes (Fig. 25–14A–D). The feet were described as paddle-shaped.

Radiographically, thinning of calvaria, pointed mastoids and spinous processes of the cervical vertebrae, kyphoscoliosis, abnormal ribs, cortical thickening of femoral shafts, and thinning of the superior pubic rami were noted. The mandible had a widened angle, increased

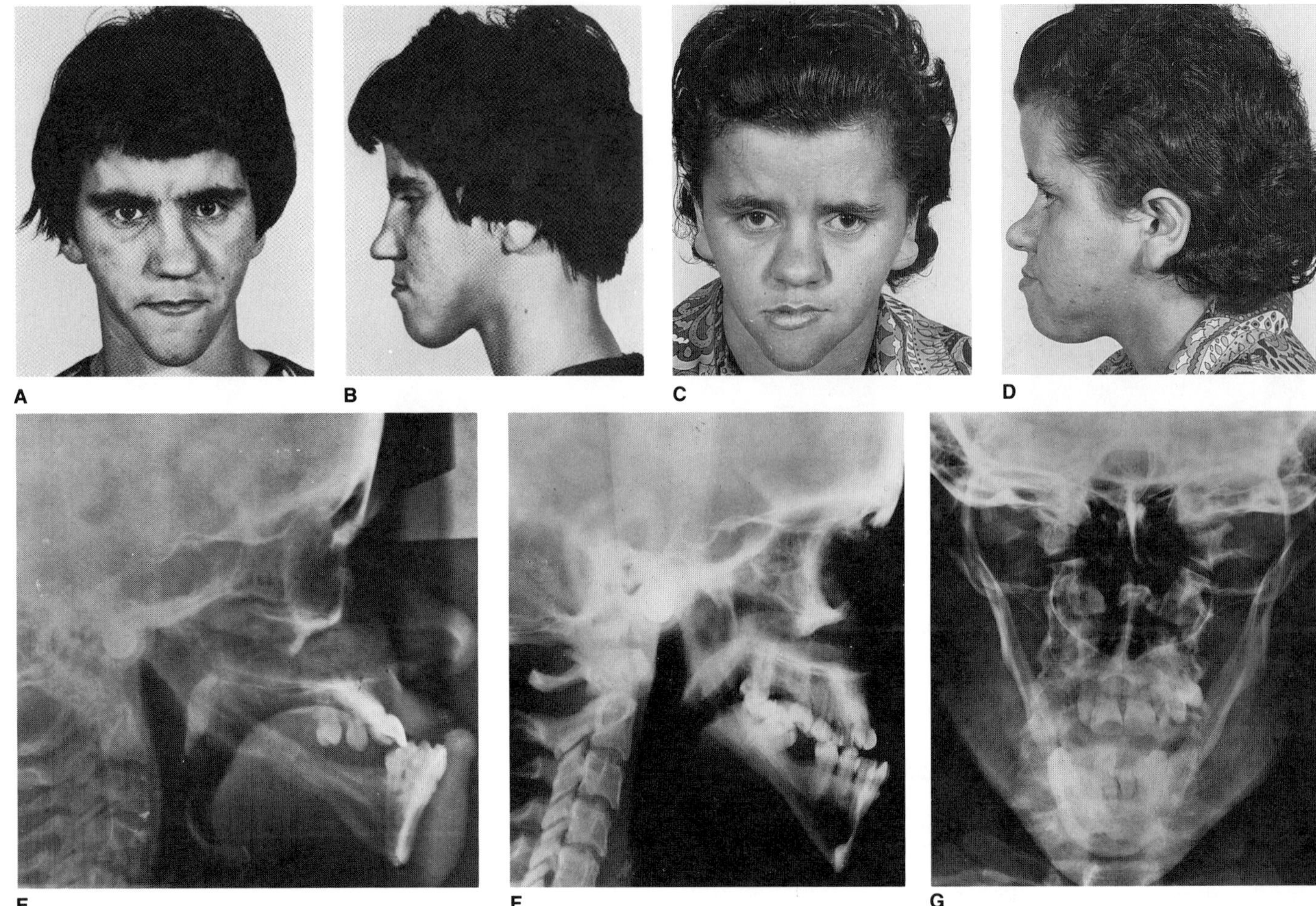


Fig. 25–14. *Familial osteodysplasia (Anderson syndrome).* (A,B) Triangular face with pointed chin and hypoplastic midface. (C,D) Sib of woman in A and B. (E) Mandible with widened angle, increased body length, reduced ramus height. (F,G) Note pointed symphysial area and marked thinning of body and ramus of mandible. (From *LG Anderson et al,* JAMA **220**:1687, 1972.)

body length, reduced ramus height, and severely decreased bigonial width (2) (Fig. 25–14E–G).

We have termed the naming of the disorder "unfortunate," because there are many different "osteodysplasias" and, in particular, *Melnick–Needles syndrome* has been called "familial osteodysplastia." The confusion has been compounded by Schendel and Delaire (3) who employed the same term to report an autosomal dominant condition in which craniosynostosis is a variable feature but which, in our opinion, has little in common with the Anderson syndrome.

### References [Familial osteodysplasia (Anderson syndrome)]

1. Anderson LG et al: Familial osteodysplasia. *JAMA* **220**:1687–1693, 1972.
2. Buchignani JS et al: Roentgenographic findings in familial osteodysplasia. *Am J Roentgenol* **116**:602–608, 1972.
3. Schendel S, Delaire J: Familial osteodysplasia. *Head Neck Surg* **4**:335–343, 1982.

## Flexion and extension deformities of the hands and unusual facies (Emery–Nelson syndrome)

Mild mental and somatic retardation, increased US/LS ratio, deformities of the first three metacarpophalangeal joints of both thumbs, clawed toes, and an unusual facies (long philtrum, flat midface) were described in a mother and daughter by Emery and Nelson (1) (Fig. 25–15A–C. RJ Gorlin has seen the same anomalies in a mother and her two sons.

### Reference [Flexion and extension deformities of the hands and unusual facies (Emery–Nelson syndrome)]

1. Emery AEH, Nelson MM: A familial syndrome of short stature, deformities of the hands and feet and an unusual facies. *J Med Genet* **7**:379–382, 1970.

## Hemimaxillofacial dysplasia

In 1987, Miles et al (2) reported a disorder in two patients consisting of unilateral enlargement of maxillary alveolar bone and gingiva associated with hypoplastic teeth, and facial asymmetry.One patient had hypertrichosis and missing maxillary premolars on the affected side (Figs. 25–16 to 25–18). The sporadically occurring disorder was detected at birth in both instances. Asymmetry was nonprogressive, and no other abnormalities were noted.

Melrose et al (1) subsequently reported eight cases and provided an excellent analysis of the radiographic and biopsy material. Because their patients had no facial hypertrichosis, they suggested a closely related variant which they named "segmental odontomaxillary dysplasia." Examination of their material clearly shows that it is a the same condition as hemimaxillofacial dysplasia because facial hypertrichosis is not an obligatory abnormality. While the term is descriptive, it does

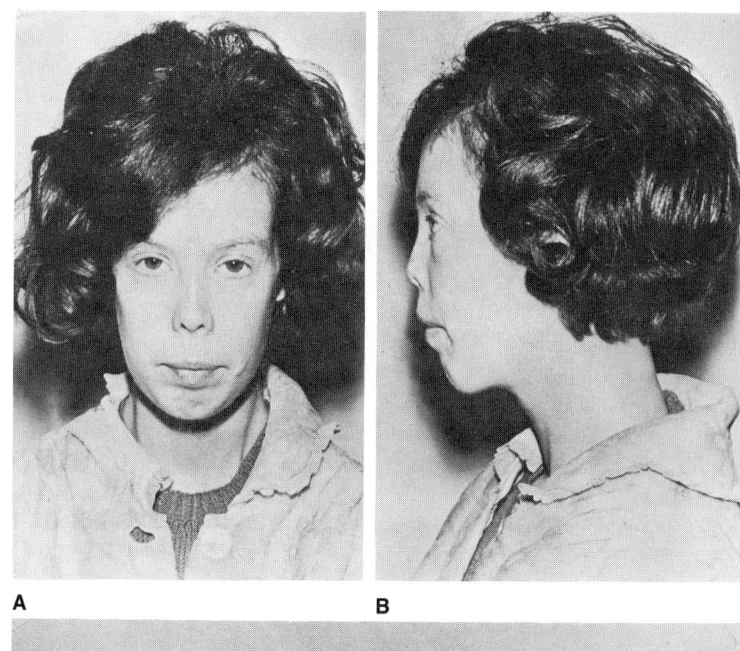


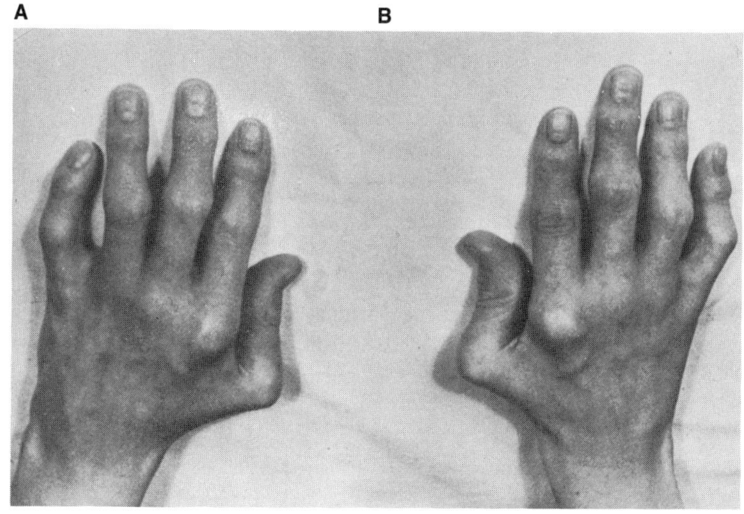


Fig. 25–15. *Flexion and extension deformities of the hands and unusual facies (Emery–Nelson syndrome).* (A,B) Flat midface with long philtrum. (C) Flexion deformity of first three metacarpophalangeal joints, extension deformity of interphalangeal joints of thumbs. (From *AEH Emery* and *MM Nelson,* J Med Genet **7**:379, 1970.)

Fig. 25–16. *Hemimaxillofacial dysplasia.* Facial asymmetry and unilateral hypertrichosis. (From *DA Miles et al,* Oral Surg Oral Med Oral Pathol **64**:445, 1987.)

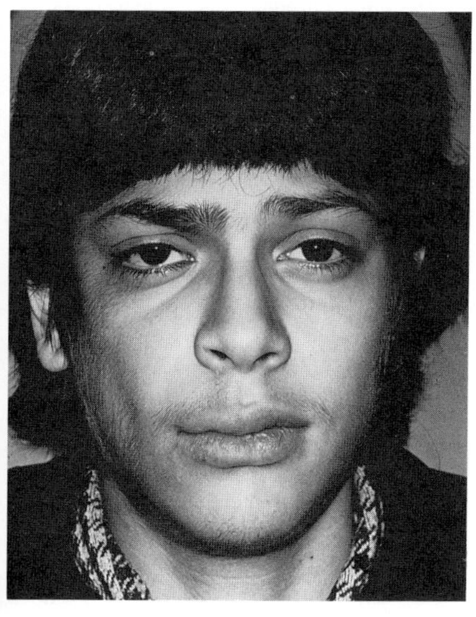

Fig. 25–17. *Hemimaxillofacial dysplasia.* Close-up showing hypertrichosis. (From *DA Miles et al,* Oral Surg Oral Med Oral Pathol **64**:445, 1987.)

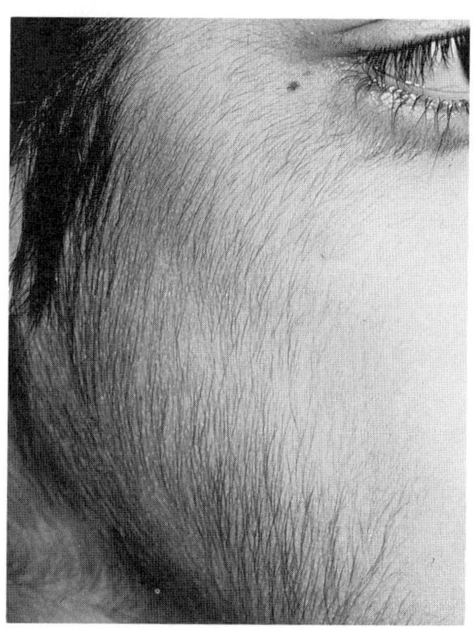

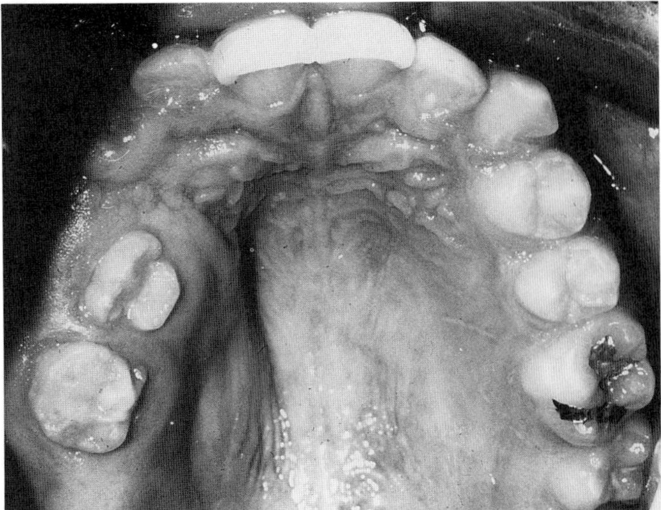

Fig. 25–18. *Hemimaxillofacial dysplasia.* Unilateral maxillary enlargement in association with hypoplastic and missing teeth. (From *DA Miles et al,* Oral Surg Oral Med Oral Pathol **64**:445, 1987.)

not convey the facial asymmetry that allows early recognition by the pediatrician.

### References (Hemimaxillofacial dysplasia)


1. Melrose R et al: Segmental odontomaxillary dysplasia: A variant of hemimaxillofacial dysplasia. *Oral Surg Oral Med Oral Pathol* (in press).
2. Miles DA et al: Hemimaxillofacial dysplasia: A newly recognized disorder of facial asymmetry, hypertrichosis of the facial skin, unilateral enlargement of the maxilla, and hypoplastic teeth in two patients. *Oral Surg Oral Med Oral Pathol* **64**:445–448, 1987.


## Oto-onycho-peroneal syndrome

Pfeiffer (2) described two male sibs with minor craniofacial malformations, multiple contractures, dysplastic pinnae, hypoplasia of nails, and hypoplasia of fibulae. The condition appears similar to that reported by Leiba et al (1), although the skeletal changes are somewhat different.

The craniofacies was characterized by dolichocephaly, flaring of temporal regions, flat facial contour, upward slanting palpebral fissures, large low-set posteriorly angulated pinnae with unfolded helix, prominent anthelix and superior crus, and hypoplastic lobule (Fig. 25–19A,B).

The thumb, index, and middle fingers were stiff with flattened knuckle creases and absent to rudimentary nails. The hallux and second toes were similarly affected (Fig. 25–20A,B). The thighs were short with abnormally shaped lower limbs. Pes calcaneovalgus was evident. The fibulae were hypoplastic to absent.

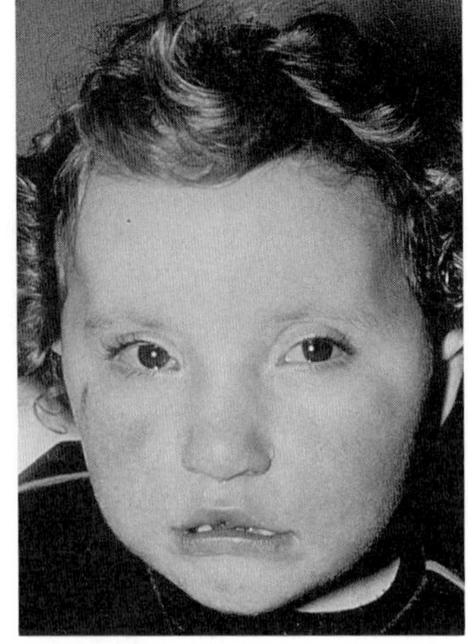

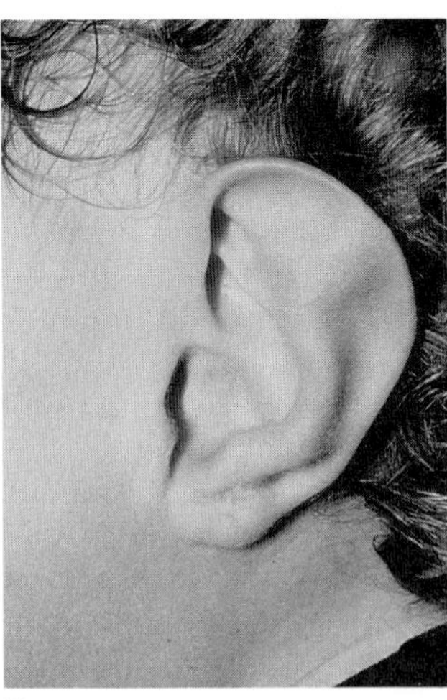

A B

Fig. 25–19. *Oto-onycho-peroneal syndrome.* (A) Mildly abnormal facies characterized by flaring of temporal regions, flat facial contour, upward-slanting palpebral fissures. (B) Pinna with unfolded helix, prominent anthelix and superior crus, hypoplastic lobules. (Courtesy of *R Pfeiffer,* Erlangen, West Germany.)

### References (Oto-onycho-peroneal syndrome)


1. Leiba S et al: Oculootonasal malformations associated with osteoonychodysplasia. *Birth Defects* **11**(2):67–73, 1975.
2. Pfeiffer RA: The oto-onycho-peroneal syndrome. *Eur J Pediatr* **138**:317–320, 1982.


## Short stature, characteristic facies, mental retardation, skeletal anomalies, and macrodontia (KBG syndrome)

In 1975, Herrmann et al (2) described an autosomal dominantly inherited syndrome of mild mental retardation, short stature, characteristic facies, macrodontia, and various skeletal anomalies. Similarly affected kindreds were described by several authors (1,3,4).

Height is below the third percentile. Intelligence quotients have ranged between 40 and 65. Several patients have a speech defect.

The craniofacial appearance is characterized by brachycephaly, wide biparietal diameter, round face, telecanthus, anteverted nostrils in some instances, outstanding pinnae, wide eyebrows with synophrys, and widely arched maxilla with short alveolar ridge (Figs. 25–21 and 25–22). Several patients have manifested oligodontia. The central and lateral incisors are enlarged. Their incisal edges have supernumerary mamelons, some as many as five (Fig. 25–23). The philtrum tends to be long. The vermilion of the lips is thin and lip shape is that of a hunter's bow.

Skeletal changes include those previously described as well as Wor-

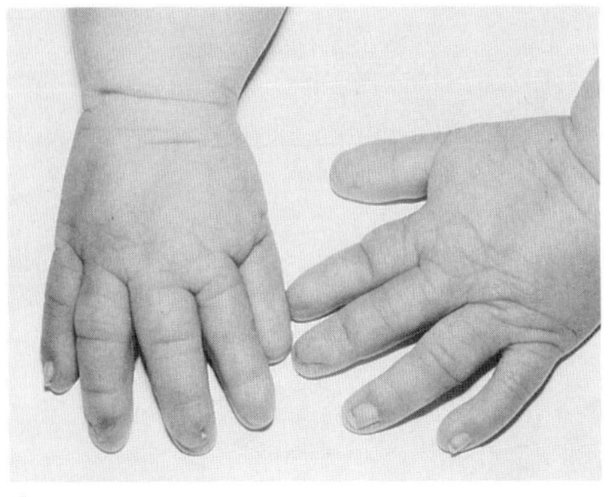

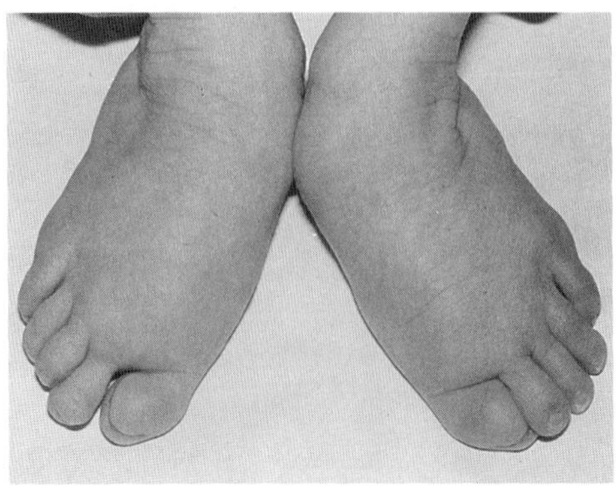

A B

Fig. 25–20. *Oto-onycho-peroneal syndrome.* (A,B) Hands and feet showing absent to rudimentary nails of middle and index fingers and thumb; similar changes in toes. Knuckle creases flattened. (Courtesy of *R Pfeiffer,* Erlangen, West Germany.)

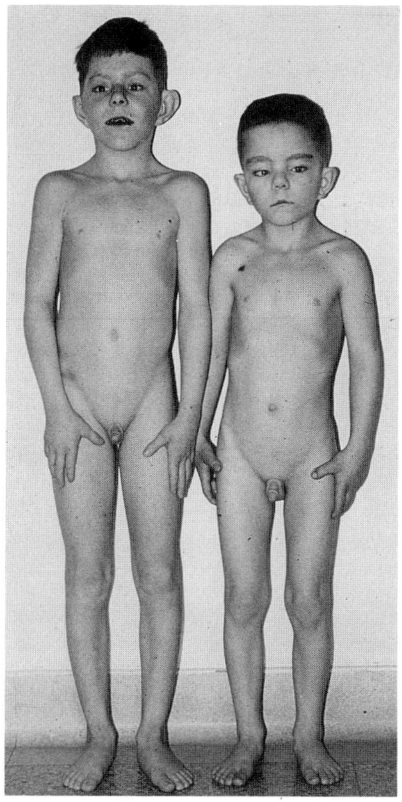

Fig. 25–21. *Short stature, characteristic facies, mental retardation, skeletal anomalies, and macrodontia (KBG syndrome).* Sibs with short stature, cryptorchidism. The facies is triangular with redundant eyebrows, telecanthus, high nasal bridge, outstanding pinnae. [From *J Herrmann et al,* Birth Defects **11**(5):7, 1975.]

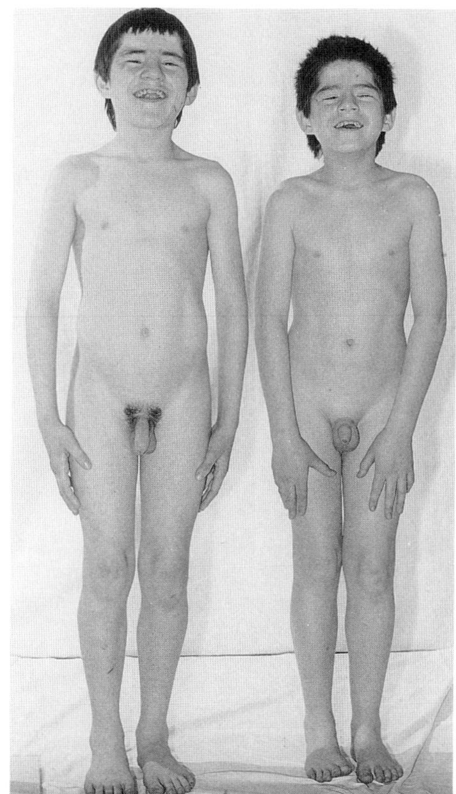

Fig. 25–22. *Short stature, characteristic facies, mental retardation, skeletal anomalies, and macrodontia (KBG syndrome).* Compare Fig. 25–21 with short mentally retarded sibs with similar facies, low implantation of frontal and temporal hair, small palpebral fissures, telecanthus. (From *C Parloir et al,* Clin Genet **12**:263, 1977.)

mian bones, cervical ribs, various vertebral anomalies (anterior notching, hypoplasia of upper and lower plates, open neural arches, block vertebrae), short tubular hand bones, short femoral neck, and delayed bone age (Fig. 25–24).

Dermatoglyphic findings consist of single palmar crease and distal axial triradius.

### References [Short stature, characteristic facies, mental retardation, skeletal anomalies, and macrodontia (KBG syndrome)]

1. Fryns JP, Haspeslagh M: Mental retardation, short stature, minor skeletal abnormalities, craniofacial dysmorphism and macrodontia in two sisters and their mother. Another variant example of the KBG syndrome. *Clin Genet* **26**:69–72, 1984.

2. Herrmann J et al: The KBG syndrome—a syndrome of short stature, characteristic facies, mental retardation, macrodontia and skeletal anomalies. *Birth Defects* **11**(5):7–18, 1975.

3. Parloir C et al: Short stature, craniofacial dysmorphism and dento-skeletal abnormalities in a large kindred. *Clin Genet* **12**:263–266, 1977.

4. Tollaro I et al: Anomalie dento-maxillo-facciali nella "Sindrome KBG." *Minerva Stomatol* **33**:437–446, 1984.

## Unusual facies, abnormal scalp hair, and diarrhea

Stankler et al (1) reported two sibs with low birthweight, unusual facies characterized by prominent eyes with flat supraorbital ridges, broad flat nose, large mouth, cleft uvula (in one), and large low-set simple pinnae. The buttocks were excoriated before the onset of diarrhea. The skin had a rubbery feel. The scalp hair was woolly and easily removed. The hair shafts exhibited budding. Persistent diarrhea began in the third week and failure to thrive became evident. Galactosuria without galactosemia, hepatic cirrhosis, renal cortical microcysts, and islet cell hyperplasia were noted on autopsy.

### Reference (Unusual facies, abnormal scalp hair, and diarrhea)

1. Stankler L et al: Unexplained diarrhea and failure to thrive in 2 siblings with unusual facies and abnormal scalp hair shafts: A new syndrome. *Arch Dis Childh* **57**:212–216, 1982.

Fig. 25–23. *Short stature, characteristic facies, mental retardation, skeletal anomalies, and macrodontia (KBG syndrome).* Cast of teeth showing macrodontia. [From *J Herrmann et al,* Birth Defects **11**(5):7, 1975.]

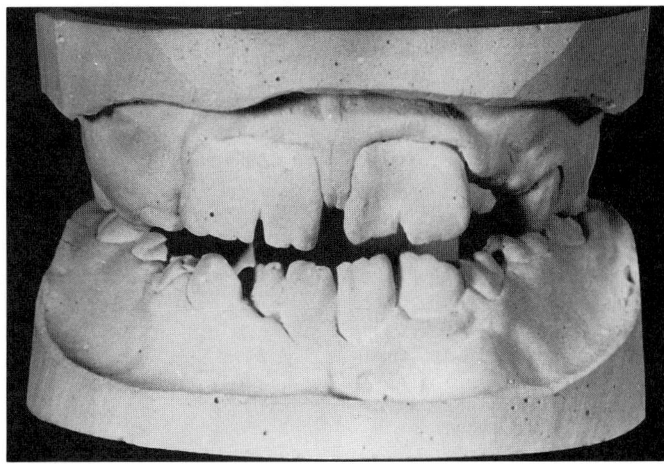

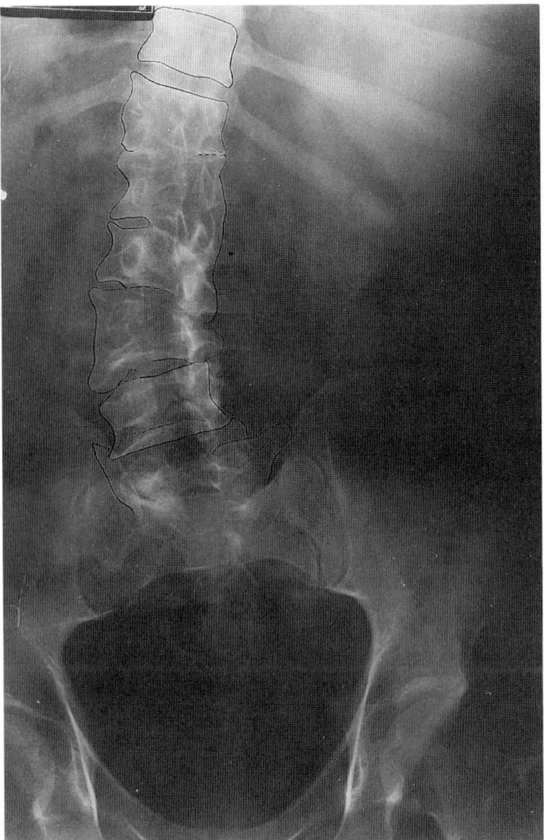

Fig. 25–24. *Short stature, characteristic facies, mental retardation, skeletal anomalies, and macrodontia (KBG syndrome).* Block vertebrae of lower spine. [From *J Herrmann et al,* Birth Defects **11**(5):7, 1975.]

## Unusual facies, congenital hypothyroidism, and severe mental retardation

Ohdo et al (1) reported siblings affected with tetra-amelia, hypotrichosis, upslanting palpebral fissures, absent lacrimal openings, hypoplastic ducts and sacs opening toward the exterior, prominent bulbous nose, large downturned mouth, preauricular pits, and developmental retardation (Fig. 25–25A–C). Inheritance is probably autosomal recessive.

#### Reference (Unusual facies, tetra-amelia, hypotrichosis, and developmental retardation)

1. Ohdo S et al: Association of tetra-amelia, ectodermal dysplasia, hypoplastic lacrimal ducts and sacs opening toward the exterior, peculiar face and developmental mental retardation. *J Med Genet* **24**:609–612, 1987.

## Unusual facies, agenesis or hypoplasia of nails and terminal phalanges, lateral clavicles, and retinal colobomas

Leiba et al (1) described a male with mild mental retardation, short stature, ptosis of eyelids, downslanting palpebral fissures and agenesis or hypoplasia of nails and terminal phalanges. The parents were consanguineous. We suspect that the condition is identical to *oto-onychoperoneal syndrome* (2), although the skeletal changes are somewhat different.

Eyelashes were on the medial half of the lids and there were bilateral colobomas of the retina. The pinnae were dysmorphic with prominent anthelix. The halluces were broad with absence of toenails. The thumbs were not broad but had dysplastic nails like those of the index fingers. The phallus was enlarged and there was hypospadias.

Radiographic examination revealed hypoplasia of malar bones and lateral clavicles. The metaphyses of the long bones were somewhat widened. The terminal phalanges of the thumbs and halluces were abbreviated. Carpal and tarsal synostosis was evident.

#### References (Unusual facies, agenesis or hypoplasia of nails and terminal phalanges, lateral clavicles, and retinal colobomas)

1. Leiba S et al: Oculootonasal malformations associated with osteoonychodysplasia. *Birth Defects* **11**(2):67–73, 1975.
2. Pfeiffer RA: The oto-onycho-peroneal syndrome. *Eur J Pediatr* **138**:317–320, 1982.

## Unusual facies characterized by short philtrum, patulous lips, ptosis, and low-set pinnae

Char (1) described autosomal dominant inheritance of short philtrum, patulous lips, ptosis, and low-set ears. Intelligence was low–normal.

The forehead was broad, the eyes slightly wide-set with ptosis of the lids and internal strabismus. The nose was short with somewhat anteverted nostrils and flat nasal bridge. The ears were low set. The mouth was large with extremely patulous lips and very short philtrum (Figs. 25–26A,B and 25–27). Only two phalanges were in each fifth finger of the proband. Patent ductus arteriosus was found in several members of the kindred.

A

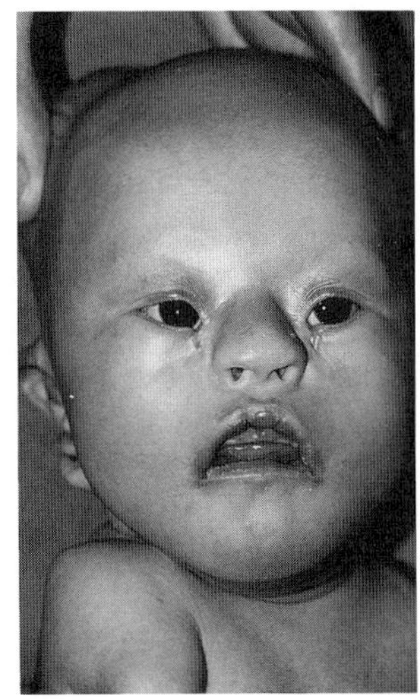

B

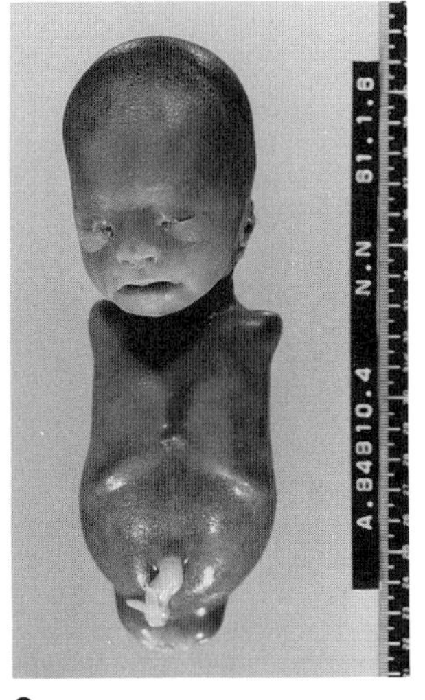

C

Fig. 25–25. *Unusual facies, tetra-amelia, hypotrichosis, and developmental retardation.* (A,B) Male child with tetra-amelia, hypotrichosis, upslanting palpebral fissures, and hypoplastic lacrimal ducts and sacs opening toward the exterior. (C) Sib of infant seen in A,B. (From *S Ohdo et al,* J Med Genet **24**:609, 1987.)

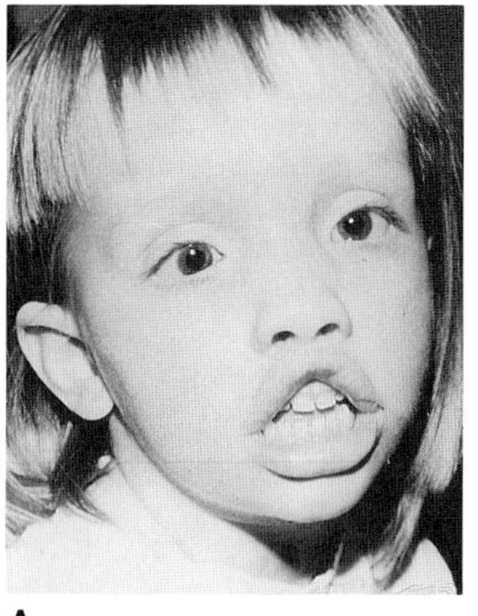

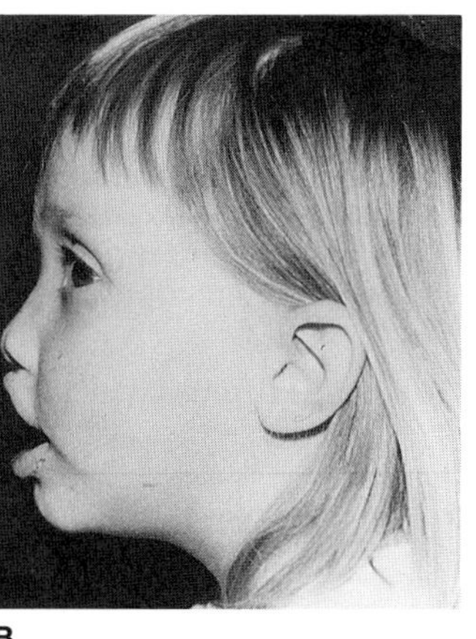

A B

Fig. 25–26. *Unusual facies characterized by short philtrum, ptosis, and low-set pinnae.* (A,B) Child has broad forehead, extremely short philtrum with patulous lips, internal strabismus, and short nose. Mother has similar appearance. (Courtesy of *CI Scott,* Wilmington, Delaware.)

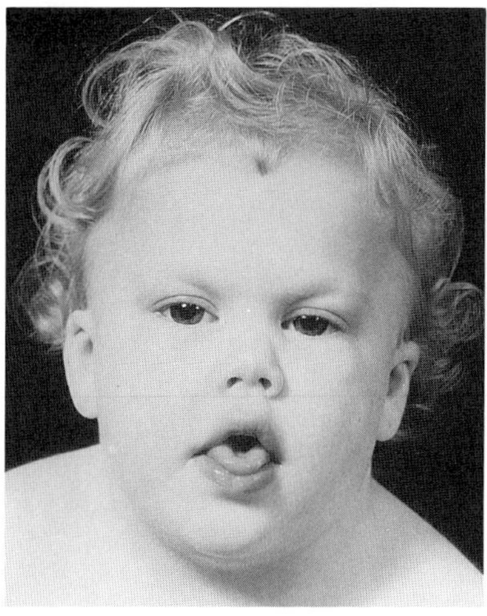

Fig. 25–27. *Unusual facies characterized by short philtrum, ptosis, and low-set pinnae.* Similar phenotype in another child. [From *F Char,* Birth Defects **14**(6B):303, 1978.]

### Reference (Unusual facies characterized by short philtrum, patulous lips, ptosis, and low-set pinnae)

1. Char F: Peculiar facies with short philtrum, duck-bill lips, ptosis, and low-set ears—a new syndrome? *Birth Defects* **14**(6B):303–305, 1978.

2. Hurst JA et al: A syndrome of mental retardation, short stature, hemolytic anemia, delayed puberty, and abnormal facial appearance: Similarities to a report of aldolase A deficiency. *Am J Med Genet* **28**:965–970, 1987.

## Unusual facies, mental retardation, short stature, hemolytic anemia, and delayed puberty

Hurst et al (2) in 1987, described male and female sibs exhibiting short stature, hypertonia, and moderate retardation with significant delay in speech and language development. In 1973, Beutler et al (1) described a similar phenotype in a male patient.

The facies was characterized by midface hypoplasia, ptosis, epicanthic folds, small mandible, cup-shaped pinnae, short neck, posterior hairline, and delayed puberty (Fig. 25–28). The male had a small penis.

### References (Unusual facies, mental retardation, short stature, hemolytic anemia, and delayed puberty)

1. Beutler E et al: Red cell aldolase deficiency and haemolytic anaemia: a new syndrome. *Trans Assoc Am Physicians* **86**:154–166, 1973.

## Unusual facies, osteosarcoma, and malformation syndrome

Schuman and Burton (1) described, in 1979, two unrelated females with frontal bossing, highly arched palate, low-set ears, small auditory canals, prominent beaked nose, malocclusion, and micrognathia. One had exotropia and bilateral cataracts; the other had severe myopia. One had sensorineural and the other had conductive hearing loss. Both had repeated urinary infections, one having deformed ureterovesical junctions and the other having duplicated right kidney and ureter. Each had osteosarcoma of an extremity.

### References (Unusual facies, osteosarcoma, and malformation syndrome)

1. Schuman SH, Burton WE: A new osteosarcoma/malformation syndrome. *Clin Genet* **15**:462–463, 1979.

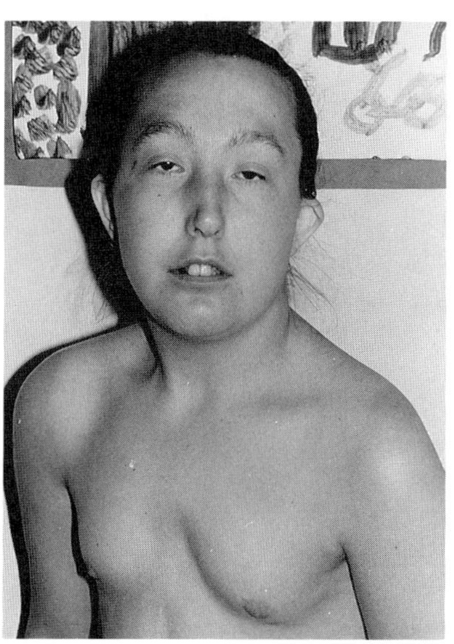

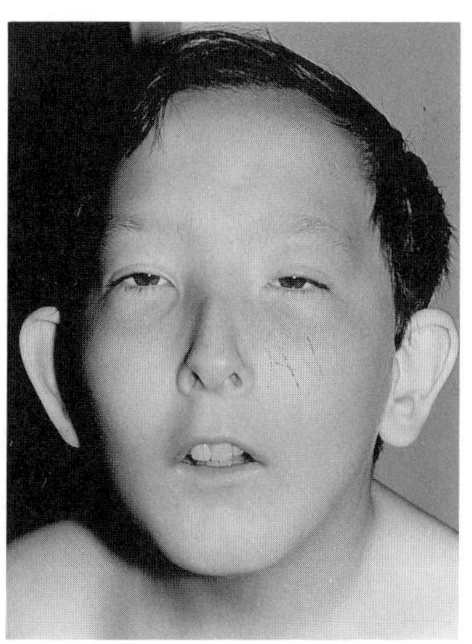

A B

Fig. 25–28. *Unusual facies, mental retardation, short stature, hemolytic anemia, and delayed puberty.* (A) Twenty-six-year-old sib exhibiting ptosis, prominent pinnae, short neck, and absent breast development. (B) Same facial appearance and anomalies as in his sister. (From *JA Hurst et al,* Am J Med Genet **28**:965, 1987.)

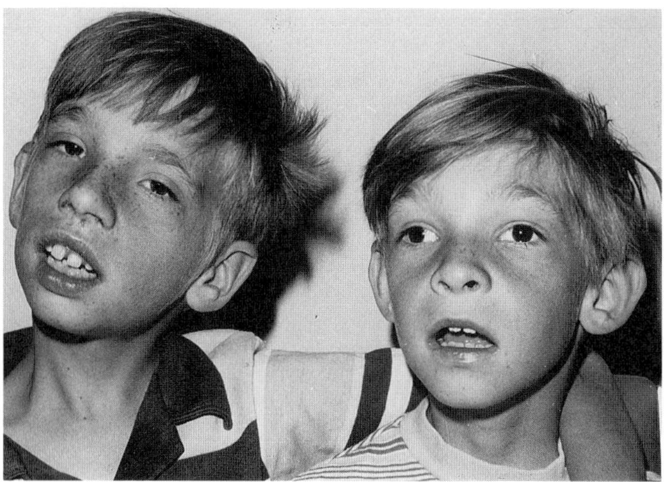

Fig. 25–29. *Unusual facies, short stature, and mental deficiency*. Microdolichocephaly, flat philtrum, prominent central maxillary incisor. (From *RD Smith et al*, Am J Med Genet **7**:5, 1980.)

## Unusual facies, short stature, and mental deficiency (Smith–Fineman–Myers syndrome)

Smith et al (1) and Stephenson and Johnson (2) reported a syndrome of unusual facial appearance, short stature, and mental retardation. Inheritance is probably autosomal recessive or X-linked recessive.

Microdolichocephaly was combined with decreased frontonasal angle, flat philtrum, prominent maxillary central incisors, and micrognathia (Fig. 25–29). Short stature, usually of prenatal onset, seizures, hypertonia, hyperreflexia, and mental retardation were noted. Other findings included chest deformity, mild foot deformities, and bridged palmar creases.

### References [Unusual facies, short stature, and mental deficiency (Smith–Fineman–Myers syndrome)]

1. Smith RD et al: Short stature, psychomotor retardation, and unusual facial appearance in two brothers. *Am J Med Genet* **7**:5–9, 1980.

2. Stephenson LD, Johnson JP: Smith–Fineman–Myers syndrome: Report of a third case. *Am J Med Genet* **22**:301–304, 1985.

## Urofacial syndrome (Ochoa syndrome)

Elejalde (1) reported a syndrome of hydronephrosis, hydroureter, and unusual facies in seven children from three families. Ochoa and Gorlin (2), in 1987, found 36 examples in 22 families. Inheritance is clearly autosomal recessive.

Facial grimacing is evident only while crying or smiling (Fig. 25–30A,B). Obstructive abnormalities of the urinary tract due to occult neuropathic bladder results in mucosal hypertrophy, trabeculation, and diverticula of the bladder, hydroureter, and hydronephrosis (Fig. 25–31). All affected males have cryptorchidism. About 65% manifest moderate to severe constipation.

Fig. 25–30. *Urofacial syndrome*. (A,B) Normal facies, except when smiling. (From *B Ochoa* and *RJ Gorlin*, Am J Med Genet **27**:661, 1987.)

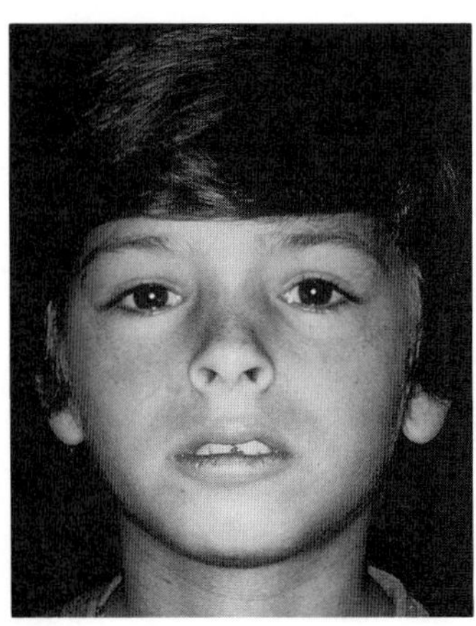

A

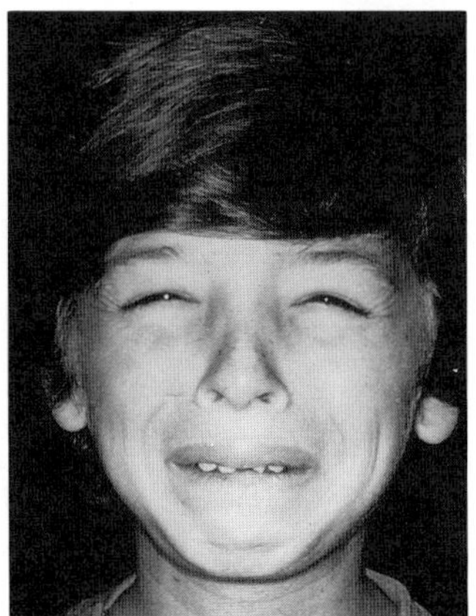

B

### References [Urofacial syndrome (Ochoa syndrome)]

1. Elejalde BR: Genetic and diagnostic considerations in three families with abnormalities of facial expression and congenital urinary obstruction: The Ochoa syndrome. *Am J Med Genet* **3**:97–108, 1979.

2. Ochoa B, Gorlin RJ: Urofacial syndrome. *Am J Med Genet* **27**:661–668, 1987.

## FACES syndrome

Friedman and Goodman (1), in 1984, described a new syndrome with unique **F**acies, **A**norexia, **C**achexia, and **E**ye and **S**kin lesions, employing the term *FACES syndrome* as an acronym. Affected were two sisters and their mother. They stemmed from a consanguineous union of Yemenite Jewish ancestry. The mode of inheritance has not been established.

The facies was characterized by mild eyelid ptosis, xanthelasma, and a nose with anteverted nostrils with bifid tip (Fig. 25–32A,B). The two daughters also exhibited nasal speech (apparently associated with a short palate) and retinitis pigmentosa. Most striking was severe muscle wasting and cachexia, the individuals appearing almost cadaverous (Fig. 25–32C). They exhibited numerous lentigines and café-au-lait spots (Fig. 25–32D). The daughters had pectus excavatum and asymmetry of the chest. All exhibited genu varum and pes planus.

Fig. 25–31. *Urofacial syndrome*. Radiograph showing hydroureter and hydronephrosis. (From *B Ochoa* and *RJ Gorlin*, Am J Med Genet **27**:661, 1987.)

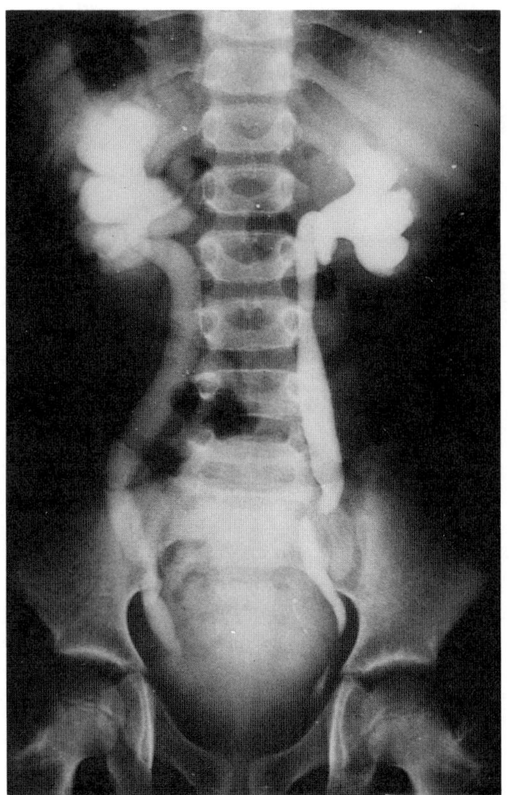

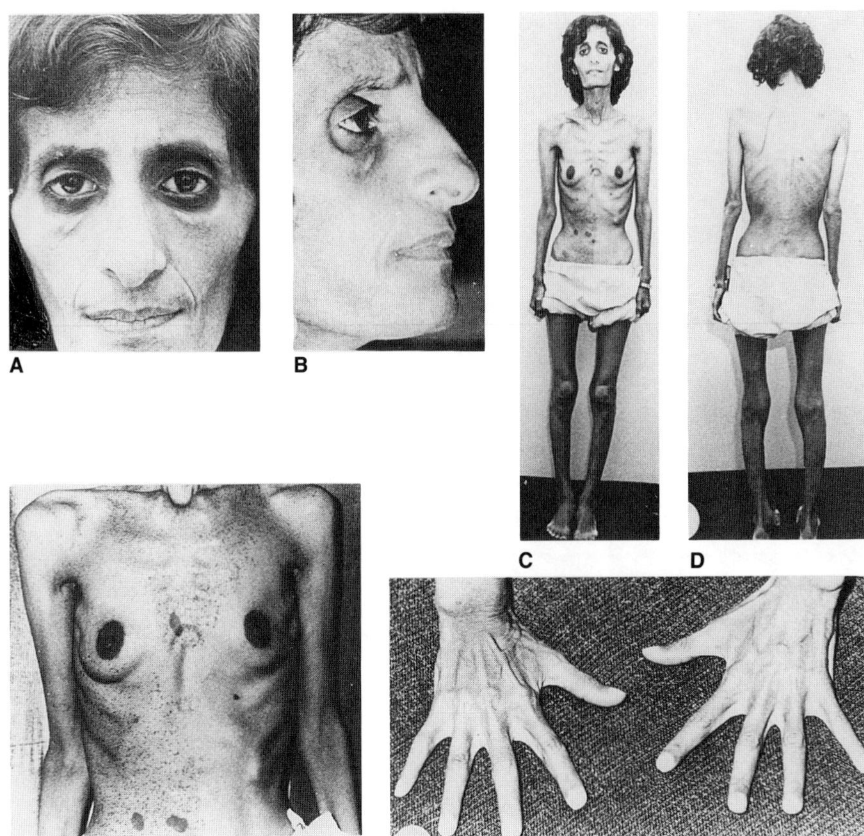


Fig. 25–32. *FACES syndrome*. (A,B) Mild ptosis, xanthelasma. (C,D) Extreme cachexia, genu varum. (E) Multiple lentigines and café-au-lait spots. (F) Mild soft tissue syndactyly. (From *E Friedman* and *RM Goodman*, J Craniofac Genet Dev Biol **4**:227, 1984.)

One daughter had papillary carcinoma of the thyroid while diffuse enlargement of the thyroid was noted in the other. Mild soft-tissue syndactyly of digits 2–4 was present in the hands and 3–4 in the feet (Fig. 25–32E,F).

### Reference (FACES syndrome)

1. Friedman E, Goodman RM: The "FACES" syndrome: A new syndrome with unique facies, anorexia, cachexia, and eye and skin lesions. *J Craniofac Genet Dev Biol* **4**:227–231, 1984.

## Unusual facies, cataract, short stature, joint contractures, and spondyloepimetaphyseal dysplasia

Borochowitz (1) described female sibs of Moroccan Jewish ancestry with unusual facies, cataract, short stature, inguinal hernia, joint contractures, and spondyloepimetaphyseal dysplasia. The parents were consanguineous. Inheritance is probably autosomal recessive.

Both had reduced height, weight, and head circumference. Ambulation was difficult due to severe knee and hip joint contractures. The facies was old appearing with a large nose and mouth and a pointed chin. Radiographic changes included spondyloepimetaphyseal dysplasia and marked precocious calcification of costochondral cartilages.

### Reference (Unusual facies, cataract, short stature, joint contractures, and spondyloepimetaphyseal dysplasia)

1. Borochowitz Z: Unusual facies, cataract, short stature, joint contractures, and spondyloepimetaphyseal dysplasia in sibs. A new premature aging-like autosomal syndrome? *David W. Smith Workshop on Malformations and Morphogenesis*, Madrid, Spain, May, 1989.

## Widow's peak, ptosis, and skeletal abnormalities

Kapur et al (1) described a large family with unusual facies and skeletal abnormalities. The patients had widow's peak, ptosis of eyelids, apparently low set angulated pinnae, difficulty in supination of forearms, recurrent dislocation of the patella, and pain in various joints.

Affected males were relatively short (162–171 cm). The patella dislocated early in life followed later in life by pain in the knees. Inability to supinate the forearm or touch the shoulder was noted in affected adult males. With age, the elbow, shoulder, wrist, hip, and interphalangeal joints show progressive pain and radiologic changes of arthritis. There is flattening of the femoral condyles. At the elbow, the capitulum of the humerus was tilted with its medial edge placed proximally at the elbow. At the wrist, there was a V-shaped angulation of radius and ulna. A lateral view of the foot showed subluxation of the navicular and cuboid on the talus and calcaneus.

Inheritance appears to be X-linked dominant. Affected females have a less marked facies but exhibit ptosis and widow's peak.

### Reference (Widow's peak, ptosis, and skeletal abnormalities)

1. Kapur S et al: Apparently previously undescribed X-linked dominant syndrome with facial and skeletal anomalies. *Am J Med Genet* **33**:357–363, 1989.

## Unusual facies, cutis laxa, skeletal abnormalities, craniosynostosis, and psychomotor retardation

Koppe et al (1), in 1989, described an X-linked recessive disorder which they called **SCARF** syndrome, an acronym for **s**keletal abnormalities, **c**utis laxa, **c**raniosynostosis, **a**mbiguous genitalia, **r**etardation and **f**acial abnormalities.

In addition to cutis laxa, the affected male cousins exhibited joint

A

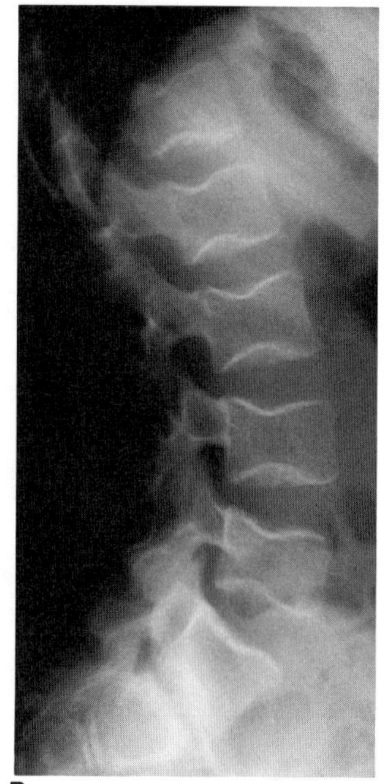

B

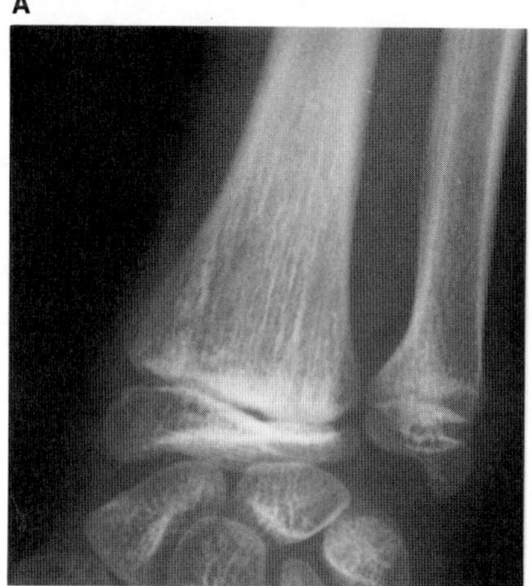

C

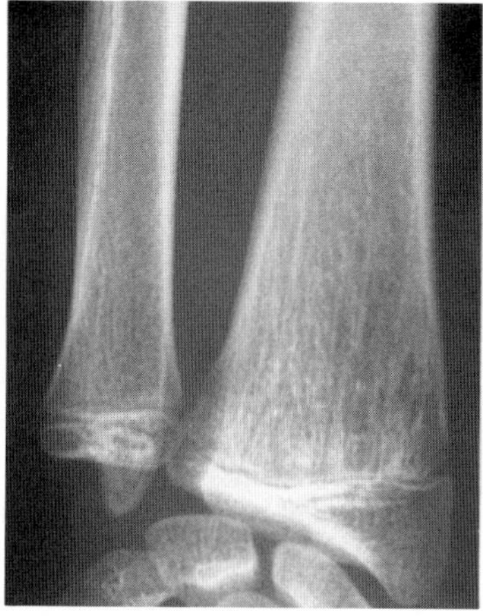

D

Fig. 25–33. *Sponastrime dysplasia.* (A) The sponastrime family with four affected girls (arrows). (B) Increased lumbar lordosis, mild deformity of vertebral body with thickened concave end plates. Mild osteoporosis. (C,D) Mildly striated metaphyses. (From *S Fanconi et al,* Helv Paediatr Acta **38**:267, 1983.)

hypermobility and umbilical and inguinal hernias. Skeletal changes included craniosynostosis, short sternum with pectus carinatum, abnormally shaped vertebral bodies, and oddly modeled tubular bones. The nipples were widely spaced. Micropenis and perineal hypospadias were noted. Both exhibited mild to moderate psychomotor retardation.

Craniofacial abnormalities included multiple hair whorls, high and broad nasal root, micrognathia, and wide webbed neck. One of the two exhibited epicanthal folds, ptosis, and lowset posteriorly rotated pinnae. The enamel was hypocalcified.

#### Reference (Unusual facies, cutis laxa, skeletal abnormalities, craniosynostosis, and psychomotor retardation)

1. Koppe R et al: Ambiguous genitalia associated with skeletal abnormalities, cutis laxa, craniosynostosis, psychomotor retardation and facial abnormalities (SCARF syndrome). *Am J Med Genet* **34**:305–312, 1989.

### Unusual facies and sponastrime dysplasia

Fanconi et al (1), in 1983, described a syndrome which they termed "sponastrime dysplasia" for **SPO**ndylar and **NA**sal changes and **STRI**ated **ME**taphyses which were characteristically altered in the disorder. The disorder was confirmed by Lachman et al (2). It may be the same as *short stature, low nasal bridge, cleft palate, and sensorineural hearing loss.*

Prenatal and postnatal growth retardation were below the fifth percentile but developmental milestones were normal.

The head appeared relatively large. The forehead was prominent and the nasal bridge sunken. These, combined with midfacial hypoplasia, produced a somewhat oriental appearance (Fig. 25–33A). Kyphoscoliosis, hyperlordosis, widened lower thorax, and prominent abdomen were evident. There was mild rotational restriction of the elbows.

Radiologic changes included vertebral alterations (kyphoscoliosis, lordosis, osteoporosis, pear-shaped vertebral bodies) and osteopathia striata of long bones (Fig. 25–33B–D).

Histopathologic studies (2) indicated a disturbance in cartilage formation with defects in collagen and proteoglycan synthesis.

Inheritance appears to be autosomal recessive. Affected sibs (1,2) and parental consanguinity (1) have been demonstrated.

#### References (Unusual facies and sponastrime dysplasia)

1. Fanconi S et al: The sponastrime dysplasia. *Helv Paediatr Acta* **38**:267–280, 1983.

2. Lachman RS et al: Sponastrime dysplasia. A radiologic-pathologic correlation. *Pediatr Radiol* **19**:417–424, 1989.

# Chapter 26
# Syndromes with Gingival/Periodontal Components

## Gingival fibromatosis and its syndromes (introduction)

Gingival fibromatosis may exist as an isolated abnormality or as part of a syndrome (1,2,4). As an isolated finding, it is often sporadic but may exhibit autosomal dominant inheritance (3) or rarely autosomal recessive inheritance (2).

### References [Gingival fibromatosis and its syndromes (introduction)]


1. Fritz ME: Idiopathic gingival fibromatosis with extensive osseous involvement in a 12-year-old boy. *Oral Surg* **30**:755–758, 1970.

2. Jorgenson RJ, Cocker ME: Variation in the inheritance and expression of gingival fibromatosis. *J Periodontol* **45**:472–477, 1974.

3. Raeste AM et al: Hereditary fibrous hyperplasia of the gingiva with varying penetrance and expressivity. *Scand J Dent Res* **86**:357–365, 1978.

4. Witkop CJ Jr: Heterogeneity in gingival fibromatosis. *Birth Defects* **7**(7):212–217, 1971.


**Gingival fibromatosis, hypertrichosis, epilepsy, and mental retardation.** This is the most commonly reported syndrome in which gingival fibromatosis is but one of the manifestations of a single gene mutation. The hypertrichosis is usually generalized, involving the eyebrows, face, ears, extremities, and sacral areas (7,8) (Fig. 26–1A–C). Epilepsy and mental retardation are variable features. The binary combination of gingival enlargement and hypertrichosis (2b,3,5,

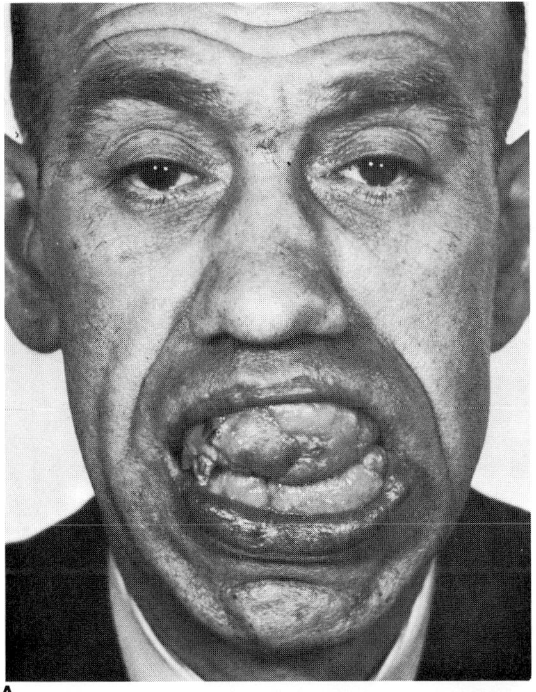

A

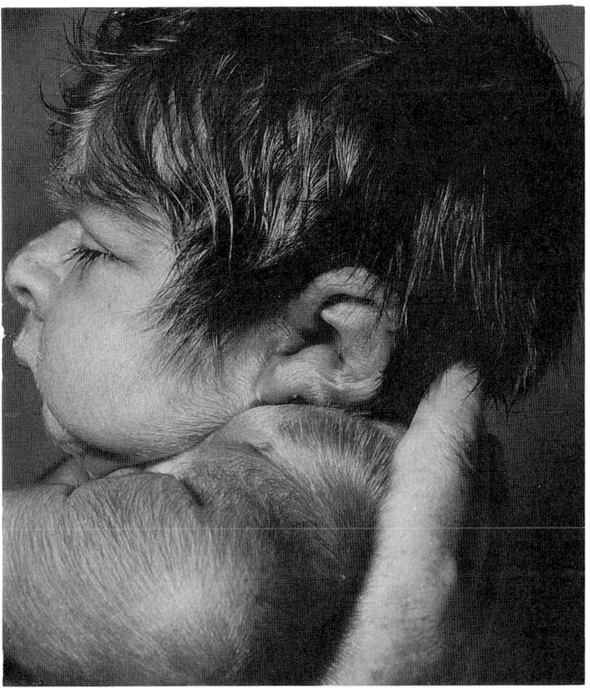

B

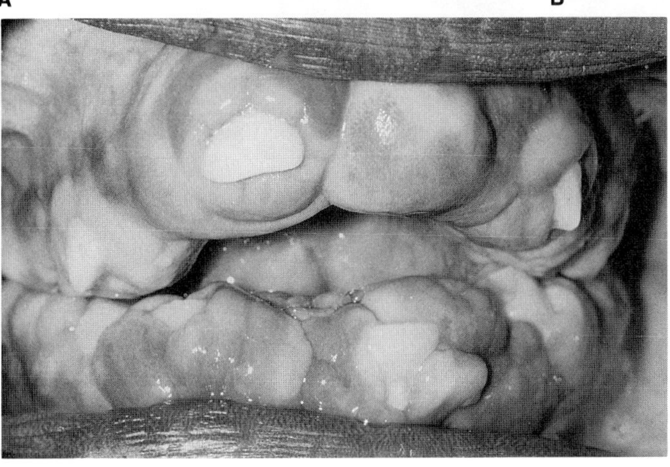

C

Fig. 26–1. *Gingival fibromatosis–hypertrichosis.* (A) Extensive gingival fibromatosis. Patient cannot close lips over hyperplastic gingival masses. (B) Note excessive overgrowth of hair on face and limbs. (C) Massive gingival overgrowth. (A from *A McIndoe* and *BO Smith,* Br J Plast Surg **11**:62, 1959. B from *F Vontobel,* Helv Pediatr Acta **38**:401, 1973. C from *MM Cohen Jr* Syndrome Ident **1**(2):12, 1974.)

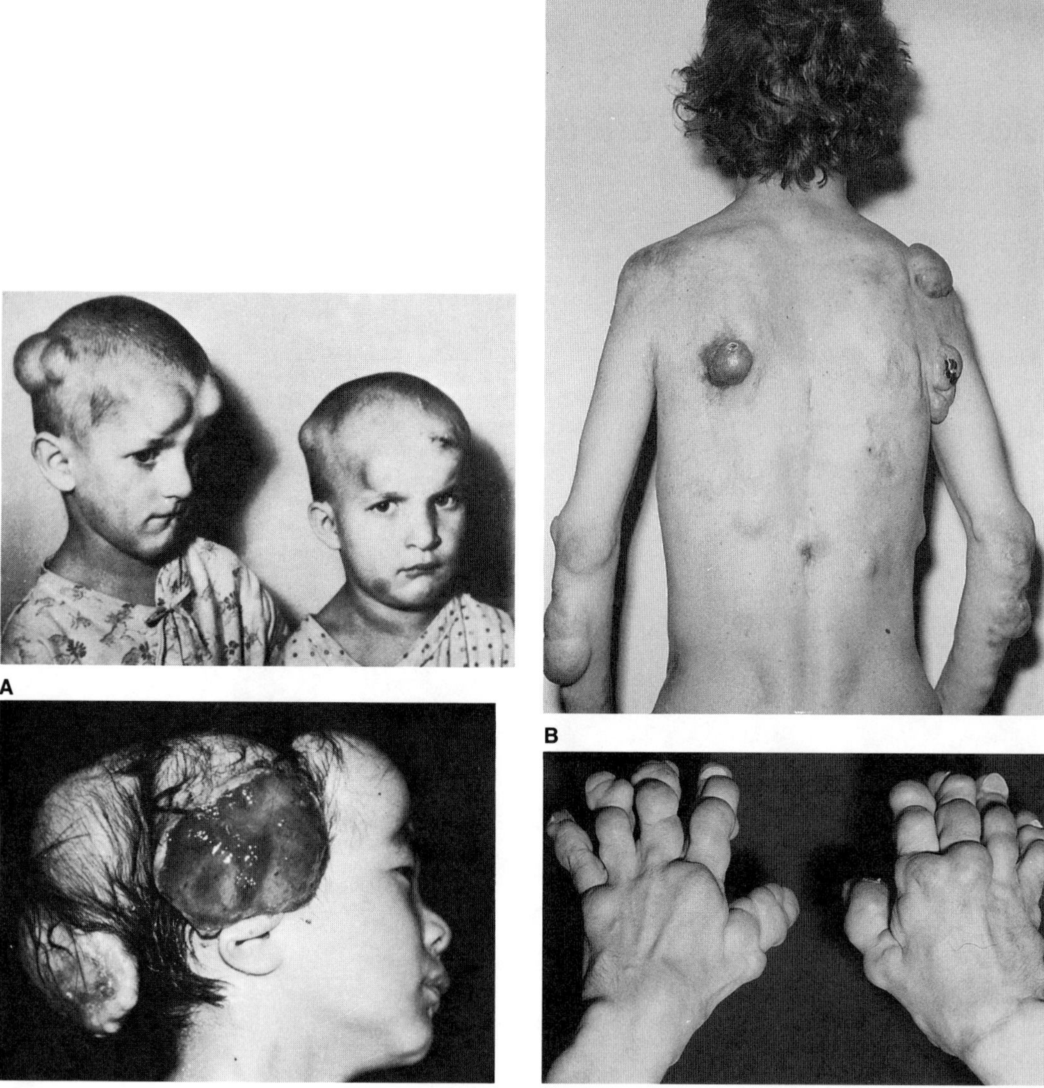


Fig. 26–2. *Gingival fibromatosis with juvenile hyaline fibromatosis (Murray–Puretić–Drescher syndrome).* (A) Tumors of the head in 5-year-old male and 4-year-old female sibs. Tumors first appeared when children were 2 years of age. (B) Note tumors of trunk and forearms. (C) Tumors of head of 7-year-old boy. (D) Extensive involvement of hands. (E) Gingival hyperplasia and tumor mass of lower lip. (F) Gingival enlargement. (G) Roentgenogram showing bone cysts and calcification of skin tumorous masses. (H) Microscopic sections showing cells lying in amorphous stroma. (A from *E Drescher et al,* J Pediatr Surg **2**:427, 1957. B,D courtesy of *S Woyke,* Szczecin, Poland. C,F from *Y Kitano et al,* Arch Dermatol **106**:877, 1972. H from *H Ishikawa* and *S Mori,* Acta Derm Venereol **53:**185, 1973.)

7–9,11,15–17) is probably much less common than is isolated gingival fibromatosis. The syndrome appears to have autosomal dominant inheritance (2b,15,16). Some cases may be autosomal recessive (1,4).

Associated mental retardation and/or epilepsy have been noted in some cases (1,1a,2,10,12,14,17), and it is possible that this combination has genetic heterogeneity, some examples possibly having autosomal recessive inheritance (4). Other possible aleatory skeletal alterations, such as narrow second cervical vertebra, long odontoid process, and abnormal first ribs, were described by Anderson (1). The gingival enlargement usually occurs at less than 5 years of age (12).

An unusual case reported by Vontobel (13) cannot be classified. Features included acromegaloid features, marked hypertrichosis, hyperconvex nails, and gingival fibromatosis. Nevin et al (6) reported male and female sibs with gingival fibromatosis, epilepsy, mental retardation, and bilateral camptodactyly of the third, fourth, and fifth fingers. Hypertrichosis was specifically denied. Inheritance was presumed autosomal recessive. Another unusual case described by Cohen et al (2a) concerns a child with hirsutism, gingival fibromatosis, prominent frontal bone, soft alar and auricular cartilages, unilateral axillary pterygium, hepatomegaly, broad ribs, and metacarpal alterations. Intelligence was normal as were fingers and toes and there was no seizure activity.

### References (Gingival fibromatosis, hypertrichosis, epilepsy, and mental retardation)

1. Anavi Y et al: Idiopathic familial gingival fibromatosis associated with mental retardation, epilepsy and hypertrichosis. *Develop Med Child Neurol* **31**:538–542, 1989.

1a. Anderson J et al: Hereditary gingival fibromatosis. *Br Med J* **3**:218–219, 1969.

2. Araiche M, Brode H: A case of fibromatosis gingivae. *Oral Surg* **12**:1307–1310, 1959.

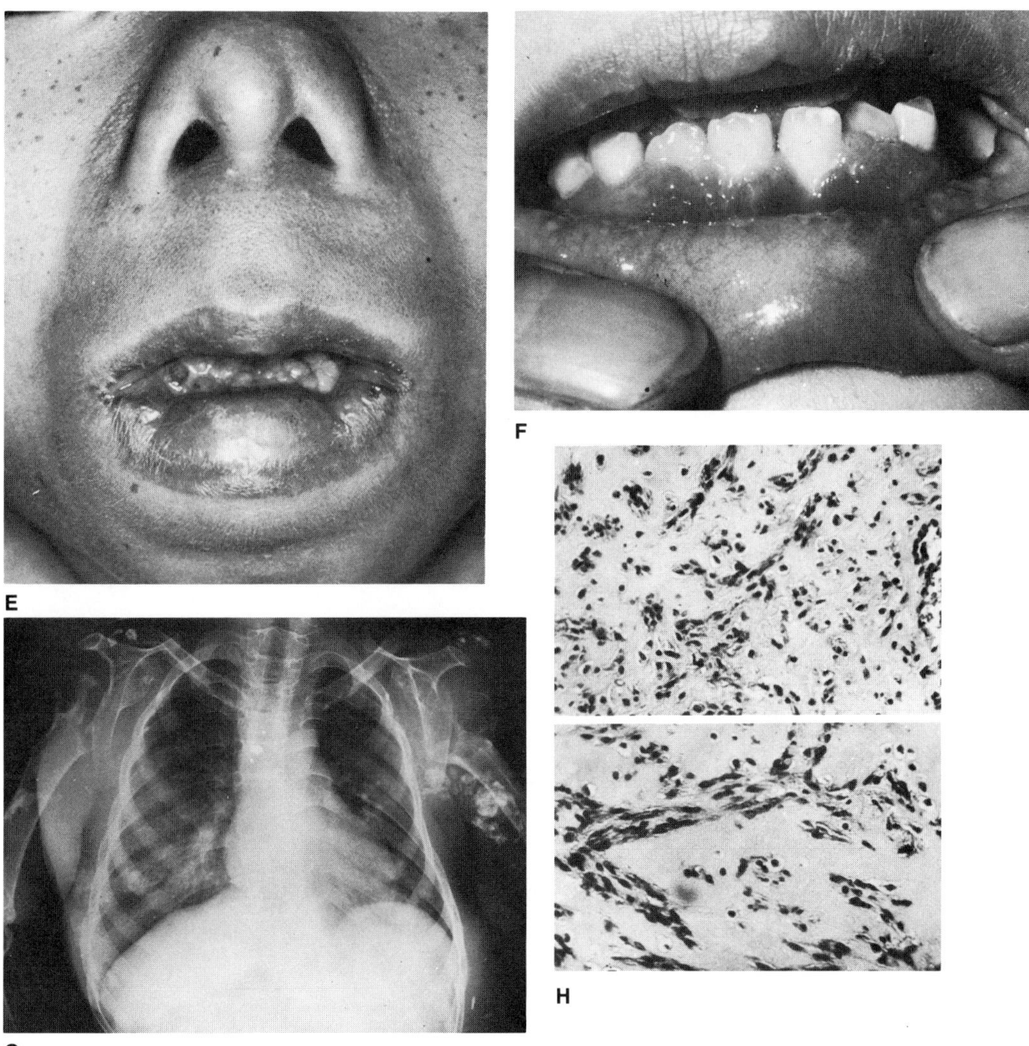


2a. Cohen MM Jr et al: Case 23. *Syndrome Ident* **1**(2):12–13, 1974.

2b. Cuestos-Carnero R, Bornancini CA: Hereditary generalized gingival fibromatosis associated with hypertrichosis: Report of five cases in one family. *J Oral Maxillofac Surg* **46**:415–420, 1988.

3. Horning GM et al: Gingival fibromatosis with hypertrichosis. *J Periodontol* **56**:344–347, 1985.

4. Jorgenson RJ, Cocker ME: Variation in the inheritance and expression of gingival fibromatosis. *J Periodontol* **45**:472–477, 1974.

5. McIndoe A, Smith BO: Congenital familial fibromatosis of the gums with the teeth as a probable etiological factor: Report of an affected family. *Br J Plast Surg* **11**:62–71, 1958.

6. Nevin NC et al: Hereditary gingival fibromatosis. *J Ment Defic Res* **15**:130–135, 1971.

7. Perkoff D: Primary generalized hypertrophy of the gums. *Lancet* **1**:1294–1297, 1929.

8. Ray AK: Hypertrichosis terminalis with simian characteristics. *J Med Genet* **3**:156, 1966.

9. Rushton MA: Hereditary or idiopathic hyperplasia of the gums. *Dent Pract Dent Rec* **7**:136–146, 1956–1957.

10. Snyder CH: Syndrome of gingival hyperplasia, hirsutism and convulsions. *J Pediatr* **67**:499–502, 1965.

11. Thoma KH, Goldman H: *Oral Pathology*, 5th ed. C.V. Mosby, St. Louis, 1960, pp 1365–1366.

12. Villa VG, Zarate AL: Extensive fibromatosis of the gingivae in the maxilla and in the mandible in a 6-year-old boy. *Oral Surg* **6**:1228–1229, 1953.

13. Vontobel F: Idiopathic gingival hyperplasia and hypertrichosis associated with acromegaloid features. *Helv Pediatr Acta* **28**:401–411, 1973.

14. Waterman T: Diseases of the jaws. *Boston Med Surg J* **81**:165–168, 1869.

15. Willert E: Ein seltener Fall einer echten Fibromatosis gingivae. *Oest Z Stomatol* **51**:663–666, 1954.

16. Winstock D: Hereditary gingivofibromatosis. *Br J Oral Surg* **2**:59–64, 1964.

17. Winter GB, Simpkiss MJ: Hypertrichosis with hereditary gingival hyperplasia. *Arch Dis Childh* **49**:394–399, 1974.

**Gingival fibromatosis with juvenile hyaline fibromatosis (Murray–Puretić–Drescher syndrome).** This condition, first reported by Murray (17) in 1873, is characterized by multiple hyaline fibromas, white papules on the skin, flexion contractures, osteolytic bone lesions, and gingival fibromatosis (4,17,18,28). Affected sibs born to normal parents and a high rate of parental consanguinity (4,5,11, 13,14a,17,24,29,31) indicate that inheritance is autosomal recessive. It is different from infantile systemic hyalinosis, which also has autosomal recessive inheritance (4a,16a). The latter may be the same as *Winchester syndrome*.

Skin. Multiple painless, freely movable, subcutaneous, hyaline, fibrous tumors of the scalp, face, chin, ears, back, digits, thighs, and legs appear between 2 months and 4 years of age (25). They slowly enlarge, sometimes to the size of a small orange and may ulcerate (4,12). Lesions have also been noted in the perianal region (9,11). With puberty, some masses may regress, and new tumors decrease (Fig. 26–2A–C). There appears to be a marked tendency to suppurative infections of the skin and mucous membranes (18). The skin undergoes poikilodermatous and sclerodermatous changes. In addition, white papules are found on the neck, pinnae, nose, nasal septum, upper lip, and behind the ears (8,11,13,24). The same material has been found in the tongue, esophagus, stomach, intestine, spleen, thymus, and lymph nodes (25). Rarely perianal lesions are mistaken for warts (14a,15,16).

On light microscopic examination, the tumors contain a homogeneous, amorphous, eosinophilic, PAS-positive ground substance that is sometimes chondroid in appearance (4,13,19,24,30). Spindle-shaped fibroblasts are present in this matrix in a streak-like pattern. The proportion of cells to ground substance varies from one tumor to another. Between the cells, argyrophilic fibers are sometimes found (Fig. 26–2H). Collagen fibers are thicker than normal. Ultrastructural studies have indicated excessive proliferation of fibroblasts, their organelles, and secretory products (13,16,24,30). Abnormalities in the Golgi apparatus have also been found (21,30). Ishikawa and Mori (12,13) and Iwata et al (14) found an increased amount of chondroitin-6-sulfate in these tumors. Electrophoretically it has been shown that an increased amount of acid mucopolysaccharides, hyaluronic acid, and dermatan sulfate is present in the urine (21,25).

Musculoskeletal. Painful progressive flexion contractures of the knees, elbows, hips, ankles, and shoulders appear within the first year of life (7,11,12,15,16,18,19,21). The temporomandibular joint may also be affected (11). Contractures were absent in a few patients (20,27). Digital tumors, osteolysis of the terminal phalanges (1a,11,21,24,25), and small cystic lesions of long bones, phalanges, and scapulae (7,8,24) have been noted in about one-half the patients, as well as generalized osteoporosis and thoracolumbar scoliosis (18) (Fig. 26–2D,G). Height and weight are reduced, and skeletal and sexual maturation are delayed. Widespread cortical hyperostosis and periosteal reactions have also been noted (7).

Psychomotor development. Mental development is normal (8,11,12,15,16,17,21).

Oral. Generalized gingival enlargement begins during the first year or so of life (6,8) and can extend onto the occlusal surfaces of the teeth (17) (Fig. 26–2E,F). Gingival biopsy has demonstrated findings similar to those of the skin (8). Patients with juvenile hyaline fibromas who do not have gingival enlargement (20,26) have been reported. The patient observed by Woyke et al (29) had skin lesions and gingival hyperplasia, but no flexion contractures. Gingival fibromatosis and flexion contractures were not mentioned in the patients reported by Enjoji et al (5).

Differential diagnosis includes congenital generalized fibromatosis, *neurofibromatosis,* and *Winchester syndrome.*

### References [Gingival fibromatosis with juvenile hyaline fibromatosis (Murray–Puretić–Drescher syndrome)]


1. Aldred MJ, Crawford PJM: Juvenile hyaline fibromatosis. *Oral Surg* **63**:71–77, 1987. (Same case as Ref. 8.)

1a. Atherton DJ et al: Juvenile hyaline fibromatosis. *Br J Dermatol* **105** (*Suppl* **19**):61–64, 1981.

2. Baptista AP et al: Fibromatose hyaline juvénile. *Ann Dermatol Venereol* **110**:751–752, 1983.

3. Costa OG et al: Fibromatosis hialina juvenil. *Med Cutan Iber Lat Am* **3**:331–340, 1975.

4. Drescher E et al: Juvenile fibromatosis in siblings (fibromatosis hyalinica multiplex juvenilis). *J Pediatr Surg* **2**:427–430, 1967.

4a. Dunger DB et al: Two cases of Winchester syndrome: With increased urinary oligosaccharide excretion. *Eur J Pediatr* **146**:615–619, 1987.

5. Enjoji M et al: Juvenile fibromatosis of the scalp in siblings. *Act Med Univ Kagoshima (Suppl)* **10**:145–151, 1968.

6. Enzinger FM, Weiss SW: *Soft Tissue Tumors.* C.V. Mosby, St. Louis, 1983, pp 83–86.

7. Familusi JB et al: Congenital generalized fibromatosis. An African case with gingival hypertrophy and other unusual features. *Am J Dis Child* **130**:1215–1217, 1976.

7a. Fayad MN et al: Juvenile hyaline fibromatosis: Two new patients and review of the literature. *Am J Med Genet* **26**:123–132, 1987.

8. Finlay AY et al: Juvenile hyaline fibromatosis. *Br J Dermatol* **108**:609–616, 1983.

9. Gianotti F et al: Fibromatose hyaline juvenile. *Ann Dermatol Venereol (Paris)* **104**:152–153, 1977.

10. Gutierrez G et al: Fibromatosis hialinica multiple juvenil. A proposito de un caso. *Med Cutan Iber Lat Am* **7**:283–286, 1973.

11. Hamada K et al: Puretić syndrome—gingival fibromatosis with hyaline fibromas. *Jpn J Hum Genet* **25**:249–255, 1980.

12. Ishikawa H, Mori S: Systemic hyalinosis or fibromatosis hyalinica multiplex juvenilis as a congenital syndrome. A new entity based on the inborn error of the acid mucopolysaccharide metabolism in connective tissue cells? *Acta Derm Venereol* **53**:185–191, 1973.

13. Ishikawa H et al: Systemic hyalinosis (juvenile hyaline fibromatosis). *Dermatol Res* **265**:195–206, 1979.

14. Iwata S et al: Systemic hyalinosis or juvenile hyaline fibromatosis. *Arch Dermatol Res* **267**:115–121, 1980.

14a. Kan AE, Rogers M: Juvenile hyaline fibromatosis: An expanded clinicopathologic spectrum. *Pediatr Dermatol* **6**:68–75, 1989.

15. Kitano Y: Juvenile hyaline fibromatosis. *Arch Dermatol* **112**:86–88, 1976.

16. Kitano Y et al: Two cases of juvenile hyaline fibromatosis: Some histological, electron microscopic, and tissue culture observations. *Arch Dermatol* **106**:877–883, 1972.

16a. Landing BH, Naddora R: Infantile systemic hyalinosis. *Pediatr Pathol* **6**:55–79, 1986.

17. Murray J: On three peculiar cases of molluscum fibrosum in children. *Med-Chir Trans London* **56**:235–238, 1873.

18. Puretić S et al: A unique form of mesenchymal dysplasia. *Br J Dermatol* **74**:8–19, 1962.

19. Puretić S, Puretić S: Clinical and histopathological observations on systemic familial mesenchymatosis. *Proc 13th Int Congr Pediatr Vienna, 1971,* **5**:373–381, 1971.

20. Quintal D, Jackson R: Juvenile hyaline fibromatosis. A 15-year follow-up. *Arch Dermatol* **121**:1062–1063, 1985.

21. Remberger K et al: Fibromatosis hyalinica multiplex (juvenile hyaline fibromatosis). Light microscopic, electron microscopic, immunohistochemical and biochemical findings. *Cancer* **56**:614–624, 1985.

22. Rimbaud P et al: Fibro-hyalinose juvenile. *Bull Soç Fr Derm Syph* **80**:435–436, 1973.

23. Schmidt BJ et al: Hemangiomatose sclerosante atypique. *Arch Fr Pédiatr* **26**:213–219, 1969.

24. Sciubba JJ, Niebloom T: Juvenile hyaline fibromatosis (Murray–Puretić–Drescher syndrome): Oral and systemic findings in siblings. *Oral Surg* **62**:397–409, 1986.

25. Stringer DA, Hall CM: Juvenile hyaline fibromatosis. *Br J Radiol* **54**:473–478, 1981.

26. Suschke J, Kunze D: Ein neuer Mucopolysaccharidose-Typ. *Dtsch Med Wochenschr* **96**:1941–1943, 1971.

27. Wang N-S, Knaack J: Fibromatosis hyalinica multiplex juvenilis. *Ultrastruct Path* **3**:153–160, 1982.

28. Whitfield A, Robinson AH: A further report on the remarkable series of cases of molluscum fibrosum in children communicated to the society by Dr. John Murray in 1873. *Med-Chir Trans London* **86**:293–301, 1903.

29. Woyke S et al: A 19-year follow-up of multiple juvenile hyaline fibromatosis. *J Pediatr Surg* **19**:302–304, 1984.

30. Woyke S et al: Ultrastructure of a fibromatosis hyalinica multiplex juvenilis. *Cancer* **26**:1157–1168, 1970.

31. Yesudian P et al: Juvenile hyaline fibromatosis. *Int J Dermatol* **23**:619–620, 1984.


**Gingival fibromatosis and corneal dystrophy (Rutherfurd syndrome).** Relatively mild gingival enlargement, failure of tooth eruption, and corneal opacities were described in three generations as an autosomal dominant trait by Rutherfurd (2). This family was followed up and reported on again in 1966 by Houston and Shotts (1), who described the corneal opacities as curtain-like, involving the superior part of the cornea. Although mental retardation and aggressive behavior have been noted, it is not certain whether these abnormalities are

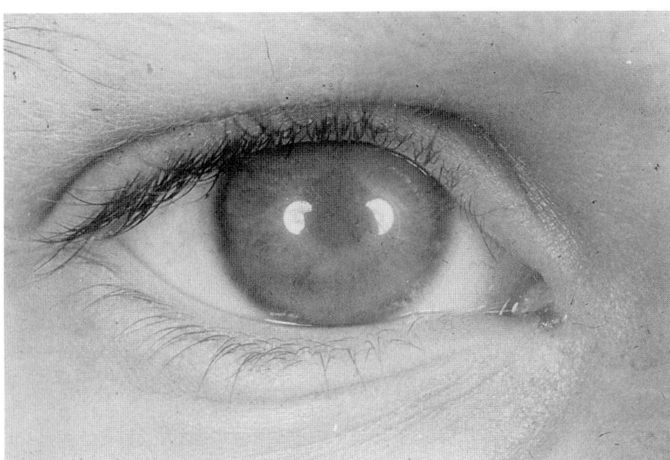

Fig. 26–3. *Gingival fibromatosis and corneal dystrophy (Rutherfurd syndrome).* Curtain-like corneal opacities. (From *I Houston* and *N Shotts,* Acta Paediatr Scand **55**:233, 1966.)

manifestations of the syndrome or are segregating independently (Fig. 26–3).

### References [Gingival fibromatosis and corneal dystrophy (Rutherfurd syndrome)]


1. Houston I, Shoots N: Rutherfurd's syndrome: A familial oculodental disorder. Acta Paediatr Scand **55**:233–238, 1966.
2. Rutherfurd ME: Three generations of inherited dental defect. *Br Med J* **2**:9–11, 1931.


**Gingival fibromatosis; ear, nose, bone, and nail defects; and hepatosplenomegaly (Laband syndrome).** In addition to gingival fibromatosis, which is manifested at birth or within the first few months of life (8), abnormalities of the nose and ears and striking hypoplastic changes in the terminal phalanges of the fingers and toes are characteristic. The syndrome is inherited as an autosomal dominant trait (1,5). Henefer and Kay (3) reported a patient with gingival fibromatosis and nasal and ear abnormalities, but the hands and feet were not described.

Facies. The nose and pinnae are large and the ears are fleshy because of the soft consistency of the cartilage (1,4–5,7,8) (Fig. 26–4A).

Skin. Mild hirsutism in the form of synophrys, hairy arms and legs, thick eyebrows and increased sacral hair has been noted (4,4a,7,8).

Skeletal alterations. The thumb nails are hypoplastic or absent. The nails and terminal phalanges on the fingers are usually rudimentary or missing. In most cases, all toes lack terminal phalanges; the first and second toes often have only one phalanx each (4,5,8). Toenails are usually absent (Fig. 26–4B). Pes cavus was noted in about half the kindred described by Laband et al (5). Joint hypermobility, especially involving the metacarpophalangeal, shoulder, and knee joints (1,5), is common. Spina bifida occulta of the fifth lumbar vertebra was described in a few cases (4,8). The third thoracic vertebra has been reported divided sagittally into two segments, with the fourth thoracic vertebra flattened (4).

Hepatosplenomegaly. Hepatomegaly (1,7) and splenomegaly (5) have been noted. Laband et al (5) suggested that splenomegaly might

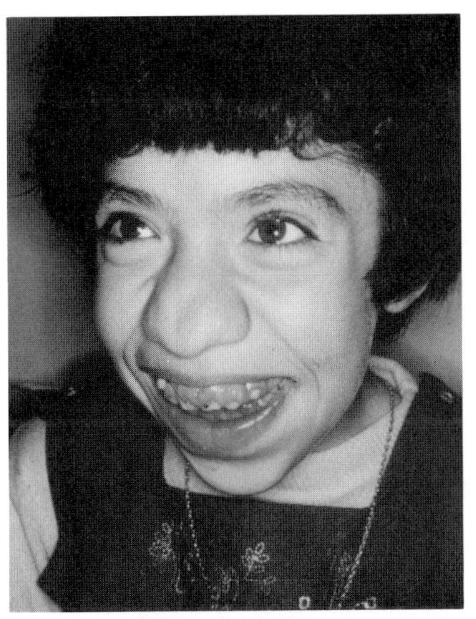

A

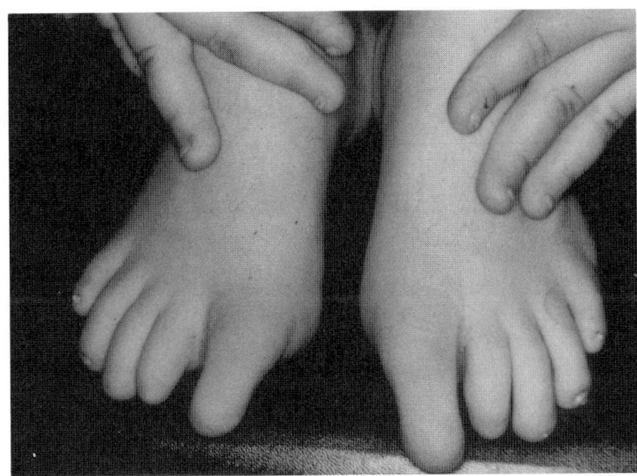

B

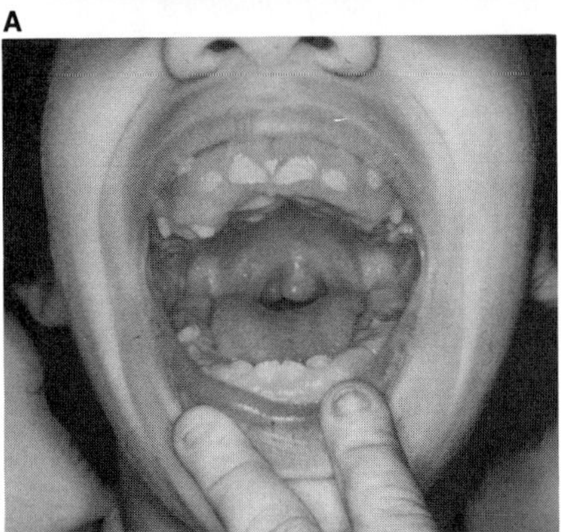

C

Fig. 26–4. *Gingival fibromatosis; ear, nose, bone, and nail defects; and hepatosplenomegaly (Laband syndrome).* (A) Nose large and poorly formed because of soft cartilage. Note marked gingival enlargement. (B) Hypoplastic nails and terminal phalanges of toes. (C) Hyperplastic gingiva. (A–C courtesy of *K Oikawa,* New York, NY.)

become more marked with age. On the other hand, enlargement of the liver and spleen has been specifically denied (7,8).

Other findings. Mental retardation has been reported in a few cases (2,5,7,8).

Oral. Gingival fibromatosis is a constant feature (Fig. 26–4C).

### References [Gingival fibromatosis; ear, nose, bone, and nail defects; and hepatosplenomegaly (Laband syndrome)]


1. Alavandar G: Elephantiasis: Report of an affected family with associated hepatomegaly, soft tissue and skeletal abnormalities. *J All India Dent Assoc* **37**:349–353, 1965.
2. Chodirker BN et al: Zimmermann–Laband syndrome and profound mental retardation. *Am J Med Genet* **25**:543–548, 1986.
3. Henefer EP, Kay LA: Congenital idiopathic gingival fibromatosis in the deciduous dentition. *Oral Surg* **24**:65–70, 1967.
4. Jacoby NM et al: Partial anonychia (recessive) with hypertrophy of the gums and multiple abnormalities of the osseous system: Report of a case. *Guy's Hosp Rep* **90**:34–40, 1940.

4a. Kiyoshi W et al: Laband syndrome. *J Oral Surg* **37**:120–122, 1979.

5. Laband PF et al: Hereditary gingival fibromatosis: Report of an affected family with associated splenomegaly and skeletal and soft-tissue abnormalities. *Oral Surg* **17**:339–351, 1964.
6. Läwen A: Über die Elephantiasis der Gingiva und ihre operative Behandlung. *Wien Med Wochenschr* **79**:607–610, 1929.
7. Oikawa K et al: Laband syndrome: Report of case. *J Oral Surg* **37**:120–122, 1975.
8. Zimmermann: Über Anomalien des Ektoderms (Cases 1 and 2). *Vjschr Zahnheilkd* **44**:419–434, 1928.


**Gingival fibromatosis, microphthalmia, mental retardation, athetosis, and hypopigmentation (Cross syndrome).** Cross et al (1) described three sibs, the product of a consanguineous union, who exhibited microphthalmia, cloudy cornea, and hypopigmented skin (Fig. 26–5). A more recent example is that of Preus et al (3). By the third month, there were signs of spasticity and mental retardation. Reduced tyrosinase-active melanosomes were demonstrated by Witkop (4,5). He noted gingival fibromatosis in two of the three sibs. The case reported by Passarge and Fuchs-Mecke (2) is doubtful. Autosomal recessive inheritance is likely.

Fig. 26–5. *Gingival fibromatosis, microphthalmia, mental retardation, athetosis, and hypopigmentation (Cross syndrome).* Note microphthalmia, hypopigmented skin, enlarged gingiva. (From *HE Cross et al,* J Pediatr **70**:398, 1967.)

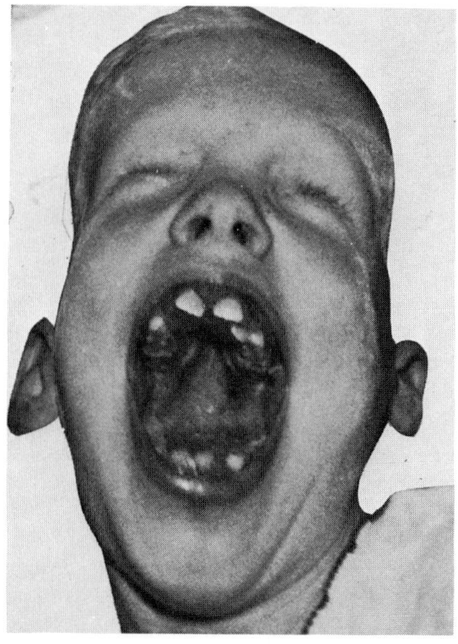

### References [Gingival fibromatosis, microphthalmia, mental retardation, athetosis, and hypopigmentation (Cross syndrome)]


1. Cross HE et al: A new oculocerebral syndrome with hypopigmentation. *J Pediatr* **70**:398–406, 1967.
2. Passarge E, Fuchs-Mecke S: Oculocerebral syndrome with hypopigmentation. *Birth Defects* **11**(2):466–467, 1975.
3. Preus M et al: An oculocerebral hypopigmentation syndrome. *J Génét Hum* **31**:323–328, 1983.
4. Witkop CJ Jr: Heterogeneity in gingival fibromatosis. *Birth Defects* **7**(7):210–221, 1971.
5. Witkop CJ Jr: Personal communication, 1975.


**Gingival fibromatosis with sensorineural hearing loss.** Jones et al (2) and Hartsfield et al (1) reported generalized gingival fibromatosis associated with hearing loss of 40–60 dB in two large kindreds. Inheritance was autosomal dominant. The hearing loss reportedly was progressive, bilateral, and manifest in the second decade of life. There was variation among affected individuals in the degree of severity and frequency.

### References (Gingival fibromatosis and sensorineural hearing loss)


1. Hartsfield JK Jr et al: Gingival fibromatosis with sensorineural hearing loss: An autosomal dominant trait. *Am J Med Genet* **22**:623–627, 1985.
2. Jones G et al: Familial gingival fibromatosis associated with progressive deafness in five generations of a family. *Birth Defects* **13**(3B):195–201, 1977.


**Gingival fibromatosis, hypertrichosis, cherubism, mental and somatic retardation, and epilepsy (Ramon syndrome).** Ramon et al (2), in 1967, and Pina-Neto et al (1), in 1986, reported affected sibs in two different families. In the first kindred, two sibs were involved; in the latter, four sibs and their second cousin.

Progressively cherubic facies associated with bony changes characteristic of cherubism (bilateral multilocular radiolucent lesions of the mandible and maxilla) became evident after the fourth year.

The gingival enlargement is a constant feature. It becomes manifest at 2 years of age, is slowly progressive, and ultimately prevents normal jaw closure. Tooth eruption is considerably delayed and incomplete. In most cases the incisors and canines erupt. The palate tends to be narrow. Gingival biopsy showed a peculiar perivascular pattern of collagen fibers that compress the blood vessels (2).

Mild to moderate mental retardation becomes evident during the first few years of life. By the fourth year, all patients experienced major seizures. Stunted growth was evident in nearly all the affected.

Hypertrichosis was present in about 50%. Pina-Neto et al (1) suggested that juvenile rheumatoid arthritis might also be a component of the disorder. It is entirely possible that this really represents polyarticular pigmented villonodular synovitis (vide infra).

Affected sibs with normal but consanguineous parents certainly suggest autosomal recessive inheritance (1,2).

It should be pointed out that cherubism has been also found in association with *neurofibromatosis, Noonan syndrome,* and a *Noonan-like syndrome with polyarticular pigmented villonodular synovitis.* Perhaps we are really dealing with a contiguous gene deletion syndrome.

### References [Gingival fibromatosis, hypertrichosis, cherubism, mental and somatic retardation, and epilepsy (Ramon syndrome)]


1. Pina-Neto JM et al: Cherubism, gingival fibromatosis, epilepsy, and mental deficiency (Ramon syndrome) plus juvenile rheumatoid arthritis. *Am J Med Genet* **25**:433–442, 1986.
2. Ramon Y et al: Gingival fibromatosis combined with cherubism. *Oral Surg* **24**:436–448, 1967.


**Gingival fibromatosis and growth hormone deficiency** Oikarinen et al (1) described five of eight children with gingival fibromatosis in association with growth hormone deficiency due to lack of growth hormone release factor.

Inheritance appears to be autosomal recessive.

#### Reference (Gingival fibromatosis and growth hormone deficiency)

1. Oikarinen K et al: Familial gingival fibromatosis associated with growth hormone deficiency. Unpublished, 1989.

**Differential diagnosis and laboratory aids.** Generalized gingival enlargement may occur as the result of inflammation, pregnancy, leukemia, *I-cell disease,* phenytoin (Dilantin), diltiazem, cyclosporin A, and nifedipine therapy (1a,1b,13–15,17). In these disorders, the gingiva is usually not as enlarged or as fibrotic as in gingival fibromatosis where it may be so severe that the affected individual may appear to have no teeth. History and physical examination will serve to distinguish these disorders from gingival fibromatosis. The presence of a complete syndrome usually gives the clinician little diagnostic difficulty.

Giansanti et al (7) reported a male with gingival fibromatosis, brachycephaly, hypertelorism, multiple skin telangiectases since birth, and café-au-lait pigmentation. The mother had multiple telangiectases only.

Hirsutism may result from many causes (6,10). Acquired hypertrichosis may occasionally be a sign of internal malignancy (8).

Extreme hypertrichosis with true anodontia or oligodontia has been described (1,3,11,12). The hair has been so profuse that some individuals have been exhibited in carnivals (2,4,5). With rare exceptions (Juliana Pastrana), these persons have not had gingival enlargement.

Other than jaw roentgenograms to determine whether teeth are present, no diagnostic aids are needed. Endocrine studies performed on these patients have not shown abnormal values.

#### References [Gingival fibromatosis and its syndromes (differential diagnosis and laboratory aids)]

1. Biegel H: Über abnorme Haarentwicklung beim Menschen. *Virchows Arch Pathol Anat* **44**:418–427, 1868.

1a. Bowman JM et al: Gingival overgrowth induced by diltiazem. *Oral Surg* **65**:183–185, 1988.

1b. Butler RT et al: Drug-induced gingival hyperplasia: Phenytoin, cyclosporine and nifedipine. *JADA* **114**:56–60, 1987.

2. Danforth CH: Studies on hair with special reference to hypertrichosis. *Arch Dermatol Syph (Chic)* **12**:380–401, 528–537, 1925.

3. Editorial: Periscope. *Dent Cosmos* 16:43–46, 1874.

4. Favelle: Un cas de pilosisme chez un jeune laôtienne. *Bull Soç Anthropol* **9**:439–488, 1886.

5. Felgenhauer WR: Hypertrichosis lanuginosa universalis. *J Génét Hum* **17**:1–44, 1969.

6. Foret J et al: Hyperplastic fibreuse idiopathique des gengives. *J Génét Hum* **13**:337–350, 1964.

7. Giansanti JS et al: Gingival fibromatosis, hypertelorism, anti-mongoloid obliquity, multiple telangiectases and café-au-lait pigmentation; a unique combination of developmental anomalies. *J Periodontol* **44**:299–302, 1973.

8. Goodfellow A et al: Hypertrichosis lanuginosa acquisita. *Br J Dermatol* **103**:431–433, 1980.

9. Miles AEW: Juliana Pastrana, the bearded lady. *Proc R Soc Med* **67**:160–164, 1974.

10. Muller SA: Hirsutism. *Am J Med* **46**:803–817, 1969.

11. Nowakowski T, Scholz A: Das Schicksal behaarter Menschen im Wandel der Geschichte. *Hautarzt* **28**:593–599, 1977.

12. Parreidt J: Über die Bezahnung bei Menschen mit abnormer Behaarung. *Dtsch Mschr Zahnheilkd* **4**:41–54, 1886.

13. Ramon Y et al: Gingival hyperplasia caused by nifedipine—a preliminary report. *Int J Cardiol* **5**:195–206, 1984.

14. Rostock MH et al: Severe gingival overgrowth associated with cyclosporine therapy. *J Periodontol* **57**:294–299, 1986.

15. Van der Wall EE et al: Gingival hyperplasia induced by nifedipine, an arterial dilating drug. *Oral Surg* **60**:38–40, 1985.

16. Virchow R: Die russischen Haarmenschen. *Berl Klin Wochenschr* **10**:337–339, 1893.

17. Wysocki GP et al: Fibrous hyperplasia of the gingiva: A side effect of cyclosporin A therapy. *Oral Surg* **55**:274–278, 1983.

### Hyperkeratosis palmoplantaris and periodontoclasia in childhood (Papillon–Lefèvre syndrome and Haim–Munk syndrome)

Papillon and Lefèvre (22), in 1924, described a syndrome consisting of hyperkeratosis of palms and soles and destruction of the supporting tissues of both primary and secondary dentitions. Over 200 cases have subsequently been described. The most comprehensive survey is that of Haneke (16).

The syndrome has autosomal recessive inheritance. Parental consanguinity has been found in about 40% (1,4,7,9,10,16,21). The frequency of the disorder is approximately 1–4/million people (13).

For a discussion of Haim–Munk syndrome, see *Differential diagnosis.*

**Skin.** Sometime between the second and fourth years of life, or on rare occasions even earlier (1,5,6), the palms and soles become diffusely red and scaly. The hyperkeratotic involvement of the palms is usually quite well demarcated, extending to the edges and over thenar eminences and to the volar wrists. The soles are usually more severely involved, the process frequently spilling over the edges, where it may be marked, extending to the Achilles tendon (Fig. 26–6A–C). Occasionally, the knees, elbows, external malleoli, tibial tuberosities, and dorsal finger and toe joints may exhibit a psoriasiform scaly redness (1,3,5,18,21). Thickening of plantar skin with resultant cracking may make walking difficult. The degree of hyperkeratosis is not severe, but normal skin markings become accentuated and involved skin may assume a parchment-like quality. The degree of involvement seems to fluctuate. Some authors have indicated that this fluctuation becomes worse during winter (1,5). The skin apparently improves somewhat with age but some degree of palmoplantar hyperkeratosis remains throughout life. Nails are rarely, if ever, involved (see *Differential diagnosis*). Relapsing pyoderma is seen in 20% (10).

**Other findings.** Calcium deposits in the attachment of the tentorium and choroid plexus (5–7,13,18,24,26) have been reported, but it is uncertain whether this finding is significant.

Increased susceptibility to infection has been suggested, but its specificity needs confirmation (8,16,17,19,28) as the same finding may occur in juvenile periodontitis.

**Oral manifestations.** The development and eruption of the deciduous teeth proceed normally, but almost simultaneously with the appearance of palmar and plantar hyperkeratosis, the gingiva swell, bleed, and become boggy. Marked halitosis develops. Destruction of the periodontium follows almost immediately the eruption of the last primary molar tooth. The teeth are involved in roughly the same order in which they erupt. Deep periodontal pocket formation precedes the exfoliation of teeth. By the age of 4 years, nearly all primary teeth have been lost. After exfoliation, the inflammation subsides and the gingiva resumes its normal appearance. The mouth then appears normal until the secondary (permanent) dentition erupts when the process is repeated in essentially the same manner. Most teeth are lost by 14 years. In some cases, the third molars do not exfoliate. The alveolar process is often completely destroyed. Even during active periodontal breakdown, the rest of the oral tissue appears perfectly normal (Fig. 26–7A,B).

**Differential diagnosis.** All disorders of diffuse palmoplantar hyperkeratosis must be excluded. However, none but Papillon–Lefèvre syndrome is associated with premature periodontal destruction with the possible exception of Haim-Munk syndrome (vide infra).

Haim and Munk (15) and others (1a,14,25,26) reported an unusual syndrome in several members of three related, inbred Jewish families from Cochin China, India. They exhibit congenital palmoplantar ker-

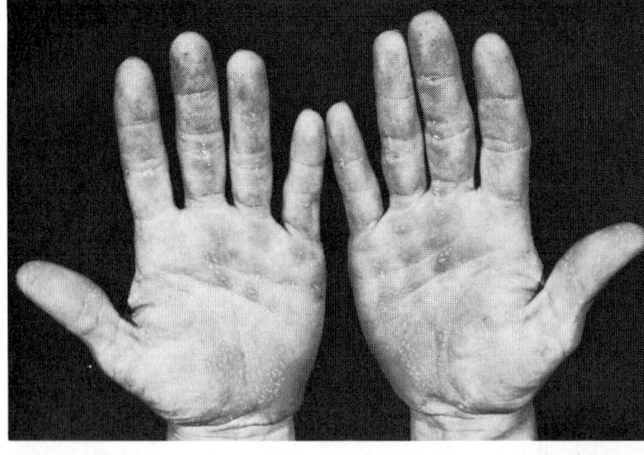

A

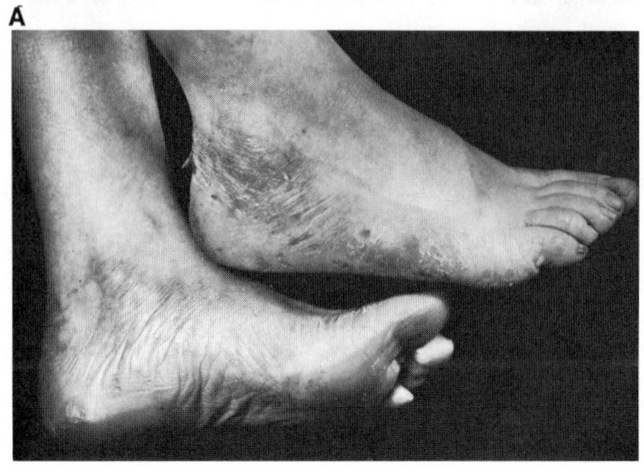

C

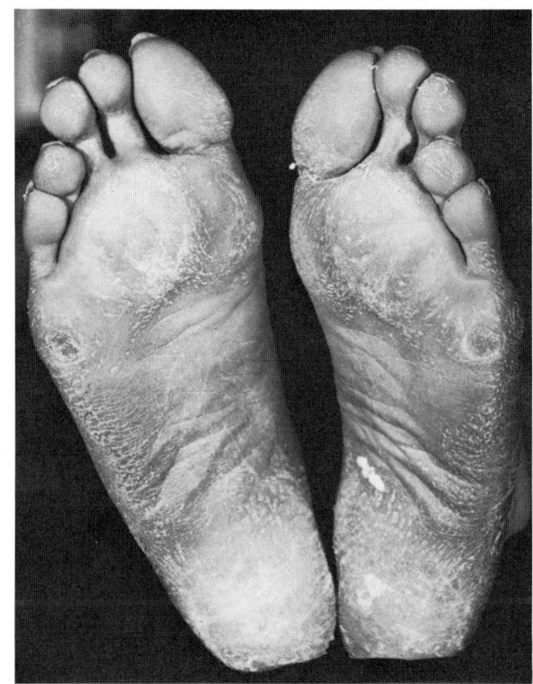

B

Fig. 26–6. *Hyperkeratosis palmoplantaris and periodontoclasia in childhood (Papillon–Lefèvre syndrome).* (A–C) Sharply demarcated hyperkeratosis of hands and feet. (C from *RK Hall,* Aust Dent J **9**:185, 1963.)

Fig. 26–7. *Hyperkeratosis palmoplantaris and periodontoclasia in childhood (Papillon–Lefèvre syndrome).* (A) Severe periodontal disease in 7-year-old child. (B) Roentgenogram showing extensive periodontal destruction. (From *JS Giansanti et al,* Oral Surg **35**:30, 1973.)

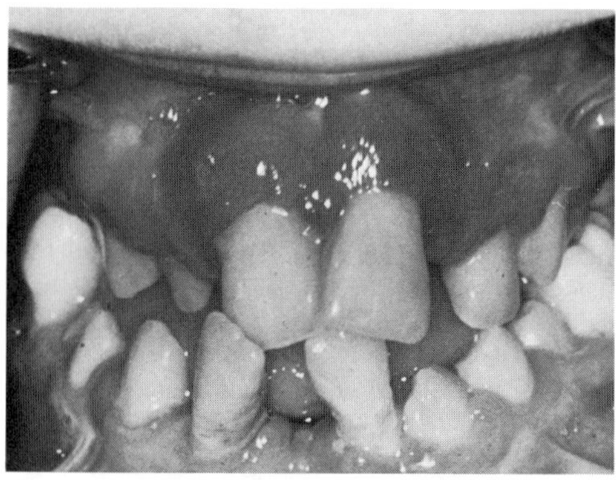

A

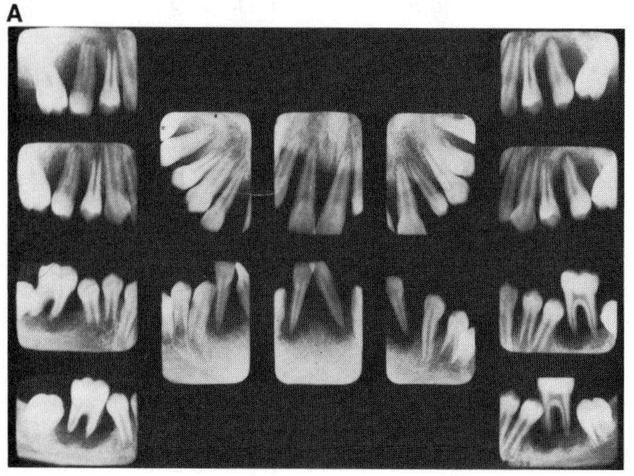

B

atosis, progressive periodontal destruction, pes planus, recurrent pyogenic skin infections, arachnodactyly, and a peculiar radiographic deformity of the fingers, consisting of tapered pointed phalangeal ends and a claw-like volar curve (Figs. 26–8A,B). In contrast to Papillon–Lefèvre syndrome, the skin manifestations were more severe and extensive and there was later onset. The periodontium also was less severely affected.

Premature loss of deciduous and/or permanent teeth is seen in trauma, acrodynia, histiocytosis X, *hypophosphatasia,* leukemia, various neutropenias, acatalasia, Chediak–Higashi syndrome, and juvenile periodontitis (11,12,20).

*Hypophosphatasia,* a condition transmitted as an autosomal recessive trait because of a deficiency in alkaline phosphatase, may be associated with a rickets-like condition. In addition to genua valga, bowing of the femurs and tibias, enlarged wrists, and other signs, the teeth are prematurely shed, are often hypoplastic, and are deficient in cementum. Increased amounts of phophoethanolamine are present in the urine (2).

Acatalasia is transmitted as an autosomal recessive trait and rarely has been observed outside Japan or Korea. It is characterized by progressive gangrenous lesions involving the gingiva and alveolar bone, resulting in exfoliation of teeth (28).

Hyperkeratosis palmaris has been seen in binary association in a family with a dentinogenesis imperfecta-like condition (30) and with a host of other disorders (11,13). It may also occur in the syndrome of *hyperkeratosis palmoplantaris and attached gingival hyperkeratosis.*

Autosomal dominantly inherited palmoplantar hyperkeratosis has been associated with short stature, unusual facies, hearing loss, seizures, clinodactyly, dysplastic nails, and oligodontia (26a).

**Laboratory aids.** Various immunologic defects have been reported. These have included impaired *in vitro* reactivity to T and B cell mitogens (19,29), a chemotactic defect, and reduced intracellular killing of *S. aureus* and *C. albicans* (8).

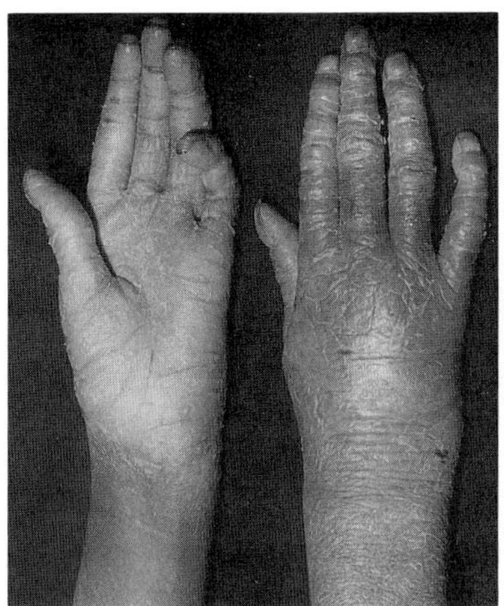

A

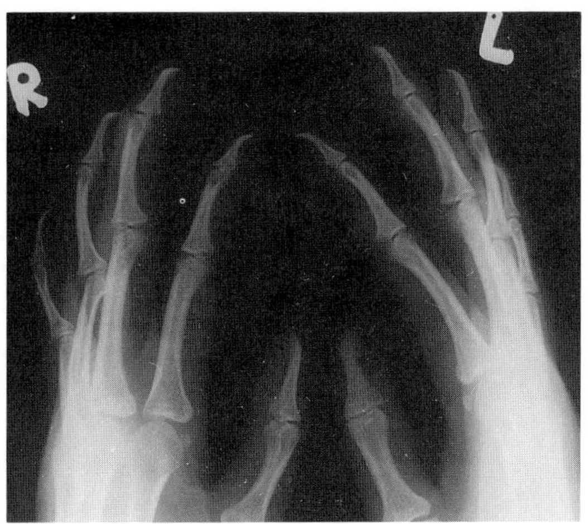


B

Fig. 26–8. *Hyperkeratosis palmoplantaris and periodontoclasia in childhood (Haim–Munk syndrome).* (A,B) Severe hyperkeratosis of hands with arachnodactyly. Note tapered pointed terminal phalanges exhibiting claw-like volar curve. (From *S Haim* and *J Munk,* Br J Dermatol **77**:42, 1965.)

**References [Hyperkeratosis palmoplantaris and periodontoclasia in childhood (Papillon–Lefèvre syndrome and Haim–Munk syndrome)]**

1. Bach JM, Levan NE: Papillon–Lefèvre syndrome. *Arch Dermatol* **97**:154–158, 1968.

1a. Bergman R, Friedman-Birnbaum R: Papillon–Lefèvre syndrome: A study of the long term clinical course, recurrent pyogenic infections and the effects of etretinate treatment. *Br J Dermatol* **119**:731–736, 1988.

2. Beumer J et al: Childhood hypophosphatasia and the premature loss of teeth. *Oral Surg* **35**:631–640, 1973.

3. Bork K, Lost C: Extrapalmoplantare Hautsymptome and weitere klinische und ätiologische, inbesondere immunologische Gesichtspunkte beim Papillon-Lefèvre-Syndrom. *Hautarzt* **31**:179–183, 1980.

4. Bravo-Piris J et al: Papillon–Lefèvre syndrome: Report of two familiar cases. *Dermatologica* **152**:168–176, 1967.

5. Brownstein MH, Soklnik P: Papillon–Lefèvre syndrome. *Arch Dermatol* **106**:533–534, 1972.

6. Corson EE: Keratosis palmaris et plantaris with dental alteration. *Arch Dermatol Syph* **40**:639, 1939.

7. Dekker G, Jansen LH: Periodontosis in a child with hyperkeratosis palmoplantaris. *J Periodontol* **29**:266–271, 1958.

8. Djawari D: Deficient phagocytic function in Papillon–Lefèvre syndrome. *Dermatologica* **150**:189–192, 1978.

9. El Darouti MA et al: Papillon–Lefèvre syndrome. Successful treatment with oral retinoids in three patients. *Int J Dermatol* **27**:63–66, 1988.

10. Gelmetti C et al: Long-term preservation of permanent teeth in a patient with Papillon–Lefèvre syndrome treated with etretinate. *Pediatr Dermatol* **6**:222–225, 1989.

11. Giansanti JS et al: Palmar-plantar hyperkeratosis and concomitant periodontal destruction (Papillon–Lefèvre syndrome). *Oral Surg* 35:30–48, 1973.

12. Gorlin RJ, Chaudhry AP: The oral manifestations of cyclic neutropenia. *Arch Dermatol* **82**:344–348, 1960.

13. Gorlin RJ et al: The syndrome of palmar-plantar hyperkeratosis and premature periodontal destruction of the teeth. *J Pediatr* **65**:895–908, 1964.

14. Hacham-Zadeh S et al: A genetic analysis of the Papillon–Lefèvre syndrome—a Jewish family from Cochin. *Am J Med Genet* **2**:153–157, 1978.

15. Haim S, Munk J: Keratosis palmo-plantaris congenita, with periodontosis, arachnodactyly and peculiar deformity of the terminal phalanges. *Br J Dermatol* **77**:42–54, 1965.

16. Haneke E: The Papillon–Lefèvre syndrome: Keratosis palmoplantaris with periodontopathy. *Hum Genet* **15**:1–35, 1979.

17. Haneke E et al: Increased susceptibility to infection in the Papillon–Lefèvre syndrome. *Dermatologica* **150**:283–286, 1975.

18. Jansen LH, Dekker G: Hyperkeratosis palmo-plantaris with periodontosis (Papillon–Lefèvre). *Dermatologica* **113**:207–219, 1956.

19. Levo Y et al: Immunological study of patients with the Papillon–Lefèvre syndrome. *Clin Exp Immunol* **40**:407–410, 1980.

20. Manson JD, Lehner T: Clinical features of juvenile periodontitis (periodontosis). *J Periodontol* **45**:636–640, 1974.

21. Naik DN et al: Papillon–Lefèvre syndrome. *Oral Surg* **25**:19–23, 1968.

22. Papillon, Lefèvre P: Deux cas de kératodermie palmaire et plantaire symetrique familiale (maladie de Meleda) chez le frère et al soeur. Coexistence dans les deux cas d'altérations dentaires graves. *Bull Soç Fr Dermatol Syphiligr* **31**:82–84, 1924.

23. Picarelli A: Le parodontopatie giovanili in corso d'ipercheratosi palmoplantare e la sindrome di Papillon–Lefèvre. *Min Stomatol* **17**:587–601, 1968.

24. Piquet B et al: Keratodermie palmaire et plantaire (Papillon–Lefèvre syndrome). *Rev Stomatol* **70**:446–459, 1969.

25. Puliyel JM, Sridharan Iyer KS: A syndrome of keratosis palmo-plantaris congenita, pes planus, onychogryphosis, periodontitis, arachnodactyly, and a peculiar acro-osteolysis. *Br J Dermatol* **115**:243–248, 1986.

26. Rosenthal SL: Periodontosis in a child resulting in exfoliation of the teeth. *J Periodontol* **22**:101–104, 1951.

26a. Seow WK: Palmoplantar hyperkeratosis with short stature, facial dysmorphism, and hypodontia in a new syndrome? *Pediatr Dent* **11**:145–150, 1989.

27. Takahara S: Acatalasemia and hypocatalasemia in the Orient. *Sem Hematol* **8**:397–416, 1971.

28. Tosti A et al: Is etretinate dangerous in Papillon–Lefèvre syndrome? *Dermatologica* **176**:148–150, 1988.

29. Van Dyke TE et al: Papillon–Lefèvre syndrome: Neutrophil dysfunction with severe periodontal disease. *Clin Immunol Immunopathol* **31**:419–429, 1984.

## Hyperkeratosis palmoplantaris and attached gingival hyperkeratosis

Fred et al (1), in 1964, reported the combination of hyperkeratosis palmoplantaris and attached gingival hyperkeratosis in a kindred. In 1974 while visiting Athens, Greece, RJ Gorlin had the opportunity to see another family with the same disorder (5). Other families have been reported (3,4,6,10). Another possible kindred has been noted by Roth et al (8).

Autosomal dominant inheritance is indicated by transmission of the disorder through several generations (1,3–5,10). There was male-to-male transmission in all kindreds.

**Skin and skin appendages.** Focal to widespread hyperkeratosis of the soles is more marked over the weight-bearing areas: heels, toe pads, and metatarsal heads (Fig. 26–9). These may be painful, inhibiting ambulation (10). Hyperkeratosis of the palms also seems trauma related (Fig. 26–10). Hyperhidrosis is noted in the hyperkeratotic areas. The hyperkeratotic areas appear around puberty in most patients and progress in severity with age. The finger- and toenails exhibit subungual and circumungual keratin deposits. First involved are the toes at 4–5 years of age followed by fingernail changes at 8–9 years. Follicular keratosis of the sebaceous areas of the face is common.

**Oral manifestations.** Sharply marginated hyperkeratosis involves the labial and lingual attached gingiva (Fig. 26–11). The hard palate, beneath denture-bearing areas, and the lateral areas or dorsum of the

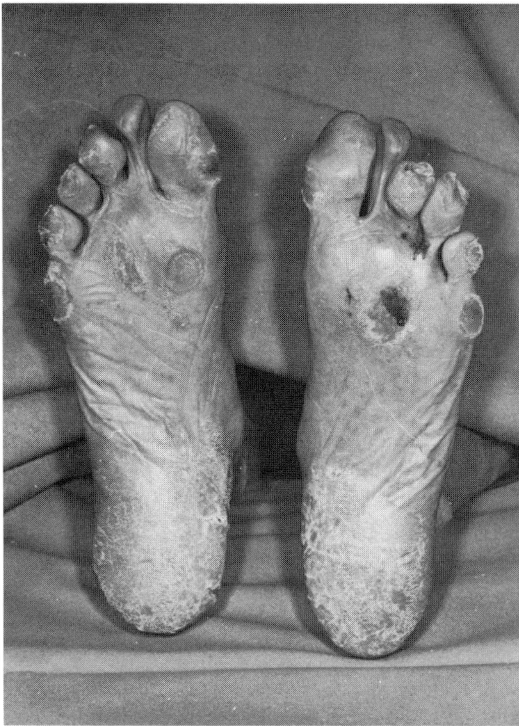

Fig. 26–9. *Hyperkeratosis palmoplantaris and attached gingival hyperkeratosis.* Numerous focal hyperkeratoses of soles, especially over pressure points.

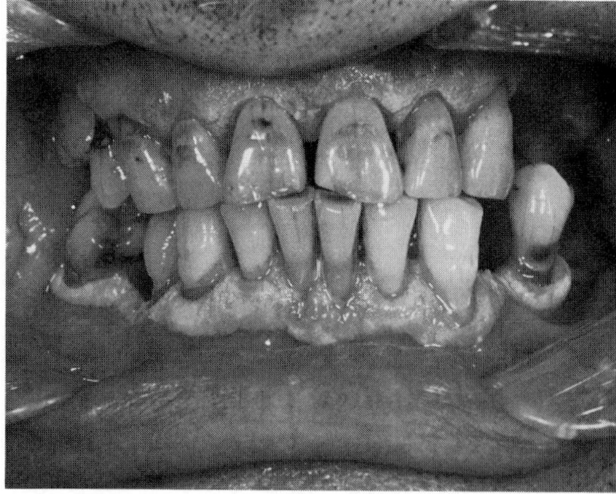

Fig. 26–11. *Hyperkeratosis palmoplantaris and attached gingival hyperkeratosis.* Hyperkeratosis limited to fixed gingiva.

### References (Hyperkeratosis palmoplantaris and attached gingival hyperkeratosis)


1. Fred HI et al: Keratosis palmaris et plantaris. *Arch Intern Med* **113**:866–871, 1974.
2. Gibbs RC, Costello MJ: *The Palms and Soles in Medicine.* Charles C. Thomas, Springfield, Illinois, 1967.
3. Gorlin RJ: Focal palmoplantar and marginal gingival hyperkeratosis in a syndrome. *Birth Defects* **12**(5):239–242, 1976.
4. James P, Beggs D: Tylosis: A case report. *Br J Oral Surg* **11**:143–145, 1973.


tongue may be also affected. The hyperkeratotic areas appear in early childhood and increase in severity with age.

**Differential diagnosis.** There are many disorders that may be associated with hyperkeratosis palmoplantaris (2). Generalized oral hyperkeratosis, especially that of the buccal mucosa, has been noted in patients with autosomal dominant tylosis and carcinoma as well as with congenital strictures and squamous cell carcinoma of the esophagus (7–9).

**Laboratory aids.** Paranuclear bodies are seen in the spinous and granular cell layers of the keratinocytes of the gingival epithelium. Ultrastructure changes show these to be condensed tonofilaments (10).

Fig. 26–10. *Hyperkeratosis palmoplantaris and attached gingival hyperkeratosis.* Focal hyperkeratoses of palms.

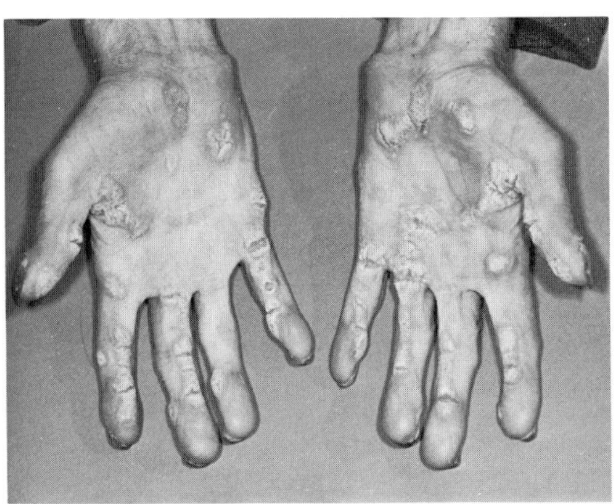

Fig. 26–12. *Oral and conjunctival amyloidosis and mental retardation.* (A) Atrophy of bulb, amyloid deposits in conjunctiva. (B) Amyloid deposits in gingiva, imparting appearance of icing. (From *J Hornová* and *O Dluhosová,* Oral Surg **25**:457, 1968.)

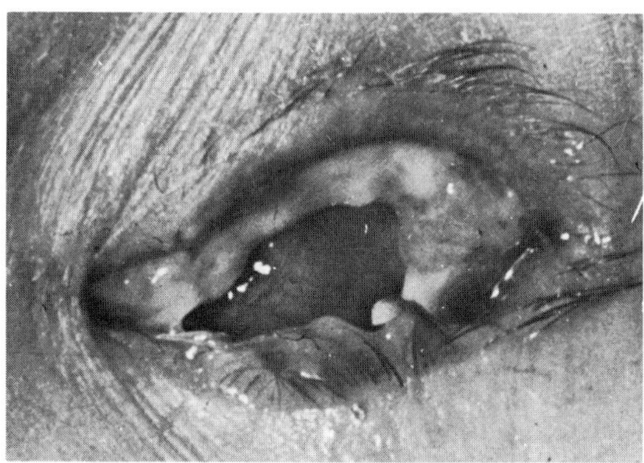

A

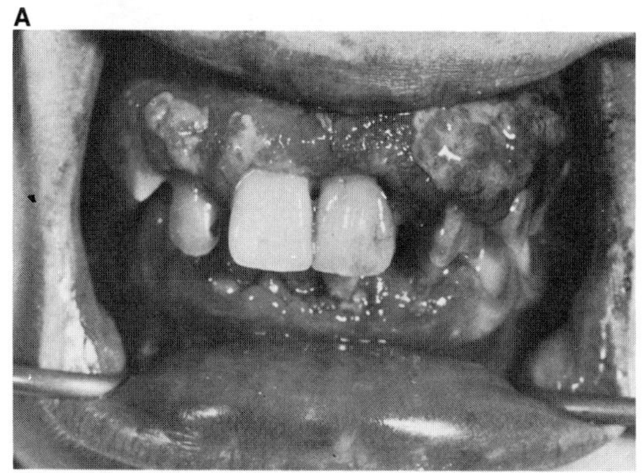

B

5. Laskaris G et al: Focal palmoplantar and oral mucosa hyperkeratosis syndrome: A report concerning five members of a family. *Oral Surg* **50**:250–253, 1980.
6. Raphael AL et al: Hyperkeratosis of gingival and plantar surfaces. *Periodontics* **6**:118–120, 1968.
7. Ritter SB, Peterson G: Esophageal cancer, hyperkeratosis and oral leukoplasia: Follow-up family study. *JAMA* **236**:1844–1845, 1976.
8. Roth W et al: Hereditary painful callosities. *Arch Dermatol* **114**:591–592, 1978.
9. Tyldesley WR, Kempson SA: Ultrastructure of the oral epithelium in leukoplakia associated with tylosis and esophageal carcinoma. *J Oral Pathol* **4**:49–58, 1975.
10. Young WG et al: Focal palmoplantar and gingival hyperkeratosis syndrome: Report of a family, with cytologic, ultrastructural and histochemical findings. *Oral Surg* **53**:473–482, 1982.

## Oral and conjunctival amyloidosis and mental retardation (Hornová–Dluhosová syndrome)

Hornová and Dluhosová (1) reported a boy and his mentally retarded sister. Within the first year of life, the eyelids were noted to be swollen, with nodular deposits of amyloid in the conjunctiva. Congenital cataracts, atrophy of ocular bulb, and amaurosis were also present. Amyloid deposits found in the enlarged boggy gingiva imparted the appearance of icing (Fig. 26–12A,B).

### Reference [Oral and conjunctival amyloidosis and mental retardation (Hornová–Dluhosová syndrome)]

1. Hornová J, Dluhosová O: Primary amyloidosis of gingiva and conjunctiva and mental disorder in a brother and sister. *Oral Surg* **25**:457–464, 1968.

# Chapter 27
# Syndromes with Unusual Dental Findings

## Rieger syndrome (hypodontia and primary mesodermal dysgenesis of the iris)

Hypodontia was described as occurring in combination with malformation of part of the anterior chamber of the eye as early as 1883 by Vossius (27). However, the condition was not recognized as a heritable syndrome until Rieger's report in 1935 (21). The syndrome has been expanded (5,14,24). The major features are absent maxillary incisor teeth, malformations of the anterior chamber of the eye, and umbilical anomalies. Its frequency has been estimated as 1/200,000 population (2).

The syndrome has autosomal dominant inheritance (2–4, 7,11,12,14,18,26,30) with almost complete penetrance and variable expressivity (2). Various karyotypic abnormalities have been found with Rieger anomaly (19,19a,21a,23a,25) but the locus has not been verified.

**Facies.** Some patients have broad flat nasal root, prominent supraorbital ridges, and relative prognathism due to underdevelopment of the maxilla and/or to loss of vertical height because of hypodontia (3a,5,15,24) (Fig. 27–1A,B).

**Eyes.** Iridogoniodysgenesis, that is, abnormal development of the anterior chamber of the eye, is characteristic. Anterior displacement and thickening of Schwalbe's line (4,5) are hallmarks, as is stromal hypoplasia. Pupils may be normal but are frequently displaced and nonspherical (dyscoria) or even slitlike. Polycoria is common and may be progressive, occasionally leading to secondary aniridia (IH Maumenee, personal communication, 1986). Iris strands to Schwalbe's line are present in 90%. Absence of the anterior iris layer in the presence of a prominent Schwalbe's line is diagnostic (IH Maumenee, personal communication, 1986) (Fig. 27–2A,B). Microcornea, megalocornea, corneal opacity, aniridia, and strabismus also occur (21a). Waring et al (28) noted onset of increased intraocular pressure in 50% by age 20. An additional 10–15% per decade developed increased intraocular pressure thereafter. Shields (24) also found glaucoma in about half the patients. The glaucoma is difficult to control and may lead to significant damage to the optic nerve head and to blindness (4,5,24).

**Umbilical abnormalities.** The only other consistent anomaly is failure of the periumbilical skin to involute (4,7,14,24) (Fig. 27–3). Reddihough et al (20) found exomphalos in two patients.

**Other abnormalities.** Hypospadias (4,14,15), inguinal hernia (2,4,7), anal stenosis (9), and Meckel's diverticulum (16,20) have been found. A wide spectrum of other abnormalities has been tabulated by Alkemade (2) and by Fitch and Kaback (9), but they formed no pattern. Psychomotor retardation has been described (2,9,15). Koshino et al (15a) reported prominent posterior clinoid process, pseudoepiphyses of fifth metacarpals and metatarsals, and irregular ossification of femoral heads and distal femoral and distal tibial epiphyses.

**Oral manifestations.** The premaxillary area is relatively underdeveloped (2,3,19), and a reduced number of teeth is frequent, more often in the upper jaw (3,3a,7,9a,14,18,30). The maxillary deciduous

Fig. 27–1. *Rieger syndrome.* (A,B) Note midfacial hypoplasia with relative mandibular prognathism. [From *E Frandsen,* Acta Ophthalmol (Kbh) **41**:757, 1963.]

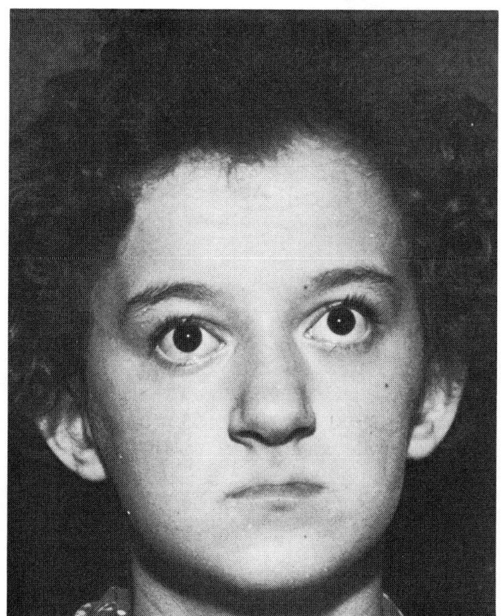

A

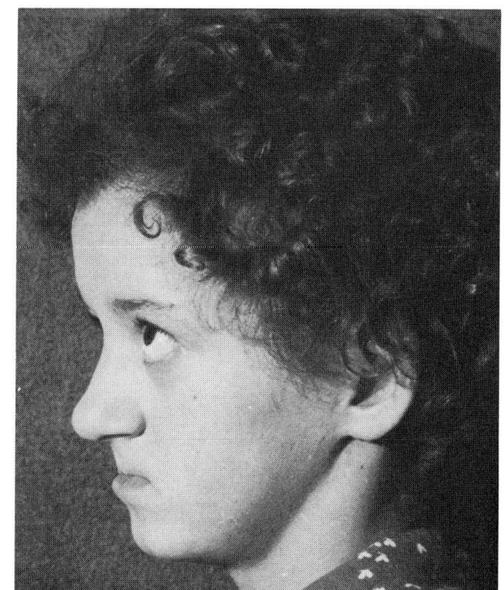

B

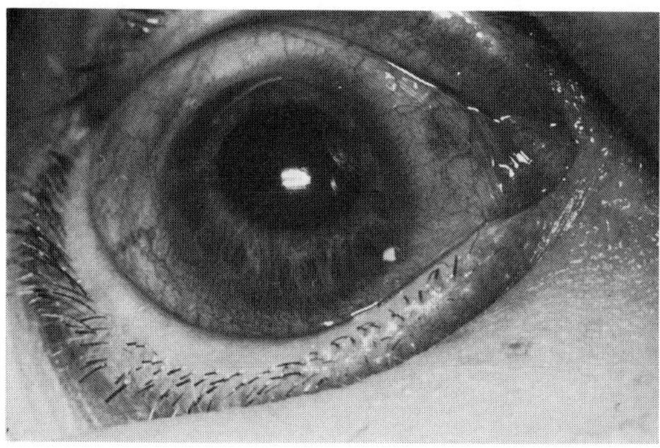

A

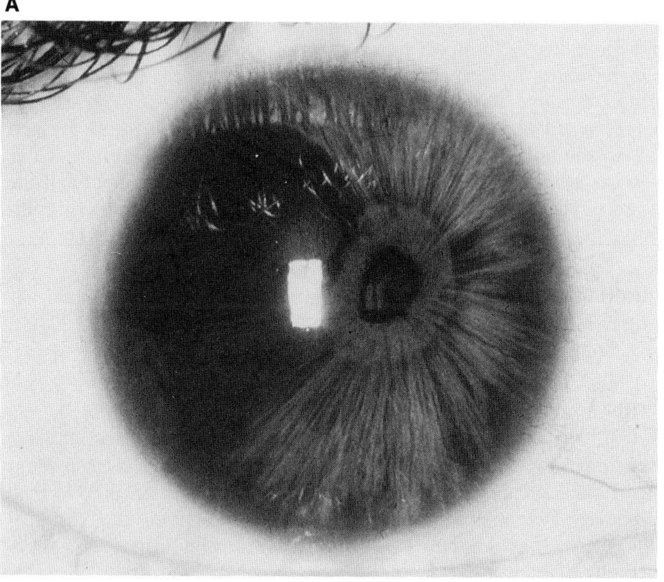

B

Fig. 27–2. *Rieger syndrome.* (A) Vertical slit pupil and polycoria are evident. (From *W Lemmingson,* Klin Monatsbl Augenheilkd **138**:96, 1961.) (B) Congenital hypoplasia of anterior iris layer. The patient had prominence of Schwalbe's line and the dental and umbilical changes of Rieger syndrome. (Courtesy of *IH Maumenee,* Baltimore, Maryland.)

and permanent incisors and second premolars are most commonly missing (14,15). Conical crown form of anterior teeth has been recorded by several authors (6,7,14,19,27,30) (Fig. 27–4). Cleft palate has been described by Fitch and Kaback (9).

**Differential diagnosis.** Congenital absence of teeth and teeth with conical crowns are seen in *hypohidrotic ectodermal dysplasia, Ellis–van Creveld syndrome, incontinentia pigmenti, hypodontia and nail dysgenesis,* and *acrodental dysostosis.*

Goniodysgenesis has been found in binary combination with anal atresia, arachnodactyly, hearing loss, glaucoma, and myopathy (13,14). Gorlin et al (10) described brothers with *SHORT syndrome* (goniodysgenesis, severe growth retardation, inguinal hernia, joint hypermobility, deep-set eyes, and delayed teething). Sensenbrenner et al (23) reported a similarly affected child. The disorder has autosomal recessive inheritance. A similar disorder was reported by Sedeghi-Najar and Senior (22) and Aarskog et al (1), but inheritance was autosomal dominant.

Mesodermal dysgenesis of the iris has been reported by De Hauwere et al (6) in association with orbital hypertelorism, psychomotor retardation, moderate sensorineural hearing loss, dilatation of the cerebral ventricles, generalized hypotonia, and pelvic anomalies. No dental abnormalities were found. Inheritance was autosomal dominant. Jerndal (13) reported a dominantly inherited syndrome of goniodysgenesis and glaucoma.

Dysgenesis of the anterior chamber angle is also found with Peter's anomaly, congenital hypoplasia of iris stroma, *rubella embryopathy,* Norrie syndrome, *Marfan syndrome,* and in some chromosomal aberrations.

Fig. 27–3. *Rieger syndrome.* Umbilical defect with characteristic projection of periumbilical skin. (From *DS Reddihough et al,* Aust Paediatr J **18**:130, 1982.)

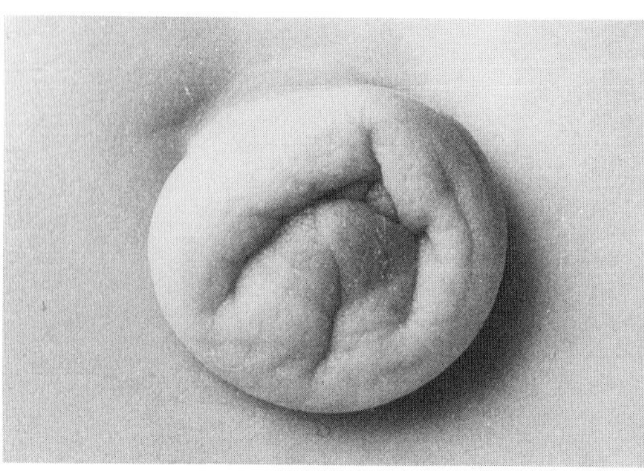

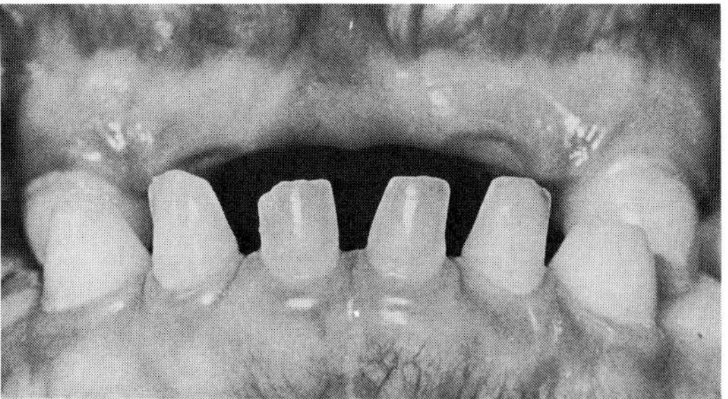

Fig. 27–4. *Rieger syndrome.* Absence of maxillary incisors and peg-shaped crowns of lower incisors and canines. (From *M Feingold et al,* Pediatrics **44**:564, 1969.)

### References [Rieger syndrome (hypodontia and primary mesodermal dysgenesis of the iris)]

1. Aarskog D et al: Autosomal dominant partial lipodystrophy associated with Rieger anomaly, short stature, and insulinopenic diabetes. *Am J Med Genet* **15**:29–38, 1983.

2. Alkemade PPH: *Dysgenesis Mesodermalis of the Iris and the Cornea.* Van Gorcum, The Netherlands, 1969.

3. Busch G et al: Dysgenesis mesodermalis et ectodermalis Rieger oder Rieger'sche Krankheit. *Klin Mbl Augenheilkd* **136**:512–523, 1960.

3a. Childers NK, Wright JT: Dental and craniofacial anomalies of Axenfeld–Rieger syndrome. *J Oral Pathol* **15**:534–539, 1986.

4. Chisholm IA, Chudley AE: Autosomal dominant iridogoniodysgenesis with associated somatic anomalies: Four-generation family with Rieger's syndrome. *Br J Ophthalmol* **67**:529–534, 1983.

5. Cross H et al: The Rieger syndrome: An autosomal dominant disorder with ocular, dental and systemic abnormalities. *Perspect Ophthalmol* **3**:3–16, 1979. (Same patients reported in Ref. 14.)

6. De Hauwere RC et al: Iris dysplasia, orbital hypertelorism, and psychomotor retardation: A dominantly inherited developmental syndrome. *J Pediatr* **82**:679–681, 1973.

7. Drum MA et al: Oral manifestations of the Rieger syndrome: Report of case. *J Am Dent Assoc* **110**:343–346, 1985.

8. Feingold M et al: Rieger's syndrome. *Pediatrics* **44**:564–569, 1969.

9. Fitch N, Kaback M: The Axenfeld syndrome and the Rieger syndrome. *J Med Genet* **15**:30–34, 1978.

9a. Fleischer-Peters A, Lang GE: Fehlende ober Schneiderzahne als Leitsymptom des Rieger-Syndroms. *Dtsch Zahnärztl Z* **44**:228–231, 1989.

10. Gorlin RJ et al: A selected miscellany: Rieger anomaly and growth retardation (the S-H-O-R-T syndrome). *Birth Defects* **11**(2):46–48, 1975.

11. Heckenlively JR et al: The Rieger syndrome, a heritable disorder associated with glaucoma. *Johns Hopkins Med J* **151**:351–355, 1982.

12. Henkind P et al: Mesodermal dysgenesis of the anterior segment: Rieger's anomaly. *Arch Ophthalmol* **73**:810–817, 1965.
13. Jerndal T: Congenital glaucoma due to dominant goniodysgenesis: A new concept of the heredity of glaucoma. *Am J Hum Genet* **35**:645–651, 1983.
14. Jorgenson RJ et al: The Rieger syndrome. *Am J Med Genet* 2:307–318, 1978. (Same patients reported in Ref. 5.)
15. Judisch GF et al: Rieger's syndrome. A case report with a 15-year follow-up. *Arch Ophthalmol* **97**:2120–2122, 1979.
15a. Koshino T et al: Bone and joint manifestations of Rieger's syndrome: A report of a family. *J Pediatr Orthoped* **9**:224–230, 1989.
16. Krespi YP, Pertsemlidis D: Rieger's syndrome associated with a large Meckel's diverticulum. *Am J Gastroenterol* **71**:608–610, 1979.
17. Langdon JD: Rieger's syndrome. *Oral Surg* **30**:788–795, 1970.
18. Lemmingson W, Riethe P: Beobachtungen bei Dysgenesis mesodermalis cornea et iridis in Kombination mit Oligodontie. *Klin Mbl Augenheilkd* **133**:877–891, 1958.
19. Ligutic I et al: Interstitial deletion of 4q and Rieger syndrome. *Clin Genet* **20**:323–327, 1981.
19a. Nielsen F, Tranebjaerg L: A case of partial monosomy 21q22.2 associated with Rieger syndrome. *J Med Genet* **21**:218–221, 1984.
20. Reddihough DS et al: Rieger syndrome with exomphalos. *Aust Paediatr J* **18**:130–131, 1981.
21. Rieger HH: Beiträge zur Kenntnis seltener Missbildungen der Iris. *Albrecht von Graefes Arch Klin Ophthalmol* **133**:602–635, 1935.
21a. Rogers CR: Rieger syndrome. *Proc Greenwood Genet Ctr* **7**:9–13, 1988.
22. Sedeghi-Najar A, Senior B: Autosomal dominant transmission of isolated growth hormone deficiency in iris-dental dysplasia (Rieger's syndrome). *J Pediatr* **85**:644–648, 1974.
23. Sensenbrenner JA et al: A low-birthweight syndrome; Rieger syndrome. *Birth Defects* **11**(2):423–426, 1975.
23a. Shiang R et al: Mapping of ADH3, EGF, and IL2 in a patient with Rieger-like phenotype and 4q23-q27 deletion. *Am J Hum Genet* **41**:185A, 1987.
24. Shields MB: Axenfeld–Rieger syndrome: A theory of mechanism and distinctions from the iridocorneal endothelial syndrome. *Trans Am Ophthalmol Soc* **81**:736–784, 1983.
25. Stathacopoulos RA et al: The Rieger syndrome and a chromosome 13 deletion. *J Pediatr Ophthalmol Strab* **24**:19–20, 1987.
26. Unger L: Beitrag zur sogen. Dysgenesis mesodermalis corneae et iridis (Rieger). *Ophthalmologia* **132**:27–35, 1956.
27. Vossius A: Kongenitale Anomalie der Iris. *Klin Mbl Augenheilkd* **21**:233–237, 1883.
28. Waring GO III et al: Anterior chamber cleavage syndrome: A stepladder classification. *Surv Ophthalmol* **20**:3–17, 1975.
29. Wesley RK et al: Rieger's syndrome (oligodontia and primary mesodermal dysgenesis of the iris). Clinical features and report of an isolated case. *J Pediatr Ophthalmol Strab* **15**:67–70, 1978.
30. Wilson JP: A case of partial anodontia. *Br Dent J* **99**:199–200, 1955.

## Tricho-dento-osseous syndrome and tricho-onycho-dental syndrome

Robinson et al (25) and Lichtenstein, Jorgenson, and co-workers (14,18,20) described a syndrome of amelogenesis imperfecta, taurodontism, curly hair, and, at times, sclerotic bones. Inheritance is autosomal dominant. Other reports have described binary combination of amelogenesis imperfecta and taurodontism (3–5,7,9,21,22,31,33) that may represent the same condition. However, it appears to stand on its own as a distinct entity (4,32). Although Gage (7) suggested X-linked inheritance in a family he reported, autosomal dominant inheritance is evident in several kindred (5).

There may be two types of tricho-dento-osseous (TDO) syndrome. In the classic form (TDO-I), the scalp hair is curly and the lashes are long (Fig. 27–5). The hair tends to straighten by the second or third decade. The teeth have pitted thin hypoplastic yellowish-brown enamel and are prone to abscess formation (Fig. 27–6). Pulp chambers are enlarged. The roots tend to be short and open. The deciduous and permanent molars are taurodont (Fig. 27–7). The teeth may be delayed in eruption and, not uncommonly, several permanent teeth are impacted. The long bones and calvaria have mild to moderately increased density (Fig. 27–8A,B). Electron microscopic study has shown a thin enamel layer with randomly distributed depressions and pits. The mineral content of the enamel is close to that of the underlying dentition (20).

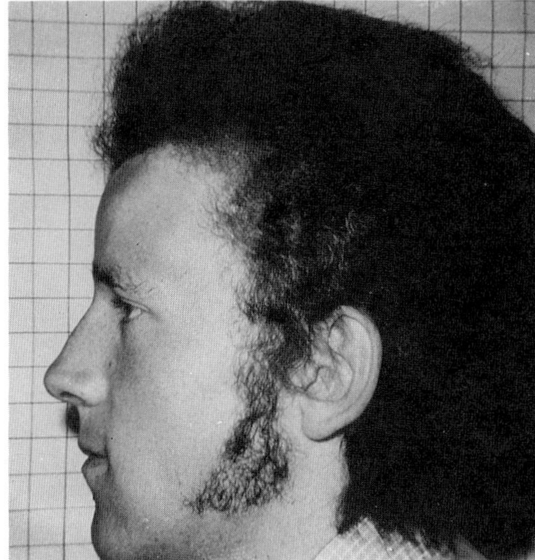

Fig. 27–5. *Tricho-dento-osseous syndrome.* Curly hair. (Courtesy of *R Jorgenson,* San Antonio, Texas.)

There may be premature closure of the sagittal suture resulting in dolichocephaly. The mastoids are poorly pneumatized. The skull base has increased density and thickness. Skeletal maturation is delayed and some patients exhibit short ramus, square jaw, and obtuse mandibular angle (19).

TDO-II, described by Shapiro et al (27) and Quattromani et al (23), is also manifest by curly hair and the same dental changes seen in TDO-I although there is no tendency toward discoloration of the enamel. The long bones do not appear to have increased density. The clavicles are undertubulated. The calvaria manifest not only increased density but increased thickness and obliterated diplöe. Frontal sinuses and mastoids tend to be obliterated. The base of the skull shows both increased density and thickness. In contrast to TDO-I, patients with TDO-II have macrocephaly. Whether these two forms represent true heterogeneity is not known. Rivas et al (24) found evidence for linkage between TDO-I locus and ABO, Kell, and Gc loci in the family reported by Lichtenstein et al (19). One would anticipate linkage studies in TDO-II families.

Fig. 27–6. *Tricho-dento-osseous syndrome.* Deciduous incisors showing generalized enamel pitting, most pronounced near the junction of the crown and the root. (From *J Lichtenstein et al,* Am J Hum Genet **24**:569, 1972.)

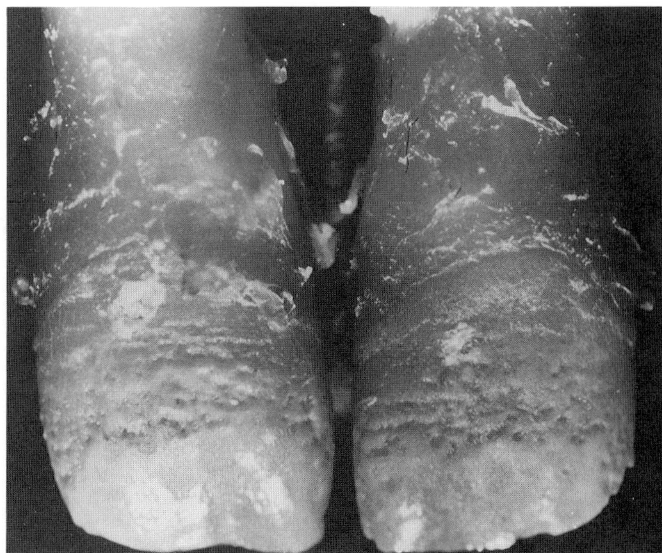

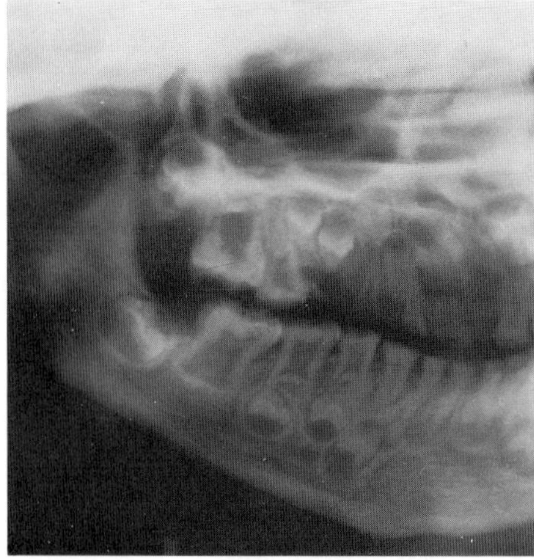
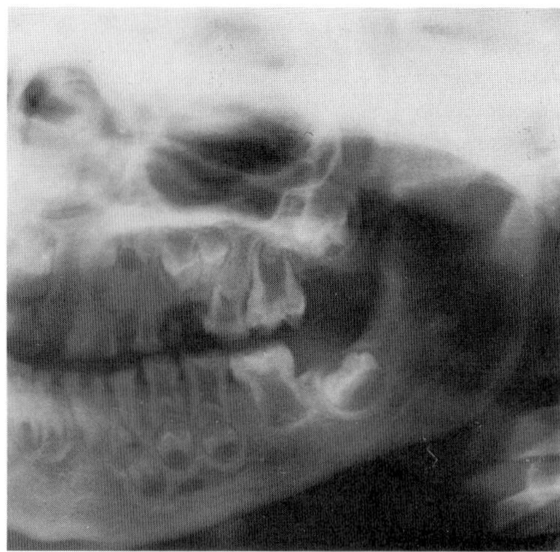

Fig. 27–7. *Tricho-dento-osseous syndrome.* Panoramic radiograph of deciduous dentition and developing permanent dentition. Deciduous molars and first permanent molars are taurodont, and the enamel is thin. There is some duplication of structures in the midline of the radiograph. (From *RJ Jorgenson* and *RW Warson,* Oral Surg **36**:693, 1973.)

Tricho-onycho-dental syndrome, described by Leisti and Sjöblom (17) and others (16,28,30) is characterized by curly hair which is relatively sparse and easily detachable. There is decreased facial, axillary, and pubic hair. The enamel is hard and thin and has indistinct enamel rods. The dentin is dysplastic and fills the pulp, being similar to that seen in dentin dysplasia, Type I. There is some tendency toward precocious eruption of teeth, taurodontism, and short open roots but no tendency for the permanent teeth to be impacted. The teeth are lost early. The long bones and calvaria have increased density but the calvaria is also increased in thickness. There is narrowing of ear canals among affected males and a tendency toward mandibular prognathism.

Taurodontism may occur as an isolated finding in 0.5–4% of the population (15) and in many syndromal associations (12,13): *taurodontism, oligodontia, and sparse hair;* microcephalic dwarfism, microdontia and diminished anterior root formation (8,26); *trisomy 21 syndrome* (11); *X-chromosome aneuploidy* (1,4a,12,13,29); osteoporosis (6); *Ackerman syndrome, taurodontism, microdontia, and dens invaginatus,* and apparent *monosomy 21* (36). There are many forms of amelogenesis imperfecta (34). The trichodental syndrome (10) consists of autosomal dominant inheritance of short hair and oligodontia. A superb discussion of taurodontism is that of Witkop et al (35).

The disorder described by Congleton and Burkes (2) is a distinct form of amelogenesis imperfecta with taurodontism but is not TDO syndrome (33a). About 33% of patients with hypodontia have at least one permanent molar with taurodontism (26a).

### References (Tricho-dento-osseous syndrome and tricho-onycho-dental syndrome)


1. Cichon JC, Pack RS: Taurodontism: Review of literature and report of case. *J Am Dent Assoc* **111**:453–455, 1985.

2. Congleton J, Burkes EJ Jr: Amelogenesis imperfecta with taurodontism. *Oral Surg* **48**:540–544, 1979.

3. Crawford JL: Concomitant taurodontism and amelogenesis imperfecta in the American Caucasian. *J Dent Child* **37**:171–175, 1970.

4. Crawford PJM et al: Amelogenesis imperfecta: Autosomal dominant hypomaturation-hypoplasia type with taurodontism. *Br Dent J* **164**:71–73, 1988.

4a. Darbyshire PA et al: Prepubertal diagnosis of Klinefelter syndrome in a patient with taurodont teeth. *Pediatr Dent* **11**:225–226, 1989.

5. Elzay RP, Chamberlain DH: Differential diagnosis of enlarged dental pulp chambers: A case report of amelogenesis imperfecta with taurodontism. *J Dent Child* **53**:388–389, 1986.

6. Fuks AB et al: Multiple taurodontism associated with osteoporosis. *J Pedodont* **7**(1):68–74, 1982.

7. Gage JP: Taurodontism and enamel hypomaturation associated with X-linked abnormalities. *Clin Genet* **14**:159–164, 1978 (Case 3).

8. Gardner DG, Girgis SS: Taurodontism, short roots, and external resorption associated with short stature and a small head. *Oral Surg* **44**:271–273, 1977.

9. Gulmen S et al: Tricho-dento-osseous syndrome. *J Endodont* **2**:117–120, 1976.

10. Keasey PJW: Tricho-dental syndrome: A disorder with a short hair cycle. *Br J Dermatol* **116**:259–263, 1987.

11. Jaspers MT, Witkop CJ Jr: Taurodontism in the Down syndrome. *Oral Surg* **51**:632–636, 1981.


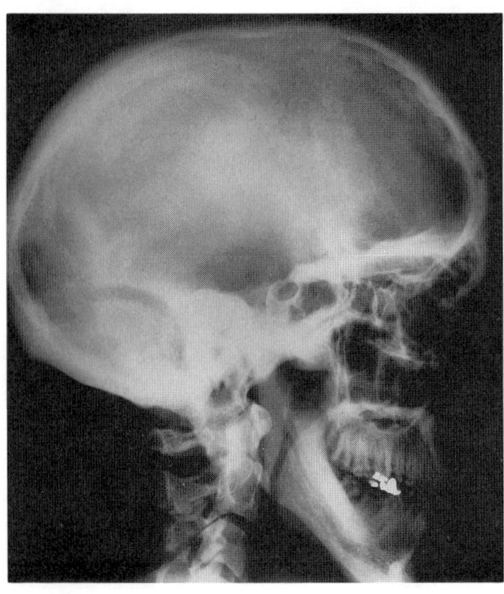
A

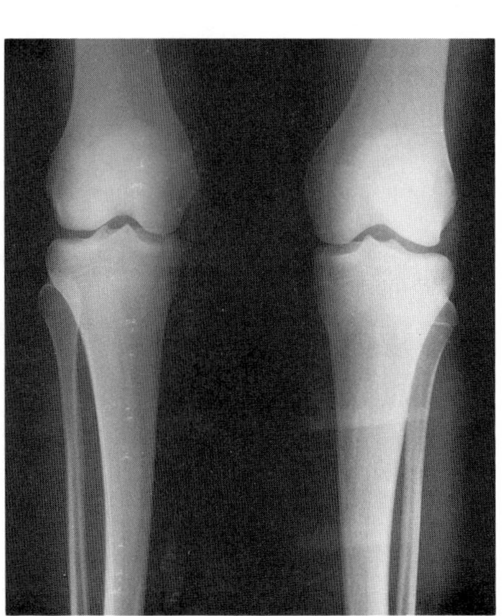
B

Fig. 27–8. *Tricho-dento-osseous syndrome.* (A) Skull radiograph showing increased bone density and thickening of chondrocranium. The mandibular angles are more oblique than normal. (B) Increased density of long bones. (A from *J Lichtenstein et al,* Am J Hum Genet **24**:569, 1972. B courtesy of *R Jorgenson,* San Antonio, Texas.)

12. Jaspers MT, Witkop CJ Jr: Taurodontism, an isolated trait associated with syndromes and X-chromosomal aneuploidy. *Am J Hum Genet* **32**:396–412, 1980.
13. Jorgenson RJ: The conditions manifesting taurodontism. *Am J Med Genet* **11**:435–442, 1982.
14. Jorgenson RJ, Warson RW: Dental abnormalities in the tricho-dento-osseous syndrome. *Oral Surg* **36**:693–700, 1973.
15. Jorgenson RJ et al: The prevalence of taurodontism in a select population. *J Craniofac Genet Div Biol* **2**:125–135, 1982.
16. Koshiba H et al: Clinical, genetic, and histologic features of the trichoonychodental (TOD) syndrome. *Oral Surg* **46**:376–385, 1978.
17. Leisti J, Sjöblom SM: A new type of autosomal dominant tricho-dento-osseous syndrome. *March of Dimes Birth Defects Conference,* San Francisco, June 11–14, 1978, p 58.
18. Lichtenstein JR, Warson RW: Syndrome of dental anomalies, curly hair and sclerotic bones. *Birth Defects* **7**(7):308–311, 1971.
19. Lichtenstein JR et al: The tricho-dento-osseous (TDO) syndrome. *Am J Hum Genet* **24**:569–582, 1972.
20. Melnick M et al: Tricho-dento-osseous syndrome: A scanning electron microscopic analysis. *Clin Genet* **12**:17–27, 1977.
21. Mjör IA: The structure of taurodont teeth. *J Dent Child* **39**:459–463, 1972.
22. Parker JL et al: Hypoplastic-hypomaturation amelogenesis imperfecta with taurodontism. *J Dent Child* **42**:379–383, 1975.
23. Quattromani F et al: Clinical heterogeneity in the tricho-dento-osseous syndrome. *Hum Genet* **64**:116–121, 1983.
24. Rivas ML et al: Possible linkage between the loci for the trichodentoosseous (TDO) syndrome and the ABO blood group system: Genetic and clinical implications. *Birth Defects* **10**(10):255–262, 1974.
25. Robinson GC et al: Hereditary enamel hypoplasia: Its association with characteristic hair structure. *Pediatrics* **37**:498–502, 1966.
26. Sauk JJ Jr, Delany JR: Taurodontism, diminished root formation and microcephalic dwarfism. *Oral Surg* **36**:231–235, 1973.
26a. Seow WK, Lai PY: Association of taurodontism with hypodontia: A controlled study. *Pediatr Dent* **11**:214–216, 1989.
27. Shapiro SD et al: Tricho-dento-osseous syndrome: Heterogeneity or clinical variability. *Am J Med Genet* **16**:225–236, 1983.
28. Siirila HS, Heikinheimo O: Odontogenesis imperfecta. *Suom Hammaslääk Toim* **58**:35–47, 1962.
29. Stewart RE: Taurodontism in X-chromosome aneuploid syndrome. *Clin Genet* **6**:341–344, 1974.
30. Westerholm N: On koinsidens vid dentogingivala och kutana patologiska tellstand och utvecklings hämmeeingar. *Suom Hammaslääk Toim* **55**:569–580, 1947.
31. Widerman FH, Serene TP: Endodontic therapy involving a taurodont tooth. *Oral Surg* **32**:618–620, 1971. (Same as Ref. 3).
32. Winter GB, Brook AH: Enamel hypoplasia and anomalies of the enamel. *Dent Clin N Am* **19**:3–24, 1975.
33. Winter GB et al: Hereditary amelogenesis imperfecta: A rare autosomal dominant type. *Br Dent J* **127**:157–164, 1969.
33a. Witkop CJ Jr: Amelogenesis imperfecta, dentinogenesis imperfecta, and dentin dysplasia revisited: Problems in classification. *J Oral Pathol* **17**:547–553, 1989.
34. Witkop CJ Jr, Rao S: Inherited defects in tooth structure. *Birth Defects* **7**(7):153–184, 1971.
35. Witkop CJ Jr et al: Taurodontism: An anomaly of teeth reflecting developmental homeostasis. *Am J Med Genet* (Suppl) **4**:85–98, 1988.
36. Wyandt HE et al: Study of a patient with apparent monosomy 21 owing to translocation: 45,XX,21-,t(18q+). *Cytogenet* **10**:413–426, 1971.

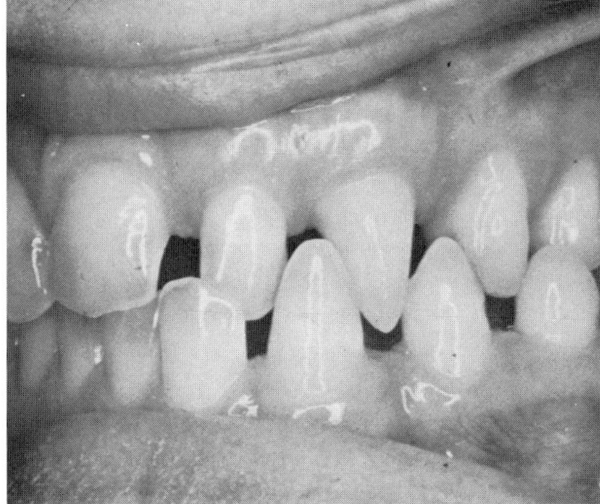

A

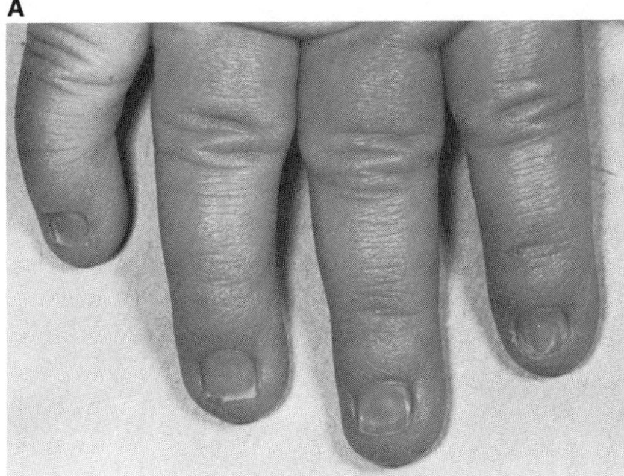

B

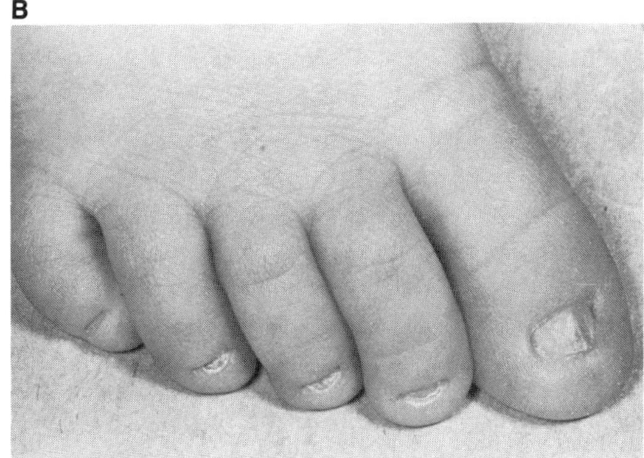

C

Fig. 27–9. *Witkop tooth–nail syndrome.* (A) Oligodontia and conical tooth crown form. (B,C) Hypoplasia of fingernails and toenails. (A from *JS Giansanti,* Oral Surg **37**:576, 1974.)

## Witkop tooth–nail syndrome (hypodontia and nail dysgenesis)

Hypodontia and nail dysgenesis (''tooth and nail'' syndrome) was first delineated by Witkop (18) although an earlier case was described (17). It is characterized chiefly by congenitally missing teeth and nail dysplasia. Evidence suggests that the syndrome, at least in some families, has autosomal dominant inheritance (4,6,10). We cannot be certain regarding classification of the disorder documented by Langsteger and Hartwagner (11). Several authors (1–3,6,9,14) reported isolated patients with hypodontia and nail anomalies but without family history. Heterogeneity cannot be excluded (8,10).

**Oral findings.** Many permanent teeth are congenitally missing or have coniform crowns (6,10) (Fig. 27-9A). In some patients, deciduous teeth are also coniform or congenitally absent (8,10). The teeth most often missing are mandibular incisors, second molars, and maxillary canines. Taurodontia is apparently not a common finding in this syndrome, although one patient depicted by Hudson and Witkop (10) had taurodont first permanent molars. Deciduous teeth had normal pulp chambers and root canals.

**Nails.** Fingernails and/or toenails are hypoplastic, spoon-shaped, and slow growing (6,10) (Fig. 27-9B,C).

**Other findings.** Although some patients have fine, thin, slow growing hair (6,10), this abnormality has not been documented in all patients with the disorder. Microscopic examination of the hair has not revealed any obvious abnormality (7,13). Sweating was reported as normal (6,10).

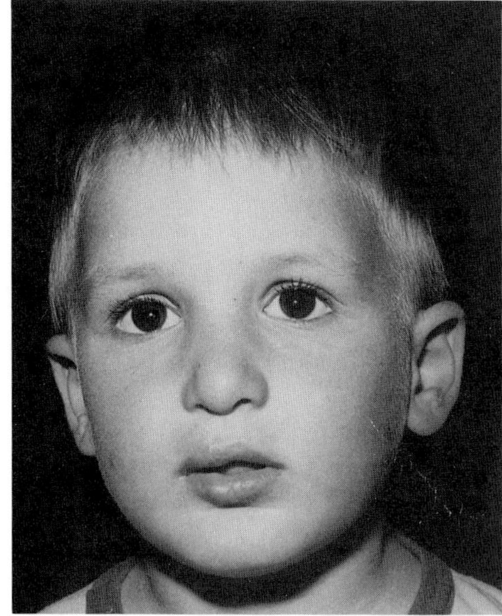

A

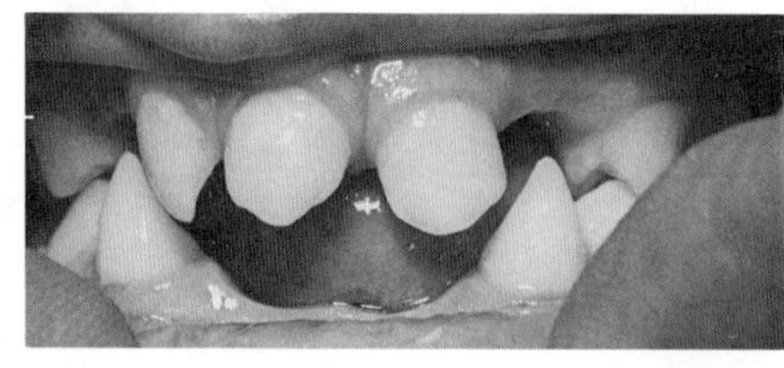

B

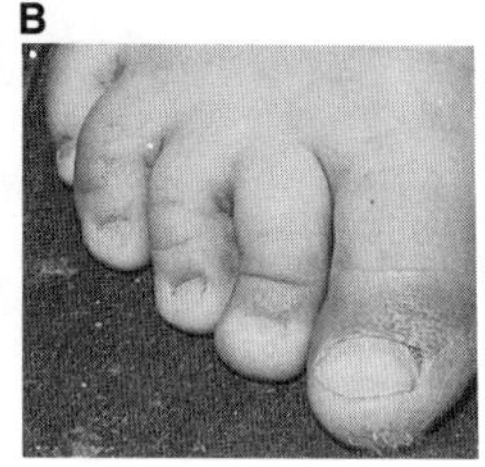

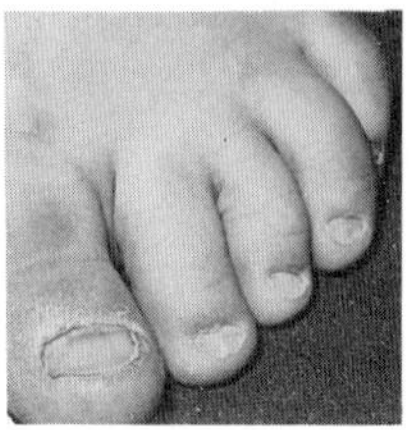

C

Fig. 27–10. *Fried syndrome.* (A) One of two sibs with fine slow-growing scalp hair and scanty eyebrows. (B) Conical crown forms and oligodontia. (C) Thin concave toenails. (From *K Fried,* J Med Genet **14**:137, 1977.)

**Differential diagnosis.** Differential diagnosis includes X-linked and autosomal recessive *hypohidrotic ectodermal dysplasia.* However, patients with this disorder have a typical facies, thin, sparse, slow growing hair, decreased sweating, and lack most teeth. Nails are usually not dysplastic. Patients with hypodontia and taurodontia have been reported (16); however, nails and hair were not described. Hypodontia and hypoplastic nails may occur with hearing loss (12).

LS Levin has seen two sibs with hypodontia, taurodontia, and sparse, slow growing, blond hair. Fingernails and toenails were spoon-shaped, thin, and slow growing. The sibs lacked deciduous maxillary incisors and molars, and the anterior deciduous teeth were pegged. The deciduous molars were taurodont and nearly all permanent teeth were missing. The parents were normal. Inheritance may be autosomal recessive. Four sibs with a similar constellation of abnormalities were reported by Stenvik et al (15), but the parents were not examined. Fried (5) reported first cousins, each born to consanguineous couples, who had congenitally absent and conically shaped deciduous teeth, fine, slow growing scalp hair, scanty eyebrows, thin fingernails, small thin concave toenails, and normal sweating. The status of the permanent teeth was unknown (Fig. 27-10A–C). This has been called *Fried syndrome.*

#### References [Witkop tooth–nail syndrome (hypodontia and nail dysgenesis)]


1. Brain RT: Familial ectodermal defect. *Proc R Soc Med* **31**:69–70, 1937.
2. Carabok JI, Pigott KLM: A case of partial anodontia, with associated ectodermal dysplasia. *Br Dent J* **100**:311–312, 1956.
3. Cohen MM Sr, Wagner R: Ectodermal dysplasia with partial anodontia. *Am J Dis Child* **68**:333–334, 1944.
4. Ellis J, Dawber RPR: Ectodermal dysplasia: A family study. *Clin Exp Dermatol* **5**:295–304, 1980.
5. Fried K: Autosomal recessive hidrotic ectodermal dysplasia. *J Med Genet* **14**:137–139, 1977.
6. Giansanti JS et al: The "tooth and nail" type of autosomal dominant ectodermal dysplasia. *Oral Surg* **37**:576–582, 1974.
7. Gorlin RJ et al: A selected miscellany: Oligodontia, taurodontia and sparse hair growth. *Birth Defects* **11**(2):39–40, 1975. (Same Case in Ref. 13.)
8. Gorlin RJ et al: A selected miscellany: Tooth-nail syndrome. *Birth Defects* **11**(2):45–46, 1975. (Same as Family B in Ref. 10.)
9. Hinrichsen CFL: Ectodermal dysplasia. *Aust Dent J* **8**:101–102, 1963.
10. Hudson CD, Witkop CJ Jr: Autosomal dominant hypodontia with nail dysgenesis. *Oral Surg* **39**:409–422, 1975. (Same as Family B in Ref. 8.)
11. Langsteger W, Hartwanger A: Ein Beitrag zum sogenannten Zahn- und Nagelsyndrom. *Z Hautkr* **57**:1802–1808, 1982.
12. Lowry RB et al: Hereditary ectodermal dysplasia: Symptoms, inheritance patterns, differential diagnosis, management. *Clin Pediatr* **5**:395–402, 1965.
13. Moller KT et al: Oligodontia, taurodontia and sparse hair growth—a syndrome. *J Speech Hear Disord* **38**:268–271, 1973. (Same as Ref. 7.)
14. Redpath TH, Winter GS: Autosomal dominant ectodermal dysplasia with significant dental defects. *Br Dent J* **126**:123–128, 1969 (Cases 12 and 16).
15. Stenvik A et al: Taurodontism and concomitant hypodontia in siblings. *Oral Surg* **33**:841–845, 1972.
16. Stoy PJ: Taurodontism associated with other dental abnormalities. *Dent Practit* **10**:202–205, 1960.
17. Weech AA: Hereditary ectodermal dysplasia (congenital ectodermal defect). *Am J Dis Child* **37**:766–790, 1929 (Case 2).
18. Witkop CJ Jr: Genetic diseases of the oral cavity, in *Oral Pathology,* Tiecke RW (ed), McGraw-Hill, New York, 1965.


## Enamel agenesis and nephrocalcinosis

MacGibbon (2) and Lubinsky et al (1) independently reported male and female sibs with amelogenesis imperfecta, nephrocalcinosis, enuresis, and polyuria. The enuresis and polyuria began at about 2 years of age. The nephrocalcinosis became apparent radiographically at 5 years. Urinary tract infections and pyelonephritis developed in late childhood and early adulthood with malignant hypertension and uremia eventuating during the third decade.

Both primary and secondary dentitions completely lacked enamel; the primary teeth erupted on time, but several permanent teeth remained unerupted with resorption of crowns within the alveolus. Large follicles were noted around the crowns of developing teeth, and multiple pulp calcifications, frequently dagger-shaped, were present in the pulp chambers.

Lubinsky et al (1) found abnormalities of calcium-binding proteins with increased levels of osteocalcin, reduced urinary carboxyglutamic acid, and reduced urinary excretion of calcium and phosphate.

Inheritance is probably autosomal recessive.

#### References (Enamel agenesis and nephrocalcinosis)


1. Lubinsky M et al: Syndrome of amelogenesis imperfecta, nephrocalcinosis, impaired renal function, and possible abnormality of calcium metabolism. *Am J Med Genet* **20**:233–243, 1985.
2. MacGibbon D: Generalized enamel hypoplasia and renal dysfunction. *Aust Dent J* **17**:61–63, 1962.

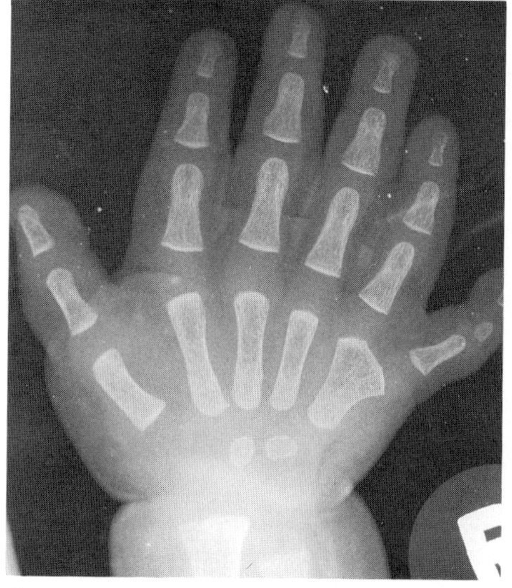


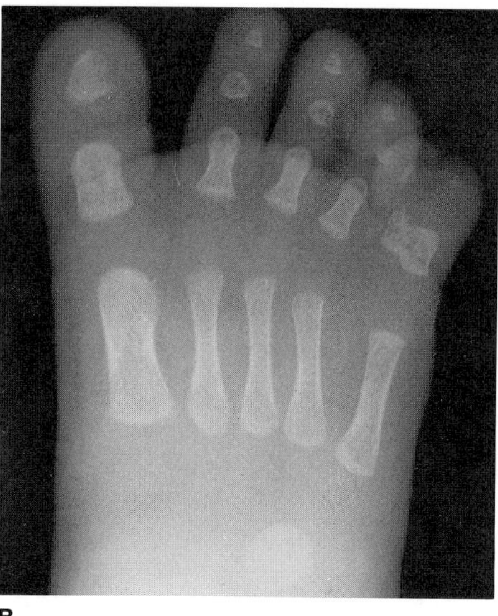


Fig. 27–11. *Acrodental dysostosis (Weyers syndrome, Curry–Hall syndrome).* (A) Postaxial hexadactyly with fusion of metacarpals V and VI. (B) Postaxial hexadactyly of feet. (From *M Roubicek* and *J Spranger*, Clin Genet **26**:585, 1984.)

## Acrodental dysostosis (Weyers syndrome, Curry–Hall syndrome)

Weyers (4–6) described several unrelated children with postaxial hexadactyly, bony cleft of the mandibular symphysis, and anomalies of the incisors and oral vestibule. Curry and Hall (1) reported a large family, several members of which had postaxial polydactyly, conical teeth, nail dysplasia, and short limbs. Additional affected kindreds have been described by Shapiro et al (3) and Roubicek and Spranger (2). Autosomal dominant transmission has been suggested (1–3,6).

Somatic growth is somewhat retarded. Shortening of limbs is very mild. Postaxial hexadactyly of the hands and feet is variable in expression. There may be fusion of the fifth and sixth metacarpals. Rarely, hexadactyly is limited to a postminimus digit. In some cases there has been complete or partial fusion of the proximal phalanges of the fifth and sixth toes. (Fig. 27-11A,B).

Fingernails and toenails may be dysplastic and discolored with thick vertical striations and splitting. The thumb and hallucal nails and the fifth toenails are most severely affected.

The facies, apart from the short somewhat retroussé nose with anteverted nostrils, is not unusual. The alae are rounded, giving the end of the nose a somewhat bulbous appearance. Although failure of fusion of the mandibular symphysis was a constant finding in Weyers' patients, all were under 1 year of age. We suspect there was only delay in fusion. Incisors in both dentitions have been described as small with conical crown form but, in some cases, as absent (Fig. 27-12). Occasionally, there are mucosal bands that obliterate the oral vestibule anteriorly, similar to those seen in *Ellis–van Creveld syndrome*.

Fig. 27–12. *Acrodental dysostosis (Weyers syndrome, Curry–Hall syndrome).* Reduced crown form and absent incisor in lower jaw. (From *M Roubicek* and *J Spranger*, Clin Genet **26**:585, 1984.)

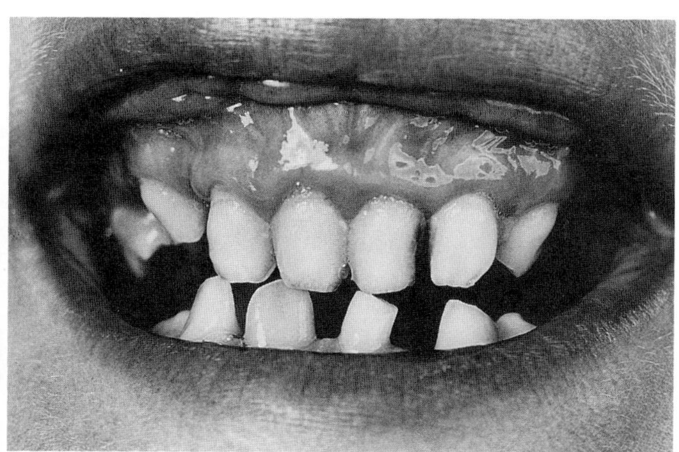

### References [Acrodental dysostosis (Weyers syndrome, Curry–Hall syndrome)]


1. Curry CJR, Hall BD: Polydactyly, conical teeth, nail dysplasia, and short limbs: A new autosomal dominant malformation syndrome. *Birth Defects* **15**(5B):253–263, 1979.
2. Roubicek M, Spranger J: Weyers acrodental dysostosis in a family. *Clin Genet* **26**:587–590, 1984.
3. Shapiro SD et al: Curry–Hall syndrome. *Am J Med Genet* **17**:579–583, 1984.
4. Weyers H: Über eine korrelierte Missbildung der Kiefer and Extremitätenakren dysostosis acro-facialis). *Fortschr Roentgenstr* **77**:562–567, 1952.
5. Weyers H: Hexadactylie, Unterkieferspalt und Oligodontie-ein neuer Symptomen-komplex: Dysostosis acrofacialis. *Ann Paediatr (Basel)* **181**:45–60, 1953.
6. Weyers H: Zur Kenntnis der Chondroektodermaldysplasie (Ellis–van Creveld). *Z Kinderheilkd* **78**:111–129, 1956.


## Amelogenesis imperfecta, epilepsy, and mental deterioration (Kohlschütter syndrome)

Kohlschütter et al (2) reported five male sibs with a sudden generalized seizure disorder that had its onset between 1 and 4 years. The infants had normal development until the appearance of epilepsy; subsequently mental deterioration followed, the degree being correlated with the frequency and severity of the seizures. Death occurred between 4 and 9 years. All affected males had marked amelogenesis imperfecta. Associated findings, established only in the survivor, included hypohidrosis, mildly elevated sodium and chloride, but markedly increased potassium in the sweat, and myopia. Inheritance is autosomal recessive. Another kindred was reported by Christodolou et al (1).

### References [Amelogenesis imperfecta, epilepsy, and mental deterioration (Kohlschütter syndrome)]


1. Christodolou J et al: A syndrome of epilepsy, dementia, and amelogenesis imperfecta: Genetic and clinical features. *J Med Genet* **25**:827–830, 1988.

2. Kohlschütter A et al: Familial epilepsy and yellow teeth—a disease of the central nervous system. Associated with enamel hypoplasia. *Helv Paediatr Acta* **29**:283–294, 1974.

## Amelogenesis imperfecta and terminal onycholysis (ameloonychohypohidrotic dysplasia)

Witkop and colleagues (1–3) described a kindred exhibiting hypoplastic hypocalcified enamel, multiple unerupted permanent teeth that underwent resorption, terminal onycholysis, seborrheic dermatitis of scalp, xerosis with keratosis pilaris of buttocks, and functional hypohidrosis (Fig. 27-13A,B). Inheritance was autosomal dominant.

### References [Amelogenesis imperfecta and terminal onycholysis (ameloonychohypohidrotic dysplasia)]

1. Witkop CJ Jr: Genetics. *Schweiz Mschr Zahnheilkd* **82**:917–941, 1972.
2. Witkop CJ Jr, Sauk JJ Jr: Heritable disorders of enamel, in *Oral Facial Genetics*, Stewart RW, Prescott GH (eds), C.V. Mosby, St. Louis, 1976.
3. Witkop CJ Jr et al: Hypoplastic enamel, onycholysis and hypohidrosis inherited as an autosomal dominant trait. *Oral Surg* **39**:71–86, 1975.

## Ankylosed teeth, maxillary hypoplasia, and fifth finger clinodactyly

Pelias and Kinnebrew (1) reported a syndrome of ankylosed teeth, bilateral clinodactyly of the fifth fingers, and mild relative mandibular prognathism due to maxillary hypoplasia in four generations. Ankylosis of the teeth, usually an isolated finding, has also been reported in association with enamel defects (2) and with multiple missing teeth (3) (Fig. 27-14).

Fig. 27–13. *Amelogenesis imperfecta and terminal onycholysis.* (A) Enamel, largely missing on some teeth, is hypomineralized. (B) Terminal onycholysis. (From *CJ Witkop Jr,* Schweiz Monatsschr Zahnheilkd **82**:917, 1972.)

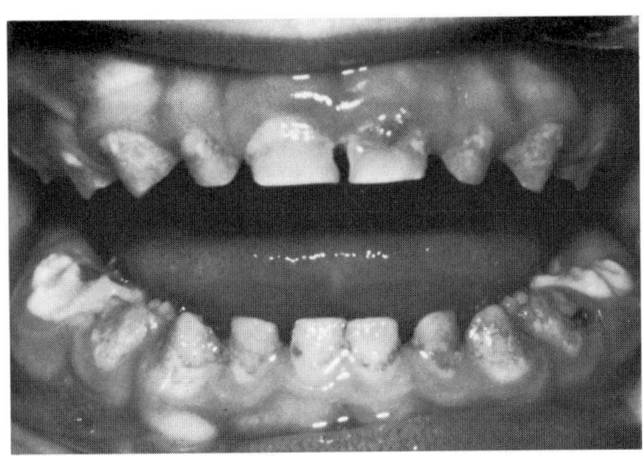

A

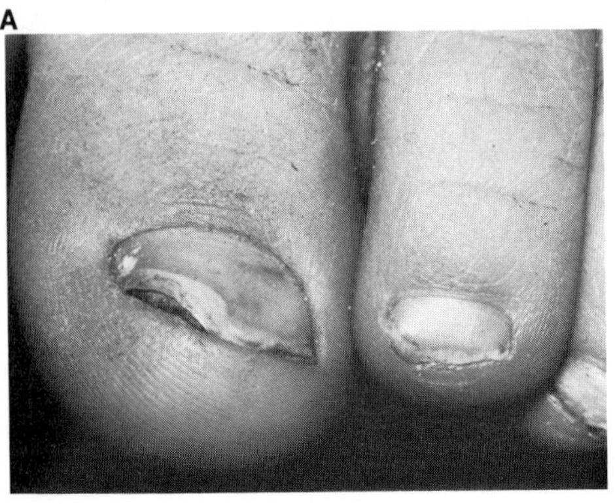

B

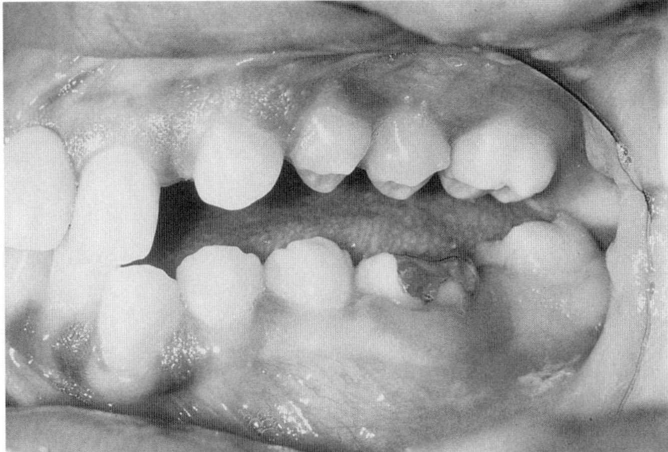

Fig. 27–14. *Ankylosed teeth, maxillary hypoplasia, and fifth finger clinodactyly.* Note failure of premolars and molars to meet at the occlusal plane. (From *MZ Pelias* and *MC Kinnebrew,* Clin Genet **27**:496, 1985.)

### References (Ankylosed teeth, maxillary hypoplasia, and fifth finger clinodactyly)

1. Pelias MZ, Kinnebrew MC: Autosomal dominant transmission of ankylosed teeth, abnormalities of the jaws, and clinodactyly: A four-generation study. *Clin Genet* **27**:496–500, 1985.
2. Rule JT et al: The relationship between ankylosed primary molars and multiple enamel defects. *J Dent Child* **39**:29–35, 1972.
3. Stewart RE, Hansen RW: Ankylosis and partial anodontia in twins. *J Calif Dent Assoc* **2**:50–52, 1974.

## Cataracts and dental abnormalities (Nance–Horan syndrome)

The syndrome of screwdriver-shaped incisors and congenital posterior cataracts was first described in detail by Nance et al (5) and Horan and Billson (3). However, Walsh and Wegman (9) may have, in reality, described the disorder even earlier, although they did not evaluate the dentition. The disorder is inherited as an X-linked trait. Several other families have been published (1,2,4,6,8). A third affected son has been born to the family reported by Goldberg and Hardy (2). The condition has also been seen by LS Levin in a black woman and in her two sons. The gene has been mapped to Xp21.1-p22.3 (7).

**Eyes.** Cataracts are congenital in both males and females. Microphthalmia and/or microcornea are common (Fig. 27-15). Affected males have total cataracts that, given the small anterior segment, are surgically difficult to remove (I Maumenee, personal communication, 1986). Carrier females have posterior Y sutural cataracts and punctate opacities (Fig. 27-16). The anterior Y suture may also be affected (I Maumenee, personal communication, 1986). Although females usually have normal vision, they frequently have mild microcornea and may be as severely affected as males with the condition (5). Cataracts in affected females may be progressive and, with age, total cataracts may develop (8).

**Other findings.** Pinnae may be anteverted with folded lobes. Shortening of the fourth metacarpal has been reported. Marfanoid habitus, with dolichocephaly, straight back, mandibular prognathism, and prolapse of the posterior leaflet of the mitral valve were noted in the brothers reported by Levin and Cortin (4); however, their affected relatives were not Marfanoid in appearance (LS Levin, unpublished). Intelligence is normal.

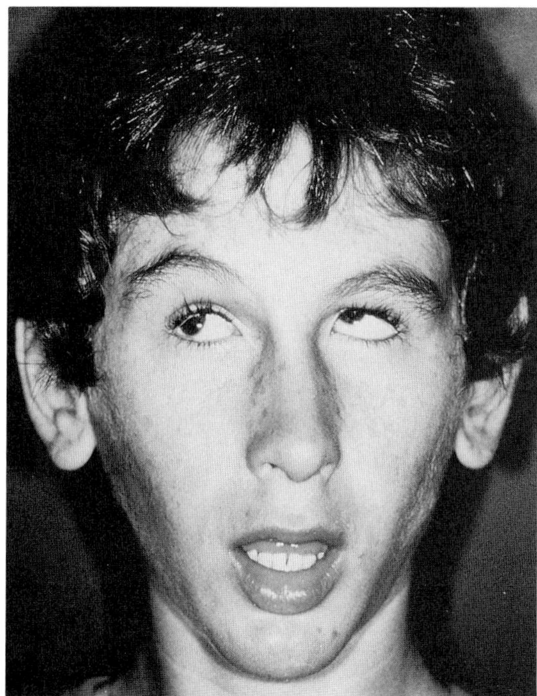

Fig. 27–15. *Cataracts and dental abnormalities.* Facies showing microphthalmia, strabismus, large anteverted pinnae. (From *WK Seow et al,* Pediatr Dent **7**:307, 1985.)

Fig. 27–16. *Cataracts and dental abnormalities.* Y-sutural cataract in a carrier female. The cataracts in affected males are total at birth.

**Oral manifestations.** Affected males have characteristic dental anomalies that include tapered, screwdriver-shaped incisors, resulting in diastemas between the teeth (5) (Fig. 27-17A–C). Molars and premolar cusps may also be tapered, the cervical width being greater than the occlusal width (LS Levin, unpublished). The molars may have a mulberry-like form (3). Deciduous and permanent teeth are affected. Maxillary supernumerary incisors (mesiodens) are also noted on dental

Fig. 27–17. *Cataracts and dental abnormalities.* (A) Deciduous dentition of an affected male. The maxillary deciduous incisors are tapered. (B) Permanent dentition of an affected male. The permanent incisors are tapered. Notches in incisal edges are mamelons. (C) Panoramic radiograph of permanent dentition showing tapered molar and incisor crown form. (D) Heterozygote dentition showing diastemas and screwdriver-shaped incisors. (D courtesy of *D Bixler,* Indianapolis, Indiana.)

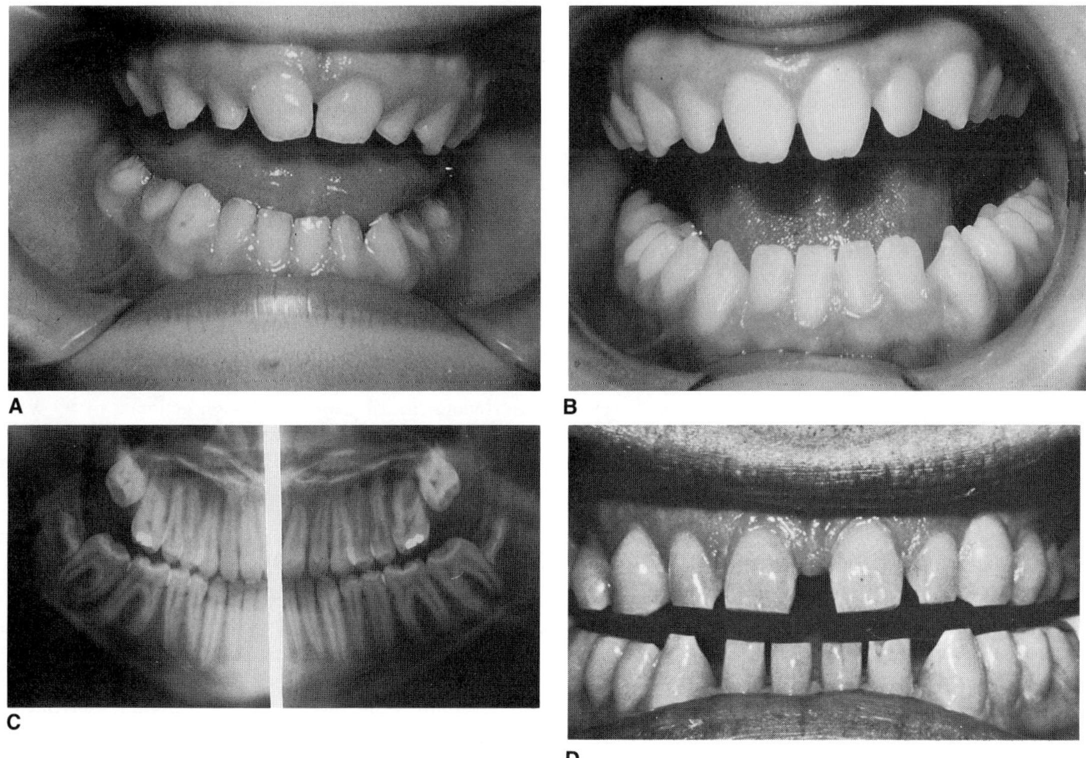

radiographs in some patients. LS Levin has evaluated an affected male in the kindred reported by Goldberg and Hardy (2) who had a supernumerary tooth in the anterior mandible. Carrier females have similarly, although less severely affected, dentitions (Fig. 27-17D). Bixler et al (1) found bilateral absence of the mandibular canines.

There is some overlap with *Lenz microphthalmia syndrome*.

#### References [Cataracts and dental abnormalities (Nance–Horan syndrome)]

1. Bixler D et al: The Nance–Horan syndrome: A rare X-linked ocular-dental trait with expression in heterozygous females. *Clin Genet* **26**:30–35, 1984.

2. Goldberg MF, Hardy J: E–X-linked cataract: Two pedigrees. *Birth Defects* **7**(3):164–165, 1971.

3. Horan MB, Billson FA: X-linked cataract and Hutchinsonian teeth. *Aust Paediatr J* **10**:98–102, 1974.

4. Levin LS, Cortin P: X-linked cataract, microcornea and dental anomalies. *Program and Abstracts of the 1976 March of Dimes Birth Defects Conference, Vancouver, B.C. Canada*, p 246a.

5. Nance WE et al: Congenital X-linked cataract, dental anomalies and brachymetacarpalia. *Birth Defects* **10**(4):285–291, 1974.

6. Seow KW et al: The Nance–Horan syndrome of dental anomalies, congenital cataracts, microphthalmia and anteverted pinnae. *Pediatr Dent* **7**:307–311, 1985.

7. Stambolian D et al: Mapping of X-linked cataract–dental syndrome. *March of Dimes Clinical Genetics Conference*, Boston, July 9–12, 1989.

8. Van Dorp DB, Delleman JW: A family with X-chromosomal recessive congenital cataract, microphthalmia, a peculiar form of the ear and dental anomalies. *J Pediatr Ophthalmol Strab* **16**:166–171, 1979.

9. Walsh FB, Wegman ME: Pedigree of hereditary cataract, illustrating sex-limited type. *Bull Johns Hopkins Hosp* **61**:125–135, 1937.

## Hyperhidrosis, premature graying of the hair, and premolar hypodontia (PHC syndrome, Böök syndrome)

Böök (1), in 1950, reported the PHC syndrome: P—premolar aplasia, H—hyperhidrosis, and C—canities prematura. The syndrome has autosomal dominant inheritance with complete penetrance.

Canities prematura (early, diffuse whitening of the hair) was the most common sign. It appeared usually between the sixth and twenty-third years of life and progressed slowly. About one-third of the patients presented whitening of the hair in other parts of the body. About 65% had marked functional hyperhidrosis. Although often palmoplantar, it involved the axillas and forehead less severely. One or more premolars were absent, with corresponding retention of the deciduous precursors.

#### Reference [Hyperhidrosis, premature graying of the hair, and premolar hypodontia (PHC syndrome, Böök syndrome)]

1. Böök JA: Clinical and genetical studies of hypodontia: I. Premolar aplasia, hyperhidrosis, and canities prematura: A new hereditary syndrome in man. *Am J Hum Genet* **2**:240–263, 1950.

## Hypodontia, sensorineural hearing loss, and dizziness

Lee et al (1) described two sibs, male and female, with profound, possibly congenital, hearing loss. Dizziness was noted at about 2 years of age and has been progressive in the presence of normal vestibular function. Missing teeth included the maxillary permanent incisors and canines in one child and missing permanent maxillary lateral incisors in the other.

#### Reference (Hypodontia, sensorineural hearing loss and dizziness)

1. Lee M et al: Autosomal recessive sensorineural hearing impairment, dizziness, and hypodontia. *Arch Otolaryngol* **104**:292–293, 1978.

## Lacrimo-auriculo-dento-digital syndrome (LADD syndrome, Levy–Hollister syndrome)

The syndrome, probably first reported by Levy (5) in 1967, was defined by Hollister et al (3) in 1973. The condition has autosomal dominant inheritance. Subsequently, over a dozen cases have been reported (1,4,6–8). Hennekam (2) suggested that the syndrome represents incomplete expression of the EEC syndrome. An excellent review is that of Wiedemann and Drescher (8). The lacrimal anomalies consist of unilateral or bilateral diminished or absent tears or recurrent or chronic tearing, recurrent or chronic conjunctivitis or dacryocystitis, nasolacrimal duct fistulas, and absence or hypoplasia of lacrimal glands or punctata, canaliculae, tear sacs, or nasal ducts. The pinnae are cup-shaped with or without sensorineural hearing loss in about 70% (Fig. 27-18A,B). The incisors are peg-shaped (Fig. 27-18F). In a few reports, patients were edentulous (3,6,7). Agenesis of salivary glands, noted in about 50%, is manifest by xerostomia and/or absence of Stensen's papillae or ducts.

Digital abnormalities consist of fingerlike, bifid, or hypoplastic thumb, syndactyly of second, third, and/or fourth digits, clinodactylous fifth digits, and hypoplastic hypothenar areas (Fig. 27-18C,E,G). Radioulnar synostosis or radial aplasia has been reported in a few cases (3–6). Calabro et al (1) described a characteristic metacarpophalangeal pattern.

Genitourinary anomalies (hypospadias, nephrosclerosis, unilateral renal agenesis) have also been documented (3,6).

#### References [Lacrimo-auriculo-dento-digital syndrome (LADD syndrome, Levy–Hollister syndrome)]

1. Calabro A et al: Lacrimo-auriculo-dento-digital (LADD) syndrome. *Eur J Pediatr* **146**:536–537, 1987.

2. Hennekam RCM: LADD syndrome. *J Med Genet* **24**:94–95, 1987.

3. Hollister DW et al: The lacrimo-auriculo-dento-digital syndrome. *J Pediatr* **83**:438–444, 1973.

4. Kreutz JM, Hoyme HE: Levy–Hollister syndrome. *Pediatrics* **82**:96–99, 1988.

5. Levy WJ: Mesoectodermal dysplasia: A new combination of anomalies. *Am J Ophthalmol* **63**:978–982, 1967.

6. LiShiang E, Holmes LB: The lacrimo-auriculo-dento-digital syndrome. *Pediatrics* **59**:927–930, 1977.

7. Thompson E et al: Phenotype variation in LADD syndrome. *J Med Genet* **22**:382–385, 1985.

8. Wiedemann H-R, Drescher J: LADD syndrome: Report of new cases and review of the clinical spectrum. *Eur J Pediatr* **144**:579–582, 1986.

## Multiple anterior dens invaginatus and sensorineural hearing loss

A female child seen by RJ Gorlin had multiple anterior dens invaginatus combined with fusion that resulted in extremely large and bizarre-shaped maxillary incisors. She was missing two lower right permanent incisors. There was dens invaginatus of the lower left central and lateral incisors (Fig. 27-19).

Facial appearance was normal except for some fullness at the bridge of the nose and somewhat dilated nostrils. She has bilateral, moderately severe sensorineural hearing loss. Although estimate of IQ was approximately 75, this may well have been in error due to a combination of hearing loss and shyness (Fig. 27-20).

## Natal teeth and steatocystoma multiplex

There have been two reports of the binary combination of natal teeth and steatocystoma multiplex. This was first noted by McDonald and Reed (2) in 1976. A five generation kindred with large numbers of affected has been reported by King and Lee (1).

This obviously overlaps with *pachyonychia congenita*, but in no case were the nails involved.

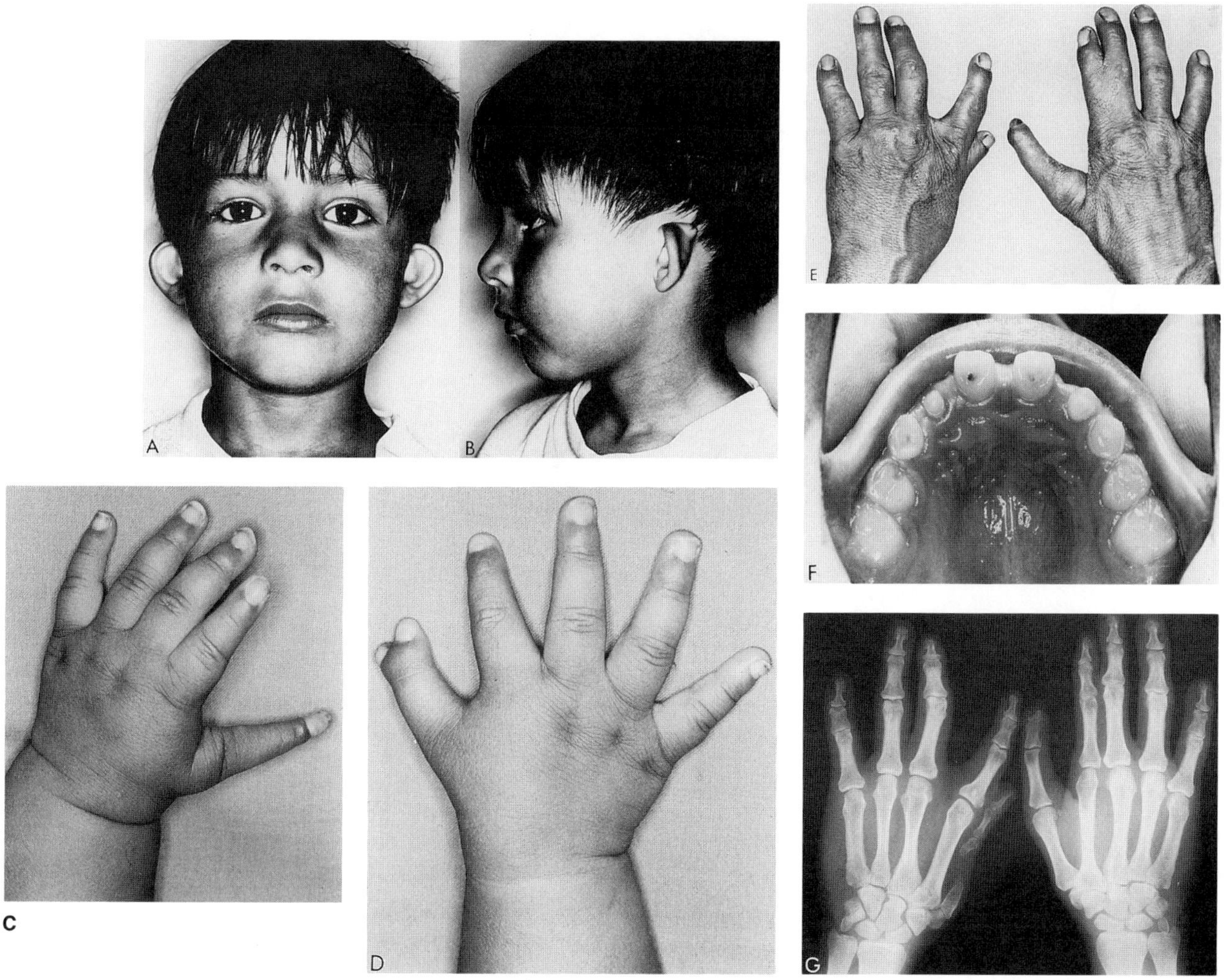

Fig. 27–18. *Lacrimo-auriculo-dento-digital syndrome.* (A,B) Prominent cup-shaped ears extend at right angles from the side of the head. The eyes have a watery, glistening appearance, and on the right there is an overflow of tears from the outer canthus. (C,D) Long, tapering thumbs with large nail, bifid thumb tip with extra ectopic nail and tapering of second and third digits bilaterally. (E) Right hand has long, tapering thumb with ectopic nail and syndactyly. Left hand has rudimentary thumb fused to index finger, prominent interdigital cleft between second and third fingers, and ulnar deviation of third digit. (F) Maxillary lateral incisors have conical crown form. Premolar cusp patterns are abnormal and show excessive wear. (G) Radiograph of hands seen in E. Tip of right thumb has extra tuftlike ossification center. Left hand shows hypoplasia of greater multangular and phalanges of thumb. Metacarpal is represented by two widely separated ossification centers. (From *D Hollister et al,* J Pediatr **84**:438, 1972.)

Fig. 27–19. *Multiple anterior dens invaginatus and sensorineural hearing loss.* Bizarre anterior teeth having dens invaginatus, fusion, unusual crown form.

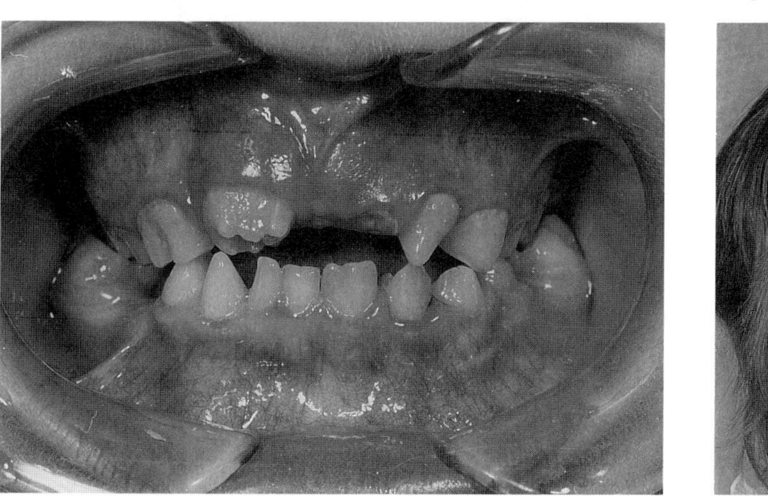

Fig. 27–20. *Multiple anterior dens invaginatus and sensorineural hearing loss.* Nine-year-old female with sensorineural hearing loss. Note mild widening of nasal bridge, dilated nostrils.

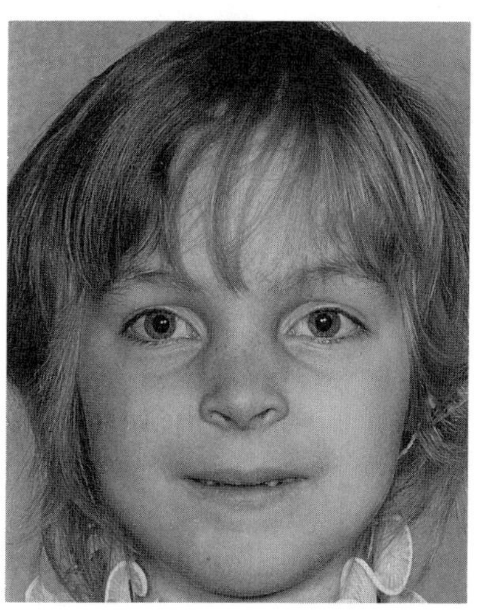

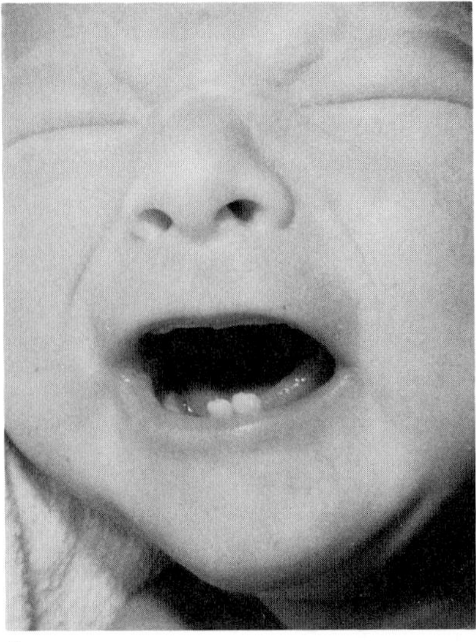

A

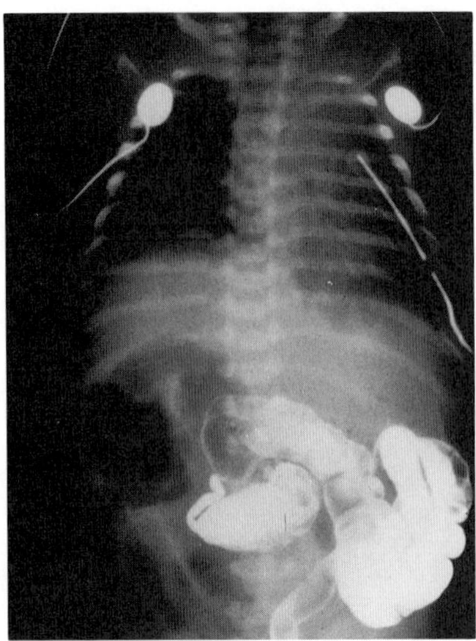

B

Fig. 27–21. *Natal teeth, patent ductus arteriosus, and intestinal pseudo-obstruction.* (A) Natal teeth. (B) Barium contrast showing microcolon and malrotation. (A from *DJ Harris et al,* Clin Genet **9**:479, 1976. B courtesy of *DJ Harris,* Kansas City, Missouri.)


#### References (Natal teeth and steatocystoma multiplex)

1. King NM, Lee AMP: Natal teeth in steatocystoma multiplex—a new syndrome. *J Craniofac Genet Dev Biol* **7**:311–317, 1987.
2. McDonald RM, Reed WB: Natal teeth in steatocystoma multiplex complicated by hidradenitis suppurativa. *Arch Dermatol* **112**:1132–1134, 1976.


## Natal teeth, patent ductus arteriosus, and intestinal pseudo-obstruction

Harris et al (1) reported brothers with dilatation and hypomobility of the small bowel and short or microcolon without anatomic obstruction. Incomplete rotation of the midgut was also noted. Both had congenital mandibular central incisors and patent ductus arteriosus, and both died within the first few months of life (Fig. 27-21A,B). The authors suggested X-linked recessive inheritance.


#### Reference (Natal teeth, patent ductus arteriosus, and intestinal pseudo-obstruction)

1. Harris DJ et al: Natal teeth, patent ductus arteriosus and intestinal pseudo-obstruction: A lethal syndrome in the newborn. *Clin Genet* **9**:479–482, 1976.


## Oligodontia, hypotrichosis, palmoplantar hyperkeratosis, and apocrine hidrocystomas of eyelid margins (Schöpf syndrome)

Schöpf et al (5) described two sisters with oligodontia, palmoplantar hyperkeratosis, and hyperhidrosis that began at puberty with hair loss occurring at 25 years, rosacea at 50 years, and 1–3 mm beaded cysts of both upper and lower eyelid margins at 60 years (Figs. 27-22A–C and 27-23A,B). The deciduous teeth were lost late and only a few rudimentary permanent teeth developed. The nails were brittle and

Fig. 27–22. *Oligodontia, hypotrichosis, palmoplantar hyperkeratosis, and apocrine hidrocystomas of eyelid margins.* (A–C) Proposita had oligodontia, sparse scalp hair, eyelid margin cysts, hyperkeratosis of soles. (Courtesy of *E Schöpf,* Heidelberg, Germany.)

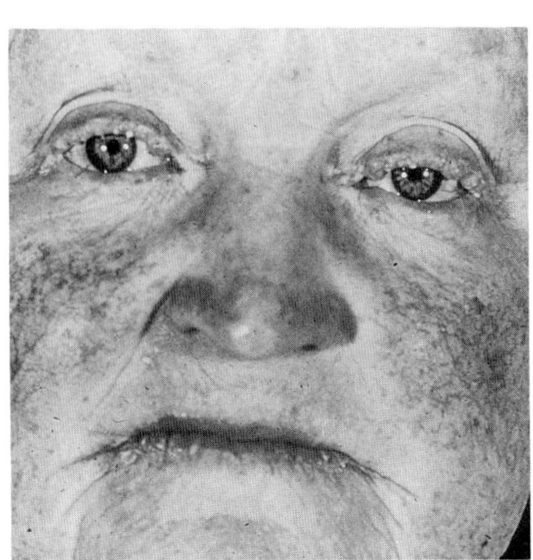

A

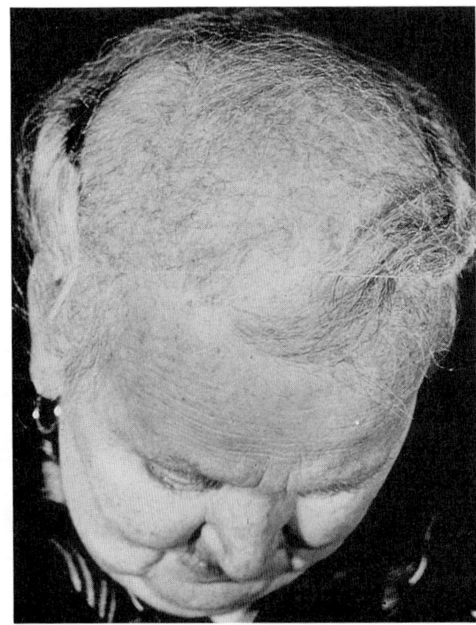

B

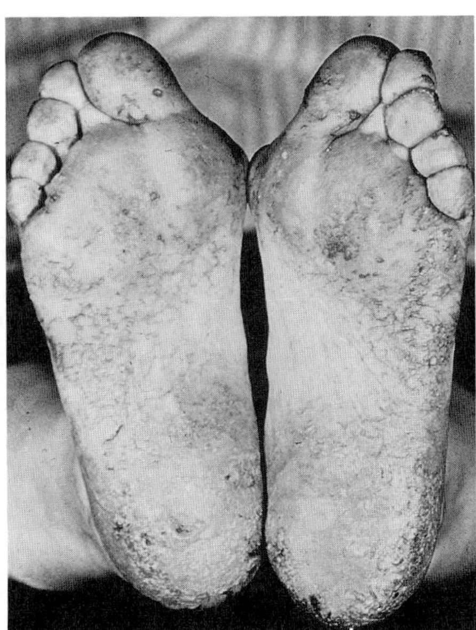

C

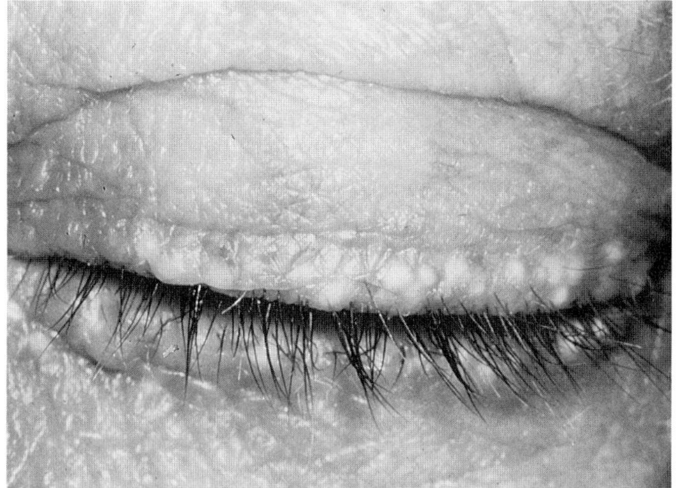

A

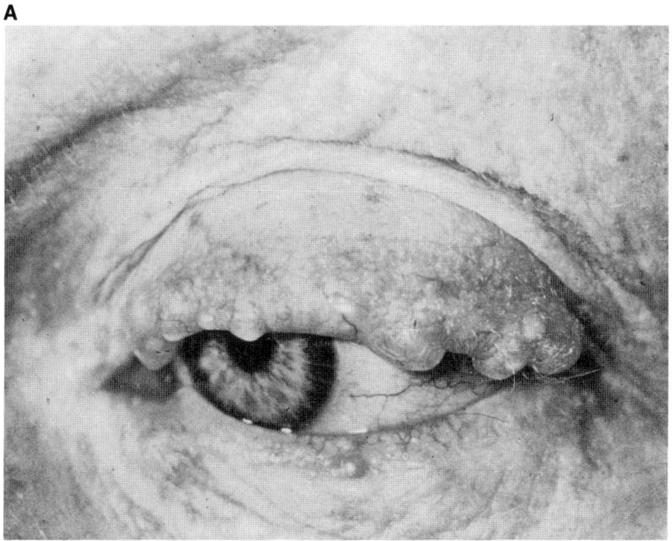

B

Fig. 27–23. *Oligodontia, hypotrichosis, palmoplantar hyperkeratosis, and apocrine hidrocystomas of eyelid margins.* (A,B) Beaded eyelid margins. (A from *RL Font et al, Arch Ophthalmol* **104**:1811, 1986. B courtesy of *E Schöpf,* Heidelberg, Germany.)

exhibited longitudinal furrowing. Photophobia, basal cell carcinoma, and eccrine poroma have been found in some patients (4).

The eyelid cysts are lined by two layers of cells: an outer myoepithelial layer and an inner layer of cuboidal cells with acidophilic cytoplasm. There were papillary infoldings. The cysts were interpreted as apocrine hidrocystomas arising from the glands of Moll.

Affected sibs (2,4,5) and consanguinity (5) suggest autosomal recessive inheritance; but because a large family (3) was found in which the disorder had variable expressivity and occurred in several generations, Hebert et al (3) opined autosomal dominant inheritance.

### References [Oligodontia, hypotrichosis, palmoplantar hyperkeratosis, and apocrine hidrocystomas of eyelid margins (Schöpf syndrome)]

1. Burket JM et al: Eyelid cysts, hypodontia and hypotrichosis. *J Am Acad Dermatol* **10**:922–925, 1984.
2. Font RL et al: Apocrine hidrocystomas of the lids, hypodontia, palmoplantar hyperkeratosis, and onychodystrophy. *Arch Ophthalmol* **104**:1811–1813, 1986.
3. Hebert SA et al: Expression and inheritance of the Schöpf syndrome. *Am J Med Genet,* in press.
4. Nordin H et al: Familial occurrence of eccrine tumours in a family with ectodermal dysplasia. *Acta Derm Venereol (Stockh)* **68**:523–530, 1988.
5. Schöpf E et al: Syndrome of cystic eyelids, palmo-plantar keratosis, hypodontia and hypotrichosis as a possible autosomal recessive trait. *Birth Defects* **7**(8):219–221, 1971.

## Oligodontia, keratitis, skin ulceration, and arthroosteolysis

Kozlova et al (1) described five sibs with a syndrome of recurrent skin ulceration, arthralgias, fever, fistulous osteolysis around joints, nail dystrophy, and keratitis that eventuated in blindness. The fingers became clawed. Involvement of the growth plates of the tibia and femur led to asymmetric shortness with secondary scoliosis (Fig. 27-24).

Several teeth were missing; others had reduced crown form.

Inheritance is autosomal recessive.

### Reference (Oligodontia, keratitis, skin ulceration, and arthroosteolysis)

1. Kozlova SI et al: Self-limited autosomal recessive syndrome of skin ulceration, arthroosteolysis with pseudoacromegaly, keratitis and oligodontia in a Kirghizian family. *Am J Med Genet* **15**:205–210, 1983.

## Oligodontia, microcephaly, short stature, and characteristic facies

Bankier et al (1) reported autosomal dominant inheritance of a syndrome of microcephaly, low normal intelligence, short stature, and dental anomalies.

Fig. 27–24. *Oligodontia, keratitis, skin ulceration, and arthroosteolysis.* Note severe scoliosis, shortening of left leg, enlargement and deformity of knees and ankles, and cicatrized fistulous scars about knees. (From *SI Kozlova et al,* Am J Med Genet **15**:205, 1983.)

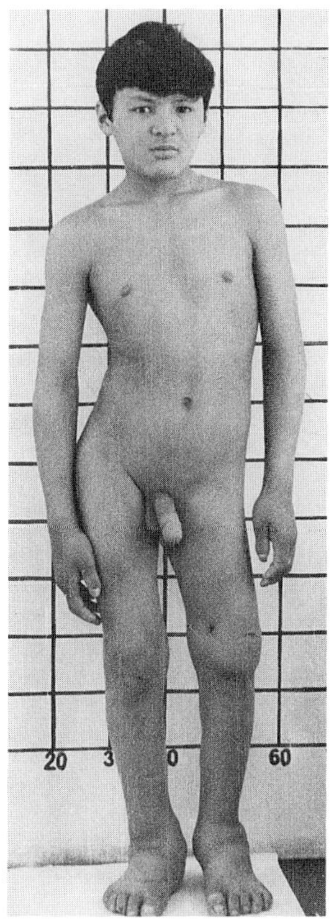

The facies was characterized by a broad forehead, long narrow nose, and flat recessed midface. Scalp hair tended to be fine and wispy. There was marked deficiency of teeth and those that were present often had conical crown form and/or diminished crown size.

#### Reference (Oligodontia, microcephaly, short stature, and characteristic facies)

1. Bankier A et al: Dental dysplasia, microcephaly, and diminutive build—a new autosomal dominant syndrome. *Am J Med Genet*, in press.

### Otodental syndrome

In 1972, Levin and Jorgenson (6) described a syndrome of dental anomalies and sensorineural hearing loss. To date, several unrelated families have been described (1–7,9,10); one case report is doubtful (8). Inheritance is clearly autosomal dominant with variable expressivity (2).

The incisors of both dentitions are spared. The crowns of the canines and posterior teeth are enlarged, bulbous, and malformed with multiple prominent lobules. The deciduous dentition is more severely involved. The relation between cusps and major grooves is eliminated. An enamel defect is frequently noted on the facial surface of the canine teeth. Premolar teeth are frequently missing or small in size. There is often delayed eruption of the deciduous malformed teeth or even of the permanent posterior teeth (1,2,9). One can observe duplicated pulp chambers with denticle formation and a longitudinal dental septum and early pulpal obliteration. The molar teeth have a tendency toward conical or taurodont root form. Complex and/or compound odontomas have also been described in the posterior maxilla and mandible (1,9) (Figs. 27-25 to 27-27).

Sensorineural hearing loss is found at all frequencies, but is more pronounced about 1000 Hz. It usually plateaus by the fourth decade (3). The age of onset of the hearing loss ranges from early childhood to middle age (3,5,7), which may complicate making the diagnosis of the disorder.

#### References (Otodental syndrome)

1. Beck-Mannagetta J et al: Odontome and pantonale Hörstörung bei otodentalem Syndrom. *Dtsch Zahnärtzl Z* **39**:232–241, 1984.

2. Chen RJ et al: "Otodental" dysplasia. *Oral Surg Oral Med Oral Pathol* **66**:353–358, 1988.

3. Cook RA et al: Otodental dysplasia: A five year study. *Ear Hear* **2**:90–94, 1981.

4. Gundlach KKH, Witkop CJ Jr: Globodontie—eine neue erbliche Zahnformanomalie. *Dtsch Zahnärztl Z* **32**:194–196, 1977.

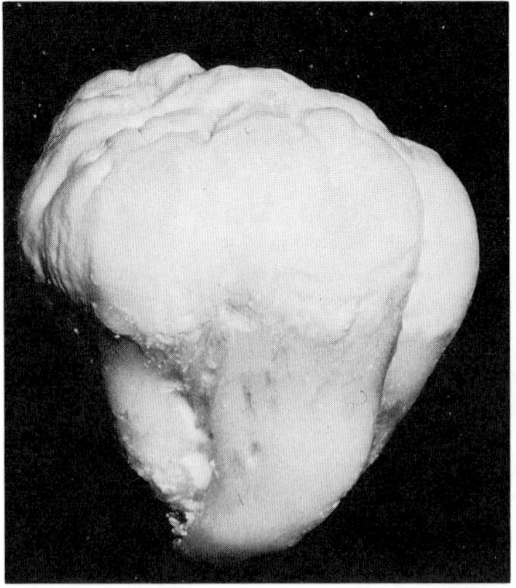

Fig. 27–26. *Otodental syndrome*. Globodontia of 6-year-molar. (From *J Beck-Mannagetta et al*, Dtsch Zahnärztl Z **39**:232, 1984.)

5. Jorgenson RJ et al: Otodental dysplasia. *Birth Defects* **11**(5):115–119, 1975.

6. Levin LS, Jorgenson RJ: Familial otodental dysplasia: A "new" syndrome. *Am J Hum Genet* **24**:61a, 1972.

7. Levin LS et al: Otodental syndrome. A new ectodermal dysplasia. *Clin Genet* **8**:136–144, 1975.

8. Stewart DJ, Kinirons MJ: Globodontia. *Br Dent J* **152**:287–288, 1982.

9. Winter GB: The association of ocular defects with the otodental syndrome. *J Int Assoc Dent Child* **14**:83–87, 1983.

10. Witkop CJ Jr: Globodontia in the otodental syndrome. *Oral Surg* **41**:472–483, 1976.

### Postaxial polydactyly–dental–vertebral syndrome

Rogers et al (1) and Temtamy and McKusick (2) reported a syndrome of bilateral broad halluces, postaxial polydactyly of hands and feet, short middle phalanges of hands, short pointed distal phalanges, abnormal vertebral bodies (variable fusion, hemivertebrae, butterfly form, small size), congenital torticollis, narrow spinal canal, kyphoscoliosis, and pectus (Figs. 27-28A,B and 29A,B). The pinnae were malformed, the neck was webbed with low posterior hairline, and the mandible was prognathic. Dental anomalies included dens invaginatus, fusion,

Fig. 27–25. *Otodental syndrome*. Occlusal view of teeth showing globodontia. Normal sized maxillary incisors contrast sharply with bulbous canine and molar teeth. (Courtesy of *CJ Witkop Jr*, Minneapolis, Minnesota.)

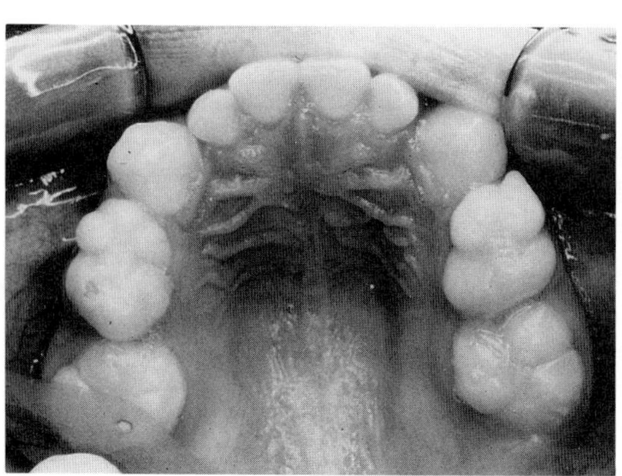

Fig. 27–27. *Otodental syndrome*. Panoramic radiograph showing 5-year-old with normal anterior teeth, globodontia of canines and molars. (From *J Beck-Mannagetta et al*, Dtsch Zahnärztl Z **39**:232, 1984.)

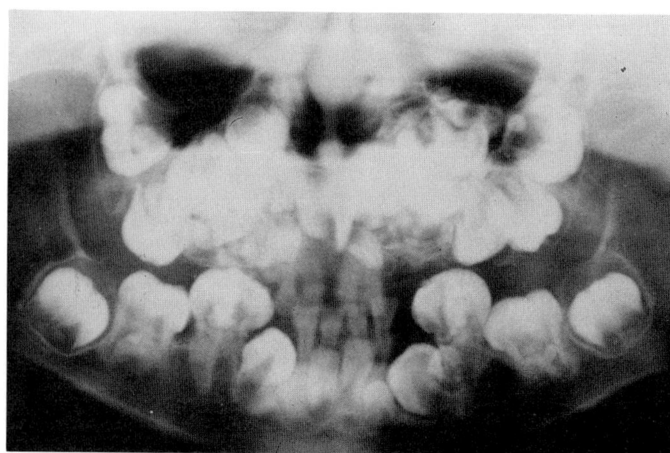

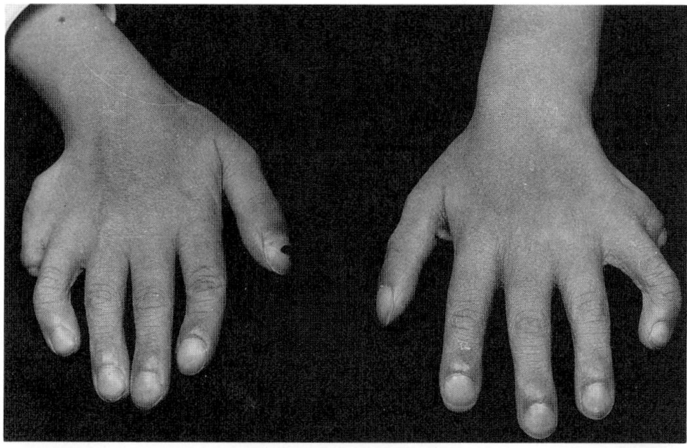

A

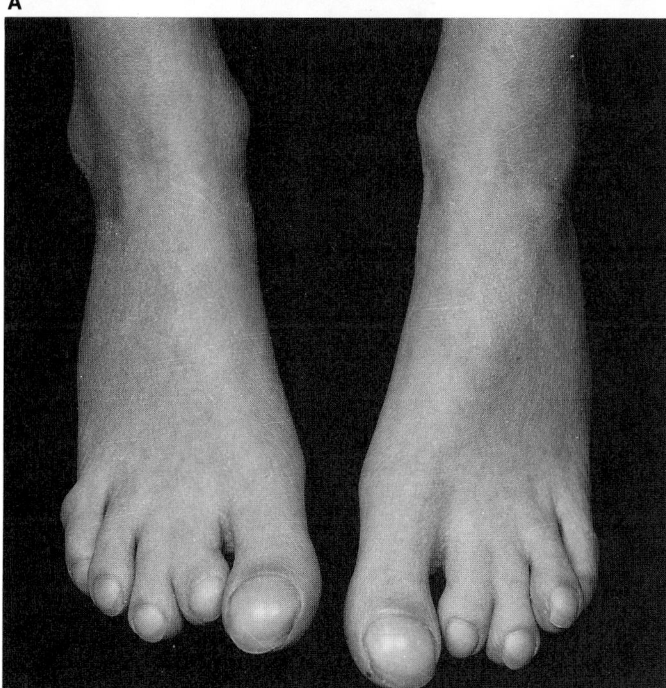

B

Fig. 27–28. *Postaxial polydactyly–dental–vertebral syndrome.* (A) Postaxial polydactyly. (B) Broad halluces. (From *JG Rogers et al,* J Pediatr **90**:230, 1977.)

macrodontia, hypodontia, enamel dysplasia, and short roots (Fig. 27-29C).

Inconstant features included high forehead, hydrocele, phimosis, agenesis of lower ribs, and congenital heart defects.

Inheritance appears to be autosomal recessive (1).

### References (Postaxial polydactyly–dental–vertebral syndrome)

1. Rogers JG et al: A postaxial polydactyly-dental-vertebral syndrome. *J Pediatr* **90**:230–235, 1977.

2. Temtamy S, McKusick V: The genetics of hand malformations. *Birth Defects* **14**(3):411–413, 1978.

## Short stature and delayed dental eruption

Arvystas (1) described autosomal dominant inheritance of short stature, prolonged retention of deciduous dentition, and delayed eruption of secondary teeth. There was frontal bossing, midface hypoplasia, Wormian bones in the lambdoidal sutures, widely open anterior fontanel in infancy, and delayed skeletal maturation.

### Reference (Short stature and delayed dental eruption)

1. Arvystas MG: Familial generalized delayed eruption of the dentition with short stature. *Oral Surg* **41**:235–243, 1976.

## Short stature and short thin dilacerated dental roots

Witkop and Jaspers (1) reported a family with short stature and teeth with short, thin, dilacerated roots in four generations. Inheritance was compatible with either autosomal or X-linked dominant inheritance.

The tooth crowns were somewhat bulbous with cervical constriction and drastically abbreviated short thin dilacerated (bent) roots. The enamel was normal (Fig. 27–30).

Height did not exceed 155 cm. Wormian bones were observed in the lambdoidal and coronal sutures.

### Reference (Short stature and short thin dilacerated dental roots)

1. Witkop CJ Jr, Jaspers MT: Teeth with short, thin, dilacerated roots in patients with short stature: A dominantly inherited trait. *Oral Surg* **54**: 553–559, 1982.

## Solitary maxillary central incisor and short stature (monosuperocentroincisivodontic dwarfism)

Rappaport et al (6,7), in 1976–1977, described the association between single maxillary central incisor and isolated growth hormone deficiency (Fig. 27–31). Vanelli et al (10) confirmed the finding. RJ Gorlin has seen this combination with congenital cataracts. In several cases, hormone levels have not been ascertained (3,8).

It is not rare to have a single maxillary central incisor, normal growth, and normal hormone levels (5,8,9,11). Single central incisor has also been reported in association with *del(18p)* (2) and with short stature and ocular coloboma (4).

The syndrome should easily be differentiated from single maxillary central incisor as incomplete expression of autosomal dominant *holoprosencephaly,* and it should be noted that single maxillary central incisor occurs much more frequently in alobar holoprosencephaly with severe facial dysmorphism than it does as a "microform" in autosomal dominant holoprosencephaly (1b). Winter et al (13) described solitary maxillary central incisor with precocious puberty and hypothalamic hamartoma. *Hypophyseal short stature and cleft lip* is discussed in Chapter 23. Single maxillary central incisor has been found in association with iris coloboma and hypomelanosis of Ito (1) and with unusual forms of ectodermal dysplasia (1a,12). P Fleming (personal communication) reported that he had seen a single central incisor in *Klippel–Feil anomaly.*

### References [Solitary maxillary central incisor and short stature (monosuperocentroincisivodontic dwarfism)]

1. Bartholomew DW et al: Single maxillary central incisor and coloboma in hypomelanosis of Ito. *Clin Genet* **32**:370–373, 1987.

1a. Buntinx I, Baraitser M: A single maxillary incisor as a manifestation of an ectodermal dysplasia. *J Med Genet* **26**:648–650, 1989.

1b. Cohen MM Jr: Perspectives on holoprosencephaly. Part III. Spectra, distinctions, continuities, and discontinuities. *Am J Med Genet* **34**:271–288, 1989.

2. Dolan LM et al: 18p− syndrome with a single central maxillary incisor. *J Med Genet* **18**:396–398, 1981.

3. Hayward JR: Observations on midline deformity and the solitary maxillary incisor. *J Hosp Dent Pract* **13**:113–114, 1979.

4. Liberfarb RM et al: Ocular coloboma associated with a solitary maxillary central incisor and growth failure: Manifestations of holoprosencephaly. *Ann Opththalmol* **19**:226–227, 1987.

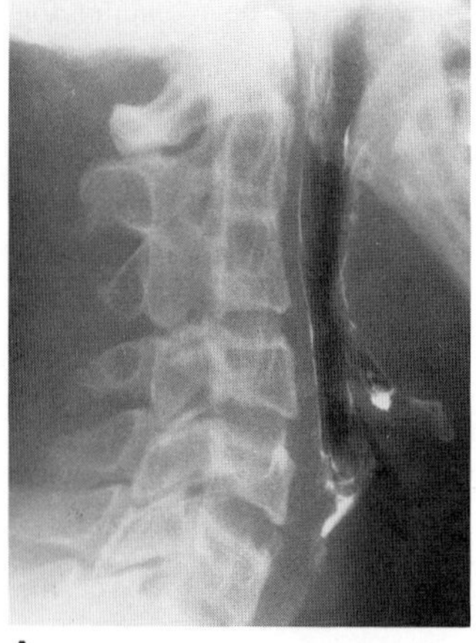
A

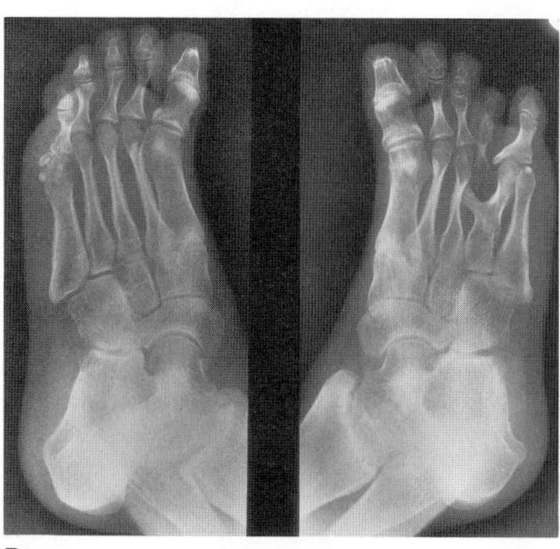
B

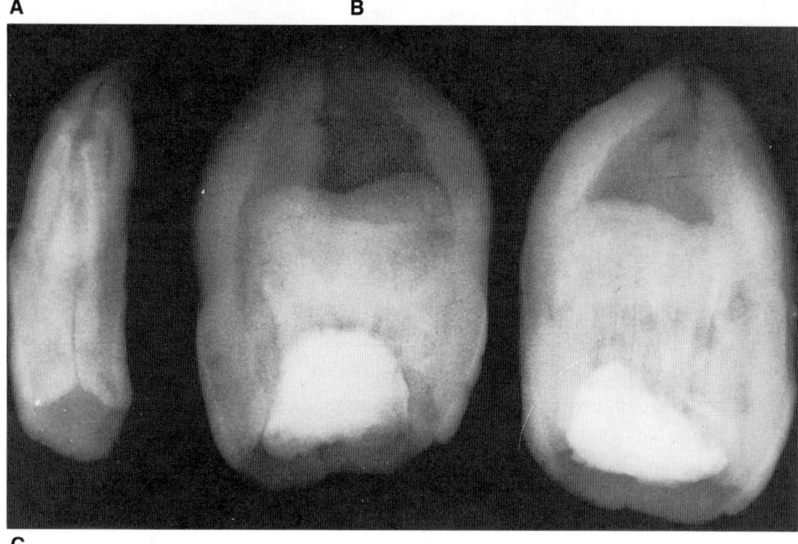
C

Fig. 27–29. *Postaxial polydactyly–dental–vertebral syndrome.* (A) Hypoplasia and fusion of cervical vertebrae. (B) Talus, calcaneus, and first and second cuneiforms are fused. Distal end of right fourth metatarsal is bifurcated. (C) Radiograph of extracted teeth showing grossly widened incisors with dens invaginatus, obliterated pulp chambers and canals, short roots. (From *JG Rogers et al,* J Pediatr **90**:230, 1977.)

5. Parker PR, Vann WF Jr: Solitary maxillary incisor. *Pediatr Dent* **7**:134–136, 1985.

6. Rappaport EB et al: Monosuperocentroincisivodontic dwarfism. *Birth Defects* **12**(5):243–245, 1976.

7. Rappaport EB et al: Solitary central incisor and short stature. *J Pediatr* **91**:924–928, 1977.

8. Santoro FP, Wesley RK: Clinical evaluation of two patients with a single maxillary central incisor. *J Dent Child* **50**:379–381, 1983.

9. Scott DC: Absence of upper central incisor. *Br Dent J* **104**:247–248, 1958.

10. Vanelli M et al: Incisive supèrieure unique et déficit en STH. *Arch Fr Pédiatr* **37**:321–322, 1980.

11. Wesley RK et al: Solitary maxillary central incisor and normal stature. *Oral Surg* **46**:837–842, 1978.

12. Winter RM et al: Sparse hair, short stature, hypoplastic thumbs, single upper central incisor, and abnormal skin pigmentation: A possible new form of ectodermal dysplasia. *Am J Med Genet* **29**:209–216, 1988.

13. Winter WE et al: Solitary central maxillary incisor associated with precocious puberty and hypothalamic hamartoma. *J Pediatr* **101**:965–967, 1982.

## Taurodontism, microdontia, and dens invaginatus

Casamassimo et al (1) described a family in which there was simultaneous occurrence of generalized microdontia, taurodontism of the first permanent molars, and teeth with multiple dens invaginatus (Figs. 27-32 and 27-33). The pedigree contained five affected males in five

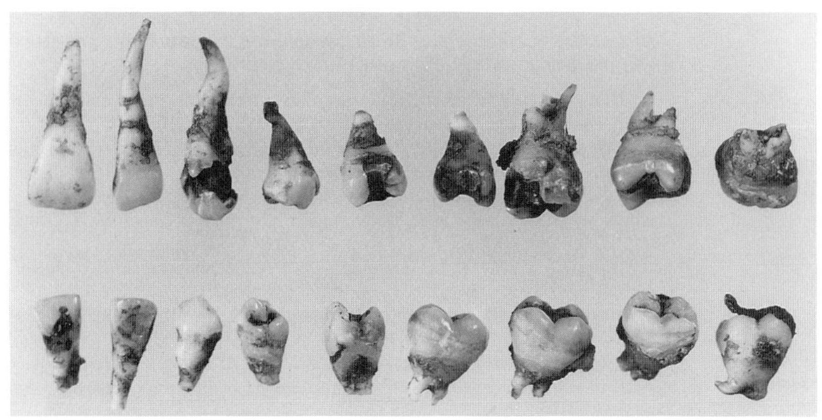

Fig. 27–30. *Short stature and short thin dilacerated dental roots.* Bulbous crowns and short dilacerated roots. (From *CJ Witkop Jr* and *MT Jaspers,* Oral Surg **54**:553, 1982.)

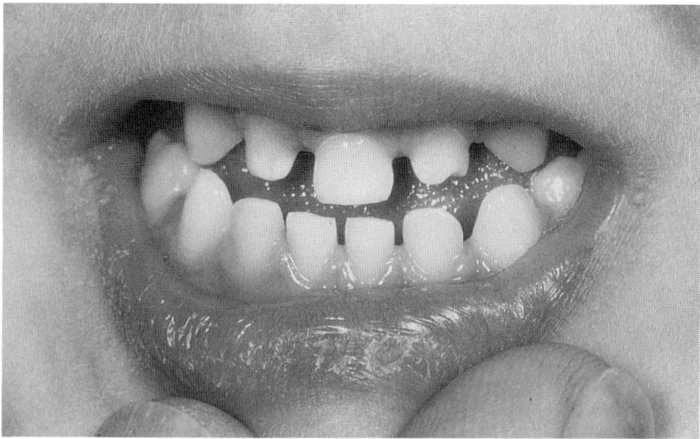

Fig. 27–31. *Solitary maxillary central incisor and short stature.* Note single central incisor.

generations. There is good reason to suspect X-linked recessive inheritance.

Ireland et al (2) described taurodontism, multiple dens invaginatus, and short tooth roots in male and female sibs. Additional children of the father had similar anomalies. They suggested autosomal dominant inheritance.

### References (Taurodontism, microdontia, and dens invaginatus)

1. Casamassimo PS et al: An unusual triad: Microdontia, taurodontism and dens invaginatus. *Oral Surg* **45**:107–112, 1978.
2. Ireland EJ et al: Short roots, taurodontia and multiple dens invaginatus. *J Pedodont* **11**:164–175, 1987.

## Taurodontism, oligodontia, and sparse hair

Stoy (6) appears to be the first author to describe a syndrome of taurodontism, oligodontia, and sparse hair growth (Fig. 27-34A,B). Several authors subsequently reported cases (1–5). Affected sibs were noted by Stenvik et al (5). All other examples were isolated cases. Care must be taken to exclude females heterozygous for *hypohidrotic ectodermal dysplasia.* Some examples may represent *Witkop tooth-nail syndrome.*

### References (Taurodontism, oligodontia, and sparse hair)

1. Davidson LE, Woolass KF: Severe hypodontia in an eight-year-old child. *Br Dent J* **158**:215–217, 1985.
2. Gorlin RJ: A selected miscellany: Oligodontia, taurodontia, and sparse hair growth. *Birth Defects* **11**(2):39–50, 1975. (Same as Ref. 4.)
3. Haunfelder D: Ein Beitrag zu den Molaren mit prismatischen Wurzeln (sog. Taurodontismus). *Dtsch Zahnärztbl* **21**:419–423, 1967.
4. Moller KT et al: Oligodontia, taurodontia and sparse hair: A syndrome. *J Speech Hear Dis* **38**:268–271, 1973.
5. Stenvik A et al: Taurodontism and concomitant hypodontia in siblings. *Oral Surg* **33**:841–845, 1972.
6. Stoy PJ: Taurodontism associated with other dental abnormalities. *Dent Pract Dent Rec* **10**:202–205, 1960.

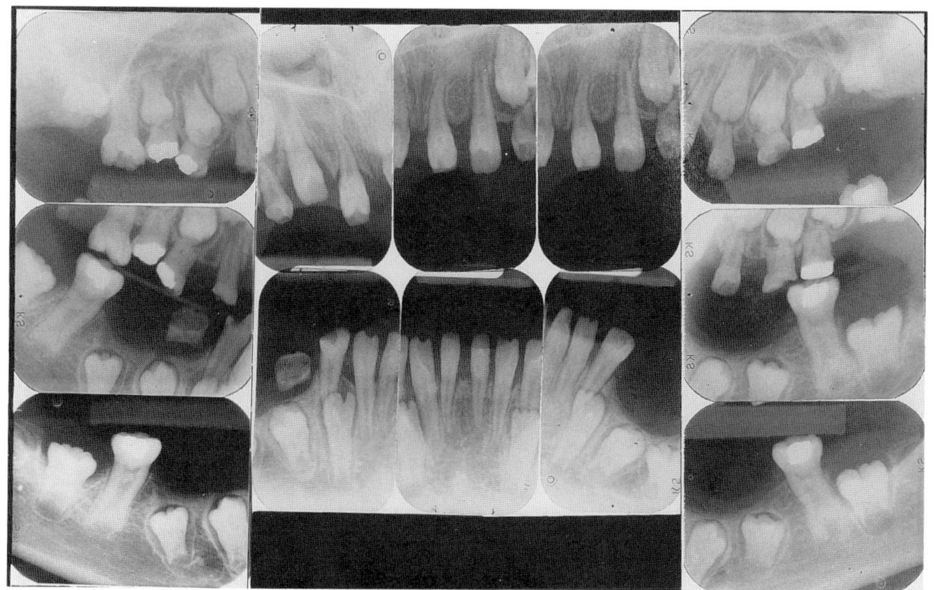

Fig. 27–32. *Taurodontism, microdontia, and dens invaginatus.* Periapical radiographs showing generalized microdontia, taurodontism of permanent molars, and several teeth with multiple dens invaginatus. (From *PS Casamassimo et al,* Oral Surg **45**:107, 1978.)

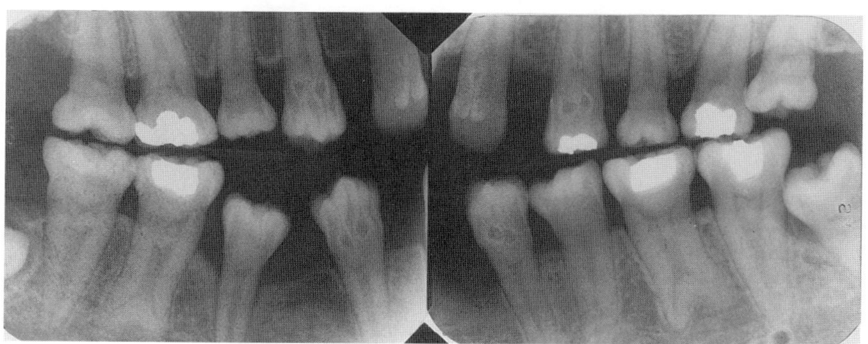

Fig. 27–33. *Taurodontism, microdontia, and dens invaginatus.* Periapical views showing taurodontism and multiple invaginations. (From *PS Casamassimo et al,* Oral Surg **45**:107, 1978.)

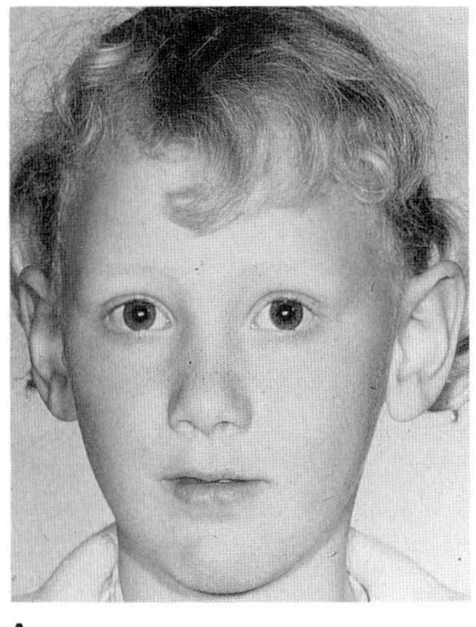

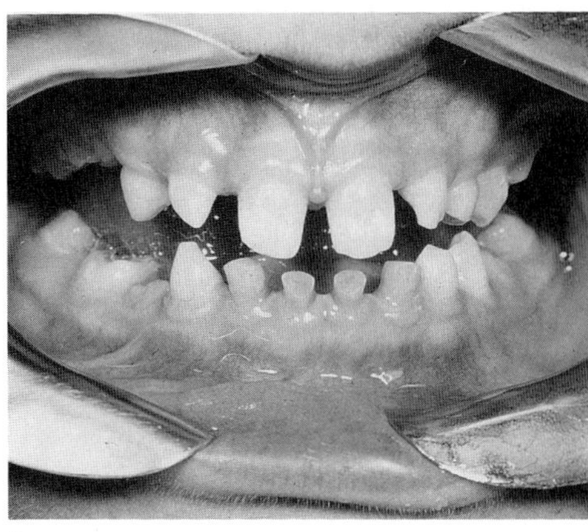

A B

Fig. 27–34. *Taurodontism, oligodontia, and sparse hair.* (A,B) Sparse hair, oligodontia.

## Taurodontism, pyramidal and fused molar roots, juvenile glaucoma, and unusual morphology of upper lip (Ackerman syndrome)

Ackerman et al (1) described a family in which several members in two generations exhibited pyramidal, taurodont, or fused molar roots with a single root canal, a finding noted on both sides of the kindred. Three of six sibs had only pyramidal molar roots.

Juvenile glaucoma was present in two of three sibs, and all had sparse body hair. The upper lip was full without cupid's bow, and there were thickening and widening of the philtrum. One of the sibs also had entropion of both lower eyelids, 3–4 soft tissue syndactyly of one hand, indurated hyperpigmented skin over the interphalangeal joints of the fingers, and clinodactyly of the fifth fingers (Fig. 27-35A–C).

### Reference [Taurodontism, pyramidal and fused molar roots, juvenile glaucoma, and unusual morphology of upper lip (Ackerman syndrome)]

1. Ackerman JL et al: Taurodont, pyramidal, and fused molar roots associated with other anomalies in a kindred. *Am J Phys Anthropol* **38**:681–694, 1973.

Fig. 27–35. *Taurodontism, pyramidal and fused molar roots, juvenile glaucoma, and unusual morphology of upper lip.* (A) Widened philtrum. (B) Full upper lip, lack of cupid's bow. (C) Pyramidal molar roots. (From *JL Ackerman et al,* Am J Phys Anthropol **38**:681, 1973.)

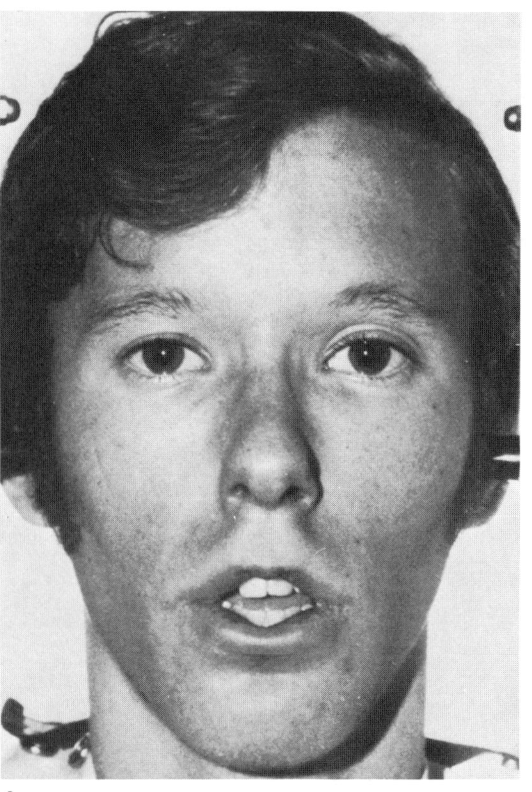

A

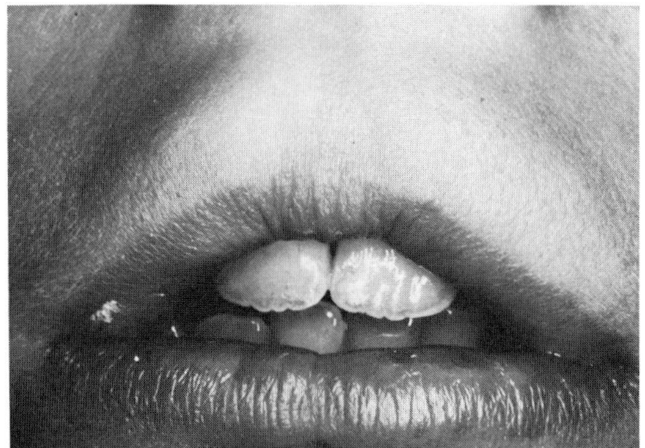

B

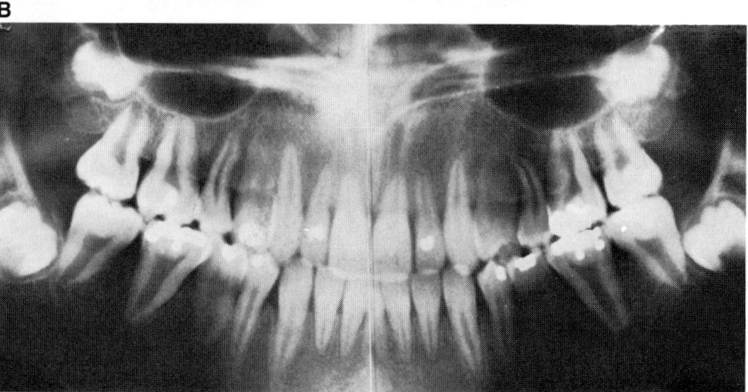

C

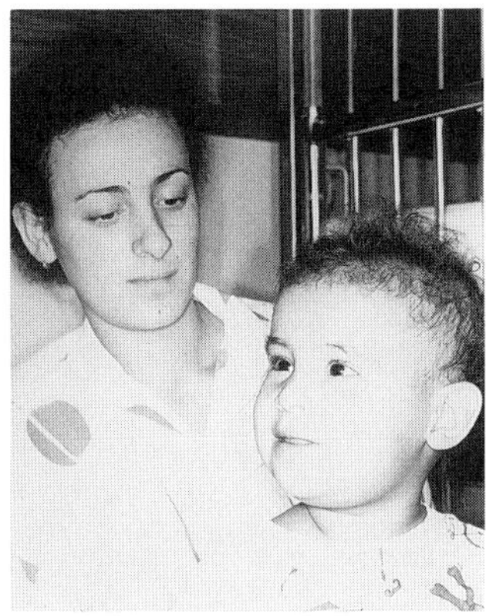

A

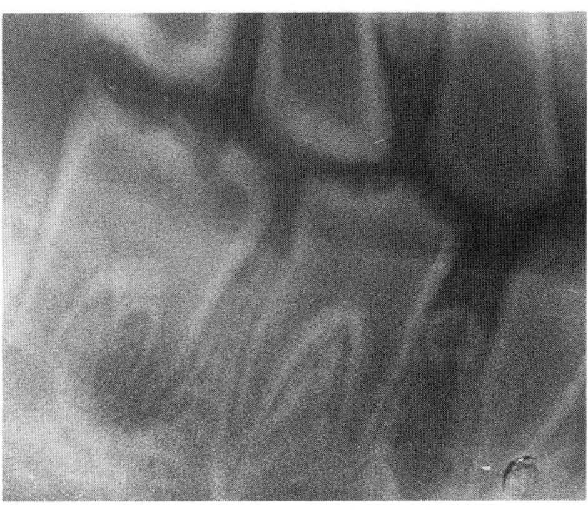

B

Fig. 27–36. *Trichodental dysplasia.* (A) Mother and child, both with short, brittle scalp hair, curly eyelashes and eyebrows. (B) Generalized shell teeth. (From *DJ Eteson* and *RD Clark,* March of Dimes Birth Defects Conference, Baltimore, 10–13 July 1988.)

## Trichodental dysplasia

Eteson and Clark (1) reported a three generation family with generalized shell teeth, sparse short slow-growing, brittle scalp hair, and curly eyelashes and eyebrows (Fig. 27-36A,B).

### Reference (Trichodental dysplasia)

1. Eteson DJ, Clark RD: A new autosomal dominant tricho-dental dysplasia. *March of Dimes Birth Defects Conference,* Baltimore, 10–13 July 1988.

## Canine radiculomegaly and congenital cataracts

Hayward (1), in 1980, reported a patient with gigantism of permanent canine teeth and congenital cataracts. AH Marashi and RJ Gorlin have observed the syndrome in a female whose brother had the same dental anomaly but normal vision and in an isolated female (2) (Fig. 27–37).

### References (Canine radiculomegaly and congenital cataracts)

1. Hayward JR: Cuspid gigantism. *Oral Surg Oral Med Oral Pathol* **49**:500–501, 1980.
2. Marashi AH, Gorlin RJ: Canine radiculomegaly and congenital cataracts—a syndrome? (in preparation).

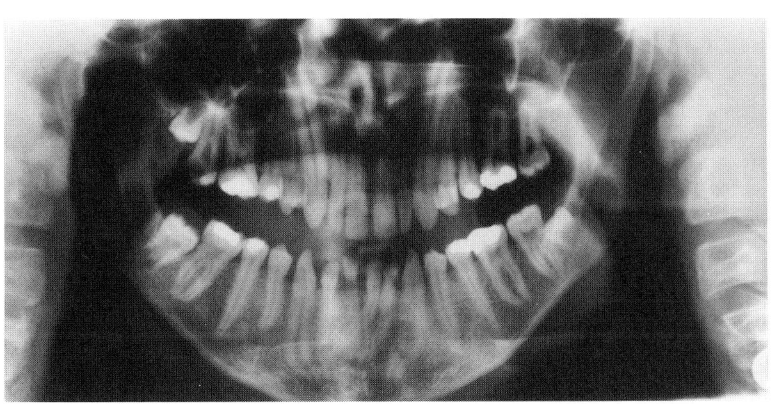

Fig. 27–37. *Canine radiculomegaly and congenital cataracts.* Note remarkable length (>45 mm) of all four canine teeth. Normal length of canines is about 26 mm.

# Chapter 28
# Well-Known Miscellaneous Syndromes

## Arteriohepatic dysplasia (Alagille syndrome)

In 1973, Watson and Miller (31) reported the association between intrahepatic cholestasis and pulmonary arterial stenosis. Alagille and co-workers (2), in 1975, added characteristic facies, mental, somatic, and sexual retardation, and vertebral and cardiac malformations. An earlier reference to the same condition is that of Vermassen and Boddaert (30). Low birthweight and growth retardation occur in about half the patients, being more common in those with vertebral anomalies. Mueller et al (17) noted that about 25% die prior to the age of 5 years due to cardiovascular or hepatic complications, but the disorder is usually benign (10,13,24). At least 90 cases have been reported.

In spite of an initial report suggesting autosomal recessive inheritance (2), many more recent reports indicate autosomal dominant inheritance with reduced penetrance and very variable expressivity (9a,13,14,17,24,25,27,31). This can be explained by a deletion at 20p11.22 found in several patients (4b,13a,26a). Genetic heterogeneity has been suggested (20). The frequency has been estimated at 1/70,000 live births (6).

**Facies.** Earlier than 1 year of age, the facies is not especially distinctive. The forehead is prominent, the eyes deeply set, and the nose long and straight with a flattened tip. The frontonasal angle is straight and there are flat midface, mild hypertelorism, and prominent chin. The pinnae are often outstanding (2,3,7,19,23,24). The postpubertal male has sparse facial hair (Fig. 28-1A–C). Sokol et al (29) pointed out that the facies is not specific for Alagille syndrome but is a general feature of congenital intrahepatic cholestatic liver disease.

**Eyes.** Bilateral posterior embryotoxon (white line showing 2 mm over the limbus), pigmentary retinopathy, and strabismus are seen in about 65% (Fig. 28-2). Less frequent are ectopic pupil, band keratopathy, Axenfeld anomaly, choroidal folds, anomalous optic disc, and infantile myopia (22,23,25).

**Cardiovascular.** Peripheral pulmonary artery stenosis is found in 85% (3). This may be an isolated finding (55%) but it may be found in combination with ASD, VSD, coarctation of the aorta, and tetralogy of Fallot (9,17,24). In about 15%, there are isolated examples of congenital heart anomalies without peripheral pulmonary artery stenosis. The stenosis may occur at a single site or at multiple sites.

**Liver.** Chronic cholestasis due to intrahepatic duct hypoplasia with associated pruritus becomes evident during the first few months of life and persists (11). However, perhaps 25% do not manifest jaundice until later in infancy (14). Acholic stools and xanthomas may be present. There is hepatosplenomegaly, but no hepatic failure or portal hypertension. Cirrhosis eventuates in 10–15% (12). Hepatocarcinoma has been reported (1,12), perhaps being related to the loss of the gene on chromosome 20.

**Musculoskeletal.** Vertebral anomalies of shape and/or segmentation (butterfly vertebrae, hemivertebrae, vertebral arch defects, reduced interpediculate distances, and plate irregularities) are present in about 85% (3,4,4a,16a,18). A pointed anterior process on C-1 has been noted (24,26). Shortened distal phalanges and ulnae are common (4a,15,23,24,26). Eleven pairs of ribs and fused ribs have also been

Fig. 28–1. *Arteriohepatic dysplasia.* (A) Note prominent forehead, deeply set eyes, upward-slanting palpebral fissures, and straight nose. (B) Compare facies in 17-year-old. Note sparse facial hair and outstanding pinnae. (C) Compare facies in child. Note sunken eyes. (A from *D Alagille et al,* J Pediatr **86:**63, 1975. C from *RF Mueller,* J Med Genet **24:**621, 1987.)

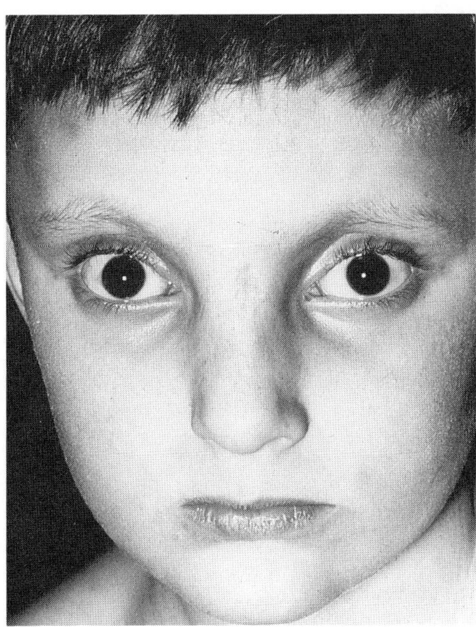

A

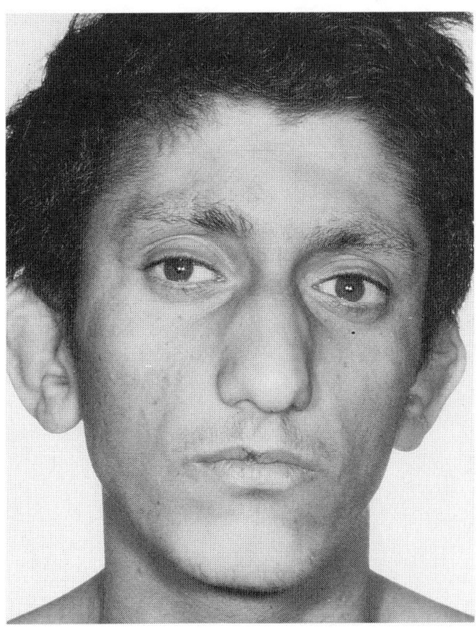

B

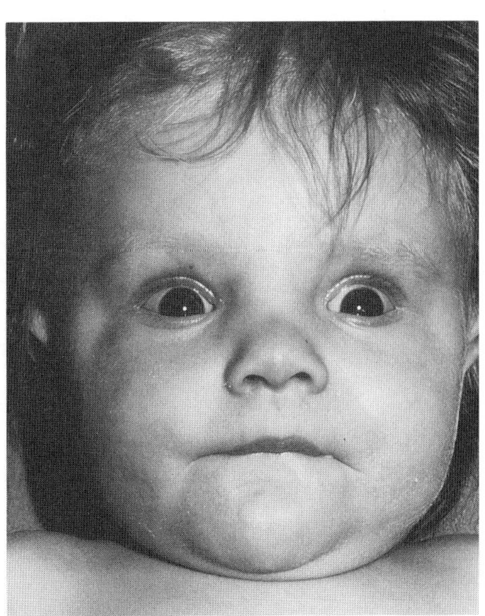

C

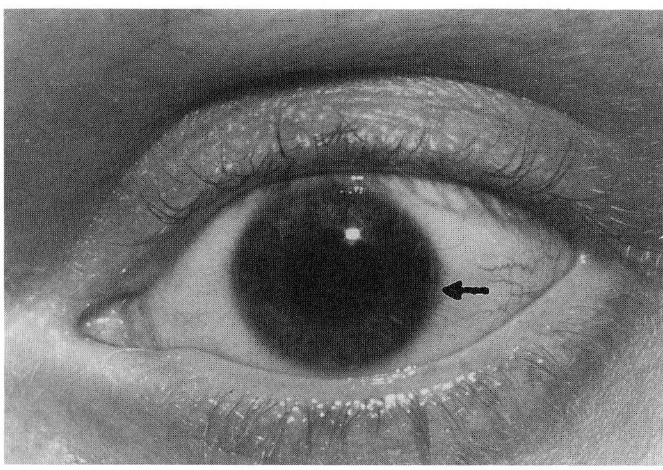

Fig. 28–2. *Arteriohepatic dysplasia.* Posterior embryotoxin. Arrow points to white line that extends about 2 mm over the limbus. (From *RF Mueller,* J Med Genet **24:**621, 1987.)

observed. Osteopenia and retarded bone age are common (5,24,31). Radioulnar synostosis and narrow lumbar spine have also been reported (4a).

**Skin.** Palmar erythema and small telangiectases are frequent on the face and trunk. The skin of the proximal phalanges may be somewhat redundant (Fig. 28-3).

**Genitourinary.** Facial and body hair is sparse but normal sexual function in males is common (10). Small kidneys, agenesis of a kidney, bifid kidney and its pelvis, mesangiolipidosis, and medullary cystic disease have been noted in 75% (3,21,24,31).

**Central nervous system.** Minimal cerebral dysfunction has been described in about 60%. Poor school performance and misdemeanors are common. Reflexes may be absent or diminished. Moderate mental retardation has been reported in about 15% (2,3,21,31). The voice is often high pitched (3).

**Pathology.** Liver biopsy shows cholestasis and continued loss to the point of absence of interlobular bile ducts without fibrosis or cirrhosis (4,5,17). The common bile duct is hypoplastic but patent.

**Differential diagnosis.** Prolonged neonatal cholestasis must be differentiated from extrahepatic biliary atresia, giant cell hepatitis, and various infections (rubella embryopathy). Arteriohepatic dysplasia must be differentiated from other forms of familial intrahepatic cholestasis such as Byler syndrome (a syndrome of abnormal bile acid synthesis), *Zellweger syndrome,* and $\alpha_1$-antitrypsin deficiency (30). Prominent Schwalbe lines (posterior embryotoxon) are seen in 8–15% of normal individuals, in *Rieger syndrome* and other anterior chamber cleavage syndromes, and in *Bannayan–Riley–Ruvalcaba syndrome.* Pigmentary retinopathy is seen in a host of disorders. Riely et al (23,24) have exhaustively discussed differential diagnosis.

Peripheral pulmonary stenosis can be seen in myriad other conditions: rubella embryopathy, *Down syndrome, Williams syndrome, thalidomide embryopathy,* and various forms of nanism (16).

Fig. 28–3. *Arteriohepatic dysplasia.* Note enlarged proximal fingers.

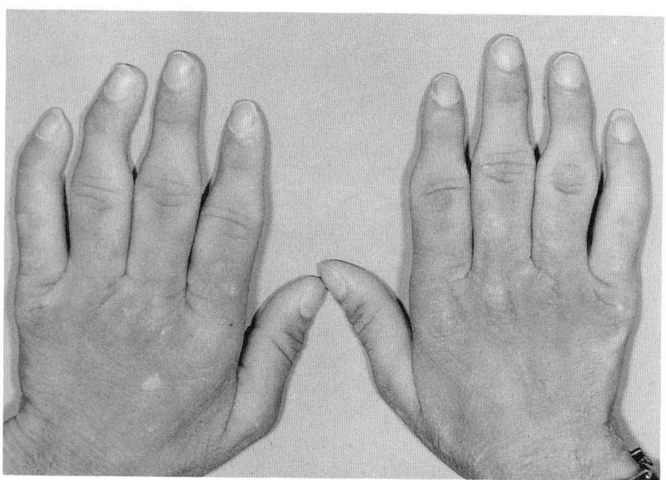

**Laboratory aids.** Atrophy of the interlobular bile ducts leads to defective secretion of bile. During the first 3 months of life there is moderate elevation of serum bilirubin and bile acids. The urine is dark and the stool is clay-colored. There is mild malabsorptioin of fat in children but not in adults. After 6 months of age, serum triglycerides and cholesterol become markedly elevated but this elevation also gradually disappears. Due to defective lipid absorption, conversion of plasma fatty acids to eicosanoids by platelets is inhibited (8).

Hepatobiliary scintigraphy or operative cholangiography may be used to exclude obstructive causes of jaundice (e.g., biliary atresia). Ultrasound in older patients shows diffuse increases in echogenicity and loss of fine detail consistent with parenchymal liver disease (28). Kidney biopsy has shown mesangiolipidosis in about 70% (3).

The pulmonary artery stenosis can be demonstrated by cardiac catheterization or inferred from Doppler echocardiography.

### References [Arteriohepatic dysplasia (Alagille syndrome)]


1. Adams PC: Hepatocellular carcinoma in association with arteriohepatic dysplasia. *Dig Dis Sci* **31**:438–442, 1986.

2. Alagille D et al: Hepatic ductular hypoplasia associated with characteristic facies, vertebral malformations, retarded physical, mental and sexual development, and cardiac murmur. *J Pediatr* **86**:63–71, 1975.

3. Alagille D et al: Syndromic paucity of interlobular bile ducts (Alagille syndrome of arteriohepatic dysplasia): Review of 80 cases. *J Pediatr* **110**:195–200, 1987.

4. Berman MD et al: Syndromatic hepatic ductular hypoplasia (arteriohepatic dysplasia): A clinical and hepatic histologic study of three patients. *Dig Dis Sci* **26**:485–497, 1981.

4a. Brunelle F et al: Skeletal anomalies in Alagille's syndrome: Radiographic study in eighty cases. *Ann Radiol* **29**:687–690, 1986.

4b. Bryne JLB et al: Del(20p) with manifestation of arteriohepatic dysplasia. *Am J Med Genet* 24:673–678, 1986.

5. Dahms BB et al: Arteriohepatic dysplasia in infancy and childhood: A longitudinal study of six patients. *Hepatology* **2**:350–358, 1982.

6. Danks DM et al: Studies of the aetiology of neonatal hepatitis and biliary atresia. *Arch Dis Childh* **52**:360–367, 1977.

7. Devloo-Blancquaert A et al: Supravalvular pulmonic and aortic stenoses and intrahepatic bile duct hypoplasia. *Acta Paediatr Belg* **33**:95–103, 1980.

8. Dupont J et al: Eicosanoid synthesis in Alagille syndrome. *N Engl J Med* **314**:718, 1986.

9. Greenwood RD et al: Syndrome of intrahepatic biliary dysgenesis and cardiovascular malformations. *Pediatrics* **58**:243–247, 1976.

9a. Grosse KP: Die arteriohepatische Dysplasie. *Leber Magen Darm* **9**:247–252, 1979.

10. Henriksen NT et al: Hereditary cholestasis combined with peripheral pulmonary stenosis and other anomalies. *Acta Paediatr Scand* **66**:7–15, 1977.

11. Kahn EI et al: Arteriohepatic dysplasia. *Hepatology* **3**:77–84, 1983.

12. Kaufman SS: Hepatoma in a child with the Alagille syndrome. *Am J Dis Child* **141**:698–700, 1987.

13. LaBrecque DR et al: Four generations of arteriohepatic dysplasia. *Hepatology* **2**:467–474, 1982.

13a. Legius E et al: Alagille syndrome (arteriohepatic dysplasia) and del(20)(p11.2). *David W. Smith Workshop on Malformations and Morphogenesis,* Madrid, Spain, 23–29 May, 1989.

14. Levin SE: Arteriohepatic dysplasia. *Pediatrics* **68**:464–465, 1981.

15. Levin SE et al: Arteriohepatic dysplasia: Association of liver disorders with pulmonary arterial stenosis as well as facial and skeletal abnormalities. *Pediatrics* **66**:876–883, 1980.

16. Löser H et al: Peripheral Pulmonalstenosen: mögliche Ursachen und syndromale Zusammenhänge. *Klin Pädiatr* **189**:137–145, 1977.

16a. Mueller RF: The Alagille syndrome. *J Med Genet* **24**:621–626, 1987.
17. Mueller RF et al: Arteriohepatic dysplasia: Phenotypic features and family studies. *Clin Genet* **25**:323–331, 1984.
18. Oestreich AE et al: Renal abnormalities in arteriohepatic dysplasia and nonsyndromic intrahepatic biliary hypoplasia. *Ann Radiol* **26**:203–209, 1983.
19. Oldigs HD, Müller-Wiefel DE: Zum Krankheitsbild der arteriohepatischen Dysplasie. *Mschr Kinderheilkd* **124**:382–383, 1976.
20. Osswald P, Löser H: Arterio-hepatische Dysplasie: ein einheitsliches Krankheitsbild? *Mschr Kinderheilkd* **125**:553–555, 1977.
21. Pokorny W et al: Gallengangshypoplasie—Syndrom mit characterischer Facies, Pulmonalgefässanomalien und facultativ anderen Missbildungen. *Klin Pädiatr* **188**:255–262, 1976.
22. Puklin JE: Anterior segment and retinal pigmentary abnormalities in arteriohepatic dysplasia. *Ophthalmology* **88**:337–347, 1981.
23. Riely CA et al: A father and son with cholestasis and peripheral pulmonic stenosis. A distinct form of intrahepatic cholestasis. *J Pediatr* **92**:405–411, 1978.
24. Riely CA et al: Arteriohepatic dysplasia: A benign syndrome of intrahepatic cholestasis with multiple organ involvement. *Ann Intern Med* **91**:520–527, 1979.
25. Romanchuk KG et al: Ocular findings in arteriohepatic dysplasia (Alagille's syndrome). *Can J Ophthalmol* **16**:94–99, 1981.
26. Rosenfield NS et al: Arteriohepatic dysplasia: Radiological features of a new syndrome. *Am J Roentgenol* **135**:1217–1223, 1980.
26a. Schnittger S et al: Molecular and cytogenetic analysis of an interstitial deletion associated with syndromal intrahepatic ductular hypoplasia (Alagille syndrome). *Hum Genet* **83**:239–244, 1989.
27. Shulman SA et al: Arteriohepatic dysplasia (Alagille syndrome): Extreme variability among affected family members. *Am J Med Genet* **19**:325–332, 1984.
28. Singcharoen T et al: Arteriohepatic dysplasia. *Br J Radiol* **59**:509–511, 1986.
29. Sokol RH et al: Intrahepatic "cholestatic facies": Is it specific for Alagille syndrome? *J Pediatr* **103**:205–208, 1983; **104**:487–488, 1984.
30. Vermassen A, Boddaert J: Congenitale agenesis van de intrahepatische galwegen. *Maand Kindergeneesk* **30**:56–58, 1962.
31. Watson GH, Miller V: Arteriohepatic dysplasia. Familial pulmonary arterial stenosis with neonatal liver diseases. *Arch Dis Childh* **48**:459–466, 1973.

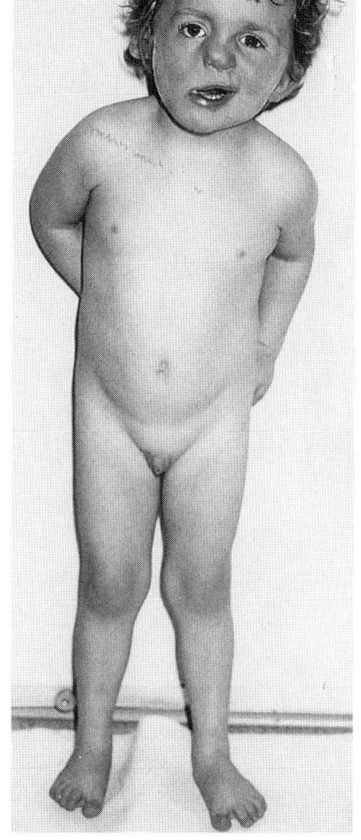
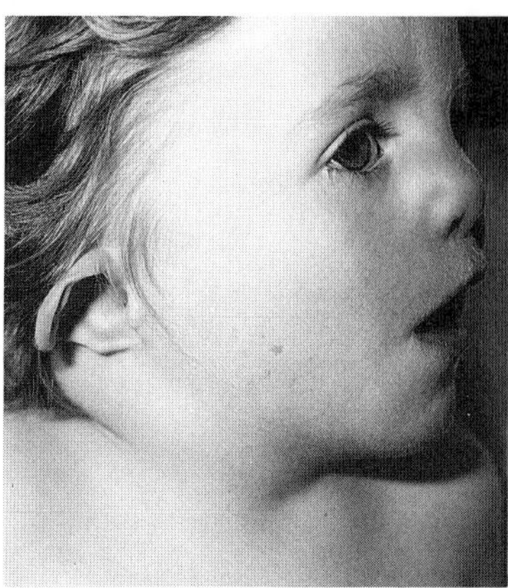

**A** **B**

Fig. 28–4. *CHARGE association.* (A,B) Mentally retarded child with choanal atresia, repaired cleft lip–palate, dysmorphic right ear with conduction hearing loss, right-sided facial palsy, coloboma of choroid and optic nerve, micropenis, cryptorchidism, pulmonary stenosis, and ASD. (Courtesy of *HH Kramer,* Düsseldorf, West Germany.)

## CHARGE association

Although the association was described earlier by a number of authors (1,1a,4,10,12,13,23), Pagon et al (17), in 1981, employed the mnemonic CHARGE (C—coloboma, H—heart defects, A—atresia choanae, R—retarded growth and development and/or CNS anomalies, G—genital hypoplasia, and E—ear anomalies and/or deafness). Several excellent reviews are available (7,8,10,13,14,16,17,20). At least 150 cases have been reported. The CHARGE association is not a specific disorder *sui generis* nor is it a diagnosis. Its main utility as a designation is to alert clinicians to search for other possible components of the constellation in a patient with one or more anomalies. The association is etiologically heterogeneous. The association may possibly result from abnormalities in development, migration, or interaction of cephalic neural crest cells (22). Although most instances are sporadic, affected parent and child, affected sibs, and monozygous twins have been reported (3,7,11,14–17).

Polyhydramnios has been noted in about 50% of the cases and in nearly all of those with bilateral choanal atresia. Approximately 35% die during the first 3 months of life. Those with bilateral choanal atresia and congenital heart disease have a higher mortality rate.

If ascertainment is done by using choanal atresia as a constant feature (100%),* usually mild mental retardation and postnatal growth retardation are the most constant abnormalities. If this bias is not introduced, about 65% have choanal atresia. Birthweight and length are usually normal but postnatal growth is poor in 50%. Seen in 40–60% are low-set, short, wide, asymmetric, outstanding, cup-shaped, simplified or lop ears, square face with malar flattening, pinched nostrils, long philtrum, prominent columella, cardiac defects (PDA, VSD, ASD, endocardial cushion defect, coarctation of aorta, vascular ring, hypoplastic left heart, tetralogy of Fallot), unilateral or bilateral ocular coloboma of iris, retina, and/or disc, hypogenitalism in males (microphallus, poorly developed scrotum, cryptorchidism), and mental retardation (6a,7a,13,24). Facial asymmetry, congenital microcephaly, high nasal bridge, malar hypoplasia, small mouth, cleft palate, swallowing difficulties, and congenital unilateral facial palsy have been noted in 40–50% (Figs. 28-4A,B and 28-5). About 10–15% exhibit postnatal microcephaly, micrognathia, and short neck. Mixed, progressive hearing loss, characterized by a wedge-shaped audiogram, is a frequent finding (24). Histologic examination of temporal bones has revealed the Mondini defect of the pars inferioris with complete absence of the pars superioris (9,18,21). Less common anomalies (10–30%) include tracheoesophageal fistula, esophageal atresia, and minor renal malformations (5,7,16).

Choanal atresia is bilateral in 65% and it occurs more often on the left side (4,10,19). The ocular colobomas range from iris coloboma to clinical anophthalmia. Not uncommonly the coloboma is limited to the choroid and/or optic nerve. Occasionally microphthalmus is noted (25). However, it should be pointed out that if colobomatous microphthalmia is used as the constant feature, the overlap of the two associations is remarkable (11,12). Some patients exhibit the *DiGeorge sequence* (6). Hypopituitarism has been reported (2).

CHARGE association exhibits some overlap with *trisomy 13, del(4p), dup(4p), cat-eye syndrome, trisomy 22, del(9q), del(11q), del(13q),* and VATER association (7,23). Choanal atresia is seen in combination with numerous syndromes (13). Undoubtedly many of the cases cited under our heading *cleft lip–palate, eye colobomas, vertebral anomalies, and mental retardation* are examples of CHARGE association.

*It should be pointed out that there is considerable bias in these data. About 40% of cases of choanal atresia are accompanied by other malformations but the uncomplicated cases are less likely to be reported (12). The preponderance of affected males suggests that micropenis makes identification easier in that sex.

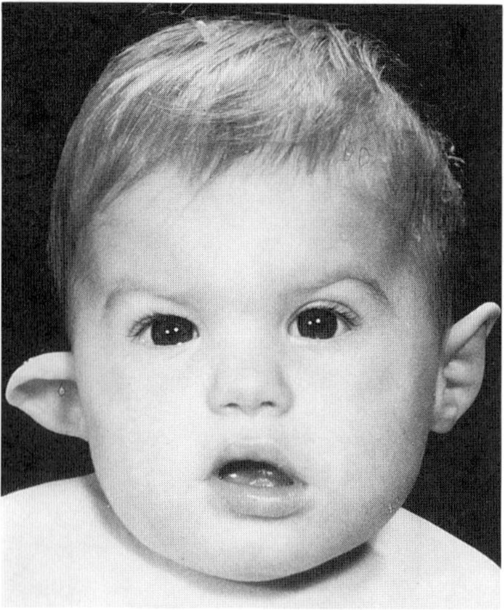

Fig. 28–5. *CHARGE association.* Rather typical face. Notice unusual pinnae. Mild degree of facial palsy. (From *CA Oley et al,* J Med Genet **25:**147, 1988.)

### References (CHARGE association)


1. Allouche C et al: L'association CHARGE. *Pédiatrie* **44**:391–395, 1989.

1a. Angelman H: Syndrome of coloboma with multiple congenital anomalies in infancy. *Br Med J* **1**:1212–1214, 1961.

2. August GP et al: Hypopituitarism and the CHARGE association. *J Pediatr* **103**:424–425, 1983.

3. Awrich PD et al: CHARGE association anomalies in siblings. *Am J Hum Genet* **34**:80A, 1982.

4. Bergstrom L, Owens O: Posterior choanal atresia: A syndromal disorder. *Laryngoscope* **94**:1273–1276, 1984.

5. Buckfield PM et al: Bilateral congenital choanal atresia association with anomalies of the foregut. *Aust Paediat J* **7**:37–44, 1971.

6. Conley ME et al: The spectrum of the DiGeorge syndrome. *J Pediatr* **94**:883–890, 1979.

6a. Cyran SE et al: Spectrum of congenital heart disease in CHARGE association. *J Pediatr* **110**:576–577, 1987.

7. Davenport SLH et al: The spectrum of clinical features in CHARGE syndrome. *Clin Genet* **29**:298–310, 1986.

7a. Davenport SLH et al: CHARGE syndrome. Part I. External ear anomalies. *Int. J Pediatr Otorhinolaryngol* **12**:127–143, 1986.

8. Goldson E et al: The CHARGE association. *Am J Dis Child* **140**:918–921, 1986.

9. Guyot JP et al: The temporal bone anomaly in CHARGE association. *Arch Otolaryngol Head Neck Surg* **113**:321–324, 1987.

10. Hall BD: Choanal atresia and associated multiple anomalies. *J Pediatr* **95**:395–398, 1979.

11. Hittner HM et al: Colobomatous microphthalmus, heart disease, hearing loss and mental retardation. *J Pediat Ophthalmol Strab* **16**:122–128, 1979.

12. Ho CK et al: Ocular colobomata, cardiac defect and other anomalies: A study of seven cases including two sibs. *J Med Genet* **12**:289–293, 1975.

13. Koletzko B, Majewski F: Congenital anomalies in patients with choanal atresia: CHARGE-association. *Eur J Pediatr* **142**:271–275, 1984.

14. Metlay LA et al: Familial CHARGE syndrome: Clinical report with autopsy findings. *Am J Med Genet* **26**:577–581, 1987.

15. Mitchell JA et al: Dominant CHARGE association. *Ophthalmol Pediatr Genet* **6**:31–36, 1985.

16. Oley CA et al: A reappraisal of the CHARGE association. *J Med Genet* **25**:147–156, 1988.

17. Pagon RA et al: Coloboma, congenital heart disease and choanal atresia with multiple anomalies: CHARGE association. *J Pediatr* **99**:223–227, 1981.

18. Paparella MM: Mondini's deafness—review of the histopathology. *Ann Otol Rhinol Laryngol* **89**(*Suppl* **67**, pt.3):1–10, 1980.

19. Rabaeus M et al: Examen diagnostique et conseil génétique dans certains cas d'atresie choanale avec malformations multiples (association CHARGE). *J Génét Hum* **31**:67–72, 1983.

20. Rogers RC, Hamilton T: CHARGE association. *Proc Greenwood Genet Ctr* **6**:3–4, 1987.

21. Sekhar HK, Sachs M: Mondini defect in association with multiple congenital anomalies. *Laryngoscope* **86**:117–125, 1976.

22. Siebert JR et al: Pathologic features of the CHARGE association: Support for involvement of the neural crest. *Teratology* **31**:331–336, 1985.

23. Stool SE, Kemper BI: Choanal atresia and/or cardiac disease. *Pediatrics* **42**:525–528, 1968.

24. Thelin JW et al: CHARGE syndrome. Part II. Hearing loss. *Int. J Pediatr Otorhinolaryngol* **12**:145–163, 1986.


## FG syndrome (unusual facies, mental retardation, congenital hypotonia, and imperforate anus)

Opitz and Kaveggia (9), in 1974, reported a syndrome consisting of mental retardation, imperforate anus, unusual facies, and other anomalies. Opitz et al (8,10) subsequently expanded the syndrome to include characteristic personality, megalencephaly, joint contractures, and broad thumbs and/or halluces. About 45 patients have been reported (1–3,6,7,11,13–17). We are skeptical about inclusion of one report (4).

Inheritance is clearly X-linked with about 30% of heterozygotes manifesting stigmata (15,16).

Stature is variable. Initial size is short but there is catch-up growth. In others, birth size is normal but there is postnatal growth failure, possibly because of severe illness. About 30% of the patients have died during the first few weeks of life from congenital heart disease or amniotic fluid aspiration. There also appears to be a second peak at about 2 years when death results from pneumonia.

**Facies.** The facies is quite variable. The head, usually normal size at birth, becomes relatively macrocephalic and plagiocephalic in about 70%. Craniosynostosis has been documented in one instance (9). The anterior fontanel tends to remain open for a prolonged period. The forehead is high and broad. The frontal hair is upswept (cowlick) in 90% and its quality is soft, silky, and sparse in about 60%. There may be several hair whorls. The palpebral fissures frequently slant downward, less often upward, and the eyes appear deep set. The medial eyebrows flare. Hypertelorism, divergent strabismus (60%), epicanthic folds, and large corneae are common. The mouth is often kept open (facial hypotonia). The philtrum is long and the upper lip is thin. The lower lip is prominent and often there is drooling. Mild micrognathia and a high narrowly arched or cleft palate (1) are reported in 5% (Figs. 28-6A,B and 28-7). The teeth are usually crowded. Minor anomalies of the auricles are seen in about 60%: small simplified occasionally protruding pinnae with mild posterior angulation and over-

Fig. 28–6. *FG syndrome.* (A) Broad forehead, downslanting palpebral fissures, strabismus, prominent lower lip, micrognathia. (B) Tall forehead, wide nasal bridge, downward slanting short palpebral fissures, long philtrum, open mouth. (From *EM Thompson et al,* Clin Genet **27:**582, 1985.)

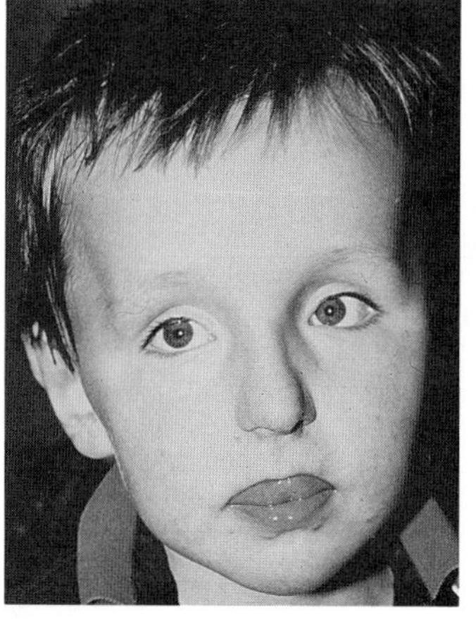

A

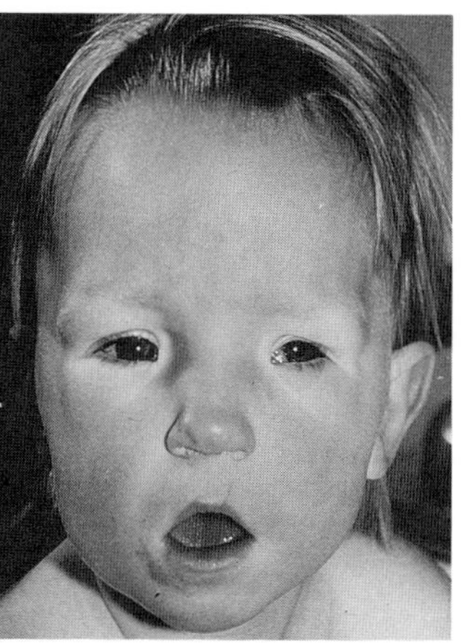

B

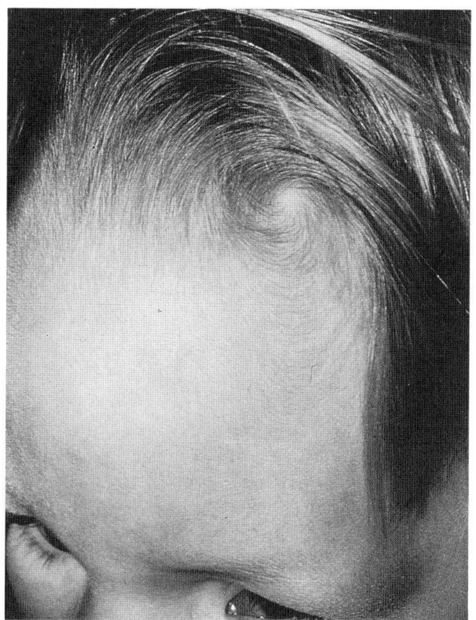

Fig. 28–7. *FG syndrome*. Frontal hair whorl. (From *EM Thompson et al*, Clin Genet **27**:582, 1985.)

folded helix. Sensorineural hearing loss has been documented in 30% (7).

**Central nervous system.** Mental retardation is constant and ranges from severe to mild but generally is moderate. Electroencephalographic disturbances are common but seizures are rare. The affect is one of hyperactivity (50%), affability, short attention span, and, occasionally, temper tantrums or aggression. Voice and speech are characteristic.

Partial absence, complete absence, or thinning of the corpus callosum has been reported (9,15–17). Thompson et al (17) reported necropsy findings consisting of simple convolutional patterning and large cavum septi pellucidi. Opitz et al (10) described midline fusion of mammillary bodies, heterotopia of neuroglial tissue in the seventh and eight nerves, ependymal cell replacement by neuroglial tissue, and diffuse defect of neuronal cell migration.

In 90% there is severe congenital hypotonia, with secondary effects such as delayed motor development, feeding difficulties, and chronic bronchopulmonary problems. Other anomalies secondary to the central nervous system disorder include squint, ptosis, open mouth, malocclusion, variable degrees of congenital joint contractures (40%), winged scapulae, clinodactyly, pes planus, and genua recurvata.

**Intestines.** The most frequent intestinal and anogenital anomalies include severe constipation (70%), anal defect [imperforate, stenotic, anteriorly displaced, perianal skin tags, (60%)], sacral dimple, malrotation of the bowel, small bowel atresia, absence of the mesentery, pyloric stenosis (10%), cryptorchidism (35%), hypospadias (25%), and inguinal and/or umbilical hernia (55%).

**Other findings.** There is a high fingertip whorl count (14) and there are fetal toe pads (Fig. 28-8). Broad thumbs and halluces are seen in 80% (11). Congenital heart defects are uncommon.

**Differential diagnosis.** The dominantly inherited *Townes–Brocks syndrome* (imperforate anus, abnormalities of hands and feet, and sensorineural hearing loss) should be excluded (12,19). There is some overlap with the neurofaciodigitorenal syndrome (severe mental retardation, megalencephaly, high forehead, cowlick, groove of nasal tip, acrorenal field defect) (5). The *Simpson–Golabi–Behmel syndrome* also needs to be excluded. Thompson et al (18) reported an X-linked mental retardation syndrome in which there was overlap with the FG syndrome but without macrocephaly and with a different facies. Fetal toe pads are also seen in *Kabuki make-up syndrome*.

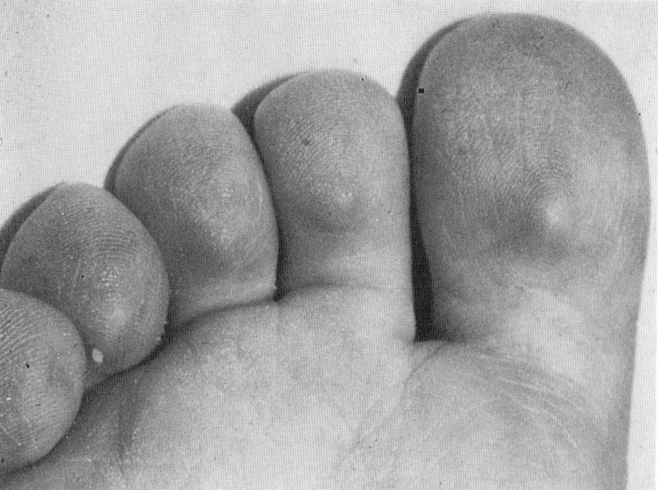

Fig. 28–8. *FG syndrome*. Toe padding. (From *EM Thompson et al*, Clin Genet **27**:582, 1985.)

#### References [FG syndrome (unusual facies, mental retardation, congenital hypotonia, and imperforate anus)]


1. Bianchi DW: FG syndrome in a premature male. *Am J Med Genet* **19**:383–386, 1984.
2. Burn J, Martin N: Two retarded male cousins with odd facies, hypotonia and severe constipation: Possible examples of the X-linked FG syndrome. *J Med Genet* **20**:97–99, 1983.
3. Cohen MM Jr: The FG syndrome. *J Pediatr* **89**:687, 1976.
4. Dallapiccola B et al: Diagnostic definition of the FG syndrome. *Am J Med Genet* **19**:379–382, 1984.
5. Freire-Maia N et al: The neurofaciodigitorenal (NFDR) syndrome. *Am J Med Genet* **11**:329–336, 1982.
6. Keller MA et al: A new syndrome of mental deficiency with craniofacial, limb and anal abnormalities. *J Pediatr* **88**:589–591, 1976.
7. Neri G et al: Sensorineural deafness in the FG syndrome: Report on four new cases. *Am J Med Genet* **19**:369–378, 1984.
8. Opitz JM: The FG syndrome. *J Pediatr* **89**:687, 1976.
9. Opitz JM, Kaveggia EG: The FG syndrome: An X-linked recessive syndrome of multiple congenital anomalies and mental retardation. *Z Kinderheilkd* **117**:1–18, 1974.
10. Opitz JM et al: The FG syndrome. Further studies on three affected individuals from the FG family. *Am J Med Genet* **12**:147–154, 1982.
11. Opitz JM et al: FG syndrome update 1988: Note of 5 new patients and bibliography. *Am J Med Genet* **30**:309–328, 1988.
12. de Pina-Neto JM: Phenotype variability in Townes–Brocks syndrome. *Am J Med Genet* **18**:147–152, 1984.
13. Riccardi VM et al: The FG syndrome: Further characterization: Report of a third family and of a sporadic case. *Am J Med Genet* **1**:47–58, 1977.
14. Richiera-Costa A: FG syndrome in a Brazilian child with additional, previously unreported signs. *Am J Med Genet* (Suppl) **2**:247–254, 1986.
15. Thompson EM, Baraitser M: The FG syndrome. *J Med Genet* **24**:139–143, 1987.
16. Thompson EM et al: The FG syndrome: 7 new cases. *Clin Genet* **27**:582–594, 1985.
17. Thompson EM et al: Necropsy finding in a child with FG syndrome. *J Med Genet* **23**:372–373, 1986.
18. Thompson EM et al: X-linked mental retardation: A family with a separate syndrome? *J Med Genet* 26:373–378, 1989.
19. Walpole IR, Hockey A: Syndromes of imperforate anus, abnormalities of hands and feet, and sensorineural deafness. *J Pediatr* **100**:250–252, 1982.


## Kartagener syndrome (immotile cilia syndrome)

The relationship between situs inversus viscerum and bronchiectasis was pointed out by Siewert (37) in 1904. Kartagener (20,21), in a series of papers from 1933 through 1935, added chronic rhinosinusi-

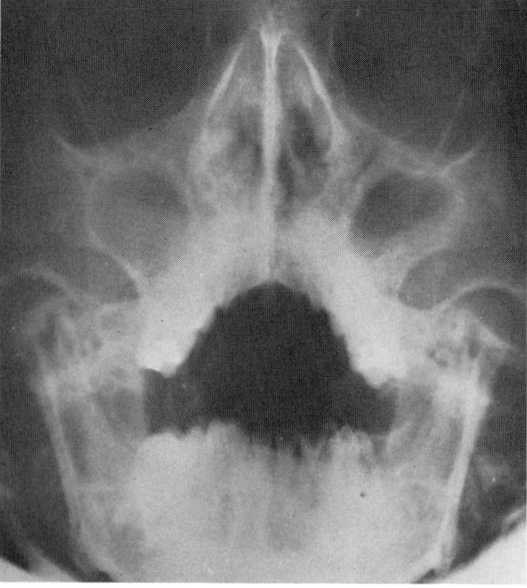

Fig. 28–9. *Kartagener syndrome.* Bilateral maxillary sinusitis, more pronounced on right side.

tis, that is, nasal polyposis, chronic hyperplastic rhinitis, and ethmoidomaxillary sinusitis. Numerous investigators have also noted the absence or hypoplasia of the frontal sinuses. More than 400 cases have been recorded.

The syndrome exhibits autosomal recessive inheritance (1,27,42). Consanguinity has been found in about 25%. There may be genetic heterogeneity (7,13). There is no racial predilection (42). The frequency of immotile cilia syndrome is about 1/16,000 population (34,42).

Abnormal cilia are considered the underlying cause of the anomalies observed in Kartagener syndrome (33). Most studies have found partial or total absence of dynein side arms (8,22,47) that would inhibit mucociliary transport. Transposed ciliary microtubules, missing radial spokes, and supernumerary microtubules have also been found (6a,7a,12,23a,32,39,40,41,46). It has been suggested that the ciliary defects may develop with age (15,31) since children may have normal nasal ultrastructure. An animal model has been found (9a).

**Respiratory tract.** Sinusitis is a frequent finding (Fig. 28-9). Similarly, nasal polyps, found in about 20%, have been reported with situs inversus in the absence of bronchiectasis (44). Agenesis or diminution of the frontal sinuses is common (28,29) but also is rather frequently seen in bronchiectasis and sinusitis without situs inversus, indicating a relationship between the sinuses and the rest of the respiratory tract, but not necessarily with visceral inversion.

In early infancy, nasal discharge, frequent colds, and chronic bronchitis are observed, as well as recurrent bouts of pneumonia. Nasal catarrh and anosmia soon intervene and are followed in most cases by chronic cough productive of foul-smelling phlegm. Nasal mucosal cilia are defective (45). Bronchiectasis is found in about 30% of older children and adults. In some persons, however, bronchiectasis seems to precede involvement of the upper part of the respiratory tract. Occasionally, there are accompanying asthma, hemoptysis, and pulmonary osteoarthropathy. Life expectancy is normal.

Among cases of situs inversus viscerum (see below), bronchiectasis is present in from 15 to 25% (28). This is far higher than the incidence among the general population (0.25–0.50%) (21).

The bronchiectasis is generally tubular or varicose, rather than cystic. There is still disagreement as to whether the bronchiectasis is of the acquired or congenital type (Fig. 28-10B).

**Situs inversus viscerum.** As an isolated phenomenon, situs inversus viscerum occurs once in about 8,000 to 10,000 births (9,29,44). There is no unanimity of opinion concerning the relationship between visceral inversion and bronchiectasis, but possibly some factor predisposes the neonate with situs inversus to defective postnatal lung growth and atelectasis. About 50% of patients with immotile cilia syndrome manifest situs inversus (13,42). Rott (34) suggests that ciliary beats in the early embryo determine the type of laterality. Without the ciliary action, laterality develops randomly (10).

Dextrocardia, without further evidence of visceral inversion, has been frequently detected (45) (Fig. 28-10A), and corrected transposition has been reported once (38). Furthermore, investigators (11a,35)

Fig. 28–10. *Kartagener syndrome.* (A) Dextrocardia. Atelectasis of left middle lobe and right basal infiltration. (B) Bronchography showing bronchiectasis of left middle and both lower lobes. (From *A Glay,* J Can Assoc Radiol **12:**22, 1961.)

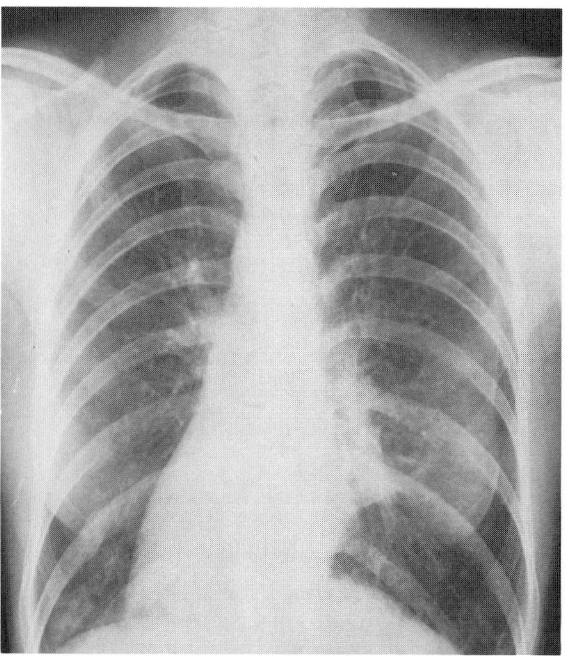



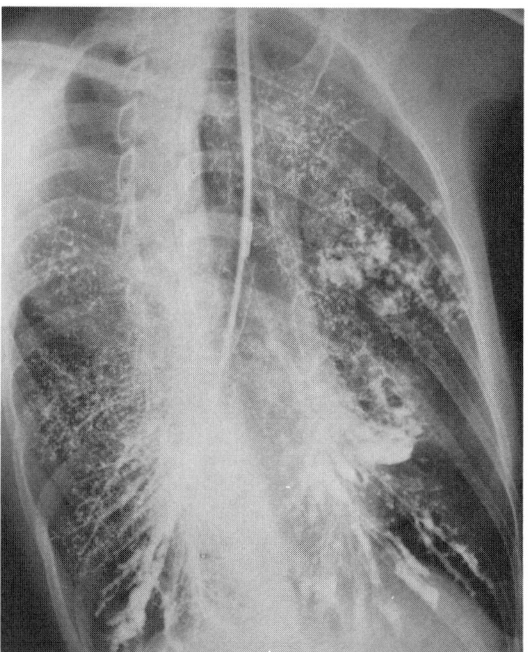

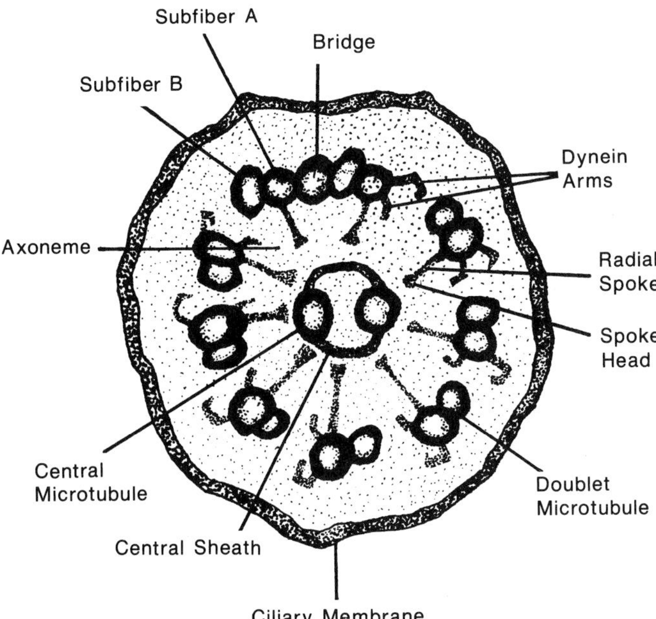


Fig. 28–11. *Kartagener syndrome.* Diagram of cross section of a cilium. Note nine outer microtubular doublets and two central microtubules held together by three kinds of connections: dynein arms, nexin links, and radial spokes. (From *DJ Imbrie,* Am J Otolaryngol **2:**215, 1981.)

have reported patients with Kartagener syndrome and features of the polysplenia field defect, suggesting a pathogenesis common to the two entities.

**Fertility.** Most males with Kartagener syndrome are infertile (3), due to an ultrastructural defect of sperm tails similar to that of respiratory cilia (10). There have been case reports of males with normal fertility (19). Female patients were originally considered to be normally fertile (2,6). However, Afzelius and Eliasson (3) found evidence that female fertility is also impaired, the Fallopian cilia being deficient in dynein arms (6,30).

**Other findings.** Rheumatoid arthritis has been reported in a few instances (24). Other cases have been described with anomalous subclavian artery, malformation of retinal vessels, cardiac and renal anomalies, turricephaly, absence of the xiphoid process, thyrotoxicosis (5), spinal arteriovenous malformation (14), Paterson–Brown–Kelly syndrome (postcricoid web dysphagia, anemia, glossitis, cheilosis, and koilonychia) (43), mesangiocapillary glomerulonephritis (9), polysplenia (11a,35) and extrahepatic biliary atresia (11a).

Various associated eye anomalies have been reviewed by Segal et al (36). Chronic otitis media and hearing loss have been reported in several patients (18); Fischer et al (11) demonstrated that middle ear cilia were abnormal.

**Differential diagnosis.** Thick tenacious sinus and bronchial secretions, nasal polyposis, chronic sinusitis, and bronchiectasis also occur in cystic fibrosis. The histologic appearance of the respiratory epithelium, the bronchial plugging, and the absence of other stigmata (situs inversus viscerum) clearly distinguish cystic fibrosis from Kartagener syndrome (17). Ultrastructural defects may be seen in many individuals with repeated upper and lower respiratory infections (7a).

**Laboratory aids.** Holmes et al (17) described transient immunoglobulin deficiencies during childhood, that is, low serum IgA levels, in this syndrome, a finding not supported by Miller and Divertie (25).

Diagnosis can be achieved by studying mucociliary clearance (4) or ultrastructural examination of ciliary structures from samples obtained by nasal and bronchial brushing (26) (Fig. 28-11). Deficiencies in dynein arms on the outer microtubular doublets of cilia are most often evident. However, a host of other ultrastructural abnormalities have been reported (complete or incomplete absence of inner or outer arms, radial spoke defect, microtubular transposition, lack of central core structures, etc.) (1,40,42) (Fig. 28-12A–E). Occasionally, no ciliary abnormalities are found (13).

### References [Kartagener syndrome (immotile cilia syndrome)]


1. Afzelius BA: Genetical and ultrastructural aspects of immotile cilia syndrome. *Am J Hum Genet* **33**:852–864, 1981.

2. Afzelius BA et al: On the function of cilia in the female reproductive tract. *Fertil Steril* **29**:72–74, 1978.

3. Afzelius BA, Eliasson R: Male and female infertility problems in the immotile-cilia syndrome. *Eur J Resp Dis* (*Suppl* **127**):144–147, 1983.

4. Antonelli M et al: Immotile cilia syndrome: Radial spokes deficiency in a patient with Kartagener's triad. *Acta Paediatr Scand* **70**:571–573, 1981.

5. Bhugra D: Kartagener's syndrome. *Br J Clin Pract* **39**:38–39, Jan 1985.

6. Bleau G et al: Absence of dynein arms in cilia of endocervical cells in a fertile woman. *Fertil Steril* **30**:362–363, 1978.

6a. Canciani M et al: The association of supernumerary microtubules and immotile cilia syndrome and defective neutrophil chemotaxis. *Acta Paediatr Scand* **77**:606–607, 1988.

7. Chao J et al: Genetic heterogeneity among dynein deficient cilia. *Am Rev Respir Dis* **126**:302, 1982.

7a. Corbeel L et al: Ultrastructural abnormalities of bronchial cilia in children with recurrent airway infections and bronchiectasis. *Arch Dis Childh* **56**:929–933, 1981.

8. Cornillie FJ et al: Atypical bronchial cilia in children with recurrent respiratory tract infections. *Path Res Pract* **178**:595–604, 1984.

9. Egbert BM et al: Kartagener syndrome: Report of a case with mesangiocapillary glomerulonephritis. *Arch Pathol Lab Med* **101**:95–99, 1972.

9a. Edwards DF et al: Familial immotile-cilia syndrome in English springer spaniel dogs. *Am J Med Genet* **33**:290–298, 1989.

10. Eliasson R et al: The immotile-cilia syndrome. *N Engl J Med* **297**:1–6, 1977.

11. Fischer TJ et al: Middle ear ciliary defect in Kartagener's syndrome. *Pediatrics* **62**:443–445, 1978.

11a. Gershoni-Baruch R et al: Immotile cilia syndrome including polysplenia, situs inversus and extrahepatic biliary atresia. *Am J Med Genet* **33**:330–333, 1989.

12. Goldman AS et al: The discovery of defects in the immotile cilia syndrome. *J Pediatr* **96**:244–247, 1980.

13. Greenstone M et al: Primary ciliary dyskinesia: Cytological and clinical features. *QJ Med* **67**:405–430, 1988.

14. Hayakawa T et al: Spinal arteriovenous malformation with Kartagener's syndrome. *Surg Neurol* **13**:413–417, 1980.

15. Herzon FS, Murphy S: Normal ciliary ultrastructure in children with Kartagener syndrome. *Ann Otol* **89**:81–83, 1980.

16. Heuckenkamp PU et al: Das Kartagener-Syndrom. *Dtsch Med Wochenschr* **97**:1458–1461, 1972.

17. Holmes LB et al: A reappraisal of Kartagener's syndrome. *Am J Med Sci* **255**:13–28, 1968.

18. Imbrie DJ: Kartagener's syndrome: A genetic defect affecting the function of cilia. *Am J Otolarygnol* **2**:215–222, 1981.

19. Jonsson MS et al: Kartagener's syndrome with motile spermatozoa. *New Engl J Med* **307**:1131–1133, 1982.

20. Kartagener M: Zur Pathogenese der Bronchiektasien. Bronchiektasien bei Situs viscerum inversus. *Beitr Klin Tuberk* **83**:489–501, 1933.

21. Kartagener M, Horlacher A: Zur Pathogenese der Bronchiektasien. Situs viscerum inversus und Polyposis nasi in einem familiärer Bronchiektasien. *Beitr Klin Tuberk* **87**:331–333, 1935.

22. Katz SM et al: Kartagener's syndrome and abnormal cilia. *N Engl J Med* **297**:1011–1012, 1977.

23. Knox G et al: A family with Kartagener's syndrome: Linkage data. *Ann Hum Genet* **24**:137–140, 1960.

23a. Lupin AJ, Misko GJ: Kartagener syndrome with abnormalities of cilia. *J Otolaryngol* **7**:95–102, 1978.

24. Mastaglia GL, Goatcher P: Rare association of rheumatoid arthritis and Kartagener's syndrome: A common denominator? *Med J Aust* **2**:400, 1980.

25. Miller RD: Divertie MB: Kartagener's syndrome. *Chest* **62**:130–135, 1972.

26. Moreau MF et al: Transposed ciliary microtubules in Kartagener's syndrome. *Acta Cytol* **29**:248–253, 1985.

27. Moreno A, Murphy EA: Inheritance of Kartagener syndrome. *Am J Med Genet* **8**:305–313, 1981.

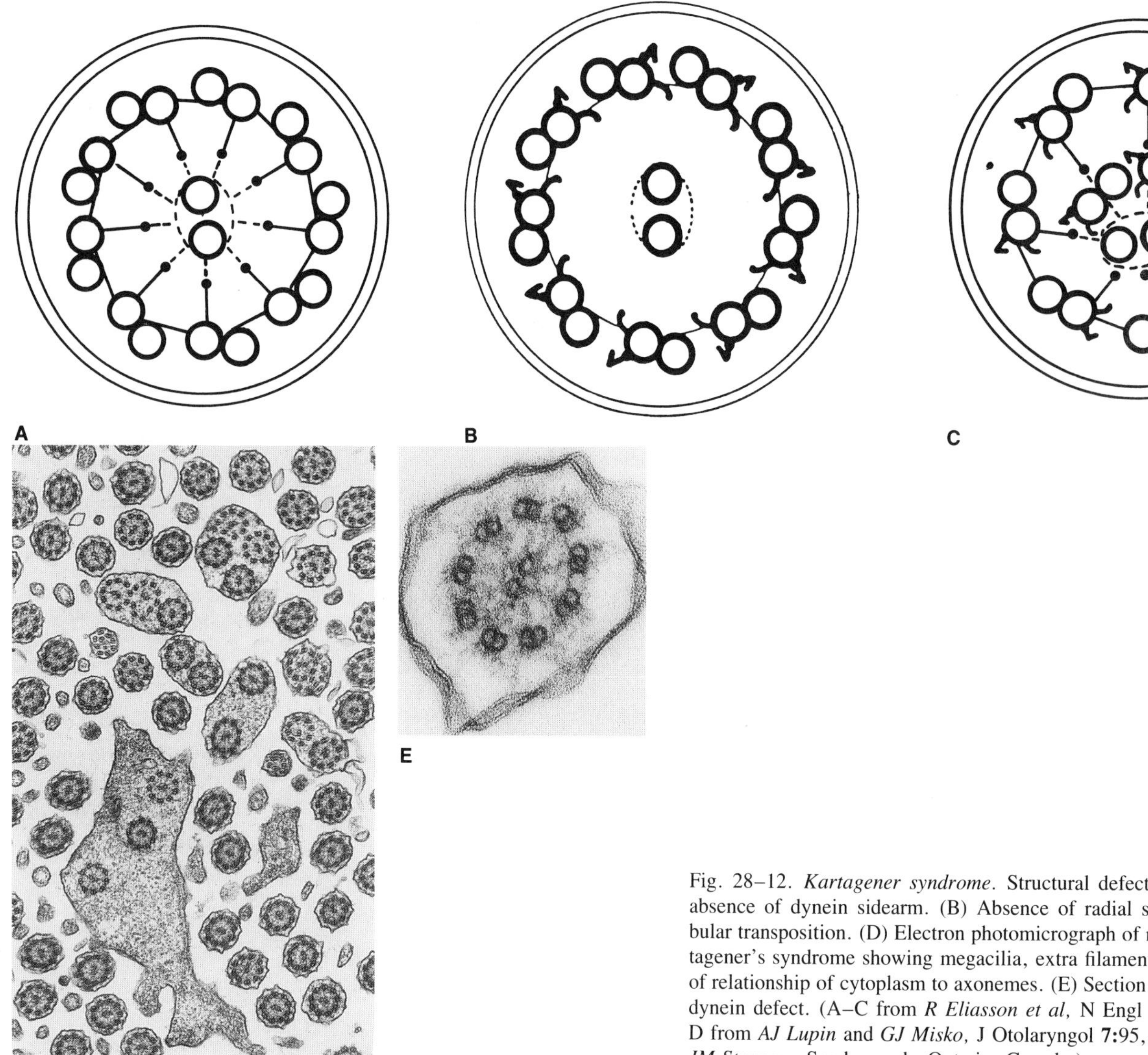


Fig. 28–12. *Kartagener syndrome*. Structural defects of cilia. (A) Total absence of dynein sidearm. (B) Absence of radial spokes. (C) Microtubular transposition. (D) Electron photomicrograph of nasal mucosa in Kartagener's syndrome showing megacilia, extra filaments, and disproportion of relationship of cytoplasm to axonemes. (E) Section illustrating complete dynein defect. (A–C from *R Eliasson et al,* N Engl J Med **297:**1, 1977. D from *AJ Lupin* and *GJ Misko,* J Otolaryngol **7:**95, 1978. E courtesy of *JM Sturgess,* Scarborough, Ontario, Canada.)

28. Nadel HR et al: The immotile cilia syndrome: Radiological manifestations. *Radiology* **154**:651–655, 1985.

29. Overholt EL, Bauman DF: Variants of Kartagener's syndrome in the same family. *Ann Intern Med* **48**:574–579, 1958.

30. Pedersen H: Absence of dynein arms in endometrial cilia: Cause of infertility? *Acta Obstet Gynecol Scand* **62**:625–627, 1983.

31. Popper H et al: Problems in the differential diagnosis of Kartagener's syndrome and ATP-ase deficiency. *Path Res Pract* **180**:481–485, 1985.

32. Pysher TJ, Neustein HB: Ciliary dysmorphology. *Persp Pediatr Pathol* **8**:101–131, 1984.

33. Rossman CM et al: Immotile cilia syndrome in persons with and without Kartagener's syndrome. *Am Rev Resp Dis* **121**:1011–1016, 1980.

34. Rott HD: Kartagener's syndrome and the syndrome of immotile cilia. *Hum Genet* **46**:249–261, 1979.

35. Schidlow DV et al: Polysplenia and Kartagener syndrome in a sibship: Association with abnormal respiratory cilia. *J Pediatr* **100**:401–403, 1982.

36. Segal P et al: Kartagener's syndrome with familial eye changes. *Am J Ophthalmol* **55**:1043–1049, 1963.

37. Siewert AK: Über einen Fall von Bronchiectasie bei einem Patienten mit Situs inversus viscerum. *Berl Klin Wochenschr* **41**:139–141, 1904.

38. Solomon MH et al: Kartagener's syndrome with corrected transposition. *Chest* **69**:677–680, 1976.

39. Sturgess JM, Turner JAP: Ultrastructural pathology in the immotile cilia syndrome. *Persp Pediatr Pathol* **8**:133–161, 1984.

40. Sturgess JM et al: Cilia with defective radial spokes: A cause of human respiratory disease. *N Engl J Med* **300**:53–57, 1979.

41. Sturgess JM et al: Transposition of ciliary microtubules: Another cause of impaired ciliary motility. *N Engl J Med* **303**:318–319, 1981.

42. Sturgess JM et al: Genetic aspects of immotile cilia syndrome. *Am J Med Genet* **25**:149–160, 1986.

43. Todd NW Jr, Yodaiken RE: A patient with Kartagener and Paterson–Brown–Kelly syndromes. *JAMA* **234**:1248–1250, 1975.

44. Torgersen J: Transposition of viscera, bronchiectasis and nasal polyps: Genetical analysis and contribution to the problem of constitution. *Acta Radiol (Stockh)* **28**:17–24, 1947.

45. Turner JAP et al: Clinical expressions of immotile cilia syndrome. *Pediatrics* **67**:805–810, 1985.

46. Veerman AJP et al: The immotile cilia syndrome: Phase contrast light microscopy, scanning and transmission electron microscopy. *Pediatrics* **65**:698–702, 1980.

47. Walter RJ et al: Cell motility and microtubules in cultured fibroblasts from patients with Kartagener syndrome. *Cell Motil* **3**:185–197, 1983.

## Klippel–Feil anomaly

Klippel–Feil anomaly is characterized by fusion of two or more cervical vertebrae and, in its most severe form, consists of massive cervical vertebral fusion, short neck, limitation of head movement, and low posterior hairline. The condition is morphologically and etiologically heterogeneous. Over 300 papers have been written on the subject

(12), approximately 350 patients have been recorded to date (12,25), and extensive reviews are available (22,25).

Vertebral fusion is actually an inaccurate term, the condition resulting from failure of normal segmentation, which can be traced back to the third embryonic week when segmentation of mesodermal somites takes place (50). Although iniencephaly has been considered to be a severe form of Klippel–Feil anomaly (7,18), it has also been classified with anencephaly with retroflexion (44). Furthermore, little is gained by identifying iniencephaly with Klippel–Feil anomaly as the latter is neither a morphologic nor an etiologic entity (50).

Klippel–Feil anomaly was described as early as the sixteenth century, and involvement of the second and third cervical vertebrae has even been found in an Egyptian mummy of about 500 B.C. (20,37). In 1912, Klippel and Feil (28) described the postmortem findings of a 46-year-old French tailor who had a short immobile neck with massive fusion of cervical and upper thoracic vertebrae. Feil (15), in 1919, added 13 cases, 12 of which had been published previously. He defined three morphologic subtypes. Type I consisted of massive fusion of the cervical and upper thoracic vertebrae. In Type II, fusion occurred at only one or two interspaces, but various anomalies such as hemivertebrae, occipitoatlantal fusion, and other defects could be present. Type III had both cervical and lower thoracic or lumbar fusion.

For Type I, birth prevalence estimates have varied from a minimum of 0.025–0.03/1000 (19,30) to a maximum of 0.16/1000 (25). Sex predilection for females has been recorded by several authors (13,18,38), and Helmi and Pruzansky (25) have estimated the male:female ratio to be 1:1.3. Type II occurs more commonly, with a prevalence of 7.3/1000 (5,25), the most frequent sites of cervical fusion being at C2–3 and C5–6.

Etiologic heterogeneity is evident and Gunderson et al (22) used Feil's (15) classification. For Type I, almost all cases have been sporadic, although there have been a few familial examples (17,22). Autosomal recessive inheritance has been postulated (22), although among sporadic cases there has been female predilection, as indicated. Type II is transmitted as an autosomal dominant disorder and, in some instances, atlantooccipital fusion or other abnormalities may be observed (22). It has been suggested that Type II is genetically heterogeneous with an autosomal recessive form also (27). For Type III, an autosomal recessive mode of inheritance has been suggested (22). Da Silva (12) reported 12 cases of a newly recognized autosomal recessive disorder in an inbred kindred; because patients had variable fusions of cervical, upper thoracic, and/or lumbar vertebrae, some presenting Type I, others Type III patterns. Raas–Rothschild et al (42a) have suggested that Klippel–Feil anomaly and sacral agenesis constitute a Type IV. This and other examples have been represented by isolated cases (3a,20,25a,50).

Warkany (50), noting a large number of teratologic experiments in animals that resulted in vertebral malformations, suggested the possibility of Klippel–Feil anomaly occurring on a teratogenic basis in humans. In this connection, Gunderson et al (22) noted a case of Type I fusion in which the mother had attempted abortion with oral and parenteral agents on three occasions during the first trimester. The possibility of Klippel–Feil anomaly occurring on the basis of subclavian artery supply disruption has also been raised (4).

*Wildervanck syndrome* consists of Klippel–Feil anomaly, abducens paralysis with retracted bulb, and sensorineural or conduction deafness (14). The appearance is striking, the head seeming to sit directly on the trunk with frequent facial asymmetry and/or torticollis. On medial gaze, the adducted globe becomes retracted as the eye slit narrows, a phenomenon known as Duane syndrome. Involvement may be unilateral or bilateral. Mental retardation has been noted on occasion. An overwhelming preponderance of females has been reported.

The MURCS association consists of **Mü**llerian duct aplasia, **R**enal aplasia, and **C**ervicothoracic **S**omite dysplasia. Vertebral defects occur especially from C5 to T1. Anomalies include Klippel–Feil anomaly, absent vagina, absence or hypoplasia of the uterus, renal agenesis, and a variety of low-frequency defects (1,11,32,33,41). This is the same as the Rokitansky–Küster–Hauser syndrome.

Chemke et al (8) described a patient with Klippel–Feil anomaly, torticollis, preauricular appendages, cleft palate, Sprengel's deformity, absent ulna, and flexion contractures of the affected upper extremity.

Fragoso et al (16) reported a female with Klippel–Feil anomaly, ocular hypertelorism, broad nose, Sprengel's deformity, and postaxial polydactyly. The patient may represent an example of craniofrontonasal dysplasia. However, no craniosynostosis was evident.

JW Hanson and MM Cohen Jr (personal observation, 1975) noted two male sibs with Klippel–Feil anomaly, mild ocular hypertelorism, downslanting palpebral fissures, prominent nasal bridge, micrognathia, submucous cleft palate, dislocated radial heads, positional foot deformities, and scoliosis.

Various authorities differ in their interpretation of what constitutes Klippel–Feil anomaly. The following description is confined to massive cervical vertebral fusion (Type I) only.

**Head and neck.** The head seems to sit directly on the thorax, without an interposing neck. The flaring trapezius muscles extend from the mastoid areas to the shoulders, producing a pterygium-like effect. Occasionally, there is facial asymmetry, with one eye situated lower than the other (35). Posteriorly, the hairline extends to the shoulders (Fig. 28-13A–D).

Associated abnormalities (Table 28-1) have been especially well reviewed by Helmi and Pruzansky (25). The most common eye finding has been convergent strabismus (3). Less frequently found are horizontal nystagmus and chorioretinal atrophy. From 25 to 50% have exhibited hearing loss that may be sensorineural, mixed, or conductive (10,13,25,31,34,40,45,48a,52); some cases may have represented Wildervanck syndrome. Mondini type dysplasia has been found in some cases (52). Cleft palate is present in about 17% (1,9,13,24,25,37–39,47,48). Frontonasal malformation has been found (16).

**Musculoskeletal system.** Characteristically, several, occasionally all, cervical vertebrae are fused into a solid mass. In some instances upper thoracic vertebrae may be involved. Scoliosis, cervical ribs, and/or the Sprengel's deformity have been noted in about 30% (19,20,23,28,46). Other common findings include spina bifida occulta, fusion of the atlas with the occipital bone, cleft vertebrae, and hemivertebrae (19,45).

**Central nervous system.** A large number of associated neurologic disturbances have been recorded: spasticity or hyperreflexia, bimanual synkinesis or mirror movements (2,3,13,21), syringomyelia or syringobulbia (35,46), hemiplegia, paraplegia, triplegia, quadriplegia, and others. The neurological findings have been thoroughly reviewed by Mosberg (35). A case with neuroschisis of the cervical spinal cord has been noted (36).

**Other findings.** Congenital heart defects, usually ventricular septal defect, are occasionally noted (34,38,49).

**Differential diagnosis.** The flaring trapezius muscles give the patient an appearance that may simulate that in *Turner syndrome* or *Noonan syndrome*. The position of the head on the thorax may resemble that seen in tuberculosis of the cervical spine or in *Morquio syndrome*. Spondylothoracic dysplasia is genetically heterogeneous, consisting of dominant and recessive forms. The entire vertebral column is involved in fusions, hemivertebrae, scoliosis, etc. (6,26,42,51). In the *nevoid basal cell carcinoma syndrome*, cervical or upper thoracic vertebral fusion is observed in about 40% (19a). In the *fetal alcohol syndrome*, cervical fusion occurs in approximately 40%, especially involving C2–3 (46a). In *Crouzon syndrome*, approximately 30% have cervical fusion most commonly involving C2–3 (29). With *Apert syndrome* and with *Binder syndrome*, fusion of C5–6 (58%) or involvement of three or more vertebrae has been reported (29). Cervical vertebral anomalies and other defects of the spine may occur with *oculo-auriculo-vertebral spectrum*. Mirror movements as an isolated finding may be inherited as an autosomal dominant trait (43). In addition, one must exclude the associations described in the introduction to this chapter

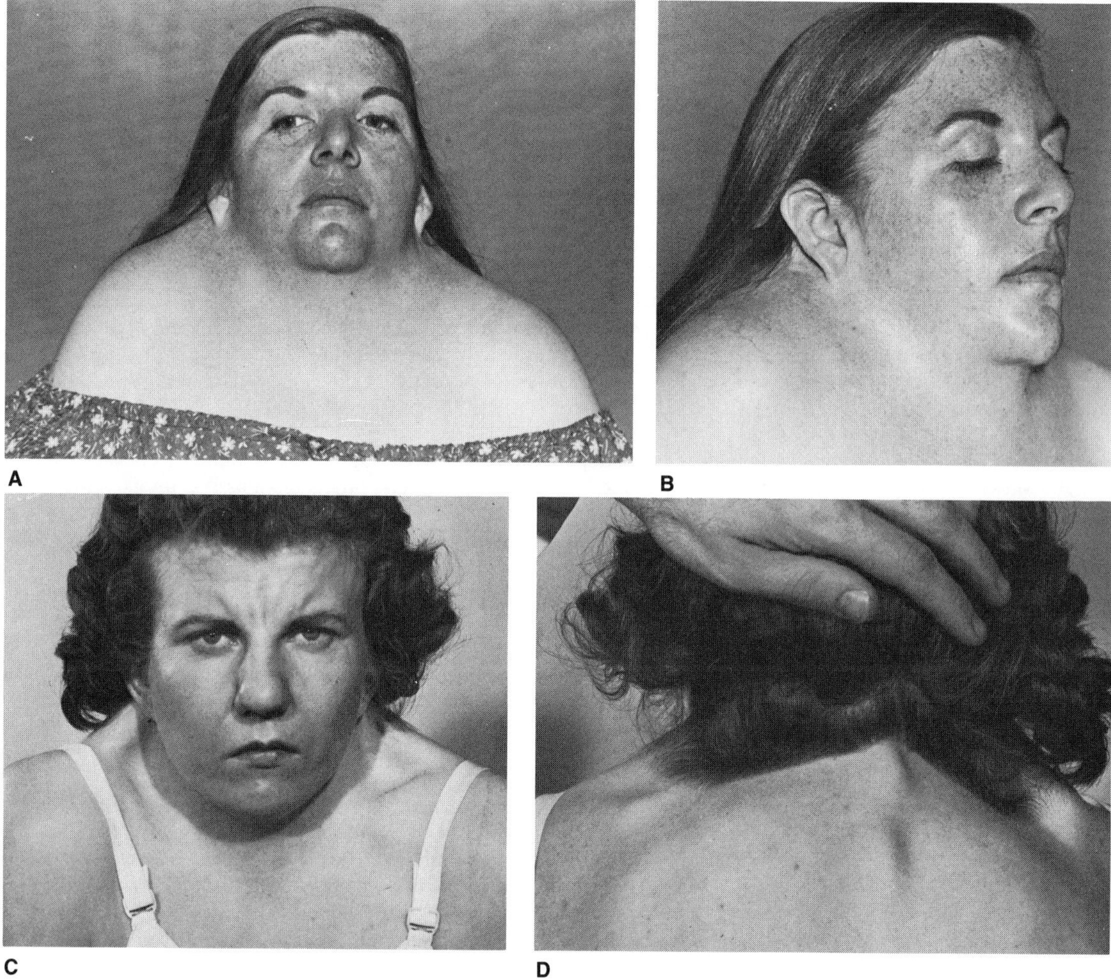


Fig. 28–13. *Klippel–Feil anomaly.* (A–D) Because of fusion of cervical vertebrae, the head appears to rest directly on the thorax, without an intervening neck. The pterygium-like structures are actually the trapezius muscles. Note pulled-down slanted auricles in B and low-set posterior hairline in D.

Table 28–1. Klippel–Feil. Associated abnormalities[a]

| Abnormalities | Percentage ($n = 160$) |
|---|---|
| Neurological abnormalities | 49 |
| Mirror movements | (16) |
| Paralysis and palsy | (9) |
| Astereognosis | (4) |
| Syringomyelia | (1) |
| Paresthesia | (4) |
| Mental retardation | (9) |
| Meningocele | (4) |
| Occipital encephalocele | (4) |
| Hydrocephaly | (2) |
| Other neurological problems | (16) |
| Hearing deficits | 24 |
| Speech defects | 17 |
| Ocular abnormalities | 21 |
| Cleft palate | 17 |
| Congenital heart defects | 9 |

[a]Based on C Helmi and S Pruzansky, *Cleft Palate J* **17**:65, 1980. Percentages are rounded off to nearest whole number. Percentages sum to greater than 100% because several abnormalities frequently occur in the same patient.

such as *Wildervanck syndrome,* MURCS association, etc. For further discussion, see *cervico-oculo-acoustic syndrome.*

### References (Klippel–Feil anomaly)

1. Baird PA, Lowry RB: Absent vagina and the Klippel–Feil anomaly. *Am J Obstet Gynecol* **118**:290–291, 1974.

2. Baird PA et al: Klippel–Feil syndrome: A study of mirror movements detected by electromyography. *Am J Dis Child* **113**:546–551, 1967.

3. Bauman GI: Absence of the cervical spine: Klippel–Feil syndrome. *JAMA* **98**:129–132, 1932.

3a. Blumel J et al: Partial and complete agenesis or malformation of the sacrum with associated anomalies. *J Bone Jt Surg* **41A**:497–518, 1959.

4. Brill CB et al: Isolation of the right subclavian artery with subclavian steal in a child with Klippel–Feil anomaly: An example of the subclavian artery supply disruption sequence. *Am J Med Genet* **26**:933–940, 1987.

5. Brown MW et al: The incidence of acquired and congenital fusions in the cervical spine. *Am J Roentgenol* **92**:1255–1259, 1964.

6. Cantu JM et al: Evidence for autosomal recessive inheritance of costovertebral dysplasia. *Clin Genet* **2**:149–154, 1971.

7. Chawla S, Bery K: Iniencephalus: Prenatal diagnosis. *Br J Radiol* **42**:29–33, 1969.

8. Chemke J et al: Absent ulna in the Klippel–Feil syndrome: An unusual associated malformation. *Clin Genet* **17**:167–170, 1980.

9. Cohney BC: The association of cleft palate with the Klippel–Feil syndrome. *Plast Reconstr Surg* **31**:179, 1963.

10. Daniilidis J et al: Stapes gusher and Klippel–Feil syndrome. *Laryngoscope* **88**:1178–1183, 1978.

11. Duncan PA: Embryologic pathogenesis of renal agenesis associated with cervical vertebral anomalies (Klippel–Feil phenotype). *Birth Defects* **13**(3D):91–101, 1977.

12. Da Silva EO: Autosomal recessive Klippel–Feil syndrome. *J Med Genet* **19**:130–134, 1982.

13. Erskine CA: An analysis of the Klippel–Feil syndrome. *Arch Pathol* **41**:269–281, 1946.

14. Everberg G et al: Wildervanck's syndrome: Klippel–Feil syndrome associated with deafness and retraction of the eyeball. *Br J Radiol* **36**:562–567, 1963.

15. Feil A: *L'absence et la diminution des vertèbres cervicales.* Thesis, Libraire Littéraire et Médicale, Paris, 1919.

16. Fragoso R et al: Frontonasal dysplasia in the Klippel–Feil syndrome: A new associated malformation. *Clin Genet* **22**:270–273, 1982.

17. Gienapp R von: Zur Erbbiologie der Klippel-Feilschen Krankheit. *Nervenarzt* **21**:74–81, 1950.

18. Gilmour JR: The essential identity of the Klippel–Feil syndrome and iniencephaly. *J Pathol Bacteriol* **53**:117–131, 1941.

19. Gjørup PA, Gjørup L: Klippel–Feil syndrome. *Dan Med Bull* **11**:50–53, 1964.

19a. Gorlin RJ: Nevoid basal-cell carcinoma syndrome. *Medicine* **66**:98–113, 1987.

20. Gray JW et al: Congenital fusion of the cervical vertebrae. *Surg Gynecol Obstet* **118**:373–385, 1964.

21. Gunderson CH, Solitaire GB: Mirror movements in patients with the Klippel–Feil syndrome. *Arch Neurol* **18**:675–679, 1968.

22. Gunderson CH et al: The Klippel–Feil syndrome: Genetic and clinical reevaluation of cervical fusion. *Medicine* **46**:491–512, 1967.

23. Hadley LA: Roentgenographic studies of the cervical spine. *Am J Roentgenol* **52**:173–195, 1944.

24. Heiner H: Über die Kombination von medianer Gaumenspalte und Klippel–Feil Syndrom. *Dtsch Stomatol* **10**:92–98, 1960.

25. Helmi C, Pruzansky S: Craniofacial and extracranial malformations in the Klippel–Feil syndrome. *Cleft Palate J* **17**:65–88, 1980.

25a. Hensinger RN et al: Klippel–Feil syndrome: A constellation of associated anomalies. *J Bone Jt Surg* **56**:1246–1253, 1974.

26. Jarcho S, Levin PM: Hereditary malformation of the vertebral bodies. *Bull Johns Hopkins Hosp* **62**:216–226, 1938.

27. Juberg RC, Gershank JJ: Cervical vertebral fusion (Klippel–Feil) syndrome with consanguineous parents. *J Med Genet* **13**:246–249, 1976.

28. Klippel M, Feil A: Un cas d'absence des vertèbres cervicales. *Nouv Iconogr Salpet* **25**:223–250, 1912.

29. Kreiborg S et al: Apert's and Crouzon's syndromes contrasted: Qualitative craniofacial x-ray findings, in *Craniofacial Surgery,* Marchac D (ed), Springer-Verlag, Berlin, 1987.

30. Luftman II, Weintraub S: Klippel–Feil syndrome in a full-term newborn infant. *NY State J Med* **51**:2035–2038, 1951.

31. McLay K, Maran AGD: Deafness and the Klippel–Feil syndrome. *J Laryngol* **83**:175–184, 1969.

32. Mecklenburg RS, Krueger PM: Extensive genitourinary anomalies associated with Klippel–Feil syndrome. *Am J Dis Child* **128**:92–93, 1974.

33. Moore WB et al: Genitourinary anomalies associated with Klippel–Feil syndrome. *J Bone Jt Surg* **57A**:355–357, 1975.

34. Morrison SG et al: Congenital brevicollis (Klippel–Feil syndrome). *Am J Dis Child* **115**:614–620, 1968.

35. Mosberg WH: The Klippel–Feil syndrome: Etiology and treatment of neurologic signs. *J Nerv Ment Dis* **117**:479–491, 1953.

36. Nagib MG et al: Neuroschisis of the cervical spinal cord in a patient with Klippel–Feil syndrome. *Neurosurg* **20**:629–631, 1987.

37. Nobel TP, Frawley JM: Klippel–Feil syndrome: Numerical reduction of cervical vertebrae. *Ann Surg* **82**:728–734, 1925.

38. Nora JJ et al: Klippel–Feil syndrome with congenital heart disease. *Am J Dis Child* **102**:858–864, 1961.

39. Nylen B, Wahlin A: Post-operative complications in pharyngeal flap surgery. *Cleft Palate J* **3**:347–356, 1966.

40. Palant DI, Carter BL: Klippel–Feil syndrome and deafness: A study with polytomography. *Am J Dis Child* **123**:218–221, 1972.

41. Park IJ, Jones HW Jr: A new syndrome in two unrelated females: Klippel–Feil deformity, conductive deafness and absent vagina. *Birth Defects* **7**(6):311–317, 1971.

42. Pochaczevsky R et al: Spondylothoracic dysplasia. *Radiology* **98**:53–58, 1971.

42a. Raas-Rothschild A et al: Klippel–Feil anomaly with sacral agenesis: An additional subtype, type IV. *J Craniofac Genet Dev Biol* **8**:297–301, 1988.

43. Regli F et al: Hereditary mirror movements. *Arch Neurol* **16**:620–623, 1967.

44. Schneider A et al: Iniencephaly and thoraco-lumbosacral meningomyelocele in a sibship. *Am J Hum Genet (Suppl)* **41**:A82, 1987.

45. Schwarze K: Zur Frage des Klippel–Feilschen Fehlers der Wirbelsäule. *Arch Orthop Unfallchir* **41**:47–63, 1941.

46. Shoul MJ, Ritvo M: Clinical and roentgenological manifestations of Klippel–Feil syndrome (congenital fusion of the cervical vertebrae brevicollis): Report of 8 additional cases and review of the literature. *Am J Roentgenol* **68**:369–385, 1952.

46a. Smith DF et al: Intrinsic defects in the fetal alcohol syndrome: Studies on 76 cases from British Columbia and the Yukon territory. *Neurobehav Toxicol Teratol* **3**:145–152, 1981.

47. Sommerfeld RM, Schweiger JW: Cleft palate with Klippel–Feil syndrome. *Oral Surg* **27**:737–739, 1969.

48. Stadnicki G, Rassumowski D: The association of cleft palate with the Klippel–Feil syndrome. *Oral Surg* **33**:335–340, 1972.

48a. Stewart EJ, O'Reilly BF: Klippel–Feil syndrome and conductive deafness. *J Laryngol Otol* **103**:947–949, 1989.

49. Tveter KJ, Kluge T: Cor triloculare biatrium associated with Klippel–Feil syndrome. *Acta Paediatr Scand* **54**:489–496, 1965.

50. Warkany J: *Congenital Malformations: Notes and Comments.* Year Book Medical Publishers, Chicago, 1971.

51. Weber V et al: Pathogenetische und genetische Fragen beim Klippel–Feil Syndrom-vergleichende Betrachtungen bei zwei Vettern. *Fortschr Roentgenstrahl* **119**:209–215, 1973.

52. Windle-Taylor PC et al: Ear deformities associated with the Klippel–Feil syndrome. *Ann Otol Rhinol Laryngol* **90**:210–216, 1981.

## Opitz trigonocephaly syndrome (C syndrome)

In 1969, Opitz and co-workers (6) reported a newly recognized syndrome in two sibs, naming it the "C syndrome of multiple congenital anomalies." The condition has subsequently been known as Opitz trigonocephaly syndrome (2,5). Characteristic features include trigonocephaly, unusual facial features, especially wide alveolar ridges, multiple frenula, limb defects, visceral anomalies, redundant skin, mental deficiency, and hypotonia.

To date, at least 15 cases have been recorded [1,2,5,6,8(Case 2),10]. It is unlikely that case 13 reported by Pfeiffer (7) also represents the disorder in two affected sibs. Normal chromosomes, normal parents with multipli-affected offspring, the equal sex ratio of affected individuals, and consanguineous matings support autosomal recessive inheritance.

**Clinical features.** The clinical findings are summarized in Table 28-2. Trigonocephaly, upslanting palpebral fissures, strabismus, hypoplastic nasal root, anomalous and posteriorly angulated ears, loose skin, and joint contractures and dislocations are common, being observed in over 70% (Figs. 28-14 to 28-16).

Wide alveolar ridges and a deep midline palatal furrow are also characteristic features in many cases. Cleft palate has been reported in one case (7). Other consistent extracranial findings include genital abnormalities and hemangiomas. Neurologic findings include hypotonia, poor sucking reflex, and severe mental retardation. Seizures have also been observed in some cases (1).

Approximately half of the patients die within the first year of life. The patient reported by Golabi et al (4) was alive at 3 years. Some abnormalities may change with age. Head size, for example, is normal at birth, but there is a tendency to develop microcephaly postnatally. In some cases, other sutures may be involved in addition to the metopic suture. Oberklaid and Danks (5) noted incomplete fusion of the sagittal suture. Flatz et al (2) observed craniosynostosis involving the coronal, sagittal, and lambdoid sutures.

Low-frequency findings include hypoplasia of middle and distal phalanges, and stenosis, renal cortical cysts, and pectus (1,2,5,6).

**Differential diagnosis and laboratory aids.** Patients with Opitz trigonocephaly or C syndrome should be studied especially carefully for possible cytogenetic abnormalities. In two instances, patients originally diagnosed as having C syndrome were found to have abnormal-

Table 28–2. Findings in the Opitz trigonocephaly syndrome (C syndrome)[a]

| Findings | Frequency |
|---|---|
| Performance | |
| Hypotonia | 13/13 |
| Mental deficiency | Constant[b] |
| Cranial | |
| Trigonocephaly | 13/13 |
| Postnatally developing microcephaly | Constant[b] |
| Cowlick | 7/13 |
| Facial | |
| Upslanting palpebral fissures | 11/13 |
| Epicanthic folds | 13/13 |
| Strabismus | 9/13 |
| Hypoplastic nasal root | 13/13 |
| Long philtrum | 10/13 |
| Wide mouth | 8/13 |
| Wide alveolar ridges | 11/13 |
| Deep midline palatal furrow | 10/13 |
| Multiple frenula | 5/13 |
| Micrognathia | 9/13 |
| Anomalous, posteriorly angulated ears | 12/13 |
| Neck | |
| Short neck | 11/13 |
| Limbs | |
| Polydactyly | 4/13 |
| Bridged palmar creases | 8/13 |
| Ulnar deviation of fingers | 4/13 |
| Short limbs | 5/13 |
| Hip dysplasia | 4/13 |
| Dislocated knees | 3/13 |
| Equinovarus or valgus deformity | 5/13 |
| Contractures | 6/13 |
| Skin | |
| Redundant skin | 11/13 |
| Hemangiomas | 9/13 |
| Deep sacral dimple | 5/13 |
| Cardiovascular | |
| Congenital heart anomaly | 10/13 |
| Genital | |
| Cryptorchidism | 6/6 |

[a]From MM Cohen Jr (ed): *Craniosynostosis: Diagnosis, Evaluation and Management,* Raven Press, New York, 1986.
[b]Approximately one-half of all patients die within the first year of life. Survivors develop microcephaly and mental deficiency.

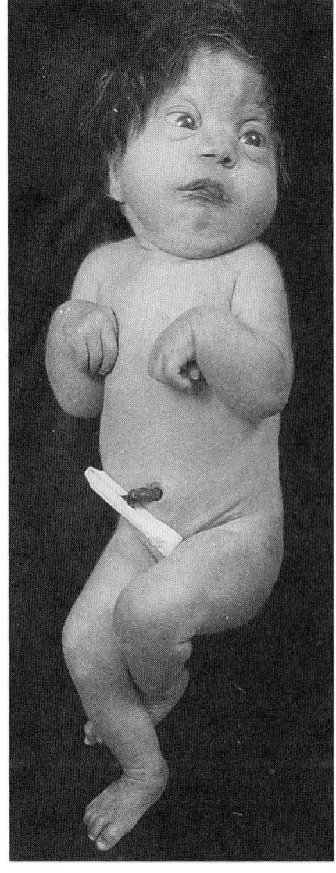

Fig. 28–14. *Opitz trigonocephaly syndrome.* Trigonocephaly, long upper lip, contractures at wrist. (From *F Oberklaid* and *DM Danks,* Am J Dis Child **129:**1348, 1975.)

ities involving chromosome 3. One of the two patients reported by Preus et al [8 (case 1)] was later found to have dup(3)(q23→ter); del(p25→pter) (9) and a patient studied by Sargent et al (10) had 46,XY,del(3)(pter→q27:). Preus et al (9) suggested that at least one other C syndrome patient from the literature had a phenotype compatible with dup(3q) syndrome. The C syndrome patient reported by Golabi et al (4) had 47/XXY karyotype.

Differential diagnosis also includes other chromosomal syndromes with trigonocephaly such as *del(9p), del(11q),* and *dup(13q)* (10). Monogenic syndromes with trigonocephaly include *Frydman trigonocephaly syndrome* (3) and *Say–Meyer trigonocephaly syndrome* (11).

### References [Opitz trigonocephaly syndrome (C syndrome)]

1. Antley RM et al: Further delineation of the C (trigonocephaly) syndrome. *Am J Med Genet* **9**:147–163, 1981.
2. Flatz SD et al: Opitz trigonocephaly syndrome: Report of two cases. *Eur J Pediatr* **141**:183–185, 1984.
3. Frydman M et al: Trigonocephaly: A new familial syndrome. *Am J Med Genet* **18**:55–59, 1984.
4. Golabi M et al: C syndrome: Report of a new case with disproportionate short stature, long survival and karyotype of 47,XXY. *David W. Smith Workshop on Malformations and Morphogenesis,* Burlington, Vermont, August 11–14, 1986.
5. Oberklaid F, Danks DM: The Opitz trigonocephaly syndrome. *Am J Dis Child* **129**:1348–1349, 1975.
6. Opitz JM et al: The C syndrome of multiple congenital anomalies. *Birth Defects* **5**(2):161–166, 1969.
7. Pfeiffer RA: Associated deformities of the head and hands. *Birth Defects* **5**(3):18–34, 1969.
8. Preus M et al: The C syndrome. *Birth Defects* **11**(2):58–62, 1975.
9. Preus M et al: Diagnosis of chromosome 3 duplication q23→qter, deletion p25→pter in a patient with C (trigonocephaly) syndrome. *Am J Med Genet* **23**:935–943, 1986.
10. Sargent C et al: Trigonocephaly and the Opitz C syndrome. *J Med Genet* **22**:39–45, 1985.
11. Say B, Meyer J: Familial trigonocephaly associated with short stature and developmental delay. *Am J Dis Child* **135**:711–712, 1981.

## Smith–Lemli–Opitz syndromes

The syndrome is characterized by microcephaly, growth and mental retardation, soft-tissue syndactyly of the second and third toes, and genital abnormalities. Since its first description in 1964 (47), over 120 patients have been documented (1,5,19,20,26). The genital abnormalities in males are probably responsible for the preponderance of reported male patients. A comprehensive bibliography is that of Opitz and Howe (38).

The relative nonspecificity of cardinal features and lack of a biochemical marker make this relatively difficult to diagnosis with certainty. Various chromosomal abnormalities have recently been reported (6,16).

Genetic heterogeneity, the classic SLO Type I and the more severe SLO Type II, has been proposed. Both have autosomal recessive inheritance and it is possible that Type II represents an allelic form (2,5,40). We present them as separate disorders.

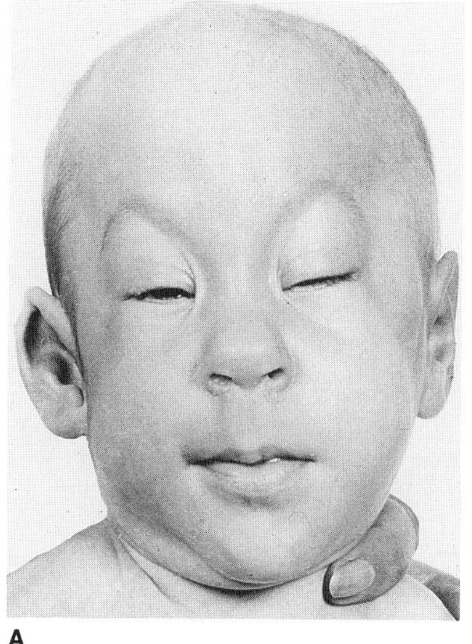

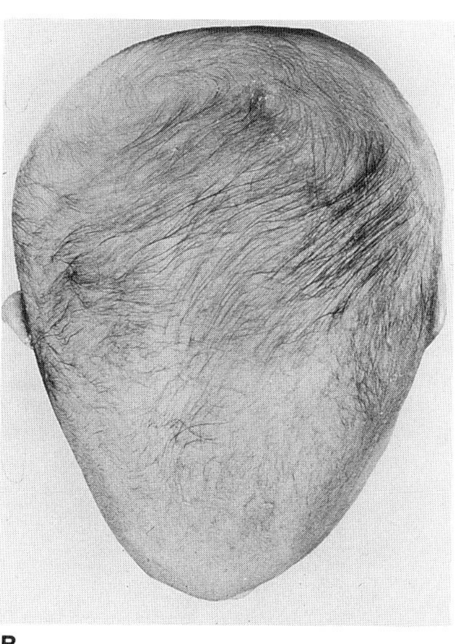

Fig. 28–15. *Opitz trigonocephaly syndrome.* (A) Striking facies with metopic ridging, small head circumference, epicanthal folds, hypoplastic nose with anteverted nares. (B) Trigonocephaly. (From *C Sargent et al,* J Med Genet **22:**39, 1985.)

**SLO Type I syndrome.** Decreased fetal movements and breech presentation may be encountered. Birthweight is less than 2500 g in approximately one-third of cases. Prematurity has been noted in 25%. Neonatal feeding difficulties, irritability, and failure to thrive are common. Prolonged high-pitched screaming has been described in 20%. Short stature is usual. Survival to adulthood is rare (14).

Facies. The facial appearance has been somewhat variable. Mild microcephaly (95%), ptosis (85%), strabismus (40%), epicanthal folds (40%), wide nasal bridge (50%), anteverted nares (75%), low-set and/or retroverted or outstanding pinnae (60%), micrognathia (80%), and short neck are reported (13,37,47) (Fig. 28-17A,B). Facial capillary hemangiomas are common.

Central nervous system. Microcephaly is found in over 90% with 40% having abnormal skull shape. Patients have a wide range of developmental defects of the brain (ventricular dilatation, seizures, partial agenesis of cerebellar vermis and/or corpus callosum, hypoplasia of frontal lobes), generally resulting in severe retardation (9,18,37), but a few have been reported with borderline normal intelligence (34). During the neonatal period, hypotonia is noted in about 50% while, later in life, hypertonia is seen in 30%. Reflexes are brisk.

Eyes. Ptosis (60%), convergent strabismus (40%), and epicanthus (40%) have been documented (18,22). Cataracts are noted in 10% (5,26). For miscellaneous eye abnormalities, see Gold and Pfaffenbach (21).

Limbs. Postaxial polydactyly occurs in about 25% (5,25,26). Soft-tissue syndactyly of the second and third toes has been noted in about 65% (Fig. 28-18). Positional foot abnormalities are common: metatarsus adductus, pes equinovarus, and metarsus varus. Other findings have included clinodactyly of fingers, hallucal hammertoe, proximally placed thumbs, and long tapered fingers (37).

Genitalia. In males, the genitalia range from relative normality to genital ambiguity (perineoscrotal hypospadias with perineal urethral opening); most common are bilateral cryptorchidism (50%) and hy-

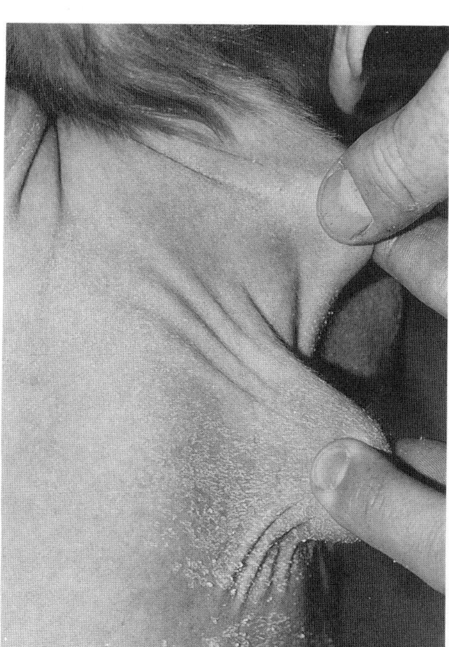

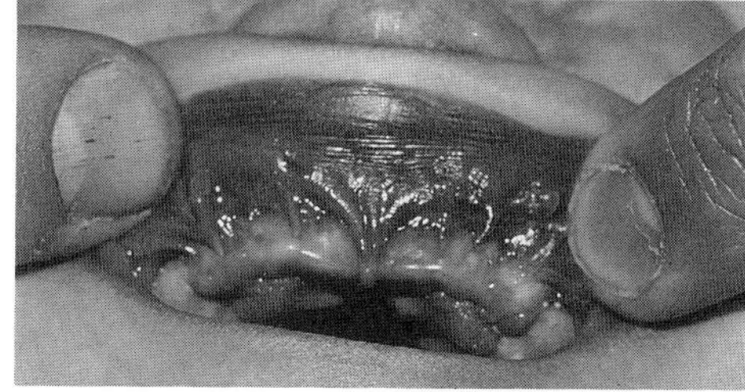

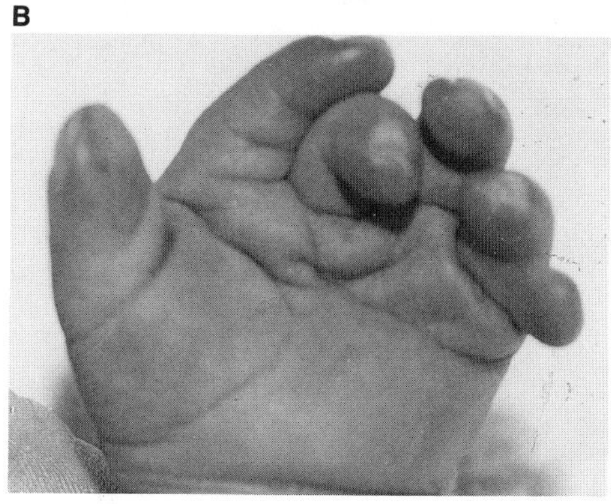

Fig. 28–16. *Opitz trigonocephaly syndrome.* (A) Redundant skin. (B) Multiple frenula. (C) Postaxial polydactyly. [From *JM Opitz et al,* Birth Defects **5**(2):161, 1969.]

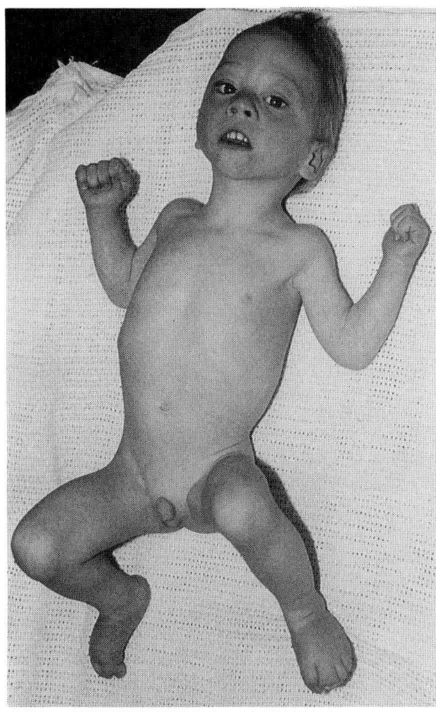

A

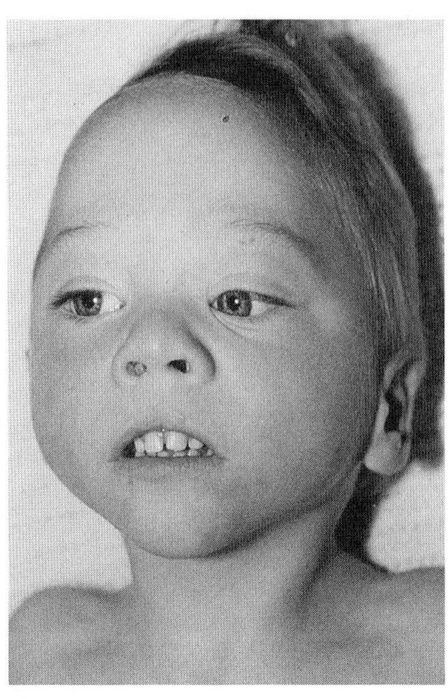

B

Fig. 28–17. *Smith–Lemli–Opitz syndrome, Type I.* (A,B) Note microcephaly, small upturned nose, epicanthal folds, abnormal genitalia, soft tissue syndactyly of second or third toes. (Courtesy of *D Donnai,* Manchester, England.)

pospadias (50%). Microphallus and/or chordee have been documented in 20% (Fig. 28-19).

Miscellaneous. Various congenital heart anomalies (PDA, VSD, tetralogy of Fallot, aberrant right subclavian artery, ASD) have been reported in about 20% (9,10,13,37,43). Minor renal changes have been reported in 45%: hypoplasia, cystic and dysplastic alterations. Incomplete lobation of the lungs has been described (28,36).

Fig. 28–18. *Smith–Lemli–Opitz syndrome, Type I.* Soft-tissue syndactyly of second and third toes. Note short fifth toes with tibial clinodactyly. (From *CG Judge et al,* Med J Aust **2:**145, 1971.)

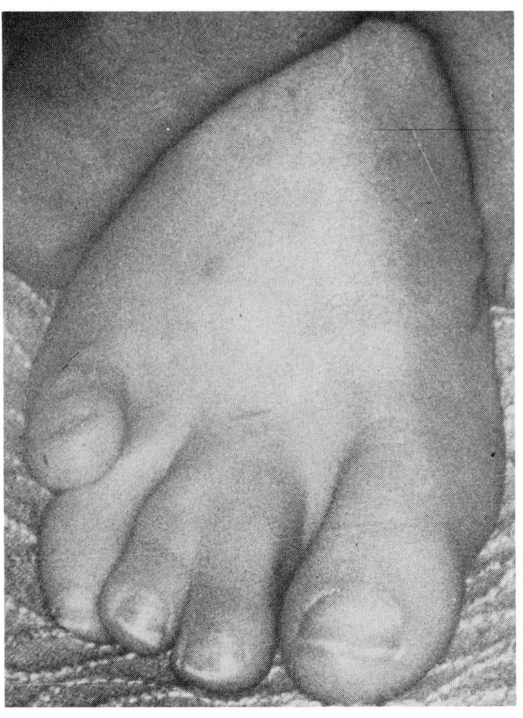

Oral manifestations. Cleft palate has been found in about 40% (5) and broad maxillary alveolar ridges in over 60% (26). The tongue tends to be small.

**SLO Type II syndrome (Lowry–Miller–MacLean syndrome, acrodysgenital dwarfism).** Donnai et al (15) and Curry et al (12) drew attention to infants who share common findings with the syndrome as originally described, but whose features individually and as a group are distinctive with marked failure of masculinization of the external genitals in XY males or even sex reversal, multiple visceral abnormalities, and frequent neonatal lethality. The term ''lethal acrodysgenital dwarfism'' has been applied to this disorder (31).

At least 50 cases of the severe form have been documented

Fig. 28–19. *Smith–Lemli–Opitz syndrome, Type I.* Hypospadias and hypoplastic scrotum. (From *CG Judge et al,* Med J Aust **2:**145, 1971.)

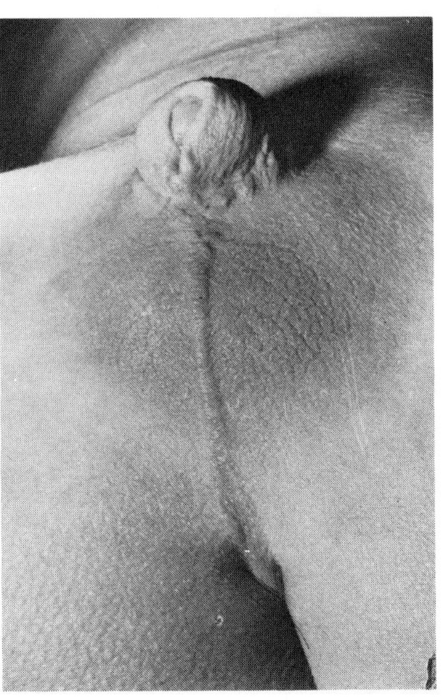

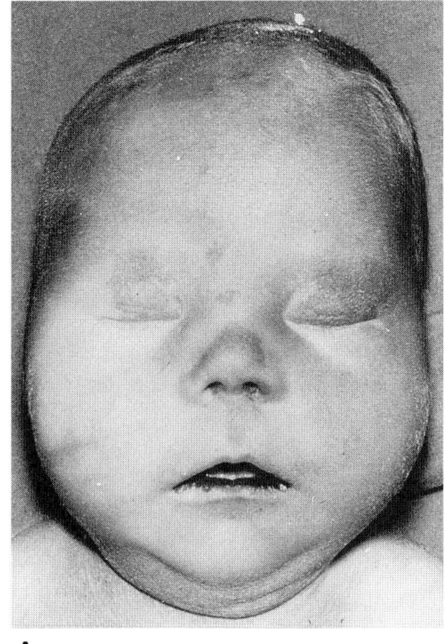


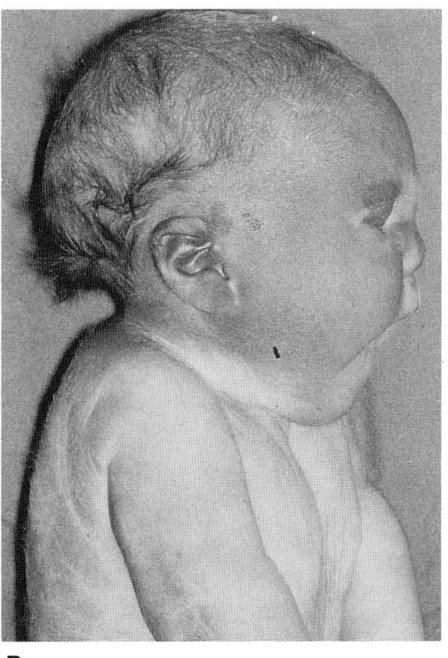


Fig. 28–20. *Smith–Lemli–Opitz syndrome, Type II.* (A,B) Facial angiomas, short nose, micrognathia, short neck. (Courtesy of *D Donnai,* Manchester, England.)

(1,5,11,12,15,17,23,27–29,31,33,38,39,41,49). SLO Type II syndrome also appears to have autosomal recessive inheritance (12,28, 31,33,37,41).

Intrauterine growth retardation (birthweight $<$ 2500 g), oligohydramnios, and lack of fetal movement have been documented by ultrasound. Breech presentation has occurred in approximately 50% and birth asphyxia is extremely common. Mild prenatal growth deficiency was seen in about half of the patients (12). Three of 41 reported patients died as a consequence of pulmonary hypoplasia secondary to renal agenesis or cystic dysplasia. Feeding difficulties and irritability are consistent findings. Almost all have died in the neonatal period with failure to thrive and hepatic dysfunction.

The occasional findings of renal cysts, large adrenals, and hepatic dysfunction in Type II are suggestive of a possible peroxisomal defect. However, investigations of long chain fatty acid levels and plasmalogen synthesis in three patients revealed no abnormalities (43). Further investigations for a metabolic defect in this syndrome seem warranted.

**Facies.** The face tends to be round with broad nasal bridge, ptosis, epicanthal folds, anteverted nares, and, occasionally, a prominent nevus flammeus over the glabella (Fig. 28-20A,B). Short neck with redundant skin folds is common. Micrognathia has been a constant feature.

**Eyes.** Cataracts have been found in 45% (12,15,29,31).

**Central nervous system.** Head circumference has been less than −2 SD for gestational age in over 50%. Hydrocephaly and agenesis of the corpus callosum have been noted in several patients (9).

**Limbs.** Postaxial polydactyly of the hands and/or feet is more common than in Type I, being found in 75% (12,31). Soft-tissue 2–3 syndactyly of the toes has been reported in about 80%. The thumbs are often short. There are short proximal limbs without evidence of primary bone dysplasia (Fig. 28-21A,B).

A short first metacarpal with secondary short thumb has been seen. A peculiar short first metatarsal and valgus foot positioning are frequent. About 50% of the infants have dislocated hips (Fig. 28–21).

Transverse palmar creases are frequent (31).

**Genitalia.** Severe genital ambiguity (male pseudohermaphroditism) is characteristic in males with complete "sex reversal" being relatively common. No male has had normal genitalia; the majority have been assigned female gender and, among the rest, sex assignment has been unclear. Features noted in XY males have included hypoplastic labia majora, absent labia, scrotalization of the labia, prominent clitoris, third degree hypospadias, micropenis, and bifid scrotum (Fig. 28-22A,B). Autopsies in severely ambiguous XY infants have revealed intraabdominal or inguinal testes, and, in most, Müllerian duct remnants. A few XY infants have had hypoplastic vagina and uterus (12,31). Occasionally, males have normal female genitalia (23,29,39).

**Other findings.** At autopsy, unusual findings have included pancreatic islet cell "giant cell" hyperplasia, unilobated lungs, pyloric stenosis, and Hirschsprung's disease (23,31,39). Renal cysts (15,23,27,28,32,33) are frequent and hypoplastic kidneys are relatively common. Cardiac anomalies of various types, most often VSD, ASD, and PDA, have been found in nearly all cases (23,29,39,45).

Fig. 28–21. *Smith–Lemli–Opitz syndrome, Type II.* Valgus deformity of feet, short halluces, polysyndactyly. (Courtesy of *D Donnai,* Manchester, England.)

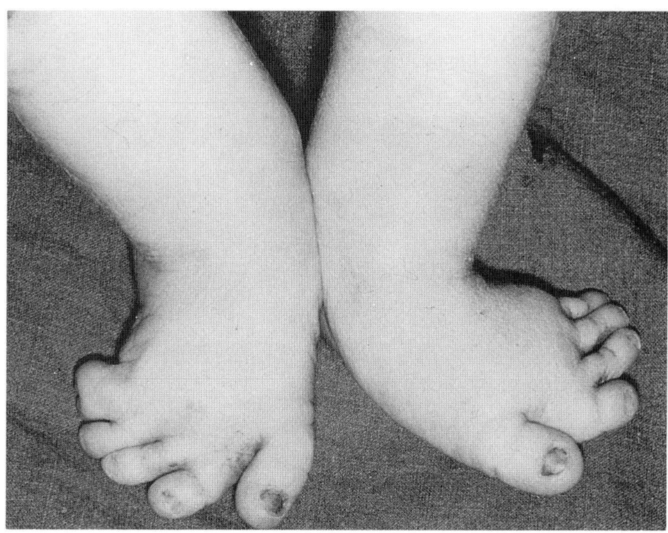

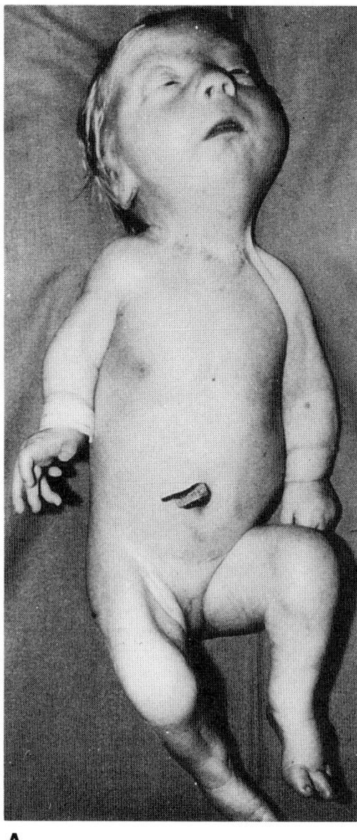

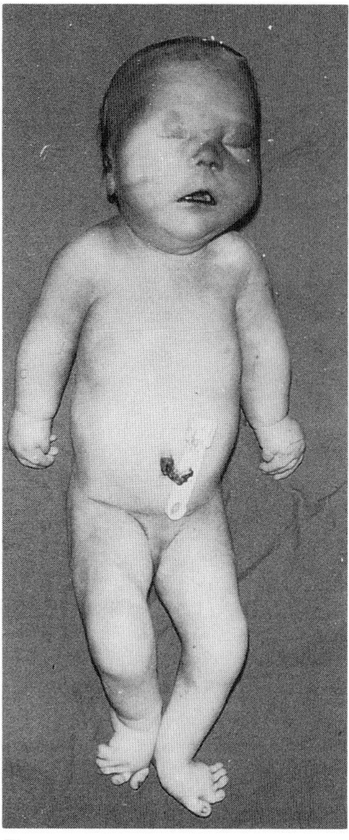

A B

Fig. 28–22. *Smith–Lemli–Opitz syndrome, Type II.* (A,B) Unusual facies, superficial angiomas, limb shortening, inguinal skin creases, postaxial polydactyly. Note apparent female genitalia in 46,XY patients. (A from *CJR Curry et al,* Am J Med Genet **28:**45, 1987. B courtesy of *D Donnai,* Manchester, England.)

Oral manifestations. Oral findings have included short tongue and redundant sublingual tissue (Fig. 28-23). Tongue cysts have been present in a few cases (31). The alveolar ridges may be broad. Cleft palate has been documented in at least 60% (5,12,23,28,31,39).

Differential diagnosis. Differential diagnosis of the Smith–Lemli–Opitz syndrome(s) can be quite extensive. A chromosome anomaly should clearly be ruled out. In Type II patients, major diagnostic confusion can exist with *Meckel syndrome.* Some patients initially diagnosed and published as examples of the Meckel syndrome likely represent Type II cases (24). Sibs published by Casamassima et al (7) appear to have features of both SLO and Meckel syndromes. There are many similarities to *Pallister–Hall syndrome.* However, the presence of a hypothalamic hamartoblastoma appears to be unique in the latter syndrome. The *hydrolethalus syndrome* may present occasional diagnostic difficulty in the hydrocephalic patient. Some infants with so-called Ullrich–Feichtiger syndrome (24,30,36) may have SLO, Type II syndrome, *trisomy 13,* or *Meckel syndrome.* In Meckel syndrome, there is a characteristic cystic kidney dysplasia. SLO, Type II syndrome should be distinguished from other syndromes with sex reversal and multiple congenital anomalies *(genito-palato-cardiac syndrome)* such as those described by Beemer and Ertbruggen (3), Robinow et al (42), Silengo et al (46), and Greenberg et al (22). In SLO, Type II, the testes are present; in *Gardner–Silengo–Wachtel syndrome,* there is gonadal dysgenesis. We are uncertain how to classify the disorder reported by Rutledge et al (44).

Fig. 28–23. *Smith–Lemli–Opitz syndrome, Type II.* Palatal cleft, small tongue, redundant sublingual tissue. (From *CJR Curry* et al, Am J Med Genet **28:**45, 1987.)

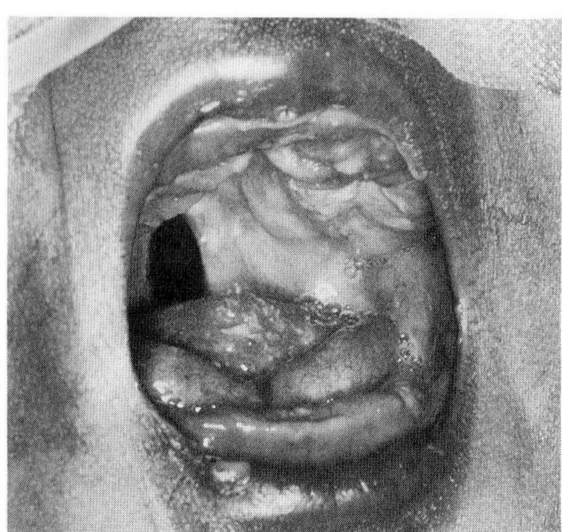

Laboratory aids. Prenatal diagnosis in SLO, Type II has not been accomplished to date but should be feasible. Amniocentesis can be utilized for fetal sexing. Ultrasonography can also be useful in fetal sexing as well as evaluation of cardiac anatomy, fetal growth and fetal movement, and assessment of amniotic fluid volume. Since Donnai et al (15) and Curry et al (12) have reported finding low estriols in late pregnancy, further prenatal investigation of adrenal status in at-risk pregnancies may be of interest.

### References (Smith–Lemli–Opitz syndromes)

1. Akl KF et al: The Smith–Lemli–Opitz syndrome. *Clin Pediatr* **16**:665–668, 1977.
2. Baraitser M, Thompson E: Nosology of Smith–Lemli–Opitz syndrome. *Am J Med Genet* **28**:733–734, 1987.
3. Beemer FA, Ertbruggen IV: Peculiar facial appearance, hydrocephalus, double-outlet right ventricle, genital anomalies and dense bones with lethal outcome. *Am J Med Genet* **19**:391–394, 1984.
4. Belmont JW et al: Two cases of severe lethal Smith–Lemli–Opitz syndrome. *Am J Med Genet* **16**:65–67, 1987.
5. Bialer MG et al: Female external genitalia and Müllerian duct derivatives in a 46,XY infant with the Smith–Lemli–Opitz syndrome. *Am J Med Genet* **28**:723–731, 1987.
6. Bundey S, Smyth HG: Three sisters with the Smith–Lemli–Opitz syndrome. *J Ment Defic Res* **18**:51, 1974.
7. Casamassima AC et al: A new syndrome with features of the Smith–Lemli–Opitz and Meckel–Gruber syndromes in a sibship with cerebellar defects. *Am J Med Genet* **26**:321–336, 1987.
8. Chakanovskis JE, Sutherland GR: The Smith–Lemli–Opitz syndrome in a profoundly retarded epileptic boy. *J Ment Def Res* **15**:153–162, 1971.
9. Cherstvoy ED et al: The pathological anatomy of the Smith–Lemli–Opitz syndrome. *Clin Genet* **7**:382–387, 1975.
10. Cotlier E, Rice P: Cataracts in the Smith–Lemli–Opitz syndrome. *Am J Ophthalmol* **72**:955–959, 1971.
11. Cruveiller JM et al: Nanisme de Smith–Lemli–Opitz: A propos de quatre observations. *Ann Pédiatr* **24**:843–851, 1977 (Cases 1,4).
12. Curry CJR et al: Smith–Lemli–Opitz syndrome Type II: Multiple congenital anomalies with male pseudohermaphroditism and frequent early lethality. *Am J Med Genet* **28**:45–57, 1987.
13. Dallaire L: Syndrome of retardation with urogenital and skeletal anomalies (Smith–Lemli–Opitz syndrome): Clinical features, mode of inheritance. *J Med Genet* **6**:113–120, 1969.
14. Deaton JG, Mendoza LD: Smith–Lemli–Opitz syndrome in a 23-year-old man. *Arch Intern Med* **132**:422–426, 1973.
15. Donnai D et al: The lethal multiple congenital anomaly syndrome of polydactyly, sex reversal, renal hypoplasia and unilobular lungs. *J Med Genet* **23**:64–71, 1986.
16. Donnenfeld A et al: A brief clinical report of an unbalanced translocation masquerading as Smith–Lemli–Opitz syndrome. Unpublished.
17. Dzarlieva R et al: The Smith–Lemli–Opitz syndrome in a five months old child. *God Ab Med Fak Skopie* **23**:253–261, 1977.
18. Fierro M et al: Smith–Lemli–Opitz syndrome: Neuropathological and ophthalmological observation. *Dev Med Child Neurol* **19**:57–62, 1977.
19. Freeman RA, Baum JL: Postlenticular membrane associated with Smith–Lemli–Opitz syndrome. *Am J Ophthalmol* **87**:675–677, 1979.
20. Gill DG, Blair AL: Smith–Lemli–Opitz syndrome: Two further cases. *J Irish Med Assoc* **68**:33–36, 1975.

21. Gold JD, Pfaffenbach DD: Ocular abnormalities in Smith–Lemli–Opitz syndrome. *J Pediatr Ophthalmol* **12**:228–234, 1975.

22. Greenberg F et al: The Gardner–Silengo–Wachtel or genito-palato-cardiac syndrome: Male pseudohermaphroditism with micrognathia, cleft palate, and conotruncal cardiac defect. *Am J Med Genet* **26**:59–64, 1987.

23. Greene C et al: The Smith–Lemli–Opitz syndrome in two 46,XY infants with female external genitalia. *Clin Genet* **25**:366–372, 1984.

24. Hövels O, Mullereisert F: Der "Typus Rostockiensis" als Charakteristisches Kombinationsbild multipler Missbildungen (Polydaktylie, Hypospadie, Kryptorchismus and Mikrognathie). *Z Kinderheilkd* **77**:454–467, 1955.

25. Jeanty P et al: Smith–Lemli–Opitz syndrome without failure to thrive. *Acta Paediatr Belg* **30**:175–178, 1977.

26. Johnson VP: Smith–Lemli–Opitz syndrome: Review and report of two affected siblings. *Z Kinderheilkd* **119**:221–234, 1975.

27. Kim EH et al: Smith–Lemli–Opitz syndrome associated with Hirschsprung disease, 46,XY female karyotype and total anomalous pulmonary venous drainage. *J Pediatr* **156**:861, 1985.

28. Kohler HG: Familial neonatally lethal syndrome of hypoplastic left heart, absent pulmonary lobation, polydactyly, and talipes, probably Smith–Lemli–Opitz (RSH) syndrome. *Am J Med Genet* **14**:423–428, 1983.

29. Kretzer FL: Ocular manifestations of the Smith–Lemli–Opitz syndrome. *Arch Ophthalmol* **99**:2000–2006, 1981.

30. Kunze J: Das Ullrich-Feichtiger-Syndrom. *Arch Kinderheilkd* **179**:182–186, 1969.

31. Le Merrer M et al: Lethal acrodysgenital dwarfism: A severe lethal condition resembling Smith–Lemli–Opitz syndrome. *J Med Genet* **25**:8–95, 1988.

32. Lipson A, Hayes A: Smith–Lemli–Opitz syndrome and Hirschsprung disease. *J Pediatr* **105**:177, 1984.

33. Lowry RB et al: Micrognathia, polydactyly and cleft palate. *J Pediatr* **72**:859–861, 1968.

34. Lowry RB, Yong SL: Borderline normal intelligence in the Smith–Lemli–Opitz (RSH) syndrome. *Am J Med Genet* **5**:137–143, 1980.

35. Meinecke P et al: Smith–Lemli–Opitz syndrome. *Am J Med Genet* **28**:735–739, 1987.

36. Meyer-Schwickerath G, Grüterich E: Mikrophthalmus mit Dyskranie und Dysphalangie. *Acta Genet Stat Med* **7**:277–279, 1957.

37. Opitz JM: The RSH syndrome. *Birth Defects* **5**(2):43–52, 1969.

38. Opitz JM, Howe JJ: The Meckel syndrome (dysencephalia splanchnocystica, the Gruber syndrome). *Birth Defects* **5**(2):167–179, 1969.

39. Patterson K et al: Hirschsprung disease in a 46,XY phenotypic infant girl with Smith–Lemli–Opitz syndrome. *J Pediatr* **103**:425–427, 1983.

40. Penchazadeh VB: The nosology of the Smith–Lemli–Opitz syndrome. *Am J Med Genet* **28**:719–721, 1987.

41. Pfeiffer RA, Slavaykoff H: Gibt es ein Syndrom nach Ullrich und Feichtiger? *Klin Pädiatr* **187**:176–180, 1975.

42. Robinow M et al: Costal dysgenesis and male pseudohermaphroditism: A new syndrome? *J Clin Dysmorphol* **2**:23–26, 1984.

43. Robinson CD et al: Smith–Lemli–Opitz syndrome with cardiovascular abnormality. *Pediatrics* **47**:844–847, 1971.

44. Rutledge JC et al: A "new" lethal multiple congenital anomaly syndrome: Joint contractures, cerebellar hypoplasia, renal hypoplasia, urogenital anomalies, tongue cysts, shortness of limbs, eye abnormalities, defects of the heart, gallbladder agenesis and ear malformations. *Am J Med Genet* **19**:255–264, 1984.

45. Scarborough PR et al: An additional case of Smith–Lemli–Opitz syndrome in a 46,XY infant with female external genitalia. *J Med Genet* **23**:174–175, 1986.

46. Silengo M et al: A 46,XY infant with uterus, dysgenetic gonads and multiple anomalies. *Hum Genet* **25**:65–68, 1974.

47. Smith DW et al: A newly recognized syndrome of multiple congenital anomalies. *J Pediatr* **64**:210–217, 1964.

48. Wilson G et al: Investigation of peroxisome function in three patients with Smith–Lemli–Opitz syndrome. Personal communication, 1986.

49. Zizka J: Intestinal aganglionosis in the Smith–Lemli–Opitz syndrome. *Acta Paediatr Scand* **72**:141–143, 1983.

# Chapter 29
# Other Miscellaneous Syndromes

## Acro-renal-mandibular syndrome

Halal et al (1) reported two female sibs whose parents were consanguineous. Both exhibited severe split hand and foot malformations, renal (polycystic kidneys, renal agenesis, and absent ureters) and genital (septate uterus, unicornuate uterus, and single tube) abnormalities and very severe micrognathia (Fig. 29-1A,B). Although split hand and foot occurs in the classic acro-renal syndrome, no patient has had micrognathia as severe as that observed in Halal's patients. Moreover, all known instances of acro-renal syndrome have been isolated males with no history of parental consanguinity. They have had unilateral renal agenesis, crossed renal ectopy, hypospadias, and/or cryptorchidism.

### Reference (Acro-renal-mandibular syndrome)


1. Halal F et al: Acro-renal-mandibular syndrome. *Am J Med Genet* **5**:277–284, 1980.


## Agenesis of salivary and lacrimal glands

There are a considerable number of case reports in which major salivary glands and lacrimal glands have been reported to be aplastic or severely hypoplastic. Duct orifices are often missing, but ducts may be present (10). The lacrimal points may be missing or atretic (1,2,4,5,7,14).

The resultant lack of secretions causes the oral and conjunctival mucosae to be dry. The lips are often cracked. The teeth decay and fracture at the neck. Mastication is extremely difficult. There are several reports of the disorder affecting two or more generations (3,4,9,11,16).

Early cases of isolated absence of glands have been reviewed (11,14,15). There are also numerous reports of dominantly inherited isolated deficiency of lacrimal glands or atresia of the lacrimal puncta. The literature has been reviewed by Caccamise and Townes (3) and McDonald et al (6). Radionuclide dacryocystography has been used in diagnosis (13).

### References (Agenesis of salivary and lacrimal glands)


1. Benedetti L: Un raro caso de xerostomia congenitale. *Riv Ital Stomatol* **5**:224–225, 1936.
2. Blackmar FB: Congenital atresia of all lacrimal puncta with absence of salivary glands. *Am J Ophthalmol* **8**:139–140, 1925.
3. Caccamise WC, Townes PL: Congenital absence of lacrimal puncta associated with alacrima and aptyalism. *Am J Ophthalmol* **89**:62–65, 1980.
4. Cortada X et al: Alacrima-aptyalism: An autosomal dominant condition. *Am J Hum Genet* **39**:57A, 1986.
5. Losch PK, Weisberger D: High caries susceptibility in diminished salivation. *Am J Orthodont* **26**:1102–1104, 1940.
6. McDonald FG et al: Salivary gland aplasia—an ectodermal disorder. *J Oral Pathol* (in press).
7. Mutch JR: Congenital absence of weeping associated with absence of salivation. *Br J Ophthalmol* **28**:333–334, 1944.
8. Raison J: Absence congénitale des parotides. *Rev Stomatol (Paris)* **27**:340, 1925.
9. Ramsay WR: A case of hereditary congenital absence of the salivary glands. *Am J Dis Child* **28**:440, 1924.
10. Sharp HS: A case of absence of salivary secretion. *J Laryngol Otol* **52**:177–178, 1937.
11. Smith NJD, Smith PB: Congenital absence of major salivary glands. *Br Dent J* **142**:259–260, 1977.
12. Steggerda FR: Observations on the water intake in an adult man with dysfunctioning salivary glands. *Am J Physiol* **132**:517–521, 1941.
13. Sucupira MS et al: Salivary gland imaging and radionuclide dacryocys-


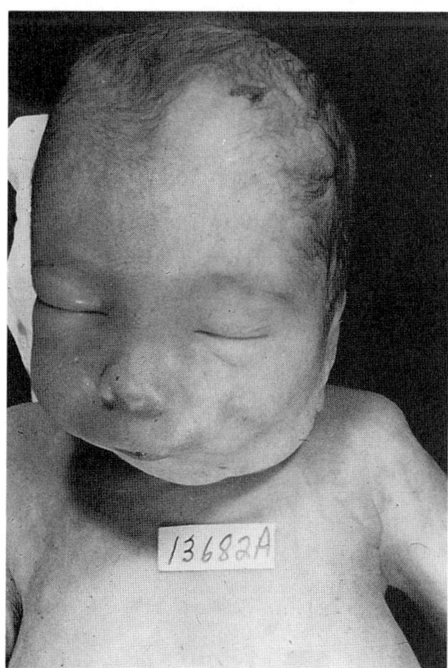


A

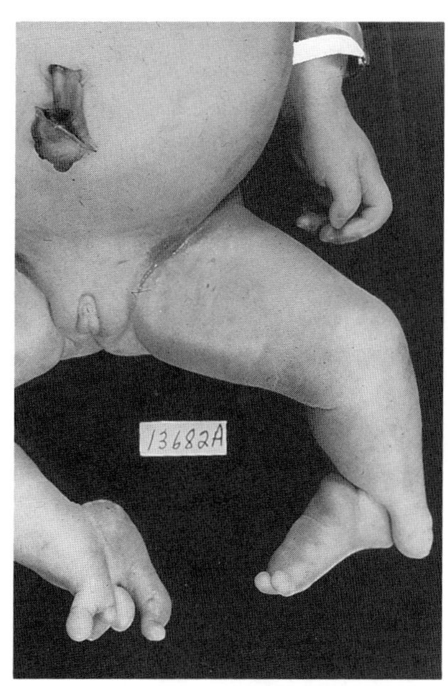


B

Fig. 29–1. *Acro-renal-mandibular syndrome.* (A) Note severe mandibular hypoplasia and mild Potter facies. (B) Bilateral split-foot anomaly. Cleft involves lower third of leg. Note hypoplastic toes. Sib was similarly affected. (From *F Halal et al,* Am J Med Genet **5:**277, 1980.)

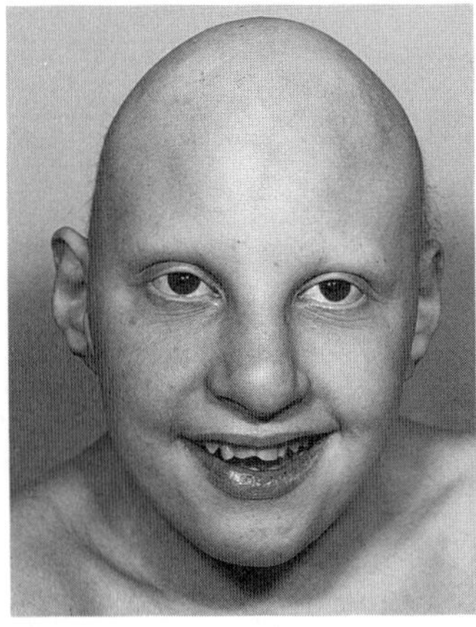

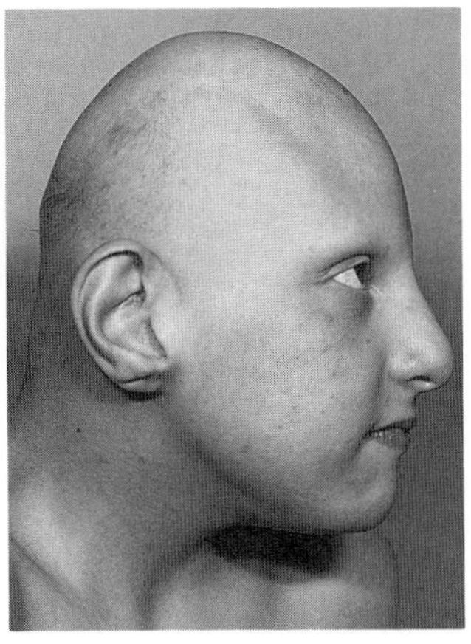

A B

Fig. 29–2. *Atrichia, somatic and mental retardation, and skeletal anomalies.* (A,B) Microbrachycephaly, baldness, absence of eyebrows and lashes, large prominent nose, short philtrum, relatively large dysmorphic pinnae. (From *A Schinzel*, Helv Paediatr Acta **35:**243, 1980.)

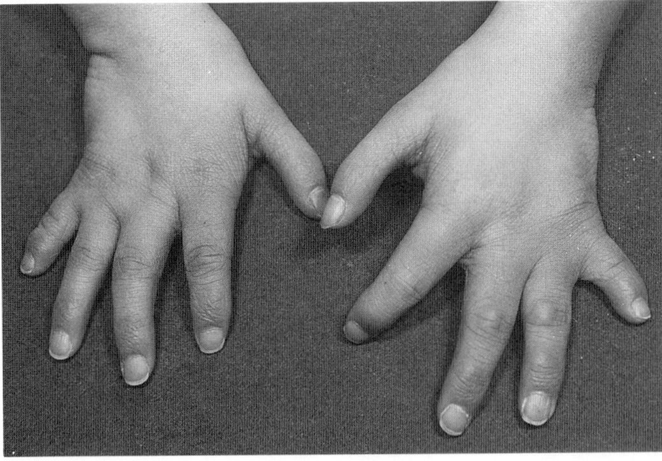

Fig. 29–3. *Atrichia, somatic and mental retardation, and skeletal anomalies.* Short index and little fingers. (From *A Schinzel,* Helv Paediatr Acta **35:**243, 1980.)

tography in agenesis of salivary glands. *Arch Otolaryngol* **109**:197–198, 1983.

14. Vogel C, Reichart P: Aplasie der Glandulae parotides und submandibularis mit Atresie der Canaliculi lacrimales. *Dtsch Zahnärztl Z* **33**:417–421, 1941.

15. Wiesenfeld D et al: Bilateral parotid gland aplasia. *Br J Oral Surg* **21**:175–178, 1983.

16. Wiesenfeld D et al: Familial parotid gland aplasia. *J Oral Med* **40**:84–85, 1985.

## Atrichia, somatic and mental retardation, and skeletal anomalies

Schinzel (6) described a female patient with severe somatic and mental retardation, almost complete absence of body hair, ichthyosiform hyperkeratosis of the skin of the lower legs, nail dysplasia, small hands and feet with short fifth fingers and toes, and complete cutaneous syndactyly of toes 4 and 5 (Figs. 29-2A,B to 29-4A).

Radiographic studies revealed fusion of several vertebral bodies, humeroradial ankylosis, fusion between talus and navicular, lunate and triquetral, proximal metacarpals 4 and 5, and hip dislocation (Figs. 29-4B,C).

Pfeiffer and Volklein (4) and Mosavy (3) reported sibs with absence of body hair, short stature, microcephaly, and mental retardation. Del Castillo et al (2) described sibs with atrichia, mental retardation, and papular lesions (follicular cysts) of the skin. Similarly affected sibs were documented by Damste and Prakken (1). Atrichia and epilepsy were reported in sibs by Richieri-Costa and Frota-Pessoa (5).

Fig. 29–4. *Atrichia, somatic and mental retardation, and skeletal anomalies.* (A) Hyperkeratosis, left hammertoe, wide space between hallux and rest of toes, 4–5 cutaneous syndactyly. (B) Lunate-triquetral fusion, proximal Y-shaped synostosis of metacarpals 4–5, short middle phalanges, especially of fingers 2 and 5. (C) Radiohumeral synostosis. (From *A Schinzel,* Helv Paediatr Acta **35:**243, 1980).

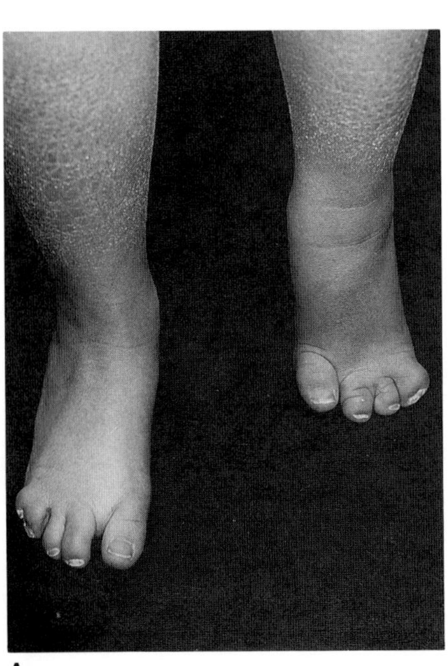

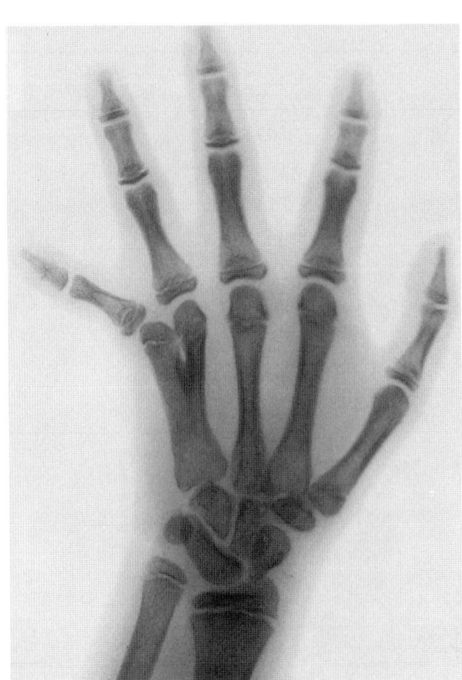

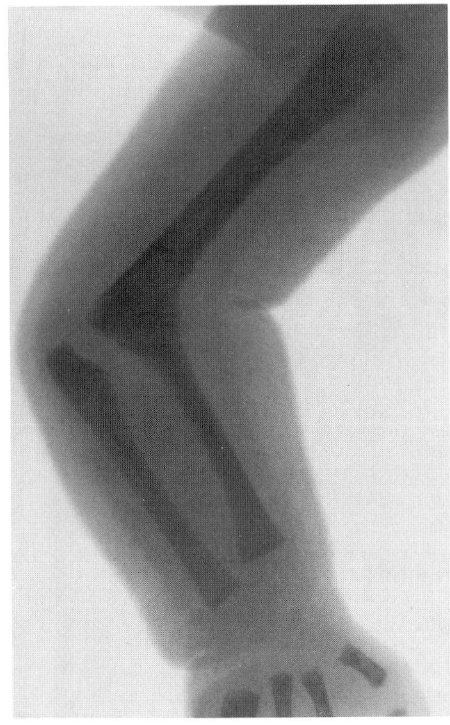

A B C

### References (Atrichia, somatic and mental retardation, and skeletal anomalies)


1. Damste TJ, Prakken JR: Atrichia with papular lesions: A variant of ectodermal dysplasia. *Dermatologica* **108**:114–121, 1954.
2. Del Castillo V et al: Atrichia with papular lesions and mental retardation in two sisters. *Int J Dermatol* **13**:261–265, 1974.
3. Mosavy SH: Universal alopecia and microcephaly in 4 siblings. *S Afr Med J* **49**:172, 1975.
4. Pfeiffer RA, Volklein J: Congenital universal alopecia, mental deficiency and microcephaly in two sibs. *J Med Genet* **19**:388–389, 1982.
5. Richiera-Costa A, Frota-Pessoa O: Atrichia, abnormal EEG, epilepsy and mental retardation in two sisters. *Hum Hered* **29**:293–297, 1979.
6. Schinzel A: A case of multiple skeletal anomalies, ectodermal dysplasia and severe growth and mental retardation. *Helv Paediatr Acta* **35**:243–251, 1980.


## Noonan-like syndrome, cherubism, and polyarticular pigmented villonodular synovitis

Falace (3) described the combination of polyarticular pigmented villonodular synovitis, "jaw cysts," cardiac defects, mild developmental delay, and distinctive facies. Review of the literature revealed several additional cases (1a,5,6,8–10). In our view, this syndrome is the same as that described by Chuong et al (1), Hoyer and Neukam (4), Dunlap et al (2), Salinas et al (7a), and others (see *Noonan syndrome*).

Facial features include mild hypertelorism, oculomotor defect (9,10), and low-set ears (9,10). In one family, multiple lentigines were found (10). Although jaw lesions were stated to be fibrous dysplasia (10) and giant cell granuloma (6), we believe that they really represented cherubism. In addition, cavernous hemangioma of the lip (5) and orbit (J Hoza, personal communication, 1989) has been documented.

As in *Noonan syndrome* and *LEOPARD syndrome,* pulmonic stenosis (5,6,9), mitral valve prolapse (10), mild growth retardation, peripheral lymphedema, and developmental delay have been evident (6,8,9). A few patients exhibited pectus (9,10).

In contrast to monoarticular villonodular synovitis which occurs in adults in their 3rd-5th decades, in this syndrome the villonodular synovitis is polyarticular, and patients are less than 10 years old. Most often the knees and ankles are involved, less often the wrists or elbows. Although a constant feature in the series of Falace (3), we suspect that this represents bias of ascertainment.

Inheritance may be autosomal dominant (10).

Cherubism usually is a lone finding with autosomal dominant inheritance. It may also be found with *Noonan syndrome* and in *neurofibromatosis* in binary combination and with *gingival fibromatosis, hypertrichosis, mental and somatic retardation and epilepsy (Ramon syndrome).* Pina-Neto et al (7) suggested that some members of the family they investigated had juvenile rheumatoid arthritis. Could this have been polyarticular pigmented villonodular synovitis? RJ Gorlin believes that these associations are examples of contiguous gene syndromes.

### References (Noonan-like syndrome, cherubism, and polyarticular pigmented villonodular synovitis)


1. Chuong R et al: Central giant cell lesions of the jaws: A clinicopathologic study. *J Oral Maxillofac Surg* **44**:708–713, 1986.
1a. Cohen MM Jr et al: Case report 16. *Syndrome Ident* **2**(1):14–17, 1974.
2. Dunlap CL et al: The Noonan syndrome–cherubism association. *Oral Surg Oral Med Oral Pathol* **67**:698–705, 1989.
3. Falace PB: Polyarticular pigmented villonodular synovitis, jaw cysts, cardiac defects, mild developmental delay and distinctive facies: A recognizable syndrome. *David W. Smith Workshop on Malformations and Morphogenesis,* Madrid, Spain, 23–29 May, 1989.
4. Hoyer PF, Neukam FW: Cherubism-eine osteofibröse Kiefererkrankung im Kindesalter. *Klin Pädiatr* **194**:128–131, 1982.
5. Leszczynski J: Pigmented villonodular synovitis in multiple joints. Occurrence in a child with cavernous haemangioma of lip and pulmonary stenosis. *Ann Rheum Dis* **34**:269–272, 1975.
6. Lindenbaum BL, Hunt T: An unusual presentation of pigmented villonodular synovitis. *Clin Orthop* **122**:263–267, 1977.
7. Pina-Neto JM et al: Cherubism, gingival fibromatosis, epilepsy, and mental deficiency (Ramon syndrome) plus juvenile rheumatoid arthritis. *Am J Med Genet* **25**:433–442, 1986.
7a. Salinas CF et al: Cherubism associated with rib anomalies. *Proc Greenwood Genet Ctr* **2**:129–130, 1983 (Case 1).
8. Wagner ML et al: Polyarticular pigmented villonodular synovitis. *Am J Roentgenol* **136**:821–823, 1981.
9. Walls JP, Nogi J: Multifocal pigmented villonodular synovitis in a child. *J Pediatr Orthop* **5**:229–231, 1985.
10. Wendt RG et al: Polyarticular pigmented villonodular synovitis in children: Evidence for a genetic condition. *J Rheumatol* **13**:921–926, 1986.


## Facio-audio-symphalangism syndrome (symphalangism–brachydactyly syndrome)

Maroteaux et al (4), in 1972, reported a syndrome having characteristic facies, proximal symphalangism of fingers and toes, various other skeletal abnormalities, and progressive conductive hearing loss. Other examples have been documented (1–3,5). Inheritance is autosomal dominant.

The nose appears hemicylindrical due to a broad base and a lack of alar flare. The upper lip is usually narrow. Occasionally there is internal strabismus (Fig. 29-6A,B). Conductive hearing loss is a constant feature.

There is proximal symphalangism of digits 2–4 with variable cutaneous syndactyly of the same fingers and toes with hypoplasia/aplasia of middle and distal phalanges and nails. The creases over the PIP joints are absent, less so the creases over the DIP joints (Fig. 29-7). The feet, upper arms, and legs tend to be short. The halluces are usually abbreviated. There is cubitus valgus. The costochondral junctions are prominent and pectus excavatum is frequent. Radiographic changes include overtubulation of tibia and fibula, dislocated radial heads, fusion of carpal and tarsal bones and tarsometatarsal bones, short broad first metacarpals and metatarsals, proximal symphalangism 2–5, distal symphalangism of 5, and absence or hypoplasia of phalanges (Fig. 29-8A–C). The ribs are broadened at the costochondral junction.

### References [Facio-audio-symphalangism syndrome (symphalangism–brachydactyly syndrome)]


1. Herrmann J: Symphalangism and brachydactyly syndrome. *Birth Defects* **10**(5):23–53, 1974.
2. Higashi K, Inoue S: Conductive deafness, symphalangism, and facial abnormalities: The WL syndrome in a Japanese family. *Am J Med Genet* **16**:105–109, 1983.
3. Hurvitz SA et al: The facio-audio-symphalangism syndrome: Report of a case and review of the literature. *Clin Genet* **28**:61–68, 1985.
4. Maroteaux P et al: La maladie des synostoses multiples. *Nouv Presse Méd* **1**:3041–3047, 1972.
5. Pedersen JC et al: Multiple synostosis syndrome. *Eur J Pediatr* **134**:273–279, 1980.


## Facio-auriculo-radial dysplasia

Harding et al (1), in 1982, described the association of unusual facies, asymmetrical radial dysplasia, abnormal pinnae, and conductive hearing loss in a mother and daughter. They pointed out that a similar family had been reported by Stoll et al (2) in 1974. There was midfacial hypoplasia. The pinnae were dysplastic. The philtrum was long and prominent. Aplasia of thumbs and index fingers, hypoplastic thumbs, aplastic or hypoplastic radii, radioulnar synostosis, somewhat hypoplastic fibulae, and minor changes in the vertebrae were found (Fig. 29-9A–D).

Inheritance is autosomal dominant with variable expression. There is overlap with the VATER association and with *oculo-auriculo-vertebral spectrum.*

### References (Facio-auriculo-radial dysplasia)


1. Harding AE et al: Autosomal dominant asymmetrical radial dysplasia, dysmorphic facies and conductive hearing loss (facioauriculoradial dysplasia). *J Med Genet* **19**:110–115, 1982.

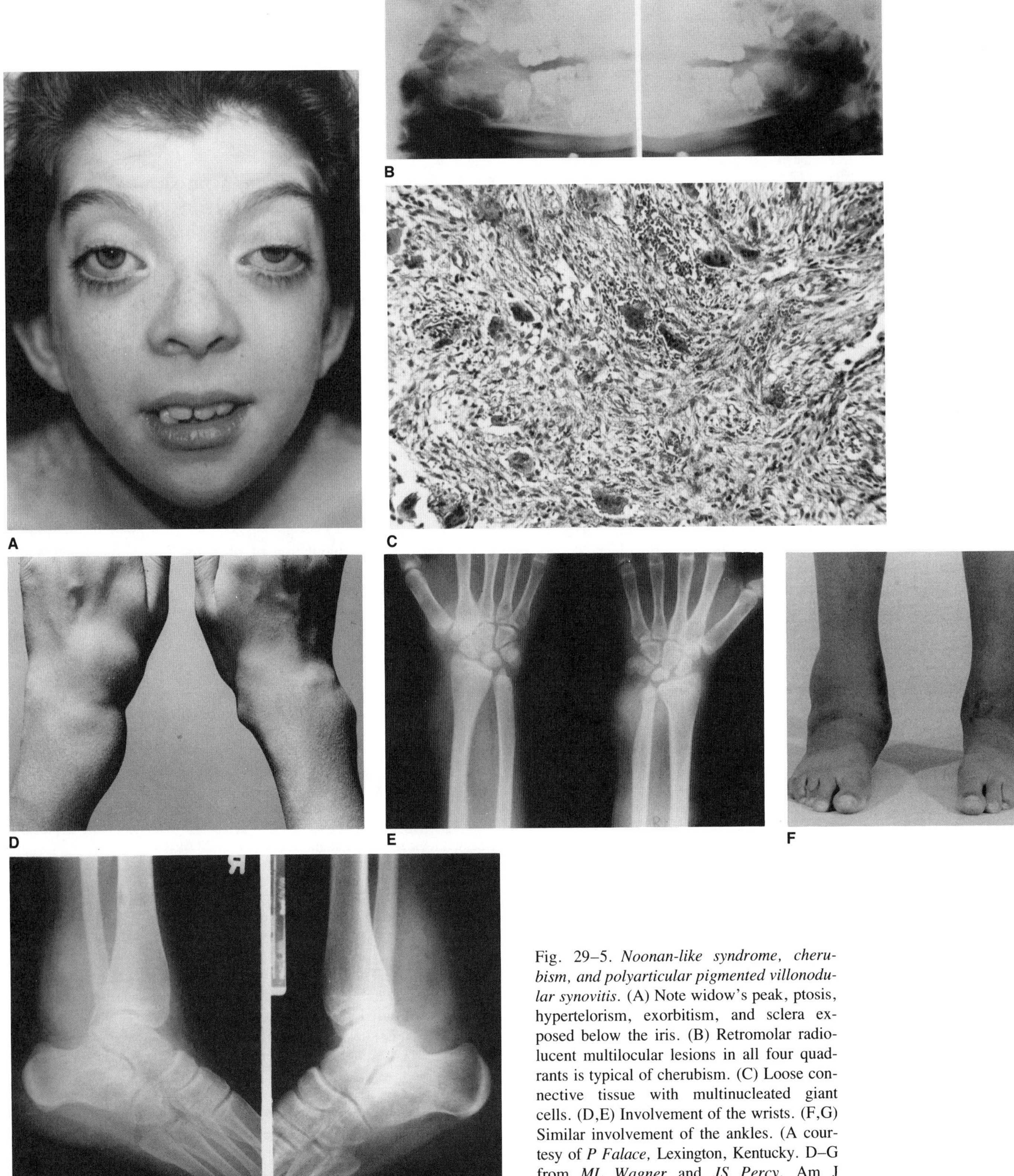

Fig. 29–5. *Noonan-like syndrome, cherubism, and polyarticular pigmented villonodular synovitis.* (A) Note widow's peak, ptosis, hypertelorism, exorbitism, and sclera exposed below the iris. (B) Retromolar radiolucent multilocular lesions in all four quadrants is typical of cherubism. (C) Loose connective tissue with multinucleated giant cells. (D,E) Involvement of the wrists. (F,G) Similar involvement of the ankles. (A courtesy of *P Falace,* Lexington, Kentucky. D–G from *ML Wagner* and *JS Percy,* Am J Roentgenol **136**:821, 1981.

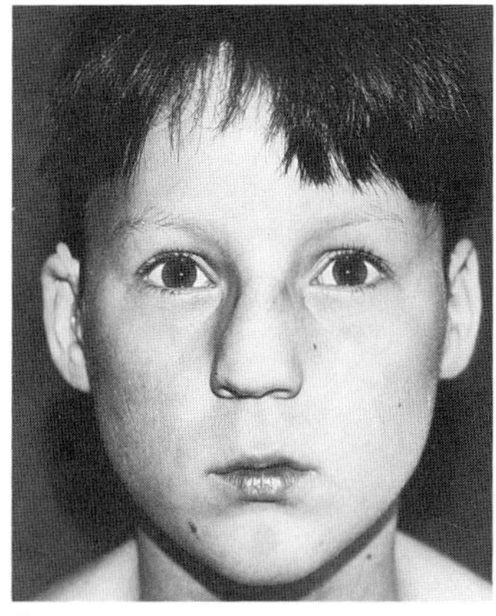
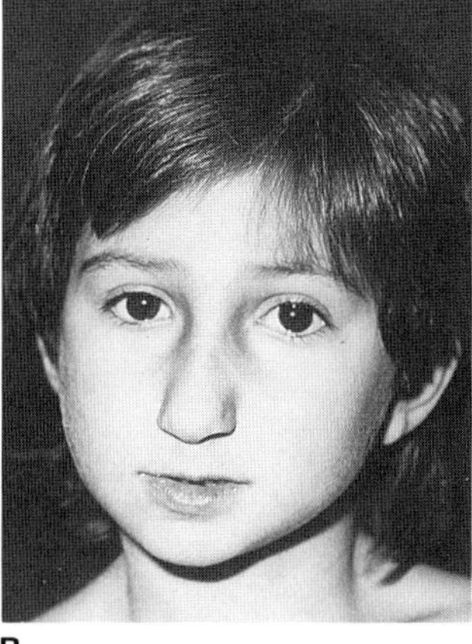

A B

Fig. 29–6. *Facio-audio-symphalangism syndrome*. (A,B) Facies characterized by thin upper lip, broad cylindrical nose. [A from *J Herrmann*, Birth Defects **10**(5):23, 1974. B from *SA Hurvitz et al*, Clin Genet **28:**61, 1985.]

2. Stoll C et al: L'association phocomelie-ectrodactylie, malformations, des oreilles avec surdité, arythmie sinusale. *Arch Fr Pédiatr* **31**:669–680, 1974.

## Facio-cardio-renal syndrome

Eastman and Bixler (1) reported three sibs with characteristic facies, severe mental retardation, congenital heart anomalies, and horseshoe kidneys. There was plagiocephaly, broad nasal root, malar hypoplasia, stiff outstanding pinnae, hypoplastic philtrum, nasal alae, and thin vermilion. The antegonial notch of the mandible was prominent. One child had cleft palate (Fig. 29-10A,B).

Various heart anomalies, including endocardial fibroelastosis, were found in all three children. All were subject to cyanotic and respiratory problems at birth and, later, with bronchopneumonia. The musculature was poorly developed. The heel cords were tight, the gait waddling, and there was limitation of motion in ankles and knees. The halluces were broad and the thumbs were clinodactylous. The nails were hypoplastic.

Inheritance is probably autosomal recessive.

### Reference (Facio-cardio-renal syndrome)

1. Eastman JR, Bixler D: Facio-cardio-renal syndrome. *Clin Genet* **11**:424–430, 1977.

Fig. 29–7. *Facio-audio-symphalangism syndrome*. Clinodactyly, mild syndactyly, brachydactyly, and absence of digital creases suggesting symphalangism. (From *SA Hurvitz et al*, Clin Genet **28:**61, 1985.)

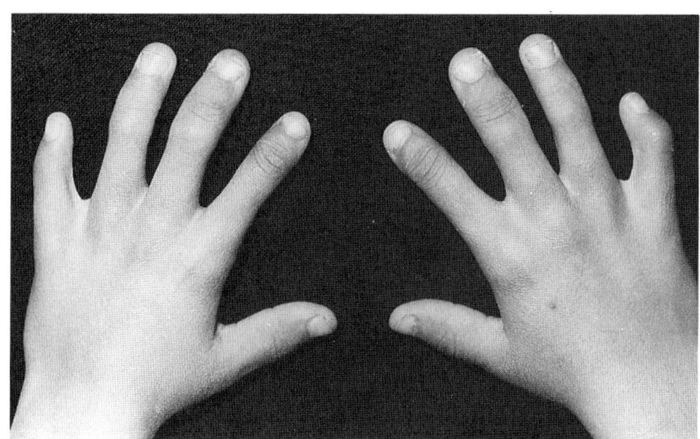

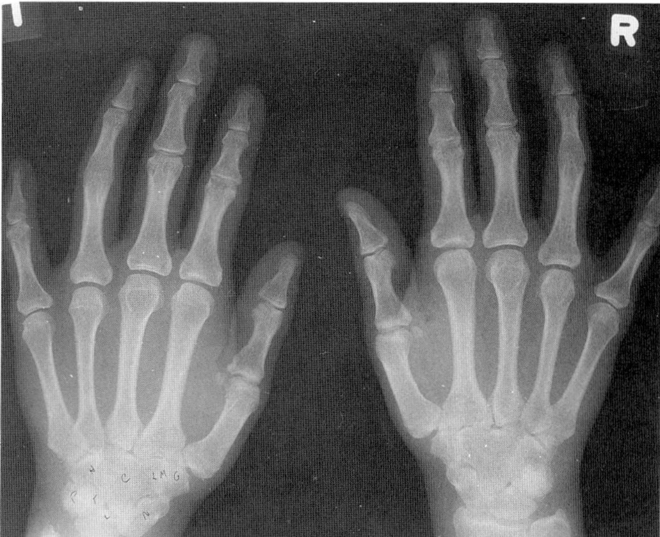


A

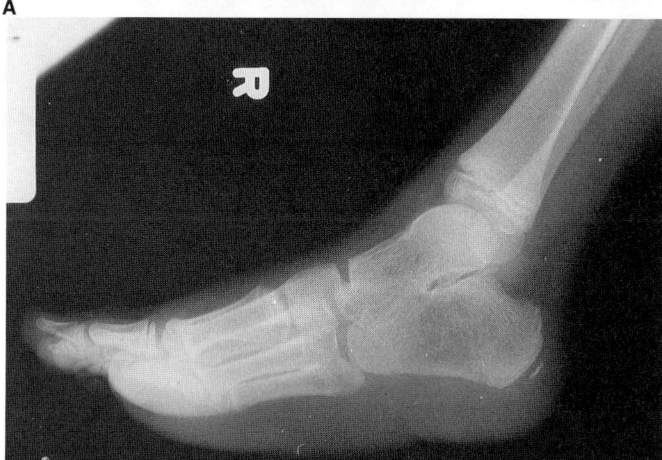


B

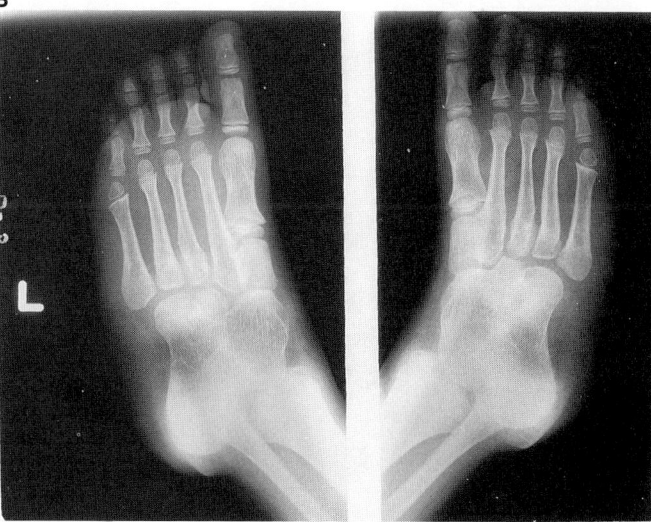


C

Fig. 29–8. *Facio-audio-symphalangism syndrome*. Radiographs showing (A) symphalangism and extensive fusion of carpal bones, bilateral brachydactyly of the middle phalanx of the fifth digit with clinodactyly, and (B,C) fusion of calcaneus, cuboid, talus, and navicular bones, reduction in number of cuneiform bones.

A

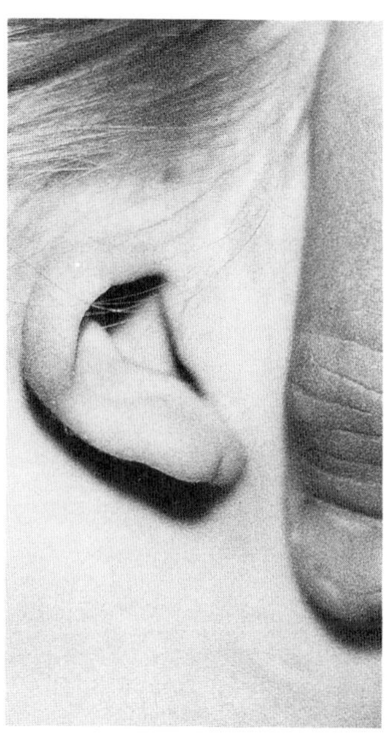

B

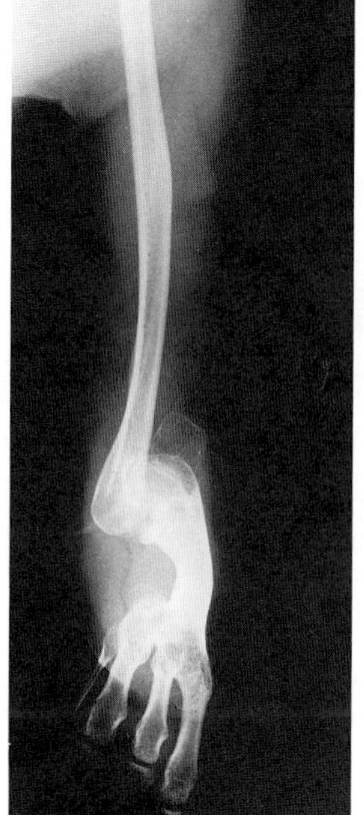

C

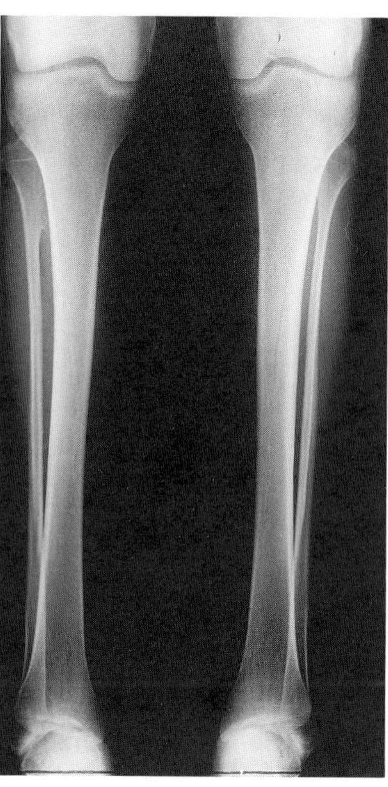

D

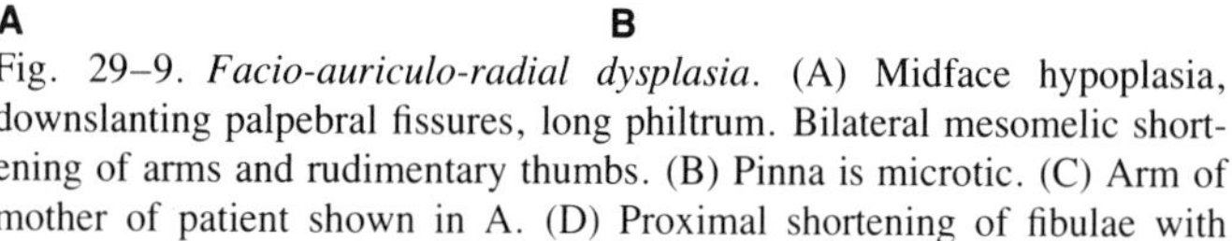

Fig. 29–9. *Facio-auriculo-radial dysplasia.* (A) Midface hypoplasia, downslanting palpebral fissures, long philtrum. Bilateral mesomelic shortening of arms and rudimentary thumbs. (B) Pinna is microtic. (C) Arm of mother of patient shown in A. (D) Proximal shortening of fibulae with exostosis at one metaphysis. Note absent radius, shortening and bowing of ulna, absent thumb, rudimentary second metacarpal. (From *AE Harding et al,* J Med Genet **19:**110, 1982.)

## Fistulas of lateral soft palate and associated anomalies

A number of cases of bilateral symmetrical defects of the soft palate have been reported (1–19). The reader is referred to the paper of Campbell (3) for review of early cases. Several examples have been found in association with other anomalies such as absence or hypoplasia of one or both palatine tonsils (3,5,7,9,16), preauricular fistulas (9), hearing loss (4), and strabismus (3).

Fig. 29–10. *Facio-cardio-renal syndrome.* (A,B) Facies of two of three affected sibs. In addition to severe mental retardation, all had plagiocephaly, long face with malar hypoplasia, broad nasal root, poorly developed philtrum, and prominent relatively inflexible pinnae. (From *JR Eastman* and *D Bixler,* Clin Genet **11:**424, 1977.)

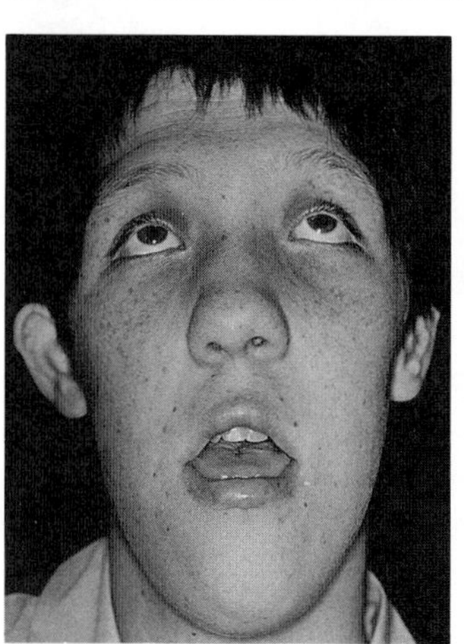

A

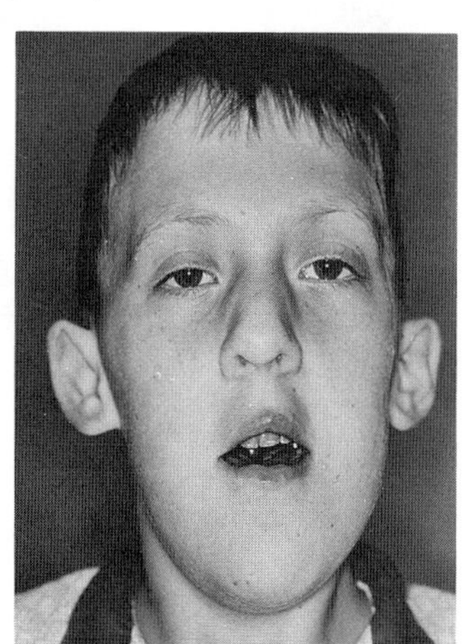

B

Though usually bilateral, the defects may be unilateral (3,7,9,16). The disorder may be familial, having been seen in two brothers (9) (Fig. 29-11).

The defect appears to be related to irregularity in development of the second pharyngeal pouch (12). Several early cases were attributed to infection, either primary or secondary to surgical complications.

### References (Fistulas of lateral soft palate and associated anomalies)

1. Broeckaert: Note sur une anomalie congénitale du voile du palais, *Rev Laryngol Otol Rhinol* **13**:577–581, 1883.

Fig. 29–11. *Fistulas of lateral soft palate and associated anomalies.* Arrows point to bilateral fistulas, often associated with agenesis of one, or both, of the palatine tonsils. (From *O Neuss,* Z Laryngol Rhinol **35:**411, 1956.)

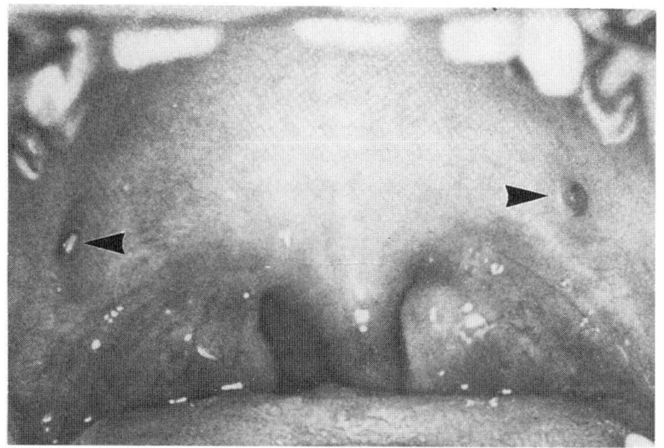

2. Bumba J: Symmetrische Defekte in den vorderen Gaumenbögen. *Z Hals Nas Ohrenheilkd* **1**:245–247, 1922.
3. Campbell EH: Perforation of faucial pillars. *Arch Otolaryngol* **1**: 503–509, 1925.
4. Chiari O: Symmetrische Defecte in den vorderen Gaumenbögen. *Mschr Ohrenheilkd* **18**:141–143, 1884.
5. Claiborne JH Jr: Hiatus in the anterior pillar of the fauces of the right side with congenital absence of the tonsil on either side. *Am J Med Sci* **89**: 490–491, 1885.
6. Cohen JS: A case of separate investment of palato-glossi muscles. *Med Rec* **14**:44–45, 1878.
7. Fridenberg P: Congenital detachment of faucial pillars and isolation of palatoglossus muscle. *Laryngoscope* **18**:567–571, 1908.
8. Lefferts GM: Cases of rare congenital deformity of the posterior nares and throat. *Med News* **40**:12, 1882.
9. Levinstein O: Unvollständige innere Halskiemenfistel in Verbindung mit doppelsseitiger Fistula preauricularis congenita. *Arch Laryngol Rhinol* **23**: 128–142, 1910.
10. Meijes WP: Un cas d'anomalie congénitale unilaterale du voile du palais. *Rev Laryngol Otol Rhinol* **14**:269, 1884.
11. Miller AS et al: Lateral soft palate fistula. *Arch Otolaryngol* **91**:200, 1970.
12. Neuss O: Anatomische Varianten und Fehlbildungen der Mundhöhle. *Z Laryngol Rhinol* **35**:411–413, 1956.
13. Newcomb JE: Anatomical defects in the faucial pillars. *Laryngoscope* **2**:220–224, 1897.
14. Rouget J: Perforations multiple des pilliers du voile du palais. *Arch Intern Laryngol* **2**:643–644, 1923.
15. Saito T: Congenital fistulas of soft palate. *J Jpn Cleft Palate Assoc* **7**:189–193, 1982.
16. Schapringer A: Ein weiterer Fall von symmetrischen Defekten in den vorderen Gaumenbögen. *Mschr Ohrenheilkd* **18**:204, 1884.
17. Seifert: Perte de substance congénitale d'un pilier palatin. *Rev Laryngol Otol Rhinol* **14**:97–102, 1894.
18. Töplitz M: Symmetrical congenital defects in the anterior pillars of the fauces. *Arch Otol* **21**:88–90, 1892.
19. Wolters: Ein ungewöhnlicher Fall von Missbildung am weichen Gaumen. *Z Rat Med* **7**:156–157, 1859.

## Pallister–Hall syndrome (hypothalamic hamartoblastoma syndrome)

In 1980, Hall, Pallister, and others (6) described a syndrome of hypothalamic hamartoblastoma, craniofacial anomalies, postaxial polydactyly, cardiac and renal defects, and endocrine dysfunction. To date, 13 examples have been reported (2,4–6,8,9). The male:female ratio is 8:5. Most cases have been sporadic, although Graham et al (4) reported a proband together with a possibly affected maternal aunt.

Most patients were at or below the tenth percentile for length and weight. The condition has been lethal in the neonatal period, although survivors have been reported (2,9). A patient reported by Iafolla et al (9) was living at 3½ years and had mild mental deficiency. The major cause of mortality in the neonatal period is hypoadrenalism (5,9). Pulmonary anomalies undoubtedly contribute to neonatal respiratory distress. The clinical and pathological findings in 13 cases are summarized in Table 29-1.

**Central nervous system.** The most striking and unique abnormality is congenital hypothalamic tumor, which consists of normal neuronal elements in abnormal mixture (1) (Fig. 29-12). The tumor probably arises from the embryonic hypothalamic plate. While undifferentiated elements may be observed, mature neural tissue was reported in one case by Iafolla et al (9), lending credence to the hypothesis that such tumors retain the ability to mature over time (1,5).

Since facial anomalies are associated with hypothalamic hamartoblastoma, embryonic disruption by the tumor may upset stereotactic relationships, thereby altering migration of embryonic structures. This results in craniofacial anomalies. Variation in timing and degree of hamartoblastomatous disruption could result in a range of phenotypic expression. Early insult resulted in holoprosencephaly, noted in three cases (1,9). Disruption during the fifth week could lead to the milder

Table 29–1. Clinical and pathological findings in Pallister–Hall syndrome

| Finding | Frequency |
|---|---|
| Central nervous system | |
| Hypothalamic hamartoma | 11/11 |
| Holoprosencephaly | 1/11 |
| Craniofacial anomalies | |
| Flat nasal bridge | 10/13 |
| Short nose | 10/13 |
| Ear anomalies | 9/13 |
| Microglossia/micrognathia | 5/13 |
| Buccal frenula | 8/13 |
| Palatal anomalies | 8/13 |
| Bifid epiglottis/cleft larynx | 7/11 |
| Limb anomalies | |
| Postaxial polydactyly/oligodactyly | 11/13 |
| Nail dysplasia/hypoplasia | 9/13 |
| Short limbs | 9/13 |
| Syndactyly | 7/13 |
| Pathological findings | |
| Renal ectopia/dysplasia | 8/13 |
| Congenital heart defects[a] | 6/13 |
| Imperforate/anteriorly placed anus | 7/13 |
| Lung segmentation anomalies | 6/11 |
| Endocrine findings[b] | |
| Adrenal hypoplasia/dysplasia hypoadrenalism | 8/10 |
| Pituitary aplasia/dysplasia panhypopituitarism | 9/12 |
| Testicular hypoplasia/cryptorchidism with micropenis in males | 7/8 |
| Thyroid aplasia/dysplasia hypothyroidism | 4/13 |

[a]PDA, VSD, endocardial cushion defect, mitral and aortic valve defects, and proximal aortic coarctation.
[b]Endocrine findings reflect premorbid endocrinologic studies and autopsy cases may be underreported.

Fig. 29–12. *Pallister–Hall syndrome.* Hypothalamic hamartoblastoma. (From *SK Clarren et al,* Am J Med Genet **7**:75, 1980.)

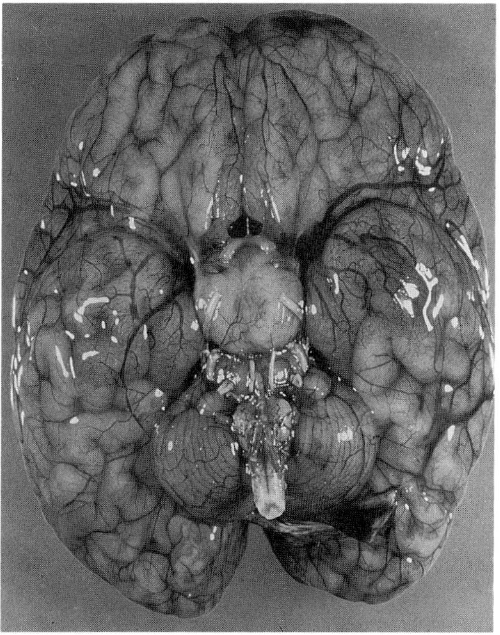

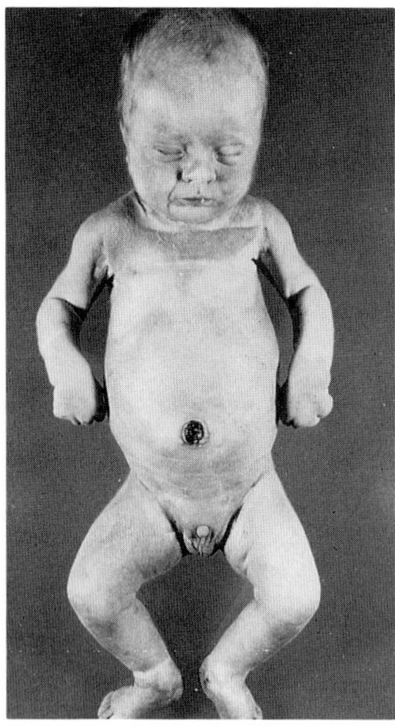
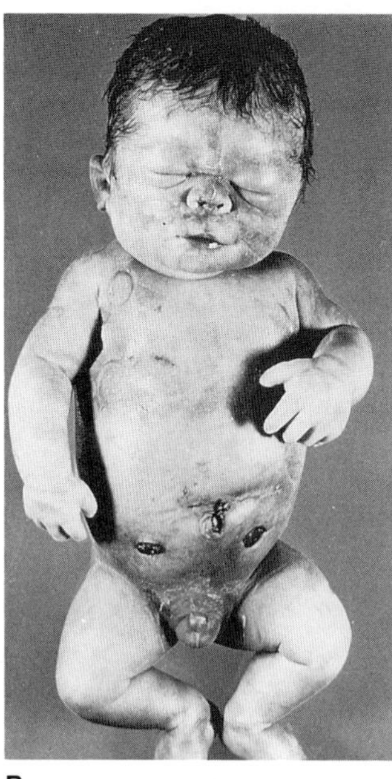

A B

Fig. 29–13. *Pallister–Hall syndrome.* (A,B) Short midface, anteverted nostrils, long philtrum, short hands, microphallus, underdeveloped scrotum. (From *JG Hall et al,* Am J Med Genet **4**:47, 1980.)

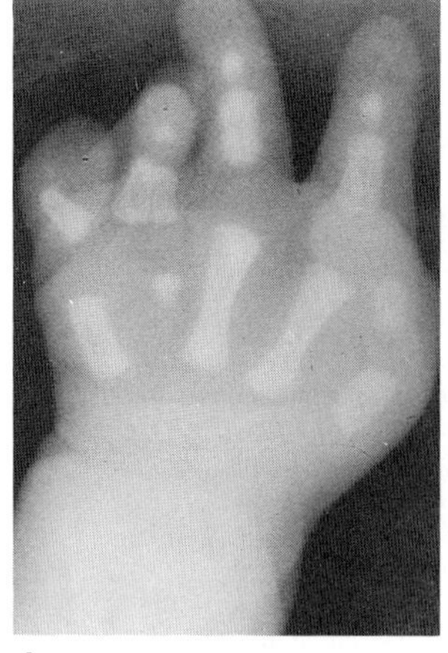
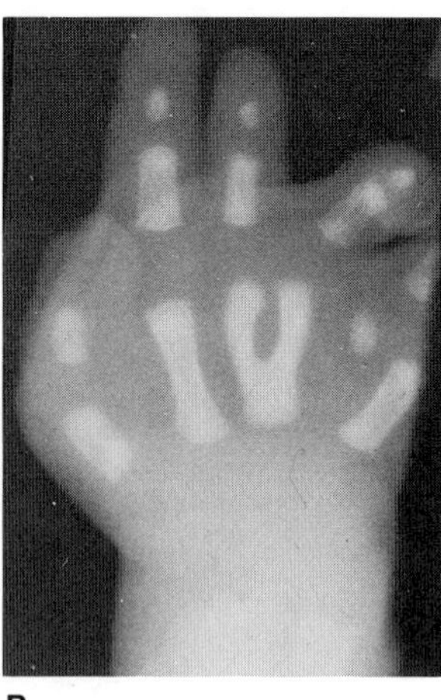

A B

Fig. 29–15. *Pallister–Hall syndrome.* (A,B) Various bony anomalies of hands (fused metacarpals, hypoplastic metacarpals, fused proximal phalanges, agenesis of distal phalanges). (From *JG Hall et al,* Am J Med Genet **4**:47, 1980.)

craniofacial anomalies usually found in the Pallister–Hall syndrome. Instances of hypothalamic hamartoma and precocious puberty without facial anomalies (9) might represent a disruptive insult that occurred even later. Since solitary central maxillary incisor may be the mildest manifestation of holoprosencephaly, its association with precocious puberty and hypothalamic hamartoma (11) helps to bridge the gap between very early hypothalamic tumors with severe craniofacial anomalies and late hypothalamic tumors without associated anomalies. Hypothalamic hamartoblastoma and its associated disruptive craniofacial features may occur alone (2a,9.10,11) or together with other embryonically noncontiguous anomalies making up a true malformation syndrome like the Pallister–Hall syndrome.

**Craniofacial anomalies.** Short midface, short nose with flat nasal bridge, anteverted nostrils, and asymmetric cranium with large fontanels are common. The ears are frequently low set and posteriorly angulated. They protrude with small or absent lobules (2,5,6,8,9) (Fig. 29-13A,B).

Oral anomalies include micrognathia, microglossia, and multiple frenula between the alveolar ridge and buccal mucosa. Natal teeth, cysts of the gingiva, cleft palate, and cleft uvula have also been noted in some patients. Mandibular ridges are often hypoplastic (2,5,6,8,9).

**Limb abnormalities.** Postaxial polydactyly, syndactyly, and nail dysplasia, involving both toes and fingers, the left side being more commonly involved than the right, are observed in most patients. The marked polydactyly is associated with short dysplastic fourth metacarpals and small underdeveloped fourth fingers. Occasionally, a bifid third metacarpal is seen (Figs. 29-14A and 29-15A,B). In the feet, the fourth and fifth digits are similarly involved (Fig. 29-14B). Subluxation of the radial head has been found in approximately half the patients (2,5,6,8,9).

**Congenital heart defects.** A wide variety of congenital heart defects have been reported in 50%, including PDA, VSD, endocardial cushion defect, mitral and aortic valve defects, and proximal aortic coarctation (5,6,8).

**Lungs.** Laryngeal cleft, aplastic, hypoplastic, or bifid epiglottis; dysplastic tracheal cartilages; and absence or hypoplasia of the lung and/or abnormal lung lobulation are common (2,5,6,8,9).

**Endocrine findings.** Severe multiple endocrine dysfunction usually includes hypopituitarism, hypothyroidism, hypoglycemia, and hypoadrenalism (5,6).

**Other findings.** Various genitourinary anomalies have been reported including absent, hypoplastic, dysplastic, or ectopic kidneys. Micropenis and cryptorchidism have also been noted. The anus is fre-

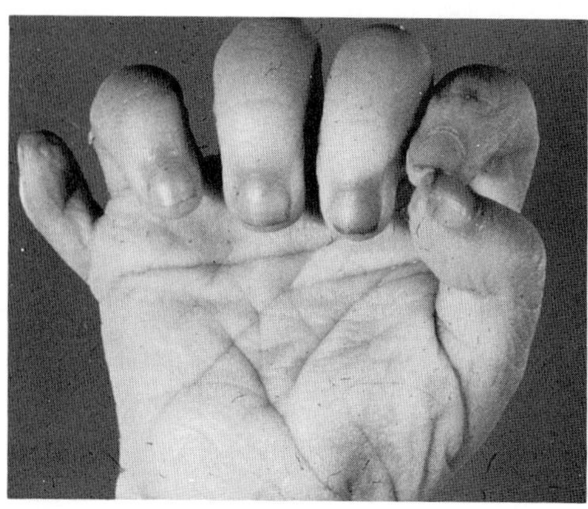
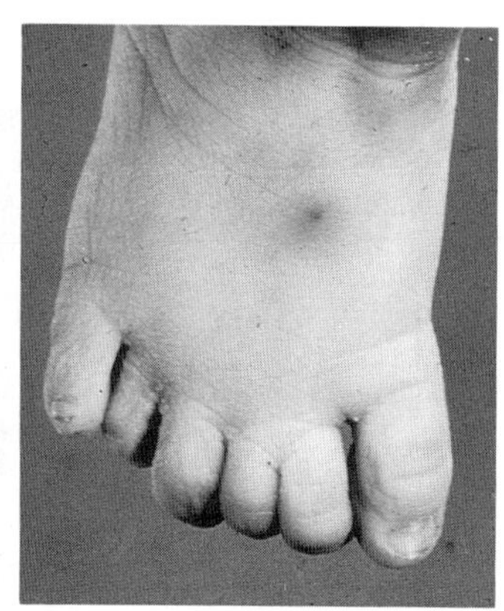

A B

Fig. 29–14. *Pallister–Hall syndrome.* (A) Postaxial polydactyly of hand. (B) Postaxial polydactyly of foot. (From *JG Hall et al,* Am J Med Genet **4**:47, 1980.)

quently imperforate or anteriorly placed. Several patients have had narrow cervical vertebrae (2,5,6,8,9).

**Differential diagnosis.** Natal teeth, multiple frenula, nail dysplasia, congenital heart defects, and postaxial polydactyly are observed in *Ellis–van Creveld syndrome*. Multiple frenula are also found in *Opitz trigonocephaly (C) syndrome*. Kaufman–McKusick syndrome, consisting of hydrometrocolpos, postaxial polydactyly, and congenital heart defect (3), should also be excluded. Marcuse et al (10), in a review of hypothalamic hamartoma unassociated with the syndrome, noted one patient with cleft palate and another with tetralogy of Fallot. Hennekam et al (7) reported congenital hypothalamic hamartoma, *frontonasal malformation,* frontal midline lipoma, and complex congenital heart defect. Gitlin and Behar (2a) reported a patient with a large hypothalamic hamartoblastoma, agenesis of the corpus callosum, absent olfactory tracts and bulbs, and bilateral nasal proboscides. Winter et al (12) reported a solitary central maxillary incisor with precocious puberty and hypothalamic hamartoma. The combination of holoprosencephaly and polydactyly may be seen in *pseudo-trisomy 13 syndrome* and *Meckel syndrome*. There is some overlap with the *oral-facial-digital syndromes* and *Smith–Lemli–Opitz syndrome*.

**Laboratory aids.** Neuroradiological detection of hypothalamic hamartoblastoma can be difficult. Magnetic resonance imaging is the procedure of choice for evaluating suprasellar tumors, but may be difficult to obtain in infants (9).

Low maternal estriols prior to birth and large fontanels suggest endocrine hypofunction and should prompt immediate evaluation for hypoadrenalism and hypothyroidism.

Prenatal diagnosis may be possible by (1) measurement of maternal estriol levels, which are decreased in the presence of hypoplastic fetal adrenals, (2) prenatal ultrasound for detection of renal, adrenal, and brain anomalies, and (3) measurement of disaccharidases in amniotic fluid that may be decreased with imperforate anus (5,6).

### References [Pallister–Hall syndrome (hypothalamic hamartoblastoma syndrome)]


1. Clarren SK et al: Congenital hypothalamic hamartoblastoma, hypopituitarism, imperforate anus, and postaxial polydactyly in a new syndrome? II. Neuropathological considerations. *Am J Med Genet* **7**:75–83, 1980.

2. Culler FL, Jones KL: Hypopituitarism in association with postaxial polydactyly. *J Pediatr* **103**:881–884, 1984.

2a. Gitlin G, Behar AJ: Meningeal angiomatosis, arhinencephaly, agenesis of the corpus callosum, and large hamartoblastoma of the brain, with neoplasia, in an infant with bilateral nasal proboscis. *Acta Anat* **41**:56–79, 1960.

3. Goecke T et al: Hydrometrocolpos, postaxial polydactyly, congenital heart diseases, and anomalies of the gastrointestinal and genitourinary tracts: A rare autosomal recessive syndrome. *Eur J Pediatr* **136**:297–305, 1981.

4. Graham JM Jr et al: Apparent familial recurrence of hypothalamic hamartoblastoma syndrome. *Proc Greenwood Genet Ctr* **2**:117–118, 1983.

5. Graham JM Jr et al: A cluster of Pallister–Hall syndrome cases, (congenital hypothalamic hamartoblastoma syndrome). *Am J Med Genet Suppl* **2**:53–63, 1986.

6. Hall JG et al: Congenital hypothalamic hamartoblastoma, hypopituitarism, imperforate anus, and post-axial polydactyly—a new syndrome? Clinical causal and pathogenetic considerations. *Am J Med Genet* **7**:47–74, 1980.

7. Hennekam RCM et al: Congenital hypothalamic hamartoma associated with severe midline defect: A developmental field defect. Report of a case. *Am J Med Genet Suppl* **2**:45–52, 1986.

8. Huff DS, Fernandez M: Two cases of congenital hypothalamic hamartoblastoma, polydactyly and other congenital anomalies (Pallister–Hall syndrome). *N Engl J Med* **306**:430–431, 1982.

9. Iafolla AK et al: Case report and delineation of the congenital hypothalamic hamartoblastoma syndrome (Pallister–Hall syndrome). *Am J Med Genet* **33**:489–499, 1989.

10. Marcuse PM et al: Hamartoma of the hypothalamus: Report of two cases with associated developmental defects. *J Pediatr* **43**:301–308, 1953.

11. Pallister PD et al: Three additional cases of the congenital hypothalamic "hamartoblastoma" (Pallister–Hall) syndrome. *Am J Med Genet* **33**:500–501, 1989.

12. Winter WE et al: Solitary central maxillary incisor associated with precocious puberty and hypothalamic hamartoma. *J Pediatr* **101**:965–967, 1982.


## Keutel syndrome

Keutel et al (3), in 1972, reported sibs with unusual facies (small depressed nose, mild midface hypoplasia), brachytelephalangy, pulmonary stenosis, and diffuse calcification of the cartilages of the pinnae, larynx, trachea, and bronchi, mixed hearing loss, and peripheral pulmonary stenosis (Figs. 29-16 to 29-18). Several additional cases have been documented (1,2,4–7). Walbaum (6) found no pulmonary stenosis but Fryns et al (2) documented VSD and infundibular pulmonary stenosis. Mental retardation has been noted (1,5).

Affected sibs (3,4) and parental consanguinity (1,3–5) suggest autosomal recessive inheritance.

### References (Keutel syndrome)


1. Cormode EJ et al: Keutel syndrome: Clinical report and literature review. *Am J Med Genet* **24**:289–294, 1986.

2. Fryns JP et al: Calcification of cartilages, brachytelephalangy and peripheral pulmonary stenosis: Confirmation of the Keutel syndrome. *Eur J Pediatr* **142**:201–203, 1984.

3. Keutel J et al: A new autosomal recessive syndrome: Peripheral pulmonary stenoses, brachytelephalangism, neural hearing loss and abnormal calcifications/ossification. *Birth Defects* **8**(5):60–68, 1972.

4. Krosroshahi HE et al: Abnormal ossification of cartilages with pulmonary artery branch stenosis: A report of four cases. *J Pediatr,* in press.

5. Say B et al: Unusual calcium deposition in cartilage associated with short stature and peculiar facial features. *Pediatr Radiol* **1**:127–129, 1973.

6. Walbaum R: Les brachydactylies. *J Génét Hum* **31**:167–181, 1983.

7. Walbaum R et al: Le syndrome de Keutel. *Ann Pédiatr* **22**:461, 1975.


## Macrocephaly, short stature, and mental retardation

Atkin et al (1), in 1985, reported a new mental retardation syndrome of macrocephaly, short stature, and coarse facial features. Additional families have been documented (2–4).

The facies is characterized by macrocephaly, prominent forehead, heavy supraorbital ridges, hypertelorism, broad nasal tip, anteverted nostrils, and thick lips.

Intelligence varied from moderate to severe.

Skeletal changes were mostly minor: hyperextensible joints, tapering fingers, genua valga and recurvata, scoliosis, and hyperkyphosis.

Inheritance is either autosomal dominant or X-linked.

### References (Macrocephaly, short stature, and mental retardation)


1. Atkin JF et al: A new X-linked mental retardation syndrome. *Am J Med Genet* **21**:697–705, 1985.


Fig. 29–16. *Keutel syndrome.* (A,B) Sibs with syndrome. Facies is essentially normal except for calcified pinnae. (Courtesy of *G Jörgensen,* Göttingen, West Germany.)

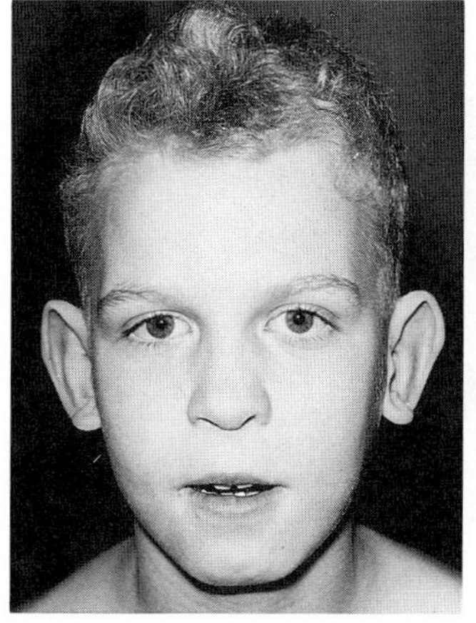

A

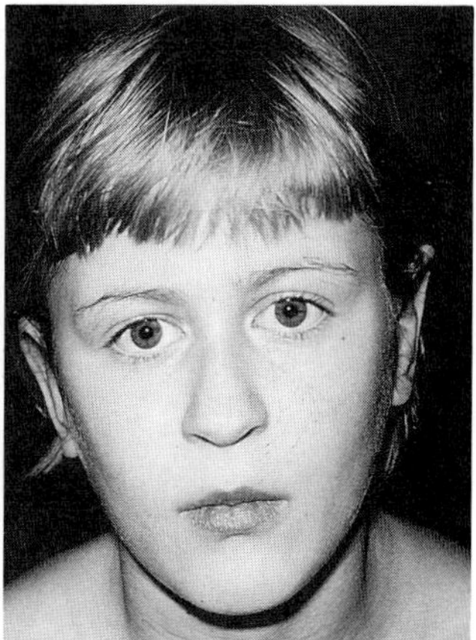

B

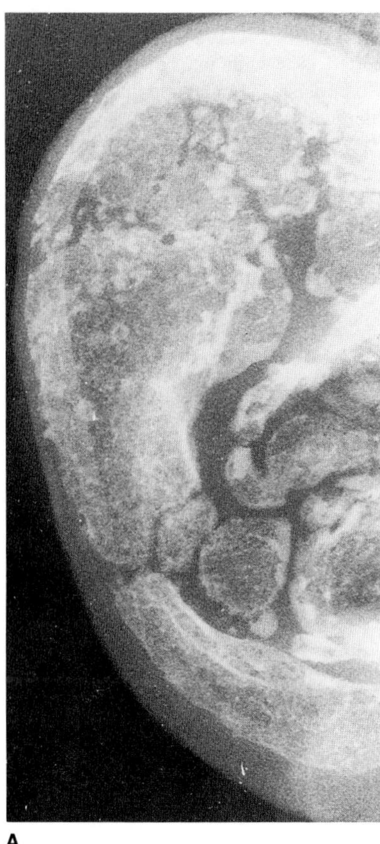

A

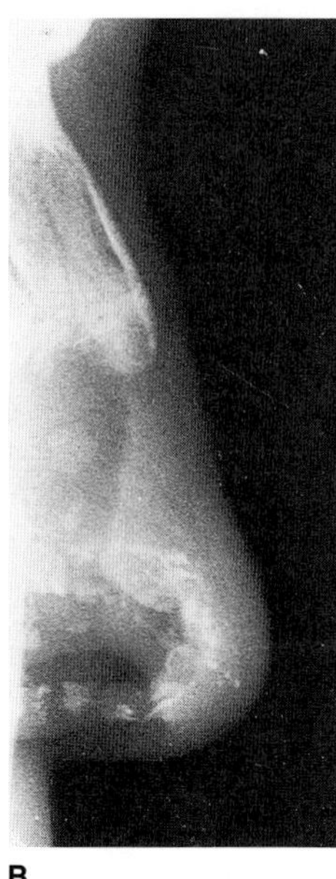

B

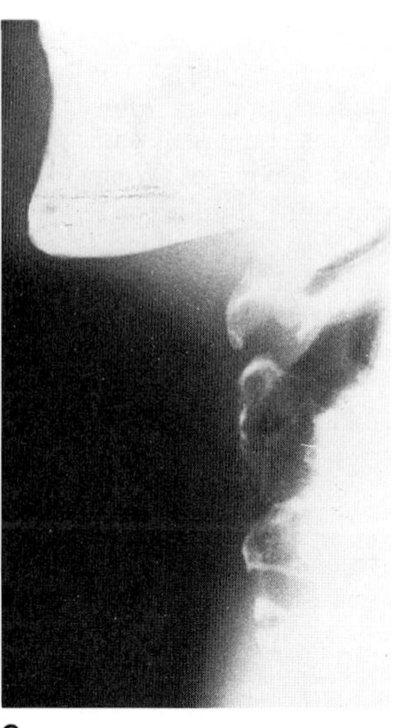

C

Fig. 29–17. *Keutel syndrome.* (A–C) Calcification of cartilages of ear, nose, and trachea. (From *G Jörgensen,* Arch Klin Exp Ohren Nasen Kehlkopfheilkd **202:**1, 1972.)

Fig. 29–18. *Keutel syndrome.* Radiograph showing abbreviation of terminal phalanges of several fingers. (Courtesy of *R Walbaum,* Roubaix, France.)

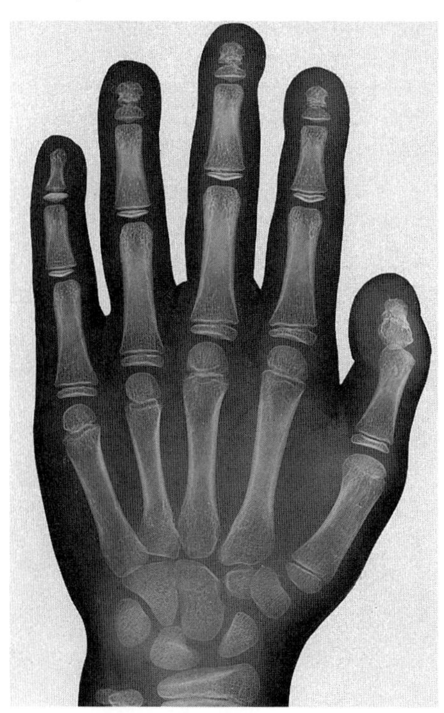

2. Clark RD, Baraitser M: A new X-linked mental retardation syndrome. *Am J Med Genet* **26**:13–15, 1987.

3. Fryns JP et al: Mental retardation, short stature and craniofacial dysmorphism in three sisters. *Clin Genet* **33**:293–298, 1988.

4. Gragg GW: Familial megalencephaly. *Birth Defects* **7**(1):228–230, 1971.

## Martsolf syndrome

In 1978, Martsolf et al (3) reported two brothers with unusual facies, severe mental retardation, short stature, cataracts, and hypogonadism.

Sanchez et al (6) described male sibs with similar problems. Hennekam et al (2) and Habord et al described an affected brother and sister. Another possible example is that of Strisciuglio et al (7). Inheritance appears to be autosomal recessive.

In addition to the severe mental retardation, frequently there are feeding problems in infancy. Height, weight, and head circumference are usually below the third percentile. The facies is characterized by an aged appearance, brachycephaly, cataracts, epicanthal folds, low nasal bridge, broad nasal tip, maxillary retrusion, short philtrum, pouting lips, and micrognathia (Fig. 29-19A,B).

Various other anomalies have included prominent nipples, lumbar hyperlordosis, thin limbs, broad fingertips, short palms, lax finger joints, various feet anomalies including abnormal toenails, hypogonadism, and cryptorchidism.

There are several similarities to *Cohen syndrome* and Mirhosseini–Holmes–Walton (4,5) syndrome, a disorder of short stature, mental retardation, microcephaly, maxillary hypoplasia, cataracts, mottled retinal pigmentation, short philtrum, narrow hands and feet with long digits, delayed puberty, and hyperextensible joints.

### References (Martsolf syndrome)

1. Habord MG et al: Microcephaly, mental retardation, cataracts, and hypogonadism in sibs: Martsolf's syndrome. *J Med Genet* **26:**397–406, 1989.

2. Hennekam RCM et al: Martsolf syndrome in a brother and sister: Clinical features and pattern of inheritance. *Eur J Pediatr* **147**:539–543, 1988.

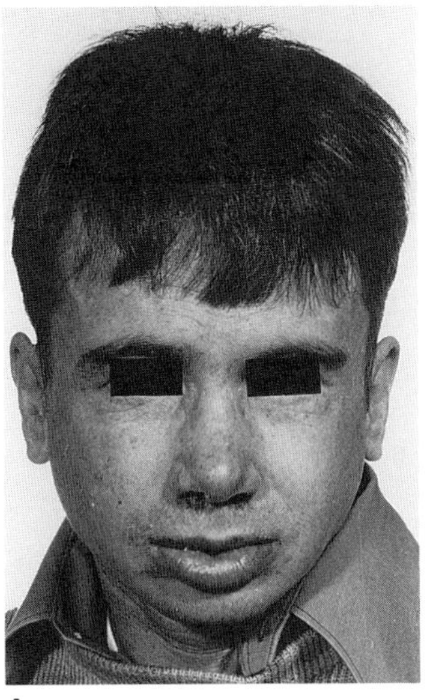
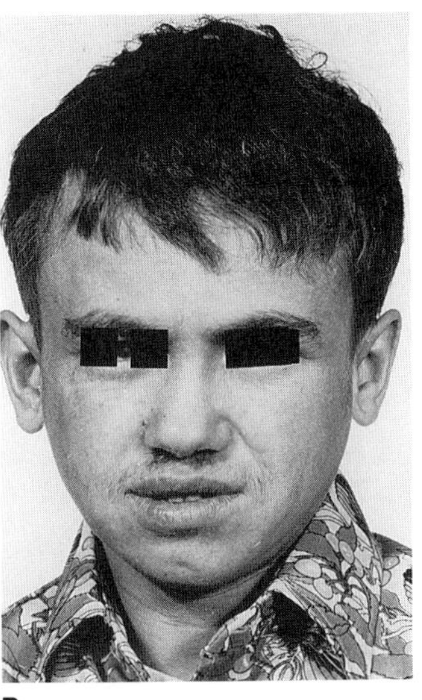

A B

Fig. 29–19. *Martsolf syndrome*. (A,B) Prominent antitragi, mild maxillary hypoplasia, short philtrum, pouty lower lip, sparse facial hair in 28- and 25-year-old brothers. (From *JT Martsolf et al,* Am J Med Genet **1**:291, 1978.)

3. Martsolf JT et al: Severe mental retardation, cataracts, short stature, and primary hypogonadism in two brothers. *Am J Med Genet* **1**:291–299, 1978.
4. Mendez HM et al: The syndrome of retinal pigmentary degeneration, microcephaly and severe mental retardation (Mirhosseini–Holmes–Walton syndrome): Report of two patients. *Am J Med Genet* **22**:223–228, 1985.
5. Mirhosseini SA et al: Syndrome of pigmentary retinal degeneration, cataract, microcephaly and severe mental retardation. *J Med Genet* **9**:193–196.
6. Sanchez JM et al: Two brothers with Martsolf's syndrome. *J Med Genet* **22**:308–310, 1985.
7. Strisciuglio P et al: Martsolf's syndrome in a non-Jewish boy. *J Med Genet* **25**:267–269, 1988.

## Midface hypoplasia, corneal clouding, subvalvular aortic stenosis, and mental and somatic retardation

Fryns and Van den Berghe (1) described male and female sibs with microcornea with diffuse clouding (Peters anomaly), glaucoma, and internal strabismus. The face was round and hypoplastic. The nose was relatively small with anteversion of the nostrils and a broad nasal bridge (Fig. 29-20).

Membranous subvalvular aortic stenosis was detected during infancy. Hands and feet were short and stubby. Mental and growth retardation were more marked in the sisters.

Haney and Falls (2) reported Peters anomaly in association with hypertelorism, upslanting palpebral fissures, brachydactyly, and mental and somatic retardation in two sibs, but we believe that this represents a different disorder.

### References (Midface hypoplasia, corneal clouding, subvalvular aortic stenosis, and mental and somatic retardation)

1. Fryns JP, Van den Berghe H: Corneal clouding, subvalvular aortic stenosis and midfacial hypoplasia associated with mental deficiency and growth retardation—a new syndrome? *Eur J Pediatr* **131**:179–183, 1979.
2. Haney WP, Falls HF: The occurrence of congenital keratoconus posterior circumscriptus. *Am J Ophthalmol* **52**:53–57, 1961.

## Mietens–Weber syndrome

Four of six children of consanguineous normal parents were reported by Mietens and Weber (1) to have a syndrome of marked growth retardation (−3 SD) and mild mental retardation (IQ 70–80). A child of unrelated parents was documented by Waring and Rodrigues (2). There was bilateral microsclerocornea and absent left fibula. Intelligence was normal. Resemblance to the kindred reported by Mietens and Weber is otherwise marked.

Clinical appearance was striking, the facies being marked by convergent strabismus and a small, narrow, pointed nose. Disc-like central opacities of the cornea, located mainly in the superficial layers, were noted bilaterally. Short forearms and flexion of the elbows with proliferation and contraction of the connective tissues on the volar aspect were also marked. Calf muscles were atrophic, and pes valgus planus was evident (Fig. 29-21A–C).

Roentgenographically, the ulna and radius were abbreviated. The head of the radius was dislocated and its epiphysis was absent bilaterally.

The disorder has autosomal recessive inheritance.

### References (Mietens–Weber syndrome)

1. Mietens C, Weber H: A syndrome characterized by corneal opacity, nystagmus, flexion contraction of the elbows, growth failure and mental retardation. *J Pediatr* **69**:624–629, 1966.
2. Waring GO III, Rodrigues MM: Ultrastructure and successful keratoplasty of sclerocornea in Mietens' syndrome. *Am J Ophthalmol* **90**:469–475, 1980.

## Oculo-cerebro-facial syndrome (Kaufman syndrome)

Kaufman et al (3) documented four of seven sibs with mental retardation, microbrachycephaly, long narrow face, sparse eyebrows, preauricular tags, upward-slanting palpebral fissures, microcornea, strabismus, pale optic disks, myopia, and micrognathia. The columella was especially low (Fig. 29-22). The hands and feet were long and thin. There were respiratory difficulties during the newborn period. Possible additional examples were reported by Jurenka and Evans (2) and Garcia-Cruz et al (1). Autosomal recessive inheritance appears likely.

### References [Oculo-cerebro-facial syndrome (Kaufman syndrome)]

1. Garcia-Cruz, J et al: Kaufman oculocerebrofacial syndrome: A corroborative report. *Dysmorphol Clin Genet* **1**:152–154, 1988.

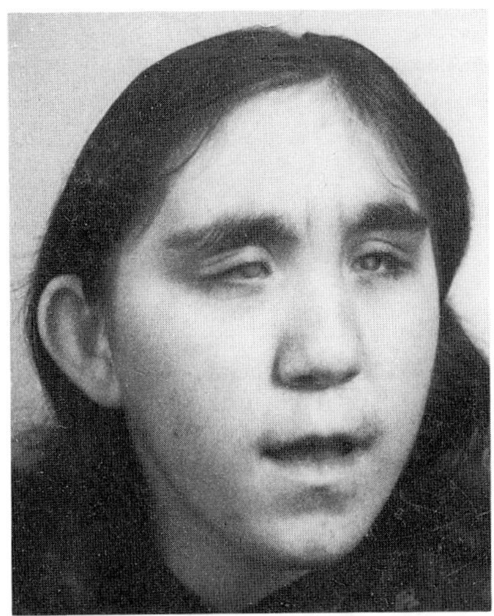

Fig. 29–20. *Midface hypoplasia, corneal clouding, subvalvular aortic stenosis, and mental and somatic retardation.* Note round and hypoplastic face, bilateral microcornea with diffuse clouding, internal strabismus. (From *JP Fryns* and *H Van den Berghe,* Eur J Pediatr **131**:179, 1979.)

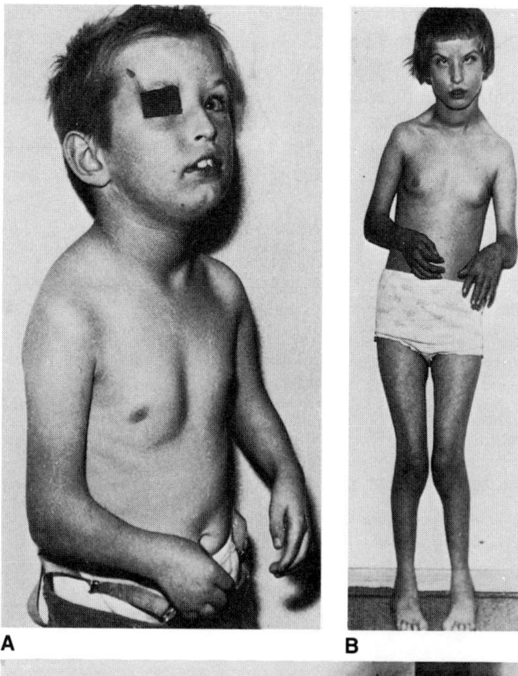



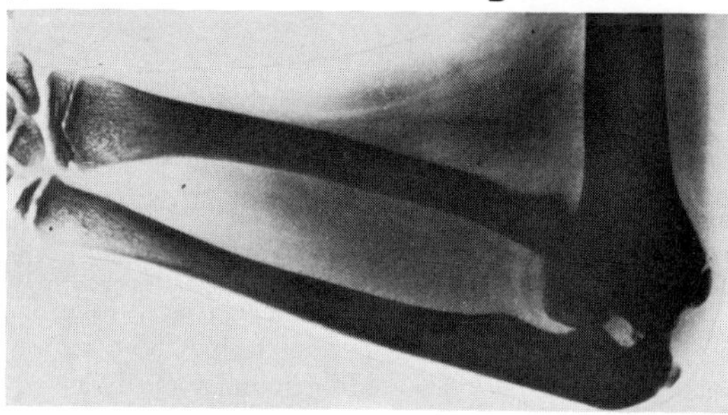



Fig. 29–21. *Mietens–Weber syndrome.* (A,B) Two of four affected sibs with convergent strabismus, narrow pointed nose, short forearms, contraction at elbows, atrophic calves. (C) Short radius and ulna; radial head dislocated, its epiphysis absent. (From *C Mietens* and *H Weber,* J Pediatr **69:**624, 1966.)

2. Jurenka SB, Evans J: Kaufman oculocerebrofacial syndrome: *Am J Med Genet* **3**:15–19, 1979.

3. Kaufman RL et al: An oculocerebrofacial syndrome. *Birth Defects* **7**(1):135–138, 1971.

## Ophthalmo-mandibulo-melic dysplasia (Pillay syndrome)

Pillay (1) in 1964, reported a father and his son and daughter with corneal opacities, short forearms due to radiohumeral and proximal radioulnar dislocations, and aplasia of lateral humeral condyle, head of radius, and lower third of ulna. The radius was bowed and short, articulating solely with the lunate. There were temporomandibular fusion, absent coronoid process, and obtuse mandibular angle (Fig. 29-23A–D).

### Reference [Ophthalmo-mandibulo-melic dysplasia (Pillay syndrome)]

1. Pillay VK: Ophthalmo-mandibulo-melic dysplasia: An hereditary syndrome. *J Bone Jt Surg* **46A**:858–862, 1964.

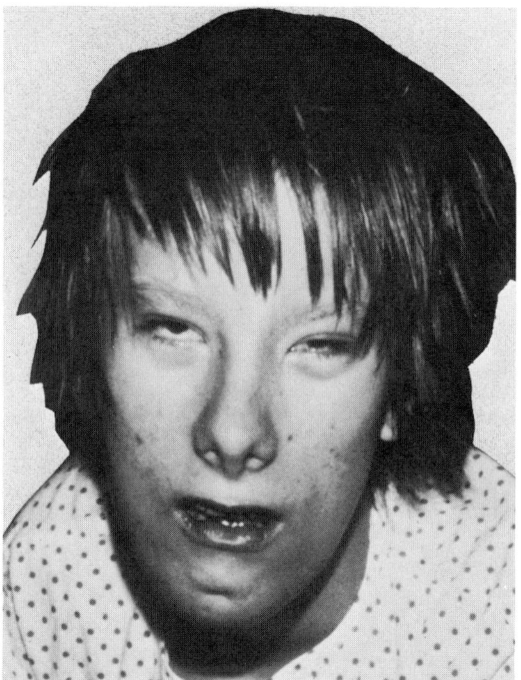

Fig. 29–22. *Oculo-cerebro-facial syndrome.* Note microcornea and unusual nasal configuration. [From *RL Kaufman et al,* Birth Defects **7**(1):135, 1971.]

## Ptosis–aortic coarctation syndrome

Cornel et al (1), in 1987, reported autosomal dominant inheritance of eyelid ptosis with coarctation of the aorta. Noted in a four generation pedigree, the syndrome is characterized by congenital bilateral ptosis, sensorineural hearing deficit, coarctation of the aorta, and asthma. Larned et al (2) noted the nonspecific association of congenital ptosis and congenital heart defects. The combination of ptosis and congenital heart defects may also be observed with *Turner syndrome* and *Noonan syndrome*.

### References (Ptosis-aortic coarctation syndrome)

1. Cornel G et al: Familial coarctation of the aortic arch with bilateral ptosis: A new syndrome? *J Pediatr Surg* **22**:724–726, 1987.

2. Larned DC et al: The association of congenital ptosis and congenital heart disease. *Ophthalmology* **93**:492–494, 1986.

## Scott syndrome (craniodigital syndrome)

In 1971, Scott et al (2) reported a newly recognized craniodigital syndrome in three affected brothers who were said to resemble patients with the *Saethre–Chotzen syndrome* but who differed significantly by having growth deficiency, moderate to severe mental retardation, and brachycephaly with absence of craniosynostosis. Other features included hirsutism, prominent eyebrows, long, dark lashes, startled facial expression, small pointed nose, micrognathia, soft-tissue syndactyly between the second, third, and fourth fingers, and soft-tissue syndactyly between the second and third toes (Fig. 29-24). The mode of inheritance is probably X-linked. A similarly affected male child was described by Lorenz et al (1).

### References [Scott syndrome (craniodigital syndrome)]

1. Lorenz P et al: Scott's craniodigital syndrome—report of a second family. *Am J Med Genet* (in press).

2. Scott CR et al: A new craniodigital syndrome with mental retardation. *J Pediatr* **78**:658–663, 1971.

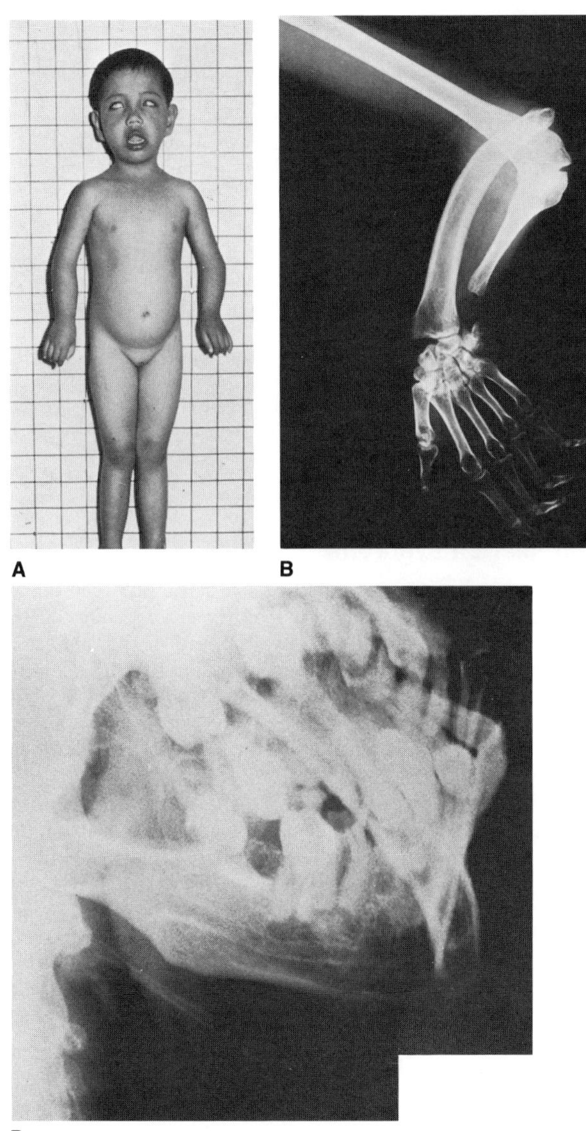


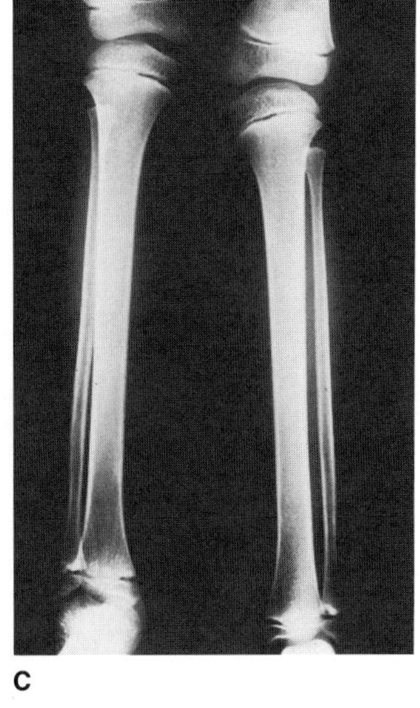


Fig. 29–23. *Ophthalmo-mandibulo-melic dysplasia.* (A) Seven-year-old girl with bilateral complete corneal opacities, shortened forearms. Father and brother similarly affected. (B) Lateral humeral condyle aplastic, abnormal trochlea, olecranon. Coronoid processes absent, as are radial heads. Both radial shaft heads end in points that are located posterolateral to lower end of humerus. Radius bowed and short, ulna short with distal third absent. Articulation of radius only with lunate. Distal interphalangeal fusion. (C) Lateral malleolus higher than medial malleolus. Fibula shorter than tibia. (D) Temporomandibular joint ankylosis, lack of mandibular angle, and absence of coronoid process. (From *VK Pillay,* J Bone Jt Surg **46A:**858, 1964.)

Fig. 29–24. *Scott syndrome (craniodigital syndrome).* Brachycephaly, pointed nose, and startled expression. (From *CR Scott et al,* J Pediatr **78:**658, 1971.)

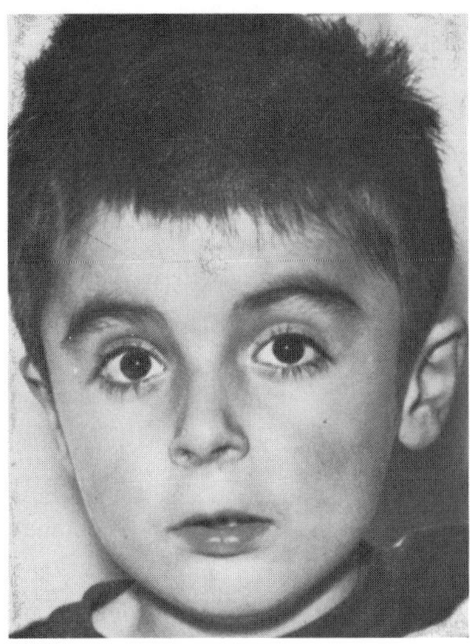

## Unusual facies, elevated IgE level, and leukocyte dysfunction (Job's syndrome, hyperimmunoglobulin E syndrome)

There are many syndromes in which there is defective neutrophil and monocyte chemotaxis (9,10,14). The most common syndrome has been called Job's syndrome due to his affliction with "sore boils" (6). About 150 patients have been described. Extensive reviews are those of Donabedian and Gallin (7), Belohradsky et al (1), and Leung and Geha (12). There is no sex or race predilection. It has been estimated that Job's syndrome constitutes about 1% of those with a primary immune defect (1).

Eczematous dermatitis, usually presenting during the first few weeks of life, often involves more than 50% of the body. There is a predilection for flexural surfaces such as popliteal and antecubital fossae and neck. From there on, nearly all patients exhibit increased susceptibility to recurrent pyogenic infections. There are abundant furuncles and carbuncles of the chest, back and axillae. However, "cold" abscesses are found in only 30% and cellulitis occurs in only 10%.

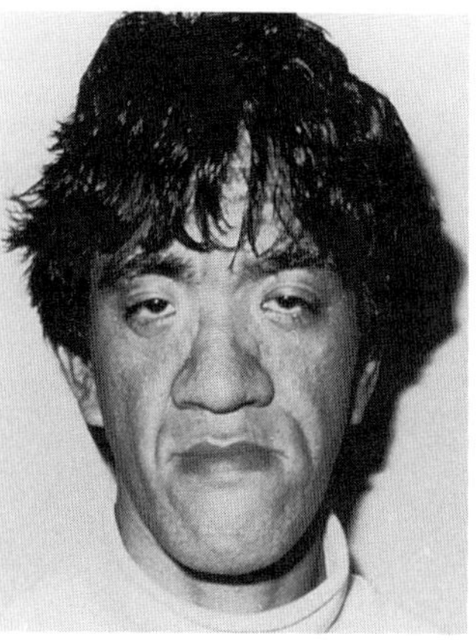

Fig. 29–25. *Unusual facies, elevated IgE level, and leukocyte dysfunction (Job's syndrome).* Patient has coarse facies typical of patients with Job's syndrome. The midface is hypoplastic and there is hypertelorism. (Courtesy of *PG Quie,* Minneapolis, Minnesota.)

Sinopulmonary infections include otitis media (45%), otitis externa (10%), sinusitis (15%), tonsillitis (12%), mastoiditis (5%), pneumonia (70%), lung abscess (40%), and empyema (10%). Gingivitis, periodontitis, and aphthae have been noted in 70% (4). Conjunctivitis and/or corneal ulceration occur in 20%. Bones, joints, and viscera are involved in less than 5% of the cases. *Staphylococcus aureus* has been isolated in 90%. Oral and vaginal candidosis and candidal nail dystrophy are evident in about 40%.

Quie and Cates (13) first noted the coarse facies characterized by hypoplastic midface, broad nasal bridge, and prominent nose. It has been estimated that the facies is seen in 50–60% (Fig. 29-25) and may be present at birth (11). Craniosynostosis has been occasionally reported (5,8,10a,16) and these cases are discussed further in Chapter 15 under *hyper-IgE syndrome.* No clear inheritance pattern is evident although there is a strong family history of atopy.

The chemotactic defect, present at about 50% of the mean normal range, is variable and often intermittent. Most patients have a low grade eosinophilia (about 10%) (7). Serum IgE levels are at least 10 times normal, i.e., greater than 2000 IU/ml, in some patients being as high as 100-fold normal values (1–3,15,17). Erythrocyte sedimentation rates are elevated in all patients.

#### References [Unusual facies, elevated IgE level, and leukocyte dysfunction (Job's syndrome, hyperimmunoglobulin E syndrome)]


1. Belohradsky BH et al: Das Hyper IgE-Syndrom (Buckley-Hiob-Syndrom). *Ergeb Inn Med Kinderheilkd* **25**:1–40, 1987.

2. Buckley RH, Becker WG: Abnormalities in the regulation of human IgE synthesis. *Immunol Rev* **41**:288–314, 1978.

3. Buckley RH et al: Extreme hyperimmunoglobulin E and undue susceptibility to infection. *Pediatrics* **49**:59–70, 1972.

4. Charon JA: et al: Gingivitis and oral ulceration in patients with neutrophil dysfunction. *J Oral Pathol* **14**:150–155, 1985.

5. Däumling S et al: Das Buckley-Syndrom: Rezidivierende, schwere Staphylokokkeninfektionen, Ekzem, und Hyperimmunoglobulinämie E. *Infection* (Suppl 3) **8**:248–254, 1980.

6. Davis SD et al: Job's syndrome: Recurrent "cold" staphylococcal abscesses. *Lancet* **1**:1013–1015, 1966.

7. Donabedian H, Gallin JI: The hyperimmunoglobulin E-recurrent infection (Job's syndrome). *Medicine (Baltimore)* **62**:195–208, 1983.

8. Fanconi S et al: Oral chloramphenicol therapy for multiple liver abscesses in hyperimmunoglobulin E syndrome. *Eur J Pediatr* **142**:292–295, 1984.

9. Gallin JI: Disorders of phagocyte chemotaxis. *Ann Intern Med* **92**:520–538, 1980.

10. Gallin JI: Abnormal phagocyte chemotaxis: Pathophysiology, clinical manifestations and management of patients. *Rev Inf Dis* **33**:1196–1220, 1981.

10a. Hörger PH et al: Craniosynostosis in hyper-IgE syndrome. *Eur J Pediatr* **144**:414–417, 1985.

11. Kamei R, Honig PJ: Neonatal Job's syndrome featuring a vesicular eruption. *Pediatr Dermatol* **5**:75–82, 1988.

12. Leung PDYM, Geha RS: Clinical and immunologic aspects of the hyperimmunoglobulin E syndrome. *Hematol Oncol Clin N Am* **2**:81–100, 1988.

13. Quie PG, Cates KL: Clinical manifestations of disorders of neutrophil chemotaxis, in *Leukocyte Chemotaxis: Methods, Physiology and Clinical Implications,* Gallin JI, Quie PG (eds), Raven Press, New York, 1978, pp 307–328.

14. Rotrosen D, Gallin JI: Disorders of phagocyte function. *Annu Rev Immunol* **5**:127–150, 1987.

15. Shuttleworth D et al: Hyperimmunoglobulin E syndrome (treatment with isotretinoin). *Br J Dermatol* **119**:93–99, 1988.

16. Smithwick EM et al: Cranial synostosis in Job's syndrome. *Lancet* **1**:826, 1978.

17. Wyre HW Jr, Johnson WT: Clinical syndrome of chemotaxis defect, infections, and hyperimmunoglobulinemia E. *Arch Dermatol* **114**:74–77, 1978.


## Singleton–Merten syndrome

A syndrome consisting of calcification of the aortic arch and aortic valve, hypoplastic tooth buds, and osteoporosis and widening of the metacarpals, carpals, and phalanges was described in two unrelated female children by Singleton and Merten (3). Additional patients were reported by McLoughlin et al (2) and Gay and Kuhn (1).

The deciduous teeth become carious at an early age and are lost prematurely. The permanent teeth are dysplastic, some never develop, and others erupt late. (Fig. 29-26A,B).

Cardiac enlargement with calcification of the thoracic aorta was seen in all children in mid or later childhood (Fig. 29-26C). Death results from calcific aortic stenosis leading to heart failure usually in the teen years.

Severe to moderate muscular weakness becomes manifest in the first or second year of life following a febrile illness. There are secondary hip and foot deformities. Stature is small. Osteoporosis of the cranial vault and all long bones is significant, being most marked in the hand bones (Fig. 29-26D). A psoriasiform skin eruption was reported by Gay and Kuhn (1).

#### References (Singleton–Merten syndrome)


1. Gay BB Jr, Kuhn JP: A syndrome of widened medullary cavities of bone, aortic calcification, abnormal dentition and muscular weakness (the Singleton–Merten syndrome). *Radiology* **118**:389–395, 1976.

2. McLoughlin MJ et al: Idiopathic calcification of the ascending aorta and aortic valve in two young women. *Br Heart J* **36**:96–100, 1974.

3. Singleton EB, Merten DF: An unusual syndrome of widened medullary cavities of the metacarpals and phalanges, aortic calcification and abnormal dentition. *Pediatr Radiol* **1**:2–7, 1973.


## Unusual facies, congenital corneal anesthesia with retinal abnormalities, sensorineural hearing loss, patent ductus arteriosus, and mental retardation

Ramos-Arroyo et al (1) described two sibs with failure to thrive and deficient stature and weight. In addition to mental retardation and persistent ductus arteriosus, the facial and ocular findings included broad flat face, upslanting palpebral fissures, ocular hypertelorism, frontal bossing, depressed nasal root and bridge, and midface hypoplasia. Keratitis, absence of peripapillary choriocapillaris, absence of retinal pigment epithelium, and poor visual acuity were noted. Both exhibited sensorineural hearing loss and patent ductus arteriosus (Fig. 29-27). The mother had mild to moderate sensorineural hearing loss, retinal changes, and similar facial features. Inheritance is assumed to be autosomal dominant.

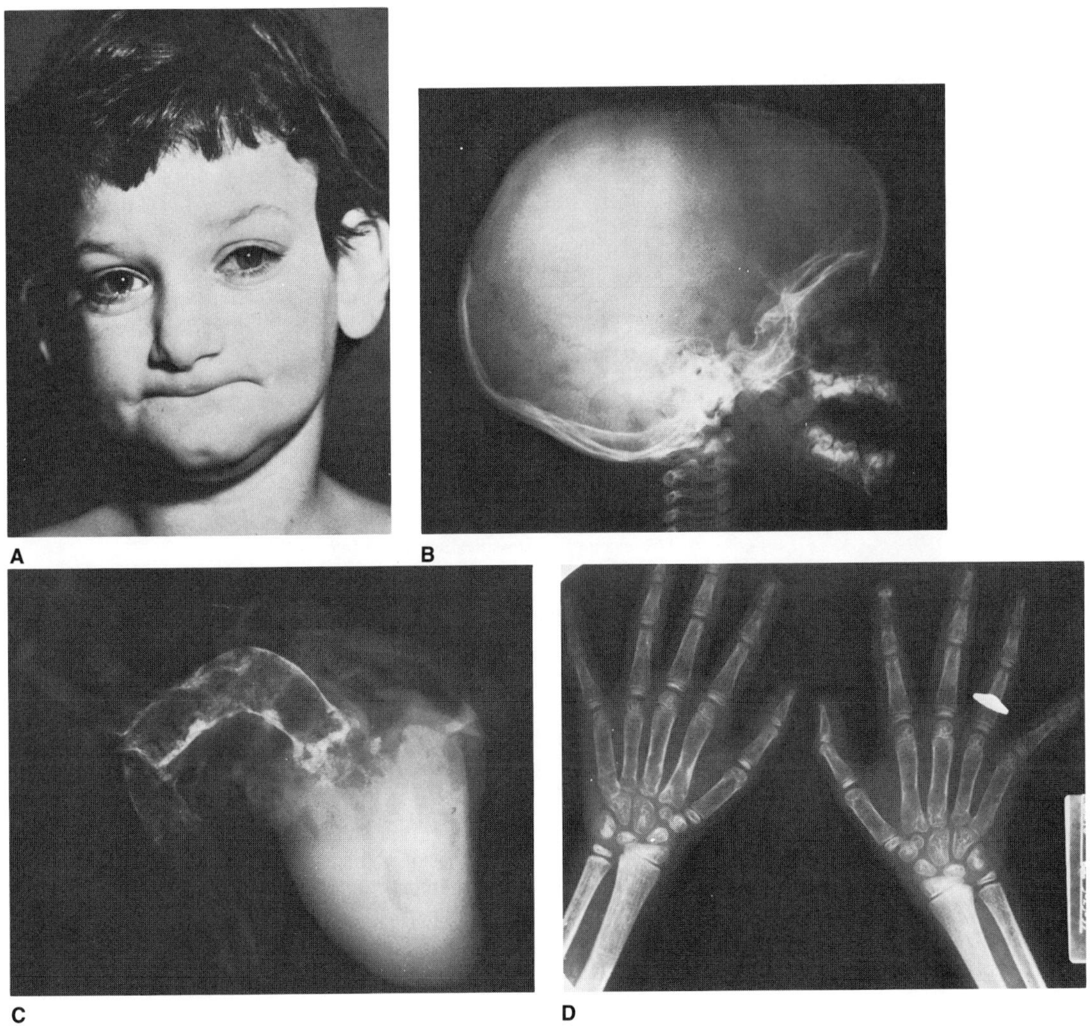

Fig. 29–26. *Singleton–Merten syndrome.* (A) Loss of vertical height in lower face due to unerupted teeth makes child look older than 10 years of age. (B) Radiograph shows mild maxillary hypoplasia, mild diffuse osteoporosis, and multiple unerupted teeth. (C) Extensive calcification of aortic arch and aortic valve. (D) Marked osteoporosis of hand bones, subungual calcification of second finger. (From *EB Singleton* and *DF Merten,* Pediatr Radiol **1**:2, 1973.)

Fig. 29–27. *Unusual facies, congenital corneal anesthesia with retinal abnormalities, sensorineural hearing loss, patent ductus arteriosus, and mental retardation.* Affected sibs with frontal bossing, hypertelorism, depressed nasal bridge, midfacial hypoplasia. Female has tarsorrhaphies. (Courtesy of *ME Hodes,* Indianapolis, Indiana.)

#### Reference (Unusual facies, congenital corneal anesthesia with retinal abnormalities, sensorineural hearing loss, patent ductus arteriosus, and mental retardation)

1. Ramos-Arroyo MA et al: Congenital corneal anesthesia with retinal abnormalities, deafness, unusual facies, persistent ductus arteriosus, and mental retardation: A new syndrome? *Am J Med Genet* **26**:345–354, 1987.

## Unusual facies, mental retardation, cataracts, muscle wasting, and skeletal abnormalities

Primrose (4), in 1982, described a slowly progressive degenerative disorder characterized by mental deficiency, wasting of limb musculature, bone abnormalities, and ossification of the pinnae. A virtually identical example was described by Collacott et al (1).

Relative macrocephaly, large calcified pinnae, sensorineural hearing loss, cataracts, and torus palatinus were found (Fig. 29-28A,B). Body build in the two males was markedly feminine. There was distal muscle wasting with contractures of fingers, hips, and knees (Fig. 29-28C).

Radiographically, generalized osteoporosis was noted. The skull was large with basilar impression and midface hypoplasia. The chest was narrow with downward-sloping ribs. The lumbar spine showed kyphosis with endplate irregularities, posterior scalloping of vertebral

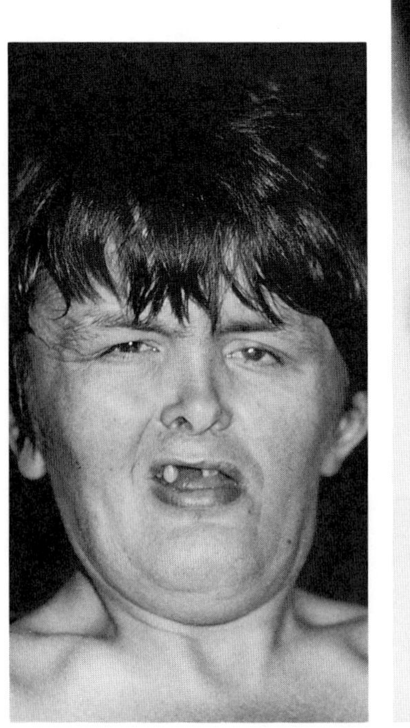
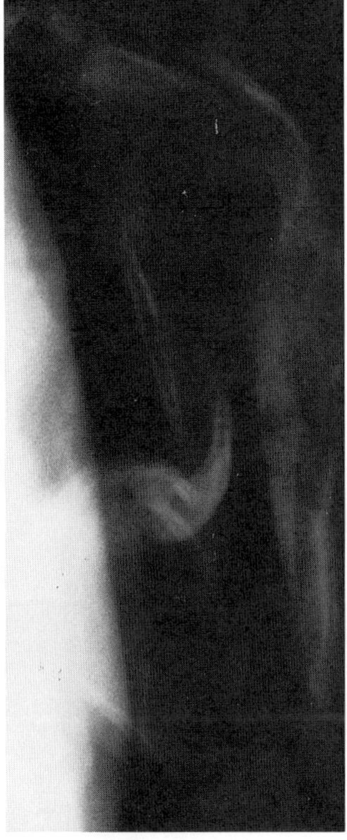
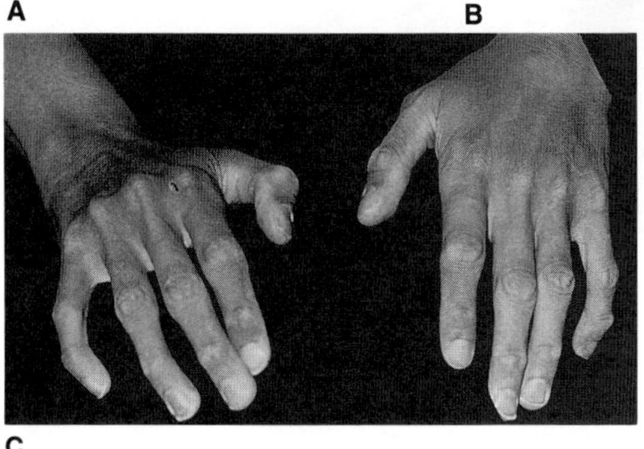

Fig. 29–28. *Unusual facies, mental retardation, cataracts, muscle wasting, and skeletal abnormalities.* (A) Macrocephaly, large calcified pinnae, and cataracts. (B) Radiograph showing calcification of pinnae. (C) Distal muscle wasting with contracture of fingers. (From *RA Collacott et al,* J Ment Defic Res **30:**301, 1986.)

bodies, and collapse-wedging of lumbar vertebrae. There were well-defined radiolucencies in the proximal femoral and humeral shafts and in both patellae. The iliac wings were narrow. The terminal phalanges were short and there were erosions at the bases of the proximal phalanges of the index and middle fingers.

The disorder appears to be different from *Keutel syndrome* (2,3) in which patients have normal or low normal intelligence, brachytelephalangy, diffuse cartilage calcification, and peripheral pulmonary stenosis. Calcified pinnae may result from trauma.

### References (Unusual facies, mental retardation, cataracts, muscle wasting, and skeletal abnormalities)

1. Collacott RA et al: The syndrome of mental handicap, cataracts, muscle wasting and skeletal abnormalities. Report of a second case. *J Ment Defic Res* **30**:301–308, 1986.

2. Fryns JP et al: Calcifications of cartilages, brachytelephalangy and peripheral pulmonary stenosis. Conformation of Keutel syndrome. *Eur J Pediatr* **42**:201–203, 1984.

3. Keutel J et al: A new autosomal recessive syndrome: Peripheral pulmonary stenosis, brachytelephalangism, neural hearing loss and abnormal cartilage calcifications/ossification. *Birth Defects* **8**(5):60–68, 1972.

4. Primrose DA: A slowly progressive degenerative condition characterized by mental deficiency, wasting of limb musculature and bone abnormalities including ossification of the pinnae. *J Ment Defic Res* **26**:101–106, 1982.

## Unusual facies, mental retardation, and intrauterine growth retardation (Pitt–Rogers–Danks syndrome)

Pitt et al (3) reported two sibs and two isolated children with intrauterine growth retardation and an unusual facies characterized by wide mouth, short upper lip with flat philtrum, prominent eyes, mild telecanthus, and mildly upslanting palpebral fissures. Two had a somewhat beaked nose. Additional examples were described by Donnai (1) and Oorthuys and Bleeker–Wagemakers (2).

Microcephaly, epilepsy, hyperactive behavior, and mental retardation ranging from moderate to severe (IQ 35–49) were found in all cases (Fig. 29-29A–C).

### References [Unusual facies, mental retardation, and intrauterine growth retardation (Pitt–Rogers–Danks syndrome)]

1. Donnai D: A further patient with the Pitt–Rogers–Danks syndrome of mental retardation, unusual face and intrauterine growth retardation. *Am J Med Genet* **24**:29–32, 1986.

2. Oorthuys JWE, Bleeker-Wagemakers EM: A girl with the Pitt–Rogers–Danks syndrome. *Am J Med Genet* **32**:140–141, 1989.

3. Pitt DB et al: Mental retardation, unusual facies and intrauterine growth retardation—a new recessive syndrome? *Am J Med Genet* **19**:307–313, 1984.

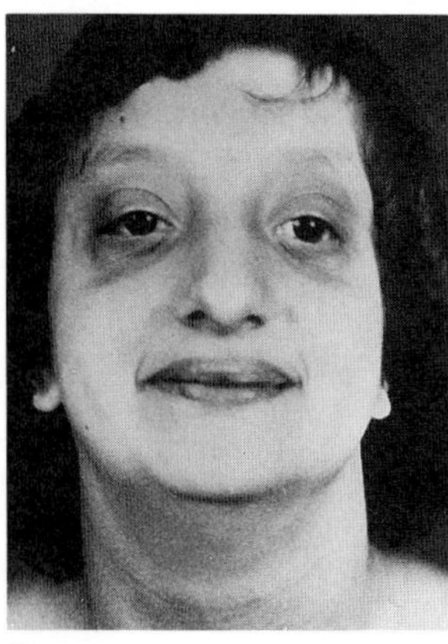
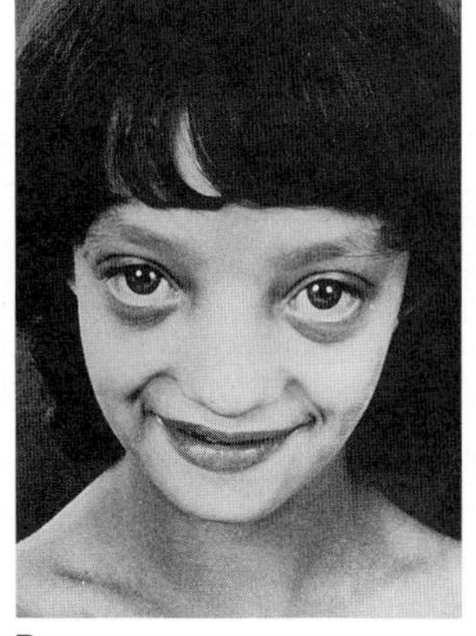
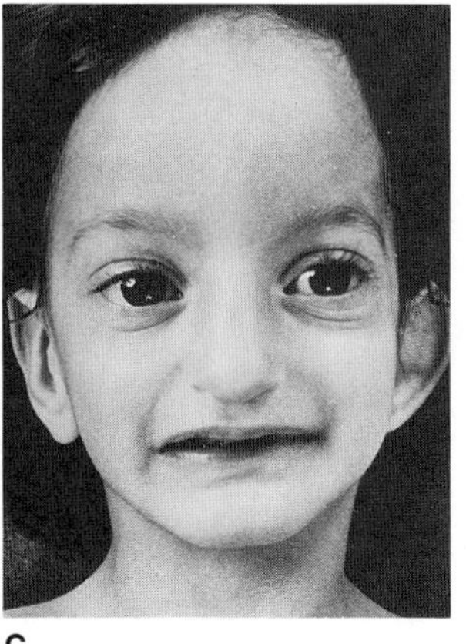

Fig. 29–29. *Unusual facies, mental retardation, and intrauterine growth retardation.* (A–C) Two sibs and isolated child with reduced head circumference, beaked nose, wide mouth, short upper lip with flat philtrum, and prominent eyes. Intelligence quotients ranged from 35 to 49. (From *DB Pitt et al,* Am J Med Genet **19:**307, 1984.)

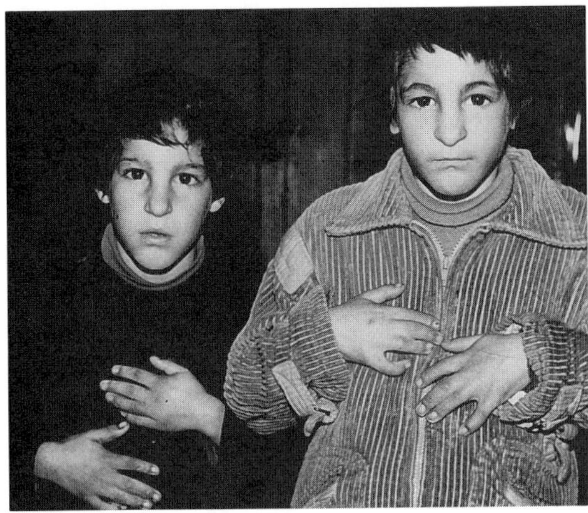

A

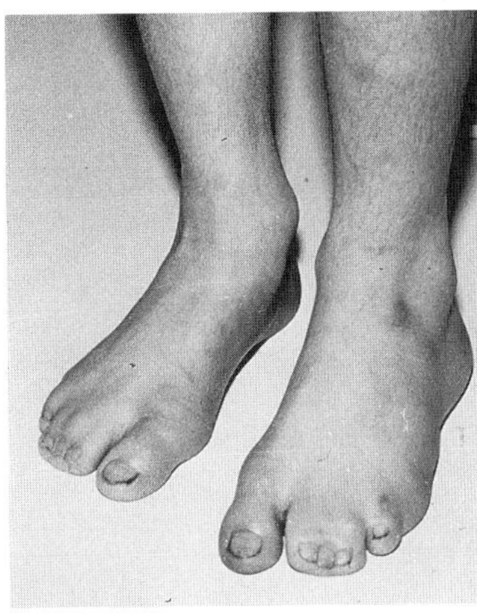

B

Fig. 29–30. *Unusual facies, microcephaly, growth and mental retardation, and syndactyly.* (A) In addition to microcephaly and moderate mental retardation, the sibs have similar facies. Note broad nasal root, lack of alar flare, and syndactyly. (B) Both brothers had syndactyly of fingers 3–4 and toes 2–3–4. (From *G Filippi,* Am J Med Genet **22:**821, 1985.)

## Unusual facies, microcephaly, growth and mental retardation, and syndactyly

Filippi (1), in 1985, described three sibs, two males, one female, with unusual facial appearance, retarded somatic and mental development, inability to speak, and syndactyly of fingers and toes. The parents were nonconsanguineous. Inheritance is probably autosomal recessive.

The facies was characterized by a somewhat protruding forehead, broad and prominent nasal root, and diminished alar flare. There was bilateral syndactyly of the third and fourth fingers with clinodactyly of the fifth fingers and bilateral syndactyly of toes 2–3–4. Both boys had cryptorchidism (Fig. 29-30A,B).

There was some overlap with the facies seen in the *blepharo-naso-facial syndrome* and *KBG syndrome.* However, the patients lacked both eye and dental findings.

### Reference (Unusual facies, microcephaly, growth and mental retardation, and syndactyly)

1. Filippi G: Unusual facial appearance, microcephaly, growth and mental retardation, and syndactyly. A new syndrome? *Am J Med Genet* **22**:821–824, 1985.

Fig. 29–31. *Unusual facies, short limbs, and congenital heart disease.* (A) Rhizomelic shortening of upper limbs more severe than of lower limbs, micrognathia, short neck, talipes equinus. (B,C) Shortened and thickened long bones. Femora and humeri especially shortened. (From *M Barrow* and *JS Fitzsimmons,* Am J Med Genet **18:**431, 1984.)

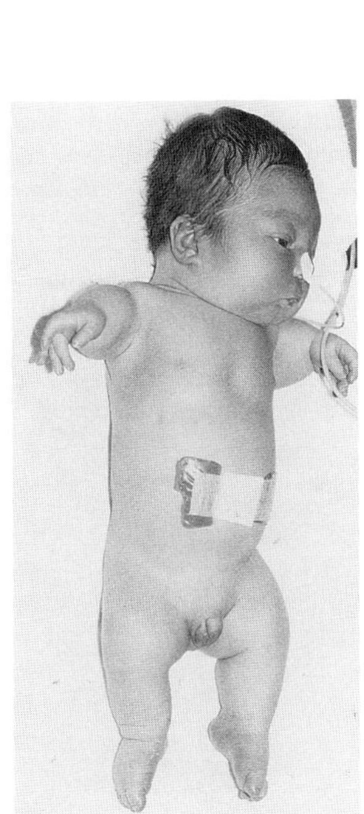

A

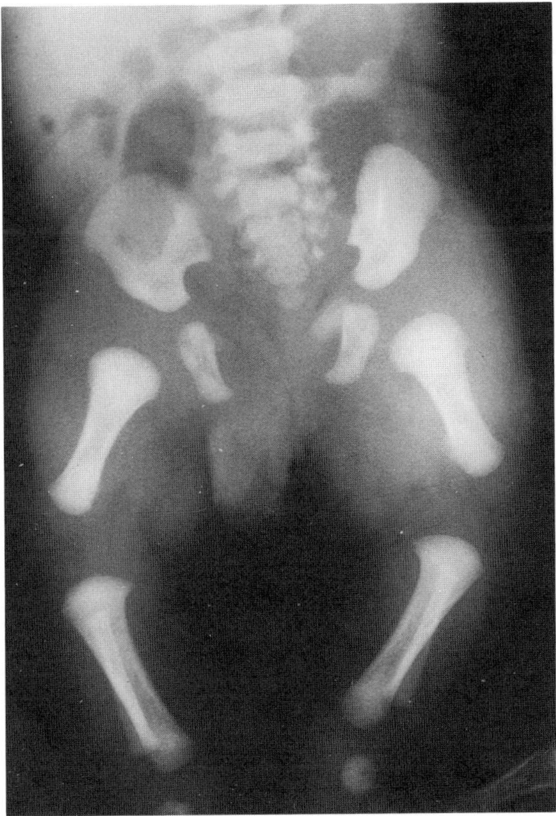

B

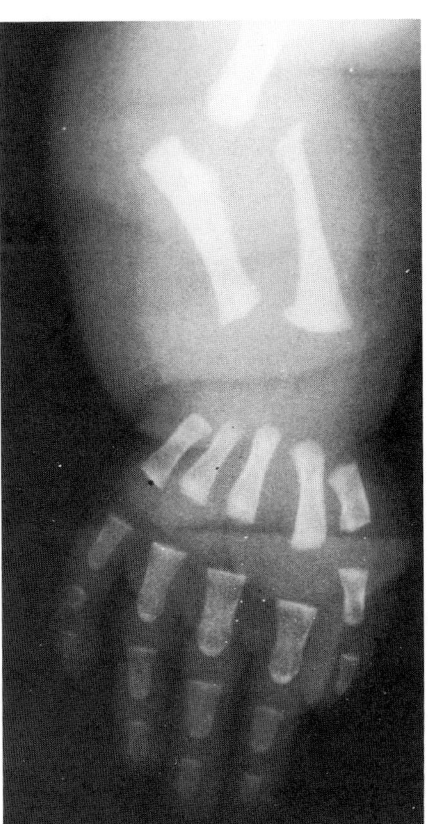

C

## Unusual facies, short limbs, and congenital heart disease

Barrow and Fitzsimmons (1) described an apparently unique lethal disorder in a male infant. The child had severe rhizomelic shortening of the limbs, more marked in the upper extremities, and mild talipes equinus.

There were posterior sloping of the forehead, prominent nasal bridge, epicanthal folds, narrow palpebral fissures with mild upslant, capillary hemangiomas involving the glabellar area and both lips, thin vermillion, micrognathia, and short neck (Fig. 29-31A–C).

Postmortem examination revealed a single ventricle leading into the truncus arteriosus that divided into right and left pulmonary arteries and continued as the aorta. The testes were intraabdominal.

### Reference (Unusual facies, short limbs, and congenital heart disease)

1. Barrow M, Fitzsimmons JS: Short limbs, abnormal facial appearance and congenital heart defect. *Am J Med Genet* **18**:431–433, 1984.

## Unusual facies, short stature, and hypoplastic penis (Floating–Harbor syndrome, Pelletier–Leisti syndrome)

Pelletier and Feingold (3) and Leisti et al (1,2) independently reported males with strikingly similar phenotype. Several additional cases were reported by Robinson et al (4). We cannot accept as an example the patient reported by Zabransky (5). There is no sex or racial predilection, and all have been isolated examples. D Donnai (personal communication, 1989) has informed us that paternal age is advanced, suggesting autosomal dominant inheritance.

Birthweight and length are usually at or below the 5th percentile. There is short stature (usually −4 to −6 SD), triangular facies, prominent occiput, deep-set eyes, long eyelashes, large bulbous nose, posteriorly angulated pinnae, thin lips, short philtrum, wide downturned mouth, short neck, hirsutism, low posterior hair line, brachydactyly, clinodactyly, loose joints, hypoplastic penis, mild mental retardation, delayed bone age, and significant speech delay (Fig. 29-32A,B). Pseudoarthrosis of the right clavicle was seen in the patient of Leisti et al (1) and in another seen by D. Donnai (personal communication, 1989).

The name Floating-Harbor syndrome is derived from one example seen at Boston Floating Hospital, the other being reported from Harbor General Hospital.

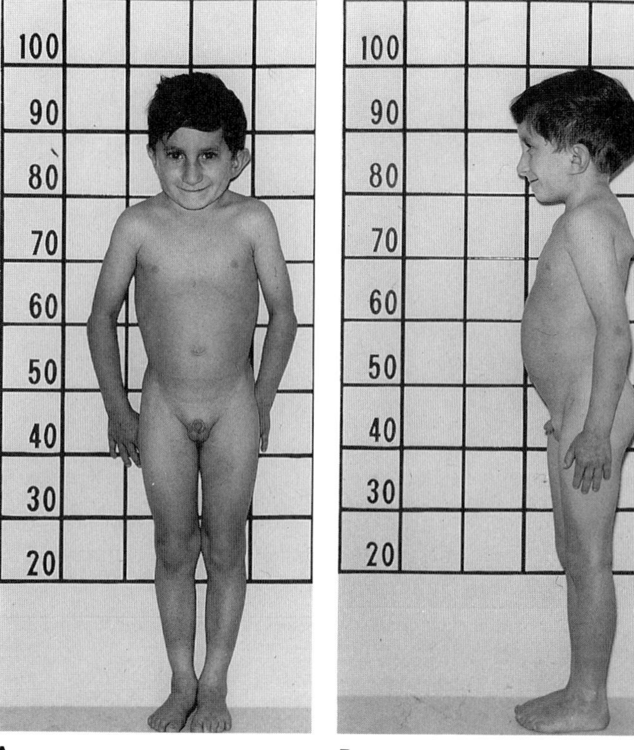


A B

Fig. 29–32. *Unusual facies, short stature, and hypoplastic penis.* (A,B) Patient at 6 years, 8 months. Notice proportionate short stature, dolichocephaly, triangular facies, deep-set eyes, large nose, thin lips, short philtrum, low hair line, short neck, large and wide nose, low-set and posteriorly rotated ears, small penis, and short fingers. [From *J Leisti et al,* Syndrome Ident **2**(1):3, 1974.]

### References [Unusual facies, short stature, and hypoplastic penis (Floating–Harbor syndrome, Pelletier–Leisti syndrome)]

1. Leisti J et al: Case report 12. *Syndrome Ident* **2**(1):3–5, 1974.
2. Leisti J et al: The Floating–Harbor syndrome. *Birth Defects* **11**(5):305–309, 1975.
3. Pelletier G, Feingold M: Case report 1. *Syndrome Ident* **1**(1):8–9, 1973.
4. Robinson PL et al: A unique association of short stature, dysmorphic features, and speech impairment (Floating–Harbor syndrome). *J Pediatr* **113**:703–706, 1988.
5. Zabransky S: Das Floating-Harbor-Homburg Syndrom *Akt Endokr Stoffw* **6**:25–29, 1985.

## Unusual facies, vitiligo, canities, and progressive spastic paraplegia

Lison et al (1) reported three affected sibs with a distinct facies, premature graying, vitiligo, and progressive spastic paraplegia.

The cutaneous findings included diffuse lentigines, graying of scalp and body hair, vitiligo, and hyperpigmentation of exposed areas (Fig. 29-33A,B).

All exhibited decreased tendon reflexes, positive Babinski reflex, foot drop, and spastic gait. Thoracic scoliosis, lumbar lordosis, and loss of muscle mass of the lower limbs were evident (Fig. 29-34A,B).

The face tended to be small and thin, producing moderately sharp features including pointed nose and chin and high cheek bones.

Inheritance is autosomal recessive.

Fig. 29–33. *Unusual facies, vitiligo, canities, and progressive spastic paraplegia.* (A,B) Premature graying of scalp hair, eyebrows, eyelashes. Note midfacial vitiligo, sharpened features. (From *M Lison et al,* Am J Med Genet **9:**351, 1981.)

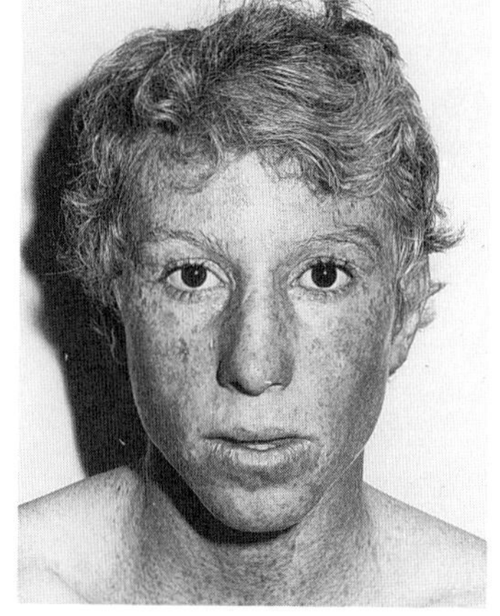

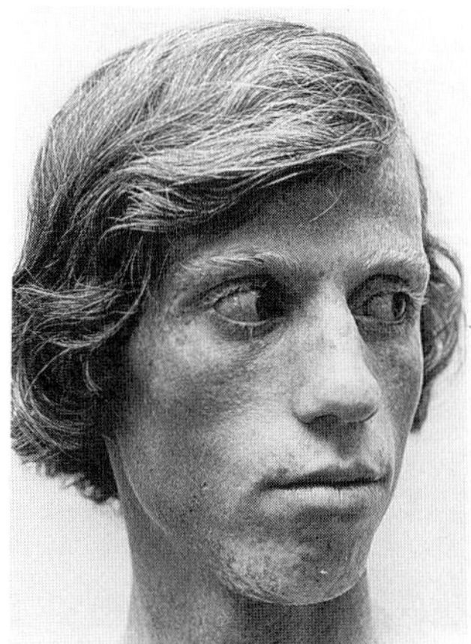

A B

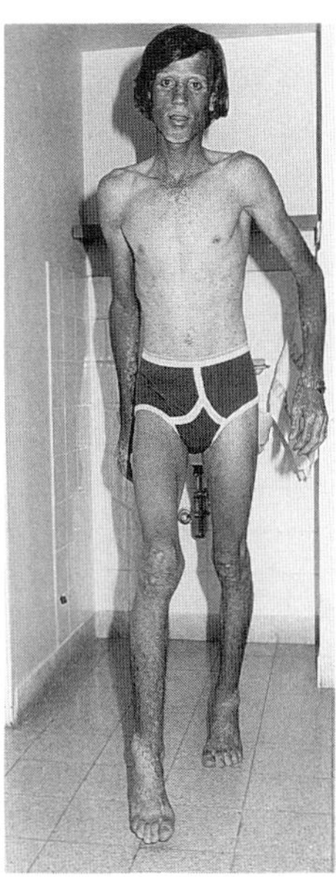

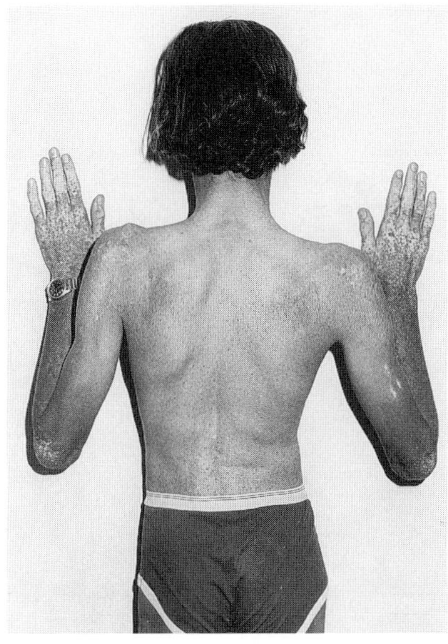

A B

Fig. 29–34. *Unusual facies, vitiligo, canities, and progressive spastic paraplegia.* (A,B) Vitiligo, lentigines, mild scoliosis, muscle wasting, and tip-toe gait. (From *M Lison et al,* Am J Med Genet **9**:351, 1981.)

#### Reference (Unusual facies, vitiligo, canities, and progressive spastic paraplegia)

1. Lison M et al: Progressive spastic paraparesis, vitiligo, premature graying, and distinct facial appearance: A new genetic syndrome in 3 sibs. *Am J Med Genet* **9**:351–357, 1981.

### Unusual facies, lymphedema, intestinal lymphangiectasia, and mental retardation (Hennekam syndrome)

Hennekam et al (1), in 1988, described three sibs and their cousin with mental retardation, lymphedematous facies, marked edema of the lower extremities, and gastrointestinal lymphangiectasia.

Inheritance is autosomal recessive. The family came from a highly inbred kindred from a small fishing town on the Zuider Zee.

Although the patients were of Dutch extraction, the facies appeared oriental due to periorbital edema. Hypertelorism, frontal upsweep, heavy eyebrows, epicanthal folds, and depressed nasal bridge gave the face a flat appearance. Some had dysmorphic pinnae, microstomia, narrow palate, retrognathia, and unusual teeth: oligodontia and incisors with conical crown form (Fig. 29-35A,B).

Lymphedema, present before birth, causes the characteristic facies. After birth, the swelling progresses and causes massive enlargement of the lower extremities and genitalia. Lymph may ooze from the swollen area and become infected, resulting in erysipelas. Intestinal lymphangiectasia of the small bowel was demonstrated in all affected, causing protein-losing enteropathy and attendant complications, such as mild growth retardation.

Mental retardation was mild to moderate in all affected and three of the four exhibited seizures.

There is some overlap with the autosomal recessive disorder described by Mücke et al (2).

Fig. 29–35. *Unusual facies, lymphedema, intestinal lymphangiectasia, and mental retardation.* (A) This boy shows congenital generalized edema but especially that of the lower extremities. (B) Hypertelorism, periorbital edema, epicanthal folds, and depressed nasal bridge impart a flat appearance to the face. (Courtesy of *RCM Hennekam,* Utrecht, The Netherlands.)

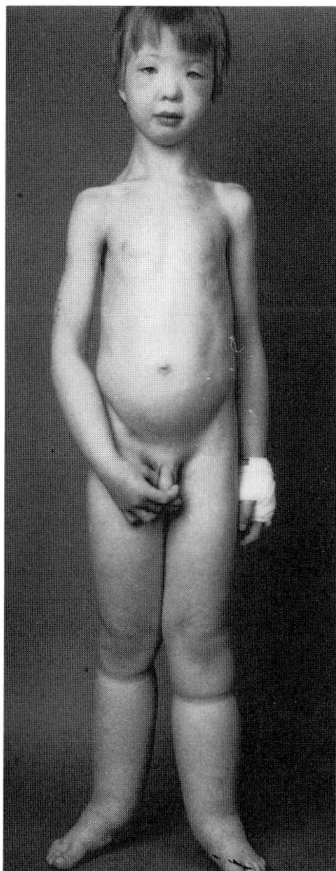

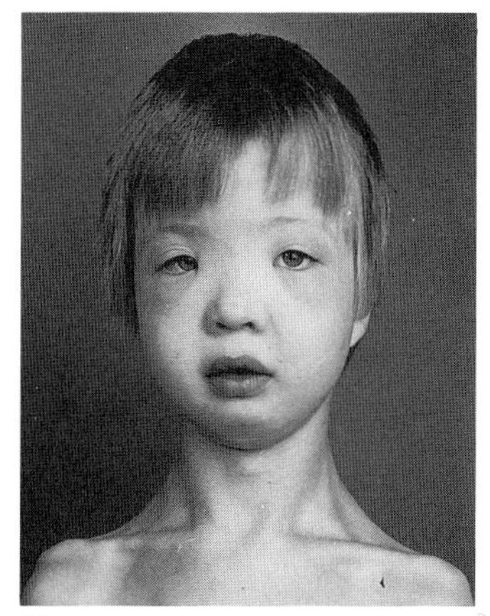

A B

#### References [Unusual facies, lymphedema, intestinal lymphangiectasia, and mental retardation (Hennekam syndrome)]

1. Hennekam RCM et al: Autosomal recessive intestinal lymphangiectasia and lymphedema with facial anomalies and mental retardation. *Am J Med Genet* **34**:593–600, 1989.

2. Mücke J et al: Early onset lymphoedema, recessive form—a new form of genetic lymphoedema syndrome. *Eur J Pediatr* **145**:195–198, 1986.

### Unusual facies–serpentine fibula–polycystic kidney syndrome

Dereymaeker et al (1) and Exner (2) each reported a female child with short stature, coarse hair, hirsutism of the forehead and neck, heavy eyebrows, and marked micrognathia. Both girls had irregular serpentine fibulae, bilateral pes valgus, and metatarsus adductus. The child reported by Exner had prominent tibial anterior tendon. The neck was short and the voice was hoarse (Fig. 29-36A–D). Polycystic kidney was found in the patient of Exner. The condition should be differentiated from *Melnick–Needles syndrome.*

#### References (Unusual facies–serpentine fibula–polycystic kidney syndrome)

1. Dereymaeker AM et al: Melnick–Needles syndrome (osteodysplasty). *Helv Paediatr Acta* **41**:339–351, 1986 (Case 1).

2. Exner GU: Serpentine fibula-polycystic kidney syndrome. A variant of the Melnick–Needles syndrome or a distinct entity? *Eur J Pediatr* **147**: 544–546, 1988.

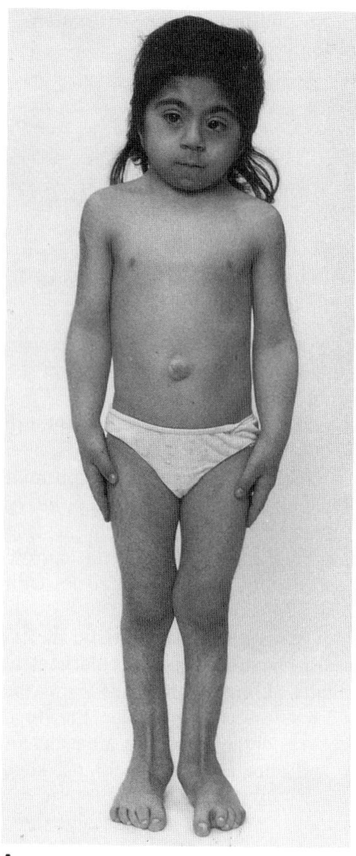
A

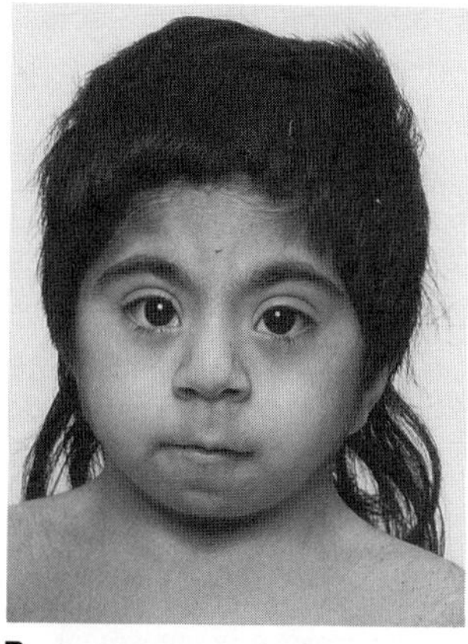
B

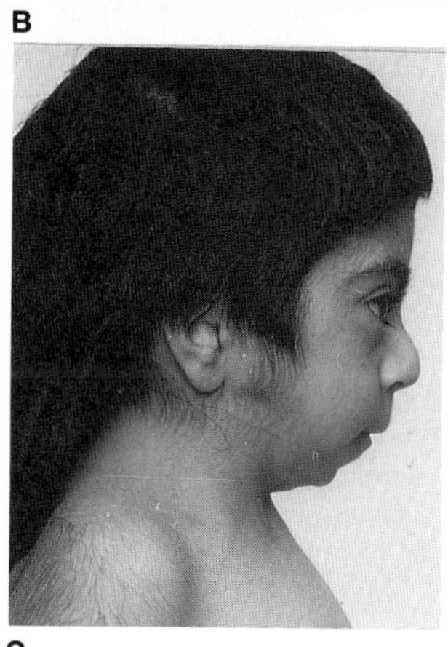
C

D

Fig. 29–36. *Unusual facies–serpentine fibula–polycystic kidney syndrome.* (A–C) Five-year-old female with unusual facial appearance characterized by short neck, hirsute forehead and neck, thick eyebrows, micrognathia, pes valgus, metatarsus adductus, and prominent tibialis anterior tendon. (D) Bilateral serpentine fibulae. (A–D from *GU Exner,* Eur J Pediatr **147:**544, 1988.)

## Unusual facies, cleft palate, microcephaly, hypoplastic nose, osteosclerosis, exophthalmos, gingival hyperplasia, and low set ears

Raine et al (1), in 1989, described a small female infant at full term with microcephaly, large open anterior and posterior fontanels, marked exophthalmos, maxillary hypoplasia with sunken nasal bridge and small nose, gingival hyperplasia, cleft soft palate, posteriorly rotated pinnae, and diffuse sclerosis of the bones.

### Reference (Unusual facies, cleft palate, microcephaly, hypoplastic nose, osteosclerosis, exophthalmos, gingival hyperplasia, and low set ears)

1. Raine J et al: Microcephaly, hypoplastic nose, exophthalmos, gum hyperplasia, cleft palate, low set ears, and osteosclerosis. *J Med Genet* **26**:786–788, 1989.

## Unusual facies, congenital hypothyroidism, and severe mental retardation

Young and Simpson (3), in 1987, reported a syndrome of unusual facies, congenital hypothyroidism and severe global retardation. Similar examples have subsequently been described (1,2). The craniofacies was characterized by microcephaly, blepharophimosis, bulbous nose, thin upper lip, low set posteriorly rotated pinnae, and micrognathia. Inconstant features have been congenital heart anomaly (3) and postaxial polydactyly of hands and feet (1).

### References (Unusual facies, congenital hypothyroidism, and severe mental retardation)

1. Cavalcanti DP: Unknown syndrome: Abnormal facies, hypothyroidism, postaxial polydactyly and severe retardation: A third patient. *J Med Genet* **26**:785–786, 1989.

2. Fryns JP, Moerman P: Unknown syndrome: Abnormal facies, hypothyroidism and severe retardation: A second patient. *J Med Genet* **25**:498–499, 1988.

3. Young ID, Simpson K: Unknown syndrome: Abnormal facies, congenital heart defects, hypothyroidism, and severe retardation. *J Med Genet* **24**:715–716, 1987.

## Unusual facies, lipodystrophy, and joint contractures

Hoepffner et al (1) described a Werner syndrome-like disorder in three male siblings. Scleroderma-like alterations of the skin, sharp facies, and joint contractures were noted. One had insulin resistant diabetes mellitus. Impaired function of three related peptide growth factors was found at the post-receptor level. The related peptides were insulin, insulin-like growth factor and epidermal growth factor.

### Reference (Unusual facies, lipodystrophy, and joint contractures)

1. Hoepffner HJ et al: A new familial syndrome with impaired function of three related peptide growth factors. *Hum Genet* **83**:209–216, 1989.

# Appendix

Documentation of dysmorphic features is an essential part of clinical evaluation. Although some features can only be visualized, others require quantification to decide whether the feature in question deviates from the norm and, if so, by how much. Objective measurements often belie subjective clinical impression. For example, the patient with Down syndrome is sometimes described as having a highly arched palate. However, metric studies show that palatal height is normal, but palatal length is dramatically shorter than normal. Some features such as epicanthic folds cannot be precisely measured, but can be graded as being mild, moderate, or severe. Other features such as Brushfield spots are of the presence–absence type. Because some dysmorphic features defy measurement and even convincing verbal description, photographic documentation is very important in syndromology; phenotypic features may change with time. In some conditions, such as the de Lange syndrome, the phenotypic features, especially of the face, are usually constant enough at any age to impart the correct diagnostic impression, and the striking resemblance of affected individuals transcends the racial background of the patient. Occasionally, mild expression may render clinical diagnosis difficult, especially at birth. In some instances, for example, the characteristic phenotype of the de Lange syndrome may evolve with time. In other conditions, such as del(5p) syndrome, the phenotypic features tend to become less distinct with time (3).

This appendix provides a number of measurements of use in evaluating the face and cranium. The more practical and efficient but less accurate measures are most commonly used. Soft-tissue measurements take precedence over radiographic ones and the use of simple rather than sophisticated techniques is the rule. In this appendix, anthropometric measurements of the face and cranium are provided, but in addition there are a large number of radiographic and cephalometric standards. In some instances, several standards for the same measurement are provided. For example, inner and outer canthal distances suffice for clinical measurement of hypertelorism. However, when craniofacial surgery is contemplated, radiographic standards for interorbital distance are mandatory. An excellent study of bony interorbital distance has been provided by Costaras et al (5,6) and is highly recommended; their standards are not found here because their use of PA cephalometric landmarks and the indices they derive are too complex to present in this appendix. Finally, the canthal index is useful for evaluating possible hypertelorism in relatives of a proband who are not available for study but whose photographs are available.

The best general group of growth references for dysmorphic conditions is found in the atlas of Saul et al (13). The reader is referred to the following references for special coverage: facial anthropometry (7), craniofacial anthropometric measurements in the newborn (27–41 gestational weeks) (1,8,10), anthropometric measurements of various syndromes and genetic conditions (7a,13–15), cephalometric atlases (2,11,12), cephalometric studies of syndromes and genetic conditions (4), and limitations of published metric studies for clinical use (3).

### References (Appendix)


1. Birholz JC, Farrell EE: Fetal ear length. *Pediatrics* **81**:555–558, 1988.
2. Broadbent RB et al: *Bolton Standards at Dentofacial Developmental Growth.* CV Mosby, St. Louis, 1975.
3. Cohen MM Jr: *The Child with Multiple Birth Defects.* Raven Press, New York, 1982.
4. Cohen MM Jr: A critical review of cephalometric studies of dysmorphic syndromes. Koski Festscrift. *Proc Finn Dent Soc* **77**:17–25, 1981.
5. Costaras M, Pruzansky S: Bony interorbital distance (BIOD), head size, and level of the cribriform plate relative to orbital height: II. Possible pathogenesis of orbital hypertelorism. *J Craniofac Genet Dev Biol* **2**:19–34, 1982.
6. Costaras M et al: Bony interorbital distance (BIOD), head size, and level of the cribriform plate relative to orbital height: I. Normal standards for age and sex. *J Craniofac Genet Dev Biol* **2**:5–18, 1982.
7. Farkas LG: *Anthropometry of the Hand and Face in Medicine.* Elsevier, Amsterdam, 1981.

7a. Farkas LG et al: Orbital measurements in 63 hyperteloric patients: Differences between anthropometric and cephalometric findings. *J Cranio-Max-Fac Surg* **17**:249–254, 1989.

8. Keen DV, Pearse RG: Weight, length and head circumference curves for boys and girls of between 20 and 42 weeks gestation. *Arch Dis Childh* **63**:1170–1172, 1988.
9. Meany FJ, Farrer LA: Clinical anthropometry and medical genetics: A compilation of body measurements in genetic and congenital disorders. *Am J Med Genet* 25:343–359, 1986.
10. Merlob P et al: Anthropometric measurements of the newborn infant (27–41 gestational weeks). *Birth Defects* **20**(7):1–32, 1984.
11. Riolo ML et al: *An Atlas of Craniofacial Growth: Cephalometric Standards from the University School Growth Study, The University of Michigan.* Monograph No. 2, Craniofacial Growth Series, Center for Human Growth and Development, University of Michigan, 1974.
12. Saksena SS et al: *A Clinical Atlas of Roentgencephalometry in Norma Lateralis.* Alan R. Liss, New York, 1987
13. Saul RA et al: Growth References from Conception to Adulthood. *Proc Greenwood Genet Ctr,* Suppl No 1, 1988.
14. Stengel-Rutkowski S et al: Anthropometric definitions of dysmorphic facial signs. *Hum Genet* **67**:272–295, 1984.
15. Ward RE: Facial morphology as determined by anthropometry: Keeping it simple. *J Craniofac Genet Dev Biol* **9**:45–60, 1989.

# Appendix Figures and Tables

## Cranial measurements



## Facial measurements



## Cephalometric measurements



### Angular measurements



### Linear measurements



## Intraoral measurements

## CRANIAL MEASUREMENTS

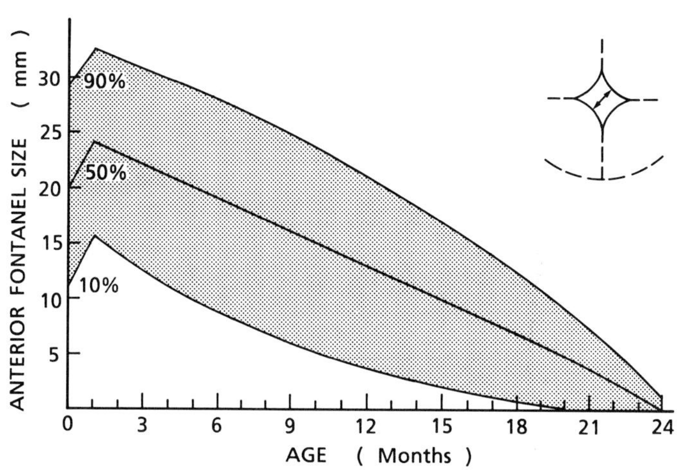


Fig. A–1. Anterior fontanel size. Percentiles from term to 24 months of age for both sexes. Inset, measurement of oblique diameter. (From *G Duc* and *RH Largo,* Pediatrics **78:**904, 1986.)

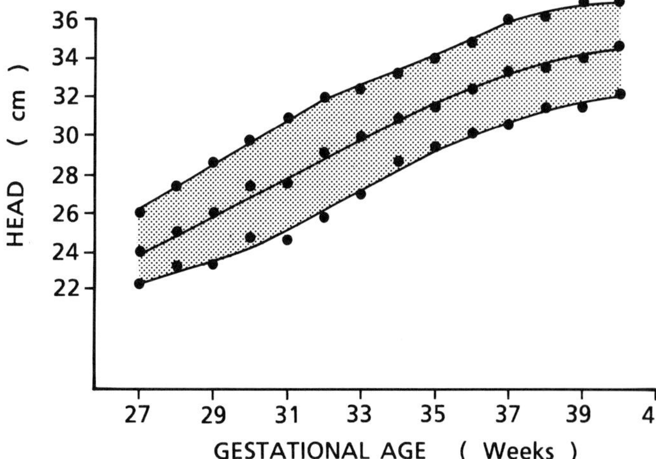


Fig. A–2. Head circumference from 27 to 41 gestational weeks. [From *P Merlob et al,* Birth Defects **20**(7):1, 1984.]

Table A–1. Fontanel and suture closure[a]

| Closure | Time |
|---|---|
| Anterior fontanel | 1 year ± 4 months |
| Posterior fontanel | Birth ± 2 months |
| Anterolateral fontanel | By third month |
| Posterolateral fontanel | During second year |
| Metopic suture | By third year (10%, never) |
| Clinical closure of sutures | 6–12 months |
| Anatomic closure of sutures | By thirtieth year |

[a]From RJ Gorlin and JJ Pindborg, *Syndromes of the Head and Neck,* 1st ed. McGraw-Hill, New York, 1964.

Table A–2. Detailed suture closure[a]

| Cranial suture | Closure begins (years) | Facial suture | Closure begins (years) |
|---|---|---|---|
| Interfrontal (metopic)[b] | 2 | Intermaxillary | |
| Interparietal (midsagittal) | 22 | (midpalatal) | 30–35 |
| Frontoparietal (coronal) | 24 | Frontomaxillary | 68–71 |
| Occipitoparietal (lambdoidal) | 26 | Frontonasal | 68 |
| Masto-occipital | 26–30 | Frontozygomatic | 72 |
| Sphenotemporal | 28–32 | Zygomaticotemporal | 70–71 |
| Temporoparietal (squamosal) | 35–39 | Zygomaticomaxillary | 70–72 |
| Sphenoparietal | 29 | Nasomaxillary | 68 |
| Sphenofrontal | 22 | | |

[a]From VC Kokich, The biology of sutures, in *Craniosynostosis: Diagnosis, Evaluation, and Management,* MM Cohen Jr (ed), Raven Press, New York, 1986, p 94.

[b]Usually obliterated by third year; persists throughout life in 10%.

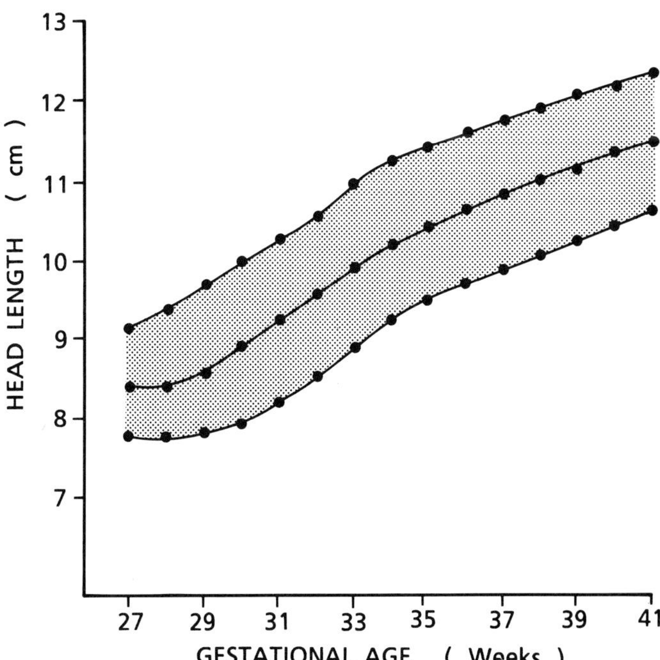


Fig. A–3. Head length from 27 to 41 gestational weeks. [From *P Merlob et al,* Birth Defects **20**(7):1, 1984.]

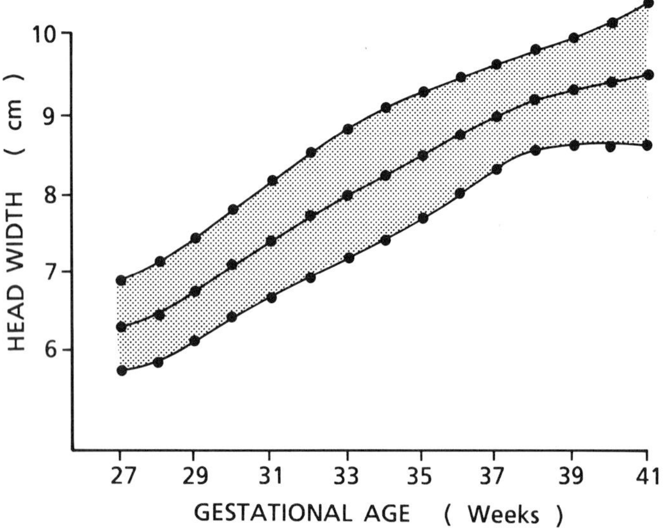


Fig. A–4. Head width from 37 to 41 gestational weeks. [From *P Merlob et al,* Birth Defects **20**(7):1, 1984.]

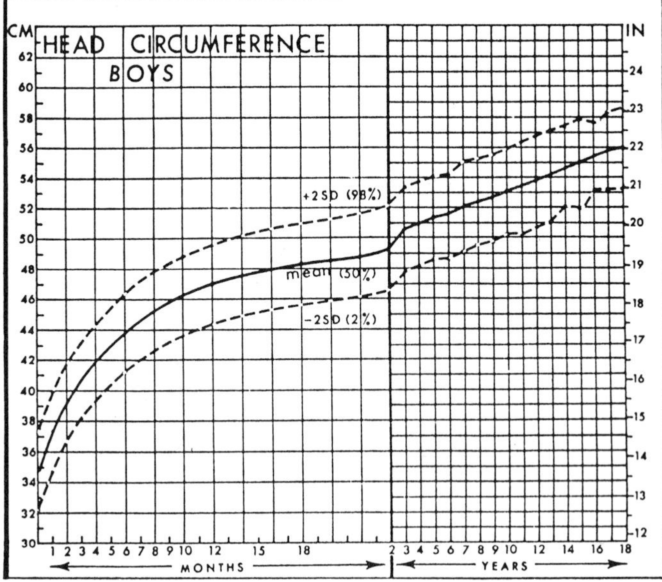


Fig. A–5. Head circumference to 18 years (males). (From *G Nelhaus,* Pediatrics **41:**106, 1968.)

Table A–3. Cranial measurements and volume (males)[a,b]

| Age | *N* | Head breadth Mean | Head breadth SD | Head length Mean | Head length SD | Head height Mean | Head height SD | Mean cranial volume |
|---|---|---|---|---|---|---|---|---|
| 7 days | 19 | 9.6 | 0.32 | 11.9 | 0.57 | 8.5 | 0.52 | 410 |
| 1 mo | 24 | 10.1 | 0.45 | 12.7 | 0.70 | 9.2 | 0.57 | 522 |
| 2 mo | 22 | 10.3 | 0.54 | 13.7 | 0.61 | 9.8 | 0.62 | 609 |
| 4 mo | 20 | 11.3 | 0.50 | 14.4 | 0.81 | 10.5 | 0.60 | 766 |
| 6 mo | 21 | 11.8 | 0.59 | 15.2 | 0.91 | 11.0 | 0.46 | 847 |
| 9 mo | 19 | 12.0 | 0.71 | 15.4 | 0.93 | 11.1 | 0.59 | 896 |
| 12 mo | 19 | 12.5 | 0.62 | 16.4 | 0.62 | 11.3 | 0.39 | 1010 |
| 18 mo | 20 | 12.9 | 0.50 | 16.8 | 0.71 | 11.8 | 0.69 | 1078 |
| 2 yr | 20 | 13.1 | 0.46 | 17.1 | 0.86 | 11.9 | 0.53 | 1129 |
| 3 yr | 29 | 13.5 | 0.51 | 17.8 | 0.61 | 12.2 | 0.47 | 1206 |
| 4 yr | 29 | 13.7 | 0.58 | 17.7 | 0.71 | 12.4 | 0.44 | 1267 |
| 5 yr | 30 | 13.9 | 0.48 | 17.8 | 0.61 | 12.4 | 0.66 | 1281 |
| 6 yr | 22 | 14.0 | 0.54 | 18.1 | 0.66 | 12.4 | 0.45 | 1309 |
| 7–8 yr | 30 | 14.4 | 0.49 | 18.2 | 0.99 | 12.3 | 0.42 | 1354 |
| 9–10 yr | 44 | 14.4 | 0.54 | 18.7 | 0.65 | 12.3 | 0.58 | 1384 |
| 11–12 yr | 49 | 14.5 | 0.57 | 18.9 | 0.64 | 12.5 | 0.44 | 1424 |
| 13–14 yr | 35 | 14.6 | 0.49 | 19.2 | 0.66 | 12.7 | 0.51 | 1480 |
| 15–16 yr | 27 | 14.9 | 0.59 | 19.3 | 0.87 | 12.8 | 0.42 | 1492 |
| 17–18 yr | 46 | 15.3 | 0.62 | 19.5 | 0.73 | 12.8 | 0.44 | 1533 |
| 19–20 yr | 30 | 15.3 | 0.52 | 19.5 | 0.71 | 12.8 | 0.44 | 1548 |

[a]From AS Dekaban, *Ann Neurol* **2**:285, 1977.
[b]Linear measurements in centimeters, volume in milliliters.

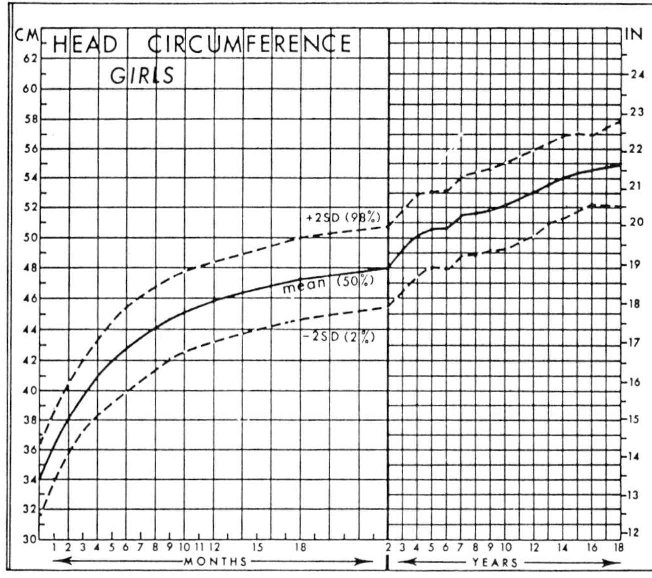


Fig. A–6. Head circumference to 18 years (females). (From G *Nelhaus*, Pediatrics **41:**106, 1968.)

Table A–5. Cephalic index[a,b]

| Index | Skull type |
|---|---|
| $x$–75.9 | Dolichocephaly |
| 76.0–80.9 | Mesocephaly |
| 81.0–85.4 | Brachycephaly |
| 85.5–$x$ | Hyperbrachycephaly |

[a]From MFA Montagu, *An Introduction to Physical Anthropology,* 3d ed. Thomas, Springfield, 1960, pp 570–571.
[b]Cephalic index = [(maximum head breadth)/(maximum head length)](100).

Table A–4. Cranial measurements and volume (females)[a,b]

| Age | N | Head breadth Mean | Head breadth SD | Head length Mean | Head length SD | Head height Mean | Head height SD | Mean cranial volume |
|---|---|---|---|---|---|---|---|---|
| 7 days | 19 | 9.3 | 0.41 | 11.6 | 0.30 | 8.3 | 0.57 | 398 |
| 1 mo | 21 | 8.8 | 0.57 | 12.8 | 0.56 | 9.2 | 0.58 | 510 |
| 2 mo | 20 | 10.3 | 0.68 | 13.4 | 0.88 | 9.5 | 0.57 | 529 |
| 4 mo | 19 | 10.8 | 0.41 | 14.0 | 0.74 | 10.0 | 0.62 | 721 |
| 6 mo | 23 | 11.4 | 0.45 | 14.7 | 0.56 | 10.4 | 0.44 | 748 |
| 9 mo | 20 | 12.0 | 0.41 | 15.3 | 0.74 | 10.8 | 0.55 | 845 |
| 12 mo | 19 | 12.3 | 0.65 | 15.6 | 0.59 | 11.0 | 0.50 | 913 |
| 18 mo | 21 | 12.7 | 0.57 | 16.3 | 0.84 | 11.4 | 0.66 | 1012 |
| 2 yr | 18 | 12.9 | 0.41 | 16.5 | 0.71 | 11.4 | 0.44 | 1033 |
| 3 yr | 21 | 13.1 | 0.57 | 17.0 | 0.65 | 11.9 | 0.45 | 1148 |
| 4 yr | 23 | 13.5 | 0.48 | 17.1 | 0.63 | 12.0 | 0.40 | 1180 |
| 5 yr | 25 | 13.6 | 0.48 | 17.3 | 0.58 | 11.9 | 0.35 | 1198 |
| 6 yr | 22 | 13.6 | 0.45 | 17.5 | 0.47 | 12.0 | 0.55 | 1218 |
| 7–8 yr | 29 | 13.8 | 0.76 | 17.9 | 0.62 | 12.0 | 1.17 | 1260 |
| 9–10 yr | 39 | 14.1 | 0.49 | 18.2 | 0.62 | 12.1 | 0.50 | 1310 |
| 11–12 yr | 31 | 14.3 | 0.50 | 18.2 | 0.69 | 12.2 | 0.58 | 1339 |
| 13–14 yr | 35 | 14.4 | 0.44 | 18.5 | 0.81 | 12.2 | 0.52 | 1371 |
| 15–16 yr | 31 | 14.5 | 0.48 | 18.6 | 0.67 | 12.2 | 0.51 | 1386 |
| 17–18 yr | 30 | 14.7 | 1.00 | 18.6 | 0.50 | 12.9 | 0.51 | 1402 |
| 19–20 yr | 37 | 14.7 | 0.39 | 18.6 | 0.95 | 12.4 | 0.42 | 1425 |

[a]From AS Dekaban, *Ann Neurol* **2**:285, 1977.
[b]Linear measurements in centimeters, volume in milliliters.

## FACIAL MEASUREMENTS

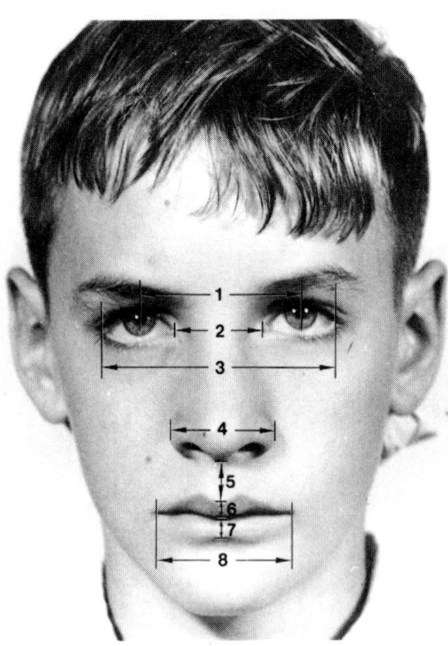


Fig. A–7. Common facial measurements. (1) Interpupillary distance, (2) inner canthal distance, (3) outer canthal distance, (4) interalar distance, (5) philtral length, (6) upper lip thickness, (7) lower lip thickness, and (8) intercommissural distance.

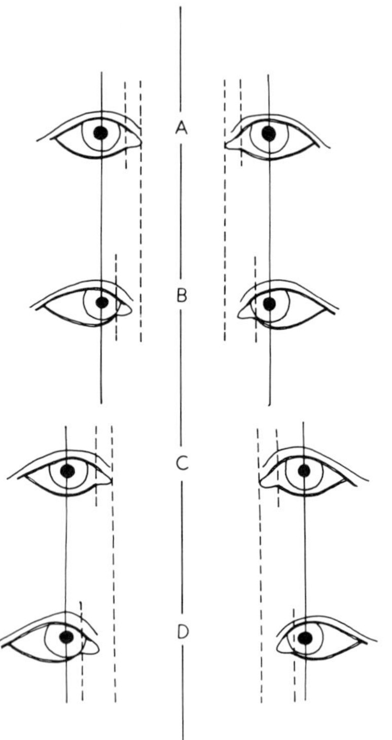


Fig. A–8. Primary telecanthus, secondary telecanthus, and hypertelorism. (A) Normal interocular distance. (B) Primary telecanthus. The inner canthi are far apart, although the outer canthi are normally spaced. Note how vertical line through lacrimal punctum cuts cornea. (C) True ocular hypertelorism. Both inner and outer canthi are abnormally far apart. (D) True ocular hypertelorism together with secondary telecanthus. Note how vertical line through lacrimal punctum cuts cornea. [From *MM Cohen Jr*, Malformation syndromes, in Surgical Correction of Dentofacial Deformities, WH Bell, WR Proffit, and RP White (eds), WB Saunders, Philadelphia, 1980, pp 7–44.]

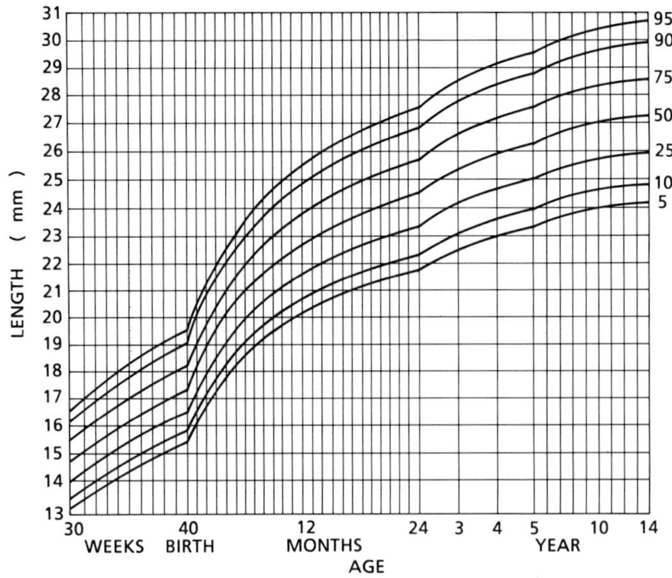


Fig. A–9. Palpebral fissure length. (From *IT Thomas et al,* J Pediatr **111**:267, 1987.)

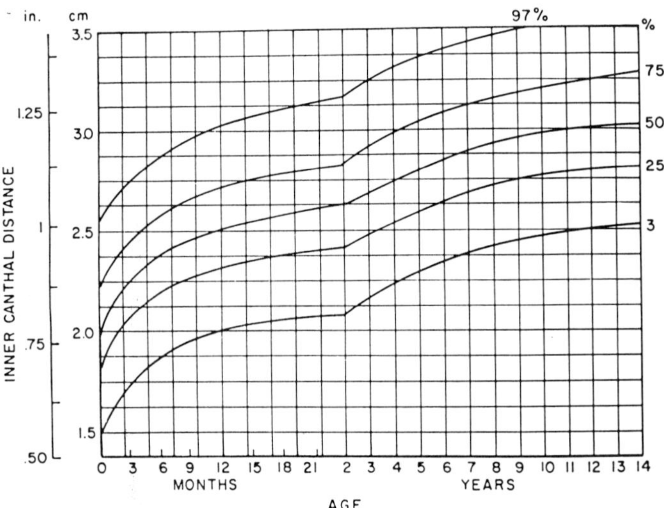


Fig. A–10. Inner canthal distance. [From *M Feingold* and *WH Bossert,* Birth Defects **10**(13):1, 1974.]

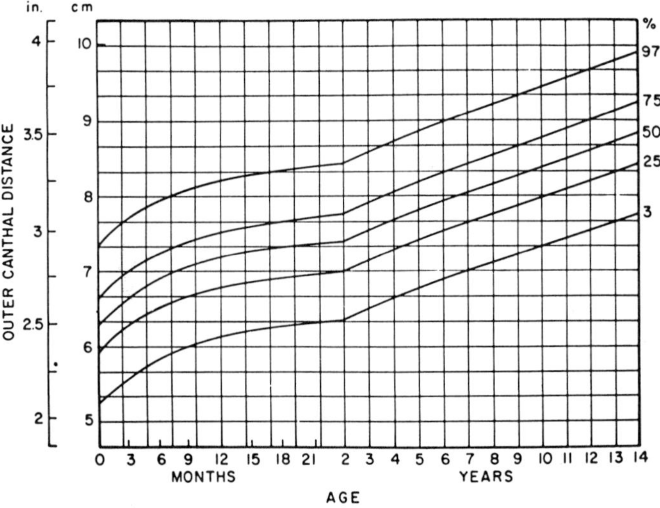


Fig. A–11. Outer canthal distance. [From *M Feingold* and *WH Bossert,* Birth Defects **10**(13):1, 1974.]

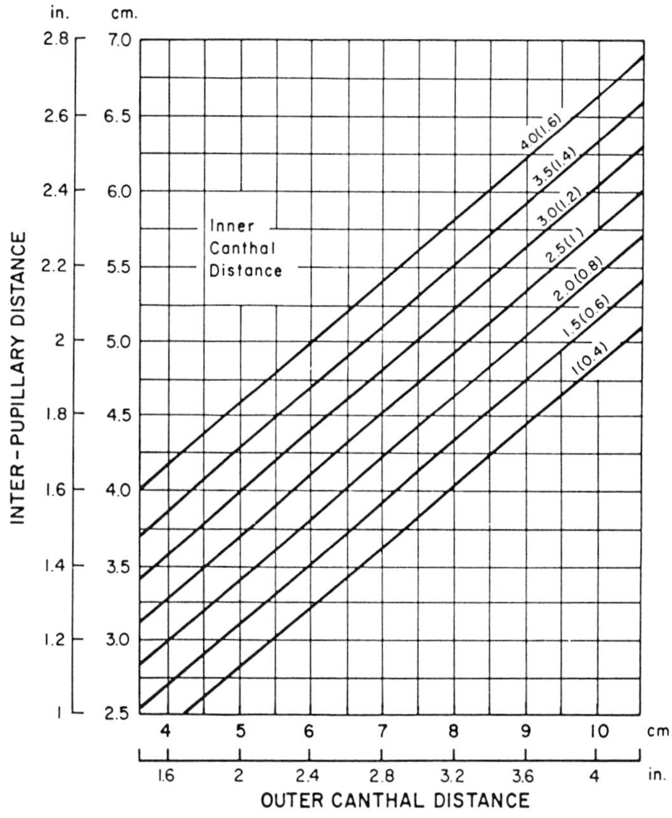


Fig. A–12. Interpupillary distance derived from inner and outer canthal distances. [From *M Feingold* and *WH Bossert,* Birth Defects **10**(13):1, 1974.]

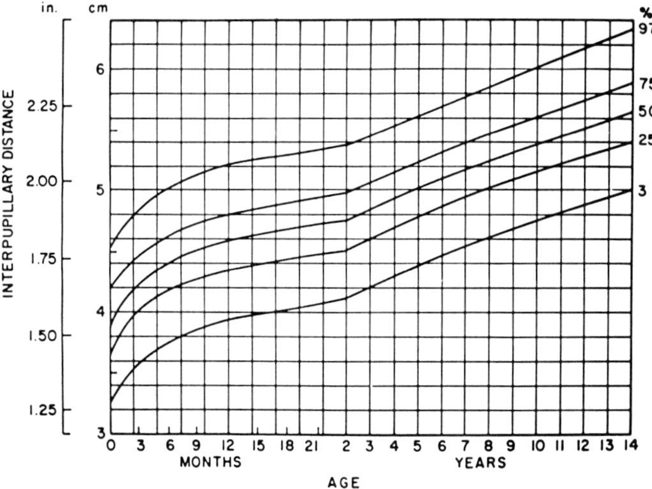


Fig. A–13. Direct interpupillary distance. [From *M Feingold* and *WH Bossert,* Birth Defects **10**(13):1, 1974.]

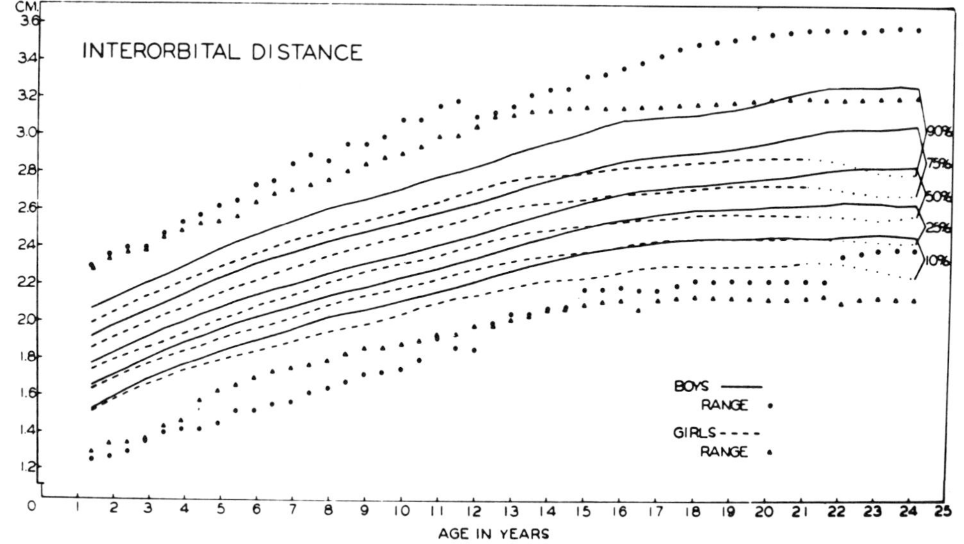


Fig. A–14. Bony interorbital distance. (From *CF Hansman,* Radiology **86**:87, 1966.)

Table A–6. Canthal index[a]

| Definitions | Use |
|---|---|
| CI = canthal index<br>IC = inner canthal distance<br>OC = outer canthal distance | For approximate distance between the eyes from photographs by using a ratio measure |

| Canthal index formula | Measurement | |
|---|---|---|
| $CI = \frac{IC}{OC} \times 100$ | 38 | Upper limit of normal |
| | 38–42 | Euryopia |
| | 42 and above | Hypertelorism |

[a]From H Günther, *Virchows Arch Pathol Anat* **290**:373, 1933.

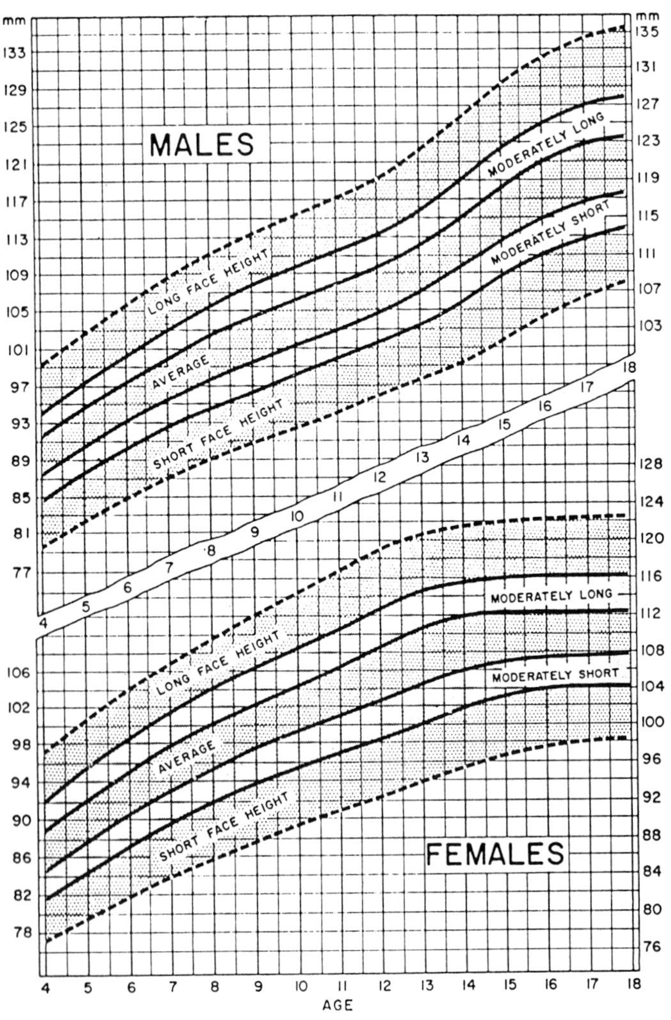


Fig. A–15. Face height based on North American white children. [From *HV Meredith,* Body size and form in childhood, with emphasis on the face, in *The Nature of Orthodontic Diagnosis,* SL Horowitz and EH Hixon (eds), Mosby, St. Louis, 1966, pp 1–30.]

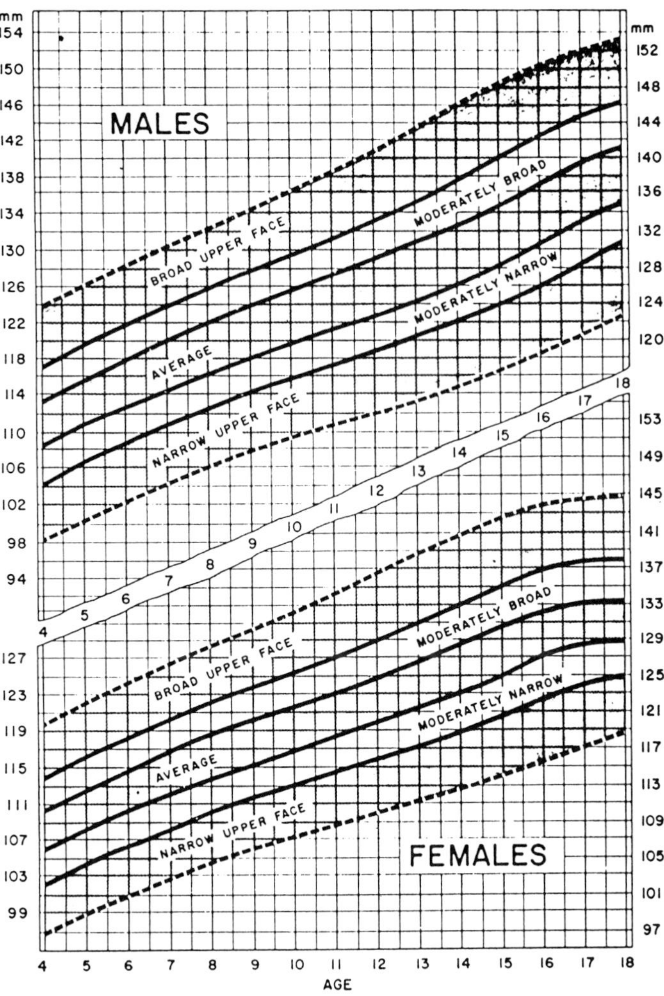


Fig. A–16. Face width based on North American white children. [From *HV Meredith,* Body size and form in childhood, with emphasis on the face, in The Nature of Orthodontic Diagnosis, SL Horowitz and EH Hixon (eds), CV Mosby, St. Louis, 1966, pp 1–30.]

Table A–7. Philtrum length and intercommissural distance from 29 to 42 weeks gestation[a]

| Gestational age (weeks) | *N* | Philtral length (mm) Mean | ± 2 SD | Oral intercommissural distance (mm) Mean | ± 2 SD |
|---|---|---|---|---|---|
| 28–29 | 7 | 6.0 | 0.3 | 21.4 | 4.0 |
| 30 | 11 | 7.3 | 2.4 | 20.3 | 1.8 |
| 31 | 12 | 6.7 | 0.6 | 22.8 | 2.9 |
| 32 | 10 | 7.3 | 1.4 | 23.7 | 4.2 |
| 33 | 18 | 7.8 | 1.8 | 24.4 | 2.8 |
| 34 | 19 | 8.3 | 2.0 | 24.8 | 4.0 |
| 35 | 13 | 8.1 | 1.6 | 24.5 | 3.7 |
| 36 | 15 | 8.5 | 0.8 | 25.0 | 5.1 |
| 37 | 20 | 9.0 | 1.5 | 26.2 | 4.0 |
| 38 | 20 | 9.1 | 1.8 | 29.0 | 5.6 |
| 39 | 28 | 9.5 | 1.1 | 27.5 | 3.2 |
| 40 | 41 | 9.8 | 1.4 | 27.9 | 3.1 |
| 41 | 25 | 9.6 | 1.2 | 28.0 | 4.2 |
| 42 | 13 | 9.5 | 2.0 | 28.9 | 5.2 |

[a]From K Mehes, *J Craniofacial Genet Dev Biol* **1**:213, 1981.

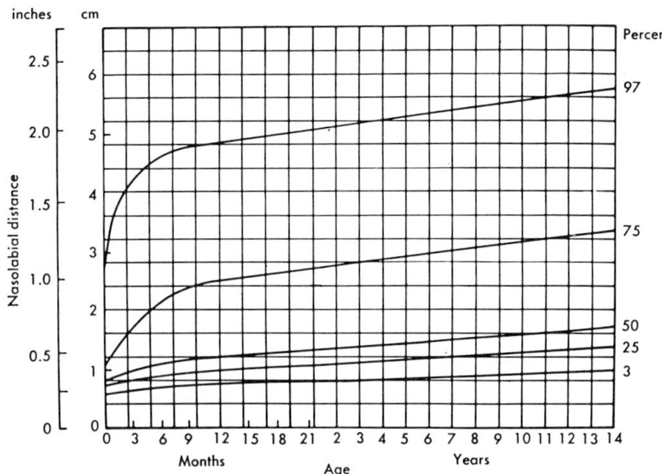


Fig. A–17. Philtrum length to age 14. (Courtesy of *M Feingold,* Boston.)

Table A–8. Intercommissural distance from birth to adulthood[a]

| Age (years) | Males (mm) | Females (mm) |
|---|---|---|
| 0–1 | 32 | 27 |
| 2–3 | 35 | 30 |
| 4–5 | 39 | 36 |
| 6–7 | 42 | 40 |
| 8–9 | 44 | 42 |
| 10–11 | 46 | 43 |
| 12–13 | 48 | 45 |
| 14–15 | 50 | 47 |
| Adult | 55 | 52 |

[a]From J Cervenka et al, *Am J Dis Child* **117**:434, 1969.

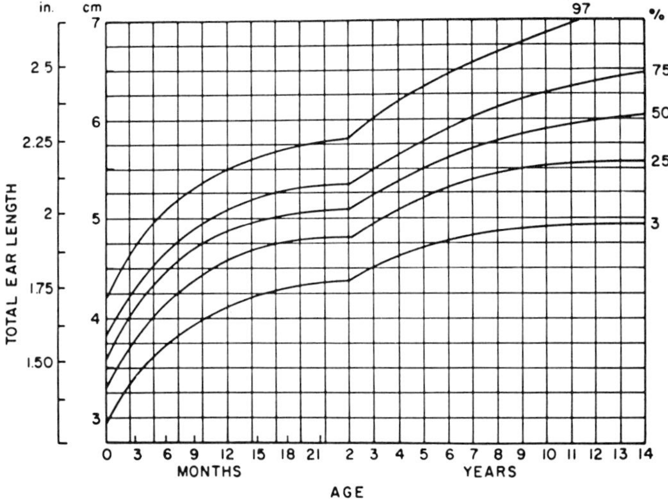


Fig. A–18. Ear length. [From *M Feingold* and *WH Bossert,* Birth Defects **10**(13):1, 1974.]

## CEPHALOMETRIC MEASUREMENTS

Table A–9. Definitions of cephalometric landmarks[a]

| Abbreviation | Definition |
|---|---|
| A | A Point: The most posterior point on the curve of the maxilla between the anterior nasal spine and supradentale |
| ANS | Anterior Nasal Spine: The tip of the median, sharp bony process of the maxilla at the lower margin of the anterior nasal opening |
| AR | Articulare (Articulare Posterior): The point of intersection of the inferior cranial base surface and the averaged posterior surfaces of the mandibular condyles |
| B | B Point: The point most posterior to a line from Infradentale to Pogonion on the anterior surface of the symphyseal outline of the mandible. B point should lie within the apical third of the incisor roots. When there is no curvature in this region and determination of B point is not possible by the above method, it is chosen with the aid of preceding or succeeding films because erupting teeth obscure mandibular concavity on occasion |
| BA | Basion: The most inferior, posterior point on the anterior margin of foramen magnum |
| GN | Gnathion: The most anterior-inferior point on the contour of the body chin symphysis. Determined by bisecting the angle formed by the mandibular plane and a line through Pogonion and Nasion |
| GO | Gonion: The midpoint of the angle of the mandible. Found by bisecting the angle formed by the mandibular plane and plane through Articulare, Posterior, and along the portion of the mandibular ramus inferior to it |
| ME | Menton: The most inferior point on the symphyseal outline |
| N | Nasion: The junction of the frontonasal suture at the most posterior point on the curve at the bridge of the nose |
| OP | Opisthion: The posterior midsagittal point on the posterior margin of foramen magnum |
| PG | Pogonion: The most anterior point on the contour of the bony chin. Determined by a tangent through Nasion |
| PNS | Posterior Nasal Spine: The most posterior point at the sagittal plane on the bony hard palate |
| S | Sella Turcica: The center of the pituitary fossa of the sphenoid bone. Determined by inspection |
| SE | Ethmoid Registration Point: Intersection of the sphenoidal plane with the averaged greater sphenoid wings |
| SYMP | Lingual Symphyseal Point: A constructed point used to determine symphyseal width at Pogonion. The SYMP point is located at the intersection of a construction line through Pogonion and parallel with the mandibular plane with the posterior border of the mandibular symphysis |

[a]From ML Riolo et al, *An Atlas of Craniofacial Growth: Cephalometric Standards from the University School Growth Study, The University of Michigan.* Center for Human Growth and Development, The University of Michigan, Ann Arbor, Michigan, 1974.

## Angular measurements

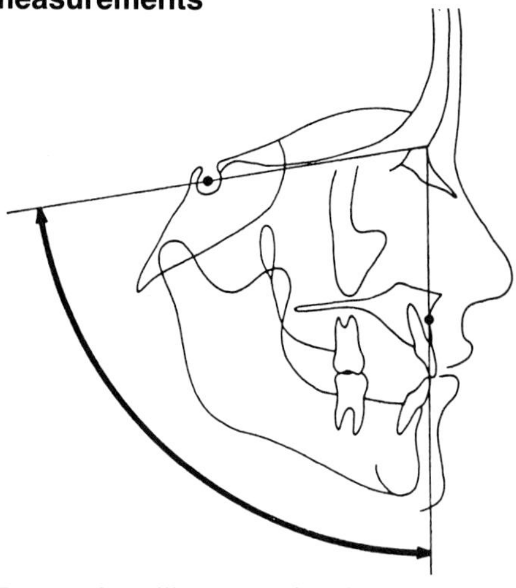

Fig. A–19. Degree of maxillary protrusion (S–N–A). (From *ML Riolo et al,* Center for Human Growth and Development, University of Michigan, Ann Arbor, 1974.)

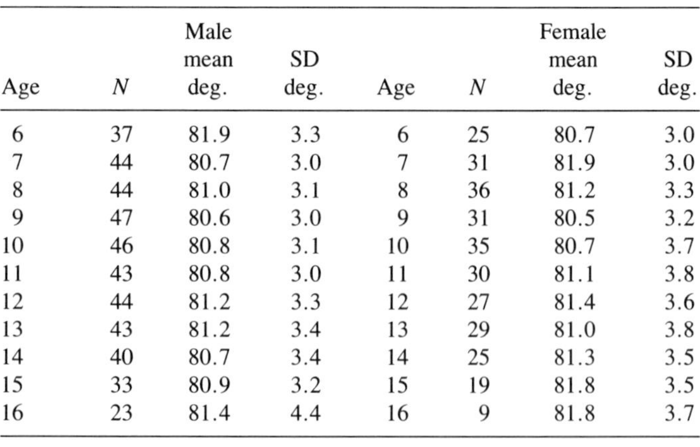

| Age | *N* | Male mean deg. | SD deg. | Age | *N* | Female mean deg. | SD deg. |
|---|---|---|---|---|---|---|---|
| 6 | 37 | 81.9 | 3.3 | 6 | 25 | 80.7 | 3.0 |
| 7 | 44 | 80.7 | 3.0 | 7 | 31 | 81.9 | 3.0 |
| 8 | 44 | 81.0 | 3.1 | 8 | 36 | 81.2 | 3.3 |
| 9 | 47 | 80.6 | 3.0 | 9 | 31 | 80.5 | 3.2 |
| 10 | 46 | 80.8 | 3.1 | 10 | 35 | 80.7 | 3.7 |
| 11 | 43 | 80.8 | 3.0 | 11 | 30 | 81.1 | 3.8 |
| 12 | 44 | 81.2 | 3.3 | 12 | 27 | 81.4 | 3.6 |
| 13 | 43 | 81.2 | 3.4 | 13 | 29 | 81.0 | 3.8 |
| 14 | 40 | 80.7 | 3.4 | 14 | 25 | 81.3 | 3.5 |
| 15 | 33 | 80.9 | 3.2 | 15 | 19 | 81.8 | 3.5 |
| 16 | 23 | 81.4 | 4.4 | 16 | 9 | 81.8 | 3.7 |

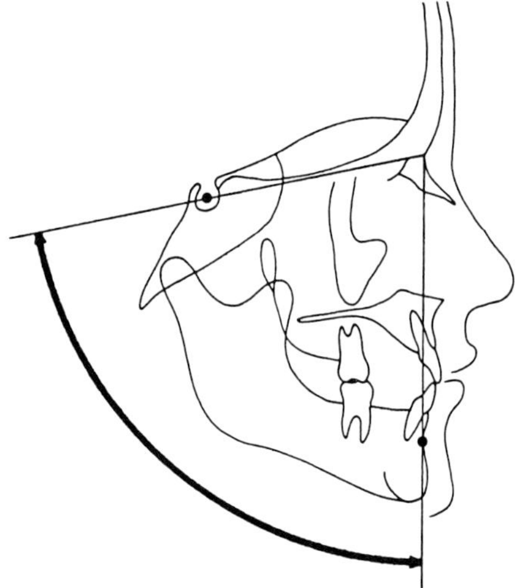

Fig. A–20. Degree of mandibular protrusion (S–N–B). (From *ML Riolo et al,* Center for Human Growth and Development, University of Michigan, Ann Arbor, 1974.)

| Age | *N* | Male mean deg. | SD deg. | Age | *N* | Female mean deg. | SD deg. |
|---|---|---|---|---|---|---|---|
| 6 | 37 | 76.5 | 2.6 | 6 | 25 | 76.0 | 3.5 |
| 7 | 44 | 75.7 | 2.8 | 7 | 31 | 76.3 | 3.1 |
| 8 | 44 | 76.3 | 2.8 | 8 | 36 | 76.7 | 3.3 |
| 9 | 46 | 76.4 | 2.5 | 9 | 31 | 76.5 | 3.4 |
| 10 | 45 | 76.5 | 2.5 | 10 | 35 | 76.7 | 3.5 |
| 11 | 43 | 76.5 | 2.6 | 11 | 30 | 77.3 | 3.9 |
| 12 | 44 | 77.3 | 2.7 | 12 | 27 | 77.7 | 3.4 |
| 13 | 43 | 77.5 | 3.0 | 13 | 29 | 77.5 | 3.9 |
| 14 | 40 | 77.3 | 3.1 | 14 | 25 | 77.9 | 3.8 |
| 15 | 33 | 77.6 | 3.0 | 15 | 19 | 78.9 | 3.9 |
| 16 | 23 | 78.2 | 3.9 | 16 | 9 | 79.2 | 2.3 |

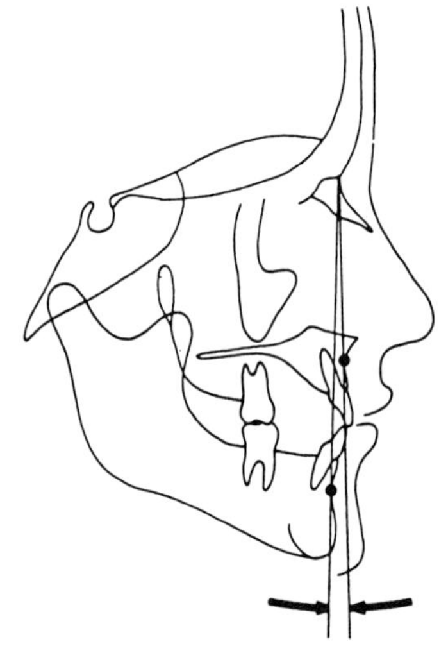

Fig. A–21. Difference between maxillary and mandibular protrusion (A–N–B). (From *ML Riolo et al,* Center for Human Growth and Development, University of Michigan, Ann Arbor, 1974.)

| Age | *N* | Male mean deg. | SD deg. | Age | *N* | Female mean deg. | SD deg. |
|---|---|---|---|---|---|---|---|
| 6 | 37 | 5.3 | 2.2 | 6 | 25 | 4.7 | 2.2 |
| 7 | 44 | 5.0 | 2.3 | 7 | 31 | 5.7 | 2.7 |
| 8 | 44 | 4.8 | 2.2 | 8 | 36 | 4.6 | 2.4 |
| 9 | 46 | 4.2 | 1.9 | 9 | 31 | 4.0 | 2.5 |
| 10 | 45 | 4.3 | 2.0 | 10 | 35 | 4.0 | 2.7 |
| 11 | 43 | 4.3 | 1.9 | 11 | 30 | 3.8 | 2.2 |
| 12 | 44 | 3.9 | 2.1 | 12 | 27 | 3.7 | 2.4 |
| 13 | 43 | 3.7 | 2.0 | 13 | 29 | 3.5 | 2.4 |
| 14 | 40 | 3.4 | 2.0 | 14 | 25 | 3.4 | 2.5 |
| 15 | 33 | 3.3 | 2.1 | 15 | 19 | 2.9 | 2.7 |
| 16 | 23 | 3.2 | 2.3 | 16 | 9 | 2.6 | 2.4 |

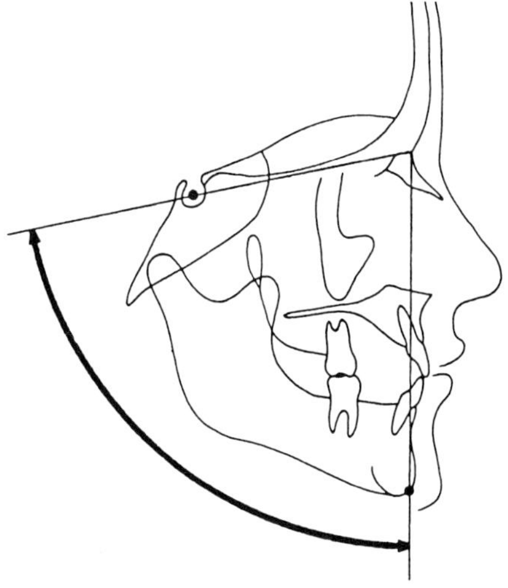

| Age | *N* | Male mean deg. | SD deg. | Age | *N* | Female mean deg. | SD deg. |
|---|---|---|---|---|---|---|---|
| 6 | 37 | 74.8 | 2.4 | 6 | 25 | 74.4 | 3.7 |
| 7 | 44 | 74.2 | 2.6 | 7 | 31 | 74.0 | 3.1 |
| 8 | 44 | 75.2 | 2.7 | 8 | 36 | 75.5 | 3.3 |
| 9 | 46 | 75.5 | 2.5 | 9 | 31 | 75.4 | 3.3 |
| 10 | 45 | 75.7 | 2.4 | 10 | 35 | 75.9 | 3.5 |
| 11 | 43 | 76.0 | 2.5 | 11 | 30 | 78.6 | 3.9 |
| 12 | 44 | 76.7 | 2.6 | 12 | 27 | 77.1 | 3.4 |
| 13 | 43 | 77.1 | 2.9 | 13 | 29 | 77.1 | 4.0 |
| 14 | 40 | 77.2 | 3.0 | 14 | 25 | 77.7 | 4.2 |
| 15 | 33 | 77.5 | 3.0 | 15 | 19 | 78.6 | 4.2 |
| 16 | 23 | 78.3 | 3.6 | 16 | 9 | 79.1 | 2.5 |

Fig. A–22. Facial angle. (From *ML Riolo et al,* Center for Human Growth and Development, University of Michigan, Ann Arbor, 1974.)

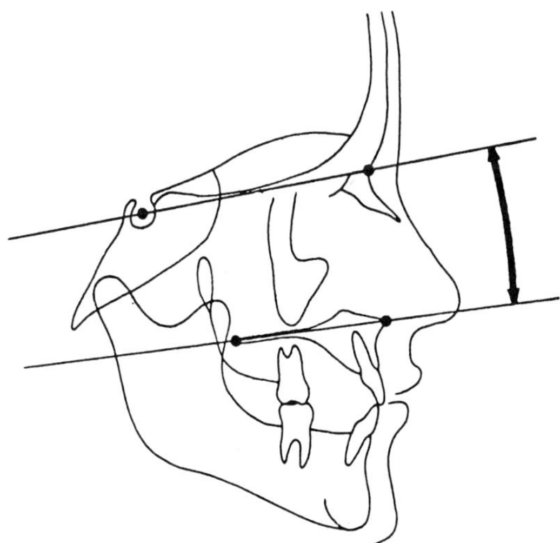

| Age | *N* | Male mean deg. | SD deg. | Age | *N* | Female mean deg. | SD deg. |
|---|---|---|---|---|---|---|---|
| 6 | 37 | 5.2 | 2.4 | 6 | 24 | 7.0 | 2.6 |
| 7 | 44 | 5.9 | 2.8 | 7 | 31 | 6.6 | 2.3 |
| 8 | 44 | 5.9 | 2.6 | 8 | 36 | 7.0 | 2.5 |
| 9 | 47 | 6.4 | 2.3 | 9 | 31 | 7.7 | 2.3 |
| 10 | 46 | 6.1 | 2.6 | 10 | 35 | 7.5 | 2.8 |
| 11 | 43 | 6.5 | 3.0 | 11 | 30 | 7.7 | 2.7 |
| 12 | 44 | 6.5 | 3.0 | 12 | 27 | 8.3 | 2.4 |
| 13 | 43 | 7.1 | 3.2 | 13 | 29 | 8.2 | 2.9 |
| 14 | 39 | 7.3 | 3.5 | 14 | 25 | 8.1 | 1.8 |
| 15 | 33 | 6.9 | 3.4 | 15 | 19 | 7.8 | 2.4 |
| 16 | 23 | 7.0 | 3.0 | 16 | 9 | 8.0 | 2.2 |

Fig. A–23. Anterior cranial base/palatal plane angle (N–S/ANS–PNS). (From *ML Riolo et al,* Center for Human Growth and Development, University of Michigan, Ann Arbor, 1974.)

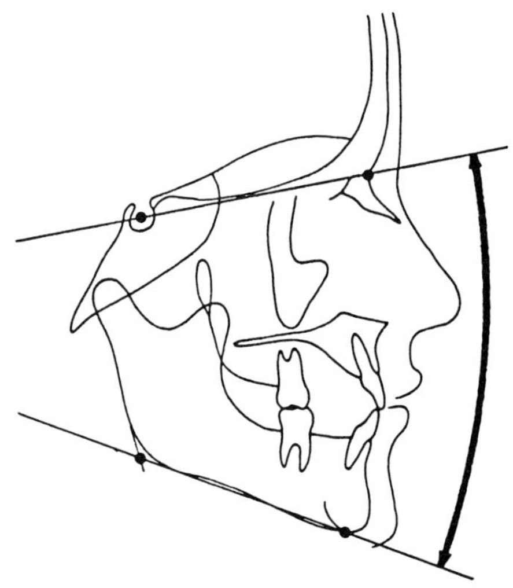

| Age | *N* | Male mean deg. | SD deg. | Age | *N* | Female mean deg. | SD deg. |
|---|---|---|---|---|---|---|---|
| 6 | 37 | 35.4 | 4.0 | 6 | 25 | 36.5 | 5.7 |
| 7 | 44 | 36.0 | 4.9 | 7 | 31 | 36.7 | 4.9 |
| 8 | 44 | 35.1 | 4.5 | 8 | 36 | 35.4 | 5.0 |
| 9 | 46 | 34.7 | 4.6 | 9 | 31 | 35.3 | 5.3 |
| 10 | 45 | 34.7 | 4.7 | 10 | 35 | 35.3 | 5.1 |
| 11 | 43 | 34.7 | 4.7 | 11 | 30 | 34.8 | 5.6 |
| 12 | 44 | 33.8 | 4.9 | 12 | 27 | 34.1 | 5.3 |
| 13 | 42 | 33.2 | 5.0 | 13 | 29 | 34.3 | 5.8 |
| 14 | 39 | 33.2 | 5.1 | 14 | 25 | 33.7 | 6.2 |
| 15 | 32 | 33.2 | 5.2 | 15 | 19 | 32.4 | 5.8 |
| 16 | 23 | 32.9 | 5.2 | 16 | 9 | 31.2 | 3.0 |

Fig. A–24. Anterior cranial base/mandibular plane angle (N–S/ME–GO). (From *ML Riolo et al,* Center for Human Growth and Development, University of Michigan, Ann Arbor, 1974.)

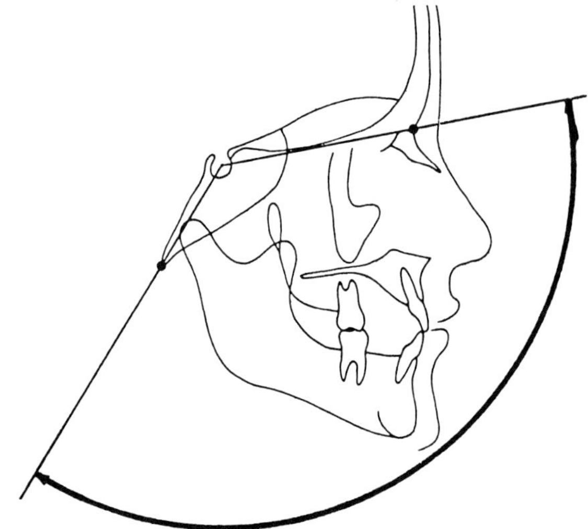

| Age | *N* | Male mean deg. | SD deg. | Age | *N* | Female mean deg. | SD deg. |
|---|---|---|---|---|---|---|---|
| 6 | 37 | 129.3 | 5.0 | 6 | 25 | 129.6 | 5.0 |
| 7 | 44 | 129.7 | 4.9 | 7 | 31 | 130.2 | 4.6 |
| 8 | 44 | 129.0 | 4.8 | 8 | 30 | 130.0 | 4.8 |
| 9 | 47 | 129.6 | 4.6 | 9 | 31 | 129.8 | 4.6 |
| 10 | 46 | 129.2 | 4.7 | 10 | 35 | 129.7 | 4.5 |
| 11 | 43 | 128.9 | 4.8 | 11 | 30 | 129.9 | 4.8 |
| 12 | 44 | 129.3 | 4.8 | 12 | 27 | 130.4 | 5.2 |
| 13 | 43 | 129.2 | 5.3 | 13 | 29 | 130.3 | 4.7 |
| 14 | 40 | 129.1 | 5.2 | 14 | 25 | 130.4 | 4.8 |
| 15 | 33 | 129.2 | 5.4 | 15 | 19 | 130.3 | 4.0 |
| 16 | 23 | 128.9 | 5.9 | 16 | 9 | 131.1 | 4.1 |

Fig. A–25. Cranial base angle (N–S–BA). (From *ML Riolo et al,* Center for Human Growth and Development, University of Michigan, Ann Arbor, 1974.)

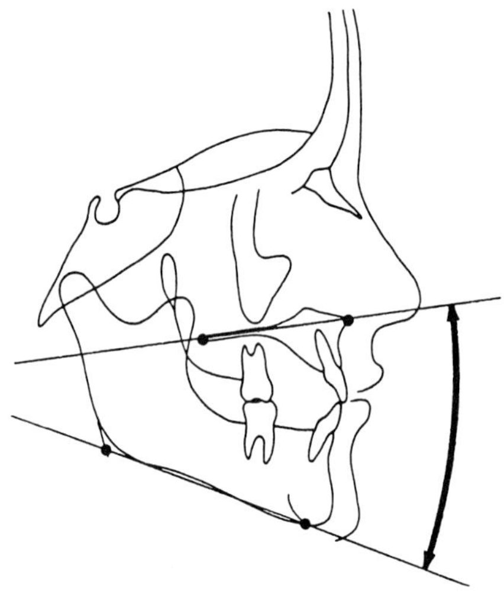

| Age | *N* | Male mean deg. | SD deg. | Age | *N* | Female mean deg. | SD deg. |
|---|---|---|---|---|---|---|---|
| 6 | 37 | 30.1 | 3.8 | 6 | 24 | 29.7 | 5.6 |
| 7 | 44 | 30.1 | 4.8 | 7 | 31 | 30.1 | 5.0 |
| 8 | 44 | 29.1 | 4.4 | 8 | 36 | 28.4 | 4.5 |
| 9 | 46 | 28.2 | 4.3 | 9 | 31 | 27.6 | 5.2 |
| 10 | 45 | 28.5 | 4.7 | 10 | 35 | 27.8 | 4.9 |
| 11 | 43 | 28.2 | 4.6 | 11 | 30 | 27.0 | 5.1 |
| 12 | 44 | 27.3 | 4.9 | 12 | 27 | 25.8 | 5.1 |
| 13 | 42 | 26.0 | 5.3 | 13 | 29 | 26.1 | 5.4 |
| 14 | 39 | 25.9 | 4.9 | 14 | 25 | 25.6 | 6.1 |
| 15 | 32 | 26.2 | 4.9 | 15 | 19 | 24.6 | 5.7 |
| 16 | 23 | 25.9 | 5.1 | 16 | 9 | 23.2 | 3.7 |

Fig. A–26. Palatal plane/mandibular plane angle (ANS–PNS/ME–GO). (From *ML Riolo et al,* Center for Human Growth and Development, University of Michigan, Ann Arbor, 1974.)

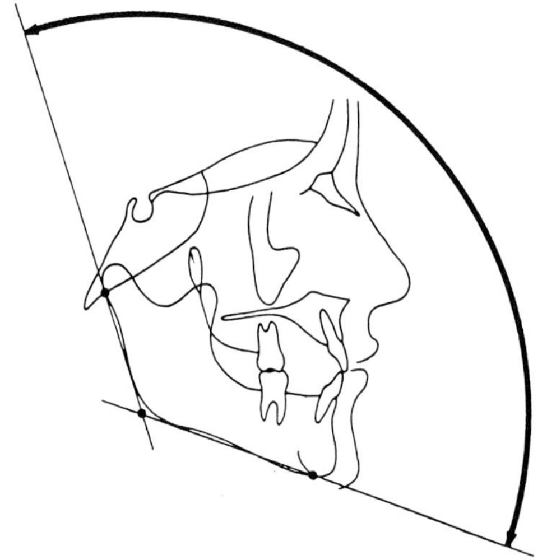

| Age | *N* | Male mean deg. | SD deg. | Age | *N* | Female mean deg. | SD deg. |
|---|---|---|---|---|---|---|---|
| 6 | 37 | 132.0 | 4.8 | 6 | 25 | 129.8 | 5.1 |
| 7 | 44 | 130.5 | 4.7 | 7 | 31 | 130.0 | 4.3 |
| 8 | 44 | 129.5 | 4.6 | 8 | 36 | 128.6 | 4.7 |
| 9 | 47 | 128.5 | 4.7 | 9 | 31 | 127.3 | 4.7 |
| 10 | 46 | 128.0 | 4.9 | 10 | 35 | 127.5 | 4.5 |
| 11 | 43 | 127.2 | 5.0 | 11 | 30 | 126.9 | 4.6 |
| 12 | 44 | 126.5 | 5.2 | 12 | 27 | 126.2 | 4.2 |
| 13 | 42 | 125.8 | 5.0 | 13 | 29 | 126.1 | 5.9 |
| 14 | 39 | 124.0 | 5.3 | 14 | 25 | 125.0 | 4.7 |
| 15 | 32 | 123.8 | 5.3 | 15 | 19 | 124.0 | 4.2 |
| 16 | 23 | 123.5 | 5.9 | 16 | 9 | 122.2 | 4.2 |

Fig. A–27. Mandibular angle (AR–GO–ME). (From *ML Riolo et al,* Center for Human Growth and Development, University of Michigan, Ann Arbor, 1974.)

## Linear measurements

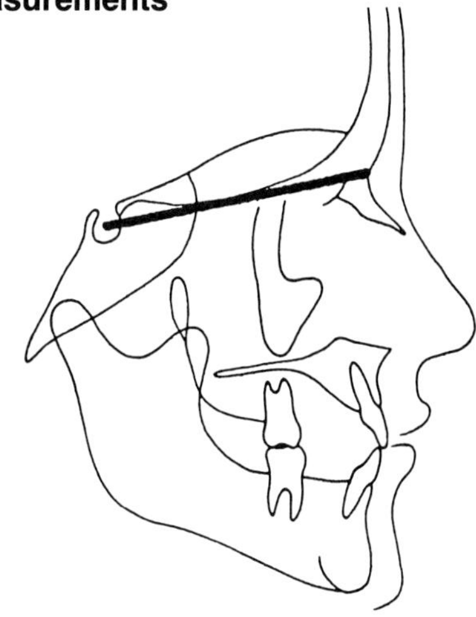

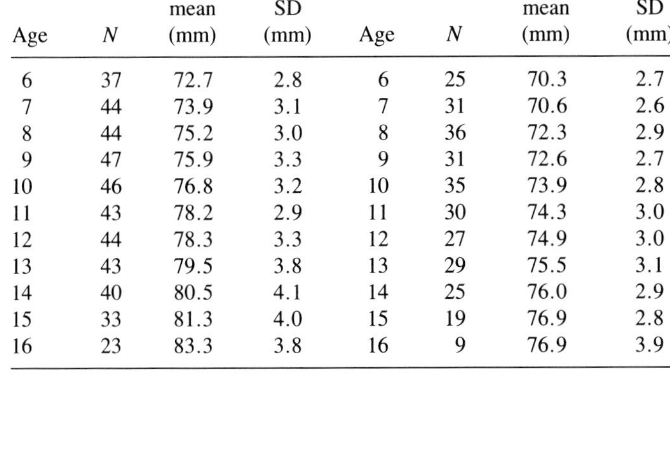

| Age | N | Male mean (mm) | SD (mm) | Age | N | Female mean (mm) | SD (mm) |
|---|---|---|---|---|---|---|---|
| 6 | 37 | 72.7 | 2.8 | 6 | 25 | 70.3 | 2.7 |
| 7 | 44 | 73.9 | 3.1 | 7 | 31 | 70.6 | 2.6 |
| 8 | 44 | 75.2 | 3.0 | 8 | 36 | 72.3 | 2.9 |
| 9 | 47 | 75.9 | 3.3 | 9 | 31 | 72.6 | 2.7 |
| 10 | 46 | 76.8 | 3.2 | 10 | 35 | 73.9 | 2.8 |
| 11 | 43 | 78.2 | 2.9 | 11 | 30 | 74.3 | 3.0 |
| 12 | 44 | 78.3 | 3.3 | 12 | 27 | 74.9 | 3.0 |
| 13 | 43 | 79.5 | 3.8 | 13 | 29 | 75.5 | 3.1 |
| 14 | 40 | 80.5 | 4.1 | 14 | 25 | 76.0 | 2.9 |
| 15 | 33 | 81.3 | 4.0 | 15 | 19 | 76.9 | 2.8 |
| 16 | 23 | 83.3 | 3.8 | 16 | 9 | 76.9 | 3.9 |

Fig. A–28. Anterior cranial base length (S–N). (From *ML Riolo et al,* Center for Human Growth and Development, University of Michigan, Ann Arbor, 1974.)

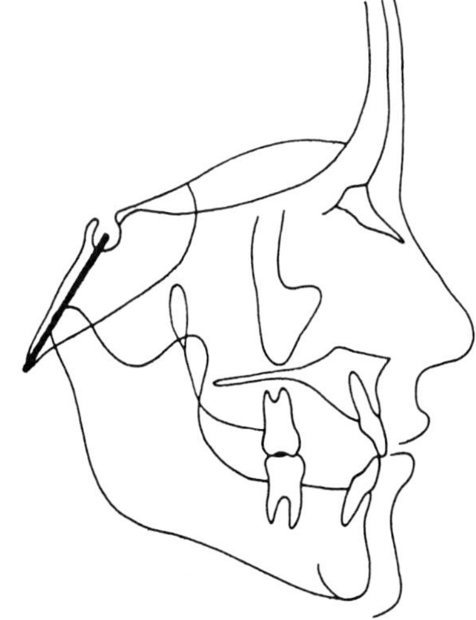

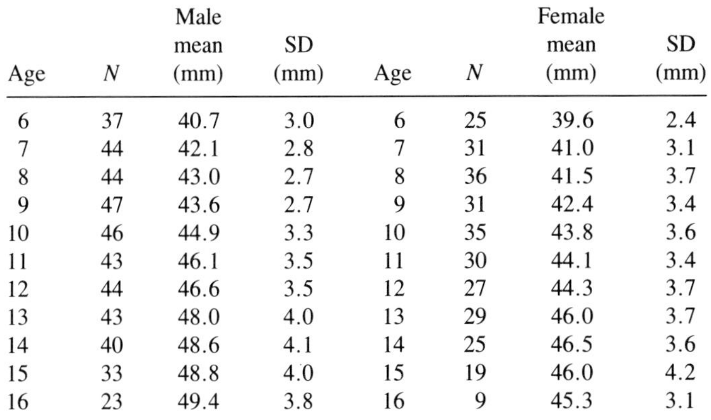

| Age | N | Male mean (mm) | SD (mm) | Age | N | Female mean (mm) | SD (mm) |
|---|---|---|---|---|---|---|---|
| 6 | 37 | 40.7 | 3.0 | 6 | 25 | 39.6 | 2.4 |
| 7 | 44 | 42.1 | 2.8 | 7 | 31 | 41.0 | 3.1 |
| 8 | 44 | 43.0 | 2.7 | 8 | 36 | 41.5 | 3.7 |
| 9 | 47 | 43.6 | 2.7 | 9 | 31 | 42.4 | 3.4 |
| 10 | 46 | 44.9 | 3.3 | 10 | 35 | 43.8 | 3.6 |
| 11 | 43 | 46.1 | 3.5 | 11 | 30 | 44.1 | 3.4 |
| 12 | 44 | 46.6 | 3.5 | 12 | 27 | 44.3 | 3.7 |
| 13 | 43 | 48.0 | 4.0 | 13 | 29 | 46.0 | 3.7 |
| 14 | 40 | 48.6 | 4.1 | 14 | 25 | 46.5 | 3.6 |
| 15 | 33 | 48.8 | 4.0 | 15 | 19 | 46.0 | 4.2 |
| 16 | 23 | 49.4 | 3.8 | 16 | 9 | 45.3 | 3.1 |

Fig. A–29. Posterior cranial base length (S–BA). (From *ML Riolo et al,* Center for Human Growth and Development, University of Michigan, Ann Arbor, 1974.)

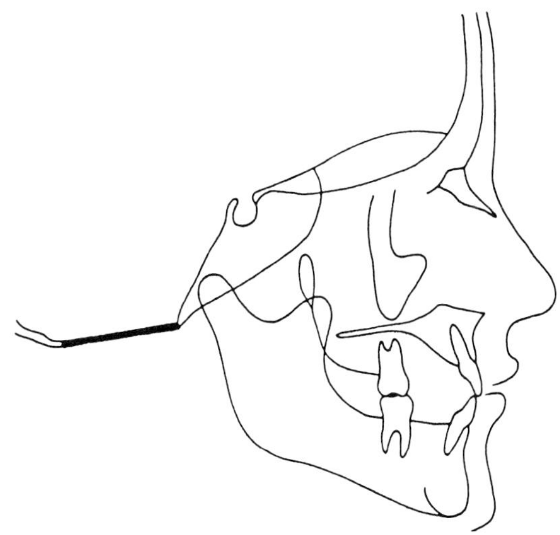

| Age | N | Male mean (mm) | SD (mm) | Age | N | Female mean (mm) | SD (mm) |
|---|---|---|---|---|---|---|---|
| 6 | 37 | 36.5 | 3.6 | 6 | 25 | 34.3 | 3.3 |
| 7 | 44 | 37.3 | 3.6 | 7 | 31 | 35.1 | 3.4 |
| 8 | 44 | 37.7 | 3.6 | 8 | 36 | 35.0 | 2.9 |
| 9 | 47 | 37.9 | 3.4 | 9 | 31 | 36.2 | 2.8 |
| 10 | 46 | 37.9 | 3.1 | 10 | 35 | 36.9 | 3.3 |
| 11 | 43 | 38.4 | 3.4 | 11 | 30 | 37.0 | 3.4 |
| 12 | 44 | 38.1 | 3.1 | 12 | 27 | 36.6 | 2.7 |
| 13 | 43 | 38.6 | 3.2 | 13 | 29 | 38.1 | 2.9 |
| 14 | 40 | 38.7 | 3.6 | 14 | 25 | 38.7 | 3.1 |
| 15 | 33 | 39.1 | 3.2 | 15 | 19 | 39.1 | 3.0 |
| 16 | 23 | 38.8 | 3.4 | 16 | 9 | 37.2 | 3.2 |

Fig. A–30. Foramen magnum diameter (BA–OP). (From *ML Riolo et al,* Center for Human Growth and Development, University of Michigan, Ann Arbor, 1974.)

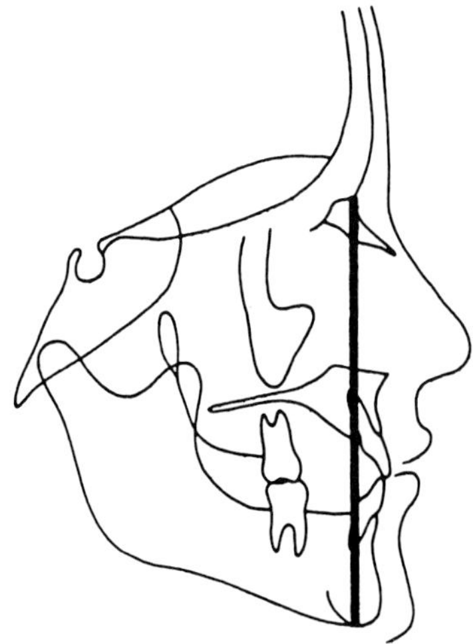

| Age | N | Male mean (mm) | SD (mm) | Age | N | Female mean (mm) | SD (mm) |
|---|---|---|---|---|---|---|---|
| 6 | 37 | 106.7 | 5.2 | 6 | 25 | 105.0 | 5.1 |
| 7 | 44 | 110.7 | 5.8 | 7 | 31 | 107.8 | 5.3 |
| 8 | 44 | 113.6 | 5.6 | 8 | 36 | 109.5 | 5.4 |
| 9 | 46 | 115.9 | 5.4 | 9 | 31 | 112.1 | 5.7 |
| 10 | 45 | 118.7 | 5.7 | 10 | 35 | 115.1 | 6.7 |
| 11 | 43 | 121.5 | 6.0 | 11 | 30 | 116.2 | 6.4 |
| 12 | 44 | 123.3 | 6.3 | 12 | 27 | 118.3 | 6.0 |
| 13 | 42 | 126.6 | 7.0 | 13 | 29 | 120.7 | 5.8 |
| 14 | 39 | 130.3 | 7.9 | 14 | 25 | 122.3 | 5.9 |
| 15 | 32 | 133.8 | 7.8 | 15 | 19 | 122.7 | 6.4 |
| 16 | 23 | 136.8 | 7.9 | 16 | 9 | 123.2 | 5.1 |

Fig. A–31. Anterior facial height (N–ME). (From *ML Riolo et al,* Center for Human Growth and Development, University of Michigan, Ann Arbor, 1974.)

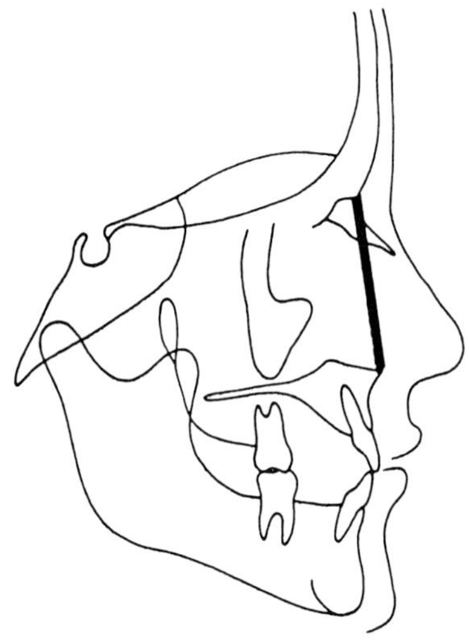

| Age | N | Male mean (mm) | SD (mm) | Age | N | Female mean (mm) | SD (mm) |
|---|---|---|---|---|---|---|---|
| 6 | 37 | 45.9 | 2.8 | 6 | 25 | 45.9 | 3.1 |
| 7 | 44 | 47.9 | 3.2 | 7 | 31 | 47.1 | 2.9 |
| 8 | 44 | 49.5 | 2.8 | 8 | 36 | 48.6 | 3.0 |
| 9 | 47 | 51.0 | 3.0 | 9 | 31 | 50.4 | 3.1 |
| 10 | 46 | 52.3 | 3.1 | 10 | 35 | 52.1 | 3.7 |
| 11 | 43 | 53.8 | 3.4 | 11 | 30 | 52.7 | 3.3 |
| 12 | 44 | 54.6 | 3.6 | 12 | 27 | 54.0 | 3.7 |
| 13 | 43 | 56.8 | 3.7 | 13 | 29 | 54.7 | 3.3 |
| 14 | 40 | 58.2 | 4.1 | 14 | 25 | 55.3 | 2.7 |
| 15 | 33 | 59.0 | 4.0 | 15 | 19 | 55.3 | 3.2 |
| 16 | 23 | 59.7 | 3.9 | 16 | 9 | 55.7 | 2.1 |

Fig. A–32. Anterior upper facial height (nasal height) (N–ANS). (From *ML Riolo et al,* Center for Human Growth and Development, University of Michigan, Ann Arbor, 1974.)

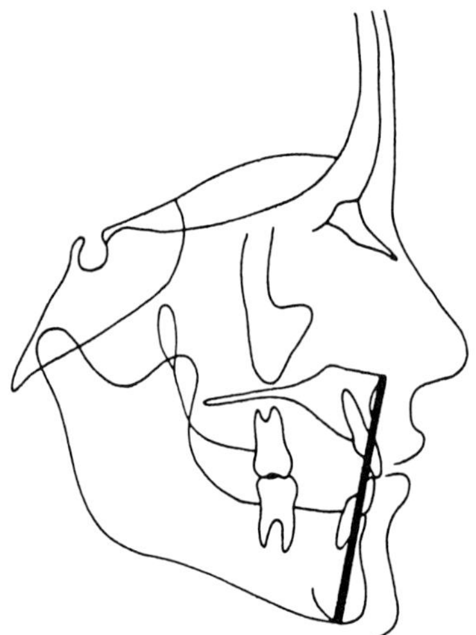

| Age | N | Male mean (mm) | SD (mm) | Age | N | Female mean (mm) | SD (mm) |
|---|---|---|---|---|---|---|---|
| 6 | 37 | 63.7 | 4.0 | 6 | 25 | 61.6 | 3.9 |
| 7 | 44 | 65.5 | 4.4 | 7 | 31 | 63.6 | 4.1 |
| 8 | 44 | 66.6 | 4.4 | 8 | 36 | 63.5 | 4.2 |
| 9 | 46 | 67.3 | 4.3 | 9 | 31 | 64.1 | 4.6 |
| 10 | 45 | 68.9 | 4.9 | 10 | 35 | 65.3 | 4.9 |
| 11 | 43 | 70.3 | 4.8 | 11 | 30 | 65.8 | 4.6 |
| 12 | 44 | 71.1 | 5.1 | 12 | 27 | 66.5 | 3.8 |
| 13 | 42 | 72.0 | 5.6 | 13 | 29 | 68.1 | 4.5 |
| 14 | 39 | 74.3 | 5.8 | 14 | 25 | 69.1 | 5.0 |
| 15 | 32 | 76.7 | 6.4 | 15 | 19 | 69.5 | 5.3 |
| 16 | 23 | 79.5 | 6.2 | 16 | 9 | 69.3 | 5.2 |

Fig. A–33. Anterior lower facial height (ANS–ME). (From *ML Riolo et al,* Center for Human Growth and Development, University of Michigan, Ann Arbor, 1974.)

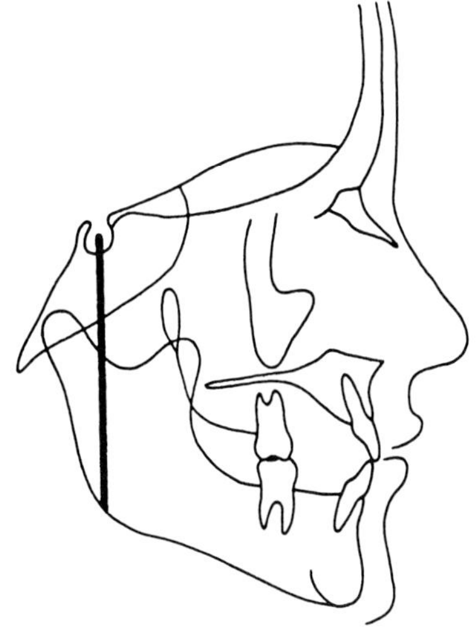

Fig. A–34. Posterior facial height (S–GO). (From *ML Riolo et al,* Center for Human Growth and Development, University of Michigan, Ann Arbor, 1974.)

| Age | *N* | Male mean (mm) | SD (mm) | Age | *N* | Female mean (mm) | SD (mm) |
|---|---|---|---|---|---|---|---|
| 6 | 37 | 65.9 | 4.7 | 6 | 25 | 62.8 | 4.3 |
| 7 | 44 | 67.4 | 4.6 | 7 | 31 | 64.6 | 4.4 |
| 8 | 44 | 70.0 | 4.3 | 8 | 36 | 66.4 | 4.0 |
| 9 | 46 | 71.9 | 4.4 | 9 | 31 | 68.5 | 4.4 |
| 10 | 45 | 73.6 | 4.6 | 10 | 35 | 70.2 | 4.1 |
| 11 | 43 | 75.4 | 4.9 | 11 | 30 | 71.1 | 4.4 |
| 12 | 44 | 77.6 | 5.3 | 12 | 27 | 73.7 | 5.1 |
| 13 | 43 | 80.3 | 6.1 | 13 | 29 | 74.8 | 4.4 |
| 14 | 40 | 82.9 | 5.6 | 14 | 25 | 76.6 | 5.5 |
| 15 | 33 | 85.4 | 5.6 | 15 | 19 | 78.2 | 5.1 |
| 16 | 23 | 88.2 | 5.9 | 16 | 9 | 79.1 | 4.3 |

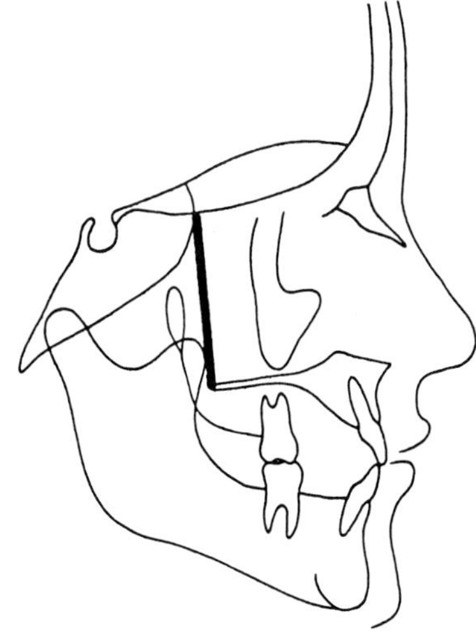

Fig. A–35. Posterior upper facial height (posterior nasal height) (SE–PNS). (From *ML Riolo et al,* Center for Human Growth and Development, University of Michigan, Ann Arbor, 1974.)

| Age | *N* | Male mean (mm) | SD (mm) | Age | *N* | Female mean (mm) | SD (mm) |
|---|---|---|---|---|---|---|---|
| 6 | 37 | 42.4 | 2.8 | 6 | 24 | 42.0 | 2.6 |
| 7 | 44 | 44.1 | 2.9 | 7 | 31 | 43.3 | 2.8 |
| 8 | 44 | 45.5 | 2.9 | 8 | 36 | 44.4 | 2.8 |
| 9 | 47 | 46.5 | 2.7 | 9 | 31 | 45.3 | 3.0 |
| 10 | 46 | 47.9 | 3.0 | 10 | 35 | 46.9 | 3.4 |
| 11 | 43 | 49.1 | 3.4 | 11 | 30 | 47.5 | 3.3 |
| 12 | 44 | 49.8 | 3.7 | 12 | 27 | 48.0 | 3.0 |
| 13 | 43 | 51.2 | 3.5 | 13 | 29 | 49.7 | 2.6 |
| 14 | 39 | 52.7 | 3.5 | 14 | 25 | 50.3 | 2.4 |
| 15 | 33 | 54.0 | 3.8 | 15 | 19 | 49.6 | 2.8 |
| 16 | 23 | 54.7 | 4.4 | 16 | 9 | 49.6 | 3.3 |

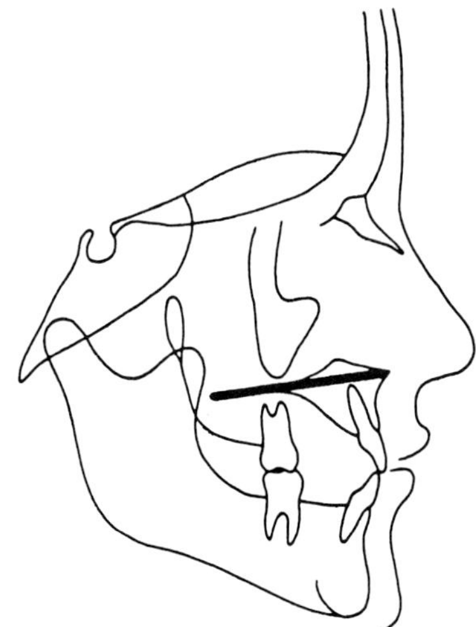

Fig. A–36. Maxillary length (ANS–PNS). (From *ML Riolo et al,* Center for Human Growth and Development, University of Michigan, Ann Arbor, 1974.)

| Age | *N* | Male mean (mm) | SD (mm) | Age | *N* | Female mean (mm) | SD (mm) |
|---|---|---|---|---|---|---|---|
| 6 | 37 | 50.2 | 2.2 | 6 | 24 | 48.9 | 2.3 |
| 7 | 44 | 51.4 | 2.5 | 7 | 31 | 50.5 | 2.9 |
| 8 | 44 | 52.1 | 2.9 | 8 | 36 | 51.2 | 2.5 |
| 9 | 47 | 53.3 | 2.9 | 9 | 31 | 51.2 | 3.2 |
| 10 | 46 | 54.4 | 2.6 | 10 | 35 | 53.1 | 3.1 |
| 11 | 43 | 56.0 | 2.4 | 11 | 30 | 53.9 | 4.0 |
| 12 | 44 | 56.7 | 3.1 | 12 | 27 | 54.1 | 3.2 |
| 13 | 43 | 57.8 | 3.1 | 13 | 29 | 55.3 | 3.0 |
| 14 | 39 | 58.7 | 3.6 | 14 | 25 | 56.7 | 2.9 |
| 15 | 33 | 59.6 | 3.6 | 15 | 19 | 57.1 | 2.7 |
| 16 | 23 | 61.6 | 3.7 | 16 | 9 | 57.0 | 4.4 |

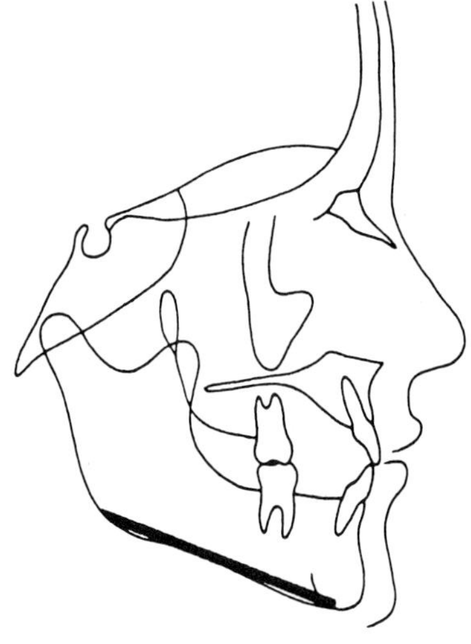

Fig. A–37. Mandibular body length (GO–ME). (From *ML Riolo et al,* Center for Human Growth and Development, University of Michigan, Ann Arbor, 1974.)

| Age | *N* | Male mean (mm) | SD (mm) | Age | *N* | Female mean (mm) | SD (mm) |
|---|---|---|---|---|---|---|---|
| 6 | 37 | 60.7 | 3.2 | 6 | 25 | 61.5 | 4.0 |
| 7 | 44 | 63.3 | 3.0 | 7 | 31 | 62.7 | 4.0 |
| 8 | 44 | 65.7 | 3.1 | 8 | 36 | 65.4 | 4.3 |
| 9 | 47 | 67.7 | 3.1 | 9 | 31 | 66.5 | 4.4 |
| 10 | 46 | 69.6 | 3.0 | 10 | 35 | 69.0 | 4.2 |
| 11 | 43 | 71.5 | 3.3 | 11 | 30 | 70.6 | 3.9 |
| 12 | 44 | 73.1 | 3.5 | 12 | 27 | 71.5 | 4.0 |
| 13 | 42 | 75.3 | 3.8 | 13 | 29 | 73.4 | 4.4 |
| 14 | 39 | 77.4 | 3.9 | 14 | 25 | 74.8 | 4.4 |
| 15 | 32 | 79.4 | 4.0 | 15 | 19 | 75.9 | 4.4 |
| 16 | 23 | 80.9 | 3.6 | 16 | 9 | 77.6 | 3.4 |

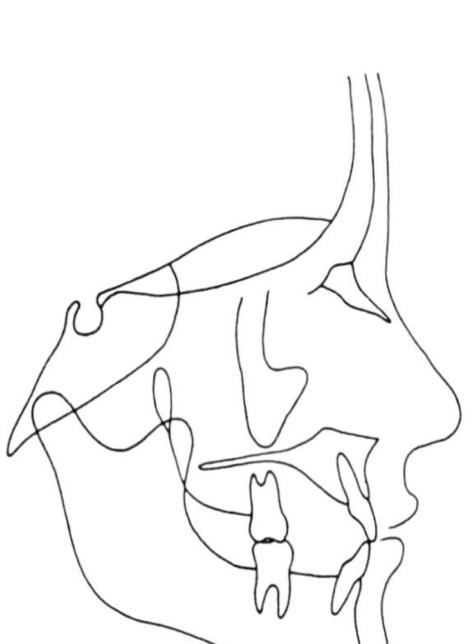

Fig. A–38. Chin prominence (PO–SYMP). (From *ML Riolo et al,* Center for Human Growth and Development, University of Michigan, Ann Arbor, 1974.)

| Age | *N* | Male mean (mm) | SD (mm) | Age | *N* | Female mean (mm) | SD (mm) |
|---|---|---|---|---|---|---|---|
| 6 | 37 | 15.2 | 1.1 | 6 | 25 | 14.6 | 1.8 |
| 7 | 44 | 15.7 | 1.3 | 7 | 31 | 14.9 | 1.8 |
| 8 | 44 | 15.9 | 1.3 | 8 | 36 | 15.1 | 1.6 |
| 9 | 47 | 16.0 | 1.4 | 9 | 31 | 15.2 | 1.8 |
| 10 | 46 | 16.3 | 1.3 | 10 | 35 | 15.6 | 1.9 |
| 11 | 43 | 16.4 | 1.5 | 11 | 30 | 15.7 | 1.8 |
| 12 | 44 | 16.7 | 1.4 | 12 | 27 | 15.8 | 1.8 |
| 13 | 43 | 16.9 | 1.6 | 13 | 29 | 16.2 | 2.0 |
| 14 | 40 | 17.6 | 1.8 | 14 | 25 | 16.3 | 1.8 |
| 15 | 33 | 17.6 | 1.7 | 15 | 19 | 16.0 | 1.7 |
| 16 | 23 | 18.1 | 1.6 | 16 | 9 | 16.0 | 1.8 |

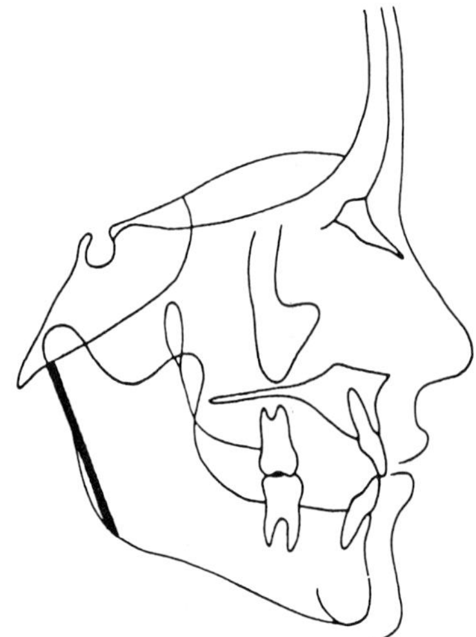

Fig. A–39. Mandibular ramus height (posterior lower facial height) (AR–GO). (From *ML Riolo et al,* Center for Human Growth and Development, University of Michigan, Ann Arbor, 1974.)

| Age | *N* | Male mean (mm) | SD (mm) | Age | *N* | Female mean (mm) | SD (mm) |
|---|---|---|---|---|---|---|---|
| 6 | 37 | 40.4 | 3.3 | 6 | 25 | 37.9 | 3.7 |
| 7 | 44 | 40.5 | 3.3 | 7 | 31 | 38.9 | 3.4 |
| 8 | 44 | 42.2 | 3.4 | 8 | 36 | 40.2 | 3.2 |
| 9 | 47 | 43.4 | 3.3 | 9 | 31 | 41.2 | 3.5 |
| 10 | 46 | 44.2 | 3.4 | 10 | 35 | 41.6 | 3.2 |
| 11 | 43 | 45.0 | 3.9 | 11 | 30 | 42.3 | 3.1 |
| 12 | 44 | 46.7 | 3.8 | 12 | 27 | 44.9 | 3.9 |
| 13 | 43 | 48.0 | 4.6 | 13 | 29 | 44.7 | 3.8 |
| 14 | 40 | 49.8 | 4.6 | 14 | 25 | 46.4 | 4.6 |
| 15 | 33 | 51.4 | 4.6 | 15 | 19 | 48.1 | 4.1 |
| 16 | 23 | 54.3 | 4.1 | 16 | 9 | 49.6 | 3.9 |

## INTRAORAL MEASUREMENTS

Table A–10. Mean palatal dimensions of Minnesota children and adults[a]

| Sex | Age group | Number of measure-ments | Mean height | SD[b] | Mean width | SD | Mean length | SD | Mean index | SD |
|---|---|---|---|---|---|---|---|---|---|---|
| Male | 6–7 | 60 | 8.0 | 1.38 | 30.8 | 2.26 | 43.6 | 2.41 | 26.3 | 4.65 |
| Female | 6–7 | 68 | 7.9 | 1.47 | 29.9 | 2.37 | 43.0 | 2.63 | 26.4 | 5.03 |
| Male | 8–9 | 124 | 8.7 | 1.48 | 31.7 | 2.83 | 45.3 | 2.87 | 27.5 | 5.27 |
| Female | 8–9 | 98 | 8.4 | 1.56 | 30.5 | 2.06 | 44.9 | 2.65 | 27.6 | 5.27 |
| Male | 10–11 | 78 | 9.5 | 1.83 | 32.5 | 2.33 | 48.1 | 2.38 | 29.4 | 5.97 |
| Female | 10–11 | 71 | 9.3 | 1.61 | 32.1 | 2.62 | 46.9 | 3.08 | 29.0 | 5.29 |
| Male | 12–13 | 67 | 11.0 | 2.24 | 32.4 | 3.03 | 50.5 | 3.30 | 34.2 | 7.41 |
| Female | 12–13 | 58 | 10.1 | 2.01 | 32.3 | 2.84 | 48.3 | 3.11 | 31.5 | 6.22 |
| Male | 14–15 | 116 | 12.3 | 2.36 | 33.9 | 2.60 | 51.4 | 3.25 | 36.5 | 7.54 |
| Female | 14–15 | 105 | 11.2 | 1.90 | 31.8 | 2.46 | 49.0 | 2.79 | 35.3 | 6.31 |
| Male | 16–18 | 130 | 13.8 | 2.64 | 33.7 | 2.92 | 51.6 | 3.03 | 41.0 | 8.78 |
| Female | 16–18 | 123 | 12.1 | 2.24 | 32.3 | 2.69 | 49.8 | 3.10 | 37.7 | 7.31 |
| Male | Adult | 101 | 14.9 | 2.93 | 34.6 | 3.03 | 51.8 | 3.28 | 43.4 | 9.68 |
| Female | Adult | 123 | 12.7 | 2.45 | 32.6 | 2.80 | 49.6 | 3.03 | 39.0 | 7.04 |

[a]From BL Shapiro et al, Measurement of normal and reportedly malformed palatal vaults. I. Normal adult measurements. *J Dent Res* **42:**1039, 1963; RS Redman et al, Measurements of normal and reportedly malformed palatal vaults. II. Normal juvenile measurements. *J Dent Res* **45:**266–269, 1966.
[b]SD, standard deviation. All measurements were made to the nearest millimeter. Each index was computed from the formula: height/width × 100.

Table A–11. Normal chronologic development of primary teeth[a]

| Tooth | Initiation, week *in utero* | Calcification begins, week *in utero* | Crown completed, month | Eruption, month | Root completed, year | Root resorption begins, year | Tooth shed, year |
|---|---|---|---|---|---|---|---|
| Central incisor | 7 | 14 (13–16) | 1–3 | 6–9 | 1½–2 | 5–6 | 7–8 |
| Lateral incisor | 7 | 16 (14½–16½) | 2–3 | 7–10 | 1½–2 | 5–6 | 7–9 |
| Canine | 7½ | 17 (15–18) | 9 | 16–20 | 2½–3¼ | 6–7 | 10–12 |
| First molar | 8 | 15½ (14½–17) | 6 | 12–16 | 2–2½ | 4–5 | 9–11 |
| Second molar | 10 | 18½ (16–23½) | 10–12 | 20–30 | 3 | 4–5 | 11–12 |

[a]From WHG Logan and R Kronfeld, Development of the human jaws and surrounding structures from birth to the age of 15 years. *J Am Dent Assoc* **20:**379–427, 1933; RC Lunt and DB Law, A review of the chronology of calcification of the deciduous teeth. *J Am Dent Assoc* **89:**599–606, 1974; I Schour and M Massler, Studies in tooth development. The growth pattern of human teeth. *J Am Dent Assoc* **27:**1918–1931, 1940.

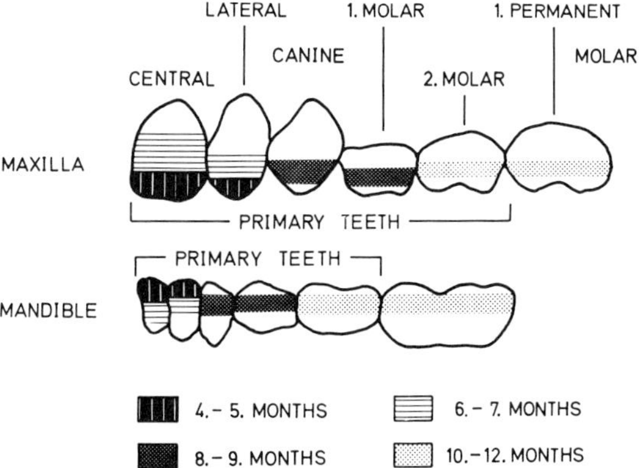


Fig. A–40. Chronology: onset of pre- and postnatal enamel hypoplasia. (From *WF Via* and *GA Churchill,* JADA **59:**702, 1959.)

Table A–12. Normal chronologic development of secondary teeth[a]

| Tooth | Initiation, month | Calcification begins | Crown completed, year | Eruption, year | Root completed, year |
|---|---|---|---|---|---|
| Maxilla | | | | | |
| Central incisor | 5–5¼ *in utero* | 3–4 months | 4–5 | 7–8 | 10 |
| Lateral incisor | 5–5¼ *in utero* | 1 year | 4–5 | 8–9 | 11 |
| Canine | 5½–6 *in utero* | 4–5 months | 6–7 | 11–12 | 13–15 |
| First premolar | Birth | 1½–1¾ years | 5–6 | 10–11 | 12–13 |
| Second premolar | 7½–8 | 2–2½ years | 6–7 | 10–12 | 12–14 |
| First molar | 3½–4 *in utero* | Birth | 2½–3 | 6–7 | 9–10 |
| Second molar | 8½–9 | 2½–3 years | 7–8 | 12–13 | 14–16 |
| Third molar | 3½–4 (yr) | 7–9 years | 12–16 | 17–25 | 18–25 |
| Mandible | | | | | |
| Central incisor | 5–5¼ *in utero* | 3–4 months | 4–5 | 6–7 | 9 |
| Lateral incisor | 5–5¼ *in utero* | 3–4 months | 4–5 | 7–8 | 10 |
| Canine | 5½–6 *in utero* | 4–5 months | 6–7 | 9–11 | 12–14 |
| First premolar | Birth | 1¾–2 years | 5–6 | 10–12 | 12–13 |
| Second premolar | 7½–8 | 2¼–2½ years | 6–7 | 11–12 | 13–14 |
| First molar | 3½–4 *in utero* | Birth | 2½–3 | 6–7 | 9–10 |
| Second molar | 8½–9 | 2½–3 years | 7–8 | 11–13 | 14–15 |
| Third molar | 3½–4 (yr) | 8–10 years | 12–16 | 17–25 | 18–25 |

[a]From WHG Logan and R Kronfeld, Development of the human jaws and surrounding structures from birth to the age of 15 years. *J Am Dent Assoc* **20:**379–427, 1933; I Schour and M Massler, Studies in tooth development. The growth pattern of human teeth. *J Am Dent Assoc* **27:**1918–1931, 1940.

Table A–13. Dental anomalies and their frequencies

| Teeth | Percentage |
|---|---|
| Congenitally missing, primary, caucasian (10, 19, 22)[a] | 0.1–0.7 |
| Congenitally missing, secondary, caucasian | |
| Maxillary central incisors (24, 27) | 0.05 |
| Maxillary lateral incisors (7, 8, 10, 20, 24, 25, 27) | 1.0–2.0 |
| Maxillary canines (7, 24, 27) | 0.3 |
| Maxillary first premolars (7, 24, 27) | 0.3 |
| Maxillary second premolars (7, 8, 10, 24, 27) | 1.0–2.0 |
| Maxillary first molars (24, 27) | 0.3 |
| Maxillary second molars (7, 24, 27) | 0.1 |
| Mandibular central incisors (7, 24, 27) | 0.3 |
| Mandibular lateral incisors (7, 24, 27) | 0.3 |
| Mandibular canines (24) | 0.1 |
| Mandibular first premolars (7, 24, 27) | 0.3 |
| Mandibular second premolars (7, 8, 10, 24, 27) | 1.2–2.5 |
| Mandibular first molars (24, 27) | 0.3 |
| Mandibular second molars (7, 24, 27) | 0.3 |
| Secondary teeth, other than third molars | |
| Caucasian (4, 5, 7, 8, 10, 27) | 3.0–7.5 |
| Japanese (21, 26) | 5.8–9.2 |
| Third molars, one or more absent (10, 12, 14) | 2.5–35.0 |
| One absent | 7–10 |
| Two absent | 10–12 |
| Three absent | 4–6 |
| Four absent | 4–7 |
| Defects, hereditary | |
| Dentin (31) | 1 : 8,000 |
| Enamel (31) | 1 : 15,000 |
| Dens invaginatus | |
| Maxillary lateral incisors (1, 9, 13, 28) | 1.2–6.6 |
| Maxillary central incisors (13) | 0.6 |
| Double formations (fusion or gemination) | |
| Primary dentition | |
| Caucasian (5, 11, 17, 19, 22) | 0.2–0.5 |
| Japanese (21, 26) | 2.5 |
| Secondary dentition | |
| Caucasian (27) | 0.2 |
| Impacted | |
| Teeth, primary (6, 18) | 17.0 |
| Maxillary canines, secondary (6) | 0.9 |
| Natal teeth (2) | 0.03–0.05 |
| Pegged | |
| Primary teeth (26) | 0.2 |
| Secondary maxillary lateral incisors | |
| Caucasian (27) | 1.0–2.0 |
| Japanese (30) | 6.2 |
| Shovel-shaped incisors | |
| Amerindian, Inuit, Oriental (15) | 60–75 |
| Supernumerary teeth | |
| Primary dentition (10, 16, 19, 22) | 0.3–0.8 |
| Secondary dentition | |
| Caucasian (4, 5, 16, 23, 29) | 1.0–3.5 |
| Japanese (21) | 2.2–5.3 |
| Maxillary incisors, secondary (2, 3, 23) | 0.3–0.5 |
| Fourth molars, secondary (29) | 0.2 |

[a] References:

1. ER Amos, Incidence of the small dens in dente. *J Am Dent Assoc* **51**:31–33, 1955.
2. J Bodenhoff, Dentitio connatalis et neonatalis. *Odontol Tidskr* **67**:645–695, 1959.
3. PJ Boyne, Supernumerary maxillary incisors. *Oral Surg* **7**:901–905, 1954.
4. CR Castaldi et al, Incidence of congenital anomalies in permanent teeth of a group of Canadian children aged 6–9. *J Can Dent Assoc* **32**:154–159, 1966.
5. JM Clayton, Congenital dental anomalies occurring in 3,557 children. *J Dent Child* **23**:206–208, 1956.
6. SF Dachi and FV Howell, A survey of 3,874 routine full mouth radiographs. II. A study of impacted teeth. *Oral Surg* **14**:1165–1169, 1961.
7. E Dolder, Deficient dentition. *Dent Rec* **57**:142–143, 1937.
8. FB Glenn, Incidence of congenitally missing permanent teeth in a private pedodontic practice. *J Dent Child* **28**:317–320, 1961.
9. H Grahnén et al, Dens invaginatus. I. A clinical roentgenological and genetical study of permanent upper lateral incisors. *Odont Revy* **10**:115–137, 1959.
10. H Grahnén and LE Granath, Numerical variation in primary dentition and their correlation with the permanent dentition. *Odont Revy* **4**:348–357, 1961.
11. H Grahnén and B Lindahl, Supernumerary teeth in the permanent dentition. A frequency study. *Odont Revy* **12**:290–294, 1961.
12. JF Gravely, A radiographic survey of third molar development. *Br Dent J* **119**:397–401, 1965.
13. GEM Hallett, The incidence, nature and clinical significance of palatal invaginations in maxillary incisor teeth. *Proc Roy Soc Med* **46**:491–499, 1953.
14. M Hellman, Our third molar teeth: Their eruption, presence and absence. *Dent Cosm* **78**:750–762, 1936.
15. A Hrdlička, Shovel-shaped teeth. *Am J Phys Anthropol* **3**:429–465, 1920.
16. JR Luten, The prevalence of supernumerary teeth in primary and mixed dentitions. *J Dent Child* **34**:346–353, 1967.
17. DR McKibben and LJ Brearley, Radiographic determination of the prevalence of selected dental anomalies in children. *J Dent Child* **38**:390–398, 1971.
18. SV Mead, Incidence of impacted teeth. *Int J Orthodont* **16**:885–890, 1930.
19. LF Menczer, Anomalies of the primary dentition. *J Dent Child* **22**:57–62, 1955.
20. TP Miller et al, A survey of congenitally missing permanent teeth. *J Am Dent Assoc* **81**:101–107, 1970.
21. JD Niswander and C Sajuku, Congenital anomalies of teeth in Japanese children. *Am J Phys Anthropol* **21**:569–574, 1963.
22. J Plaetschke, Okklusionsanomalien im Milchgebiss. *Deutsche Zahn Mund Kieferheilk* **5**:435–451, 1938.
23. JJ Ravin and LA Nielsen, An ortopantomografisk undersøgelse af overtal og aplasier hos 1,530 Kobenhavnske skolebørn. *Tandlaegebladet* **77**:12–22, 1973.
24. M Ringqvist and B Thilander, The frequency of hypodontia in an orthodontic material. *Svensk Tandläk Tidskr* **62**:535–541, 1969.
25. JS Rose, A survey of congenitally missing teeth, excluding third molars, in 6,000 orthodontic patients. *Dent Pract (Bristol)* **17**:107–111, 1966.
26. T Saito, A genetic study on the degenerative anomalies of deciduous teeth. *Jpn J Hum Genet* **4**:27–53, 1959.
27. C Schulze, Developmental abnormalities of the teeth and jaws, in RJ Gorlin and HM Goldman (eds), *Thoma's Oral Pathology,* 6th ed., CV Mosby, St. Louis, 1970.
28. WC Shafer, Dens in dente. *NY Dent J* **19**:220–223, 1953.
29. EC Stafne, Supernumerary teeth. *Dent Cosm* **74**:653–659, 1932.
30. Y Sumiya, Statistical study on dental anomalies in the Japanese. *J Anthropol Soc Nippon* **67**:171–172, 1959.
31. CJ Witkop and S Rao, Inherited defects in tooth structure. *Birth Defects* **7**(7):153–184, 1971.

Table A–14. Nondental anomalies and their frequencies

| Anomaly | Frequency in ostensibly normal population, approximate percentage |
|---|---|
| Cheek biting (39)[a] | 2 |
| Cleft | |
| Lip–palate | |
| Caucasian (12, 14) | 0.06–0.15 |
| Afro-American (3, 4, 10, 24) | 0.04 |
| Japanese (20, 33) | 0.17 |
| Amerindian (33, 44) | 0.3 |
| Palate, isolated | |
| Caucasian (10, 12) | 0.04–0.05 |
| Oriental (33) | 0.06–0.07 |
| Afro-American (3, 4) | 0.02–0.04 |
| Uvula | |
| Amerindian (8, 40) | 10–20 |
| Inuit (16) | 3–6 |
| Japanese (16) | 1–2 |
| Caucasian (30, 43) | 1–2 |
| Afro-American (36, 38) | 0.25–0.50 |
| Cysts, eruption (12) | 0.2 |
| Fordyce granules (sebaceous glands) | |
| Lips | |
| Caucasian (15, 25, 39a) | 50–80 |
| Buccal mucosa | |
| Caucasian (15, 25, 39a) | 60–95 |
| Afro-American (29) | 55 |
| Lip pits | |
| Paramedian (45) | 1:200,000 |
| Commissural | |
| Afro-American (5, 38) | 20 |
| Caucasian (5) | 12 |
| Oriental (5) | 6 |
| Amerindian (8) | 9 |
| Inuit (19) | 8 |
| Tongue | |
| Ability to touch nose with (personal observation) | 7–10 |
| Ankyloglossia (7, 28, 47) | 1:300 to 1 per 2000 |
| Fissured | |
| 5–18 years (9, 35) | |
| Overall (1, 15, 47) | 2–5 |
| Folding of tip | |
| Within mouth | 1–2 |
| Outside mouth (46) | 1 per 600 |
| Geographic | |
| Caucasian (9, 15) | 1.0–2.5 |
| Median rhomboid "glossitis" | |
| Caucasian (15, 47) | 0.2–0.3 |
| Inuit (19) | 0.6 |
| Rolling or tubing | |
| Caucasian (13, 41) | 60–75 |
| Afro-American (23) | 80 |
| Japanese (22) | 25–30 |
| Thyroid gland inclusion | |
| Microscopic evidence (6, 37) | 10 |
| Clinical evidence (6) | 1 per 10,000 |
| Torus mandibularis | |
| Caucasian and Afro-American (17, 21, 42) | 3–16 |
| Inuit (17, 18, 19, 26, 27) | 10–80 |
| Oriental (38) | 20–30 |
| Torus palatinus | |
| Caucasian and Afro-American (21, 31, 38, 42) | 20 (F-2, M-15) |
| Amerindian and Inuit (17, 19, 32) | 25–75 |
| Oriental (2, 38) | 40–90 |

[a]References:

1. V Aboyans and A Ghaemmaghami, The incidence of fissured tongue among 4,009 Iranian dental outpatients. *Oral Surg* **36:**34–38, 1973.
2. E Akabori, Torus mandibularis. *J Shanghai Sci Inst* **4:**239–255, 1939.
3. LA Altemus and AD Ferguson, Comparative incidence of birth defects in negro and white children. *Pediatrics* **36:**56–61, 1965.
4. LA Altemus, The incidence of cleft lip and palate among North American negroes. *Cleft Palate J* **3:**357–361, 1966.
5. WR Baker, Pits of the lip commissures in caucasoid males. *Oral Surg* **21:**56–60, 1966.
6. RA Baughman, Lingual thyroid and lingual thyroglossal tract remnants. *Oral Surg* **34:**781–799, 1972.
7. FI Catlin and V De Haas, Tongue-tie. *Arch Otolaryngol* **94:**548–557, 1971.
8. J Cervenka, Unpublished data, 1970.
9. A Chosack et al, The prevalence of scrotal tongue and geographic tongue in 70,359 Israeli school children. *Community Dent Oral Epidemiol* **2:**253–257, 1974.
10. CE Chung and NC Myrianthopoulos, Racial and prenatal factors in major congenital malformations. *Am J Hum Genet* **20:**44–60, 1968.
11. CA Clark, A survey of eruption cysts in the newborn. *Oral Surg* **15:**917, 1962.
12. CM Drillien et al, *The Causes and Natural History of Cleft Lip and Palate*. E. & S. Livingston, Ltd., Edinburgh, 1966.
13. EE Gahres, Tongue rolling and tongue folding and other hereditary movements of the tongue. *J Hered* **43:**221–225, 1952.
14. JC Greene et al, Epidemiologic study of cleft lip and cleft palate in four states *J Am Dent Assoc* **68:**387–404, 1964.
15. V Halperin et al, The occurrence of Fordyce spots, benign migratory glossitis, median rhomboid glossitis and fissured tongue in 2,478 dental patients. *Oral Surg* **6:**1072–1077, 1953.
16. GM Heathcote, The prevalence of cleft uvula in an Inuit population. *Am J Phys Anthropol* **41:**433–438, 1974.
17. EA Hooton, On certain Eskimoidal characters in Icelandic skulls. *Am J Phys Anthropol* **1:**53–76, 1918.
18. A Hrdlička, Mandibular and maxillary hyperostoses. *Am J Phys Anthropol* **27:**1–55, 1940.
19. A Jarvis and RJ Gorlin, Minor orofacial anomalies in an Eskimo population. *Oral Surg* **33:**417–427, 1972.
20. YA Kobayashi, A genetic study of harelip and cleft palate. *Jpn J Hum Genet* **3:**73–107, 1958.
21. S Kolas et al, The occurrence of torus palatinus and torus mandibularis in 2,478 dental patients. *Oral Surg* **6:**1134–1141, 1953.
22. T Komai, Notes on lingual gymnastics. *J Hered* **42:**293–297, 1951.
23. JW Lee, Tongue-folding and tongue-rolling in an American negro population sample. *J Hered* **46:**289–291, 1955.
24. CG Longenecker et al, Cleft lip and cleft palate, incidence at a large charity hospital. *Plast Reconstr Surg* **35:**548–550, 1965.
25. EA Martin and RT Wales, Sebaceous glands in the mouth. *Irish J Med Sci* **6:**481–486, 1964.
26. JT Mayhall, Torus mandibularis in an Alaskan Eskimo population. *Am J Phys Anthropol* **33:**57–60, 1970.
27. JT Mayhall and MF Mayhall, Torus mandibularis in two Northwest Territories villages. *Am J Phys Anthropol* **34:**143–148, 1971.
28. ET McEnery and FP Gaines, Tongue-tie in infants and children. *J Pediatr* **18:**252–255, 1941.
29. RC McGoodwin, Fordyce's granules in pigmented oral mucosa. *J Dent Res* **43:**773, 1964.
30. LH Meskin et al, The prevalence of cleft uvula. *Cleft Palate J* **1:**342–346, 1964.
31. SC Miller and H Roth, Torus palatinus: A statistical study. *J Am Dent Assoc* **27:**1950–1957, 1940.
32. CFA Moorrees, The dentition as a criterion of race with special reference to the Aleut. *J Dent Res* **30:**815–821, 1951.

33. JV Neel, A study of major congenital defects in Japanese infants. *Am J Hum Genet* **10:**398–445, 1958.
34. J Niswander and MS Adams, Oral clefts in the American Indians. *Public Health Rep* **82:**807–812, 1967.
35. RS Redman, The prevalence of geographic tongue, fissured tongue, median rhomboid glossitis and hairy tongue among 3,611 Minnesota school children. *Oral Surg* **30:**390–395, 1970.
36. ER Richardson, Cleft uvula in negroes. *Cleft Palate J* **7:**669–672, 1970.
37. JJ Sauk Jr, Ectopic lingual thyroid. *J Pathol* **102:**239–243, 1970.
38. BF Schaumann et al, Minor craniofacial anomalies among a negro population. *Oral Surg* **29:**566–575, 729–734, 1970.
39. I Sewerin, The prevalence of morsicatio buccarum/labiorum in 841 Copenhagen school children. *Tandlaegebladet* **77:**861–864, 1973.
39a. I Sewerin, The sebaceous glands in the vermilion border of the lips and in the oral mucosa of man. *Acta Odontol Scand* **33:**Suppl **68:**1–226, 1975.
40. BL Shapiro et al, Cleft uvula: A microform of facial clefts and its genetic basis. *Birth Defects* **7**(7):80–82, 1971.
41. AH Sturtevant, A new inherited character in man. *Proc Nat Acad Sci (Wash)* **26:**100–102, 1940.
42. CJ Summers, Prevalence of tori. *J Oral Surg* **26:**718–720, 1968.
43. M Tolarova et al, Distribution of signs considered to be a microform of cleft lip and/or cleft palate in a population of normal 18- to 21-year-old subjects. *Acta Chir Plast (Praha)* **9:**1–14, 1967.
44. VE Tretsven, Incidence of cleft lip and palate in Montana Indians. *J Speech Hearing Dis* **28:**52–57, 1963.
45. A Van der Woude, Fistula labii inferioris congenita and its association with cleft lip and palate. *Am J Hum Genet* **6:**244–256, 1954.
46. DD Whitney, Tongue tip overfolding. *J Hered* **40:**19–21, 1949.
47. CJ Witkop and L Barros, Oral and genetic studies of Chileans. Oral anomalies. *Am J Phys Anthropol* **21:**15–24, 1963.

# Index